CHILTON®

ASIAN SERVICE MANUAL 2010 EDITION VOLUME V SCION TOYOTA

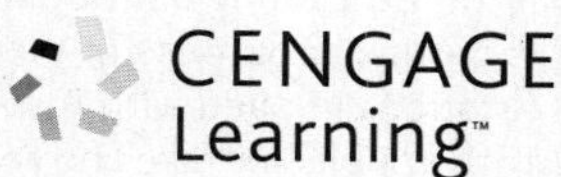

Australia • Brazil • Japan • Korea • Mexico • Singapore • Spain • United Kingdom • United States

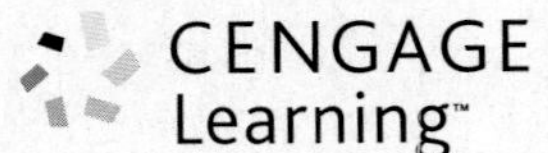

CHILTON®
Asian Service Manual
2010 Edition
Volume V
Scion, Toyota

Vice President, Technology Professional Business Unit:
Gregory L. Clayton

Publisher, Technology Professional Business Unit:
David Koontz

Director of Marketing:
Beth A. Lutz

Director Education Production:
Carolyn Miller

Marketing Manager:
Jennifer Barbic

Marketing Coordinator:
Rachael Torres

Chilton Content Specialist:
Paula Baillie

Graphical Designer:
Melinda Possinger

Art Director:
Benjamin Gleeksman

Sr. Content Project Manager:
Elizabeth C. Hough

Senior Editor:
Christine L. Sheeky

Editors:
Dennis Bailey
Nick D'Andrea
Eugene F. Hannon Jr., A.S.E.
Kyla Nyjordet
Lance Williams

ISBN-13: 978-1-1110-3768-0
ISBN-10: 1-1110-3768-X
ISSN: 1939-621X

Chilton
5 Maxwell Drive
Clifton Park, NY 12065-2919
USA

Chilton products are represented in Canada by Nelson Education, Ltd.

NOTICE TO THE READER

Publisher does not warrant or guarantee any of the products described herein or perform any independent analysis in connection with any of the product information contained herein. Publisher does not assume, and expressly disclaims, any obligation to obtain and include information other than that provided to it by the manufacturer.

The reader is expressly warned to consider and adopt all safety precautions that might be indicated by the activities described herein and to avoid all potential hazards. By following the instructions contained herein, the reader willingly assumes all risks in connection with such instructions.

The publisher makes no representations or warranties of any kind, including but not limited to, the warranties of fitness for particular purpose or merchantability, nor are any such representations implied with respect to the material set forth herein, and the publisher takes no responsibility with respect to such material. The publisher shall not be liable for any special, consequential, or exemplary damages resulting, in whole or part, from the readers' use of, or reliance upon, this material.

Printed in the United States of America
1 2 3 4 5 6 7 15 14 13 12 11

Contents

Sections

Model Index

USING THIS INFORMATION

Organization

To find where a particular model section or procedure is located, look in the Table of Contents. Main topics are listed with the page number on which they may be found. Following the main topics is an alphabetical listing of all of the procedures within the section and their page numbers.

Manufacturer and Model Coverage

This product covers 2009–2010 Asian models that are produced in sufficient quantities to warrant coverage, and which have technical content available from the vehicle manufacturers before our publication date. Although this information is as complete as possible at the time of publication, some manufacturers may make changes which cannot be included here. While striving for total accuracy, the publisher cannot assume responsibility for any errors, changes, or omissions that may occur in the compilation of this data.

Part Numbers and Special Tools

Part numbers and special tools are recommended by the publisher and vehicle manufacturer to perform specific jobs. Before substituting any part or tool for the one recommended, you must be completely satisfied that neither your personal safety, nor the performance of the vehicle will be endangered.

ACKNOWLEDGEMENT

Portions of materials contained herein have been reprinted under license from Toyota Motor Sales, U.S.A., Inc., License Agreement TMS1005.

All information contained herein about Toyota, Lexus, and Scion vehicles is based on the latest product information available at the time of publication, is provided "as is" without warranty of any kind, and is intended for service providers and other interested parties in Canada, Mexico, and the United States of America, including Guam, Puerto Rico, and the U.S. Virgin Islands.

Specifications and procedures are subject to change without notice. This information is provided expressly for the purpose of use by professional automobile technicians who have special techniques and certifications. Repair or service by non-specialized or uncertified technicians using only this information, or without proper equipment or tools, may cause severe injury to the individual or other individuals and could possibly cause damage to the vehicle. Certain procedures or content elements may make reference to Toyota Warranty policy or practice - these policies or practices are only applicable to Toyota Lexus, or Scion dealers.

PRECAUTIONS

Before servicing any vehicle, please be sure to read all of the following precautions, which deal with personal safety, prevention of component damage, and important points to take into consideration when servicing a motor vehicle:

- Always wear safety glasses or goggles when drilling, cutting, grinding or prying.
- Steel-toed work shoes should be worn when working with heavy parts. Pockets should not be used for carrying tools. A slip or fall can drive a screwdriver into your body.
- Work surfaces, including tools and the floor should be kept clean of grease, oil or other slippery material.
- When working around moving parts, don't wear loose clothing. Long hair should be tied back under a hat or cap, or in a hair net.
- Always use tools only for the purpose for which they were designed. Never pry with a screwdriver.
- Keep a fire extinguisher and first aid kit handy.
- Always properly support the vehicle with approved stands or lift.
- Always have adequate ventilation when working with chemicals or hazardous material.
- Carbon monoxide is colorless, odorless and dangerous. If it is necessary to operate the engine with vehicle in a closed area such as a garage, always use an exhaust collector to vent the exhaust gases outside the closed area.
- When draining coolant, keep in mind that small children and some pets are attracted by ethylene glycol antifreeze, and are quite likely to drink any left in an open container, or in puddles on the ground. This will prove fatal in sufficient quantity. Always drain the coolant into a sealable container.
- To avoid personal injury, do not remove the coolant pressure relief cap while the engine is operating or hot. The cooling system is under pressure; steam and hot liquid can come out forcefully when the cap is loosened slightly. Failure to follow these instructions may result in personal injury. The coolant must be recovered in a suitable, clean container for reuse. If the coolant is contaminated it must be recycled or disposed of correctly.
- When carrying out maintenance on the starting system be aware that heavy gauge leads are connected directly to the battery. Make sure the protective caps are in place when maintenance is completed. Failure to follow these instructions may result in personal injury.
- Do not remove any part of the engine emission control system. Operating the engine without the engine emission control system will reduce fuel economy and engine ventilation. This will weaken engine performance and shorten engine life. It is also a violation of Federal law.
- Due to environmental concerns, when the air conditioning system is drained, the refrigerant must be collected using refrigerant recovery/recycling equipment. Federal law requires that refrigerant be recovered into appropriate recovery equipment and the process be conducted by qualified technicians who have been certified by an approved organization, such as MACS, ASI, etc. Use of a recovery machine dedicated to the appropriate refrigerant is necessary to reduce the possibility of oil and refrigerant incompatibility concerns. Refer to the instructions provided by the equipment manufacturer when removing refrigerant from or charging the air conditioning system.
- Always disconnect the battery ground when working on or around the electrical system.

• Batteries contain sulfuric acid. Avoid contact with skin, eyes, or clothing. Also, shield your eyes when working near batteries to protect against possible splashing of the acid solution. In case of acid contact with skin or eyes, flush immediately with water for a minimum of 15 minutes and get prompt medical attention. If acid is swallowed, call a physician immediately. Failure to follow these instructions may result in personal injury.

• Batteries normally produce explosive gases. Therefore, do not allow flames, sparks or lighted substances to come near the battery. When charging or working near a battery, always shield your face and protect your eyes. Always provide ventilation. Failure to follow these instructions may result in personal injury.

• When lifting a battery, excessive pressure on the end walls could cause acid to spew through the vent caps, resulting in personal injury, damage to the vehicle or battery. Lift with a battery carrier or with your hands on opposite corners. Failure to follow these instructions may result in personal injury.

• Observe all applicable safety precautions when working around fuel. Whenever servicing the fuel system, always work in a well-ventilated area. Do not allow fuel spray or vapors to come in contact with a spark, open flame, or excessive heat (a hot drop light, for example). Keep a dry chemical fire extinguisher near the work area. Always keep fuel in a container specifically designed for fuel storage; also, always properly seal fuel containers to avoid the possibility of fire or explosion. Do not smoke or carry lighted tobacco or open flame of any type when working on or near any fuel-related components.

• Fuel injection systems often remain pressurized, even after the engine has been turned OFF. The fuel system pressure must be relieved before disconnecting any fuel lines. Failure to do so may result in fire and/or personal injury.

• The evaporative emissions system contains fuel vapor and condensed fuel vapor. Although not present in large quantities, it still presents the danger of explosion or fire. Disconnect the battery ground cable from the battery to minimize the possibility of an electrical spark occurring, possibly causing a fire or explosion if fuel vapor or liquid fuel is present in the area. Failure to follow these instructions can result in personal injury.

• The EPA warns that prolonged contact with used engine oil may cause a number of skin disorders, including cancer! You should make every effort to minimize your exposure to used engine oil. Protective gloves should be worn when changing oil. Wash your hands and any other exposed skin areas as soon as possible after exposure to used engine oil. Soap and water, or waterless hand cleaner should be used.

• Some vehicles are equipped with an air bag system, often referred to as a Supplemental Restraint System (SRS) or Supplemental Inflatable Restraint (SIR) system. The system must be disabled before performing service on or around system components, steering column, instrument panel components, wiring and sensors. Failure to follow safety and disabling procedures could result in accidental air bag deployment, possible personal injury and unnecessary system repairs.

• Always wear safety goggles when working with, or around, the air bag system. When carrying a non-deployed air bag, be sure the bag and trim cover are pointed away from your body. When placing a non-deployed air bag on a work surface, always face the bag and trim cover upward, away from the surface. This will reduce the motion of the module if it is accidentally deployed.

• Electronic modules are sensitive to electrical charges. The ABS module can be damaged if exposed to these charges.

• Brake pads and shoes may contain asbestos, which has been determined to be a cancer-causing agent. Never clean brake surfaces with compressed air. Avoid inhaling brake dust. Clean all brake surfaces with a commercially available brake cleaning fluid.

• When replacing brake pads, shoes, discs or drums, replace them as complete axle sets.

• When servicing drum brakes, disassemble and assemble one side at a time, leaving the remaining side intact for reference.

• Brake fluid often contains polyglycol ethers and polyglycols. Avoid contact with the eyes and wash your hands thoroughly after handling brake fluid. If you do get brake fluid in your eyes, flush your eyes with clean, running water for 15 minutes. If eye irritation persists, or if you have taken brake fluid internally, immediately seek medical assistance.

• Clean, high quality brake fluid from a sealed container is essential to the safe and proper operation of the brake system. You should always buy the correct type of brake fluid for your vehicle. If the brake fluid becomes contaminated, completely flush the system with new fluid. Never reuse any brake fluid. Any brake fluid that is removed from the system should be discarded. Also, do not allow any brake fluid to come in contact with a painted or plastic surface; it will damage the paint.

• Never operate the engine without the proper amount and type of engine oil; doing so will result in severe engine damage.

• Timing belt maintenance is extremely important! Many models utilize an interference- type, non freewheeling engine. If the timing belt breaks, the valves in the cylinder head may strike the pistons, causing potentially serious (also time-consuming and expensive) engine damage.

• Disconnecting the negative battery cable on some vehicles may interfere with the functions of the on-board computer system(s) and may require the computer to undergo a relearning process once the negative battery cable is reconnected.

• Steering and suspension fasteners are critical parts because they affect performance of vital components and systems and their failure can result in major service expense. They must be replaced with the same grade or part number or an equivalent part if replacement is necessary. Do not use a replacement part of lesser quality or substitute design. Torque values must be used as specified during reassembly.

SCION

tC • xB • xD

1

SPECIFICATIONS AND MAINTENANCE CHARTS

ENGINE AND VEHICLE IDENTIFICATION

Engine							Model Year	
Code ①	Liters (cc)	Cu. In.	Cyl.	Fuel Sys.	Engine Type	Eng. Mfg.	Code ②	Year
2AZ-FE	2.4 (2362)	144.2	4	SFI	DOHC	Toyota	9	2009
2ZR-FE	1.8 (1798)	109.7	4	SFI	DOHC	Toyota	A	2010

SFI: Sequential Fuel Injection

DOHC: Double Overhead Camshaft

① Stamped on the left side of the engine block

② 10th digit of the Vehicle Identification Number (VIN)

3768X_SCIO_C0001

GENERAL ENGINE SPECIFICATIONS

Year	Model	Engine Displacement Liters	Engine Series ID	Net Horsepower @ rpm	Net Torque @ rpm (ft. lbs.)	Bore x Stroke (in.)	Com-pression Ratio	Oil Pressure @ rpm
2009	xB	2.4	2AZ-FE	161@6000	163@4000	3.48x3.78	9.8:1	36-78@3000
	tC	2.4	2AZ-FE	161@6000	163@4000	3.48x3.78	9.8:1	36-78@3000
	xD	1.8	2ZR-FE	128@6000	125@4400	3.17x3.48	10.0:1	36-78@3000
2010	xB	2.4	2AZ-FE	161@6000	163@4000	3.48x3.78	9.8:1	36-78@3000
	tC	2.4	2AZ-FE	161@6000	163@4000	3.48x3.78	9.8:1	36-78@3000
	xD	1.8	2ZR-FE	128@6000	125@4400	3.17x3.48	10.0:1	36-78@3000

3768X_SCIO_C0002

ENGINE TUNE-UP SPECIFICATIONS

Year	Engine Displacement Liters	Engine ID	Spark Plug Gap (in.)	Ignition Timing (deg.)	Fuel Pump (psi)	Idle Speed (rpm)	Valve Clearance Intake	Valve Clearance Exhaust
2009	1.8	2ZR-FE	0.043	8-12B	44-50	①	0.0060-0.0100	0.0100-0.0140
	2.4	2AZ-FE	0.043	5-12B	44-50	①	0.0075-0.0114	0.0150-0.0189
2010	1.8	2ZR-FE	0.043	8-12B	44-50	①	0.0060-0.0100	0.0100-0.0140
	2.4	2AZ-FE	0.043	5-12B	44-50	①	0.0075-0.0114	0.0150-0.0189

NOTE: The Vehicle Emission Control Information label often reflects specification changes made during production. The label figures must be used if they differ from those in this chart.

B: Before top dead center

① M/T: 600-700
A/T: 650-750

3768X_SCIO_C0003

CAPACITIES

Year	Model	Engine Displacement Liters	Engine ID	Engine Oil with Filter (qts.)	Transmission (pts.)		Front Drive Axle (pts.)	Fuel Tank (gal.)	Cooling System (qts.)
					Manual	Auto			
2009	xD	1.8	2ZR-FE	3.9	4.0	6.2	①	11.9	②
	xB	2.4	2AZ-FE	3.9	4.0	6.2	①	11.9	②
	tC	2.4	2AZ-FE	4.0	5.2	7.4	①	14.5	②
2010	xD	1.8	2ZR-FE	3.9	4.0	6.2	①	11.9	②
	xB	2.4	2AZ-FE	3.9	4.0	6.2	①	11.9	②
	tC	2.4	2AZ-FE	4.0	5.2	7.4	①	14.5	②

① Included in transmission capacity

② With Manual Transmission: 4.7 qts.
With Automatic Transmission: 4.5 qts.

3768X_SCIO_C0004

FLUID SPECIFICATIONS

Year	Model	Engine Displ. Liters	Engine Oil	Man. Trans.	Auto. Trans.	Drive Axle		Transfer Case	Power Steering Fluid	Brake Master Cylinder	Cooling System
						Front	Rear				
2009	All	1.8, 2.4	ILSAC	①	②	NA	NA	NA	③	④	⑤
2010	All	1.8, 2.4	ILSAC	①	②	NA	NA	NA	③	④	⑤

DOT: Department Of Transpotation

NA: Not Available

① ILSAC multi-grade engine oil

② API GL-4 SAE 75W-90

③ ATF Dexron II or III

④ SAE J1703 or FMVSS No. 116 DOT3

⑤ Toyota Super Long Life Coolant

3768X_SCIO_C0013

VALVE SPECIFICATIONS

Year	Engine Displacement Liters	Engine ID	Seat Angle (deg.)	Face Angle (deg.)	Spring Test Pressure (lbs. @ in.)	Spring Installed Height (in.)	Stem-to-Guide Clearance (in.)		Stem Diameter (in.)	
							Intake	Exhaust	Intake	Exhaust
2009	1.8	2ZR-FE	NA	NA	NA	NA	0.0010-0.0024	0.0012-0.0026	0.2154-0.2159	0.2152-0.2157
	2.4	2AZ-FE	45	45	NA	NA	0.0010-0.0024	0.0012-0.0026	0.2154-0.2159	0.2152-0.2158
2010	1.8	2ZR-FE	NA	NA	NA	NA	0.0010-0.0024	0.0012-0.0026	0.2154-0.2159	0.2152-0.2157
	2.4	2AZ-FE	45	45	NA	NA	0.0010-0.0024	0.0012-0.0026	0.2154-0.2159	0.2152-0.2158

NA: Not Available

3768X_SCIO_C0005

CAMSHAFT AND BEARING SPECIFICATIONS CHART

All measurements are given in inches.

Year	Engine Displ. Liters	Engine ID/VIN	Journal Dia.	Brg. Oil Clearance	Shaft End-play	Runout	Lobe Height	
							Intake	Exhaust
2009	1.8	2ZR-FE	①	0.0012-0.0024	0.0016-0.0037	0.0016	1.6857-1.6896	1.7455-1.7494
	2.4	2AZ-FE	②	③	④	0.0012	1.8654-1.8664	1.8104-1.8143
2010	1.8	2ZR-FE	①	0.0012-0.0024	0.0016-0.0037	0.0016	1.6857-1.6896	1.7455-1.7494
	2.4	2AZ-FE	②	③	④	0.0012	1.8654-1.8664	1.8104-1.8143

① No. 1 Journal: 1.3563-1.3569 in.
All Others: 0.9035-0.9041 in.

② No. 1 Journal: 1.4162-1.4167 in.
All Others: 0.9039-0.9045 in.

③ No. 1 Journal: 0.0003-0.0015 in.
All Others: 0.0010-0.0024 in.

④ Intake: 0.0016-0.0037 in.
Exhaust: 0.0032-0.0053 in.

3768X_SCIO_C0007

CRANKSHAFT AND CONNECTING ROD SPECIFICATIONS

All measurements are given in inches.

Year	Engine Displacement Liters	Engine ID	Crankshaft				Connecting Rod		
			Main Brg. Journal Dia.	Main Brg. Oil Clearance	Shaft End-play	Thrust on No.	Journal Diameter	Oil Clearance	Side Clearance
2009	1.8	2ZR-FE	1.8893-1.8898	0.0006-0.0015	NA	3	1.8504-1.8513	0.0012-0.0024	0.0063-0.0135
	2.4	2AZ-FE	2.1649-2.1654	0.0007-0.0016	0.0016-0.0095	3	1.8894-1.8898	0.0009-0.0019	0.0063-0.0143
2010	1.8	2ZR-FE	1.8893-1.8898	0.0006-0.0015	NA	3	1.8504-1.8513	0.0012-0.0024	0.0063-0.0135
	2.4	2AZ-FE	2.1649-2.1654	0.0007-0.0016	0.0016-0.0095	3	1.8894-1.8898	0.0009-0.0019	0.0063-0.0143

3768X_SCIO_C0006

PISTON AND RING SPECIFICATIONS

All measurements are given in inches.

Year	Engine Displ. Liters	Engine ID	Piston Clearance	Ring Gap Top Comp.	Ring Gap Bottom Comp.	Ring Gap Oil Control	Ring Side Clearance Top Comp.	Ring Side Clearance Bottom Comp.	Ring Side Clearance Oil Control
2009	1.8	2ZR-FE	0.0004-0.0017	0.0079-0.0118	0.0138-0.0197	0.0039-0.0157	0.0008-0.0028	0.0008-0.0024	0.0008-0.0026
	2.4	2AZ-FE	0.0008-0.0017	0.0094-0.0122	0.0130-0.0169	0.0040-0.0119	0.0008-0.0028	0.0008-0.0024	0.0008-0.0028
2010	1.8	2ZR-FE	0.0004-0.0017	0.0079-0.0118	0.0138-0.0197	0.0039-0.0157	0.0008-0.0028	0.0008-0.0024	0.0008-0.0026
	2.4	2AZ-FE	0.0008-0.0017	0.0094-0.0122	0.0130-0.0169	0.0040-0.0119	0.0008-0.0028	0.0008-0.0024	0.0008-0.0028

3768X_SCIO_C0008

TORQUE SPECIFICATIONS

All readings in ft. lbs.

Year	Engine Displacement Liters	Engine ID	Cylinder Head Bolts	Main Bearing Bolts	Rod Bearing Bolts	Crankshaft Damper Bolts	Flywheel Bolts	Manifold Intake	Manifold Exhaust	Spark Plugs	Oil Pan Drain Plug
2009	1.8	2ZR-FE	①	②	③	140	④	21	32	15	27
	2.4	2AZ-FE	⑤	⑥	⑦	133	⑧	22	27	18	18
2010	1.8	2ZR-FE	①	②	③	140	④	21	32	15	27
	2.4	2AZ-FE	⑤	⑥	⑦	133	⑧	22	27	18	18

① Step 1: 36 ft. lbs.
Step 2: Plus 90 degrees
Step 3: Plus 45 degrees

② Step 1: 30 ft. lbs.
Step 2: Plus 90 degrees

③ Step 1: 15 ft. lbs.
Step 2: Plus 90 degrees

④ Manual Trans.Step 1: 38 ft. lbs.
Step 2: Plus 90 degrees
Auto Trans.: 65 ft. lbs.

⑤ Step 1: 52 ft. lbs.
Step 2: Plus 90 degrees

⑥ Step 1: 30 ft. lbs.
Step 2: Plus 90 degrees

⑦ Step 1: 18 ft. lbs.
Step 2: Plus 90 degrees

⑧ Auto Trans.: 72 ft. lbs.
Manual Trans.: 96 ft. lbs.

3768X_SCIO_C0009

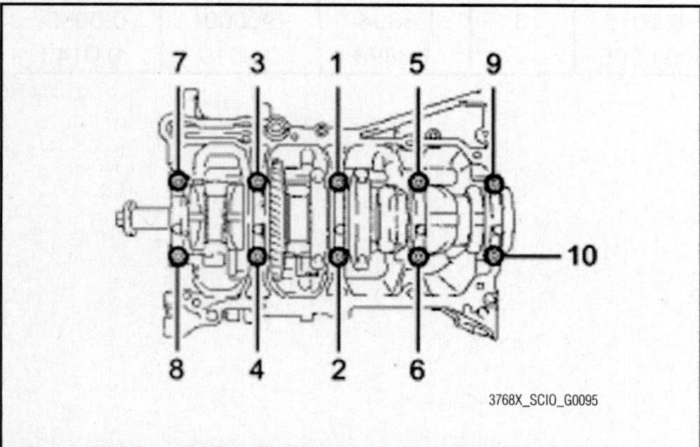

Fig. 1 Main bearing cap bolt torque sequence—2.4L engines

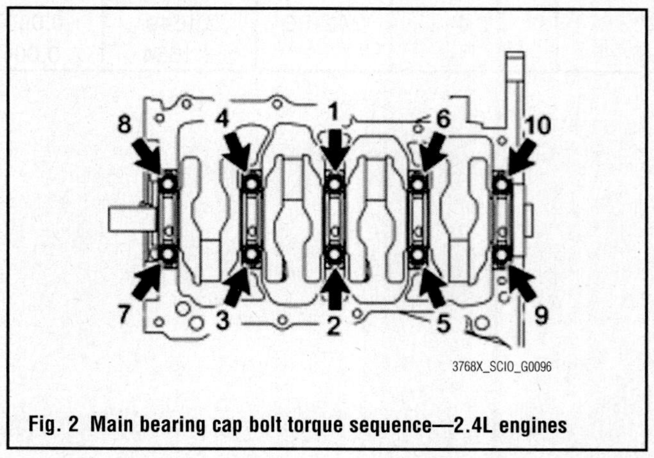

Fig. 2 Main bearing cap bolt torque sequence—2.4L engines

WHEEL ALIGNMENT

Year	Model		Caster Range (+/-Deg.)	Caster Preferred Setting (Deg.)	Camber Range (+/-Deg.)	Camber Preferred Setting (Deg.)	Toe-in (in.)
2009	xD	F	0.75	+4.85	0.75	-0.18	0.06+/-0.08
		R	—	—	0.5	-0.95	0.13+/-0.11
	xB	F	0.75	+5.75	0.75	-0.16	0.08+/-0.08
		R	—	—	0.5	-1.42	0.08+/-0.11
	tC	F	0.75	+3.03	0.75	-0.52	0+/-0.08
		R	—	—	0.5	-0.90	0.12+/-0.08
2010	xD	F	0.75	+4.85	0.75	-0.18	0.06+/-0.08
		R	—	—	0.5	-0.95	0.13+/-0.11
	xB	F	0.75	+5.75	0.75	-0.16	0.08+/-0.08
		R	—	—	0.5	-1.42	0.08+/-0.11
	tC	F	0.75	+3.03	0.75	-0.52	0+/-0.08
		R	—	—	0.5	-0.90	0.12+/-0.08

F: Front

R: Rear

3768X_SCIO_C0010

TIRE, WHEEL AND BALL JOINT SPECIFICATIONS

Year	Model	OEM Tires Standard	OEM Tires Optional	Tire Pressures (psi) Front	Tire Pressures (psi) Rear	Wheel Size	Ball Joint Inspection	Lug Nut Torque (ft. lbs.)
2009	xD	P195/60R16	N/A	33	33	5.5-JJ	44 in. ①	76
	xB	P205/55R16	N/A	35	32	6.0-JJ	44 in. ①	76
	tC	P215/45ZR17	205/55R16	32	29	7.0-JJ	44 in. ①	76
2010	xD	P195/60R16	N/A	33	33	5.5-JJ	44 in. ①	76
	xB	P205/55R16	N/A	35	32	6.0-JJ	44 in. ①	76
	tC	P215/45ZR17	205/55R16	32	29	7.0-JJ	44 in. ①	76

OEM: Original Equipment Manufacturer

PSI: Pounds Per Square Inch

STD: Standard

OPT: Optional

① Torque required (in inch lbs.) to rotate ball joint when removed from the knuckle

3768X_SCIO_C0011

BRAKE SPECIFICATIONS

All measurements in inches unless noted

			Brake Disc			Brake Drum				Brake Caliper	
Year	Model		Original Thickness	Minimum Thickness	Maximum Run-out	Original Inside Diameter	Max. Wear Limit	Maximum Machine Diameter	Minimum Lining Thickness	Bracket Bolts (ft. lbs.)	Mounting Bolts (ft. lbs.)
2009	xD	F	0.866	0.748	0.0020	—	—	—	0.039	79	25
		R	—	—	—	9.000	—	9.039	0.039	—	—
	xB	F	0.984	0.866	0.0020	—	—	—	0.039	130	26
		R	0.394	0.335	0.0020	—	—	—	0.039	42	25
	tC	F	0.984	0.906	0.0020	—	—	—	0.039	79	25
		R	0.354	0.295	0.0059	—	—	—	0.039	34	29
2010	xD	F	0.866	0.748	0.0020	—	—	—	0.039	79	25
		R	—	—	—	9.000	—	9.039	0.039	—	—
	xB	F	0.984	0.866	0.0020	—	—	—	0.039	130	26
		R	0.394	0.335	0.0020	—	—	—	0.039	42	25
	tC	F	0.984	0.906	0.0020	—	—	—	0.039	79	25
		R	0.354	0.295	0.0059	—	—	—	0.039	34	29

F: Front

R: Rear

3768X_SCIO_C0012

SCHEDULED MAINTENANCE INTERVALS

SCION - xB, xD and tC

TO BE SERVICED	TYPE OF SERVICE	VEHICLE MILEAGE INTERVAL (x1000)												
		5	10	15	20	25	30	35	40	45	50	55	60	65
Air cleaner filter	R						✓						✓	
Transmission fluid	S/I						✓						✓	
Ball joints & dust covers	S/I			✓			✓			✓			✓	
Bolts & nuts on chassis & body	S/I													
Brake line pipes & hoses	S/I			✓			✓			✓			✓	
Brake pads & discs/linings & drums (front & rear)	S/I	✓	✓	✓	✓	✓	✓	✓	✓	✓	✓	✓	✓	✓
Drive belts	S/I												✓	
Driveshaft boots	S/I			✓			✓			✓			✓	
Engine coolant	S/I			✓			✓			✓			✓	
Engine coolant	R	Replace at 100,000 miles												
Engine oil & filter	R	✓	✓	✓	✓	✓	✓	✓	✓	✓	✓	✓	✓	✓
Exhaust pipes & mountings	S/I			✓			✓			✓			✓	
Fuel lines & connections	S/I						✓						✓	
Propeller shaft bolt	S/I			✓			✓			✓			✓	
Radiator core & condenser	S/I			✓			✓			✓			✓	
Front differential fluid	S/I						✓						✓	
Rotate Tires	S/I	✓	✓	✓	✓	✓	✓	✓	✓	✓	✓	✓	✓	✓
Spark plugs (tC)	R	Replace at 120,000 miles												
Spark plugs (xA & xB)	R						✓						✓	
Steering linkage & gear box	S/I			✓			✓			✓			✓	

R: Replace S/I: Service or Inspect

Drivebelts: After initial inspection at 60,000 miles, inspect every 15,000 miles thereafter.

FREQUENT OPERATION MAINTENANCE (SEVERE SERVICE)

If a vehicle is operated under any of the following conditions it is considered severe service:

- Desert/Extremely dusty areas.

- Trailer towing usage.

Air cleaner filter: service or inspect every 5000 miles

Ball joints & dust covers: service or inspect every 5000 miles.

Bolts & nuts on chassis & body: service or inspect every 5000 miles.

Driveshaft boots: service or inspect every 5000 miles.

Steering linkage: service or inspect every 5000 miles.

Transmission and Front differential fluid: replace every 30,000 miles.

3768X_SCIO_C0014

BRAKES — INFORMATION AND PRECAUTIONS

ANTI-LOCK SYSTEMS

- Certain components within the ABS system are not intended to be serviced or repaired individually.
- Do not use rubber hoses or other parts not specifically specified for and ABS system. When using repair kits, replace all parts included in the kit. Partial or incorrect repair may lead to functional problems and require the replacement of components.
- Lubricate rubber parts with clean, fresh brake fluid to ease assembly. Do not use shop air to clean parts; damage to rubber components may result.
- Use only DOT 3 brake fluid from an unopened container.
- If any hydraulic component or line is removed or replaced, it may be necessary to bleed the entire system.
- A clean repair area is essential. Always clean the reservoir and cap thoroughly before removing the cap. The slightest amount of dirt in the fluid may plug an orifice and impair the system function. Perform repairs after components have been thoroughly cleaned; use only denatured alcohol to clean components. Do not allow ABS components to come into contact with any substance containing mineral oil; this includes used shop rags.
- The Anti-Lock control unit is a microprocessor similar to other computer units in the vehicle. Ensure that the ignition switch is **OFF** before removing or installing controller harnesses. Avoid static electricity discharge at or near the controller.
- If any arc welding is to be done on the vehicle, the control unit should be unplugged before welding operations begin.

DISC AND DRUM SYSTEMS

CAUTION

Dust and dirt accumulating on brake parts during normal use may contain asbestos fibers from production or aftermarket brake linings. Breathing excessive concentrations of asbestos fibers can cause serious bodily harm. Exercise care when servicing brake parts. Do not sand or grind brake lining unless equipment used is designed to contain the dust residue. Do not clean brake parts with compressed air or by dry brushing. Cleaning should be done by dampening the brake components with a fine mist of water, then wiping the brake components clean with a dampened cloth. Dispose of cloth and all residue containing asbestos fibers in an impermeable container with the appropriate label. Follow practices prescribed by the Occupational Safety and Health Administration (OSHA) and the Environmental Protection Agency (EPA) for the handling, processing, and disposing of dust or debris that may contain asbestos fibers.

BRAKES — BLEEDING THE BRAKE SYSTEM

BLEEDING PROCEDURE

BLEEDING PROCEDURE

Brake Lines

See Figure 3.

CAUTION

Read the "Precautions" in this section before beginning any repair work.

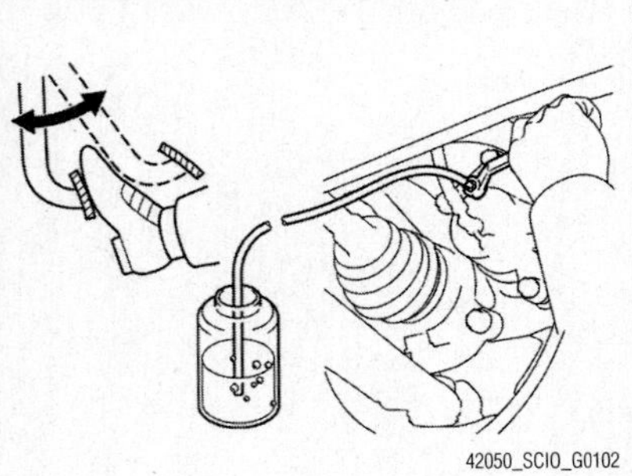

42050_SCIO_G0102

Fig. 3 Connect a vinyl tube to either one of the bleeder plugs. Depress the pedal several times, and then loosen the bleeder plug with the pedal depressed. When fluid stops coming out, immediately tighten the bleeder plug and release the pedal

1. Fill the brake master cylinder reservoir with brake fluid. Use SAE J1703 or FMVSS No. 116 DOT3 brake fluid.
2. Remove the bleeder plug cap.
3. Connect a vinyl tube to either one of the bleeder plugs.
4. Depress the pedal several times, and then loosen the bleeder plug with the pedal depressed.
5. When fluid stops coming out, immediately tighten the bleeder plug. Then release the pedal.
6. Repeat steps (3) and (4) until all the air in the fluid is gone.
7. Tighten the bleeder plug to 73 inch lbs. (8.3 Nm).
8. Install the cap.
9. Repeat the procedure to bleed air from the brake line for each wheel.

Master Cylinder

See Figure 4.

CAUTION

Read the "Precautions" in this section before beginning any repair work.

➡If the master cylinder has been disassembled or if the reservoir becomes empty, bleed air from the master cylinder.

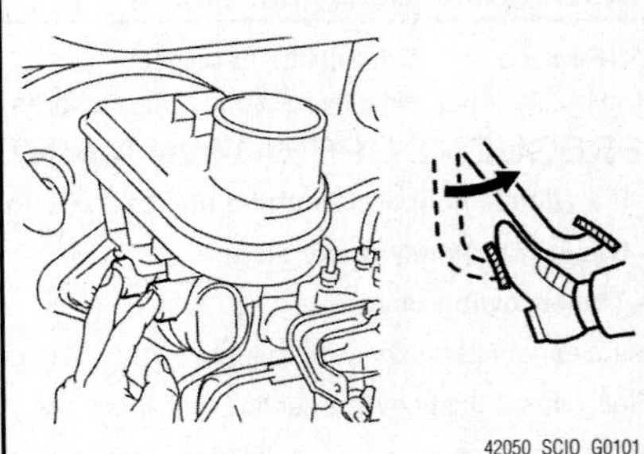

42050_SCIO_G0101

Fig. 4 Slowly depress and hold the brake pedal. Cover the outer holes with your fingers, and release the pedal

1. Fill the brake master cylinder reservoir with brake fluid. Use SAE J1703 or FMVSS No. 116 DOT3 brake fluid.
2. Using SST 09023-00101, or equivalent, disconnect the brake lines from the master cylinder.
3. Slowly depress and hold the brake pedal.
4. Cover the outer holes with your fingers, and release the pedal.
5. Repeat steps **3** and **4** several times.

➡Use a torque wrench with a fulcrum length of 30 cm (11.81 in.) to tighten the brake lines.

6. Connect the brake lines to the master cylinder and tighten to 11 ft. lbs. (15 Nm) without the special tool, or to 10 ft. lbs. (14 Nm) with the special tool.

Actuator Assembly (With Vehicle Stability Control)

See Figure 5.

✻✻ CAUTION

Read the "Precautions" in this section before beginning any repair work.

➡After bleeding air from the brake system, the height and/or feel of the pedal may still be awkward. If so, use the hand-held tester to bleed air from the actuator (see below).

Perform the air bleeding by following the steps displayed on the hand-held tester. Make sure that the fluid in the reservoir does not become empty.

1. Fill the brake master cylinder reservoir with brake fluid. Use SAE J1703 or FMVSS No. 116 DOT3 brake fluid.
2. With the engine stopped, depress the pedal more than 20 times.
3. Connect the hand-held tester to the DLC3, and then turn the ignition switch **ON**.

✻✻ WARNING

Do NOT start the engine.

4. Select "AIR BLEEDING" on the hand-held tester.

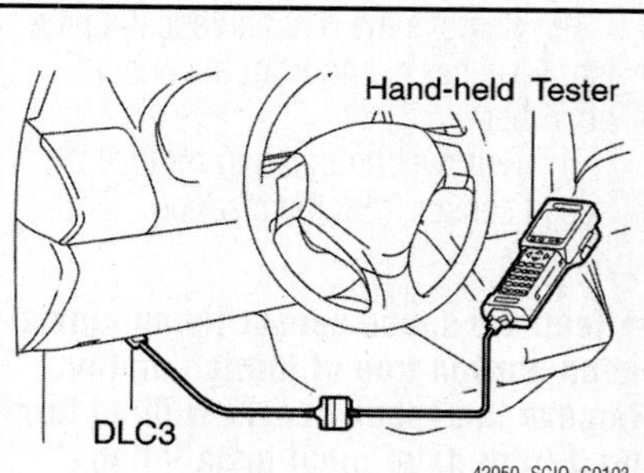

Fig. 5 Connect a hand-held tester to the DLC, turn the ignition ON, but do NOT start the engine

➡Please refer to the hand-held tester operator's manual for further operating instructions.

5. Bleed air from the brake line in "Step 1: Increase" on the hand-held tester display:
   a. Remove the bleeder plug cap.
   b. Connect a vinyl tube to either one of the bleeder plugs.
   c. Depress the pedal several times, and then loosen the bleeder plug with the pedal depressed.
   d. When fluid stops coming out, immediately tighten the bleeder plug. Then release the pedal.
   e. Repeat steps the previous 2 steps until all the air in the brake line is gone.
   f. Tighten the bleeder plug to 73 inch lbs. (8.3 Nm).
   g. Install the cap.
   h. Repeat the procedure to bleed the air from the brake line at each wheel.
6. Bleed air from the suction line in "Step 2: Inhalation" on the hand-held tester display.
   a. Remove the bleeder plug cap.
   b. Connect a vinyl tube to the bleeder plug at the right front wheel or the right rear wheel and loosen the bleeder plug.
   c. Using the hand-held tester, operate the actuator to bleed the air.

➡The operation stops automatically in 4 seconds. At this time, be sure to release the pedal.

   d. View the hand-held tester display and check that the operation has stopped.
   e. Repeat the previous steps until all the air in the fluid is gone.
   f. Tighten the bleeder plug to 73 inch lbs. (8.3 Nm).
   g. Install the cap.
   h. Repeat the procedure to bleed air from the brake line for each wheel.
7. Bleed air from the pressure reduction line in "Step 3: Decrease" on the hand-held tester display:
   a. Remove the bleeder plug cap.
   b. Connect a vinyl tube to either one of the bleeder plugs.
   c. Loosen the bleeder plug.
   d. Using the hand-held tester, operate the actuator. At the same time, completely depress the pedal and maintain this position.

➡The operation stops automatically in 4 seconds. When performing this procedures continuously, an interval of at least 20 seconds is required. When the operation is completed, the pedal slightly goes down. This is normal and is caused when the solenoid opens. During this procedure, the pedal may seem heavy, but completely depress it so that the fluid comes out from the bleeder plug. Be sure to keep depressing the pedal. Never depress and release the pedal repeatedly.

   e. Tighten the bleeder plug to 73 inch lbs. (8.3 Nm), and then release the pedal.
   f. Repeat the previous 3 steps until all the air in the fluid is gone.
   g. Install the cap.
   h. Repeat the procedure to bleed the air from the brake line at each wheel.
8. Bleed air from the brake line again in "Step 4: Increase" on the hand-held tester display:
   a. Remove the bleeder plug cap.
   b. Connect a vinyl tube to either one of the bleeder plugs.
   c. Depress the pedal several times, and then loosen the bleeder plug with the pedal depressed.
   d. When fluid stops coming out, immediately tighten the bleeder plug. Then release the pedal.
   e. Repeat the steps until all the air in the fluid is gone.
   f. Tighten the bleeder plug to 73 inch lbs. (8.3 Nm).
   g. Install the cap.
   h. Repeat the procedure to bleed air from the brake line at each wheel.
9. Check the fluid level in the brake fluid reservoir and add fluid if necessary.

WHEEL SPEED SENSORS

REMOVAL & INSTALLATION

tC Models

See Figure 6.

✲✲ CAUTION

Read the "Precautions" in this section before beginning any repair work.

1. Raise and safely support the vehicle.
2. Remove the front wheel and tire assembly.
3. Remove front fender liner.
4. Disengage the clamp and disconnect the wheel speed sensor connector.
5. Remove the 2 bolts and separate the front speed sensor wiring from the body.
6. Remove the bolt, then remove the front speed sensor.

✲✲ CAUTION

Do not let any foreign matter or debris attach to the sensor tip.

To install:

7. Install the front speed sensor with the bolt and tighten to 80 inch lbs. (9 Nm).
8. Install the speed sensor wiring with the 2 bolts and tighten as follows:
    a. Bolt a: 80 inch lbs. (9.0 Nm)
    b. Bolt b: 21 ft. lbs. (29 Nm)
9. Connect the connector and engage the clamp.
10. Install front fender liner.
11. Install front wheel.
12. Carefully lower the vehicle.

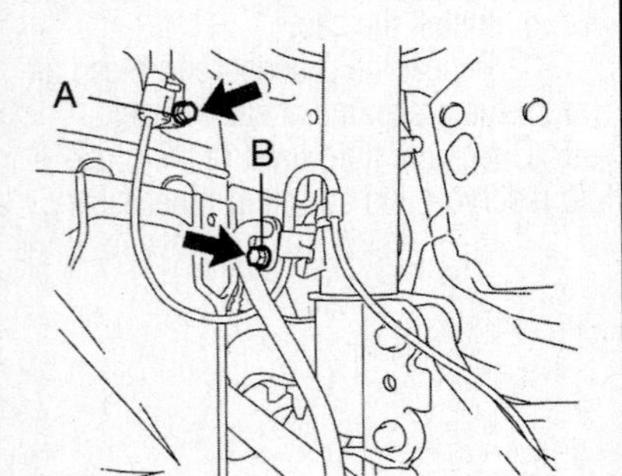

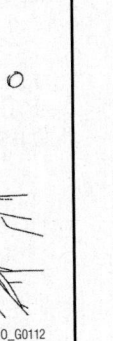

Fig. 6 View of the speed sensor wiring and attaching bolts (a & b)

xB Models

See Figures 7 and 8.

1. Raise and safely support the vehicle.
2. Remove the front wheel and tire assembly.
3. Disconnect the negative, then the positive battery cables, then remove the battery, if necessary for access.
4. Disconnect the speed sensor connector.
5. Remove the 2 clamp bolts holding the speed sensor wire harness from the body and the shock absorber.
6. Disconnect the resin clip and the speed sensor wire harness from the shock absorber.
7. Remove the bolt, then remove the front speed sensor.

✲✲ CAUTION

Do not let any foreign matter or debris attach to the sensor tip.

To install:

8. Install the front speed sensor with the bolt and tighten to 71 inch lbs. (8 Nm).
9. Install the speed sensor wire harness with the 2 bolts on the body and the shock absorber. Tighten as follows:
    a. Bolt a: 71 inch lbs. (8.0 Nm)
    b. Bolt b: 22 ft. lbs. (29 Nm)

➡Do not twist the sensor wire when installing the sensor.

10. Connect the resin clip and the speed sensor wire harness to the shock absorber.
11. Attach the speed sensor connector.
12. Install the front wheel and tire assembly.
13. Install battery, then connect the positive and negative battery cables.

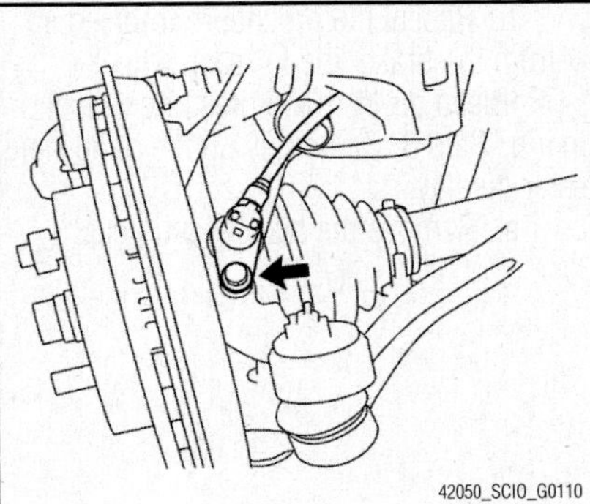

Fig. 7 Unfasten the mounting bolt, then remove the front speed sensor

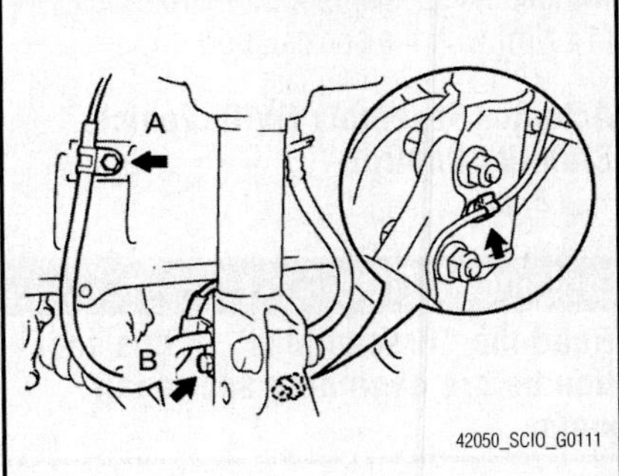

Fig. 8 View of the speed sensor wire harness and 2 mounting bolts (a & b)

xD Models

See Figure 9.

1. Raise and safely support the vehicle.
2. Remove the front wheel.
3. Remove the rocker panel molding cover.
4. Remove the side mudguard.
5. Remove the front fender liner.
6. Remove the front speed sensor.
    a. Disengage the clamp A from the body.
    b. Disconnect the speed sensor connector.
    c. Disengage the clamps B, C and D from the body.
    d. Remove the bolt and separate the No. 2 sensor clamp from the body.
    e. Remove the bolt and separate the No. 1 sensor clamp from the shock absorber.
    f. Remove the bolt and remove the speed sensor from the steering knuckle.

➡Keep the speed sensor tip and installation portion free of foreign matter. Remove the speed sensor without turning it from its original installation angle.

To install:

7. Install the front speed sensor as follows:
    a. Install the speed sensor onto the steering knuckle with the bolt. Torque: 75 inch lbs. (8.5 Nm).

➡Check that the speed sensor tip and installation portion are free of foreign matter. Install the speed sensor without turning it from its original installation angle.

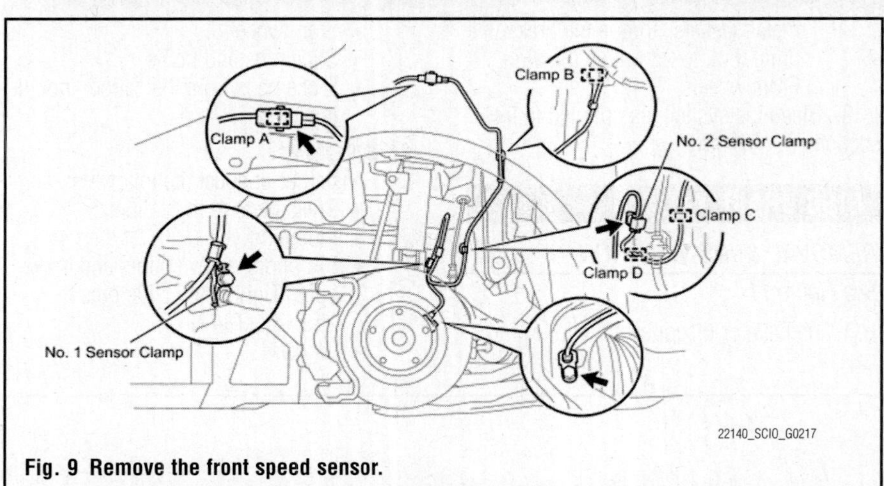

Fig. 9 Remove the front speed sensor.

b. Install the No. 1 sensor clamp onto the shock absorber with the bolt. Torque: 22 ft. lbs. (29 Nm).

c. Install the No. 2 sensor clamp onto the body with the bolt. Torque: 53 inch lbs. (6.0 Nm).

d. Engage the clamps D, C and B onto the body.

e. Connect the speed sensor connector.

f. Engage the clamp A onto the body.

8. 2. Install the front fender liner.
9. 3. Install the side mudguard.
10. 4. Install the rocker panel molding cover.
11. 5. Install the front wheel.

BRAKES

FRONT DISC BRAKES

BRAKE CALIPER

REMOVAL & INSTALLATION

See Figure 10.

✱✱ CAUTION

Read the "Precautions" in this section before beginning any repair work.

1. Remove or disconnect the following:
   - Front wheels
   - Brake line at the caliper
   - Brake pads
   - Brake pad support plate

29 (296, 21)
FRONT FLEXIBLE HOSE
● GASKET
FRONT DISC
FRONT DISC BRAKE CYLINDER SLIDE PIN
DISC BRAKE CYLINDER ASSEMBLY
● FRONT DISC BRAKE BUSH DUST BOOT
x2
34 (350, 25)
x2
107 (1089, 79)
FRONT DISC BRAKE CYLINDER SLIDE PIN
● FRONT DISC BRAKE CYLINDER SLIDE BUSH
● FRONT DISC BRAKE BUSH DUST BOOT
FRONT DISC BRAKE CYLINDER MOUNTING
FRONT NO. 1 DISC BRAKE PAD SUPPORT PLATE
FRONT NO. 2 DISC BRAKE PAD SUPPORT PLATE
N*m (kgf*cm, ft.*lbf) : Specified torque
● Non-reusable part
⬅ Lithium soap base glycol grease
22140_SCIO_G0218

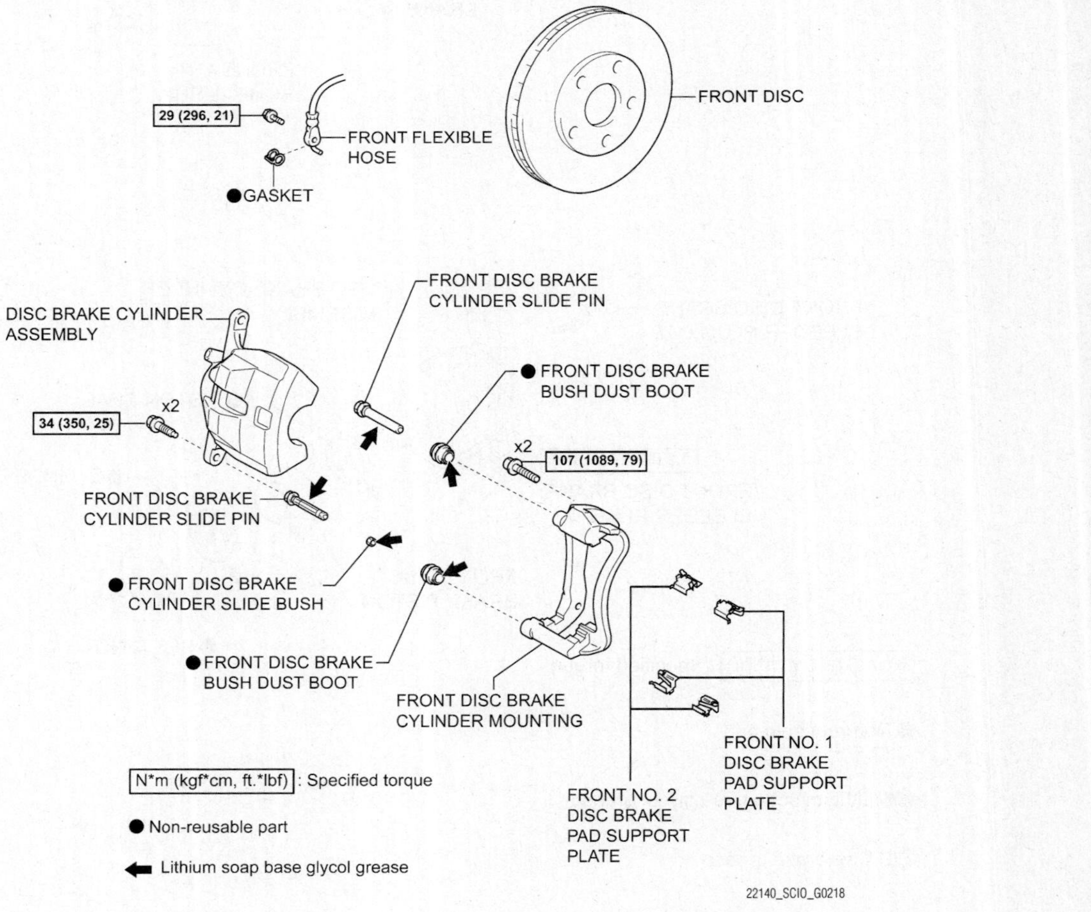

Fig. 10 Exploded view of the front brake components

- Pin and sleeve boots
- Caliper bolts
- Caliper

To install:

2. Install or connect the following:
   - Caliper onto its mounting and install the lower mounting bolt. Torque: 79 ft. lbs. (107 Nm).
   - Pin boots, sleeve boots and brake pads. Tighten the slide pins to 25 ft. lbs. (34 Nm).
   - Brake line to the caliper with 2 new metal gaskets. Torque the brake line union bolt to 22 ft. lbs. (30 Nm).
   - Front wheels
3. Bleed the system, as outlined in this section.

DISC BRAKE PADS

REMOVAL & INSTALLATION

See Figure 11.

1. Remove or disconnect the following:
   - Front wheels
   - Slide pins and boots
   - Brake pads from the caliper mounting

To install:

2. Install or connect the following:
   - Brake pads to the caliper mounting
   - Pin boots, sleeve boots and brake pads. Tighten the slide pins to 25 ft. lbs. (34 Nm).
   - Wheels

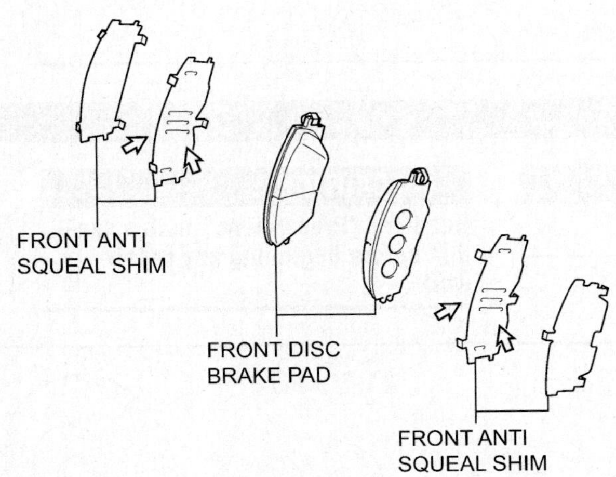

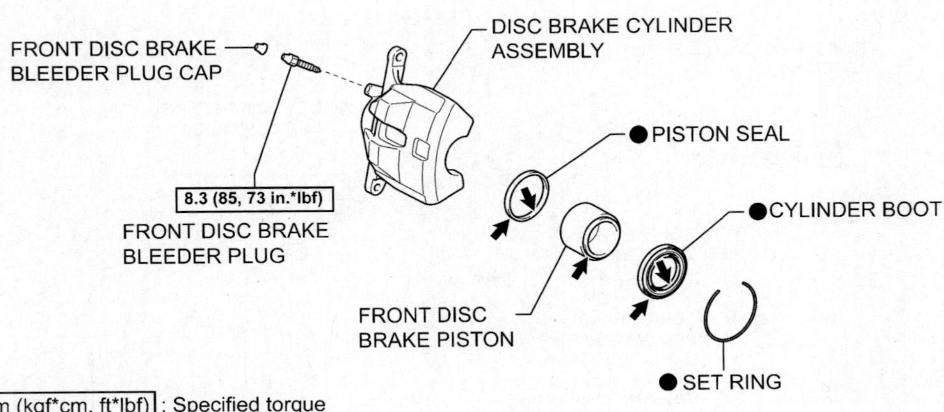

Fig. 11 Exploded view of the front brake pad components

BRAKE CALIPER

REMOVAL & INSTALLATION

See Figure 12.

⁂ CAUTION

Read the "Precautions" in this section before beginning any repair work.

1. Remove or disconnect the following:
   - Wheel
   - Brake hose
   - Caliper assembly mounting bolts
   - Caliper

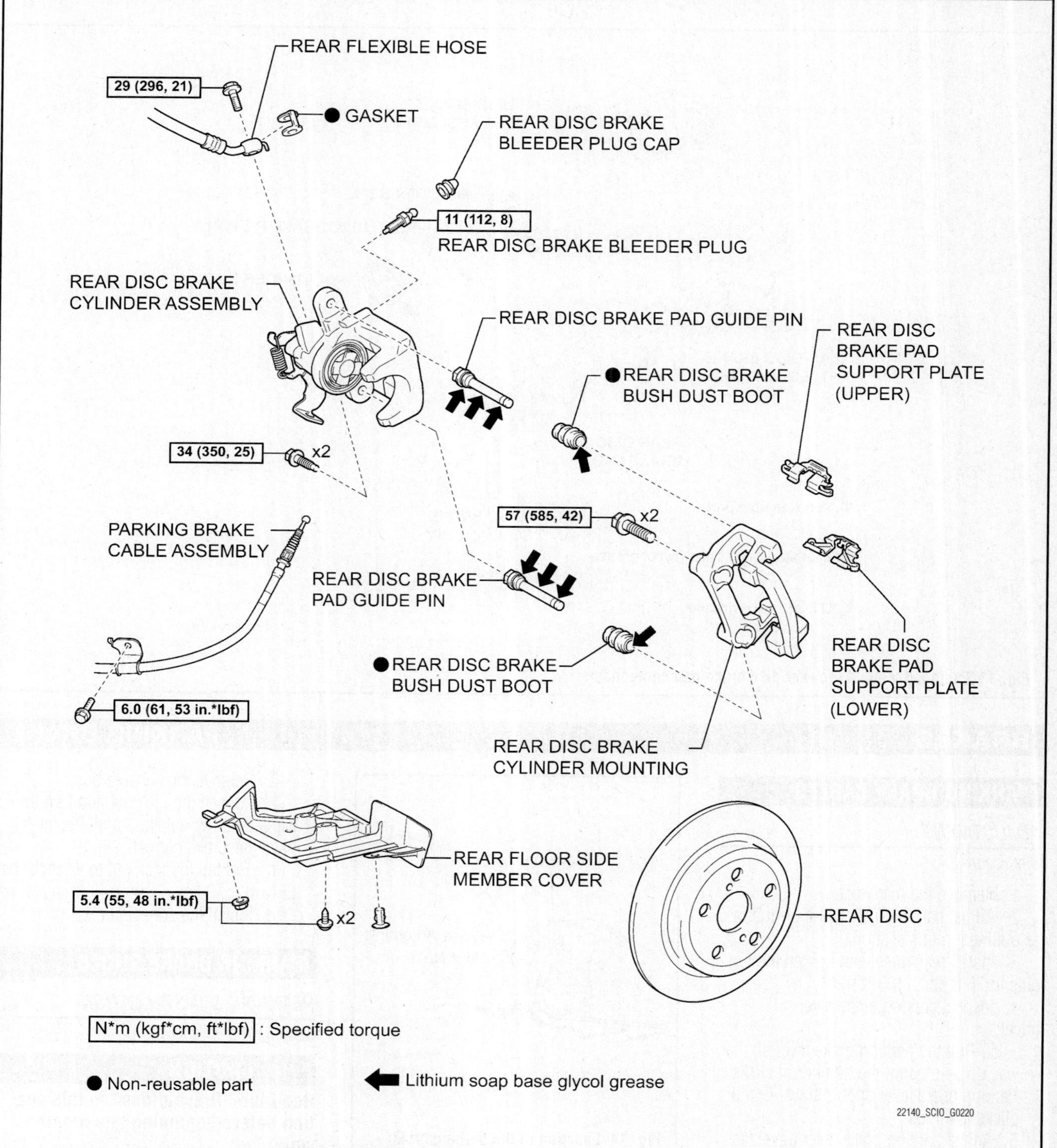

Fig. 12 Exploded view of the rear disc brake components

To install:

2. Install or connect the following:
   - Caliper. Tighten the mounting bolts to 29 ft. lbs. (39 Nm).
   - Brake hose
   - Wheel
3. Bleed the system, as outlined in this section.

DISC BRAKE PADS

REMOVAL & INSTALLATION

See Figure 13.

1. Remove or disconnect the following:
   - Front wheels
   - Slide pins and boots
   - Brake pads from the caliper mounting

To install:

2. Install or connect the following:
   - Brake pads to the caliper mounting
   - Pin boots, sleeve boots and brake pads. Tighten the slide pins to 29 ft. lbs. (39 Nm).
   - Wheels

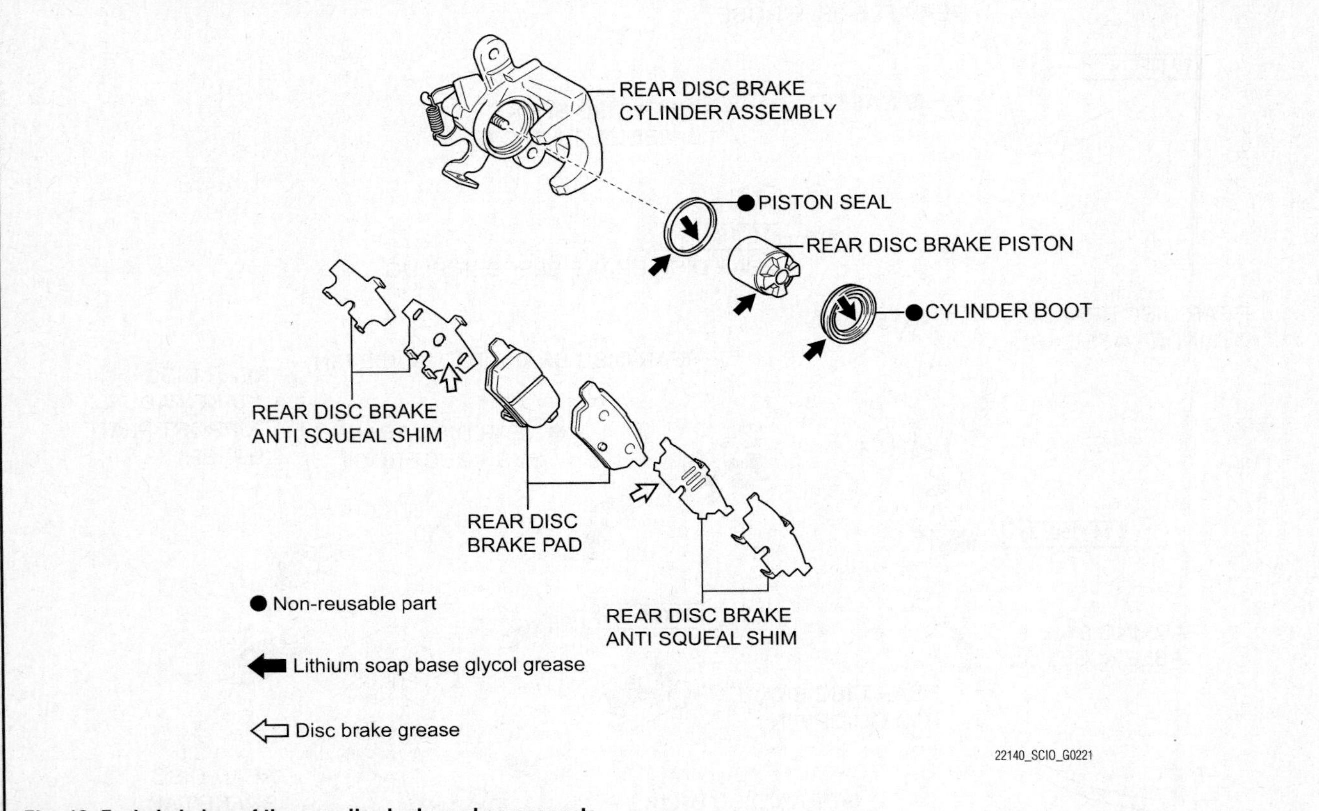

Fig. 13 Exploded view of the rear disc brake pad components

BRAKES

PARKING BRAKE

PARKING BRAKE CABLES

ADJUSTMENT

See Figure 14.

1. Remove the rear wheel.
2. Adjust parking brake shoe clearance, as outlined in this section.
3. Install rear wheel and tighten the lug nuts to 76 ft. lbs. (103 Nm).
4. Inspect parking brake lever travel:
   a. Pull the parking brake lever all the way up, and count the number of clicks. Parking brake lever should travel 6 to 9 clicks at 44 lbs.
5. Adjust parking brake lever travel, as follows:

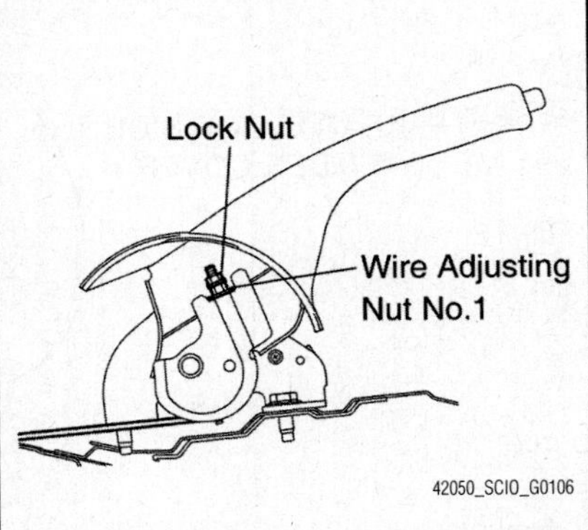

Fig. 14 Location of the parking brake lever adjusting nut

   a. Remove the console box.
   b. Loosen the lock nut and turn the wire adjusting nut No.1 until the lever travel becomes correct.
   c. Tighten the lock nut to 44 inch lbs. (5 Nm).
   d. Install the console box.

PARKING BRAKE SHOES

REMOVAL & INSTALLATION

See Figures 15 and 16.

⁂ CAUTION

Read the "Precautions" in this section before beginning any repair work.

1. Make sure the parking brake is fully released.
2. Raise and safely support the vehicle.
3. Remove the rear wheel.
4. Remove the brake caliper and rotor.

➡Matchmark the disc (rotor) and the axle hub. If the rotor cannot be removed easily, turn the shoe adjuster until the wheel turns freely.

5. Remove parking brake shoe strut, as follows:
   a. Using needle-nose pliers, remove the 2 parking brake shoe return tension springs on the upper side of the shoe.
   b. Remove the parking brake shoe strut.
6. Remove parking brake shoe adjusting screw, as follows:
   a. Using needle-nose pliers, remove the anchor side parking brake shoe return tension spring.
   b. Remove the parking brake shoe adjusting screw set.
7. Remove parking brake shoe no.2 as follows:
   a. Using SST 09718-00010, or equivalent, remove the parking brake shoe hold-down compression spring, parking brake shoe hold- down spring pin and parking brake shoe no.1.
   b. Using the special tool, remove the parking brake shoe hold-down compression spring and parking brake shoe hold-down spring pin.
   c. Disconnect the parking brake shoe LH from the parking brake shoe lever LH, and remove the parking brake shoe LH.
8. Inspect parking brake shoe lining thickness:
   a. Using a ruler, measure the thickness of the shoe lining as follows:
   - Standard thickness: 3.2 mm (0.126 in.)
   - Minimum thickness: 1.0 mm (0.039 in.)
9. Inspect brake disc and parking brake shoe lining for proper contact. Apply chalk to the inside surface of the disc, then grind down the brake shoe lining to fit. If the contact between the brake disc and the shoe lining is improper, repair it using a brake shoe grinder or replace the brake shoe assembly.

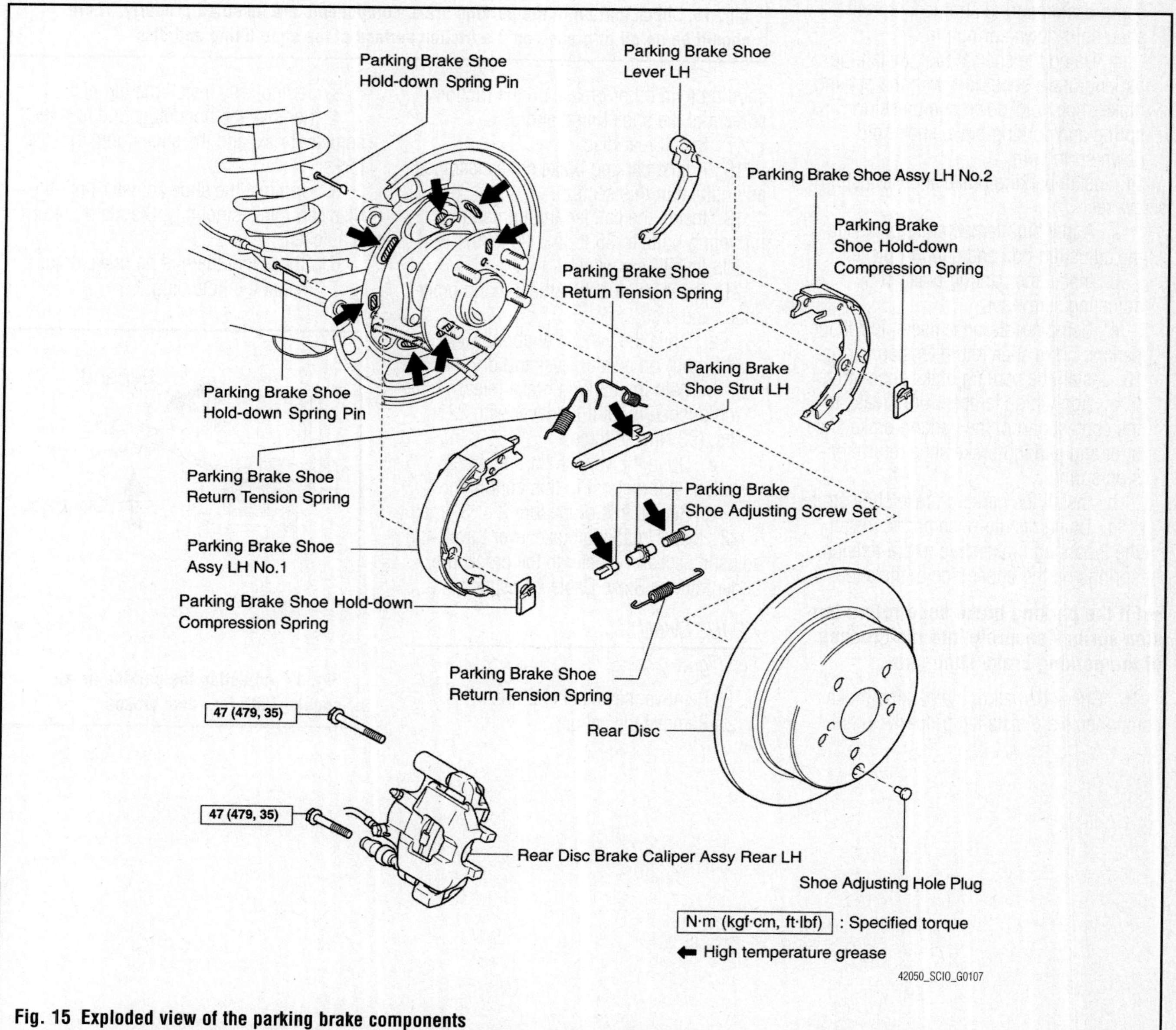

Fig. 15 Exploded view of the parking brake components

10. Remove the parking brake shoe lever:
    a. Using needle-nose pliers, disconnect the parking brake cable no.3 and remove the parking brake shoe lever.

To install:

11. Apply high temperature grease to the shoe attachment surfaces of the backing plate.
12. Using needle-nose pliers, connect the parking brake cable no.3 to the parking brake shoe lever.
13. Install parking brake shoe no.2:
    a. Apply high temperature grease to the contact part of the parking brake shoe and parking brake shoe lever.
    b. Using SST 09718-00010, or equivalent, install the parking brake shoe no.2 with the parking brake shoe hold-down compression spring and parking brake shoe hold-down spring pin.
    c. Using the special tool, install the parking brake shoe no.1 with the parking brake shoe hold-down compression spring and parking brake shoe hold-down spring pin.
14. Install parking brake shoe adjusting screw set:
    a. Apply high temperature grease to the adjusting bolt and support piece.
    b. Install the parking brake shoe adjusting screw set.
    c. Using needle-nose pliers, install the parking brake shoe return tension spring.
15. Install the parking brake shoe strut:
    a. Apply high temperature grease to the contact part of the parking brake shoe strut and parking brake shoe return tension spring.
    b. Install the parking brake shoe strut.
    c. Using needle-nose pliers, install the 2 parking brake shoe return tension springs on the upper side of the shoe.

➡Fit the parking brake shoe return tension springs securely into the grooves of the parking brake shoe strut.

16. Check that all of the parking brake components are installed properly. There should be no oil or grease on the friction surface of the shoe lining and disc.

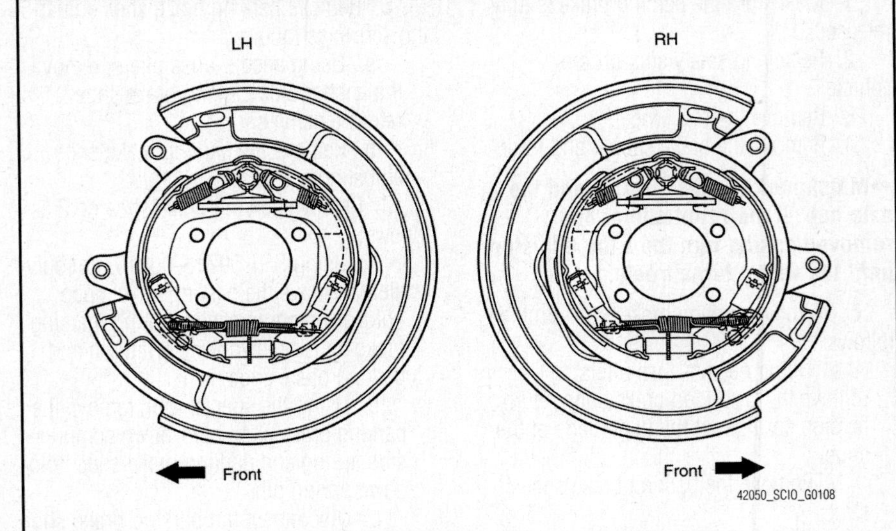

Fig. 16 Check that all of the parking brake components are installed properly. There should be no oil or grease on the friction surface of the shoe lining and disc

17. Install rear disc
18. Adjust parking brake shoe clearance, as outlined in this section.
19. Install the caliper and tighten the mounting bolts to 35 ft. lbs. (47 Nm).
20. Install rear wheel
21. Break in, or settle, the parking brake shoes and disc:
    a. Drive the vehicle at about 50 km/h (31 mph) on a safe, level and dry road.
    b. With the parking brake release button pushed in, pull the lever with 22 ft. lbs. (98 Nm) of force.
    c. Drive the vehicle for about 0.25 miles (400 meters) in this condition.
    d. Repeat this procedure 2 or 3 times.
22. Inspect parking brake lever travel, and adjust if necessary. Refer to the procedure under Parking Brake Cable Adjustment.

ADJUSTMENT

See Figure 17.

1. Remove the wheel and tire.
2. Remove the caliper.
3. Temporarily install the lug nuts.
4. Remove the hole plug, and turn the adjuster to expand the shoes until the disc locks.
5. Contract the shoe adjuster until the disc can rotate smoothly. Standard : return 8 notches
6. Check that there is no brake drag.
7. Install the hole plug.

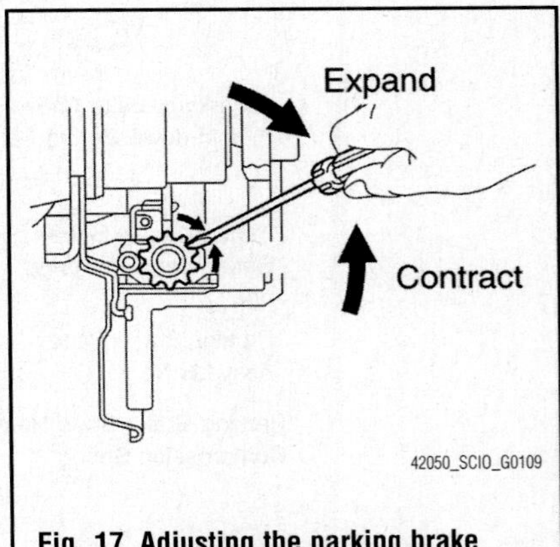

Fig. 17 Adjusting the parking brake shoes—with rear disc brakes

CHASSIS ELECTRICAL

AIR BAG (SUPPLEMENTAL RESTRAINT SYSTEM)

GENERAL INFORMATION

⁂ CAUTION

These vehicles are equipped with an air bag system. The system must be disarmed before performing service on, or around, system components, the steering column, instrument panel components, wiring and sensors. Failure to follow the safety precautions and the disarming procedure could result in accidental air bag deployment, possible injury and unnecessary system repairs.

SERVICE PRECAUTIONS

⁂ CAUTION

Disconnect and isolate the battery negative cable before beginning any airbag system component diagnosis, testing, removal, or installation procedures. Wait at least 90 seconds after the ignition switch is turned off and the negative (-) terminal cable is disconnected from the battery before starting the operation. The SRS is equipped with a backup power source, so if work is started within 90 seconds after disconnecting the negative (-) terminal cable from the battery, the SRS may be deployed. Failure to disable the airbag system may result in accidental airbag deployment, personal injury, or death.

DISARMING THE SYSTEM

Disconnect and isolate the negative battery cable. Wait 90 seconds for the system capacitor to discharge before performing any service.

ARMING THE SYSTEM

Reconnect the negative battery cable.

DRIVE TRAIN

CLUTCH

ADJUSTMENTS

Clutch Release Point

See Figure 18.

1. Pull the parking brake lever and install wheel chocks.
2. Start the engine and run it at idle.
3. Without depressing the clutch pedal, slowly move the shift lever into reverse until the gears contact.
4. Gradually depress the clutch pedal and measure the stroke distance from the point that the gear noise stops (release point) up to the full stroke end position. Standard distance is 0.98 in. (25 mm) or more.

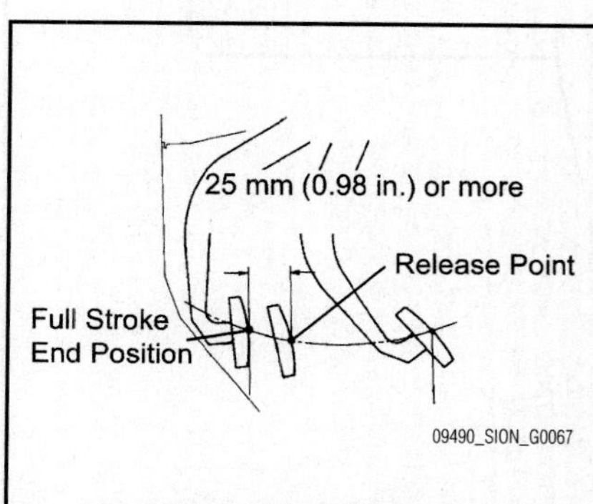

Fig. 18 Measuring the clutch release point.

Pedal Free Play

See Figure 19.

1. Depress the pedal until clutch resistance begins to be felt. Standard pedal free play is 0.197–0.591 in. (5–15 mm).

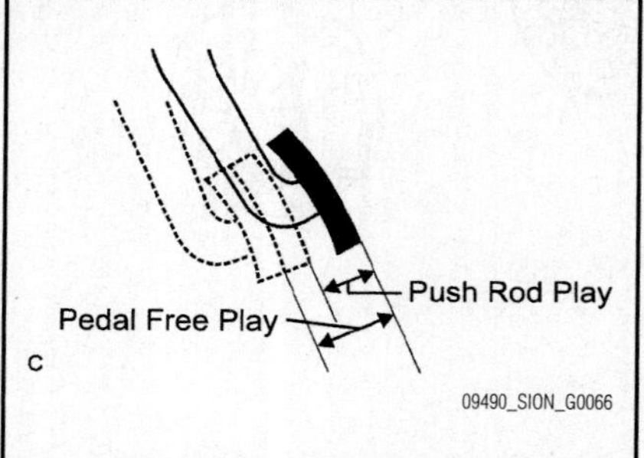

Fig. 19 Measuring pedal and push rod free play.

2. Gently depress the pedal until the resistance begins to increase a little. Standard push rod play at pedal top is 0.039–0.197 in. (1–5 mm).
3. Loosen the lock nut and turn the push rod until correct free play and push rod play are obtained. Tighten the lock nut to 9 ft. lbs. (12 Nm).

➡After adjusting the pedal free play, check the pedal height.

Pedal Height

See Figure 20.

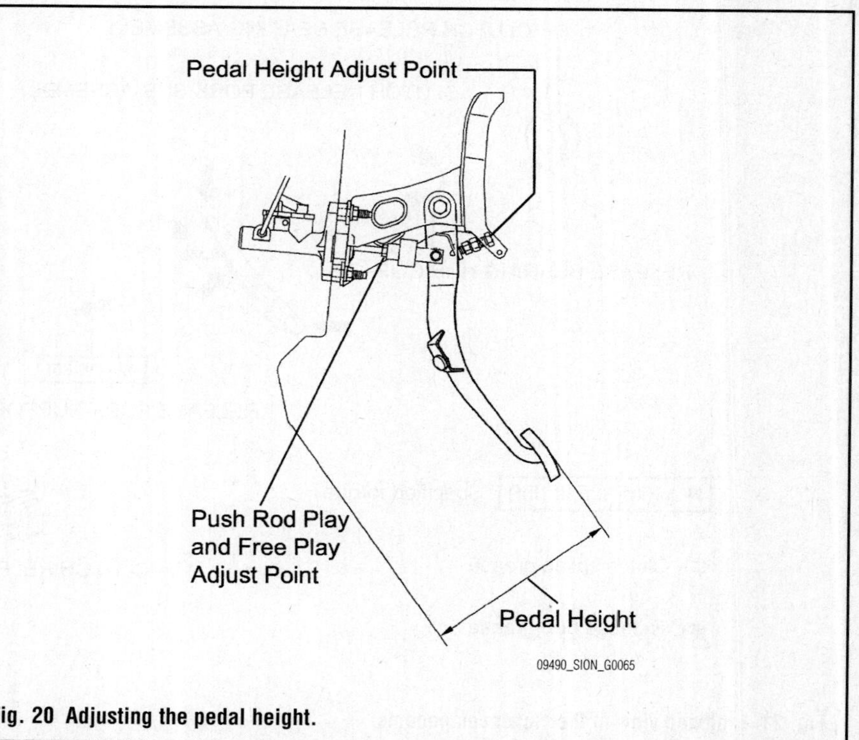

Fig. 20 Adjusting the pedal height.

1. Loosen the lock nut and turn the stopper bolt until the pedal height is 5.287–5.681 in. (134.3–144.3 mm).

2. Tighten the lock nut to 13 ft. lbs. (18 Nm).

REMOVAL & INSTALLATION

tC & xB Models

See Figures 21 and 22.

1. Remove or disconnect the following:
   - Transaxle
   - Release fork and boot
   - Release bearing assembly
   - Release fork support
2. Matchmark the clutch cover and flywheel. Loosen the clutch cover bolts one turn at a time until the spring tension is released.
3. Remove the clutch cover.
4. Remove the clutch disc.

To install:

5. Install the clutch disc on the flywheel.
6. Align the matchmarks on the clutch cover and flywheel.
7. Evenly tighten the mounting bolts in several steps to 14 ft. lbs. (19 Nm) by following the order shown.
8. Install or connect the following:
   - Release fork
   - Release bearing hub clip

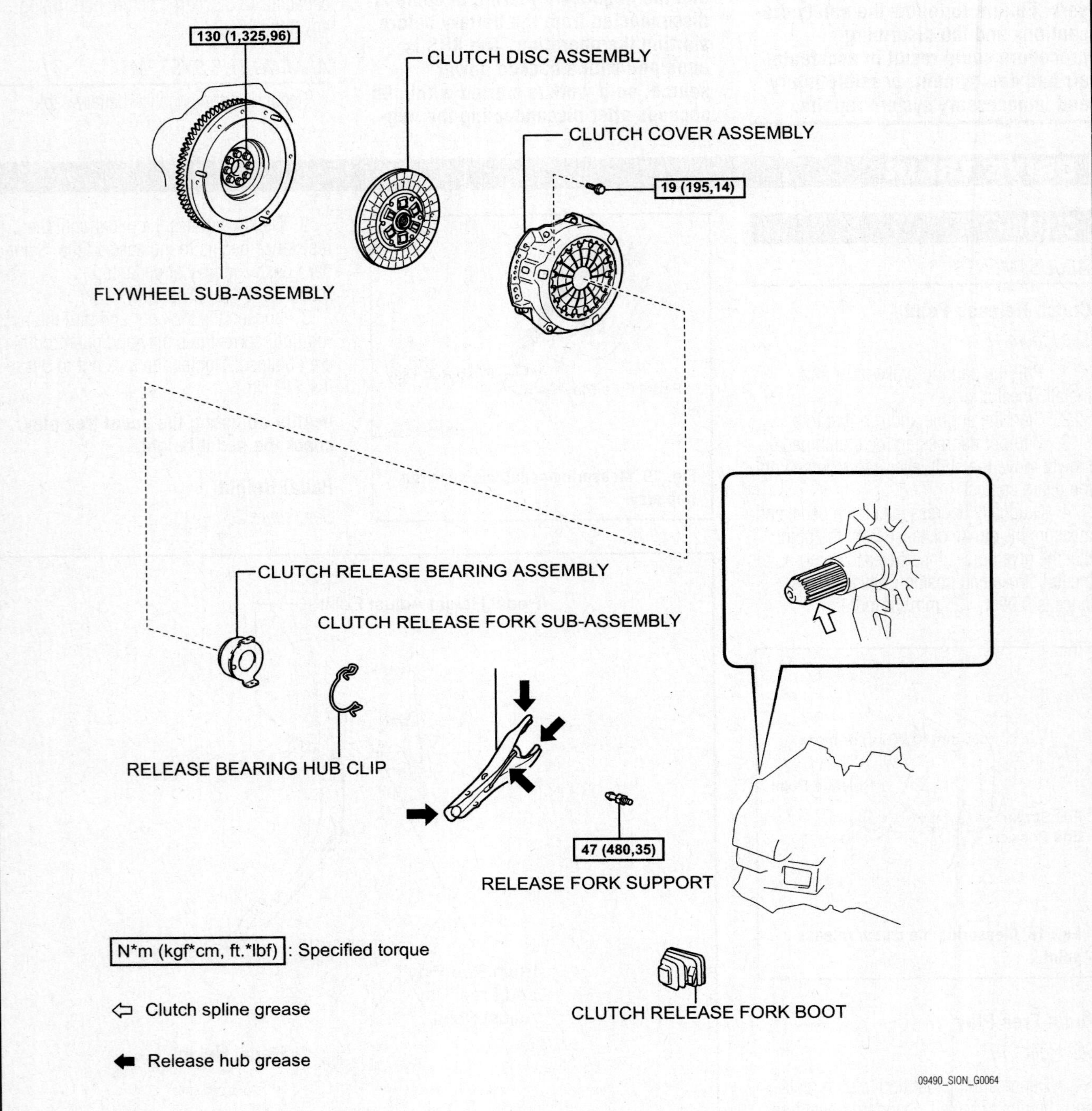

Fig. 21 Exploded view of the clutch components.

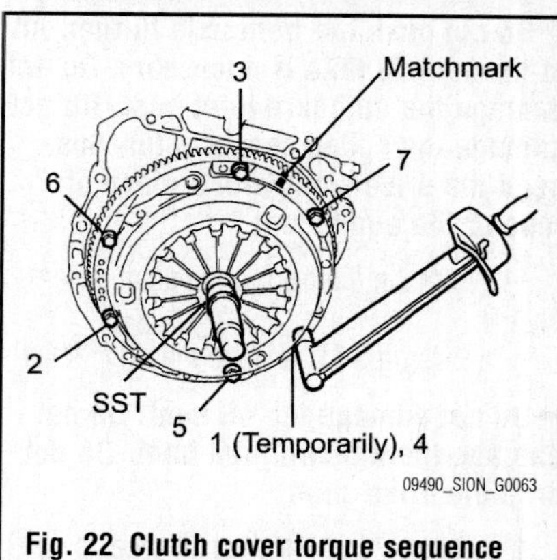

Fig. 22 Clutch cover torque sequence

- Release bearing assembly
- Release fork boot
- Transaxle

xD Models

See Figure 23.

1. Remove the manual transaxle assembly.
2. Remove the clutch release fork sub-assembly.
   a. Remove the clutch release fork with the clutch release bearing from the manual transaxle.
3. Remove the clutch release fork boot.
4. Remove the clutch release bearing assembly.
   a. Remove the clutch release bearing from the clutch release fork.
5. Remove the release bearing hub clip.
6. Remove the release fork support.
   a. Remove the release fork support from the manual transaxle.
7. Remove the clutch cover assembly.
   a. Put the matchmarks on the clutch cover assembly and the flywheel.
   b. Loosen each set bolt one turn at a time until the spring tension is released.
   c. Remove the set bolts and pull off the clutch cover.

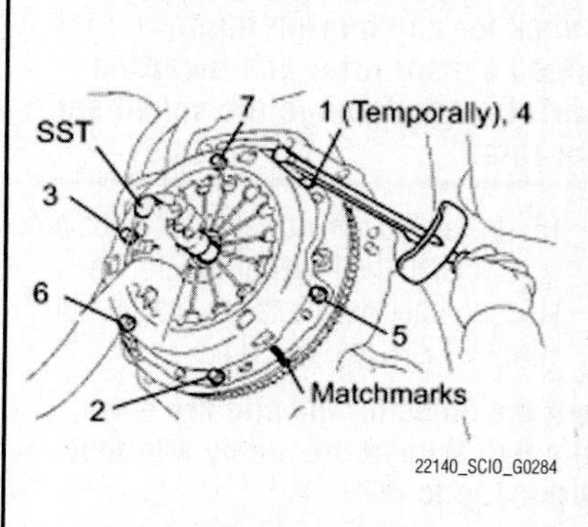

Fig. 23 Tighten the 6 bolts in order

⁂ CAUTION

Do not drop the clutch disc.

8. Remove the clutch disc assembly.

To install:

9. Install the clutch disc assembly.
   a. Insert SST into the clutch disc assembly, then insert them both into the flywheel sub-assembly.

⁂ CAUTION

Insert clutch disc assembly in the correct direction.

10. Install the clutch cover assembly.
    a. Align the matchmark on the clutch cover assembly with that on the flywheel sub-assembly.
    b. Following the procedures shown in the illustration, tighten the 6 bolts in order, starting with the bolt located near the knock pin at the top. Torque: 14 ft. lbs. (19 Nm).

➡Following the order in the illustration, tighten the bolts evenly one at a time. Move SST up and down, right and left lightly after checking that the disc is in the center, and tighten the bolts.

11. Inspect and adjust the clutch cover assembly.
    a. Using a dial indicator with a roller instrument, check the diaphragm spring tip alignment.

➡Maximum non-alignment: 0.5 mm (0.020 in.)

    b. If the alignment is not as specified, using SST, adjust the diaphragm spring tip alignment.
12. Install the release fork support.
    a. Install the release fork support onto the transaxle assembly. Torque: 27 ft. lbs. (37 Nm).
13. Install the clutch release fork boot.
14. Install the release bearing hub clip.
15. Install the clutch release fork sub-assembly.
    a. Apply release hub grease to the contact surfaces of the release fork and release bearing assembly, release fork and push rod, and release fork and fork support.
    b. Install the release fork onto the release bearing assembly.
16. Install the clutch release bearing assembly.
    a. Apply clutch spline grease to the input shaft spline.
    b. Install the clutch release bearing with release fork onto the transaxle assembly.

➡After the installation, move the fork forward and backward to check that the release bearing slides smoothly.

17. Install the manual transaxle assembly.

HALFSHAFTS

REMOVAL & INSTALLATION

tC & xB Models

See Figures 24 and 25.

1. Before servicing the vehicle, refer to the Precautions Section.
2. Drain the transaxle fluid.
3. Remove or disconnect the following:
   - Front wheel
   - Engine undercover
   - Hub nut
   - Wheel speed sensor
   - Front stabilizer bar
   - Lower control arm
   - Tie rod end
4. Using Special Tool 09520-01010, tap out the left halfshaft.

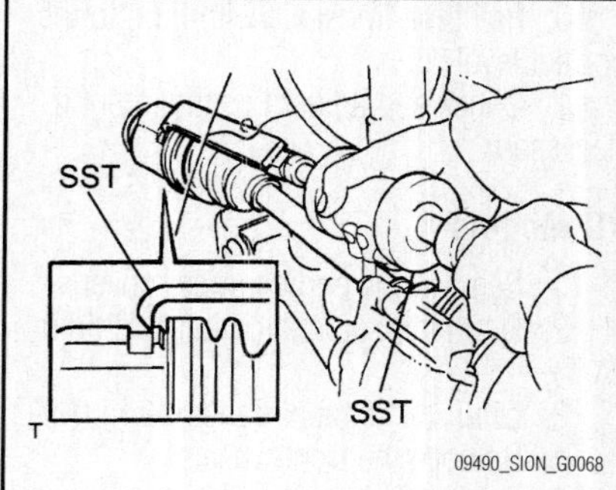

Fig. 24 Use Special Tool 09520-01010 to remove the left halfshaft—xB

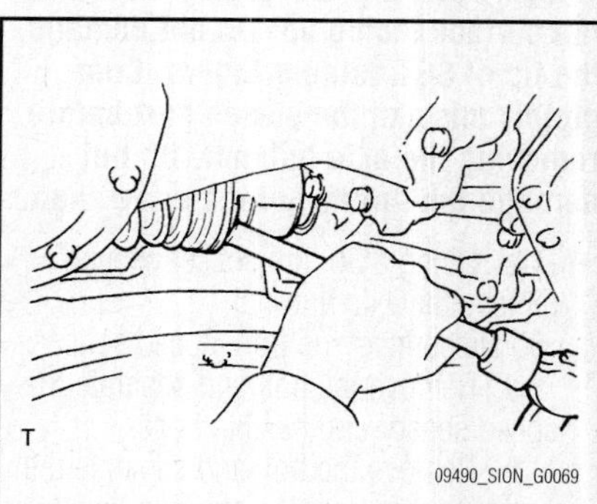

Fig. 25 Tap out the right halfshaft with a brass bar and hammer

5. Remove the right halfshaft as follows:

a. xB: Using a brass bar and hammer, tap out the right halfshaft.

b. tC: Remove the two mounting bolts and remove the halfshaft from the transaxle.

To install:

6. Coat the splines of the inboard joint shaft with gear oil (M/T) or ATF (A/T).

7. Align the shaft splines and tap in the left halfshaft with a brass bar and hammer.

8. Install the right halfshaft as follows:

a. xB: Align the shaft splines and tap in the halfshaft with a brass bar and hammer.

b. tC: Align the shaft splines and install the halfshaft to the transaxle. Tighten bolts to 47 ft. lbs. (64 Nm).

9. Install or connect the following:

- Tie rod end. Tighten nut to 36 ft. lbs. (49 Nm).
- Lower control arm
- Front stabilizer arm. xB: Tighten nut to 13 ft. lbs. (18 Nm). tC: Tighten nut to 55 ft. lbs. (74 Nm).
- Wheel speed sensor
- New hub nut. Tighten to 159 ft. lbs. (216 Nm).
- Engine undercover
- Front wheel

10. Refill the transaxle with fluid to the correct level.

11. Check and adjust the alignment if necessary.

xD Models

1. Remove the engine under cover.

2. Drain the automatic transaxle fluid (A/T).

3. Drain the manual transaxle oil (M/T).

4. Remove the front wheels.

5. Remove the front axle hub nut.

a. Using SST and a hammer, release the staked part of the axle hub nut.

➡Insert SST into the groove with the flat surface facing up. Do not damage the tip of SST using grinders. Completely unstake the staked part before removing the axle hub nut. Do not damage the threads of the drive shaft.

b. Using a 30 mm socket wrench, remove the axle hub nut.

6. Disconnect the front speed sensor.

a. Remove the bolt and separate the speed sensor and flexible hose.

b. Remove the bolt and separate the speed sensor from the steering knuckle.

➡Keep the speed sensor tip and installation portion free of foreign matter. Remove the speed sensor without turning it from its original installation angle.

7. Separate the front stabilizer link assembly.

a. Remove the nut and separate the stabilizer link from the shock absorber.

8. Separate the tie rod end sub-assembly.

a. Remove the cotter pin and castle nut.

b. Install SST to the threaded section of the tie rod end.

➡Make sure the upper ends of the threaded section of the tie rod end and SST (spacer B) are aligned.

c. Using SST, separate the tie rod end from the front axle assembly.

➡Make sure to tie the string of SST to the vehicle to prevent SST from dropping. Install SST so that A and B are parallel. Be sure to place the wrench on the part indicated in the illustration. Do not damage the ball joint dust cover. Do not damage the front disc brake dust cover.

9. Separate the front lower suspension arm.

a. Remove the clip and castle nut.

b. Install SST (spacer B) to the threaded section of the lower ball joint.

➡Make sure the upper ends of the threaded section of the lower ball joint and SST are aligned.

c. Using SST, separate the front lower suspension arm from the front axle assembly.

➡Make sure to tie the string of SST to the vehicle to prevent SST from dropping. Install SST so that A and B are parallel. Be sure to place the wrench on the part indicated in the illustration. Do not damage the lower ball joint dust cover. Do not damage the drive shaft outboard joint boots. Do not damage the front disc brake dust cover.

10. Separate the front axle assembly.

a. Using a plastic hammer, tap the end of the drive shaft and disengage the fitting between the drive shaft and front axle.

➡If it is difficult to disengage the fitting, tap the end of the drive shaft with a brass bar and hammer.

b. Push the front axle out of the vehicle to remove the drive shaft from the front axle.

➡Do not push the front axle further out of the vehicle than is necessary. Do not damage the outboard joint boot. Do not damage the speed sensor rotor. Suspend the drive shaft with a piece of rope or the equivalent.

11. Remove the front drive shaft assembly LH.

a. Using SST, remove the drive shaft.

➡Do not damage the oil seal. Do not damage the inboard joint boot. Do not drop the drive shaft.

12. Remove front drive shaft assembly RH.

a. Using a screwdriver and hammer, remove the drive shaft.

To install:

13. Install the front drive shaft assembly LH.

a. Coat the spline of the inboard joint with gear oil.

b. Align the inboard joint splines and install the drive shaft with a screwdriver and hammer.

⁂ CAUTION

Face the cut area of the front drive shaft hole snap ring downward. Do not damage the oil seal. Do not damage the inboard joint boot. Confirm whether the drive shaft is securely driven in by checking the reaction force and sound.

14. Install the front drive shaft assembly RH.

a. Use the same procedure as for the LH side.

15. Install the front axle assembly.

a. Push the front axle out of the vehicle to align the spline of the drive shaft with the front axle and insert the front axle.

⁂ CAUTION

Do not push the front axle further out of the vehicle than is necessary. Do not damage the outboard joint boot. Check for any foreign matter on the speed sensor rotor and insertion part. Do not damage the speed sensor rotor.

16. Install the front lower suspension arm.

a. Install the lower arm onto the steering knuckle with a new castle nut. Torque: 72 ft. lbs. (98 Nm).

➡If the holes for the clip are not aligned, tighten the nut by a further turn of up to 60°.

b. Install a new clip.

17. Install the tie rod end sub-assembly.
a. Install the tie rod end onto the steering knuckle with a new castle nut. Torque: 36 ft. lbs. (49 Nm).

➡**If the holes for the clip are not aligned, tighten the nut by a further turn of up to 60°.**

b. Install a new cotter pin.
18. Install the front stabilizer link assembly.
a. Install the stabilizer link with the nut. Torque: 55 ft. lbs. (74 Nm).
19. Install the front speed sensor.
a. Install the speed sensor onto the steering knuckle with the bolt.

➡**Check that the speed sensor tip and installation portion are free of foreign matter. Install the speed sensor without turning it from its original installation angle.**

b. Install the flexible hose and speed sensor with the bolt. Torque: 22 ft. lbs. (29 Nm).
c. Install the flexible hose and speed sensor without twisting them.
20. Install the front axle hub nut.
a. Clean the threaded parts on the drive shaft and axle hub nut using a non-residue solvent.

➡**Be sure to perform this work for a new drive shaft. Keep the threaded parts free of oil and foreign objects.**

b. Using a 30 mm socket wrench, install a new axle hub nut. Torque: 160 ft. lbs. (216 Nm).
c. Using a chisel and hammer, stake the axle hub nut.
21. Install the front wheels.
22. Add automatic transaxle fluid (A/T).
23. Inspect A/T fluid leak (A/T).
24. Add manual transaxle oil (M/T).
25. Inspect M/T oil leak (M/T).
26. Inspect and adjust front wheel alignment.
27. Install the engine under cover.

ENGINE COOLING

ENGINE FAN

REMOVAL & INSTALLATION

➡**For engine fan removal and installation, see "Radiator" in this section.**

RADIATOR

REMOVAL & INSTALLATION

tC Models

See Figures 26 and 27.

1. Disconnect the negative battery cable.
2. Drain the engine coolant into a suitable container.
3. Remove the right side engine under cover.
4. Disconnect the radiator inlet and outlet hoses.
5. If equipped with an automatic transaxle, disconnect the oil cooler inlet and outlet tubes.
6. Remove radiator , as follows:
a. Detach the 2 cooling fan motor connectors.
b. Disconnect the 2 clamps for the engine room main wire.
c. Disconnect the radiator reserve tank hose.
d. Remove the 2 bolts and the 2 radiator upper supports, then remove the radiator from the vehicle.
7. Remove the radiator lower support.
8. Remove fan shroud, as follows:
a. Remove the 2 bolts.
b. Remove the 2 pawl fittings, then remove the fan shroud together with motor.

To install:

9. Install fan shroud and tighten the retainers to 44 inch lbs. (5 Nm).
10. Install the radiator lower support.
11. Install the radiator , as follows:
a. Install the radiator with the 2 bolts and 2 radiator upper supports. Tighten to 14 ft. lbs. (19 Nm).
b. Connect the radiator reserve tank hose.
c. Connect the 2 clamps of the engine room main wire.
d. Connect the 2 cooling fan motor connectors.
12. Connect oil cooler outlet and inlet tubes, if equipped with an automatic transaxle.
13. Connect the radiator hoses
14. Refill the engine cooling system to the correct level.
15. Connect the negative battery cable.
16. Start the engine and check for leaks.
17. Install the engine under cover.

xB Models

See Figures 28 through 32.

1. Disconnect the negative battery cable.
2. Drain the engine coolant into a suitable container.
3. Disconnect the inlet and outlet radiator hoses.
4. If equipped with an automatic transaxle, disconnect the oil cooler inlet and outlet hoses.
5. Remove the radiator grille.
6. Remove the front bumper cover.

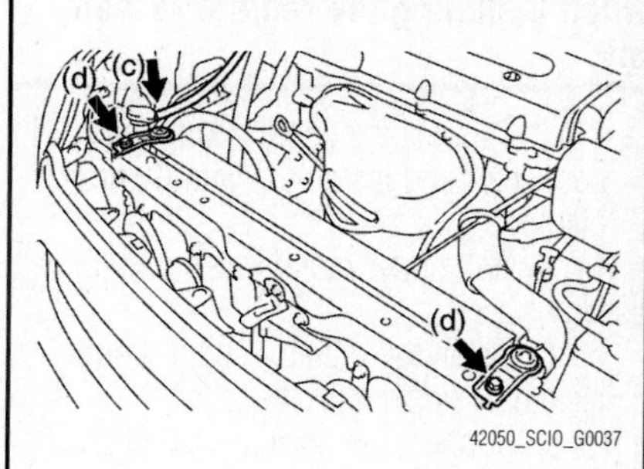

Fig. 26 Detach the reserve tank hose (c), then remove the 2 bolts and upper supports (d)

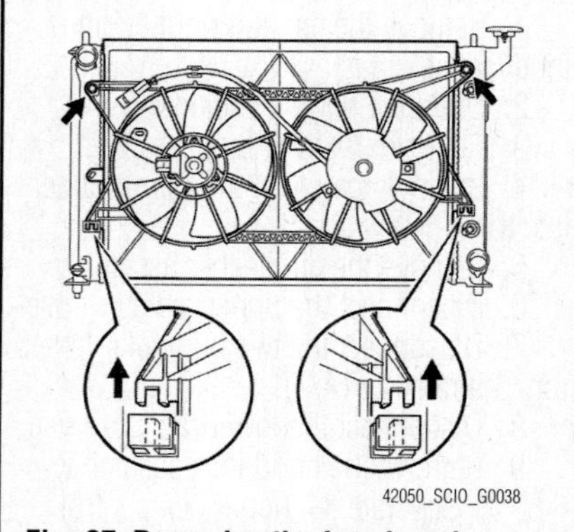

Fig. 27 Removing the fan shroud

Fig. 28 Remove the hood lock assembly—xB

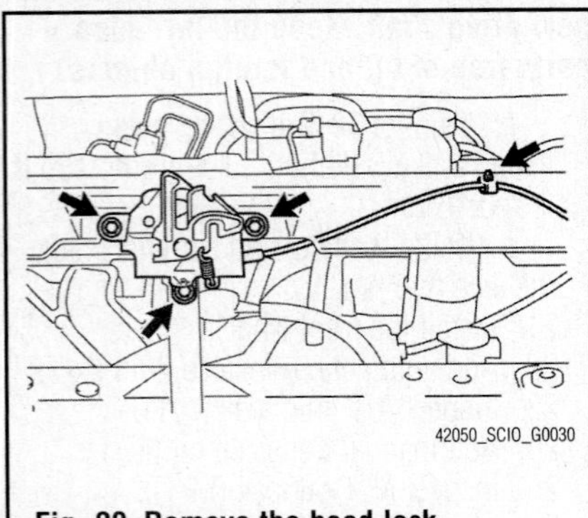

Fig. 29 Remove the hood lock assembly—xB

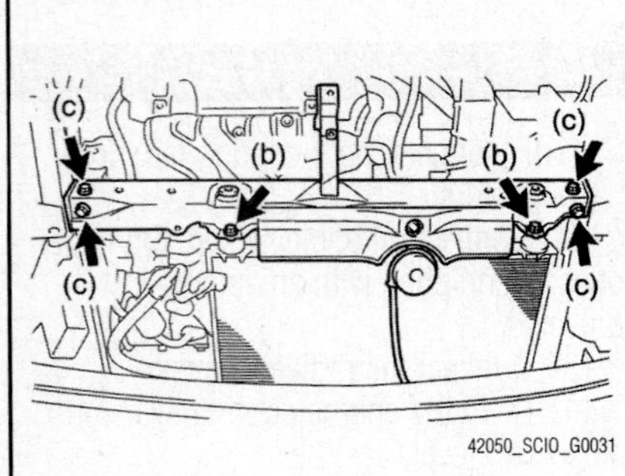

Fig. 30 Removing the upper radiator support—xB

7. Remove the 3 bolts, clip and hood lock assembly.
8. Remove the 5 bolts, and then remove the hood lock support brace.
9. Remove the 2 clips and radiator support opening cover.
10. Remove the upper radiator support, as follows:
   a. Disconnect the connector from the horn .
   b. Remove the 2 bolts and separate the condenser from the upper radiator support.
   c. Remove the 4 bolts and the upper radiator support.
11. Remove the radiator , as follows:
   a. Disconnect the connector from the cooling fan motor, and then remove the radiator from the engine compartment.
   b. Remove the 3 bolts and detach the fan shroud with fan.
   c. Remove the 2 bolts and water filler.
   d. Remove the 2 radiator support cushions and radiator support lower.

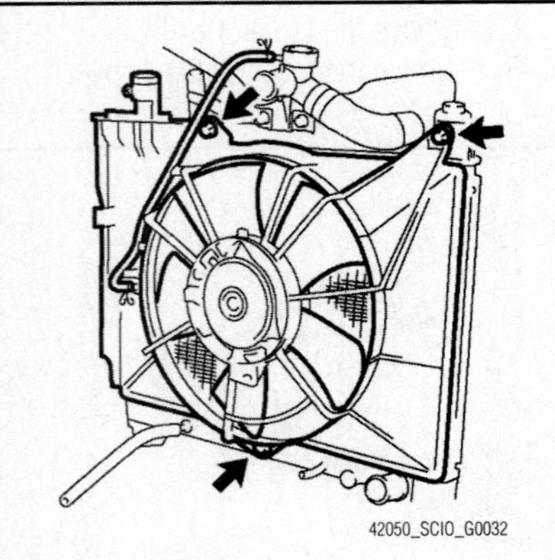

Fig. 31 Disconnect the connector from the cooling fan motor, and then remove the radiator from the engine compartment

To install:

12. Install the radiator , as follows:
   a. Install the 2 radiator support cushions and radiator support lower onto the radiator.
   b. Install the water filler with the 2 bolts and tighten to 66 inch lbs. (7.5 Nm).
   c. Install the fan shroud with fan with the 3 bolts and tighten to 66 inch lbs. (7.5 Nm).
   d. Install the radiator , and attach the connector to the cooling fan motor.
13. Install the upper radiator support:
   a. Install the radiator support upper with the 4 bolts and tighten to 44 inch lbs. (5 Nm).
   b. Install the condenser onto the radiator support upper with the 2 bolts and tighten to 7.2 ft. lbs. (9.8 Nm).
   c. Attach the connector to the horn .
14. Install the hood lock support brace with the 5 bolts and tighten to 44 inch lbs. (5 Nm).
15. Temporarily install the hood lock with the 3 bolts, then the clip.
16. Adjust the hood.
17. Install the front bumper cover.
18. Add the proper type and amount of engine coolant.
19. Start the engine and check for leaks.

xD Models

1. Remove the left and right engine under cover.
2. Drain the engine coolant.
3. Remove the front bumper cover.
4. Remove the radiator support upper absorber.
5. Remove the air cleaner assembly.
6. Disconnect the upper radiator hose.
7. Disconnect the two oil cooler hoses from the radiator (A/T).
8. Disconnect the lower radiator hose.
9. Remove the hood lock assembly.
   a. Separate the hood lock control cable assembly from the 2 clamps.

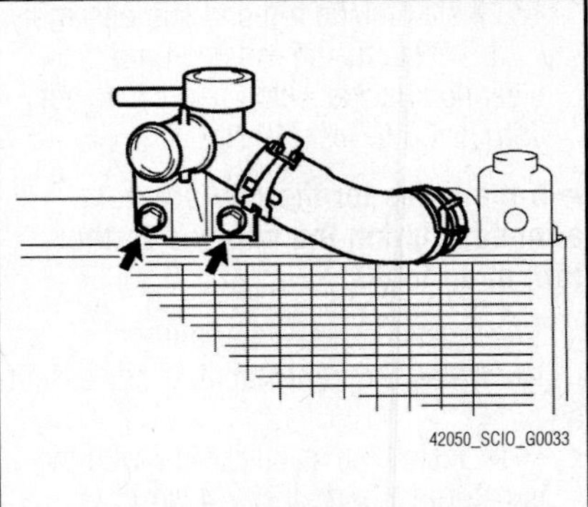

Fig. 32 Unfasten the bolts and remove the water filler

   b. Remove the 3 bolts and remove the hood lock assembly.
10. Remove the No. 1 cooler cover.
   a. Remove the 2 clips and remove the No. 1 cooler cover.
11. Remove the upper radiator support sub-assembly.
   a. Disconnect the horn connector.
   b. Remove the 5 bolts and remove the radiator support sub-assembly upper.
12. Remove the radiator assembly.
   a. Disconnect the cooling fan motor connector and separate the wire harness clamp.
   b. Disengage the 2 claws and remove the radiator assembly from the vehicle.

CAUTION

Do not apply excessive force to the cooler condenser assembly or piping when removing the radiator assembly.

To install:

13. Install radiator assembly.
   a. Engage the 2 claws and install the radiator assembly onto the vehicle.

CAUTION

Do not apply excessive force to the cooler condenser assembly or piping when installing the radiator assembly.

   b. Connect the cooling fan motor connector and install the wire harness clamp.
14. Install upper radiator support sub-assembly.
   a. Install the radiator support sub-assembly upper with the 5 bolts.
   b. Connect the horn assembly connector.
15. Install No. 1 cooler cover.
   a. Install the No. 1 cooler cover with the 2 clips.

16. Install hood lock assembly.
   a. Install the hood lock assembly with the 3 bolts.
   b. Install the hood lock control cable assembly with the 2 clamps.
17. Connect lower radiator hose.
18. Connect the two oil cooling hoses to the radiator (A/T).
19. Connect upper radiator hose.
20. Install air cleaner assembly.
21. Install radiator support upper absorber.
22. Install front bumper cover.
23. Add engine coolant.
24. Adjust hood lock assembly.
   a. Loosen the 3 bolts.
   b. Adjust the hood lock position so that the striker can enter it smoothly.
   c. Tighten the 3 bolts after the adjustment.
25. Inspect for engine coolant leak.

THERMOSTAT

REMOVAL & INSTALLATION

See Figure 33.

1. Drain the engine coolant to a level below the thermostat.
2. Remove the 2 nuts, then remove the water inlet from the cylinder block.
3. Remove the thermostat. Thoroughly clean the gasket mating surfaces.

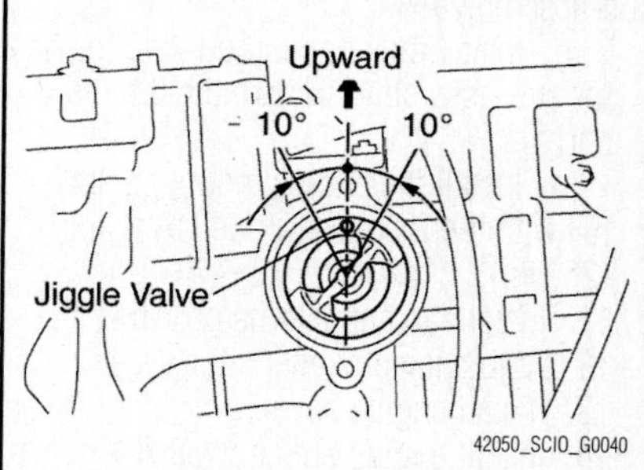

Fig. 33 Proper position of the jiggle valve when installing the thermostat—2.4L engine

To install:

4. Position a new gasket onto the thermostat, then install the thermostat with the jiggle valve facing upward.

➡The jiggle valve may be set within 10° of either side as shown in the accompanying illustration.

5. Install the water inlet and tighten to 80 inch lbs. (9 Nm).
6. Refill the engine cooling system to the correct level.
7. Start the engine and check for leaks.

WATER PUMP

REMOVAL & INSTALLATION

tC & xB Models

See Figure 34.

1. Drain the cooling system.
2. Remove or disconnect the following:
   - Negative battery cable
   - Right-hand front fender apron
   - Engine undercover
   - Accessory drive belt
   - Alternator
3. Using Special Tool 09960-10010, remove the water pump pulley.
4. Remove the water pump assembly.

To install:

5. Install the water pump with a new gasket. Tighten to 80 inch lbs. (9 Nm).
6. Using Special Tool 09960-10010, install the water pump pulley and tighten to 19 ft. lbs. (26 Nm).
7. Install or connect the following:
   - Alternator
   - Accessory drive belt
   - Engine undercover
   - Right-hand front fender apron
   - Negative battery cable
8. Refill the engine cooling system to the correct level.
9. Start the engine and check for leaks.

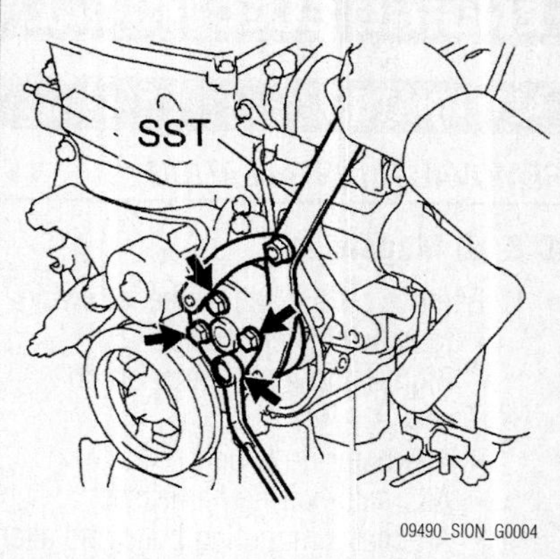

Fig. 34 Use Special Tool 09960-10010 to hold the water pump pulley while removing the mounting bolts.

xD Models

See Figure 34.

1. Drain the cooling system.
2. Remove or disconnect the following:
   - Negative battery cable
   - Right-hand front fender apron
   - Engine undercover
   - Accessory drive belt
   - Alternator
3. Using Special Tool 09960-10010, remove the water pump pulley.
4. Remove the water pump assembly.

To install:

5. Install the water pump with a new gasket. Tighten to 80 inch lbs. (9 Nm).
6. Using Special Tool 09960-10010, install the water pump pulley and tighten to 19 ft. lbs. (26 Nm).
7. Install or connect the following:
   - Alternator
   - Accessory drive belt
   - Engine undercover
   - Right-hand front fender apron
   - Negative battery cable
8. Refill the engine cooling system to the correct level.
9. Start the engine and check for leaks.

ALTERNATOR

REMOVAL & INSTALLATION

tC & xB Models

1. Remove or disconnect the following:
   - Negative battery cable
   - Right-hand front fender apron seal
   - Alternator drive belt
   - Alternator wiring harnesses
   - Alternator mounting bolts and alternator

To install:

➡Confirm the Crankshaft Position (CKP) sensor wiring harnesses is secured in the clamp bracket on the timing chain cover.

2. Install or connect the following:
   - Alternator. Tighten upper bolt to 16 ft. lbs. (21 Nm) and lower bolt to 38 ft. lbs. (52 Nm).
   - Alternator wiring harnesses
   - Alternator drive belt
   - Right-hand front fender apron seal
   - Negative battery cable

xD Models

See Figures 35 and 36.

1. Disconnect the cable from negative battery terminal
2. Remove the engine under cover RH
3. Remove the engine cover.
4. Remove the fan and alternator V-belt.
5. Remove the engine mounting insulator sub-assembly RH
   a. Remove 2 bolts and remove the engine mounting stay RH .
   b. Place a wooden block on a jack underneath the engine.

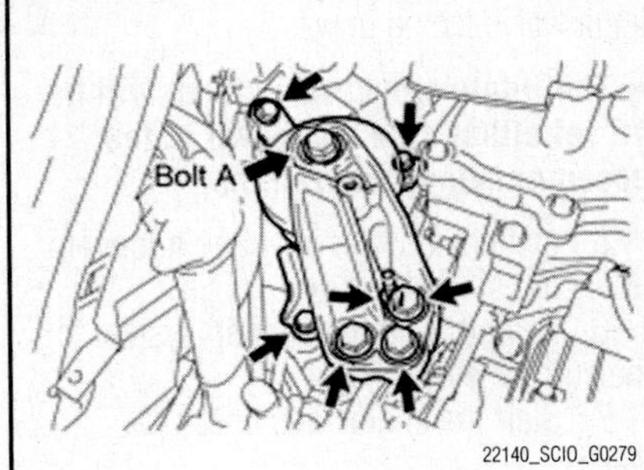

Fig. 35 Remove the 6 bolts and nut and remove the engine mounting insulator RH.

Fig. 36 Tighten the bolt A to the specified torque.

   c. Remove the 6 bolts and nut and remove the engine mounting insulator sub-assembly RH.

➡Do not remove bolt A.

6. Remove the transverse engine mounting bracket.
   a. Remove the 3 bolts and remove the engine mounting bracket.
7. Remove the alternator assembly.
   a. Remove the 2 bolts and remove the fan belt adjusting bar.
   b. Remove the terminal cap.
   c. Remove the nut and remove terminal B.
   d. Disconnect the connector and harness clamp.
   e. Remove the bolt and remove the alternator assembly.
   f. Remove the bolt and remove the wire harness bracket.

To install:

8. Install alternator assembly.
   a. Install the wire harness bracket with the bolt.
   b. Provisionally install the alternator assembly with the bolt.
   c. Provisionally install the fan belt adjusting bar and alternator assembly with the 2 bolts.
   d. Tighten the bolt A to the specified torque. Torque: 14 ft. lbs. (19 Nm).
   e. Connect the connector and wire harness clamp.
   f. Install terminal B with the nut.
   g. Install the terminal cap.
9. Install transverse engine mounting bracket.
   a. Install the engine mounting bracket with the 3 bolts. Torque: 38 ft. lbs. (51 Nm).
10. Install engine mounting insulator sub-assembly RH.
   a. Install the engine mounting insulator sub-assembly RH with the 6 bolts and nut.
   b. Install the engine mounting stay RH with the 2 bolts. Torque: 19 ft. lbs. (26 Nm).
11. Install fan and alternator V-belt.
12. Adjust fan and alternator V-belt.
13. Install engine cover.
14. Install engine under cover RH
15. Connect cable to negative battery terminal.

ENGINE ELECTRICAL

DISTRIBUTORLESS IGNITION SYSTEM

FIRING ORDERS

Firing order: 1–3–4–2

IGNITION COIL

REMOVAL & INSTALLATION

See Figures 37 and 38.

1. Disconnect the negative battery cable.
2. Remove the engine cover

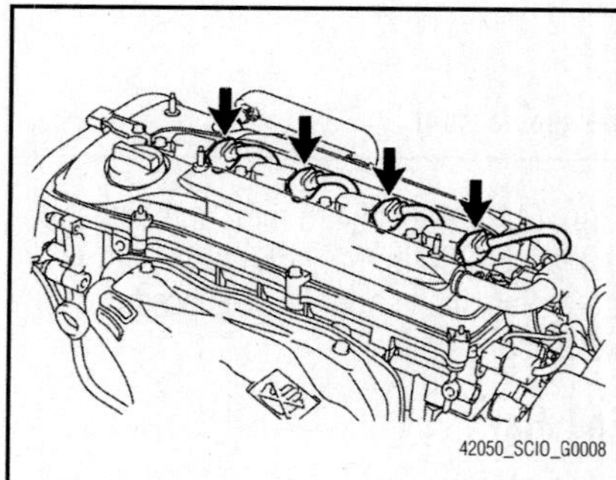

Fig. 37 Detach the ignition coil electrical connectors

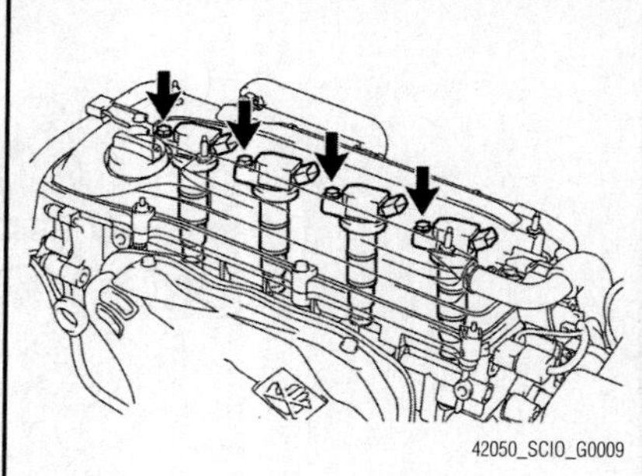

Fig. 38 Each ignition coil is secured with one mounting bolt

3. Detach the 4 ignition coil connectors.
4. Remove the 4 bolts (one bolt for each coil), and remove the 4 ignition coils.

To install:

5. Install the ignition coils and secure with the bolts. Tighten to 80 inch lbs. (9 Nm).
6. Attach the 4 ignition coil connectors.
7. Install the engine cover.
8. Connect the negative battery cable.

IGNITION TIMING

ADJUSTMENT

The ignition timing is controlled by the Powertrain Control Module (PCM). No adjustment is necessary or possible.

SPARK PLUGS

REMOVAL & INSTALLATION

1. Disconnect the negative battery cable.
2. Remove the engine cover.
3. Remove the ignition coils, as outlined in this section.
4. Use a spark plug wrench to remove the spark plugs.
5. Inspect the spark plugs, as outlined in this section.

To install:

6. Use a spark plug wrench to install and tighten the spark plugs to 14 ft. lbs. (19 Nm).
7. Install the ignition coil. Tighten the retainers to 80 inch lbs. (9 Nm). Refer to "Ignition Coil" in this section for more details.
8. Install the engine cover.
9. Connect the negative battery cable.

ENGINE ELECTRICAL

STARTING SYSTEM

STARTER

REMOVAL & INSTALLATION

tC & xB Models

See Figure 39.

1. Remove or disconnect the following:
   - Negative battery cable
   - Starter electrical connections
   - Starter mounting bolts
   - Starter

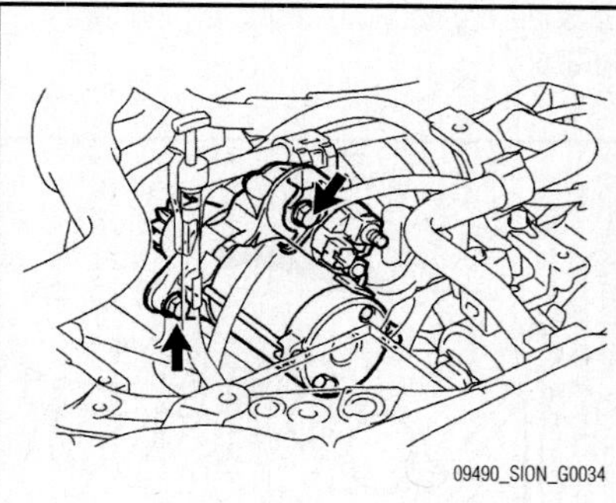

Fig. 39 Location of the starter mounting bolts—2.4L Engine

To install:

2. Install or connect the following:
   - Starter. Tighten the mounting bolts to 27 ft. lbs. (37 Nm).
   - Starter electrical connections. Tighten the starter wire nut to 7 ft. lbs. (10 Nm).
   - Negative battery cable

xD Models

1. Disconnect the cable from negative battery terminal.
2. Remove the engine under covers LH and RH.
3. Remove the flywheel housing side cover.
   a. Disengage the claw by pulling it outward and remove the flywheel housing side cover.
4. Remove the starter assembly.
   a. Remove the terminal cap.
   b. Remove the nut and disconnect terminal 30.
   c. Disconnect the connector.
   d. Remove the 2 bolts and remove the starter assembly.

To install:

5. 1. Install starter assembly.
   a. Install the starter assembly with the 2 bolts. Torque: 27 ft. lbs. (37 Nm).
   b. Connect the connector.
   c. Connect terminal 30 with the nut.
   d. Close the terminal cap.
6. Install flywheel housing side cover.
   a. Insert the protruding portion into the end of the cylinder block and while pushing it along the cylinder block, fit the claw into the cylinder block.

➡Make sure that the claw makes a click sound, indicating that it fits tightly. Replace the claw with a new one if it does not fit tightly or is deformed.

7. Install engine under cover RH.
8. Install engine under cover LH.
9. Connect cable to negative battery terminal.

ENGINE MECHANICAL

ACCESSORY DRIVE BELTS

ACCESSORY BELT ROUTING

tC & xB Models

See Figure 42.

xD Models

See Figure 40.

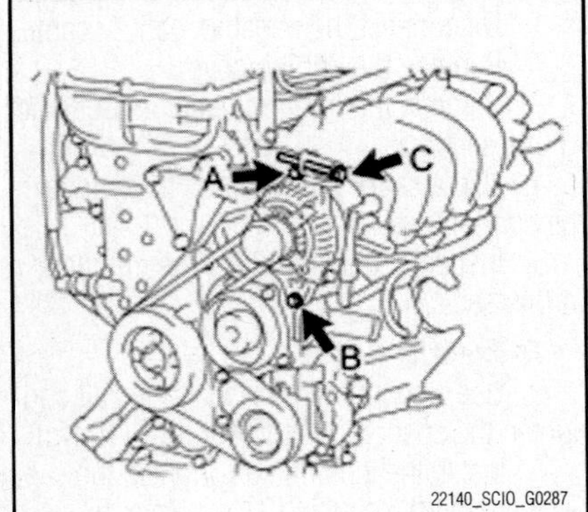

Fig. 40 Accessory belt routing—1.8L Engine

ADJUSTMENT

➡The accessory belt uses an automatic tensioner, so no manual adjustment is possible.

INSPECTION

Inspect the drive belt for signs of glazing or cracking. A glazed belt will be perfectly smooth from slippage, while a good belt will have a slight texture of fabric visible. Cracks will usually start at the inner edge of the belt and run outward. All worn or damaged drive belts should be replaced immediately.

REMOVAL & INSTALLATION

See Figure 41.

1. Remove the front right side fender apron seal.
2. Using SST 09249-63010, or equivalent tool, loosen the V-ribbed belt tensioner arm clockwise, then remove the fan-alternator belt.

➡Be sure to connect SST and the tools so they align using. When retracting the tensioner, turn it clockwise slowly in 3 second or more. Be sure to not apply force rapidly. After the tensioner is retracted all the way, do not apply and more force than necessary.

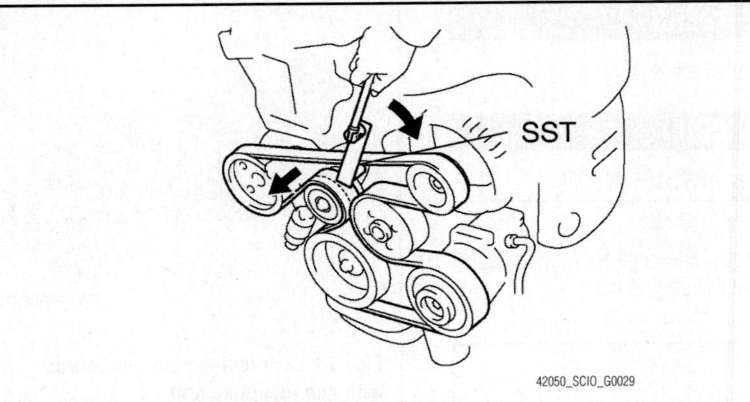

Fig. 41 Removing and installing the belt using the special tool

To install:

3. Using the special tool, loosen the V-ribbed belt tensioner arm clockwise, then install the fan and alternator V-belt.
4. Install the front right side fender apron seal.

CAMSHAFT & VALVE LIFTERS

INSPECTION

Cam Height

TC & XB Models

1. Remove the camshafts.
2. Measure the cam height with a micrometer.
3. The intake camshaft should measure as follows:
    a. 2.4L Engine: between 1.8305–1.8345 inches (46.495–46.595 mm) and not less than 1.8262 inches (46.385 mm).
4. The exhaust camshaft should measure as follows:
    a. 2.4L Engine: between 1.8104–1.8143 inches (45.983–46.083 mm) and not less than 1.8060 inches (45.873 mm).
5. Camshaft should be replaced if it exceeds the limit.

XD Models

1. Remove the camshafts.
2. Measure the cam height with a micrometer.
3. The camshaft should measure as follows:
    a. No. 1 camshaft: 1.6857—1.6896 in. (42.816—42.916 mm) and not less than 1.6798 in. (42.666 mm)
    b. No. 2 camshaft: 1.7455—1.7494 in. (44.336—44.436 mm) and not less than 1.7396 in. (44.186 mm)
4. Camshaft should be replaced if it exceeds the limit.

End Play

TC & XB Models

See Figure 42.

1. Install a dial indicator in the thrust direction on the front end of the camshaft. Measure the end play of the dial indicator when the camshaft is moved back and forth. The dial indicator should measure as follows:
    a. 2.4L Engine: Intake camshaft between 0.0016–0.0037 inches (0.040–0.095 mm) and not exceed 0.0043 inches (0.11 mm). Exhaust camshaft between 0.0032–0.0053 inches (0.080–0.135 mm) and not exceed 0.0059 inches (0.15 mm).
2. Replace the cylinder head assembly if the measurement is exceeded. Replace the camshaft if damage is found on the thrust surfaces.

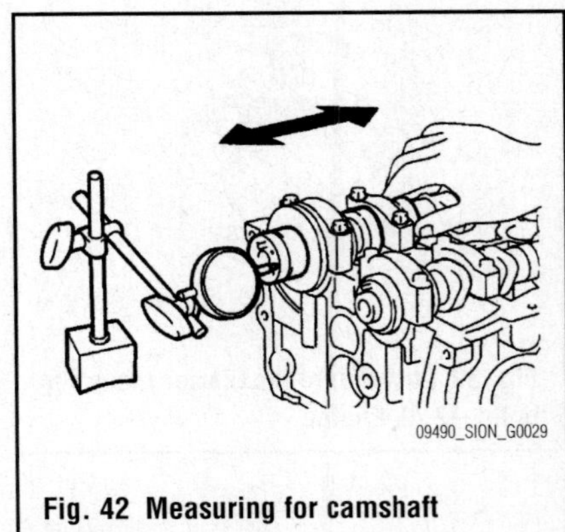

Fig. 42 Measuring for camshaft endplay

XD Models

1. Install a dial indicator in the thrust direction on the front end of the camshaft. Measure the end play of the dial indicator when the camshaft is moved back and forth. The dial indicator should measure as follows:
   a. Both camshafts: 0.06– 0.155 mm (0.0024– 0.0061 in.)
2. Replace the cylinder head assembly if the measurement is exceeded. Replace the camshaft if damage is found on the thrust surfaces.

Journal Oil Clearance

TC & XB Models

1. Remove the camshafts.
2. Clean the 10 bearing caps and camshaft journals.
3. Place the 2 camshafts on the cylinder head.
4. Lay a strip of Plastigage® across each of the camshaft journals.
5. Install the 10 bearing caps.

➡Do not turn the camshaft.

6. Remove the 10 bearing caps.
7. Measure the Plastigage® at its widest point.
8. Oil clearance maximum should not exceed:
   - Camshaft No. 1 journals: 0.0033 in. (0.085 mm)
   - Camshaft other journals: 0.0035 in. (0.09 mm)

XD Models

1. Remove the camshafts.
2. Clean the 10 bearing caps and camshaft journals.
3. Place the 2 camshafts on the cylinder head.
4. Lay a strip of Plastigage® across each of the camshaft journals.
5. Install the 10 bearing caps.

➡Do not turn the camshaft.

6. Remove the 10 bearing caps.
7. Measure the Plastigage® at its widest point.
8. Oil clearance should measure as follows:
   a. 2.4L Engine: No. 1 intake journal should measure between 0.0003–0.0015 inches (0.007–0.038 mm). No. 1 exhaust journal should measure between 0.0016–0.0031 inches (0.040–0.079 mm). All other journals should be between 0.0010–0.0024 inches (0.025–0.062 mm). If any clearance measurement is more than 0.0028 inches (0.07 mm) for the No. 1 intake journal, or more than 0.0039 inches (0.10 mm) for all other journals, the camshaft or cylinder head sub-assembly (or both) needs to be replaced. If the No. 1 journal clearance is greater than the maximum, replace the bearing cap.

Runout

TC & XB Models

1. Remove the camshafts.
2. Place the camshaft on a V-block, on a precise flat table.
3. Set the dial indicator to center journal.
4. Turn the camshaft to one direction by hand and measure the camshaft runout.
5. Runout should measure less than 0.0012 inches (0.03 mm).
6. Camshaft should be replaced if it exceeds the limit.

XD Models

1. Remove the camshafts.
2. Place the camshaft on a V-block, on a precise flat table.
3. Set the dial indicator to center journal.
4. Turn the camshaft to one direction by hand and measure the camshaft runout.
5. Maximum circle runout should not exceed: 0.0016 in. (0.04 mm)
6. Camshaft should be replaced if it exceeds the limit.

REMOVAL & INSTALLATION

tC & xB Models

See Figures 43 through 50.

1. Drain the cooling system.
2. Drain the engine oil.
3. Relieve the fuel system pressure.
4. Remove or disconnect the following:
   - Negative battery cable
   - Hood
   - Right-hand front wheel
   - Engine undercover
   - Right-hand front fender apron
   - Engine appearance cover
   - Wiper arm assembly
   - Top cowl seal
   - Left-hand top cowl ventilator louver
   - Wiper linkage assembly
   - Top outer cowl panel
   - Left-hand cowl body mounting bracket
   - Air intake assembly
   - Throttle body
   - Fuel rail
   - Intake manifold
   - Front exhaust pipe
   - Oil dipstick guide
   - Exhaust manifold
   - Accessory drive belts
   - Alternator
   - Power steering pump
5. Support the engine with a suitable jack.
6. Remove or disconnect the following:
   - Right-hand engine mount
   - Ignition coil
   - Cylinder head cover
   - Accessory drive belt tensioner
   - Crankshaft Position (CKP) sensor
   - Oil pan
7. Turn the crankshaft pulley until its groove and the timing mark on the front cover are aligned to set the No. 1 cylinder to TDC.
8. Matchmark the timing chain and camshaft sprockets.
9. Holding the exhaust camshaft with a wrench, loosen the camshaft timing set bolt.
10. Using several steps, loosen the bearing cap bolts in the sequence shown. Remove the bearing caps.

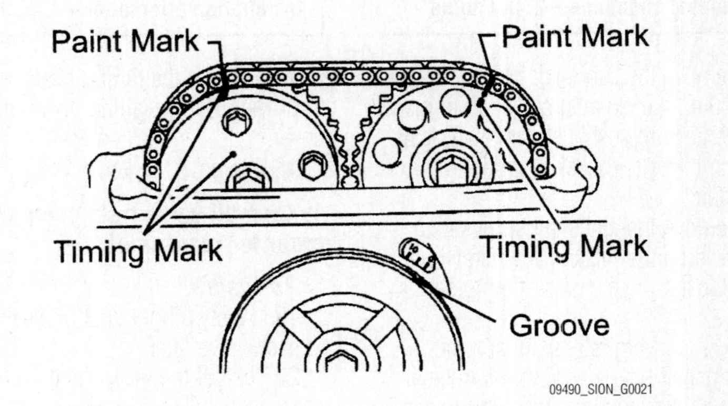

Fig. 43 Match the groove on the crankshaft pulley to the front cover timing mark. Place matchmarks on the camshaft sprocket and timing chain—2.4L Engine

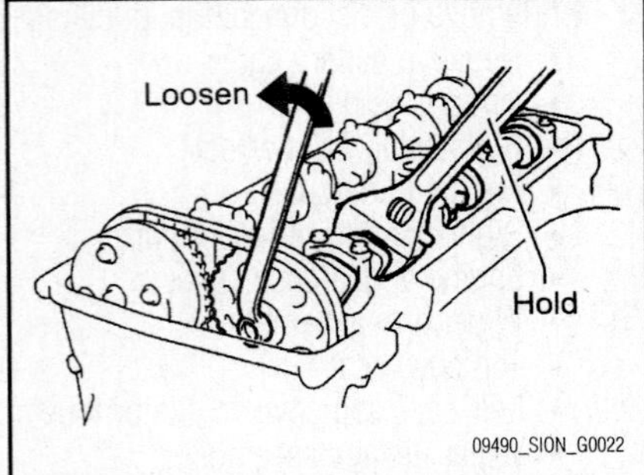

Fig. 44 Secure the camshaft with a wrench when removing the set bolt—2.4L Engine

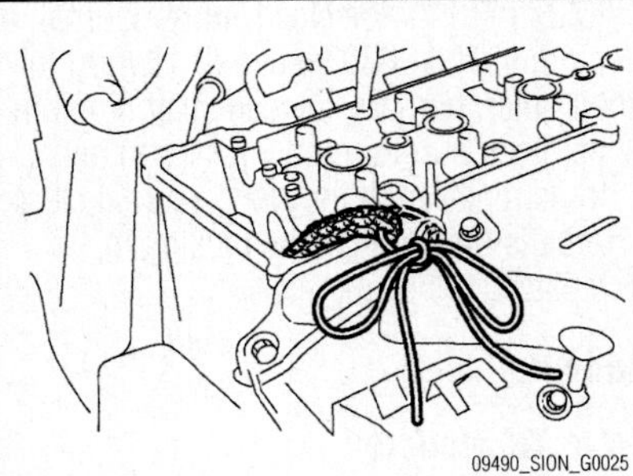

Fig. 47 Secure the timing chain with string after camshafts are removed—2.4L Engine

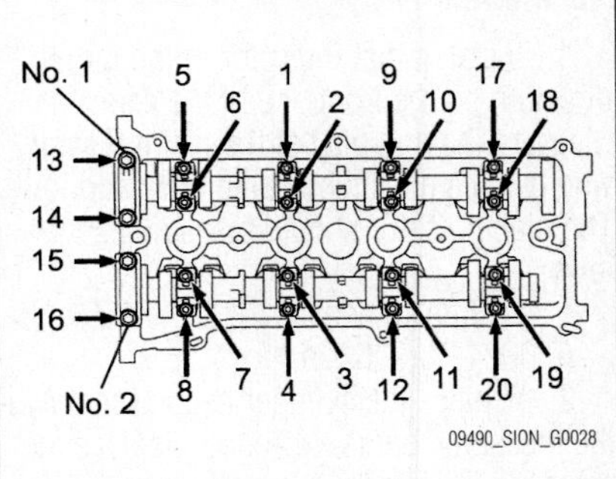

Fig. 50 Camshaft bearing cap bolt torque sequence—2.4L Engine

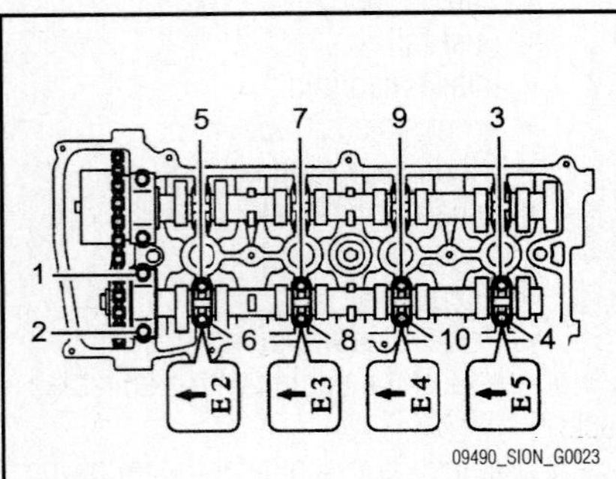

Fig. 45 Exhaust camshaft bearing cap bolt removal sequence—2.4L Engine

Fig. 48 Ensure the No. 1 cam lobes are facing the correct direction during installation—2.4L Engine

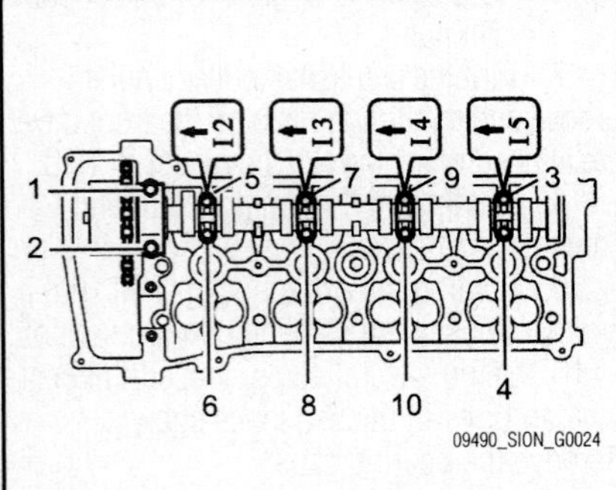

Fig. 46 Intake camshaft bearing cap bolt removal sequence—2.4L Engine

Fig. 49 Compare the markings on the camshafts to the illustration for correct installation orientation—2.4L Engine

11. Remove the camshaft timing set bolt while holding the exhaust camshaft in place.
12. Remove the exhaust camshaft, leaving the camshaft sprocket wrapped in the timing chain.
13. Remove the camshaft sprocket.
14. Holding the intake camshaft with a wrench, loosen the camshaft timing set bolt.
15. Using several steps, loosen the intake camshaft bearing cap bolts in the sequence shown. Remove the bearing caps.
16. Remove the intake camshaft, with sprocket attached, while holding the timing chain by hand.
17. Secure the timing chain with string to prevent it from falling down into the front cover.
18. Remove the valve lifters.

➡Keep all valve train components in order for reassembly.

To install:

19. Apply a light coat of clean engine oil to each valve lifter.
20. Install the valve lifters in their original places.
21. Apply a light coat of clean engine oil to the journals of the camshafts.
22. Install the camshafts on the cylinder head with the No. 1 cam lobes facing the directions shown.
23. Examine the camshaft markings to ensure correct orientation of the camshafts for installation.
24. Apply a light coat of clean engine oil to the threads and under the heads of the bearing cap bolts.
25. Tighten the bearing cap bolts in sequence using the following torque values:
    a. No. 1 and No. 2 bearing cap to 22 ft. lbs. (30 Nm).
    b. Remaining bolts to 80 inch lbs. (9 Nm).
26. The remainder of the installation is the reverse order of removal.
27. Refill the engine with oil to the correct level.
28. Refill the cooling system to the correct level.
29. Start the engine and check for leaks.

xD Models

See Figures 51 through 62.

1. Discharge fuel system pressure.
2. Remove battery and tray.
3. Remove front wheels.
4. Remove engine under covers.
5. Drain engine coolant.
6. Drain engine oil.
7. Remove front wiper arm head cap.
8. Remove both wiper arm and blade assemblies
9. Remove hood to cowl top seal.
10. Remove cowl top ventilator louver sub-assembly.
11. Remove LH cowl top ventilator louver.
12. Remove front wiper motor and link.
13. Remove RH front air shutter seal.
14. Remove outer cowl top panel.
15. Remove air cleaner assembly and bracket.
16. Remove battery carrier.
17. Remove No. 2 cylinder head cover.
18. Remove fan and generator V-belt.

19. Remove No. 1 and No. 2 radiator hoses.
20. Remove front bumper cover.
21. Disconnect No. 1 and No. 2 oil cooler outlet hoses (A/T).
22. Remove hood lock assembly.
23. Remove No. 1 cooler cover.
24. Remove upper radiator support absorber.
25. Remove radiator assembly.
26. Separate, with pulley attached, the compressor assembly. Set aside.
27. Separate transmission control cable assembly.
28. Separate union to check valve hose.
29. Separate No. 1 fuel vapor feed hose.
30. Disconnect engine wire.
31. Disconnect heater water outlet hose A (from heater unit).
32. Disconnect heater water hose inlet A.
33. Disconnect fuel tube sub-assembly.
34. Separate clutch release cylinder assembly (M/T).
35. Remove column hole cover silencer sheet.
36. Separate steering sliding yoke sub-assembly.
37. Separate No. 1 steering column hole cover sub-assembly.
38. Remove shift lever knob sub-assembly (M/T).
39. Remove console panel upper.
40. Remove console box cover rear.
41. Remove console box carpet.
42. Remove rear console box assembly.
43. Remove front console box.
44. Remove front floor brace center.
45. Remove front exhaust pipe assembly.
46. Remove front axle shaft left and right side nuts.
47. Separate both front speed sensors.
48. Separate both tie rod end sub-assemblies.
49. Separate both front stabilizer link assemblies.
50. Separate both front lower suspension arm sub-assemblies.
51. Separate both front axle assemblies.
52. Remove engine assembly with transaxle.
53. Remove front suspension cross-member sub-assembly.
54. Remove No. 1 oil cooler inlet tube (A/T).
55. Remove No. 1 oil cooler outlet tube (A/T).
56. Remove No. 2 oil cooler tube clamp (A/T).
57. Remove transmission oil level gauge and tube sub-assembly (A/T).
58. Remove intake manifold.
59. Remove oil level dipstick.
60. Remove oil level gage guide.
61. Remove alternator assembly.
62. Remove fan belt adjusting bar.
63. Loosen the clip and separate the water inlet hose.
64. Loosen the clip and separate the No. 3 water by-pass hose.
65. Remove water inlet.
66. Remove thermostat.
67. Remove ignition coil assembly.
68. Remove radio setting condenser.
69. Remove No. 2 ventilation hose.
70. Remove air tube assembly.
71. Remove cylinder head cover sub-assembly.
72. Set No. 1 cylinder to TDC on compression.
73. Remove crankshaft pulley.
74. Remove No. 1 chain tensioner assembly.
75. Remove the timing chain and related components. See "Timing Chain & Sprockets" in this section.
76. Remove camshaft timing gear assembly as follows:
   a. Check the lock of the camshaft timing gear.
   b. Release the lock pin.

➡Before removing the camshaft timing gear assembly, make sure that the lock pin has been released.

   c. After cleaning and degreasing the VVT oil hole on the intake side of the No. 1 camshaft bearing cap, completely seal the oil hole with adhesive tape or equivalent as shown in the illustration to prevent air from leaking.

➡Be sure to cover the oil hole completely because air leaks due to insufficient sealing will prevent the lock pin from being released.

   d. Prick a hole in the tape covering the oil hole as shown in the illustration.
   e. Apply approximately 22 psi (150 kPa) of air pressure to the hole pricked in the tape to release the lock pin.

➡If air leaks out, reattach the adhesive tape.

   f. Cover the oil hole with a shop rag or piece of cloth when applying air pressure to prevent oil from spraying.

➡Do not lock the camshaft timing gear assembly. If it is locked, release the lock pin again.

➡The camshaft timing gear assembly may be turned in the advance direction without applying any force.

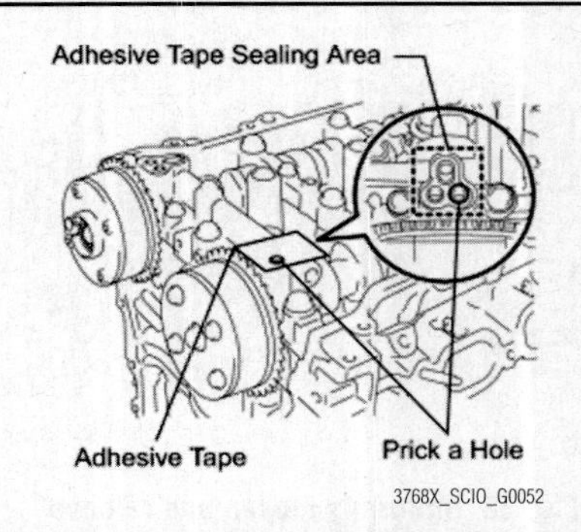

Fig. 51 Prick a hole in the tape covering the oil hole as shown in the illustration

➡If enough air pressure cannot be applied because of air leakage from the port, releasing the lock pin may be difficult.

   g. Remove the adhesive tape from the No. 1 camshaft bearing cap.
   h. Remove the flange bolt while holding the hexagonal portion of the camshaft, and then remove the camshaft timing gear assembly.

➡Before removing the camshaft timing gear, make sure that the lock pin has been released. Be sure not to remove the other 4 bolts.

   i. Keep the camshaft timing gear assembly horizontal while removing it from the camshaft.

77. Remove the flange bolt while holding the hexagonal portion of the camshaft, and then remove the camshaft timing exhaust gear assembly.

➡Be sure not to remove the other 4 bolts.

   a. Keep the camshaft timing exhaust gear assembly horizontal while removing it from the camshaft.

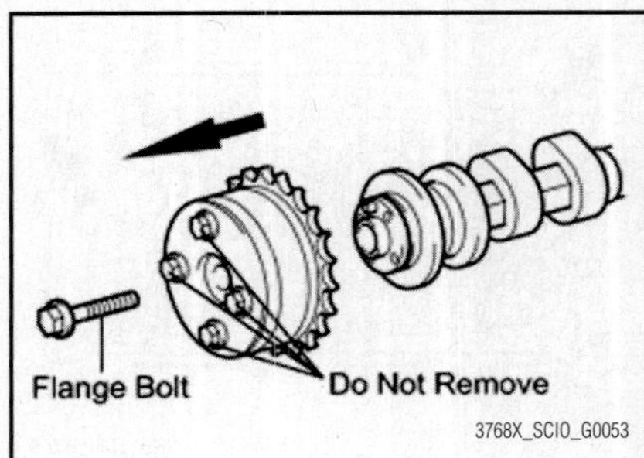

Fig. 52 Remove the flange bolt while holding the hexagonal portion of the camshaft, and then remove the camshaft timing gear assembly.

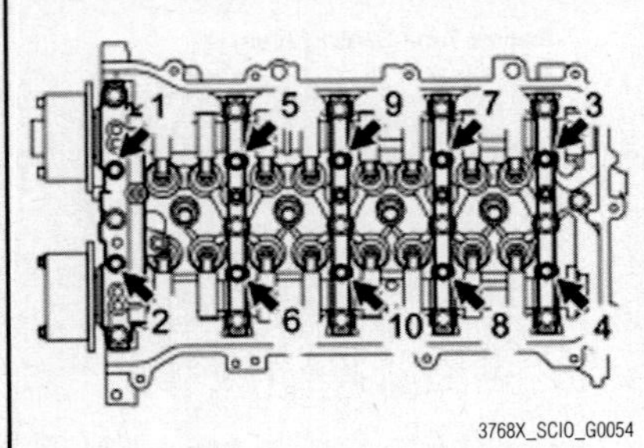

Fig. 53 Uniformly loosen and remove the 10 bearing cap bolts in the sequence shown in the illustration.

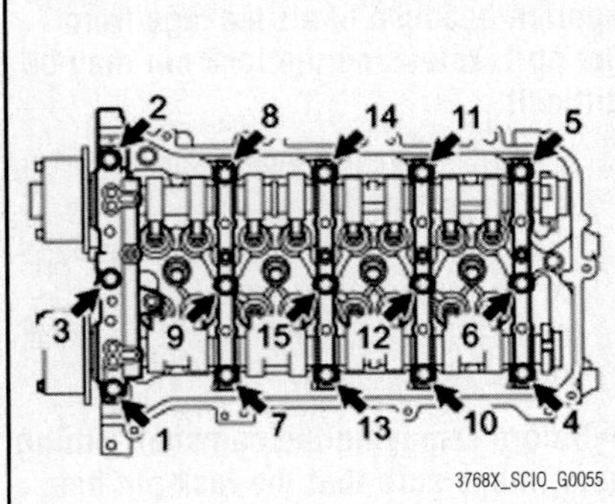

Fig. 54 Uniformly loosen and remove the 15 bearing cap bolts in the sequence shown in the illustration.

78. Remove camshaft bearing cap as follows:

a. Uniformly loosen and remove the 10 bearing cap bolts in the sequence shown in the illustration.

b. Uniformly loosen and remove the 15 bearing cap bolts in the sequence shown in the illustration.

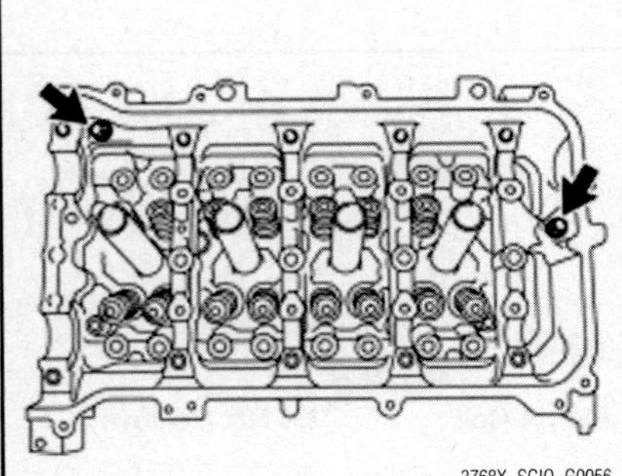

Fig. 55 Remove the 2 bolts and remove the camshaft housing by prying between the cylinder head and camshaft housing with a screwdriver.

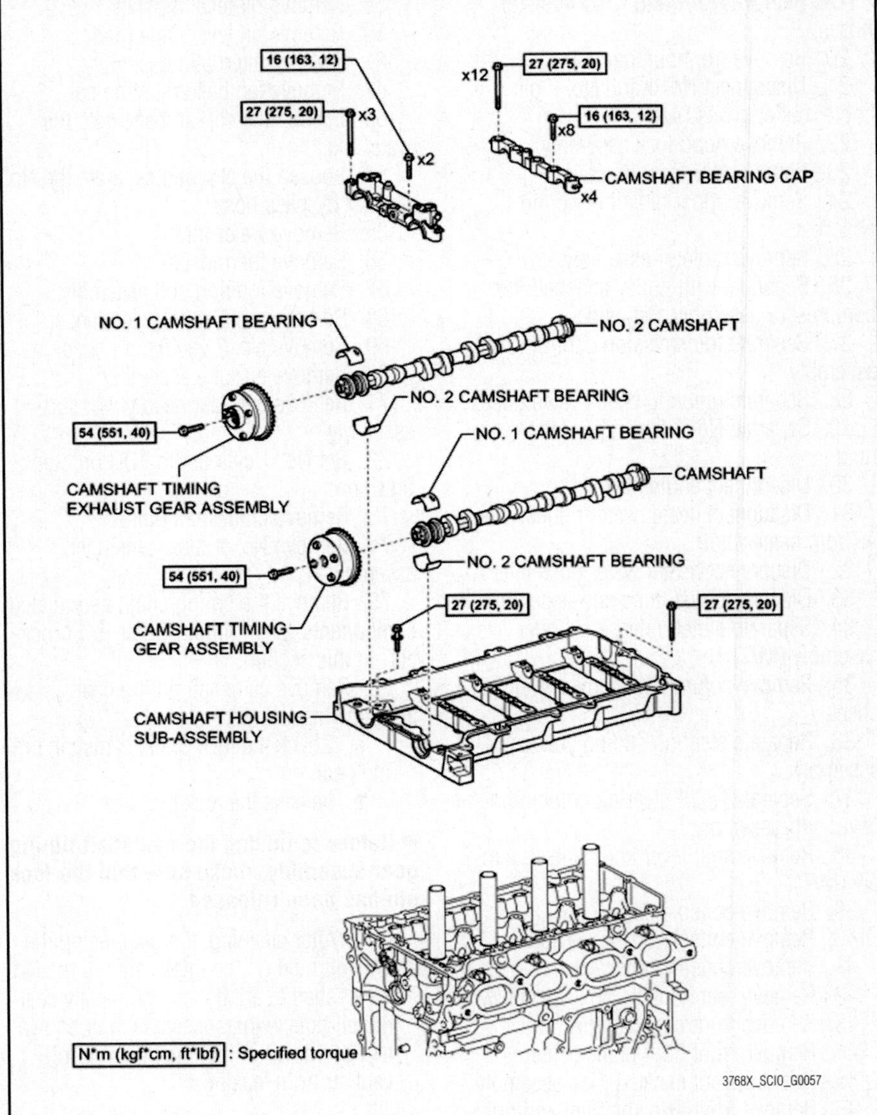

Fig. 56 Exploded view of the cylinder head and camshaft assemblies

➡If the camshaft bearing cap bolts have been loosened, reapply seal packing between the camshaft housing and cylinder head.

c. Remove the 5 bearing caps.

d. Arrange the removed parts in the correct order.

e. Remove the camshaft and No. 2 camshaft.

79. Remove the camshaft.

80. Remove the No. 2 camshaft.

81. Remove the 2 No. 1 camshaft bearings.

82. Remove the 2 No. 2 camshaft bearings.

83. Remove the 2 bolts and remove the camshaft housing by prying between the cylinder head and camshaft housing with a screwdriver.

⁂ CAUTION

Be careful not to damage the contact surfaces of the cylinder head and camshaft housing.

To install:

84. Clean the both surfaces of the bearing, then install the 2 No. 1 camshaft bearings.

85. Using calipers, measure the distance between the bearing cap's edge and the camshaft bearing's edge. Measurement (A–B) should be 0.0276 in. (0.7 mm) or less.

➡Position the bearing to the center of the bearing cap by measuring dimension A–B.

86. Install the No. 2 camshaft bearing as follows:

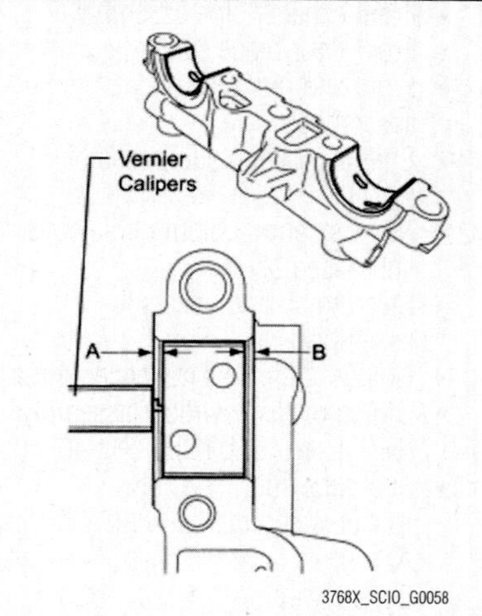

Fig. 57 Using calipers, measure the distance between the bearing cap's edge and the camshaft bearing's edge. Measurement (A–B) should be 0.0276 in. (0.7 mm) or less.

a. Clean both surfaces of the bearing.

b. Install the 2 No. 2 camshaft bearings.

c. Using calipers, measure the distance between the bearing cap's edge and the camshaft bearing's edge. Dimension A should be 0.042–0.068 in. (1.05–1.75 mm).

➡Position the bearing to the center of the bearing cap by measuring dimension A.

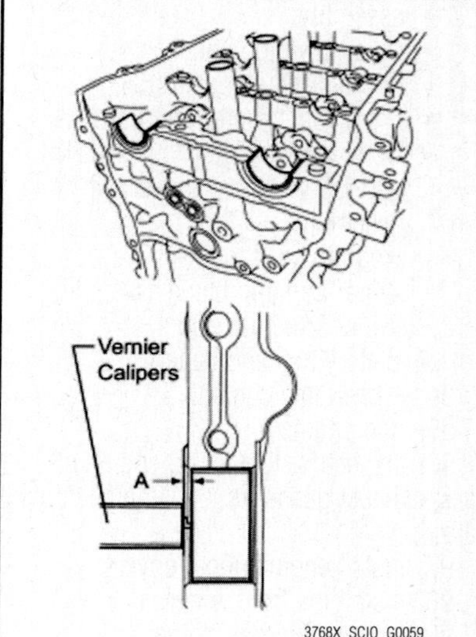

Fig. 58 Using calipers, measure the distance between the bearing cap's edge and the camshaft bearing's edge. Dimension A should be 0.042–0.068 in. (1.05–1.75 mm).

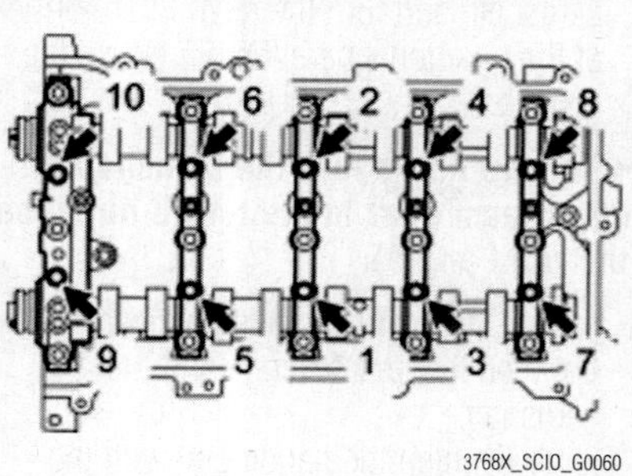

Fig. 59 Tighten the 10 bolts in the order shown in the illustration to 12 ft. lbs. (16 Nm).

87. Install the No. 2 camshaft as follows:

a. Clean the camshaft journals.

b. Apply a light coat of engine oil to the camshaft journals, camshaft housings and bearing caps.

c. Install the No. 2 camshaft to the camshaft housing.

88. Install the other camshaft as follows:

a. Clean the camshaft journals.

b. Apply a light coat of engine oil to the camshaft journals, camshaft housings and bearing caps.

c. Install the camshaft to the camshaft housing.

89. Install the camshaft bearing caps as follows:

a. Apply engine oil to the camshaft journals, camshaft housing and bearing caps.

b. Check the marks and numbers on the camshaft bearing caps and place them in the proper position and direction.

➡Make sure that the knock pin of the camshaft is positioned as shown in the illustration.

c. Tighten the 10 bolts in the order shown in the illustration to 12 ft. lbs. (16 Nm).

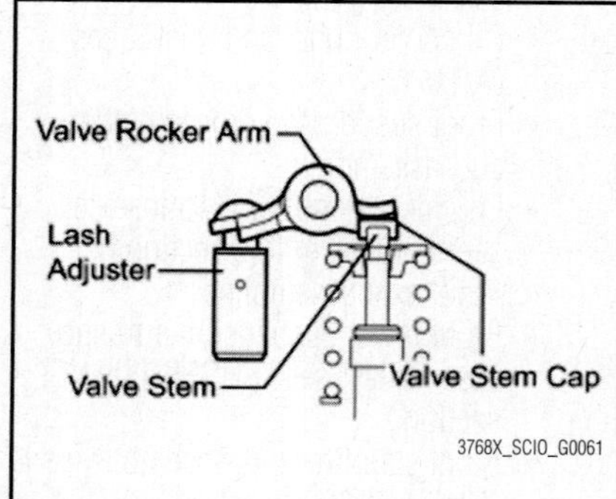

Fig. 60 Make sure that the valve rocker arm is installed as shown in the illustration.

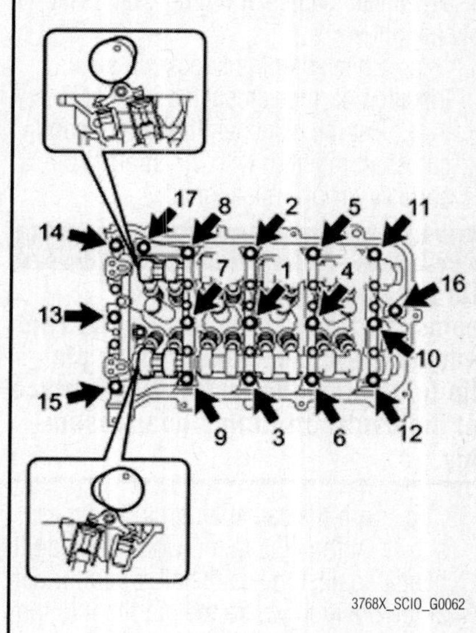

Fig. 61 Set the camshaft and No. 2 camshaft as shown in the illustration.

90. Install the camshaft housing sub-assembly as follows:

a. Make sure that the valve rocker arm is installed as shown in the illustration.

b. Apply seal packing in a continuous line around the outer edge.

➡Remove any oil from the contact surface. Install the camshaft housing sub-assembly RH within 3 minutes of applying the seal packing. Do not start the engine for at least 2 hours after installing.

c. Set the camshaft and No. 2 camshaft as shown in the illustration.

d. Install the camshaft housing and tighten the 17 bolts in the order shown in the illustration. Torque to 20 ft. lbs. (27 Nm).

➡After installing the camshaft housing, make sure that the cam lobes are positioned as shown in the illustration.

➡If any of the bolts are loosened during installation, remove the camshaft housing, clean the installation surfaces, and reapply seal packing.

➡If the camshaft housing is removed because any of the bolts are loosened during installation, make sure that the previously applied seal packing does not enter any oil passages.

e. After installing the camshaft housing, wipe off any seal packing that seeped out from between the housing and the cylinder head.

91. Install camshaft timing gear assembly as follows:

a. Check that the knock pin is installed on the camshaft.

b. Put the camshaft timing gear and camshaft together with the straight pin and key groove misaligned.

CAUTION

Do not forcefully push in the camshaft timing gear assembly. This may cause the camshaft knock pin tip to damage the installation surface of the camshaft timing gear assembly.

c. Turn the camshaft timing gear as shown in the illustration while pushing it gently against the camshaft. Push further at the position where the pin fits into the groove.

➡Do not turn the camshaft timing gear in the retard direction (the right angle).

d. Measure the clearance between the gear and the camshaft. Clearance should be 0.004–0.016 in. (0.1–0.4 mm).

e. Tighten the flange bolt with the camshaft timing gear fixed in place. Torque to 40 ft. lbs. (54 Nm).

f. Check that the camshaft timing gear can move to the retard angle side (the right direction) and is locked in the most retarded position.

92. Install camshaft timing exhaust gear assembly as follows:

a. Check that the knock pin is installed on the camshaft.

b. Put the camshaft timing exhaust gear and camshaft together by aligning the key groove and straight pin.

c. Lightly press the gear against the camshaft, and turn the gear. Push further at the position where the pin enters the groove.

➡Be sure not to turn the camshaft timing exhaust gear in the retard direction (the right angle).

d. Check that there is no clearance between the gear's flange and the camshaft.

e. Tighten the flange bolt with the camshaft timing exhaust gear fixed to 40 ft. lbs. (54 Nm).

f. Check the camshaft timing exhaust gear lock.

g. Make sure that the camshaft timing exhaust gear is locked.

93. Install or connect the following components:

- Crankshaft timing gear or sprocket
- No. 1 chain vibration damper
- Timing chain assembly. See "timing chain & sprockets" in this section.
- Cylinder head cover sub-assembly
- Air tube assembly
- No. 2 ventilation hose
- Radio setting condenser
- Ignition coil assembly
- Thermostat
- Water inlet
- No. 3 water by-pass hose
- Water inlet hose
- Fan belt adjusting bar
- Alternator assembly; see "engine electrical" section.
- Oil level gauge guide and dipstick
- Intake manifold; see "intake manifold" in this section.
- Transmission oil filler tube sub-assembly (A/T)
- Transmission oil level gauge sub-assembly (A/T)
- Oil cooler outlet and inlet tubes (A/T)
- Front suspension crossmember sub-assembly
- Engine assembly with transaxle; see "engine" in this section.
- Front axle assemblies
- Front lower suspension arm sub-assemblies; see "SUSPENSION" section.
- Front stabilizer link assemblies; see "SUSPENSION" section.
- Tie rod end sub-assemblies.
- Front wheel speed sensors; see "BRAKES" section.
- Front axle shaft nut.
- Front exhaust pipe assembly
- Front floor brace center
- Front and rear console assemblies
- Shift lever knob sub-assembly (M/T)
- No. 1 steering column hole cover sub-assembly
- Steering sliding yoke sub-assembly
- Column hole cover silencer sheet
- Clutch release cylinder assembly (M/T); see "CLUTCH" section.
- Fuel tube sub-assembly
- Heater water inlet and outlet hoses A
- Engine wire
- No. 1 fuel vapor feed hose
- Union to check valve hose
- Transmission control cable assembly
- Compressor assembly.
- Radiator assembly; see "ENGINE COOLING" section.
- Upper radiator support sub-assembly and support absorber
- No. 1 cooler cover
- Hood lock assembly
- No. 2 oil cooler inlet hose and no. 1 cooler outlet hose (A/T)
- No. 1 oil cooler outlet hose (A/T)
- Front bumper cover
- No. 1 and No. 2 radiator hoses
- Fan and generator V-belt
- No. 2 cylinder head cover
- Battery carrier
- Air cleaner bracket and air cleaner assembly
- Outer cowl top panel
- Front air shutter seal (RH)
- Front wiper motor and link
- Cowl top ventilator louver (LH) and top ventilator louver sub-assembly
- Hood to cowl top seal
- Wiper arm and blade assemblies
- Wiper arm head cap
- Battery tray and battery

94. Add engine coolant.

95. Add engine oil.

96. Inspect for fuel leaks, engine oil leaks, exhaust gas leaks, and engine coolant leaks.

97. Install engine under covers.

98. Install the front wheels.

99. Inspect ignition timing.

100. Inspect engine idling speed.

101. Inspect the emissions level.

102. Inspect front wheel alignment.

103. Inspect the ABS sensor signal.

104. Inspect the VSC warning.

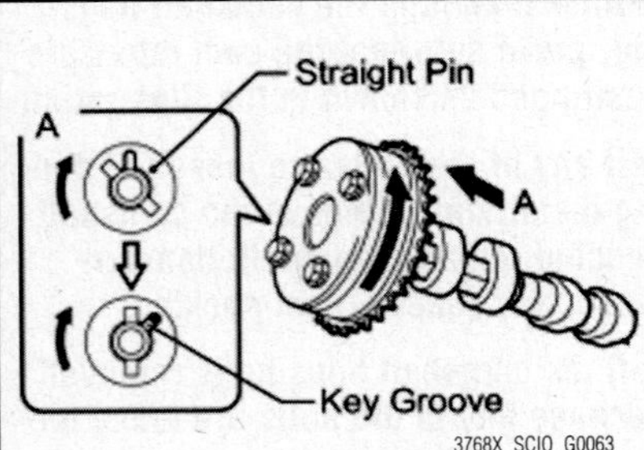

Fig. 62 Turn the camshaft timing gear as shown in the illustration while pushing it gently against the camshaft. Push further at the position where the pin fits into the groove.

CRANKSHAFT DAMPER

REMOVAL & INSTALLATION

➡See "Crankshaft Front Seal" for procedure.

CRANKSHAFT FRONT SEAL

REMOVAL & INSTALLATION

tC & xB Models

➡Manufacturer does not provide specific removal and installation procedures for this component.

xD Models

See Figures 63 and 64.

1. Remove the RH engine under cove.
2. Remove the engine cover.
3. Remove the fan and alternator v-belt.
   a. Loosen bolts A and B.
   b. Turn adjusting bolt C to release the tension and remove the V-belt from the pulleys.
   c. Remove the fan and alternator V-belt.
4. Remove the crankshaft pulley.
   a. Using SST, hold the pulley in place and loosen the pulley bolt.
   b. Using SST, remove the pulley bolt and pulley.
5. Remove the timing chain or belt cover oil seal
   a. Using a knife, cut off the oil seal lip.
   b. Using a screwdriver with its tip taped, pry out the oil seal.

➡After the removal, check the crankshaft for damage. If it is damaged, smooth the surface with 400-grit sandpaper.

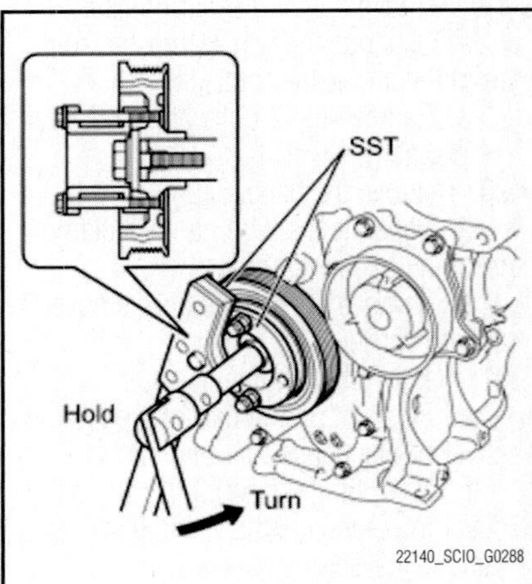

Fig. 63 Using SST, hold the pulley in place and loosen the pulley bolt.

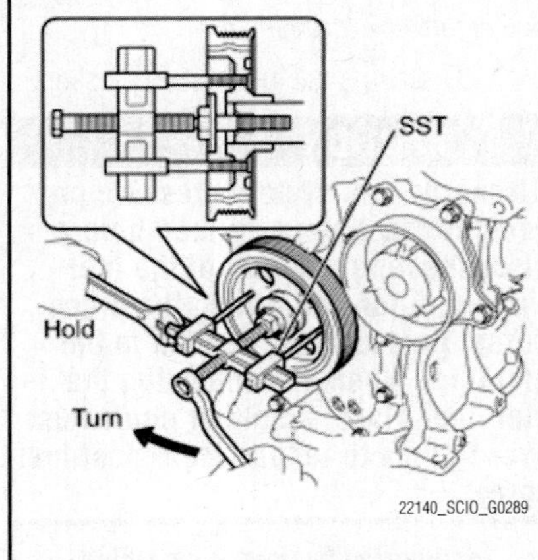

Fig. 64 Using SST, remove the pulley bolt and pulley.

To install:

6. Install timing chain or belt cover oil seal.
   a. Apply MP grease to a new oil seal lip.
   b. Using SST and a hammer, tap the oil seal until its surface is flush with the timing chain cover edge.

➡Do not tap the oil seal at an angle. Keep the lip free of foreign matter.

7. Install crankshaft pulley.
   a. Align the pulley set key with the key groove of the pulley.
   b. Using SST, hold the pulley in place and tighten the bolt. Torque: 140 ft. lbs. (190 Nm).
8. Install fan and alternator V-belt.
9. Adjust fan and alternator V-belt.
10. Inspect fan and alternator V-belt.
11. Install Engine cover.
12. Inspect for engine oil leak.
13. Install engine under cover RH.

CYLINDER HEAD

REMOVAL & INSTALLATION

tC & xB Models

See Figures 65 and 66.

1. Before servicing the vehicle, refer to the Precautions Section.
2. Drain the cooling system.
3. Drain the engine oil.
4. Relieve the fuel system pressure.
5. Remove or disconnect the following:
   - Negative battery cable
   - Hood
   - Right-hand front wheel
   - Engine undercover
   - Right-hand front fender apron
   - Engine appearance cover
   - Wiper arm assembly
   - Top cowl seal
   - Left-hand top cowl ventilator louver
   - Wiper linkage assembly
   - Top outer cowl panel
   - Left-hand cowl body mounting bracket
   - Air intake assembly
   - Throttle body
   - Fuel rail
   - Intake manifold
   - Front exhaust pipe
   - Oil dipstick guide
   - Exhaust manifold
   - Accessory drive belts
   - Alternator
   - Power steering pump
6. Support the engine with a suitable jack.
7. Remove or disconnect the following:
   - Right-hand engine mount
   - Ignition coil
   - Cylinder head cover
   - Accessory drive belt tensioner
   - Crankshaft Position (CKP) sensor
   - Oil pan
8. Turn the crankshaft pulley until its groove and the timing mark on the front cover are aligned to set the No. 1 cylinder to TDC.
9. Remove or disconnect the following:
   - Camshafts
   - Crankshaft pulley
   - Front cover
   - Timing chain
   - Timing chain vibration damper
   - Camshaft timing oil control valve
   - Radiator inlet hose
   - All remaining sensors connectors
   - Ground wire
10. Loosen the cylinder head bolts in the sequence shown and remove the cylinder head and gasket.

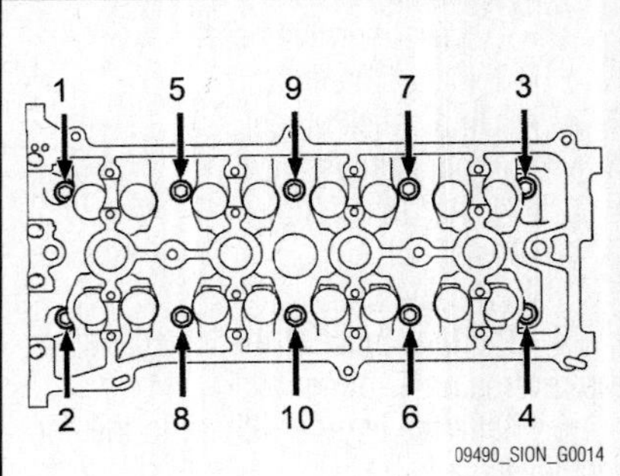

Fig. 65 Cylinder head bolt removal sequence—2.4L Engine

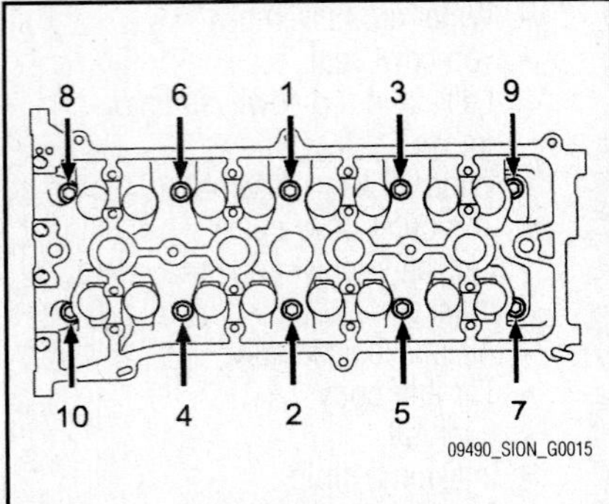

Fig. 66 Cylinder head bolt torque sequence—2.4L Engine

To install:

11. Install the cylinder head with a new gasket. The Lot Number on the gasket should face upward.
12. Apply a light coat of new engine oil to the threads of the cylinder head bolts. Tighten the bolts in the sequence shown as follows:
    a. Step 1: Tighten to 58 ft. lbs. (79 Nm).
    b. Step 2: Plus 90 degrees.
13. Install or connect the following:
    - Ground wire
    - Sensor connectors to the cylinder head
    - Radiator inlet hose
    - Camshafts
    - Camshaft timing oil control valve
    - Exhaust manifold
    - Oil dipstick guide
    - Timing chain vibration damper
    - Timing chain
    - Front cover
    - Crankshaft pulley. Tighten to 133 ft. lbs. (180 Nm).
    - Oil pan
    - Crankshaft Position (CKP) sensor
    - Accessory drive belt tensioner
    - Cylinder head cover
    - Ignition coil
    - Right-hand engine mount
    - Power steering pump
    - Alternator
    - Accessory drive belts
    - Exhaust manifold
    - Oil dipstick guide
    - Front exhaust pipe
    - Intake manifold
    - Fuel rail
    - Throttle body
14. The remainder of the installation is the reverse order of removal.
15. Refill the engine with oil to the correct level.
16. Refill the cooling system to the correct level.
17. Start the engine and check for leaks.

xD Models

See Figures 67 through 76.

1. Discharge the fuel system pressure.

⁂ CAUTION

Discharge fuel system pressure procedures must be performed before disconnecting any part of the fuel system. After performing the procedures, pressure will remain in the fuel line. When disconnecting the fuel line, place a cloth or equivalent over fittings to reduce the risk of fuel spray.

   a. Remove the rear seat cushion assembly LH
   b. Remove the rear floor service hose cover.
   c. Disconnect the connector from the fuel pump assembly.
   d. Start the engine. After the engine stops naturally, turn the ignition switch OFF.
   e. Start the engine again and make sure that engine does not start.
   f. Remove the fuel tank cap, and let the air out of the fuel tank.
   g. Connect the connector.
   h. Install the rear floor service hole cover with new butyl tape.
   i. Install the rear seat cushion assembly LH.
2. Remove the battery.
   a. Disconnect the cable from the battery terminal.
   b. Loosen the nut and remove the battery clamp.
   c. Remove the battery.
3. Remove the battery tray.
4. Remove the front wheels.
5. Remove the engine under cover LH.
6. Remove the engine under cover RH.
7. Drain the engine coolant.
8. Drain the engine oil.
9. Remove front wiper arm head cap.
   a. Using a screwdriver with its tip wrapped in protective tape, disengage the 2 claws and remove the 2 front wiper arm caps.
10. Remove front wiper arm and blade assembly LH.
    a. Operate the wiper, then stop the wiper motor in the automatic stop position.
    b. Remove the nut and front wiper arm and blade assembly LH.
11. Remove front wiper arm and blade assembly RH.

➡Use the same procedure as for the LH side.

12. Remove hood to cowl top seal.
    a. Disengage the 7 clips and remove the hood to cowl top seal.
13. Remove cowl top ventilator louver sub-assembly.
    a. Remove the clip.
    b. Disengage the 4 claws and 4 guides and remove the cowl top ventilator louver RH.
14. Remove cowl top ventilator louver LH.
    a. Remove the clip.
    b. Disengage the 5 claws and 5 guides and remove the cowl top ventilator louver LH.
15. Remove front wiper motor and link.
    a. Disengage the clamp.
    b. Remove the 2 bolts.
    c. Slide the front wiper motor and link. Disengage the link from the rubber pin, then disconnect the connector and remove the front wiper motor and link.
16. Remove the front air shutter seal RH.
    a. Disengage the 2 claws and remove the front air shutter seal RH.
17. Remove the outer cowl top panel.
    a. Disengage the wire harness clamp.
    b. Remove the 9 bolts and remove the outer cowl top panel.
18. Remove the air cleaner assembly.
    a. Separate the intake air flow meter connector and 2 wire harness clamps.
    b. Separate the ventilation hose from the air cleaner hose.
    c. Unfasten the 2 clamps.
    d. Remove the air cleaner cap sub-assembly with air cleaner hose.
    e. Remove the air cleaner element.
    f. Separate the wire harness clamp from the air cleaner case.
    g. Remove the 2 bolts and remove the air cleaner case with air cleaner inlet.
19. Remove the air cleaner bracket.
    a. Separate the wire harness clamp from the air cleaner bracket.
    b. Remove the 2 bolts and remove the air cleaner bracket.
20. Remove the battery carrier.
    a. Separate the wire harness clamp from the battery carrier.
    b. Remove the 5 bolts and remove the battery carrier.
21. Remove the engine cover.
    a. First, lift up the rear of the Engine cover to disengage the 2 fittings
    b. Next, lift the front of the Engine cover to disengage the 2 fittings. Then remove the Engine cover.

➡Ensure that the rubber grommets remain attached to the Engine cover. If

any grommets are attached to the bolts, move them to the Engine cover.

22. Remove the fan and alternator V-belt.
   a. Loosen bolts A and B.
   b. Turn adjusting bolt C to release the tension and remove the V-belt from the pulleys.
   c. Remove the fan and alternator V-belt.
23. Remove the lower radiator hose.
   a. Loosen the 2 clips and remove the No. 2 radiator hose.
24. Remove the front bumper cover.
   a. Apply protective tape, as shown in the illustration.
   b. Remove the 8 screws.
   c. Remove the 9 clips.
   d. Disengage the 6 claws and remove the front bumper cover.
25. Disconnect the transaxle cooling lines.
26. Remove the hood lock assembly.
27. Remove the hood lock assembly.
   a. Separate the hood lock control cable assembly from the 2 clamps.
   b. Remove the 3 bolts and remove the hood lock assembly.
28. Remove the No. 1 cooler cover.
   a. Remove the 2 clips and remove the No. 1 cooler cover.
29. Remove the upper radiator support absorber.
   a. Using a screwdriver with its tip wrapped in protective tape, disengage the 6 claws and remove the radiator support upper absorber.
30. Remove the upper radiator support sub-assembly.
   a. Disconnect the horn connector.
   b. Remove the 5 bolts and remove the upper radiator support sub-assembly.
31. Remove the radiator assembly.
   a. Disconnect the cooling fan motor connector and separate the wire harness clamp.

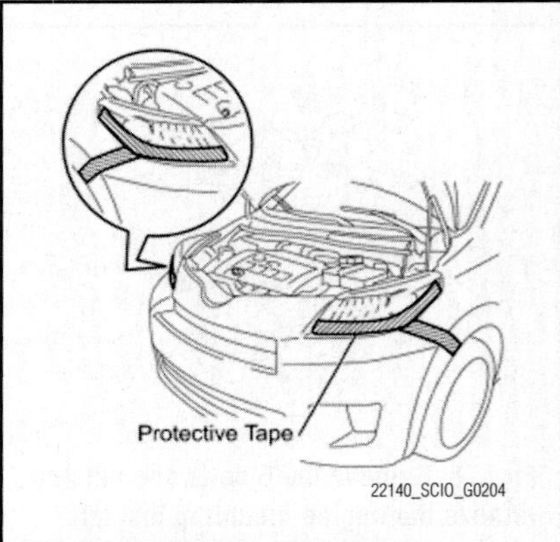

Fig. 67 Apply protective tape

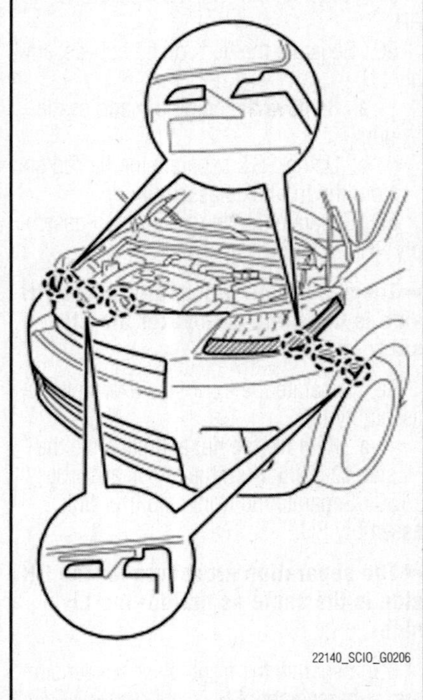

Fig. 68 Disengage the 6 claws and remove the front bumper cover.

   b. Disengage the 2 claws and remove the radiator assembly from the vehicle.

CAUTION

Do not apply excessive force to the cooler condenser assembly or piping when removing the radiator assembly.

32. Separate with pulley compressor assembly.
33. Separate the transmission control cable assembly (A/T).
   a. Remove the nut and disconnect the transmission control cable assembly from the control shaft lever.
34. Separate the transmission control cable assembly (M/T).
   a. Remove the 2 clips and the washers, and disconnect the 2 cables from the transaxle.
   b. Remove the 2 clips and disconnect the 2 cables from the control cable bracket.
35. Separate the union to check valve hose.
   a. Loosen the clip and separate the union to check valve hose.
36. Separate No. 1 fuel vapor feed hose.
   a. Separate the No. 1 fuel vapor feed hose from the vacuum switching valve assembly.
37. Disconnect the engine wire.
   a. Remove the bolt and wire harness clamp and separate the transaxle ground wire (A/T).
   b. Remove the bolt and 2 wire harness clamp and separate the transaxle ground wire (M/T).
   c. Disconnect the 2 connectors.
   d. Pull up the lever and disconnect the connector from the engine control computer.
   e. Separate the 2 wire harness clamps.
   f. Remove the 2 connectors and clamp from the engine room junction block and disconnect the 2 wire harness clamps.
   g. Disconnect all wire harnesses and connectors.

CAUTION

Make sure that no wire harness is connected between the body and engine.

38. Disconnect the upper and lower heater hoses (from heater unit).
39. Disconnect the fuel tube sub-assembly.
   a. Remove the No. 1 fuel pipe clamp.
   b. Pinch the retainer, then pull the fuel tube connector out of the pipe.

➡Remove any dirt and foreign matter from the fuel tube connector before performing this work. Do not allow any scratches or foreign matter on the parts when disconnecting, as the fuel tube connector has the O-rings that seal the pipe. Perform this work by hand. Do not use any tools. Do not forcibly bend, twist or turn the nylon tube. Protect the disconnected parts by covering them with vinyl bags after disconnecting the fuel tube. If the fuel tube connector and pipe are stuck, push and pull them to release them.

40. Separate the clutch release cylinder assembly (M/T).
   a. Remove the 4 bolts, then separate the clutch release cylinder.

➡Suspend the clutch release cylinder with a piece of rope so as not to overload the clutch pipe.

41. Remove the column hole cover silencer sheet.
   a. Remove the floor carpet and 2 clips and remove the column hole cover silencer.
42. Separate the steering sliding yoke sub-assembly.

a. Use a seat belt to fix the steering wheel assembly, in order to avoid breakage of the spiral cable.

b. Place matchmarks on the sliding yoke of the steering intermediate shaft assembly and the power steering.

c. Loosen bolt A, remove bolt B and separate the steering intermediate shaft assembly.

43. Separate the No. 1 steering column hole cover sub-assembly.

a. Remove clip A, separate clip B from the body and separate No. 1 steering column hole cover.

CAUTION

Do not damage clip B.

44. Remove the shift lever knob sub-assembly (M/T).

a. Remove the shift lever knob by turning the knob counterclockwise.

45. Remove the upper console panel.

a. Disengage the 2 clips and the 6 claws and remove the upper console panel.

46. Remove the rear console box cover.

a. Disengage the 9 claws and remove the rear console box cover.

47. Remove the console box carpet.

48. Remove the rear console box assembly.

a. Remove the 3 screws.

b. Disengage the 4 claws and remove the rear console box.

49. Remove the front console box.

a. Remove the 2 screws.

b. Disengage the clip and 2 claws and remove the front console box.

50. Remove the front floor center brace.

a. Remove the 2 bolts and remove the front floor center brace.

51. Remove the front exhaust pipe assembly.

a. Disconnect the heated oxygen sensor connector.

b. Remove the grommet and pull the sensor connector out of the cabin through the floor panel.

c. Remove the 2 bolts and 2 springs and separate the front exhaust pipe assembly from the exhaust tail pipe assembly.

d. Remove the 2 bolts and 2 springs and separate the front exhaust pipe assembly from the exhaust manifold.

e. Remove the 3 exhaust pipe supports and remove the front exhaust pipe assembly.

52. Remove the front axle shaft LH nut.

53. Remove the front axle shaft RH nut.

54. Separate the speed sensor front LH.

55. Separate the speed sensor front RH.

56. Separate the tie rod end sub-assembly LH.

a. Remove the cotter pin and castle nut.

b. Using SST, separate the tie rod end from the front axle assembly.

57. Separate the tie rod end sub-assembly RH

➡The separation procedure for the RH side is the same as that for the LH side.

58. Separate the front stabilizer link assembly LH.

a. Remove the nut and separate the stabilizer link from the shock absorber.

59. Separate the front stabilizer link assembly RH.

➡The separation procedure for the RH side is the same as that for the LH side.

60. Separate the front lower suspension arm sub-assembly LH.

a. Remove the clip and castle nut.

b. Install SST to the threaded section of the lower ball joint.

➡Make sure the upper ends of the threaded section of the lower ball joint and SST (spacer B) are aligned.

c. Using SST, separate the front lower suspension arm from the front axle assembly.

61. Separate the front lower suspension arm sub-assembly RH.

➡The separation procedure for the RH side is the same as that for the LH side.

62. Separate the front axle assembly LH.

a. Using a plastic hammer, tap the end of the drive shaft and disengage the fitting between the drive shaft and front axle.

➡If it is difficult to disengage the fitting, tap the end of the drive shaft with a brass bar and hammer.

b. Push the front axle out of the vehicle to remove the drive shaft from the front axle.

➡Do not push the front axle further out of the vehicle than is necessary. Do not damage the outboard joint boot. Do not damage the speed sensor rotor. Suspend the drive shaft with a piece of rope or the equivalent.

63. Separate the front axle assembly RH.

➡The separation procedure for the RH side is the same as that for the LH side.

64. Remove the engine assembly with transaxle.

a. Set the engine lifter.

b. Remove 2 bolts and remove the engine mounting stay RH.

c. Remove the 6 bolts and nut and remove the engine mounting insulator RH.

➡Be sure not to remove the other bolt A.

d. Remove the through bolt and nut and separate the transverse engine mounting insulator.

e. Remove the 6 bolts and remove the engine assembly with transaxle and front suspension crossmember from the vehicle.

65. Disconnect the transaxle oil cooler inlet and outlet hoses from the radiator (A/T).

66. Remove the transmission oil level gage sub-assembly (A/T)

67. Separate the transmission oil filler tube sub-assembly (A/T)

68. Remove the ignition coil assembly.

a. Disconnect the 4 ignition coil assembly connectors.

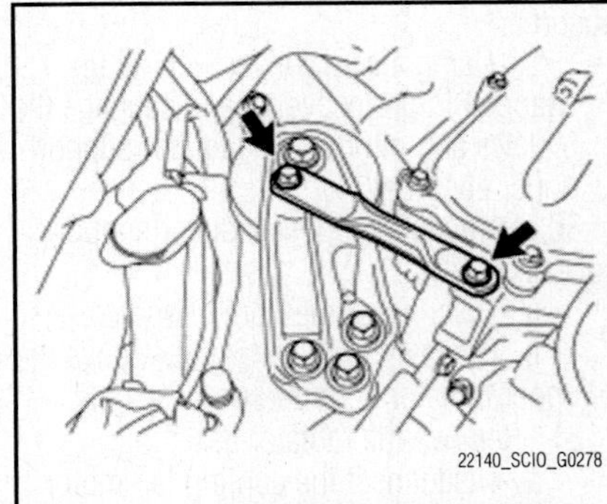

Fig. 69 Remove 2 bolts and remove the engine mounting stay RH.

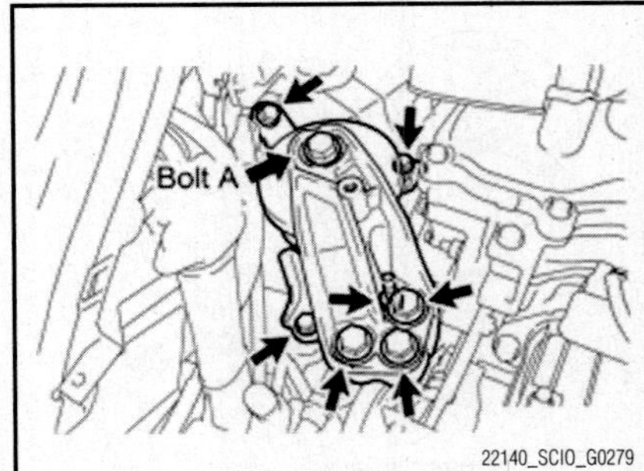

Fig. 70 Remove the 6 bolts and nut and remove the engine mounting insulator RH.

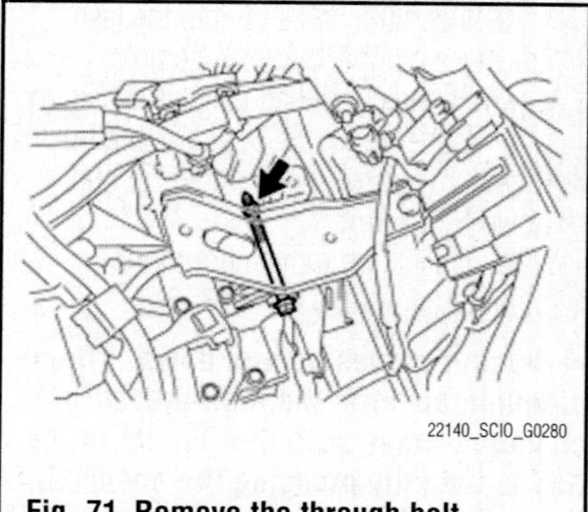

Fig. 71 Remove the through bolt and nut

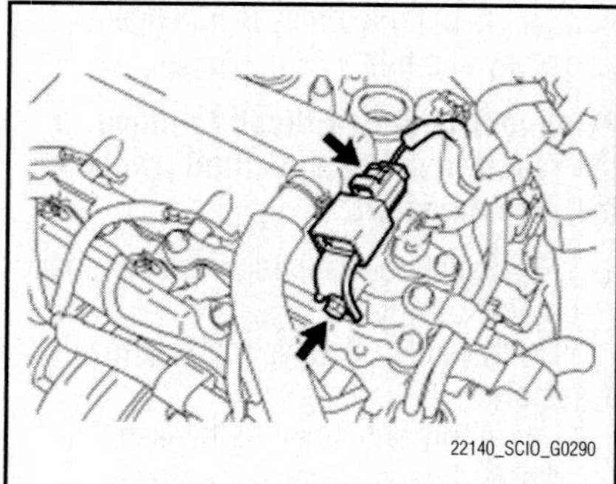

Fig. 72 Remove the radio setting condenser.

b. Remove the 4 bolts and remove the 4 ignition coil assemblies.

69. Remove the radio setting condenser.

a. Disconnect the radio setting condenser connector.

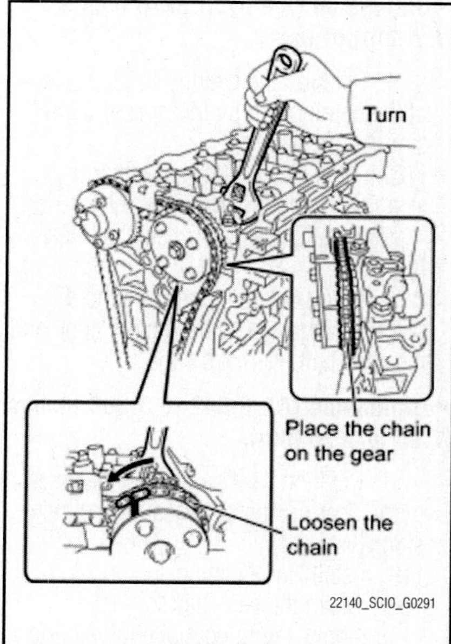

Fig. 73 Remove the chain sub-assembly.

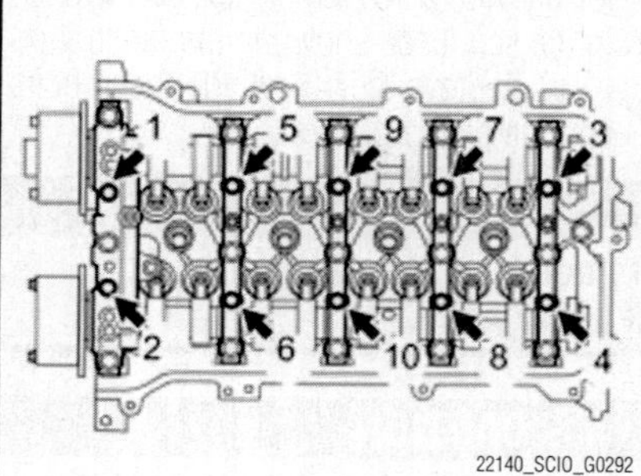

Fig. 74 Remove the 10 bearing cap bolts in the sequence

b. Remove the bolt and remove the radio setting condenser.

70. Remove the oil breather hose.

71. Remove the heater water hoses (from heater unit).

72. Remove the upper radiator hose.

73. Remove the intake manifold.

a. Loosen the 2 clips and separate the 2 water by-pass hoses.

b. Separate the No. 1 fuel vapor feed hose from the throttle body.

c. Loosen the clip and separate the No. 1 vacuum transmitting hose.

d. Separate the ventilation hose.

e. Separate the wire harness clamp from the intake manifold.

f. Remove the 4 bolt and 2 nuts and remove the intake manifold and manifold stay.

g. Remove the gasket from the intake manifold.

74. Remove the fuel delivery pipe sub-assembly.

75. Remove the fuel delivery pipe spacers.

76. Remove the injector vibration insulators.

77. Remove the fuel injector assembly.

a. Pull the 4 fuel injectors out of the fuel delivery pipe.

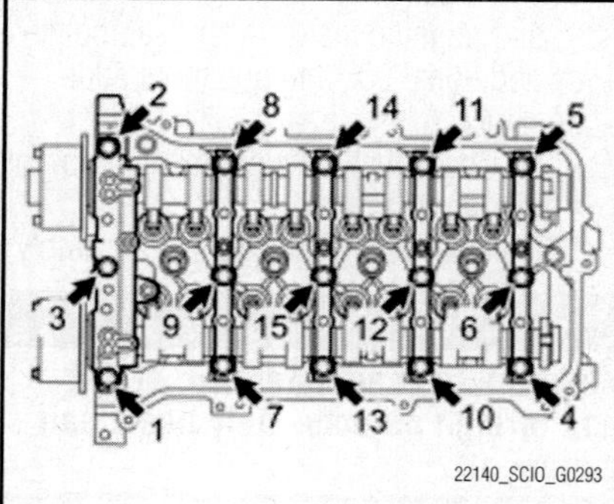

Fig. 75 Remove the 15 bearing cap bolts in the sequence

b. Place the fuel injector in a plastic bag to prevent foreign matter from entering.

78. Remove alternator assembly.

79. Remove the fan belt adjusting bar.

80. Remove the oil level dipstick

81. Remove the oil level gage guide.

82. Remove the water by-pass hose.

83. Remove the water inlet hose.

84. Remove the water by-pass pipe.

85. Remove the vacuum tube assembly.

86. Remove the exhaust manifold heat insulator.

87. Remove the manifold stay.

88. Remove the exhaust manifold.

89. Remove the wire harness clamp bracket.

90. Remove the water inlet.

91. Remove the thermostat.

92. Remove the cylinder head cover sub-assembly.

93. Set No. 1 cylinder to TDC / compression.

94. Remove the crankshaft pulley.

95. Remove the No. 1 chain tensioner assembly.

96. Remove the timing chain cover sub-assembly.

97. Remove the timing chain or belt cover oil seal.

98. Remove the No. 2 chain vibration damper.

99. Remove the chain tensioner slipper.

100. Remove the No. 1 chain vibration damper.

101. Remove the chain sub-assembly.

a. Hold the hexagonal portion of the camshaft with a wrench and turn the camshaft timing gear assembly counterclockwise to loosen the chain on the camshaft timing gears.

b. With the chain loosened, release the chain from the camshaft timing gear assembly and rest it on the camshaft timing gear assembly.

CAUTION

Be sure to release the chain from the sprocket completely.

c. Turn the camshaft clockwise to return it to the original position and remove the chain.

102. Remove the crankshaft timing gear sprocket.

103. Remove the camshaft bearing cap.

a. Uniformly loosen and remove the 10 bearing cap bolts in the sequence shown in the illustration.

b. Uniformly loosen and remove the 15 bearing cap bolts in the sequence shown in the illustration.

c. Uniformly loosen the bolts while keeping the camshaft level.

➡If the camshaft bearing cap bolts have been loosened, reapply seal packing between the camshaft housing and cylinder head.

d. Remove the 5 bearing caps.

104. Remove the camshaft.

a. Remove the camshaft from the camshaft housing.

105. Remove the No. 2 camshaft.

a. Remove the No. 2 camshaft from the camshaft housing.

106. Remove the No. 1 valve rocker arm sub-assembly.

a. Remove the 16 valve rocker arms.

107. Remove the valve lash adjuster assembly.

a. Remove the 16 valve lash adjusters from the cylinder head.

108. Remove the No. 1 camshaft bearing.

a. Remove the 2 No. 1 camshaft bearings.

109. Remove the No. 2 camshaft bearing.

a. Remove the 2 No. 2 camshaft bearings.

110. Remove the camshaft housing sub-assembly.

a. Remove the 2 bolts.

b. Remove the camshaft housing by prying between the cylinder head and camshaft housing with a screwdriver.

CAUTION

Be careful not to damage the contact surfaces of the cylinder head and camshaft housing. Tape the screwdriver tip before use.

111. Remove the cylinder head sub-assembly.

a. Using several steps, loosen and remove the 10 cylinder head bolts uniformly with a 10 mm bi-hexagon wrench in the sequence shown in the illustration.

b. Remove the 10 cylinder head bolts and the plate washers.

CAUTION

Do not drop the washers into the cylinder head.

CAUTION

Head warpage or cracking could result from removing bolts in the wrong order.

c. Remove the cylinder head sub-assembly.

112. Remove the cylinder head gasket.

a. Remove the cylinder head gasket from the cylinder block.

To install:

113. Install cylinder head gasket

a. Place a new cylinder head gasket on the cylinder block with the Lot No. stamp facing upward.

b. Remove any oil from the contact surface.

➡Pay attention to the mounting orientation of the cylinder head gasket.

CAUTION

Do not damage the cylinder gasket when installing the cylinder head onto the cylinder block.

114. Install cylinder head sub-assembly

a. The cylinder head bolts are tightened in 2 successive steps.

b. Apply a light coat of engine oil to the threads of the cylinder head bolts.

c. Using several steps, install and tighten the 10 cylinder head bolts and plate washers uniformly in the sequence shown in the illustration. Torque: 36 ft. lbs. (49 Nm).

d. Mark the front of the cylinder head bolt with paint.

e. Retighten the cylinder head bolts by additional 90° and one more additional 45°.

f. Check that the paint mark is now at a 135° angle from the front.

115. Install valve lash adjuster assembly

CAUTION

Keep the lash adjuster free of dirt and foreign objects. Only use clean engine oil.

a. Place the lash adjuster into a container filled with engine oil.

b. Insert the SST's tip into the lash adjuster's plunger and use the tip to press down on the check ball inside the plunger.

c. Squeeze the SST and lash adjuster together to move the plunger up and down 5 or 6 times.

d. Check the movement of the plunger and bleed the air.

➡Plunger moves up and down. When bleeding air from the high-pressure chamber, make sure that the tip of the SST is actually pressing the check ball. If the check ball is not pressed, air will not bleed.

e. After bleeding the air, remove the SST. Then, try to press the plunger quickly and firmly with a finger.

➡Plunger is very difficult to move. If the result is not as specified, replace the lash adjuster.

f. Install the lash adjusters into their original positions.

116. Install No. 1 valve rocker arm sub-assembly

a. Apply engine oil to the lash adjuster tip and valve stem cap end.

b. Install the valve rocker arms as shown in the illustration.

117. Install No. 1 camshaft bearing

a. Clean the both surfaces of the bearing.

b. Install the 2 No. 1 camshaft bearings.

c. Using calipers, measure the distance between the bearing cap's edge and the camshaft bearing's edge.

➡Dimension (A—B): 0.0276 inches (0.7 mm) or less.

d. Position the bearing to the center of the bearing cap by measuring dimension A—B.

118. Install No. 2 camshaft bearing

a. Clean both surfaces of the bearing.

b. Install the 2 No. 2 camshaft bearings.

c. Using calipers, measure the distance between the bearing cap's edge and the camshaft bearing's edge.

➡Dimension (A): 0.042 to 0.068 inches (1.05 to 1.75 mm).

d. Position the bearing to the center of the bearing cap by measuring dimension A.

119. Install No. 2 camshaft

a. Clean the camshaft journals.

b. Apply a light coat of engine oil to the camshaft journals, camshaft housings and bearing caps.

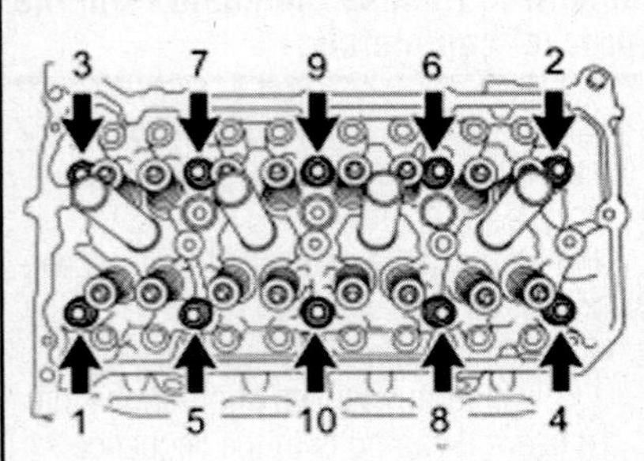

Fig. 76 Remove the 10 cylinder head bolts

c. Install the No. 2 camshaft to the camshaft housing.

120. Install camshaft as follows:

a. Clean the camshaft journals.

b. Apply a light coat of engine oil to the camshaft journals, camshaft housings and bearing caps.

c. Install the camshaft to the camshaft housing.

121. Install camshaft bearing cap

a. Apply engine oil to the camshaft journals, camshaft housing and bearing caps.

b. Check the marks and numbers on the camshaft bearing caps and place them in the proper position and direction.

c. Make sure that the knock pin of the camshaft is positioned as shown in the illustration.

d. Tighten the 10 bolts in the order shown in the illustration. Torque: 12 ft. lbs. (16 Nm).

122. Install camshaft housing sub-assembly

a. Make sure that the valve rocker arm is installed as shown in the illustration.

b. Apply seal packing in a continuous line as shown in the illustration.

➡Seal diameter: 0.138 to 0.158 inches (3.5 to 4.0 mm).

c. Remove any oil from the contact surface.

d. Install the camshaft housing sub-assembly RH within 3 minutes of applying the seal packing.

e. Do not start the engine for at least 2 hours after installing.

f. Set the camshaft and No. 2 camshaft.

g. Install the camshaft housing and tighten the 17 bolts in the order shown in the illustration. Torque: 20 ft. lbs. (27 Nm).

➡After installing the camshaft housing, make sure that the cam lobes are positioned as shown in the illustration. If any of the bolts are loosened during installation, remove the camshaft housing, clean the installation surfaces, and reapply seal packing. If the camshaft housing is removed because any of the bolts are loosened during installation, make sure that the previously applied seal packing does not enter any oil passages. After installing the camshaft housing, wipe off any seal packing that seeped out from between the housing and the cylinder head.

123. Install crankshaft timing gear or sprocket

124. Install No. 1 chain vibration damper

125. Install chain sub-assembly

126. Install chain tensioner slipper

127. Install No. 2 chain vibration damper

128. Install timing chain or belt cover oil seal

129. Install timing chain or belt cover sub-assembly

130. Install crankshaft pulley

131. Install No. 1 chain tensioner assembly

132. Install cylinder head cover sub-assembly

133. Install thermostat

134. Install water inlet

135. Install wire harness clamp bracket

136. Install exhaust manifold

137. Install manifold stay

138. Install No. 1 exhaust manifold heat insulator

139. Install air tube assembly

140. Install No. 1 water by-pass pipe

141. Install water inlet hose

142. Install water by-pass hose

143. Install oil level gage guide

144. Install oil level dipstick

145. Install fan belt adjusting bar

146. Install alternator assembly

147. Install fuel injector

148. Install injector vibration insulator

149. Install No. 1 delivery pipe spacer

150. Install fuel delivery pipe sub-assembly

151. Install intake manifold

152. Install No. 1 radiator hose

153. Install heater water inlet hose a

154. Install heater water outlet hose a (from heater unit)

155. Install No. 2 ventilation hose

156. Install radio setting condenser

157. Install ignition coil assembly

158. Install transmission oil filler tube sub-assembly (A/T)

159. Install transmission oil level gage sub-assembly (A/T)

160. Install No. 1 oil cooler outlet tube (A/T)

161. Install No. 1 oil cooler inlet tube (A/T)

162. Install No. 2 oil cooler tube clamp (A/T)

163. Install front suspension crossmember sub-assembly

164. Install engine assembly with transaxle

165. Install front axle assembly LH

166. Install front axle assembly RH

167. Install front lower suspension arm sub-assembly LH

168. Install front lower suspension arm sub-assembly RH

169. Install front stabilizer link assembly LH

170. Install front stabilizer link assembly RH

171. Install tie rod end sub-assembly LH

172. Install tie rod end sub-assembly RH

173. Install speed sensor front LH

174. Install speed sensor front RH

175. Install front axle shaft LH nut

176. Install front axle shaft RH nut

177. Install front exhaust pipe assembly

178. Install front floor brace center

179. Install front console box

180. Install rear console box assembly

181. Install console box carpet

182. Install console box rear cover

183. Install console panel upper

184. Install shift lever knob sub-assembly (M/T)

185. Install No. 1 steering column hole cover sub-assembly

186. Install steering sliding yoke sub-assembly

187. Install column hole cover silencer sheet

188. Install clutch release cylinder assembly (M/T)

189. Connect fuel tube sub-assembly

190. Connect heater water inlet hose a

191. Connect heater water outlet hose a (from heater unit)

192. Connect engine wire

193. Install No. 1 fuel vapor feed hose

194. Install union to check valve hose

195. Install transmission control cable assembly (A/T)

196. Install transmission control cable assembly (M/T)

197. Install with pulley compressor assembly

198. Install radiator assembly

199. Install upper radiator support sub-assembly

200. Install upper radiator support absorber

201. Install No. 1 cooler cover

202. Install hood lock assembly

203. Connect No. 1 radiator hose

204. Connect No. 2 oil cooler inlet hose (A/T)

205. Connect No. 1 oil cooler outlet hose (A/T)

206. Install No. 2 radiator hose

207. Install front bumper cover

208. Install fan and alternator v-belt

209. Adjust fan and alternator v-belt

210. Inspect fan and alternator v-belt

211. Install Engine cover

212. Install battery carrier

213. Install air cleaner bracket

214. Install air cleaner assembly

215. Install outer cowl top panel

216. Install front air shutter seal RH
217. Install front wiper motor and link
218. Install cowl top ventilator louver LH
219. Install cowl top ventilator louver sub-assembly
220. Install hood to cowl top seal
221. Install front wiper arm and blade assembly LH
222. Install front wiper arm and blade assembly RH
223. Install front wiper arm head cap
224. Install battery tray
225. Install battery
226. Add engine coolant
227. Add engine oil
228. Inspect for fuel leak
229. Inspect for engine oil leak
230. Inspect for exhaust gas leak
231. Inspect for engine coolant leak
232. Install engine under cover RH
233. Install engine under cover LH
234. Install front wheels

EXHAUST MANIFOLD

REMOVAL & INSTALLATION

tC & xB Models

See Figures 77 and 78.

1. Before servicing the vehicle, refer to the Precautions Section.
2. Drain the cooling system.
3. Relieve the fuel system pressure.
4. Remove or disconnect the following:
   - Negative battery cable
   - Engine appearance cover
   - Air intake assembly
   - Throttle body
   - Fuel rail
   - Intake manifold
   - Front exhaust pipe
   - Oil dipstick guide
   - Exhaust manifold stays
   - Exhaust manifold heat shield
   - Exhaust manifold and gasket

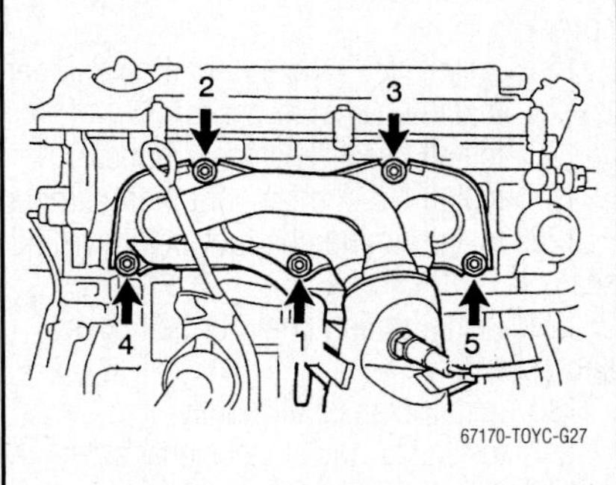

Fig. 78 Exhaust manifold torque sequence—2.4L Engine

To install:

5. Install the exhaust manifold with a new gasket. Tighten the bolts in sequence to 27 ft. lbs. (37 Nm).
6. Install or connect the following:
   - Exhaust manifold heat shield. Tighten to 9 ft. lbs. (12 Nm).
   - Exhaust manifold stays. Tighten to 32 ft. lbs. (44 Nm).
   - Oil dipstick guide. Tighten to 80 inch lbs. (9 Nm).
   - Front exhaust pipe. Tighten to 32 ft. lbs. (44 Nm).
   - Intake manifold
   - Fuel rail
   - Throttle body
   - Air intake assembly
   - Engine appearance cover
   - Negative battery cable

xD Models

See Figures 79 and 80.

1. Remove the air fuel ratio sensor.
   a. Disconnect the air fuel ratio sensor connector and clamp.
   b. Using SST, remove the air fuel ratio sensor.
2. Remove the engine under cover RH.
3. Remove the front exhaust pipe assembly.
4. Drain the automatic transaxle fluid (A/T).
5. Drain the manual transaxle oil (M/T).
6. Remove the front wheels.
7. Remove the front axle shaft RH nut
8. Remove the front speed sensor RH.
9. Remove the tie rod end sub-assembly RH.
10. Remove the front stabilizer link assembly RH.
11. Remove the No. 1 front lower suspension arm sub-assembly RH.
    a. Remove the clip and castle nut.
    b. Install SST to the threaded section of the lower ball joint.

➡Make sure the upper ends of the threaded section of the lower ball joint and SST (spacer B) are aligned.

    c. Using SST, separate the front lower suspension arm from the front axle assembly.
12. Separate the front axle assembly RH.
13. Remove the drive shaft assembly RH.
14. Remove the No. 1 exhaust manifold heat insulator.
    a. Remove the 4 bolts and remove the No. 1 exhaust manifold heat insulator.
15. Remove the manifold stay.
    a. Remove the 3 bolts and remove the manifold stay.
16. Remove the exhaust manifold.
    a. Remove the 5 nuts and remove the exhaust manifold.

To install:

17. Install exhaust manifold.
    a. Install a new exhaust manifold gasket and exhaust manifold with the 5 nuts. Torque: 16 ft. lbs. (21 Nm).
18. Install manifold support bracket.

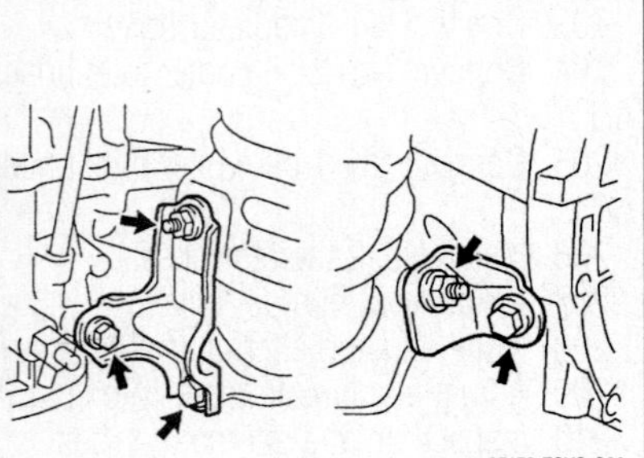

Fig. 77 Exhaust manifold stays—2.4L Engine

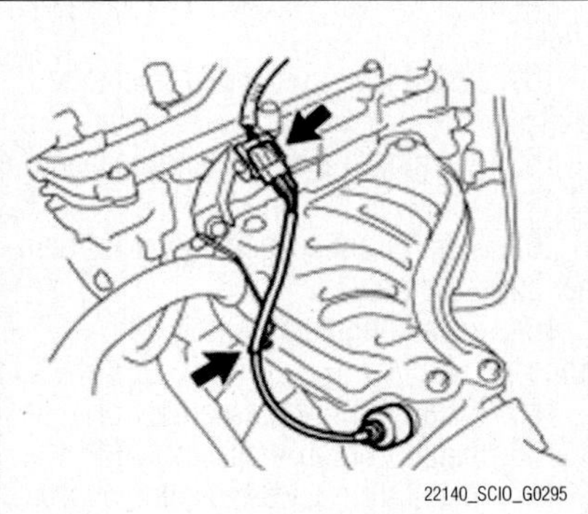

Fig. 79 Remove the air fuel ratio sensor.

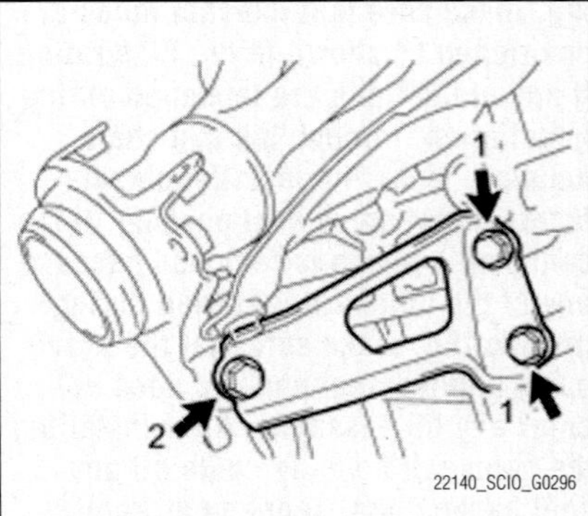

Fig. 80 Install the manifold support bracket with the 3 bolts in the sequence

a. Install the manifold support bracket with the 3 bolts in the sequence. Torque: 32 ft. lbs. (43 Nm).

19. Install No. 1 exhaust manifold heat insulator.

a. Install the No. 1 exhaust manifold heat insulator with the 4 bolts. Torque: 9 ft. lbs. (12 Nm).

20. Install drive shaft assembly RH.
21. Install front axle assembly RH.
22. Install No. 1 front suspension arm sub-assembly RH.
23. Install front stabilizer link assembly RH.
24. Install tie rod end sub-assembly RH.
25. Install front speed sensor RH.
26. Install front axle shaft RH nut.
27. Install front wheel.
28. Install front exhaust pipe assembly.
29. Install air fuel ratio sensor.
30. Add automatic transaxle fluid (A/T).
31. Inspect automatic transaxle fluid (A/T).
32. Add manual transaxle oil (M/T).
33. Inspect manual transaxle oil (M/T).
34. Inspect A/T fluid leak (A/T).
35. Inspect M/T oil leak.
36. Inspect for exhaust gas leak.
37. Inspect and adjust front wheel alignment.
38. Install engine under cover RH.

INTAKE MANIFOLD

REMOVAL & INSTALLATION

tC & xB Models

See Figures 81 and 82.

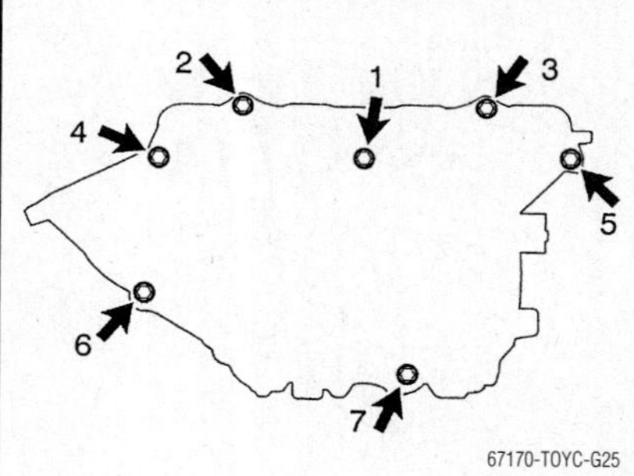

Fig. 82 Intake manifold fastener torque sequence—2.4L Engine

1. Before servicing the vehicle, refer to the Precautions Section.
2. Drain the cooling system.
3. Relieve the fuel system pressure.
4. Remove or disconnect the following:
   - Negative battery cable
   - Hood
   - Engine appearance cover
   - Wiper arm assembly
   - Top cowl seal
   - Left-hand top cowl ventilator louver
   - Wiper linkage assembly
   - Top outer cowl panel
   - Left-hand cowl body mounting bracket
   - Air intake assembly
   - Throttle body
   - Fuel rail
   - Water by-pass hoses from the throttle body
   - Intake manifold and gasket

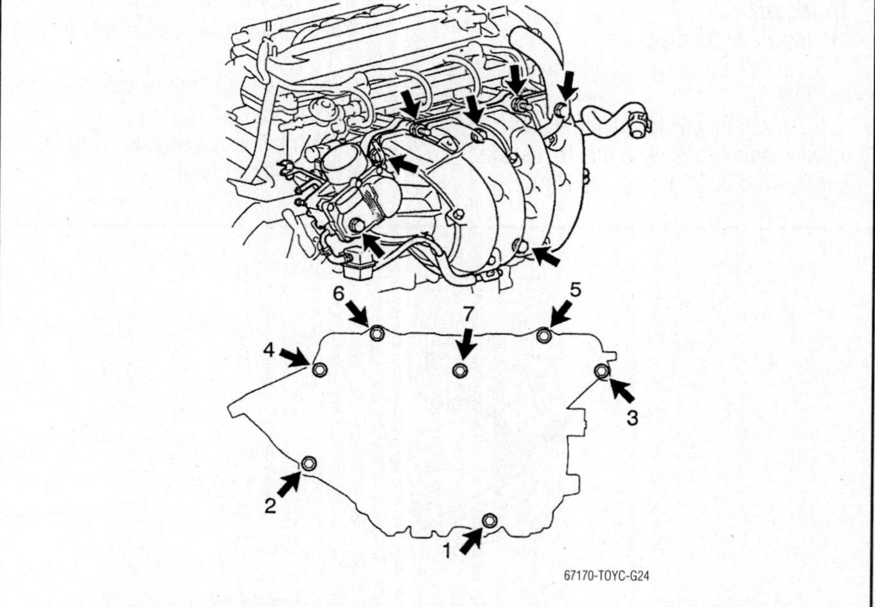

Fig. 81 Intake manifold fastener location and loosening sequence—2.4L Engine

To install:

5. Install a new gasket into the intake manifold.
6. Install the intake manifold and tighten fasteners in sequence to 22 ft. lbs. (30 Nm).
7. Install or connect the following:
   - Water bypass hoses from the throttle body
   - Fuel rail
   - Throttle body. Tighten bolts to 22 ft lbs. (30 Nm).
   - Air intake assembly
8. The remainder of the installation is the reverse order of removal.
9. Refill the cooling system to the correct level.
10. Start the engine and check for leaks.

xD Models

See Figures 83 through 87.

1. Discharge the fuel system pressure.

✻✻ CAUTION

The discharge fuel system pressure procedures must be performed before disconnecting any part of the fuel system. After performing the procedures, pressure will remain in the fuel line. When disconnecting the fuel line, place a cloth or equivalent over fittings to reduce the risk of fuel spray.

a. Remove the rear seat cushion assembly LH

b. Remove the rear floor service hose cover.

c. Disconnect the connector from the fuel pump assembly.

d. Start the engine. After the engine stops naturally, turn the ignition switch OFF.

e. Start the engine again and make sure that engine does not start.

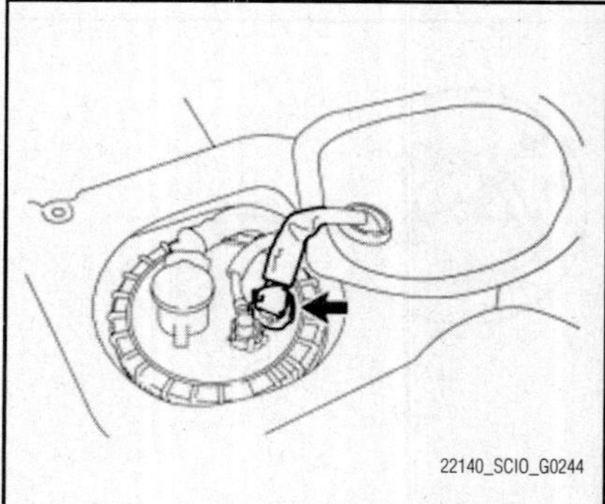

Fig. 83 Disconnect the connector from the fuel pump assembly.

f. Remove the fuel tank cap, and let the air out of the fuel tank.

g. Connect the connector.

h. Install the rear floor service hole cover with new butyl tape.

i. Install the rear seat cushion assembly LH.

2. Disconnect the cable from negative battery terminal.

3. Remove the engine under cover LH.

4. Remove the engine under cover RH.

5. Drain the engine coolant.

6. Remove the engine cover.

7. Remove the air cleaner cap sub-assembly with hose.

a. Disconnect the wire harness clamp and Mass Air Flow (MAF) meter connector.

b. Disconnect the No. 2 ventilation hose.

c. Loosen the hose clamp, unlock the 2 clamps and remove air cleaner cap sub-assembly with hose.

8. Separate the engine wire harness.

a. Remove the 3 bolts, harness clamp and 4 fuel injector connectors and separate the engine wire harness.

9. Remove the harness bracket.

10. Disconnect the No. 1 fuel vapor feed hose.

11. Disconnect the vacuum hose.

12. Remove the fuel pipe clamp.

13. Disconnect the fuel tube sub-assembly.

14. Remove fuel the delivery pipe sub-assembly.

15. Remove the No. 1 delivery pipe spacer.

16. Remove the injector vibration insulator.

17. Remove the starter assembly.

18. Remove the oil cooler tube.

a. Using a union nut wrench, separate the inlet No. 1 oil cooler tube while holding the oil cooler tube union with a wrench.

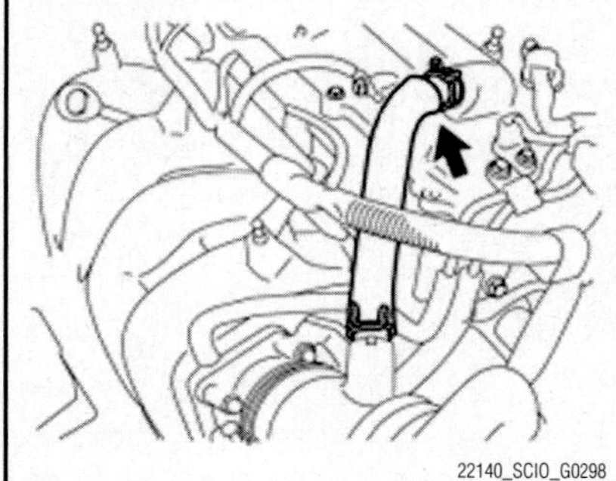

Fig. 85 Disconnect the No. 2 ventilation hose.

b. Using a union nut wrench, separate the outlet No. 1 oil cooler tube while holding the oil cooler tube union with a wrench.

c. Remove the bolt and remove the oil cooler tube clamp.

19. Remove the transmission oil level gauge sub-assembly.

a. Remove the transmission oil level dipstick.

b. Disconnect the breather hose.

c. Remove the bolt and remove the transmission oil level gauge sub-assembly.

20. Remove the engine oil level dipstick.

21. Remove the throttle body assembly.

22. Remove the intake manifold.

a. Disconnect the 4 wire harness clamps.

b. Disconnect the ventilation hose.

c. Remove the 4 bolts and 2 nuts and remove the intake manifold and intake manifold stay.

To install:

23. Install intake manifold

a. Install a new gasket onto the intake manifold.

b. Install the intake manifold and intake manifold stay with the 4bolts and 2 nuts. Torque: 21 ft. lbs. (28 Nm).

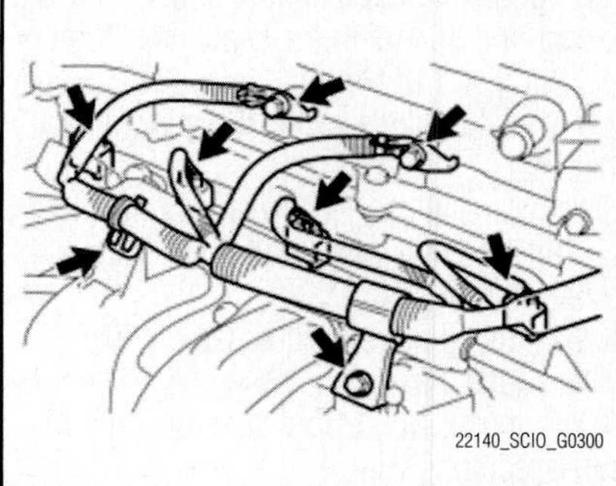

Fig. 87 Separate the engine wire harness.

c. Connect the ventilation hose.

d. Connect the 4 wire harness clamps.

24. Install throttle body assembly

25. Install transmission oil level gauge sub-assembly

a. Install the transmission oil level gauge sub-assembly with the bolt.

b. Connect the breather hose.

c. Install the transmission oil level dipstick.

26. Install oil cooler tube

a. Install the oil cooler tube clamp.

b. Provisionally install the outlet No. 1 oil cooler tube into the oil cooler tube union.

c. Provisionally install the inlet No. 1 oil cooler tube into the oil cooler tube union.

d. Install the No. 2 oil cooler tube clamp onto the transmission oil filler tube sub-assembly with the bolt.

e. Using a union nut wrench, install the inlet No. 1 oil cooler tube while holding the oil cooler tube union with a spanner.

- Without union nut wrench - Torque: 25 ft. lbs. (34 Nm).
- With union nut wrench - Torque: 24 ft. lbs.32 Nm).

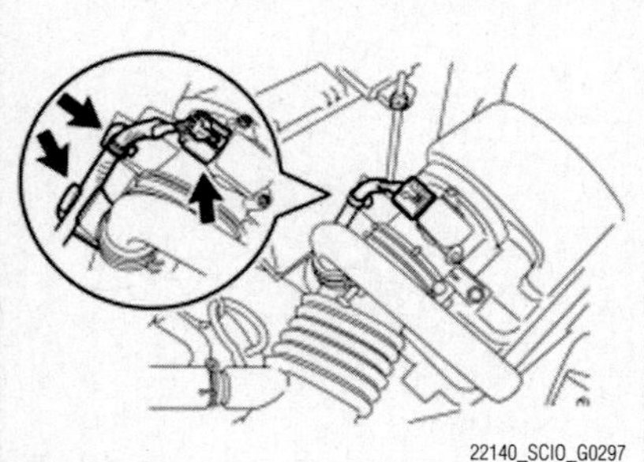

Fig. 84 Disconnect the wire harness clamp and Mass Air Flow (MAF) meter connector.

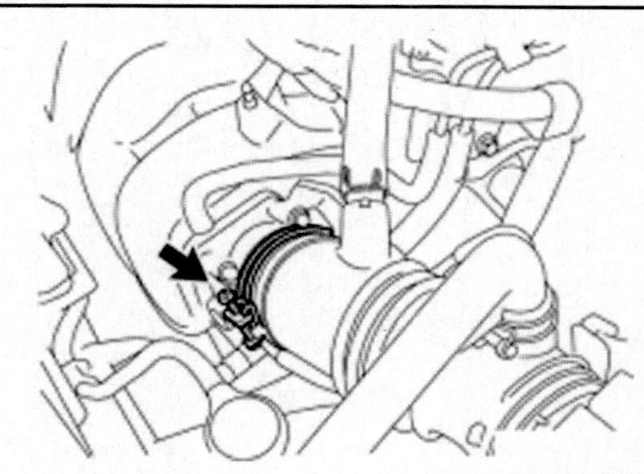

Fig. 86 Remove air cleaner cap sub-assembly with hose

➡This torque value can be obtained by using a torque wrench with a fulcrum length of 13.58 inches (345 mm) and a union nut wrench with a fulcrum length of 1.18 inches (30 mm). This torque value is effective when union nut wrench is parallel to a torque wrench.

27. Install engine oil level dipstick
28. Install starter assembly
29. Install injector vibration insulator
30. Install No. 1 delivery pipe spacer
31. Install fuel delivery pipe sub-assembly
32. Connect fuel tube sub-assembly
33. Install fuel pipe clamp
34. Connect vacuum hose
35. Connect No. 1 fuel vapor feed hose
36. Install harness bracket
37. Install engine wire harness
38. Install air cleaner cap sub-assembly with hose
39. Connect cable to negative battery terminal
40. Add engine coolant
41. Inspect for coolant leak
42. Inspect for fuel leak
43. Install engine cover
44. Install engine under cover RH
45. Install engine under cover LH

OIL PAN

REMOVAL & INSTALLATION

See Figure 88.

1. Before servicing the vehicle, refer to the Precautions Section.
2. Drain the engine oil.
3. Remove the engine assembly from the vehicle and secure to a suitable chain block and sling device.
4. Remove the oil pan mounting bolts.
5. Using Special Tool 09032-00100, cut off the sealant between the front cover, cylinder block and oil pan.
6. Remove the oil pan.

To install:

7. Remove any old sealant from the oil pan.
8. Apply a continuous bead of sealant to the contact surfaces of the oil pan.

➡Do not expose the sealant to engine oil for at least 2 hours after installation.

9. Install the oil pan. Tighten the bolts in sequence to 80 inch lbs. (9 Nm).
10. Install the engine assembly.

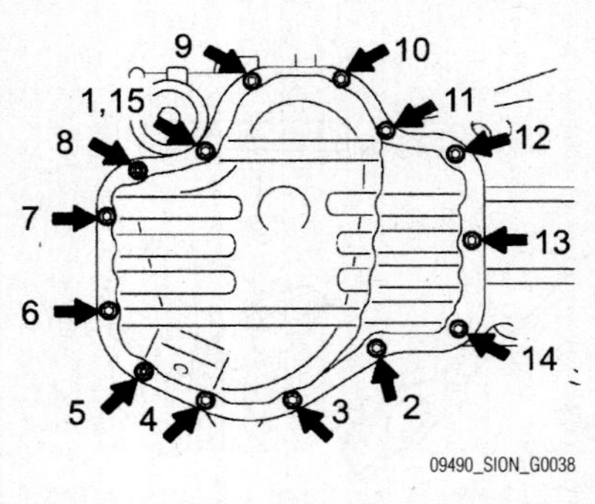

Fig. 88 Oil pan torque sequence—2.4L Engine

11. Fill the engine with oil to the correct level.
12. Start the engine and check for leaks.

OIL PUMP

REMOVAL & INSTALLATION

tC & xB Models

See Figures 89 and 90.

1. Before servicing the vehicle, refer to the Precautions Section.
2. Drain the engine oil.
3. Remove or disconnect the following:
   - Negative battery cable
   - Front cover
   - Timing chain
   - Oil pump assembly

To install:

4. Install the oil pump assembly with a new gasket. Tighten the mounting bolts to 14 ft. lbs. (19 Nm).
5. Install or connect the following:
   - Timing chain
   - Front cover
   - Negative battery cable
6. Refill the engine with oil to the correct level.
7. Start the engine and check for leaks.

xD Models

➡The engine must be removed from the vehicle to access the oil pump. Refer to Engine Assembly R&I.

1. Remove the engine assembly with transaxle.
2. Remove the front suspension cross-member sub-assembly.
   a. Remove the bolt and remove the air fuel ratio sensor bracket.
   b. Install the engine hangers with the bolts. Torque: 32 ft. lbs. (43 Nm).
   c. Using an engine sling device and a chain block, suspend the engine assembly with transaxle and front suspension crossmember.
   d. Remove the through bolt from the engine moving control rod and remove the front suspension crossmember.
3. Remove the No. 1 oil cooler inlet tube (A/T).
4. Remove the No. 1 oil cooler outlet tube (A/T).
5. Remove the No. 2 oil cooler tube clamp (A/T).
6. Remove the transmission oil level gage sub-assembly (A/T).
7. Separate the transmission oil filler tube sub-assembly (A/T).
8. Remove the intake manifold.
9. Remove the oil level dipstick.
10. Remove the oil level gage guide.
11. Remove the alternator assembly.
12. Remove the fan belt adjusting bar.
13. Separate the water inlet hose.
14. Separate the No. 3 water by-pass hose.

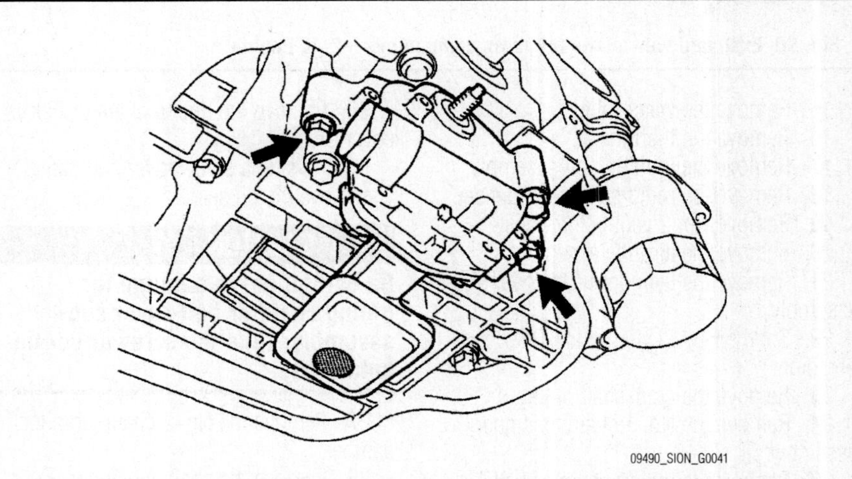

Fig. 89 Remove the three mounting bolts to remove the oil pump assembly—2.4L Engine

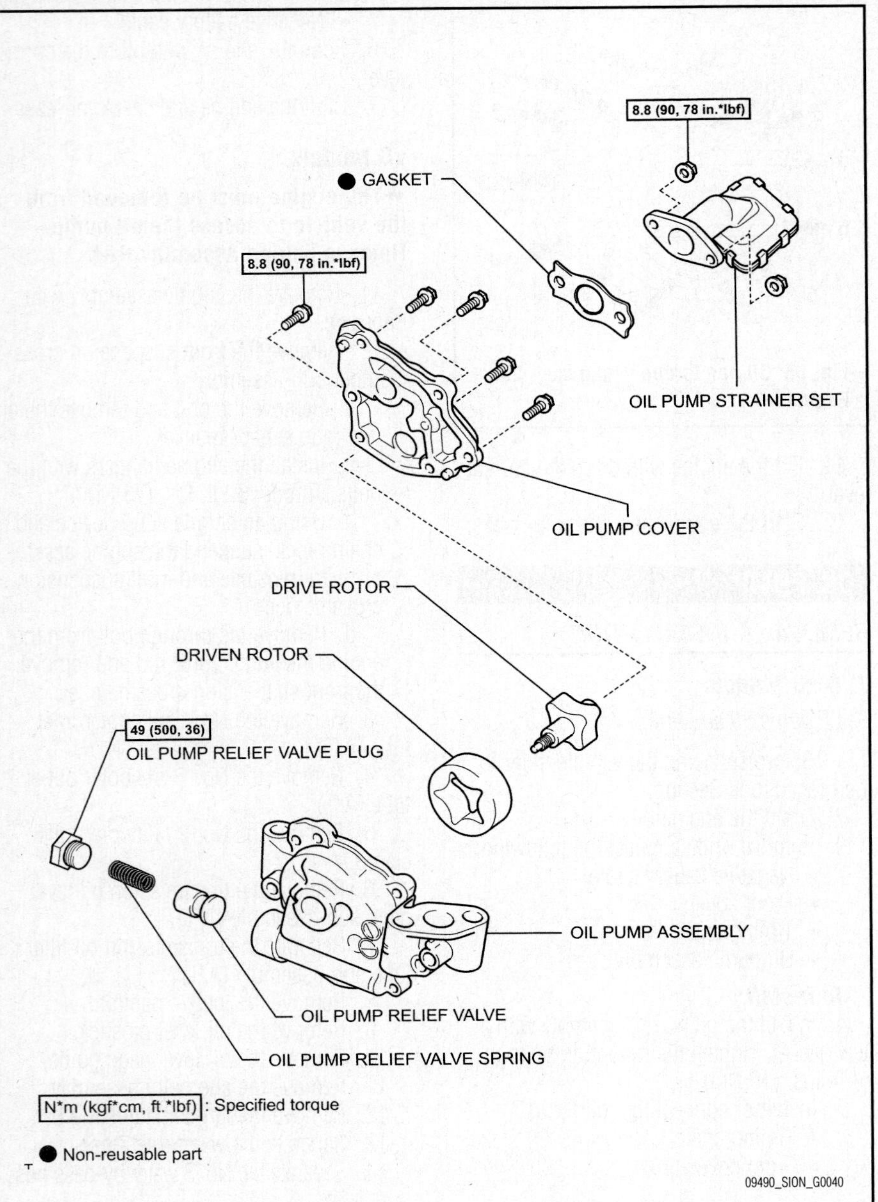

Fig. 90 Exploded view of the oil pump components—2.4L Engine

15. Remove the water inlet.
16. Remove the thermostat.
17. Remove the ignition coil assembly.
18. Remove the radio setting condenser.
19. Remove No. 2 ventilation hose.
20. Remove the air tube assembly.
21. Remove the cylinder head cover sub-assembly.
22. Set the No. 1 cylinder to TDC / compression.
23. Remove the crankshaft pulley.
24. Remove the No. 1 chain tensioner assembly.
25. Remove the timing chain or belt cover sub-assembly.
26. Remove the timing chain or belt cover oil seal.
    a. Using a screwdriver and hammer, remove the oil seal

CAUTION

Be careful not to damage the timing chain or belt cover sub-assembly. Tape the screwdriver tip before use.

27. Remove the No. 2 chain vibration damper.
28. Remove the chain tensioner slipper.
29. Remove the No. 1 chain vibration damper.
30. Remove the chain sub-assembly.
31. Remove the crankshaft timing gear or sprocket.
32. Remove the No. 2 chain sub-assembly.
33. Remove the No. 2 oil pan sub-assembly.
34. Remove the oil pump assembly.
    a. Remove the 3 bolts and remove the oil pump assembly.
35. Installation is the reverse of the removal procedure.

PISTON AND RING

POSITIONING

tC & xB Models

See Figures 91 and 92.

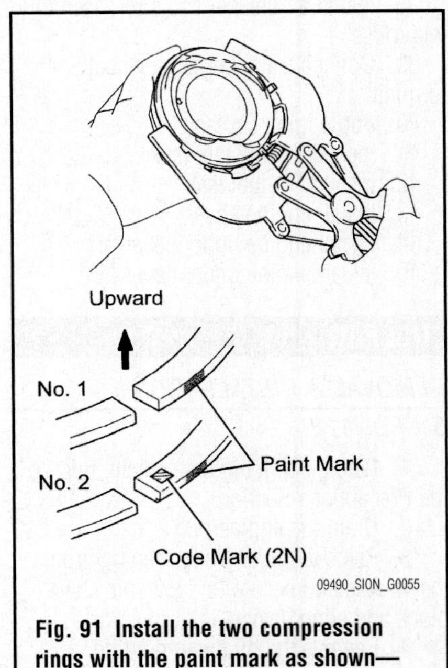

Fig. 91 Install the two compression rings with the paint mark as shown—2.4L Engine

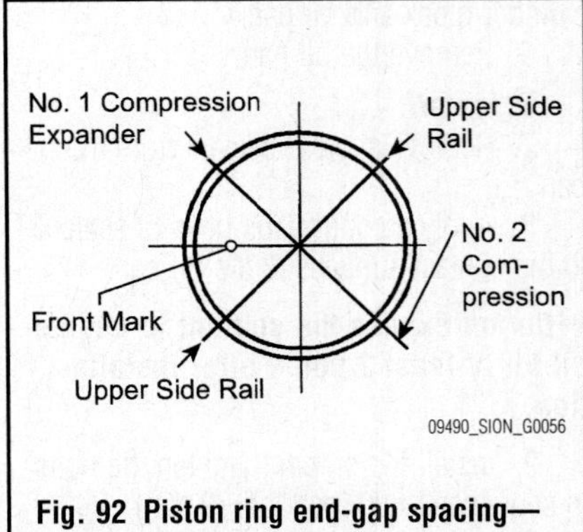

Fig. 92 Piston ring end-gap spacing—2.4L Engine

xD Models

See Figures 93 and 94.

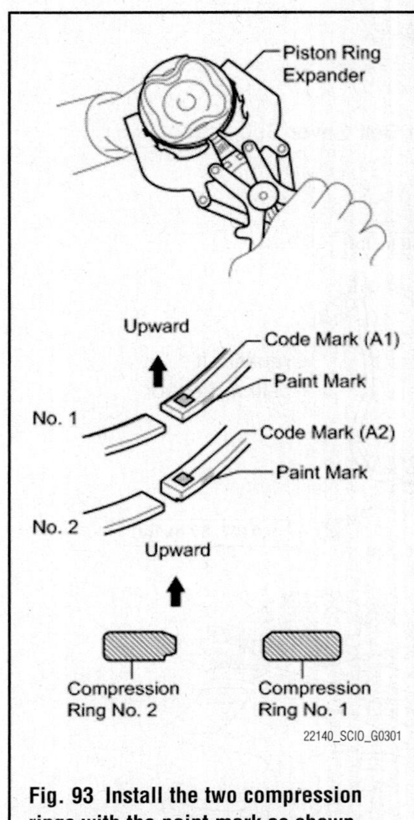

Fig. 93 Install the two compression rings with the paint mark as shown—1.8L Engine

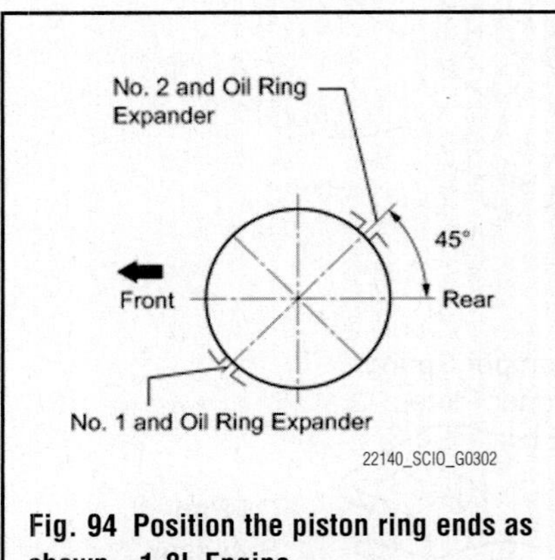

Fig. 94 Position the piston ring ends as shown—1.8L Engine

REAR MAIN SEAL

REMOVAL & INSTALLATION

1. Before servicing the vehicle, refer to the Precautions Section.
2. Remove or disconnect the following:
   - Transaxle
   - Flywheel/Driveplate
   - Oil seal

To install:

3. Using Special Tool 09223-56010 or suitable seal installer, tap in the oil until its surface is flush with the seal retainer edge.
4. Install or connect the following:
   - Flywheel/Driveplate
   - Transaxle
5. Start the engine check for leaks.

TIMING CHAIN & SPROCKETS

REMOVAL & INSTALLATION

tC & xB Models

See Figures 95 through 103.

1. Before servicing the vehicle, refer to the Precautions Section.
2. Drain the engine oil.
3. Drain the cooling system.
4. Support the engine with a suitable jack.
5. Remove or disconnect the following:
   - Negative battery cable
   - Hood
   - Right-hand front wheel
   - Engine undercover
   - Front fender apron
   - Engine appearance cover
   - Front exhaust pipe
   - Accessory drive belt
   - Alternator
   - Power steering pump and reservoir
   - Right-hand engine mount
   - Ignition coil
   - Cylinder head cover
   - Accessory drive belt tensioner

➡Lift the engine with a suitable jack to gain access to the tensioner mounting bolts.

   - Crankshaft Position (CKP) sensor
   - Oil pan
6. Set the No. 1 cylinder to TDC.
7. Using Special Tool 09213-54015, remove the crankshaft pulley bolt.
8. Remove the crankshaft pulley. Use Special Tool 09950-50013 and the pulley bolt if necessary.
9. Remove or disconnect the following:
   - Timing chain tensioner
   - Front cover
   - Crankshaft Position (CKP) sensor plate
   - Timing chain guide
   - Timing chain tensioner slipper
   - Upper timing chain
   - Timing chain vibration damper
   - Crankshaft sprocket
10. Remove the lower timing chain as follows:
    a. Turn the crankshaft by 90° counterclockwise to align the adjusting hole of the oil pump drive shaft gear with the groove of the oil pump.
    b. Insert a 4 mm diameter bar into the adjusting hole of the oil pump drive shaft gear to lock in position, then remove the nut.
    c. Remove the bolt, then remove the chain tensioner plate and spring.
    d. Remove the oil pump drive gear, oil pump drive shaft gear and lower timing chain.

To install:

11. Set the crankshaft key in the left horizontal position. Turn the cutout of the drive shaft to the top.
12. Align the yellow mark links with the timing marks of the each gear as shown in the illustration.
13. Install the gears onto the crankshaft and oil pump shaft with the lower chain wrapped.
14. Temporarily tighten the oil pump drive shaft gear with the nut.
15. Insert the damper spring into the adjusting hole, then install the chain tensioner plate with the bolt. Tighten to 9 ft. lbs. (12 Nm).
16. Align the adjusting hole of the oil pump drive shaft gear with the groove of the oil pump.
17. Insert a 4 mm diameter bar into the adjusting hole of the oil pump drive shaft gear to lock in position, then tighten the nut to 22 ft. lbs. (30 Nm).
18. Turn the crankshaft clockwise by 90° to position the crankshaft key upward.
19. Install the crankshaft sprocket.
20. Turn the camshafts with a wrench on the hexagonal lobe to align the timing marks of the camshaft timing gear with each timing mark located on the No. 1 and No. 2 bearing caps as shown in the illustration.
21. Using the crankshaft pulley bolt, turn the crankshaft to position the key on the crankshaft upward.
22. Install the upper timing chain onto the crankshaft timing gear with the gold or orange mark link aligned with the timing mark on the crankshaft sprocket.
23. Using Special Tool 09309-37010, tap in the crankshaft timing gear.
24. Align the gold or yellow mark links with each timing mark located on the camshaft timing gears, then install the upper timing chain.

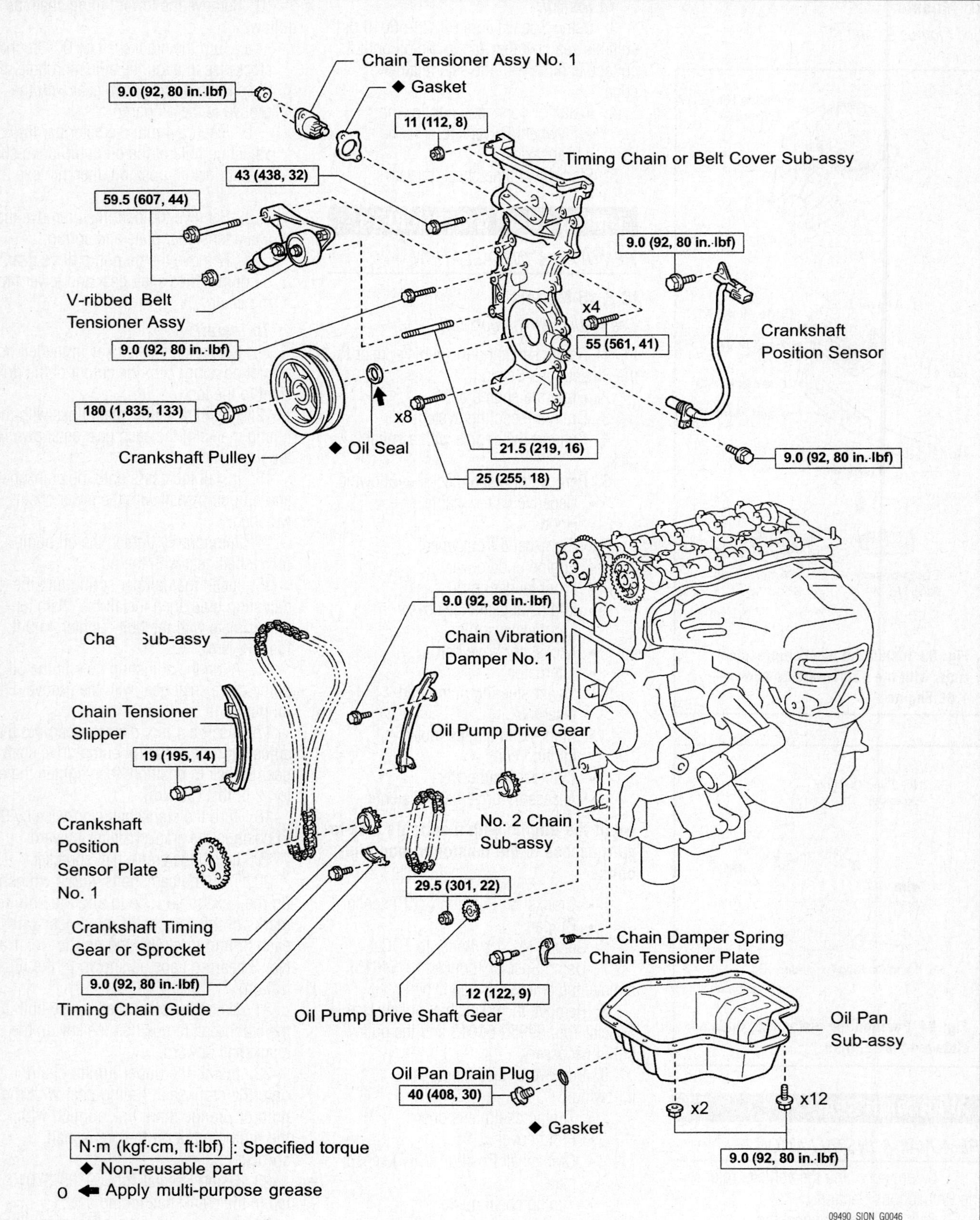

Fig. 95 Exploded view of the timing chain components—2.4L Engine

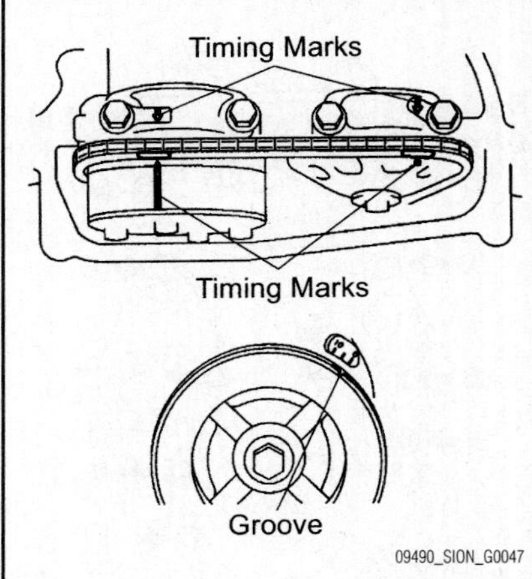

Fig. 96 Align the timing marks on the camshaft timing gears and Nos. 1 and 2 camshaft bearing caps to ensure the No. 1 cylinder is at TDC compression.

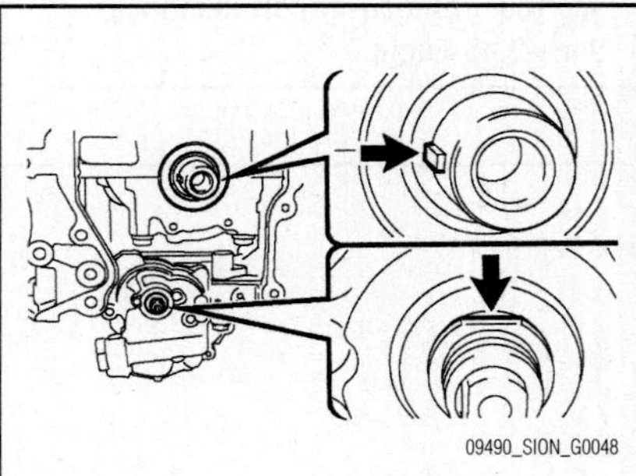

Fig. 97 Set the crankshaft key and drive shaft cutout as shown—2.4L Engine

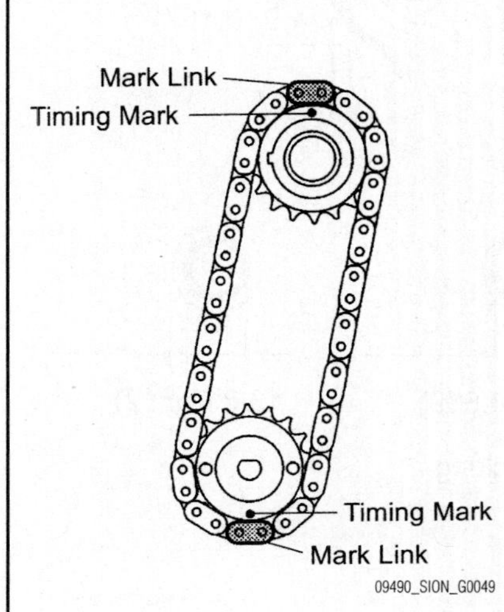

Fig. 98 Align the marked links with the timing gear marks—2.4L Engine

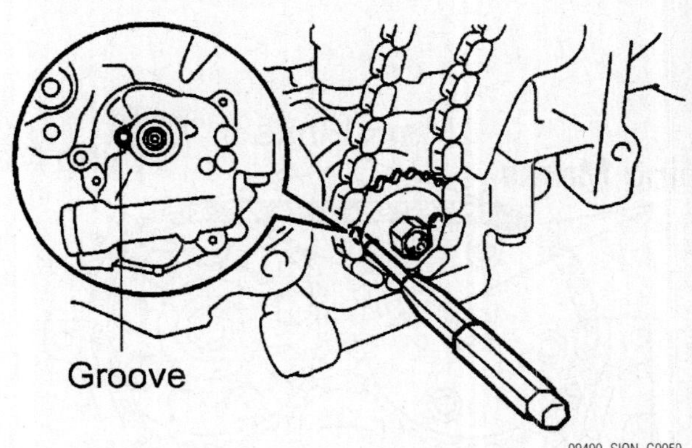

Fig. 99 Insert a bar into the adjusting hole to lock the drive shaft gear in position—2.4L Engine

25. Install or connect the following:
    - Timing chain tensioner slipper. Tighten to 14 ft. lbs. (19 Nm).
    - Timing chain guide. Tighten to 80 inch lbs. (9 Nm).
    - Crankshaft Position (CKP) sensor plate
26. Remove any old sealant from the front cover.
27. Apply a continuous bead of sealant to the front cover contact surfaces as shown.
28. Install the front cover. Tighten the bolts as follows:
    a. Bolt A to 80 inch lbs. (9 Nm).
    b. Bolts B to 18 ft. lbs. (25 Nm).
    c. Bolts C to 41 ft. lbs. (55 Nm).
    d. Bolt D to 32 ft. lbs. (43 Nm).
    e. Nuts to 8 ft. lbs. (11 Nm).
29. Install or connect the following:
    - Timing chain tensioner. Tighten to 80 inch lbs. (9 Nm).
    - Crankshaft pulley. Tighten to 133 ft. lbs. (180 Nm).
    - Accessory drive belt tensioner. Tighten to 44 ft. lbs. (60 Nm).
    - Cylinder head cover
    - Ignition coil
    - Right-hand engine mount
    - Power steering pump and reservoir
    - Alternator
    - Accessory drive belts
    - Front exhaust pipe
    - Engine undercover
    - Right-hand front fender apron
    - Front wheel
    - Negative battery cable
30. Refill the engine with oil to the correct level.
31. Refill the cooling system to the correct level.
32. Start the engine and check for leaks.

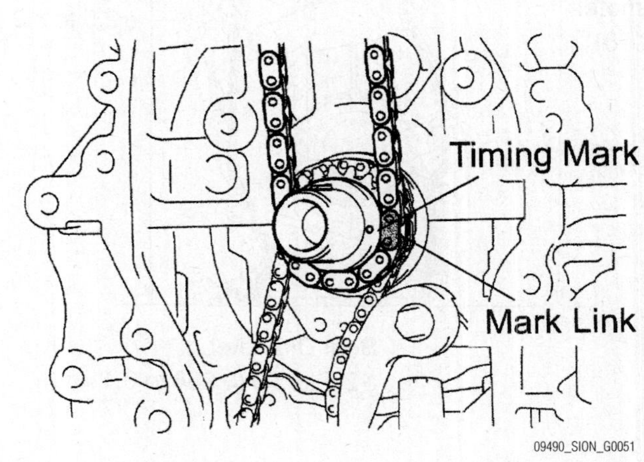

Fig. 100 Align the timing marks installing the upper timing chain on the crankshaft sprocket—2.4L Engine

Mark Links
Timing Marks
Timing Mark
09490_SION_G0052

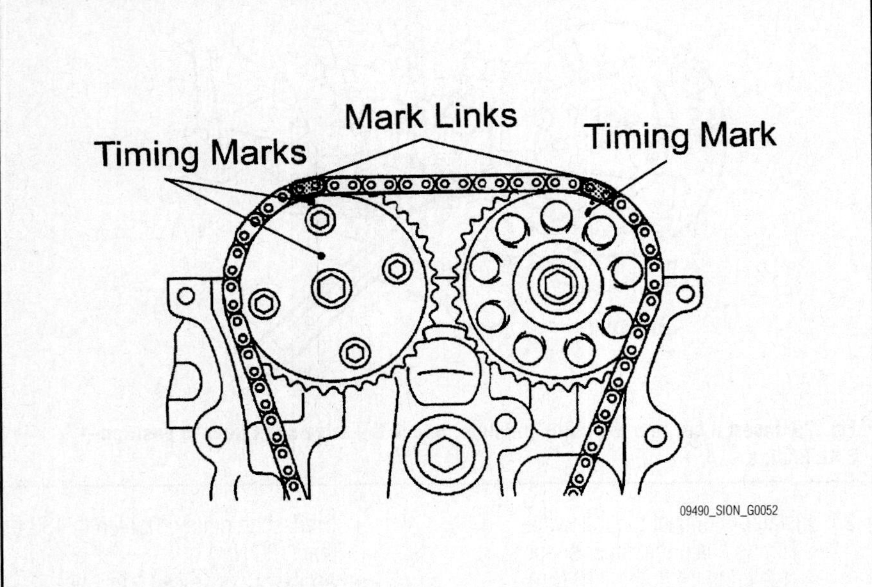

Fig. 101 Align the timing marks with the camshaft timing gears when installing the upper timing chain—2.4L Engine

Nut
D
Nut
C
C
C
C
A
B
B
B
B
B
B
B
B
09490_SION_G0054

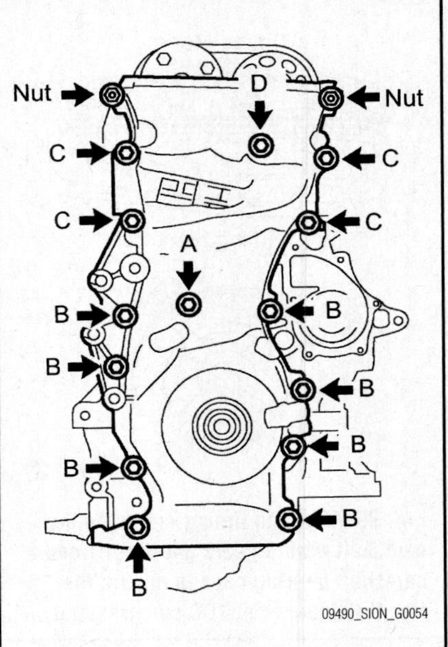

Fig. 103 Front cover bolt identification—2.4L Engine

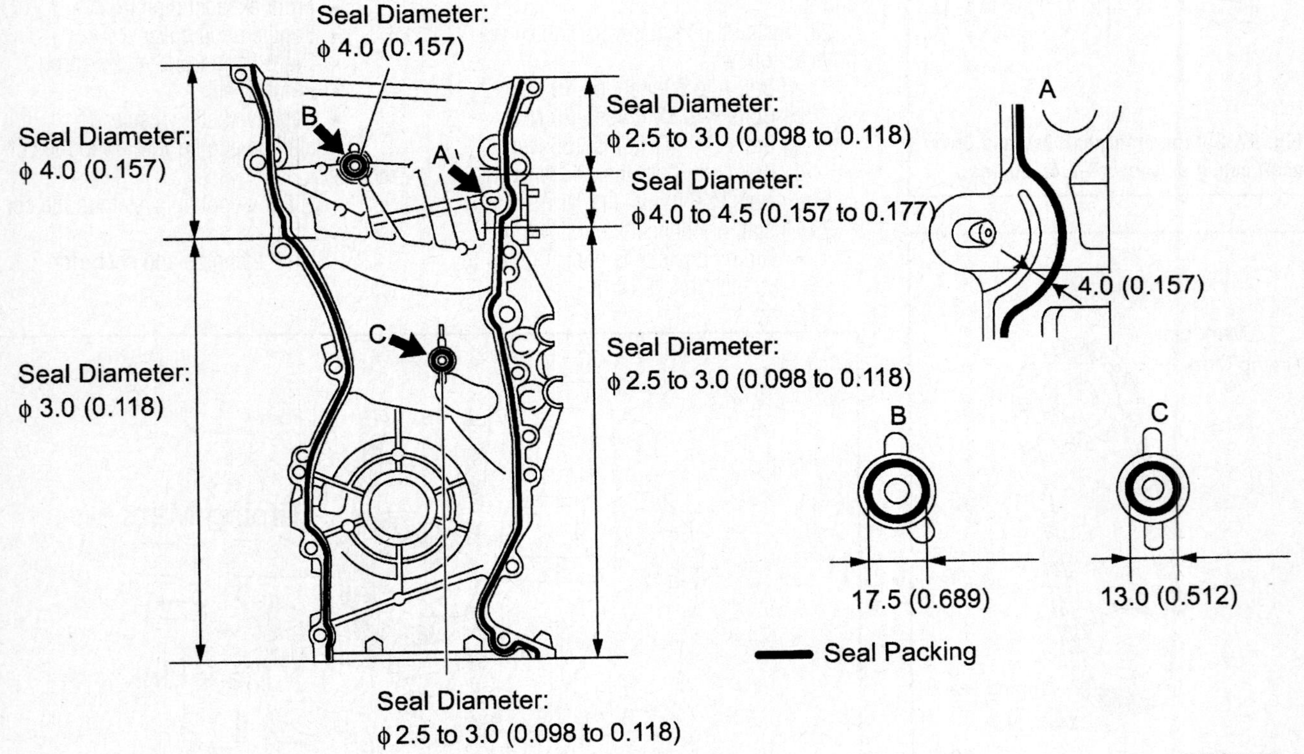

Seal Diameter:
ϕ 4.0 (0.157)
Seal Diameter:
ϕ 4.0 (0.157)
B
A
Seal Diameter:
ϕ 2.5 to 3.0 (0.098 to 0.118)
Seal Diameter:
ϕ 4.0 to 4.5 (0.157 to 0.177)
A
4.0 (0.157)
Seal Diameter:
ϕ 3.0 (0.118)
C
Seal Diameter:
ϕ 2.5 to 3.0 (0.098 to 0.118)
B
C
17.5 (0.689)
13.0 (0.512)
Seal Packing
Seal Diameter:
ϕ 2.5 to 3.0 (0.098 to 0.118)
09490_SION_G0053

Fig. 102 Apply sealant to the front cover contact surfaces shown—2.4L Engine

xD Models

See Figures 104 through 116.

➡Engine must be removed from vehicle. Refer to "Engine" in this section.

➡Refer to "Timing Chain Cover & Seal" in this section.

1. Remove the 3 O-rings from the cylinder head and cylinder block.
2. Remove the 3 bolts and the water pump assembly.
3. Remove the water inlet housing.
4. Remove the alternator bracket.
5. Remove the No. 2 chain vibration damper.
6. Remove the chain tensioner slipper.
7. Remove the chain vibration damper.
8. Remove the chain sub-assembly.
   a. Hold the hexagonal portion of the camshaft with a wrench and turn the camshaft timing gear assembly counterclockwise to loosen the chain on the camshaft timing gears.
   b. With the chain loosened, release the chain from the camshaft timing gear assembly and rest it on the camshaft timing gear assembly.

➡Be sure to release the chain from the sprocket completely.

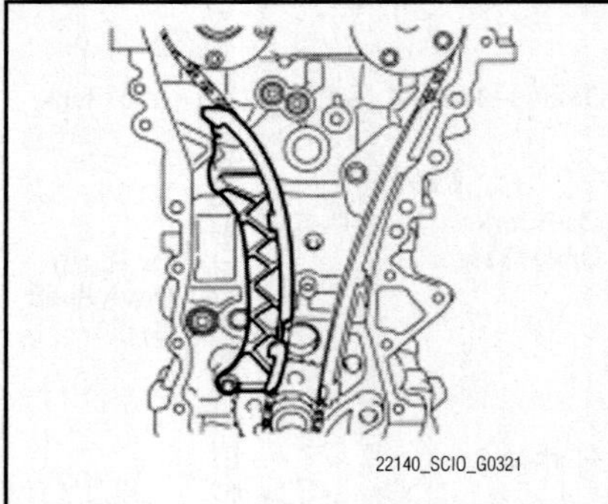

22140_SCIO_G0321

Fig. 106 Remove the chain tensioner slipper.

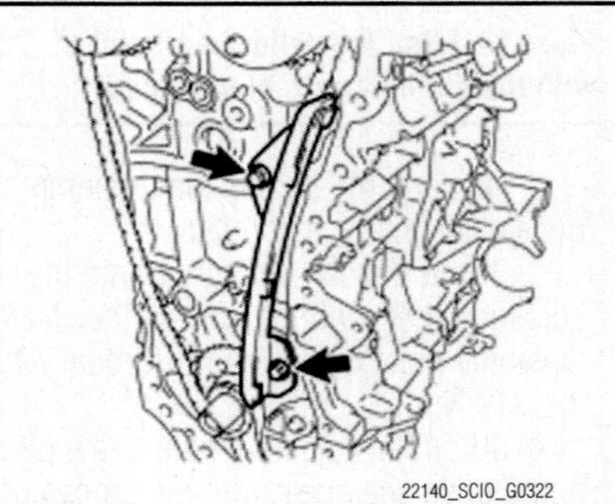

22140_SCIO_G0322

Fig. 107 Remove the chain vibration damper.

   c. Turn the camshaft clockwise to return it to the original position and remove the chain.
9. Remove the crankshaft timing gear or sprocket.
10. Remove the No. 2 chain sub-assembly.
    a. Turn the crankshaft 90° clockwise to align the adjusting hole of the oil pump drive shaft gear with the groove of the oil pump.
    b. Insert a 4 mm diameter bar into the adjusting hole of the oil pump drive shaft gear to lock the gear in position, and then remove the nut.
    c. Remove the bolt, chain tensioner plate and spring.
    d. Remove the oil pump drive shaft gear, oil pump drive gear and chain.

To install:

11. Install No. 2 chain sub-assembly.
    a. Set the crankshaft key as shown.
    b. Turn the drive shaft so that the cutout faces right.
    c. Align the yellow mark links with the timing marks of each gear.
    d. Install the gears onto the crankshaft and oil pump shaft with the chain on the gears.

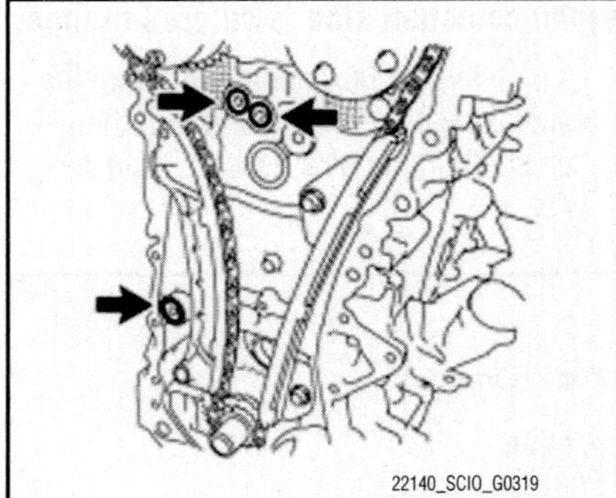

22140_SCIO_G0319

Fig. 104 Remove the 3 O-rings from the cylinder head and cylinder block.

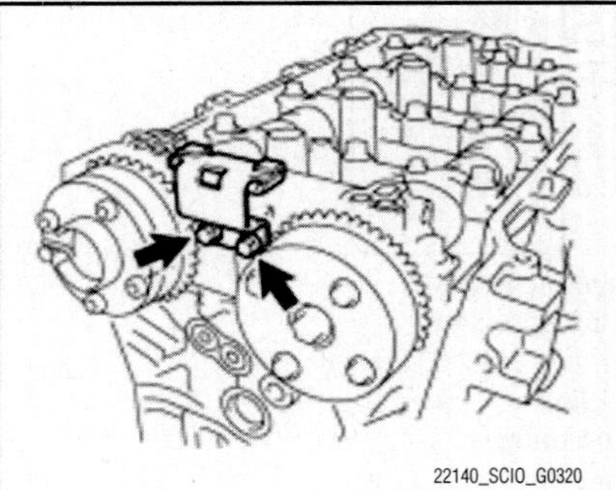

22140_SCIO_G0320

Fig. 105 Remove the No. 2 chain vibration damper.

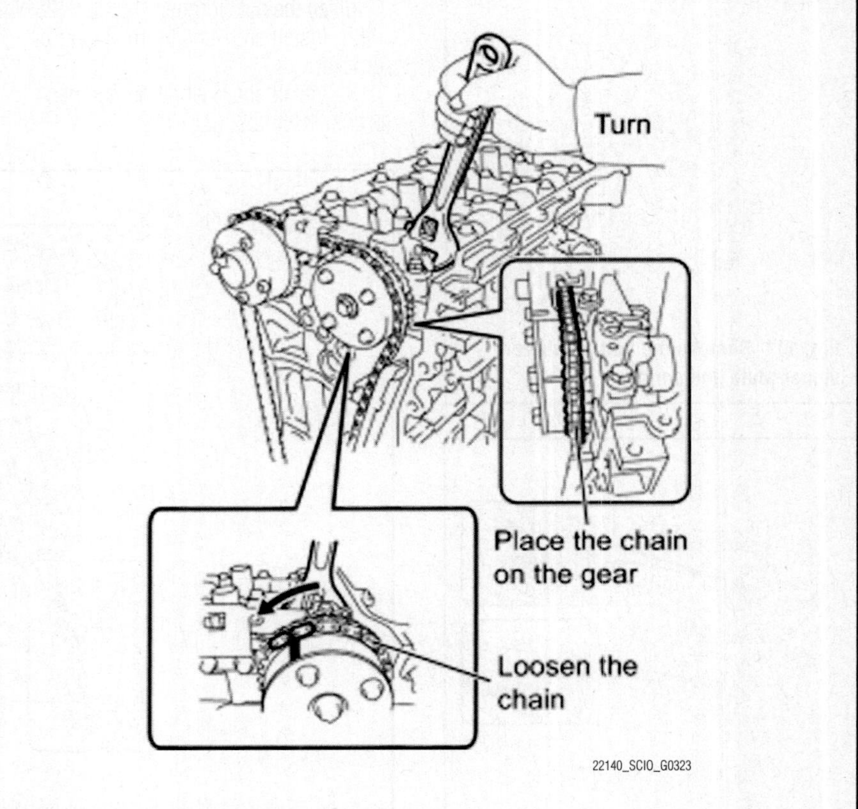

22140_SCIO_G0323

Fig. 108 Remove the chain sub-assembly.

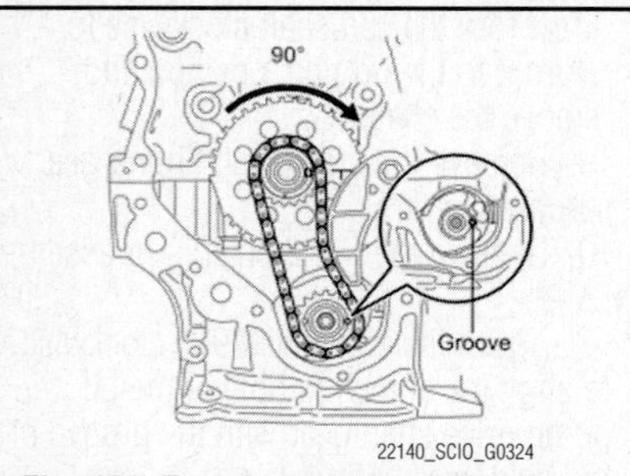

Fig. 109 Turn the crankshaft 90° clockwise to align the adjusting hole of the oil pump drive shaft gear with the groove of the oil pump.

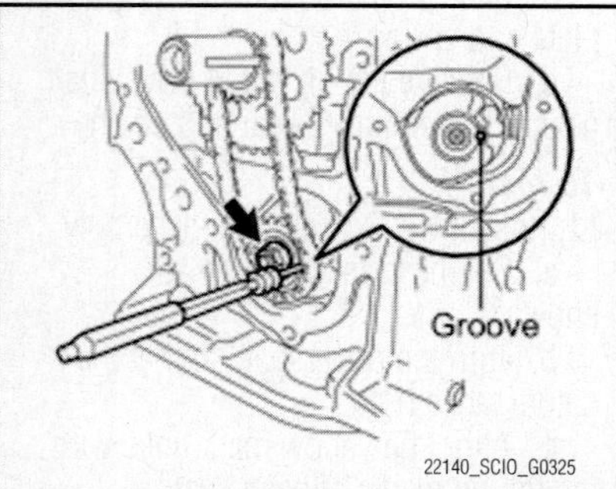

Fig. 110 Insert a 4 mm diameter bar into the adjusting hole of the oil pump drive shaft gear

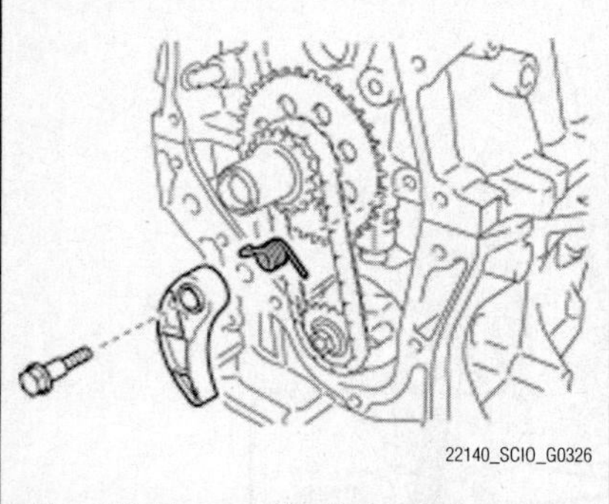

Fig. 111 Remove the bolt, chain tensioner plate and spring.

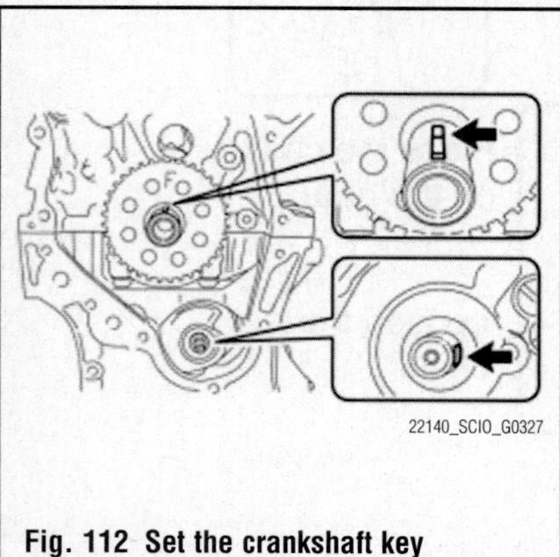

Fig. 112 Set the crankshaft key

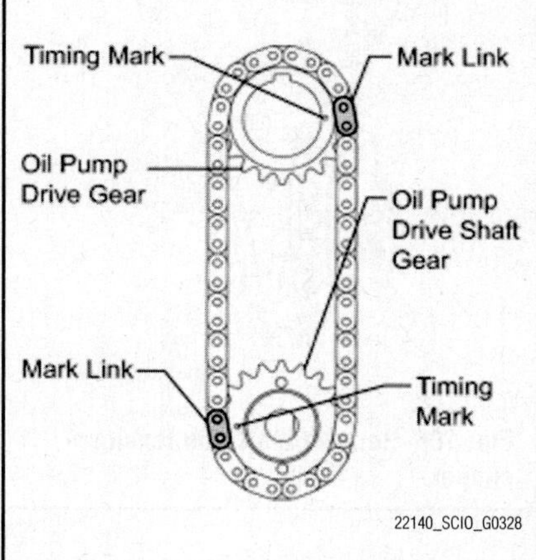

Fig. 113 Align the yellow mark links with the timing marks of each gear

e. Temporarily tighten the oil pump drive shaft gear with the nut.

f. Insert the damper spring into the adjusting hole, and then install the chain tensioner plate with the bolt. Torque: 7 ft. lbs. (10 Nm).

g. Align the adjusting hole of the oil pump drive shaft gear with the groove of the oil pump.

h. Insert a 4 mm diameter bar into the adjusting hole of the oil pump drive shaft gear to lock the gear in position, and then tighten the nut. Torque: 21 ft. lbs. (28 Nm).

12. Install crankshaft timing gear or sprocket.

13. Install chain vibration damper. Torque: 16 ft. lbs. (21 Nm).

14. Install chain sub-assembly.

a. Check the No. 1 cylinder TDC/compression.

b. Temporarily tighten the crankshaft pulley bolt.

c. Turn the crankshaft counterclockwise until the timing gear key is facing up.

d. Remove the crankshaft pulley bolt.

e. Check the timing mark on each camshaft timing gear.

f. Align the mark plate (orange) with the timing mark and install the chain.

➡Be sure to position the mark plate at the front of the engine. The mark plate on the camshaft side is colored orange.

g. Do not pass the chain around the sprocket of the camshaft timing gear assembly. Only place it on the sprocket.

h. Pass the chain through the No. 1 vibration damper.

i. Place the chain on the crankshaft without passing it around the shaft.

j. Hold the hexagonal portion of the camshaft with a wrench and turn the camshaft timing gear assembly counterclockwise to align the mark plate (orange) and timing mark.

➡Be sure to position the mark plate at the front of the engine. The mark plate on the camshaft side is colored orange.

k. Hold the hexagonal portion of the camshaft with a wrench and turn the camshaft timing gear assembly clockwise.

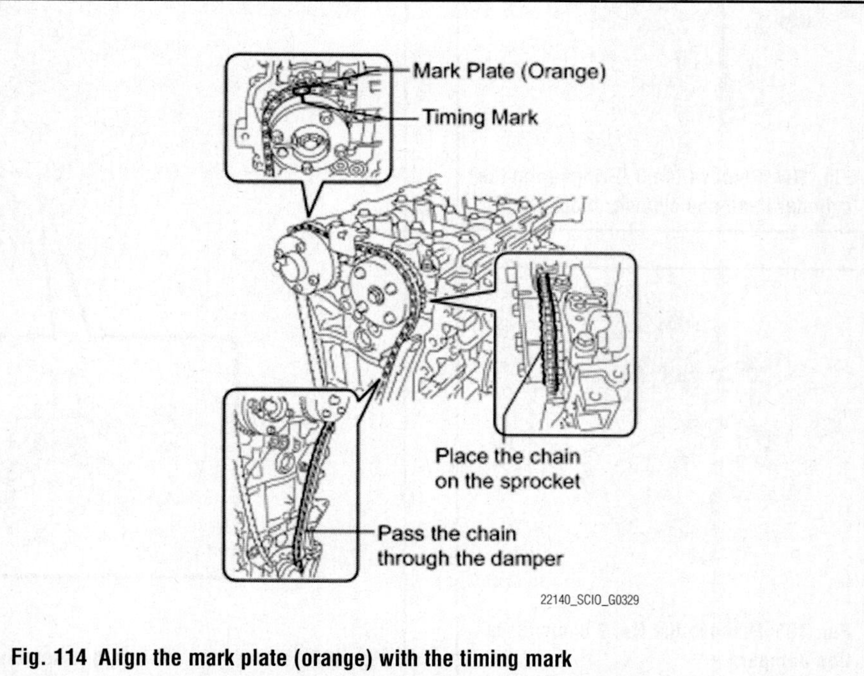

Fig. 114 Align the mark plate (orange) with the timing mark

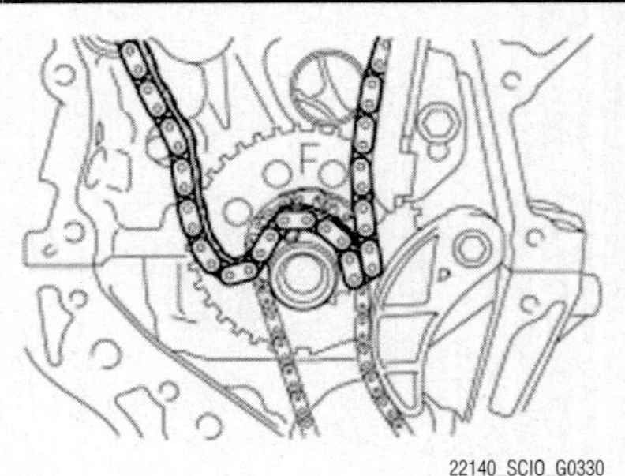

Fig. 115 Place the chain on the crankshaft without passing it around the shaft.

l. When tensioning the chain, turn the camshaft timing gear assembly clockwise slowly to prevent the chain from being misaligned.

m. Align the mark plate (yellow) and timing mark and install the chain to the crankshaft timing gear. The mark plate on the crankshaft side is colored yellow.

n. Recheck each timing mark at TDC/compression.

15. Install chain tensioner slipper.

16. Install No. 2 chain vibration damper. Torque: 7 ft. lbs. (10 Nm).

17. Install alternator bracket. Torque: 16 ft. lbs. (21 Nm).

18. Install water inlet housing. Torque: 16 ft. lbs. (21 Nm).

19. Refer to Timing Chain Cover R&I.

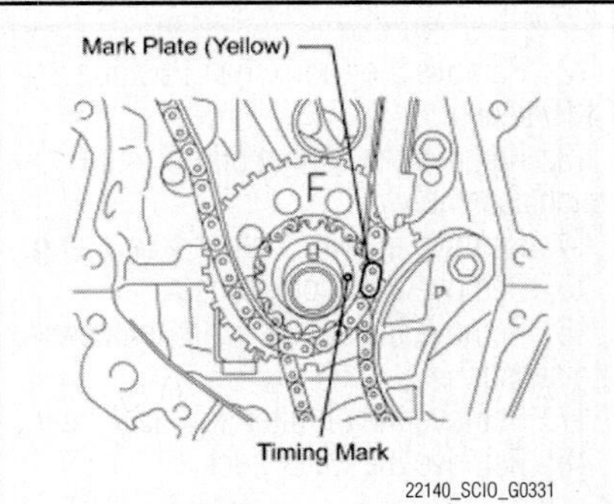

Fig. 116 Align the mark plate (yellow) and timing mark

TIMING CHAIN COVER & SEAL

REMOVAL & INSTALLATION

tC & xB Models

See Figures 95 through 103.

1. Before servicing the vehicle, refer to the Precautions Section.
2. Drain the engine oil.
3. Drain the cooling system.
4. Support the engine with a suitable jack.
5. Remove or disconnect the following:
   - Negative battery cable
   - Hood
   - Right-hand front wheel
   - Engine undercover
   - Front fender apron
   - Engine appearance cover
   - Front exhaust pipe
   - Accessory drive belt
   - Alternator
   - Power steering pump and reservoir
   - Right-hand engine mount
   - Ignition coil
   - Cylinder head cover
   - Accessory drive belt tensioner

➡Lift the engine with a suitable jack to gain access to the tensioner mounting bolts.

   - Crankshaft Position (CKP) sensor
   - Oil pan

6. Set the No. 1 cylinder to TDC.
7. Using Special Tool 09213-54015, remove the crankshaft pulley bolt.
8. Remove the crankshaft pulley. Use Special Tool 09950-50013 and the pulley bolt if necessary.
9. Remove or disconnect the following:
   - Timing chain tensioner
   - Front cover
   - Crankshaft Position (CKP) sensor plate
   - Timing chain guide
   - Timing chain tensioner slipper
   - Upper timing chain
   - Timing chain vibration damper
   - Crankshaft sprocket
10. Remove the lower timing chain as follows:

    a. Turn the crankshaft by 90° counterclockwise to align the adjusting hole of the oil pump drive shaft gear with the groove of the oil pump.

    b. Insert a 4 mm diameter bar into the adjusting hole of the oil pump drive shaft gear to lock in position, then remove the nut.

    c. Remove the bolt, then remove the chain tensioner plate and spring.

    d. Remove the oil pump drive gear, oil pump drive shaft gear and lower timing chain.

To install:

11. Set the crankshaft key in the left horizontal position. Turn the cutout of the drive shaft to the top.
12. Align the yellow mark links with the timing marks of the each gear as shown in the illustration.
13. Install the gears onto the crankshaft and oil pump shaft with the lower chain wrapped.
14. Temporarily tighten the oil pump drive shaft gear with the nut.
15. Insert the damper spring into the adjusting hole, then install the chain tensioner plate with the bolt. Tighten to 9 ft. lbs. (12 Nm).
16. Align the adjusting hole of the oil pump drive shaft gear with the groove of the oil pump.
17. Insert a 4 mm diameter bar into the adjusting hole of the oil pump drive shaft gear to lock in position, then tighten the nut to 22 ft. lbs. (30 Nm).
18. Turn the crankshaft clockwise by 90° to position the crankshaft key upward.
19. Install the crankshaft sprocket.
20. Turn the camshafts with a wrench on the hexagonal lobe to align the timing marks of the camshaft timing gear with each timing mark located on the No. 1 and No. 2 bearing caps as shown in the illustration.
21. Using the crankshaft pulley bolt, turn the crankshaft to position the key on the crankshaft upward.
22. Install the upper timing chain onto the crankshaft timing gear with the gold or orange mark link aligned with the timing mark on the crankshaft sprocket.
23. Using Special Tool 09309-37010, tap in the crankshaft timing gear.
24. Align the gold or yellow mark links with each timing mark located on the camshaft timing gears, then install the upper timing chain.
25. Install or connect the following:
    - Timing chain tensioner slipper. Tighten to 14 ft. lbs. (19 Nm).
    - Timing chain guide. Tighten to 80 inch lbs. (9 Nm).
    - Crankshaft Position (CKP) sensor plate
26. Remove any old sealant from the front cover.
27. Apply a continuous bead of sealant to the front cover contact surfaces as shown.
28. Install the front cover. Tighten the bolts as follows:

    a. Bolt A to 80 inch lbs. (9 Nm).

    b. Bolts B to 18 ft. lbs. (25 Nm).

    c. Bolts C to 41 ft. lbs. (55 Nm).

    d. Bolt D to 32 ft. lbs. (43 Nm).

    e. Nuts to 8 ft. lbs. (11 Nm).
29. Install or connect the following:
    - Timing chain tensioner. Tighten to 80 inch lbs. (9 Nm).

- Crankshaft pulley. Tighten to 133 ft. lbs. (180 Nm).
- Accessory drive belt tensioner. Tighten to 44 ft. lbs. (60 Nm).
- Cylinder head cover
- Ignition coil
- Right-hand engine mount
- Power steering pump and reservoir
- Alternator
- Accessory drive belts
- Front exhaust pipe
- Engine undercover
- Right-hand front fender apron
- Front wheel
- Negative battery cable

30. Refill the engine with oil to the correct level.
31. Refill the cooling system to the correct level.
32. Start the engine and check for leaks.

xD Models

See Figures 117 through 132.

➡Engine must be removed from vehicle. Refer to Engine Assembly R&I.

1. Remove the engine hanger.
2. Remove the oil filler cap.
3. Remove the oil filler cap gasket.
4. Remove the spark plugs.
5. Remove the 2 Camshaft Position (CMP) sensors.
6. Remove the camshaft timing oil control valve assembly.
   a. Remove the 2 bolts, 2 O-rings, bracket and 2 oil control valves.
7. Remove the cylinder head cover sub-assembly.
   a. Remove the 13 bolts, seal washer and cylinder head cover.
   b. Remove the 3 gaskets from the camshaft bearing cap.

✱✱ CAUTION

When removing the cylinder head cover, some of the gaskets may stick to it so be careful not to drop any of the gaskets into the engine.

8. Remove the cylinder head cover gasket.
9. Set the No. 1 cylinder to TDC / compression.
   a. Turn the crankshaft pulley, and align its timing notch with the "0" timing mark on the timing chain cover sub-assembly.
   b. Check that timing marks on both the camshaft timing exhaust gear and camshaft timing gear are facing upward. If not, turn the crankshaft 1 complete revolution (360°) and align the marks as above.
10. Remove the crankshaft pulley.
   a. Using SST, hold the pulley in place and loosen the pulley bolt.
   b. Using SST, remove the pulley bolt and pulley.
11. Remove the No. 1 chain tensioner assembly.

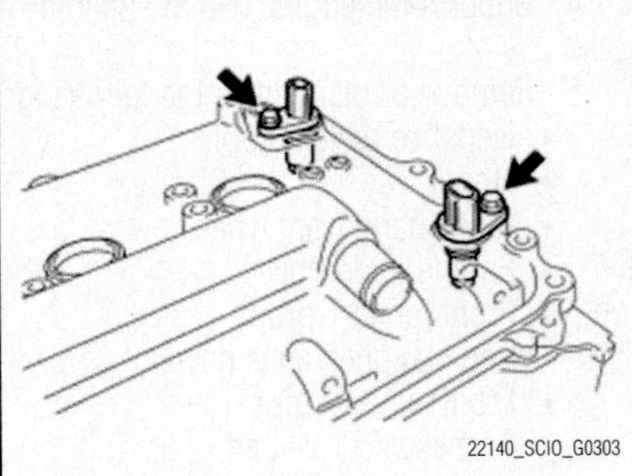

Fig. 117 Remove the 2 Camshaft Position (CMP) sensors.

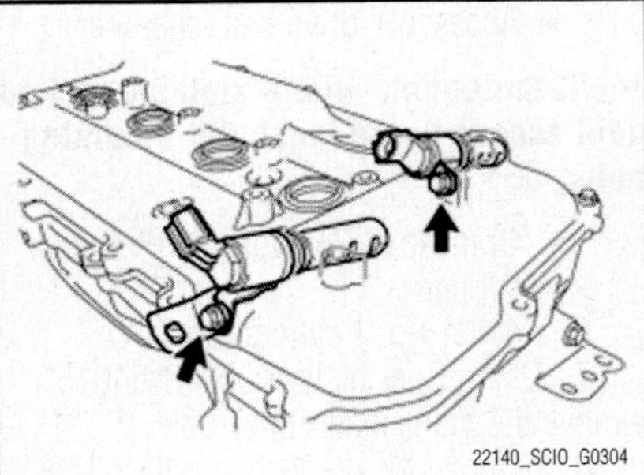

Fig. 118 Remove the camshaft timing oil control valve assembly.

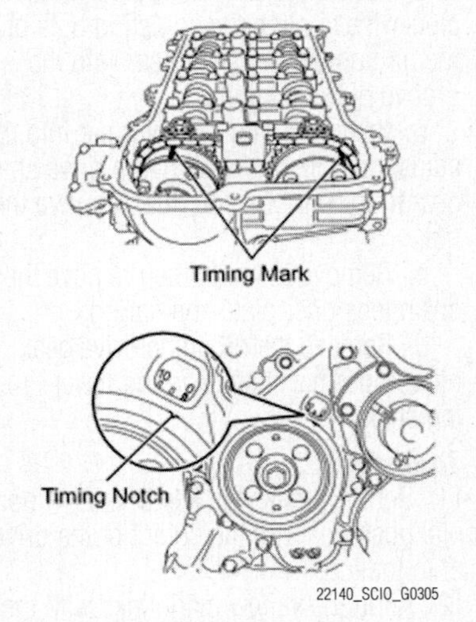

Fig. 119 Set the No. 1 cylinder to TDC / compression.

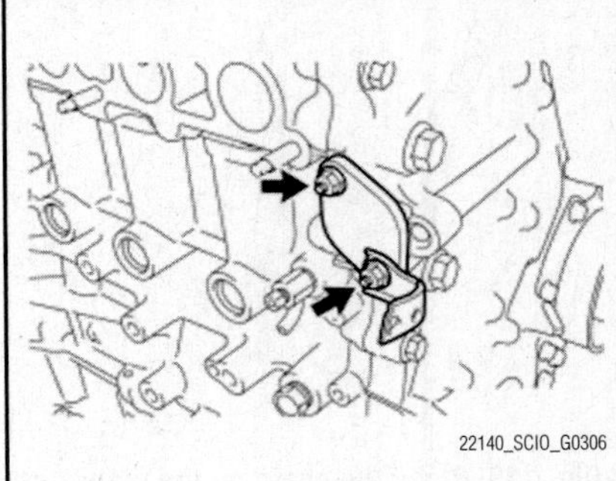

Fig. 120 Remove the No. 1 chain tensioner assembly.

Fig. 121 Remove the Crankshaft Position (CKP) sensor.

✱✱ CAUTION

Do not turn the crankshaft after you have removed the chain tensioner.

12. Remove the Crankshaft Position (CKP) sensor.
13. Remove the engine oil pressure switch assembly.
14. Remove the No. 1 taper screw plug.
15. Remove the knock sensor.
16. Remove the engine water temperature sensor.
17. Remove the oil filter sub-assembly.
18. Remove the water inlet.

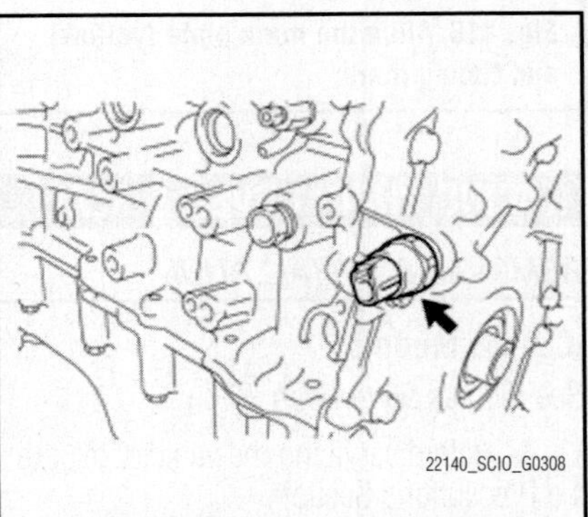

Fig. 122 Remove the engine oil pressure switch assembly.

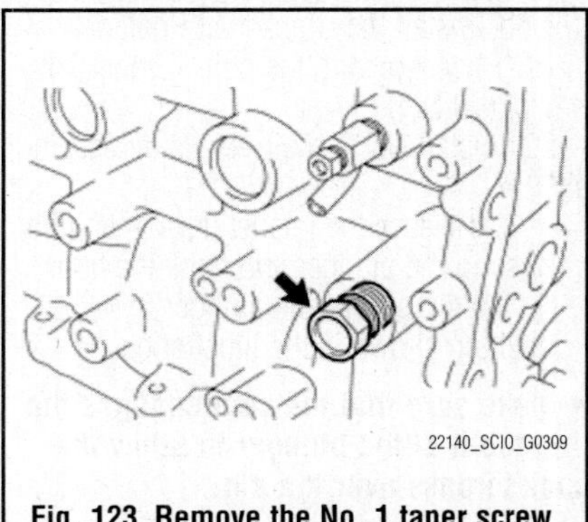
Fig. 123 Remove the No. 1 taper screw plug.

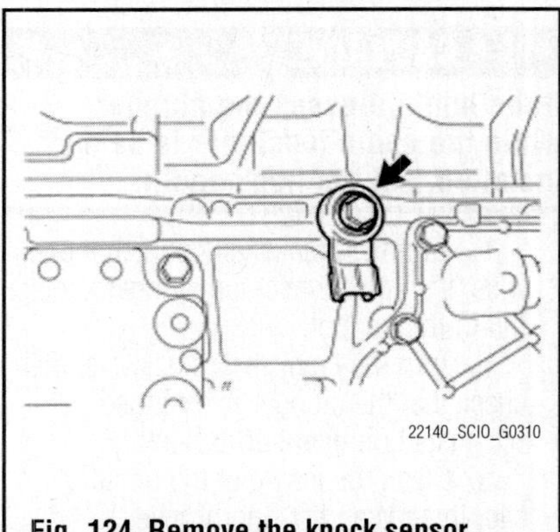
Fig. 124 Remove the knock sensor.

19. Remove the thermostat.
20. Remove the timing chain cover sub-assembly.
a. Remove the 3 bolts and remove the engine mounting bracket RH.
b. Remove the 4 bolts and remove the oil filter bracket.
c. Remove the 2 O-rings from the timing chain cover sub-assembly.
d. Remove the 19 bolts.
e. Remove the timing chain cover sub-assembly by prying between the timing chain cover sub-assembly and cylinder head or cylinder block

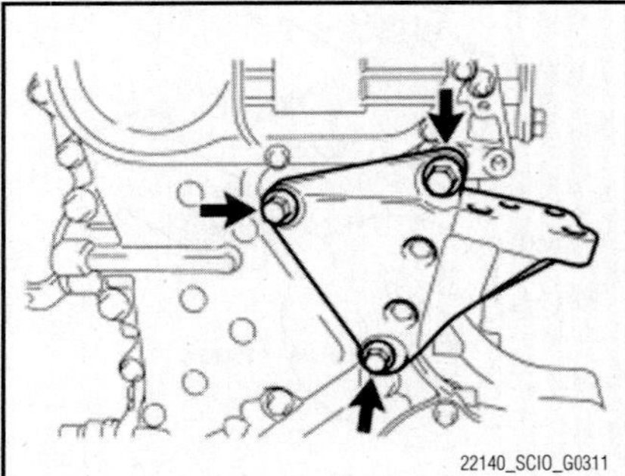
Fig. 125 Remove the 3 bolts and remove the engine mounting bracket RH.

Fig. 126 Remove the 4 bolts and remove the oil filter bracket.

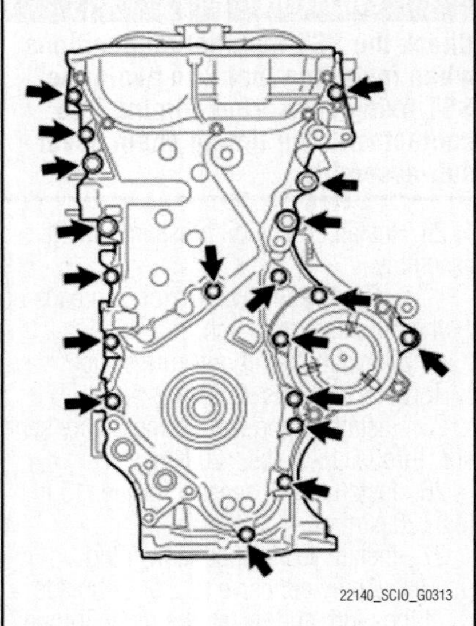
Fig. 127 Remove the 19 bolts.

with a screwdriver wrapped in protective tape.

To install:

21. Install timing chain cover sub-assembly.
a. Remove any old sealant material and be careful not to drop any oil on the contact surfaces of the timing chain cover, cylinder head and cylinder block.
b. Apply sealant as shown in the illustration.
c. Apply sealant in a continuous line to the timing chain cover.
d. Install the chain cover within 3 minutes and tighten the bolts within 15 minutes of applying the sealant.

CAUTION

Do not start the engine for at least 2 hours after installing.

e. Install a new gasket.
f. Install 2 new O-rings.
g. Install the timing chain cover with the 19 bolts as shown in the illustration.
h. Bolt A and C - Torque: 19 ft. lbs. (26 Nm)
i. Bolt B and D - Torque: 38 ft. lbs. (51 Nm)
j. Bolt E - Torque: 7 ft. lbs. (10 Nm)
k. Tighten bolt D through the engine mounting bracket, pressing it against the timing chain cover sub-assembly.
l. Tighten bolt C through the oil filter bracket, pressing it against the timing chain cover sub-assembly. Note the bolt lengths:

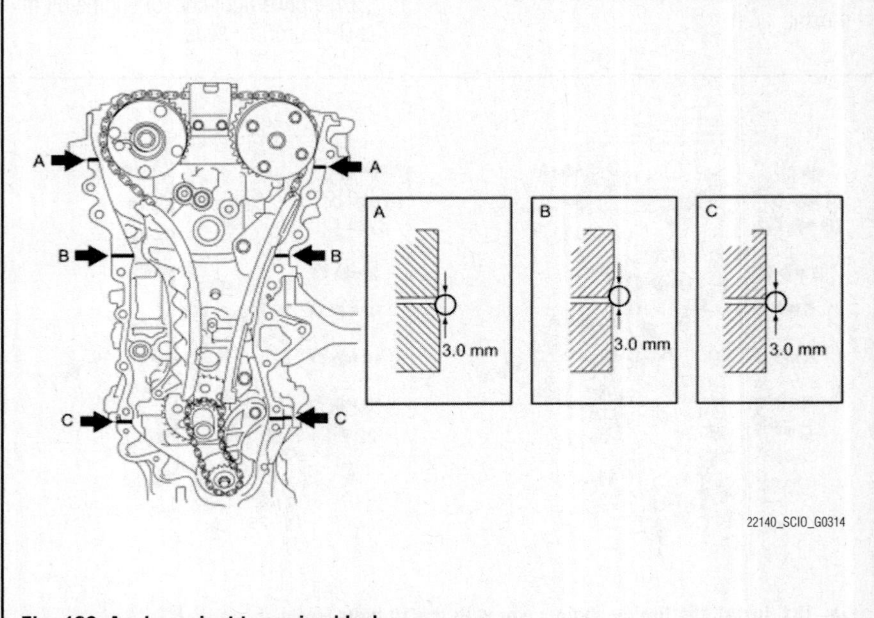

Fig. 128 Apply sealant to engine block.

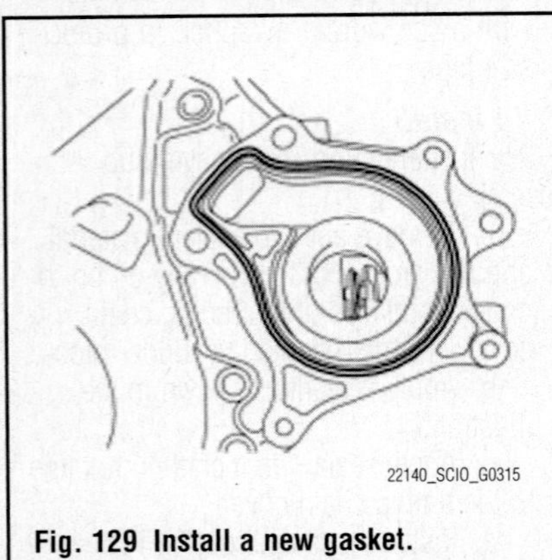

Fig. 129 Install a new gasket.

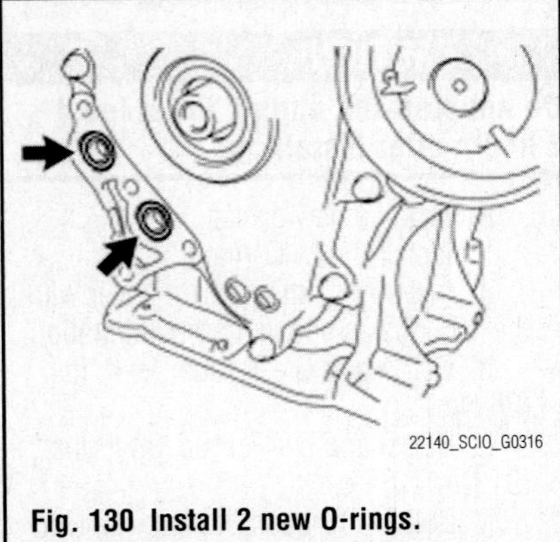

Fig. 130 Install 2 new O-rings.

- Bolt A and C: 1.38 inches (35 mm)
- Bolt B: 2.16 inches (55 mm)
- Bolt D: 3.15 inches (80 mm)
- Bolt E: 1.57 inches (40 mm)

m. Install the water pump with the 3 bolts. Torque: 18 ft. lbs. (24 Nm).

22. Install thermostat.

a. Install a new gasket onto the thermostat.

b. Install the thermostat with the jiggle valve facing upward.

➡The jiggle valve may be set within 10° on either side of the top.

c. Install water inlet. Torque: 7 ft. lbs. (10 Nm).

23. Install crankshaft pulley.

a. Align the pin hole in the crankshaft pulley with the pin position and install the crankshaft pulley.

b. Provisionally install the bolt.

c. Using SST, tighten the bolt while holding the crankshaft pulley. Torque: 140 ft. lbs. (190 Nm).

**** CAUTION**

Check the SST installation positions when installing them, to avoid the SST fixing bolts from coming into contact with the timing chain cover sub-assembly.

24. Install engine oil pressure switch assembly.

a. Apply adhesive to 2 or 3 threads of the oil pressure switch.

b. Install the oil pressure switch. Torque: 11 ft. lbs. (15 Nm).

25. Install engine water temperature sensor. Torque: 15 ft. lbs. (20 Nm).

26. Install knock sensor. Torque: 15 ft. lbs. (20 Nm).

27. Install No. 1 taper screw plug.

a. Apply adhesive to 2 or 3 threads of the plug, and install the plug. Torque: 32 ft. lbs. (43 Nm).

28. Install Crankshaft Position (CKP) sensor.

a. Apply a light coat of engine oil to the O-ring of the sensor.

b. Install the Crankshaft Position (CKP) sensor with the bolt. Torque: 7 ft. lbs. (10 Nm).

29. Install No. 1 chain tensioner assembly.

a. Release the ratchet pawl, then fully push in the plunger and hook the hook on the pin so that the plunger is in the position shown in the illustration.

➡Make sure that the cam engages the first tooth of the plunger to allow the hook to pass over the pin.

b. Install a new gasket, bracket and the No. 1 chain tensioner with the 2 nuts. Torque: 7 ft. lbs. (10 Nm).

**** CAUTION**

If the hook releases the plunger while the chain tensioner is being installed, set the hook again.

c. Turn the crankshaft counterclockwise, then disconnect the plunger knock pin from the hook.

d. Turn the crankshaft clockwise, then check that the plunger is extended.

30. Install oil filter sub-assembly.

a. Clean the inside of the oil filter cap, threads and O-ring groove.

b. Apply a small amount of engine oil to a new O-ring and install it onto the oil filter cap.

c. Set a new oil filter element into the oil filter cap.

d. Apply a small amount of engine oil

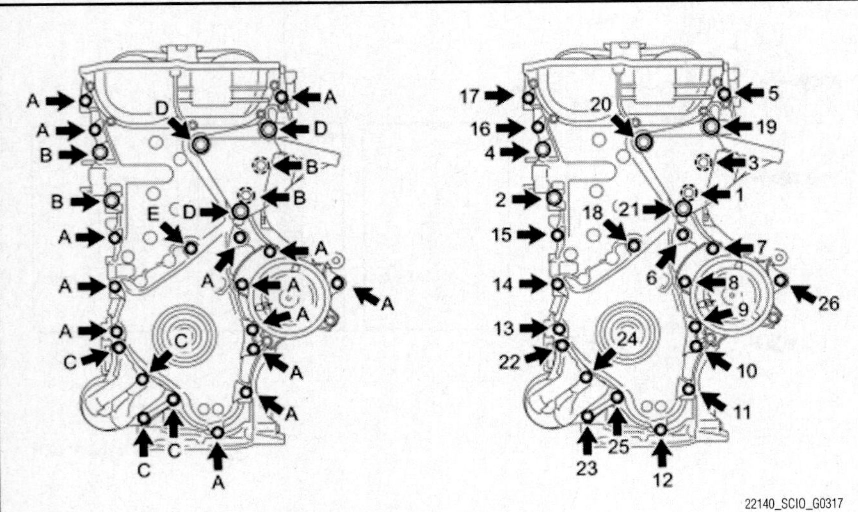

Fig. 131 Install the timing chain cover with the 19 bolts

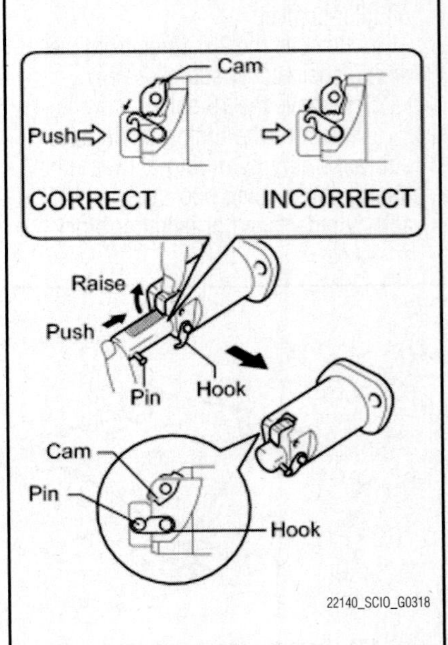

Fig. 132 Install No. 1 chain tensioner assembly.

to the O-ring again and install the oil filter cap.

e. Using SST, tighten the oil filter cap. Torque: 18 ft. lbs. (25 Nm).

31. Install cylinder head cover gasket.
32. Install cylinder head cover sub-assembly.

a. Install 3 new gaskets onto the No. 1 camshaft bearing cap.

b. Install the cylinder head cover with a new seal washer and the 13 bolts. Torque: 7 ft. lbs. (10 Nm).

33. Install camshaft timing oil control valve assembly.

a. Apply a light coat of engine oil to 2 new O-rings, then install each of them onto the camshaft timing oil control valve.

b. Install the 2 camshaft timing oil control valves and bracket with the 2 bolts. Torque: 7 ft. lbs. (10 Nm).

34. Install Camshaft Position (CMP) sensor.

a. Apply a light coat of engine oil to the O-ring of the sensor.

b. Install the 2 sensors with the 2 bolts. Torque: 7 ft. lbs. (10 Nm).

35. Install spark plugs.
36. Install oil filler cap gasket.
37. Install oil filler cap sub-assembly.
38. Install engine hangers.

VALVE LASH

ADJUSTMENT

tC & xB Models

See Figures 133 and 134.

1. Remove or disconnect the following:
   - Negative battery cable
   - Right-hand front wheel
   - Engine undercover
   - Engine appearance cover
   - Accessory drive belt
   - Power steering pump
   - Right-hand engine mount
   - Ignition coil
   - Cylinder head cover
2. Turn the crankshaft to set the No. 1 cylinder at TDC compression.

➡**Inspect the valve clearance when the engine is cold.**

3. Turn the crankshaft pulley to set the No. 1 cylinder at TDC.

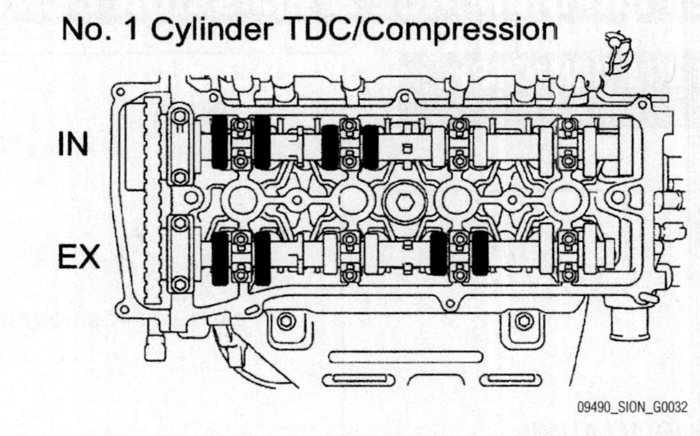

Fig. 133 Measure indicated valve lifters with No. 1 cylinder at TDC—2.4L Engine

4. Using a feeler gauge, measure the valve clearance shown in the illustration. Record the measurements.
5. Turn the crankshaft 1 complete revolution and set the No. 4 cylinder to TDC.
6. Using a feeler gauge, measure the valve clearance shown in the illustration. Record the measurements.

a. Intake valve clearance is 0.0075–0.0114 inches (0.19–0.29 mm).

b. Exhaust valve clearance is 0.0118–0.0158 inches (0.30–0.40 mm).

7. To adjust the valve clearance:
8. Remove the camshafts.
9. Remove the valve lifters.

a. Using a micrometer, measure the thickness of the removed lifter.

10. Calculate the thickness of a new lifter so that the valve clearance comes within the specified value.
    - Intake: New lifter thickness = Used lifter thickness + Recorded valve clearance measurement – 0.0095 in. (0.24 mm)
    - Exhaust: New lifter thickness = Used lifter thickness + Recorded valve clearance measurement – 0.0138 in. (0.35 mm)
11. Select a new lifter with the thickness as close to the calculated value as possible.
12. Reinstall the camshafts and check the valve clearance.
13. Install the cylinder head cover and remaining components.
14. Start the engine and check for leaks.

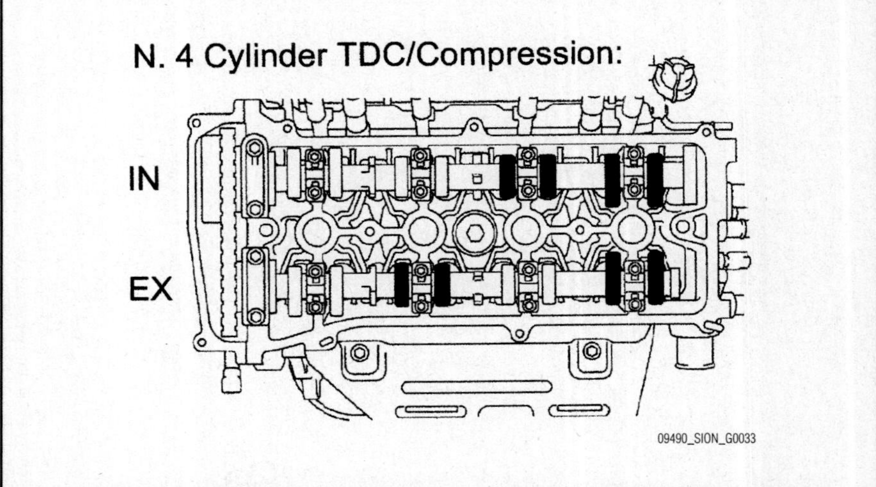

Fig. 134 Measure indicated valve lifters with No. 4 cylinder at TDC—2.4L Engine

ENGINE PERFORMANCE & EMISSION CONTROLS

CAMSHAFT POSITION (CMP) SENSOR

LOCATION

tC & xB Models

See Figure 135.

xD Models

See Figure 136.

REMOVAL & INSTALLATION

tC & xB Models

1. Disconnect the cable from the negative battery terminal. Wait at least 90 seconds after disconnecting the cable from the negative (-) battery terminal to prevent airbag and seat belt pretensioner activation.
2. Remove the engine cover.
3. Remove the air cleaner cap.
4. Disconnect the CMP sensor connector.
5. Remove the bolt and CMP sensor.
6. Installation is the reverse of the removal procedure.
7. Apply a thin coat of engine oil to the O-ring during installation.

xD Models

1. Remove the engine cover.
2. Remove the CMP sensor as follows:

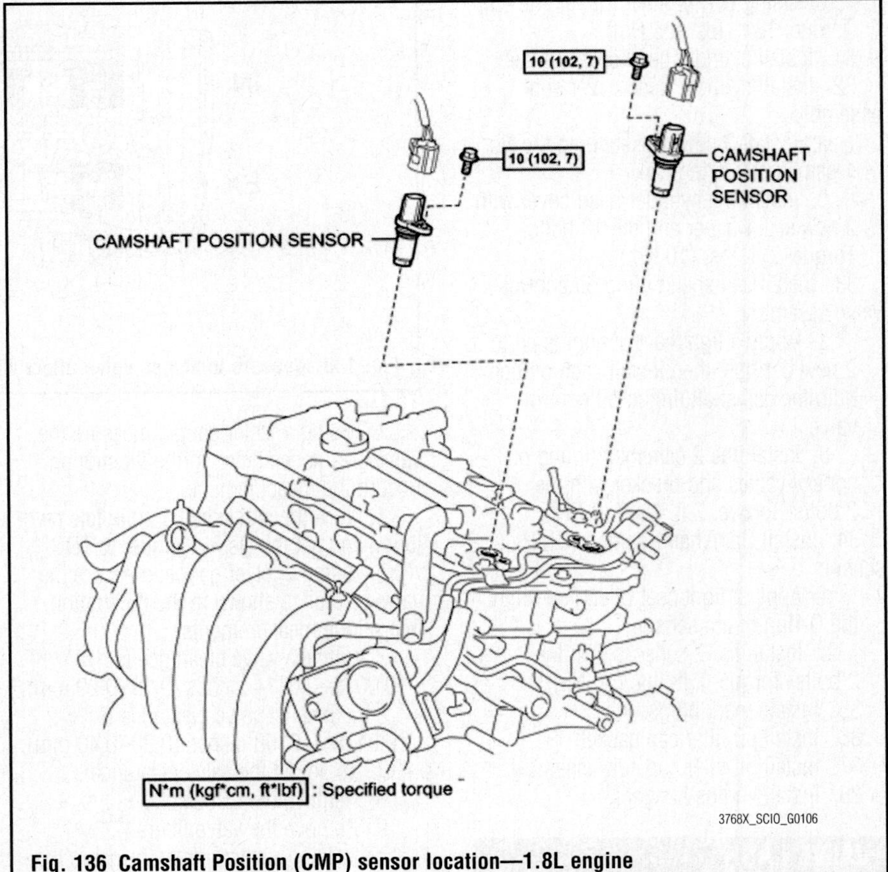

Fig. 136 Camshaft Position (CMP) sensor location—1.8L engine

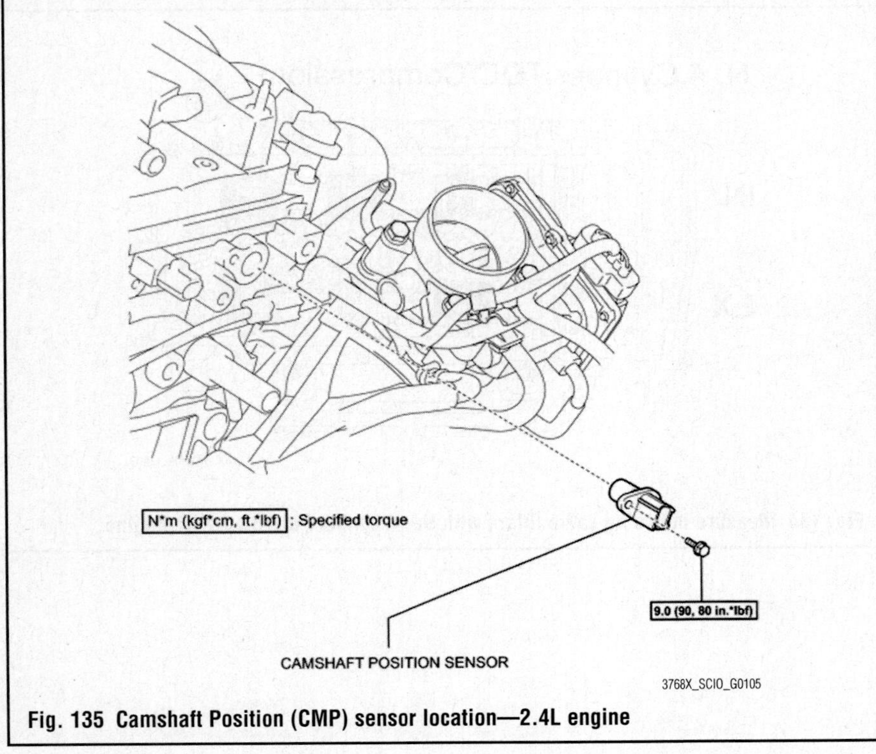

Fig. 135 Camshaft Position (CMP) sensor location—2.4L engine

a. Disconnect the duty vacuum switching valve connector and 3 engine wire harness clamps (for exhaust camshaft side).

b. Remove the bolt and remove the Camshaft Position (CMP) sensor (for exhaust camshaft side).

c. Remove the bolt and remove the Camshaft Position (CMP) sensor (for intake camshaft side).

3. Installation is the reverse of removal.

CRANKSHAFT POSITION (CKP) SENSOR

LOCATION

tC & xB Models

See Figure 137.

xD Models

See Figure 138.

REMOVAL & INSTALLATION

tC & xB Models

1. Disconnect the cable from the negative battery terminal. Wait at least 90 sec-

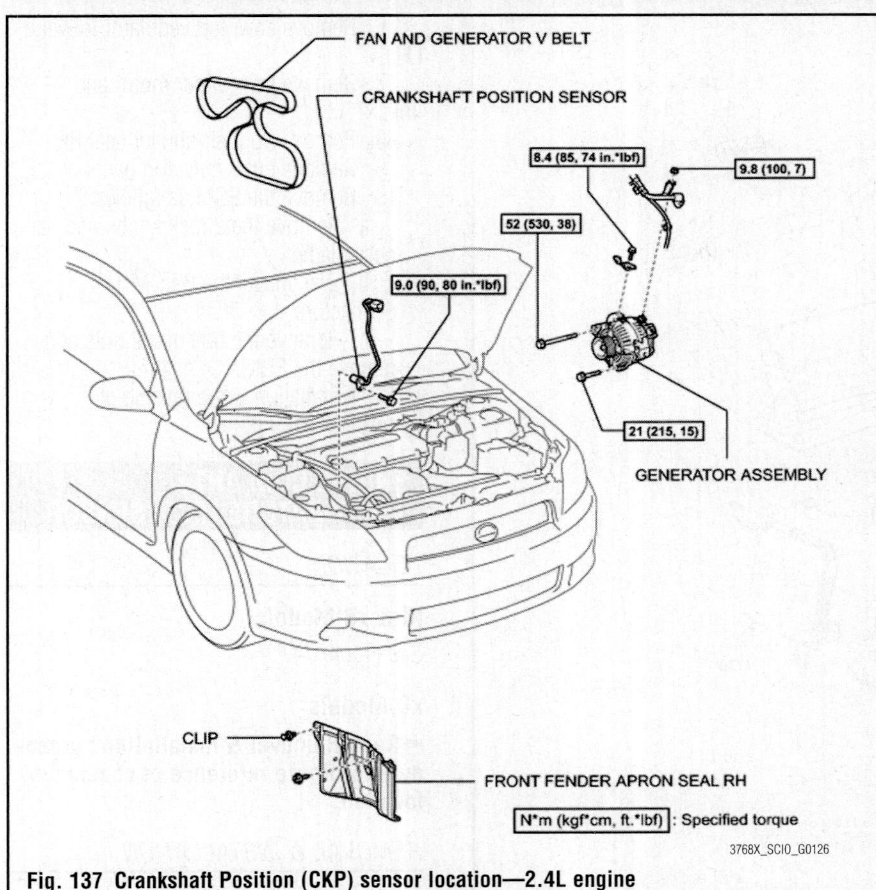

Fig. 137 Crankshaft Position (CKP) sensor location—2.4L engine

onds after disconnecting the cable from the negative (-) battery terminal to prevent airbag and seat belt pretensioner activation.

2. Remove the RH front fender apron seal.
3. Remove the fan and generator V-belt.
4. Remove the alternator assembly. See "ENGINE ELECTRICAL" section.
5. Remove the Crankshaft Position (CKP) sensor as follows:
   a. Disconnect the sensor connector.
   b. Remove the connector clamp.
   c. Remove the wire harness from the wire harness clamp bracket.
   d. Remove the wire harness clamp.
   e. Remove the bolt and sensor.

To install:

6. Installation is the reverse of the removal procedure.
7. Apply thin coat of engine oil to the O-ring during installation.
8. Check for oil leaks.
9. Perform initialization.

➡Certain systems need to be initialized after disconnecting and reconnecting the cable from the negative (-) battery terminal.

xD Models

1. Remove the RH engine under cover.
2. Remove the CKP sensor as follows:
   a. Disconnect the Crankshaft Position (CKP) sensor connector.
   b. Remove the bolt and remove the Crankshaft Position (CKP) sensor.
3. Installation is the reverse of removal.
4. Perform initialization.

➡Certain systems need to be initialized after disconnecting and reconnecting the cable from the negative (-) battery terminal.

ELECTRONIC CONTROL MODULE (ECM)

LOCATION

tC & xB Models

See Figure 139.

xD Models

See Figure 140.

REMOVAL & INSTALLATION

tC & xB Models

1. Disconnect the cable from the negative battery terminal. Wait at least 90 seconds

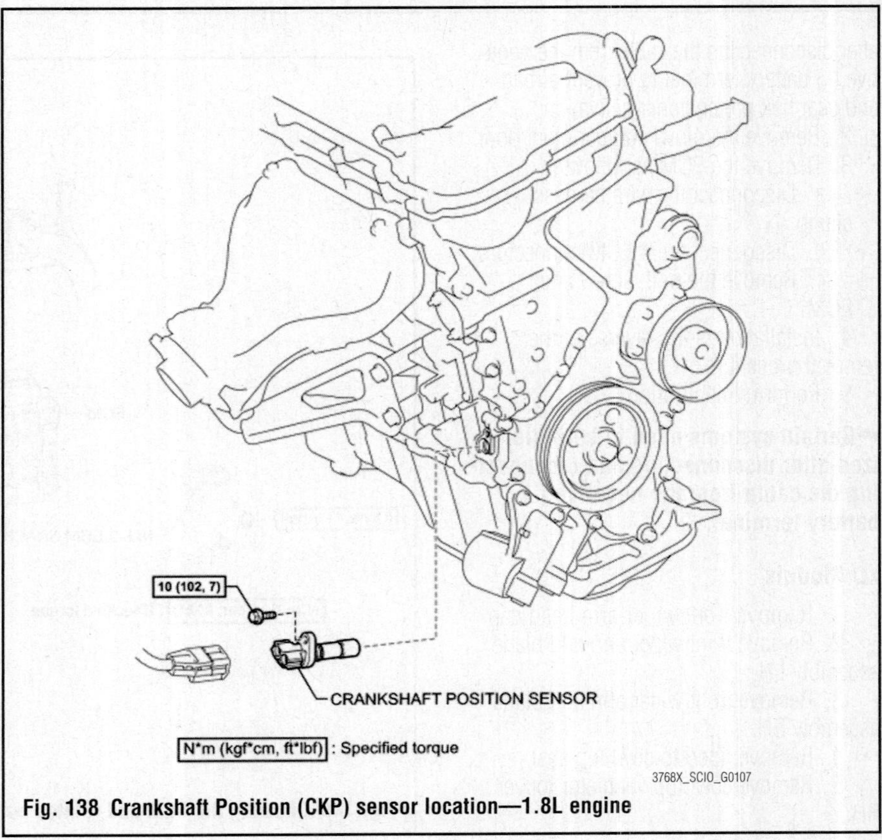

Fig. 138 Crankshaft Position (CKP) sensor location—1.8L engine

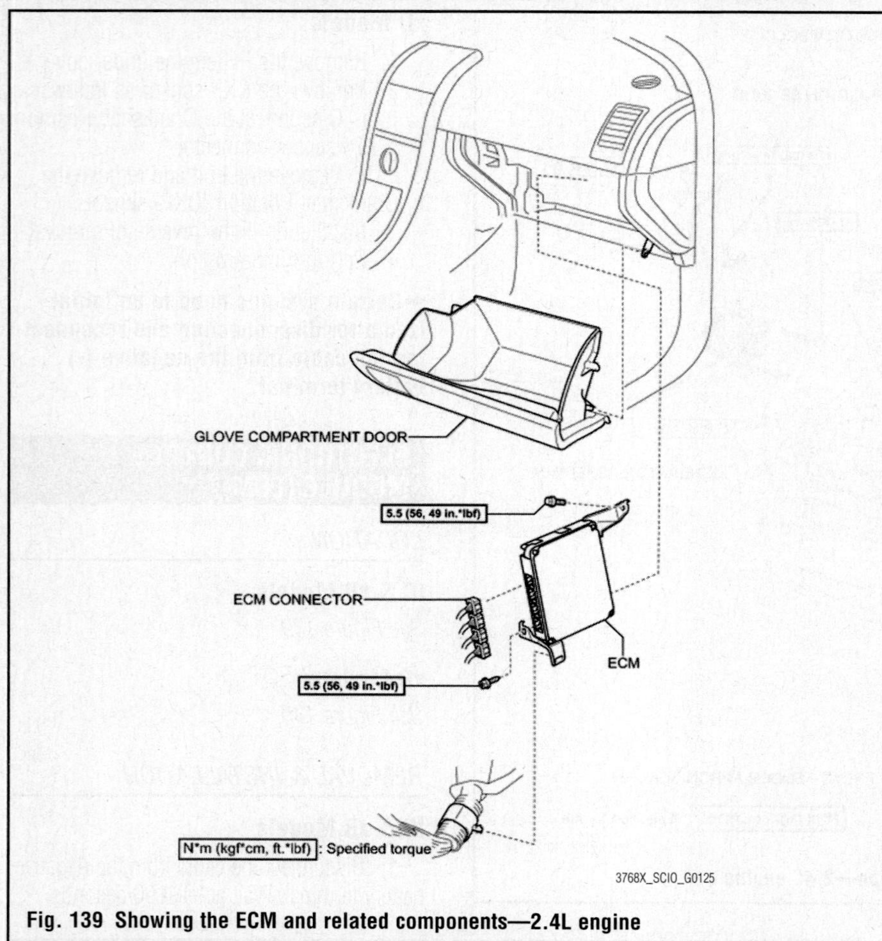

Fig. 139 Showing the ECM and related components—2.4L engine

after disconnecting the cable from the negative (-) battery terminal to prevent airbag and seat belt pretensioner activation.

2. Remove the glove compartment door.
3. Remove the ECM as follows:
   a. Disconnect the wire harness clamp.
   b. Disconnect the 4 ECM connectors.
   c. Remove the bolt, screw and ECM.
4. Installation is the reverse of the removal procedure.
5. Perform initialization.

➡Certain systems need to be initialized after disconnecting and reconnecting the cable from the negative (-) battery terminal.

xD Models

1. Remove front wiper arm head cap.
2. Remove front wiper arm and blade assembly LH.
3. Remove front wiper arm and blade assembly RH.
4. Remove hood to cowl top seal.
5. Remove cowl top ventilator louver RH.
6. Remove cowl top ventilator louver LH.
7. Remove front wiper motor and link.
8. Remove front air shutter seal RH.
9. Remove outer cowl top panel.
10. Remove the ECM as follows:
    a. Remove the 2 lock knobs and harness clamp.
    b. Disconnect the 2 ECM connectors.
    c. Remove the bolt and 2 nuts and remove the ECM.
11. Installation is the reverse of removal.

ENGINE COOLANT TEMPERATURE (ECT) SENSOR

LOCATION

tC & xB Models

See Figure 141.

xD Models

➡See "Removal & Installation" procedure for figure reference of component location.

REMOVAL & INSTALLATION

tC & xB Models

See Figures 142 through 144.

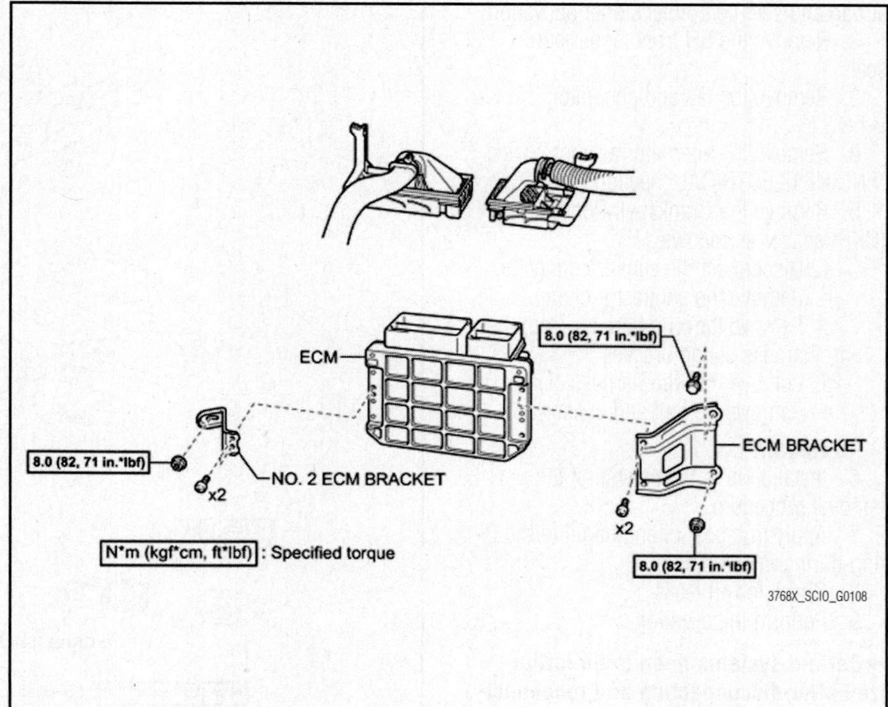

Fig. 140 Showing the ECM and related components—1.8L engine

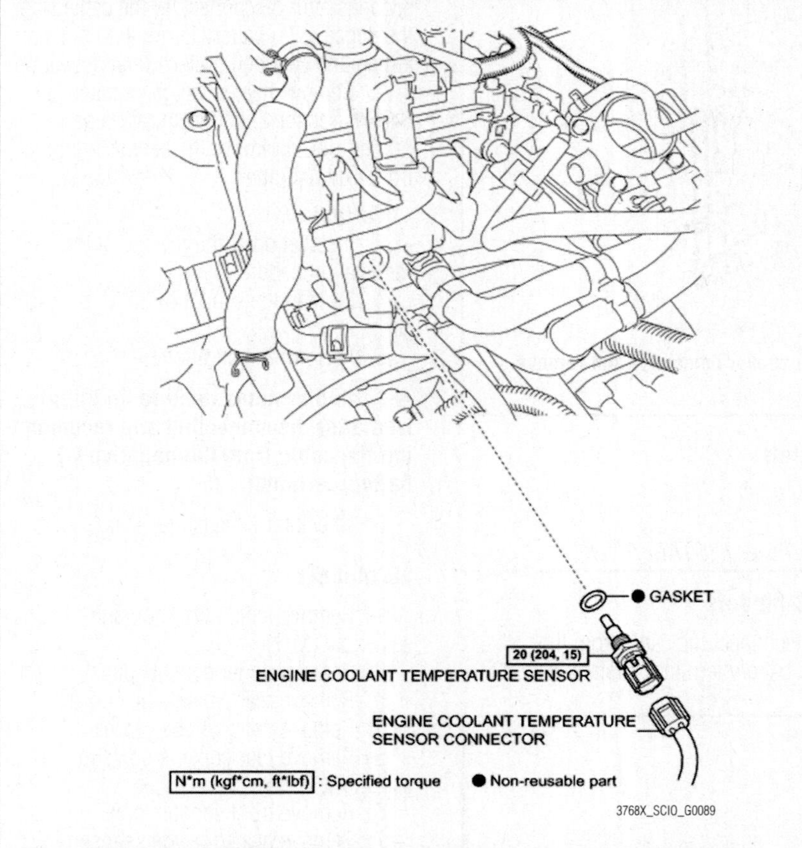

Fig. 141 Showing the location of the Engine Coolant Temperature (ECT) sensor—2.4L engines

1. Drain engine coolant
2. Remove air cleaner assembly as follows:
   a. Disconnect the Mass Air Flow (MAF) meter connector.
   b. Separate the wire harness.
   c. Separate the No. 1 vacuum switching valve.
   d. Disconnect the ventilation hose.
   e. Separate the No. 2 fuel vapor feed hose.
   f. Loosen the No. 1 air cleaner hose clamp, unlock the air cleaner assembly clamp and remove the air cleaner cap sub-assembly with No. 1 air cleaner hose.
   g. Remove the air cleaner filter element sub-assembly.
   h. Separate the wire harness.
   i. Remove the 3 bolts and remove the air cleaner case sub-assembly.
3. Remove Engine Coolant Temperature (ECT) sensor as follows:

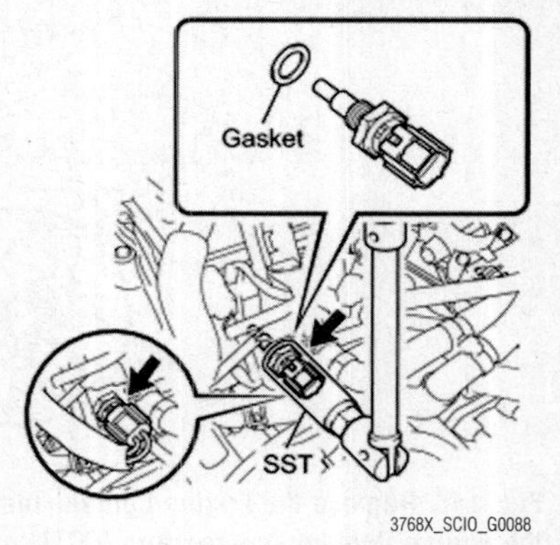

Fig. 144 Using a proper removal tool, remove the Engine Coolant Temperature (ECT) sensor and gasket.

   a. Disconnect the Engine Coolant Temperature (ECT) sensor connector.
   b. Using a proper removal tool, remove the Engine Coolant Temperature (ECT) sensor and gasket.

To install:

4. Installation is the reverse of the removal procedure.
5. Tighten the sensor to 14 ft. lbs. (20 Nm).

xD Models

See Figure 145.

1. Drain engine coolant
2. Remove the No. 2 cylinder head cover.
3. Remove the Engine Coolant Temperature (ECT) sensor connector and remove the Engine Coolant Temperature (ECT) sensor.
4. Installation is the reverse of the removal procedure.
5. Tighten the sensor to 14 ft. lbs. (20 Nm).

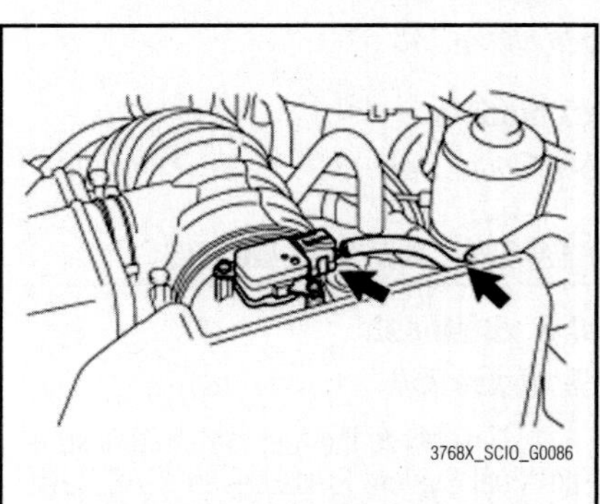

Fig. 142 Disconnect the Mass Air Flow (MAF) meter connector.

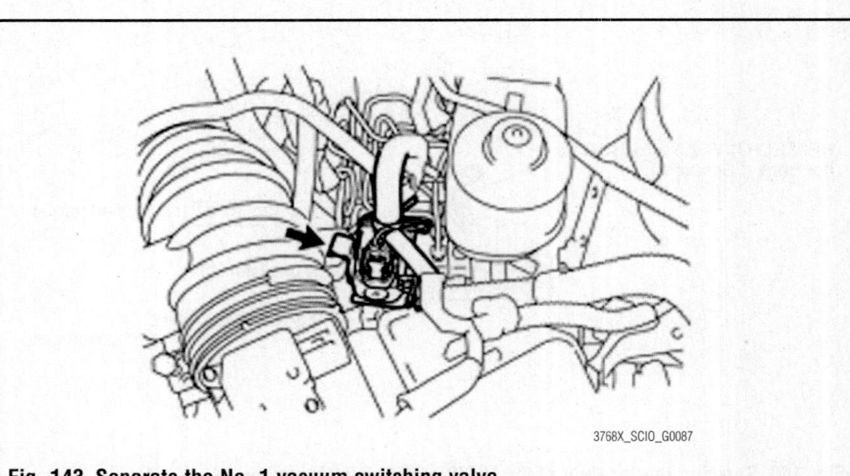

Fig. 143 Separate the No. 1 vacuum switching valve.

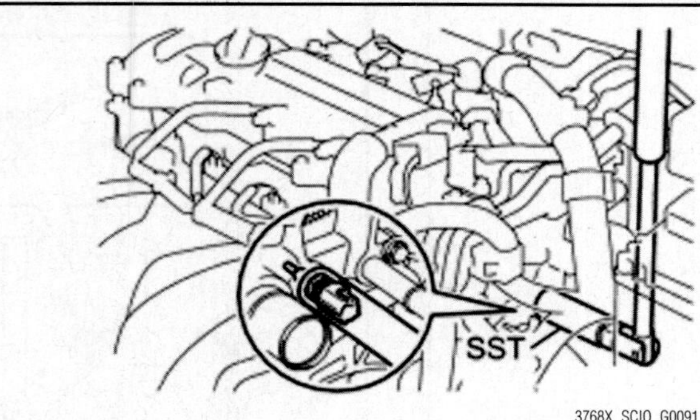

Fig. 145 Remove the Engine Coolant Temperature (ECT) sensor connector and remove the Engine Coolant Temperature (ECT) sensor.

HEATED OXYGEN SENSOR (HO2S)

LOCATION

tC & xB Models

See Figure 146.

xD Models

See Figure 147.

REMOVAL & INSTALLATION

tC & xB Models

1. Disconnect the cable from the negative battery terminal. Wait at least 90 seconds after disconnecting the cable from the negative (-) battery terminal to prevent airbag and seat belt pretensioner activation.
2. Remove heated oxygen sensor (for Bank 1 Sensor 2) connector and use a proper tool to remove the sensor from the front exhaust pipe.

To install:

3. Installation is the reverse of the removal procedure.
4. Tighten the sensor to 30 ft. lbs. (40 Nm).
5. Perform initialization.

➡Certain systems need to be initialized after disconnecting and reconnecting the cable from the negative (-) battery terminal.

6. Check for exhaust gas leaks.

Fig. 146 Showing the location of the HO2S sensor—2.4L engine

xD Models

1. Remove shift lever knob sub-assembly (M/T).
2. Remove upper console panel.
3. Remove rear console box cover.
4. Remove console box carpet.
5. Remove rear console box sub-assembly.
6. Remove front console box.
7. Remove heated oxygen sensor.
   a. Disconnect the heated oxygen sensor connector.
   b. Remove the grommet and pull the sensor connector out of the cabin through the floor panel.
   c. Remove the wire harness clamp bracket.
   d. Using SST, remove heated oxygen sensor.
8. Installation is the reverse of removal.

KNOCK SENSOR (KS)

LOCATION

tC & xB Models

See Figure 148.

xD Models

See Figure 149.

REMOVAL & INSTALLATION

tC & xB Models

See Figure 150.

1. Discharge the fuel system pressure. See "Fuel System Pressure" in "GASOLINE FUEL INJECTION SYSTEM" section.
2. Drain coolant

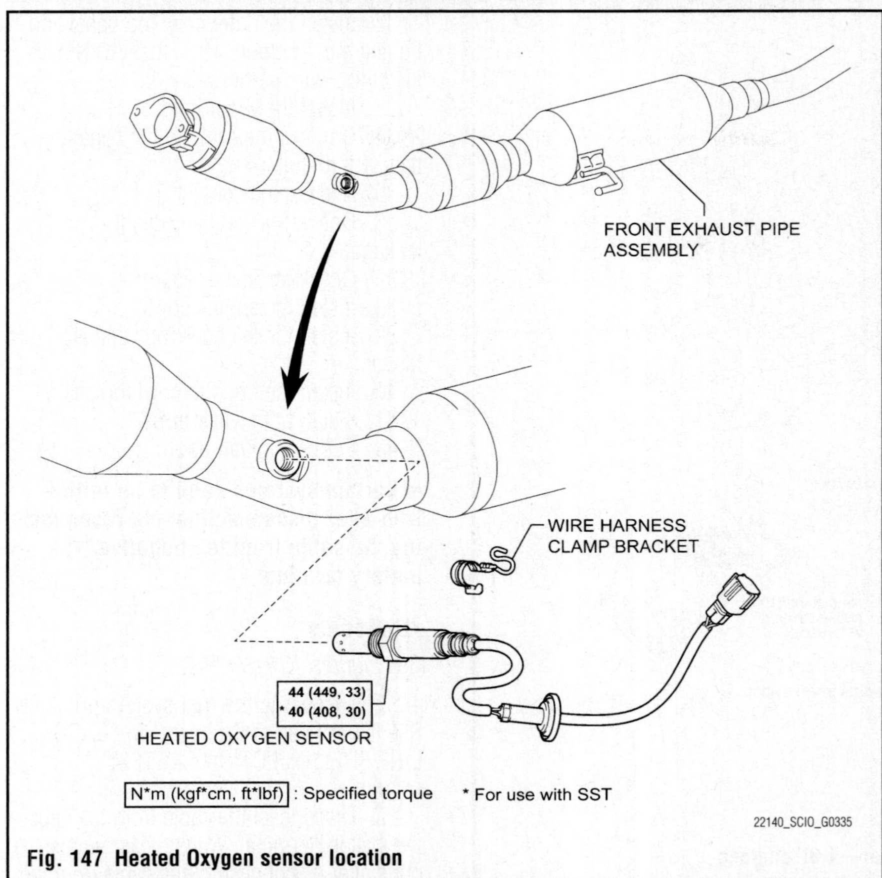

Fig. 147 Heated Oxygen sensor location

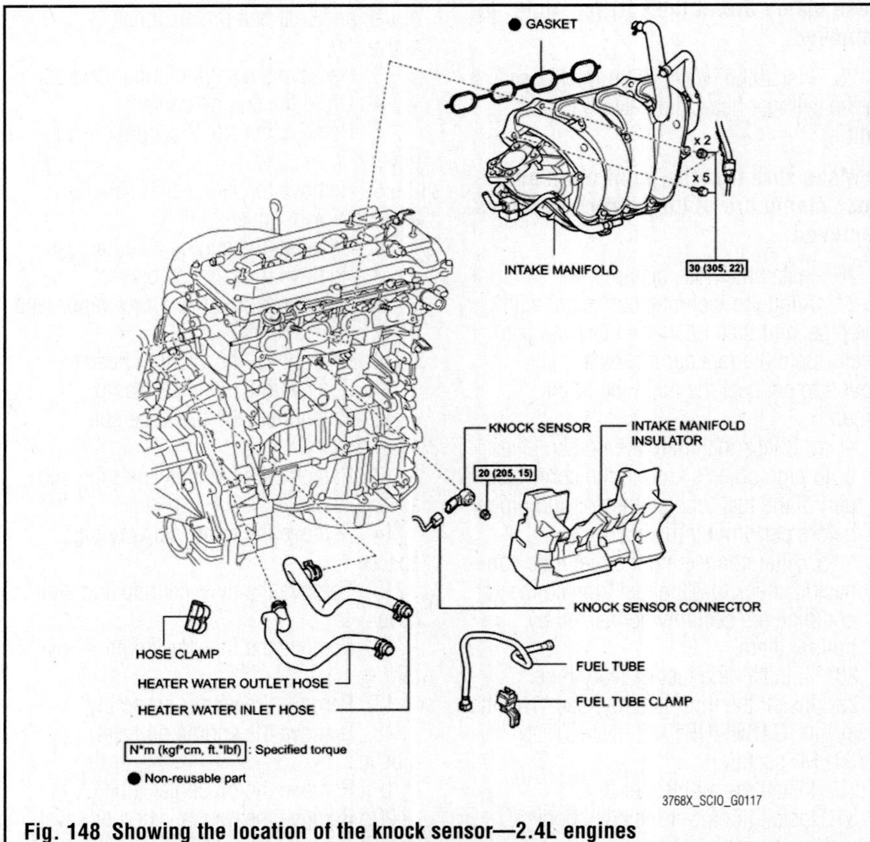

Fig. 148 Showing the location of the knock sensor—2.4L engines

3. Disconnect the cable from the negative battery terminal. Wait at least 90 seconds after disconnecting the cable from the negative (-) battery terminal to prevent airbag and seat belt pretensioner activation.
4. Remove the front wiper arms.
5. Remove the hood to cowl top seal.
6. Remove the left cowl top ventilator louver.
7. Remove the wiper link.
8. Remove the outer cowl top panel.
9. Remove the left lower cowl body mounting bracket (bolt, clip and bracket).
10. Remove the engine cover.
11. Remove the air cleaner cap.
12. Remove the throttle body. See "Throttle Body" in "GASOLINE FUEL INJECTION SYSTEM" section.
13. Remove the fuel delivery pipe.

➡**Remove any dirt and foreign objects on the fuel tube connector before performing this work. When disconnecting the connector, do not damage the connector and pipe, and do not allow foreign matter to contact the connector and pipe. Damage or foreign matter may damage the O-ring that seals the pipe and lead to leaks.**

➡**Perform this work by hand. Do not use any tools.**

CAUTION

Do not forcibly bend, twist or turn the nylon tube.

14. Remove the fuel tube. Protect the disconnected part by covering it with a plastic bag after disconnecting the fuel tube. If the fuel tube connector and pipe are stuck, push and pull to release them.
15. Remove the hose clamp. Remove the heater water inlet hose from the heater radiator unit and cylinder head.
16. Remove the heater water outlet hose from the heater radiator unit and water bypass pipe.
17. Remove the intake manifold. See "Intake Manifold" in "ENGINE MECHANICAL" section.
18. Remove the intake manifold insulator from the cylinder block.
19. Disconnect the knock sensor connector, remove the nut and the knock sensor.

To install:

20. Install the knock sensor with the nut. Torque to 15 ft. lbs. (20 Nm).

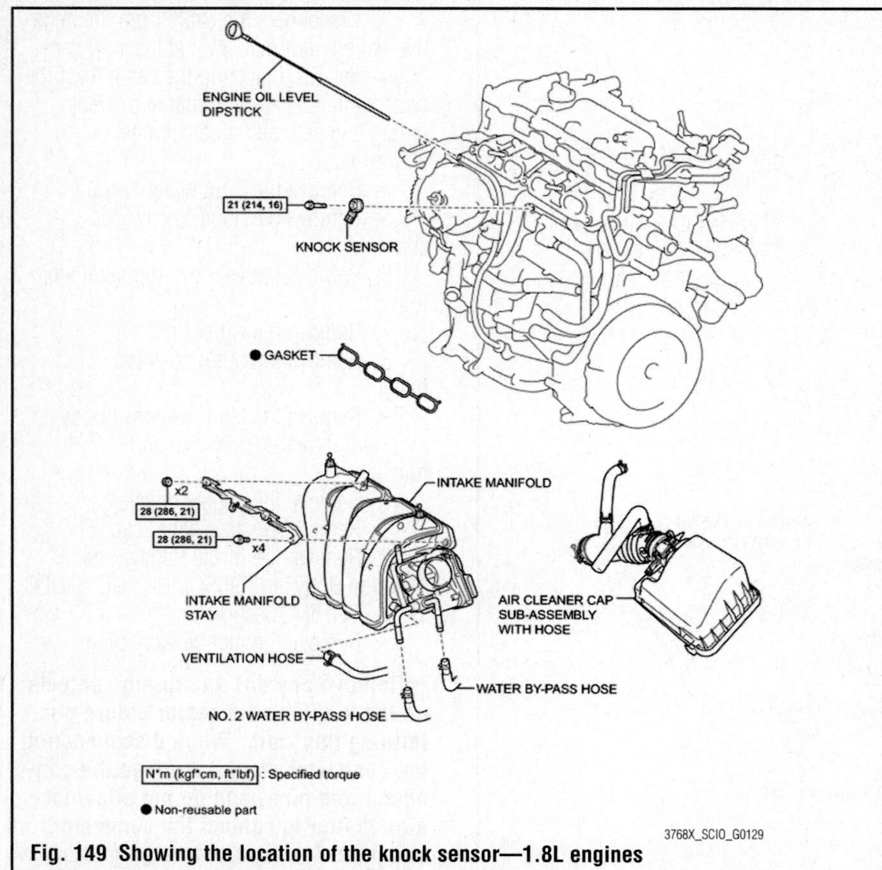

Fig. 149 Showing the location of the knock sensor—1.8L engines

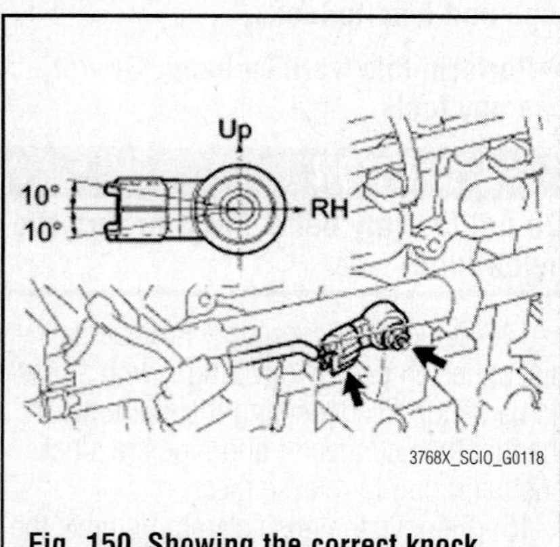

Fig. 150 Showing the correct knock sensor installed direction

➡Make sure that the knock sensor is in the correct direction.

21. Connect the sensor connector.
22. Install the intake manifold insulator onto the cylinder block.
23. Install the intake manifold. See "Intake Manifold" in "ENGINE MECHANICAL" section.
24. Install the heater water outlet hose to the water by-pass pipe and heater radiator unit.

➡Make sure that the paint mark and hose clamp are at the correct angle, as removed.

25. Install the heater water inlet hose to the cylinder head and heater radiator unit.

➡Make sure that the paint mark and hose clamp are at the correct angle, as removed.

26. Install the hose clamp.
27. Align the fuel tube connector with the pipe, and then push the fuel tube connector until the retainer makes a "click" sound to connect the fuel tube to the pipe.
    a. Check that there are no scratches or foreign objects around the connected part of the fuel tube connector and pipe before performing this work.
    b. After connecting the fuel tube connector, check that the fuel tube connector and pipe are securely connected by pulling them.
28. 7. Install the fuel delivery pipe.
29. Install the throttle body. See "Throttle Body" in "GASOLINE FUEL INJECTION SYSTEM" section.
30. Install the air cleaner cap.
31. Install the lower left cowl body mounting bracket.
32. Install the outer cowl top panel and tighten the 7 bolts to 10 ft. lbs. (13 Nm). Install the wire harness clamp.
33. Install the wiper link.
34. Connect the cable to the negative battery terminal.
35. Add engine coolant.
36. Check for engine coolant leakage.
37. Check for fuel leakage
38. Install the engine cover.
39. Install the left cowl top ventilator louver.
40. Install the hood to cowl top seal.
41. Install both wiper arms.
42. Perform initialization.

➡Certain systems need to be initialized after disconnecting and reconnecting the cable from the negative (-) battery terminal.

xD Models

See Figures 151 and 152.

1. Discharge the fuel system pressure. See "Fuel System Pressure" in "GASOLINE FUEL INJECTION SYSTEM" section.
2. Disconnect the cable from the negative battery terminal. Wait at least 90 seconds after disconnecting the cable from the negative (-) battery terminal to prevent airbag and seat belt pretensioner activation.
3. Remove the engine under covers.
4. Drain the engine coolant.
5. Remove the No. 2 cylinder head cover.
6. Remove the air cleaner cap sub-assembly with hose.
7. Separate the engine wire harness.
8. Remove the harness bracket.
9. Disconnect the No. 1 fuel vapor feed hose.
10. Disconnect the vacuum hose.
11. Remove the fuel pipe clamp.
12. Disconnect the fuel tube sub-assembly.
13. Remove the fuel delivery pipe sub-assembly.
14. Remove the No. 1 delivery pipe spacer.
15. Remove the flywheel housing side cover.
16. Remove the injector vibration insulator.
17. Remove the starter assembly.
18. Remove the engine oil level dipstick.
19. Remove the oil cooler tube (A/T).
20. Remove the transmission oil level gauge sub-assembly (A/T).

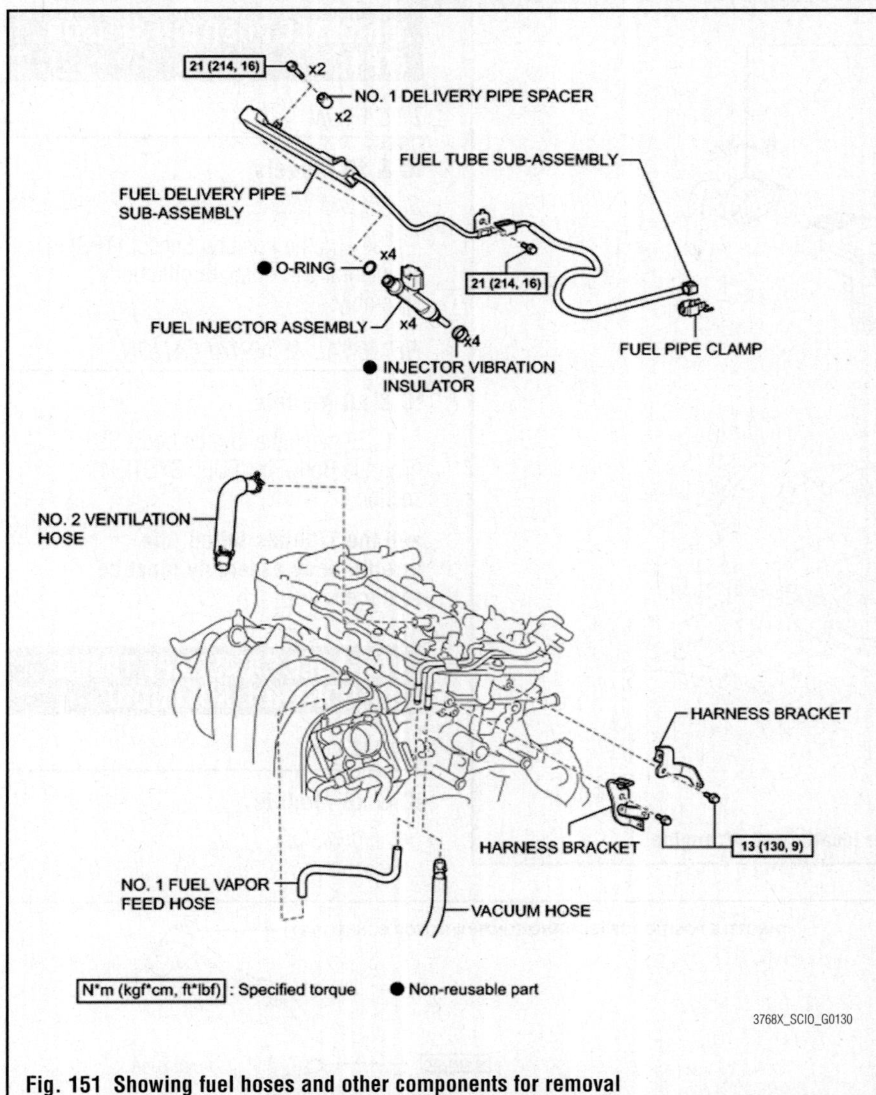

Fig. 151 Showing fuel hoses and other components for removal

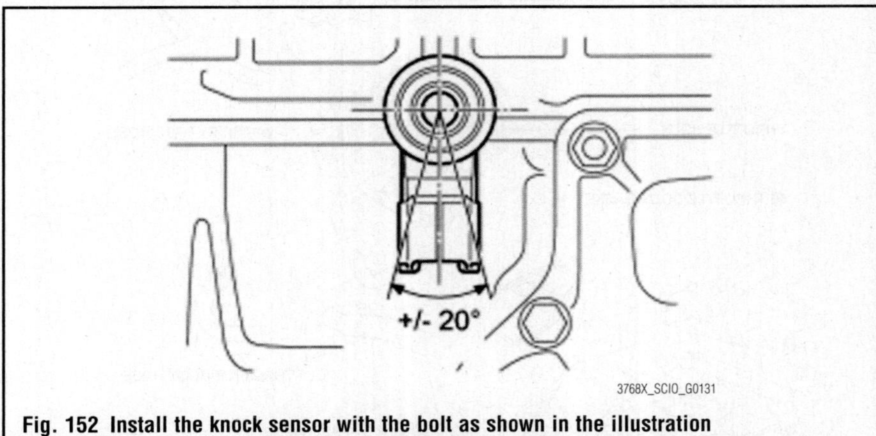

Fig. 152 Install the knock sensor with the bolt as shown in the illustration

21. Remove the intake manifold. See "Intake Manifold" in "ENGINE MECHANICAL" section.
22. Disconnect the knock sensor connector.
23. Remove the bolt and remove the knock sensor.

To install:

24. Install the knock sensor with the bolt as shown in the illustration. Torque the sensor to 16 ft. lbs. (21 Nm). Connect the knock sensor connector.
25. Install the intake manifold. See "Intake Manifold" in "ENGINE MECHANICAL" section.
26. Install the transmission oil level gauge sub-assembly (A/T).
27. Install the oil cooler tube (A/T).
28. Install the engine oil level dipstick.
29. Install the starter assembly.
30. Install the flywheel housing side cover.
31. Install the injector vibration insulator.
32. Install the No. 1 delivery pipe spacer.
33. Install the fuel delivery pipe sub-assembly.
34. Connect the fuel tube sub-assembly.
35. Install the fuel pipe clamp.
36. Connect the vacuum hose.
37. Connect the No. 1 fuel vapor feed hose.
38. Install the harness bracket.
39. Install the engine wire harness.
40. Install the air cleaner cap sub-assembly with hose.
41. Connect the cable to negative battery terminal.
42. Add the engine coolant.
43. Inspect for coolant leak.
44. Inspect for fuel leak.
45. Install the No. 2 cylinder head cover.
46. Install the engine under covers.

MASS AIR FLOW (MAF) METER

LOCATION

tC & xB Models

See Figure 153.

xD Models

See Figure 154.

REMOVAL & INSTALLATION

1. Disconnect the cable from the negative battery terminal. Wait at least 90 seconds after disconnecting the cable from the negative (-) battery terminal to prevent airbag and seat belt pretensioner activation.
2. Disconnect the Mass Air Flow (MAF) meter connector. Remove the 2 screws and mass air flow meter.
3. Installation is the reverse of the removal procedure.

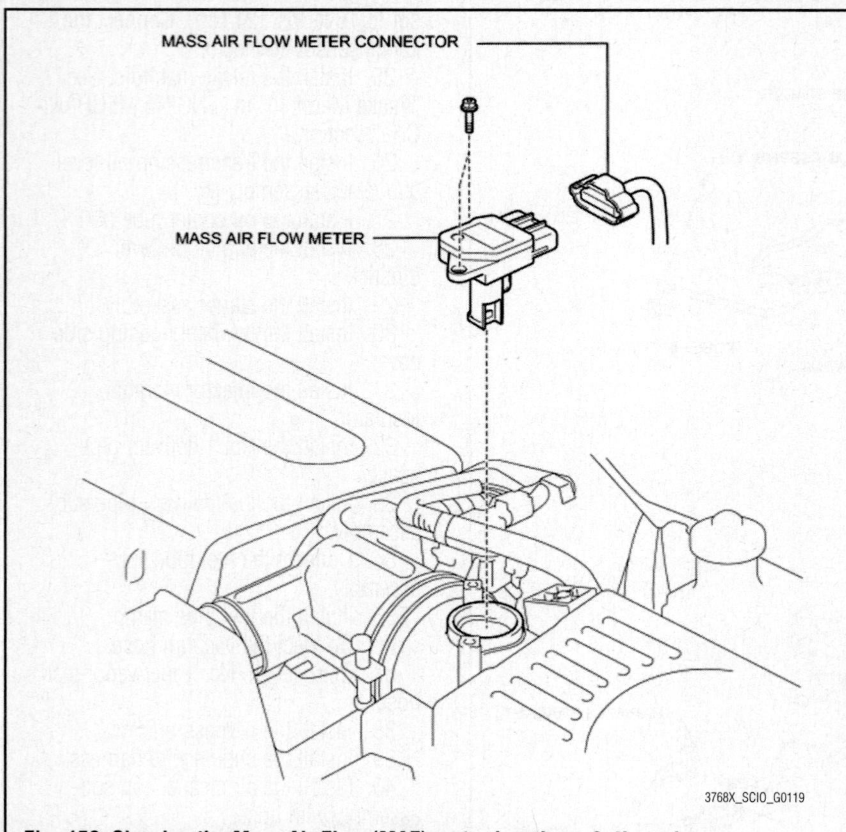

Fig. 153 Showing the Mass Air Flow (MAF) meter location—2.4L engine

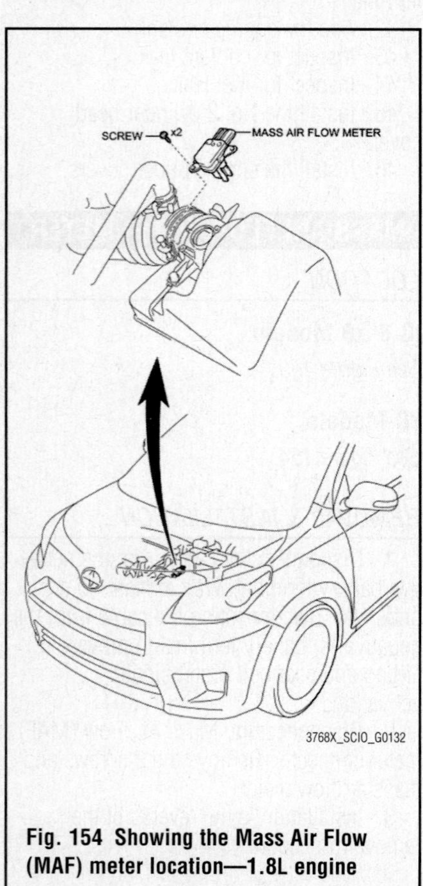

Fig. 154 Showing the Mass Air Flow (MAF) meter location—1.8L engine

THROTTLE POSITION SENSOR (TPS)

LOCATION

tC & xB Models

See Figure 155.

The Throttle Position Sensor (TPS) is an integral part of the throttle body assembly

REMOVAL & INSTALLATION

tC & xB Models

1. Remove the throttle body. See "Throttle Body" in "FUEL SYSTEM" section.

➡If the TPS has failed, the throttle body assembly must be replaced.

VARIABLE CAMSHAFT TIMING OIL CONTROL SOLENOID

LOCATION

tC & xB Models

See Figure 156.

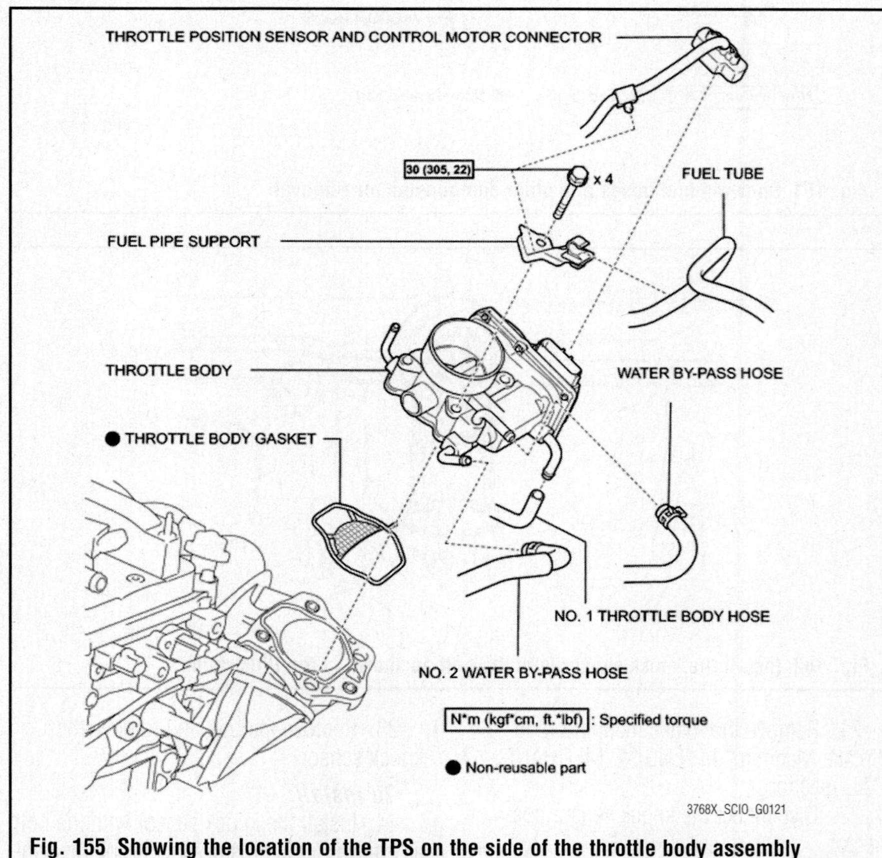

Fig. 155 Showing the location of the TPS on the side of the throttle body assembly

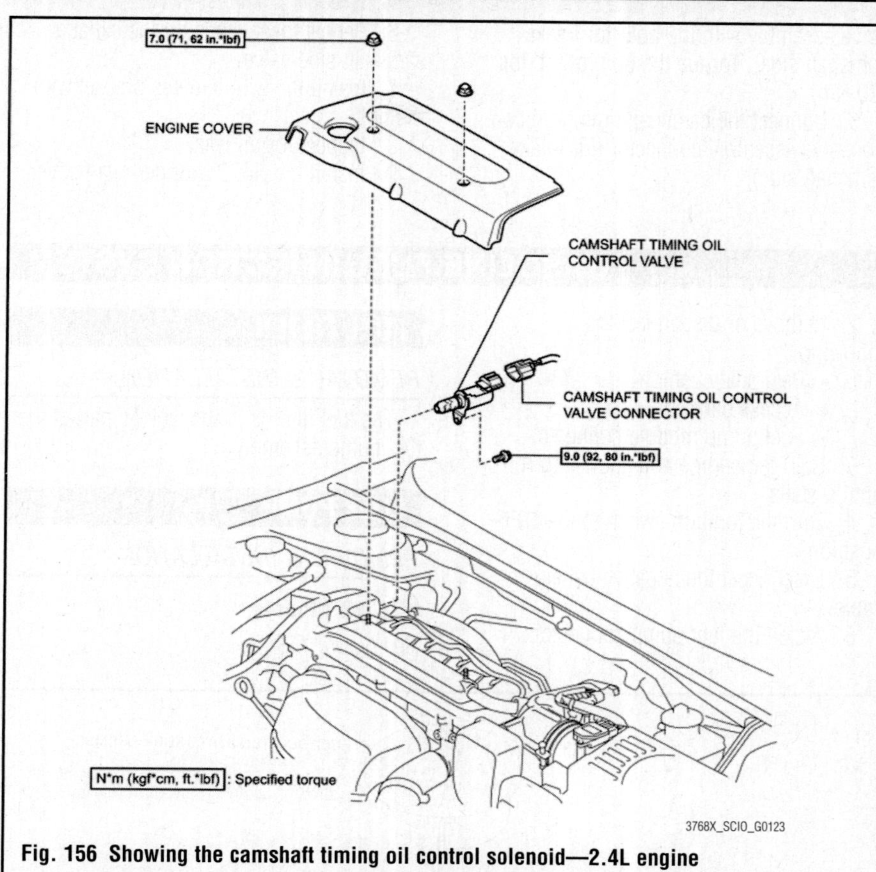

Fig. 156 Showing the camshaft timing oil control solenoid—2.4L engine

xD Models

See Figure 157.

REMOVAL & INSTALLATION

tC & xB Models

1. Disconnect the cable from the negative battery terminal. Wait at least 90 seconds after disconnecting the cable from the negative (-) battery terminal to prevent airbag and seat belt pretensioner activation.
2. Remove the engine cover.
3. Remove the camshaft timing oil control valve assembly as follows:
   a. Disconnect the oil control valve connector.
   b. Remove the bolt and oil control valve.

To install:

4. Installation is the reverse of the removal procedure.
5. Apply a thin coat of engine oil to the valve O-ring.
6. Reconnect the battery cable.
7. Check for oil leaks.
8. Perform initialization.

➡Certain systems need to be initialized after disconnecting and reconnecting the cable from the negative (-) battery terminal.

xD Models

1. Remove the No. 2 cylinder head cover.
2. Remove the camshaft timing oil control valve assembly as follows:
   a. Remove the bolt and the wire harness bracket.
   b. Disconnect the camshaft timing oil control valve assembly connector (for exhaust camshaft side).
   c. Remove the bolt and remove the camshaft timing oil control valve assembly (for exhaust camshaft side).
   d. Remove the O-ring from the camshaft timing oil control valve assembly (for exhaust camshaft side).
   e. Disconnect the camshaft timing oil control valve assembly connector (for intake camshaft side).
   f. Remove the bolt and remove the camshaft timing oil control valve assembly (for intake camshaft side).
   g. Remove the O-ring from the camshaft timing oil control valve assembly (for intake camshaft side).

To install:

3. Apply a light coat of engine oil to a new O-ring and install it onto the camshaft

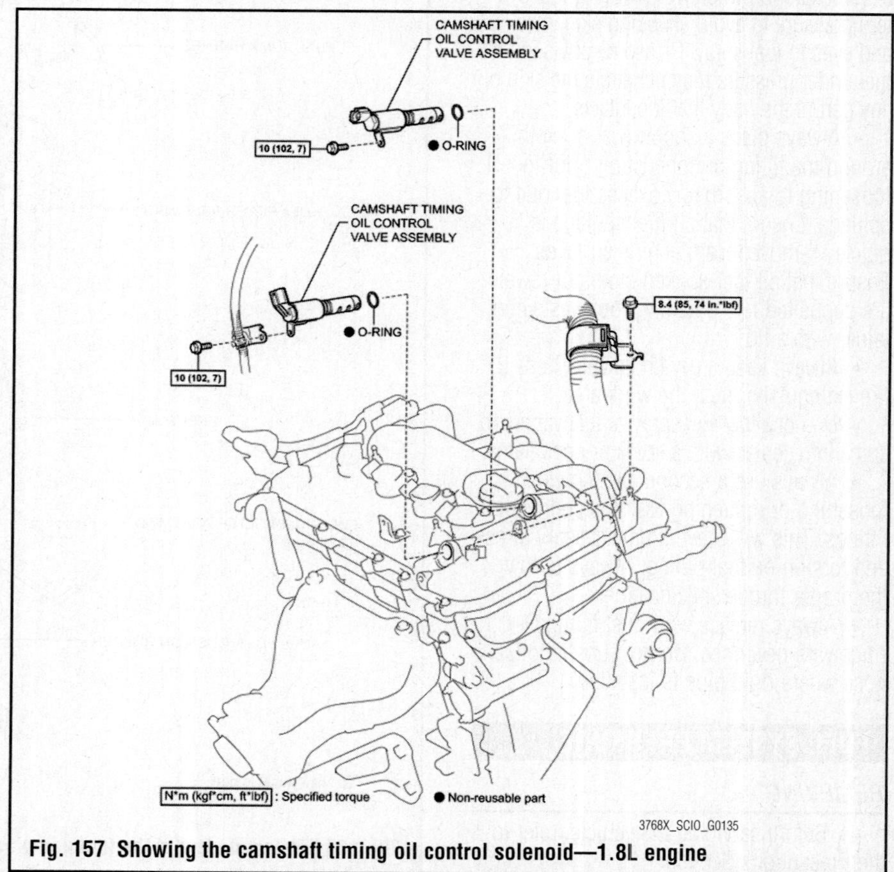

Fig. 157 Showing the camshaft timing oil control solenoid—1.8L engine

timing oil control valve assembly (for intake camshaft side).

CAUTION

Do not twist the O-ring.

4. Install the camshaft timing oil control valve assembly with the bolt (for intake camshaft side). Torque the bolt to 7 ft. lbs. (10 Nm).
5. Connect the camshaft timing oil control valve assembly connector (for intake camshaft side).
6. Repeat these steps for the exhaust camshaft side valve.
7. Install the wire harness bracket with the bolt.
8. Inspect for oil leak.
9. Install the No. 2 cylinder head cover.

FUEL

GASOLINE FUEL INJECTION SYSTEM

FUEL SYSTEM SERVICE PRECAUTIONS

Safety is the most important factor when performing not only fuel system maintenance, but any type of maintenance. Failure to conduct maintenance and repairs in a safe manner may result in serious personal injury or death. Work on a vehicle's fuel system components can be accomplished safely and effectively by adhering to the following rules and guidelines.

- To avoid the possibility of fire and personal injury, always disconnect the negative battery cable unless the repair or test procedure requires that battery voltage be applied.
- Always relieve the fuel system pressure prior to disconnecting any fuel system component (injector, fuel rail, pressure regulator, etc.) fitting or fuel line connection. Exercise extreme caution whenever relieving fuel system pressure to avoid exposing skin, face and eyes to fuel spray. Please be advised that fuel under pressure may penetrate the skin or any part of the body that it contacts.
- Always place a shop towel or cloth around the fitting or connection prior to loosening to absorb any excess fuel due to spillage. Ensure that all fuel spillage is quickly removed from engine surfaces. Ensure that all fuel-soaked cloths or towels are deposited into a flame-proof waste container with a lid.
- Always keep a dry chemical (Class B) fire extinguisher near the work area.
- Do not allow fuel spray or fuel vapors to come into contact with a spark or open flame.
- Always use a second wrench when loosening or tightening fuel line connection fittings. This will prevent unnecessary stress and torsion on fuel piping. Always follow the proper torque specifications.
- Always replace worn fuel fitting O-rings with new ones. Do not substitute fuel hose where rigid pipe is installed.

FUEL SYSTEM PRESSURE

RELIEVING

1. Before servicing the vehicle, refer to the Precautions Section.
2. Remove or disconnect the following:
   - Rear seat cushion
   - Access panel
   - Fuel pump module connector
3. Start the engine and allow it to run until it stalls.
4. Turn the ignition switch to the **OFF**-position.
5. Disconnect the negative battery cable.
6. Attach the fuel pump connector.

FUEL FILTER

REMOVAL & INSTALLATION

The fuel filter is in the tank as part of the fuel pump assembly.

FUEL PUMP

REMOVAL & INSTALLATION

tC & xB Models

See Figure 158.

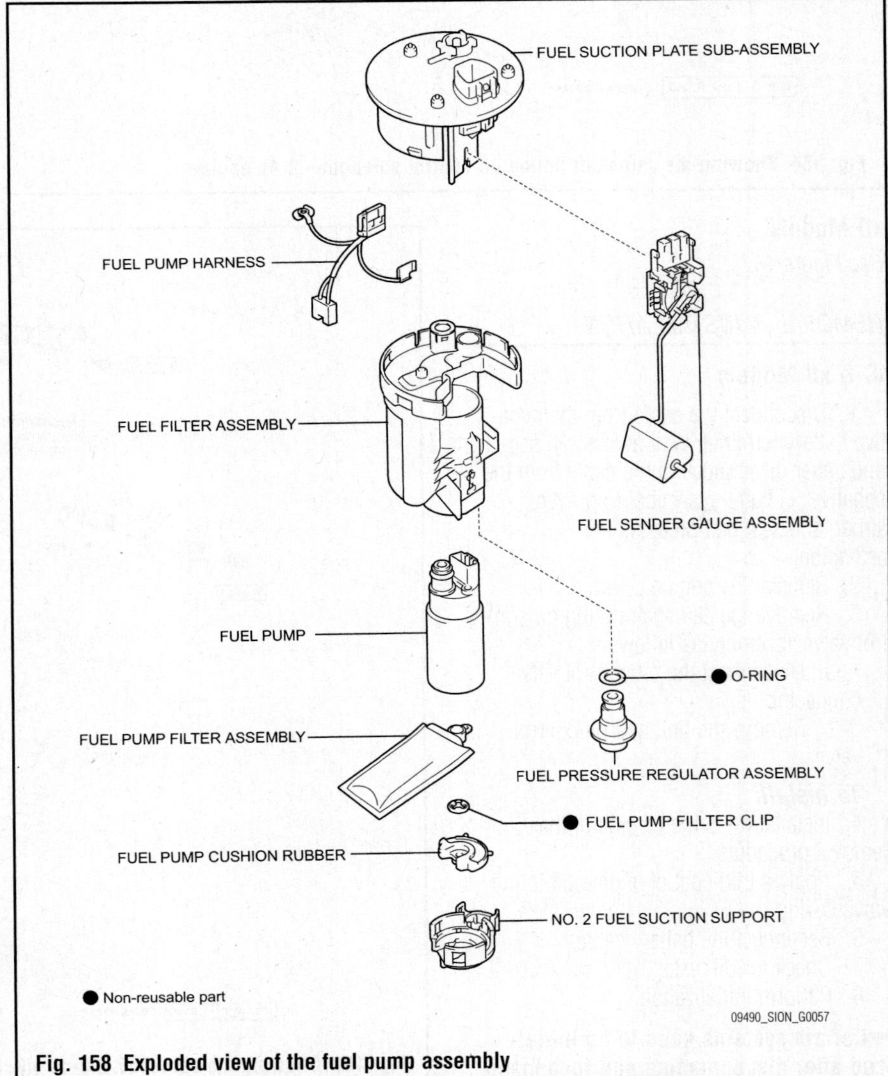

Fig. 158 Exploded view of the fuel pump assembly

1. Remove or disconnect the following:
   - Negative battery cable
   - Rear seat cushion assembly
   - Rear seat cushion support bracket (xB only)
   - Rear floor service hole cover
   - Fuel pump connector
   - Fuel supply and vent hoses
   - Fuel tank vent tube set plate
   - Fuel pump assembly

To install:

2. Install or connect the following:
   - Fuel pump assembly with a new gasket.
   - Fuel tank vent tube set plate. Tighten bolts to 31 inch lbs. (4 Nm) for and xB models. 53 inch lbs. (6 Nm) for tC models.
   - Fuel supply and vent hoses
   - Fuel pump connector
   - Rear floor service hole cover
   - Rear seat cushion support bracket
   - Rear seat cushion
   - Negative battery cable
3. Start the engine and check for leaks.

xD Models

See Figure 159.

1. Remove the No. 1 rear seat leg cover.
2. Remove the No. 2 seat leg cover.
3. Remove the rear seat assembly.
4. Remove the rear floor service hole cover.
5. Discharge the fuel system pressure. See "Relieving Fuel System Pressure" in "GASOLINE FUEL INJECTION SYSTEM" section.
6. Disconnect fuel tank main tube sub-assembly as follows:
   a. Spread the tip of the tube joint clip and pull the clip off.
   b. Disconnect the fuel tank main tube.

⁂ CAUTION

Keep the O-ring free of any foreign matter, as it becomes contaminated easily. Do not use any tools in this procedure. Do not forcefully bend or twist the tube.

   c. Put the tube in a plastic bag to prevent damage and contamination.
   d. If the fuel suction plate and tube are stuck together, pinch the tube and turn it carefully to disconnect them.
   e. If a clip is damaged, replace it.
7. Disconnect the fuel tank vent hose.

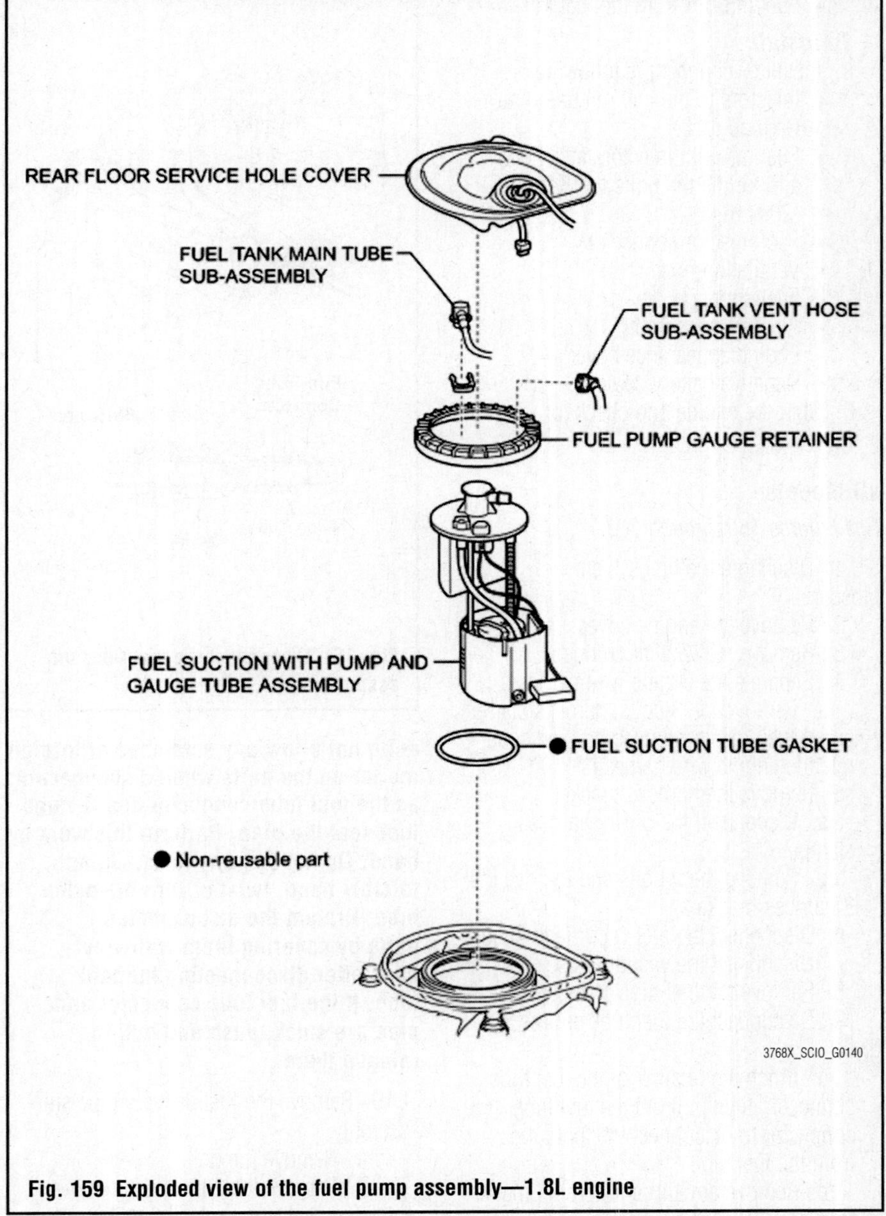

Fig. 159 Exploded view of the fuel pump assembly—1.8L engine

8. Using SST 09808-14020, or equivalent, remove the fuel pump gauge retainer.

➡Align the claws of the fuel pump gauge retainer with the tips of SST.

9. Remove the fuel suction with pump and gauge tube.

⁂ CAUTION

Do not bend the arm of the sender gauge.

10. Remove the fuel suction tube gasket from the fuel tank.
11. Installation is the reverse of the removal procedure.

FUEL RAIL & INJECTORS

REMOVAL & INSTALLATION

tC & xB Models

1. Before servicing the vehicle, refer to the Precautions Section.
2. Relieve the fuel system pressure.
3. Remove or disconnect the following:
   - Negative battery cable
   - Engine appearance cover
   - Air intake assembly
   - Fuel supply hose
   - Ventilation hose
   - Fuel injector connectors
   - Fuel rail with the injectors attached

- Fuel injector from the fuel rail

To install:

4. Install or connect the following:
   - Injectors to the fuel rail using new O-rings
   - Fuel rail with injectors attached and torque the bolts to 15 ft. lbs. (20 Nm).
   - Fuel injector connectors
   - Ventilation hose
   - Fuel supply hose
   - Air intake assembly
   - Engine appearance cover
   - Negative battery cable
5. Start the engine and check for leaks.

xD Models

See Figures 160 through 163.

1. Discharge the fuel system pressure.
2. Remove the engine cover.
3. Remove the ventilation hose.
4. Separate the engine wire harness.
   a. Remove the 3 bolts, harness clamp and 4 fuel injector connectors and separate the engine wire harness.
5. Remove the harness bracket.
   a. Disconnect the 5 wire harness clamps.
   b. Remove the 2 bolts and remove the 2 harness brackets.
6. Disconnect the fuel vapor feed hose.
7. Disconnect the vacuum hose.
8. Remove the fuel pipe clamp.
9. Disconnect the fuel tube sub-assembly.
   a. Pinch the retainer of the fuel tube connector, then pull out the fuel tube connector to disconnect the fuel tube from the fuel pipe.
   b. Remove any dirt and foreign matter from the fuel tube connector before performing this work.

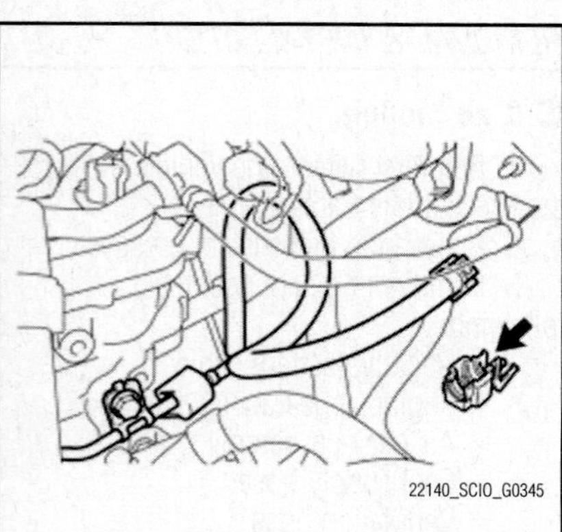

Fig. 160 Remove the fuel pipe clamp.

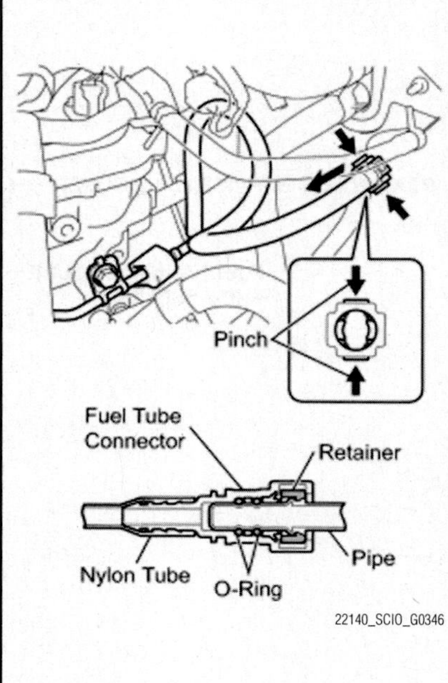

Fig. 161 Disconnect the fuel tube sub-assembly.

➡Do not allow any scratches or foreign matter on the parts when disconnecting as the fuel tube connector has O-rings that seal the pipe. Perform this work by hand. Do not use any tools. Do not forcibly bend, twist or turn the nylon tube. Protect the disconnected parts by covering them with vinyl bags after disconnecting the fuel tube. If the fuel tube connector and pipe are stuck, push and pull to release them.

10. Remove the fuel delivery pipe sub-assembly.
    a. Remove the bolt.
    b. Remove the 2 bolts and remove the fuel delivery pipe with the 4 fuel injectors.

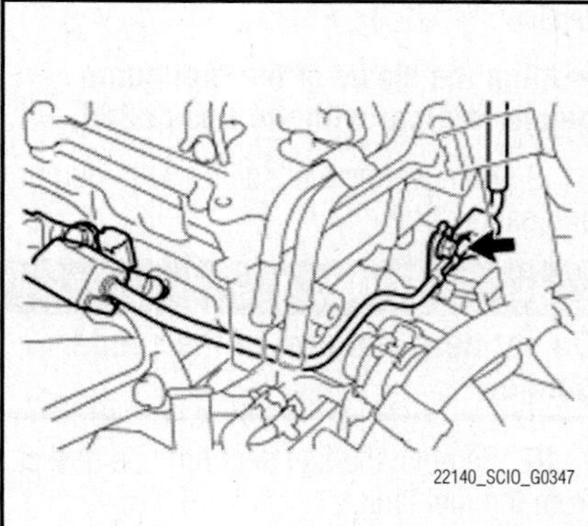

Fig. 162 Remove the fuel delivery pipe sub-assembly.

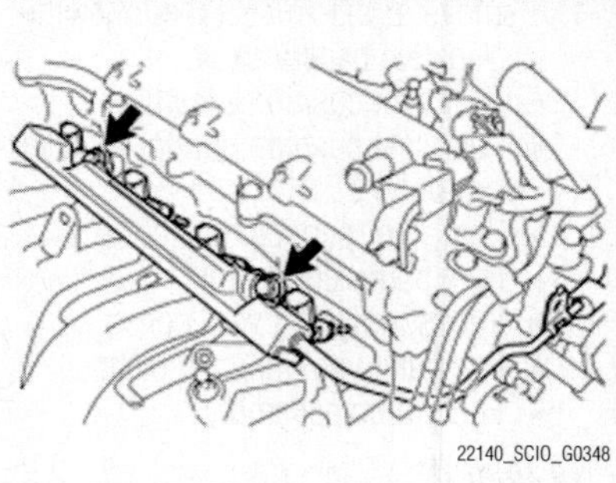

Fig. 163 Remove the 2 bolts and remove the fuel delivery pipe with the 4 fuel injectors.

✱✱ CAUTION

Do not drop the fuel injectors when removing the fuel delivery pipe.

11. Remove the 2 delivery pipe spacers.
12. Remove the 4 injector vibration insulator.
13. Remove the fuel injector assembly.
    a. Pull the 4 fuel injectors out of the fuel delivery pipe.
    b. Place the fuel injector in a plastic bag to prevent foreign matter from entering.

To install:

14. Install fuel injector assembly.
    a. Apply a light coat of gasoline to new O-rings, then install one onto each fuel injector.
    b. Apply a light coat of gasoline to the contact surfaces of the fuel delivery pipe and the O-ring of the fuel injector.
    c. While turning the fuel injector left and right, install it onto the fuel delivery pipe.

✱✱ CAUTION

Do not twist the O-ring.

    d. After installing the fuel injectors, check that they turn smoothly. If they do not, replace the O-ring with a new one.
15. Install 4 new injector vibration insulators onto the cylinder head.
16. Install the 2 No. 1 delivery pipe spacers onto the cylinder head.

➡Install the No. 1 delivery pipe spacers in the correct direction.

17. Install fuel delivery pipe sub-assembly.
    a. Install the fuel delivery pipe with the 4 fuel injectors, then provisionally install the 3 bolts.

✱✱ CAUTION

Do not drop the fuel injectors when installing the fuel delivery pipe.

b. Check that the fuel injectors rotate smoothly after installing the fuel delivery pipe.

c. Tighten the 3 bolts to the specified torque. Torque: 16 ft. lbs. (21 Nm).

18. Connect fuel tube sub-assembly.

a. Insert the fuel tube connector into the fuel pipe until a click sound can be heard.

b. Check that there are no scratches or foreign matter around the disconnected parts of the fuel tube connector and pipe before performing this work.

c. After connecting the fuel tube, check that the fuel tube connector and pipe are securely connected by pulling them.

19. Install the fuel pipe clamp.
20. Connect the vacuum hose.
21. Connect the fuel vapor feed hose.
22. Install harness bracket.

a. Install the 2 harness brackets with the 2 bolts. Torque: 9 ft. lbs. (13 Nm).

b. Connect the 5 wire harness clamps.

23. Install engine wire harness.

a. Install the engine wire harness with 3 bolts and harness clamp.

b. Connect the 4 fuel injector connectors.

24. Install the ventilation hose.
25. Inspect for fuel leak.
26. Install Engine cover.

FUEL TANK

REMOVAL & INSTALLATION

tC Models

See Figure 164.

1. Remove the rear seat cushion assembly.
2. Remove the rear floor service hole cover.
3. Disconnect the cable from the negative battery terminal. Wait at least 90 seconds after disconnecting the cable from the negative (-) battery terminal to prevent airbag and seat belt pretensioner activation.
4. Remove the fuel suction with pump and gauge tube assembly.
5. Drain the fuel
6. Remove the center exhaust pipe assembly.
7. Remove the No. 1 fuel tank protector 5 bolts and fuel tank protector.
8. Remove the EVAP canister assembly. See "Evaporative Emissions (EVAP) Canister" in "ENGINE PERFORMANCE & ENGINE CONTROLS" section.

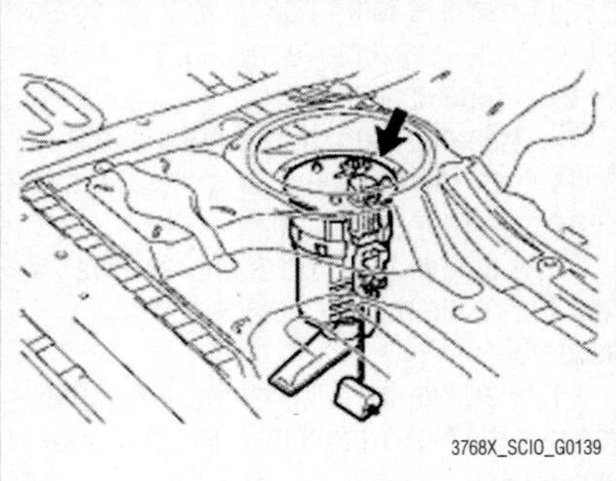

Fig. 164 Remove the fuel suction with pump and gauge tube assembly.

9. Remove the lower LH and RH rear suspension braces.
10. Remove the fuel tank assembly as follows:

a. Disconnect the fuel tank main tube. Pinch the tab of the retainer of the fuel tube connector to remove the lock claws and push it down.

b. Pull the fuel tank main tube out of the pipe.

CAUTION

Remove any dirt and foreign objects on the fuel tube connector before performing this work. Do not allow any scratches or foreign objects on the parts when disconnecting, as the fuel tube connector has the O-ring that seals the pipe.

➡Perform this work by hand. Do not use any tools. Do not forcibly bend, twist or turn the nylon tube.

c. Protect the disconnected part by covering it with a plastic bag after disconnecting the fuel tank main tube.

d. If the fuel tube connector and pipe are stuck, push and pull to release them.

e. Disconnect the fuel tank breather tube from the fuel tank.

f. Pinch the retainer, then pull the fuel tube connector out of the pipe.

g. Separate the claw fitting, then remove the fuel hose protector.

h. Loosen the hose clamp bolt, then disconnect the fuel tank to filler pipe hose.

i. Hold the fuel tank using a transmission jack.

j. Remove the 4 bolts and 2 fuel tank bands.

k. Operate the transmission jack, then remove the fuel tank.

To install:

11. Install the fuel tank assembly as follows:

a. Set the fuel tank onto a transmission jack.

b. Operate the transmission jack, then install the fuel tank into the vehicle.

c. Install the 2 fuel tank bands with the 4 bolts. Torque to 29 ft. lbs. (39 Nm).

d. Connect the fuel tank to filler pipe hose to the fuel tank, then tighten the hose clamp. Make sure that the mark and hose clamp are at the correct angle when installing.

e. Install the fuel hose protector.

f. Connect the fuel tank breather tube to the fuel tank. Align the fuel tube connector with the pipe, then push the fuel tube connector until the retainer makes a "click" sound to connect the fuel tube breather tube to the pipe.

➡Check that there are no scratches or foreign objects around the connected part of the fuel tube connector and pipe before performing this work.

g. After connecting the fuel tube connector, check that the fuel tube connector and pipe are securely connected by pulling them.

h. Connect the fuel tank main tube. Align the fuel tube connector with the pipe, then push the fuel tube connector until the it comes into contact with the seat to connect the fuel tank main tube to the pipe. Then push the retainer up until the claws lock.

i. After connecting the fuel tank main tube, check that the fuel tank main tube is securely connected by pulling the fuel tube connector and pipe.

12. Install lower rear suspension braces. Torque to 47 ft. lbs. (63 Nm).
13. Install the EVAP canister assembly. See "Evaporative Emissions (EVAP) Canister" in "ENGINE PERFORMANCE & ENGINE CONTROLS" section.
14. Install the No. 1 fuel tank protector.
15. Install the center exhaust pipe assembly.
16. Add fuel to the tank.
17. Install the fuel suction with pump and gauge tube assembly.
18. Install rear floor service hole cover.
19. Connect the cable to the negative battery terminal.
20. Perform initialization.

➡Certain systems need to be initialized after disconnecting and reconnecting the cable from the negative (-) battery terminal.

21. Inspect for fuel leak.
22. Install the rear seat cushion assembly

xB Models

See Figures 165 through 168.

CAUTION

Avoid getting any scratches or foreign matter on the parts when disconnecting, as the quick connector has an O-ring that seals the plug. Perform this work by hand. Do not use any tools. Do not forcibly bend, twist or turn the nylon tube.

➡Protect the disconnected parts by covering them with a plastic bag.

1. Remove the rear seat cushion with cover pad sub-assembly.
2. Discharge the fuel system pressure. See "Fuel System Pressure" in this section.
3. Disconnect the fuel tank main tube sub-assembly.
4. Remove the fuel pump gauge retainer.
5. Remove the fuel suction with pump and gauge tube assembly.
6. Drain the fuel.
7. Remove the EVAP charcoal canister assembly.

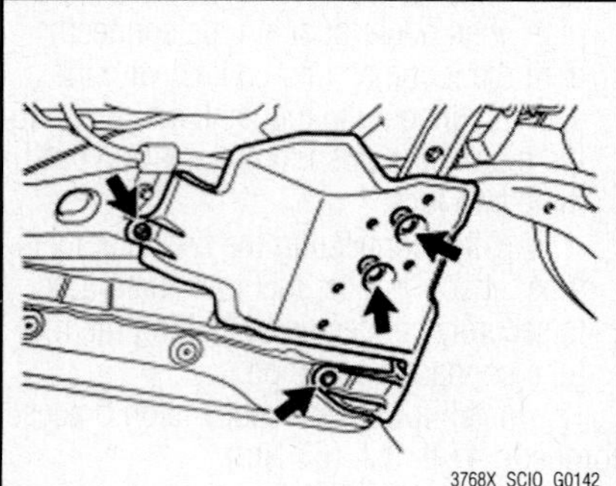

Fig. 165 Remove the nut, grommet and 2 bolts, then remove the rear floor side member cover from each side—RH shown; LH similar

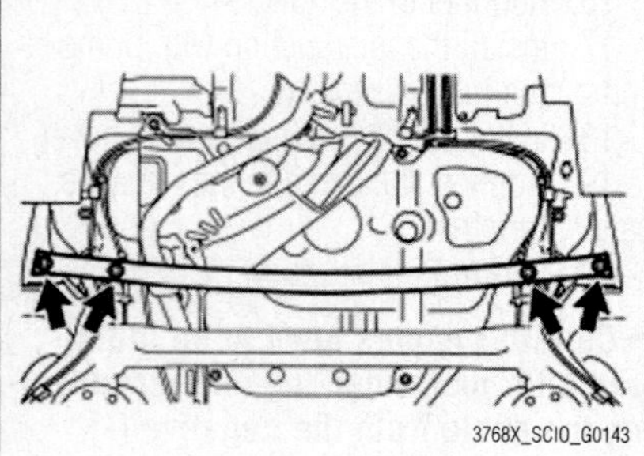

Fig. 166 Remove the 4 bolts and remove the rear floor side member brace sub-assembly.

8. Remove the 4 bolts and 2 compression spring and remove the No. 2 center exhaust pipe assembly.
9. Remove the nut, grommet and 2 bolts, then remove the rear floor side member cover from each side.
10. Remove the 4 bolts and remove the rear floor side member brace sub-assembly.
11. Remove the 3 bolts and a nut, then remove the No. 1 fuel tank protector sub-assembly.
12. Release the lock, then pull and remove the fuel tank main tube.
13. Remove any dirt and foreign matter from the clip before performing this work.
14. If the connector and pipe are stuck, disconnect the nylon tube by turning it by hand to release it.
15. Disconnect the fuel tank to filler pipe hose.
16. Release the lock, then pull and remove the No. 6 fuel tank breather tube.
17. Remove any dirt and foreign matter from the clip before performing this work.
18. Remove the 2 bolts and separate the parking brake cable assembly.
19. Set the jack under the fuel tank.
20. Remove the 4 bolts and remove the fuel tank assembly.
21. Remove the No. 1 fuel tank cushion and 3 No. 2 fuel tank cushions.
22. Remove the fuel tank main tube sub-assembly from the fuel tank.

To install:

23. Position the fuel tank, after assembling the tube sub-assembly and tank cushions, using the jack.

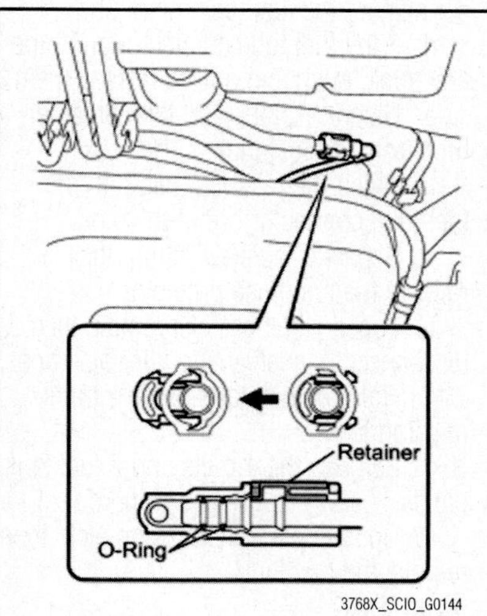

Fig. 167 Release the lock, then pull and remove the fuel tank main tube.

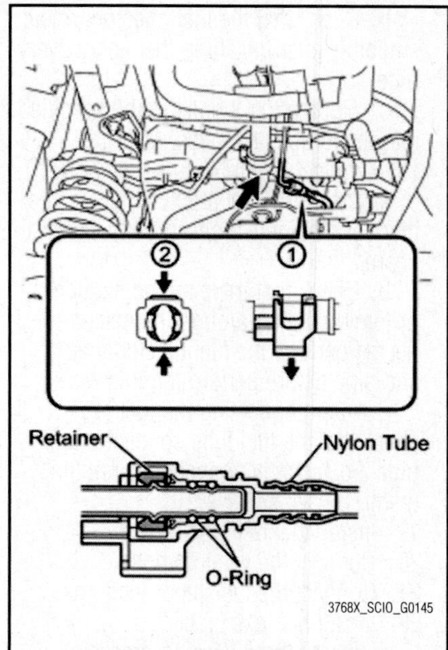

Fig. 168 Release the lock, then pull and remove the No. 6 fuel tank breather tube.

24. Install the 4 retaining bolts. Torque to 29 ft. lbs. (39 Nm).
25. Install or connect the following:
    - Parking brake cable
    - No. 6 fuel tank breather tube
    - Filler pipe
    - Fuel tank main tube
    - Fuel tank protector plate
    - Rear floor side member brace sub-assembly
    - Rear floor side member cover
    - No. 2 center exhaust pipe assembly
    - EVAP charcoal canister assembly
    - Fuel suction with pump and gauge tube assembly
    - Fuel pump gauge retainer
    - Fuel tank main tube sub-assembly
26. Fill the tank with gasoline.
27. Start the engine and check for fuel leaks.
28. Replace the rear seat assembly.

xD Models

See Figure 169.

1. Remove the rear seat.
2. Remove the rear floor service hole cover.
3. Discharge the fuel system pressure.
4. Disconnect the fuel tank main tube sub-assembly.
5. Disconnect the fuel tank vent hose.
6. Remove the fuel pump gauge retainer.
7. Remove fuel suction with pump and gauge tube assembly.

8. Drain the fuel.
9. Remove the front floor heat insulator
10. Remove the fuel tank protector.
11. Disconnect the fuel tank main tube sub-assembly.
a. Release the lock then pull and remove the fuel tank main tube.
b. Remove any dirt and foreign matter from the clip before performing this work.
c. Avoid any scratches or foreign matter on the parts when disconnecting them, as the quick connector has an O-ring that seals the plug.

➡**Perform this work by hand. Do not use any tools. Do not forcibly bend, twist or turn the nylon tube. Protect the disconnected parts by covering them with a plastic bag. If the connector and pipe are stuck, disconnect the nylon tube by turning it by hand to release them.**

12. Disconnect the fuel tank vent hose.
a. Disconnect the fuel tank vent hose from the charcoal canister assembly.
13. Disconnect the fuel tank breather hose.
a. Release the lock then pull and remove the fuel tank breather hose.
b. Remove any dirt and foreign matter from the clip before performing this work.
c. Avoid any scratches or foreign matter on the parts when disconnecting them, as the quick connector has an O-ring that seals the plug.

➡**Perform this work by hand. Do not use any tools. Do not forcibly bend, twist or turn the nylon tube. Protect the disconnected parts by covering them with a plastic bag. If the connector and pipe are stuck, disconnect the nylon tube by turning it by hand to release them.**

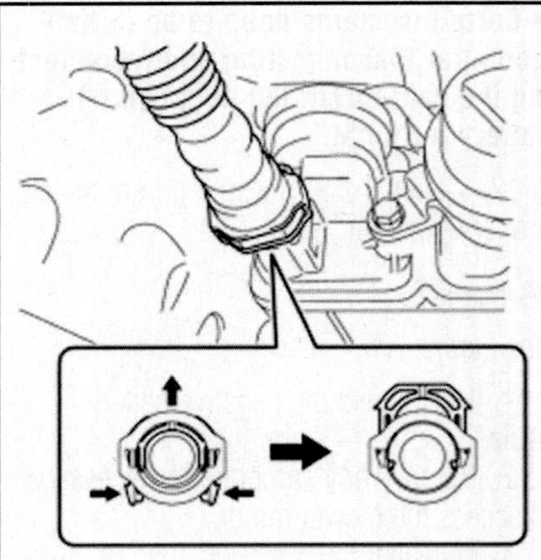

Fig. 169 Disconnect the fuel tank vent hose.

14. Disconnect the lower fuel tank filler pipe sub-assembly.
15. Remove the fuel tank assembly.
a. Remove the 4 bolts and remove the fuel tank.
b. Remove the fuel tank main tube from the fuel tank.
c. Remove the fuel tank vent hose from the fuel tank.

To install:

16. Install fuel tank assembly.
a. Install the fuel vent hose onto the fuel tank.
b. Install the fuel tank main tube onto the fuel tank.
c. Clean the bolt hole and remove any grease.
d. Install the fuel tank with 4 new bolts. Torque: 10 ft. lbs. (14 Nm).
17. Connect lower fuel tank filler pipe sub-assembly.
18. Connect fuel tank breather tube.
19. Connect fuel tank vent hose.
a. Connect the fuel tank vent hose to the charcoal canister assembly.
20. Connect fuel tank main tube sub-assembly.
a. Align the fuel tube connector with the pipe, push the fuel tube connector in until the retainer makes a click sound, then lock the cover of the connector.
21. Install fuel tank protector.
22. Install front floor heat insulator.
23. Install fuel suction with pump and gauge tube assembly.
24. Install fuel pump gauge retainer.
25. Connect fuel tank vent hose sub-assembly.
26. Connect fuel tank main tube sub-assembly.
27. Connect rear floor service hole cover.
28. Install rear seat cushion with cover pad sub-assembly LH.
29. Install rear seat.
30. Inspect for fuel leak.

IDLE SPEED

ADJUSTMENT

Idle speed is maintained by the Powertrain Control Module (PCM). No adjustment is necessary or possible.

THROTTLE BODY

REMOVAL & INSTALLATION

xD Models

See Figures 170 and 171.

1. Drain the engine coolant.
2. Remove the engine cover.
3. Remove the air cleaner cap sub-assembly with hose.
a. Disconnect the wire harness clamp and Mass Air Flow (MAF) meter connector.
b. Disconnect the 2 ventilation hose.
c. Loosen the hose clamp, unlock the 2 clamps and remove air cleaner cap sub-assembly with hose.
4. Remove the throttle body assembly.
a. Disconnect the throttle body connector.
b. Disconnect the water by-pass hoses.
c. Remove the 2 bolts and 2 nuts and remove the throttle body assembly.
d. Remove the gasket from the intake manifold.

To install:

5. Install throttle body assembly
a. Install a new gasket onto the intake manifold.

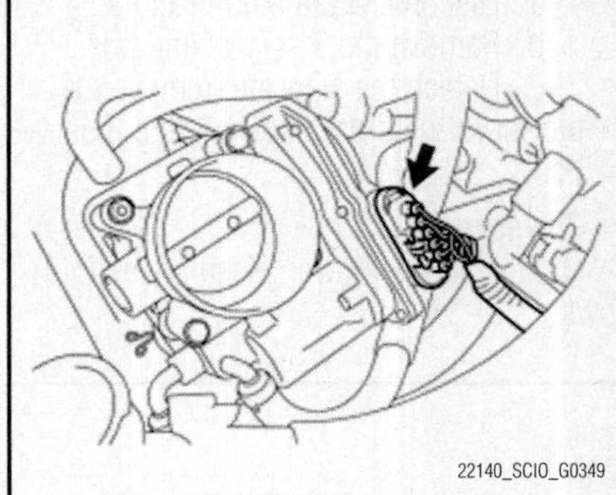

Fig. 170 Disconnect the throttle body connector.

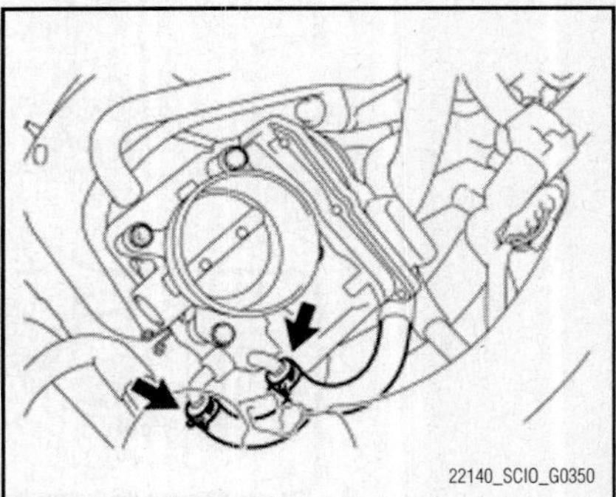

Fig. 171 Disconnect the water by-pass hoses.

b. Install the throttle body assembly with the 2 bolts and 2 nuts. Torque: 7 ft. lbs. (10 Nm).

c. Connect the water by-pass hoses.

d. Connect the throttle body assembly connector.

6. Install air cleaner cap sub-assembly with hose

a. Install air cleaner cap sub-assembly with hose and lock the 2 clamps.

b. Tighten the hose clamp to the specified torque.

c. Connect the ventilation hose.

d. Connect the wire harness clamp and Mass Air Flow (MAF) meter connector.

7. Add engine coolant.

8. Inspect for coolant leak.

9. Install engine cover.

HEATING & AIR CONDITIONING SYSTEM

BLOWER MOTOR

REMOVAL & INSTALLATION

tC Models

See Figures 172 through 174.

1. Disconnect the cable from the negative battery terminal. Wait at least 90 seconds after disconnecting the cable from the negative (-) battery terminal to prevent airbag and seat belt pretensioner activation.

2. Remove the upper and lower instrument panel sub-assemblies.

3. Remove the ECM. See "Engine Control Module (ECM)" in "ENGINE PERFORMANCE & EMISSION CONTROLS" section.

4. Remove the air duct.

5. Remove the blower assembly as follows:

a. Disconnect the 3 connectors.

b. Remove the 3 screws and bolt.

c. Detach the claw and remove the blower in the order shown in the illustration.

To install:

6. Install the blower assembly as follows:

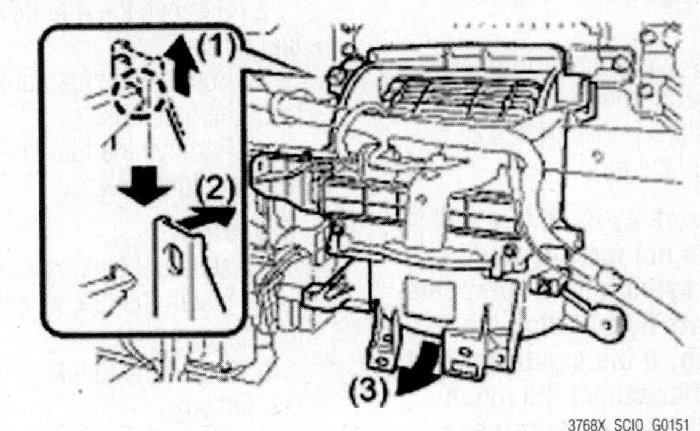

Fig. 173 Detach the claw and remove the blower in the order shown in the illustration

a. Install the blower assembly with the claw in the order shown in the illustration.

b. Install the blower assembly with the bolt and 3 screws.

c. Connect the 3 connectors.

7. Install the air duct.

8. Install the ECM. See "Engine Control Module (ECM)" in "ENGINE PERFORMANCE & EMISSION CONTROLS" section.

9. Install the instrument panel assemblies.

10. Connect the cable to the negative battery terminal.

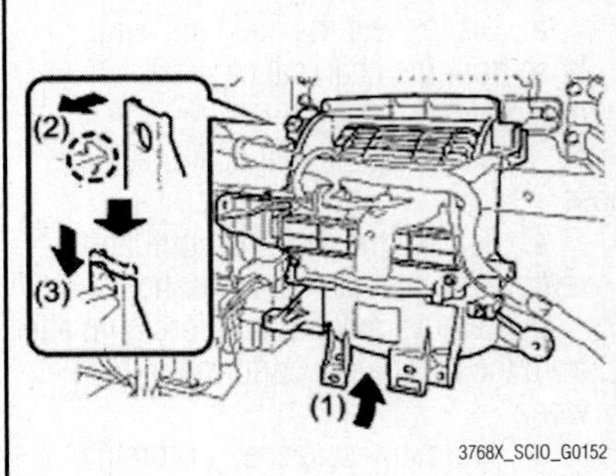

Fig. 174 Install the blower assembly with the claw in the order shown in the illustration.

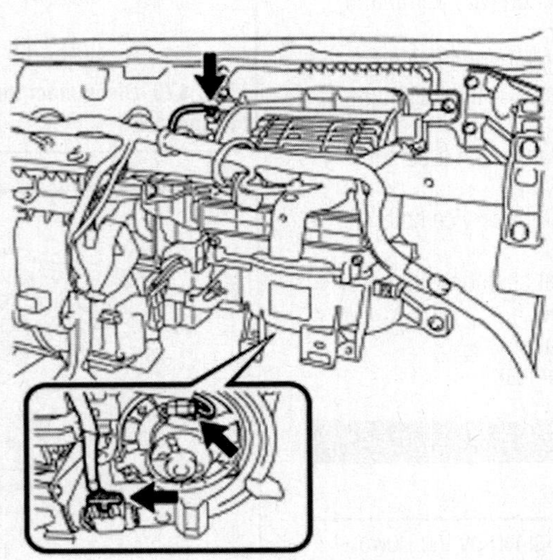

Fig. 172 Disconnect the 3 connectors

11. Perform initialization.

➡Certain systems need to be initialized after disconnecting and reconnecting the cable from the negative (-) battery terminal.

12. Check the SRS warning light for proper operation.

xB Models

See Figure 175.

1. Disconnect the negative battery cable.

2. Remove any components necessary to access the blower motor.

3. Unfasten the 3 screws, then remove the blower motor.

4. Installation is the reverse of the removal procedure.

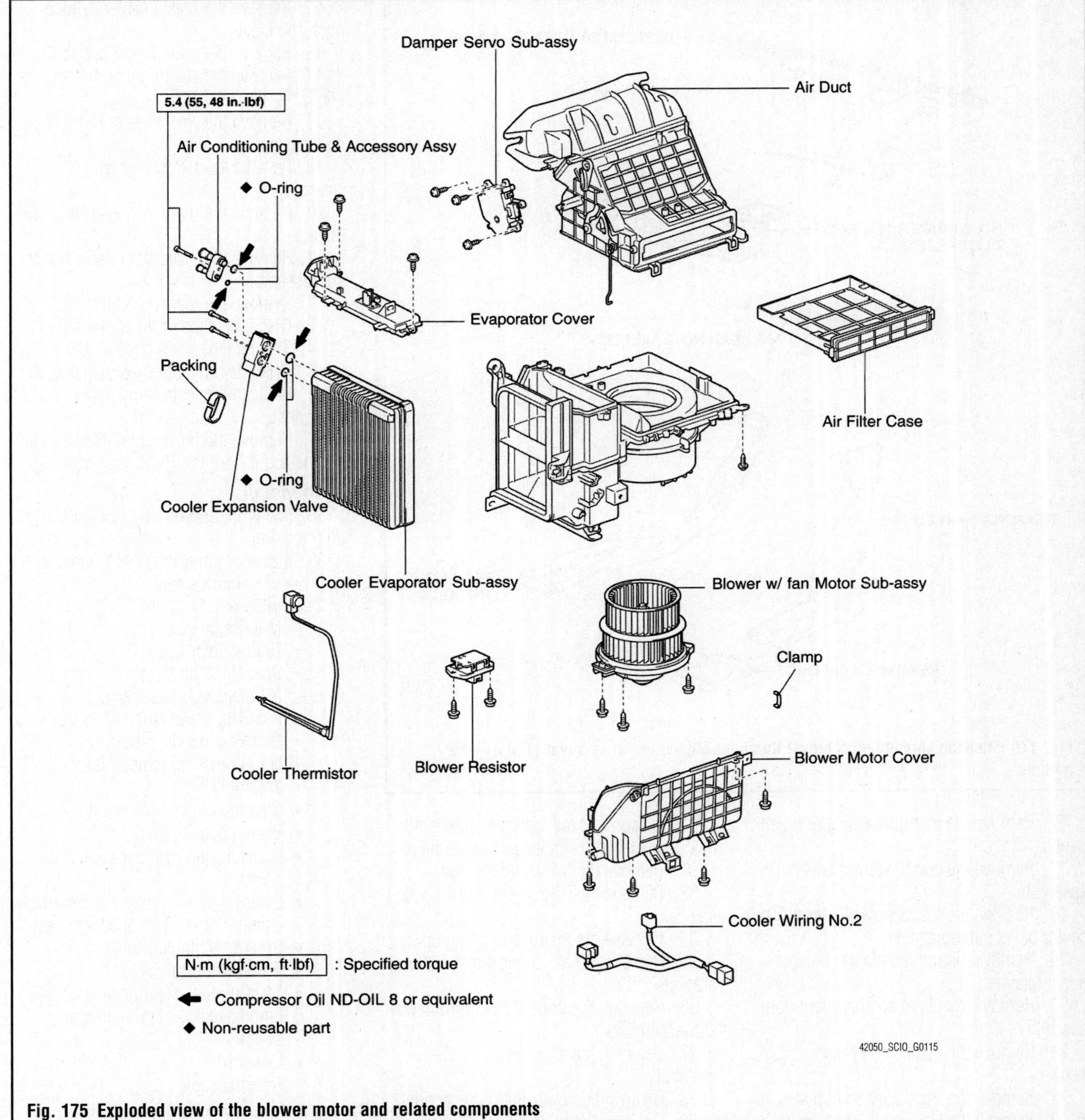

Fig. 175 Exploded view of the blower motor and related components

xD Models

See Figures 176 through 179.

1. Disconnect the cable from the negative battery terminal. Wait at least 90 seconds after disconnecting the cable from the negative (-) battery terminal to prevent airbag and seat belt pretensioner activation.
2. Discharge the refrigerant from the refrigeration system.
3. Drain the engine coolant.
4. Remove the front wiper arm and blade assemblies.
5. Remove the hood to cowl top seal.
6. Remove the cowl top ventilator louver sub-assembly.
7. Remove the LH cowl top ventilator louver.
8. Remove the front wiper motor and link.
9. Remove the RH front air shutter seal.
10. Remove the outer cowl top panel.
11. Disconnect the A/C suction tube sub-assembly and the A/C liquid tube sub-assembly from the firewall connections.
12. Disconnect heater water outlet hose a (from heater unit)
13. Disconnect the heater water hose inlet A.
14. Separate the front door opening trim weatherstrips.
15. Remove the front pillar garnishes.

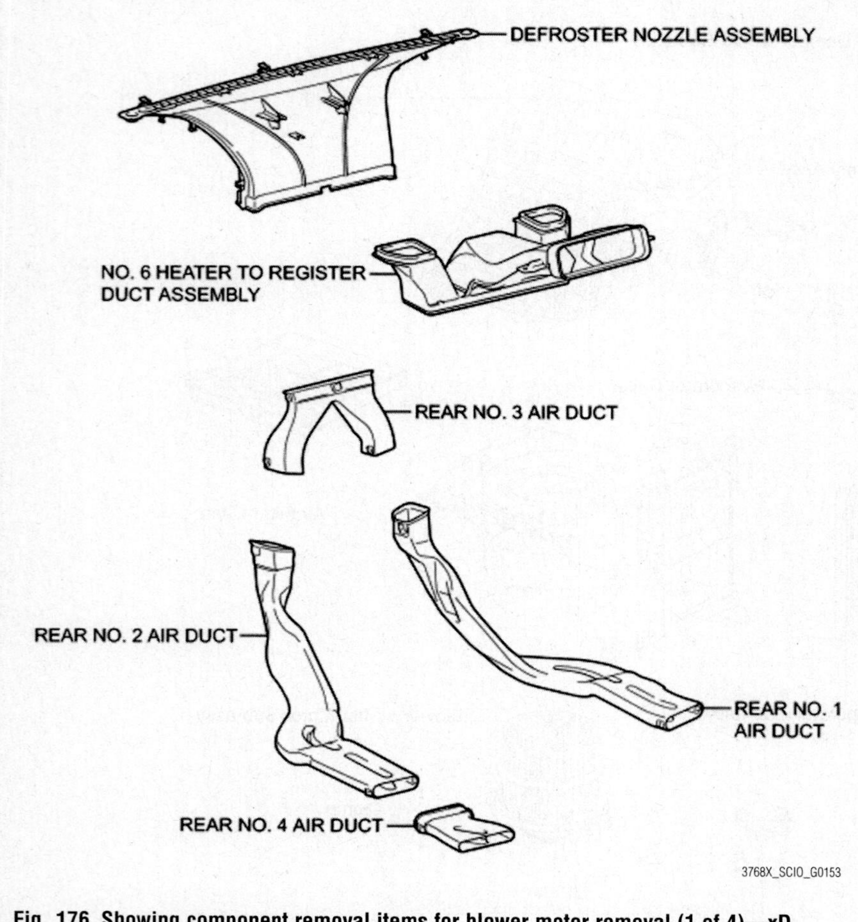

Fig. 176 Showing component removal items for blower motor removal (1 of 4)—xD models

16. Remove the instrument cluster finish panel.
17. Remove the combination meter assembly.
18. Remove the instrument cluster finish center panel sub-assembly.
19. Remove instrument cluster finish panel retainer.
20. Remove the glove compartment door assembly.
21. Remove the upper instrument panel.
22. Remove the front door scuff plates.
23. Remove the No. 2 and No. 1 instrument panel under cover sub-assemblies.
24. Remove the cowl side trim boards.
25. Remove the shift lever knob sub-assembly (M/T).
26. Remove the upper console panel.
27. Remove the rear console box cover.
28. Remove the console box carpet.
29. Remove the rear console box sub-assembly.
30. Separate the air conditioning panel assembly.
31. Disconnect the defroster, inlet, and mix damper control cable sub-assemblies.
32. Remove the front console box.
33. Remove the instrument panel under tray.
34. Remove the instrument panel box.
35. Remove the No. 2 and No. 1 radio brackets.
36. Separate the hood lock control lever sub-assembly.
37. Separate the antenna cord sub-assembly.
38. Remove the lower instrument panel.
39. Position the front wheels facing straight ahead.
40. Remove steering pad, steering wheel, column covers, and combination switch.
41. Disconnect the power steering Electronic Control Unit (ECU).
42. Remove the instrument panel sub reinforcement.
43. Remove the column hole cover silencer sheet.
44. Separate the steering sliding yoke sub-assembly.
45. Remove the brake master cylinder push rod clevis.
46. Remove the brake pedal support.
47. Remove the steering column assembly.
48. Remove the No. 6 heater to register duct assembly.
49. Remove the defroster nozzle assembly.
50. Remove the rear No. 1 and No. 2 air ducts.
51. Remove the instrument panel brace sub-assembly.
52. Remove the rear No. 3 air duct.
53. Disconnect the drain cooler hose.
54. Separate the main body ECU.
55. Separate the instrument panel wire.
56. Remove the instrument panel reinforcement.
57. Remove the blower unit (3 screws) and detach it from the HVAC unit housing.

To install:

58. Install the blower unit to the HVAC unit housing.
59. Install or connect the following:
    - I/P reinforcement
    - I/P wire
    - Main body ECU
    - Drain cooler hose
    - Rear No. 3 air duct
    - I/P brace sub-assembly
    - Rear No. 1 and No. 2 air ducts
    - Defroster nozzle assembly
    - No. 6 heater to register duct assembly
    - Steering column assembly
    - Brake pedal support
    - Brake master cylinder push rod clevis
    - Steering sliding yoke sub-assembly
    - Column hole cover silencer sheet
    - I/P sub reinforcement
    - Power steering ECU
    - Steering pad, steering wheel, column covers, and combination switch
    - Lower I/P
    - Antenna cable
    - Hood lock control lever sub-assembly
    - No. 2 and No. 1 radio brackets
    - I/P box and under tray
    - Front console box
    - Defroster, inlet, and mix damper control cable sub-assemblies
    - A/C panel
    - Rear console box sub-assembly, carpet and box cover
    - Upper console panel
    - Shift lever knob
    - Cowl side trim boards

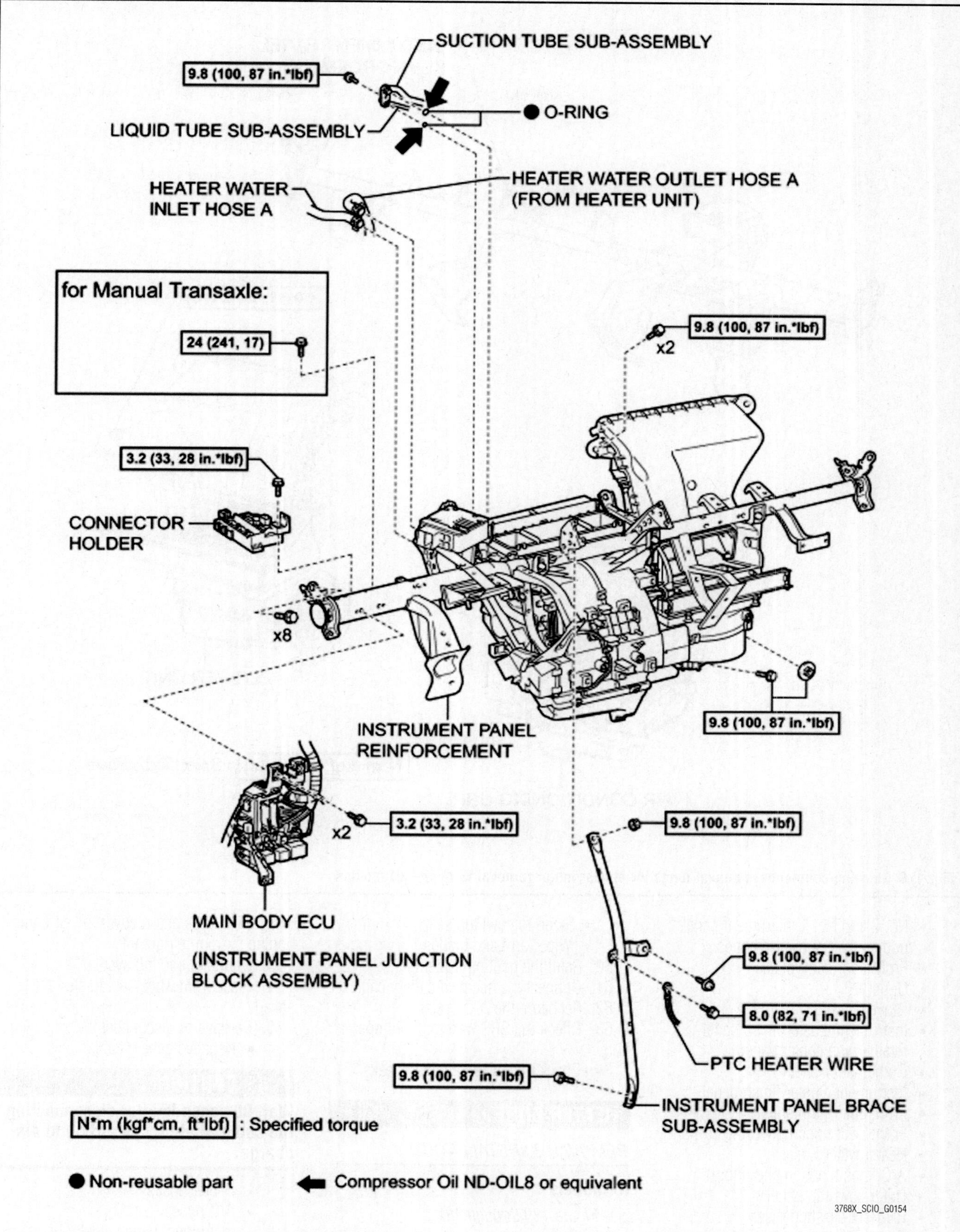

Fig. 177 Showing component removal items for blower motor removal (2 of 4)—xD models

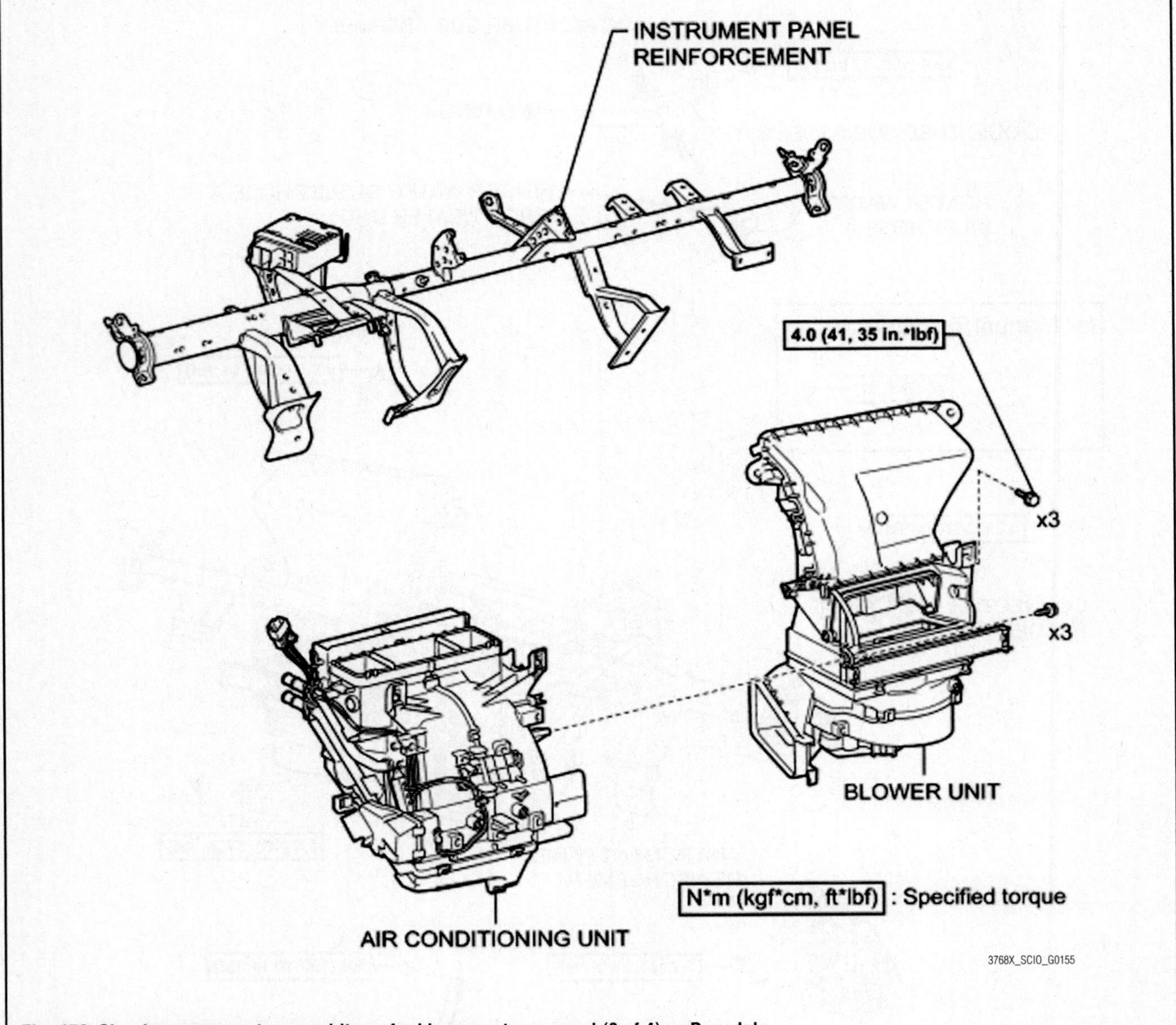

Fig. 178 Showing component removal items for blower motor removal (3 of 4)—xD models

- No. 2 and No. 1 instrument panel under cover sub-assemblies
- Front door scuff plates
- Upper I/P
- Glove box door
- Instrument cluster finish panel retainer and center panel
- Combination meter
- Instrument cluster finish panel
- Front pillar garnishes
- Front door opening weatherstrips
- Heater water hoses
- A/C lines (suction and liquid)
- Outer cowl top panel
- RH front air shutter seal
- Front wiper motor and link
- Cowl top ventilator louver
- Hood to cowl top seal
- Wiper arm and blades

60. Refill the cooling system.
61. Connect the negative battery cable.
62. Recharge the A/C system.
63. Check the SRS warning light operation.
64. Check for A/C and coolant leaks.

HEATER CORE

REMOVAL & INSTALLATION

tC Models

See Figure 180 through 182.

1. Before servicing the vehicle, refer to the Precautions Section.
2. Discharge and recover the air conditioning system refrigerant.
3. Drain the cooling system.
4. Place front wheel facing straight ahead.
5. Remove or disconnect the following:
   - Negative battery cable

CAUTION

Wait 90 seconds after disconnecting the cable to allow the airbag to discharge.

   - Heater hoses
   - A/C hoses
   - Instrument gauge cluster hood panel

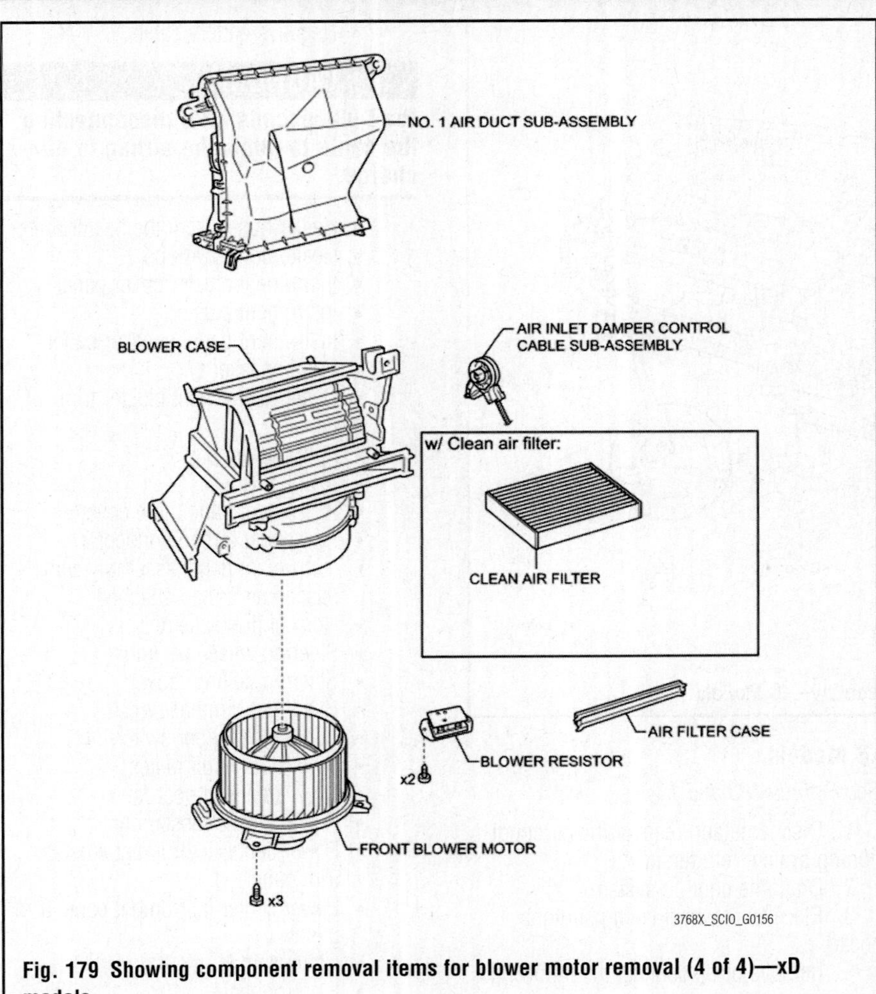

Fig. 179 Showing component removal items for blower motor removal (4 of 4)—xD models

- Instrument gauge cluster
- Shift lever knob, if equipped with manual transmission
- Upper front floor console panel
- Heater control assembly
- Heater control assembly
- Center cluster module knob
- Lower heater control base
- Air conditioner amplifier assembly
- Radio support brackets
- Glove compartment door
- Glove compartment door stopper
- A-pillar trim

➡If equipped with side curtain airbags, cover the airbag with a protective cover as soon as the trim pieces are removed.

- Passenger airbag connector
- Instrument panel assembly with passenger airbag attached.
- Steering wheel assembly
- Steering column cover
- Headlight dimmer switch
- Windshield wiper switch assembly
- Floor console top panel
- Floor console
- Front door scuff plates
- Cowl side trim panels
- Center heater to register duct
- Lower instrument panel assembly as follows:
- ECM
- Center heater to register duct
- Lower defroster nozzle assembly

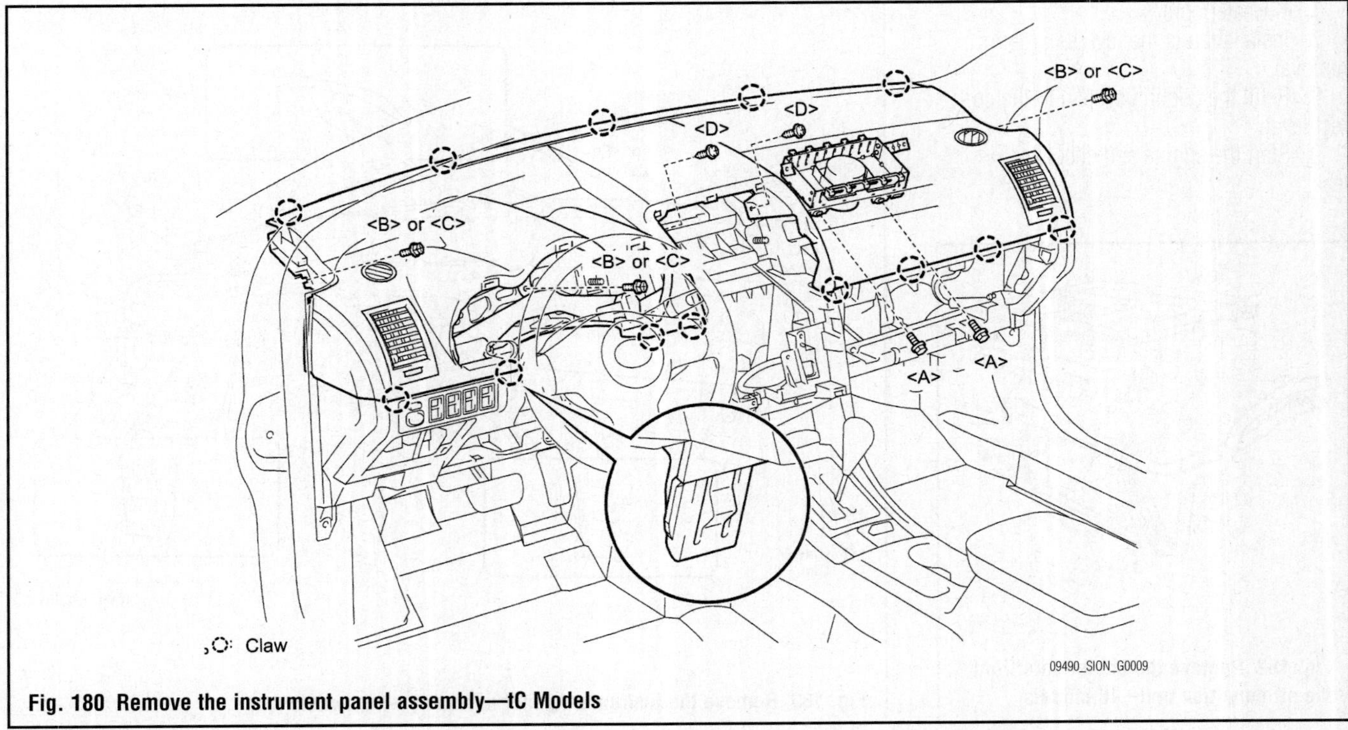

Fig. 180 Remove the instrument panel assembly—tC Models

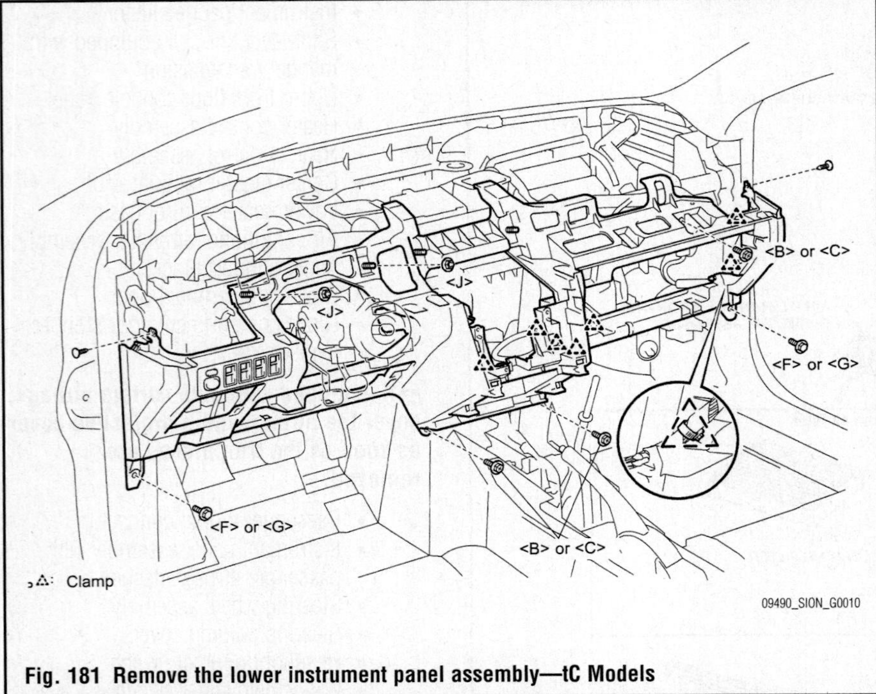

Fig. 181 Remove the lower instrument panel assembly—tC Models

- Instrument panel-to-cowl brace assembly
- Blower assembly
- Air conditioning unit assembly

6. Remove the following from the air conditioning unit:
   - Drain hose
   - Air duct
   - Air outlet control motor
   - Air-mix control motor
   - Heater piping cover
   - Heater core
7. Installation is the reverse order of removal.
8. Refill the cooling system to the correct level.
9. Start the engine and check for leaks.

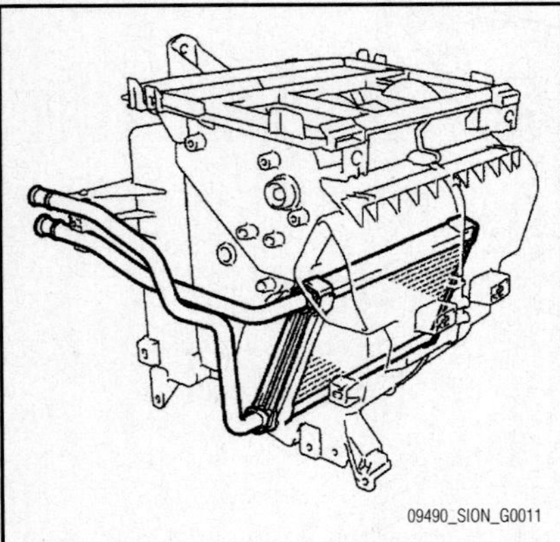

Fig. 182 Remove the heater core from the air condition unit—tC models

xB Models

See Figures 183 and 184.

1. Discharge and recover the air conditioning system refrigerant.
2. Drain the cooling system.
3. Place front wheel facing straight ahead.
4. Remove or disconnect the following:
   - Negative battery cable

CAUTION

Wait 90 seconds after disconnecting the cable to allow the airbag to discharge.

- Heater hoses from the heater unit
- Heater control knobs
- Instrument cluster cover panel
- Instrument cluster
- Instrument panel speaker panel
- Front speaker
- Center instrument cluster trim panel
- Radio assembly
- A-pillar trim
- Instrument panel hole cover
- Passenger airbag connector
- Instrument panel assembly with passenger airbag attached
- Horn button assembly
- Steering wheel assembly
- Steering column cover
- Headlight dimmer switch
- Windshield wiper switch
- Front door scuff plates
- Cowl side trim boards
- Heater control assembly
- Lower center instrument cluster trim panel
- Lower center instrument panel trim panel
- Glove compartment door
- Floor console

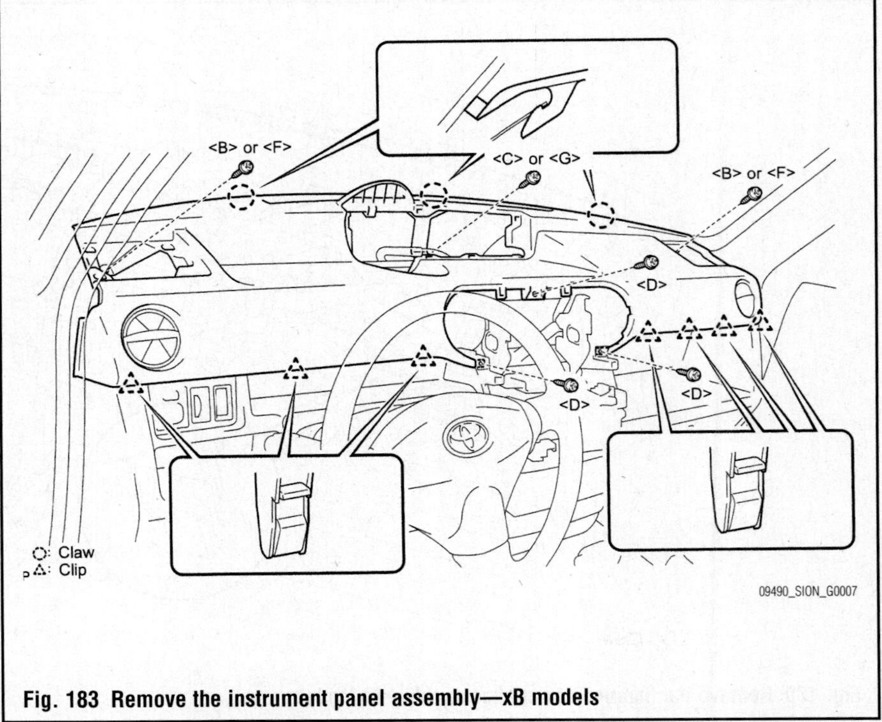

Fig. 183 Remove the instrument panel assembly—xB models

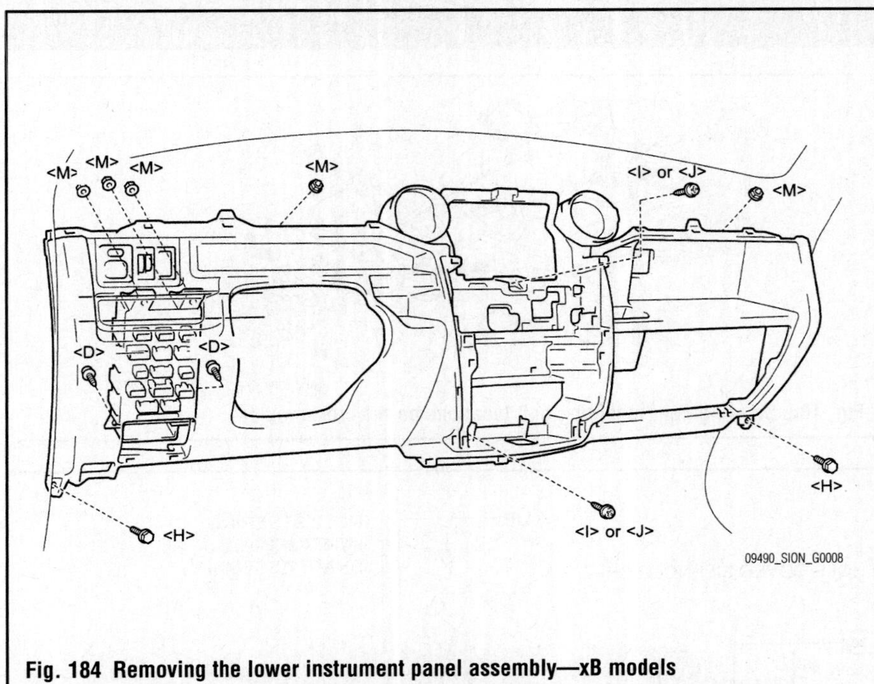

Fig. 184 Removing the lower instrument panel assembly—xB models

- Floor parking brake cable assembly, if equipped with automatic transmission
- Steering column hole cover plate
- Intermediate steering shaft assembly
- Steering column assembly
- Antenna cord
- Lower instrument panel assembly with panel reinforcement attached
- Defroster nozzle assembly
- ECM
- Air conditioner blower assembly
- Defroster damper control cable
- Air-mix damper control cable
- Rear air ducts
- Heater unit assembly
- Air duct assembly from the heater unit
- Heater core cover
- Thermistor assembly
- Heater core

To install:

5. Installation is the reverse order of removal.
6. Refill the cooling system to the correct level.
7. Start the engine and check for leaks.

xD Models

⁂ WARNING

SRS components are located in this area. Read and observe all precautions and service procedures in "AIR BAG (SUPPLEMENTAL RESTRAINT SYSTEM)" section before proceeding.

1. Disconnect the cable from negative battery terminal. Wait for at least 90 seconds after disconnecting the cable to prevent the airbag from working.
2. Remove or disconnect the following:
    - Front door opening trim weatherstrip
    - Front pillar garnish
    - Instrument cluster finish panel
    - Combination meter assembly
    - Instrument cluster finish center panel sub-assembly
    - Instrument cluster finish panel retainer
    - Glove compartment door assembly
    - Upper instrument panel
    - Front door scuff plate
    - Instrument panel under cover sub-assembly
    - Cowl side trim board
    - Shift lever knob sub-assembly (M/T)
    - Upper console panel
    - Rear console box cover
    - Console box carpet
    - Rear console box sub-assembly
    - Air conditioning panel assembly
    - Defroster damper control cable sub-assembly
    - Air inlet damper control cable sub-assembly
    - Air mix damper control cable sub-assembly
    - Front console box
    - Instrument panel under tray
    - Instrument panel box
    - Radio bracket
    - Hood lock control lever sub-assembly
    - Antenna cord sub-assembly
    - Lower instrument panel
    - Steering pad
    - Steering wheel assembly
    - Steering column lower cover
    - Combination switch assembly
    - Power steering ECU
    - Instrument panel sub reinforcement
    - Column hole cover silencer sheet
    - Steering sliding yoke sub-assembly
    - Brake master cylinder push rod clevis
    - Brake pedal support
    - Steering column assembly
    - Rear No. 2 air duct
    - Instrument panel brace sub-assembly
3. Remove No. 1 air duct
    a. Disengage the 3 claws and remove the air duct.
4. Remove heater assembly
    a. Disconnect the connector and 2 clamps.
    b. Remove the 2 screws and heater from the air conditioning unit.

To install:

5. Installation is the reverse order of removal.
6. Refill the cooling system to the correct level.
7. Start the engine and check for leaks.

STEERING

POWER RACK & PINION STEERING GEAR

REMOVAL & INSTALLATION

tC Models

See Figures 185 through 189.

1. Face the front wheels straight ahead.
2. Drain the power steering fluid.
3. Remove or disconnect the following:
   - Steering column hole cover plate
   - Intermediate steering shaft

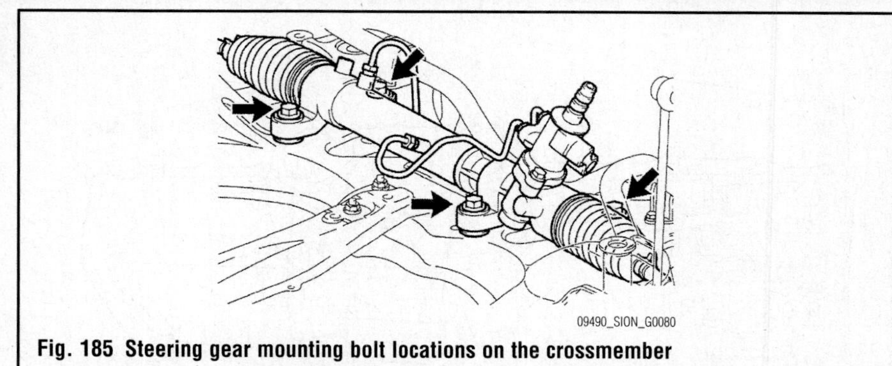

Fig. 185 Steering gear mounting bolt locations on the crossmember

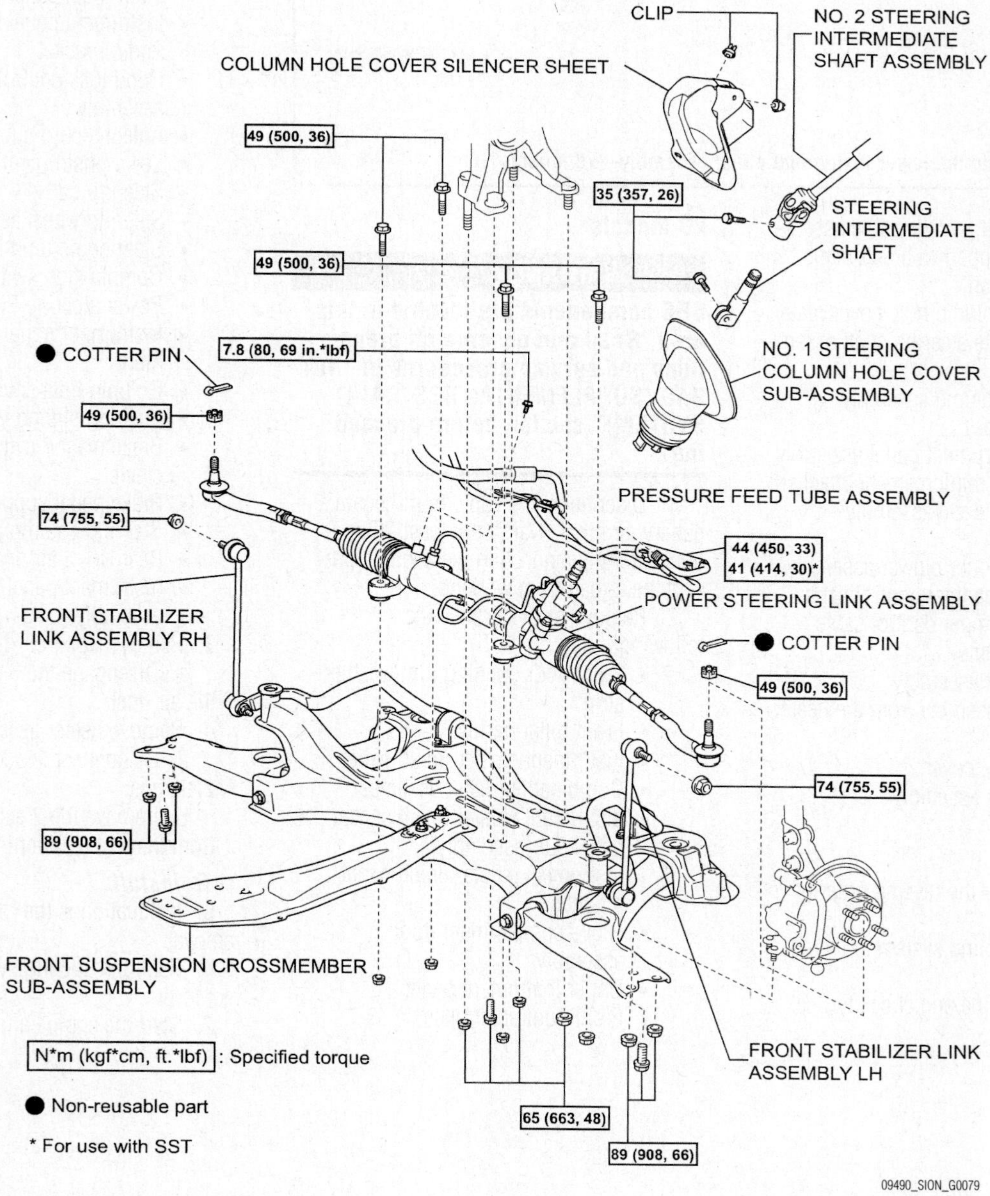

Fig. 186 Exploded view of the power steering assembly—tC

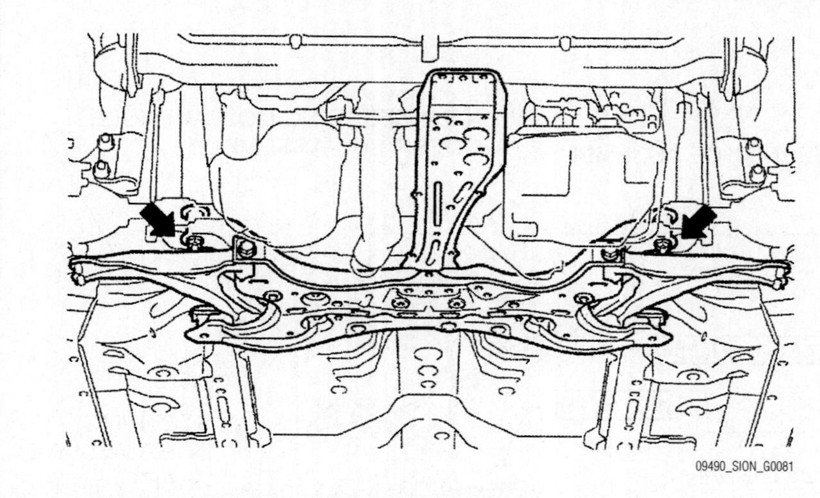

Fig. 187 Crossmember front nuts to 98 ft. lbs. (133 Nm)—tC

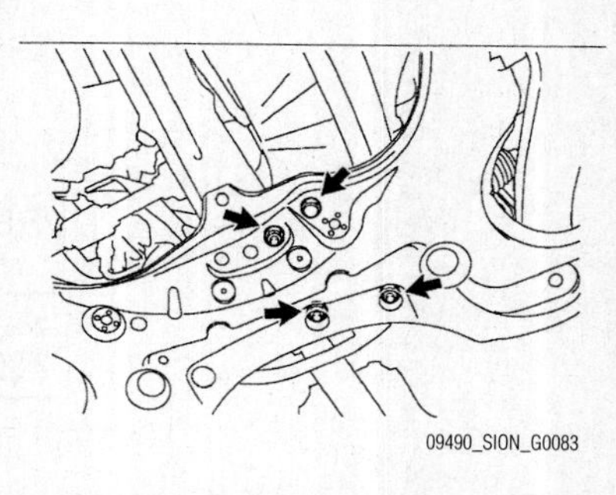

Fig. 189 Rear engine mounting insulator bolts and nuts—tC

- Front wheel
- Engine undercover
- Tie rod ends
- Power steering hoses
- Front stabilizer links
- Lower control arms
- Front floor panel brace
- Center exhaust pipe
- Hood
- Engine appearance cover

4. Attach an engine crane to support the engine.

5. Remove or disconnect the following:
   - Front suspension crossmember
   - Steering gear assembly, attached to the crossmember, from the vehicle
   - Steering gear assembly from the crossmember

To install:

6. Install the power steering gear assembly to the front crossmember. Tighten the mounting bolts to 36 ft. lbs. (49 Nm).

7. Using a suitable jack, lift the front crossmember into the vehicle. Tighten as follows:
   a. Two front nuts to 98 ft. lbs (133 Nm)
   b. Crossmember bracket: Bolt A to 98 ft. lbs. (133 Nm). Bolts B to 59 ft. lbs. (80 Nm)
   c. Rear engine mounting insulator bolt and nuts to 48 ft. lbs. (65 Nm).

8. Install or connect the following:
   - Lower control arms
   - Stabilizer links
   - Power steering hoses
   - Tie rod ends
   - Center exhaust pipe
   - Front floor panel brace
   - Front wheels
   - Steering intermediate shaft. Tighten the bolts to 26 ft. lbs. (35 Nm).
   - Steering column hole cover
   - Engine appearance cover
   - Hood
   - Engine undercover

9. Refill the power steering reservoir to the correct level.

10. Bleed the air from the power steering system.

11. Start the engine and check for leaks.

12. Check and adjust the alignment if necessary.

xB & xD Models

See Figures 190 and 191.

1. Face the front wheels straight ahead.
2. Drain the power steering fluid.
3. Remove or disconnect the following:
   - Wiper arms
   - Top cowl ventilator louvers
   - Wiper motor and linkage
   - Steering column hole cover plate
4. Matchmark the steering sliding yoke and intermediate shaft.
5. Disconnect the sliding yoke from the intermediate shaft.

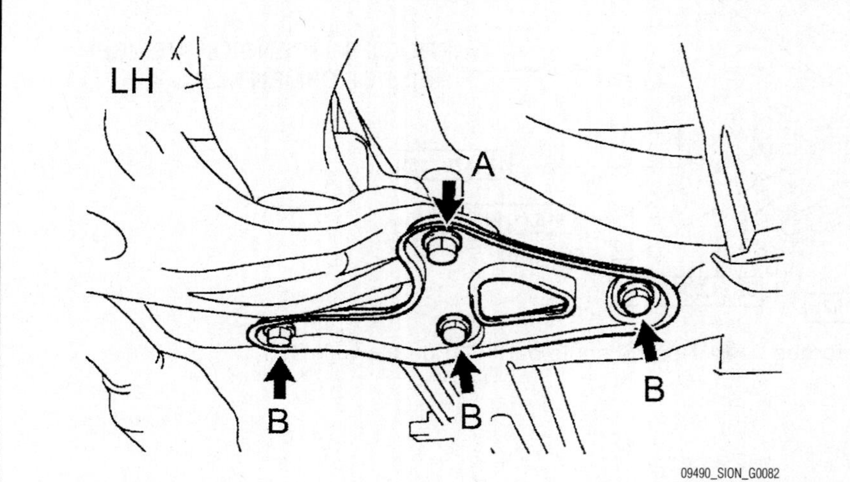

Fig. 188 Crossmember bracket bolt identification, right-hand similar—tC

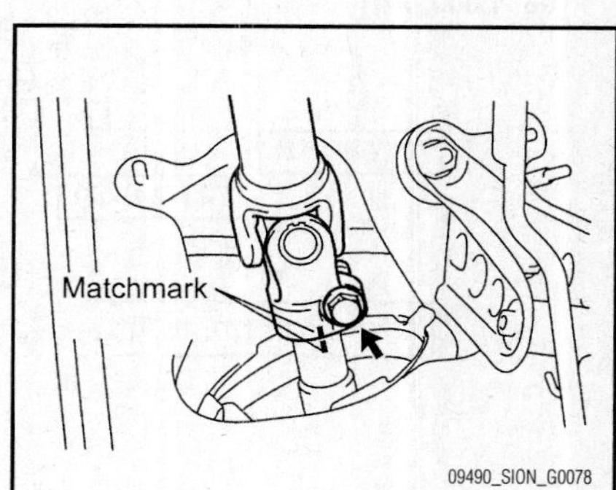

Fig. 190 Place matchmarks on the yoke and intermediate shaft.

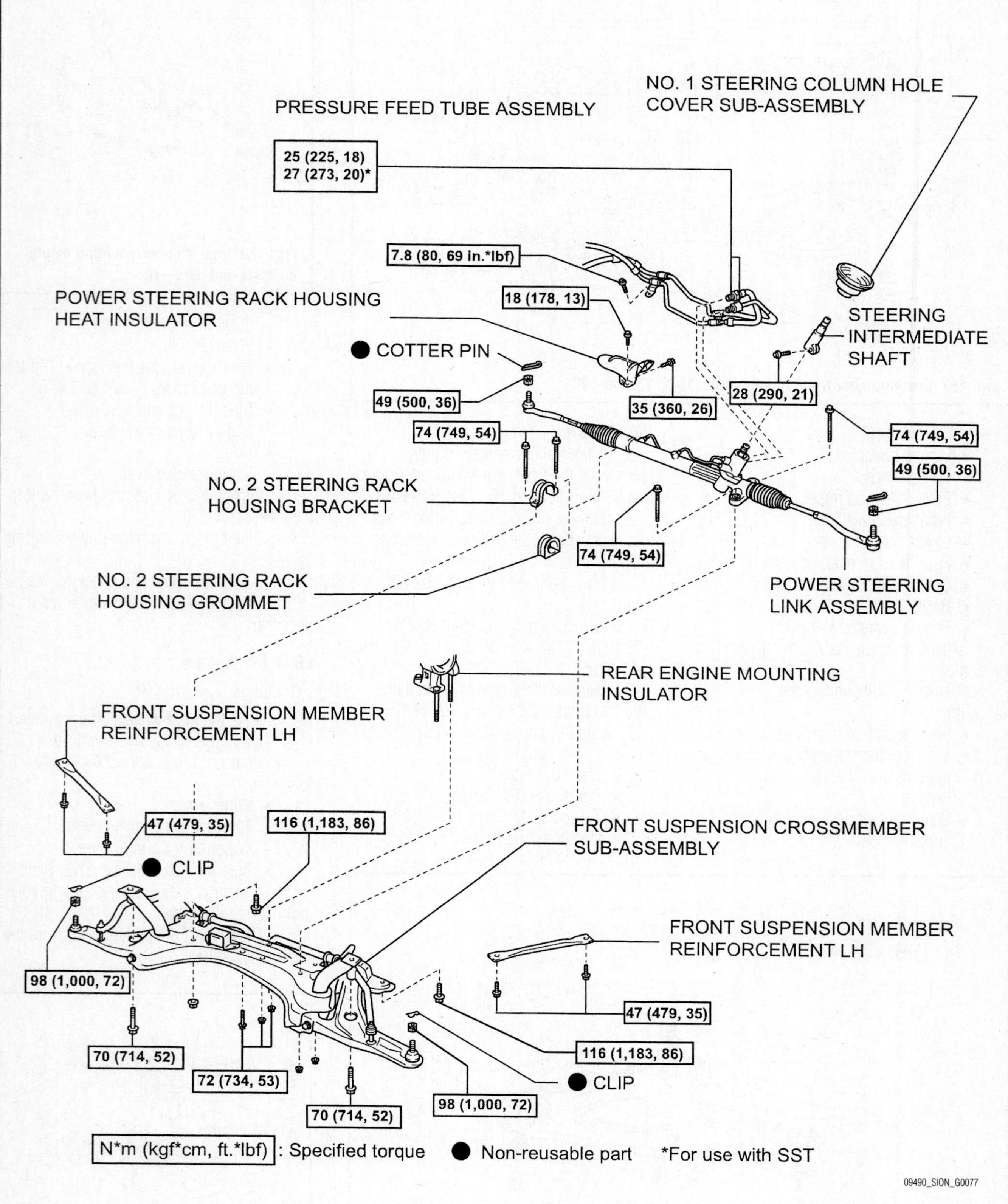

Fig. 191 Exploded view of the steering linkage components

6. Remove or disconnect the following:
- Hood
- Front wheels
- Tie rod ends
- Engine undercover
- Power steering hoses
- Lower control arms

7. At this point, attach an engine crane to support the engine.
8. Remove or disconnect the following:
- Front suspension crossmember
- Power steering rack heat insulator
- Intermediate shaft
- Steering gear assembly attached to the front crossmember.

9. Remove the steering rack clamps and remove the steering gear from the crossmember.

To install:

10. Install the steering gear to the cross-member. Tighten to the nuts 54 ft. lbs. (74 Nm).
11. Install or connect the following:
- Intermediate shaft
- Power steering rack heat insulator. Tighten to 26 ft. lbs. (35 Nm).
- Front suspension crossmember. Tighten the front outside bolts to 52 ft. lbs. (70 Nm). Tighten the rear outside bolts to 86 ft. lbs. (116 Nm). Tighten to inside nuts to 53 ft. lbs. (72 Nm).
- Lower control arms
- Power steering hoses
- Tie rod ends
- Front wheels
- Sliding yoke to the intermediate shaft
- Hood
- Engine undercover
- Steering column hole cover plate
- Wiper motor and linkage
- Top cowl ventilator louvers
- Wiper arms

12. Refill the power steering reservoir to the correct level.
13. Bleed the air from the power steering system.
14. Start the engine and check for leaks.
15. Check and adjust the alignment if necessary.

POWER STEERING PUMP

REMOVAL & INSTALLATION

tC & xB Models

See Figures 192 and 193.

1. Raise and safely support the vehicle.
2. Remove the right front wheel and tire assembly.
3. Drain the power steering fluid into a suitable container.
4. Remove the front right fender apron seal.
5. Remove the fan-alternator belt, as outlined in the Engine Mechanical Section.
6. Remove the clip and disconnect the oil reservoir to pump hose no.1.

⁂ WARNING

Do not spill fluid on the belt.

7. Disconnect the pressure feed tube as follows:
   a. Remove the bolt and separate the pressure feed tube from the pump bracket rear.
   b. Using a wrench (27 mm) to hold the pressure port union, remove the union bolt and gasket.
8. Remove the power steering pump as follows:
   a. Disconnect the connector from the oil pressure switch.
   b. Using SST 09249-63010, or equivalent, and a deep socket (14 mm), loosen the 2 bolts.
   c. Remove the power steering pump assembly.

To install:

9. Install the power steering pump as follows:
   a. Temporarily install the 2 bolts to the power steering pump assembly.
   b. Install the power steering pump with the 2 bolts.
   c. Using SST 09249-63010, or equivalent, and a deep socket (14 mm), fully tighten the bolts. To 25 ft. lbs. (34 Nm).

➡Use a torque wrench with a fulcrum length of 300 mm (11.81 in.). This torque value is effective when the special tool is parallel to the torque wrench.

   d. Attach the connector to the oil pressure switch.
10. Connect the pressure feed tube as follows:
   a. Install the pressure feed tube and a new gasket to the power steering pump with the union bolt.

➡Make sure the stopper of the pressure feed tube touches the power steering pump body.

   b. Using a 27mm wrench to hold the pressure port union, tighten the union bolt to 38 ft. lbs. (52 Nm).
11. Install the pressure feed tube with the bolt to the rear pump bracket and tighten to 69 inch lbs. (7.8 Nm).

Fig. 192 Using the special tool and a deep socket (14 mm), loosen the 2 bolts, then remove the power steering pump

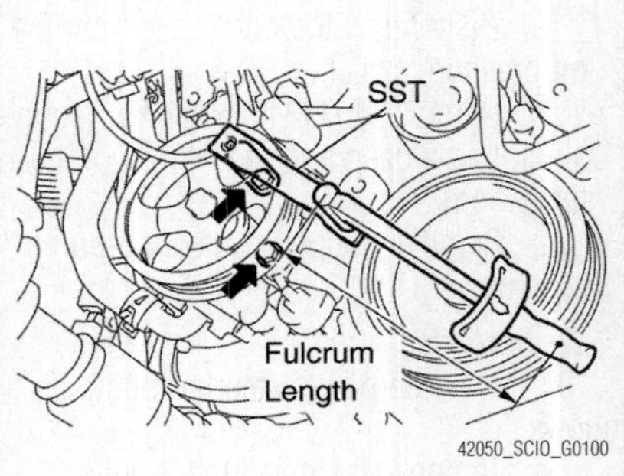

Fig. 193 Using SST 09249-63010, or equivalent, and a deep socket (14 mm), fully tighten the bolts. To 25 ft. lbs. (34 Nm). Use a torque wrench with a fulcrum length of 300 mm (11.81 in.). This torque value is effective when the special tool is parallel to the torque wrench

12. Connect the oil reservoir to pump hose no.1 with the clip.

⁂ WARNING

Do not spill fluid on the belt.

13. Install the fan-alternator belt.
14. Install the front fender apron seal and front wheel and tire.
15. Add power steering fluid and bleed the system.
16. Check for fluid leaks.

xD Models

See Figures 192 and 193.

1. Raise and safely support the vehicle.
2. Remove the right front wheel and tire assembly.
3. Drain the power steering fluid into a suitable container.

4. Remove the front right fender apron seal.
5. Remove the fan-alternator belt.
6. Remove the clip and disconnect the oil reservoir to pump hose no.1.

CAUTION

Do not spill fluid on the belt.

7. Disconnect the pressure feed tube as follows:
   a. Remove the bolt and separate the pressure feed tube from the pump bracket rear.
   b. Using a wrench (27 mm) to hold the pressure port union, remove the union bolt and gasket.
8. Remove the power steering pump as follows:
   a. Disconnect the connector from the oil pressure switch.
   b. Using SST 09249-63010, or equivalent, and a deep socket (14 mm), loosen the 2 bolts.
   c. Remove the power steering pump assembly.

To install:

9. Install the power steering pump as follows:
   a. Temporarily install the 2 bolts to the power steering pump assembly.
   b. Install the power steering pump with the 2 bolts.
   c. Using SST 09249-63010, or equivalent, and a deep socket (14 mm), fully tighten the bolts. To 25 ft. lbs. (34 Nm).

➡Use a torque wrench with a fulcrum length of 300 mm (11.81 in.). This torque value is effective when the special tool is parallel to the torque wrench.

   d. Attach the connector to the oil pressure switch.
10. Connect the pressure feed tube as follows:
   a. Install the pressure feed tube and a new gasket to the power steering pump with the union bolt.

➡Make sure the stopper of the pressure feed tube touches the power steering pump body.

   b. Using a 27mm wrench to hold the pressure port union, tighten the union bolt to 38 ft. lbs. (52 Nm).
11. Install the pressure feed tube with the bolt to the rear pump bracket and tighten to 69 inch lbs. (7.8 Nm).
12. Connect the oil reservoir to pump hose no.1 with the clip.

WARNING

Do not spill fluid on the belt.

13. Install the fan-alternator belt.
14. Install the front fender apron seal and front wheel and tire.
15. Add power steering fluid and bleed the system, as outlined in this section.
16. Check for fluid leaks.

BLEEDING

1. Check the fluid level in the power steering reservoir and add if necessary.
2. Jack up the front of the vehicle and support it with stands.
3. With the engine stopped, turn the wheel slowly from lock to lock several times.
4. Lower the vehicle.
5. Start the engine.
6. Run the engine at idle for a few minutes.
7. With the engine idling, turn the wheel to the left or right full lock position and keep it there for 2 to 3 seconds, then turn the wheel to the opposite full lock

   position and keep it there for 2 to 3 seconds.
8. Repeat the previous steps several times.
9. Stop the engine.
10. Check for foaming or emulsification. Especially, if the system has to be bled twice because of foaming or emulsification, check for fluid leaks in the system.
11. Check the fluid level and add if necessary.
   a. Keep the vehicle level.
   b. With the engine stopped, check the fluid level in the oil reservoir.
   c. If necessary, add fluid. These models use Dexron® II or II ATF fluid.

➡When hot, check that the fluid level is within the HOT range on the oil reservoir. If the fluid is cold, check that it is within the COLD range.

12. Start the engine and run at idle.
13. Turn the steering wheel from lock to lock several times to raise fluid temperature. The Fluid temperature should be: 167–176°F (75–80°C).
14. Check for foaming or emulsification. If foaming or emulsification is found, bleed the power steering system again.
15. With the engine idling, measure the fluid level in the oil reservoir.
16. Stop the engine.
17. Wait a few minutes and recheck the fluid level in the oil reservoir. The maximum fluid level rise is : 5 mm (0.20 in.)
18. If a problem is found, bleed the power steering system again.
19. Check the fluid level and add if necessary.

LOWER BALL JOINT

REMOVAL & INSTALLATION

tC Models

See Figure 194.

1. Remove or disconnect the following:
   - Front wheel
   - Hub nut
   - Wheel speed sensor
   - Tie rod end
   - Lower control arm mounting bolts
2. Disconnect the halfshaft from the axle hub.
3. Remove the lower shock absorber mounting bolts.
4. Using Special Tool 09628-62011 or suitable puller, remove the lower ball joint from the steering knuckle.

To install:

➡Use a new split pin for assembly.

5. Install the ball joint and torque the mounting nut to 76 ft. lbs. (103 Nm)
6. Install the halfshaft to the front axle. Tighten the steering knuckle mounting bolts to 76 ft. lbs. (103 Nm).
7. Install or connect the following :
   - Lower control arm mounting bolts. Tighten to 66 ft. lbs. (89 Nm).
   - Tie rod ends. Tighten to 36 ft. lbs. (49 Nm).
   - Hub nut. Tighten to 159 ft. lbs. (216 Nm).
   - Front wheel
8. Check and/or adjust the wheel alignment.

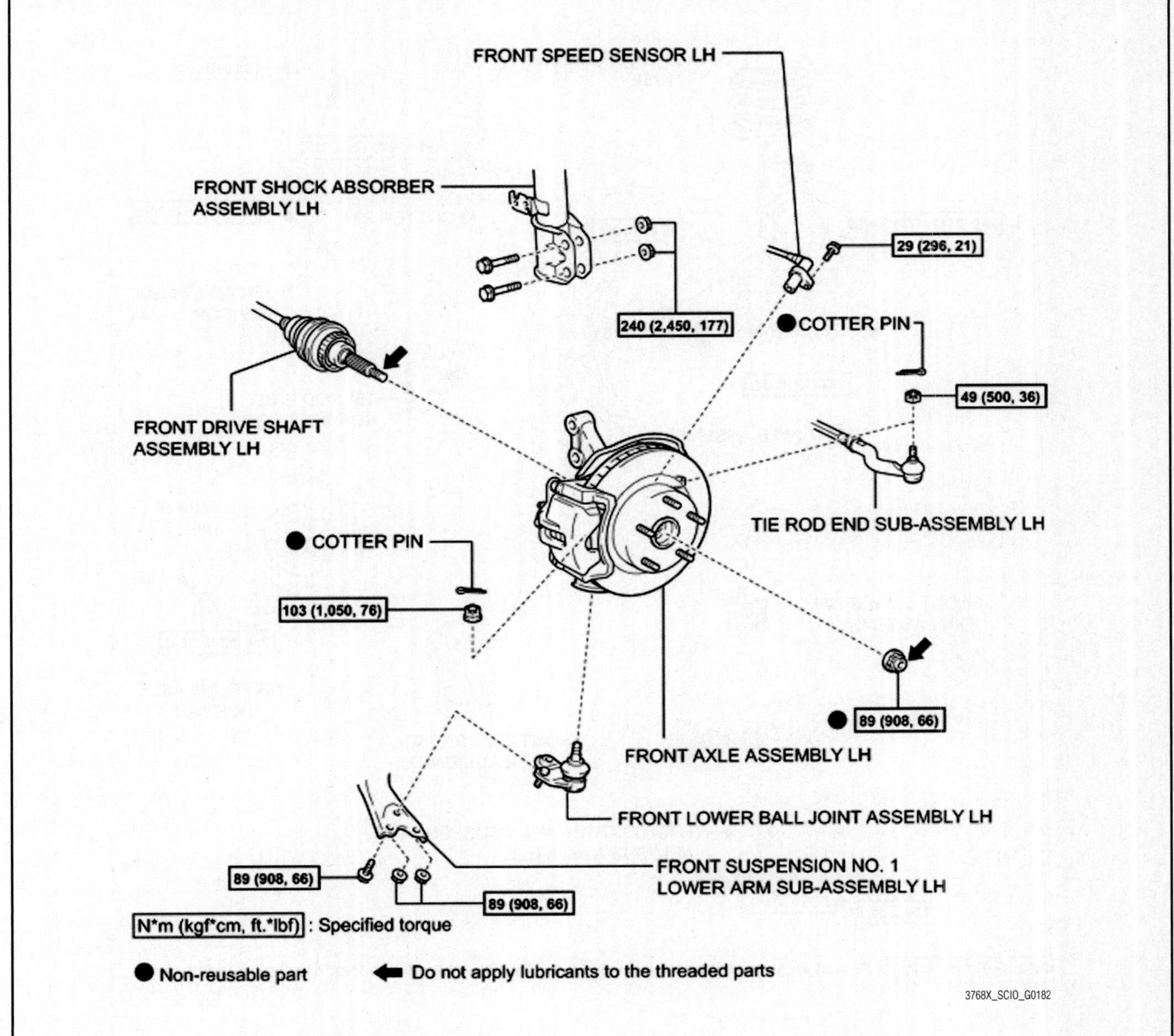

Fig. 194 Exploded view of the lower ball joint and related components—tC models

xB Models

See Figure 195.

1. Remove the front wheel.
2. Remove the front axle hub nut.
3. Separate the front speed sensor.
4. Separate the tie rod end sub-assembly.
5. Separate the front disc brake caliper assembly.
6. Remove the front brake disc.
7. Separate the front lower No. 1 suspension arm sub-assembly.
8. Remove the front axle assembly.
9. Remove the front lower ball joint assembly as follows:
   a. Secure the front axle assembly in a vise.
   b. Remove the cotter pin and nut.
   c. Install the removal tool to the front lower ball joint.
   d. Check that the clearance measurement between SST and the front axle assembly is 0.04 in. (1 mm).
   e. Using proper tools (SST 09960-20010 or equivalent), remove the front lower ball joint from the front axle assembly.

CAUTION

Do not damage the front lower ball joint dust cover.

To install:

10. Install the front lower ball joint assembly as follows:
    a. Secure the front axle assembly in a vise.
    b. Install the front lower ball joint onto the front axle assembly with the nut. Torque to 98 ft. lbs. (133 Nm).
    c. Install a new cotter pin.

➡Further tighten the nut up to 60° if the holes for the cotter pin are not aligned; do not overtighten.

11. Install the front axle assembly.
12. Connect the front lower No. 1 suspension arm sub-assembly.

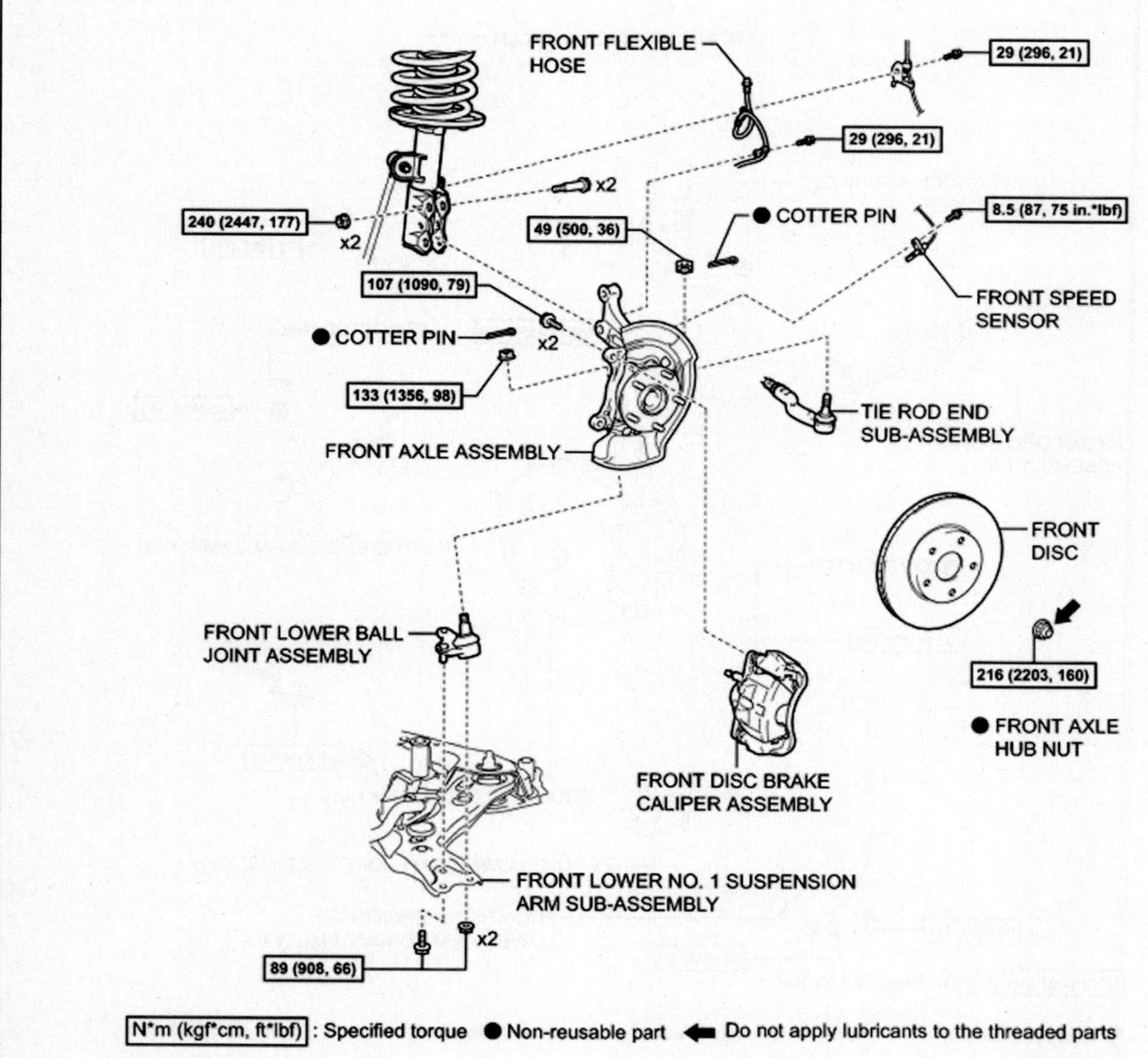

Fig. 195 Exploded view of the front suspension, showing the lower ball joint and related components—xB models

13. Connect the tie rod end sub-assembly.
14. Install the front disc.
15. Install the front disc brake caliper assembly.
16. Install the front speed sensor.
17. Install the front axle hub nut as follows:
    a. Clean the threaded parts on the driveshaft and axle hub nut using a non-residue solvent.

➡Keep the threaded parts free of oil and foreign objects.

    b. Using a socket wrench (30 mm), install a new axle hub nut. Torque to 160 ft. lbs. (220 Nm).
    c. Stake the nut after inspecting for looseness.
18. Install the front wheels.
19. Inspect and adjust the front wheel alignment.
20. Check for VSC sensor signal.

xD Models

See Figures 196 through 198.

1. Remove the front wheels.
2. Remove the front lower suspension arm sub-assembly LH (M/T).
    a. Remove the clip and castle nut.
    b. Install SST (spacer B) to the threaded section of the lower ball joint.

➡Make sure the upper ends of the threaded section of the lower ball joint and SST (spacer B) are aligned.

    c. Using SST, separate the lower arm.
    d. Remove the 2 bolts and lower arm.
3. Remove the hood sub-assembly.
4. Remove the front wiper arm head cap.
5. Remove the front wiper arm and blade assembly LH.
6. Remove the front wiper arm and blade assembly RH.
7. Remove the hood to cowl top seal.
8. Remove the cowl top ventilator louver RH.

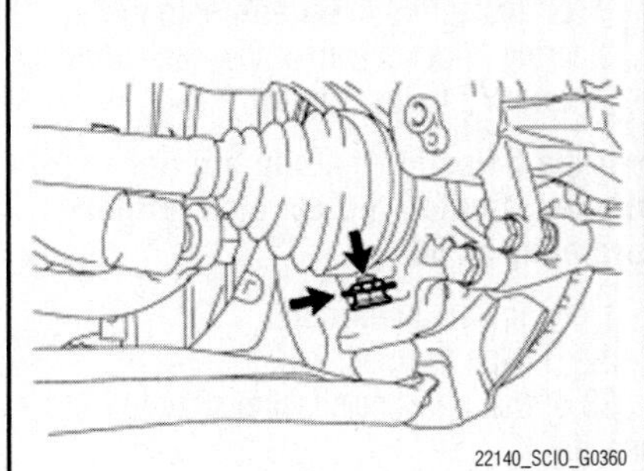

Fig. 196 Remove the front lower suspension arm sub-assembly LH

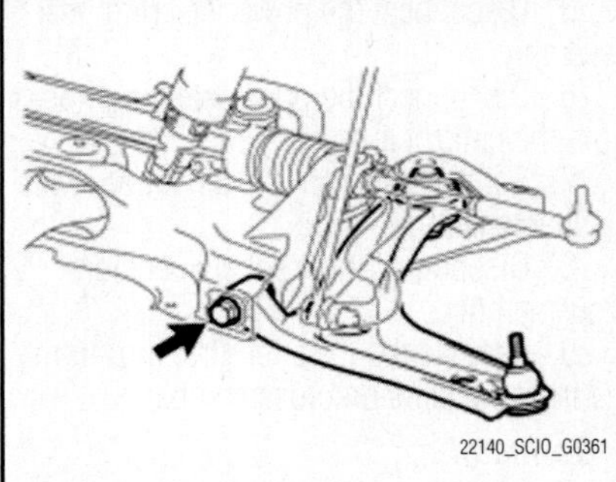

Fig. 197 Remove the front lower suspension arm sub-assembly

9. Remove the cowl top ventilator louver LH.
10. Remove the front wiper motor and link.
11. Remove the front air shutter seal RH.
12. Remove the outer cowl top panel.
13. Position the wheels facing straight ahead.
14. Remove the column hole cover silencer sheet.
15. Remove the steering sliding yoke sub-assembly.
16. Remove the steering column hole cover sub-assembly.
17. Separate the tie rod end sub-assembly LH and RH.
18. Separate the front lower suspension arm sub-assembly LH and RH.
19. Separate the front stabilizer link assembly LH and RH.
20. Suspend the engine assembly.
21. Remove the front suspension cross-member sub-assembly.
22. Remove the front lower suspension arm sub-assembly LH and RH.
    a. Remove the bolt and lower arm.

To install:

23. Temporarily tighten front lower suspension arm sub-assembly.

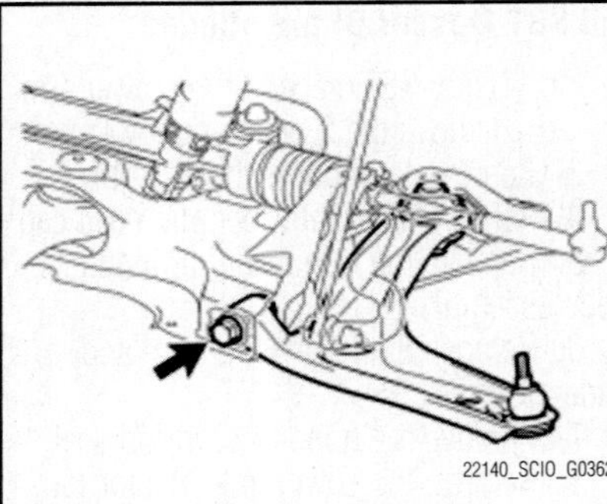

Fig. 198 Fully tighten front lower suspension arm sub-assembly

    a. Install the lower arm onto the crossmember and provisionally tighten the bolt.
24. Install front suspension crossmember sub-assembly.
25. Install front stabilizer link assembly.
26. Install front lower suspension arm sub-assembly.
27. Install tie rod end sub-assembly.
28. Install steering column hole cover sub-assembly.
29. Install steering sliding yoke sub-assembly.
30. Install column hole cover silencer sheet.
31. Temporarily tighten front lower suspension arm sub-assembly LH (M/T).
    a. Provisionally tighten the lower arm with the 2 bolts.
    b. Install the lower arm onto the steering knuckle with a new castle nut. Torque: 72 ft. lbs. (98 Nm).

➡If the holes for the clip are not aligned, tighten the nut by a further turn of up to 60°.

    c. Install a new clip.
32. Install front wheels.
33. Position wheels facing straight ahead.
34. Stabilize suspension.
    a. Lower the vehicle from the jack.
    b. Bounce the vehicle up and down several times to stabilize the suspension.
35. Fully tighten front lower suspension arm sub-assembly.
    a. Fully tighten the 2 bolts:
        - Bolt A: Torque: 101 ft. lbs. (137 Nm)
        - Bolt B: Torque: 118 ft. lbs. (160 Nm).
36. Install outer cowl top panel.
37. Install front air shutter seal.
38. Install front wiper motor and link.
39. Install cowl top ventilator louver.
40. Install hood to cowl top seal.
41. Install front wiper arm and blade assembly.
42. Install front wiper arm head cap.
43. Inspect hood sub-assembly.
44. Adjust hood sub-assembly.
45. Inspect and adjust front wheel alignment.

LOWER CONTROL ARM

REMOVAL & INSTALLATION

tC Models

1. Face the front wheels straight ahead.
2. Drain the power steering fluid.
3. Remove or disconnect the following:
    - Steering column hole cover plate
    - Intermediate steering shaft
    - Front wheel

- Engine undercover
- Tie rod ends
- Power steering hoses
- Front stabilizer links
- Lower control arms
- Front floor panel brace
- Center exhaust pipe
- Hood
- Engine appearance cover

4. Attach an engine crane to support the engine.
5. Remove or disconnect the following:
   - Front suspension crossmember
   - Lower control arm from the cross-member

To install:

6. Install the lower control arm to the crossmember, and temporarily tighten the bolts.
7. Using a suitable jack, lift the front crossmember into the vehicle. Tighten as follows:
   a. Two front nuts to 98 ft. lbs (133 Nm)
   b. Crossmember bracket: Bolt A to 98 ft. lbs. (133 Nm). Bolts B to 59 ft. lbs. (80 Nm)
   c. Rear engine mounting insulator bolt and nuts to 48 ft. lbs. (65 Nm).
8. Install or connect the following:
   - Lower control arms to the steering knuckle. Tighten to 66 ft. lbs. (89 Nm).
   - Stabilizer links. Tighten to 55 ft. lbs. (74 Nm).
   - Power steering hoses
   - Tie rod ends
   - Center exhaust pipe
   - Front floor panel brace
   - Front wheels
   - Steering intermediate shaft. Tighten the bolts to 26 ft. lbs. (35 Nm).
   - Steering column hole cover
   - Engine appearance cover
   - Hood
   - Engine undercover
9. Start the engine and check for leaks.
10. Check and adjust the alignment if necessary.

xB Models

1. Remove or disconnect the following:
   - Front wheel
   - Hood
2. Attach an engine crane to support the engine.
3. Disconnect the lower control arm from the steering knuckle.
4. Disconnect the front stabilizer bar from the lower control arm.
5. Disconnect the power steering gear assembly.
6. Disconnect the power steering hoses from the control arm.
7. Support the front crossmember with a suitable jack.
8. Disconnect the front suspension crossmember.
9. Remove the lower control arm from the front suspension crossmember.

To install:

10. Install the lower control arm to the crossmember and temporarily tighten the bolts.
11. Using a suitable jack, lift the front crossmember into the vehicle. Tighten as follows:
   a. Two front nuts to 98 ft. lbs (133 Nm)
   b. Crossmember bracket: Bolt A to 98 ft. lbs. (133 Nm). Bolts B to 59 ft. lbs. (80 Nm)
   c. Rear engine mounting insulator bolt and nuts to 48 ft. lbs. (65 Nm).
12. Install or connect the following:
   - Power steering hoses
   - Steering gear assembly
   - Front stabilizer bar. Tighten to 13 ft. lbs. (18 Nm).
   - Lower control arm to the steering knuckle. Tighten to 72 ft. lbs. (98 Nm).
   - Front wheel
13. Fully tighten the lower control arm mounting on the crossmember to 97 ft. lbs. (132 Nm).

xD Models

See Figures 196 through 198.

1. Remove the front wheels.
2. Remove the front lower suspension arm sub-assembly LH (M/T).
   a. Remove the clip and castle nut.
   b. Install SST (spacer B) to the threaded section of the lower ball joint.

➡Make sure the upper ends of the threaded section of the lower ball joint and SST (spacer B) are aligned.

   c. Using SST, separate the lower arm.
   d. Remove the 2 bolts and lower arm.
3. Remove the hood sub-assembly.
4. Remove the front wiper arm head cap.
5. Remove the front wiper arm and blade assembly LH.
6. Remove the front wiper arm and blade assembly RH.
7. Remove the hood to cowl top seal.
8. Remove the cowl top ventilator louver RH.
9. Remove the cowl top ventilator louver LH.
10. Remove the front wiper motor and link.
11. Remove the front air shutter seal RH.
12. Remove the outer cowl top panel.
13. Position the wheels facing straight ahead.
14. Remove the column hole cover silencer sheet.
15. Remove the steering sliding yoke sub-assembly.
16. Remove the steering column hole cover sub-assembly.
17. Separate the tie rod end sub-assembly LH and RH.
18. Separate the front lower suspension arm sub-assembly LH and RH.
19. Separate the front stabilizer link assembly LH and RH.
20. Suspend the engine assembly.
21. Remove the front suspension cross-member sub-assembly.
22. Remove the front lower suspension arm sub-assembly LH and RH.
   a. Remove the bolt and lower arm.

To install:

23. Temporarily tighten front lower suspension arm sub-assembly.
   a. Install the lower arm onto the crossmember and provisionally tighten the bolt.
24. Install front suspension crossmember sub-assembly.
25. Install front stabilizer link assembly.
26. Install front lower suspension arm sub-assembly.
27. Install tie rod end sub-assembly.
28. Install steering column hole cover sub-assembly.
29. Install steering sliding yoke sub-assembly.
30. Install column hole cover silencer sheet.
31. Temporarily tighten front lower suspension arm sub-assembly LH (M/T).
   a. Provisionally tighten the lower arm with the 2 bolts.
   b. Install the lower arm onto the steering knuckle with a new castle nut. Torque: 72 ft. lbs. (98 Nm).

➡If the holes for the clip are not aligned, tighten the nut by a further turn of up to 60°.

   c. Install a new clip.
32. Install front wheels.
33. Position wheels facing straight ahead.
34. Stabilize suspension.
   a. Lower the vehicle from the jack.
   b. Bounce the vehicle up and down several times to stabilize the suspension.

35. Fully tighten front lower suspension arm sub-assembly.
a. Fully tighten the 2 bolts. Bolt A: Torque: 101 ft. lbs. (137 Nm); Bolt B: Torque: 118 ft. lbs. (160 Nm).
36. Install outer cowl top panel.
37. Install front air shutter seal.
38. Install front wiper motor and link.
39. Install cowl top ventilator louver.
40. Install hood to cowl top seal.
41. Install front wiper arm and blade assembly.
42. Install front wiper arm head cap.
43. Inspect hood sub-assembly.
44. Adjust hood sub-assembly.
45. Inspect and adjust front wheel alignment.

MACPHERSON STRUT

REMOVAL & INSTALLATION

tC Models

1. Remove or disconnect the following:
   - Front wheel
   - Wiper arms
   - Top cowl ventilator louvers
   - Stabilizer links
   - Wheel speed sensor
2. Remove the strut assembly as follows:
   a. Remove the support dust cover
   b. Loosen the top center lock nut
   c. Lower mounting nuts
   d. 3 upper mounting nuts
   e. Remove the strut assembly

To install:

3. Install the strut assembly as follows:
   a. Upper mounting nuts to 38 ft. lbs. (52 Nm)
   b. Lower mounting bolts to 177 ft. lbs. (240 Nm)
   c. Center lock nut to 35 ft. lbs. (47 Nm)
   d. Apply multipurpose grease to the center lock nut well and install the dust cover.
4. The remainder of the installation is the reverse of removal.
5. Check and adjust the alignment if necessary.

xB Models

See Figure 199.

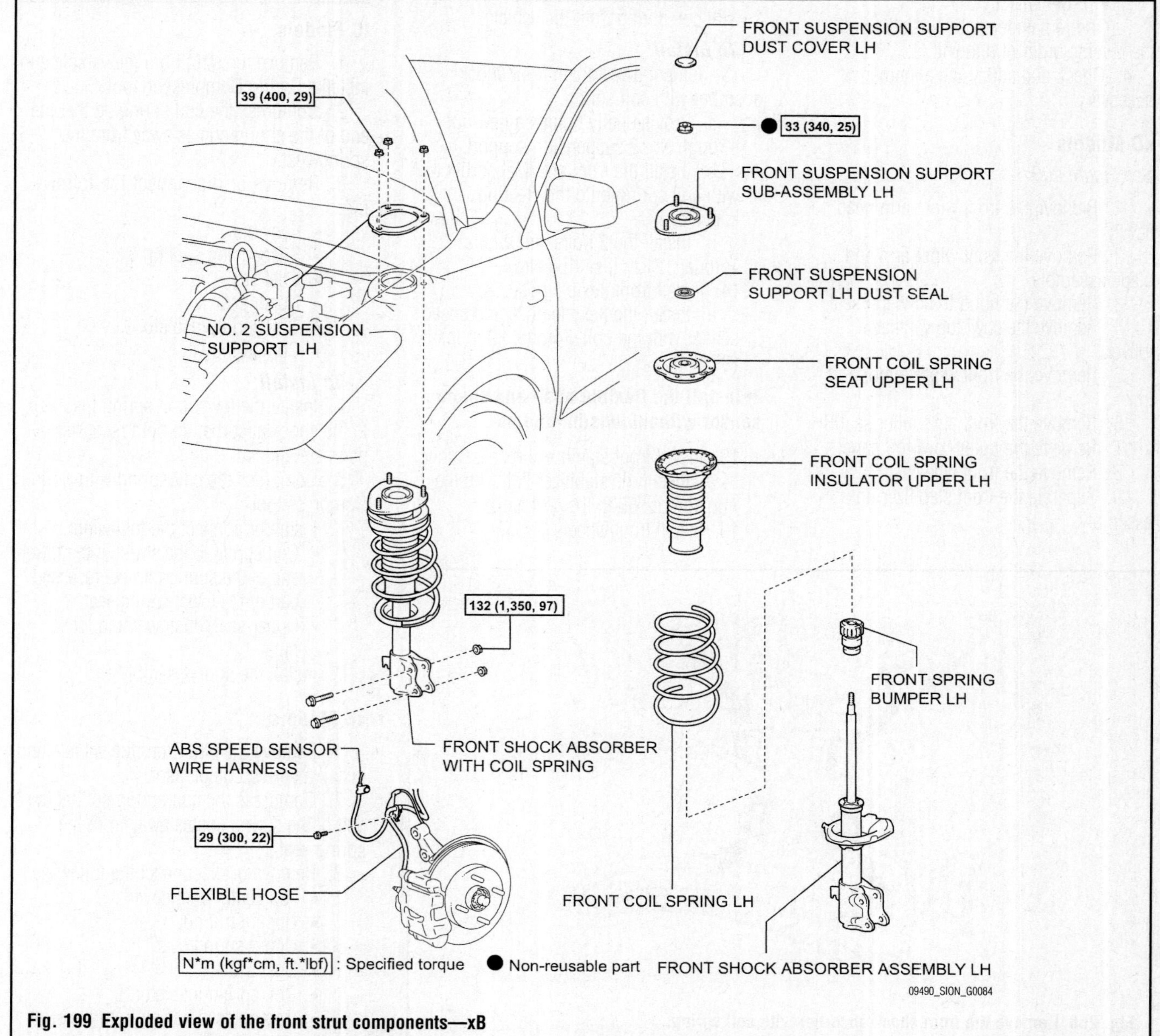

Fig. 199 Exploded view of the front strut components—xB

1. Remove or disconnect the following:
   - Wiper arms
   - Top cowl ventilator louvers
   - Wiper link assembly
   - Outer top cowl panel
   - Front wheel
   - Brake hose and wheel speed sensor
   - Lower mounting bolts
   - Upper mounting nuts

To install:

2. Install or connect the following:
   - Strut assembly. Tighten the upper mounting nuts to 29 ft. lbs. (39 Nm). Lower mounting bolts to 97 ft. lbs. (132 Nm).
   - Brake hose and wheel speed sensor
   - Front wheel

3. The remainder of the installation is the reverse order of removal.

4. Check and adjust the alignment if necessary.

xD Models

See Figure 200.

1. Remove the front wiper arm head cap.
2. Remove the front wiper arm and blade assembly.
3. Remove the hood to cowl top seal.
4. Remove the cowl top ventilator louver.
5. Remove the front wiper motor and link.
6. Remove the front air shutter seal RH.
7. Remove the outer cowl top panel.
8. Remove the front wheels.
9. Separate the front stabilizer link assembly.
   a. Remove the nut and separate the stabilizer link from the shock absorber.
10. Separate the front flexible hose.
    a. Remove the bolt and separate the speed sensor and flexible hose.
11. Remove the front suspension support dust cover.
12. Remove the front shock absorber with coil spring.
    a. Remove the 2 nuts and 2 bolts and separate the shock absorber with coil spring from the steering knuckle.
    b. Using a socket hexagon wrench, fix the shock absorber rod and remove the nut.
    c. Remove the No. 2 suspension support.
    d. Remove the front shock absorber with coil spring from the vehicle.

To install:

13. Temporarily tighten front shock absorber with coil spring.
    a. Provisionally tighten a new nut through No. 2 suspension support.
    b. Install the front shock absorber with coil spring onto the steering knuckle.
    c. Install the 2 bolts and 2 nuts. Torque: 121 ft. lbs. (164 Nm).
14. Install front flexible hose.
    a. Install the flexible hose and speed sensor with the bolt. Torque: 22 ft. lbs. (29 Nm).

➡Install the flexible hose and speed sensor without twisting them.

15. Install front stabilizer link assembly
    a. Install the stabilizer link with the nut. Torque: 55 ft. lbs. (74 Nm).
16. Install front wheels.
17. Fully tighten front shock absorber with coil spring.
    a. Using a socket hexagon wrench 6, fix the shock absorber rod and tighten the nut. Torque: 41 ft. lbs. (55 Nm).
18. Install front suspension support dust cover.
19. Install outer cowl top panel.
20. Install front air shutter seal RH.
21. Install front wiper motor and link.
22. Install cowl top ventilator louver.
23. Install hood to cowl top seal.
24. Install front wiper arm and blade assembly.
25. Install front wiper arm head cap.
26. Inspect and adjust front wheel alignment.

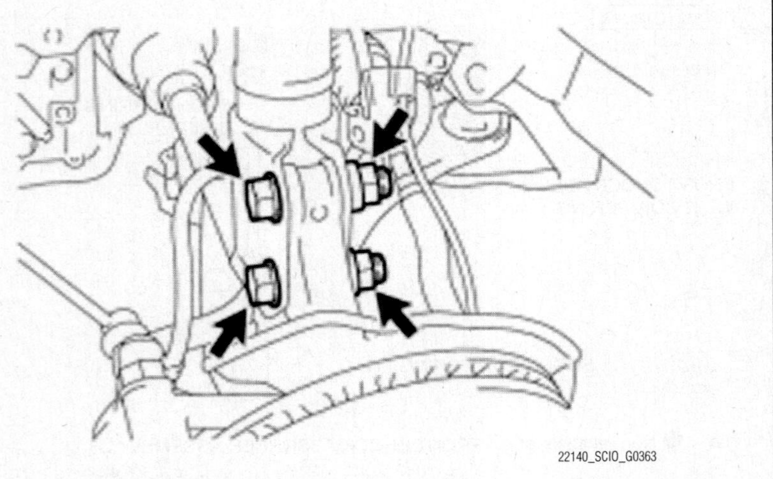

Fig. 200 Remove the front shock absorber with coil spring.

OVERHAUL

tC Models

1. Remove the strut from the vehicle and install a Spring Compressor Tool.
2. Compress the coil spring so that the end of the spring comes away from the spring seat.
3. Remove or disconnect the following:
   - Lock nut
   - Suspension support
   - Dust seal
   - Upper seat
   - Coil spring insulator
   - Coil spring

To install:

4. Install the lower coil spring insulator to the shock absorber so both recessed parts are aligned.
5. Compress the coil spring with Spring Compress tool.
6. Install or connect the following:
   - Coil spring to the strut. Fit the lower end of the spring into the recessed part of the lower spring seat.
   - Upper seat, dust seal and lock nut.
7. Install the strut assembly.

xB Models

1. Remove the strut from the vehicle and install a spring compressor.
2. Compress the coil spring so that the end of the spring comes away from the spring seat.
3. Remove or disconnect the following:
   - Top center dust cover
   - Upper strut nut
   - Upper spring seat
   - Dust seal
   - Coil spring insulator
   - Compressed spring from the strut

- Spring from the spring compressor

To install:

4. Compress the spring and install it on the strut.
5. Install or connect the following:
    - Coil spring insulator
    - Dust seal
    - Upper spring seat and the upper strut mount. Torque the nut to 25 ft. lbs. (33 Nm)
    - Strut to the vehicle
6. Check and/or adjust the wheel alignment.

xD Models

1. Remove the front suspension support sub-assembly.
2. Remove the front support to front shock absorber nut.
    a. Using a socket hexagon wrench 6, fix the shock absorber rod and loosen the nut.
    b. Using SST, compress the coil spring.
    c. Remove the nut.
3. Remove the front coil spring upper seat.
    a. Remove the coil spring seat upper with the strut mounting bearing and spring bumper from the shock absorber.
4. Remove the front spring bumper.
    a. Remove the spring bumper from the coil spring seat upper.
5. Remove the front coil spring upper insulator.
6. Remove the front coil spring.
7. Remove the strut mounting bearing.
    a. Using a brass bar and press, remove the strut mounting bearing from the coil spring seat upper.

To install:

8. Install the strut mounting bearing.
    a. Using a brass bar and press, install the strut mounting bearing onto the coil spring seat upper.

➡The strut mounting bearing must be securely installed.

9. Install the front coil spring.
    a. Using SST, compress the coil spring.
    b. Install the coil spring onto the shock absorber.

➡A spring of a smaller diameter should be installed in the upward direction.

➡Fit the lower end of the coil spring into the gap of the absorber lower seat.

10. Install the front coil spring upper insulator.
11. Install the front spring bumper.
    a. Install the spring bumper onto the coil spring seat upper with the air discharge groove of the spring bumper and the rib portion of the coil spring seat upper aligned.
    b. Securely insert the spring bumper into the coil spring seat upper.
12. Install the front coil spring upper seat.
    a. Install the coil spring seat upper with the strut mounting bearing and spring bumper onto the shock absorber.
13. Install the front support to front shock absorber nut.
    a. Provisionally tighten a new nut.
    b. Remove SST from the front coil spring.
    c. Using a socket hexagon wrench 6, fix the shock absorber rod and tighten the nut. Torque: 25 ft. lbs. (33 Nm).
14. Install the front suspension support sub-assembly.

STABILIZER BAR & LINKS

REMOVAL & INSTALLATION

tC Models

See Figures 201 through 204.

1. Place front wheels facing straight ahead
2. Remove the column hole cover silencer sheet.
3. Separate the steering intermediate shaft.
4. Separate the steering column hole cover.
5. Remove the front wheels
6. Remove the left and right engine under covers.

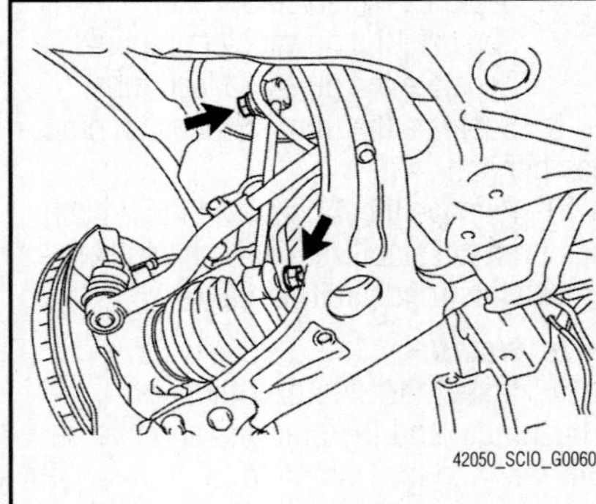

Fig. 201 Remove the stabilizer link

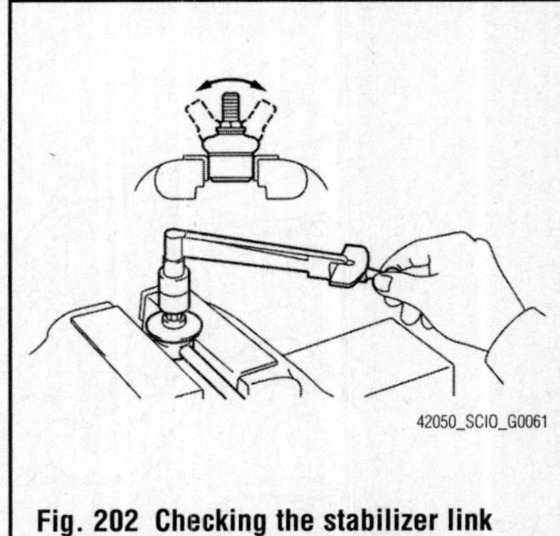

Fig. 202 Checking the stabilizer link

7. Remove front stabilizer links:
    a. Remove the 2 nuts and front stabilizer link.

➡Use a hexagon wrench (6 mm) to hold the stud if the ball joint turns together with the nut.

8. Inspect the stabilizer link as follows:
    a. Secure the front stabilizer link in a vise.
    b. Install the nut to the stud bolt.
    c. Flip the ball joint back and forth 5 times or more.
    d. Use a torque wrench to turn the nut continuously at a rate of 3 to 5 seconds per turn. Take the torque reading on the 5th turn. The turning torque should be 8.9 inch lbs. (1 Nm) or less. If not, replace the link.
    e. Check the dust boots for cracks or grease leakage.
9. Separate the tie rod ends.
10. Disconnect the pressure feed tube.
11. Separate the lower control arms.
12. Remove the front floor panel brace.
13. Remove the center exhaust pipe.
14. Remove the hood.
15. Remove the engine cover.
16. Suspend the engine assembly with the suitable equipment.
17. Remove the no.1 hook.
18. Remove the front suspension crossmember.
19. Remove the stabilizer bracket bolts and the brackets.
20. Remove the 2 front stabilizer bar bushings from the stabilizer bar.
21. Remove the stabilizer bar from the vehicle.

To install:

22. Install the stabilizer bar.
23. Install front stabilizer bar bushings:
    a. Install the bushing to the outer side of the stopper on the stabilizer bar.

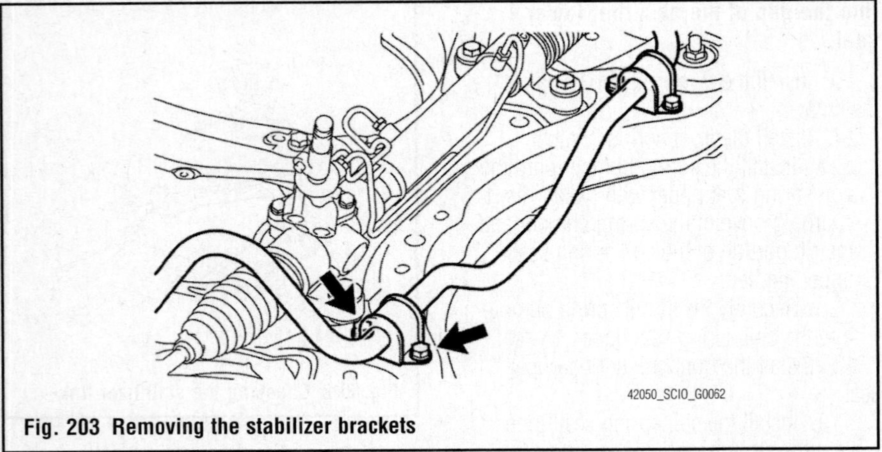

Fig. 203 Removing the stabilizer brackets

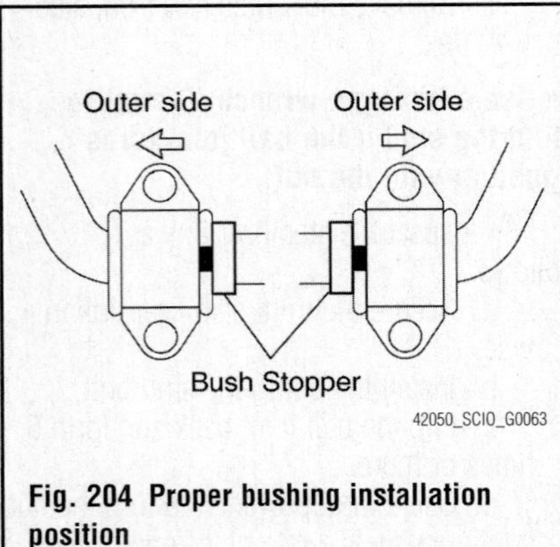

Fig. 204 Proper bushing installation position

➡Place the cutout of the stabilizer bushing facing the rear side.

24. Install the stabilizer bracket to the front suspension crossmember with the 2 bolts, and tighten to 14 ft. lbs. (19 Nm).
25. Install the stabilizer brackets.
26. Install the front suspension crossmember.
27. Install the no.1 hook.
28. Connect the lower control arms.
29. Install the stabilizer links with the retaining nuts and tighten to 55 ft. lbs. (74 Nm).

➡Use a 6mm hex wrench to hold the stud if the ball joint turns together with the nut.

30. Connect pressure feed tube.
31. Connect the tie rod end.
32. Install the center exhaust pipe.
33. Install the front floor panel brace.
34. Install the front wheels
35. Connect the steering column hole cover.
36. Connect the steering intermediate shaft.
37. Install the column hole cover silencer sheet
38. Install the engine cover.
39. Add power steering fluid, then bleed the power steering system, as outlined in this section.
40. Check power steering fluid level in reservoir and add if necessary.
41. Check for fluid leaks
42. Install the hood. Check to make sure the hood is aligned properly.
43. Check and adjust front wheel alignment.
44. Install the engine under covers.

xB Models

⁂ WARNING

This procedure requires you to suspend the engine using the proper engine lifting and holding equipment. Do not attempt this procedure unless you have the proper equipment.

1. Raise and safely support the vehicle.
2. Remove the front wheels
3. Remove the hood.
4. Suspend the engine assembly.
5. Separate the lower control arms. Refer to the procedure in this section for details.
6. Separate the stabilizer bar.
7. Separate the power steering link.
8. Separate the pressure feed tube.
9. Separate the front suspension crossmember sub-.
10. Remove the 4 bolts and the 2 bushings from the stabilizer bar bracket, then remove the front stabilizer bar.

To install:

11. Install the stabilizer bar, the 2 bushings, and the brackets and secure with the 4 bolts. Tighten to 27 ft. lbs. (327 Nm).

➡Install the bushings with the slit facing to the rear side of the vehicle. Install the bushings on the outside of each paint line.

12. Connect the front suspension crossmember.
13. Connect the pressure feed tube.
14. Connect the power steering link.
15. Connect the stabilizer bar.
16. Connect the lower control arms.
17. Install front wheel, then carefully lower the vehicle.
18. Install the hood, making sure it is aligned correctly.
19. Inspect and adjust the front wheel alignment.

xD Models

1. Remove the hood sub-assembly.
2. Remove the front wiper arm head cap.
3. Remove the front wiper arm and blade assembly.
4. Remove the hood to cowl top seal.
5. Remove the cowl top ventilator louver.
6. Remove the front wiper motor and link.
7. Remove the front air shutter seal RH.
8. Remove the outer cowl top panel.
9. Position the wheels facing straight ahead.
10. Remove the front wheel.
11. Remove the column hole cover silencer sheet.
12. Remove the steering sliding yoke sub-assembly.
13. Remove the steering column hole cover sub-assembly.
14. Separate the tie rod end sub-assembly.
15. Separate the front lower suspension arm sub-assembly.
16. Remove the front stabilizer link assembly.
    a. Remove the 2 nuts and stabilizer link
17. Suspend the engine assembly.
18. Remove the front suspension crossmember sub-assembly.
19. Remove the power steering gear.
20. Remove the front stabilizer bracket.
    a. Remove the 2 bolts and the stabilizer bracket.
21. Remove the front stabilizer bar.
22. Remove the front stabilizer bar bushings..

To install:

23. Install front stabilizer bar bushings.
    a. Install the stabilizer bar bush onto the stabilizer bar.

➡Install the bush onto the stabilizer so that the bush stopper of the stabilizer bar faces the outside of the vehicle.

b. Install the bushing with its cutout facing the front of the vehicle.

24. Install front stabilizer bar
    a. Install the stabilizer onto the cross-member with the paint mark on the left side of the vehicle.
25. Install front stabilizer bracket.
    a. Provisionally tighten bolt A.
    b. Tighten the bolts to the specified torque, in the order of B then A. Torque: 35 ft. lbs. (47 Nm).
26. Install power steering gear.
27. Install front suspension crossmember sub-assembly.
28. Install front stabilizer link assembly.
    a. Install the stabilizer link with the 2 nuts. Torque: 55 ft. lbs. (74 Nm).
29. Install front lower suspension arm sub-assembly.
30. Install tie rod end sub-assembly.
31. Install steering column hole cover sub-assembly.
32. Install steering sliding yoke sub-assembly.
33. Install column hole cover silencer sheet.
34. Install front wheels.
35. Position wheels facing straight ahead.
36. Install outer cowl top panel
37. Install front air shutter seal RH.
38. Install front wiper motor and link.
39. Install cowl top ventilator louver.
40. Install hood to cowl top seal.
41. Install front wiper arm and blade assembly.
42. Install front wiper arm head cap.
43. Inspect hood sub-assembly.
44. Adjust hood sub-assembly.
45. Inspect and adjust front wheel alignment.

WHEEL BEARINGS

REMOVAL & INSTALLATION

See Figures 205 through 207.

1. Remove or disconnect the following:
    - Front wheel
    - Hub nut
    - Wheel speed sensor
    - Brake caliper
    - Front disc
    - Tie rod ends
    - Lower control arm
    - Axle from the hub assembly
    - Steering knuckle from strut
    - Lower ball joint
    - Axle hub snap ring

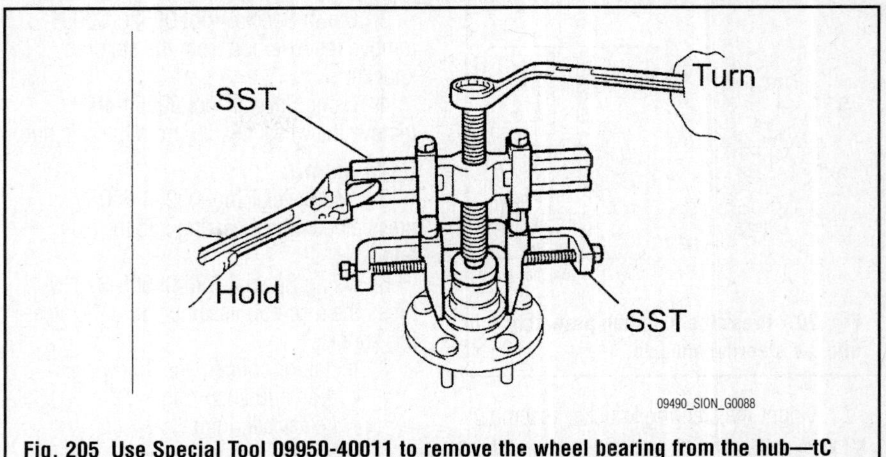

Fig. 205 Use Special Tool 09950-40011 to remove the wheel bearing from the hub—tC

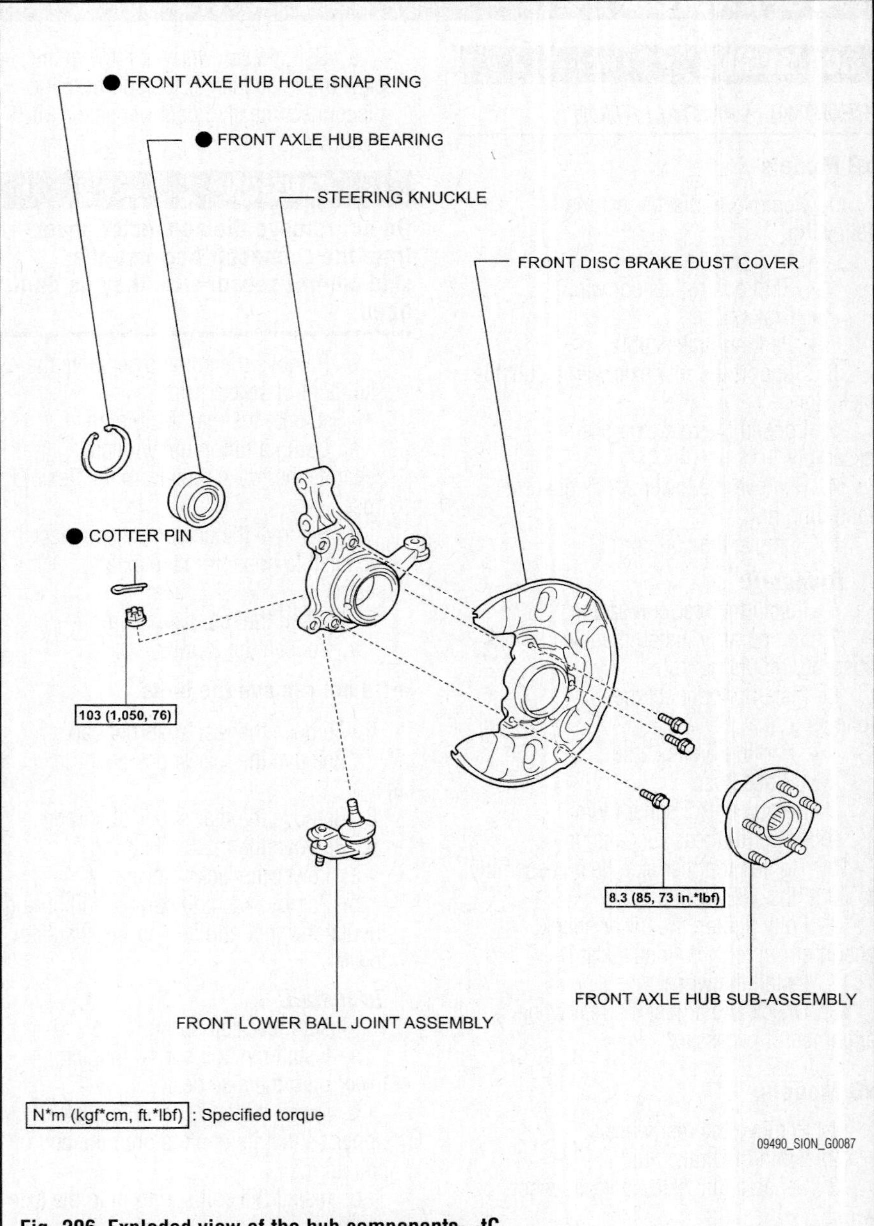

Fig. 206 Exploded view of the hub components—tC

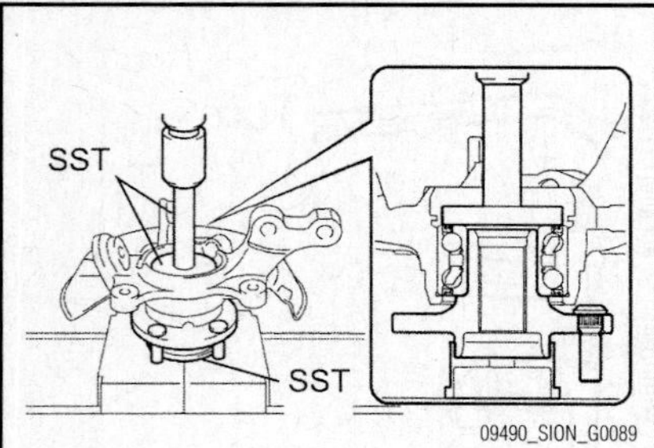

Fig. 207 Press the axle hub assembly into the steering knuckle.

2. Mount the steering knuckle assembly in a vise.
3. Using Special Tool 09520-00031, remove the axle hub from the steering knuckle.
4. Using Special Tool 09950-40011, remove the wheel bearing from the axle hub.

To install:

5. Use Special Tool 09950-60020 to press a new wheel bearing into the axle hub.
6. Using Special Tool 09608-32010, press the axle hub assembly into the steering knuckle.
7. Install or connect the following:
   - Axle hub snap ring
   - Lower ball joint
   - Steering knuckle mounting bolts
   - Axle into the hub assembly
   - Lower control arm
   - Tie rod ends
   - Front disc
   - Brake caliper
   - Wheel speed sensor
   - Hub nut
   - Front wheel

ADJUSTMENT

➡Wheel bearings are an integral, sealed unit in the hub assembly, with no allowance for adjustment.

SUSPENSION

REAR SUSPENSION

COIL SPRING

REMOVAL & INSTALLATION

xB Models

1. Remove or disconnect the following:
   - Rear wheel
   - Skid control sensor wire
   - Brake hose
   - Parking brake cable
2. Support the rear axle with a suitable jack stand.
3. Loosen the rear axle beam assembly.
4. Remove the lower shock absorber mounting nut.
5. Remove the coil spring.

To install:

6. Install the rear coil spring.
7. Temporarily install the lower shock absorber mounting nut.
8. Install or connect the following:
   - Parking brake cable
   - Brake hose
   - Skid control sensor wire
9. Lower the rear suspension.
10. Tighten the rear axle beam assembly to 60 ft. lbs. (82 Nm).
11. Fully tighten the lower shock absorber nut to 36 ft. lbs (49 Nm).
12. Install the wheel.
13. Check and adjust the rear wheel alignment if necessary.

xD Models

1. Remove the rear wheels.
2. Drain the brake fluid.
3. Separate the skid control sensor wire.
   a. Using a screwdriver, remove the claw of the connector lock portion and disconnect the skid control sensor wire connector.

⁂ CAUTION

Do not remove the connector cover from the connector because the skid control sensor wire may be damaged.

   b. Remove the nut and separate the skid control sensor wire.
4. Separate the rear flexible hose.
   a. Using a union nut wrench, separate the brake tube from the flexible hose.
   b. Remove the clip and disconnect the flexible hose from the axle beam.
5. Loosen the rear axle beam.
   a. Loosen the 2 bolts.

➡Do not remove the bolts.

6. Remove the rear absorber cap.
7. Remove the rear shock absorber cap.
8. Remove the rear shock absorber.
9. Remove the rear coil spring.
   a. Lower the jacks slowly.
   b. Remove the coil spring, coil spring insulator upper and coil spring insulator lower.

To install:

10. Install rear coil spring.
    a. Install the coil spring insulator lower onto the axle beam.
    b. Install the coil spring insulator upper so that its gap fits onto the end of coil spring.
    c. Install the coil spring onto the axle beam.

➡The paint mark of the coil spring should be towards the underside and rear side of the vehicle.

11. Temporarily tighten rear shock absorber.
12. Install rear shock absorber cap.
13. Install rear absorber cap.
14. Install rear flexible hose.
    a. Connect the flexible hose onto the axle beam with a new clip.
    b. Using a union nut wrench, install the brake tube.
    c. Without union nut wrench - Torque: 11 ft. lbs. (15 Nm).
    d. With union nut wrench - Torque: 10 ft. lbs.14 Nm).

➡This torque value can be obtained by using a torque wrench with a fulcrum length of 11.8 inches (300 mm) and a union nut wrench with a fulcrum length of 0.866 inches (22 mm). This torque value is effective when union nut wrench is parallel to a torque wrench.

15. Install skid control sensor wire.
    a. Install the skid control sensor wire onto the axle beam with the nut.
    b. Connect the skid control sensor wire connector.
16. Install rear wheels.
17. Stabilize suspension.
    a. Lower the vehicle from the jack.
    b. Bounce the vehicle up and down several times to stabilize the suspension.
18. Fully tighten rear axle beam.
19. Fully tighten rear shock absorber.
20. Fill reservoir with brake fluid.
21. Bleed master cylinder.
22. Bleed brake line.
23. Bleed brake actuator (w/ VSC).

24. Check fluid level in reservoir.
25. Check for brake fluid leakage.
26. Inspect rear wheel alignment

MACPHERSON STRUTS

REMOVAL & INSTALLATION

tC Models

See Figure 208.

1. Remove or disconnect the following:
   - Tonneau cover assembly
   - Rear deck board assembly
   - Rear floor board
   - Rear seat cushion assembly
   - Rear seat back assembly
   - Side trim assembly
   - Rear wheel
   - Skid control sensor wire
   - Rear stabilizer link
2. Support the lower control arm with a suitable jack.
3. Remove or disconnect the following:
   - Strut lower mounting bolt
   - Upper mounting nuts
   - Lower suspension brace
   - Parking brake cable
4. Lower the jack and remove the strut assembly.

To install:

5. Install the strut assembly. Tighten the upper mounting nuts to 59 ft. lbs. (80 Nm). Temporarily tighten the lower mounting bolt.
6. Install or connect the following:
   - Parking brake cable
   - Lower suspension brace. Tighten to 47 ft. lbs. (64 Nm).
   - Rear stabilizer link
   - Skid control sensor wire
   - Rear wheel
   - Lower strut mounting bolt. Fully tighten to 103 ft. lbs. (140 Nm).
7. The remainder of the installation is the reverse order of removal.
8. Check and adjust the rear wheel alignment if necessary.

OVERHAUL

1. Remove the strut from the vehicle and install a Spring Compressor Tool.
2. Compress the coil spring so that the end of the spring comes away from the spring seat.
3. Remove or disconnect the following:
   - Upper mounting nut
   - Shock cushion washer
   - Suspension support
   - Front spring bracket
   - Spring bumper
   - Upper spring insulator
   - Coil spring

To install:

4. Compress the coil spring with Spring Compress tool.
5. Install the coil spring on the strut assembly.
6. Install or connect the following:
   - Spring bumper
   - Suspension support
   - Upper spring insulator
   - Shock cushion washer
   - Upper mounting nut. Tighten to 41 ft. lbs. (56 Nm)

SHOCK ABSORBER

REMOVAL & INSTALLATION

xB Models

See Figure 209.

1. Remove the rear wheels.
2. Remove the rear floor service hole cover.
3. Remove the rear shock absorber cushion retainer as follows:
   a. Support the spring seat of the rear axle beam assembly using a jack and wooden block.

➡Do not excessively jack up the rear axle beam assembly.

   b. Support the rear shock absorber at a position where it compresses by approximately 0.78–1.18 in. (20–30 mm).
   c. Using a socket hexagon wrench (6 mm), secure the rear shock absorber rod and remove the lock nut.
   d. Remove the rear shock absorber cushion retainer.
4. Remove the rear suspension support.
5. Remove the rear shock absorber assembly.
6. Remove the rear No. 1 spring bumper.

To install:

7. Installation is the reverse of the removal procedure.
8. Install but do not tighten the lower shock absorber mounting bolt.
9. Tighten the upper shock absorber mounting bolt to 18 ft. lbs. (25 Nm).
10. Install the rear wheels. Torque the nuts to 76 ft. lbs. (103 Nm).
11. Settle the rear suspension by bouncing the vehicle. Repeat several times.
12. Tighten the lower shock absorber mounting nut to 67 ft. lbs. (90 Nm).

xD Models

1. Remove the rear wheels.
2. Remove the rear absorber cap.
3. Remove the rear shock absorber cap.
4. Remove the rear shock absorber.
   a. Support the axle beam with a jack. Insert a wooden block between the jack and the rear axle spring seat to prevent damage.
   b. Remove the 2 nuts while keeping the piston rod from rotating.
   c. Remove the cushion retainer and suspension support.
   d. Remove the bolt while keeping the nut from rotating and remove the shock absorber.

➡Remove the nut from the bolt side because the one on the lower side is a jam nut.

5. Remove the rear suspension support stopper.
6. Remove the rear suspension support assembly.

To install:

7. Install rear suspension support assembly.
8. Install rear suspension support stopper.
9. Temporarily tighten rear shock absorber.
   a. Support the axle beam with a jack. Insert a wooden block between the jack and the rear axle spring seat to prevent damage.
   b. Jack up the axle beam slowly, and provisionally install the shock absorber (lower side) with the bolt and nut onto the axle beam.
   c. Install the suspension support and cushion retainer.
   d. While holding the piston rod, install a new nut (lower nut).
   e. While holding the piston rod, tighten a new nut (upper nut). Torque: 18 ft. lbs. (25 Nm).
10. Install rear shock absorber cap.
11. Install rear absorber cap.
12. Install rear wheels.
13. Stabilize suspension.
   a. Lower the vehicle from the jack.
   b. Bounce the vehicle up and down several times to stabilize the suspension.
14. Fully tighten rear shock absorber.
   a. Fully tighten the shock absorber (lower side) with the bolt. Torque: 36 ft. lbs. (49 Nm).
15. Inspect rear wheel alignment.

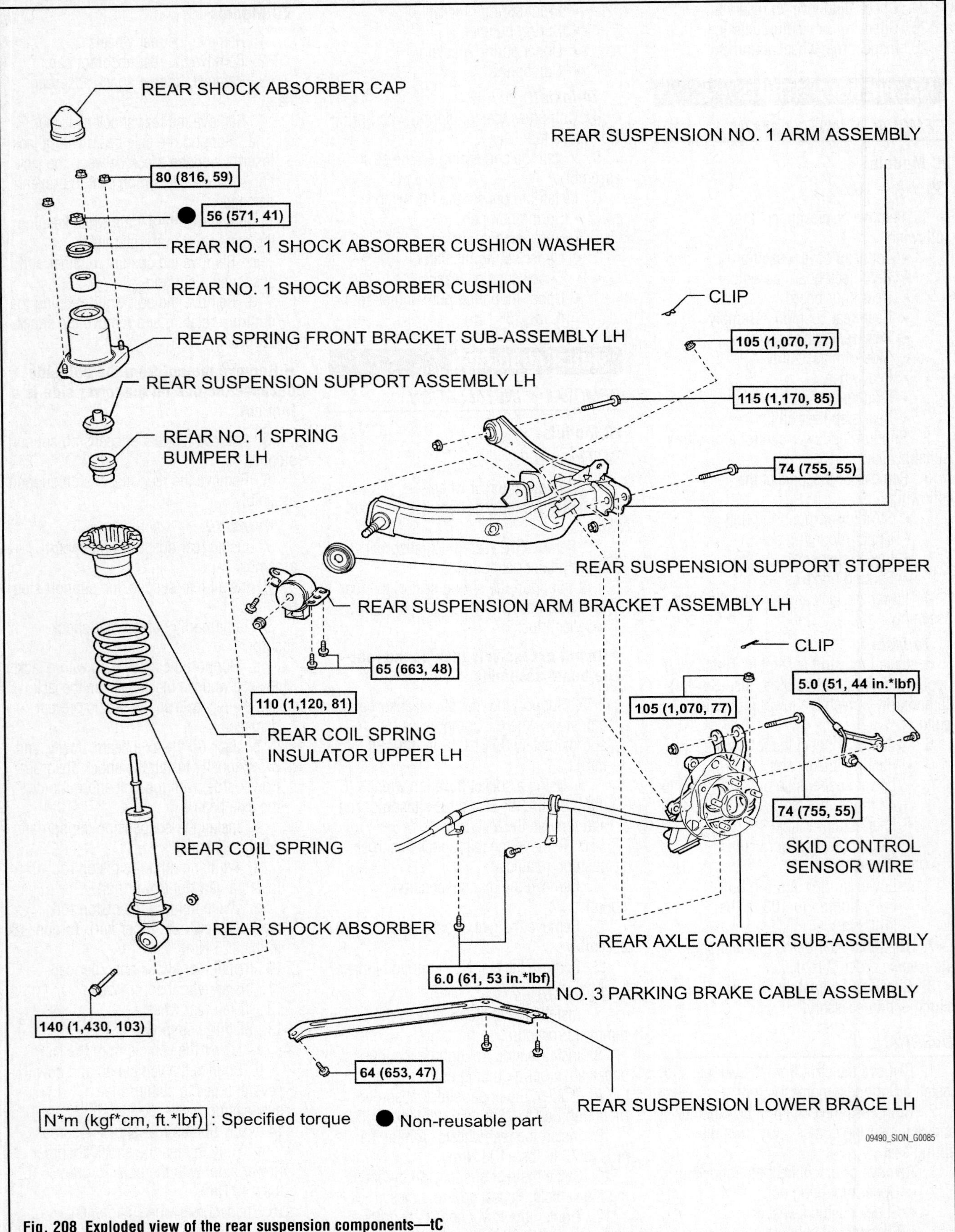

Fig. 208 Exploded view of the rear suspension components—tC

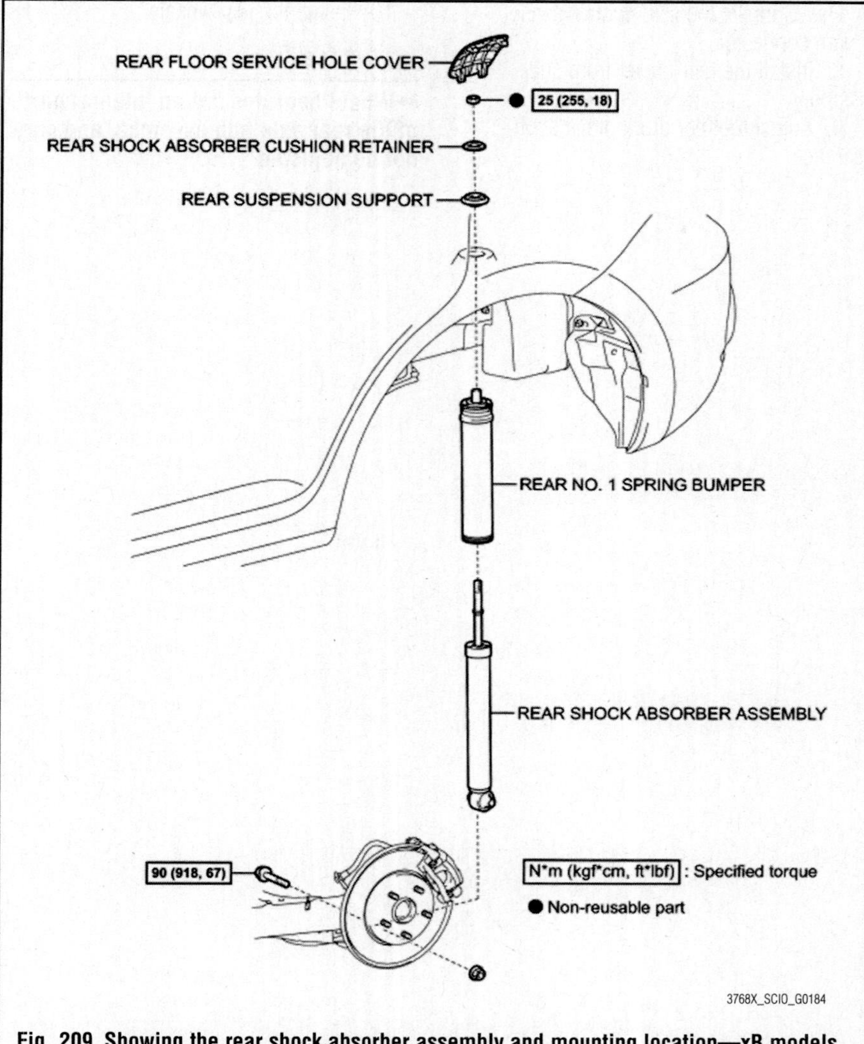

Fig. 209 Showing the rear shock absorber assembly and mounting location—xB models

STABILIZER BAR & LINKS

REMOVAL & INSTALLATION

tC Models

See Figure 210.

1. Raise and safely support the vehicle.
2. Remove the 2 nuts and the stabilizer link assemblies.

➡If the ball joint turns together with the nut, use a 5mm hex wrench to hold the stud.

3. Remove the stabilizer link assemblies.
4. Inspect the stabilizer link assembly, as follows:
   a. Before installing the nut, flip the ball joint stud back and Forth 5 times as shown in the illustration.
   b. Using a torque wrench, continuously turn the nut for 2 to 4 seconds per 1 turn with a torque of 8.7 inch lbs. (1 Nm) or less,, and take the torque reading on the 5th turn.
   c. Check that neither unusual drag nor rattle occurs during the rotation.
   d. Check that neither crack nor grease leakage exists on the dust cover.

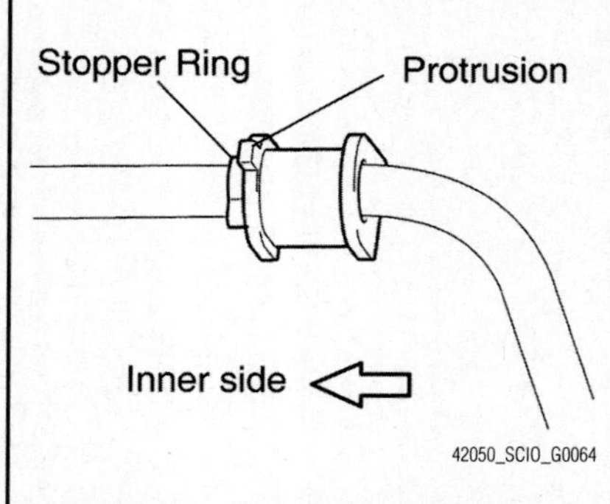

Fig. 210 Rear stabilizer bar bushing installation

5. Make sure that the stabilizer link is not deformed.
6. Remove the 2 bolts, 2 nuts and 2 stabilizer bar brackets from the rear suspension member.
7. Remove the 2 stabilizer bushings from the stabilizer bar.

To install:

8. Install the 2 stabilizer bushings to the stabilizer bar.

➡Install the stabilizer bush rear to the outer side of the stopper ring on the stabilizer bar. Install the stabilizer bush with the protrusion facing towards the inside of the vehicle.

9. Install the stabilizer bar and 2 rear stabilizer bar Bracket no.3 with 2 bolts and 2 nuts. Tighten to 26 ft. lbs. (35 Nm).
10. Install the rear stabilizer link assembly with the 2 nuts and tighten to 32 ft. lbs. (44 Nm).

➡If the ball joint turns together with the nut, use a 5mm hex wrench to hold the stud.

11. Install rear stabilizer links.
12. Inspect and adjust rear wheel alignment.

WHEEL BEARINGS

REMOVAL & INSTALLATION

1. Remove the rear wheels.
2. Remove the rear brake drum sub-assembly.
3. Disconnect the speed sensor wire.
   a. Using a screwdriver, remove the claw of the connector lock portion and disconnect the skid control sensor wire connector.

✱✱ CAUTION

Do not remove the connector cover from the connector because the skid control sensor wire may be damaged.

4. Remove the rear axle hub and bearing assembly.
   a. Remove the 4 bolts and remove the axle hub and bearing from the axle beam.
   b. Suspend the backing plate with a piece of rope.

To install:

5. Install the rear axle hub and bearing assembly.

a. Install the axle hub and bearing onto the axle beam with the 4 bolts.
Torque: 67 ft. lbs. (90 Nm)

6. Inspect the rear axle hub bearing.
7. Connect the speed sensor wire.
   a. Connect the skid control sensor wire connector.
8. Install the rear brake drum sub-assembly.
9. Adjust the rear drum brake shoe clearance.
10. Install the rear wheels.

ADJUSTMENT

➡Wheel bearings are an integral part of the rear axle hub assembly and cannot be adjusted.

SCION

Diagnostic Trouble Codes

2

DIAGNOSTIC TROUBLE CODES

OBD II VEHICLE APPLICATIONS

SCION

tC
2009–2010
- 2.4L I4 .... 2AZ-FE

xB
2009–2010
- 2.4L I4 .... 2AZ-FE

xD
2009–2010
- 1.8L I4 .... 2ZR-FE

OBD II Trouble Code List (P0XXX Codes)

DTC	Trouble Code Title, Conditions & Possible Causes
DTC: P0010 **1T ECM, MIL: Yes** **Year:** 2009, 2010 **Model:** tC, xD **Engine:** 1.8L L4 VIN U, 2.4L L4 **Transmission:** All	**Camshaft Position "A" Actuator Circuit (Bank 1):** Starter: OFF Engine Switch: On (IG) Time after turning engine switch off to on (IG): 0.5 seconds or more **Possible Causes:** • Open or short in Oil Control Valve (OCV) for intake camshaft (Bank 1) circuit • OCV for intake camshaft (Bank 1) • ECM
DTC: P0010 **2T ECM, MIL: Yes** **Year:** 2009, 2010 **Model:** xB **Engine:** 2.4L L4 **Transmission:** All	**VVT Oil Control Circuit Malfunction (Bank 1):** Key on or engine running; and the PCM detected an unexpected voltage condition on the VVT Oil Control Valve Bank 1 circuit. The VVT system controls the intake camshaft in order to provide optimal valve timing during all conditions based signals from the ECT, IAT and TP sensor. The VVT regulates the intake camshaft angle using oil pressure through the Oil Control Valve. This results in the relative position between the camshaft and crankshaft to become optimal. The result is higher torque, better fuel economy and low emissions. **Possible Causes:** • OCV assembly connector is damaged or loose • OCV assembly control circuit is open or shorted to ground • OCV assembly is damaged or has failed • ECM has failed
DTC: P0011 **1T ECM, MIL: Yes** **Year:** 2009, 2010 **Model:** tC, xD **Engine:** 1.8L L4 VIN U, 2.4L L4 **Transmission:** All	**Camshaft Position "A" - Timing Over-Advanced or System Performance (Bank 1):** Valve timing is not adjusted in valve timing advance. Battery Voltage: 11V or more Engine: 500-4000 rpm ECT: 167-212 degrees F (75-100 degrees C) **Possible Causes:** • Valve timing • Oil Control Valve (OCV) for intake camshaft (Bank 1) • OCV filter (Bank 1) • Intake camshaft timing gear assembly (Bank 1) • ECM
DTC: P0011 **2T** **Year:** 2009, 2010 **Model:** xB **Engine:** 2.4L L4 **Transmission:** All	**Camshaft Position 'A' Over-Advanced Or System Performance (Bank 1):** Engine started, ECT sensor more than 158°F, vehicle driven at an engine speed of 400-4000 rpm, and the PCM detected the valve timing did not change from the "current" valve timing, or the valve timing remain fixed during testing. The VVT system controls the intake camshaft in order to provide optimal valve timing during all conditions based signals from the ECT, IAT and TP sensor. The VVT regulates the intake camshaft angle using oil pressure through the Oil Control Valve. This results in the relative position between the camshaft and crankshaft to become optimal. The result is better engine torque, fuel economy and lower emissions. **Possible Causes:** • Engine valve timing malfunction • Camshaft timing oil control valve unit is damaged or has failed • PCM or VVT ECM has failed
DTC: P0012 **2T ECM, MIL: Yes** **Year:** 2009, 2010 **Model:** tC, xD **Engine:** 1.8L L4 VIN U, 2.4L L4 **Transmission:** All	**Camshaft Position "A" - Timing Over-Retarded (Bank 1):** Valve timing is not adjusted in valve timing retard range. Battery Voltage: 11V or more Engine: 500-4000 rpm ECT: 167-212 degrees F (75-100 degrees C) **Possible Causes:** • Valve timing • OCV for intake camshaft (Bank 1) • OCV filter (Bank 1) • Intake camshaft timing gear assembly (Bank 1) • ECM

DTC	Trouble Code Title, Conditions & Possible Causes
DTC: P0013 **1T ECM, MIL: Yes** **Year:** 2009, 2010 **Model:** xD **Engine:** 1.8L L4 VIN U **Transmission:** All	**Camshaft Position "B" Actuator Circuit / Open (Bank 1):** All of following conditions met, starter OFF, ignition switch ON. Time after ignition switch OFF to ON 0.5 seconds or more. One of following conditions met 1). All of following conditions met: Battery voltage 11 to 13 V, CPU commanded duty less than 70%, current cut status Not Cut, 2). All of following conditions met: Battery voltage 13 V or more, Target duty ratio less than 80%, current cut status Not Cut **Possible Causes:** • Open or short in OCV (for exhaust camshaft) circuit • OCV (for exhaust camshaft) • ECM
DTC: P0014 **2T ECM, MIL: Yes** **Year:** 2009, 2010 **Model:** xD **Engine:** 1.8L L4 VIN U **Transmission:** All	**Camshaft Position "B" - Timing Over-Advanced or System Performance (Bank 1):** Battery voltage 11 V or more, engine RPM 500 to 4000 rpm, engine coolant temperature 75 to 100°C (167 to 212°F). The valve timing is not adjusted in exhaust valve timing advance range. **Possible Causes:** • Valve timing • Camshaft Timing Oil Control Valve (OCV) (for exhaust camshaft) • OCV filter • Camshaft timing gear assembly (for exhaust camshaft) • ECM
DTC: P0015 **1T ECM, MIL: Yes** **Year:** 2009, 2010 **Model:** xD **Engine:** 1.8L L4 VIN U **Transmission:** All	**Camshaft Position "B" - Timing Over-Retarded (Bank 1):** Battery voltage 11 V or more, engine RPM 500 to 4000 rpm, engine coolant temperature 75 to 100°C (167 to 212°F). The valve timing is not adjusted in exhaust valve timing retard range. **Possible Causes:** • Valve timing • Camshaft Timing Oil Control Valve (OCV) (for exhaust camshaft) • OCV filter • Camshaft timing gear assembly (for exhaust camshaft) • ECM
DTC: P0016 **2T ECM, MIL: Yes** **Year:** 2009, 2010 **Model:** tC, xD **Engine:** 1.8L L4 VIN U, 2.4L L4 **Transmission:** All	**Crankshaft Position - Camshaft Position Correlation (Bank 1 Sensor A):** Deviations in crankshaft and camshaft position sensor signals. Engine RPM: 500-1000 RPM **Possible Causes:** • Valve timing • Camshaft timing oil control valve for intake camshaft • Camshaft timing oil control valve filter • Intake camshaft timing gear assembly • ECM
DTC: P0017 **2T ECM, MIL: Yes** **Year:** 2009, 2010 **Model:** xD **Engine:** 1.8L L4 VIN U **Transmission:** All	**Crankshaft Position - Camshaft Position Correlation (Bank 1 Sensor B):** Engine RPM 500 to 1000 rpm. A deviation in crankshaft position sensor signal and camshaft position sensor (for exhaust camshaft) signal is detected. **Possible Causes:** • Mechanical system (Timing chain has jumped tooth or chain stretched) • Camshaft Timing Oil Control Valve (OCV) (for exhaust camshaft) • Camshaft timing gear assembly (for exhaust camshaft) • ECM
DTC: P0031 **1T ECM, MIL: Yes** **Year:** 2009, 2010 **Model:** xB **Engine:** 2.4L L4 **Transmission:** All	**Oxygen (A/F) Sensor Heater Control Circuit Low (Bank 1 Sensor 1):** The battery voltageis 10.5 V or more, A/F sensor heater duty-cycle ratio is 50% or more. Time after engine start 10 seconds or more. The Air Fuel Ratio (A/F) sensor heater current less than 0.8 A. **Possible Causes:** • Open in Air Fuel Ratio (A/F) sensor heater circuit • A/F sensor heater (sensor 1) • ECM power source circuit • ECM

DTC	Trouble Code Title, Conditions & Possible Causes
DTC: P0031 **1T ECM, MIL: Yes** **Year:** 2009, 2010 **Model:** xD **Engine:** 1.8L L4 VIN U **Transmission:** All	**Oxygen (A/F) Sensor Heater Control Circuit Low (Bank 1 Sensor 1):** Battery voltage 10.5 V or more, time after heater ON 5 seconds or more. Heater output duty 10 % or more. The Air Fuel Ratio (A/F) sensor heater current is less than 0.8 A. **Possible Causes:** • Open in A/F sensor heater circuit • A/F sensor heater (sensor 1) • ECM power source circuit • ECM
DTC: P0032 **1T ECM, MIL: Yes** **Year:** 2009, 2010 **Model:** tC, xB, xD **Engine:** 1.8L L4 VIN U, 2.4L L4 **Transmission:** All	**Oxygen (A/F) Sensor Heater Control Circuit High (Bank 1 Sensor 1):** Battery voltage: 10.5 V or more Heater output duty: More than 50% Time after engine start: 10 seconds or more Active heater off control: Not operating Active heater on control: Not operating **Possible Causes:** • Short in A/F sensor heater circuit • A/F sensor heater • A/F sensor heater relay • ECM
DTC: P0037 **1T ECM, MIL: Yes** **Year:** 2009, 2010 **Model:** tC, xB, xD **Engine:** 1.8L L4 VIN U, 2.4L L4 **Transmission:** All	**Oxygen Sensor Heater Control Circuit Low (Bank 1 Sensor 2):** Battery voltage: 10.5-20V. Heated Oxygen (HO2) sensor (sensor 2) heater current less than 0.3 A. **Possible Causes:** • Open in HO2 sensor heater circuit • HO2 sensor heater • Engine room junction block (EFI relay) • ECM
DTC: P0038 **1T ECM, MIL: Yes** **Year:** 2009, 2010 **Model:** tC, xB, xD **Engine:** 1.8L L4 VIN U, 2.4L L4 **Transmission:** All	**Oxygen Sensor Heater Control Circuit High (Bank 1 Sensor 2):** (Case 1) Battery voltage: 10.5 V or more Engine: Running Starter: OFF (Case 2) Battery voltage: 10.5-20V **Possible Causes:** • Short in HO2 sensor heater circuit • HO2 sensor heater • Engine room junction block (EFI relay) • ECM
DTC: P0100 **1T ECM, MIL: Yes** **Year:** 2009, 2010 **Model:** tC **Engine:** 2.4L L4 **Transmission:** All	**Mass or Volume Air Flow Circuit:** Monitor runs whenever following DTCs are not present: None MAF meter voltage: Less than 0.2 V, or more than 4.9 V **Possible Causes:** • Open or short in MAF meter circuit • MAF meter • ECM
DTC: P0101 **2T ECM, MIL: Yes** **Year:** 2009, 2010 **Model:** tC, xB, xD **Engine:** 1.8L L4 VIN U, 2.4L L4 **Transmission:** All	**Mass Air Flow Circuit Range / Performance Problem:** TP (Throttle position) sensor voltage: 0.24 to 2 V Engine: Running Battery voltage: 10.5 V or more ECT: 158°F (70°C) or more Estimated load: 30 to 70% **Possible Causes:** • Mass Air Flow (MAF) meter • Air induction system • PCV hose connections

DTC	Trouble Code Title, Conditions & Possible Causes
DTC: P0102 **1T ECM, MIL: Yes** **Year:** 2009, 2010 **Model:** tC, xB, xD **Engine:** 1.8L L4 VIN U, 2.4L L4 **Transmission:** All	**Mass or Volume Air Flow Circuit Low Input:** Monitor runs whenever following DTCs are not present Mass air flow meter voltage: Less than 0.2 V for 3 seconds. **Possible Causes:** • Open in MAF meter circuit • Short in MAF meter circuit • MAF meter • ECM
DTC: P0103 **1T ECM, MIL: Yes** **Year:** 2009, 2010 **Model:** tC, xB, xD **Engine:** 1.8L L4 VIN U, 2.4L L4 **Transmission:** All	**Mass or Volume Air Flow Circuit High Input:** MAF meter voltage is more than 4.9 V for 3 seconds. **Possible Causes:** • Open in Mass Air Flow (MAF) meter power source circuit • Open or short in VG circuit • MAF meter • ECM
DTC: P0110 **1T ECM, MIL: Yes** **Year:** 2009, 2010 **Model:** tC **Engine:** 2.4L L4 **Transmission:** All	**Intake Air Temperature Circuit Malfunction:** Open or short in IAT sensor circuit for 0.5 seconds. **Possible Causes:** • Open or short in IAT sensor circuit • IAT sensor (built into MAF meter) • ECM
DTC: P0111 **2T ECM, MIL: Yes** **Year:** 2009, 2010 **Model:** tC, xB, xD **Engine:** 1.8L L4 VIN U, 2.4L L4 **Transmission:** All	**Intake Air Temperature Sensor Gradient Too High:** After Engine Stop: Time after engine start: 10 seconds or more Battery voltage: 10.5 V or more ECT change since engine stop: -40°F (-40°C) or more Accumulated MAF amount before engine stop: 2033 g or more Key-off duration: 30 minutes After Cold Engine Start: Key-off duration:5 hours Time after engine start: 10 seconds or more ECT: 158° F (70° C) or more Accumulated MAF amount: 2033 g or more One of the following conditions 1 or 2 is met: 1. Duration while engine load is low: 120 seconds or more 2. Duration while engine load is high: 10 seconds or more **Possible Causes:** • MAF Meter
DTC: P0112 **1T ECM, MIL: Yes** **Year:** 2009, 2010 **Model:** tC, xB, xD **Engine:** 1.8L L4 VIN U, 2.4L L4 **Transmission:** All	**Intake Air Temperature Circuit Low Input:** Monitor runs whenever following DTCs are not present: None Short in intake air temperature sensor circuit for 0.5 seconds. Battery voltage: 8 V or more Engine switch: On (IG) Starter: OFF **Possible Causes:** • Short in IAT sensor circuit • IAT sensor (built into MAF meter) • ECM
DTC: P0113 **1T ECM, MIL: Yes** **Year:** 2009, 2010 **Model:** tC, xB, xD **Engine:** 1.8L L4 VIN U, 2.4L L4 **Transmission:** All	**Intake Air Temperature Circuit High Input:** Open in IAT sensor circuit for 0.5 seconds. **Possible Causes:** • Open in IAT sensor circuit • IAT sensor (inside MAF meter) • ECM

DTC	Trouble Code Title, Conditions & Possible Causes
DTC: P0115 **1T ECM, MIL: Yes** **Year:** 2009, 2010 **Model:** tC, xB, xD **Engine:** 1.8L L4 VIN U, 2.4L L4 **Transmission:** All	**Engine Coolant Temperature Circuit Malfunction:** Monitor runs whenever following DTCs are not present: None Engine coolant temperature sensor voltage: Less than 0.14 V, or more than 4.91 V **Possible Causes:** • Open or short in ECT sensor circuit • ECT sensor • ECM
DTC: P0116 **2T ECM, MIL: Yes** **Year:** 2009, 2010 **Model:** tC, xB, xD **Engine:** 1.8L L4 VIN U, 2.4L L4 **Transmission:** All	**Engine Coolant Temperature Circuit Range / Performance Problem:** Monitor runs whenever following DTCs are not present: None ECT sensor cold start monitor: Battery voltage: 10.5 V or more Time after engine start: 1 second or more ECT at engine start: Less than 140°F (60°C) Soak time: 0 second Accumulated MAF:1241 g or more Fuel cut: OFF Difference between ECT at engine start and IAT: Less than 104°F (40°C) ECT sensor soak monitor: Battery voltage: 10.5 V or more Engine: Running Soak time: 5 hours or more ECT at engine start: 140°F (60°C) or more Accumulated MAF: 2334 g or more **Possible Causes:** • Thermostat • ECT sensor
DTC: P0117 **1T ECM, MIL: Yes** **Year:** 2009, 2010 **Model:** tC, xB, xD **Engine:** 1.8L L4 VIN U, 2.4L L4 **Transmission:** All	**Engine Coolant Temperature Circuit Low Input:** Monitor runs whenever following DTCs are not present: None Engine coolant temperature sensor voltage: Less than 0.14 V **Possible Causes:** • Short in ECT sensor circuit • ECT sensor • ECM
DTC: P0118 **1T ECM, MIL: Yes** **Year:** 2009, 2010 **Model:** tC, xB, xD **Engine:** 1.8L L4 VIN U, 2.4L L4 **Transmission:** All	**Engine Coolant Temperature Circuit High Input:** Monitor runs whenever following DTCs are not present: None Engine coolant temperature sensor voltage: More than 4.91 V **Possible Causes:** • Open in ECT sensor circuit • ECT sensor • ECM
DTC: P011B **2T ECM, MIL: Yes** **Year:** 2009, 2010 **Model:** tC, xB, xD **Engine:** 1.8L L4 VIN U, 2.4L L4 **Transmission:** All	**Engine Coolant Temperature / Intake Air Temperature Correlation:** Monitor runs whenever following DTCs are not present: None All of following conditions met: Conditions 1 and 2 1. All of the following conditions are met: Conditions (a), (b), (c) and (d) (a) After engine switch on (IG) and engine not running time: Less than 20 seconds (b) Soak Time: 7 hours or more (c) Battery voltage: 10.5 V or more (d) Time after engine start: 15 seconds or more 2. Either of the following conditions are met: Condition (a) and (b) (a) Minium IAT after engine start: 14° F (-10° C) or more (b) ECT before engine start: 14° F (-10° C) or more **Possible Causes:** • IAT sensor • ECT sensor • ECM

DTC	Trouble Code Title, Conditions & Possible Causes
DTC: P0120 **1T ECM, MIL: Yes** **Year:** 2009, 2010 **Model:** tC, xB, xD **Engine:** 1.8L L4 VIN U, 2.4L L4 **Transmission:** All	**Throttle / Pedal Position Sensor / Switch "A" Circuit Malfunction:** Monitor runs whenever following DTCs are not present: None Either of the following conditions A or B is met: A. Engine switch on (IG): 0.012 seconds or more B. Electronic throttle actuator power: ON **Possible Causes:** • Throttle position sensor (built into throttle body) • ECM
DTC: P0121 **1T ECM, MIL: Yes** **Year:** 2009, 2010 **Model:** tC, xB, xD **Engine:** 1.8L L4 VIN U, 2.4L L4 **Transmission:** All	**Throttle / Pedal Position Sensor / Switch "A" Circuit Range / Performance Problem:** Monitor runs whenever following DTCs are not present: None Either of the following conditions A or B is met: A. Engine switch: On (IG) B. Electric throttle motor power: ON Throttle position sensor malfunction (P0120, P0122, P0123, P0220, P0222, P0223, P2135): not detected **Possible Causes:** • Throttle position sensor (built into throttle body) • Throttle position sensor circuit • ECM
DTC: P0122 **1T ECM, MIL: Yes** **Year:** 2009, 2010 **Model:** tC, xB, xD **Engine:** 1.8L L4 VIN U, 2.4L L4 **Transmission:** All	**Throttle / Pedal Position Sensor / Switch "A" Circuit Low Input:** Monitor runs whenever following DTCs are not present: None Either of the following conditions A or B is met: A. Engine switch on (IG): 0.012 seconds or more B. Electronic throttle actuator power: ON **Possible Causes:** • Throttle position sensor (built into throttle body) • Short in VTA1 circuit • Open in VC circuit • ECM
DTC: P0123 **1T ECM, MIL: Yes** **Year:** 2009, 2010 **Model:** tC, xB, xD **Engine:** 1.8L L4 VIN U, 2.4L L4 **Transmission:** All	**Throttle / Pedal Position Sensor / Switch "A" Circuit High Input:** Monitor runs whenever following DTCs are not present: None Either of the following conditions A or B is met: A. Engine switch on (IG): 0.012 seconds or more B. Electronic throttle actuator power: ON **Possible Causes:** • Throttle position sensor (built into throttle body) • Open in VTA1 circuit • Open in E2 circuit • Short between VC and VTA1 circuits • ECM
DTC: P0125 **T ECM, MIL: Yes** **Year:** 2009, 2010 **Model:** tC, xB, xD **Engine:** 1.8L L4 VIN U, 2.4L L4 **Transmission:** All	**Insufficient Coolant Temperature for Closed Loop Fuel Control:** Monitor runs whenever following DTCs are not present: MAF sensor circuit fail (P0102, P0103) IAT sensor circuit fail (P0112, P0113) ECT sensor circuit fail (P0115, P0117, P0118) Thermostat fail (P0128) **Possible Causes:** • Engine coolant temperature sensor • Cooling system • Thermostat

DTC	Trouble Code Title, Conditions & Possible Causes
DTC: P0128 **2T ECM, MIL: Yes** **Year:** 2009, 2010 **Model:** tC, xB, xD **Engine:** 1.8L L4 VIN U, 2.4L L4 **Transmission:** All	**Coolant Thermostat (Coolant Temperature Below Thermostat Regulating Temperature):** Battery voltage: 11V or more Either following condition 1 or 2 is met: 1. All of the following conditions are met: *ECT at engine start - IAT at engine start: 5-44.6°F (-15-7°C) *ECT at engine start: 14-132.8°F (-10-56°C) *IAT at engine start: 14-132.8°F (-10-56°C) 2. All of following conditions are met: *ECT at engine start - IAT at engine start: More than 44.6°F (7°C) *ECT at engine start: 132.8°F (56°C) or less *IAT at engine start: 14°F (-10°C) or more Accumulated time that vehicle speed is 80 mph (128 km/h) or more speed: Less than 20 seconds **Possible Causes:** • Thermostat • Cooling system • ECT sensor • ECM
DTC: P0136 **2T ECM, MIL: Yes** **Year:** 2009, 2010 **Model:** tC, xB, xD **Engine:** 1.8L L4 VIN U, 2.4L L4 **Transmission:** All	**Oxygen Sensor Circuit Malfunction (Bank 1 Sensor 2):** Active A/F control: Performing Battery voltage: 11 V or more ECT: 167°F (75°C) or more Idle: OFF Engine rpm: Less than 3200 rpm A/F sensor status: Activatted Fuel system status: Closed loop Fuel cut: OFF Engineload: 10 to 70% Shift position: 3rd or more Estimated rear HO2S temperature: Less than 1292°F (700°C) ECM monitor: Completed P0607: Not set **Possible Causes:** • HO2 sensor (sensor 2) • HO2 sensor heater (sensor 2) • Air-Fuel Ratio (A/F) sensor (sensor 1) • Gas leakage from exhaust system • Fuel pressure • PCV valve and hose • Air induction system

DTC	Trouble Code Title, Conditions & Possible Causes
DTC: P0137 **2T ECM, MIL: Yes** **Year:** 2009, 2010 **Model:** tC, xB, xD **Engine:** 1.8L L4 VIN U, 2.4L L4 **Transmission:** All	**Oxygen Sensor Circuit Low Voltage (Bank 1 Sensor 2):** Active A/F control: Performing Battery voltage: 11 V or more ECT: 167°F (75°C) or more Idle: OFF Engine rpm: Less than 3200 rpm A/F sensor status: Activatted Fuel system status: Closed loop Fuel cut: OFF Engineload: 10 to 70% Shift position: 4rd or more Battery voltage: 11 V or more Estimated rear HO2S temperature: 842-1382°F (450-750°C) P0607: Not preset **Possible Causes:** • Open in HO2 sensor (sensor 2) circuit • HO2 sensor (sensor 2) • HO2 sensor heater (sensor 2) • Air-Fuel Ratio (A/F) sensor (sensor 1) • Gas leakage from exhaust system
DTC: P0138 **2T ECM, MIL: Yes** **Year:** 2009, 2010 **Model:** tC, xB, xD **Engine:** 1.8L L4 VIN U, 2.4L L4 **Transmission:** All	**Oxygen Sensor Circuit High Voltage (Bank 1 Sensor 2):** Active A/F control: Performing Battery voltage: 11 V or more ECT: 167°F (75°C) or more Idle: OFF Engine rpm: Less than 3200 rpm A/F sensor status: Activatted Fuel system status: Closed loop Fuel cut: OFF Engineload: 10 to 70% Shift position: 4rd or more Battery voltage: 11 V or more Time after engine start: 2 seconds or more **Possible Causes:** • Short in HO2 sensor (sensor 2) circuit • HO2 sensor (sensor 2) • ECM internal circuit malfunction • A/F sensor (sensor 1)
DTC: P0139 **2T ECM, MIL: Yes** **Year:** 2009, 2010 **Model:** tC, xB, xD **Engine:** 1.8L L4 VIN U, 2.4L L4 **Transmission:** All	**Oxygen Sensor Circuit Slow Response (Bank 1 Sensor 2):** ECT: 167°F (75°C) or more Estimated catalyst temperature: 752°F (400°C) or more Fuel cut: ON **Possible Causes:** • Short in HO2 sensor (sensor 2) • HO2 sensor (sensor 2) • ECM internal circuit malfunction
DTC: P0141 **2T ECM, MIL: Yes** **Year:** 2009, 2010 **Model:** tC, xB, xD **Engine:** 1.8L L4 VIN U, 2.4L L4 **Transmission:** All	**Oxygen Sensor Heater Circuit Malfunction (Bank 1 Sensor 2):** One of the following conditions is met: Condition A or B A. All of the following conditions are met: Conditions 1, 2, 3, 4 and 5 1. Battery voltage: 10.5 V or more 2. Fuel cut: OFF 3. Time after fuel cut ON to OFF: 30 seconds or more 4. Accumulated heater ON time: 100 seconds or more 5. Learned heater OFF current operation: Completed B. Duration that rear heated oxygen sensor impedance is less than 15 kΩ: 2 seconds or more **Possible Causes:** • Open or short in HO2 sensor heater circuit • HO2 sensor heater (sensor 2) • EFI relay • ECM

DTC	Trouble Code Title, Conditions & Possible Causes
DTC: P0171 **2T ECM, MIL: Yes** **Year:** 2009, 2010 **Model:** tC, xB, xD **Engine:** 1.8L L4 VIN U, 2.4L L4 **Transmission:** All	**System Too Lean (Bank 1):** Monitor runs whenever following DTCs are not present: P0010, P0020 (OCV bank 1, 2) P0011, P0021 (VVT system bank 1, 2 - advance) P0012, P0022 (VVT system bank 1, 2 - retard) P0013, P0023 (Exhaust OCV bank 1, 2) P0014, P0024 (Exhaust VVT system bank 1, 2 - advance) P0015, P0025 (Exhaust VVT system bank 1, 2 - retard) P0016, P0018 (VVT system bank 1, 2 - misalignment) P0017, P0019 (Exhaust VVT system bank 1, 2 - misalignment) P0031, P0032, P0051, P0052 (A/F sensor heater) P0102, P0103 (MAF meter) P0115, P0117, P0118 (ECT sensor) P0120, P0121, P0122, P0123, P0220, P0222, P0223, P2135 (TP sensor) P0125 (Insufficient ECT for closed loop) P0335 (CKP sensor) P0340, P0342, P0343, P0345, P0347, P0348 (VVT sensor) P0351, P0352, P0353, P0354, P0355, P0356 (Igniter) P0365, P0367, P0368, P0390, P0392, P0393 (Exhaust VVT sensor) P0500 (Vehicle speed sensor) Fuel system status: Closed loop Battery voltage: 11 V or more Either following condition is met: 1. Engine RPM: less than 1100 rpm 2. Intake air amount per revolution: 0.22 g/rev or more Catalyst monitor: Not executed **Possible Causes:** • Air induction system • Injector blockage • MAF meter • ECT sensor • Fuel pressure • Gas leakage from exhaust system • Open or short in A/F sensor (bank sensor 1) circuit • A/F sensor (bank 1 sensor 1) • A/F sensor heater (bank 1 sensor 1) • A/F sensor heater relay • A/F sensor heater and A/F sensor heater relay circuits • PCV valve and hose • PCV hose connections • ECM

DTC	Trouble Code Title, Conditions & Possible Causes
DTC: P0172 **2T ECM, MIL: Yes** **Year:** 2009, 2010 **Model:** tC, xB, xD **Engine:** 1.8L L4 VIN U, 2.4L L4 **Transmission:** All	**System Too Rich (Bank 1):** Monitor runs whenever following DTCs are not present: P0010, P0020 (OCV bank 1, 2) P0011, P0021 (VVT system bank 1, 2 - advance) P0012, P0022 (VVT system bank 1, 2 - retard) P0013, P0023 (Exhaust OCV bank 1, 2) P0014, P0024 (Exhaust VVT system bank 1, 2 - advance) P0015, P0025 (Exhaust VVT system bank 1, 2 - retard) P0016, P0018 (VVT system bank 1, 2 - misalignment) P0017, P0019 (Exhaust VVT system bank 1, 2 - misalignment) P0031, P0032, P0051, P0052 (A/F sensor heater) P0102, P0103 (MAF meter) P0115, P0117, P0118 (ECT sensor) P0120, P0121, P0122, P0123, P0220, P0222, P0223, P2135 (TP sensor) P0125 (Insufficient ECT for closed loop) P0335 (CKP sensor) P0340, P0342, P0343, P0345, P0347, P0348 (VVT sensor) P0351, P0352, P0353, P0354, P0355, P0356 (Igniter) P0365, P0367, P0368, P0390, P0392, P0393 (Exhaust VVT sensor) P0500 (Vehicle speed sensor) Fuel system status: Closed loop Battery voltage: 11 V or more Either following condition is met: 1. Engine RPM: less than 1100 rpm 2. Intake air amount per revolution: 0.22 g/rev or more Catalyst monitor: Not executed **Possible Causes:** • Injector leakage or blockage • MAF meter • ECT sensor • Ignition system • Fuel pressure • Gas leakage from exhaust system • Open or short in A/F sensor (sensor 1) circuit • A/F sensor (sensor 1) • A/F sensor heater (sensor 1) • A/F sensor heater relay • A/F sensor heater and A/F sensor heater relay circuits • ECM
DTC: P0220 **1T ECM, MIL: Yes** **Year:** 2009, 2010 **Model:** tC, xB, xD **Engine:** 1.8L L4 VIN U, 2.4L L4 **Transmission:** All	**Throttle / Pedal Position Sensor / Switch "B" Circuit:** Monitor runs whenever following DTCs are not present: None Either of the following conditions A or B is met: A. Engine switch on (IG): 0.012 seconds or more B. Electronic throttle actuator power: ON **Possible Causes:** • Throttle position sensor (built into throttle body) • ECM
DTC: P0222 **1T ECM, MIL: Yes** **Year:** 2009, 2010 **Model:** tC, xB **Engine:** 2.4L L4 **Transmission:** All	**Throttle / Pedal Position Sensor / Switch "B" Circuit Low Input:** Monitor runs whenever following DTCs are not present: None Either of the following conditions A or B is met: A. Engine switch on (IG): 0.012 seconds or more B. Electronic throttle actuator power: ON **Possible Causes:** • Throttle position sensor (built into throttle body) • Short in VTA2 circuit • Open in VC circuit • ECM

DTC	Trouble Code Title, Conditions & Possible Causes
DTC: P0223 **1T ECM, MIL: Yes** **Year:** 2009, 2010 **Model:** tC, xB, xD **Engine:** 1.8L L4 VIN U, 2.4L L4 **Transmission:** All	**Throttle / Pedal Position Sensor / Switch "B" Circuit High Input:** Monitor runs whenever following DTCs are not present: None Either of the following conditions A or B is met: A. Engine switch on (IG): 0.012 seconds or more B. Electronic throttle actuator power: ON **Possible Causes:** • Throttle position sensor (built into throttle body) • Open in VTA2 circuit • Open in E2 circuit • Short between VC and VTA2 circuits • ECM
DTC: P0300 **2T ECM, MIL: Yes** **Year:** 2009, 2010 **Model:** tC, xB, xD **Engine:** 1.8L L4 VIN U, 2.4L L4 **Transmission:** All	**Random / Multiple Cylinder Misfire Detected:** Battery voltage: 8 V or more VVT system: Not operated by scan tool Engine RPM: 450-6500 rpm Either of the following conditions (a) or (b) is met: (a) Engine coolant temperature at engine start: More than 19.4°F (-7°C) (b) Engine coolant temperature: More than 68°F (20°C) Fuel cut: OFF Monitor period of emission-related misfire: First 1000 revolutions after engine start, or Check Mode: Crankshaft 1000 revolutions Except above: Crankshaft 1000 revolutions X 4 Monitor period of catalyst-damaged misfire (MIL blinks): All of the following conditions 1, 2 and 3 are met: Crankshaft 200 revolutions 1. Driving cycles: 1st 2. Check mode: OFF 3. Enine RPM: Less than 2300 rpm Except above: Crankshaft 200 revolutions X 3 **Possible Causes:** • Open or short in engine wire harness • Connector connections • Vacuum hose connections • Ignition system • Injector • Fuel pressure • MAF meter • ECT sensor • Compression pressure • Valve timing • PCV valve and hose • PCV hose connections • Air induction system • ECM

DTC	Trouble Code Title, Conditions & Possible Causes
DTC: P0301 **2T ECM, MIL: Yes** **Year:** 2009, 2010 **Model:** tC, xB, xD **Engine:** 1.8L L4 VIN U, 2.4L L4 **Transmission:** All	**Cylinder 1 Misfire Detected:** Battery voltage: 8 V or more VVT system: Not operated by scan tool Engine RPM: 450-6500 rpm Either of the following conditions (a) or (b) is met: (a) Engine coolant temperature at engine start: More than 19.4°F (-7°C) (b) Engine coolant temperature: More than 68°F (20°C) Fuel cut: OFF Monitor period of emission-related misfire: First 1000 revolutions after engine start, or Check Mode: Crankshaft 1000 revolutions Except above: Crankshaft 1000 revolutions X 4 Monitor period of catalyst-damaged misfire (MIL blinks): All of the following conditions 1, 2 and 3 are met: Crankshaft 200 revolutions 1. Driving cycles: 1st 2. Check mode: OFF 3. Enine RPM: Less than 2300 rpm Except above: Crankshaft 200 revolutions X 3 **Possible Causes:** • Open or short in engine wire harness • Connector connections • Vacuum hose connections • Ignition system • Injector • Fuel pressure • MAF meter • ECT sensor • Compression pressure • Valve timing • PCV valve and hose • PCV hose connections • Air induction system • ECM

DTC	Trouble Code Title, Conditions & Possible Causes
DTC: P0302 **2T ECM, MIL: Yes** **Year:** 2009, 2010 **Model:** tC, xB, xD **Engine:** 1.8L L4 VIN U, 2.4L L4 **Transmission:** All	**Cylinder 2 Misfire Detected:** Battery voltage: 8 V or more VVT system: Not operated by scan tool Engine RPM: 450-6500 rpm Either of the following conditions (a) or (b) is met: (a) Engine coolant temperature at engine start: More than 19.4°F (-7°C) (b) Engine coolant temperature: More than 68°F (20°C) Fuel cut: OFF Monitor period of emission-related misfire: First 1000 revolutions after engine start, or Check Mode: Crankshaft 1000 revolutions Except above: Crankshaft 1000 revolutions X 4 Monitor period of catalyst-damaged misfire (MIL blinks): All of the following conditions 1, 2 and 3 are met: Crankshaft 200 revolutions 1. Driving cycles: 1st 2. Check mode: OFF 3. Enine RPM: Less than 2300 rpm Except above: Crankshaft 200 revolutions X 3 **Possible Causes:** • Open or short in engine wire harness • Connector connections • Vacuum hose connections • Ignition system • Injector • Fuel pressure • MAF meter • ECT sensor • Compression pressure • Valve timing • PCV valve and hose • PCV hose connections • Air induction system • ECM

DTC	Trouble Code Title, Conditions & Possible Causes
DTC: P0303 **2T ECM, MIL: Yes** **Year:** 2009, 2010 **Model:** tC, xB, xD **Engine:** 1.8L L4 VIN U, 2.4L L4 **Transmission:** All	**Cylinder 3 Misfire Detected:** Battery voltage: 8 V or more VVT system: Not operated by scan tool Engine RPM: 450-6500 rpm Either of the following conditions (a) or (b) is met: (a) Engine coolant temperature at engine start: More than 19.4°F (-7°C) (b) Engine coolant temperature: More than 68°F (20°C) Fuel cut: OFF Monitor period of emission-related misfire: First 1000 revolutions after engine start, or Check Mode: Crankshaft 1000 revolutions Except above: Crankshaft 1000 revolutions X 4 Monitor period of catalyst-damaged misfire (MIL blinks): All of the following conditions 1, 2 and 3 are met: Crankshaft 200 revolutions 1. Driving cycles: 1st 2. Check mode: OFF 3. Enine RPM: Less than 2300 rpm Except above: Crankshaft 200 revolutions X 3 **Possible Causes:** • Open or short in engine wire harness • Connector connections • Vacuum hose connections • Ignition system • Injector • Fuel pressure • MAF meter • ECT sensor • Compression pressure • Valve timing • PCV valve and hose • PCV hose connections • Air induction system • ECM

DTC	Trouble Code Title, Conditions & Possible Causes
DTC: P0304 **2T ECM, MIL: Yes** **Year:** 2009, 2010 **Model:** tC, xB, xD **Engine:** 1.8L L4 VIN U, 2.4L L4 **Transmission:** All	**Cylinder 4 Misfire Detected:** Battery voltage: 8 V or more VVT system: Not operated by scan tool Engine RPM: 450-6500 rpm Either of the following conditions (a) or (b) is met: (a) Engine coolant temperature at engine start: More than 19.4°F (-7°C) (b) Engine coolant temperature: More than 68°F (20°C) Fuel cut: OFF Monitor period of emission-related misfire: First 1000 revolutions after engine start, or Check Mode: Crankshaft 1000 revolutions Except above: Crankshaft 1000 revolutions X 4 Monitor period of catalyst-damaged misfire (MIL blinks): All of the following conditions 1, 2 and 3 are met: Crankshaft 200 revolutions 1. Driving cycles: 1st 2. Check mode: OFF 3. Enine RPM: Less than 2300 rpm Except above: Crankshaft 200 revolutions X 3 **Possible Causes:** • Open or short in engine wire harness • Connector connections • Vacuum hose connections • Ignition system • Injector • Fuel pressure • MAF meter • ECT sensor • Compression pressure • Valve timing • PCV valve and hose • PCV hose connections • Air induction system • ECM
DTC: P0327 **1T ECM, MIL: Yes** **Year:** 2009, 2010 **Model:** tC, xB, xD **Engine:** 1.8L L4 VIN U, 2.4L L4 **Transmission:** All	**Knock Sensor 1 Circuit Low Input (Bank 1 or Single Sensor):** Monitor runs whenever following DTCs are not present: None Battery voltage: 10.5 V or more Time after engine start: 5 seconds or more Engine switch: On (IG) Starter: OFF **Possible Causes:** • Short in knock sensor circuit • Knock sensor • ECM
DTC: P0328 **1T ECM, MIL: Yes** **Year:** 2009, 2010 **Model:** tC, xB, xD **Engine:** 1.8L L4 VIN U, 2.4L L4 **Transmission:** All	**Knock Sensor 1 Circuit High Input (Bank 1 or Single Sensor):** Monitor runs whenever following DTCs are not present: None Battery voltage: 10.5 V or more Time after engine start: 5 seconds or more Engine switch: On (IG) Starter: OFF **Possible Causes:** • Open in knock sensor circuit • Knock sensor • ECM

DTC	Trouble Code Title, Conditions & Possible Causes
DTC: P0335 **1T ECM, MIL: Yes** **Year:** 2009, 2010 **Model:** tC, xB, xD **Engine:** 1.8L L4 VIN U, 2.4L L4 **Transmission:** All	**Crankshaft Position Sensor "A" Circuit:** Monitor runs whenever following DTCs are not present: None CKP Sensor Range Check/Rationality: Time after starter OFF to ON: 3 seconds or more Battery voltage: 7 V or more Minimum battery voltage while starter ON: Less than 11 V Number of VVT sensor signal pulse: 6 times Camshaft position sensor circuit fail (P0340,P0342,P0343): Not detected CKP Sensor Verify Pulse Input (Case 1): Engine speed: 600 rpm or less Starter: OFF Time after starter ON to OFF: 3 seconds or more CKP Sensor Verify Pulse Input (Case 2): Starter: ON Minimum battery voltage starter ON: Below 11 V **Possible Causes:** • Open or short in CKP sensor circuit • CKP sensor • CKP sensor plate • ECM
DTC: P0339 **1T ECM, MIL: Yes** **Year:** 2009, 2010 **Model:** tC, xB, xD **Engine:** 1.8L L4 VIN U, 2.4L L4 **Transmission:** All	**Crankshaft Position Sensor "A" Circuit Intermittent:** Monitor runs whenever following conditions are met: Engine speed 1,000 rpm or more. No Crankshaft Position (CKP) sensor signal for 0.05 seconds or more. 3 seconds or more have elapsed since starter signal switched from ON to OFF. **Possible Causes:** • Open or short in CKP sensor circuit • CKP sensor • CKP sensor plate • ECM
DTC: P0340 **2T ECM, MIL: Yes** **Year:** 2009, 2010 **Model:** tC, xB, xD **Engine:** 1.8L L4 VIN U, 2.4L L4 **Transmission:** All	**Camshaft Position Sensor "A" Circuit (Bank 1 or Single Sensor):** Monitor runs whenever following DTCs are not present: None Camshaft Position Sensor Range Check: Stater: ON Minimal battery voltage while starter ON: Less than 11 V Camshaft Position/Crankshaft Position Misalignment: Engine speed: 600 rpm or more Starter: OFF **Possible Causes:** • Open or short in CMP sensor circuit • CMP sensor • Camshaft • Jumped tooth of timing chain • ECM
DTC: P0342 **1T ECM, MIL: Yes** **Year:** 2009, 2010 **Model:** xD **Engine:** 1.8L L4 VIN U **Transmission:** All	**Camshaft Position Sensor "A" Circuit Low Input (Bank 1 or Single Sensor):** Starter OFF, ignition switch ON. Time after ignition switch OFF to ON 2 seconds. The output voltage of CMP sensor is less than 0.3 V for 4 seconds. **Possible Causes:** • Open or short in CMP sensor circuit for intake camshaft • CMP sensor for intake camshaft • Camshaft timing gear for intake camshaft • Jumped tooth of timing chain for intake camshaft • ECM

DTC	Trouble Code Title, Conditions & Possible Causes
DTC: P0343 **1T ECM, MIL: Yes** **Year:** 2009, 2010 **Model:** xD **Engine:** 1.8L L4 VIN U **Transmission:** All	**Camshaft Position Sensor "A" Circuit High Input (Bank 1 or Single Sensor):** Starter OFF, ignition switch ON. Time after ignition switch OFF to ON 2 seconds. The output voltage of the CMP sensor is more than 4.7 V for 4 seconds. **Possible Causes:** • Open or short in CMP sensor circuit for intake camshaft • CMP sensor for intake camshaft • Camshaft timing gear for intake camshaft • Jumped tooth of timing chain for intake camshaft • ECM
DTC: P0351 **1T ECM, MIL: Yes** **Year:** 2009, 2010 **Model:** xB **Engine:** 2.4L L4 **Transmission:** All	**Ignition Coil "A" Primary / Secondary Circuit:** No IGF signal to ECM while engine running . **Possible Causes:** • Ignition system • Open or short in IGF1 or IGT circuit (1 to 4) between ignition coil and ECM • No. 1 to No. 4 ignition coils • ECM
DTC: P0351 **1T ECM, MIL: Yes** **Year:** 2009, 2010 **Model:** xD **Engine:** 1.8L L4 VIN U **Transmission:** All	**Ignition Coil "A" Primary / Secondary Circuit:** No IGF signal to ECM while engine is running, the ECM does not receive any IGF signal despite ECM sending IGT signal to igniter. **Possible Causes:** • Ignition system • Open or short in IGF1 or IGT circuit (1 to 4) between ignition coil and ECM • No. 1 to No. 4 ignition coils • ECM
DTC: P0352 **1T ECM, MIL: Yes** **Year:** 2009, 2010 **Model:** tC, xB, xD **Engine:** 1.8L L4 VIN U, 2.4L L4 **Transmission:** All	**Ignition Coil "B" Primary / Secondary Circuit:** Monitor runs whenever following DTCs are not present: None Either of the following condition A or B is met: A. Engine RPM: 1500 rpm or less B. Starter: OFF Either of the following condition C or D is met: C. Both of the following conditions are met: (a) Engine speed: 500 rpm or less (b) Battery voltage: 6 V or more D. All of the following conditions are met: (a) Engine speed: More than 500 rpm (b) Battery voltage: 10 V or more (c) Number of sparks after CPU reset: 5 sparks or more **Possible Causes:** • Ignition system • Open or short in IGF1 or IGT2 circuit (1 to 4) or (1 to 6) between ignition coil and ECM • No. 1 to No. 4 ignition coils or No. 1 to No. 6 ignition coils • ECM

DTC	Trouble Code Title, Conditions & Possible Causes
DTC: P0353 **1T ECM, MIL: Yes** **Year:** 2009, 2010 **Model:** tC, xB, xD **Engine:** 1.8L L4 VIN U, 2.4L L4 **Transmission:** All	**Ignition Coil "C" Primary / Secondary Circuit:** Monitor runs whenever following DTCs are not present: None Either of the following condition A or B is met: A. Engine RPM: 1500 rpm or less B. Starter: OFF Either of the following condition C or D is met: C. Both of the following conditions are met: (a) Engine speed: 500 rpm or less (b) Battery voltage: 6 V or more D. All of the following conditions are met: (a) Engine speed: More than 500 rpm (b) Battery voltage: 10 V or more (c) Number of sparks after CPU reset: 5 sparks or more **Possible Causes:** • Ignition system • Open or short in IGF1 or IGT3 circuit (1 to 4) or (1 to 6) between ignition coil and ECM • No. 1 to No. 4 ignition coils or No. 1 to No. 6 ignition coils • ECM
DTC: P0354 **1T ECM, MIL: Yes** **Year:** 2009, 2010 **Model:** tC, xB, xD **Engine:** 1.8L L4 VIN U, 2.4L L4 **Transmission:** All	**Ignition Coil "D" Primary / Secondary Circuit:** Monitor runs whenever following DTCs are not present: None Either of the following condition A or B is met: A. Engine RPM: 1500 rpm or less B. Starter: OFF Either of the following condition C or D is met: C. Both of the following conditions are met: (a) Engine speed: 500 rpm or less (b) Battery voltage: 6 V or more D. All of the following conditions are met: (a) Engine speed: More than 500 rpm (b) Battery voltage: 10 V or more (c) Number of sparks after CPU reset: 5 sparks or more **Possible Causes:** • Ignition system • Open or short in IGF1 or IGT4 circuit (1 to 4) or (1 to 6) between ignition coil and ECM • No. 1 to No. 4 ignition coils or No. 1 to No. 6 ignition coils • ECM
DTC: P0365 **1T ECM, MIL: Yes** **Year:** 2009, 2010 **Model:** xD **Engine:** 1.8L L4 VIN U **Transmission:** All	**Camshaft Position Sensor "B" Circuit (Bank 1):** When one of following conditions is met: Misfiring exhaust camshaft position sensor signal signal for 5 seconds at engine speed of 600 rpm or more (1 trip detection logic). Input voltage to ECM remains less than 0.3 V, or more than 4.7 V for 4 seconds when 2 or more seconds have elapsed after turning ignition switch ON (1 trip detection logic). **Possible Causes:** • Open or short in exhaust camshaft position sensor circuit • Exhaust camshaft position sensor • Camshaft timing gear for exhaust camshaft • Jumped tooth of timing chain for exhaust camshaft • ECM

DTC	Trouble Code Title, Conditions & Possible Causes
DTC: P0367 **1T ECM, MIL: Yes** **Year:** 2009, 2010 **Model:** xD **Engine:** 1.8L L4 VIN U **Transmission:** All	**Camshaft Position Sensor "B" Circuit Low Input (Bank 1):** Starter OFF, ignition switch ON. Time after ignition switch OFF to ON 2 seconds. The output voltage of exhaust camshaft position sensor is less than 0.3 V for 4 seconds. **Possible Causes:** • Open or short in exhaust camshaft position sensor circuit • Exhaust camshaft position sensor • Camshaft timing gear for exhaust camshaft • Jumped tooth of timing chain for exhaust camshaft • ECM
DTC: P0368 **1T ECM, MIL: Yes** **Year:** 2009, 2010 **Model:** xD **Engine:** 1.8L L4 VIN U **Transmission:** All	**Camshaft Position Sensor "B" Circuit High Input (Bank 1):** Starter OFF, ignition switch ON. Time after ignition switch OFF to ON 2 seconds. The output voltage of exhaust camshaft position sensor is more than 4.7 V for 4 seconds. **Possible Causes:** • Open or short in exhaust camshaft position sensor circuit • Exhaust camshaft position sensor • Camshaft timing gear for exhaust camshaft • Jumped tooth of timing chain for exhaust camshaft • ECM

DTC	Trouble Code Title, Conditions & Possible Causes
DTC: P0420 **2T ECM, MIL: Yes** **Year:** 2009, 2010 **Model:** tC, xB, xD **Engine:** 1.8L L4 VIN U, 2.4L L4 **Transmission:** All	**Catalyst System Efficiency Below Threshold (Bank 1):** Monitor runs whenever following DTCs are not present: P0010, P0020 (OCV bank 1, 2) P0011, P0021 (VVT system bank 1, 2 - advance) P0012, P0022 (VVT system bank 1, 2 - retard) P0013, P0023 (Exhaust OCV bank 1, 2) P0014, P0024 (Exhaust VVT system bank 1, 2 - advance) P0015, P0025 (Exhaust VVT system bank 1, 2 - retard) P0016, P0018 (VVT system bank 1, 2 - misalignment) P0017, P0019 (Exhaust VVT system bank 1, 2 - misalignment) P0031, P0032, P0051, P0052 (A/F sensor heater) P0037, P0038, P0057, P0058 (HO2 sensor heater) P0102, P0103 (MAF meter) P0115, P0117, P0118 (ECT sensor) P0120, P0121, P0122, P0123, P0220, P0222, P0223, P2135 (TP sensor) P0125 (Insufficient ECT for closed loop) P0136, P0137, P0138, P0139 (HO2 sensor bank 1) P014C, P014D, P014E, P014F (Air fuel ratio sensor - slow response) P015A, P015B, P015C, P015D (Air fuel ratio sensor - delayed response) P0156, P0157, P0158, P0159 (HO2 sensor bank 2) P0171, P0172, P0174, P0175 (Fuel system) P0301, P0302, P0303, P0304, P0305, P0306 (Misfire) P0335 (CKP sensor) P0340, P0342, P0343, P0345, P0347, P0348 (VVT sensor) P0351, P0352, P0353, P0354, P0355, P0356 (Igniter) P0365, P0367, P0368, P0390, P0392, P0393 (Exhaust VVT sensor) P0500 (Vehicle speed sensor) P0607 (HO2 sensor - sensor 2) P2195, P2196, P2197, P2198 (A/F sensor - rationality) P2237, P2240 (A./F sensor - open) P2238, P2241, P2252, P2255 (A/F sensor - low impedance) P2239, P2242, P2253, P2256 (A/F sensor - high impedance) Battery voltage: 11 V or more IAT: 14°F (-10°C) or more ECT: 167°F (75°C) or more Atmospheric pressure: 570 mmHg (76 kPa) or more Idle: OFF Enigne RPM: Less than 3200 rpm A/F Sensor: Activated Fuel system status: Closed loop Engine load: 10-70% All of the following conditions 1, 2 and 3 are met: 1. MAF: 5-60 g/sec. 2. Front catalyst tempperature (estimated): 1112-1472°F (600-800°C) 3. Rear catalyst temperature (estimated): 212-1652°F (100-900°C) Shift position: 4rd or higher **Possible Causes:** • Gas leakage from exhaust system • A/F sensor (bank 1 sensor 1) • HO2 sensor (bank 1 sensor 2) • Exhaust manifold sub-assembly RH (TWC: Front catalyst) • Front exhaust pipe assembly (TWC: Rear catalyst)

DTC	Trouble Code Title, Conditions & Possible Causes
DTC: P043E **1T ECM, MIL: Yes** **Year:** 2009, 2010 **Model:** tC, xB, xD **Engine:** 1.8L L4 VIN U, 2.4L L4 **Transmission:** All	**Evaporative Emission System Reference Orifice Clog Up:** Reference orifice clogged. Reference orifice high-flow. Leak detection pump OFF malfunction. Leak detection pump ON malfunction. Vent valve ON (close) malfunction. **Possible Causes:** • Canister pump module • Connector/wire harness (Canister pump module - ECM) • ECM
DTC: P043F **1T ECM, MIL: Yes** **Year:** 2009, 2010 **Model:** tC, xB, xD **Engine:** 1.8L L4 VIN U, 2.4L L4 **Transmission:** All	**Evaporative Emission System Reference Orifice High Flow:** Reference orifice clogged. Reference orifice high-flow. Leak detection pump OFF malfunction. Leak detection pump ON malfunction. Vent valve ON (close) malfunction. **Possible Causes:** • Canister pump module • Connector/wire harness (Canister pump module - ECM) • ECM
DTC: P0441 **2T ECM, MIL: Yes** **Year:** 2009, 2010 **Model:** tC, xB, xD **Engine:** 1.8L L4 VIN U, 2.4L L4 **Transmission:** All	**Evaporative Emission Control System Incorrect Purge Flow:** Purge VSV (Vacuum Switching Valve) stuck open: Leak detection pump creates negative pressure (vacuum) in EVAP system and EVAP system pressure is measured. 0.02 inch leak pressure standard is measured at start and at end of leak check. If stabilized pressure is higher than [second 0.02 inch leak pressure standard x 0.2], ECM determines that Purge VSV is stuck open. Purge VSV stuck closed: After EVAP leak check is performed, Purge VSV is turned ON (open), and atmospheric air is introduced into EVAP system. 0.02 inch leak pressure standard is measured at start and at end of the check. If pressure does not return to near atmospheric pressure, ECM determines that purge valve is stuck closed. Purge flow: While engine is running, following conditions are successively met: * Negative pressure is not created in EVAP system when Purge VSV is turned ON (open) * EVAP system pressure change is less than 0.5 kPa (3.75 mmHg) when vent valve is turned ON (closed) * Atmospheric pressure change before and after purge flow monitor is less than 0.1 kPa (0.75 mmHg) **Possible Causes:** • Purge VSV (Vacuum Switching Valve) stuck open: • Purge VSV • Connector/wire harness (Purge VSV - ECM) • ECM • Canister pump module • Leakage from EVAP system • Purge VSV stuck closed: • Purge VSV • Connector/wire harness (Purge VSV - ECM) • ECM • Canister pump module • Leakage from EVAP system • Purge flow: • Purge VSV • Connector/wire harness (Purge VSV - ECM) • Leakage from EVAP line (Purge VSV - Intake manifold) • ECM

DTC	Trouble Code Title, Conditions & Possible Causes
DTC: P0450 **1T , MIL: Yes** **Year:** 2009, 2010 **Model:** tC **Engine:** 2.4L L4 **Transmission:** All	**Evaporative Emission Control System Pressure Sensor / Switch:** Monitor runs whenever following DTCs are not present: None Either of following conditions is met: (a) or (b) (a) Ignition switch: ON (b) Soak timer: ON Battery voltage: 8 V or more Starter: OFF **Possible Causes:** • Canister pump module • EVAP system hose (pipe from air inlet port to canister pump module, canister filter, fuel tank vent hose) • ECM
DTC: P0451 **2T ECM, MIL: Yes** **Year:** 2009, 2010 **Model:** tC, xB, xD **Engine:** 1.8L L4 VIN U, 2.4L L4 **Transmission:** All	**Evaporative Emission Control System Pressure Sensor Range / Performance:** Canister pressure sensor noise: Atmospheric pressure: 70 kPa (525 mmHg) to 110 kPa (825 mmHg) Battery voltage: 10.5 V or more Intake air temperature: 40 to 95°F (4.4 to 35°C) EVAP pressure sensor malfunction (P0452, P0453): Not detected Either of following conditions is met: 1. Engine: Running 2. Time after key off: 5, 7 or 9.5 hours **Possible Causes:** • Canister pressure sensor noising: • Canister pump module • Connector/wire harness (Canister pump module - ECM) • ECM • Canister pressure sensor stuck: • Canister pump module • Connector/wire harness (Canister pump module - ECM) • ECM
DTC: P0452 **1T ECM, MIL: Yes** **Year:** 2009, 2010 **Model:** xD **Engine:** 1.8L L4 VIN U **Transmission:** All	**Evaporative Emission Control System Pressure Sensor / Switch Low Input:** Ignition switch ON, EVAP monitoring (ignition switch OFF). The EVAP pressure is less than 42.1 kPa for 0.5 seconds. **Possible Causes:** • Canister pump module • Connector/wire harness (Canister pump module - ECM) • EVAP system hose (pipe from air inlet port to canister pump module, canister filter, fuel tank vent hose) • ECM
DTC: P0452 **1T ECM, MIL: Yes** **Year:** 2009, 2010 **Model:** xB **Engine:** 2.4L L4 **Transmission:** All	**Evaporative Emission Control System Pressure Sensor / Switch Low Input:** With the ignition switch ON, EVAP monitoring. (ignition switch OFF) The EVAP pressure is less than 42.1 kPa for 0.5 seconds. **Possible Causes:** • Canister pump module • Connector/wire harness (Canister pump module to ECM) • EVAP system hose (pipe from air inlet port to canister pump module, canister filter, fuel tank vent hose) • ECM
DTC: P0453 **1T ECM, MIL: Yes** **Year:** 2009, 2010 **Model:** tC, xB, xD **Engine:** 1.8L L4 VIN U, 2.4L L4 **Transmission:** All	**Evaporative Emission Control System Pressure Sensor / Switch High Input:** Monitor runs whenever following DTCs are not present: None Battery voltage: 8 V or more Starter: OFF Engine switch: ON **Possible Causes:** • Canister pump module • Connector/wire harness (Canister pump module - ECM) • ECM

DTC	Trouble Code Title, Conditions & Possible Causes
DTC: P0455 **2T ECM, MIL: Yes** **Year:** 2009, 2010 **Model:** tC, xB, xD **Engine:** 1.8L L4 VIN U, 2.4L L4 **Transmission:** All	**Evaporative Emission Control System Leak Detected (Gross Leak):** EVAP key-off monitor runs when all of following conditions are met: Atmospheric pressure: 70 to 110 kPa (525 to 825 mmHg) Battery voltage: 10.5 V or more Vehicle speed: Below 2.5 mph (4 km/h) Engine switch: OFF Time after key off: 5, 7 or 9.5 hours EVAP pressure sensor malfunction (P0451, P0452 and P0453): Not detected Purge VSV: Not operated by scan tool Vent valve: Not operated by scan tool Leak detection pump: Not operated by scan tool Both of following conditions are met before key OFF: Conditions 1 and 2 1. Duration that vehicle is being driven: 5 minutes or more 2. EVAP purge operation: Performed ECT: 4.4 to 35°C (40 to 95°F): P0456 4.4 to 50°C (40 to 122°F): P0455 IAT:4.4 to 35°C (40 to 95°F): P0456 4.4 to 50°C (40 to 122°F): P0455 **Possible Causes:** • Fuel tank cap (loose) • Leakage from EVAP line (Canister - Fuel tank) • Leakage from EVAP line (Purge VSV - Canister) • Canister pump module • Leakage from fuel tank • Leakage from canister
DTC: P0456 **2T ECM, MIL: Yes** **Year:** 2009, 2010 **Model:** tC, xB, xD **Engine:** 1.8L L4 VIN U, 2.4L L4 **Transmission:** All	**Evaporative Emission Control System Leak Detected (Very Small Leak):** EVAP key-off monitor runs when all of following conditions are met: Atmospheric pressure: 70 to 110 kPa (525 to 825 mmHg) Battery voltage: 10.5 V or more Vehicle speed: Below 2.5 mph (4 km/h) Engine switch: OFF Time after key off: 5, 7 or 9.5 hours EVAP pressure sensor malfunction (P0451, P0452 and P0453): Not detected Purge VSV: Not operated by scan tool Vent valve: Not operated by scan tool Leak detection pump: Not operated by scan tool Both of following conditions are met before key OFF: Conditions 1 and 2 1. Duration that vehicle is being driven: 5 minutes or more 2. EVAP purge operation: Performed ECT: 4.4 to 35°C (40 to 95°F): P0456 IAT:4.4 to 35°C (40 to 95°F): P0456 **Possible Causes:** • Fuel tank cap (loose) • Leakage from EVAP line (Canister - Fuel tank) • Leakage from EVAP line (Purge VSV - Canister) • Canister pump module • Leakage from fuel tank • Leakage from canister

DTC	Trouble Code Title, Conditions & Possible Causes
DTC: P0500 **1T ECM, MIL: Yes** **Year:** 2009, 2010 **Model:** tC, xB, xD **Engine:** 1.8L L4 VIN U, 2.4L L4 **Transmission:** All	**Vehicle Speed Sensor "A":** The monitor will run whenever these DTCs are not present: None Time after engine switch off to on: 3 seconds or more Engine: Running Battery voltage: 8 V or more Engine switch: ON Starter: OFF Either of the following conditions are met: Condition 1 or 2 1. All fo the following conditions are met: Condition (a), (b) or (c) (a) ECT: 60°F (20°C) or more (b) ECT fail: Not detected (c) Time after park/neutral position switch ON to OFF: 10 seconds or more 2. All of the following conditions are met: Either (a) or (b) is set (a) ECT: Less than 68°F (20°C) (b) ECT fail: Detected (c) Time after park/neutral position switch ON to OFF: 30 seconds or more **Possible Causes:** • Vehicle speed sensor • Vehicle speed sensor signal circuit • Combination meter assembly • ECM • Skid control ECU • Main Body ECU • Tire pressure warning ECU • Windshield wiper relay • TCM • Multi-display assembly • Stereo component amplifier
DTC: P0504 **1T ECM, MIL: Yes** **Year:** 2009, 2010 **Model:** tC, xB, xD **Engine:** 1.8L L4 VIN U, 2.4L L4 **Transmission:** All	**Brake Switch "A" / "B" Correlation:** The monitor will run whenever these DTCs are not present: None Engine switch: ON Starter: OFF Battery voltage: 8 V or more GO (Vehicle speed is 18.65 mph (30 km/h) or more): Once STOP (Vehicle speed is less than 1.86 mph (3 km/h)): Once **Possible Causes:** • Short in stop light switch signal circuit • STOP fuse • IGN fuse • Stop light switch • ECM

DTC	Trouble Code Title, Conditions & Possible Causes
DTC: P0505 **2T ECM, MIL: Yes** **Year:** 2009, 2010 **Model:** tC, xB, xD **Engine:** 1.8L L4 VIN U, 2.4L L4 **Transmission:** All	**Idle Control System Malfunction:** Monitor will run whenever these DTCs are not present: P0010, P0020 (OCV bank 1, 2) P0011, P0021 (VVT system bank 1, 2 - advance) P0012, P0022 (VVT system bank 1, 2 - retard) P0013, P0023 (Exhaust OCV bank 1, 2) P0014, P0024 (Exhaust VVT system bank 1, 2 - advance) P0015, P0025 (Exhaust VVT system bank 1, 2 - retard) P0016, P0018 (VVT system bank 1, 2 - misalignment) P0017, P0019 (Exhaust VVT system bank 1, 2 - misalignment) P0031, P0032, P0051, P0052 (A/F sensor heater) P0102, P0103 (MAF meter) P0115, P0117, P0118 (ECT sensor) P0120, P0121, P0122, P0123, P0220, P0222, P0223, P2135 (TP sensor) P0125 (Insufficient ECT for closed loop) P0171, P0172, P0174, P0175 (Fuel system) P0301, P0302, P0303, P0304, P0305, P0306 (Misfire) P0335 (CKP sensor) P0340, P0342, P0343, P0345, P0347, P0348 (VVT sensor) P0351, P0352, P0353, P0354, P0355, P0356 (Igniter) P0365, P0367, P0368, P0390, P0392, P0393 (Exhaust VVT sensor) P0451, P0452, P0452 (EVAP system) P0500 (Vehicle speed sensor) P2195, P2196, P2197, P2198 (A/F sensor - rationality) P2237, P2240 (A/F sensor - open) P2238, P2241, P2252, P2255 (A/F sensor - low impedance) P2239, P2242, P2253, P2256 (A/F sensor - high impedance) P2A00, P2A03 (A/F sensor - slow response) Engine: Running **Possible Causes:** • ETCS (Electronic Throttle Control System) • Air induction system • PCV hose connection • ECM

DTC	Trouble Code Title, Conditions & Possible Causes
DTC: P050A **2T ECM, MIL: Yes** **Year:** 2009, 2010 **Model:** tC, xB, xD **Engine:** 1.8L L4 VIN U, 2.4L L4 **Transmission:** All	**Cold Start Idle Air Control System Performance:** Monitor runs whenever following DTCs are not present: P0010, P0020 (OCV bank 1, 2) P0011, P0021 (VVT system bank 1, 2 - advance) P0012, P0022 (VVT system bank 1, 2 - retard) P0013, P0023 (Exhaust OCV bank 1, 2) P0014, P0024 (Exhaust VVT system bank 1, 2 - advance) P0015, P0025 (Exhaust VVT system bank 1, 2 - retard) P0016, P0018 (VVT system bank 1, 2 - misalignment) P0017, P0019 (Exhaust VVT system bank 1, 2 - misalignment) P0102, P0103 (MAF meter) P0115, P0117, P0118 (ECT sensor) P0120, P0121, P0122, P0123, P0220, P0222, P0223, P2135 (TP sensor) P0125 (Insufficient ECT for closed loop) P0171, P0172, P0174, P0175 (Fuel system) P0301, P0302, P0303, P0304, P0305, P0306 (Misfire) P0335 (CKP sensor) P0340, P0342, P0343, P0345, P0347, P0348 (VVT sensor) P0351, P0352, P0353, P0354, P0355, P0356 (Igniter) P0365, P0367, P0368, P0390, P0392, P0393 (Exhaust VVT sensor) P0500 (Vehicle speed sensor) P2195, P2196, P2197, P2198 (A/F sensor - rationality) P2237, P2240 (A/F sensor - open) P2238, P2241, P2252, P2255 (A/F sensor - low impedance) P2239, P2242, P2253, P2256 (A/F sensor - high impedance) P2A00, P2A03 (A/F sensor - slow response) Battery voltage: 8 V or more Time after engine start: 3 seconds or more Starter: OFF ECT at engine start: −10°C (14°F) or more ECT: -10°C to 50°C (14°F to 122°F) Engine idling time: 3 seconds or more Fuel-cut: OFF Vehicle speed: 1.875 mph (3.01 km/h) or less Time after shift position changed: 1 second or more Atmospheric pressure: 76 kPa (570 mmHg) or more **Possible Causes:** • Throttle body assembly • MAF meter • Air induction system • PCV hose connections • VVT system • Air cleaner filter element • ECM
DTC: P050B **2T ECM, MIL: Yes** **Year:** 2009, 2010 **Model:** xD **Engine:** 1.8L L4 VIN U **Transmission:** All	**Cold Start Ignition Timing Performance:** Battery voltage 8 V or more, time after engine start 3 seconds or more. Starter OFF, ECT at engine start -10°C (14°F) or more, ECT -10 to 50°C (14 to 122°F). Engine idling time 3 seconds or more, Fuel-cut OFF. Vehicle speed less than 3 km/h (1.875 mph). Insufficient ignition timing retard at cold start. **Possible Causes:** • Throttle body assembly • Mass air flow meter • PCV system • Air cleaner filter element • Air induction system • VVT system • ECM

DTC	Trouble Code Title, Conditions & Possible Causes
DTC: P0560 **1T ECM, MIL: Yes** **Year:** 2009, 2010 **Model:** tC, xB, xD **Engine:** 1.8L L4 VIN U, 2.4L L4 **Transmission:** All	**System Voltage:** Monitor runs whenever following DTCs are not present: None Stand by RAM: Initialize **Possible Causes:** • Faulty battery • Open in back-up power source circuit • EFI fuse • ECM
DTC: P0604 **1T ECM, MIL: Yes** **Year:** 2009, 2010 **Model:** tC, xB, xD **Engine:** 1.8L L4 VIN U, 2.4L L4 **Transmission:** All	**Internal Control Module Random Access Memory (RAM) Error:** The ECM continuously monitors its internal memory status. This self-check ensures that the ECM is functioning properly. It is diagnosed by internal "mirroring" of the main CPU and sub CPU to detect the Random Access Memory (RAM) errors. If outputs from these CPUs are different and deviate from the standards, the ECM will illuminate the MIL and set a DTC immediately. Monitor will run whenever this DTC is not present: None **Possible Causes:** • ECM
DTC: P0606 **1T ECM, MIL: Yes** **Year:** 2009, 2010 **Model:** tC, xB, xD **Engine:** 1.8L L4 VIN U, 2.4L L4 **Transmission:** All	**ECM/PCM Processor:** Monitor will run whenever this DTC is not present: None With the engine running. Estimated A/F sensor temperature 450 to 800°C (842 to 1,472°F). Estimated HO2S temperature 450 to 80. ECM CPUs malfunction. A/F sensor transistors malfunction. HO2S transistors malfunction. **Possible Causes:** • Exhaust gas leak • HO2 sensor • ECM
DTC: P0607 **T ECM, MIL: Yes** **Year:** 2009, 2010 **Model:** tC, xB, xD **Engine:** 1.8L L4 VIN U, 2.4L L4 **Transmission:** All	**Control Module Performance:** Monitor runs whenever the following DTCs are not present: None Engine: Running **Possible Causes:** • Exhaust gas leak • HO2 sensor • ECM
DTC: P060A **T ECM, MIL: Yes** **Year:** 2009, 2010 **Model:** xD **Engine:** 1.8L L4 VIN U **Transmission:** All	**Internal Control Module Monitoring Processor Performance:** ECM sub CPU error, the MIL operation is immediate. **Possible Causes:** • ECM
DTC: P060A **T ECM, MIL: Yes** **Year:** 2009, 2010 **Model:** xB **Engine:** 2.4L L4 **Transmission:** All	**Internal Control Module Monitoring Processor Performance:** When either of the following conditions below are met: Condition: 1. CPU reset 1 time or more, Learned TP - Learned APP 0.4 V or more, Electronic throttle actuator is OFF Condition: 2. CPU reset 2 times or more. **Possible Causes:** • ECM
DTC: P060D **T ECM, MIL: Yes** **Year:** 2009, 2010 **Model:** xB, xD **Engine:** 1.8L L4 VIN U, 2.4L L4 **Transmission:** All	**Internal Control Module Accelerator Pedal Position Performance:** When the difference of the main APP and sub APP are 0.3 V or more. for mor than 1 second. MIL operation is immediate. **Possible Causes:** • ECM

DTC	Trouble Code Title, Conditions & Possible Causes
DTC: P060E **T ECM, MIL: Yes** **Year:** 2009, 2010 **Model:** xB, xD **Engine:** 1.8L L4 VIN U, 2.4L L4 **Transmission:** All	**Internal Control Module Throttle Position Performance:** When one of following conditions is met: Condition 1 or 2 1. Difference of main TP and sub TP 0.3 V or more. 2. Difference of main brake switch signal and sub brake switch signal. The MIL operation is immediate. **Possible Causes:** • ECM
DTC: P0617 **1T ECM, MIL: Yes** **Year:** 2009, 2010 **Model:** tC, xB, xD **Engine:** 1.8L L4 VIN U, 2.4L L4 **Transmission:** All	**Starter Relay Circuit High:** Monitor runs whenever this DTC is not present: None Battery voltage: 10.5 V or more Vehicle speed: 12.43 mph (20 km/h) or more Engine speed: 1000 rpm or more **Possible Causes:** • Park/Neutral Position (PNP) switch • Starter relay circuit • Cranking holding function circuit • ECM
DTC: P062F **T ECM, MIL: Yes** **Year:** 2009, 2010 **Model:** xB, xD **Engine:** 1.8L L4 VIN U, 2.4L L4 **Transmission:** All	**Internal Control Module EEPROM Error:** Time after engine start 10 seconds or more, battery voltage 8 V or higher, ignition switch ON, starter OFF. MIL operation is immediate. Mismatch (3 times or more) **Possible Causes:** • ECM
DTC: P0630 **T ECM, MIL: Yes** **Year:** 2009, 2010 **Model:** tC, xB, xD **Engine:** 1.8L L4 VIN U, 2.4L L4 **Transmission:** All	**VIN not Programmed or Mismatch - ECM / PCM:** Battery voltage: 8 V or more Engine switch: ON Starter: OFF **Possible Causes:** • ECM
DTC: P0657 **T ECM, MIL: Yes** **Year:** 2009, 2010 **Model:** tC, xB, xD **Engine:** 1.8L L4 VIN U, 2.4L L4 **Transmission:** All	**Actuator Supply Voltage Circuit / Open:** Monitor will run whenever this DTC is not present: None Engine switch: Front ON to OFF **Possible Causes:** • ECM
DTC: P0705 **2T ECM, MIL: Yes** **Year:** 2009, 2010 **Model:** tC, xB, xD **Engine:** 1.8L L4 VIN U, 2.4L L4 **Transmission:** All	**Transmission Range Sensor Circuit Malfunction (PRNDL Input):** The monitor will run whenever this DTC is not present: None Engine switch: ON Battery voltage: 10.5 V or more Condition (C) One of the following conditions is met: Condition (a) or (b) (a) Park Neutral position switch: ON (b) R range positions switch: ON **Possible Causes:** • Open or Short in park/neutral position switch circuit • Park/neutral position switch • TCM/ECM

DTC	Trouble Code Title, Conditions & Possible Causes
DTC: P0711 **2T ECM, MIL: Yes** **Year:** 2009, 2010 **Model:** tC, xB, xD **Engine:** 1.8L L4 VIN U, 2.4L L4 **Transmission:** All	**Transmission Fluid Temperature Sensor "A" Performance:** The monitor will run whenever this DTC is not present (Not circuit malfunction) P0712, P0713 (ATF temperature sensor circuit (TFT sensor)) P0115, P0117, P0118 (ECT sensor circuit) P0112, P0113 (IAT sensor circuit) P0715, P0717 (Turbine speed sensor circuit) P0791, P0793 (Intermediate shaft speed sensor circuit) P0748 (Shift solenoid valve SL1 circuit) P0778 (Shift solenoid valve SL2 circuit) P0798 (Shift solenoid valve SL3 circuit) P2810 (Shift solenoid valve SL4 circuit) P0327, P0328, P0332, P0333 (KCS sensor circuit) P0120, P0121, P0122, P0123, P0220, P0222, P0223, P0604, P0606, P060A, P060B, P060D, P060E, P0657, P1607, P2102, P2103, P2111, P2112, P2118, P2119, P2135 ((ETCS) Electronic throttle control system) U0100 (CAN communication system) TFT (Transmission fluid temperature) sensor circuit: No circuit malfunction ECT (Engine coolant temperature) sensor circuit: No circuit malfunction IAT (Intake air temperature) sensor circuit: No circuit malfunction Turbine speed sensor circuit: No circuit malfunction Intermediate shaft speed sensor: No circuit malfunction Intermediate shaft speed sensor: No circuit malfunction Shift solenoid valve SL1 circuit: No circuit malfunction Shift solenoid valve SL2 circuit: No circuit malfunction Shift solennoid valve SL3 circuit: No circuit malfunction Shift solenoid valve SL4 circuit: No circuit malfunction (KCS) Knock control sensor circuit: No circuit malfunction (ETCS) Electronic throttle control system: No system down CAN communication system: Not system down Time after engine start: 16 min. and 40 sec. or more ECT (Engine Coolant Temperature): 5°F (-15°C) or more **Possible Causes:** • Transmission wire • ATF temperature sensor • TCM/ECM
DTC: P0712 **1T ECM, MIL: Yes** **Year:** 2009, 2010 **Model:** tC, xB **Engine:** 2.4L L4 **Transmission:** All	**Transmission Fluid Temperature Sensor "A" Circuit Low Input:** The monitor will run whenever this DTC is not present: None The typical enabling condition is not available. **Possible Causes:** • Short in ATF temperature sensor circuit • Transmission wire • ATF temperature sensor • TCM/ECM
DTC: P0713 **1T ECM, MIL: Yes** **Year:** 2009, 2010 **Model:** tC, xB, xD **Engine:** 1.8L L4 VIN U, 2.4L L4 **Transmission:** All	**Transmission Fluid Temperature Sensor "A" Circuit High Input:** The monitor will run whenever this DTC is not present: None The typical enabling condition is not available. **Possible Causes:** • Open in ATF temperature sensor circuit • Transmission wire • ATF temperature sensor • TCM/ECM
DTC: P0717 **1T ECM, MIL: Yes** **Year:** 2009, 2010 **Model:** xB, xD **Engine:** 1.8L L4 VIN U, 2.4L L4 **Transmission:** All	**Input Speed Sensor Circuit No Signal:** ECM detects conditions (a), (b) and (c) continuously for 5 seconds or more. (a) Output shaft speed: 1000 rpm or more (b) Park/Neutral position switch (R) is OFF (c) Speed sensor NT: Less than 300 rpm **Possible Causes:** • Open or short in speed sensor NT circuit • Speed sensor NT • ECM

DTC	Trouble Code Title, Conditions & Possible Causes
DTC: P0724 **2T TCM, MIL: Yes** **Year:** 2009, 2010 **Model:** tC, xB, xD **Engine:** 1.8L L4 VIN U, 2.4L L4 **Transmission:** All	**Brake Switch "B" Circuit High:** The monitor will run whenever this DTC is not present: None GO: (Vehicle speed is 18.63 mph (30 km/h) or more): 18.7 mph (30 km/h) or more STOP: (Vehicle speed is less than 1.86 mph (3 km/h)): Less than 1.86 mph (3 km/h) Starter: OFF Battery voltage: 8 V or more Engine switch: ON **Possible Causes:** • Short in stop light switch signal circuit • Stop light switch • TCM
DTC: P0741 **2T ECM, MIL: Yes** **Year:** 2009, 2010 **Model:** tC, xB **Engine:** 2.4L L4 **Transmission:** All	**Torque Converter Clutch Solenoid Performance (Shift Solenoid Valve DSL):** Lock-up does not occur when driving in lock-up range (normal driving at 80 km/h [50 mph]), or lock-up remains ON in lock-up OFF range (2 trip detection logic) **Possible Causes:** • Shift solenoid valve DSL remains open or closed • Valve body is blocked • Torque converter clutch • Automatic transaxle (clutch, brake or gear etc.) • Line pressure is too low
DTC: P0746 **2T TCM, MIL: Yes** **Year:** 2009, 2010 **Model:** tC, xB **Engine:** 2.4L L4 **Transmission:** All	**Pressure Control Solenoid "A" Performance (Shift Solenoid Valve SL1):** The monitor will run whenever this DTC is not present. (Not circuit malfunction): P0115, P0117, P0118 (ECT sensor circuit) P0715, P0717 (Turbine speed sensor circuit) P0791, P0793 (Intermediate shaft speed sensor circuit) P0748 (Shift solenoid valve SL1 circuit) P0778 (Shift solenoid valve SL2 circuit) P0798 (Shift solenoid valve SL3 circuit) P2810 (Shift solenoid valve SL4 circuit) P0327, P0328, P0332, P0333 (KCS sensor circuit) P0120, P0121, P0122, P0123, P0220, P0222, P0223, P0604, P0606, P060A, P060B, P060D, P060E, P0657, P1607, P2102, P2103, P2111, P2112, P2118, P2119, P2135 ((ETCS) Electronic throttle control system) U0100 (CAN communication system) Transmission range: "D" TFT (Transmission fluid temperature): -10°C (14°F) or more TFT (Transmission fluid temperature) sensor circuit: No circuit malfunction ECT (Engine coolant temperature) sensor circuit: No circuit malfunction Turbine speed sensor circuit: No circuit malfunction Intermediate shaft speed sensor circuit: No circuit malfunction Shift solenoid valve SL1 circuit: No circuit malfunction Shift solenoid valve SL2 circuit: No circuit malfunction Shift solenoid valve SL3 circuit: No circuit malfunction Shift solenoid valve SL4 circuit: No circuit malfunction (KCS) Knock control sensor circuit: No circuit malfunction (ETCS) Electronic throttle control system: Not system down CAN communication system: Not system down **Possible Causes:** • Shift solenoid valve SL1 (open or closed) • Valve body (blocked) • Automatic transaxle (clutch, brake or gear etc.)

DTC	Trouble Code Title, Conditions & Possible Causes
DTC: P0748 **1T ECM, MIL: Yes** **Year:** 2009, 2010 **Model:** tC, xB **Engine:** 2.4L L4 **Transmission:** All	**Pressure Control Solenoid "A" Electrical (Shift Solenoid Valve SL1):** The monitor will run whenever this DTC is not present: None Engine switch: ON Starter: OFF Condition (A): Battery voltage: 12 V or more Condition (B): Battery voltage: 10 V or more and less than 12 V Target current: Less than 0.75 A Condition (C): Battery voltage: 8 V or more Target current: 0.25 A or more **Possible Causes:** • Open or short in shift solenoid valve SL1 circuit • Shift solenoid valve SL1 • Automatic transaxle (clutch, brake or gear etc.)
DTC: P0751 **2T ECM, MIL: Yes** **Year:** 2009, 2010 **Model:** xD **Engine:** 1.8L L4 VIN U **Transmission:** All	**Shift Solenoid "A" Performance (Shift Solenoid Valve S1) :** Transmission shift position D, time after shifting N to D 4.5 seconds or more ECT (Engine Coolant Temperature) 60°C (140°F) or more. The gear required by the ECM does not match the actual gear when driving. **Possible Causes:** • Shift solenoid valve S1 remains open or closed • Valve body is blocked • Shift solenoid valve S1 • Automatic transaxle (clutch, brake or gear etc.)
DTC: P0756 **2T ECM, MIL: Yes** **Year:** 2009, 2010 **Model:** xD **Engine:** 1.8L L4 VIN U **Transmission:** All	**Shift Solenoid "B" Performance (Shift Solenoid Valve S2):** Transmission shift position D, time after shifting N to D 4.5 seconds or more. ECT (Engine Coolant Temperature) 60°C (140°F) or more. The gear required by the ECM does not match the actual gear when driving. **Possible Causes:** • Shift solenoid valve S2 remains open or closed • Valve body is blocked • Shift solenoid valve S2 • Automatic transaxle (clutch, brake or gear etc.)
DTC: P0776 **2T ECM, MIL: Yes** **Year:** 2009, 2010 **Model:** tC, xB **Engine:** 2.4L L4 **Transmission:** All	**Pressure Control Solenoid "B" Performance (Shift Solenoid Valve SL2) :** The gear required by the ECM does not match the actual gear when driving. **Possible Causes:** • Shift solenoid valve SL2 remains open or closed • Valve body is blocked • Shift solenoid valve SL2 • Automatic transaxle (clutch, brake or gear etc.) • ECM
DTC: P0778 **1T ECM, MIL: Yes** **Year:** 2009, 2010 **Model:** tC, xB **Engine:** 2.4L L4 **Transmission:** All	**Pressure Control Solenoid "B" Electrical (Shift Solenoid Valve SL2):** ECM checks for an open or short circuit in shift solenoid valves SL2 (1-trip detection logic) Hybrid IC for solenoid indicates fail. **Possible Causes:** • Open or short in shift solenoid valve SL2 circuit • Shift solenoid valve SL2 • ECM
DTC: P0787 **1T ECM, MIL: Yes** **Year:** 2009, 2010 **Model:** xD **Engine:** 1.8L L4 VIN U **Transmission:** All	**Shift / Timing Solenoid Low (Shift Solenoid Valve ST):** Solenoid ON, time after solenoid OFF to ON more than 0.008 seconds. Battery voltage 8 V or more, starter OFF. The ECM detects short in solenoid valve ST circuit 2-4 times when solenoid valve ST is operated. **Possible Causes:** • Short in shift solenoid valve ST circuit • Shift solenoid valve ST • ECM

DTC	Trouble Code Title, Conditions & Possible Causes
DTC: P0788 **1T ECM, MIL: Yes** **Year:** 2009, 2010 **Model:** xD **Engine:** 1.8L L4 VIN U **Transmission:** All	**Shift / Timing Solenoid High (Shift Solenoid Valve ST):** Solenoid OFF, time after solenoid OFF to ON more than 0.008 seconds. Battery voltage 8 V or more. Starter OFF. The ECM detects open in solenoid valve ST circuit 2 times when solenoid valve ST is not operated. **Possible Causes:** • Open in shift solenoid valve ST circuit • Shift solenoid valve ST • ECM
DTC: P0793 **1T ECM, MIL: Yes** **Year:** 2009, 2010 **Model:** tC, xB **Engine:** 2.4L L4 **Transmission:** All	**Intermediate Shaft Speed Sensor "A":** ECM detects conditions (a), (b) and (c) continuously for 5 sec. or more: (1-trip detection logic) (a) Vehicle speed: 50 km/h (31 mph) or more (b) Park/neutral position switch (STAR) is OFF (c) Speed sensor (NC): less than 300 rpm **Possible Causes:** • Open or short in transmission revolution sensor NC (speed sensor NC) circuit • Transmission revolution sensor NC (speed sensor NC) • ECM
DTC: P0973 **1T ECM, MIL: Yes** **Year:** 2009, 2010 **Model:** xD **Engine:** 1.8L L4 VIN U **Transmission:** All	**Shift Solenoid "A" Control Circuit Low (Shift Solenoid Valve S1):** Solenoid ON, time after solenoid OFF to ON more than 0.008 seconds. The ECM detects short in solenoid valve S1 circuit 2 times when solenoid valve S1 is operated. **Possible Causes:** • Short in shift solenoid valve S1 circuit • Shift solenoid valve S1 • ECM
DTC: P0974 **1T ECM, MIL: Yes** **Year:** 2009, 2010 **Model:** xD **Engine:** 1.8L L4 VIN U **Transmission:** All	**Shift Solenoid "A" Control Circuit High (Shift Solenoid Valve S1):** Solenoid OFF, time after solenoid OFF to ON more than 0.008 seconds. The ECM detects open in solenoid valve S1 circuit 2 times when solenoid valve S1 is not operated. **Possible Causes:** • Open in shift solenoid valve S1 circuit • Shift solenoid valve S1 • ECM
DTC: P0976 **1T TCM, MIL: Yes** **Year:** 2009, 2010 **Model:** xD **Engine:** 1.8L L4 VIN U **Transmission:** All	**Shift Solenoid "B" Control Circuit Low (Shift Solenoid Valve S2):** The monitor will run whenever the following DTCs are not stored: None Battery voltage: 8 V or higher Ignition switch: ON Starter: OFF Shift solenoid valve S2: ON **Possible Causes:** • Short in shift solenoid valve S2 circuit • Shift solenoid valve S2 • ECM
DTC: P0977 **1T TCM, MIL: Yes** **Year:** 2009, 2010 **Model:** xD **Engine:** 1.8L L4 VIN U **Transmission:** All	**Shift Solenoid "B" Control Circuit High (Shift Solenoid Valve S2):** The monitor will run whenever the following DTCs are not stored: None Battery voltage: 8 V or higher Ignition switch: ON Starter: OFF Shift solenoid valve S2: OFF **Possible Causes:** • Open in shift solenoid valve S2 circuit • Shift solenoid valve S2 • ECM
DTC: P0982 **1T ECM, MIL: Yes** **Year:** 2009, 2010 **Model:** tC, xB **Engine:** 2.4L L4 **Transmission:** All	**Shift Solenoid "D" Control Circuit Low (Shift Solenoid Valve S4):** ECM detects short in solenoid valve S4 circuit 2 times when solenoid valve S4 is operated. **Possible Causes:** • Short in shift solenoid valve S4 circuit • Shift solenoid valve S4 • ECM

DTC	Trouble Code Title, Conditions & Possible Causes
DTC: P0983 **1T ECM, MIL: Yes** **Year:** 2009, 2010 **Model:** tC, xB **Engine:** 2.4L L4 **Transmission:** All	**Shift Solenoid "D" Control Circuit High (Shift Solenoid Valve S4):** ECM detects open in solenoid valve S4 circuit 2 times when solenoid valve S4 is not operated. **Possible Causes:** • Open in shift solenoid valve S4 circuit • Shift solenoid valve S4 • ECM

OBD II Trouble Code List (P1XXX Codes)

DTC	Trouble Code Title, Conditions & Possible Causes
DTC: P101D **1T ECM, MIL: Yes** **Year:** 2009, 2010 **Model:** xD **Engine:** 1.8L L4 VIN U **Transmission:** All	**A/F Sensor Heater Circuit Performance Bank 1 Sensor 1 Stuck ON:** Battery voltage 10.5 V or higher, time after heater ON 5 seconds or more. A/F sensor heater duty-cycle 10 to 60%, A/F sensor heater current 0.8 A or higher. The heater current is higher than the specified value while the heater is not operating. **Possible Causes:** • ECM
DTC: P102D **1T ECM, MIL: Yes** **Year:** 2009, 2010 **Model:** xD **Engine:** 1.8L L4 VIN U **Transmission:** All	**O2 Sensor Heater Circuit Performance Bank 1 Sensor 2 Stuck ON:** Monitor runs whenever following DTCs are not present: None Battery voltage: 10.5 V or more Engine: Running Starter: OFF Catalyst active air fuel ratio control: Not operating Time after heater ON: 10 seconds or more Learned heater OFF current operation completed flag: ON Heated oxygen sensor heater OFF current: More than 3.5 A Hybrid IC high current limiter monitor input: Fail Heated oxygen sensor heater high current fail (P0038): Not detected **Possible Causes:** • Hybrid vehicle control ECU
DTC: P1603 **1T ECM, MIL: Yes** **Year:** 2009, 2010 **Model:** xB, xD **Engine:** 1.8L L4 VIN U, 2.4L L4 **Transmission:** All	**Engine Stall History:** After 5 seconds or more elapse after starting the engine, with the engine running, the engine stops (the engine speed drops to 200 rpm or less) for 0.5 seconds or more, without the ignition switch or engine switch being operated. **Possible Causes:** • If any DTCs other than P1603 are output, troubleshoot those DTCs first. • Check freeze frame data

DTC	Trouble Code Title, Conditions & Possible Causes
DTC: P1604 **1T ECM, MIL: Yes** **Year:** 2009, 2010 **Model:** xD **Engine:** 1.8L L4 VIN U **Transmission:** All	**Startability Malfunction:** The engine speed is below 500 rpm with the STA signal on for a certain amount of time. After the engine starts (engine speed is 500 rpm or more), the engine speed drops to 200 rpm or less within approximately 2 seconds. **Possible Causes:** • Engine assembly (excess friction, compression loss) • Starter • Crankshaft position sensor • Camshaft position sensor • Engine coolant temperature sensor • Fuel pump • Fuel pump control system • Fuel pipes • Fuel injector • Throttle body • Pressure regulator • Battery • Drive plate/Flywheel • Spark plug • Ignition coil circuit • Intake system • Camshaft timing oil control valve • Mass air flow meter • Air fuel ratio sensor • Valve timing • Fuel • Purge VSV • Intake valve • Exhaust valve • ECM
DTC: P1605 **T ECM, MIL: Yes** **Year:** 2009, 2010 **Model:** xB, xD **Engine:** 1.8L L4 VIN U, 2.4L L4 **Transmission:** All	**Rough Idling:** 5 seconds or more elapse after starting the engine and the engine is running. The engine speed drops to 400 rpm or less. **Possible Causes:** • If any DTCs other than P1605 are output, troubleshoot those DTCs first. • Check freeze frame data.
DTC: P1607 **T ECM, MIL: Yes** **Year:** 2009, 2010 **Model:** xB, xD **Engine:** 1.8L L4 VIN U, 2.4L L4 **Transmission:** All	**Cruise Control Input Processor:** The ECM continuously monitors its main and sub CPUs for the cruise control. This self-check ensures that the ECM is functioning properly. If outputs from the CPUs are different and deviate from the standards, the ECM will illuminate the MIL and set the DTC immediately. **Possible Causes:** • ECM

OBD II Trouble Code List (P2XXX Codes)

DTC	Trouble Code Title, Conditions & Possible Causes
DTC: P2102 **1T ECM, MIL: Yes** **Year:** 2009, 2010 **Model:** tC, xB, xD **Engine:** 1.8L L4 VIN U, 2.4L L4 **Transmission:** All	**Throttle Actuator Control Motor Circuit Low:** Monitor runs whenever following DTCs are not present: None Throttle motor: 80 % or more Throttle actuator power supply: 8 V or more Current motor current-Motor current before 0.016 sec.: Less than 0.2A **Possible Causes:** • Open in throttle actuator circuit • Throttle actuator • ECM

DTC	Trouble Code Title, Conditions & Possible Causes
DTC: P2103 **T ECM, MIL: Yes** **Year:** 2009, 2010 **Model:** tC, xB, xD **Engine:** 1.8L L4 VIN U, 2.4L L4 **Transmission:** All	**Throttle Actuator Control Motor Circuit High:** Monitor runs whenever following DTCs are not present: None Throttle motor: ON Either following condition 1 or 2 is met: 1. Throttle actuator power supply: 8 V or more 2. Throttle actuator power: ON Battery voltage: 8 V or more Starter: OFF **Possible Causes:** • Short in throttle actuator circuit • Throttle actuator • Throttle valve • Throttle body assembly • ECM
DTC: P2109 **5T ECM, MIL: Yes** **Year:** 2009, 2010 **Model:** xB, xD **Engine:** 1.8L L4 VIN U, 2.4L L4 **Transmission:** All	**Throttle / Pedal Position Sensor "A" Minimum Stop Performance:** The engine is started with a coolant temperature of 45°C (113°F) or lower, and after the engine has warmed up, the ISC learning conditions are met, or the engine has been running for at least an hour after ignition switch is ON, the engine is warmed up, and the ISC learning conditions are met. Vehicle has been driven at a speed of 30 km/h (18.65 mph) or more. Mass air flow meter is operating correctly. Atmospheric Pressure is 85 kPa (637.5 mmHg) or more (typically when altitude is 1400 m or lower). **Possible Causes:** • Dirty or faulty throttle body
DTC: P2111 **1T ECM, MIL: Yes** **Year:** 2009, 2010 **Model:** tC, xB, xD **Engine:** 1.8L L4 VIN U, 2.4L L4 **Transmission:** All	**Throttle Actuator Control System - Stuck Open:** Monitor runs whenever following DTCs are not present: None All of the following conditions are met: System guard*:ON Throttle actuator current: 2 A or more Duty cycle to close throttle: 80% or more * System guard is ON when the following conditions are met Throttle actuator: ON Throttle actuator duty calculation: Executing Throttle position sensor: Fail determined Throttle actuator current-cut operation: Not executing Throttle actuator power supply: 4 V or more Throttle actuator: Fail determined **Possible Causes:** • Throttle actuator • Throttle body assembly • Throttle valve
DTC: P2112 **1T ECM, MIL: Yes** **Year:** 2009, 2010 **Model:** tC, xB, xD **Engine:** 1.8L L4 VIN U, 2.4L L4 **Transmission:** All	**Throttle Actuator Control System - Stuck Closed:** Monitor runs whenever following DTCs are not present: None All of the following conditions are met: System guard*:ON Throttle actuator current: 2 A or more Duty cycle to open throttle: 80% or more * System guard is ON when the following conditions are met Throttle actuator: ON Throttle actuator duty calculation: Executing Throttle position sensor: Fail determined Throttle actuator current-cut operation: Not executing Throttle actuator power supply: 4 V or more Throttle actuator: Fail determined **Possible Causes:** • Throttle actuator • Throttle body assembly • Throttle valve

DTC	Trouble Code Title, Conditions & Possible Causes
DTC: P2118 **1T ECM, MIL: Yes** **Year:** 2009, 2010 **Model:** tC, xB, xD **Engine:** 1.8L L4 VIN U, 2.4L L4 **Transmission:** All	**Throttle Actuator Control Motor Current Range / Performance:** Monitor runs whenever following DTCs are not present: None Battery voltage: 8 V or more Throttle actuator power: ON **Possible Causes:** • Open in ETCS power source circuit • ETCS fuse • ECM
DTC: P2119 **1T ECM, MIL: Yes** **Year:** 2009, 2010 **Model:** tC, xB, xD **Engine:** 1.8L L4 VIN U, 2.4L L4 **Transmission:** All	**Throttle Actuator Control Throttle Body Range / Performance:** Monitor runs whenever following DTCs are not present: None System guard*: ON *System guard is set when following conditions are met: Throttle actuator: ON Throttle actuator duty calculation: Executing TP sensor: Fail determined Throttle actuator current cut operation: Not executing Throttle actuator power supply: 4 V or more Throttle actuator: Fail determined **Possible Causes:** • ETCS • ECM
DTC: P2120 **1T ECM, MIL: Yes** **Year:** 2009, 2010 **Model:** tC, xB, xD **Engine:** 1.8L L4 VIN U, 2.4L L4 **Transmission:** All	**Throttle / Pedal Position Sensor / Switch "D" Circuit:** VPA fluctuates rapidly beyond upper and lower malfunction thresholds for 0.5 seconds or more, No other DTC's are present. **Possible Causes:** • Accelerator Pedal Position (APP) sensor • ECM
DTC: P2121 **1T ECM, MIL: Yes** **Year:** 2009, 2010 **Model:** tC, xB, xD **Engine:** 1.8L L4 VIN U, 2.4L L4 **Transmission:** All	**Throttle / Pedal Position Sensor / Switch "D" Circuit Range / Performance:** Monitor runs whenever following DTCs are not present: None Either of the following conditions met: Condition (a) or (b) (a) Engine switch: ON (b) Throttle actuator power: ON **Possible Causes:** • Accelerator Pedal Position (APP) sensor • ECM
DTC: P2122 **1T ECM, MIL: Yes** **Year:** 2009, 2010 **Model:** tC, xB, xD **Engine:** 1.8L L4 VIN U, 2.4L L4 **Transmission:** All	**Throttle / Pedal Position Sensor / Switch "D" Circuit Low Input:** Monitor runs whenever following DTCs are not present: None VPA 0.4 V or less for 0.5 seconds or more when accelerator pedal depressed. **Possible Causes:** • APP sensor • Open in VCP1 circuit • Open or ground short in VPA circuit • ECM
DTC: P2123 **1T ECM, MIL: Yes** **Year:** 2009, 2010 **Model:** tC, xB, xD **Engine:** 1.8L L4 VIN U, 2.4L L4 **Transmission:** All	**Throttle / Pedal Position Sensor / Switch "D" Circuit High Input:** Monitor runs whenever following DTCs are not present: None VPA 4.8 V or more for 2.0 seconds or more. **Possible Causes:** • APP sensor • Open in EPA circuit • ECM
DTC: P2125 **1T ECM, MIL: Yes** **Year:** 2009, 2010 **Model:** tC, xB, xD **Engine:** 1.8L L4 VIN U, 2.4L L4 **Transmission:** All	**Throttle / Pedal Position Sensor / Switch "E" Circuit:** Monitor runs whenever following DTCs are not present: None VPA2 1.2 V or less for 0.5 seconds or more when accelerator pedal depressed. **Possible Causes:** • APP sensor • ECM

DTC	Trouble Code Title, Conditions & Possible Causes
DTC: P2127 **1T ECM, MIL: Yes** **Year:** 2009, 2010 **Model:** tC, xB, xD **Engine:** 1.8L L4 VIN U, 2.4L L4 **Transmission:** All	**Throttle / Pedal Position Sensor / Switch "E" Circuit Low Input:** Monitor runs whenever following DTCs are not present: None **Possible Causes:** • APP sensor • Open in VCP2 circuit • Open or ground short in VPA2 circuit • ECM
DTC: P2128 **1T ECM, MIL: Yes** **Year:** 2009, 2010 **Model:** tC, xB, xD **Engine:** 1.8L L4 VIN U, 2.4L L4 **Transmission:** All	**Throttle / Pedal Position Sensor / Switch "E" Circuit High Input:** Monitor runs whenever following DTCs are not present: None With the ignition switch ON for 0.012 seconds or more and throttle actuator power ON. Conditions (a) and (b) continue for 2.0 seconds or more: (a) VPA2 4.8 V or more (b) VPA between 0.4 V and 3.45 V **Possible Causes:** • APP sensor • Open in EPA2 circuit • ECM
DTC: P2135 **1T ECM, MIL: Yes** **Year:** 2009, 2010 **Model:** tC, xB, xD **Engine:** 1.8L L4 VIN U, 2.4L L4 **Transmission:** All	**Throttle / Pedal Position Sensor / Switch "A" / "B" Voltage Correlation:** Monitor runs whenever following DTCs are not present: None Either of the following conditions A or B is met: A. Engine switch on (IG): 0.012 seconds or more B. Electronic throttle actuator power: ON **Possible Causes:** • Short between VTA1 and VTA2 circuits • Throttle position sensor (built into throttle body) • ECM
DTC: P2138 **1T ECM, MIL: Yes** **Year:** 2009, 2010 **Model:** tC, xB, xD **Engine:** 1.8L L4 VIN U, 2.4L L4 **Transmission:** All	**Throttle / Pedal Position Sensor / Switch "D" / "E" Voltage Correlation:** Monitor runs whenever following DTCs are not present: None **Possible Causes:** • Short between VPA and VPA2 circuits • APP sensor • ECM

DTC	Trouble Code Title, Conditions & Possible Causes
DTC: P2195 **2T ECM, MIL: Yes** **Year:** 2009, 2010 **Model:** tC, xB, xD **Engine:** 1.8L L4 VIN U, 2.4L L4 **Transmission:** All	**Oxygen (A/F) Sensor Signal Stuck Lean (Bank 1 Sensor 1):** Monitor runs whenever following DTCs are not present: P0016, P0018 (VVT system bank 1, 2 - misalignment) P0017, P0019 (Exhaust VVT system bank 1, 2 - misalignment) P0031, P0032, P0051, P0052 (A/F sensor heater) P0102, P0103 (MAF meter) P0110, P0112, P0113 (IAT sensor) P0115, P0117, P0118 (ECT sensor) P0120, P0121, P0122, P0123, P0220, P0222, P0223, P2135 (TP sensor) P0125 (Insufficient ECT for closed loop) P0128 (Thermostat) P0171, P0172, P0174, P0175 (Fuel system) P0301, P0302, P0303, P0304, P0305, P0306 (Misfire) P0335 (CKP sensor) P0451, P0452, P0452 (EVAP system) P0500 (Vehicle speed sensor) P0505 (Idle speed control) Sensor voltage detection monitor (Lean side malfunction) Time after engine start: 30 seconds or more Fuel system status: Closed loop Sensor current detection monitor (High and low side malfunction): Battery voltage: 11 V or more Atmospheric pressure: 76 kPa (570 mmHg) or more A/F sensor status: Activated Continuous time of fuel cut: 3-10 seconds ECT: 167°F (75°C) or more **Possible Causes:** • Open or short in A/F sensor (bank 1 sensor 1) circuit • A/F sensor (bank 1 sensor 1) • A/F sensor (bank 1 sensor 1) heater • A/F sensor heater relay • A/F sensor heater and relay circuits • Air induction system • Injector • ECM

DTC	Trouble Code Title, Conditions & Possible Causes
DTC: P2196 **2T ECM, MIL: Yes** **Year:** 2009, 2010 **Model:** tC, xB, xD **Engine:** 1.8L L4 VIN U, 2.4L L4 **Transmission:** All	**Oxygen (A/F) Sensor Signal Stuck Rich (Bank 1 Sensor 1):** Monitor runs whenever following DTCs are not present: P0016, P0018 (VVT system bank 1, 2 - misalignment) P0017, P0019 (Exhaust VVT system bank 1, 2 - misalignment) P0031, P0032, P0051, P0052 (A/F sensor heater) P0102, P0103 (MAF meter) P0110, P0112, P0113 (IAT sensor) P0115, P0117, P0118 (ECT sensor) P0120, P0121, P0122, P0123, P0220, P0222, P0223, P2135 (TP sensor) P0125 (Insufficient ECT for closed loop) P0128 (Thermostat) P0171, P0172, P0174, P0175 (Fuel system) P0301, P0302, P0303, P0304, P0305, P0306 (Misfire) P0335 (CKP sensor) P0451, P0452, P0452 (EVAP system) P0500 (Vehicle speed sensor) P0505 (Idle speed control) Sensor voltage detection monitor (Rich side malfunction): Time after engine start: 30 seconds or more Fuel system status: Closed loop Sensor current detection monitor (High and low side malfunction): Battery voltage: 11 V or more Atmospheric pressure: 76 kPa (570 mmHg) or more A/F sensor status: Activated Continuous time of fuel cut: 3-10 seconds ECT: 167°F (75°C) or more **Possible Causes:** • Open or short in A/F sensor (bank 1 sensor 1) circuit • A/F sensor (bank 1 sensor 1) • A/F sensor (bank 1 sensor 1) heater • A/F sensor heater relay • A/F sensor heater and relay circuits • Air induction system • Injector • ECM
DTC: P2237 **2T ECM, MIL: Yes** **Year:** 2009, 2010 **Model:** tC, xB, xD **Engine:** 1.8L L4 VIN U, 2.4L L4 **Transmission:** All	**Oxygen (A/F) Sensor Pumping Current Circuit / Open (Bank 1 Sensor 1):** Monitor runs whenever following DTCs are not present: P0016, P0018 (VVT system bank 1, 2 - misalignment) P0017, P0019 (Exhaust VVT system bank 1, 2 - misalignment) P0031, P0032, P0051, P0052 (A/F sensor heater) P0102, P0103 (MAF meter) P0110, P0112, P0113 (IAT sensor) P0115, P0117, P0118 (ECT sensor) P0120, P0121, P0122, P0123, P0220, P0222, P0223, P2135 (TP sensor) P0125 (Insufficient ECT for closed loop) P0128 (Thermostat) P0171, P0172, P0174, P0175 (Fuel system) P0301, P0302, P0303, P0304, P0305, P0306 (Misfire) P0335 (CKP sensor) P0340 (CMP sensor) P0451, P0452, P0452 (EVAP system) P0500 (Vehicle speed sensor) P0505 (Idle speed control) Open circuit between AF+ and AF-: Estimated sensor temperature: 842-1022°F (450-550°C) **Possible Causes:** • Open or short in A/F sensor (bank 1 sensor 1) circuit • Air fuel ratio sensor (bank 1 sensor 1) • ECM

DTC	Trouble Code Title, Conditions & Possible Causes
DTC: P2238 **2T ECM, MIL: Yes** **Year:** 2009, 2010 **Model:** xB **Engine:** 2.4L L4 **Transmission:** All	**Oxygen (A/F) Sensor Pumping Current Circuit Low (Bank 1 Sensor 1):** Air Fuel Ratio (A/F) sensor output drops while engine is running. Voltage at terminal A1A+ is 0.5 V or less. Voltage difference between terminals A1A+ and A1A- is 0.1 V or less. **Possible Causes:** • Open or short in Air Fuel Ratio (A/F) sensor (sensor 1) circuit • A/F sensor (sensor 1) • ECM
DTC: P2239 **2T ECM, MIL: Yes** **Year:** 2009, 2010 **Model:** tC, xB, xD **Engine:** 1.8L L4 VIN U, 2.4L L4 **Transmission:** All	**Oxygen (A/F) Sensor Pumping Current Circuit High (Bank 1 Sensor 1):** Monitor runs whenever following DTCs are not present: P0016, P0018 (VVT system bank 1, 2 - misalignment) P0017, P0019 (Exhaust VVT system bank 1, 2 - misalignment) P0031, P0032, P0051, P0052 (A/F sensor heater) P0102, P0103 (MAF meter) P0110, P0112, P0113 (IAT sensor) P0115, P0117, P0118 (ECT sensor) P0120, P0121, P0122, P0123, P0220, P0222, P0223, P2135 (TP sensor) P0125 (Insufficient ECT for closed loop) P0128 (Thermostat) P0171, P0172, P0174, P0175 (Fuel system) P0301, P0302, P0303, P0304, P0305, P0306 (Misfire) P0335 (CKP sensor) P0340 (CMP sensor) P0451, P0452, P0452 (EVAP system) P0500 (Vehicle speed sensor) P0505 (Idle speed control) Battery voltage: 11 V or more Engine switch: ON Time after engine switch OFF to ON: 5 seconds or more **Possible Causes:** • Open or short in A/F sensor (bank 1 sensor 1) circuit • A/F sensor (bank 1 sensor 1) • A/F sensor heater • A/F sensor heater relay • A/F sensor heater and relay circuits • ECM

DTC	Trouble Code Title, Conditions & Possible Causes
DTC: P2252 **2T ECM, MIL: Yes** **Year:** 2009, 2010 **Model:** tC, xB, xD **Engine:** 1.8L L4 VIN U, 2.4L L4 **Transmission:** All	**Oxygen (A/F) Sensor Reference Ground Circuit Low (Bank 1 Sensor 1):** Monitor runs whenever following DTCs are not present: P0016, P0018 (VVT system bank 1, 2 - misalignment) P0017, P0019 (Exhaust VVT system bank 1, 2 - misalignment) P0031, P0032, P0051, P0052 (A/F sensor heater) P0102, P0103 (MAF meter) P0110, P0112, P0113 (IAT sensor) P0115, P0117, P0118 (ECT sensor) P0120, P0121, P0122, P0123, P0220, P0222, P0223, P2135 (TP sensor) P0125 (Insufficient ECT for closed loop) P0128 (Thermostat) P0171, P0172, P0174, P0175 (Fuel system) P0301, P0302, P0303, P0304, P0305, P0306 (Misfire) P0335 (CKP sensor) P0340 (CMP sensor) P0451, P0452, P0452 (EVAP system) P0500 (Vehicle speed sensor) P0505 (Idle speed control) Battery voltage: 11 V or more Engine switch: ON Time after engine switch OFF to ON: 5 seconds or more **Possible Causes:** • Open or short in A/F sensor (bank sensor 1) circuit • A/F sensor (bank 1 sensor 1) • A/F sensor heater • A/F sensor heater relay • A/F sensor heater and relay circuits • ECM
DTC: P2253 **2T ECM, MIL: Yes** **Year:** 2009, 2010 **Model:** tC, xB, xD **Engine:** 1.8L L4 VIN U, 2.4L L4 **Transmission:** All	**Oxygen (A/F) Sensor Reference Ground Circuit High (Bank 1 Sensor 1):** Monitor runs whenever following DTCs are not present: P0016, P0018 (VVT system bank 1, 2 - misalignment) P0017, P0019 (Exhaust VVT system bank 1, 2 - misalignment) P0031, P0032, P0051, P0052 (A/F sensor heater) P0102, P0103 (MAF meter) P0110, P0112, P0113 (IAT sensor) P0115, P0117, P0118 (ECT sensor) P0120, P0121, P0122, P0123, P0220, P0222, P0223, P2135 (TP sensor) P0125 (Insufficient ECT for closed loop) P0128 (Thermostat) P0171, P0172, P0174, P0175 (Fuel system) P0301, P0302, P0303, P0304, P0305, P0306 (Misfire) P0335 (CKP sensor) P0340 (CMP sensor) P0451, P0452, P0452 (EVAP system) P0500 (Vehicle speed sensor) P0505 (Idle speed control) Battery voltage: 11 V or more Engine switch: ON Time after engine switch OFF to ON: 5 seconds or more **Possible Causes:** • Open or short in A/F sensor (bank 1 sensor 1) circuit • A/F sensor (bank 1 sensor 1) • A/F sensor heater • A/F sensor heater relay • A/F sensor heater and relay circuits • ECM

DTC	Trouble Code Title, Conditions & Possible Causes
DTC: P2401 **2T ECM, MIL: Yes** **Year:** 2009, 2010 **Model:** tC, xB, xD **Engine:** 1.8L L4 VIN U, 2.4L L4 **Transmission:** All	**Evaporative Emission System Leak Detection Pump Control Circuit Low:** Reference orifice clogged. Reference orifice high-flow. Leak detection pump OFF malfunction. Leak detection pump ON malfunction. Vent valve ON (close) malfunction. **Possible Causes:** • Canister pump module (Reference orifice, leak detection pump, vent valve) • Connector/wire harness (Canister pump module - ECM) • EVAP system hose (pipe from air inlet port to canister pump module, canister filter, fuel tank vent hose) • ECM
DTC: P2402 **2T ECM, MIL: Yes** **Year:** 2009, 2010 **Model:** tC, xB, xD **Engine:** 1.8L L4 VIN U, 2.4L L4 **Transmission:** All	**Evaporative Emission System Leak Detection Pump Control Circuit High:** Reference orifice clogged. Reference orifice high-flow. Leak detection pump OFF malfunction. Leak detection pump ON malfunction. Vent valve ON (close) malfunction. **Possible Causes:** • Canister pump module (Reference orifice, leak detection pump, vent valve) • Connector/wire harness (Canister pump module - ECM) • EVAP system hose (pipe from air inlet port to canister pump module, canister filter, fuel tank vent hose) • ECM
DTC: P2419 **2T ECM, MIL: Yes** **Year:** 2009, 2010 **Model:** tC, xB, xD **Engine:** 1.8L L4 VIN U, 2.4L L4 **Transmission:** All	**Evaporative Emission System Switching Valve Control Circuit Low:** Reference orifice clogged. Reference orifice high-flow. Leak detection pump OFF malfunction. Leak detection pump ON malfunction. Vent valve ON (close) malfunction. **Possible Causes:** • Canister pump module (Reference orifice, leak detection pump, vent valve) • Connector/wire harness (Canister pump module - ECM) • EVAP system hose (pipe from air inlet port to canister pump module, canister filter, fuel tank vent hose) • ECM
DTC: P2420 **2T ECM, MIL: Yes** **Year:** 2009, 2010 **Model:** tC, xB, xD **Engine:** 1.8L L4 VIN U, 2.4L L4 **Transmission:** All	**Evaporative Emission System Switching Valve Control Circuit High:** The monitor will run whenever these DTCs are not present: None Key-off monitor is run when all of the following conditions are met: Atmospheric pressure: 70 to 110 kPa (525 to 825 mmHg) Battery voltage: 10.5 V or more Vehicle speed: 2.5 mph (4 km/h) or less Engine switch: OFF Time after key-off: 5 or 7 or 9.5 hours EVAP pressure sensor malfunction (P0452, P0453): Not detected EVAP canister purge valve: Not operated by scan tool EVAP canister vent valve: Not operated by scan tool EVAP leak detection pump: Not operated by scan tool Both of the following conditions 1 and 2 are met before key-off: 1. Duration that vehicle has been driven: 5 minutes or more 2. EVAP purge operation: Performed ECT: 40-95°F (4.4-35°C) IAT: 40-95°F (4.4-35°C) **Possible Causes:** • Canister pump module (0.02 inch orifice, leak detection pump, vent valve) • Connector / wire harness (Canister pump module - ECM) • ECM

DTC	Trouble Code Title, Conditions & Possible Causes
DTC: P2610 **2T ECM, MIL: Yes** **Year:** 2009, 2010 **Model:** tC, xB, xD **Engine:** 1.8L L4 VIN U, 2.4L L4 **Transmission:** All	**ECM / PCM Internal Engine Off Timer Performance:** Case 1: Engine switch: ON Engine: Running Battery voltage: 8 V or more Starter: OFF Case 2: Internal engine OFF timer (elapsed time from engine stop): 10-30 minutes Battery voltage: 8 V or more Engine switch: ON Starter: OFF Case 3: Internal engine OFF timer (elapsed time from engine stop): 40 minutes Battery voltage: 8 V or more Engine switch: ON Starter: OFF **Possible Causes:** • ECM
DTC: P2714 **2T , MIL: Yes** **Year:** 2009, 2010 **Model:** tC, xB, xD **Engine:** 1.8L L4 VIN U, 2.4L L4 **Transmission:** All	**Pressure Control Solenoid "D" Performance (Shift Solenoid Valve SLT):** The monitor will run whenever this DTC is not present. (Not circuit malfunction): P0712, P0713 (ATF temperature sensor circuit (TFT sensor)) P0115, P0117, P0118 (ECT sensor circuit) P0715, P0717 (Turbine speed sensor circuit) P0791, P0793 (Intermediate shaft speed sensor circuit) P0748 (Shift solenoid valve SL1 circuit) P0778 (Shift solenoid valve SL2 circuit) P0798 (Shift solenoid valve SL3 circuit) P2810 (Shift solenoid valve SL4 circuit) P2716 (Shift solenoid valve SLT circuit) P0327, P0328, P0332, P0333 (KCS sensor circuit) P0120, P0121, P0122, P0123, P0220, P0222, P0223, P0604, P0606, P060A, P060B, P060D, P060E, P0657, P1607, P2102, P2103, P2111, P2112, P2118, P2119, P2135 ((ETCS) Electronic throttle control system) U0100 (CAN communication system) Transmission range: "D" Duration time from shifting "N" to "D": 4 sec. or more TFT (Transmission fluid temperature): -10°C (14°F) or more TFT (Transmission fluid temperature) sensor circuit: No circuit malfunction ECT (Engine coolant temperature) sensor circuit: No circuit malfunction Turbine speed sensor circuit: No circuit malfunction Intermediate shaft speed sensor circuit: No circuit malfunction Shift solenoid valve SL1 circuit: No circuit malfunction Shift solenoid valve SL2 circuit: No circuit malfunction Shift solenoid valve SL3 circuit: No circuit malfunction Shift solenoid valve SL4 circuit: No circuit malfunction Shift solenoid valve SLT circuit: No circuit malfunction (KCS) Knock control sensor circuit: No circuit malfunction (ETCS) Electronic throttle control system: Not system down CAN communication system: Not system down Engine: Starting Input turbine torque: 100 N*m or more Turbine speed: 250 rpm or more Output speed: 250 rpm or more **Possible Causes:** • Shift solenoid valve SLT (open) • Shift solenoid valve SL1, SL2, SL3 or SL4 (open or closed) • Valve body (blocked) • Torque converter clutch • Automatic transaxle (clutch, brake or gear, etc.)

DTC	Trouble Code Title, Conditions & Possible Causes
DTC: P2716 **1T ECM, MIL: Yes** **Year:** 2009, 2010 **Model:** tC, xB, xD **Engine:** 1.8L L4 VIN U, 2.4L L4 **Transmission:** All	**Pressure Control Solenoid "D" Electrical (Shift Solenoid Valve SLT):** Open or short is detected in shift solenoid valve SLT circuit for 1 second or more while driving (1-trip detecting logic). Solenoid current cut status: Not cut Battery voltage 11 V or more, ignition switch ON, starter OFF, CPU commanded duty ratio to SLT 19% or more. **Possible Causes:** • Open or short in shift solenoid valve SLT circuit • Shift solenoid valve SLT • ECM
DTC: P2757 **2T TCM, MIL: Yes** **Year:** 2009, 2010 **Model:** xD **Engine:** 1.8L L4 VIN U **Transmission:** All	**Torque Converter Clutch Pressure Control Solenoid Performance (Shift Solenoid Valve SLU):** The monitor will run whenever this DTC is not present. (Not circuit malfunction): P0712, P0713 (ATF temperature sensor circuit (TFT sensor)) P0115, P0117, P0118 (ECT sensor circuit) P0715, P0717 (Turbine speed sensor circuit) P0791, P0793 (Intermediate shaft speed sensor circuit) P0748 (Shift solenoid valve SL1 circuit) P0778 (Shift solenoid valve SL2 circuit) P0798 (Shift solenoid valve SL3 circuit) P2810 (Shift solenoid valve SL4 circuit) P2759 (Shift solenoid valve SLU circuit) P2769, P2770 (Shift solenoid valve SL circuit) P0327, P0328, P0332, P0333 (KCS sensor circuit) P0120, P0121, P0122, P0123, P0220, P0222, P0223, P0604, P0606, P060A, P060B, P060D, P060E, P0657, P1607, P2102, P2103, P2111, P2112, P2118, P2119, P2135 ((ETCS) Electronic throttle control system) U0100 (CAN communication system) ECT (Engine coolant temperature): 40°C (104°F) or more Spark advance from Max. retard timing by KCS control: 0°CA or more Transmission range: "D" TFT (Transmission fluid temperature): -10°C (14°F) or more TFT (Transmission fluid temperature) sensor circuitL: No circuit malfunction ECT (Engine coolant temperature) sensor circuit: No circuit malfunction Turbine speed sensor circuit: No circuit malfunction Intermediate shaft speed sensor circuit: No circuit malfunction Shift solenoid valve SL1 circuit: No circuit malfunction Shift solenoid valve SL2 circuit: No circuit malfunction Shift solenoid valve SL3 circuit:n: No circuit malfunction Shift solenoid valve SL4 circuit: No circuit malfunction Shift solenoid valve SLU circuit: No circuit malfunction Shift solenoid valve SL circuit: No circuit malfunction (KCS) Knock control sensor circuit: No circuit malfunction (ETCS) Electronic throttle control system: Not system down CAN communication system: Not system down TCM selected gear: Not 1st Vehicle speed: 5.5 mph (25 km/h) or more Turbine speed/Output speed (NT/NO) with 1st: 3.304 to 7.724 Turbine speed/Output speed (NT/NO) with 2nd: 1.901 to 2.340 Turbine speed/Output speed (NT/NO) with 3rd: 1.399 to 1.649 Turbine speed/Output speed (NT/NO) with 4th: 0.998 to 1.138 Turbine speed/Output speed (NT/NO) with 5th: 0.705 to 0.836 Turbine speed/Output speed (NT/NO) with 6th: 0.568 to 0.695 **Possible Causes:** • Shift solenoid valve SLU (open or closed) • Valve body (blocked) • Torque converter clutch • Automatic transaxle (clutch, brake or gear, etc.) • Line pressure is too low

DTC	Trouble Code Title, Conditions & Possible Causes
DTC: P2759 **1T ECM, MIL: Yes** **Year:** 2009, 2010 **Model:** xD **Engine:** 1.8L L4 VIN U **Transmission:** All	**Torque Converter Clutch Pressure Control Solenoid Control Circuit Electrical (Shift Solenoid Valve SLU):** The monitor will run whenever this DTC is not present: None Solenoid current cut status: Not cut CPU commanded duty: 19% or more Battery voltage: 8 V or more Engine switch: ON Starter: OFF **Possible Causes:** • Open or short in shift solenoid valve SLU circuit • Shift solenoid valve SLU • ECM
DTC: P2769 **2T ECM** **Year:** 2009, 2010 **Model:** tC, xB **Engine:** 2.4L L4 **Transmission:** All	**Torque Converter Clutch Solenoid Circuit Low (Shift Solenoid Valve DSL):** The ECM detects a short in the solenoid valve DSL circuit (0.1 sec.) when solenoid valve DSL is operated. Shift solenoid valve DSL: ON, Solenoid current cut status: Not cut, battery voltage 8 V or more, ignition switch ON, starter OFF. **Possible Causes:** • Short in shift solenoid valve DSL circuit • Shift solenoid valve DSL • ECM
DTC: P2770 **2T ECM, MIL: Yes** **Year:** 2009, 2010 **Model:** tC, xB **Engine:** 2.4L L4 **Transmission:** All	**Torque Converter Clutch Solenoid Circuit High (Shift Solenoid Valve DSL):** The ECM detects a short in the solenoid valve DSL circuit (0.1 sec.) when solenoid valve DSL is operated. Shift solenoid valve DSL: ON, Solenoid current cut status: Not cut, battery voltage 8 V or more, ignition switch ON, starter OFF. **Possible Causes:** • Open in shift solenoid valve DSL circuit • Shift solenoid valve DSL • ECM

DTC	Trouble Code Title, Conditions & Possible Causes
DTC: P2A00 **2T ECM, MIL: Yes** **Year:** 2009, 2010 **Model:** tC, xB, xD **Engine:** 1.8L L4 VIN U, 2.4L L4 **Transmission:** All	**A/F Sensor Circuit Slow Response (Bank 1 Sensor 1):** Monitor runs whenever following DTCs are not present: P0016, P0018 (VVT system bank 1, 2 - misalignment) P0017, P0019 (Exhaust VVT system bank 1, 2 - misalignment) P0031, P0032, P0051, P0052 (A/F sensor heater) P0102, P0103 (MAF meter) P0110, P0112, P0113 (IAT sensor) P0115, P0117, P0118 (ECT sensor) P0120, P0121, P0122, P0123, P0220, P0222, P0223, P2135 (TP sensor) P0125 (Insufficient ECT for closed loop) P0128 (Thermostat) P0171, P0172, P0174, P0175 (Fuel system) P0301, P0302, P0303, P0304, P0305, P0306 (Misfire) P0335 (CKP sensor) P0340 (CMP sensor) P0451, P0452, P0452 (EVAP system) P0500 (Vehicle speed sensor) P0505 (Idle speed control) Active A/F control: Performing Battery voltage: 11 V or more ECT: 167°F (75°C) or more Idle: OFF Engine rpm: Less than 4000 rpm A/F sensor status: Activated Fuel cut: OFF Engine load: 10-70% Shift position: 2 or more Catalyst monitor: Not yet MAF: 2.5-15 g/sec **Possible Causes:** • Open or short in A/F sensor (sensor 1) circuit • A/F sensor (sensor 1) • A/F sensor (sensor 1) heater • EFI relay • A/F sensor heater and EFI relay circuits • Air induction system • Fuel pressure • Injector • PCV valve and hose • PCV hose connection • ECM

OBD II Trouble Code List (U0XXX Codes)

DTC	Trouble Code Title, Conditions & Possible Causes
DTC: U0073 **T ECM, MIL: Yes** **Year:** 2009, 2010 **Model:** tC, xB, xD **Engine:** 1.8L L4 VIN U, 2.4L L4 **Transmission:** All	**Control Module Communication Bus OFF:** When any of the following is detected: 1. With the IG1 terminal voltage is between 10 and 17.4 V, after the output of data from the skid control ECU is completed, the sending continues for 5 seconds or more. 2. The condition that bus OFF state occurs once or more within 0.1 seconds occurs 10 times in succession. (Sent signals cannot be received.) 3. With the IG1 terminal voltage is between 10 and 17.4 V, a delay in receiving data from the yaw rate and acceleration sensor and steering angle sensor continues for 1 second or more. 4. With the IG1 terminal voltage is between 10 and 17.4 V, the following occurs 10 times in succession. The condition that a delay in receiving data from the yaw rate and acceleration sensor and steering angle sensor occurs more than once within 5 seconds **Possible Causes:** • CAN communication system

DTC	Trouble Code Title, Conditions & Possible Causes
DTC: U0100 **T ECM, MIL: Yes** **Year:** 2009, 2010 **Model:** tC, xB, xD **Engine:** 1.8L L4 VIN U, 2.4L L4 **Transmission:** All	**Lost Communication with ECM / PCM:** With the IG1 terminal voltage is between 10 and 17.4 V, and the vehicle speed 15 km/h (9 mph) or more, data cannot be sent to the ECM for 2 seconds or more. The IG1 terminal voltage is between 10 and 17.4 V, and the vehicle speed 15 km/h (9 mph) or more for 2 seconds or more. **Possible Causes:** • CAN communication system
DTC: U0124 **T ECM, MIL: Yes** **Year:** 2009, 2010 **Model:** tC, xB **Engine:** 2.4L L4 **Transmission:** All	**Lost Communication with Lateral Acceleration Sensor Module:** When either of the following is detected: With the IG1 terminal voltage is between 10 and 17.4 V, data from the acceleration sensor cannot be received for 1 second or more. With the IG1 terminal voltage is between 10 and 17.4 V, the following occurs 10 times in succession. The condition that data from the acceleration sensor cannot be received occurs once or more within 5 seconds. **Possible Causes:** • CAN communication system (Skid control ECU to yaw rate and acceleration sensor)

TOYOTA

4Runner

3

SPECIFICATIONS AND MAINTENANCE CHARTS

ENGINE AND VEHICLE IDENTIFICATION

Engine							Model Year	
Code ①	Liters (cc)	Cu. In.	Cyl.	Fuel Sys.	Engine Type	Eng. Mfg.	Code ②	Year
2TR-FE	2.7 (2697)	164.4	4	SFI	DOHC	Toyota	9	2009
1GR-FE	4.0 (3956)	241	6	SFI	DOHC	Toyota	A	2010
2UZ-FE	4.7 (4664)	285	8	SFI	DOHC	Toyota		

SFI: Sequential Fuel Injection

DOHC: Double Overhead Camshaft

① Stamped on the left side of the engine block

② 10th digit of the Vehicle Identification Number (VIN)

3768X_4RUN_C0001

GENERAL ENGINE SPECIFICATIONS

Year	Model	Engine Displacement Liters	Engine Series ID	Net Horsepower @ rpm	Net Torque @ rpm (ft. lbs.)	Bore x Stroke (in.)	Com-pression Ratio	Oil Pressure @ rpm
2009	4Runner	4.0	1GR-FE	236@5200	266@4000	3.70x3.74	10:01	45-65@3000
	4Runner	4.7	2UZ-FE	260@5400	306@3400	3.70x3.31	10:01	45-65@3000
2010	4Runner	4.0	1GR-FE	270@5600	278@4400	3.70x3.74	10.4:01	43-85@3000
	4Runner	2.7	2TR-FE	157@5200	178@3800	3.74x3.74	9.6:01	23-71@3000

3768X_4RUN_C0002

ENGINE TUNE-UP SPECIFICATIONS

Year	Engine Displacement Liters	Engine ID	Spark Plug Gap (in.)	Ignition Timing (deg.)*	Fuel Pump (psi)	Idle Speed (rpm)		Valve Clearance	
						MT	AT	Intake	Exhaust
2009	4.0	1GR-FE	0.043	N/A	38-44	—	650-750	0.006-0.010	0.011-0.015
	4.7	2UZ-FE	0.043	N/A	38-44	—	650-750	0.006-0.010	0.010-0.014
2010	4.0	1GR-FE	0.043	N/A	46.5-47.4	—	690-790	NA	NA
	2.7	2TR-FE	0.043	N/A	41-42	—	600-700	NA	NA

NOTE: The Vehicle Emission Control Information label often reflects specification changes made during production. The label figures must be used if they differ from those in this chart.

B: Before top dead center

* With terminals TC and E1 connected to DLC1 or for 4.7L Terminal TC and CG of DLC3 connected

3768X_4RUN_C0003

CAPACITIES

Year	Model	Engine Displacement Liters	Engine ID	Engine Oil with Filter (qts.)	Transmission (qts.) 5-Spd	Transmission (qts.) Auto.*	Transfer Case (pts.)	Drive Axle Front (pts.)	Drive Axle Rear (pts.)	Fuel Tank (gal.)	Cooling System (qts.)
2009	4Runner	4.0	1GR-FE	5.5	—	11.3	3.0	3.2	6.4	23.0	10.4
	4Runner	4.7	2UZ-FE	6.5		11.3	3.0	3.0	6.4	23.0	13.0
2010	4Runner	4.0	1GR-FE	5.5	—	11.3	3.0	3.2	6.4	23.0	11.1
	4Runner	2.7	2TR-FE	5.1		10.5	2.2	3.0	6.4	23.0	8.6

*After draining, add the following amounts, then fill to the cold full line

3768X_4RUN_C0004

FLUID SPECIFICATIONS

Year	Model	Engine Displacement Liters	Engine ID/VIN	Engine Oil	Auto. Trans.	Power Steering Fluid	Brake Master Cylinder
2009	4Runner	4.0	1GR-FE	5W-20	Toyota Genuine ATF WS	ATF Dexron II Or III	DOT 3
		4.7	2UZ-FE	5W-30	Toyota Genuine ATF WS	ATF Dexron II Or III	DOT 3
2010	4Runner	4.0	1GR-FE	5W-30	Toyota Genuine ATF WS	ATF Dexron II Or III	DOT 3
		2.7	2TR-FE	5W-30	Toyota Genuine ATF WS	ATF Dexron II Or III	DOT 3

DOT: Department Of Transpotation

NA: Not Available

3768X_4RUN_C0014

VALVE SPECIFICATIONS

Year	Engine Displacement Liters	Engine ID	Seat Angle (deg.)	Face Angle (deg.)	Spring Test Pressure (lbs. @ in.)	Spring Installed Height (in.)	Stem-to-Guide Clearance (in.) Intake	Stem-to-Guide Clearance (in.) Exhaust	Stem Diameter (in.) Intake	Stem Diameter (in.) Exhaust
2009	4.0	1GR-FE	45	44.5	41.9-46.3@ 1.311	1.311	0.0010-0.0024	0.0012-0.0026	0.2154-0.2159	0.2152-0.0158
	4.7	2UZ-FE	45	44.5	47.2-50.7@ 1.378	1.380	0.0010-0.0024	0.0012-0.0026	0.2154-0.2159	0.2152-0.2157
2010	4.0	1GR-FE	NA	NA	53-58.3@ 1.45	1.311	0.0010-0.0024	0.0012-0.0026	0.2154-0.2159	0.2152-0.0158
	2.7	2TR-FE	NA	NA	47.2-50.7@ 1.378	1.380	0.0010-0.0024	0.0012-0.0026	0.2154-0.2159	0.2152-0.2157

3768X_4RUN_C0005

CAMSHAFT AND BEARING SPECIFICATIONS CHART

All measurements are given in inches.

Year	Engine Displacement Liters	Engine ID/VIN	Journal Dia.	Brg. Oil Clearance	Shaft End-play	Runout	Journal Bore	Lobe Height Intake	Lobe Height Exhaust
2009	4.0	1GR-FE	①	②	NA	NA	NA	1.728-1.7320	1.743-1.7470
	4.7	2UZ-FE	③	④	NA	0.0031	NA	1.8624-1.8664	1.8104-1.1843
2010	4.0	1GR-FE	①	②	NA	NA	NA	1.728-1.7320	1.743-1.7470
	2.7	2TR-FE	1.415-1.4160	NA	NA	0.00118	NA	1.687-1.6910	1.687-1.6910

NA: Not Available

① No. 1 journal: 1.4152-1.4157
Other Journals: 1.0221-1.0226 in.

② No. 1 journal: 0.00126-0.00248 in.
Other journal: 0.000984-0.00244 in.
Other Maximum Journals: 0.00276 in.

③ 1.0612-1.0618 in.

④ 0.012-0.0028 in.

3768X_4RUN_C0013

CRANKSHAFT AND CONNECTING ROD SPECIFICATIONS

All measurements are given in inches.

Year	Engine Displacement Liters	Engine ID	Crankshaft Main Brg. Journal Dia.	Crankshaft Main Brg. Clearance	Crankshaft Shaft End-play	Crankshaft Thrust on No.	Connecting Rod Journal Diameter	Connecting Rod Oil Clearance	Connecting Rod Side Clearance
2009	4.0	1GR-FE	2.8342-2.8346	0.0007-0.0018	—	2	2.2044-2.2047	0.0010 0.0018	0.0059 0.0138
	4.7	2UZ-FE	2.6373-2.6378	①	0.0008-0.0087	3	2.0465-2.0472	0.0011-0.0021	0.0063-0.0138
2010	4.0	1GR-FE	2.8342-2.8346	0.0010-0.0018	—	—	2.2044-2.2047	0.0016 0.0026	0.0059 0.0138
	2.7	2TR-FE	2.361-2.3620	②	—	—	2.086-2.0870	0.0015-0.0026	0.00591-0.0138

① Nos. 1 and 2: 0.0011-0.0018
All others: 0.0016-0.0023

② No. 3: 0.00142-0.00264
All others: 0.0018-0.0024

3768X_4RUN_C0006

PISTON AND RING SPECIFICATIONS

All measurements are given in inches.

Year	Engine Displacement Liters	Engine ID	Piston Clearance	Ring Gap Top Comp.	Ring Gap Bottom Comp.	Ring Gap Oil Control	Ring Side Clearance Top Comp.	Ring Side Clearance Bottom Comp.	Ring Side Clearance Oil Control
2009	4.0	1GR-FE	0.0031-0.0051	0.0118-0.0157	0.0157-0.0197	①	0.0008-0.0028	0.0008-0.0024	0.0028 0.0060
	4.7	2UZ-FE	0.0035-0.0044	0.0118-0.0157	0.0157-0.0217	0.0051-0.0150	0.0012-0.0031	0.0012-0.0028	SNUG
2010	4.0	1GR-FE	0.000354-0.0020	0.00866-0.0126	0.0138-0.0177	0.0394-0.0157	0.0008-0.0028	0.0008-0.0024	0.0028 0.0059
	2.7	2TR-FE	0.0035-0.0044	0.0102-0.0150	0.0232-0.0280	0.0039-0.0157	0.00079-0.0030	0.00079-0.0026	0.00079-0.0028

① No 1: 0.039, No 2: 0.043, No 3: 0.039

3768X_4RUN_C0007

TORQUE SPECIFICATIONS

All readings in ft. lbs.

Year	Engine Displacement Liters	Engine ID	Cylinder Head Bolts	Main Bearing Bolts	Rod Bearing Bolts	Crankshaft Damper Bolts	Flywheel Bolts	Manifold Intake	Manifold Exhaust	Spark Plugs	Oil Pan Drain Plug
2009	4.0	1GR-FE	①	②	③	184	61	19	22	13	29
	4.7	2UZ-FE	④	⑤	⑥	181	⑦	13	33	13	29
2010	4.0	1GR-FE	①	⑧	③	184	61	15	15	13	30
	2.7	2TR-FE	⑨	⑩	⑥	181	⑦	18	27	13	28

① Step 1: 45 ft. lbs.
Step 2: Plus 180 degrees

② Step 1: 16 main cap bolts 45 ft. lbs.
Step 2: Plus 90 degr
Step 3: 8 side main cap bolts 19 ft. lbs.

③ Step 1: 18 ft. lbs.
Step 2: Plus 90 degrees

④ Step 1: 30 ft. lbs.
Step 2: Plus 90 degrees
Step 3: Plus 90 degrees

⑤ Step 1: 20 ft. lbs.
Step 2: Plus 90 degrees

⑥ Step 1: 18 ft. lbs.
Step 2: Plus 90 degrees

⑦ Step 1: 22 ft. lbs.
Step 2: Plus 90 degrees

⑧ Step 1: 45 ft. lbs.
Step 2: Plus 90 degrees

⑨ Step 1: 29 ft. lbs.
Step 2: Plus 90 degrees
Step 3: Plus 90 degrees

⑩ Bolt A: 9 ft. lbs.
Bolt B: 11 ft. lbs.

3768X_4RUN_C0008

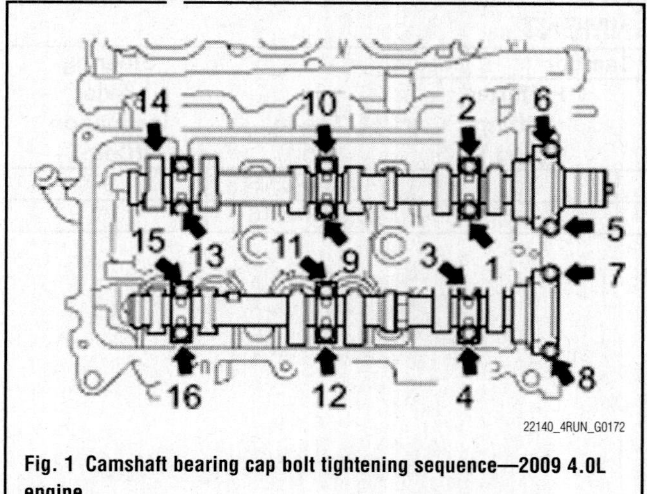

Fig. 1 Camshaft bearing cap bolt tightening sequence—2009 4.0L engine

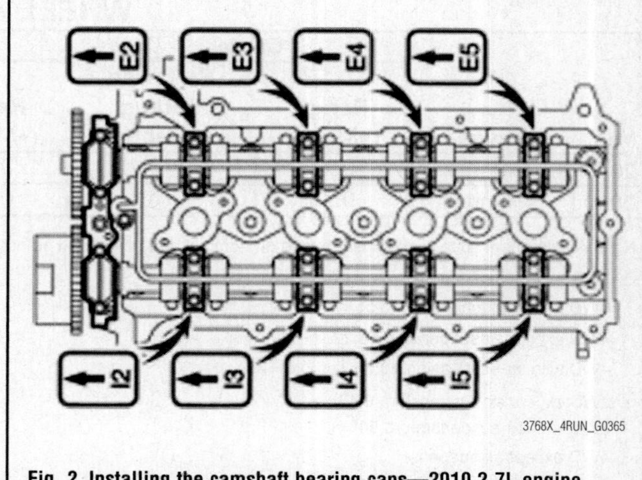

Fig. 2 Installing the camshaft bearing caps—2010 2.7L engine

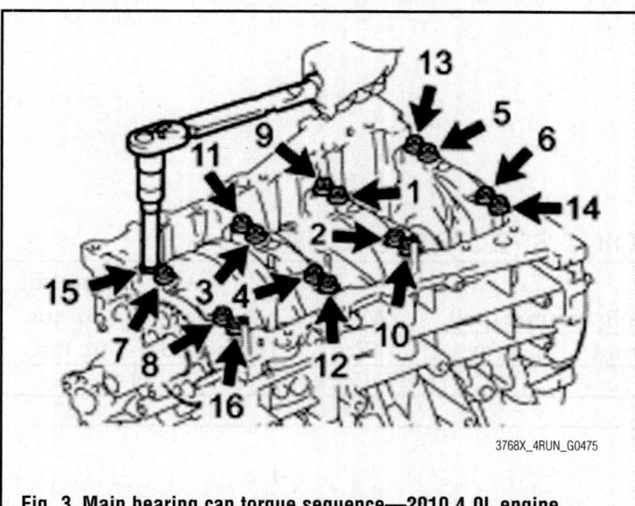

Fig. 3 Main bearing cap torque sequence—2010 4.0L engine

WHEEL ALIGNMENT

		Caster		Camber			
Year	Model	Range (+/-Deg.)	Preferred Setting (Deg.)	Range (+/-Deg.)	Preferred Setting (Deg.)	Toe-in (in.)	Steering Axis Inclination (Deg.)
2009	4Runner	0.75	①	0.75	②	0.08+/-0.16	③
2010	4Runner	0.75	①	0.75	②	0.08+/-0.16	③

Note: All alignment specifications are based on nominal ride height and standard tires

① 2WD except air suspension +3.38
2WD with air suspension +3.55
4WD except air suspension +3.22
4WD with air suspension +3.37

② 2WD except air suspension -0.47
2WD with air suspension -0.50
4WD except air suspens
4WD with air suspension -0.17

③ 2WD except air suspension 12.97+/-.075
2WD with air suspension 13.00+/-.075
4WD except air suspension 12.65+/-.075
4WD with air suspension 12.67+/-.075

22140_4RUN_C0009

TIRE, WHEEL AND BALL JOINT SPECIFICATIONS

		OEM Tires		Tire Pressures (psi)				
Year	Model	Standard	Optional	Front	Rear	Wheel Size	Ball Joint Inspection	Lug Nut Torque (ft. lbs.)
2009	4Runner	P265/70R16	P265/65R17	①	①	7-J/7.5J	②	83
2010	4Runner	P265/70R16	P265/65R17	①	①	7-J/7.5J	②	83

OEM: Original Equipment Manufacturer

① See placard on vehicle

② Upper arm ball joint turning torque: 39 inch lbs.
Lower arm ball joint turning torque: 26 inch lbs.

3768X_4RUN_C0010

BRAKE SPECIFICATIONS

All measurements in inches unless noted

			Brake Disc				Brake Caliper	
Year	Model		Original Thickness	Minimum Thickness	Maximum Runout	Minimum Lining Thickness	Bracket Bolts (ft. lbs.)	Mounting Bolts (ft. lbs.)
2009	4Runner	F	1.102	1.024	0.0020	0.039	—	91
		R	0.709	0.630	0.0079	0.039	77	65
2010	4Runner	F	1.260	1.150	0.0020	0.039	—	91
		R	0.709	0.630	0.0079	0.039	77	65

F: Front
R: Rear

3768X_4RUN_C0011

SCHEDULED MAINTENANCE INTERVALS

4Runner

TO BE SERVICED	TYPE OF SERVICE	VEHICLE MILEAGE INTERVAL (x1000)													
		5	10	15	20	25	30	35	40	45	50	55	60	90	120
Engine oil & filter	R	✓	✓	✓	✓	✓	✓	✓	✓	✓	✓	✓	✓	✓	✓
Automatic transmission fluid	S/I			✓			✓			✓			✓	✓	✓
Ball joints & dust covers	S/I			✓			✓			✓			✓	✓	✓
Bolts & nuts on chassis & body	S/I			✓			✓			✓			✓	✓	✓
Brake linings & drums	S/I	✓	✓	✓	✓	✓	✓	✓	✓	✓	✓	✓	✓	✓	✓
Brake line pipes & hoses	S/I			✓			✓			✓			✓	✓	✓
Brake pads & discs (front & rear)	S/I	✓	✓	✓	✓	✓	✓	✓	✓	✓	✓	✓	✓	✓	✓
Brake fluid	R						✓						✓	✓	✓
Rack and pinion assembly	S/I			✓			✓			✓			✓	✓	✓
Steering linkage & boots	S/I			✓			✓			✓			✓	✓	✓
Air cleaner filter	R						✓						✓	✓	✓
Spark plugs ①	R														✓
Drive belts	S/I												✓	✓	✓
Exhaust pipes & mountings	S/I			✓			✓			✓			✓	✓	✓
Fuel lines & connections	S/I						✓						✓	✓	✓
Engine coolant ②	S/I			✓			✓			✓			✓	✓	
Rear differential & transfer case oil	S/I			✓			✓			✓			✓	✓	✓
Fuel tank cap gasket	S/I						✓						✓	✓	✓
Rotate tires	S/I			✓			✓			✓			✓		✓
Clean air conditioning filter ③	S/I			✓			✓			✓			✓		✓
Axle shaft bolts	S/I			✓			✓			✓			✓	✓	✓
Brake pad thickness and rotor	S/I						✓						✓	✓	✓

R: Replace S/I: Service or Inspect

① Spark plugs are replaced at 120,000 miles

② Replace engine coolant at 100,000 miles and then inspect every 15,000 miles

③ Replace air conditioning filter every 30,000 miles

FREQUENT OPERATION MAINTENANCE (SEVERE SERVICE)

If a vehicle is operated under any of the following conditions it is considered severe service:

- Extremely dusty areas.

- 50% or more of the vehicle operation is in 32°C (90°F) or higher temperatures, or constant temperatures below 0°C (32°F).

- Prolonged idling (vehicle operation in stop and go traffic).

- Frequent short running periods (engine does not warm to normal operating temperatures).

- Police, taxi, delivery usage or trailer towing usage.

Air cleaner filter: service or inspect every 5000 miles

Rear differential & transfer case oil: replace every 15,000 miles.

Ball joints & dust covers: service or inspect every 5000 miles.

Bolts & nuts on chassis & body: service or inspect every 5000 miles.

Axle shaft bolts: service or inspect every 5000 miles.

Steering linkage: service or inspect every 5000 miles.

3768X_4RUN_C0012

BRAKES — INFORMATION AND PRECAUTIONS

ANTI-LOCK SYSTEMS

- Certain components within the ABS system are not intended to be serviced or repaired individually.
- Do not use rubber hoses or other parts not specifically specified for and ABS system. When using repair kits, replace all parts included in the kit. Partial or incorrect repair may lead to functional problems and require the replacement of components.
- Lubricate rubber parts with clean, fresh brake fluid to ease assembly. Do not use shop air to clean parts; damage to rubber components may result.
- Use only DOT 3 brake fluid from an unopened container.
- If any hydraulic component or line is removed or replaced, it may be necessary to bleed the entire system.
- A clean repair area is essential. Always clean the reservoir and cap thoroughly before removing the cap. The slightest amount of dirt in the fluid may plug an orifice and impair the system function. Perform repairs after components have been thoroughly cleaned; use only denatured alcohol to clean components. Do not allow ABS components to come into contact with any substance containing mineral oil; this includes used shop rags.
- The Anti-Lock control unit is a microprocessor similar to other computer units in the vehicle. Ensure that the ignition switch is **OFF**before removing or installing controller harnesses. Avoid static electricity discharge at or near the controller.
- If any arc welding is to be done on the vehicle, the control unit should be unplugged before welding operations begin.

DISC AND DRUM SYSTEMS

CAUTION

Dust and dirt accumulating on brake parts during normal use may contain asbestos fibers from production or aftermarket brake linings. Breathing excessive concentrations of asbestos fibers can cause serious bodily harm. Exercise care when servicing brake parts. Do not sand or grind brake lining unless equipment used is designed to contain the dust residue. Do not clean brake parts with compressed air or by dry brushing. Cleaning should be done by dampening the brake components with a fine mist of water, then wiping the brake components clean with a dampened cloth. Dispose of cloth and all residue containing asbestos fibers in an impermeable container with the appropriate label. Follow practices prescribed by the Occupational Safety and Health Administration (OSHA) and the Environmental Protection Agency (EPA) for the handling, processing, and disposing of dust or debris that may contain asbestos fibers.

BRAKES — BLEEDING THE BRAKE SYSTEM

BLEEDING PROCEDURE

➡If any work is done on the brake system or if air is suspected in the brake lines, bleed the air from the system.

➡Do not let brake fluid remain on a painted surface. Wash it off immediately.

1. Before servicing the vehicle, refer to the Precautions section.
2. Check the fluid level in the reservoir after bleeding each wheel. Add DOT3 fluid, if necessary.
3. If the hydraulic brake booster was disassembled or if the reservoir becomes empty, bleed the air from the hydraulic brake booster as follows:

➡Perform this step only if the brake booster with accumulator pump assembly is removed and/or installed.

   a. Turn the ignition switch OFF, depress the brake pedal 20 times or more to release the pressure from the accumulator.
   b. Fully depress the brake pedal 10 times.
   c. Turn the ignition switch to the ON position and start the brake booster pump.
   d. Make sure the pump operates for 8 to 14 seconds.

➡If the pump does not operate as specified, repeat the above and recheck the operating time.

4. Bleeding Front Brake Lines:
   a. Turn the ignition switch to the ON position and wait until the pump motor has stopped.
   b. Connect the vinyl tube to the brake caliper.
   c. Depress the brake pedal several times, then loosen the bleeder plug with the pedal held down.
   d. At the point when the fluid stops coming out, tighten the bleeder plug, 8 ft. lbs. (11 Nm) then release the brake pedal.
   e. Repeat procedure until all the air in the fluid has been bled out.
   f. Repeat the above procedures to bleed the other brake line.
5. Bleeding Rear Brake Lines:
   a. Turn the ignition switch to the ON position and depress the brake pedal.
   b. Connect the vinyl tube to the brake caliper.
   c. Loosen the bleeder plug and release air.

➡Brake fluid is sent through the pump, so keep the brake pedal depressed until the air is completely bled out.

   d. When the air is completely bled out of the brake fluid through the bleeder plug, tighten the bleeder plug to 8 ft. lbs. (11 Nm) then release.
   e. Repeat the above procedures to bleed the other brake line.
6. Bleeding Master Cylinder Solenoid is only possible with a Toyota proprietary scan system.

WHEEL SPEED SENSORS

REMOVAL & INSTALLATION

Front

2009 Models

➡**Replacement of RH side is same as that of LH side.**

1. Disconnect negative battery cable.
2. Remove front wheel.
3. Disconnect the speed sensor connector,
4. Using a hexagon wrench (5 mm), remove the bolt and front speed sensor from the steering knuckle

➡**Do not stick and foreign matter on the sensor tip. Do not let the foreign matter into the sensor installation hole.**

To install:

➡**Make sure the sensor tip is clean.**

5. To install, Using a hexagon wrench (5 mm), install the front speed sensor with the bolt to the steering knuckle and torque to 73 inch lbs. (8.3 Nm).
6. Connect the speed sensor connector.
7. Install front wheel and tighten to 82 ft. lbs. (112 Nm).
8. Connect negative battery cable.
9. Check speed sensor signal.

2010 Models

See Figure 4.

➡**The procedure listed below is for the LH side.**

1. Remove the front wheel.
2. Remove the front skid control sensor wire.
   a. Disconnect the connector from the front speed sensor.
   b. Remove the 2 bolts and 2 harness clamps.
   c. Detach the clip.
   d. Remove the 2 bolts and the 2 harness clamps.
   e. Disconnect the connector as follows.

LH

a. Disconnect the connector.
b. Detach the connector.

RH

a. Disconnect the connector.
b. Detach the connector from the skid control sensor clamp.
c. Detach the clip.
d. Remove the bolt and skid control sensor clamp.

3. Remove the skid control sensor clamp.
   a. Remove the bolt and skid control sensor clamp from the knuckle.
4. Remove the front speed sensor from the knuckle.

➡**Pull out the sensor while trying as much as possible not to rotate it.**

To install:

5. Install the front speed sensor. Tighten to 75 inch lbs. (8.5 Nm).

➡**Make sure there are no pieces of iron or other foreign matter attached to the sensor tip.**

➡**While inserting the speed sensor into the knuckle hole, do not strike or damage the sensor tip.**

➡**After installing the speed sensor, make sure there is no clearance or foreign matter between the sensor stay part and the knuckle.**

➡**Make sure there is no foreign matter attached to the speed sensor rotor.**

6. Install the skid control sensor clamp. Tighten the bolt to 9 ft. lbs. (13 Nm).

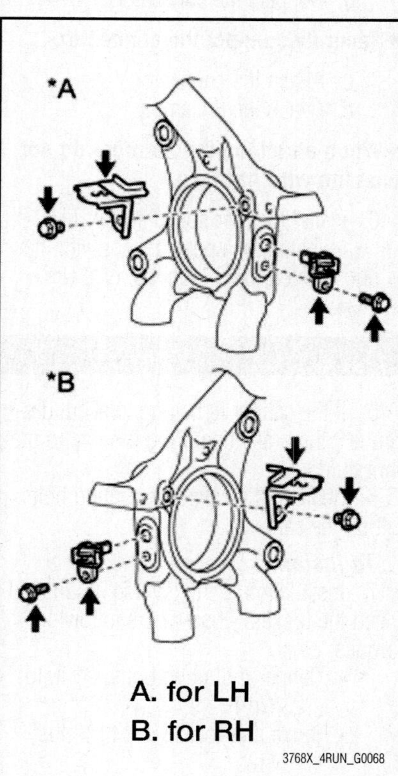

Fig. 4 Removing the skid control sensor clamp

➡**Install the clamp so that the rotation stopper touches the knuckle.**

7. Install the front skid control sensor wire.
   a. For the LH, attach the connector, and then connector the connector.
   b. For the RH, install the skid control sensor clamp with the bolt. Tighten to 44 inch lbs. (5 Nm).

➡**Make sure the clamp rotation stopper touches the installation position.**

c. For the RH, attach the connector, and then connect the connector.

➡**Securely connect the connector.**

➡**When connecting the connector, do not twist the wire harness.**

d. For the RH, attach the clip.
e. Install the 2 harness clamps with the 2 bolts. Tighten to 9 ft. lbs. (13 Nm).

➡**When installing the clamps, do not twist the wire harness.**

➡**Make sure the clamp rotation stopper touches the installation position.**

f. Install the 2 harness clamps with the 2 bolts. Tighten to 9 ft. lbs. (13 Nm).

➡**When installing the clamps, do not twist the wire harness.**

➡**Make sure the clamp rotation stopper touches the installation position.**

g. Attach the clip.
h. Connect the connector.

➡**Securely connect the connector.**

8. Install the front wheel.
9. Check the speed sensor signal.

Rear

2009 Models

➡**Replacement of RH side is same as that of LH side.**

1. Disconnect negative battery cable.
2. Remove rear wheel.
3. Disconnect the speed sensor connector.
4. Remove the bolt and front speed sensor from the axle hub.
5. Disconnect retaining clips.

➡**Do not stick and foreign matter on the sensor tip. Do not let the foreign matter into the sensor installation hole.**

To install:

➡**Make sure the sensor tip is clean.**

6. To install, install the front speed sensor with the bolt to the axle hub and torque to 73 inch lbs. (8.3 Nm).
7. Connect the speed sensor connector.
8. Connect retaining clips.
9. Install rear wheel and tighten to 82 ft. lbs. (112 Nm).
10. Connect negative battery cable.
11. Check speed sensor signal.

2010 Models

See Figures 5 and 6.

1. Remove the rear wheel.
2. Remove the rear speed sensor LH.
   a. Disconnect the speed sensor connector.
   b. Remove the nut and speed sensor.

➡Pull out the sensor while trying as much as possible not to rotate it.

3. Remove the rear speed sensor RH.
4. Remove the skid control sensor wire.
   a. Disconnect the connector.
   b. Detach the connector.

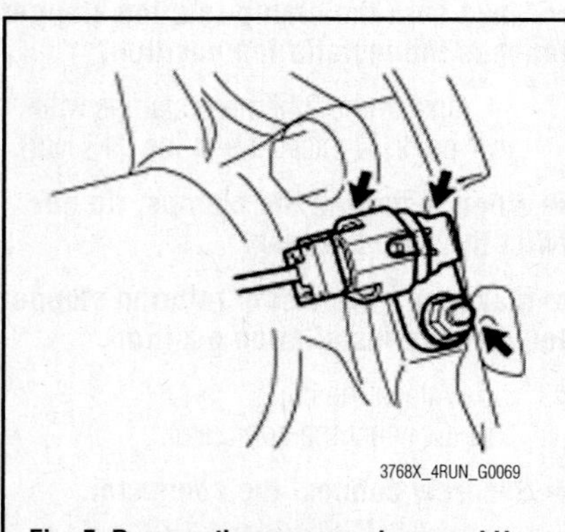

Fig. 5 Remove the rear speed sensor LH

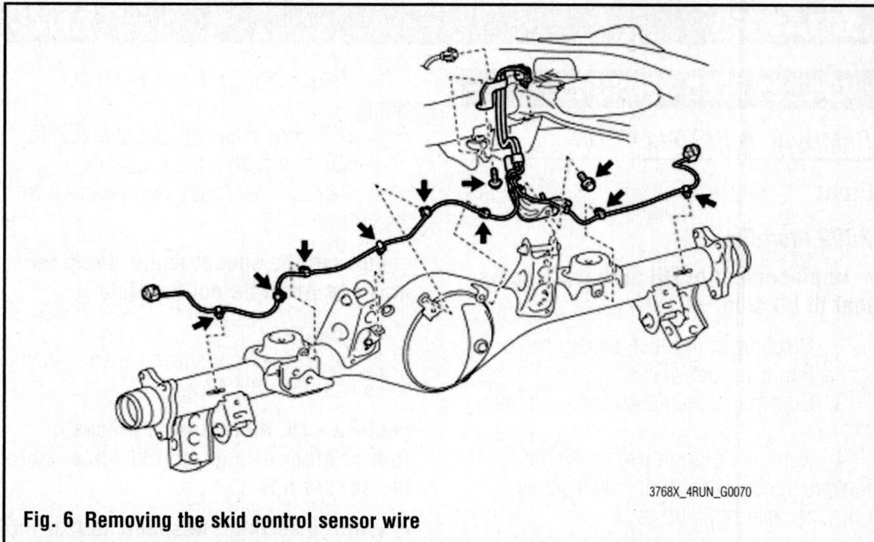

Fig. 6 Removing the skid control sensor wire

   c. Detach the 8 clamps.
   d. Remove the 2 bolts and 2 sensor clamps.

To install:

5. Install the skid control sensor wire.
   a. Install the 2 sensor clamps with the 2 bolts. Tighten to 9 ft. lbs. (13 Nm).

➡Make sure the clamp rotation stopper touches the installation position.

   b. Connect the connector.

➡Securely connect the connector.

   c. Attach the connector.
   d. Attach the 8 clamps.

➡When attaching the clamps, do not twist the wire harness.

6. Install the rear speed sensor LH.
   a. Install the speed sensor with the nut. Tighten to 69 inch lbs. (7.8 Nm).

➡Make sure there are no pieces of iron or other foreign matter attached to the sensor tip.

➡While inserting the speed sensor into the axle hole, do not strike or damage the sensor tip.

➡After installing the speed sensor, make sure there is no clearance or foreign matter between the sensor stay part and the axle.

➡Make sure there is no foreign matter attached to the speed sensor rotor.

   b. Connect the speed sensor connector.

➡Securely connect the connector.

7. Install the rear speed sensor RH.
8. Install the rear wheel.
9. Check the speed sensor signal.

BRAKES

FRONT DISC BRAKES

BRAKE CALIPER

REMOVAL & INSTALLATION

2009 Models

See Figure 7.

➡Use the same procedure for the RH and LH sides.

➡The procedure listed below is for the LH side.

1. Remove the wheel.
2. Remove the anti-rattle spring from the caliper.
3. Remove the clips and anti-rattle pins.
4. Lift out the pads and shims.
5. If the caliper is being replaced, disconnect the brake line. Plug the line to prevent fluid loss.
6. Remove the caliper mounting bolts. Lift off the caliper.

To install:

7. Installation is the reverse of removal. Bleed the brakes. Observe the following torques:
   - Caliper mounting bolts: 91 ft. lbs. (123 Nm)
   - Brake line-to-caliper: 11 ft. lbs. (15 Nm)

2010 Models

See Figures 8 and 9.

➡Use the same procedure for the RH and LH sides.

➡The procedure listed below is for the LH side.

1. Remove front wheel.
2. Drain the brake fluid.

➡Wash off the brake fluid immediately if it comes into contact with a painted surface.

3. Remove the front disc brake pad.
4. Remove the disc brake caliper assembly LH.
   a. Using a union nut wrench, disconnect the brake tube from the disc brake caliper assembly.

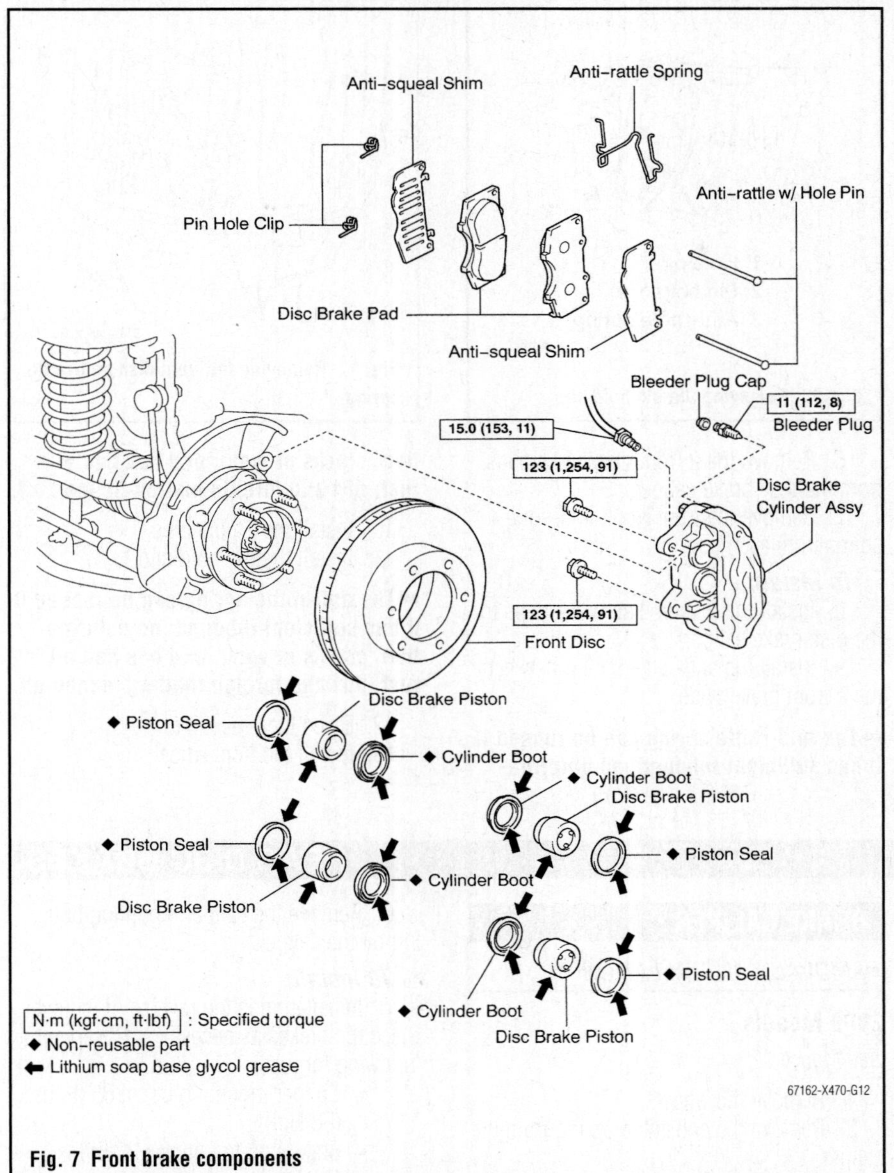

Fig. 7 Front brake components

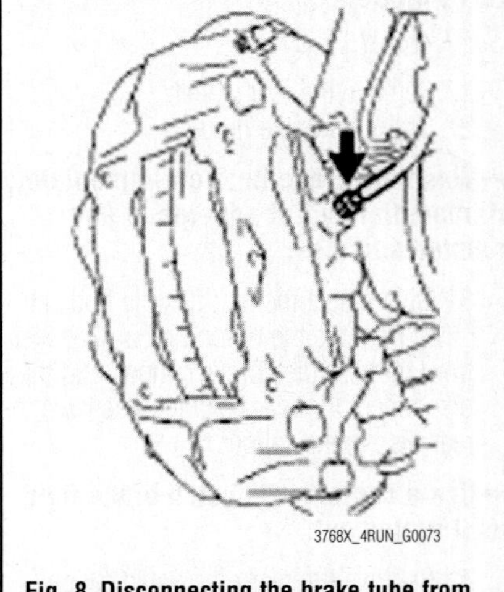

Fig. 8 Disconnecting the brake tube from the disc brake caliper

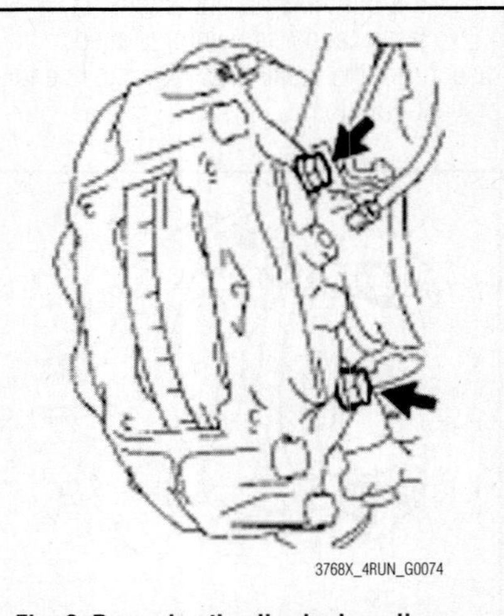

Fig. 9 Removing the disc brake caliper assembly

➡Use a container to catch the brake fluid as it drains out.

b. Remove the 2 bolts and the disc brake caliper assembly.

To install:

5. Install the disc brake caliper assembly LH.
   a. Install the disc brake caliper assembly with the 2 bolts. Tighten to 91 ft. lbs. (123 Nm).
   b. Using a union nut wrench, connect the brake tube to the disc brake caliper assembly. Tighten to 11 ft. lbs. (15 Nm).
6. Install the front No. 1 anti-squeal shim to the front disc brake pads.

➡If necessary, replace the anti-squeal shim when replacing the brake pad.

➡There should be no oil or grease on the friction surfaces of the front disc pads and the front disc.

7. Install the front disc brake pad.
8. Bleed the brake line.
9. Install the front wheel.

DISC BRAKE PADS

REMOVAL & INSTALLATION

2009 Models

See Figure 7.

1. Raise the vehicle and support it safely.
2. Remove the wheels.
3. Remove the clip, pins and anti-rattle spring.
4. Withdraw the pads and remove the anti-squeal shims.

To install:

5. Before installing the new pads, check the disc thickness and disc runout.
6. Siphon out a small amount of brake fluid from the reservoir.
7. Press in the pistons with a hammer handle or equivalent.
8. Apply disc brake grease to both sides of the inner anti-squeal shim. Install the anti-squeal shims to the new pads.
9. Install the pads.
10. Install the anti-rattle springs and pins. Install the clip.
11. Install the wheels.
12. Check and adjust the fluid level. Apply the brake pedal several times.
13. Road-test the vehicle for proper operation.

2010 Models

See Figures 10 and 11.

➡Use the same procedure for the RH and LH sides.

➡The procedure listed below is for the LH side.

1. Remove front wheel.
2. Drain the brake fluid.

➡Wash off the brake fluid immediately if it comes into contact with a painted surface.

3. Remove the pin hold clip.

➡The pin hold clip can be reused if it has sufficient rebound; no deformation or wear; and has had all rust, dirt and foreign matter cleaned off.

4. Remove the 2 hole pins.
5. Remove the front disc brake anti-rattle spring.

➡The anti-rattle spring can be reused if it has sufficient rebound; no deformation, cracks or wear; and has had rust, dirt and foreign matter cleaned.

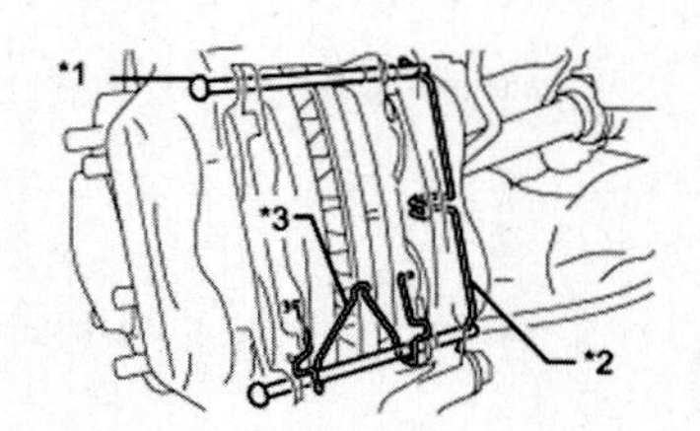

Fig. 10 Removing the pin hold clip

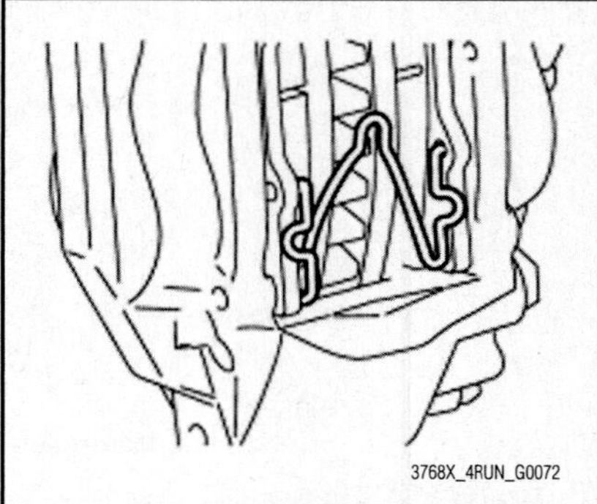

Fig. 11 Removing the front disc anti-rattle spring

6. Remove the 2 front disc brake pads from the disc brake caliper.
7. Remove the front No. 1 anti-squeal shims from each pad.

To install:

8. Install the 2 front disc brake pads to the disc brake caliper.
9. Install the anti-rattle spring between the 2 front brake pads.

➡The anti-rattle spring can be reused if it has sufficient rebound; no deformation, cracks or wear, and has had all rust, dirt and foreign matter cleaned off.

10. Install the 2 hole pins.
11. Install the pin hold clip.

➡The anti-rattle spring can be reused if it has sufficient rebound; no deformation, cracks or wear, and has had all rust, dirt and foreign matter cleaned off.

12. Bleed the brake line.
13. Install the front wheel.

BRAKES — REAR DISC BRAKES

⁂ CAUTION

Dust and dirt accumulating on brake parts during normal use may contain asbestos fibers from production or aftermarket brake linings. Breathing excessive concentrations of asbestos fibers can cause serious bodily harm. Exercise care when servicing brake parts. Do not sand or grind brake lining unless equipment used is designed to contain the dust residue. Do not clean brake parts with compressed air or by dry brushing. Cleaning should be done by dampening the brake components with a fine mist of water, then wiping the brake components clean with a dampened cloth. Dispose of cloth and all residue containing asbestos fibers in an impermeable container with the appropriate label. Follow practices prescribed by the Occupational Safety and Health Administration (OSHA) and the Environmental Protection Agency (EPA) for the handling, processing, and disposing of dust or debris that may contain asbestos fibers.

BRAKE CALIPER

REMOVAL & INSTALLATION

2009 Models

See Figure 12.

1. Remove the wheel.
2. Remove the anti-rattle spring from the caliper.
3. Remove the clips and anti-rattle pins.
4. Lift out the pads and shims.
5. If the caliper is being replaced, disconnect the brake line. Plug the line to prevent fluid loss.

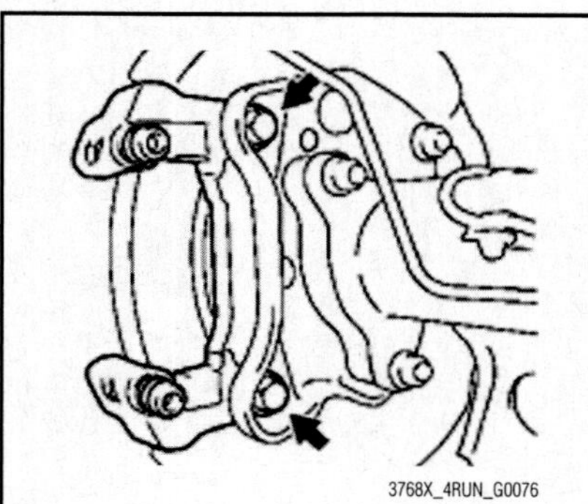

Fig. 12 Removing the rear brake caliper—2009

6. Remove the caliper mounting bolts. Lift off the caliper.

To install:

7. Installation is the reverse of removal. Bleed the brakes as necessary. Observe the following torques:
   - Caliper mounting bolts: 65 ft. lbs. (88 Nm)
   - Brake line-to-caliper: 23 ft. lbs. (31 Nm)

2010 Models

See Figure 13.

1. Remove the rear wheel.
2. Drain the brake fluid.

➡Wash the brake fluid off immediately if immediately if it adheres to any painted surfaces.

3. Disconnect the rear flexible hose LH.
   a. Remove the union bolt and gasket from the rear disc brake caliper, and then disconnect the rear flexible hose from the rear disc brake caliper.

➡Use a container to catch brake fluid as it drains out.

4. Remove the rear disc brake caliper assembly LH.

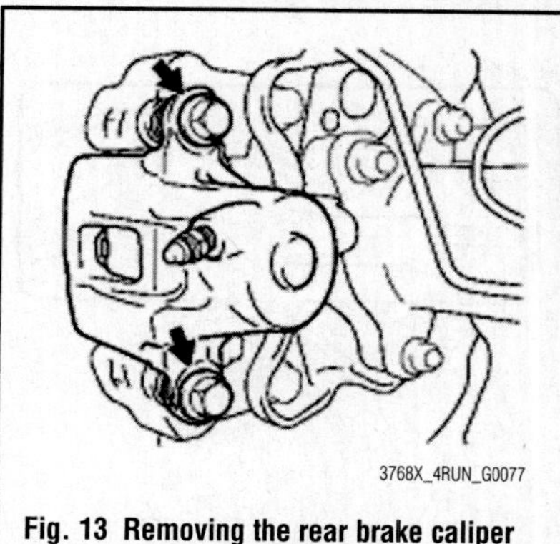

Fig. 13 Removing the rear brake caliper

a. Remove the 2 caliper slide pins from the rear disc brake caliper.

b. Remove the rear disc brake caliper from the rear disc brake caliper mounting.

To install:

5. To install, reverse the removal procedure. Apply a light coat of lithium soap base glycol grease to the sliding surfaces of the rear disc brake caliper slide pins. Tighten the bolts to 65 ft. lbs. (88 Nm).

DISC BRAKE PADS

REMOVAL & INSTALLATION

2009 Models

1. Raise the vehicle and support it safely.
2. Remove the wheels.
3. Remove the brake caliper and suspend it so the hose is not stretched.
4. Remove the brake pads, anti-squeal shim, pad support plates and wear indicators.

To install:

5. Before installing the new pads, check the disc thickness and disc runout.
6. Install the pad support plates.
7. Install the pad wear indicator plates on each pad.
8. Install the anti-squeal shim to the outer pad. Install the pads.
9. Install the brake caliper and tighten to 65 ft. lbs. (88 Nm).
10. Install the wheels.
11. Apply the brake pedal several times.
12. Road-test the vehicle for proper operation.

2010 Models

1. Remove the rear wheel.
2. Drain the brake fluid.
3. Remove the rear brake cylinder assembly LH.
4. Remove the rear disc brake pad.
   a. Remove the 2 rear disc brake pads together with the rear disc brake anti squeal shims from the rear disc brake caliper mounting.

To install:

5. To install, reverse the removal procedure.

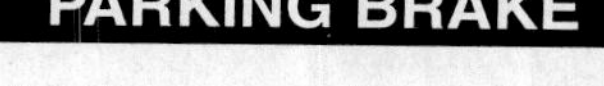

PARKING BRAKE CABLES

ADJUSTMENT

1. Remove rear wheel.
2. Adjust parking brake shoe clearance.
3. Install rear wheel an tighten to 82 ft. lbs. (112 Nm)
4. Inspect parking brake lever travel.
5. Slowly depress the parking brake lever all the way, and count the number of clicks.
   a. Parking brake lever travel at 66 ft. lbs. (294 Nm) 5 to 7 clicks.
6. Adjust parking brake lever travel by removing the console panel upper.
7. Turn the adjusting nut until the parking brake lever travel becomes correct.
8. Check whether parking brake drags or not.
9. When operating the parking brake lever, check that the parking brake lever indicator light comes on.
10. Install the console panel upper.

PARKING BRAKE SHOES

REMOVAL & INSTALLATION

2009 Models

See Figure 14.

1. Before servicing the vehicle, refer to the Precautions section.
2. Raise and safely support the vehicle.

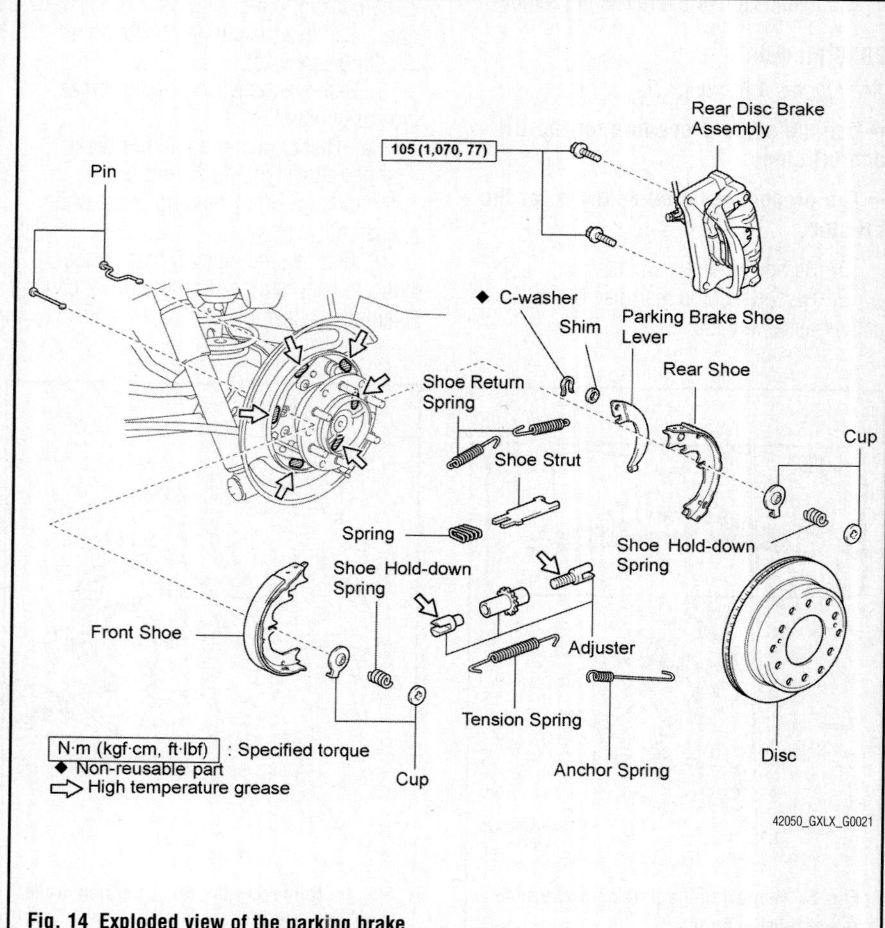

Fig. 14 Exploded view of the parking brake

3. Remove the rear wheel.

4. Remove the 2 mounting bolts and remove the disc brake assembly.

5. Suspend the disc brake securely and so the hose is not stretched.

6. Release the parking brake lever.

7. Place matchmarks on the disc and rear axle hub.

8. Remove the disc.

➡If the disc cannot be removed easily, turn the shoe adjuster until the wheel turns freely.

9. Using needle-nose pliers, remove the 2 shoe return springs.

➡At the time of reassembly, install the strut with the spring facing forward.

10. Slide the front shoe toward outside and remove the shoe adjuster.

11. Using a needle-nose pliers, disconnect the anchor spring and tension spring from the front shoe.

12. Using a needle-nose pliers, disconnect the anchor spring and tension spring from the rear shoe.

To install:

13. Installation is the reverse of removal.

2010 Models

See Figures 15 through 22.

➡Use the same procedure for the RH and LH sides.

➡The procedure listed below is for the LH side.

1. Remove the rear wheel.

2. Disconnect the rear disc brake caliper assembly LH.

   a. Remove the 2 bolts and disconnect the rear disc brake caliper.

➡Do not disconnect the flexible hose from the disc brake caliper.

➡Do not twist or bend the flexible hose.

3. Remove the rear disc.

4. Remove the parking brake shoe return tension spring.

   a. Remove the 2 parking brake shoe return tension springs.

5. Disconnect the No. 1 parking brake shoe assembly LH.

   a. Remove the parking brake shoe hold down spring cup and parking brake shoe hold down spring to disconnect the No. 1 parking brake shoe assembly from the backing plate.

6. Remove the parking brake shoe strut and parking brake shoe strut compression spring.

7. Disconnect the No. 2 parking brake shoe assembly LH.

   a. Remove the parking brake shoe hold down spring cup and parking brake shoe hold down spring to disconnect the No. 2 parking brake shoe assembly from the backing plate.

8. Remove the parking brake shoe adjusting screw set.

9. Remove the No. 1 parking brake shoe assembly LH.

   a. Disconnect the parking brake shoe return tension spring to remove the No. 1 parking brake shoe assembly.

10. Remove the parking brake shoe return tension spring from the No. 2 parking brake shoe assembly.

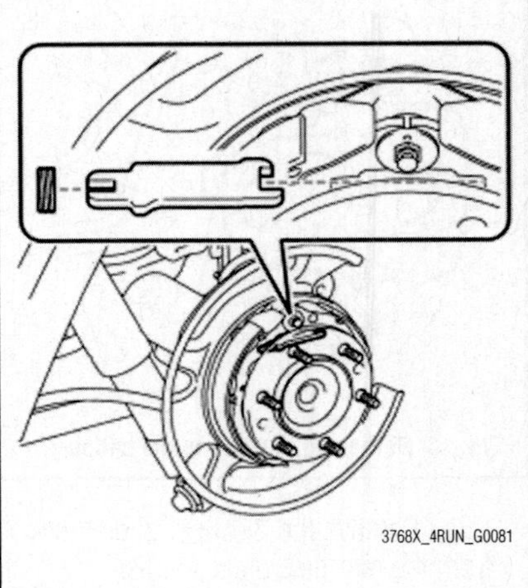

Fig. 17 Removing the parking brake shoe strut LH

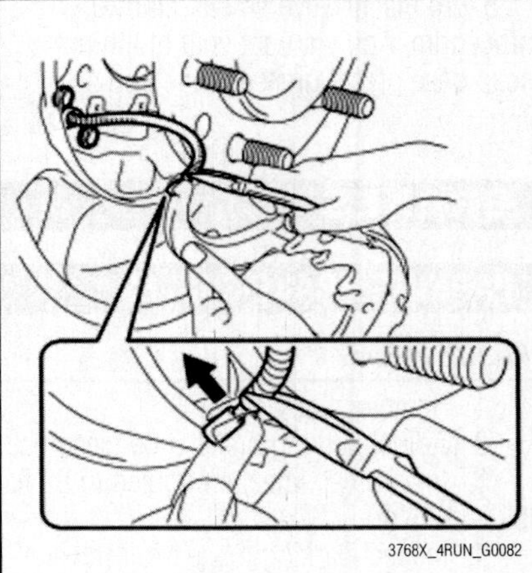

Fig. 18 Removing the No. 2 parking brake shoe assembly with the parking brake shoe lever

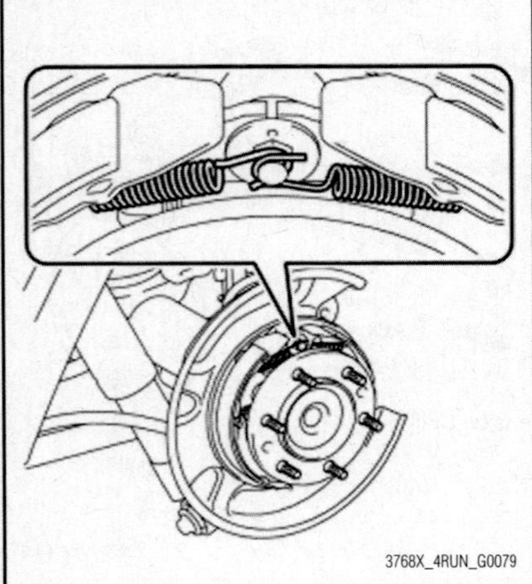

Fig. 15 Removing the parking brake shoe return tension spring

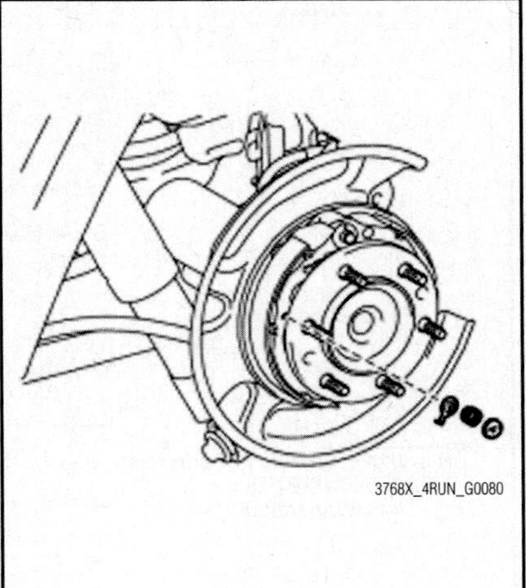

Fig. 16 Removing the No. 1 parking brake shoe assembly LH

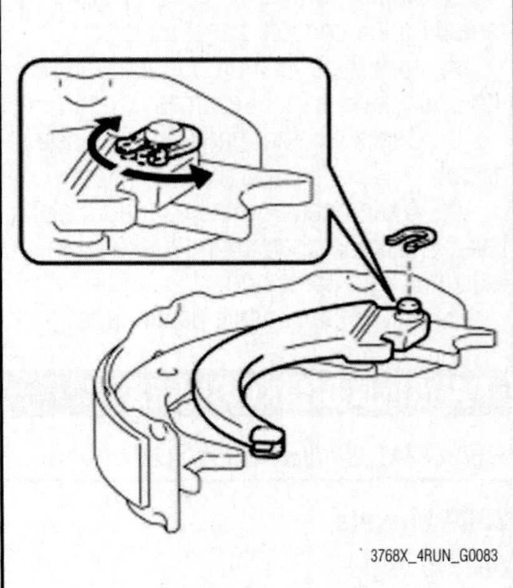

Fig. 19 Removing the parking brake shoe lever

11. Remove the No. 2 parking brake shoe assembly with parking brake shoe lever.
   a. Using needle-nose pliers, disconnect the No. 3 parking brake cable assembly from the parking brake shoe lever.

➡**Be careful not to damage the No. 3 parking brake cable assembly.**

12. Remove the parking brake shoe lever.
   a. Remove the C-washer, shim and parking brake shoe lever from the No. 2 parking brake shoe assembly.
13. Remove the parking brake shoe hold down spring pin.
   a. Remove the parking brake shoe hold down spring pin (for front side).
   b. Remove the parking brake shoe hold down spring pin (for rear side).

To install:

14. Install the parking brake shoe hold down spring.
15. Apply high temperature grease.
   a. Apply a light coat of high-temperature grease to the areas of the backing plate which make contact with the shoe.
16. Install the parking brake shoe lever.
   a. Apply a light coat of high-temperature grease to the areas of the parking brake shoe lever which make contact with the No. 2 parking brake shoe assembly.
   b. Install the parking brake shoe lever and shim to the No. 2 parking brake shoe assembly with a new C-washer.
   c. Using a feeler gauge, measure the clearance between the No. 2 parking brake shoe assembly and the parking brake lever. Standard clearance is 0.00984 inch (0.25 mm). If the clearance is not within the specification, replace the shim with one of a different thickness so that the clearance is within the specification.
17. Install the No. 2 parking brake shoe assembly with the parking brake shoe lever.

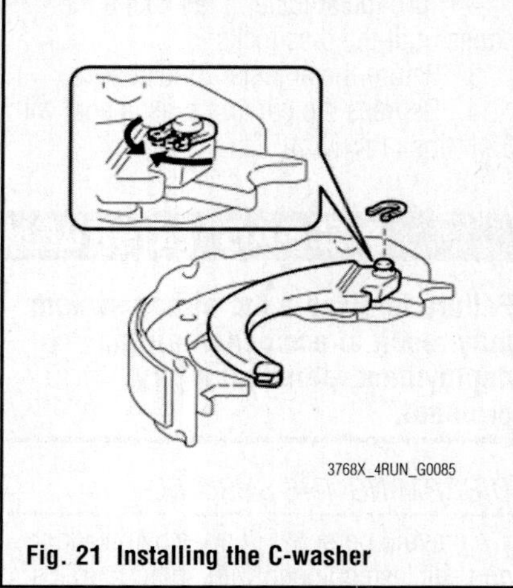

Fig. 21 Installing the C-washer

   a. Using needle-nose pliers, connect the No. 3 parking brake cable assembly to the parking brake shoe lever.

➡**Be careful not to damage the No. 3 parking brake cable assembly.**

18. Install the parking brake shoe return tension spring.
   a. Install the parking brake shoe return tension spring to the No. 2 parking b rake shoe assembly.
19. Install the No. 1 parking brake shoe assembly LH.
   a. Connect the parking brake shoe return tension spring to install the No. 1 parking brake shoe assembly.
20. Install the parking brake shoe adjusting screw set.
   a. Apply a light coat of high-temperature grease to the areas of the shoe adjusting screw set.
   b. Install the parking brake shoe adjusting screw set.
21. Install the No. 2 parking brake shoe assembly LH.
   a. Install the No. 2 parking brake shoe assembly to the backing plate with the parking brake shoe hold down spring cup and parking brake shoe hold down spring.
22. Install the parking brake shoe strut and parking brake shoe strut compression spring.
23. Install the parking brake shoe assembly LH.
   a. Install the No. 1 parking brake shoe assembly to the backing plate with the parking brake shoe hold down spring cup and parking brake shoe hold down spring.
24. Install the parking brake shoe return tension spring.

➡**First install the front side spring, and then the rear side spring.**

25. Check the parking brake installation.
   a. Check that each part is installed properly.

➡**There should be no oil or grease adhering to the friction surfaces of the shoe lining and disc.**

26. Install the rear disc.
27. Connect the rear disc brake caliper assembly LH. Tighten the 2 bolts to 77 ft. lbs. (105 Nm).
28. Adjust the parking brake shoe clearance and parking brake pedal travel.
29. Inspect the parking brake pedal travel.
30. Install the rear wheel.
31. Settle the parking brake shoe and disc.
   a. Drive the vehicle for approximately 0.25 mile (400 m). The vehicle speed must be approximately 31 mph (50 km/h) and on a safe, level and dry road. The parking brake pedal is being depressed with a force of 37.7 ldf (150 N).

➡**Set a 5 minute interval between each procedure to prevent the brake assembly from overheating.**

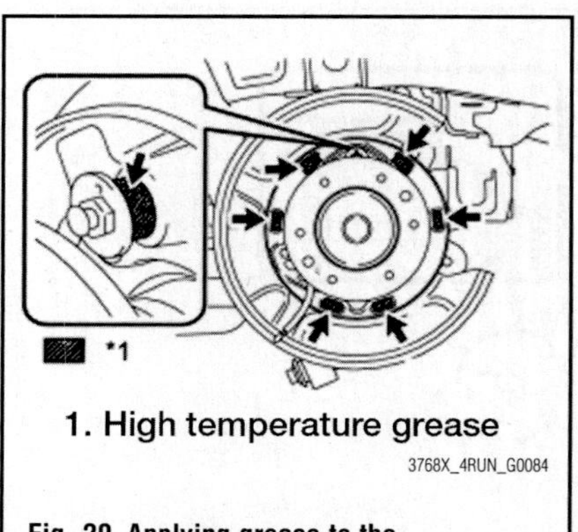

Fig. 20 Applying grease to the backing plate

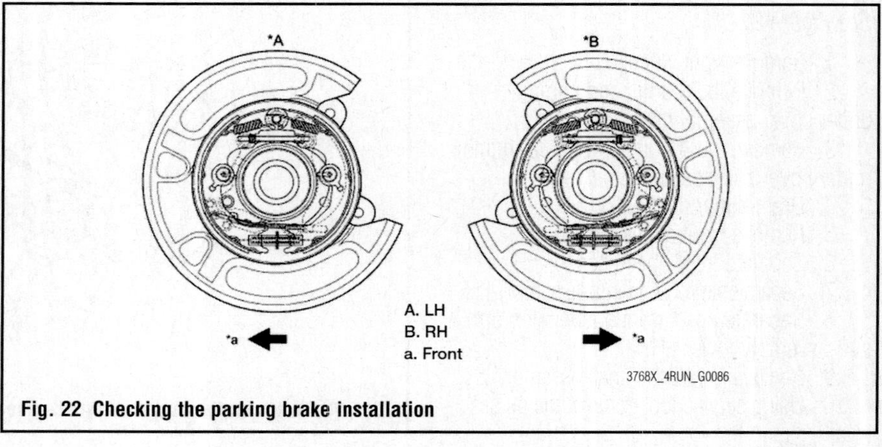

Fig. 22 Checking the parking brake installation

ADJUSTMENT

2009 Models

1. Before servicing the vehicle, refer to the Precautions section.
2. Turn the adjuster and expand the shoes until the disc locks.
3. Return the adjuster 8 notches.
4. Depress the parking brake pedal with 33 ft. lbs (147 Nm).
5. Drive the vehicle at about 50 km/h (31 mph) on a safe, level and dry road for about 400 meters (0.25 mile) in this condition.
6. Repeat this procedure 2 or 3 times.

CHASSIS ELECTRICAL

AIR BAG (SUPPLEMENTAL RESTRAINT SYSTEM)

GENERAL INFORMATION

CAUTION

These vehicles are equipped with an air bag system. The system must be disarmed before performing service on, or around, system components, the steering column, instrument panel components, wiring and sensors. Failure to follow the safety precautions and the disarming procedure could result in accidental air bag deployment, possible injury and unnecessary system repairs.

SERVICE PRECAUTIONS

CAUTION

Disconnect and isolate the battery negative cable before beginning any airbag system component diagnosis, testing, removal, or installation procedures. Allow system capacitor to discharge for two minutes before beginning any component service. This will disable the airbag system. Failure to disable the airbag system may result in accidental airbag deployment, personal injury, or death.

DISARMING THE SYSTEM

To avoid personal injury when working on vehicles equipped with an air bag, the negative battery cable must be disconnected and at least 90 seconds must elapse before working on the system. Failure to do so may result in deployment of the air bag.

ARMING THE SYSTEM

Reconnect the negative battery cable. Wait 2 minutes for performing any service on the vehicle.

CLOCKSPRING CENTERING

2009 Models

See Figure 23.

1. Check that the front wheels are facing straight ahead.
2. Check that the ignition switch is off.
3. Check that the battery negative (-) terminal is disconnected

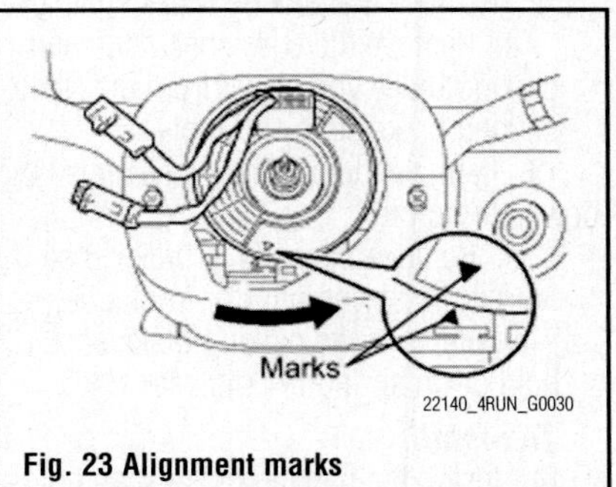

Fig. 23 Alignment marks

CAUTION

After removing the terminal, wait for at least 90 seconds before starting the operation.

4. Rotate the spiral cable clockwise slowly by hand until it feels firm.

➡Do not turn the spiral cable by the airbag wire harness

5. Rotate the spiral cable counterclockwise approximately 2.5 turns to align the marks

➡The spiral cable will rotate approximately 2.5 turns to both the left and right from the center

DRIVE TRAIN

FRONT AXLE SHAFT, BEARING & SEAL

REMOVAL & INSTALLATION

2009 Models

See Figure 24.

1. Remove front wheel.
2. Remove the 6 bolts and engine under cover assembly rear.
3. Remove the 4 bolts and no. 1 engine under cover sub-assembly.
4. Drain differential oil.
5. Remove front axle shaft LH nut.
6. Separate front speed sensor LH.
7. Separate tie rod end sub-assembly LH.
8. Separate no. 1 front suspension arm sub-assembly lower LH.
9. Remove front drive shaft assembly LH.
10. Using service tool, remove the oil seal.

To install:

11. Using SST and a hammer, install a new oil seal.
12. Coat the oil seal lip with MP grease.
13. Install front drive shaft assembly LH.
14. Connect no. 1 front suspension arm sub-assembly lower LH.
15. Connect tie rod end sub-assembly LH.
16. Connect front speed sensor LH.

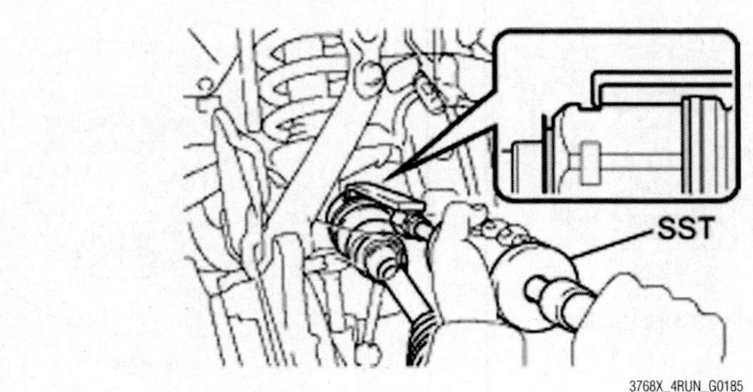

Fig. 24 Removing the front drive shaft assembly LH

17. Install front axle shaft LH nut and tighten to 173 ft. lbs. (235 Nm).
18. Fill up differential oil.
19. Install the no. 1 engine under cover sub-assembly with the 4 bolts.
20. Install the engine under cover assembly with the 6 bolts.
21. Install front wheel.
22. Inspect ABS speed sensor signal.

FRONT DRIVESHAFT

REMOVAL & INSTALLATION

2010 Models

See Figure 25.

➡Use the same procedure for the RH and LH sides.

➡The procedures listed below is for the LH side.

1. Remove the front wheel.
2. Drain the differential oil.
3. Remove the front axle hub grease cap.
4. Remove the front axle shaft nut.
5. Remove the front speed sensor.
6. Disconnect the tie rod end sub assembly LH.
7. Disconnect the front lower ball joint attachment LH.
   a. Remove the 2 bolts and disconnect the lower ball joint attachment from the steering knuckle.
8. Remove the front drive shaft assembly LH.

➡Be careful not to damage the dust cover or oil seal.

9. Keep the drive shaft level while handling it.

To install:

10. Install the front drive shaft assembly.
    a. Coat the spline of the inboard joint shaft assembly with ATF.
    b. Align the shaft splines and install the drive shaft with a brass bar and hammer.

➡Set the snap ring with the opening facing downward.

➡Be careful not to damage the oil seal, boot or dust cover.

➡Whether the inboard joint shaft is in contact with the pinion shaft or not can be confirmed from the sound or feeling when tapping in the shaft.

11. Install the front speed sensor.
12. Install the lower ball joint attachment LH.
    a. Install the lower ball joint attachment with the 2 bolts. Tighten to 118 ft. lbs. (160 Nm).
13. Connect the tie rod end sub assembly LH.
14. Install the front axle shaft nut.
15. Install the front grease hub cap.
16. Add differential oil.
17. Install the front wheel. Tighten to 83 ft. lbs. (112 Nm).
18. Check the front speed sensor signal.

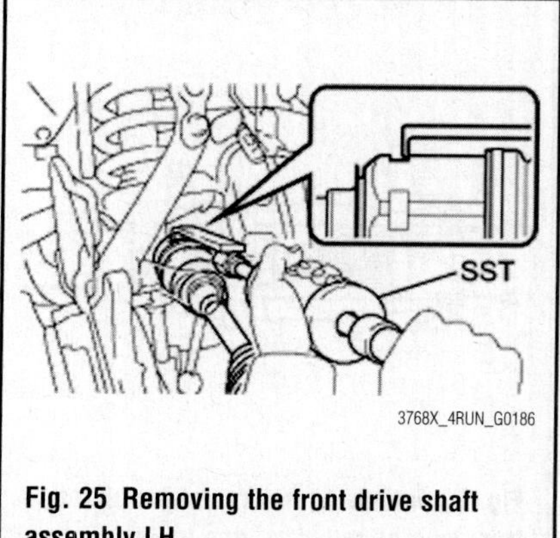

Fig. 25 Removing the front drive shaft assembly LH

FRONT HALFSHAFT

REMOVAL & INSTALLATION

2009 Models

See Figures 26 and 27.

1. Before servicing the vehicle, refer to the precautions section.
2. Remove the wheel.
3. Drain the differential oil.

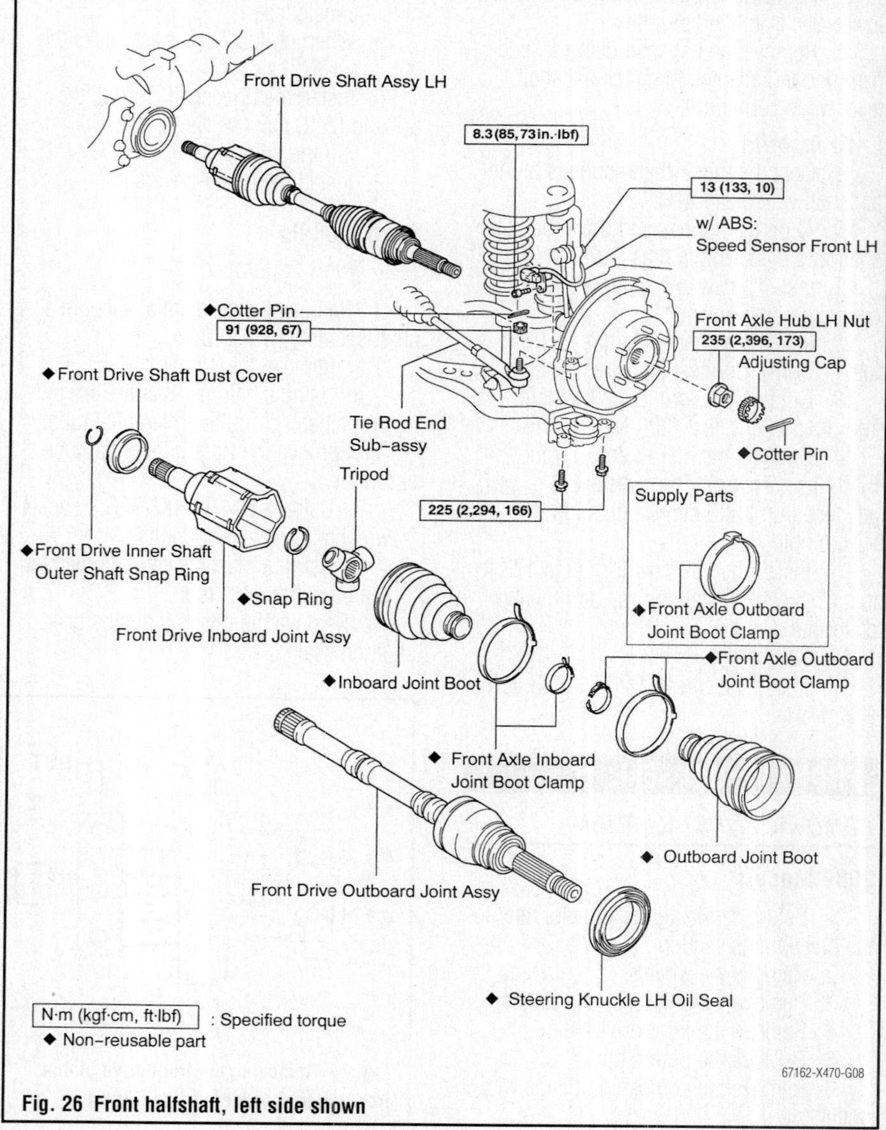

Fig. 26 Front halfshaft, left side shown

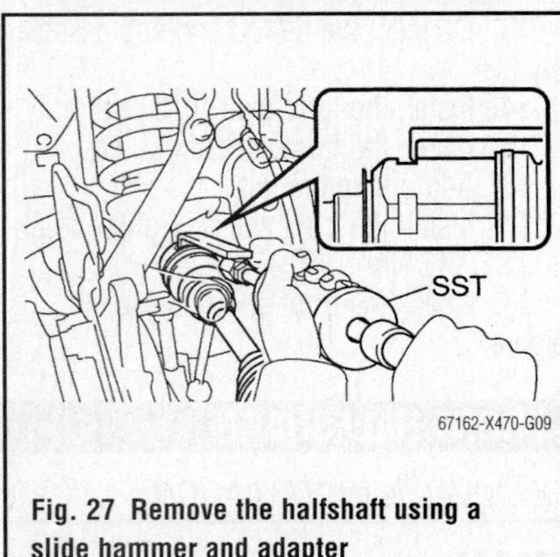

Fig. 27 Remove the halfshaft using a slide hammer and adapter

4. Remove the cotter pin and cap, then remove the hub nut.
5. Remove the speed sensor wiring harness. Remove the sensor.
6. Remove the tie rod end from the knuckle.
7. Remove the 2 bolts and separate the lower arm from the ball joint.
8. Remove the halfshaft using a slide hammer and adapter. Keep the halfshaft level when carrying it.

To install:

9. Coat the inboard end splines of the halfshaft with clean ATF.
10. Align the splines and drive the half-shaft into place with a brass drift.
11. Install a new snap ring with the opening facing down.
12. Install the sensor. Torque to 10 ft. lbs. (13 Nm). Connect the wire harness.
13. Connect the arm to the ball joint. Torque to 166 ft. lbs. (225 Nm).
14. Connect the tie rod end. Torque to 67 ft. lbs. (91 Nm). The nut can be advanced up to 60 degrees to align the cotter pin hole.
15. Install the hub nut. Torque to 173 ft. lbs. (235 Nm). Install the cap and a new cotter pin.
16. Fill the differential.
17. Install the wheel. Torque to 83 ft. lbs. (112 Nm).

FRONT PINION SEAL

REMOVAL & INSTALLATION

2009 Models

1. Before servicing the vehicle, refer to the precautions section.
2. Remove the wheels.
3. Remove the engine under-covers.
4. Remove the front driveshaft.
5. Remove the pinion nut.
6. Remove the companion flange with a puller.
7. Remove the oil seal with a seal puller.
8. Remove the oil slinger.
9. Remove the bearing with a puller.
10. Remove the oil storage ring.
11. Remove the spacer and discard it.

To install:

12. Install a new spacer.
13. Install the oil storage ring using a brass drift.
14. Install the bearing.
15. Install the slinger.
16. Using a seal driver, install the new oil seal. Drive the seal into a depth of 4.35mm +/- 0.45mm.
17. Install the companion flange. Coat the threads of a new flange nut with gear oil. Hold the flange and torque the nut to 273 ft. lbs. (370 Nm).
18. Using an inch-pound torque wrench, check the preload. Preload for a new bearing should be 9-14 inch lbs.; for a used bearing, 4.3-7 inch lbs. If not, a new spacer must be installed.
19. When preload is correct, stake the nut.
20. Install the driveshaft. Torque the bolts to 65 ft. lbs. (88 Nm).
21. Fill the differential.
22. Install the under-covers.

2010 Models

See Figures 28 through 30.

1. Remove the differential vacuum actuator assembly (w/A.D.D.).
   a. Remove the 4 bolts.
   b. Using a hammer handle, pry out the actuator from the differential tube.
2. Remove the front differential tube assembly.
   a. Using a E14 TORX®.$ socket wrench, remove the 4 bolts.
   b. Using a plastic faced hammer, tap out the differential tube.
3. Remove the differential side gear shaft oil seal.
4. Remove the differential side gear inner shaft sub assembly (w/A.D.D.).
   a. Remove the snap ring from the side gear inter shaft.
5. Remove the front differential side bearing retainer deflector (w/A.D.D.).
   a. Using a screwdriver pry out the bearing retainer deflector.

➡Tape the screwdriver tip before use.

6. Remove the front drive pinion companion flange nut.
   a. Using a special tool and a hammer, unstake the nut.
   b. Using the special tool to hold the companion flange, remove the nut.
7. Remove the front drive pinion companion flange sub assembly.
8. Remove the front differential dust deflector.
9. Remove the front differential carrier oil seal.
10. Remove the front differential drive pinion oil slinger.
    a. Remove the oil slinger from the drive pinion.
11. Remove the front drive pinion front tapered roller bearing (inner).
    a. Using the special tool, remove the front tapered roller bearing (inner) from the drive pinion.
12. Remove the front drive pinion front tapered roller bearing (outer).
    a. Using the special tool, remove the front tapered roller bearing (outer).
13. Remove the front differential oil storage ring.
    a. Using a screwdriver and hammer, tap out the oil storage ring.
14. Remove the front differential drive pinion bearing spacer.
15. Remove the differential side bearing retainer.

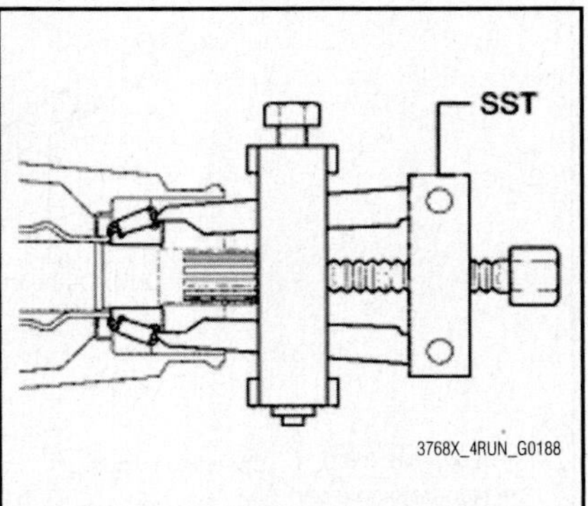

Fig. 28 Removing the front drive pinion front tapered roller bearing (inner)

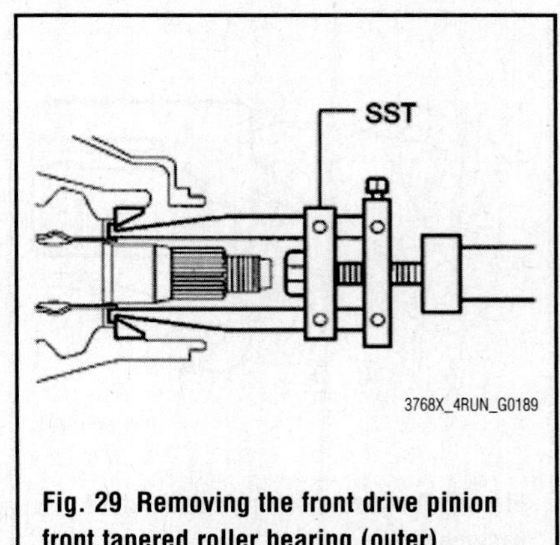

Fig. 29 Removing the front drive pinion front tapered roller bearing (outer)

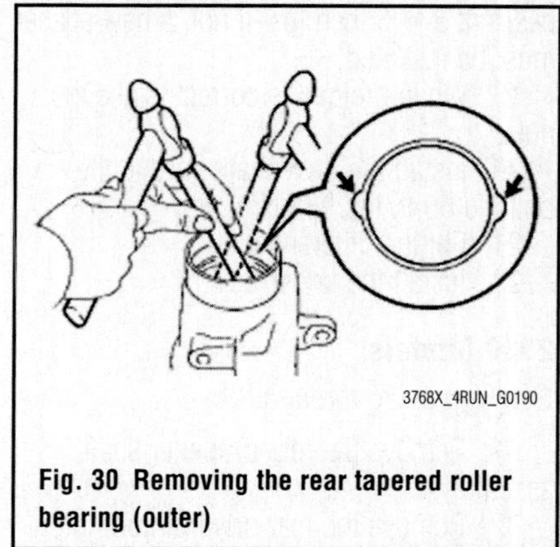

Fig. 30 Removing the rear tapered roller bearing (outer)

a. Using a screwdriver, remove the union.

b. Remove the 10 bolts and tap out the side bearing retainer with a plastic faced hammer.

16. Remove the differential case assembly.

17. Remove the differential drive pinion.

18. Remove the front drive pinion rear tapered roller bearing (inner).

a. Using the special tool and a press, remove the rear tapered roller bearing (inner) and washer from the drive pinion.

➡Do not drop the drive pinion.

➡If the drive gear or ring gear is damaged, replace them as a set.

19. Remove the front drive pinion tapered roller bearing (outer).

20. Using a brass bar and hammer, remove the rear tapered roller bearing (outer).

To install:

21. To install, reverse the removal procedure.

REAR AXLE SHAFT, BEARING & SEAL

REMOVAL & INSTALLATION

2009 Models

See Figure 31.

1. Remove the wheel.
2. Remove the speed sensor.
3. Remove the caliper.
4. Remove the rotor.
5. Remove the parking brake assembly.
6. Remove the 4 nuts and pull out the axle shaft with backing plate.
7. Remove the oil seal with a slide hammer.

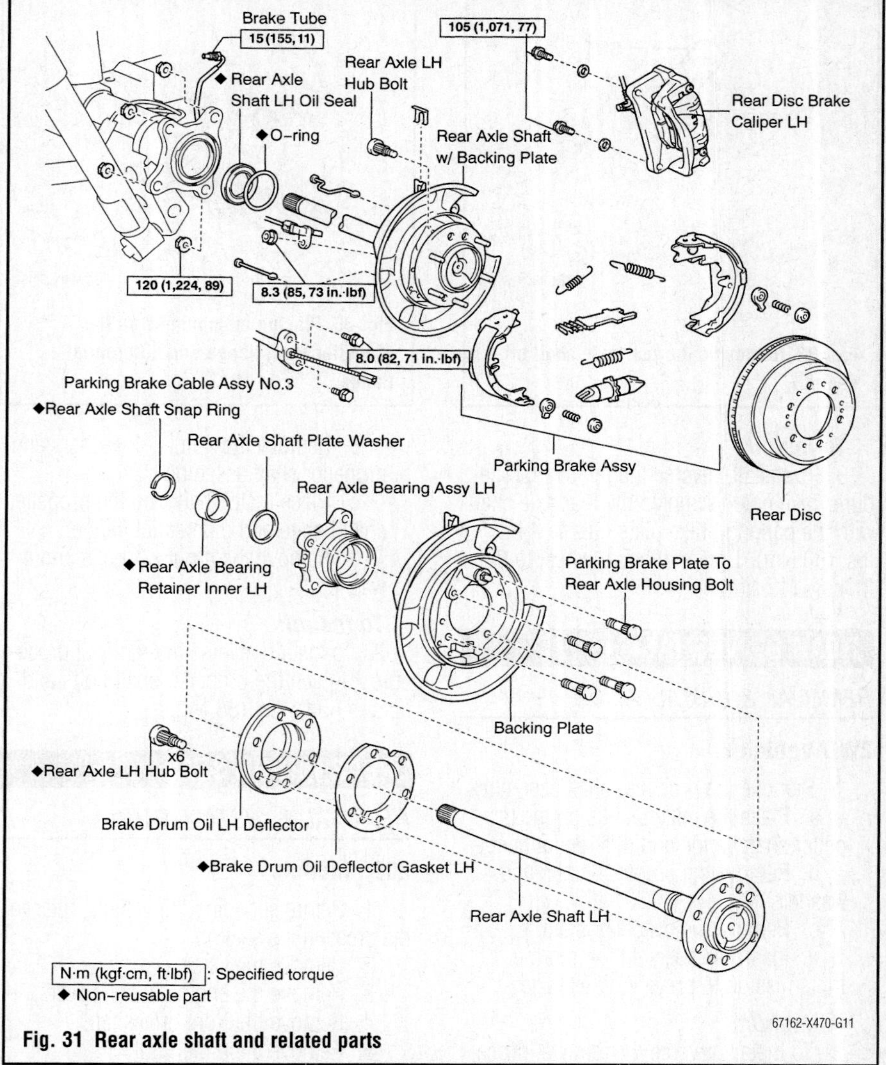

Fig. 31 Rear axle shaft and related parts

To install:

8. Installation is the reverse of removal. Torque the nuts to 89 ft. lbs. (120 Nm).

2010 Models

See Figures 32 and 33.

➡Use the same procedure for the RH and LH sides.

➡The procedure listed below is for the LH side.

1. Disconnect the cable from the negative battery terminal.

➡When disconnecting the cable, some systems need to be initialized after the cable is reconnected.

2. Remove the rear wheel.
3. Drain the brake fluid.
4. Disconnect the rear flexible hose LH.
5. Remove the rear speed sensor LH.
6. Remove the parking brake assembly.
7. Remove the rear axle shaft with the parking brake plate LH.

a. Remove the 4 nuts and rear axle shaft with parking brake plate.

b. Remove the o-ring.

8. Remove the rear axle shaft oil seal LH.

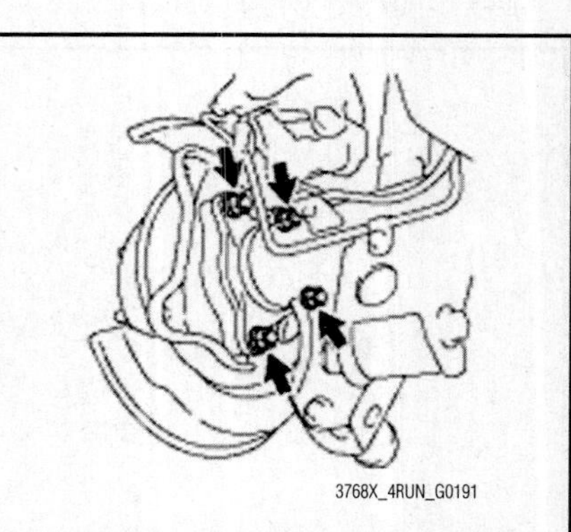

Fig. 32 Removing the rear axle shaft with the parking brake plate LH

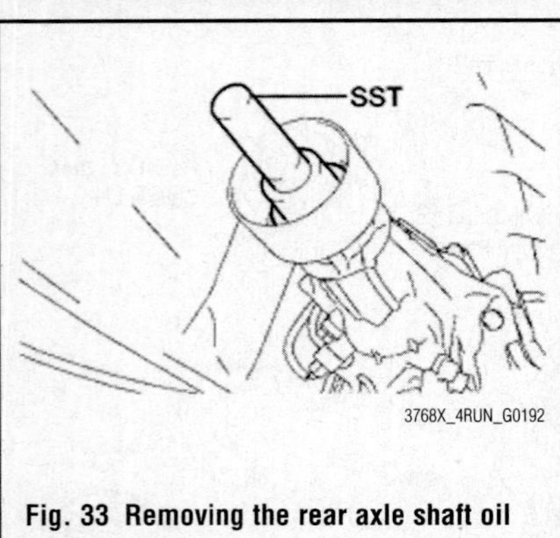

Fig. 33 Removing the rear axle shaft oil seal LH

To install:

9. To install, reverse the removal procedure. Take note to tighten the rear axle shaft with the parking brake plate nuts to 44 ft. lbs. (60 Nm). Tighten the rear wheel to 82 ft. lbs. (112 Nm).

REAR DRIVESHAFT

REMOVAL & INSTALLATION

2WD Vehicles

1. Remove the propeller shaft assembly.
   a. Place the matchmarks on the propeller shaft flange and differential flange.
   b. Remove the 4 nuts, 4 bolts and 4 washers.
   c. Remove the propeller shaft.
   d. Insert the special tool into the transmission to prevent oil leakage.

To install:

2. To install, reverse the removal procedure. Tighten the propeller shaft assembly bolts and nuts to 65 ft. lbs. (88 Nm).

4WD Vehicles

See Figures 34 and 35.

1. Remove the propeller shaft assembly.
   a. Place matchmarks on the propeller shaft flange and transfer flange.

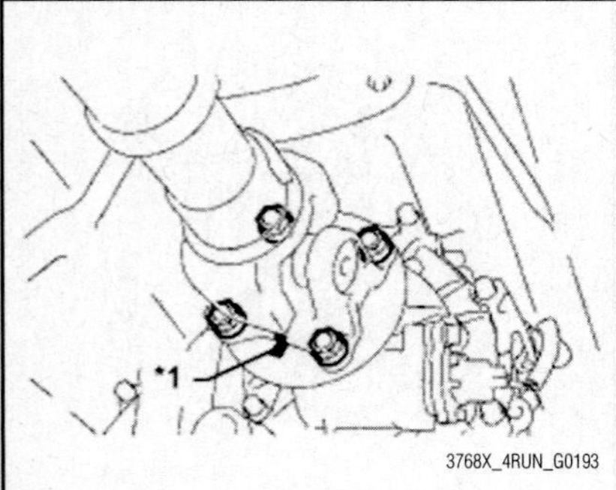

Fig. 34 Placing matchmarks on the propeller shaft flange and transfer flange

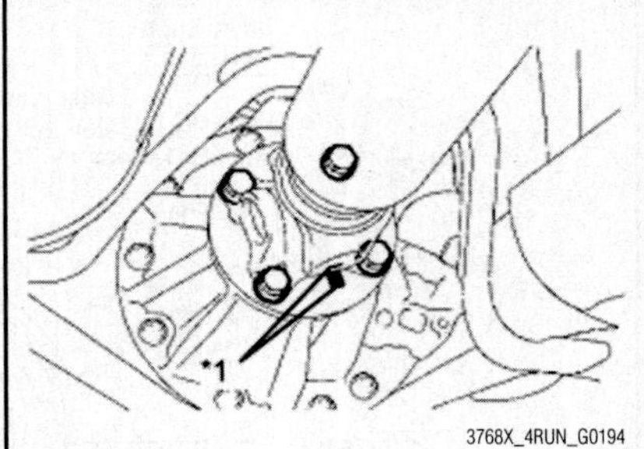

Fig. 35 Placing matchmarks on the propeller shaft flange and differential flange

   b. Remove the 4 nuts, 4 washers and propeller shaft assembly.
   c. Place matchmarks on the propeller shaft flange and differential flange.
   d. Remove the 4 nuts, 4 bolts and 4 washers.

To install:

2. To install, reverse the removal procedure. Tighten the propeller shaft bolts and nuts to 65 ft. lbs. (88 Nm).

REAR PINION SEAL

REMOVAL & INSTALLATION

2009 Models

1. Before servicing the vehicle, refer to the precautions section.
2. Remove the wheels.
3. Remove the engine under-covers.
4. Remove the front driveshaft.
5. Remove the pinion nut.
6. Remove the companion flange with a puller.
7. Remove the oil seal with a seal puller.
8. Remove the oil slinger.
9. Remove the bearing with a puller.
10. Remove the oil storage ring.
11. Remove the spacer and discard it.

To install:

12. Install a new spacer.
13. Install the oil storage ring using a brass drift.
14. Install the bearing.
15. Install the slinger.
16. Using a seal driver, install the new oil seal. Drive the seal into a depth of 1.00mm +/- 0.45mm.
17. Install the companion flange. Coat the threads of a new flange nut with gear oil. Hold the flange and torque the nut to 273 ft. lbs. (370 Nm).
18. Using an inch-pound torque wrench, check the preload. Preload for a new bearing should be 9–15 inch lbs.; for a used bearing, 5–7.5 inch lbs. If not, a new spacer must be installed.
19. When preload is correct, stake the nut.
20. Install the driveshaft. Torque the bolts to 65 ft. lbs. (88 Nm).
21. Fill the differential.
22. Install the under-covers.

2010 Models

See Figures 36 through 40.

1. Remove the rear propeller shaft assembly.
2. Remove the rear drive pinion nut.
   a. Using the special tool and a hammer, loosen the staked part of the rear drive pinion nut.

➡Be sure to use special tool with the tapered surface facing the shaft.

➡Do not grind the tip of the SST with a grinder, etc.

➡Completely loosen the staked part of the nut when removing it.

➡Do not damage the threads of the drive pinion nut.

   b. Use the special tool to hold the companion flange.

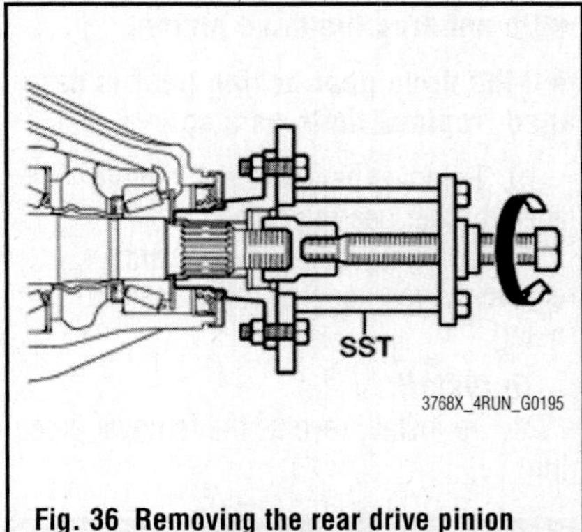

Fig. 36 Removing the rear drive pinion companion flange sub assembly

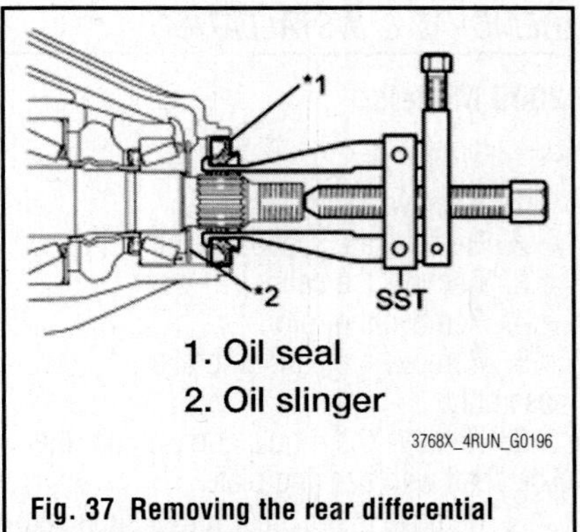

Fig. 37 Removing the rear differential carrier oil seal

c. Using a 30 mm socket wrench, remove the rear drive pinion nut.

3. Remove the rear drive pinion companion flange sub assembly.

a. Using a special tool, remove the rear drive pinion companion flange sub assembly.

➡**Before using the special tool (center bolt), apply hypoid gear oil to its threads and tip.**

4. Remove the rear differential carrier oil seal.

5. Remove the rear differential drive pinion oil slinger.

➡**Apply grease to the threads and tip of the special tool center bolt before use.**

6. Remove the rear drive pinion from tapered roller bearing.

7. Using the special tool, remove the rear drive pinion tapered roller bearing (inner).

➡**Apply grease to the threads and tip of the special tool center bolt before use.**

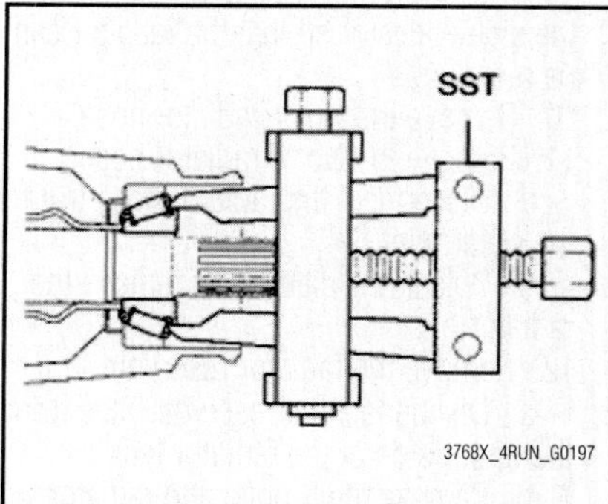

Fig. 38 Removing the rear drive pinion tapered roller bearing (inner)

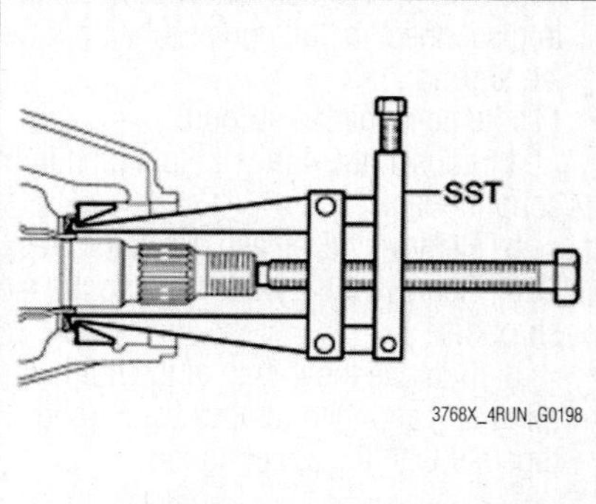

Fig. 39 Tapping out the rear drive pinion tapered roller bearing (outer)

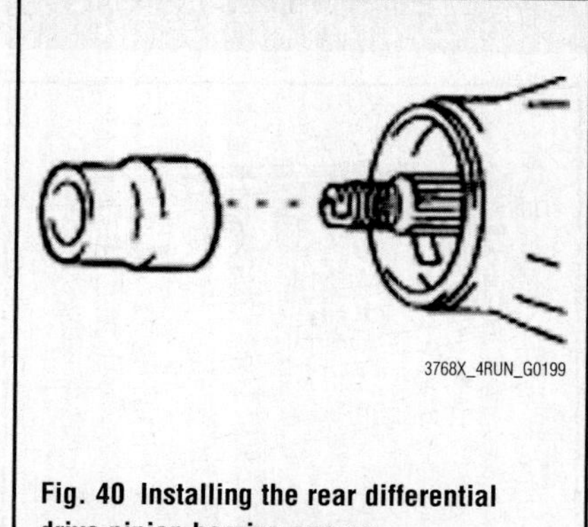

Fig. 40 Installing the rear differential drive pinion bearing spacer

a. Using the special tool, tap out the rear drive pinion tapered roller bearing (outer).

8. Remove the differential drive pinion bearing spacer.

To install:

9. Install the rear differential drive pinion bearing spacer.

➡**Install the spacer so that it is facing in the correct direction.**

10. Install the differential oil storage ring using a hammer.

➡**Be careful not to damage the oil storage ring.**

11. Install the rear drive pinion front tapered roller bearing.

a. Using the special tool and a hammer, tap in the rear drive pinion front roller bearing (outer).

b. Install the rear drive pinion front roller bearing (inner).

12. Install the front differential drive pinion oil slinger.

13. Install the rear differential carrier oil seal.

a. Apply MP grease to the lip of a new oil seal.

b. Using the special tool and a hammer, tap in the rear differential carrier oil seal.

14. Install the rear drive pinion companion flange sub assembly.

➡**Before using the special tool (center bolt), apply hypoid gear oil to its threads and tip.**

a. Using the special tool to hold the companion flange in place, install the rear drive pinion nut. Tighten to 337 ft. lbs. (457 Nm).

15. Inspect the differential drive pinion preload.

a. Using a torque wrench, measure the preload.

Standard Preload (at Starting):

- New bearing: 7.35-19.3 inch lbs. (0.83-2.18 Nm)
- Used bearing: 7.79-17.5 inch lbs. (0.88-1.98 Nm)

➡**If the results are not as specified, adjust the preload. The total preload at starting is 9.6-21.0 inch lbs. (1.08-2.38 Nm).**

16. Stake the rear drive pinion nut.

a. Using a hammer and chisel, stake the rear drive pinion nut.

17. Install the rear propeller shaft assembly.

18. Install the differential oil.

19. Check for differential oil leakage.

TRANSFER CASE ASSEMBLY

REMOVAL & INSTALLATION

2009 Models

1. Before servicing the vehicle, refer to the Precautions section.
2. Drain the fluid.
3. Remove the skid plate.
4. Remove the transmission.
5. Remove the 8 bolts and 2 clamps from transfer assembly
6. Separate the transfer case from the transmission.

To install:

7. Installation is the reverse of removal. Torque the bolts to 17 ft. lbs. (24 Nm).

2010 Models

1. Drain the transfer oil.
2. Remove the automatic transmission assembly.
3. Remove the transfer assembly.
   a. Remove the 8 bolts.
   b. Remove the transfer from the transmission.

To install:

4. Install the transfer case.
   a. Install the transfer to the transmission.
   b. Install the 8 bolts. Tighten to 18 ft. lbs. (24 Nm).
5. Install the automatic transmission assembly.
6. Add transfer oil.
7. Check for transfer oil leaks.

ENGINE COOLING

ENGINE FAN

REMOVAL & INSTALLATION

2009 Models

1. Disconnect cable from negative battery terminal.

⁂ CAUTION

Wait at least 90 seconds after disconnecting the cable from negative (-) battery terminal to prevent airbag and seat belt pretensioner activation.

2. Remove no. 1 engine under cover.
3. Drain engine coolant.
4. Remove the 4 bolts and engine under cover.
5. Remove V-bank cover.
6. Remove the 11 clips and radiator support seal upper.
7. Loosen the 4 fluid coupling bolts.
8. Remove fan and alternator v-belt.

To install:

9. Install fan and alternator v-belt.
10. Tighten the 4 fluid coupling bolts tighten bolts to 15 ft. lbs. (21 Nm).
11. Install radiator support seal upper.
12. Add engine coolant.
13. Install the V-bank cover with the 2 nuts and tighten to 66 inch lbs. (7.5 Nm).
14. Install the engine under cover with the 4 bolts and tighten bolts to 21 ft. lbs. (29 Nm).
15. Reinstall batter cable.
16. Perform initialization, if necessary.

2010 Models

Refer to RADIATOR, for removal and installation procedures.

RADIATOR

REMOVAL & INSTALLATION

2009 Models

See Figures 41 and 42.

1. Before servicing the vehicle, refer to the Precautions section.
2. Remove the engine under cover.
3. Drain the engine coolant.
4. Disconnect the radiator reservoir hose from the radiator.
5. Disconnect the upper radiator hose from radiator.
6. Disconnect the lower radiator hose from radiator.

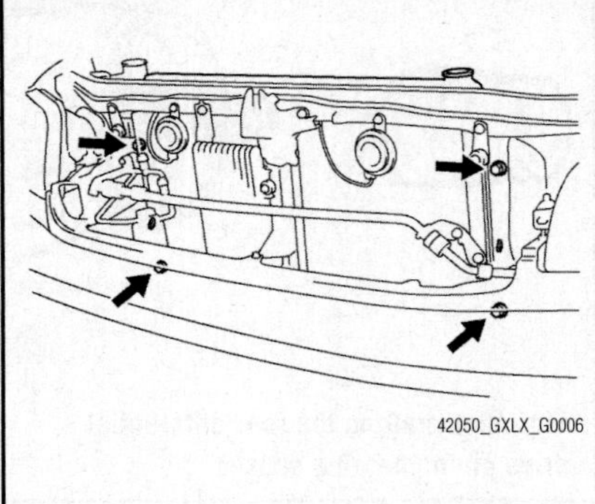

Fig. 41 Radiator attachment bolts

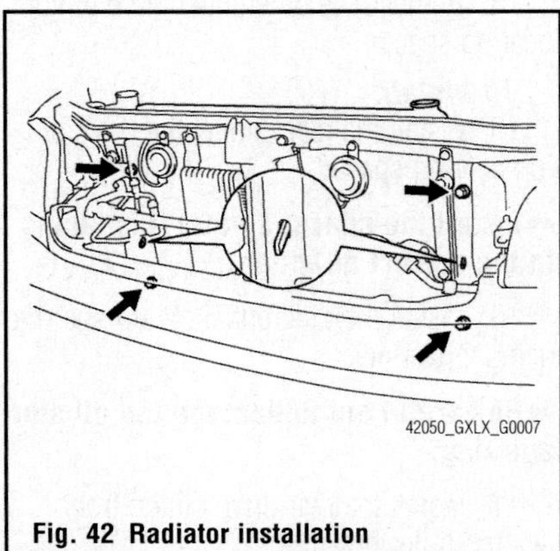

Fig. 42 Radiator installation

7. Disconnect the A/T oil cooler hoses from radiator.
8. Remove the 2 clips and No.2 fan shroud.
9. Remove the 4 bolts and radiator assembly.

To install:

10. Set the radiator bracket hooks to the radiator support holes.
11. Install the 4 bolts and tighten to 9 ft. lbs. (12 Nm)
12. Install the No.2 fan shroud with the 2 clips.
13. Connect A/T oil cooler hoses to the radiator.
14. Connect the upper radiator hose to the radiator.
15. Connect the lower radiator hose to the radiator.
16. Connect the radiator reservoir hose to the radiator.
17. Fill with engine coolant.
18. Start engine and check for engine coolant leaks.
19. Recheck engine coolant level.
20. Install engine under cover.

2010 Models

4.0L Engine

See Figures 43 through 51.

1. Remove the upper radiator support seal.
2. Remove the front bumper cover lower.
3. Remove the No. 1 engine under cover sub assembly.
4. Drain the engine coolant.
5. Remove the V-bank cover.
6. Remove the front bumper cover.
7. Remove the upper front bumper retainer.
   a. Remove the 3 bolts and retainer.
8. Remove the radiator side deflector RH.
   a. Using a clip remover, detach the 3 claws and remove the clip. Then move the side deflector so that the radiator can be removed.
9. Remove the radiator side deflector LH.
   a. Using a clip remover, detach the 3 claws and remove the clip. Then move the side deflector so that the radiator can be removed.
10. Remove the No. 1 radiator hose.
11. Remove the No. 2 radiator hose.
    a. Disconnect the radiator hose from the water inlet.
    b. Detach the clamp and remove the radiator hose.
12. Remove the radiator reservoir.
    a. Disconnect the reservoir hose from the upper side of the radiator tank.
    b. Remove the 3 bolts and radiator reservoir.
13. Disconnect the oil cooler tube.
    a. Detach the claw to open the flexible hose clamp, and then remove the 2 bolts to disconnect the oil cooler tube from the fan shroud.
14. Remove the fan shroud.
    a. Loosen the 4 nuts holding the fluid coupling fan.
    b. Remove the fan and alternator v-belt.
    c. Remove the 2 bolts holding the fan shroud.
    d. Remove the 4 nuts of the fluid coupling fan, and then remove the shroud together with the coupling fan.

➡Be careful not to damage the radiator core.

    e. Remove the fan pulley from the water pump.
15. Remove the radiator assembly.
    a. Disconnect the 2 oil cooler hoses.

b. Remove the 4 bolts and radiator.

16. Remove the No. 1 radiator to support seal.

a. Remove the seal from the radiator assembly.

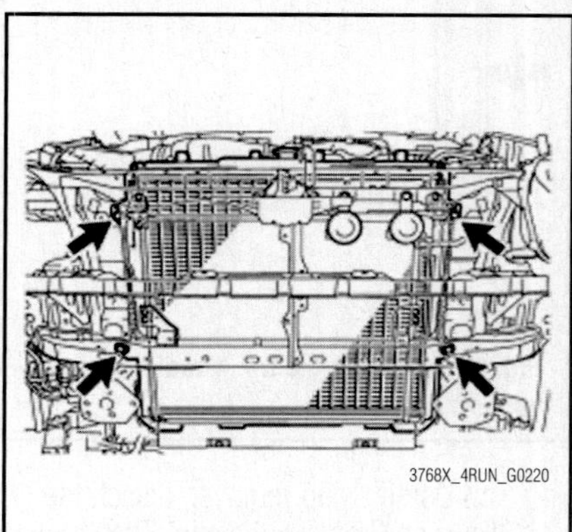

Fig. 43 Removing the radiator

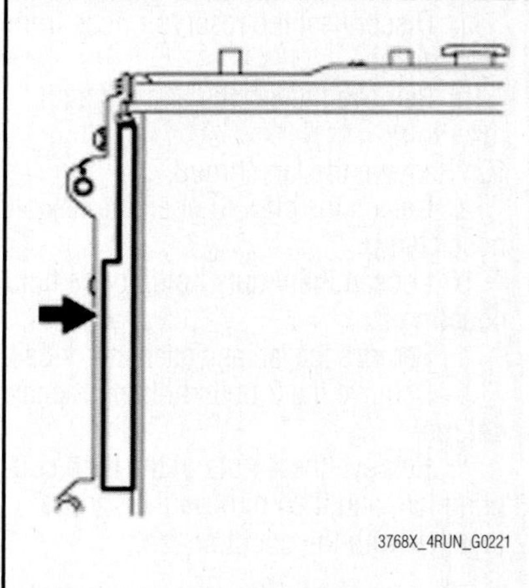

Fig. 44 Installing the No. 1 radiator to the support seal

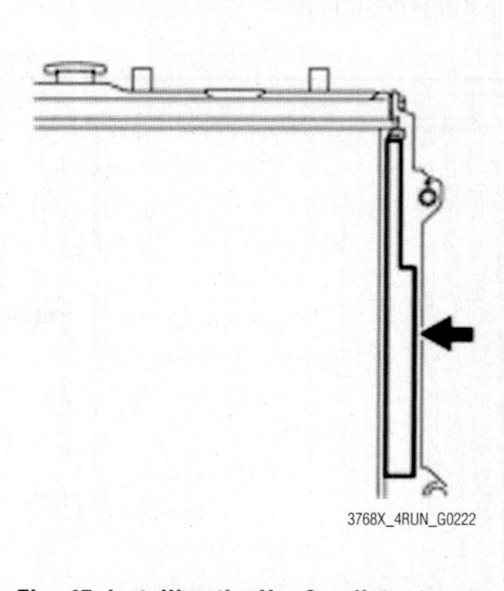

Fig. 45 Installing the No. 2 radiator to support seal

17. Remove the No. 2 radiator to support seal.

a. Remove the seal from the radiator assembly.

To install:

18. Install the No. 1 radiator to support seal.

19. Install the No. 2 radiator to support seal.

20. Install the radiator assembly.

a. Insert the radiator bracket hooks into the radiator support holes.

b. Install the radiator with the 4 bolts. Tighten the bolts to 13 ft. lbs. (18 Nm).

c. Connect the 2 oil cooler hoses.

➡Make sure the direction of the hose clamp is as shown.

21. Install the fan shroud.

a. Install the fan pulley to the water pump.

b. Place the shroud together with the coupling fan between the radiator and engine.

➡Be careful not to damage the radiator core.

c. Align the pain marks on the heads of the water pump stud bolts with the paint marks of the same color on the outer edge of the fluid coupling flange and install the fluid coupling to the water pump.

d. Temporarily install the fluid coupling fan to the water pump with the 4 nuts. Tighten the nuts as much as possible by hand.

e. Attach the claws of the shroud to the radiator.

f. Install the shroud with the 2 bolts. Tighten the bolts to 44 inch lbs. (5 Nm).

g. Install the fan and alternator v-belt.

h. Tighten the 4 nuts of the fluid coupling fan to 15 ft. lbs. (21 Nm).

22. Connect the oil cooler tube with the 2 bolts, and attach the claw to close the flexible hose clamp. Tighten to 49 inch lbs. (5.5 Nm).

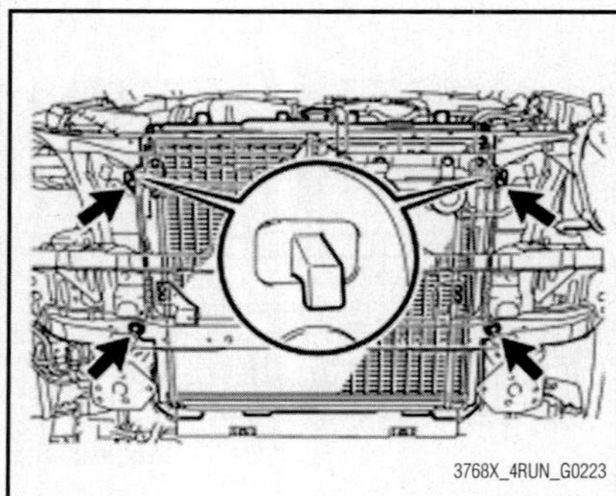

Fig. 46 Inserting the radiator bracket hooks into the radiator support holes

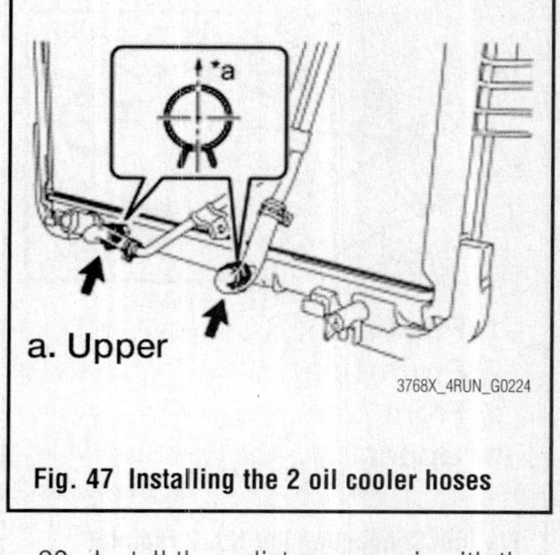

Fig. 47 Installing the 2 oil cooler hoses

23. Install the radiator reservoir with the 3 bolts. Tighten to 44 inch lbs. (5 Nm). Connect the reservoir hose to the upper side of the radiator tank upper.

24. Install the No. 2 radiator hose.

a. Connect the No. 2 radiator hose so that its paint mark aligns with the radiator protrusion.

➡Make sure the direction of the hose clamp is as shown.

b. Connect the No. 2 radiator hose so that its paint mark aligns with the water inlet housing protrusion.

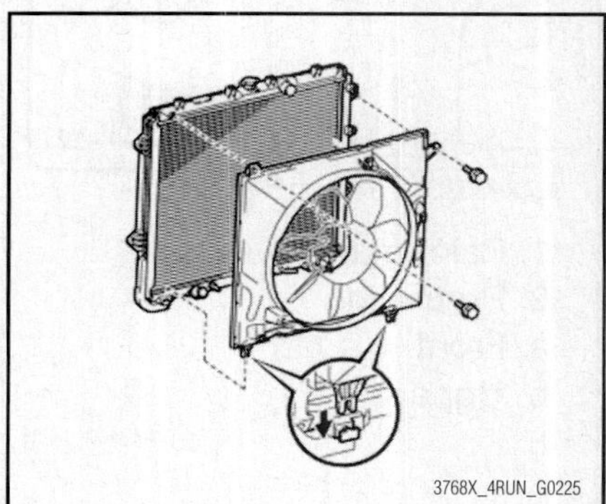

Fig. 48 Attaching the claws of the shroud to the radiator

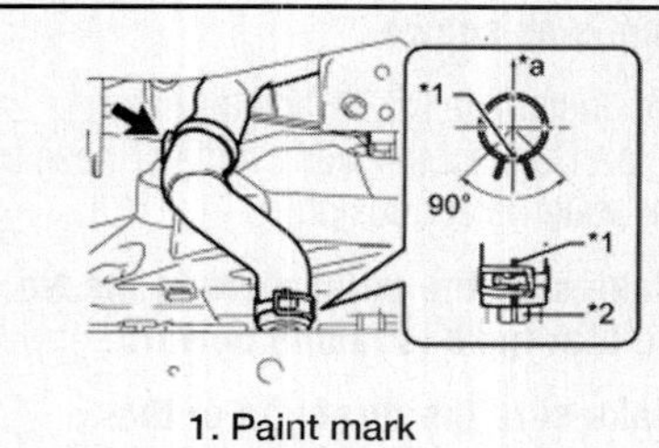

Fig. 49 Connecting the No. 2 radiator hose so that the paint marks align with the radiator protrusion

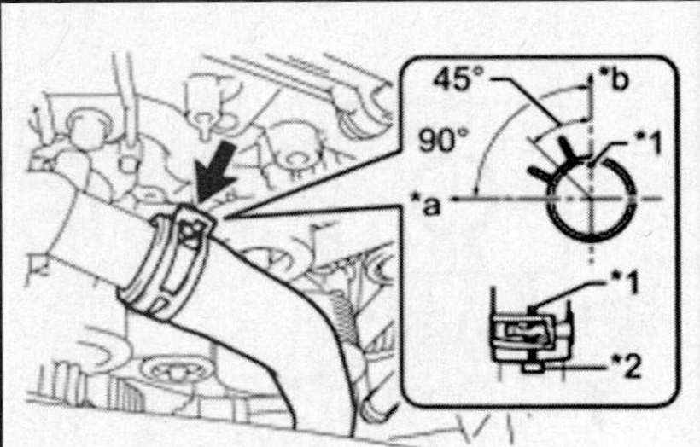

Fig. 50 Connecting the No. 2 radiator hose so that its paint mark aligns with the water inlet housing protrusion

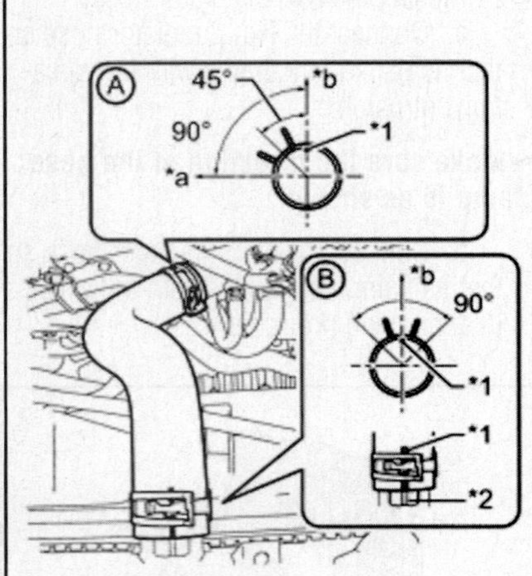

Fig. 51 Connecting the No. 1 radiator hose to the water inlet housing

➡Make sure the direction of the hose clamp is as shown.

25. Install the No. 1 radiator hose.
   a. Connect the No. 1 radiator hose to the water inlet housing.

➡Make sure the paint mark of the No. 1 radiator hose is facing upward.

➡Make sure the direction of the hose clamp is as shown in the illustration.

   b. Connect the No. 1 radiator hose so that its paint mark aligns with the radiator protrusion as shown in the illustration labeled B.

➡Make sure the paint mark of the No. 1 radiator hose is facing upward.

➡Make sure the direction of the hose clamp is as shown

26. Install the radiator side deflector RH.
   a. Attach the 3 screws.
   b. Install the deflector with the clip.
27. Install the radiator side deflector RH.
   a. Attach the 3 claws.
   b. Install the deflector with the clip.
28. Install the upper front bumper retainer.
   a. Install the retainer with the 3 bolts. Tighten to 71 inch lbs. (8 Nm).
29. Install the front bumper cover.
30. Add engine coolant.
31. Inspect for engine coolant leaks.
32. Install the No. 1 engine under cover sub assembly.
33. Install the front bumper cover lower.
34. Install the v-bank cover.
35. Install the upper radiator support seal.

2.7L Engine

See Figures 52 through 61.

1. Remove the upper radiator support seal.
2. Remove the front bumper cover lower.
3. Remove the No. 1 engine under cover sub assembly.
4. Drain the engine coolant.
5. Remove the front bumper cover.
6. Remove the upper front bumper retainer.
   a. Remove the 3 bolts and upper front bumper retainer.
7. Remove the radiator side deflector RH.
   a. Using a clip remover, detach the 3 claws and remove the clip. Then move the side deflector so that the radiator can be removed.
8. Remove the radiator side deflector LH.

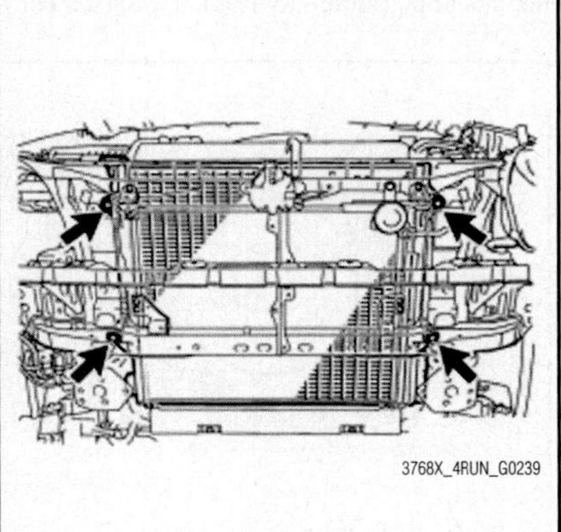

Fig. 52 Removing the radiator

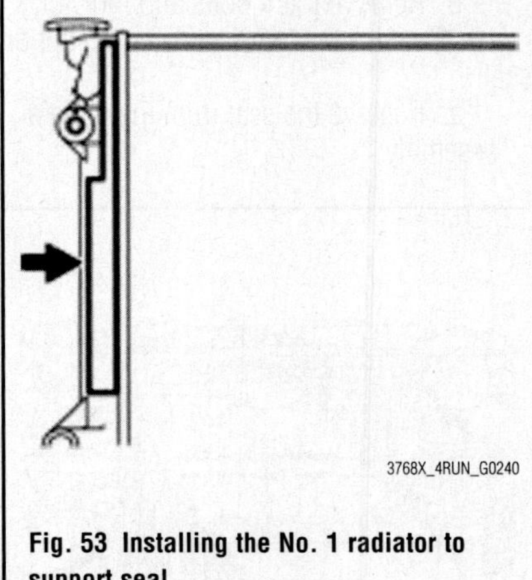

Fig. 53 Installing the No. 1 radiator to support seal

   a. Using a clip remover, detach the 3 claws and remove the clip. Then move the side deflector so that the radiator can be removed.
9. Remove the radiator reservoir.
   a. Disconnect the reservoir hose from the radiator.
   b. Remove the 3 bolts and radiator reservoir.
10. Remove the fan shroud.
   a. Detach the claw to open the flexible hose clamp.
   b. Loosen the 4 nuts holding the fluid coupling fan.
   c. Remove the fan and alternator v-belt.
   d. Remove the 2 bolts holding the fan shroud.
   e. Remove the 4 nuts of the fluid coupling fan, and then remove the shroud together with the coupling fan.

➡Be careful not to damage the radiator core.

11. Disconnect the No. 1 radiator hose.
   a. Disconnect the No. 1 radiator hose from the radiator.

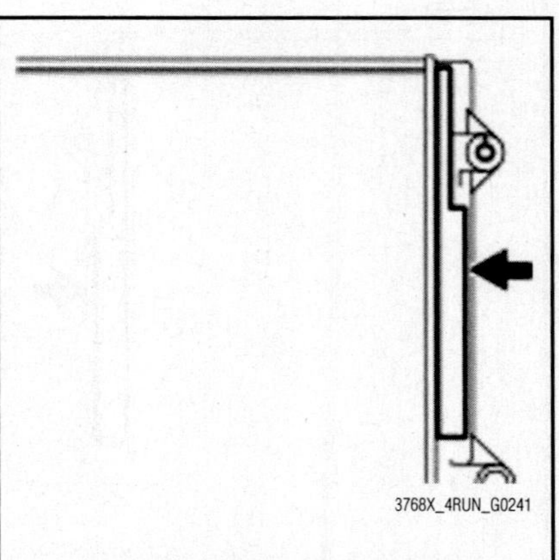

Fig. 54 Installing the No. 2 radiator to support seal

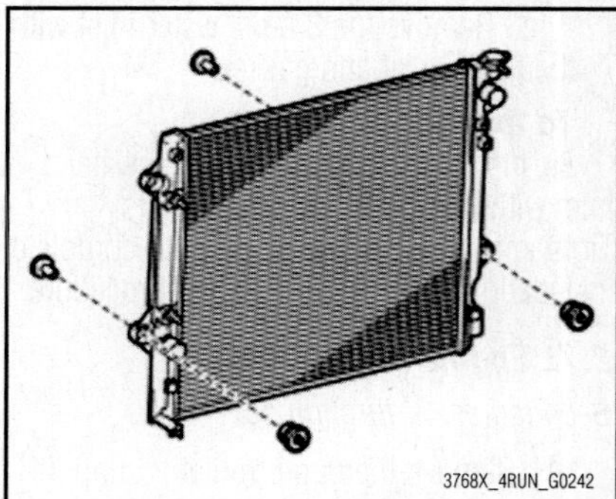

Fig. 55 Installing the 2 radiator supports and 2 No. 1 radiator support bushes

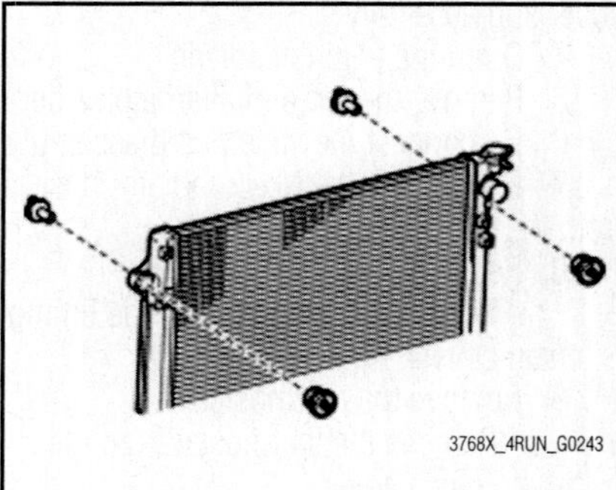

Fig. 56 Installing the No. 1 radiator support

12. Disconnect the No. 2 radiator hose from the radiator.
13. Disconnect the oil cooler inlet hose from the radiator.
14. Disconnect the oil cooler outlet hose from the radiator.
15. Remove the radiator assembly.
   a. Remove the 4 bolts and the radiator.
16. Remove the No. 1 radiator support.
   a. Remove the 2 radiator supports and 2 No. 1 radiator support bushes.

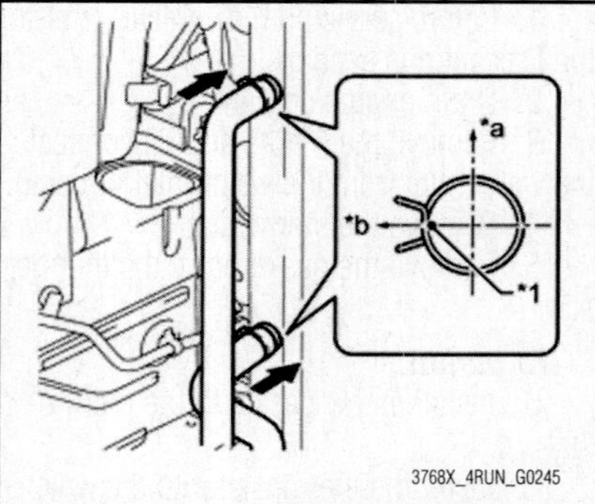

Fig. 58 Connecting the oil cooler outlet hose

17. Remove the No. 2 radiator support.
   a. Remove the 2 radiator supports and 2 No. 2 radiator support bushes.
18. Remove the No. 1 radiator to support seal from the radiator.
19. Remove the No. 2 radiator to support seal from the radiator.

To install:

20. Install the No. 1 radiator to support seal.
   a. Install a new seal to the radiator as shown.
21. Install the No. 2 radiator to support seal.
22. Install the No. 2 radiator support.
   a. Install the 2 radiator supports and 2 No. 1 radiator support bushes.
23. Install the No. 1 radiator support.
   a. Install the 2 radiator supports and 2 No. 1 radiator support bushes.
24. Install the radiator assembly.
   a. Insert the No. 1 radiator support hooks into the radiator support holes.
   b. Install the radiator with the 4 bolts. Tighten the bolts to 13 ft. lbs. (18 Nm).
25. Connect the oil cooler outlet hose.
   a. Connect the oil cooler outlet hose to the radiator.

➡Make sure the direction of the hose clamp is as shown.

26. Connect the oil cooler inlet hose to the radiator.
27. Connect the No. 2 radiator hose to the radiator.

➡Make sure the direction of the hose clamp is as shown.

28. Connect the No. 1 radiator hose to the radiator.

➡Make sure the direction of the hose clamp is as shown.

29. Install the fan shroud.
   a. Install the fan pulley to the water pump.
   b. Place the shroud together with the coupling fan between the radiator and engine.

➡Be careful not to damage the radiator core.

   c. Temporarily install the fluid coupling fan to the water pump with the 4 nuts. Tighten the nuts as much as possible by hand.
   d. Attach the claws of the shroud to the radiator.
   e. Install the shroud with the 2 bolts. Tighten to 44 inch lbs. (5 Nm).
   f. Install the fan and alternator v-belt.
   g. Tighten the 4 nuts of the fluid coupling fan. Tighten to 18 ft. lbs. (25 Nm).
   h. Attach the claw to close the flexible hose clamp.
30. Install the radiator reservoir with the 3 bolts. Tighten to 44 inch lbs. (5 Nm).

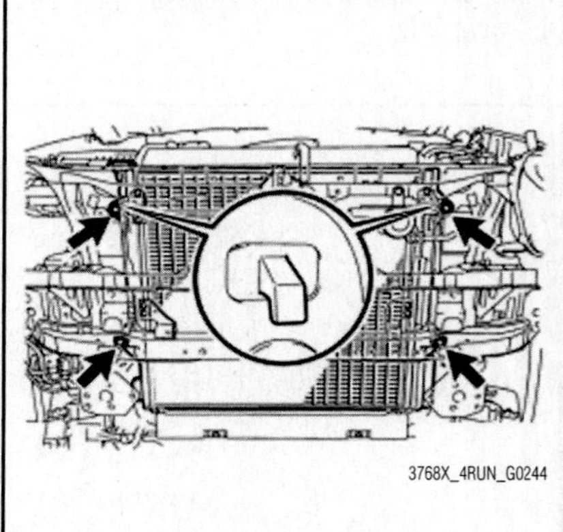

Fig. 57 Inserting the No. 1 radiator support hooks into the radiator support holes

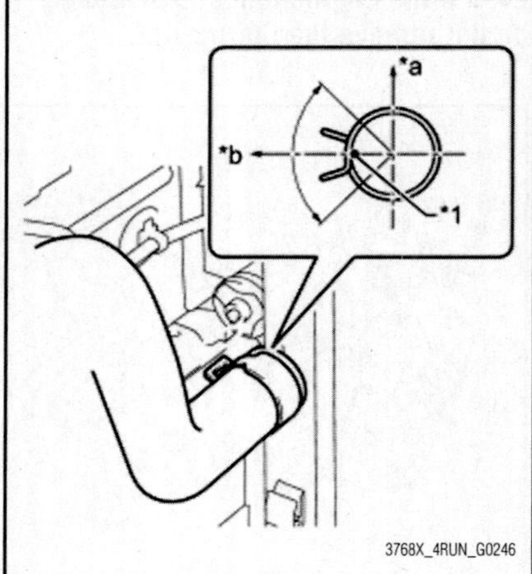

Fig. 59 Connecting the No. 2 radiator hose

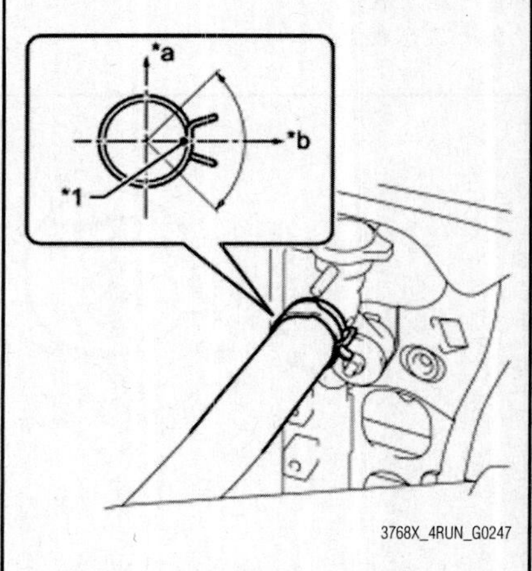

Fig. 60 Connecting the No. 1 radiator hose

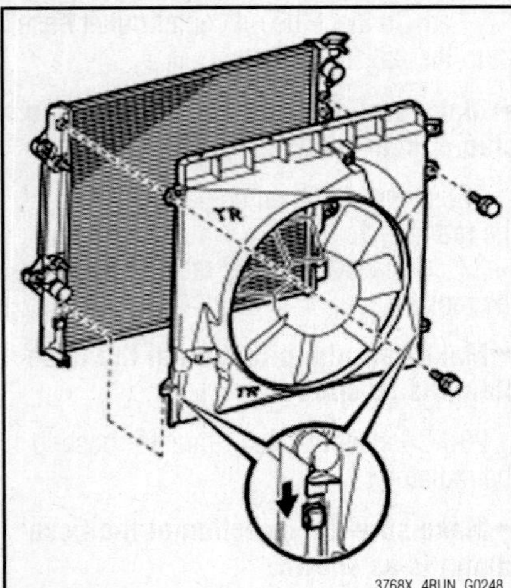

Fig. 61 Attaching the claws of the shroud to the radiator

a. Connect the reservoir hose to the radiator.
31. Install the radiator side deflector LH.
a. Attach the 3 claws.
b. Install the deflector with the clip.
32. Install the radiator side deflector RH.
33. Attach the 3 claws.
a. Install the deflector with the clip.
34. Install the upper front bumper retainer with the 3 bolts. Tighten to 71 inch lbs. (8 Nm).
35. Install the front bumper cover.
36. Add engine coolant.
37. Inspect for coolant leaks.
38. Install the No. 1 engine under cover sub assembly.
39. Install the front bumper cover lower.
40. Install the upper radiator support seal.

THERMOSTAT

REMOVAL & INSTALLATION

2009 Models

See Figure 62.

1. Before servicing the vehicle, refer to the Precautions section.
2. Drain engine coolant.
3. Remove the 3 nuts and disconnect the water inlet from the water inlet housing.
4. Remove the thermostat.
5. Remove the gasket from the thermostat.

To install:

6. Install a new gasket to the thermostat.
7. Insert the thermostat into the water inlet housing with the jiggle valve facing straight upward.

➡The jiggle valve may be set within 30° of either side of the prescribed position.

8. Install the water inlet with the 3 nuts and tighten to 19 Nm (14 ft. lbs.).
9. Fill with engine coolant.
10. Start engine and check for coolant leaks.
11. Recheck engine coolant level.

2010 Models

4.0L Engine

See Figure 63.

1. Remove the upper radiator support seal.
2. Remove the front bumper cover lower.
3. Remove the No. 1 engine under cover sub assembly.
4. Drain the engine coolant.
5. Remove the v-bank cover.
6. Disconnect the No. 2 radiator hose.
a. Disconnect the No. 2 radiator hose.
7. Remove the water inlet with the thermostat.

➡If the thermostat was not installed, cooling efficiency would decrease. Even if the engine tends to overheat, do not remove the thermostat.

a. Remove the 3 nuts, water inlet with the thermostat and gasket.

To install:

8. Install a new gasket and the water inlet with thermostat with the 3 nuts. Tighten to 80 inch lbs. (9 Nm). To complete installation, reverse the removal procedure.

2.7L Engine

See Figures 64 through 66.

1. Remove the upper radiator support seal.
2. Remove the front bumper cover lower.
3. Remove the No. 1 engine under cover sub assembly.
4. Drain the engine coolant.
5. Remove the fan and alternator v-belt.
6. Disconnect the vane pump assembly.
7. Disconnect the No. 2 radiator hose from the water inlet.
8. Remove the water inlet.
a. Remove the gasket from the timing chain cover.
9. Remove the thermostat.
a. Remove the thermostat from the timing chain cover.
b. Remove the gasket from the thermostat.

To install:

10. Install the thermostat.
a. Install a new gasket to the thermostat.
b. Install the thermostat with the jiggle valve upward.

➡The jiggle valve may be set within 10°of either side of the prescribed position.

11. Install the water inlet with a new gasket, the bolt and 2 nuts. Tighten to 21 ft. lbs. (28 Nm).
12. Connect the No. 2 radiator hose to the water inlet.

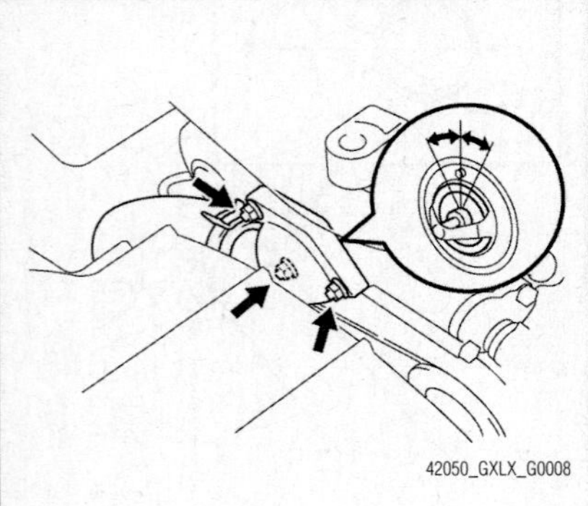

Fig. 62 Thermostat positioning and installation

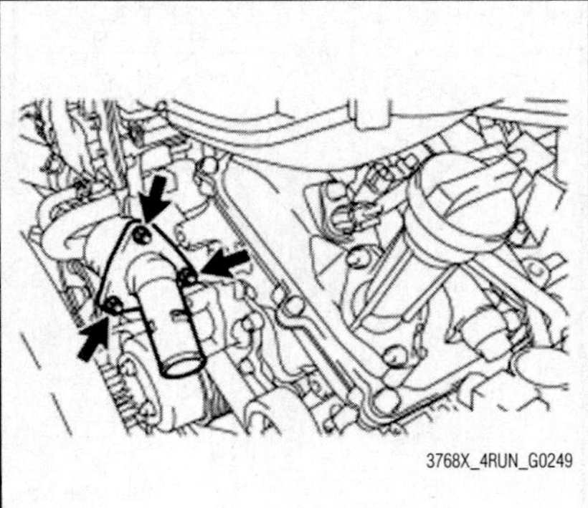

Fig. 63 Removing the thermostat—4.0L engine

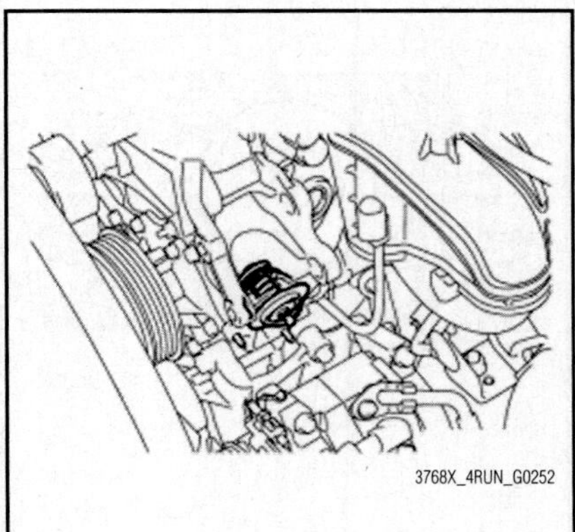

Fig. 64 Removing the thermostat

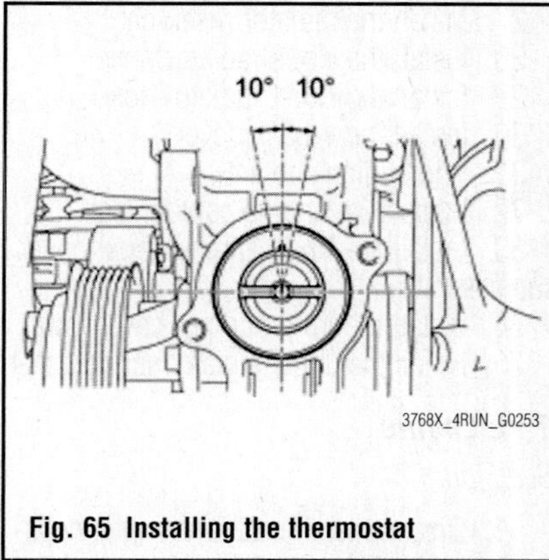

Fig. 65 Installing the thermostat

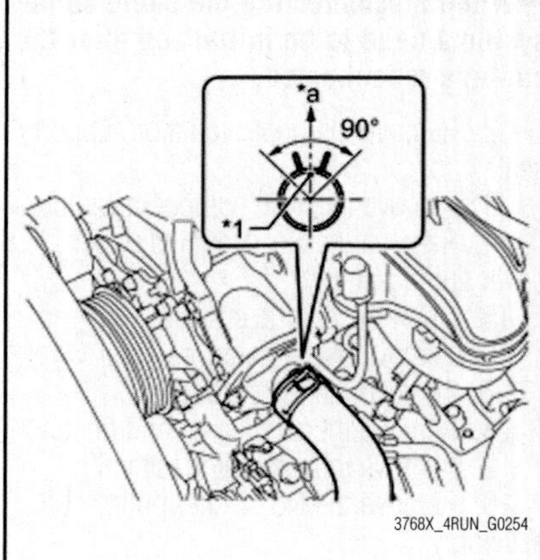

Fig. 66 Connecting the No. 2 radiator hose to the water inlet

➡Make sure the direction of the hose clamp is as shown.

13. Connect the vane pump assembly.
14. Install the fan and alternator v-belt.
15. Add engine coolant.
16. Inspect for coolant leaks.
17. Install the No. 1 engine under cover sub assembly.
18. Install the front bumper cover lower.
19. Install the upper radiator support seal.

WATER PUMP

REMOVAL & INSTALLATION

2009 Models

4.0L Engine

See Figure 67.

1. Remove the 4 bolts and engine under cover.
2. Drain engine coolant.
3. Remove v-bank cover.
4. Remove the 11 clips and radiator support seal upper.
5. Loosen the 4 fluid coupling bolts.
6. Remove fan and alternator v-belt.
7. Unfasten the hose clamp, and then separate the 2 oil cooler hoses from the fan shroud.
8. Disconnect the radiator reserve tank hose.
9. Remove the 2 fan shroud bolts and radiator reserve tank bolt.
10. Remove the 4 fluid coupling nuts, and then remove the fan with fluid coupling and fan shroud together.
11. Remove the fan pulley.
12. Disconnect ventilation hose no.2.
13. Remove air cleaner assembly.
14. Remove water inlet.
15. Disconnect the 2 oil cooler hoses (with oil cooler).
16. Disconnect the 2 radiator hoses.
17. Disconnect the 5 water by-pass hoses.
18. Remove the 5 bolts and water inlet.
19. Remove the O-ring from the water outlet pipe.
20. Remove the gasket from the water pump.
21. Remove the 2 bolts and 2 idler pulleys.
22. Remove alternator assembly.
23. Separate cooler compressor assembly.
24. Remove v-ribbed belt tensioner assembly.
25. Remove the 17 bolts, water pump and gasket.

To install:

26. Install a new gasket and the water pump with the 17 bolts and tighten 10 mm to 80 inch lbs. (9.0 Nm) 12 mm to 17 ft. lbs. (23 Nm).
27. Install v-ribbed belt tensioner assembly.
28. Install cooler compressor assembly.
29. Install alternator assembly.
30. Install the 2 idler pulleys with the 2 bolts and tighten bolts to 29 ft. lbs.
31. Install a new O-ring to the water outlet pipe.
32. Install a new gasket to the water pump.
33. Apply soapy water to the O-ring.
34. Install the water inlet with the 5 bolts and tighten bolts to 80 inch lbs. (9.0 Nm).
35. Connect the 5 water by-pass hoses.
36. Connect the 2 radiator hoses.
37. Connect the 2 oil cooler hoses (with oil cooler).
38. Put the fan pulley to the water pump.
39. Put the fan with fluid coupling and fan shroud together.
40. Temporarily install the 4 fluid coupling nuts.
41. Tighten the 2 fan shroud bolts and radiator reserve tank bolt to 44 inch lbs. (5.0 Nm).
42. Connect the reserve tank hose.
43. Fasten the 2 oil cooler hoses with the hose clamp.
44. Install fan and alternator v-belt.
45. Tighten the 4 fluid coupling bolts.
46. Install radiator support seal upper.
47. Add engine coolant.
48. Inspect engine coolant leak.
49. Install the V-bank cover with the 2 nuts and tighten to 66 inch lbs. (7.5 Nm).
50. Install the engine under cover with the 4 bolts and tighten bolts to 21 ft. lbs. (29 Nm).

4.7L Engine

See Figure 68.

1. Before servicing the vehicle, refer to the precautions section.
2. Drain the cooling system.
3. Remove or disconnect the following:
   - Negative battery cable
   - Timing belt.
   - No. 2 idler pulley
   - Radiator hose
   - Bypass hose
   - Water inlet housing assembly
   - Water pump

Fig. 67 Removing the water pump assembly

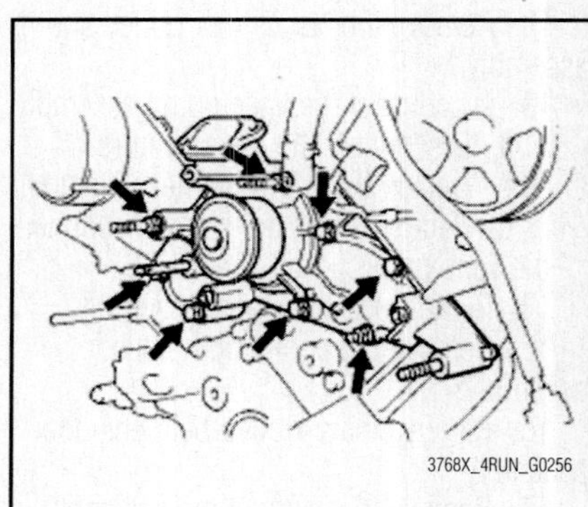

Fig. 68 Removing the water pump assembly

To install:

4. Install or connect the following:
   - Water pump. Use a new gasket and tighten the bolts to 15 ft. lbs. (21 Nm). Tighten the stud bolt and nut to 13 ft. lbs. (18 Nm).
   - Water inlet housing assembly. Use a new O-ring and apply sealant as shown. Tighten the bolts to 13 ft. lbs. (18 Nm).
   - Bypass hose
   - Radiator hose
   - No. 2 idler pulley
   - Timing belt
   - Negative battery cable
5. Fill the cooling system.
6. Start the engine and check for leaks.

2010 Models

4.0L Engine

See Figures 69 and 70.

1. Remove the upper radiator support seal.
2. Remove the front bumper cover lower.
3. Remove the No. 1 engine under cover sub assembly.
4. Drain the engine coolant.
5. Remove the v-bank cover.
6. Remove the No. 1 radiator hose.
7. Remove the No. 2 radiator hose.
8. Remove the radiator reservoir.
9. Disconnect the oil cooler tube.
10. Remove the fan shroud.
11. Remove the water inlet housing.
    a. Disconnect the throttle body connector.
    b. Disconnect the 5 water by pass hoses.
    c. Disconnect the oil cooler hose.
    d. Disconnect the No. 2 oil cooler hose.
    e. Remove the 5 bolts and water inlet.
    f. Remove the o-ring from the water outlet pipe.
    g. Remove the gasket from the water pump.
12. Remove the No. 2 idler pulley sub assembly.
13. Disconnect the vane pump assembly.
    a. Disconnect the 2 connectors.
    b. Detach the 2 wire harness clamps.
    c. Remove the 2 bolts and disconnect the vane pump.
14. Remove the alternator assembly.
15. Remove the cooler compressor assembly.
16. Remove the v-ribbed belt tensioner assembly.
17. Remove the water pump assembly.
    a. Remove the 17 bolts, water pump and gasket.

Fig. 69 Removing the water pump

To install:

18. Install the water pump assembly.
    a. Install a new gasket and the water pump with the 17 bolts. Tighten bolt A to 35 ft. lbs. (47 Nm). Tighten bolt B to 8 ft. lbs. (11 Nm). Tighten bolt C to 17 ft. lbs. (23 Nm).
19. Install the v-ribbed belt tensioner assembly.
20. Install the cooler compressor assembly.
21. Install the alternator assembly.
22. Connect the vane pump assembly. Tighten the 2 bolts to 32 ft. lbs. (43 Nm).
    a. Attach the 2 wire harness clamps.
    b. Connect the 2 connectors.
23. Install the No. 2 idler puller sub assembly. Tighten to 40 ft. lbs. (54 Nm).
24. Install the water inlet housing.
    a. Install a new O-ring to the water outlet pipe.
    b. Install a new gasket to the water pump.
    c. Apply soapy water to the gasket of the water outlet pipe.
    d. Install the water inlet with the 5 bolts. Tighten to 7 ft. lbs. (10 Nm).
    e. Connect the 5 water by-pass hoses.
    f. Connect the No. 2 oil cooler hose.
    g. Connect the oil cooler hose.
    h. Connect the throttle body connector.
25. Install the fan shroud.
26. Connect the oil cooler tube.
27. Install the radiator reservoir.
28. Install the No. 2 radiator hose.
29. Install the No. 1 radiator hose.
30. Install the v-bank cover.
31. Add engine coolant.
32. Inspect for engine coolant leak.
33. Install the No. 1 engine under cover sub assembly.
34. Install the front bumper cover lower.
35. Install the upper radiator support seal.

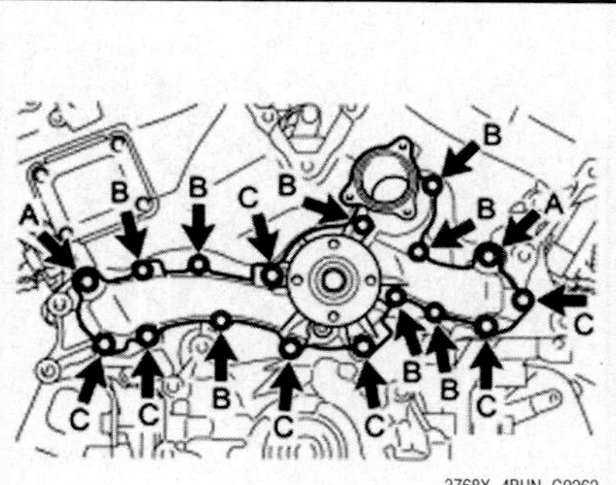

Fig. 70 Installing the water pump

2.7L Engine

See Figure 71.

1. Disconnect the cable from the negative battery terminal.

➡When disconnecting the cable some systems need to be initialized after the cable is reconnected.

2. Remove the upper radiator support seal.
3. Remove the front bumper cover lower.
4. Remove the No. 1 engine under cover sub assembly.
5. Drain the engine coolant.
6. Remove the radiator reservoir.
7. Remove the fan shroud.
8. Remove the air cleaner and hose.
9. Remove the alternator assembly.
10. Remove the No. 1 idler pulley sub assembly.
11. Remove the v-ribbed belt tensioner assembly.
12. Remove the water pump assembly.
    a. Remove the 10 bolts, water pump and gasket.

To install:

13. Install the water pump assembly.
    a. Install a new gasket and the water pump with the 10 bolts. Tighten the A bolts (black arrows) to 19 ft. lbs. (26 Nm). Tighten the B bolts (white arrows) to 79 inch lbs. (8.9 Nm).
14. To complete installation, reverse the installation procedure.

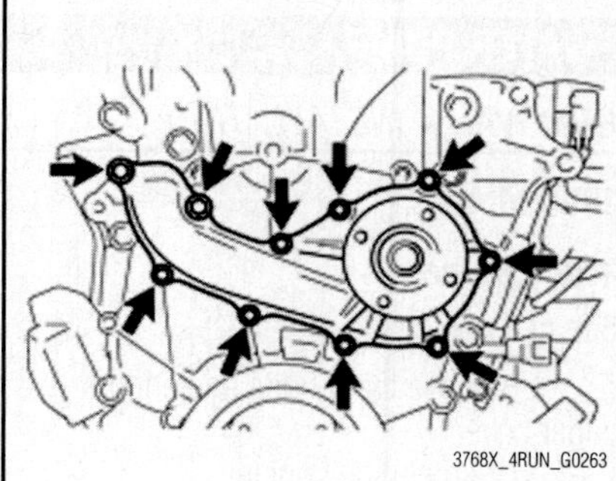

Fig. 71 Removing the water pump—2.7L engine

ALTERNATOR

REMOVAL & INSTALLATION

2009 Models

4.0L Engine

See Figure 72.

1. Remove the v-bank cover.
2. Remove the engine No. 1 under cover sub assembly.
 a. Remove the 4 bolts and engine under cover sub assembly No. 1.
3. Remove the fan and alternator v-belt.
4. Remove the battery.
5. Remove the alternator assembly.
 a. Disconnect or remove the following:
 - Bolt and wire harness stay
 - Connector from the alternator assembly
 - Terminal cap and nut
 - Wire harness from the terminal B

 b. Remove the 2 bolts and wire harness clamp bracket from the alternator assembly.
 c. Remove the 2 bolts and alternator assembly.

To install:

6. Install the alternator assembly with the 2 bolts. Tighten to 32 ft. lbs. (43 Nm).
 a. Install the wire harness clamp bracket with the 2 bolts. Tighten to 71 inch lbs. (8 Nm).
 b. Connect the wire harness to terminal B and install the nut. Tighten the nut to 7 ft. lbs. (9.8 Nm). Connect the connector to the alternator assembly. Install the wire harness stay with the bolt. Tighten to 71 inch lbs. (8 Nm).
7. Install the battery.
8. Install the fan and alternator v-belt.
9. Install the engine No. 1 under cover sub assembly.
 a. Install the engine under cover sub assembly No. 1 with the 4 bolts. Tighten the bolts to 21 ft. lbs. (29 Nm).
10. Install the v-bank cover.

4.7L Engine

See Figure 73.

1. Disconnect the battery negative terminal.
2. Remove the radiator upper support seal.
 a. Remove the 11 clips and radiator upper support seal.
3. Remove the fan and alternator v-belt.
4. Remove the fan and fluid coupling.
 a. Separate the 2 oil cooler hoses from fan shroud.
 b. Remove the 3 bolts and separate the fan shroud from radiator.
 c. Remove the 4 nuts and separate the fan w/ fluid coupling from engine.
 d. Remove the fan shroud and fan w/ fluid coupling from the vehicle at the same time.
5. Remove the vane pump assembly.
 a. Disconnect the vacuum hose.
 b. Remove the nut, 2 bolts and vane pump assembly.

➡Hold up the hoses instead of detaching.

6. Remove the alternator assembly.
 a. Disconnect the alternator connector.
 b. Remove the terminal cap and nut, and disconnect the alternator wire.
 c. Disconnect the alternator wire clamp from the cord clip on the alternator.
 d. Remove the bolt, nut and alternator.

To install:

7. Install the alternator assembly with the bolt and nut. Tighten the bolt to 29 ft. lbs. (39 Nm). Tighten the M10 nut to 29 ft. lbs. (39 Nm). Tighten the M8 nut to 11 ft. lbs. (16 Nm).
 a. Connect the alternator connector.
 b. Connect the alternator wire with the nut and rubber cap. Tighten to 86 inch lbs. (9.8 Nm).
 c. Install the alternator wire clamp to the cord clip on the alternator.
8. Install the vane pump assembly with the nut and 2 bolts. Tighten to 32 ft. lbs. (43 Nm).
 a. Connect the vacuum hose.
9. Install the fan with the fluid coupling.
 a. Put the fan with the fluid coupling and fan shroud into the engine room at the same time.
 b. Install the fan shroud with the 3 bolts. Tighten to 44 inch lbs. (5 Nm).
 c. Install the fan with the fluid coupling with the 4 bolts. Tighten to 21 ft. lbs. (29 Nm).
10. Install the fan and alternator v-belt.
11. Connect the battery negative terminal.

2010 Models

4.0L Engine

See Figure 74.

1. Disconnect the cable from the negative battery terminal.

➡When disconnecting the cable, some systems need to be initialized after the cable is reconnected.

2. Disconnect the cable from the positive battery terminal.
3. Remove the battery clamp.
 a. Loosen the 2 nuts and remove the battery clamp.
4. Remove the battery.
5. Remove the battery tray.

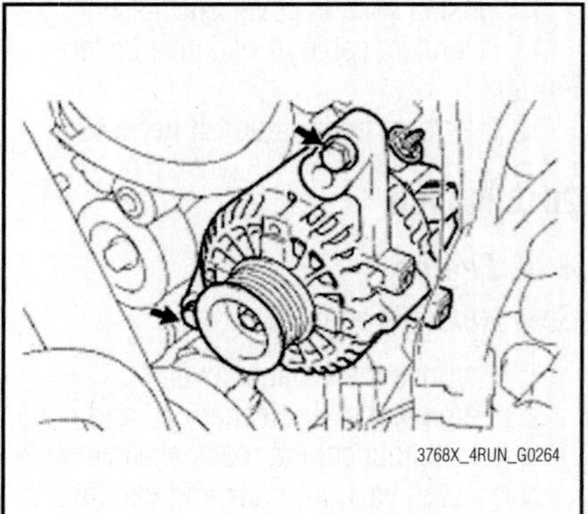

Fig. 72 Removing the alternator assembly

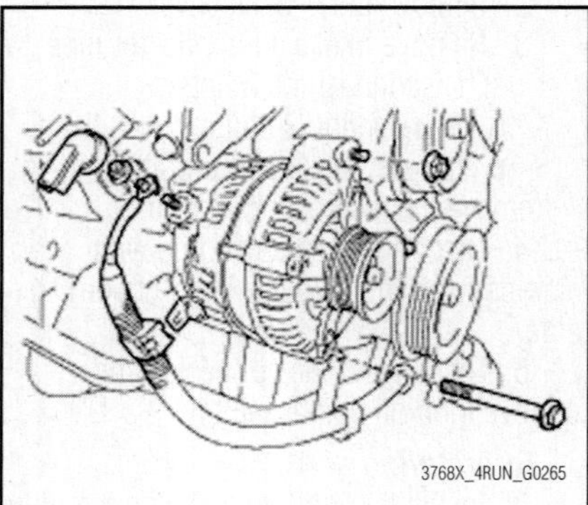

Fig. 73 Removing the alternator

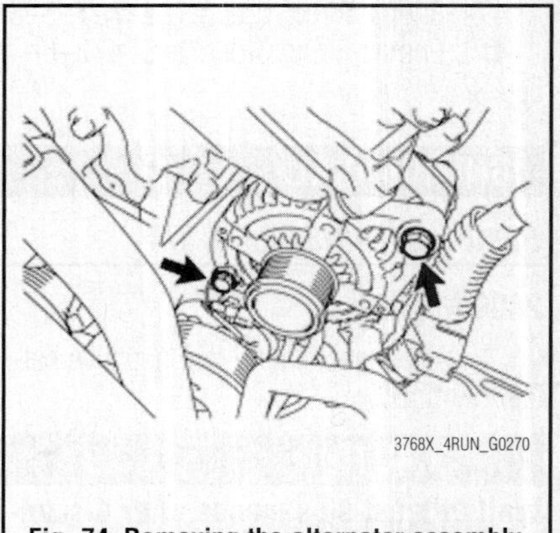

Fig. 74 Removing the alternator assembly

6. Remove the v-bank cover.
7. Remove the fan and alternator v-belt.
8. Remove the No. 2 idler pulley sub assembly.
9. Remove the battery.
10. Remove the battery tray.
11. Remove the v-bank cover.
12. Remove the fan and alternator v-belt.
13. Remove the No. 2 idler pulley sub assembly.
14. Remove the wiring harness clamp bracket.
   a. Detach the clamp.
   b. Remove the bolt and wiring harness clamp bracket.
15. Remove the No. 2 exhaust manifold heat insulator.
16. Remove the alternator assembly.
   a. Open the terminal cap.
   b. Remove the nut and disconnect the wire harness from terminal B.
   c. Disconnect the alternator connector from the alternator assembly.
   d. Remove the 2 bolts and disconnect the wire harness.
   e. Disconnect the wire harness clamp.
   f. Remove the bolt and disconnect the alternator bracket.
   g. Remove the 2 bolts and alternator assembly.
   h. Remove the bolt and alternator bracket.

To install:

17. Install the alternator assembly.
   a. Install the alternator bracket to the alternator with the bolt. Tighten to 15 ft. lbs. (20 Nm).
   b. Install the alternator with the 2 bolts. Tighten to 32 ft. lbs. (43 Nm).
   c. Install the alternator bracket with the bolt. Tighten to 15 ft. lbs. (20 Nm).
   d. Attach the wire harness clamp.
   e. Install the wire harness with the 2 bolts. Tighten to 71 inch lbs. (8 Nm).
   f. Connect the alternator connector to the alternator.
   g. Install the alternator wire with the nut. Tighten to 87 inch lbs. (9.8 Nm).
   h. Close the terminal cap.
18. Install the No. 2 exhaust manifold heat insulator.
19. Install the wiring harness clamp bracket with the bolt. Tighten to 71 inch lbs. (8 Nm).
   a. Attach the clamp.
20. Install the No. 2 idler pulley sub assembly. Tighten to 40 ft. lbs. (54 Nm).
21. Install the fan and alternator v-belt.
22. Install the v-bank cover.
23. Install the battery tray.
24. Install the battery clamp with the 2 nuts. Tighten to 53 inch lbs. (6 Nm).
25. Connect the cable to the positive battery terminal.
26. Connect the cable to the negative battery terminal.

2.7L Engine

See Figure 75.

1. Disconnect the cable from the negative battery terminal.

➡When disconnecting the cable, some systems need to be initialized after the cable is reconnected.

2. Remove the air cleaner and hose.
3. Remove the fan and alternator v-belt.
4. Remove the alternator.

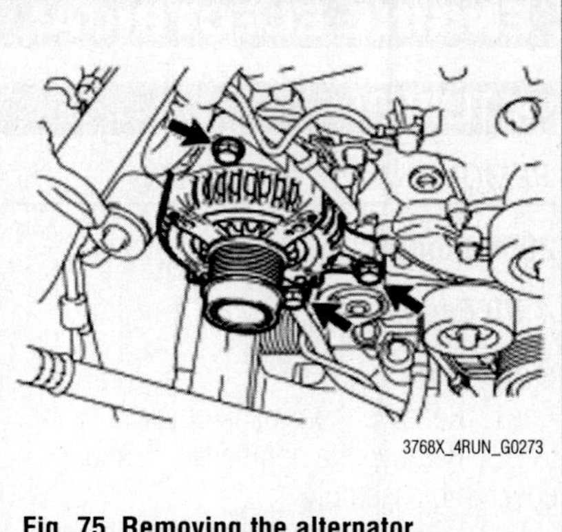

Fig. 75 Removing the alternator

   a. Disconnect the alternator connector.
   b. Remove the terminal cap.
   c. Remove the nut and disconnect the alternator wire from terminal B.
   d. Remove the 3 bolts and the alternator.

To install:

5. Install the alternator assembly.
   a. Install the alternator with the 3 bolts. Tighten to 32 ft. lbs. (43 Nm).
   b. Connect the alternator wire to terminal B with the nut. Tighten to 87 inch lbs. (9.8 Nm).
   c. Install the terminal cap.
   d. Connect the alternator connector.
6. Install the fan and alternator v-belt.
7. Install the air cleaner and hose.
8. Connect the cable to the negative battery terminal.

➡When disconnecting the cable, some systems need to be initialized after the cable is reconnected.

ENGINE ELECTRICAL — IGNITION SYSTEM

FIRING ORDER

2.7L Engine Firing order: 1–3–4–2
4.0L Engine Firing order: 1–2–3–4–5–6
4.7L Engine Firing order: 1–8–4–3–6–5–7–2

IGNITION COIL

REMOVAL & INSTALLATION

2009 Models

1. Disconnect cable from negative battery terminal.

⁂ CAUTION

Wait at least 90 seconds after disconnecting the cable from the negative (-) battery terminal to prevent airbag and seat belt pretensioner activation.

2. Remove the V-bank cover.
3. Remove intake air connector pipe
   a. Disconnect the 3 hoses.
   b. Remove the 2 bolts.
   c. Loosen the 2 clamp bolts and remove the intake air connector.
4. Disconnect the 2 engine wire clamps from the cylinder head cover LH side.
5. Disconnect the connector and remove ignition coil assembly.

To install:

6. Install the ignition coil with the bolt; tighten bolt to 66 inch lbs. (7.5 Nm)
7. Connect the connector.
8. Connect the 2 engine wire clamps to the cylinder head cover LH side.
9. Install intake air connector pipe.
10. Install v-bank cover sub-assembly.
11. Connect cable to negative battery terminal.
12. Perform initialization if necessary.

2010 Models

4.0L Engine

See Figures 76 and 77.

1. Remove the v-bank cover.
2. Remove the air cleaner cap and hose.
   a. Disconnect the mass air flow meter connector, vacuum hose and ventilation hose and detach the 4 clamps.
   b. Loosen the clamp.

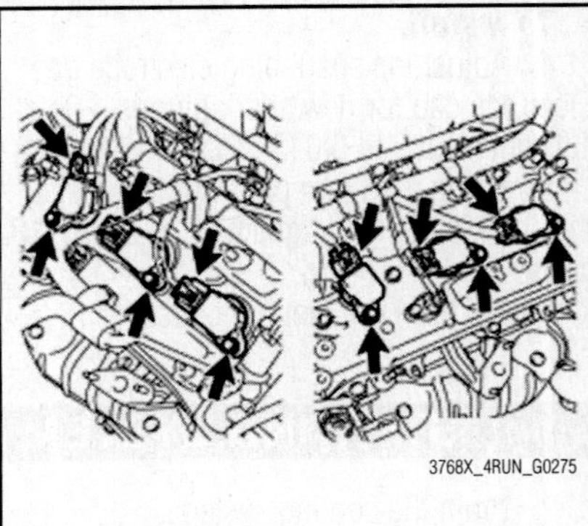

3768X_4RUN_G0275

Fig. 76 Removing the ignition coil

c. Unfasten the 4 hook clamps, and then remove the bolt and air cleaner cap and hose.

3. Remove the ignition coil assembly.

a. Disconnect the 6 ignition coil connectors.

b. Remove the 6 bolts and 6 ignition coils.

4. Remove the 6 spark plugs.

To install:

5. Install the spark plugs and tighten to 13 ft. lbs. (18 Nm).

6. Install the ignition coil assembly with the bolts. Tighten to 7 ft. lbs. (10 Nm). Connect the connectors.

7. Install the air cleaner cap and hose.

a. Install the air cleaner cap and hose with the bolt and fasten the 4 hook clamps. Tighten to 44 inch lbs. (5 Nm).

a. Top
b. Front
c. RH
d. Align cutout portion of the hose with the protrusion of the throttle
e. Paint mark

3768X_4RUN_G0276

Fig. 77 Installing the air cleaner cap and hose

b. Tighten the clamp to 44 inch lbs. (5 Nm).

c. Attach the 4 clamps and connect the ventilation hose, vacuum hose and mass air flow meter connector.

➡**The direction of the hose clamp is illustrated.**

2.7L Engine

See Figure 78.

1. Remove the air cleaner.
2. Remove the intake air connector.
3. Remove the ignition coil assembly.

a. Disconnect the 4 ignition coil connectors.

b. Remove the 4 bolts and 4 ignition coils.

4. Remove the spark plugs using a 16 mm spark plug wrench.

To install:

5. Install the spark plugs using a 16 mm spark plug wrench. Tighten to 13 ft. lbs. (18 Nm).

6. Install the ignition coil assembly.

a. Install the 4 ignition coils with the 4 bolts. Tighten to 80 inch lbs. (9 Nm).

b. Connect the 4 ignition coil connectors.

7. Install the intake air connector.

8. Install the air cleaner and hose.

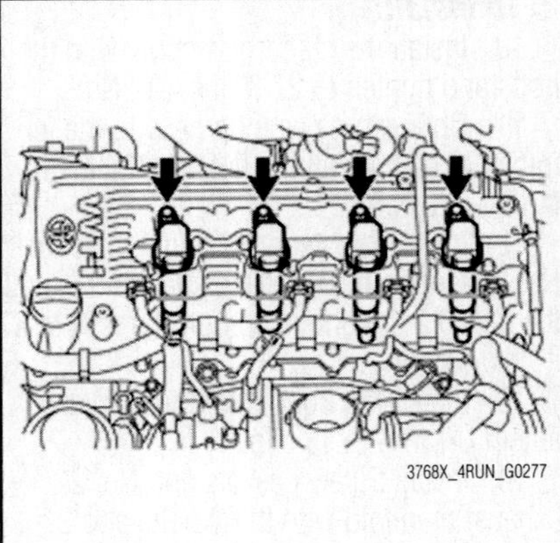

3768X_4RUN_G0277

Fig. 78 Removing the ignition coils

IGNITION TIMING

INSPECTION

See Figures 79 and 80.

1. Warm up the engine.

2. Connect the terminals 13 (TC) and 4 (CG) of the DLC3.

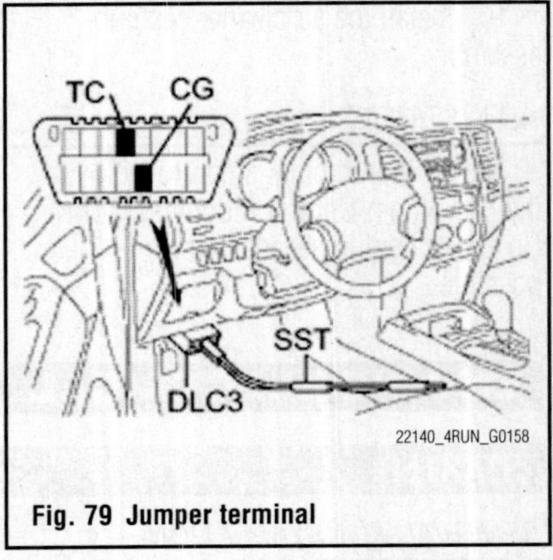

22140_4RUN_G0158

Fig. 79 Jumper terminal

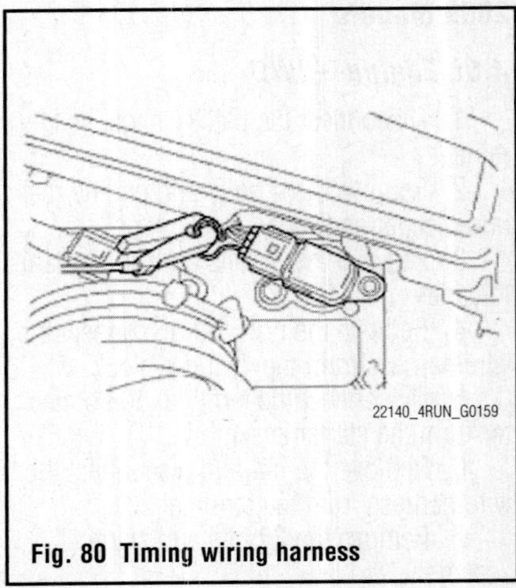

22140_4RUN_G0159

Fig. 80 Timing wiring harness

➡**Be sure not to connect the terminals wrongly. It causes breakage of the engine.**

3. Remove the air cleaner cap sub-assembly.

4. Pull out the wire harness as shown in the illustration.

5. Connect the tester probe of a timing light to the wire of the ignition coil connector for No. 1 cylinder.

➡**Use a timing light that detects the first signal. After checking, be sure to wrap the wire harness with tape.**

6. Inspect the ignition timing during idling. 8 to 12°CA BTDC during idling (Transmission in neutral position).

7. Remove the connector from the DLC3.

8. Inspect the ignition timing during idling. 7 to 24°CA BTDC during idling (Transmission in neutral position).

9. Disconnect the timing light from the engine.

10. Install the air cleaner cap sub-assembly.

ADJUSTMENT

The ignition timing is controlled by the Powertrain Control Module (PCM). No adjustment is necessary or possible.

SPARK PLUGS

REMOVAL & INSTALLATION

1. Remove the ignition coils.
2. Using a 16 mm plug wrench, remove the spark plugs.
3. Clean the spark plugs.

To install:

4. Adjust the spark plug electrode gap. Electrode gap for new spark plug is 1.0 to 1.1 mm (0.039 to 0.043 in.).
5. Using a 16 mm plug wrench, install the spark plugs and tighten to 17.5 Nm (13 ft lbs)
6. Reinstall the ignition coils.

ENGINE ELECTRICAL

STARTING SYSTEM

STARTER

REMOVAL & INSTALLATION

2009 Models

4.0L Engine—2WD

1. Disconnect the battery negative terminal.
2. Remove the 4 bolts and engine rear under cover assembly.
3. Remove the 3 bolts and No. 2 manifold stay.
4. Remove the bolt and disconnect the wire harness from the cylinder block.
5. Disconnect the terminal 50 connector from the starter assembly.
6. Remove the nut and disconnect the wire harness from the terminal 30.
7. Remove the 2 bolts and starter assembly.

To install:

8. Install the starter assembly with the 2 bolts and tighten to 27 ft. lbs. (37 Nm).
9. Connect the wire harness to the terminal 30 and install the nut Tighten to 7 ft. lbs. (9.8 Nm).
10. Connect the wire harness to the cylinder block and install the bolt. 10 ft. lbs. (13 Nm).
11. Install the No. 2 manifold stay with the 3 bolts and tighten to 30 ft. lbs. (40 Nm).
12. Install the engine rear under cover assembly with the 4 bolts and tighten to 21 ft. lbs. (29 Nm).
13. Connect the battery negative terminal.

4.0L Engine—4WD

See Figure 81.

1. Disconnect the battery negative terminal.
2. Remove the 4 bolts and engine rear under cover assembly.
3. Disconnect the heated oxygen sensor (bank 2) connector.
4. Remove the 2 bolts, 2 nuts, No. 2 front exhaust pipe assembly and 2 gaskets.
5. Remove the 5 clips and front fender rear splash shield LH.

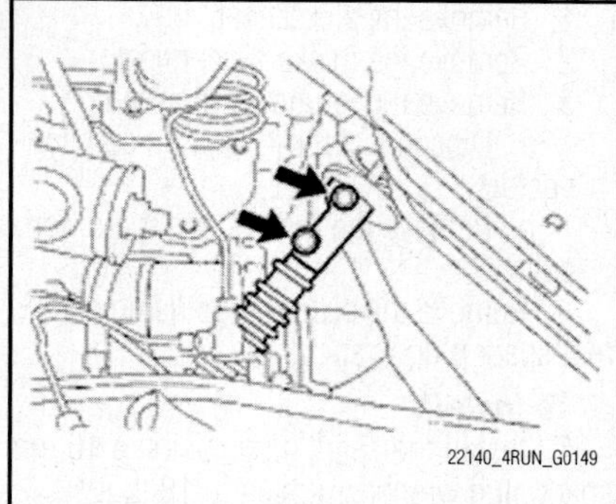

Fig. 81 Steering intermediate shaft

6. Loosen the bolt A and remove the bolt B, then separate the No. 2 steering intermediate shaft sub-assembly.
7. Disconnect the terminal 50 connector from the starter assembly.
8. Remove the nut and disconnect the wire harness from the terminal 30.
9. Remove the 2 bolts and starter assembly.

To install:

10. Install the starter assembly with the 2 bolts and tighten to 27 ft. lbs. (37 Nm).
11. Connect the wire harness to the terminal 30 and install the nut and tighten to 7 ft. lbs. (9.8 Nm).
12. Install the steering sliding yoke to No. 2 steering intermediate shaft and install the bolt B and tighten the bolt B and A to 27 ft. lbs. (36 Nm).
13. Install the front fender rear splash shield LH with the 5 clips.
14. Install 2 new gaskets and No. 2 exhaust manifold with the 2 bolts and 2 nuts and tighten nut to 46 ft. lbs. (62 Nm). Tighten bolt to 35 ft. lbs. (48 Nm).
15. Connect the heated oxygen sensor (bank 2) connector.
16. Install the engine rear under cover assembly with the 4 bolts and tighten to 21 ft. lbs. (29 Nm).
17. Connect the battery negative terminal.

4.7L Engine

1. Before servicing the vehicle, refer to the precautions section.
2. Drain the cooling system.
3. Relieve the fuel system pressure.
4. Remove or disconnect the following:
   - Negative battery cable
   - Engine appearance cover
   - Air intake tube
   - No1 & No2 fuel hoses
   - Ventilation hose
   - Purge VSV
   - Vacuum control valve
   - Engine wire harness
   - Throttle body water bypass hose
   - Intake manifold
   - Air pump and switching valve
   - Starter motor mounting bolts
   - Starter wiring connectors
   - Starter motor

To install:

5. Install or connect the following:
   - Starter motor
   - Starter wiring connectors. Tighten the cable nut to 86 inch lbs. (10 Nm).
   - Starter motor mounting bolts. Tighten the bolts to 29 ft. lbs. (39 Nm).
   - Air pump and switching valve
   - Intake manifold
   - Throttle body water bypass hose
   - Engine wire harness
   - Vacuum control valve
   - Purge VSV
   - Ventilation hose
   - No1 & No2 fuel hoses
   - Air intake tube
   - Engine appearance cover
   - Negative battery cable
6. Fill the cooling system.
7. Start the engine and check for leaks.
8. Perform initialization if necessary.

2010 Models

4.0L Engine

See Figure 82.

1. Disconnect the cable from the negative battery terminal.

➡When disconnecting the cable, some systems need to be initialized after the cable is reconnected.

Fig. 82 Removing the starter

2. Remove the exhaust manifold sub assembly LH.
3. Remove the 3 bolts and the starter cover.
4. Remove the starter assembly.
 a. Remove the bolt and disconnect the ground wire.
 b. Disconnect the starter connector.
 c. Open the terminal cap.
 d. Remove the 2 bolts and starter.
5. Remove the flywheel housing side cover.

To install:

6. Install the flywheel housing side cover.
7. Install the starter assembly with the 2 bolts. Tighten the bolts to 27 ft. lbs. (37 Nm).
 a. Connect the starter wire with the bolt and nut. Tighten the bolts to 71 inch lbs. (8 Nm). Tighten the nut to 87 inch lbs. (9.8 Nm).
 b. Close the terminal cap.
 c. Connect the starter connector.
 d. Connect the ground wire with the bolt. Tighten to 10 ft. lbs. (13 Nm).
8. Install the starter cover with the 3 bolts. Tighten to 8 ft. lbs. (12 Nm).
9. Install the exhaust manifold sub assembly LH.
10. Connect the cable to the negative battery terminal.

➡When disconnecting the cable, some systems need to be initialized after the cable is reconnected.

2.7L Engine

See Figure 83.

1. Disconnect the cable from the negative battery terminal.

➡When disconnecting the cable, some systems need to be initialized after the cable is reconnected.

2. Remove the rear engine under cover assembly.
3. Remove the transmission oil filter tube sub assembly.
4. Remove the starter assembly.
 a. Disconnect the starter connector.
 b. Remove the terminal cap.
 c. Remove the nut and disconnect the starter wire.
 d. Remove the 2 bolts and the starter.

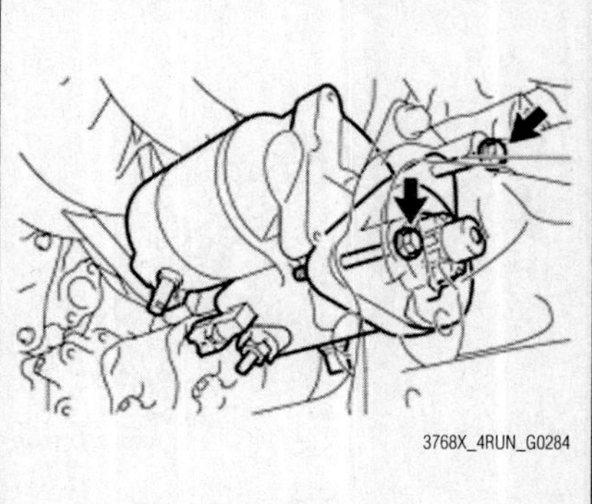

Fig. 83 Removing the starter

To install:

5. Install the starter assembly with the 2 bolts. Tighten to 27 ft. lbs. (37 Nm).
 a. Connect the starter connector.
 b. Connect the starter wire harness with the nut. Tighten to 87 inch lbs. (9.8 Nm).
 c. install the terminal cap.
6. Install the transmission oil filler tube sub assembly.
7. Install the rear engine under cover assembly.
8. Connect the cable to the negative battery terminal.

➡When disconnecting the cable, some systems need to be initialized after the cable is reconnected.

ENGINE MECHANICAL

ACCESSORY DRIVE BELTS

ACCESSORY BELT ROUTING

See Figures 84 and 85.

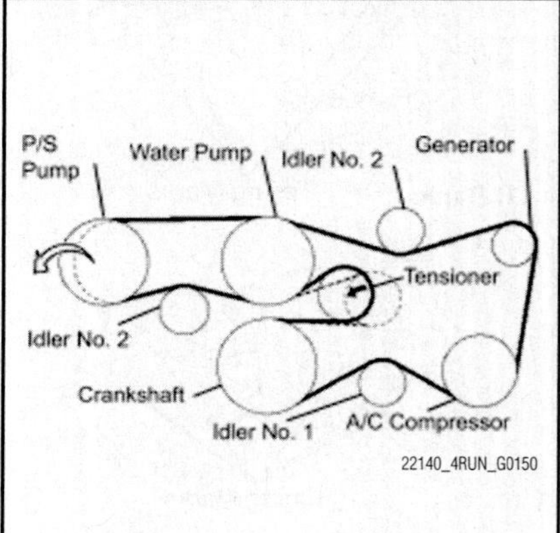

Fig. 84 Accessory drive belt routing—4.0L engine

INSPECTION

Inspect the drive belt for signs of glazing or cracking. A glazed belt will be perfectly smooth from slippage, while a good belt will have a slight texture of fabric visible. Cracks will usually start at the inner edge of the belt and run outward. All worn or damaged drive belts should be replaced immediately.

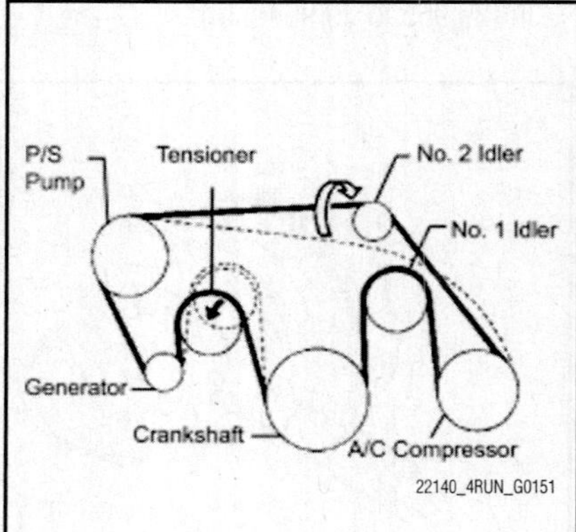

Fig. 85 Accessory drive belt routing—4.7 engine

ADJUSTMENT

Belt adjustment is automatic and non-adjustable.

REMOVAL & INSTALLATION

See Figures 84 through 86.

1. Before servicing the vehicle, refer to the Precautions section.
2. Loosen the drive belt tension by turning the drive belt tensioner counterclockwise, and remove the drive belt.

To install:

3. Installation is the reverse of removal.

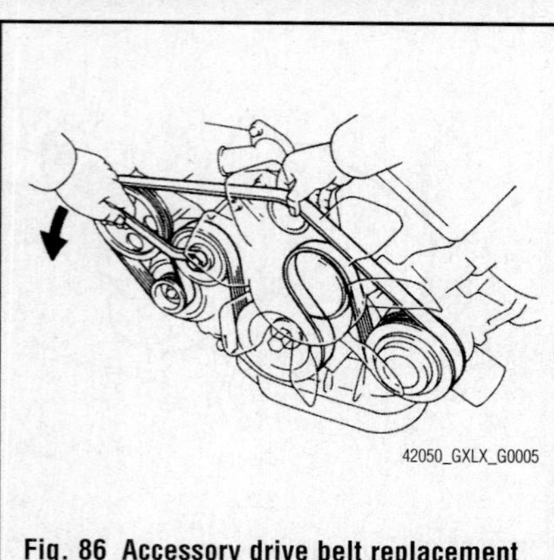

Fig. 86 Accessory drive belt replacement

CAMSHAFT AND VALVE LIFTERS

INSPECTION

2009 Models

4.0L Engine

See Figure 87.

1. Place the camshaft on V-blocks.
2. Using a dial indicator, measure the circle runout at the center journal.
    a. Maximum circle runout 0.0024 in. (0.06 mm).
3. If the circle runout is greater than maximum, replace the camshaft.
4. Using a micrometer, measure the cam lobe height.
5. Standard cam lobe height:
    a. Intake: 1.7389 to 1.7428 in. (44.168 to 44.268 mm).
    b. Exhaust: 1.7551 to 1.7591 in. (44.580 to 44.680 mm).
6. Minimum cam lobe height:
    a. Intake: 1.7330 in (44.018 mm).
    b. Exhaust: 1.7492 in. (44.430 mm).
7. If the cam lobe height is less than the minimum, replace the camshaft.
8. Using a micrometer, measure the journal diameter.
    a. Journal diameter: 1.4162 to 1.4167 in. (35.971 to 35.985 mm).
9. If the journal diameter is not as specified, check the oil clearance.

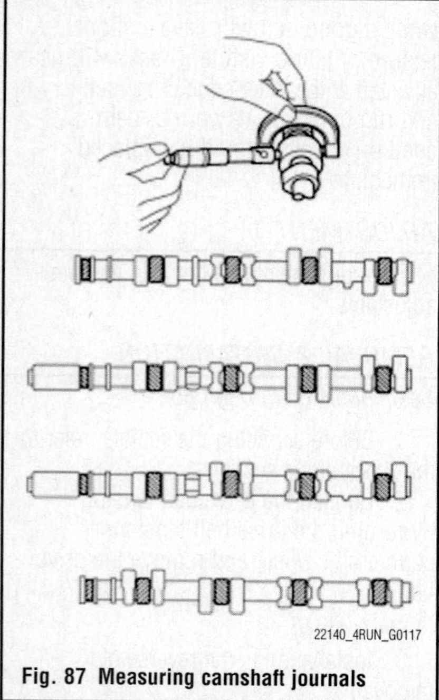

Fig. 87 Measuring camshaft journals

4.7L Engine

See Figures 87 through 89.

1. Place the camshaft on V-blocks.
2. Using a dial indicator, measure the circle runout at the center journal.
    a. Maximum circle runout 0.0031 in. (0.08 mm).
3. If the circle runout is greater than maximum, replace the camshaft.
4. Using a micrometer, measure the cam lobe height.
5. Standard cam lobe height:
    a. Intake: 1.6512 to 1.6551 in. (41.94 to 42.04 mm).
    b. Exhaust: 1.6520 to 1.6559 in. (41.96 to 42.06 mm).
6. Minimum cam lobe height:
    a. Intake: 1.6453 in (41.79 mm).
    b. Exhaust: 1.6461 in. (41.81 mm).
7. If the cam lobe height is less than the minimum, replace the camshaft.
8. Using a micrometer, measure the journal diameter.
    a. Journal diameter: 1.0612 to 1.0618 in. (26.954 to 26.970 mm).
9. If the journal diameter is not as specified, check the oil clearance.
10. Using a micrometer, measure the journal diameter.
    a. Journal diameter: 1.5730 to 1.5734 in (39.955 to 39.964 mm).

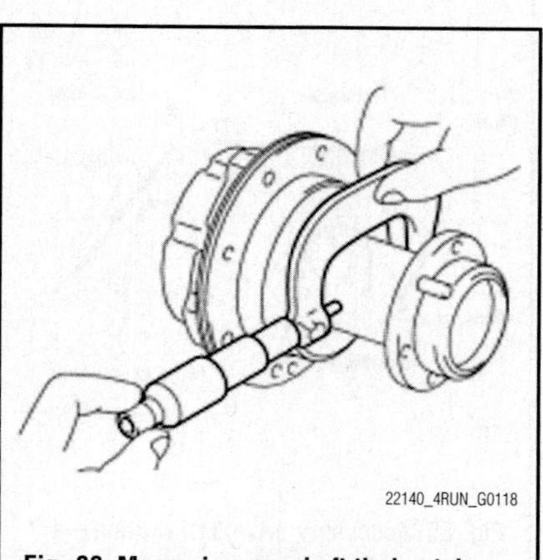

Fig. 88 Measuring camshaft timing tube

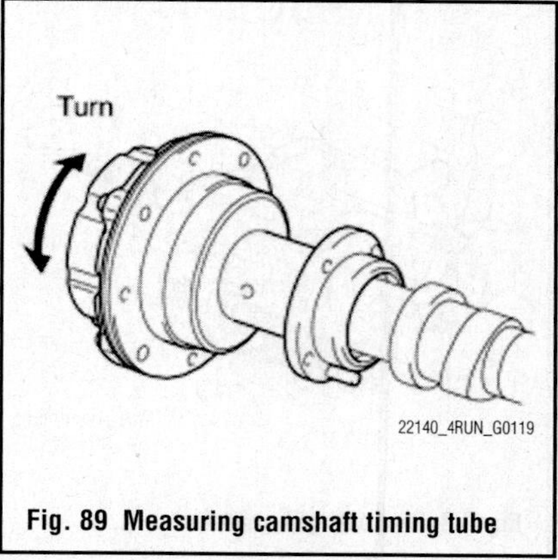

Fig. 89 Measuring camshaft timing tube

11. If the journal diameter is not as specified, check the oil clearance.
12. Install the timing tube to the intake camshaft, and check that the timing tube turns smoothly.
13. If necessary, replace the timing tube and intake camshaft.
14. Using vernier calipers, measure the gap distance of the gear spring.
    a. Gap distance: 0.717 to 0.740 in. (18.2 to 18.8 mm).
15. If the gap distance is not as specified, replace the gear spring.

REMOVAL & INSTALLATION

2009 Models

4.0L Engine

See Figures 90 through 98.

1. Drain engine coolant.
2. Remove v-bank cover.

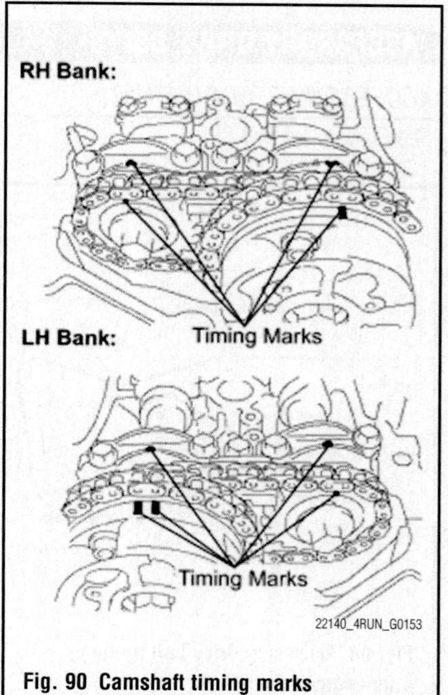

Fig. 90 Camshaft timing marks

3. Disconnect no. 2 ventilation hose.
4. Remove air cleaner assembly.
5. Disconnect the 2 water by-pass hoses.
6. Disconnect the fuel vapor feed hose.
7. Disconnect the 2 VSV connectors.
8. Disconnect the throttle body with motor connector.
9. Separate the 3 wire harness clamps and hose clamp.
10. Remove the 2 bolts and throttle body bracket.
11. Remove the bolt and oil baffle plate.
12. Remove the 4 bolts and 2 surge tank stays.
13. Remove the 2 nuts.
14. Using an 8 mm socket hexagon wrench, remove the 4 bolts, intake air surge tank and gasket.
15. Remove ignition coil assembly.
16. Remove the 10 bolts, 3 seal washers, 2 nuts, cylinder head cover and gasket.
17. Turn the crankshaft pulley, and align the notch with the timing mark "0" of the timing chain cover.
18. Check that the timing marks of the camshaft timing gears are aligned with the timing marks of the bearing caps as shown in the illustration. If not, turn the crankshaft 1 complete revolution (360°) and align the timing marks as above.
19. Place paint marks on the No. 1 chain links that correspond with the timing marks of the camshaft timing gears.

➡Never rotate the crankshaft with the chain tensioner removed. When rotating the camshaft with the timing chain removed, rotate the crankshaft counterclockwise 40° from TDC first.

20. Remove the 4 bolts, timing chain cover plate and gasket.
21. While turning the stopper plate of the tensioner clockwise, push in the plunger of the chain tensioner as shown in the illustration.
22. While turning the stopper plate of the tensioner counterclockwise, insert a bar of 3.5 mm (0.138 in.) into the holes on the stopper plate and tensioner to fix the stopper plate in place.
23. Remove the 2 bolts and chain tensioner.

➡As the thrust clearance of the camshaft is small, the camshaft must be kept level while it is being removed. If the camshaft is not kept level, the portion of the cylinder head which receives the shaft thrust may crack or be damaged, causing the camshaft to seize or break. To avoid this, the following steps should be carried out.

24. While raising up the No. 2 chain tensioner, insert a pin of 1.0 mm (0.039 in.) into the hole to fix it in place.
25. Hold the hexagonal portion of the No. 2 camshaft with a wrench, and remove the camshaft timing gear set bolt.

➡Be careful not to damage the cylinder head and valve lifter with the wrench.

26. Separate the camshaft timing gear from the No. 2 camshaft.
27. Rotate the camshaft counterclockwise using a wrench so that the cam lobes of the No. 1 cylinder face upward.
28. Using several steps, loosen and remove the 8 bearing cap bolts uniformly in the sequence shown in the illustration.
29. Remove the 4 bearing caps and No. 2 camshaft.
30. Remove the No. 2 chain tensioner bolt, and then remove the No. 2 chain tensioner and camshaft timing gear.

➡As the thrust clearance of the camshaft is small, the camshaft must be kept level while it is being removed. If the camshaft is not kept level, the portion of the cylinder head which receives the shaft thrust may crack or be damaged, causing the camshaft to seize or break. To avoid this, the following steps should be carried out.

31. Hold the hexagonal portion of the No. 1 camshaft with a wrench, and loosen the camshaft timing gear set bolt.

➡Be careful not to damage the cylinder head and valve lifter with the wrench. Do not disassemble the camshaft timing gear assembly.

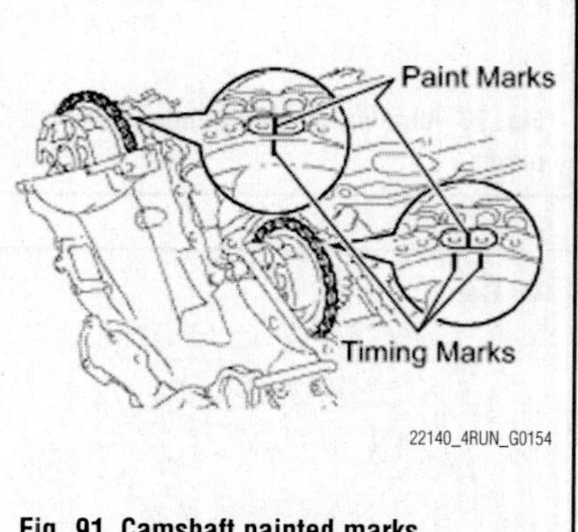

Fig. 91 Camshaft painted marks

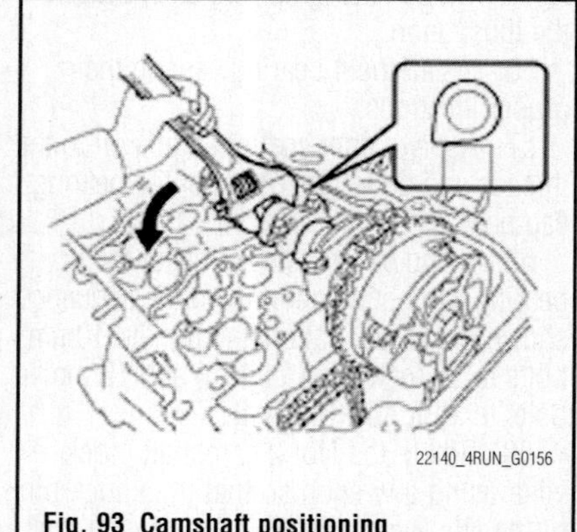

Fig. 93 Camshaft positioning

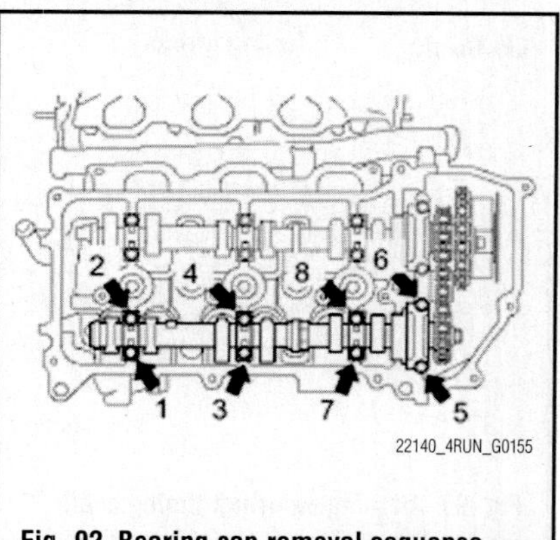

Fig. 92 Bearing cap removal sequence

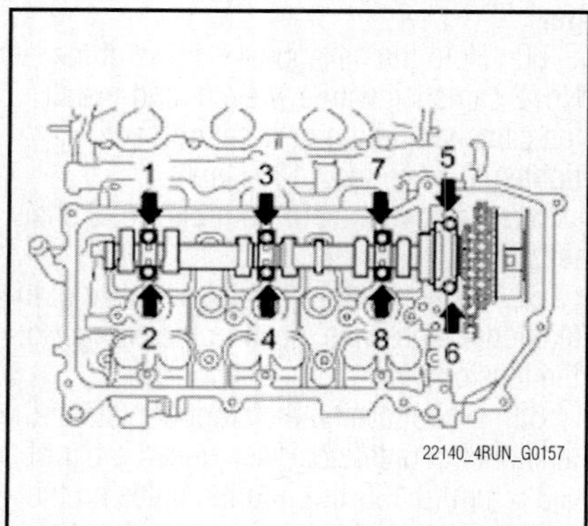

Fig. 94 Bearing cap removal sequence

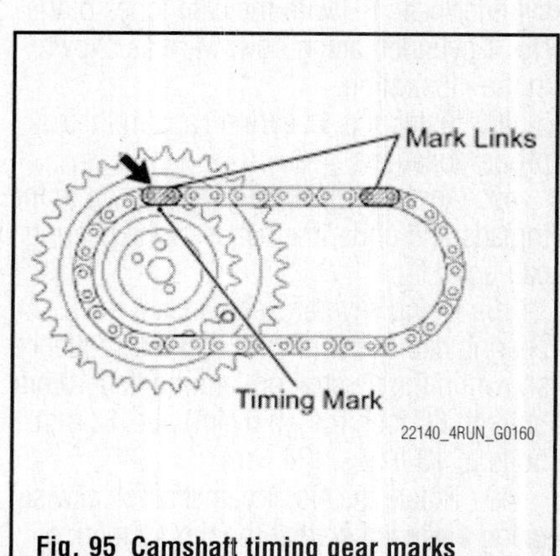

Fig. 95 Camshaft timing gear marks

32. Slide the camshaft timing gear and separate the No. 1 chain from the camshaft timing gear
33. Rotate the No. 1 camshaft counterclockwise using a wrench so that the cam lobes of the No. 1 cylinder face downward as shown in the illustration.
34. Using several steps, loosen and remove the 8 bearing cap bolts in the sequence shown in the illustration.
35. Remove the 4 bearing caps.
36. Remove the camshaft timing gear set bolt with the No. 1 camshaft lifted up, and then remove the No. 1 camshaft and camshaft timing gear with the No. 2 chain.
37. Tie the No. 1 chain with a string.

➡Be careful not to drop anything inside the timing chain cover.

38. Remove the valve lifters.

To install:

39. Install valve lifters

➡As the thrust clearance of the camshaft is small, the camshaft must be kept level while it is being installed. If the camshaft is not kept level, the portion of the cylinder head which receives the shaft thrust may crack or be damaged, causing the camshaft to seize or break. To avoid this, the following steps should be carried out.

40. Align the mark link (yellow) with the timing mark (1-dot mark) of the camshaft timing gear as shown in the illustration.
41. Apply new engine oil to the thrust portions and journals of the camshafts
42. Temporarily put the No. 1 chain on the No. 2 chain of the camshaft timing gear
43. Align the knock pin hole of the camshaft timing gear with the knock pin of the No. 1 camshaft, and insert the No. 1 camshaft into the camshaft timing gear.
44. Temporarily install the camshaft timing gear set bolt.
45. Set the No. 1 camshaft onto the cylinder head RH with the cam lobes of the No. 1 cylinder facing downward as shown in the illustration.
46. Install the 4 bearing caps in their proper locations
47. Apply a light coat of engine oil to the threads and under the heads of the bearing cap bolts
48. Using several steps, install the 8 bearing cap bolts uniformly in the sequence shown in the illustration, and tighten 10mm bolts to 80 inch lbs. (9.0 Nm) and 12 mm bolts to 18 ft. lbs. (24 Nm)
49. Rotate the No. 1 camshaft clockwise using a wrench so that the timing mark of the camshaft timing gear is aligned with the timing mark of the camshaft bearing cap
50. Align the paint mark of the No. 1 chain with the timing mark of the camshaft timing gear
51. Hold the hexagonal portion of the No. 1 camshaft with a wrench, and tighten the camshaft timing gear set bolt and tighten to 74 ft lbs. (100 Nm)
52. While pushing in the tensioner, insert a pin of 1.0 mm (0.039 in.) into the hole to fix it in place
53. Temporarily install the camshaft timing gear and No. 2 chain tensioner with the bolt and align the mark links (yellow) with the timing marks (1-dot mark) of the camshaft timing gears
54. Tighten the No. 2 chain tensioner bolt to 14 ft. lbs. (19 Nm).

➡As the thrust clearance of the camshaft is small, the camshaft must be kept level while it is being installed. If the camshaft is not kept level, the portion of the cylinder head which receives the shaft thrust may crack or be damaged, causing the camshaft to seize or break. To avoid this, the following steps should be carried out.

55. Set the No. 2 camshaft onto the cylinder head RH with the cam lobes of the No. 1 cylinder facing upward as shown in the illustration.
56. Install the 4 bearing caps in their proper locations.
57. Apply a light coat of engine oil to the threads and under the heads of the bearing cap bolts.
58. Using several steps, install the 8 bearing cap bolts uniformly in the sequence shown in the illustration and tighten 10mm bolts to 80 inch lbs. (9.0 Nm) and 12 mm bolts to 18 ft. lbs. (24 Nm).
59. Rotate the No. 2 camshaft clockwise using a wrench so that the knock pin of the No. 2 camshaft is aligned with the knock pin hole of the camshaft timing gear.
60. Hold the hexagonal portion of the No. 2 camshaft with a wrench, and install the camshaft timing gear set bolt and tighten to 74 ft. lbs. (100 Nm).
61. Remove the pin from the No. 2 chain tensioner.
62. While turning the stopper plate of the tensioner clockwise, push in the plunger of the tensioner.
63. While turning the stopper plate of the tensioner counterclockwise, insert a bar of ϕ 3.5 mm (0.138 in.) into the holes on the stopper plate and tensioner to fix the stopper plate in place.
64. Install the chain tensioner with the 2 bolts and tighten to 7 ft lbs. (10 Nm).
65. Remove the bar from the chain tensioner.
66. Install a new gasket and the timing chain cover plate with the 4 bolts and tighten to 80 inch lbs.
67. Turn the crankshaft pulley 2 complete revolutions slowly, and align the notch with the timing mark "0" of the timing chain cover.
68. Check that the timing marks of the camshaft timing gears are aligned with the timing marks of the bearing caps as shown in the illustration.
69. Set no. 1 cylinder to TDC/compression.
70. Inspect and adjust valve clearance.
71. Remove any old packing (FIPG) material and be careful not to drop any oil on the contact surfaces of the cylinder head, timing chain cover and cylinder head cover.

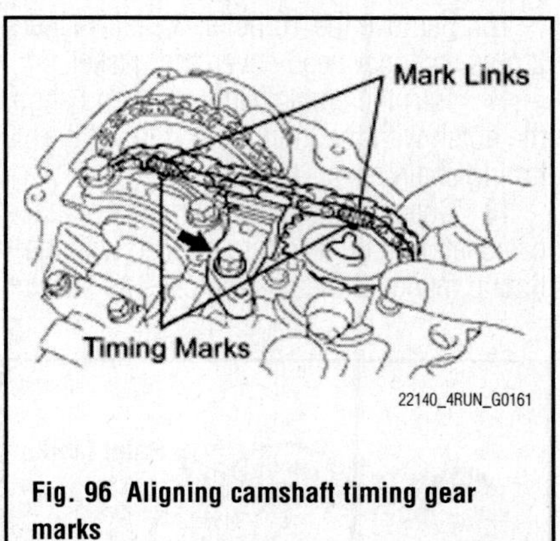

Fig. 96 Aligning camshaft timing gear marks

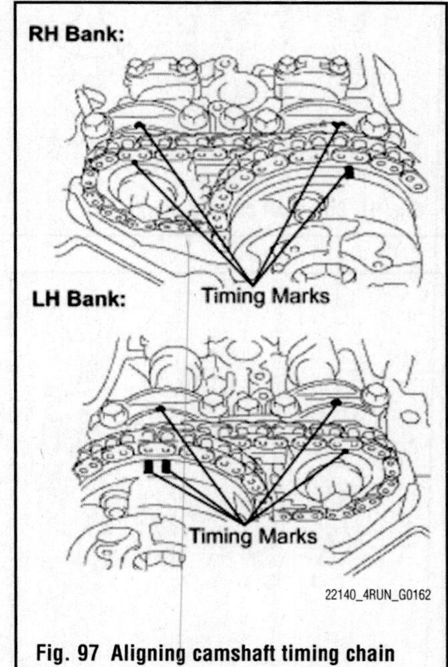

Fig. 97 Aligning camshaft timing chain marks

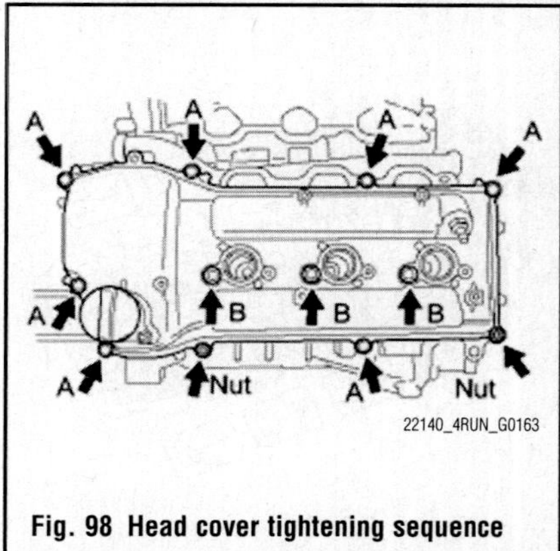

Fig. 98 Head cover tightening sequence

72. Apply seal packing (diameter: 2 to 3 mm (0.08 to 0.12 in.)) to the cylinder head and timing chain cover.

73. Install the seal washers to the bolts.

74. Install the cylinder head cover with the 10 bolts and 2 nuts. Tighten the bolts and nuts uniformly in several steps. Tighten bolt A to 7 ft. lbs. (10 Nm) and bolt B and nut to 90 inch lbs. (9.0 Nm).

75. Install the 3 ignition coils with the 3 bolts and tighten to 7 ft. lbs. (10 Nm).

76. Connect the 3 connectors to the 3 ignition coils.

77. Install a new gasket to the intake air surge tank.

78. Using an 8 mm socket hexagon wrench, install the intake air surge tank with the 4 bolts and tighten to 21 ft. lbs. (28 Nm).

79. Install the 2 intake air surge tank nuts and tighten to 21 ft. lbs. (28 Nm)

80. Install the 3 wire harness clamps and hose clamp.

81. Connect the throttle body with motor connector.

82. Connect the 2 VSV connectors.

83. Install the 2 surge tank stays with the 4 bolts.

84. Install the oil baffle plate with the bolt and tighten to 80 inch lbs. (9.0 Nm).

85. Install the throttle body bracket with the 2 bolts and tighten to 15 ft. lbs. (21 Nm).

86. Connect the ventilation hose.

87. Connect the fuel vapor feed hose.

88. Connect the 2 water by-pass hoses.

89. Install air cleaner assembly.

90. Connect no. 2 ventilation hose.

91. Add engine coolant.

92. Install V-bank cover and tighten to 66 inch lbs. (7.5 Nm).

93. Inspect ignition timing.

4.7L Engine

See Figures 99 through 112.

1. Before servicing the vehicle, refer to the precautions section.

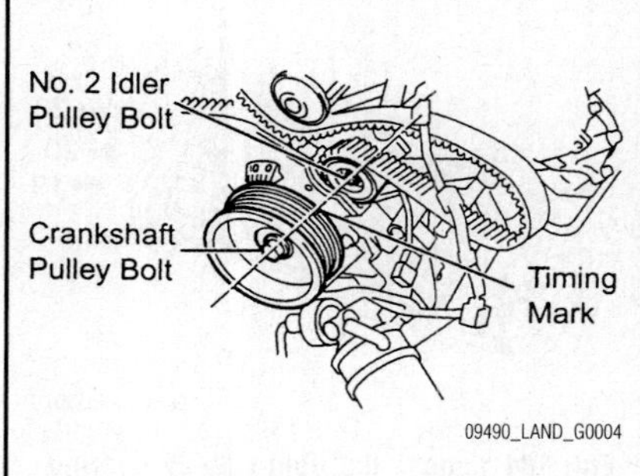

Fig. 99 Check the timing mark of the crankshaft pulley is aligned with the center(s) of the crankshaft pulley bolt and idler pulley bolt

2. Drain the cooling system.

3. Relieve the fuel system pressure.

4. Remove the V bank cover.

5. Remove the timing belt.

6. Remove the camshaft pulleys.

7. Remove the Camshaft Position (CMP) sensor.

8. Remove the power steering pump and set it aside with the lines still attached.

9. Remove the front exhaust pipe.

10. On models with an automatic transmission, remove the oil dipstick and tube.

11. Remove the ignition coils.

12. Remove the rear timing belt plates being careful not to drop anything.

13. Disconnect the fuel inlet hose.

14. Remove the intake manifold.

15. Remove the water inlet and inlet housing. Refer to water pump removal.

16. Remove the front and rear water bypass joint.

17. Remove the engine hangers and if needed the oil dipstick and tube.

18. Remove the valve covers.

➡Since the thrust level of the camshaft is small, the camshaft must be kept level during removal.

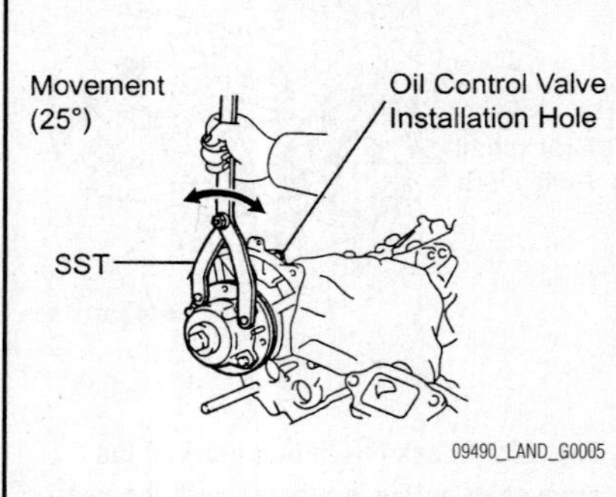

Fig. 100 Release the oil from the front bearing caps using the tool illustrated

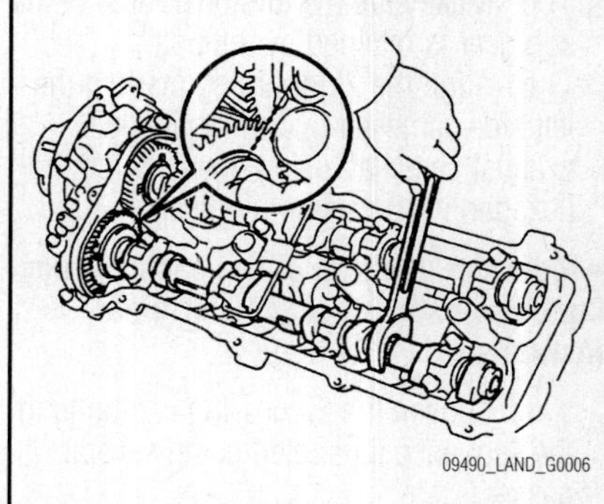

Fig. 101 Align the 2 dot timing mark of the left side camshaft by turning the left exhaust camshaft using a wrench on the hexagon head portion of the shaft

If not kept level serious damage could occur.

19. Check the timing mark of the crankshaft pulley is aligned with the center(s) of the crankshaft pulley bolt and idler pulley bolt.

➡If the crankshaft pulley is wrongly positioned, this can cause the piston to contact the head causing severe damage. Make sure the crankshaft pulley is properly positioned.

20. Release the oil from the front bearing caps using the tool illustrated. Rotate the camshaft timing tube from left to right 2 to 3 times within its VVT-I range of 25 degrees and collect the oil from the timing oil control valve installation hole using a rag.

21. Remove the left hand camshafts as follows:

a. Bring the service bolt of the sub gear up by turning the left exhaust camshaft using a wrench on the hexagon head portion of the shaft.

b. Secure the sub gear to the main gear using a 16 to 20 mm bolt with a diameter of 6mm and a thread pitch of 1mm.

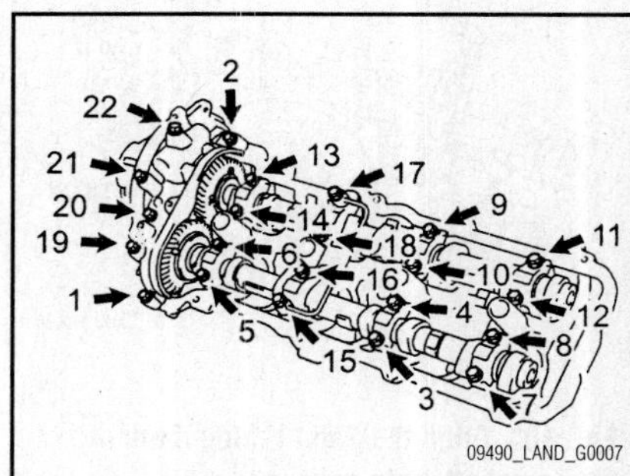

Fig. 102 Loosen the left side 22 bearing cap bolts in the sequence illustrated using several passes

c. Make sure the torsional force of the sub gear is retained by the bolt.

d. Align the 2 dot timing mark of the left side camshaft by turning the left exhaust camshaft using a wrench on the hexagon head portion of the shaft.

➡Mark the position of the caps so they can be reinstalled in their original positions.

e. Loosen the 22 bearing cap bolts in the sequence illustrated using several passes.

f. Remove the bolts, washers, oil feed pipe, bearing caps, camshaft housing plug, oil control valve filter and the camshafts.

22. Remove the right hand camshafts as follows:

a. Bring the service bolt of the sub gear up by turning the right exhaust camshaft using a wrench on the hexagon head portion of the shaft.

b. Secure the sub gear to the main gear using a 16 to 20 mm bolt with a diameter of 6mm and a thread pitch of 1mm.

c. Make sure the torsional force of the sub gear is retained by the bolt.

d. Align the 1 dot timing mark of the camshaft main gear (about 10 degrees) angle by turning the right exhaust camshaft using a wrench on the hexagon head portion of the shaft.

➡Mark the position of the caps so they can be reinstalled in their original positions.

e. Loosen the 22 bearing cap bolts in the sequence illustrated using several passes.

f. Remove the bolts, washers, oil feed pipe, bearing caps, camshaft housing plug, oil control valve filter and the camshafts.

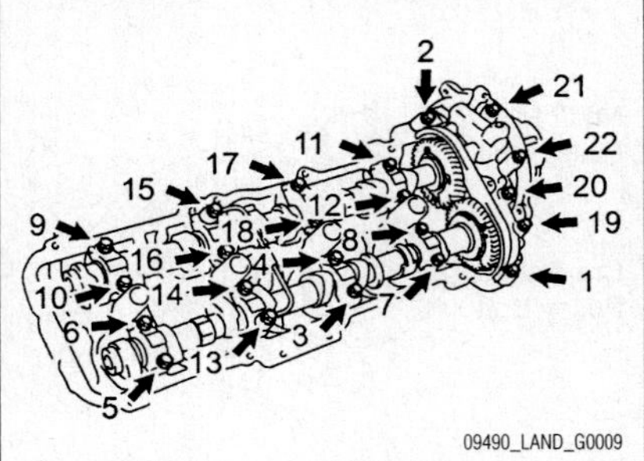

Fig. 104 Loosen the right side 22 bearing cap bolts in the sequence illustrated using several passes

To install:

23. Check the timing mark of the crankshaft pulley is aligned with the center(s) of the crankshaft pulley bolt and idler pulley bolt.

➡If the crankshaft pulley is wrongly positioned, this can cause the piston to contact the head causing severe damage. Make sure the crankshaft pulley is properly positioned.

24. Install the left side camshafts as follows:

a. Apply multipurpose grease to the thrust portion of the camshafts.

b. Align the 2 dot timing mark of the camshaft drive and driven main gears and install the camshafts.

c. Apply seal packing to the camshaft housing plug.

d. Install the camshaft housing plug on the cylinder head as illustrated. Install the strainer on the head being careful it is properly positioned.

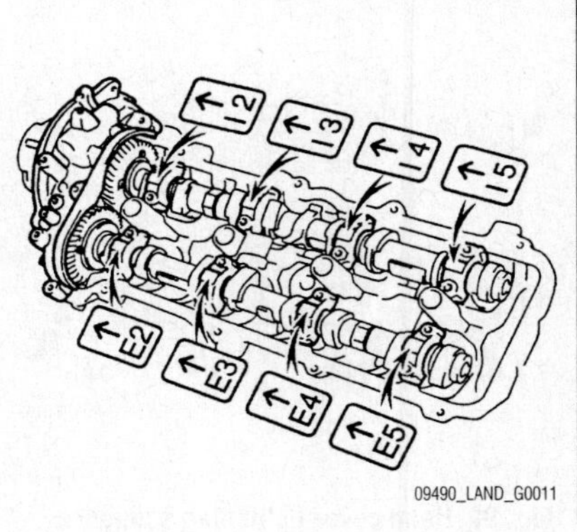

Fig. 106 Install the front bearing cap and then the other caps in the sequence illustrated on the left side camshafts

e. Apply seal packing to the front bearing cap.

f. Install the front bearing cap and then the other caps in the sequence illustrated.

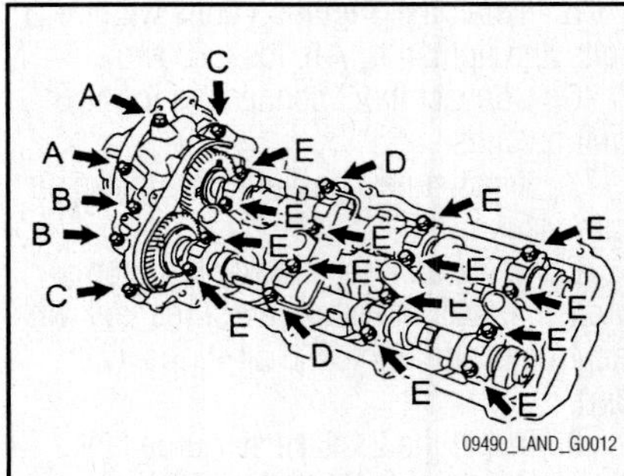

Fig. 107 Apply a light coating of clean oil to the threads and underside of the bolt heads D and E. make sure no oil gets under the heads of bolts A, B and C on the left side camshafts

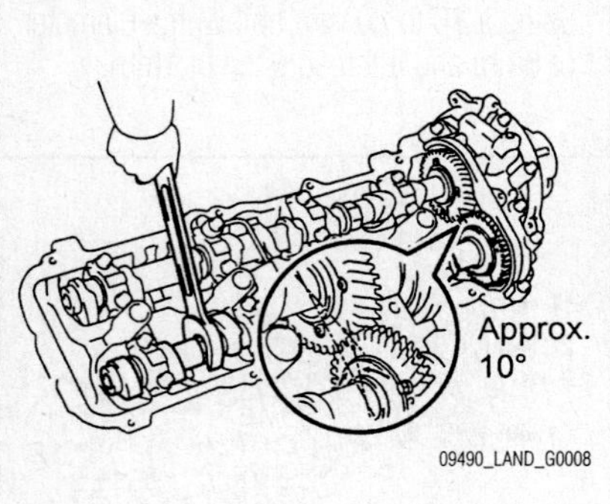

Fig. 103 Align the 1 dot timing mark of the camshaft main gear (about 10 degrees) angle by turning the right exhaust camshaft using a wrench on the hexagon head portion of the shaft

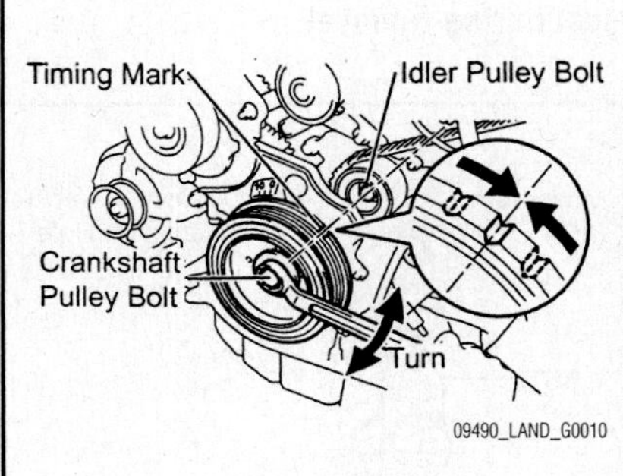

Fig. 105 Check the timing mark of the crankshaft pulley is aligned with the center(s) of the crankshaft pulley bolt and idler pulley bolt

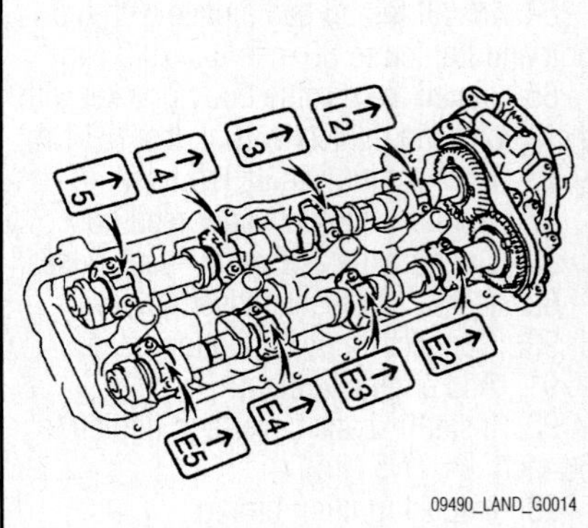

Fig. 108 Install the front bearing cap and then the other caps in the sequence illustrated on the left side camshafts

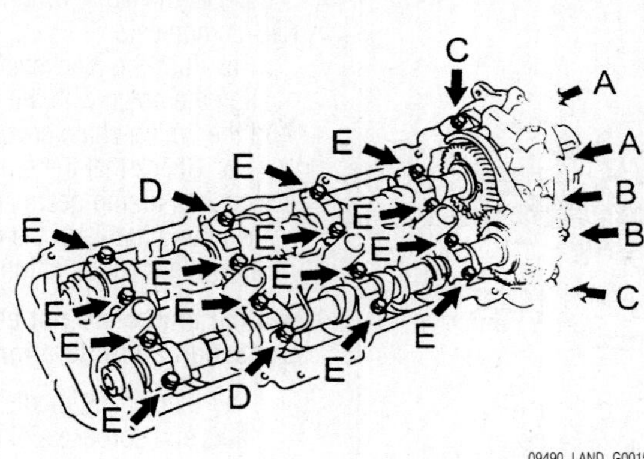

Fig. 109 Apply a light coating of clean oil to the threads and underside of the bolt heads D and E. make sure no oil gets under the heads of bolts A, B and C on the right side camshafts

g. Push in the camshaft oil seal.

h. Install 4 new seal washers to the bearing cap bolts A and B, refer to the illustration.

i. Apply a light coating of clean oil to the threads and underside of the bolt heads D and E. make sure no oil gets under the heads of bolts A, B and C.

j. The bolt lengths and positions are as follows. refer to the illustration for bolt location:

- 94mm bolts A
- 72mm bolt B
- 25mm bolt C
- 52mm bolt D
- 38mm bolts E

k. Tighten the cap bolts using several passes. Tighten bolt C to 66 inch lbs. (7.5 Nm) an the remaining bolts to 12 ft. lbs. (16 Nm).

l. Remove the service bolt.

25. Install the right side camshafts as follows:

a. Apply multi-purpose grease to the thrust portion of the camshafts.

b. Align the 1 dot timing mark of the camshaft drive and driven main gears and install the camshafts.

c. Set the 1 dot timing mark of the camshaft drive and driven gears at a 10 degree angle.

d. Apply seal packing to the camshaft housing plug.

e. Install the camshaft housing plug on the cylinder head as illustrated. Install the strainer on the head being careful it is properly positioned.

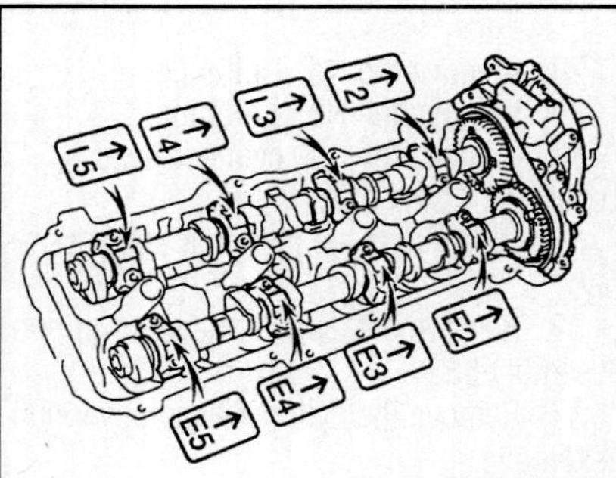

Fig. 111 Install the front bearing cap and then the other caps in the sequence illustrated on the right side camshafts

f. Apply seal packing to the front bearing cap.

g. Install the front bearing cap and then the other caps in the sequence illustrated.

h. Push in the camshaft oil seal.

i. Install 4 new seal washers to the bearing cap bolts A and B, refer to the illustration.

j. Apply a light coating of clean oil to the threads and underside of the bolt heads D and E. make sure no oil gets under the heads of bolts A, B and C.

k. The bolt lengths and positions are as follows. refer to the illustration for bolt location:

- 94mm bolts A
- 72mm bolt B
- 25mm bolt C
- 52mm bolt D
- 38mm bolts E

l. Tighten the cap bolts using several passes. Tighten bolt C to 66 inch lbs. (7.5 Nm) an the remaining bolts to 12 ft. lbs. (16 Nm).

m. Remove the service bolt.

26. Check and adjust the valve clearance.
27. Install the camshaft timing control valve.
28. Install the 4 half moon plugs onto the cylinder heads.
29. Install the valve covers and tighten to 53 inch lbs. (6 Nm).
30. Install the engine hangers and tighten to 27 ft. lbs. (37 Nm).
31. Install the VVT sensors.
32. Install the oil dipstick tube and dipstick.
33. Install the ignition coils.
34. Install the water bypass joint and tighten the retainers to 13 ft. lbs. (18 Nm).
35. Install the water inlet and housing assembly.
36. Install the intake manifold.
37. Install the timing belt rear plates, right plates first, and then left plates. Tighten the retainers to 66 inch lbs. (7 Nm).
38. Install the throttle body cover.
39. Install the front exhaust pipe, power steering pump.
40. Install the Camshaft Position (CMP) sensor and camshaft timing pulleys; tighten to 25 ft. lbs. (34 Nm).
41. Install the timing belt.
42. Fill the cooling system and perform an oil change.
43. Start the vehicle and check for leaks.

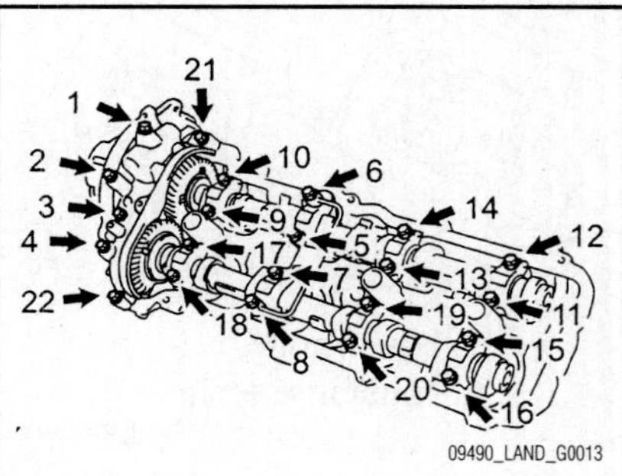

Fig. 110 Left side camshaft bolt torque sequence

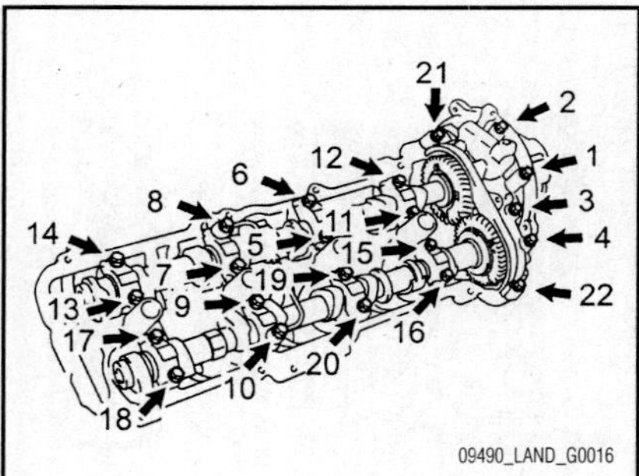

Fig. 112 Right side camshaft bolt torque sequence

2010 Models

4.0L Engine

See Figures 113 through 146.

1. Disconnect the cable from the negative battery terminal.

➡**When disconnecting the cable, some systems need to be initialized after the cable is reconnected.**

2. Remove the front bumper.
3. Remove the No. 1 engine under cover sub assembly.
4. Drain the engine oil.
5. Drain the coolant.
6. Remove the upper radiator support seal.
7. Disconnect the cable from the positive battery terminal.
8. Remove the battery hold down clamp.
9. Remove the battery.
10. Remove the battery tray.
11. Remove the v-bank cover.
12. Remove the air cleaner cap and hose.
13. Remove the air cleaner case sub assembly.
14. Remove the No. 1 radiator hose.
15. Remove the No. 2 radiator hose.
16. Remove the radiator reservoir.
17. Disconnect the oil cooler tube.
18. Remove the fan shroud.
19. Remove the intake air surge tank.
20. Remove the ignition coil assembly.
21. Disconnect the vane pump assembly.
22. Remove the No. 2 idler pulley sub assembly.
23. Remove the wiring harness clamp bracket.
24. Remove the No. 2 exhaust manifold heat insulator.
25. Remove the alternator assembly.
26. Remove the engine oil level dipstick guide.
27. Remove the water by pass pipe sub assembly.

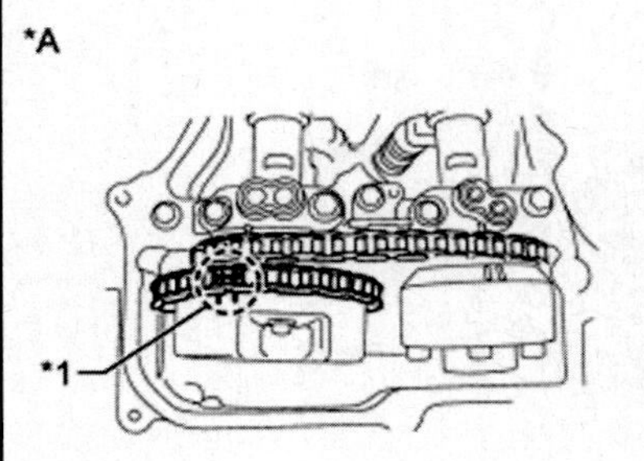

Fig. 114 Checking the timing marks of the camshaft timing gears are aligned with the timing marks of the bearing caps

28. Remove the No. 1 oil pipe.
29. Remove the No. 2 oil pipe.
30. Remove the rear cylinder head cover.
31. Disconnect the fuel pipe sub assembly.
32. Remove the cylinder head cover sub assembly LH.
33. Remove the cylinder head cover sub assembly.
34. Remove the timing chain cover plate.
    a. Remove the 4 bolts, timing chain cover plate and gasket.
35. Set the No. 1 cylinder to TDC/compression.
    a. Turn the crankshaft pulley and align the notch with the "0" timing mark of the timing chain cover.
    b. Check that the timing marks of the camshaft timing gears are aligned with the timing marks of the bearing caps as shown in the illustration.

➡**If the marks are not aligned, turn the crankshaft again to align the marks.**

    c. Place paint marks on the timing marks and sprockets of each camshaft timing gear and on the links of the No. 1 chain.

➡**Be sure to place the paint marks on 2 links of the chain and on the sprockets of the camshaft timing gears at the locations of the timing marks of the camshaft timing gears.**

36. Remove the No. 1 chain tensioner assembly.
    a. Turn the crankshaft approximately 30° counterclockwise so that there is some slack in the chain.

➡**This prevents the valves and pistons from interfering with each other.**

    b. Align the hole in the lever of the tensioner with the hole in the tensioner body as shown in the illustration, and then insert a pin with a diameter of 0.0500 inch (1.27 mm) into the hole.
    c. Turn the crankshaft clockwise and align the notch with the "0" timing mark of the timing chain cover.
    d. Remove the 2 bolts and chain tensioner.

➡**Do not drop the No. 1 chain tensioner assembly or bolts into the timing chain cover.**

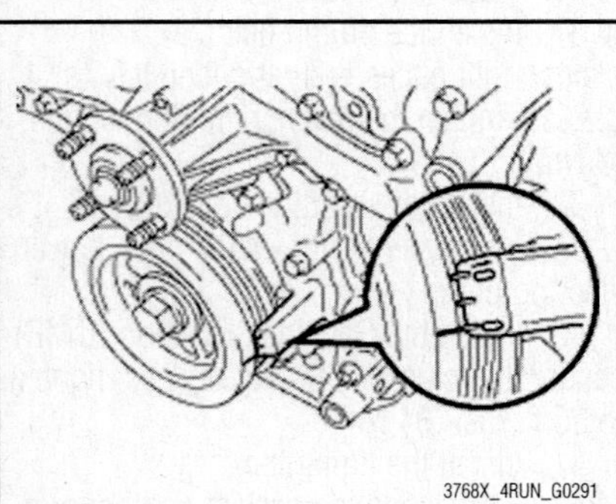

Fig. 113 Aligning the notch with the "0" timing mark of the timing chain cover

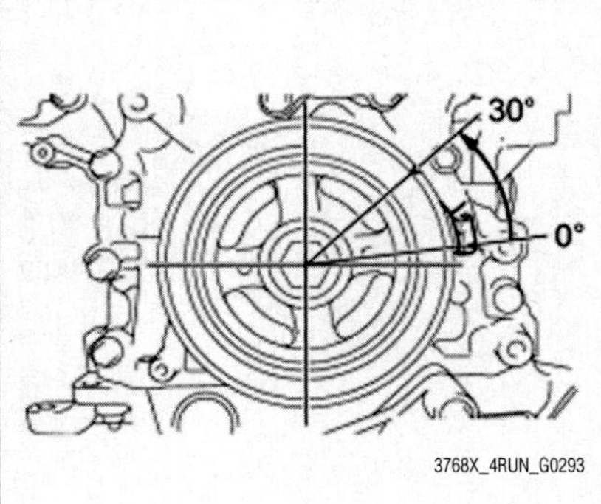

Fig. 115 Turning the crankshaft

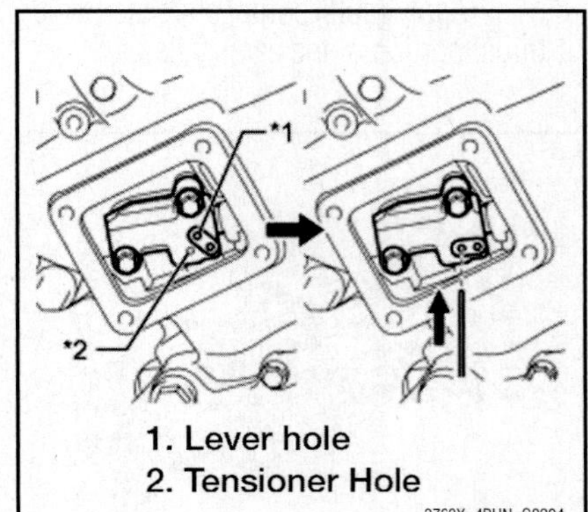

Fig. 116 Aligning the hole in the lever of the tensioner with the hole in the tensioner body

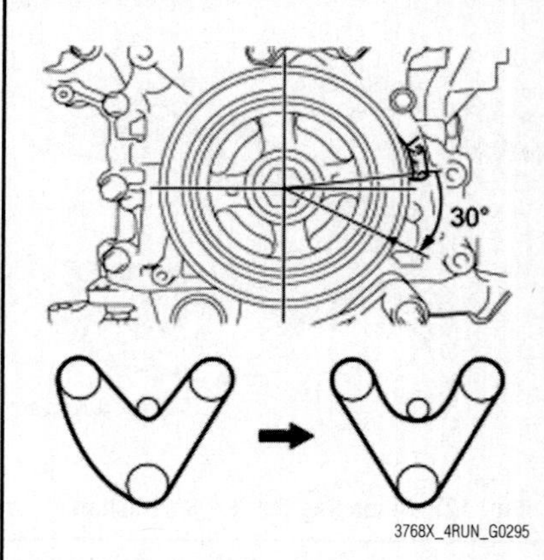

Fig. 117 Turning the crankshaft clockwise to the correct position

37. Disconnect the chain sub assembly (for Bank 1).

a. Turn the crankshaft clockwise until it is in the position shown in the illustration so that there is some slack in the chain between the banks.

➡When turning the crankshaft, engine oil may spray out of the oil holes.

CAUTION

As the camshafts turn suddenly, do not touch the camshafts or camshaft timing gears.

b. Turn the crankshaft clockwise until it is in the position shown in the illustration so that the chain can be removed easily.

➡When turning the crankshaft, engine oil may spray out of the oil holes.

c. Remove the chain from the sprocket of the camshaft timing gear and set it on the gear.

CAUTION

As the camshaft may turn suddenly and pinch your fingers when the chain is removed, pinch the chain and lift it upward to remove it from the sprocket.

38. Remove the camshaft bearing cap (for Bank 1).

a. Remove the bolts and bearing caps in the order shown in the illustration. Immediately after removing a bearing cap, install VVT bolt kit in the order shown in the illustration. Tighten to 21 ft. lbs. (28 Nm).

➡Arrange the removed parts so that they can be reinstalled in their original locations.

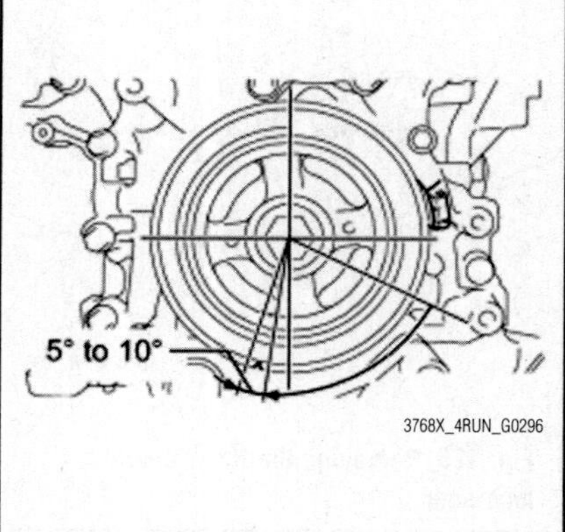

Fig. 118 Turning the crankshaft clockwise to remove the chain easily

➡Do not install the bearing caps when installing VVT bolt kit.

➡Be sure to follow the numerical order when performing this procedure.

➡Do not allow the VVT bolt kit to contact the camshaft.

➡Do not drop the VVT bolt kit into the cylinder head.

39. Remove the No. 2 camshaft.

a. Remove the bolt of the No. 2 chain tensioner assembly.

b. Remove the No. 2 chain tensioner assembly while lifting up the No. 2 camshaft.

c. While lifting up the No. 2 camshaft, pass it through the No. 2 chain and pull it out towards the front of the vehicle to remove it.

40. Remove the camshaft.

a. Lift up the rear of the camshaft so that it is at an angle.

b. Remove the chain from the camshaft timing gear and pull out the

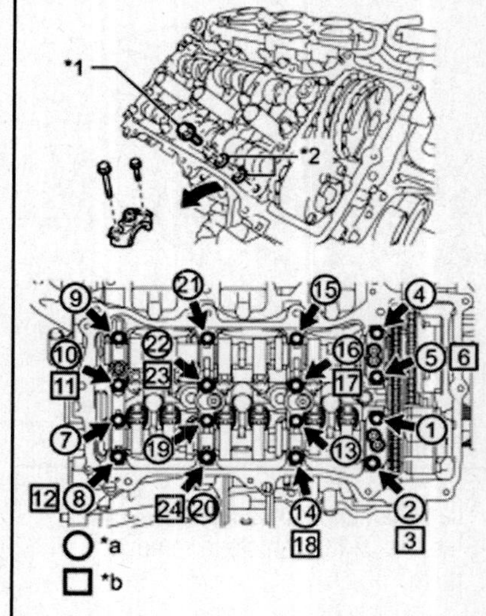

Fig. 120 Bearing cap removal and installation sequence

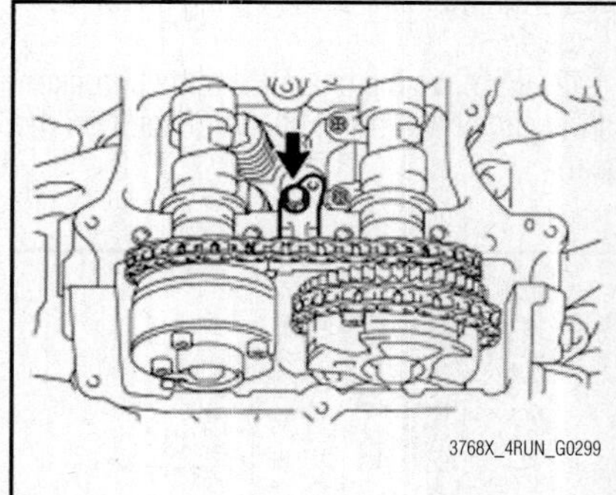

Fig. 121 Removing the No. 2 chain tensioner assembly bolt

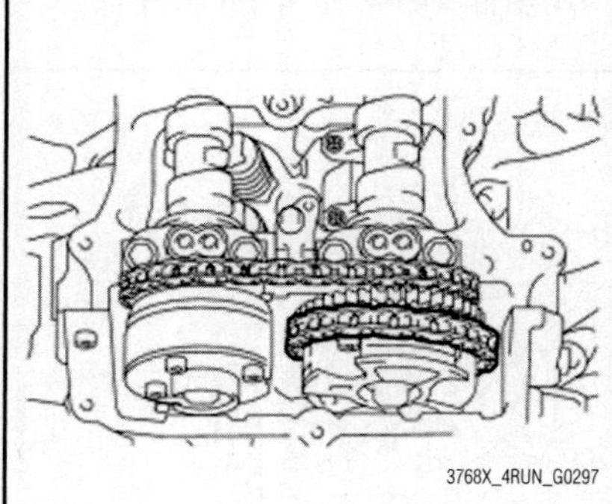

Fig. 119 Removing the chain from the sprocket of the camshaft timing gear

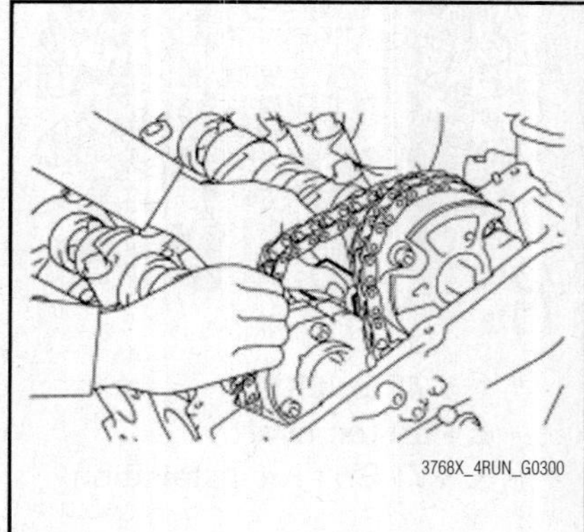

Fig. 122 Removing the No. 2 chain tensioner assembly

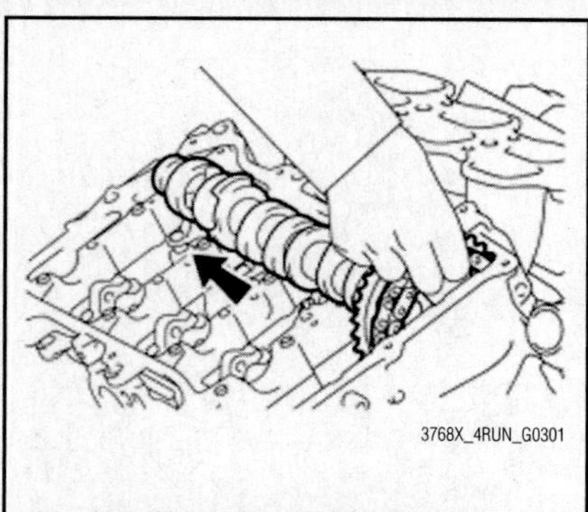

Fig. 123 Removing the chain from the camshaft timing gear

camshaft and No. 2 chain towards the rear of the vehicle to remove them.

➡**Do not drop the chain into the gap between the engine and cover.**

c. Suspend the chain with a string or equivalent.

41. Disconnect the chain sub assembly (for Bank 2).

a. Turn the crankshaft counterclockwise and align the notch with the "0" timing mark of the timing chain cover.

42. Remove the chain from the sprocket of the camshaft timing gear and set it on the gear.

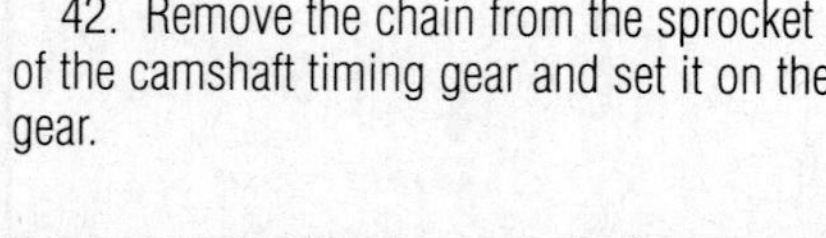

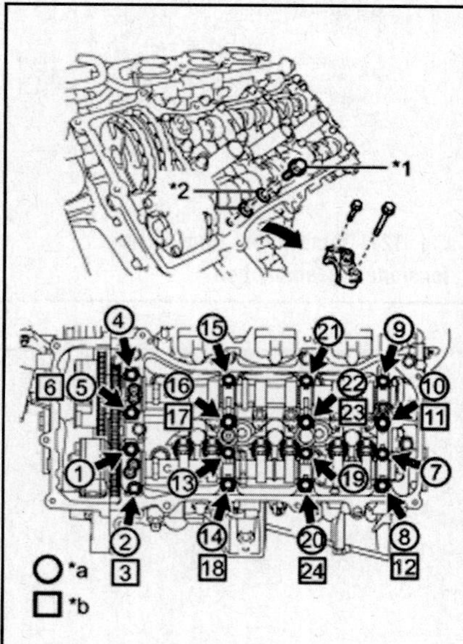

Fig. 124 Identifying the bearing cap (for Bank 2) removal and installation sequence

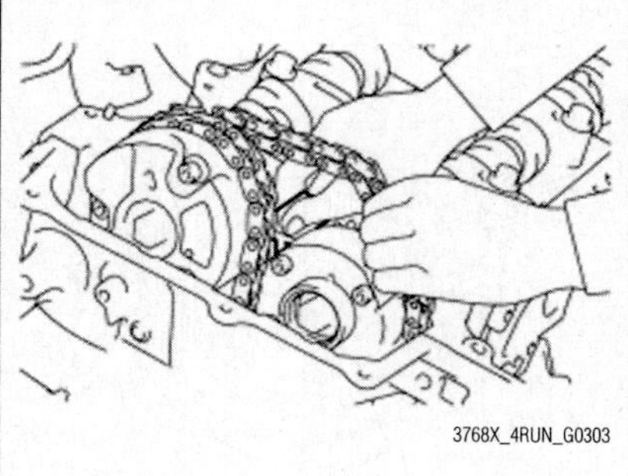

Fig. 125 Removing the No. 3 chain tensioner

⁂ CAUTION

As the camshaft may turn suddenly and pinch your fingers when the chain is removed, pinch the chain and lift it upward to remove it from the sprocket.

43. Remove the camshaft bearing cap (for Bank 2).

a. Remove the bolts and bearing caps in the order shown in the illustration. Immediately after removing a bearing cap, install VVT bolt kit in the order shown in the illustration. Tighten to 21 ft. lbs. (28 Nm).

➡**Arrange the removed parts so that they can be reinstalled in their original locations.**

➡**Do not install the bearing caps when installing VVT bolt kit.**

➡**Be sure to follow the numerical order when performing this procedure.**

➡**Do not allow the VVT bolt kit to contact the camshaft.**

➡**Do not drop the VVT bolt kit into the cylinder head.**

44. Remove the No. 4 camshaft.

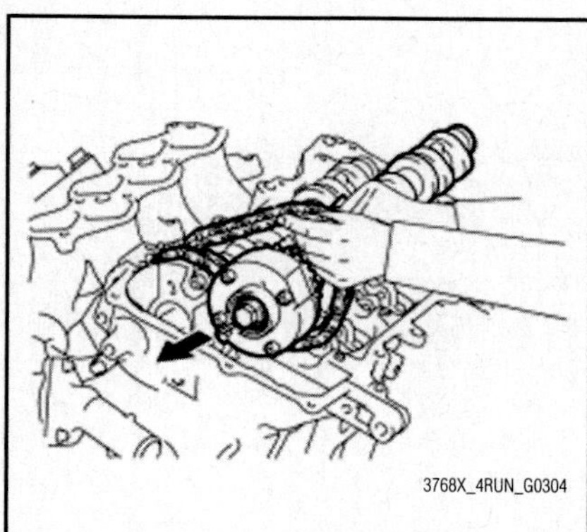

Fig. 126 Removing the No. 4 camshaft

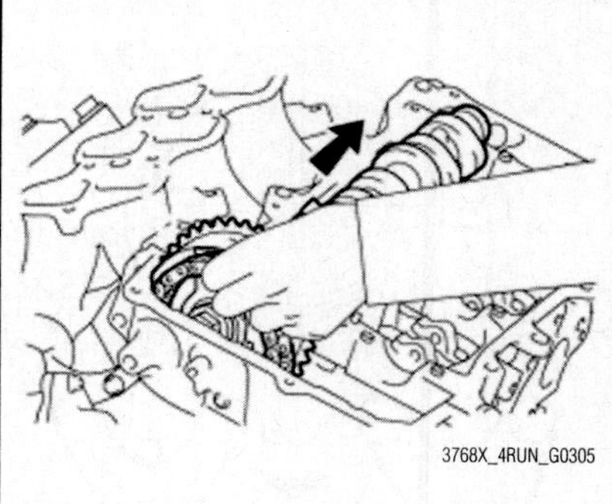

Fig. 127 Removing the No. 3 camshaft

a. Remove the bolt of the No. 3 chain tensioner assembly.

b. Remove the No. 3 chain tensioner assembly while lifting up the No. 4 camshaft.

c. While lifting up the No. 4 camshaft, pass it through the No. 2 chain and pull it out towards the front of the vehicle to remove it.

45. Remove the No. 3 camshaft.

a. Lift up the rear of the camshaft so that it is at an angle.

b. Remove the chain from the camshaft timing gear and pull out the No. 3 camshaft and No. 2 chain towards the rear of the vehicle to remove them.

➡**Do not drop the chain into the gap between the engine and cover.**

c. Suspend the chain with a string or equivalent.

46. Remove the camshaft timing gear assembly.

a. Fix the camshaft in place.

➡**Be careful not to damage the camshaft.**

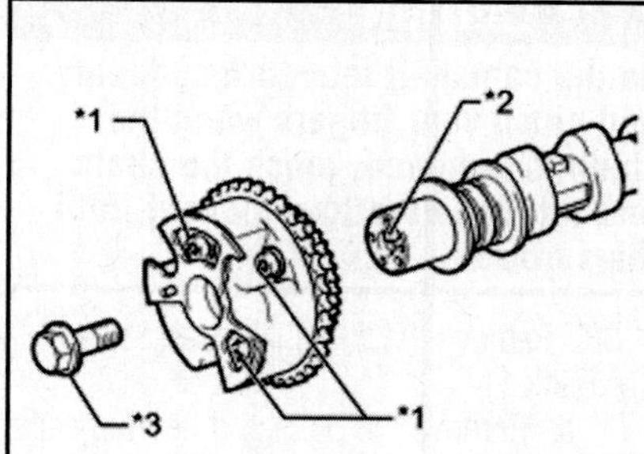

Fig. 128 Removing the camshaft timing gear assembly

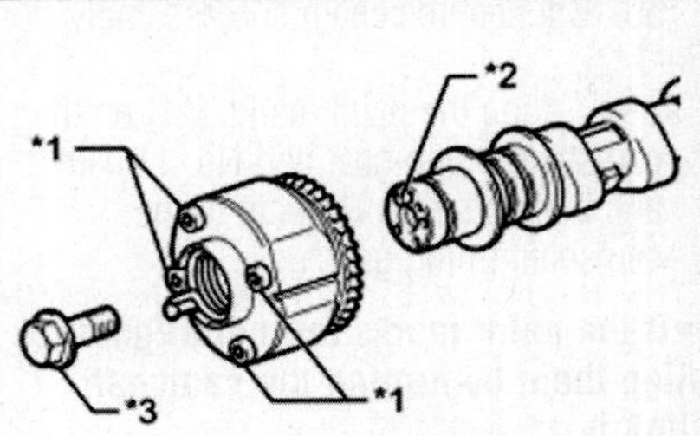

1. DO NOT REMOVE
2. Straight Pin
3. Flange Bolt

3768X_4RUN_G0307

Fig. 129 Removing the camshaft timing exhaust gear assembly

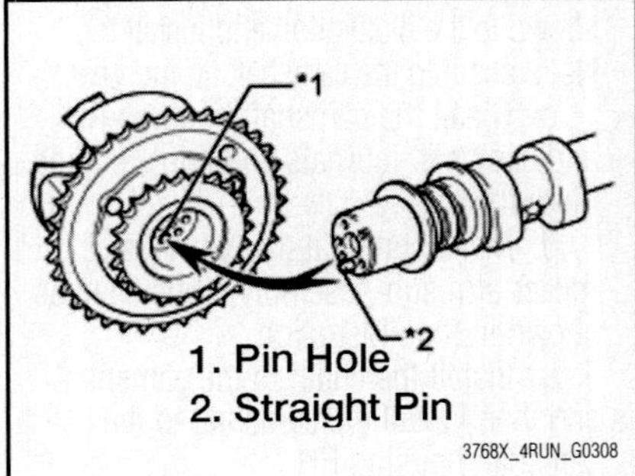

1. Pin Hole
2. Straight Pin

3768X_4RUN_G0308

Fig. 130 Putting the camshaft timing gear and camshaft together

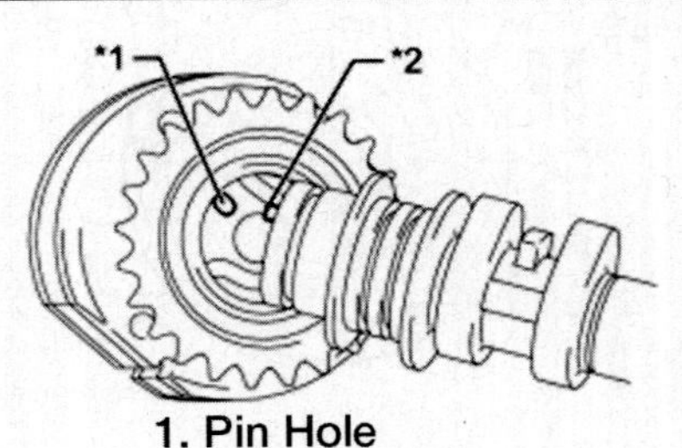

1. Pin Hole
2. Straight Pin

3768X_4RUN_G0309

Fig. 131 Putting the camshaft timing exhaust gear assembly and camshaft together

b. Remove the flange bolt and camshaft timing gear assembly.

➡Do not remove the other 3 bolts.

➡If planning to reuse the camshaft timing gear, be sure to release the straight pin lock before installing the camshaft timing gear.

47. Remove the camshaft timing exhaust gear assembly.
   a. Fix the camshaft in place.

➡Be careful not to damage the camshaft.

   b. Remove the flange bolt and camshaft timing exhaust gear assembly.

➡Be sure not to remove the other 4 bolts.

➡If planning to reuse the gear, be sure to release the straight pin lock before installing the gear.

To install:

48. Inspect the camshaft timing gear assembly.
49. Inspect the camshaft timing exhaust gear assembly.
50. Install the camshaft timing gear assembly.
   a. Fix the camshaft in place.

➡Be careful not to damage the camshaft.

   b. Put the camshaft timing gear assembly and camshaft together by aligning the pin hole and straight pin.
   c. Lightly press and turn the camshaft timing gear assembly against the camshaft, and press harder after the pin enters the hole.

➡Be sure not to turn the camshaft timing gear assembly in the retard direction.

   d. Check that there is no clearance between the camshaft timing gear assembly flange and camshaft.
   e. Install the flange bolt while holding the camshaft. Tighten to 74 ft. lbs. (100 Nm).
   f. Check the lock of the camshaft timing gear assembly.
   g. Fix the camshaft in place and confirm that the camshaft timing gear assembly is locked.

➡Be careful not to damage the camshaft.

51. Install the camshaft timing exhaust gear assembly.
   a. Fix the camshaft in place.

➡Be careful not to damage the camshaft.

   b. Put the camshaft timing exhaust gear assembly and camshaft together by aligning the pin hole and straight pin.
   c. Lightly press and turn the camshaft timing gear assembly against the camshaft, and press harder after the pin enters the hole.

➡Be sure not to turn the camshaft timing exhaust gear in the advanced direction.

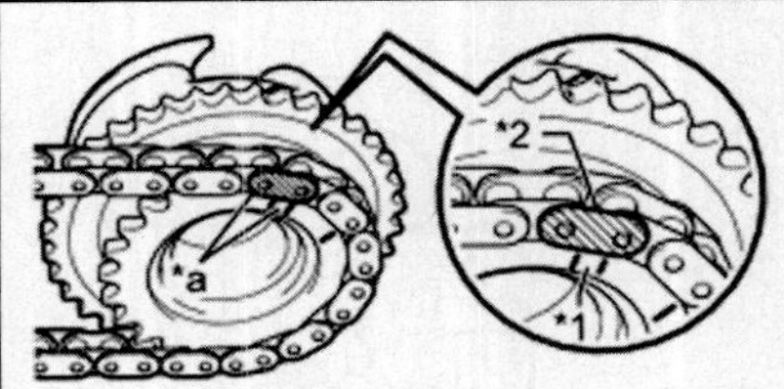

1. Timing mark
2. Mark plate (yellow)
a. Align

3768X_4RUN_G0310

Fig. 132 Aligning the mark plate with the timing mark of the camshaft timing gear and installing the No. 2 chain to the camshaft timing gear

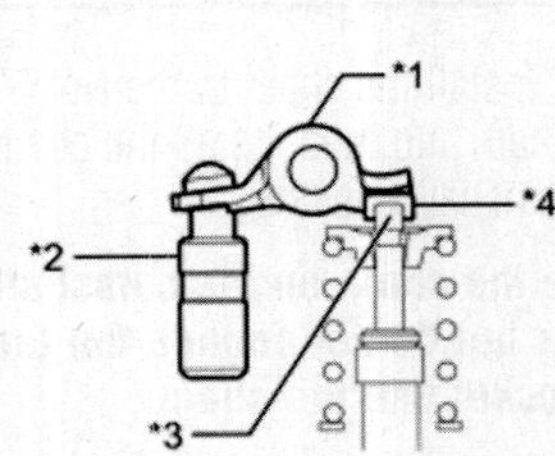

1. Valve rocker arm
2. Lash adjuster
3. Valve stem
4. Valve stem cap

3768X_4RUN_G0311

Fig. 133 Checking the No. 1 valve rocker arm sub assembly

   d. Check that there is no clearance between the gear flange and camshaft.
   e. Install the flange bolt while holding the camshaft. Tighten to 74 ft. lbs. (100 Nm).
   f. Check the camshaft timing exhaust gear lock.
   g. Make sure that the camshaft timing exhaust gear assembly locks.
52. Install the No. 3 camshaft.
   a. Check that the notch is aligned with the "0" timing mark of the timing chain cover.
   b. Align the mark plate (yellow) with the timing mark of the camshaft timing gear as shown in the illustration and install the No. 2 chain to the camshaft timing gear.
   c. Clean the camshaft housing LH and camshaft journals and apply engine oil to them.
   d. Make sure that the No. 1 valve rocker arm sub-assembly is installed as shown in the illustration.

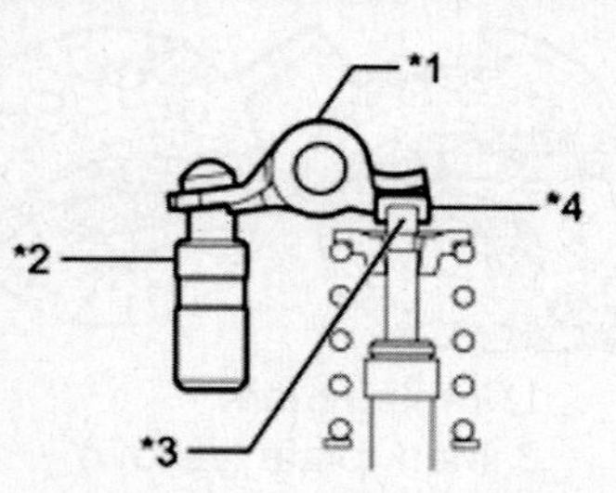

Fig. 134 Installing the No. 1 valve rocker arm sub assembly

e. Install the chain to the No. 3 camshaft, and then install the camshaft to the camshaft housing LH.

➡**Place the chain on the camshaft timing gear but do not engage the teeth of the sprocket and the chain.**

➡**Install the camshaft so that the timing mark is facing upward.**

53. Install the No. 4 camshaft.

a. Clean the camshaft housing LH and camshaft journals and apply engine oil to them.

b. Pass the No. 4 camshaft through the No. 2 chain from the front of the vehicle, align the mark plate (yellow) with the timing mark and install the No. 2 chain to the camshaft timing exhaust gear.

54. While lifting up the No. 4 camshaft, pass the No. 3 chain tensioner assembly through the No. 2 chain and set it in place.

a. Install the No. 4 camshaft to the camshaft housing LH, and then install the No. 3 chain tensioner assembly with the bolt. Tighten to 15 ft. lbs. (21 Nm).

55. Install the camshaft bearing cap (for Bank 2).

a. Clean the camshaft bearing caps and apply engine oil to them.

b. Make sure that the No. 1 valve rocker arm sub assembly.

c. Check the marks and numbers on the camshaft bearing caps, and then remove VVT bolt kit in the order shown in the illustration. Immediately after removing VVT bolt kit in the location for a bearing cap, install the bearing cap with the bolts in the order shown in the illustration. Tighten bolt A to 21 ft. lbs. (28 Nm). Tighten bolt B to 12 ft. lbs. (16 Nm).

➡**Be sure to follow the numerical order when performing this procedure.**

➡**Do not drop the VVT bolt kit into the cylinder head.**

d. Check the torque of each bolt again.

56. Connect the chain sub assembly (for Bank 2).

a. Align the paint marks (*1) on the camshaft timing gear and No. 1 chain and install the No. 1 chain to the camshaft timing gear.

➡**If the paint marks are not aligned, align them by turning the camshaft slightly.**

57. Install the camshaft.

a. Turn the crankshaft clockwise until it is in the position shown in the illustration so that the chain can be installed easily.

➡**When turning the crankshaft, engine oil may spray out of the oil holes.**

b. Align the mark plate (yellow) with the timing mark of the camshaft timing gear as shown in the illustration and install the No. 2 chain to the camshaft timing gear.

c. Clean the camshaft housing RH and camshaft journals and apply engine oil to them.

d. Make sure that the No. 1 valve rocker arm sub-assembly is installed as shown in the illustration.

e. Install the chain to the camshaft, and then install the camshaft to the camshaft housing RH.

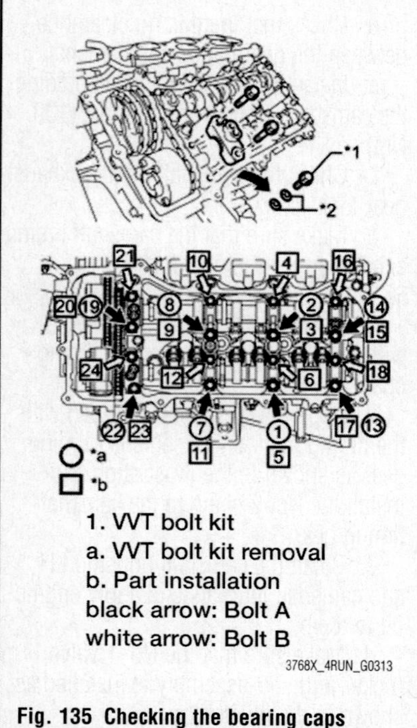

Fig. 135 Checking the bearing caps

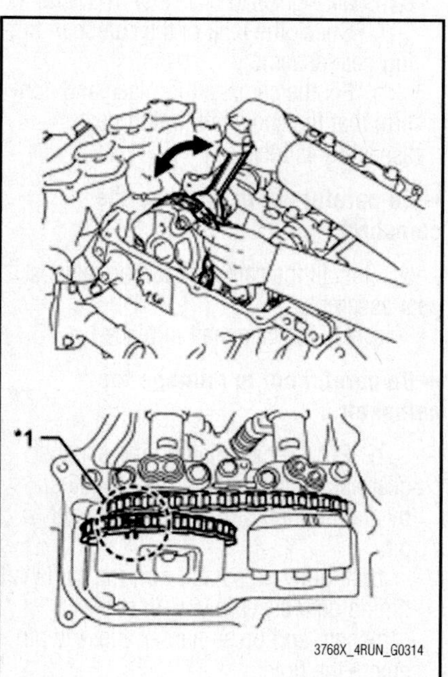

Fig. 136 Connecting the chain sub assembly (for Bank 2)

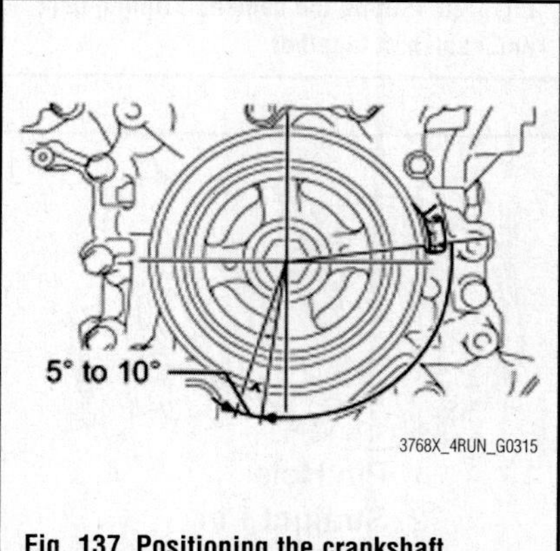

Fig. 137 Positioning the crankshaft

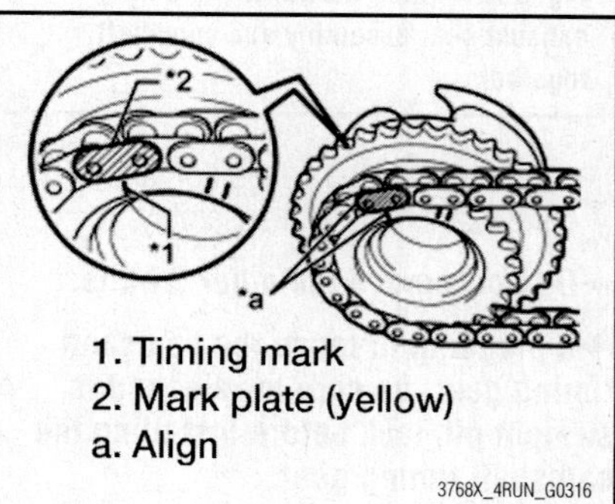

Fig. 138 Installing the No. 2 chain to the camshaft timing gear

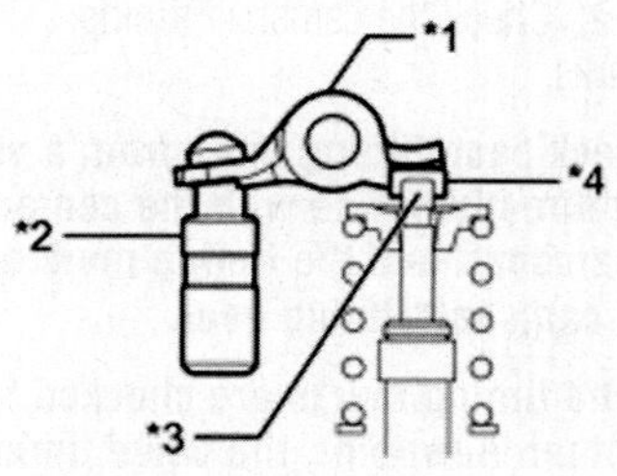

1. Valve rocker arm
2. Lash adjuster
3. Valve stem
4. Valve stem cap

3768X_4RUN_G0317

Fig. 139 Installing the No. 1 valve rocker arm

➡**Place the chain on the camshaft timing gear but do not engage the teeth of the sprocket and the chain.**

➡**Install the camshaft so that the timing mark is facing upward.**

58. Install the No. 2 camshaft.
    a. Clean the camshaft housing RH and camshaft journals and apply engine oil to them.
    b. Pass the No. 2 camshaft through the No. 2 chain from the front of the vehicle, align the mark plate (yellow) with the timing mark and install the No. 2 chain to the camshaft timing exhaust gear.
    c. While lifting up the No. 2 camshaft, pass the No. 2 chain tensioner assembly through the No. 2 chain and set it in place.
    d. Install the No. 2 camshaft to the camshaft housing RH, and then install the No. 2 chain tensioner assembly with the bolt. Tighten to 15 ft. lbs. (21 Nm).

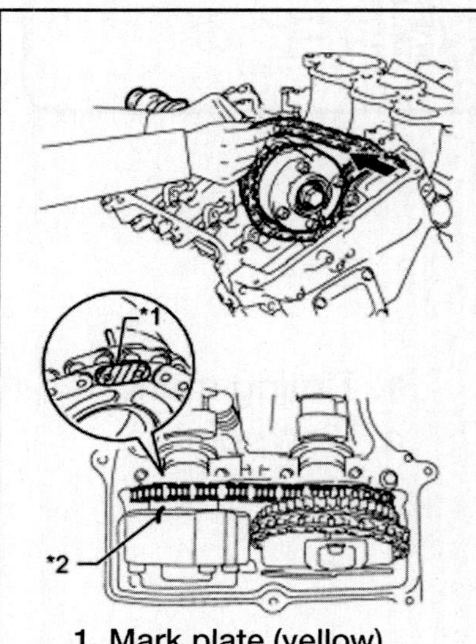

Fig. 140 Installing the No. 2 camshaft

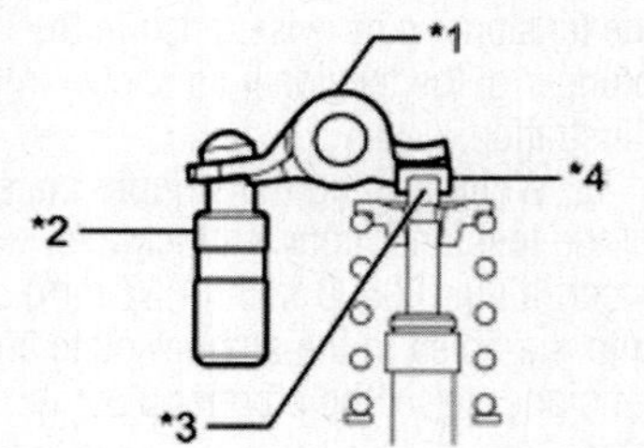

Fig. 141 Installing the No. 1 valve rocker arm sub assembly

59. Install the camshaft bearing cap (for Bank 1).
    a. Clean the camshaft bearing caps and apply engine oil to them.
    b. Make sure that the No. 1 valve rocker arm sub assembly is installed as shown in the illustration.
    c. Check the marks and numbers on the camshaft bearing caps, and then remove VVT bolt kit in the order shown in the illustration. Immediately after removing VVT bolt kit in the location for a bearing cap, install the bearing cap with the bolts in the order shown in the illustration. Tighten bolt A to 21 ft. lbs. (28 Nm). Tighten bolt B to 12 ft. lbs. (16 Nm).

➡**Be sure to follow the numerical order when performing this procedure.**

➡**Do not drop the VVT bolt kit into the cylinder head.**

    d. Check the torque of each bolt again.

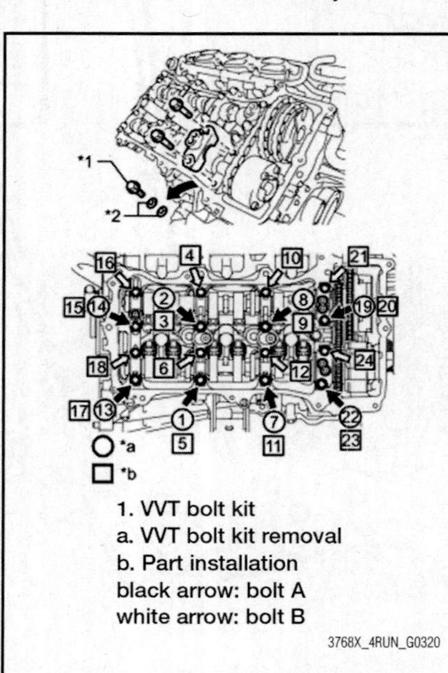

Fig. 142 Checking the bearing cap installation

60. Connect the chain sub assembly (for Bank 1).
    a. Align the paint marks on the camshaft timing gear and No. 1 chain and install the No. 1 chain to the camshaft timing gear.

➡**If the paint marks are not aligned, align them by turning the camshaft slightly.**

61. Install the No. 1 chain tensioner assembly.
    a. Turn the crankshaft counterclockwise 30°past the "0" timing mark, and then turn it clockwise to align the notch with the "0" timing mark.
    b. Turn the crankshaft slightly to eliminate the slack in the chain.

➡**Make sure there is some slack in the chain around the area where the chain tensioner is installed.**

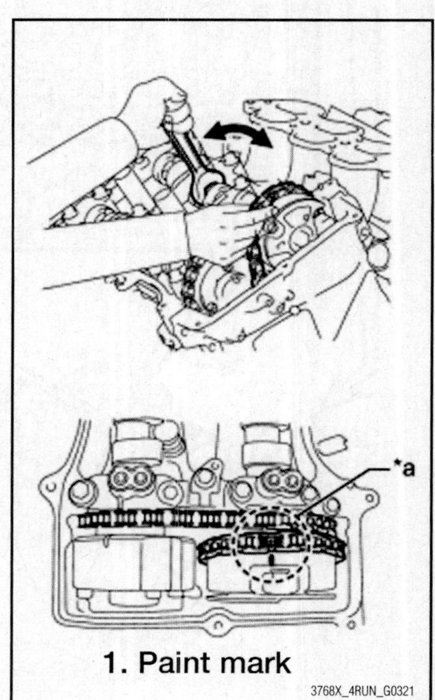

Fig. 143 Aligning the paint marks on the camshaft timing gear and No. 1 chain to the camshaft timing gear

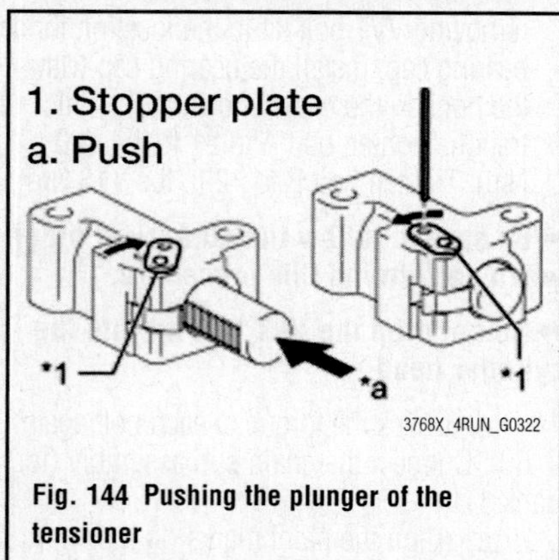

Fig. 144 Pushing the plunger of the tensioner

c. While turning the stopper plate of the tensioner clockwise, push in the plunger of the tensioner as shown in the illustration.

d. While turning the stopper plate of the tensioner counterclockwise, insert a pin 0.0500 inch (1.27 mm) into the holes in the stopper plate and tensioner to fix the stopper plate in place.

e. Install the chain tensioner with the 2 bolts. Tighten to 7 ft. lbs. (10 Nm).

f. Remove the pin from the No. 1 chain tensioner.

62. Inspect the valve timing.

a. Check the camshaft timing marks.

➡Check each timing mark from a viewpoint directly in line with the center of the camshaft and the timing mark on each camshaft timing gear.

➡If the timing marks are checked from any other viewpoint, the valve timing may appear misaligned.

b. Check that each camshaft timing mark is positioned as shown in the illustration.

1. Timing mark
a. Viewpoint

3768X_4RUN_G0323

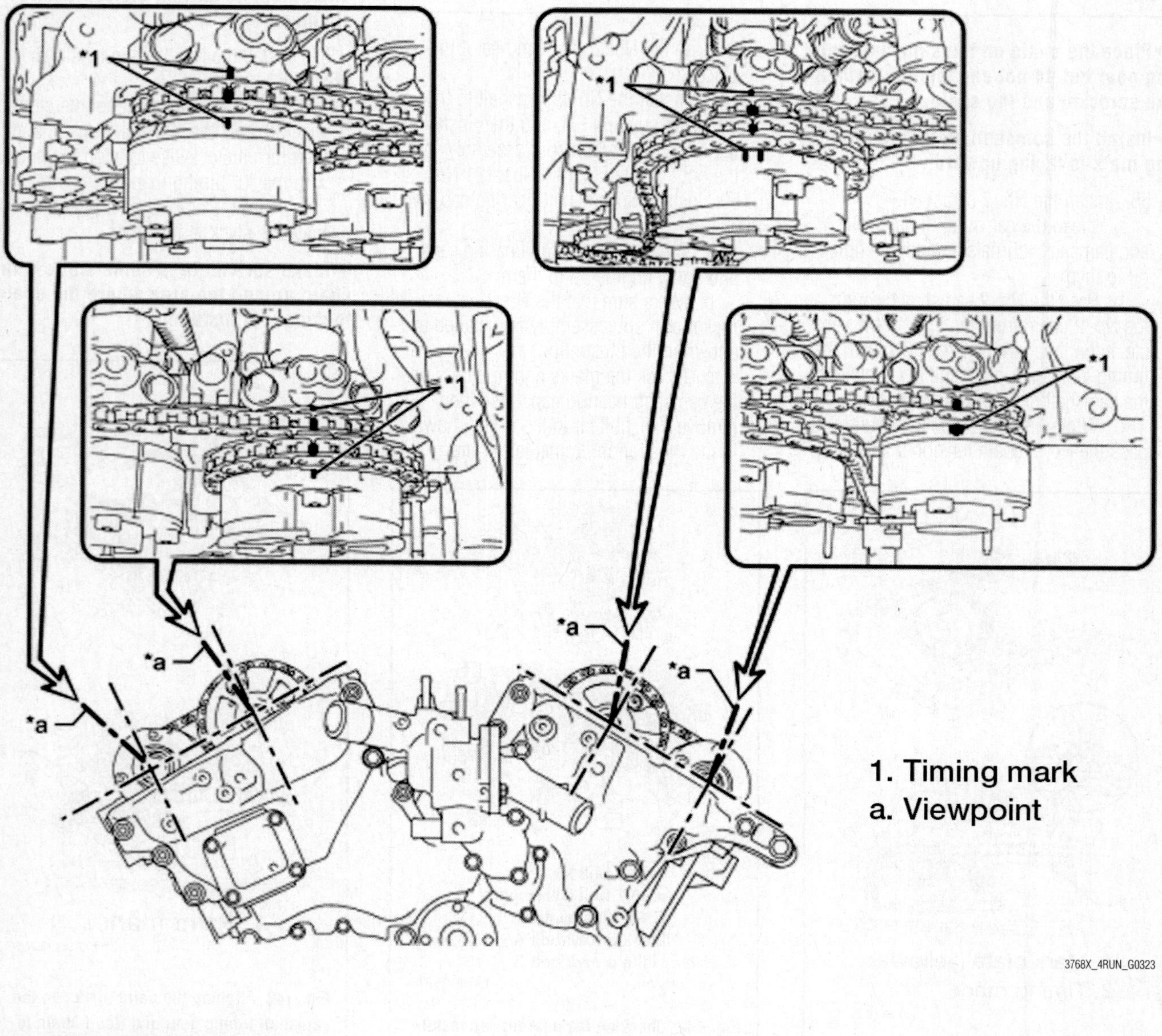

Fig. 145 Checking the valve timing

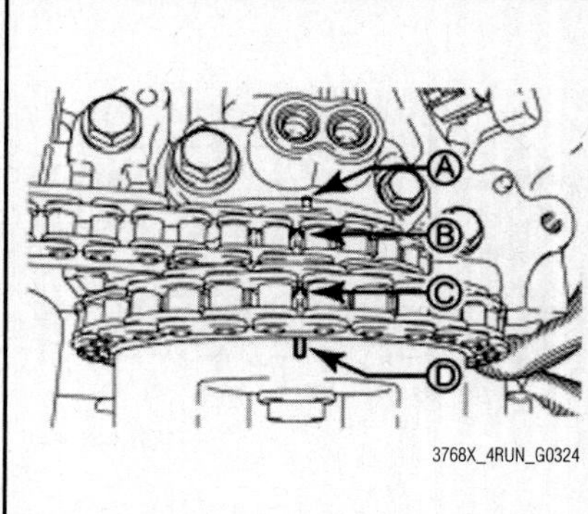

Fig. 146 Checking the intake camshaft

➡**Be sure to check mark A at the point when marks B, C and D are positioned in line. If the marks are checked from any other viewpoint, they cannot be checked correctly.**

c. If the valve timing is misaligned, reinstall the timing chain.

d. Turn the crankshaft 2 revolutions, set the No. 1 cylinder to TDC/compression and check the timing marks again.

63. Install the timing chain cover plate with a new gasket and the 4 bolts. Tighten to 80 inch lbs. (9 Nm).
64. Pour engine oil.
65. Install the cylinder head cover sub assembly.
66. Install the cylinder head cover sub assembly LH.
67. Connect the fuel pipe sub assembly.
68. Install the rear cylinder head cover.
69. Install the No. 2 oil pipe.
70. Install the No. 1 oil pipe.
71. Install the water by pass pipe sub assembly.
72. Install the engine oil level dipstick guide.
73. Install the alternator assembly.
74. Install the No. 2 exhaust manifold heat insulator.
75. Install the wiring harness clamp bracket.
76. Install the No. 2 idler pulley sub assembly.
77. Connect the vane pump assembly.
78. Install the ignition coil assembly.
79. Install the intake air surge tank.
80. Install the fan shroud.
81. Connect the oil cooler tube.
82. Install the radiator reservoir.
83. Install the No. 2 radiator hose.
84. Install the No. 1 radiator hose.
85. Install the air cleaner case sub assembly.
86. Install the air cleaner cap and hose.
87. Install the battery tray.
88. Install the battery.
89. Install the battery hold down clamp.
90. Connect the cable to the positive battery terminal.
91. Connect the cable to the negative battery terminal.

➡**When disconnecting the cable, some systems need to be initialized after the cable is reconnected.**

92. Add engine coolant.
93. Add engine oil.
94. Inspect for coolant leaks.
95. Inspect for engine oil leaks.
96. Install the v-bank cover.
97. Install the upper radiator support seal.
98. Install the No. 1 engine under cover sub assembly.
99. Install the front bumper cover lower.
100. Inspect the ignition timing.

2.7L Engine

See Figures 147 through 162.

1. Disconnect the cable from the negative battery terminal.

➡**When disconnecting the cable, some systems need to be initialized after the cable is reconnected.**

2. Remove the front bumper cover lower.
3. Remove the No. 1 engine under cover sub assembly.
4. Remove the upper radiator support seal.
5. Drain the engine oil.
6. Remove the air cleaner cap sub assembly.
7. Remove the intake air connector.
8. Remove the radiator reservoir.
9. Remove the fan shroud.
10. Remove the ignition coil assembly.
11. Remove the Camshaft Position (CMP) sensor.
12. Remove the cylinder head cover sub assembly.
13. Remove the timing chain.

a. Remove the 2 bolts, chain guide and o-ring.

14. Remove the camshaft timing sprocket.

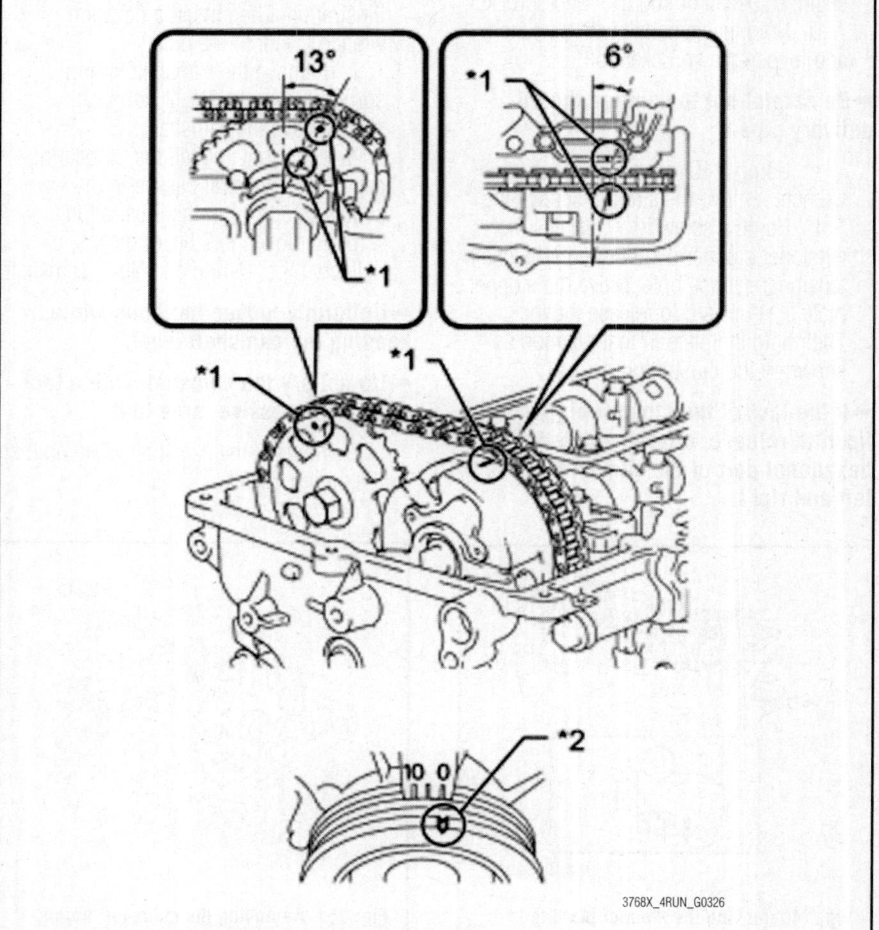

Fig. 147 Aligning the timing mark of the timing chain

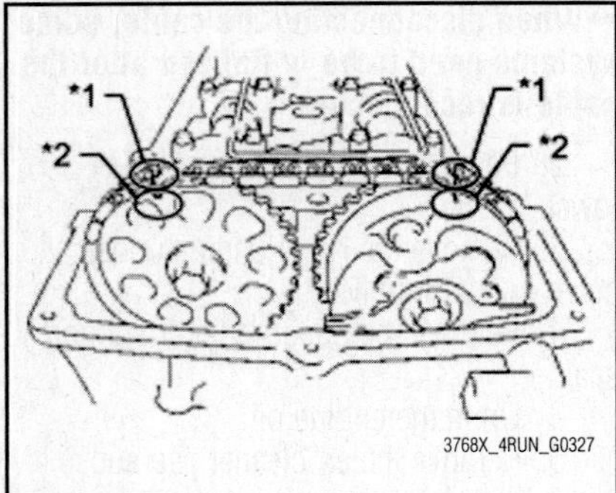

Fig. 148 Marking the timing chain, camshaft timing gear and sprocket

a. Turn the crankshaft pulley, and align its groove with the "0" timing mark of the timing chain cover.

b. Check the timing marks of the camshaft timing gear and sprocket are aligned with the timing marks of the No. 1 bearing cap.

➡**If the timing marks do not align, rotate the crankshaft clockwise again and align the timing marks.**

c. Place paint marks on the timing chain, camshaft timing gear and sprocket.

d. Hold the camshaft with a wrench and loosen the sprocket bolt.

➡**Be careful not to damage the oil delivery pipe.**

e. Using a 10 mm socket hexagon wrench, remove the straight screw plug.

f. Using a screwdriver, access the tensioner stopper plate through the chain tensioner service hole. Move the stopper plate (*1) upward to release the lock. Then hold the plate in that position as shown in the illustration.

➡**If the lock of the stopper plate is difficult to release, slightly rotate the hexagonal part of the camshaft to the left and right.**

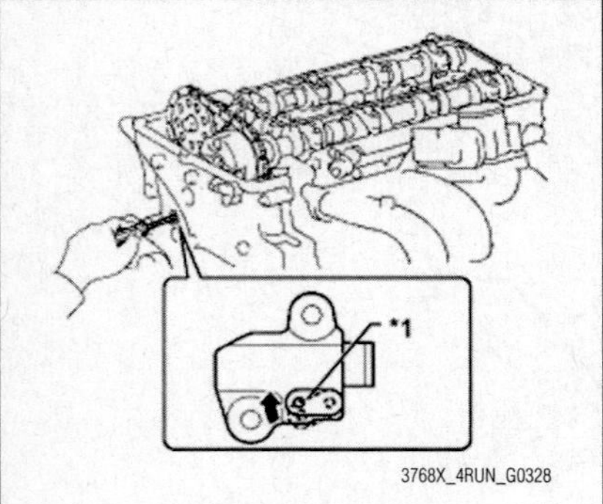

Fig. 149 Moving the stopper plate to release the lock

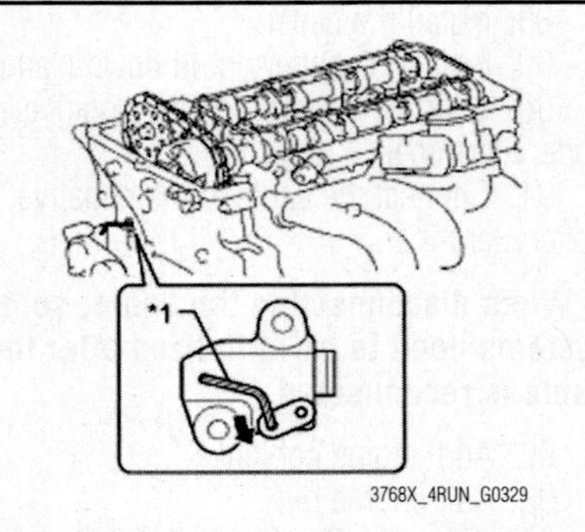

Fig. 150 Moving the stopper plate to a specified position

g. With the lock of the stopper plate released, slightly rotate the camshaft clockwise and keep it in that position.

➡**Rotating the camshaft clockwise will cause pressure to be applied to the tensioner plunger.**

➡**Be careful not to damage the oil delivery pipe.**

h. Remove the screwdriver from the chain tensioner service hole. Move the stopper plate to the position shown in the illustration. Then insert a hexagon wrench (*1) into the hole.

i. Remove the camshaft timing sprocket from the No. 2 camshaft.

15. Remove the camshaft.

a. Uniformly loosen the 21 bearing cap bolts in several passes in the sequence shown in the illustration.

b. Remove the 9 bearing caps, oil delivery pipe, O-ring and No. 2 camshaft.

➡**Uniformly loosen the bolts while keeping the camshaft level.**

➡**Do not pry the camshaft with a tool or apply excessive force to it.**

c. Remove the camshaft while holding the timing chain.

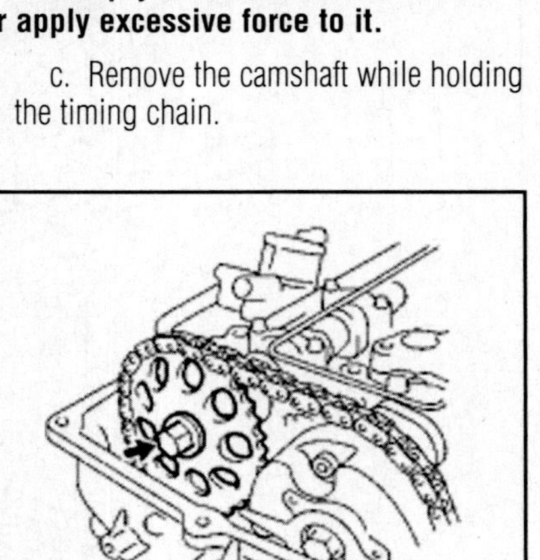

Fig. 151 Removing the camshaft timing sprocket from the No. 2 camshaft

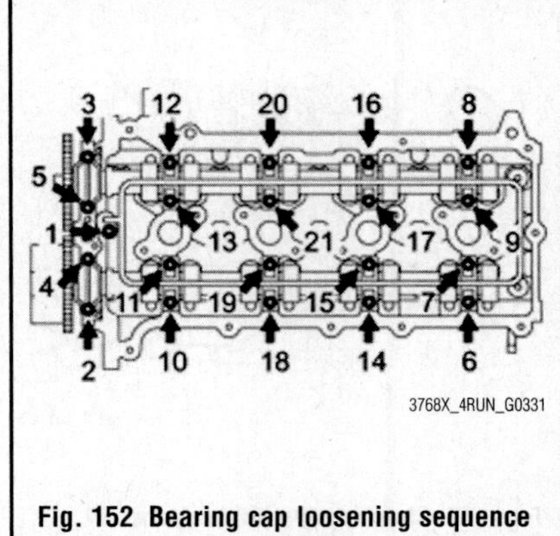

Fig. 152 Bearing cap loosening sequence

d. Secure the timing chain with a string.

16. Remove the No. 1 valve rocket arm sub assembly.

a. Remove the 16 valve rocket arms from the cylinder head.

➡**Arrange the removed parts in the correct order.**

17. Remove the valve lash adjuster assembly.

a. Remove the 16 valve lash adjusters from the cylinder head.

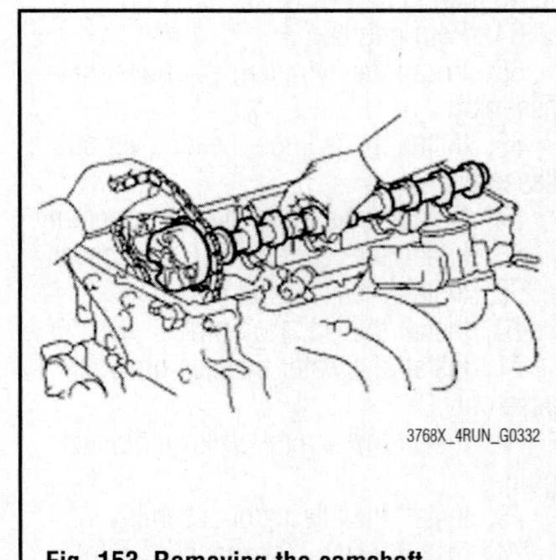

Fig. 153 Removing the camshaft

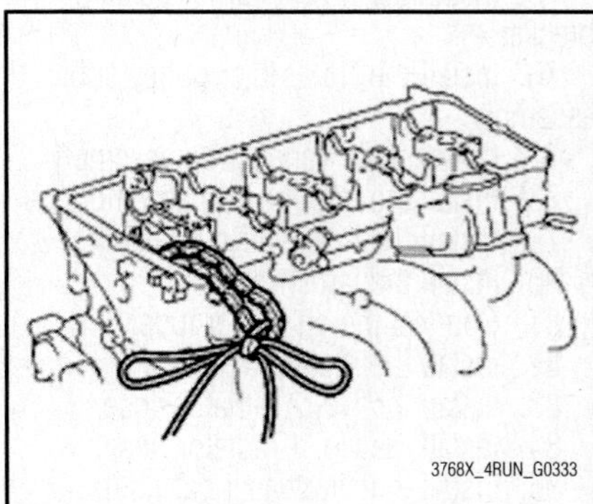

Fig. 154 Securing the timing chain with string

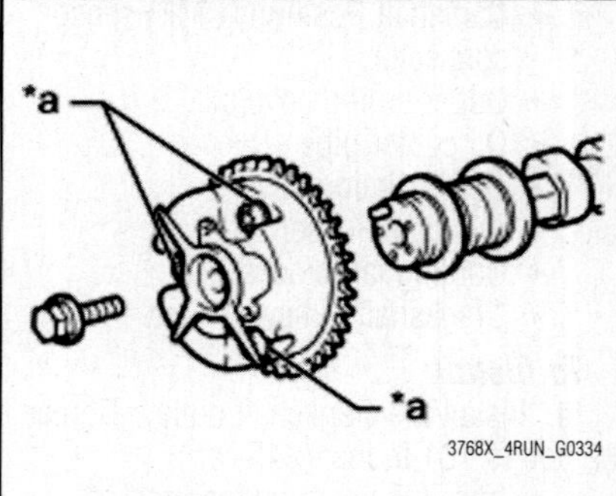

3768X_4RUN_G0334

Fig. 155 Removing the camshaft timing gear assembly

➡Arrange the removed parts in the correct order.

18. Remove the camshaft timing gear assembly.

a. Remove the flange bolt and camshaft timing gear.

➡Be sure not to remove the other 3 bolts.

➡If planning to reuse the gear, be sure to release the straight pin lock before installing the gear.

19. Inspect the valve lash adjuster assembly.

20. Inspect the camshaft timing gear assembly.

To install:

21. Install the camshaft timing gear assembly.

a. Align the pin hole and straight pin and install the camshaft timing gear to the camshaft.

b. Lightly press the gear against the camshaft and turn the gear. Push further at the position where the pin enters the groove.

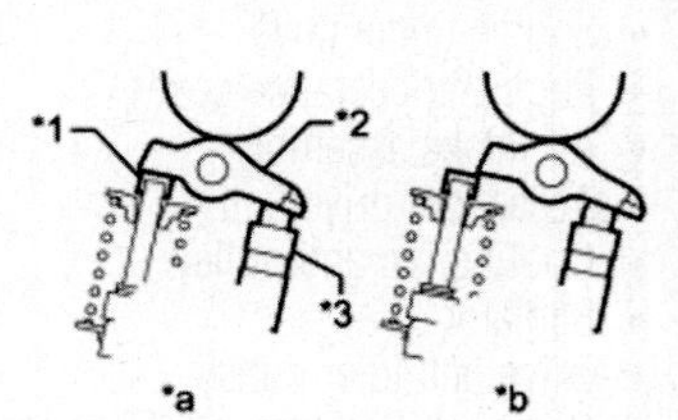

1. Valve stem cap
2. Valve rocker arm
3. Valve lash adjuster
a. Correct
b. Incorrect

3768X_4RUN_G0335

Fig. 156 Installing the No. 1 valve rocker arm sub assembly

c. Check that there is no gap between the flange of the gear and the camshaft.

d. With the camshaft timing gear fixed in place, install the flange bolt. Tighten to 58 ft. lbs. (78 Nm).

e. Check that the camshaft timing gear can move in the retard direction and becomes locked at the most retarded position.

22. Install valve lash adjuster assembly.

a. Inspect each valve lash adjuster before installing it.

b. Install the 16 valve lash adjusters to the cylinder head.

➡Install each lash adjuster to the same place it was removed from.

23. Install the No. 1 valve rocker arm sub assembly.

a. Apply clean engine oil to the valve lash adjuster tips and valve stem cap surfaces.

b. Install the 16 valve rocker arms as shown.

➡install each valve rocker arm to the same place it was removed from.

24. Install the camshaft.

a. Apply clean engine oil to the camshaft cams and cylinder head journals.

b. Install the timing chain to the camshaft timing gear with the painted mark of the link aligned with the timing mark of the camshaft timing gear.

c. Position the 2 camshafts as shown in the illustration

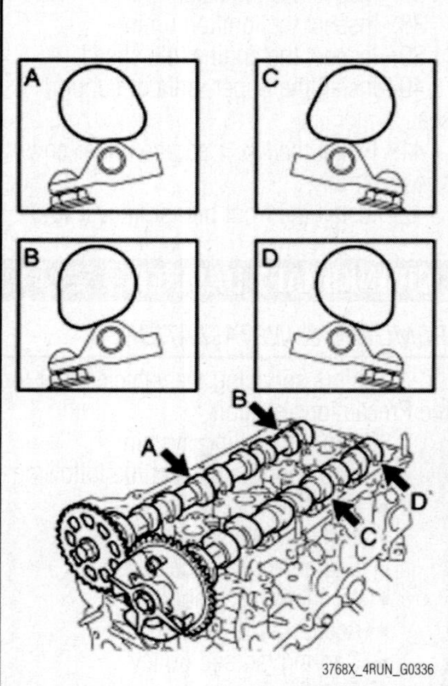

3768X_4RUN_G0336

Fig. 157 Positioning the camshafts

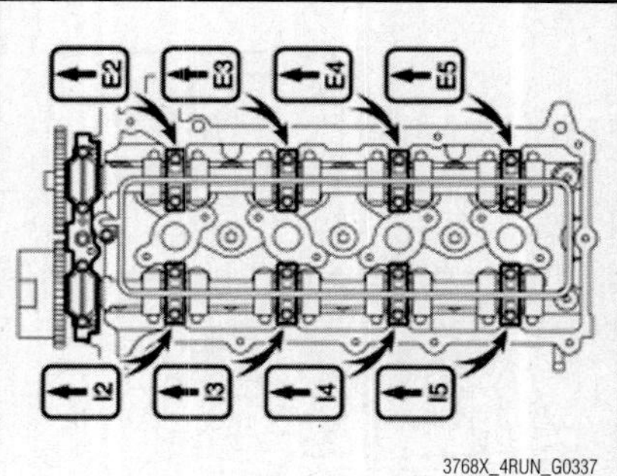

3768X_4RUN_G0337

Fig. 158 Temporarily installing the No. 1 camshaft bearing cap

➡Align the paint mark and timing mark before positioning the camshaft.

➡Before and after positioning the camshaft and No. 2 camshaft, check that the rocker arm is firmly set on the lash adjuster.

d. Temporarily install the No. 1 camshaft bearing cap.

e. Check the proper location of each camshaft bearing cap and install each one.

f. Install a new O-ring to the No. 1 camshaft bearing cap.

g. Temporarily install the oil delivery pipe.

h. Install the 21 bolts and tighten them in the order shown in the illustration. Tighten bolt A to 9 ft. lbs. (12 Nm). Tighten all other bolts to 11 ft. lbs. (16 Nm).

25. Install the camshaft timing sprocket.

a. Rotate the camshaft so that the camshaft timing mark and No. 2 camshaft knock pin are as shown in the illustration.

b. Turn the crankshaft pulley and align its groove with the "0" timing mark of the timing chain cover.

c. Install the timing chain to the camshaft timing sprocket with the paint mark aligned with the timing marks on the camshaft timing sprocket.

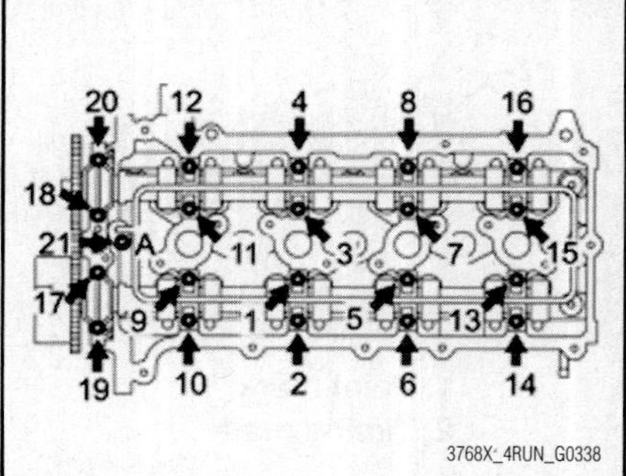

3768X_4RUN_G0338

Fig. 159 Identifying the No. 1 camshaft bearing cap bolt torque sequence

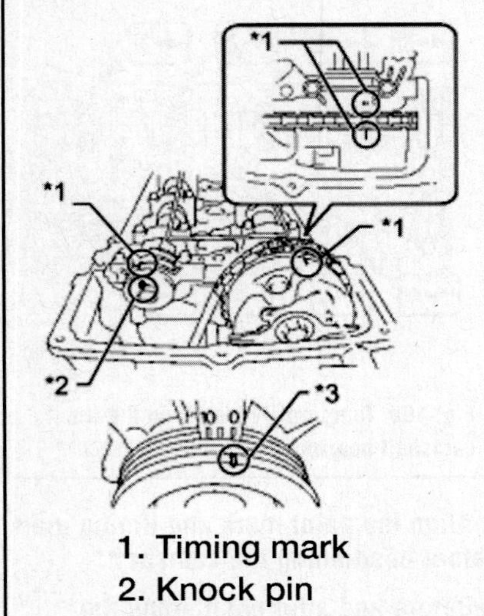

Fig. 160 Rotating the camshaft

d. Align the No. 2 camshaft knock pin and camshaft timing sprocket pin hole. Then install the camshaft timing sprocket to the No. 2 camshaft.

➡If the knock pin and pin hole are difficult to align, slightly rotate the No. 2 camshaft back and forth using the hexagonal part of the camshaft. Then attempt alignment again.

e. Hold the camshaft with a wrench and tighten the sprocket bolt. Tighten to 58 ft. lbs. (78 Nm).

f. Remove the hexagon wrench from the chain tensioner.

g. Apply adhesive to 2 or 3 threads of the straight screw plug.

➡Remove any oil from the bolt hole.

h. Using a 10 mm socket hexagon wrench, install the straight screw plug. Tighten to 12 ft. lbs. (17 Nm).

26. Install the timing chain guide.

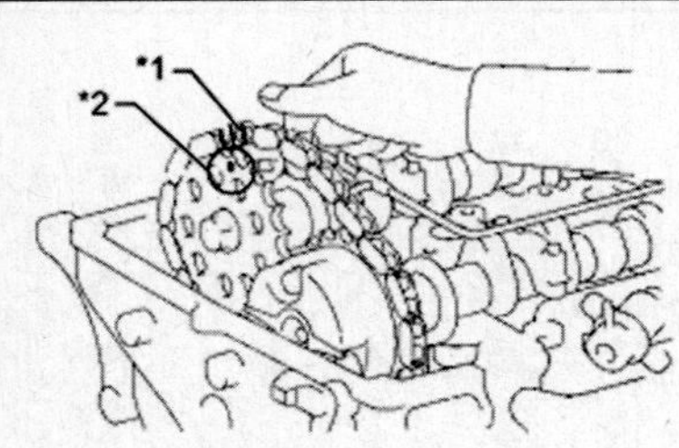

Fig. 161 Installing the timing chain to the camshaft timing sprocket

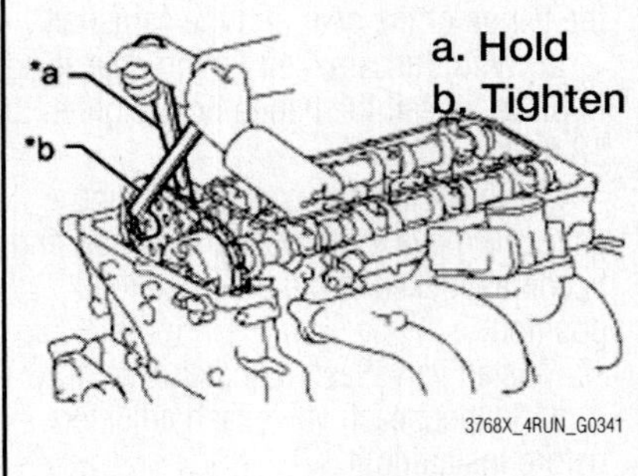

Fig. 162 Tightening the camshaft sprocket bolt

a. Install a new o-ring to the camshaft bearing cap.

b. Install the timing chain guide with the 2 bolts. Tighten to 7 ft. lbs. (10 Nm).

27. Install the cylinder head cover sub assembly.
28. Install the Camshaft Position (CMP) sensor.
29. Install the ignition coil assembly.
30. Install the fan shroud.
31. Install the radiator reservoir.
32. Install the intake air connector.
33. Install the air cleaner cap sub assembly.
34. Connect the cable to the negative battery terminal.

➡When disconnecting the cable, some systems need to be initialized after the cable is reconnected.

35. Add engine oil.
36. Inspect the engine oil level.
37. Inspect for oil leaks.
38. Inspect the ignition timing.
39. Inspect the engine idle speed.
40. Install the upper radiator support seal.
41. Install the No. 1 engine under cover sub assembly.
42. Install the front bumper cover lower.

CRANKSHAFT DAMPER

REMOVAL & INSTALLATION

1. Before servicing the vehicle, refer to the Precautions section.
2. Drain the cooling system.
3. Remove or disconnect the following:
   - Negative battery cable
   - Engine under cover
   - Engine appearance cover
   - Air intake assembly
   - Accessory drive belt
   - Cooling fan and pulley
   - Radiator
   - Drive belt idler pulley
   - Camshaft Position (CMP) sensor connector
   - Upper timing covers
   - Oil cooler pipe
   - Center timing cover
   - A/C compressor
   - Cooling fan bracket
   - Crankshaft pulley

To install:

4. Install the crankshaft pulley. Tighten the bolt to 181 ft. lbs. (245 Nm).
   - Install the cooling fan bracket. Tighten the 12mm bolts to 12 ft. lbs. (16 Nm) and the 14mm bolts to 24 ft. lbs. (32 Nm).
5. Install the A/C compressor.
6. Install the center timing cover.
7. Install the oil cooler pipe.
8. Install the upper timing covers.
9. Install the CMP sensor connector.
10. Install the drive belt idler pulley. Tighten the bolt to 27 ft. lbs. (37 Nm).
11. Install the radiator.
12. Install the cooling fan and pulley. Tighten the nuts to 16 ft. lbs. (21 Nm).
13. Install the accessory drive belt.
14. Install the air intake assembly.
15. Install the engine appearance cover.
16. Install the engine under cover.
17. Connect the negative battery cable.
18. Fill the cooling system.
19. Start the engine and check for leaks.

CRANKSHAFT FRONT SEAL

REMOVAL & INSTALLATION

2009 Models

1. Before servicing the vehicle, refer to the Precautions section.
2. Drain the cooling system.
3. Remove or disconnect the following:
   - Negative battery cable
   - Engine under cover
   - Engine appearance cover
   - Air intake assembly
   - Accessory drive belt
   - Cooling fan and pulley
   - Radiator
   - Drive belt idler pulley
   - Camshaft Position (CMP) sensor connector
   - Upper timing covers
   - Oil cooler pipe
   - Center timing cover
   - A/C compressor
   - Cooling fan bracket
   - Crankshaft pulley
   - Lower timing cover
   - Timing belt.

- Crankshaft timing sprocket
- Front crankshaft seal

To install:

4. Install the oil seal so that it is flush with the oil pump housing.
5. Install or connect the following:
   - Crankshaft timing sprocket
   - Timing belt
   - Lower timing cover
   - Crankshaft pulley. Tighten the bolt to 181 ft. lbs. (245 Nm).
   - Cooling fan bracket. Tighten the 12mm bolts to 12 ft. lbs. (16 Nm) and the 14mm bolts to 24 ft. lbs. (32 Nm).
   - A/C compressor
   - Center timing cover
   - Oil cooler pipe
   - Upper timing covers
   - CMP sensor connector
   - Drive belt idler pulley. Tighten the bolt to 27 ft. lbs. (37 Nm).
   - Radiator
   - Cooling fan and pulley. Tighten the nuts to 16 ft. lbs. (21 Nm).
   - Accessory drive belt
   - Air intake assembly
   - Engine appearance cover
   - Engine under cover
   - Negative battery cable
6. Fill the cooling system.
7. Start the engine and check for leaks.

2010 Models

4.0L Engine

See Figure 163.

1. Remove the radiator assembly.
2. Disconnect the water by pass pipe sub assembly.
3. Remove the oil filter bracket.
4. Remove the crankshaft pulley.
5. Remove the front crankshaft oil seal.
   a. Using a screwdriver, pry out the oil seal.

➡Tape the tip of the screwdriver before use.

➡Do not damage the surfaces of the oil seal press fit hole or crankshaft.

To install:

6. Install the front crankshaft oil seal.
   a. Apply MP grease to the lip of a new oil seal.
   b. Using a special tool and hammer, tap in the oil seal until its surface is flush with the timing chain cover edge.

➡Keep the lip free from foreign matter.

➡Do not tap the oil seal at an angle.

7. Install the crankshaft pulley.
8. Install the oil filter bracket.
9. Connect the water by pass pipe sub assembly.
10. Install the radiator assembly.

2.7L Engine

See Figure 164.

1. Remove the upper radiator support seal.
2. Remove the radiator reservoir.
3. Remove the fan shroud.
4. Remove the crankshaft pulley.
   a. Using a special tool, hold the crankshaft pulley and loosen the pulley bolt until 2 or 3 threads are screwed into the crankshaft.
   b. Using a special tool and pulley bolt, remove the crankshaft pulley.
5. Remove the front crankshaft oil seal.
   a. Using a screwdriver, pry out the oil seal.

➡Tape the screwdriver tip before use.

➡Do not damage the surface of the oil seal press fit hole of the crankshaft.

➡Check the crankshaft for damage after removing the oil seal. If the crankshaft is damaged, smooth the surface with 400 grit sandpaper.

To install:

6. Install the front crankshaft oil seal.
   a. Apply MP grease to the lip of a new oil seal.

➡Do not allow foreign matter to contact the lip of the oil seal.

➡Do not allow MP grease to contact the dust seal.

   b. Temporarily install the oil seal to the timing chain cover.
   c. Using the special tool and a hammer, tap in the oil seal until its surface is flush with the chin cover edge.

➡Keep the lip free from foreign matter.

➡Do not tap the oil seal at an angle.

7. Install the crankshaft pulley.
   a. Align the key groove of the pulley with the pulley set key and slide on the pulley.
   b. Using the special tool, install a new crankshaft pulley bolt. Tighten the bolt to 192 ft. lbs. (260 Nm).

➡Do not reuse the pulley bolt.

8. Install the fan shroud.
9. Install the radiator reservoir.
10. Install the upper radiator support seal.
11. Inspect for oil leaks.
12. Inspect the engine oil level.

CYLINDER HEAD

REMOVAL & INSTALLATION

2009 Models

4.0L Engine

See Figures 165 through 177.

1. Discharge fuel system pressure.
2. Drain engine coolant.
3. Drain engine oil.
4. Remove timing chain.
5. Remove the 2 bolts and cool air inlet with air cleaner hose.
6. Remove the ATF level gauge.
7. Remove the bolt and pull out the oil filler tube.
8. Remove the O-ring from the oil filler tube.
9. Remove exhaust pipe assembly.
10. Remove the 3 bolts and manifold stay.
11. Disconnect the A/F sensor connector.
12. Remove the 6 nuts, exhaust manifold and gasket.
13. Remove No. 2 exhaust front pipe assembly.
14. Remove the 3 bolts and No. 2 manifold stay.

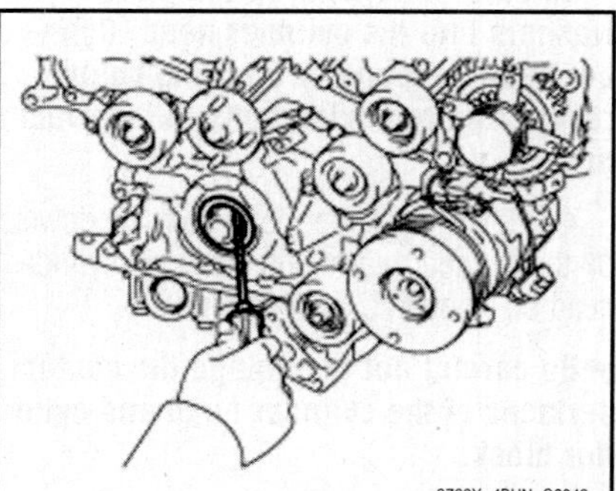

Fig. 163 Removing the front crankshaft oil seal

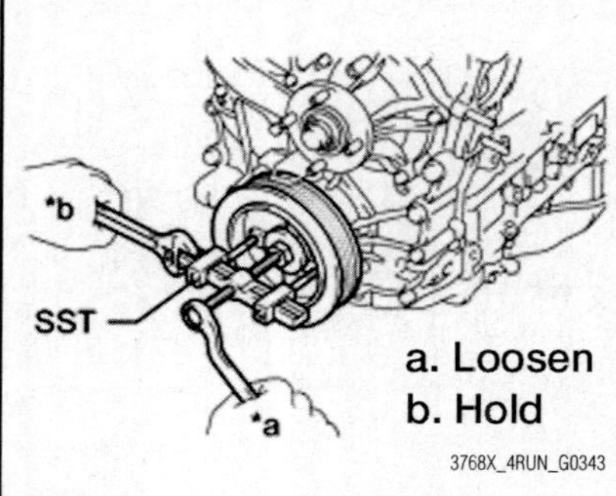

Fig. 164 Removing the crankshaft pulley

15. Disconnect the A/F sensor connector.

16. Remove the 6 nuts, exhaust manifold and gasket.

17. Disconnect No. 1 and No. 2 fuel pipe sub-assembly.

18. Disconnect the 6 fuel injector connectors.

19. Remove the 10 bolts, intake manifold and 2 gaskets.

20. Disconnect the Engine Coolant Temperature (ECT) sensor connector.

21. Disconnect the heater hose.

22. Remove the 2 bolts, 4 nuts, water by-pass joint RR and 2 gaskets.

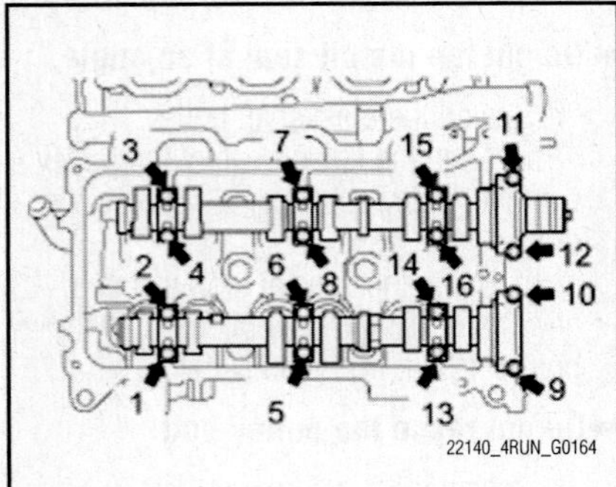

Fig. 165 Bearing cap removal sequence bank No. 1

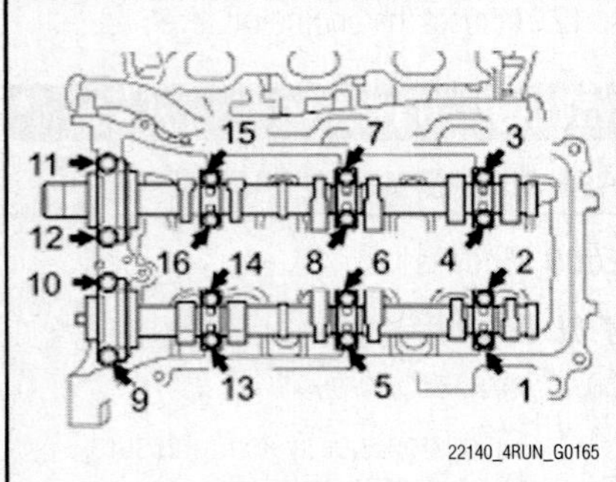

Fig. 166 Bearing cap removal sequence bank No. 2

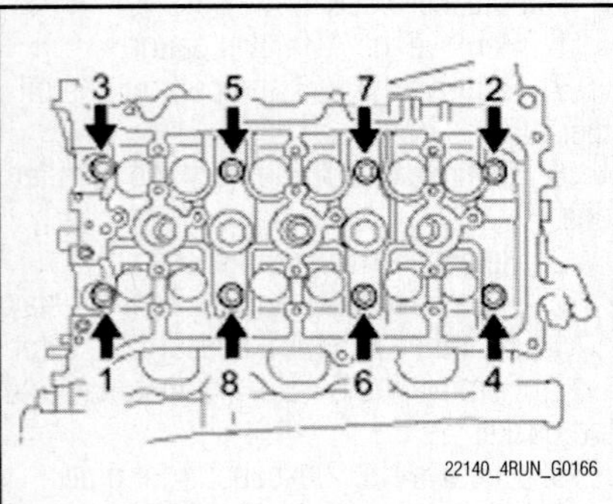

Fig. 167 Cylinder head bolt removal sequence

23. Remove the O-ring from the water outlet pipe.

24. While raising the No. 2 chain tensioner, insert a pin of 1.0 mm (0.039 in.) into the hole to fix it.

25. Hold the hexagonal portion of the camshaft with a wrench, and remove the 2 bolts, camshaft timing gear, camshaft timing gear assembly and No. 2 timing chain.

➡Be careful not to damage the cylinder head and valve lifter with the wrench. Do not disassemble the camshaft timing gear assembly.

26. Remove the bolt and No. 2 chain tensioner.

➡As the thrust clearance of the camshaft is small, the camshaft must be kept level while it is being removed. If the camshaft is not kept level, the portion of the cylinder head which receives the shaft thrust may crack or be damaged, causing the camshaft to seize or break. To avoid this, the following steps should be carried out.

27. Rotate the camshafts counterclockwise using a wrench so that the cam lobes of No. 1 cylinder face each direction.

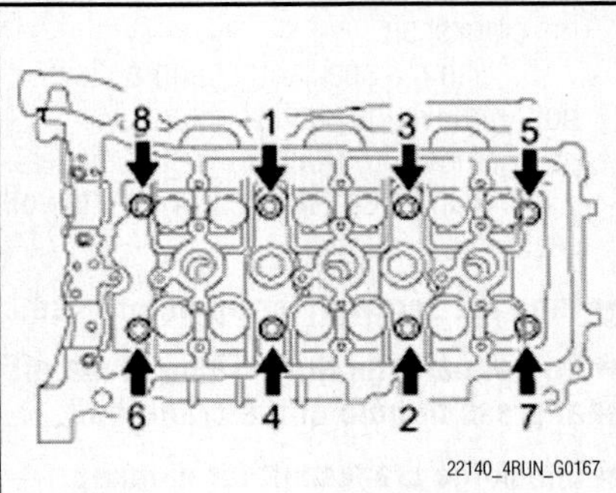

Fig. 168 Cylinder head bolt installation sequence

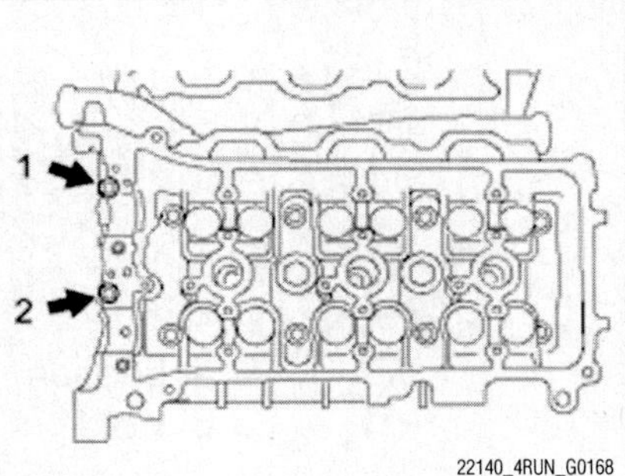

Fig. 169 Cylinder head bolt installation sequence

28. Using several steps, loosen and remove the 16 bearing cap bolts uniformly in the sequence as shown in the illustration.

29. Remove the 8 bearing caps and 2 camshafts.

30. Remove No. 2 camshaft bearing.

31. Remove the 2 bolts and No. 1 chain vibration damper.

32. While pushing down the No. 3 chain tensioner, insert a pin of 1.0 mm (0.039 in.) into the hole to fix it.

33. Hold the hexagonal portion of the camshaft with a wrench, and remove the 2 bolts, camshaft timing gear, camshaft timing gear assembly and No. 2 timing chain.

➡Be careful not to damage the cylinder head and valve lifter with the wrench. Do not disassemble the camshaft timing gear assembly.

34. Remove the bolt and No. 3 chain tensioner.

➡As the thrust clearance of the camshaft is small, the camshaft must be kept level while it is being removed. If the camshaft is not kept level, the portion of the cylinder head which receives the shaft thrust may crack or be damaged, causing the camshaft to seize or break. To avoid this, the following steps should be carried out.

35. Using several steps, loosen and remove the 16 bearing cap bolts uniformly in the sequence as shown in the illustration.

36. Remove the 8 bearing caps and 2 camshafts.

37. Remove the 2 bolts and separate the 2 ground cables.

38. Using several steps, loosen the 8 cylinder head bolts on the cylinder head uniformly with a 10 mm bi-hexagon wrench in the sequence as shown in the illustration.

39. Remove the 8 cylinder head bolts and plate washers.

➡Be careful not to drop the plate washers into the cylinder head. Cylinder head warpage or cracking could result from removing bolts in incorrect order.

40. Lift the cylinder head from the dowels on the cylinder block, and place the cylinder head on wooden blocks on a bench.

➡Be careful not to damage the contact surfaces of the cylinder head and cylinder block.

To install:

41. Remove any old packing (FIPG) material and be careful not to drop any oil

on the contact surfaces of the cylinder head and cylinder block.

42. Apply seal packing (diameter: 2.5 to 3 mm (0.098 to 0.118 in.)) to a new cylinder head gasket.

➡Install the cylinder head within 3 minutes after applying seal packing. After installing it, the cylinder head bolts must be tightened within 15 minutes. Otherwise the seal packing must be removed and reapplied.

43. Place the cylinder head gasket on the cylinder block surface with the Lot No. stamp upper side facing upward.

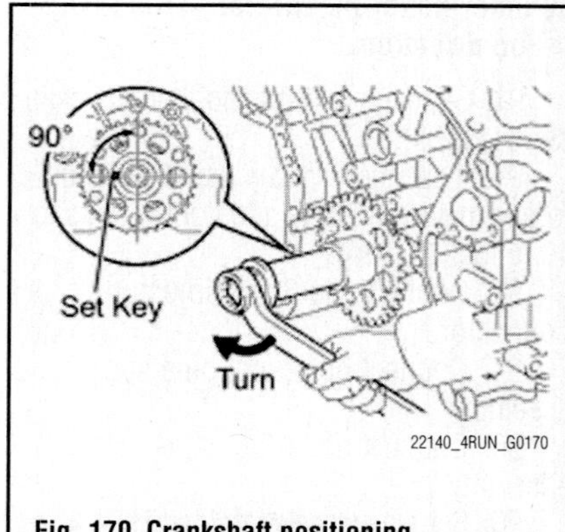

Fig. 170 Crankshaft positioning

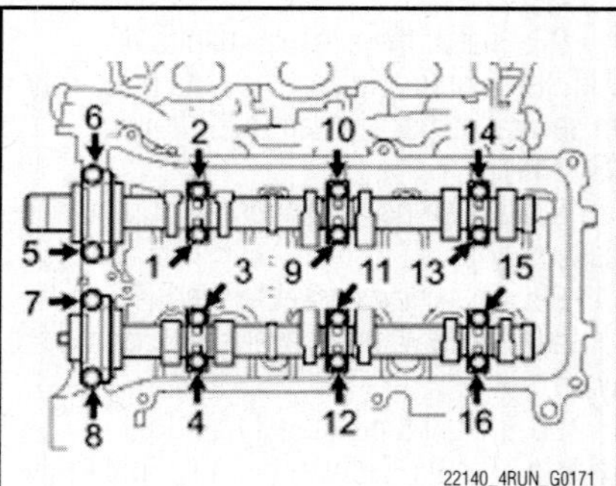

Fig. 171 Camshaft bearing cap bolt tightening sequence

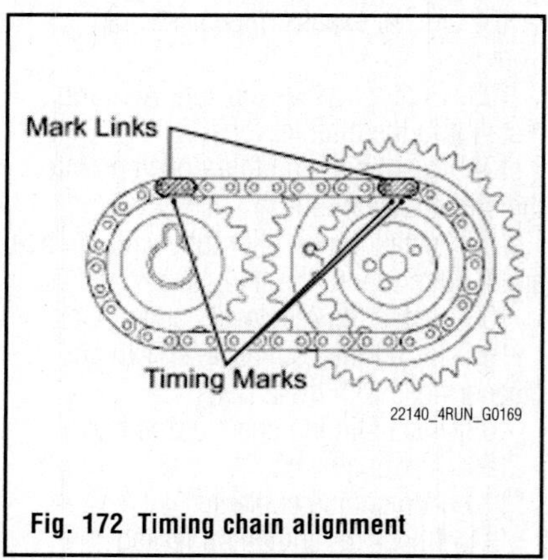

Fig. 172 Timing chain alignment

44. Place the cylinder head on the cylinder head gasket.

45. Install the 8 cylinder head bolts.

➡The cylinder head bolts are tightened in 2 successive steps. If any cylinder head bolt is broken or deformed, replace it.

46. Apply a light coat of engine oil to the threads of the cylinder head bolts.

47. Install the plate washer to the cylinder head bolt.

48. Using several steps, tighten each bolt uniformly with a 10 mm bi-hexagon wrench in the sequence. Tighten to 27 ft. lbs. (36 Nm). If any one of the cylinder head bolts does not meet the torque specification, replace the cylinder head bolt.

49. Retighten the cylinder head bolts 180°.

50. Install the front 2 cylinder head bolts.

51. Using several steps, install and tighten the 2 cylinder head bolts uniformly in the sequence to 22 ft. lbs. (30 Nm).

52. Install the ground cable with the bolt.

53. Perform the installation procedure for RH cylinder head.

54. Install camshafts for Bank 2.

➡As the thrust clearance of the camshaft is small, the camshaft must be kept level while it is being installed. If the camshaft is not kept level, the portion of the cylinder head which receives the shaft thrust may crack or be damaged, causing the camshaft to seize or break. To avoid this, the following steps should be carried out.

55. Set the crankshaft position. Install the crankshaft pulley set bolt, and turn the crankshaft, and set the crankshaft set key at the left horizontal position.

➡Setting the crankshaft at a wrong angle can cause the piston head and valve head to come into contact with each other when you install the camshaft, causing damage. So always set the crankshaft at the correct angle.

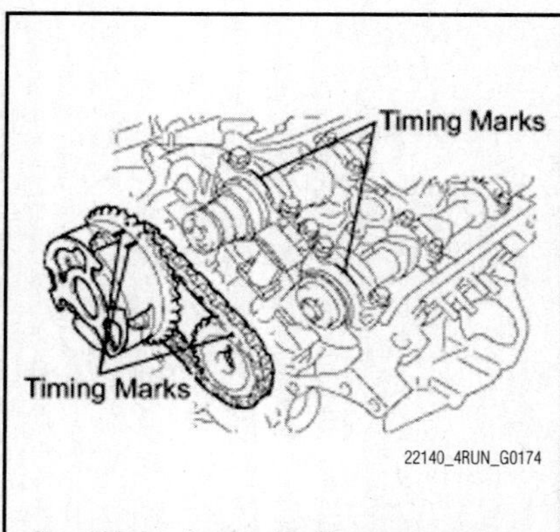

Fig. 173 Timing chain alignment marks

56. Apply new engine oil to the thrust portions and journals of the camshafts.

57. Place the 2 camshafts onto the cylinder head with the cam lobes of No. 2 cylinder facing each correct direction as shown in the illustration.

58. Install the 8 bearing caps in their proper locations.

59. Apply a light coat of engine oil to the threads and under the heads of the bearing cap bolts

60. Using several steps, install and tighten the 16 bearing cap bolts uniformly in the sequence and tighten 10mm bolts to 80 inch lbs. (9.0 Nm) and 12 mm to 18 ft. lbs. (24 Nm).

61. While pushing in the tensioner, insert a pin of 1.0 mm (0.039 in.) into the hole to fix it.

62. Install the No. 3 chain tensioner with the bolt and tighten to 14 ft. lbs. (19 Nm).

63. Align the mark links (yellow) with the timing marks (1-dot mark and 2-dot mark) of the camshaft timing gears.

64. Align the timing marks on the camshaft timing gears with the timing marks on the bearing caps, and install the camshaft timing gears with the chain to the LH camshafts.

65. Temporarily install the 2 camshaft timing gear bolts.

➡Do not push the camshaft timing gear assembly to the camshaft forcibly when installing it.

66. Hold the hexagonal portion of the camshaft with a wrench, and tighten the 2 bolts to 74 ft. lbs. 100 Nm).

67. Remove the pin from the No. 3 tensioner.

68. Install the No. 1 chain vibration damper with the 2 bolts and tighten to 14 ft. lbs. (19 Nm).

69. Install the No. 2 camshaft bearing to the cylinder head.

➡Clean the backside of the bearing and the contact surface of the cylinder head and prevent oil from adhering to them

70. Install camshafts for Bank 1.

➡As the thrust clearance of the camshaft is small, the camshaft must be kept level while it is being installed. If the camshaft is not kept level, the portion of the cylinder head

which receives the shaft thrust may crack or be damaged, causing the camshaft to seize or break. To avoid this, the following steps should be carried out.

71. Set the crankshaft position. Install the crankshaft pulley set bolt, and turn the crankshaft, and set the crankshaft set key at the left horizontal position.

➡Setting the crankshaft at a wrong angle can cause the piston head and valve head to come into contact with each other when you install the camshaft, causing damage. So always set the crankshaft at the correct angle.

72. Apply new engine oil to the thrust portions and journals of the camshafts.
73. Place the 2 camshafts onto the cylinder head with the cam lobes of No. 2 cylinder facing each correct direction as shown in the illustration.
74. Install the 8 bearing caps in their proper locations.
75. Apply a light coat of engine oil to the threads and under the heads of the bearing cap bolts.
76. Using several steps, install and tighten the 16 bearing cap bolts uniformly in the sequence and tighten 10mm bolts to 80 inch lbs. (9.0 Nm) and 12 mm to 18 ft. lbs. (24 Nm).
77. Turn the camshafts clockwise until the knock pin comes to 90°position to the cylinder head.
78. While pushing in the tensioner, insert a pin of 1.0 mm (0.039 in.) into the hole to fix it.
79. Install the No. 2 chain tensioner with the bolt and tighten to 14 ft. lbs. (19 Nm).
80. Align the mark links (yellow) with the timing marks (1-dot mark and 2-dot mark) of the camshaft timing gears.
81. Align the timing marks on the camshaft timing gears with the timing marks on the bearing caps, and install the camshaft timing gears with the chain to the RH camshafts.
82. Temporarily install the 2 camshaft timing gear bolts.

➡Do not push the camshaft timing gear assembly to the camshaft forcibly when installing it.

83. Hold the hexagonal portion of the camshaft with a wrench, and tighten the 2 bolts to 74 ft. lbs. 100 Nm).
84. Remove the pin from the No. 2 tensioner.
85. Install a new O-ring to the water outlet pipe.
86. Apply soapy water to the O-ring.
87. Install 2 new gaskets and water by-pass joint RR with the 2 bolts and 4 nuts and tighten to 80 inch lbs. (9.0 Nm).
88. Connect the heater hose.
89. Connect the Engine Coolant Temperature (ECT) sensor connector.
90. Set a new intake gasket on each cylinder head.

➡Align the port holes of the gasket and cylinder head. Be careful of the installation direction.

91. Set the intake manifold on the cylinder heads.
92. Install and tighten the 10 bolts uniformly in several steps and tighten bolts to 19 ft. lbs. (26 Nm).
93. Connect the 6 fuel injector connectors.
94. Connect no. 2 fuel pipe sub-assembly.
95. Connect no. 1 fuel pipe sub-assembly.
96. Set a new gasket to the LH cylinder head with the oval shape facing backward.
97. Install the exhaust manifold with the 6 nuts. Tighten the nuts uniformly in several steps. Tighten to 22 ft .lbs. (30 Nm).
98. Install no. 2 exhaust front pipe assembly.
99. Set a new gasket to the RH cylinder head with the oval shape facing forward.
100. Install the exhaust manifold with the 6 nuts. Tighten the nuts uniformly in several steps. Tighten to 22 ft. lbs. (30 Nm).
101. Install the manifold stay with the 3 bolts tighten to 30 ft. lbs. (40 Nm).
102. Install exhaust pipe assembly front.
103. Coat a new O-ring with ATF, and install it to the oil filler tube.
104. Push in the oil filler tube end into the oil pan lower tube.
105. Install the oil filler tube with the bolt and tighten to 9 ft. lbs. (12 Nm).
106. Install the ATF level gauge.
107. Install the cool air inlet with air cleaner hose with the 2 bolts.
108. Install timing chain assembly.
109. Add engine oil.
110. Add engine coolant.
111. Check for engine oil leakage.

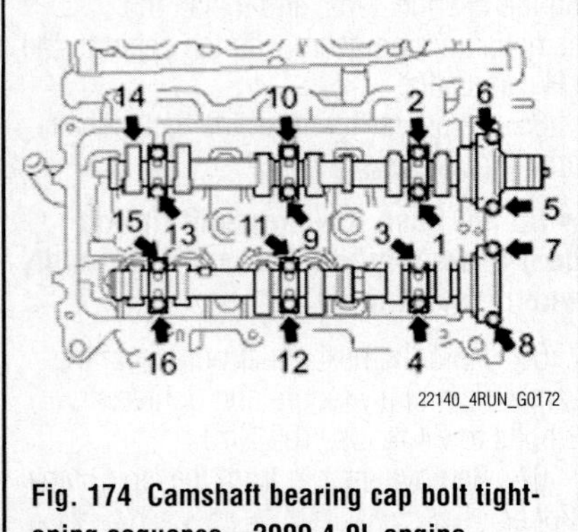

Fig. 174 Camshaft bearing cap bolt tightening sequence—2009 4.0L engine

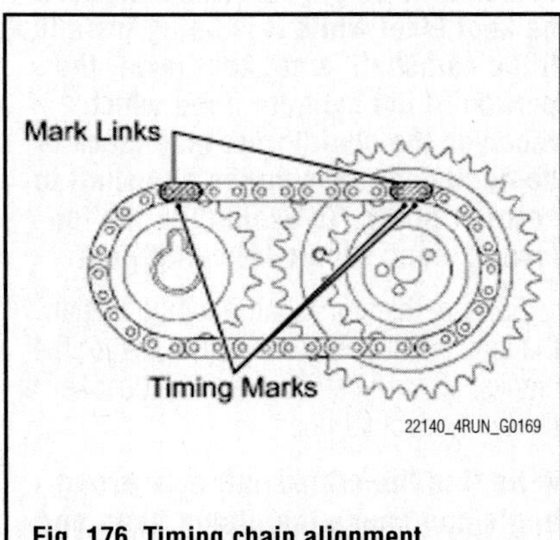

Fig. 176 Timing chain alignment

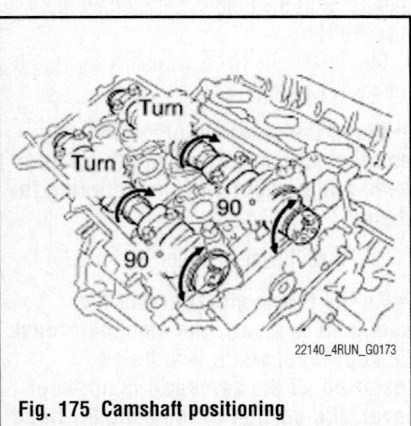

Fig. 175 Camshaft positioning

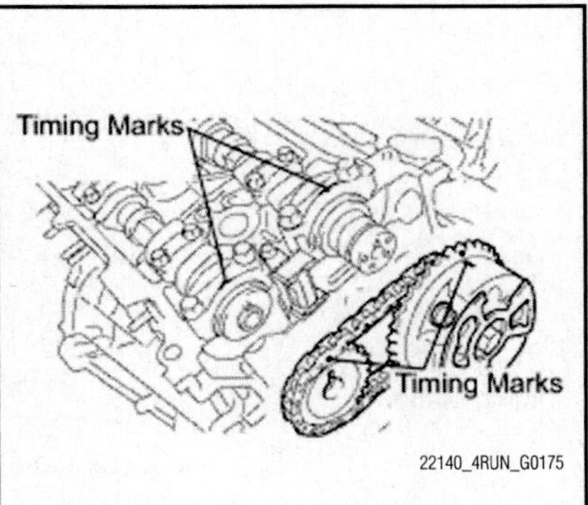

Fig. 177 Timing chain alignment marks

112. Check for engine coolant leakage.
113. Check for fuel leakage.
114. Check for exhaust gas leakage.
115. Inspect engine idle speed.

4.7L Engine

See Figure 178.

1. Before servicing the vehicle, refer to the precautions section.
2. Drain the cooling system.
3. Relieve the fuel system pressure.
4. Remove the V bank cover.
5. Remove the timing belt.
6. Remove the camshaft pulleys.
7. Remove the Camshaft Position (CMP) sensor.
8. Remove the power steering pump and set it aside with the lines still attached.
9. Remove the front exhaust pipe.
10. On models with an automatic transmission, remove the oil dipstick and tube.
11. Remove the ignition coils.
12. Remove the rear timing belt plates being careful not to drop anything.
13. Disconnect the fuel inlet hose.
14. Remove the intake manifold.
15. Remove the water inlet and inlet housing. Refer to water pump removal.
16. Remove the front and rear water bypass joint.
17. Remove the engine hangers and if needed the oil dipstick and tube.
18. Remove the valve covers.

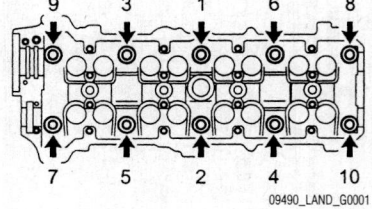

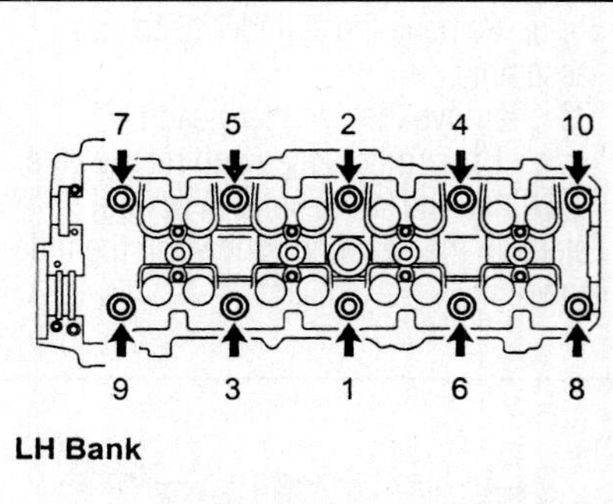

Fig. 178 Cylinder head loosening sequence—4.7L 2UZ-FE engine

➡Since the thrust level of the camshaft is small, the camshaft must be kept level during removal. If not kept level serious damage could occur.

19. Check the timing mark of the crankshaft pulley is aligned with the center(s) of the crankshaft pulley bolt and idler pulley bolt.

➡If the crankshaft pulley is wrongly positioned, this can cause the piston to contact the head causing severe damage. Make sure the crankshaft pulley is properly positioned.

20. Release the oil from the front bearing caps using the tool illustrated. Rotate the camshaft timing tube from left to right 2 to 3 times within its VVT-I range of 25 degrees and collect the oil from the timing oil control valve installation hole using a rag.
21. Remove the left hand camshafts as follows:
    a. Bring the service bolt of the sub gear up by turning the left exhaust camshaft using a wrench on the hexagon head portion of the shaft.
    b. Secure the sub gear to the main gear using a 16 to 20 mm bolt with a diameter of 6mm and a thread pitch of 1mm.
    c. Make sure the torsional force of the sub gear is retained by the bolt.
    d. Align the 2 dot timing mark of the left side camshaft by turning the left exhaust camshaft using a wrench on the hexagon head portion of the shaft.

➡Mark the position of the caps so they can be reinstalled in their original positions.

    e. Loosen the 22 bearing cap bolts in the sequence illustrated using several passes.
    f. Remove the bolts, washers, oil feed pipe, bearing caps, camshaft housing plug, oil control valve filter and the camshafts.
22. Remove the right hand camshafts as follows:
    a. Bring the service bolt of the sub gear up by turning the right exhaust camshaft using a wrench on the hexagon head portion of the shaft.
    b. Secure the sub gear to the main gear using a 16 to 20 mm bolt with a diameter of 6mm and a thread pitch of 1mm.
    c. Make sure the torsional force of the sub gear is retained by the bolt.
    d. Align the 1 dot timing mark of the camshaft main gear (about 10 degrees) angle by turning the right exhaust camshaft using a wrench on the hexagon head portion of the shaft.

➡Mark the position of the caps so they can be reinstalled in their original positions.

    e. Loosen the 22 bearing cap bolts in the sequence illustrated using several passes.
    f. Remove the bolts, washers, oil feed pipe, bearing caps, camshaft housing plug, oil control valve filter and the camshafts.
23. Loosen the cylinder head bolts in the sequence shown, using several passes.
24. Remove the cylinder heads and exhaust manifolds together as an assembly.

To install:

25. Install new gaskets and the cylinder heads
26. Tighten the bolts in sequence as follows:
    a. Step 1: 30 ft. lbs. (40 Nm).
    b. Step 2: Plus 90 degrees.
    c. Step 3: Plus 90 degrees.
27. Check the timing mark of the crankshaft pulley is aligned with the center(s) of the crankshaft pulley bolt and idler pulley bolt.

➡If the crankshaft pulley is wrongly positioned, this can cause the piston to contact the head causing severe damage. Make sure the crankshaft pulley is properly positioned.

28. Install the left side camshafts as follows:
    a. Apply multipurpose grease to the thrust portion of the camshafts.
    b. Align the 2 dot timing mark of the camshaft drive and driven main gears and install the camshafts.
    c. Apply seal packing to the camshaft housing plug.
    d. Install the camshaft housing plug on the cylinder head as illustrated. Install the strainer on the head being careful it is properly positioned.
    e. Apply seal packing to the front bearing cap.
    f. Install the front bearing cap and then the other caps in the sequence illustrated.
    g. Push in the camshaft oil seal.
    h. Install 4 new seal washers to the bearing cap bolts A and B, refer to the illustration.

i. Apply a light coating of clean oil to the threads and underside of the bolt heads D and E. make sure no oil gets under the heads of bolts A, B and C.

j. The bolt lengths and positions are as follows. refer to the illustration for bolt location:

- 94mm bolts A
- 72mm bolt B
- 25mm bolt C
- 52mm bolt D
- 38mm bolts E

k. Tighten the cap bolts using several passes. Tighten bolt C to 66 inch lbs. (7.5 Nm) an the remaining bolts to 12 ft. lbs. (16 Nm).

l. Remove the service bolt.

29. Install the right side camshafts as follows:

a. Apply multipurpose grease to the thrust portion of the camshafts.

b. Align the 1 dot timing mark of the camshaft drive and driven main gears and install the camshafts.

c. Set the 1 dot timing mark of the camshaft drive and driven gears at a 10 degree angle.

d. Apply seal packing to the camshaft housing plug.

e. Install the camshaft housing plug on the cylinder head as illustrated. Install the strainer on the head being careful it is properly positioned.

f. Apply seal packing to the front bearing cap.

g. Install the front bearing cap and then the other caps in the sequence illustrated.

h. Push in the camshaft oil seal.

i. Install 4 new seal washers to the bearing cap bolts A and B, refer to the illustration.

j. Apply a light coating of clean oil to the threads and underside of the bolt heads D and E. make sure no oil gets under the heads of bolts A, B and C.

k. The bolt lengths and positions are as follows. refer to the illustration for bolt location:

- 94mm bolts A
- 72mm bolt B
- 25mm bolt C
- 52mm bolt D
- 38mm bolts E

l. Tighten the cap bolts using several passes. Tighten bolt C to 66 inch lbs. (7.5 Nm) an the remaining bolts to 12 ft. lbs. (16 Nm).

m. Remove the service bolt.

30. Check and adjust the valve clearance.

31. Install the camshaft timing control valve.

32. Install the 4 half moon plugs onto the cylinder heads.

33. Install the valve covers and tighten to 53 inch lbs. (6 Nm).

34. Install the engine hangers and tighten to 27 ft. lbs. (37 Nm).

35. Install the VVT sensors.

36. Install the oil dipstick tube and dipstick.

37. Install the ignition coils.

38. Install the water bypass joint and tighten the retainers to 13 ft. lbs. (18 Nm).

39. Install the water inlet and housing assembly.

40. Install the intake manifold.

41. Install the timing belt rear plates, right plates first, then left plates. Tighten the retainers to 66 inch lbs. (7 Nm).

42. Install the throttle body cover.

43. Install the front exhaust pipe, power steering pump.

44. Install the Camshaft Position (CMP) sensor and camshaft timing pulleys, tighten to 25 ft. lbs. (34 Nm).

45. Install the timing belt.

46. Fill the cooling system and perform an oil change.

47. Start the vehicle and check for leaks.

2010 Models

4.0L Engine

See Figures 179 through 187.

1. Remove the timing chain cover sub assembly.

2. Set the No. 1 cylinder to TDC/compression.

3. Remove the No. 1 chain tensioner assembly.

4. Remove the chain tensioner slipper.

5. Remove the chain sub assembly.

6. Remove the No. 1 idle gear shaft.

7. Remove the No. 1 chain vibration damper.

8. Remove the No. 2 chain vibration damper.

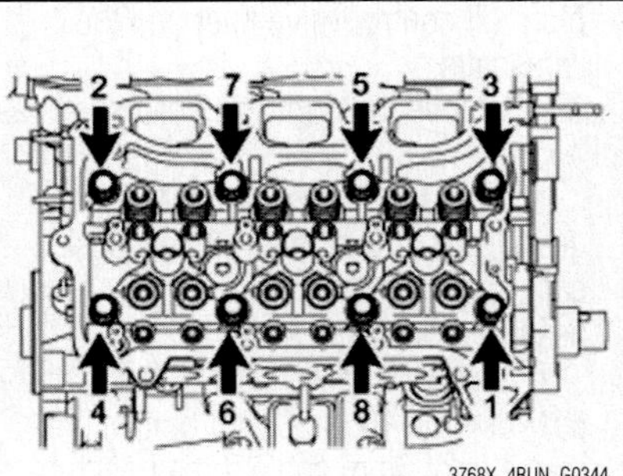

Fig. 179 Removing the cylinder head sub assembly

9. Remove the crankshaft timing sprocket.

10. Remove the camshaft timing gears and No. 2 chain (Bank 1).

11. Remove the No. 2 chain tensioner assembly.

12. Remove the camshaft bearing cap (Bank 1).

13. Remove the camshaft housing sub assembly RH.

14. Remove the camshaft timing gears and No. 2 chain (Bank 2).

15. Remove the No. 3 chain tensioner assembly.

16. Remove the camshaft bearing cap (Bank 2).

17. Remove the camshaft housing sub assembly LH.

18. Remove the No. 1 valve rocker arm sub assembly.

19. Remove the valve lash adjuster assembly.

20. Remove the valve stem cap.

21. Remove the cylinder head sub assembly.

a. Using a 10 mm bi-hexagon wrench, uniformly loosen the 8 cylinder head bolts in the sequence shown in the illustration. Remove the 8 cylinder head bolts and plate washers.

➡Be careful not to drop washers into the cylinder head sub assembly.

➡Cylinder head warpage or cracking could result from removing bolts in an incorrect order.

➡Arrange the removed parts in the correct order.

b. Remove the cylinder head sub assembly.

22. Remove the cylinder head LH.

a. Uniformly loosen and remove the 2 cylinder head set bolts in several steps in the sequence shown in the illustration.

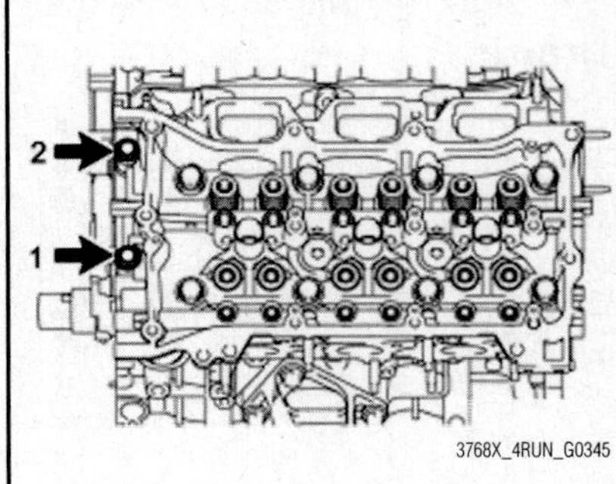

Fig. 180 Loosening the 2 cylinder head set bolts

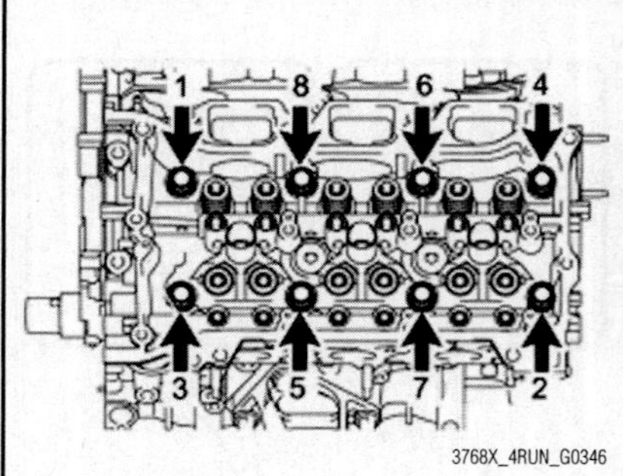

Fig. 181 Removing the 8 cylinder head bolts and plate washers

b. Using a 10 mm bi-hexagon wrench, uniformly loosen the 8 bolts in the sequence shown in the illustration. Remove the 8 cylinder head bolts and plate washers.

➡**Be careful not to drop washers into the cylinder head sub-assembly.**

➡**Cylinder head warpage or cracking could result from removing bolts in an incorrect order.**

➡**Be sure to keep the removed parts for each installation position separate.**

c. Remove the cylinder head LH.

To install:

23. Inspect the cylinder head set bolt.
24. Inspect the cylinder head sub assembly.
25. Install the cylinder head gasket.

a. Remove any old packing (FIPG) material and be careful not to drop any oil on the contact surfaces of the cylinder head or cylinder block.

b. Apply seal packing to a new cylinder head gasket as shown in the illustration.

➡**Remove any oil from the contact surface.**

➡**Install the cylinder head gasket within 3 minutes and tighten the bolts within 15 minutes after applying seal packing.**

➡**Do not add engine oil within 2 hours of installation.**

c. Place the cylinder head gasket on the cylinder block surface with the front face of the Lot No. stamp upward.

➡**Make sure that the gasket is installed facing the proper direction.**

26. Install the cylinder head sub assembly.

a. Place the cylinder head on the cylinder block.

➡**Gently place the cylinder head in order not to damage the gasket with the bottom part of the head.**

➡**Make sure that no oil is on the mounting surface of the cylinder head.**

➡**The cylinder head bolts are tightened in 3 progressive steps.**

b. Apply a light coat of engine oil to the threads and under the heads of the cylinder head bolts.

Step 1:

a. Using a 10 mm bi-hexagon wrench, install and uniformly tighten the 8 cylinder head bolts with the plate washers in several steps in the sequence shown in the illustration. Tighten to 27 ft. lbs. (36 Nm).

Step 2:

a. Mark the front side of each cylinder head bolt head with paint.

b. Tighten the cylinder head bolts another 90°.

Step 3:

a. Tighten the cylinder head bolts an additional 90°.

b. Check that the paint mark is now at a 180° angle to the front.

➡**Thoroughly wipe clean any seal packing.**

27. Install No. 2 cylinder head gasket.

a. Remove any old packing (FIPG) material and be careful not to drop any oil on the contact surfaces of the cylinder head or cylinder block.

➡**Remove any oil from the contact surface.**

➡**Install the cylinder head gasket within 3 minutes and tighten the bolts within 15 minutes after applying seal packing.**

➡**Do not add engine oil within 2 hours of installation.**

b. Place the cylinder head gasket on the cylinder block surface with the front face of the Lot No. stamp upward.

➡**Make sure the gasket is installed facing the proper direction.**

28. Install the cylinder head LH.

a. Place the cylinder head on the cylinder block.

➡**Gently place the cylinder head in order not to damage the gasket with the bottom part of the head.**

➡**Make sure that no oil is on the mounting surface of the cylinder head.**

➡**The cylinder head bolts are tightened in 3 progressive steps.**

b. Apply a light coat of engine oil to the threads and under the heads of the cylinder head bolts.

Step 1:

a. Using a 10 mm bi-hexagon wrench, install and uniformly tighten the 8 cylinder head bolts with plate washers in several steps in the sequence shown. Tighten to 27 ft. lbs. (36 Nm).

Step 2:

a. Mark the front side of each cylinder head bolt head with paint.

b. Tighten the cylinder head bolts another 90°.

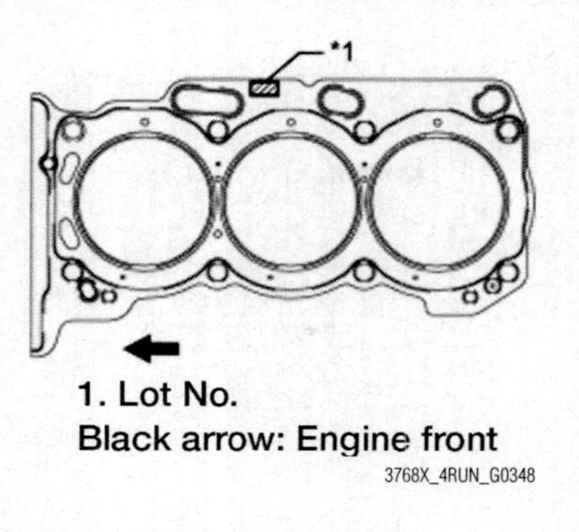

Fig. 183 Installing the gasket

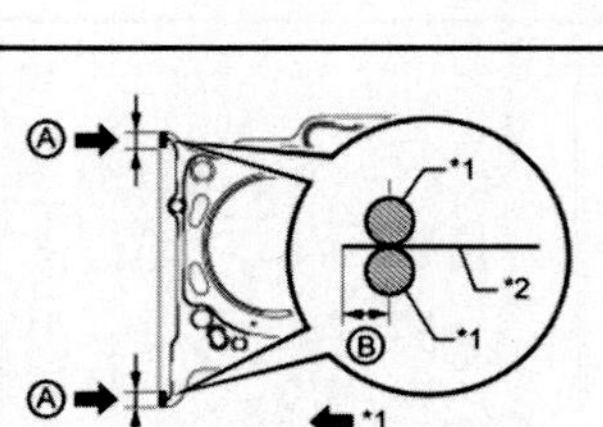

Fig. 182 Applying seal packing to a new cylinder head gasket

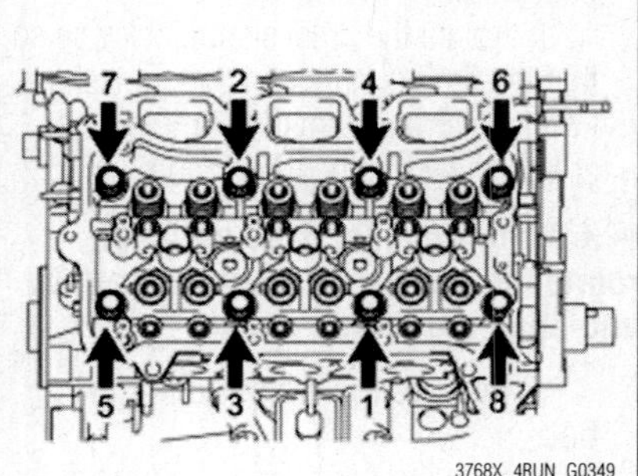

Fig. 184 Cylinder head sub assembly bolt installation sequence

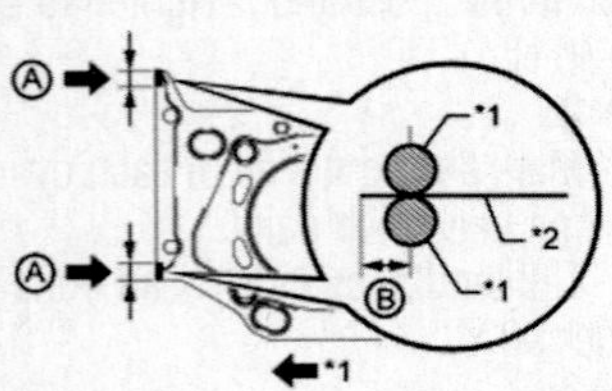

Fig. 185 Installing the No. 2 cylinder head gasket

Step 3:

29. Tighten the cylinder head bolts and additional 90°.
    a. Check that the paint mark is now at a 180°angle to the front.
    b. Tighten the 2 bolts in the order shown in the illustration. Tighten to 22 ft. lbs. (30 Nm).

➡Thoroughly wipe clean any seal packing.

30. Install the valve stem cap.
31. Install the valve lash adjuster assembly.
32. Install the No. 1 valve rocker arm sub assembly.
33. Install the camshaft bearing cap (Bank 2).
34. Install the camshaft housing sub assembly LH.
35. Install the camshaft bearing cap (Bank 1).
36. Install the camshaft housing sub assembly RH.
37. Install the No. 3 chain tensioner assembly.
38. Install the camshaft timing gears and No. 2 chain (Bank 2).
39. Install the No. 2 chain tensioner assembly.

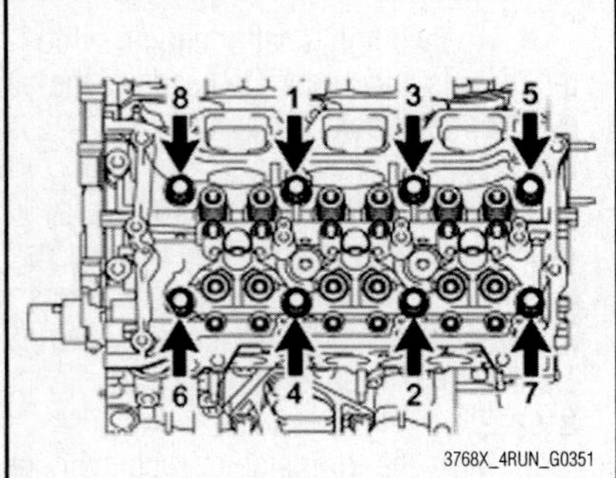

Fig. 186 LH cylinder head bolt tightening sequence

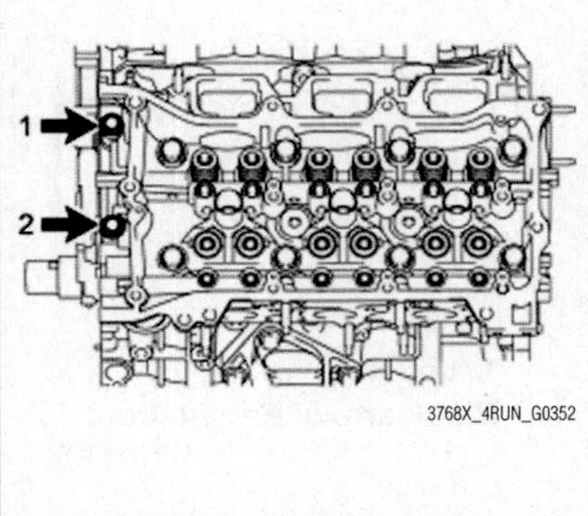

Fig. 187 Tightening the 2 remaining bolts

40. Install the camshaft timing gears and No. 2 chain (Bank 1).
41. Install the No. 1 chain vibration damper.
42. Install the No. 2 chain vibration damper.
43. Install the crankshaft timing sprocket.
44. Install No. 1 idle gear shaft.
45. Install the chain sub assembly.
46. Install the chain tensioner slipper
47. Install the No. 1 chain tensioner assembly.
48. Inspect the valve timing.
49. Install the timing chain cover sub assembly.

2.7L Engine

See Figures 188 through 202.

1. Remove the engine assembly.
2. Remove the No. 1 exhaust manifold heat insulator.
3. Remove the No. 4 intake pipe.
4. Remove the air switching valve assembly.
5. Remove the exhaust manifold.
6. Remove the timing chain cover sub assembly.
7. Remove the No. 1 cylinder to TDC/compression.
    a. Temporarily install the crankshaft pulley bolt.
    b. Rotate the crankshaft clockwise so that the timing marks on the crankshaft timing gear and camshaft timing gears are as shown in the illustration.

➡If the timing marks do not align, rotate the crankshaft clockwise again and align the timing marks.

    c. Remove the crankshaft pulley bolt.
8. Removing the timing chain guide.
9. Remove the No. 1 chain tensioner assembly.

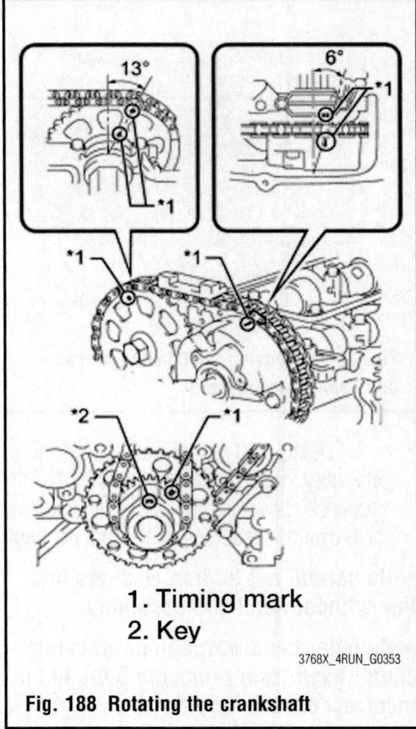

Fig. 188 Rotating the crankshaft

➡When the chain tensioner is removed, do not rotate the crankshaft.

➡When the chain is removed and the camshaft needs to be rotated, rotate the crankshaft 90°to the right.

    a. Move the stopper plate upward to release the lock and push the plunger deep into the tensioner.
    b. Move the stopper plate downward to set the lock and insert a 3.0 mm (0.118 in.) diameter bar into the stopper plate hole.
    c. Remove the bolt, nut, chain tensioner and gasket.
10. Remove the chain tensioner slipper.
11. Remove the No. 1 chain vibration damper.
12. Remove the chain sub assembly.

Fig. 189 Removing the chain tensioner

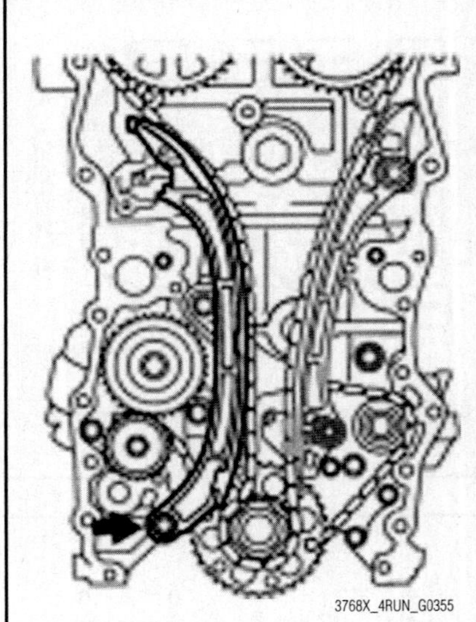

Fig. 190 Removing the chain tensioner slipper

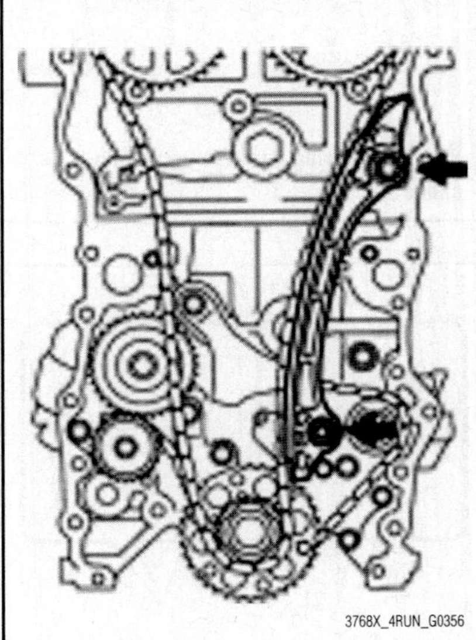

Fig. 191 Removing the No. 1 chain vibration damper

13. Remove the camshaft bearing cap.

a. Uniformly loosen and remove the 21 bearing cap bolts in the sequence shown in the illustration.

➡Uniformly loosen the bolts while keeping the camshaft level.

b. Remove the oil delivery pipe and O-ring from the bearing caps.

c. Remove the 9 bearing caps.

➡Arrange the removed parts in the correct order.

14. Remove the camshaft.
15. Remove the No. 2 camshaft.
16. Remove the No. 1 valve rocker arm sub assembly.
17. Remove the valve lash adjuster assembly.
18. Remove the valve step cap.

➡Arrange the removed parts in the correct order.

19. Remove the cylinder head sub assembly.

a. Uniformly loosen the 10 bolts in the sequence shown in the illustration. Remove the 10 cylinder head bolts and plate washers.

➡Be careful not to drop the washers into the cylinder head.

➡Head warpage or cracking could result from removing the bolts in the wrong order.

20. Remove the cylinder head gasket.
21. Inspect the cylinder head set bolt.
22. Inspect the cylinder head sub assembly.

To install:

23. Install the cylinder head gasket.

a. Place a new cylinder head gasket on the cylinder block surface with the lot No. stamp facing upward.

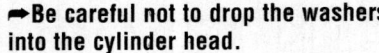

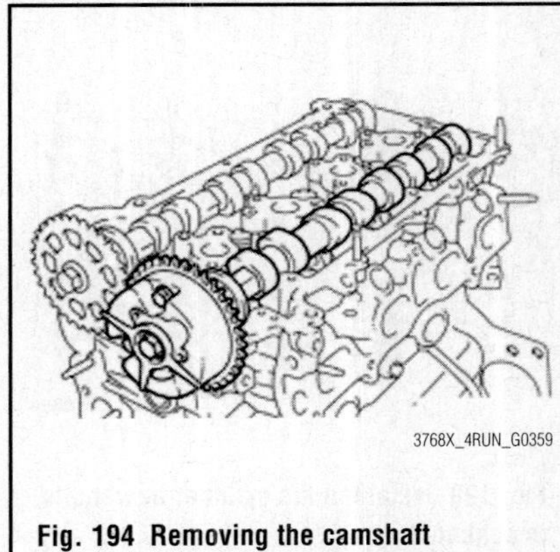

Fig. 194 Removing the camshaft

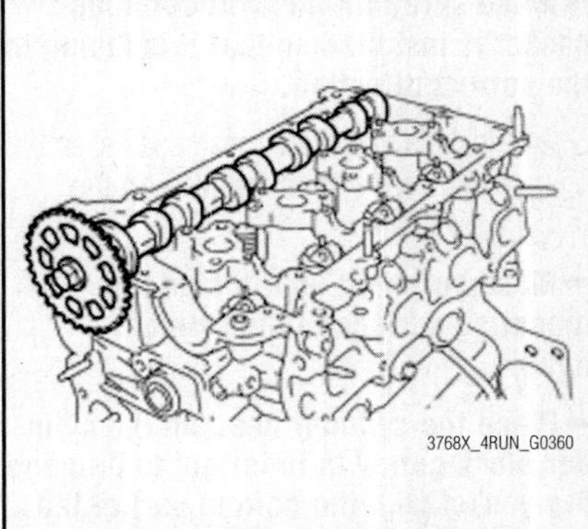

Fig. 195 Removing the No. 2 camshaft

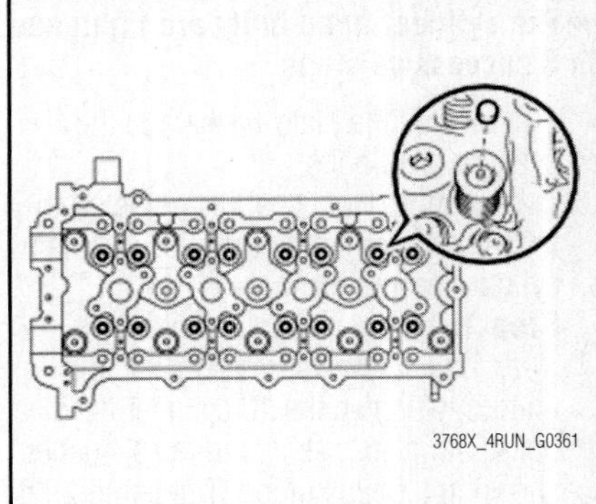

Fig. 196 Removing the valve stem caps

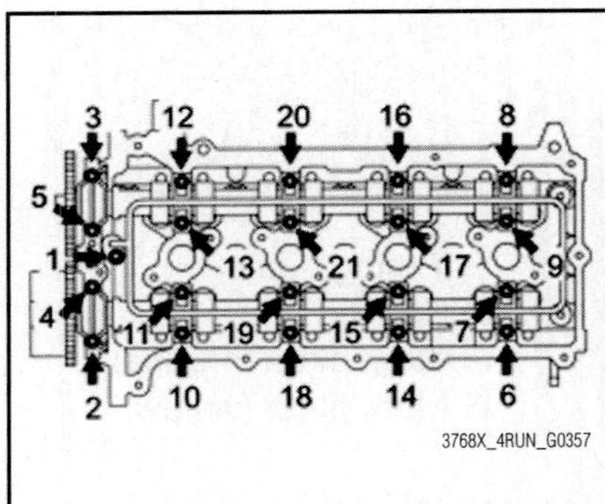

Fig. 192 Removing the camshaft bearing cap bolts in sequence

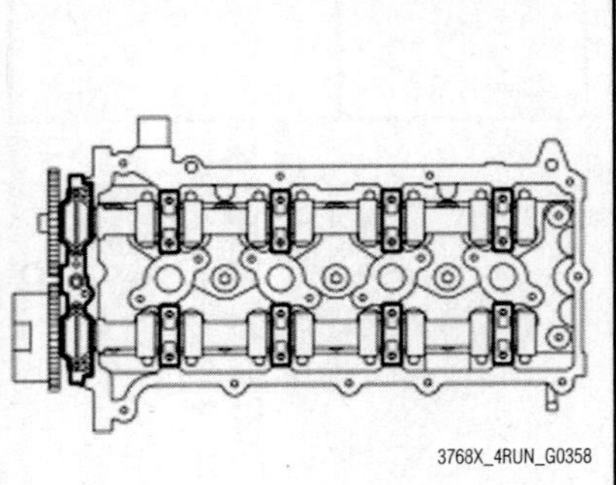

Fig. 193 Removing the bearing caps

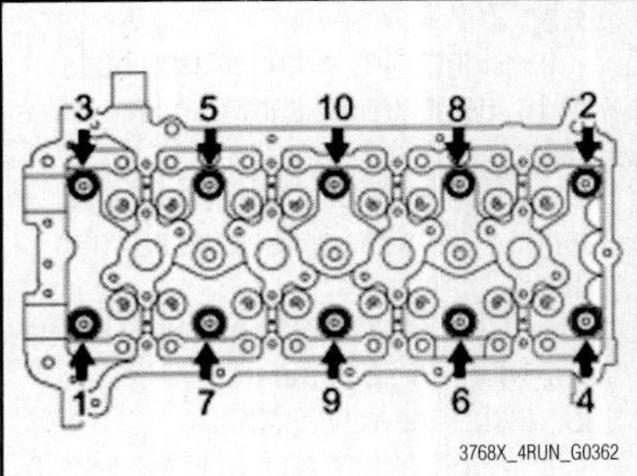

Fig. 197 Removing the cylinder head sub assembly

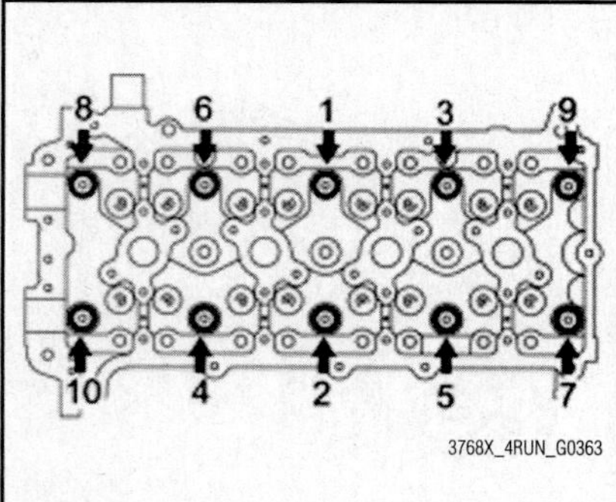

Fig. 198 Installing the cylinder head bolts in sequence

➡**Make sure that the cylinder head gasket is installed so that it is facing in the correct direction.**

24. Install the cylinder head sub assembly.
   a. Place the cylinder head on the cylinder block.

➡**Make sure that no oil is on the mounting surface of the cylinder head.**

➡**Place the cylinder head on the cylinder block gently in order not to damage the gasket with the bottom part of the head.**

   b. Install the cylinder head bolts.

➡**The cylinder head bolts are tightened in 3 successive steps.**

   c. Install the plate washers to the cylinder head bolts.
   d. Apply a light coat of engine oil to the threads and under the heads of the cylinder head bolts.

Step 1:

   a. Using several steps, install and uniformly tighten the 10 cylinder head bolts with plate washers in the sequence shown in the illustration. Tighten to 29 ft. lbs. (39 Nm).
   b. Mark the front of each cylinder head bolt head with paint.

Step 2:

   a. Tighten the cylinder head bolts 90°in the sequence shown in step 1.

Step 3:

   a. Tighten the cylinder head bolts another 90°in the sequence shown in step 1.
   b. Check that the paint marks are now at a 180c. angle to the front.

25. Install the valve step cap.
   a. Apply a light coat of engine oil to the valve stem ends.
   b. Install the 16 valve stem caps to the cylinder head.

➡**Do not drop the valve stem caps into the cylinder head.**

26. Install the valve lash adjuster assembly.
27. Install the No. 1 valve rocker arm sub assembly.
28. Install the camshaft.
   a. Apply clean engine oil to the camshaft cams and cylinder head journals.
   b. Position the camshaft and No. 2 camshaft as shown in the illustration.
29. Install the camshaft bearing cap.
   a. Temporarily install the No. 1 camshaft bearing cap.
   b. Check the proper location of each No. 2 camshaft bearing cap and install them.
   c. Install a new O-ring to the No. 1 camshaft bearing cap.
   d. Temporarily install the oil delivery pipe.
   e. Install the 21 bolts and tighten them in the order shown. Tighten bolt A to 9 ft. lbs. (12 Nm). Tighten all the remaining bolts to 11 ft. lbs. (16 Nm).
30. Install the No. 1 chain vibration damper.
   a. Install the vibration damper with the bolt and nut. Tighten the bolt to 15 ft. lbs. (21 Nm). Tighten the nut to 13 ft. lbs. (18 Nm).
31. Install the chain sub assembly.
   a. As shown in the illustration, install the chain to the sprocket and gear with the mark plates aligned with the timing marks on the sprocket and gear.

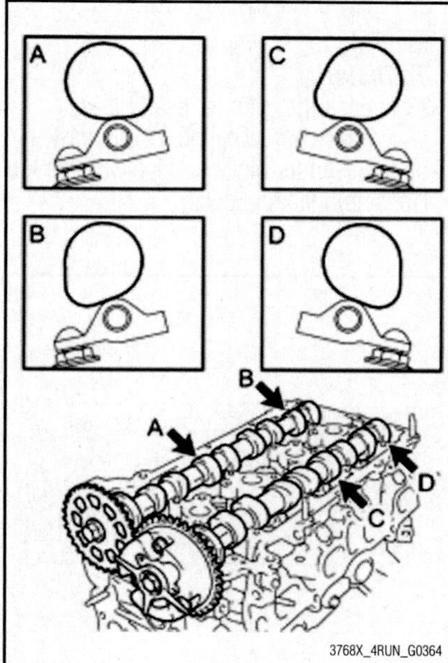

Fig. 199 Installing the camshaft

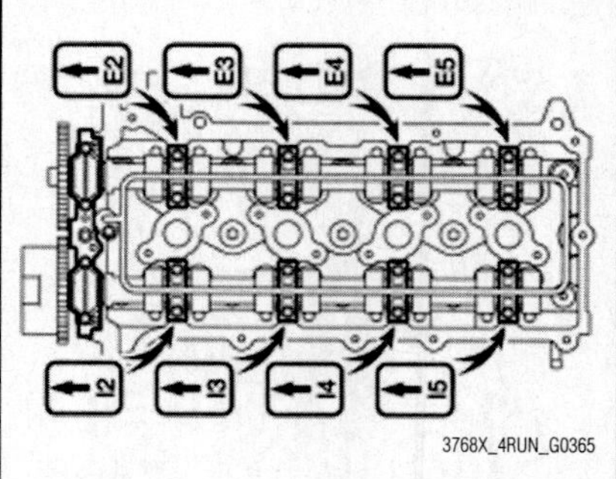

Fig. 200 Installing the camshaft bearing caps

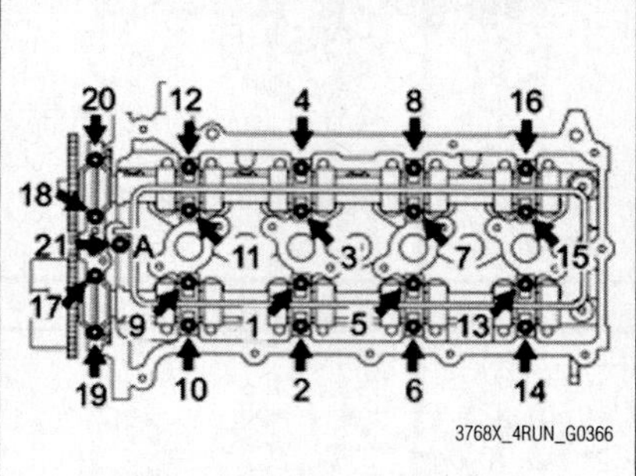

Fig. 201 Bearing cap bolt tightening sequence

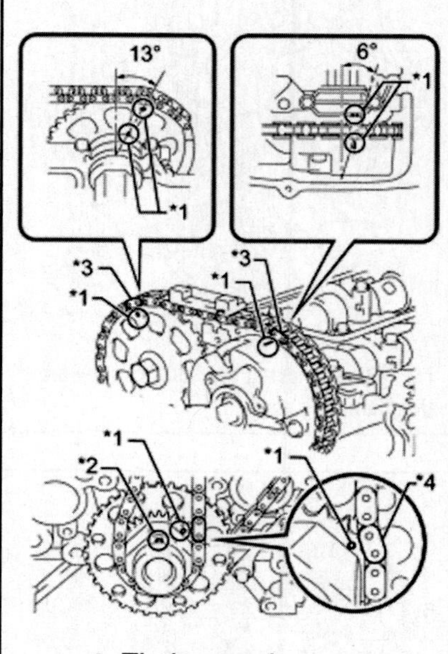

Fig. 202 Installing the chain to the sprocket and gear

➡The camshaft mark plate is orange.

➡The crankshaft mark plate is yellow.

b. Use a rope to secure the chain of the crankshaft timing sprocket. Tie the rope near the sprocket.

➡After the chain tensioner has been installed, the rope must be removed.

➡The rope is used to prevent the chain from jumping a tooth.

32. Install the chain tensioner slipper with the bolt. Tighten to 15 ft. lbs. (21 Nm).
33. Install the No. 1 chain tensioner assembly.
    a. Move the stopper plate upward to release the lock and push the plunger deep into the tensioner.
    b. Move the stopper plate downward to set the lock and insert a hexagon wrench into the hole of the stopper plate.
    c. Install a new gasket and the chain tensioner with the bolt and nut. Tighten to 7 ft. lbs. (10 Nm).
34. Install the timing chain guide.
35. Install the timing chain sub assembly.
36. Install the exhaust manifold.
37. Install the air switching valve assembly.
38. Install the No. 4 intake pipe.
39. Install the No. 1 exhaust manifold heat insulator.
40. Install the engine assembly.

EXHAUST MANIFOLD

REMOVAL & INSTALLATION

2009 Models

1. Before servicing the vehicle, refer to the precautions section.
2. Attach a hoist to the engine lifting eyes.
3. Remove or disconnect the following:
    - Negative battery cable
    - Heated Oxygen (HO_2S) sensor connectors
    - Exhaust manifold heat shield
    - Exhaust front pipe
    - Motor mount
    - Motor mount bracket
    - Exhaust manifold

To install:

➡Use new exhaust manifold nuts for assembly.

4. Install or connect the following:
    - Exhaust manifold. Tighten the nuts to 32 ft. lbs. (44 Nm).
    - Motor mount bracket. Tighten the bolts to 27 ft. lbs. (36 Nm).
    - Motor mount. Tighten the fasteners to 22 ft. lbs. (30 Nm).
    - Exhaust front pipe. Tighten the nuts to 46 ft. lbs. (62 Nm).
    - Exhaust manifold heat shield
    - HO_2S sensor connectors
    - Negative battery cable
5. Start the engine and check for leaks.

2010 Models

4.0L Engine

See Figure 203.

1. Remove the upper radiator support seal.
2. Remove the v-bank cover.
3. Remove the air cleaner cap and hose.
4. Remove the air cleaner case sub assembly.
5. Remove the front fender apron seal LH.
6. Remove the front fender apron seal RH.
7. Remove the front No. 1 fender apron to frame seal LH.
8. Remove the front No. 1 fender apron to frame seal RH.
9. Remove the front exhaust pipe assembly.
10. Remove the manifold stay.
    a. Remove the 3 bolts and manifold stay.
11. Remove the No. 1 exhaust manifold heat insulator.
12. Remove the exhaust manifold sub assembly RH.
    a. Disconnect the air fuel ratio sensor connector.
    b. Remove the 6 nuts, manifold and gasket.
13. Remove the air fuel ratio sensor (Bank 1, Sensor 1).
14. Remove the No. 2 manifold stay.
    a. Remove the 3 bolts and the No. 2 manifold stay.
15. Remove the No. 1 exhaust manifold heat insulator.
    a. Remove the 3 bolts and heat insulator.
16. Remove the exhaust manifold sub assembly LH.
    a. Disconnect the air fuel ratio sensor connector.
    b. Remove the 6 nuts, manifold and gasket.
17. Remove the air fuel ratio sensor (Bank 2, Sensor 1).

To install:

18. Install the air fuel ratio sensor (Bank 2, Sensor 1).
19. Install the exhaust manifold sub assembly LH.

➡Be careful of the installation direction.

    a. Temporarily install the manifold with 6 new nuts.
    b. Tighten the 6 nuts in the sequence shown in the illustration. Tighten to 15 ft. lbs. (21 Nm).
    c. Connect the air fuel ratio sensor connector.
20. Install the No. 2 exhaust manifold heat in the illustration. Tighten to 15 ft. lbs. (21 Nm).
    a. Connect the air fuel ratio sensor connector.
21. Install the No. 2 exhaust manifold heat insulator with the 3 bolts. Tighten to 10 ft. lbs. (13 Nm).
22. Install the manifold stay with the 3 bolts. Tighten to 30 ft. lbs. (40 Nm).
23. Install the air fuel ratio sensor (Bank 1, Sensor 1).
24. Install the exhaust manifold sub assembly RH.
    a. Install a new gasket onto the cylinder head.

➡Be careful of the installation direction.

    b. Temporarily install the manifold with 6 new nuts.
    c. Tighten the 6 nuts in the sequence shown.
    d. Connect the air fuel ratio sensor connector.
25. Install the No. 1 exhaust manifold heat insulator with the 3 bolts. Tighten to 10 ft. lbs. (13 Nm).
26. Install the No. 2 manifold stay with the 3 bolts. Tighten to 30 ft. lbs. (40 Nm).
27. Install the front exhaust pipe assembly.

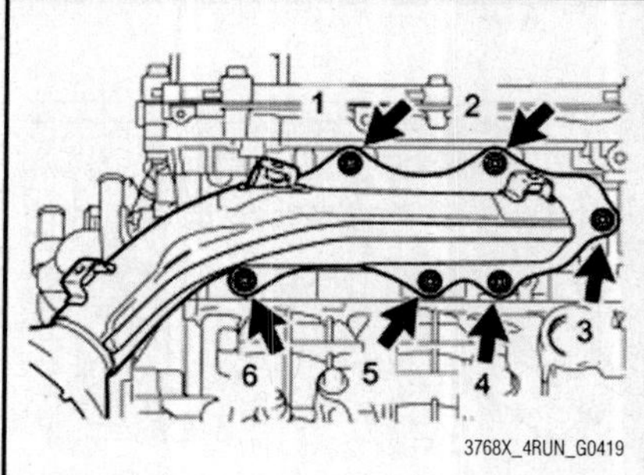

Fig. 203 Identifying the exhaust manifold sub assembly RH nut tightening sequence

28. Install the air cleaner case sub assembly.
29. install the air cleaner cap and hose.
30. Install the v-bank cover.
31. Install the radiator support seal upper.
32. Inspect for exhaust gas leaks.
33. Install the front fender apron seal LH.
34. Install the front fender apron seal RH.
35. Install the front No. 1 fender apron to frame seal LH.
36. Install the front No. 1 fender apron to frame seal RH.

2.7L Engine

See Figure 204.

1. Remove the front fender apron seal RH.
2. Remove the No. 1 fender apron to frame seal RH.
3. Remove the air cleaner and hose.
4. Remove the air cleaner case.
5. Disconnect the No. 1 air injection system hose.
6. Remove the No. 1 exhaust manifold heat insulator.
7. Remove the No. 4 intake pipe.
8. Remove the air switching valve assembly.
9. Remove the front exhaust pipe assembly.
   a. Disconnect the air fuel ratio sensor connector and detach the wire harness clamp.
   b. Disconnect the heated oxygen sensor connector.
   c. Remove the 4 bolts and 4 compression springs.
   d. Remove the front exhaust pipe from the pipe support.
   e. Remove the gasket.
10. Remove the manifold stay.
11. Remove the exhaust manifold.
   a. Remove the 8 nuts and exhaust manifold.
   b. Remove the 2 gaskets.
12. Installation is the reverse of the removal procedure.

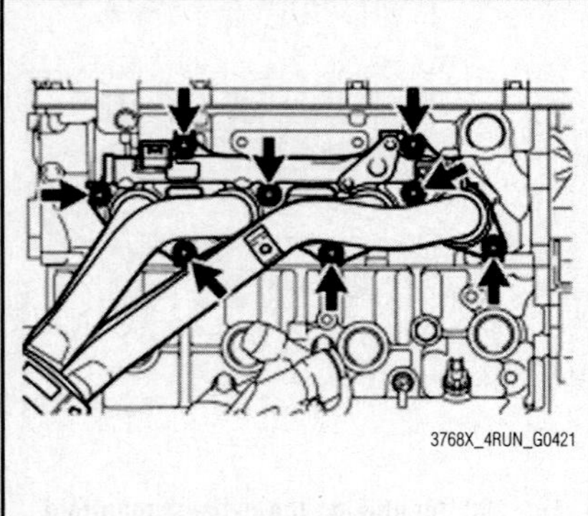

Fig. 204 Removing the exhaust manifold

INTAKE MANIFOLD

REMOVAL & INSTALLATION

2009 Models

4.0L Engine

See Figure 205.

1. Discharge fuel system pressure.
2. Drain engine coolant.
3. Drain engine oil.
4. Remove the 2 bolts and cool air inlet with air cleaner hose.
5. Remove the ATF level gauge.
6. Remove the bolt and pull out the oil filler tube.
7. Remove the O-ring from the oil filler tube.
8. Disconnect the A/F sensor connector.
9. Disconnect No. 1 and No. 2 fuel pipe sub-assembly.
10. Disconnect the 6 fuel injector connectors.
11. Remove the 10 bolts, intake manifold and 2 gaskets.

To install:

12. Set a new gasket on each cylinder head.
13. Set the intake manifold on the cylinder heads.
14. Install and tighten the 10 bolts uniformly in several steps and tighten to 19 ft. lbs. (26 Nm).
15. Connect the 6 fuel injector connectors.
16. Connect No. 2 fuel pipe sub-assembly.
17. Connect No. 1 fuel pipe sub-assembly.
18. Coat a new O-ring with ATF, and install it to the oil filler tube.
19. Push in the oil filler tube end into the oil pan lower tube.
20. Install the oil filler tube with the bolt.

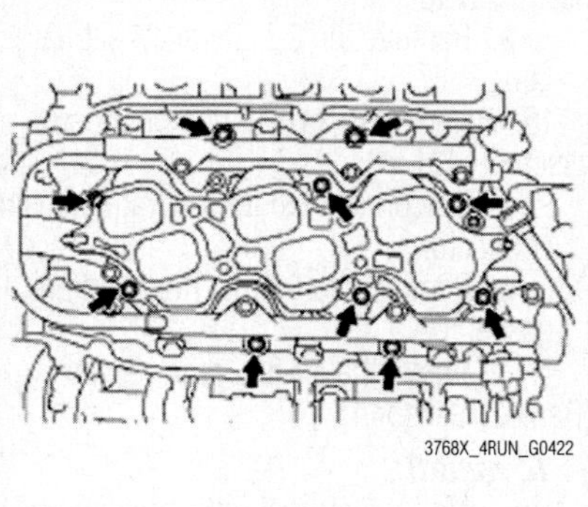

Fig. 205 Removing the intake manifold

21. Install the ATF level gauge.
22. Install the cool air inlet with air cleaner hose with the 2 bolts.
23. Add engine oil.
24. Add engine coolant.
25. Check for leakage.

4.7L Engine

1. Discharge fuel system pressure.
2. Drain engine coolant.
3. Remove the 2 nuts and throttle body cover sub-assembly.
4. Disconnect the vacuum hoses (for the power steering idle-up and fuel pressure regulator) and ventilation hose.
5. Remove the air cleaner hose assembly.
6. Disconnect fuel hose.
7. Disconnect fuel hose No.2.
8. Disconnect the throttle control connector.
9. Disconnect the purge VSV.
10. Disconnect the 8 injector connectors.
11. Disconnect the ECT sensor connector.
12. Disconnect the 2 VSV connectors for the air injection system.
13. Disconnect the 8 ignition coil connectors.
14. Disconnect the 2 air fuel ratio sensor connectors.
15. Disconnect the vacuum hose from the fuel pressure regulator.
16. Disconnect the PCV hoses from the PCV valve on the LH cylinder head.
17. Disconnect the EVAP hose (from the charcoal canister) from the purge VSV.
18. Disconnect the 2 vacuum hoses from the VSV for the air injection system.
19. Disconnect the 2 water by-pass hoses from the throttle body.
20. Disconnect the 2 wire clamps from the wire clamp bracket on the RH delivery pipe.
21. Remove the bolt and nut holding the engine wire protector from the intake manifold and cylinder head.
22. Remove the 2 bolts and ground cables from the RH and LH cylinder heads.
23. Remove the bolt and V-bank cover bracket from the intake manifold.
24. Disconnect the engine wire from the engine hanger and wire bracket.
25. Remove the bolt and wire bracket from the intake manifold.
26. Remove the 6 bolts, 4 nuts, intake manifold assembly and 2 gaskets.
27. Remove air pump assembly w/ bracket.
28. Remove the 2 nuts and 2 knock sensors.

To install:

29. Install the 2 knock sensors with the 2 nuts and tighten to 15 ft. lbs. (20 Nm).
30. Connect the 2 knock sensor connectors.
31. Place 2 new gaskets on the intake manifold.
32. Place the intake manifold on the cylinder heads.
33. Install and uniformly tighten the 6 bolts and 4 nuts in several steps and tighten to 13 ft .lbs. (18 Nm).
34. Install the V-bank cover bracket to the intake manifold.
35. Install the wire bracket to the intake manifold with the bolt.
36. Connect the engine wire to the engine hanger and wire bracket.
37. Connect the wire protector to the intake manifold and cylinder heads with the bolt and nut.
38. Install the 2 ground cables with the 2 bolts to the RH and LH cylinder heads.
39. Connect the 2 water by-pass hoses to the throttle body.
40. Connect the 2 wire clamps to the wire clamp bracket on the RH delivery pipe.
41. Connect the vacuum hose to the fuel pressure regulator.
42. Connect the PCV hose to the PCV valve on the LH cylinder head.
43. Connect the EVAP hose (from the charcoal canister) to the purge VSV.
44. Connect the 2 vacuum hoses to the VSV for the air injection system.
45. Connect the throttle control connector.
46. Connect the 2 VSV connectors for the air injection system.
47. Connect the purge VSV connector.
48. Connect the 8 injector connectors.
49. Connect the ECT sensor connector.
50. Connect the 8 ignition coil connectors.
51. Connect the 2 air fuel ratio sensor connectors.
52. Install fuel hose No.2.
53. Install fuel hose.
54. Install throttle body cover sub-assembly.
55. Add engine coolant.
56. Check for leaks.

2010 Models

4.0L Engine

See Figures 206 through 209.

1. Discharge the fuel system pressure.
2. Disconnect the cable from the negative battery terminal.

➡When disconnecting the cable, some systems need to be initialized after the cable is reconnected.

3. Remove the front bumper cover lower.
4. Remove the No. 1 engine under cover sub assembly.
5. Drain the engine coolant.
6. Remove the v-bank cover.
   a. Raise the front of the v-bank cover to detach the 2 pins. Then remove the 2 v-bank cover hooks from the bracket, then remove the v-bank cover.
7. Remove the No. 1 air cleaner hose.
   a. Disconnect the ventilation hose and vacuum hose.
   b. Detach the wire harness clamp.
   c. Remove the bolt and loosen the 2 hose clamps.
   d. remove the No. 1 air cleaner hose.
8. Remove the intake air surge tank.
   a. Disconnect the throttle body connector.
   b. Disconnect the No. 4 water by pass hose.
   c. Disconnect the No. 5 water by pass hose.
   d. Disconnect the No. 1 fuel vapor feed hose.
   e. Disconnect the No. 1 vacuum switching valve connector.
   f. Disconnect the No. 1 ventilation hose.
   g. Detach the 2 heater hose clamps.
   h. Remove the 2 bolts and throttle body bracket.
   i. Using a clip remover, detach the wire harness clamp.
   j. Remove the 2 bolts and No. 1 surge tank stay.
   k. Remove the 2 bolts and No. 2 surge tank stay.
   l. Remove the 2 nuts, 4 bolts and intake air surge tank.
   m. Remove the gasket.
9. Remove the fuel delivery pipe sub assembly.

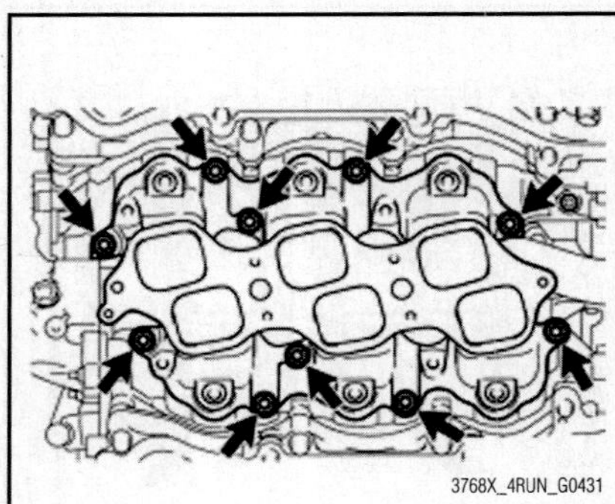

Fig. 206 Removing the intake manifold

10. Remove the intake manifold.
    a. Remove the 4 nuts, 6 bolts and 2 gaskets.

To install:

11. Install the intake manifold.
    a. Set a new gasket on each cylinder head.

➡Align the port holes of the gasket and cylinder head.

➡Be careful of the installation direction.

   b. Set the intake manifold on the cylinder heads.
   c. Install and uniformly tighten the 6 bolts and 4 nuts in several passes. Tighten to 15 ft. lbs. (21 Nm).

➡Tighten the inner installation bolts of the intake manifold before tightening the outer bolts.

12. Install the fuel delivery pipe sub assembly.

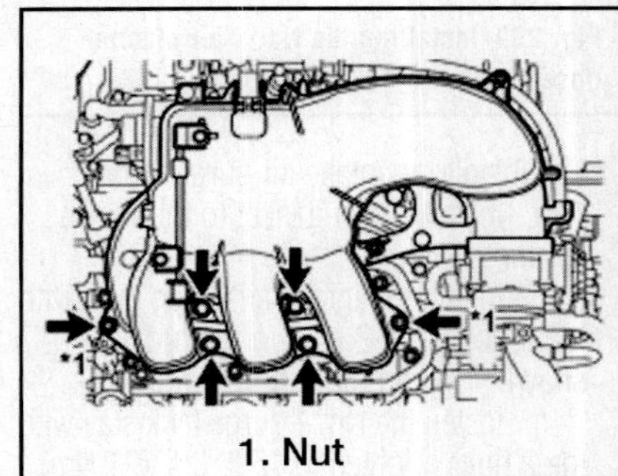

Fig. 207 Identifying the intake air surge tank nut and bolt installation sequence

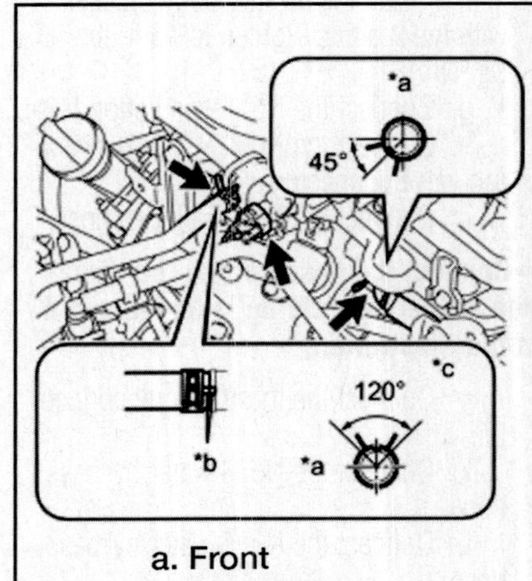

Fig. 208 Connecting the hoses

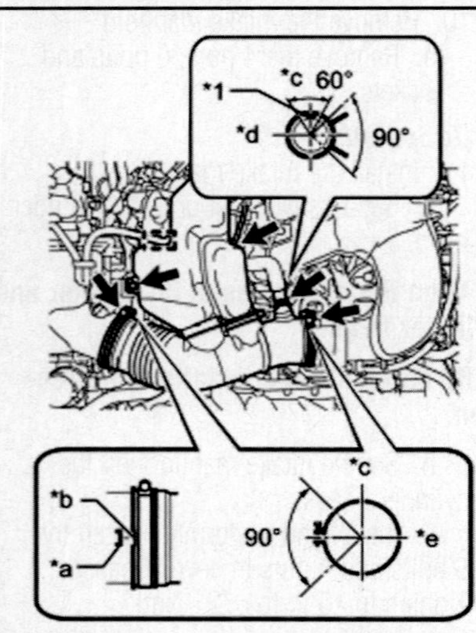

Fig. 209 Installing the No. 1 air cleaner hose

13. Install the intake air surge tank.
   a. Install a new gasket to the intake air surge tank.
   b. Install the intake air surge tank with the 4 bolts and 2 nuts in the order shown. Tighten to 21 ft. lbs. (28 Nm).
   c. Install the No. 1 surge tank stay with the 2 bolts. Tighten to 15 ft. lbs. (21 Nm).
   d. Attach the wire harness clamp.
   e. Install the No. 2 surge tank stay with the 2 bolts. Tighten to 15 ft. lbs. (21 Nm).
   f. Install the throttle body bracket with the 2 bolts. Tighten to 15 ft. lbs. (21 Nm).
   g. Connect the No. 1 ventilation hose.
   h. Connect the No. 1 vacuum switching valve connector.
   i. Connect the No.1 fuel feed hose.

➡Connect the hose so that the direction of the hose clamp is as indicated in the illustration.

   j. Connect the throttle body connector.
   k. Connect the No. 4 water by-pass hose.
   l. Connect the No. 5 water by-pass hose.

➡Connect the hose so that the direction of the hose clamp is as indicated in the illustration.

   m. Connect the 2 heater hose clamps.

14. Install the No. 1 air cleaner hose with the 2 clamps. Tighten to 44 inch lbs. (5 Nm).
   a. Install the bolt. Tighten to 44 inch lbs. (5 Nm).
   b. Connect the vacuum hose and ventilation hose.

➡The direction of the hose clamp is indicated in the illustration.

15. Connect the cable to the negative battery terminal.

➡When disconnecting the cable, some systems need to be initialized after the cable is reconnected.

16. Add engine coolant.
17. Inspect for coolant leaks.
18. Inspect for fuel leaks.
19. Install the v-bank cover.
   a. Attach the 2 V-bank cover hooks to the bracket. Then align the 2 V-bank cover grommets with the 2 pins and press down on the V-bank cover to attach the pins.
20. Install the No. 1 engine under cover sub assembly.
21. Install the front bumper cover lower.

2.7L Engine

See Figure 210.

1. Discharge the fuel system pressure.
2. Disconnect the cable from the negative battery terminal.

➡When disconnecting the cable, some systems need to be initialized after the cable is reconnected.

3. Remove the throttle body with the motor assembly.
4. Remove the fuel delivery pipe with the fuel injectors.
5. Remove the purge VSV.
6. Remove the rear engine under cover assembly.

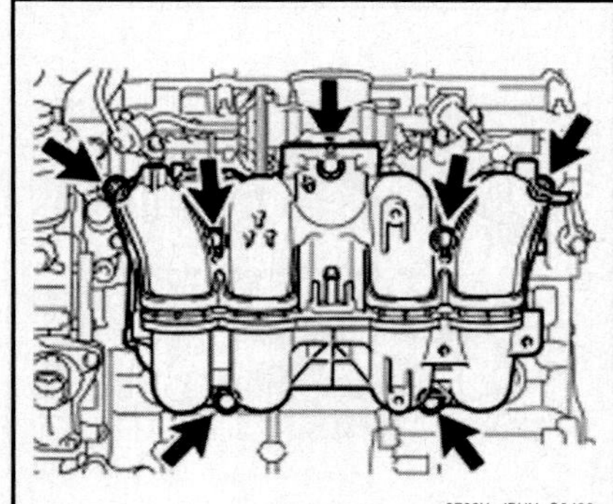

Fig. 210 Removing the intake manifold

7. Remove the transmission oil filter tube sub assembly.
8. Remove the starter assembly.
9. Remove the 5 clips and the front fender apron seal LH.
10. Remove the 5 clips and the front No. 1 fender apron to frame seal LH.
11. Remove the intake manifold.
   a. For the engine front side, detach the 2 wire harness clamps from the 2 wire harness clamp brackets.
   b. For the engine rear side, detach the No. 2 water by pass hose and disconnect the No. 3 ventilation hose from the intake manifold.
   c. Remove the 5 bolts, 2 nuts and intake manifold.
   d. Remove the gasket from the intake manifold.
   e. Remove the 2 bolts, 2 wire harness clamp brackets and purge line hose from the intake manifold.

To install:

12. Install the intake manifold.
   a. Install the purge line hose to the intake manifold.
   b. Install the 2 wire harness clamp brackets to the intake manifold with the 2 bolts. Tighten to 71 inch lbs. (8 Nm).
   c. Install a new gasket to the intake manifold.
   d. Install the intake manifold with the 5 bolts and 2 nuts. Tighten to 18 ft. lbs. (25 Nm).
   e. For the engine rear side, attach the No. 2 water by pass hose and connect the No. 3 ventilation hose to the intake manifold.
   f. For the engine front side, attach the 2 wire harness clamps to the 2 wire harness clamp brackets.
13. Install the purge VSV.
14. Install the front No. 1 fender apron to the frame seal LH with the 5 clips.
15. Install the front fender apron seal LH with the 5 clips.
16. Install the starter assembly.
17. Install the transmission oil filler tube sub assembly.
18. Install the rear engine under cover assembly.
19. Install the fuel delivery pipe with the fuel injectors.
20. Install the throttle body with the motor assembly.
21. Connect the cable to the negative battery terminal.

➡When disconnecting the cable, some systems need to be initialized after the cable is reconnected.

OIL PAN

REMOVAL & INSTALLATION

2009 Models

4.0L Engine

See Figure 211.

1. Before servicing the vehicle, refer to the precautions section.
2. Remove the engine from the vehicle and mount it on a stand.
3. Remove or disconnect the following:
   - Remove the drain plug and gasket
   - Remove the 10 bolts and 2 nuts
   - Insert the blade of oil pan seal cutter between the oil pan and No. 2 oil pan, and cut off applied sealer and remove the No. 2 oil pan

➡Be careful not to damage the contact surfaces of the oil pan and No. 2 oil pan. Be careful not to damage the No. 2 oil pan flange.

   - Remove the 2 nuts, oil strainer and gasket
   - Remove the 17 bolts and 2 nuts
   - Remove the 4 stud bolts
   - Using a screwdriver, remove the oil pan by prying between the oil pan and cylinder block

To install:

4. Remove any old packing (FIPG) material and be careful not to drop any oil on the contact surfaces of the cylinder block, rear oil seal retainer and oil pan.
5. Install the 4 stud bolts and tighten to 35 ft. lbs. (4.0 Nm).
6. Install a new O-ring to the oil pump.
7. Apply a continuous bead of seal packing (diameter: 3 to 4 mm (0.12 to 0.16 in.)) to the oil pan.

➡Install the oil pan within 3 minutes after applying seal packing. After installing it, the oil pan bolts and nuts must be tightened within 15 minutes. Otherwise the seal packing must be removed and reapplied.

8. Install the oil pan with the 17 bolts and 2 nuts. Tighten the bolts and nuts uniformly in several steps.
9. Tighten bolts as follows:
   - Bolt A 15 ft. lbs. (21 Nm)
   - Bolt B 15 ft. lbs. (21 Nm)
   - Bolt C 7 ft. lbs. (10 Nm)
   - Nut 15 ft. lbs. (21 Nm)
10. Install a new gasket and the oil strainer with the 2 nuts and tighten to 80 inch lbs. (9.0 Nm).
11. Remove any old packing (FIPG) material and be careful not to drop any oil on the contact surfaces of the oil pan and No. 2 oil pan.

➡Install the No. 2 oil pan within 3 minutes after applying seal packing. After installing it, the No. 2 oil pan bolts and nuts must be tightened within 15 minutes. Otherwise the seal packing must be removed and reapplied.

12. Install the No. 2 oil pan with the 10 bolts and 2 nuts. Tighten the bolts and nuts uniformly in several steps. Tighten bolts to 80 inch lbs. (9.0 Nm) and nuts to 7 ft. lbs. (10 Nm).
13. Install the drain plug and a new gasket.
14. Reinstall engine.
15. Start engine and check for leaks.

4.7L Engine

See Figures 212 through 214.

1. Before servicing the vehicle, refer to the precautions section.
2. Remove the engine from the vehicle and mount it on a stand.
3. Remove or disconnect the following:
   - Oil dipstick tube
   - Lower oil pan
   - Oil pan baffle
   - Upper oil pan

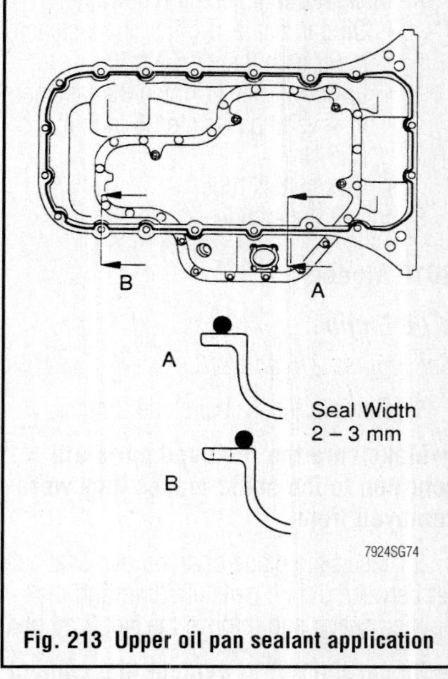

Fig. 213 Upper oil pan sealant application

To install:

4. The upper oil pan bolts are different lengths and are identified as follows:
   - A: 0.79 inch (20mm) w/10mm head
   - B: 0.98 inch (25mm) w/12mm head
   - C: 2.36 inch (60mm) w/12mm head
   - D: 1.38 inch (35mm) w/10mm head
5. Apply silicone sealant to the upper oil pan as shown.
6. Install the upper oil pan and tighten the fasteners in several passes to the following specifications:
   - 10mm: 66 inch lbs. (7.5 Nm)
   - 12mm: 21 ft. lbs. (28 Nm)

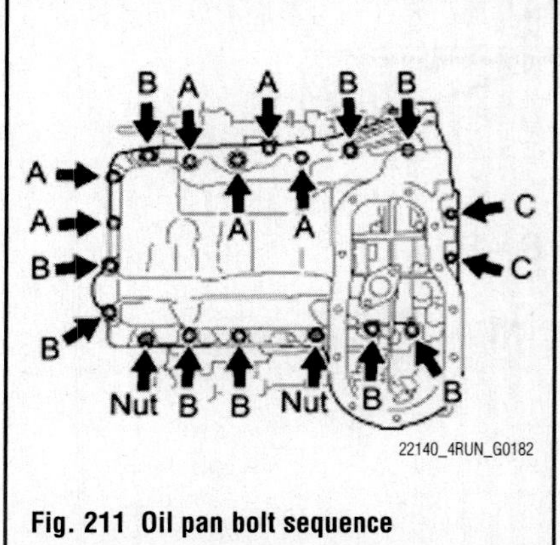

Fig. 211 Oil pan bolt sequence

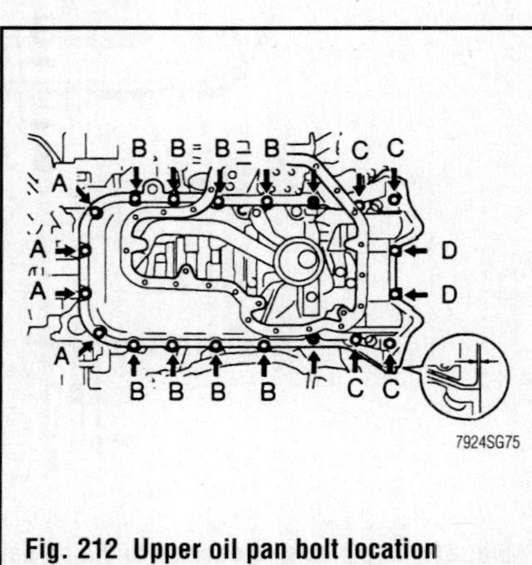

Fig. 212 Upper oil pan bolt location

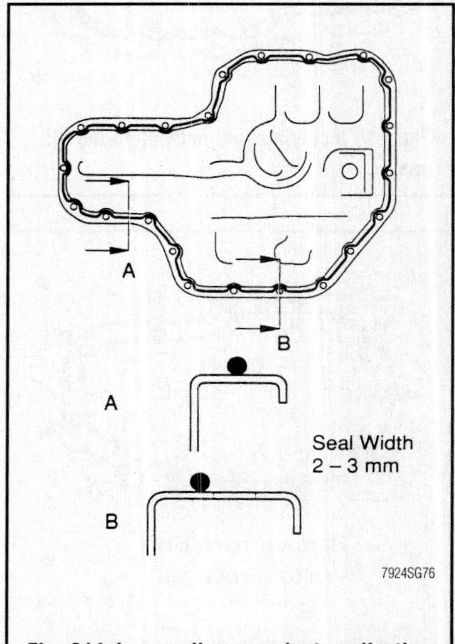

Fig. 214 Lower oil pan sealant application

7. Install or connect the following:
   - Oil pan baffle. Tighten the fasteners to 66 inch lbs. (7.5 Nm).
   - Lower oil pan. Tighten the fasteners in several passes to 66 inch lbs. (7.5 Nm).
   - Oil dipstick tube
8. Install the engine.

2010 Models

4.0L Engine

See Figures 215 and 216.

1. Remove the 10 bolts and 2 nuts.

➡Make sure the removed parts are returned to the same places they were removed from.

2. Insert the blade of an oil pan seal cutter between the oil pans. Cut through the applied sealer and remove the No. 2 oil pan.

➡Be careful not to damage the contact surfaces of the oil pans.

To install:

3. Apply seal packing (* 1) in a continuous line.

➡Remove any oil from the contact surface.

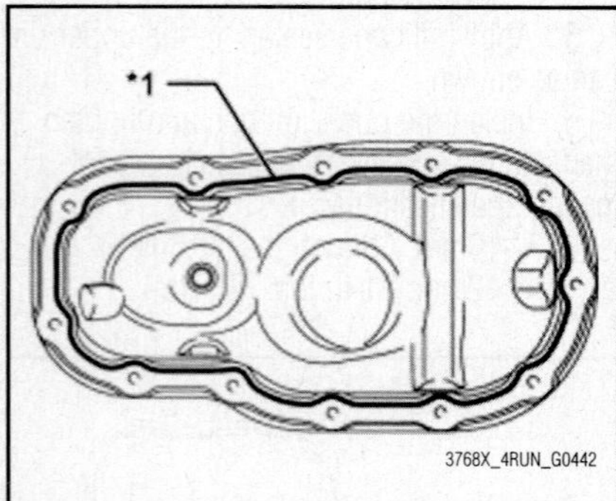

Fig. 215 Applying seal packing to the oil pan

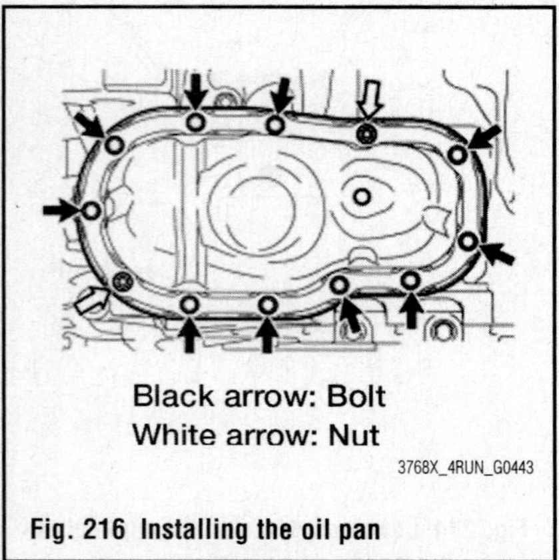

Fig. 216 Installing the oil pan

➡Install the No. 2 oil pan within 3 minutes after applying seal packing.

a. Install the No. 2 oil pan with the 10 bolts and 2 nuts. Tighten the bolts and nuts uniformly in several steps. Tighten to 7 ft. lbs. (10 Nm).

➡Tighten the nuts first. After tightening the bolts, check that the nuts and bolts are tightened to the specified torque.

➡Do not start the engine for at least 2 hours after the installation.

2.7L Engine

See Figures 217 and 218.

1. Remove the drain plug and gasket.
   a. Remove the 18 bolts and 2 nuts.
2. Insert the blade of an oil pan seal cutter between the oil pans. Cut through the applied sealer and remove the oil pan.

➡Be careful not to damage the contact surfaces of the oil pans.

To install:

3. Apply seal packing in a continuous line.

➡Remove any oil from the contact surface.

➡Install the oil pan within 3 minutes after applying seal packing.

➡Do not start the engine for at least 4 hours after the installation.

4. Temporarily install the oil pan with the 16 bolts and 2 nuts.

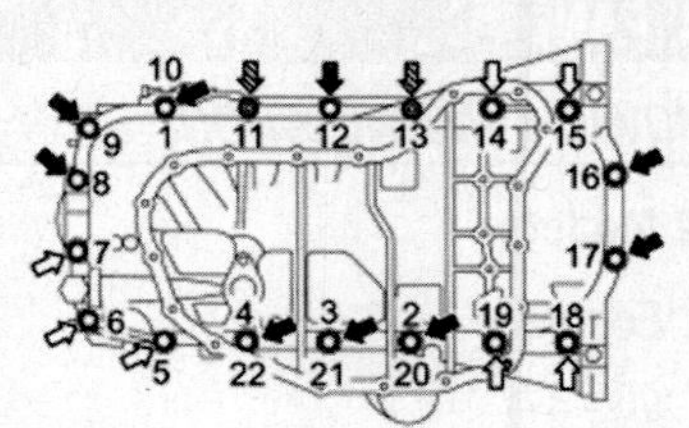

Fig. 218 Identifying the oil pan bolt and nut tightening sequence

5. Uniformly tighten the 16 bolts and 2 nuts in the order shown. Tighten to 19 ft. lbs. (26 Nm).

OIL PUMP

REMOVAL & INSTALLATION

2009 Models

4.0L Engine

See Figures 219 through 224.

1. Remove power steering link assembly.
2. Remove differential carrier assembly front.
3. Drain engine coolant.
4. Drain engine oil.
5. Remove battery.
6. Remove v-bank cover.
7. Remove the 11 clips and radiator support seal upper.

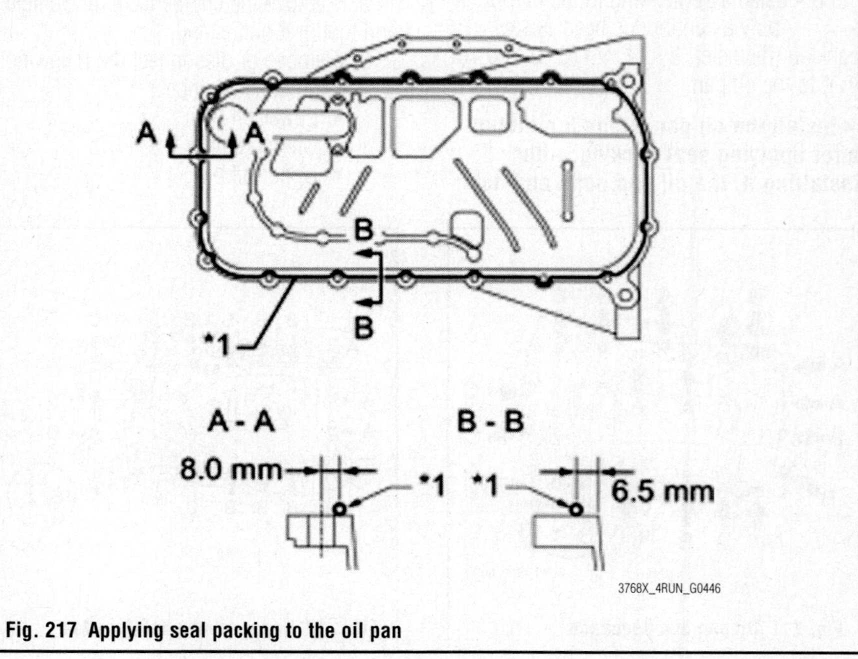

Fig. 217 Applying seal packing to the oil pan

8. Loosen fan fluid coupling assembly.
9. Remove fan and alternator v-belt.
10. Remove fluid coupling assembly.
11. Disconnect no. 2 ventilation hose.
12. Remove air cleaner assembly.
13. Remove the oil level gauge.
14. Remove the bolt and pull out the oil level gauge guide.
15. Remove the O-ring from the oil level gauge guide.
16. Remove water inlet.
17. Disconnect the P/S oil pressure switch connector.
18. Remove the 2 bolts, and separate the vane pump.

➡Do not hit the pulley to other parts when separating the vane pump.

19. Remove alternator assembly.
20. Separate cooler compressor assembly.
21. Remove the 5 bolts and V-ribbed belt tensioner.
22. Remove the 2 bolts and 2 idler pulleys.
23. Remove the bolt and idler pulley.
24. Using service tool, hold the crankshaft pulley and loosen the pulley set bolt.
25. Using the pulley set bolt and service tool, remove the crankshaft pulley.
26. Remove the 10 bolts and 2 nuts.
27. Insert the blade of oil pan seal cutter between the oil pan and No. 2 oil pan, cut off applied sealer and remove the No. 2 oil pan.

➡Be careful not to damage the contact surfaces of the oil pan and No. 2 oil pan. Be careful not to damage the No. 2 oil pan flange.

28. Remove the 2 nuts, oil strainer and gasket.
29. Remove the 4 housing bolts.
30. Remove the flywheel housing under cover.
31. Remove the 17 bolts and 2 nuts.
32. Using a screwdriver, remove the oil pan by prying between the oil pan and cylinder block.
33. Remove the O-ring from the oil pump.
34. Remove intake air surge tank.
35. Remove ignition coil assembly.
36. Remove cylinder head cover sub-assemblies.
37. Disconnect the 2 oil control valve connectors.
38. Remove the 2 bolts and 2 camshaft timing oil control valves.
39. Remove VVT sensor.
40. Remove timing chain or belt cover sub-assembly.
41. Remove timing gear case or timing chain case oil seal.
42. Remove oil pump cover and gears and relief valve.

To install:

43. Coat the relief valve with engine oil and insert the relief valve and spring into the valve hole.
44. Install the relief valve plug and tighten to 36 ft. lbs. (49 Nm).
45. Apply fresh engine oil to the drive and driven rotors.
46. Place the drive and driven rotors into the timing chain cover with the marks facing the oil pump cover side.

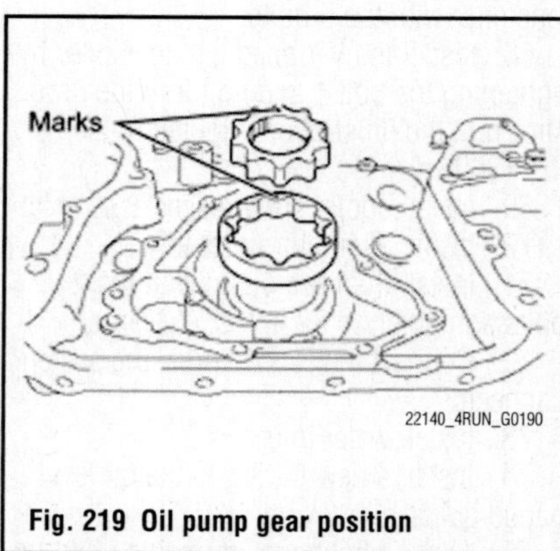

Fig. 219 Oil pump gear position

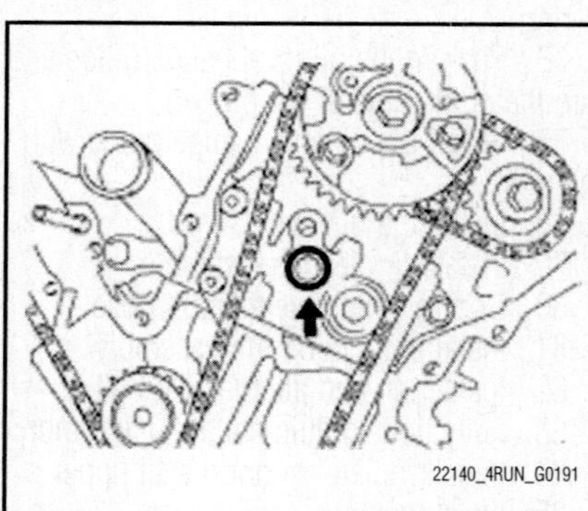

Fig. 220 Timing chain cover O-ring position

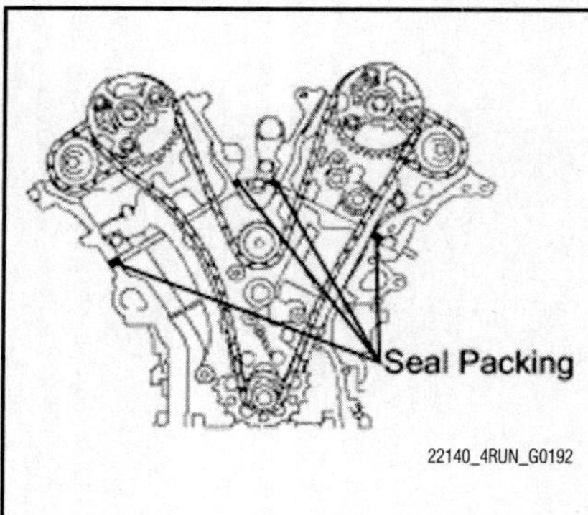

Fig. 221 Timing Chain Cover Seal Packing

47. Install the oil pump cover with the 7 bolts and tighten to 80 inch lbs. (9.0 Nm).
48. Install oil pump gears and cover.
49. Install timing gear case or timing chain case oil seal.
50. Install timing chain or belt cover sub-assembly.
    a. Install a new O-ring onto the LH cylinder head.
    b. Apply continuous beads of seal packing (diameter 3 to 4 mm (0.12 to 0.16 in.)) to the 4 locations as shown in the illustration.
    c. Apply continuous beads of seal packing (diameter 3 to 4 mm (0.12 to 0.16 in.)) to the timing chain cover as shown in the illustration.

➡Install the timing chain cover within 3 minutes of applying the seal packing. Timing chain cover bolts and nuts must be tightened within 15 minutes of installation. Otherwise, the seal packing must be removed and reapplied. Do not apply seal packing to portion A shown in the illustration.

    d. Align the key way of the oil pump drive rotor with the rectangular portion of the crankshaft timing gear, and slide the timing chain cover into place.
    e. Install the timing chain cover with the 24 bolts and 2 nuts. Tighten the bolts and nuts uniformly in several steps. Tighten to 17 ft. lbs. (23 Nm).

➡Pay attention not to wrap the chain and slipper over the timing chain cover seal line.

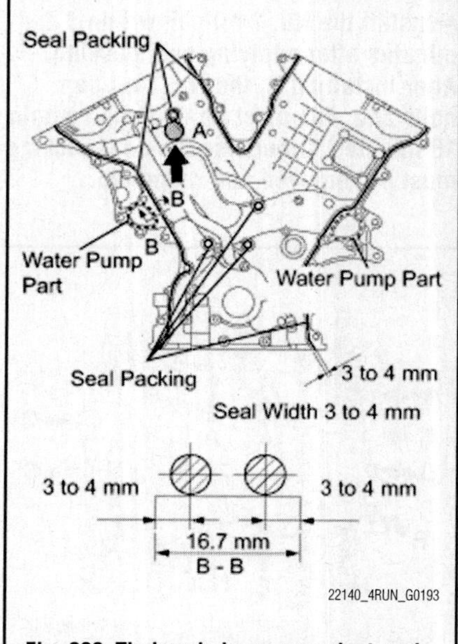

Fig. 222 Timing chain cover sealant position

51. Install VVT sensor.

52. Insert the camshaft timing oil control valves to each cylinder head, and tighten the 2 bolts to 7 ft. lbs. (10 Nm).

53. Install intake air surge tank.

54. Remove any old packing (FIPG) material and be careful not to drop any oil on the contact surfaces of the cylinder block, rear oil seal retainer and oil pan.

55. Install a new O-ring to the oil pump.

56. Apply a continuous bead of seal packing (diameter: 3 to 4 mm (0.12 to 0.16 in.)) to the oil pan.

➡Install the oil pan within 3 minutes after applying seal packing. After installing it, the oil pan bolts and nuts must be tightened within 15 minutes. Otherwise the seal packing must be removed and reapplied.

57. Install the oil pan with the 17 bolts and 2 nuts. Tighten the bolts and nuts uniformly in several steps.

58. Tighten bolts as follows:

- Bolt A 15 ft. lbs. (21 Nm)
- Bolt B 15 ft. lbs. (21 Nm)
- Bolt C 7 ft. lbs. (10 Nm)
- Nut 15 ft. lbs. (21 Nm)

59. Install the 4 housing bolts and tighten to 27 ft. lbs. (37 Nm).

60. Install the flywheel housing under cover.

61. Install a new gasket and the oil strainer with the 2 nuts and tighten to 80 inch lbs. (9.0 Nm).

62. Remove any old packing (FIPG) material and be careful not to drop any oil on the contact surfaces of the oil pan and No. 2 oil pan.

➡Install the No. 2 oil pan within 3 minutes after applying seal packing. After installing it, the No. 2 oil pan bolts and nuts must be tightened within 15 minutes. Otherwise the seal packing must be removed and reapplied.

63. Install the No. 2 oil pan with the 10 bolts and 2 nuts. Tighten the bolts and nuts uniformly in several steps. Tighten bolts to 80 inch lbs. (9.0 Nm) and nuts to 7 ft. lbs. (10 Nm).

64. Install the pulley set bolt and tighten to 184 ft. lbs. (250 Nm).

65. Install the idler pulley with the bolt and tighten to 40 ft. lbs. (54 Nm).

➡"DOUBLE" is marked on the No. 1 idler pulley to distinguish it from the No. 2 idler pulley.

66. Install the 2 idler pulleys with the 2 bolts and tighten to 29 ft. lbs. (39 Nm).

67. Temporarily install the V-ribbed belt tensioner with the 5 bolts.

68. Install the V-ribbed belt tensioner by tightening the bolt 1 and bolt 2 in the order shown in the illustration. Tighten to 27 ft. lbs. (36 Nm).

69. Install cooler compressor assembly.

70. Install alternator assembly.

71. Install the vane pump with the 2 bolts an tighten to 32 ft. lbs. (43 Nm).

72. Connect the P/S oil pressure switch connector.

73. Install water inlet.

74. Install a new O-ring to the oil level gauge guide.

75. Apply a light coat of engine oil to the O-ring.

76. Push in the oil level gauge guide end into the guide hole of the oil pan.

77. Install the oil level gauge guide with the bolt.

78. Install the oil level gauge.

79. Install air cleaner assembly.

80. Connect No. 2 ventilation hose.

81. Install fluid coupling assembly.

82. Install fan and alternator v-belt.

83. Fully tighten fluid coupling assembly.

84. Install radiator support seal upper.

85. Install battery.

86. Install differential carrier assembly front as necessary.

87. Install power steering link assembly.

88. Add engine oil.

89. Add engine coolant.

90. Check for engine leaks.

91. Install the V-bank cover with the 2 nuts.

4.7L Engine

See Figures 225 through 227.

1. Before servicing the vehicle, refer to the precautions section.

2. Remove the engine from the vehicle and mount it on a stand.

3. Remove or disconnect the following:

- Front cover
- Timing belt.
- Timing belt idler pulleys
- Crankshaft timing sprocket
- Oil dipstick tube
- Oil filter and bracket
- Crankshaft Position (CKP) sensor
- Oil pan and baffle
- Oil pump strainer
- Oil pump

To install:

4. Install a new O-ring on the engine block.

5. Apply silicone sealant to the oil pump housing as shown.

6. Install the oil pump. Tighten the bolts in several passes to the following specifications:

- 12mm: 11 ft. lbs. (15.5 Nm)
- 14mm: 22 ft. lbs. (30.5 Nm)
- 6mm Hex: 11 ft. lbs. (15.5 Nm)

7. The upper oil pan bolts are different lengths and are identified as follows:

- A: 1.38 inch (35mm) w/12mm head
- B: 1.97 inch (50mm) w/12mm head
- C: 4.17 inch (106mm) w/12mm head
- D: 1.57 inch (40mm) w/14mm head
- E: 1.18 inch (30mm) w/6mm hex head

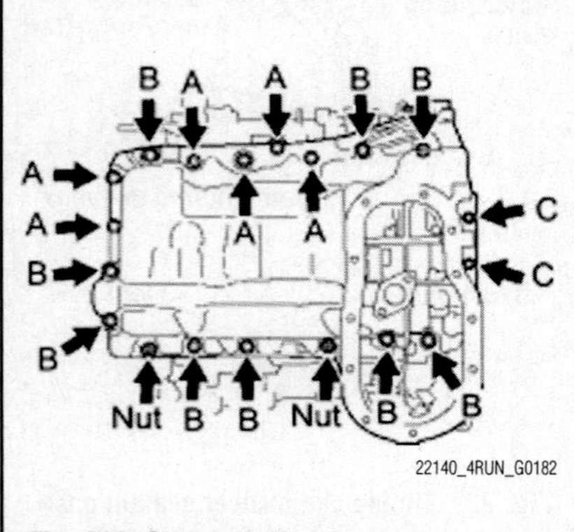

Fig. 223 Oil pan bolt sequence

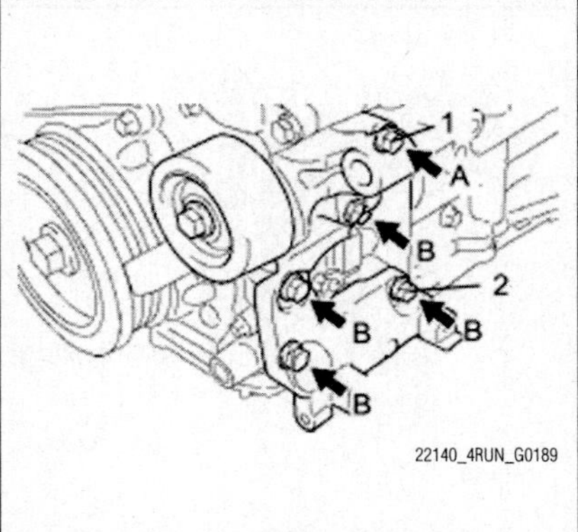

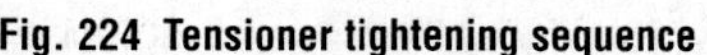

Fig. 224 Tensioner tightening sequence

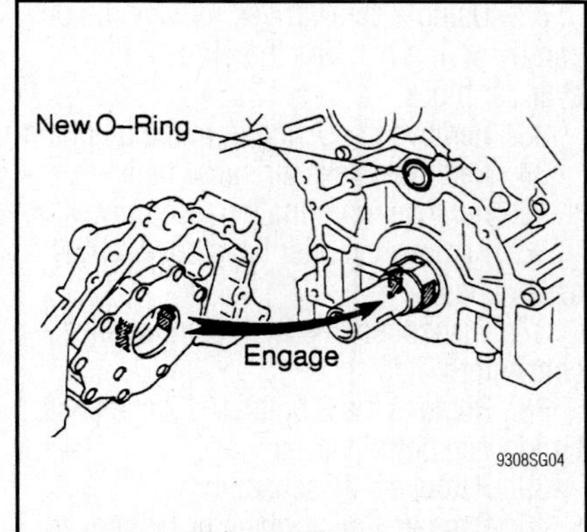

Fig. 225 Location of the O-ring seal

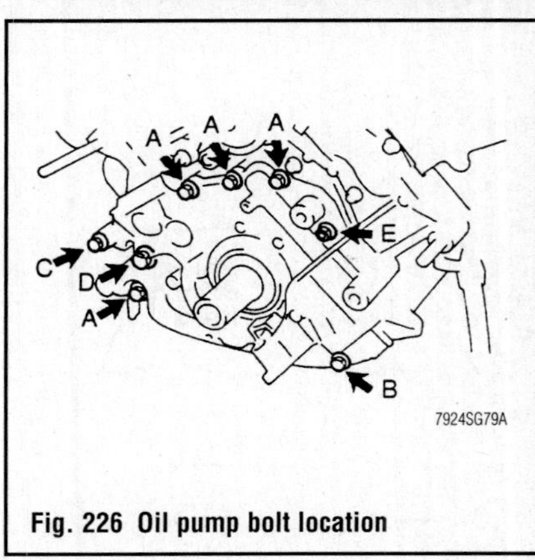

Fig. 226 Oil pump bolt location

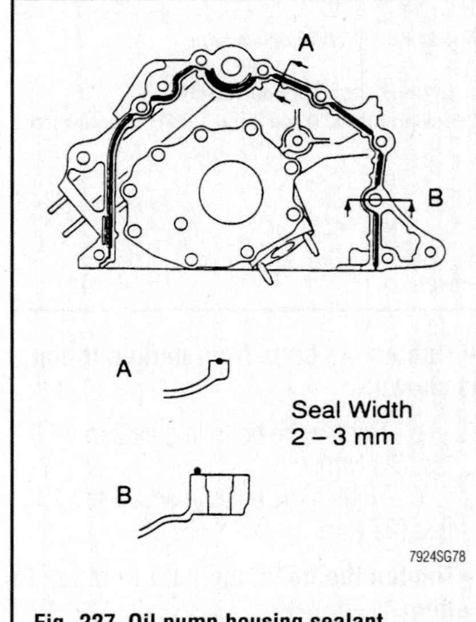

Fig. 227 Oil pump housing sealant application

8. Install or connect the following:
   - Oil pump pickup tube. Tighten the bolts to 66 inch lbs. (7.5 Nm).
   - Oil pan and baffle
   - CKP sensor
   - Oil filter and bracket. Tighten the bolts to 13 ft. lbs. (18 Nm).
   - Oil dipstick tube
   - Crankshaft timing sprocket
   - Timing belt idler pulleys
   - Timing belt
   - Front cover
9. Install the engine.

2010 Models

See Figures 228 through 237.

1. Remove the engine assembly.
2. Remove the ignition coil assembly.
3. Remove the engine oil level dipstick guide.
   a. Remove the dipstick.
   b. Remove the bolt and dipstick guide.
   c. Remove the O-ring from the dipstick guide.
4. Remove the water by pass sub assembly.
   a. Disconnect the 2 hoses.
   b. Remove the 3 bolts and water by pass pipe.
5. Remove the water inlet housing.
   a. Disconnect the 3 water by pass hoses.
   b. Remove the 5 bolts and water inlet housing.
   c. Remove the o-ring and gasket from the water outlet pipe and water pump.
   d. Remove the 3 water by pass hoses.
6. Remove the No. 1 idler pulley sub assembly.
   a. Remove the bolt and No. 1 idler pulley.
7. Remove the No. 2 idler pulley sub assembly.
8. Remove the v-ribbed belt tensioner assembly.
   a. Remove the 5 bolts and v-ribbed belt tensioner.
9. Remove the oil filter bracket.
   a. Remove the 2 nuts, bolt, oil filter bracket and gasket.
10. Remove the crankshaft pulley.
   a. Using special tool, hold the crankshaft pulley and loosen the pulley bolt. Continue to loosen the bolt until only 2 or 3 threads are screwed into the crankshaft.
   b. Using the pulley set bolt and special tool, remove the crankshaft pulley and pulley bolt.
11. Remove the No. 1 oil pipe.
   a. Remove the 2 oil pipe unions, oil control valve filter LH, 3 gaskets and the No. 1 oil pipe.

➡Do not touch the mesh when removing the oil control valve filter.

12. Remove the rear cylinder head cover.
13. Disconnect the fuel pipe sub assembly.
   a. Remove the 2 bolts and disconnect the fuel pipe.
14. Remove the cylinder head cover sub assembly LH.
   a. Remove the 12 bolts, seal washer, cylinder head cover and gasket.

➡Make sure the removed parts are returned to the same places they were removed from.

   b. Remove the 3 gaskets.
15. Remove the cylinder head cover sub assembly.
   a. Remove the 12 bolts, seal washer, cylinder head cover and gasket.

➡Make sure the removed parts are returned to the same places they were removed from.

   b. Remove the 3 gaskets.
16. Remove the No. 2 oil pan sub assembly.
   a. Remove the 10 bolts and 2 nuts.

➡Make sure the removed parts are returned to the same places they were removed from.

   b. Insert the blade of an oil pan seal cutter between the oil pans. Cut through the applied sealer and remove the No. 2 oil pan.

➡Be careful not to damage the contact surfaces of the oil pans.

17. Remove the oil strainer sub assembly.
   a. Remove the 2 nuts, oil strainer and gasket.
18. Remove the oil pan sub assembly.
   a. Remove the 17 bolts and 2 nuts.

➡Make sure the removed parts are returned to the same places they were removed from.

   b. Using the screwdriver, remove the oil pan by prying between the oil pan and cylinder block.

➡Tape the screwdriver tip before use.

➡Be careful not to damage the contact surfaces of the cylinder block and oil pan.

   c. Remove the 3 o-rings from the timing chain cover.
19. Remove the timing chain cover sub assembly.
   a. Remove the 26 bolts and the 2 nuts.

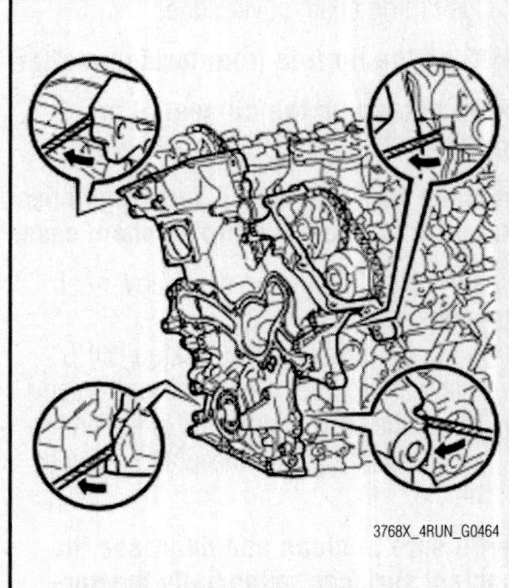

Fig. 228 Locating the pry points

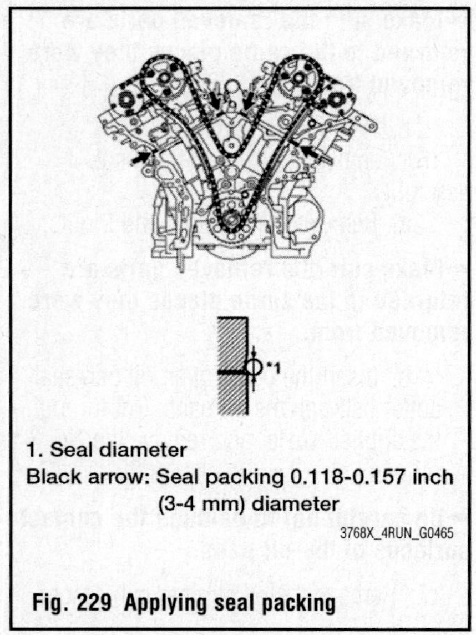

1. Seal diameter
Black arrow: Seal packing 0.118-0.157 inch (3-4 mm) diameter

3768X_4RUN_G0465

Fig. 229 Applying seal packing

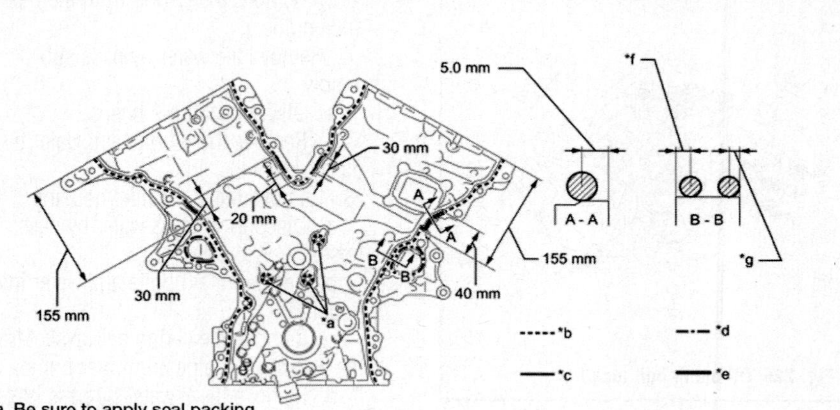

a. Be sure to apply seal packing
b. Dashed area (seal packing: Toyota Genuine Seal Packing Black, Three Bond 1207B or equivalent)
c. Continuous line area
(seal packing: Toyota Genuine Seal Packing Black, Three Bond 1207B or equivalent)
d. Alternate long and short dashed line area
(seal packing: Toyota Genuine Seal Packing 1282B, Three Bond 1282B or equivalent)
e. Diagonal line area (seal packing: Toyota Genuine Seal Packing Black, Three Bond 1207B or equivalent
f. 3-4 mm
g 2-3 mm

3768X_4RUN_G0466

Fig. 230 Applying seal packing to the timing chain cover

b. Remove the timing chain cover by prying between the timing chain cover and cylinder head or cylinder block with a screwdriver.

➡Tape the screwdriver tip before use.

➡Be careful not to damage the contact surfaces of the timing chain cover, cylinder block and cylinder head.

c. Remove the oil pump gasket from the cylinder block.

20. Remove the front crankshaft oil seal.

a. Using a screwdriver and wooden block, pry out the oil seal.

➡Tape the screwdriver tip before use.

➡Do not damage the surface of the oil seal press fit hole.

To install:

21. Install the front crankshaft oil seal.

a. Using SST and a hammer, tap in a new oil seal until its surface is flush with the timing chain cover edge.

➡Keep the lip free from foreign matter.

➡Do not tap on the oil seal at an angle.

➡Make sure that the oil seal edge does not protrude from the timing chain case.

22. Install the timing chain cover sub assembly.

a. Remove any old packing (FIPG material) and be careful not to drop any oil on the contact surfaces of the timing chain cover, cylinder head and cylinder block.

➡Be sure to clean and degrease the contact surfaces, especially the surfaces indicated in the illustration.

b. Apply a light coat of engine oil to a new oil pump gasket.

c. Install the oil pump gasket.

d. Apply seal packing as shown in the illustration.

e. Apply seal packing to the timing chain cover in a continuous line as shown in the following illustration.

➡When the contact surfaces are wet, wipe them with an oil-free cloth before applying seal packing.

➡Install the chain cover within 3 minutes and tighten the bolts within 10 minutes after applying seal packing.

f. Align the oil pump drive rotor spline and crankshaft as shown in the illustration. Install the drive rotor and chain cover to the crankshaft.

g. Install the timing chain cover with the 26 bolts, labeled A, B, C and D, and the 2 nuts. Tighten the bolts and nuts uniformly in several steps.

Standard bolt:

- Bolt A: 0.984 inch (25 mm)
- Bolt B: 2.17 inch (55 mm)
- Bolt C: 1.38 inch (35 mm)
- Bolt D: 2.56 inch (65 mm)

➡Make sure that there is no oil on the bolt threads.

a. Tighten the bolts in area 1 to 35 ft. lbs. (47 Nm).

➡Tighten the bolts from bottom to top as shown.

b. Tighten the bolts in area 2 to 17 ft. lbs. (23 Nm).

c. Tighten the bolts in area 3 to 17 ft. lbs. (23 Nm).

➡Tighten the bolts and nuts from top to bottom.

d. Tighten the bolts in area 4 to 17 ft. lbs. (23 Nm).

➡Tighten the bolts from bottom to top as shown in the illustration.

➡Do not start the engine for at least 2 hours after installation.

➡Wipe off any seal packing that seeped out around the surfaces of the oil pan and head cover and make sure that there is no seal packing seeping out around the edges.

23. Install the oil pan sub assembly.

➡Do not start the engine for at least 2 hours after installing.

24. Install the oil strainer sub assembly.

a. Install a new gasket and the oil strainer with the 2 nuts. Tighten to 80 inch lbs. (9 Nm).

25. Install the No. 2 oil pan sub assembly.

26. Install the cylinder head cover sub assembly.

a. Remove any old packing (FIPG

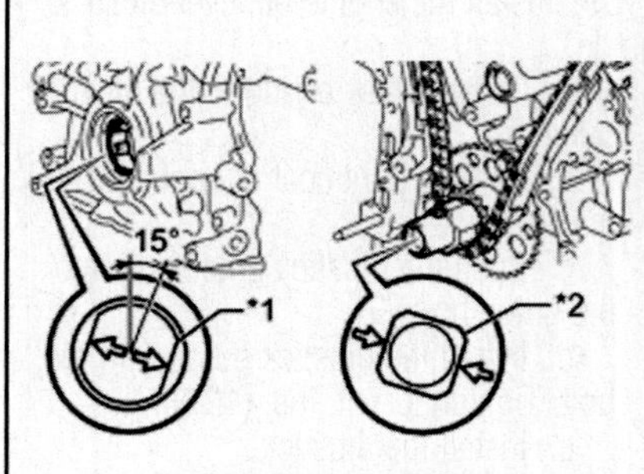

Fig. 231 Aligning the oil pump drive rotor spline and crankshaft

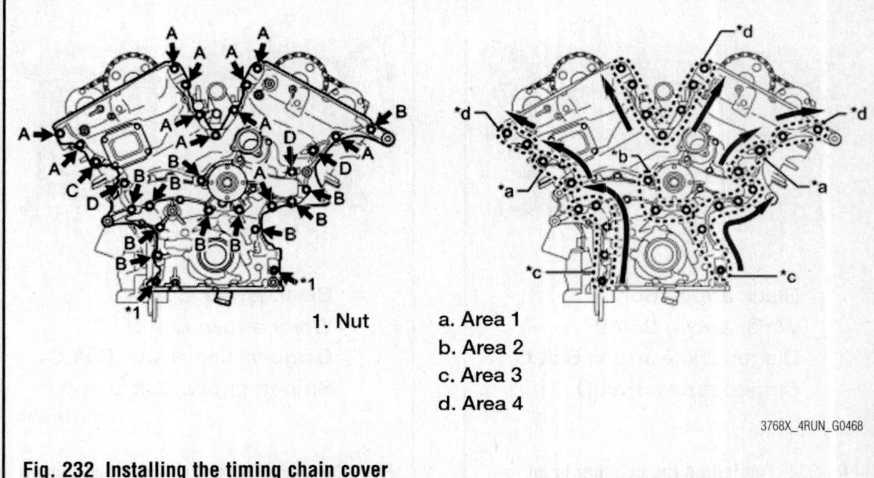

Fig. 232 Installing the timing chain cover

material) and be careful not to drop any oil on the contact surfaces of the cylinder head, timing chain cover and cylinder head cover.

b. Apply seal packing as shown in the illustration.

➡Remove any oil from the contact surface.

➡Install the cylinder head cover within 3 minutes and tighten the bolts within 15 minutes after applying seal packing.

c. Install 3 new gaskets.

d. Install a new gasket to the cylinder head cover.

e. Install the seal washers to the bolts.

f. Temporarily install the cylinder head cover with the 12 bolts. Tighten the bolts uniformly in several steps. Tighten bolts A and D to 7 ft. lbs. (10 Nm). Tighten bolts B and C to 15 ft. lbs. (21 Nm).

Standard Bolts:

- A: 0.984 inch (25 mm)
- B: 1.38 inch (35 mm)
- C: 2.56 inch (65 mm)
- D: 2.36 inch (60 mm)

➡Do not start the engine for at least 2 hours after installation.

27. Install the cylinder head cover sub assembly LH.

a. Remove any old packing (FIPG material) and be careful not to drop any oil on the contact surfaces of the cylinder head, timing chain cover and cylinder head cover.

b. Apply seal packing as shown in the illustration.

➡Remove any oil from the contact surface.

➡Install the cylinder head cover within 3 minutes and tighten the bolts within 15 minutes after applying seal packing.

c. Install 3 new gaskets.

d. Install a new gasket to the cylinder head cover.

e. Install the seal washers to the bolts.

f. Temporarily install the cylinder head cover with the 12 bolts. Tighten the bolts uniformly in several steps. Tighten bolts A and D to 7 ft. lbs. (10 Nm). Tighten bolts B and C to 15 ft. lbs. (21 Nm).

Standard Bolts:

- A: 0.984 inch (25 mm)
- B: 1.38 inch (35 mm)
- C: 2.76 inch (70 mm)
- D: 2.36 inch (60 mm)

➡Do not start the engine for at least 2 hours after installation.

28. Connect the fuel pipe sub assembly with the 2 bolts. Tighten to 80 inch lbs. (9 Nm).

29. Install the rear cylinder head cover.

30. Install the No. 2 oil pipe.

a. Make sure that there is no foreign matter on the mesh of the oil control valve filter RH.

➡Do not touch the mesh when installing the oil control valve filter.

b. Install a new gasket and temporarily install the oil pipe (on the cylinder head side) with the oil check valve bolt.

c. Install the oil control valve filter RH to the oil pipe union. Install new gaskets and temporarily install the oil pipe (on the head cover side).

d. Tighten the oil pipe union (on the cylinder head side). Tighten to 48 ft. lbs. (65 Nm).

➡If the link that connects the gaskets is broken, remove the connecting link by using side cutters or a similar tool.

e. Tighten the oil pipe union (on the head cover side). Tighten to 48 ft. lbs. (65 Nm).

31. Install the No. 1 oil pipe.

a. Make sure that there is no foreign matter on the mesh of the oil control valve filter LH.

➡Do not touch the mesh when installing the oil control valve filter.

b. Install a new gasket and temporarily install the oil pipe (on the cylinder head side) with the oil check valve bolt.

c. Install the oil control valve filter LH to the oil pipe union. Install new gaskets and temporarily install the oil pipe (on the head cover side).

d. Tighten the oil pipe union (on the cylinder head side). Tighten to 48 ft. lbs. (65 Nm).

➡If the link that connects the gaskets is broken, remove the connecting link by using side cutters or a similar tool.

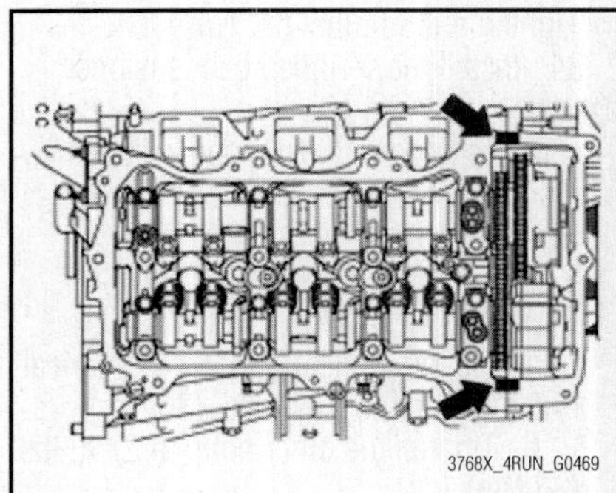

Fig. 233 Applying the seal packing to the cylinder head cover sub assembly

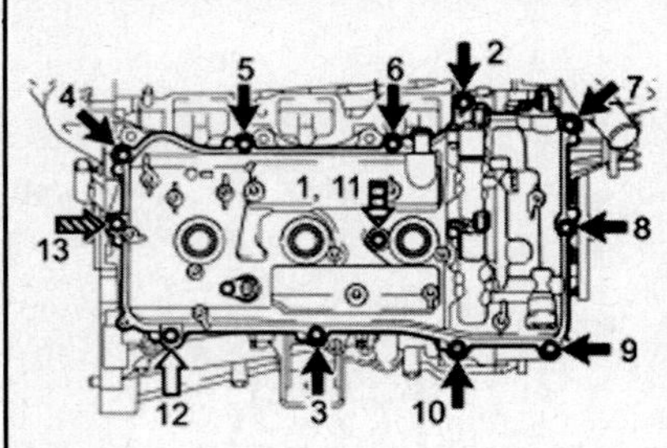

Black arrow: Bolt A
White arrow: Bolt B
Diagonal line arrow: Bolt C
Striped arrow: Bolt D

3768X_4RUN_G0470

Fig. 234 Tightening the cylinder head bolts

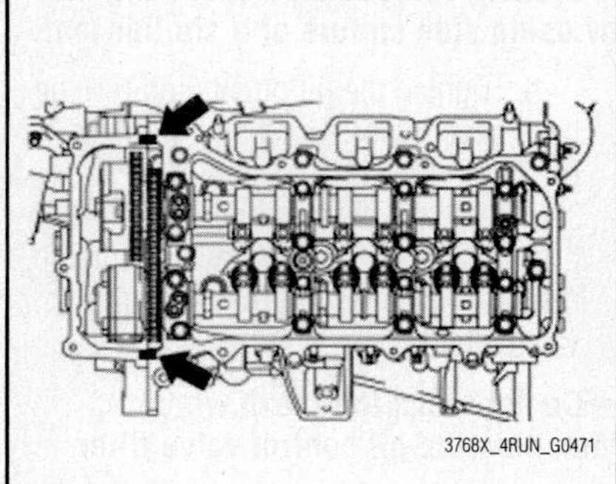

3768X_4RUN_G0471

Fig. 235 Applying seal packing to the cylinder head cover sub assembly LH

e. Tighten the oil pipe union (on the head cover side). Tighten to 48 ft. lbs. (65 Nm).

32. Install the crankshaft pulley.

a. Using the special tool, install the crankshaft pulley with the pulley set bolt. Tighten to 192 ft. lbs. (260 Nm).

33. Install the oil filter bracket.

a. Install the oil filter bracket and a new gasket with the 2 nuts and bolt. Tighten to 15 ft. lbs. (21 Nm).

34. Install the v-ribbed belt tensioner assembly.

a. Temporarily install the v-ribbed belt tensioner with the 5 bolts.

Standard bolt:

- A: 2.76 inch (70 mm)
- B: 1.30 inch (33 mm)

a. Tighten bolts 1 and 2 in numerical order. Tighten to 27 ft. lbs. (36 Nm).

b. Tighten the other bolts to 27 ft. lbs. (36 Nm).

35. Install the No. 2 idler pulley sub assembly.

36. Install the No. 1 idler pulley sub

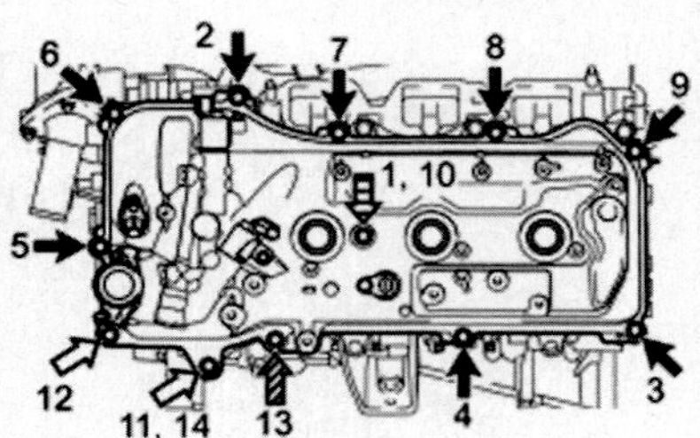

Black arrow: Bolt A
White arrow: Bolt B
Diagonal line arrow: Bolt C
Striped arrow: Bolt D

3768X_4RUN_G0472

Fig. 236 Installing the cylinder head cover LH

Black arrow: Bolt A
White arrow: Bolt B

3768X_4RUN_G0473

Fig. 237 Installing the v-ribbed belt tensioner assembly

assembly with the bolt. Tighten to 40 ft. lbs. (54 Nm).

➡"Double" is marked on the No. 1 idler pulley to distinguish it from the No. 2 idler pulley.

37. Install the water inlet housing.

a. Install the 3 water by-pass hoses.

b. Apply soapy water to a new O-ring and install the O-ring to the water outlet pipe.

c. Install a new O-ring to the water outlet pipe.

d. Install a new gasket to the water pump.

e. Install the water inlet with the 5 bolts. Tighten to 7 ft. lbs. (10 Nm).

f. Connect the 3 water by pass hoses.

38. Install the water by pass pipe sub assembly with the 3 bolts. Tighten to 7 ft. lbs. (10 Nm).

a. Connect the 2 hoses.

➡The direction of the hose clamp is indicated in the illustration.

39. Install the engine oil level dipstick guide.

a. Install a new o-ring to the dipstick guide.

b. Apply a light coat of engine oil to the O-ring.

c. Push the dipstick guide end into the guide hole.

d. Install the dipstick guide with the bolt. Tighten to 7 ft. lbs. (10 Nm).

e. Install the dipstick.

40. Install the ignition coil assembly.

41. Install the engine assembly.

PISTON AND RING

POSITIONING

2009 Models

See Figures 238 through 240.

2010 Models

See Figures 241 and 242.

REAR MAIN SEAL

REMOVAL & INSTALLATION

2009 Models

4.0L Engine

1. Remove automatic transmission assembly.
2. Remove drive plate & ring gear sub-assembly.
3. Remove the 8 bolts, front spacer, drive plate and rear spacer.
4. Using a knife, cut off the oil seal lip.

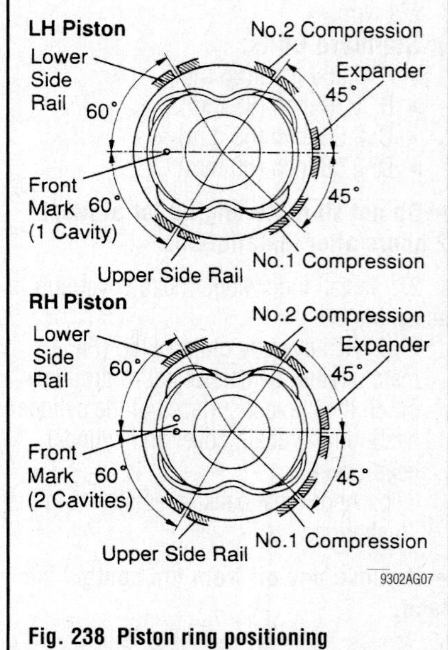

Fig. 238 Piston ring positioning

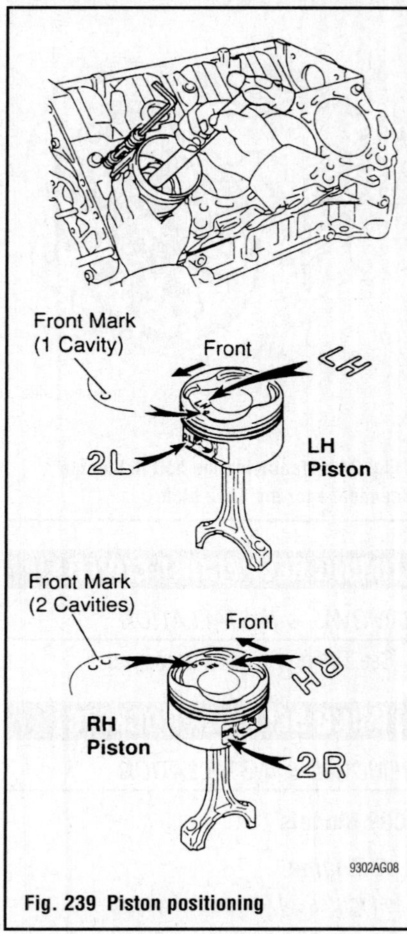

Fig. 239 Piston positioning

5. Using a screwdriver, pry out the oil seal.

➡Be careful not to damage the crankshaft. Tape the screwdriver tip.

To install:

6. Apply MP grease to a new oil seal lip.
7. Using service tool and a hammer, tap in the oil seal until its surface is flush with the rear oil seal retainer edge.
8. Apply adhesive to 2 or 3 threads of the mounting bolt end.
9. Install the front spacer, drive plate and rear spacer on the crankshaft.

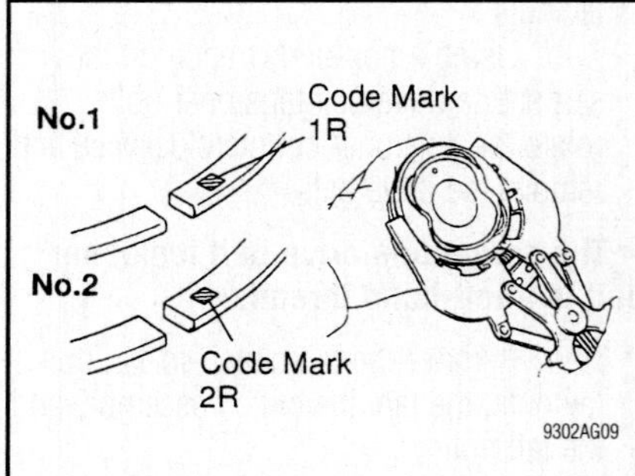

Fig. 240 Piston ring identification

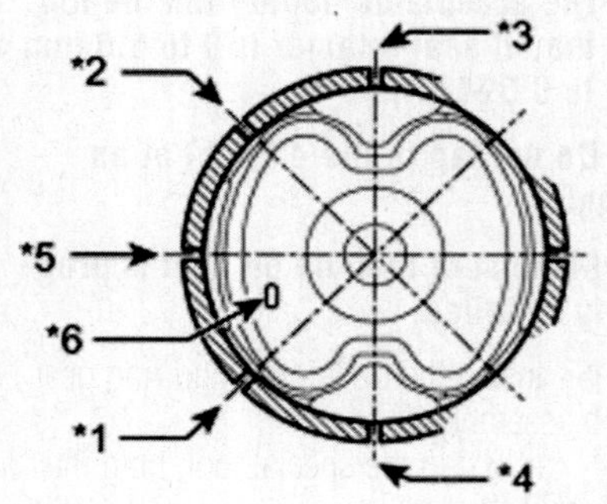

Fig. 241 Piston ring positioning—4.0L engine

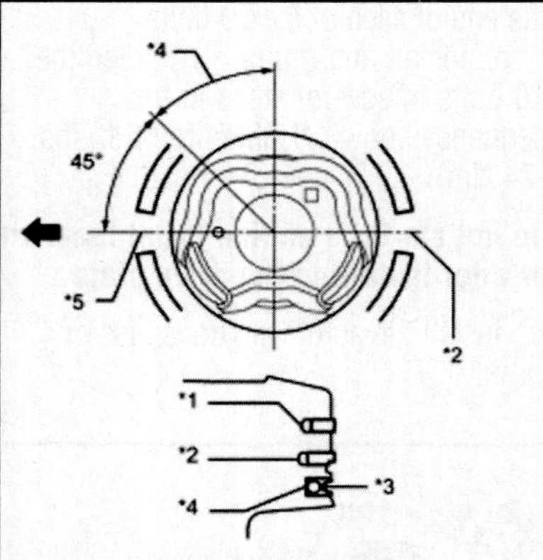

Fig. 242 Piston ring positioning—2.7L engine

10. Using several steps, install and tighten the 8 mounting bolts uniformly in the sequence as shown in the illustration and tighten to 61 ft. lbs. (83 Nm).
11. Install automatic transmission assembly.
12. Check for oil leaks.

4.7L Engine

1. Before servicing the vehicle, refer to the precautions section.
2. Remove the transmission and flywheel from the vehicle.
3. Cut off the rubber lip portion of the seal with a sharp knife.
4. Pry out the oil seal.

To install:

5. Install the rear main seal so that it is flush with the seal retainer housing.
6. Install or connect the following:
   - Flywheel/driveplate. Tighten the bolts to 35 ft. lbs. (48 Nm) plus a 90 degree turn.
   - Transmission

2010 Models

4.0L Engine

See Figure 243.

1. Remove the automatic transmission assembly.
2. Remove the drive plate and ring gear sub assembly.
3. Remove the rear crankshaft oil seal.
   a. Using a knife, cut off the lip of the oil seal.
   b. Using a screwdriver, pry out the oil seal.

➡Tape the screwdriver tip before use.

➡Do not damage the surface of the oil seal press fit hole or crankshaft.

To install:

4. Install the rear crankshaft oil seal.
   a. Apply MP grease to the lip of a new oil seal.
   b. Using the special tool and a hammer, tap in the oil seal until its surface is flush with the rear oil seal retainer edge.

➡Keep the lip free foreign matter.

➡Do not tap the oil seal at an angle.

5. Install the drive plate and ring gear sub assembly.
6. Install the automatic transmission assembly.

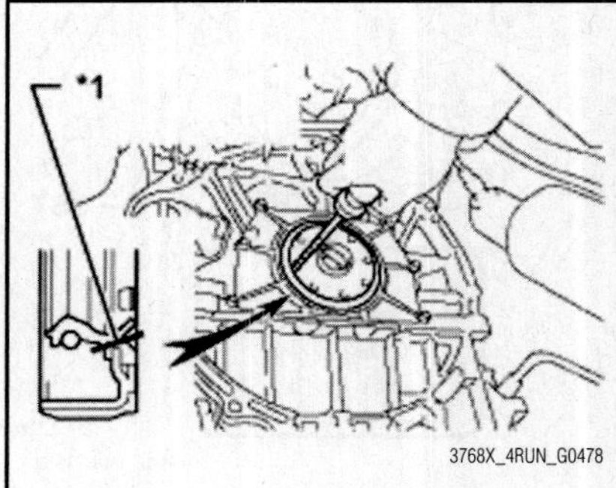

Fig. 243 Removing the rear crankshaft oil seal

2.7L Engine

See Figures 244 through 246.

1. Remove the automatic transmission assembly.
2. Remove the drive plate and ring gear sub assembly.
   a. Using the special tool, hold the crankshaft pulley.
   b. Remove the 10 bolts, rear drive plate spacer, drive plate and front drive plate spacer.
3. Remove the rear crankshaft oil seal.
   a. Using a knife, cut off the lip of the oil seal.
   b. Using a screwdriver pry out the oil seal.

➡Tape the screwdriver tip before use.

➡Do not damage the surface of the oil seal press fit hole or the crankshaft.

➡After removing the oil seal, check the crankshaft for damage. If damaged, smooth the surface with 400-grit sandpaper.

To install:

4. Install the rear crankshaft oil seal.
   a. Apply a light coat of MP grease to the lip of a new oil seal.

➡Do not allow foreign matter to contact the lip of the oil seal.

➡Do not allow MP grease to contact the dust seal.

   b. Using the special tool and a hammer, tap in the oil seal until its surface is flush with the rear oil seal retainer edge.

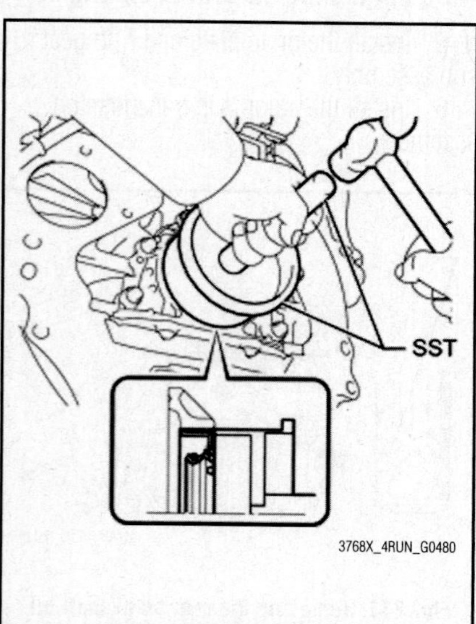

Fig. 244 Installing the crankshaft oil seal

➡The acceptable depth from the top of the oil seal retainer is 0 to 1.0 mm (0 to 0.394 in.)

➡Do not tap in the oil seal at an angle.

➡Make sure that the oil seal is properly installed.

5. Install the drive plate and ring gear sub assembly.
   a. Using the special tool, hold the crankshaft pulley.
   b. Install the front drive plate spacer, drive plate and rear drive plate spacer to the crankshaft.

➡The front drive plate spacer is reversible.

➡As the rear drive plate spacer and drive plate and ring gear are not reversible, be sure to install it so that it is facing in the direction shown in the illustration.

   c. Clean the bolts and bolt holes.
   d. Apply adhesive to 2 or 3 threads at the end of each of the 10 bolts.
   e. Install and uniformly tighten the 10 bolts in several steps in the sequence shown. Tighten to 55 ft. lbs. (74 Nm).

➡Do not start the engine for at least an hour after installing the drive plate.

6. Install the automatic transmission assembly.

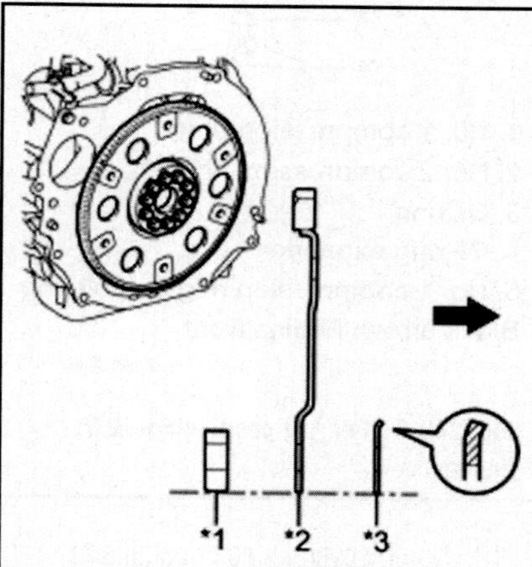

Fig. 245 Installing the front drive plate spacer, drive plate and rear drive plate spacer to the crankshaft

Fig. 246 Identifying the bolt tightening sequence for the drive plate

TIMING BELT FRONT COVER

REMOVAL & INSTALLATION

See Timing Belt and Sprockets.

TIMING BELT & SPROCKETS

REMOVAL & INSTALLATION

2009 Models

4.7L Engine

See Figures 247 through 254.

1. Disconnect the negative battery cable.
2. Raise and safely support the vehicle.
3. Remove the oil pan protector and the engine under cover.
4. Drain the cooling system and store the coolant for refilling purposes.
5. Lower the vehicle and remove the battery clamp cover.
6. From the top of the engine, remove the fuel return hose, the engine cover nuts/bolts and the cover.
7. Remove the air cleaner and the intake air connector assembly.
8. Remove the cooling fan pulley by performing the following procedures:
   a. Loosen the 4 fan clutch-to-fan pulley nuts.
   b. Using a box-end wrench on the serpentine drive belt tensioner bolt, rotate the tensioner counterclockwise and remove the drive belt.

➡The serpentine drive belt tensioner bolt is a left-hand thread.

   c. Remove the fan clutch-to-fan pulley nuts, the fan, the clutch assembly and the fan pulley.
9. Remove the radiator by performing the following procedures:

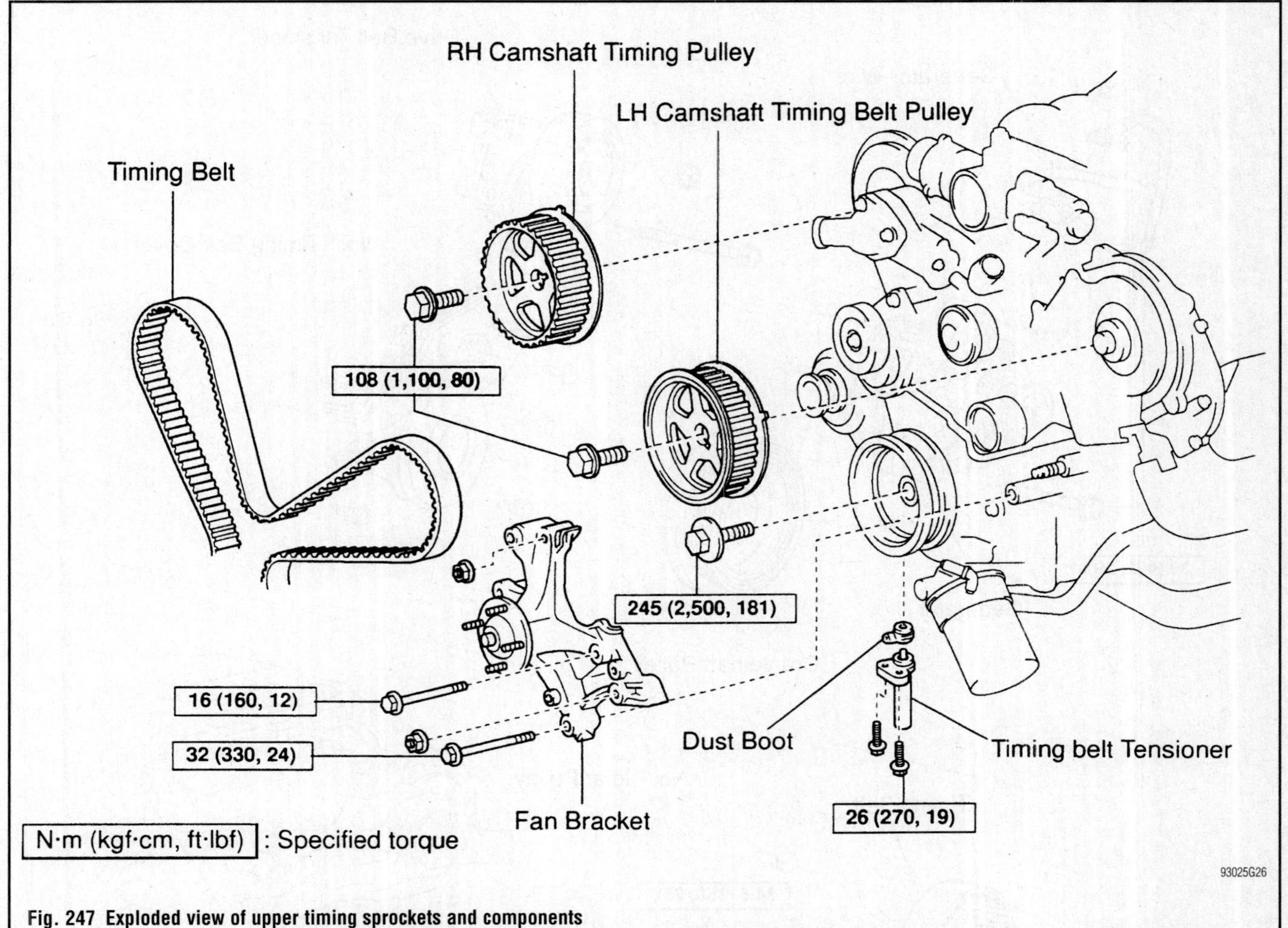

Fig. 247 Exploded view of upper timing sprockets and components

a. Disconnect the upper, lower and reservoir hoses from the radiator.

b. Disconnect and plug the automatic transmission oil cooler at the radiator. Disconnect the automatic transmission oil cooler hoses from the fan shroud clamp.

c. Remove the radiator reservoir tank.

d. Remove the fan shroud-to-radiator bolts and the shroud.

e. Remove the 2 upper radiator-to-chassis nuts.

f. Remove the middle radiator-to-chassis nut/bolts and brackets.

g. Carefully, lift the radiator from the vehicle.

10. Remove the serpentine drive belt idler pulley bolt, cover plate and pulley.

11. Remove the right side (No. 3) timing belt cover.

12. Remove the left side (No. 3) timing belt cover by performing the following procedures:

a. Disconnect the engine wire from both wire clamps.

b. Disconnect the Camshaft Position (CMP) sensor wire from the wire clamp on the left-side (No.3) timing belt cover.

c. Disconnect the sensor connector from the connector bracket.

d. Disconnect the sensor connector.

e. Remove the wire grommet from the left-side (No. 3) timing belt cover.

f. Remove the oil cooler tube bolts and tube.

13. Remove the middle (No. 2) timing belt cover bolts and cover.

14. Remove the cooling fan bracket nuts/bolts and bracket.

➡If reusing the timing belt, make sure that there are 3 installation marks on the belt; if there are none, install them.

15. Using the Crankshaft Pulley Holding tool 09213-70010, Bolt tool 90105-08076 and Companion Flange Holding tool 09330-00021, or equivalent, loosen the crankshaft pulley bolt.

16. Position the No. 1 cylinder to approximately 50 degrees After Top Dead Center (ATDC) of the compression stroke by performing the following procedures:

a. Rotate the crankshaft pulley (CLOCKWISE) to align its groove with the timing mark "0" on the lower (No. 1) timing belt cover.

b. Check that the camshaft sprocket timing marks are aligned with the rear timing belt plate marks; if not, rotate the crankshaft 1 revolution (360 degrees).

c. Rotate the crankshaft pulley approximately 50 degrees (CLOCKWISE) and align the crankshaft pulley timing mark between the centers of the crankshaft pulley bolt and the idler pulley bolt.

✻✻ WARNING

If the timing belt is disengaged, having the crankshaft pulley in the wrong angle can cause the valve to come into contact with the piston when removing the camshaft pulley.

17. Remove the crankshaft pulley bolt.

➡If reusing the timing belt and the installation marks have disappeared, place new installation marks on the

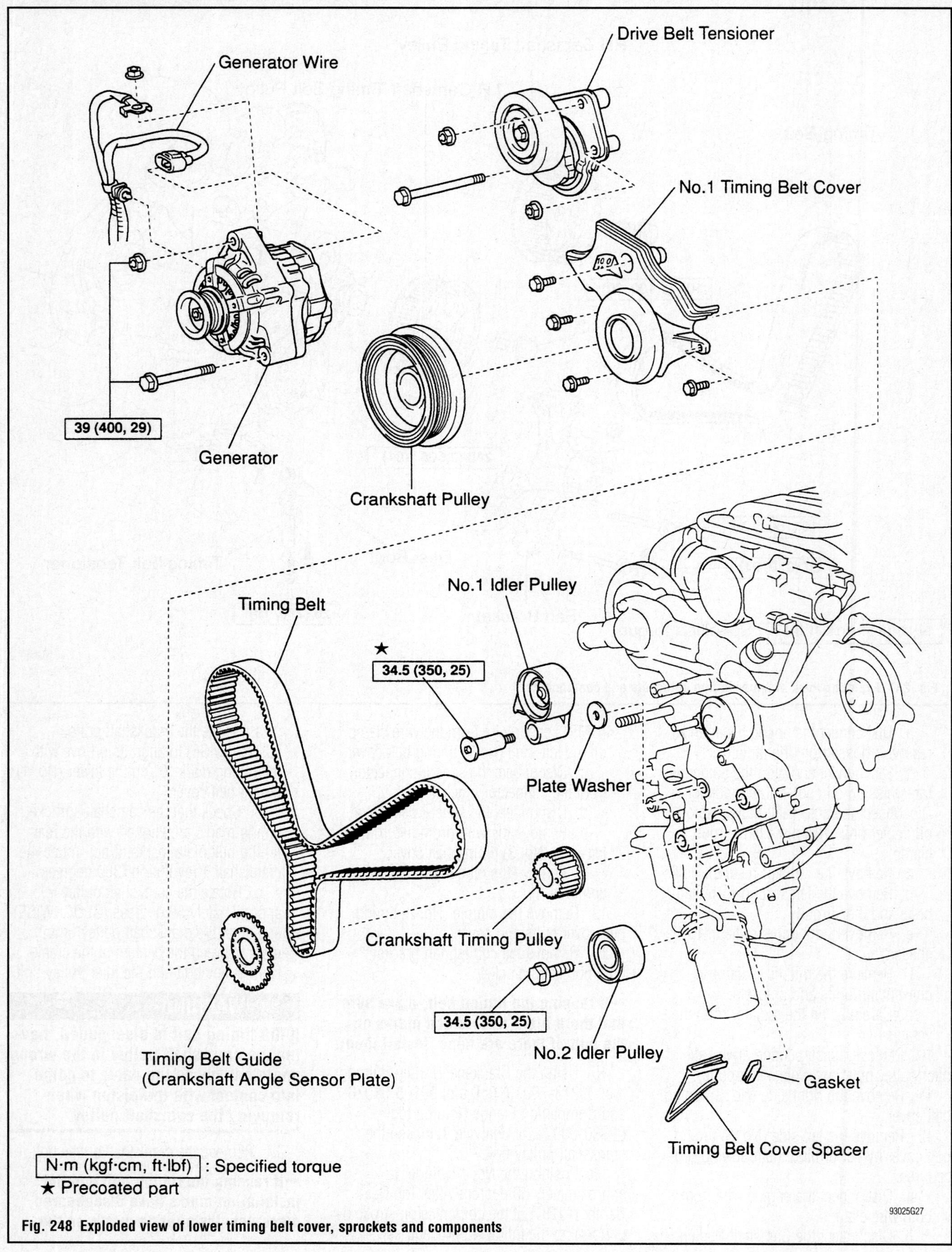

Fig. 248 Exploded view of lower timing belt cover, sprockets and components

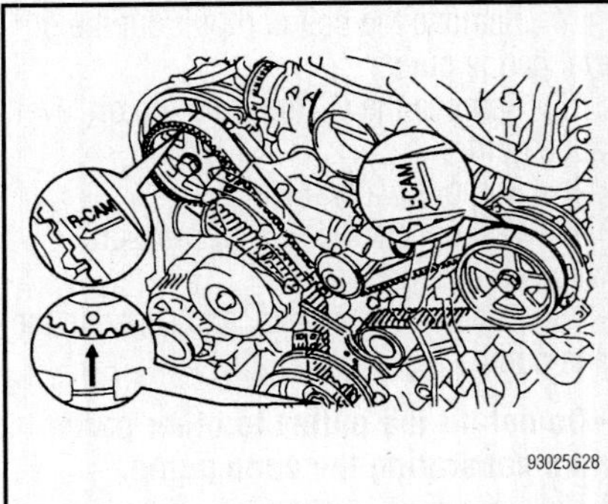

Fig. 249 Alignment of timing belt with the timing sprockets

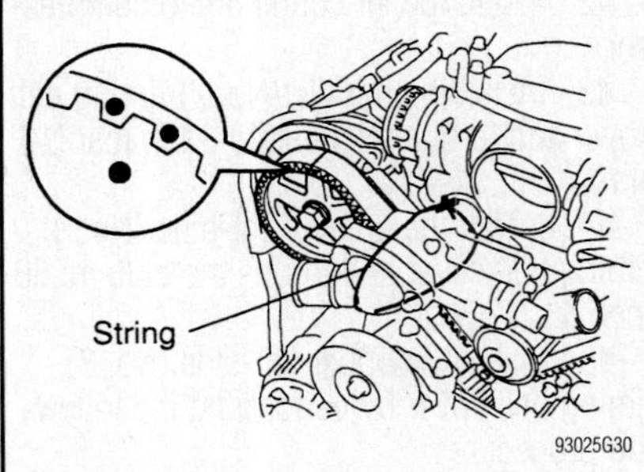

Fig. 251 Securing the timing belt with string and matchmarking the camshaft with the timing belt

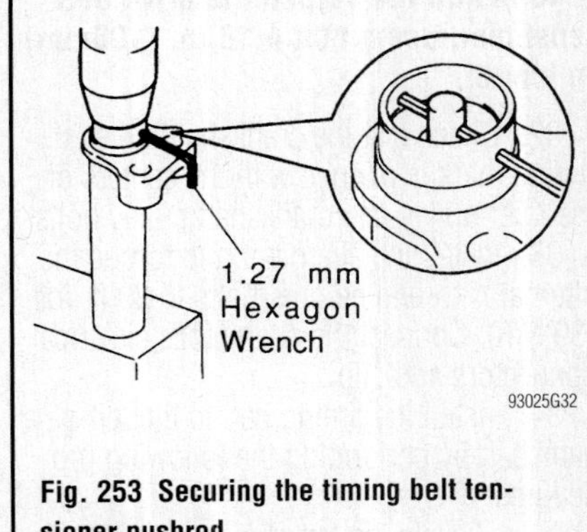

Fig. 253 Securing the timing belt tensioner pushrod

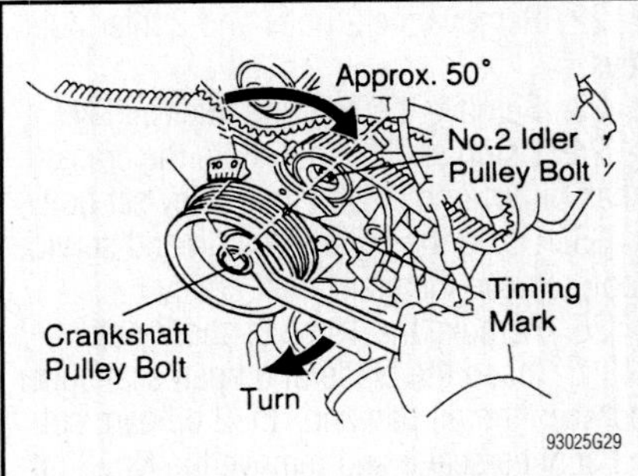

Fig. 250 Aligning of crankshaft pulley timing mark with the center line of the crankshaft pulley bolt and the idler pulley bolt

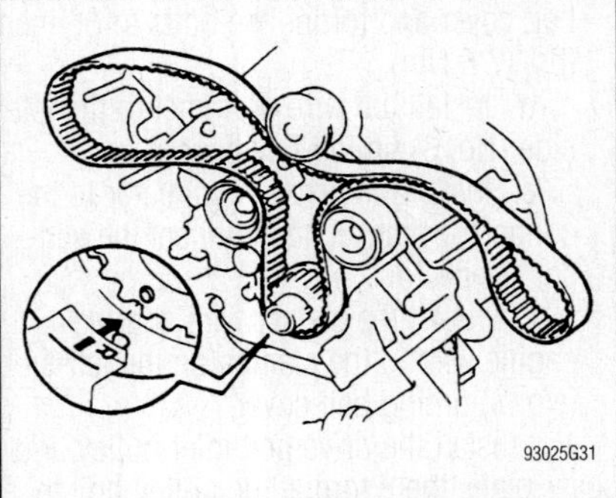

Fig. 252 Installing the timing belt on the crankshaft sprocket

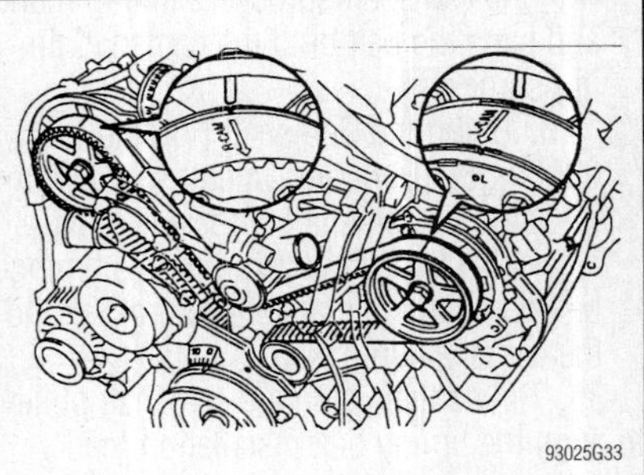

Fig. 254 Checking the TDC alignment marks after rotating the crankshaft 2 revolutions

timing belt to match the camshaft timing sprocket marks.

➡To avoid meshing the timing sprocket and the timing belt, secure one with a string; then, place matchmarks on the timing belt and the right-side camshaft timing sprocket.

18. Remove the timing belt tensioner bolts and the tensioner.
19. Using the Camshaft Holding tool 09960-10010, or equivalent, slightly turn the left-side camshaft sprocket clockwise to loosen the tension spring. Then, disconnect the timing belt from the camshaft sprockets.
20. Remove the alternator by performing the following procedures:
    a. Disconnect the electrical connector from the alternator.
    b. Remove the rubber cap/nut and disconnect the battery wire from the alternator.
    c. Disconnect the wire clamp from the alternator cord clip.
    d. Remove the alternator-to-engine nuts/bolts and the alternator.
21. Remove the serpentine drive belt tensioner nuts/bolts and the tensioner.
22. Using the Crankshaft Puller Assembly tool 09950-50012, or equivalent, press the crankshaft pulley from the crankshaft.

⁂ WARNING

DO NOT rotate the crankshaft pulley.

23. Remove the lower (No. 1) timing belt cover bolts and the cover.
24. Remove the timing belt guide, spacer and the timing belt.

To install:

➡With the timing belt removed, this is a perfect opportunity to inspect and/or replace the water pump.

25. Inspect the timing belt tensioner by performing the following procedures:
    a. Inspect the seal for leakage; if leakage is suspected, replace the tensioner.
    b. Using both hands to hold the tensioner facing upward, strongly press the pushrod against a solid surface. If the pushrod moves, replace the tensioner.

⁂ WARNING

Never hold the tensioner with the pushrod facing downward.

    c. Measure the pushrod protrusion from the housing end, it should be 0.413–0.453 in. (10.5–11.5mm). If the protrusion is not as specified, replace the tensioner.
26. Temporarily install the timing belt by performing the following procedures:
    a. Align the timing belt's installation mark with the crankshaft timing sprocket.
    b. Install the timing belt on the crankshaft timing sprocket, the No. 1 idler pulley and the No. 2 idler pulley.
27. Install the gasket to the timing belt cover spacer and install the cover spacer.
28. Install the timing belt guide with the cup side facing outward.
29. Install the lower (No. 1) timing belt cover.
30. Install the crankshaft pulley by performing the following procedures:
    a. Align the crankshaft pulley with the crankshaft key.
    b. Using the Crankshaft Installer tool 09223-46011, or equivalent, and a hammer, tap the crankshaft pulley into position.
31. Install the serpentine drive belt tensioner and torque the tensioner-to-engine bolts to 12 ft. lbs. (16 Nm).

➡To install the serpentine drive belt tensioner, use a bolt 4.18 in. (106mm) in length.

32. Check that the crankshaft pulley's timing mark is aligned with the centers of the idler pulley and crankshaft pulley bolts.
33. Install the alternator and torque the alternator-to-engine nuts/bolts to 29 ft. lbs. (39 Nm). Connect the alternator's electrical connectors and clip.
34. Install the timing belt to the left-side camshaft by performing the following procedures:
    a. Rotate the left-side camshaft pulley to align the timing belt installation mark with the camshaft sprocket's timing mark and slide the belt onto the camshaft timing sprocket.
    b. Using the Camshaft Holding tool 09960-10010, or equivalent, slightly turn the left-side camshaft sprocket counterclockwise to place tension on the timing belt between the crankshaft sprocket and the camshaft sprocket.
35. Rotate the right-side camshaft pulley to align the timing belt installation mark with the camshaft sprocket's timing mark and slide the belt onto the camshaft timing sprocket.
36. Using a vertical press, slowly press the pushrod into the housing using 200–2205 lbs. (981–9807 N) until the holes align, then, install a 1.27mm Allen®wrench to secure the pushrod and release the press. Install the dust boot on the tensioner housing.
37. Install the timing belt tensioner and torque the bolts to 19 ft. lbs. (26 Nm).
38. Using a pair of pliers, remove the Allen®wrench from the tensioner housing.
39. Check the valve timing by performing the following procedure:
    a. Temporarily install the crankshaft pulley bolt.
    b. Slowly, rotate the crankshaft pulley 2 revolutions (CLOCKWISE) and realign the TDC marks.

➡If the pulley/sprocket timing marks do not realign, remove the timing belt and reinstall it.

40. Using the Crankshaft Pulley Holding tool 09213-70010, Bolt tool 90105-08076 and Companion Flange Holding tool 09330-00021, or equivalent, torque the crankshaft pulley bolt to 181 ft. lbs. (245 Nm).
41. Install the cooling fan bracket and torque the 12mm (head size) bolt to 12 ft. lbs. (16 Nm) and the 14mm (head size) bolt to 24 ft. lbs. (32 Nm).
42. Install the air conditioning compressor.
43. Install the middle (No. 2) timing belt cover and torque the bolts to 12 ft. lbs. (16 Nm).
44. Install the upper right-side (No. 3) timing belt cover and torque the bolts to 66 inch lbs. (7.5 Nm).
45. Install the upper left-side (No. 3) timing belt cover by performing the following procedures:
    a. Install the oil cooler tube and bolt.
    b. Feed the Camshaft Position (CMP) sensor (CPS) through the left-side (No. 3) timing belt cover hole.
    c. Install the left-side (No. 3) timing belt cover and torque the bolts to 66 inch lbs. (7.5 Nm).
    d. Install the wire grommet to the left-side (No. 3) timing belt cover.
    e. Install the sensor connector to the connector bracket and connect the sensor connector.
    f. Install the sensor wire and the engine wire to the clamps on the left-side (No. 3) timing belt cover.
46. Install the drive belt idler pulley and cover plate; then, torque the pulley bolt to 27 ft. lbs. (37 Nm).
47. To complete the installation, reverse the removal procedures.
48. Refill the cooling system and connect the negative battery cable.

TIMING CHAIN & SPROCKETS

REMOVAL & INSTALLATION

2009 Models

4.0L Engine

See Figures 255 through 264.

1. Remove power steering link assembly.
2. Remove differential carrier assembly front.
3. Drain engine coolant.
4. Drain engine oil.
5. Remove battery.
6. Remove v-bank cover.
7. Remove the 11 clips and radiator support seal upper.
8. Loosen fan fluid coupling assembly.
9. Remove fan and alternator v-belt.
10. Remove fluid coupling assembly.
11. Disconnect no. 2 ventilation hose.
12. Remove air cleaner assembly.
13. Remove the oil level gauge.
14. Remove the bolt and pull out the oil level gauge guide.
15. Remove the O-ring from the oil level gauge guide.
16. Remove water inlet.
17. Disconnect the P/S oil pressure switch connector.
18. Remove the 2 bolts, and separate the vane pump.

➡Do not hit the pulley to other parts when separating the vane pump.

19. Remove alternator assembly.
20. Separate cooler compressor assembly.
21. Remove the 5 bolts and V-ribbed belt tensioner.
22. Remove the 2 bolts and 2 idler pulleys.
23. Remove the bolt and idler pulley.
24. Using service tool, hold the crankshaft pulley and loosen the pulley set bolt.
25. Using the pulley set bolt and service tool, remove the crankshaft pulley.
26. Remove the 10 bolts and 2 nuts.
27. Insert the blade of oil pan seal cutter between the oil pan and No. 2 oil pan, cut off applied sealer and remove the No. 2 oil pan.

➡Be careful not to damage the contact surfaces of the oil pan and No. 2 oil pan. Be careful not to damage the No. 2 oil pan flange.

28. Remove the 2 nuts, oil strainer and gasket.
29. Remove the 4 housing bolts.
30. Remove the flywheel housing under cover.
31. Remove the 17 bolts and 2 nuts.
32. Using a screwdriver, remove the oil pan by prying between the oil pan and cylinder block.
33. Remove the O-ring from the oil pump.
34. Remove intake air surge tank.
35. Remove ignition coil assembly.
36. Remove cylinder head cover sub-assemblies.
37. Disconnect the 2 oil control valve connectors.
38. Remove the 2 bolts and 2 camshaft timing oil control valves.
39. Remove VVT sensor.
40. Remove timing chain or belt cover sub-assembly.
41. Remove timing gear case or timing chain case oil seal.
42. Set No. 1 cylinder to TDC by installing the crankshaft pulley set bolt, and turn the crankshaft to align the crankshaft set key with the timing line of the cylinder block.

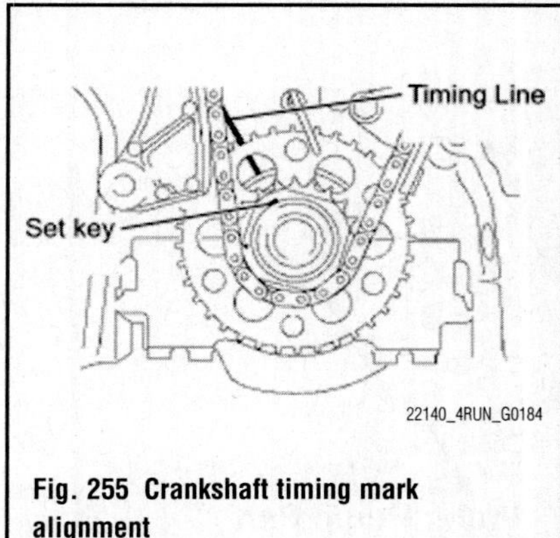

Fig. 255 Crankshaft timing mark alignment

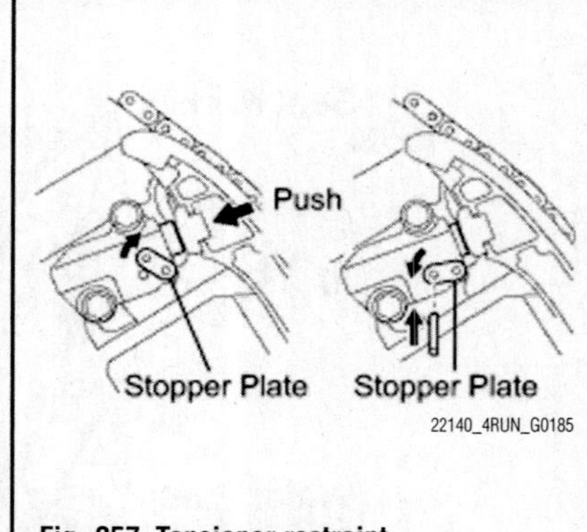

Fig. 257 Tensioner restraint

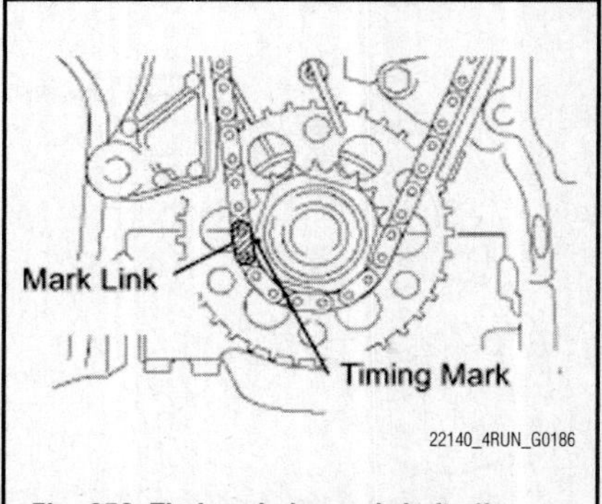

Fig. 258 Timing chain crankshaft alignment

43. Check that the timing marks of the camshaft timing gears are aligned with the timing marks of the bearing caps as shown in the illustration. If not, turn the crankshaft 1 complete revolution (360°) and align the timing marks.

➡Never rotate the crankshaft with the chain tensioner removed. When rotating the camshaft with the timing chain removed, rotate the crankshaft counterclockwise 40°from the TDC first.

44. While turning the stopper plate of the tensioner clockwise, push in the plunger of the chain tensioner.
45. While turning the stopper plate of the tensioner counterclockwise, insert a bar of 3.5 mm (0.138 in.) into the holes on the stopper plate and tensioner to fix the stopper plate.
46. Remove the 2 bolts and chain tensioner.

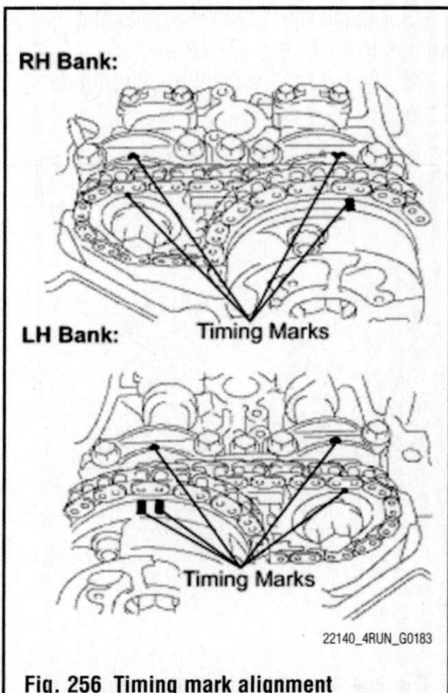

Fig. 256 Timing mark alignment

47. Remove chain tensioner slipper.
48. Using a 10 mm hexagon wrench, remove the No. 2 idle gear shaft, No. 1 idle gear and No. 1 idle gear shaft.
49. Remove the 2 No. 2 chain vibration dampers.
50. Remove chain sub-assembly.

To install:

51. Install chain tensioner slipper.
52. While turning the stopper plate of the tensioner clockwise, push in the plunger of the tensioner.
53. While turning the stopper plate of the tensioner counterclockwise, insert a bar of _3.5 mm (0.138 in.) into the holes on the stopper plate and tensioner to fix the stopper plate.
54. Install the chain tensioner with the 2 bolts and tighten to 7 ft. lbs, (10 Nm).
55. Set the No. 1 cylinder to the TDC / compression.
56. Align the timing marks of the camshaft timing gears and bearing caps.
57. Install the crankshaft pulley set bolt, and turn the crankshaft to align the crankshaft set key with the timing line of the cylinder block.
58. Align the mark link (yellow) with the timing mark of the crankshaft timing gear.
59. Align the mark links (orange) with the timing marks of the camshaft timing gears, and install the chain.
60. Install the 2 No. 2 chain vibration dampers.
61. Apply a light coat of engine oil to the rotating surface of the No. 1 idle gear shaft.
62. Temporarily install the No. 1 idle gear shaft and No. 1 idle gear with the No. 2 idle gear shaft while aligning the knock pin of the No. 1 idle gear shaft with the knock pin groove of the cylinder block.

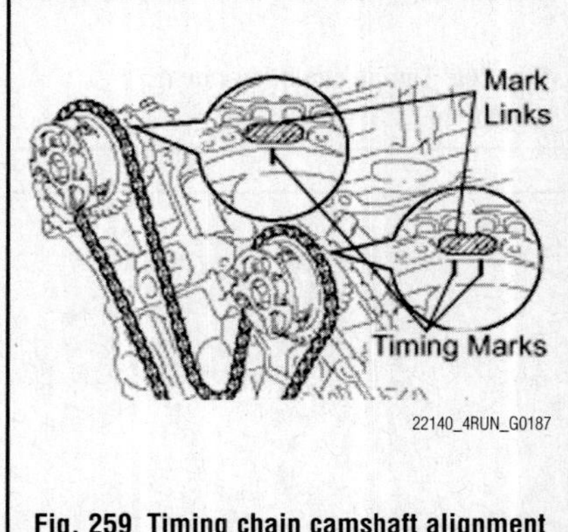

Fig. 259 Timing chain camshaft alignment

➡Be careful of the idle gear direction.

63. Using a 10 mm hexagon wrench, tighten the No. 2 idle gear shaft to 44 ft. lbs. (60 Nm).
64. Remove the bar from the chain tensioner.
65. Install timing gear case or timing chain case oil seal.
66. Install timing chain or belt cover sub-assembly.
    a. Install a new O-ring onto the LH cylinder head.
    b. Apply continuous beads of seal packing (diameter 3 to 4 mm (0.12 to 0.16 in.)) to the 4 locations as shown in the illustration.
    c. Apply continuous beads of seal packing (diameter 3 to 4 mm (0.12 to 0.16 in.)) to the timing chain cover as shown in the illustration.

➡Install the timing chain cover within 3 minutes of applying the seal packing. Timing chain cover bolts and nuts must be tightened within 15 minutes of installation. Otherwise, the seal packing must be removed and reapplied. Do not apply seal packing to portion A shown in the illustration.

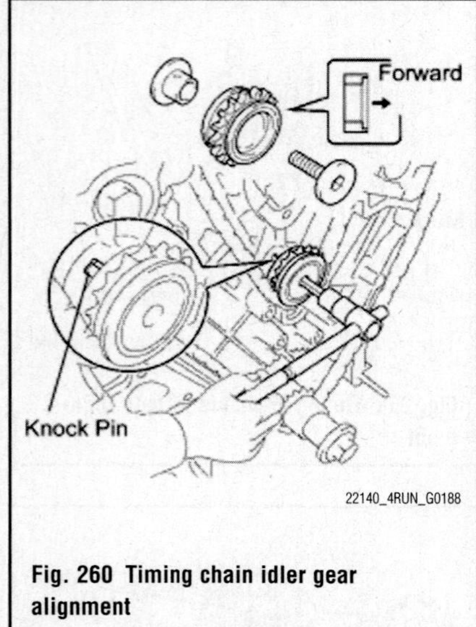

Fig. 260 Timing chain idler gear alignment

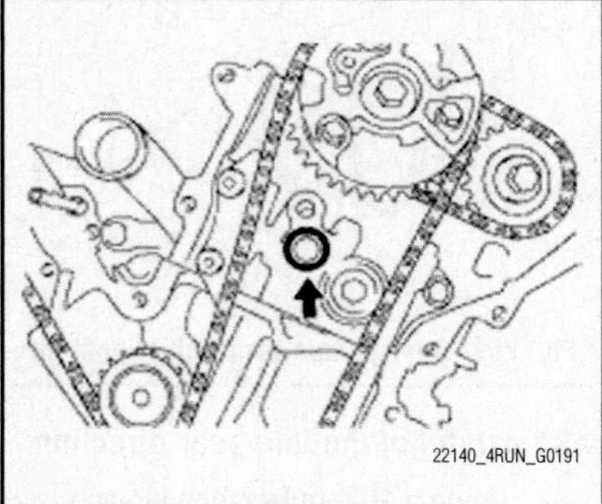

Fig. 261 Timing chain cover o-ring position

d. Align the key way of the oil pump drive rotor with the rectangular portion of the crankshaft timing gear, and slide the timing chain cover into place.

e. Install the timing chain cover with the 24 bolts and 2 nuts. Tighten the bolts and nuts uniformly in several steps. Tighten to 17 ft. lbs. (23 Nm).

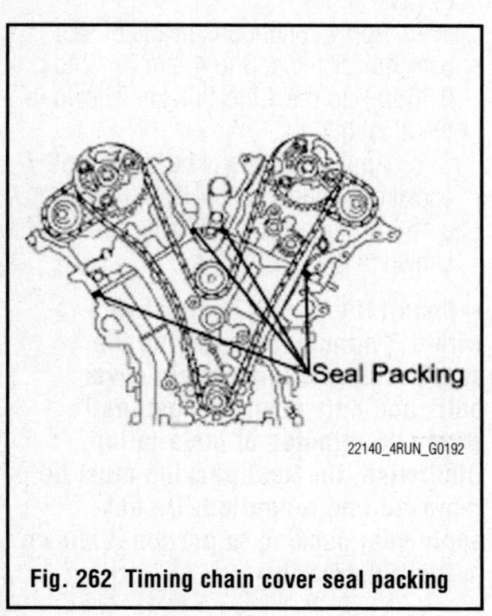

Fig. 262 Timing chain cover seal packing

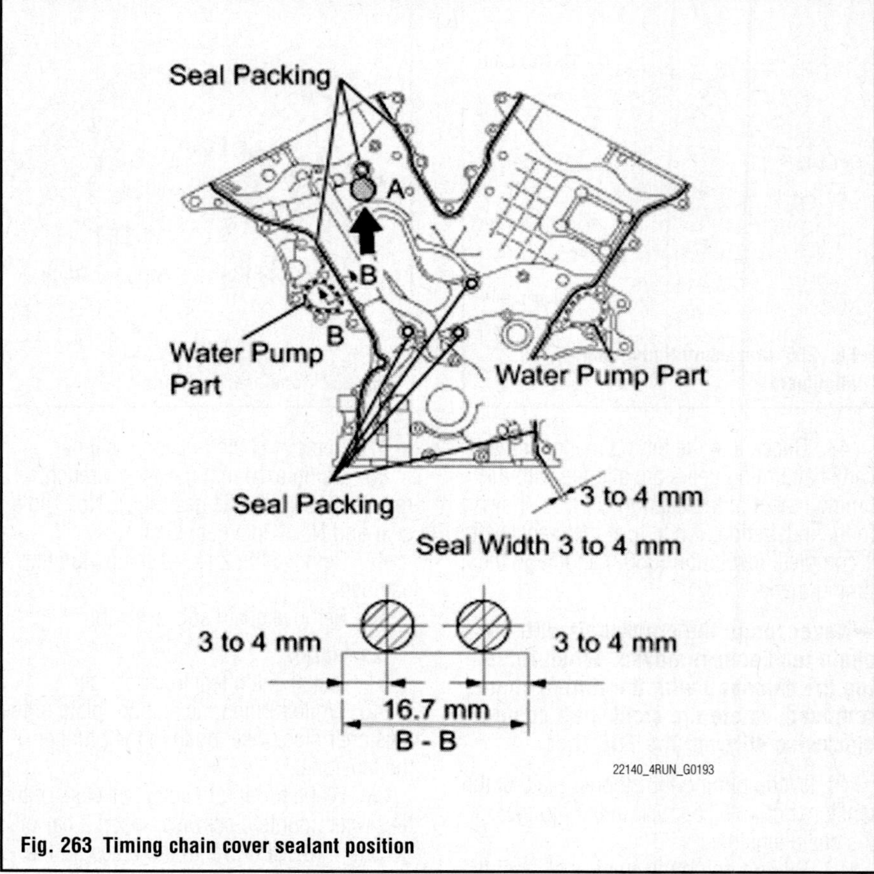

Fig. 263 Timing chain cover sealant position

➡Pay attention not to wrap the chain and slipper over the timing chain cover seal line.

67. Install VVT sensor.
68. Insert the camshaft timing oil control valves to each cylinder head, and tighten the 2 bolts to 7 ft. lbs. (10 Nm).
69. Install intake air surge tank.
70. Remove any old packing (FIPG) material and be careful not to drop any oil on the contact surfaces of the cylinder block, rear oil seal retainer and oil pan.
71. Install a new O-ring to the oil pump.
72. Apply a continuous bead of seal packing (diameter: 3 to 4 mm (0.12 to 0.16 in.)) to the oil pan.

➡Install the oil pan within 3 minutes after applying seal packing. After installing it, the oil pan bolts and nuts must be tightened within 15 minutes. Otherwise the seal packing must be removed and reapplied.

73. Install the oil pan with the 17 bolts and 2 nuts. Tighten the bolts and nuts uniformly in several steps.
74. Tighten bolts as follows (refer to the Oil Pan procedure for bolt location):
    - Bolt A 15 ft. lbs. (21 Nm)
    - Bolt B 15 ft. lbs. (21 Nm)
    - Bolt C 7 ft. lbs. (10 Nm)
    - Nut 15 ft. lbs. (21 Nm)
75. Install the 4 housing bolts and tighten to 27 ft. lbs. (37 Nm).
76. Install the flywheel housing under cover.

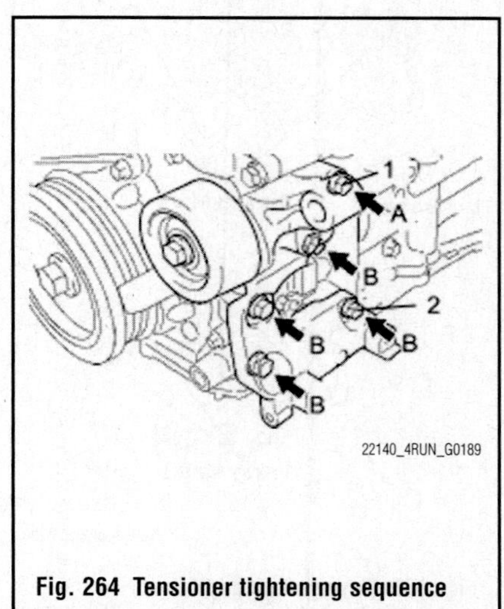

Fig. 264 Tensioner tightening sequence

77. Install a new gasket and the oil strainer with the 2 nuts and tighten to 80 inch lbs. (9.0 Nm).

78. Remove any old packing (FIPG) material and be careful not to drop any oil on the contact surfaces of the oil pan and No. 2 oil pan.

➡Install the No. 2 oil pan within 3 minutes after applying seal packing. After installing it, the No. 2 oil pan bolts and nuts must be tightened within 15 minutes. Otherwise the seal packing must be removed and reapplied.

79. Install the No. 2 oil pan with the 10 bolts and 2 nuts. Tighten the bolts and nuts uniformly in several steps. Tighten bolts to 80 inch lbs. (9.0 Nm) and nuts to 7 ft. lbs. (10 Nm).

80. Install the pulley set bolt and tighten to 184 ft. lbs. (250 Nm).

81. Install the idler pulley with the bolt and tighten to 40 ft. lbs. (54 Nm).

➡"DOUBLE" is marked on the No. 1 idler pulley to distinguish it from the No. 2 idler pulley.

82. Install the 2 idler pulleys with the 2 bolts and tighten to 29 ft. lbs. (39 Nm).

83. Temporarily install the V-ribbed belt tensioner with the 5 bolts.

84. Install the V-ribbed belt tensioner by tightening the bolt 1 and bolt 2 in the order shown in the illustration. Tighten to 27 ft. lbs. (36 Nm).

85. Install cooler compressor assembly.

86. Install alternator assembly.

87. Install the vane pump with the 2 bolts an tighten to 32 ft. lbs. (43 Nm).

88. Connect the P/S oil pressure switch connector.

89. Install water inlet.

90. Install a new O-ring to the oil level gauge guide.

91. Apply a light coat of engine oil to the O-ring.

92. Push in the oil level gauge guide end into the guide hole of the oil pan.

93. Install the oil level gauge guide with the bolt.

94. Install the oil level gauge.

95. Install air cleaner assembly.

96. Connect No. 2 ventilation hose.

97. Install fluid coupling assembly.

98. Install fan and alternator v-belt.

99. Fully tighten fluid coupling assembly.

100. Install radiator support seal upper.

101. Install battery.

102. Install differential carrier assembly front as necessary.

103. Install power steering link assembly.

104. Add engine oil.

105. Add engine coolant.

106. Check for engine leaks.

107. Install the V-bank cover with the 2 nuts.

2010 Models

Refer to CAMSHAFT for removal and installation procedures.

VALVE LASH

ADJUSTMENT

2009 Models

4.0L Engine

See Figures 265 through 268.

➡Measure valve clearance with the engine cold.

1. Before servicing the vehicle, refer to the precautions section.
2. Drain the cooling system.
3. Remove or disconnect the following:
   - Negative battery cable
   - Ignition coils
   - Valve covers
4. Turn the crankshaft pulley, and align the notch with the timing mark "0" of the timing chain cover
5. Check the valves indicated in the illustration
6. Turn the crankshaft 2/3 of a revolution (240°), and check the valves indicated in the illustration0
7. Turn the crankshaft 2/3 of a revolution (240°), and check the valves indicated in the illustration.
8. The valve clearance specifications are as follows:
   - Intake: 0.006–0.010 in. (0.15–0.25mm)
   - Exhaust: 0.011–0.015 in. (0.29–0.39mm)
9. Record the measurements for each valve.
10. When all valve clearances have been measured, remove the camshafts if necessary.
11. Remove the valve shims and measure them. Note this measurement along with the clearance measurement recorded earlier.
12. Using the valve clearance and shim thickness measurements, find replacement shims in the Adjusting Shim Selection charts.

To install:

13. Install or connect the following:
    - Replacement valve shims
    - Camshafts
    - Valve covers
    - Ignition coils
    - Negative battery cable
14. Fill the cooling system.
15. Start the engine and check for leaks.

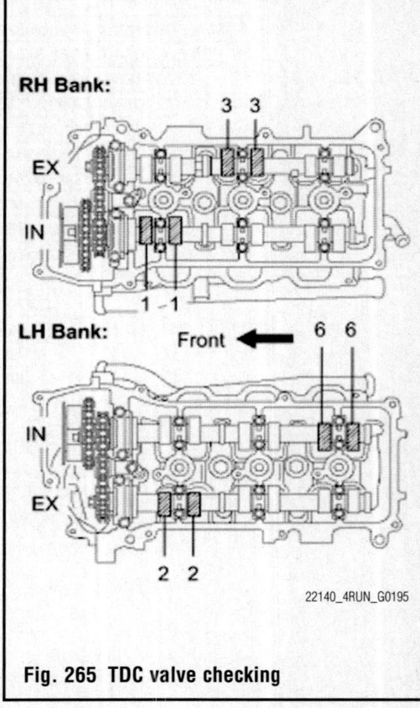

Fig. 265 TDC valve checking

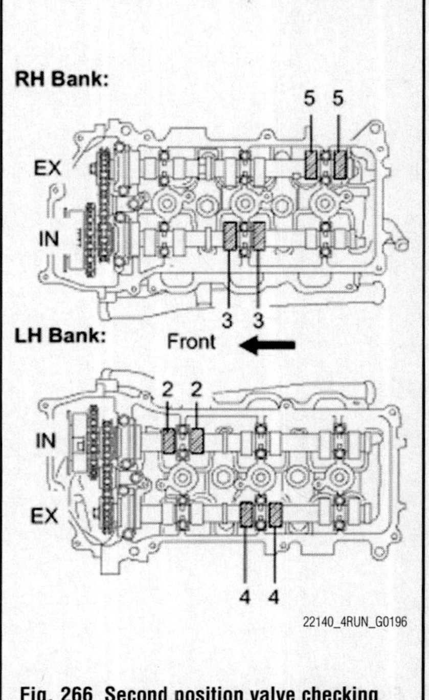

Fig. 266 Second position valve checking

Valve Lifter Selection Chart (Intake)

0.851 - 0.870 (0.0335 - 0.0343)	0.831 - 0.850 (0.0327 - 0.0335)	0.811 - 0.830 (0.0319 - 0.0327)	0.791 - 0.810 (0.0311 - 0.0319)	0.771 - 0.790 (0.0304 - 0.0311)	0.751 - 0.770 (0.0296 - 0.0303)	0.731 - 0.750 (0.0288 - 0.0295)	0.711 - 0.730 (0.0280 - 0.0287)	0.691 - 0.710 (0.0272 - 0.0280)	0.671 - 0.690 (0.0264 - 0.0272)	0.651 - 0.670 (0.0256 - 0.0264)	0.631 - 0.650 (0.0248 - 0.0256)	0.611 - 0.630 (0.0241 - 0.0248)	0.591 - 0.610 (0.0233 - 0.0240)	0.571 - 0.590 (0.0225 - 0.0232)	0.551 - 0.570 (0.0217 - 0.0224)	0.531 - 0.550 (0.0209 - 0.0217)	0.511 - 0.530 (0.0201 - 0.0209)	0.491 - 0.510 (0.0193 - 0.0201)	0.471 - 0.490 (0.0185 - 0.0193)	0.451 - 0.470 (0.0178 - 0.0185)	0.431 - 0.450 (0.0170 - 0.0177)	0.411 - 0.430 (0.0162 - 0.0169)	0.391 - 0.410 (0.0154 - 0.0161)	0.371 - 0.390 (0.0146 - 0.0154)	0.351 - 0.370 (0.0138 - 0.0146)	0.331 - 0.350 (0.0130 - 0.0138)	0.311 - 0.330 (0.0122 - 0.0130)	0.291 - 0.310 (0.0115 - 0.0122)	0.271 - 0.290 (0.0107 - 0.0114)	0.251 - 0.270 (0.0099 - 0.0106)	0.141 - 0.149 (0.0056 - 0.0059)	0.121 - 0.140 (0.0048 - 0.0055)	0.101 - 0.120 (0.0040 - 0.0047)	0.081 - 0.100 (0.0032 - 0.0039)	0.061 - 0.080 (0.0024 - 0.0031)	0.041 - 0.060 (0.0016 - 0.0024)	0.021 - 0.040 (0.0008 - 0.0016)	0.000 - 0.020 (0.0000 - 0.0008)	Measured clearance mm (in.) / Removed lifter thickness mm (in.)
72	70	68	66	64	62	60	58	56	54	52	50	48	46	44	42	40	38	36	34	32	30	28	26	24	22	20	18	16	14	12									5.060 (0.1992)
	72	70	68	66	64	62	60	58	56	54	52	50	48	46	44	42	40	38	36	34	32	30	28	26	24	22	20	18	16	14									5.080 (0.2000)
		72	70	68	66	64	62	60	58	56	54	52	50	48	46	44	42	40	38	36	34	32	30	28	26	24	22	20	18	16									5.100 (0.2008)
			72	70	68	66	64	62	60	58	56	54	52	50	48	46	44	42	40	38	36	34	32	30	28	26	24	22	20	18	06								5.120 (0.2016)
				72	70	68	66	64	62	60	58	56	54	52	50	48	46	44	42	40	38	36	34	32	30	28	26	24	22	20	08	08							5.140 (0.2024)
					72	70	68	66	64	62	60	58	56	54	52	50	48	46	44	42	40	38	36	34	32	30	28	26	24	22	10	10	08						5.160 (0.2031)
						72	70	68	66	64	62	60	58	56	54	52	50	48	46	44	42	40	38	36	34	32	30	28	26	24	12	12	10	08					5.180 (0.2039)
							72	70	68	66	64	62	60	58	56	54	52	50	48	46	44	42	40	38	36	34	32	30	28	26	14	14	12	10	08				5.200 (0.2047)
							74	72	70	68	66	64	62	60	58	56	54	52	50	48	46	44	42	40	38	36	34	32	30	28	16	14	12	10	08	06			5.210 (0.2051)
								72	70	68	66	64	62	60	58	56	54	52	50	48	46	44	42	40	38	36	34	32	30	28	16	16	14	12	10	08			5.220 (0.2055)
								74	72	70	68	66	64	62	60	58	56	54	52	50	48	46	44	42	40	38	36	34	32	30	18	16	14	12	10	08	06		5.230 (0.2059)
									72	70	68	66	64	62	60	58	56	54	52	50	48	46	44	42	40	38	36	34	32	30	18	18	16	14	12	10	08		5.240 (0.2063)
									74	72	70	68	66	64	62	60	58	56	54	52	50	48	46	44	42	40	38	36	34	32	20	18	16	14	12	10	08	06	5.250 (0.2067)
										72	70	68	66	64	62	60	58	56	54	52	50	48	46	44	42	40	38	36	34	32	20	20	18	16	14	12	10	08	5.260 (0.2071)
										74	72	70	68	66	64	62	60	58	56	54	52	50	48	46	44	42	40	38	36	34	22	20	18	16	14	12	10	08	5.270 (0.2075)
											72	70	68	66	64	62	60	58	56	54	52	50	48	46	44	42	40	38	36	34	22	22	20	18	16	14	12	10	5.280 (0.2079)
											74	72	70	68	66	64	62	60	58	56	54	52	50	48	46	44	42	40	38	36	24	22	20	18	16	14	12	10	5.290 (0.2083)
												72	70	68	66	64	62	60	58	56	54	52	50	48	46	44	42	40	38	36	24	24	22	20	18	16	14	12	5.300 (0.2087)
												74	72	70	68	66	64	62	60	58	56	54	52	50	48	46	44	42	40	38	26	24	22	20	18	16	14	12	5.310 (0.2091)
													72	70	68	66	64	62	60	58	56	54	52	50	48	46	44	42	40	38	26	26	24	22	20	18	16	14	5.320 (0.2094)
													74	72	70	68	66	64	62	60	58	56	54	52	50	48	46	44	42	40	28	26	24	22	20	18	16	14	5.330 (0.2098)
														72	70	68	66	64	62	60	58	56	54	52	50	48	46	44	42	40	28	28	26	24	22	20	18	16	5.340 (0.2102)
														74	72	70	68	66	64	62	60	58	56	54	52	50	48	46	44	42	30	28	26	24	22	20	18	16	5.350 (0.2106)
															72	70	68	66	64	62	60	58	56	54	52	50	48	46	44	42	30	30	28	26	24	22	20	18	5.360 (0.2110)
															74	72	70	68	66	64	62	60	58	56	54	52	50	48	46	44	32	30	28	26	24	22	20	18	5.370 (0.2114)
																72	70	68	66	64	62	60	58	56	54	52	50	48	46	44	32	32	30	28	26	24	22	20	5.380 (0.2118)
																74	72	70	68	66	64	62	60	58	56	54	52	50	48	46	34	32	30	28	26	24	22	20	5.390 (0.2122)
																	72	70	68	66	64	62	60	58	56	54	52	50	48	46	34	34	32	30	28	26	24	22	5.400 (0.2126)
																	74	72	70	68	66	64	62	60	58	56	54	52	50	48	36	34	32	30	28	26	24	22	5.410 (0.2130)
																		72	70	68	66	64	62	60	58	56	54	52	50	48	36	36	34	32	30	28	26	24	5.420 (0.2134)
																		74	72	70	68	66	64	62	60	58	56	54	52	50	38	36	34	32	30	28	26	24	5.430 (0.2138)
																			72	70	68	66	64	62	60	58	56	54	52	50	38	38	36	34	32	30	28	26	5.440 (0.2142)
																			74	72	70	68	66	64	62	60	58	56	54	52	40	38	36	34	32	30	28	26	5.450 (0.2146)
																				72	70	68	66	64	62	60	58	56	54	52	40	40	38	36	34	32	30	28	5.460 (0.2150)
																				74	72	70	68	66	64	62	60	58	56	54	42	40	38	36	34	32	30	28	5.470 (0.2154)
																					72	70	68	66	64	62	60	58	56	54	42	42	40	38	36	34	32	30	5.480 (0.2157)
																					74	72	70	68	66	64	62	60	58	56	44	42	40	38	36	34	32	30	5.490 (0.2161)
																						72	70	68	66	64	62	60	58	56	44	44	42	40	38	36	34	32	5.500 (0.2165)
																						74	72	70	68	66	64	62	60	58	46	44	42	40	38	36	34	32	5.510 (0.2169)
																							72	70	68	66	64	62	60	58	46	46	44	42	40	38	36	34	5.520 (0.2173)
																							74	72	70	68	66	64	62	60	48	46	44	42	40	38	36	34	5.530 (0.2177)
																								72	70	68	66	64	62	60	48	48	46	44	42	40	38	36	5.540 (0.2181)
																								74	72	70	68	66	64	62	50	48	46	44	42	40	38	36	5.550 (0.2185)
																									72	70	68	66	64	62	50	50	48	46	44	42	40	38	5.560 (0.2189)
																									74	72	70	68	66	64	52	50	48	46	44	42	40	38	5.570 (0.2193)
																										72	70	68	66	64	52	52	50	48	46	44	42	40	5.580 (0.2197)
																										74	72	70	68	66	54	52	50	48	46	44	42	40	5.590 (0.2201)
																											72	70	68	66	54	54	52	50	48	46	44	42	5.600 (0.2205)
																												72	70	68	56	56	54	52	50	48	46	44	5.620 (0.2213)
																													72	70	58	58	56	54	52	50	48	46	5.640 (0.2220)
																														72	60	60	58	56	54	52	50	48	5.660 (0.2228)
																															62	62	60	58	56	54	52	50	5.680 (0.2236)
																															64	64	62	60	58	56	54	52	5.700 (0.2244)
																															66	66	64	62	60	58	56	54	5.720 (0.2252)
																															68	68	66	64	62	60	58	56	5.740 (0.2260)

22140_4RUN_G0197

Fig. 267 Intake valve lifter chart

Valve Lifter Selection Chart (Exhaust)

Removed lifter thickness mm (in.) \ Measured clearance mm (in.)	0.991 - 1.010 (0.0390 - 0.0398)	0.971 - 0.990 (0.0382 - 0.0390)	0.951 - 0.970 (0.0374 - 0.0382)	0.931 - 0.950 (0.0367 - 0.0374)	0.911 - 0.930 (0.0359 - 0.0366)	0.891 - 0.910 (0.0351 - 0.0358)	0.871 - 0.890 (0.0343 - 0.0350)	0.851 - 0.870 (0.0335 - 0.0343)	0.831 - 0.850 (0.0327 - 0.0335)	0.811 - 0.830 (0.0319 - 0.0327)	0.791 - 0.810 (0.0311 - 0.0319)	0.771 - 0.790 (0.0304 - 0.0311)	0.751 - 0.770 (0.0296 - 0.0303)	0.731 - 0.750 (0.0288 - 0.0295)	0.711 - 0.730 (0.0280 - 0.0287)	0.691 - 0.710 (0.0272 - 0.0280)	0.671 - 0.690 (0.0264 - 0.0272)	0.651 - 0.670 (0.0256 - 0.0264)	0.631 - 0.650 (0.0248 - 0.0256)	0.611 - 0.630 (0.0241 - 0.0248)	0.591 - 0.610 (0.0233 - 0.0240)	0.571 - 0.590 (0.0225 - 0.0232)	0.551 - 0.570 (0.0217 - 0.0224)	0.531 - 0.550 (0.0209 - 0.0217)	0.511 - 0.530 (0.0201 - 0.0209)	0.491 - 0.510 (0.0193 - 0.0201)	0.471 - 0.490 (0.0185 - 0.0193)	0.451 - 0.470 (0.0178 - 0.0185)	0.431 - 0.450 (0.0170 - 0.0177)	0.411 - 0.430 (0.0162 - 0.0169)	0.391 - 0.410 (0.0154 - 0.0161)	0.281 - 0.289 (0.0111 - 0.0114)	0.261 - 0.280 (0.0103 - 0.0110)	0.241 - 0.260 (0.0095 - 0.0102)	0.221 - 0.240 (0.0087 - 0.0094)	0.201 - 0.220 (0.0079 - 0.0087)	0.181 - 0.200 (0.0071 - 0.0079)	0.161 - 0.180 (0.0063 - 0.0071)	0.141 - 0.160 (0.0056 - 0.0063)	0.121 - 0.140 (0.0048 - 0.0055)	0.101 - 0.120 (0.0040 - 0.0047)	0.081 - 0.100 (0.0032 - 0.0039)	0.061 - 0.080 (0.0024 - 0.0031)	0.041 - 0.060 (0.0016 - 0.0024)	0.021 - 0.040 (0.0008 - 0.0016)	0.000 - 0.020 (0.0000 - 0.0008)
5.060 (0.1992)	72	70	68	66	64	62	60	58	56	54	52	50	48	46	44	42	40	38	36	34	32	30	28	26	24	22	20	18	16	14	12															
5.080 (0.2000)		72	70	68	66	64	62	60	58	56	54	52	50	48	46	44	42	40	38	36	34	32	30	28	26	24	22	20	18	16	14															
5.100 (0.2008)			72	70	68	66	64	62	60	58	56	54	52	50	48	46	44	42	40	38	36	34	32	30	28	26	24	22	20	18	16															
5.120 (0.2016)				72	70	68	66	64	62	60	58	56	54	52	50	48	46	44	42	40	38	36	34	32	30	28	26	24	22	20	18	06														
5.140 (0.2024)					72	70	68	66	64	62	60	58	56	54	52	50	48	46	44	42	40	38	36	34	32	30	28	26	24	22	20	08	08													
5.160 (0.2031)						72	70	68	66	64	62	60	58	56	54	52	50	48	46	44	42	40	38	36	34	32	30	28	26	24	22	10	10	08												
5.180 (0.2039)							72	70	68	66	64	62	60	58	56	54	52	50	48	46	44	42	40	38	36	34	32	30	28	26	24	12	12	10	08											
5.200 (0.2047)								72	70	68	66	64	62	60	58	56	54	52	50	48	46	44	42	40	38	36	34	32	30	28	26	14	14	12	10	08										
5.210 (0.2051)								74	72	70	68	66	64	62	60	58	56	54	52	50	48	46	44	42	40	38	36	34	32	30	28	16	14	12	10	08	06									
5.220 (0.2055)									72	70	68	66	64	62	60	58	56	54	52	50	48	46	44	42	40	38	36	34	32	30	28	16	16	14	12	10	08									
5.230 (0.2059)									74	72	70	68	66	64	62	60	58	56	54	52	50	48	46	44	42	40	38	36	34	32	30	18	16	14	12	10	08	06								
5.240 (0.2063)										72	70	68	66	64	62	60	58	56	54	52	50	48	46	44	42	40	38	36	34	32	30	18	18	16	14	12	10	08								
5.250 (0.2067)										74	72	70	68	66	64	62	60	58	56	54	52	50	48	46	44	42	40	38	36	34	32	20	18	16	14	12	10	08	06							
5.260 (0.2071)											72	70	68	66	64	62	60	58	56	54	52	50	48	46	44	42	40	38	36	34	32	20	20	18	16	14	12	10	08							
5.270 (0.2075)											74	72	70	68	66	64	62	60	58	56	54	52	50	48	46	44	42	40	38	36	34	22	20	18	16	14	12	10	08	06						
5.280 (0.2079)												72	70	68	66	64	62	60	58	56	54	52	50	48	46	44	42	40	38	36	34	22	22	20	18	16	14	12	10	08						
5.290 (0.2083)												74	72	70	68	66	64	62	60	58	56	54	52	50	48	46	44	42	40	38	36	24	22	20	18	16	14	12	10	08	06					
5.300 (0.2087)													72	70	68	66	64	62	60	58	56	54	52	50	48	46	44	42	40	38	36	24	24	22	20	18	16	14	12	10	08					
5.310 (0.2091)													74	72	70	68	66	64	62	60	58	56	54	52	50	48	46	44	42	40	38	26	24	22	20	18	16	14	12	10	08	06				
5.320 (0.2094)														72	70	68	66	64	62	60	58	56	54	52	50	48	46	44	42	40	38	26	26	24	22	20	18	16	14	12	10	08				
5.330 (0.2098)														74	72	70	68	66	64	62	60	58	56	54	52	50	48	46	44	42	40	28	26	24	22	20	18	16	14	12	10	08	06			
5.340 (0.2102)															72	70	68	66	64	62	60	58	56	54	52	50	48	46	44	42	40	28	28	26	24	22	20	18	16	14	12	10	08			
5.350 (0.2106)															74	72	70	68	66	64	62	60	58	56	54	52	50	48	46	44	42	30	28	26	24	22	20	18	16	14	12	10	08	06		
5.360 (0.2110)																72	70	68	66	64	62	60	58	56	54	52	50	48	46	44	42	30	30	28	26	24	22	20	18	16	14	12	10	08		
5.370 (0.2114)																74	72	70	68	66	64	62	60	58	56	54	52	50	48	46	44	32	30	28	26	24	22	20	18	16	14	12	10	08	06	
5.380 (0.2118)																	72	70	68	66	64	62	60	58	56	54	52	50	48	46	44	32	32	30	28	26	24	22	20	18	16	14	12	10	08	
5.390 (0.2122)																	74	72	70	68	66	64	62	60	58	56	54	52	50	48	46	34	32	30	28	26	24	22	20	18	16	14	12	10	08	06
5.400 (0.2126)																		72	70	68	66	64	62	60	58	56	54	52	50	48	46	34	34	32	30	28	26	24	22	20	18	16	14	12	10	08
5.410 (0.2130)																		74	72	70	68	66	64	62	60	58	56	54	52	50	48	36	34	32	30	28	26	24	22	20	18	16	14	12	10	08
5.420 (0.2134)																			72	70	68	66	64	62	60	58	56	54	52	50	48	36	36	34	32	30	28	26	24	22	20	18	16	14	12	10
5.430 (0.2138)																			74	72	70	68	66	64	62	60	58	56	54	52	50	38	36	34	32	30	28	26	24	22	20	18	16	14	12	10
5.440 (0.2142)																				72	70	68	66	64	62	60	58	56	54	52	50	38	38	36	34	32	30	28	26	24	22	20	18	16	14	12
5.450 (0.2146)																				74	72	70	68	66	64	62	60	58	56	54	52	40	38	36	34	32	30	28	26	24	22	20	18	16	14	12
5.460 (0.2150)																					72	70	68	66	64	62	60	58	56	54	52	40	40	38	36	34	32	30	28	26	24	22	20	18	16	14
5.470 (0.2154)																					74	72	70	68	66	64	62	60	58	56	54	42	40	38	36	34	32	30	28	26	24	22	20	18	16	14
5.480 (0.2157)																						72	70	68	66	64	62	60	58	56	54	42	42	40	38	36	34	32	30	28	26	24	22	20	18	16
5.490 (0.2161)																						74	72	70	68	66	64	62	60	58	56	44	42	40	38	36	34	32	30	28	26	24	22	20	18	16
5.500 (0.2165)																							72	70	68	66	64	62	60	58	56	44	44	42	40	38	36	34	32	30	28	26	24	22	20	18
5.510 (0.2169)																							74	72	70	68	66	64	62	60	58	46	44	42	40	38	36	34	32	30	28	26	24	22	20	18
5.520 (0.2173)																								72	70	68	66	64	62	60	58	46	46	44	42	40	38	36	34	32	30	28	26	24	22	20
5.530 (0.2177)																								74	72	70	68	66	64	62	60	48	46	44	42	40	38	36	34	32	30	28	26	24	22	20
5.540 (0.2181)																									72	70	68	66	64	62	60	48	48	46	44	42	40	38	36	34	32	30	28	26	24	22
5.550 (0.2185)																									74	72	70	68	66	64	62	50	48	46	44	42	40	38	36	34	32	30	28	26	24	22
5.560 (0.2189)																										72	70	68	66	64	62	50	50	48	46	44	42	40	38	36	34	32	30	28	26	24
5.570 (0.2193)																										74	72	70	68	66	64	52	50	48	46	44	42	40	38	36	34	32	30	28	26	24
5.580 (0.2197)																											72	70	68	66	64	52	52	50	48	46	44	42	40	38	36	34	32	30	28	26
5.590 (0.2201)																											74	72	70	68	66	54	52	50	48	46	44	42	40	38	36	34	32	30	28	26
5.600 (0.2205)																												72	70	68	66	54	54	52	50	48	46	44	42	40	38	36	34	32	30	28
5.620 (0.2213)																													72	70	68	56	56	54	52	50	48	46	44	42	40	38	36	34	32	30
5.640 (0.2220)																														72	70	58	58	56	54	52	50	48	46	44	42	40	38	36	34	32
5.660 (0.2228)																															72	60	60	58	56	54	52	50	48	46	44	42	40	38	36	34
5.680 (0.2236)																																62	62	60	58	56	54	52	50	48	46	44	42	40	38	36
5.700 (0.2244)																																64	64	62	60	58	56	54	52	50	48	46	44	42	40	38
5.720 (0.2252)																																66	66	64	62	60	58	56	54	52	50	48	46	44	42	40
5.740 (0.2260)																																68	68	66	64	62	60	58	56	54	52	50	48	46	44	42

22140_4RUN_G0198

Fig. 268 Exhaust valve lifter chart

INSPECTION

2010 Models

See Figure 269.

➡Keep the valve lash adjuster assembly free of dirt and foreign objects.

➡Only use clean engine oil.

1. Place the valve lash adjuster assembly into a container filled with engine oil.
2. Insert the tip of special tool into the valve lash adjuster assembly plunger and use the tip to press down on the check ball inside the plunger.
3. Squeeze SST and the valve lash adjuster assembly together to move the plunger up and down 5 to 6 times.
4. Check the movement of the plunger and bleed the air. If the movement is OK, the plunger moves up and down.

➡When bleeding air from the high-pressure chamber, make sure that the tip of SST is actually pressing the check ball as shown in the illustration. If the check ball is not pressed, air will not bleed.

5. After bleeding the air, remove SST. Then try to quickly and firmly press the plunger by hand. If everything is OK, the plunger is very difficult to move. If the result is not as specified, replace the valve lash adjuster assembly.

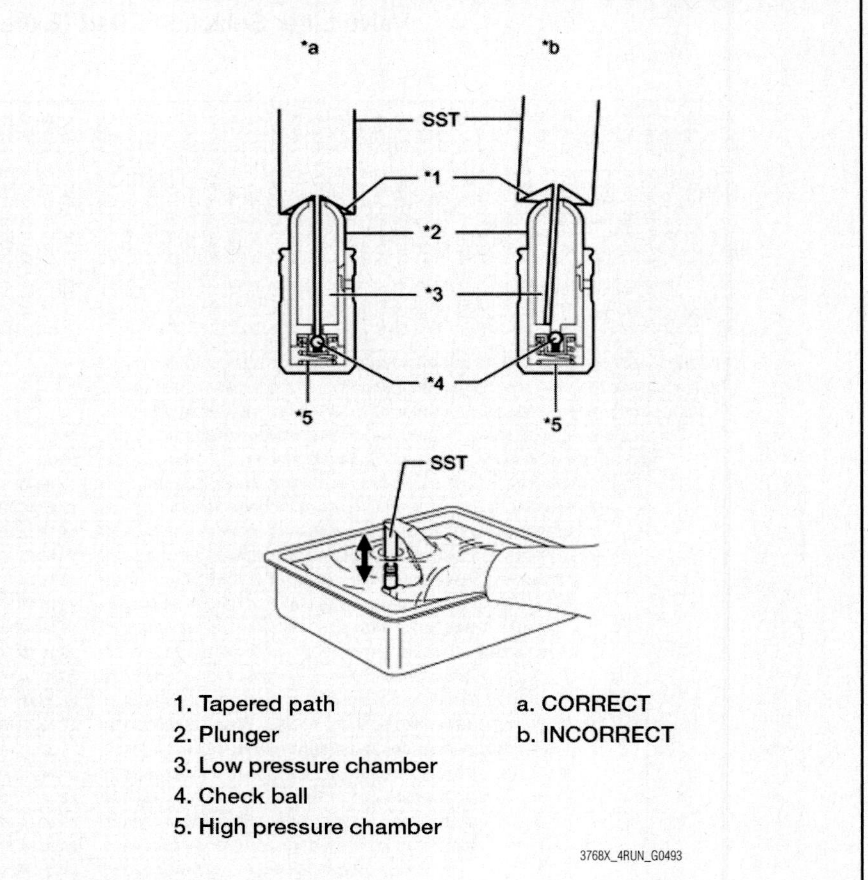

Fig. 269 Inspecting the valve lash adjuster

ENGINE PERFORMANCE & EMISSION CONTROLS

ACCELERATOR PEDAL POSITION (APP) SENSOR

LOCATION

See Figures 270 and 271.

REMOVAL & INSTALLATION

1. Disconnect the accelerator pedal connector.
2. Remove the 2 nuts and accelerator pedal assembly.

To install:

➡Be care not to give a shock to the accelerator pedal assembly. Be careful not to disassemble the accelerator pedal assembly.

3. Install the accelerator pedal assembly with the 2 nuts. Tighten nuts to 44 inch lbs. (5.0 Nm)
4. Connect the accelerator pedal connector.

CAMSHAFT POSITION (CMP) SENSOR

LOCATION

See Figures 272 through 274.

REMOVAL & INSTALLATION

2009 Models

1. Drain engine coolant.
2. Remove V-bank cover sub-assembly.
3. Remove fan and alternator v-belt.
4. Remove oil cooler pipe.
5. Remove timing belt cover sub-assembly no. 3 LH.
6. Disconnect the Camshaft Position (CMP) sensor connector.
7. Remove the bolt, stud bolt and Camshaft Position (CMP) sensor.

To install:

8. Install the Camshaft Position (CMP) sensor with the bolt and stud bolt. Tighten bolt to 66 inch lbs. (7.5 Nm).
9. Reconnect the Camshaft Position (CMP) sensor connector.
10. Install timing belt cover sub-assembly no. 3 LH.
11. Install oil cooler pipe.
12. Install fan and alternator v-belt.
13. Add engine coolant.
14. Check for engine coolant leaks.
15. Install V-bank cover sub-assembly.

2010 Models

4.0L Engine

1. Remove the v-bank cover.
2. Remove the No. 1 air cleaner hose.
3. Remove the VVT sensor (for intake side of Bank 1).
    a. Disconnect the sensor connector.
    b. Remove the bolt and sensor.
4. Remove the VVT sensor (exhaust side of Bank 1).
    a. Disconnect the sensor connector.
    b. Remove the bolt and sensor.
5. Remove the VVT sensor (intake side of Bank 2).

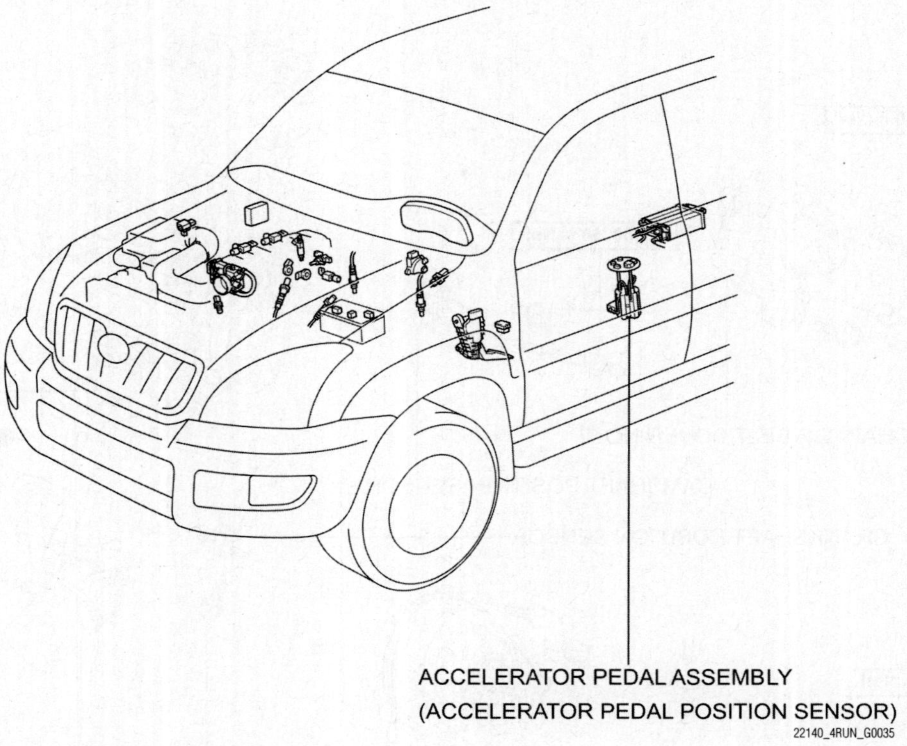

Fig. 270 Accelerator Pedal Position (APP) sensor location—2009 models

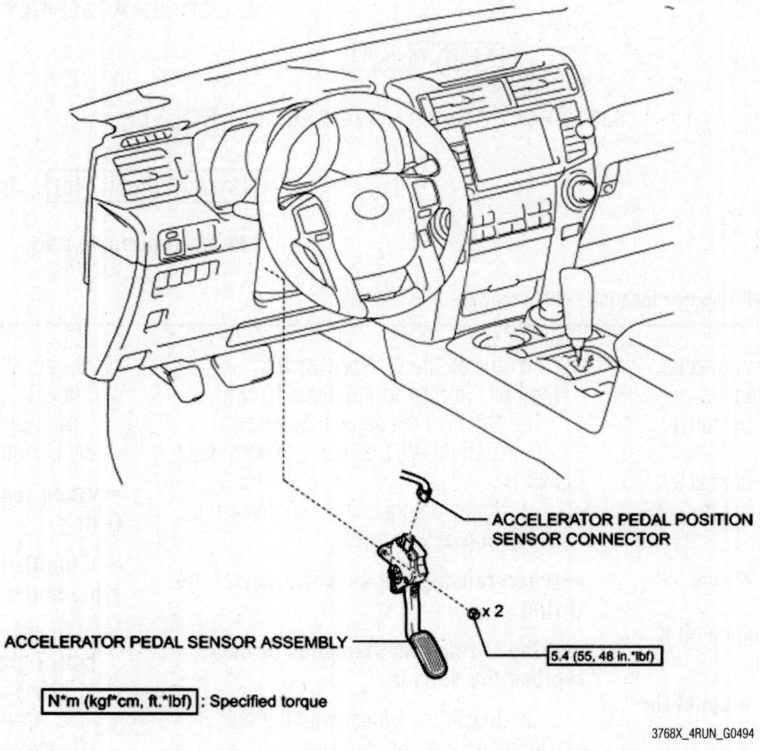

Fig. 271 Accelerator Pedal Position (APP) sensor—2010 models

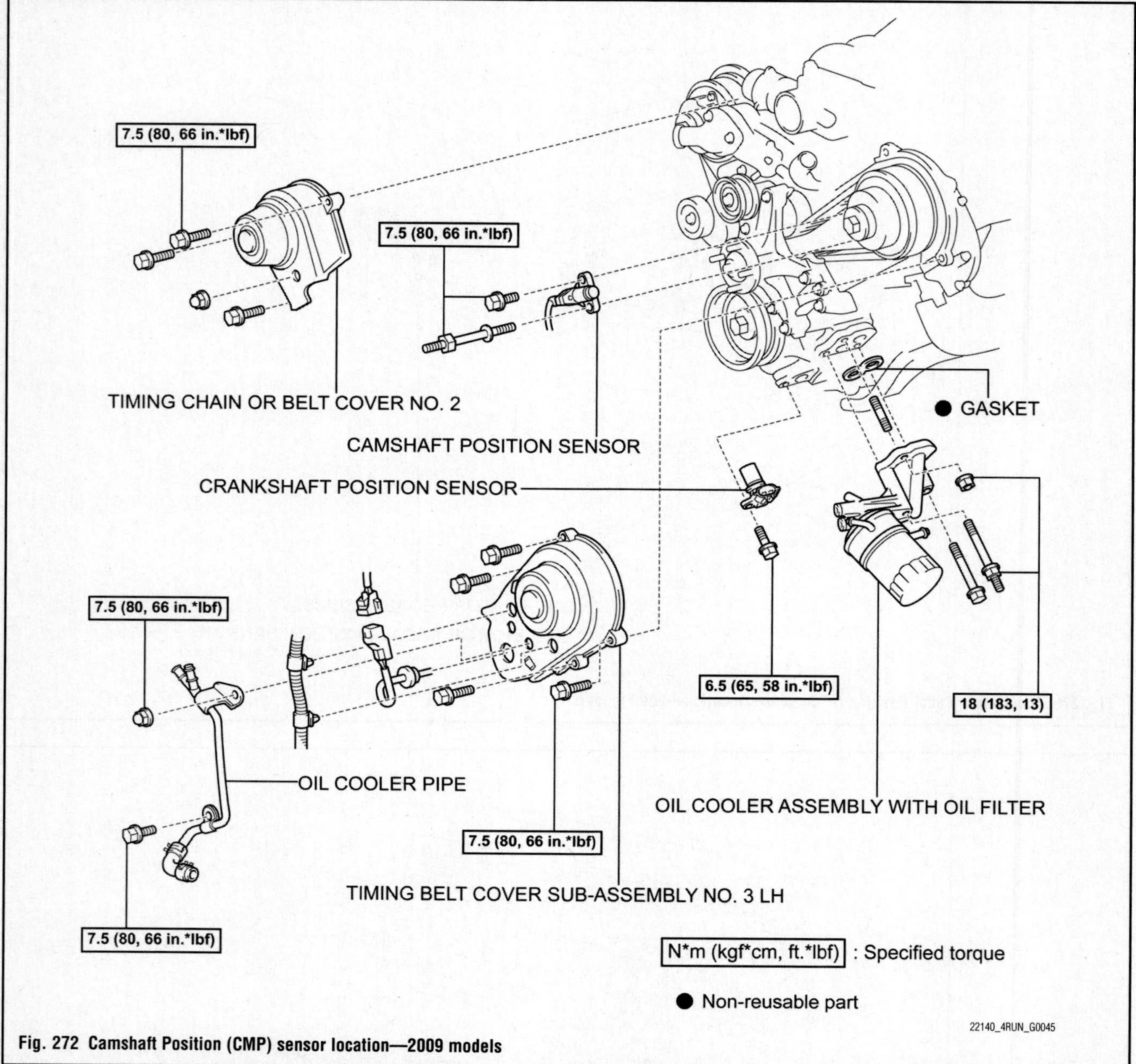

Fig. 272 Camshaft Position (CMP) sensor location—2009 models

a. Disconnect the sensor connector.
b. Remove the bolt and sensor.

6. Remove the VVT sensor (exhaust side of Bank 2).
a. Disconnect the sensor connector.
b. Remove the bolt and sensor.

To install:

7. Install the VVT sensor (exhaust side of Bank 2).
a. Apply a light coat of engine oil to the O-ring of the VVT sensor.

➡When reusing the sensor, inspect the O-ring.

➡If the O-ring has scratches or cuts, replace the sensor.

b. Install the VVT sensor with the bolt. Tighten to 7 ft. lbs. (10 Nm).
c. Connect the sensor connector.

8. Install the VVT sensor (intake side of Bank 2).
a. Apply a light coat of engine oil to the O-ring of the sensor.

➡When reusing the sensor, inspect the O-ring.

➡If the O-ring has scratches or cuts, replace the sensor.

b. Install the sensor with the bolt. Tighten to 7 ft. lbs. (10 Nm).
c. Connect the sensor connector.

9. Install the VVT sensor (exhaust side of Bank 1).
a. Apply a light coat of engine oil to the O-ring of the sensor.

➡When reusing the sensor, inspect the O-ring.

➡If the O-ring has scratches or cuts, replace the sensor.

b. Install the sensor with the bolt. Tighten to 7 ft. lbs. (10 Nm).
c. Connect the sensor connector.

10. Install the VVT sensor (intake side of Bank 1).

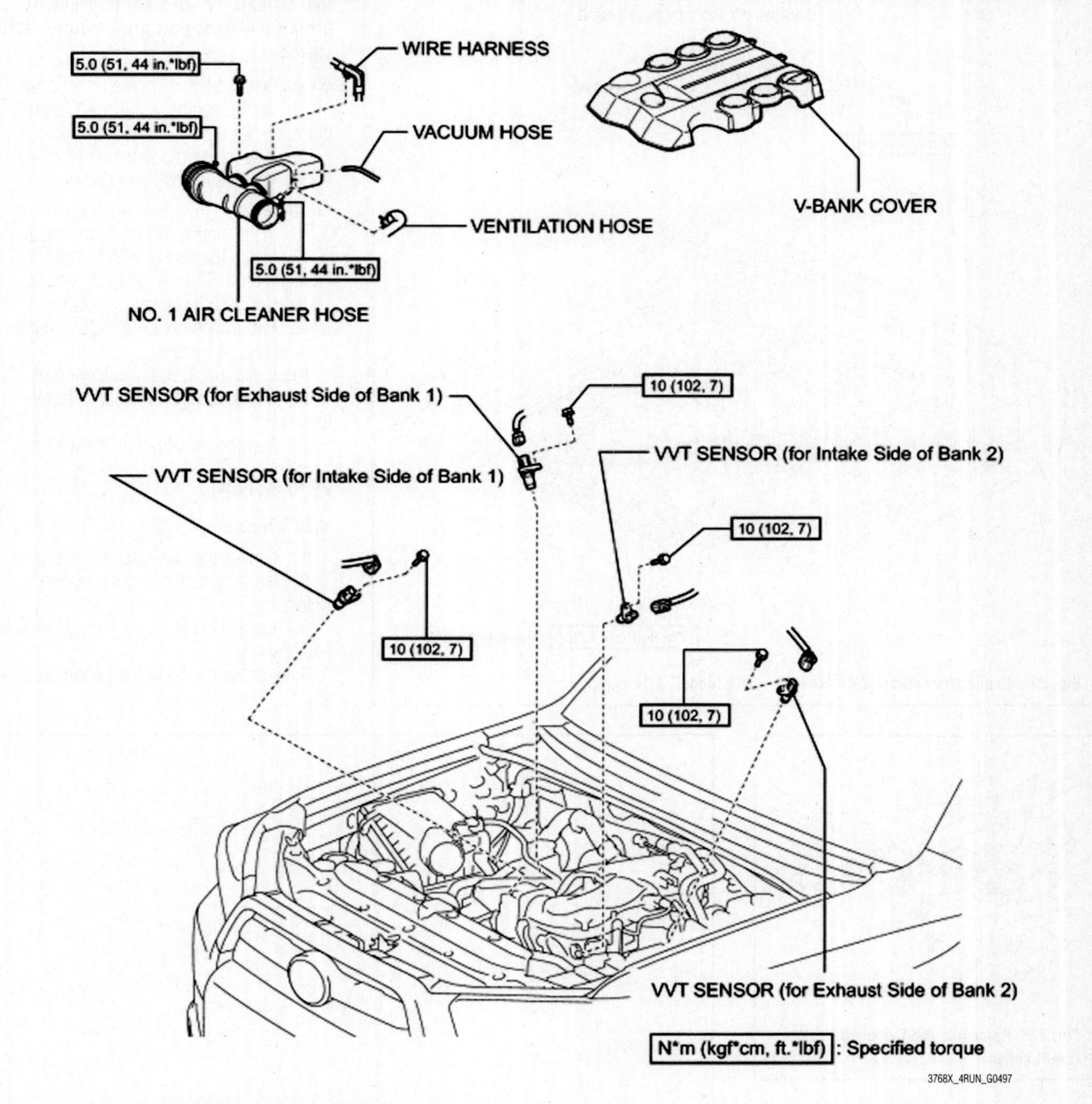

Fig. 273 Camshaft Position (CMP) sensor—2010 model, 4.0L engine

a. Apply a light coat of engine oil to the O-ring of the sensor.

➡**When reusing the sensor, inspect the O-ring.**

➡**If the O-ring has scratches or cuts, replace the sensor.**

b. Install the sensor with the bolt. Tighten to 7 ft. lbs. (10 Nm).

11. Install the No. 1 air cleaner hose.
12. Install the v-bank cover.

2.7L Engine

See Figure 275.

1. Remove the Camshaft Position (CMP) sensor.
   a. Disconnect the Camshaft Position (CMP) sensor connector.
   b. Remove the bolt and Camshaft Position (CMP) sensor.

To install:

2. Install the Camshaft Position (CMP) sensor.
   a. Apply a light coat of engine oil to the o-ring of the Camshaft Position (CMP) sensor.
   b. Install the Camshaft Position (CMP) sensor with the bolt. Tighten to 75 inch lbs. (8.5 Nm).

➡**Make sure that the o-ring is not cracked or jammed when installing it.**

   c. Connect the Camshaft Position (CMP) sensor connector.

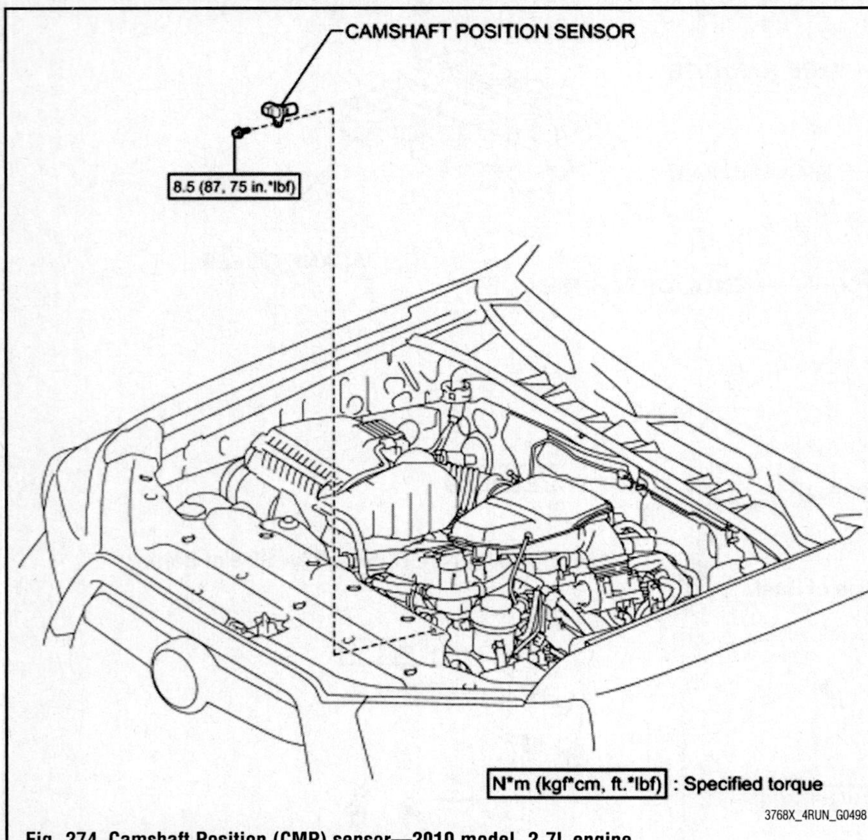

Fig. 274 Camshaft Position (CMP) sensor—2010 model, 2.7L engine

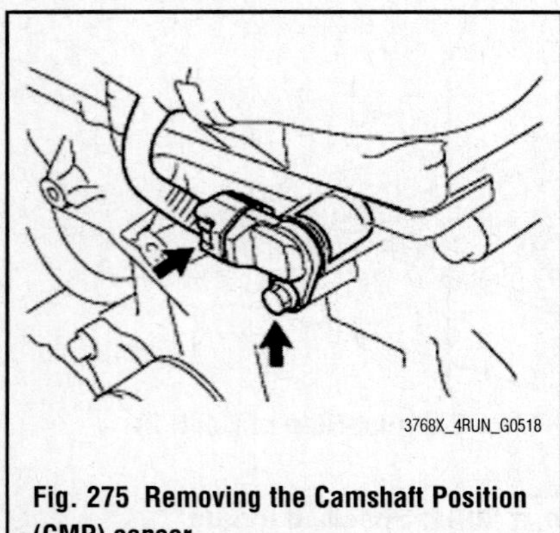

Fig. 275 Removing the Camshaft Position (CMP) sensor

CRANKSHAFT POSITION (CKP) SENSOR

LOCATION

See Figures 276 through 278.

REMOVAL & INSTALLATION

2009 Models

1. Disconnect cable from negative battery terminal.

➡Wait at least 90 seconds after disconnecting the cable from the negative (-) battery terminal to prevent airbag and seat belt pretensioner activation.

2. Remove No. 1 engine under cover.
3. For V6 engine, remove A/C compressor.
4. Disconnect the sensor connector.
5. Remove the bolt and sensor.

To install:

6. Install crankshaft position sensor.
7. Install the sensor with the bolt. Tighten bolt to 57 inch lbs. (6.5 Nm).
8. Connect the sensor connector.
9. For V6 engine, install A/C compressor and charge system.
10. Install no. 1 engine under cover.
11. Connect cable to negative battery terminal.
12. Perform initialization, if necessary. .

2010 Models

4.0L Engine

1. Remove the alternator assembly.
2. Remove the cooler compressor assembly.
3. Remove the No. 1 engine under cover sub assembly.
4. Remove the crankshaft position sensor.

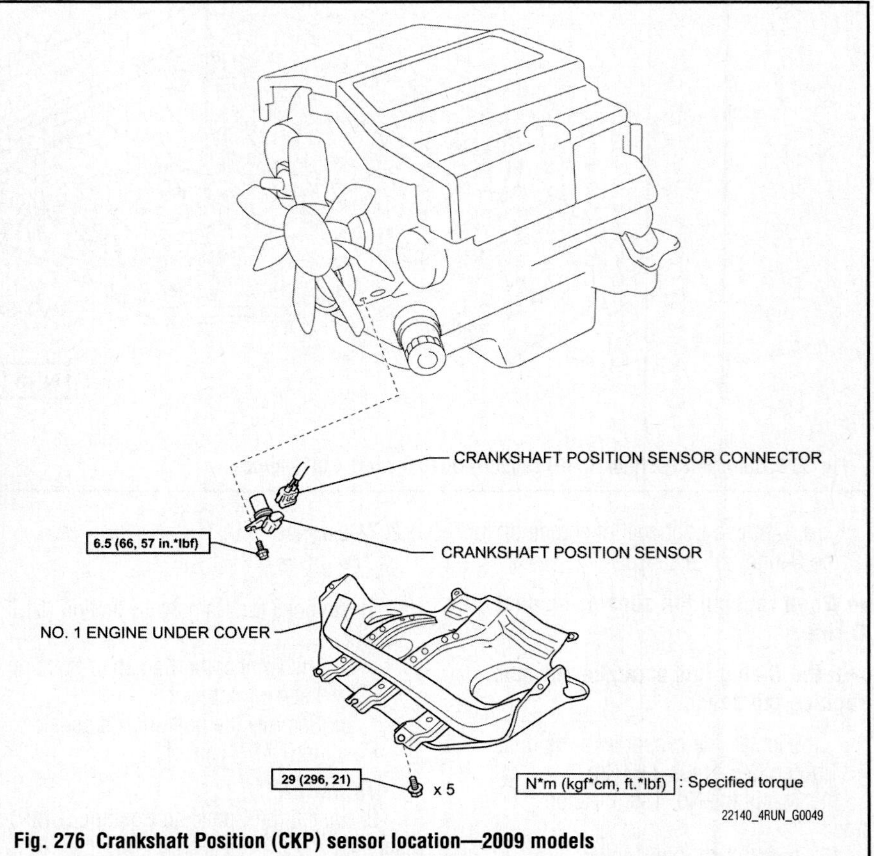

Fig. 276 Crankshaft Position (CKP) sensor location—2009 models

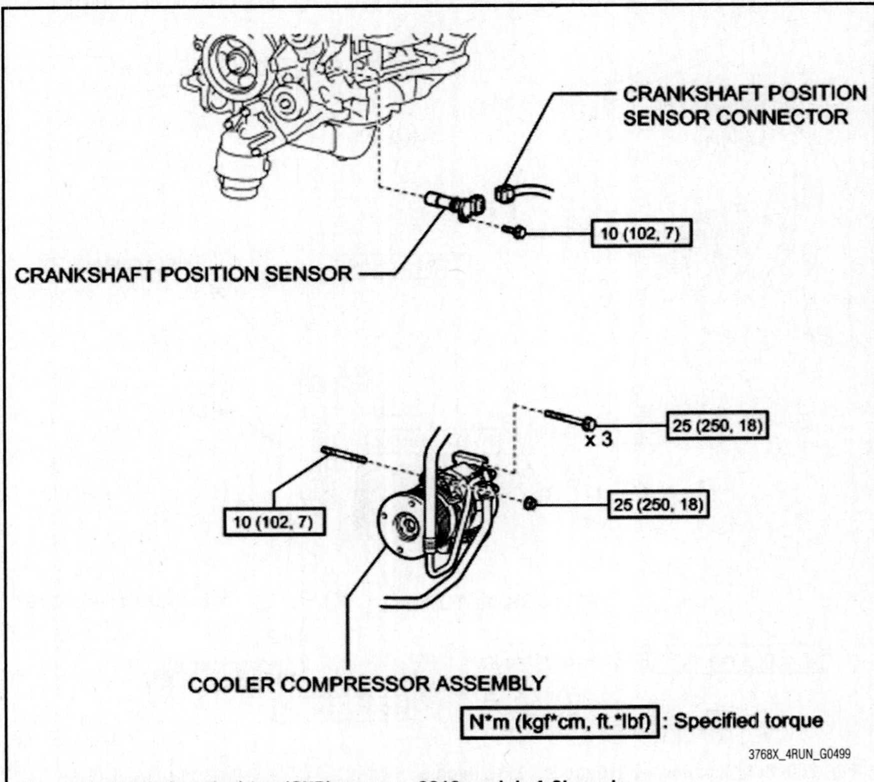

Fig. 277 Crankshaft Position (CKP) sensor—2010 model, 4.0L engine

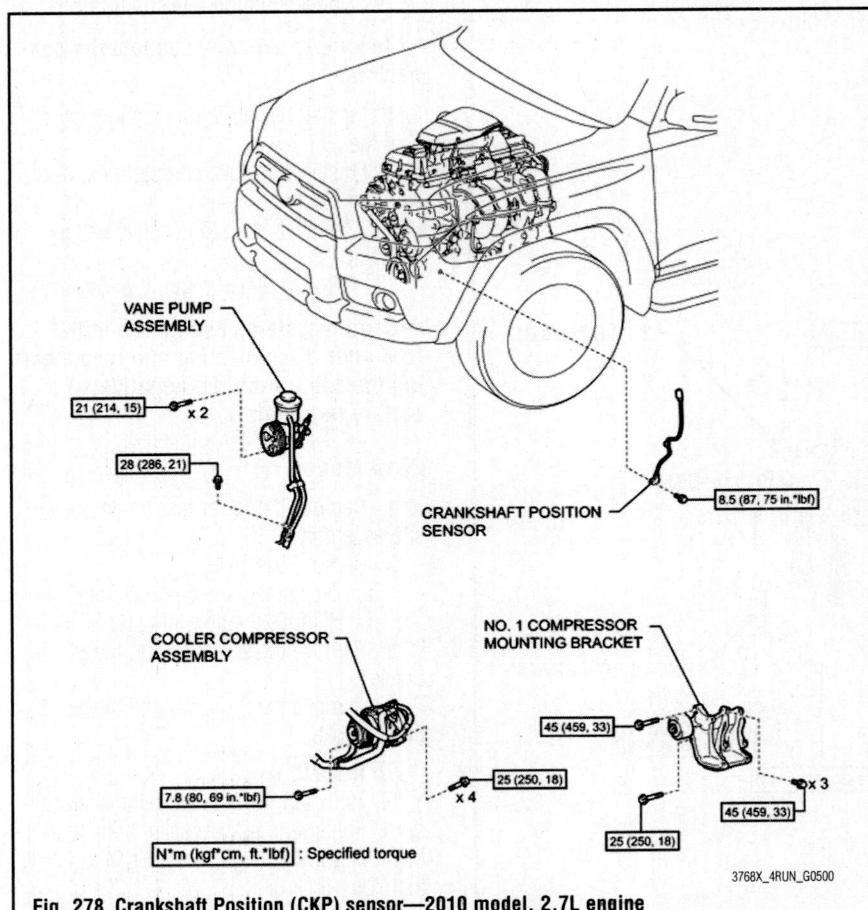

Fig. 278 Crankshaft Position (CKP) sensor—2010 model, 2.7L engine

a. Disconnect the sensor connector.

b. Remove the bolt and sensor.

To install:

5. Install the crankshaft position sensor.

a. Apply a light coat of engine oil to the o-ring of the sensor.

➡When reusing the sensor, inspect the o-ring.

➡If the o-ring has scratches or cuts, replace the sensor.

b. Install the sensor with the bolt. Tighten to 7 ft. lbs. (10 Nm).

c. Connect the sensor connector.

6. Install the No. 1 engine under cover sub assembly.

7. Install the cooler compressor assembly.

8. Install the alternator assembly.

2.7L Engine

1. Remove the fan shroud.
2. Disconnect the vane pump assembly.
3. Disconnect the cooler compressor assembly.
4. Remove the No. 1 compressor mounting bracket.
5. Remove the crankshaft position sensor.

a. Disconnect the Crankshaft Position (CKP) sensor connector and detach the wire harness clamp.

b. Remove the bolt and crankshaft position sensor.

To install:

6. Install the crankshaft position sensor.

a. Apply a light coat of engine oil to the O-ring of the crankshaft position sensor.

b. Install the Crankshaft Position (CKP) sensor with the bolt. Tighten to 75 inch lbs. (8.5 Nm).

➡Make sure that the O-ring is not cracked or jammed when installing it.

c. Connect the Crankshaft Position (CKP) sensor connector and attach the wire harness clamp.

7. Install the No. 1 compressor mounting bracket.
8. Connect the cooler compressor assembly.
9. Connect the vane pump assembly.
10. Install the fan shroud.

ELECTRONIC CONTROL MODULE (ECM)

LOCATION

See Figures 279 and 280.

REMOVAL & INSTALLATION

2009 Models

1. Disconnect cable from negative battery terminal.

➡Wait at least 90 seconds after disconnecting the cable from the negative (-) battery terminal to prevent airbag and seat belt pretensioner activation.

2. Remove glove box compartment door.
3. Remove the 2 screws and glove compartment door.
4. Remove no. 2 finish panel lower.
5. Remove the 3 screws and no. 2 finish panel lower.
6. Disconnect the 5 ECM connectors.
7. Remove the 3 screws and ECM.

To install:

8. Install the ECM with the 3 screws, tighten to 49 inch lbs. (5.5 Nm).

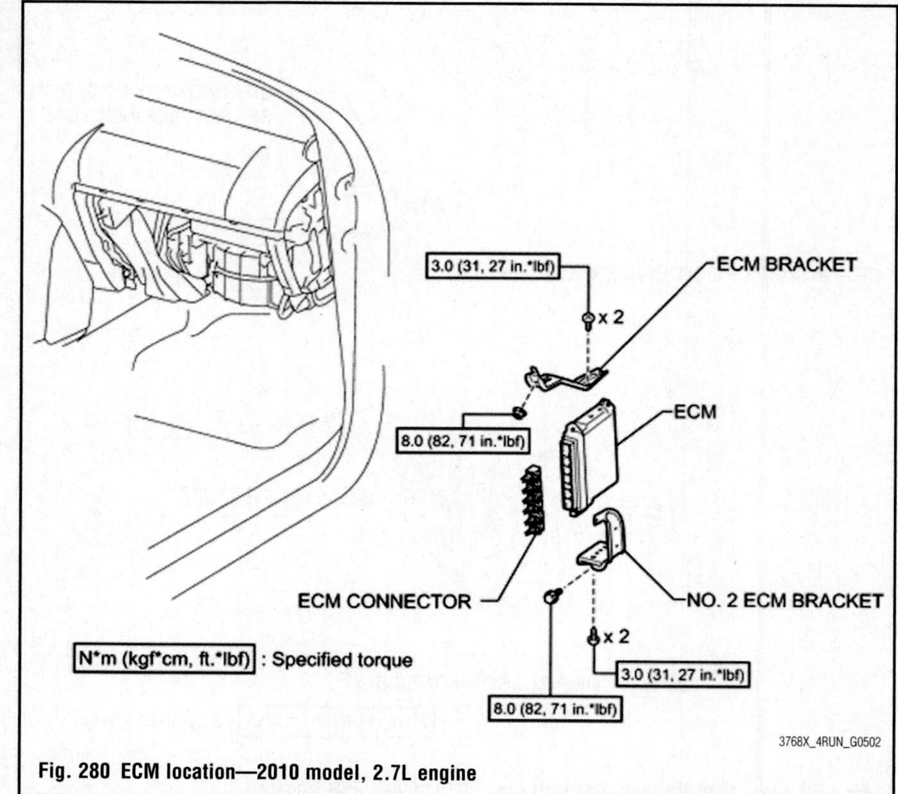

Fig. 280 ECM location—2010 model, 2.7L engine

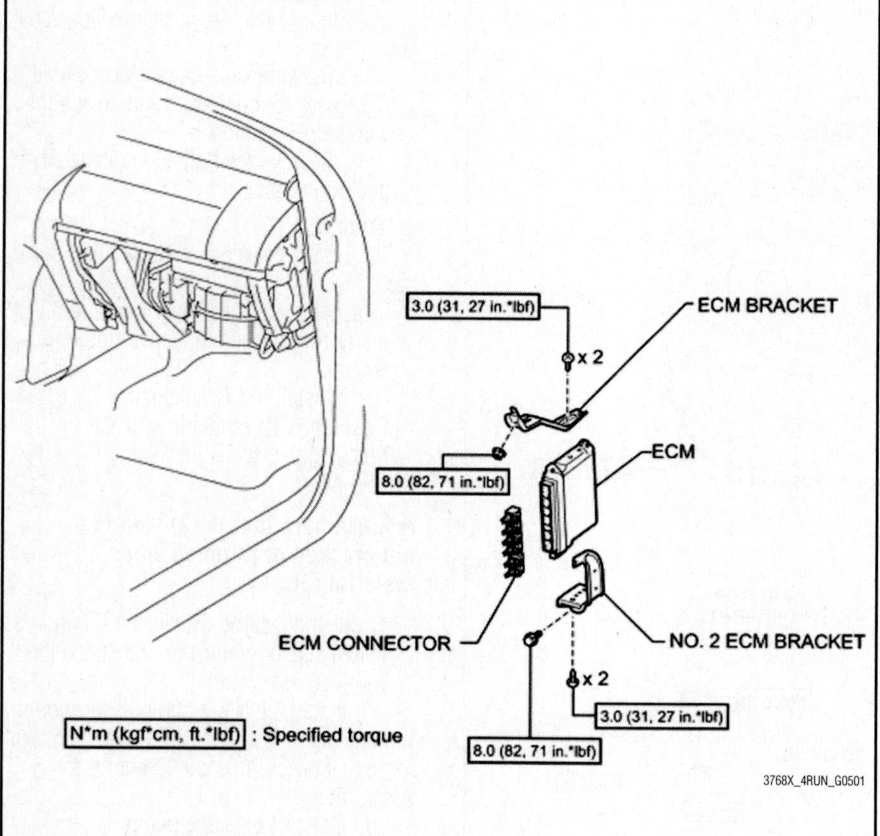

Fig. 279 ECM location—2010 model, 4.0L engine

9. Connect the 5 ECM connectors.

➡Be sure to securely connect the connectors.

10. Install the No. 2 finish panel lower with the 3 screws.
11. Install the glove compartment door with the 2 screws.
12. Connect cable to negative battery terminal.
13. Perform initialization, if necessary.

➡Certain systems need to be initialized after disconnecting and reconnecting the cable from the negative (-) battery terminal.

2010 Models

1. Remove the lower instrument panel sub assembly.
2. Remove the ECM.
    a. Disconnect the 6 connectors.
    b. Remove the bolt, nut and ECM.
3. Remove the 2 screws and the ECM bracket.
4. Remove the 2 screws and the No. 2 ECM bracket.

To install:

5. To install, reverse the removal procedure. Tighten the ECM bracket bolts to 27 inch lbs. (3 Nm). Tighten the ECM to 71 inch lbs. (8 Nm).

ENGINE COOLANT TEMPERATURE (ECT) SENSOR

LOCATION

See Figures 281 and 282.

REMOVAL & INSTALLATION

2010 Models

4.0L Engine

1. Remove the intake manifold.
2. Remove the Engine Coolant Temperature (ECT) sensor.
   a. Disconnect the sensor connector.
   b. Using a 19 mm deep socket wrench, remove the sensor.
   c. Remove the gasket from the sensor.

To install:

3. To install, reverse the removal procedure. Tighten the sensor to 14 ft. lbs. (20 Nm).

2.7L Engine

1. Remove the intake manifold.
2. Remove the Engine Coolant Temperature (ECT) sensor.
   a. Disconnect the Engine Coolant Temperature (ECT) sensor connector.
   b. Using a 19 mm deep socket wrench, remove the Engine Coolant Temperature (ECT) sensor and gasket.

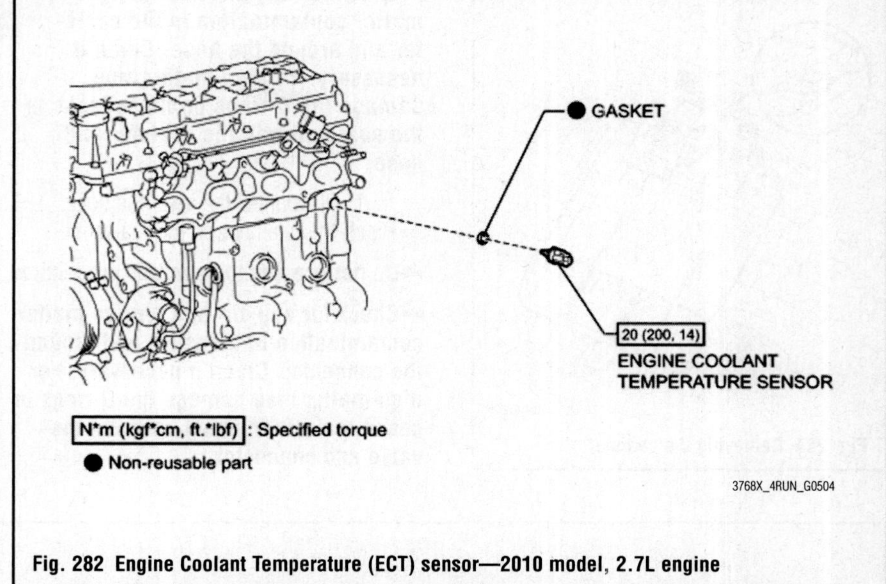

Fig. 282 Engine Coolant Temperature (ECT) sensor—2010 model, 2.7L engine

To install:

3. To install, reverse the removal procedure. Tighten the sensor to 14 ft. lbs. (20 Nm).

EVAPORATIVE EMISSIONS (EVAP) CANISTER

REMOVAL & INSTALLATION

2010 Models

See Figures 283 and 284.

1. Remove the spare tire.
2. Remove the canister assembly.
   a. Disconnect the connector.
   b. Disconnect the purge line hose.
   c. Disconnect the air inlet line hose. Push the hose firmly toward the canister side. Pinch the hose as shown in the illustration. Pull out the hose.

➡Do not use any tools in this procedure.

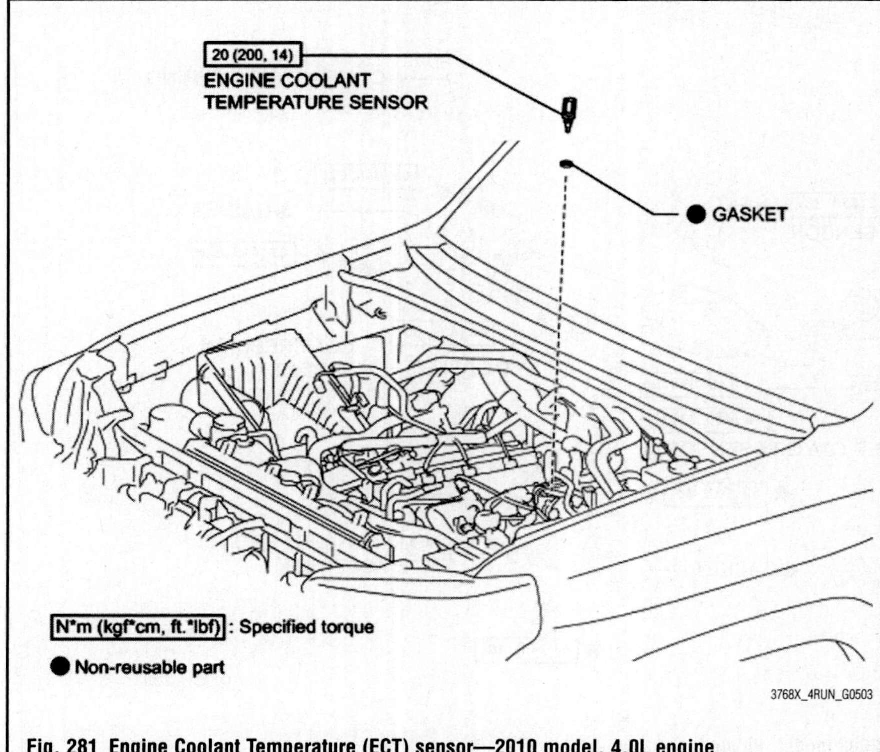

Fig. 281 Engine Coolant Temperature (ECT) sensor—2010 model, 4.0L engine

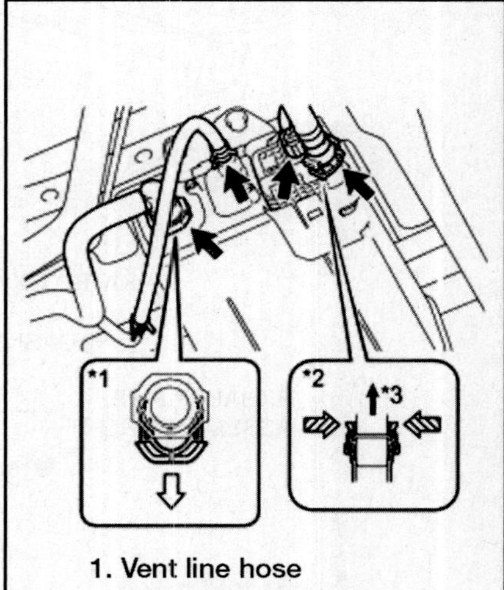

Fig. 283 Disconnecting the connectors and purge line hoses

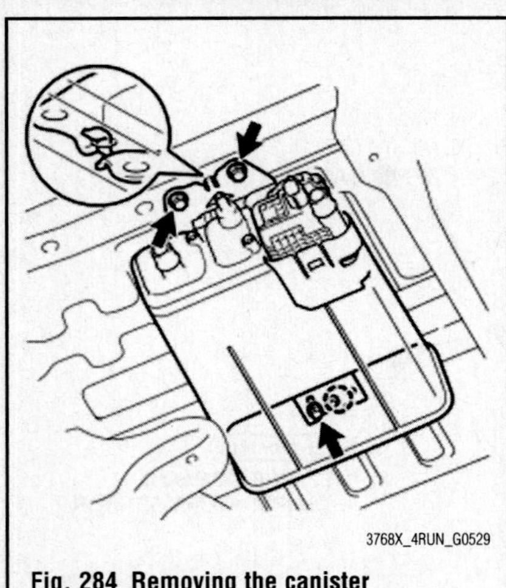

Fig. 284 Removing the canister

➡Check for any dirt and foreign matter contamination in the canister and around the hose. Clean if necessary. Foreign matter may damage the O-rings or cause leaks in the seal between the canister and hose.

d. Disconnect the vent line hose. Pinch the retainer and then raise it.

➡Do not use any tools in this procedure.

➡Check for any dirt and foreign matter contamination in the valve and around the connector. Clean if necessary. Foreign matter may damage the O-rings or cause leaks in the seal between the valve and connector.

e. Detach the claw, and remove the 3 bolts and canister.

To install:

3. To install, reverse the removal procedure. Tighten the canister bolts to 15 ft. lbs. (20 Nm).

HEATED OXYGEN SENSOR (HO2S)

LOCATION

See Figures 285 through 288.

REMOVAL & INSTALLATION

2009 Models

See Figures 285 and 286.

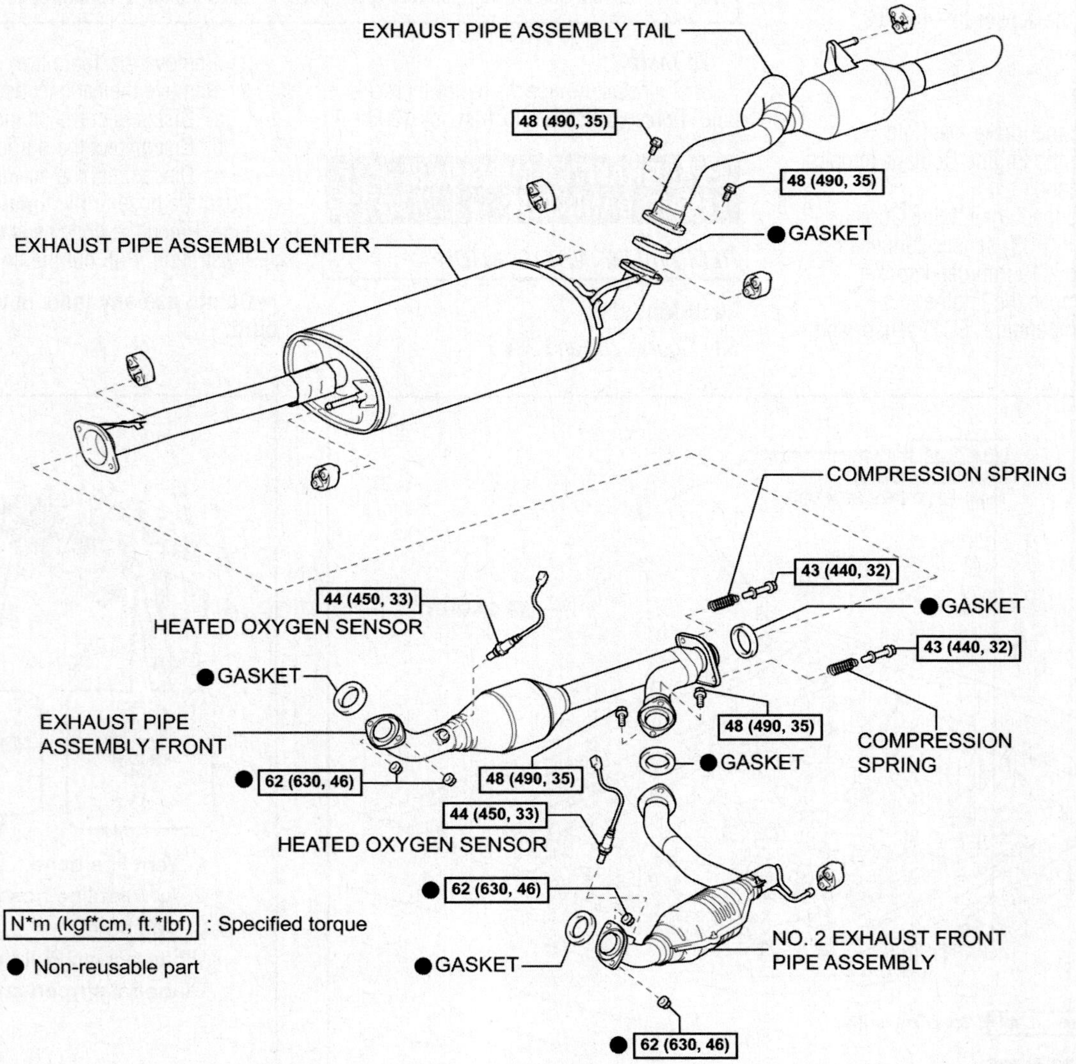

Fig. 285 Heated Oxygen Sensor (HO2S) location—2009 model, V6 engine

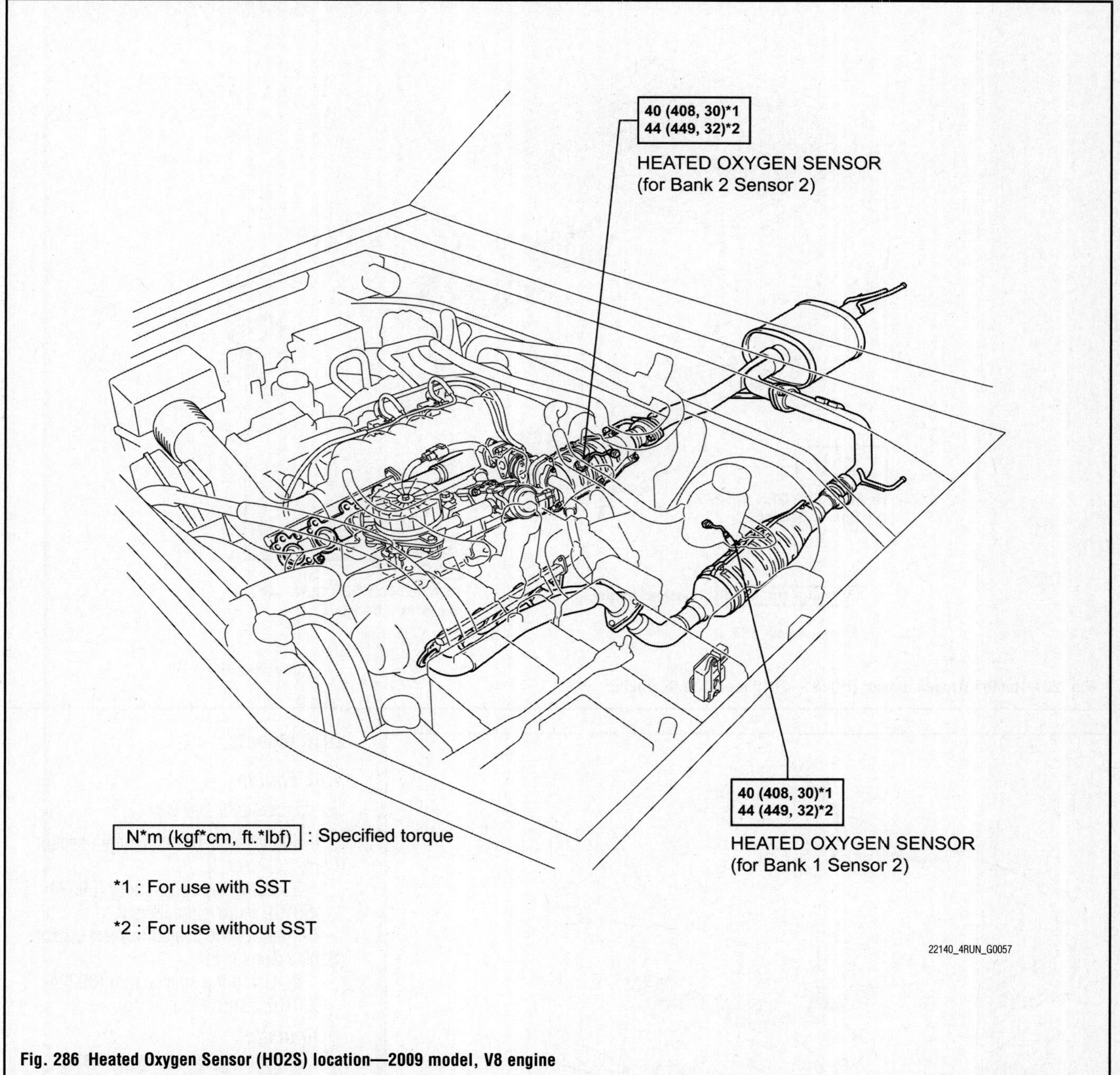

Fig. 286 Heated Oxygen Sensor (HO2S) location—2009 model, V8 engine

CAUTION

Wear protective gloves when removing the sensor. The exhaust pipe assembly is extremely hot immediately after the engine has stopped. Confirm that the exhaust pipe assembly has cooled down before removing it.

1. Disconnect cable from negative battery terminal.

CAUTION

Wait at least 90 seconds after disconnecting the cable from the negative (-) battery terminal to prevent airbag and seat belt pretensioner activation.

2. Disconnect the sensor connector and remove heated oxygen sensor (for Bank 1 Sensor 2).
3. Disconnect the sensor connector and remove heated oxygen sensor (for Bank 2 Sensor 2).

To install:

4. Install heated oxygen sensor (for Bank 1 Sensor 2) and tighten to 32 ft. lbs. (44 Nm).

➡Use a torque wrench with a fulcrum length of 11.81 inch (30 cm).

5. Connect the sensor connector.
6. Install heated oxygen sensor (for Bank 2 Sensor 2) and tighten to 32 ft. lbs. (44 Nm).

➡Use a torque wrench with a fulcrum length of 11.81 inch (30 cm).

7. Connect the sensor connector.
8. Connect cable to negative battery terminal.
9. Perform initialization, if necessary.

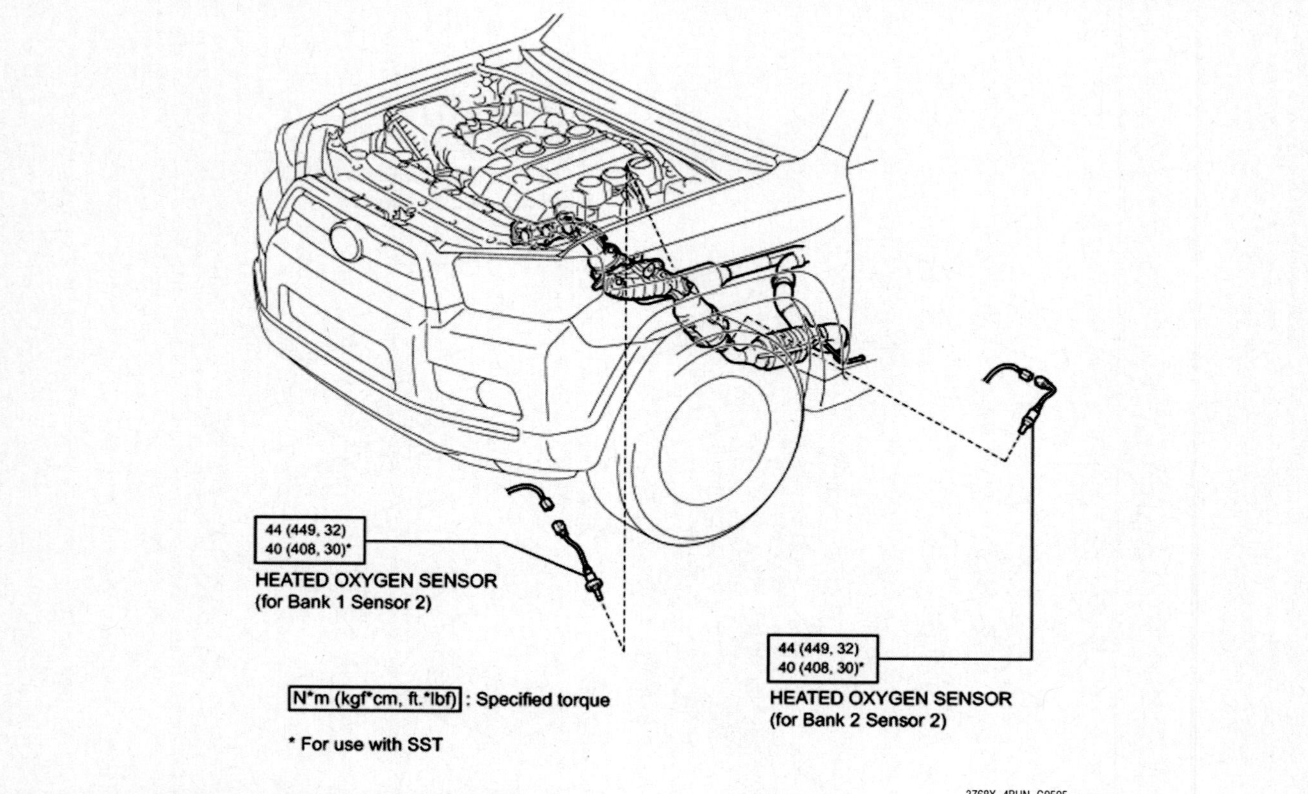

Fig. 287 Heated Oxygen Sensor (HO2S)—2010 model, 4.0L engine

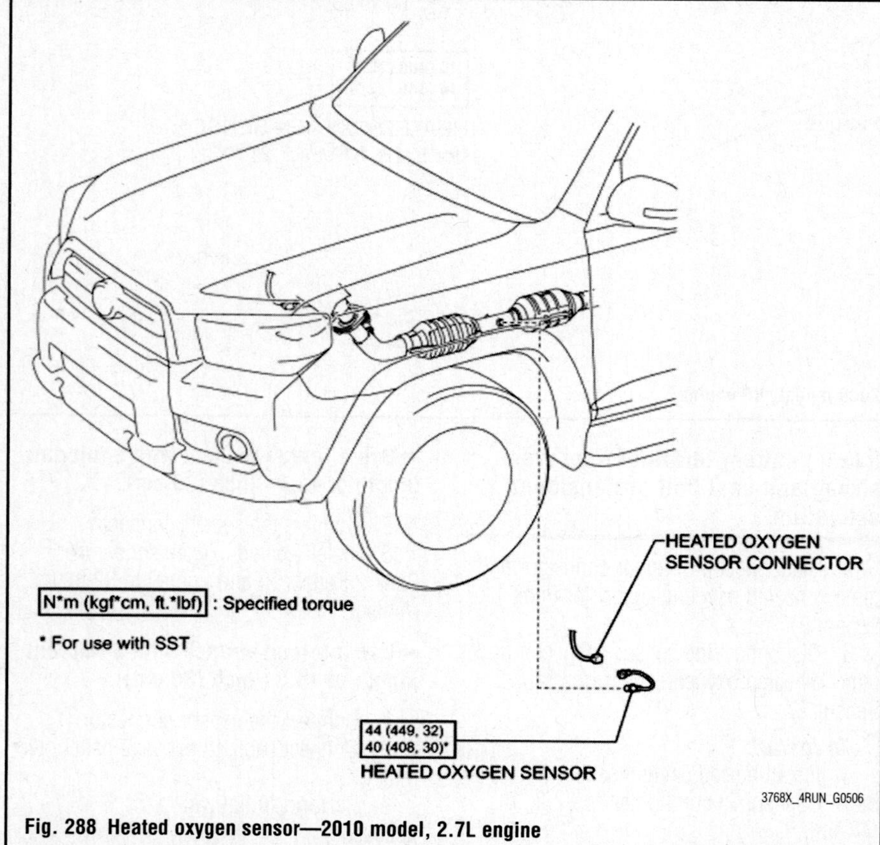

Fig. 288 Heated oxygen sensor—2010 model, 2.7L engine

2010 Models

4.0L Engine

See Figures 289 and 290.

1. Remove the heated oxygen sensor (Bank 1 Sensor 2).
   a. Using the special tool (09224-00010), remove the sensor.
2. Remove the heated oxygen sensor (Bank 2 Sensor 2).
   a. Using the special tool (09224-00010), remove the sensor.

To install:

3. Install the heated oxygen sensor (Bank 1 Sensor 2).
   a. Temporarily install the sensor to the exhaust pipe by hand.
   b. Using the special tool (09224-00010). If tightening without the special tool, tighten to 32 ft. lbs. (44 Nm). If tightening with the special tool, tighten to 30 ft. lbs. (40 Nm).

➡Use a torque wrench with a fulcrum length of 11.8 inch (30 cm). When using a torque wrench with a fulcrum length that is not 11.8 inch (30 cm), calculate the torque specification for the torque wrench and SST based on

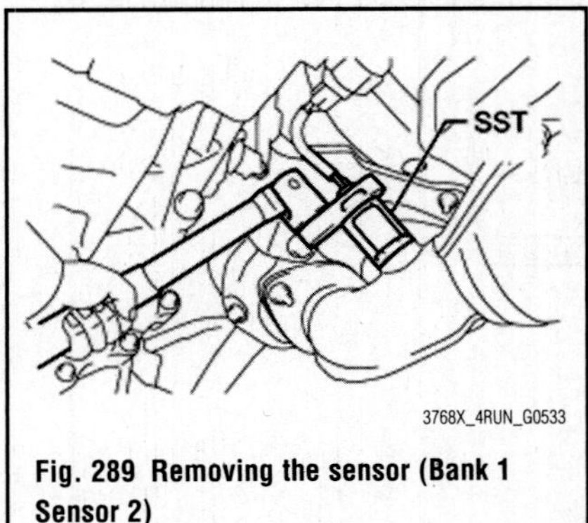

Fig. 289 Removing the sensor (Bank 1 Sensor 2)

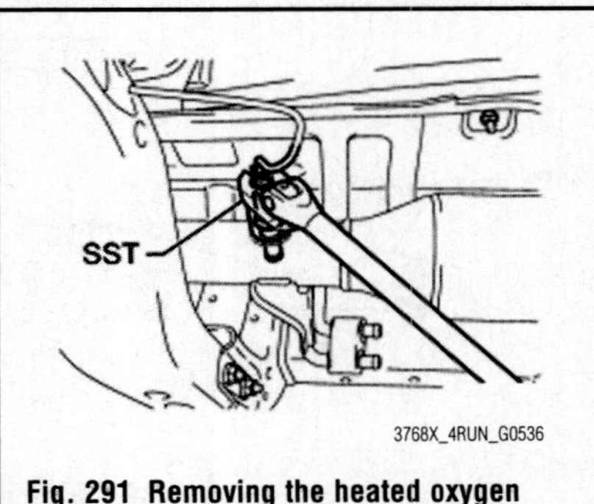

Fig. 291 Removing the heated oxygen sensor

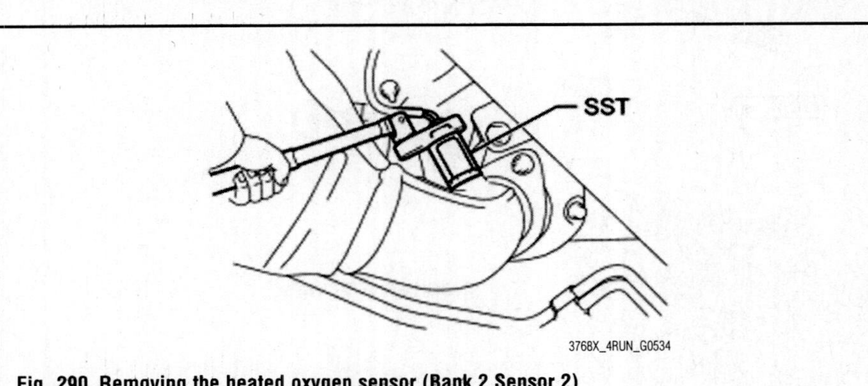

Fig. 290 Removing the heated oxygen sensor (Bank 2 Sensor 2)

the "without SST" torque specification.

➡Make sure SST and the wrench are connected in a straight line.

c. Connect the sensor connector.

4. Install the heated oxygen sensor (Bank 2 Sensor 2).

a. Temporarily install the sensor to the exhaust pipe by hand.

b. Using the special tool (09224-00010). If tightening without the special tool, tighten to 32 ft. lbs. (44 Nm). If tightening with the special tool, tighten to 30 ft. lbs. (40 Nm).

➡Use a torque wrench with a fulcrum length of 11.8 inch (30 cm). When using a torque wrench with a fulcrum length that is not 11.8 inch (30 cm), calculate the torque specification for the torque wrench and SST based on the "without SST" torque specification.

➡Make sure SST and the wrench are connected in a straight line.

c. Connect the sensor connector.

2.7L Engine

See Figure 291.

1. Remove the heated oxygen sensor.

a. Disconnect the heated oxygen sensor connector.

b. Using the special tool (09224-00010), remove the heated oxygen sensor.

To install:

2. To install, reverse the removal procedure. Tighten the heated oxygen sensor to 32 ft. lbs. (44 Nm) if using the special tool. If tightening with the special tool, tighten to 30 ft. lbs. (40 Nm).

INTAKE AIR TEMPERATURE (IAT) SENSOR

LOCATION

The Intake Air Temperature (IAT) sensor, built into the Mass Air Flow (MAF) meter.

REMOVAL & INSTALLATION

See Mass Air Flow Meter.

KNOCK SENSOR (KS)

LOCATION

See Figures 292 through 294.

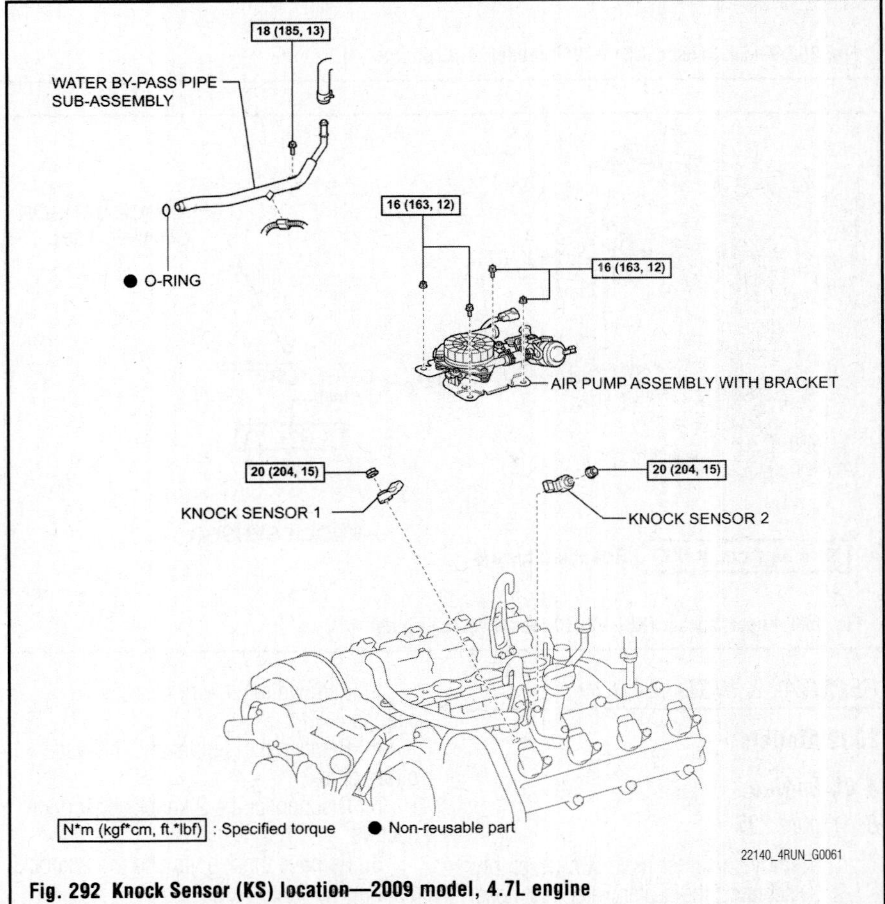

Fig. 292 Knock Sensor (KS) location—2009 model, 4.7L engine

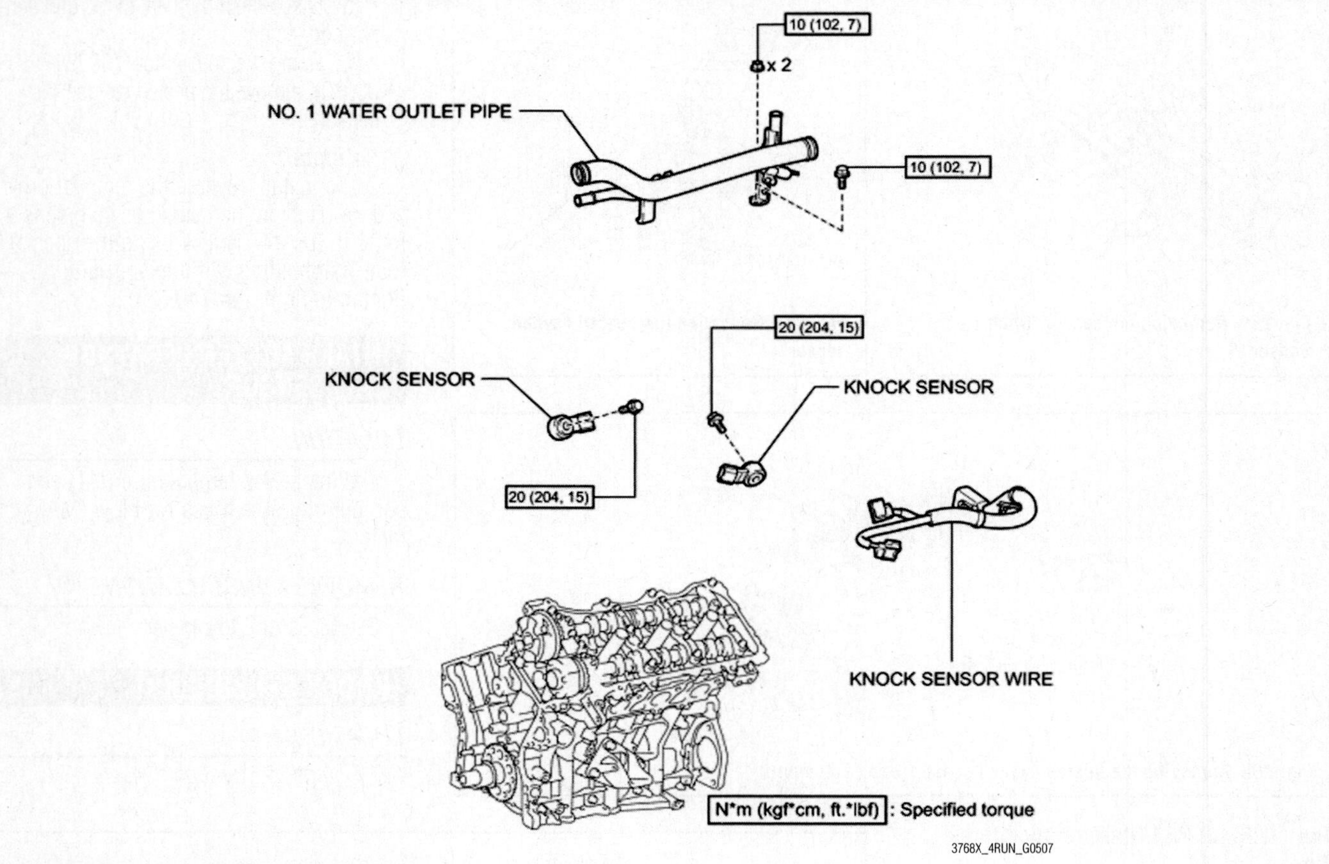

Fig. 293 Knock Sensor (KS)—2010 model, 4.0L engine

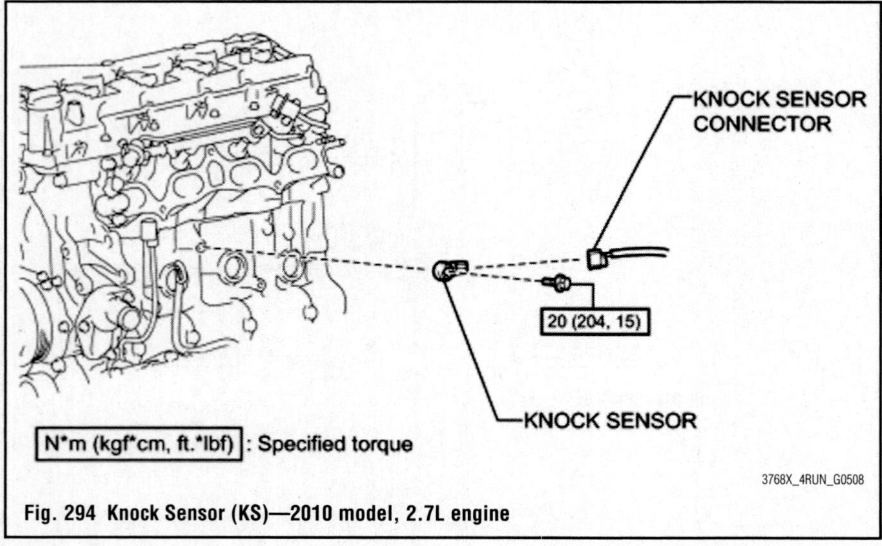

Fig. 294 Knock Sensor (KS)—2010 model, 2.7L engine

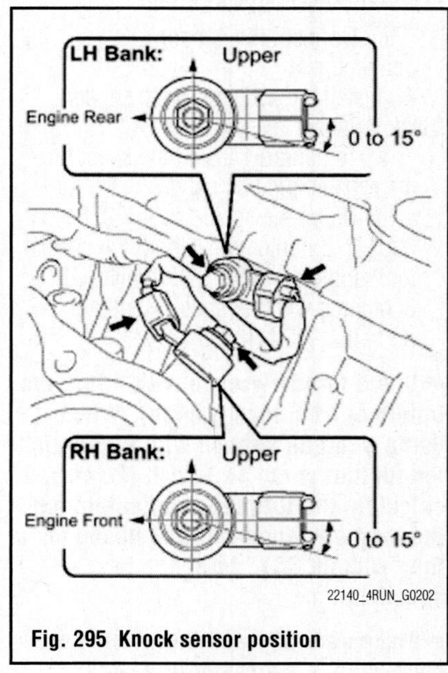

Fig. 295 Knock sensor position

REMOVAL & INSTALLATION

2009 Models

4.0L Engine

See Figure 295.

1. Remove cylinder head sub-assembly.
2. Disconnect the heater water inlet hose.
3. Remove the 4 wire harness clamps.
4. Remove the 3 bolts and the water outlet pipe.
5. Disconnect the 2 knock sensor connectors.
6. Remove the 2 bolts and the 2 knock sensors.

To install:

7. Install the 2 knock sensors with the 2 bolts and tighten to 15 ft. lbs. (20 Nm).
8. Connect the 2 knock sensor connectors.

9. Install the 3 bolts and the water outlet pipe and tighten to 7 ft. lbs. (10 Nm).
10. Install the 4 wire harness clamps.
11. Connect the heater water inlet hose.
12. Install cylinder head sub-assembly.

4.7L Engine

See Figure 296.

1. Discharge fuel system pressure.
2. Drain engine coolant.
3. Remove V-bank cover sub-assembly.
4. Disconnect the vacuum hoses (for the power steering idle-up and fuel pressure regulator) and ventilation hose.
5. Remove the air cleaner hose assembly.
6. Disconnect fuel hose.
7. Disconnect fuel hose no.2.
8. Disconnect the throttle control connector.
9. Disconnect the purge VSV connector.
10. Disconnect the 8 injector connectors.
11. Disconnect the ECT sensor connector.
12. Disconnect the 8 ignition coil connectors.
13. Disconnect the 2 VSV connectors for the air injection system.
14. Disconnect the 8 ignition coil connectors.
15. Disconnect the 2 air fuel ratio sensor connectors.
16. Disconnect the vacuum hose [A] from the fuel pressure regulator.
17. Disconnect the PCV hoses [B] from the PCV valve on the LH cylinder head.
18. Disconnect the EVAP hose (from the charcoal canister) [C] from the VSV for the EVAP.
19. Disconnect the 2 vacuum hoses [D] from the VSV for the air injection system.
20. Disconnect the 2 water by-pass hoses from the throttle body.
21. Disconnect the 2 wire clamps from the wire clamp bracket on the RH delivery pipe.
22. Remove the bolt and nut holding the engine wire protector from the intake manifold and cylinder head.
23. Remove the 2 bolts and ground cables from the RH and LH cylinder heads.
24. Remove the bolt and V-bank cover bracket from the intake manifold.
25. Disconnect the engine wire from the engine hanger and wire bracket.
26. Remove the bolt and wire bracket from the intake manifold.
27. Remove the 6 bolts, 4 nuts, intake manifold assembly and 2 gaskets.
28. Remove air pump assembly with bracket.
29. Remove knock sensor.
30. Disconnect the 2 knock sensor connectors.

To install:

31. Install the 2 knock sensors with the 2 nuts as shown in the illustration and tighten nuts to 15 ft. lbs. (20 Nm).
32. Connect the 2 knock sensor connectors.
33. Place 2 new gaskets on the intake manifold.
34. Place the intake manifold on the cylinder heads.
35. Install and uniformly tighten the 6 bolts and 4 nuts in several steps to 13 ft. lbs. (18 Nm). Refer to the Intake Manifold procedure for tightening sequence.
36. Install the V-bank cover bracket to the intake manifold.
37. Install the wire bracket to the intake manifold with the bolt.
38. Connect the engine wire to the engine hanger and wire bracket.
39. Connect the wire protector to the intake manifold and cylinder heads with the bolt and nut.
40. Install the 2 ground cables with the 2 bolts to the RH and LH cylinder heads.
41. Connect the 2 water by-pass hoses to the throttle body.
42. Connect the 2 wire clamps to the wire clamp bracket on the RH delivery pipe.
43. Connect the vacuum hose to the fuel pressure regulator.
44. Connect the PCV hose to the PCV valve on the LH cylinder head.
45. Connect the EVAP hose (from the charcoal canister) to the purge VSV.
46. Connect the 2 vacuum hoses to the VSV for the air injection system.
47. Connect the throttle control connector.
48. Connect the 2 VSV connectors for the air injection system.
49. Connect the purge VSV connector.
50. Connect the 8 injector connectors.
51. Connect the ECT sensor connector.
52. Connect the 8 ignition coil connectors.
53. Connect the 2 air fuel ratio sensor connectors.
54. Install fuel hose no.2.
55. Install fuel hose.
56. Install V-bank cover sub-assembly.
57. Add engine coolant.
58. Check for engine coolant leaks.
59. Check for fuel leaks.

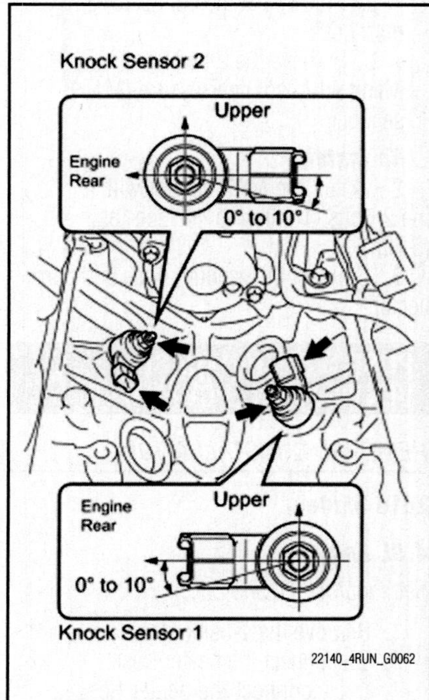

Fig. 296 Installing the knock sensor

2010 Models

4.0L Engine

1. Remove the cylinder head sub assembly (Bank 1).
2. Remove the No. 1 water outlet pipe.
   a. Detach the 3 wire harness clamps.
   b. Remove the 2 nuts, bolt and water outlet pipe.
3. Remove the knock sensor.
   a. Disconnect the 2 sensor connectors.
   b. Remove the 2 bolts and 2 sensors.

To install:

4. Install the knock sensor.
   a. Install the 2 sensors with the 2 bolts. Tighten to 15 ft. lbs. (20 Nm).
   b. Connect the 2 sensor connectors.
5. Install the No. 1 water outlet pipe.
   a. Install the water outlet pipe with the 2 nuts and bolt. Tighten the bolt to 7 ft. lbs. (10 Nm).
   b. Attach the 3 wire harness clamps.
6. Install the cylinder head sub assembly (for Bank 1).

2.7L Engine

1. Remove the intake manifold.
2. Remove the knock sensor.
   a. Disconnect the knock sensor connector.
   b. Remove the bolt and knock sensor.

To install:

3. Install the knock sensor.
   a. Install the knock sensor with the bolt. Tighten to 15 ft. lbs. (20 Nm).

➡Make sure that the knock sensor is at the correct angle when installing it.

   b. Connect the knock sensor connector.
4. Install the intake manifold.

MALFUNCTION INDICATOR LIGHT (MIL)

RESET PROCEDURE

Clearing DTC codes resets MIL.

MASS AIR FLOW (MAF) METER

LOCATION

See Figures 297 and 298.

The MAF meter is located in the air intake snorkel.

REMOVAL & INSTALLATION

1. Disconnect the electrical connector.
2. Remove attaching screws and remove MAF meter.

To install:

3. Installation is the reverse of the removal procedure.

MANIFOLD ABSOLUTE PRESSURE (MAP) SENSOR

LOCATION

See Figure 299.

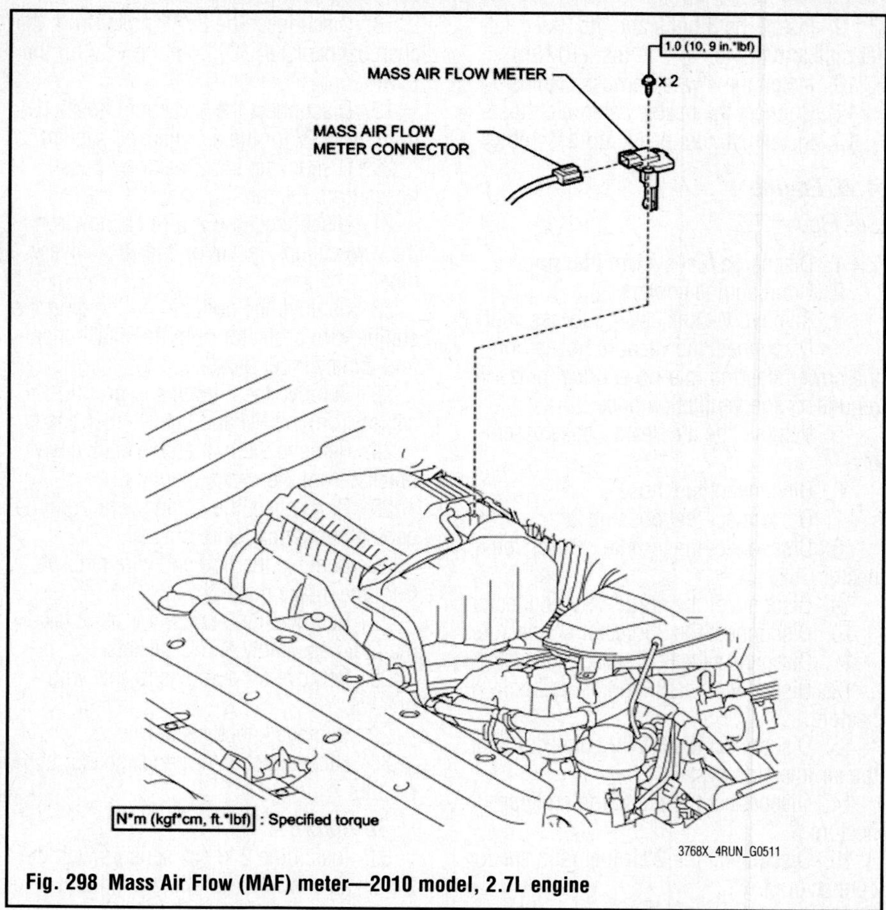

Fig. 298 Mass Air Flow (MAF) meter—2010 model, 2.7L engine

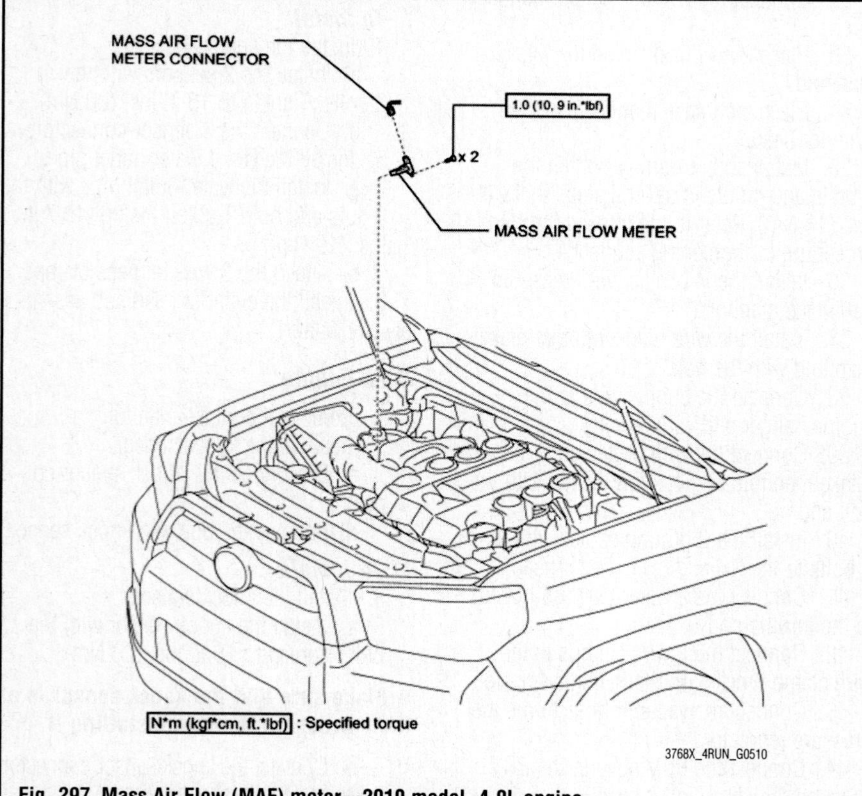

Fig. 297 Mass Air Flow (MAF) meter—2010 model, 4.0L engine

REMOVAL & INSTALLATION

1. Remove the manifold absolute pressure sensor.
   a. Disconnect the connector and vacuum hose.
   b. Remove the 2 bolts and Manifold Absolute Pressure (MAP) sensor.

To install:

2. Install the MAP sensor with the 2 bolts. Tighten to 44 inch lbs. (5 Nm).
3. Connect the vacuum hose and connector.

POSITIVE CRANKCASE VENTILATION (PCV) VALVE

REMOVAL & INSTALLATION

2010 Models

4.0L Engine

See Figures 300 and 301.

1. Remove the v-bank cover.
2. Disconnect the heater hose.
   a. Disconnect the heater hose clamp.

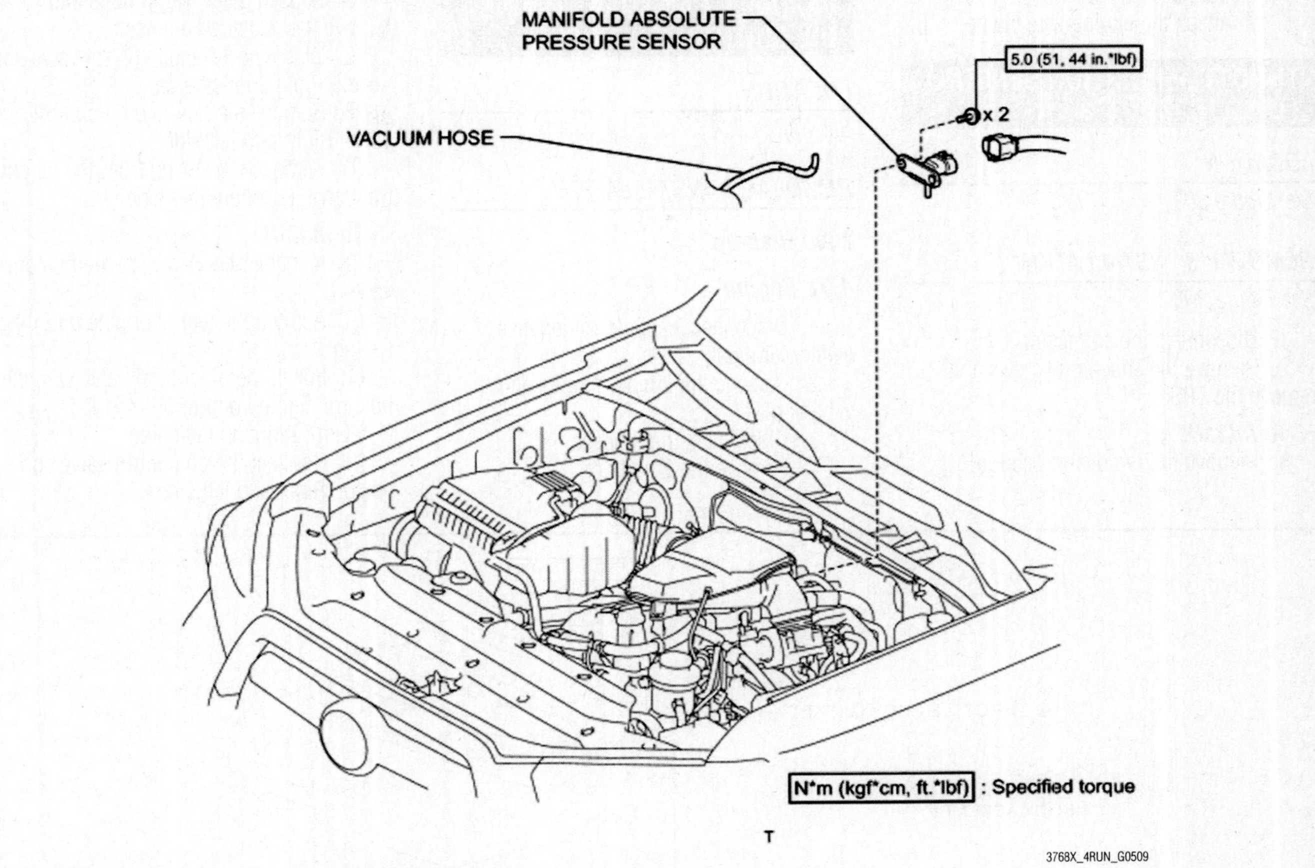

MANIFOLD ABSOLUTE PRESSURE SENSOR
5.0 (51, 44 in.*lbf)
x 2
VACUUM HOSE
N*m (kgf*cm, ft.*lbf) : Specified torque
T
3768X_4RUN_G0509

Fig. 299 Manifold Absolute Pressure (MAP) sensor—2010 model, 2.7L engine

3768X_4RUN_G0548

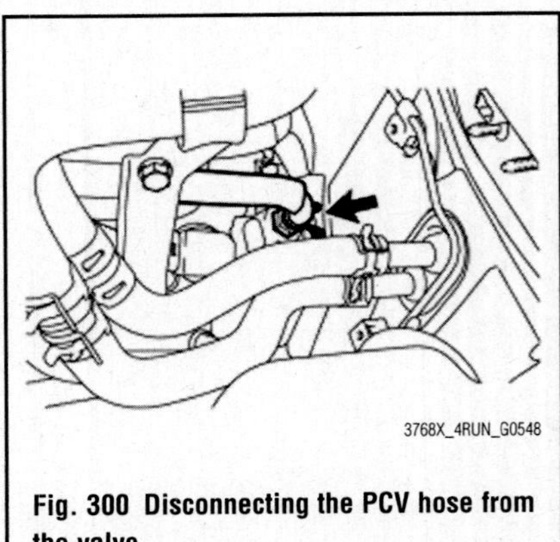

Fig. 300 Disconnecting the PCV hose from the valve

3768X_4RUN_G0549

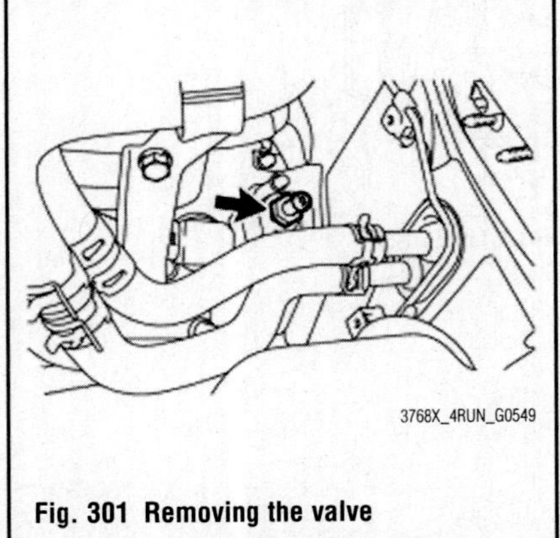

Fig. 301 Removing the valve

3. Remove the PCV valve sub assembly.
   a. Loosen the hose clamp and disconnect the PCV hose from the valve.
   b. Remove the valve.

To install:

4. Install the PCV valve sub assembly:
   a. Apply adhesive to 2 or 3 threads of the PCV valve.
   b. Install the PCV valve. Tighten to 20 ft. lbs. (27 Nm).
   c. Connect the PCV hose to the valve.
   d. Secure the hose with the clamp.
5. Connect the heater hose with the clamp.
6. Install the v-bank cover.

2.7L Engine

See Figure 302.

1. Remove the PCV valve sub assembly.
   a. Detach the engine wire clamp.
   b. Disconnect the PCV hose from the ventilation valve.
   c. Using a 22 mm ball joint lock nut wrench, remove the PCV valve.

To install:

2. Install the PCV valve sub assembly.
   a. Apply adhesive to 2 or 3 threads of the valve.
   b. Using a 22 mm ball joint lock nut wrench, install the PCV valve. Tighten to 44 inch lbs. (5 Nm).

3768X_4RUN_G0550

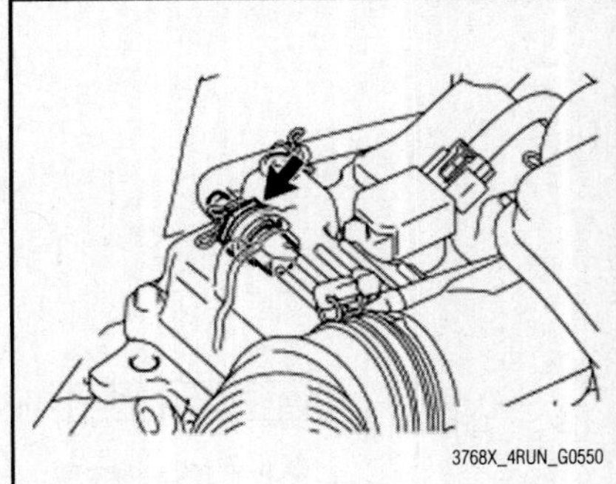

Fig. 302 Disconnecting the PCV hose from the ventilation valve

c. Connect the PCV hose.
d. Attach the engine wire clamp.

THROTTLE POSITION SENSOR (TPS)

LOCATION

See Figure 303.

REMOVAL & INSTALLATION

See Figure 303.

1. Disconnect the connector.
2. Remove the attaching screws and remove the TPS.

To install:

3. Remove the removal procedure.

VARIABLE CAMSHAFT TIMING OIL CONTROL SOLENOID

LOCATION

See Figure 304.

REMOVAL & INSTALLATION

2009 Models

4.7L Engine

1. Disconnect cable from negative battery terminal.
2. Remove the 2 nuts, 2 bolts and V-bank cover.
3. Disconnect the 3 hoses.
4. Remove the 2 bolts.
5. Loosen the 2 clamp bolts and remove the intake air connector.
6. Disconnect the oil control valve connector. Right and left side.
7. Remove the bolt and oil control valve. Right and left side.
8. Remove the O-ring from the oil control valve. Right and left side.

To install:

9. Install a new O-ring to the oil control valve.
10. Apply light coat of engine oil to the O-ring.
11. Install the oil control valve with the bolt and tighten bolt to 66 inch lbs. (7.5 Nm). Right and left side.
12. Connect the oil control valve connector. Right and left side.

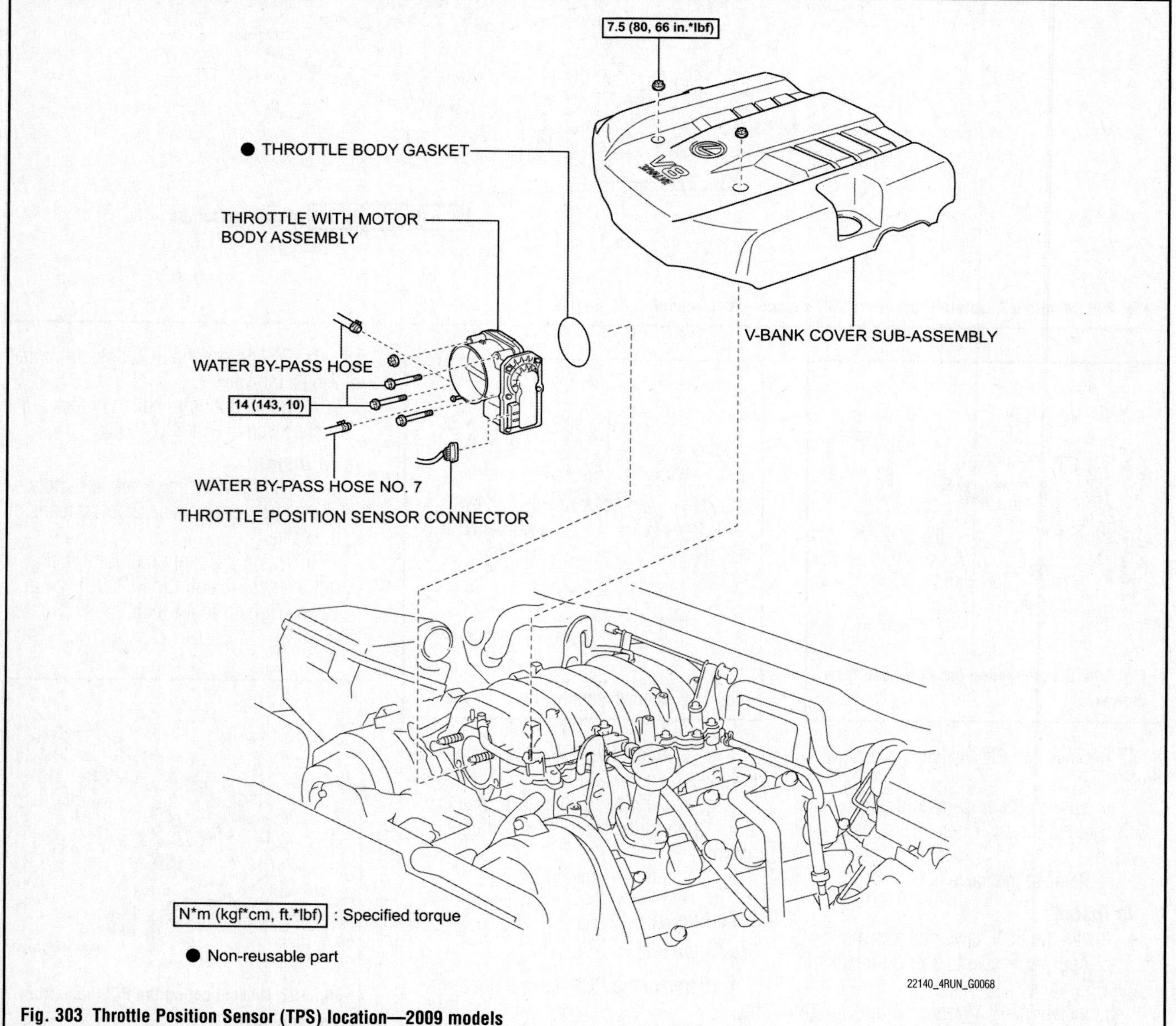

Fig. 303 Throttle Position Sensor (TPS) location—2009 models

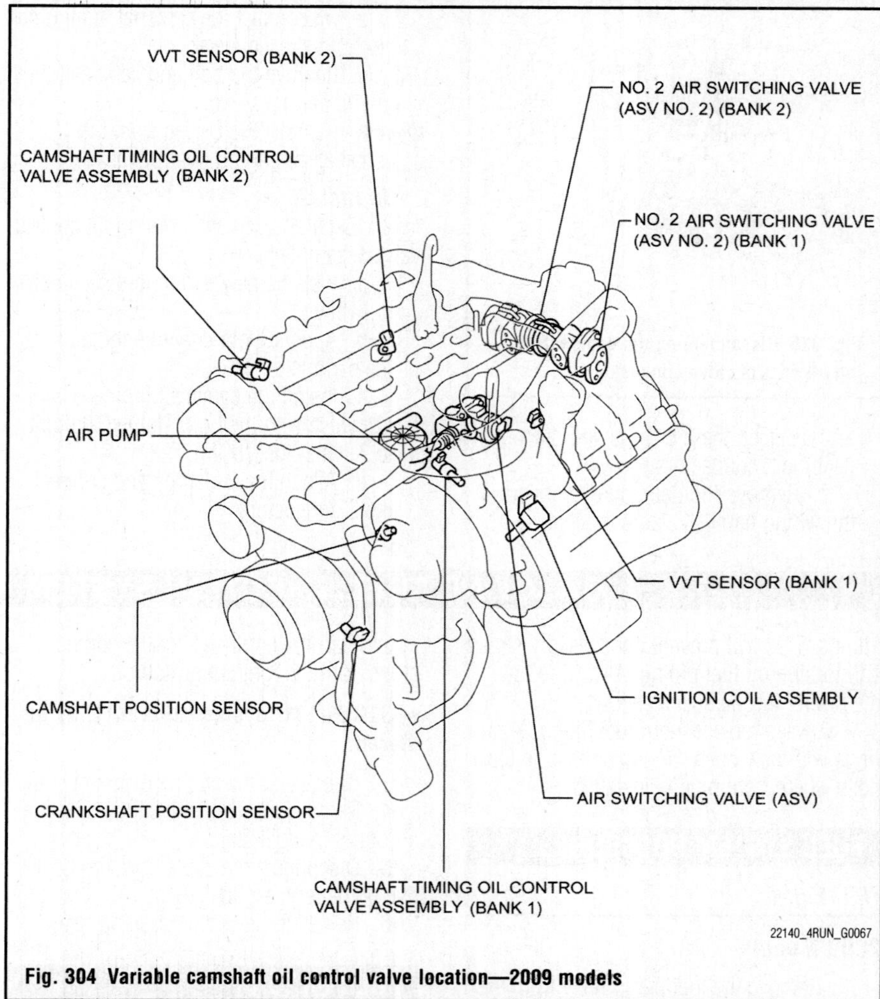

Fig. 304 Variable camshaft oil control valve location—2009 models

2010 Models

4.0L Engine

See Figures 305 through 308.

1. Remove the v-bank cover.
2. Remove the No. 1 air cleaner hose.
3. Remove the camshaft timing oil control valve assembly (for intake side of Bank 1).
   a. Disconnect the oil control valve connector.
   b. Remove the bolt and oil control valve.
   c. Remove the O-ring from the oil control valve.
4. Remove the camshaft timing oil control valve assembly (for exhaust side of Bank 1).
   a. Disconnect the oil control valve connector.

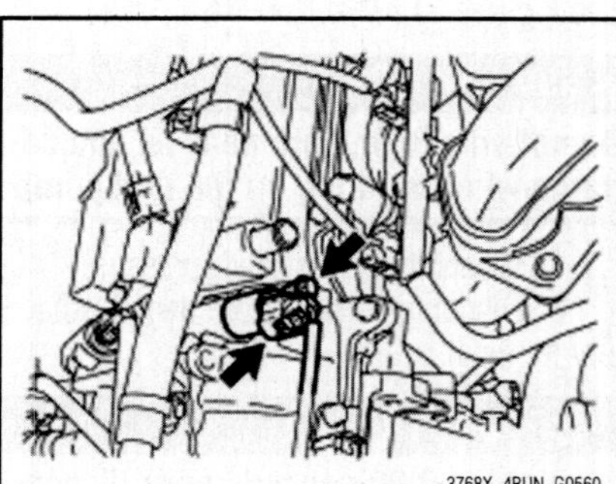

Fig. 305 Removing the camshaft timing oil control valve assembly (intake side of Bank 1)

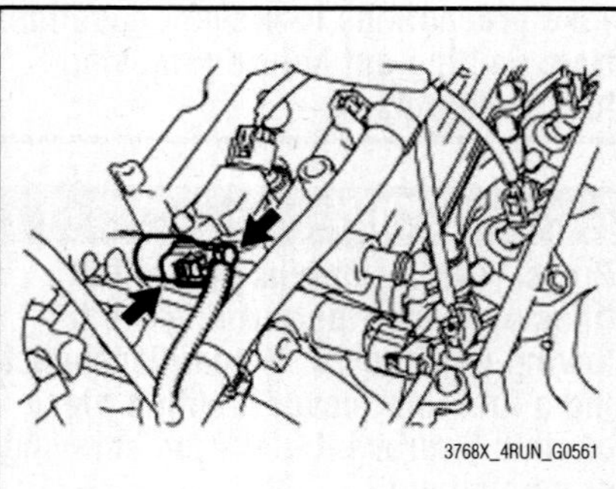

Fig. 306 Removing the camshaft timing oil control valve assembly

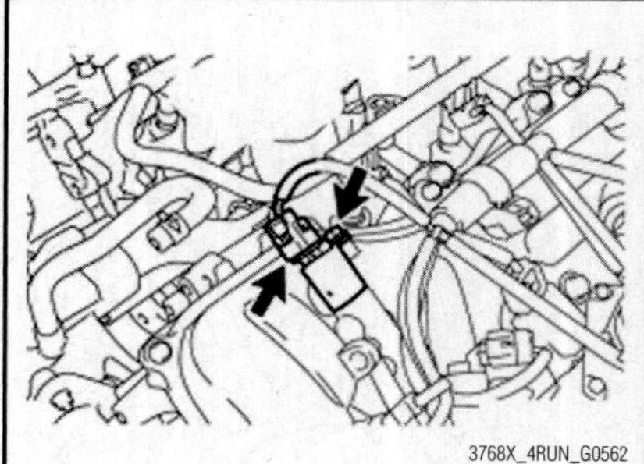

Fig. 307 Removing the camshaft timing oil control valve assembly (intake side of Bank 2)

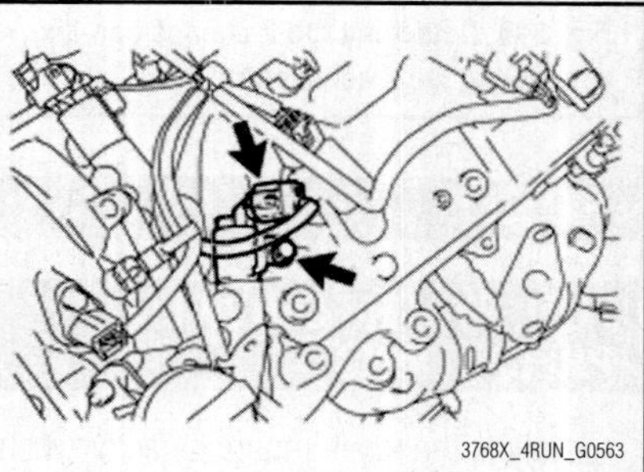

Fig. 308 Removing the camshaft timing oil control valve assembly (exhaust side of Bank 2)

   b. Remove the bolt and oil control valve.
   c. Remove the O-ring from the oil control valve.
5. Remove the camshaft timing oil control valve assembly (for intake side of Bank 2).
   a. Disconnect the oil control valve connector.
   b. Remove the bolt and oil control valve.
   c. Remove the O-ring from the oil control valve.
6. Remove the camshaft timing oil control valve assembly (for exhaust side of Bank 2).
   a. Disconnect the oil control valve connector.
   b. Remove the bolt and oil control valve.
   c. Remove the O-ring from the oil control valve.

To install:

7. To install, reverse the removal procedure. Apply a light coat of engine oil to new o-rings. Tighten the camshaft timing oil control valve assemblies to 7 ft. lbs. (10 Nm).

2.7L Engine

See Figures 309 and 310.

1. Remove the camshaft timing oil control valve assembly.

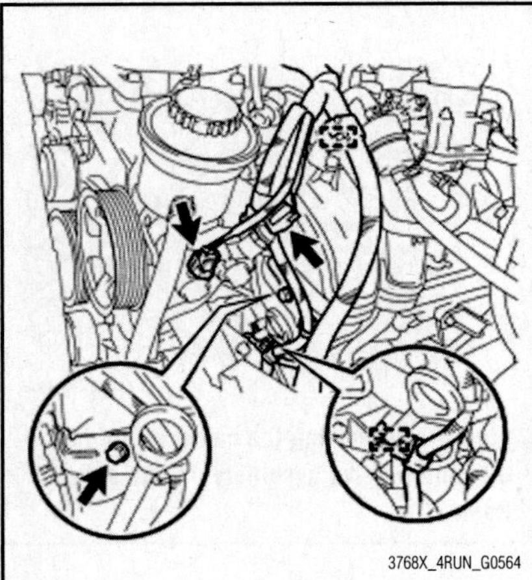

Fig. 309 Detaching the 2 clamps and disconnecting the 2 connectors

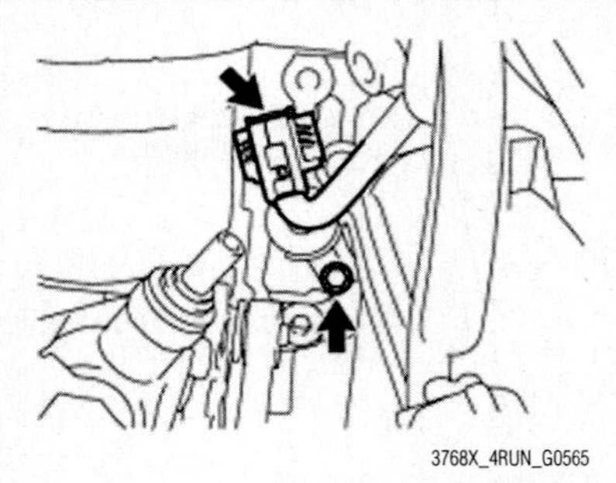

Fig. 310 Disconnecting the camshaft timing oil control valve connector

a. Detach the 2 clamps and disconnect the 2 connectors.

b. Remove the bolt and disconnect the wiring harness clamp bracket.

c. Disconnect the camshaft timing oil control valve connector.

d. Remove the bolt and camshaft timing oil control valve.

e. Remove the O-ring from the camshaft timing oil control valve.

To install:

2. Install the camshaft timing oil control valve assembly.

a. Install a new o-ring to the oil control valve.

b. Apply a light coat of engine oil to the o-ring.

c. Install the camshaft timing oil control valve with the bolt. Tighten the bolt to 71 inch lbs. (8 Nm).

d. Attach the 2 clamps and connect the 2 connectors.

FUEL — GASOLINE FUEL INJECTION SYSTEM

FUEL SYSTEM SERVICE PRECAUTIONS

Safety is the most important factor when performing not only fuel system maintenance, but any type of maintenance. Failure to conduct maintenance and repairs in a safe manner may result in serious personal injury or death. Work on a vehicle's fuel system components can be accomplished safely and effectively by adhering to the following rules and guidelines.

- To avoid the possibility of fire and personal injury, always disconnect the negative battery cable unless the repair or test procedure requires that battery voltage be applied.
- Always relieve the fuel system pressure prior to disconnecting any fuel system component (injector, fuel rail, pressure regulator, etc.) fitting or fuel line connection. Exercise extreme caution whenever relieving fuel system pressure to avoid exposing skin, face and eyes to fuel spray. Please be advised that fuel under pressure may penetrate the skin or any part of the body that it contacts.
- Always place a shop towel or cloth around the fitting or connection prior to loosening to absorb any excess fuel due to spillage. Ensure that all fuel spillage is quickly removed from engine surfaces. Ensure that all fuel-soaked cloths or towels are deposited into a flame-proof waste container with a lid.
- Always keep a dry chemical (Class B) fire extinguisher near the work area.
- Do not allow fuel spray or fuel vapors to come into contact with a spark or open flame.
- Always use a second wrench when loosening or tightening fuel line connection fittings. This will prevent unnecessary stress and torsion on fuel piping. Always follow the proper torque specifications.
- Always replace worn fuel fitting O-rings with new ones. Do not substitute fuel hose where rigid pipe is installed.

FUEL SYSTEM PRESSURE

RELIEVING

2009 Models

1. Remove the fuel pump relay from the engine compartment relay block.
2. Start the engine and let it run until it shuts off.
3. Turn the ignition to OFF.
4. Try to start the engine and make sure it won't start.
5. Disconnect the negative battery cable.
6. Install the relay.

2010 Models

**** CAUTION**

Take precautions to prevent gasoline from spilling out before removing fuel system parts.

**** CAUTION**

Pressure still remains in the fuel lines even after performing the following procedures. When disconnecting a fuel line, cover it with a piece of cloth to prevent fuel from spraying or coming out.

1. Remove the circuit opening relay from the engine room relay block.
2. Start the engine. After the engine stops, turn the ignition switch off.

➡DTC P0171 (system to lean) may be stored.

3. Check that the engine does not start.
4. Remove the fuel tank cap and let the air out of the fuel tank.
5. Disconnect the cable from the negative (-) battery terminal.
6. Reinstall the circuit opening relay.

FUEL FILTER

REMOVAL & INSTALLATION

See Figure 311.

The fuel filter is part of the fuel pump module unit and is not a normally replaced item.

FUEL LEVEL SENDING UNIT

REMOVAL & INSTALLATION

2009 Models

See Figures 312 through 314.

**** CAUTION**

Do not smoke or work near an open flame when working on the fuel pump.

1. Discharge fuel system pressure.
2. Disconnect cable from negative battery terminal.

**** CAUTION**

Wait at least 90 seconds after disconnecting the cable from the negative (-) battery terminal to prevent airbag and seat belt pretensioner activation.

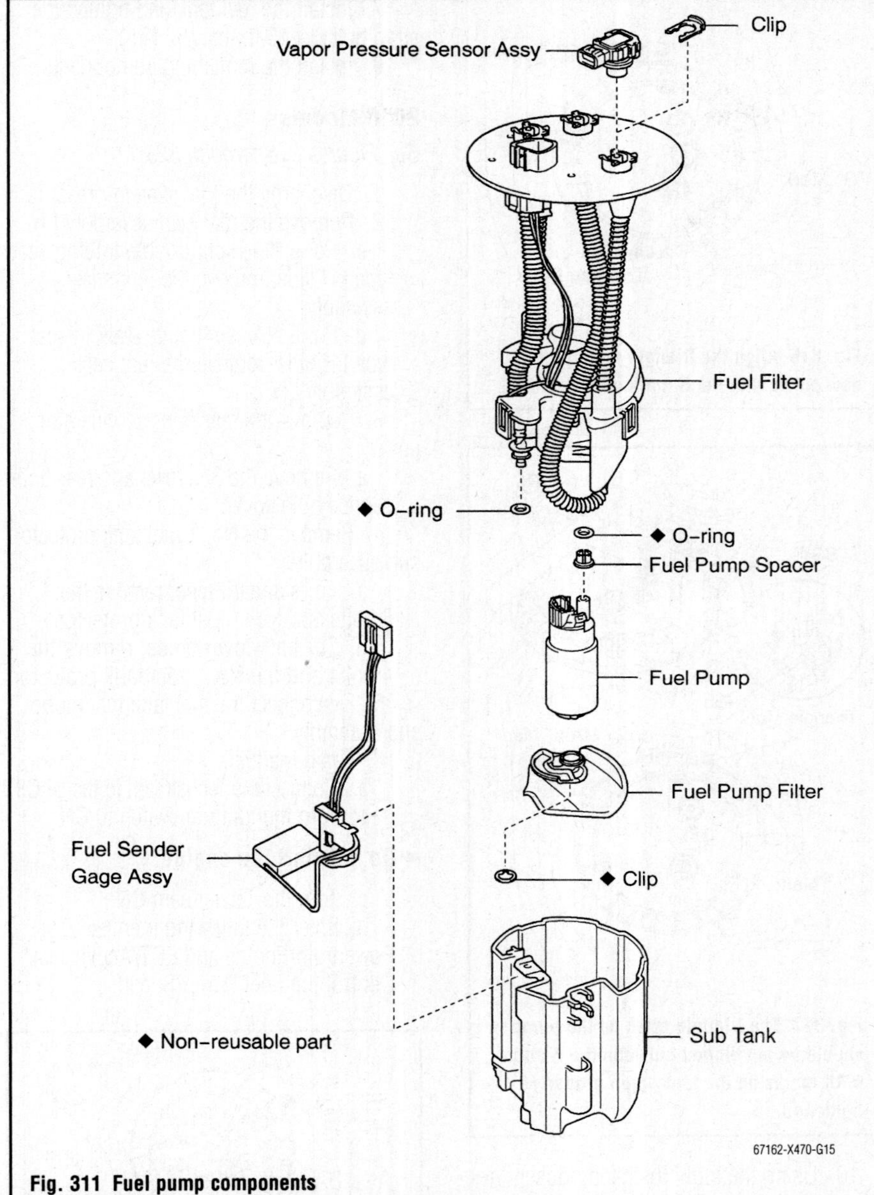

Fig. 311 Fuel pump components

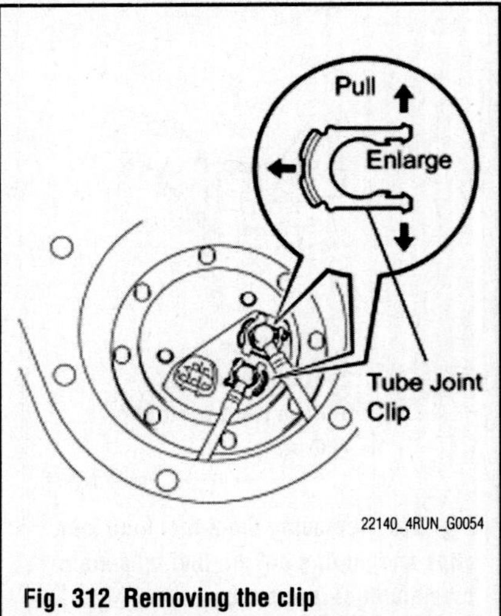

Fig. 312 Removing the clip

3. Remove the 2 rear seats.
4. Remove rear door scuff plate.
5. Remove step plate.
6. Remove rear seat lock cover.
7. Take off the front and rear floor carpets.
8. Remove the 2 screws and floor service hole cover.

CAUTION

Prevent the retained pressure in the fuel line from splashing inside the vehicle compartment. When sealing the tube and suction plates with the O-ring of the quick connector, be careful not to damage any contact surfaces or allow foreign matter to contact any surface. Be sure to perform the disconnection by hand. Do not use tools. Do not bend or turn the nylon tube by force.

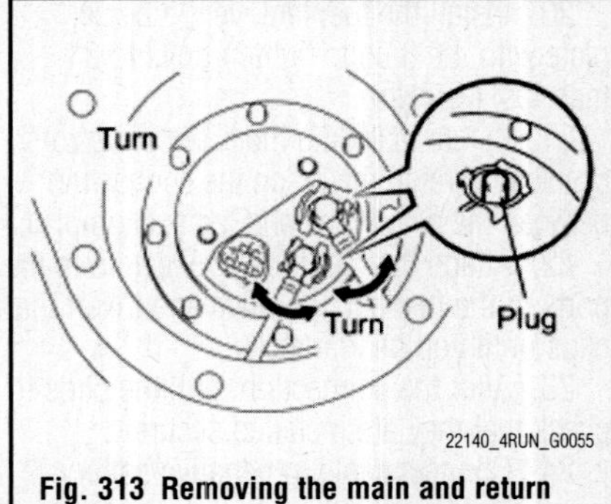

Fig. 313 Removing the main and return tubes

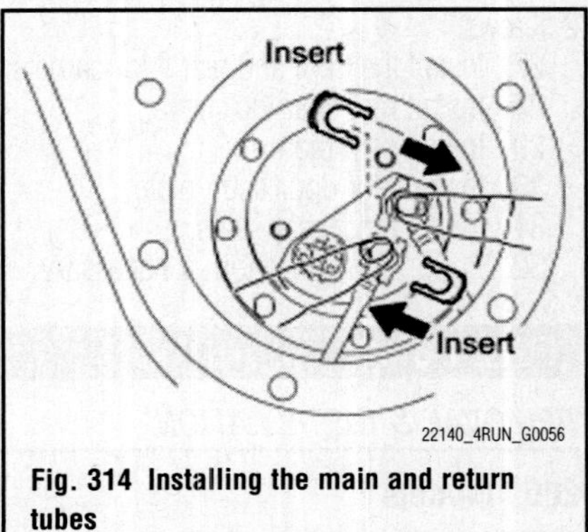

Fig. 314 Installing the main and return tubes

9. Disconnect fuel main tube and return tube.
10. Before the operation, remove foreign matter or dirt sticking to the tube joint clips.
11. Widen the tip of the clips with your fingers and pull them out for disconnection.
12. Pull out the fuel main tube and the return tube. If the nylon tube and the suction plate stick together, turn the nylon tube with your fingers and pull it out for disconnection.
13. After the disconnection, protect the connector with a plastic bag.
14. Remove the 8 bolts.
15. Pull out the fuel pump and sender gauge assembly.

➡Do not damage the fuel pump filter. Be careful that the arm of the sender gauge is not bent.

16. Remove sending unit.

To install:

17. Install sending unit.
18. Install a new gasket to the fuel suction plate.
19. Insert the fuel pump and sender gauge assembly into the fuel tank.

20. Install the fuel tank vent tube set plate with the 8 bolts tighten bolts to 31 inch lbs. (3.5 Nm).

21. Before installing the tube connectors, check for foreign matter on the connection between the nylon tube and the suction plate.

22. Attach the fuel tube connectors to the ports of the fuel suction plate and insert the clips until you hear a click.

23. After the connection, pull the clips to check that they are installed securely.

24. Connect cable to negative battery terminal.

25. Check for fuel leaks.

26. Install the service hole cover with the 2 screws.

27. Install the front and rear floor carpets.

28. Install rear seat lock cover.

29. Install step plate.

30. Install rear door scuff plate.

31. Install the 2 rear seats.

32. Perform initialization, if necessary.

FUEL PUMP

REMOVAL & INSTALLATION

2009 Models

See Figures 315 through 317.

1. Before servicing the vehicle, refer to the precautions section.
2. Relieve the fuel system pressure.
3. Remove the spare tire.
4. Disconnect the fuel pump connector and remove the fuel tank protector.
5. Disconnect the main and fuel return tubes.
6. Disconnect the fuel tank vent hose.
7. Disconnect the inlet and breather hoses.
8. Support the fuel tank with a jack, loosen the tank strap bolts remove the straps and lower the tank.
9. Disconnect any necessary hoses and wiring from the pump.

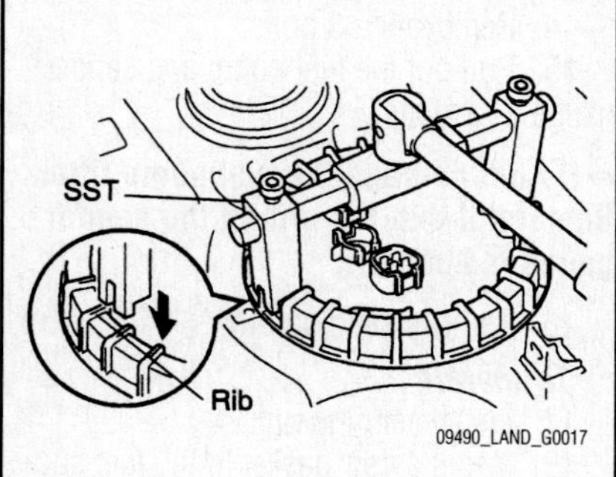

Fig. 315 Use the tool illustrated to remove the fuel pump retainer

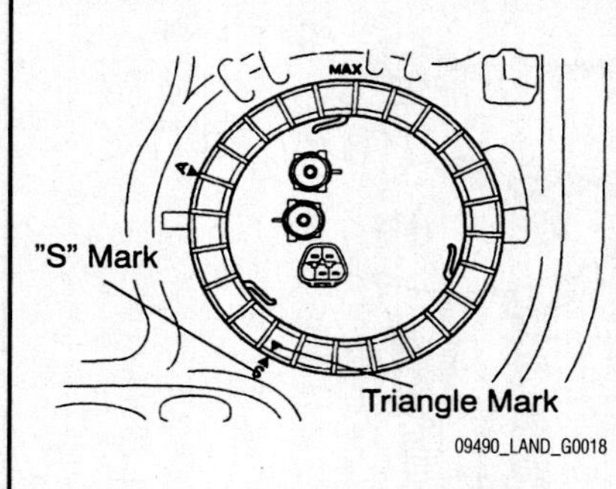

Fig. 316 Align the triangle mark on the new pump with the S mark on the tank

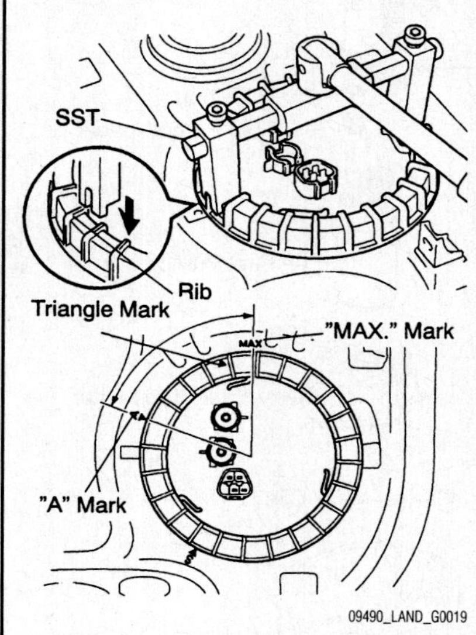

Fig. 317 The triangle mark on the pump should be positioned between the A and MAX marks on the tank when properly tightened

10. Using the tool illustrated, loosen the pump retainer.
11. Remove the pump and gasket.

To install:

12. Install a new gasket and the pump. Make sure to align the keyway of the suction tube with the key of the suction plate No. 1.
13. Apply a multipurpose grease to the whole surface of the pump retainer.
14. Align the triangle mark on the new pump retainer with the S mark on the tank while pushing the suction tube down and attach the gauge retainer.
15. Using the same tool used to remove the pump retainer, tighten the retainer 1 1/2 times. The triangle mark on the pump should be positioned between the A and MAX marks on the tank.
16. Attach any electrical connections and hoses.
17. Install the fuel tank and tighten the strap bolts to 45 ft. lbs. (62 Nm).
18. Install the remaining components.

2010 Models

See Figures 318 through 323.

1. Discharge the fuel system pressure.
2. Remove the rear seat assembly LH.
   a. For a 60/40 split double folding seat type LH side, remove the rear seat assembly.
   b. For a 60/40 split slide walk in seat type LH side, remove the rear seat assembly LH.
3. Remove the rear floor service hole cover.
   a. Remove the 3 screws and rear floor service hole cover.
4. Remove the No. 1 fuel tank protector sub assembly.
   a. For standard type, remove the 6 bolts and No. 1 fuel tank protector.
   b. For half cover types, remove the 4 bolts and the No. 1 fuel tank protector.
5. Disconnect the fuel tank main tube sub assembly.
6. Drain the fuel.
   a. Connect the Techstream to the DLC3.
   b. Turn the ignition switch to ON.

➡Do not start the engine.

   c. Turn the Techstream ON.
   d. Enter the following menus: Powertrain/Engine and ECT/Active Test/ Control the Fuel Pump/Speed.

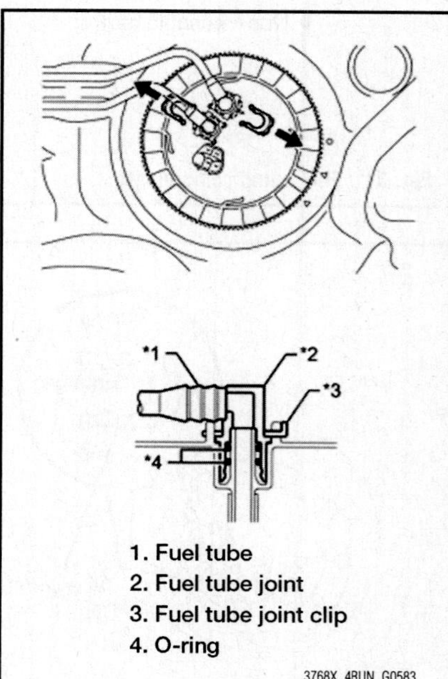

Fig. 318 Removing the 2 fuel tube joint clips and pulling out the fuel tank main tube and fuel return tube

Fig. 319 Removing the fuel tank main tube and fuel return tube from the fuel tank

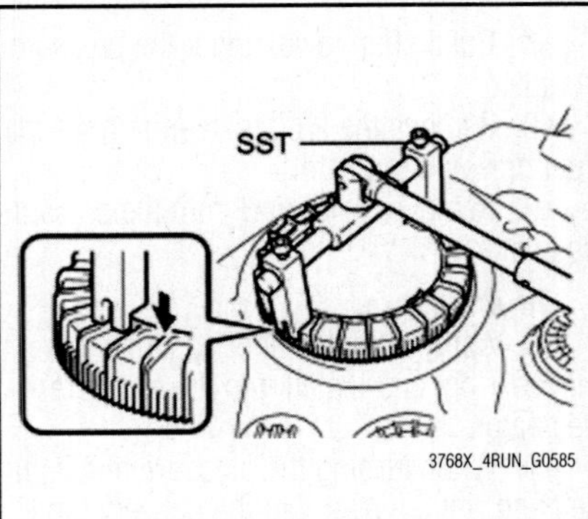

Fig. 320 Loosening the retainer

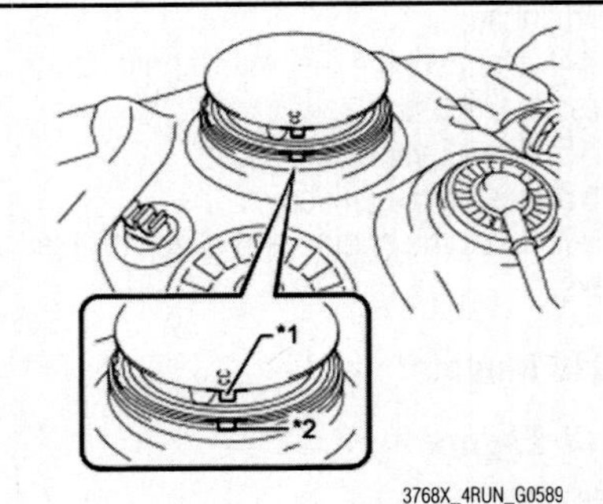

Fig. 321 Installing the fuel suction with the pump and gauge tube assembly

e. Operate the fuel pump and drain the fuel from the fuel tank.

CAUTION

Do not smoke or be near an open flame when working on the fuel system.

CAUTION

Secure good ventilation.

CAUTION

Keep gasoline away from rubber or leather parts.

f. Disconnect the fuel pump and sender gauge connector.

7. Disconnect the cable from the negative battery terminal.

➡When disconnecting the cable, some systems need to be initialized after the cable is reconnected.

8. Disconnect the fuel return tube sub assembly.

a. Disconnect the fuel return tube.

9. Disconnect the fuel tank vent hose sub assembly.

a. Disconnect the fuel tank vent hose.

10. Disconnect the fuel tank breather tube sub assembly.

a. Disconnect the fuel tank breather tube.

11. Disconnect the fuel tank to the filler pipe hose.

12. Remove the fuel tank sub assembly.

a. Place a transmission jack under the fuel tank.

b. Remove the 2 bolts, 2 clips, 2 pins and 2 fuel tank bands.

c. Slowly lower the transmission jack slightly.

13. Remove the fuel tank cushion.

a. Remove the No. 1, No. 2 and No. 3 fuel tank cushions from the fuel tank.

14. Remove the No. 1 fuel tank protector.

15. Remove the fuel tank breather tube sub assembly.

16. Remove the fuel tank main tube sub assembly and fuel return tube sub assembly.

SST
*1
*1
*2
A

1. Fuel tank side mark
2. Retainer side mark

3768X_4RUN_G0590

Fig. 322 Tightening the retainer

a. Remove the 2 fuel tube joint clips and pull out the fuel tank main tube and fuel return tube.

➡Remove any dirt and foreign matter on the fuel tube joint before performing this work.

➡Do not allow any scratches or foreign matter on the parts when disconnecting them, as the fuel tube joint contains the O-rings that seal the plug.

➡Perform this work by hand. Do not use any tools.

➡Do not forcibly bend, twist or turn the nylon tube.

➡Protect the disconnected part by covering it with a plastic bag and tape after disconnecting the fuel tubes.

b. Remove the fuel tank main tube and fuel return tube from the fuel tank.

17. Remove the fuel suction with pump and gauge tube assembly.

a. Using the special tool (09808-14020), loosen the retainer.

➡Fit the tips of the special tool onto the ribs of the retainer.

➡When the retainer is loosened, be careful as the pump and gauge tube will spring upward from the force of the spring.

➡Remove any foreign matter around the fuel suction with pump and gauge tube before this operation.

b. Remove the retainer.

c. Remove the fuel suction with pump and gauge tube assembly from the fuel tank.

➡Be careful not to bend the arm of the fuel sender gauge.

d. Remove the gasket from the fuel tank.

To install:

18. Install the fuel suction with the pump and gauge tube assembly.

a. Apply a light coat of gasoline or grease to a new gasket and install the gasket to the fuel tank.

b. Install the fuel suction with pump and gauge tube into the fuel tank.

➡Align the protrusion of the fuel suction with pump and gauge tube with the groove of the fuel tank.

➡Be careful not to bend the arm of the fuel sender gauge.

c. Put the new retainer on the fuel tank. While holding the fuel suction with pump and gauge tube, tighten the retainer one complete turn by hand.

d. Using special tool (09808-14020), tightens the retainer until the mark on the retainer is within range A on the fuel tank, as shown in the illustration.

➡**Fit the tips of special tool onto the ribs of the retainer.**

19. Install the fuel tank main tube sub assembly and fuel return tube sub assembly.

a. Install the fuel tank main tube and fuel return tube with the 2 fuel tube joint clips.

➡**Check that there are no scratches or foreign objects on the connecting parts.**

➡**Check that the fuel tube joints are inserted securely.**

➡**Check that the fuel tube joint clips are on the collars of the fuel tube joints.**

➡**After installing the fuel tube joint clips, check that the fuel tube joints cannot be pulled off.**

b. Install the fuel tank main tube and fuel return tube to the fuel tank.

20. Install the fuel tank breather tube sub assembly.

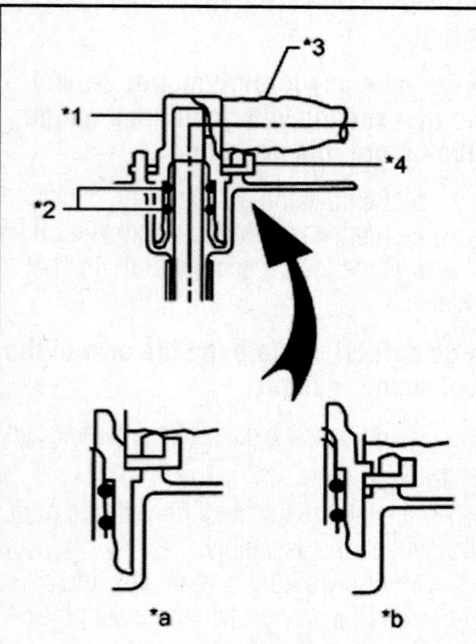

Fig. 323 Installing the fuel tank main tube sub assembly and fuel return tube sub assembly

21. Install the No. 1 fuel tank protector to the fuel tank.

22. Install the fuel tank cushion.

a. Install the 3 new fuel tank cushions to the fuel tank.

23. Install the fuel tank sub assembly.

a. Set the fuel tank on a transmission jack and lift up the transmission jack.

➡**Do not allow the fuel tank to contact the vehicle, especially the differential.**

b. Connect the 2 fuel tank bands with the 2 bolts. Tighten to 30 ft. lbs. (40 Nm).

24. Connect the fuel tank to filler pipe hose to the filler pipe.

25. Connect the fuel tank breather tube sub assembly.

26. Connect the fuel tank vent hose sub assembly.

27. Connect the fuel return tube sub assembly.

28. Connect the fuel tank main tube sub assembly.

29. Install the No. 1 fuel tank protector sub assembly. Tighten to 15 ft. lbs. (20 Nm).

30. Install the rear floor service hole cover.

a. Connect the fuel pump and fuel sender gauge connector.

b. Install the rear floor service hole cover with the 3 screws.

31. Install the rear seat assembly LH.

32. Connect the cable to the negative battery terminal.

➡**When disconnecting the cable, some systems need to be initialized after the cable is reconnected.**

33. Inspect for fuel leaks.

FUEL PRESSURE REGULATOR

REMOVAL & INSTALLATION

2009 Models

See Figure 324.

➡**For V6 engines, the following procedure applies to the fuel pressure regulator. For V8 engines, the primary fuel pressure regulator is part of the fuel pump assembly. The information below is related to the fuel pulsation dampener assembly.**

1. Before servicing the vehicle, refer to the Precautions section.
2. Relieve the fuel system pressure.
3. Remove V-bank cover
4. Disconnect the no. 2 ventilation hose
5. Remove air cleaner assembly
6. Disconnect No. 2 fuel pipe sub-assembly

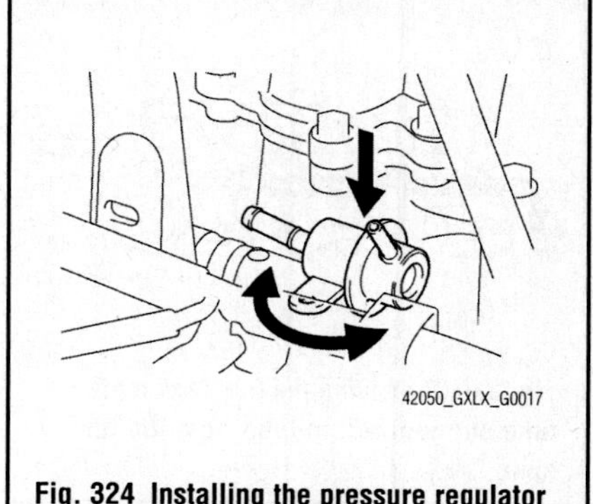

Fig. 324 Installing the pressure regulator

7. Put a shop towel under the pressure regulator.
8. Remove the 2 bolts, then remove the fuel pressure regulator
9. Remove the O-ring from the pressure regulator.

To install:

10. Apply a light coat of gasoline to a new O-ring, and install it to the pressure regulator.
11. While turning the pressure regulator left and right, install it to the delivery pipe.
12. Install the pressure regulator with the 2 bolts and tighten to 7.5 Nm (66 in. lbs.).
13. Connect the vacuum hose to intake air resonator.
14. Connect the fuel return hose to the pressure regulator.
15. Install ventilation hose
16. Install V-bank cover
17. Start the engine and check for fuel leaks.

2010 Models

4.0L Engine

See Figure 325.

1. Discharge the fuel system pressure.
2. Disconnect the cable from the negative battery terminal.

Fig. 325 Removing the fuel pressure regulator

➡When disconnecting the cable, some systems need to be initialized after the cable is reconnected.

3. Remove the intake air surge tank.
4. Disconnect the No. 2 fuel pipe sub assembly.
5. Remove the fuel pressure regulator assembly.
   a. Disconnect the vacuum hose.
   b. Remove the 2 bolts and fuel pressure regulator.
   c. Remove the o-ring from the fuel pressure regulator.

To install:

6. Install the fuel pressure regulator assembly.
   a. Apply a light coat of spindle oil or gasoline to a new O-ring and install the O-ring to the fuel pressure regulator.
   b. Install the fuel pressure regulator with the 2 bolts. Tighten to 80 inch lbs. (9 Nm).
   c. Connect the vacuum hose.
7. Connect the No. 2 fuel pipe sub assembly.
8. Install the intake air surge tank.
9. Connect the cable to the negative battery terminal.

➡When disconnecting the cable, some systems need to be initialized after the cable is reconnected.

10. Inspect for fuel leaks.

2.7L Engine

See Figure 326.

1. Discharge the fuel system pressure.
2. Disconnect the cable from the negative battery terminal.

➡When disconnecting the cable, some systems need to be initialized after the cable is reconnected.

3. Remove the throttle body with the motor assembly.
4. Disconnect the No. 2 fuel hose.
5. Remove the fuel pressure regulator assembly.
   a. Disconnect the purge VSV connector.
   b. Detach the wire harness clamp.
   c. Remove the bolt and wire harness clamp bracket.
   d. Remove the 3 bolts and fuel pressure regulator.

To install:

6. Install the fuel pressure regulator assembly.
   a. Apply a light coat of gasoline or spindle oil to the o-ring.
   b. Install the fuel pressure regulator with the 3 bolts. Tighten to 75 inch lbs. (8.5 Nm).
   c. install the wire harness clamp bracket with the bolt. Tighten to 71 inch lbs. (8 Nm).
   d. Attach the wire harness clamp.
   e. Connect the purge VSV connector.
7. Connect the No. 2 fuel hose.
8. Install the throttle body with the motor assembly.
9. Connect the cable to the negative battery terminal.

➡When disconnecting the cable, some systems need to be initialized after the cable is reconnected.

10. Inspect for fuel leaks.

FUEL RAIL AND INJECTOR

REMOVAL & INSTALLATION

2010 Models

4.0L Engine

See Figures 327 through 329.

1. Remove the intake air surge tank.
2. Remove the rear cylinder head cover:
   a. Remove the 3 bolts and cover.
3. Disconnect the No. 1 fuel pipe sub assembly.
4. Disconnect the No. 2 fuel pipe sub assembly.
5. Remove the fuel delivery pipe sub assembly.
   a. Disconnect the 6 fuel injector connectors.
   b. Remove the 4 bolts and the fuel delivery pipe together with the 6 fuel injectors.

➡Be careful not to drop the injectors when removing the fuel delivery pipe.

6. Remove the fuel injector assembly.
   a. Remove the 6 fuel injectors from the fuel delivery pipe.
   b. Remove the o-ring and injector vibration insulator from each fuel injector.

To install:

7. Install the fuel injector assembly.
   a. Install a new insulator to each fuel injector.
   b. Apply a light coat of spindle oil or gasoline to new O-rings and install one to each fuel injector.
   c. Install the 6 injectors. While turning each fuel injector left and right, install it to the fuel delivery pipe. Position the fuel injectors with the connectors facing outward.
8. Install the fuel delivery pipe sub assembly.

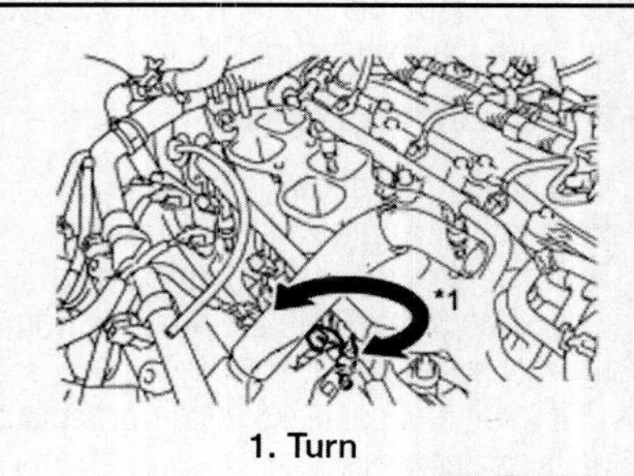

Fig. 328 Installing the fuel delivery pipe sub assembly

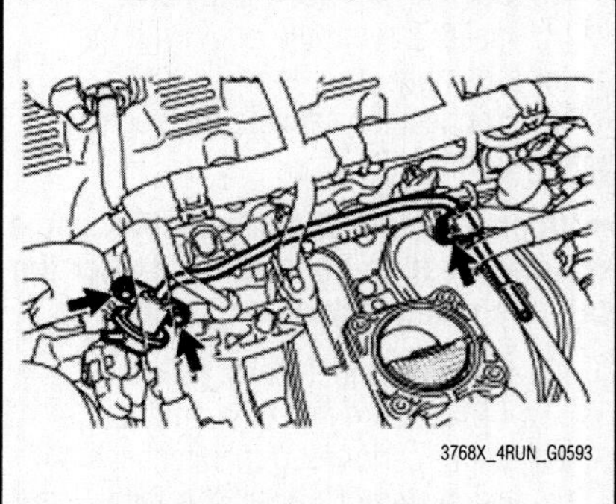

Fig. 326 Removing the fuel pressure regulator

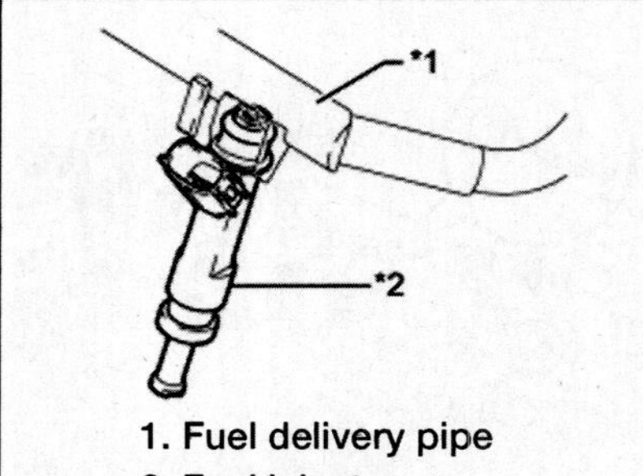

Fig. 327 Removing the fuel injector assembly

Fig. 329 Identifying the rear cylinder head cover bolt tightening sequence

a. Place the fuel delivery pipe together with the 6 fuel injectors on the intake manifold.

b. Temporarily install the 4 bolts, which are used to hold the fuel delivery pipe in place, to the intake manifold.

c. Check that the fuel injectors rotate smoothly.

➡**If the fuel injectors do not rotate smoothly, replace the O-ring of any injector that does not rotate smoothly.**

d. Position the fuel injectors with the connectors facing outward.

e. Tighten the 6 bolts. Tighten to 15 ft. lbs. (21 Nm).

f. Connect the 6 fuel injector connectors.

9. Connect the No. 2 fuel pipe sub assembly.

10. Connect the No. 1 fuel pipe sub assembly.

11. Install the rear cylinder head cover with the 3 bolts.

a. Tighten the 3 bolts in the sequence shown in the illustration. Tighten to 80 inch lbs. (9 Nm).

12. Install the air surge tank.

13. Inspect for fuel leaks.

a. Make sure that there are no fuel leaks after performing maintenance on the fuel system. Connect the Techstream to the DLC3. Turn the ignition switch to ON and turn the Techstream ON.

➡**Do not start the engine.**

b. Enter the following menus: Powertrain / Engine and ECT / Active Test / Control the Fuel Pump / Speed.

c. Check that there are no leaks from the fuel system.

If there are fuel leaks, repair or replace parts as necessary.

d. Turn the ignition switch off.

e. Disconnect the Techstream from the DLC3.

2.7L Engine

See Figures 330 and 331.

1. Discharge the fuel system pressure.

2. Disconnect the cable from the negative battery terminal.

➡**When disconnecting the cable, some systems need to be initialized after the cable is reconnected.**

3. Remove the throttle body with the motor assembly.

4. Disconnect the fuel hose.

5. Disconnect the No. 2 fuel hose.

6. Remove the fuel pressure pulsation damper assembly.

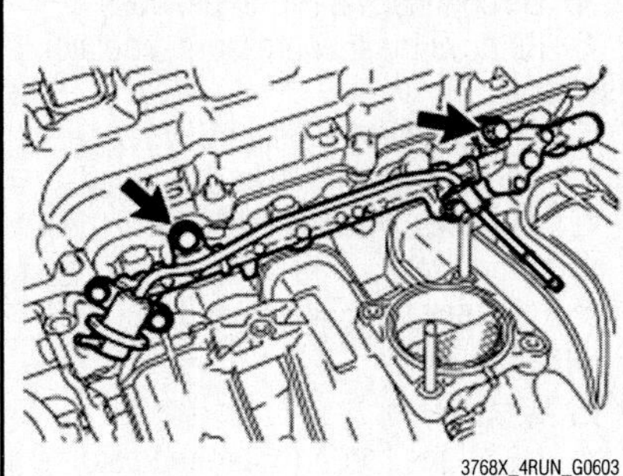

Fig. 330 Removing the fuel delivery pipe with the injectors

7. Remove the fuel delivery pipe with the fuel injector.

a. Disconnect the 4 injector connectors.

b. Disconnect the purge VSV connector.

c. Detach the wire harness clamp.

d. Remove the bolt and wire harness clamp bracket.

e. Remove the bolt and disconnect the purge VSV.

f. Remove the 2 bolts and fuel delivery pipe together with the fuel injectors.

➡**Be careful not to drop the fuel injectors when removing the fuel delivery pipe.**

g. Remove the 2 No. 1 delivery pipe spacers.

h. Remove the 4 injector vibration insulators.

i. Using a screwdriver, remove the 4 spacers.

8. Remove the fuel injector assembly.

a. Remove the 4 fuel injectors from the fuel delivery pipe.

To install:

9. Install the fuel injector assembly.

a. Apply a light coat of gasoline or spindle oil to new O-rings, and then install one to each fuel injector.

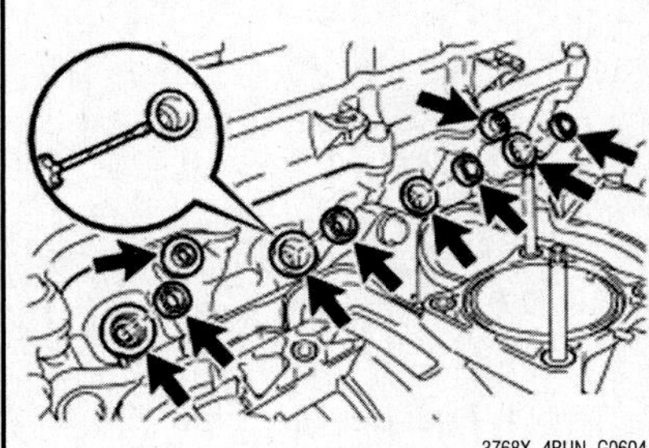

Fig. 331 Removing the delivery pipe spacers, injector vibration insulators and spacers

b. Apply a light coat of gasoline or spindle oil to the contact surfaces of the fuel delivery pipe and the O-ring of the fuel injector.

c. Apply a light coat of gasoline or spindle oil to the O-ring again, and then install the fuel injector to the fuel delivery pipe by turning it right and left.

➡**Make sure that the O-ring is not cracked or jammed when installing.**

d. Check that the fuel injector rotates smoothly. If the fuel injector does not rotate smoothly, replace the O-ring.

10. Install the fuel delivery pipe with the fuel injector.

a. Apply a light coat of gasoline or spindle oil to new O-rings, and then install one to each spacer.

b. Install the 4 spacers to the cylinder head.

c. Install 4 new injector vibration insulators to the cylinder head.

d. Install the 2 No. 1 delivery pipe spacers to the cylinder head.

e. Install the fuel delivery pipe together with the 4 fuel injectors, and then temporarily install the fuel delivery pipe with the 2 bolts.

➡**Do not drop the fuel injectors when installing the fuel delivery pipe.**

f. Check that the fuel injectors rotate smoothly. If any fuel injector does not rotate smoothly, replace its O-ring.

g. Tighten the 2 bolts to 9 ft. lbs. (12 Nm).

h. Install the purge VSV with the bolt. Tighten to 80 inch lbs. (9 Nm).

i. Install the harness clamp bracket with the bolt. Tighten to 71 inch lbs. (8 Nm).

j. Attach the wire harness clamp.

k. Connect the purge VSV connector.

l. Connect the 4 injector connectors.

11. Install the fuel pressure pulsation damper assembly.

12. Connect the fuel hose.

13. Connect the No. 2 fuel hose.

14. Install the throttle body with the motor assembly.

15. Connect the cable to the negative battery terminal.

➡**When disconnecting the cable, some systems need to be initialized after the cable is reconnected.**

16. Inspect for fuel leaks.

a. Make sure that there are no fuel leaks after performing maintenance on the fuel system. Connect the Techstream to the DLC3. Turn the ignition switch to ON and turn the Techstream on.

➡**Do not start the engine.**

b. Enter the following menus: Powertrain / Engine and ECT / Active Test / Control the Fuel Pump / Speed.

c. Check that there are no leaks from the fuel system.

If there are fuel leaks, repair or replace parts as necessary.

d. Turn the ignition switch off.

e. Disconnect the Techstream from the DLC3.

FUEL TANK

DRAINING

1. Connect the Techstream to the DLC3.
2. Turn the ignition switch to ON.

➡**Do not start the engine.**

3. Turn the Techstream on.
4. Enter the following menus: Powertrain / Engine and ECT / Active Test / Control the Fuel Pump / Speed.
5. Operate the fuel pump and drain the fuel from the fuel tank main tube.

CAUTION

Do not smoke or be near an open flame when working on the fuel system.

CAUTION

Secure good ventilation.

CAUTION

Keep gasoline away from rubber or leather parts.

6. Disconnect the fuel pump and sender gauge connector.

REMOVAL & INSTALLATION

2009 Models

See Figures 332 through 334.

1. Discharge fuel system pressure.
2. Disconnect cable from negative battery terminal.
3. Disconnect vent line tube.
4. Remove the bolt and bracket from the fuel tank band.
5. Remove the bolt and bracket from the body.
6. Disconnect the fuel main tube, return tube and fuel tube.

a. With Fuel hose connector cover type disengage the lock claw by lifting up the cover, as shown in the illustration.

b. Check for dirt or mud on the pipe and around the connector before disconnection. Clean if necessary.

c. Disconnect the connector and pipe by hand.

d. If the connector and the pipe stuck, pinch the connector, and push and pull the pipe to disconnect it.

➡**Do not use any tools.**

e. Check for dirt or mud on the seal surface of the disconnected pipe. Clean if necessary.

f. To protect the disconnected pipe and connector from damage and contamination, cover it with a plastic bag.

7. Loosen the bolt of the clamp and disconnect the fuel inlet hose from the fuel inlet pipe.
8. Set up a transmission jack under the fuel tank.
9. Remove the 2 bolts and disconnect the 2 fuel tank bands from the fuel tank.
10. Slightly lower the mission jack so that the fuel pump and sender gauge connector and 2 clamps can be removed.

➡**Do not lower the mission jack excessively as this may damage the connector.**

11. Operate the transmission jack and remove the fuel tank.
12. Remove fuel pump and sender gauge assembly.
13. Remove fuel inlet hose.
14. Remove fuel hose.

To install:

15. Install fuel hose.
16. Install fuel inlet hose.
17. Install fuel pump and sender gauge assembly.
18. Install fuel tank assembly.
19. Operate the transmission jack so that the fuel pump and sender gauge connector and 2 clamps can be installed. Then raise the transmission jack again to install the fuel tank.

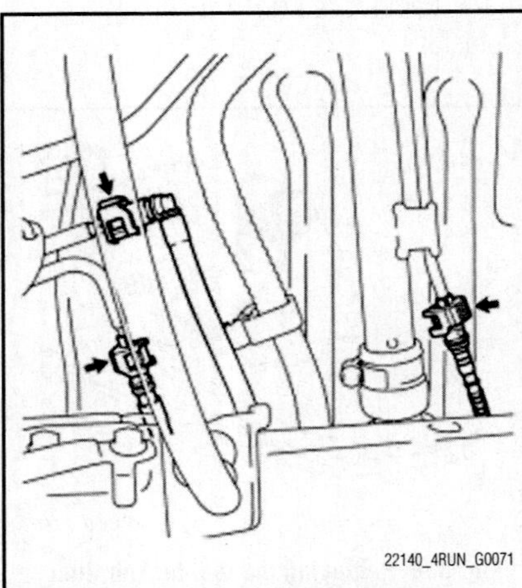

Fig. 332 Disconnecting the fuel lines

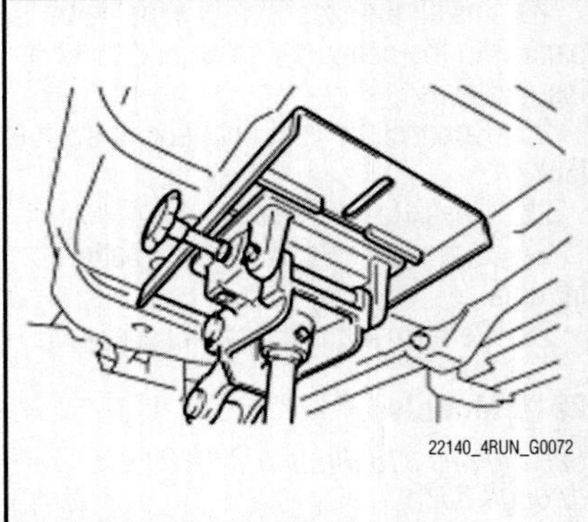

Fig. 333 Removing the fuel tank

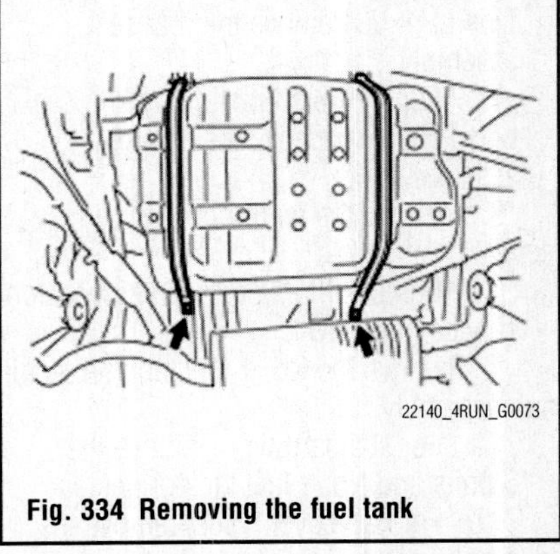

Fig. 334 Removing the fuel tank

20. Install the 2 fuel tank bands with the 2 bolts and tighten to 30 ft. lbs. (40 Nm).
21. Connect the fuel main tube, return tube and fuel hose.

a. Check that there is no damage or contamination in the connected part of the pipe.

b. Align the axis of the connector with the axis of the pipe. Push the pipe into the connector until the connector makes a "click" sound. If the connection is tight, apply a little amount of fresh engine oil on the tip of the pipe.

c. After having finished the connection, try to pull apart the pipe and the connector and confirm that they are securely connected.

d. With fuel hose connector cover type attach the lock claw by lifting up the cover, as shown in the illustration.

22. Connect the fuel inlet hose to the fuel inlet pipe and tighten the bolt of the clamp to 66 inch lbs. (7.5 Nm).
23. Install the bracket to the body with the bolt and tighten bolt to 11 ft. lbs. (15 Nm).

24. Install the bracket to the fuel tank band with the bolt and tighten bolt to 11 ft. lbs. (15 Nm).
25. Connect the vent line tube to the fuel tank.
26. Check for fuel leaks.
27. Connect cable to negative battery terminal.
28. Perform initialization, if necessary.

2010 Models

See Figures 318 through 323, 335 through 337.

1. Discharge the fuel system pressure.
2. Remove the rear seat assembly LH.
   a. For a 60/40 split double folding seat type LH side, remove the rear seat assembly.
   b. For a 60/40 split slide walk in seat type LH side, remove the rear seat assembly LH.
3. Remove the rear floor service hole cover.
   a. Remove the 3 screws and rear floor service hole cover.
4. Remove the No. 1 fuel tank protector sub assembly.
   a. For standard type, remove the 6 bolts and No. 1 fuel tank protector.
   b. For half cover types, remove the 4 bolts and the No. 1 fuel tank protector.
5. Disconnect the fuel tank main tube sub assembly.
6. Drain the fuel.
   a. Connect the Techstream to the DLC3.
   b. Turn the ignition switch to ON.

➡Do not start the engine.

   c. Turn the Techstream ON.
   d. Enter the following menus: Powertrain/Engine and ECT/Active Test/ Control the Fuel Pump/Speed.
   e. Operate the fuel pump and drain the fuel from the fuel tank.

CAUTION

Do not smoke or be near an open flame when working on the fuel system.

CAUTION

Secure good ventilation.

CAUTION

Keep gasoline away from rubber or leather parts.

   f. Disconnect the fuel pump and sender gauge connector.
7. Disconnect the cable from the negative battery terminal.

➡When disconnecting the cable, some systems need to be initialized after the cable is reconnected.

8. Disconnect the fuel return tube sub assembly.
   a. Disconnect the fuel return tube.
9. Disconnect the fuel tank vent hose sub assembly.
   a. Disconnect the fuel tank vent hose.
10. Disconnect the fuel tank breather tube sub assembly.
    a. Disconnect the fuel tank breather tube.
11. Disconnect the fuel tank to the filler pipe hose.
12. Remove the fuel tank sub assembly.
    a. Place a transmission jack under the fuel tank.
    b. Remove the 2 bolts, 2 clips, 2 pins and 2 fuel tank bands.
    c. Slowly lower the transmission jack slightly.
13. Remove the fuel tank cushion.
    a. Remove the No. 1, No. 2 and No. 3 fuel tank cushions from the fuel tank.
14. Remove the No. 1 fuel tank protector.
15. Remove the fuel tank breather tube sub assembly.
16. Remove the fuel tank main tube sub assembly and fuel return tube sub assembly.
    a. Remove the 2 fuel tube joint clips and pull out the fuel tank main tube and fuel return tube.

➡Remove any dirt and foreign matter on the fuel tube joint before performing this work.

➡Do not allow any scratches or foreign matter on the parts when disconnecting them, as the fuel tube joint contains the O-rings that seal the plug.

➡Perform this work by hand. Do not use any tools.

➡Do not forcibly bend, twist or turn the nylon tube.

➡Protect the disconnected part by covering it with a plastic bag and tape after disconnecting the fuel tubes.

    b. Remove the fuel tank main tube and fuel return tube from the fuel tank.
17. Remove the fuel suction with pump and gauge tube assembly.
    a. Using the special tool (09808-14020), loosen the retainer.

➡Fit the tips of the special tool onto the ribs of the retainer.

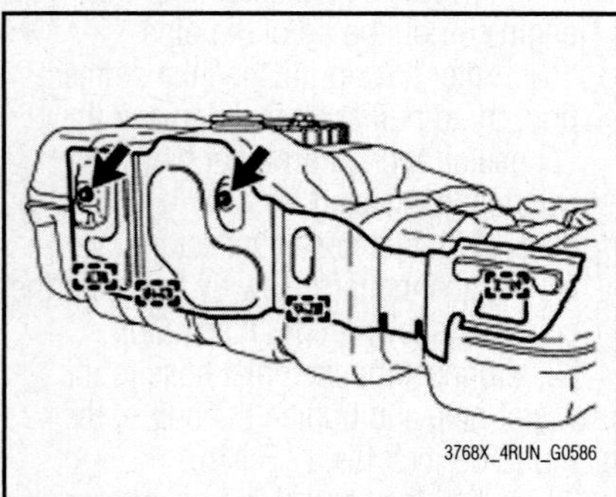

Fig. 335 Removing the No. 3 fuel tank protector

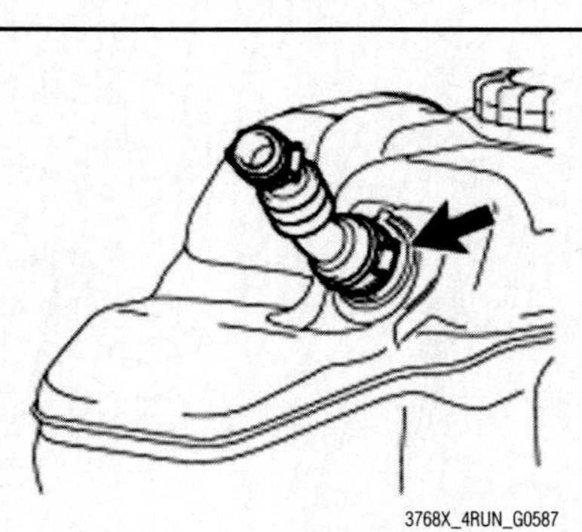

Fig. 336 Removing the fuel tank to filler pipe hose

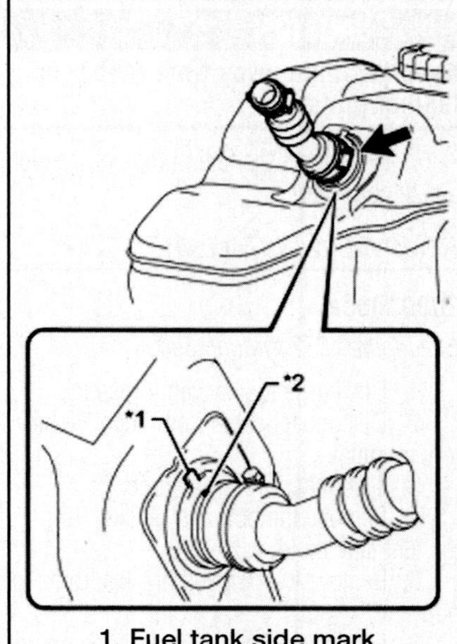

Fig. 337 Installing the fuel tank to the filler pipe hose

➡When the retainer is loosened, be careful as the pump and gauge tube will spring upward from the force of the spring.

➡Remove any foreign matter around the fuel suction with pump and gauge tube before this operation.

b. Remove the retainer.
c. Remove the fuel suction with pump and gauge tube assembly from the fuel tank.

➡Be careful not to bend the arm of the fuel sender gauge.

d. Remove the gasket from the fuel tank.
18. Remove the No. 3 fuel tank protector.
a. Remove the 2 bolts.
b. Detach the 4 clamps and remove the No. 3 fuel tank protector.
19. Remove the fuel tank to filler pipe hose.
a. Remove the fuel tank to filler pipe hose from the fuel tank.

To install:

20. Install the fuel tank to filler pipe hose.

➡Align the fuel tank side mark with the hose side mark when installing the hose.

a. Install the 2 bolts and tighten to 44 inch lbs. (5 Nm).
21. Install the fuel suction with the pump and gauge tube assembly.
a. Apply a light coat of gasoline or grease to a new gasket and install the gasket to the fuel tank.
b. Install the fuel suction with pump and gauge tube into the fuel tank.

➡Align the protrusion of the fuel suction with pump and gauge tube with the groove of the fuel tank.

➡Be careful not to bend the arm of the fuel sender gauge.

c. Put the new retainer on the fuel tank. While holding the fuel suction with pump and gauge tube, tighten the retainer one complete turn by hand.
d. Using special tool (09808-14020), tighten the retainer until the mark on the retainer is within range A on the fuel tank, as shown in the illustration.

➡Fit the tips of special tool onto the ribs of the retainer.

22. Install the fuel tank main tube sub assembly and fuel return tube sub assembly.
a. Install the fuel tank main tube and fuel return tube with the 2 fuel tube joint clips.

➡Check that there are no scratches or foreign objects on the connecting parts.

➡Check that the fuel tube joints are inserted securely.

➡Check that the fuel tube joint clips are on the collars of the fuel tube joints.

➡After installing the fuel tube joint clips, check that the fuel tube joints cannot be pulled off.

b. Install the fuel tank main tube and fuel return tube to the fuel tank.
23. Install the fuel tank breather tube sub assembly.
24. Install the No. 1 fuel tank protector to the fuel tank.
25. Install the fuel tank cushion.
a. Install the 3 new fuel tank cushions to the fuel tank.
26. Install the fuel tank sub assembly.
a. Set the fuel tank on a transmission jack and lift up the transmission jack.

➡Do not allow the fuel tank to contact the vehicle, especially the differential.

b. Connect the 2 fuel tank bands with the 2 bolts. Tighten to 30 ft. lbs. (40 Nm).
27. Connect the fuel tank to filler pipe hose to the filler pipe.
28. Connect the fuel tank breather tube sub assembly.
29. Connect the fuel tank vent hose sub assembly.
30. Connect the fuel return tube sub assembly.
31. Connect the fuel tank main tube sub assembly.
32. Install the No. 1 fuel tank protector sub assembly. Tighten to 15 ft. lbs. (20 Nm).
33. Install the rear floor service hole cover.
a. Connect the fuel pump and fuel sender gauge connector.
b. Install the rear floor service hole cover with the 3 screws.
34. Install the rear seat assembly LH.
35. Connect the cable to the negative battery terminal.

➡When disconnecting the cable, some systems need to be initialized after the cable is reconnected.

36. Inspect for fuel leaks.

IDLE SPEED

ADJUSTMENT

Idle speed is maintained by the Powertrain Control Module (PCM). No adjustment is necessary or possible.

THROTTLE BODY

REMOVAL & INSTALLATION

2009 Models

4.0L Engine

1. Before servicing the vehicle, refer to the Precautions section.
2. Disconnect cable to negative battery terminal.
3. Remove the 2 nuts, then remove the V-bank cover.
4. Drain engine coolant.
5. Disconnect the ventilation hose No. 2.
6. Disconnect the vacuum hose.
7. Disconnect the mass air flow meter connector.
8. Remove the 2 wire harness clamps.
9. Loosen the 2 hose clamps.
10. Remove the 2 bolts, then remove the air cleaner.
11. Disconnect the water by-pass hose No. 5.
12. Disconnect the water by-pass hose No. 4.
13. Disconnect the throttle motor connector.
14. Remove the 4 bolts, then remove the throttle w/ motor body and gasket.

To install:

15. Install a new gasket and the throttle with motor body with the 4 bolts.
16. Connect the throttle motor connector.
17. Connect the water by-pass hose No. 4.
18. Connect the water by-pass hose No. 5.
19. Install the air cleaner with the 2 bolts.
20. Tighten the 2 hose clamps.
21. Install the 2 wire harness clamps.
22. Connect the mass air flow meter connector.
23. Connect the vacuum hose.
24. Connect the ventilation hose No. 2.
25. Connect cable to negative battery terminal.
26. Add engine coolant.
27. Check for engine coolant leakage.
28. Install the V-bank cover with the 2 nuts.

4.7L Engine

1. Before servicing the vehicle, refer to the Precautions section.
2. Remove the 2 nuts, then remove the V-bank cover.
3. Remove throttle body cover.
4. Drain engine coolant.
5. Remove intake air connector.
6. Disconnect the throttle control connector.
7. Disconnect the 2 water bypass hoses from the throttle body.
8. Remove the nut and 3 bolts, and remove the throttle body from the intake manifold.

To install:

9. Install the throttle body with the nut and 3 bolts. Tighten them to 14 Nm (10 ft. lbs.).
10. Connect the 2 water bypass hoses to the throttle body.
11. Connect the throttle control connector.
12. Install intake air connector.
13. Fill with engine coolant.
14. Start engine and check for engine coolant leaks.
15. Install throttle body cover.

2010 Models

4.0L Engine

1. Remove the upper radiator support seal.
2. Remove the front bumper cover lower.
3. Remove the No. 1 engine under cover sub assembly.
4. Drain the engine coolant.
5. Remove the v-bank cover.
6. Remove the No. 1 air cleaner hose.
7. Remove the throttle body with the motor assembly.
   a. Disconnect the No. 5 water by pass hose.
   b. Disconnect the No. 4 water by pass hose.
   c. Disconnect the throttle position sensor and throttle control motor connector.
   d. Remove the 4 bolts, throttle body with motor and gasket.

To install:

8. Install the throttle body with the motor assembly with the 4 bolts. Tighten to 7 ft. lbs. (10 Nm).
   a. Connect the throttle position sensor and throttle control motor connector.
   b. Connect the No. 4 water by pass hose.
   c. Connect the No. 5 water by pass hose.
9. Install the No. 1 air cleaner hose.
10. Add engine coolant.
11. Inspect for engine coolant leak.
12. Install the No. 1 engine under cover sub assembly.
13. Install the front bumper cover lower.
14. Install the v-bank cover.
15. Install the upper radiator support seal.
16. Perform the initialization.

➡Be sure to perform this procedure after reassembling the throttle body or removing and reinstalling any throttle body component.

➡Perform the following procedure after replacing the ECM, throttle body or any throttle body components. The following procedure should also be performed if the throttle body is cleaned.

➡Be sure to perform this procedure after replacing the ECM and reconnecting the battery cable.

   a. Disconnect the EFI and ETCS fuses at the same time. Wait at least 60 seconds, and then reconnect the fuses.
   b. Turn the ignition switch to ON without operating the accelerator pedal.

➡If the accelerator pedal is operated, perform the above steps again.

   c. Connect the Techstream to the DLC3 and clear the DTCs.
   d. Start the engine and check that the MIL is not illuminated and that the idle speed is within the specified range when the A/C is switched off after the engine is warmed up.

The standard idle speed is 690 to 790 rpm.

➡Be sure to perform this step with all accessories off.

➡Make sure that the shift lever is in neutral.

   e. Enter the following menus: Powertrain / Engine and ECT / Data List / All Data / Throttle Sensor Position. Fully depress the accelerator pedal and check that the value is 60% or more.
   f. Perform a road test and confirm that there are no abnormalities.

2.7L Engine

See Figures 338 and 339.

1. Remove the front bumper cover lower.
2. Remove the No. engine under cover sub assembly.

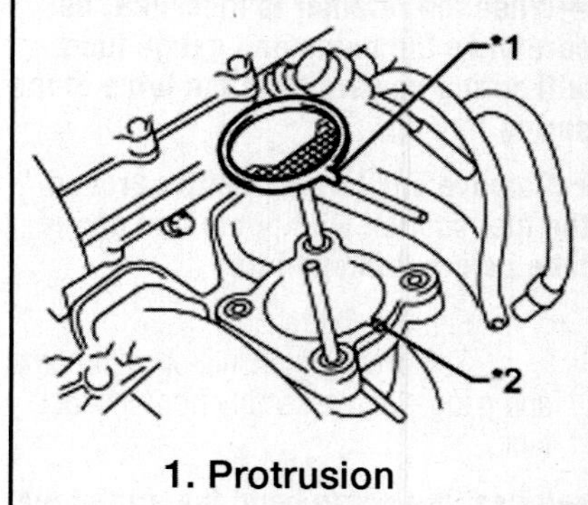

Fig. 338 Aligning the protrusion of the new gasket with the groove of the intake manifold

3. Drain the engine coolant.
4. Remove the air cleaner and hose.
   a. Detach the 3 clamps and disconnect the mass air flow meter connector.
   b. Detach the 4 clamps.
   c. Loosen the hose clamp and remove the air cleaner and hose.
5. Remove the intake air connector.
   a. Disconnect the vacuum hose from the fuel pressure regulator.
   b. Disconnect the No. 2 ventilation hose.
   c. Detach the wire harness clamp.
   d. Disconnect the connector.
   e. Disconnect the vacuum hose from the manifold absolute pressure sensor.
   f. Detach the vacuum hose.
   g. Loosen the clamp.
   h. Remove the 3 bolts and intake air connector.
6. Remove the throttle body with the motor assembly.
   a. Disconnect the water by pass hose.
   b. Disconnect the No. 2 water bypass hose.
   c. Disconnect the throttle position sensor and throttle control motor connector.
   d. Remove the 2 bolts, 2 nuts and throttle body with motor.
   e. Remove the gasket from the intake manifold.

To install:

7. Install the throttle body with the motor assembly.
   a. Align the protrusion of a new gasket with the groove of the intake manifold.
   b. Install a new gasket to the intake manifold.
   c. Install the throttle body with motor

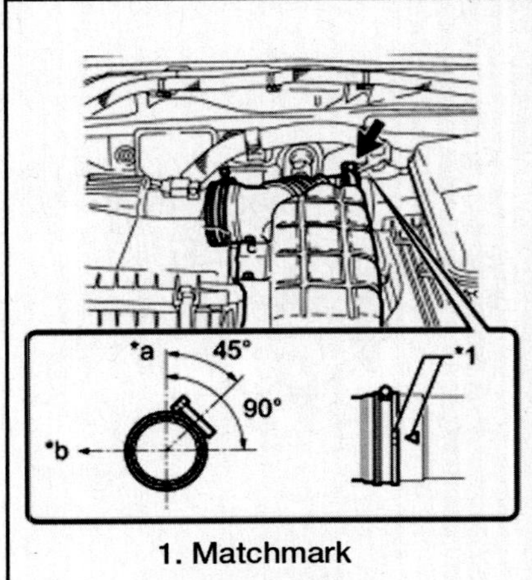

Fig. 339 Installing the air cleaner

with the 2 bolts and 2 nuts. Tighten to 80 inch lbs. (9 Nm).

d. Connect the water by-pass hose.

e. Connect the No. 2 water by-pass hose.

f. Connect the throttle position sensor and throttle control motor connector.

8. Install the intake air connector.

a. Install the intake air connector with the 3 bolts. Tighten to 71 inch lbs. (8 Nm).

b. Tighten the hose clamp. Tighten to 44 inch lbs. (5 Nm).

c. Attach the vacuum hose.

d. Connect the vacuum hose to the manifold absolute pressure sensor.

e. Connect the connector.

f. Attach the wire harness clamp.

g. Connect the No. 2 ventilation hose.

h. Connect the vacuum hose to the fuel pressure regulator.

9. Install the air cleaner and hose.

a. Install the air cleaner and hose, align its matchmark with the matchmark of the air cleaner cap as shown in the illustration.

b. Tighten the hose clamp. Tighten to 44 inch lbs. (5 Nm).

c. Attach the 4 clamps.

d. Attach the 3 clamps and connect the mass air flow meter connector.

10. Add engine coolant.

11. Inspect for coolant leak.

12. Install the No. 1 engine under cover sub assembly.

13. Install the front bumper cover lower.

14. Perform the initialization.

➡Be sure to perform this procedure after reassembling the throttle body assembly or removing and reinstalling any throttle body component.

➡Perform the following procedure after replacing the ECM, throttle body assembly or any throttle body components. The following procedure should also be performed if the throttle body is cleaned.

➡Be sure to perform this procedure after reconnecting the battery cable and after replacing the ECM.

a. Disconnect the cable from the negative (-) battery terminal. Wait at least 60 seconds and reconnect the cable.

b. Turn the ignition switch to ON without operating the accelerator pedal.

➡If the accelerator pedal is operated, perform the above steps again.

c. Connect the Techstream to the DLC3 and clear the DTCs.

d. Start the engine, and check that the MIL is not illuminated and that the idle speed is within the specified range when the A/C is switched off after the engine is warmed up. Engine idle speed with the A/C switched off should be 600-700 rpm.

➡Be sure to perform this step with all accessories off.

➡Make sure that the shift lever is in neutral.

e. Enter the following menus: Powertrain / Engine and ECT / Data List / Throttle Sensor Position. Fully depress the accelerator pedal and check that the value is 60% or more.

f. Perform a road test and confirm that there are no abnormalities.

HEATING & AIR CONDITIONING SYSTEM

BLOWER MOTOR

REMOVAL & INSTALLATION

1. Before servicing the vehicle, refer to the Precautions section.

2. Disconnect the connector.

3. Remove the three screws and the blower motor.

To install:

4. Install the blower motor with three screws.

5. Connect the connector.

HEATER CORE

REMOVAL & INSTALLATION

See Figures 340 through 347.

1. Discharge refrigerant from refrigeration system.

2. Disconnect cooler refrigerant suction pipe c.

3. Disconnect cooler refrigerant liquid pipe c.

4. Using pliers, grip the claws of the clip and slide the clip and disconnect the heater water outlet hose.

5. Disconnect heater water outlet hose.

6. Remove instrument panel safety pad sub-assembly.

7. Remove air conditioning amplifier assembly.

8. Remove the 2 clips.

9. Release the 2 claw fittings and remove the side defroster nozzle duct No. 1.

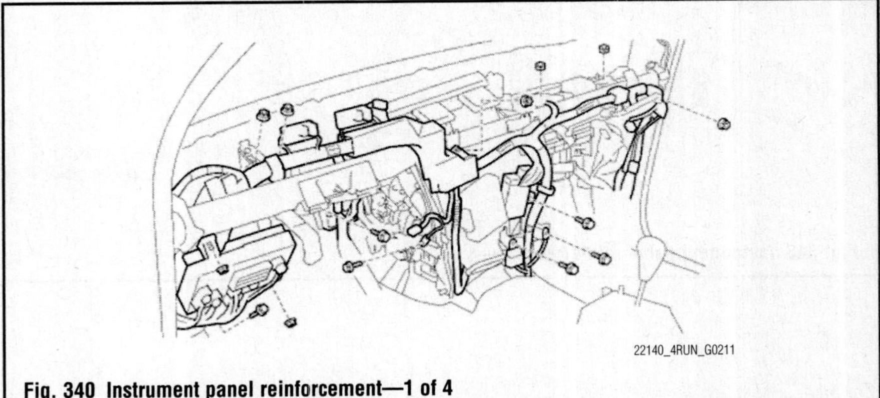

Fig. 340 Instrument panel reinforcement—1 of 4

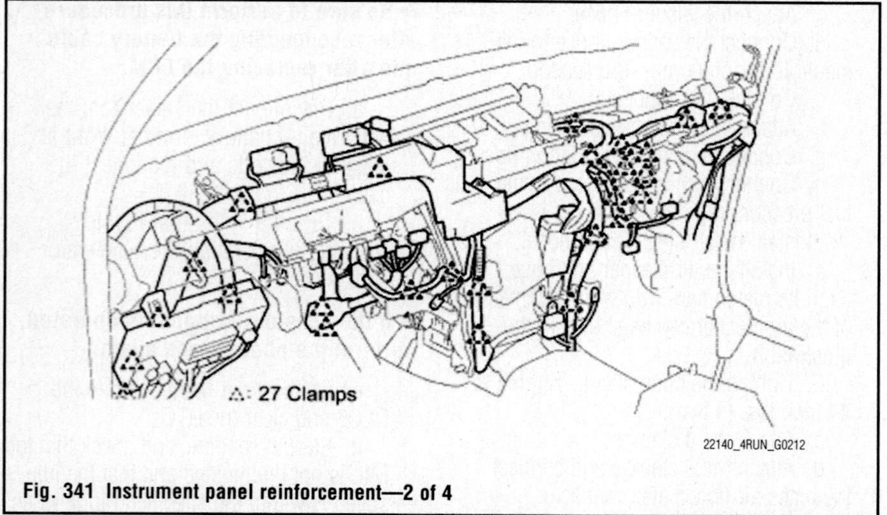

Fig. 341 Instrument panel reinforcement—2 of 4

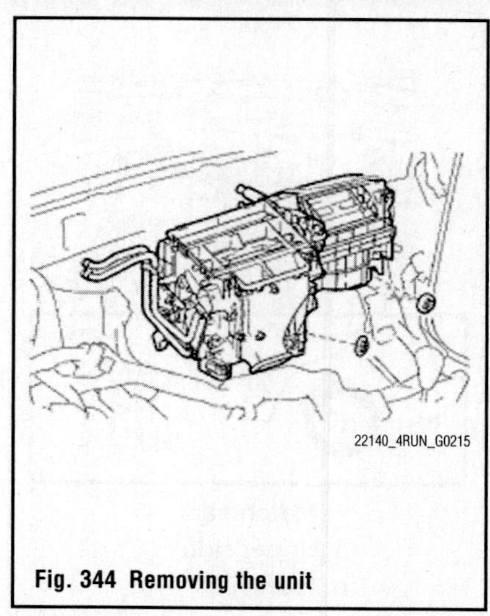

Fig. 344 Removing the unit

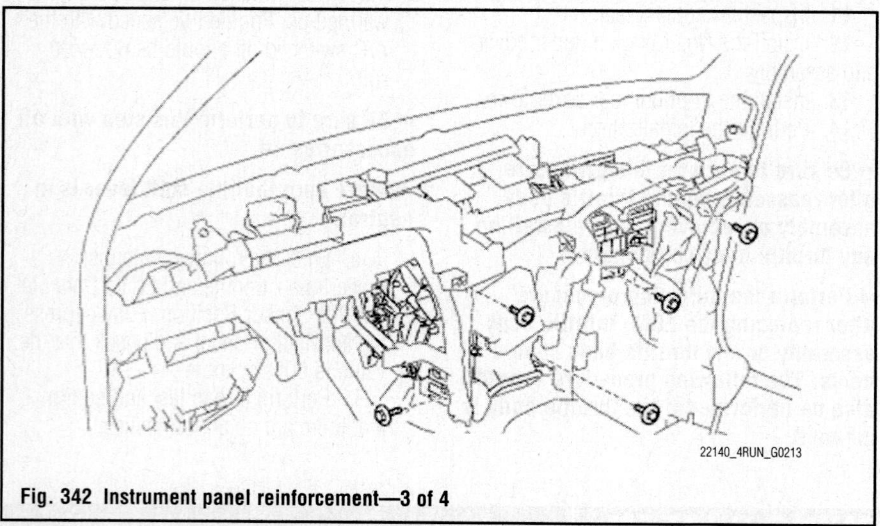

Fig. 342 Instrument panel reinforcement—3 of 4

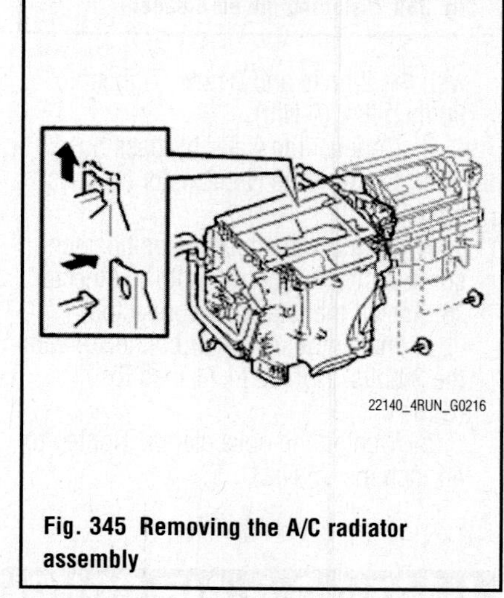

Fig. 345 Removing the A/C radiator assembly

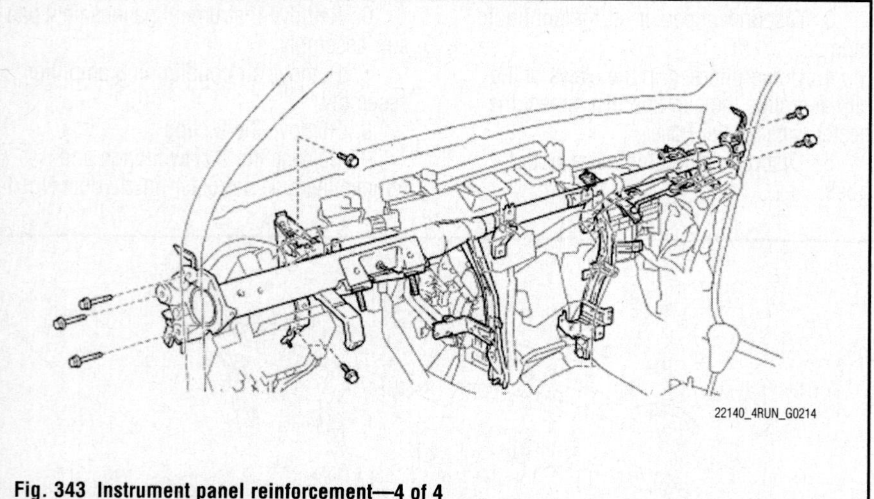

Fig. 343 Instrument panel reinforcement—4 of 4

10. Remove the 2 clips.
11. Release the 2 claw fittings and remove the side defroster nozzle duct No. 2.
12. Remove the screw.
13. Release the 2 pin fittings and remove the heater to register duct No. 1.
14. Remove the screw.
15. Release the 2 pin fittings and remove the heater to register duct No. 3.
16. Release the 6 claw fittings and 3 clamps and remove the air duct rear No. 2.
17. Release the 6 claw fittings and 3 clamps and remove the air duct rear No. 1.
18. Remove the clip and disconnect the console box duct No. 1.
19. Release the 3 claw fittings and remove the air duct No. 1.
20. Release the 3 claw fittings and remove the air duct No. 2.

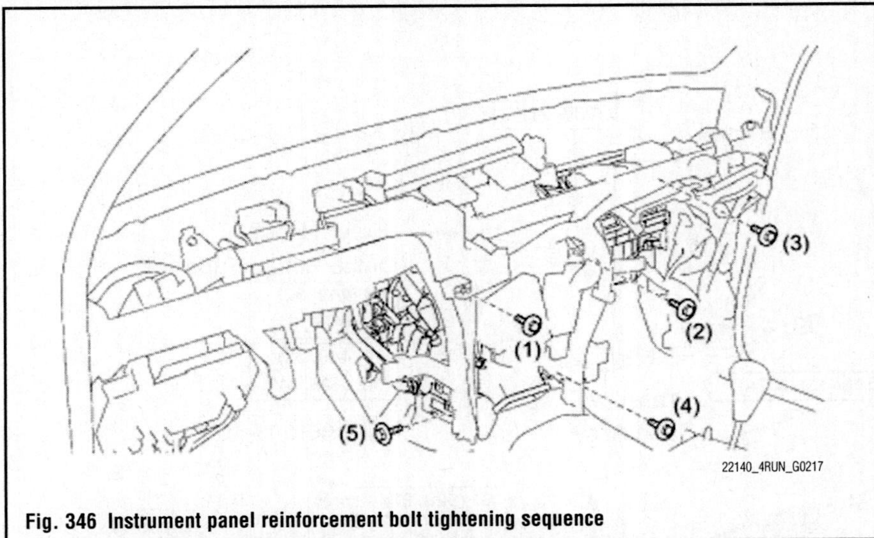

Fig. 346 Instrument panel reinforcement bolt tightening sequence

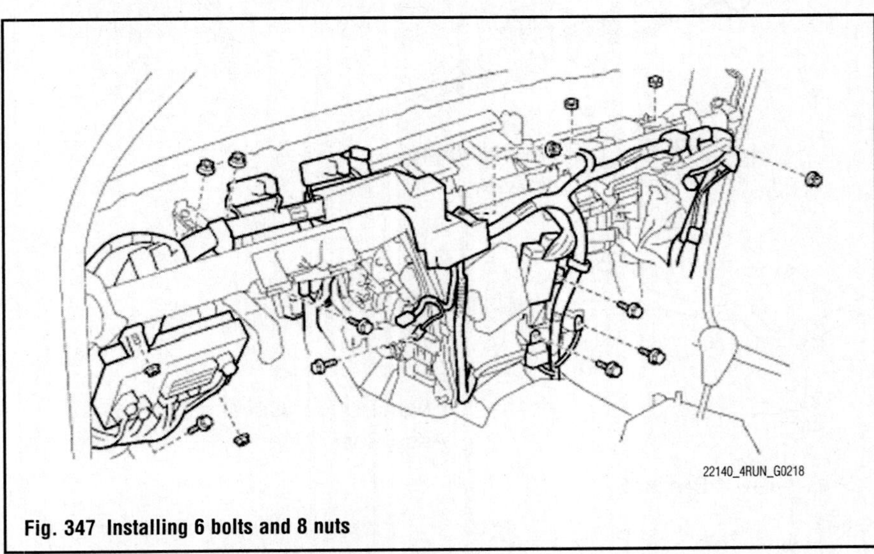

Fig. 347 Installing 6 bolts and 8 nuts

21. Remove the bolt, nut and instrument panel brace mounting brackets.
22. Release the 3 clamps and disconnect the connector.
23. Remove the 4 nuts and disconnect the steering column assembly.
24. Remove instrument panel reinforcement.
    a. Remove the 6 bolts and 8 nuts.
    b. Release the 27 clamps.
    c. Disconnect the connectors.
    d. Remove the 5 bolts.
    e. Remove the 7 bolts and instrument panel reinforcement.
25. Release the 4 claw fittings and remove the heater to register duct center.
26. Release the 4 claw fittings and remove the defroster nozzle assembly lower.
27. Disconnect the connectors.
28. Remove the 2 nut and air conditioner unit assembly.
29. Remove the 2 screws and air conditioning radiator assembly.

To install:

30. Installation is reverse of removal.
31. Tighten bolts and nuts to the following torque:
    - A/C unit nuts 48 inch lbs. (5.4 Nm)
    - (7) Instrument panel reinforcement bolts 87 inch lbs. (9.8 Nm)
    - (5) Instrument panel reinforcement bolts 87 inch lbs. (9.8 Nm) in the order of the illustration
    - 6 bolts and 8 nuts as shown in illustration
    - Steering column and tighten to 19 ft. lbs. (26 Nm)

STEERING

POWER RACK & PINION STEERING GEAR

REMOVAL & INSTALLATION

See Figure 348.

1. Before servicing the vehicle, refer to the precautions section.
2. Disconnect the battery ground cable.
3. Place the front wheels in the straight ahead position.
4. Remove the horn pad.
5. Remove the steering wheel.
6. Remove the lower steering column cover.
7. Remove the turn signal switch.
8. Remove the spiral cable assembly.
9. Remove the front wheels.
10. Remove the engine under-covers.
11. Remove the stabilizer bar.
12. Remove the tie rod ends from the knuckle.
13. Remove the steering intermediate shaft.
14. Disconnect the pressure and return lines.
15. Remove the 2 bolts and remove the steering gear assembly.

To install:

16. Position the gear and install the 2 bolts. Torque to 74 ft. lbs. (100 Nm).

➡The nuts have detents. Never turn the nuts, just the bolts.

17. Install the stabilizer bar. Torque the end links to 52 ft. lbs. (70 Nm); the clamp bolts to 30 ft. lbs. (40 Nm).
18. Connect the return line. Use a torque wrench with SST 09023-12700, or equivalent. The torque wrench should have a fulcrum length of 300mm. Torque to 31 ft. lbs. (42 Nm).
19. Connect the pressure line at the subframe. Torque to 21 ft. lbs. (28 Nm).
20. Connect the pressure line to the gear. Use a torque wrench with SST 09023-12700, or equivalent. The torque wrench should have a fulcrum length of 300mm. Torque to 31 ft. lbs. (42 Nm).
21. Connect the intermediate shaft. Torque to 26 ft. lbs. (36 Nm).
22. Connect the tie rod ends. Torque to 67 ft. lbs. (91 Nm).
23. Install the under-covers.
24. The remainder of installation is the reverse of removal.

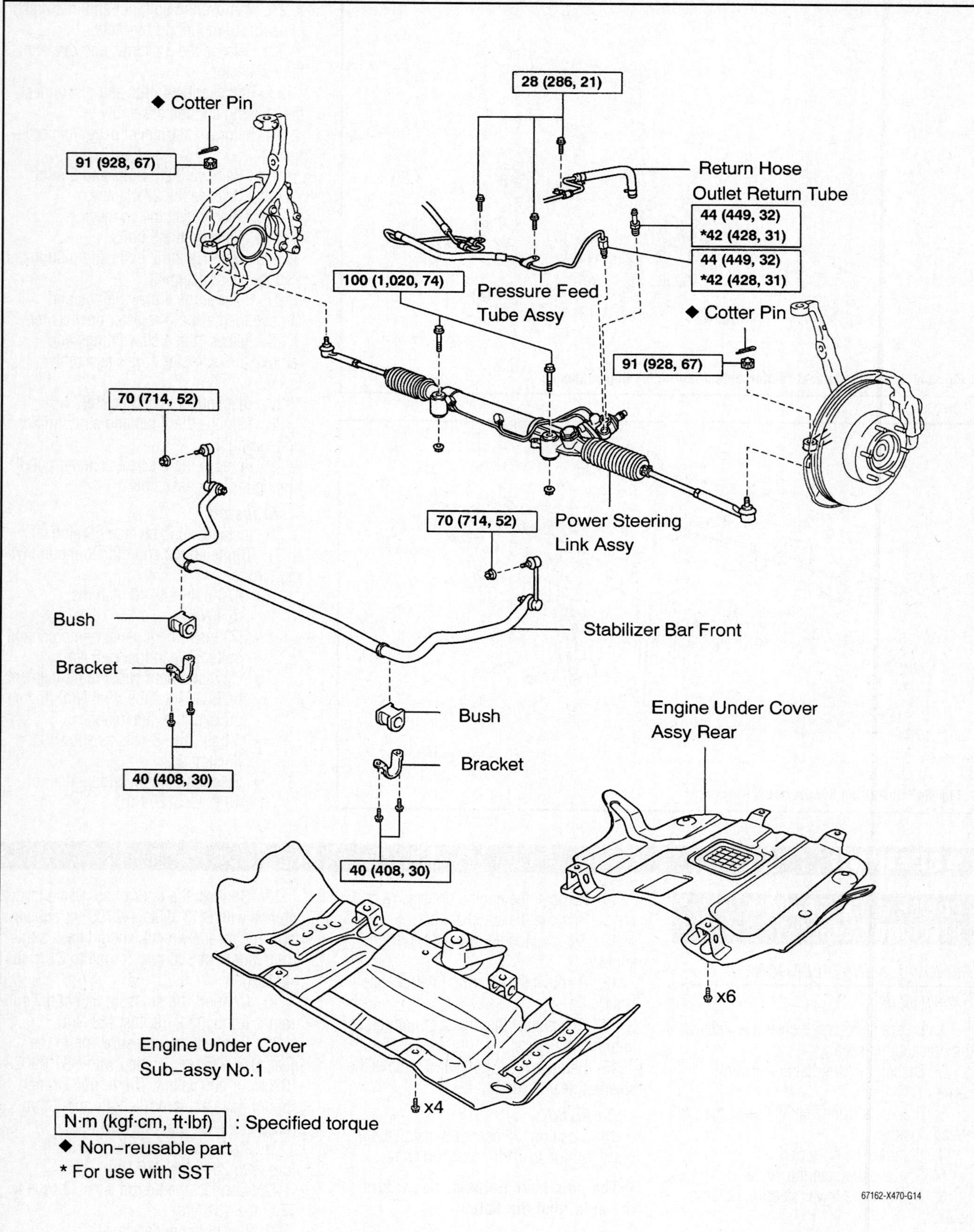
28 (286, 21)
◆ Cotter Pin
91 (928, 67)
Return Hose
Outlet Return Tube
44 (449, 32)
*42 (428, 31)
44 (449, 32)
*42 (428, 31)
100 (1,020, 74)
Pressure Feed
Tube Assy
◆ Cotter Pin
91 (928, 67)
70 (714, 52)
70 (714, 52)
Power Steering
Link Assy
Bush
Stabilizer Bar Front
Bracket
Bush
Engine Under Cover
Assy Rear
Bracket
40 (408, 30)
40 (408, 30)
x6
Engine Under Cover
Sub-assy No.1
x4
N·m (kgf·cm, ft·lbf) : Specified torque
◆ Non-reusable part
* For use with SST
67162-X470-G14

Fig. 348 Steering gear and related parts

POWER STEERING PUMP

REMOVAL & INSTALLATION

See Figure 349.

1. Before servicing the vehicle, refer to the Precautions section.
2. Disconnect the MAF meter connector.
3. Disconnect the hoses.
4. Remove the clamp.
5. Remove the 3 bolts and air cleaner assembly with air cleaner hose connected.
6. Loosen the drive belt tension by turning the drive belt tensioner counterclockwise, and remove the drive belt.
7. Remove the 2 clips and disconnect the 2 vacuum hoses.
8. Remove the clip and disconnect the return hose.
9. Remove the union bolt and gasket, disconnect the pressure feed tube.
10. Remove the 2 bolts, nut, stud bolt and power steering pump assembly.

To install:

11. Install the power steering pump assembly with the stud bolt.
12. Tighten the stud bolt to 22 Nm (16 ft. lbs.).
13. Install the 2 bolts and nut and tighten them to 44 Nm (33 ft. lbs.).

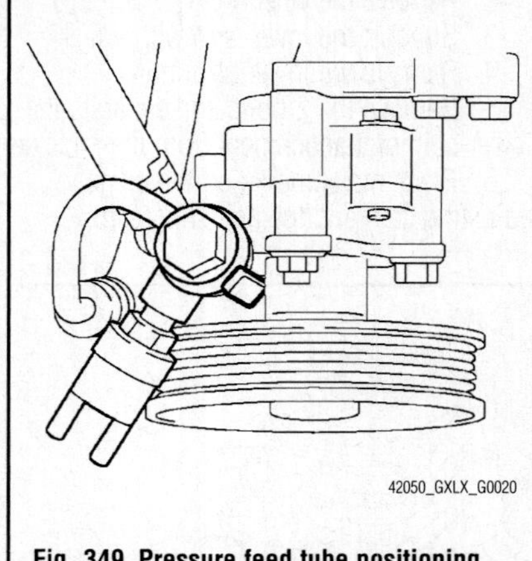

Fig. 349 Pressure feed tube positioning

14. Install a new gasket and the union bolt on the pressure feed tube.

➡Make sure that the stopper of the pressure feed tube contacts the power steering pump body as shown in the illustration.

15. Tighten the union bolt to 46.5 Nm (34 ft. lbs.).
16. Connect the return hose with the clip.
17. Connect the 2 vacuum hoses and install the 2 clips.
18. Loosen the drive belt tension by turning the drive belt tensioner counterclockwise, and install the belt.
19. Install the air cleaner assembly with air cleaner hose and the 3 bolts.
20. Install the clamp.
21. Connect the MAF meter connector.
22. Fill with power steering fluid and bleed the system.

BLEEDING

1. Before servicing the vehicle, refer to the Precautions section.
2. Check fluid level.
3. Jack up front of vehicle and support it with stands.
4. With the engine stopped, turn the wheel slowly from lock to lock several times.
5. Lower the vehicle.
6. Start the engine and run at idle for a few minutes.
7. With the engine idling, turn the wheel left or right to the full lock position and keep it there for 2 to 3 seconds, then turn the wheel to the opposite full lock position and keep it there for 2 to 3 seconds. Repeat several times.
8. Stop the engine.
9. Check for foaming or emulsification of the power steering fluid.
10. If the system has to be bled twice specifically because of foaming or emulsification, check for fluid leaks in the system.
11. Check fluid level.

SUSPENSION

FRONT SUSPENSION

COIL SPRING

REMOVAL & INSTALLATION

See Figure 350.

1. Remove the strut.
2. Place the strut in a compressor, such as SST 09727-30021, and compress the spring.

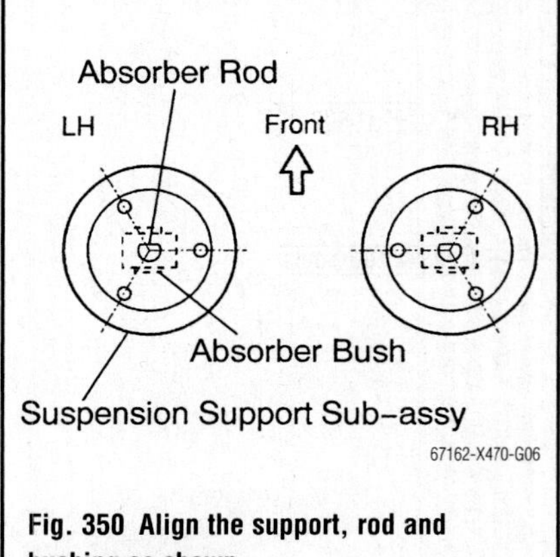

Fig. 350 Align the support, rod and bushing as shown

3. Hold the rod and remove the nut.

➡Don't use an impact wrench.

4. Remove the bushing retainer
5. Remove the upper bushing.
6. Remove the support.
7. Remove the lower bushing retainer.
8. Remove the spring.
9. Remove the lower bushing.

To install:

10. Install the new lower bushing.
11. Compress the spring and install it.
12. Install the bushing retainer.
13. Install the suspension support.
14. Install the upper bushing.
15. Install the retainer.
16. Align the support, rod and bushing as shown. Install the locknut and torque to 18 ft. lbs. (25 Nm).
17. Release the spring from the compressor and check the alignment of the parts.
18. Install the strut.

CONTROL LINKS

REMOVAL & INSTALLATION

Non KDSS

1. Remove front disc wheel.
2. Remove the 2 nuts and the front stabilizer link assembly.

➡If the ball joint turns together with the nut, use a hexagon (6 mm) wrench to hold the stud.

3. Remove front stabilizer link assembly.

To install:

Installation is reverse of removal. Tighten nuts to 52 ft. lbs. (70 Nm).

LOWER BALL JOINT

REMOVAL & INSTALLATION

The lower ball joint is serviced with the lower control arm as an assembly.

LOWER CONTROL ARM

REMOVAL & INSTALLATION

See Figure 351.

1. Before servicing the vehicle, refer to the precautions section.
2. Remove the wheel.
3. Support the lower arm with a jack.
4. Remove the lower strut bolt.
5. Remove the 2 bolts and separate the lower ball joint attachment from the knuckle.
6. Place matchmarks on the camber adjusting cam and toe adjusting cam.
7. Remove the 2 nuts and remove the arm along with the cams.

To install:

8. Installation is the reverse of removal. Align all matchmarks. Use new nuts and cotter pins. Don't fully tighten the control arm bolts until the vehicle is on the ground

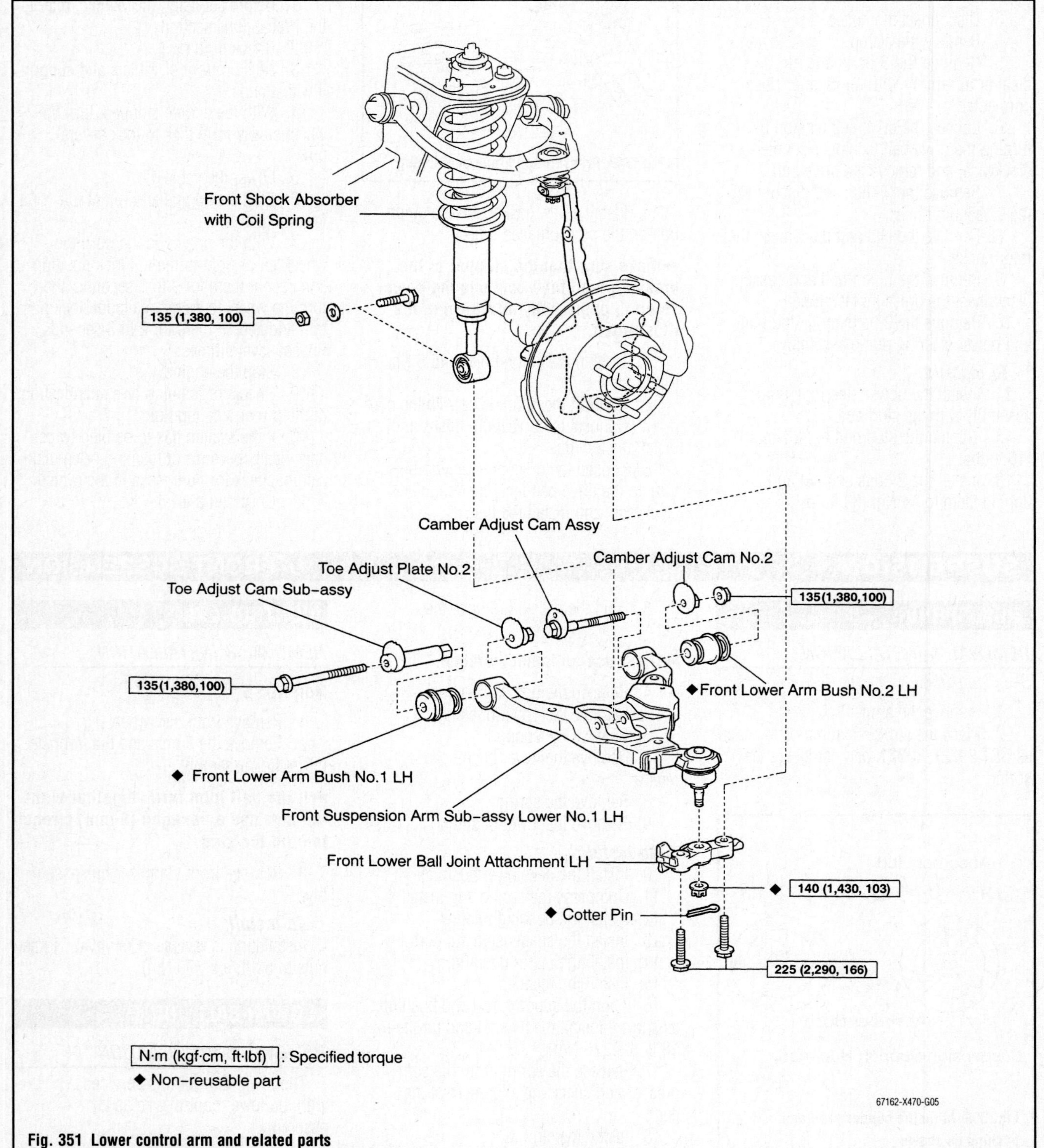

Fig. 351 Lower control arm and related parts

and the suspension jounced a few times. Observe the following torques:

- Lower ball joint stud: 103 ft. lbs. (140 Nm)
- Lower ball joint attachment bolts: 166 ft. lbs. (225 Nm)
- Lower arm bolts: 100 ft. lbs. (135 Nm)

MACPHERSON STRUT

REMOVAL & INSTALLATION

Non REAS Suspension

See Figure 352.

1. Before servicing the vehicle, refer to the precautions section.
2. Remove the wheel.
3. Remove the stabilizer bar.
4. Remove the clamps and connector.
5. Remove the wire bracket.
6. Remove the lower strut bolt.
7. Remove the 3 upper strut nuts.
8. Remove the strut.

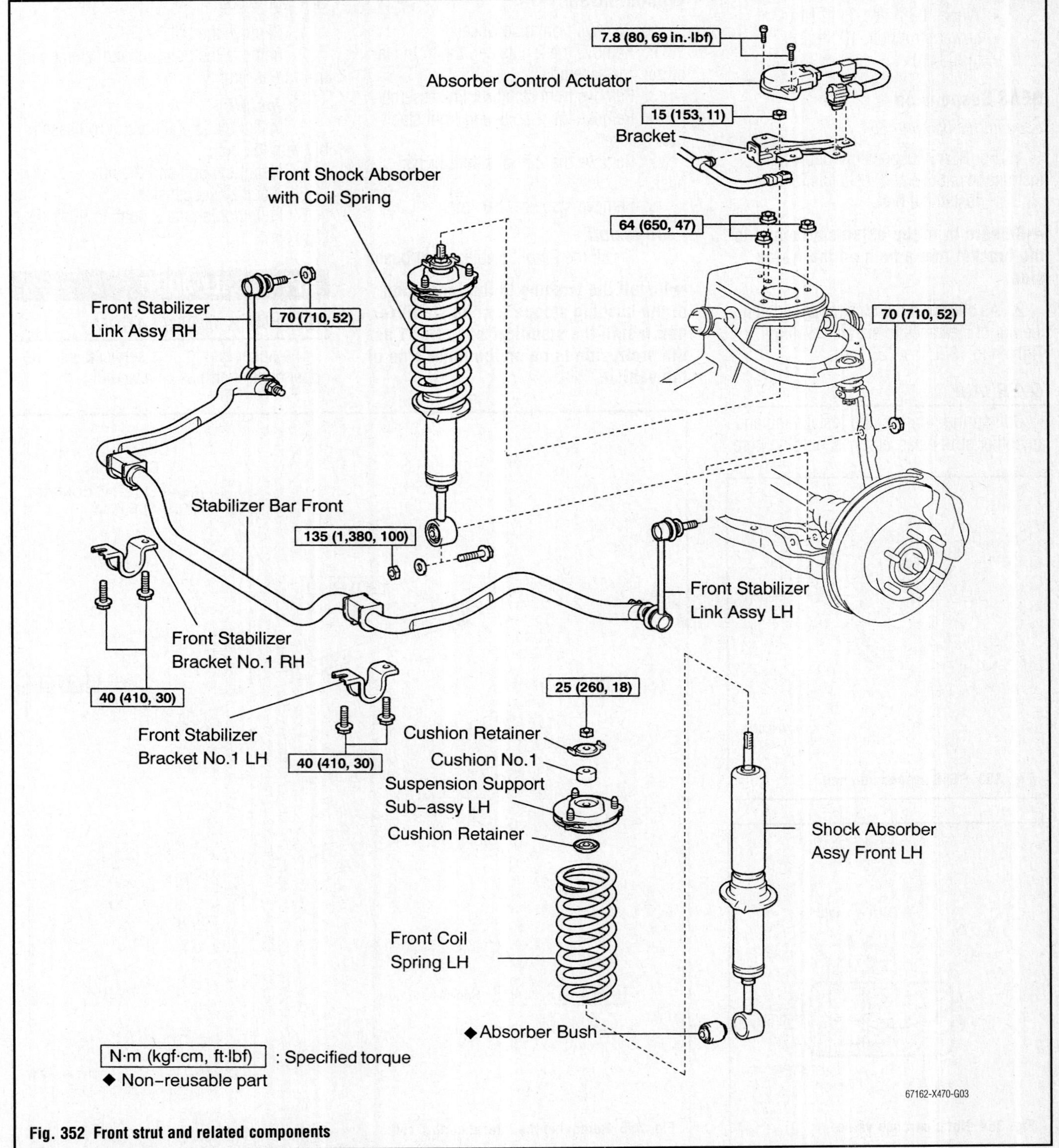

Fig. 352 Front strut and related components

To install:

9. Installation is the reverse of removal. Do not fully tighten the lower strut bolt until the vehicle is resting on the ground and the suspension has been jounced a few times. Observe the following torques:

- Upper nuts: 47 ft. lbs. (64 Nm)
- Bracket nut: 11 ft. lbs. (15 Nm)
- Stabilizer bar links: 52 ft. lbs. (70 Nm)
- Wheel: 83 ft. lbs. (112 Nm)
- Lower strut bolt: 100 ft. lbs. (135 Nm)

REAS Suspension

See Figures 353 and 354.

1. For REAS suspension follow the instructions above and install as follows:
   a. Install the bolt.

➡Be sure to fit the detents attached to the bracket into a hole on the frame side.

2. As shown in the illustration, tighten the nut of clearance to standard value. Tighten to 18 ft. lbs. (25 Nm).

OVERHAUL

Coil spring removal and installation and any other strut disassembly should go here

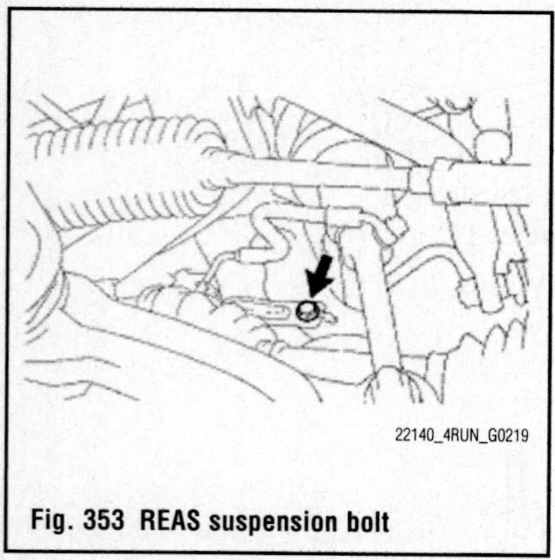

Fig. 353 REAS suspension bolt

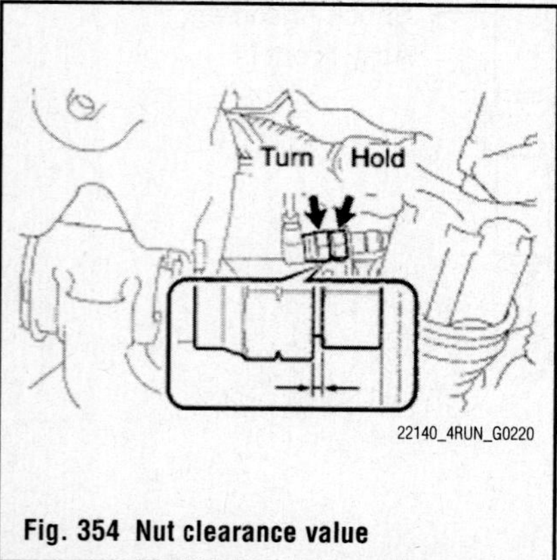

Fig. 354 Nut clearance value

SHOCK ABSORBERS

REMOVAL & INSTALLATION

See MacPherson Strut.

STABILIZER BAR & LINKS

REMOVAL & INSTALLATION

Without KDSS

1. Remove front disc wheel.
2. Remove the 2 nuts and the front stabilizer link assembly.
3. Remove front stabilizer link assembly.
4. Remove the 2 bolts and front stabilizer bracket
5. Remove the 2 front stabilizer bar bush.
6. Remove stabilizer bar front.

To install:

7. Install the 2 front stabilizer bar bush.

➡Install the bushing to the inner side of the bushing stopper on the stabilizer bar. Install the stabilizer bush No. 1 as the protrusion to be on the inner side of the vehicle.

8. Install the front stabilizer bracket and tighten to 30 ft. lbs. (40 Nm).
9. Install the front stabilizer link assemblies.
10. Install front wheels.

TRACK BAR

REMOVAL & INSTALLATION

See Figure 355.

1. Remove the bolt.
2. Remove the bolt, nut and lateral control rod assembly.

To install:

3. Install the lateral control rod assembly with the bolt.
4. Install the bolt and the nut.
5. Stabilize suspension.
6. Fully tighten the 2 bolts to 96 ft. lbs. (130 Nm).

UPPER BALL JOINT

REMOVAL & INSTALLATION

The upper ball joint is serviced with the upper control arm as an assembly.

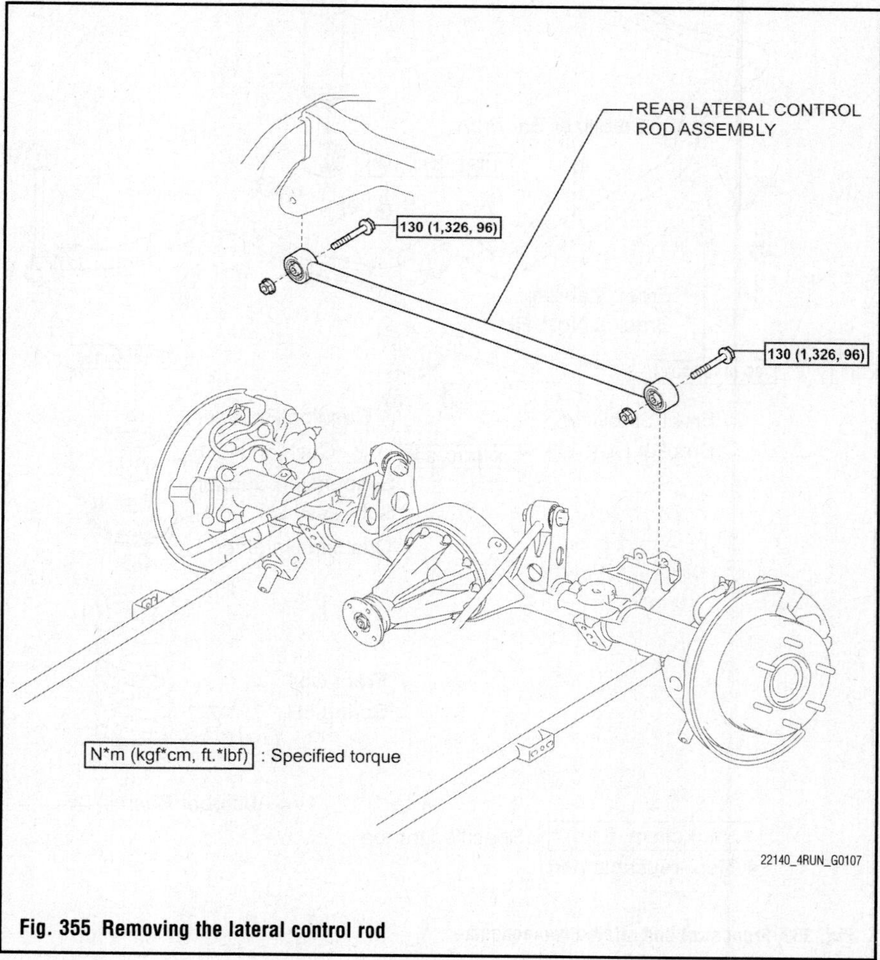

Fig. 355 Removing the lateral control rod

UPPER CONTROL ARM

REMOVAL & INSTALLATION

See Figure 356.

1. Before servicing the vehicle, refer to the precautions section.
2. Remove the wheel.
3. Disconnect the skid control wire.
4. Support the lower arm with a jack.
5. Remove the cable bracket.
6. Disconnect the ball joint from the knuckle.
7. Remove the through-bolt, washers and nut.
8. Remove the arm.

To install:

9. Installation is the reverse of removal. Don't fully tighten the through-bolt until the vehicle is on the ground and the suspension is jounced a few times.
   - Ball joint nut: 81 ft. lbs. (110 Nm)
   - Through-bolt: 85 ft. lbs. (115 Nm)

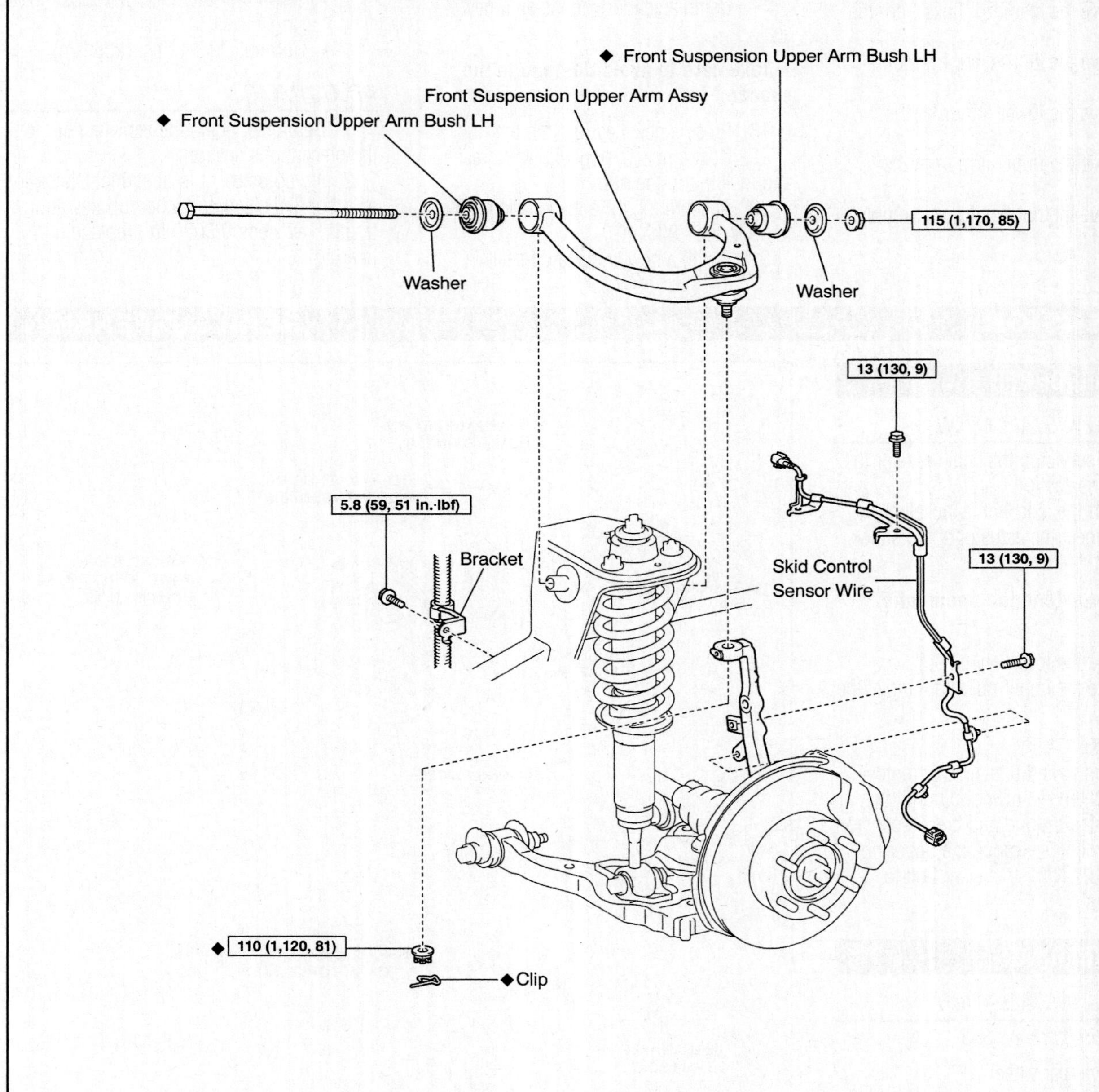

Fig. 356 Upper control arm and related parts

WHEEL BEARINGS

REMOVAL & INSTALLATION

1. Remove the wheel.
2. Remove the caliper.
3. Remove the hub grease cap.
4. Remove the cotter pin.
5. Remove the hub nut.
6. Remove the speed sensor.
7. Remove the stabilizer links from the knuckles.
8. Remove the tie rod end from the knuckle.
9. Remove the lower arm from the knuckle.
10. Remove the upper arm from the knuckle.
11. Remove the hub/knuckle assembly from the shaft.
12. Mount the assembly in a vise.
13. Remove the knuckle oil seal.
14. Remove the 4 bolts and remove the hub assembly from the knuckle.
15. Using SST 09710-30021 and its components, remove the bearing from the hub.
16. Remove the oil seal.

To install:

17. Using a seal driver, install a new seal.

➡Take care to avoid damage to the spacer.

18. Press a new bearing into the hub.
19. Coat a new O-ring with MP grease and install it in the hub.
20. Attach the hub to the knuckle. Torque to 59 ft. lbs. (80 Nm).
21. Install a new knuckle oil seal.
22. The remainder of installation is the reverse of removal. Observe the following torques:
    - Upper arm ball stud nut: 81 ft. lbs. (110 Nm)
    - Lower arm ball joint attachment bolts: 166 ft. lbs. (225 Nm)
    - Tie rod end ball stud nut: 67 ft. lbs. (91 Nm)
    - Stabilizer end links: 52 ft. lbs. (70 Nm)
    - Hub nut: 173 ft. lbs. (235 Nm)

ADJUSTMENT

1. Before servicing the vehicle, refer to the precautions section.
2. No adjustment is possible. Check for axle hub backlash and axle hub deviation. If either exceeds 0.0020 in., replace the bearing.

SUSPENSION — REAR SUSPENSION

SHOCK ABSORBER

REMOVAL & INSTALLATION

1. Before servicing the vehicle, refer to the precautions section.
2. Support the axle with a jackstand.
3. Disconnect the actuator at the shock absorber.

➡Don't over-extend the pneumatic shock.

4. Remove the lower shock bolt.
5. Remove the upper nut and remove the shock.

To install:

6. Installation is the reverse of removal. Don't fully tighten the lower bolt until the vehicle is on the ground and the suspension jounced a few times. Torque the upper nut to 18 ft. lbs. (25 Nm); the lower bolt to 72 ft. lbs. (98 Nm).

WHEEL BEARINGS

REMOVAL & INSTALLATION

See Figures 357 through 363.

1. Remove rear wheel.
2. Remove rear speed sensor (w/ ABS).
3. Separate rear disc brake caliper assembly.
4. Remove rear disc.
5. Remove parking brake shoe return tension spring.
6. Remove parking brake shoe strut compression spring.
7. Remove parking brake shoe strut.
8. Remove parking brake shoe.

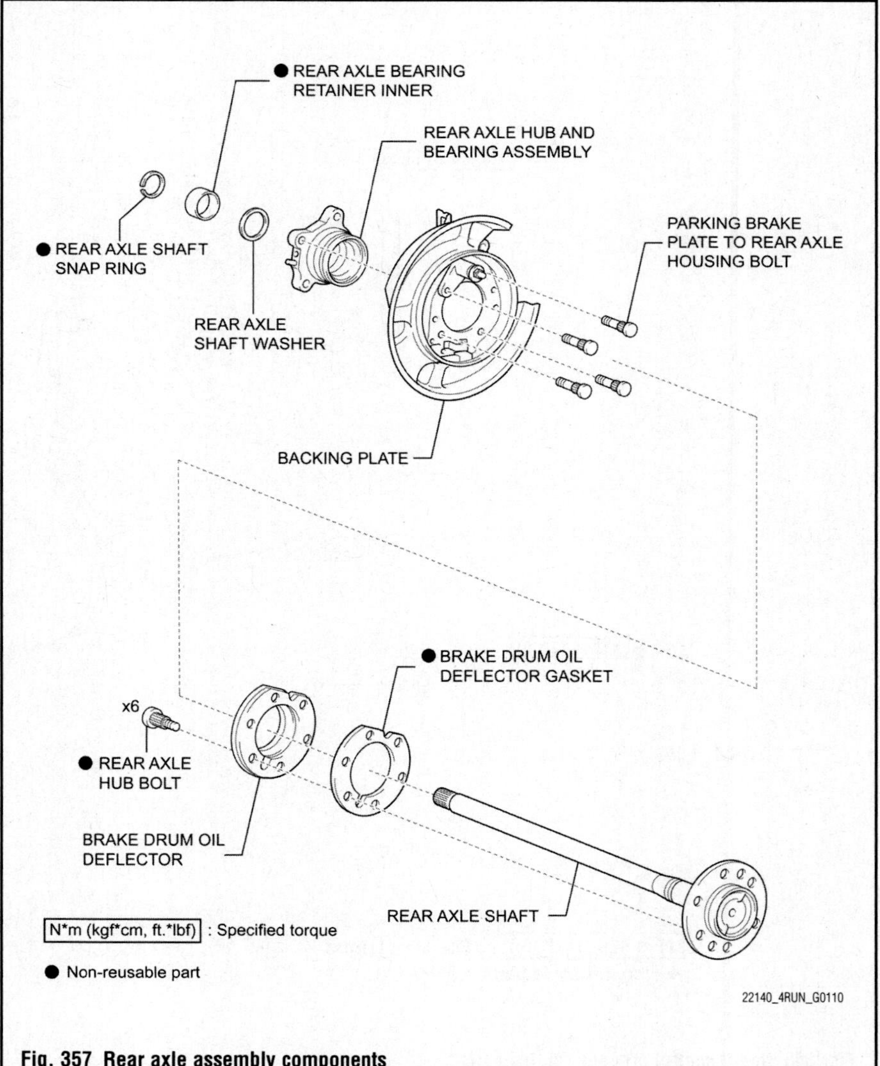

Fig. 357 Rear axle assembly components

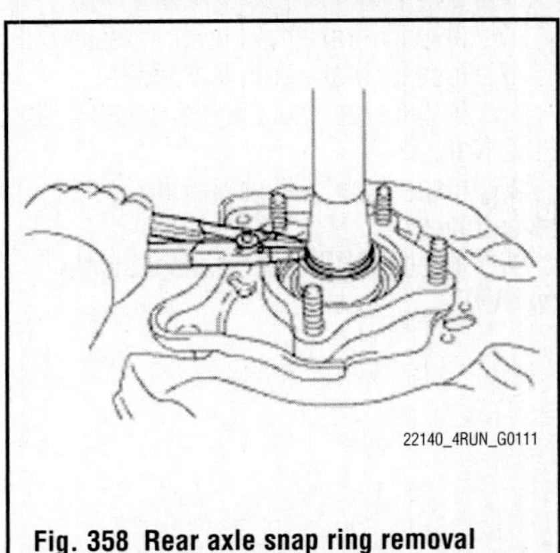
Fig. 358 Rear axle snap ring removal

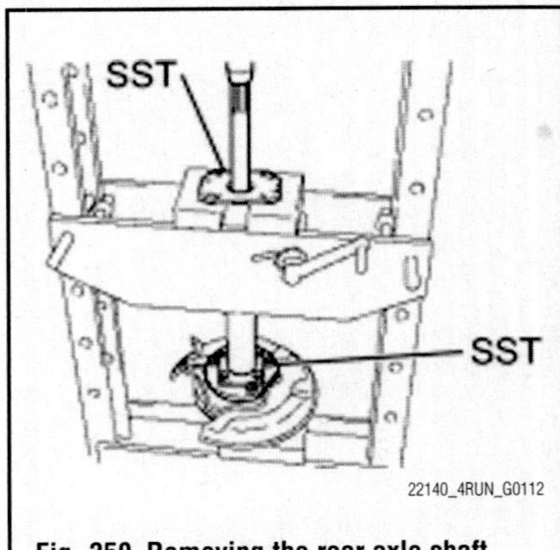

Fig. 359 Removing the rear axle shaft from the bearing

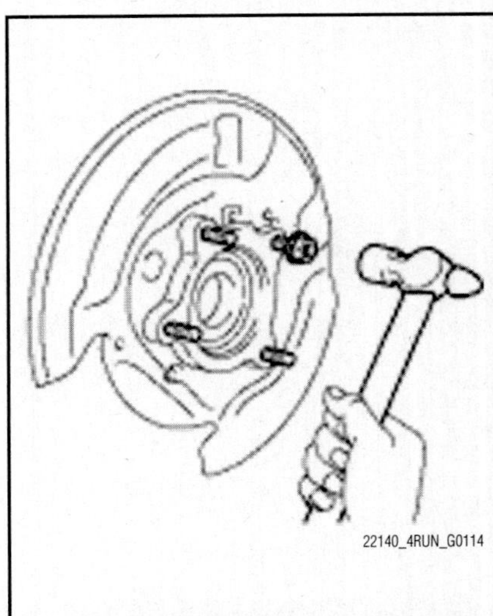
Fig. 360 Removing the parking brake plate

9. Remove rear axle shaft with backing plate.
10. Remove the 4 nuts and rear axle shaft with backing plate.
11. Remove the O-ring.
12. Using a snap ring expander, remove the snap ring.
13. Remove the rear axle shaft from bearing.

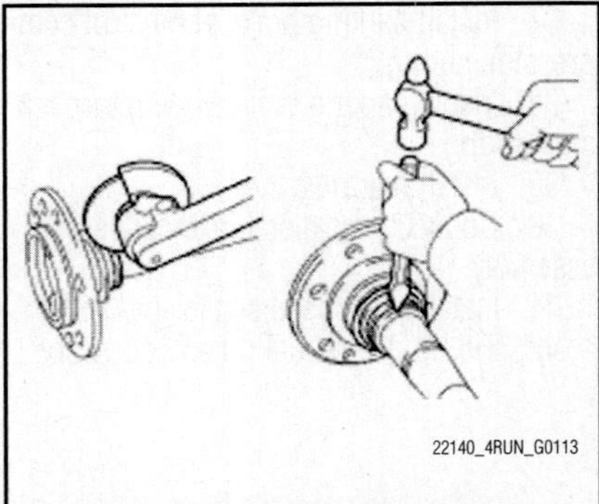
Fig. 361 Removing the rear axle bearing inner race

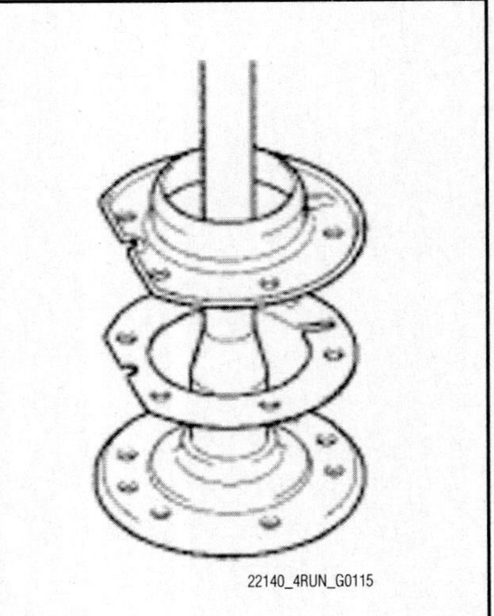
Fig. 362 Installing the rear axle deflector gaskets

14. Remove the rear axle bearing retainer inner from the rear axle bearing assembly.
15. Remove the rear axle shaft washer from the rear axle bearing assembly.
16. Attach the 4 nuts to the parking brake plate to rear axle housing bolts.
17. Using a hammer, remove the 4 parking brake plate to rear axle housing bolts and rear axle bearing assembly.

➡Do not reuse the nuts previously removed from the vehicle.

18. Remove the 6 hub bolts.
19. Remove brake drum oil deflector.
20. Remove brake drum oil deflector gasket.
21. Remove rear axle bearing oil seal.
22. Grind the rear axle bearing inner race surface using a grinder, then chisel them out with a chisel.
23. Remove the rear axle shaft oil seal from the rear axle shaft.

To install:

24. Install a new deflector gasket and deflector to the rear axle shaft.

➡Align the 2 notches.

25. Pass the 6 bolts through the axle hub and install.
26. Install rear axle hub and bearing assembly.
27. Install the rear axle shaft plate washer onto the rear axle shaft.
28. Position the backing plate on a rear axle bearing assembly, and install the 4 parking brake plate to rear axle housing bolts using 2 socket wrenches and a press.
29. Install the rear axle shaft plate washer onto the rear axle shaft.
30. Install a new rear axle bearing retainer inner to the rear axle shaft.
31. With and press and the appropriate tool, install the rear axle shaft to the rear axle bearing assembly.

➡Do not damage the speed sensor rotor.

32. Using a snap ring expander, install a new rear axle shaft snap ring.
33. Install a new O-ring.

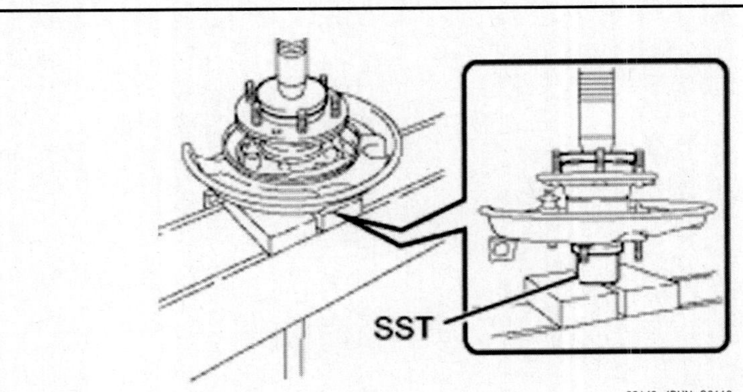

Fig. 363 Assembling the rear axle shaft to axle bearing assembly

34. Install the rear axle shaft with backing plate with the 4 nuts, tighten to 89 ft. lbs. (120 Nm).

➡Do not damage the speed sensor rotor. Inspect no damage and no foreign matter at the speed sensor rotor.

35. Install parking brake shoe.
36. Install parking brake shoe strut.
37. Install parking brake shoe strut compression spring.
38. Install parking brake shoe return tension spring.
39. Install rear disc.
40. Connect rear disc brake caliper assembly.
41. Install rear speed sensor (w/ ABS).
42. Fill up differential oil as necessary.
43. Inspect brake fluid level in reservoir.
44. Inspect brake fluid leakage.
45. Install rear wheel tighten to 83 ft. lbs. (112 Nm).
46. Inspect and adjust parking brake lever travel.
47. Inspect ABS speed sensor signal (w/ ABS).

TOYOTA

Avalon

4

SPECIFICATIONS AND MAINTENANCE CHARTS

ENGINE AND VEHICLE IDENTIFICATION

Engine							Model Year	
Code ①	Liters (cc)	Cu. In.	Cyl.	Fuel Sys.	Engine Type	Eng. Mfg.	Code ②	Year
2GR-FE	3.5 (3456)	210	6	SFI	DOHC	Toyota	9	2009
2GR-FE	3.5 (3456)	210	6	SFI	DOHC	Toyota	A	2010

SFI: Sequential Fuel Injection

DOHC: Double Overhead Camshaft

NA: Information not available

① Stamped on the left side of the engine block

② 10th digit of the Vehicle Identification Number (VIN)

3768X_AVAL_C0001

GENERAL ENGINE SPECIFICATIONS

Year	Model	Engine Displacement Liters	Engine Series ID	Net Horsepower @ rpm	Net Torque @ rpm (ft. lbs.)	Bore x Stroke (in.)	Com-pression Ratio	Oil Pressure @ rpm
2009	Avalon	3.5	2GR-FE	270@6200	251@4700	3.70x3.27	10.8:1	36-78@3000
2010	Avalon	3.5	2GR-FE	270@6200	251@4700	3.70x3.27	10.8:1	36-78@3000

3768X_AVAL_C0002

ENGINE TUNE-UP SPECIFICATIONS

Year	Engine Displacement Liters	Engine ID	Spark Plug Gap (in.)	Ignition Timing (deg.)*	Fuel Pump (psi)	Idle Speed (rpm)	Valve Clearance Intake	Valve Clearance Exhaust
2009	3.5	2GR-FE	0.039-0.043	N/A	44-50	650-750	NA	NA
2010	3.5	2GR-FE	0.039-0.043	N/A	44-50	650-750	NA	NA

NOTE: The Vehicle Emission Control Information label often reflects specification changes made during production.

The label figures must be used if they differ from those in this chart.

NA: Not available

3768X_AVAL_C0003

CAPACITIES

Year	Model	Engine Displacement Liters	Engine ID	Engine Oil with Filter (qts.)	Transmission (qts.)		Transfer Case (pts.)	Drive Axle		Fuel Tank (gal.)	Cooling System (qts.)
					5-Spd	Auto.*		Front (pts.)	Rear (pts.)		
2009	Avalon	3.5	2GR-FE	6.4	—	6.8	—	—	—	18.5	8.8
2010	Avalon	3.5	2GR-FE	6.4	—	6.8	—	—	—	18.5	8.8

3768X_AVAL_C0004

FLUID SPECIFICATIONS

Year	Model	Engine Displacement Liters	Engine ID/VIN	Engine Oil	Auto. Trans.	Drive Axle	Power Steering Fluid	Brake Master Cylinder
2009	Avalon	3.5	2GR-FE	5W-30	NA	—	ATF Dexron II Or III	DOT 3
2010	Avalon	3.5	2GR-FE	5W-30	NA	—	ATF Dexron II Or III	DOT 3

DOT: Department Of Transportation

NA: Not Available

3768X_AVAL_C0013

VALVE SPECIFICATIONS

Year	Engine Displacement Liters	Engine ID	Seat Angle (deg.)	Face Angle (deg.)	Spring Test Pressure (lbs. @ in.)	Spring Installed Height (in.)	Stem-to-Guide Clearance (in.)		Stem Diameter (in.)	
							Intake	Exhaust	Intake	Exhaust
2009	3.5	2GR-FE	45	44.5	NA	NA	0.0010-0.0024	0.0012-0.0026	0.2154-0.2159	0.2151-0.2157
2010	3.5	2GR-FE	45	44.5	NA	NA	0.0010-0.0024	0.0012-0.0026	0.2154-0.2159	0.2151-0.2157

NA: Information not available

3768X_AVAL_C0005

CAMSHAFT AND BEARING SPECIFICATIONS CHART

All measurements are given in inches.

Year	Engine Displ. Liters	Engine ID/VIN	Journal Dia.	Brg. Oil Clearance	Shaft End-play	Runout	Journal Bore	Lobe Height Intake	Lobe Height Exhaust
2009	3.5	2GR-FE	①	②	NA	0.0016	NA	1.7447-1.7487	1.7426-1.7465
2010	3.5	2GR-FE	①	②	NA	0.0016	NA	1.7447-1.7487	1.7426-1.7465

NA: Not Available

① No. 1 journal: 1.4152-1.4157
Other Journals: 1.0220-1.0226 in.

② No. 1 journal: 0.0016-0.0031
Other Journals: 1.0010-1.0024 in.
Maximum No.1 Journal: 0.0039 in.
Other Maximum Journals: 0.0035 in.

3768X_AVAL_C0014

CRANKSHAFT AND CONNECTING ROD SPECIFICATIONS

All measurements are given in inches.

Year	Engine Displacement Liters	Engine ID	Crankshaft Main Brg. Journal Dia.	Crankshaft Main Brg. Oil Clearance	Crankshaft Shaft End-play	Crankshaft Thrust on No.	Connecting Rod Journal Diameter	Connecting Rod Oil Clearance	Connecting Rod Side Clearance
2009	3.5	2GR-FE	2.4011-2.4016	0.0010-0.0019	0.0016-0.0095	2	2.0863-2.0866	0.0018-0.0026	0.0059-0.0157
2010	3.5	2GR-FE	2.4011-2.4016	0.0010-0.0019	0.0016-0.0095	2	2.0863-2.0866	0.0018-0.0026	0.0059-0.0157

3768X_AVAL_C0006

PISTON AND RING SPECIFICATIONS

All measurements are given in inches.

Year	Engine Displ. Liters	Engine ID	Piston Clearance	Ring Gap Top Comp.	Ring Gap Bottom Comp.	Ring Gap Oil Control	Ring Side Clearance Top Comp.	Ring Side Clearance Bottom Comp.	Ring Side Clearance Oil Control
2009	3.5	2GR-FE	0.0018-0.0020	0.0098-0.0138	0.0197-0.0236	0.0039-0.0157	0.0008-0.0028	0.0008-0.0024	0.0028-0.0059
2010	3.5	2GR-FE	0.0018-0.0020	0.0098-0.0138	0.0197-0.0236	0.0039-0.0157	0.0008-0.0028	0.0008-0.0024	0.0028-0.0059

3768X_AVAL_C0007

TORQUE SPECIFICATIONS

All readings in ft. lbs.

Year	Engine Displacement Liters	Engine ID	Cylinder Head Bolts	Main Bearing Bolts	Rod Bearing Bolts	Crankshaft Damper Bolts	Flywheel Bolts	Manifold Intake	Manifold Exhaust	Spark Plugs	Oil Pan Drain Plug
2009	3.5	2GR-FE	①	②	③	184	61	15	15	13	30
2010	3.5	2GR-FE	①	②	③	184	61	15	15	13	30

① Step 1: 10 mm bolts to 27 ft. lbs.

Step 2: 10mm point cap bolts plus 90 degrees

Step 3: 10mm point cap bolts plus 90 degrees

Step 4: Front bolts to 22 ft. lbs.

② Step 1: 16 cap bolts to 45 ft. lbs.

Step 2: 16 cap bolts plus 90 degrees

Step 3: 8 side bolts to 38 ft. lbs.

③ Step 1: 18 ft. lbs.

Step 2: Plus 90 degrees

3768X_AVAL_C0008

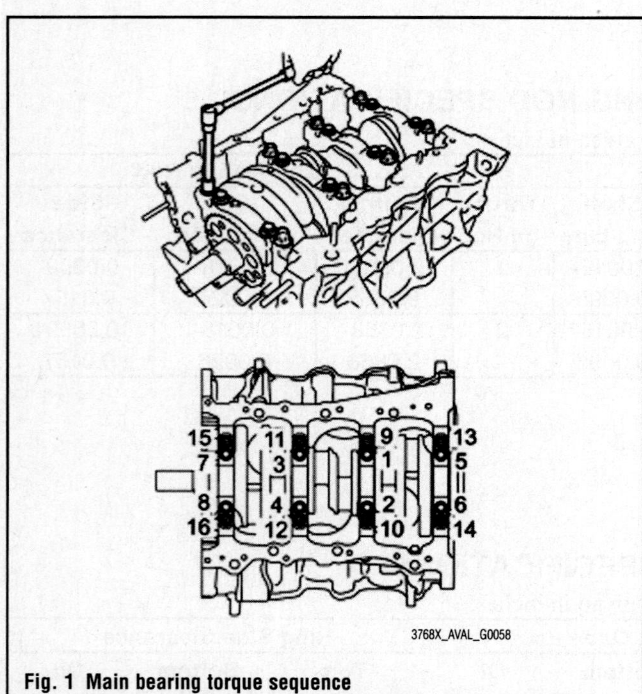

Fig. 1 Main bearing torque sequence

WHEEL ALIGNMENT

Year	Model		Caster Range (+/-Deg.)	Caster Preferred Setting (Deg.)	Camber Range (+/-Deg.)	Camber Preferred Setting (Deg.)	Toe-in (in.)	Steering Axis Inclination (Deg.)
2009	Avalon XL	Front	0.75	2.65	0.75	-0.67	0+/-0.04	12.25+/-0.75
		Rear	—	—	0.75	-1.15	0.16+/-0.08	—
	Avalon Touring	Front	0.75	2.72	0.75	-0.72	0+/-0.04	12.37+/-0.75
		Rear	—	—	0.75	-1.22	0.16+/-0.08	—
	Avalon XLS	Front	0.75	2.70	0.75	-0.72	0+/-0.04	12.33+/-0.75
		Rear	—	—	0.75	-1.22	0.16+/-0.08	—
	Avalon Limited	Front	0.75	2.80	0.75	-0.72	0+/-0.04	12.33+/-0.75
		Rear	—	—	0.75	-1.25	0.16+/-0.08	—
2010	Avalon XL	Front	0.75	2.65	0.75	-0.67	0+/-0.04	12.25+/-0.75
		Rear	—	—	0.75	-1.15	0.16+/-0.08	—
	Avalon Touring	Front	0.75	2.72	0.75	-0.72	0+/-0.04	12.37+/-0.75
		Rear	—	—	0.75	-1.22	0.16+/-0.08	—
	Avalon XLS	Front	0.75	2.70	0.75	-0.72	0+/-0.04	12.33+/-0.75
		Rear	—	—	0.75	-1.22	0.16+/-0.08	—
	Avalon Limited	Front	0.75	2.80	0.75	-0.72	0+/-0.04	12.33+/-0.75
		Rear	—	—	0.75	-1.25	0.16+/-0.08	—

3768X_AVAL_C0009

TIRE, WHEEL AND BALL JOINT SPECIFICATIONS

Year	Model	OEM Tires Standard	OEM Tires Optional	Tire Pressures (psi) Front	Tire Pressures (psi) Rear	Wheel Size	Ball Joint Inspection	Lug Nut Torque (ft. lbs.)
2009	Avalon	P215/60R16	P215/55R17	29	29	6.5-JJ	①	76
2010	Avalon	P215/60R16	P215/55R17	29	29	6.5-JJ	①	76

OEM: Original Equipment Manufacturer

PSI: Pounds Per Square Inch

STD: Standard

OPT: Optional

① Replace if any measurable movement is found.

3768X_AVAL_C0010

BRAKE SPECIFICATIONS

All measurements in inches unless noted

Year	Model		Brake Disc Original Thickness	Brake Disc Minimum Thickness	Brake Disc Maximum Runout	Minimum Lining Thickness	Brake Caliper Bracket Bolts (ft. lbs.)	Brake Caliper Mounting Bolts (ft. lbs.)
2009	Avalon	F	1.102	0.983	0.0020	0.039	79	25
		R	0.390	0.334	0.0059	0.039	46	32
2010	Avalon	F	1.102	0.983	0.0020	0.039	79	25
		R	0.390	0.334	0.0059	0.039	46	32

F: Front

R: Rear

3768X_AVAL_C0011

SCHEDULED MAINTENANCE INTERVALS

TOYOTA—AVALON

TO BE SERVICED	TYPE OF SERVICE	VEHICLE MILEAGE INTERVAL (x1000)													
		5	10	15	20	25	30	35	40	45	50	55	60	90	120
Engine oil & filter	R	✓	✓	✓	✓	✓	✓	✓	✓	✓	✓	✓	✓	✓	✓
Automatic transmission fluid	S/I			✓			✓			✓			✓	✓	✓
Ball joints & dust covers	S/I			✓			✓			✓			✓	✓	✓
Bolts & nuts on chassis & body	S/I			✓			✓			✓			✓	✓	✓
Brake linings & drums	S/I	✓	✓	✓	✓	✓	✓	✓	✓	✓	✓	✓	✓	✓	✓
Brake line pipes & hoses	S/I			✓			✓			✓			✓	✓	✓
Brake pads & discs (front & rear)	S/I	✓	✓	✓	✓	✓	✓	✓	✓	✓	✓	✓	✓	✓	✓
Brake fluid	R						✓						✓	✓	✓
Rack and pinion assembly	S/I			✓			✓			✓			✓	✓	✓
Steering linkage & boots	S/I			✓			✓			✓			✓	✓	✓
Air cleaner filter	R						✓						✓	✓	✓
Spark plugs ①	R														✓
Drive belts	S/I												✓	✓	✓
Exhaust pipes & mountings	S/I			✓			✓			✓			✓	✓	✓
Fuel lines & connections	S/I						✓						✓	✓	✓
Engine coolant ②	S/I			✓			✓			✓			✓	✓	
Fuel tank cap gasket	S/I						✓						✓	✓	✓
Rotate tires	S/I			✓			✓			✓			✓		✓
Clean air conditioning filter ③	S/I			✓			✓			✓			✓		✓
Axle shaft bolts	S/I			✓			✓			✓			✓	✓	✓
Brake pad thickness and rotor runout	S/I						✓						✓	✓	✓

R: Replace S/I: Service or Inspect

① Spark plugs are replaced at 120,000 miles

② Replace engine coolant at 100,000 miles and then inspect every 15,000 miles

③ Replace air conditioning filter every 30,000 miles

FREQUENT OPERATION MAINTENANCE (SEVERE SERVICE)

If a vehicle is operated under any of the following conditions it is considered severe service:

- Extremely dusty areas.

- 50% or more of the vehicle operation is in 32°C (90°F) or higher temperatures, or constant temperatures below 0°C (32°F).

- Prolonged idling (vehicle operation in stop and go traffic).

- Frequent short running periods (engine does not warm to normal operating temperatures).

- Police, taxi, delivery usage or trailer towing usage.

Air cleaner filter: service or inspect every 5000 miles

Rear differential & transfer case oil: replace every 15,000 miles.

Ball joints & dust covers: service or inspect every 5000 miles.

Bolts & nuts on chassis & body: service or inspect every 5000 miles.

Axle shaft bolts: service or inspect every 5000 miles.

Steering linkage: service or inspect every 5000 miles.

3768X_AVAL_C0012

BRAKES — INFORMATION AND PRECAUTIONS

ANTI-LOCK SYSTEMS

- Certain components within the ABS system are not intended to be serviced or repaired individually.
- Do not use rubber hoses or other parts not specifically specified for and ABS system. When using repair kits, replace all parts included in the kit. Partial or incorrect repair may lead to functional problems and require the replacement of components.
- Lubricate rubber parts with clean, fresh brake fluid to ease assembly. Do not use shop air to clean parts; damage to rubber components may result.
- Use only DOT 3 brake fluid from an unopened container.
- If any hydraulic component or line is removed or replaced, it may be necessary to bleed the entire system.
- A clean repair area is essential. Always clean the reservoir and cap thoroughly before removing the cap. The slightest amount of dirt in the fluid may plug an orifice and impair the system function. Perform repairs after components have been thoroughly cleaned; use only denatured alcohol to clean components. Do not allow ABS components to come into contact with any substance containing mineral oil; this includes used shop rags.
- The Anti-Lock control unit is a microprocessor similar to other computer units in the vehicle. Ensure that the ignition switch is **OFF** before removing or installing controller harnesses. Avoid static electricity discharge at or near the controller.
- If any arc welding is to be done on the vehicle, the control unit should be unplugged before welding operations begin.

DISC AND DRUM SYSTEMS

CAUTION

Dust and dirt accumulating on brake parts during normal use may contain asbestos fibers from production or aftermarket brake linings. Breathing excessive concentrations of asbestos fibers can cause serious bodily harm. Exercise care when servicing brake parts. Do not sand or grind brake lining unless equipment used is designed to contain the dust residue. Do not clean brake parts with compressed air or by dry brushing. Cleaning should be done by dampening the brake components with a fine mist of water, then wiping the brake components clean with a dampened cloth. Dispose of cloth and all residue containing asbestos fibers in an impermeable container with the appropriate label. Follow practices prescribed by the Occupational Safety and Health Administration (OSHA) and the Environmental Protection Agency (EPA) for the handling, processing, and disposing of dust or debris that may contain asbestos fibers.

BRAKES — BLEEDING THE BRAKE SYSTEM

ACTUATOR ASSEMBLY

BLEEDING PROCEDURE

➡After bleeding the air from the brake system, if the height or feel of the brake pedal cannot be obtained, perform air bleeding in the brake actuator assembly with a hand-held tester by following the procedures below.

1. Depress the brake pedal more than 20 times with the engine off.
2. Connect the hand-held tester to the DLC3, then turn the ignition switch to the **ON** position, but do NOT start the engine.
3. Select "AIR BLEEDING" on the hand-held tester.

➡Refer to the hand-held tester operator's manual for more details.

4. Bleed the air out of the regular brake line when "Step 1: Increase" appears on the hand-held tester display, as follows:

➡Bleed the air by following the steps displayed on the hand-held tester. Make sure that the brake fluid in the master cylinder reservoir tank does not become empty.

a. Connect the vinyl tube to either one of the bleeder plugs.

b. Have an assistant depress the brake pedal several times, and then loosen the bleeder plug connected to the vinyl tube with the pedal depressed.

c. When fluid stops coming out, tighten the bleeder plug and release the brake pedal.

d. Repeat the previous 2 steps until all the air in the fluid is completely bled out.

e. Tighten the bleeder plug completely to 73 inch lbs. (8.3 Nm).

f. Repeat the above procedures for each wheel to bleed the air out of the brake line.

5. Bleed the air out of the suction line when "Step 2: Inhalation" appears on the hand-held tester display, as follows:

➡Bleed the air by following the steps displayed on the hand-held tester. Make sure that the brake fluid in the master cylinder reservoir tank does not become empty.

a. Connect the vinyl tube to the bleeder plug at the right front wheel or the right rear wheel and loosen the bleeder plug.

b. Operate the brake actuator assembly to perform air bleeding from the suction line using the hand-held tester.

➡This operation stops automatically after 4 seconds. At this time, be sure to release the brake pedal.

c. Check if the operation has stopped by referring to the hand-held tester display and tighten the bleeder plug.

d. Repeat the previous 2 steps until all air in the fluid is completely bled out.

e. Tighten the bleeder plug completely to 73 inch lbs. (8.3 Nm).

f. Repeat the above procedures to bleed the air out of the brake line for each wheel.

6. Bleed the air out of the pressure reduction line when "Step 3: Decrease" appears on the hand-held tester display, as follows:

➡Bleed the air by following the steps displayed on the hand-held tester. Make sure that the brake fluid in the master cylinder reservoir tank does not become empty.

a. Connect a vinyl tube to either one of the bleeder plugs.

b. Loosen the bleeder plug.

c. Using the hand-held tester, operate the brake actuator assembly, completely depress the brake pedal and keep it depressed.

➡The operation stops automatically after 4 seconds. When performing this procedure continuously, set an interval of at least 20 seconds. When the operation is complete, the brake pedal goes down slightly. This is a normal phenomenon caused when the solenoid opens. During this procedure, the pedal will feel heavy, but completely depress it so that the brake fluid comes out from the bleeder plug. Be sure to keep depressing the brake pedal. Do not depress and release the pedal repeatedly.

d. Tighten the bleeder plug, then release the brake pedal.

e. Repeat the previous 2 steps until all the air in the fluid is completely bled out.

f. Tighten the bleeder plug completely to 73 inch lbs. (8.3 Nm).

g. Repeat the above procedures for each wheel to bleed the air out of the brake line.

7. Bleed the air out of the brake line again when "Step 4: Increase" appears on the hand-held tester display, as follows:

➡Bleed the air by following the steps displayed on the hand-held tester. Make sure that the brake fluid in the master cylinder reservoir tank does not become empty.

a. Connect the vinyl tube to either one of the bleeder plugs.

b. Depress the brake pedal several times, then loosen the bleeder plug connected to the vinyl tube with the pedal depressed.

8. When fluid stops coming out, tighten the bleeder plug, then release the brake pedal.

a. Repeat the previous 2 steps until all the air in the fluid is completely bled out.

b. Tighten the bleeder plug completely to 73 inch lbs. (8.3 Nm).

c. Repeat the above procedures for each wheel to bleed the air out of the brake line.

d. Finish "AIR BLEEDING" on the hand-held tester and turn the hand-held tester off.

e. Disconnect the hand-held tester from the DLC3 from the DLC3.

f. Turn the ignition switch off.

g. Inspect for fluid leak.

9. Check the fluid level and add fluid if necessary. Use SAE J1703 or FMVSS No. 116 DOT3 Brake fluid.

BRAKE LINES

BLEEDING PROCEDURE

➡Bleed air from the brake line of the wheel farthest from the master cylinder.

1. Raise and safely support the vehicle.
2. Connect a vinyl tube to the bleeder plug.
3. Have an assistant depress the brake pedal several times, then loosen the bleeder plug while the pedal is depressed.
4. When fluid stops coming out, tighten the bleeder plug, then release the brake pedal.
5. Repeat steps 3 and 4 until all the air in the fluid has been bled out.
6. Tighten the brake bleeder plug to 73 inch lbs. (8.3 Nm).
7. Repeat the above steps to bleed the air out of the brake line for each wheel.

MASTER CYLINDER

BLEEDING PROCEDURE

If the master cylinder is reinstalled or if the reservoir becomes empty, bleed the air from the master cylinder. To prevent brake fluid from adhering, cover nearly painted surfaces with a shop rag or a piece of cloth.

1. Using a union nut wrench (10 mm), disconnect the 2 brake lines from the master cylinder, using a suitable brake line wrench.
2. Have an assistant slowly depress the brake pedal and hold it.
3. Cover the 2 outer holes with your fingers, and have your assistant release the brake pedal.
4. Repeat the previous 2 steps 3 or 4 times.
5. Using a union nut wrench (10 mm), connect the 2 the brake lines to the master cylinder and tighten to 11 ft. lbs. (15 Nm).

➡Use a torque wrench with a fulcrum length of 250 mm (9.84 in.).

➡This torque value is effective when the union nut wrench is parallel to the torque wrench.

BRAKES — ANTI-LOCK BRAKE SYSTEM (ABS)

WHEEL SPEED SENSORS

REMOVAL & INSTALLATION

Front

1. Remove the front wheel.
2. Remove the front fender liner.
3. Remove the front speed sensor, as follows:

a. Disconnect the speed sensor connector and clamp.

b. Remove the 2 bolts and separate the speed sensor harness from the body and shock absorber assembly.

c. Remove the clamp from the steering knuckle.

d. Remove the bolt and the front speed sensor.

➡Do not allow foreign matter to attach to the sensor tip.

➡Clean the installation hole and surface for the speed sensor every time the speed sensor is removed.

To install:

4. Install the front speed sensor with the bolt and tighten to 71 inch lbs. (8 Nm).

➡Do not allow foreign matter to attach to the sensor tip.

5. Install the sensor harness brackets with the 2 bolts to the body and shock absorber assembly and tighten to 44 inch lbs. (5 Nm), and 14 ft. lbs. (19 Nm).

a. Connect the clamp to the steering knuckle.

➡Do not twist the sensor wire when installing the sensor.

6. Connect the speed sensor connector and clamp.
7. Install the front fender liner.
8. Install the front wheel and tighten the lug nuts to 76 ft. lbs. (103 Nm).
9. Check ABS speed sensor signal.

Rear

See Figure 2.

1. Remove the rear wheel.
2. Disconnect the connector from the skid control sensor.
3. Separate the rear disc brake caliper assembly.
4. Remove the rear disc.
5. Remove the 4 bolts and rear axle hub and bearing assembly.
6. Remove the rear speed sensor, as follows:

a. Mount the rear axle hub in a vise between aluminum plates.

⁂ WARNING

Replace the hub and bearing assembly if it is dropped or receives a strong shock.

b. Using a pin punch and hammer, drive out the 2 pins and remove the 2 attachments from SST.

c. Using SST (SST: 09520-00031, SST: 09521-00020, SST: 09950-00020 or equivalent) and 2 bolts (Diameter: 12 mm, pitch: 1.5 mm), remove the rear speed sensor from the rear axle hub.

➡ **If the sensor rotor is damaged, replace the hub and bearing assembly.**

➡ **Do not scratch the contact surface of the axle hub and speed sensor.**

To install:

7. Install the rear speed sensor, as follows:

a. Clean the contact surface of the axle hub and a new skid control sensor.

➡ **Make sure the sensor rotor is clean.**

b. Place the skid control sensor on the rear axle hub so that the connector is positioned as shown in the illustration.

c. Using SST (SST: 09830-36010, 09950-60010, 09950-70010 or equivalent) and a press, install the skid control sensor to the axle hub.

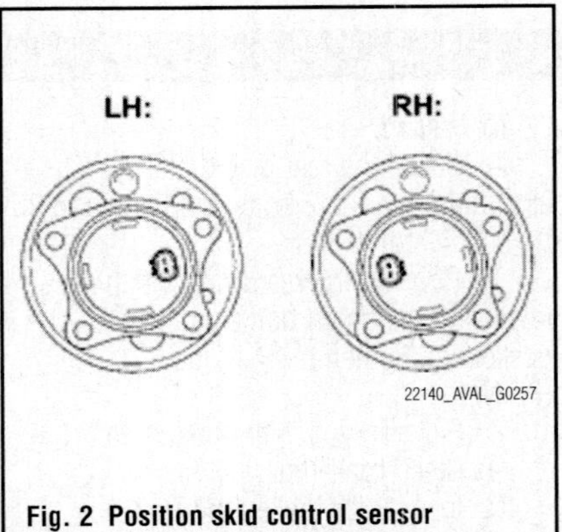

Fig. 2 Position skid control sensor

➡ **Do not tap the skid control sensor directly with a hammer.**

➡ **Check that the skid control sensor detection part is clean.**

➡ **Press in the skid control sensor straight and slowly.**

8. Install the rear axle hub and bearing assembly with the 4 bolts and tighten to 59 ft. lbs. (80 Nm).
9. Install the rear disc.
10. Install the rear disc brake caliper assembly.
11. Connect the connector to the skid control sensor.
12. Install the rear wheel and tighten the lug nuts to 76 ft. lbs. (103 Nm).
13. Inspect and adjust rear wheel alignment.
14. Check ABS speed sensor signal.

BRAKES

FRONT DISC BRAKES

BRAKE CALIPER

REMOVAL & INSTALLATION

See Figure 3.

1. Before servicing the vehicle, refer to the Precautions Section.
2. Remove the front wheel.
3. Drain brake fluid.

⁂ WARNING

Do not let brake fluid sit on painted surfaces, as it will eat through the paint. Wash it off immediately.

4. Remove the union bolt and gasket from the disc brake cylinder assembly, then disconnect the front brake flexible hose.

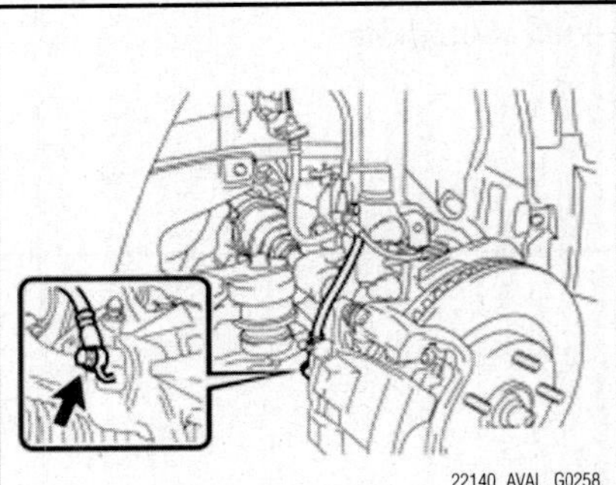

Fig. 3 Remove the union bolt and gasket

5. Hold the front disc brake cylinder slide pin and remove the 2 bolts and disc brake cylinder assembly.

➡ **Remove the disc brake cylinder assembly while holding both of the brake pads or the anti-squeal springs may fall off the brake pads.**

To install:

6. Install the disc brake cylinder assembly with the 2 bolts and tighten to 25 ft. lbs. (34 Nm).

➡ **Be sure that the anti-squeal springs are installed to the front disc brake pads.**

7. Connect the flexible hose with the union bolt and a new gasket and tighten to 21 ft. lbs. (29 Nm).

➡ **Install the front brake flexible hose lock securely in the lock hole in the disc brake cylinder.**

8. Fill reservoir with brake fluid.
9. Bleed brake line.
10. Inspect for brake fluid leak.
11. Inspect brake fluid level in reservoir.
12. Install the front wheel and tighten the lug nuts to 76 ft. lbs. (103 Nm).

DISC BRAKE PADS

REMOVAL & INSTALLATION

See Figures 4 and 5.

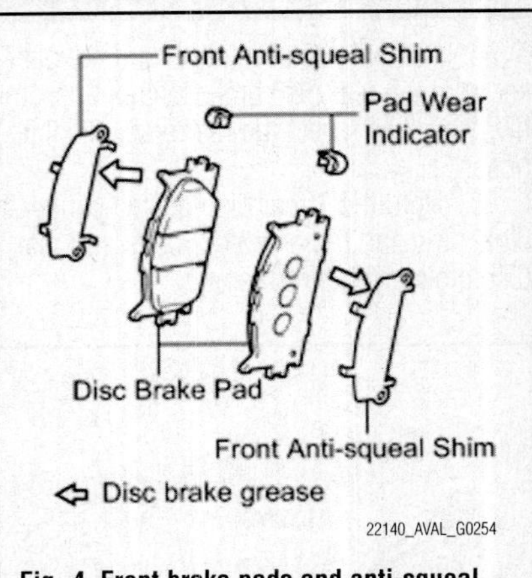

Fig. 4 Front brake pads and anti-squeal shims

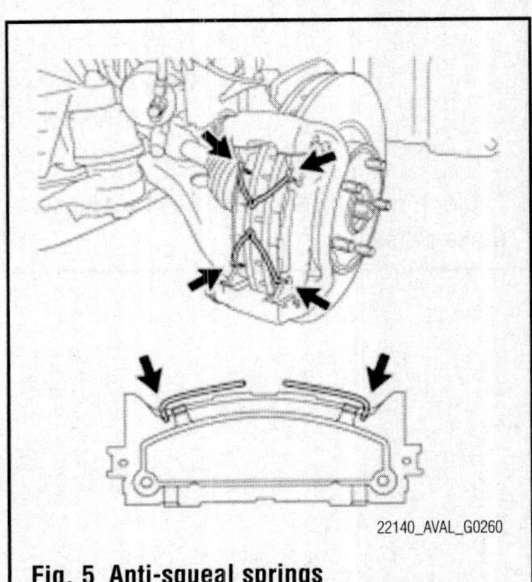

Fig. 5 Anti-squeal springs

1. Before servicing the vehicle, refer to the Precautions Section.
2. Remove the 2 front disc brake cylinder slide pins (upper and lower) from the front disc brake cylinder mounting.
3. Remove brake cylinder.
4. Remove the 2 anti-squeal springs.
5. Remove the 2 brake pads from the front disc brake cylinder mounting.

To install:

6. Install the 2 brake pads with front anti-squeal shims to the front disc brake cylinder mounting.

➡When replacing worn pads, the front anti-squeal springs must be replaced at the same time.

➡Be sure to install the anti-squeal springs into the front disc brake pad installation holes as far as they will go.

7. Install the 2 front disc brake cylinder slide pins (upper and lower) from the front disc brake cylinder mounting.
8. Install the brake cylinder.

BRAKES

REAR DISC BRAKES

BRAKE CALIPER

REMOVAL & INSTALLATION

See Figure 6.

1. Before servicing the vehicle, refer to the Precautions Section.
2. Remove the rear wheel.
3. Drain brake fluid.

WARNING

Do not let brake fluid sit on painted surfaces, as it will eat through the paint. Wash it off immediately.

4. Remove the union bolt and the gasket from the rear disc brake cylinder assembly, then disconnect the rear brake flexible hose.
5. Hold the 2 rear disc brake cylinder slide pins and remove the 2 bolts and rear disc brake cylinder assembly.

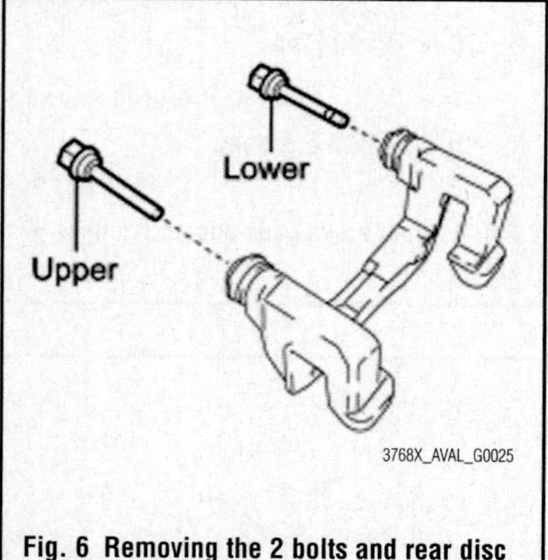

Fig. 6 Removing the 2 bolts and rear disc brake cylinder

To install:

6. Install the rear disc brake cylinder assembly with the 2 bolts and tighten to 20 ft. lbs. (77 Nm).
7. Connect the rear brake flexible hose with the union bolt and a new gasket and tighten to 24 ft. lbs. (33 Nm).
8. Fill reservoir with brake fluid.
9. Bleed brake line.
10. Inspect for brake fluid leak.
11. Inspect brake fluid level in reservoir.
12. Install the rear wheel and tighten the lug nuts to 76 ft. lbs. (103 Nm).

DISC BRAKE PADS

REMOVAL & INSTALLATION

See Figure 7.

1. Before servicing the vehicle, refer to the Precautions Section.
2. Remove the 2 rear disc brake cylinder slide pins (upper and lower) from the rear disc brake cylinder mounting.
3. Remove brake cylinder.
4. Remove the 2 brake pads with the rear anti-squeal shims.

To install:

5. Installation is the reverse of removal procedure.

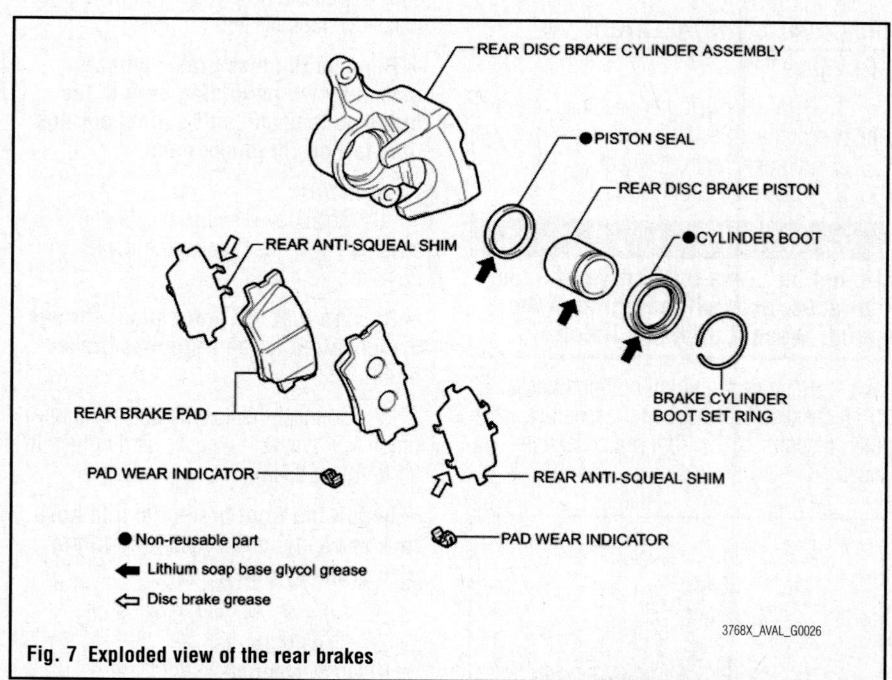

Fig. 7 Exploded view of the rear brakes

PARKING BRAKE CABLES

ADJUSTMENT

1. Inspect the parking brake pedal travel, as follows:
 a. Fully depress the parking brake pedal and release it to engage the parking brake.
 b. Depress the pedal to the floor again, and release it to disengage the parking brake.
 c. Slowly depress the parking brake pedal to the floor, and count the number of clicks. Parking brake pedal travel: 9 to 11 notches at 67.5 lbs. (300 N).
2. Adjust the parking brake pedal travel, as follows:
 a. Depress the parking brake pedal. Hold the wire adjusting nut No. 1 using a wrench and loosen the lock nut.
 b. Release the parking brake pedal.
 c. Turn the No. 1 wire adjusting nut until the parking brake pedal travel meets the above specification.
 d. Hold the No. 1 wire adjusting nut using a wrench or equivalent tool and tighten the lock nut to 48 inch lbs. (5.4 Nm).
 e. Count the number of clicks after depressing and releasing the parking brake pedal 3 or 4 times. Parking brake pedal travel: 9 to 11 notches at 67.5 lbs. (300 N).
 f. Check whether the parking brake drags or not.
 g. When operating the parking brake pedal, check that the parking brake indicator light comes on.

PARKING BRAKE SHOES

REMOVAL & INSTALLATION

See Figures 8 through 12.

1. Remove the rear wheel.
2. Remove the 2 bolts and separate the rear disc brake caliper assembly. Do not disconnect the flexible hose from the disc brake caliper assembly.
3. Remove the parking brake shoe adjusting hole plug from the rear disc.
4. Release the parking brake and place the matchmarks on the rear disc and the axle hub.
5. Remove the rear disc.

➡If the disc cannot be removed easily, turn the shoe adjuster until the disc turns freely.

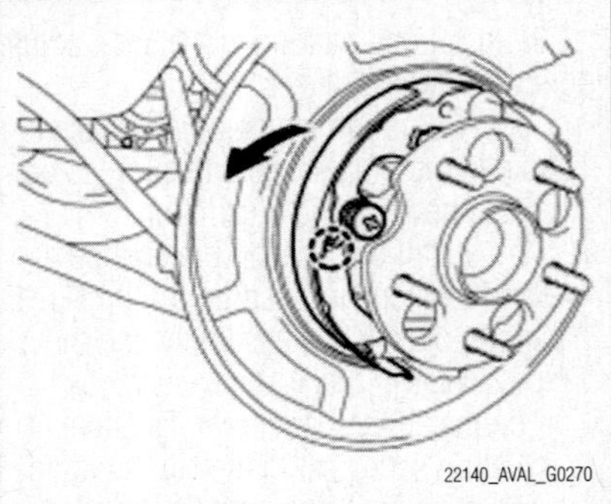

Fig. 8 Remove the No. 1 parking brake shoe assembly

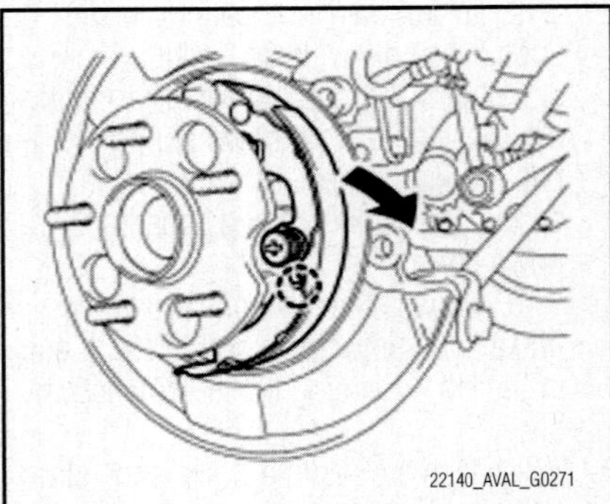

Fig. 9 Remove the No. 2 parking brake shoe assembly

6. Using needle-nose pliers, remove the 2 parking brake shoe return tension No. 1 springs.
7. Remove the parking brake shoe strut and the parking brake shoe strut compression spring.
8. Remove the No. 1 parking brake shoe assembly, as follows:
 a. Release the claw of the parking brake shoe hold down spring No. 2 cup.
 b. Remove the No. 1 parking brake shoe assembly.
 c. Remove the parking brake shoe hold down spring No. 1 cup, the parking brake shoe hold down spring, the parking brake shoe hold down spring No. 2 cup, and the parking brake shoe hold down spring No. 1 pin.
9. Remove the parking brake shoe adjusting screw set.
10. Remove the parking brake shoe return tension No. 2 spring.
11. Remove the No. 2 parking brake shoe assembly, as follows:
 a. Release the claw of the parking brake shoe hold down spring No. 2 cup.
 b. Remove the No. 2 parking brake shoe assembly.
 c. Remove the parking brake shoe hold down spring No. 1 cup, the parking brake shoe hold down spring, the parking brake shoe hold down spring No. 2 cup, and the parking brake shoe hold down spring No. 2 pin.
 d. Using needle-nose pliers, disconnect the No. 3 parking brake cable assembly from the parking brake shoe lever.

➡Be careful not to damage the No. 3 parking brake cable assembly.

12. Using a screwdriver, remove the C-washer, shim and the parking brake shoe lever.
13. Remove the parking brake shoe guide plate set bolt and the parking brake shoe guide plate.

To install:

14. Apply high temperature grease to the backing plate where it contacts the shoe.
15. Apply adhesive (Toyota Genuine Adhesive 1344, Three Bond 1344 or equivalent)
 to the threads of the parking brake shoe guide plate set bolt.
16. Install the parking brake shoe guide plate with the parking brake shoe guide plate set bolt and tighten to 13 ft. lbs. (18 Nm).
17. Install the parking brake shoe lever and shim to the No. 2 parking brake shoe assembly with a new C-washer.
18. Using a feeler gauge, measure the clearance between the No. 2 parking brake

Shim Thickness	Shim Thickness
0.3 mm (0.012 in.)	0.9 mm (0.035 in.)
0.6 mm (0.024 in.)	-

22140_AVAL_G0272

Fig. 10 Shim thickness

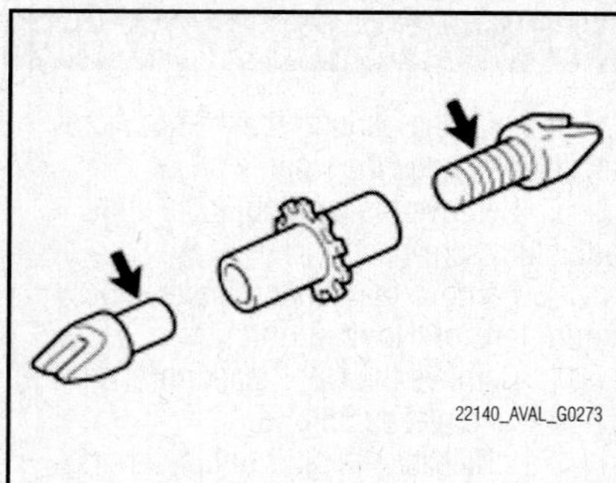

Fig. 11 Apply high temperature grease to the parking brake shoe adjusting screw set

shoe assembly and parking brake shoe lever. Standard clearance: Less than 0.014 inch (0.35 mm). If the clearance is not as specified, replace the shim with one of the correct size.

19. Install the No. 2 parking brake shoe assembly as follows:

a. Using needle-nose pliers, connect the No. 3 parking brake cable assembly to the parking brake shoe lever.

b. Install the No. 2 parking brake shoe assembly with the parking brake shoe hold down spring No. 2 pin, the parking brake shoe hold down spring No. 2 cup, the parking brake shoe hold down spring and the parking brake shoe hold down spring No. 1 cup.

c. Engage the claw of the parking brake shoe hold down spring No. 2 cup to the No. 2 parking brake shoe assembly.

20. Install the parking brake shoe adjusting screw set, as follows:

a. Apply high temperature grease to the parking brake shoe adjusting screw set as shown in the illustration.

b. Install the parking brake shoe return tension No. 2 spring to the No. 1 parking brake shoe assembly and the No. 2 parking brake shoe assembly.

c. Install the parking brake shoe adjusting screw set to the No. 1 parking brake shoe assembly and the No. 2 parking brake shoe assembly.

21. Install the No. 1 parking brake shoe assembly as follows:

a. Install the No. 1 parking brake shoe assembly with the parking brake shoe hold down spring No. 1 pin, parking brake shoe hold down spring No. 2 cup, parking brake shoe hold down spring and parking brake shoe hold down spring No. 1 cup.

b. Engage the claw of the parking brake shoe hold down spring No. 2 cup to the No. 1 parking brake shoe assembly.

22. Attach the parking brake shoe strut and the parking brake shoe strut compression spring to the No. 1 parking brake shoe assembly and No. 2 parking brake shoe assembly.

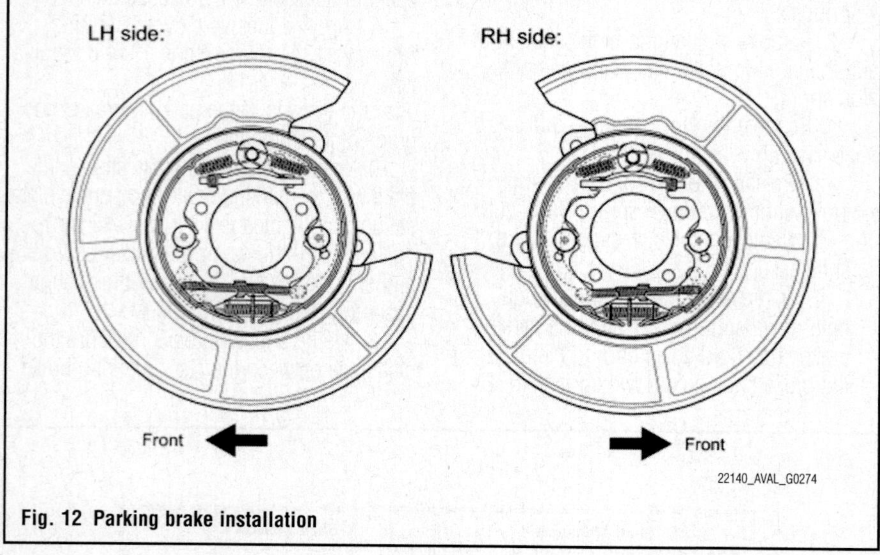

Fig. 12 Parking brake installation

23. Using needle-nose pliers, install the 2 parking brake shoe return tension No. 1 springs. First install the front side spring and then the rear side spring.

24. Inspect parking brake installation and check that each part is installed properly.

➡There should be no oil or grease on the friction surfaces of the shoe linings and discs.

25. Install the rear disc.

26. Install the parking brake shoe adjusting hole plug.

27. Adjust parking brake shoe clearance.

28. Install the rear disc brake caliper assembly with the 2 bolts and tighten to 46 ft. lbs. (62 Nm).

29. Install the rear wheel.

30. Adjust the parking brake pedal travel.

31. Bed in parking brake shoes to discs, as follows:

a. Drive the vehicle at about 31 mph (50 km/h) on a safe, level and dry road.

b. Depress the parking brake pedal with 34 lbs. (150 N) of force.

32. Drive the vehicle about 0.25 miles (400 m) in this condition.

a. Repeat this procedure 3 times using 5-minute intervals between each procedure to prevent the parking brake assembly from overheating.

33. Remove the rear wheel.

34. Adjust parking brake shoe clearance.

35. Adjust the parking brake pedal travel.

36. Install the rear wheel and tighten the lug nuts to 76 ft. lbs. (103 Nm).

ADJUSTMENT

1. Adjust parking brake shoe clearance, as follows:

a. Temporarily install the hub nuts.

b. Remove the shoe adjusting hole plug, turn the adjuster and expand the shoes until the disc locks.

c. Contract the shoe adjuster until the disc rotates smoothly. Standard: returns 8 notches

d. Check that the disc has no brake drag.

e. Install the shoe adjusting hole plug.

CHASSIS ELECTRICAL

AIR BAG (SUPPLEMENTAL RESTRAINT SYSTEM)

GENERAL INFORMATION

CAUTION

The vehicle is equipped with a Supplemental Restraint System (SRS). It consists of a driver airbag, front passenger airbag, driver side knee airbag, front seat side airbag and curtain shield airbag. Failure to carry out service operations in the correct sequence could cause the SRS to unexpectedly deploy during servicing, possibly leading to a serious accident. Further, if a mistake is made in servicing the SRS, it is possible that the SRS may fail to operate when required. Before performing servicing (including removal or installation of parts, inspection or replacement), be sure to read the following items carefully, then follow the correct procedures indicated in the repair manual.

SERVICE PRECAUTIONS

CAUTION

Disconnect and isolate the battery negative cable before beginning any airbag system component diagnosis, testing, removal, or installation procedures. Wait at least 90 seconds after the ignition switch is turned off and the negative (-) terminal cable is disconnected from the battery before starting the operation. The SRS is equipped with a backup power source, so if work is started within 90 seconds after disconnecting the negative (-) terminal cable from the battery, the SRS may be deployed. Failure to disable the airbag system may result in accidental airbag deployment, personal injury, or death.

DISARMING THE SYSTEM

To avoid personal injury when working on vehicles equipped with an air bag, the negative battery cable must be disconnected and at least 90 seconds must elapse before working on the system. Failure to do so may result in deployment of the air bag.

ARMING THE SYSTEM

To arm the system after service is finished, connect the negative battery cable.

CLOCKSPRING CENTERING

See Figures 13 and 14.

1. Before servicing the vehicle, refer to the Precautions Section.
2. Check that the ignition switch is **OFF**.
3. Check that the battery negative (-) terminal is disconnected.

CAUTION

After removing the terminal, wait for at least 90 seconds before starting the operation.

4. Rotate the spiral cable counterclockwise slowly by hand until it feels firm.

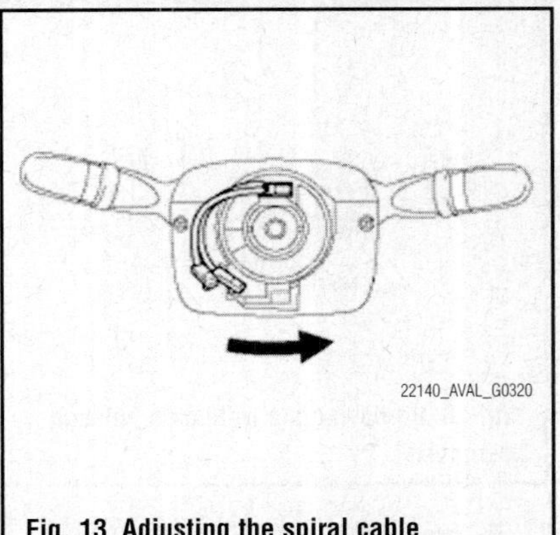

Fig. 13 Adjusting the spiral cable

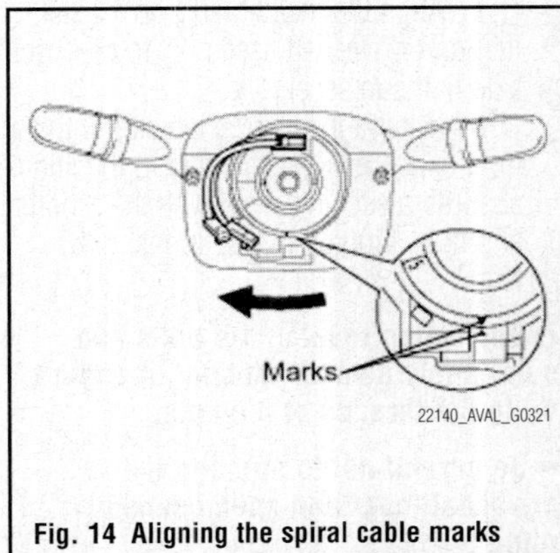

Fig. 14 Aligning the spiral cable marks

5. Rotate the spiral cable clockwise approximately 2.5 turns to align the marks.

➡Do not turn the spiral cable by the airbag wire harness.

➡The spiral cable will rotate approximately 2.5 turns to both the left and right from the center.

DRIVE TRAIN

FRONT AXLE SHAFT, BEARING & SEAL

REMOVAL & INSTALLATION

See Figures 15 and 16.

1. Remove the front wheel.
2. Remove the front axle shaft nuts.
3. Separate the speed sensor.
4. Separate the front disc brake caliper assembly.
    a. Remove the 2 bolts and separate the front disc brake caliper assembly from the steering knuckle.

➡Use a wire or an equivalent to keep the brake caliper from hanging down by the flexible hose.

5. Remove the front disc.
6. Separate the tie rod end sub assembly.
7. Separate the front suspension arm sub assembly No. 1.
8. Remove the front axle assembly.
    a. Put matchmarks on the front driveshaft assembly and the front axle hub sub assembly.
    b. Using a plastic hammer, separate the front driveshaft assembly from the front axle hub sub assembly.

➡Be careful not to damage the driveshaft boot and speed sensor.

    c. Remove the 2 bolts, nuts and steering knuckle with the front axle hub sub assembly.

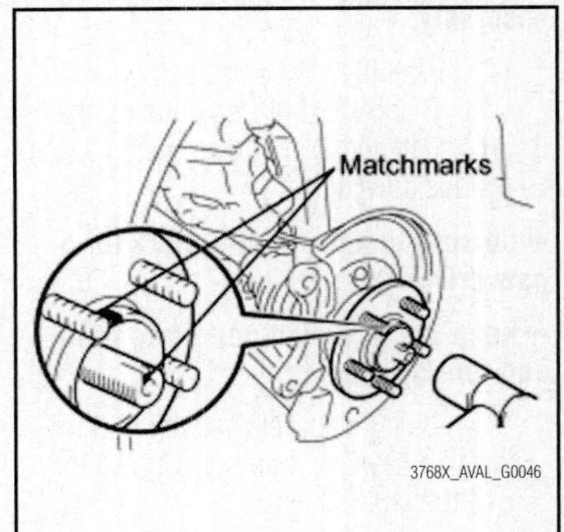

Fig. 15 Removing the front axle assembly

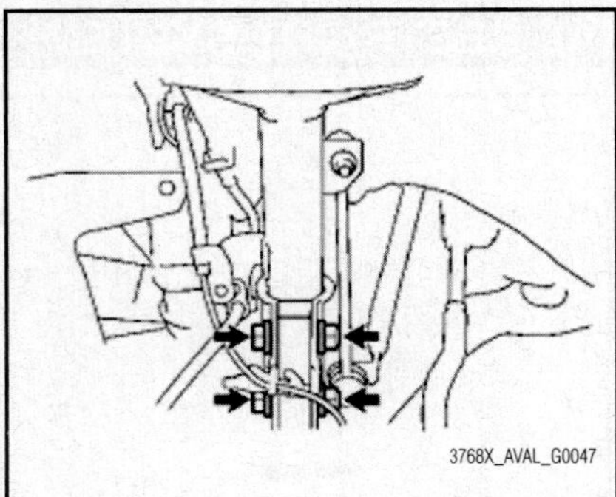

Fig. 16 Removing the front axle hub sub assembly

To install:

9. Install the front axle assembly.
   a. Align the matchmarks and install the front driveshaft assembly to the front axle hub sub assembly.
   b. Install the steering knuckle with the front axle hub sub assembly to the shock absorber assembly front with the 2 bolts and nuts. Tighten to 155 ft. lbs. (210 Nm).

➡Only when reusing the bolts and nuts, apply a small amount of engine oil to the threads of the nuts.

➡Be careful not to damage the driveshaft boot and speed sensor rotor.

10. Install the front suspension arm sub assembly lower No. 1.
11. Install the tie rod end sub assembly.
12. Install the front disc.
13. Install the front disc brake caliper assembly.
    a. Install the front disc brake caliper assembly to the steering knuckle with the 2 bolts. Tighten to 79 ft. lbs. (107 Nm).

➡Do not twist the brake hose when installing the front disc brake caliper assembly.

14. Install the front axle shaft nut.
    a. Clean the threaded parts on the driveshaft and axle hub nut using a non residue solvent.

➡Be sure to perform this work for a new driveshaft.

➡Keep the threaded parts free of oil and foreign objects.

   b. Using a socket wrench (30 mm), install a new axle hub nut. Tighten to 217 ft. lbs. (294 Nm).

➡Stake the nut after inspecting for looseness and runout through the following steps.

15. Separate the front disc brake caliper assembly.
    a. Remove the 2 bolts and separate the front disc brake caliper assembly from the steering knuckle.

➡Use a wire or an equivalent to keep the brake caliper from hanging down by the flexible hose.

16. Remove the front disc.
17. Inspect the front axle hub bearing for looseness.
18. Inspect the front axle hub runout.
19. Install the front disc.
20. Install the front disc brake caliper assembly.
    a. Install the front disc brake caliper assembly with the 2 bolts to the steering knuckle. Tighten to 79 ft. lbs. (107 Nm).

➡Do not twist the brake hose when installing the front disc brake caliper assembly.

21. Install the speed sensor front.
22. Install the front axle shaft nut.
    a. Using a chisel and hammer, stake the axle hub nut.
23. Install the front wheel and tighten to 76 ft. lbs. (103 Nm).
24. Inspect and adjust the front wheel alignments.
25. Check the speed sensor signal.

FRONT HALFSHAFT

REMOVAL & INSTALLATION

See Figures 17 through 22.

1. Remove the engine under cover.
2. Remove the drain plug and gasket, and then drain the automatic transaxle fluid.
3. Install a new gasket and drain plug and tighten to 36 ft. lbs. (49 Nm).
4. Remove front wheel.
5. Using SST (SST: 09930-00010) and hammer, release the staked part of the front axle hub nut.

➡Loosen the staked part of the nut completely, otherwise the screw of the driveshaft may be damaged.

6. While applying the brakes, remove the front axle hub nut.
7. Remove the nut and separate the front stabilizer link assembly.

➡If the ball joint turns together with the nut, use a hexagon wrench (6mm) to hold the stud.

8. Remove the bolt and clip, and separate the speed sensor wire and flexible hose from the shock absorber.

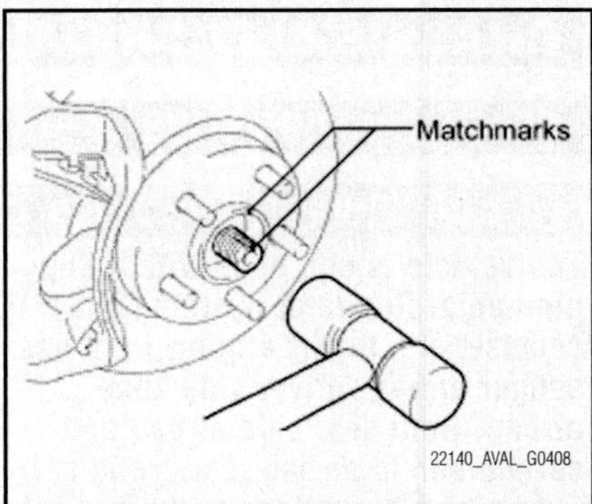

Fig. 17 Put matchmarks on the front driveshaft assembly and the axle hub

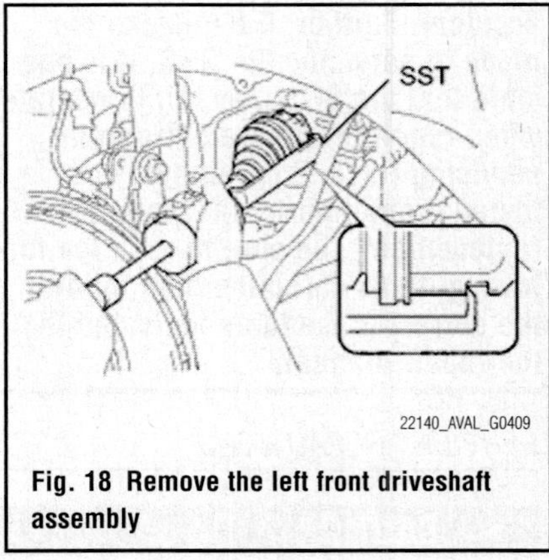

Fig. 18 Remove the left front driveshaft assembly

9. Remove the bolt and separate the front speed sensor from the steering knuckle.

➡Do not allow foreign matter to adhere to the speed sensor. Be careful not to damage the speed sensor.

10. Separate tie rod end sub-assembly, as follows:
    a. Remove the cotter pin and nut.
    b. Using SST (SST: 09628-62011) or equivalent, separate the tie rod end sub-assembly from the steering knuckle.

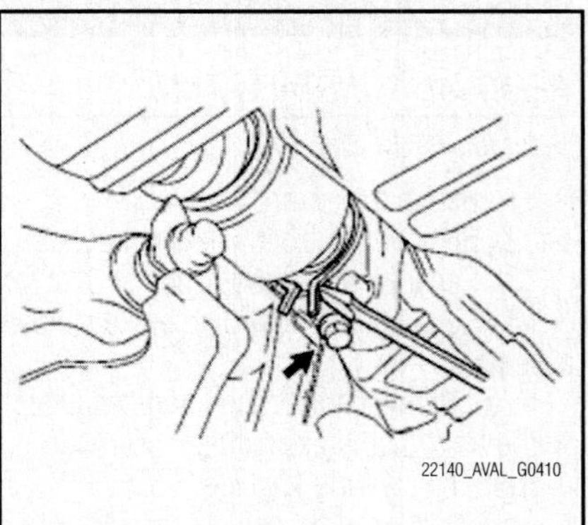

Fig. 19 Remove the right front driveshaft assembly

➡**Do not damage the ball joint dust cover.**

11. Remove the bolt and 2 nuts, and separate the lower No. 1 front suspension arm sub-assembly from the lower ball joint.
12. Put matchmarks on the front driveshaft assembly and the axle hub.
13. Using a plastic hammer, separate the front driveshaft assembly from the front axle hub sub-assembly.

➡**Be careful not to damage the driveshaft boot and speed sensor rotor.**

14. Remove the front driveshaft assembly(s), as follows:
    a. For left front driveshaft, use SST (SST: 09520-01010, SST: 09520-24010) or equivalent, and remove the left front driveshaft assembly.

➡**Be careful not to damage the driveshaft dust cover, boot and oil seal. Be careful not to drop the driveshaft assembly.**

    b. For right front driveshaft, use a screwdriver and remove the bearing bracket hole snap ring.
    c. Remove the bolt and right front driveshaft assembly from the driveshaft bearing bracket.

➡**Do not damage the boot and oil seal.**

15. Secure the front axle hub bearing. The hub bearing could be damaged if it is subjected to the vehicle's full weight, such as moving the vehicle with the driveshaft removed. If it is necessary to place the vehicle's weight on the hub bearing, first support it with SST (SST: 09608-16042).

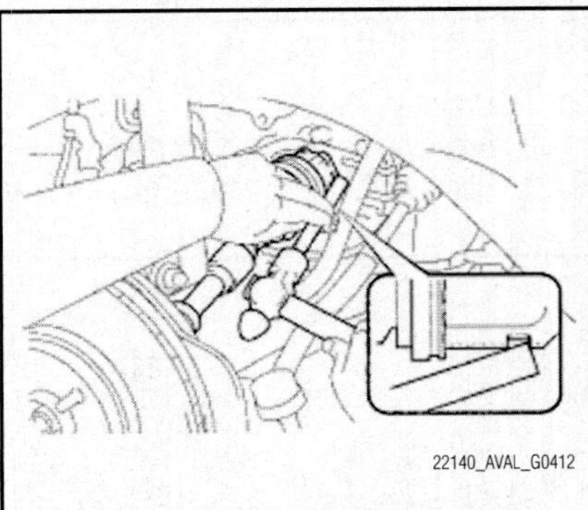

Fig. 20 Install the left front driveshaft assembly

To install:

16. Install the front driveshaft assembly(s), as follows:
    a. Coat the spline of the inboard joint shaft assembly with automatic transaxle fluid.
    b. For the left front driveshaft, align the shaft splines and install the driveshaft assembly with a brass bar and hammer.

➡**Set the shaft snap ring with the opening side facing down.**

    c. For the right front driveshaft, install the driveshaft and use a screwdriver to install a new bearing bracket hole snap ring, and install a new bolt tightened to 24 ft. lbs. (32 Nm).

➡**Be careful not to damage the driveshaft dust cover, boot and oil seal.**

➡**Move the driveshaft assembly while keeping it level.**

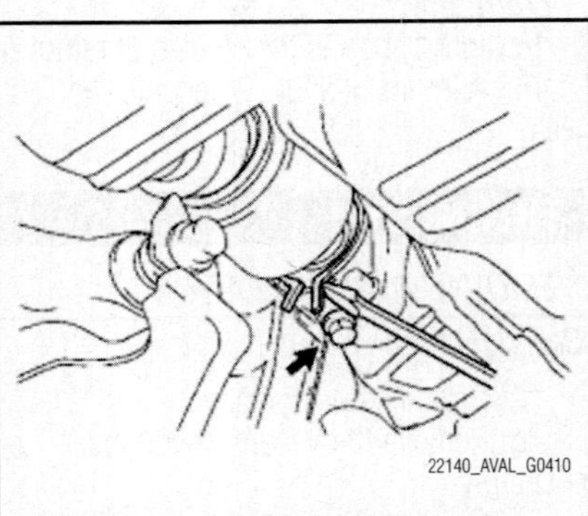

Fig. 21 Install the right front driveshaft assembly

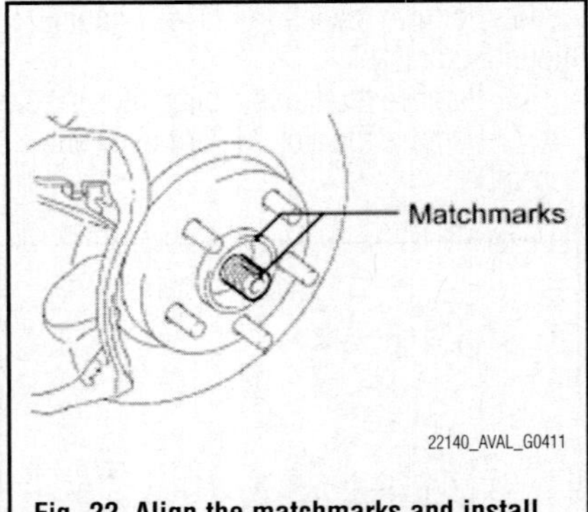

Fig. 22 Align the matchmarks and install the front driveshaft assembly

17. Align the matchmarks and install the front driveshaft assembly to the front axle hub sub-assembly.

➡**Be careful not to damage the driveshaft boot and speed sensor rotor.**

18. Install the lower ball joint to the lower No. 1 front suspension arm sub-assembly with the bolt and 2 nuts and tighten to 55 ft. lbs. (75 Nm).
19. Install the tie rod end sub-assembly to the steering knuckle with the nut and tighten to 36 ft. lbs. (49 Nm).
20. Install a new cotter pin. If the holes for the cotter pin are not aligned, tighten the nut up to 60° further.
21. Install the front speed sensor to the steering knuckle with the bolt and tighten to 71 inch lbs. (8 Nm).
22. Install the flexible hose and the speed sensor to the shock absorber with the bolt and set the sensor clip on the knuckle and tighten to 14 ft. lbs. (19 Nm).

➡**Be careful not to damage the speed sensor. Do not allow foreign matter to adhere to the speed sensor. Do not twist the sensor wire when installing the speed sensor.**

23. Install the stabilizer link assembly with the nut and tighten to 55 ft. lbs. (74 Nm).

➡**If the ball joint turns together with the nut, use a hexagon wrench (6 mm) to hold the stud.**

24. Clean the threaded parts on the driveshaft and axle hub nut using a non-residue solvent.

➡**Be sure to perform this work for a new driveshaft. Keep the threaded parts free of oil and foreign objects.**

25. Using a socket wrench (30 mm), install a new axle hub nut and tighten to 217 ft. lbs. (294 Nm).
26. Using a chisel and hammer, stake the front axle hub nut.
27. Install front wheel.
28. Add automatic transaxle fluid.
29. Inspect automatic transaxle fluid.
30. Inspect and adjust front wheel alignment.
31. Install the engine under cover.
32. Check the ABS speed sensor signal.

ENGINE COOLING

ENGINE FAN

REMOVAL & INSTALLATION

Refer to the Radiator removal and installation procedure.

RADIATOR

REMOVAL & INSTALLATION

See Figures 23 and 24.

1. Drain engine coolant.
2. Remove battery.
3. Remove the No. 2air cleaner inlet.
4. Remove the No. 1air cleaner inlet.
5. Separate the radiator reserve tank hose from the radiator assembly.
6. Separate the radiator hose inlet from the radiator assembly.
7. Separate the radiator hose outlet from the radiator assembly.
8. Separate the No. 1 oil cooler inlet tube from the radiator assembly.
9. Separate the No. 1 oil cooler outlet tube from the radiator assembly.
10. Remove the radiator support upper, as follows:
    a. Remove the hood lock nut cap from the hood lock assembly.
    b. Remove the 3 bolts and separate the hood lock assembly from the upper radiator support.
    c. Remove the 5 bolts and radiator support upper.
11. Remove the radiator assembly as follows:
    a. Disconnect the fan motor connector.
    b. Remove the 4 bolts and separate the condenser assembly from the radiator assembly.
    c. Remove the radiator assembly from the body.

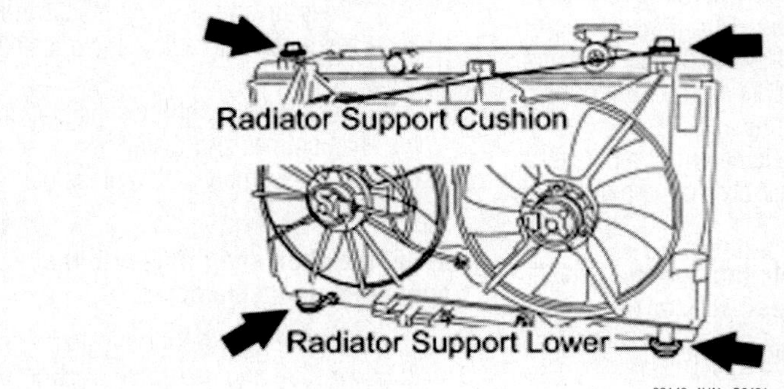

Fig. 24 Remove the 2 radiator support cushions

12. Remove the 2 radiator support cushions from the radiator assembly.
13. Remove the 2 radiator support lowers from the radiator assembly.
14. Disassemble the 3 snap fits and lift the fan assembly w/motor from the radiator.

To install:

15. Installation is the reverse of removal.
16. After installation, inspect for coolant leak.

THERMOSTAT

REMOVAL & INSTALLATION

See Figure 25.

1. Drain engine coolant.
2. Remove the V-bank cover sub-assembly.
3. Remove the engine moving control rod.
4. Remove the engine mounting control bracket.
5. Remove the left front No. 1 engine mounting bracket.
6. Remove the fan and alternator V-belt.
7. Remove the No. 2 idler pulley sub-assembly.
8. Separate the radiator hose outlet.
9. Remove the 2 bolts and water inlet.
10. Remove the thermostat.

To install:

11. Install a new gasket to the thermostat.
12. Install the thermostat with the jiggle valve facing up.

➡The jiggle valve may be set within 10°of either side of the prescribed position.

13. Install the water inlet and tighten to 7 ft. lbs. (10 Nm).
14. The remainder of installation is the reverse of removal.
15. After installation, inspect for coolant leak.

WATER PUMP

REMOVAL & INSTALLATION

See Figures 26 through 29.

1. Drain engine coolant.
2. Remove the right front wheel.
3. Remove the right engine under cover.

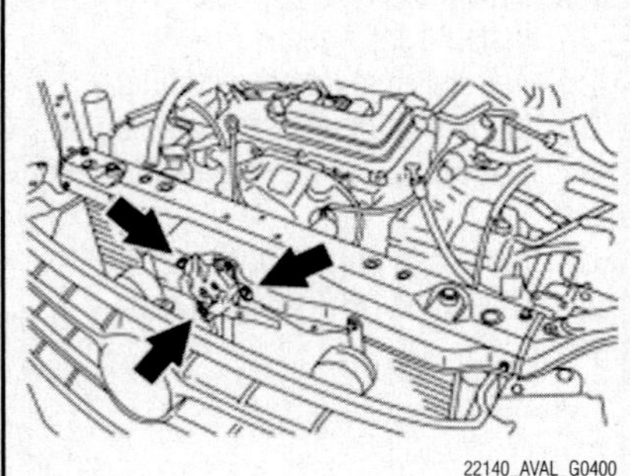

Fig. 23 Separate the hood lock assembly from the upper radiator support

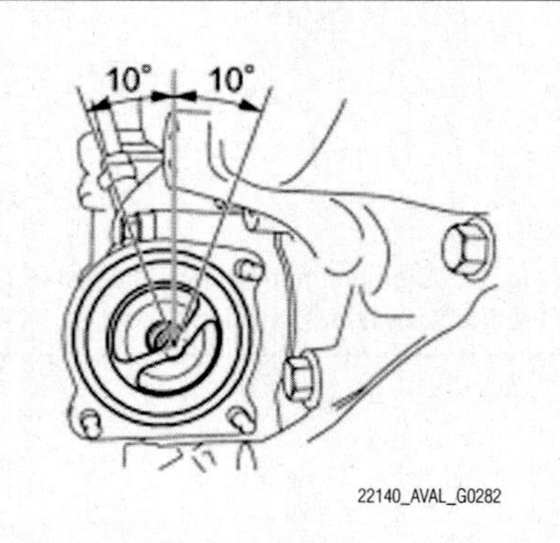

Fig. 25 Radiator jiggle valve

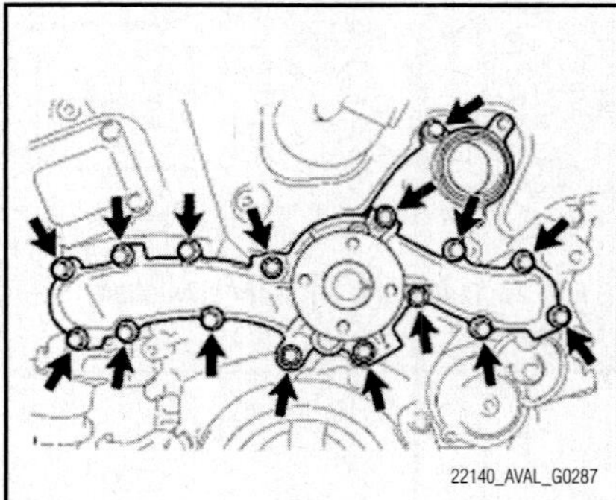

Fig. 26 Remove the 16 bolts, water pump assembly and water pump gasket

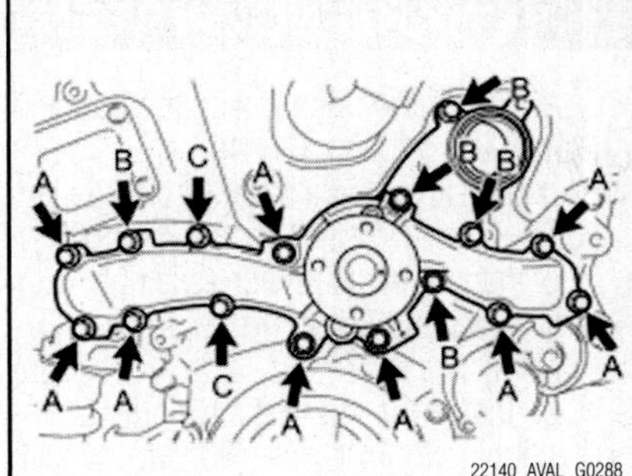

Fig. 27 Water pump assembly bolt position

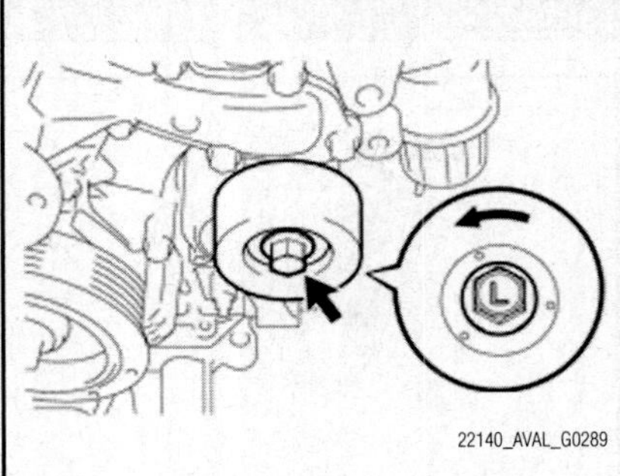

Fig. 28 Install the idler pulley with the bolt

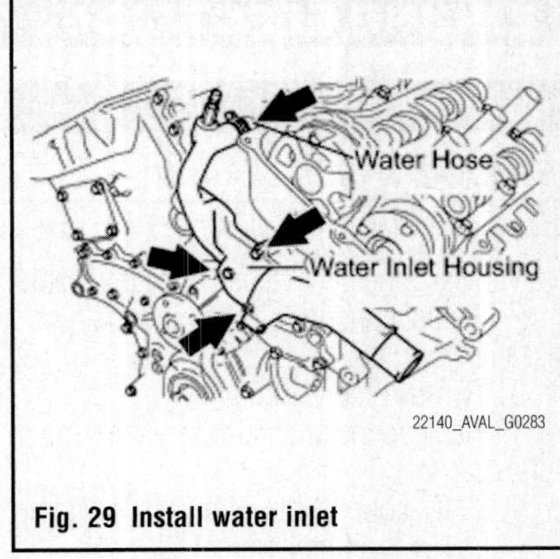

Fig. 29 Install water inlet

4. Remove the V-bank cover sub-assembly.
5. Remove the engine moving control rod.
6. Remove the engine mounting control bracket.
7. Remove the left front No. 1 engine mounting bracket.
8. Remove the fan and alternator V-belt.
9. Separate the radiator hose outlet.
10. Remove water inlet housing, as follows:
    a. Separate the water hose.
    b. Remove the 2 bolts, nut and water inlet housing.
    c. Remove the water inlet housing gasket No. 1 and water outlet pipe O-ring.
11. Remove crankshaft pulley.
12. Using SST (SST: 09960-10010), hold the water pump pulley.
13. Remove the 4 bolts and water pump pulley.
14. Remove the No. 2 idler pulley sub-assembly, as follows:
    a. Remove the bolt, No. 2 idler pulley cover plate and No. 2 idler pulley sub-assembly.
    b. Remove the bolt and idler pulley.

➡Be careful when loosening the bolt because it is left-hand threaded.

15. Separate the vane pump assembly.
16. Remove the water pump assembly, as follows:
    a. Remove the 16 bolts, water pump assembly and water pump gasket.

To install:

17. Install a new water pump gasket and the water pump assembly with the 16 bolts and tighten to 15 ft. lbs. (21 Nm), and 81 inch lbs. (9.1 Nm).

➡Make sure that there is no oil on the threads of the A bolts.

➡Be sure to replace the 2 C bolts with new ones or reuse them after applying adhesive (Part No. 08833-00080, three bond 1344 or equivalent).

18. Install vane pump assembly.
19. Install the idler pulley with the bolt and tighten to 32 ft. lbs. (43 Nm).

➡Be careful when tightening the bolt because it is left-hand threaded.

20. Install the No. 2 idler pulley cover plate and No. 2 idler pulley sub-assembly with the bolt tighten to 32 ft. lbs. (43 Nm).
21. Install the water pump pulley, as follows:
22. Temporarily install the water pump pulley with the 4 bolts.
    a. Using SST (SST: 09960-10010), hold the water pump pulley.
    b. Tighten the 4 bolts to 15 ft. lbs. (21 Nm).
23. Install crankshaft pulley.
24. Install water inlet housing, as follows:
    a. Install a new No. 1 water inlet housing gasket and water outlet pipe O-ring.
    b. Install the water inlet with the 2 bolts and nut and tighten to 7 ft. lbs. (10 Nm).

➡Be careful not to allow the O-ring to get caught between the parts.

    c. Install the water hose.
25. Install the radiator hose outlet.
26. Install the fan and alternator V-belt.
27. Install the left front No. 1 engine mounting bracket.
28. Install the engine mounting control bracket.
29. Install the engine moving control rod.
30. Install the V-bank cover sub-assembly.
31. Add engine coolant.
32. Inspect for coolant leak.
33. Install the right engine under cover.
34. Install the right front wheel.

ENGINE ELECTRICAL — CHARGING SYSTEM

ALTERNATOR

REMOVAL & INSTALLATION

See Figure 30.

1. Disconnect the negative battery cable.
2. Remove the V-bank cover sub-assembly.
3. Remove the V-ribbed belt.
4. Remove the alternator assembly, as follows:
    a. Disconnect the wire harness clamp.
    b. Remove the terminal cap.
    c. Remove the nut and disconnect the wire harness from terminal B.
    d. Disconnect the alternator connector from the alternator assembly.
    e. Remove the nut from the cylinder block.
    f. Remove the 2 bolts and alternator assembly.

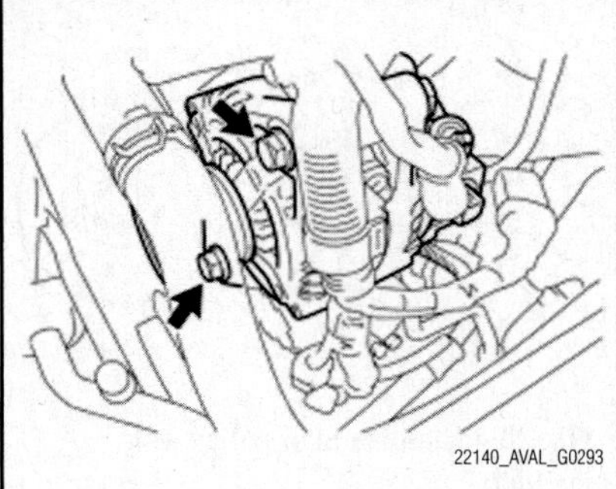

Fig. 30 Remove the 2 bolts and alternator assembly

    g. Remove the bolt and wire harness clamp stay.
    h. Remove the bolt and bracket.

To install:

5. Install the alternator assembly, as follows:
    a. Install the bracket with the bolt and tighten to 15 ft. lbs. (20 Nm).
    b. Install the wire harness clamp stay and tighten to 74 inch lbs. (8.4 Nm).
    c. Install the alternator assembly with the 2 bolts and tighten to 32 ft. lbs. (43 Nm).
    d. Install the nut to the cylinder block and tighten to 15 ft. lbs. (20 Nm).
    e. Connect the alternator connector to the alternator assembly.
    f. Install the alternator wire with the nut and tighten to 87 inch lbs. (9.8 Nm).
    g. Install the terminal cap.
    h. Connect the wire harness clamp.
6. Install the V-ribbed belt.
7. Install the V-bank cover sub-assembly.
8. Connect the negative battery cable.
9. Perform initialization.

ENGINE ELECTRICAL — IGNITION SYSTEM

FIRING ORDER

Firing order for 3.5L engine: 1–2–3–4–5–6

IGNITION COIL

REMOVAL & INSTALLATION

See Figure 31.

1. Disconnect the negative battery cable.
2. Remove the V-bank cover.
3. Remove the intake air surge tank.
4. Disconnect the 6 ignition coil connectors.
5. Remove the 6 bolts and 6 ignition coils.

To install:

6. Installation is the reverse of removal. Torque the ignition coils to 66 inch lbs. (7.5 Nm).

IGNITION TIMING

INSPECTION

1. Warm up the engine.
2. Using SST: 09843-18040, connect terminals 13 (TC) and 4 (CG) of the DLC3.

WARNING

Confirm the terminal numbers before connecting them. Connecting the wrong terminals can damage the engine.

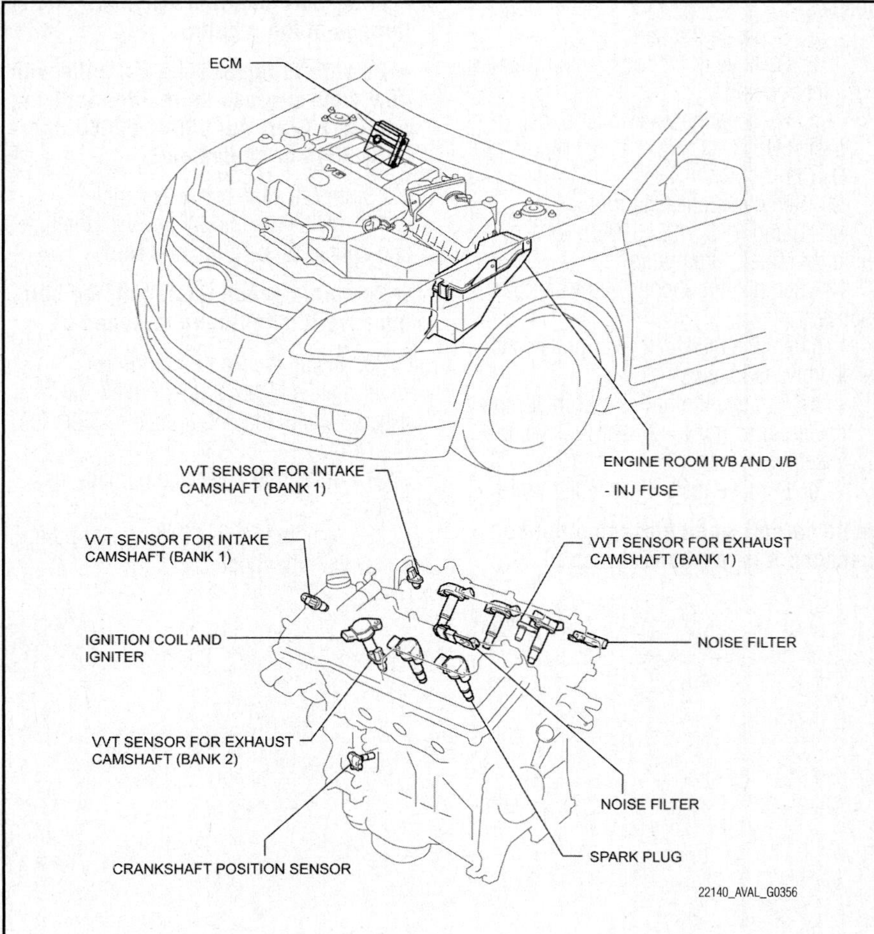

Fig. 31 Ignition coil, spark plugs and related components

➡**Turn off all electrical systems before connecting the terminals.**

➡**Perform this inspection after the cooling fan motor is turned off.**

3. Remove the v-bank cover.
4. Pull out the red lead wire harness.
5. Connect the tester terminal of the timing light to the red lead wire.

➡**Use a timing light which can detect the first signal.**

6. Check the ignition timing at idle. Standard ignition timing: 8 to 12°BTDC at idle.

➡**When checking the ignition timing, the transmission should be in the neutral position.**

➡**Run the engine at 1000 to 1300 rpm for 5 seconds, and then check that the engine rpm returns to idle speed.**

7. Disconnect terminals 13 (TC) and 4 (CG) of the DLC3.
8. Check the ignition timing at idle. Standard ignition timing: 7 to 24°BTDC at idle.
9. Confirm that the ignition timing advances immediately when the engine rpm is increased.
10. Remove the timing light from the engine.

ADJUSTMENT

All engines are equipped with a Distributorless Ignition System (DIS). No timing adjustment is possible.

SPARK PLUGS

REMOVAL & INSTALLATION

See Figure 31.

1. Disconnect the negative battery cable.
2. Remove the V-bank cover.
3. Remove the intake air surge tank.
4. Disconnect the 6 ignition coil connectors.
5. Remove the 6 bolts and 6 ignition coils.
6. Using a 16 mm (0.63 in.) plug wrench, remove the spark plugs.

To install:

7. Installation is the reverse of removal, noting the following:
   a. Torque the ignition coils to 66 inch lbs. (7.5 Nm) and the spark plugs to 13 ft. lbs (18 Nm).

ENGINE ELECTRICAL

STARTING SYSTEM

STARTER

REMOVAL & INSTALLATION

See Figure 32.

1. Disconnect the negative battery cable.
2. Remove the air cleaner assembly with hose.
3. Remove the No. 1 air cleaner inlet.
4. Remove the starter assembly, as follows:
   a. Disconnect terminal 50 of the connector from the starter assembly.
   b. Open the terminal cap, remove the nut and disconnect the wire harness from terminal 30.
   c. Remove the 2 bolts and the starter assembly.

To install:

5. Install the starter assembly with the 2 bolts and tighten to 26 ft. lbs. (37 Nm).
6. Connect the wire harness to terminal 30 and install the nut and tighten to 87 inch lbs. (9.8 Nm).
7. Cover the nut with the cap.
8. Connect terminal 50 to the starter assembly.
9. Install the No. 1 air cleaner inlet.
10. Install the air cleaner assembly with hose.
11. Connect the negative battery cable.
12. Perform initialization.

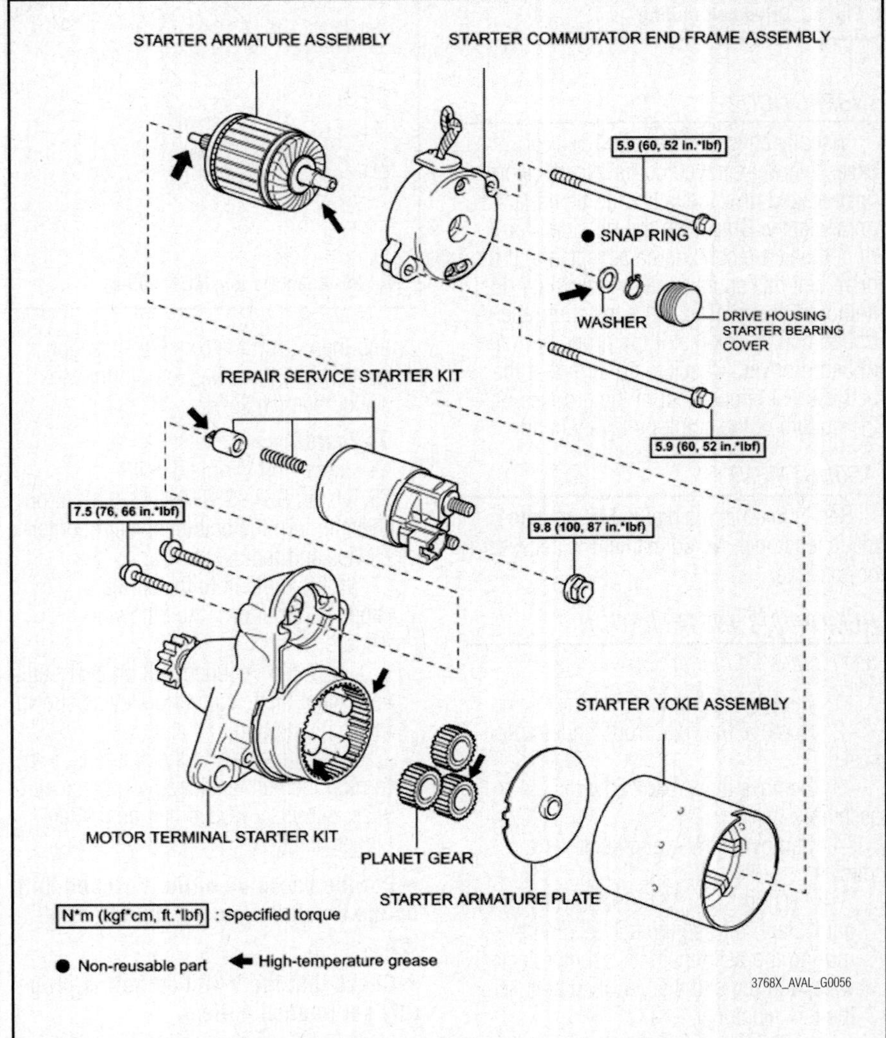

Fig. 32 Starter and related component locations

ENGINE MECHANICAL

ACCESSORY DRIVE BELTS

ACCESSORY BELT ROUTING

See Figure 33.

Refer to the accompanying illustration for Drive Belt routing.

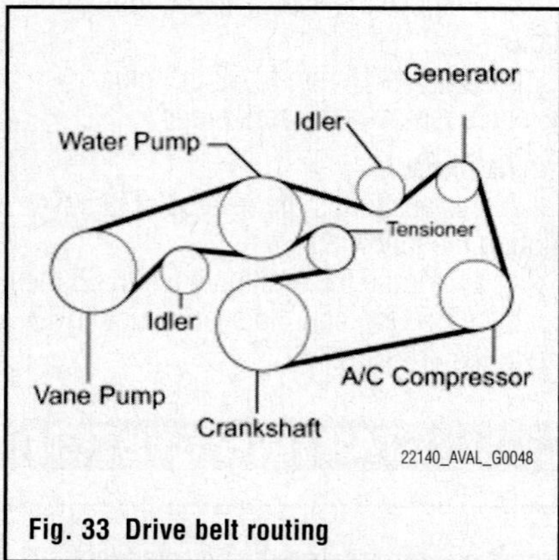

Fig. 33 Drive belt routing

INSPECTION

Visually check the V-ribbed belt for excessive wear, frayed cords, etc. All worn or damaged drive belts should be replaced immediately. Cracks on the rib side of a V-ribbed belt are considered acceptable, If the drive belt has chunks missing from its ribs, it should be replaced. After installing the V-ribbed belt, check that it fits properly in the ribbed grooves. Check to confirm that the belt has not slipped out of the grooves on the bottom of the crank pulley by hand.

ADJUSTMENT

Belt tension is maintained by an automatic tensioner. No adjustment is necessary or possible.

REMOVAL & INSTALLATION

See Figures 34 through 36.

1. Remove the right front wheel.
2. Remove the right front fender apron seal.
3. Remove the V-bank cover sub-assembly.
4. Remove the V-ribbed belt, as follows:
   a. Using SST (SST: 09249-63010) or equivalent, release the belt tension by turning the belt tensioner counterclockwise, and remove the V-ribbed belt from the belt tensioner.
   b. While turning the belt tensioner counterclockwise, align with its holes,

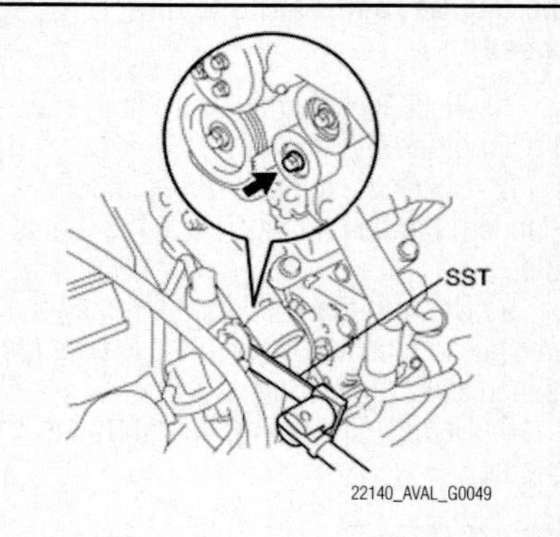

Fig. 34 Release the belt tension by turning the belt tensioner counterclockwise

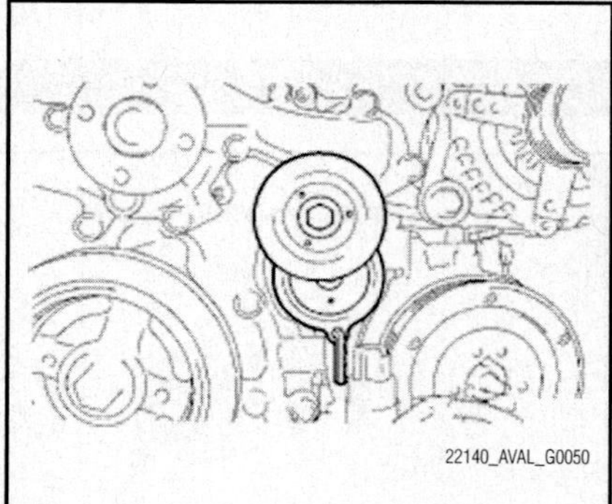

Fig. 35 Align the tensioner holes

and then insert the 5 mm bi-hexagon wrench into the holes to hold the V-ribbed belt tensioner.

To install:

5. Install the V-ribbed belt.
6. Using SST (SST: 09249-63010) or equivalent, turn the belt tensioner counterclockwise and remove the bar.
7. If it is difficult to install the V-ribbed belt, perform the following procedure:
   a. Put the V-ribbed belt on every pulley except the tensioner pulley as shown in the illustration.
   b. While releasing the belt tension by turning the belt tensioner counterclockwise, put the V-ribbed belt on the tensioner pulley.

➡Put the backside of the V-ribbed belt on the tensioner pulley and idler pulley.

➡Check that the V-ribbed belt is properly set to each pulley.

8. Install the V-bank cover sub-assembly.

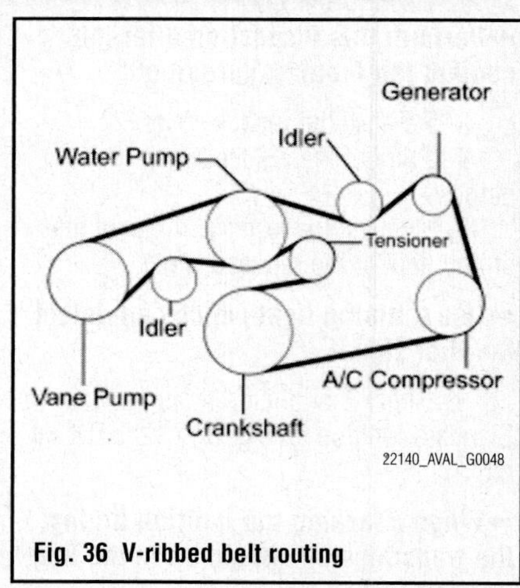

Fig. 36 V-ribbed belt routing

9. Install the right front fender apron seal.
10. Install the right front wheel and tighten the lug nuts to 76 ft. lbs. (103 Nm).

CAMSHAFT AND VALVE LIFTERS

INSPECTION

See Figures 37 through 42.

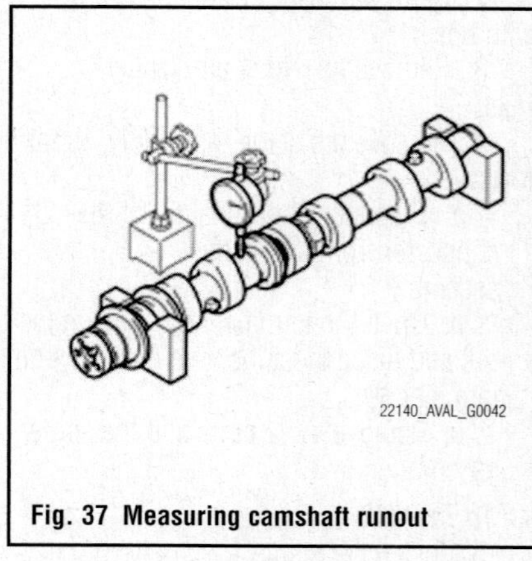

Fig. 37 Measuring camshaft runout

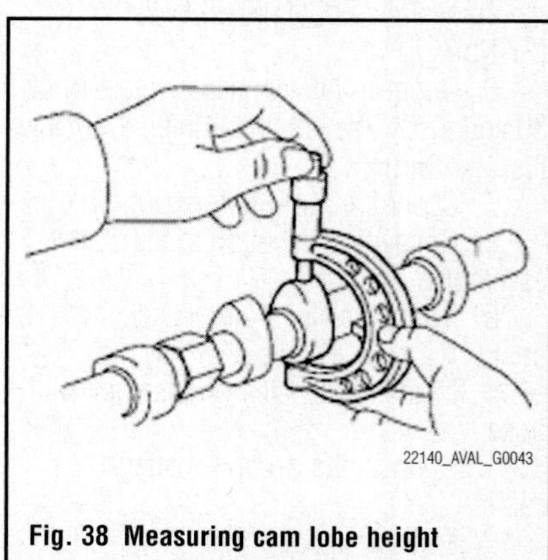

Fig. 38 Measuring cam lobe height

Item	Specification
Intake	44.316 to 44.416 mm (1.7447 to 1.7487 in.)
Exhaust	44.262 to 44.362 mm (1.7426 to 1.7465 in.)

22140_AVAL_G0044

Fig. 39 Standard cam lobe height

Item	Specification
Intake	44.166 mm (1.7388 in.)
Exhaust	44.112 mm (1.7367 in.)

22140_AVAL_G0045

Fig. 40 Maximum cam lobe height

1. To inspect the camshaft, place the camshaft on V-blocks.
   a. Using a dial indicator, measure the circle runout at the center journal.
   b. Maximum runout: 0.0016 inches (0.04 mm). If the runout is greater than the maximum, replace the camshaft.

➡Check the oil clearance after replacing the camshaft.

2. Using a micrometer, measure the cam lobe height.
3. Using a micrometer, measure the journal diameter.
   a. If the journal diameter is not as specified, check the oil clearance.

4. The valve lash adjuster assembly can be inspected, as follows:

➡Keep the lash adjuster free of dirt and foreign objects. Only use clean engine oil.

   a. Place the lash adjuster into a container filled with engine oil.
   b. Insert the SST's tip (SST: 09276-75010) into the lash adjuster's plunger and use the tip to press down on the check ball inside the plunger.
   c. Squeeze the SST and lash adjuster together to move the plunger up and down 5 to 6 times.
   d. Check the movement of the plunger and bleed the air. OK: Plunger moves up and down.

➡When bleeding air from the high-pressure chamber, make sure that the tip of the SST is actually pressing the check ball as shown in the illustration. If the check ball is not pressed, air will not bleed.

   e. After bleeding the air, remove the SST. Then, try to press the plunger quickly and firmly with a finger. OK: Plunger is very difficult to move. If the result is not as specified, replace the lash adjuster.
   f. Install the lash adjusters. Install the lash adjuster to the same place where it was removed from.

Item	Specification
No. 1 journal	35.946 to 35.960 mm (1.4152 to 1.4157 in.)
Other journal	25.959 to 25. 975 mm (1.0220 to 1.0226 in.)

22140_AVAL_G0046

Fig. 41 Standard journal diameter

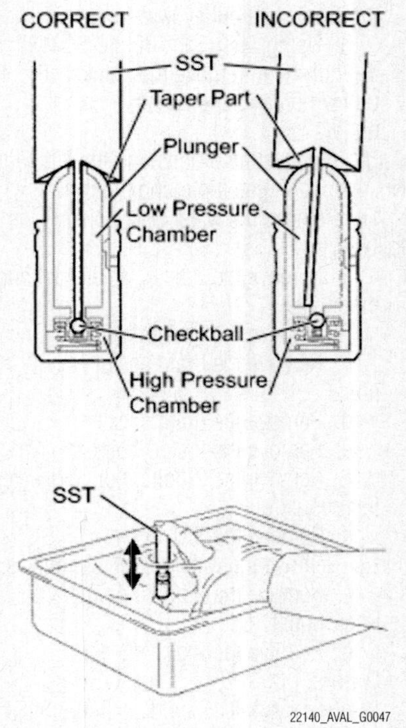

CORRECT
INCORRECT
SST
Taper Part
Plunger
Low Pressure Chamber
Checkball
High Pressure Chamber
SST

22140_AVAL_G0047

Fig. 42 Valve lash adjuster

REMOVAL & INSTALLATION

See Figures 43 through 68.

1. Before servicing the vehicle, refer to the Precautions Section.
2. Remove the engine assembly.
3. Install on engine stand.
4. Remove the oil filler cap and gasket.
5. Remove the spark plugs and ignition coil assembly.
6. Remove the drain plug and gasket.
7. Remove the ventilation valve.
8. Remove the 4 bolts and 4 camshaft position sensors.
9. Remove the 4 bolts and 4 camshaft timing oil control valves.
10. Remove the bolt and crankshaft position sensor.
11. Remove the No. 1 oil pipe.
12. Remove the oil pipe.
13. Remove the cylinder block water drain cock sub-assembly, as follows:
    a. Remove the water drain cocks from the cylinder block.
    b. Remove the water drain cock plugs from the water drain cocks.
14. Remove the oil filter.
15. Remove the crankshaft pulley, as follows:

a. Using SST (SST: 09213-70011, SST: 09330-00021) or equivalent, loosen the crankshaft pulley bolt.

b. Using SST (SST: 09950-50013) or equivalent, remove the crankshaft pulley bolt and crankshaft pulley.

16. Remove the 6 bolts and the left hand No. 1 front engine mounting bracket.

17. Remove the water inlet housing, as follows:

a. Remove the 2 nuts, water inlet and thermostat.

b. Remove the gasket.

c. Remove the drain cock plug.

d. Remove the drain cock.

e. Remove the 2 stud bolts.

f. Remove the 2 bolts, nut, and water inlet housing.

g. Remove the 2 O-rings.

18. Remove the water outlet, as follows:

a. Remove the 2 bolts, 4 nuts and water outlet.

b. Remove the 2 gaskets and O-ring.

19. Remove the 12 bolts, cylinder head cover sub-assembly (for Bank 1), and gasket.

20. Remove the 12 bolts, cylinder head cover sub-assembly (for Bank 2), and gasket.

21. Remove the No. 2 oil pan sub-assembly.

22. Remove the oil strainer sub-assembly.

23. Remove the oil pan sub-assembly.

24. Remove the No. 1 oil pan baffle plate.

25. Remove the engine rear oil seal, as follows:

a. Remove the 6 bolts.

b. Using a screwdriver with the tip taped, pry out the oil seal retainer. Be careful not to damage the oil seal retainer.

Fig. 43 Crankshaft timing gear

26. Remove the water pump assembly.

27. Remove the timing chain cover sub-assembly.

28. Set the No. 1 cylinder to TDC/ compression.

29. Remove the No. 1 chain tensioner assembly.

30. Remove the chain tensioner slipper.

31. Remove the chain sub-assembly.

32. Remove the idle sprocket assembly.

33. Remove the No. 1 chain vibration damper.

34. Remove the No. 2 chain vibration damper.

35. Remove the crankshaft timing gear, as follows:

a. Remove the pulley set bolt.

b. Remove the crankshaft timing gear from the crankshaft.

c. Remove the 2 pulley set keys from the crankshaft.

36. Remove the camshaft timing gears and No. 2 chain (for right-hand bank), as follows:

a. While raising the No. 2 chain tensioner, insert a pin of 0.039 inch (1.0 mm) into the hole to hold the No. 2 chain tensioner.

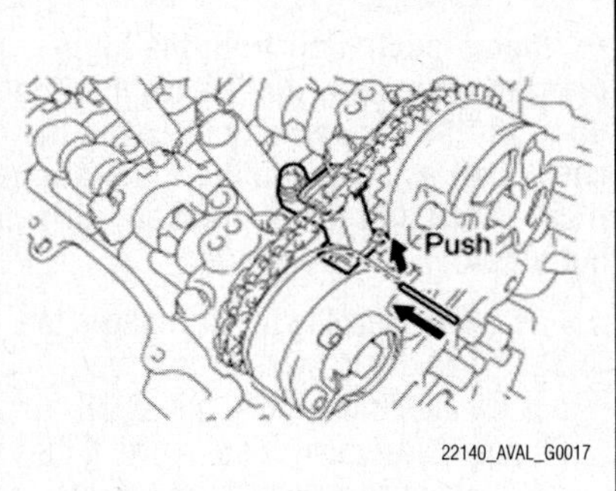

Fig. 44 Pinning No. 2 tensioner—right-hand bank

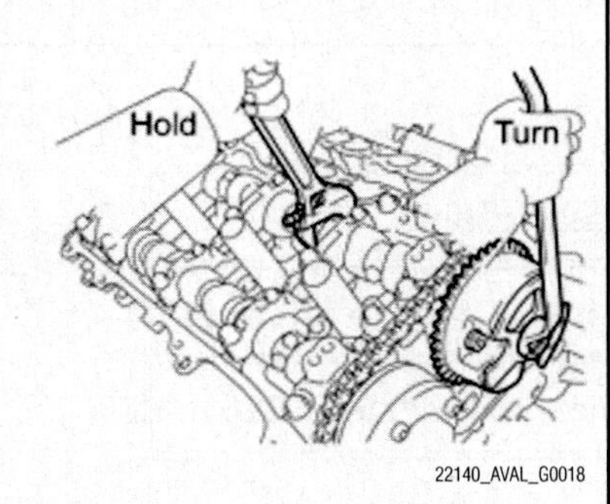

Fig. 45 Removing gear assemblies—right-hand bank

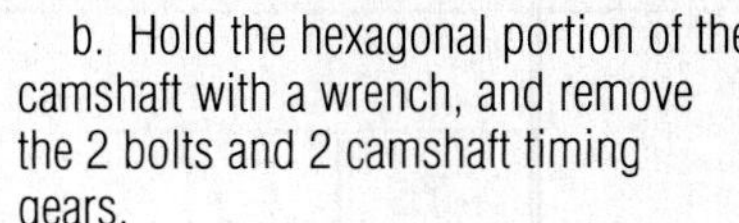

b. Hold the hexagonal portion of the camshaft with a wrench, and remove the 2 bolts and 2 camshaft timing gears.

➡Be careful not to damage the cylinder head with the wrench.

➡Do not disassemble the camshaft timing gear assemblies.

c. Remove the No. 2 chain.

37. Remove the bolt and No. 2 chain tensioner assembly.

38. Remove the camshaft bearing cap, as follows:

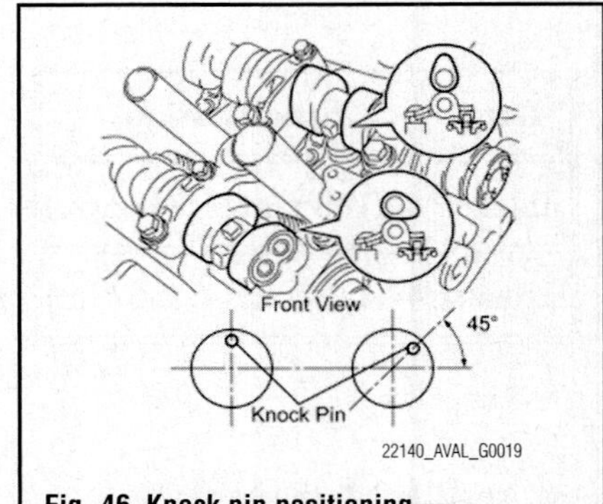

Fig. 46 Knock pin positioning—right-hand bank

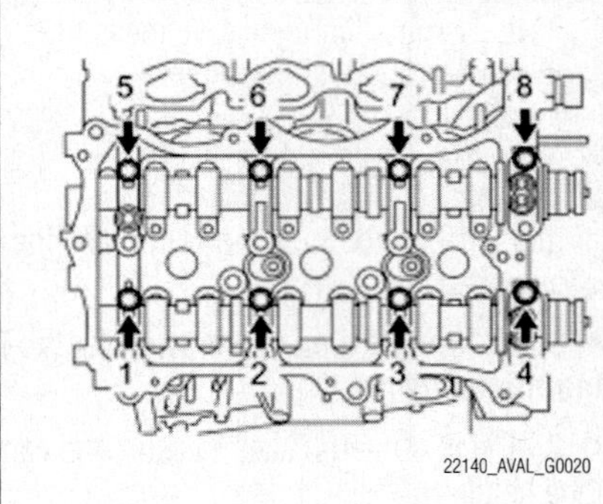

Fig. 47 Bearing cap 8 bolt removal sequence—right-hand bank

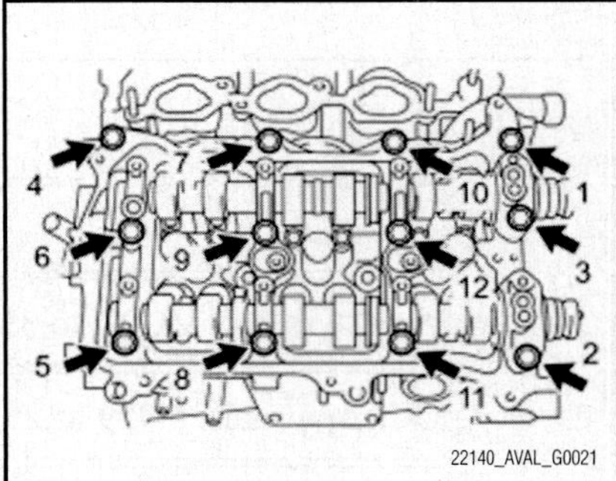

Fig. 48 Bearing cap 12 bolt removal sequence—right-hand bank

a. Remove the 3 gaskets.

b. Make sure that the knock pin of the camshaft is positioned as shown in the illustration.

c. Uniformly loosen and remove the 8 bearing cap bolts in the sequence shown.

d. Uniformly loosen and remove the 12 bearing cap bolts in the sequence shown. Loosen the bolts while keeping the camshaft level

e. Remove the 5 camshaft bearing caps.

39. Remove the No. 1 camshaft.

40. Remove the No. 2 camshaft.

41. Remove the right hand camshaft housing sub-assembly by prying between the cylinder head and the camshaft housing with a screwdriver with the tip taped.

➡Be careful not to damage the contact surfaces of the cylinder head and the camshaft housing.

42. Remove the camshaft timing gears and No. 2 chain (for left-hand bank), as follows:

a. While pushing down the No. 3 chain tensioner, insert a pin of 1.0 mm (0.039 in.) into the hole to hold the No. 3 chain tensioner.

Fig. 49 Pinning No. 3 tensioner—left-hand bank

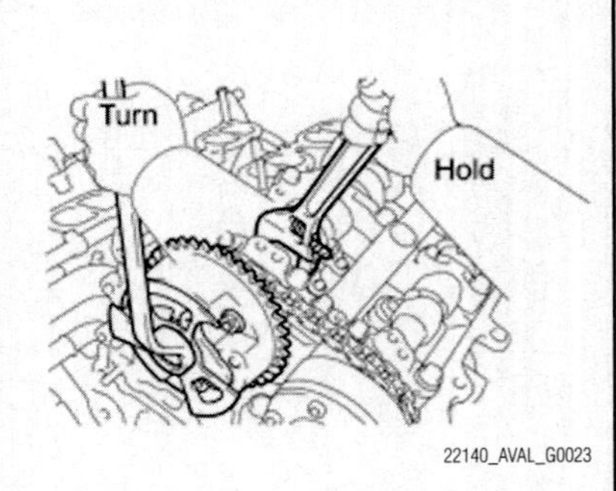

Fig. 50 Removing gear assemblies—left-hand bank

b. Hold the hexagonal portion of the camshaft with a wrench, and remove the 2 bolts and 2 camshaft timing gears.

➡Be careful not to damage the cylinder head with the wrench.

➡Do not disassemble the camshaft timing gear assemblies.

c. Remove the No. 2 chain.

43. Remove the bolt and No. 3 chain tensioner.

44. Remove the camshaft bearing cap, as follows:

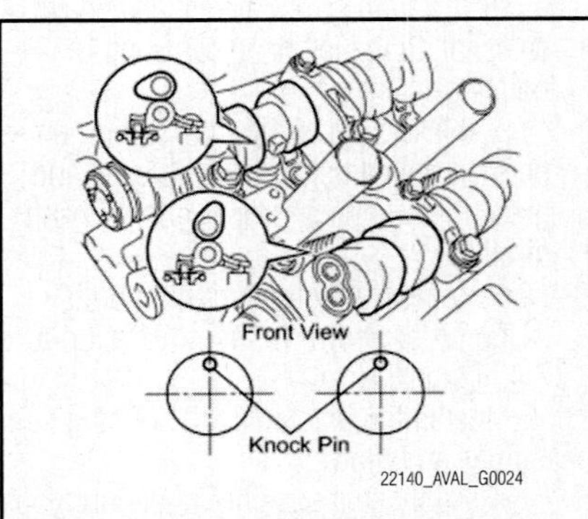

Fig. 51 Knock pin positioning—left-hand bank

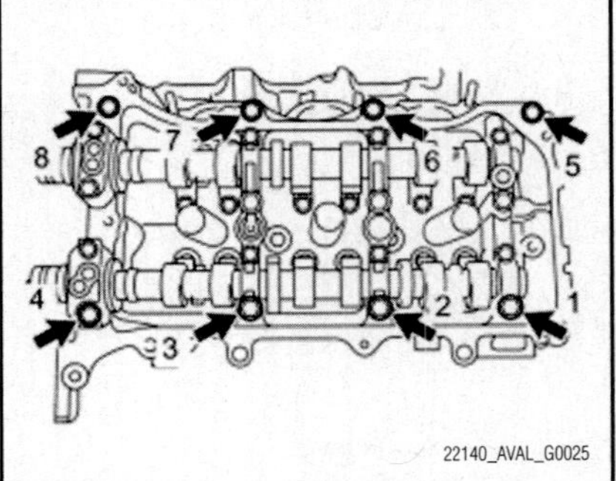

Fig. 52 Bearing cap 8 bolt removal sequence—left-hand bank

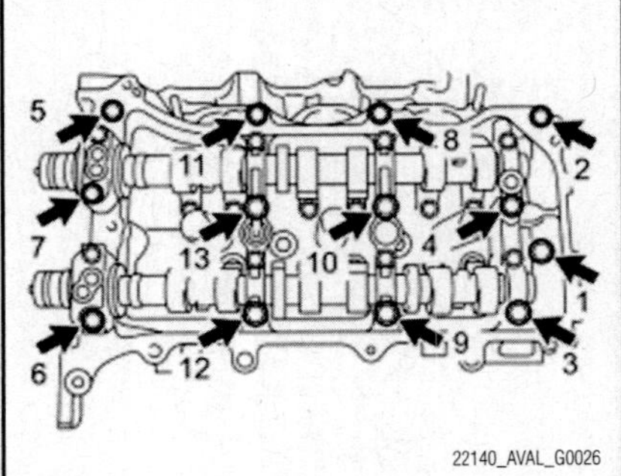

Fig. 53 Bearing cap 12 bolt removal sequence—left-hand bank

a. Remove the 3 gaskets.

b. Make sure that the knock pin of the camshaft is positioned as shown in the illustration.

c. Uniformly loosen and remove the 8 bearing cap bolts in the sequence shown in the illustration.

d. Uniformly loosen and remove the 13 bearing cap bolts in the sequence shown in the illustration. Loosen the bolts while keeping the camshaft level

e. Remove the 5 camshaft bearing caps.

45. Remove the No. 4 camshaft.

46. Remove the No. 3 camshaft.

47. Remove the left hand camshaft housing sub-assembly by prying between the cylinder head and the camshaft housing with a screwdriver with the tip taped.

➡Be careful not to damage the contact surfaces of the cylinder head and the camshaft housing.

48. Remove the No. 1 valve rocker arm sub-assembly, as follows:

a. Remove the 24 valve rocker arms.

➡Arrange the removed parts in the correct order.

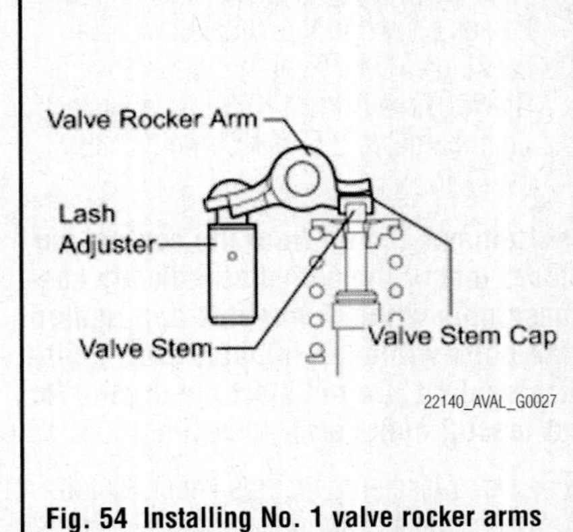

Fig. 54 Installing No. 1 valve rocker arms

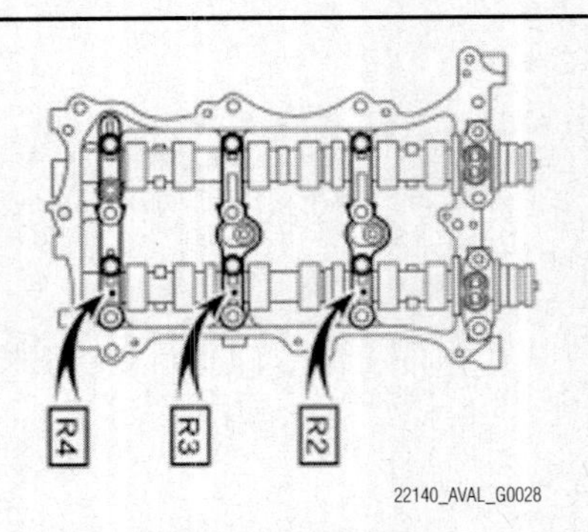

Fig. 55 Camshaft bearing caps placement—right-hand

49. Remove the valve lash adjuster assembly, as follows:

a. Remove the 24 valve lash adjusters from the cylinder head.

➡Arrange the removed parts in the correct order.

To install:

50. Install the valve lash adjuster assembly.

51. Install the No. 1 valve rocker arm sub-assembly, as follows:

a. Apply engine oil to the lash adjuster tips and valve stem cap ends.

b. Make sure that the valve rocker arms are installed as shown in the illustration.

52. Install the right-hand camshaft bearing cap, as follows:

a. Apply engine oil to the camshaft journals, camshaft housing and bearing caps.

b. Install the No. 1camshaft and No. 2 camshaft to the right camshaft housing.

c. Make sure of the marks and numbers on the camshaft bearing caps and place them in each proper position and direction.

d. Temporarily tighten the 8 bearing cap bolts to 7 ft. lbs. (10 Nm) in the order shown in the illustration.

53. Install the right hand camshaft housing sub-assembly, as follows:

a. Apply seal packing in a continuous line as shown in the illustration. Seal packing: Toyota Genuine Seal Packing Black, Three Bond 1207B or equivalent. Seal diameter: 3.5 to 4.5 mm (0.138 to 0.177 in.).

➡Remove any oil from the contact surface. Install the camshaft housing sub-assembly within 3 minutes and tighten the bolts within 15 minutes after applying sealant. Do not start the engine for at least 2 hours after installing.

b. Make sure that the knock pins of the camshafts are positioned as shown. Install the right-hand camshaft housing and tighten the 12 bolts in the order shown in the illustration to 21 ft. lbs. (28 Nm).

c. Tighten the 8 bolts to 12 ft. lbs. (16 Nm) in the order shown in the illustration.

➡Thoroughly wipe clean any seal packing.

d. Install 3 new gaskets.

54. Install the camshaft bearing cap, as follows:

a. Apply engine oil to the camshaft journals, camshaft housing and bearing caps.

b. Install the No. 3 camshaft and No. 4 camshaft to the left hand camshaft housing.

c. Make sure of the marks and numbers on the camshaft bearing caps and place them in each proper position and direction.

d. Temporarily tighten the 8 bolts in the order shown in the illustration to 7 ft. lbs. (10 Nm).

55. Install the left camshaft housing sub-assembly, as follows:

a. Apply seal packing in a continuous line as shown in the illustration. Seal packing: Toyota Genuine Seal Packing Black, Three Bond 1207B or equivalent. Seal diameter: 3.5 to 4.5 mm (0.138 to 0.177 in.).

➡Remove any oil from the contact surface. Install the camshaft housing sub-assembly within 3 minutes and tighten the bolts within 15 minutes after applying sealant. Do not start the engine for at least 2 hours after installing.

b. Make sure that the knock pins of the camshafts are positioned as shown. Install the left hand camshaft housing and tighten the 13 bolts in the order shown in the illustration to 21 ft. lbs. (28 Nm).

c. Tighten the 8 bolts to 12 ft. lbs. (16 Nm) in the order shown in the illustration.

➡Thoroughly wipe clean any sealant.

d. Install 3 new gaskets.

56. Install the No. 2 chain tensioner assembly with the bolt and tighten to 15 ft. lbs. (21 Nm).

57. While pushing in the tensioner, insert a pin of 0.039 inch (1.0 mm) diameter into the hole to hold it.

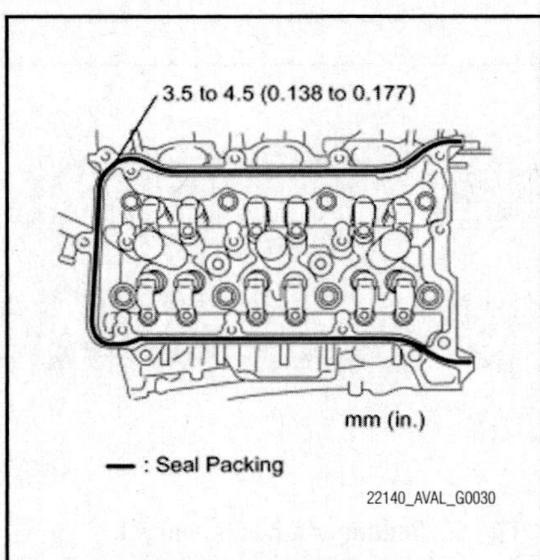
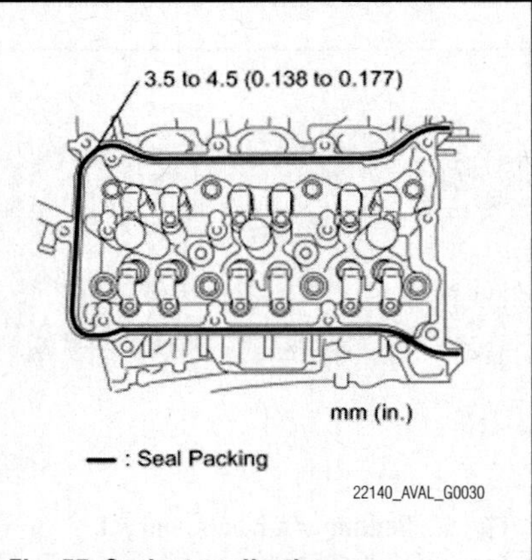

Fig. 57 Sealant application

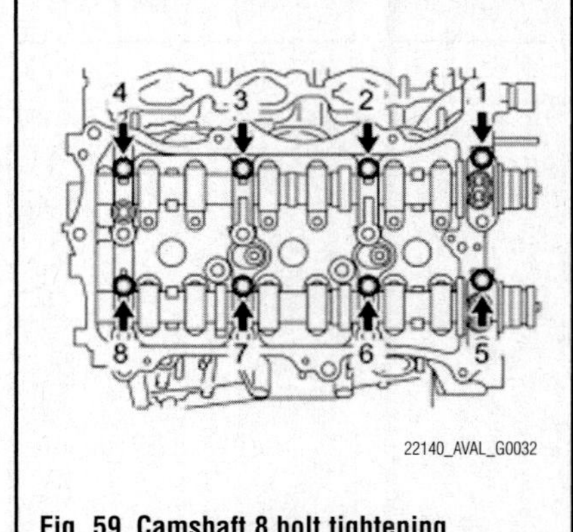

Fig. 59 Camshaft 8 bolt tightening sequence—right-hand

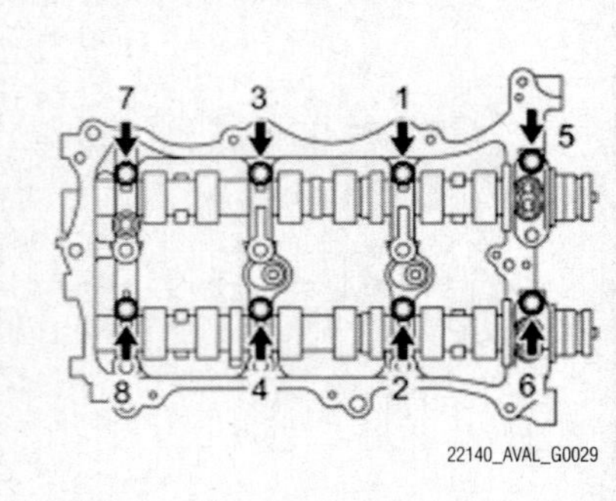

Fig. 56 Camshafts bearing cap tightening sequence—right-hand

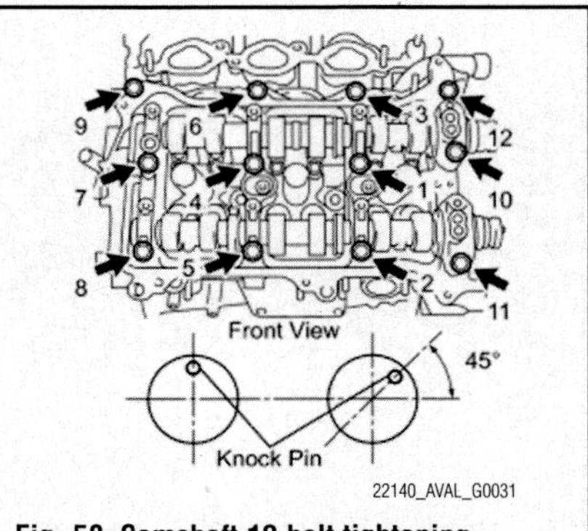

Fig. 58 Camshaft 12 bolt tightening sequence—right-hand

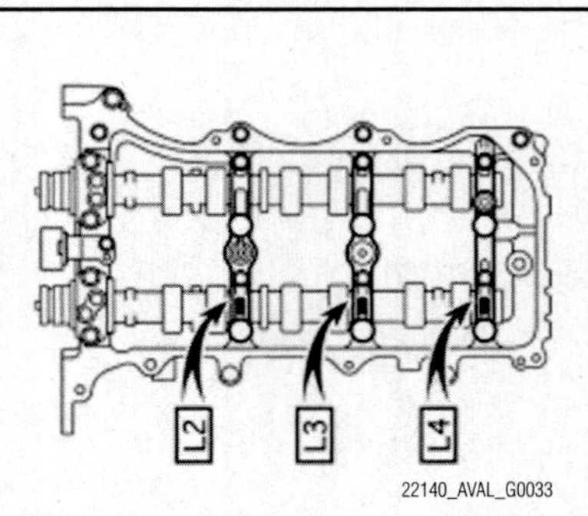

Fig. 60 Camshaft bearing cap positioning—left-hand

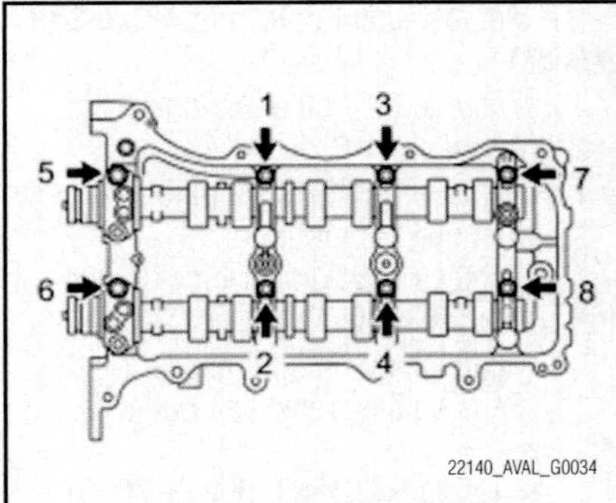

Fig. 61 Camshaft 8 bolt tightening sequence—left-hand

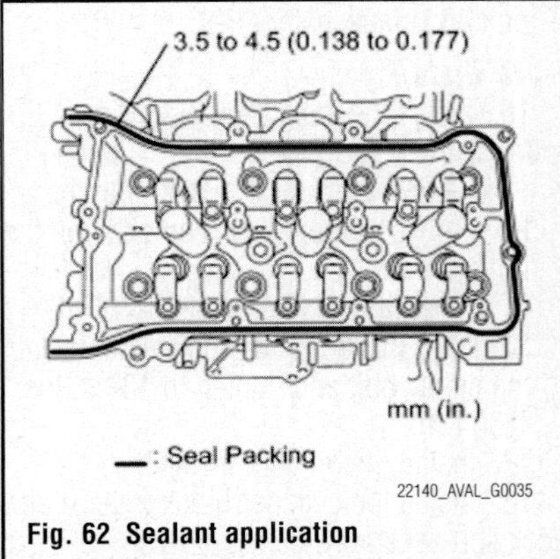

Fig. 62 Sealant application

58. Install the camshaft timing gears and No. 2 chain (for Right-hand Bank).

59. Install the No. 3 chain tensioner assembly with the bolt and tighten to 15 ft. lbs. (21 Nm).

60. While pushing in the tensioner, insert a pin of 0.039 inch (1.0 mm) diameter into the hole to hold it.

61. Install the camshaft timing gears and No. 2 chain (for Left-hand Bank).

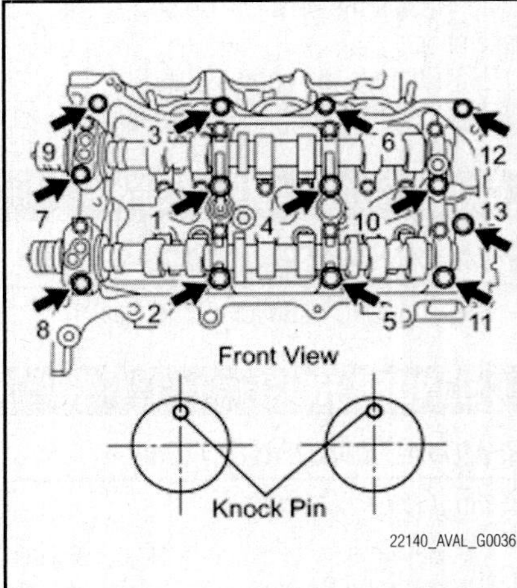

Fig. 63 Camshaft housing 13 bolt tightening sequence—left-hand

62. Install the No 1 chain vibration damper with the 2 bolts and tighten to 17 ft. lbs. (23 Nm).

63. Install the No 2 chain vibration damper.

64. Install the timing gear set keys and timing gear as shown in the illustration.

65. Install the idle sprocket assembly.

66. Install the chain sub-assembly.

67. Install the chain tensioner slipper.

68. Install the No. 1 chain tensioner assembly.

69. Install the water pump assembly.

70. Install the timing chain cover sub-assembly.

71. Install the water inlet housing.

72. Install the No. 1 left front engine mounting bracket, as follows:

a. Install the No. 1 left front engine mounting bracket with the 6 bolts and tighten to 40 ft. lbs. (54 Nm).

➡Install the water inlet and mounting bracket within 15 minutes after installing the chain cover. Do not start the engine for at least 2 hours after installation.

73. Install the No. 1 oil pan baffle plate with the 7 bolts and tighten to 7 ft. lbs. (10 Nm).

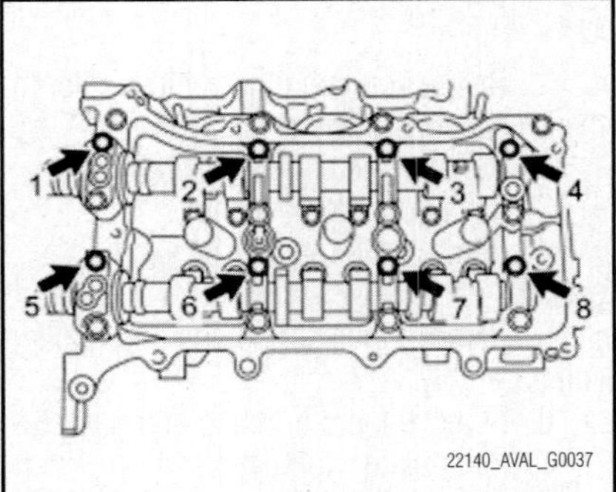

Fig. 64 Camshaft housing 8 bolt tightening sequence—left-hand

Fig. 65 Crankshaft timing gear installation

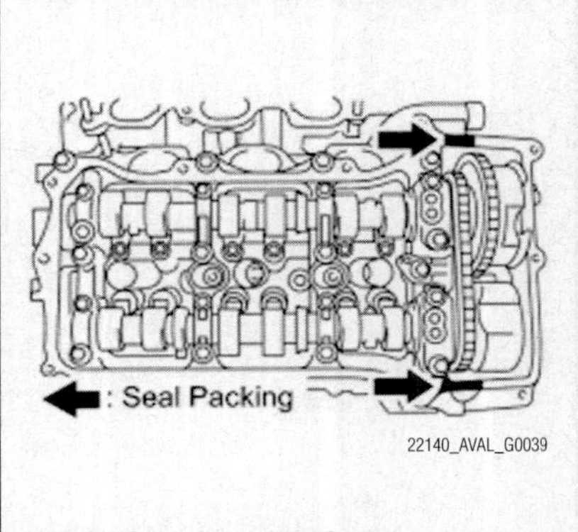

Fig. 66 Sealant application

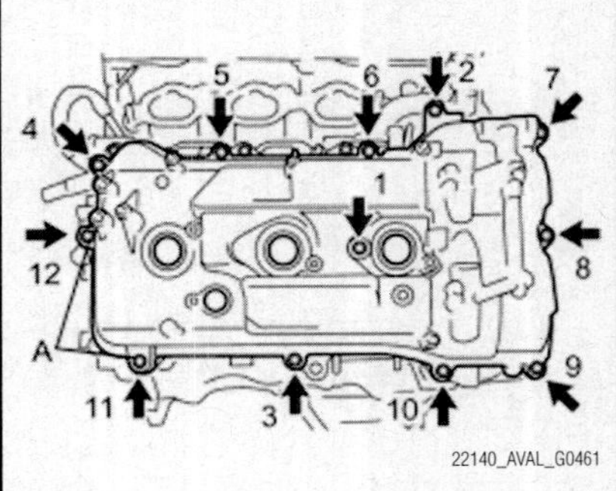

Fig. 67 Cylinder head cover bolt tightening sequence

74. Install the oil pan sub-assembly.

75. Install the oil strainer sub-assembly.

76. Install the No. 2 oil pan sub-assembly.

77. Install a new gasket and oil pan drain plug and tighten to 30 ft. lbs. (40 Nm).

78. Install the cylinder head cover sub-assembly, as follows:

a. Apply seal packing (Toyota Genuine Seal Packing Black, Three Bond 1207B or equivalent) as shown in the illustration.

➡Remove any oil from the contact surface. Install the crankcase within 3 minutes after applying seal packing. Do not start the engine for at least 2 hours after installation.

b. Install the gasket to the head cover.

c. Install the head cover with the 12 bolts. Tighten bolt A to 15 ft. lbs. (21 Nm), and other bolts to 7 ft. lbs. (10 Nm). Be certain to tighten bolt 1.

79. Install the left-hand cylinder head cover sub-assembly, as follows:

a. Apply seal packing (Toyota Genuine Seal Packing Black, Three Bond 1207B or equivalent) as shown.

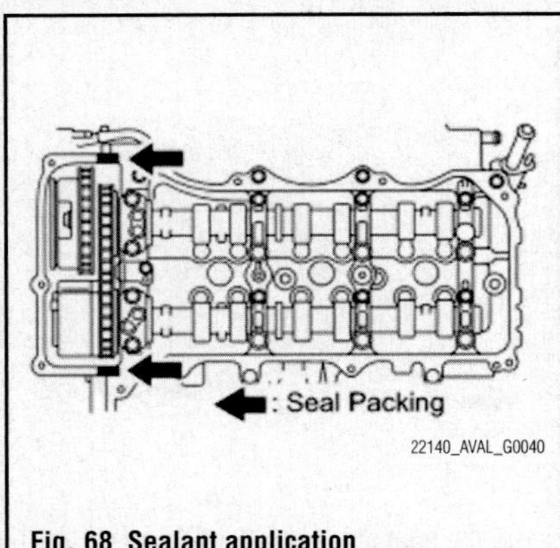

Fig. 68 Sealant application

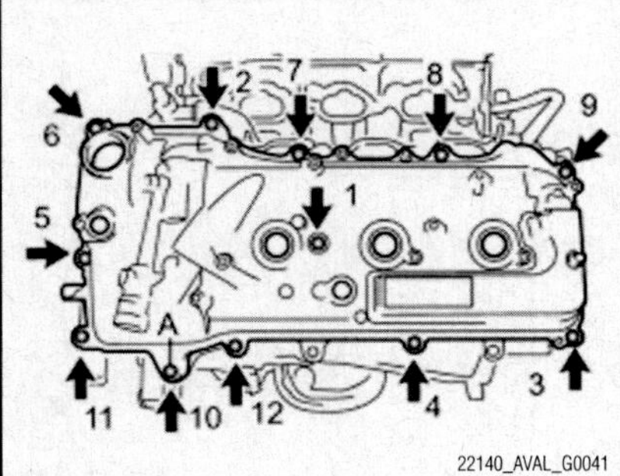

Fig. 69 Left cylinder head cover bolt tightening sequence

➡Remove any oil from the contact surface. Install the crankcase within 3 minutes after applying seal packing. Do not start the engine for at least 2 hours after installation.

b. Install the gasket to the head cover.

c. Install the head cover with the 14 bolts. Tighten bolt A to 15 ft. lbs. (21 Nm), and other bolts to 7 ft. lbs. (10 Nm). Be certain to tighten bolts 1 and 10.

80. Install water outlet.
81. Install the crankshaft pulley.
82. Install the oil filter element.
83. Install the cylinder block water drain cock sub-assembly, as follows:

a. Apply adhesive around the drain cocks. Adhesive: Toyota Genuine Adhesive 1324, Three Bond 1324 or Equivalent.

b. Install the water drain cocks and tighten to 18 ft. lbs. (25 Nm). Do not rotate the drain cocks more than 1 revolution (360°) after tightening the drain cocks with the specified torque. Do not loosen after setting correctly.

c. Install the water drain cock plug to the water drain cocks and tighten to 9 ft. lbs. (13 Nm).

84. Install the No. 1 oil pipe.
85. Install the oil pipe.
86. Install the crankshaft position sensor with the bolt and tighten to 7 ft. lbs. (10 Nm).
87. Install the 4 camshaft timing oil control valves with the 4 bolts and tighten to 7 ft. lbs. (10 Nm).
88. Install the 4 camshaft position sensors with the 4 bolts and tighten to 7 ft. lbs. (10 Nm).
89. Install the ventilation valve sub-assembly, as follows:

a. Apply adhesive (Toyota Genuine Adhesive 1324, Three Bond 1324 or equivalent) around the ventilation valve.

b. Install the ventilation valve and tighten to 20 ft. lbs. (27 Nm).

90. Install the 6 spark plugs and the ignition coil assembly.
91. Install the oil filler cap sub-assembly.
92. Remove the engine stand.
93. Install the engine assembly.

CRANKSHAFT DAMPER (BALANCER)

REMOVAL & INSTALLATION

See Figure 70.

1. Before servicing the vehicle, refer to the Precautions Section.
2. Remove the engine assembly.
3. Install on engine stand.
4. Remove the oil filler cap and gasket.
5. Remove the spark plugs and ignition coil assembly.
6. Remove the drain plug and gasket.
7. Remove the ventilation valve.
8. Remove the 4 bolts and 4 camshaft position sensors.

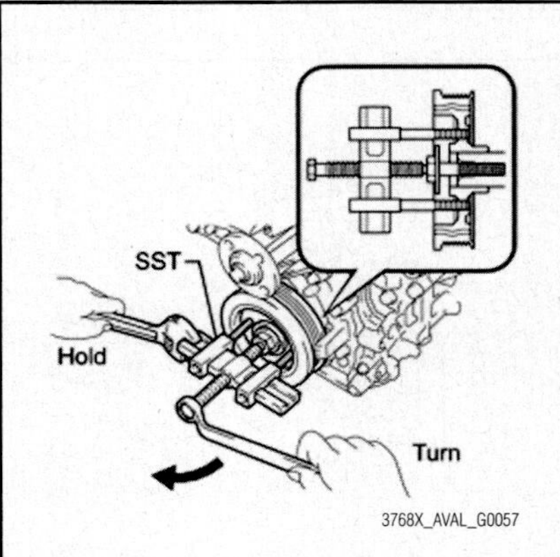

Fig. 70 Removing the crankshaft pulley

9. Remove the 4 bolts and 4 camshaft timing oil control valves.
10. Remove the bolt and crankshaft position sensor.
11. Remove the No. 1 oil pipe.
12. Remove the oil pipe.
13. Remove the cylinder block water drain cock sub-assembly.
14. Remove the oil filter element.
15. Remove the crankshaft pulley, as follows:

a. Using SST (SST: 09213-70011, SST: 09330-00021) or equivalent, loosen the crankshaft pulley bolt.

b. Using SST (SST: 09950-50013) or equivalent, remove the crankshaft pulley bolt and crankshaft pulley.

To install:

16. Install the crankshaft pulley, as follows:

a. Align the pulley set key with the key groove of the pulley, and slide on the pulley.

b. Using SST (SST: 09213-70011, SST: 09330-00021) or equivalent, install the pulley bolt and tighten to 184 ft. lbs. (250 Nm).

17. Install the oil filter element.
18. Install the cylinder block water drain cock sub-assembly.
19. Install the No. 1 oil pipe.
20. Install the oil pipe.
21. Install the crankshaft position sensor.
22. Install the camshaft timing oil control valve assembly.
23. Install the camshaft position sensor.
24. Install the ventilation valve sub-assembly.
25. Install the spark plugs and the ignition coil assembly.
26. Install the oil filler cap sub-assembly.
27. Remove the engine stand.
28. Install the engine assembly.

CRANKSHAFT FRONT SEAL

REMOVAL & INSTALLATION

See Timing Chain Cover and Seal.

CYLINDER HEAD

REMOVAL & INSTALLATION

See Figures 71 through 79.

1. Before servicing the vehicle, refer to the Precautions Section.
2. Remove the engine assembly with transaxle.

3. Secure engine.
4. Remove the oil filler cap sub-assembly.
5. Remove the spark plugs and ignition coil assembly.
6. Remove the oil pan drain plug and gasket.
7. Remove the ventilation valve sub-assembly.
8. Remove the camshaft position sensor.
9. Remove the camshaft timing oil control valve assembly.
10. Remove crankshaft position sensor.
11. Remove the No. 1 oil pipe.
12. Remove the oil pipe.
13. Remove the cylinder block water drain cock sub-assembly.
14. Remove the oil filter.
15. Remove the crankshaft pulley.
16. Remove the left hand No. 1 front engine mounting bracket.
17. Remove the water inlet housing.
18. Remove the water outlet.
19. Remove the cylinder head covers and gaskets.
20. Remove the No. 2 oil pan sub-assembly.
21. Remove the oil strainer sub-assembly.
22. Remove the oil pan sub-assembly.
23. Remove the No. 1 oil pan baffle plate.
24. Remove the engine rear oil seal.
25. Remove the water pump assembly.
26. Remove the timing chain cover.
27. Set the No. 1 cylinder to TDC/compression.
28. Remove the No. 1 chain tensioner assembly.
29. Remove the chain tensioner slipper.
30. Remove the chain sub-assembly.
31. Remove the idle sprocket assembly.
32. Remove the No. 1 and 2 chain vibration damper.
33. Remove the left hand cylinder head sub-assembly, as follows:
   a. Uniformly loosen and remove the 2 bolts in the sequence shown.
   b. Using a 10 mm bi-hexagon wrench, uniformly loosen the 8 bolts in the sequence shown. Remove the 8 cylinder head bolts and plate washers.

⁂ WARNING

Be careful not to drop washers into the cylinder head.

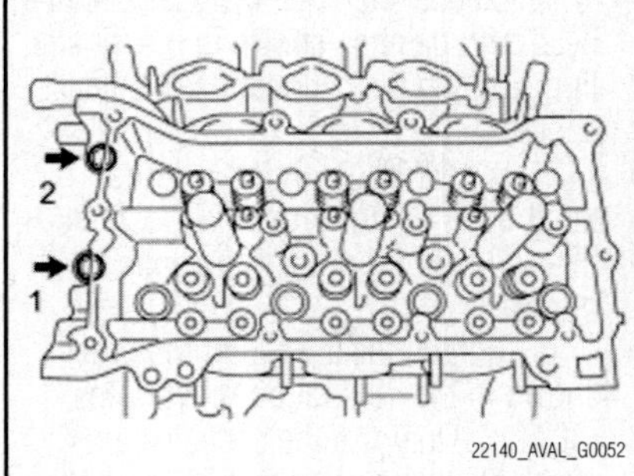

Fig. 71 Left hand cylinder head 2 bolt removal sequence

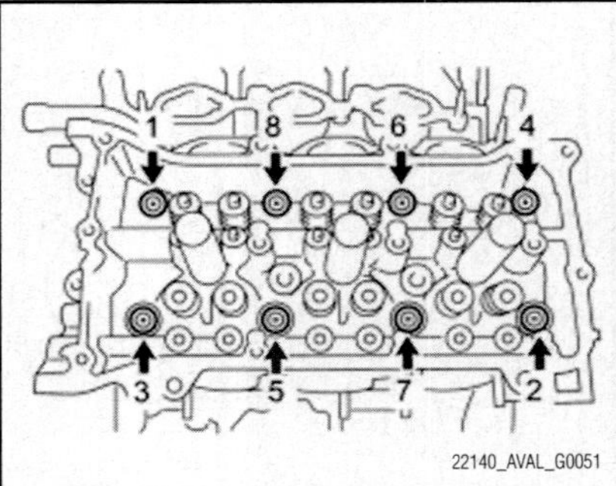

Fig. 72 Left hand cylinder head 8 bolt removal sequence

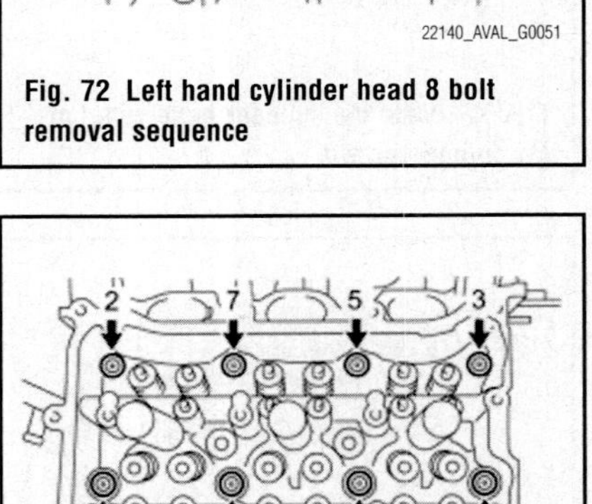

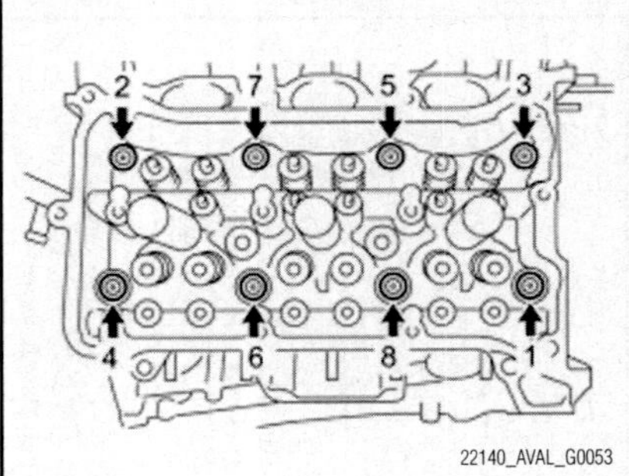

Fig. 73 Right hand cylinder head bolt removal sequence

⁂ WARNING

Cylinder head warpage or cracking could result from removing bolts in an incorrect order.

➡Be sure to keep separate the removed parts for each installation position.

   c. Remove the cylinder head and gasket.

34. Remove the right hand cylinder head sub-assembly, as follows:
   a. Using a 10 mm bi-hexagon wrench, uniformly loosen the 8 bolts in the sequence shown. Remove the 8 cylinder head bolts and plate washers.

⁂ WARNING

Be careful not to drop washers into the cylinder head.

⁂ WARNING

Cylinder head warpage or cracking could result from removing bolts in an incorrect order.

➡Be sure to keep separate the removed parts for each installation position.

   b. Remove the cylinder head and gasket.

To install:

35. Place the right hand cylinder head gasket on the cylinder block surface with the front face of the Lot No. stamp upward.

➡Be careful of the installation direction.

⁂ WARNING

Gently place the cylinder head in order not to damage the gasket with the bottom part of the head.

36. Place the cylinder head on the cylinder block.

➡Do not allow oil to adhere to the mounting surface of the cylinder head.

37. Apply a light coat of engine oil to the threads and under the heads of the cylinder head bolts.
38. The cylinder head bolts are tightened in 3 progressive steps:
   a. Step 1: Using a 10 mm bi-hexagon wrench, install and uniformly tighten the 8 cylinder head bolts with the plate washers in several steps and in the sequence shown in the illustration. Tighten to 27 ft. lbs. (36 Nm).
   b. Step 2: Mark the cylinder head bolt head with paint as shown in the illustration. Tighten the cylinder head bolts another 90°.
   c. Step 3: Tighten the cylinder head bolts an additional 90°. Check that the painted mark is now facing rearward.
   d. Seal packing will seep out on the engine's front side. Thoroughly wipe clean any seal packing.
39. Place the left hand cylinder head gasket on the cylinder block surface with the front face of the Lot No. stamp upward.

➡Be careful of the installation direction.

➡Gently place the cylinder head in order not to damage the gasket with the bottom part of the head.

40. Place the cylinder head on the cylinder block.

➡Do not allow oil to adhere to the mounting surface of the cylinder head.

41. Apply a light coat of engine oil to the threads and under the heads of the cylinder head bolts.

42. The cylinder head bolts are tightened in 3 progressive steps:

a. Step 1: Using a 10 mm bi-hexagon wrench, install and uniformly tighten the 8 cylinder head bolts with the plate washers in several steps and in the sequence shown in the illustration. Tighten to 27 ft. lbs. (36 Nm).

b. Step 2: Mark the cylinder head bolt head with paint as shown in the illustration. Tighten the cylinder head bolts another 90°.

c. Step 3: Tighten the cylinder head bolts an additional 90°. Check that the painted mark is now facing rearward.

d. Tighten the 2 bolts in the order shown in the illustration to 22 ft. lbs. (30 Nm). Only use the specifications stated above when tightening the bolts 1 and 2 shown in the illustration.

e. Seal packing will seep out on the engine's front side. Thoroughly wipe clean any seal packing.

43. Install the No. 2 chain tensioner assembly.

44. Install the No. 3 chain tensioner.

45. Install the No. 1 and 2 chain vibration damper.

46. Install the idle sprocket assembly.

47. Install the chain sub-assembly.

48. Install the chain tensioner slipper.

49. Install the No. 1 chain tensioner assembly.

50. Install the water pump assembly.

51. Install the timing chain cover.

52. Install the water inlet housing.

53. Install the left hand No. 1 front engine mounting bracket.

54. Install the No. 1 oil pan baffle plate.

55. Install the oil pan sub-assembly.

56. Install the oil strainer sub-assembly.

57. Install the No. 2 oil pan sub-assembly.

58. Install the oil pan drain plug and gasket.

59. Install the cylinder head covers and gaskets

60. Install the water outlet.

61. Install the crankshaft pulley.

62. Install the oil filter.

63. Install the cylinder block water drain cock sub-assembly.

64. Install the No. 1 oil pipe.

65. Install the oil pipe.

66. Install the crankshaft position sensor.

67. Install the camshaft timing oil control valve assembly.

68. Install the camshaft position sensor.

69. Install the ventilation valve sub-assembly.

70. Install the spark plugs and ignition coil assembly.

71. Install the oil filler cap sub-assembly.

72. Install the engine assembly with transaxle.

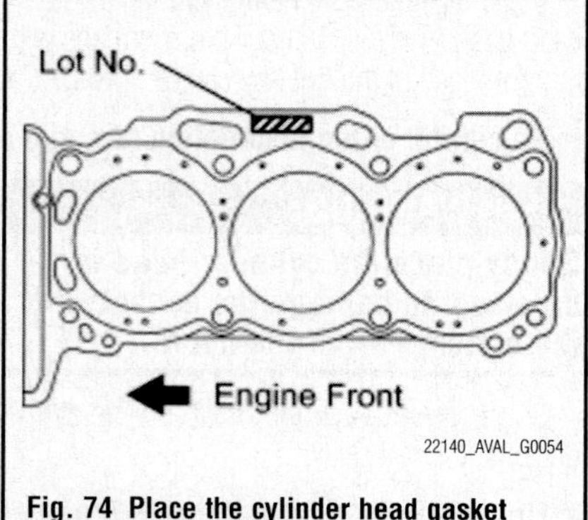

Fig. 74 Place the cylinder head gasket with Lot No. stamp upward

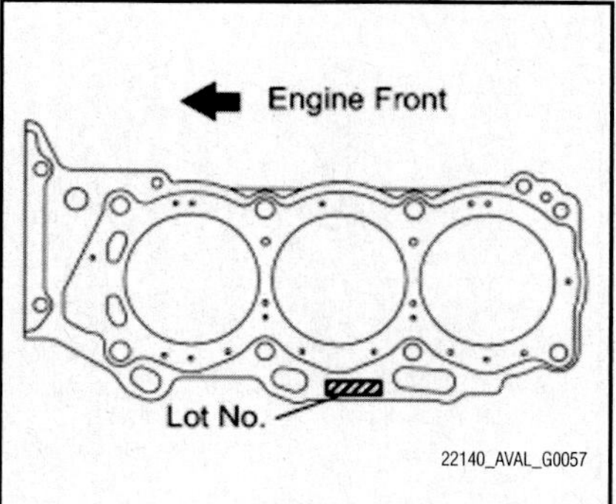

Fig. 77 Place the cylinder head with Lot No. stamp upward

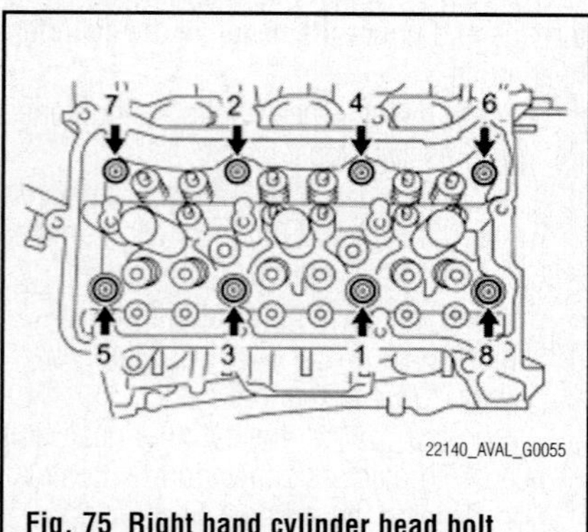

Fig. 75 Right hand cylinder head bolt tightening sequence

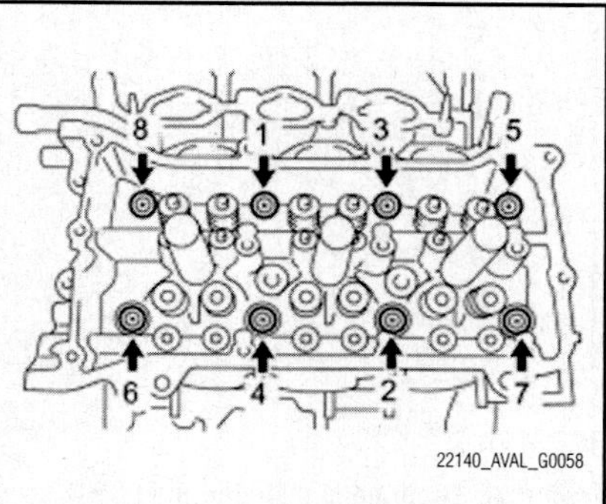

Fig. 78 Left hand cylinder head 8 bolt tightening sequence

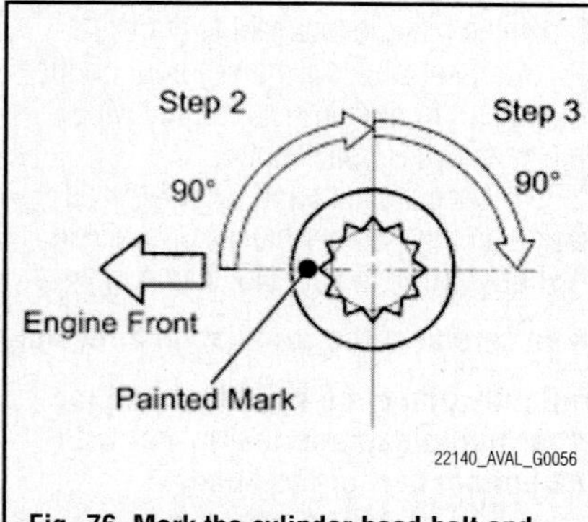

Fig. 76 Mark the cylinder head bolt and tighten another 90°

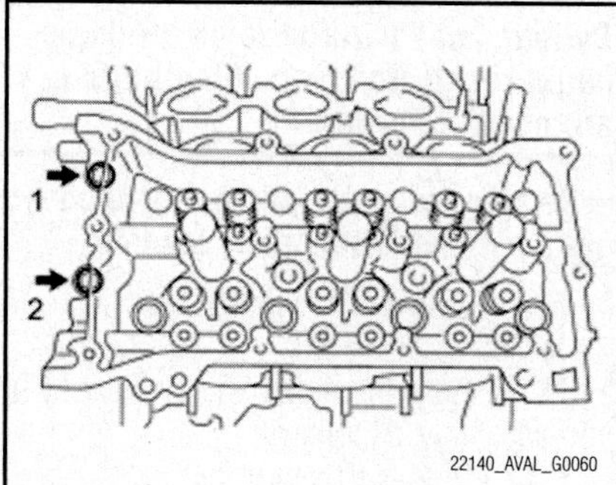

Fig. 79 Left hand cylinder head 2 bolt tightening sequence

EXHAUST MANIFOLD

REMOVAL & INSTALLATION

See Figures 80 through 86.

1. Before servicing the vehicle, refer to the Precautions Section.

2. Remove the ignition coil assembly.

3. Remove the right No. 2 engine mounting stay.

4. Remove the intake manifold.

5. Remove the right exhaust manifold sub-assembly, as follows:

a. Uniformly loosen and remove the 6 nuts.

b. Remove the manifold and gasket.

6. Remove the oil level gauge guide sub-assembly.

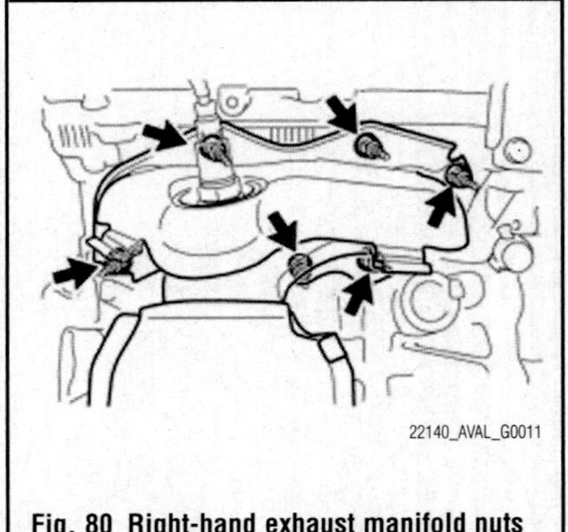

Fig. 80 Right-hand exhaust manifold nuts

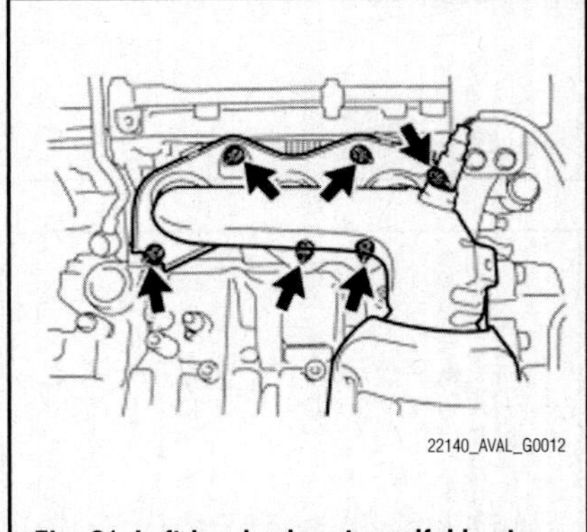

Fig. 81 Left-hand exhaust manifold nuts

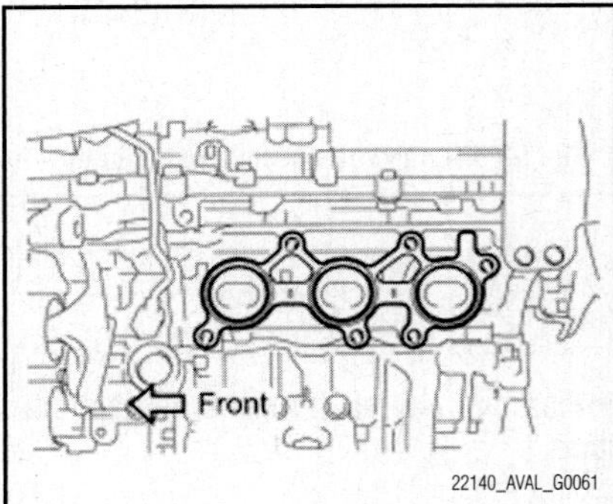

Fig. 82 Install a new gasket

7. Remove the bolt, nut and No. 2 manifold stay.

8. Remove the 3 bolts and No. 2 exhaust manifold heat insulator.

9. Remove the left exhaust manifold sub-assembly, as follows:

a. Uniformly loosen and remove the 6 nuts.

b. Remove the manifold and gasket.

To install:

10. Install the left exhaust manifold sub-assembly, as follows:

Fig. 83 Left-hand exhaust manifold nuts

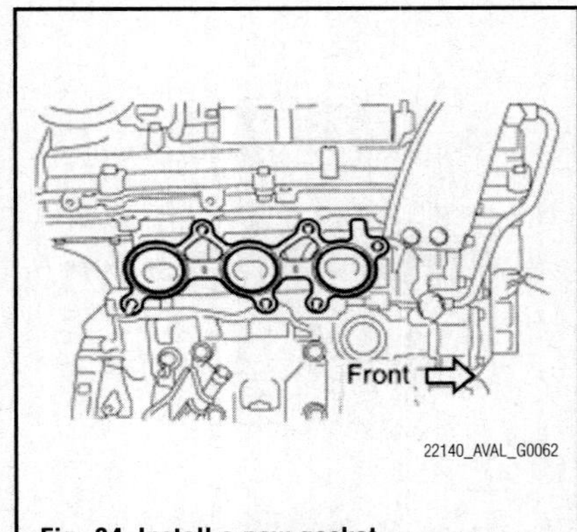

Fig. 84 Install a new gasket

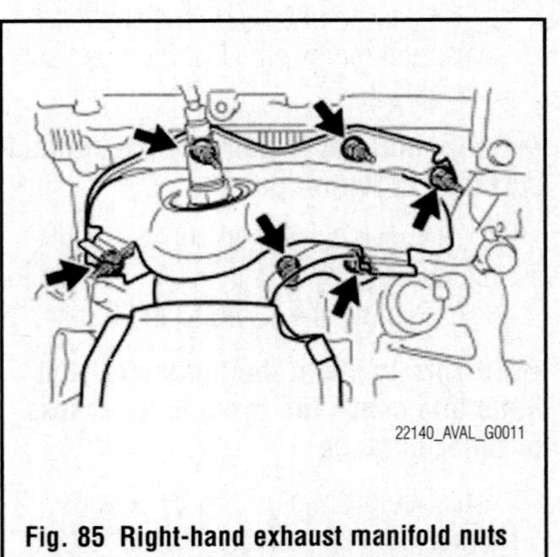

Fig. 85 Right-hand exhaust manifold nuts

a. Install a new gasket as shown in the illustration.

b. Install the left exhaust manifold sub-assembly with the 6 nuts and tighten to 15 ft. lbs. (21 Nm).

11. Install the No. 2 exhaust manifold heat insulator with the 3 bolts and tighten to 75 inch lbs. (8.5 Nm).

12. Install the No. 2 manifold stay with the bolt and nut and tighten to 25 ft. lbs. (34 Nm).

13. Install the oil level gauge guide sub-assembly.

14. Install the right exhaust manifold sub-assembly, as follows:

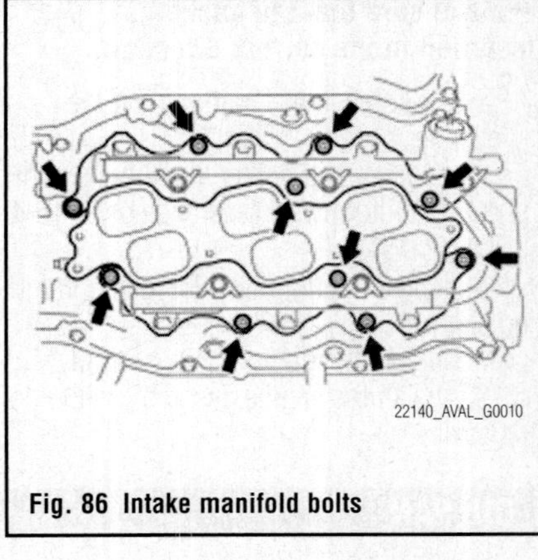

Fig. 86 Intake manifold bolts

a. Install a new gasket as shown in the illustration.

b. Install the right exhaust manifold sub-assembly with the 6 nuts and tighten to 15 ft. lbs. (21 Nm).

15. Install the intake manifold.

16. Install the right No. 2 engine mounting stay.

17. Install the ignition coil assembly.

18. Install the engine assembly with transaxle.

INTAKE MANIFOLD

REMOVAL & INSTALLATION

See Figure 86.

1. Before servicing the vehicle, refer to the Precautions Section.

2. Remove the engine assembly with transaxle.

3. Secure the engine.

4. Remove the ignition coil assembly.

5. Remove the right No. 2 engine mounting stay.

6. Remove the intake manifold, as follows:

a. Uniformly loosen and remove the 6 bolts and 4 nuts.

b. Remove the intake manifold and 2 gaskets.

To install:

7. Install the intake manifold, as follows:

**** WARNING**

DO NOTapply oil to the intake manifold and cylinder head sub-assembly bolts.

a. Set a new gasket on each cylinder head.

➡Align the port holes of the gasket and cylinder head.

➡Make sure that the gasket is installed in the correct direction.

b. Set the intake manifold on the cylinder heads.

c. Install and tighten the 6 bolts and 4 nuts uniformly in several steps to 15 ft. lbs. (21 Nm).

8. Install the right No. 2 engine mounting stay.
9. Install the ignition coil assembly.
10. Install the engine assembly with transaxle.

OIL PAN

REMOVAL & INSTALLATION

See Figures 87 through 94.

1. Before servicing the vehicle, refer to the Precautions Section.
2. Drain the engine oil.
3. Remove the engine assembly with transaxle.
4. Secure the engine.
5. Remove the oil filler cap and gasket.
6. Remove the oil pan drain plug and gasket.
7. Remove the No. 1 oil pipe, as follows:
   a. Remove the 2 oil pipe unions and oil pipe.
   b. Remove the left hand oil control valve filter and gaskets.
8. Remove the oil pipe, as follows:
   a. Remove the bolt.
   b. Remove the 2 oil pipe unions and oil pipe.
   c. Remove the right oil control valve filter and gaskets.
9. Remove the oil filter element, as follows:
   a. Remove the drain plug. Do not remove the O-ring.
   b. Connect the hose to the pipe.
   c. Insert the pipe with the hose into the oil filter cap.
   d. Make sure that the oil is completely drained and remove the pipe and O-ring.
   e. Using SST (SST: 09228-06501) or equivalent, remove the oil filter cap.
   f. Remove the oil filter element and O-ring from the oil filter cap. Do not use any tools when removing the O-ring to prevent the O-ring groove from being damaged.
10. Remove the No. 2 oil pan sub-assembly, as follows:
   a. Remove the 16 bolts and 2 nuts.
   b. Insert the blade of SST (SST: 09032-00100) or equivalent tool between

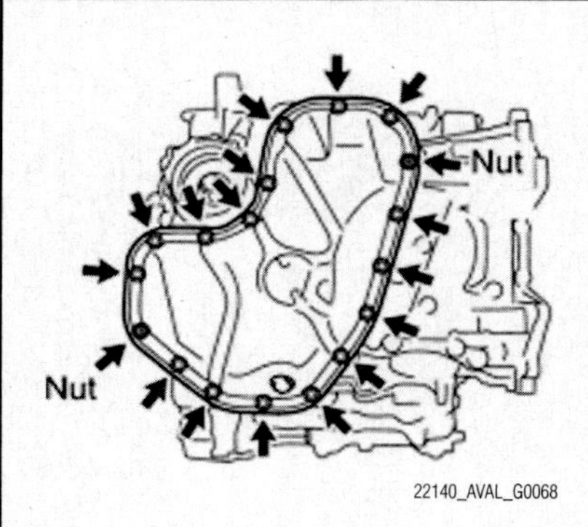

Fig. 87 No. 2 oil pan bolts and nuts

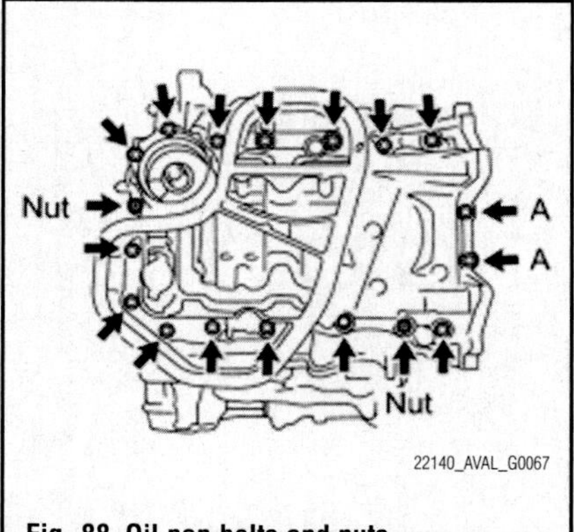

Fig. 88 Oil pan bolts and nuts

the oil pans. Cut through the applied sealer and remove the No. 2 oil pan sub-assembly.

➡Be careful not to damage the contact surfaces of the oil pans.

11. Remove the oil pan sub-assembly, as follows:
   a. Remove the 16 bolts and 2 nuts.

➡Be sure to clean the bolts and stud bolts and check the threads for cracks or other damage.

   b. Remove the oil pan by prying between the oil pan and cylinder block with a taped screwdriver.

➡Be careful not to damage the contact surfaces of the cylinder block and oil pan.

   c. Remove the 2 O-rings.

To install:

12. Install the oil pan sub-assembly, as follows:
   a. Using an E8 Torx® socket wrench, install the stud bolts as shown in the illustration. Tighten to 7 ft. lbs (10 Nm).
   b. Apply seal packing (Toyota Genuine Seal Packing Black, Three Bond

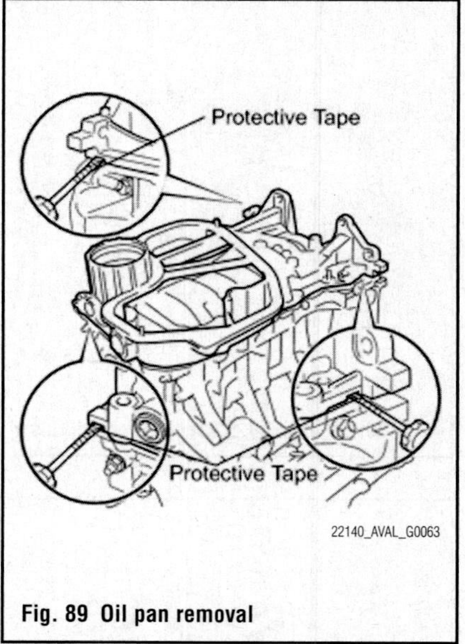

Fig. 89 Oil pan removal

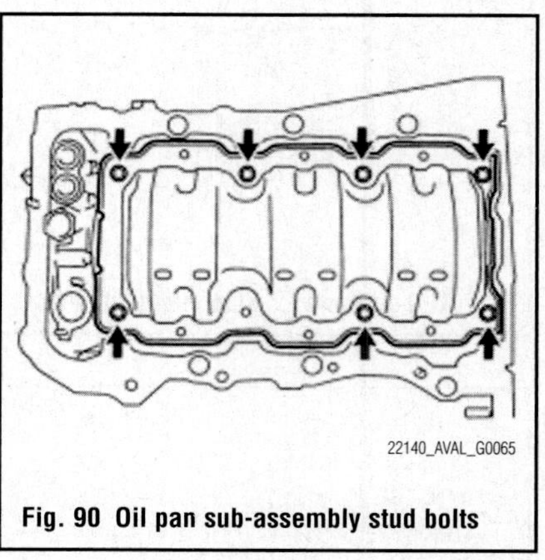

Fig. 90 Oil pan sub-assembly stud bolts

1207B or equivalent) in a continuous line as shown in the illustration. Seal diameter: 0.118 to 0.156 inch (3.0 to 4.0 mm).

➡Remove any oil from the contact surface.

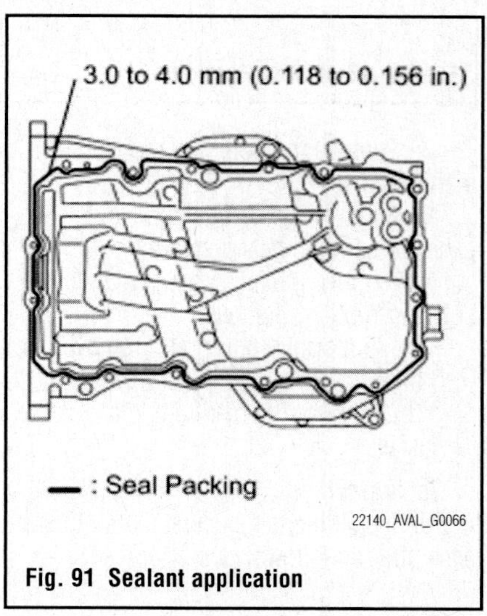

Fig. 91 Sealant application

➡Install the oil pan within 3 minutes after applying seal packing.

➡Do not start the engine for at least 2 hours after installing.

c. Install 2 new O-rings.

d. Install the oil pan with the 16 bolts and 2 nuts and tighten to 7 ft. lbs (10 Nm), and 15 ft. lbs (21 Nm).

13. Install the No. 2 oil pan sub-assembly, as follows:

a. Apply seal packing (Toyota Genuine Seal Packing Black, Three Bond 1207B or equivalent) in a continuous line as shown in the illustration. Seal diameter: 0.118 to 0.156 inch (3.0 to 4.0 mm).

➡Remove any oil from the contact surface.

➡Install the No. 2 oil pan within 3 minutes after applying seal packing.

➡Do not start the engine for at least 2 hours after installing.

b. Using an E6 Torx® socket wrench, install the stud bolts as shown in the illustration and tighten to 35 inch lbs (4 Nm).

c. Install the No. 2 oil pan with the 16 bolts and 2 nuts and tighten to 7 ft. lbs (10 Nm).

14. Install the oil pan drain plug and a new gasket. Tighten to 30 ft. lbs (40 Nm).

15. Install the oil filter element, as follows:

a. Clean the inside of the oil filter cap, the threads and O-ring groove.

b. Apply a small amount of engine oil to a new O-ring and install it to the oil filter cap.

c. Set a new oil filter element to the oil filter cap.

d. Remove dirt or foreign matter from the installation surface and inside of the engine.

e. Apply a small amount of engine oil to the O-ring again and install the oil filter cap.

➡Be careful that the O-ring does not get caught between the parts. The O-ring must not be twisted on the groove.

f. Using SST (SST: 09228-06501) or equivalent, install the oil filter cap and tighten to 18 ft. lbs (25 Nm). Make sure that the oil filter is installed securely as shown.

g. Apply a light coat of engine oil to a new O-ring and install it to the oil filter cap. Remove all dirt and foreign matter from the installation surface.

h. Install the oil filter drain plug to the oil filter cap and tighten to 9 ft. lbs (13 Nm). Make sure that the O-ring does not get caught between the parts.

16. Install the oil filler cap sub-assembly.

17. Install the engine assembly with transaxle.

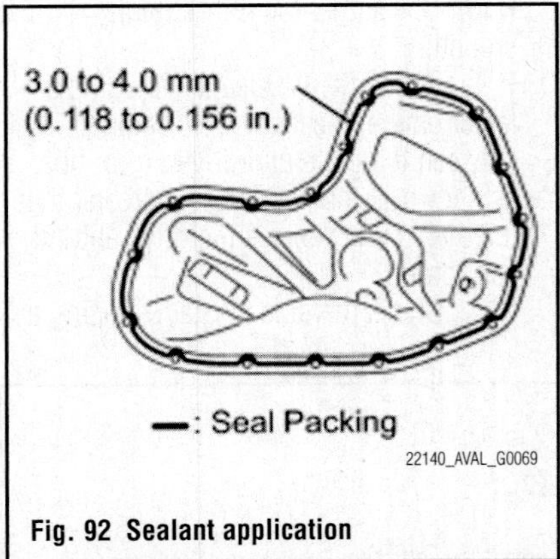

Fig. 92 Sealant application

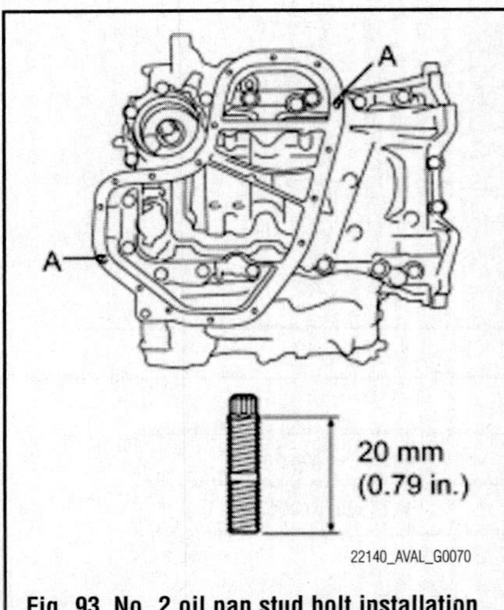

Fig. 93 No. 2 oil pan stud bolt installation

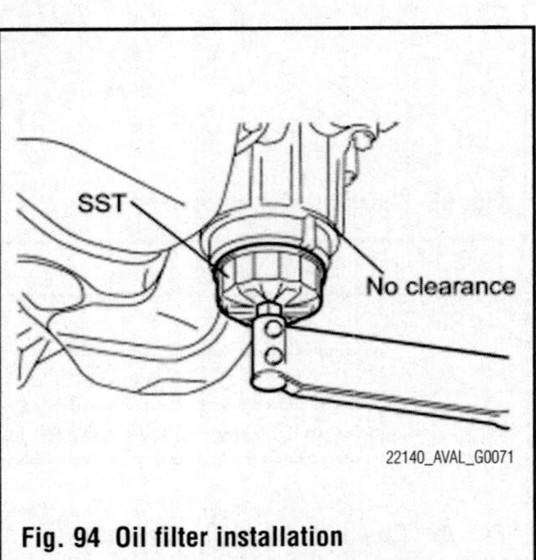

Fig. 94 Oil filter installation

OIL PUMP

REMOVAL & INSTALLATION

See Figures 95 and 96.

1. Before servicing the vehicle, refer to the Precautions Section.
2. Remove the engine assembly with transaxle.
3. Secure engine.
4. Remove the engine wire.
5. Remove the front frame assembly.
6. Remove the starter assembly.
7. Remove the automatic transaxle assembly.
8. Remove the oil level gauge guide sub-assembly.
9. Remove the right and left exhaust manifold sub-assemblies.
10. Remove the drive plate and ring gear sub-assembly.
11. Remove the No. 2 idler pulley sub-assembly.
12. Remove the V-ribbed belt tensioner assembly.
13. Remove the water pump pulley.
14. Remove the water inlet housing.
15. Remove the crankshaft pulley.
16. Remove the No. 2 oil pan sub-assembly.
17. Remove the oil strainer sub-assembly.
18. Remove the oil pan sub-assembly.
19. Remove the intake air surge tank assembly.
20. Remove the ignition coil assembly.
21. Remove the No. 1 and 2 oil pipes.
22. Remove the right and left cylinder head cover sub-assemblies.
23. Remove the timing chain cover sub-assembly. Refer to Timing Chain Cover and Seal procedures for instructions.

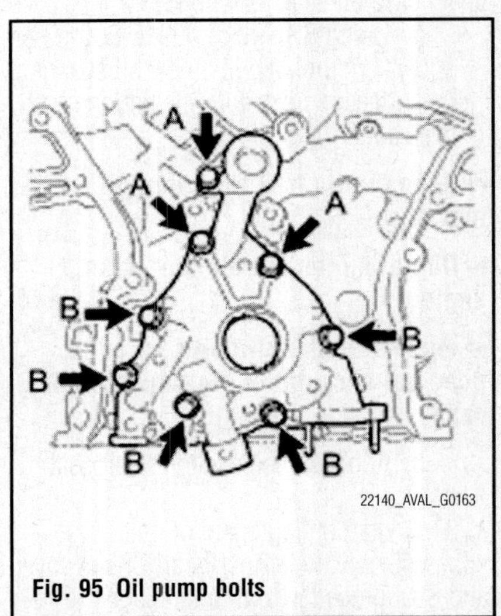

Fig. 95 Oil pump bolts

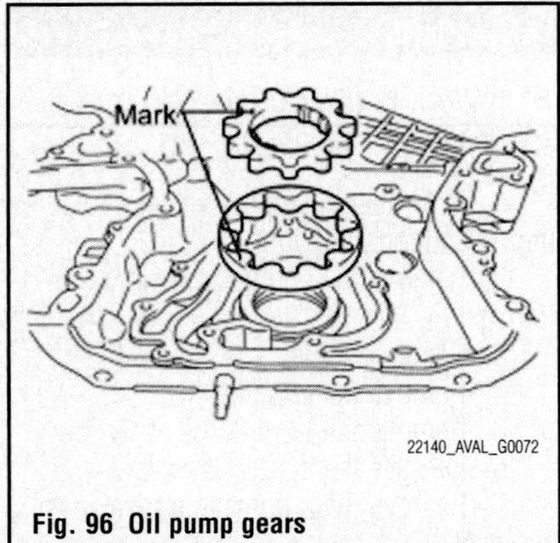

Fig. 96 Oil pump gears

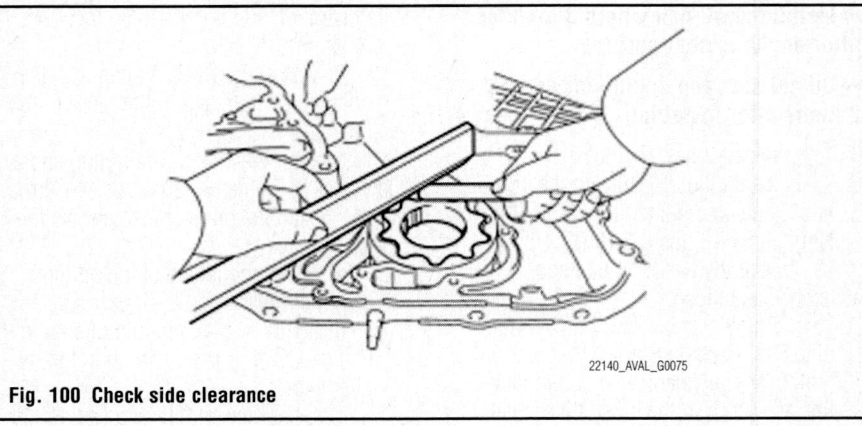

Fig. 100 Check side clearance

24. Using a screwdriver with the tip taped, pry out the timing gear case or timing chain case oil seal.
25. Using a 27 mm socket wrench, remove the relief valve plug.
26. Remove the valve spring and oil pump relief valve.
27. Remove the 8 bolts, oil pump cover, drive rotor and driven rotor.

To install:

28. Coat the drive and driven rotors with engine oil and place them into the timing chain cover with the marks facing outward (oil pump cover side). Check that the rotors revolve smoothly.
29. Install the oil pump cover with the 8 bolts and tighten to 81 inch lbs. (9.1 Nm). Bolt length: 0.87 in. (22 mm) for bolt A, 1.58 in. (40 mm) for bolt B.
30. Coat the oil pump relief valve with engine oil.
31. Insert the relief valve and relief valve spring into the oil pump cover hole.
32. Using a 27 mm socket wrench, install the plug and tighten to 36 ft. lbs. (49 Nm).
33. Install timing gear case or timing chain case oil seal, as follows:
    a. Using SST (SST: 09316-60011) or equivalent tool, tap in a new oil seal until its surface is flush with the timing chain case edge.

➡Keep the lip free from foreign matter.

➡Do not tap on the oil seal at an angle.

➡Make sure that the oil seal edge does not stick out of the timing chain case.

    b. Apply MP grease to the oil seal lip.

34. Install timing chain or belt cover sub-assembly. Refer to Timing Chain Cover and Seal procedures for instructions.
35. The remainder of installation is the reverse of removal.

INSPECTION

See Figures 97 through 103.

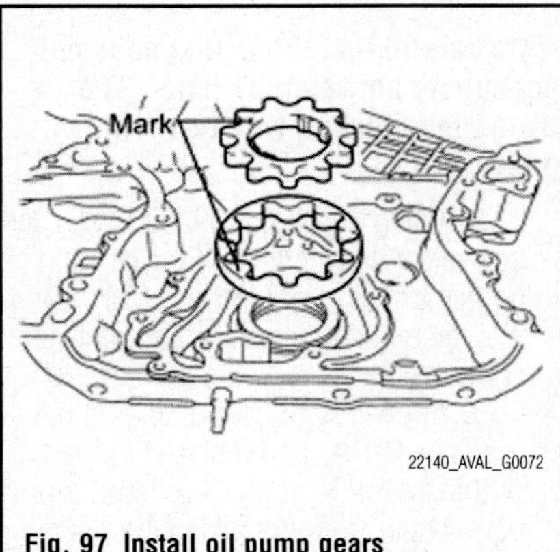

Fig. 97 Install oil pump gears

1. Inspect the oil pump relief valve, as follows:
    a. Coat the relief valve with engine oil and check that it falls smoothly into the valve hole by its own weight. If the valve does not fall smoothly, replace the relief valve. If necessary, replace the oil pump assembly.
2. Inspect the oil pump rotor set, as follows:
    a. Install the rotors to the timing chain cover with the rotors' marks outward. Check that the rotors rotate smoothly.
    b. Check the tip clearance: using a feeler gauge, measure the clearance between the drive and driven rotor tips, as shown. If the clearance is greater than the maximum, replace the drive and driven rotors.
    c. Check the side clearance: using a

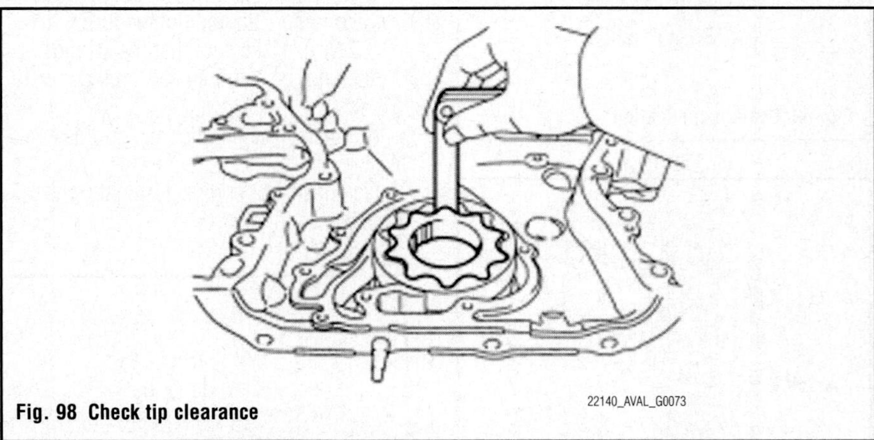

Fig. 98 Check tip clearance

Standard	Maximum
0.060 to 0.160 mm (0.0024 to 0.0063 in.)	0.16 mm (0.0063 in.)

22140_AVAL_G0074

Fig. 99 Tip clearance specifications

Standard	Maximum
0.030 to 0.090 mm (0.0012 to 0.0035 in.)	0.090 mm (0.0035 in.)

22140_AVAL_G0076

Fig. 101 Side clearance specifications

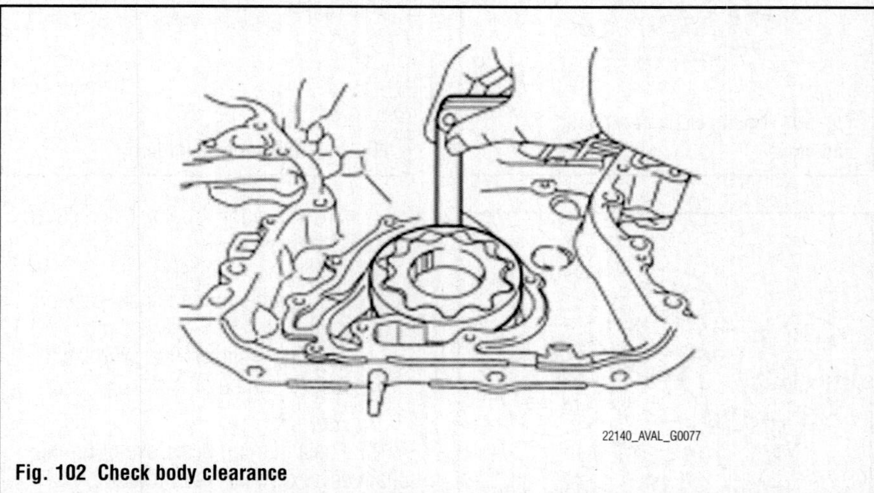

Fig. 102 Check body clearance

Standard	Maximum
0.250 to 0.325 mm (0.0098 to 0.0128 in.)	0.325 mm (0.0128 in.)

22140_AVAL_G0078

Fig. 103 Body clearance specifications

feeler gauge and precision straightedge, measure the clearance between the rotors and precision straightedge, as shown in the illustration. If the side clearance is greater than the maximum, replace the timing chain cover sub-assembly.

d. Check the body clearance: using a feeler gauge, measure the clearance between the timing chain cover and driven rotor, as shown in the illustration. If the body clearance is greater than the maximum, replace the timing chain cover sub-assembly.

PISTON AND RING

POSITIONING

See Figure 104.

Refer to the graphic provided for piston ring positioning.

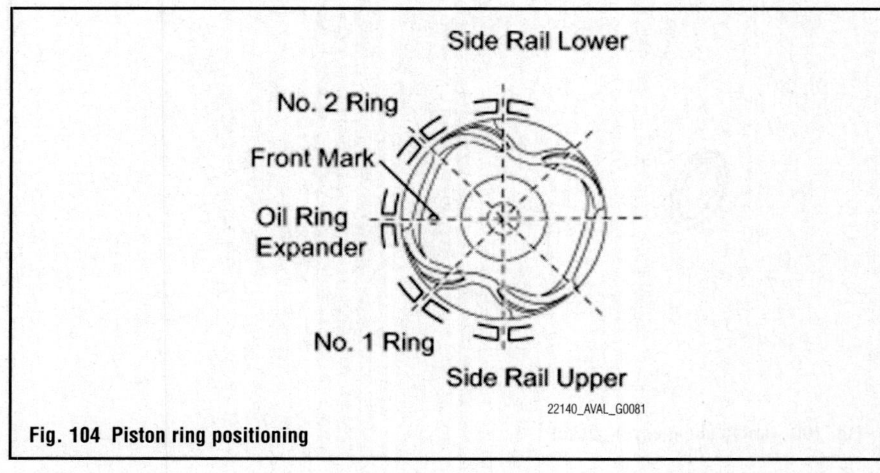

Fig. 104 Piston ring positioning

REAR MAIN SEAL

REMOVAL & INSTALLATION

See Figures 105 and 106.

1. Before servicing the vehicle, refer to the Precautions Section.
2. Remove the automatic transaxle assembly.
3. Remove the drive plate and ring gear sub-assembly.
4. Remove the rear main seal, as follows:
   a. Using a knife, cut off the oil seal lip.
   b. Using a screwdriver with the tip taped, pry out the oil seal.

➡Be careful not to damage the crankshaft.

To install:

5. Apply MP grease to a new oil seal lip.
6. Using SST (SST: 09223-15030, SST: 09950-70010) or equivalent and a hammer, tap in the oil seal. Oil seal tap in depth: -0.020 to 0.020 inch (-0.5 to 0.5 mm).
7. Install the drive plate and ring gear sub-assembly.
8. Install automatic transaxle assembly.

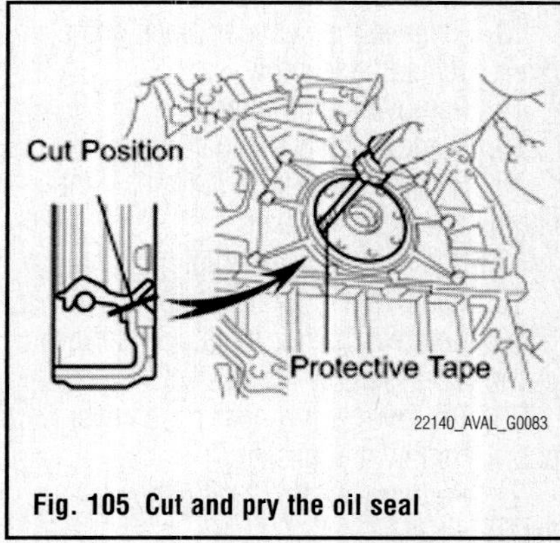

Fig. 105 Cut and pry the oil seal

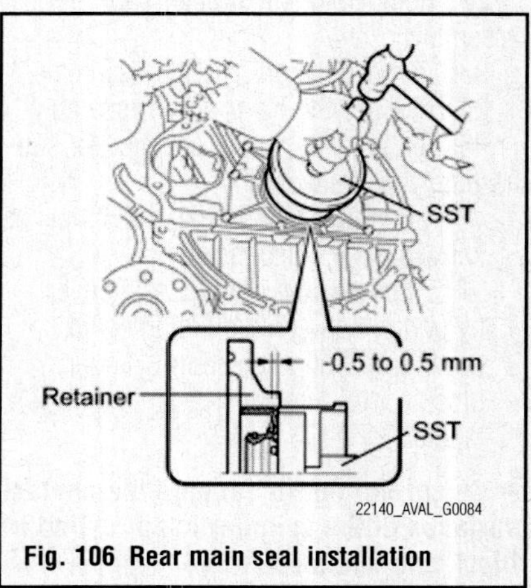

Fig. 106 Rear main seal installation

ROCKER ARMS

REMOVAL & INSTALLATION

See Camshaft and Valve Lifters.

TIMING CHAIN COVER & SEAL

REMOVAL & INSTALLATION

See Figures 107 through 116.

1. Before servicing the vehicle, refer to the Precautions Section.
2. Remove the engine assembly with transaxle.
3. Secure engine.
4. Remove the oil filler cap sub-assembly.
5. Remove the spark plugs and ignition coil assembly.
6. Remove the oil pan drain plug and gasket.
7. Remove the ventilation valve sub-assembly.
8. Remove the camshaft position sensor.
9. Remove the camshaft timing oil control valve assembly.
10. Remove crankshaft position sensor.
11. Remove the No. 1 oil pipe.
12. Remove the oil pipe.
13. Remove the cylinder block water drain cock sub-assembly.
14. Remove the oil filter.
15. Remove the crankshaft pulley.
16. Remove the left hand No. 1 front engine mounting bracket.
17. Remove the water inlet housing.
18. Remove the water outlet.
19. Remove the left-hand cylinder head cover sub-assembly and gasket.
20. Remove the cylinder head cover sub-assembly and gasket.
21. Remove the No. 2 oil pan sub-assembly.
22. Remove the oil strainer sub-assembly.
23. Remove the oil pan sub-assembly.
24. Remove the water pump assembly.
25. Remove the timing chain cover sub-assembly, as follows:
   a. Remove the 15 bolts and 2 nuts as shown in the illustration.
   b. Remove the timing chain cover by prying between the timing chain cover and cylinder head or cylinder block with a screwdriver with the tip taped.

➡Be careful not to damage the contact surfaces of the cylinder head, cylinder block and chain cover.

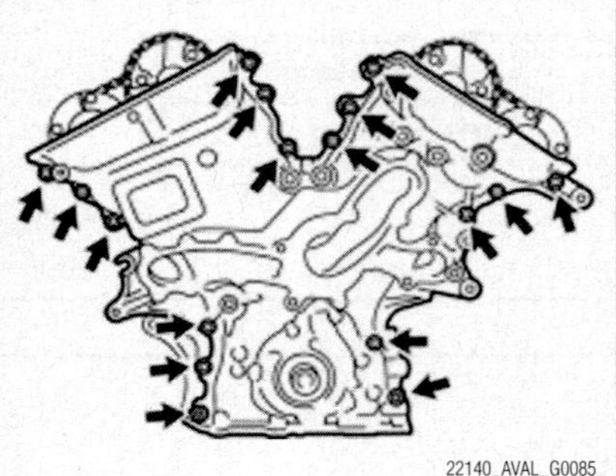

Fig. 107 Timing chain cover bolts and nuts

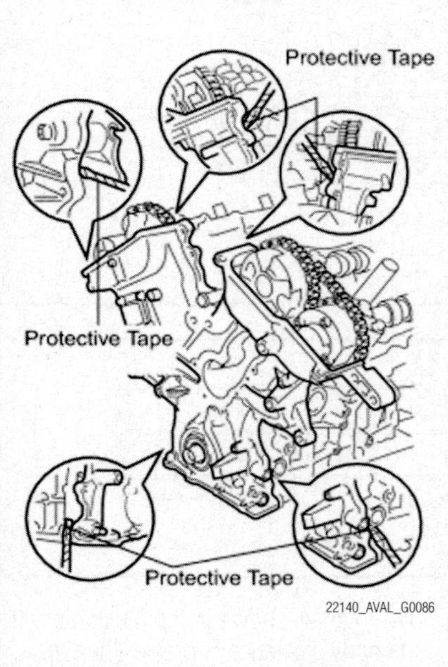

Fig. 108 Timing chain cover removal

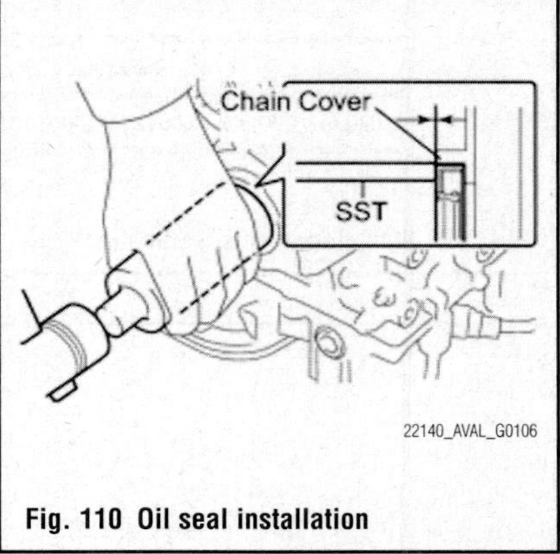

Fig. 110 Oil seal installation

   c. Remove the 4 bolts, chain cover plate and gasket.
   d. Remove the gasket.
26. Remove the timing chain case oil seal, as follows:
   a. Using a screwdriver with the tip taped, pry out the oil seal.

To install:

27. Install timing gear case or timing chain cover oil seal, as follows:
   a. Apply MP grease to a new oil seal lip.
   b. Using SST (SST: 09316-60011) and a hammer, tap in the oil seal until its surface is flush with the timing chain cover edge.

➡Keep the lip free from foreign matter.

➡Do not tap on the oil seal at an angle.

➡Make sure that the oil seal edge does not stick out of the timing chain cover.

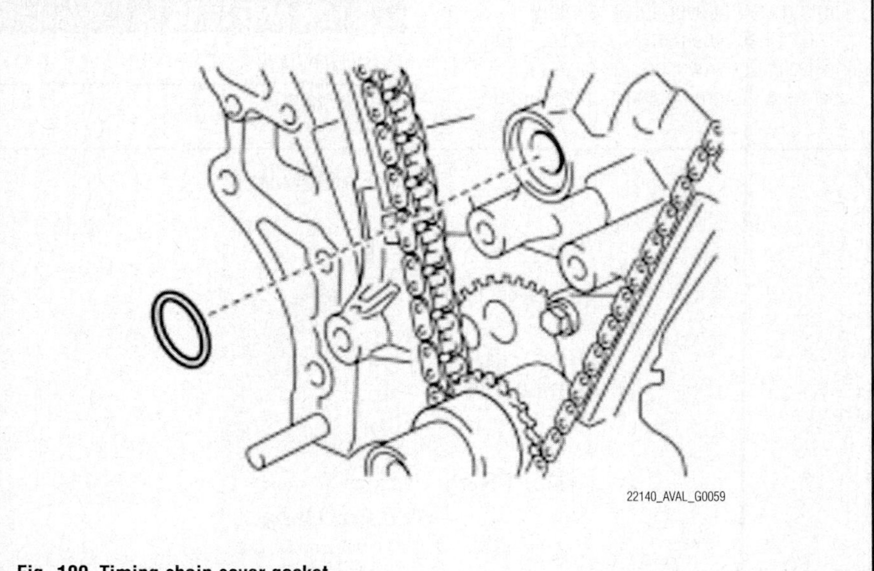

Fig. 109 Timing chain cover gasket

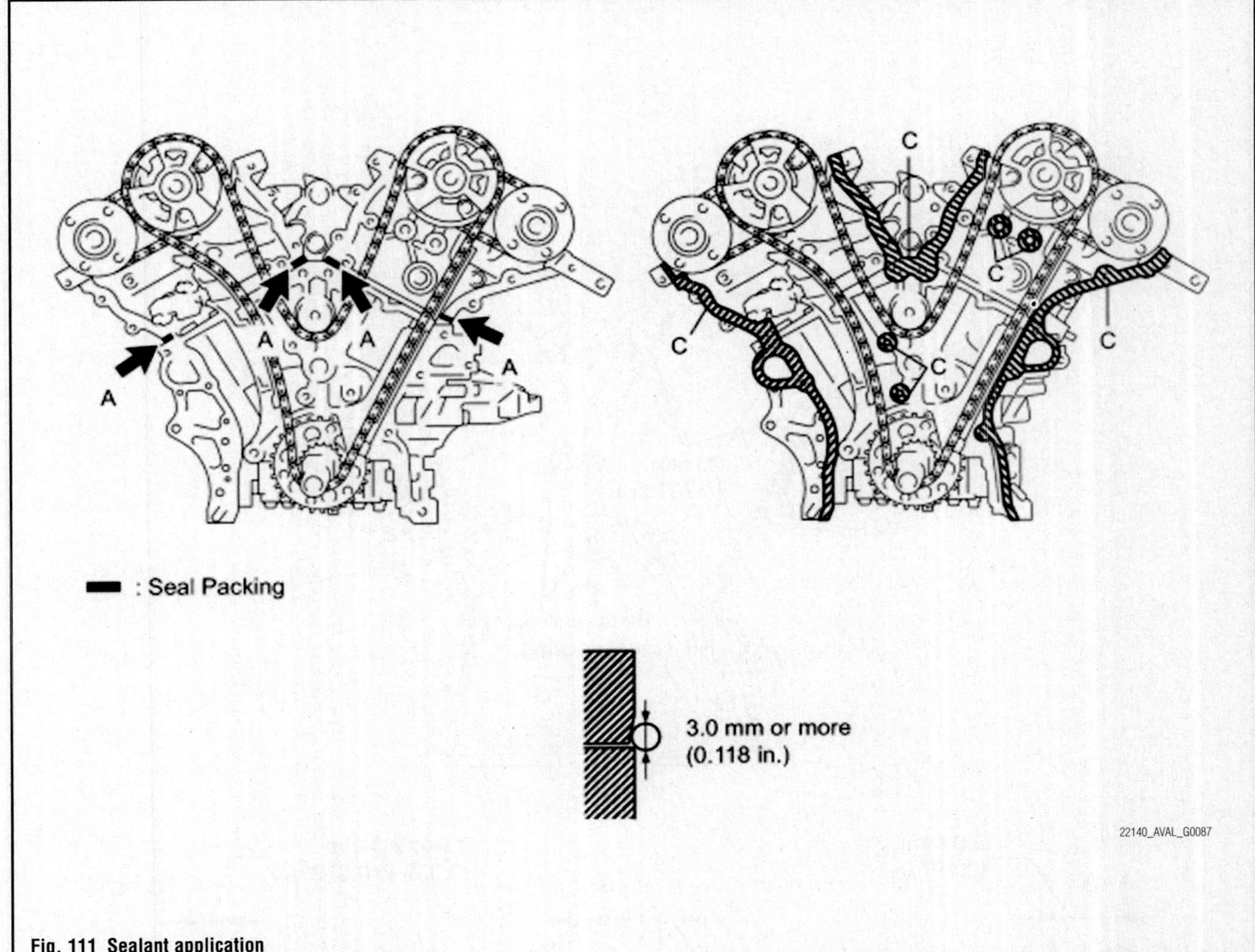

Fig. 111 Sealant application

28. Install the timing chain cover sub-assembly, as follows:
    a. Install a new gasket and the chain cover plate with the 4 bolts and tighten to 81 inch lbs. (9.1 Nm).
    b. Apply seal packing (Toyota Genuine Seal Packing Black, Three Bond 1207B or equivalent) in a continuous line to the engine unit as shown in the illustration. Seal diameter: 3.0 mm (0.118 in.).

➡Be sure to clean and degrease the contact surfaces, especially the surfaces indicated by C in the illustration.

➡When the contact surfaces are wet, wipe them with an oil-free cloth before applying seal packing.

➡Install the chain cover within 3 minutes after applying seal packing.

➡Do not start the engine for at least 2 hours after installing.

    c. Apply seal packing in a continuous line to the timing chain cover as shown in the following illustration. Seal packing: Toyota Genuine Seal Packing Black, Three Bond 1207B or equivalent, Toyota Genuine Seal Packing Black, Three Bond 1282B, Three Bond 1282B or equivalent.

➡When the contact surfaces are wet, wipe them with an oil-free cloth before applying seal packing.

➡Install the crankcase within 3 minutes and tighten the bolts within 15 minutes after applying seal packing.

➡Do not start the engine for at least 2 hours after installing.

    d. Install a new gasket.
    e. Align the oil pump's drive rotor spline and the crankshaft as shown in the illustration. Install the spline and chain cover to the crankshaft.
    f. Loosely install the timing chain cover with the 23 bolts and 2 nuts, but do not tighten the bolts and 2 nuts yet.

⁂ WARNING

Make sure that there is no oil on the bolt and nut threads.

    g. Fully tighten the bolts in this order: Area 1 and Area 2, tighten to 15 ft. lbs. (21 Nm).
    h. Fully tighten the bolts in Area 3 to 15 ft. lbs. (21 Nm). Tighten the bolts and nuts in the order of upper to lower as shown in the illustration.
    i. Fully tighten the bolts in Area 4 to 32 ft. lbs. (43 Nm), and to 15 ft. lbs. (21 Nm). Tighten the bolts and nuts in the order of lower to upper as shown in the illustration.
29. Install the water pump assembly.
30. Install the water inlet housing.
31. Install the left hand No. 1 front engine mounting bracket.

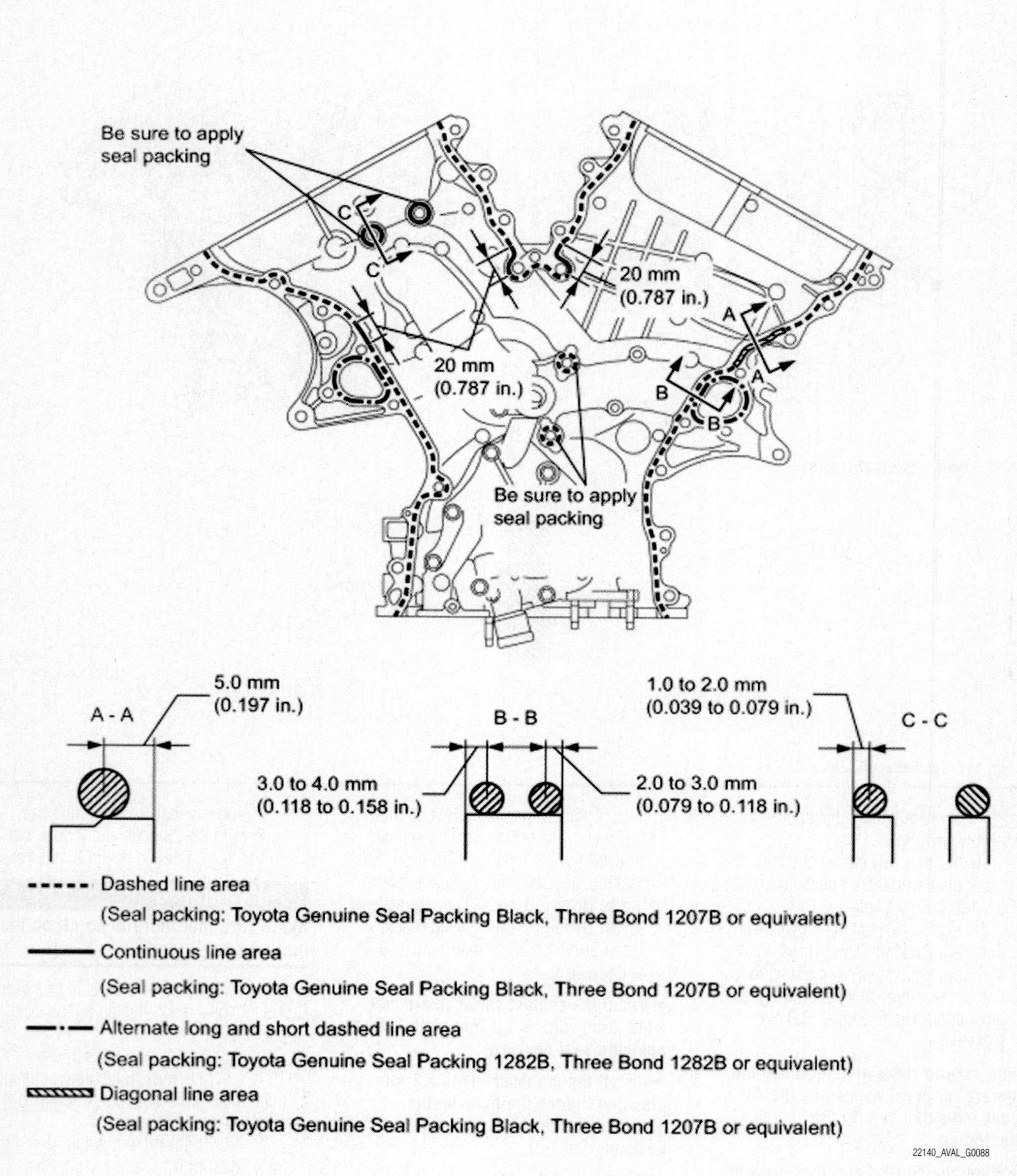

Fig. 112 Timing chain cover sealant application

Area	Seal Packing Diameter	Application Position from Inside Seal Line
Dashed line area	3.5 mm or more (0.138 in.)	3.0 to 4.0 mm (0.118 to 0.158 in.)
Continuous line area	4.5 mm or more (0.177 in.)	3.0 to 4.0 mm (0.118 to 0.158 in.)
Alternate long and short dashed line area	3.5 mm or more (0.138 in.)	2.0 to 3.0 mm (0.079 to 0.118 in.)
Diagonal line area	6.0 mm or more (0.236 in.)	5.0 mm (0.197 in.)

22140_AVAL_G0089

Fig. 113 Sealant application diameter and position

32. Install the oil pan sub-assembly.
33. Install the oil strainer sub-assembly.
34. Install the No. 2 oil pan sub-assembly.
35. Install the oil pan drain plug and gasket.
36. Install the cylinder head cover sub-assembly.
37. Install the left-hand cylinder head cover sub-assembly.
38. Install the water outlet.
39. Install the crankshaft pulley.
40. Install the oil filter.
41. Install the cylinder block water drain cock sub-assembly.
42. Install the No. 1 oil pipe.
43. Install the oil pipe.
44. Install the crankshaft position sensor.
45. Install the camshaft timing oil control valve assembly.
46. Install the camshaft position sensor.
47. Install the ventilation valve sub-assembly.
48. Install the spark plugs and ignition coil assembly.
49. Install the oil filler cap sub-assembly.
50. Install the water pump assembly.
51. Install the engine assembly with transaxle.

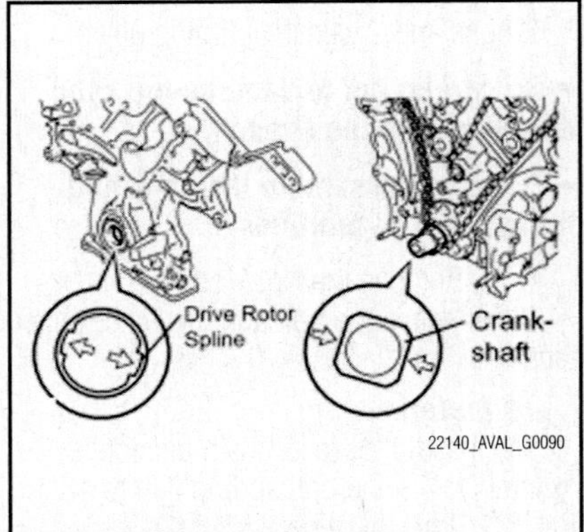

Fig. 114 Oil pump alignment

Item	Length
Bolt A	40 mm (1.57 in.)
Bolt B	55 mm (2.17 in.)
Bolt C	25 mm (0.98 in.)

22140_AVAL_G0092

Fig. 116 Bolt length

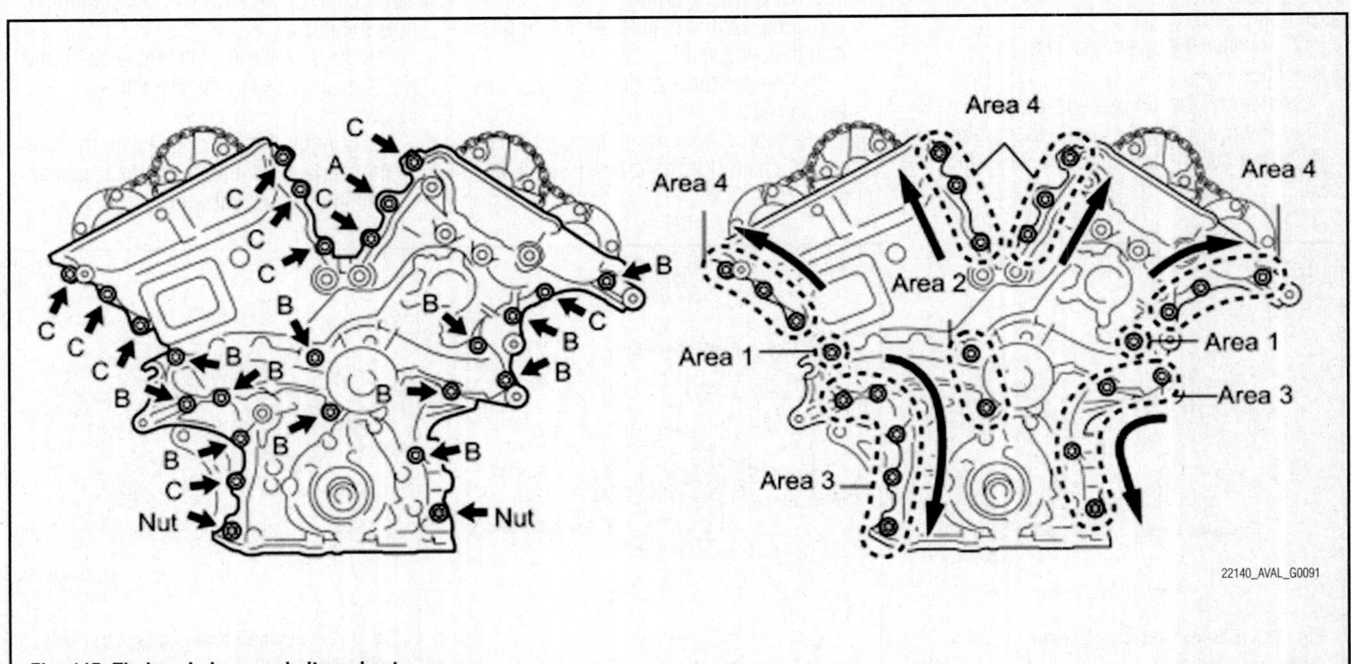

Fig. 115 Timing chain cover bolts and nuts

TIMING CHAIN & SPROCKETS

REMOVAL & INSTALLATION

See Figures 117 through 129.

1. Before servicing the vehicle, refer to the Precautions Section.
2. Remove the engine assembly with transaxle.
3. Secure engine.
4. Remove the oil filler cap sub-assembly.
5. Remove the spark plugs and ignition coil assembly.
6. Remove the oil pan drain plug and gasket.
7. Remove the ventilation valve sub-assembly.
8. Remove the camshaft position sensor.
9. Remove the camshaft timing oil control valve assembly.
10. Remove crankshaft position sensor.
11. Remove the No. 1 oil pipe.
12. Remove the oil pipe.
13. Remove the cylinder block water drain cock sub-assembly.
14. Remove the oil filter.
15. Remove the crankshaft pulley.
16. Remove the left hand No. 1 front engine mounting bracket.
17. Remove the water inlet housing.
18. Remove the water outlet.
19. Remove the left-hand cylinder head cover sub-assembly and gasket.
20. Remove the cylinder head cover sub-assembly and gasket.
21. Remove the No. 2 oil pan sub-assembly.
22. Remove the oil strainer sub-assembly.
23. Remove the oil pan sub-assembly.
24. Remove the water pump assembly.
25. Remove the timing chain cover and seal.
26. Remove the No. 1 chain tensioner assembly, as follows:
    a. Move the stopper plate upward to release the lock, and push the plunger deep into the tensioner.
    b. Move the stopper plate downward to set the lock, and insert a hexagon wrench into the stopper plate's hole.
    c. Remove the 2 bolts and chain tensioner.
27. Remove the chain tensioner slipper.
28. Remove the chain sub-assembly, as follows:
    a. Turn the crankshaft counterclockwise 10°to loosen the chain of the crankshaft timing sprocket.
    b. Remove the chain from the crankshaft timing sprocket and place it on the crankshaft.
    c. Turn the camshaft timing gear assembly on the right hand bank clockwise (approximately 60°) and set it as shown in the illustration. Be sure to loosen the chain between the banks.
    d. Remove the chain.
29. Remove the idle sprocket assembly, as follows:
    a. Using a 10 mm hexagon wrench, remove the No. 2 idle gear shaft, sprocket and No. 1 idle gear shaft.
30. Remove the 2 bolts and the No. 1 chain vibration damper.
31. Remove the No. 2 chain vibration damper.
32. Remove the crankshaft timing sprocket, as follows:
    a. Remove the pulley set bolt.
    b. Remove the crankshaft timing gear from the crankshaft.
    c. Remove the 2 pulley set keys from the crankshaft.
33. Remove the camshaft timing gears and No. 2 chain (for Right-hand Bank), as follows:
    a. While raising up the No. 2 chain tensioner, insert a pin of 1.0 mm (0.039 in.) into the hole to hold it.
    b. Hold the hexagonal portion of the camshaft with a wrench, and remove the 2 bolts and 2 camshaft timing gears.

➡Be careful not to damage the cylinder head with the wrench.

➡Do not disassemble the camshaft timing gear assemblies.

    c. Remove the No. 2 chain.
34. Remove the bolt and No. 2 chain tensioner.
35. Remove the camshaft timing gears and No. 2 chain (for Left-hand Bank), as follows:
    a. While pushing down on the No. 3 chain tensioner, insert a pin of 1.0 mm (0.039 in.) into the hole to hold it.
    b. Hold the hexagonal portion of the camshaft with a wrench, and remove the 2 bolts and 2 camshaft timing gears.

➡Be careful not to damage the cylinder head with the wrench.

➡Do not disassemble the camshaft timing gear assemblies.

    c. Remove the No. 2 chain.
36. Remove the bolt and the No. 3 chain tensioner.

To install:

37. Install the No. 2 chain tensioner assembly with the bolt and tighten to 15 ft. lbs. (21 Nm).
38. While pushing in the tensioner, insert a pin of 1.0 mm (0.039 in.) into the hole to hold it.
39. Install the camshaft timing gears and No. 2 chain (for Right-hand Bank), as follows:
    a. Align the mark plate with the timing marks (1-dot mark) of the camshaft timing gears as shown.

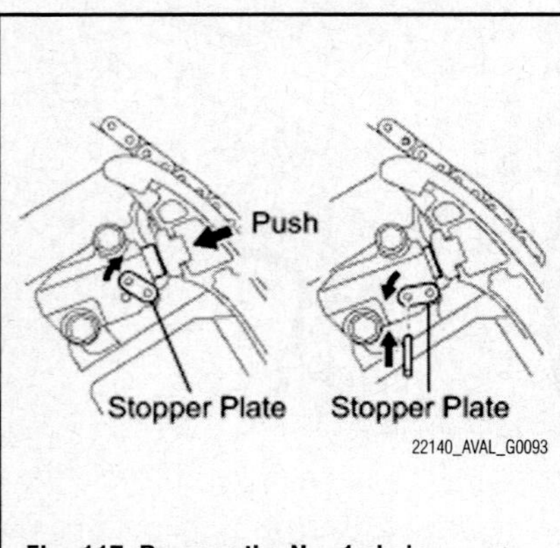

Fig. 117 Remove the No. 1 chain tensioner assembly

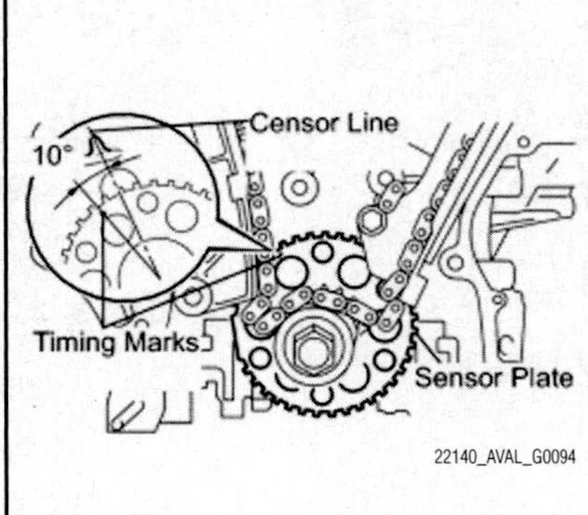

Fig. 118 Crankshaft timing sprocket

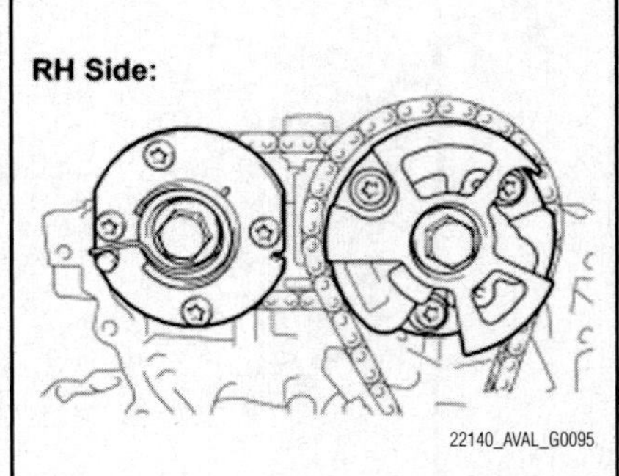

Fig. 119 Camshaft timing gear assembly positioning

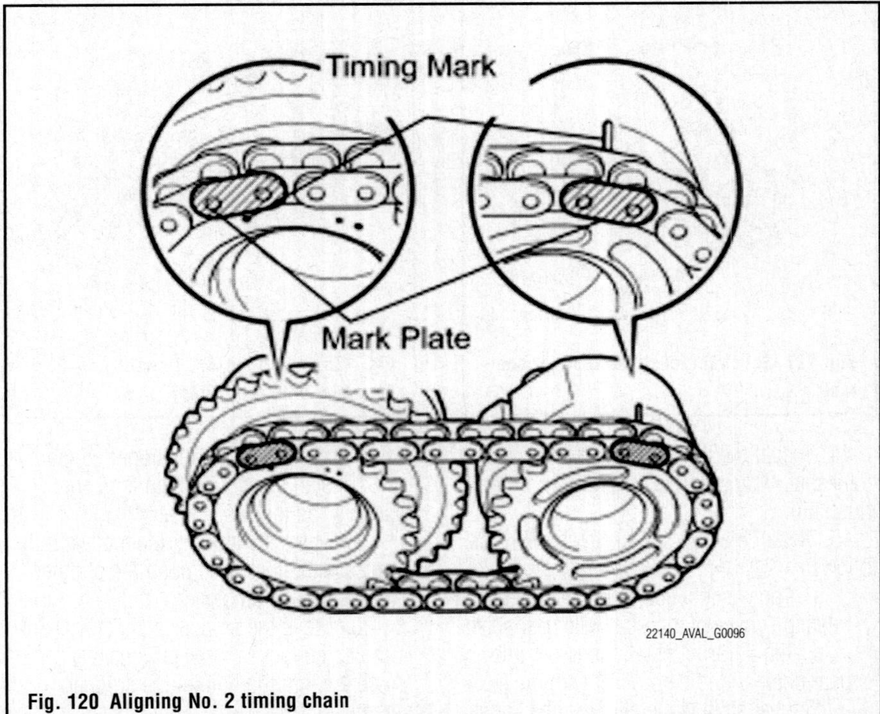

Fig. 120 Aligning No. 2 timing chain

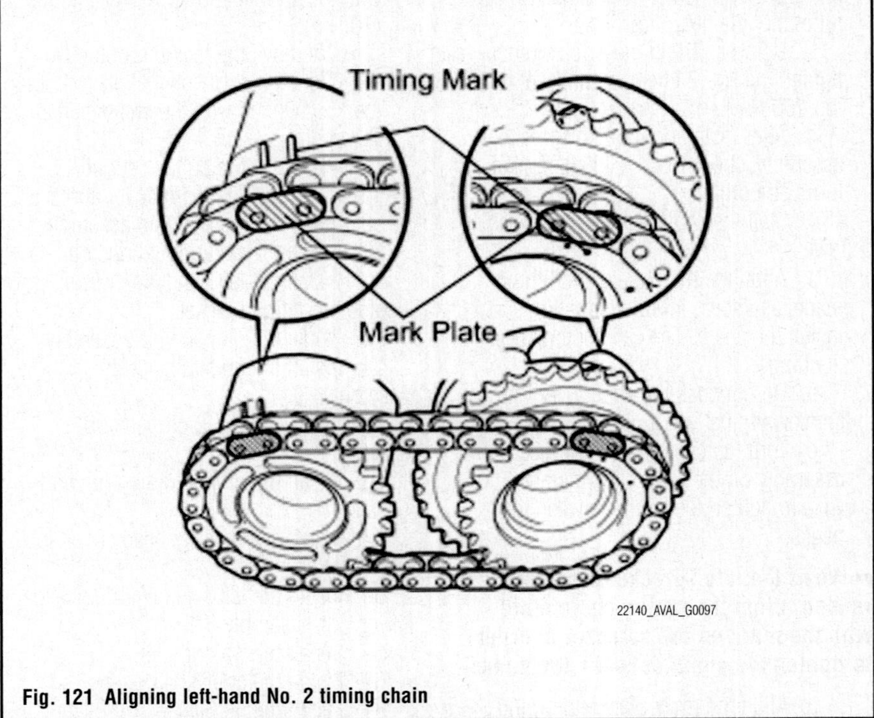

Fig. 121 Aligning left-hand No. 2 timing chain

b. Apply a light coat of engine oil to the bolt threads and bolt-seating surface.

c. Align the knock pin of the camshaft with pin hole of the camshaft timing gear. Install the camshaft timing gear and the right camshaft timing exhaust gear with the No. 2 chain installed.

d. Hold the hexagonal portion of the camshaft with the wrench and tighten the two bolts to 74 ft. lbs. (100 Nm).

e. Remove the pin from the No. 2 chain tensioner.

40. Install the No. 3 chain tensioner assembly with the bolt and tighten to 15 ft. lbs. (21 Nm).

41. While pushing in the tensioner,

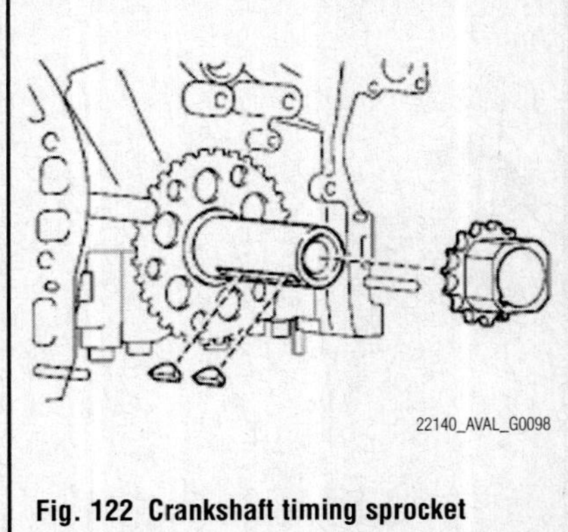

Fig. 122 Crankshaft timing sprocket

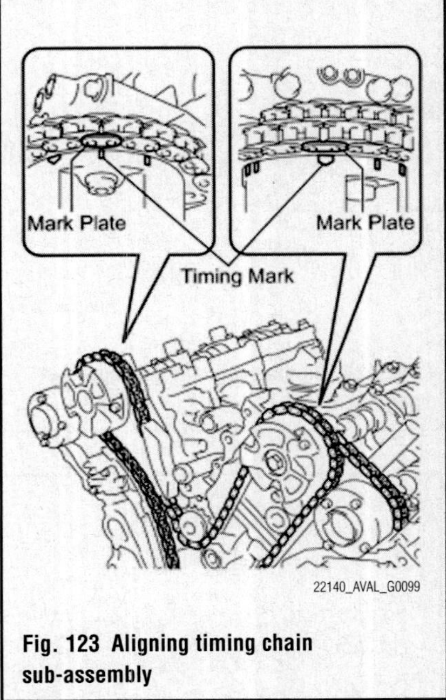

Fig. 123 Aligning timing chain sub-assembly

insert a pin of 1.0 mm (0.039 in.) into the hole to hold it.

42. Install the camshaft timing gears and No. 2 chain (for Left-hand Bank), as follows:

a. Align the mark plate (yellow) with the timing marks (2-dot mark) of the camshaft timing gears as shown.

b. Apply a light coat of engine oil to the bolt threads and bolts seating surface.

c. Align the knock pin of the camshaft with pin hole of the camshaft timing gear. Install the camshaft timing gear and the left camshaft timing exhaust gear with the No. 2 chain installed.

d. Hold the hexagonal portion of the camshaft with the wrench and tighten the two bolts to 74 ft. lbs. (100 Nm).

e. Remove the pin from the No 2 chain tensioner.

43. Install the No. 1 and 2 chain vibration dampers.

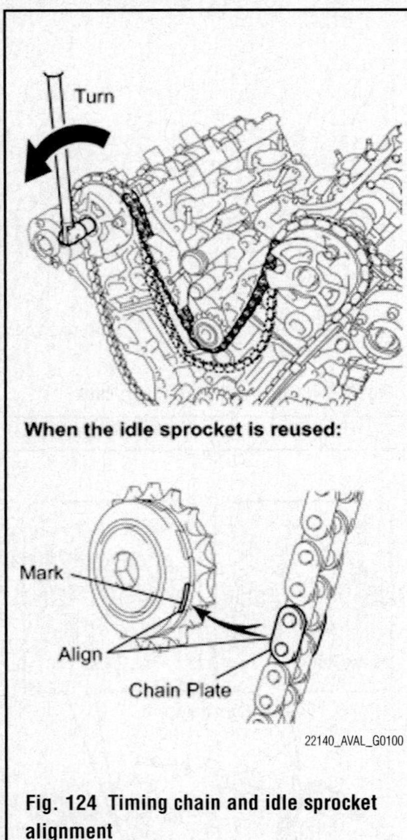

Fig. 124 Timing chain and idle sprocket alignment

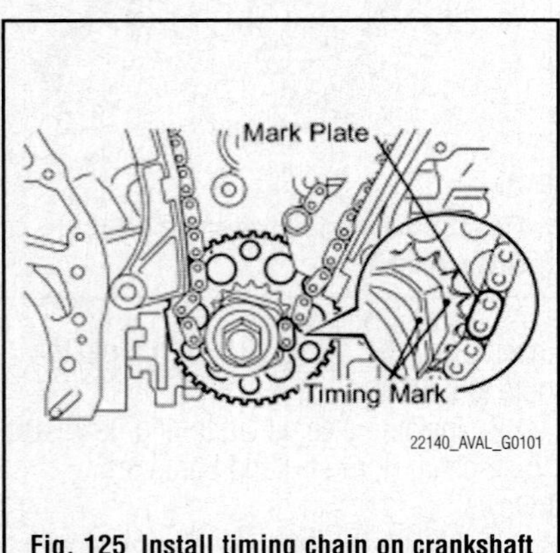

Fig. 125 Install timing chain on crankshaft

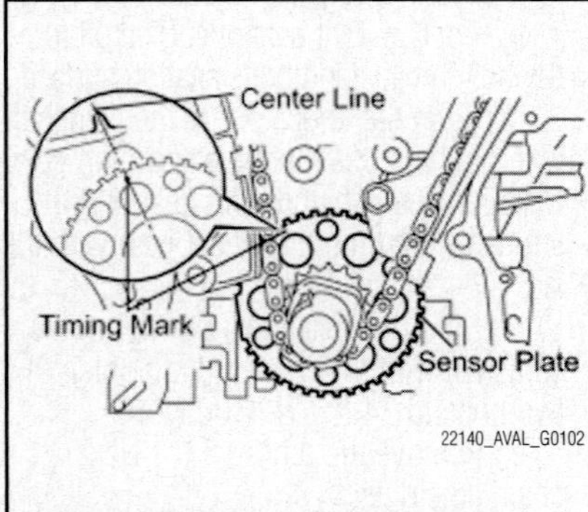

Fig. 126 Timing chain and crankshaft alignment

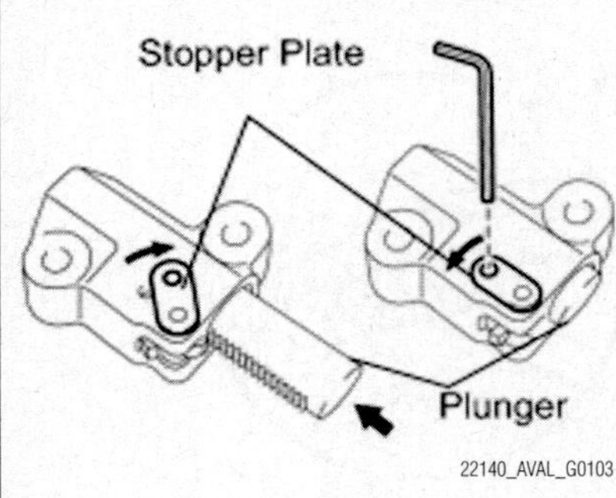

Fig. 127 Set chain tensioner plunger position

44. Install the timing gear set keys and crankshaft timing sprocket as shown in the illustration.

45. Install the idle sprocket assembly, as follows:

   a. Apply a light coat of engine oil to the rotating surface of the No. 1 idle gear shaft.

   b. Temporarily install the No. 1 idle gear shaft and idle sprocket with the No. 2 idle gear shaft while aligning the knock pin of the No. 1 idle gear with the knock pin groove of the cylinder block. Be careful of the idle gear direction.

   c. Using a 10 mm hexagon wrench, tighten the No. 2 idle gear shaft to 44 ft. lbs. (60 Nm).

   d. After installing the idle sprocket assembly, check that the idle sprocket turns smoothly.

46. Install the chain sub-assembly, as follows:

   a. Align the mark plate and timing marks as shown in the illustration and install the chain. The camshaft mark plate is orange.

   b. Do not pass the chain over the crankshaft, just put it on.

   c. Turn the camshaft timing gear assembly on the right bank counterclockwise to tighten the chain between the banks.

➡When the idle sprocket assembly is reused, align the timing chain plate with the mark on the sprocket in order to tighten the chain between the banks.

   d. Align the mark plate and timing marks as shown in the illustration and install the chain onto the crankshaft timing sprocket. The crankshaft to mark plate is yellow.

   e. Temporarily tighten the pulley set bolt.

   f. Turn the crankshaft clockwise to set it to the right-hand block bore more centerline. (TDC/compression).

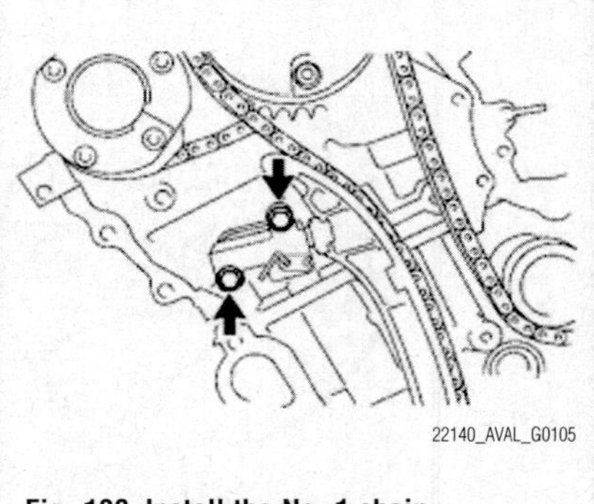

Fig. 128 Install the No. 1 chain tensioner with the 2 bolts

47. Install the chain tensioner slipper.

48. Install the No. 1 chain tensioner assembly, as follows:

   a. Move the stopper plate upward to release the lock, and push the plunger deep into the tensioner.

   b. Move the stopper plate downward to set the lock, and insert a hexagon wrench into the hole of the stopper plate.

   c. Install the No. 1 chain tensioner with the 2 bolts and tighten to 7 ft. lbs. (10 Nm).

   d. Remove the lock pin of the No. 1 chain tensioner. Check that each timing mark is aligned with the crankshaft at TDC/compression.

   e. Remove the pulley set bolt.

49. Install timing chain cover and seal.

50. Install the water pump assembly.

51. Install the water inlet housing.

52. Install the left hand No. 1 front engine mounting bracket.

53. Install the oil pan sub-assembly.

54. Install the oil strainer sub-assembly.

55. Install the No. 2 oil pan sub-assembly.

56. Install the oil pan drain plug and gasket.

57. Install the cylinder head cover sub-assembly.

58. Install the left-hand cylinder head cover sub-assembly.

59. Install the water outlet.

60. Install the crankshaft pulley.

61. Install the oil filter.

62. Install the cylinder block water drain cock sub-assembly.

63. Install the No. 1 oil pipe.

64. Install the oil pipe.

65. Install the crankshaft position sensor.

66. Install the camshaft timing oil control valve assembly.

67. Install the camshaft position sensor.

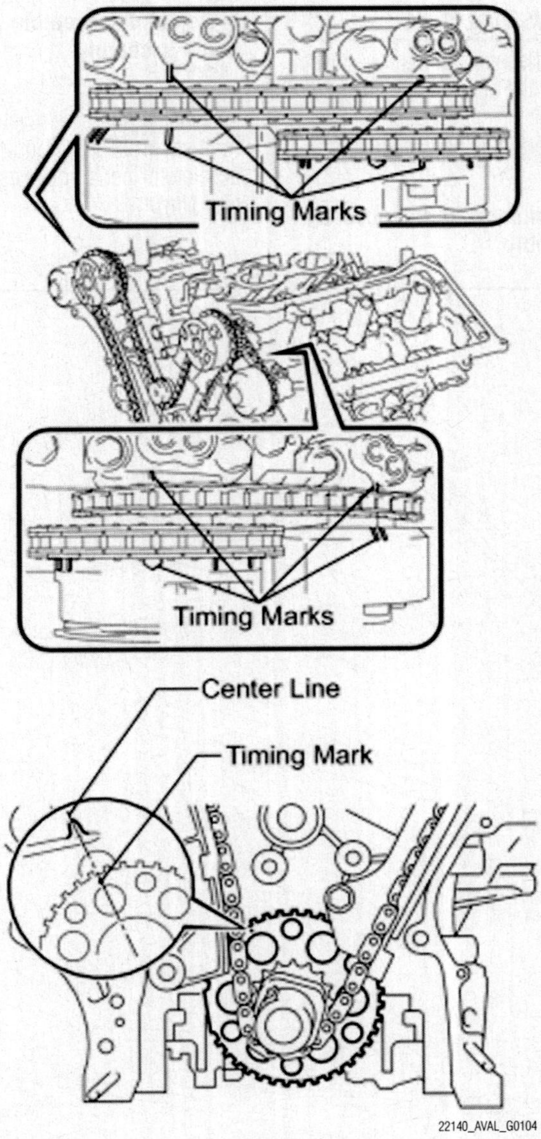

Fig. 129 Aligning timing marks

68. Install the ventilation valve sub-assembly.
69. Install the spark plugs and ignition coil assembly.
70. Install the oil filler cap sub-assembly.
71. Install the engine assembly with transaxle.

VALVE LASH

ADJUSTMENT

See Figure 130.

➡**Keep the lash adjuster free of dirt and foreign objects.**

➡**Only use clean engine oil.**

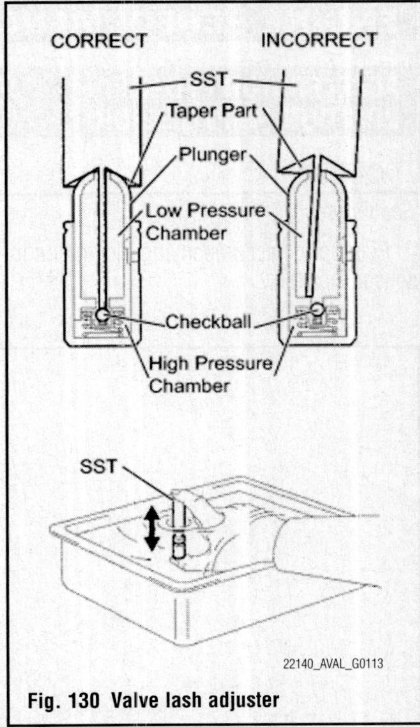

Fig. 130 Valve lash adjuster

1. Place the lash adjuster into a container filled with engine oil.
2. Insert the SST's (SST: 09276-75010) tip into the lash adjuster's plunger and use the tip to press down on the check ball inside the plunger.
3. Squeeze the SST and lash adjuster together to move the plunger up and down 5 to 6 times.
4. Check the movement of the plunger and bleed the air. OK: Plunger moves up and down.

➡**When bleeding air from the high-pressure chamber, make sure that the tip of the SST is actually pressing the check ball as shown in the illustration. If the check ball is not pressed, air will not bleed.**

5. After bleeding the air, remove the SST. Then, try to press the plunger quickly and firmly with a finger. OK: Plunger is very difficult to move. If the result is not as specified, replace the lash adjuster.
6. Install the lash adjusters.

➡**Install the lash adjuster to the same place where it was removed from.**

ENGINE PERFORMANCE & EMISSION CONTROLS

ACCELERATOR PEDAL POSITION (APP) SENSOR

LOCATION

See Figure 131.

Refer to the accompanying illustration for sensor location.

REMOVAL & INSTALLATION

1. Remove left center floor carpet cover.
2. Disconnect the accelerator pedal connector.
3. Remove the 2 nuts and accelerator pedal assembly.

➡Avoid physical shock to the accelerator pedal assembly.

➡Do not disassemble the accelerator pedal assembly.

To install:

Installation is the reverse of the removal procedure. Tighten the accelerator pedal rod nuts to 43 inch lbs. (5.4 Nm).

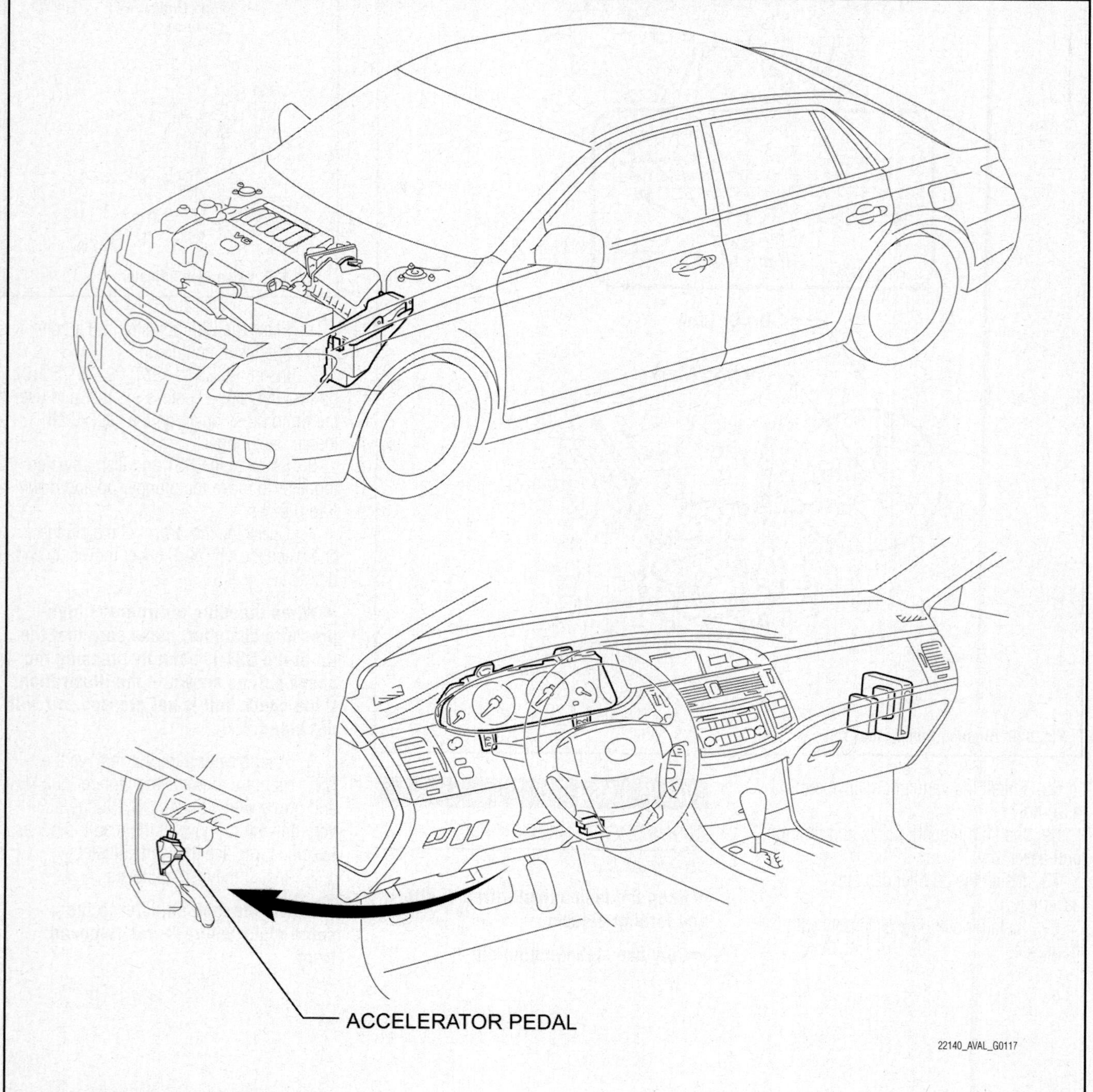

Fig. 131 Accelerator pedal assembly location

CAMSHAFT POSITION (CMP) SENSOR

LOCATION

See Figure 132.

Refer to the accompanying illustration for sensor location.

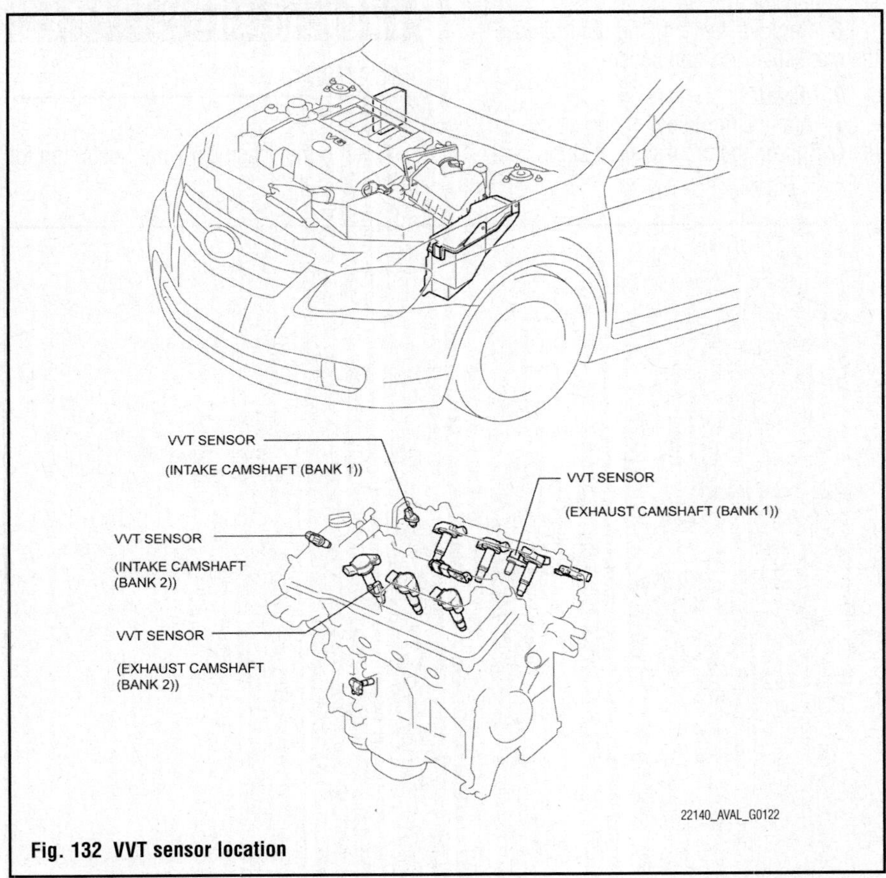

Fig. 132 VVT sensor location

REMOVAL & INSTALLATION

See Figures 133 through 136.

1. Remove the V-bank cover sub-assembly.

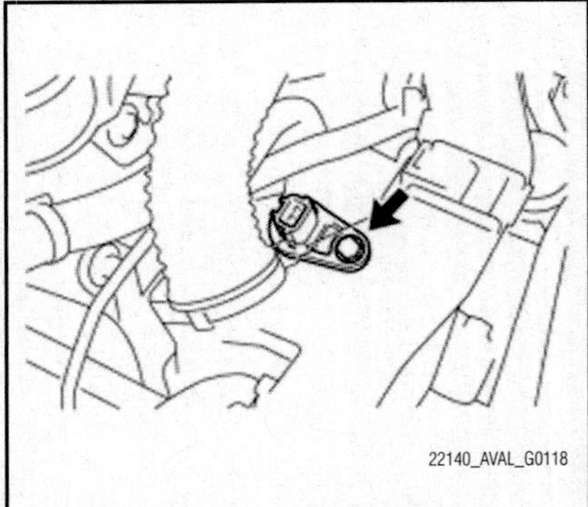

Fig. 133 Bank 1 intake camshaft VVT sensor

2. Remove the Intake camshaft VVT sensor (Bank 1), as follows:
   a. Disconnect the VVT sensor connector.
   b. Remove the bolt and VVT sensor.
3. Remove the Exhaust camshaft VVT sensor (Bank 1), as follows:
   a. Remove the windshield wiper link assembly.
   b. Remove the front cowl top outside panel.
   c. Disconnect the VVT sensor connector.
   d. Remove the bolt and VVT sensor.
4. Remove the Exhaust camshaft VVT sensor (Bank 2), as follows:
   a. Disconnect the VVT sensor connector.
   b. Remove the bolt and VVT sensor.
5. Remove the Intake camshaft VVT sensor (Bank 2), as follows:
   a. Disconnect the VVT sensor connector.
   b. Remove the bolt and VVT sensor.

To install:

6. Install the VVT sensor.
   a. For the intake camshaft (Bank 1), install the sensor with the bolt. Tighten the bolt to 7 ft. lbs. (10 Nm). Connect the VVT sensor connector.
   b. For the exhaust camshaft (Bank 1), install the VVT sensor with the bolt. Tighten the bolt to 7 ft. lbs. (10 Nm). Connect the connector. Install the cowl top outside panel front. Install the windshield wiper link assembly.
   c. For the exhaust camshaft (Bank 2), install the VVT sensor with the bolt. Tighten the bolt to 7 ft. lbs. (10 Nm). Connect the VVT sensor connector.

Fig. 134 Bank 1 exhaust camshaft VVT sensor

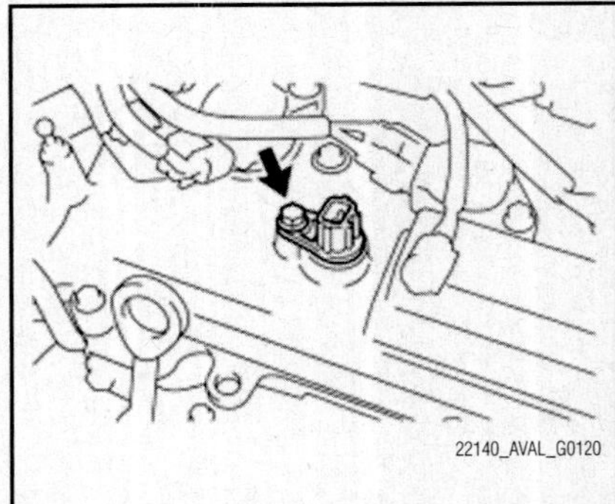

Fig. 135 Bank 2 intake camshaft VVT sensor

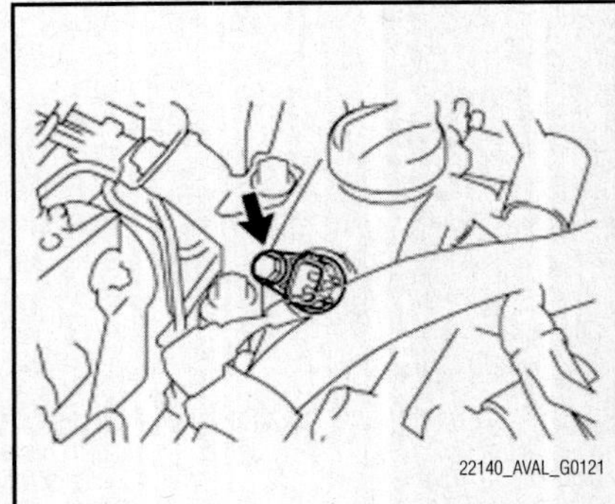

Fig. 136 Bank 2 exhaust camshaft VVT sensor

d. For the intake camshaft (Bank 2), install the VVT sensor with the bolt. Tighten the bolt to 7 ft. lbs. (10 Nm). Connect the VVT sensor connector.

7. Install the V-bank cover sub assembly.

CRANKSHAFT POSITION (CKP) SENSOR

LOCATION

See Figure 137.

Refer to the accompanying illustration for sensor location.

REMOVAL & INSTALLATION

1. Disconnect the negative battery terminal.
2. Remove the V-ribbed belt.
3. Remove alternator assembly.
4. Disconnect the cooler compressor assembly.
5. Remove the crankshaft position sensor connector.
6. Remove the bolt, and then remove the crankshaft position sensor.

To install:

7. Apply a light coat of engine oil to the O-ring on the crankshaft position sensor.
8. Install the crankshaft position sensor with the bolt and tighten to 7 ft. lbs. (10 Nm).
9. Connect the crankshaft position sensor connector.
10. The remainder of installation is the reverse of the removal procedure.

ELECTRONIC CONTROL MODULE (ECM)

LOCATION

See Figure 138.

Refer to the accompanying illustration for ECM location.

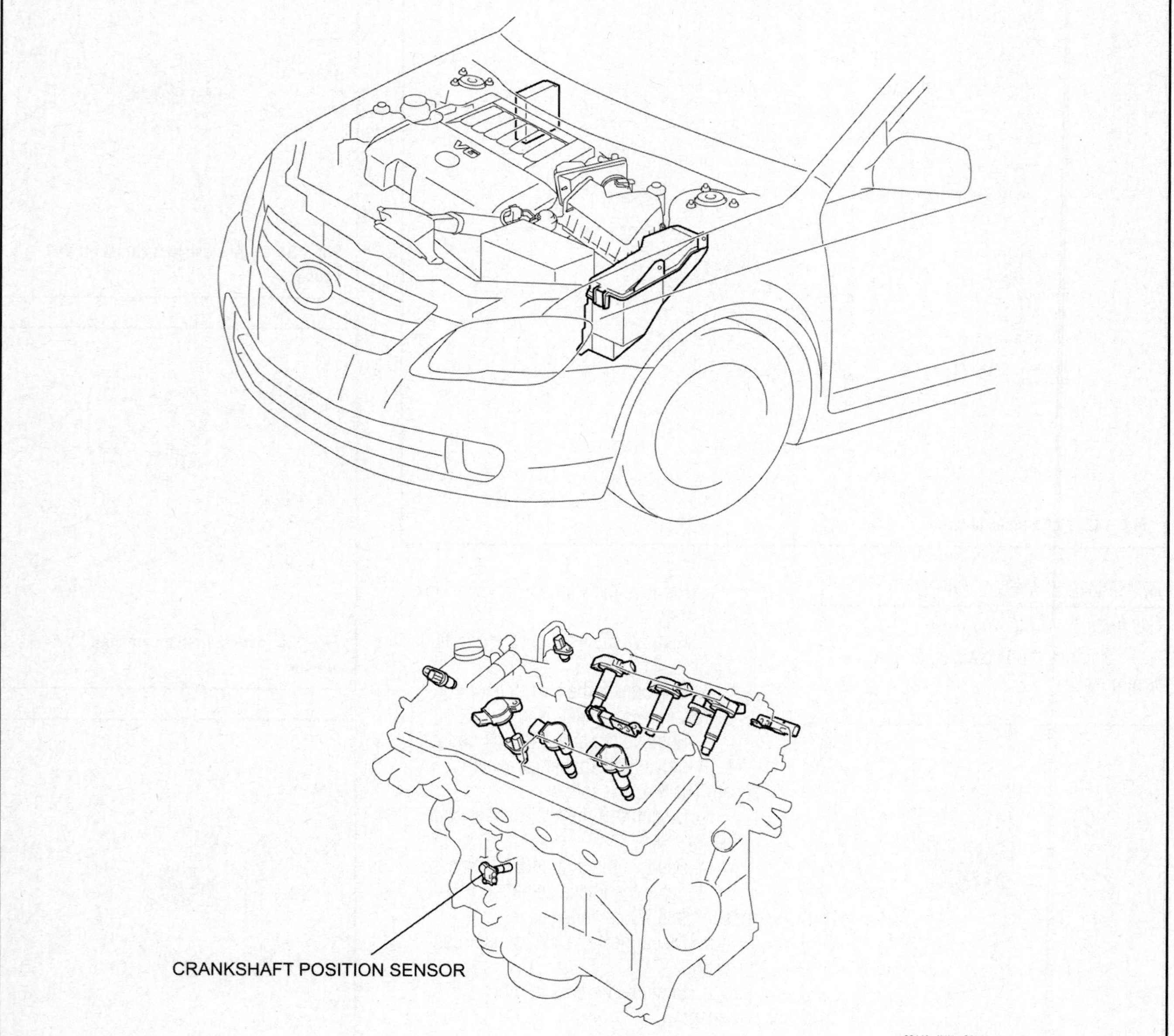

Fig. 137 Crankshaft position sensor location

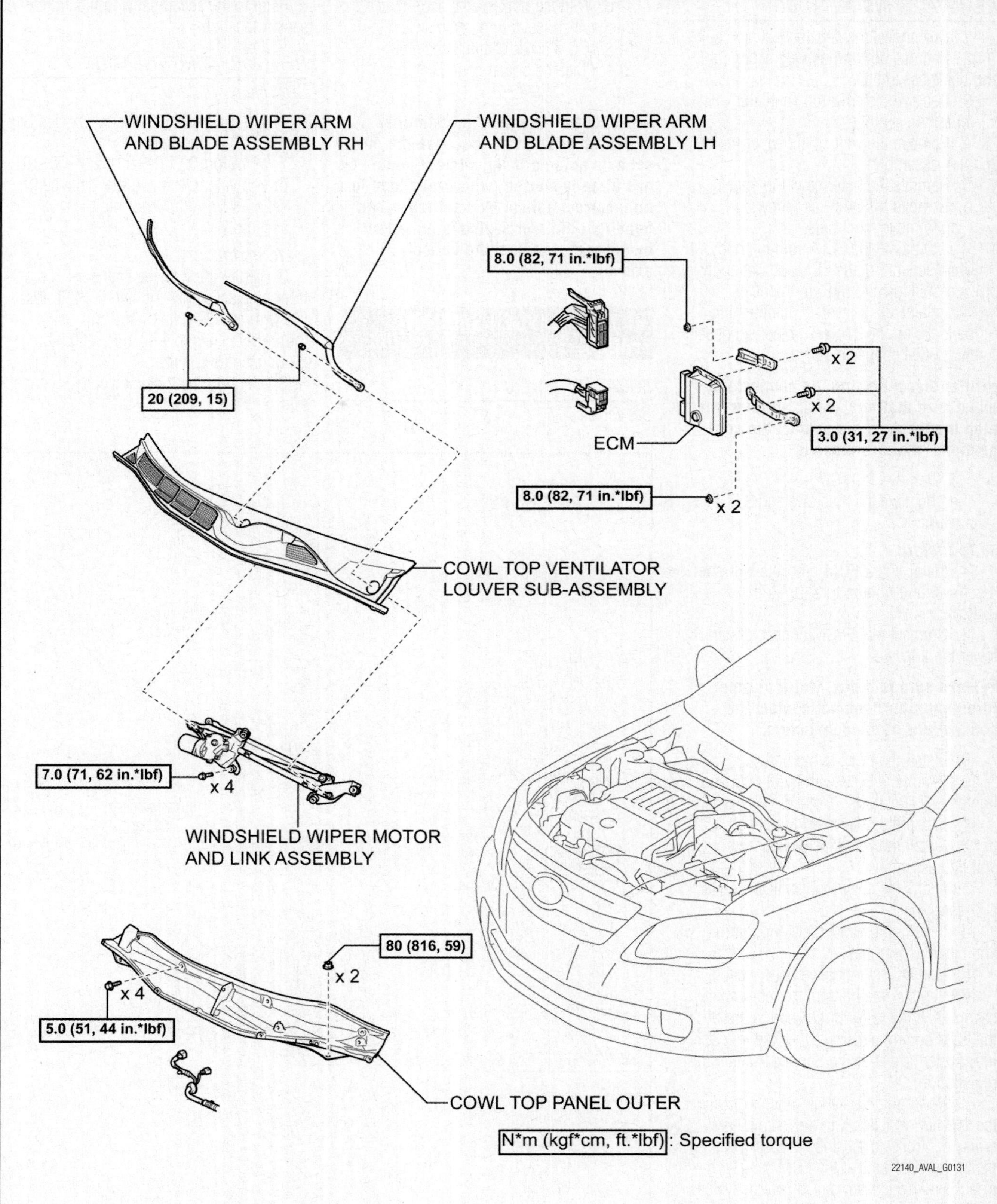

Fig. 138 ECM location

REMOVAL & INSTALLATION

1. Disconnect the negative battery cable.
2. Remove both windshield wiper arm and blade assemblies.
3. Remove the cowl top ventilator louver sub-assembly.
4. Remove the windshield wiper motor and link assembly.
5. Remove the outer cowl top panel.
6. Remove the ECM, as follows:
   a. Remove the 3 nuts.
   b. Separate the ECM from the body. When separating the ECM, do not apply excessive force to the wire harness.
   c. Raise the 2 levers while pushing the locks on the 2 levers, and disconnect the 2 ECM connectors.

➡After disconnecting the connectors, make sure that dirt, water or other foreign matter does not contact the connections of the connectors.

   d. Remove the ECM.
   e. Remove the 4 screws and 2 ECM brackets.

To install:

7. Install the 2 ECM brackets with the 4 screws, and tighten to 27 inch lbs. (3 Nm).
8. Connect the 2 ECM connectors and lower the 2 levers.

➡Make sure that dirt, water or other foreign matter does not contact the connections of the connectors.

9. Install the ECM to the body.
10. Attach the ECM with the 3 nuts and tighten to 71 inch lbs. (8 Nm).
11. Install the outer cowl top panel.
12. Install the windshield wiper motor and link assembly.
13. Install the cowl top ventilator louver sub-assembly.
14. Install both windshield wiper arm and blade assemblies.
15. Connect the negative battery cable.
16. Register the immobilizer communication ID. If the ECM is replaced, register the ECM communication ID for the immobilizer system (refer to the Service Bulletin for registration).
17. Perform initialization. After replacing the ECM on vehicles with a dynamic laser cruise control system, it is necessary to initialize the ECM so that the ECM can recognize the dynamic laser cruise control system.
18. Be sure to perform the following procedure after replacing the ECM:
   a. Turn the ignition switch on (IG).
   b. Turn the cruise control main switch on.
   c. With the brake pedal depressed, push the cruise control main switch to RES/ACC 3 times within 3 seconds. Check that the buzzer sounds at this time.

➡Do not turn the headlight dimmer switch on at this time because the optical axis automatic adjustment mode has already started, which may lead to an incorrect optical axis setting. If the headlight dimmer switch is turned on by mistake, readjust the optical axis.

HEATED OXYGEN SENSOR (HO2S)

LOCATION

See Figure 139.

Refer to the accompanying illustration for sensor location.

REMOVAL & INSTALLATION

See Figures 140 and 141.

1. Disconnect the 2 Heated Oxygen Sensor (HO2S) connectors.
   a. Using SST (SST: 09224-00010) or equivalent, remove the 2 oxygen sensors from the front pipe assembly.

To install:

2. Install the 2 oxygen sensors to the front pipe assembly. Tighten to 32 ft. lbs. (44 Nm) and 30 ft. lbs. (40 Nm). Use a torque wrench with a fulcrum length of 11.81 inch (300 mm).
3. Connect the 2 oxygen sensor connectors.

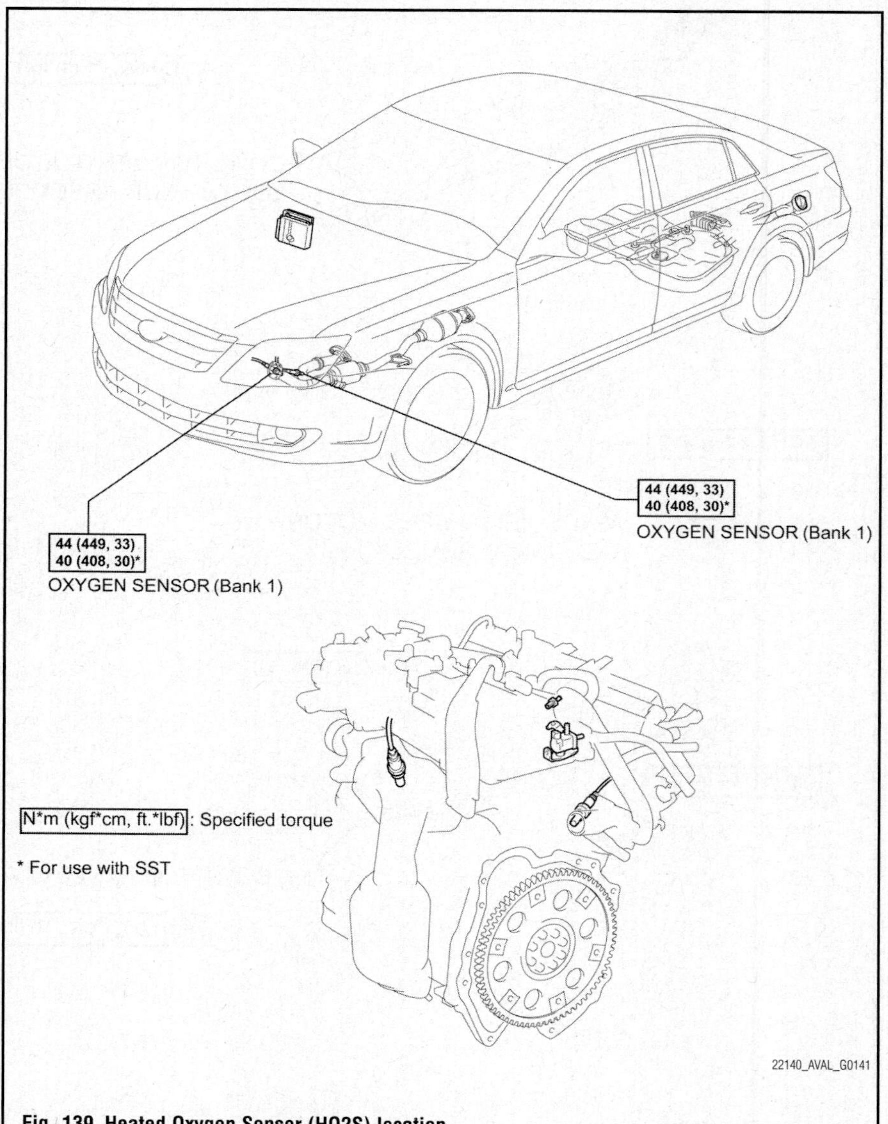

Fig. 139 Heated Oxygen Sensor (HO2S) location

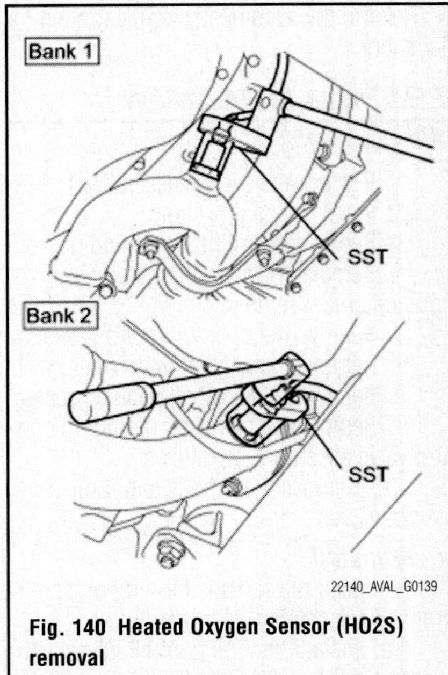

Fig. 140 Heated Oxygen Sensor (HO2S) removal

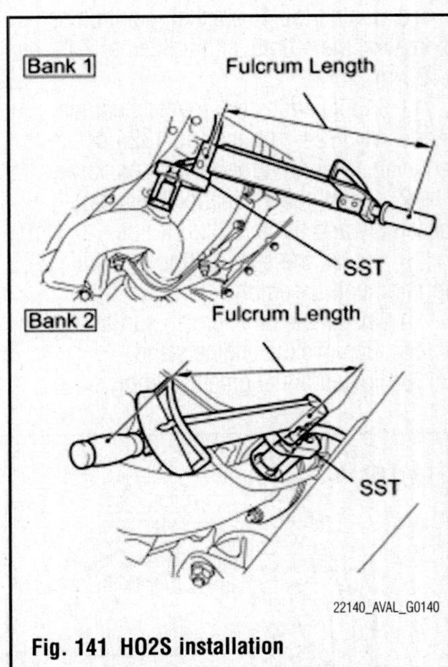

Fig. 141 HO2S installation

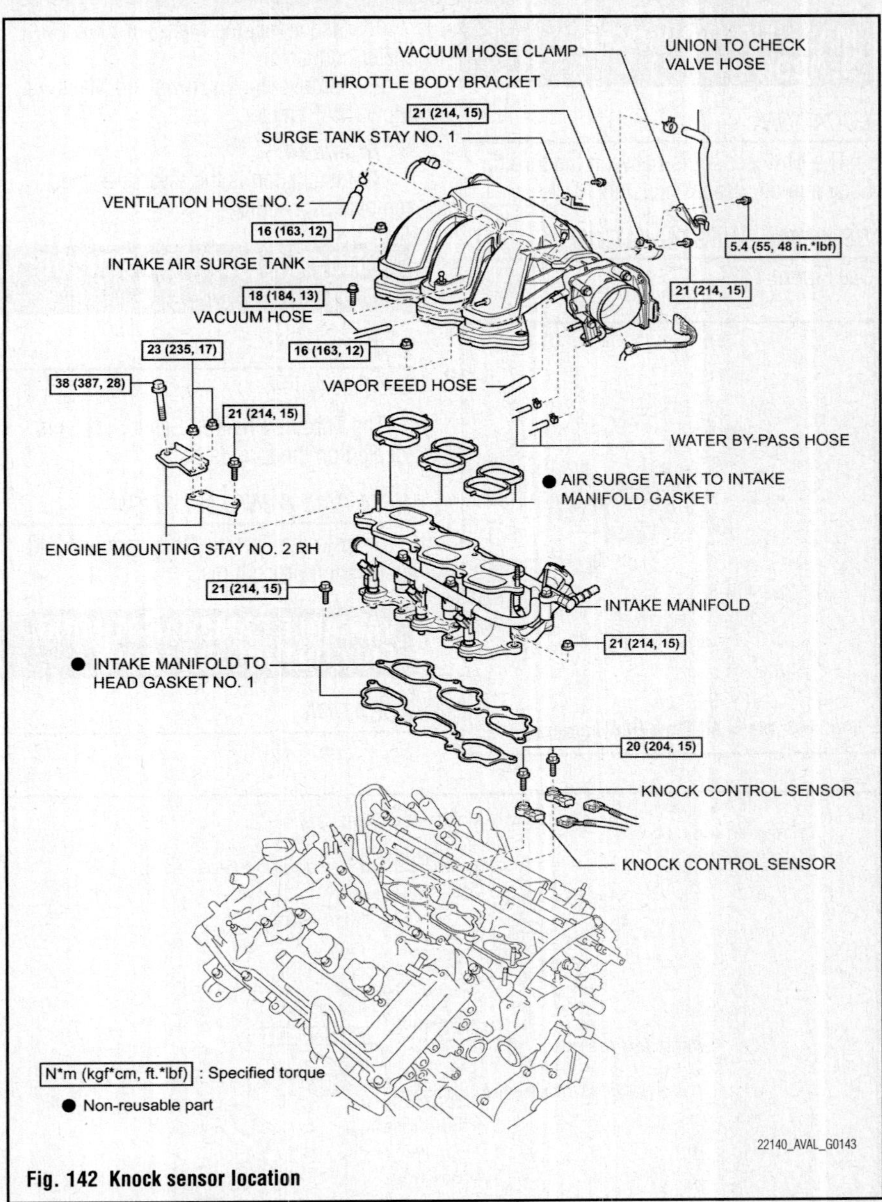

Fig. 142 Knock sensor location

INTAKE AIR TEMPERATURE (IAT) SENSOR

LOCATION

The Intake Air Temperature (IAT) sensor is mounted on the Mass Air Flow (MAF) meter.

REMOVAL & INSTALLATION

See Mass Air Flow (MAF) meter.

KNOCK SENSOR (KS)

LOCATION

See Figure 142.

Refer to the accompanying illustration for sensor location.

REMOVAL & INSTALLATION

1. Properly discharge the fuel system pressure.
2. Disconnect battery negative cable.
3. Drain engine coolant.
4. Remove the windshield wiper motor and link assembly.
5. Remove the outer cowl top panel.
6. Remove the V-bank cover sub-assembly.
7. Remove the air cleaner cap with air cleaner hose.
8. Remove the intake air surge tank.
9. Remove the intake manifold.
10. Disconnect the 2 knock control sensor connectors.
11. Remove the 2 bolts and 2 knock control sensors.

To install:

12. Install the 2 knock control sensors with the 2 bolts as shown in the illustration and tighten to 15 ft. lbs. (20 Nm).
13. Connect the 2 knock control sensor connectors.
14. The remainder of installation is the reverse of the removal procedure.
15. Inspect for fuel leak and check the function of throttle body.

MALFUNCTION INDICATOR LIGHT (MIL)

RESET PROCEDURE

Clear the DTC codes.

MASS AIR FLOW (MAF) SENSOR

LOCATION

The MAF sensor is between the throttle body and air cleaner housing.

REMOVAL & INSTALLATION

See Figure 143.

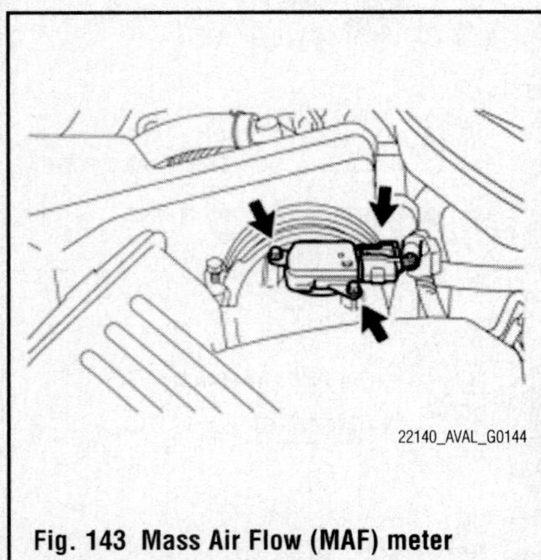

Fig. 143 Mass Air Flow (MAF) meter

1. Disconnect the Mass Air Flow (MAF) meter connector.
2. Remove the 2 screws and Mass Air Flow (MAF) meter.

To install:

3. Installation is the reverse of the removal procedure.

THROTTLE POSITION SENSOR (TPS)

LOCATION

See Figure 144.

The Throttle Position Sensor (TPS) is located on the throttle body.

REMOVAL & INSTALLATION

Refer to the Throttle Body removal and installation procedures.

VARIABLE CAMSHAFT TIMING OIL CONTROL SOLENOID

LOCATION

See Figure 145.

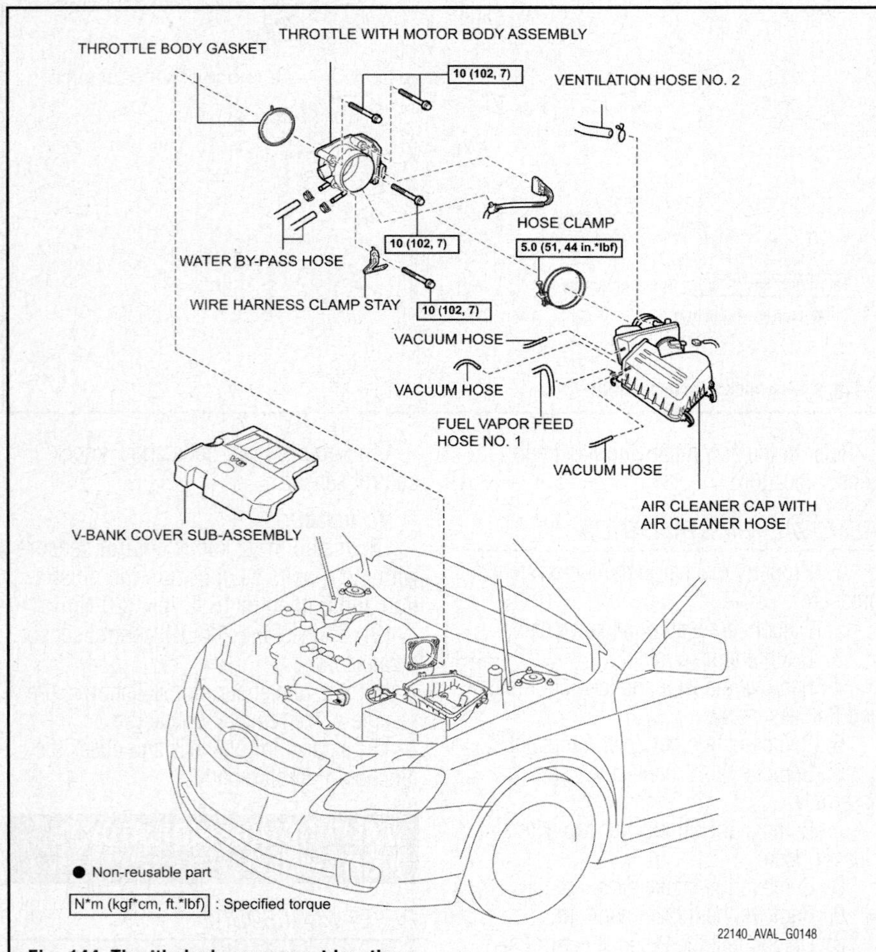

Fig. 144 Throttle body component locations

Refer to the accompanying illustration for location.

REMOVAL & INSTALLATION

See Figure 145.

1. Remove the engine assembly.
2. Install on engine stand.
3. Remove the oil filler cap and gasket.
4. Remove the spark plugs and ignition coil assembly.
5. Remove the drain plug and gasket.
6. Remove the ventilation valve.
7. Remove the camshaft position sensor.
8. Remove the camshaft timing oil control valve assembly, as follows:
    a. Remove the 4 bolts and 4 oil control valves.

To install:

9. Install the camshaft timing oil control valve assembly, as follows:
    a. Install the 4 oil control valves with the 4 bolts and tighten to 7 ft. lbs. (10 Nm).
10. Install the 4 camshaft position sensors with the 4 bolts and tighten to 7 ft. lbs. (10 Nm).
11. Apply adhesive (Toyota Genuine Adhesive 1324, Three Bond 1324 or equivalent) around the ventilation valve.
12. Install the ventilation valve sub-assembly and tighten to 20 ft. lbs. (27 Nm).
13. Install the spark plugs and the ignition coil assembly.
14. Install the oil filler cap sub-assembly.
15. Remove the engine stand.
16. Install the engine assembly.

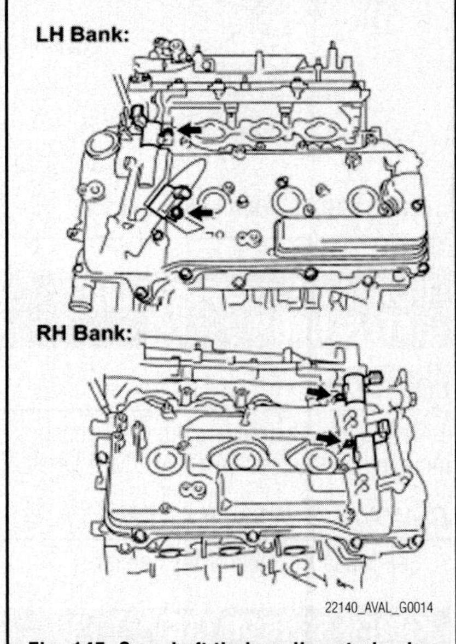

Fig. 145 Camshaft timing oil control valve assembly (Variable camshaft timing oil control solenoid) location

FUEL SYSTEM SERVICE PRECAUTIONS

Safety is the most important factor when performing not only fuel system maintenance, but any type of maintenance. Failure to conduct maintenance and repairs in a safe manner may result in serious personal injury or death. Work on a vehicle's fuel system components can be accomplished safely and effectively by adhering to the following rules and guidelines.

- To avoid the possibility of fire and personal injury, always disconnect the negative battery cable unless the repair or test procedure requires that battery voltage be applied.
- Always relieve the fuel system pressure prior to disconnecting any fuel system component (injector, fuel rail, pressure regulator, etc.) fitting or fuel line connection. Exercise extreme caution whenever relieving fuel system pressure to avoid exposing skin, face and eyes to fuel spray. Please be advised that fuel under pressure may penetrate the skin or any part of the body that it contacts.
- Always place a shop towel or cloth around the fitting or connection prior to loosening to absorb any excess fuel due to spillage. Ensure that all fuel spillage is quickly removed from engine surfaces. Ensure that all fuel-soaked cloths or towels are deposited into a flame-proof waste container with a lid.
- Always keep a dry chemical (Class B) fire extinguisher near the work area.
- Do not allow fuel spray or fuel vapors to come into contact with a spark or open flame.
- Always use a second wrench when loosening or tightening fuel line connection fittings. This will prevent unnecessary stress and torsion on fuel piping. Always follow the proper torque specifications.
- Always replace worn fuel fitting O-rings with new ones. Do not substitute fuel hose where rigid pipe is installed.

FUEL SYSTEM PRESSURE

RELIEVING

CAUTION

Perform the following procedures to prevent fuel from spilling out before removing any fuel system parts.

CAUTION

Pressure will still remain in the fuel line even after performing the following procedures. When disconnecting the fuel line, cover with a shop rag or a piece of cloth to prevent fuel from spraying or spewing out.

1. Disconnect the fuel pump connector.
   a. Remove the rear seat cushion assembly.
   b. Remove the rear floor service hole cover.
   c. Disconnect the fuel pump connector.
   d. Start the engine.
   e. After the engine stops, turn the ignition switch off.

➡DTC P0171/25 (fuel problem) may be detected.

   f. Crank the engine again. Check that the engine does not start.
   g. Remove the fuel tank cap to discharge pressure from the fuel tank.
   h. Disconnect the cable from the negative (-) battery terminal.
   i. Reconnect the fuel pump connector.
   j. Install the rear floor service hole cover.
   k. Install the rear seat.

FUEL FILTER

REMOVAL & INSTALLATION

See Figure 146.

1. Before servicing the vehicle, refer to the precautions section.

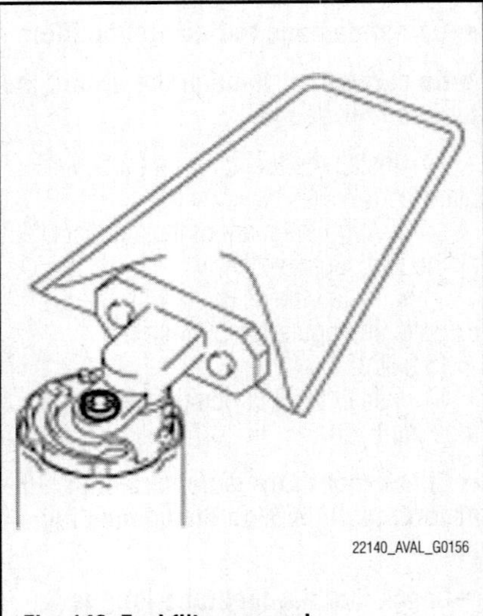

Fig. 146 Fuel filter removal

2. Remove the fuel pump from the vehicle.
3. Remove the fuel pump filter, as follows:
   a. Using a screwdriver, pry out the clip.
   b. Pull out the fuel pump filter from the fuel pump.

To install:

4. Install the fuel pump filter with a new clip.
5. Install the fuel pump.

FUEL LEVEL SENDING UNIT

REMOVAL & INSTALLATION

See Figures 147 and 148.

1. Before servicing the vehicle, refer to the precautions section.
2. Remove the fuel pump from the vehicle.
3. Remove the fuel sender gauge assembly, as follows:
   a. Disconnect the fuel sender gauge connector.
   b. Unlock the fuel sender gauge, and slide it to remove.

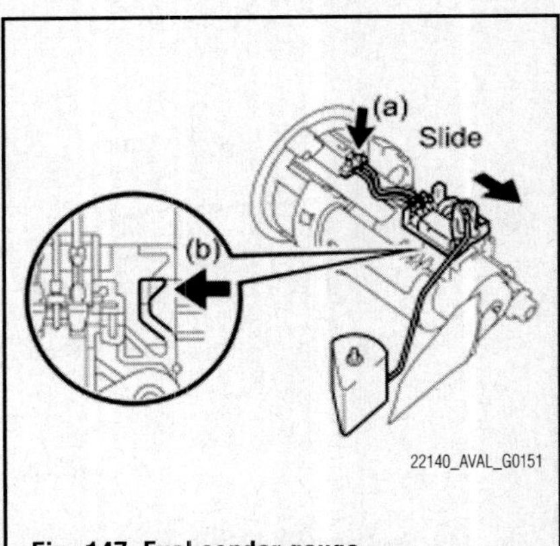

Fig. 147 Fuel sender gauge

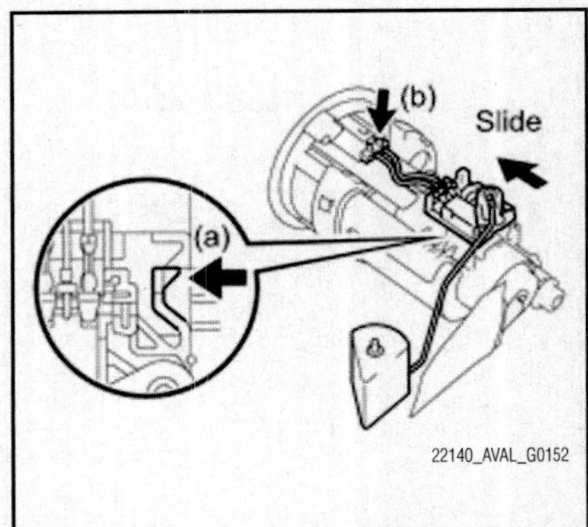

Fig. 148 Fuel sender gauge installation

To install:

4. Slide the fuel sender gauge to engage with the claw.
   a. Connect the fuel sender gauge connector.
5. Install the fuel pump.

FUEL PUMP

REMOVAL & INSTALLATION

See Figures 149 and 150.

1. Before servicing the vehicle, refer to the precautions section.
2. Discharge fuel system pressure.
3. Disconnect battery negative cable.
4. Remove the rear seat cushion assembly.
   a. Remove the rear floor service hole cover.
   b. Disconnect the fuel pump connector.

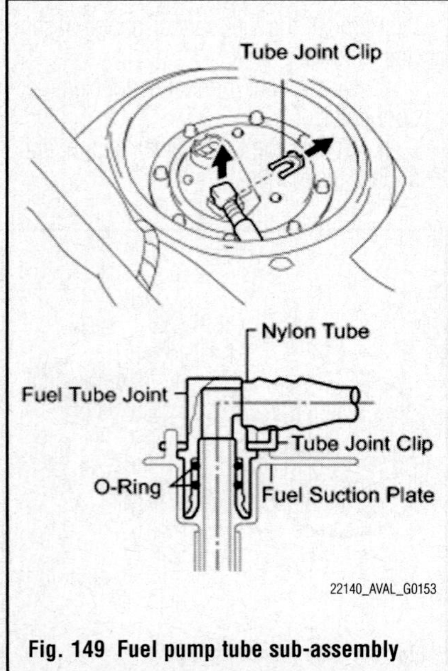

Fig. 149 Fuel pump tube sub-assembly

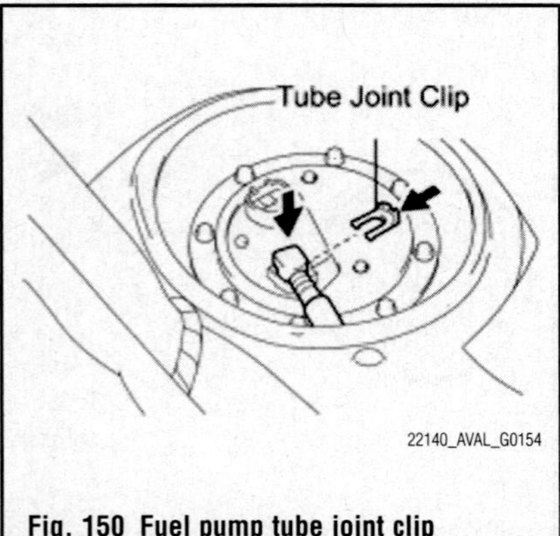

Fig. 150 Fuel pump tube joint clip

5. Separate the fuel pump tube sub-assembly, as follows:
   a. Remove the tube joint clip, and pull out the fuel pump tube.

➡Check if there is any dirt or mud around the connector before this operation and remove the dirt as necessary.

➡Be careful of mud because the quick connector has an O-ring which seals the pipe and connector that can be contaminated.

➡Do not use any tools in this operation.

➡Do not bend or twist the nylon tube. Cover the fuel tube joint with a plastic bag.

➡When the fuel tube joint and fuel suction plate are stuck, pinch the fuel tank tube between fingers, and turn it carefully to release it. Disconnect the fuel tank tube.

6. Remove fuel tank vent tube set plate, as follows:
   a. Remove the 8 bolts and set plate.
7. Remove fuel suction tube assembly with pump and gauge, as follows:
   a. Pull out the fuel suction tube from the fuel tank.

➡Do not damage the fuel pump filter.

➡Be careful not to bend the arm of the fuel sender gauge.

   b. Remove the gasket from the fuel suction tube.

To install:

8. Install a new gasket to the fuel suction tube.
9. Install the fuel suction tube.

➡Do not damage the fuel pump filter.

➡Be careful not to bend the arm of the fuel sender gauge.

10. Install the fuel tank vent tube set plate, as follows:
    a. Align the mark of the set plate with the fuel suction tube.
    b. Install the set plate with the 8 bolts and tighten to 52 inch lbs. (5.9 Nm).
11. Install the fuel pump tube with the tube joint clip.

➡Check that there is no scratches or foreign objects on the connecting part.

➡Check that the fuel tube joint is inserted securely.

➡Check that the tube joint clip is on the collar of the fuel tube joint.

➡After installing the tube joint clip, check that the fuel tube joint is pulled off.

12. Connect battery negative cable.
13. Inspect for fuel leak.
14. Install the rear floor service hole cover.
15. Install the rear seat cushion assembly.

FUEL PRESSURE REGULATOR

REMOVAL & INSTALLATION

See Figure 151.

1. Before servicing the vehicle, refer to the precautions section.
2. Remove the fuel pump from the vehicle.
3. Remove the fuel pump filter.
4. Pull out the fuel pressure regulator from the fuel tank fuel filter.
5. Remove the O-ring from the fuel pressure regulator.

Fig. 151 Fuel pressure regulator

To install:

6. Apply a light coat of spindle oil or gasoline to a new O-ring, and install it to the fuel pressure regulator.
7. Push in the fuel pressure regulator to the fuel tank fuel filter.
8. Install the fuel pump filter.
9. Install the fuel pump.

FUEL RAIL (SUPPLY MANIFOLD) & INJECTORS

REMOVAL & INSTALLATION

See Figures 152 through 154.

1. Properly discharge the fuel system pressure.

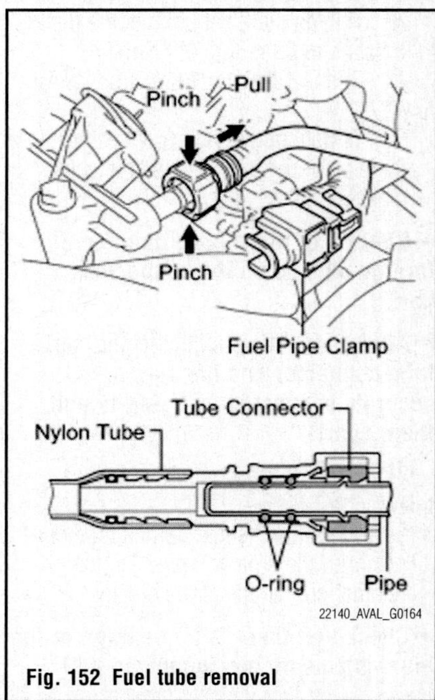

Fig. 152 Fuel tube removal

2. Disconnect battery negative cable.
3. Drain engine coolant.
4. Remove both windshield wiper arm and blade assemblies.
5. Remove the right cowl top ventilator louver.
6. Remove the windshield wiper motor and link assembly.
7. Remove the outer cowl top panel.
8. Remove the V-bank cover sub-assembly.
9. Remove the air cleaner cap with air cleaner hose.
10. Remove the intake air surge tank.
11. Disconnect the fuel tube sub-assembly, as follows:
   a. Remove the No. 2 fuel pipe clamp.
   b. Pinch the tube connector and then pull out the fuel pipe.

➡Check that there is no dirt or other foreign objects around the connector before removing fuel tube, and clean the connector as necessary.

➡It is necessary to prevent mud or dirt from entering the connector. If mud or dirt gets in the connector, the O-rings may not seal properly.

➡Do not use any tools in this operation.

➡Do not bend, kink or twist the nylon tube. Protect the connector by covering it with a plastic bag.

➡When the pipe and connector are stuck, push and pull the connector to release and pull the connector out carefully.

12. Remove the fuel injector assembly, as follows:
   a. Disconnect the 6 fuel injector connectors.
   b. Remove the 5 bolts and fuel delivery pipe together with the 6 fuel injectors.

➡Be careful not to drop the fuel injectors when removing the fuel delivery pipe.

   c. Remove the 6 insulators from the intake manifold.
   d. Pull out the fuel injector from the fuel delivery pipe.

To install:

13. Install the fuel injector assembly, as follows:
   a. Apply a light coat of spindle oil or gasoline to new O-rings, and install one to each injector.
   b. Apply a light coat of spindle oil or gasoline where the fuel delivery pipe contacts the O-ring.
   c. Push the fuel injector while twisting it back and forth to install it in the fuel delivery pipe.
   d. Position the fuel injector connector outward.

➡Be careful not to twist the O-ring.

➡After installing the fuel injector, check that it turns smoothly. If not, reinstall it with a new O-ring.

   e. Install 6 new insulators to the intake manifold.
   f. Place the fuel delivery pipe and the 6 fuel injectors together to the intake manifold.

➡Be careful not to drop the fuel injectors when installing the fuel delivery pipe.

   g. Temporarily install the 6 bolts which are used to hold the fuel delivery pipe to the intake manifold.

➡After installing the fuel injector, check that it turns smoothly. If not, reinstall it with a new O-ring.

   h. Tighten the 5 bolts which are used to hold the fuel delivery pipe to the intake manifold to 15 ft. lbs. (21 Nm).

14. Push in the tube connector to the pipe until the tube connector makes a "click" sound.

➡Before connecting the tube, make sure that it is not damaged. Make sure that there is no dirt present on the connecting surfaces.

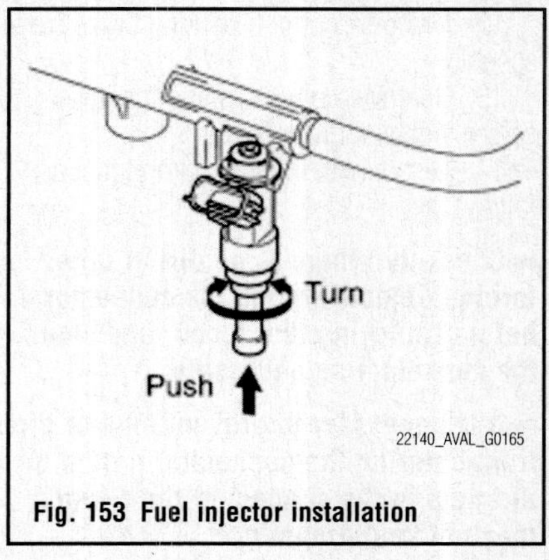

Fig. 153 Fuel injector installation

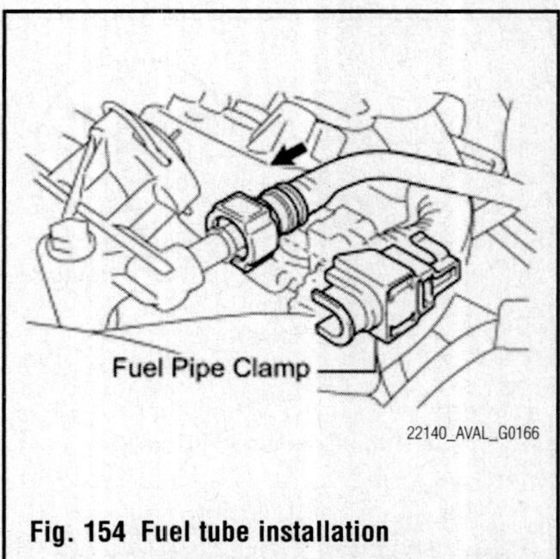

Fig. 154 Fuel tube installation

➡After connecting, check if the fuel tube connector and the pipe are securely connected by pulling on them.

15. Install the No. 2 fuel pipe clamp.
16. The remainder of installation is the reverse of the removal procedure.
17. Check for coolant leak and fuel leak.

FUEL TANK

REMOVAL & INSTALLATION

See Figures 155 through 157.

1. Before servicing the vehicle, refer to the precautions section.
2. Discharge fuel system pressure.
3. Disconnect battery negative cable.
4. Remove rear seat cushion assembly.
5. Remove rear floor service hole cover.
6. Separate fuel pump tube sub-assembly.
7. Remove fuel tank vent tube set plate.
8. Remove fuel suction tube assembly with pump and gauge.
9. Drain fuel.
10. Remove center exhaust pipe assembly.
11. Disconnect no. 2 parking brake cable assembly.

12. Disconnect no. 3 parking brake cable assembly.

13. Remove the 4 bolts and the lower center fuel tank protector.

14. Disconnect the fuel pump tube, as follows:

➡Check that there is no dirt or other foreign objects around the connector before removing fuel tubes, and clean the connector as necessary.

➡It is necessary to prevent mud or dirt from entering the connector. If mud or dirt gets in the connector, the O-rings may not seal properly.

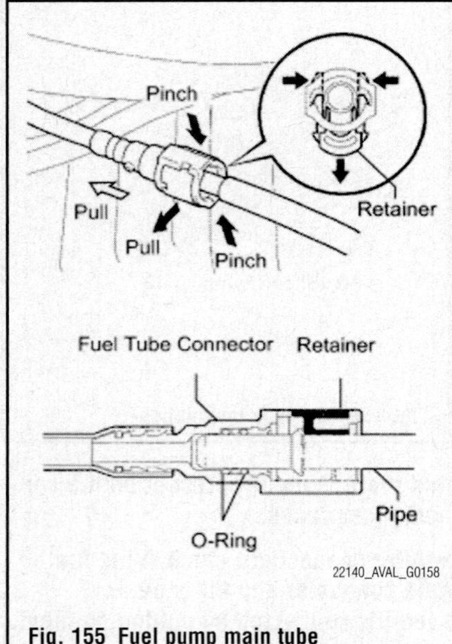

Fig. 155 Fuel pump main tube

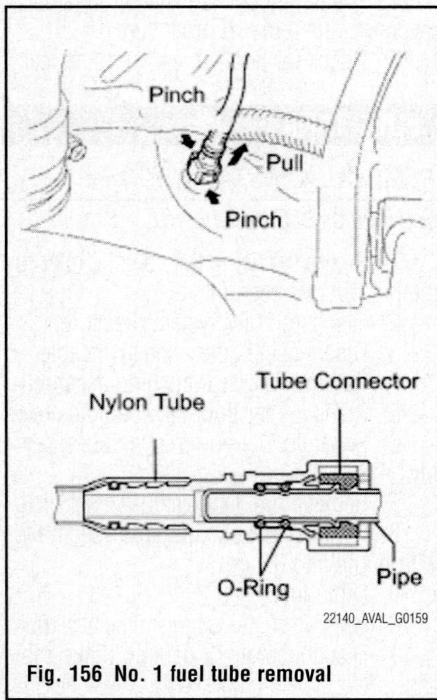

Fig. 156 No. 1 fuel tube removal

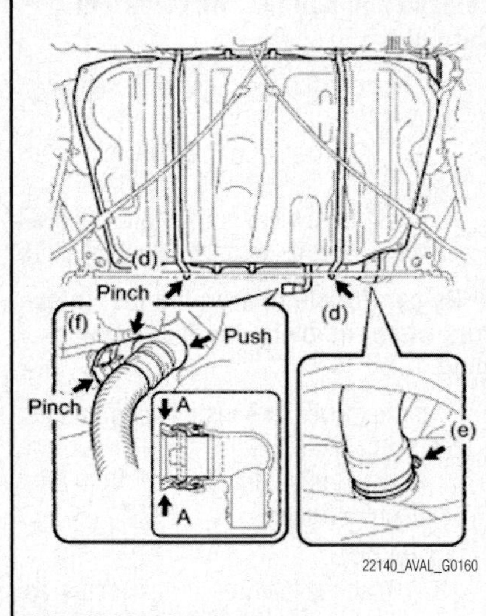

Fig. 157 Fuel tank removal

➡Do not use any tools in these operations.

➡Do not bend, kink or twist the nylon tubes. Protect the connector by covering it with a plastic bag.

➡When the pipe and connector are stuck, push and pull the connector to release and pull the connector out carefully.

a. Pinch the tabs of the retainer to remove the lock claws and pull it down as shown in the illustration.

b. Pull out the fuel tank main tube.

15. Pinch the tube connector and then pull out the No. 1 fuel tube.

16. Set up a transmission jack underneath the fuel tank.

17. Remove the 2 set bolts of the fuel tank bands.

18. Remove the hose clamp and disconnect the fuel tank to filter pipe hose.

19. Disconnect the fuel tank vent hose from the charcoal canister, as follows:

a. Push the connector deep into the charcoal canister to release the locking pin.

b. Pinch portion A.

c. Pull out the connector.

20. Remove the 2 pins and 2 fuel tank bands as shown in the illustration.

21. Remove the 4 clip nuts.

To install:

22. Install the 4 clip nuts.

23. Install the 2 fuel tank bands with the 2 pins.

24. Connect the fuel tank vent hose.

25. Connect the fuel tank inlet pipe with the fuel filter pipe clamp.

26. Tighten the 2 set bolts of the fuel tank bands to 29 ft. lbs. (39 Nm).

27. Connect the No. 1 fuel tube, as follows:

a. Push the fuel tube connector into the pipe until the fuel tube connector makes a "click" sound.

➡Check that there is no damage or foreign objects on the connected part.

➡After connecting, check if the fuel tube connector and the pipe are securely connected by trying to pull them apart.

28. Connect the fuel pump tube, as follows:

a. Push in the fuel pumps tube connector to the pipe and push up the retainer so that the claws engage.

➡Check that there is no damage or foreign objects on the connected part.

➡After connecting, check if the fuel tube connector and the pipe are securely connected by trying to pull them apart.

29. Install the lower center fuel tank protector and tighten to 48 inch lbs. (5.4 Nm).

30. Install the No. 3 parking brake cable assembly with the bolt and nut and tighten to 53 inch lbs. (6 Nm), and 75 inch lbs. (8.5 Nm).

31. Install the No. 2 parking brake cable assembly with the bolt and nut and tighten to 53 inch lbs. (6 Nm), and 75 inch lbs. (8.5 Nm).

32. Install the center exhaust pipe assembly.

33. Install the fuel suction tube assembly with pump and gauge.

34. Install the fuel tank vent tube set plate.

35. Connect the fuel pump tube sub-assembly.

36. Add fuel.

37. Connect battery negative cable.

38. Inspect for fuel leak and exhaust gas leak.

39. Install the rear floor service hole cover.

40. Install the rear seat cushion assembly.

IDLE SPEED

ADJUSTMENT

Idle speed is maintained by the ECM. No adjustment is necessary or possible.

THROTTLE BODY

REMOVAL & INSTALLATION

1. Before servicing the vehicle, refer to the precautions section.
2. Disconnect battery negative cable.
3. Drain engine coolant.
4. Remove the V-bank cover sub-assembly.
5. Remove the air cleaner cap with air cleaner hose.
6. Disconnect the 2 water by-pass hoses from the throttle w/ motor body assembly.
7. Disconnect the throttle w/ motor body assembly connector.
8. Remove the 4 bolts and throttle w/ motor body assembly from the intake air surge tank.
9. Remove the throttle body gasket from the intake air surge tank.

To install:

10. Install a new throttle body gasket to the intake air surge tank.
11. Install the throttle w/ motor body assembly and wire harness clamp stay to the intake air surge tank with the 4 bolts and tighten to 7 ft. lbs. (10 Nm).
12. Connect the throttle w/ motor body assembly connector.
13. The remainder of installation is the reverse of the removal procedure.
14. Check for coolant leak.
15. Check the function of the throttle body.

HEATING, VENTILATION & AIR CONDITIONING

BLOWER MOTOR

REMOVAL & INSTALLATION

See Figure 158.

1. Remove the glove box assembly.
2. Disconnect the connector from the blower motor.
3. Remove the 3 screws and the blower motor.

To install:

4. Replace the blower motor and the three screws.
5. Connect the electrical connector.
6. Replace the glove box assembly.

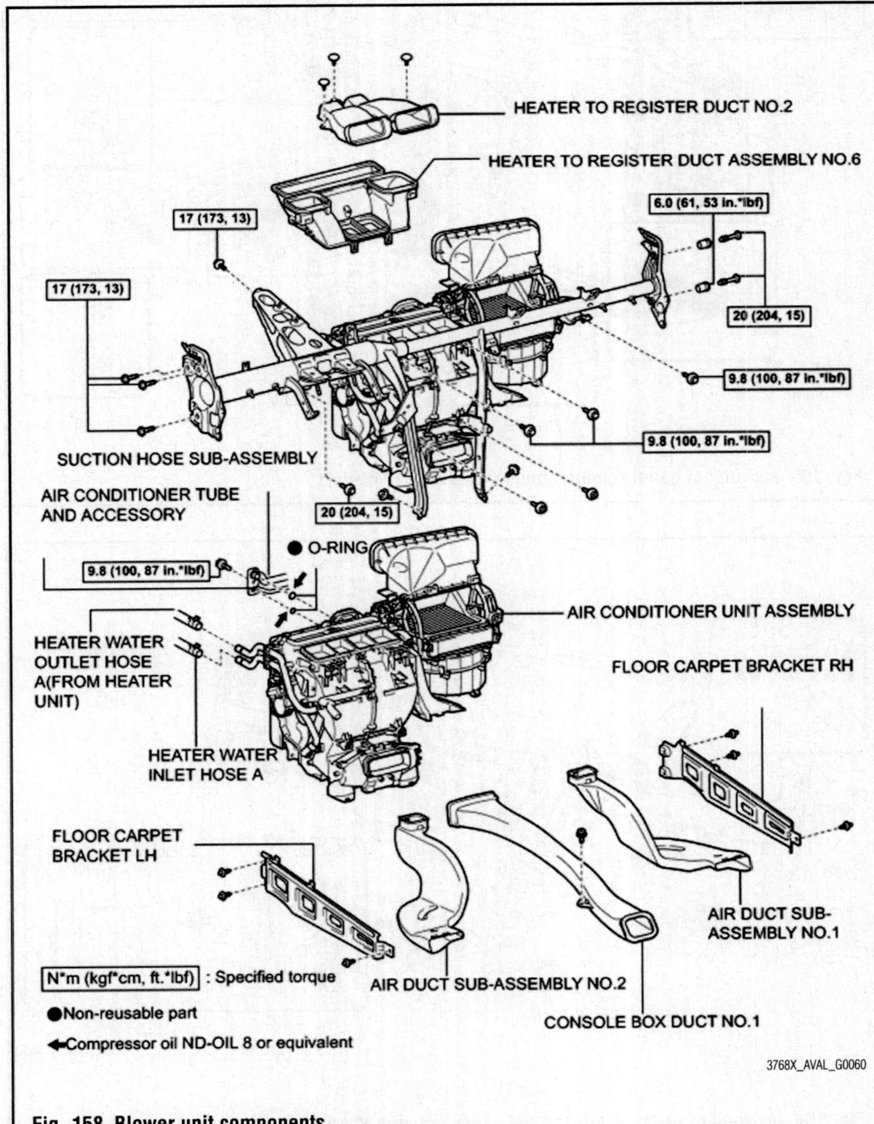

Fig. 158 Blower unit components

HEATER CORE

REMOVAL & INSTALLATION

See Figures 159 through 169.

1. Before servicing the vehicle, refer to the precautions section.
2. Disconnect battery negative terminal.

CAUTION

Wait for 90 seconds after disconnecting the cable to prevent the airbag from deploying.

3. Drain engine coolant.
4. Remove both windshield wiper arm and blade assemblies.
5. Remove the right cowl top ventilator louver.
6. Remove the windshield wiper motor and link.
7. Remove the outer front cowl top panel.
8. Discharge refrigerant from refrigeration system.

Fig. 159 Heater water outlet hoses

9. Disconnect the suction hose sub-assembly, as follows:
   a. Remove the bolt, and slide the hook connector.
   b. Disconnect the suction hose sub-assembly.
   c. Remove the O-ring from the suction hose sub-assembly.

➡Seal the openings of the disconnected parts using vinyl tape to prevent moisture and foreign matter from entering.

10. Disconnect the air conditioner tube and accessory.

11. Remove the O-ring from the air conditioner tube and accessory.

12. Slide the clip and disconnect the heater water outlet hose A.

13. Slide the clip and disconnect the heater water inlet hose A.

➡Do not apply excessive force to the heater water hoses

➡Prepare a drain pan or cloth in case the cooling water leaks.

14. Remove the instrument panel safety pad sub-assembly.

➡Refer to the removal procedures for the instrument panel safety pad sub-assembly w/ front passenger airbag assembly.

15. Remove the 3 clips and No. 2 heater to register duct.

16. Disengage the 4 claws and then remove the No. 6 heater to register duct.

17. Remove the clip and No. 1 console box duct.

18. Remove the left and right hand floor carpet brackets, as follows:

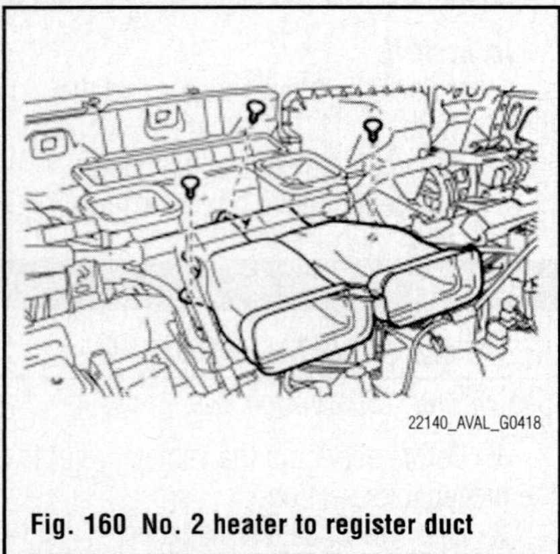

Fig. 160 No. 2 heater to register duct

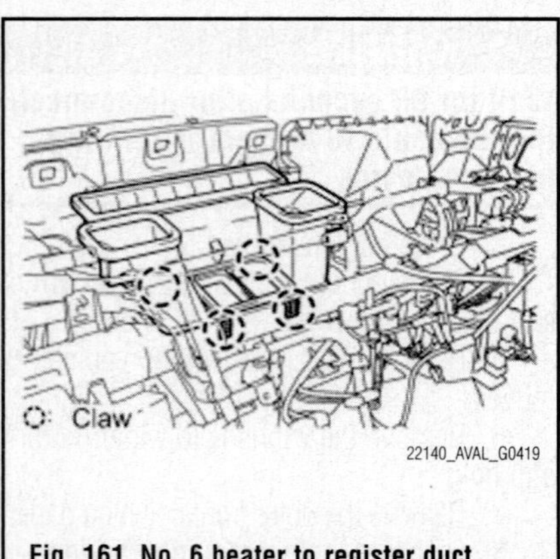

Fig. 161 No. 6 heater to register duct

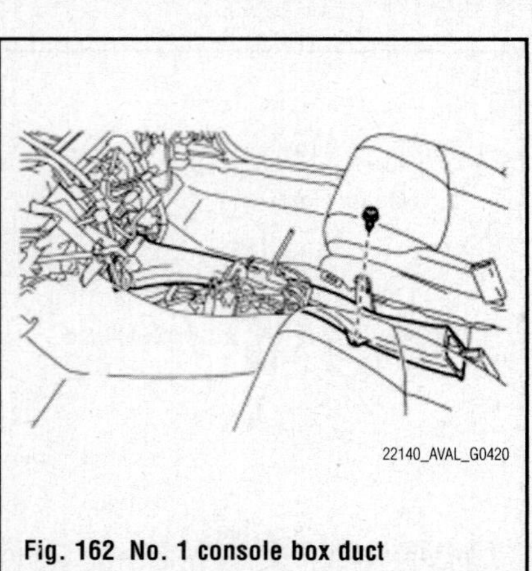

Fig. 162 No. 1 console box duct

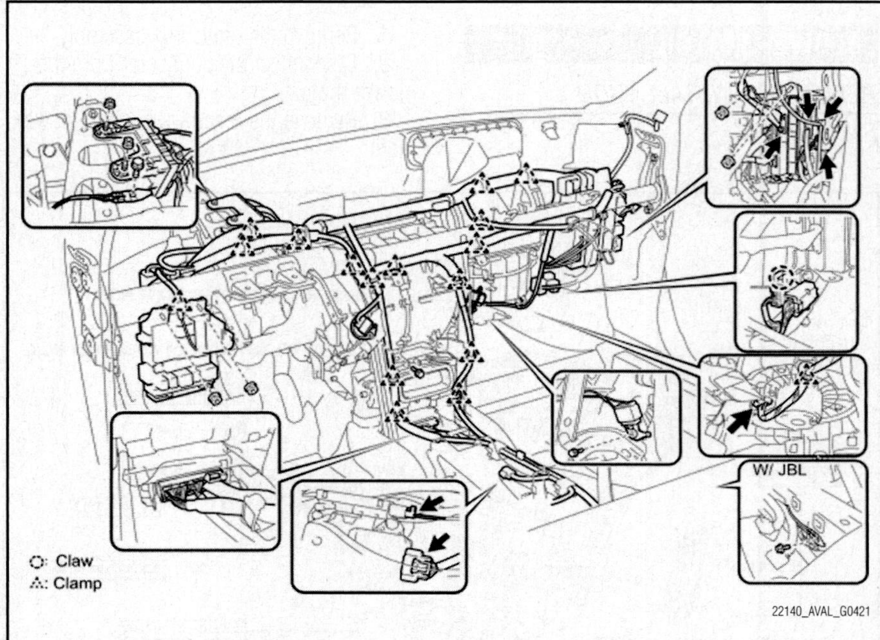

Fig. 163 Instrument panel reinforcement assembly connectors

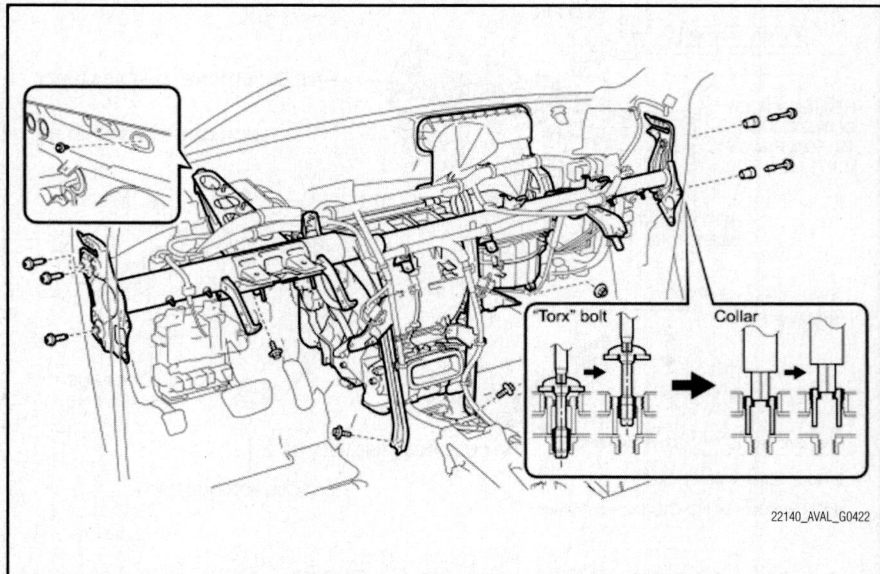

Fig. 164 Instrument panel reinforcement assembly with the air conditioner unit assembly

a. Release the clamps (2 from the left and 3 from the right).
b. Turn back the floor carpet.
c. Remove the 3 clips.
d. Remove the floor carpet brackets.

19. Release the 2 claws and remove the rear No. 2 air duct.
20. Release the 2 claws and remove the rear No. 1 air duct.
21. Remove the No. 1 air duct sub-assembly.
22. Separate the steering intermediate shaft assembly.
23. Remove the steering column assembly.
24. Remove instrument panel reinforcement assembly, as follows:

a. Disconnect each connector and remove each clamp. Disconnect the wire harness.
b. Remove the 6 nuts and 3 bolts.
c. Remove the 3 bolts and nut.
d. Using a Torx® socket wrench (T40), remove the 5 Torx® bolts.

➡The Torx® bolts on the passenger side can be removed with the collar or for adjustment.

e. Using a hexagon wrench 12 mm, remove the 2 collars and instrument panel reinforcement assembly with the air conditioner unit assembly.
f. Remove the 3 bolts, 2 screws and instrument panel reinforcement assembly.

25. Remove the No. 2 air duct sub-assembly.
26. Remove the blower assembly.
27. Remove the air outlet control servo motor.
28. Remove the air mix control servo motor.
29. Remove the heater radiator unit sub-assembly, as follows:

a. Remove the screw and clamp.
b. Release the 4 claws and remove the clamp.
c. Remove the heater radiator unit sub-assembly from the air conditioning radiator assembly.

➡Prepare a drain pan or cloth in case the cooling water leaks.

To install:

30. Install the heater radiator unit sub-assembly.

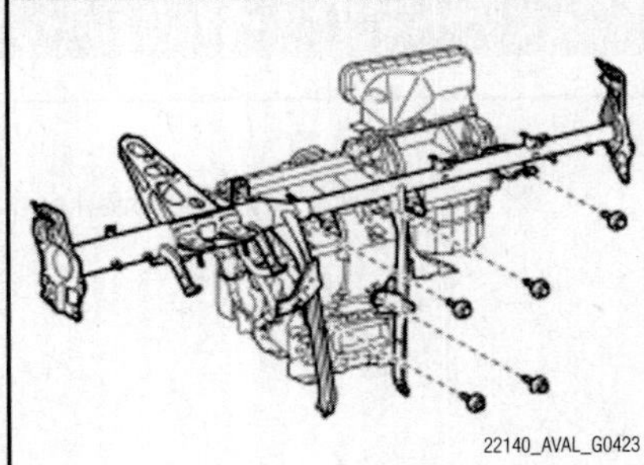

Fig. 165 Instrument panel reinforcement assembly with the air conditioner unit assembly

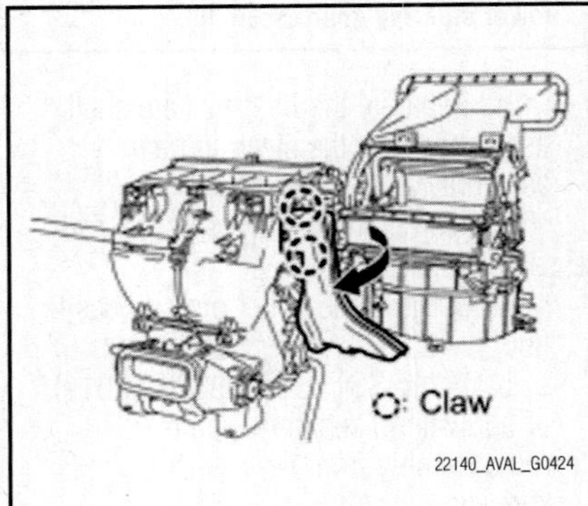

Fig. 166 No. 2 air duct sub-assembly

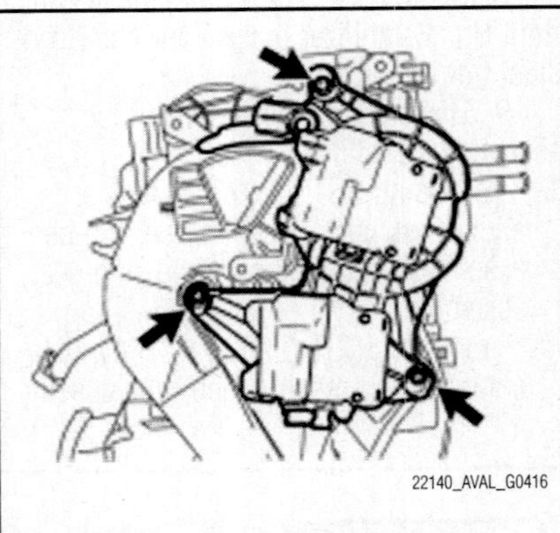

Fig. 167 Air outlet control servomotor

31. Install the air mix control servo motor.
32. Install the air outlet control servo motor.
33. Install the blower assembly.
34. Install the No. 2 air duct sub-assembly.
35. Install the air conditioner unit assembly to the instrument panel reinforcement assembly with the 2 screws and 3 bolts and tighten in the order shown to 87 inch lbs. (9.8 Nm).

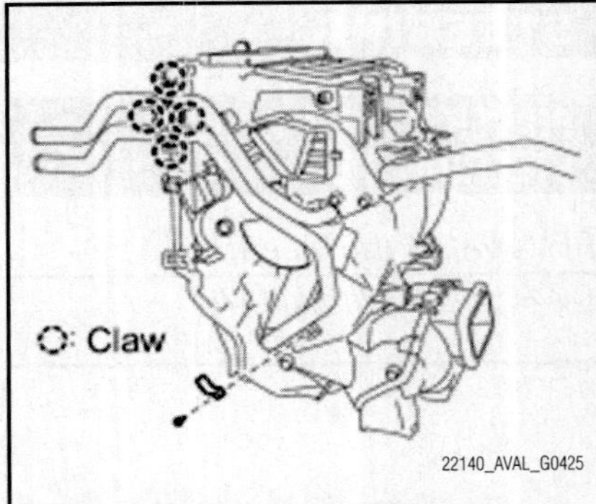

Fig. 168 Heater radiator unit sub-assembly

36. Driver seat:

a. Using a Torx® socket wrench (T40), install the instrument panel reinforcement assembly with the 3 Torx® bolts and tighten to 13 ft. lbs. (17 Nm).

37. Passenger seat:

a. Using a hexagon wrench 12 mm, install the instrument panel reinforcement assembly with the 2 bolts and tighten to 53 inch lbs. (6 Nm).
b. Using a Torx® socket wrench (T40), install the instrument panel reinforcement assembly with the 2 Torx® bolts and tighten to 15 inch lbs. (20 Nm).
c. Install the 3 bolts and nut and tighten to 87 inch lbs. (9.8 Nm) and 15 ft. lbs. (20 Nm).
d. Connect the connectors and clamps.
e. Install the 6 nuts and 3 bolts.

38. The remainder of installation is the reverse of removal.
39. Initialize system.
40. After adding engine coolant and charging refrigerant, check for coolant and refrigerant leaks.

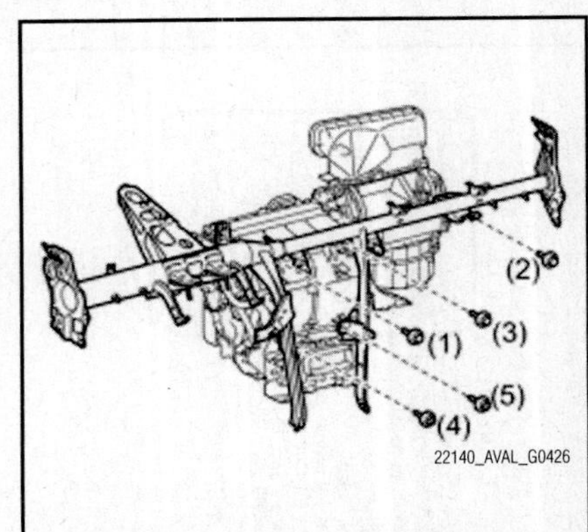

Fig. 169 A/C unit bolt tightening sequence

STEERING

POWER RACK & PINION STEERING GEAR

REMOVAL & INSTALLATION

See Figures 170 through 178.

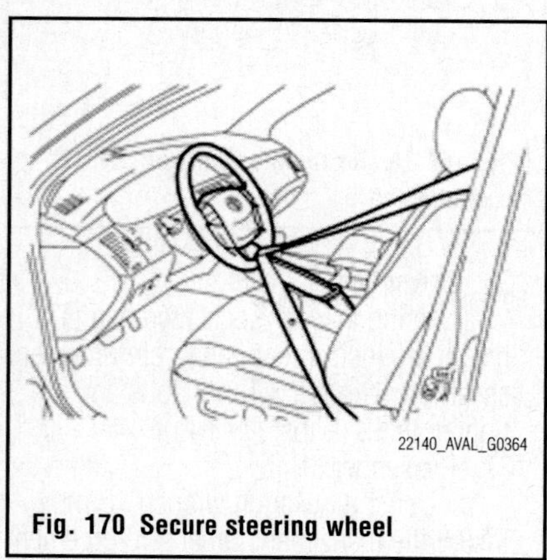
Fig. 170 Secure steering wheel

1. Before servicing the vehicle, refer to the precautions section.
2. Position the front wheels straight ahead.
3. Remove front wheel.
4. Remove right and left front fender apron seals.
5. Separate steering intermediate shaft assembly, as follows:
   a. Secure the steering wheel with the seat belt in order to prevent rotation. This will help prevent damage to the spiral cable.
   b. Remove the bolt.
   c. Put matchmarks on the steering intermediate shaft assembly and the power steering gear assembly.

Fig. 171 Remove the bolt

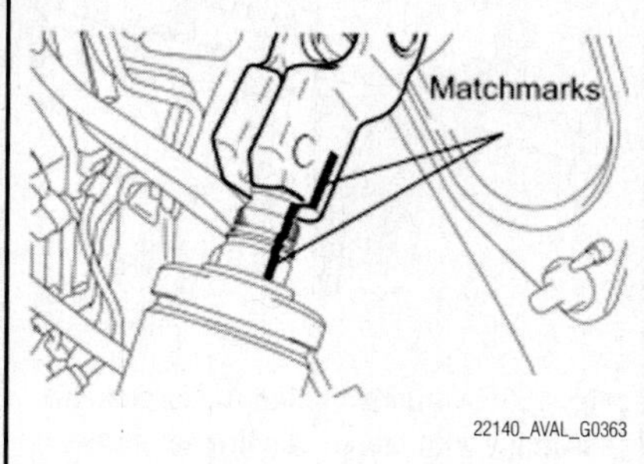

Fig. 172 Matchmarks on the steering intermediate shaft assembly and the power steering gear assembly

   d. Separate the intermediate shaft assembly from the steering gear assembly.
6. Separate the left and right tie rod assemblies, as follows:
   a. Remove the cotter pin and castle nut.
   b. Using SST (SST: 09628-62011) or equivalent, separate the tie rod assembly from the steering knuckle.
7. Remove the right and left front stabilizer link assemblies.
8. Remove the 2 bolts and the left and right No. 1 stabilizer bracket and the No. 1 stabilizer bar bush.
9. Disconnect pressure feed tube assembly, as follows:
   a. Using SST (09023-12701) or equivalent, disconnect the return tube assembly from the power steering gear assembly.
   b. Using SST (09023-12701), disconnect the pressure feed tube assembly from the power steering gear assembly.
   c. Remove the 2 bolts and separate the pressure feed tube clamp from the power steering gear assembly.
10. Remove the 2 bolts, nuts and the power steering gear assembly.

➡Do not turn the nut because the nut has its own stopper. Loosen the bolt with the nut secured.

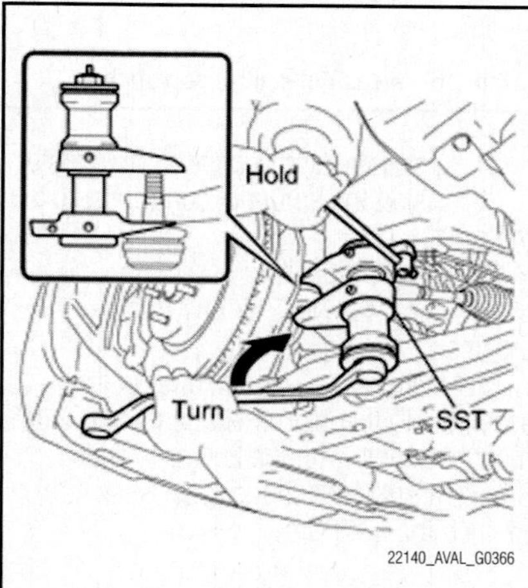

Fig. 173 Separate the tie rod assembly from the steering knuckle

To install:

11. Install the power steering gear assembly with the 2 bolts and 2 nuts and tighten to 52 ft. lbs. (70 Nm).

➡Do not turn the nut because the nut has its own stopper. Tighten the bolt with the nut secured.

➡For the next 2 steps, use a torque wrench with a fulcrum length of 300 mm (11.81 in.). These torque values are effective when SST is parallel to the torque wrench.

12. Using SST (09023-12701) or equivalent, connect the pressure feed tube

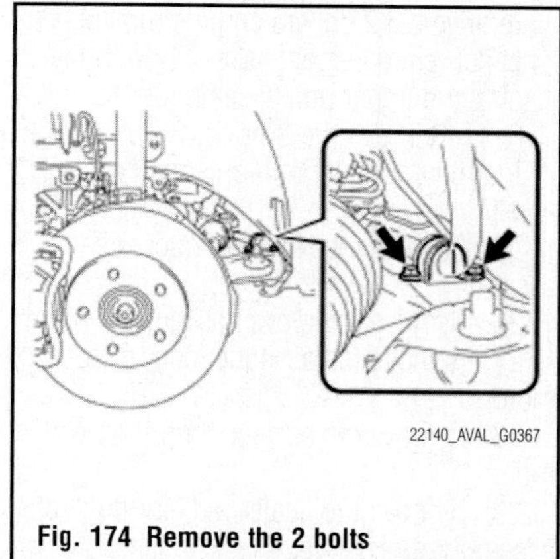
Fig. 174 Remove the 2 bolts

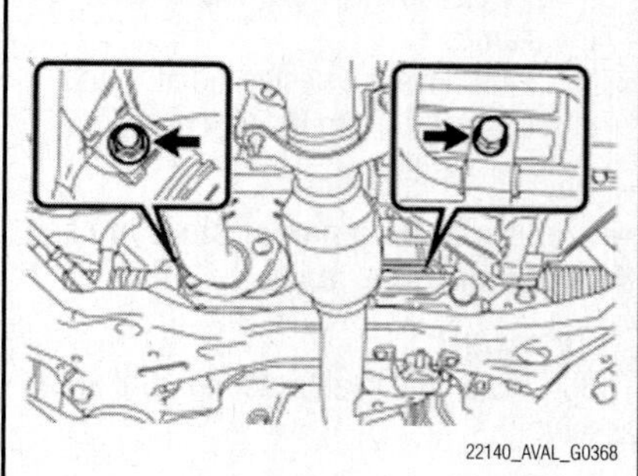
Fig. 175 Remove the 2 bolts and separate the pressure feed tube clamp

assembly to the power steering gear assembly and tighten to 16 ft. lbs. (22 Nm).

13. Using SST (09023-12701) or equivalent, connect the return tube assembly to the power steering gear assembly and tighten to 16 ft. lbs. (22 Nm).

14. Install the pressure feed tube clamp with the 2 bolts and tighten to 87 inch lbs. (9.8 Nm).

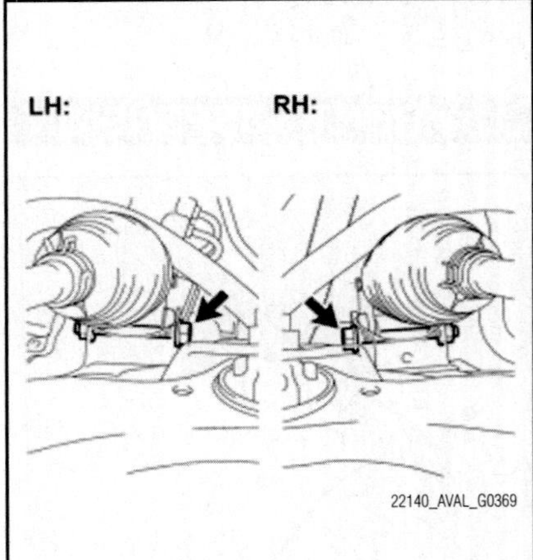

Fig. 176 Install the power steering gear assembly

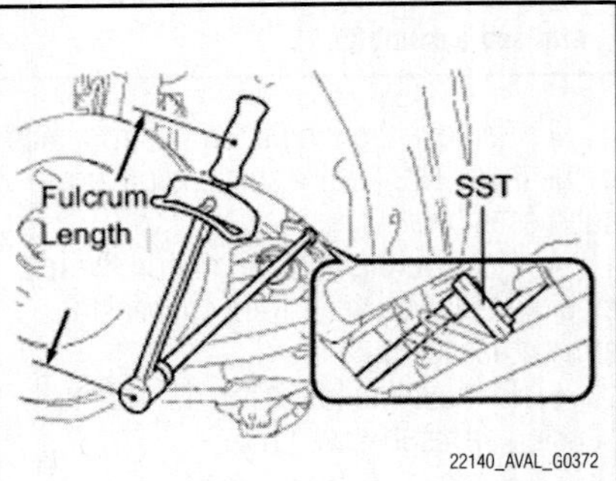

Fig. 177 Connect the pressure feed tube assembly

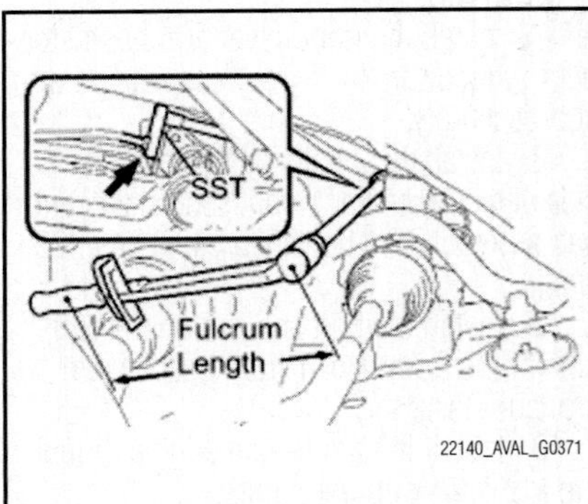

Fig. 178 Connect the return tube assembly

15. Install the left and right front stabilizer brackets, as follows:
   a. Install the No. 1 stabilizer bar bush to the stabilizer bar.
   b. Install the No. 1 front stabilizer bracket with the 2 bolts and tighten to 20 ft. lbs. (27 Nm).

16. Install the right and left front stabilizer link assemblies.
   a. Connect the left and right tie rod assemblies, as follows:
   b. Connect the tie rod assembly to the steering knuckle with the castle nut and tighten to 36 ft. lbs. (49 Nm).
   c. Install a new cotter pin. If the holes for the cotter pin are not aligned, tighten the nut up to 60° further.

17. Align the matchmarks on the steering intermediate shaft assembly and the power steering gear assembly.
   a. Install the bolt and tighten to 26 ft. lbs. (35 Nm).

18. Bleed power steering fluid.

19. Inspect for power steering fluid.

20. Install the left and right front fender apron seals.

21. Install the front wheel.

22. Inspect the front wheels.

23. Inspect the front wheel alignment.

POWER STEERING PUMP

REMOVAL & INSTALLATION

See Figures 179 and 180.

1. Before servicing the vehicle, refer to the precautions section.

➡Do not over tighten when using a vise.

➡When installing, coat the parts indicated by arrows with power steering fluid. Take care not to spill power steering fluid on the V-belt.

2. Remove the right front wheel.
3. Drain power steering fluid.
4. Remove the right front fender apron seal.
5. Remove fan and generator V-belt.
6. Remove the clip and disconnect the No. 1 oil reservoir to pump hose.
7. Remove the union bolt and disconnect the pressure feed tube assembly from the vane pump assembly.
8. Remove the bolt and separate the pressure feed tube assembly clamp.
9. Remove the gasket from the pressure feed tube assembly.
10. Remove the vane pump assembly, as follows:
   a. Disconnect the connector from the power steering oil pressure switch.
   b. Remove the 2 bolts and the vane pump assembly.

To install:

11. Install the vane pump assembly with the 2 bolts and tighten to 32 ft. lbs. (43 Nm).

12. Connect the connector to the power steering oil pressure switch.

13. Install a new gasket to the pressure feed tube assembly.

14. Connect the pressure feed tube assembly to the vane pump assembly with the union bolt and tighten to 38 ft. lbs. (52 Nm).

➡Make sure that the stopper of the pressure feed tube assembly contacts the vane pump assembly as shown in the illustration, then tighten the union bolt.

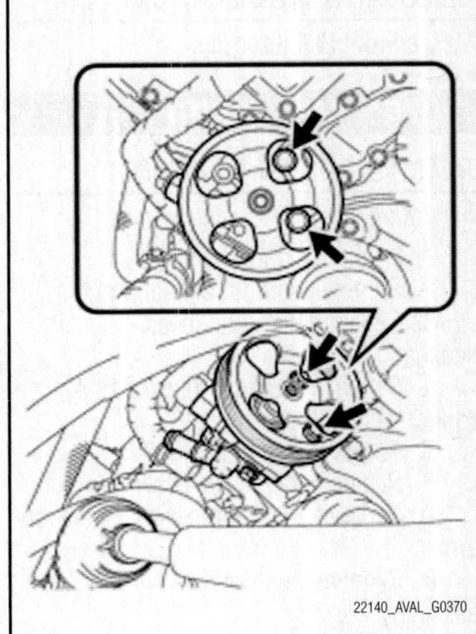

Fig. 179 Remove the 2 bolts and the vane pump assembly

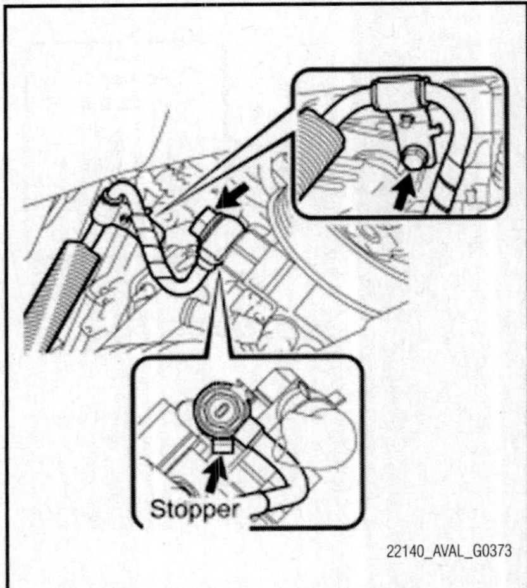

Fig. 180 Connect the pressure feed tube assembly

15. Install the pressure feed tube clamp with the bolt and tighten to 87 inch lbs. (9.8 Nm).
16. Connect the No. 1 oil reservoir to pump hose with the clip.
17. Install the fan and generator V-belt.
18. Install the right front fender apron seal.
19. Install the right front wheel.
20. Bleed the power steering fluid.
21. Inspect for power steering fluid.

BLEEDING

1. Before servicing the vehicle, refer to the precautions section.
2. Check the fluid level.
3. Jack up the front of the vehicle and support it with stands.
4. With the engine stopped, turn the wheel slowly from lock to lock several times.
5. Lower the vehicle.
6. Start the engine.
7. Run the engine at idle for a few minutes.
8. With the engine idling, turn the wheel left or right to the full lock position and keep it there for 2 to 3 seconds, then turn the wheel to the opposite full lock position and keep it there for 2 to 3 seconds.
9. Repeat the above steps several times.
10. Stop the engine.
11. Check for foaming or emulsification. If the system has to be bled twice because of foaming or emulsification, check for fluid leaks in the system.
12. Check the fluid level.

SUSPENSION

FRONT SUSPENSION

CONTROL LINKS

REMOVAL & INSTALLATION

See Front Stabilizer Bar.

LOWER BALL JOINT

REMOVAL & INSTALLATION

See Figure 181.

1. Remove the front wheel.
2. Remove the front axle hub nut.
3. Separate the front speed sensor.
4. Separate the front disc the brake caliper assembly.
5. Remove front disc.
6. Separate the tie rod assembly.
7. Separate the front lower suspension arm.
8. Remove the front axle assembly.
9. Remove the front lower ball joint assembly, as follows:

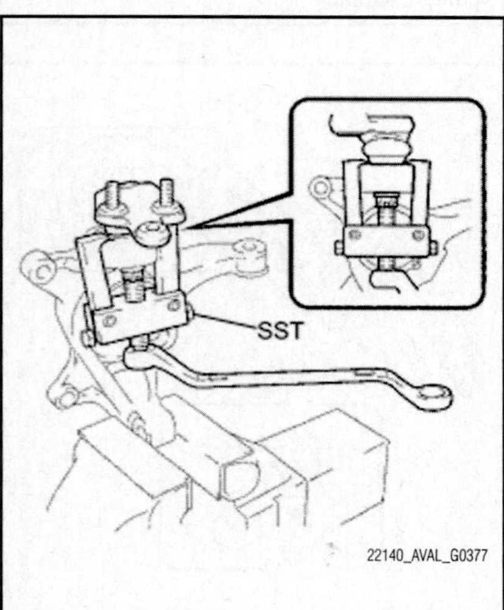

Fig. 181 Remove the front lower ball joint assembly

a. Remove the cotter pin and castle nut.
b. Using SST (SST: 09628-62011) or equivalent, remove the front lower ball joint assembly.

To install:

10. Installation is the reverse of the removal procedure, noting the following:
a. Install the front lower ball joint assembly to the steering knuckle with the castle nut and tighten to 91 ft. lbs. (123 Nm). Further tighten the nut up to 60°if the holes for the cotter pin are not aligned.
b. Inspect and adjust the front wheel alignment.
c. Inspect the ABS speed sensor signal.

LOWER CONTROL ARM

REMOVAL & INSTALLATION

See Figures 182 and 183.

1. Remove the engine assembly with transaxle.
2. Remove the 3 nuts and transverse engine mounting insulator.

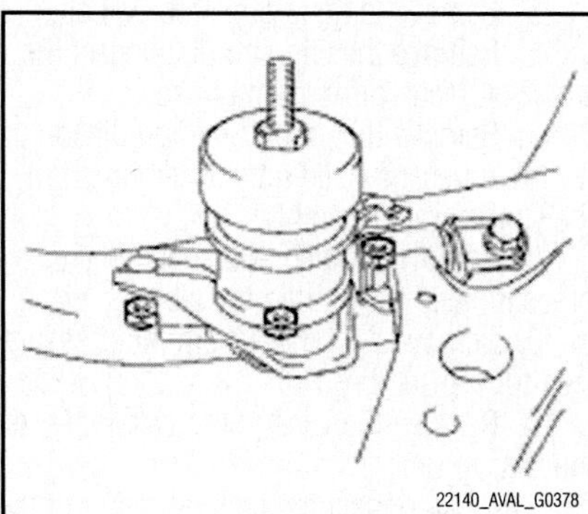

Fig. 182 Transverse engine mounting insulator with 3 nuts

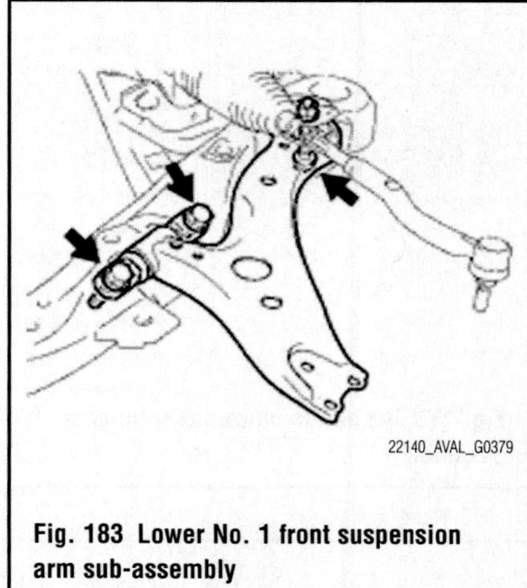

Fig. 183 Lower No. 1 front suspension arm sub-assembly

3. Remove the 2 bolts on the front side of the lower No. 1 front suspension arm sub-assembly.
4. Remove the bolt and nut on the rear side of the lower No. 1 front suspension arm sub-assembly.
5. Remove the lower No. 1 front suspension arm sub-assembly.
6. Remove the front lower arm bush stopper from the lower No. 1 front suspension arm sub-assembly.

To install:

7. Install the front lower arm bush stopper to the lower No. 1 front suspension arm sub-assembly.
8. Install the 2 bolts on the front side of the lower No. 1 front suspension arm sub-assembly and tighten to 148 ft. lbs. (200 Nm).
9. Install the bolt and nut on the rear side of the lower No. 1 front suspension arm sub-assembly.
10. Install the transverse engine mounting insulator with the 3 nuts.
11. Install the engine assembly with transaxle.

SHOCK ABSORBERS

REMOVAL & INSTALLATION

See Figures 184 through 186.

1. Remove the front wheel.
2. Remove the nut and disconnect the front stabilizer link assembly from the front shock absorber assembly.

➡Use a hexagon (6 mm) wrench to hold the stud if the ball joint turns together with the nut.

3. Remove the front shock absorber with coil spring, as follows:
    a. Loosen the lock nut. If not disassembling the shock absorber it is not necessary to loosen the nut.
    b. Remove the bolt and disconnect the No. 1 front flexible hose and front speed sensor wire harness.
    c. Remove the 2 nuts on the lower side of the front shock absorber with coil spring. Keep the bolts inserted.
    d. Remove the 3 nuts on the upper side of the front shock absorber with coil spring.
    e. Remove the 2 bolts on the lower side of the front shock absorber and front shock absorber with coil spring.

➡Be careful not to drop the collar in the case that there is front suspension upper brace center.

To install:

4. Install the front shock absorber with coil spring.
5. Install the 3 nuts to the upper side of the front shock absorber with coil spring and tighten to 63 ft. lbs. (85 Nm).

➡Be careful not to drop the collar in the case that there is front suspension upper brace center.

6. Install the 2 bolts and 2 nuts to the lower side of the front shock absorber with coil spring and tighten to 155 ft. lbs. (210 Nm).

➡Keep the bolts from rotating and torque the 2 nuts when installing the 2 nuts.

➡Insert the bolts from the front side of the vehicle.

7. Fully tighten the lock nut to 51 ft. lbs. (70 Nm).
8. Install the No. 1 front flexible hose and front speed sensor with the bolt and tighten to 14 ft. lbs. (19 Nm).
9. Install the front stabilizer link assembly with the nut and tighten to 55 ft. lbs. (74 Nm).

➡Use a hexagon (6 mm) wrench to hold the stud if the ball joint turns together with the nut.

10. Install the front wheel.
11. Inspect and adjust front wheel alignment.

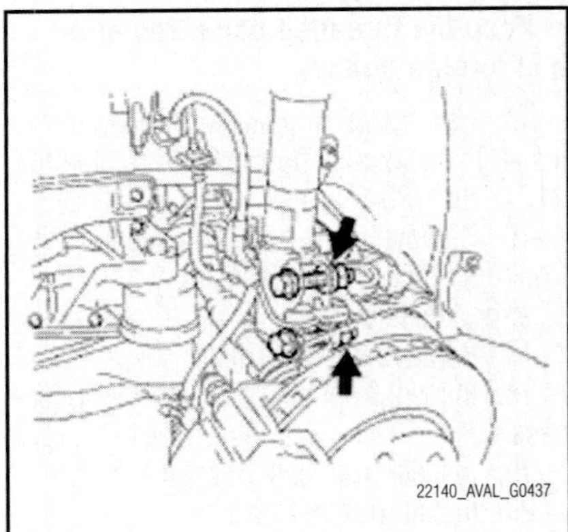

Fig. 184 Remove the 2 nuts on the lower side of the front shock absorber

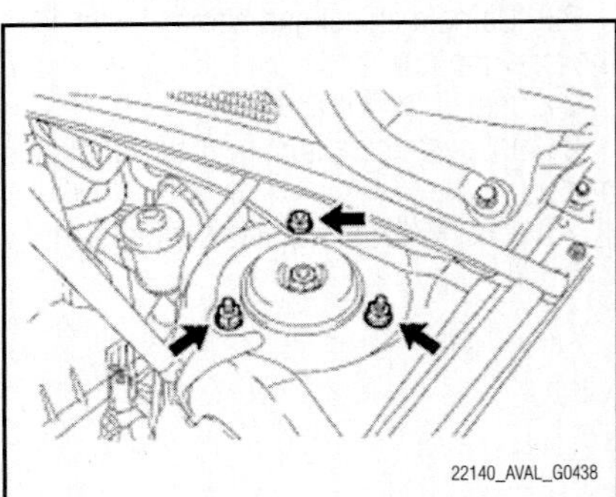

Fig. 185 Remove the 3 nuts on the upper side of the front shock absorber

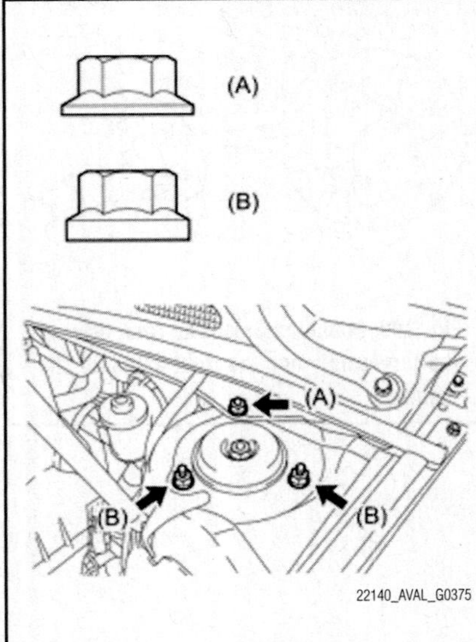

Fig. 186 Install the 3 nuts to the upper side of the front shock absorber

STEERING KNUCKLE

REMOVAL & INSTALLATION

See Wheel Hub and Bearing.

STABILIZER BAR

REMOVAL & INSTALLATION

See Figures 187 through 189.

1. Remove the front wheel.
2. Remove the nuts and the right and left front stabilizer link assemblies.

➡Use a hexagon (6 mm) wrench to hold the stud if the ball joint turns together with the nut.

3. Remove the engine assembly with transaxle.
4. Remove the bolts and the left and right No. 1 stabilizer brackets.
5. Remove the 2 bushes from the stabilizer.

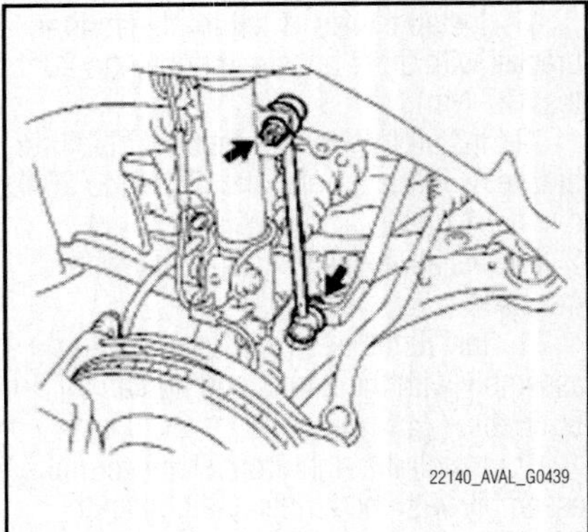

Fig. 187 Remove the 2 nuts and front stabilizer link assembly (left hand shown)

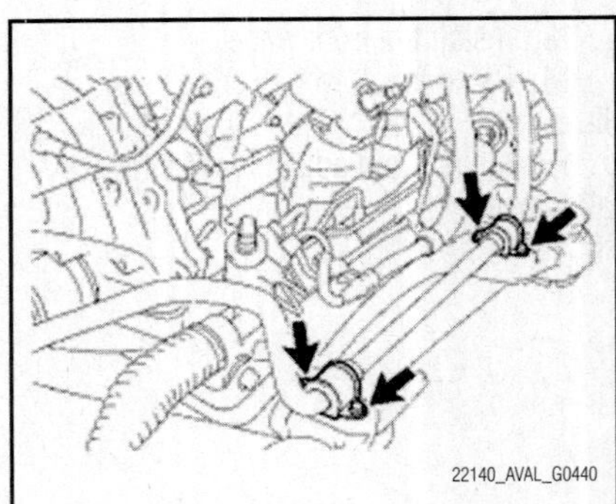

Fig. 188 Remove the 2 bolts and No. 1 stabilizer bracket (left hand shown)

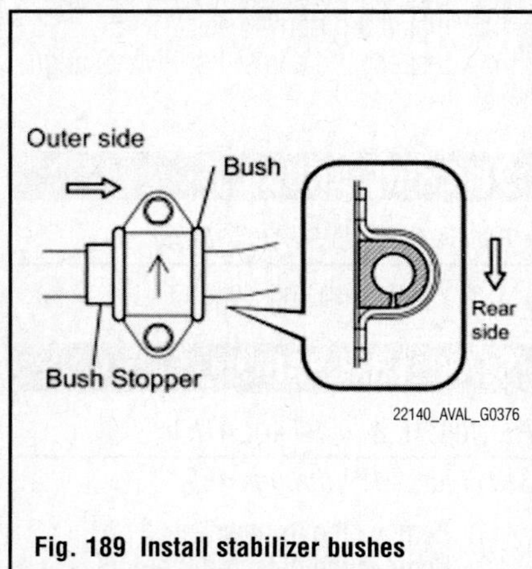

Fig. 189 Install stabilizer bushes

6. Remove the front stabilizer bar from the vehicle.

To install:

7. Install the stabilizer bar front to the vehicle.

➡Install the bushes as to the outer side of each bush stopper on the stabilizer bar.

➡Place the cutout of the stabilizer bushes as facing the rear side as shown in the illustration.

8. Install the No. 1 left front stabilizer bracket with the 2 bolts and tighten to 20 ft. lbs. (27 Nm).
9. Install the No. 1 right front stabilizer bracket with the 2 bolts and tighten to 20 ft. lbs. (27 Nm).
10. Install the engine assembly with transaxle.
11. Install the left front stabilizer link assembly with the 2 nuts and tighten to 55 ft. lbs. (74 Nm).
12. Install the right front stabilizer link assembly with the 2 nuts and tighten to 55 ft. lbs. (74 Nm).

➡Use a hexagon (6 mm) wrench to hold the stud if the ball joint turns together with the nut.

13. Install the front wheel.
14. Bleed the power steering fluid.
15. Inspect and adjust the front wheel alignment.

WHEEL BEARINGS

REMOVAL & INSTALLATION

See Figure 190.

1. Remove the front wheel.
2. Remove the front axle shaft nut.
3. Separate the front speed sensor.
4. Remove the 2 bolts and separate the front disc brake caliper assembly from the steering knuckle.
5. Remove the front disc.
6. Separate the tie rod end sub-assembly.
7. Separate the No. 1 front suspension arm sub-assembly.
8. Remove the front axle assembly, as follows:
   a. Put matchmarks on the front driveshaft assembly and the front axle hub sub-assembly.
   b. Using a plastic hammer, separate the front driveshaft assembly from the front axle hub sub-assembly.

⁂ WARNING

Be careful not to damage the boot and ABS speed sensor rotor.

   c. Remove the 2 bolts, nuts and steering knuckle with the front axle hub sub-assembly.

To install:

9. Install the front axle assembly, as follows:

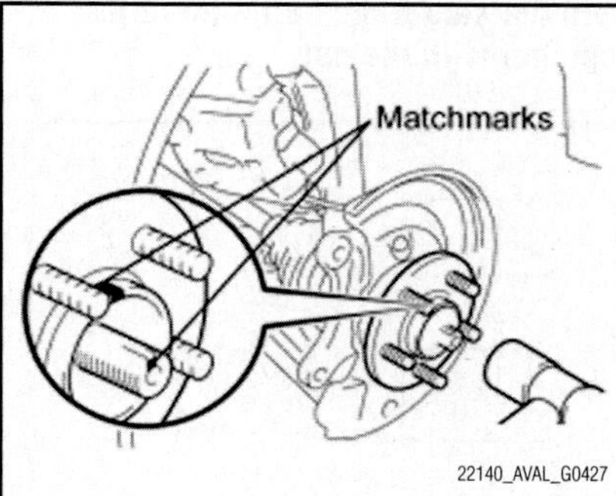

Fig. 190 Matchmarks on the front driveshaft assembly and the front axle hub sub-assembly

   a. Align the matchmarks and install the front driveshaft assembly to the front axle hub sub-assembly.
   b. Install the steering knuckle with the front axle hub sub-assembly to the front shock absorber assembly with the 2 bolts and 2 nuts and tighten to 155 ft. lbs. (210 Nm).

➡Only when reusing the bolts and nuts, apply the small amount of engine oil to the screw part of the nuts.

⁂ WARNING

Be careful not to damage the driveshaft boot or speed sensor rotor.

10. Install the lower No. 1 front suspension arm sub-assembly.
11. Install the tie rod end sub-assembly.
12. Install the front disc.
13. Install the front disc brake caliper assembly with the 2 bolts to the steering knuckle and tighten to 79 ft. lbs. (107 Nm).
14. Clean the threaded parts on the driveshaft and axle hub nut using a non-residue solvent.

➡Be sure to perform this work for a new driveshaft.

➡Keep the threaded parts free of oil and foreign objects.

15. Using a 30 mm socket wrench, install the front axle hub nut and tighten to 217 ft. lbs. (294 Nm).
16. Remove the 2 bolts and separate the front disc brake caliper assembly from the steering knuckle.
17. Remove the front disc.
18. Inspect front axle hub bearing looseness.
19. Inspect front axle hub runout.
20. Install the front disc.
21. Install the front disc brake caliper assembly with the 2 bolts to the steering knuckle and tighten to 79 ft. lbs. (107 Nm).
22. Install the front speed sensor.
23. Using a chisel and hammer, stake the axle hub nut.
24. Install the front wheel.
25. Inspect and adjust front wheel alignment.
26. Check ABS speed sensor signal.

CONTROL ARMS/LINKS

REMOVAL & INSTALLATION

No. 1 Suspension Arm

See Figures 191 through 204.

1. Remove the rear wheel.
2. Remove the center exhaust pipe assembly.
3. Remove the rear stabilizer bar.
4. Remove the rear strut rod.
5. The rear height control sensor sub-assembly (with discharge heard light) is on the right side only. Remove the nut and separate the right rear height control sensor sub-assembly.
6. Remove the bolt, nut and the left No. 2 rear suspension arm (outer side) from the rear axle carrier. When removing the bolt, keep the nut from rotating.
7. Remove the right No. 2 rear suspension arm using the same procedure as for the left side.
8. Remove the bolt, nut and the left and right No. 1 rear suspension arm (outer side) from the rear axle carrier. When removing the bolt, keep the nut from rotating.

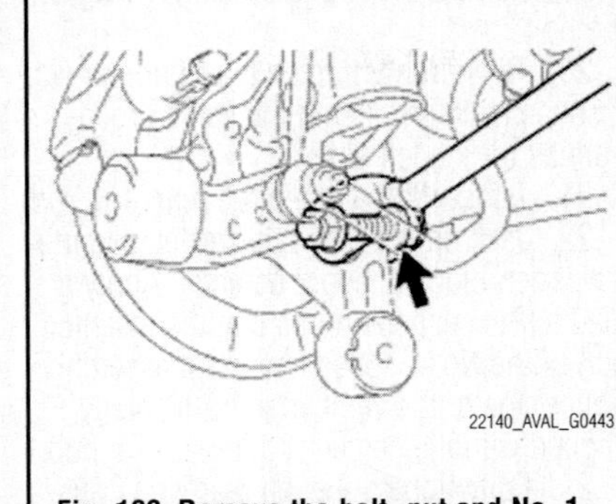

Fig. 193 Remove the bolt, nut and No. 1 rear suspension arm

9. Support the rear suspension member with a jack.
10. Remove the 4 nuts, 2 bolts and 4 retainers from the rear suspension member.
11. Lower the rear suspension member.
12. Remove the bolt and No. 1 rear suspension arm.

To install:

13. Install the No. 1 rear suspension arm with the bolt, and temporarily tighten the bolt.

➡Install the No. 1 rear suspension arm so that the bracket leans toward the front side of the vehicle, as shown in the illustration.

➡Ensure that the paint mark faces the rear side of the vehicle.

14. Set the No. 1 rear suspension arm in the position shown in the illustration, and fully tighten the bolt to 74 ft. lbs. (100 Nm).

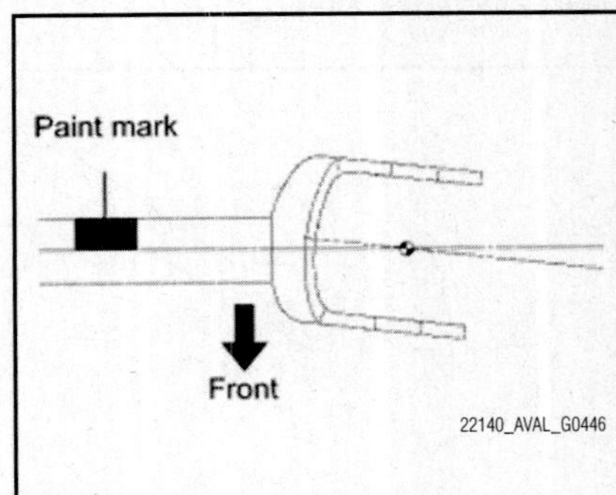

Fig. 196 Ensure that the paint mark faces the rear side of the vehicle

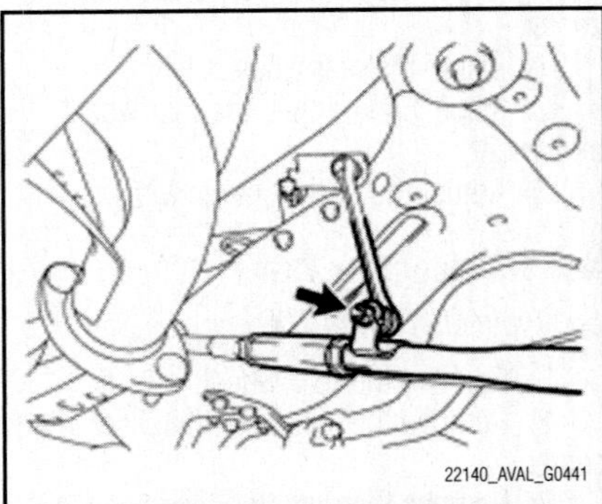

Fig. 191 Remove the nut and separate the right rear height control sensor sub-assembly

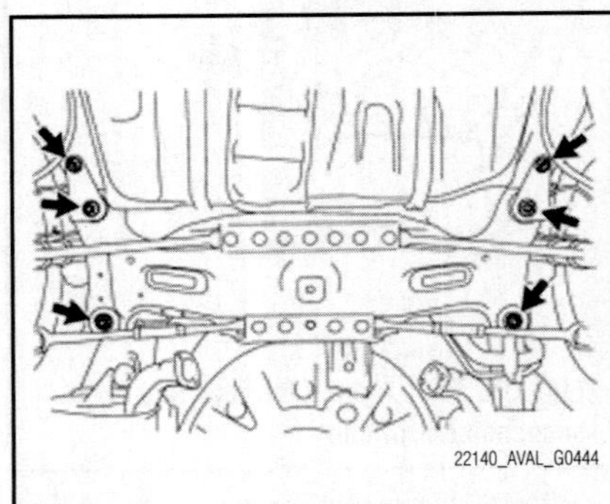

Fig. 194 Remove the 4 nuts, 2 bolts and 4 retainers from the rear suspension member

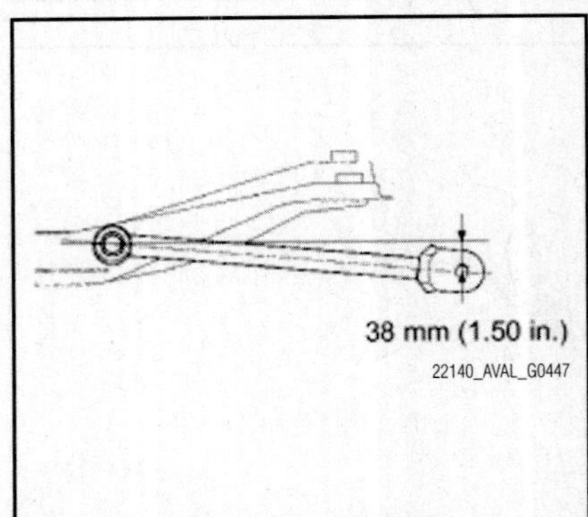

Fig. 197 Set the No. 1 rear suspension arm

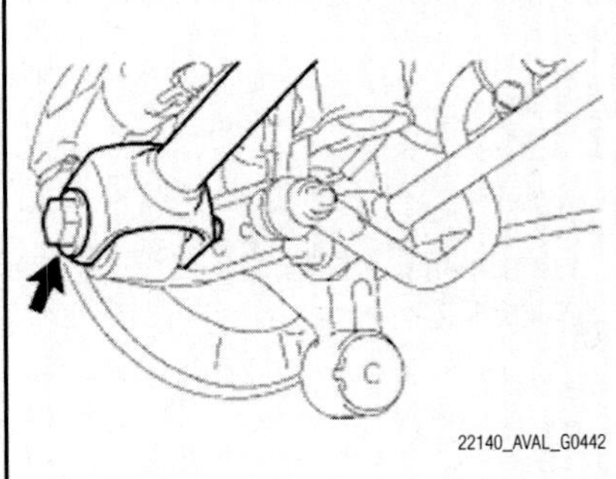

Fig. 192 Remove the bolt, nut and the left No. 2 rear suspension arm

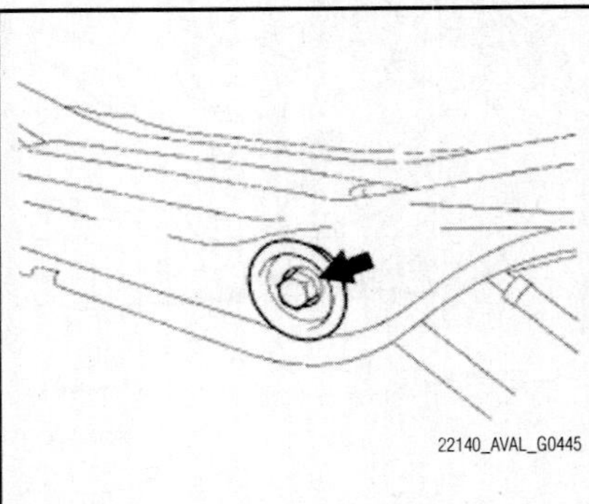

Fig. 195 Remove the bolt and No. 1 rear suspension arm

Fig. 198 Install the rear suspension member

15. Raise the rear suspension member with a jack.

16. Install the rear suspension member with the 4 nuts, 2 bolts and 4 retainers and tighten to 41 ft. lbs. (55 Nm), and 28 ft. lbs. (38 Nm).

17. Connect the left No. 1 rear suspension arm (outer side) to the rear axle carrier with the bolt and nut and temporarily tighten. Insert the bolt from the front side of the vehicle and temporarily install it.

18. Temporarily tighten the right rear No. 1 suspension arm using the same procedure as for the left side.

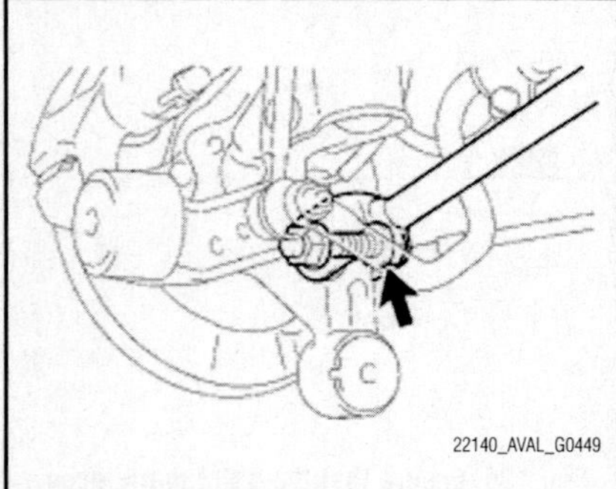

Fig. 199 Connect the left No. 1 rear suspension arm

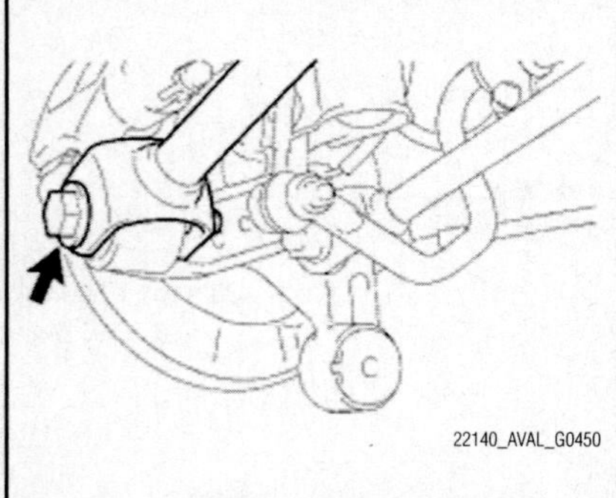

Fig. 200 Connect the left No. 2 rear suspension arm

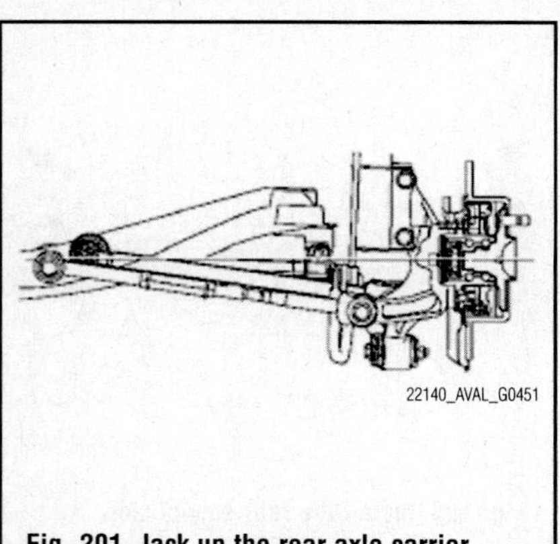

Fig. 201 Jack up the rear axle carrier

19. Connect the left No. 2 rear suspension arm (outer side) to the rear axle carrier with the bolt and nut and temporarily tighten the bolt.

20. Temporarily tighten the right rear No. 2 suspension arm using the same procedure as for the left side.

21. Temporarily tighten the rear strut rod.

22. Jack up the rear axle carrier, placing a wooden block to avoid damage. Apply a load to the suspension so that the installed bolt of the No. 1 suspension arm assembly (inner side of the vehicle) is horizontally aligned with the center of the rear axle hub.

23. Fully tighten the left rear No. 1 suspension arm bolt to 74 ft. lbs. (100 Nm). When installing the bolt, hold the nut and tighten the bolt.

24. Fully tighten the right rear No. 1 suspension arm using the same procedure as for the left side.

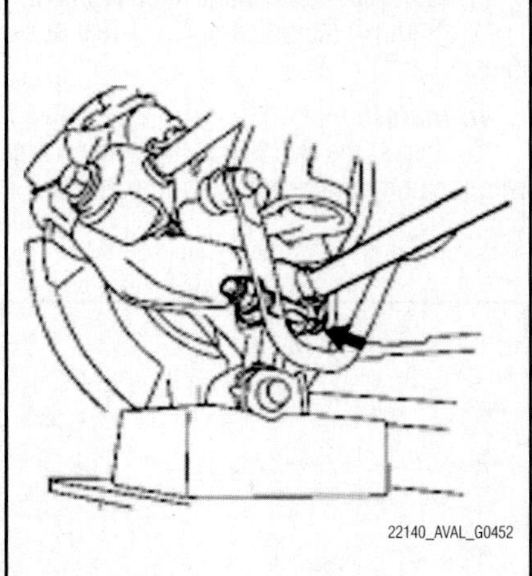

Fig. 202 Fully tighten the rear No. 1 suspension arm bolt

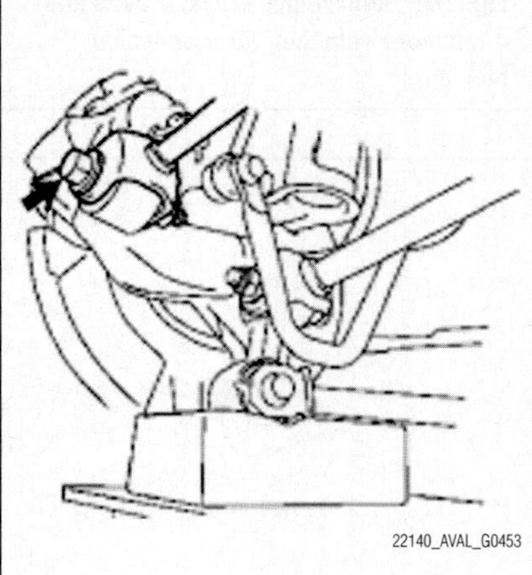

Fig. 203 Fully tighten the rear No. 2 suspension arm bolt

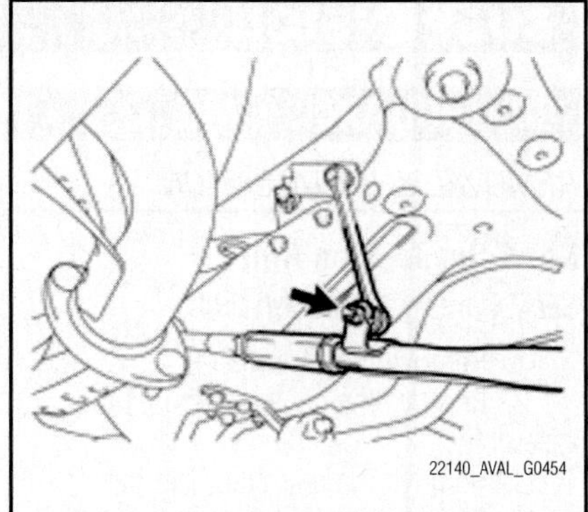

Fig. 204 Install the right rear height control sensor

25. Fully tighten the left rear No. 2 suspension arm bolt to 74 ft. lbs. (100 Nm). When installing the bolt, hold the nut and tighten the bolt.

26. Fully tighten the right rear No. 2 suspension arm using the same procedure as for the left side.

27. Fully tighten the rear strut rod.

28. Install the rear stabilizer bar.

29. Install the right rear height control sensor sub-assembly (with discharge head light) with the nut and tighten to 48 inch lbs. (5.4 Nm).

30. Install the exhaust pipe assembly center.

31. Install the rear wheel.

32. Inspect and adjust the rear wheel alignment.

33. Adjust the headlight aiming.

No. 2 Suspension Arm

See Figures 205 and 206.

1. Remove the rear wheel.
2. Remove the exhaust pipe assembly center.
3. Remove the rear stabilizer bar.
4. Remove the rear strut rod.
5. Separate the rear height control sensor sub-assembly.

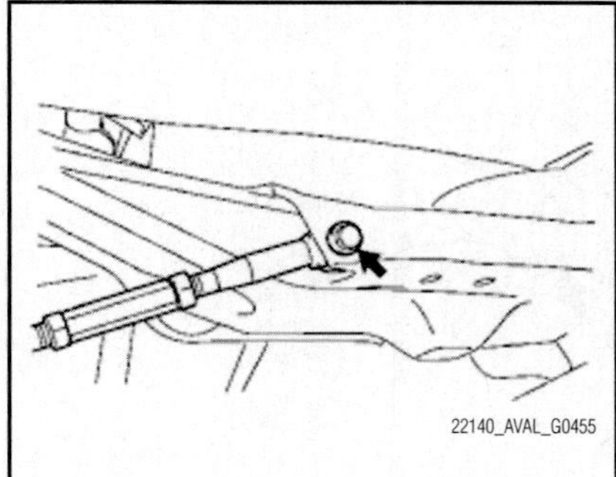

Fig. 205 Remove the bolt, and disconnect the rear No. 2 suspension arm

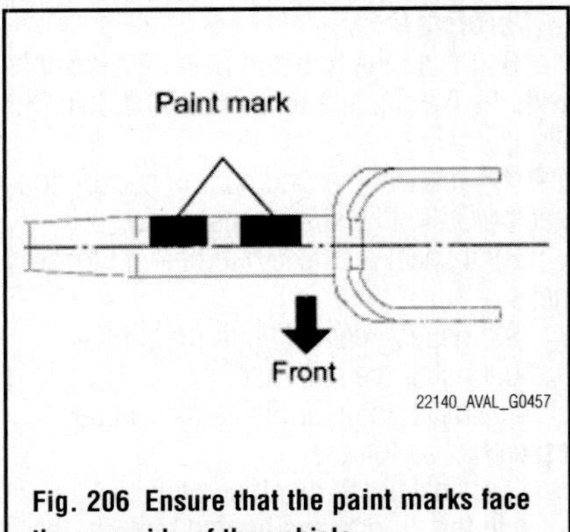

Fig. 206 Ensure that the paint marks face the rear side of the vehicle

6. Separate the left and right rear No. 1 suspension arms.
7. Separate the left and right rear No. 2 suspension arms.
8. Remove the rear suspension member sub-assembly.
9. Remove the bolt, and disconnect the rear No. 2 suspension arm (inner side).

To install:

10. Install the rear No. 2 suspension arm (inner side) with the bolt and tighten to 74 ft. lbs. (100 Nm).

➡Ensure that the paint marks face the rear side of the vehicle.

11. Install the rear suspension member sub-assembly.
12. Temporarily tighten the left and right rear No. 1 suspension arms.
13. Temporarily tighten the left and right rear No. 2 suspension arms.
14. Temporarily tighten the rear strut rod.
15. Stabilize the suspension.
16. Fully tighten the left and right rear No. 1 suspension arms.
17. Fully tighten the left and right rear No. 2 suspension arms.
18. Fully tighten the rear strut rod.
19. Install the rear stabilizer bar.
20. Install the rear height control sensor sub-assembly.
21. Install the exhaust pipe assembly center.
22. Install the rear wheel.
23. Inspect and adjust the rear wheel alignment.
24. Adjust the headlight aiming.

SHOCK ABSORBER

REMOVAL & INSTALLATION

See Figures 207 through 210.

1. Remove the rear seat cushion assembly.

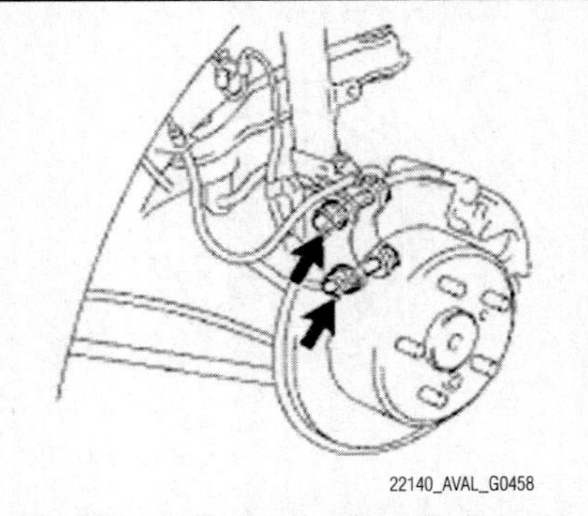

Fig. 207 Loosen the 2 nuts on the lower side of the shock absorber

Fig. 208 Remove the 3 nuts

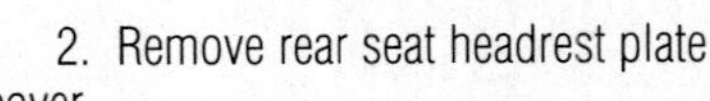

2. Remove rear seat headrest plate cover.
3. Remove rear seat headrest assembly.
4. Remove the rear seatback assembly.
5. Remove the rear wheel.
6. Separate rear stabilizer link assembly, as follows:
   a. Support the rear axle carrier with a jack.
   b. Remove the nut, and disconnect the stabilizer link from the shock absorber.

➡If the ball joint turns together with the nut, use a hexagon wrench (5 mm) to hold the stud.

7. Remove the 2 bolts, and disconnect the flexible hose and skid control sensor wire from the shock absorber.
8. Remove the rear shock absorber with coil spring, as follows:
   a. Loosen the 2 nuts on the lower side of the shock absorber. Do not remove the 2 bolts and 2 nuts.
   b. Remove the No. 1 rear suspension support cover.
   c. Loosen the support suspension center nut. Do not remove the nut. It is not necessary to loosen the nut if the shock absorber is not being disassembled.

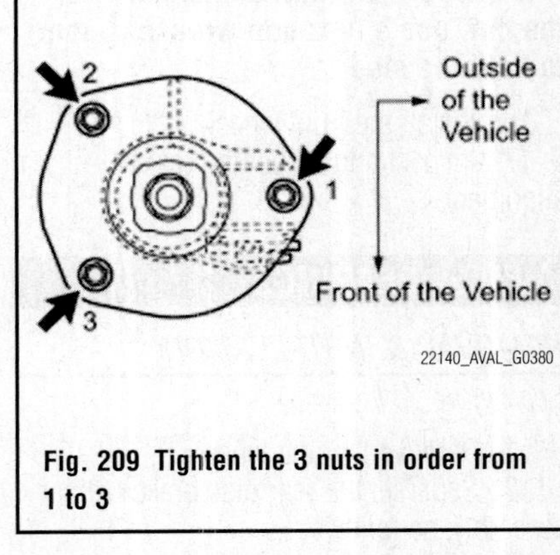

Fig. 209 Tighten the 3 nuts in order from 1 to 3

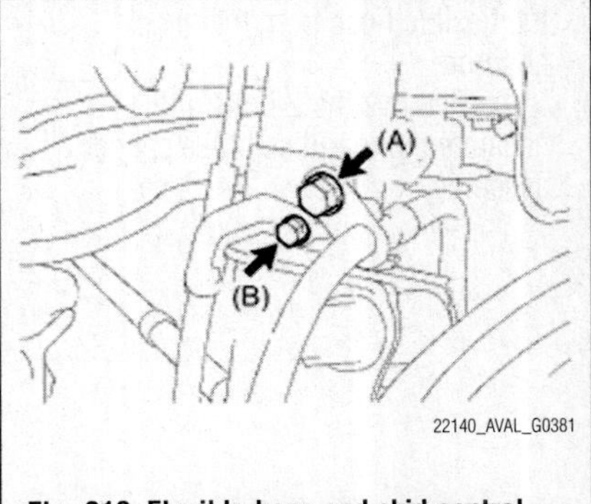

Fig. 210 Flexible hose and skid control sensor wire bolt tightening sequence

   d. Remove the 3 nuts.
   e. Lower the rear axle carrier, and remove the 2 nuts and 2 bolts on the lower side of the shock absorber.
   f. Remove the shock absorber with coil spring.

To install:

9. Temporarily install the 3 nuts to the upper side of the rear shock absorber with coil spring.
10. Fully tighten the 3 nuts in order from 1 to 3 to 29 ft. lbs. (39 Nm).
11. Install the 2 bolts and 2 nuts to the shock absorber with coil spring tighten to 133 ft. lbs. (180 Nm).
12. Fully tighten the nut installed on the top of the shock absorber with coil spring and tighten to 41 ft. lbs. (55 Nm). If the shock absorber has not been disassembled, it is not necessary to torque the nut.
13. Install the No. 1 rear suspension support cover.
14. Install the flexible hose and skid control sensor wire with the 2 bolts and tighten to 14 ft. lbs. (19 Nm), 49 inch lbs. (5.5 Nm).
15. Install the stabilizer link to the shock absorber with the nut and tighten to 29 ft. lbs. (39 Nm).

➡**If the ball joint turns together with the nut, use a hexagon wrench (5 mm) to hold the stud.**

16. Install the front wheel.
17. Inspect and adjust the rear wheel alignment.

WHEEL BEARINGS

REMOVAL & INSTALLATION

See Figure 211.

1. Remove the rear wheel.
2. Separate the rear disc brake caliper assembly, as follows:
   a. Remove the bolt and separate the flexible hose from the shock absorber.
   b. Remove the 2 bolts and separate the rear disc brake caliper assembly.

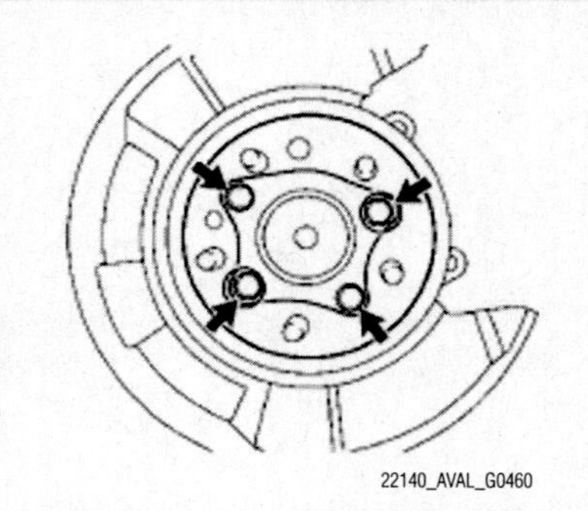

Fig. 211 Remove the 4 bolts and the rear axle hub and bearing assembly

3. Remove the rear disc.
4. Disconnect the skid control sensor connector.
5. Remove the 4 bolts and the rear axle hub and bearing assembly.

To install:

6. Install the hub and bearing assembly with the 4 bolts and tighten to 59 ft. lbs. (80 Nm).
7. Connect the skid control sensor connector. Do not twist the sensor wire.
8. Inspect rear axle hub bearing looseness.
9. Inspect rear axle hub runout.
10. Install the rear disc.
11. Install the rear disc brake caliper assembly, as follows:
    a. Install the rear disc brake caliper with the 2 bolts and tighten to 46 ft. lbs. (62 Nm).
    b. Install the flexible hose with the bolt and tighten to 14 ft. lbs. (19 Nm).
12. Install the rear wheel.
13. Inspect and adjust the rear wheel alignment.
14. Check ABS speed sensor signal.

TOYOTA

Camry

5

SPECIFICATIONS AND MAINTENANCE CHARTS

ENGINE AND VEHICLE IDENTIFICATION

Engine							Model Year	
Code ①	Liters (cc)	Cu. In.	Cyl.	Fuel Sys.	Engine Type	Eng. Mfg.	Code ②	Year
2AZ-FE	2.4 (2362)	144	4	SFI	DOHC	Toyota	9	2009
2AR-FE	2.5 (2494)	152	4	SFI	DOHC	Toyota	A	2010
2GR-FE	3.5 (3456)	210	6	SFI	DOHC	Toyota		

SFI: Sequential Fuel Injection

DOHC: Double Overhead Camshaft

NA: Information not available

① Stamped on the left side of the engine block

② 10th digit of the Vehicle Identification Number (VIN)

3768X_CAMR_C0001

GENERAL ENGINE SPECIFICATIONS

All measurements are given in inches.

Year	Model	Engine Displacement Liters	Engine Series VIN	Net Horsepower @ rpm	Net Torque @ rpm (ft. lbs.)	Bore x Stroke (in.)	Com-pression Ratio	Oil Pressure @ rpm
2009	Camry	2.4	2AZ-FE	155@6000	158@4000	3.48x3.78	9.8:1	55@3000
		3.5	2GR-FE	268@6200	248@4700	3.70x3.27	10.8:1	36-78@3000
2010	Camry	2.5	2AR-FE	179@6000	171@4000	3.54x3.86	10.4:1	55@3000
		3.5	2GR-FE	268@6200	248@4700	3.70x3.27	10.8:1	36-78@3000

3768X_CAMR_C0002

GASOLINE ENGINE TUNE-UP SPECIFICATIONS

Year	Engine Displacement Liters	Engine VIN	Spark Plug Gap (in.)	Ignition Timing (deg.)	Fuel Pump (psi)	Idle Speed (rpm)	Valve Clearance (in.)	
							Intake	Exhaust
2009	2.4	2AZ-FE	0.039-0.043	NA	44-50	①	0.0075-0.0114	0.0150-0.0189
	3.5	2GR-FE	0.039-0.043	NA	44-50	650-750	NA	NA
2010	2.5	2AR-FE	0.039-0.043	NA	44-50	①	0.0075-0.0114	0.0150-0.0189
	3.5	2GR-FE	0.039-0.043	NA	44-50	650-750	NA	NA

NOTE: The Vehicle Emission Control Information label often reflects specification changes made during production.

The label figures must be used if they differ from those in this chart.

NA: Not available

① Manual transmission: 650 to 750 rpm, Automatic transmission: 610 to 710 rpm

3768X_CAMR_C0003

CAPACITIES

Year	Model	Engine Displacement Liters	Engine VIN	Engine Oil with Filter (qts.)	Transmission (pts.) 5-Spd	Transmission (pts.) Auto.	Transfer Case (pts.)	Drive Axle Front (pts.)	Drive Axle Rear (pts.)	Fuel Tank (gal.)	Cooling System (qts.)
2009	Camry	2.4	2AZ-FE	4.5	—	3.7	—	—	—	18.5	6.6
		3.5	2GR-FE	6.4	—	6.8	—	—	—	18.5	8.8
2010	Camry	2.5	2AR-FE	4.6	—	3.7	—	—	—	18.5	7.5
		3.5	2GR-FE	6.4	—	6.8	—	—	—	18.5	9.5

NOTE: All capacities are approximate. Add fluid gradually and check to be sure a proper fluid level is obtained.

3768X_CAMR_C0004

FLUID SPECIFICATIONS

Year	Model	Engine Displacement Liters	Engine ID/VIN	Engine Oil	Auto. Trans.	Drive Axle	Power Steering Fluid	Brake Master Cylinder
2009	Camry	2.4	2AZ-FE	5W-20	NA	—	ATF Dexron II Or III	DOT 3
		3.5	2GR-FE	5W-30	NA	—	ATF Dexron II Or III	DOT 3
2010	Camry	2.5	2AR-FE	0W-20	NA	—	ATF Dexron II Or III	DOT 3
		3.5	2GR-FE	5W-30	NA	—	ATF Dexron II Or III	DOT 3

DOT: Department Of Transpotation

NA: Not Available

3768X_CAMR_C0005

VALVE SPECIFICATIONS

Year	Engine Displacement Liters	Engine VIN	Seat Angle (deg.)	Face Angle (deg.)	Spring Test Pressure (lbs. @ in.)	Spring Installed Height (in.)	Stem-to-Guide Clearance (in.) Intake	Stem-to-Guide Clearance (in.) Exhaust	Stem Diameter (in.) Intake	Stem Diameter (in.) Exhaust
2009	2.4	2AZ-FE	45	44.5	NA	NA	0.0010-0.0031	0.0012-0.0039	0.2154-0.2159	0.2151-0.2157
	3.5	2GR-FE	45	44.5	NA	NA	0.0010-0.0024	0.0012-0.0026	0.2154-0.2159	0.2151-0.2157
2010	2.5	2AR-FE	45	44.5	NA	NA	0.0010-0.0031	0.0012-0.0039	0.2154-0.2159	0.2151-0.2157
	3.5	2GR-FE	45	44.5	NA	NA	0.0010-0.0024	0.0012-0.0026	0.2154-0.2159	0.2151-0.2157

NA: Not Available

3768X_CAMR_C0006

CAMSHAFT AND BEARING SPECIFICATIONS CHART

All measurements are given in inches.

Year	Engine Displ. Liters	Engine ID/VIN	Journal Dia.	Brg. Oil Clearance	Shaft End-play	Runout	Journal Bore	Lobe Height Intake	Lobe Height Exhaust
2009	2.4	2AZ-FE	①	NA	NA	NA	NA	1.8624-1.8664	1.8104-1.1843
	3.5	2GR-FE	②	③	NA	0.0016	NA	1.7447-1.7487	1.7426-1.7465
2010	2.5	2AR-FE	④	NA	NA	0.0012	NA	1.7387-1.7443	1.7379-1.7435
	3.5	2GR-FE	②	③	NA	0.0016	NA	1.7447-1.7487	1.7426-1.7465

NA: Not Available

① Mark 1, 2 and 3: 1.4162-1.4167

② No. 1 journal: 1.4152-1.4157
Other Journals: 1.0220-1.0226 in.

③ No. 1 journal: 0.0016-0.0031
Other Journals: 1.0010-1.0024 in.
Maximum No.1 Journal: 0.0039 in.
Other Maximum Journals: 0.0035 in.

④ No 1 journal: 1.3562-1.3568
Other journals: 0.9039-0.9045

3768X_CAMR_C0007

CRANKSHAFT AND CONNECTING ROD SPECIFICATIONS

All measurements are given in inches.

Year	Engine Displacement Liters	Engine VIN	Crankshaft Main Brg. Journal Dia.	Crankshaft Main Brg. Oil Clearance	Crankshaft Shaft End-play	Crankshaft Thrust on No.	Connecting Rod Journal Diameter	Connecting Rod Oil Clearance	Connecting Rod Side Clearance
2009	2.4	2AZ-FE	2.0863-2.0866	0.0007-0.0016	0.0016-0.0095	3	1.8894-1.8898	0.0013-0.0025	0.0063-0.0143
	3.5	2GR-FE	2.4011-2.4016	0.0010-0.0019	0.0016-0.0095	2	2.0863-2.0866	0.0018-0.0026	0.0059-0.0157
2010	2.5	2AR-FE	2.1649-2.1654	0.0006-0.0015	0.0016-0.0095	3	1.8894-1.8898	0.0012-0.0025	0.0063-0.0202
	3.5	2GR-FE	2.4011-2.4016	0.0010-0.0019	0.0016-0.0095	2	2.0863-2.0866	0.0018-0.0026	0.0059-0.0157

3768X_CAMR_C0008

PISTON AND RING SPECIFICATIONS

All measurements are given in inches.

	Engine			Ring Gap			Ring Side Clearance		
Year	Displ. Liters	Engine VIN	Piston Clearance	Top Compression	Bottom Compression	Oil Control	Top Compression	Bottom Compression	Oil Control
2009	2.4	2AZ-FE	0.0020-0.0029	0.0094-0.0122	0.0130-0.0169	0.0040-0.0119	0.0008-0.0028	0.0008-0.0024	0.0008-0.0028
	3.5	2GR-FE	0.0018-0.0020	0.0098-0.0138	0.0197-0.0413	0.0039-0.0157	0.0008-0.0028	0.0008-0.0024	0.0028-0.0059
2010	2.5	2AR-FE	0.0020-0.0029	0.0094-0.0122	0.0130-0.0169	0.0040-0.0119	0.0008-0.0028	0.0021-0.0037	0.0023-0.0085
	3.5	2GR-FE	0.0018-0.0020	0.0098-0.0138	0.0197-0.0413	0.0039-0.0157	0.0008-0.0028	0.0008-0.0024	0.0028-0.0059

3768X_CAMR_C0009

TORQUE SPECIFICATIONS

All readings in ft. lbs.

Year	Engine Displacement Liters	Engine VIN	Cylinder Head Bolts	Main Bearing Bolts	Rod Bearing Bolts	Crankshaft Damper Bolts	Flywheel Bolts	Manifold Intake	Manifold Exhaust	Spark Plugs	Oil Pan Drain Plug
2009	2.4	2AZ-FE	①	②	③	125	④	22	27	13	18
	3.5	2GR-FE	⑤	⑥	⑦	184	61	15	15	13	30
2010	2.5	2AR-FE	⑧	②	⑨	125	④	22	27	13	18
	3.5	2GR-FE	⑤	⑥	⑦	184	61	15	15	13	30

① Step 1: 52 ft. lbs.
Step 2: plus 90 degrees

② Step 1: 15 ft. lbs.
Step 2: 30 ft. lbs.
Step 3: Plus 90 degrees

③ Step 1: Cap bolts to 18 ft. lbs.
Step 2: cap bolts plus 90 degrees

④ Auto driveplate: 72 ft. lbs.
Manual Flywheel : 96 ft. lbs.

⑤ Step 1: 10mm point cap bolts to 27 ft. lbs.
Step 2: 10mm point cap bolts plus 90 degrees
Step 3: 10mm point cap bolts plus 90 degrees
Step 4: Front bolts to 22 ft. lbs.

⑥ Step 1: 18 ft. lbs.
Step 2: Plus 90 degrees

⑦ Step 1: 16 cap bolts to 45 ft. lbs.
Step 2: 16 cap bolts plus 90 degrees
Step 3: 8 side bolts to 38 ft. lbs.

⑧ Step 1: 27 ft. lbs.
Step 2: 27 ft. lbs.
Step 3: Plus 90 degrees
Step 3: Plus 90 degrees

⑨ Step 1: 30 ft. lbs.
Step 2: Plus 90 degrees

3768X_CAMR_C0010

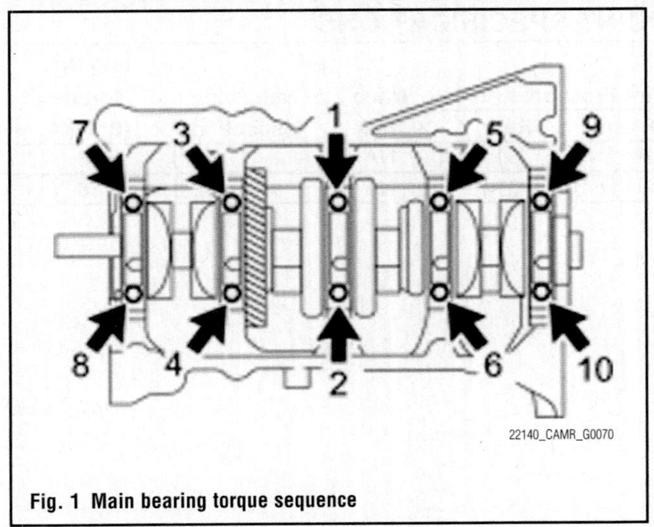

Fig. 1 Main bearing torque sequence

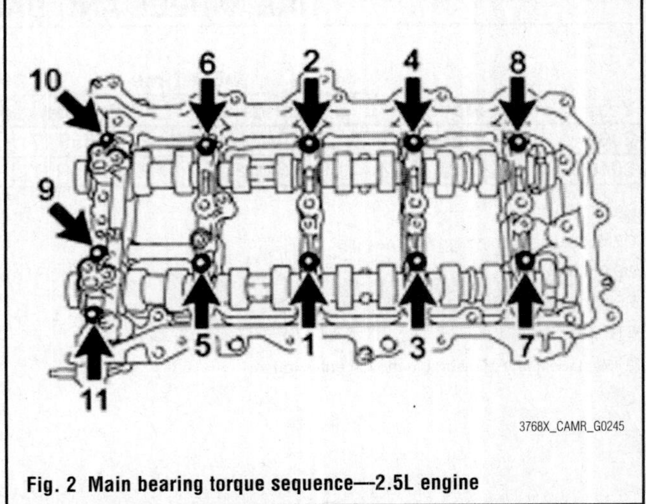

Fig. 2 Main bearing torque sequence—2.5L engine

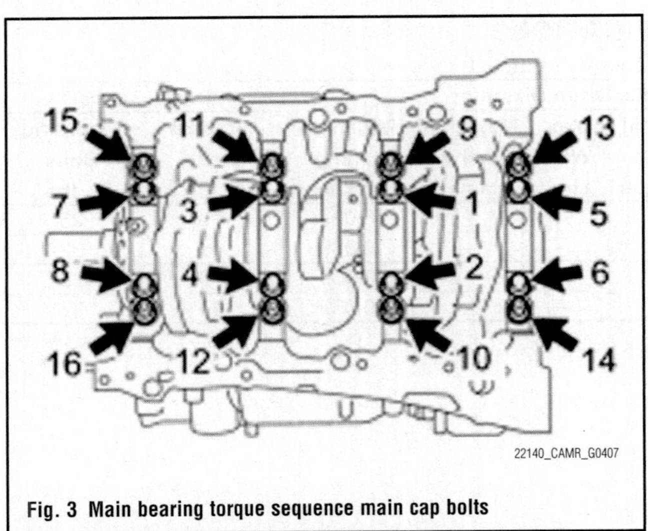

Fig. 3 Main bearing torque sequence main cap bolts

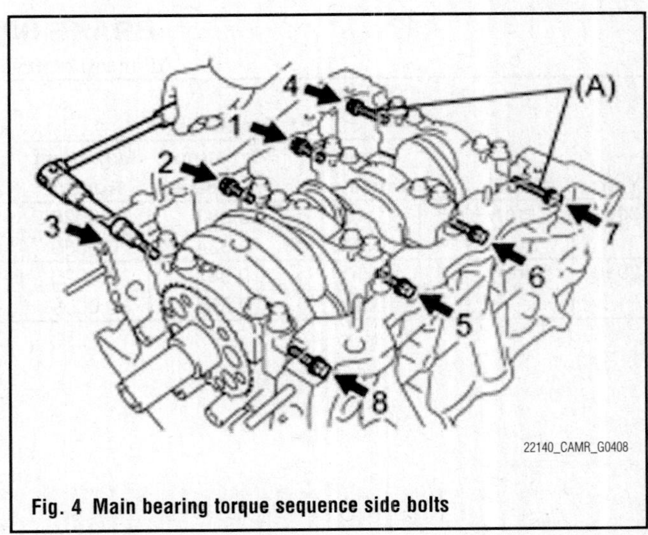

Fig. 4 Main bearing torque sequence side bolts

WHEEL ALIGNMENT

Year	Model		Caster Range (+/-Deg.)	Caster Preferred Setting (Deg.)	Camber Range (+/-Deg.)	Camber Preferred Setting (Deg.)	Toe-in (in.)	Steering Axis Inclination (Deg.)
2009	Camry	Front	0.75	2.65	0.75	-0.67	0+/-0.04	12.25+/-0.75
		Rear	—	—	0.75	-1.15	0.16+/-0.08	—
2010	Camry	Front	0.75	2.92	0.75	-0.67	0+/-0.08	12.25+/-0.75
		Rear	—	—	0.75	-1.15	0.16+/-0.08	—

3768X_CAMR_C0011

TIRE, WHEEL AND BALL JOINT SPECIFICATIONS

		OEM Tires		Tire Pressures (psi)		Wheel	Ball Joint	Lug Nut Torque
Year	Model	Standard	Optional	Front	Rear	Size	Inspection	(ft. lbs.)
2009	Camry	P215/60R16	P215/55R17	31	31	NA	①	76
2010	Camry	P215/60R16	P215/55R17	31	31	NA	①	76

NA: Not Available

OEM: Original Equipment Manufacturer

OPT: Optional

PSI: Pounds Per Square Inch

STD: Standard

① Replace if any measurable movement is found.

3768X_CAMR_C0012

BRAKE SPECIFICATIONS

All measurements in inches unless noted

			Brake Disc			Brake Drum Diameter				Brake Caliper	
Year	Model		Original Thickness	Minimum Thickness	Maximum Runout	Original Inside Diameter	Max. Wear Limit	Maximum Machine Diameter	Minimum Lining Thickness	Bracket Bolts (ft. lbs.)	Mounting Bolts (ft. lbs.)
2009	Camry	F	1.102	0.983	0.0020	—	—	—	0.039	79	25
		R	0.390	0.334	0.0059	—	—	—	0.039	46	32
2010	Camry	F	1.102	0.983	0.0020	—	—	—	0.039	79	25
		R	0.390	0.334	0.0059	—	—	—	0.039	46	32

3768X_CAMR_C0013

SCHEDULED MAINTENANCE INTERVALS

TOYOTA—CAMRY

TO BE SERVICED	TYPE OF SERVICE	VEHICLE MILEAGE INTERVAL (x1000)													
		5	10	15	20	25	30	35	40	45	50	55	60	90	120
Engine oil & filter	R	✓	✓	✓	✓	✓	✓	✓	✓	✓	✓	✓	✓	✓	✓
Automatic transmission fluid	S/I			✓			✓			✓			✓	✓	✓
Ball joints & dust covers	S/I			✓			✓			✓			✓	✓	✓
Bolts & nuts on chassis & body	S/I			✓			✓			✓			✓	✓	✓
Brake linings & drums	S/I	✓	✓	✓	✓	✓	✓	✓	✓	✓	✓	✓	✓	✓	✓
Brake line pipes & hoses	S/I			✓			✓			✓			✓	✓	✓
Brake pads & discs (front & rear)	S/I	✓	✓	✓	✓	✓	✓	✓	✓	✓	✓	✓	✓	✓	✓
Brake fluid	R						✓						✓	✓	✓
Rack and pinion assembly	S/I			✓			✓			✓			✓	✓	✓
Steering linkage & boots	S/I			✓			✓			✓			✓	✓	✓
Air cleaner filter	R						✓						✓	✓	✓
Spark plugs ①	R														✓
Drive belts	S/I												✓	✓	✓
Exhaust pipes & mountings	S/I			✓			✓			✓			✓	✓	✓
Fuel lines & connections	S/I												✓	✓	✓
Engine coolant ②	S/I			✓			✓			✓			✓	✓	✓
Fuel tank cap gasket	S/I						✓						✓	✓	✓
Rotate tires	S/I	✓	✓	✓	✓	✓	✓	✓	✓	✓	✓	✓	✓	✓	✓
Clean air conditioning filter ③	S/I			✓			✓			✓			✓	✓	✓
Axle shaft bolts	S/I			✓			✓			✓			✓	✓	✓
Brake pad thickness and rotor runout	S/I						✓						✓	✓	✓

R: Replace S/I: Service or Inspect

① Spark plugs are replaced at 120,000 miles

② Replace engine coolant at 100,000 miles and then inspect every 15,000 miles

③ Replace air conditioning filter every 30,000 miles

FREQUENT OPERATION MAINTENANCE (SEVERE SERVICE)

If a vehicle is operated under any of the following conditions it is considered severe service:

- Extremely dusty areas.

- 50% or more of the vehicle operation is in 32°C (90°F) or higher temperatures, or constant temperatures below 0°C (32°F).

- Prolonged idling (vehicle operation in stop and go traffic).

- Frequent short running periods (engine does not warm to normal operating temperatures).

- Police, taxi, delivery usage or trailer towing usage.

Air cleaner filter: service or inspect every 5000 miles

Rear differential & transfer case oil: replace every 15,000 miles.

Ball joints & dust covers: service or inspect every 5000 miles.

Bolts & nuts on chassis & body: service or inspect every 5000 miles.

Axle shaft bolts: service or inspect every 5000 miles.

Steering linkage: service or inspect every 5000 miles.

3768X_CAMR_C0014

BRAKES — INFORMATION AND PRECAUTIONS

ANTI-LOCK SYSTEMS

- Certain components within the ABS system are not intended to be serviced or repaired individually.
- Do not use rubber hoses or other parts not specifically specified for and ABS system. When using repair kits, replace all parts included in the kit. Partial or incorrect repair may lead to functional problems and require the replacement of components.
- Lubricate rubber parts with clean, fresh brake fluid to ease assembly. Do not use shop air to clean parts; damage to rubber components may result.
- Use only DOT 3 brake fluid from an unopened container.
- If any hydraulic component or line is removed or replaced, it may be necessary to bleed the entire system.
- A clean repair area is essential. Always clean the reservoir and cap thoroughly before removing the cap. The slightest amount of dirt in the fluid may plug an orifice and impair the system function. Perform repairs after components have been thoroughly cleaned; use only denatured alcohol to clean components. Do not allow ABS components to come into contact with any substance containing mineral oil; this includes used shop rags.
- The Anti-Lock control unit is a microprocessor similar to other computer units in the vehicle. Ensure that the ignition switch is **OFF**before removing or installing controller harnesses. Avoid static electricity discharge at or near the controller.
- If any arc welding is to be done on the vehicle, the control unit should be unplugged before welding operations begin.

DISC AND DRUM SYSTEMS

CAUTION

Dust and dirt accumulating on brake parts during normal use may contain asbestos fibers from production or aftermarket brake linings. Breathing excessive concentrations of asbestos fibers can cause serious bodily harm. Exercise care when servicing brake parts. Do not sand or grind brake lining unless equipment used is designed to contain the dust residue. Do not clean brake parts with compressed air or by dry brushing. Cleaning should be done by dampening the brake components with a fine mist of water, then wiping the brake components clean with a dampened cloth. Dispose of cloth and all residue containing asbestos fibers in an impermeable container with the appropriate label. Follow practices prescribed by the Occupational Safety and Health Administration (OSHA) and the Environmental Protection Agency (EPA) for the handling, processing, and disposing of dust or debris that may contain asbestos fibers.

BRAKES — BLEEDING THE BRAKE SYSTEM

BLEEDING PROCEDURE

Actuator Assembly

➡After bleeding the air from the brake system, if the height or feel of the brake pedal cannot be obtained, perform air bleeding in the brake actuator assembly with a hand-held tester by following the procedures below.

1. Depress the brake pedal more than 20 times with the engine off.
2. Connect the hand-held tester to the DLC3, then turn the ignition switch to the **ON** position, but do NOT start the engine.
3. Select "AIR BLEEDING" on the hand-held tester.

➡Refer to the hand-held tester operator's manual for more details.

4. Bleed the air out of the regular brake line when "Step 1: Increase" appears on the hand-held tester display, as follows:

➡Bleed the air by following the steps displayed on the hand-held tester. Make sure that the brake fluid in the master cylinder reservoir tank does not become empty.

a. Connect the vinyl tube to either one of the bleeder plugs.
b. Have an assistant depress the brake pedal several times, then loosen the bleeder plug connected to the vinyl tube with the pedal depressed.
c. When fluid stops coming out, tighten the bleeder plug and release the brake pedal.
d. Repeat the previous 2 steps until all the air in the fluid is completely bled out.
e. Tighten the bleeder plug completely to 73 inch lbs. (8.3 Nm).
f. Repeat the above procedures for each wheel to bleed the air out of the brake line.

5. Bleed the air out of the suction line when "Step 2: Inhalation" appears on the hand-held tester display, as follows:

➡Bleed the air by following the steps displayed on the hand-held tester. Make sure that the brake fluid in the master cylinder reservoir tank does not become empty.

a. Connect the vinyl tube to the bleeder plug at the right front wheel or the right rear wheel and loosen the bleeder plug.
b. Operate the brake actuator assembly to perform air bleeding from the suction line using the hand-held tester.

➡This operation stops automatically after 4 seconds. At this time, be sure to release the brake pedal.

c. Check if the operation has stopped by referring to the hand-held tester display and tighten the bleeder plug.
d. Repeat the previous 2 steps until all air in the fluid is completely bled out.
e. Tighten the bleeder plug completely to 73 inch lbs. (8.3 Nm).
f. Repeat the above procedures to bleed the air out of the brake line for each wheel.

6. Bleed the air out of the pressure reduction line when "Step 3: Decrease" appears on the hand-held tester display, as follows:

➡Bleed the air by following the steps displayed on the hand-held tester. Make sure that the brake fluid in the master cylinder reservoir tank does not become empty.

a. Connect a vinyl tube to either one of the bleeder plugs.
b. Loosen the bleeder plug.
c. Using the hand-held tester, operate the brake actuator assembly, completely depress the brake pedal and keep it depressed.

➡The operation stops automatically after 4 seconds. When performing this procedure continuously, set an interval of at least 20 seconds. When the operation is complete, the brake pedal goes

down slightly. This is a normal phenomenon caused when the solenoid opens. During this procedure, the pedal will feel heavy, but completely depress it so that the brake fluid comes out from the bleeder plug. Be sure to keep depressing the brake pedal. Do not depress and release the pedal repeatedly.

d. Tighten the bleeder plug, then release the brake pedal.

e. Repeat the previous 2 steps until all the air in the fluid is completely bled out.

f. Tighten the bleeder plug completely to 73 inch lbs. (8.3 Nm).

g. Repeat the above procedures for each wheel to bleed the air out of the brake line.

7. Bleed the air out of the brake line again when "Step 4: Increase" appears on the hand-held tester display, as follows:

➡Bleed the air by following the steps displayed on the hand-held tester. Make sure that the brake fluid in the master cylinder reservoir tank does not become empty.

a. Connect the vinyl tube to either one of the bleeder plugs.

b. Depress the brake pedal several times, then loosen the bleeder plug connected to the vinyl tube with the pedal depressed.

8. When fluid stops coming out, tighten the bleeder plug, then release the brake pedal.

a. Repeat the previous 2 steps until all the air in the fluid is completely bled out.

b. Tighten the bleeder plug completely to 73 inch lbs. (8.3 Nm).

c. Repeat the above procedures for each wheel to bleed the air out of the brake line.

d. Finish "AIR BLEEDING" on the hand-held tester and turn the hand-held tester off.

e. Disconnect the hand-held tester from the DLC3 from the DLC3.

f. Turn the ignition switch off.

g. Inspect for fluid leak.

9. Check the fluid level and add fluid if necessary. Use SAE J1703 or FMVSS No. 116 DOT3 Brake fluid.

Brake Lines

➡Bleed air from the brake line of the wheel farthest from the master cylinder.

1. Raise and safely support the vehicle.
2. Connect a vinyl tube to the bleeder plug.
3. Have an assistant depress the brake pedal several times, then loosen the bleeder plug while the pedal is depressed.
4. When fluid stops coming out, tighten the bleeder plug, then release the brake pedal.
5. Repeat steps 3 and 4 until all the air in the fluid has been bled out.
6. Tighten the brake bleeder plug to 73 inch lbs. (8.3 Nm).
7. Repeat the above steps to bleed the air out of the brake line for each wheel.

Master Cylinder

If the master cylinder is reinstalled or if the reservoir becomes empty, bleed the air from the master cylinder. To prevent brake fluid from adhering, cover nearly painted surfaces with a shop rag or a piece of cloth.

1. Using a union nut wrench (10 mm), disconnect the 2 brake lines from the master cylinder, using a suitable brake line wrench.
2. Have an assistant slowly depress the brake pedal and hold it.
3. Cover the 2 outer holes with your fingers, and have your assistant release the brake pedal.
4. Repeat the previous 2 steps 3 or 4 times.
5. Using a union nut wrench (10 mm), connect the 2 the brake lines to the master cylinder and tighten to 11 ft. lbs. (15 Nm).

➡Use a torque wrench with a fulcrum length of 250 mm (9.84 in.).

➡This torque value is effective when the union nut wrench is parallel to the torque wrench.

BRAKES — ANTI-LOCK BRAKE SYSTEM (ABS)

WHEEL SPEED SENSORS

REMOVAL & INSTALLATION

1. Before servicing the vehicle, refer to the Precautions Section.
2. Remove the front wheel.
3. Remove the 3 screws and the front wheel opening extension pad
4. Remove the front fender liner.
5. Remove the front speed sensor, as follows:

a. Disconnect the speed sensor connector and clamp.

b. Remove the 2 bolts and separate the speed sensor harness from the body and shock absorber assembly.

c. Disengage the 2 claws from the steering knuckle.

d. Remove the bolt and the front speed sensor.

➡Do not allow foreign matter to attach to the sensor tip.

➡Clean the installation hole and surface for the speed sensor every time the speed.

To install:

➡Do not twist the wire harness for the front speed sensor when installing the speed sensor.

➡The bolt B tightens the brake flexible hose and front speed sensor together. Make sure that the flexible hose is positioned over the front speed sensor.

➡Install the stopper firmly to the hole in the body.

6. To install, reverse removal procedure.
7. Tighten the front speed sensor with the bolt to 71 inch lbs. (8 Nm).

➡Do not allow foreign matter to attach to the sensor tip.

8. Tighten the sensor harness brackets with the 2 bolts to the body and shock absorber assembly and 44 inch lbs. (5 Nm), and then to 14 ft. lbs. (19 Nm).
9. Install the front wheel and tighten the lug nuts to 76 ft. lbs. (103 Nm).
10. Check ABS speed sensor signal.

BRAKE CALIPER

REMOVAL & INSTALLATION

See Figure 5.

1. Before servicing the vehicle, refer to the Precautions Section.
2. Remove the front wheel.

➡Do not let brake fluid sit on painted surfaces, as it will eat through the paint. Wash it off immediately.

3. Drain brake fluid.
4. Remove the union bolt and gasket from the disc brake caliper assembly, then disconnect the flexible hose.

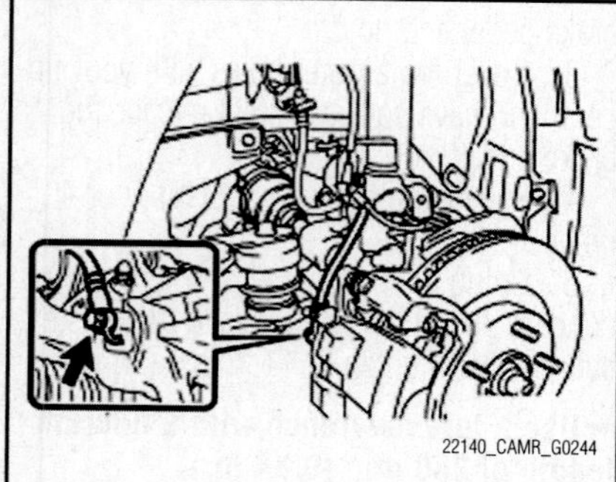

Fig. 5 Removing the union bolt and gasket

➡Remove the disc brake caliper assembly while holding both of the brake pads or the anti-squeal springs may fall off the brake pads.

5. Hold the front disc brake caliper slide pin and remove the 2 bolts and disc brake caliper assembly.

To install:

6. Install the disc brake caliper assembly with the 2 bolts and tighten to 25 ft. lbs. (34 Nm).
7. Check the installation of the anti-squeal springs. Visually check for any clearance between the brake pad and front disc brake pad support plates.

➡If the anti-squeal springs are installed correctly, there will be no clearance between the brake pad and the front disc brake pad support plates. If there is a clearance, the anti-squeal springs may not be installed properly.

➡Check all 4 contact surfaces between the brake pad and the front disc brake pad support plates.

8. Connect the flexible hose with the union bolt and a new gasket and tighten to 21 ft. lbs. (29 Nm).

➡Install the front brake flexible hose lock securely in the lock hole in the disc brake caliper.

9. Fill reservoir with brake fluid.
10. Bleed master cylinder.
11. Bleed brake line.
12. Bleed brake actuator assembly.
13. Inspect for brake fluid leak.
14. Inspect brake fluid level in reservoir.
15. Install the front wheel and tighten the lug nuts to 76 ft. lbs. (103 Nm).

DISC BRAKE PADS

REMOVAL & INSTALLATION

See Figure 6.

1. Before servicing the vehicle, refer to the Precautions Section.
2. Remove the 2 front disc brake caliper slide pins (upper and lower) from the front disc brake caliper mounting.
3. Remove brake cylinder.
4. Remove the 2 anti-squeal springs.
5. Remove the 2 brake pads from the front disc brake caliper mounting.

To install:

6. Install the 2 brake pads with front anti-squeal shims to the front disc brake caliper mounting.

➡When replacing worn pads, the front anti-squeal springs must be replaced at the same time.

➡Be sure to install the anti-squeal springs into the front disc brake pad installation holes as far as they will go.

7. Install the 2 front disc brake caliper slide pins (upper and lower) from the front disc brake caliper mounting.
8. Install the brake cylinder.

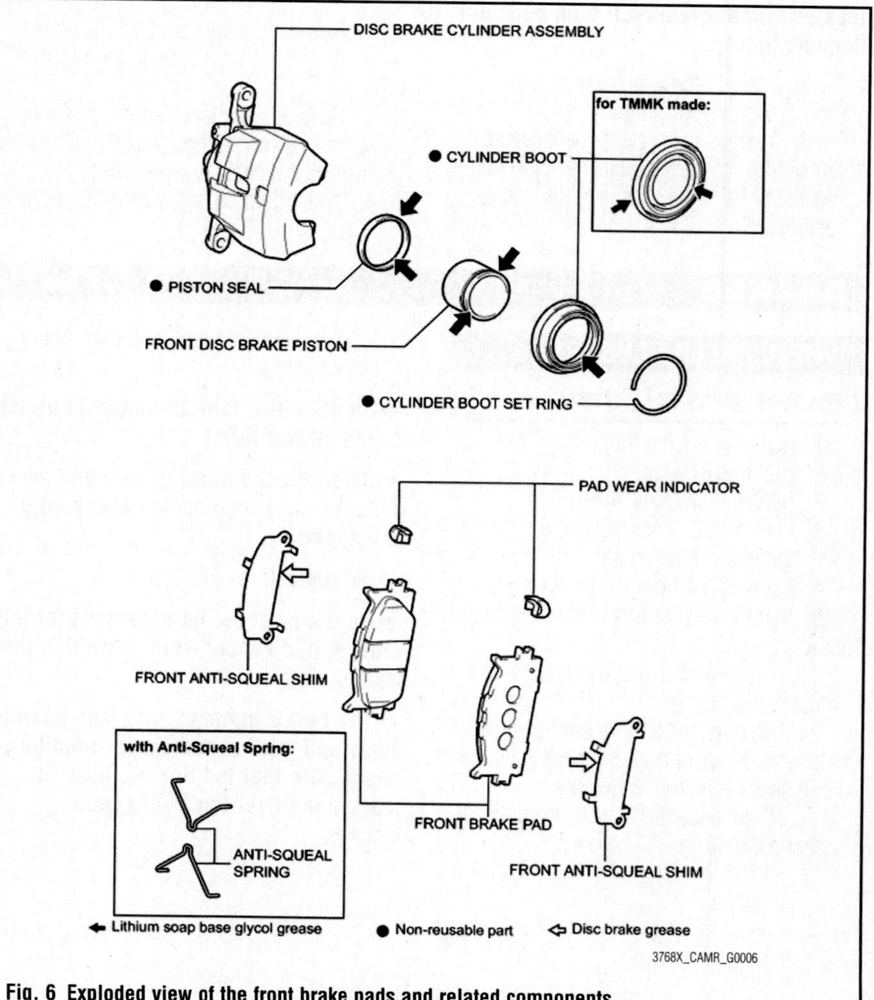

Fig. 6 Exploded view of the front brake pads and related components

BRAKES

REAR DISC BRAKES

BRAKE CALIPER

REMOVAL & INSTALLATION

See Figure 7.

1. Before servicing the vehicle, refer to the Precautions Section.
2. Remove the rear wheel.
3. Drain brake fluid.

⁂ WARNING

Do not let brake fluid sit on painted surfaces, as it will eat through the paint. Wash it off immediately.

4. Remove the union bolt and the gasket from the rear disc brake caliper assembly, then disconnect the rear brake flexible hose.
5. Hold the 2 rear disc brake caliper slide pins and remove the 2 bolts and rear disc brake caliper assembly.

To install:

6. Install the rear disc brake caliper assembly with the 2 bolts and tighten to 20 ft. lbs. (77 Nm).
7. Connect the rear brake flexible hose with the union bolt and a new gasket and tighten to 24 ft. lbs. (33 Nm).
8. Fill reservoir with brake fluid.

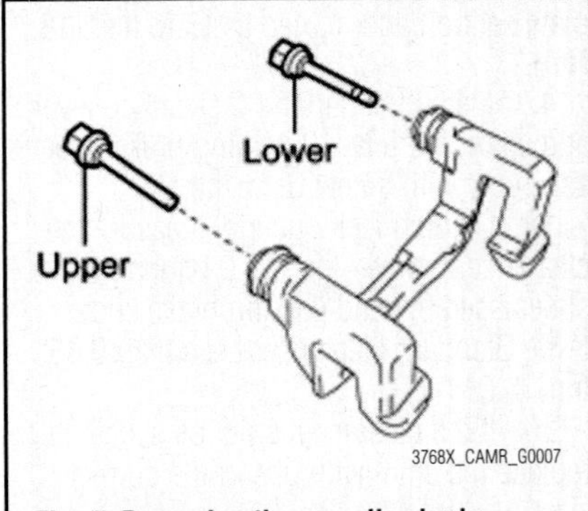

Fig. 7 Removing the rear disc brake caliper

9. Bleed brake line.
10. Inspect for brake fluid leak.
11. Inspect brake fluid level in reservoir.
12. Install the rear wheel and tighten the lug nuts to 76 ft. lbs. (103 Nm).

DISC BRAKE PADS

REMOVAL & INSTALLATION

See Figure 8.

1. Before servicing the vehicle, refer to the Precautions Section.
2. Remove the 2 rear disc brake caliper slide pins (upper and lower) from the rear disc brake caliper mounting.
3. Remove brake cylinder.
4. Remove the 2 brake pads with the rear anti-squeal shims.

To install:

5. Installation is the reverse of removal procedure.

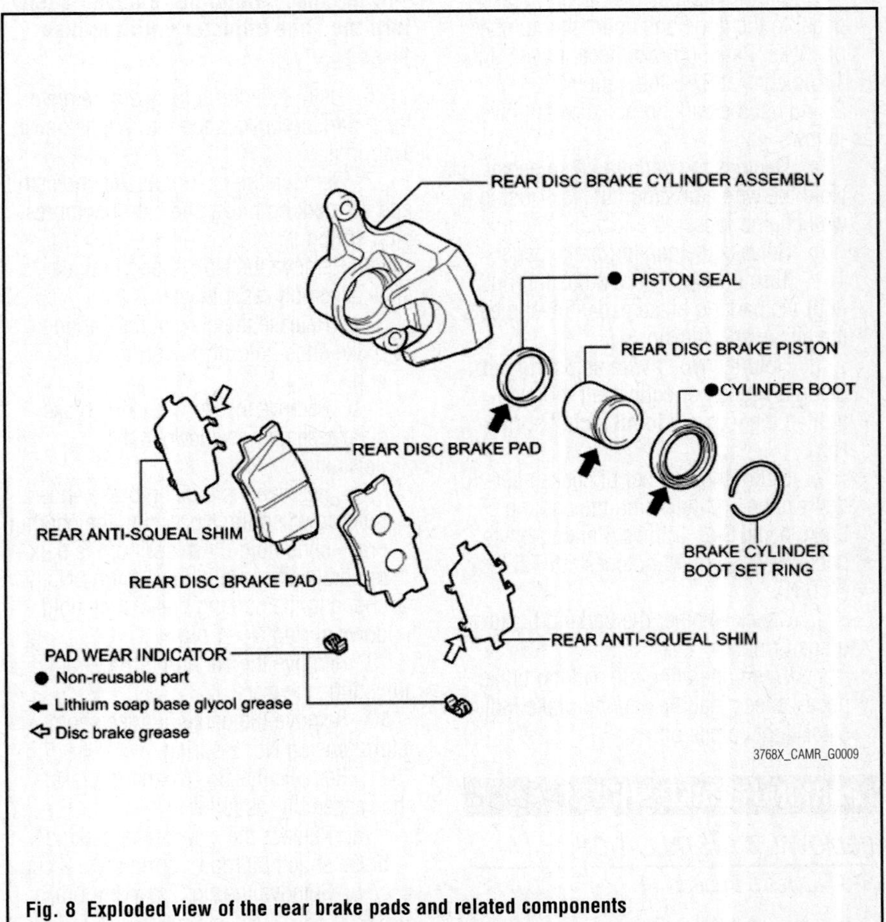

Fig. 8 Exploded view of the rear brake pads and related components

PARKING BRAKE CABLES

ADJUSTMENT

1. Inspect the parking brake pedal travel, as follows:
   a. Fully depress the parking brake pedal and release it to engage the parking brake.
   b. Depress the pedal to the floor again, and release it to disengage the parking brake.
   c. Slowly depress the parking brake pedal to the floor, and count the number of clicks. Parking brake pedal travel: 9 to 11 notches at 67.5 lbs. (300 N).
2. Adjust the parking brake pedal travel, as follows:
   a. Depress the parking brake pedal. Hold the wire adjusting nut No. 1 using a wrench and loosen the lock nut.
   b. Release the parking brake pedal.
   c. Turn the No. 1 wire adjusting nut until the parking brake pedal travel meets the above specification.
   d. Hold the No. 1 wire adjusting nut using a wrench or equivalent tool and tighten the lock nut to 48 inch lbs. (5.4 Nm).
   e. Count the number of clicks after depressing and releasing the parking brake pedal 3 or 4 times. Parking brake pedal travel: 9 to 11 notches at 67.5 lbs. (300 N).
   f. Check whether the parking brake drags or not.
   g. When operating the parking brake pedal, check that the parking brake indicator light comes on.

PARKING BRAKE SHOES

REMOVAL & INSTALLATION

See Figures 9 through 11.

1. Remove the rear wheel.
2. Remove the 2 bolts and separate the rear disc brake caliper assembly. Do not disconnect the flexible hose from the disc brake caliper assembly.
3. Remove the parking brake shoe adjusting hole plug from the rear disc.
4. Release the parking brake and place the matchmarks on the rear disc and the axle hub.
5. Remove the rear disc.

➡If the disc cannot be removed easily, turn the shoe adjuster until the disc turns freely.

6. Using needle-nose pliers, remove the 2 parking brake shoe return tension No. 1 springs.
7. Remove the parking brake shoe strut and the parking brake shoe strut compression spring.
8. Remove the No. 1 parking brake shoe assembly, as follows:
   a. Release the claw of the parking brake shoe hold down spring No. 2 cup.
   b. Remove the No. 1 parking brake shoe assembly as shown in the illustration.
   c. Remove the parking brake shoe hold down spring No. 1 cup, the parking brake shoe hold down spring, the parking brake shoe hold down spring No. 2 cup, and the parking brake shoe hold down spring No. 1 pin.
9. Remove the parking brake shoe adjusting screw set.
10. Remove the parking brake shoe return tension No. 2 spring.
11. Remove the No. 2 parking brake shoe assembly, as follows:
    a. Release the claw of the parking brake shoe hold down spring No. 2 cup.
    b. Remove the No. 2 parking brake shoe assembly as shown in the illustration.
    c. Remove the parking brake shoe hold down spring No. 1 cup, the parking brake shoe hold down spring, the parking brake shoe hold down spring No. 2 cup, and the parking brake shoe hold down spring No. 2 pin.
    d. Using needle-nose pliers, disconnect the No. 3 parking brake cable assembly from the parking brake shoe lever.

➡Be careful not to damage the No. 3 parking brake cable assembly.

12. Using a screwdriver, remove the C-washer, shim and the parking brake shoe lever.
13. Remove the parking brake shoe guide plate set bolt and the parking brake shoe guide plate.

To install:

14. Apply high temperature grease to the backing plate where it contacts the shoe.
15. Apply adhesive (Toyota Genuine Adhesive 1344, Three Bond 1344 or equivalent) to the threads of the parking brake shoe guide plate set bolt.
16. Install the parking brake shoe guide plate with the parking brake shoe guide plate set bolt and tighten to 13 ft. lbs. (18 Nm).
17. Install the parking brake shoe lever and shim to the No. 2 parking brake shoe assembly with a new C-washer.
18. Using a feeler gauge, measure the clearance between the No. 2 parking brake shoe assembly and parking brake shoe lever. Standard clearance: Less than 0.35 mm (0.014 in.).
19. If the clearance is not as specified, replace the shim with one of the correct size.
20. Install the No. 2 parking brake shoe assembly as follows:
    a. Using needle-nose pliers, connect the No. 3 parking brake cable assembly to the parking brake shoe lever.
    b. Install the No. 2 parking brake shoe assembly with the parking brake shoe hold down spring No. 2 pin, the parking brake shoe hold down spring No. 2 cup, the parking brake shoe hold down spring and the parking brake shoe hold down spring No. 1 cup.
    c. Engage the claw of the parking brake shoe hold down spring No. 2 cup to the No. 2 parking brake shoe assembly.

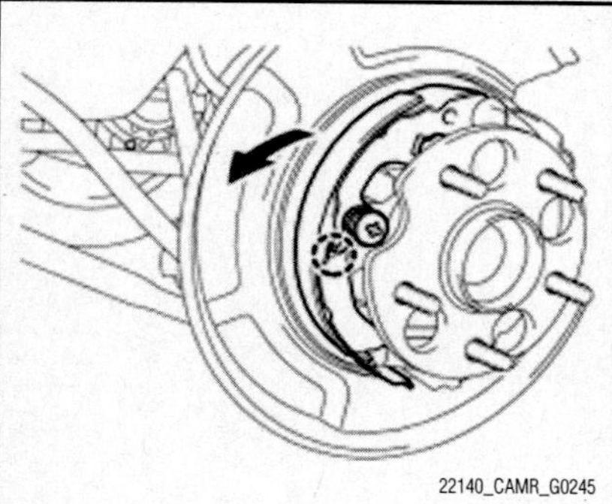

Fig. 9 Remove the No. 1 parking brake shoe assembly

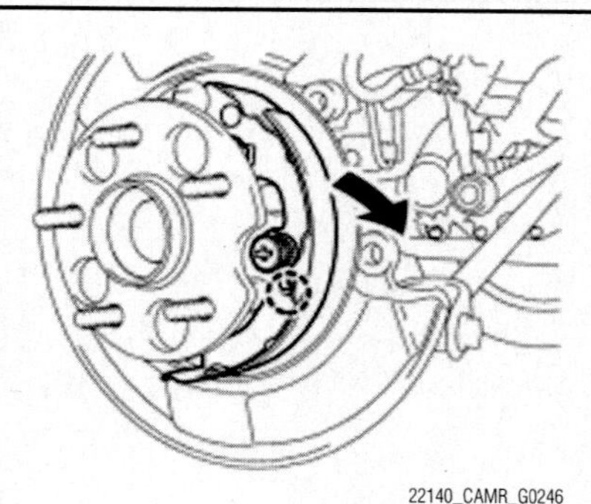

Fig. 10 Remove the No. 2 parking brake shoe assembly

21. Install the parking brake shoe adjusting screw set, as follows:

a. Apply high temperature grease to the parking brake shoe adjusting screw set as shown in the illustration.

b. Install the parking brake shoe return tension No. 2 spring to the No. 1 parking brake shoe assembly and the No. 2 parking brake shoe assembly.

c. Install the parking brake shoe adjusting screw set to the No. 1 parking brake shoe assembly and the No. 2 parking brake shoe assembly.

22. Install the No. 1 parking brake shoe assembly as follows:

a. Install the No. 1 parking brake shoe assembly with the parking brake shoe hold down spring No. 1 pin, parking brake shoe hold down spring No. 2 cup, parking brake shoe hold down spring and parking brake shoe hold down spring No. 1 cup.

b. Engage the claw of the parking brake shoe hold down spring No. 2 cup to the No. 1 parking brake shoe assembly.

23. Attach the parking brake shoe strut and the parking brake shoe strut compression spring to the No. 1 parking brake shoe assembly and No. 2 parking brake shoe assembly.

24. Using needle-nose pliers, install the 2 parking brake shoe return tension No. 1 springs. First install the front side spring and then the rear side spring.

25. Inspect parking brake installation and check that each part is installed properly.

➡There should be no oil or grease on the friction surfaces of the shoe linings and discs.

26. Install the rear disc.

27. Install the parking brake shoe adjusting hole plug.

28. Adjust parking brake shoe clearance.

29. Install the rear disc brake caliper assembly with the 2 bolts and tighten to 46 ft. lbs. (62 Nm).

30. Install the rear wheel.

31. Adjust the parking brake pedal travel.

32. Bed in parking brake shoes to discs, as follows:

a. Drive the vehicle at about 31 mph (50 km/h) on a safe, level and dry road.

b. Depress the parking brake pedal with 34 lbs. (150 N) of force.

33. Drive the vehicle about 0.25 miles (400 m) in this condition.

a. Repeat this procedure 3 times using 5-minute intervals between each procedure to prevent the parking brake assembly from overheating.

34. Remove the rear wheel.

35. Adjust parking brake shoe clearance.

36. For A/T vehicles, adjust the parking brake pedal travel.

37. For M/T vehicles, adjust the parking brake pedal travel.

38. Install the rear wheel and tighten the lug nuts to 76 ft. lbs. (103 Nm).

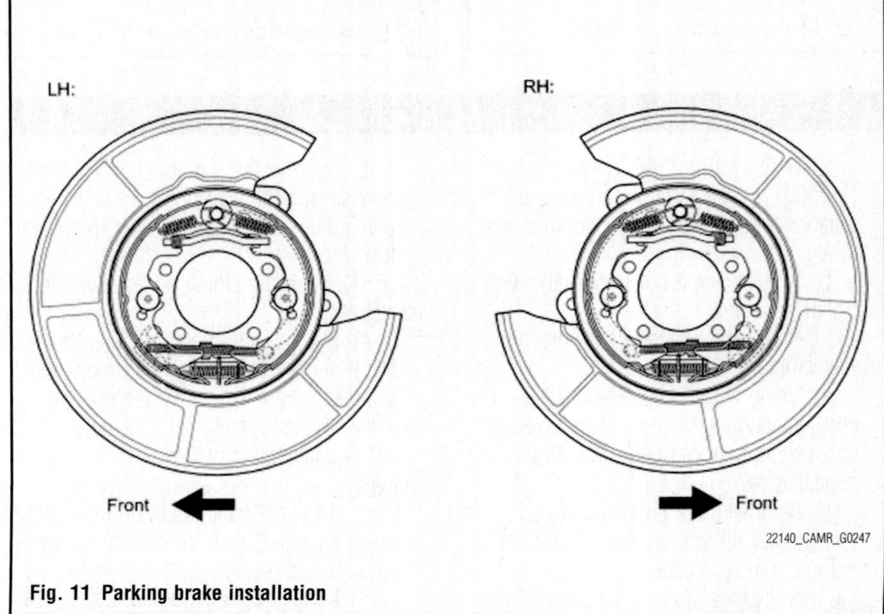

Fig. 11 Parking brake installation

ADJUSTMENT

1. Adjust parking brake shoe clearance, as follows:

a. Temporarily install the hub nuts.

b. Remove the shoe adjusting hole plug, turn the adjuster and expand the shoes until the disc locks.

c. Contract the shoe adjuster until the disc rotates smoothly. Standard: returns 8 notches

d. Check that the disc has no brake drag.

e. Install the shoe adjusting hole plug.

Parking Brake Lever Travel Adjustment (M/T)

1. Pull up the parking brake lever. Hold the No. 1 wire adjusting nut using a wrench and loosen the lock nut.
2. Release the parking brake lever.
3. Turn the No. 1 wire adjusting nut until the parking brake lever travel meets the above specification.
4. Hold the No. 1 wire adjusting nut using a wrench or an equivalent tool and tighten the lock nut to 44 inch lbs. (5 Nm).
5. Count the number of clicks after depressing and releasing the parking brake lever 3 or 4 times.
6. Check whether the parking brake drags.
7. When operating the parking brake lever, check that the parking brake indicator light comes on.

Parking Brake Lever Travel Adjustment (A/T)

1. Depress the parking brake pedal. Hold the No. 1 wire adjusting nut using a wrench and loosen the lock nut.
2. Release the parking brake pedal.
3. Turn the No. 1 wire adjusting nut until the parking brake pedal travel meets the above specification.
4. Hold the No. 1 wire adjusting nut using a wrench or an equivalent tool and tighten the lock nut to 48 inch lbs. (5.4 Nm).
5. Count the number of clicks after depressing and releasing the parking brake pedal 3 or 4 times.
6. Check whether the parking brake drags.
7. When operating the parking brake pedal, check that the parking brake indicator light comes on.

CHASSIS ELECTRICAL

AIR BAG (SUPPLEMENTAL RESTRAINT SYSTEM)

GENERAL INFORMATION

This vehicle is equipped with a Supplemental Restraint System (SRS). It consists of a driver airbag, front passenger airbag, driver side knee airbag, front seat side airbag and curtain shield airbag. Failure to carry out service operations in the correct sequence could cause the SRS to unexpectedly deploy during servicing, possibly leading to a serious accident. Further, if a mistake is made in servicing the SRS, it is possible that the SRS may fail to operate when required. Before performing servicing (including removal or installation of parts, inspection or replacement), be sure to read the following carefully, then follow the correct procedures indicated in the repair manual.

SERVICE PRECAUTIONS

**** CAUTION**

Disconnect and isolate the battery negative cable before beginning any airbag system component diagnosis, testing, removal, or installation procedures. Wait at least 90 seconds after the ignition switch is turned off and the negative (-) terminal cable is disconnected from the battery before starting the operation. The SRS is equipped with a backup power source, so if work is started within 90 seconds after disconnecting the negative (-) terminal cable from the battery, the SRS may be deployed. Failure to disable the airbag system may result in accidental airbag deployment, personal injury, or death.

DISARMING THE SYSTEM

To avoid personal injury when working on vehicles equipped with an air bag, the negative battery cable must be disconnected and at least 90 seconds must elapse before working on the system. Failure to do so may result in deployment of the air bag.

ARMING THE SYSTEM

To arm the system after service is finished, connect the negative battery cable.

CLOCKSPRING CENTERING

See Figures 12 and 13.

1. Before servicing the vehicle, refer to the Precautions Section.

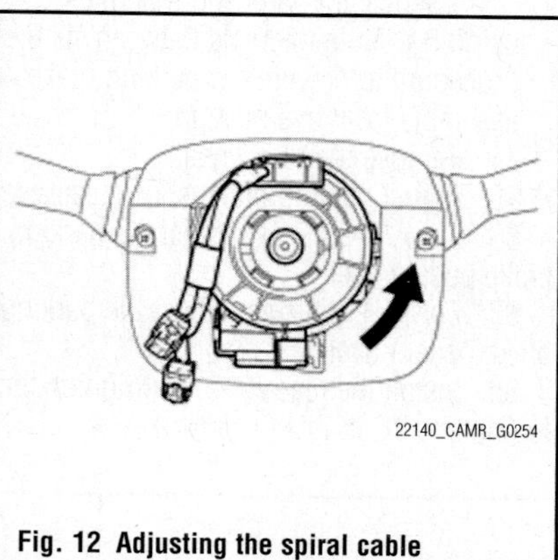

Fig. 12 Adjusting the spiral cable

2. Check that the ignition switch is **OFF**.
3. Check that the battery negative (-) terminal is disconnected.

**** CAUTION**

After removing the terminal, wait for at least 90 seconds before starting the operation.

4. Rotate the spiral cable counterclockwise slowly by hand until it feels firm.
5. Rotate the spiral cable clockwise approximately 2.5 turns to align the marks.

➡Do not turn the spiral cable by the airbag wire harness.

➡The spiral cable will rotate approximately 2.5 turns to both the left and right from the center.

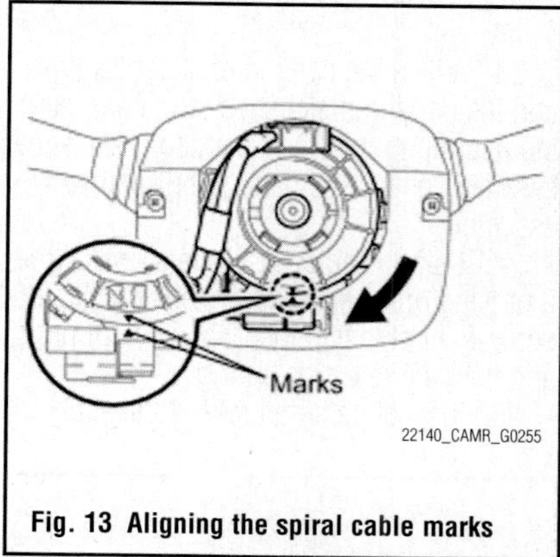

Fig. 13 Aligning the spiral cable marks

DRIVE TRAIN

CLUTCH

REMOVAL & INSTALLATION

See Figures 14 through 17.

➡When the transaxle is removed, be sure to use a new clutch release with bearing cylinder and new installation bolts. Removal of the transaxle allows the compressed clutch release with bearing cylinder to return to its original position, and dust from the moving section could damage the seal of the clutch release with bearing cylinder, possibly causing clutch fluid leaks.

1. Remove the manual transaxle assembly.
2. Remove the clutch orifice assembly.
   a. Using a union nut wrench (10 mm), remove the clutch release bleeder tube and orifice to flexible hose tube.
   b. Remove the 2 bolts and the clutch orifice assembly.
3. Remove the clutch release bleeder sub assembly.
   a. Using a union nut wrench (10 mm), separate the clutch release bleeder sub assembly from the bleeder clutch release tube.
   b. Remove the 2 bolts and clutch release bleeder sub assembly from the manual transaxle case.
4. Remove the clutch release with the bearing cylinder assembly.
   a. Remove the clutch tube boot from the front transaxle case.
   b. Remove the 3 bolts and the clutch release with the bearing cylinder assembly and clutch release cylinder to bleeder tube.
5. Remove the clutch release cylinder to bleeder tube.
   a. Using a union nut wrench (10 mm), remove the clutch release cylinder to bleeder tube from the clutch release with the bearing cylinder assembly.
6. Remove the clutch cover assembly.
   a. Align the matchmark on the clutch cover assembly with the one on the flywheel sub assembly.
   b. Loosen each set bolt one turn at a time until the spring tension is released.
   c. Remove the set bolts and pull off the clutch cover.

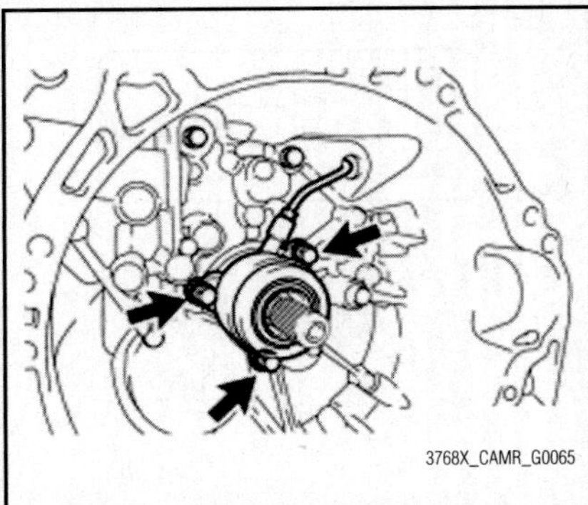

Fig. 14 Removing the clutch release with bearing cylinder assembly

➡Do not drop the clutch disc.

7. Remove the clutch disc assembly.

To install:

➡When the transaxle is removed, be sure to use a new clutch release with bearing cylinder and new installation bolts. Removal of the transaxle allows the compressed clutch release with bearing cylinder to return to its original position, and dust from the moving section could damage the seal of the clutch release with bearing cylinder, possibly causing clutch fluid leaks.

8. Install the clutch disc assembly.
   a. Insert the special tool into the clutch disc assembly, then insert them both into the flywheel sub assembly.

➡Insert the clutch disc assembly in the correct direction.

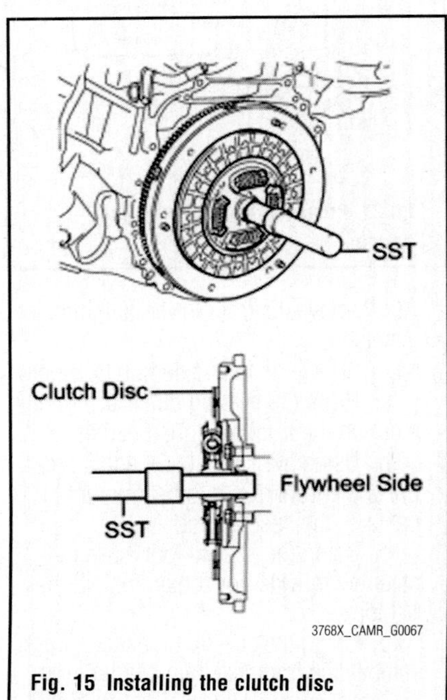

Fig. 15 Installing the clutch disc assembly

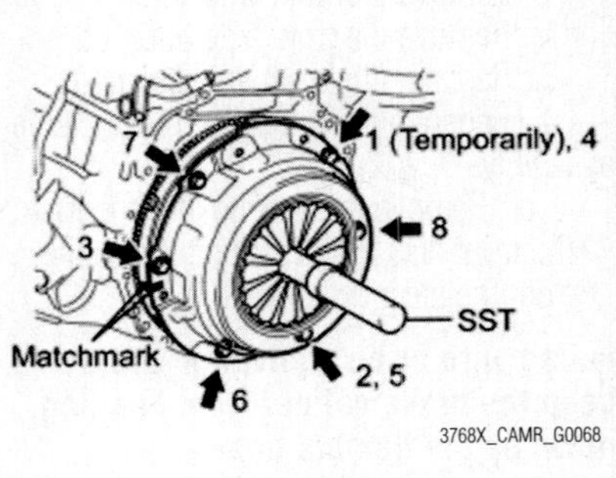

Fig. 16 Installing the clutch cover assembly

9. Install the clutch cover assembly.
   a. Align the matchmark on the clutch cover assembly with that on the flywheel sub assembly.
   b. Following the procedure shown in the illustration, tighten the 6 bolts in order, starting with the bolt located near the knock pin at the top. Tighten to 14 ft. lbs. (19 Nm).

➡Following the order in the illustration, tighten the bolts evenly one at a time.

➡Move SST up and down, right and left lightly after checking that the disc is in the center, and tighten the bolts.

10. Inspect and adjust the clutch cover assembly.
    a. Using a dial indicator with a roller instrument, check the diaphragm spring tip alignment. Maximum non-alignment is 0.0197 inch (0.5 mm).
    b. If the alignment is not as specified, using a special tool, adjust the diaphragm spring tip alignment.
11. Install the clutch release with bearing cylinder assembly.
    a. Temporarily tighten the clutch release cylinder to bleeder tube onto a new clutch release with bearing cylinder assembly.
    b. Clean and degrease all installation surfaces for clutch release with bearing cylinder assembly.
    c. Install the clutch release with bearing cylinder assembly with 3 new bolts. Tighten to 17 ft. lbs. (23 Nm).

➡The clutch release with bearing cylinder and installation bolts cannot be reused and must be replaced with new ones.

➡Clean and degrease all installation surfaces and make sure that the clutch release with bearing cylinder fits securely with the transaxle during installation. The first bolt should be tightened by hand while holding the clutch release with bearing cylinder.

➡Ensure that none of the clutch disc spline grease adheres to the clutch release with bearing cylinder.

➡The clutch release with bearing cylinder cannot be disassembled.

    d. Install the clutch tube boot onto the transaxle case.
    e. Temporarily tighten the clutch release cylinder to bleeder tube onto the clutch release bleeder sub-assembly.
    f. Temporarily tighten the 2 bolts and install the clutch release bleeder.
    g. Using a union nut wrench (10 mm), install the clutch release cylinder to bleeder tube. Tighten to 10 ft. lbs. (14 Nm). If tightening without a union nut wrench tighten to 11 ft. lbs. (15 Nm).

➡Use a torque wrench with a fulcrum length of 9.84 inch (250 mm).

➡This torque value is effective when the union nut wrench is parallel to the torque wrench.

    h. Apply clutch spline grease to the input shaft splines.
12. Remove the clutch release bleeder sub assembly.
    a. Separate the clutch release cylinder to bleeder tube from the clutch release bleeder.
    b. Remove the 2 bolts and the clutch release bleeder.
13. Inspect the clutch pipe line.
    a. Using the special tool, apply pressure of 0.1 MPa to the clutch pipe location shown, and confirm that pressure is maintained for 15 seconds or more.
14. Install the clutch release bleeder sub assembly.

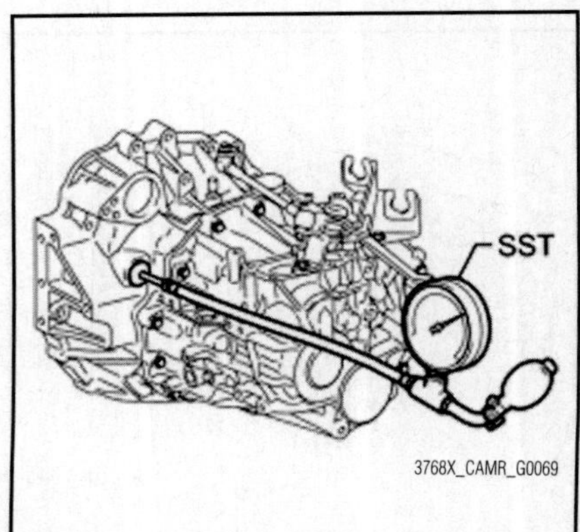

Fig. 17 Inspecting the clutch pipe line

a. Temporarily tighten the clutch release cylinder to bleeder tube onto the clutch release bleeder.

b. Install the 2 bolts and clutch release bleeder. Tighten to 12 ft. lbs. (17 Nm).

c. Using a union nut wrench (10 mm), install the clutch release cylinder to bleeder tube. Tighten to 10 ft. lbs. (14 Nm). If tightening without a union nut wrench, tighten to 11 ft. lbs. (15 Nm).

➡Use a torque wrench with a fulcrum length of 9.84 inches (250 mm).

➡This torque value is effective when the union nut wrench is parallel to the torque wrench.

15. Install the clutch orifice assembly.

a. Install the 2 bolts and clutch orifice assembly. Tighten to 9 ft. lbs. (12 Nm).

b. Using a union nut wrench (10 mm), install the clutch release bleeder tube and orifice to flexible hose tube. Tighten to 10 ft. lbs. (14 Nm). If tightening without the union nut wrench, tighten to 11 ft. lbs. (15 Nm).

➡Use a torque wrench with a fulcrum length of 9.84 inches (250 mm).

➡This torque value is effective when the union nut wrench is parallel to the torque wrench.

16. Install the manual transaxle assembly.

FRONT AXLE SHAFT, BEARING & SEAL

REMOVAL & INSTALLATION

See Figures 18 through 22.

➡Use the same procedures for the RH side as the LH side.

➡The procedures listed below are for the LH side.

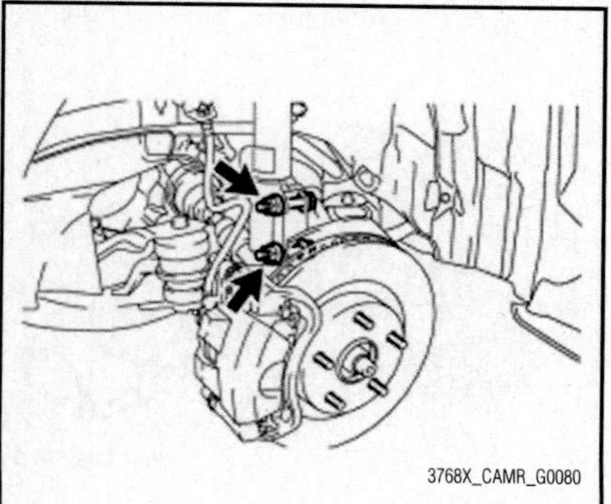

Fig. 18 Removing the front axle assembly

1. Remove the front wheel.
2. Remove the front axle hub nut.
3. Separate the front speed sensor.
4. Separate the front disc brake caliper assembly.

a. Remove the 2 bolts and separate the front disc brake caliper assembly from the steering knuckle.

➡Use wire or an equivalent tool to keep the brake caliper from hanging down by the flexible hose.

5. Remove the front disc.
6. Separate the tie rod end sub assembly.
7. Separate the front suspension lower No. 1 arm.
8. Remove the front axle assembly.

a. Put matchmarks on the front driveshaft assembly and the front axle hub sub assembly.

b. Using a plastic hammer, separate the front driveshaft assembly from the front axle assembly.

➡Be careful not to damage the driveshaft boot and speed sensor rotor.

c. Remove the 2 bolts, 2 nuts, and the front axle assembly.

9. Remove the No. 1 front wheel bearing dust deflector.

a. Using a screwdriver with its tip wrapped with vinyl tape, remove the No. 1 front wheel bearing dust deflector.

➡Be careful not to damage the steering knuckle.

10. Remove the front axle hub hole snap ring.

a. Using the snap ring pliers, remove the front axle hub hole snap ring.

11. Remove the front axle hub sub assembly.

a. Hole the front axle assembly between aluminum plates in a vise.

➡Do not over tighten the vise.

b. Using the special tool, remove the front axle hub sub assembly.

c. Using the special tool and a press, remove the bearing inner race (outside) from the front axle hub sub assembly.

➡Be careful not to drop the front axle hub sub assembly.

12. Remove the front disc brake dust cover.

a. Remove the 4 bolts and the disc brake dust cover from the steering knuckle.

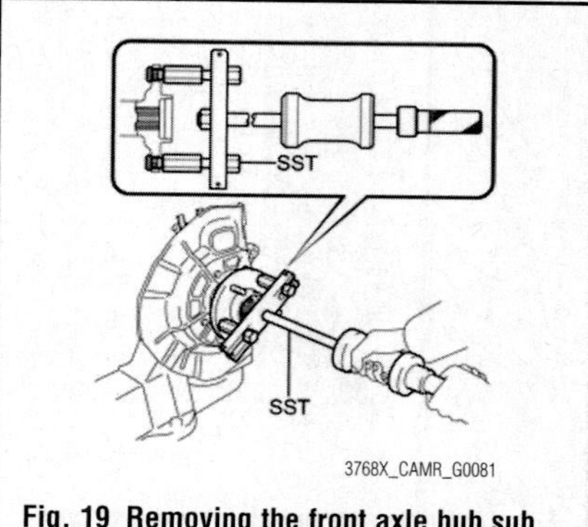

Fig. 19 Removing the front axle hub sub assembly

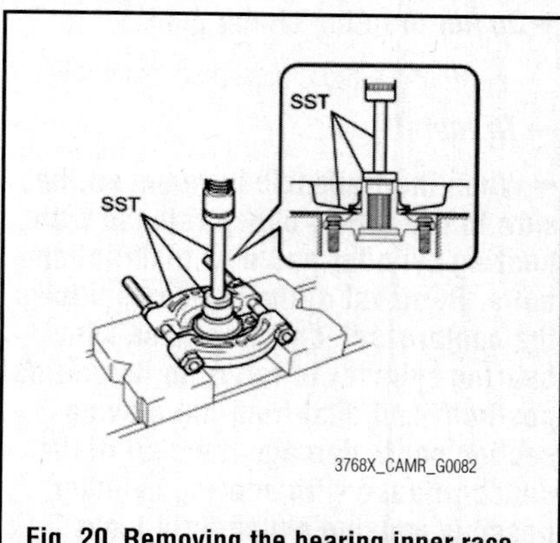

Fig. 20 Removing the bearing inner race

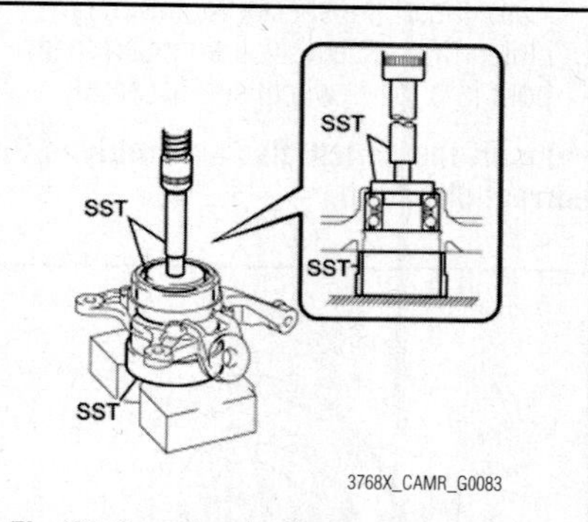

Fig. 21 Pressing the front axle hub bearing

13. Remove the front lower ball joint assembly.
14. Remove the front axle hub bearing.

a. Place the bearing inner race (outside) on the front axle hub bearing.

b. Using the special tool and a press, press the front axle hub bearing until it contacts the special tool.

c. Using the special tool make the steering knuckle horizontal, fix it to the V-block.

d. Using the special tool and a press, remove the front axle hub bearing from the steering knuckle.

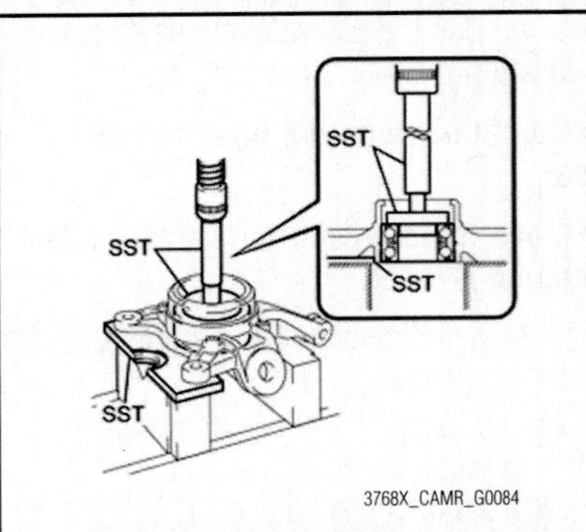

Fig. 22 Fixing the steering knuckle to the V-block

To install:

15. Install the front axle hub bearing.
 a. Using the special tool and a press, install a new front axle hub bearing to the steering knuckle.
16. Install the front lower ball joint assembly.
17. Install the front disc brake dust cover to the steering knuckle with the 4 bolts. Tighten the bolts to 73 inch lbs. (8.3 Nm)
18. Install the front axle hub sub assembly.
 a. Using the special tool and a press, install the front axle hub sub assembly.
19. Install the front axle hub hole snap ring.
 a. Using snap ring pliers, install a new front axle hub hole snap ring.
20. Install the No. 1 front wheel bearing dust deflector.
 a. Using the special tool and a hammer, install a new No. 1 front wheel bearing dust deflector.

➡Align the hole for the speed sensor in the No. 1 front wheel bearing dust deflector with the steering knuckle.

21. Install the front axle assembly.
 a. Align the matchmarks and install the front driveshaft assembly to the front axle hub sub assembly.
 b. Install the steering knuckle with the front axle assembly to the front shock absorber with the 2 bolts and 2 nuts. Tighten to 155 ft. lbs. (210 Nm).

➡Only when reusing the bolts and nuts, apply a small amount of engine oil to the threads of the nuts.

➡Be careful not to damage the driveshaft boot and speed sensor rotor.

22. Install the front suspension lower No. 1 arm.
23. Install the tie rod end sub assembly.
24. Install the front disc.
25. Install the front disc brake caliper assembly.
 a. Install the front disc brake caliber assembly to the steering knuckle with the 2 bolts. Tighten to 79 ft. lbs. (107 Nm).

➡Do not twist the brake hose when installing the front disc brake caliper assembly.

26. Install the front axle hub nut.
 a. Clean the threaded parts on the driveshaft and axle hub nut using a non residue solvent.

➡Be sure to perform this work for a new driveshaft.

➡Keep the threaded parts free of oil and foreign objects.

 b. Using a socket wrench (30 mm), install a new axle hub nut. Tighten to 217 ft. lbs. (294 Nm).

➡Take the nut after inspecting for looseness and runout in the following steps.

27. Separate the front disc brake caliper assembly.
 a. Remove the 2 bolts and separate the front disc brake caliper assembly from the steering knuckle.

➡Use wire or an equivalent tool to keep the brake caliper from hanging down by the flexible hose.

28. Remove the front disc.
29. Inspect the front axle hub bearing looseness.
30. Inspect the front axle hub runout.
31. Install the front disc.
32. Install the front disc brake caliper assembly.
 a. Install the front disc brake caliper assembly with the 2 bolts to the steering knuckle. Tighten to 79 ft. lbs. (107 Nm).

➡Do not twist the brake hose when installing the front disc brake caliper assembly.

33. Install the front speed sensor.
34. Install the front axle hub nut.
 a. Using a chisel and hammer, stake the axle hub nut.
35. Install the front wheel. Tighten to 76 ft. lbs. (103 Nm).
36. Adjust the front wheel alignment.
37. Check the ABS speed sensor signal.

FRONT DRIVESHAFT

REMOVAL & INSTALLATION

TMC Made

See Figures 23 through 26.

➡Use the same procedures for the RH side and LH side.

➡The procedures listed below are for the LH side.

1. Drain the automatic transaxle fluid.
2. Remove the front wheel.
3. Remove the front axle hub nut.
 a. Using the special tool and a hammer, release the staked part of the front axle hub nut.

➡Loosen the staked part of the nut completely, otherwise the thread of the driveshaft may be damaged.

 b. While applying the brakes, remove the front axle hub nut.
4. Separate the front stabilizer link assembly.
 a. Remove the nut and separate the front stabilizer link assembly.
5. Separate the front speed sensor.
 a. Remove the bolt and clip, and separate the speed sensor wire and flexible hose from the shock absorber.
 b. Remove the bolt and separate the speed sensor from the steering knuckle.

➡Prevent foreign matter from adhering to the speed sensor.

➡Be careful not to damage the speed sensor.

6. Separate the tie rod end sub assembly.
 a. Remove the cotter pin and nut.
 b. Using the special tool, separate the tie rod end sub assembly from the steering knuckle.

➡Make sure that the string of the SST is securely tied to the vehicle.

➡Be careful not to damage the ball joint dust cover.

➡Be careful not to damage the steering knuckle.

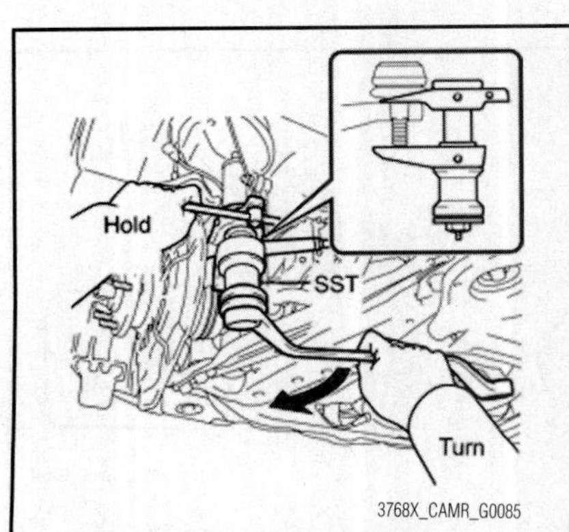

Fig. 23 Separating the tie rod end sub assembly

➡Be careful not to damage the front disc brake dust cover.

7. Separate the front suspension lower No. 1 arm.
   a. Remove the bolt and 2 nuts, and separate the front suspension lower No. 1 arm from the lower ball joint.
8. Separating the front axle assembly.
   a. Put matchmarks on the front driveshaft assembly and the axle hub.
   b. Using a plastic hammer, separate the front driveshaft assembly from the front axle assembly.

➡Be careful not to damage the driveshaft boot and speed sensor rotor.

9. Remove the front driveshaft assembly LH.
   a. Using the special tool, remove the front driveshaft assembly LH.

➡Be careful not to damage the driveshaft dust cover, boot or seal.

➡Be careful not to drop the driveshaft assembly.

10. Remove the front driveshaft assembly RH.
   a. Using a screwdriver, remove the bearing bracket hole snap ring.
   b. Remove the bolt and the front driveshaft assembly RH from the driveshaft bearing bracket.

➡Do not damage the boot or oil seal.

11. Secure the front axle hub sub assembly.
   a. Secure the front axle hub bearing.

➡The hub bearing may be damaged if it is subjected to the vehicle's full weight, such as moving the vehicle with the driveshaft removed. If it is necessary to place the vehicle's weight on the hub bearing, first support it with SST.

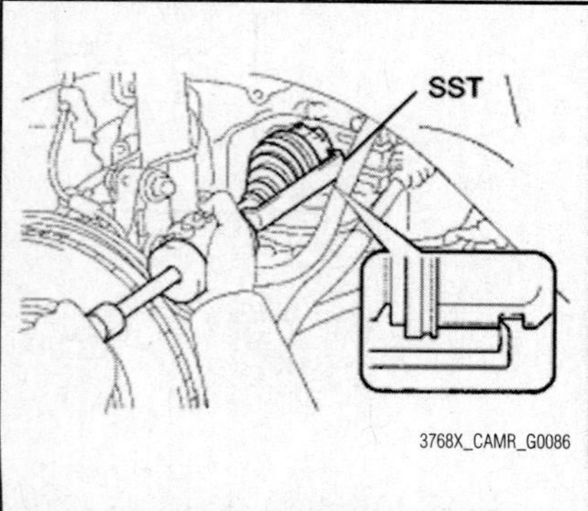

Fig. 24 Removing the front driveshaft assembly LH

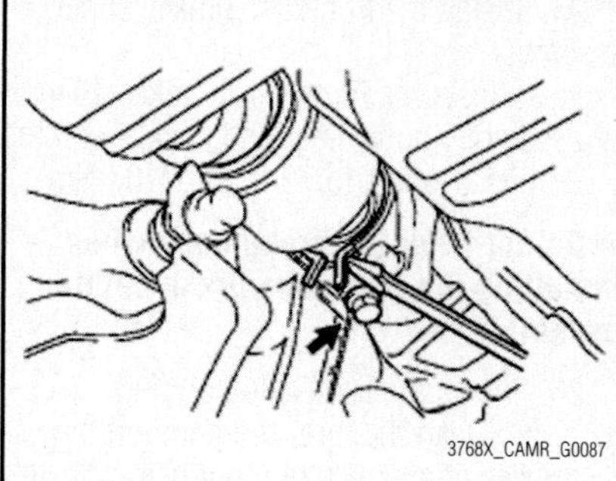

Fig. 25 Removing the front driveshaft assembly RH

To install:

12. Install the front driveshaft assembly LH.
   a. Coat the spline of the inboard joint shaft assembly with ATF.
   b. Align the shaft splines and install the driveshaft assembly LH with a brass bar and a hammer.

➡Set the shaft snap ring with the opening side facing down.

➡Be careful not to damage the driveshaft dust cover, boot, or oil seal.

➡Move the driveshaft assembly while keeping it level.

13. Install the front driveshaft assembly RH.
   a. Coat the spline of the inboard joint shaft assembly with ATF.
   b. Install the front driveshaft assembly RH.

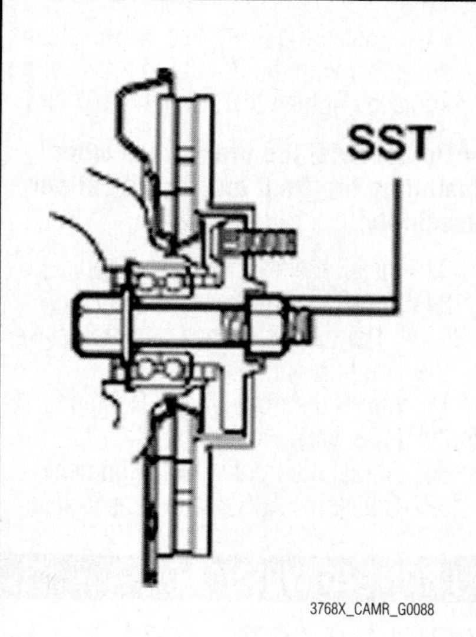

Fig. 26 Securing the front axle hub sub assembly

   c. Using a screwdriver, install a new bearing bracket hole snap ring.

➡Do not damage the boot and oil seal.

➡Move the driveshaft assembly while keeping it level.

   d. Install a new bolt and tighten to 24 ft. lbs. (32 Nm).
14. Install the front axle assembly.
   a. Align the matchmarks and install the front driveshaft assembly to the front axle hub sub assembly.

➡Be careful not to damage the driveshaft boot or speed sensor rotor.

15. Install the front suspension lower No. 1 arm.
   a. Install the lower ball joint to the front suspension lower No. 1 arm with the bolt and 2 nuts. Tighten to 55 ft. lbs. (75 Nm).
16. Install the tie rod end sub assembly.
   a. Install the tie rod end sub assembly to the steering knuckle with the nut. Tighten to 36 ft. lbs. (49 Nm).
   b. Install a new cotter pin.

➡If the holes for the cotter pin are not aligned, tighten the nut up to 60°further.

17. Install the front speed sensor.
   a. Install the front speed sensor to the steering knuckle with the bolt. Tighten to 71 inch lbs. (8 Nm).

➡Prevent foreign matter from adhering to the speed sensor.

➡Be careful not to damage the speed sensor.

   b. Install the flexible hose and the speed sensor to the shock absorber with the bolt and the set sensor clip on the knuckle. Tighten to 14 ft. lbs. (19 Nm).

➡Be careful not to damage the speed sensor.

➡Prevent foreign matter from adhering to the speed sensor.

➡Do not twist the sensor wire when installing the speed sensor.

18. Install the front stabilizer link assembly.
   a. Install the stabilizer link assembly with the nut. Tighten to 55 ft. lbs. (74 Nm).
19. Install the front axle hub nut.
   a. Clean the threaded parts on the driveshaft and axle hub nut using a non residue solvent.

➡**Be sure to perform this work for a new driveshaft.**

➡**Keep the threaded parts free of oil and foreign matter.**

b. Using a socket wrench (30 mm), install a new axle hub. Tighten to 217 ft. lbs. (294 Nm).

c. Using a chisel and hammer, stake the front axle hub nut.

20. Install the front wheel. Tighten to 76 ft. lbs. (103 Nm).
21. Add automatic transaxle fluid.
22. Inspect the automatic transaxle fluid.
23. Adjust the front wheel alignment.
24. Check the ABS speed sensor signal.

TMMK Made

See Figures 27 and 28.

➡**Use the same procedures for the RH side and the LH side.**

➡**The procedures listed below are for the LH side.**

1. Drain the automatic transaxle fluid.
2. Drain the manual transaxle oil (MT).
3. Remove the front wheel.
4. Remove the front axle hub nut.
5. Separate the front stabilizer link assembly.
6. Separate the front speed sensor.
7. Separate the front suspension lower No. 1 arm.
8. Separate the front axle assembly.
9. Remove the front driveshaft assembly LH.

➡**Be careful not to damage the driveshaft dust cover, boot and oil seal.**

➡**Be careful not to drop the driveshaft assembly.**

10. Remove the front driveshaft assembly RH.

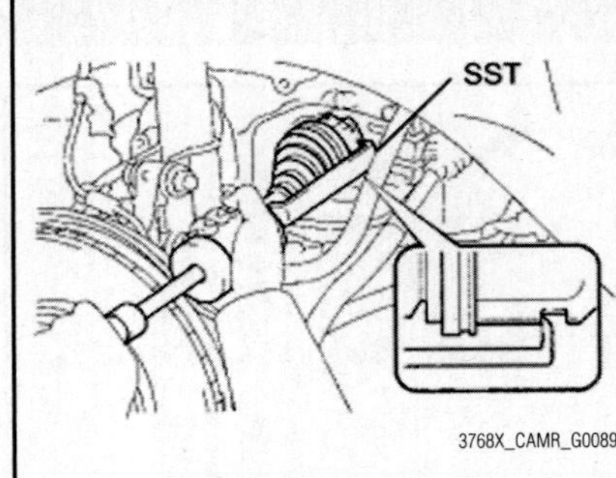

Fig. 27 Removing the front driveshaft assembly LH

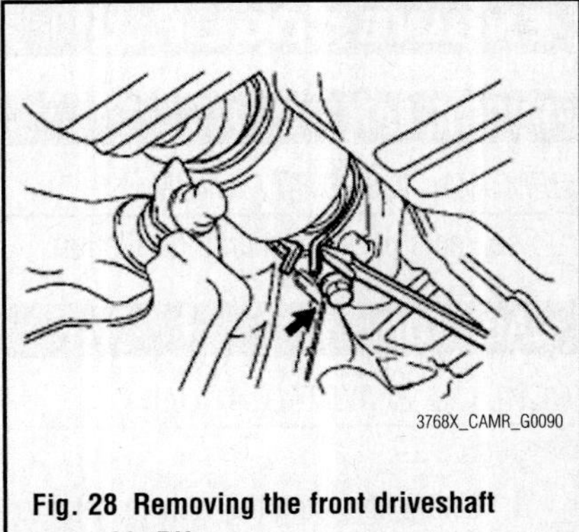

Fig. 28 Removing the front driveshaft assembly RH

a. Using a screwdriver, remove the bearing bracket hole snap ring.

b. Remove the bolt and front driveshaft assembly RH front the driveshaft bearing bracket.

➡**Do not damage the boot and oil seal.**

11. Secure the front axle hub sub assembly.
12. Inspect the front driveshaft assembly.

To install:

13. Install the front driveshaft assembly LH.

a. Coat the spline of the inboard joint shaft assembly with ATF.

b. Align the shaft splines and install the driveshaft assembly LH with a brass bar and hammer.

➡**Set the shaft snap ring with the opening side facing down.**

➡**Be careful not to damage the driveshaft dust cover, boot, and oil seal.**

➡**Move the driveshaft assembly while keeping it level.**

14. Install the front driveshaft assembly RH.

a. Coat the spline of the inboard joint shaft assembly with ATF.

b. Install the front driveshaft assembly RH.

c. Using a screwdriver, install a new bearing bracket hole snap ring.

➡**Do not damage the boot and oil seal.**

➡**Move the driveshaft assembly while keeping it level.**

d. Install a new bolt. Tighten to 24 ft. lbs. (32 Nm).

15. Install the front axle assembly.
16. Install the front suspension lower No. 1 arm.
17. Install the tie rod end sub assembly.
18. Install the front speed sensor.
19. Install the front stabilizer link assembly.
20. Install the front axle hub nut.
21. Install the front wheel. Tighten to 76 ft. lbs. (103 Nm).
22. Add automatic transaxle fluid (AT).
23. Add manual transaxle oil (MT).
24. Inspect the automatic transaxle fluid.
25. Inspect the manual transaxle oil.
26. Adjust the front wheel alignment.
27. Check the ABS speed sensor signal.

ENGINE COOLING

ENGINE FAN

REMOVAL & INSTALLATION

See Radiator removal and installation.

RADIATOR

REMOVAL & INSTALLATION

See Figure 29.

1. Before servicing the vehicle, refer to the Precautions Section.
2. Drain engine coolant.
3. Remove both front wheel opening extension pads.
4. Remove both engine under cover.
5. For 2.4L and 2.5L engines, perform the following:
   a. Remove air cleaner cap sub-assembly.
   b. Remove air cleaner inlet assembly.
6. For 3.5L engines, perform the following:
   a. Remove v-bank cover sub-assembly.
   b. Remove cool air intake duct seal.
   c. Remove air cleaner cap sub-assembly.
   d. Remove air cleaner inlet sub-assembly.
   e. Remove no. 1 air cleaner inlet.
7. For all vehicles, remove front bumper assembly.
8. Remove front bumper energy absorber.
9. Separate the radiator reserve tank hose from the radiator assembly.
10. Disconnect the inlet and outlet radiator hose from the radiator assembly.
11. Disconnect the oil from the radiator assembly.
12. For vehicles with A/T, disconnect the oil cooler inlet hose and outlet hose from the radiator assembly.
13. For 2.5L engines, remove the hood lock assembly.
14. Remove the upper radiator support, as follows:
    a. Disconnect the horn connector.
    b. Remove the 3 bolts and separate the hood lock assembly from the radiator support upper.
    c. Remove the clamp and separate the hood lock control cable from the radiator support upper.
    d. Remove the 5 bolts and radiator support upper.
15. Remove the radiator assembly, as follows:

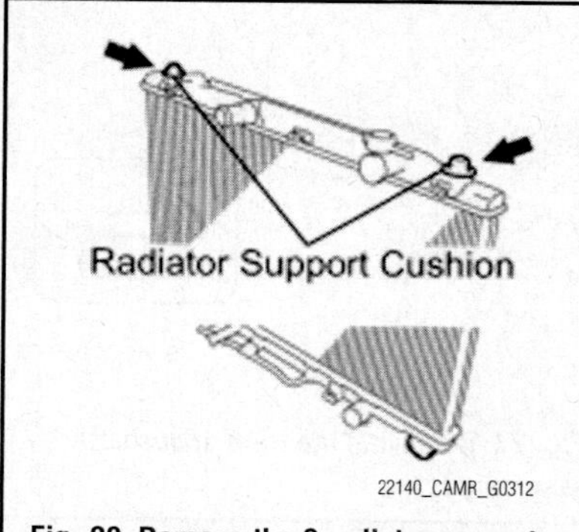

Fig. 29 Remove the 2 radiator support cushions

   a. Remove the 3 clamps and 2 connectors.
   b. Remove the 4 bolts and separate the condenser assembly from the radiator assembly.
   c. Remove the radiator assembly from the body by pulling upwards.
16. Release the 3 snap fits and lift the fan assembly with motor from the radiator.
17. Remove the 2 radiator support cushions from the radiator assembly.
18. Remove the 2 radiator support lowers from the radiator assembly.

To install:

19. Installation is the reverse of removal.
20. After installation, inspect for coolant leak.

THERMOSTAT

REMOVAL & INSTALLATION

2.4L and 3.5L Engines

See Figures 30 through 33.

1. Before servicing the vehicle, refer to the Precautions Section.

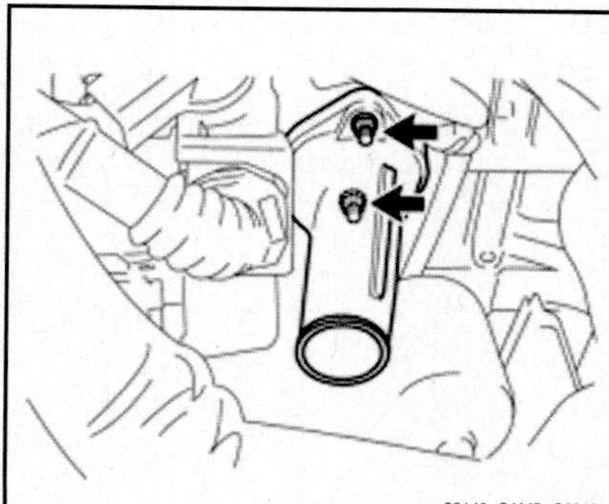

Fig. 30 Removing the 2 thermostat nuts—2.4L engine

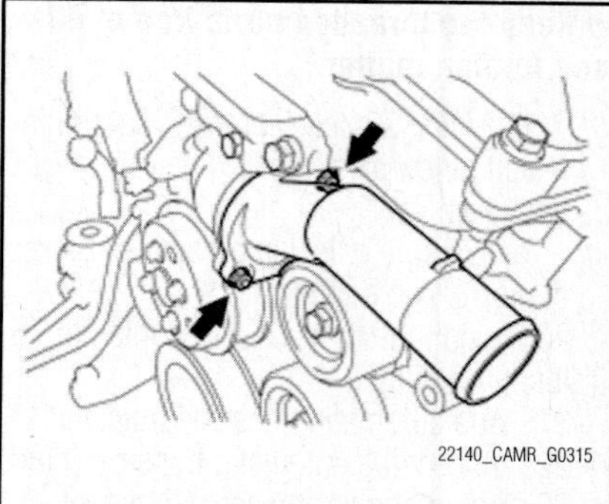

Fig. 31 Removing the 2 thermostat nuts—3.5L engine

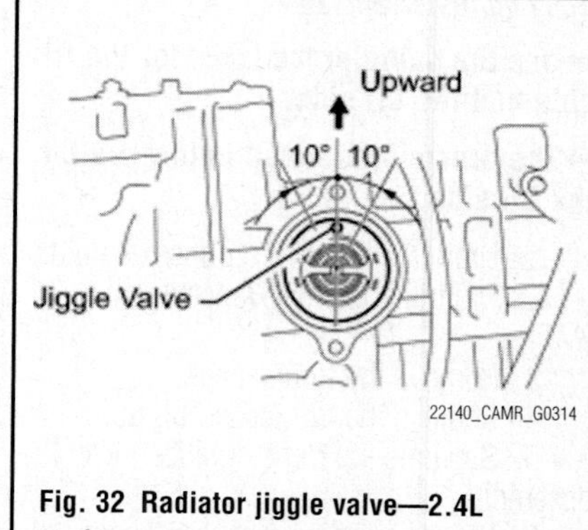

Fig. 32 Radiator jiggle valve—2.4L engine

2. For 2.4L engines, perform the following:
   a. Remove both front wheel opening extension pads.
   b. Remove both engine under cover.
   c. Drain engine coolant.
3. For 3.5L engines, perform the following:
   a. Drain engine coolant.
   b. Remove the V-bank cover sub-assembly.
   c. Remove the RH front fender apron seal.
   d. Remove the RH No. 2 engine mounting stay.
   e. Remove the drive belt.
   f. Remove No. 2 idler pulley sub-assembly.
4. Separate the radiator hose outlet.
5. Remove the 2 nuts and disconnect the water inlet from the cylinder block.
6. Remove the thermostat and the gasket from the thermostat.

To install:

7. Install a new gasket to the thermostat.

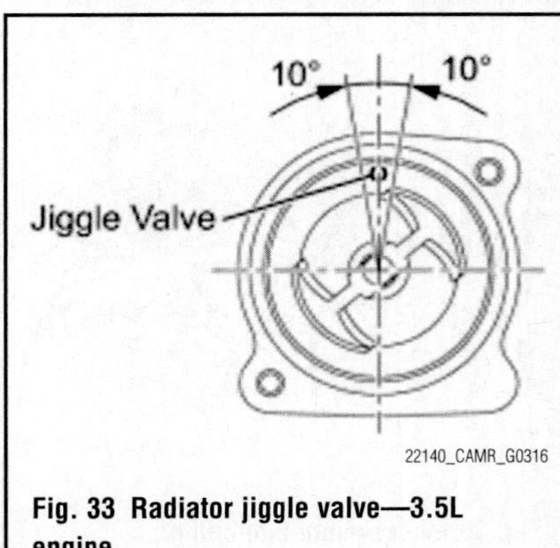

Fig. 33 Radiator jiggle valve—3.5L engine

8. Install a new gasket to the thermostat.

➡**The jiggle valve may be set within 10°of either side of the prescribed position.**

9. Install the thermostat with the jiggle valve facing up.
10. Install the water inlet and tighten to 7 ft. lbs. (10 Nm).
11. The remainder of installation is the reverse of removal.
12. After installation, inspect for coolant leak.

2.5L Engine

See Figure 34.

1. Disconnect the cable from the negative battery terminal.
2. Remove the front wheel RH.
3. Remove the front wheel opening extension pad RH.
4. Remove the front wheel opening extension pad LH.
5. Remove the engine under cover RH and LH.
6. Remove the front fender apron seal RH.
7. Remove the No. 1 engine cover sub assembly.
8. Drain the engine coolant.
9. Remove the v-ribbed belt.
10. Remove the alternator assembly.
11. Disconnect the outlet radiator hose.
12. Remove the water inlet.
    a. Remove the 2 nuts and water inlet.
13. Remove the thermostat.
    a. Remove the thermostat.
    b. Remove the gasket from the thermostat.

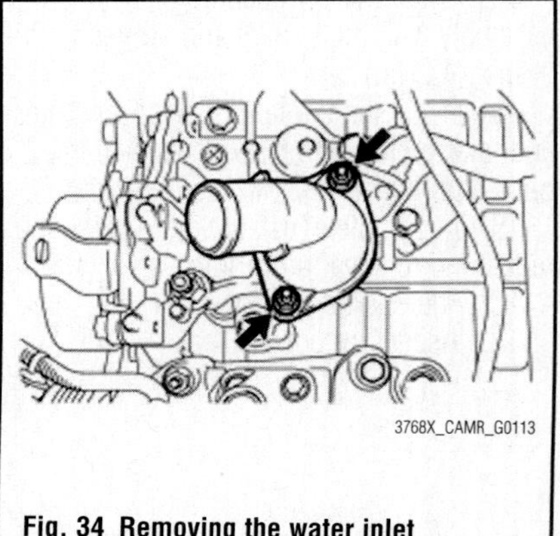

Fig. 34 Removing the water inlet

WATER PUMP

REMOVAL & INSTALLATION

2.4L Engine

See Figures 35 and 36.

1. Before servicing the vehicle, refer to the Precautions Section.
2. Disconnect the negative battery cable.
3. Remove both front wheel opening extension pads.
4. Remove both engine under covers.
5. Drain and recycle the engine coolant.
6. Remove RH front fender apron seal.
7. Remove RH No. 2 engine mounting stay.
8. Remove engine moving control rod sub-assembly.
9. Remove RH No. 2 engine mounting bracket
10. Remove the drive belt.
11. Remove alternator assembly.
12. Using SST: 09960-10010, remove the 4 bolts and water pump pulley.
13. Remove water pump assembly by performing the following:
    a. Remove the clamp of the Crankshaft Position (CKP) sensor from the water pump.
    b. Disconnect the wire of the Crankshaft Position (CKP) sensor from the clamp bracket.

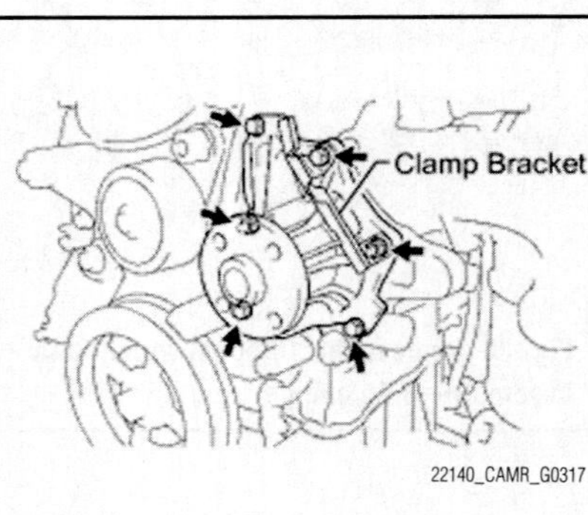

Fig. 35 Removing the 4 bolts, 2 nuts and clamp bracket—2.4L engine

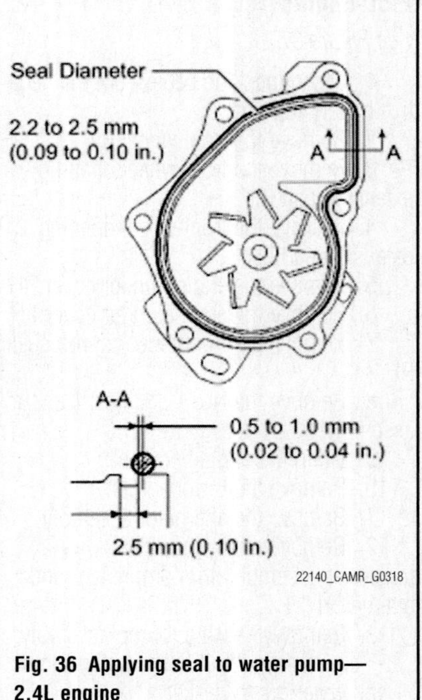

Fig. 36 Applying seal to water pump—2.4L engine

    c. Remove the 4 bolts, 2 nuts and clamp bracket.

➡**Be careful not to damage the contact surfaces of the water pump and cylinder block.**

    d. Using a screwdriver, pry between the water pump and cylinder block, and then remove the water pump. Tape the screwdriver tip before use.

To install:

14. Install water pump assembly:
    a. Remove any old seal packing material from the water pump assembly contact surface.

➡**Remove any oil from the contact surface. The parts must be set within 3 minutes after applying seal packing. Otherwise, the material must be removed and reapplied.**

    b. Apply a continuous line of seal packing. Seal packing: Toyota Genuine Seal Packing Black, Three Bond 1207B or Equivalent. Standard seal diameter: 0.09 to 0.10 in. (2.2 to 2.5 mm)
15. Install the water pump and clamp bracket with the 4 bolts and 2 nuts and tighten to 80 inch lbs. (9 Nm).
16. To complete installation, reverse removal procedure.
17. Tighten the water pump bolts to 19 ft. lbs. (26 Nm).
18. Add engine coolant.
19. Inspect for coolant leak.

2.5L Engine

See Figure 37.

1. Disconnect the cable from the negative battery terminal.
2. Remove the front wheel RH.
3. Remove the front wheel opening extension pad RH.
4. Remove the front wheel opening extension pad LH.
5. Remove the engine under cover RH.
6. Remove the engine under cover LH.
7. Remove the front fender apron seal RH.
8. Remove the No. 1 engine cover sub assembly.
9. Drain the engine coolant.
10. Remove the v-ribbed belt.
11. Remove the alternator assembly.
12. Remove the v-ribbed belt tensioner assembly (for compressor and alternator).
13. Remove the water pump assembly.
    a. Remove the 7 bolts, water pump and water pump gasket.

To install:

14. Install the water pump assembly.
    a. Install a new gasket and the water pump with the 7 bolts. Tighten to 15 ft. lbs. (21 Nm).
15. Install the v-ribbed belt tensioner assembly (for compressor and alternator).
16. Install the alternator assembly.
17. Install the v-ribbed belt.
18. Connect the cable to the negative battery terminal.
19. Add engine coolant.
20. Inspect for coolant leaks.
21. Install the No. 1 engine cover sub assembly.

3768X_CAMR_G0114

Fig. 37 Removing the water pump—2.5L engine

22. Install the front fender apron seal RH.
23. Install the engine under cover RH.
24. Install the engine under cover LH.
25. Install the front wheel opening extension pad RH.
26. Install the front wheel opening extension pad LH.
27. Install the front wheel RH.

3.5L Engine

See Figures 38 and 39.

1. Before servicing the vehicle, refer to the Precautions Section.
2. Remove engine assembly and transaxle. Secure the engine stand.
3. Remove RH front No. 1 engine mounting bracket.
4. Remove the No. 2 idler pulley sub-assembly, as follows:
    a. Remove the 2 bolts, 2 idler pulley cover plates and 2 idler pulley sub-assemblies.
5. Remove the 5 bolts and V-ribbed belt tensioner assembly.
6. Using SST: 09960-10010, hold the water pump pulley. Remove the 4 bolts and water pump pulley.
7. Remove water inlet housing, as follows:
    a. Separate the water hose.
    b. Remove the 2 bolts, nut and water inlet housing.
    c. Remove the water inlet housing gasket and water outlet pipe O-ring.
8. Remove the 16 bolts, water pump assembly and water pump gasket.

To install:

➡Make sure that there is no oil on the threads of the A bolts.

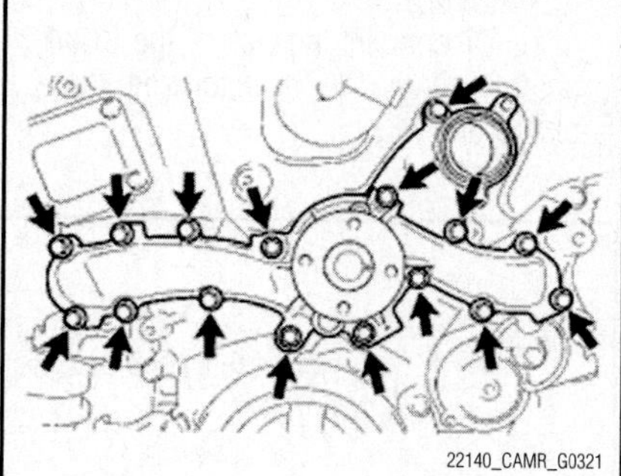

Fig. 38 Remove the 16 bolts, water pump assembly—3.5L engine

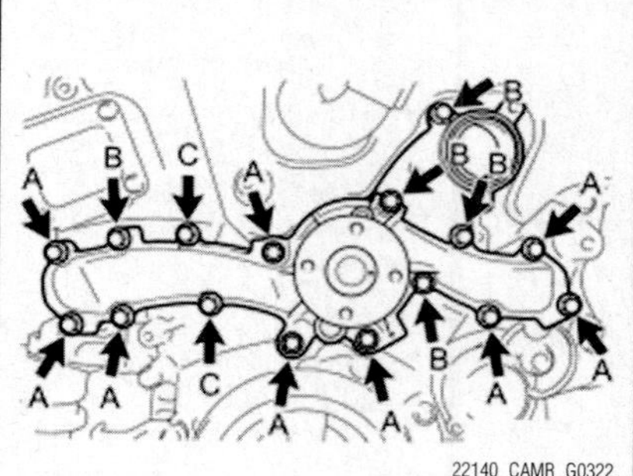

Fig. 39 Water pump tightening sequence—3.5L engine

➡Be sure to replace the 2 C bolts with new ones or reuse them after applying adhesive (Part No. 08833-00080, three bond 1344 or equivalent).

9. Install a new water pump gasket and the water pump assembly with the 16 bolts and tighten to:
    a. Bolt A: 15 ft. lbs. (21 Nm).
    b. Bolt B: 81 inch lbs. (9.1 Nm).
    c. Bolt C: 81 inch lbs. (9.1 Nm).
10. Install a new water inlet housing No. 1 gasket and water outlet pipe O-ring.

➡Be careful not to allow the O-ring to get caught between the parts.

11. Install water inlet housing, as follows:
    a. Install a new No. 1 water inlet housing gasket and water outlet pipe O-ring.
    b. Install the water inlet with the 2 bolts and nut and tighten to 7 ft. lbs. (10 Nm).
12. Temporarily install the water pump pulley with the 4 bolts.
    a. Using SST (SST: 09960-10010) or equivalent, hold the water pump pulley.
    b. Tighten the 4 bolts to 15 ft. lbs. (21 Nm).
13. Install the V-ribbed belt tensioner assembly with the 5 bolts and tighten to 32 ft. lbs. (43 Nm).
14. Install the 2 idler pulley cover plates and idler pulley sub-assemblies with the 2 bolts and tighten to 32 ft. lbs. (43 Nm).
15. To complete installation, reverse the remaining removal procedure.
16. Add engine coolant.
17. Inspect for coolant leak.

ALTERNATOR

REMOVAL & INSTALLATION

2.4L Engine

See Figure 40.

1. Disconnect the negative battery cable.
2. Remove the front wheel RH.
3. Remove the engine front cover RH.
4. Remove the front fender apron seal RH.
5. Remove the V-bank cover sub-assembly.
6. Remove the V-ribbed belt.
7. Remove the alternator assembly, as follows:
    a. Disconnect the alternator connector.
    b. Remove the nut and disconnect the wire harness from terminal B.
    c. Remove the bolt and wire harness clamp bracket.
    d. Remove the wire harness clamps.
    e. Remove the 2 bolts and alternator assembly.

To install:

8. Install the alternator assembly, as follows:
    a. Install the bracket with the bolt and tighten to 15 ft. lbs. (20 Nm).
    b. Install the wire harness clamp stay and tighten to 74 inch lbs. (8.4 Nm).
    c. Install the alternator assembly with the 2 bolts and tighten to 32 ft. lbs. (43 Nm).
    d. Install the nut to the cylinder block and tighten to 15 ft. lbs. (20 Nm).
    e. Connect the alternator connector to the alternator assembly.
    f. Install the alternator wire with the nut and tighten to 87 inch lbs. (9.8 Nm).
    g. Install the terminal cap.
    h. Connect the wire harness clamp.
9. Install the V-ribbed belt.
10. Install the V-bank cover sub-assembly.
11. Connect the negative battery cable.
12. Perform initialization.

2.5L Engine

See Figure 41.

1. Disconnect the cable from the negative battery terminal.
2. Remove the No. 1 engine cover sub assembly.
3. Remove the v-ribbed belt.
4. Remove the alternator assembly.
    a. Disconnect the alternator connector.
    b. Turn back the terminal cap.
    c. Remove the nut and disconnect the alternator wire.
    d. Remove the bolt and wire harness clamp bracket.
    e. Remove the 2 bolts and alternator.

To install:

5. Install the alternator assembly.
    a. Install the alternator with the 2 bolts. Tighten to 38 ft. lbs. (52 Nm).
    b. Install the wire harness clamp bracket with the bolt. Tighten to 74 inch lbs. (8.4 Nm).
    c. Connect the alternator wire with the nut. Tighten to 87 inch lbs. (9.8 Nm).
    d. Install the terminal cap.
    e. Connect the alternator connector.
6. Install the v-ribbed belt.
7. Install the No. 1 engine cover sub assembly.
8. Connect the cable to the negative battery terminal.

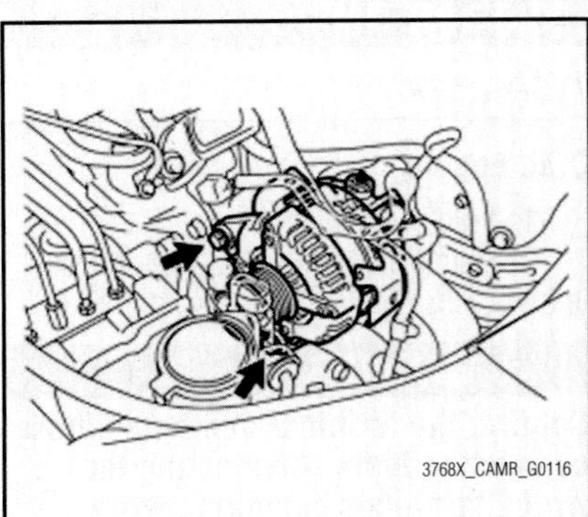

Fig. 40 Removing the alternator

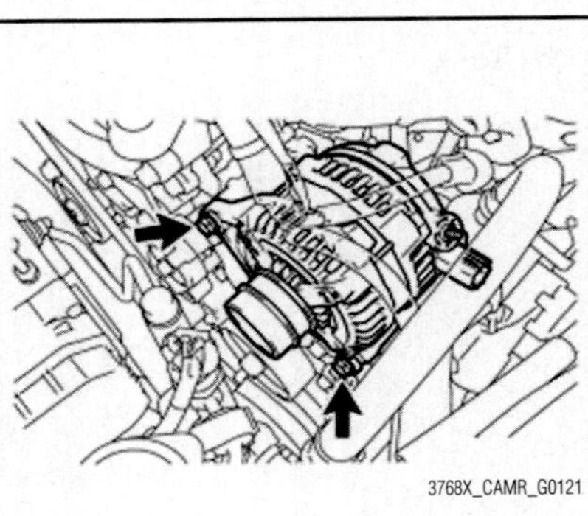

Fig. 41 Removing the alternator

3.5L Engine

1. Disconnect the cable from the negative battery terminal.
2. Remove the front wheel.
3. Remove the front fender apron seal RH.
4. Remove the front wheel opening extension pad RH.
5. Remove the front wheel opening extension pad LH.
6. Remove the engine under cover RH.
7. Remove the engine under cover LH.
8. Drain the engine coolant.
9. Remove the v-bank cover sub assembly.
10. Remove the cool air intake duct seal.
11. Remove the air cleaner inlet assembly.
12. Remove the No. 1 air cleaner inlet.
13. Remove the front bumper assembly.
14. Remove the front bumper energy absorber.
15. Separate the radiator reserve tank hose.
16. Separate the radiator inlet hose.
17. Separate the radiator outlet hose.
18. Separate the No. 1 oil cooler inlet hose.
19. Separate the No. 1 oil cooler outlet hose.
20. Remove the radiator support upper.
21. Remove the fan shroud.
22. Remove the radiator assembly.
23. Remove the v-ribbed belt.
24. Remove the alternator assembly.
    a. Remove the terminal cap.
    b. Remove the nut and disconnect the wire harness from terminal B.
    c. Disconnect the alternator connector from the alternator assembly.
    d. Disconnect the connector from the compressor and magnetic clutch.
    e. Disconnect the 2 wire harness clamps.
    f. Remove the 2 bolts.
    g. Remove the bolt from the cylinder block.
    h. Disconnect the wire harness clamp and remove the alternator assembly.
    i. Remove the bolt and wire harness clamp stay.
    j. Remove the bolt and bracket.

To install:

25. Install the alternator assembly.
    a. Install the bracket with the bolt. Tighten to 15 ft. lbs. (20 Nm).
    b. Install the wire harness clamp stay and tighten to 74 inch lbs. (8.4 Nm).

c. Connect the wire harness clamp.

d. Install the alternator assembly to the cylinder block with the bolt. Tighten to 15 ft. lbs. (20 Nm).

e. Install the 2 bolts. Tighten to 32 ft. lbs. (43 Nm).

f. Connect the alternator connector to the alternator assembly.

g. Install the alternator wire with the nut and tighten to 87 inch lbs. (9.8 Nm).

h. Install the terminal cap.

i. Connect the 2 wire harness clamps.

j. Connect the magnetic clutch connector to the compressor and magnetic clutch.

26. Install the v-ribbed belt.
27. Install the radiator assembly.
28. Install the fan shroud.
29. Install the radiator support upper.
30. Connect the No. 1 oil cooler outlet tube.
31. Connect the No. 1 oil cooler inlet tube.
32. Connect the radiator reserve tank hose.
33. Install the front bumper energy absorber.
34. Install the front bumper assembly.
35. Install the No. 1 air cleaner inlet.
36. Install the air cleaner cap sub assembly.
37. Install the air cleaner inlet assembly.
38. Connect the cable to the negative battery terminal.
39. Add engine coolant.
40. Check for engine coolant leaks.
41. Install the v-bank cover sub assembly.
42. Install the cool air intake duct seal.
43. Install the front fender apron seal RH.
44. Install the engine under cover RH.
45. Install the engine under cover LH.
46. Install the front wheel opening extension pad RH.
47. Install the front wheel opening extension pad LH.
48. Install the front wheel and tighten to 76 ft. lbs. (103 Nm).

ENGINE ELECTRICAL

IGNITION SYSTEM

FIRING ORDER

Firing order for the 2.4L and 2.5L engines are 1–3–4–2.

Firing order for 3.5L engine: 1–2–3–4–5–6

IGNITION COIL

REMOVAL & INSTALLATION

2.4L Engine

See Figure 42.

1. Before servicing the vehicle, refer to the Precautions Section.
2. Disconnect the negative battery cable.
3. Remove engine cover(s).
4. Disconnect the 4 ignition coil connectors. Remove the 4 bolts and 4 ignition coils.

To install:

5. To install, reverse removal procedure. Tighten ignition coil bolts to 80 inch lbs. (9 Nm).

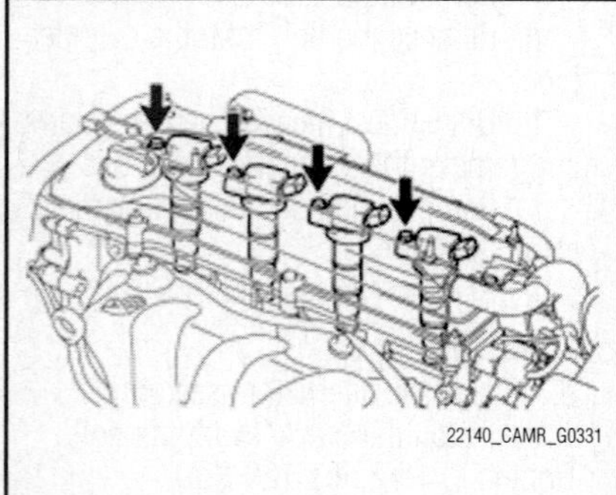

Fig. 42 Removing ignition coils—2.4L engines

2.5L Engine

For ignition coil removal and installation refer to SPARK PLUG.

3.5L Engine

See Figure 43.

1. Before servicing the vehicle, refer to the Precautions Section.
2. Disconnect the negative battery cable.
3. Drain and recycle the engine coolant.
4. Remove windshield wiper link assembly.
5. Remove cowl top panel outer sub-assembly.
6. Remove v-bank cover sub-assembly.
7. Remove air cleaner cap sub-assembly.
8. Remove intake air surge tank assembly.
9. Remove No. 1 surge tank stay by performing the following:
   a. Remove the bolt and disconnect the harness clamp.
   b. Remove the bolt and No. 1 surge tank stay.
10. Disconnect the 6 ignition coil connectors.
11. Remove the 6 bolts and 6 ignition coils.

To install:

12. To install, reverse removal procedure.
13. Tighten the following to specification:
   a. 6 ignition coil bolt: 10 ft. lbs. (10 Nm).
   b. No. 1 surge tank stay bolt: 15 ft. lbs. (21 Nm).
   c. No. 1 surge tank stay bolt and clamp: 62 inch lbs. (7 Nm).

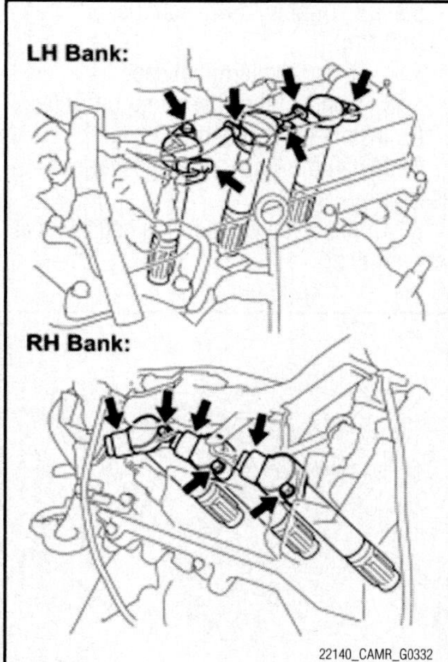

Fig. 43 Removing ignition coils—3.5L engines

IGNITION TIMING

INSPECTION

3.5L Engine

1. Warm up the engine.
2. Using SST: 09843-18040, connect terminals 13 (TC) and 4 (CG) of the DLC3.

⁂ WARNING

Confirm the terminal numbers before connecting them. Connecting the wrong terminals can damage the engine.

➡Turn off all electrical systems before connecting the terminals.

➡Perform this inspection after the cooling fan motor is turned off.

3. Remove the v-bank cover.
4. Pull out the red lead wire harness.
5. Connect the tester terminal of the timing light to the red lead wire.

➡Use a timing light which can detect the first signal.

6. Check the ignition timing at idle. Standard ignition timing: 8 to 12°BTDC at idle.

➡When checking the ignition timing, the transmission should be in the neutral position.

➡Run the engine at 1000 to 1300 rpm for 5 seconds, and then check that the engine rpm returns to idle speed.

7. Disconnect terminals 13 (TC) and 4 (CG) of the DLC3.
8. Check the ignition timing at idle. Standard ignition timing: 7 to 24°BTDC at idle.
9. Confirm that the ignition timing advances immediately when the engine rpm is increased.
10. Remove the timing light from the engine.

ADJUSTMENT

All engines are equipped with a Distributorless Ignition System (DIS). No timing adjustment is possible.

SPARK PLUGS

REMOVAL & INSTALLATION

2.4L Engine

1. Before servicing the vehicle, refer to the Precautions Section.
2. Remove the plastic engine cover.
3. Disconnect the 4 ignition coil connectors and remove the 4 bolts and ignition coils.
4. Using a 16 mm (0.63 in.) spark plug wrench, remove the 4 spark plugs.

To install:

5. Installation is the reverse of removal.
   a. Torque the ignition coils to 80 inch lbs. (9 Nm) and the spark plugs to 13 ft. lbs (18 Nm).

2.5L Engine

1. Remove the No. 1 engine cover sub assembly.
2. Remove the ignition coil assembly.
   a. Disconnect the 4 ignition coil assembly connectors.
   b. Remove the 4 bolts and 4 ignition coil assemblies.
3. Remove the spark plug.
   a. Using a spark plug wrench, remove the 4 spark plugs.

To install:

4. Install the spark plugs.
   a. Using a spark plug wrench, install the 4 spark plugs. Tighten to 18 ft. lbs. (25 Nm).
5. Install the ignition coil assembly.
   a. Install the 4 ignition coil assemblies with the 4 bolts. Tighten to 7 ft. lbs. (10 Nm).

3.5L Engine

1. Before servicing the vehicle, refer to the Precautions Section.
2. Remove the V-bank cover.
3. Remove the intake air surge tank.
4. Disconnect the 6 ignition coil connectors.
5. Remove the 6 bolts and 6 ignition coils.
6. Using a 16 mm (0.63 in.) plug wrench, remove the spark plugs.

To install:

7. Installation is the reverse of removal, noting the following:
   a. Torque the ignition coils to 66 inch lbs. (7.5 Nm) and the spark plugs to 13 ft. lbs (18 Nm).

ENGINE ELECTRICAL

STARTING SYSTEM

STARTER

REMOVAL & INSTALLATION

2.4L Engine

See Figures 44 and 45.

1. Disconnect the cable from the negative battery terminal.
2. Remove the air cleaner inlet assembly.
3. Remove the air cleaner cap sub assembly.
4. Remove the air cleaner case sub assembly.
5. Remove the starter assembly (manual transaxle):
   a. Disconnect the terminal 50 connector from the starter assembly.
   b. Remove the nut and disconnect the wire harness from terminal 30.
   c. Remove the 3 bolts, clutch flexible hose bracket and starter assembly.
6. Remove the starter assembly (automatic transaxle):
   a. Disconnect the terminal 50 connector from the starter assembly.
   b. Remove the nut and disconnect the wire harness from terminal 30.
   c. Remove the 2 bolts and starter assembly.

To install:

7. Install the starter assembly (manual transaxle).
   a. Install the starter assembly and clutch flexible hose bracket with the 3 bolts. Tighten bolt A to 28 ft. lbs. (37 Nm). Tighten bolt B to 9 ft. lbs. (12 Nm).
   b. Connect the wire harness to terminal 30 and install the nut. Then, attach the terminal cap. Tighten to 87 inch lbs. (9.8 Nm).
   c. Connect the terminal 50 connector to the starter assembly.
8. Install the starter assembly (automatic transaxle).
   a. Install the starter assembly with the 2 bolts. Tighten to 28 ft. lbs. (37 Nm).

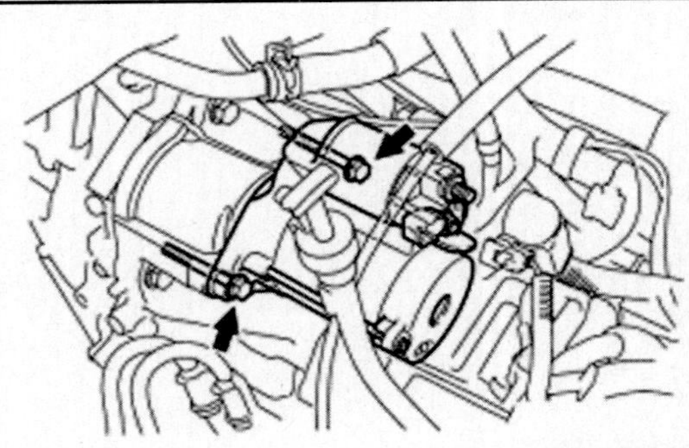

Fig. 44 Removing the starter assembly (2.4L engine, automatic transaxle)

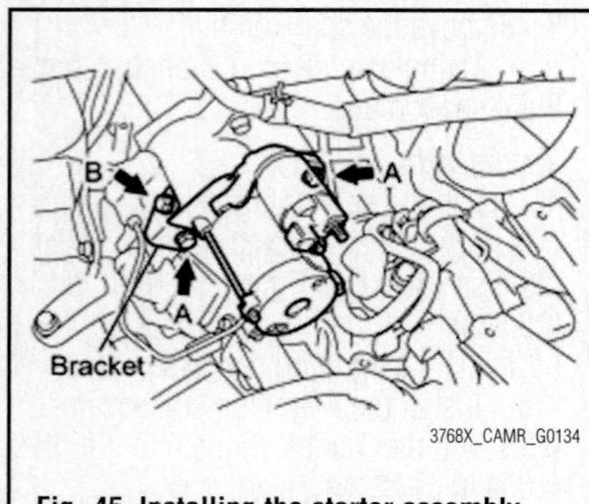

Fig. 45 Installing the starter assembly (manual transaxle)

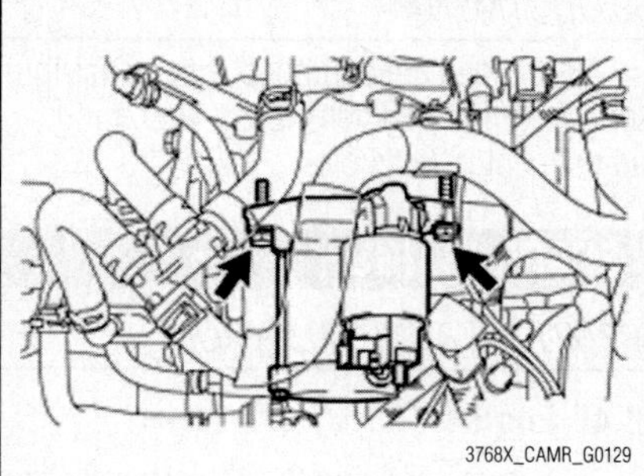

Fig. 46 Removing the starter—2.5L engine

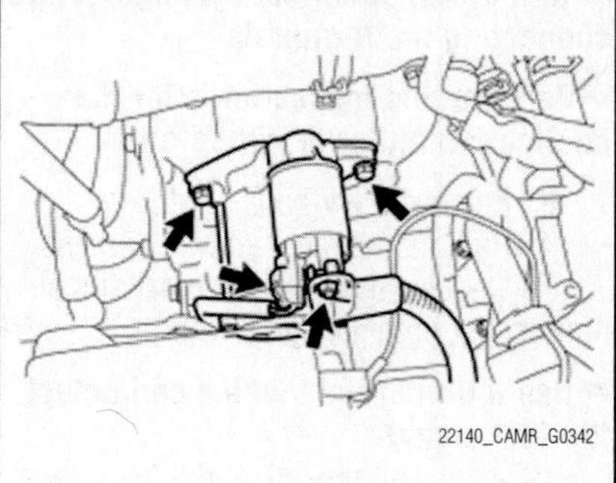

Fig. 47 Removing starter assembly—3.5L engine

b. Connect the wire harness to terminal 30 and install the nut. Then, attach the terminal cap. Tighten to 87 inch lbs. (9.8 Nm).

c. Connect the terminal 50 connector to the starter assembly.

9. Install the air cleaner case sub assembly.
10. Install the air cleaner cap sub assembly.
11. Install the air cleaner inlet assembly.
12. Connect the cable to the negative battery terminal.

2.5L Engine

See Figure 46.

1. Disconnect the cable from the negative battery terminal.
2. Remove the No. 1 engine cover sub assembly.
3. Remove the inlet air cleaner assembly.
4. Remove the air cleaner cap sub assembly.
5. Remove the air cleaner filter element.
6. Remove the air cleaner case sub assembly.
7. Remove the starter assembly.

a. Disconnect the terminal 50 connector from the starter assembly.

b. Turn back the terminal cap.

c. Remove the nut and disconnect the wire harness from terminal 30.

d. Remove the 2 bolts and starter assembly

To install:

8. Install the starter assembly.

a. Install the starter assembly with the 2 bolts. Tighten to 27 ft. lbs. (37 Nm).

b. Connect the wire harness to terminal 30 and install the nut. Then, attach the terminal cap. Tighten to 87 inch lbs. (9.8 Nm).

c. Connect the terminal 50 connector to the starter assembly.

9. Install the air cleaner case sub assembly.
10. Install the air cleaner filter element.
11. Install the air cleaner cap sub assembly.
12. Install the air cleaner assembly.
13. Install the No. 1 engine cover sub assembly.
14. Connect the cable to the negative battery terminal.

3.5L Engine

See Figure 47.

1. Before servicing the vehicle, refer to the Precautions Section.
2. Disconnect the negative battery cable.
3. Remove cool air intake duct seal.
4. Remove v-bank cover sub-assembly.
5. Remove air cleaner inlet assembly.
6. Remove air cleaner cap sub-assembly.
7. Remove air cleaner case sub-assembly.
8. Remove No. 1 air cleaner inlet.
9. Disconnect the terminal 50 connector from the starter assembly.
10. Remove the nut and disconnect the wire harness from terminal 30.
11. Remove the 2 bolts and starter assembly.

To install:

12. Install the starter assembly with the 2 bolts and tighten to 26 ft. lbs. (37 Nm).
13. Connect the wire harness to terminal 30 and install the nut and tighten to 87 inch lbs. (9.8 Nm).
14. Cover the nut with the cap.
15. Connect terminal 50 to the starter assembly.
16. To complete installation, reverse removal procedure.

ENGINE MECHANICAL

ACCESSORY DRIVE BELTS

ACCESSORY BELT ROUTING

See Figures 48 through 50.

INSPECTION

See Figure 51.

Visually check the V-ribbed belt for excessive wear, frayed cords, etc. If any defect has been found, replace the V-ribbed belt.

- Cracks on the rib side of a belt are considered acceptable. If the belt has chunks missing from the ribs, it should be replaced

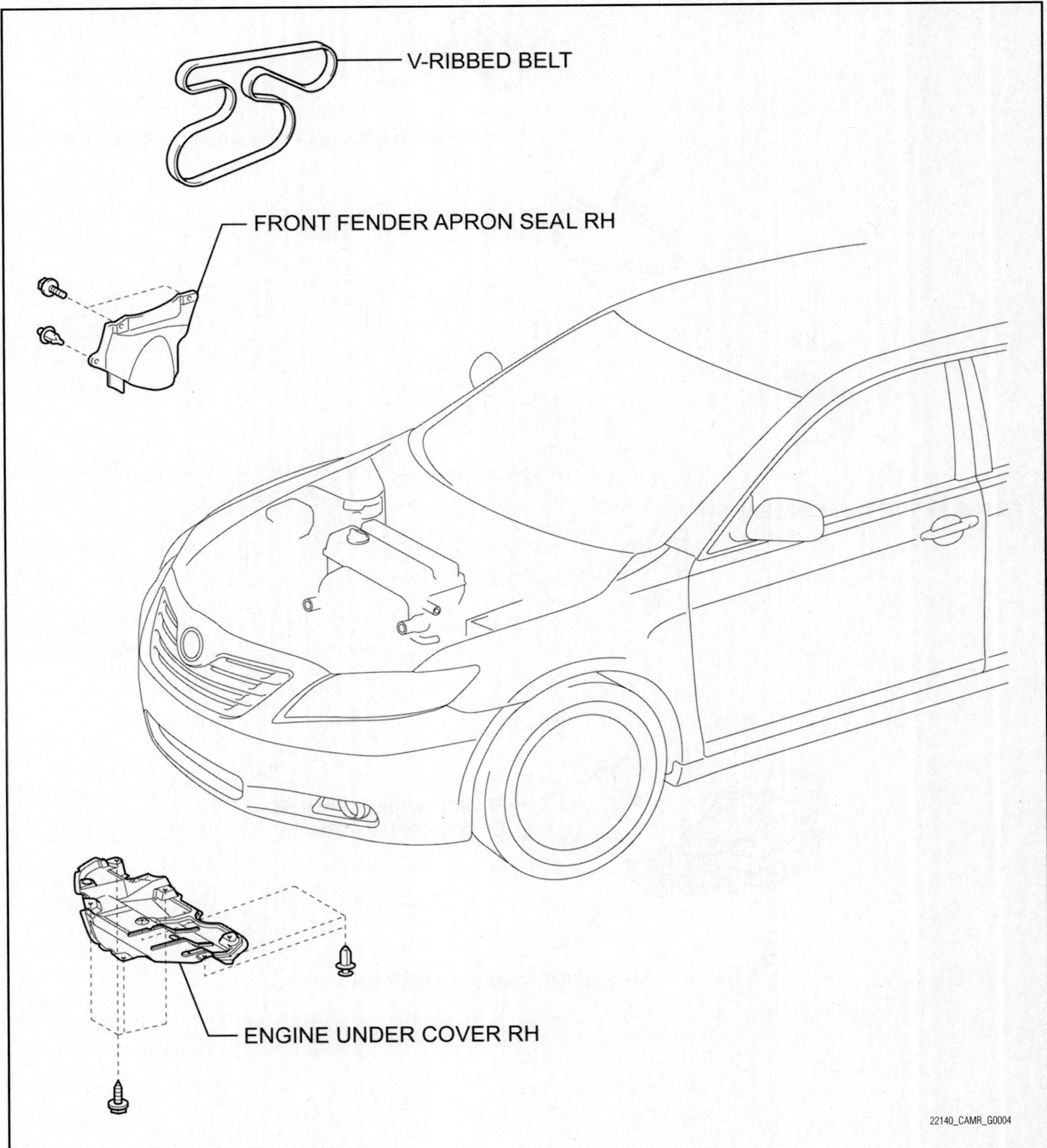

Fig. 48 Locating drive belt components—2.4L engine

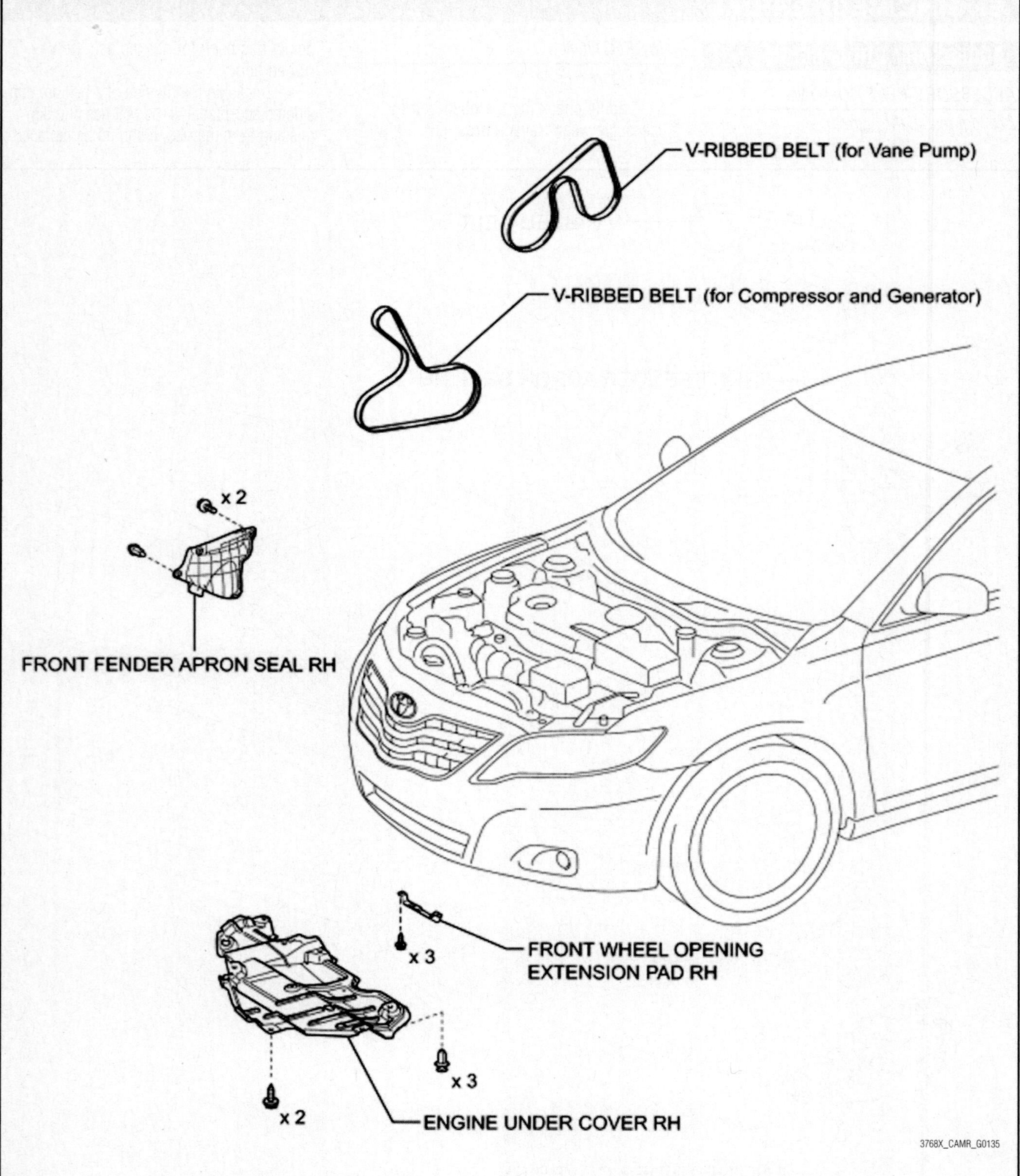

Fig. 49 Locating drive belt components—2.5L engine

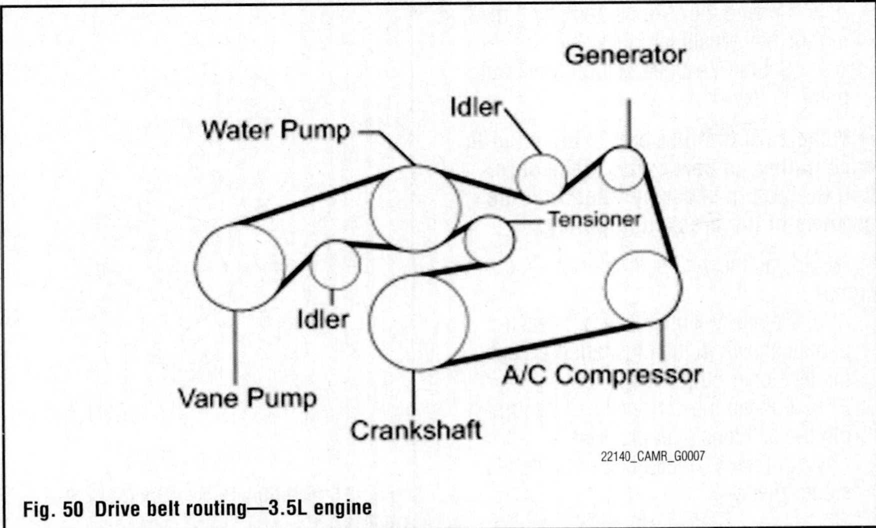

Fig. 50 Drive belt routing—3.5L engine

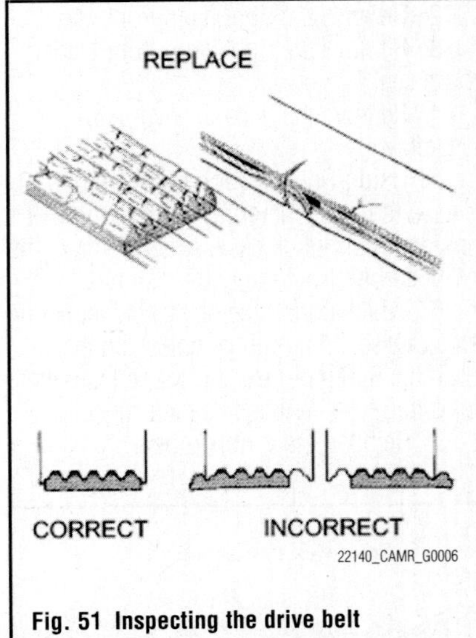

Fig. 51 Inspecting the drive belt

- A "new belt" is a belt which has been used for less than 5 minutes with the engine running
- A "used belt" is a belt which has been used for 5 minutes or more with the engine running

ADJUSTMENT

This vehicle is equipped with an auto-tensioner and cannot be adjusted.

REMOVAL & INSTALLATION

2.4L Engine

See Figure 48.

1. Before servicing the vehicle, refer to the Precautions Section.
2. Remove the right hand front wheel.
3. Remove the right hand engine under cover.
4. Remove the right hand front fender apron seal.

➡Before removing, take note of the following:

- Be sure to connect Special Tool: 09216—42010 and the tools so that they are in line during use
- When retracting the tensioner, turn it clockwise slowly for 3 seconds or more. Do not apply force rapidly
- After the tensioner is fully retracted, do not apply force any more than necessary

5. Using the Special Tool and a 19 mm socket wrench, loosen the v-ribbed belt tensioner arm clockwise, then remove the v-ribbed belt.
6. Remove the v-ribbed belt.

To install:

7. To install, reverse removal procedure.
8. After installing the V-ribbed belt, check that it fits properly in the ribbed grooves. Check to confirm that the belt has not slipped out of the grooves on the bottom of the crank pulley by hand.
9. Tighten the right hand front wheel to: 76 ft. lbs. (103 Nm).

2.5L Engine

See Figures 52 through 55.

➡The high strength drive belt is used for the alternator side drive belt.

➡When replacing the drive belt, use Toyota genuine drive belt or equivalent high strength drive belt. If the high strength drive belt is not used, durability of the belt may become less than expected. The high strength drive belt is a belt with Aramid core which has higher strength compared to usually available belts with PET or PEN core.

1. Remove the front wheel RH.
2. Remove the front wheel opening extension pad RH.
3. Remove the engine under cover RH.
4. Remove the front fender apron seal RH.
5. Remove the v-ribbed belt (for vane pump).
   a. Attach a wrench to the hexagonal portion of the belt tensioner as shown in the illustration, rotate the belt tensioner clockwise, and remove the V-ribbed belt.
6. Remove the v-ribbed belt (for compressor and alternator).
   a. Attach a wrench to the hexagonal portion of the belt tensioner as shown in the illustration, rotate the belt tensioner clockwise, and remove the V-ribbed belt.

To install:

7. Install the v-ribbed belt (for compressor and alternator).
   a. Set the v-ribbed belt onto each part as shown, except the water pump pulley.

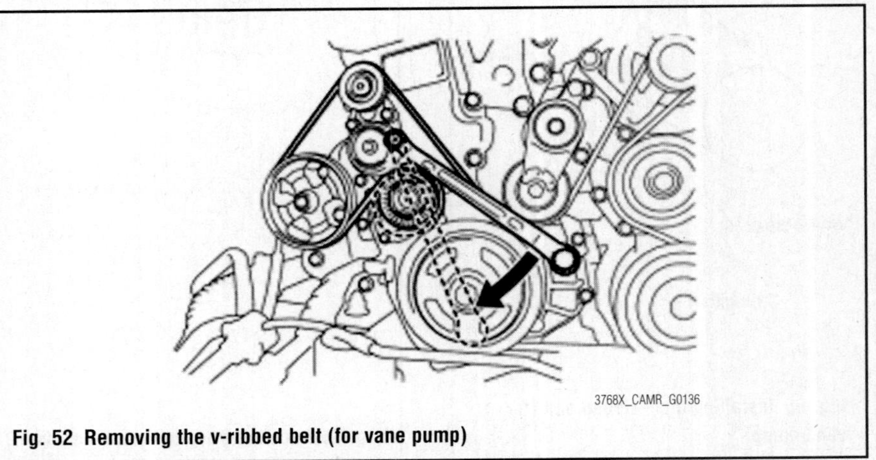

Fig. 52 Removing the v-ribbed belt (for vane pump)

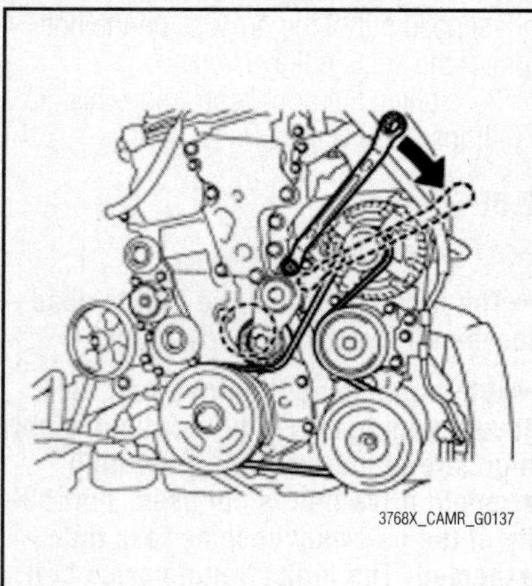

Fig. 53 Removing the v-ribbed belt (for compressor and alternator)

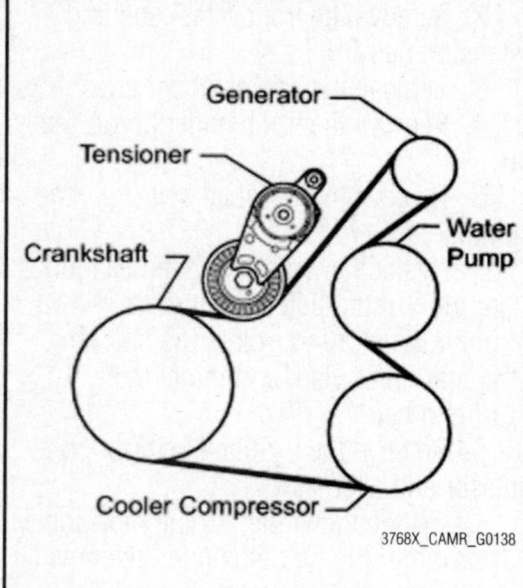

Fig. 54 Installing the v-ribbed belt (for compressor and alternator)

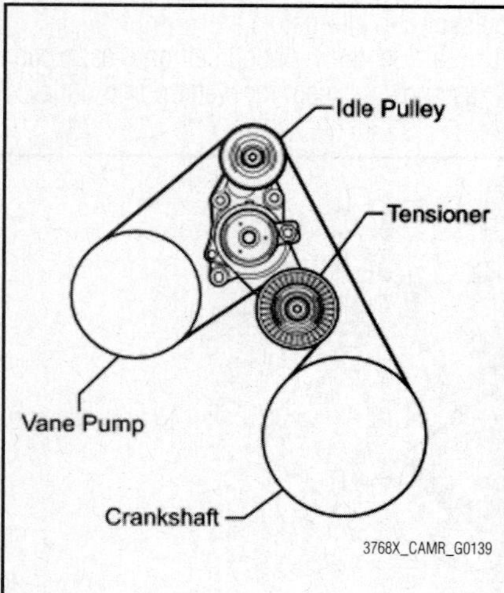

Fig. 55 Installing the v-ribbed belt (for vane pump)

b. Loosen the V-ribbed belt by turning the belt tensioner clockwise.

c. Set the V-ribbed belt onto the water pump pulley.

➡Make sure that the belt is attached to each pulley. In particular, make sure that the belt is securely fitted into the grooves of the crankshaft pulley.

8. Install the v-ribbed belt (for vane pump).

a. Set the V-ribbed belt onto each part as shown in the illustration except the tensioner pulley.

b. Loosen the V-ribbed belt by turning the belt tensioner clockwise.

c. Set the V-ribbed belt onto the tensioner pulley.

➡Make sure that the belt is attached to each pulley. In particular, make sure that the belt is securely fitted into the grooves of the crankshaft pulley.

9. Install the front fender apron seal RH.
10. Install the engine under cover RH.
11. Install the front wheel opening extension pad RH.
12. Install the front wheel RH. Tighten to 76 ft. lbs. (103 Nm).

3.5L Engine

See Figures 56 and 57.

1. Before servicing the vehicle, refer to the Precautions Section.

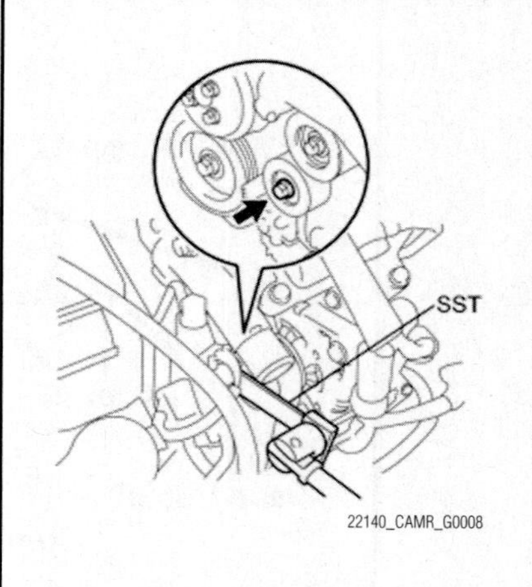

Fig. 57 Installing Special Tool: 09249—63010

2. Remove the right hand front wheel.
3. Remove the right hand front fender apron seal.
4. Remove the V-bank cover sub-assembly.
5. Using Special Tool: 09249—63010, release the belt tension by turning the belt tensioner counterclockwise, and remove the V-ribbed belt from the belt tensioner.
6. While turning the belt tensioner counterclockwise, align with its holes and then insert the 5 mm bi-hexagon wrench into the holes to fix the V-ribbed belt tensioner.
7. Remove the v-ribbed belt.

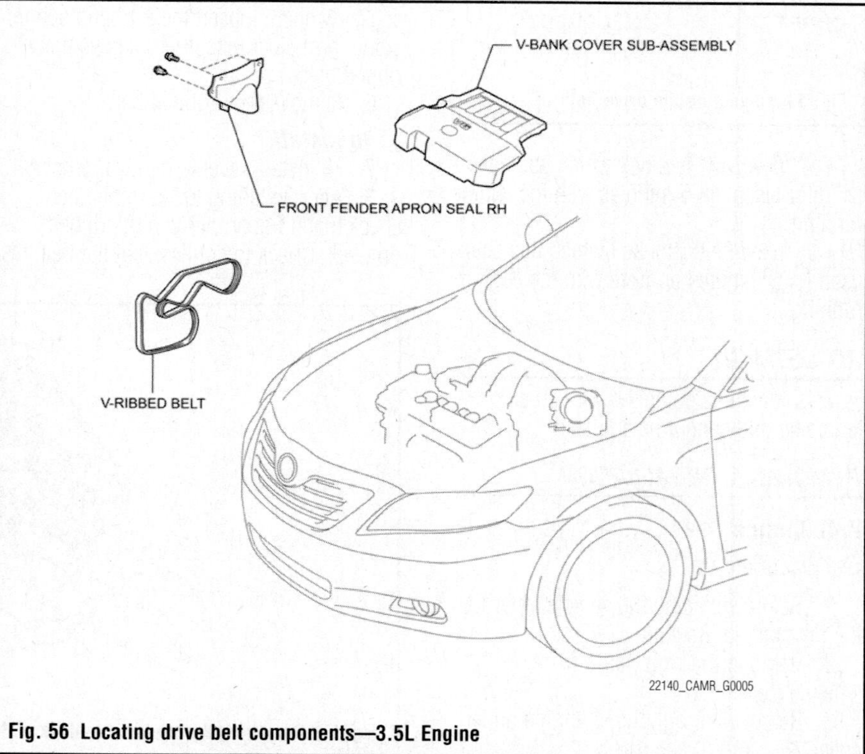

Fig. 56 Locating drive belt components—3.5L Engine

To install:

8. To install, reverse removal procedure.
9. If it is difficult to install the V-ribbed belt, perform the following procedure:
   a. Put the V-ribbed belt on every pulley except the tensioner pulley.
   b. While releasing the belt tension by turning the belt tensioner counterclockwise, put the V-ribbed belt on the tensioner pulley.

➡Put the backside of the V-ribbed belt on the tensioner pulley and idler pulley. Check that the V-ribbed belt is properly set to each pulley.

10. After installing the V-ribbed belt, check that it fits properly in the ribbed grooves. Check to confirm that the belt has not slipped out of the grooves on the bottom of the crank pulley by hand.
11. Tighten the right hand front wheel to: 76 ft. lbs. (103 Nm).

BALANCE SHAFT

REMOVAL & INSTALLATION

2.4L Engine

See Figures 58 through 64.

1. Before servicing the vehicle, refer to the Precautions Section.
2. Connect the negative battery cable.
3. Drain the engine oil.
4. Remove the oil pump. Refer to Oil Pump below for removal procedure.
5. Remove the No. 1 and No. 2 balance shaft sub-assembly. Remove the eight bolts in sequence.
6. Remove the No. 1 and No. 2 balance shafts.
7. Remove the balance shaft bearings if necessary.

To install:

➡Do not apply engine oil to the bearings and the contact surfaces.

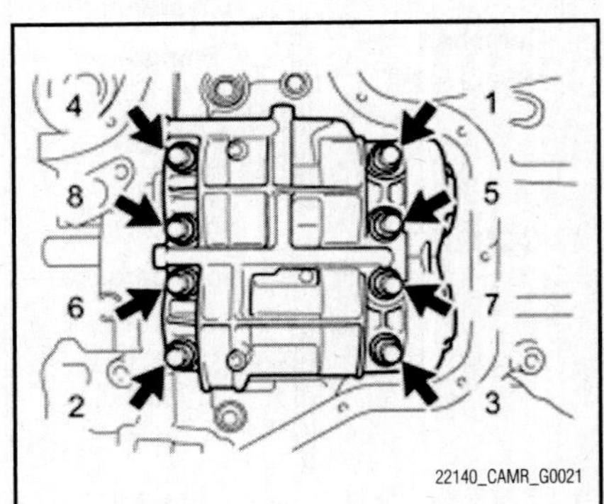

Fig. 58 Sub-assembly bolt removal sequence

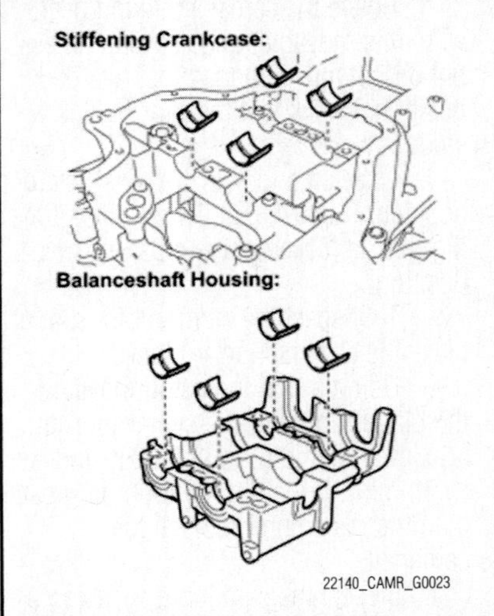

Fig. 59 Identifying and removing the balance shaft bearings

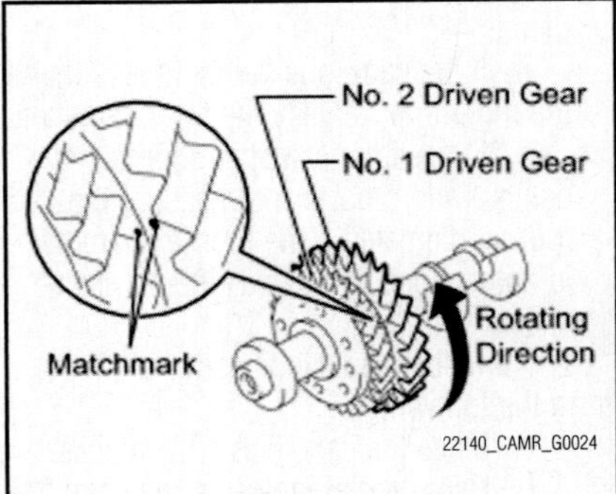

Fig. 60 Rotate the driven gear No. 1 of balance shaft No. 1

8. Install the bearings in the crankcase and balance shaft housing.
9. Apply a light coat of engine oil to the bearings.

➡Confirm that the match marks on driven gears No. 1 and No. 2 are matched.

10. Install No. 1 and No. 2 balance shaft sub-assembly. Rotate the driven gear No. 1 of balance shaft No. 1 in the rotating direction until it hits the stopper.
11. Align the alignment marks of the No. 1 and No. 2 balance shafts as shown.
12. Place the No. 1 and No. 2 balance shafts on the crankcase.
13. Apply a light coat of engine oil under the heads of the balance shaft housing bolts.
14. Install the balance shaft housing bolts. The balance shaft housing bolts should be tightened in 2 progressive steps as follows:

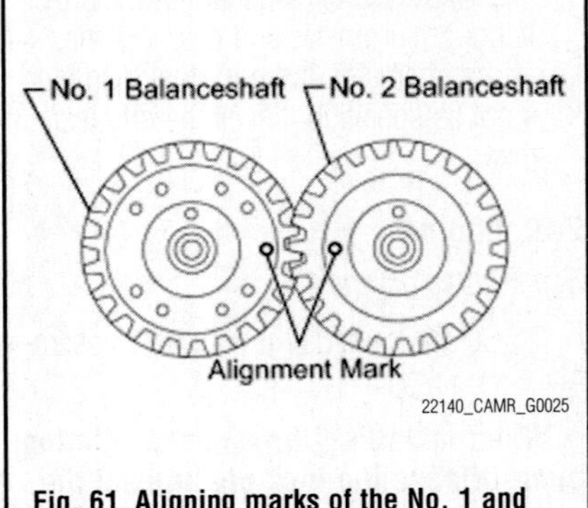

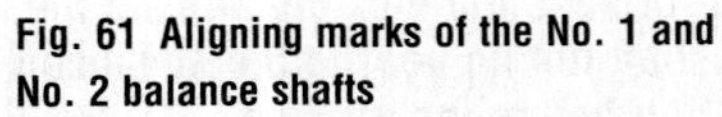

Fig. 61 Aligning marks of the No. 1 and No. 2 balance shafts

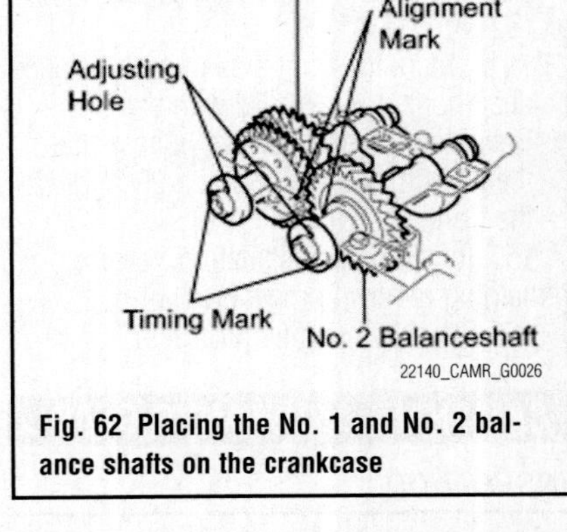

Fig. 62 Placing the No. 1 and No. 2 balance shafts on the crankcase

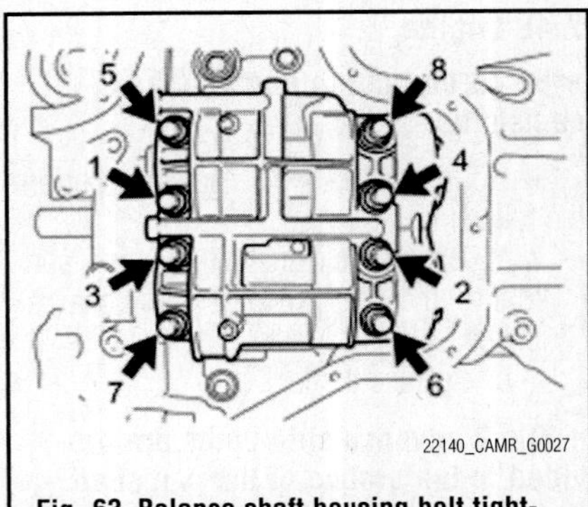

Fig. 63 Balance shaft housing bolt tightening sequence

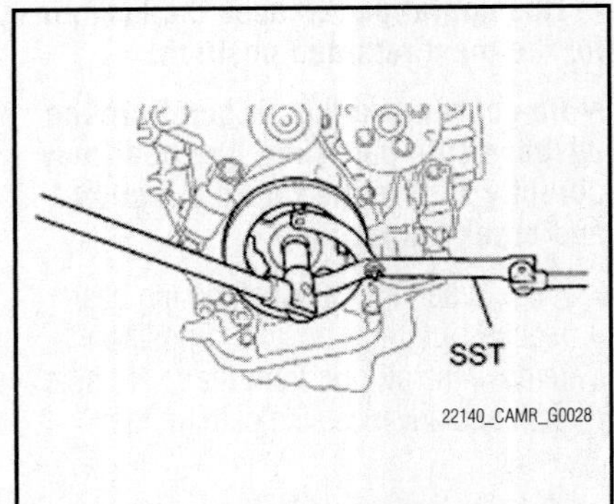

Fig. 64 Marking the bolt head and tightening procedure

a. Tighten the eight balance shaft housing bolts in sequence to: 16 ft. lbs. (22 Nm).

b. Mark the front side of each balance shaft housing bolt head with paint. Retighten the bolts by 90°. Check that the paint marks are now at a 90°angle to the front.

15. To complete installation, reverse remaining removal procedure.

16. Check the engine for leaks.

CAMSHAFT AND TIMING GEAR

INSPECTION

2.4L Engine

➡Be careful not to damage the camshafts.

1. To inspect the No.1 camshaft, perform the following:

a. Clamp the camshaft in a vise, and confirm that the camshaft timing gear is locked.

b. Release the lock pin.

➡The 2 advance side paths are provided in the groove of the camshaft. Plug one of the paths with a rubber piece.

c. Cover the 4 oil paths of the cam journal with vinyl tape.

d. Break through the tape of the advance side path and the retard side path on the opposite side to the hole of the advance side path.

➡Cover the paths with a piece of cloth when applying pressure to keep oil from splashing.

e. Apply approximately 28 psi of air pressure to the two broken paths.

f. Check that the camshaft timing gear revolves in the advance direction when reducing the air pressure of the retard side path.

➡This operation releases the lock pin for the most retarded position.

➡Do not remove the air gun from the advance side path first. The gear may abruptly shift in the retard direction and break the lock pin.

g. When the camshaft timing gear reaches the most advanced position, remove the air gun from the retard side path and advance side path, in that order.

➡Do not use an air gun to check for smooth operation.

h. Rotate the camshaft timing gear within its movable range several times, but do not turn it to the most retarded position. Check that the gear rotates smoothly.

i. Check the lock in the most retarded position. Confirm that the camshaft timing gear is locked at the most retarded position.

j. To inspect the camshaft for runout, place the camshaft on V-blocks.

k. Using a dial indicator, measure the circle runout at the center journal. Maximum circle runout: 0.0012 inches (0.03 mm). If the circle runout is greater than the maximum, replace the camshaft.

l. Inspect the cam lobes by using a micrometer and measure the cam lobe height. Standard cam lobe height: 1.8624 to 1.8664 inches. (47.306 to 47.406 mm). Minimum cam lobe height: 1.8581 inches (47.196 mm).

m. If the cam lobe height is less than the minimum, replace the No.1 camshaft.

n. Inspect the camshaft journals by using a micrometer and measure the journal diameter. If the journal diameter is not as specified, check the oil clearance.

2. To inspect the No.2 Camshaft, perform the following:

a. Place the camshaft on V-blocks.

b. Using a dial indicator, measure the circle runout at the center journal. Maximum circle runout: 0.0012 inches. (0.03 mm). If the circle runout is greater than the maximum, replace the No. 2 camshaft.

c. Inspect the cam lobes by using a micrometer and measure the cam lobe height. Standard cam lobe height: 1.8104 to 1.8143 inches. (45.983 to 46.083 mm). Minimum cam lobe height: 1.8060 inches (45.873 mm).

d. If the cam lobe height is less than the minimum, replace the No.2 camshaft.

e. Inspect the camshaft journals by using a micrometer and measure the journal diameter. If the journal diameter is not as specified, check the oil clearance.

2.5L Engine

See Figures 65 and 66.

1. Set the camshaft timing gear assembly.

➡When installing the camshaft timing gear, release the lock pin and set the camshaft timing gear to the advanced position before installation.

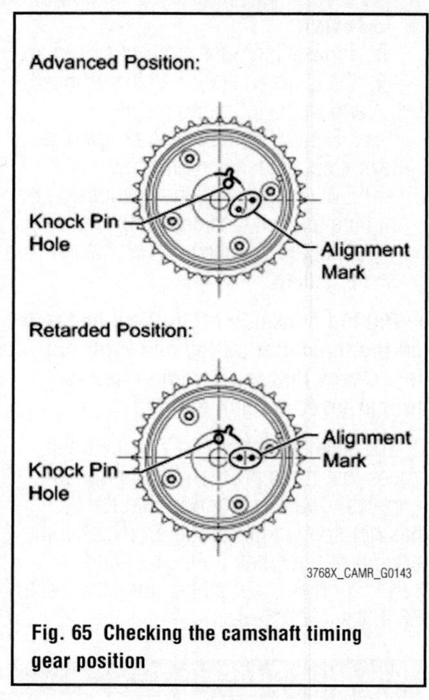

Fig. 65 Checking the camshaft timing gear position

a. Check the camshaft timing gear position.

➡If the camshaft timing gear is set to the advanced position, do not let the camshaft timing gear rotate clockwise during installation.

➡If the camshaft timing gear rotates to the retarded position, release the lock pin and set the camshaft timing gear to the advanced position.

b. Align and attach the knock pin of the No. 1 camshaft with the pin hole of the camshaft timing gear.

c. Check that there is no clearance between the camshaft timing gear and camshaft flange.

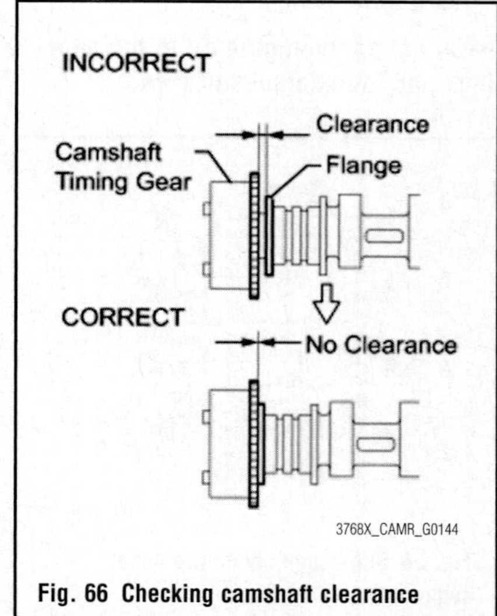

Fig. 66 Checking camshaft clearance

3.5L Engine

1. To inspect the camshaft, place the camshaft on V-blocks.
2. Using a dial indicator, measure the circle runout at the center journal.
    a. Maximum circle runout: 0.0016 inches (0.04 mm). If the circle runout is greater than the maximum, replace the camshaft.

➡**Check the oil clearance after replacing the camshaft.**

3. Inspect the cam lobes by using a micrometer and measure the cam lobe height.
    a. Standard cam lobe height:
    - Intake: 1.7447 to 1.7487 inches (44.316 to 44.416 mm)
    - Exhaust: 1.7426 to 1.7465 inches. (44.262 to 44.362 mm)

    b. Maximum cam lobe height:
    - 1.7388 inches (44.166 mm)
    - 1.7367 inches (44.112 mm)
4. Inspect the camshaft journals by using a micrometer and measure the journal diameter. If the journal diameter is not as specified, check the oil clearance.
    a. No. journal: 1.4152 to 1.4157 inches (35.946 to 35.960 mm).
    b. Other journals: 1.0220 to 1.0226 inches (25.959 to 25. 975 mm).

REMOVAL & INSTALLATION

2.4L Engine

See Figures 67 through 78.

1. Before servicing the vehicle, refer to the Precautions Section.
2. Disconnect the negative battery cable.
3. Loosen the lug nuts on the front right hand wheel.
4. Apply the parking brake, block the rear wheels, then raise and safely support the front of the vehicle securely on jackstands.

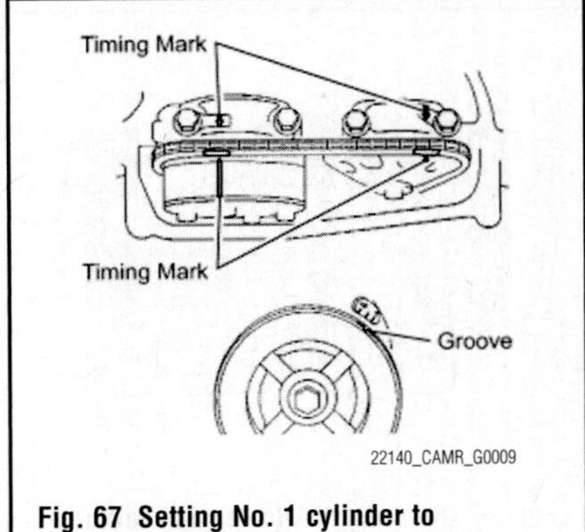

Fig. 67 Setting No. 1 cylinder to TDC/Compression

5. Remove the front right hand wheel.
6. Remove the left and right hand under cover.
7. Remove the No.1 engine cover sub-assembly and 2 nuts.
8. Remove the ignition coil assembly.
9. Remove the cylinder head cover sub-assembly
10. Set No.1 Cylinder to TDC/Compression by performing the following:
    - Turn the crankshaft pulley until its groove and the timing mark "0" of the timing chain cover are aligned
    - Check that each timing mark of the camshaft timing gear and sprocket is aligned with each timing mark located on the No. 1 and No. 2 bearing caps as shown in the illustration. If not, turn the crankshaft by 1 revolution (360°) to align the timing marks

➡**Do not turn the crankshaft without the chain tensioner.**

11. Remove No.1 Chain tensioner assembly. Remove the 2 nuts, tensioner and gasket.
12. Remove the No. 2 camshaft by performing the following:
    a. While holding the camshaft with a wrench, loosen the camshaft timing set bolt.
    b. Using several steps, uniformly loosen and remove the 10 bearing cap bolts in the sequence shown in the illustration.
    c. While holding the No. 2 camshaft by hand, remove the camshaft timing sprocket set bolt.
    d. Remove the camshaft timing sprocket from the No. 2 camshaft with the timing chain wrapped on the sprocket.
    e. Remove the camshaft timing sprocket from the timing chain.
13. Remove the No. 1 camshaft by performing the following:
    a. Using several steps, uniformly loosen and remove the 10 bearing cap bolts in the sequence shown in the illustration.
    b. Remove the 5 bearing caps.
    c. Remove the camshaft and camshaft timing gear while holding the timing chain by hand.

➡**Be careful not to drop anything inside the timing chain cover.**

    d. Tie the timing chain with a string.
14. Remove the camshaft timing gear assembly by performing the following:
    a. Clamp the camshaft in a vise, and make sure that the camshaft timing gear does not rotate.
    b. Cover all the oil ports except the advance side port shown in the illustration with vinyl tape

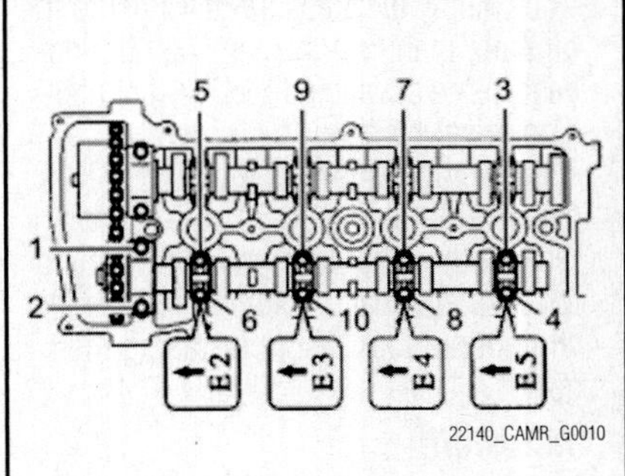

Fig. 68 No. 2 camshaft bearing cap bolt removal sequence

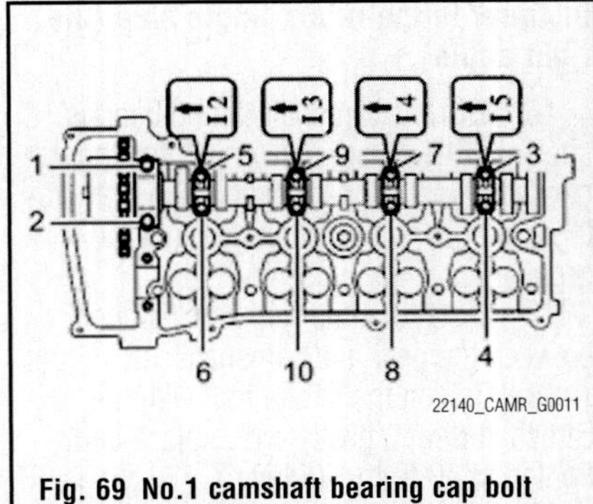

Fig. 69 No.1 camshaft bearing cap bolt removal sequence

➡**Cover the paths with a shop rag or piece of cloth to avoid oil splashes.**

➡**Depending on the air pressure, the camshaft timing gear will turn to the advance angle side without applying force by hand. Also, if the pressure is difficult to apply because of air leakage from the port, the lock may be difficult to release.**

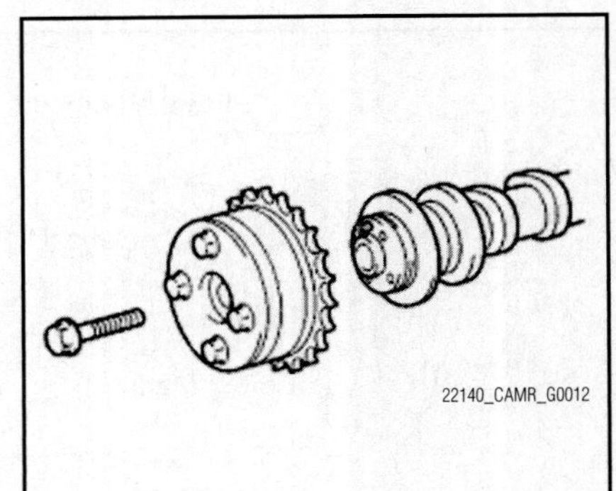

Fig. 70 Removing the flange bolt from the camshaft timing gear

c. Apply air pressure of 14 psi to the oil path, then turn the camshaft timing gear in the advance direction (counter-clockwise) by hand.

d. Remove the flange bolt of the camshaft timing gear. Be sure not to remove the other four bolts. If planning to reuse the gear, be sure to release the straight pin lock before installing the gear.

To install:

15. Put the camshaft timing gear and camshaft together with the straight pin and key groove misaligned.

➡Be sure not to turn the camshaft timing gear to the retard angle side (the right angle).

16. Turn the camshaft timing gear as shown in the illustration while pushing it gently against the camshaft. Push further at the position where the pin fits into the groove.

17. Check that there is no clearance between the gear and camshaft.

18. Tighten the flange bolt with the camshaft timing gear fixed in place and tighten to 40 ft. lbs. (54 Nm).

19. Check that the camshaft timing gear can move to the retard angle side (the right direction) and is locked in the most retarded position.

20. To install the No.1 camshaft, perform the following:

a. Apply a light coat of engine oil to the journal portion of the camshaft.

b. Install the timing chain onto the camshaft timing gear with the paint mark aligned with the timing mark in the camshaft timing gear.

c. Examine the front marks and numbers, and check that the order is as shown in the illustration below. Then install the bearing caps into the cylinder head.

d. Apply a light coat of engine oil to the threads and under the heads of the bearing cap bolts.

e. Using several steps, uniformly tighten the 10 bearing cap bolts in the sequence shown in the illustration. Tighten the following to specification:

- No. 1 Bearing cap: 22 ft. lbs. (30 Nm)
- No. 3 Bearing cap: 80 inch lbs. (9 Nm)

21. To install camshaft No. 2, perform the following:

a. Apply a light coat of engine oil to the journal portion of the No. 2 camshaft.

b. Put the No. 2 camshaft on the cylinder head with the paint mark of the chain aligned with the timing mark on the camshaft timing sprocket.

c. While holding the No. 2 camshaft by hand, temporarily tighten the camshaft timing sprocket set bolt.

d. Examine the front marks and numbers, and check that the order is as shown in the illustration. Then install the bearing caps onto the cylinder head.

e. Apply a light coat of engine oil to the threads and under the heads of the bearing cap bolts.

f. Using several steps, uniformly tighten the 10 bearing cap bolts in the sequence shown in the illustration. Tighten the following to specification:

- No. 1 Bearing cap: 22 ft. lbs. (30 Nm)
- No. 3 Bearing cap: 80 inch lbs. (9 Nm)

g. While holding the camshaft with a wrench, tighten the camshaft timing sprocket set bolt and tighten to 40 ft. lbs. (54 Nm).

h. Check that the paint marks on the chain are aligned with the timing marks

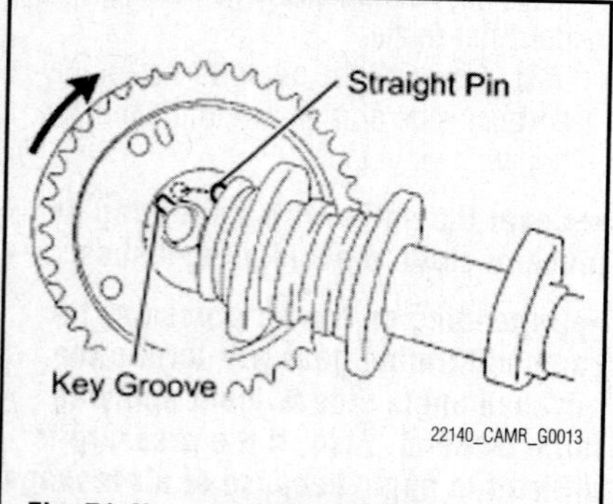

Fig. 71 No. 1 camshaft straight pin and key groove misaligned

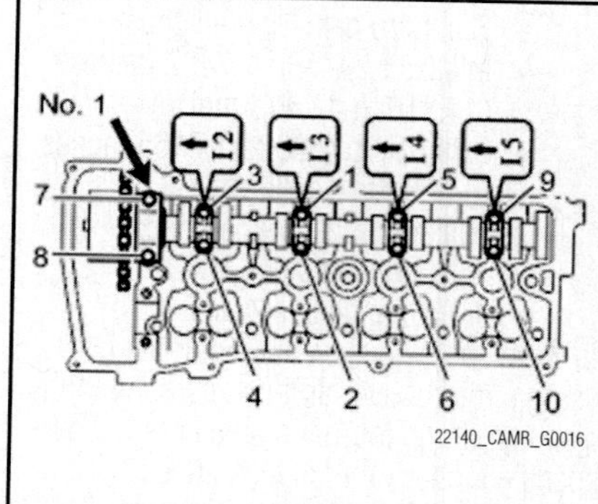

Fig. 74 No. 1 camshaft bearing cap tightening sequence

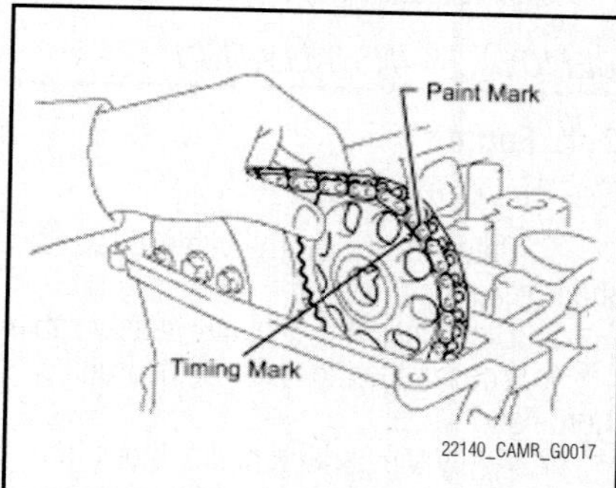

Fig. 75 Aligning No. 2 camshaft paint mark with the timing mark

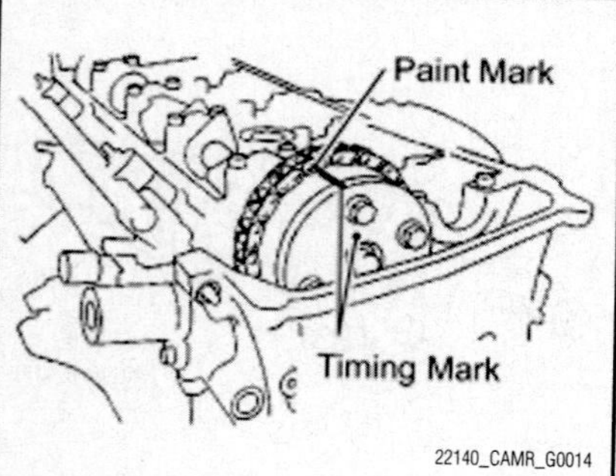

Fig. 72 Aligning No. 1 camshaft paint mark with the timing mark

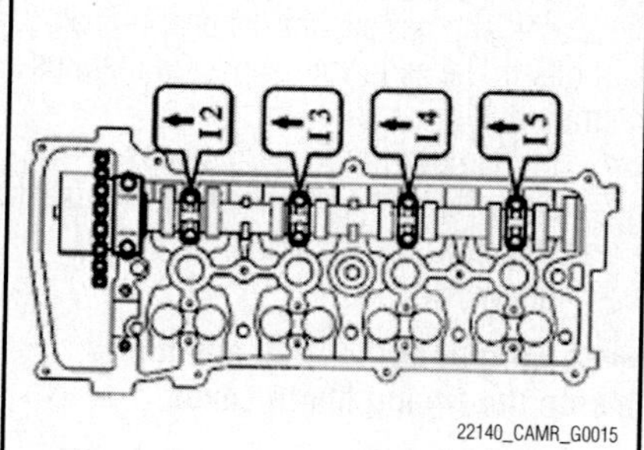

Fig. 73 Checking the No. 1 camshaft bearing cap front marks, numbers and order

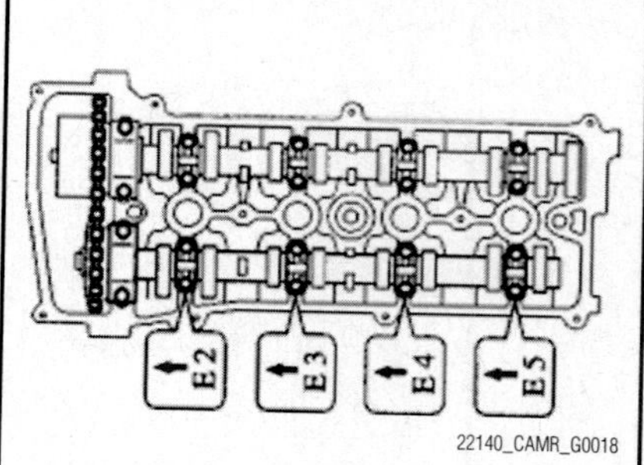

Fig. 76 Checking the No. 2 camshaft bearing cap front marks, numbers and order

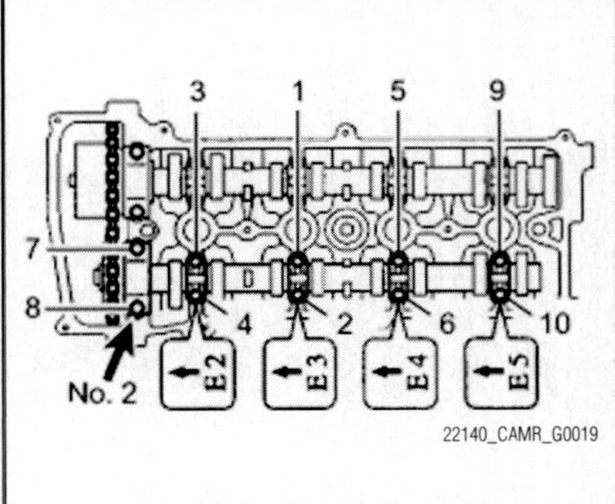

Fig. 77 No. 2 camshaft bearing cap tightening sequence

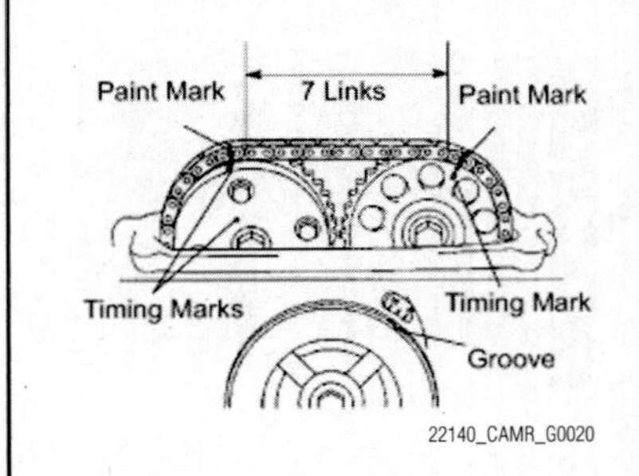

Fig. 78 Aligning paint marks on the chain with the timing marks on the camshaft timing gear and camshaft timing sprocket

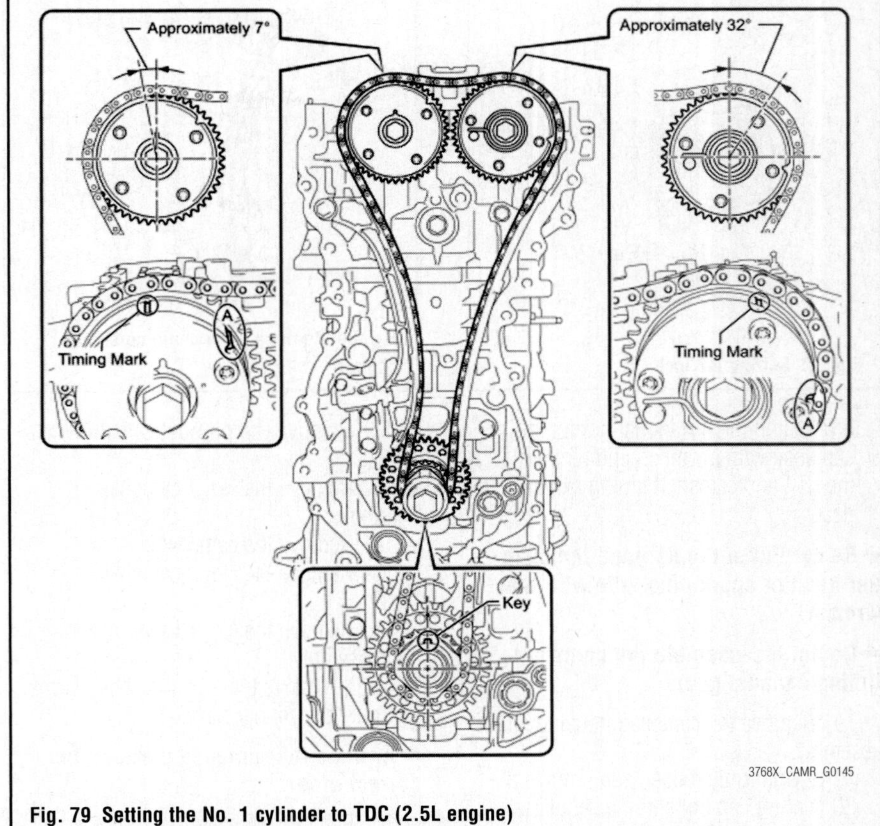

Fig. 79 Setting the No. 1 cylinder to TDC (2.5L engine)

on the camshaft timing gear and camshaft timing sprocket. Also, check that the crankshaft pulley groove is aligned with the timing mark "0" of the timing chain cover.

22. To complete installation, reverse remaining removal procedure.

23. Check engine for oil leaks.

24. Connect the negative battery cable.

2.5L Engine

See Figures 79 through 105.

1. Remove the timing chain cover sub assembly.

2. Set the No. 1 cylinder to TDC/Compression.

a. Temporarily install the crankshaft pulley bolt.

➡"A" is not a timing mark.

b. Rotate the crankshaft clockwise so that the timing marks on the crankshaft timing gear and camshaft timing gears are as shown in the illustration.

➡If the timing marks do not align, rotate the crankshaft clockwise again and align the timing marks.

c. Remove the crankshaft pulley bolt.

3. Remove the timing chain guide.

4. Remove the No. 1 chain tensioner assembly.

a. Allow the plunger to extend slightly, and then rotate the stopper plate counterclockwise to release the lock. Once the lock is released, push the plunger into the tensioner.

b. Move the stopper plate clockwise to set the lock, and insert a pin into the stopper plate hole.

c. Remove the 2 bolts, chain tensioner and the gasket.

5. Remove the chain tensioner slipper.

6. Remove the chain sub assembly.

7. Remove the 2 bolts and No. 1 chain vibration damper.

8. Remove the camshaft timing gear assembly.

a. Hold the hexagonal portion of the camshaft with a wrench and remove the bolt and camshaft timing gear.

➡Be careful not to damage the cylinder head or spark plug tube with the wrench.

➡Do not disassemble the camshaft timing gear.

9. Remove the camshaft timing exhaust gear assembly.

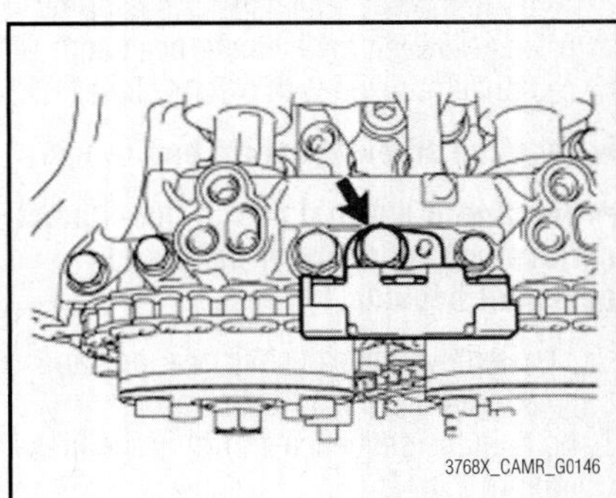

Fig. 80 Removing the timing chain bolt and guide (2.5L engine)

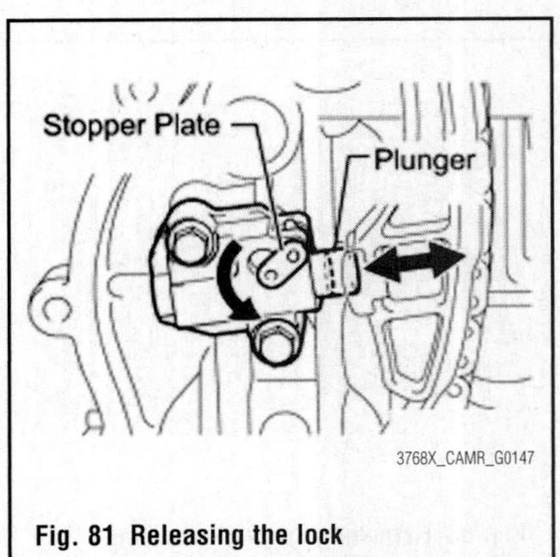

Fig. 81 Releasing the lock

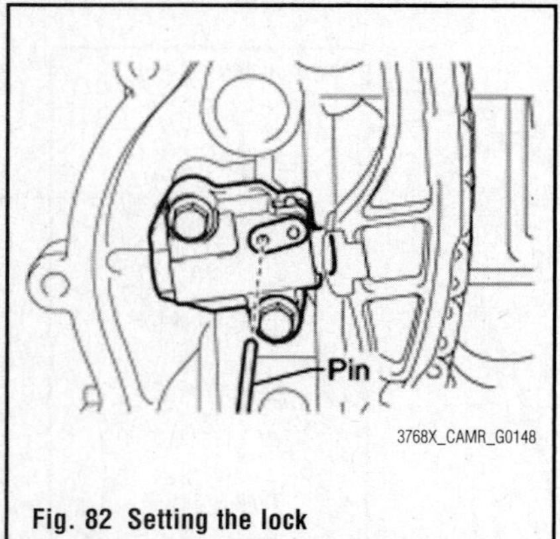

Fig. 82 Setting the lock

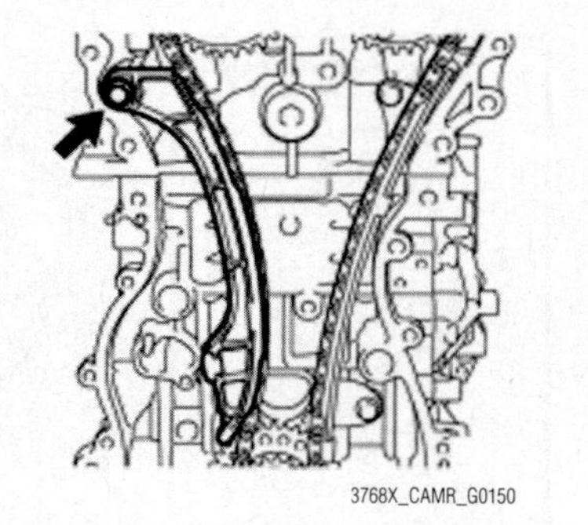

Fig. 84 Removing the chain tensioner slipper

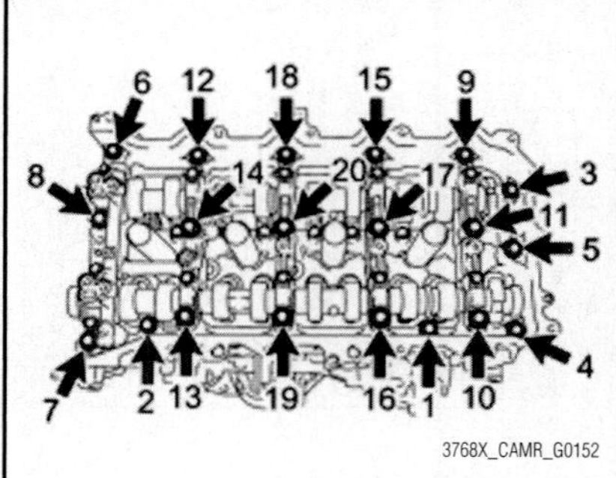

Fig. 86 Removing the camshaft bearing cap bolts in sequence

a. Hold the hexagonal portion of the camshaft with a wrench and remove the bolt and camshaft timing exhaust gear.

➡Be careful not to damage the cylinder head or spark plug tube with the wrench.

➡Do not disassemble the camshaft timing exhaust gear.

10. Remove the camshaft housing sub assembly.
 a. Uniformly loosen and remove the 20 bearing cap bolts in the sequence shown in the illustration.
 b. Remove the camshaft housing by prying between the cylinder head and camshaft housing with a screwdriver.

➡Tape the screwdriver tip before use.

➡Be careful not to damage the contact surfaces of the cylinder head and camshaft housing.

11. Remove the camshaft bearing cap.
 a. Remove the 11 bearing cap bolts in the sequence shown in the illustration.
 b. Removing the 5 bearing caps.

➡Arrange the removed parts in the correct order.

12. Remove the oil control valve filter.
13. Remove the No. 1 camshaft bearing.
14. Remove the camshafts.
15. Remove the No. 2 camshaft bearing.
16. Remove the No. 1 valve rocker arm sub assembly.
 a. Remove the 16 valve rocker arms from the cylinder head.

➡Arrange the removed parts in the correct order.

17. Remove the valve lash adjuster assembly.
 a. Remove the 16 valve lash adjusters from the cylinder head.

➡Arrange the removed parts in the correct order.

To install:

18. Set the camshaft timing gear assembly.

➡When installing the camshaft timing gear, release the lock pin and set the camshaft timing gear to the advanced position before installation.

 a. Check the camshaft timing gear position.

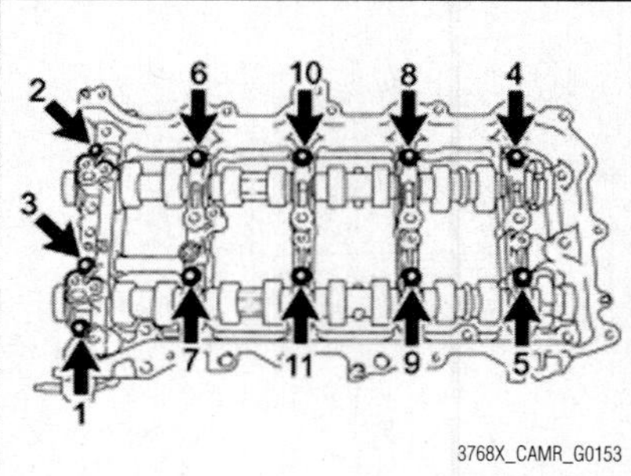

Fig. 87 Removing the 11 bearing cap bolts in sequence

➡If the camshaft timing gear is set to the advanced position, do not let the camshaft timing gear rotate clockwise during installation.

➡If the camshaft timing gear rotates to the retarded position, release the lock pin and set the camshaft timing gear to the advanced position.

 b. Align and attach the knock pin of the No. 1 camshaft with the pin hole of the camshaft timing gear.
 c. Check that there is no clearance between the camshaft timing gear and camshaft flange.
 d. Secure the camshaft in place by hand, and then install the installation bolt of the camshaft timing gear by hand.

➡Do not use any tools to install the bolt. If the bolt is installed using a tool, the lock pin will be damaged.

 e. Release the lock pin. Clean the camshaft journal with non-residue solvent. Cover the 4 oil paths of the cam journal with vinyl tape as shown in the illustration.

➡There are 4 oil paths in the grooves of the camshaft. Plug three of the paths with pieces of rubber.

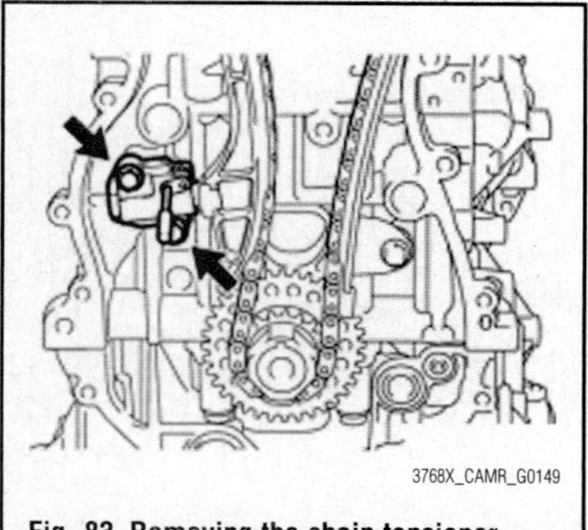

Fig. 83 Removing the chain tensioner

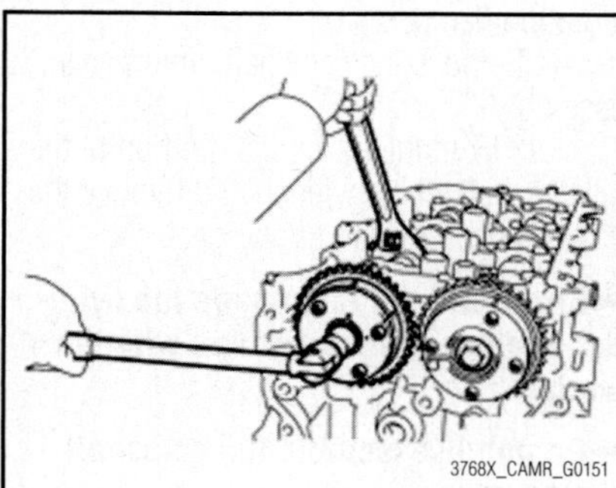

Fig. 85 Removing the camshaft timing gear assembly

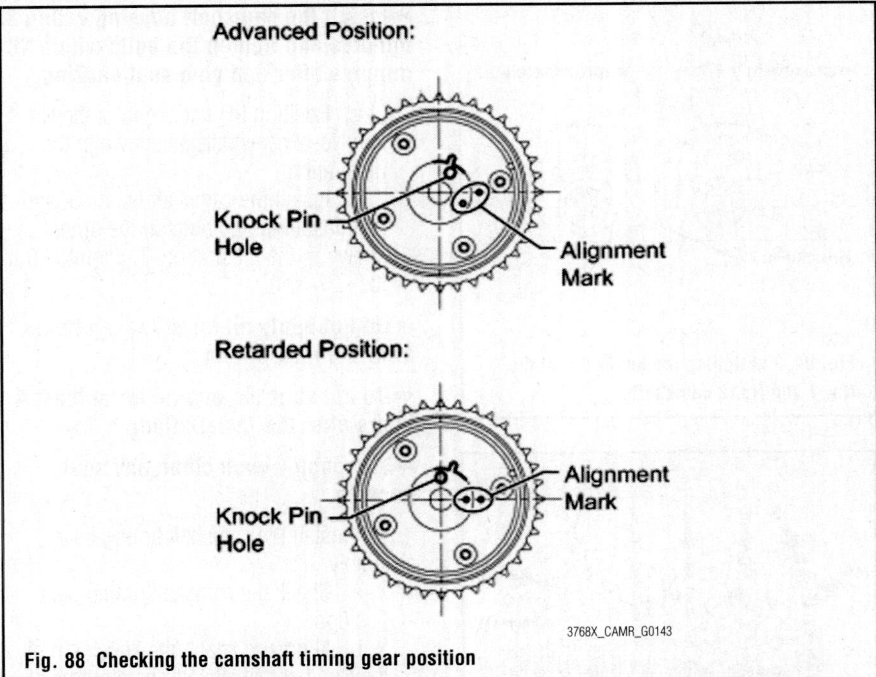

Fig. 88 Checking the camshaft timing gear position

f. Open a hole at port A shown in the illustration.

g. While applying approximately 29 psi (200 kPa) of air pressure to the oil path, forcibly turn the camshaft timing gear assembly in the advance direction (counterclockwise).

CAUTION

Cover the paths with a piece of cloth when applying pressure to keep oil from splashing.

➡Do not allow the camshaft timing gear assembly to lock. If it locks, release the lock pin again.

➡The camshaft timing gear assembly may be turned in the advance direction without applying any force.

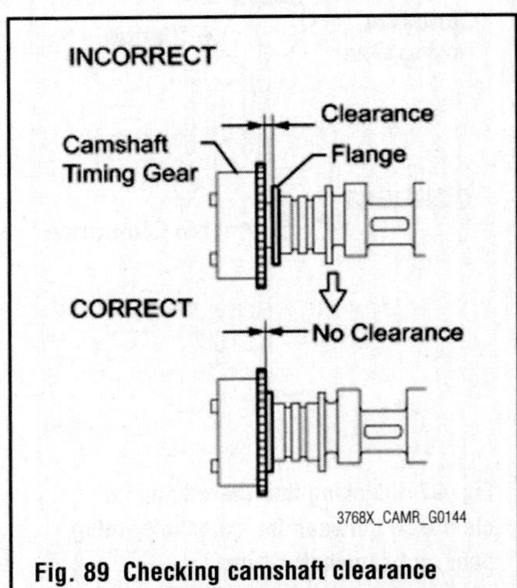

Fig. 89 Checking camshaft clearance

➡If enough air pressure cannot be applied because of air leakage from the port, releasing the lock pin may be difficult.

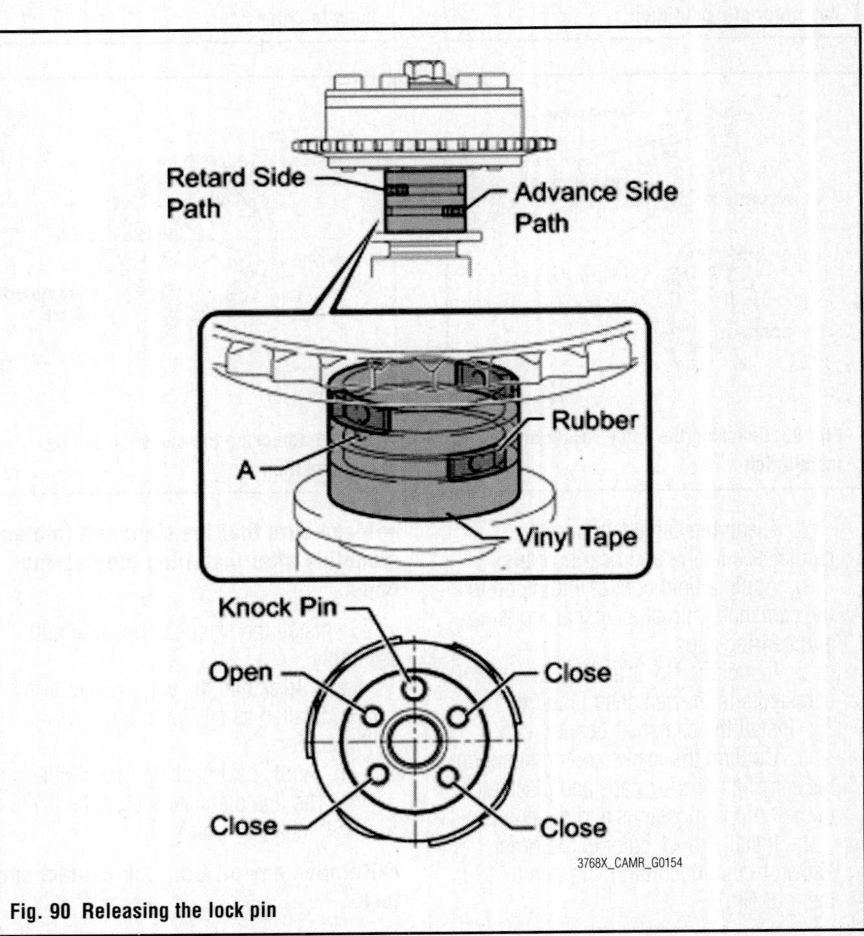

Fig. 90 Releasing the lock pin

h. Remove the vinyl tape and rubber pieces from the camshaft.

i. Remove the bolt and camshaft timing gear.

➡Do not allow the camshaft timing gear assembly to lock. If it locks, release the lock pin again.

19. Install the valve lash adjuster assembly.

a. Inspect the valve lash adjuster before installing it.

b. Inspect the 16 lash adjusters to the cylinder head.

➡Install the lash adjuster to the same place it was removed from.

20. Install the No. 1 valve rocker arm sub assembly.

a. Apply engine oil to the lash adjuster tips and valve stem caps.

b. Install the 16 valve rocker arms as shown.

21. Install the No. 2 camshaft bearing.

22. Install the No. 1 camshaft bearing.

23. Install the oil control valve filter.

24. Install the camshaft.

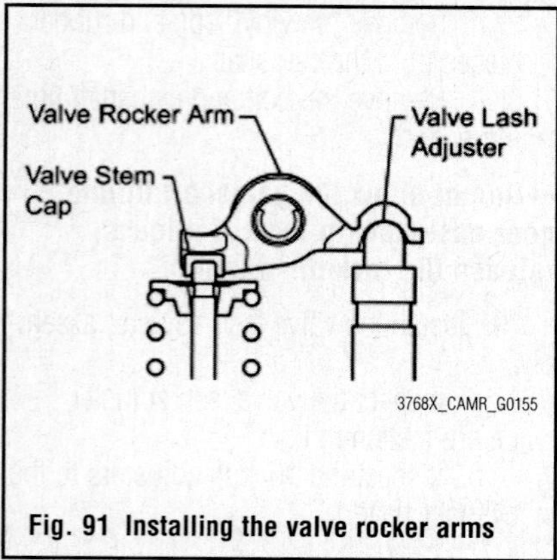

Fig. 91 Installing the valve rocker arms

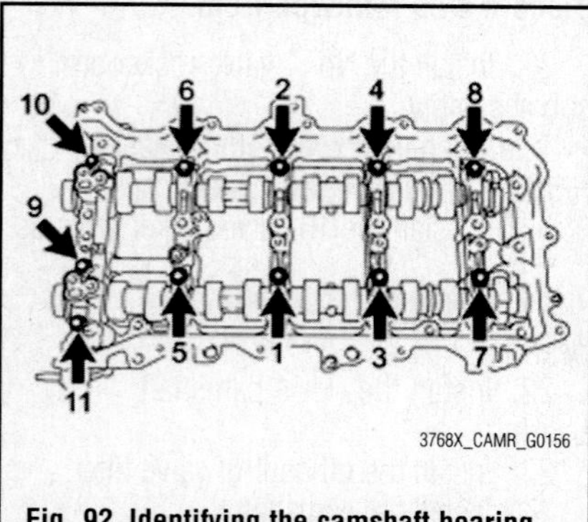

Fig. 92 Identifying the camshaft bearing cap order of installation

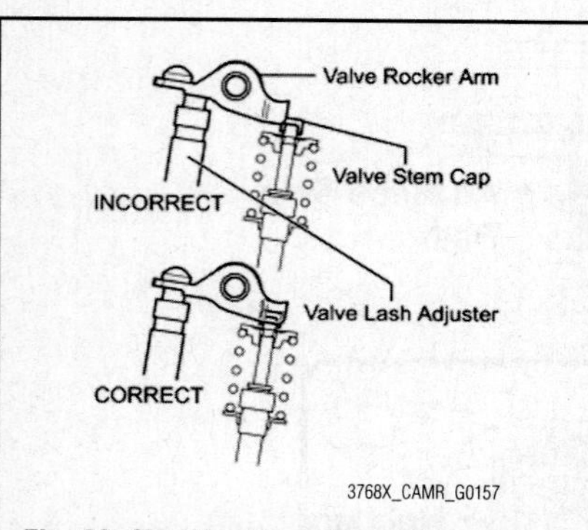

Fig. 93 Checking the valve rocker arm installation

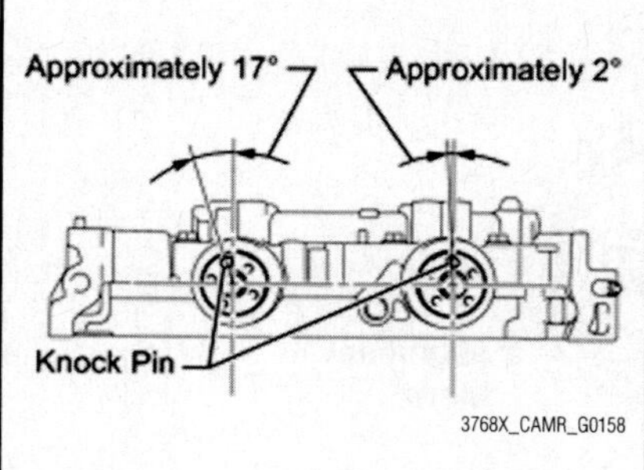

Fig. 94 Positioning the knock pin of the No. 1 and No. 2 camshafts

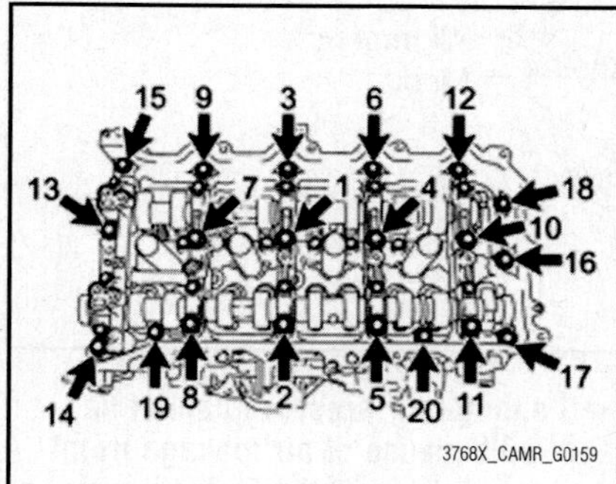

Fig. 95 Installing the camshaft housing bolts in order

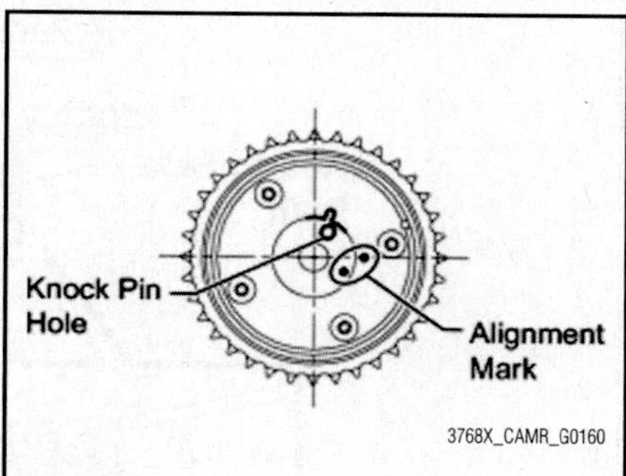

Fig. 96 Checking the camshaft timing gear position

a. Clean the camshaft journals, camshaft housing and bearing caps.

b. Apply a light coat of engine oil to the camshaft journal, camshaft housing and bearing caps.

c. Install the No. 1 and No. 2 camshafts to the camshaft housing.

25. Install the camshaft bearing cap.

a. Confirm the marks and numbers on the camshaft bearing caps and place them in their proper positions and directions.

b. Install the 11 bolts in the order shown in the illustration. Tighten to 12 ft. lbs. (16 Nm).

➡Make sure that the camshaft rotates smoothly after installing the bearing caps.

26. Install the camshaft housing sub assembly.

a. Check that the valve rocker arms are installed as shown in the illustration.

b. Apply seal packing in a continuous line. The line diameter should be 0.118-0.157 inch (3-4 mm).

➡Remove any oil from the contact surface.

➡Install the camshaft housing within 3 minutes and tighten the bolts within 10 minutes after applying seal packing.

c. Position the knock pin of the No. 1 and No. 2 camshafts as shown in the illustration.

d. Install the camshaft housing, and then install the 20 bolts in the order shown in the illustration. Tighten to 20 ft. lbs. (27 Nm).

➡Do not apply oil for at least 4 hours after the installation.

➡Do not start the engine for at least 4 hours after the installation.

➡Thoroughly wipe clean any seal packing.

27. Install the camshaft timing gear assembly.

a. Check the camshaft timing gear position.

b. Align and attach the knock pin of the No. 1 camshaft with the pin hole of the camshaft timing gear.

c. Check that there is no clearance between the camshaft timing gear and camshaft flange.

d. Using a wrench to hold the hexagonal portion of the No. 1 camshaft, install the bolt. Tighten to 63 ft. lbs. (85 Nm).

➡Be careful not to damage the cylinder head or spark plug tube with the wrench.

➡Do not disassemble the camshaft timing gear

28. Install the camshaft timing exhaust gear assembly.

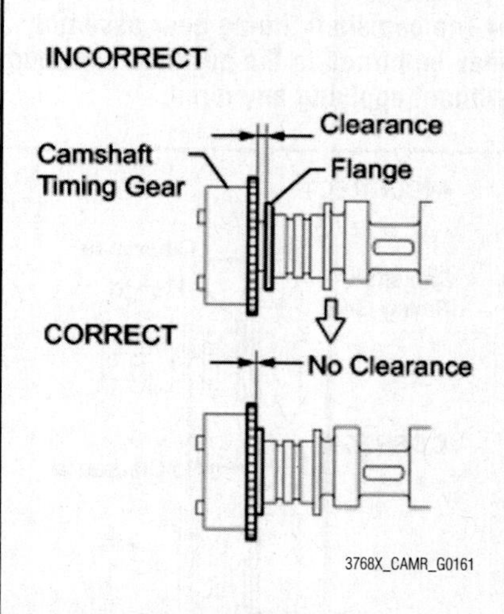

Fig. 97 Checking that there is no clearance between the camshaft timing gear and camshaft flange

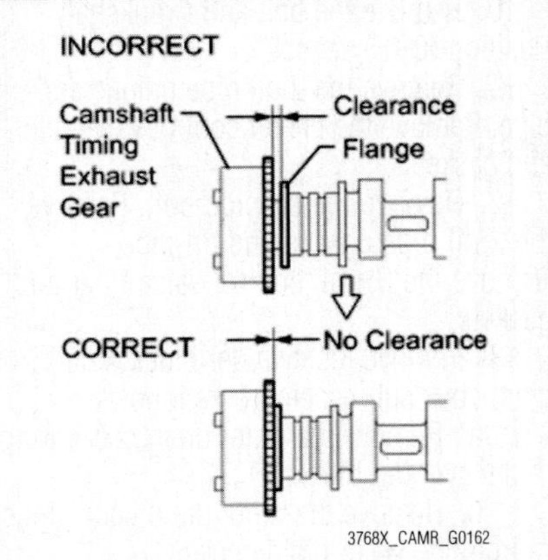

Fig. 98 Checking that there is no clearance between the camshaft timing exhaust gear and the camshaft flange

a. Align and attach the knock pin of the No. 2 camshaft with the pin hole of the camshaft timing exhaust gear.

b. Check that there is no clearance between the camshaft timing exhaust gear and camshaft flange.

c. Using a wrench to hold the hexagonal portion of the No. 2 camshaft, install the bolt. Tighten to 63 ft. lbs. (85 Nm).

Fig. 99 Adding engine oil

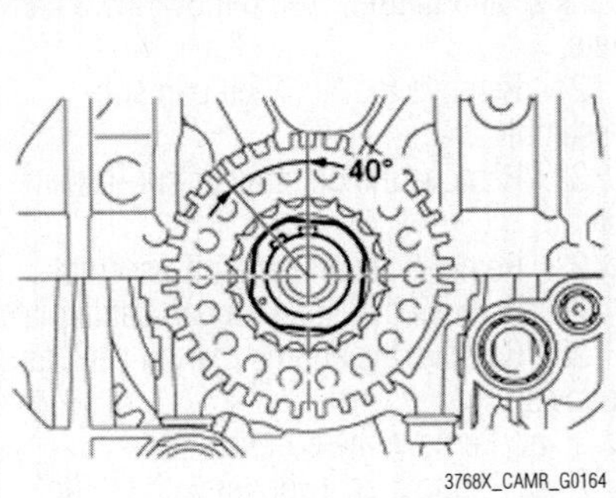

Fig. 100 Rotating the crankshaft 40°

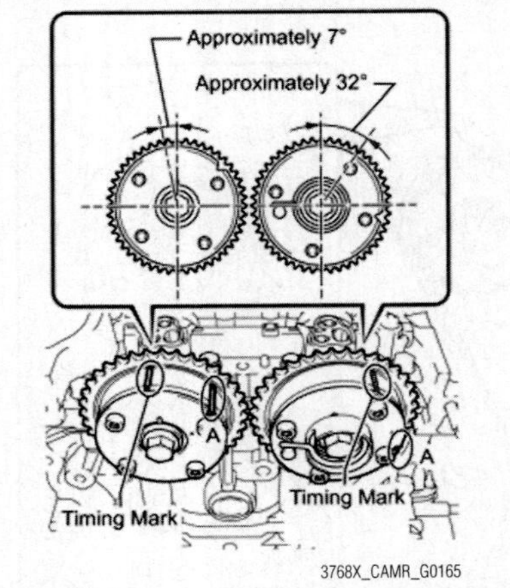

Fig. 101 Checking the camshaft timing gear timing marks

➡Be careful not to damage the cylinder head or spark plug tube with the wrench.

➡Do not disassemble the camshaft timing exhaust gear.

29. Add engine oil.
    a. Add 3.1 cu. Inch (50 cc) of engine oil into the oil hole.

➡Oil must be added if the lash adjusters were removed.

➡Make sure that the low pressure chamber and oil paths of the lash adjusters are full of engine oil.

30. Set No. 1 cylinder to TDC/Compression.
    a. Temporarily install the crankshaft pulley bolt.
    b. Rotate the crankshaft 40°counterclockwise to position the crankshaft pulley key as shown in the illustration.
    c. Check that the timing marks of the camshaft timing gears are as shown in the illustration.

➡"A" is not a timing mark.

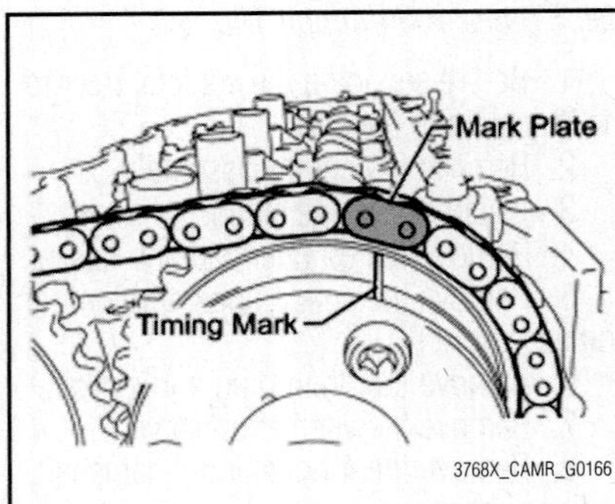

Fig. 102 Aligning the mark plate with the timing mark

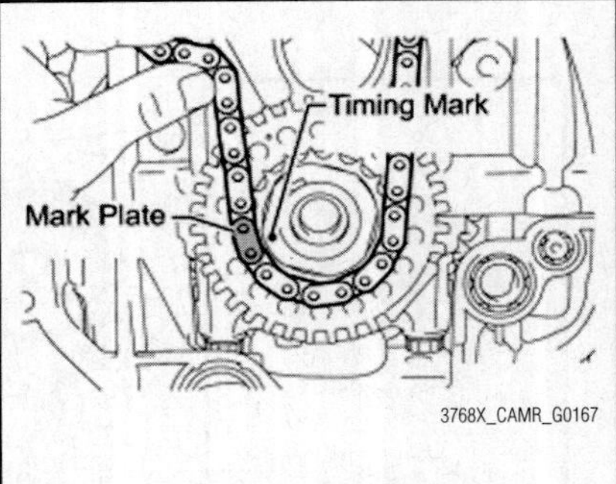

Fig. 103 Aligning the mark plate of the chain with the timing mark of the sprocket

31. Install the No. 1 chain vibration damper with the 2 bolts. Tighten to 15 ft. lbs. (21 Nm).
32. Install the chain sub assembly.
33. Place the chain onto the camshaft timing gears and crankshaft timing sprocket.

➡Make sure the mark plate of the chain faces away from the engine.

➡It is not necessary to install the chain to the teeth of the gears and sprocket.

a. Align the mark plate (yellow or gold) of the chain with the timing mark of the camshaft timing exhaust gear and install the chain to the camshaft timing exhaust gear.

b. Align the mark plate (pink or gold) of the chain with the timing mark of the crankshaft timing sprocket and install the chain to the crankshaft timing sprocket.

c. Tie a string above the crankshaft timing sprocket so that the chain is secure.

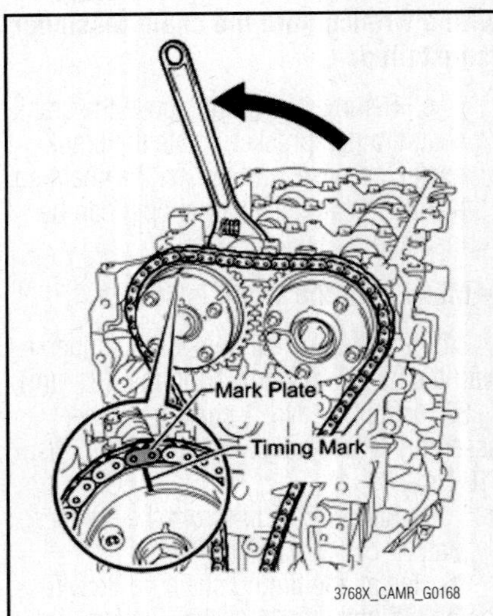

Fig. 104 Rotating the intake camshaft to align the timing marks

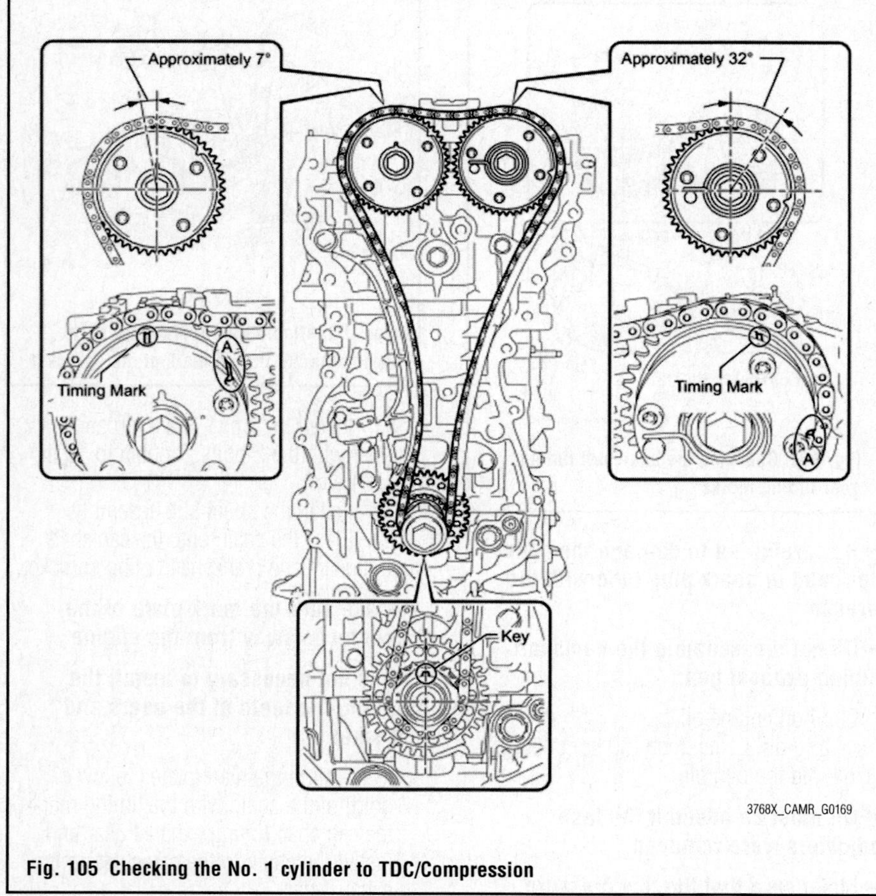

Fig. 105 Checking the No. 1 cylinder to TDC/Compression

d. Using the hexagonal portion of the intake camshaft, rotate the intake camshaft counterclockwise with a wrench, align the timing mark of the camshaft timing gear with the mark plate (yellow or gold) of the chain and install the chain to the camshaft timing gear.

➡Hold the intake camshaft in place with a wrench until the chain tensioner is installed.

e. Remove the string above the crankshaft timing sprocket, rotate the crankshaft clockwise, and loosen the chain so that the chain tensioner slipper can be installed.

➡Make sure the chain is secure.

34. Install the chain tensioner slipper with the bolt. Tighten to 15 ft. lbs. (21 Nm).
35. Install the No. 1 chain tensioner assembly with 2 bolts. Tighten to 7 ft. lbs. (10 Nm).
    a. Remove the pin from the stopper plate.
36. Install the timing chain guide with the bolt. Tighten to 15 ft. lbs. (21 Nm).
37. Check the No. 1 cylinder to TDC/Compression.
    a. Temporarily install the crankshaft pulley bolt.
    b. Rotate the crankshaft clockwise, and check that the timing marks on the crankshaft timing sprocket and camshaft timing gears are as shown in the illustration.

➡"A" is not a timing mark.

    c. Remove the crankshaft pulley bolt.
38. Install the timing chain cover sub assembly.

3.5L Engine

See Figures 106 through 125.

1. Before servicing the vehicle, refer to the Precautions Section.
2. Remove the engine assembly.
3. Install on engine stand.
4. Remove the oil filler cap and gasket.
5. Remove the spark plugs and ignition coil assembly.
6. Remove the drain plug and gasket.
7. Remove the ventilation valve.
8. Remove the 4 bolts and 4 camshaft position sensors.
9. Remove the 4 bolts and 4 camshaft timing oil control valves.
10. Remove the bolt and Crankshaft Position (CKP) sensor.
11. Remove the 2 oil pipe unions and oil pipe. Remove the LH oil control valve filter and gaskets.
12. Remove the oil pipe bolt. Remove the 2 oil pipe unions and oil pipe. Remove the RH oil control valve filter and gaskets.
13. Remove the cylinder block water drain cock sub-assembly, as follows:
    a. Remove the water drain cocks from the cylinder block.
    b. Remove the water drain cock plugs from the water drain cocks.
14. Remove the oil filter.
15. Remove the crankshaft pulley, as follows:
    a. Using SST (SST: 09213-70011, SST: 09330-00021) or equivalent, loosen the crankshaft pulley bolt.
    b. Using SST (SST: 09950-50013) or equivalent, remove the crankshaft pulley bolt and crankshaft pulley.
16. Remove the 6 bolts and the left hand No. 1 front engine mounting bracket. Using Torx®socket wrench E8, remove the 2 stud bolts.
17. Remove the water inlet housing, as follows:
    a. Remove the 2 nuts, water inlet and thermostat.
    b. Remove the gasket.
    c. Remove the drain cock plug.
    d. Remove the drain cock.
    e. Remove the 2 stud bolts.
    f. Remove the 2 bolts, nut, and water inlet housing.
    g. Remove the 2 O-rings.
18. Remove the water outlet, as follows:
    a. Remove the 2 bolts, 4 nuts and water outlet.
    b. Remove the 2 gaskets and O-ring.
19. Remove the 12 bolts, valve cover (for Bank 1) and gasket, and then remove the 3 gaskets.
20. Remove the 12 bolts, valve cover (for Bank 2) and gasket, then remove the 3 gaskets.
21. Remove the No. 2 oil pan sub-assembly.
22. Remove the oil strainer sub-assembly.
23. Remove the oil pan sub-assembly.
24. Remove the No. 1 oil pan baffle plate.
25. Remove the engine rear oil seal, as follows:
    a. Remove the 6 bolts.
    b. Using a screwdriver with the tip taped, pry out the oil seal retainer. Be careful not to damage the oil seal retainer.

➡Be careful not to damage the engine rear oil seal retainer.

➡Tape the screwdriver tip before use.

26. Place the oil seal retainer on wooden blocks. Using a screwdriver and a hammer, tap out the oil seal.
27. Remove the water pump assembly.
28. Remove the timing chain cover sub-assembly.
29. Remove timing chain case oil seal.
30. Set the No. 1 cylinder to TDC/compression.
31. Remove the No. 1 chain tensioner assembly.
32. Remove the chain tensioner slipper.
33. Remove the chain sub-assembly.
34. Remove the idle sprocket assembly.
35. Remove the No. 1 chain vibration damper.
36. Remove crankshaft timing sprocket. Remove the 2 pulley set keys from the crankshaft.
37. Remove camshaft timing gears and No. 2 chain (for Bank 1), as follows:
    a. While raising the No. 2 chain tensioner, insert a pin of 0.039 in (1.0 mm) into the hole to fix the No. 2 chain tensioner.

➡Be careful not to damage the cylinder head with the wrench.

➡Do not disassemble the camshaft timing gear assemblies.

    b. Hold the hexagonal portion of the camshaft with a wrench, and remove the 2 bolts and 2 camshaft timing gears.
    c. Remove the No. 2 chain.
38. Remove the bolt and No. 2 chain tensioner assembly.
39. Remove camshaft bearing cap (for Bank 1), as follows:
    a. Check that the camshafts are positioned as shown in the illustration.
    b. Uniformly loosen and remove the 8 bearing cap bolts in the sequence shown in the illustration.
    c. Uniformly loosen and remove the 12 bearing cap bolts in the sequence shown in the illustration. Uniformly loosen the bolts while keeping the camshaft level.
    d. Remove the 5 bearing caps.
40. Remove the camshaft.
41. Remove the No. 2 camshaft.
42. If necessary, remove the right hand camshaft housing sub-assembly by prying between the cylinder head and the camshaft housing with a screwdriver with the tip taped.

➡Be careful not to damage the contact surfaces of the cylinder head and the camshaft housing.

43. Remove the camshaft timing gears and No. 2 chain (for Left-hand Bank), as follows:
    a. While pushing down the No. 3 chain tensioner, insert a pin of 1.0 mm (0.039 in.) into the hole to fix the No. 3 chain tensioner.
    b. Hold the hexagonal portion of the camshaft with a wrench, and remove the 2 bolts and 2 camshaft timing gears.

➡Be careful not to damage the cylinder head with the wrench.

➡Do not disassemble the camshaft timing gear assemblies.

    c. Remove the No. 2 chain.
44. Remove the bolt and No. 3 chain tensioner.
45. Remove the camshaft bearing cap (bank 2), as follows:
    a. Make sure that the knock pin of the camshaft is positioned as shown in the illustration.
    b. Uniformly loosen and remove the 8 bearing cap bolts in the sequence shown in the illustration.
    c. Uniformly loosen and remove the 13 bearing cap bolts in the sequence shown in the illustration. Loosen the bolts while keeping the camshaft level.
    d. Remove the 5 camshaft bearing caps.
46. Remove the No. 3 camshaft.
47. Remove the No. 4 camshaft.
48. If necessary, remove the left hand camshaft housing sub-assembly by prying between the cylinder head and the camshaft

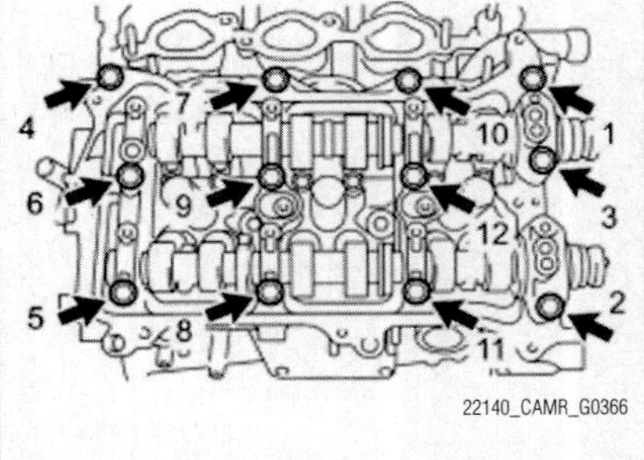

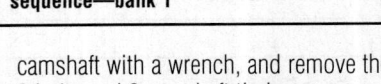

Fig. 109 12 bearing cap bolts removal sequence—bank 1

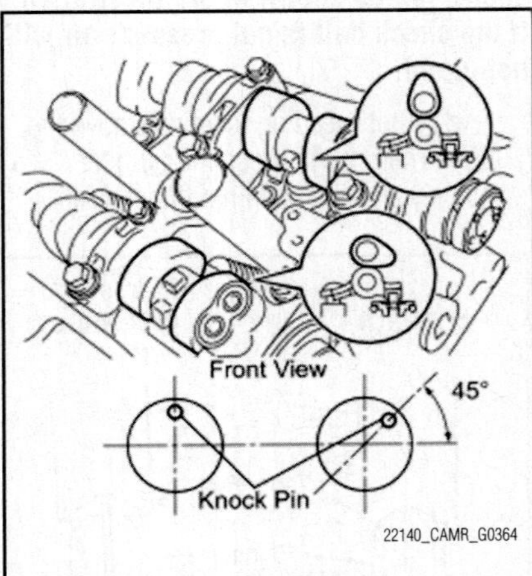

Fig. 107 Camshafts knock pin positioning—bank 1

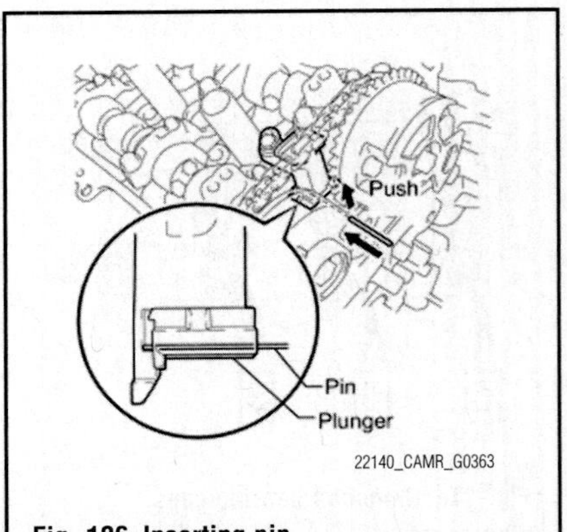

Fig. 106 Inserting pin

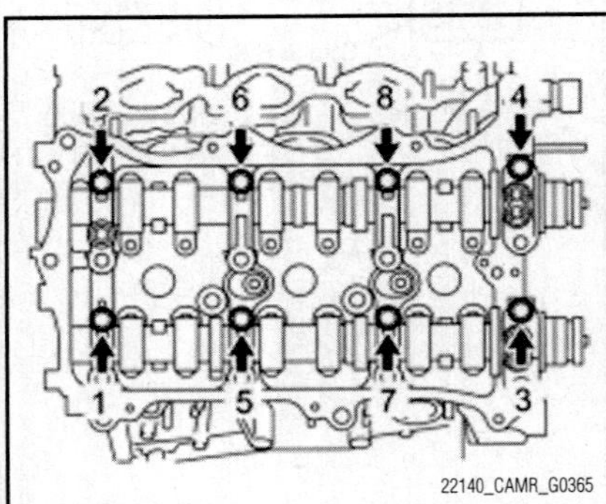

Fig. 108 8 bearing cap bolts removal sequence—bank 1

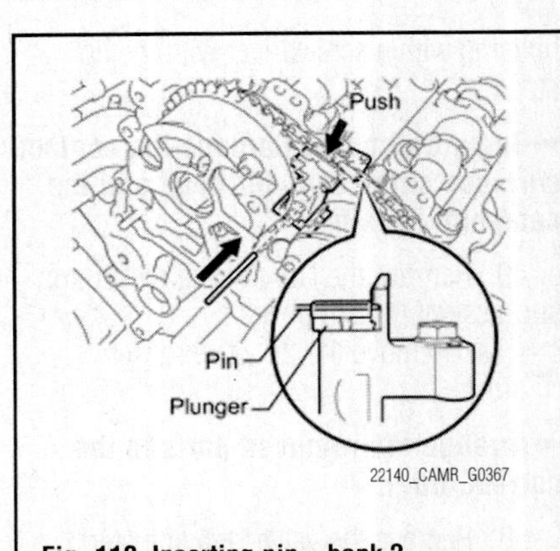

Fig. 110 Inserting pin—bank 2

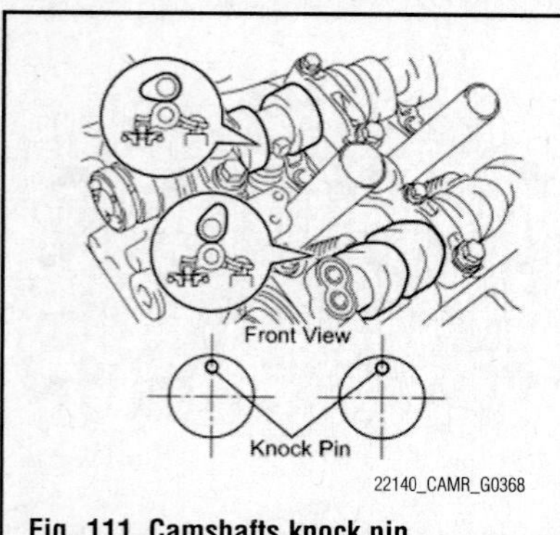

Fig. 111 Camshafts knock pin positioning—bank 2

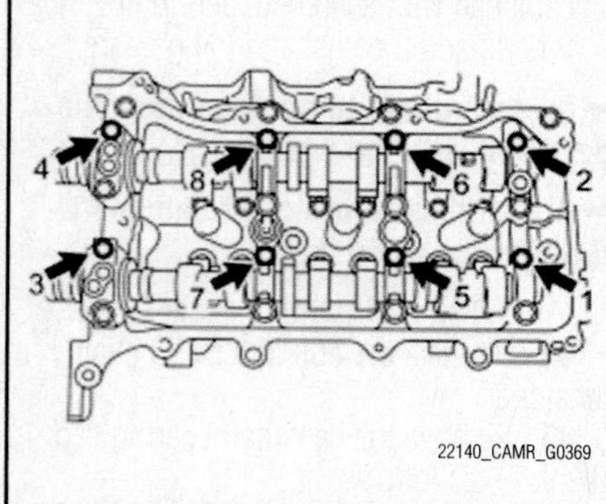

Fig. 112 8 bearing cap bolts removal sequence—bank 1

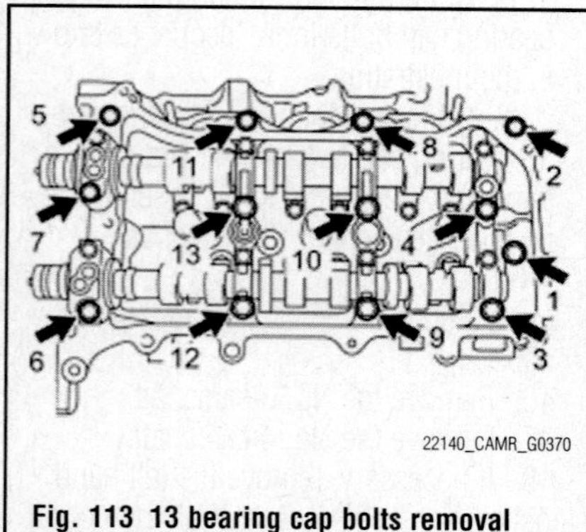

Fig. 113 13 bearing cap bolts removal sequence—bank 1

housing with a screwdriver with the tip taped.

➡**Be careful not to damage the contact surfaces of the cylinder head and the camshaft housing.**

49. Remove the No. 1 valve rocker arm sub-assembly, as follows:

a. Remove the 24 valve rocker arms.

➡**Arrange the removed parts in the correct order.**

50. Remove the valve lash adjuster assembly, as follows:

a. Remove the 24 valve lash adjusters from the cylinder head.

➡**Arrange the removed parts in the correct order.**

To install:

➡**Keep the lash adjuster free of dirt and foreign objects. Only use clean engine oil.**

51. Install the valve lash adjuster assembly, as follows:

a. Place the lash adjuster into a container filled with engine oil. Insert the SST 09276-75010 tip into the lash adjuster's plunger and use the tip to press down on the check ball inside the plunger.

b. Squeeze the SST and lash adjuster together to move the plunger up and down 5 to 6 times.

c. Check the movement of the plunger and bleed the air. Make sure the plunger moves up and down.

➡**When bleeding air from the high-pressure chamber, make sure that the tip of the SST is actually pressing the check ball as shown in the illustration. If the check ball is not pressed, air will not bleed.**

d. After bleeding the air, remove the SST. Then, try to press the plunger quickly and firmly with a finger. Make sure the plunger is very difficult to move. If the result is not as specified, replace the lash adjuster.

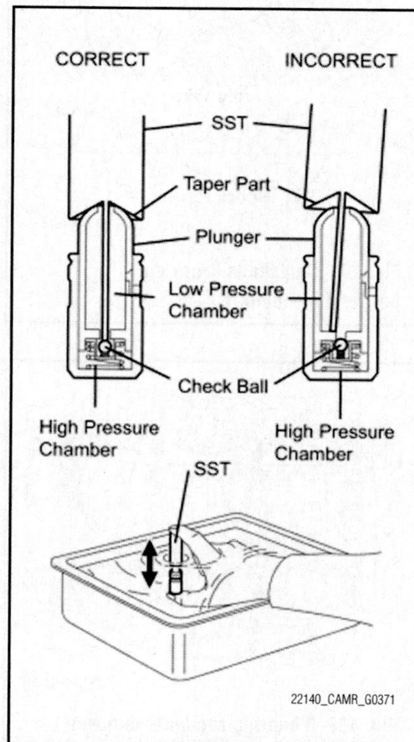

Fig. 114 Installing lash adjuster assembly

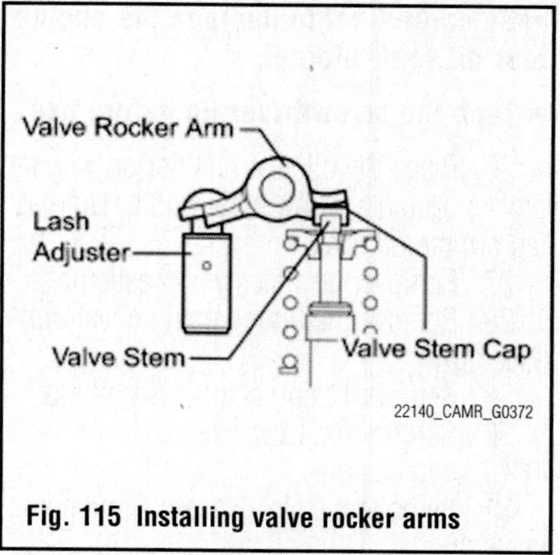

Fig. 115 Installing valve rocker arms

➡**Install the lash adjuster to the same place it was removed from.**

e. Install the lash adjusters.

52. Install the No. 1 valve rocker arm sub-assembly, as follows:

a. Apply engine oil to the lash adjuster tips and valve stem cap ends.

b. Make sure that the valve rocker arms are installed as shown in the illustration.

53. Install the right hand camshaft bearing cap, as follows:

a. Apply engine oil to the camshaft journals, camshaft housing and bearing caps.

b. Install the camshaft and No. 2 camshaft to the right camshaft housing.

c. Make sure of the marks and numbers on the camshaft bearing caps and place them in each proper position and direction.

d. Temporarily tighten the 8 bearing cap bolts to 7 ft. lbs. (10 Nm) in the order shown in the illustration.

54. Install the right hand camshaft housing sub-assembly, as follows:

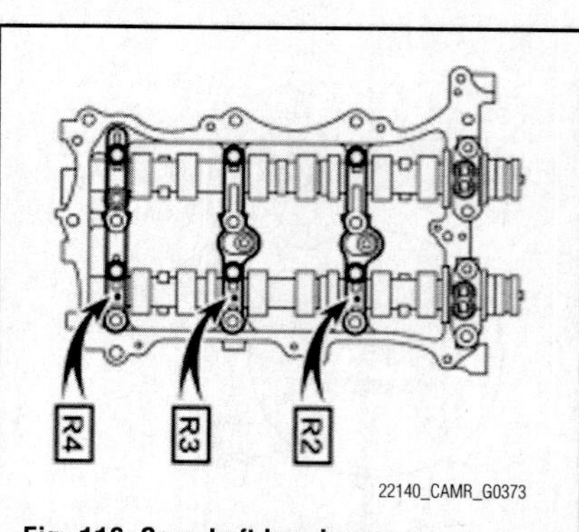

Fig. 116 Camshaft bearing caps placement

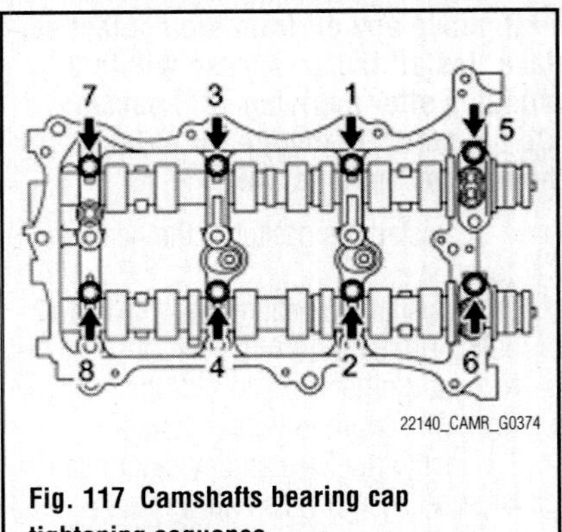

Fig. 117 Camshafts bearing cap tightening sequence

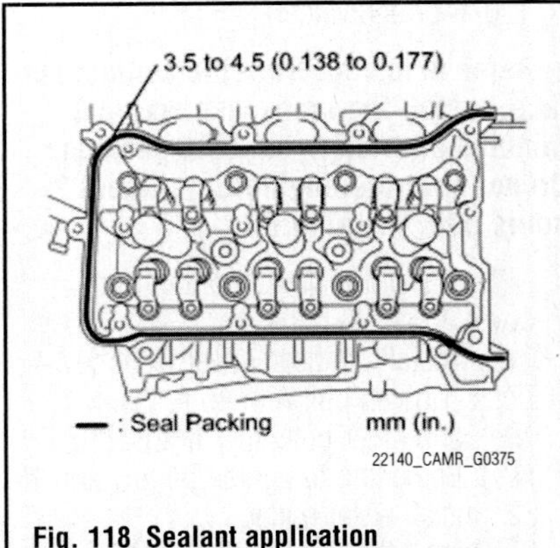

Fig. 118 Sealant application

a. Apply seal packing in a continuous line as shown in the illustration. Seal packing: Toyota Genuine Seal Packing Black, Three Bond 1207B or equivalent. Seal diameter: 3.5 to 4.5 mm (0.138 to 0.177 in.).

➡Remove any oil from the contact surface. Install the camshaft housing sub-assembly within 3 minutes and tighten the bolts within 15 minutes after applying sealant. Do not start the engine for at least 2 hours after installing.

b. Make sure that the knock pins of the camshafts are positioned as shown. Install the right hand camshaft housing and tighten the 12 bolts in the order shown in the illustration to 21 ft. lbs. (28 Nm).

➡When installing the camshaft housing RH, it is necessary to correctly position the camshafts as shown in the illustration. Failure to correctly position these parts may result in damage due to contact between the pistons and valves. If a camshaft is rotated with a piston at TDC, valve contact will occur.

➡If any of the bolts are loosened during installation, remove the camshaft

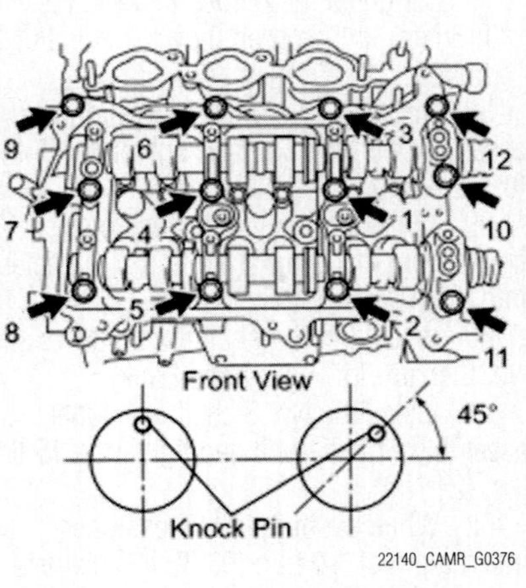

Fig. 119 Camshaft bolt tightening sequence and knock pin position

housing, clean the installation surfaces, and reapply seal packing.

➡If the camshaft housing is removed because any of the bolts are loosened during installation, make sure that the previously applied seal packing does not enter any oil passages.

c. Tighten the 8 bolts to 12 ft. lbs. (16 Nm) in the order shown in the illustration.

55. Install the camshaft bearing cap (bank 2), as follows:

a. Apply engine oil to the camshaft journals, camshaft housing and bearing caps.

b. Install the No. 3 camshaft and No. 4 camshaft to the left hand camshaft housing.

c. Make sure of the marks and numbers on the camshaft bearing caps and place them in each proper position and direction.

d. Temporarily tighten the 8 bolts in the order shown in the illustration to 7 ft. lbs. (10 Nm).

56. Make sure that the valve rocker arm is installed.

57. Install the left camshaft housing sub-assembly, as follows:

a. Apply seal packing in a continuous line as shown in the illustration. Seal packing: Toyota Genuine Seal Packing Black, Three Bond 1207B or equivalent. Seal diameter: 3.5 to 4.5 mm (0.138 to 0.177 in.).

➡Remove any oil from the contact surface. Install the camshaft housing sub-assembly within 3 minutes and tighten the bolts within 15 minutes after applying sealant. Do not start the engine for at least 2 hours after installing.

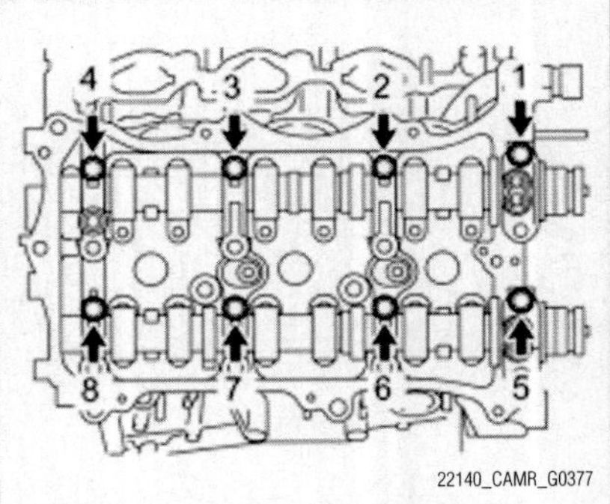

Fig. 120 Camshaft bolt tightening sequence

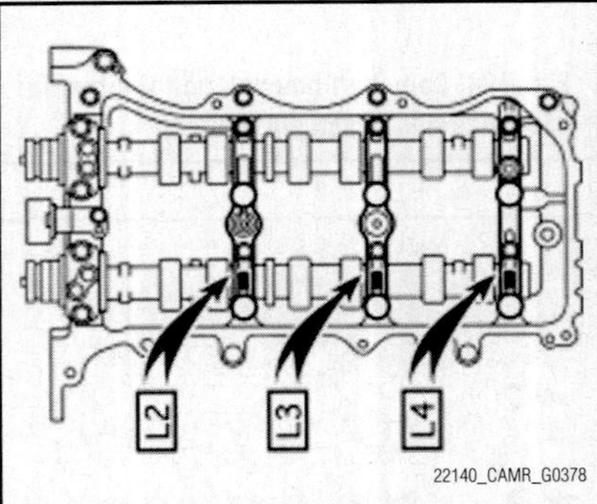

Fig. 121 Camshaft bearing cap positioning

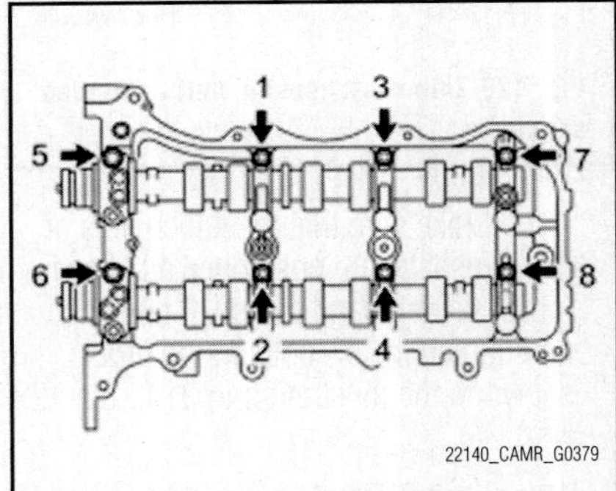

Fig. 122 Camshaft bolt tightening sequence

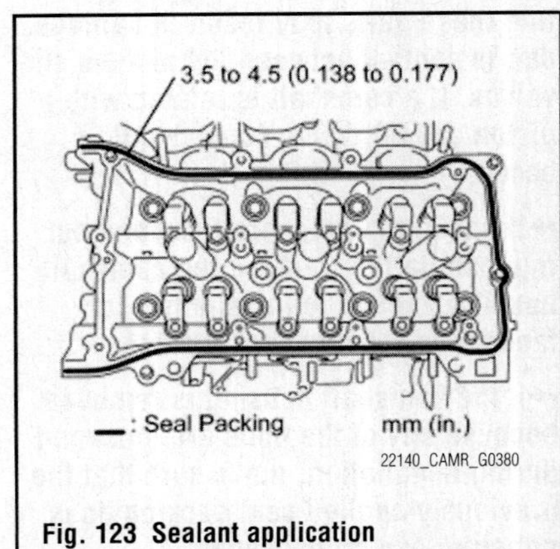

Fig. 123 Sealant application

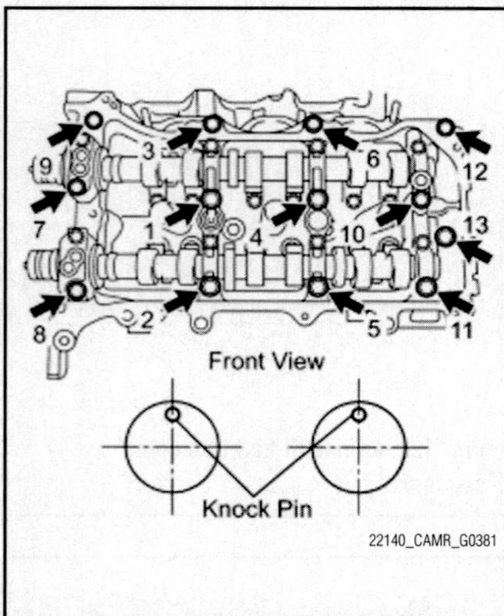

Fig. 124 Camshaft housing bolt tightening sequence and knock pin position

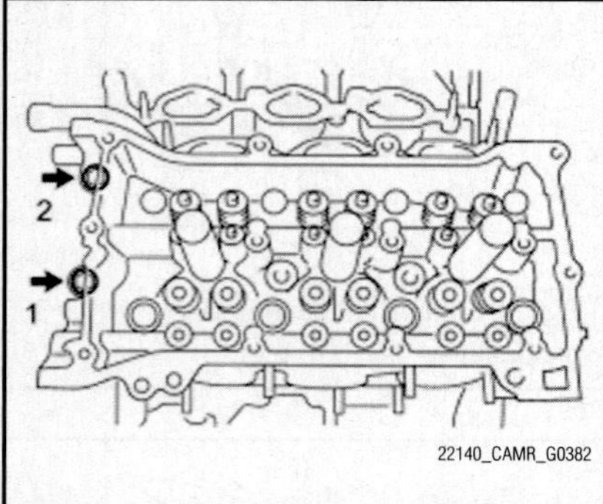

Fig. 125 Camshaft housing bolt tightening sequence

b. Make sure that the knock pins of the camshafts are positioned as shown. Install the left hand camshaft housing and tighten the 13 bolts in the order shown in the illustration to 21 ft. lbs. (28 Nm).

➡**When installing the camshaft housing LH, it is necessary to correctly position the camshafts as shown in the illustration. Failure to correctly position these parts may result in damage due to contact between the pistons and valves. If a camshaft is rotated with a piston at TDC, valve contact will occur.**

➡**If any of the bolts are loosened during installation, remove the camshaft housing, clean the installation surfaces, and reapply seal packing.**

➡**If the camshaft housing is removed because any of the bolts are loosened during installation, make sure that the previously applied seal packing does not enter any oil passages.**

c. Tighten the 8 bolts to 12 ft. lbs. (16 Nm) in the order shown in the illustration.

58. Install the No. 2 chain tensioner assembly with the bolt and tighten to 15 ft. lbs. (21 Nm).
59. While pushing in the tensioner, insert a pin of 1.0 mm (0.039 in.) diameter into the hole to hold it.
60. Install the camshaft timing gears and No. 2 chain (for Right-hand Bank).
61. Install the No. 3 chain tensioner assembly with the bolt and tighten to 15 ft. lbs. (21 Nm).
62. While pushing in the tensioner, insert a pin of 1.0 mm (0.039 in.) diameter into the hole to hold it.
63. Install the camshaft timing gears and No. 2 chain (for Left-hand Bank).
64. Install the No 1 chain vibration damper with the 2 bolts and tighten to 17 ft. lbs. (23 Nm).
65. Install the No 2 chain vibration damper.
66. Install the timing gear set keys and timing gear as shown in the illustration.
67. Install the idle sprocket assembly.
68. Install the chain sub-assembly.
69. Install the chain tensioner slipper.
70. Install the No. 1 chain tensioner assembly.
71. Install the water pump assembly.
72. Install the timing chain cover sub-assembly.
73. Install the water inlet housing.
74. Install the No. 1 left front engine mounting bracket, as follows:

a. Install the No. 1 left front engine mounting bracket with the 6 bolts and tighten to 40 ft. lbs. (54 Nm).

➡**Install the water inlet and mounting bracket within 15 minutes after installing the chain cover. Do not start the engine for at least 2 hours after installation.**

75. Install the No. 1 oil pan baffle plate with the 7 bolts and tighten to 7 ft. lbs. (10 Nm).
76. Install the oil pan sub-assembly.
77. Install the oil strainer sub-assembly.
78. Install the No. 2 oil pan sub-assembly.
79. Install a new gasket and oil pan drain plug and tighten to 30 ft. lbs. (40 Nm).
80. Install the cylinder head cover sub-assembly, as follows:

a. Apply seal packing (Toyota Genuine Seal Packing Black, Three Bond 1207B or equivalent).

➡**Remove any oil from the contact surface. Install the crankcase within 3 minutes after applying seal packing. Do not start the engine for at least 2 hours after installation.**

b. Install the gasket to the head cover.

c. Install the head cover with the 12 bolts. Tighten bolt A to 15 ft. lbs. (21 Nm), and other bolts to 7 ft. lbs. (10 Nm). Be certain to tighten bolt 1.

81. Install the left-hand cylinder head cover sub-assembly, as follows:

a. Apply seal packing (Toyota Genuine Seal Packing Black, Three Bond 1207B or equivalent).

➡**Remove any oil from the contact surface. Install the crankcase within 3 minutes after applying seal packing. Do not start the engine for at least 2 hours after installation.**

b. Install the gasket to the head cover.

c. Install the head cover with the 14 bolts. Tighten bolt A to 15 ft. lbs. (21 Nm), and other bolts to 7 ft. lbs. (10 Nm). Be certain to tighten bolts 1 and 10.

82. Install water outlet.
83. Install the crankshaft pulley.
84. Install the oil filter element.
85. Install the cylinder block water drain cock sub-assembly, as follows:

a. Apply adhesive around the drain cocks. Adhesive: Toyota Genuine Adhesive 1324, Three Bond 1324 or Equivalent.

b. Install the water drain cocks and tighten to 18 ft. lbs. (25 Nm). Do not rotate the drain cocks more than 1 revolution (360°) after tightening the drain cocks with the specified torque. Do not loosen after setting correctly.

c. Install the water drain cock plug to the water drain cocks and tighten to 9 ft. lbs. (13 Nm).

86. Install the No. 1 oil pipe.
87. Install the oil pipe.
88. Install the Crankshaft Position (CKP) sensor with the bolt and tighten to 7 ft. lbs. (10 Nm).
89. Install the 4 camshaft timing oil control valves with the 4 bolts and tighten to 7 ft. lbs. (10 Nm).
90. Install the 4 camshaft position sensors with the 4 bolts and tighten to 7 ft. lbs. (10 Nm).
91. Install the ventilation valve sub-assembly, as follows:

a. Apply adhesive (Toyota Genuine Adhesive 1324, Three Bond 1324 or equivalent) around the ventilation valve.

b. Install the ventilation valve and tighten to 20 ft. lbs. (27 Nm).

92. Install the 6 spark plugs and the ignition coil assembly.

93. Install the oil filler cap sub-assembly.

94. Remove the engine stand.

95. Install the engine assembly.

CRANKSHAFT DAMPER (BALANCER)

REMOVAL & INSTALLATION

2.4L and 3.5L Engine

See Figure 126.

1. Before servicing the vehicle, refer to the Precautions Section.
2. Remove the engine assembly.
3. Install on engine stand.
4. Remove the oil filler cap and gasket.
5. Remove the spark plugs and ignition coil assembly.
6. Remove the drain plug and gasket.
7. Remove the ventilation valve.
8. Remove the 4 bolts and 4 camshaft position sensors.
9. Remove the 4 bolts and 4 camshaft timing oil control valves.
10. Remove the bolt and Crankshaft Position (CKP) sensor.
11. Remove the No. 1 oil pipe.
12. Remove the oil pipe.
13. Remove the cylinder block water drain cock sub-assembly.
14. Remove the oil filter element.
15. Remove the crankshaft pulley, as follows:
    a. Using SST (SST: 09213-70011, SST: 09330-00021) or equivalent, loosen the crankshaft pulley bolt.
    b. Using SST (SST: 09950-50013) or equivalent, remove the crankshaft pulley bolt and crankshaft pulley.

To install:

16. Install the crankshaft pulley, as follows:
    a. Align the pulley set key with the key groove of the pulley, and slide on the pulley.
    b. Using SST (SST: 09213-70011, SST: 09330-00021) or equivalent, install the pulley bolt and tighten to 184 ft. lbs. (250 Nm).
17. Install the oil filter element.
18. Install the cylinder block water drain cock sub-assembly.
19. Install the No. 1 oil pipe.
20. Install the oil pipe.
21. Install the Crankshaft Position (CKP) sensor.
22. Install the camshaft timing oil control valve assembly.
23. Install the camshaft position sensor.
24. Install the ventilation valve sub-assembly.
25. Install the spark plugs and the ignition coil assembly.
26. Install the oil filler cap sub-assembly.
27. Remove the engine stand.
28. Install the engine assembly.

2.5L Engine

See Figures 127 through 129.

1. Remove the engine cover joint.
2. Remove the spark plug.
3. Remove the camshaft timing oil control valve assembly (intake and exhaust side).
4. Remove the camshaft position sensors.
5. Remove the oil filler cap sub assembly.
    a. Remove the oil filler cap from the cylinder head.
    b. Remove the gasket from the oil filler cap.
6. Remove the Crankshaft Position (CKP) sensor.
7. Remove the ventilation valve sub assembly.
8. Remove the ventilation case sub assembly.
    a. Remove the 8 bolts and 2 nuts.
    b. Remove the ventilation case by prying between the ventilation case and cylinder block with a screwdriver.

➡Tape the tip of the screwdriver before use.

➡Be careful not to damage the contact surfaces of the cylinder block ventilation case.

9. Remove the separator case.
10. Remove the No. 1 water by pass pipe.
11. Remove the inlet water.
12. Remove the thermostat.
13. Remove the v-ribbed belt tensioner assembly (for compressor and alternator).
14. Remove the water pump assembly.
15. Remove the water drain cock.
16. Remove the oil cooler assembly.
17. Remove the inlet water housing.
    a. Remove the 4 bolts, nut, inlet water housing and gasket.
18. Remove the crankshaft pulley.
    a. Using the special tool, hold the crankshaft pulley and loosen the pulley bolt. Further loosen the bolt until 2 or 3 threads are screwed into the crankshaft.
    b. Using the special tool and the pulley bolt, remove the crankshaft pulley.

➡Apply a lubricant to the threads and end of SST.

To install:

19. Install the crankshaft pulley.
    a. Align the pulley set key with the key groove of the crankshaft pulley.
    b. Using SST, hold the crankshaft pulley and install the pulley bolt. Tighten to 192 ft. lbs. (260 Nm).
20. Install the Crankshaft Position (CKP) sensor.
21. Install the inlet water housing.
    a. Install a new gasket ad the inlet water housing with the 4 bolts and nut. Tighten to 32 ft. lbs. (43 Nm).
22. Install the oil cooler assembly.
23. Install the water pump assembly.
24. Install the v-ribbed belt tensioner assembly (for compressor and alternator). Tighten to 15 ft. lbs. (21 Nm).
25. Install the thermostat.
26. Install the inlet water.
27. Install the water drain cock.

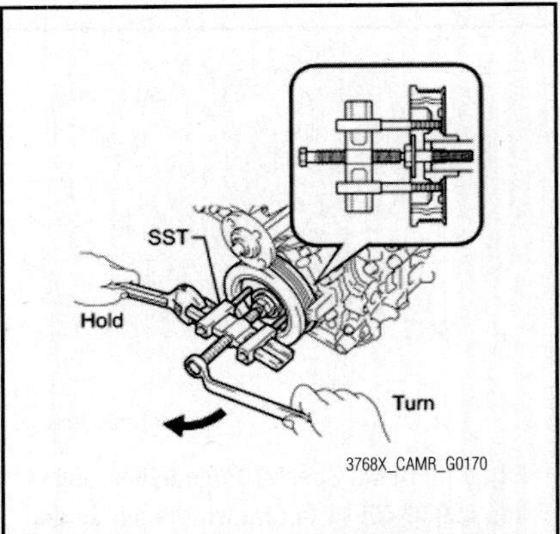

Fig. 126 Removing the crankshaft pulley

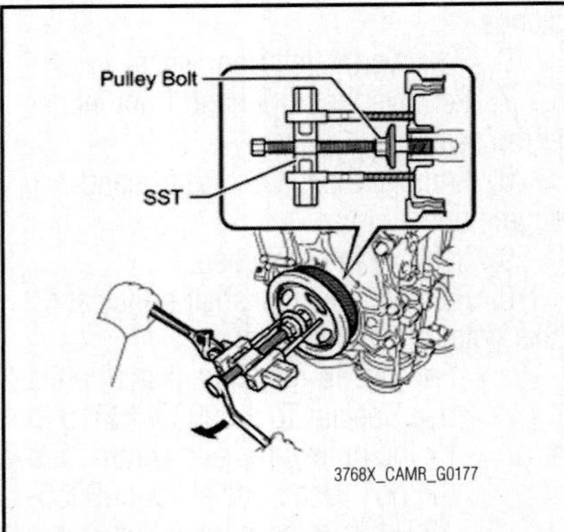

Fig. 127 Removing the crankshaft pulley

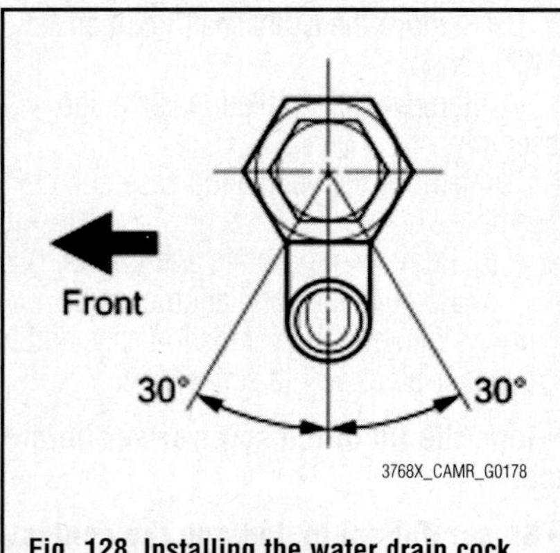

Fig. 128 Installing the water drain cock

a. Apply adhesive to 2 or 3 threads of the drain cock.

b. Install the water drain cock as shown. Tighten to 15 ft. lbs. (20 Nm).

➡Do not rotate the drain cock more than 1 revolution (360°) after tightening the drain cock to the specified torque.

➡Do not loosen the drain cock to adjust it. If an adjustment is necessary, remove the drain cock and reinstall it.

c. Install the water drain cock plug to the water drain cock. Tighten to 9 ft. lbs. (13 Nm).

28. Install the No. 1 water by pass pipe.

a. Install a new gasket and the water by pass pipe with the 2 nuts and bolt. Tighten to 7 ft. lbs. (10 Nm).

29. Install the separator case.

a. Apply a light coat of engine oil to a new gasket.

b. Install the gasket to the separator case.

c. Install the separator case with the 2 bolts. Tighten to 7 ft. lbs. (10 Nm).

30. Install the ventilation case sub assembly.

a. Apply seal packing in a continuous line with a diameter of 0.0984-0.138 inch (2.5-3.5 mm).

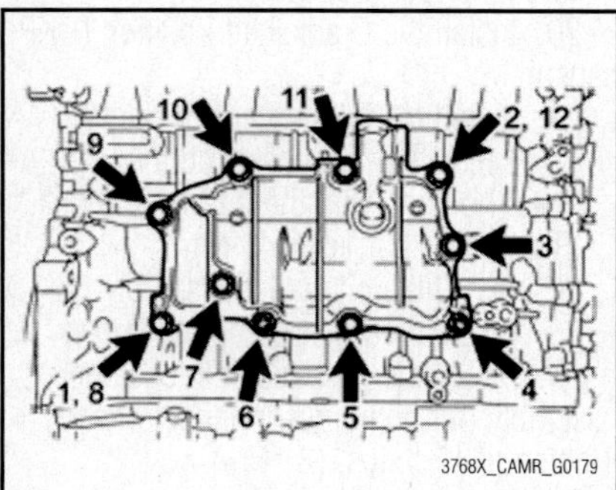

Fig. 129 Identifying the ventilation case bolt tightening sequence

➡Remove any oil from the contact surface.

➡Install the ventilation case within 3 minutes and tighten the bolts and nuts within 15 minutes after applying seal packing.

b. Install the ventilation case, and install the 8 bolts and 2 nuts in the order shown. Tighten to 15 ft. lbs. (21 Nm).

➡Bolt A is tightened twice.

31. Install the ventilation valve sub assembly.

32. Install the camshaft position sensors (intake and exhaust side).

33. Install the camshaft timing oil control valve assemblies (intake and exhaust side).

34. Install the oil filler cap sub assembly.

a. Install a new gasket to the oil filler cap.

b. Install the oil filler cap to the cylinder head.

35. Install the spark plug.

36. Install the engine cover joints. Tighten to 7 ft. lbs. (10 Nm).

CRANKSHAFT FRONT SEAL

REMOVAL & INSTALLATION

2.4L Engine

See Figures 130 through 134.

1. Before servicing the vehicle, refer to the Precautions Section.
2. Disconnect the negative battery cable.
3. Drain the engine oil
4. Loosen the lug nuts on the left front wheel.
5. Apply the parking brake, block the rear wheels, then raise and safely support the front of the vehicle securely on jack stands.
6. Remove the left front wheel.
7. Remove the right hand front fender apron seal.
8. Remove the left and right hand engine under cover.
9. Remove the drive belt.
10. Remove the crankshaft pulley and take note of the following:
    - For TMMK made crankshaft pulley, use Special Tool 09960-10010 to fix the pulley in place and to loosen the bolt. Use Special Tool 09950-40011 to remove the pulley and bolt.

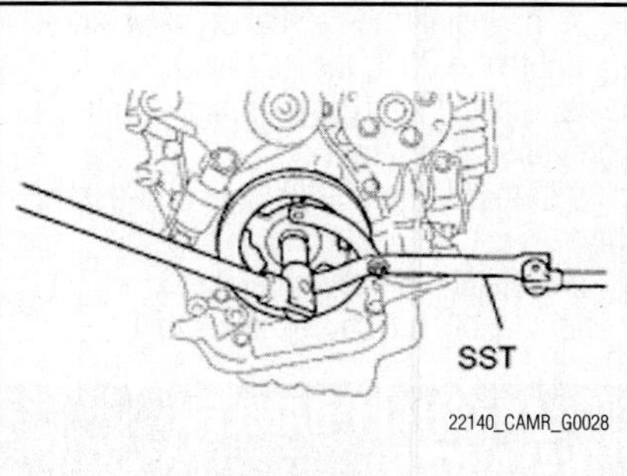

Fig. 130 Using Special Tool 09960-10010, fix the pulley in place and loosen the bolt—TMMK Pulley

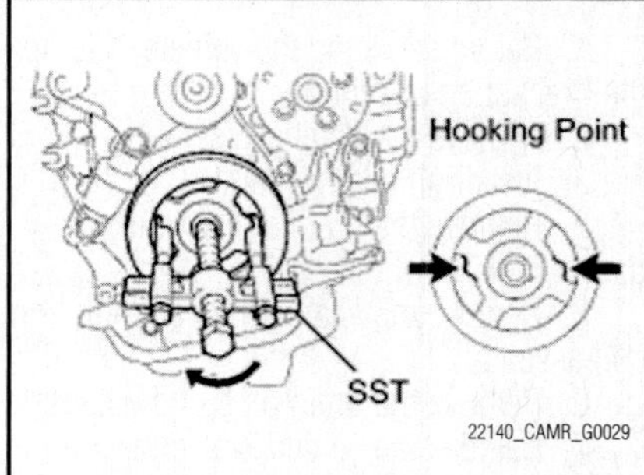

Fig. 131 Using Special Tool 09950-40011 to remove the pulley and bolt—TMMK Pulley

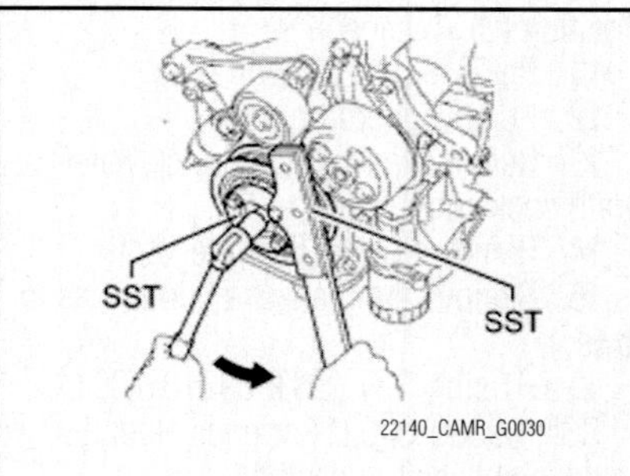

Fig. 132 Using Special Tools 09213-54015 and 09330-00021 to fix the pulley in place and to loosen the bolt—TMC Pulley

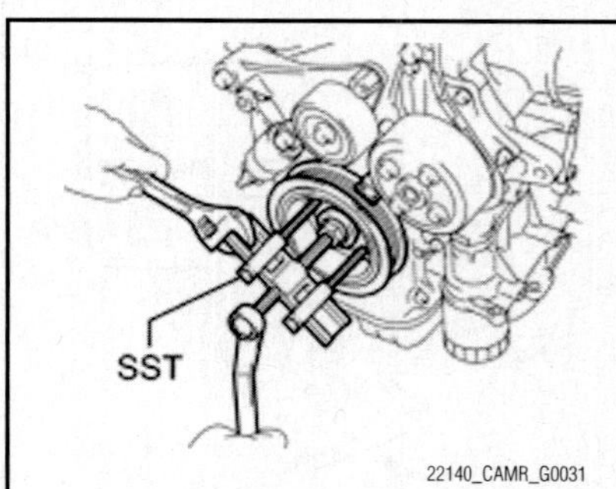

Fig. 133 Using Special Tools 09950-50013 and 09950-40011 to remove the pulley and bolt—TMC Pulley

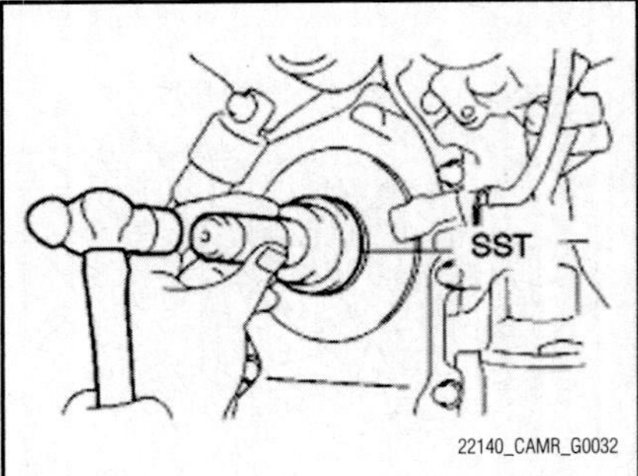

Fig. 134 Using Special Tool 09223-22010 and hammer to install new oil seal

- For TMC made crankshaft pulley, use Special Tools 09213-54015 and 09330-00021 to fix the pulley in place and to loosen the bolt. Use Special Tools 09950-50013 and 09950-40011 to remove the pulley and bolt.

11. Using a knife, cut off the oil seal lip. Using a screwdriver with the tip taped, pry out the oil seal.
12. After the removal, check the crankshaft for damage. If it is damaged, smooth the surface with 400-grit sandpaper.

To install:

13. Apply MP grease to a new oil seal lip. Keep the lip free from foreign matter.
14. Using Special Tool 09223-22010 and a hammer, tap in the oil seal until its surface is flush with the rear oil seal retainer edge.
15. For TMMK made pulleys, align the pulley set key with the key groove of the pulley. Using Special Tool 09960-10010, keep the pulley in place and tighten the bolt to: 125 ft. lbs. (170 Nm).
16. For TMC made pulleys, align the pulley set key with the key groove of the pulley. Using Special Tools 09213-54015 and 09330-00021, keep the pulley in place and tighten the bolt to 133 ft. lbs. (180 Nm).
17. To complete installation, reverse remaining removal procedure.
18. Check engine for oil leaks.

2.5L Engine

See Figure 135.

1. Remove the front wheel RH.
2. Remove the front wheel opening extension pad RH.
3. Remove the engine under cover RH.
4. Remove the front fender apron seal RH.
5. Remove the v-ribbed belts.
6. Remove the crankshaft pulley.

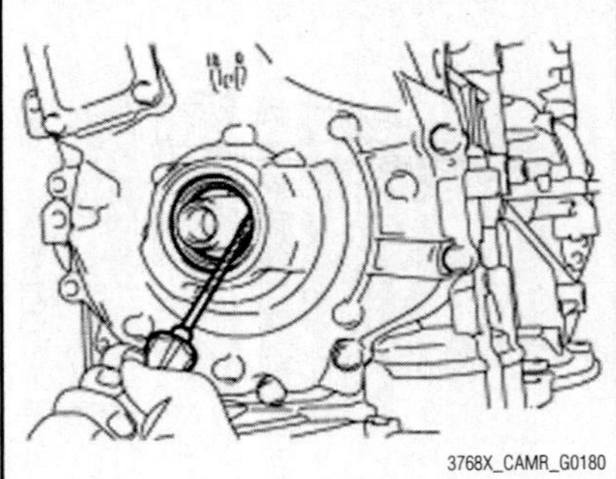

Fig. 135 Removing the timing chain coil seal

7. Remove the timing chain cover oil seal.
   a. Using a screwdriver, pry out the oil seal.

➡Tape the screwdriver tip before use.

➡Do not damage the surface of the oil seal press fit hole or the crankshaft.

To install:

8. Install the timing chain cover oil seal.
   a. Apply MP grease to the lip of a new oil seal.

➡Do not allow foreign matter to contact the lip of the oil seal.

➡Do not allow MP grease to contact the dust seal.

   b. Using the special tool and a hammer, tap in the oil seal until its surface is flush with the timing chain cover edge.

➡Keep the lip of the oil seal free from foreign matte.

➡Do not tap in the oil seat at an angle.

9. Install the crankshaft pulley.
   a. Align the pulley set key with the key groove of the crankshaft pulley.
   b. Using the special tool, hold the crankshaft pulley and install the pulley bolt. Tighten to 192 ft. lbs. (260 Nm).
10. Install the v-ribbed belts.
11. Install the front fender apron seal RH.
12. Install the engine under cover RH.
13. Install the front wheel opening extension pad RH.
14. Install the front wheel RH. Tighten to 76 ft. lbs. (103 Nm).

3.5L Engine

See Timing Chain Cover and Seal.

CYLINDER HEAD

REMOVAL & INSTALLATION

2.4L Engine

See Figures 136 through 142.

1. Discharge the fuel system pressure.
2. Disconnect the cable from the negative battery terminal.
3. Remove the engine under covers.
4. Remove the front fender apron seal RH.
5. Remove the No. 1 engine cover sub assembly.
6. Drain the engine coolant.
7. Drain the engine oil.
8. Remove the windshield wiper link assembly.
9. Remove the cowl top panel outer sub assembly.
10. Remove the air cleaner inlet assembly.
11. Remove the air cleaner cap sub assembly.
12. Remove the air cleaner case sub assembly.
13. Remove the battery.
14. Remove the throttle body assembly.
15. Disconnect the fuel tube sub assembly.
16. Remove the fuel delivery pipe with the injectors.
17. Remove the intake manifold.
18. Remove the intake air control valve (PZEV).
19. Remove the No. 1 intake manifold insulator.
20. Remove the front exhaust pipe assembly.
21. Remove the No. 2 engine mounting stay RH.
22. Remove the engine moving control rod sub assembly.
23. Remove the No. 2 engine mounting bracket RH.
24. Remove the v-ribbed belt.
25. Remove the alternator assembly.
26. Remove the oil level gauge sub assembly.
27. Remove the oil level gauge guide.
28. Remove the manifold stay.
29. Remove the No. 2 manifold stay.
30. Remove the exhaust manifold converter sub assembly.
31. Remove the No. 2 camshaft.
32. Remove the camshaft.
33. Remove the camshaft timing oil control valve assembly.
34. Disconnect the radiator hose inlet.
35. Disconnect the engine wire.

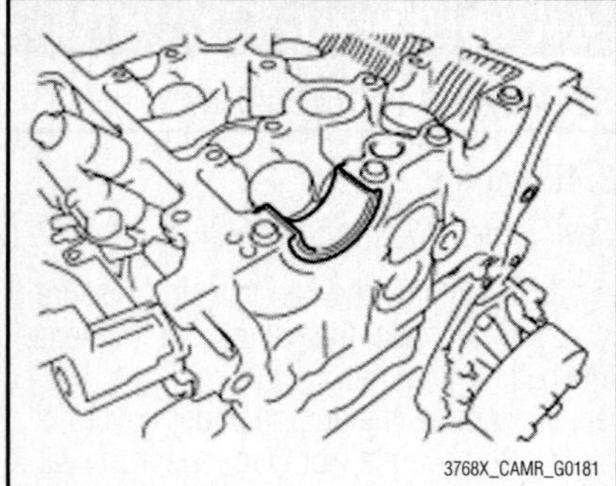

Fig. 136 Removing the No. 2 camshaft bearing

Fig. 138 Removing the cylinder head gasket

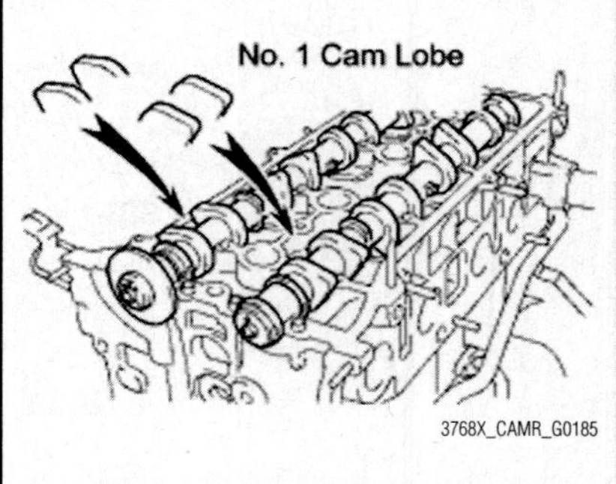

Fig. 140 Placing the camshafts on the cylinder head

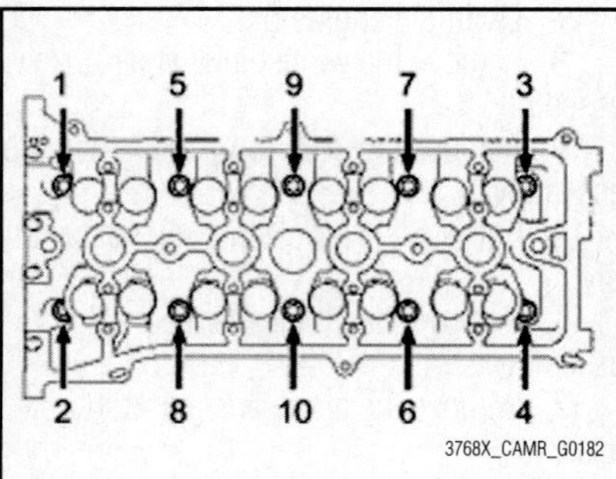

Fig. 137 Identifying the cylinder head bolt and plate washer removal sequence

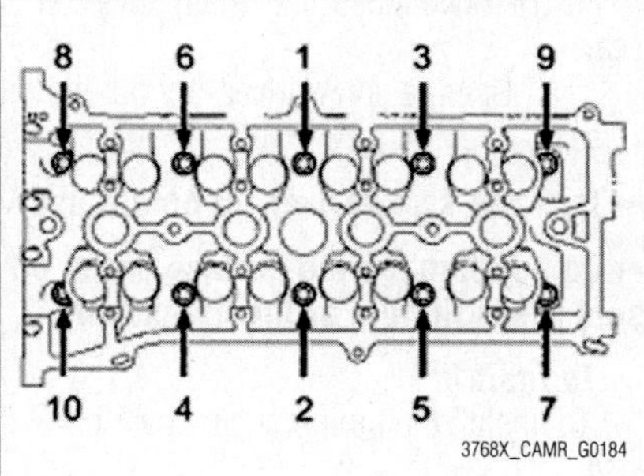

Fig. 139 Identifying the cylinder head set bolts and plate washer installation order

Fig. 141 Examining front marks and numbers

a. Disconnect the radio setting condenser connector.

b. Disconnect the engine oil pressure switch connector.

c. Disconnect the Engine Coolant Temperature (ECT) sensor connector.

d. Disconnect the camshaft position sensor connector.

e. Remove the bolt and ground cable.

36. Remove the No. 2 camshaft bearing.

37. Remove the cylinder head sub assembly.

a. Using several steps, uniformly loosen and remove the 10 cylinder head bolts and 10 plate washers with a 10 mm bi-hexagon wrench in the sequence shown in the illustration.

➡Head warpage or cracking could result from removing the bolts in the wrong order.

b. Using a screwdriver with its tip wrapped with tape, pry between the cylinder head and cylinder block, and remove the cylinder head.

➡Be careful not to damage the contact surfaces of the cylinder head and cylinder block.

38. Remove the cylinder head gasket.

To install:

39. Install the cylinder head gasket.

a. Place a new gasket on the cylinder block surface with the Lot No. stamp facing upward.

➡Remove any oil from the contact surfaces.

➡Make sure that the gasket is installed in the correct direction.

40. Install the cylinder head sub assembly.

a. Install the cylinder head on the cylinder block.

➡The cylinder head bolts are tightened in 2 progressive steps.

b. Apply a light coat of engine oil to the bolt threads and the area beneath the bolt heads that come in contact with the washers.

c. Install the bolts and plate washers to the cylinder head.

➡Do not drop the washers into the cylinder head.

d. Using several steps, uniformly install and tighten the 10 cylinder head set bolts and plate washers with a 10 mm bi-hexagon wrench in the order shown in the illustration. Tighten to 52 ft. lbs. (70 Nm).

e. Mark the front side of the cylinder head bolt with paint.

f. Retighten the cylinder head bolts 90°in the same sequence provided before.

g. Check that the paint mark is now at a 90°angle to the front.

41. Connect the engine wire.

a. Connect the ground cable with the bolt and tighten to 74 inch lbs. (8.4 Nm).

b. Connect the camshaft position sensor connector.

c. Connect the Engine Coolant Temperature (ECT) sensor connector.

d. Connect the engine oil pressure switch connector.

e. Connect the radio setting condenser connector.

42. Connect the radiator hose inlet.

43. Install the No. 2 camshaft bearing.

44. Install the camshaft timing oil control valve assembly.

45. Install the camshafts.

a. Apply a light coat of engine oil to the camshaft journals.

b. Place the 2 camshafts on the cylinder head with the No. 1 cam lobes facing the directions shown in the illustration.

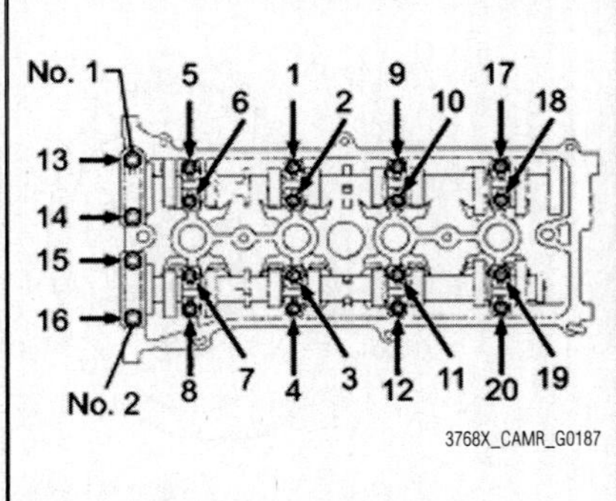

Fig. 142 Bearing cap installation sequence

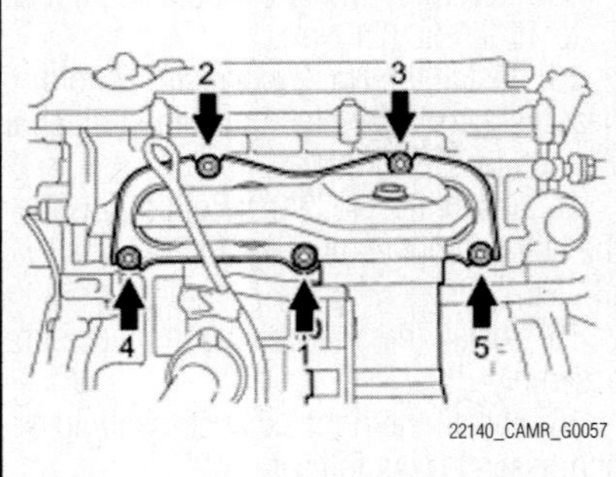

Fig. 144 Exhaust manifold tightening sequence—PZEV vehicles

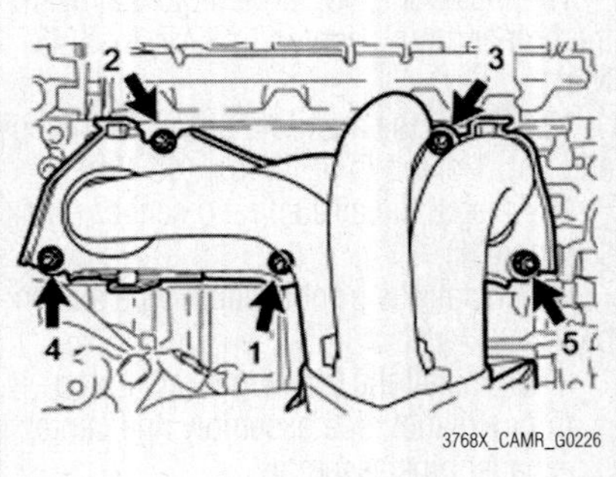

Fig. 145 Installing the exhaust manifold converter sub assembly

c. Examine the front marks and numbers, and check that the order is as shown in the illustration. Then install the bearing caps onto the cylinder head.

d. Apply a light coat of engine oil to the threads and under the heads of the bearing cap bolts.

e. Using several steps, uniformly tighten the 20 bearing cap bolts in the sequence shown in the illustration. Tighten the No. 1 and No. 2 bearing caps to 22 ft. lbs. (30 Nm). Tighten the No. 3 bearing cap to 80 inch lbs. (9 Nm).

46. To complete installation, reverse the remaining removal procedure.

EXHAUST MANIFOLD

REMOVAL & INSTALLATION

2.4L Engine

See Figures 143 and 144.

1. Before servicing the vehicle, refer to the Precautions Section.
2. Disconnect the negative battery cable.
3. Remove engine cover.
4. Remove air cleaner assembly.
5. Remove manifold stays.

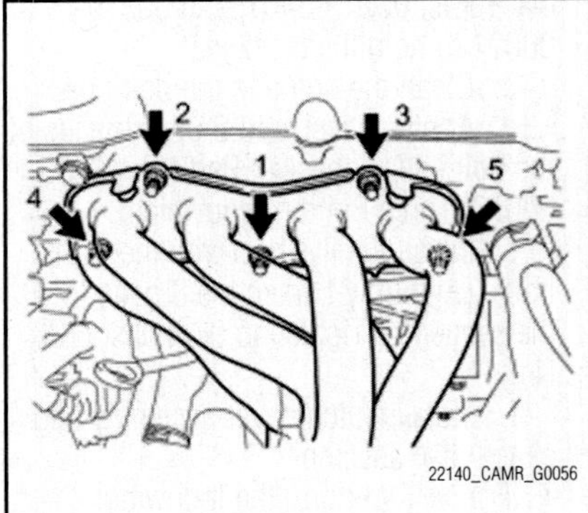

Fig. 143 Exhaust manifold tightening sequence—except PZEV vehicles

6. Disconnect the air-fuel ratio sensor connector.
7. Remove or disconnect remaining components from the exhaust manifold.
8. For PZEV vehicles, perform the following:
   a. Remove the five nuts, manifold converter and gasket.
9. For non-PZEV vehicles, perform the following:
   a. Remove the four bolts and insulator.
   b. Remove the five nuts, manifold converter and gasket.
10. Remove exhaust manifold from catalytic converter.

To install:

11. To install, reverse removal procedure.
12. Install new gaskets for exhaust manifolds.
13. For non-PZEV vehicles, tighten exhaust manifold bolts in sequence to: 27 ft. lbs. (37 Nm). Tighten the exhaust manifold heat insulator bolts to: 9 ft. lbs. (12 Nm).
14. For PZEV vehicles, tighten exhaust manifold bolts in sequence to: 27 ft. lbs. (37 Nm).
15. Tighten the exhaust manifold stays to 32 ft. lbs. (44 Nm).

2.5L Engine

See Figure 145.

1. Remove the front wheel opening extension pad RH.
2. Remove the front wheel opening extension pad LH.
3. Remove the engine under cover LH.
4. Remove the engine under cover RH.
5. Remove the front exhaust pipe assembly.
   a. Disconnect the connector.
   b. Remove the 2 bolts and support bracket.
   c. Remove the 2 nuts and bracket.
   d. Remove the 4 nuts, 2 bolts and front exhaust pipe assembly.
   e. Remove the 2 gaskets from the front exhaust pipe assembly and center exhaust pipe assembly.
6. Remove the air fuel ratio sensor (bank 1, sensor 1).
7. Remove the manifold stay.
8. Remove the No. 2 manifold stay.
9. Remove the No. 1 exhaust manifold heat insulator.
   a. Remove the 4 bolts and No. 1 exhaust manifold heat insulator.
10. Remove the exhaust manifold converter sub assembly.
    a. Remove the 5 nuts and exhaust manifold converter sub assembly.
11. Remove the No. 2 exhaust manifold heat insulator.
    a. Remove the 2 bolts and No. 2 exhaust manifold heat insulator.
12. Remove the No. 1 manifold converter insulator.
    a. Remove the 4 bolts and the No. 1 manifold converter insulator.

To install:

13. Install the No. 1 manifold converter insulator.
    a. Install the No. 1 manifold converter insulator with the 4 bolts. Tighten to 9 ft. lbs. (12 Nm).
14. Install the No. 2 exhaust manifold heat insulator with the 2 bolts. Tighten to 9 ft. lbs. (12 Nm).
15. Install the exhaust manifold converter sub assembly.
    a. Install a new gasket onto the cylinder head.
    b. Temporarily install the exhaust manifold converter sub assembly with the 5 nuts.
    c. Tighten the 5 nuts in the sequence shown. Tighten to 26 ft. lbs. (35 Nm).
16. Install the No. 1 exhaust manifold heat insulator with the 4 bolts. Tighten to 9 ft. lbs. (12 Nm).

17. Install the No. 2 manifold stay with the bolt and nut. Tighten to 32 ft. lbs. (43 Nm).
18. Install the manifold stay with the bolt and nut. Tighten to 32 ft. lbs. (43 Nm).
19. Install the air fuel ratio sensor (bank 1, sensor 1).
20. Install the front exhaust pipe assembly.
  a. Install the 2 new gaskets to the front exhaust pipe assembly and center exhaust pipe assembly.
  b. Install the front exhaust pipe with the 2 nuts to the exhaust manifold converter sub assembly. Tighten to 40 ft. lbs. (55 Nm).
  c. Install the front exhaust pipe with the 2 nuts and 2 bolts to the center exhaust pipe assembly. Tighten to 36 ft. lbs. (49 Nm).
  d. Install the bracket with the 2 nuts. Tighten to 24 ft. lbs. (33 Nm).
  e. Install the support bracket with the 2 bolts. Tighten to 24 ft. lbs. (33 Nm).
  f. Connect the connector.
21. Inspect for exhaust gas leaks.
22. Install the engine under covers.
23. Install the front wheel opening extension pads.

3.5L Engine

1. Before servicing the vehicle, refer to the Precautions Section.
2. Remove the engine assembly with transaxle.
3. Secure the engine.
4. Remove the ignition coil assembly.
5. Remove the right No. 2 engine mounting stay.
6. Remove the intake manifold.
7. Remove the right exhaust manifold sub-assembly, as follows:
  a. Uniformly loosen and remove the 6 nuts.
  b. Remove the manifold and gasket.
8. Remove the oil level gauge guide sub-assembly.
9. Remove the bolt, nut and No. 2 manifold stay.
10. Remove the 3 bolts and No. 2 exhaust manifold heat insulator.
11. Remove the left exhaust manifold sub-assembly, as follows:
  a. Uniformly loosen and remove the 6 nuts.
  b. Remove the manifold and gasket.

To install:

12. Install the left exhaust manifold sub-assembly, as follows:
  a. Install a new gasket.
  b. Install the left exhaust manifold sub-assembly with the 6 nuts and tighten to 15 ft. lbs. (21 Nm).
13. Install the No. 2 exhaust manifold heat insulator with the 3 bolts and tighten to 75 inch lbs. (8.5 Nm).
14. Install the No. 2 manifold stay with the bolt and nut and tighten to 25 ft. lbs. (34 Nm).
15. Install the oil level gauge guide sub-assembly.
16. Install the right exhaust manifold sub-assembly, as follows:
  a. Install a new gasket.
  b. Install the right exhaust manifold sub-assembly with the 6 nuts and tighten to 15 ft. lbs. (21 Nm).
17. Install the intake manifold.
18. Install the right No. 2 engine mounting stay.
19. Install the ignition coil assembly.
20. Install the engine assembly with transaxle.

FLYWHEEL/FLEXPLATE

REMOVAL & INSTALLATION

See Figures 146 through 149.

1. Before servicing the vehicle, refer to the Precautions Section.
2. Disconnect the negative battery cable.
3. To gain access to the flywheel or flexplate, remove the transaxle.
4. For A/T, perform the following:
  a. For TMMK made flexplate: use Special Tool: 09960-10010 to hold the crankshaft.
  b. For TMC made flexplate: use Special Tools: 09213-54015and 09330-00021 to hold the crankshaft.
  c. Remove the eight bolts, rear spacer, drive plate and front spacer.
5. For M/T, perform the following:
  a. Remove the clutch cover assembly and clutch disc assembly.

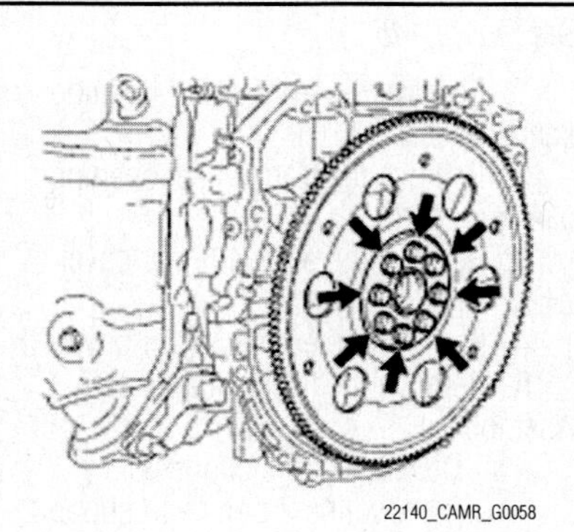

Fig. 146 Removing A/T flexplate bolts

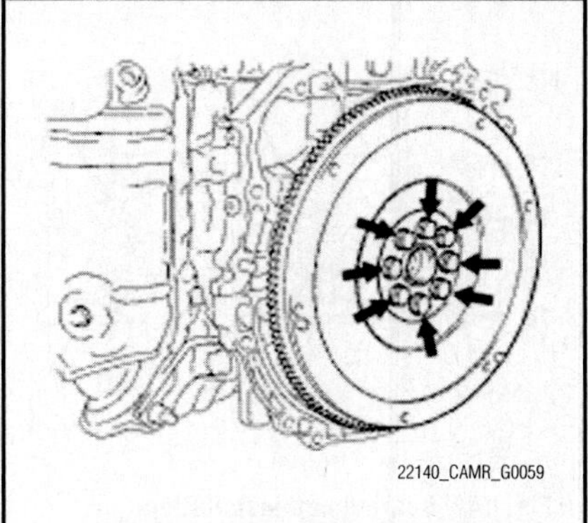

Fig. 147 Removing M/T flywheel bolts

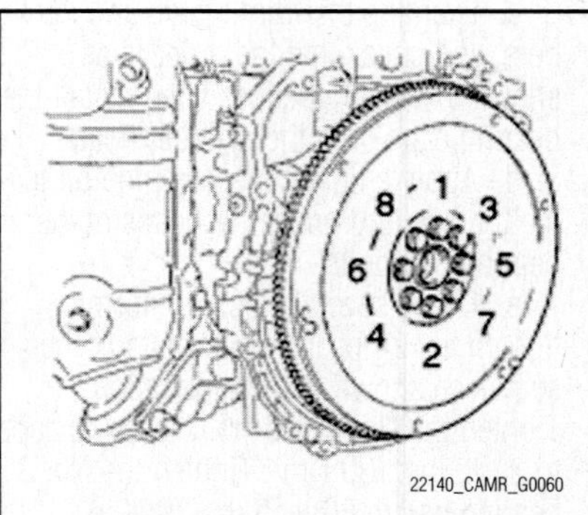

Fig. 148 M/T flywheel bolt tightening sequence

  b. For TMMK made flywheels: use Special Tool: 09960-10010 to hold the crankshaft.
  c. For TMC made flywheels: use Special Tools: 09213-54015and 09330-00021 to hold the crankshaft.
  d. Remove the eight bolts and flywheel.

To install:

6. For M/T, perform the following:
  a. For TMMK made flywheels: use Special Tool: 09960-10010 to hold the crankshaft.
  b. For TMC made flywheels: use Special Tools: 09213-54015and 09330-00021 to hold the crankshaft.
  c. Clean the bolt and the bolt hole.
  d. Apply adhesive to 2 or 3 threads of the bolt end. Adhesive: Part No. 08833-00070, Three Bond or equivalent.
  e. Install the flywheel with the eight bolts. Uniformly tighten the 8 bolts in the sequence. Tighten to 96 ft. lbs. (130 Nm).
  f. Install clutch cover assembly and clutch disc assembly.
7. For M/T, perform the following:
  a. For TMMK made flywheel: use Special Tool: 09960-10010 to hold the crankshaft.

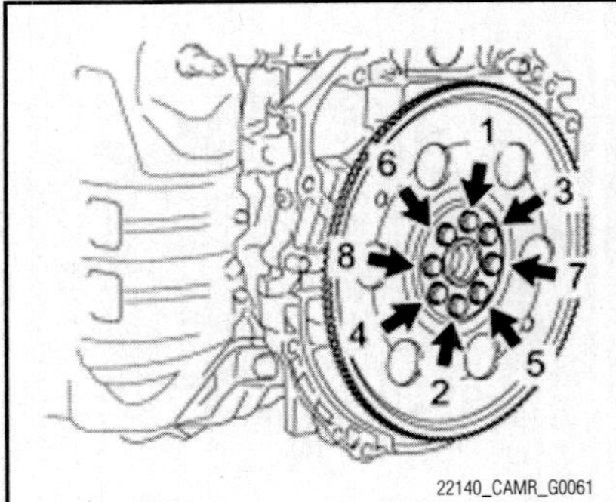

Fig. 149 A/T flywheel bolt tightening sequence

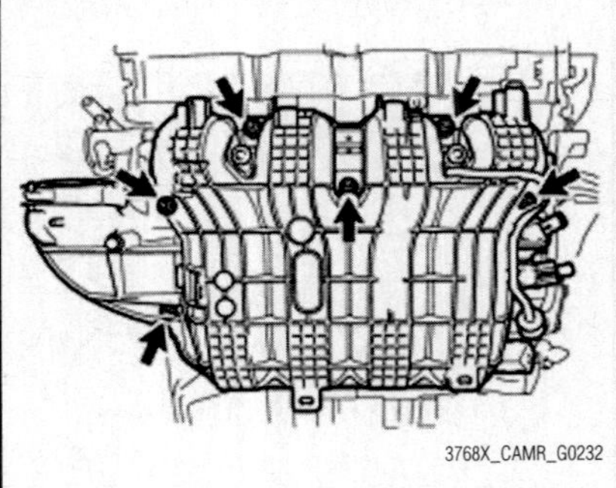

Fig. 151 Removing the intake manifold bolts

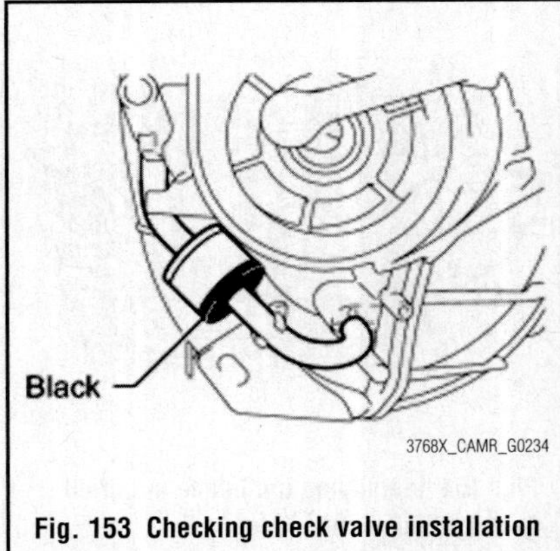

Fig. 153 Checking check valve installation

b. For TMC made flywheels: use Special Tools: 09213-54015and 09330-00021 to hold the crankshaft.

c. Clean the bolt and the bolt hole.

d. Apply adhesive to 2 or 3 threads of the bolt end. Adhesive: Part No. 08833-00070, Three Bond or equivalent.

e. Install the front spacer, drive plate and rear spacer with the 8 bolts. Uniformly tighten the eight bolts in the sequence. Tighten 72 ft. lbs. (98 Nm).

INTAKE MANIFOLD

REMOVAL & INSTALLATION

2.5L Engine

See Figures 150 through 154.

1. Drain engine coolant.
2. Remove the fuel delivery pipe sub assembly.
3. Remove the throttle body assembly.
4. Remove the vacuum switching valve assembly.
   a. Disconnect the 2 vacuum hoses, connector and clamp.
   b. Remove the bolt and vacuum switching valve assembly.
5. Disconnect the bolt and vacuum switching valve assembly.

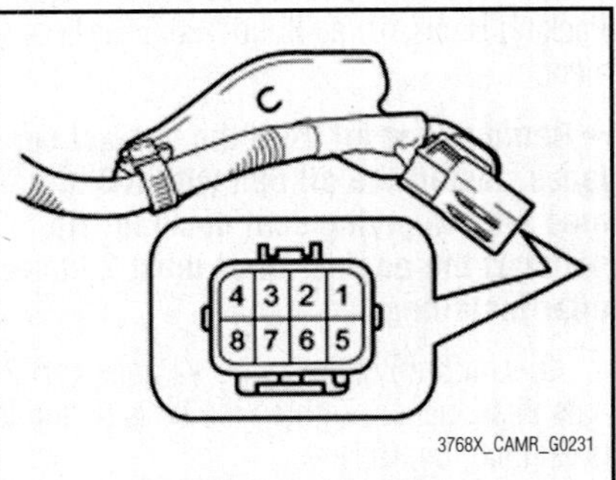

Fig. 150 Applying battery voltage to the terminals of the connector to close the tumble control valves

6. Disconnect the No. 2 ventilation hose from the intake manifold.
7. Remove the union to connector tube hose from the intake manifold.
8. Remove the No. 1 vacuum switching valve assembly (PZEV).
9. Remove the intake manifold.
   a. Disconnect the fuel vapor feed hose, 3 connectors and 8 clamps.
   b. Remove the 2 bolts and 2 wire harness brackets.
   c. Remove the bolt and wire harness bracket.
   d. Apply battery voltage to the terminals of the connector to close the tumble control valves.

Positive (+) battery voltage applied to terminal 8 (M-), and negative (-) battery voltage applied to terminal 4 (M+): Open → Closed

➡If this procedure is not performed, the valves may be damaged when the intake manifold is removed.

➡Apply battery voltage for 1 to 3 seconds.

➡If battery voltage is applied for more than 3 seconds, the actuator may be damaged.

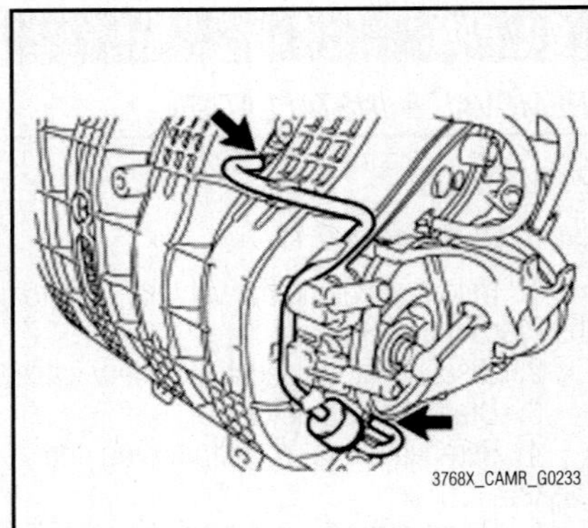

Fig. 152 Removing the check valve

➡Do not allow the lead wires to contact the other terminals.

a. Remove the bolt and separate the wire harness.

b. Detach the 2 clamps from the intake manifold and bracket.

c. Disconnect the intake air control valve actuator connector.

d. Remove the 6 bolts and intake manifold.

➡The tumble control valves may be damaged if they are not closed before installing the intake manifold.

➡Connect the battery to the terminals of the actuators to operate the motor and close the valves.

e. Remove the intake manifold gasket from the intake manifold.

10. Remove the check valve.
    a. Disconnect the 2 vacuum hoses from the intake manifold and remove the check valve.
11. Remove the wiring harness clamp bracket.

To install:

12. Install the wiring harness clamp bracket with the bolt. Tighten to 7 ft. lbs. (10 Nm).
13. Install the check valve.
    a. Connect the 2 vacuum hoses to the intake manifold to install the check valve.
    b. Check that the check valve is installed properly.
14. Install the intake manifold.
    a. Close the tumble control valves.

➡The tumble control valves may be damaged if they are not closed before installing the intake manifold.

➡Connect the battery to the terminals of the actuator to operate the motor and close the valves.

b. Install a new gasket to the intake manifold.

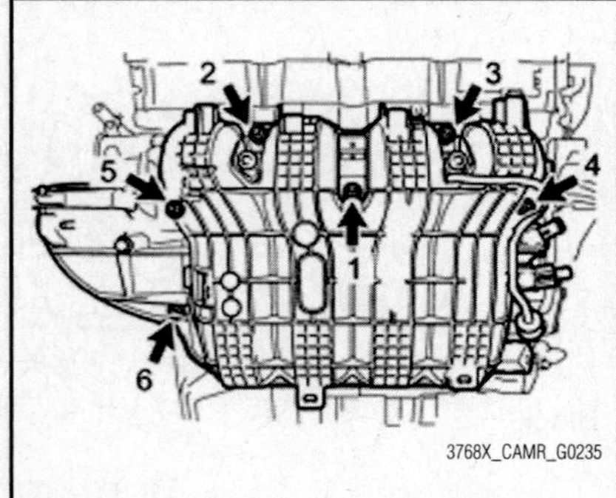

Fig. 154 Identifying the intake manifold bolt tightening sequence

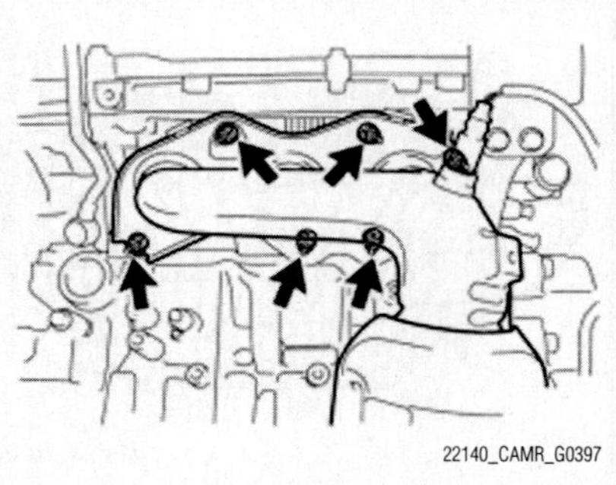

Fig. 155 Locating intake manifold bolts and nuts

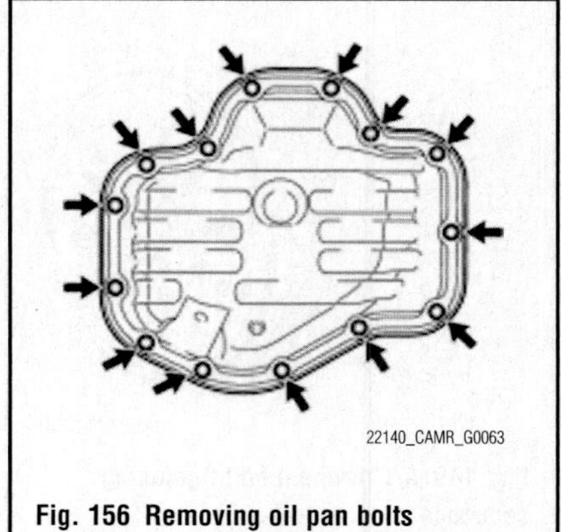

Fig. 156 Removing oil pan bolts

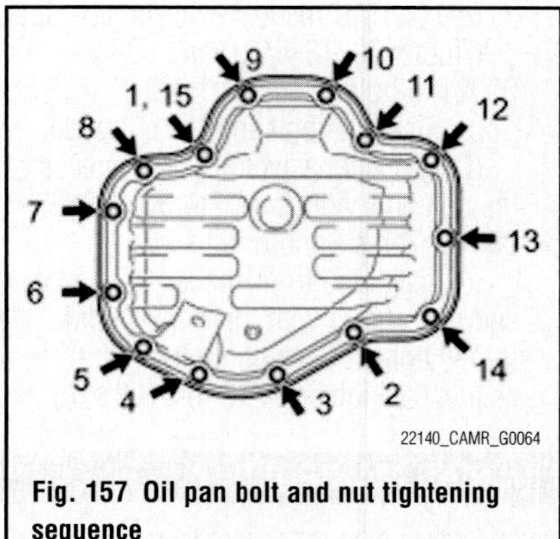

Fig. 157 Oil pan bolt and nut tightening sequence

c. Install the intake manifold by tightening the 6 bolts in the sequence shown. Tighten to 15 ft. lbs. (21 Nm).

d. Connect the intake air control actuator connector.

e. Attach the 2 clamps to the intake manifold and bracket.

f. Install the wire harness with the bolt. Tighten the bolt to 7 ft. lbs. (10 Nm).

g. Install the 2 wire harness brackets with the 2 bolts. Tighten the bolts to 7 ft. lbs. (10 Nm).

h. Connect the fuel vapor feed hose, 3 connectors and 8 clamps.

15. Install the No. 1 vacuum switching valve assembly (PZEV).
16. Install the union to the connector tube hose to the intake manifold.
17. Connect the No. 2 ventilation hose to the intake manifold.
18. Install the vacuum switching valve assembly (ACIS).

a. Install the vacuum switching valve with the bolt. Tighten to 80 inch lbs. (9 Nm).

b. Connect the 2 vacuum hoses, connector and clamp.

19. Install the throttle body assembly.
20. Install the fuel delivery pipe sub assembly.
21. Add engine coolant.
22. Inspect for coolant leaks.

3.5L Engine

See Figure 155.

1. Before servicing the vehicle, refer to the Precautions Section.
2. Remove the engine assembly with transaxle.
3. Secure the engine.
4. Remove the ignition coil assembly.
5. Remove the right No. 2 engine mounting stay.
6. Remove the intake manifold, as follows:

a. Uniformly loosen and remove the 6 bolts and 4 nuts.

b. Remove the intake manifold and 2 gaskets.

To install:

7. Install the intake manifold, as follows:

WARNING

DO NOTapplies oil to the intake manifold and cylinder head sub-assembly bolts.

a. Set a new gasket on each cylinder head.

➡Align the port holes of the gasket and cylinder head.

➡Make sure that the gasket is installed in the correct direction.

b. Set the intake manifold on the cylinder heads.

c. Install and tighten the 6 bolts and 4 nuts uniformly in several steps to 15 ft. lbs. (21 Nm).

8. Install the right No. 2 engine mounting stay.
9. Install the ignition coil assembly.
10. Install the engine assembly with transaxle.

OIL PAN

REMOVAL & INSTALLATION

2.4L Engine

See Figures 156 and 157.

1. Before servicing the vehicle, refer to the Precautions Section.
2. Disconnect the negative battery cable.
3. Drain the engine oil.
4. Remove the oil pan drain plug and gasket.
5. Remove the oil pan 12 bolts and 2 nuts.

➡Be careful not to damage the contact surfaces of the crankcase, chain cover and oil pan.

6. Insert the blade of Special Tool: 09032-00100 between the crankcase and oil pan. Cut through the sealer and remove the oil pan.

To install:

7. Remove any old packing material and be careful not to drop any oil on the contact surfaces of the cylinder block and oil pan.
8. Apply a continuous bead of seal packing (Diameter 0.118 to 0.157 inches (3.0 to 4.0 mm)). Use Toyota Genuine Seal Packing Block, Three Bond 1207B or Equivalent.

➡Remove any oil from the contact surfaces. Install the oil pan within 3 minutes after applying seal packing. Do not start the engine for at least 2 hours after installing.

9. Uniformly tighten the 12 bolts and 2 nuts in sequence. Tighten the bolts and nuts to 80 inch lbs. (9 Nm).
10. To complete installation, reverse remaining removal procedure.
11. Add oil and check for leaks.

2.5L Engine

See Figures 158 and 159.

1. Remove the oil filter cap assembly.
2. Remove the oil pan sub assembly.
   a. Remove the 11 bolts and 2 nuts.
   b. Insert the blade of an oil pan seal cutter between the oil and stiffening crankcase, cut off the applied sealer and remove the oil pan.

To install:

3. Install the oil pan sub assembly.
   a. Apply seal packing in a continuous line. Use Toyota Genuine Seal Packing Black, Three Bond 1207B or equivalent. The seal diameter should ne 0.0984-0.138 inch (2.5-3.5 mm).

➡Remove any foreign oil from the contact surface.

➡Install the oil pan within 3 minutes and tighten the bolts and nuts within 10 minutes after applying seal packing.

➡Do not apply oil for at least 4 hours after the installation.

   b. Install the oil pan with the 11 bolts and 2 nuts in several steps, in the sequence shown in the illustration. Tighten to 7 ft. lbs. (10 Nm).

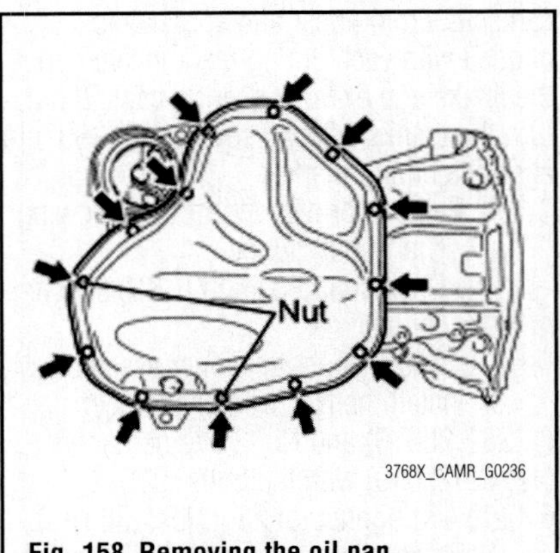

Fig. 158 Removing the oil pan

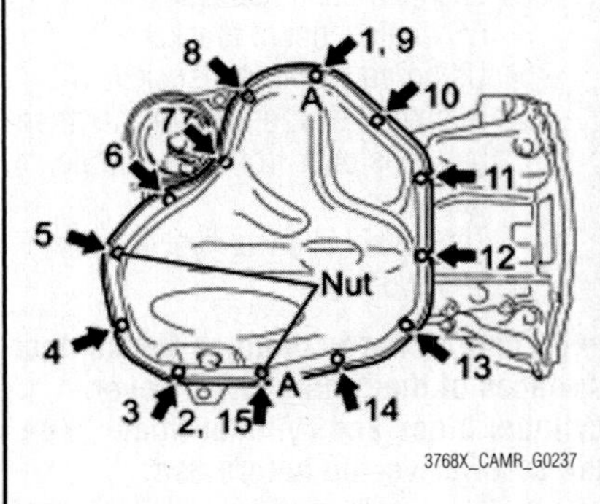

Fig. 159 Identifying the oil pan bolt installation pattern

➡Bolt A and nut A are tightened twice.

4. Install the oil filter cap assembly.

3.5L Engine

See Figures 160 through 162.

1. Before servicing the vehicle, refer to the Precautions Section.
2. Drain the engine oil.
3. Remove the engine assembly with transaxle.
4. Secure the engine.
5. Remove the oil filler cap and gasket.
6. Remove the oil pan drain plug and gasket.
7. Remove the oil pan drain plug and gasket.
8. Remove the No. 1 oil pipe, as follows:
   a. Remove the 2 oil pipe unions and oil pipe.
   b. Remove the left hand oil control valve filter and gaskets.
9. Remove the oil pipe, as follows:
   a. Remove the bolt.
   b. Remove the 2 oil pipe unions and oil pipe.
   c. Remove the right oil control valve filter and gaskets.
10. Remove the oil filter element, as follows:
   a. Remove the drain plug. Do not remove the O-ring.
   b. Connect the hose to the pipe.
   c. Insert the pipe with the hose into the oil filter cap.
   d. Make sure that the oil is completely drained and remove the pipe and O-ring.
   e. Using SST (SST: 09228-06501) or equivalent, remove the oil filter cap.
   f. Remove the oil filter element and O-ring from the oil filter cap. Do not use any tools when removing the O-ring to prevent the O-ring groove from being damaged.
11. Remove the No. 2 oil pan sub-assembly, as follows:
   a. Remove the 16 bolts and 2 nuts.
   b. Insert the blade of SST (SST: 09032-00100) or equivalent tool between the oil pans. Cut through the applied sealer and remove the No. 2 oil pan sub-assembly.

➡Be careful not to damage the contact surfaces of the oil pans.

12. Remove the oil pan sub-assembly, as follows:
   a. Remove the 16 bolts and 2 nuts.

➡Be sure to clean the bolts and stud bolts and check the threads for cracks or other damage.

   b. Remove the oil pan by prying between the oil pan and cylinder block with a taped screwdriver.

➡Be careful not to damage the contact surfaces of the cylinder block and oil pan.

   c. Remove the 2 O-rings.

To install:

13. Install the oil pan sub-assembly, as follows:
   a. Using an E8 Torx® socket wrench, install the stud bolts as shown in the illustration. Tighten to 7 ft. lbs (10 Nm).
   b. Install 2 new O-rings.

➡Remove any oil from the contact surface.

➡Install the oil pan within 3 minutes after applying seal packing.

➡Do not start the engine for at least 2 hours after installing.

   c. Apply seal packing (Toyota Genuine Seal Packing Black, Three Bond 1207B or equivalent) in a continuous line as shown in the illustration. Seal diameter: 3.0 to 4.0 mm (0.118 to 0.156 in.).

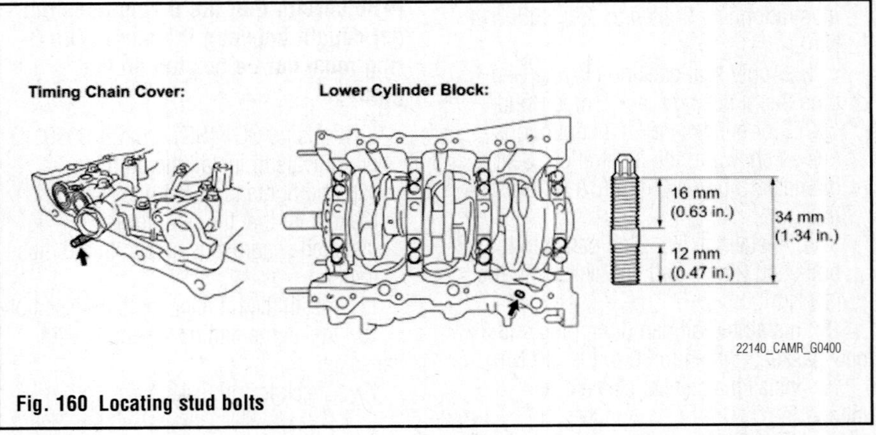

Fig. 160 Locating stud bolts

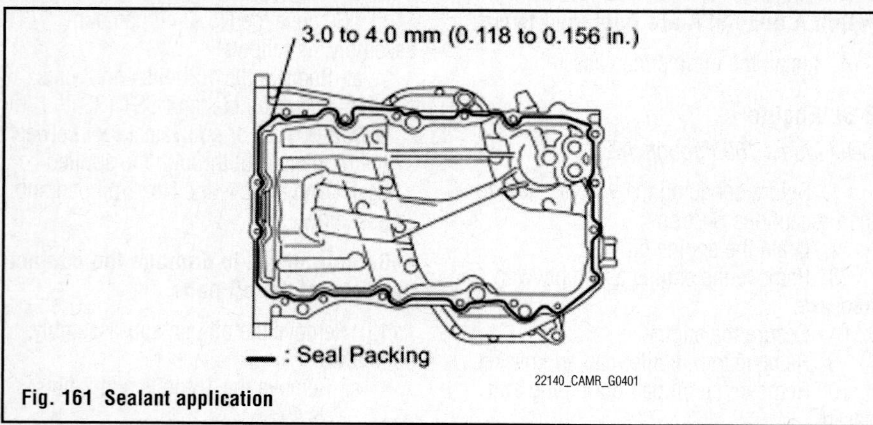

Fig. 161 Sealant application

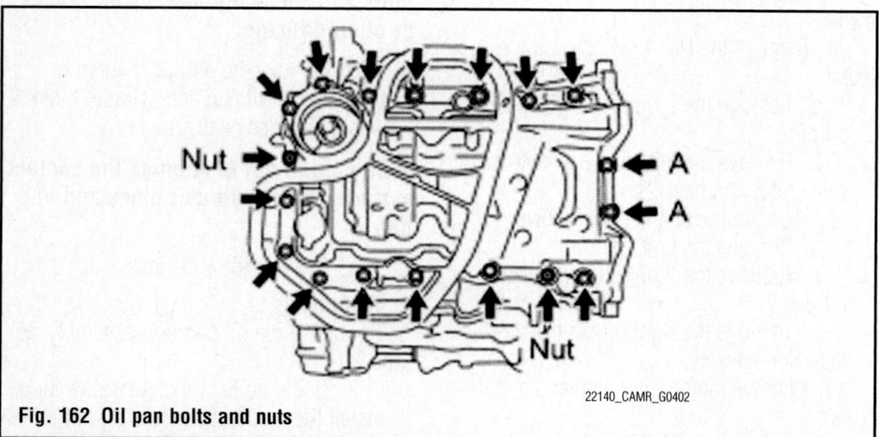

Fig. 162 Oil pan bolts and nuts

d. Install the oil pan with the 16 bolts and 2 nuts and tighten to 7 ft. lbs (10 Nm), and 15 ft. lbs (21 Nm).

14. Install the No. 2 oil pan sub-assembly, as follows:

➡**Remove any oil from the contact surface.**

➡**Install the No. 2 oil pan within 3 minutes after applying seal packing.**

➡**Do not start the engine for at least 2 hours after installing.**

a. Using an E6 Torx® socket wrench, install the stud bolts as shown in the illustration and tighten to 35 inch lbs (4 Nm).

b. Apply seal packing (Toyota Genuine Seal Packing Black, Three Bond 1207B or equivalent) in a continuous line as shown in the illustration. Seal diameter: 3.0 to 4.0 mm (0.118 to 0.156 in.).

c. Install the No. 2 oil pan with the 16 bolts and 2 nuts and tighten to 7 ft. lbs (10 Nm).

15. Install the oil pan drain plug and a new gasket. Tighten to 30 ft. lbs (40 Nm).

16. Install the oil filter element, as follows:

a. Clean the inside of the oil filter cap, the threads and O-ring groove.

b. Apply a small amount of engine oil to a new O-ring and install it to the oil filter cap.

c. Set a new oil filter element to the oil filter cap.

d. Remove dirt or foreign matter from the installation surface and inside of the engine.

e. Apply a small amount of engine oil to the O-ring again and install the oil filter cap.

➡**Be careful that the O-ring does not get caught between the parts. The O-ring must not be twisted on the groove.**

f. Using SST (SST: 09228-06501) or equivalent, install the oil filter cap and tighten to 18 ft. lbs (25 Nm). Make sure that the oil filter is installed securely as shown in the illustration.

17. Install the oil filler cap sub-assembly.

18. Install the engine assembly with transaxle.

19. Check for oil leaks.

OIL PUMP

REMOVAL & INSTALLATION

2.4L Engine

See Figure 163.

1. Before servicing the vehicle, refer to the Precautions Section.
2. Disconnect the negative battery cable.
3. Drain the engine oil.
4. Remove or disconnect the following:
   - Plastic engine cover and two nuts
   - Front right wheel
   - LH and RH engine under cover
   - RH front fender apron seal
   - Front exhaust pipe assembly
   - RH engine mount stay
   - Engine mount
   - RH engine mounting bracket
   - Drive belt
   - Alternator assembly
   - Vane pump
   - Ignition coil assembly
   - Ventilation hoses
   - Valve cover
5. Turn the crankshaft pulley until its groove and the timing mark "0" of the timing chain cover are aligned.
6. Check that each timing mark of the camshaft timing gear and sprocket is aligned with each timing mark located on the intake and exhaust bearing caps. If not, turn the crankshaft by 1 revolution (360°) to align the timing marks.
7. Remove or disconnect the following:
   - Crankshaft pulley
   - Crankshaft Position (CKP) sensor
   - Oil pan
   - Chain upper tensioner assembly
8. Install the No. 1 engine hanger (12281-28010) and No. 2 engine hanger (12282-28010) with the bolts (91512-61020) and tighten to: 28 ft. lbs. (38 Nm).
9. Remove or disconnect the following:
   - Drive belt tensioner
   - Engine mount insulator
   - RH engine mount bracket
   - Using an E10 Torx® socket, remove the stud bolt for the drive belt tensioner from the cylinder block
   - Timing chain cover twelve bolts and two nuts

➡**Be careful not to damage the contact surfaces of the timing chain cover, cylinder block and cylinder head. Tape the screwdriver tip before use.**

   - Timing chain cover by prying between the timing chain cover and

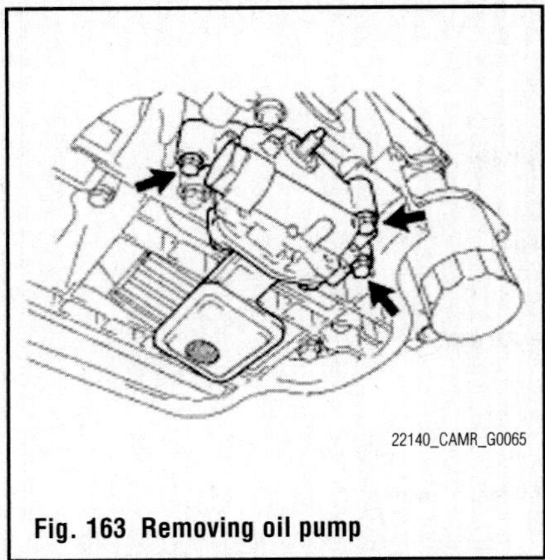
Fig. 163 Removing oil pump

cylinder head or cylinder block with a screwdriver

➡Tape the screwdriver tip before use.

- Using a screwdriver and a hammer, tap out the timing chain case oil seal
- Crankshaft Position (CKP) sensor plate
- Chain tensioner slipper
- Chain vibration damper
- Timing chain guide
- Upper chain assembly
- Crankshaft timing sprocket
- Lower chain assembly
- Three bolts, oil pump and gasket

To install:

10. Install a new gasket and the oil pump with the 3 bolts. Tighten the bolts 14 ft. lbs. (19 Nm).
11. To complete installation, reverse removal procedure. Refer to the appropriate sections to install components correctly.

2.5L Engine

See Figures 164 through 167.

➡Do not remove the oil pump or oil pump relief valve from the timing chain cover sub assembly.

1. Remove the engine and transaxle.
2. Remove the engine wire.
3. Remove the ignition coil assembly.
4. Remove the cylinder head cover sub assembly.
5. Remove the Crankshaft Position (CKP) sensor.
6. Remove the crankshaft pulley.
7. Remove the No. 2 timing chain cover.
   a. Remove the 2 bolts and No. 2 timing chain cover.
8. Remove the 5 bolts engine mounting bracket RH.

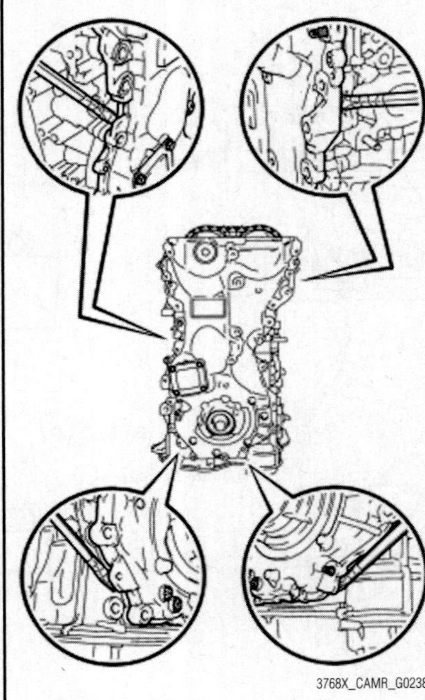
Fig. 164 Removing the timing chain cover

9. Remove the v-ribbed belt tensioner assembly (compressor and alternator).
10. Remove the v-ribbed belt tensioner assembly (vane pump).
11. Remove the timing chain cover sub assembly.
    a. Remove the 17 bolts and 2 nuts.
    b. Remove the timing chain cover by prying between the timing chain cover and the cylinder head, camshaft housing, cylinder block and stiffening crankcase with a screwdriver.

➡Be careful not to damage the contact surfaces of the cylinder head, camshaft housing, cylinder block, stiffening crankcase or chain cover.

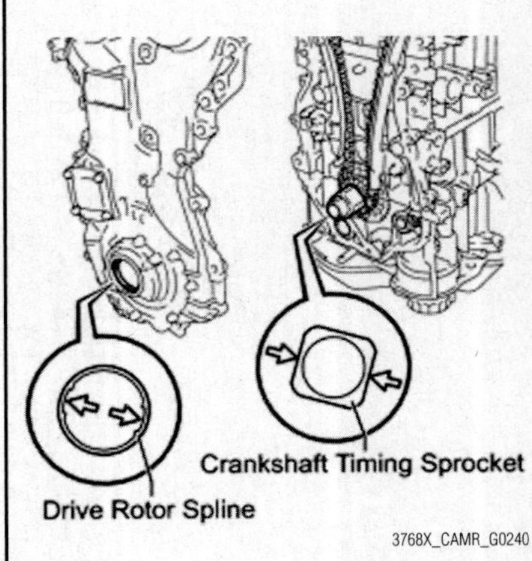

Fig. 166 Aligning the drive rotor spline and the crankshaft timing sprocket

➡Tape the screwdriver tip before use.

   c. Remove the 3 gaskets from the stiffening crankcase.

12. Remove the timing chain cover oil seal.
    a. Using a screwdriver and wooden block, pry out the oil seal.

➡Do not damage the surface of the oil seal press fit hole.

13. Tape the screwdriver tip before use.

To install:

14. Install the timing chain cover sub assembly.
    a. Apply a light coat of engine oil to 3 new gaskets.
    b. Install the 3 gaskets to the stiffening crankcase.
    c. Align the drive rotor spline and the crankshaft timing sprocket.
    d. Apply seal packing in a line to the timing chain cover.

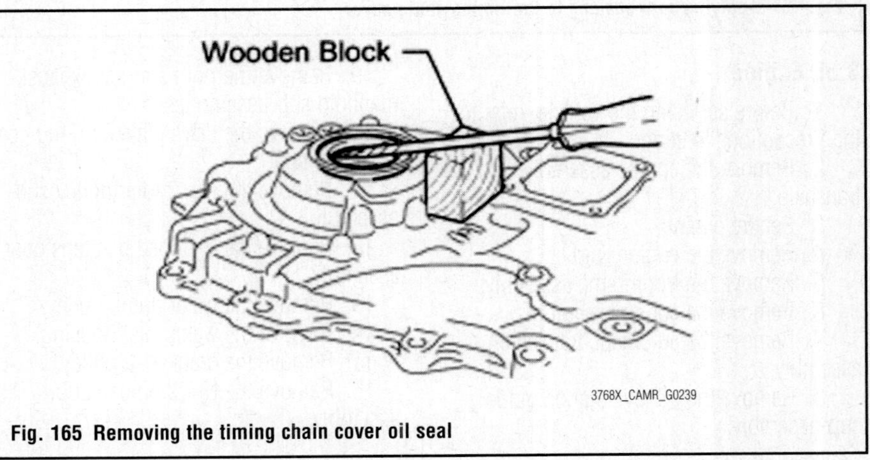

Fig. 165 Removing the timing chain cover oil seal

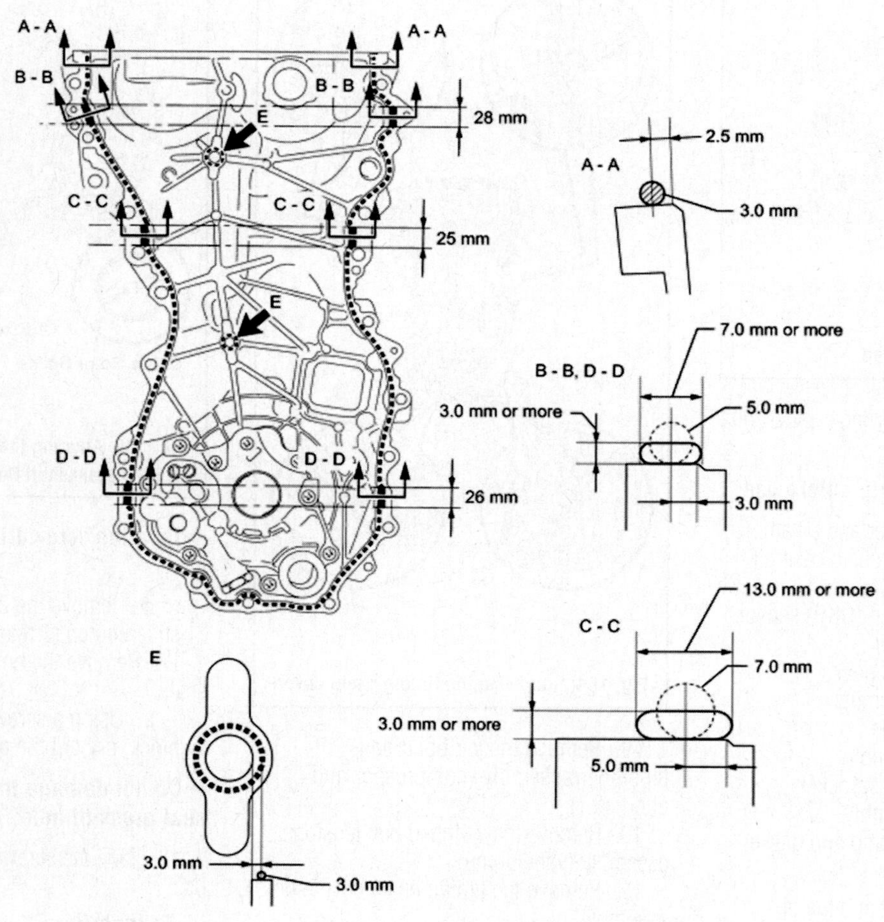

Area: Seal Packing Diameter (round)/Seal Pacing Dimension (flat)
Distance From Edge Of Cover to Center of Seal Packing: Seal Packing Application Length

Dashed line: 0.118 inch (3 mm)/ -: 0.0984 inch (2.5 mm): -
A-A: 0.118 inch (3 mm)/ -: 0.984 inch (2.5 mm): -
B-B: 0.197 inch (5 mm)/0.276 inch (7 mm) or more wide and 0.118 inch (3 mm) or more thick: 0.118 inch (3 mm): 1.10 inch (28 mm)
C-C: 0.276 inch (7 mm)/0.512 inch (13 mm) or more wide and 0.118 inch (3 mm) or more thick: 0.197 inch (5 mm): 0.984 inch (25 mm)
D-D: 0.197 inch (5 mm)/0.276 inch (7 mm) or more wide and 0.118 inch (3 mm) or more thick: 0.118 inch (3 mm): 1.02 inch (26 mm)
E: 0.118 inch (3 mm)/ -: 0.118 inch (3 mm): -

3768X_CAMR_G0241

Fig. 167 Applying seal packing to the timing chain cover

3.5L Engine

1. Before servicing the vehicle, refer to the Precautions Section.
2. Remove the engine assembly with transaxle.
3. Secure engine.
4. Remove the engine wire.
5. Remove the front frame assembly.
6. Remove the starter assembly.
7. Remove the automatic transaxle assembly.
8. Remove the oil level gauge guide sub-assembly.
9. Remove the right and left exhaust manifold sub-assemblies.
10. Remove the drive plate and ring gear sub-assembly.
11. Remove the No. 2 idler pulley sub-assembly.
12. Remove the V-ribbed belt tensioner assembly.
13. Remove the water pump pulley.
14. Remove the water inlet housing.
15. Remove the crankshaft pulley.
16. Remove the No. 2 oil pan sub-assembly.
17. Remove the oil strainer sub-assembly.
18. Remove the oil pan sub-assembly.
19. Remove the intake air surge tank assembly.
20. Remove the ignition coil assembly.
21. Remove the No. 1 and 2 oil pipes.
22. Remove the right and left cylinder head cover sub-assemblies.
23. Remove the timing chain or belt cover sub-assembly.
24. Remove the timing gear case or timing chain case oil seal, as follows:

a. Using a screwdriver with the tip taped, pry out the oil seal.

To install:

25. Install timing gear case or timing chain case oil seal, as follows:

a. Using SST (SST: 09316-60011) or equivalent tool, tap in a new oil seal until its surface is flush with the timing chain case edge.

➡Keep the lip free from foreign matter.

➡Do not tap on the oil seal at an angle.

➡Make sure that the oil seal edge does not stick out of the timing chain case.

b. Apply MP grease to the oil seal lip.

26. Install timing chain or belt cover sub-assembly.

27. Install the right and left cylinder head cover sub-assemblies.

28. Install the No. 1 and 2 oil pipes.

29. Install the ignition coil assembly.

30. Install the intake air surge tank assembly.

31. Install the oil pan sub-assembly.

32. Install the oil strainer sub-assembly.

33. Install the No. 2 oil pan sub-assembly.

34. Install the crankshaft pulley.

35. Install the water inlet housing.

36. Install the water pump pulley.

37. Install the V-ribbed belt tensioner assembly.

38. Install the No. 2 idler pulley sub-assembly.

39. Install the drive plate and ring gear sub-assembly.

40. Install the right and left exhaust manifold sub-assemblies.

41. Install the oil level gauge guide sub-assembly.

42. Install the automatic transaxle assembly.

43. Install the starter assembly.

44. Install the front frame assembly.

45. Install the engine wire.

46. Install the engine assembly with transaxle.

PISTON AND RING

POSITIONING

See Figure 168.

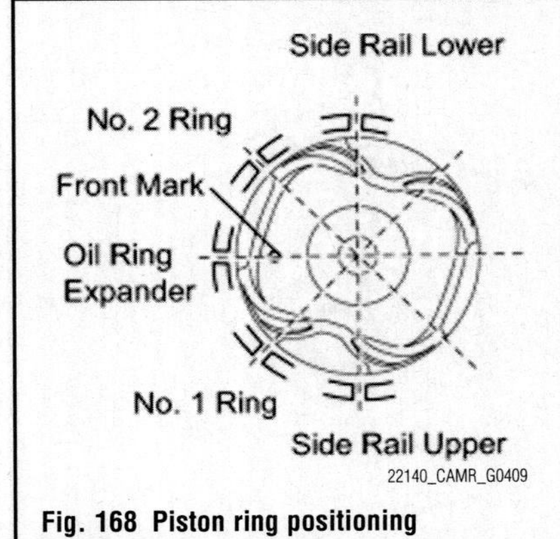

Fig. 168 Piston ring positioning

REAR MAIN SEAL

REMOVAL & INSTALLATION

2.4L Engine

See Figures 169 and 170.

1. Before servicing the vehicle, refer to the Precautions Section.

2. Disconnect the negative battery cable.

3. For A/T vehicles, separate the transaxle. Refer to the Drive Train section.

4. For M/T vehicles, separate the transaxle. Refer to the Drive Train section.

5. For A/T vehicles, remove the flex-plate. Refer to Flexplate and Flywheel above.

6. For M/T vehicles, remove the fly-wheel. Refer to Flexplate and Flywheel above.

7. Using a knife, cut through the oil seal lip.

8. Using a screwdriver with its tip taped, pry out the oil seal.

9. After the removal, check the crank-shaft for damage. If it is damaged, smooth the surface with 400-grit sandpaper.

To install:

➡Keep the lip free from foreign matter.

10. Apply MP grease to a new oil seal lip.

11. Using Special Tools: 09223-15030, 09950-70010 and a hammer, tap in the oil seal until its surface is flush with the rear oil seal retainer edge. Wipe off extra grease from the crankshaft.

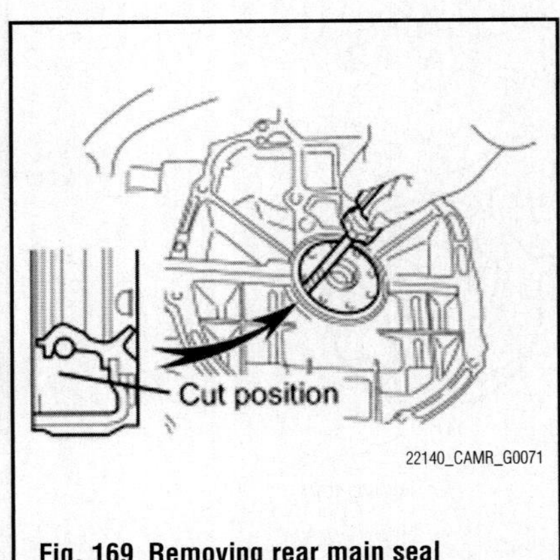

Fig. 169 Removing rear main seal

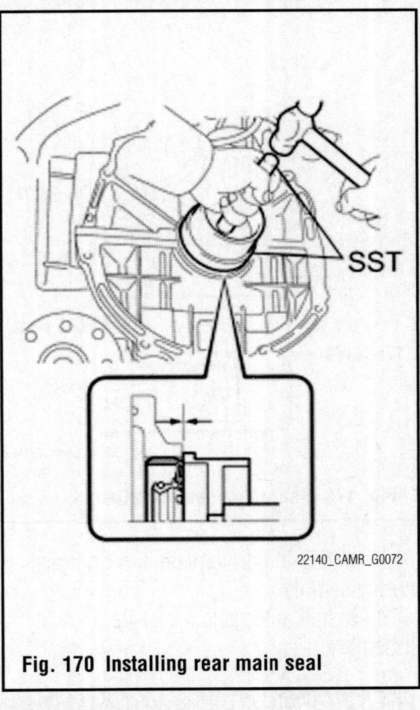

Fig. 170 Installing rear main seal

12. To complete installation, reverse remaining removal procedure.

3.5L Engine

See Figures 171 and 172.

1. Before servicing the vehicle, refer to the Precautions Section.

2. Remove the automatic transaxle assembly.

3. Remove the drive plate and ring gear sub-assembly.

4. Remove the rear main seal, as follows:

a. Using a knife, cut off the oil seal lip.

To install:

5. Apply MP grease to a new oil seal lip.

6. Using SST (SST: 09223-15030, SST: 09950-70010) or equivalent and a hammer, tap in the oil seal. Oil seal tap in depth: -0.020 to 0.020 in. (-0.5 to 0.5 mm)

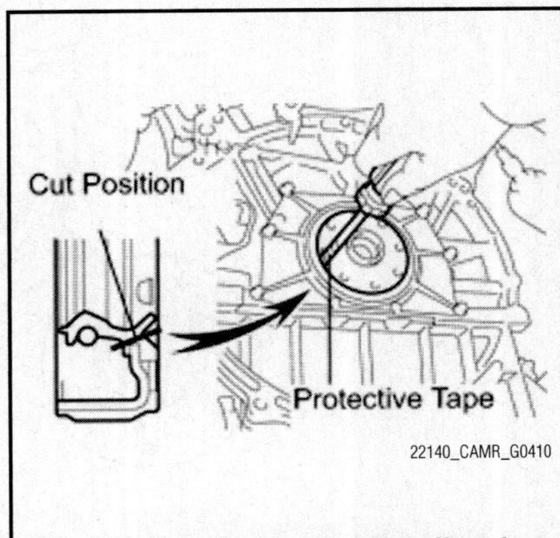

Fig. 171 Cut off and pry out the oil seal

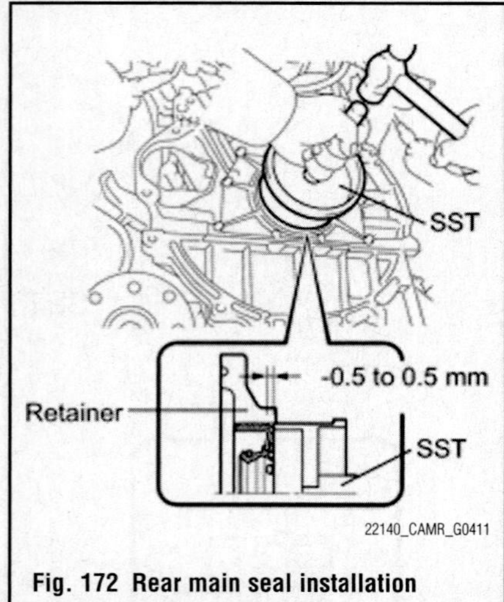

Fig. 172 Rear main seal installation

7. Install the drive plate and ring gear sub-assembly.
8. Install automatic transaxle assembly.

ROCKER ARMS

REMOVAL & INSTALLATION

For 3.5L engines, see Camshafts.

TIMING CHAIN FRONT COVER

REMOVAL & INSTALLATION

2.4L Engine

See Figures 173 through 177

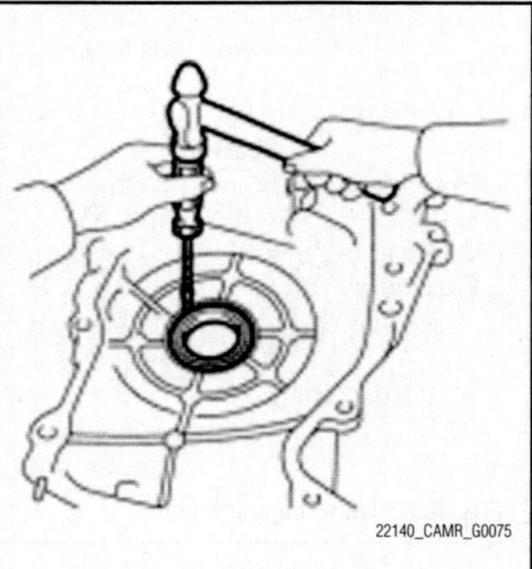

Fig. 174 Removing timing chain case oil seal

1. Before servicing the vehicle, refer to the Precautions Section.
2. Using a E10 Torx® socket, remove the stud bolt for the drive belt tensioner from the cylinder block.
3. Remove the 12 bolts and 2 nuts.

➡Be careful not to damage the contact surfaces of the timing chain cover, cylinder block and cylinder head.

4. Remove the timing chain cover by prying between the timing chain cover and cylinder head or cylinder block with a screwdriver. Tape the screwdriver tip before use.
5. Using a screwdriver and a hammer, tap out the oil seal.

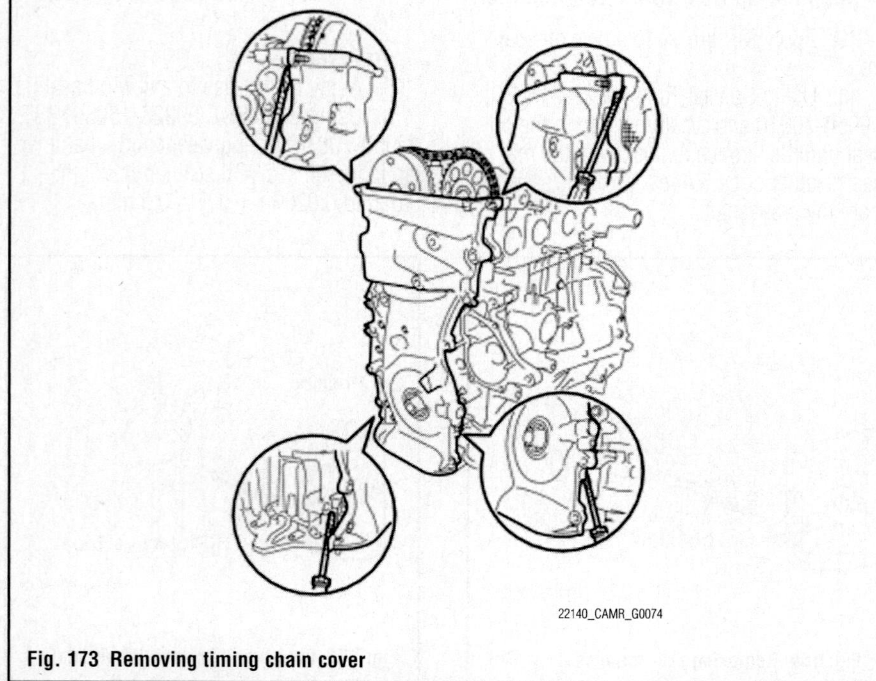

Fig. 173 Removing timing chain cover

To install:

➡Keep the gap between the timing chain cover edge and the oil seal free of foreign matter.

6. Using Special Tool: 09223-22010, tap in a new oil seal until its surface is flush with the timing chain cover edge. Apply a light coat of MP grease to the lip of the oil seal.
7. Remove any old packing (FIPG) material and be careful not to drop any oil on the contact surfaces of the timing chain cover, cylinder head and cylinder block.
8. Apply Toyota Genuine Seal Packing Black, Three Bond 1207B or Equivalent seal packing in a diameter of 0.157 to 0.177 inches (4.0 to 4.5 mm).
9. Apply seal packing in a continuous bead as shown in the illustration below.

➡Remove any oil from the contact surface. Install the chain cover within 3 minutes after applying seal packing. Do not start the engine for at least 2 hours after installing.

10. Install the timing chain cover with the twelve bolts and two nuts in sequence to the following torque specification:
    - Bolt A length: 1.18 inches (30 mm) for 10 mm head: 80 inch lbs. (9 Nm)
    - Bolt B length: 1.18 inches (30 mm) for 12 mm head: 18 ft. lbs. (25 Nm)
    - Bolt C length: 1.57 inches (40 mm) for 14 mm head: 41 ft. lbs. (55 Nm)
    - Nut: 8 ft. lbs. (11 Nm)
11. Using a E10 Torx® socket, install the stud bolt to the drive belt tensioner and tighten to 16 ft. lbs. (22 Nm).

2.5L Engine

For removal and installation procedures, refer to TIMING CHAIN & SPROCKETS.

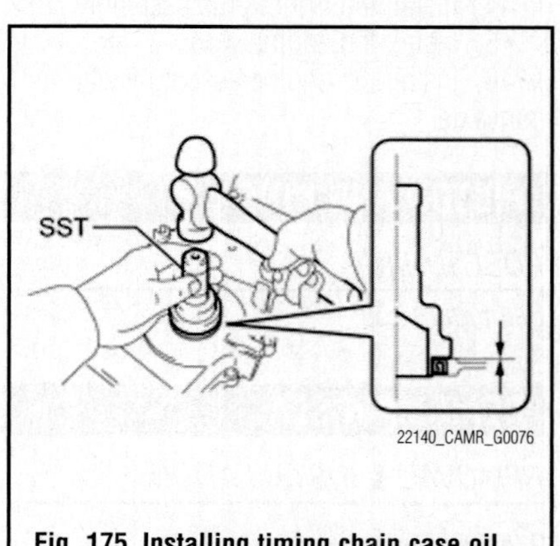

Fig. 175 Installing timing chain case oil seal

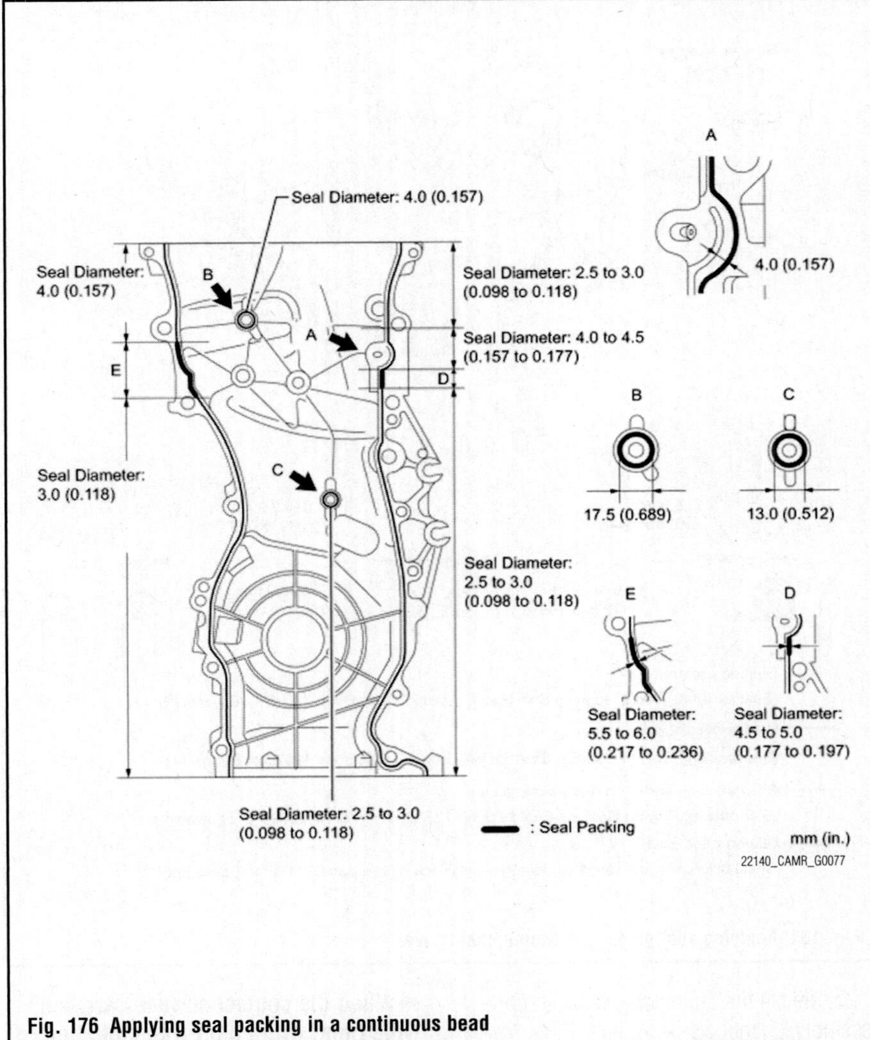

Fig. 176 Applying seal packing in a continuous bead

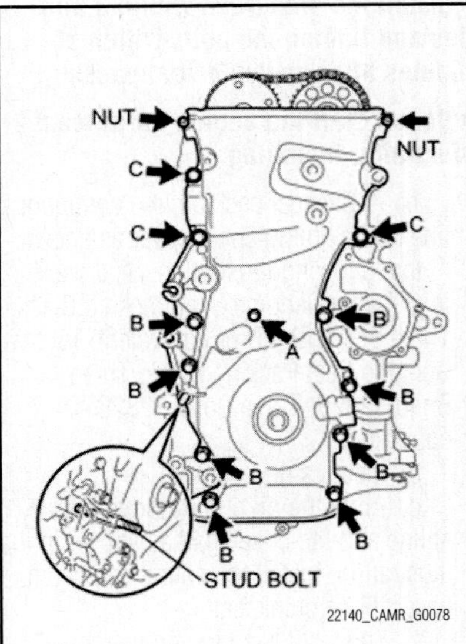

Fig. 177 Installing the twelve bolts and two nuts

3.5L Engine

See Figures 178 through 183.

1. Before servicing the vehicle, refer to the Precautions Section.
2. Remove the engine assembly with transaxle.
3. Secure engine.
4. Remove the oil filler cap sub-assembly.
5. Remove the spark plugs and ignition coil assembly.
6. Remove the oil pan drain plug and gasket.
7. Remove the ventilation valve sub-assembly.
8. Remove the camshaft position sensor.
9. Remove the camshaft timing oil control valve assembly.
10. Remove Crankshaft Position (CKP) sensor.
11. Remove the No. 1 oil pipe.
12. Remove the oil pipe.
13. Remove the cylinder block water drain cock sub-assembly.
14. Remove the oil filter.
15. Remove the crankshaft pulley.
16. Remove the left hand No. 1 front engine mounting bracket.
17. Remove the water inlet housing.
18. Remove the water outlet.
19. Remove the left-hand cylinder head cover sub-assembly and gasket.
20. Remove the cylinder head cover sub-assembly and gasket.
21. Remove the No. 2 oil pan sub-assembly.
22. Remove the oil strainer sub-assembly.
23. Remove the oil pan sub-assembly.
24. Remove the water pump assembly.
25. Remove the timing chain cover sub-assembly, as follows:
    a. Remove the 15 bolts and 2 nuts as shown in the illustration.

➡Be careful not to damage the contact surfaces of the cylinder head, cylinder block and chain cover.

    b. Remove the timing chain cover by prying between the timing chain cover and cylinder head or cylinder block with a screwdriver with the tip taped.
    c. Remove the 4 bolts, chain cover plate and gasket.
    d. Remove the gasket.

To install:

26. Install timing gear case or timing chain cover oil seal, as follows:

➡Keep the lip free from foreign matter.

➡Do not tap on the oil seal at an angle.

➡Make sure that the oil seal edge does not stick out of the timing chain cover.

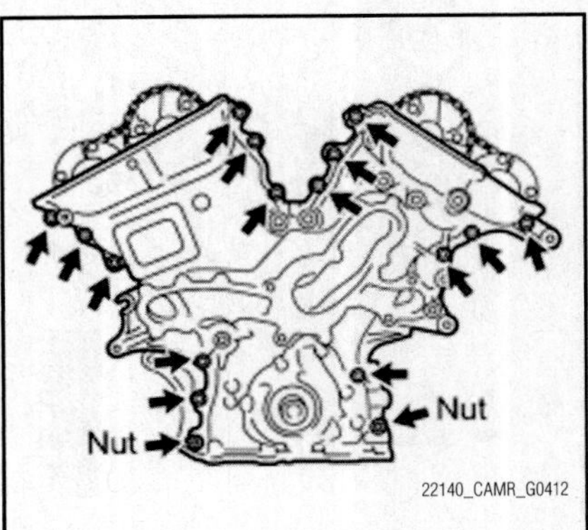

Fig. 178 Locating timing chain cover sub-assembly bolts and nuts

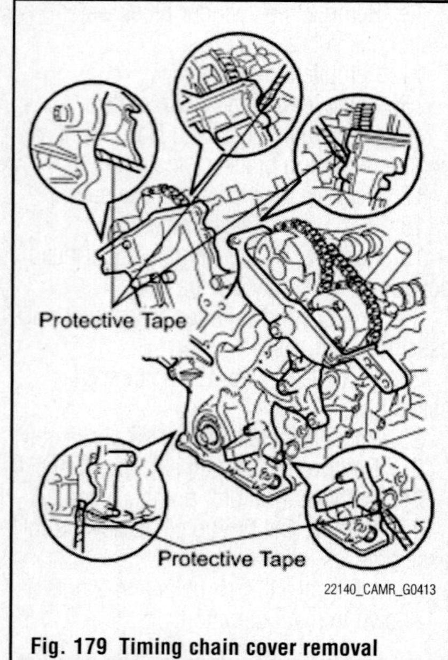

Fig. 179 Timing chain cover removal

a. Apply MP grease to a new oil seal lip.

b. Using SST (SST: 09316-60011) and a hammer, tap in the oil seal until its surface is flush with the timing chain cover edge.

➡**Be sure to clean and degrease the contact surfaces, especially the surfaces indicated by C in the illustration.**

➡**When the contact surfaces are wet, wipe them with an oil-free cloth before applying seal packing.**

➡**Install the chain cover within 3 minutes after applying seal packing.**

➡**Do not start the engine for at least 2 hours after installing.**

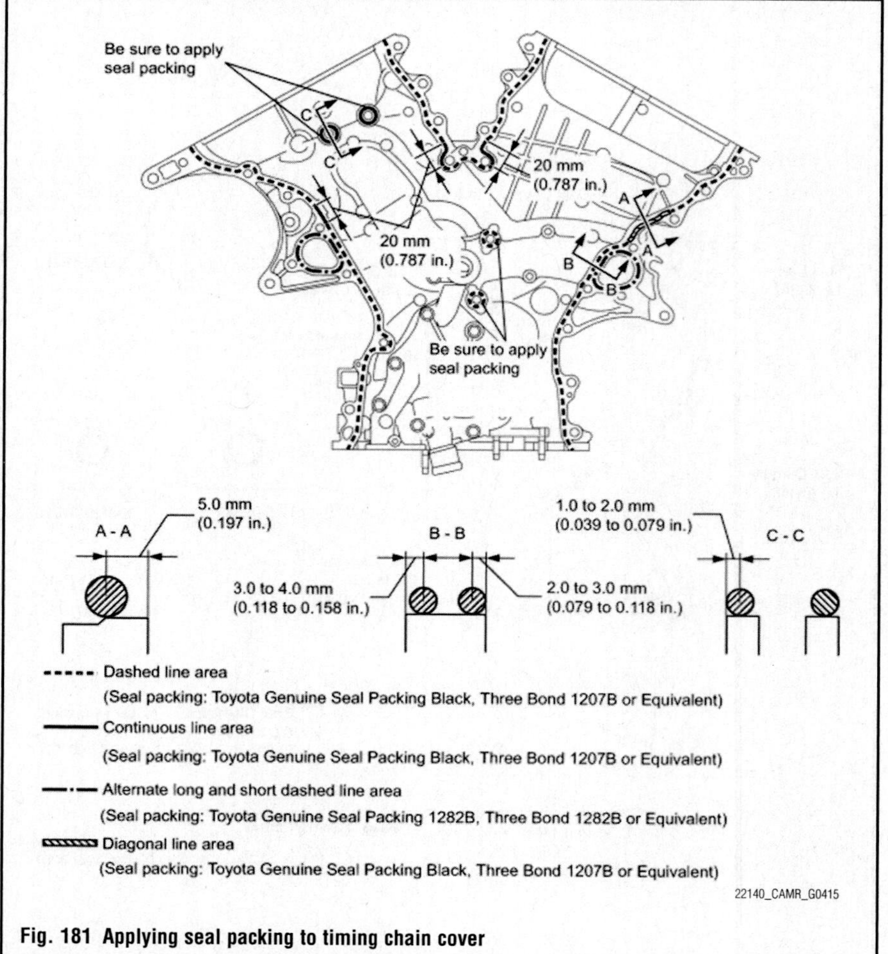

Fig. 181 Applying seal packing to timing chain cover

27. Install the timing chain cover sub-assembly, as follows:

a. Apply seal packing (Toyota Genuine Seal Packing Black, Three Bond 1207B or equivalent) in a continuous line to the engine unit as shown in the illustration. Seal diameter: 3.0 mm (0.118 in.).

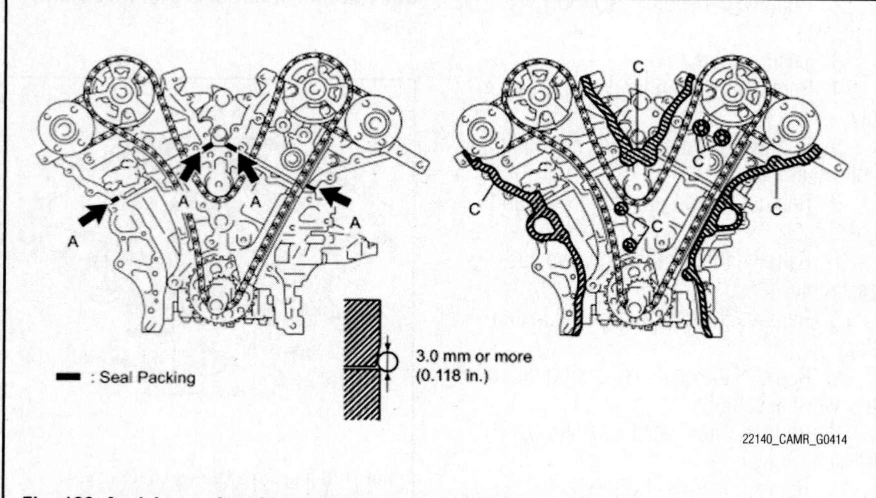

Fig. 180 Applying seal packing to timing chain cover sub-assembly

➡**When the contact surfaces are wet, wipe them with an oil-free cloth before applying seal packing.**

➡**Install the crankcase within 3 minutes and tighten the bolts within 15 minutes after applying seal packing.**

➡**Do not start the engine for at least 2 hours after installing.**

b. Apply seal packing in a continuous line to the timing chain cover as shown in the following illustration. Seal packing: Toyota Genuine Seal Packing Black, Three Bond 1207B or equivalent, Toyota Genuine Seal Packing Black, Three Bond 1282B, Three Bond 1282B or equivalent.

c. Install a new gasket.

d. Align the oil pump's drive rotor spline and the crankshaft as shown in the illustration. Install the spline and chain cover to the crankshaft.

e. Loosely install the timing chain cover with the 23 bolts and 2 nuts, but do not tighten the bolts and 2 nuts yet.

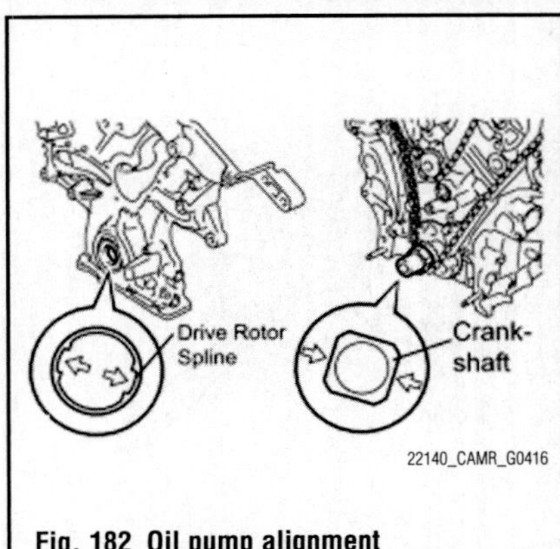

Fig. 182 Oil pump alignment

⁂ CAUTION

Make sure that there is no oil on the bolt and nut threads.

f. Fully tighten the bolts in this order: Area 1 and Area 2, tighten to 15 ft. lbs. (21 Nm).

g. Fully tighten the bolts in Area 3 to 15 ft. lbs. (21 Nm). Tighten the bolts and nuts in the order of upper to lower as shown in the illustration.

h. Fully tighten the bolts in Area 4 to 32 ft. lbs. (43 Nm), and to 15 ft. lbs. (21 Nm). Tighten the bolts and nuts in the order of lower to upper as shown in the illustration.

- Bolt A: 1.57 inches (40 mm)
- Bolt B: 2.17 inches (55 mm)
- Bolt C: 0.98 inches (25 mm)

28. Install the water pump assembly.
29. Install the water inlet housing.
30. Install the left hand No. 1 front engine mounting bracket.
31. Install the oil pan sub-assembly.
32. Install the oil strainer sub-assembly.
33. Install the No. 2 oil pan sub-assembly.
34. Install the oil pan drain plug and gasket.
35. Install the cylinder head cover sub-assembly.
36. Install the left-hand cylinder head cover sub-assembly.
37. Install the water outlet.
38. Install the crankshaft pulley.
39. Install the oil filter.
40. Install the cylinder block water drain cock sub-assembly.
41. Install the No. 1 oil pipe.
42. Install the oil pipe.
43. Install the Crankshaft Position (CKP) sensor.
44. Install the camshaft timing oil control valve assembly.
45. Install the camshaft position sensor.
46. Install the ventilation valve sub-assembly.
47. Install the spark plugs and ignition coil assembly.
48. Install the oil filler cap sub-assembly.
49. Install the water pump assembly.
50. Install the engine assembly with transaxle.

TIMING CHAIN & SPROCKETS

REMOVAL & INSTALLATION

2.4L Engine

See Figures 184 through 202.

1. Before servicing the vehicle, refer to the Precautions Section.
2. Remove the Crankshaft Position (CKP) sensor plate.
3. Remove the bolt and chain tensioner slipper.
4. Remove the two bolts and chain vibration damper.
5. Remove the bolt and timing chain guide.
6. Remove the upper chain assembly.
7. Remove the crankshaft timing sprocket.

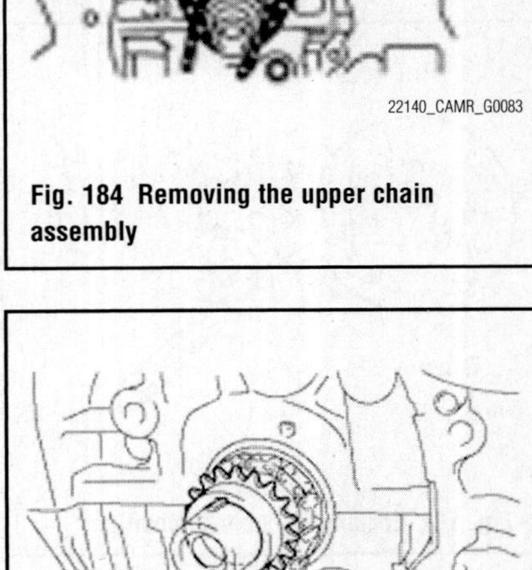

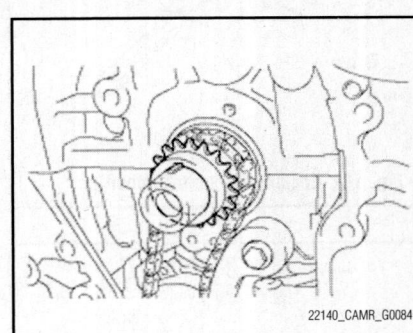

Fig. 184 Removing the upper chain assembly

Fig. 185 Removing the crankshaft timing sprocket

8. To remove the lower chain assembly. Turn the crankshaft by 90°counterclockwise to align the adjusting hole of the oil pump driveshaft sprocket with the groove of the oil pump.
9. Insert a 4 mm diameter bar into the adjusting hole of the oil pump driveshaft sprocket to lock the gear in position, and then remove the nut.
10. Remove the bolt, chain tensioner plate and spring.
11. Remove the chain tensioner, oil pump driven sprocket and chain.

To install:

12. Set the crankshaft key into the left horizontal position.
13. Turn the driveshaft so that the cutout faces upward.
14. Align the yellow mark links with the timing marks of each gear.

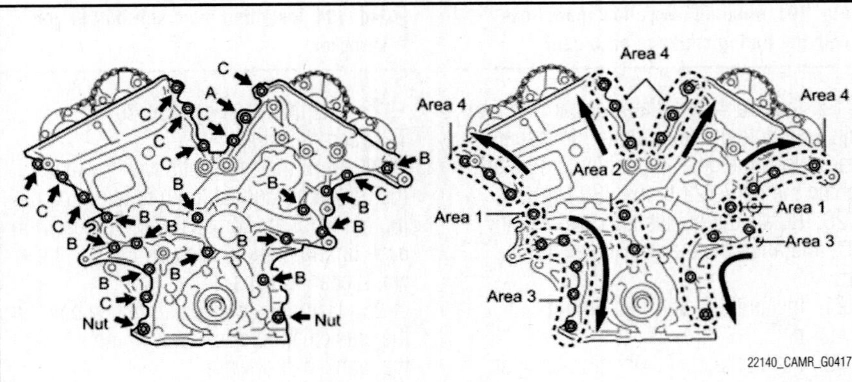

Fig. 183 Timing chain cover bolts and nuts tightening sequence

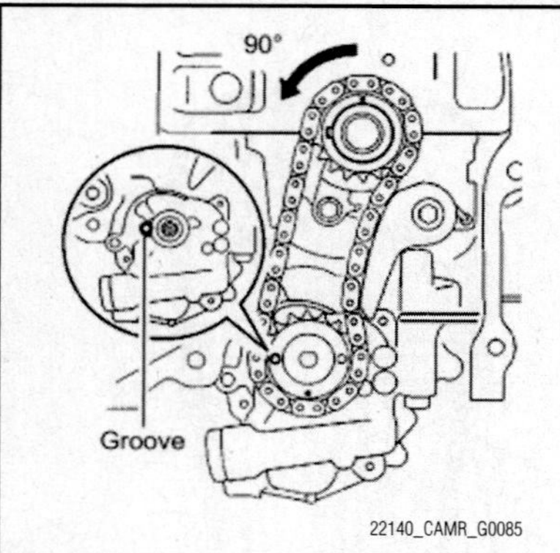

Fig. 186 Turning the crankshaft by 90°counterclockwise to align the adjusting hole

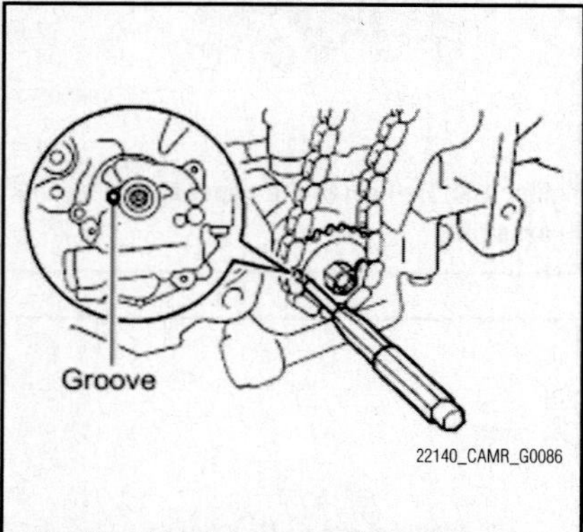

Fig. 187 Locking the gear in position

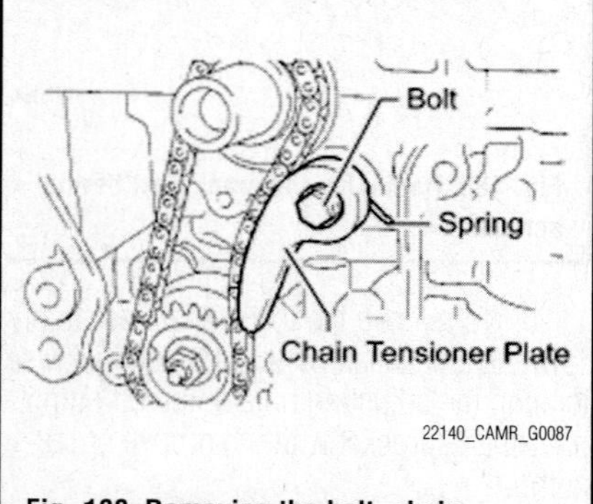
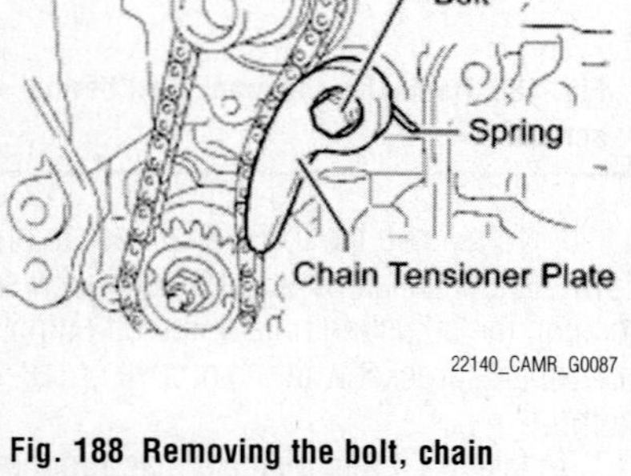

Fig. 188 Removing the bolt, chain tensioner plate and spring

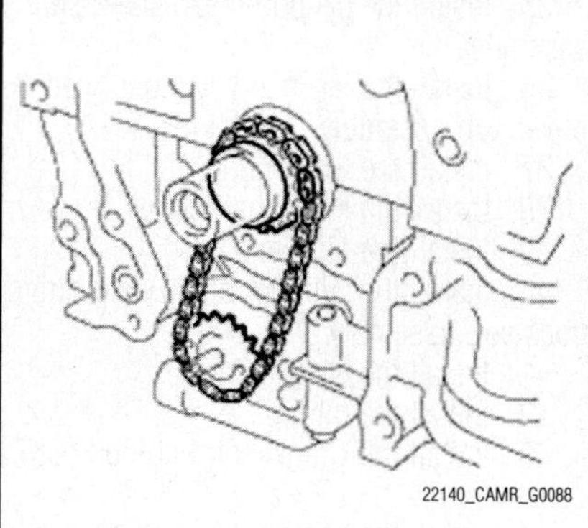

Fig. 189 Removing the chain tensioner, oil pump driven sprocket and chain

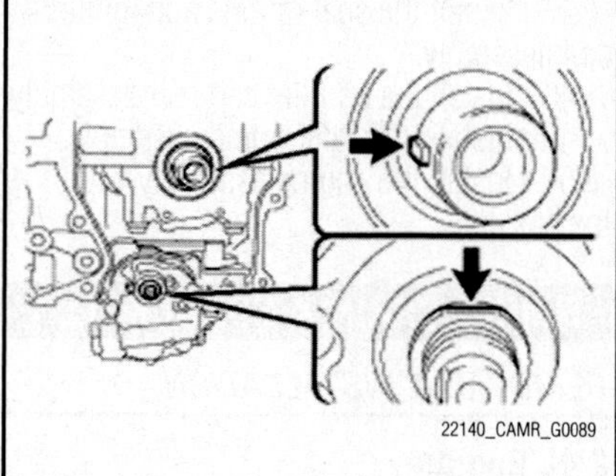

Fig. 190 Setting the crankshaft key and cutout face

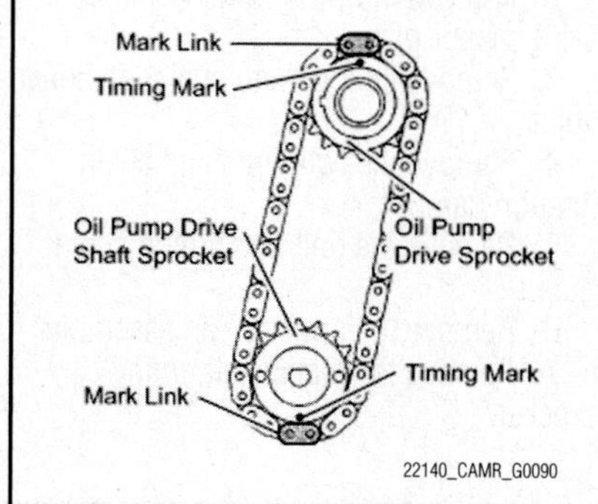

Fig. 191 Aligning the yellow mark links with the timing marks of each gear

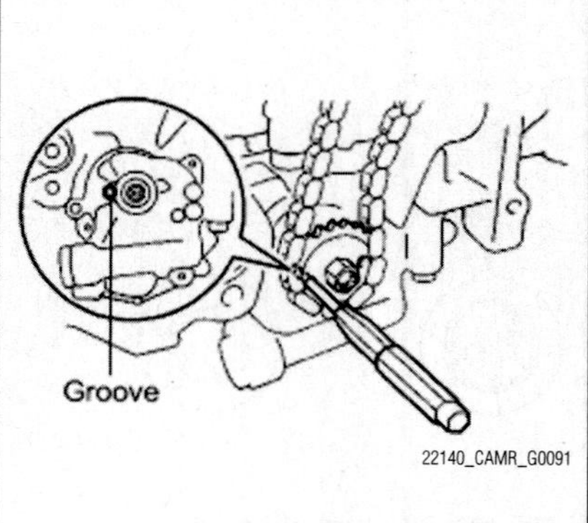

Fig. 192 Aligning the adjusting hole of the oil pump driveshaft sprocket

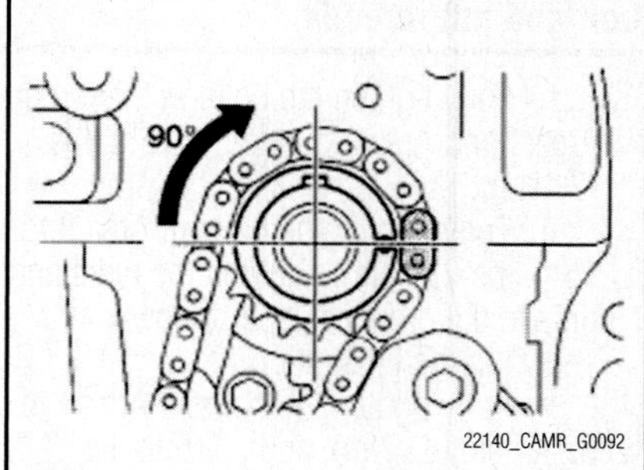

Fig. 193 Rotating the crankshaft clockwise by 90°

Fig. 194 Installing the crankshaft timing sprocket

15. Install the sprockets onto the crankshaft and oil pump shaft with the chain wrapped on the gears
16. Temporarily tighten the oil pump driveshaft sprocket with the nut.
17. Insert the damper spring into the adjusting hole, and then install the chain tensioner plate with the bolt and tighten to 9 ft. lbs. (12 Nm).
18. Align the adjusting hole of the oil pump driveshaft sprocket with the groove of the oil pump.
19. Insert a 4 mm diameter bar into the adjusting hole of the oil pump driveshaft gear to lock the gear in position, and then tighten the nut to 22 ft. lbs. (30 Nm).
20. Rotate the crankshaft clockwise by 90°, and align the crankshaft key to the top.
21. Install the crankshaft timing sprocket.
22. Install the chain vibration damper with the 2 bolts and tighten to 80 inch lbs. (9 Nm).
23. Set the No. 1 cylinder to TDC/compression.
24. Turn the camshafts with a wrench (using the hexagonal lobe) to align the timing marks of the camshaft timing gear with each timing mark located on the No. 1 and No. 2 bearing caps.
25. Using the crankshaft pulley bolt, turn the crankshaft to position with the key on the crankshaft upward.
26. Install the chain onto the crankshaft timing sprocket with the gold or pink mark

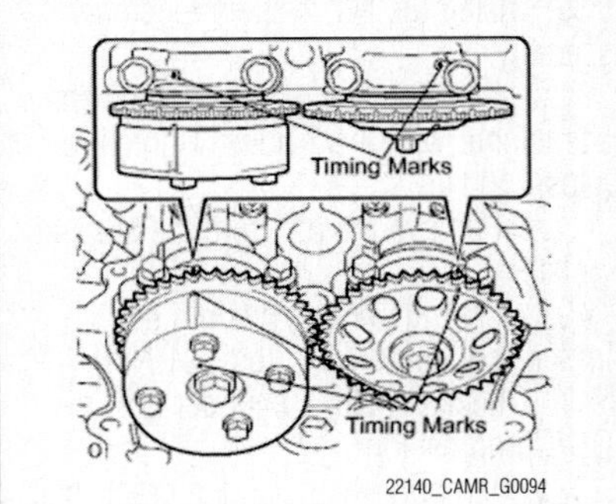

Fig. 195 Setting the No. 1 cylinder to TDC/compression

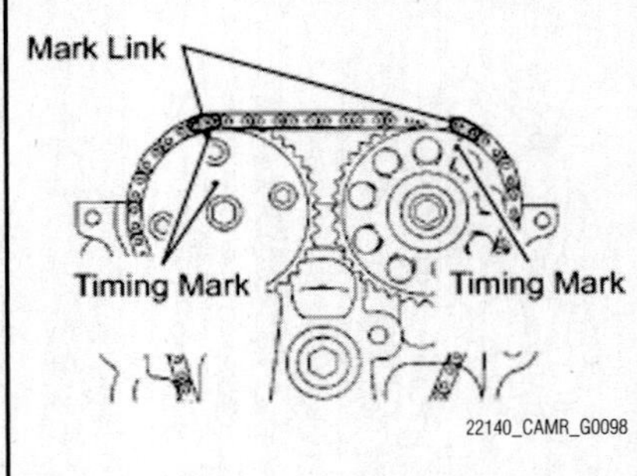

Fig. 199 Aligning the gold or yellow link with each timing mark located on the camshaft timing gear and sprocket

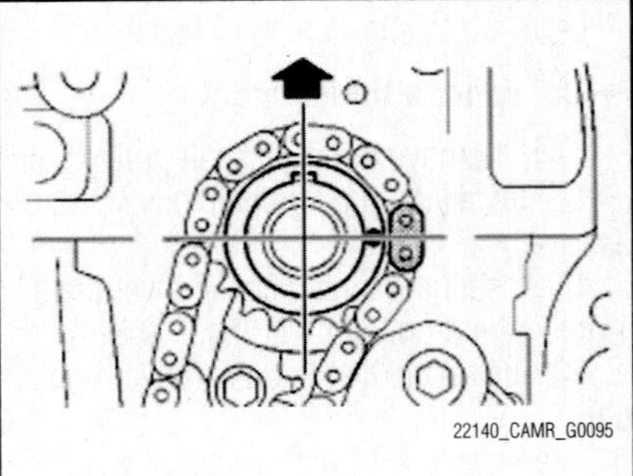

Fig. 196 Turning the crankshaft to position with the key on the crankshaft upward

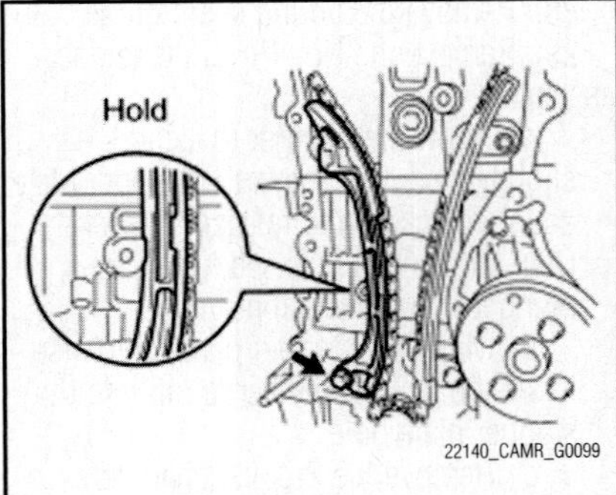

Fig. 200 Installing the chain tensioner slipper and bolt

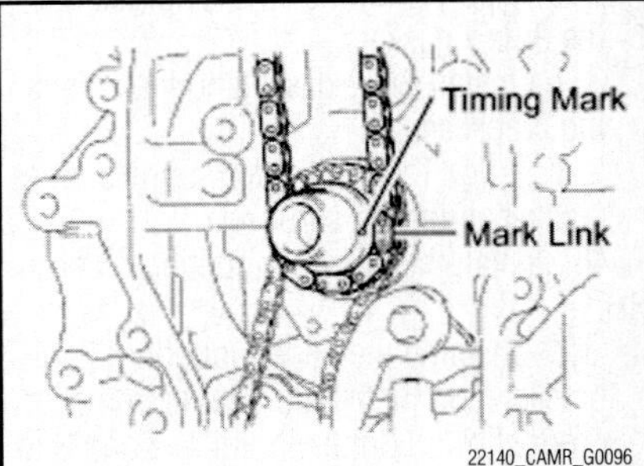

Fig. 197 Installing the chain onto the crankshaft timing sprocket with the gold or pink mark link

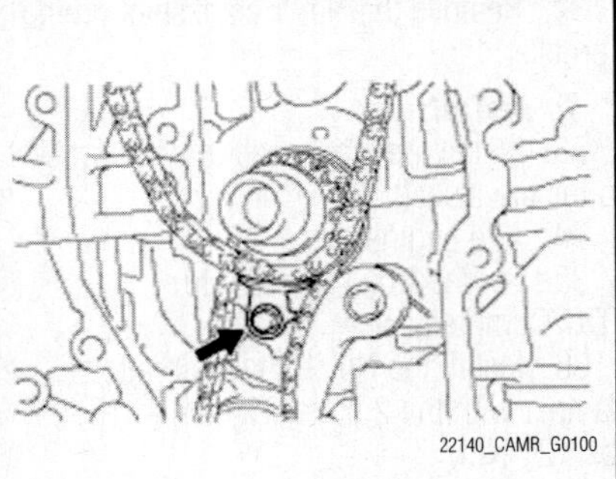

Fig. 201 Installing the timing chain guide and bolt

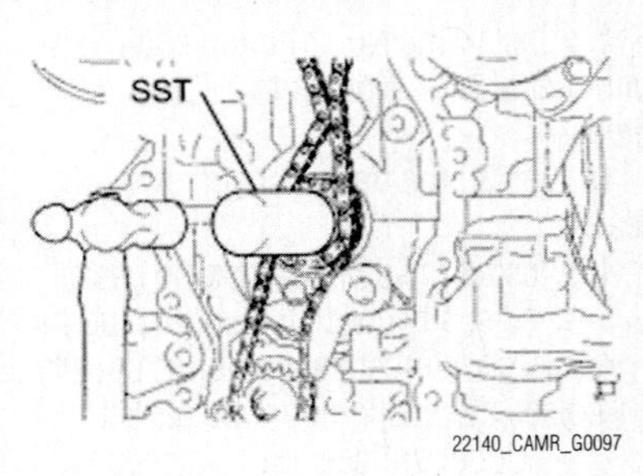

Fig. 198 Using Special Tool: 09309-37010 and a hammer to install the crankshaft timing sprocket

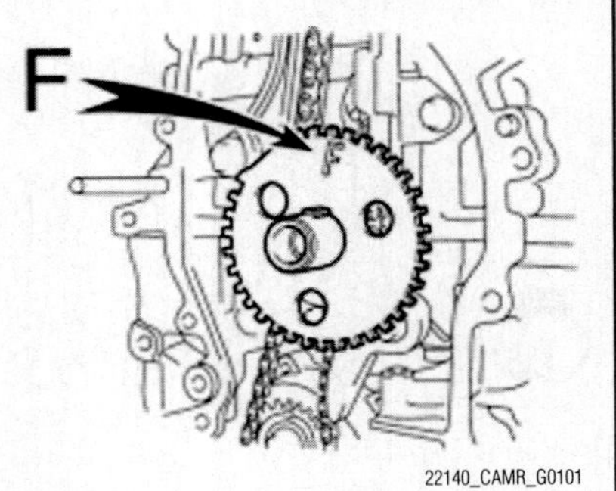

Fig. 202 Installing the crankshaft sensor plate with the "F" mark facing forward

link aligned with the timing mark on the crankshaft.

27. Using Special Tool: 09309-37010 and a hammer, tap in the crankshaft timing sprocket.
28. Align the gold or yellow link with each timing mark located on the camshaft timing gear and sprocket, then install the chain.
29. Install the chain tensioner slipper with the bolt and tighten to: 14 ft. lbs. (19 Nm).
30. Install the timing chain guide with the bolt and tighten to: 80 inch lbs. (9 Nm).
31. Install the crankshaft sensor plate with the "F" mark facing forward.

2.5L Engine

See Figures 203 through 214.

1. Remove the engine cover joints.
2. Remove the spark plug.
3. Remove the camshaft timing oil control valve assembly (intake and exhaust sides).
4. Remove the camshaft position sensors (intake and exhaust side).
5. Remove the oil filler cap sub assembly.
   a. Remove the oil filler cap from the cylinder head.
   b. Remove the gasket from the oil filler cap.
6. Remove the Crankshaft Position (CKP) sensor.
7. Remove the ventilation valve sub assembly.
8. Remove the ventilation case sub assembly.
   a. Remove the 8 bolts and 2 nuts.
   b. Remove the ventilation case by prying between the ventilation case and cylinder block with a screwdriver.

➡Tape the screwdriver tip before use.

➡Be careful not to damage the contact surfaces of the cylinder block and ventilation case.

9. Remove the separator case.
   a. Remove the 2 bolts, separator case and gasket.
10. Remove the No. 1 water by pass pipe and gasket.
11. Remove the inlet water.
12. Remove the thermostat.
13. Remove the v-ribbed belt tensioner assembly (compressor and alternator).
14. Remove the water pump assembly.
15. Remove the water drain cock plug from the water drain cock.
   a. Remove the water drain cock from the cylinder block.

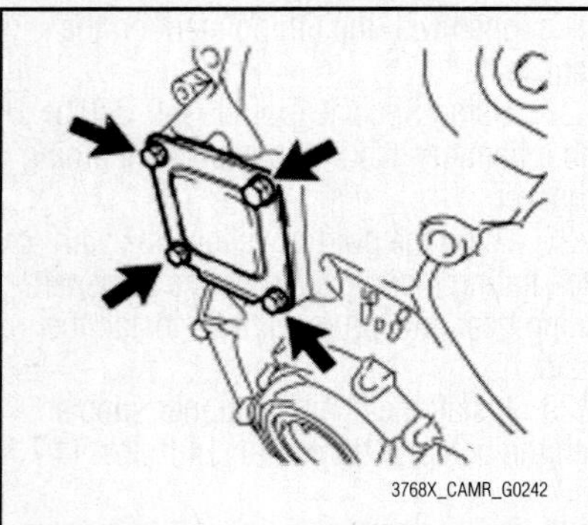

Fig. 203 Removing the timing chain cover plate

16. Remove the oil cooler assembly (with the oil cooler).
17. Remove the inlet water housing.
    a. Remove the 4 bolts, nut, inlet water housing and gasket.
18. Remove the crankshaft pulley.
    a. Using the special tool, hold the crankshaft pulley ad loosen the pulley bolt. Further loosen the bolt until 2 or 3 threads are screwed into the crankshaft.
    b. Using the special tool and the pulley bolt, remove the crankshaft pulley.

➡Apply a lubricant to the threads and end of the special tool.

19. Remove the cylinder head cover sub assembly and 3 gaskets from the camshaft bearing caps.
20. Remove the spark plug tube gasket.
    a. Using a screwdriver, pry out the 4 plug tube gaskets.

➡Be careful not to damage the cylinder head cover.

➡Tape the screwdriver tip before use.

21. Remove the No. 2 timing chain cover.
22. Remove the engine mounting bracket RH.
23. Remove the timing chain cover sub assembly.
24. Remove the timing chain cover tight plug.
    a. Using a 14 mm hexagon wrench, remove the plug and gasket.
25. Remove the timing chain cover plate.
    a. Remove the 4 bolts, timing chain cover plate and gasket.
26. Remove the timing chain cover oil seal.
27. Set the No. 1 cylinder to TDC/Compression.

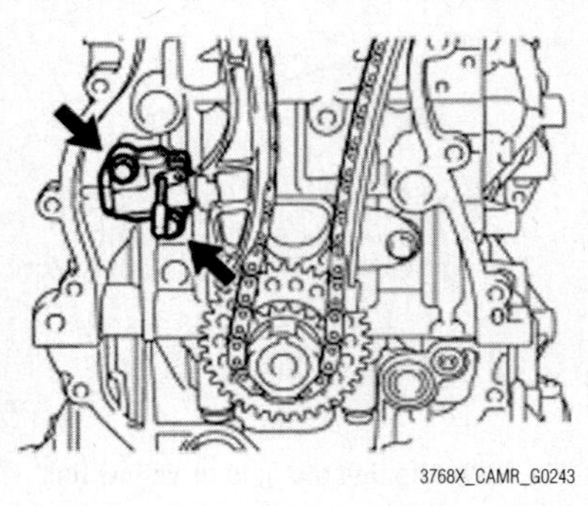

Fig. 204 Removing the chain tensioner

28. Remove the timing chain guide.
29. Remove the No. 1 chain tensioner assembly.
    a. Allow the plunger to extend slightly, and then rotate the stopper plate counterclockwise to release the lock. Once the lock is released, push the plunger into the tensioner.
    b. Move the stopper plate clockwise to set the lock, and insert a pin into the stopper plate hole.
    c. Remove the 2 bolts, chain tensioner and gasket.
30. Remove the chain tensioner slipper.
31. Remove the chain sub assembly.
32. Remove the No. 1 crankshaft timing sprocket.

To install:

33. Install the crankshaft timing sprocket to the crankshaft.
34. Add engine oil.
35. Set the No. 1 cylinder to TDC/Compression.
36. Install the No. 1 chain vibration damper with the 2 bolts. Tighten to 15 ft. lbs. (21 Nm).
37. Install the chain sub assembly.
38. Install the chain tensioner slipper. Tighten to 15 ft. lbs. (21 Nm).

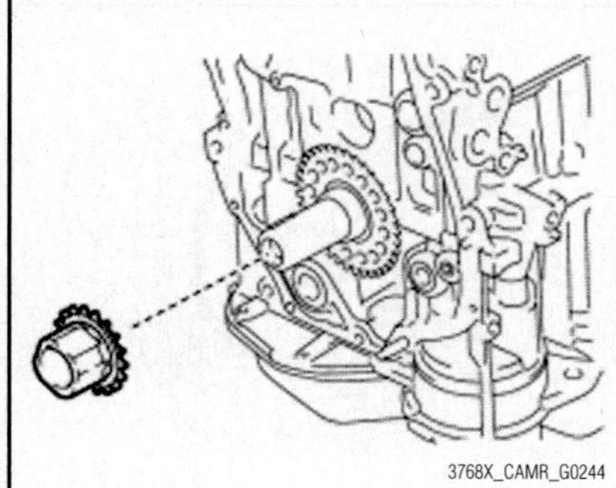

Fig. 205 Removing the No. 1 crankshaft timing sprocket

39. Install the No. 1 chain tensioner assembly.
    a. Install a new gasket and the chain tensioner with the 2 bolts. Tighten to 7 ft. lbs. (10 Nm).
    b. Remove the pin from the stopper plate.
40. Install the timing chain guide. Tighten the bolt to 15 ft. lbs. (21 Nm).
41. Check the No. 1 cylinder to TDC/Compression.
    a. Temporarily install the crankshaft pulley bolt.
    b. Rotate the crankshaft clockwise, and check that the timing marks on the crankshaft timing sprocket and camshaft timing gears are as shown in the illustration.

➡"A" is not a timing mark.

    c. Remove the crankshaft pulley bolt.
42. Install the timing chain cover sub assembly.
43. Install the timing chain cover plate. Tighten the 4 bolts to 7 ft. lbs. (10 Nm).
44. Install the timing chain cover tight plug.
    a. Using a 14 mm hexagon wrench, install a new gasket and the plug. Tighten to 15 ft. lbs. (20 Nm).
45. Install the timing chain cover sub assembly.
    a. Apply a light coat of engine oil to the 3 new gaskets.
    b. Install the 3 gaskets to the stiffening crankcase.
    c. Align the drive rotor spline and the crankshaft timing sprocket.
46. Install the engine mounting bracket RH.
    a. Install the engine mounting bracket, and install the 5 bolts in the order shown. Tighten bolts 1, 2 and 3 to 41 ft. lbs. (55 Nm). Tighten bolts 4 and 5 to 16 ft. lbs. (21 Nm).

➡After applying seal packing to the timing chain cover install the engine mounting bracket within 10 minutes.

47. Install the No. 2 timing chain cover with the 2 bolts. Tighten to 7 ft. lbs. (10 Nm).
48. Install the timing chain cover oil seal.
49. Install the spark plug tube gasket.
    a. Visually check the spark plug tube gasket. If the results are not as specified, replace the spark plug tube gasket.
    - Upper surface: no scratches or deformation
    - Outer lip: no scratches or deformation
    - Inner lip: No scratches

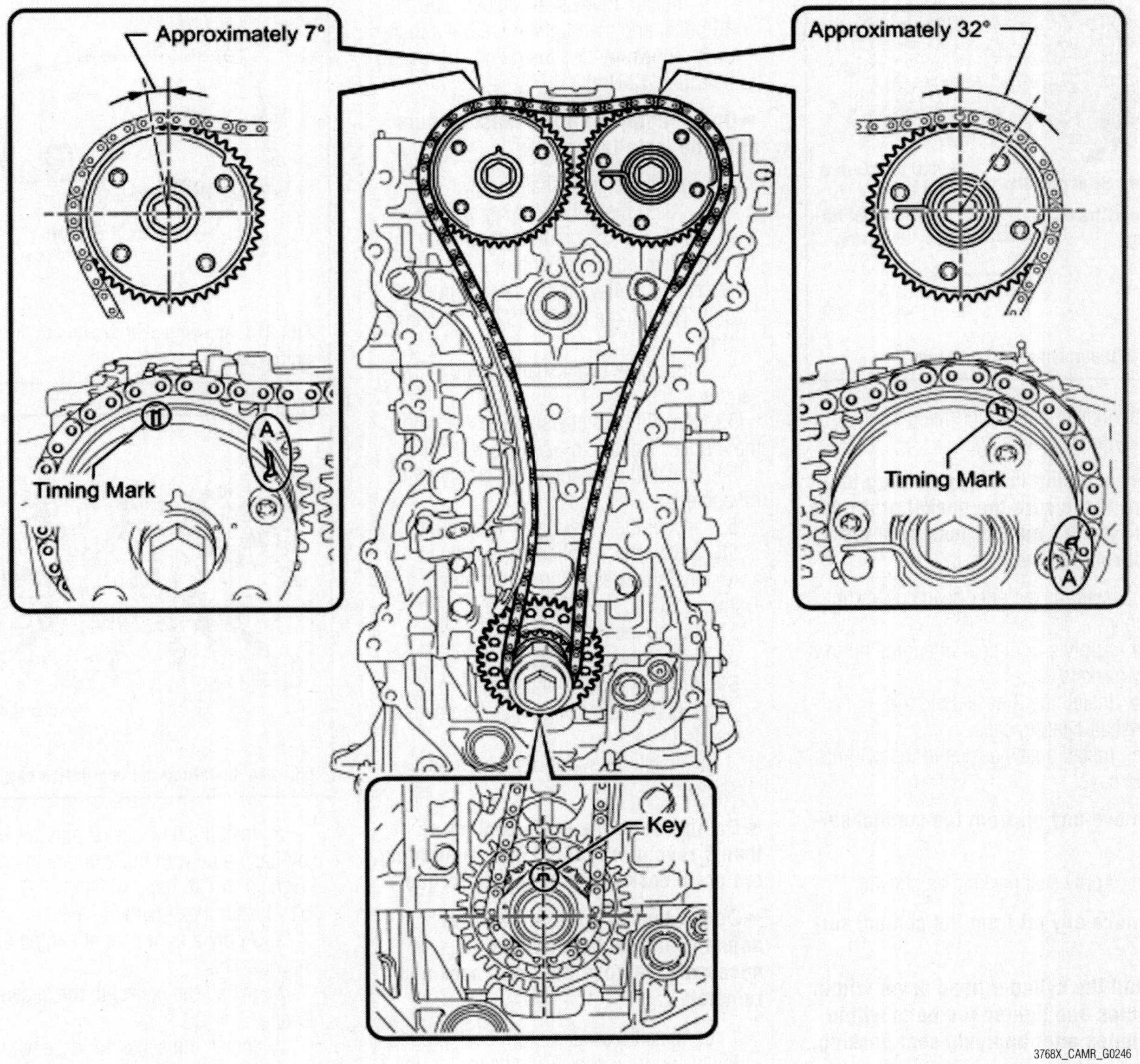

Approximately 7°
Approximately 32°
Timing Mark
Timing Mark
A
A
Key
3768X_CAMR_G0246

Fig. 206 Checking the timing marks

Crankshaft Timing Sprocket
Drive Rotor Spline
3768X_CAMR_G0247

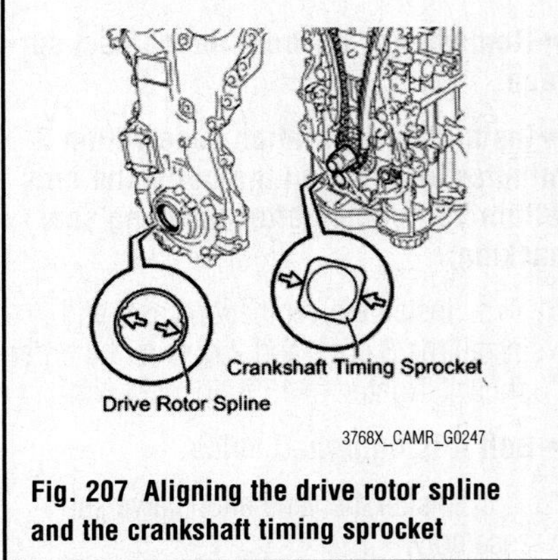

Fig. 207 Aligning the drive rotor spline and the crankshaft timing sprocket

5
2
3
4
1
3768X_CAMR_G0248

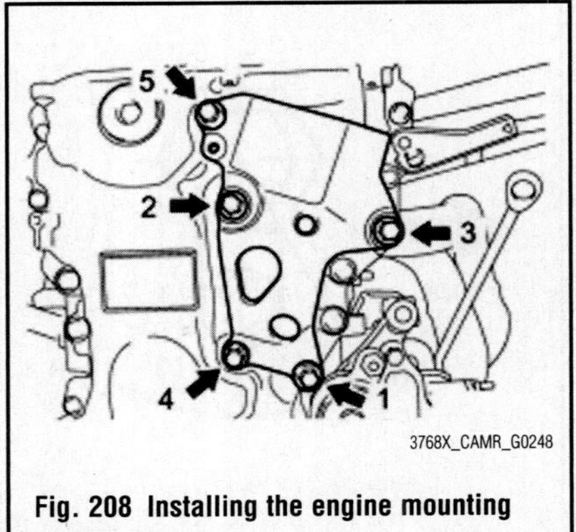

Fig. 208 Installing the engine mounting bracket RH

Upper Surface
Inner Lip
Outer Lip
3768X_CAMR_G0249

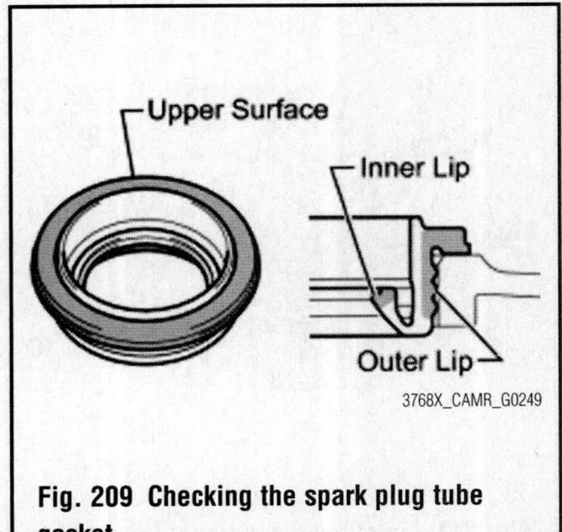

Fig. 209 Checking the spark plug tube gasket

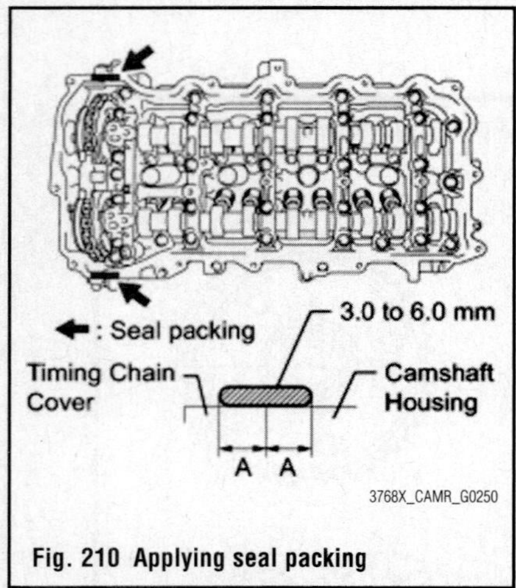

Fig. 210 Applying seal packing

b. Install the 4 plug tube gaskets to the cylinder head cover.

➡After pressing in the spark plug tube gasket, make sure the gasket protrudes 0.0394 inch (1 mm) or less from the cylinder head cover.

50. Install the cylinder head cover sub assembly.
 a. Apply a light coat of engine oil to 3 new gaskets.
 b. Install the 3 gaskets to the camshaft bearing caps.
 c. Install a new gasket to the cylinder head cover.

➡Remove any oil from the contact surface.

 d. Apply seal packing as shown.

➡Remove any oil from the contact surface.

➡Install the cylinder head cover within 3 minutes and tighten the bolts within 15 minutes after applying seal packing.

 e. Align the cylinder head cover with pin A. Then align the cylinder head cover with pin B and install the cylinder head cover.
 f. Install 3 new seal washers and the 16 bolts, and then tighten the bolts in the order shown in the illustration. Tighten to 9 ft. lbs. (12 Nm).

➡Do not apply oil for at least 4 hours after the installation.

51. Install the crankshaft pulley.
 a. Align the pulley set key with the key groove of the crankshaft pulley.
 b. Using the special tool, hold the crankshaft pulley and install the pulley bolt. Tighten the bolt to 192 ft. lbs. (260 Nm).
52. Install the Crankshaft Position (CKP) sensor.
53. Install the inlet water housing with the 4 bolts. Tighten to 32 ft. lbs. (43 Nm).
54. Install the oil cooler assembly (with oil cooler).
55. Install the water pump assembly.
56. Install the v-ribbed belt tensioner assembly (compressor and alternator). Tighten to 15 ft. lbs. (21 Nm).
57. Install the thermostat.
58. Install the inlet water.
59. Install the water drain cock.
 a. Apply adhesive to 2 or 3 threads of the drain cock.
 b. Install the water drain cock as shown.

➡Do not rotate the drain cock more than 1 revolution (360°) after tightening the drain cock to the specified torque.

➡Do not loosen the drain cock to adjust it. If an adjustment is necessary, remove the drain cock and reinstall it.

 c. Install the water drain cock plug to the water drain cock. Tighten to 9 ft. lbs. (13 Nm).
60. Install the No. 1 water by pass pipe.

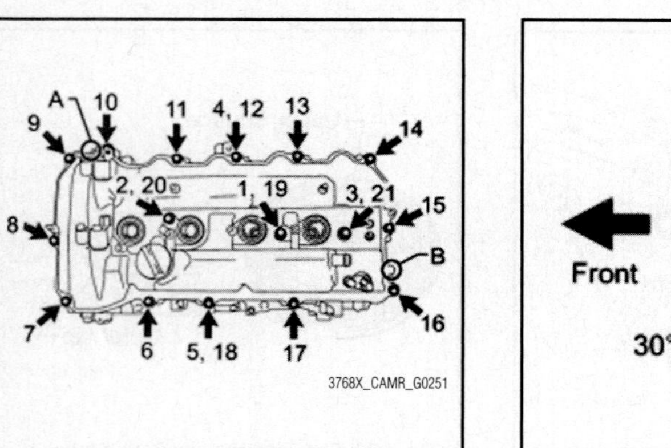

Fig. 211 Installing the cylinder head cover

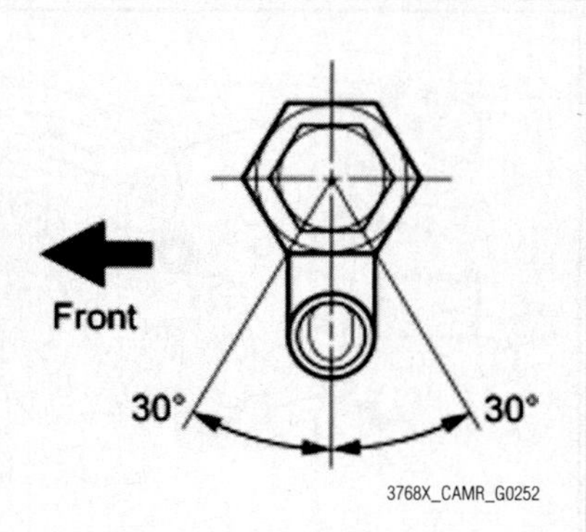

Fig. 212 Installing the water drain cock

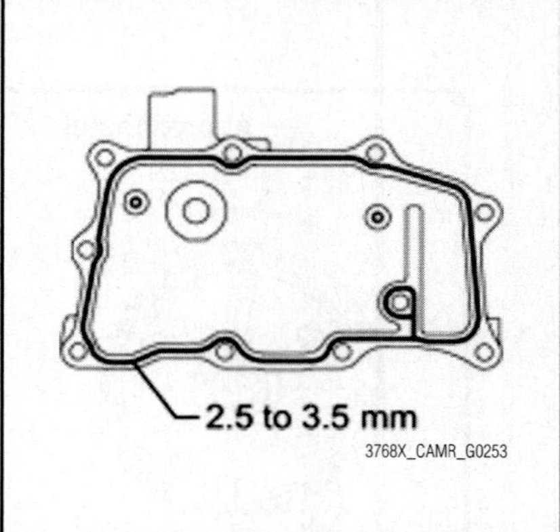

Fig. 213 Applying seal packing to the separator case

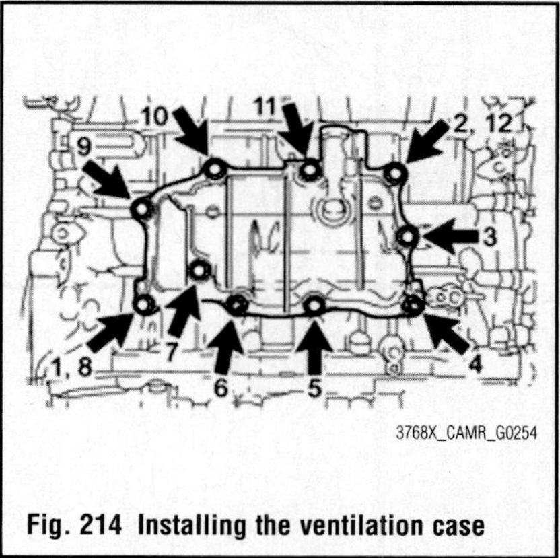

Fig. 214 Installing the ventilation case

 a. Install a new gasket and the water by-pass pipe with the 2 nuts and bolt. Tighten to 7 ft. lbs. (10 Nm).
61. Install the separator case.
 a. Apply a light coat of engine oil to a new gasket.
 b. Install the gasket to the separator case.
 c. Install the separator case with the 2 bolts. Tighten to 7 ft. lbs. (10 Nm).
62. Install the ventilation case sub assembly.
 a. Apply seal packing in a continuous line as shown.

➡Remove any oil from the contact surface.

➡Install the ventilation case within 3 minutes and tighten the bolts and nuts within 15 minutes after applying seal packing.

 b. Install the ventilation case, and install the 8 bolts and 2 nuts in the order shown. Tighten to 15 ft. lbs. (21 Nm).

➡Bolt A is tightened twice.

 c. Install the ventilation valve sub assembly.

63. Install the camshaft position sensor (intake and exhaust side).

64. Install the camshaft timing oil control valve assembly (intake and exhaust side).

65. Install the oil filler cap sub assembly.

a. Install a new gasket to the filler cap.

b. Install the oil filler cap to the cylinder head.

66. Install the spark plug.

67. Install the engine cover joints. Tighten to 7 ft. lbs. (10 Nm).

3.5L Engine

See Figures 215 through 226.

1. Before servicing the vehicle, refer to the Precautions Section.

2. Remove the engine assembly with transaxle.

3. Secure engine.

4. Remove the oil filler cap sub-assembly.

5. Remove the spark plugs and ignition coil assembly.

6. Remove the oil pan drain plug and gasket.

7. Remove the ventilation valve sub-assembly.

8. Remove the camshaft position sensor.

9. Remove the camshaft timing oil control valve assembly.

10. Remove Crankshaft Position (CKP) sensor.

11. Remove the No. 1 oil pipe.

12. Remove the oil pipe.

13. Remove the cylinder block water drain cock sub-assembly.

14. Remove the oil filter.

15. Remove the crankshaft pulley.

16. Remove the left hand No. 1 front engine mounting bracket.

17. Remove the water inlet housing.

18. Remove the water outlet.

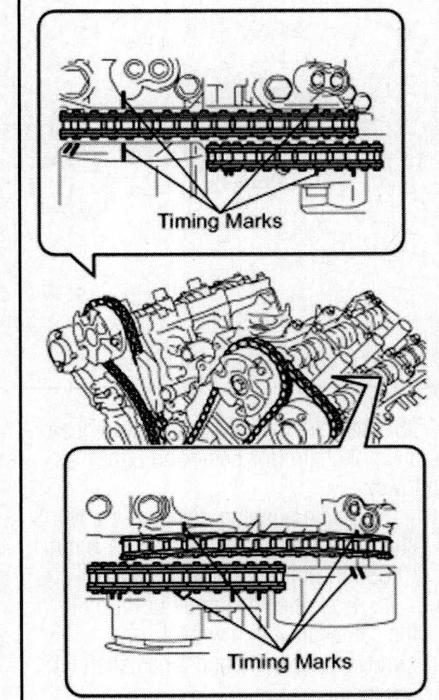

Fig. 216 Check that the timing marks of the camshaft timing gears

19. Remove the left-hand cylinder head cover sub-assembly and gasket.

20. Remove the cylinder head cover sub-assembly and gasket.

21. Remove the No. 2 oil pan sub-assembly.

22. Remove the oil strainer sub-assembly.

23. Remove the oil pan sub-assembly.

24. Remove the water pump assembly.

25. Remove the timing chain cover and seal.

26. Set no. 1 cylinder to TDC/compression, as follows:

a. Temporarily tighten the pulley set bolt. Set the timing mark on the crank angle sensor plate to the RH block bore center line (TDC/compression).

b. Check that the timing marks of the camshaft timing gears are aligned with the timing marks of the bearing cap as shown in the illustration. If not, turn the crankshaft 1 revolution (360°) and align the timing marks as above.

27. Remove the No. 1 chain tensioner assembly, as follows:

a. Move the stopper plate upward to release the lock, and push the plunger deep into the tensioner.

b. Move the stopper plate downward to set the lock, and insert a pin of 1.27 mm (0.05 in.) into the stopper plate's hole.

c. Remove the 2 bolts and chain tensioner.

28. Remove the chain tensioner slipper.

29. Remove the chain sub-assembly, as follows:

a. Turn the crankshaft counterclockwise 10° to loosen the chain of the crankshaft timing sprocket.

b. Remove the pulley set bolt.

c. Remove the chain from the crankshaft timing sprocket and place it on the crankshaft.

d. Turn the camshaft timing gear assembly on the right hand bank clockwise (approximately 60°) and set it as shown in the illustration. Be sure to loosen the chain between the banks.

e. Remove the chain.

30. Remove the idle sprocket assembly, as follows:

a. Using a 10 mm hexagon wrench, remove the No. 2 idle gear shaft, sprocket and No. 1 idle gear shaft.

31. Remove the 2 bolts and the No. 1 chain vibration damper.

32. Remove the No. 2 chain vibration damper.

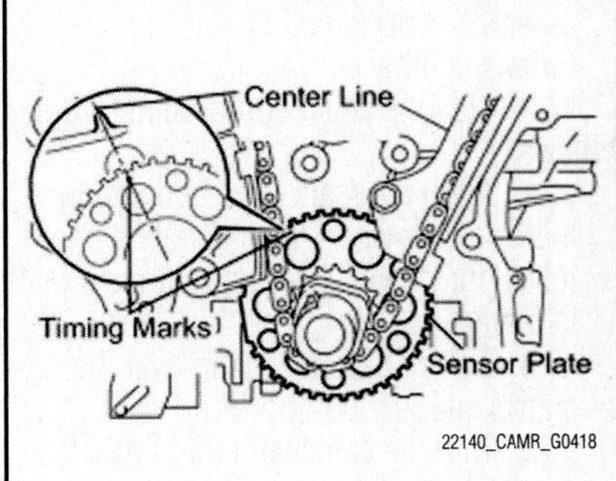

Fig. 215 Set the timing mark on the crank angle sensor plate

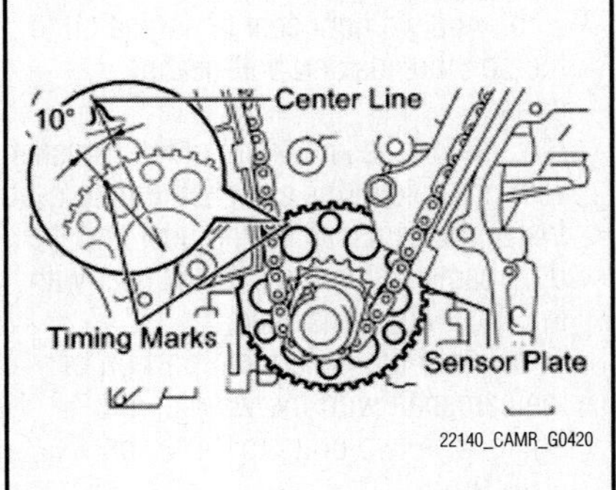

Fig. 217 Turning the crankshaft counterclockwise 10°

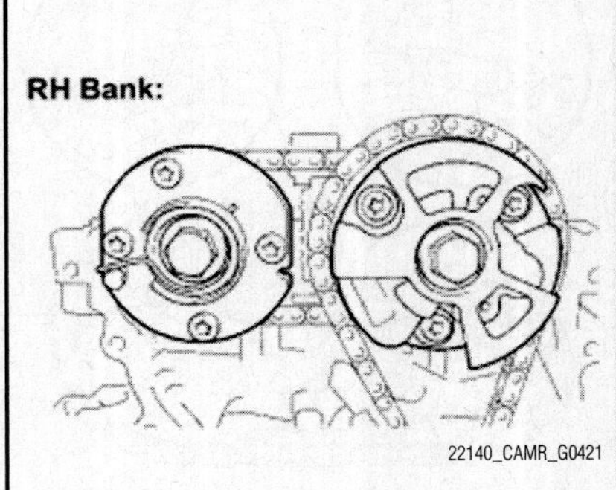

Fig. 218 Camshaft timing gear assembly positioning

33. Remove the crankshaft timing sprocket, as follows:

a. Remove the pulley set bolt.

b. Remove the crankshaft timing gear from the crankshaft.

c. Remove the 2 pulley set keys from the crankshaft.

34. Remove the camshaft timing gears and No. 2 chain (for Right-hand Bank), as follows:

a. While raising up the No. 2 chain tensioner, insert a pin of 1.0 mm (0.039 in.) into the hole to hold it.

b. Hold the hexagonal portion of the camshaft with a wrench, and remove the 2 bolts and 2 camshaft timing gears.

➡Be careful not to damage the cylinder head with the wrench.

➡Do not disassemble the camshaft timing gear assemblies.

c. Remove the No. 2 chain.

35. Remove the bolt and No. 2 chain tensioner.

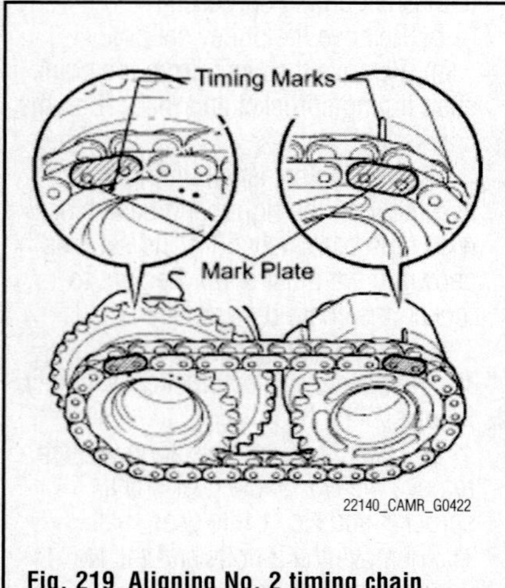

Fig. 219 Aligning No. 2 timing chain

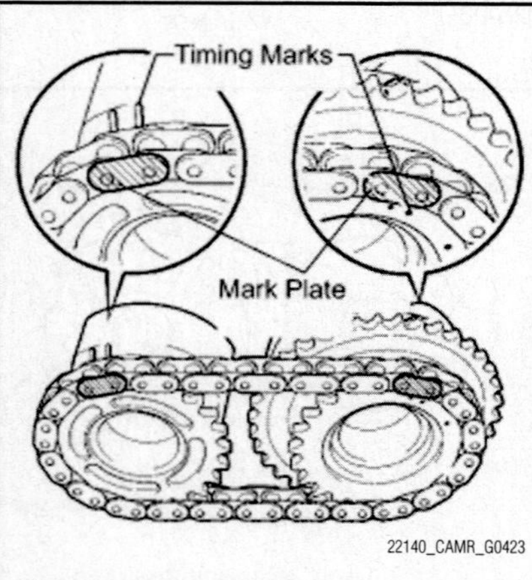

Fig. 220 Aligning No. 2 timing chain (bank 2)

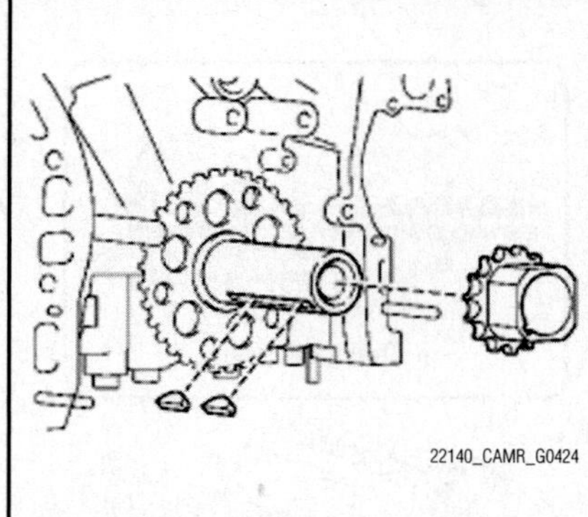

Fig. 221 Crankshaft timing sprocket

36. Remove the camshaft timing gears and No. 2 chain (for Left-hand Bank), as follows:

a. While pushing down on the No. 3 chain tensioner, insert a pin of 1.0 mm (0.039 in.) into the hole to hold it.

b. Hold the hexagonal portion of the camshaft with a wrench, and remove the 2 bolts and 2 camshaft timing gears.

➡Be careful not to damage the cylinder head with the wrench.

➡Do not disassemble the camshaft timing gear assemblies.

c. Remove the No. 2 chain.

37. Remove the bolt and the No. 3 chain tensioner.

To install:

38. Install the No. 2 chain tensioner assembly with the bolt and tighten to 15 ft. lbs. (21 Nm).

39. While pushing in the tensioner, insert a pin of 1.0 mm (0.039 in.) into the hole to hold it.

40. Install the camshaft timing gears and No. 2 chain (for Right-hand Bank), as follows:

a. Align the mark plate with the timing marks (1-dot mark) of the camshaft timing gears as shown.

b. Apply a light coat of engine oil to the bolt threads and bolt-seating surface.

c. Align the knock pin of the camshaft with pin hole of the camshaft timing gear. Install the camshaft timing gear and the right camshaft timing exhaust gear with the No. 2 chain installed.

d. Hold the hexagonal portion of the camshaft with the wrench and tighten the two bolts to 74 ft. lbs. (100 Nm).

e. Remove the pin from the No. 2 chain tensioner.

41. Install the No. 3 chain tensioner assembly with the bolt and tighten to 15 ft. lbs. (21 Nm).

42. While pushing in the tensioner, insert a pin of 1.0 mm (0.039 in.) into the hole to hold it.

43. Install the camshaft timing gears and No. 2 chain (for Left-hand Bank), as follows:

a. Align the mark plate (yellow) with the timing marks (2-dot mark) of the camshaft timing gears as shown.

b. Apply a light coat of engine oil to the bolt threads and bolts seating surface.

c. Align the knock pin of the camshaft with pin hole of the camshaft timing gear. Install the camshaft timing gear and the left camshaft timing exhaust gear with the No. 2 chain installed.

d. Hold the hexagonal portion of the camshaft with the wrench and tighten the two bolts to 74 ft. lbs. (100 Nm).

e. Remove the pin from the No 2 chain tensioner.

44. Install the No. 1 and 2 chain vibration dampers.

45. Install the timing gear set keys and crankshaft timing sprocket as shown in the illustration.

46. Install the idle sprocket assembly, as follows:

a. Apply a light coat of engine oil to the rotating surface of the No. 1 idle gear shaft.

b. Temporarily install the No. 1 idle gear shaft and idle sprocket with the No. 2 idle gear shaft while aligning the knock pin of the No. 1 idle gear with the knock pin groove of the cylinder block. Be careful of the idle gear direction.

c. Using a 10 mm hexagon wrench, tighten the No. 2 idle gear shaft to 44 ft. lbs. (60 Nm).

d. After installing the idle sprocket assembly, check that the idle sprocket turns smoothly.

47. Install the chain sub-assembly, as follows:

a. Align the mark plate and timing marks as shown in the illustration and install the chain. The camshaft mark plate is orange.

b. Do not pass the chain over the crankshaft, just put it on.

c. Turn the camshaft timing gear assembly on the right bank counterclockwise to tighten the chain between the banks.

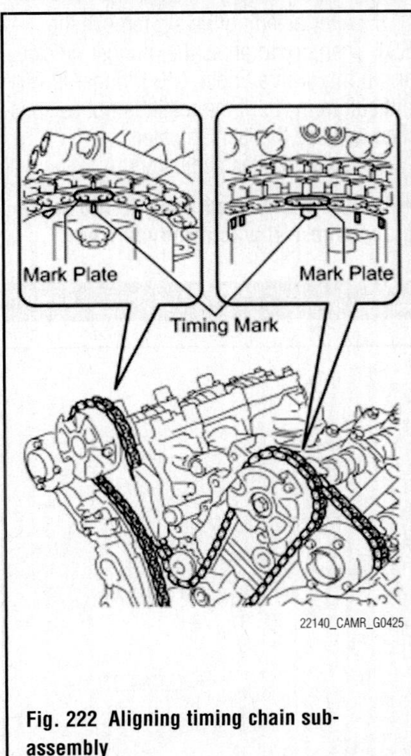

Fig. 222 Aligning timing chain sub-assembly

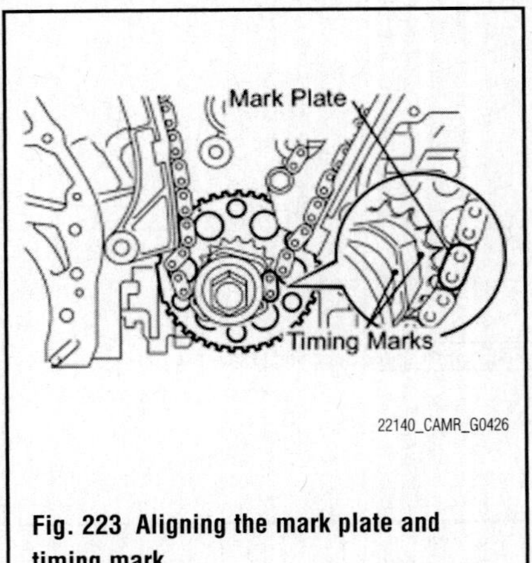

Fig. 223 Aligning the mark plate and timing mark

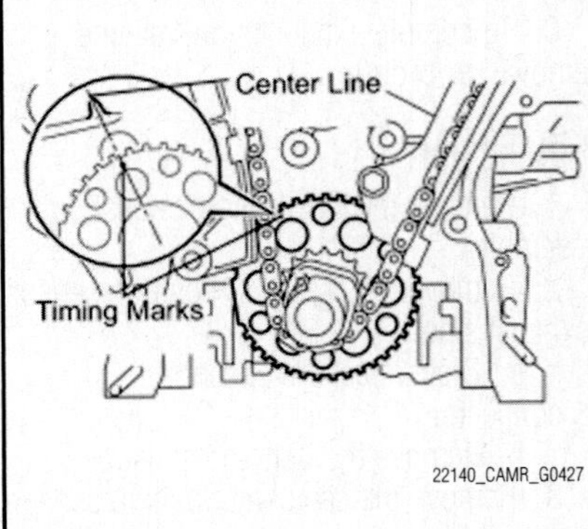

Fig. 224 Turning the crankshaft clockwise (TDC/compression)

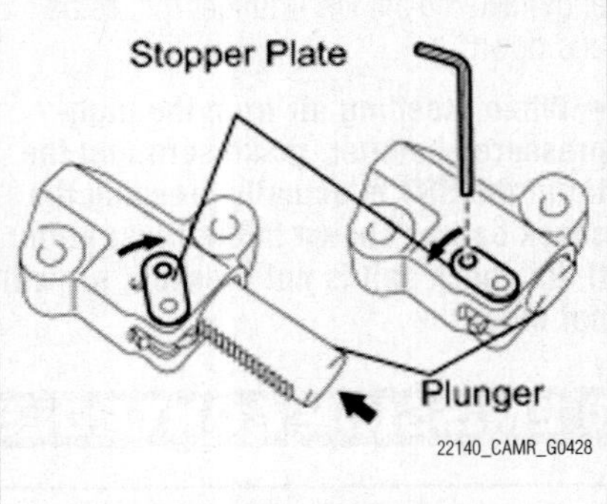

Fig. 225 Set chain tensioner plunger position

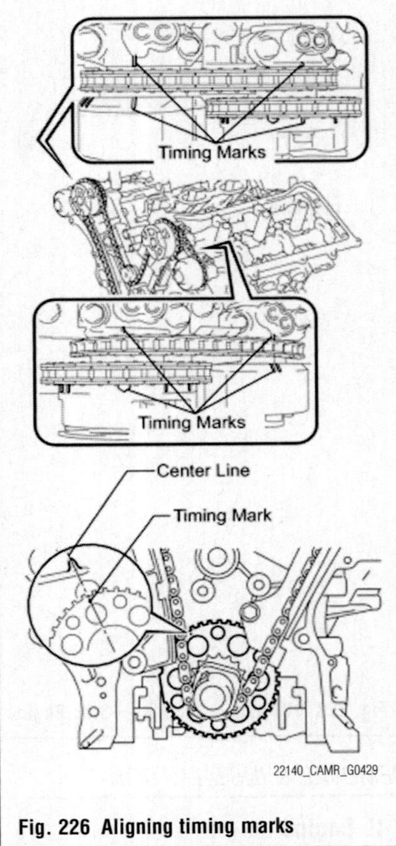

Fig. 226 Aligning timing marks

➡When the idle sprocket assembly is reused, align the timing chain plate with the mark on the sprocket in order to tighten the chain between the banks.

d. Align the mark plate and timing marks as shown in the illustration and install the chain onto the crankshaft timing sprocket. The crankshaft to mark plate is yellow.

e. Temporarily tighten the pulley set bolt.

f. Turn the crankshaft clockwise to set it to the right-hand block bore more centerline. (TDC/compression).

48. Install the chain tensioner slipper.
49. Install the No. 1 chain tensioner assembly, as follows:
    a. Move the stopper plate upward to release the lock, and push the plunger deep into the tensioner.
    b. Move the stopper plate downward to set the lock, and insert a hexagon wrench into the hole of the stopper plate.
    c. Install the No. 1 chain tensioner with the 2 bolts and tighten to 7 ft. lbs. (10 Nm).
    d. Remove the lock pin of the No. 1 chain tensioner. Check that each timing mark is aligned with the crankshaft at TDC/compression.
    e. Remove the pulley set bolt.
50. Install timing chain cover and seal.
51. Install the water pump assembly.
52. Install the water inlet housing.
53. Install the left hand No. 1 front engine mounting bracket.
54. Install the oil pan sub-assembly.
55. Install the oil strainer sub-assembly.
56. Install the No. 2 oil pan sub-assembly.
57. Install the oil pan drain plug and gasket.
58. Install the cylinder head cover sub-assembly.
59. Install the left-hand cylinder head cover sub-assembly.
60. Install the water outlet.
61. Install the crankshaft pulley.
62. Install the oil filter.
63. Install the cylinder block water drain cock sub-assembly.
64. Install the No. 1 oil pipe.
65. Install the oil pipe.
66. Install the Crankshaft Position (CKP) sensor.
67. Install the camshaft timing oil control valve assembly.
68. Install the camshaft position sensor.
69. Install the ventilation valve sub-assembly.
70. Install the spark plugs and ignition coil assembly.
71. Install the oil filler cap sub-assembly.
72. Install the engine assembly with transaxle.

VALVE LASH

ADJUSTMENT

➡Keep the lash adjuster free of dirt and foreign objects.

➡Only use clean engine oil.

1. Place the lash adjuster into a container filled with engine oil.

2. Insert the SST's (SST: 09276-75010) tip into the lash adjuster's plunger and use the tip to press down on the check ball inside the plunger.

3. Squeeze the SST and lash adjuster together to move the plunger up and down 5 to 6 times.

4. Check the movement of the plunger and bleed the air. OK: Plunger moves up and down.

➡**When bleeding air from the high-pressure chamber, make sure that the tip of the SST is actually pressing the check ball as shown in the illustration. If the check ball is not pressed, air will not bleed.**

5. After bleeding the air, remove the SST. Then, try to press the plunger quickly and firmly with a finger. OK: Plunger is very difficult to move. If the result is not as specified, replace the lash adjuster.

6. Install the lash adjusters.

➡**Install the lash adjuster to the same place where it was removed from.**

ENGINE PERFORMANCE & EMISSION CONTROLS

ACCELERATOR PEDAL POSITION (APP) SENSOR

LOCATION

See Figure 227.

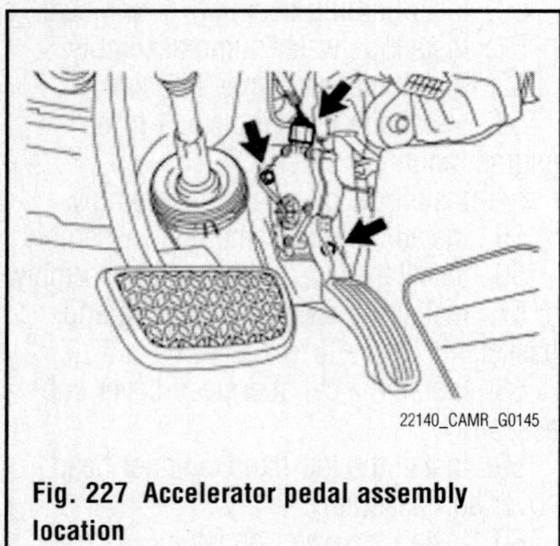

Fig. 227 Accelerator pedal assembly location

REMOVAL & INSTALLATION

See Figure 227.

1. Before servicing the vehicle, refer to the Precautions Section.

2. Remove left center floor carpet cover.

3. Disconnect the accelerator pedal connector.

4. Remove the 2 nuts and accelerator pedal assembly.

➡**Avoid physical shock to the accelerator pedal assembly.**

➡**Do not disassemble the accelerator pedal assembly.**

To install:

5. Installation is the reverse of the removal procedure. Tighten the accelerator pedal rod nuts to 43 inch lbs. (5.4 Nm).

CAMSHAFT POSITION (CMP) SENSOR

LOCATION

See Figure 228.

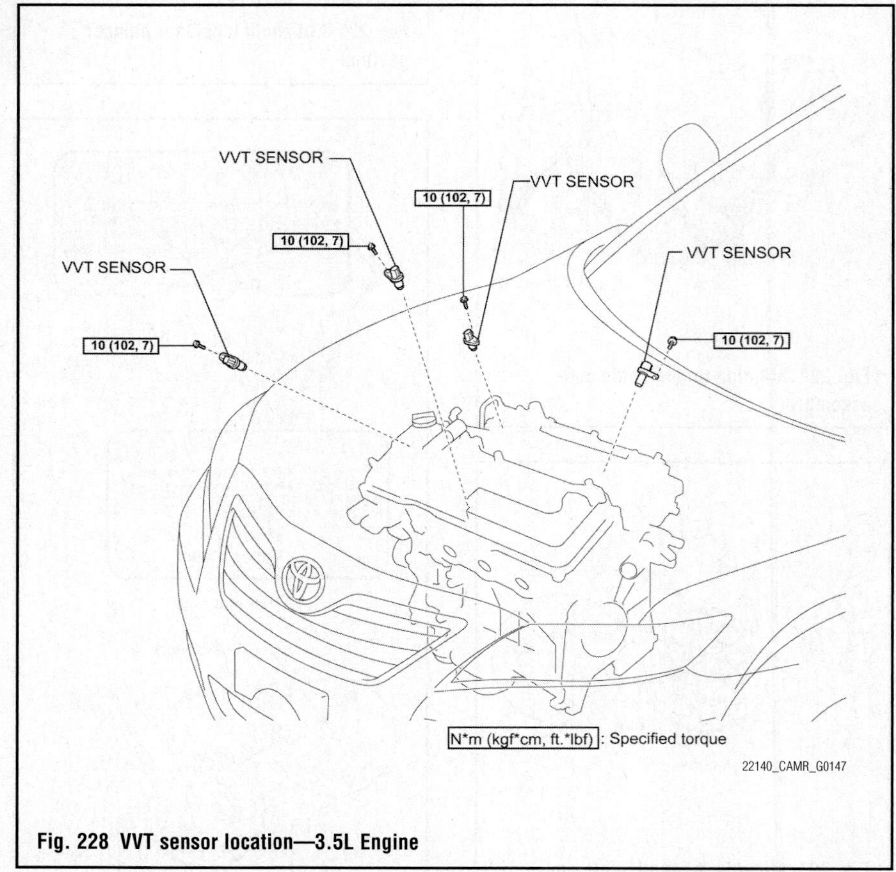

Fig. 228 VVT sensor location—3.5L Engine

REMOVAL & INSTALLATION

2.4L Engine

1. Before servicing the vehicle, refer to the Precautions Section.

2. Remove the plastic engine cover.

3. Remove air cleaner cap sub-assembly.

4. Disconnect the camshaft position sensor connector.

5. Remove the bolt and camshaft position sensor.

To install:

➡**Make sure that the O-ring is not cracked or jammed when installing it.**

6. Apply a light coat of engine oil to the O-ring of the sensor.

7. Install the camshaft position sensor with the bolt and tighten to: 80 inch lbs. (9 Nm).

8. To complete installation, reverse removal procedure.

2.5L Engine

1. Remove the No. 1 engine cover sub assembly.

2. Remove the camshaft position sensor (exhaust side).

a. Disconnect the sensor connector.

b. Remove the bolt and sensor.

3. Remove the camshaft position sensor (intake side).

a. Disconnect the connector.

b. Remove the bolt and sensor.

To install:

4. Install the camshaft position sensor (exhaust side).
   a. Apply a light coat of engine oil to the O-ring of the sensor.

➡Make sure that the O-ring is not cracked or jammed when installing the sensor.

   b. Apply adhesive to 2 or 3 threads of the bolt.
   c. Install the sensor with the bolt. Tighten to 7 ft. lbs. (10 Nm).
   d. Connect the sensor connector.
5. Repeat the above steps for the camshaft position sensor (intake side).

3.5L Engine

1. Before servicing the vehicle, refer to the Precautions Section.
2. Drain and recycle the engine coolant.
3. Disconnect the negative battery cable.
4. Remove the V-bank cover sub-assembly.
5. Remove the windshield wiper link assembly.
6. Remove the front cowl top outside panel.
7. Remove the Intake camshaft VVT sensor (Bank 1), as follows:
   a. Disconnect the VVT sensor connector.
   b. Remove the bolt and VVT sensor.
8. Remove the Exhaust camshaft VVT sensor (Bank 1), as follows:
   a. Disconnect the VVT sensor connector.
   b. Remove the bolt and VVT sensor.
9. Remove the Exhaust camshaft VVT sensor (Bank 2), as follows:
   a. Disconnect the VVT sensor connector.
   b. Remove the bolt and VVT sensor.
10. Remove the Intake camshaft VVT sensor (Bank 2), as follows:
   a. Disconnect the VVT sensor connector.
   b. Remove the bolt and VVT sensor.

To install:

11. Install the VVT sensors and tighten to 7 ft. lbs. (10 Nm).
   a. Connect the VVT sensor connectors.
12. To complete installation, reverse the remaining removal procedure.

CRANKSHAFT POSITION (CKP) SENSOR

LOCATION

See Figures 229 through 231.

Fig. 229 Crankshaft Position (CKP) sensor location—2.4L engine

Fig. 231 Crankshaft Position (CKP) sensor location—3.5L engine

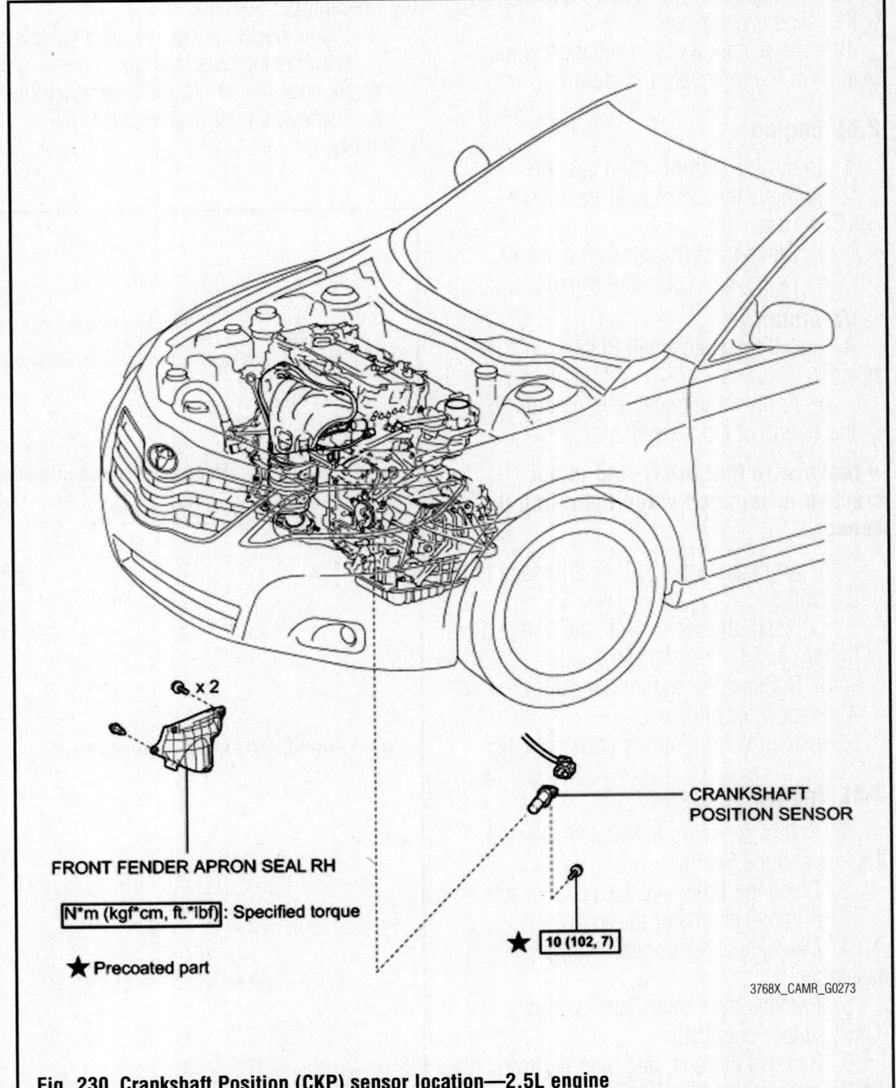

Fig. 230 Crankshaft Position (CKP) sensor location—2.5L engine

REMOVAL & INSTALLATION

2.4L Engine

1. Before servicing the vehicle, refer to the Precautions Section.
2. Disconnect the negative battery cable.
3. Remove the front right wheel.
4. Remove the right hand front fender apron.
5. Remove the drive belt.
6. Remove the alternator assembly.

7. Disconnect the Crankshaft Position (CKP) sensor connector.
8. Remove the connector clamp and wire harness clamp.
9. Remove the wire harness clamp bracket from the wire harness.
10. Remove the bolt, then remove the Crankshaft Position (CKP) sensor.

To install:

11. Apply a light coat of engine oil to the O-ring on the Crankshaft Position (CKP) sensor.
12. Install the Crankshaft Position (CKP) sensor with the bolt and tighten to 7 ft. lbs. (10 Nm).
13. Connect the Crankshaft Position (CKP) sensor connector.
14. The remainder of installation is the reverse of the removal procedure.

2.5L Engine

1. Remove the front apron seal RH.
2. Remove the Crankshaft Position (CKP) sensor.
   a. Disconnect the sensor connector.
   b. Remove the bolt and sensor.

To install:

3. Install the Crankshaft Position (CKP) sensor.
   a. Apply a light coat of engine oil to the O-ring of the sensor.

➡Make sure that the O-ring is not cracked or jammed when installing the sensor.

   b. Apply adhesive to 2 or 3 threads of the bolt.
   c. Install the sensor with the bolt and tighten to 7 ft. lbs. (10 Nm).
   d. Connect the sensor connector.
4. Inspect for oil leaks.
5. Install the front fender apron seal RH.

3.5L Engine

1. Before servicing the vehicle, refer to the Precautions Section.
2. Disconnect the negative battery cable.
3. Remove alternator assembly.
4. Disconnect the cooler compressor assembly.
5. Remove the Crankshaft Position (CKP) sensor connector.
6. Remove the bolt, and then remove the Crankshaft Position (CKP) sensor.

To install:

7. Apply a light coat of engine oil to the O-ring on the Crankshaft Position (CKP) sensor.
8. Install the Crankshaft Position (CKP) sensor with the bolt and tighten to 7ft. lbs. (10 Nm).
9. Connect the Crankshaft Position (CKP) sensor connector.
10. The remainder of installation is the reverse of the removal procedure.

ELECTRONIC CONTROL MODULE (ECM)

LOCATION

See Figures 232 through 234.

REMOVAL & INSTALLATION

2.4L Engine

1. Before servicing the vehicle, refer to the Precautions Section.
2. Disconnect the negative battery cable.
3. Remove the plastic engine cover.
4. Remove the air cleaner inlet assembly.
5. Remove the air cleaner cap sub-assembly.
6. Remove the air cleaner case sub-assembly.
7. Remove the two bolts and air cleaner bracket.
8. Disconnect the two ECM connectors by raising the two levers. While pushing the locks on the two levers, disconnect the two ECM connectors.

➡After disconnecting the connectors, make sure that dirt, water or other foreign matter does not contact the connections of the connectors.

   a. Remove the ECM with the bracket and three bolts.
   b. Remove the four screws and ECM brackets.

To install:

9. Install the bracket to the ECM with the 4 screws, and tighten to 27 inch lbs. (3 Nm).
10. Attach the ECM with the three nuts and tighten to 71 inch lbs. (8 Nm).

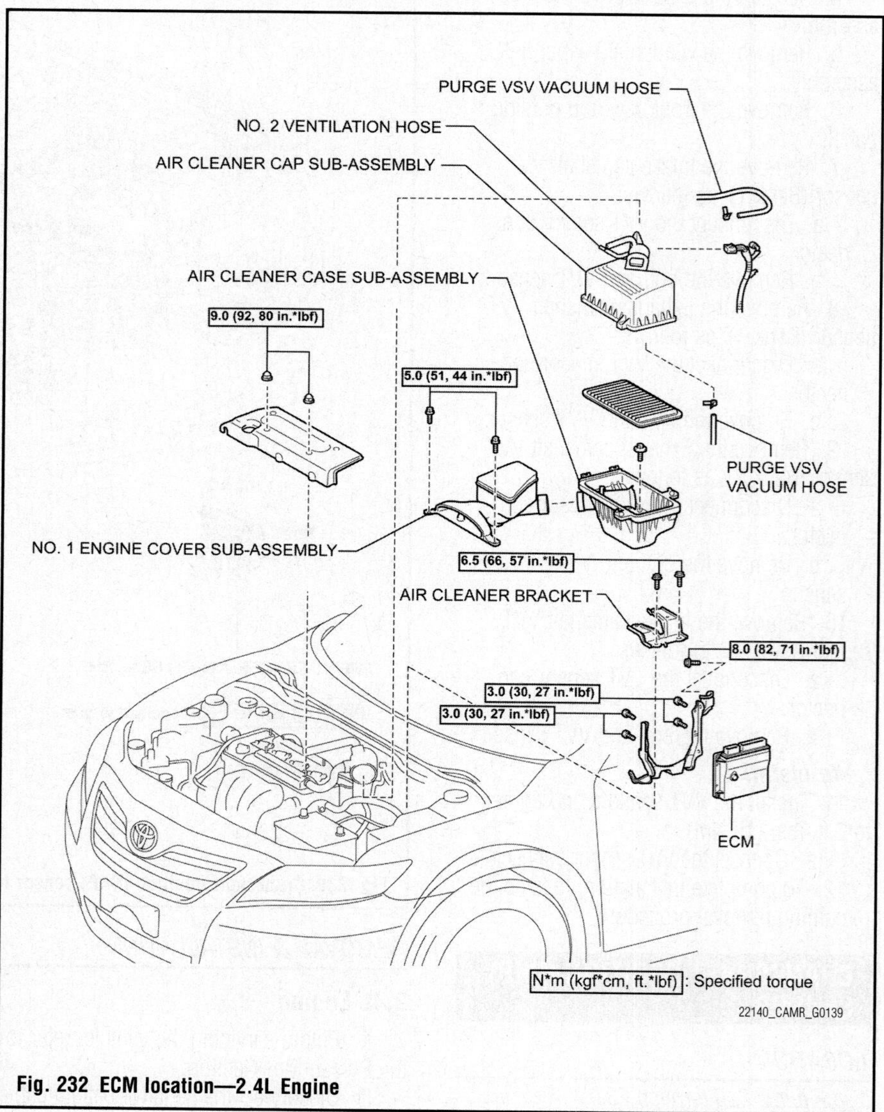

Fig. 232 ECM location—2.4L Engine

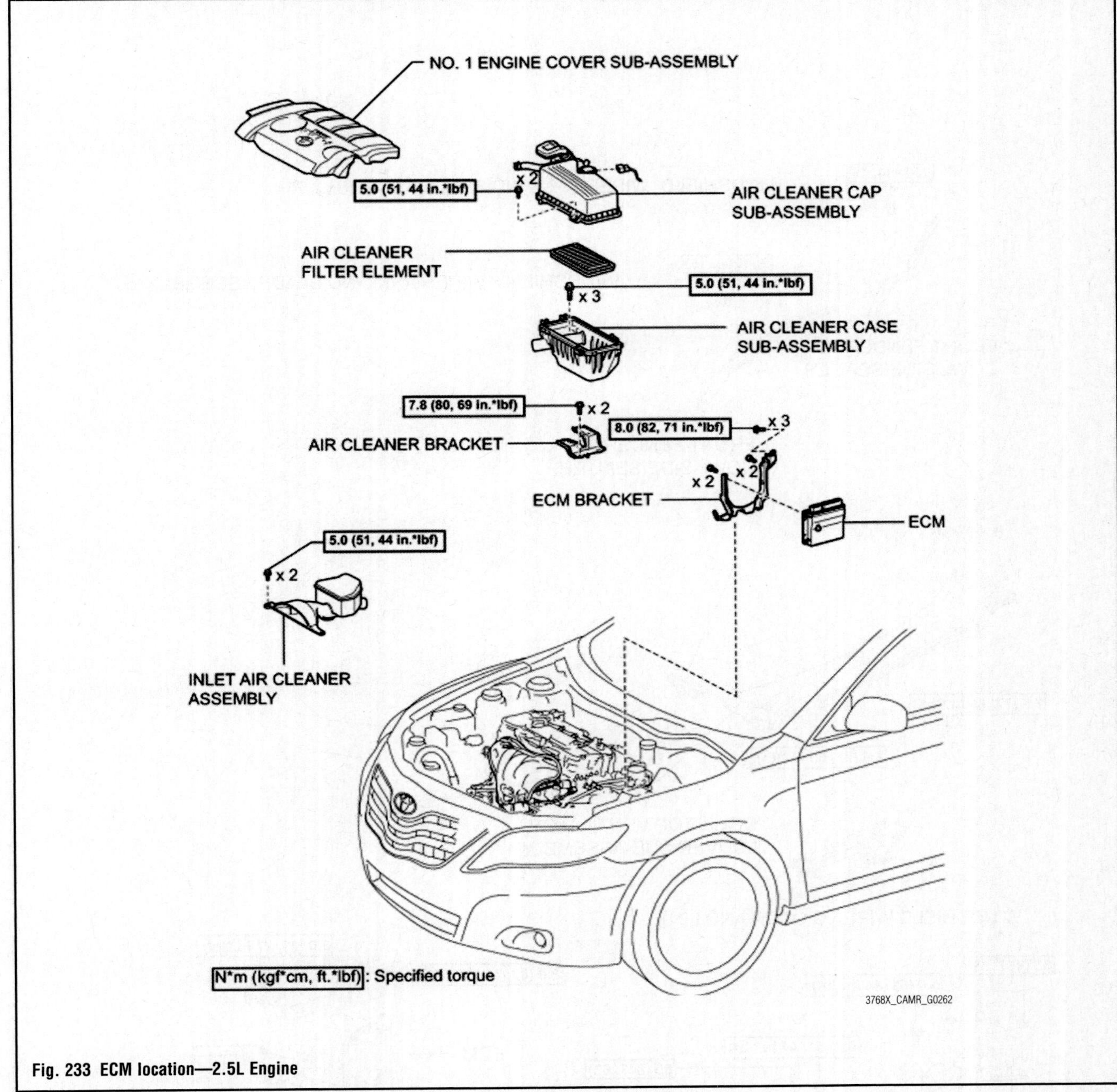

Fig. 233 ECM location—2.5L Engine

➡Make sure that dirt, water or other foreign matter does not contact the connections of the connectors.

11. Connect the two ECM connectors and lower the two levers.
12. Install the ECM to the body.
13. Install the two bolts and air cleaner bracket.
14. Install the air cleaner case sub-assembly.
15. Install the air cleaner cap sub-assembly.
16. Install the air cleaner inlet assembly.
17. Remove the plastic engine cover.
18. Connect the negative battery cable.
19. Register the immobilizer communication ID. If the ECM is replaced, register the ECM communication ID for the immobilizer system (refer to the Service Bulletin for registration).
20. Perform initialization. After replacing the ECM on vehicles with a dynamic laser cruise control system, it is necessary to initialize the ECM so that the ECM can recognize the dynamic laser cruise control system.
21. Be sure to perform the following procedure after replacing the ECM:
    a. Turn the ignition switch on (IG).
    b. Turn the cruise control main switch on.
    c. With the brake pedal depressed, push the cruise control main switch to RES/ACC 3 times within 3 seconds. Check that the buzzer sounds at this time.

➡Do not turn the headlight dimmer switch on at this time because the optical axis automatic adjustment mode has already started, which may lead to an incorrect optical axis setting. If the headlight dimmer switch is turned on by mistake, readjust the optical axis.

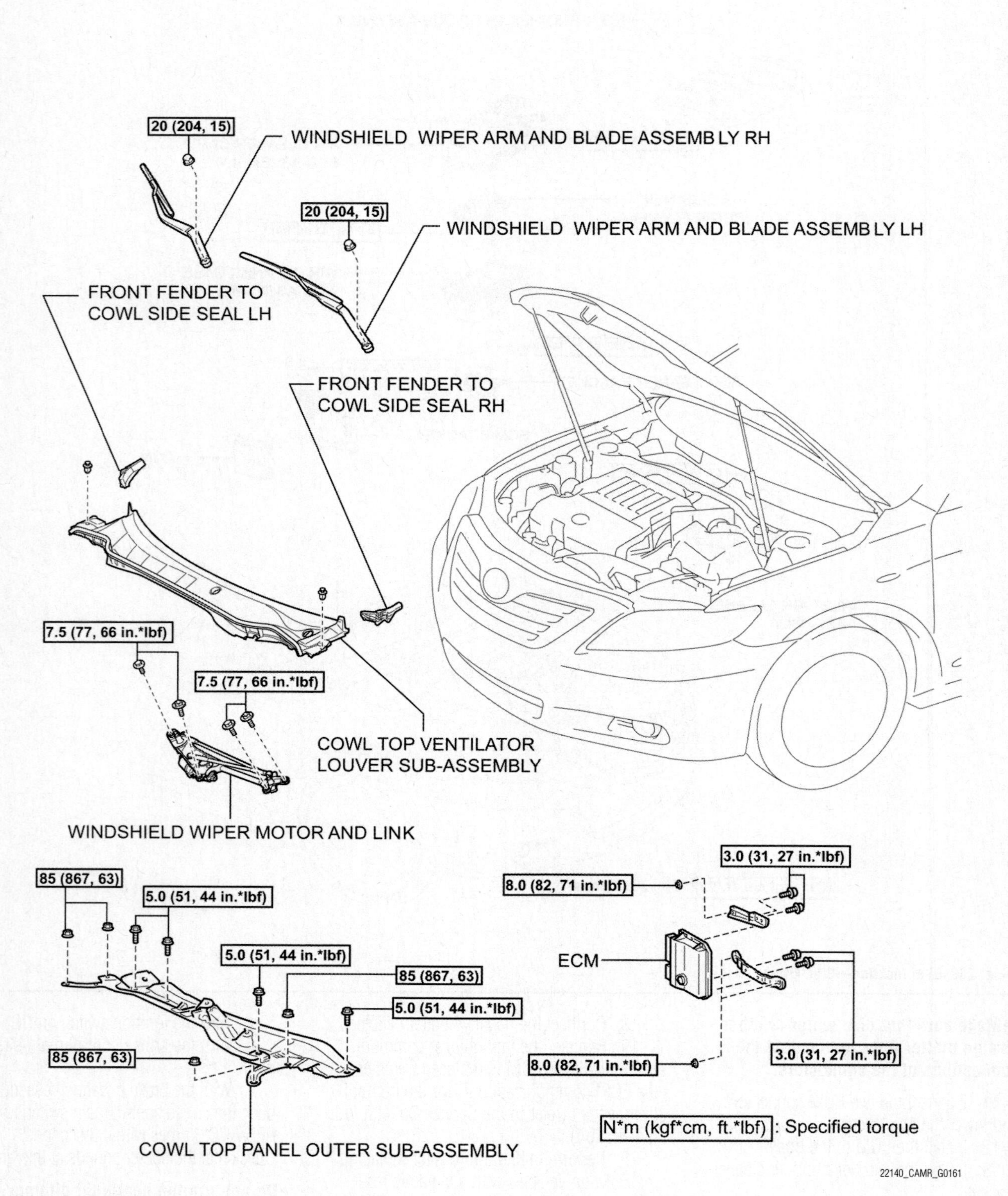

Fig. 234 ECM location—3.5L Engine

2.5L Engine

1. Disconnect the cable from the negative battery cable terminal.
2. Remove the No. 1 engine cover sub assembly.
3. Remove the inlet air cleaner assembly.
4. Remove the air cleaner cap sub assembly.
5. Remove the air cleaner filter element.
6. Remove the air cleaner case sub assembly.
7. Remove the air cleaner bracket.
   a. Remove the 2 bolts and the air cleaner bracket.
8. Disconnect the ECM connector.
   a. Separate the 2 wire harness clamps.
   b. Raise the levers while pushing the locks on the levers, and disconnect the 2 ECM connectors.

➡After disconnecting the connector, make sure that dirt, water and other foreign matter does not contact the connecting part of the connector.

9. Remove the ECM.
   a. Remove the 3 bolts of the ECM bracket.
   b. Remove the ECM with the bracket.
   c. Remove the 4 screws and the ECM bracket.

To install:

10. Install the ECM.
   a. Install the bracket to the ECM with the 4 screws.
   b. Install the ECM with the 3 bolts. Tighten to 71 inch lbs. (8 Nm).
11. Connect the connector.
   a. Connect the 2 ECM connectors and lower the 2 levers.

➡When connecting the connectors, make sure that dirt, water and other foreign matter does not become stuck between the connectors and other parts.

➡Make sure that the 2 levers are securely lowered.

   b. Install the 2 wire harness clamps.
12. Install the air cleaner bracket with the 2 bolts. Tighten to 69 inch lbs. (7.8 Nm).
13. Instll the air cleaner case sub assembly.
14. Install the air cleaner filter element.
15. Install the air cleaner cap sub assembly.
16. Install the air cleaner assembly,
17. Install the No. 1 engine cover sub assembly.
18. Perform registration.

➡Perform VIN registration when replacing the ECM.

➡Registration cannot be completed by only disconnecting and reconnecting the cable of the negative battery terminal.

18. Perform initialization.

➡Initialization cannot be completed by only disconnecting and reconnecting the cable of the negative battery terminal.

3.5L Engine

1. Before servicing the vehicle, refer to the Precautions Section.
2. Disconnect the negative battery cable.
3. Remove both windshield wiper arm and blade assemblies.
4. Remove the cowl top ventilator louver sub-assembly.
5. Remove the windshield wiper motor and link assembly.
6. Remove the outer cowl top panel.
7. Remove the ECM, as follows:
   a. Remove the 3 nuts.
   b. Separate the ECM from the body. When separating the ECM, do not apply excessive force to the wire harness.
   c. Raise the 2 levers while pushing the locks on the 2 levers, and disconnect the 2 ECM connectors.

➡After disconnecting the connectors, make sure that dirt, water or other foreign matter does not contact the connections of the connectors.

   d. Remove the ECM.
   e. Remove the 4 screws and 2 ECM brackets.

To install:

8. Install the 2 ECM brackets with the 4 screws, and tighten to 27 inch lbs. (3 Nm).
9. Connect the 2 ECM connectors and lower the 2 levers.

➡Make sure that dirt, water or other foreign matter does not contact the connections of the connectors.

10. Install the ECM to the body.
11. Attach the ECM with the 3 nuts and tighten to 71 inch lbs. (8 Nm).
12. Install the outer cowl top panel.
13. Install the windshield wiper motor and link assembly.
14. Install the cowl top ventilator louver sub-assembly.
15. Install both windshield wiper arm and blade assemblies.
16. Connect the negative battery cable.
17. Register the immobilizer communication ID. If the ECM is replaced, register the ECM communication ID for the immobilizer system (refer to the Service Bulletin for registration).
18. Perform initialization. After replacing the ECM on vehicles with a dynamic laser cruise control system, it is necessary to initialize the ECM so that the ECM can recognize the dynamic laser cruise control system.
19. Be sure to perform the following procedure after replacing the ECM:
   a. Turn the ignition switch on (IG).
   b. Turn the cruise control main switch on.
   c. With the brake pedal depressed, push the cruise control main switch to RES/ACC 3 times within 3 seconds. Check that the buzzer sounds at this time.

➡Do not turn the headlight dimmer switch on at this time because the optical axis automatic adjustment mode has already started, which may lead to an incorrect optical axis setting. If the headlight dimmer switch is turned on by mistake, readjust the optical axis.

RESET

➡The Vehicle Identification Number (VIN) must be input into the replacement ECM.

➡The VIN is of a 17-digit alphanumeric vehicle identification number. The Techstream is required to register the VIN.

Description

Read VIN: Explains the VIN reading process in a flowchart. This process allows the VIN stored in the ECM to be read in order to confirm that the 2 VINs, provided with the vehicle and stored in the vehicle ECM, are the same.

Write VIN: Explains the VIN writing process in a flowchart. This process allows the VIN to be input into the ECM. If the ECM is replaced, or the ECM VIN and vehicle VIN do not match, the VIN can be registered, or overwritten in the ECM by following this procedure.

Read VIN

1. Confirm the vehicle VIN.
2. Connect the Techstream to the DLC3.
3. Turn the ignition switch to ON.
4. Turn the Techstream on.
5. Enter the following menus: Powertrain / Engine / Utility / VIN / VIN Read.

Write VIN

6. Confirm the vehicle VIN.
7. Connect the Techstream to the DLC3.

8. Turn the ignition switch to ON.
9. Turn the Techs ream on.
10. Enter the following menus: Powertrain / Engine / Utility / VIN / VIN Write.

ENGINE COOLANT TEMPERATURE (ECT) SENSOR

LOCATION

The ECT sensor is connected to the air cleaner case sub assembly.

REMOVAL & INSTALLATION

2.4L Engine

See Figure 235.

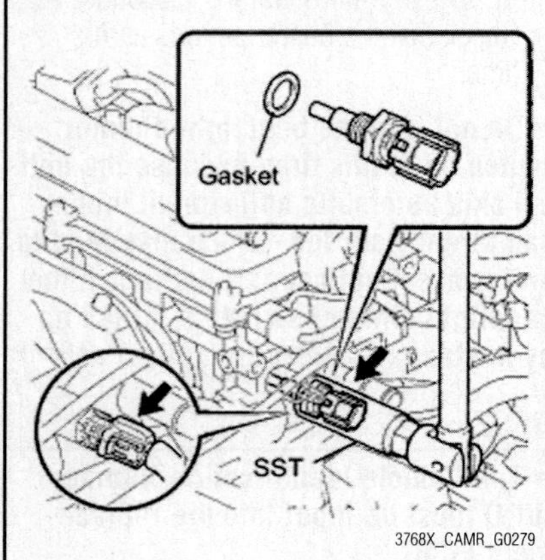

Fig. 235 Removing the ECT sensor—2.4L engine

1. Drain the engine coolant.
2. Remove the No. 1 engine cover sub assembly.
3. Remove the air cleaner inlet assembly.
4. Remove the air cleaner cap sub assembly.
5. Remove the air cleaner case sub assembly.
6. Remove the Engine Coolant Temperature (ECT) sensor.
   a. Disconnect the ECT sensor connector.
   b. Using the special tool, remove the ECT sensor and gasket.

To install:

7. To install, reverse the removal procedure. Tighten the ECT sensor to 15 ft. lbs. (20 Nm).

2.5L Engine

See Figure 236.

1. Remove the front wheel opening extension pads.

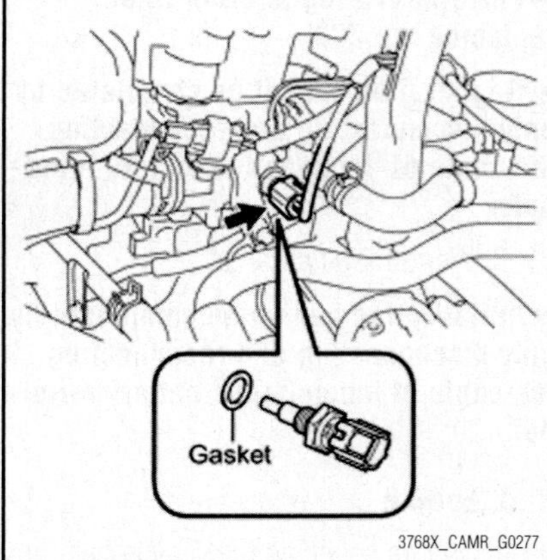

Fig. 236 Removing the Engine Coolant Temperature (ECT) sensor—2.5L engine

2. Remove the engine under covers.
3. Drain the engine coolant.
4. Remove the No. 1engine cover sub assembly.
5. Remove the inlet air cleaner assembly.
6. Remove the air cleaner cap sub assembly.
7. Remove the air cleaner filter element.
8. Remove the air cleaner case sub assembly.
9. Remove the Engine Coolant Temperature (ECT) sensor:
   a. Disconnect the Engine Coolant Temperature (ECT) sensor connector.
   b. Remove the Engine Coolant Temperature (ECT) sensor and gasket.

To install:

10. To install, reverse the removal procedure. Tighten the ECT sensor to 15 ft. lbs. (20 Nm).

3.5L Engine

See Figure 237.

Fig. 237 Removing the ECT sensor—3.5L engine

1. Remove the engine under covers.
2. Drain the engine coolant.
3. Remove the v-bank cover sub assembly.
4. Remove the air cleaner inlet assembly.
5. Remove the air cleaner cap sub assembly.
6. Remove the air cleaner case sub assembly.
7. Remove the No. 1 air cleaner inlet.
8. Remove the Engine Coolant Temperature (ECT) sensor.
   a. Remove the ECT sensor connector.
   b. Remove the Engine Coolant Temperature (ECT) sensor.

To install:

9. To install, reverse the removal procedure. Tighten the ECT sensor to 15 ft. lbs. (20 Nm).

EVAPORATIVE EMISSIONS (EVAP) CANISTER

REMOVAL & INSTALLATION

2.4L Engine

See Figure 238.

1. Remove the fuel tank.
2. Remove the charcoal canister assembly.
   a. Disconnect the fuel tank vent hose from the charcoal canister. Push the connector deep inside. Pinch portion A. Pull out the connector.
   b. Disconnect the charcoal canister filter sub assembly from the charcoal canister. Push the connector deep inside. Pinch portion A. Pull out the connector.
   c. Disconnect the vapor pressure sensor connector.
   d. Disconnect the wire harness clamp.
   e. Disconnect the purge line hose from the charcoal canister.

Fig. 238 Removing the charcoal canister

f. Remove the 2 bolts, clip and charcoal canister.

To install:

3. To install, reverse the removal procedure. Tighten the charcoal canister to 29 ft. lbs. (39 Nm).

2.5L Engine

See Figures 239 through 242.

1. Remove the fuel tank.
2. Remove the charcoal canister assembly (except PZEV).
    a. Disconnect the air inlet tube.

Fig. 239 Removing the charcoal canister assembly (except PZEV)

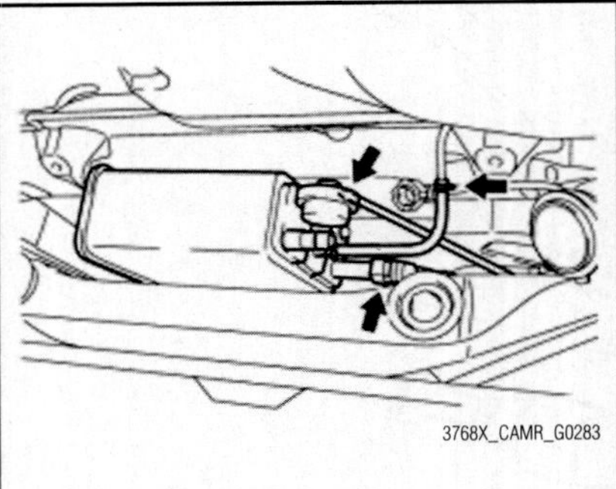

Fig. 240 Disconnecting the vacuum hose, purge line hose and vent hose (PZEV)

Fig. 241 Removing the charcoal canister assembly (PZEV)

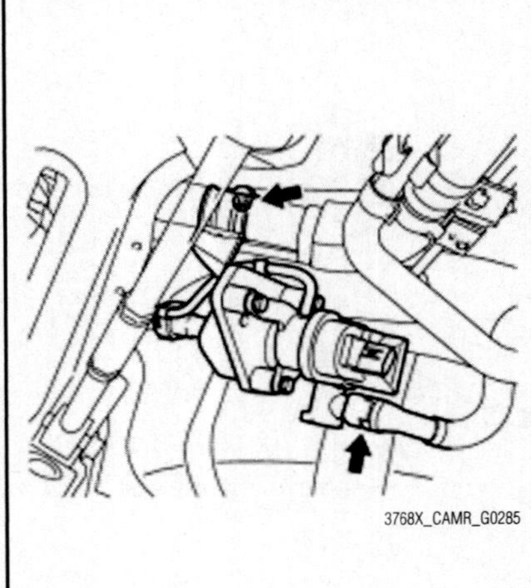

Fig. 242 Removing the trap canister with outlet valve assembly (PZEV)

    b. Disconnect the connector, clamp and purge line hose.
    c. Remove the 2 bolts, clip and charcoal canister assembly.
3. Remove the charcoal canister assembly (PZEV).
    a. Disconnect the vacuum hose, purge line hose and vent hose.
    b. Remove the 2 bolts, clip and charcoal canister assembly.
4. Remove the trap canister with outlet valve assembly (PZEV).
    a. Disconnect the air inlet tube.
    b. Remove the nut and trap canister with outlet valve assembly.

To install:

5. Install the trap canister with the outlet valve assembly (PZEV) with the nut. Tighten to 66 inch lbs. (7.5 Nm). Connect the air inlet tube.
6. Install the charcoal canister (PZEV).
    a. Install the charcoal canister assembly with the 2 bolts and clip. Tighten to 29 ft. lbs. (39 Nm).
    b. Connect the vacuum hose, air inlet tube and vent hose.
7. Install the charcoal canister assembly (except PZEV).
    a. Install the charcoal canister assembly with the 2 bolts and clip. Tighten to 29 ft. lbs. (39 Nm).
    b. Connect the purge line hose, connector and clamp.
    c. Connect the air inlet tube.
8. Install the fuel tank assembly.

3.5L Engine

See Figure 243.

1. Remove the fuel tank.

Fig. 243 Removing the charcoal canister

2. Remove the charcoal canister assembly.
    a. Disconnect the fuel tank vent hose from the charcoal canister. Push the connector deep inside. Pinch portion A. Pull out the connector.
    b. Disconnect the charcoal canister filter sub assembly from the charcoal canister. Push the connector deep inside. Pinch portion A. Pull out the connector.
    c. Disconnect the vapor pressure sensor connector.
    d. Disconnect the wire harness clamp.
    e. Disconnect the purge line hose from the charcoal canister.
    f. Remove the 2 bolts, clip and the charcoal canister.

To install:

3. Install the charcoal canister assembly.
    a. Install the 2 bolts, clip and charcoal canister. Tighten to 29 ft. lbs. (39 Nm).
    b. Connect the purge line hose to the charcoal canister.
    c. Connect the wire harness clamp.
    d. Connect the vapor pressure sensor connector.
    e. Connect the charcoal canister filter sub-assembly to the charcoal canister.
    f. Connect the fuel tank vent hose to the charcoal canister.
4. Install the fuel tank.

HEATED OXYGEN SENSOR (HO2S)

LOCATION

See Figures 244 through 246.

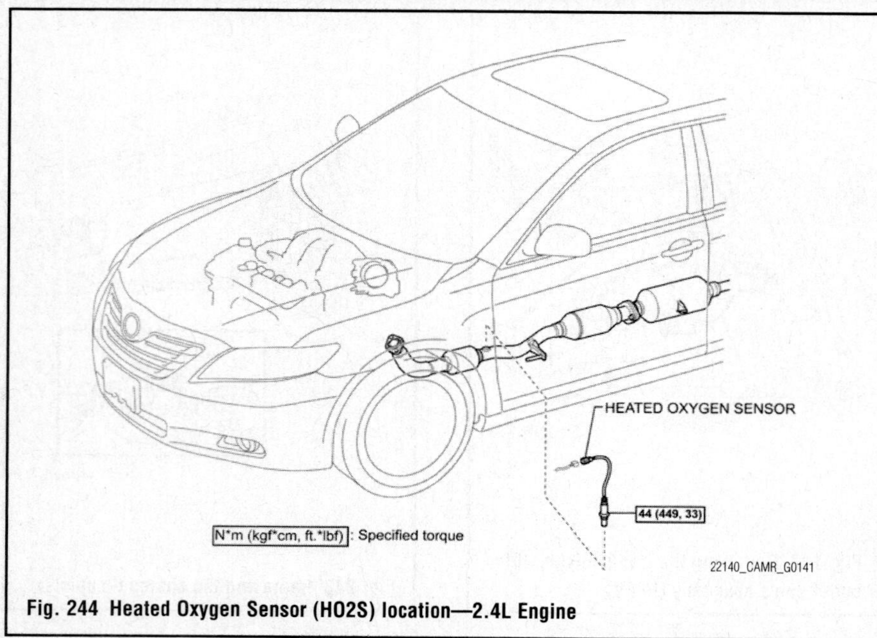

Fig. 244 Heated Oxygen Sensor (HO2S) location—2.4L Engine

REMOVAL & INSTALLATION

2.4L Engine

See Figures 247 and 248.

1. Before servicing the vehicle, refer to the Precautions Section.
2. Disconnect the oxygen sensor connectors.

➡Do not damage the Heated Oxygen Sensor (HO2S).

3. Using Special Tool: 09224-00010 or equivalent, remove the two HO2S from the front pipe assembly.

To install:

4. Install the oxygen sensor to the front pipe assembly. Tighten to 32 ft. lbs. (44 Nm). Use a torque wrench with a fulcrum length of 300 mm (11.81 in.).
5. Connect the oxygen sensor connector.

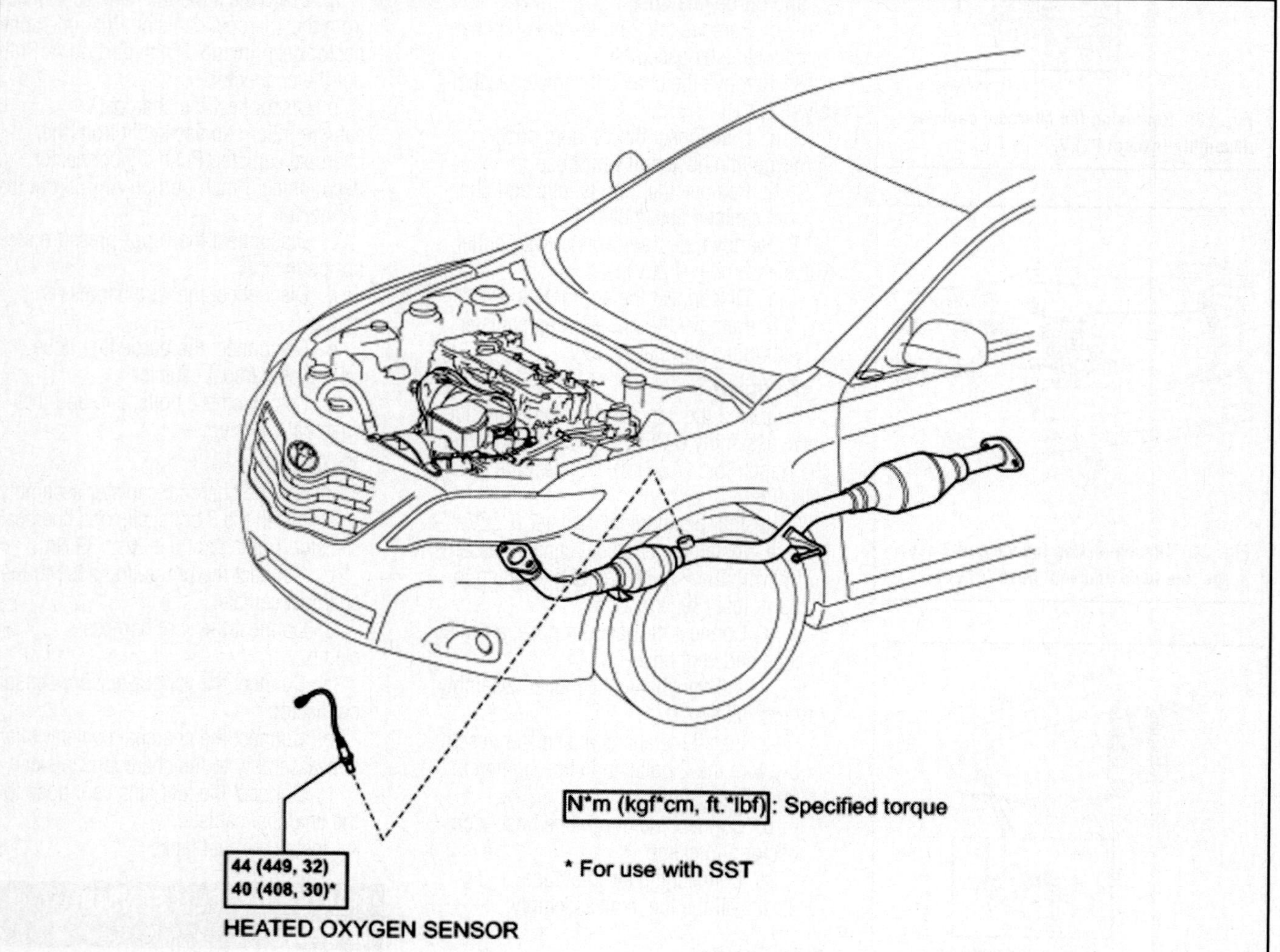

Fig. 245 Heated Oxygen Sensor (HO2S) location—2.5L Engine

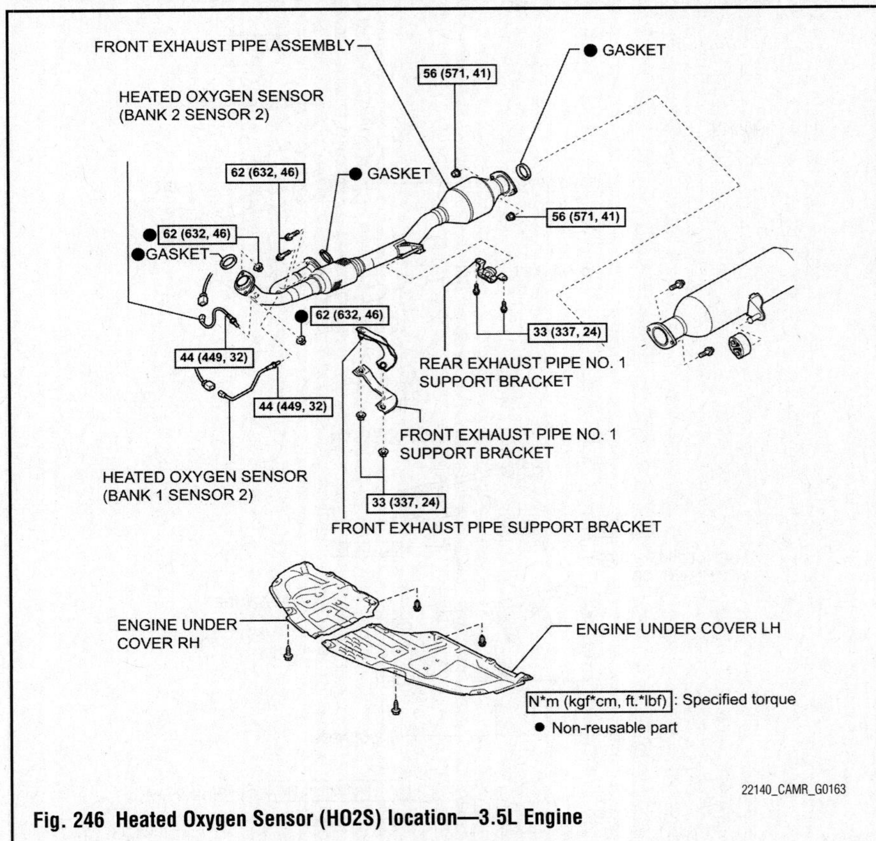

Fig. 246 Heated Oxygen Sensor (HO2S) location—3.5L Engine

Fig. 247 Removing oxygen sensor—2.4L Engine

2.5L Engine

See Figure 249.

1. Remove the Heated Oxygen Sensor (HO2S).
   a. Disconnect the Heated Oxygen Sensor (HO2S) connector.
   b. Using the special tool, remove the Heated Oxygen Sensor (HO2S) from the front exhaust pipe.

To install:

2. To install, reverse the removal procedure. If using the special tool tighten the Heated Oxygen Sensor (HO2S) to 30 ft. lbs. (40 Nm). If tightening without the special tool, tighten to 32 ft. lbs. (44 Nm). Inspect for exhaust gas leaks.

➡The "with special tool" torque value is effective when using SST with a fulcrum length of 1.18 inch (30 mm).

➡The "with special tool" torque value is effective when using a torque wrench with a fulcrum length of 11.81 inch (300 mm).

3.5L Engine

See Figures 250 and 251.

1. Before servicing the vehicle, refer to the Precautions Section.
2. Remove the front exhaust pipe assembly.
3. Disconnect the 2 Heated Oxygen Sensor (HO2S) connectors.
4. Using Special Tool: 09224-00010 or equivalent, remove the 2 HO2S from the front pipe assembly.

To install:

5. Install the 2 HO2S to the front pipe assembly. Tighten to 32 ft. lbs. (44 Nm) and 30 ft. lbs. (40 Nm). Use a torque wrench with a fulcrum length of 300 mm (11.81 in.).
6. Connect the 2 sensor connectors.

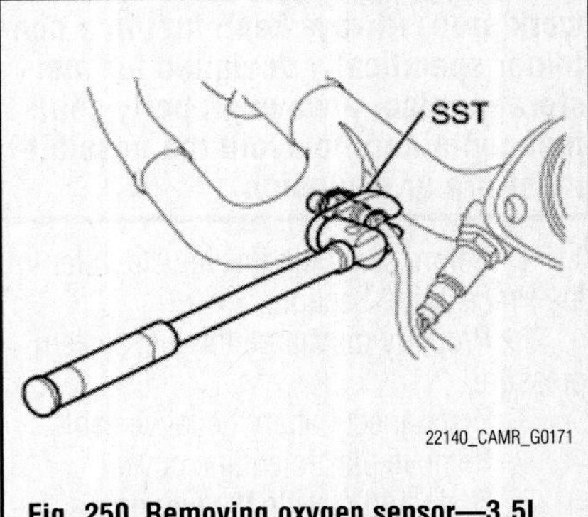

Fig. 250 Removing oxygen sensor—3.5L Engine

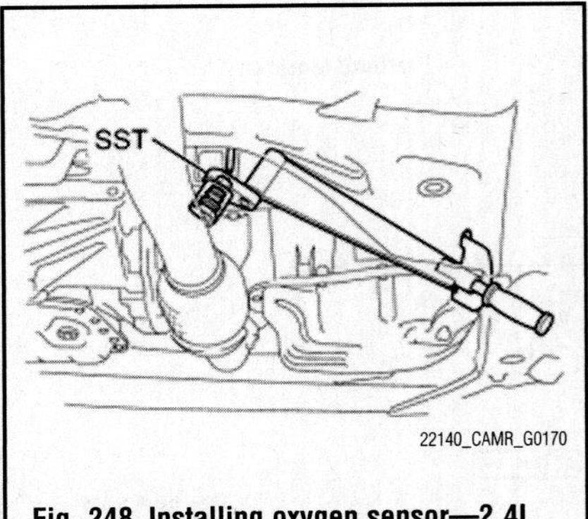

Fig. 248 Installing oxygen sensor—2.4L Engine

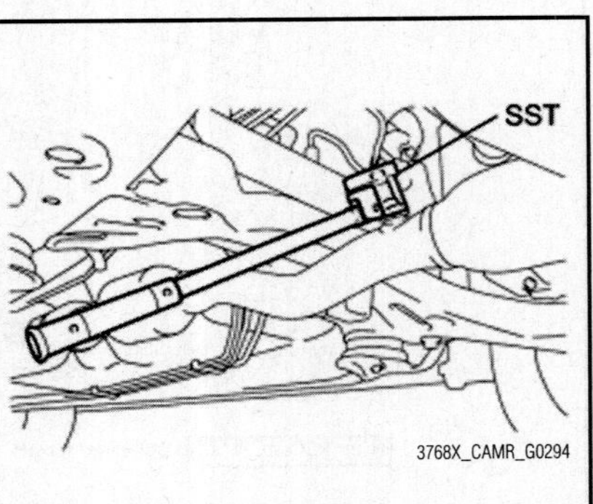

Fig. 249 Removing the Heated Oxygen Sensor (HO2S)

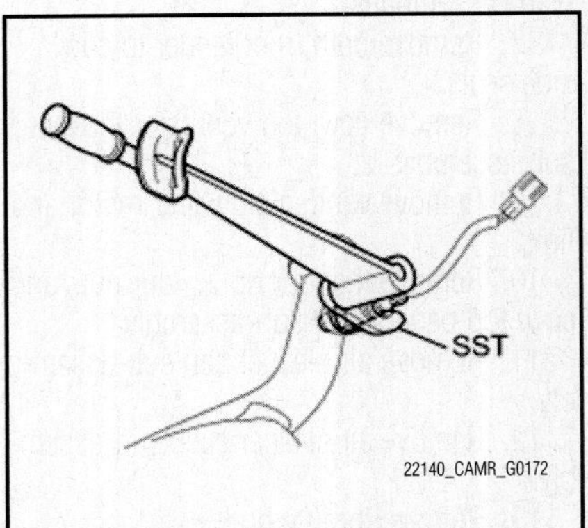

Fig. 251 Installing HO2S—3.5L Engine

7. Remove the front exhaust pipe assembly.

INTAKE AIR TEMPERATURE (IAT) SENSOR

LOCATION

The Intake Air Temperature (IAT) sensor is mounted on the Mass Air Flow (MAF) meter.

REMOVAL & INSTALLATION

See Mass Air Flow (MAF) meter.

KNOCK SENSOR (KS)

LOCATION

See Figures 252 through 254.

REMOVAL & INSTALLATION

2.4L Engine

CAUTION

Observe all applicable safety precautions when working around fuel. Whenever servicing the fuel system, always work in a well ventilated area. Do not allow fuel spray or vapors to come in contact with a spark or open flame. Keep a dry chemical fire extinguisher near the work area. Always keep fuel in a container specifically designed for fuel storage; also, always properly seal fuel containers to avoid the possibility of fire or explosion.

1. Before servicing the vehicle, refer to the Precautions Section.
2. Properly discharge the fuel system pressure.
3. Disconnect battery negative cable.
4. Remove plastic engine cover.
5. Drain and recycle the engine coolant.
6. Remove windshield wiper arms and blade assemblies.
7. Remove both front fender to cowl side seals.
8. Remove cowl top ventilator louver sub-assembly.
9. Remove windshield wiper motor and link.
10. Remove the four bolts, four nuts and cowl top panel outer sub-assembly.
11. Remove air cleaner cap sub-assembly.
12. Remove air cleaner case sub-assembly.
13. Remove throttle body.
14. Disconnect fuel tube.

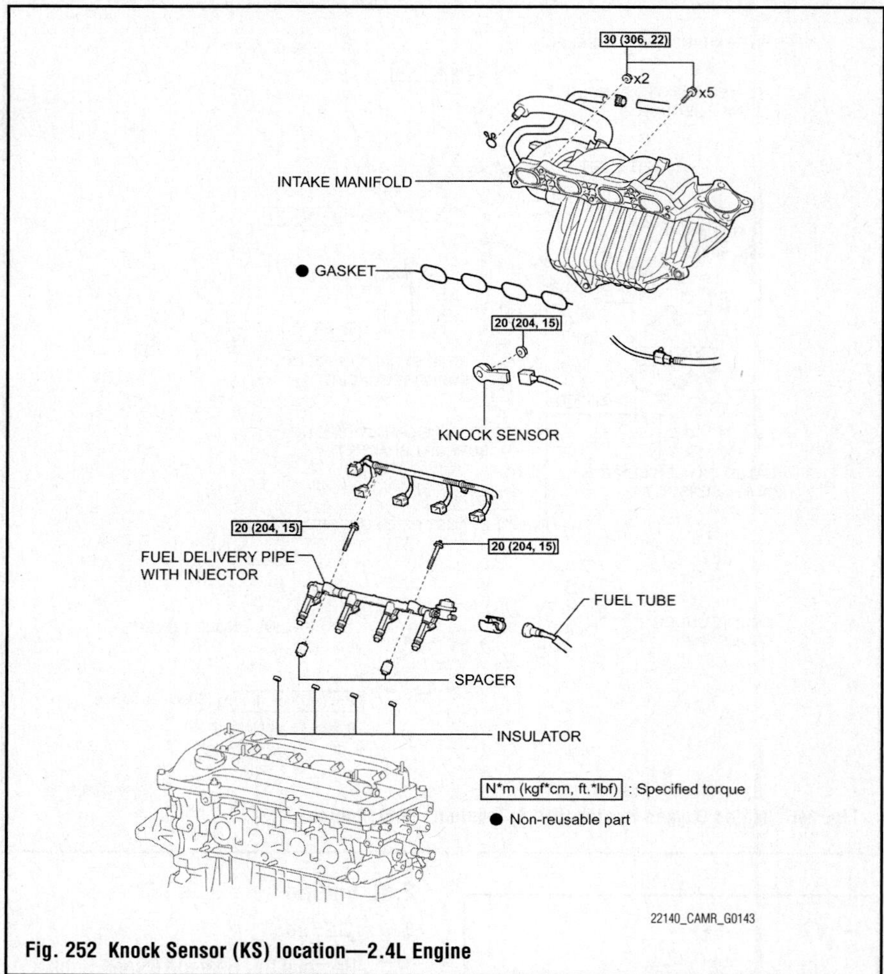

Fig. 252 Knock Sensor (KS) location—2.4L Engine

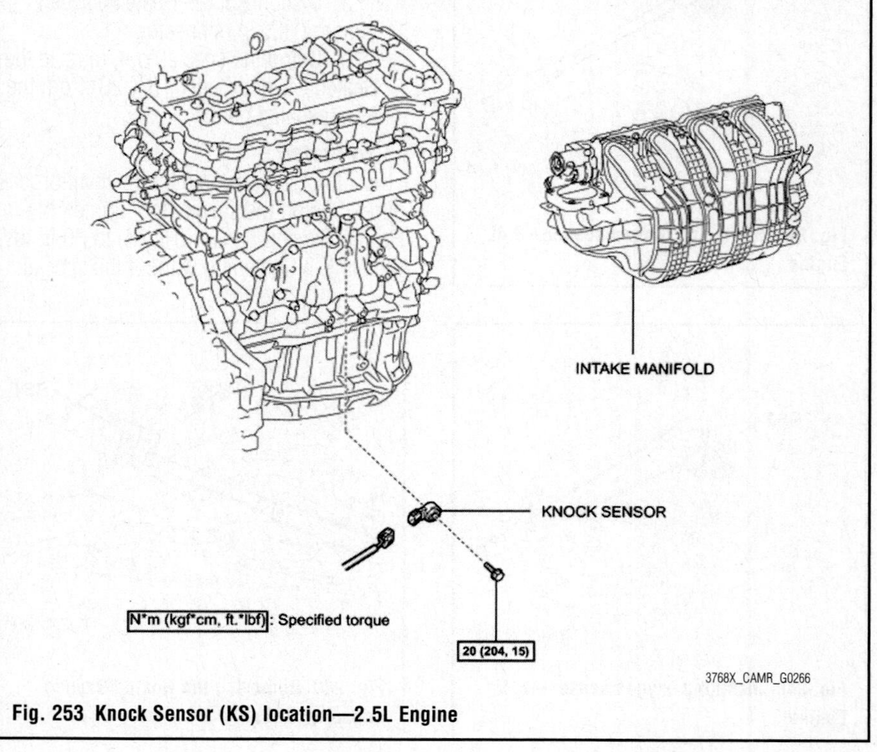

Fig. 253 Knock Sensor (KS) location—2.5L Engine

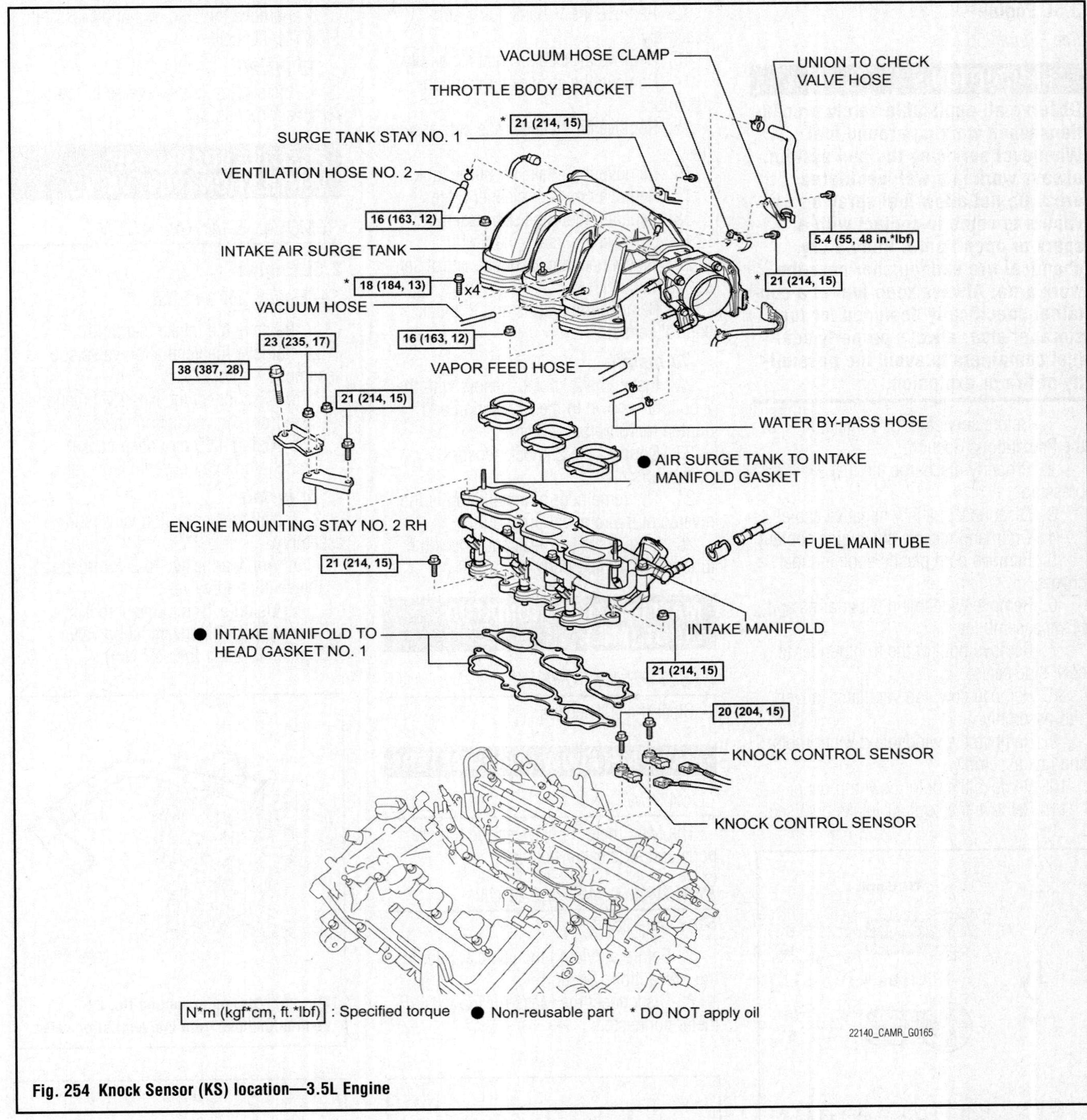

Fig. 254 Knock Sensor (KS) location—3.5L Engine

15. Remove fuel delivery pipe with injector.
16. Disconnect the union to check valve hose from the brake booster.
17. Disconnect the camshaft timing oil control valve connector.
18. Remove the wire harness clamp.
19. Remove the union to check valve hose from the vacuum hose clamp.
20. Remove the 5 bolts, 2 nuts and intake manifold. Remove the gasket from the intake manifold.
21. Disconnect the Knock Sensor (KS) connector. Remove the nut and knock sensor.

To install:

➡Make sure that the Knock Sensor (KS) is in the correct position.

22. Install the knock sensor with the nut and tighten to 15 ft. lbs. (20 Nm).
23. Connect the KS connector.
24. The remainder of installation is the reverse of the removal procedure.
25. Inspect for fuel leak and check the function of throttle body.

2.5L Engine

1. Remove the intake manifold.
2. Remove the Knock Sensor (KS).
   a. Disconnect the sensor connector.
   b. Remove the bolt and sensor.

To install:

3. To install, reverse the removal procedure. Tighten the sensor to 15 ft. lbs. (20 Nm).

3.5L Engine

See Figure 255.

CAUTION

Observe all applicable safety precautions when working around fuel. Whenever servicing the fuel system, always work in a well ventilated area. Do not allow fuel spray or vapors to come in contact with a spark or open flame. Keep a dry chemical fire extinguisher near the work area. Always keep fuel in a container specifically designed for fuel storage; also, always properly seal fuel containers to avoid the possibility of fire or explosion.

1. Before servicing the vehicle, refer to the Precautions Section.
2. Properly discharge the fuel system pressure.
3. Disconnect battery negative cable.
4. Drain and recycle the engine coolant.
5. Remove both plastic engine under covers.
6. Remove windshield wiper arms and blade assemblies.
7. Remove both of the front fender to cowl side seals.
8. Remove cowl top ventilator louver sub-assembly.
9. Remove the windshield wiper motor and link assembly.
10. Remove the outer cowl top panel.
11. Remove the cool air intake duct seal.
12. Remove the V-bank cover sub-assembly.
13. Remove air cleaner inlet assembly.
14. Remove the air cleaner cap sub-assembly.
15. Remove air cleaner case sub-assembly.
16. Remove the intake air surge tank.
17. Remove no. 1 air cleaner inlet.
18. Separate fuel tube sub-assembly.
19. Remove the intake manifold.
20. Disconnect the 2 knock control sensor connectors.
21. Remove the 2 bolts and 2 knock control sensors.

To install:

22. Install the 2 knock sensors with the 2 bolts as shown in the illustration and tighten to 15 ft. lbs. (20 Nm).
23. Connect the 2 knock sensor connectors.
24. The remainder of installation is the reverse of the removal procedure.
25. Inspect for fuel leak and check the function of throttle body.

Fig. 255 Installing Knock sensor—3.5L Engine

MALFUNCTION INDICATOR LIGHT (MIL)

RESET PROCEDURE

Clear the DTC codes.

MASS AIR FLOW (MAF) METER

LOCATION

The MAF meter is between the throttle body and air cleaner housing.

REMOVAL & INSTALLATION

See Figure 256.

1. Before servicing the vehicle, refer to the Precautions Section.
2. Disconnect the Mass Air Flow (MAF) meter connector.
3. Remove the 2 screws and Mass Air Flow (MAF) meter.

To install:

4. Installation is the reverse of the removal procedure.

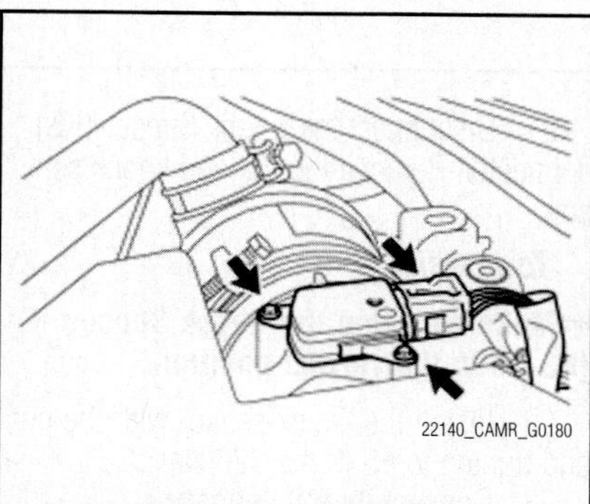

Fig. 256 Mass Air Flow (MAF) meter—2.4L shown, 2.5L and 3.5L are similar

POSITIVE CRANKCASE VENTILATION (PCV) VALVE

REMOVAL & INSTALLATION

2.5L Engine

See Figures 257 and 258.

1. Remove the intake manifold.
2. Remove the ventilation valve sub assembly.
   a. Disconnect the No. 2 ventilation hose from the ventilation valve.
   b. Using a 19 mm deep socket wrench, remove the ventilation valve.

To install:

3. Install the ventilation valve sub assembly.
   a. Apply adhesive to 2-3 threads of the ventilation valve.
   b. Using a 19 mm deep socket wrench, install the ventilation valve. Tighten to 20 ft. lbs. (27 Nm).

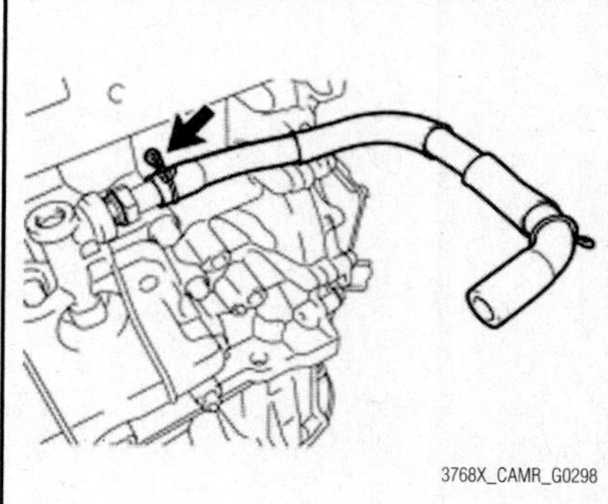

Fig. 257 Disconnecting the No. 2 ventilation hose from the ventilation valve

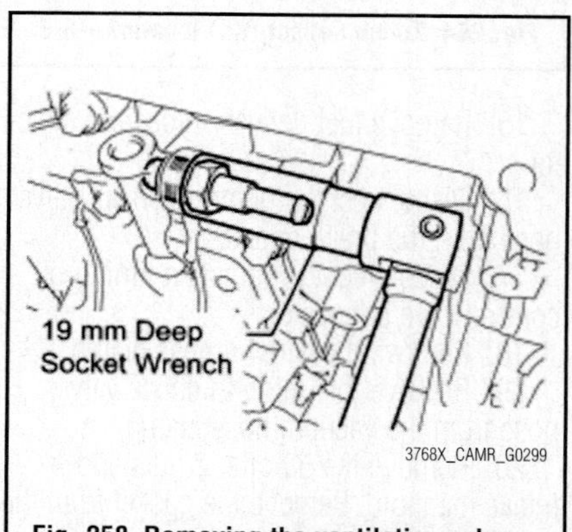

Fig. 258 Removing the ventilation valve

3.5L Engine

See Figures 259 and 260.

1. Remove the v-bank cover sub assembly.
2. Disconnect the ventilation hose.
   a. Disconnect the ventilation hose from the ventilation valve.
3. Remove the ventilation valve.

To install:

4. Install the ventilation valve.

Fig. 259 Disconnecting the ventilation hose from the ventilation valve

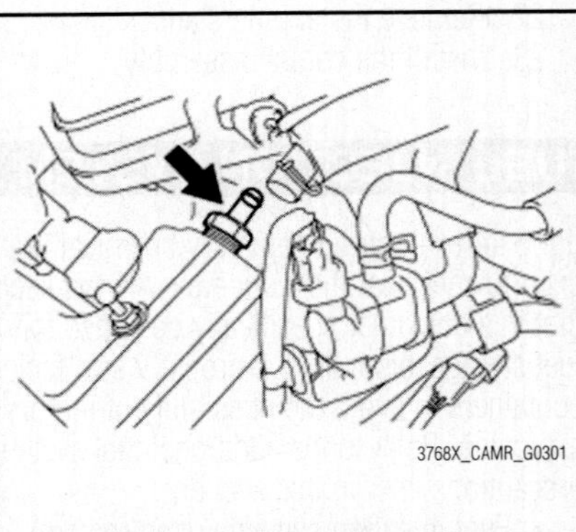

Fig. 260 Removing the ventilation valve

   a. Apply adhesive to 2 or 3 threads.
   b. Install the ventilation valve. Tighten to 20 ft. lbs. (27 Nm).
5. Connect the ventilation hose to the ventilation hose.
6. Install the v-bank cover sub assembly.

THROTTLE POSITION SENSOR (TPS)

LOCATION

The Throttle Position Sensor (TPS) is mounted to the throttle body.

REMOVAL & INSTALLATION

Refer to the Throttle Body removal and installation procedures.

VARIABLE CAMSHAFT TIMING OIL CONTROL SOLENOID

REMOVAL & INSTALLATION

2.4L Engine

See Figure 261.

1. Before servicing the vehicle, refer to the Precautions Section.
2. Remove the plastic engine cover.

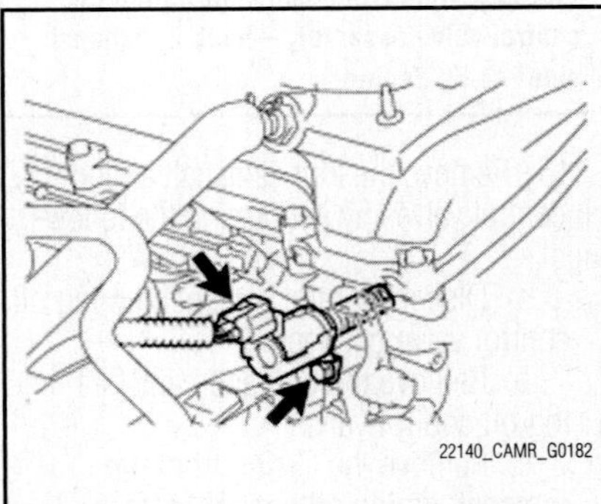

Fig. 261 Removing the camshaft timing oil control valve assembly—2.4L Engine

3. Remove the bolt and disconnect the vacuum hose clamp.
4. Remove the clip nut.
5. Disconnect the camshaft timing oil control valve assembly connector.
6. Remove the bolt and camshaft timing oil control valve assembly.

To install:

7. Apply a light coat of engine oil to an O-ring of the camshaft timing oil control valve assembly sensor.
8. Install the camshaft timing oil control valve assembly with the bolt and tighten to 80 inch lbs. (9 Nm).
9. To complete installation, reverse removal procedure.

2.5L Engine

See Figures 262 and 263.

1. Remove the No. 1 engine cover sub assembly.
2. Remove the camshaft timing oil control valve assembly (exhaust side).
   a. Disconnect the oil control valve connector.
   b. Remove the bolt and oil control valve.
   c. Remove the O-ring from the oil control valve.
3. Remove the camshaft timing oil control valve assembly (intake side).
   a. Disconnect the oil control valve connector,

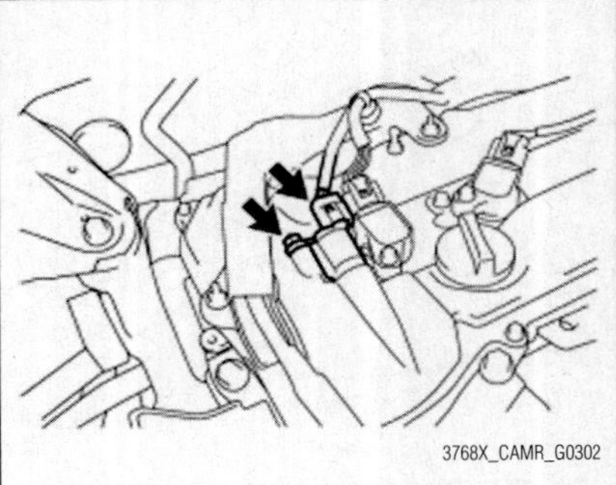

Fig. 262 Disconnecting the oil control valve connector (exhaust side)

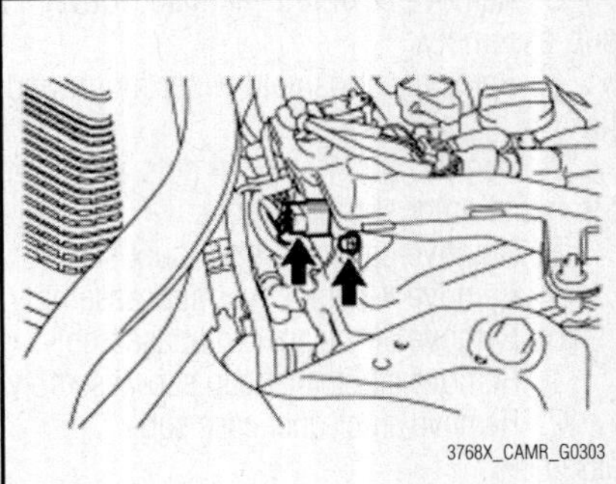

Fig. 263 Disconnecting the oil control valve connector (intake side)

To install:

4. To install, reverse the removal procedure. Tighten the camshaft timing oil control valve assembly to 7 ft. lbs. (10 Nm).

3.5L Engine

See Figures 264 through 267.

1. Remove engine under cover.
2. Drain and recycle the engine coolant.
3. Remove windshield wiper arm and blade assembly.
4. Remove front fender to cowl side seal.

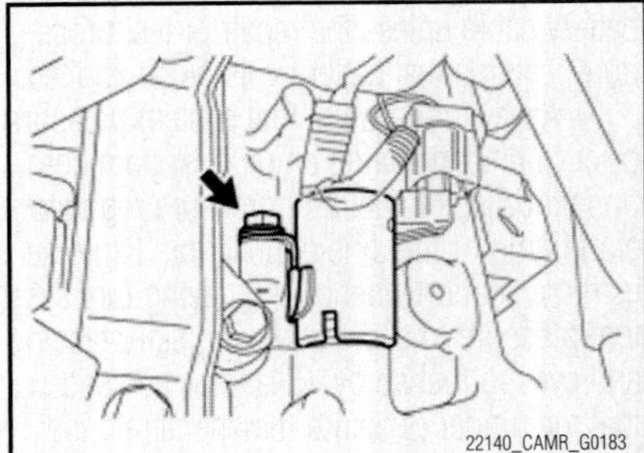

Fig. 264 Removing camshaft timing oil control valve assembly—bank 1, exhaust side—3.5L Engine

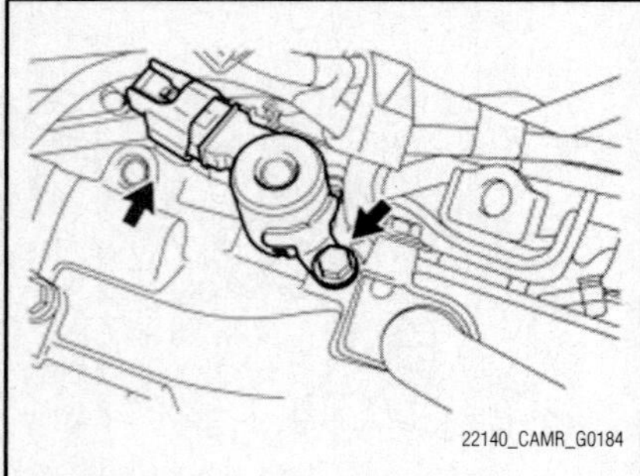

Fig. 265 Removing camshaft timing oil control valve assembly—bank 1, intake side—3.5L Engine

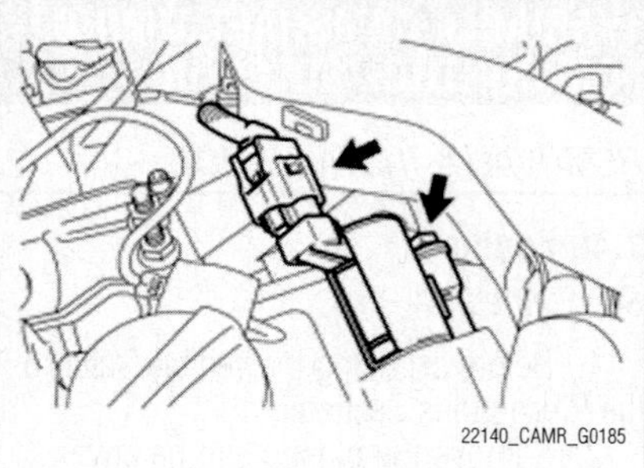

Fig. 266 Removing camshaft timing oil control valve assembly—bank 2, exhaust side—3.5L Engine

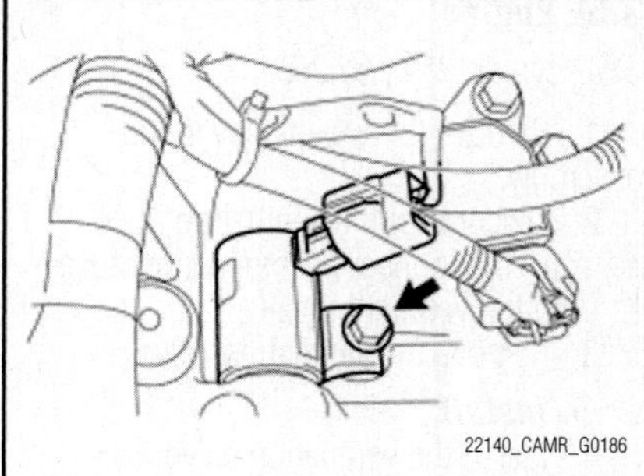

Fig. 267 Removing camshaft timing oil control valve assembly—bank 2, intake side—3.5L Engine

5. Remove cowl top ventilator louver sub-assembly.
6. Remove windshield wiper motor and link.
7. Remove the 4 bolts, 4 nuts and cowl top panel outer sub-assembly.
8. Remove cool air intake duct seal.
9. Remove v-bank cover sub-assembly.
10. Remove air cleaner inlet assembly.
11. Remove air cleaner cap sub-assembly.
12. Remove air cleaner case sub-assembly.
13. Remove no. 1 air cleaner inlet.
14. Remove intake air surge tank.
15. Remove the four camshaft timing oil control valves by performing the following:
   a. Disconnect the camshaft timing oil control valve connector.
   b. Remove the bolt and camshaft timing oil control valve.
   c. Remove the O-ring from the camshaft timing oil control valve.

To install:

16. Install the camshaft timing oil control valve assembly, as follows:
   a. Install the 4 oil control valves with the 4 bolts and tighten to 7 ft. lbs. (10 Nm).
17. Install the 4 camshaft position sensors with the 4 bolts and tighten to 7 ft. lbs. (10 Nm).
18. Apply adhesive (Toyota Genuine Adhesive 1324, Three Bond 1324 or equivalent) around the ventilation valve.
19. Install the ventilation valve sub-assembly and tighten to 20 ft. lbs. (27 Nm).
20. Install the spark plugs and the ignition coil assembly.
21. Install the oil filler cap sub-assembly.
22. Remove the engine stand.
23. Install the engine assembly.

FUEL — GASOLINE FUEL INJECTION SYSTEM

FUEL SYSTEM SERVICE PRECAUTIONS

Safety is the most important factor when performing not only fuel system maintenance, but any type of maintenance. Failure to conduct maintenance and repairs in a safe manner may result in serious personal injury or death. Work on a vehicle's fuel system components can be accomplished safely and effectively by adhering to the following rules and guidelines.

- To avoid the possibility of fire and personal injury, always disconnect the negative battery cable unless the repair or test procedure requires that battery voltage be applied.
- Always relieve the fuel system pressure prior to disconnecting any fuel system component (injector, fuel rail, pressure regulator, etc.) fitting or fuel line connection. Exercise extreme caution whenever relieving fuel system pressure to avoid exposing skin, face and eyes to fuel spray. Please be advised that fuel under pressure may penetrate the skin or any part of the body that it contacts.
- Always place a shop towel or cloth around the fitting or connection prior to loosening to absorb any excess fuel due to spillage. Ensure that all fuel spillage is quickly removed from engine surfaces. Ensure that all fuel-soaked cloths or towels are deposited into a flame-proof waste container with a lid.
- Always keep a dry chemical (Class B) fire extinguisher near the work area.
- Do not allow fuel spray or fuel vapors to come into contact with a spark or open flame.
- Always use a second wrench when loosening or tightening fuel line connection fittings. This will prevent unnecessary stress and torsion on fuel piping. Always follow the proper torque specifications.
- Always replace worn fuel fitting O-rings with new ones. Do not substitute fuel hose where rigid pipe is installed.

Before servicing any vehicle, please be sure to read all of the following precautions, which deal with personal safety, prevention of component damage, and important points to take into consideration when servicing a motor vehicle:

- Observe all applicable safety precautions when working around fuel. Whenever servicing the fuel system, always work in a well-ventilated area. Do not allow fuel spray or vapors to come in contact with a spark, open flame, or excessive heat (a hot drop light, for example). Keep a dry chemical fire extinguisher near the work area. Always keep fuel in a container specifically designed for fuel storage; also, always properly seal fuel containers to avoid the possibility of fire or explosion. Refer to the additional fuel system precautions later in this section.
- Fuel injection systems often remain pressurized, even after the engine has been turned **OFF**. The fuel system pressure must be relieved before disconnecting any fuel lines. Failure to do so may result in fire and/or personal injury.
- All new vehicles are now equipped with an air bag system, often referred to as a Supplemental Restraint System (SRS) or Supplemental Inflatable Restraint (SIR) system. The system must be disabled before performing service on or around system components, steering column, instrument panel components, wiring and sensors. Failure to follow safety and disabling procedures could result in accidental air bag deployment, possible personal injury and unnecessary system repairs.
- Disconnecting the negative battery cable on some vehicles may interfere with the functions of the on-board computer system(s) and may require the computer to

undergo a relearning process once the negative battery cable is reconnected.

Before inspecting and repairing the fuel system, disconnect the cable from the negative (-) battery terminal.

Keep gasoline away from rubber or leather parts.

1. Check that there are no fuel leaks from the fuel system after doing any maintenance or repairs.

FUEL SYSTEM PRESSURE

RELIEVING

CAUTION

Perform the following procedures to prevent fuel from spilling out before removing any fuel system parts.

CAUTION

Pressure will still remain in the fuel line even after performing the following procedures. When disconnecting the fuel line, cover it with a shop rag or a piece of cloth to prevent fuel from spraying or coming out.

1. Disconnect the fuel pump connector:
   a. Remove the rear seat cushion assembly.
   b. Remove the rear floor service whole cover.
   c. Disconnect the fuel pump connector.
   d. Start the engine.
   e. After the engine stops, turn the ignition switch off.

➡DTC P0171/25 (fuel problem) may be detected.

   f. Crank the engine again. Check that the engine does not start.
   g. Remove the fuel tank cap to discharge pressure from the fuel tank.
   h. Disconnect the cable from the negative (-) battery terminal.
   i. Reconnect the fuel pump connector.
   j. Install the rear floor service hole cover.
   k. Install the rear seat.
2. Check that there are no fuel leaks from the fuel system after doing any maintenance or repairs.

FUEL FILTER

REMOVAL & INSTALLATION

1. Before servicing the vehicle, refer to the precautions section.
2. Remove the fuel pump from the vehicle.
3. Remove the fuel pump filter, as follows:

➡Do not damage the fuel pump filter. Do not remove the suction filter.

   a. Using a screwdriver, pry out the clips.
   b. Pull out the fuel pump filter from the fuel pump.

To install:

4. Install the fuel pump filter with a new clip.
5. Install the fuel pump.

FUEL LEVEL SENDING UNIT

REMOVAL & INSTALLATION

1. Before servicing the vehicle, refer to the precautions section.
2. Remove the fuel pump from the vehicle.
3. Remove the fuel sender gauge assembly, as follows:
   a. Disconnect the fuel sender gauge connector.
   b. Unlock the fuel sender gauge, and slide it to remove.

To install:

4. Slide the fuel sender gauge to engage with the claw.
   a. Connect the fuel sender gauge connector.
5. Install the fuel pump.

FUEL PUMP

REMOVAL & INSTALLATION

See Figures 268 and 269.

1. Before servicing the vehicle, refer to the precautions section.
2. Discharge fuel system pressure.
3. Disconnect battery negative cable.
4. Remove the rear seat cushion assembly.
5. Remove the rear floor service hole cover.
6. Disconnect the fuel pump connector.
7. Separate the fuel pump tube sub-assembly, as follows:

➡Check if there is any dirt or mud around the connector before this operation and remove the dirt as necessary.

➡Be careful of mud because the quick connector has an O-ring which seals the pipe and connector that can be contaminated.

➡Do not use any tools in this operation.

➡Do not bend or twist the nylon tube.

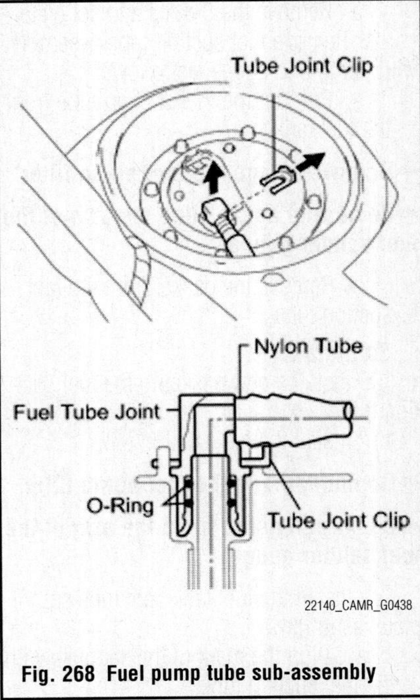

Fig. 268 Fuel pump tube sub-assembly

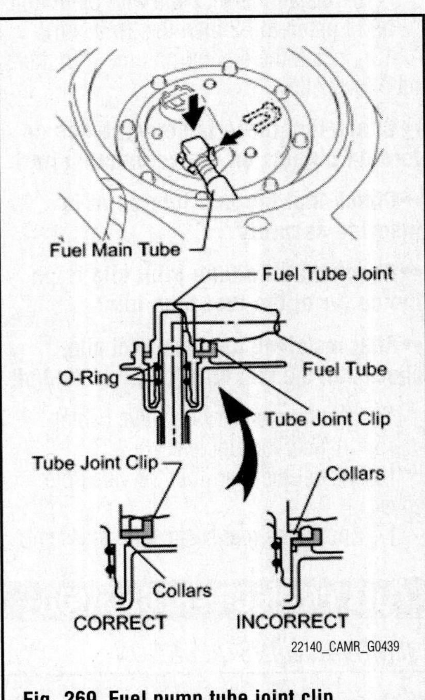

Fig. 269 Fuel pump tube joint clip

Cover the fuel tube joint with a plastic bag.

➡When the fuel tube joint and fuel suction plate are stuck, pinch the fuel tank tube between fingers, and turn it carefully to release it. Disconnect the fuel tank tube.

   a. Remove the tube joint clip, and pull out the fuel pump tube.
8. Remove fuel tank vent tube set plate, as follows:

a. Remove the 8 bolts and set plate.

9. Remove fuel suction tube assembly with pump and gauge, as follows:

a. Pull out the fuel suction tube from the fuel tank.

➡Do not damage the fuel pump filter.

➡Be careful not to bend the arm of the fuel sender gauge.

b. Remove the gasket from the fuel suction tube.

To install:

10. Install a new gasket to the fuel suction tube.

11. Install the fuel suction tube.

➡Do not damage the fuel pump filter.

➡Be careful not to bend the arm of the fuel sender gauge.

12. Install the fuel tank vent tube set plate, as follows:

a. Align the mark of the set plate with the fuel suction tube.

b. Install the set plate with the 8 bolts and tighten to 52 inch lbs. (5.9 Nm).

13. Install the fuel pump tube with the tube joint clip.

➡Check that there is no scratches or foreign objects on the connecting part.

➡Check that the fuel tube joint is inserted securely.

➡Check that the tube joint clip is on the collar of the fuel tube joint.

➡After installing the tube joint clip, check that the fuel tube joint is pulled off.

14. Connect battery negative cable.

15. Inspect for fuel leak.

16. Install the rear floor service hole cover.

17. Install the rear seat cushion assembly.

FUEL PRESSURE REGULATOR

REMOVAL & INSTALLATION

2.4L Engine

See Figure 270.

1. Before servicing the vehicle, refer to the precautions section.

2. Remove the fuel pump from the vehicle.

3. Using a screwdriver with its tip wrapped in protective tape, remove the fuel pressure regulator from the fuel filter.

4. Remove the 2 O-rings from the fuel pressure regulator.

To install:

5. Apply a light coat of spindle oil or gasoline to a new O-ring, and install it to the fuel pressure regulator.

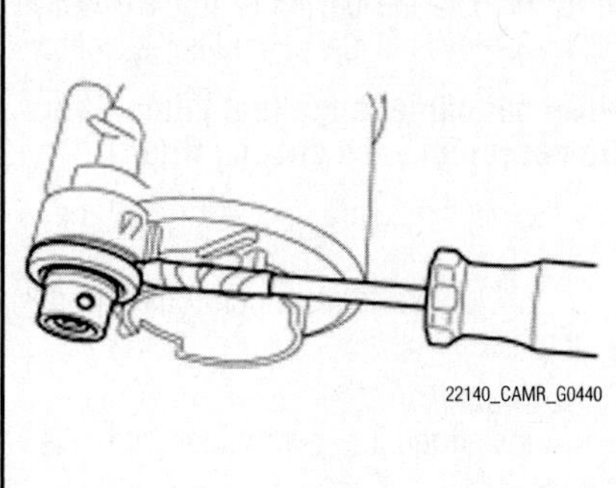

Fig. 270 Removing the fuel pressure regulator from the fuel filter—2.4L

6. Push in the fuel pressure regulator to the fuel tank fuel filter.

7. Install the fuel pump filter.

8. Install the fuel pump.

2.5L Engine

See Figures 271 through 274.

1. Before servicing the vehicle, refer to the precautions section.

2. Remove the fuel suction tube assembly with the pump and gauge.

3. Remove the fuel sender gauge assembly.

4. Remove the No. 1 fuel tank band bracket.

a. Disconnect the fuel pump harness connector from the fuel suction plate.

b. Using needle nose pliers, remove the e-ring.

c. Disengage the 4 claws of the No. 1 fuel suction support and remove the No. 1 fuel tank band bracket from the fuel suction plate with the fuel filter.

d. Remove the spring from the fuel suction plate.

➡Do not disconnect the tube shown in the illustration when disassembling the fuel suction tube assembly with pump and gauge. Doing so will cause reassembly of the fuel suction tube assembly with pump and gauge to be impossible as the tube is welded to the plate.

5. Remove the No. 1 fuel suction support.

a. Using a screwdriver with the tip taped, disengage the claw and remove the No. 1 fuel suction support.

6. Remove the fuel pump.

a. Using a screwdriver with its tip wrapped in protective tape, disengage the 5 claws and pull out the fuel pump from the fuel filter assembly.

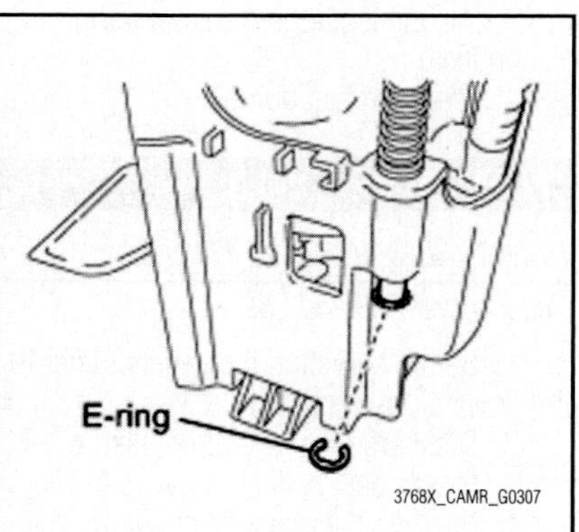

Fig. 271 Removing the e-ring

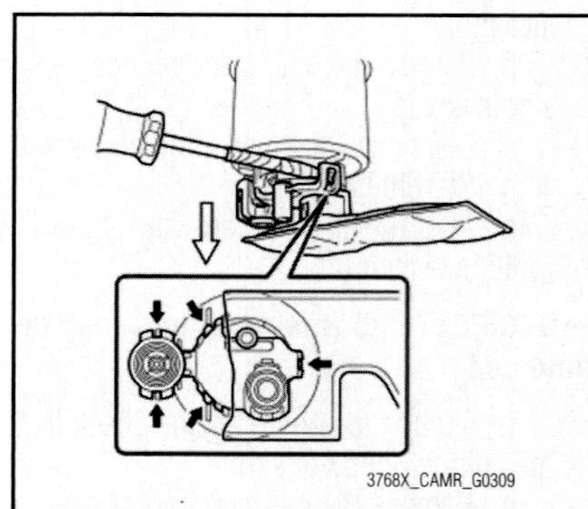

Fig. 273 Disengaging the fuel pump from the fuel filter assembly

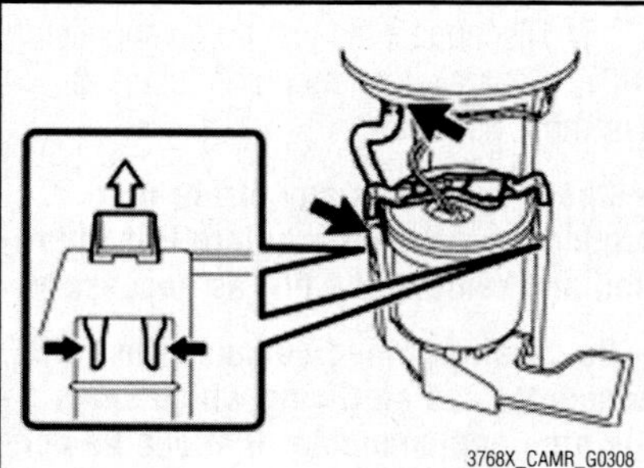

Fig. 272 Removing the No. 1 fuel tank band bracket from the fuel suction plate with the fuel filter

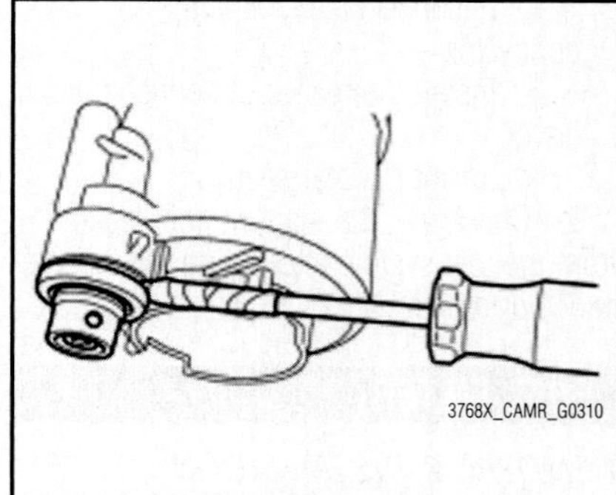

Fig. 274 Removing the fuel pressure regulator from the fuel filter

➡**Do not damage the fuel pump filter.**

➡**Do not remove the suction filter.**

b. Remove the fuel pump harness connector.

c. Remove the O-ring from the fuel pump.

7. Remove the fuel pressure regulator assembly.

a. Using a screwdriver with its tip wrapped in protective tape, remove the fuel pressure regulator from the fuel filter.

b. Remove the 2 O-rings from the fuel pressure regulator.

To install:

8. To install, reverse the removal procedure.

3.5L Engine

See Figure 275.

1. Before servicing the vehicle, refer to the precautions section.
2. Remove the fuel pump from the vehicle.
3. Remove the fuel pump filter.
4. Pull out the fuel pressure regulator from the fuel tank fuel filter.
5. Remove the O-ring from the fuel pressure regulator.

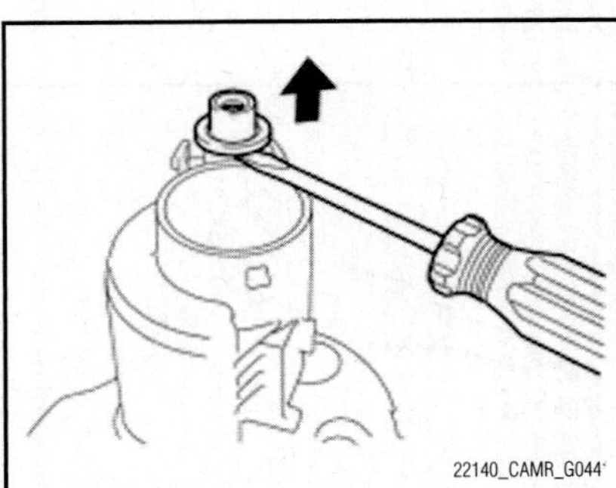

Fig. 275 Removing the fuel pressure regulator from the fuel filter—3.5L engine

To install:

6. Apply a light coat of spindle oil or gasoline to a new O-ring, and install it to the fuel pressure regulator.
7. Push in the fuel pressure regulator to the fuel tank fuel filter.
8. Install the fuel pump filter.
9. Install the fuel pump.

FUEL RAIL (SUPPLY MANIFOLD) AND INJECTOR

REMOVAL & INSTALLATION

2.4L Engine

See Figures 276 through 278.

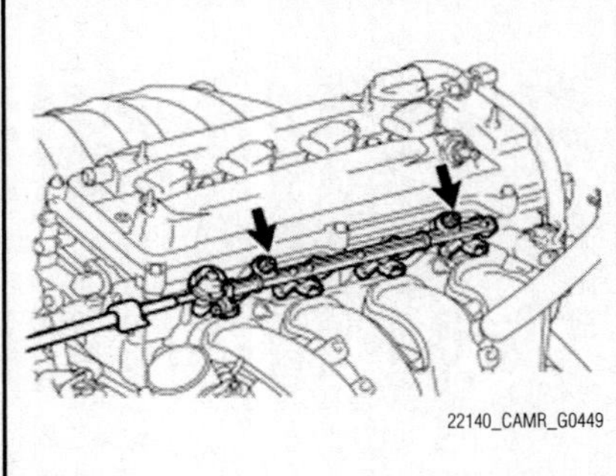

Fig. 276 Removing the fuel delivery pipe together with the 4 fuel injectors

1. Before servicing the vehicle, refer to the precautions section.
2. Discharge fuel system pressure.
3. Disconnect battery negative cable.
4. Remove air cleaner cap sub-assembly.
5. Remove No. 1 engine cover sub-assembly.

➡**Check for foreign matter on the pipe and around the connector before disconnecting the quick connector. Clean the connector if necessary.**

6. Disconnect fuel tube sub-assembly, as follows:

a. Remove the No. 1 fuel pipe clamp.

CAUTION

Do not use any tools in this following procedure. Check for foreign matter on the sealing surface of the disconnected pipe. Clean it if necessary.

b. If the connector and pipe are stuck, pinch the connector, and push and pull the pipe to disconnect them.

c. Separate the fuel tube from the fuel hose clamp.

7. Disconnect the No. 2 ventilation hose from the ventilation valve.
8. Remove fuel delivery pipe with injector as follows:

a. Remove the 2 wire harness clamps.

CAUTION

Be careful not to drop the fuel injectors when removing the fuel delivery pipe.

b. Remove the 2 bolts, then remove the fuel delivery pipe together with the 4 fuel injectors.

c. Remove the 2 delivery pipe spacers from the cylinder head.

d. Remove the 4 insulators from the cylinder head.

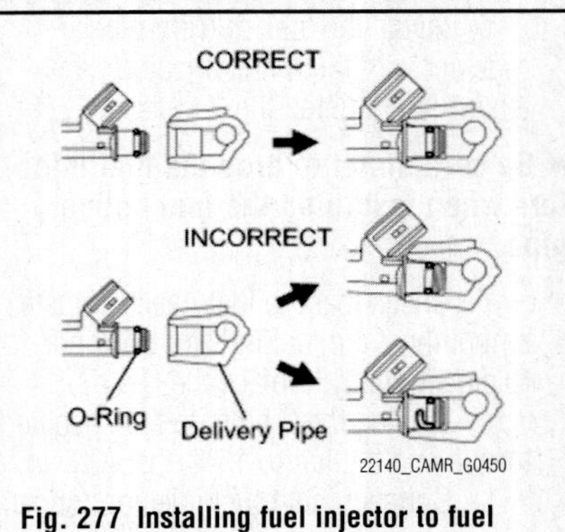

Fig. 277 Installing fuel injector to fuel tube

9. Pull out the 4 injectors from the delivery pipe.
10. Remove the 4 O-rings from the injectors.

To install:

11. Install the fuel injector assembly, as follows:

a. Apply a light coat of spindle oil or gasoline to new O-rings, and install one to each injector.

b. Apply a light coat of spindle oil or gasoline where the fuel delivery pipe contacts the O-ring.

c. Apply a light coat of gasoline or spindle oil to the O-ring again, then install the right and left fuel injectors onto the fuel delivery pipe.

➡**Make sure that the O-ring is not cracked or jammed before installing the injector.**

d. Check that the fuel injector rotates smoothly. If the fuel injector does not rotate, replace the O-ring.

12. Install fuel delivery pipe with injector as follows:

a. Install 4 new insulators into the cylinder head.

b. Install the 2 delivery pipe spacers onto the cylinder head.

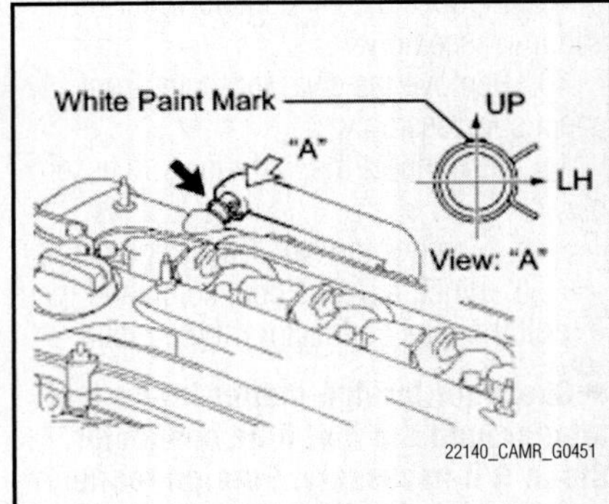

Fig. 278 Positioning paint mark and hose clamp

c. Install the fuel delivery pipe together with the 4 fuel injectors, then temporarily tighten the 2 bolts.

➡Be careful not to drop the fuel injectors when installing the fuel delivery pipe.

d. Check that the fuel injector rotates smoothly. If the fuel injector does not rotate smoothly, replace the O-ring.

e. Tighten the 2 bolts to the specified torque and tighten to 15 ft. lbs. (20 Nm).

f. Connect the 4 fuel injector connectors.

g. Install the 2 wire harness clamps.

13. Connect the No. 2 ventilation hose to the ventilation valve.
14. Install the fuel tube to the fuel hose clamp.
15. Push the fuel tube connector until it makes a "click" sound.
16. Install the No. 1 fuel pipe clamp.
17. Install air cleaner cap sub-assembly.
18. Connect cable to negative battery terminal.
19. Check for fuel leaks.
20. Install No. 1 engine cover sub-assembly.

2.5L Engine

See Figures 279 through 282.

1. Discharge the fuel system pressure.
2. Disconnect the cable from the negative battery terminal.
3. Remove the No. 1 engine cover sub assembly.
4. Remove the air cleaner cap sub assembly.
5. Remove the air cleaner filter element sub assembly.
6. Remove the front wiper arm and blade assemblies.
7. Remove the front fender to cowl side seals.
8. Remove the cowl top ventilator louver sub assembly.
9. Remove the windshield wiper motor and link assembly.
10. Remove the cowl top outer front panel sub assembly.
11. Disconnect the fuel tube sub assembly.

a. Remove the No. 1 fuel pipe clamp.

b. Pinch the tube connector, and then pull the tube connector off the pipe.

➡Check for foreign matter in the fuel tube around the fuel tube connector. Clean it if necessary. Foreign matter can affect the ability of the O-ring to seal the connector and fuel pipe.

Fig. 279 Removing the fuel delivery pipe with the injectors

➡Do not use any tools to separate the connector and pipe.

➡Do not forcefully bend, kink or twist the hose.

➡Keep the connector and pipe free from foreign matter.

➡If the connector and pipe are stuck together, pinch the connector and turn it carefully to disconnect it.

➡Put the connector in a plastic bag to prevent damage and contamination.

c. Remove the fuel tube sub assembly from the fuel hose clamp.

12. Disconnect the wire harness.

a. Disconnect the 4 fuel injector connectors.

b. Disconnect the 3 connectors.

c. Remove the 2 bolts and 2 wire harness brackets.

d. Detach the 2 clamps to disconnect the wire harness.

13. Remove the vacuum switching valve assembly (ACIS).

a. Remove the 2 bolts, and then remove the fuel delivery pipe together with the 4 fuel injectors.

➡Be careful not to drop the fuel injectors when removing the fuel delivery pipe.

b. Remove the 2 fuel delivery spacers from the cylinder head.

c. Remove the 4 injector vibration insulators from the cylinder head.

14. Remove the fuel injector assembly.

a. Pull the 4 fuel injectors out of the fuel delivery pipe.

To install:

15. Install the fuel injector assembly.

a. Apply a light coat of gasoline or spindle oil to the new O-rings, and then install one onto each fuel injector.

b. Apply a light coat of gasoline or spindle oil to the part of the fuel delivery pipe which comes into contact with the O-ring of the fuel injector.

Fig. 280 Removing the fuel delivery spacers and injector vibration insulators from the cylinder head

c. Apply a light coat of gasoline or spindle oil to the O-ring again, and then install the fuel injectors onto the fuel delivery pipe.

➡Make sure that the O-ring is not cracked or jammed when installing the injector.

d. Check that the fuel injector rotates smoothly. If the fuel injector does not rotate, replace the O-ring.

16. Install the fuel delivery pipe sub assembly.

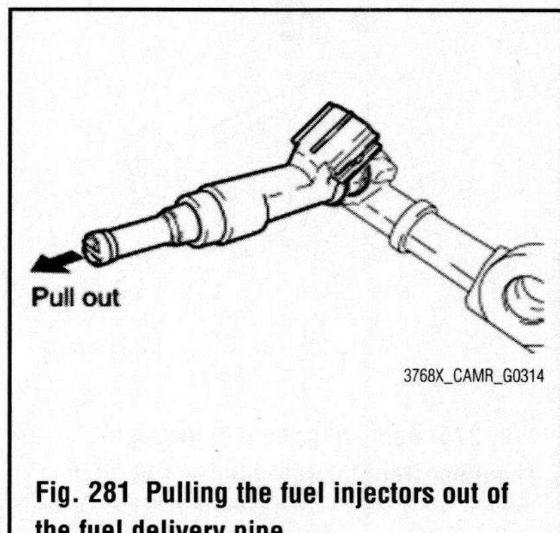

Fig. 281 Pulling the fuel injectors out of the fuel delivery pipe

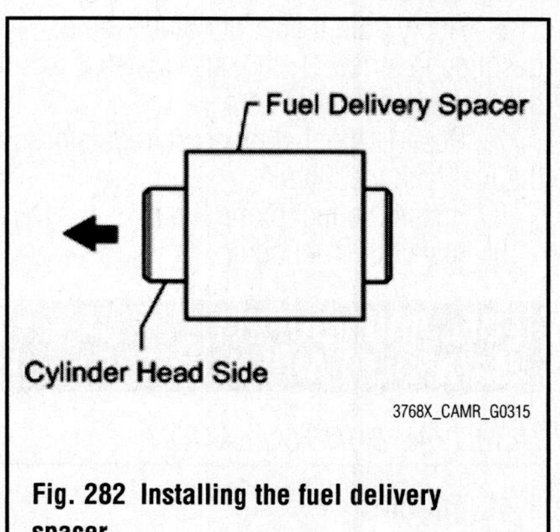

Fig. 282 Installing the fuel delivery spacer

a. Install 4 new injector vibration insulators to the cylinder head.

b. Install the 2 fuel delivery spacers onto the cylinder head.

➡Install the fuel delivery spacer so that the longer protrusion is on the cylinder head side.

c. Install the fuel delivery pipe together with the 4 fuel injectors to the cylinder head, and then temporarily install the 2 bolts.

➡Be careful not to drop the fuel injectors when installing the fuel delivery pipe.

d. Check that the fuel injector rotates smoothly. If the fuel injector does not rotate, replace the O-ring.

e. Tighten the 2 bolts to 15 ft. lbs. (21 Nm).

17. Connect the wire harness.

a. Install the 2 wire harness brackets with the 2 bolts. Tighten to 7 ft. lbs. (10 Nm).

b. Connect the 3 connectors.

c. Connect the 4 fuel injector connectors.

d. Attach the 2 clamps to connect the wire harness.

18. Install the vacuum switching valve assembly (ACIS).

19. Connect the fuel tube sub assembly.

a. Push the tube connector to the pipe until the tube connector makes a "click" sound.

➡Before connecting the connector and fuel pipe, check that there is no damage or foreign matter on the connecting part of the fuel pipe.

➡After connecting the fuel tube connector and pipe, check that they are securely connected by trying to pull them apart.

b. Install the No. 1 fuel pipe clamp.

c. Install the fuel tube sub-assembly to the fuel hose clamp.

20. Install the cowl top outer front panel sub assembly.

21. Install the windshield wiper motor and link assembly.

22. Install the cowl top ventilator louver sub assembly.

23. Install the front fender to cowl side seals.

24. Install the front wiper arm and blade assemblies.

25. Install the air cleaner cap sub assembly.

26. Install the No. 1 engine cover sub assembly.

27. Connect the cable to the negative battery terminal.

28. Inspect for fuel leaks.

3.5L Engine

See Figures 283 and 284.

1. Properly discharge the fuel system pressure.
2. Disconnect battery negative cable.
3. Drain engine coolant.
4. Remove both windshield wiper arm and blade assemblies.
5. Remove the right cowl top ventilator louver.
6. Remove the windshield wiper motor and link assembly.
7. Remove front fender to cowl side seal.
8. Remove the cowl top ventilator louver sub-assembly.
9. Remove windshield wiper motor and link assembly.
10. Remove cowl top panel outer sub-assembly.
11. Remove the V-bank cover sub-assembly.
12. Remove the air cleaner cap with air cleaner hose.
13. Remove the intake air surge tank.
14. Disconnect the fuel tube sub-assembly, as follows:

a. Remove the No. 2 fuel pipe clamp.

b. Pinch the tube connector and then pull out the fuel pipe.

➡Check that there is no dirt or other foreign objects around the connector before removing fuel tube, and clean the connector as necessary.

➡It is necessary to prevent mud or dirt from entering the connector. If mud or dirt gets in the connector, the O-rings may not seal properly.

➡Do not use any tools in this operation.

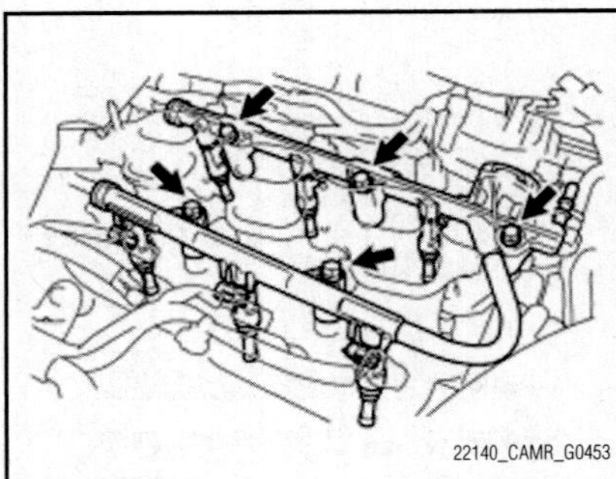

Fig. 283 Removing the 5 bolts and fuel delivery pipe together with the 6 fuel injectors

➡Do not bend, kink or twist the nylon tube. Protect the connector by covering it with a plastic bag.

➡When the pipe and connector are stuck, push and pull the connector to release and pull the connector out carefully.

15. Remove the fuel injector assembly, as follows:

a. Disconnect the 6 fuel injector connectors.

b. Remove the 5 bolts and fuel delivery pipe together with the 6 fuel injectors.

➡Be careful not to drop the fuel injectors when removing the fuel delivery pipe.

c. Remove the 6 insulators from the intake manifold.

d. Pull out the fuel injector from the fuel delivery pipe.

e. Remove the 6 O-rings from the injectors.

To install:

16. Install the fuel injector assembly, as follows:

a. Apply a light coat of spindle oil or gasoline to new O-rings, and install one to each injector.

b. Apply a light coat of spindle oil or gasoline where the fuel delivery pipe contacts the O-ring.

➡Be careful not to twist the O-ring.

➡After installing the fuel injector, check that it turns smoothly. If not, reinstall it with a new O-ring.

c. Push the fuel injector while twisting it back and forth to install it in the fuel delivery pipe.

d. Position the fuel injector connector outward.

e. Install 6 new insulators to the intake manifold.

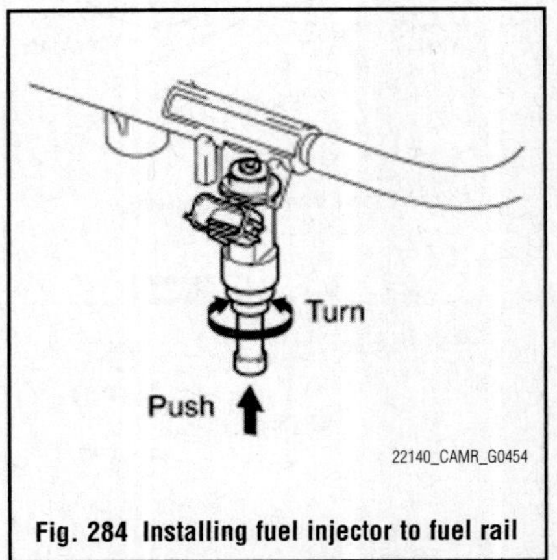

Fig. 284 Installing fuel injector to fuel rail

f. Place the fuel delivery pipe and the 6 fuel injectors together to the intake manifold.

➡**Be careful not to drop the fuel injectors when installing the fuel delivery pipe.**

g. Temporarily install the 6 bolts which are used to hold the fuel delivery pipe to the intake manifold.

➡**After installing the fuel injector, check that it turns smoothly. If not, reinstall it with a new O-ring.**

h. Tighten the 5 bolts which are used to hold the fuel delivery pipe to the intake manifold to 15 ft. lbs. (21 Nm).

17. Push in the tube connector to the pipe until the tube connector makes a "click" sound.

➡**Before connecting the tube, make sure that it is not damaged. Make sure that there is no dirt present on the connecting surfaces.**

➡**After connecting, check if the fuel tube connector and the pipe are securely connected by pulling on them.**

18. Install the No. 2 fuel pipe clamp.
19. Te remainder of installation is the reverse of the removal procedure.
20. Check for coolant leak and fuel leak.

FUEL TANK

REMOVAL & INSTALLATION

See Figures 285 through 290.

1. Before servicing the vehicle, refer to the precautions section.
2. Discharge fuel system pressure.
3. Disconnect battery negative cable.
4. Remove rear seat cushion assembly.
5. Remove rear floor service whole cover.
6. Separate fuel pump tube sub-assembly.
7. Remove fuel tank vent tube set plate.
8. Remove fuel suction tube assembly with pump and gauge.
9. Drain fuel.
10. Remove center exhaust pipe assembly.
11. Disconnect no. 2 parking brake cable assembly.
12. Disconnect no. 3 parking brake cable assembly.
13. Remove rear stabilizer BAR No. 1 bracket.
14. Remove lower center fuel tank protector, as follows:

a. Remove the 4 bolts and the (Except SE grade).

b. Remove the 4 bolts and 2 clips (for SE Grade).

c. Remove the fuel tank protector (for SE Grade).

➡**Check that there is no dirt or other foreign objects around the connector before removing fuel tubes, and clean the connector as necessary.**

➡**It is necessary to prevent mud or dirt from entering the connector. If mud or dirt gets in the connector, the O-rings may not seal properly.**

➡**Do not use any tools in these operations.**

➡**Do not bend, kink or twist the nylon tubes. Protect the connector by covering it with a plastic bag.**

➡**When the pipe and connector are stuck, push and pull the connector to release and pull the connector out carefully.**

15. Disconnect the fuel pump tube, as follows:

a. Pinch the tabs of the retainer to remove the lock claws and pull it down as shown in the illustration.

b. Pull out the fuel tank main tube.

16. Pinch the tube connector and then pull out the No. 1 fuel tube.
17. Set up a transmission jack underneath the fuel tank.
18. Remove the 2 set bolts of the fuel tank bands.

➡**Check that there is no dirt or other foreign objects around the connector before removing fuel tubes, and clean the connector as necessary.**

➡**It is necessary to prevent mud or dirt from entering the connector. If mud or dirt gets in the connector, the O-rings may not seal properly.**

➡**Do not use any tools in these operations.**

➡**Do not bend, kink or twist the nylon tubes. Protect the connector by covering it with a plastic bag.**

➡**When the pipe and connector are stuck, push and pull the connector to release and pull the connector out carefully.**

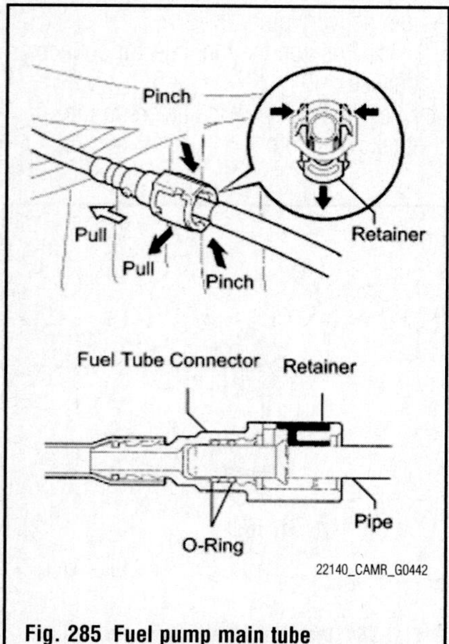

Fig. 285 Fuel pump main tube

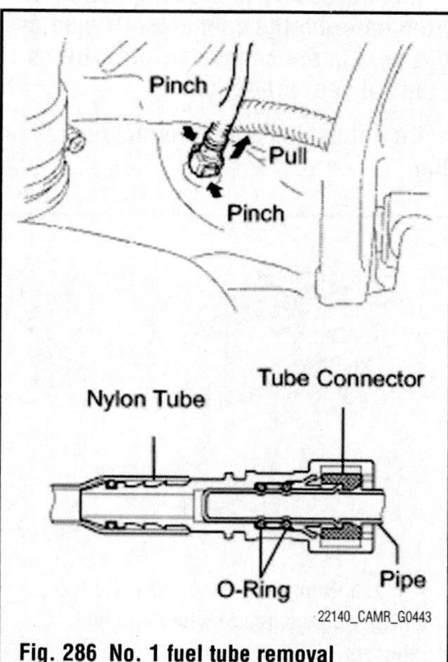

Fig. 286 No. 1 fuel tube removal

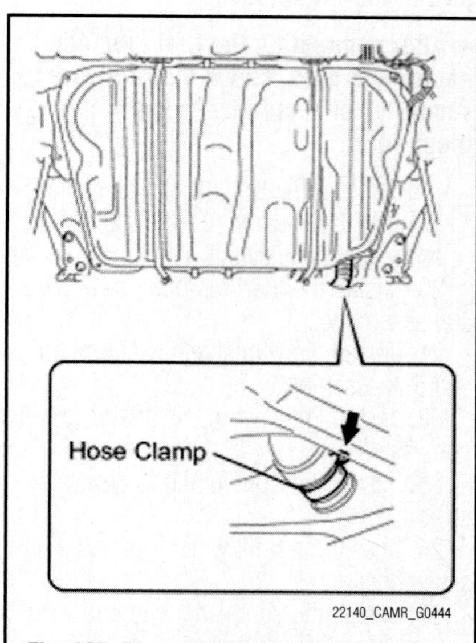

Fig. 287 Disconnecting the fuel tank to filter pipe hose—except PZEV

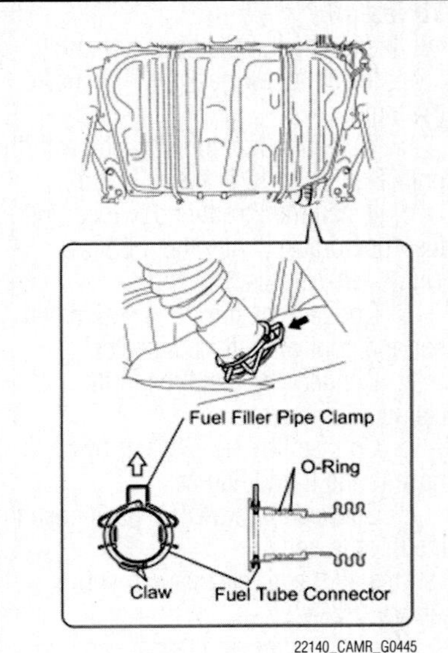

Fig. 288 Disconnecting the fuel tank to filter pipe hose—PZEV

19. Remove the hose clamp and disconnect the fuel tank to filter pipe hose (except PZEV).
20. Remove the clamp and disconnect the fuel tank to filter pipe hose (for PZEV).
21. Slightly lower the transmission jack.
22. Disconnect the fuel tank vent hose from the charcoal canister, as follows:
    a. Push the connector deep into the charcoal canister to release the locking pin.
    b. Pinch portion A.
    c. Pull out the connector.
23. Remove the 2 pins and 2 fuel tank bands as shown in the illustration.
24. Remove the 4 clip nuts.

To install:

25. Install the 4 clip nuts.
26. Install the 2 fuel tank bands with the 2 pins.
27. Connect the fuel tank vent hose.
28. Connect the fuel tank inlet pipe with the fuel filter pipe clamp.

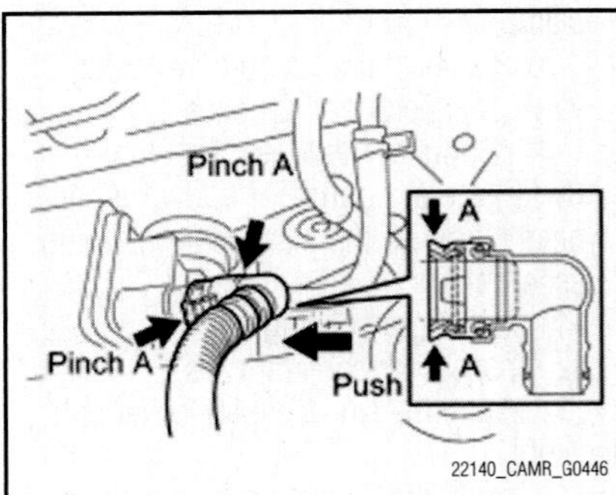

Fig. 289 Disconnect the fuel tank vent hose from the charcoal canister

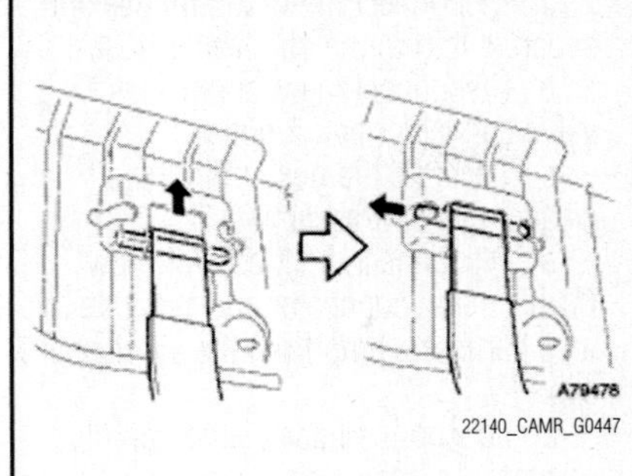

Fig. 290 Removing the 2 pins and 2 fuel tank bands

29. Tighten the 2 set bolts of the fuel tank bands to 29 ft. lbs. (39 Nm).
30. Connect the No. 1 fuel tube, as follows:
    a. Push the fuel tube connector into the pipe until the fuel tube connector makes a "click" sound.

➡Check that there is no damage or foreign objects on the connected part.

➡After connecting, check if the fuel tube connector and the pipe are securely connected by trying to pull them apart.

31. Connect the fuel pump tube, as follows:
    a. Push in the fuel pump tube connector to the pipe and push up the retainer so that the claws engage.

➡Check that there is no damage or foreign objects on the connected part.

➡After connecting, check if the fuel tube connector and the pipe are securely connected by trying to pull them apart.

32. Install the lower center fuel tank protector and tighten to 48 inch lbs. (5.4 Nm).
33. Install the No. 3 parking brake cable assembly with the bolt and nut and tighten to 53 inch lbs. (6 Nm), and 75 inch lbs. (8.5 Nm).
34. Install the No. 2 parking brake cable assembly with the bolt and nut and tighten to 53 inch lbs. (6 Nm), and 75 inch lbs. (8.5 Nm).
35. Install the center exhaust pipe assembly.
36. Install the fuel suction tube assembly with pump and gauge.
37. Install the fuel tank vent tube set plate.
38. Connect the fuel pump tube sub-assembly.
39. Add fuel.
40. Connect battery negative cable.
41. Inspect for fuel leak and exhaust gas leak.
42. Install the rear floor service hole cover.
43. Install the rear seat cushion assembly.

IDLE SPEED

ADJUSTMENT

Idle speed is maintained by the ECM. No adjustment is necessary or possible.

THROTTLE BODY

REMOVAL & INSTALLATION

2.4L Engine

1. Drain engine coolant.
2. Remove No. 1 engine cover sub-assembly.
3. Remove air cleaner cap sub-assembly, as follows:
    a. Disconnect the Mass Air Flow (MAF) meter connector (1).
    b. Disconnect the purge VSV connector (2).
    c. Disconnect the 2 purge VSV vacuum hoses (3).
    d. Disconnect the purge line hose from the clamp (4).
    e. Disconnect the No. 2 ventilation hose from the air cleaner hose.
    f. Lock the No. 1 air cleaner hose clamp, and then disconnect the No. 1 air cleaner hose from the throttle body.
    g. Remove the 2 bolts and air cleaner cap.
    h. Remove the air cleaner filter element from the air cleaner case.
4. Remove air cleaner case sub-assembly.
5. Disconnect the throttle position sensor connector and wire harness clamp.
6. Remove the 4 bolts, and then remove the fuel pipe support and throttle body.
7. Disconnect the purge line hose from the throttle body.
8. Disconnect the water by-pass hose from the throttle body.
9. Disconnect the No. 2 water by-pass hose from the throttle body.
10. Disconnect the No. 1 throttle body hose from the throttle body.
11. Remove the gasket from the intake manifold.

To install:

12. Install a new gasket onto the intake manifold.
13. Connect the purge line hose to the throttle body.
14. Connect the water by-pass hose to the throttle body.
15. Connect the No. 2 water by-pass hose to the throttle body.
16. Connect the No. 1 throttle body hose to the throttle body.
17. Install the throttle body and fuel pipe clamp with the 4 bolts and tighten to 22 ft. lbs. (30 Nm).
18. Connect the fuel tube into the clamp.
19. Connect the throttle position sensor connector.
20. Connect the wire harness clamp.
21. Install air cleaner case sub-assembly.
22. Install air cleaner cap sub-assembly, as follows:
    a. Install the air cleaner filter element onto the air cleaner case.
    b. Insert the hinges. Install the air cleaner cap sub-assembly with the 2 bolts.
    c. Align the matchmarks of the No. 1 air cleaner hose and throttle body, and then connect the air cleaner hose No. 1 to the throttle body and unfasten the No. 1 air cleaner hose clamp.

➡Make sure that the hose clamp is at the correct angle.

    d. Connect the No. 2 ventilation hose to the air cleaner hose.
    e. Connect the purge line hose to the clamp.
    f. Connect the 2 purge VSV vacuum hoses.
    g. Connect the purge VSV connector.
    h. Connect the Mass Air Flow (MAF) meter connector.
23. Install air cleaner inlet assembly.
24. Add engine coolant.
25. Check for engine coolant leaks.
26. Install No. 1 engine cover sub-assembly.

2.5L Engine

See Figure 291.

1. Remove the front wheel opening extension pads.
2. Remove the engine under covers.
3. Drain the engine coolant.
4. Remove the No. 1 engine cover sub assembly.
5. Remove the air cleaner cap sub assembly (PZEV).
    a. Disconnect the vacuum hose and separate it from the air cleaner hose.
    b. Disconnect the vacuum switching valve connector and 2 hoses.
    c. Separate the hose from the air cleaner cap sub-assembly
    d. Disconnect the Mass Air Flow (MAF) meter connector and separate the wire harness clamp from the air cleaner cap.
    e. Loosen the hose clamp and disconnect the hose.
    f. Loosen the hose clamp and disconnect the air cleaner hose.
    g. Remove the 2 bolts and the air cleaner cap sub-assembly.
6. Remove the air cleaner cap sub assembly (except PZEV).
    a. Disconnect the Mass Air Flow (MAF) meter connector and separate the wire harness clamp from the air cleaner cap.
    b. Disconnect the vacuum switching valve connector and 2 hoses.
    c. Separate the hose from the air cleaner cap sub-assembly.
    d. Loosen the hose clamp and disconnect the hose.
    e. Loosen the hose clamp and disconnect the air cleaner hose.
    f. Remove the 2 bolts and the air cleaner cap sub-assembly.
7. Remove the throttle body.
    a. Disconnect the water by pass hose from the throttle body.
    b. Disconnect the No. 2 water by-pass hose from the throttle body.
    c. Disconnect the fuel tube from the clamp.
    d. Disconnect the throttle position sensor and control motor connector.
    e. Remove the 4 bolts and the throttle body with fuel tube bracket.
    f. Remove the bolt and fuel tube bracket.
    g. Remove the gasket from the intake manifold.

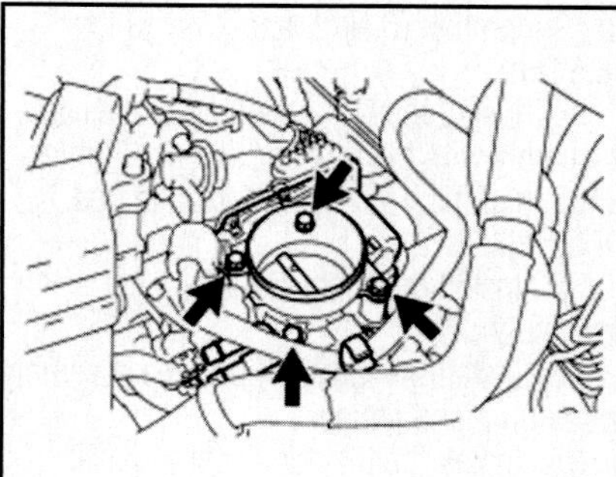

Fig. 291 Removing the throttle body with the fuel tube bracket

To install:

8. Install the throttle body assembly.
    a. Install a new gasket to the intake manifold.
    b. Install the fuel tube bracket with the bolt. Tighten to 10 ft. lbs. (13 Nm).
    c. Install the throttle body (with the fuel tube bracket) with the 4 bolts. Tighten to 7 ft. lbs. (10 Nm).
    d. Connect the throttle position sensor and control motor connector.
    e. Connect the fuel tube to the clamp.
    f. Connect the No. 2 water by-pass hose to the throttle body.
    g. Connect the water by-pass hose to the throttle body.
9. Install the air cleaner cap sub assembly (PZEV).
    a. Connect the air cleaner cap sub assembly with the hose clamp.
    b. Install the air cleaner cap sub assembly with the 2 bolts. Tighten to 44 inch lbs. (5 Nm).
    c. Connect the hose with the hose clamp.
    d. Engage the hose with the hose clamp.
    e. Connect the vacuum switching valve connector and 2 hoses.
    f. Connect the Mass Air Flow (MAF) meter connector and wire harness clamp to the air cleaner cap sub assembly.
    g. Connect the hose and engage the hose to the clips on the air cleaner hose.
10. Install the air cleaner cap sub assembly (except PZEV).
    a. Connect the air cleaner cap sub assembly with the hose clamp.
    b. Install the air cleaner cap sub assembly with the 2 bolts. Tighten to 44 inch lbs. (5 Nm).
    c. Connect the hose with the hose clamp.
    d. Engage the hose with the hose clamp.
    e. Connect the vacuum switching valve connector and 2 hoses.
    f. Connect the Mass Air Flow (MAF) meter connector and wire harness clamp to the air cleaner cap sub assembly.
11. Add engine coolant.
12. Inspect for coolant leaks.
13. Install the No. 1 engine cover sub assembly.
14. Install the engine under covers.
15. Install the front wheel opening extension pads.
16. Perform the initialization.

➡**Be sure to perform this procedure after reassembling the throttle body assembly or removing and reinstalling any throttle body component.**

➡**Perform the following procedure after replacing the ECM, throttle body assembly or any throttle body components. The following procedure should also be performed if the throttle body is cleaned.**

➡**Be sure to perform this procedure after reconnecting the battery cable or replacing the ECM.**

a. Disconnect the cable of the negative (-) battery terminal. Wait at least 60 seconds and reconnect the cable of the negative (-) battery terminal.

b. Turn the ignition switch to ON without operating the accelerator pedal.

➡**If the accelerator pedal is operated, perform the above steps again.**

c. Connect the Techstream to the DLC3 and clear the DTCs.

d. Start the engine and check that the MIL is not illuminated. After the engine is warmed up, check that the idle speed is within the specified range when the A/C is switched off.

➡**Standard idle speed for an automatic transaxle with the A/C switched off is 610-710 rpm. For a manual transaxle the standard idle speed with the A/C switched off is 650-750 rpm.**

➡**Be sure to perform this step with all accessories off.**

➡**Make sure that the shift lever is in neutral.**

e. Enter the following menus: Powertrain / Engine / Data List / Throttle Sensor Position. Fully depress the accelerator pedal and check that the value is 60%or more.

f. Perform a road test and confirm that there are no abnormalities.

3.5L Engine

See Figure 292.

1. Before servicing the vehicle, refer to the precautions section.
2. Disconnect battery negative cable.
3. Drain engine coolant.
4. Remove cool air intake duct seal.
5. Remove the V-bank cover sub-assembly.
6. Remove air cleaner inlet assembly.
7. Remove air cleaner cap sub-assembly, as follows:
   a. Disconnect the 3 vacuum hoses.
   b. Disconnect the Mass Air Flow (MAF) meter connector.
   c. Disconnect the No. 2 ventilation hose.
   d. Disconnect the hose band.
   e. Disconnect the 3 bands, and remove the air cleaner cap sub-assembly.
8. Remove air cleaner case sub-assembly.
9. Remove No. 1 air cleaner inlet.

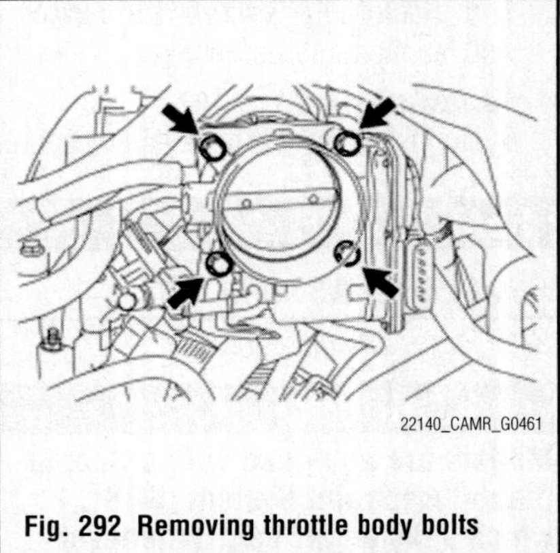

Fig. 292 Removing throttle body bolts

10. Disconnect the throttle body connector and clamp.
11. Disconnect the 2 water by-pass hoses from the throttle body.
12. Remove the 4 bolts and throttle body.
13. Remove the throttle body gasket from the intake air surge tank.

To install:

14. Install a new throttle body gasket to the intake air surge tank.
15. Install the throttle w/ motor body assembly and wire harness clamp stay to the intake air surge tank with the 4 bolts and tighten to 7 ft. lbs. (10 Nm).
16. Connect the throttle w/ motor body assembly connector.
17. Te remainder of installation is the reverse of the removal procedure.
18. Check for coolant leak.
19. Check the function of the throttle body.

HEATING & AIR CONDITIONING SYSTEM

BLOWER MOTOR

REMOVAL & INSTALLATION

See Figures 293 and 294.

1. Drain and recycle the engine coolant.
2. Disconnect the negative battery cable.
3. Remove instrument panel.
4. For TMC made:
   a. Disconnect the connector.
   b. Remove the 2 screws and blower assembly.
   c. Remove the 3 screws and blower with fan motor sub-assembly.
5. For TMMK made:
   a. Remove cooler expansion valve.
   b. Remove the connector and clamp, and disconnect the wire harness.
   c. Remove the 6 screws and then the blower assembly with the cooler evaporator sub-assembly.

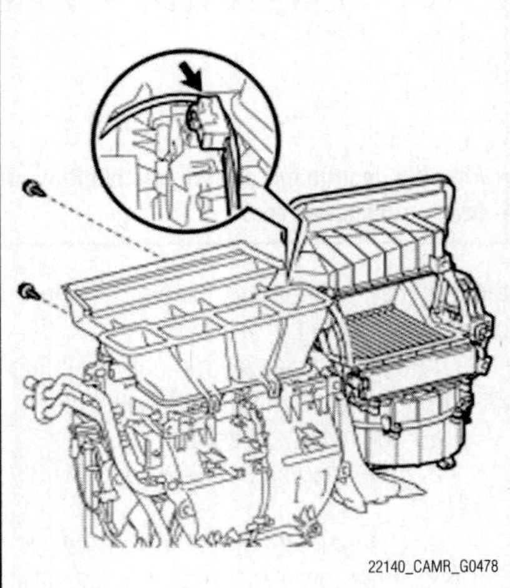

Fig. 293 Removing the connector, clamp and wire harness—TMMK

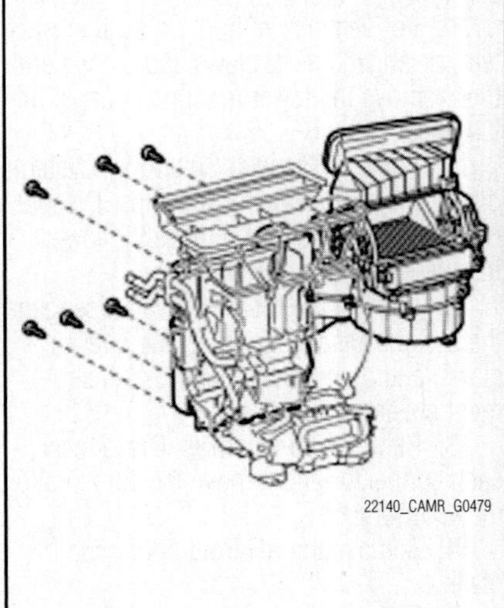

Fig. 294 Removing the 6 screws and blower assembly—TMMK

d. Remove the 3 screws and blower with fan motor sub-assembly.

To install:

6. To install, reverse removal procedure.

HEATER CORE

REMOVAL & INSTALLATION

See Figures 295 through 302.

⁂ CAUTION

Models are equipped with a Supplemental Restraint System (SRS), which uses an air bag. Whenever working near any of the SRS components, such as the impact sensors, the air bag module, steering column and instrument panel, disable the SRS.

1. Before servicing the vehicle, refer to the Precautions Section.

⁂ CAUTION

Wait for 90 seconds after disconnecting the cable to prevent airbag deployment.

2. Disconnect battery negative terminal.
3. Remove the lower No. 2 and No. 3 steering wheel covers.
4. Remove the steering pad.
5. Remove the steering wheel assembly.
6. Remove the LH front door scuff plate.
7. Remove the LH cowl side trim sub-assembly.
8. Remove steering column cover.
9. Remove the turn signal switch assembly.
10. For vehicles without Smart Key System, disengage the 2 claws and 2 clips and then remove the lower instrument panel finish panel.
11. For vehicles with Smart Key System, disengage the 2 claws and 2 clips. Disconnect the connector and remove the lower instrument panel finish panel.
12. Using a molding remover, disengage the 2 clips. Disengage the guide and 4 claws, and then remove the No. 1 instrument cluster finish panel.
13. Remove the 4 screws. Disconnect each connector and remove the combination meter assembly.
14. Remove the RH front door scuff plate.
15. Remove cowl side trim sub-assembly.
16. Disengage the 4 claws. Disengage the 2 guides and remove the No. 2 under cover sub-assembly.
17. Remove lower instrument panel sub-assembly by performing the following:
    a. Remove the 4 screws.
    b. Disengage the 3 claws and the 3 clips.
    c. Disconnect the connector and remove the lower instrument panel sub-assembly.
18. Turn the shift lever knob counter-clockwise and remove the shift lever knob sub-assembly.

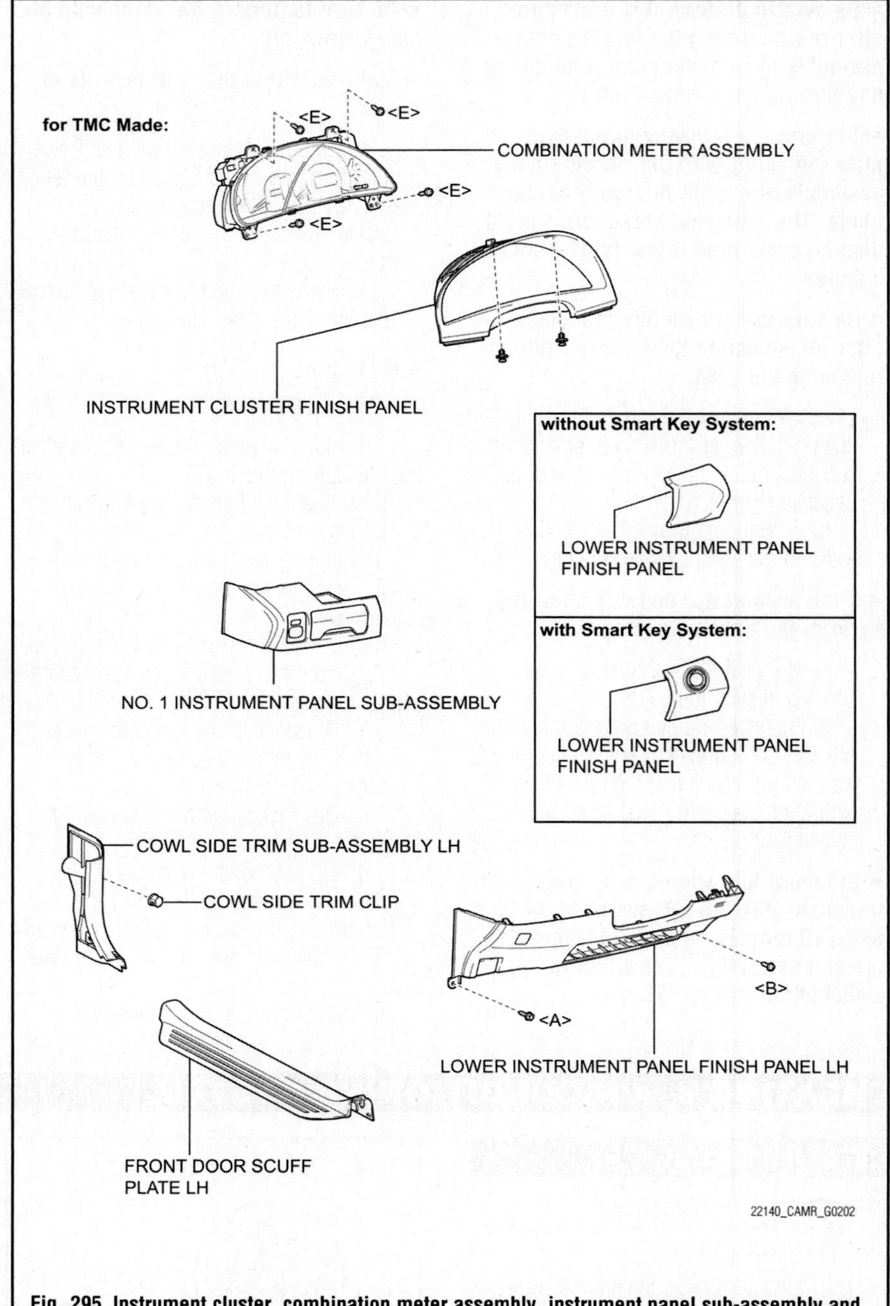

Fig. 295 Instrument cluster, combination meter assembly, instrument panel sub-assembly and lower trip sub-assembly

19. Disengage the 2 clips and remove the No. 1 instrument cluster finish panel garnish.
20. Disengage the 2 clips and remove the No. 2 instrument cluster finish panel garnish.
21. For A/T vehicles, Disengage the 6 claws and the 3 clips, and then remove the floor shift position indicator housing sub-assembly. If equipped with Seat Heater System, disconnect each connector.
22. For M/T vehicles, open the lid of the upper console panel. Apply protective tape

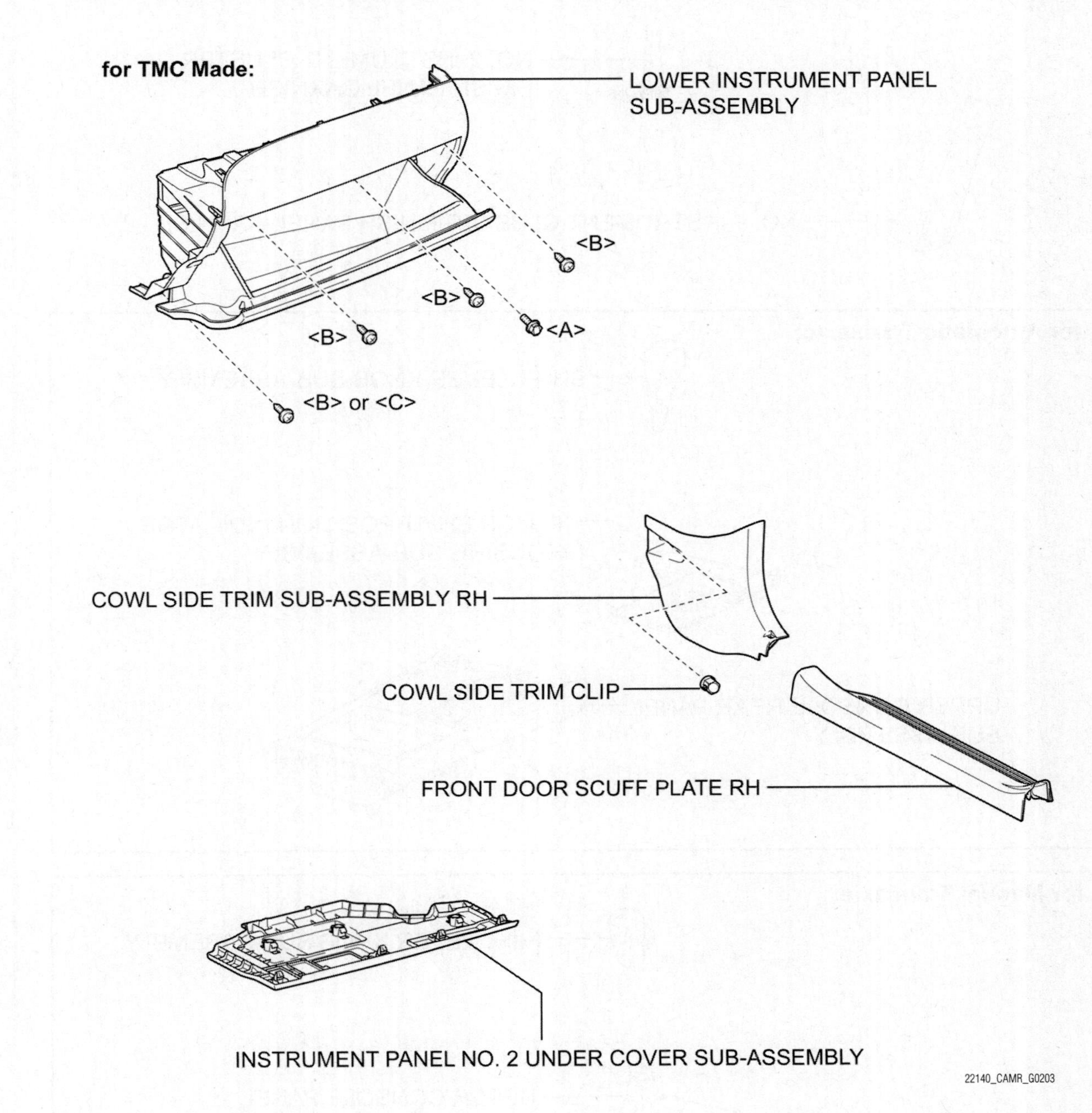

Fig. 296 Lower instrument panel sub-assembly and No. 2 under cover sub-assembly

to the area. Using a moulding remover, disengage the 2 claws and the 5 clips, and then remove the upper console panel.

23. Disengage the 3 claws and the 5 clips. Disconnect the connector and remove the upper console rear panel sub-assembly.

24. Remove instrument panel no. 2 register assembly by performing the following:

a. Apply protective tape to the areas.

b. Using a moulding remover, disengage the 3 clips.

c. Using a moulding remover, disengage the 4 clips.

d. Disconnect the connector and remove the instrument panel No. 2 register assembly.

25. Remove radio receiver with heater control panel assembly.

26. Remove the console box pocket

27. Remove the console box pocket.

28. Remove the console box assembly by performing the following:

a. Remove the 2 screws.

b. Disengage the clamp.

c. Remove the 2 bolts and the console box assembly.

29. Remove both of the front console box inserts by performing the following:

a. Remove the 3 screws.

b. Disengage the clip and remove the front console box insert.

30. Remove the LH front pillar garnish.

31. Disengage the 4 clips and remove the instrument panel No. 1 register assembly.

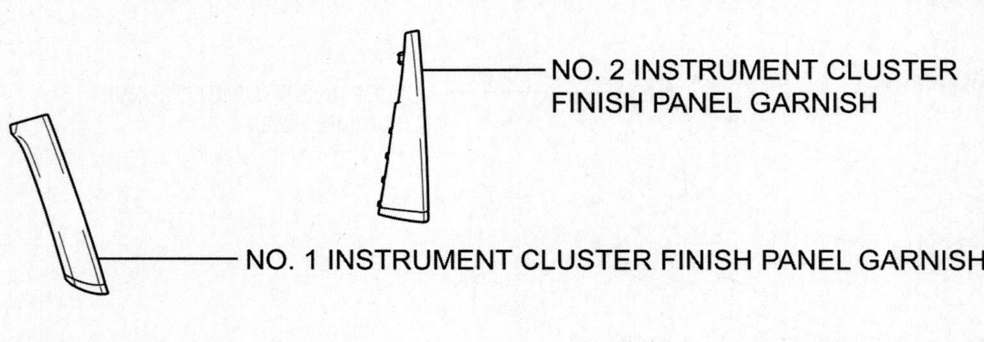

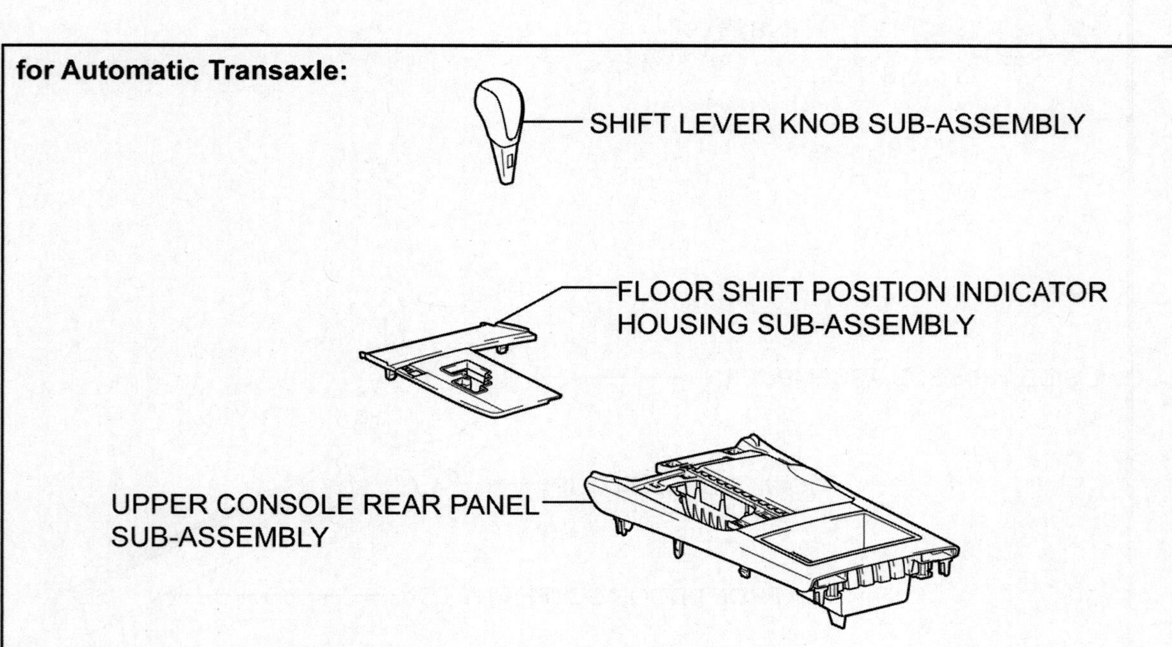

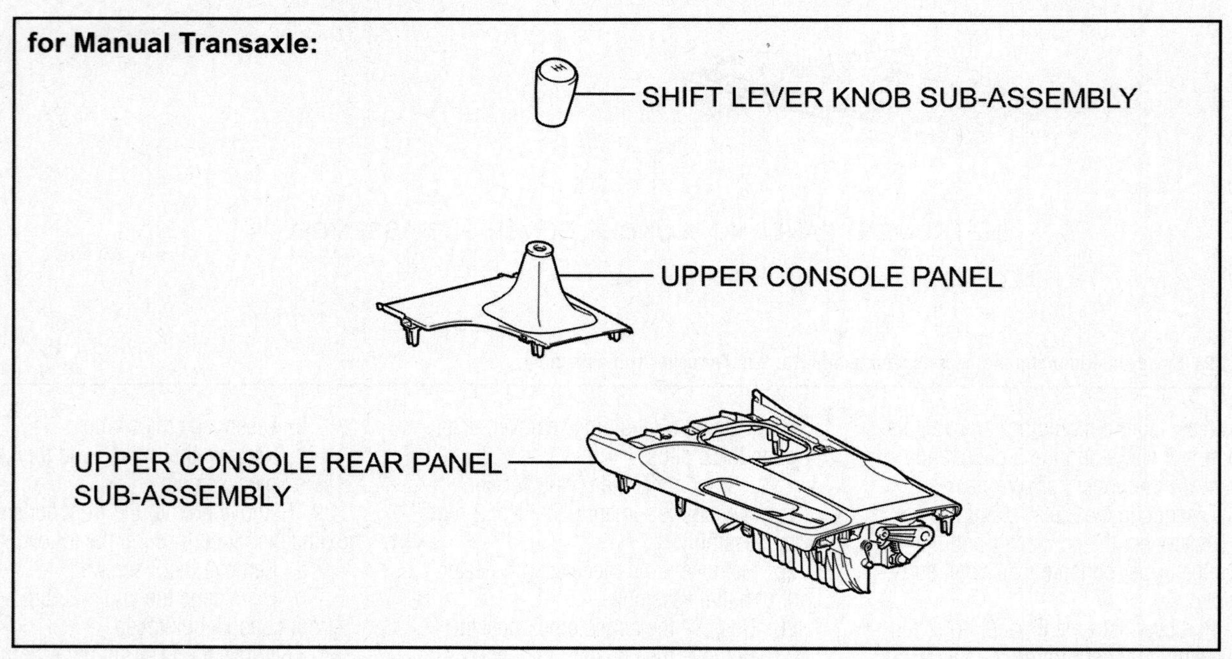

Fig. 297 Instrument cluster panel garnish, floor housing sub-assembly and upper console rear panel sub-assembly

for TMC Made:

without Navigation System:

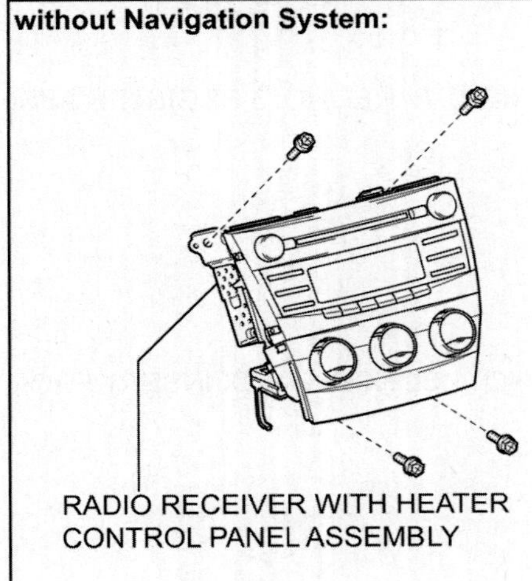

with Navigation System:

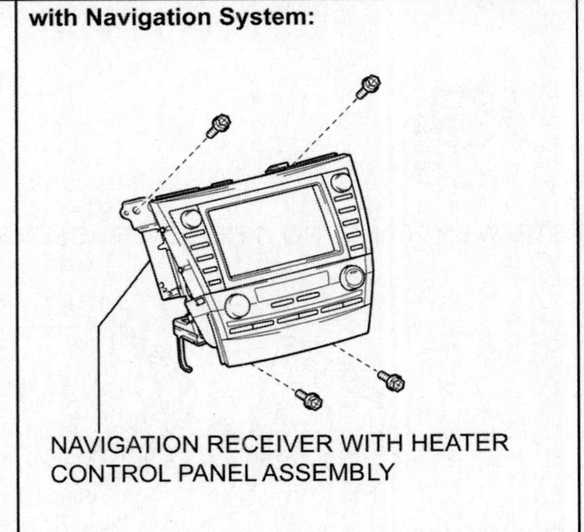

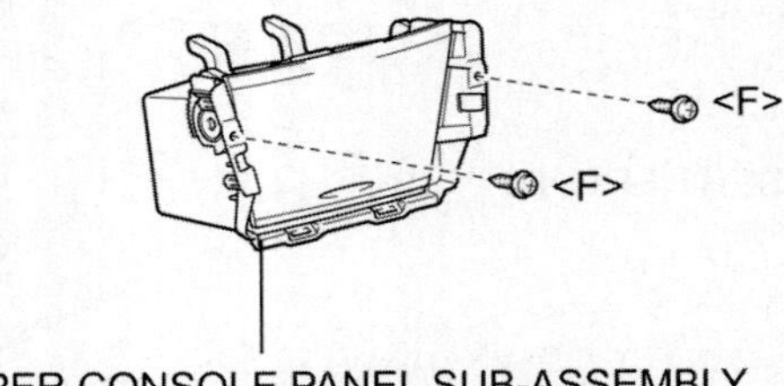

Fig. 298 Instrument panel no. 2 register assembly, control panel assembly and upper console sub-assembly

32. Remove instrument panel no. 1 speaker panel sub-assembly by performing the following:
    a. Disengage the 6 claws and the 2 clips.
    b. Disengage the 2 guides and remove the instrument panel No. 1 speaker panel sub-assembly.
33. Remove RH front no. 2 speaker assembly.
34. Remove the RH front pillar garnish.
35. Disengage the 4 clips and remove the instrument panel No. 3 register assembly.
36. Remove instrument panel no. 2 speaker panel sub-assembly by performing the following:
    a. Disengage the 6 claws and the 2 clips.
    b. Disengage the 2 guides and remove the instrument panel No. 2 speaker panel sub-assembly.
37. Remove LH and RH front no. 2 speaker assemblies.
38. Remove no. 1 defroster nozzle garnish by performing the following:
    a. Disengage the 8 clips and the 4 guides.
    b. Disconnect each connector and remove the No. 1 defroster nozzle garnish.
39. Disconnect instrument panel wire assembly.

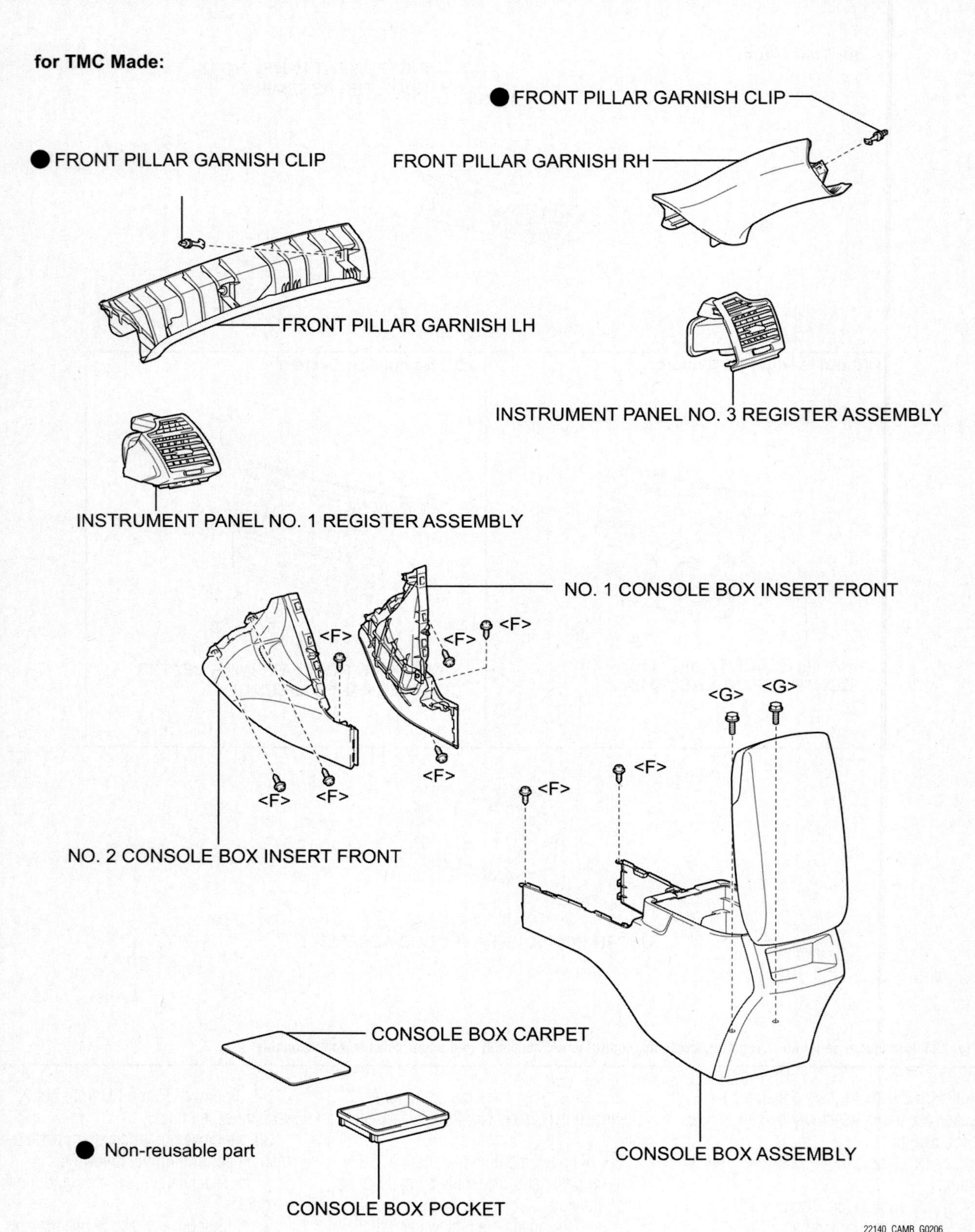

Fig. 299 Console box assembly, front console box inserts, LH front pillar garnish and instrument panel No. 1 and No. 3 register assembly

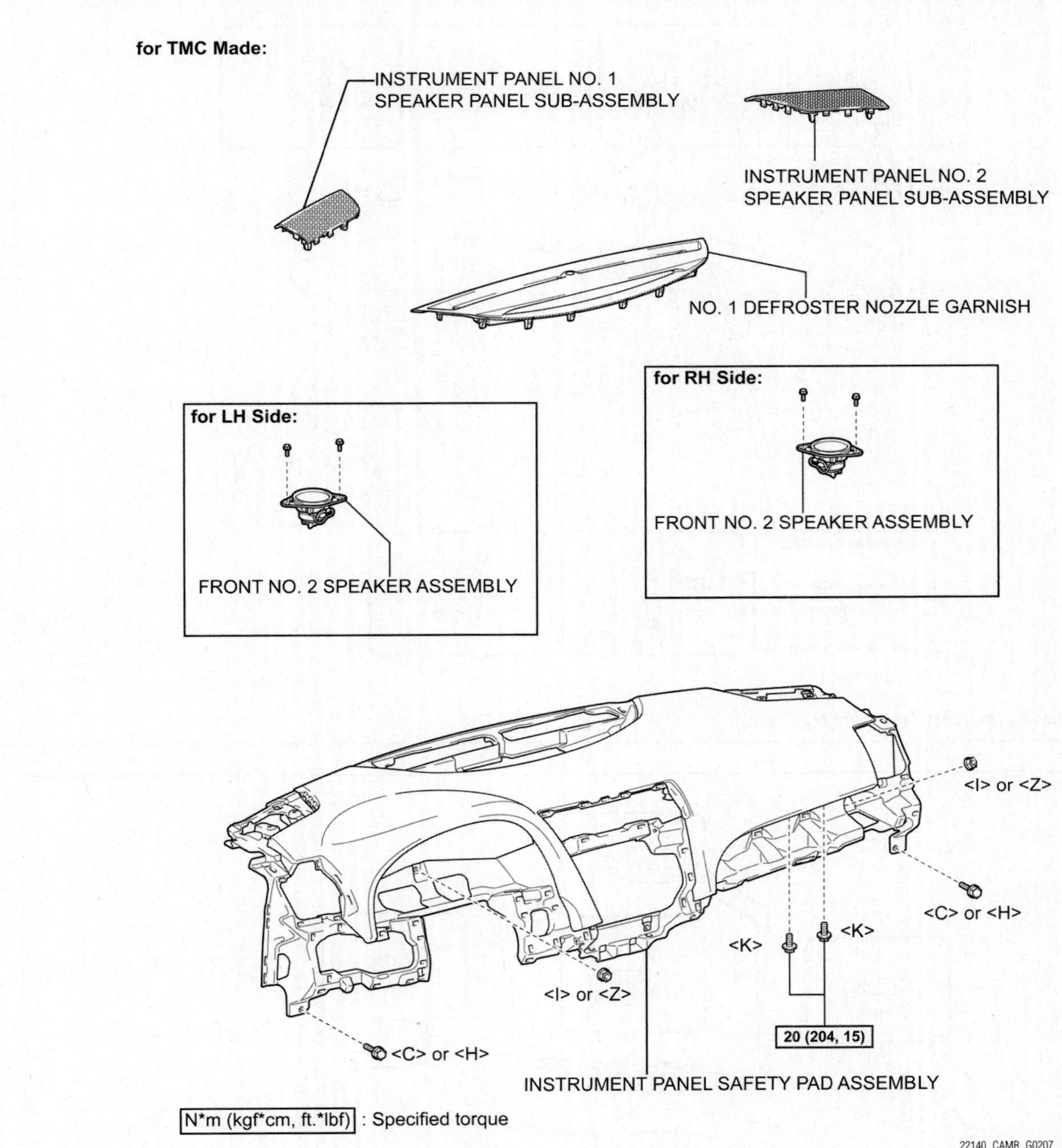

Fig. 300 Speakers, speaker panel sub-assembly and No. 1 defroster nozzle garnish

40. Remove instrument panel safety pad assembly by performing the following:
    a. Disengage each clamp.
    b. Disconnect each connector.
    c. Remove the bolt (J).

CAUTION

Models are equipped with a Supplemental Restraint System (SRS), which uses an air bag. Whenever working near any of the SRS components, such as the impact sensors, the air bag module, steering column and instrument panel, disable the SRS.

    d. Remove the 2 passenger airbag bolts (K).
    e. If equipped with Plasmacluster, disconnect the connector
    f. Disconnect the connector
    g. Remove the 2 bolts (C) or (H).
    h. Disengage the 5 claws and remove the instrument panel safety pad assembly.
    i. Disengage the claw and remove the 5 instrument panel stays.
41. Remove heater core as necessary.

To install:

42. Installation is the reverse of the removal procedure.
43. Perform initialization.
44. Inspect the steering pad.
45. Inspect the SRS warning light.

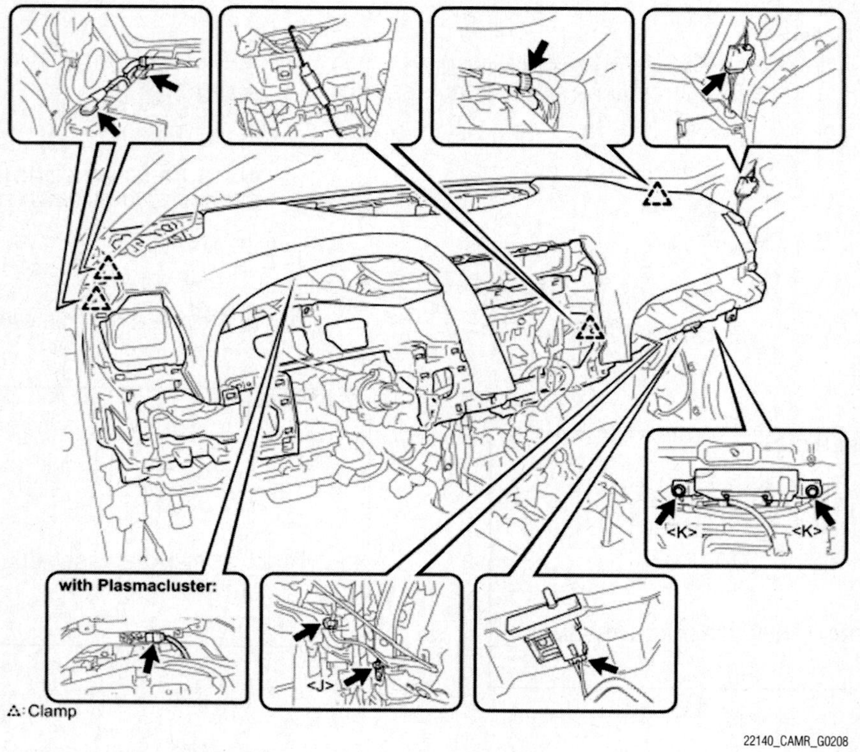

Fig. 301 Instrument panel safety pad assembly (1 of 2)

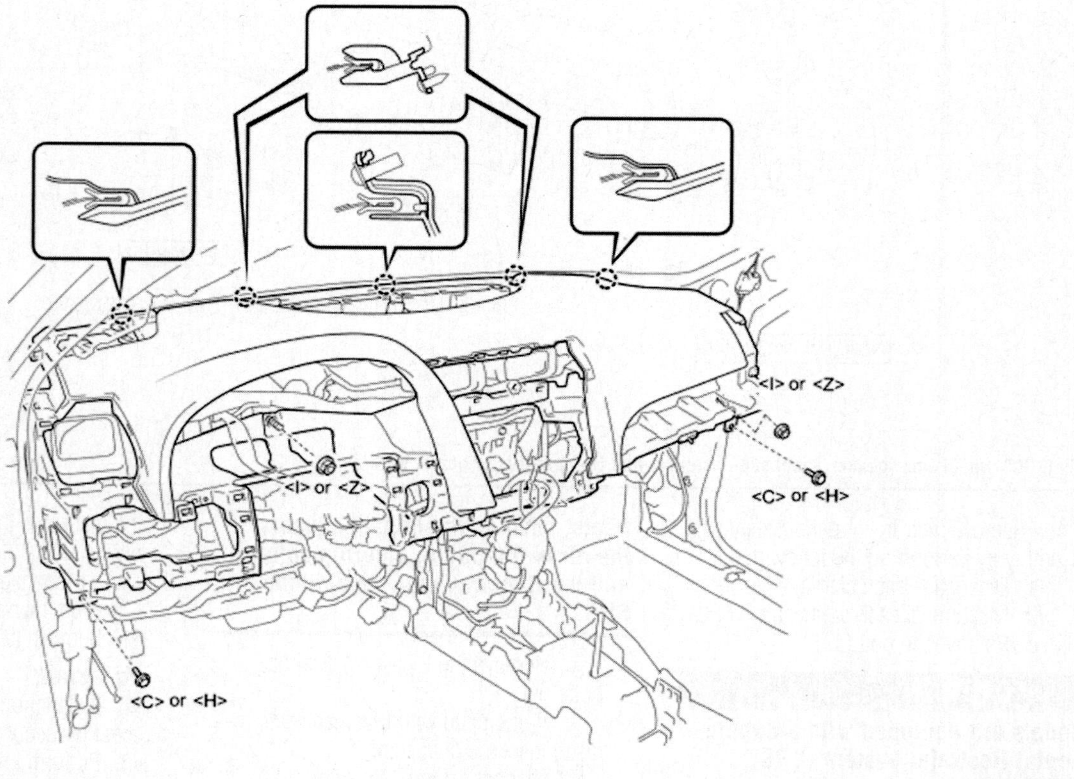

Fig. 302 Instrument panel safety pad assembly (2 of 2)

STEERING

POWER RACK & PINION STEERING GEAR

REMOVAL & INSTALLATION

See Figures 303 through 306.

CAUTION

Models are equipped with a Supplemental Restraint System (SRS), which uses an air bag. Whenever working near any of the SRS components, such as the impact sensors, the air bag module, steering column and instrument panel, disable the SRS.

➡**Be sure to turn the front wheels straight ahead when removing and installing the power steering link assembly.**

➡**If disconnecting the steering sliding yoke and the pinion shaft of the power steering link assembly, be sure to put matchmarks before starting the operation.**

1. Before servicing the vehicle, refer to the precautions section.
2. Place front wheels facing straight ahead.
3. Disconnect the negative battery cable.
4. Remove front wheels.
5. Separate steering sliding yoke, as follows:
   a. Secure the steering wheel with the seat belt in order to prevent rotation. This operation is useful to prevent damage to the spiral cable.
   b. Remove the bolt and slide the steering sliding yoke. Do not separate the steering sliding yoke from the power steering link assembly.
   c. Put matchmarks on the steering sliding yoke and the power steering link assembly.
   d. Separate the steering sliding yoke from the power steering link assembly.
6. Separate both tie rod assemblies, as follows:
   a. Remove the cotter pin and the nut.
   b. Using SST: 09628-00011, separate the tie rod assembly LH from the steering knuckle.
7. Remove engine assembly with transaxle.
8. Disconnect pressure feed tube assembly, as follows:

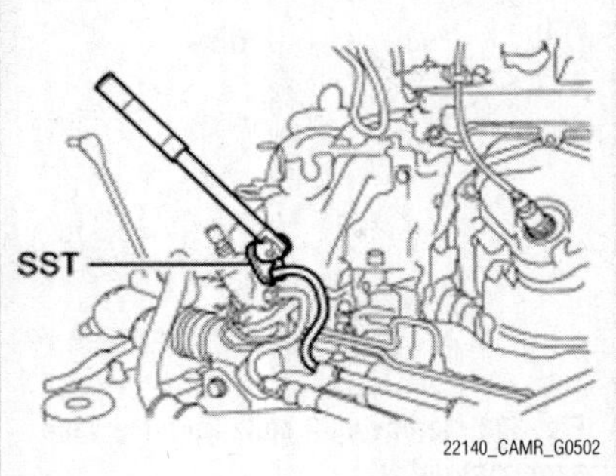

Fig. 303 Disconnect the pressure feed tube assembly—return tube side

   a. Using SST: 09023-12701, disconnect the pressure feed tube assembly (return tube side) from the power steering link assembly.
   b. Using SST: 09023-12701, disconnect the pressure feed tube assembly (pressure feed tube side) from the power steering link assembly.
   c. Remove the 2 bolts and separate the pressure feed tube clamp.

➡**Because the nut has its own stopper, do not turn the nut. Loosen the bolt with the nut fixed.**

9. Remove the 2 bolts, 2 nuts, and the power steering link assembly.
10. If equipped, remove the power steering rack housing heat insulator from the power steering link assembly.

To install:

11. If equipped, install the power steering rack housing heat insulator to the power steering link assembly.
12. Install the power steering link assembly with the 2 bolts and 2 nuts. Tighten to 52 ft. lbs. (70 Nm).
13. Connect pressure feed tube assembly, as follows:

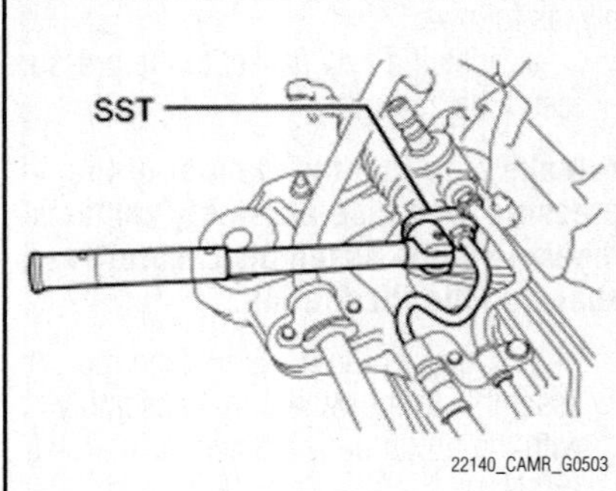

Fig. 304 Disconnect the pressure feed tube assembly—pressure feed tube side

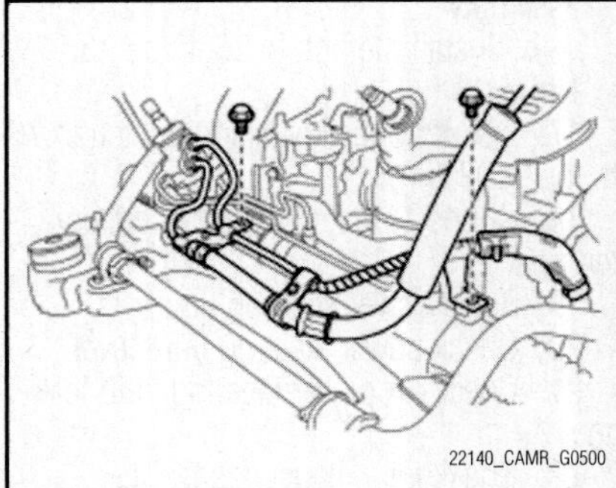

Fig. 305 Removing the 2 bolts and pressure feed tube clamp

Fig. 306 Removing power steering link assembly

   a. Temporarily connect the pressure feed tube assembly to the power steering link assembly.
   b. Install the pressure feed tube assembly clamp with the 2 bolts and tighten to 87 inch lbs. (9.8 Nm).

➡**Use a torque wrench with a fulcrum length of 11.81 inches (300 mm).**

➡**This torque value is effective when SST is parallel to the torque wrench.**

   c. Using SST: 09023-12701, tighten the pressure feed tube assembly (pressure feed tube side) to 16 ft. lbs. (22 Nm).
   d. Using SST: 09023-12701, tighten the pressure feed tube assembly (return tube side) to 16 ft. lbs. (22 Nm).
14. Install engine assembly with transaxle.
15. Connect both of the tie rod assemblies to the steering knuckle with the nut. Tighten to 36 ft. lbs. (49 Nm). Install a new cotter pin. Further tighten the nut up to 60°if the holes for the cotter pin are not aligned.
16. Connect steering sliding yoke, as follows:

a. Align the matchmarks on the steering sliding yoke and the steering link assembly.

b. Install the bolt to 26 ft. lbs. (35 Nm).

17. Install front wheels and tighten to 76 ft. lbs. (103 Nm).
18. Connect cable to negative battery terminal.
19. Bleed power steering fluid.
20. Check power steering fluid level.
21. Check for power steering fluid leakage.
22. Check for exhaust gas leaks.
23. Place front wheels facing straight ahead.
24. Inspect and adjust front wheel alignment.

POWER STEERING PUMP

REMOVAL & INSTALLATION

2.4L Engine

See Figures 307 and 308.

1. Before servicing the vehicle, refer to the precautions section.
2. Drain power steering fluid.
3. Remove RH engine under cover.
4. Remove RH front fender apron seal.
5. Remove fan and alternator v belt.
6. Slide the clip and disconnect the No. 1 fluid reservoir to pump hose from the vane pump assembly.
7. Disconnect pressure feed tube assembly, as follows:

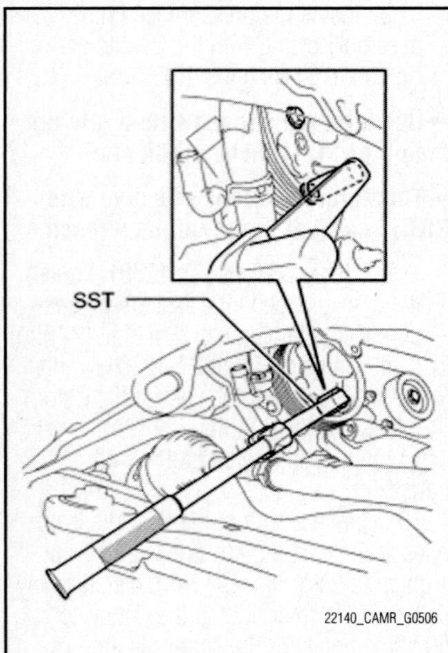

Fig. 307 Removing the vane pump assembly

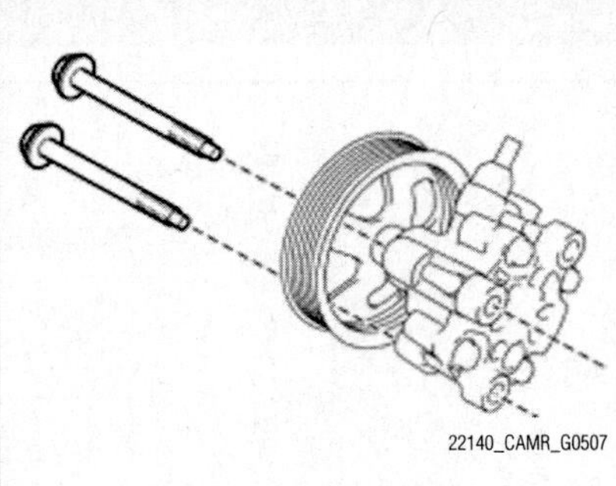

Fig. 308 Removing 2 bolts from the vane pump assembly

a. Remove the union bolt and disconnect the pressure feed tube assembly from the vane pump assembly.

b. Remove the gasket from the pressure feed tube assembly.

8. Disconnect the power steering fluid pressure switch connector.
9. Remove vane pump assembly, as follows:

a. Using SST: 09249-63010, loosen the 2 bolts and remove the vane pump assembly.

b. Remove the 2 bolts from the vane pump assembly.

To install:

10. Install vane pump assembly, as follows:

a. Temporarily install the 2 bolts to the vane pump assembly.

b. Install the vane pump assembly.

➡Use a torque wrench with a fulcrum length of 11.81 inches (300 mm).

➡This torque value is effective when SST is parallel to the torque wrench.

c. Using SST: 09249-63010, tighten the 2 bolts to 32 ft. lbs. (43 Nm).

11. Connect the connector to the power steering fluid pressure switch.
12. Connect pressure feed tube assembly, as follows:

a. Install a new gasket to the pressure feed tube assembly.

➡Make sure that the stopper of the pressure feed tube assembly contacts the vane pump assembly securely as shown in the illustration.

b. Connect the pressure feed tube assembly to the vane pump assembly with the union bolt. Tighten to 37 ft. lbs. (50 Nm).

13. Connect No. 1 fluid reservoir to pump hose, as follows:

➡Connect the No. 1 oil reservoir to pump hose with the paint mark facing toward the rear of the vehicle.

➡Push the No. 1 oil reservoir to pump hose as far as it will go.

➡Install the clip at the position specified in the illustration.

a. Connect the No. 1 fluid reservoir to pump hose to the vane pump assembly with the clip.

14. To complete installation, reverse removal procedure.

2.5L Engine

See Figures 309 and 310.

1. Drain the power steering fluid.
2. Remove the front wheel RH.
3. Remove the front fender apron seal RH.
4. Remove the v-ribbed belt.

a. Attach a wrench to the hexagonal portion of the belt tensioner as shown in the illustration, rotate the belt tensioner clockwise, and remove the V-ribbed belt.

5. Disconnect the No. 1 fluid reservoir to pump hose.

a. Slide the clip and disconnect the No. 1 fluid reservoir to pump hose from the vane pump assembly.

6. Disconnect the pressure feed tube assembly.

a. Remove the union bolt and disconnect the pressure feed tube assembly from the vane pump assembly.

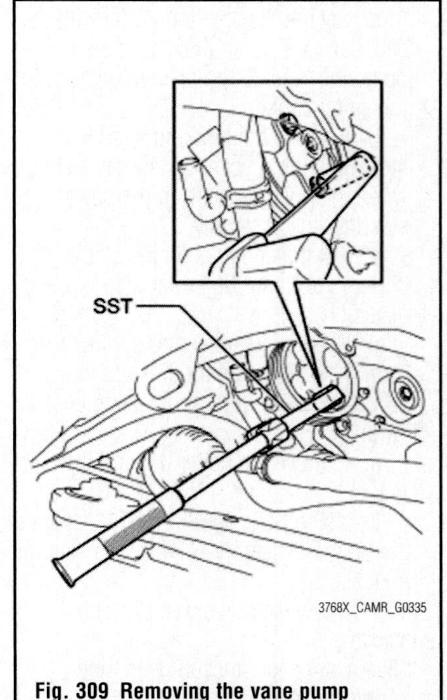

Fig. 309 Removing the vane pump

b. Remove the union bolt and disconnect the pressure feed tube assembly from the vane pump assembly.

7. Disconnect the power steering fluid pressure switch connector.

8. Remove the vane pump assembly.

a. Using the special tool, loosen the 2 bolts and remove the vane pump assembly.

b. Remove the 2 bolts from the vane pump assembly.

To install:

9. Install the vane pump.

a. Temporarily install the 2 bolts to the vane pump assembly.

b. Install the vane pump assembly.

c. Using the special tool, tighten to 2 bolts. Using the special tool, tighten to 21 ft. lbs. (29 Nm). Tightening without the special tool, tighten to 32 ft. lbs. (43 Nm).

10. Connect the power steering fluid pressure switch connector.

11. Connect the pressure feed tube assembly and a new gasket. Tighten to 37 ft. lbs. (50 Nm).

➡Make sure that the stopper of the pressure feed tube assembly contacts the vane pump assembly securely.

12. Connect the No. 1 fluid reservoir to pump hose with the clip.

➡Connect the No. 1 fluid reservoir to pump hose with the paint mark facing the rear of the vehicle.

➡Push the No. 1 fluid reservoir to pump hose as far as it will go as shown in the illustration.

➡Install the clip at the position specified in the illustration.

13. Install the v-ribbed belt.

a. Set the v-ribbed belt onto each part.

b. Loosen the v-ribbed belt by turning the belt tensioner clockwise.

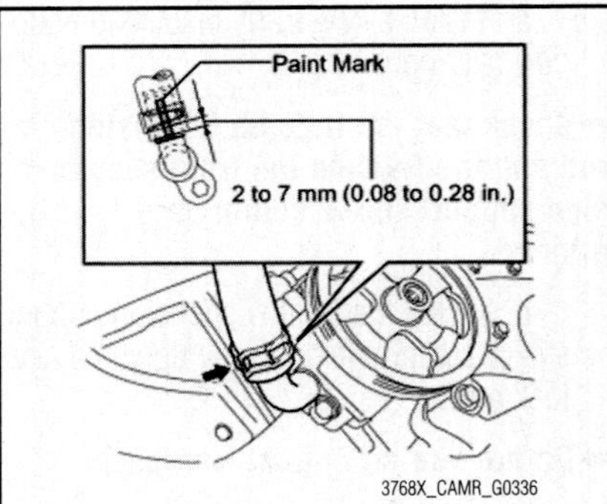

Fig. 310 Connecting the No. 1 fluid reservoir to the pump hose

c. Set the v-ribbed belt onto the tensioner pulley.

➡Make sure that the belt is attached to each pulley. In particular, make sure that the belt is securely fitted into the grooves of the crankshaft pulley.

14. Add power steering fluid.

15. Bleed the power steering fluid.

16. Check the power steering fluid level.

17. Inspect for power steering fluid leaks.

18. Install the front fender apron seal RH.

19. Install the front wheel RH. Tighten to 76 ft. lbs. (103 Nm).

3.5L Engine

See Figures 311 and 312.

1. Before servicing the vehicle, refer to the precautions section.

2. Drain power steering fluid.

3. Remove RH engine under cover.

4. Remove RH front fender apron seal.

5. Remove v-bank cover sub-assembly.

6. Remove fan and alternator v belt.

7. Slide the clip and disconnect the No. 1 fluid reservoir to pump hose from the vane pump assembly.

8. Disconnect pressure feed tube assembly, as follows:

a. Remove the union bolt and disconnect the pressure feed tube assembly from the vane pump assembly.

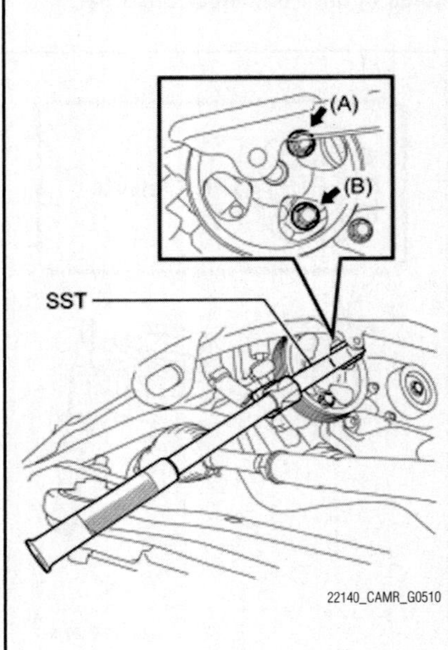

Fig. 311 Removing the vane pump assembly

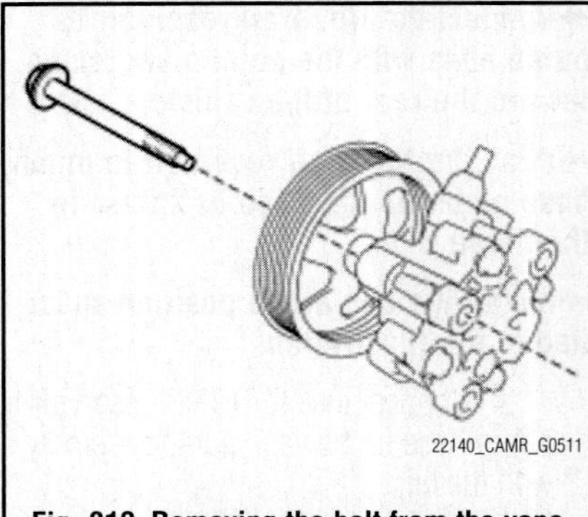

Fig. 312 Removing the bolt from the vane pump assembly

b. Remove the bolt and separate the pressure feed tube clamp.

c. Remove the gasket from the pressure feed tube assembly.

9. Disconnect the power steering fluid pressure switch connector.

10. Using SST: 09249-63010, loosen bolt (A) and remove bolt (B), and then remove the vane pump assembly.

11. Remove the bolt from the vane pump assembly.

To install:

12. Install vane pump assembly, as follows:

a. Temporarily install the bolt to the vane pump assembly.

b. Install the vane pump assembly.

➡Use a torque wrench with a fulcrum length of 11.81 inches (300 mm).

➡This torque value is effective when SST is parallel to the torque wrench.

c. Using SST: 09249-63010, tighten the 2 bolts to 32 ft. lbs. (43 Nm).

13. Connect the connector to the power steering fluid pressure switch.

14. Connect pressure feed tube assembly, as follows:

a. Install a new gasket to the pressure feed tube assembly.

b. Temporarily connect the pressure feed tube assembly to the vane pump assembly with the union bolt.

c. Install the pressure feed tube assembly clamp with the bolt. Tighten to 87 ft. lbs. (10 Nm).

d. Fully tighten the union bolt and tighten to 37 ft. lbs. (50 Nm).

➡Make sure that the stopper of the pressure feed tube assembly contacts the vane pump assembly securely.

15. Connect No. 1 fluid reservoir to pump hose, as follows:

➡**Connect the No. 1 oil reservoir to pump hose with the paint mark facing toward the rear of the vehicle.**

➡**Push the No. 1 oil reservoir to pump hose as far as it will go as shown in the illustration.**

➡**Install the clip at the position specified in the illustration.**

a. Connect the No. 1 fluid reservoir to pump hose to the vane pump assembly with the clip.

16. To complete installation, reverse removal procedure.

BLEEDING

1. Before servicing the vehicle, refer to the precautions section.
2. Check the fluid level.
3. Jack up the front of the vehicle and support it with stands.
4. With the engine stopped, turn the wheel slowly from lock to lock several times.
5. Lower the vehicle.
6. Start the engine.
7. Run the engine at idle for a few minutes.
8. With the engine idling, turn the wheel left or right to the full lock position and keep it there for 2 to 3 seconds, then turn the wheel to the opposite full lock position and keep it there for 2 to 3 seconds.
9. Repeat the above steps several times.
10. Stop the engine.
11. Check for foaming or emulsification. If the system has to be bled twice because of foaming or emulsification, check for fluid leaks in the system.
12. Check the fluid level.

SUSPENSION

COIL SPRING

REMOVAL & INSTALLATION

See Figures 313 through 318.

1. Before servicing the vehicle, refer to the precautions section.
2. Remove the front shock absorber.
3. As shown in the illustration, secure the front shock absorber with coil spring in a vise using aluminum plates by clamping onto a double nutted bolt affixed to the bracket at the bottom of the absorber.

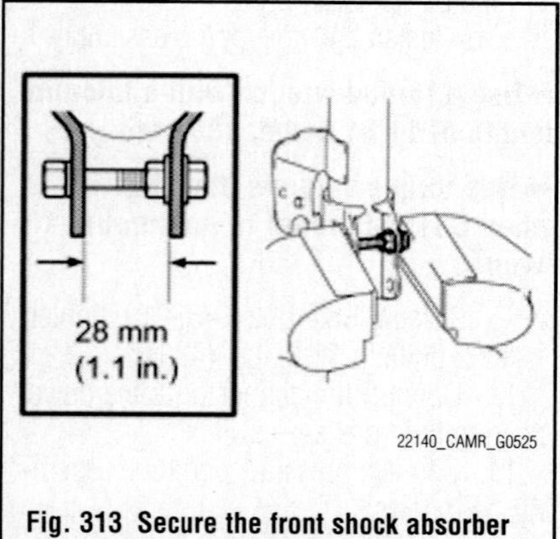

Fig. 313 Secure the front shock absorber

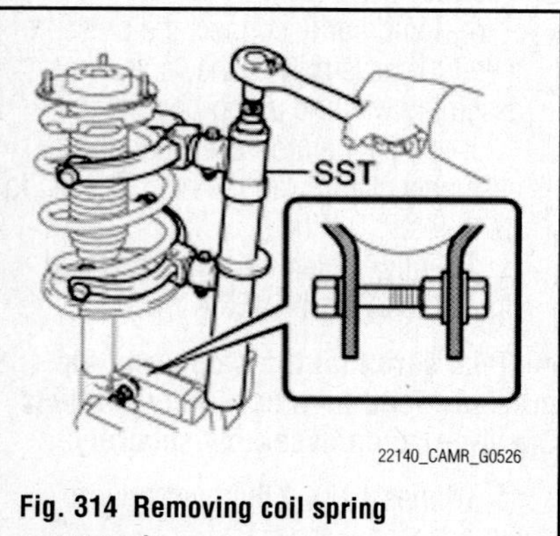

Fig. 314 Removing coil spring components

➡**Do not use an impact wrench.**

➡**If the front coil spring is compressed at an angle, using 2 SST will make the work easier.**

4. Using SST: 09727-30021, compress the front coil spring.
5. Remove the front suspension support sub-assembly, front suspension support bearing, front coil spring upper seat, front coil spring upper insulator, front coil spring, front spring bumper, and front coil spring lower insulator from the front shock absorber.

To install:

6. Install front coil spring as follows:

a. Install the front spring bumper to the piston rod.

➡**Align the 2 protrusions of the front coil spring lower insulator and the 2 holes in the front shock absorber.**

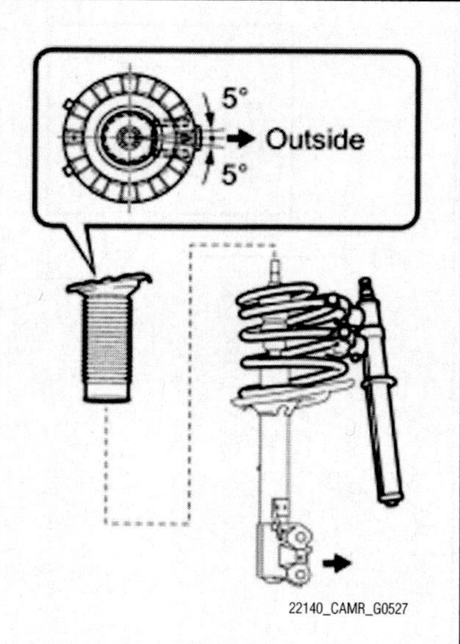

Fig. 315 Installing the front coil spring upper insulator

➡**Do not use an impact wrench.**

b. Install the front coil spring lower insulator onto the front shock absorber.

c. Using SST: 09727-30021, compress the front coil spring.

➡**The smaller diameter end of the front coil spring must face upward.**

➡**Fit the lower end of the front coil spring into the gap of the insulator.**

d. Install the front coil spring to the front shock absorber.

➡**Any misalignment between the front shock absorber lower bracket and the matchmark must be +/-5°.**

e. Install the front coil spring upper insulator as shown in the illustration.

➡**Any misalignment between the front shock absorber lower bracket and the matchmark must be +/-5°.**

7. Install the front coil spring upper seat with the mark facing to the outside of the vehicle.

➡**If there is foreign matter inside the front suspension support bearing, replace it with a new one.**

a. Install a new front suspension support bearing.

➡**Check that the flats on the piston rod and the flats on the front suspension support sub-assembly are aligned.**

b. Install the front suspension support sub-assembly. Temporarily tighten a new lock nut.

➡**Do not use an impact wrench.**

➡**Any misalignment between the front shock absorber lower bracket and the matchmark must be +/-5°.**

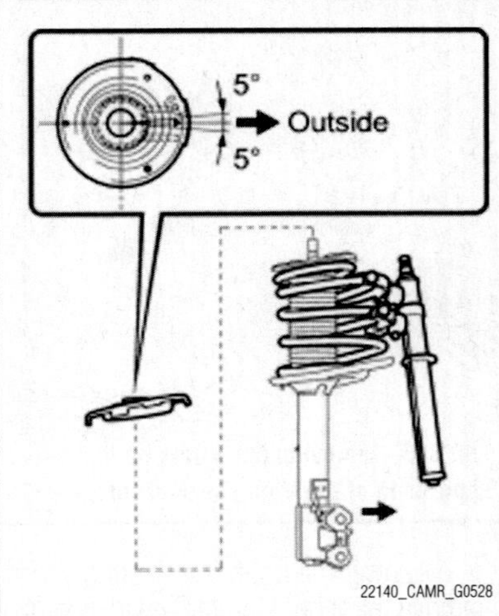

Fig. 316 Installing the front coil spring upper seat

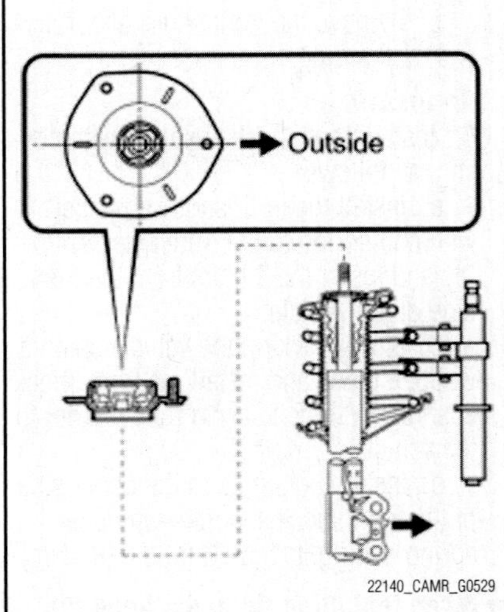

Fig. 317 the front suspension support sub-assembly

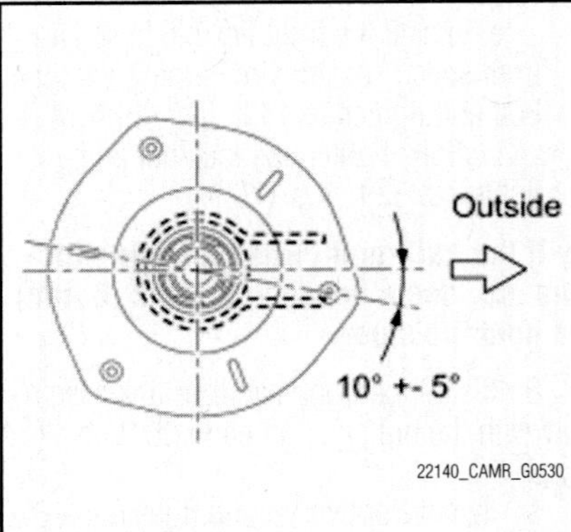

Fig. 318 Aligning the front shock absorber lower bracket and arrows

c. Remove the SST slowly in order to release the coil spring.

CONTROL LINKS

REMOVAL & INSTALLATION

See Front Stabilizer Bar.

LOWER BALL JOINT

REMOVAL & INSTALLATION

See Figure 319.

1. Before servicing the vehicle, refer to the precautions section.
2. Remove the front wheel.
3. Remove the front axle hub nut.
4. Separate the front speed sensor.
5. Separate the front disc the brake caliper assembly.
6. Remove front disc.
7. Separate the tie rod assembly.
8. Separate the No. 1 front lower suspension arm.
9. Remove the front axle assembly.
10. Remove front wheel No. 1 bearing dust deflector.
11. Remove front axle hub hole snap ring.
12. Remove front axle hub.
13. Remove front disc brake dust cover.
14. Remove the front lower ball joint assembly, as follows:
    a. Secure the steering knuckle in a vise using aluminum plates.
    b. Remove the cotter pin and castle nut.

➡Do not damage the dust cover of the ball joint.

➡Do not damage the steering knuckle.

c. Using SST (SST: 09628-62011) or equivalent, remove the front lower ball joint assembly.

To install:

15. Installation is the reverse of the removal procedure, noting the following:
    a. Install the front lower ball joint assembly to the steering knuckle with the castle nut and tighten to 91 ft. lbs. (123 Nm). Further tighten the nut up to 60°if the holes for the cotter pin are not aligned.
    b. Inspect and adjust the front wheel alignment.
    c. Inspect the ABS speed sensor signal.

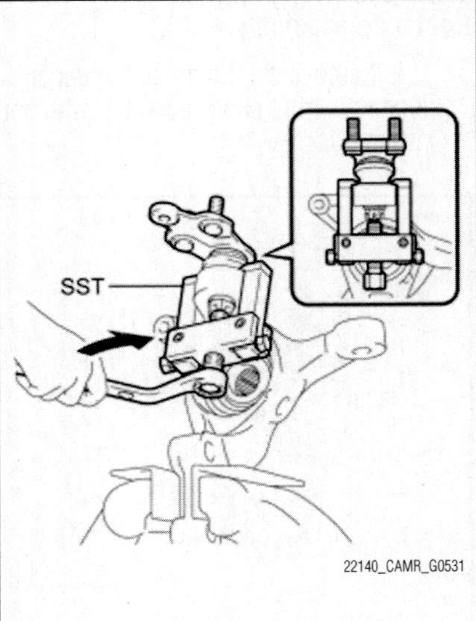

Fig. 319 Remove the front lower ball joint assembly

LOWER CONTROL ARM

REMOVAL & INSTALLATION

See Figures 320 through 322.

1. Before servicing the vehicle, refer to the precautions section.
2. Remove the engine assembly with transaxle.

➡Use the same procedures for the RH side and the LH side. The procedures listed below are for the LH side.

3. Remove the 3 nuts and the engine mounting insulator.
4. Remove the 3 bolts and the nut on the front suspension lower No. 1 arm and remove it from the front frame assembly.
5. Remove the front lower arm bushing stopper.

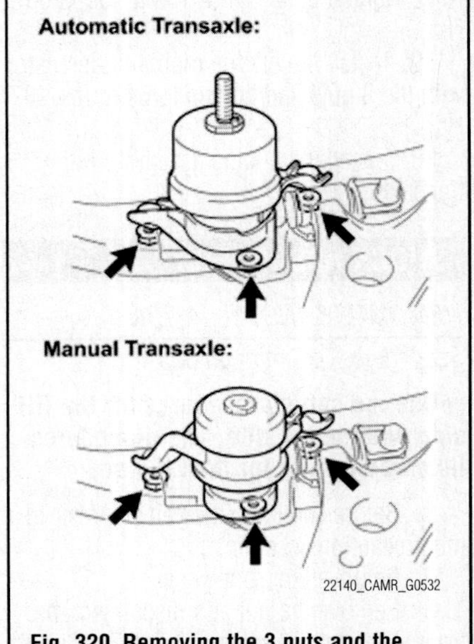

Fig. 320 Removing the 3 nuts and the engine mounting insulator

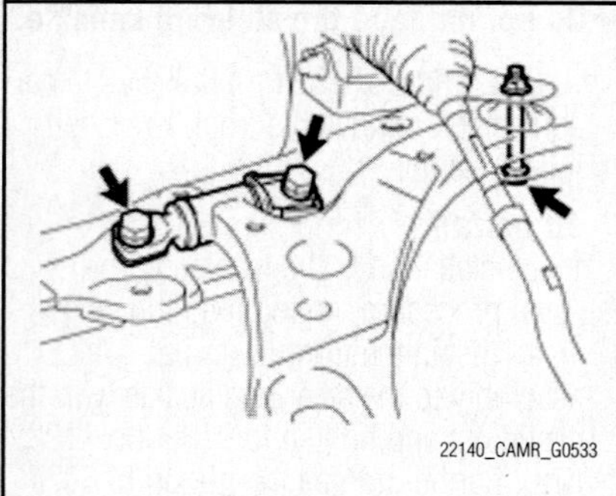

Fig. 321 Removing the 3 bolts and the nut on the front suspension lower No. 1 arm

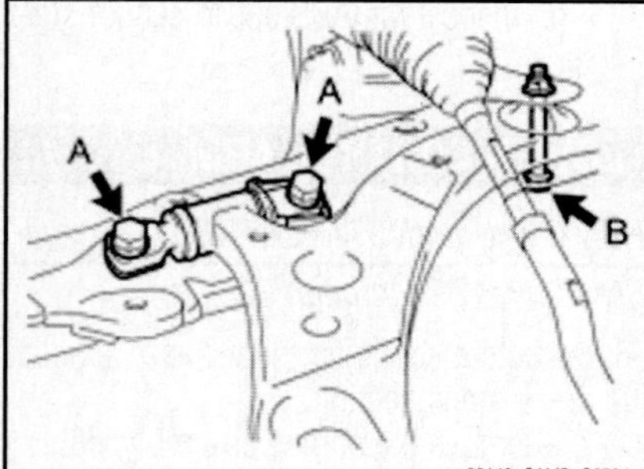

Fig. 322 Tightening the 3 bolts and the nut on the front suspension lower No. 1 arm

To install:

6. Install the front lower arm bushing stopper.

7. Install the front suspension lower No. 1 arm to the front frame assembly with the 3 bolts and the nut, but do not tighten them yet.

8. Tighten bolts "A" to 148 ft. lbs. (200 Nm). Tighten bolts "B" to 152 ft. lbs. (206 Nm).

9. Install the engine mounting insulator with the 3 nuts and tighten to 64 ft. lbs. (87 Nm).

10. Install the engine assembly with transaxle.

SHOCK ABSORBERS

REMOVAL & INSTALLATION

See Figures 323 through 325.

➡Use the same procedures for the RH side and the LH side. The procedures listed below are for the LH side.

1. Before servicing the vehicle, refer to the precautions section.

2. Remove the front wheel.

3. Remove the nut and disconnect the front stabilizer link assembly from the front shock absorber assembly.

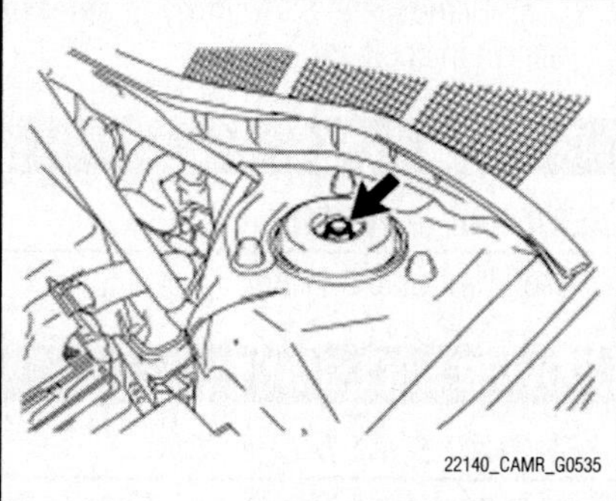

Fig. 323 Loosening the lock nut of the front shock absorber arm

4. Remove front shock absorber with coil spring, as follows:

a. Loosen the lock nut of the front shock absorber with coil spring.

➡Do not remove the lock nut.

➡Only loosen the nut when disassembling the front shock absorber with coil spring.

b. Remove the bolt and disconnect the front flexible hose and front speed sensor wire harness from the front shock absorber with coil spring.

➡Be sure to remove the front speed sensor from the front shock absorber with coil spring.

c. Remove the 2 nuts on the lower side of the front shock absorber with coil spring.

➡When removing the nuts, keep the bolts from rotating.

➡Keep the bolts inserted to secure the front axle assembly.

d. Remove the 3 nuts on the upper side of the front shock absorber with coil spring.

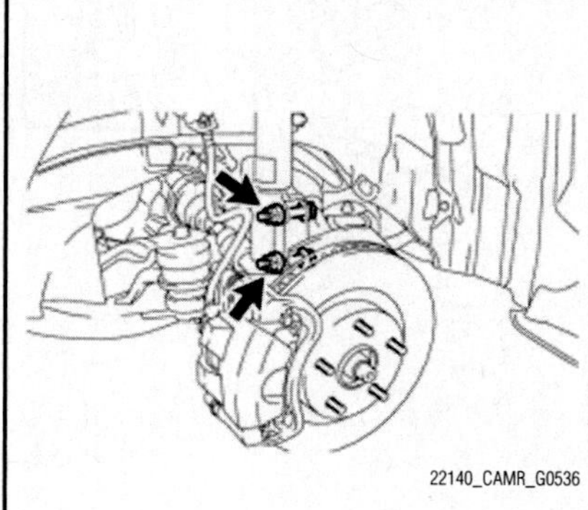

Fig. 324 Remove the 2 nuts on the lower side of the front shock absorber

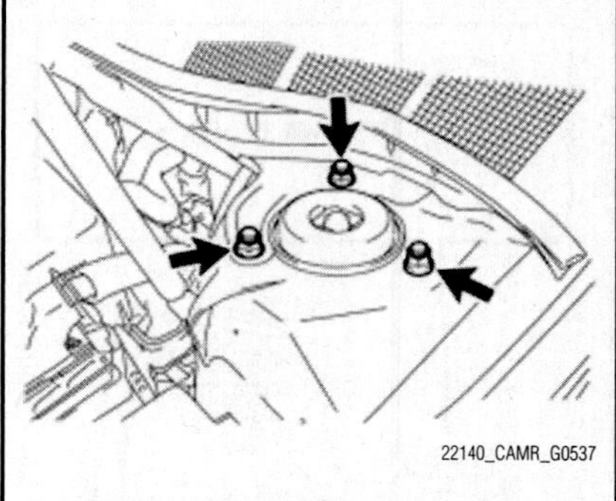

Fig. 325 Removing the 3 nuts on the upper side of the front shock absorber

e. Lower the front axle assembly, and remove the 2 bolts on the lower side of the front shock absorber.

➡Make sure that the front speed sensor is disconnected from the front shock absorber with coil spring.

f. Remove the front shock absorber with coil spring.

To install:

5. Install front shock absorber with coil spring, as follows:

a. Install the front shock absorber with coil spring to the front axle assembly and insert the 2 bolts from the front side of the vehicle.

b. Slowly jack up the vehicle using a wooden block and install the front shock absorber with coil spring (upper side) to the vehicle.

c. Install the 3 nuts to the upper side of the front shock absorber with coil spring and tighten to 63 ft. lbs. (85 Nm).

➡When installing the nuts, keep the bolts from rotating.

d. Install the 2 nuts to the lower side of the front shock absorber with coil spring and tighten to 155 ft. lbs. (210 Nm).

e. Install the front flexible hose and front speed sensor wire harness with the bolt and tighten to 14 ft. lbs. (19 Nm).

f. Fully tighten the lock nut and tighten to 52 ft. lbs. (70 Nm).

➡If the ball joint turns together with the nut, use a hexagon wrench (6 mm) to hold the stud.

6. Install the front stabilizer link assembly with the nut and tighten to 55 ft. lbs. (74 Nm).

7. Install front wheel and tighten to 76 ft. lbs. (103 Nm).

8. Inspect and adjust front wheel alignment.

STEERING KNUCKLE

REMOVAL & INSTALLATION

See Wheel Hub and Bearing.

STABILIZER BAR

REMOVAL & INSTALLATION

See Figures 326 through 328.

1. Before servicing the vehicle, refer to the precautions section.
2. Remove the front wheels.
3. Separate steering intermediate shaft assembly.
4. Separate tie rod end sub-assembly.

➡If the ball joint turns together with the nut, use a hexagon wrench (6 mm) to hold the stud.

5. Remove the 2 nuts and the front stabilizer link assembly
6. Remove the engine assembly with transaxle.
7. Remove the bolts and the left and right No. 1 stabilizer brackets.
8. Remove the engine assembly with transaxle.
9. Remove the bolts and the left and right No. 1 stabilizer brackets.

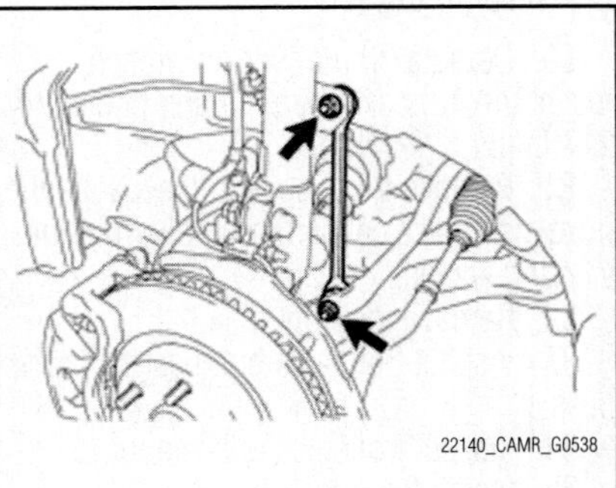

Fig. 326 Remove the 2 nuts and front stabilizer link assembly (left hand shown)

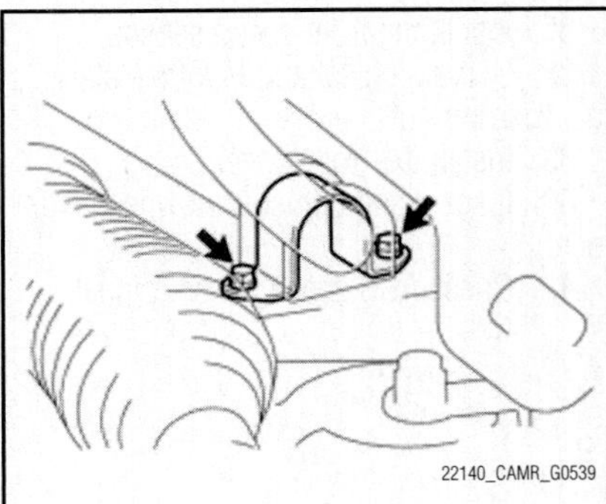

Fig. 327 Remove the 2 bolts and No. 1 stabilizer bracket (left hand shown)

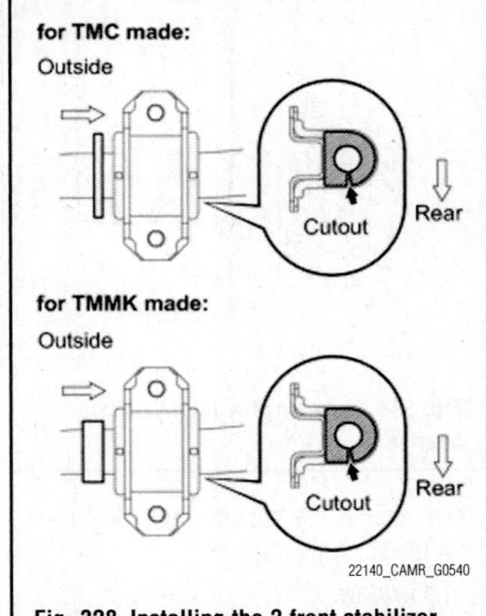

Fig. 328 Installing the 2 front stabilizer bar bushings

10. Remove the 2 front No. 1 stabilizer bar bushings from the front stabilizer bar.
11. Remove the front stabilizer bar from the vehicle.

To install:

➡Make sure that the cutout of the front stabilizer bar bushing No. 1 faces the rear side as shown in the illustration.

12. Install the 2 front stabilizer bar bushings No. 1 to the outside of the bushing stopper on the front stabilizer bar.
13. Install the No. 1 left front stabilizer bracket with the 2 bolts and tighten to 20 ft. lbs. (27 Nm).
14. Install the No. 1 right front stabilizer bracket with the 2 bolts and tighten to 20 ft. lbs. (27 Nm).
15. Install the engine assembly with transaxle.
16. Install the left front stabilizer link assembly with the 2 nuts and tighten to 55 ft. lbs. (74 Nm).
17. Install the right front stabilizer link assembly with the 2 nuts and tighten to 55 ft. lbs. (74 Nm).
18. To complete installation, reverse removal procedure.
19. Inspect and adjust the front wheel alignment.

WHEEL BEARINGS

REMOVAL & INSTALLATION

See Figures 329 through 334.

1. Before servicing the vehicle, refer to the Precautions Section.
2. Remove front wheel.
3. Remove front axle hub nut.

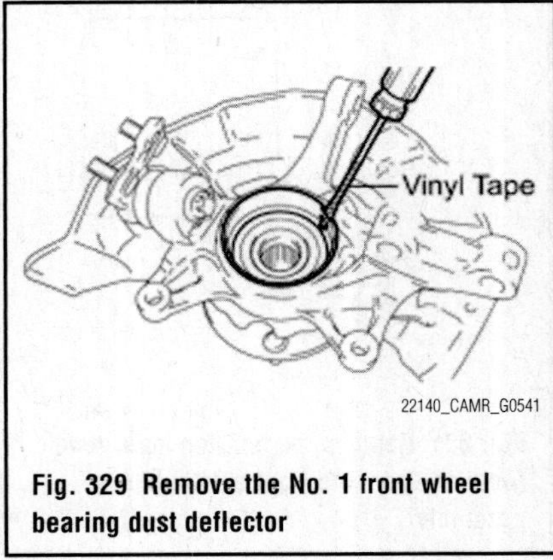

Fig. 329 Remove the No. 1 front wheel bearing dust deflector

4. Separate front speed sensor.
5. Remove the 2 bolts and separate the front disc brake caliper assembly from the steering knuckle. Use wire or an equivalent tool to keep the brake caliper from hanging down by the flexible hose.
6. Remove front disc.
7. Separate tie rod end sub-assembly.
8. Separate front suspension lower no. 1 arm.
9. Remove front axle assembly.
10. Using a screwdriver with its tip wrapped with vinyl tape, remove the No. 1 front wheel bearing dust deflector. Be careful not to damage the steering knuckle.
11. Using snap ring pliers, remove the front axle hub hole snap ring.
12. Remove front axle hub sub-assembly by performing the following:
    a. Hold the front axle assembly between aluminum plates in a vise.

➡Do not overtighten the vise.

    b. Using SST 09520-00031, remove the front axle hub sub-assembly.

➡Be careful not to drop the front axle hub sub-assembly.

    c. Using SST 09555-55010, SST: 09950-60010 and SST: 09950-70010

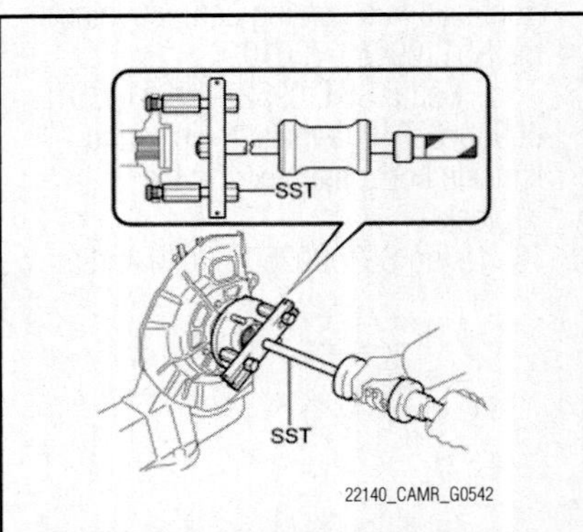

Fig. 330 Remove the front axle hub sub-assembly

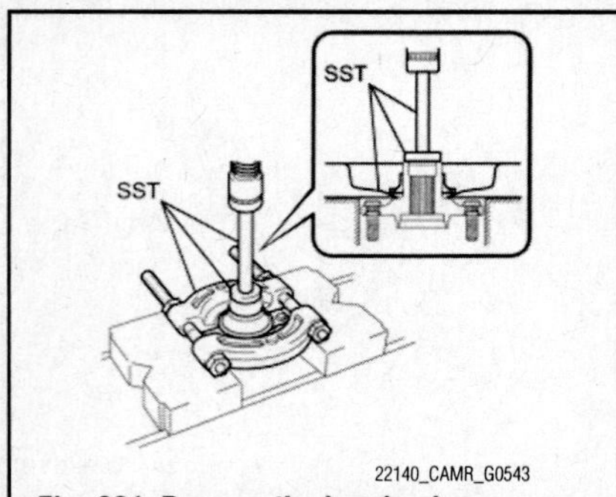

Fig. 331 Remove the bearing inner race (outside) from the front axle hub sub-assembly

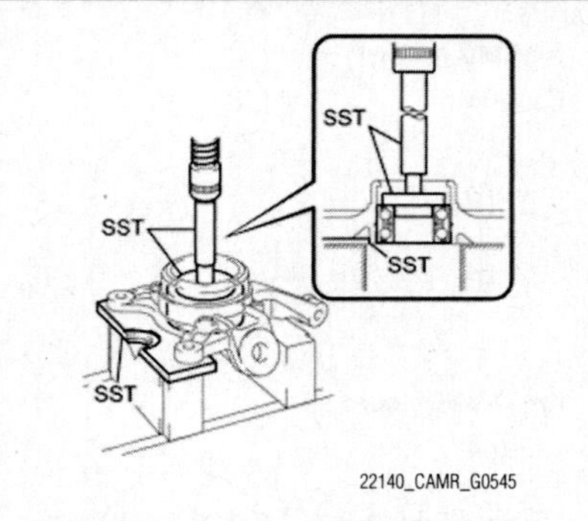

Fig. 333 Removing the front axle hub bearing

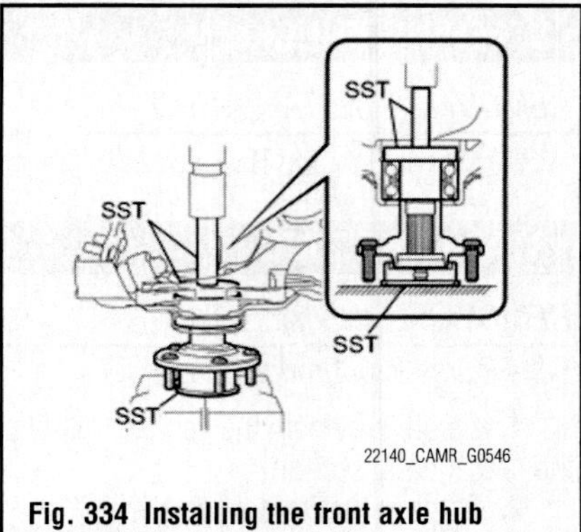

Fig. 334 Installing the front axle hub sub-assembly

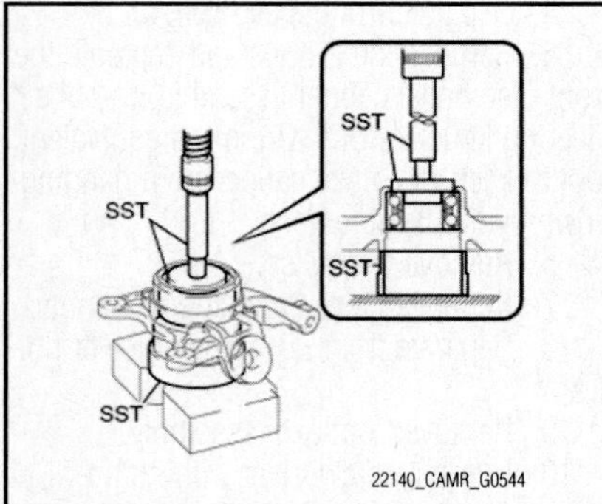

Fig. 332 Pressing the front axle hub bearing

and a press, remove the bearing inner race (outside) from the front axle hub sub-assembly.

13. Remove the 4 bolts and disc brake dust cover from the steering knuckle.
14. Remove front lower ball joint assembly.
15. Remove front axle hub bearing by performing the following:
    a. Place the bearing inner race (outside) on the front axle hub bearing.
    b. Using SST 09527-17011, SST: 09950-60010 and a press, press the front axle hub bearing until it contacts the SST: 09950-70010.
    c. Using SST: 09527-20011, SST: 09950-60010 to make the steering knuckle horizontal, fix it to the V-block.
16. Using SST: 09950-70010 and a press, remove the front axle hub bearing from the steering knuckle.

To install:

17. Using SST's: 09950-60020, 09950-70010 and a press, install a new front axle hub bearing to the steering knuckle.
18. Install front lower ball joint assembly.
19. Install the disc brake dust cover to the steering knuckle with the 4 bolts and tighten to 73 inch lbs. (8.3 Nm).
20. Using SST's: 09608-32010, 09950-60020, 09950-70010 and a press, install the front axle hub sub-assembly.
21. Using snap ring pliers, install a new front axle hub hole snap ring.

➡Align the hole for the speed sensor in the No. 1 front wheel bearing dust deflector with the steering knuckle.

22. Using SST's: 09316-60011, 09608-32010 and a hammer, install a new No. 1 front wheel bearing dust deflector.

➡Only when reusing the bolts and nuts, apply the small amount of engine oil to the screw part of the nuts.

➡Be careful not to damage the driveshaft boot or speed sensor rotor.

23. Align the matchmarks and install the front driveshaft assembly to the front axle hub sub-assembly.
24. Install the steering knuckle with the front axle hub sub-assembly to the front shock absorber assembly with the 2 bolts and 2 nuts and tighten to 155 ft. lbs. (210 Nm).
25. Install the lower No. 1 front suspension arm sub-assembly.
26. Install the tie rod end sub-assembly.
27. Install the front disc.
28. Install the front disc brake caliper assembly with the 2 bolts to the steering knuckle and tighten to 79 ft. lbs. (107 Nm).
29. Clean the threaded parts on the driveshaft and axle hub nut using a non-residue solvent.

➡Be sure to perform this work for a new driveshaft.

➡Keep the threaded parts free of oil and foreign objects.

30. Using a 30 mm socket wrench, install the front axle hub nut and tighten to 217 ft. lbs. (294 Nm).
31. Remove the 2 bolts and separate the front disc brake caliper assembly from the steering knuckle.
32. Remove the front disc.
33. Inspect front axle hub bearing looseness.
34. Inspect front axle hub runout.
35. Install the front disc.
36. Install the front disc brake caliper assembly with the 2 bolts to the steering knuckle and tighten to 79 ft. lbs. (107 Nm).
37. Install the front speed sensor.
38. Using a chisel and hammer, stake the axle hub nut.
39. Install the front wheel.
40. Inspect and adjust front wheel alignment.
41. Check ABS speed sensor signal.

COIL SPRING

REMOVAL & INSTALLATION

See Figures 335 through 337.

1. Before servicing the vehicle, refer to the precautions section.
2. Secure the rear shock absorber with coil spring in a vise using aluminum plates by closing the vise onto the double nutted bolt affixed to the bracket at the bottom of the absorber.

➡Do not use an impact wrench.

➡If the rear coil spring is compressed at an angle, using 2 SST will make the work easier.

3. Using SST: 09727-30021, compress the rear coil spring
4. Remove the nut, rear shock absorber collar and rear suspension support assembly.
5. Remove the rear coil spring, rear No. 1 spring bumper, and rear coil spring lower insulator.

To install:

6. Install the rear No. 1 spring bumper to the piston rod.
7. Install the rear coil spring lower insulator onto the rear shock absorber.

➡Do not use an impact wrench.

8. Using SST: 09727-30021, compress the rear coil spring.

➡The smaller diameter end must face upward.

➡Fit the lower end of the rear coil spring into the gap of the lower seat.

➡If the front coil spring is compressed at an angle, using 2 SST will make the work easier.

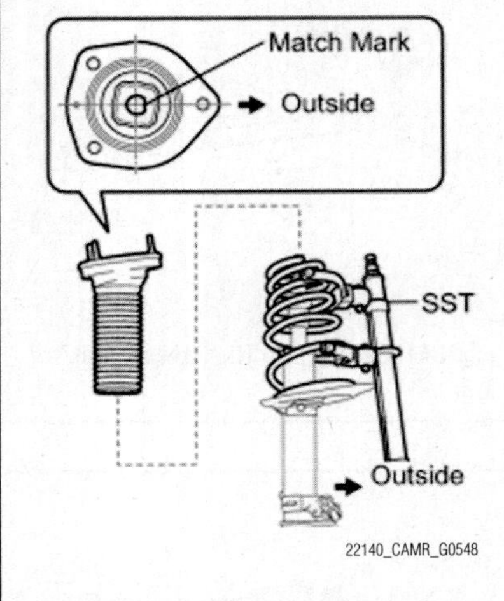

Fig. 336 Align the notches of the shock absorber

9. Install the rear coil spring to the rear shock absorber.

➡Align the notches of the piston rod and the rear suspension support assembly as shown in the illustration before installing the rear suspension support assembly.

10. Install the rear suspension support assembly.
11. Align the notches of the shock absorber with the notch of the rear suspension support assembly so that the notches face the outside of the vehicle.
12. Install the rear shock absorber collar.
13. Loosely tighten a new lock nut to the rear suspension piston rod.

➡Do not use an impact wrench.

➡When lining up the rear suspension support assembly's stud bolts at the middle point between the two sides of the bracket, the maximum permissible degree of error is plus or minus 5°.

14. Release the spring while adjusting the rear suspension support assembly to the position shown in the illustration, and remove the SST from the rear coil spring.

CONTROL ARMS/LINKS

REMOVAL & INSTALLATION

No. 1 Suspension Arm

See Figures 338 through 343.

1. Before servicing the vehicle, refer to the precautions section.

➡Check if an old gasket still remains on the pipe. If so, remove it. Also, check if any bolts or nuts are rusted. If so, replace them.

2. Remove rear wheel.
3. Remove center exhaust pipe assembly.
4. Remove tail exhaust pipe assembly.
5. Separate both rear stabilizer link assemblies.
6. Remove rear stabilizer bar no. 2 and no. 1 bracket.
7. Remove rear stabilizer bar.
8. Remove rear stabilizer bushing.
9. Separate rear strut rod.

➡When removing the bolt, keep the nut from rotating.

10. Remove the bolt, nut and separate the rear suspension No. 2 arm (outer side) from the rear axle carrier.

➡When removing the bolt, keep the nut from rotating.

11. Remove the bolt, nut and the rear No. 1 suspension arm (outer side) from the rear axle carrier.

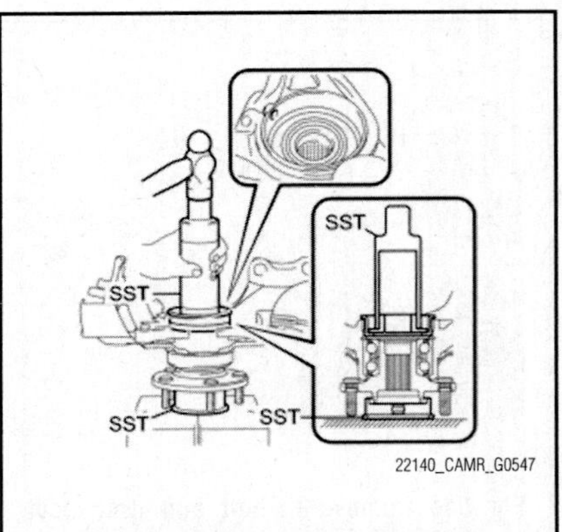

Fig. 335 Remove the rear coil spring

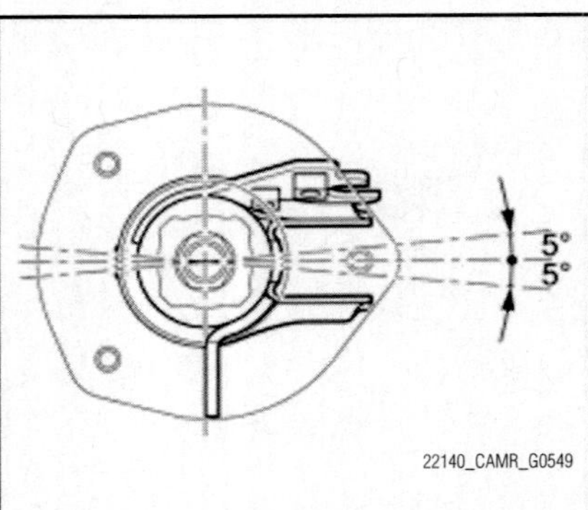

Fig. 337 Lining up the rear suspension support assembly's stud bolts

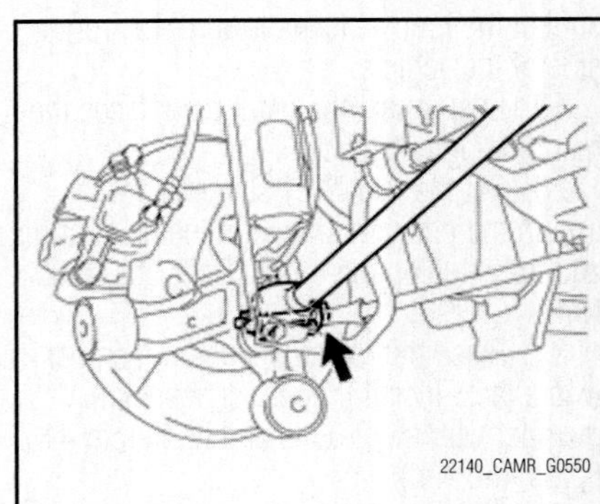

Fig. 338 Removing the bolt, nut and the rear No. 1 suspension arm—LH shown

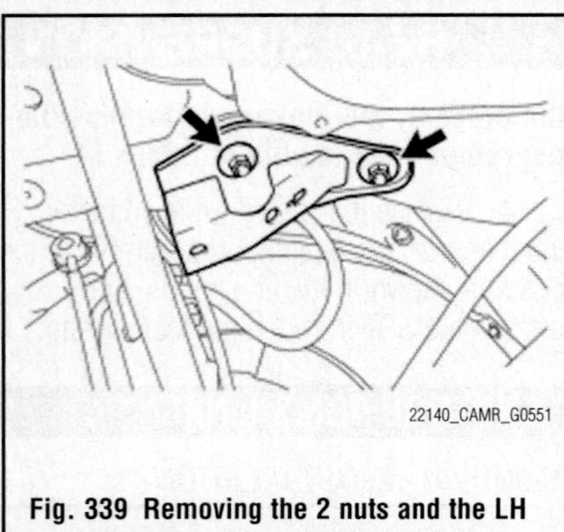

Fig. 339 Removing the 2 nuts and the LH rear suspension member lower stopper

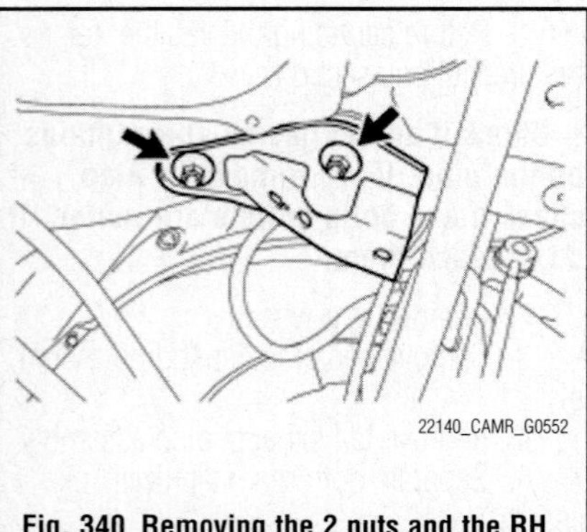

Fig. 340 Removing the 2 nuts and the RH rear suspension member lower stopper

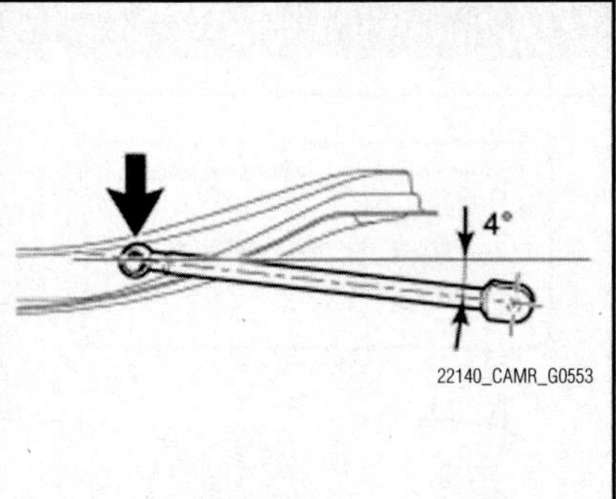

Fig. 341 Set the rear No.1 suspension arm

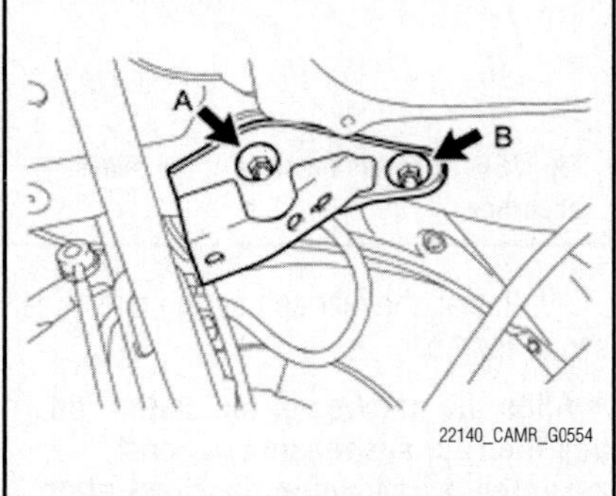

Fig. 342 LH rear suspension member lower stopper tightening sequence

12. Remove the 2 nuts and the LH rear suspension member lower stopper.
13. Remove the 2 nuts and the RH rear suspension member lower stopper.
14. Support the rear suspension member with a jack.
15. Remove the 2 bolts, and the rear suspension member sub-assembly.
16. Remove the bolt and rear No. 1 suspension arm assembly.

To install:

17. Install the No. 1 rear suspension arm (inner side) with the bolt, and temporarily tighten the bolt.
18. Install the rear No. 1 suspension arm so that the bracket leans toward the front side of the vehicle.
19. Ensure that the paint mark faces the rear side of the vehicle.
20. Set the rear No.1 suspension arm in the position shown in the illustration, and fully tighten the bolt to 74 ft. lbs. (100 Nm).
21. Raise the rear suspension member with a jack. Install the rear suspension member with the 2 bolts and tighten to 41 ft. lbs. (56 Nm).
22. Install both the rear suspension member lower stoppers with the 2 nuts and tighten to:
    a. Nut A: 41 ft. lbs. (55 Nm).
    b. Nut B: 28 ft. lbs. (38 Nm).

➡Insert the bolt from the front of the vehicle and temporarily install the bolt.

23. Connect the rear No.1 suspension arm (outer side) to the rear axle carrier with the bolt and nut and temporarily tighten the bolt and nut. When temporarily tightening the bolt, keep the nut from rotating.

➡Insert the bolt from the inside of the vehicle and temporarily install the bolt.

24. Connect the strut rod assembly rear to the axle carrier with the bolt and nut and temporarily tighten the bolt. When temporarily tightening the bolt, keep the nut from rotating.
25. Jack up the rear axle carrier, placing a wooden block to avoid damage. Apply load to the suspension so that the installed bolt of the rear No. 1 suspension arm (inner side) is horizontally aligned with the center of the rear axle hub.
26. Fully tighten rear No. 1 suspension arm and tighten the bolt to 74 ft. lbs. (100 Nm).
27. Fully tighten rear No. 2 suspension arm and tighten the bolt to 74 ft. lbs. (100 Nm).
28. To complete installation, reverse removal procedure.

No. 2 Suspension Arm

See Figures 344 and 345.

1. Before servicing the vehicle, refer to the precautions section.
2. Remove the rear wheel.
3. Remove the bolt, and disconnect the rear No. 2 suspension arm (inner side).

➡When removing the bolt, keep the nut from rotating.

4. Remove the bolt, nut and the rear No. 2 suspension arm (outer side) from the rear axle carrier.

To install:

➡Ensure that the paint mark faces to the rear of the vehicle.

5. Install the rear No. 2 suspension arm (inner side) with the bolt, and temporarily tighten the bolt.

➡When temporarily tightening the bolt, keep the nut from rotating.

6. Connect the rear No. 2 suspension arm (outer side) to the rear axle carrier with the bolt and nut, and temporarily tighten the bolt.

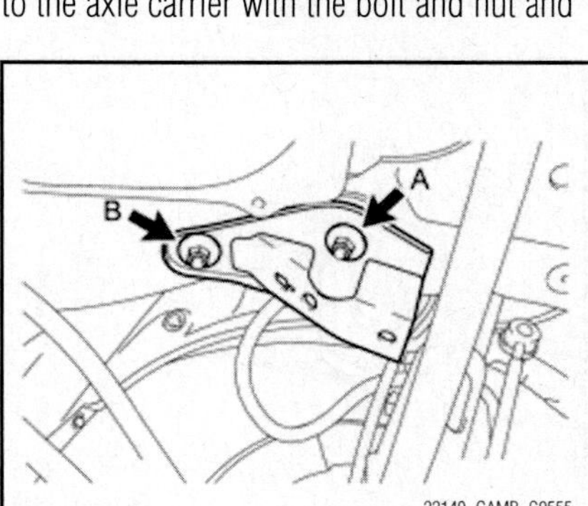

Fig. 343 LH rear suspension member lower stopper tightening sequence

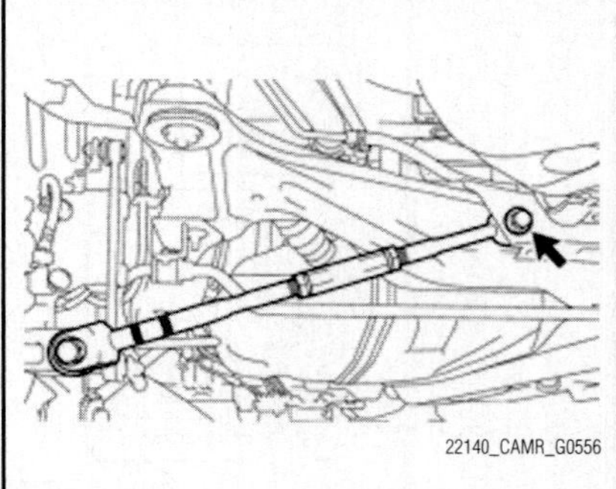

Fig. 344 Remove the bolt, and disconnect the rear No. 2 suspension arm (inner side)

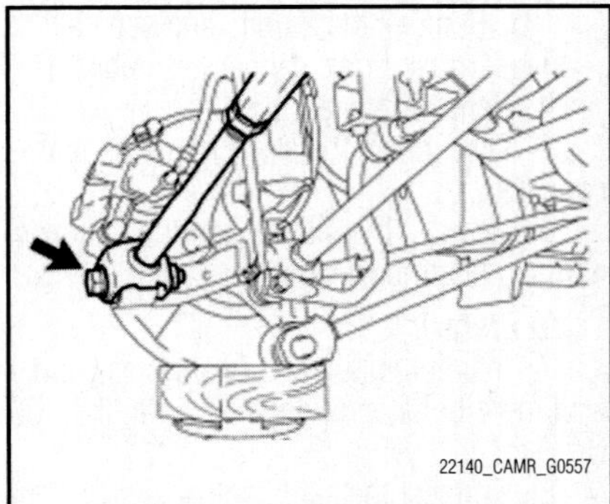

Fig. 345 Removing the bolt, nut and the rear No. 2 suspension arm (outer side)

7. Stabilize suspension.
8. Fully tighten the rear No. 2 suspension arm bolt (inner side) to 74 ft. lbs. (100 Nm).
9. Fully tighten the rear No. 2 suspension arm bolt (outer side) to 74 ft. lbs. (100 Nm).
10. Install the rear wheel.
11. Inspect and adjust the rear wheel alignment.

SHOCK ABSORBER

REMOVAL & INSTALLATION

See Figures 346 and 347.

1. Before servicing the vehicle, refer to the precautions section.
2. Remove the rear seat cushion assembly.
3. Remove rear seat headrest plate cover.
4. Remove rear seat headrest assembly.
5. Remove the rear seatback assembly.
6. Remove the rear wheel.
7. Separate LH rear stabilizer link assembly.
8. Remove the 2 bolts, and disconnect the rear brake flexible hose and rear speed sensor from the rear shock absorber with coil spring and rear axle carrier.

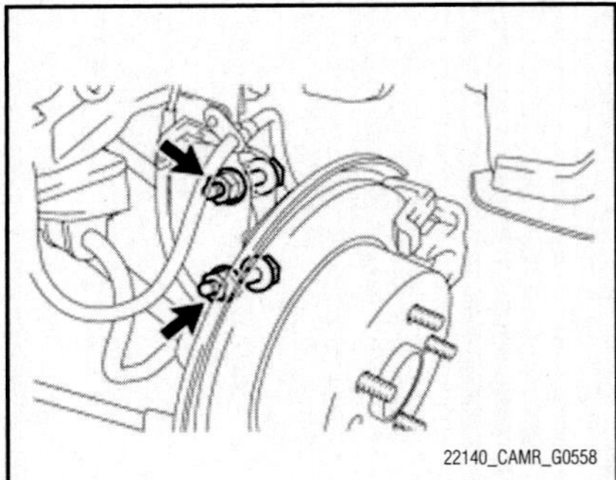

Fig. 346 Loosen the 2 nuts on the lower side of the shock absorber

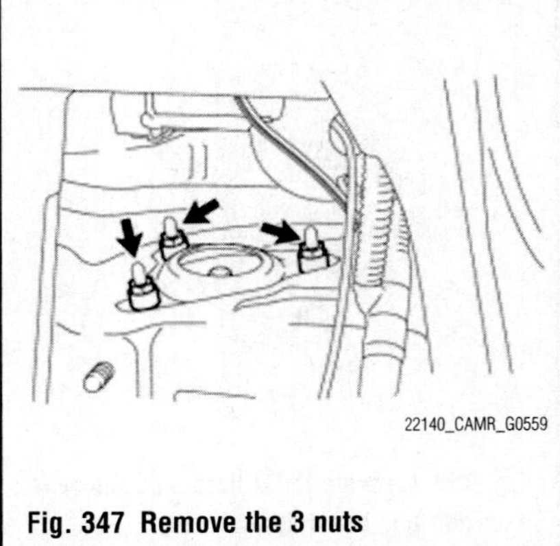

Fig. 347 Remove the 3 nuts

9. Remove the 4 claws and the rear suspension support No. 1 cover.
10. Remove the rear shock absorber with coil spring, as follows:

➡Do not remove the lock nut.

➡Only loosen the nut when disassembling the rear shock absorber with coil spring.

a. Loosen the lock nut of the rear shock absorber with coil spring.

➡When removing the nuts, keep the bolts from rotating.

➡Keep one bolt inserted to secure the hub and disc rotor.

b. Remove the 2 nuts and 2 bolts on the lower side of the rear shock absorber with coil spring.
c. Remove the 3 nuts on the upper side of the rear shock absorber with coil spring.

➡Make sure that the rear speed sensor is disconnected from the rear shock absorber with coil spring.

d. Lower the rear axle carrier, and remove the 2 bolts on the lower side of the rear shock absorber with coil spring.

To install:

11. Install the rear shock absorber with coil spring to the rear axle carrier assembly and insert the 2 bolts from the rear of the vehicle.
12. Slowly jack up the vehicle using a wooden block and install the rear shock absorber with coil spring (upper side) to the vehicle.
13. Install the 3 nuts to the upper side of the rear shock absorber with coil spring and tighten to 29 ft. lbs. (39 Nm).

➡When installing the nuts, keep the bolts from rotating.

14. Install the 2 nuts and 2 bolts to the lower side of the rear shock absorber with coil spring and tighten 133 ft. lbs. (180 Nm).
15. Fully tighten the lock nut to 41 ft. lbs. (55 Nm).
16. Connect rear speed sensor.
17. Install the LH rear stabilizer link assembly.
18. Engage the 4 claws and install the rear suspension support No. 1 cover.
19. To complete installation, reverse remaining removal.
20. Check abs speed sensor signal.
21. Inspect and adjust rear wheel alignment.

STABILIZER BAR

REMOVAL & INSTALLATION

See Figures 348 and 349.

1. Before servicing the vehicle, refer to the precautions section.
2. Remove rear wheels.
3. For 2.4L and 2.5L engines, remove center exhaust pipe assembly.
4. For 3.5L engines, remove tail exhaust pipe assembly.
5. For 3.5L engines, center exhaust pipe assembly.
6. Remove rear stabilizer link assembly.
7. Remove the 2 bolts and rear stabilizer bar No. 2 bracket.
8. Remove the 2 bolts and rear stabilizer bar No. 1 bracket.
9. Remove the 2 rear stabilizer bushings from the rear stabilizer bar.
10. Remove rear stabilizer bar.

To install:

11. Install the 2 rear stabilizer bushings to the outside of the stopper ring on the stabilizer bar.
12. Install the rear stabilizer bar No. 2 and No. 1 bracket.

Fig. 348 Removing the 2 bolts and No. 2 bracket

Fig. 349 Removing the 2 bolts and No. 1 bracket

13. Install the rear stabilizer bar with the 2 bolts and tighten to 23 ft. lbs. (31 Nm).
14. Install rear stabilizer link assembly.
15. To complete installation, reverse remaining removal.
16. Check abs speed sensor signal.
17. Inspect and adjust rear wheel alignment.

WHEEL BEARINGS

REMOVAL & INSTALLATION

See Figure 350.

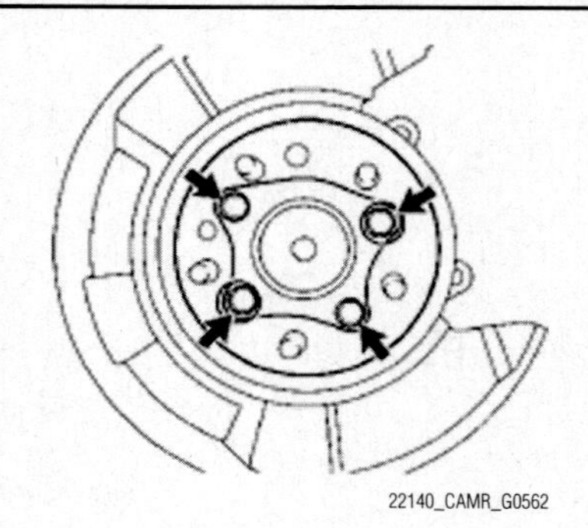

Fig. 350 Remove the 4 bolts and the rear axle hub and bearing assembly

➡Use the same procedures for the RH side and LH side.

➡The procedures listed below are for the LH side.

1. Before servicing the vehicle, refer to the precautions section.
2. Remove the rear wheel.
3. Separate the rear disc brake caliper assembly, as follows:
   a. Remove the bolt and separate the flexible hose from the shock absorber.
   b. Remove the 2 bolts and separate the rear disc brake caliper assembly.
4. Remove the rear disc.
5. Disconnect the skid control sensor connector.
6. Remove the 4 bolts and the rear axle hub and bearing assembly.

To install:

7. Install the hub and bearing assembly with the 4 bolts and tighten to 59 ft. lbs. (80 Nm).
8. Connect the skid control sensor connector. Do not twist the sensor wire.
9. Inspect rear axle hub bearing looseness.
10. Inspect rear axle hub runout.
11. Install the rear disc.
12. Install the rear disc brake caliper assembly, as follows:
    a. Install the rear disc brake caliper with the 2 bolts and tighten to 46 ft. lbs. (62 Nm).
    b. Install the flexible hose with the bolt and tighten to 14 ft. lbs. (19 Nm).
13. Install the rear wheel.
14. Inspect and adjust the rear wheel alignment.
15. Check ABS speed sensor signal.

TOYOTA

Camry Hybrid

6

SPECIFICATIONS AND MAINTENANCE CHARTS

ENGINE AND VEHICLE IDENTIFICATION

Engine						
Code ①	Liters (cc)	Cu. In.	Cyl.	Fuel Sys.	Engine Type	Eng. Mfg.
2AZ-FXE	2.4 (2362)	144	4	SFI	DOHC	Toyota

Model Year	
Code ②	Year
9	2009
A	2010

SFI: Sequential Fuel Injection

DOHC: Double Overhead Camshaft

① Stamped on the left side of the engine block

② 10th digit of the Vehicle Identification Number (VIN)

3768X_CMRH_C0001

GENERAL ENGINE SPECIFICATIONS

All measurements are given in inches.

Year	Model	Engine Displacement Liters	Engine Series VIN	Net Horsepower @ rpm	Net Torque @ rpm (ft. lbs.)	Bore x Stroke (in.)	Com-pression Ratio	Oil Pressure @ rpm
2009	Camry HV	2.4	2AZ-FXE	155@6000	158@4000	3.48x3.78	9.8:1	55@3000
2010	Camry HV	2.4	2AZ-FXE	155@6000	158@4000	3.48x3.78	9.8:1	55@3000

3768X_CMRH_C0002

GASOLINE ENGINE TUNE-UP SPECIFICATIONS

Year	Engine Displacement Liters	Engine VIN	Spark Plug Gap (in.)	Ignition Timing (deg.)	Fuel Pump (psi)	Idle Speed (rpm)	Valve Clearance (in.)	
							Intake	Exhaust
2009	2.4	2AZ-FXE	0.039-0.043	N/A	44-50	①	0.0075-0.0114	0.0150-0.0189
2010	2.4	2AZ-FXE	0.039-0.043	N/A	44-50	①	0.0075-0.0114	0.0150-0.0189

NOTE: The Vehicle Emission Control Information label often reflects specification changes made during production.

The label figures must be used if they differ from those in this chart.

NA: Not available

① Manual transmission: 650 to 750 rpm, Automatic transmission: 610 to 710 rpm

3768X_CMRH_C0003

CAPACITIES

Year	Model	Engine Displacement Liters	Engine VIN	Engine Oil with Filter (qts.)	Transmission (pts.) 5-Spd	Transmission (pts.) Auto.	Transfer Case (pts.)	Drive Axle Front (pts.)	Drive Axle Rear (pts.)	Fuel Tank (gal.)	Cooling System (qts.)
2009	Camry HV	2.4	2AZ-FXE	4.5	—	3.7	—	—	—	18.5	6.6
2010	Camry HV	2.4	2AZ-FXE	4.5	—	3.7	—	—	—	18.5	6.6

NOTE: All capacities are approximate. Add fluid gradually and check to be sure a proper fluid level is obtained.

3768X_CMRH_C0004

FLUID SPECIFICATIONS

Year	Model	Engine Displacement Liters	Engine ID/VIN	Engine Oil	Auto. Trans. ①	Drive Axle	Power Steering Fluid	Engine Coolant	Brake Master Cylinder
2009	Camry HV	2.4	2AZ-FXE	5W-20	H V fluid	NA	N/S	②	DOT 3
2010	Camry HV	2.4	2AZ-FXE	5W-20	H V fluid	NA	N/S	②	DOT 3

DOT: Department Of Transpotation

NA: Not Available

N/S: Not Specified

① Hybrid Transaxle fluid.

② Toyota Super long life Coolant (SLLC)

3768X_CMRH_C0005

VALVE SPECIFICATIONS

Year	Engine Displacement Liters	Engine VIN	Seat Angle (deg.)	Face Angle (deg.)	Spring Test Pressure (lbs. @ in.)	Spring Installed Height (in.)	Stem-to-Guide Clearance (in.) Intake	Stem-to-Guide Clearance (in.) Exhaust	Stem Diameter (in.) Intake	Stem Diameter (in.) Exhaust
2009	2.4	2AZ-FXE	45	44.5	NA	NA	0.0010-0.0031	0.0012-0.0039	0.2154-0.2159	0.2151-0.2157
2010	2.4	2AZ-FXE	45	44.5	NA	NA	0.0010-0.0031	0.0012-0.0039	0.2154-0.2159	0.2151-0.2157

NA: Not Available

3768X_CMRH_C0006

CAMSHAFT AND BEARING SPECIFICATIONS CHART

All measurements are given in inches.

Year	Engine Displ. Liters	Engine ID/VIN	Journal Dia.	Brg. Oil Clearance	Shaft End-play	Runout	Journal Bore	Lobe Height	
								Intake	Exhaust
2009	2.4	2AZ-FXE	①	NA	NA	NA	NA	1.8624-1.8664	1.8104-1.1843
2010	2.4	2AZ-FXE	①	NA	NA	NA	NA	1.8624-1.8664	1.8104-1.1843

NA: Not Available

① Mark 1, 2 and 3: 1.4162-1.4167

② No. 1 journal: 1.4152-1.4157

Other Journals: 1.0220-1.0226 in.

③ No. 1 journal: 0.0016-0.0031

Other Journals: 1.0010-1.0024 in.

Maximum No.1 Journal: 0.0039 in.

Other Maximum Journals: 0.0035 in.

3768X_CMRH_C0007

CRANKSHAFT AND CONNECTING ROD SPECIFICATIONS

All measurements are given in inches.

Year	Engine Displacement Liters	Engine VIN	Crankshaft				Connecting Rod		
			Main Brg. Journal Dia.	Main Brg. Oil Clearance	Shaft End-play	Thrust on No.	Journal Diameter	Oil Clearance	Side Clearance
2009	2.4	2AZ-FXE	2.0863-2.0866	0.0007-0.0016	0.0016-0.0095	3	1.8894-1.8898	0.0013-0.0025	0.0063-0.0143
2010	2.4	2AZ-FXE	2.0863-2.0866	0.0007-0.0016	0.0016-0.0095	3	1.8894-1.8898	0.0013-0.0025	0.0063-0.0143

3768X_CMRH_C0008

PISTON AND RING SPECIFICATIONS

All measurements are given in inches.

Year	Engine Displ. Liters	Engine VIN	Piston Clearance	Ring Gap			Ring Side Clearance		
				Top Compression	Bottom Compression	Oil Control	Top Compression	Bottom Compression	Oil Control
2009	2.4	2AZ-FXE	0.0020-0.0029	0.0094-0.0122	0.0130-0.0169	0.0040-0.0119	0.0008-0.0028	0.0008-0.0024	0.0008-0.0028
2010	2.4	2AZ-FXE	0.0020-0.0029	0.0094-0.0122	0.0130-0.0169	0.0040-0.0119	0.0008-0.0028	0.0021-0.0037	0.0023-0.0085

3768X_CMRH_C0009

TORQUE SPECIFICATIONS

All readings in ft. lbs.

Year	Engine Displacement Liters	Engine VIN	Cylinder Head Bolts	Main Bearing Bolts	Rod Bearing Bolts	Crankshaft Damper Bolts	Flywheel Bolts	Manifold Intake	Manifold Exhaust	Spark Plugs	Oil Pan Drain Plug
2009	2.4	2AZ-FXE	①	②	③	125	④	22	27	13	18
2010	2.4	2AZ-FXE	①	②	③	125	④	22	27	13	18

① Step 1: 52 ft. lbs.
Step 2: plus 90 degrees

② Step 1: 15 ft. lbs.
Step 2: 30 ft. lbs.
Step 3: Plus 90 degrees

③ Step 1: Cap bolts to 18 ft. lbs.
Step 2: Cap bolts plus 90 degrees

④ Auto driveplate: 72 ft. lbs.
Manual Flywheel : 96 ft. lbs.

3768X_CMRH_C0010

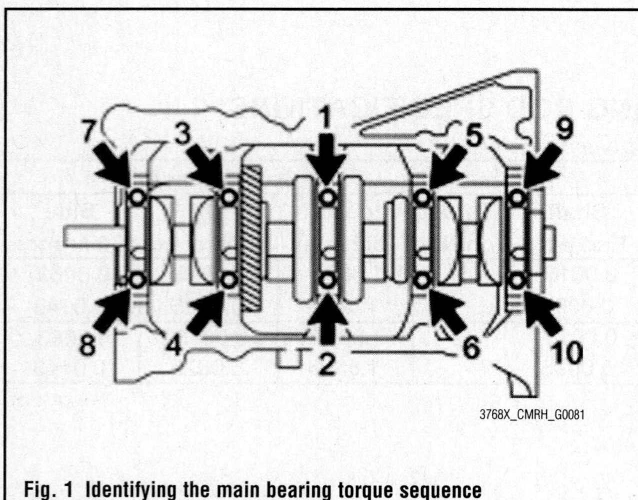

Fig. 1 Identifying the main bearing torque sequence

WHEEL ALIGNMENT

Year	Model		Caster Range (+/-Deg.)	Caster Preferred Setting (Deg.)	Camber Range (+/-Deg.)	Camber Preferred Setting (Deg.)	Toe-in (in.)	Steering Axis Inclination (Deg.)
2009	Camry HV	Front	0.75	2.65	0.75	-0.67	0+/-0.04	12.25+/-0.75
		Rear	—	—	0.75	-1.15	0.16+/-0.08	—
2010	Camry HV	Front	0.75	2.65	0.75	-0.67	0+/-0.04	12.25+/-0.75
		Rear	—	—	0.75	-1.15	0.16+/-0.08	—

3768X_CMRH_C0011

TIRE, WHEEL AND BALL JOINT SPECIFICATIONS

Year	Model	OEM Tires Standard	OEM Tires Optional	Tire Pressures (psi) Front	Tire Pressures (psi) Rear	Wheel Size	Ball Joint Inspection	Lug Nut Torque (ft. lbs.)
2009	Camry HV	P215/60R16	P215/55R17	31	31	NA	①	76
2010	Camry HV	P215/60R16	P215/55R17	31	31	NA	①	76

NA: Not Available

OEM: Original Equipment Manufacturer

OPT: Optional

PSI: Pounds Per Square Inch

STD: Standard

① Replace if any measurable movement is found.

3768X_CMRH_C0012

BRAKE SPECIFICATIONS

All measurements in inches unless noted

Year	Model		Brake Disc Original Thickness	Brake Disc Minimum Thickness	Brake Disc Maximum Runout	Minimum Lining Thickness	Brake Caliper Bracket Bolts (ft. lbs.)	Brake Caliper Mounting Bolts (ft. lbs.)
2009	Camry HV	Front	1.102	0.983	0.0020	0.039	79	25
		Rear	0.390	0.334	0.0059	0.039	46	32
2010	Camry HV	Front	1.102	0.983	0.0020	0.039	79	25
		Rear	0.390	0.334	0.0059	0.039	46	32

3768X_CMRH_C0013

SCHEDULED MAINTENANCE INTERVALS

TOYOTA—CAMRY HV

TO BE SERVICED	TYPE OF SERVICE	VEHICLE MILEAGE INTERVAL (x1000)													
		5	10	15	20	25	30	35	40	45	50	55	60	90	120
Engine oil & filter	R	✓	✓	✓	✓	✓	✓	✓	✓	✓	✓	✓	✓	✓	✓
Automatic transmission fluid	S/I			✓			✓			✓			✓	✓	✓
Ball joints & dust covers	S/I			✓			✓			✓			✓	✓	✓
Bolts & nuts on chassis & body	S/I			✓			✓			✓			✓	✓	✓
Brake linings & drums	S/I	✓	✓	✓	✓	✓	✓	✓	✓	✓	✓	✓	✓	✓	✓
Brake line pipes & hoses	S/I			✓			✓			✓			✓	✓	✓
Brake pads & discs (front & rear)	S/I	✓	✓	✓	✓	✓	✓	✓	✓	✓	✓	✓	✓	✓	✓
Brake fluid	R						✓						✓	✓	✓
Rack and pinion assembly	S/I			✓			✓			✓			✓	✓	✓
Steering linkage & boots	S/I			✓			✓			✓			✓	✓	✓
Steering gear box	S/I			✓			✓			✓			✓	✓	✓
Spark plugs ①	R														✓
Drive belts	S/I												✓	✓	✓
Exhaust pipes & mountings	S/I			✓			✓			✓			✓	✓	✓
Fuel lines & connections	S/I						✓						✓	✓	✓
Engine/inverter coolant ②	S/I			✓			✓			✓			✓	✓	
Radiator, condenser and/or intercooler	S/I			✓			✓			✓			✓	✓	
Fuel tank cap gasket	S/I						✓						✓	✓	✓
Rotate tires	S/I	✓	✓	✓	✓	✓	✓	✓	✓	✓	✓	✓	✓	✓	✓
Clean air conditioning filter ③	S/I			✓			✓			✓			✓	✓	✓
Axle shaft bolts	S/I			✓			✓			✓			✓	✓	✓
Brake pad thickness and rotor runout	S/I						✓						✓	✓	✓

R: Replace S/I: Service or Inspect

① Spark plugs are replaced at 120,000 miles

② Replace engine coolant at 100,000 miles and then inspect every 15,000 miles

③ Replace air conditioning filter every 30,000 miles

FREQUENT OPERATION MAINTENANCE (SEVERE SERVICE)

If a vehicle is operated under any of the following conditions it is considered severe service:

- Extremely dusty areas.

- 50% or more of the vehicle operation is in 32°C (90°F) or higher temperatures, or constant temperatures below 0°C (32°F).

- Prolonged idling (vehicle operation in stop and go traffic).

- Frequent short running periods (engine does not warm to normal operating temperatures).

- Police, taxi, delivery usage or trailer towing usage.

Air cleaner filter: service or inspect every 5000 miles

Rear differential & transfer case oil: replace every 15,000 miles.

Ball joints & dust covers: service or inspect every 5000 miles.

Bolts & nuts on chassis & body: service or inspect every 5000 miles.

Axle shaft bolts: service or inspect every 5000 miles.

Steering linkage: service or inspect every 5000 miles.

3768X_CMRH_C0014

BRAKES — INFORMATION AND PRECAUTIONS

ANTI-LOCK SYSTEMS

- Certain components within the ABS system are not intended to be serviced or repaired individually.
- Do not use rubber hoses or other parts not specifically specified for and ABS system. When using repair kits, replace all parts included in the kit. Partial or incorrect repair may lead to functional problems and require the replacement of components.
- Lubricate rubber parts with clean, fresh brake fluid to ease assembly. Do not use shop air to clean parts; damage to rubber components may result.
- Use only DOT 3 brake fluid from an unopened container.
- If any hydraulic component or line is removed or replaced, it may be necessary to bleed the entire system.
- A clean repair area is essential. Always clean the reservoir and cap thoroughly before removing the cap. The slightest amount of dirt in the fluid may plug an orifice and impair the system function. Perform repairs after components have been thoroughly cleaned; use only denatured alcohol to clean components. Do not allow ABS components to come into contact with any substance containing mineral oil; this includes used shop rags.
- The Anti-Lock control unit is a microprocessor similar to other computer units in the vehicle. Ensure that the ignition switch is **OFF** before removing or installing controller harnesses. Avoid static electricity discharge at or near the controller.
- If any arc welding is to be done on the vehicle, the control unit should be unplugged before welding operations begin.

DISC AND DRUM SYSTEMS

CAUTION

Dust and dirt accumulating on brake parts during normal use may contain asbestos fibers from production or aftermarket brake linings. Breathing excessive concentrations of asbestos fibers can cause serious bodily harm. Exercise care when servicing brake parts. Do not sand or grind brake lining unless equipment used is designed to contain the dust residue. Do not clean brake parts with compressed air or by dry brushing. Cleaning should be done by dampening the brake components with a fine mist of water, then wiping the brake components clean with a dampened cloth. Dispose of cloth and all residue containing asbestos fibers in an impermeable container with the appropriate label. Follow practices prescribed by the Occupational Safety and Health Administration (OSHA) and the Environmental Protection Agency (EPA) for the handling, processing, and disposing of dust or debris that may contain asbestos fibers.

BRAKES — BLEEDING THE BRAKE SYSTEM

BLEEDING PROCEDURE

BLEEDING PROCEDURE

Actuator Assembly

➡After bleeding the air from the brake system, if the height or feel of the brake pedal cannot be obtained, perform air bleeding in the brake actuator assembly with a hand-held tester by following the procedures below.

1. Depress the brake pedal more than 20 times with the engine off.
2. Connect the hand-held tester to the DLC3, then turn the ignition switch to the **ON** position, but do NOT start the engine.
3. Select "AIR BLEEDING" on the hand-held tester.

➡Refer to the hand-held tester operator's manual for more details.

4. Bleed the air out of the regular brake line when "Step 1: Increase" appears on the hand-held tester display, as follows:

➡Bleed the air by following the steps displayed on the hand-held tester. Make sure that the brake fluid in the master cylinder reservoir tank does not become empty.

a. Connect the vinyl tube to either one of the bleeder plugs.

b. Have an assistant depress the brake pedal several times, then loosen the bleeder plug connected to the vinyl tube with the pedal depressed.

c. When fluid stops coming out, tighten the bleeder plug and release the brake pedal.

d. Repeat the previous 2 steps until all the air in the fluid is completely bled out.

e. Tighten the bleeder plug completely to 73 inch lbs. (8.3 Nm).

f. Repeat the above procedures for each wheel to bleed the air out of the brake line.

5. Bleed the air out of the suction line when "Step 2: Inhalation" appears on the hand-held tester display, as follows:

➡Bleed the air by following the steps displayed on the hand-held tester. Make sure that the brake fluid in the master cylinder reservoir tank does not become empty.

a. Connect the vinyl tube to the bleeder plug at the right front wheel or the right rear wheel and loosen the bleeder plug.

b. Operate the brake actuator assembly to perform air bleeding from the suction line using the hand-held tester.

➡This operation stops automatically after 4 seconds. At this time, be sure to release the brake pedal.

c. Check if the operation has stopped by referring to the hand-held tester display and tighten the bleeder plug.

d. Repeat the previous 2 steps until all air in the fluid is completely bled out.

e. Tighten the bleeder plug completely to 73 inch lbs. (8.3 Nm).

f. Repeat the above procedures to bleed the air out of the brake line for each wheel.

6. Bleed the air out of the pressure reduction line when "Step 3: Decrease" appears on the hand-held tester display, as follows:

➡Bleed the air by following the steps displayed on the hand-held tester. Make sure that the brake fluid in the master cylinder reservoir tank does not become empty.

a. Connect a vinyl tube to either one of the bleeder plugs.

b. Loosen the bleeder plug.

c. Using the hand-held tester, operate the brake actuator assembly, completely depress the brake pedal and keep it depressed.

➡**The operation stops automatically after 4 seconds. When performing this procedure continuously, set an interval of at least 20 seconds. When the operation is complete, the brake pedal goes down slightly. This is a normal phenomenon caused when the solenoid opens. During this procedure, the pedal will feel heavy, but completely depress it so that the brake fluid comes out from the bleeder plug. Be sure to keep depressing the brake pedal. Do not depress and release the pedal repeatedly.**

d. Tighten the bleeder plug, then release the brake pedal.

e. Repeat the previous 2 steps until all the air in the fluid is completely bled out.

f. Tighten the bleeder plug completely to 73 inch lbs. (8.3 Nm).

g. Repeat the above procedures for each wheel to bleed the air out of the brake line.

7. Bleed the air out of the brake line again when "Step 4: Increase" appears on the hand-held tester display, as follows:

➡**Bleed the air by following the steps displayed on the hand-held tester. Make sure that the brake fluid in the master cylinder reservoir tank does not become empty.**

a. Connect the vinyl tube to either one of the bleeder plugs.

b. Depress the brake pedal several times, then loosen the bleeder plug connected to the vinyl tube with the pedal depressed.

8. When fluid stops coming out, tighten the bleeder plug, then release the brake pedal.

a. Repeat the previous 2 steps until all the air in the fluid is completely bled out.

b. Tighten the bleeder plug completely to 73 inch lbs. (8.3 Nm).

c. Repeat the above procedures for each wheel to bleed the air out of the brake line.

d. Finish "AIR BLEEDING" on the hand-held tester and turn the hand-held tester off.

e. Disconnect the hand-held tester from the DLC3 from the DLC3.

f. Turn the ignition switch off.

g. Inspect for fluid leak.

9. Check the fluid level and add fluid if necessary. Use SAE J1703 or FMVSS No. 116 DOT3 Brake fluid.

Brake Lines

➡**Bleed air from the brake line of the wheel farthest from the master cylinder.**

1. Raise and safely support the vehicle.
2. Connect a vinyl tube to the bleeder plug.
3. Have an assistant depress the brake pedal several times, then loosen the bleeder plug while the pedal is depressed.
4. When fluid stops coming out, tighten the bleeder plug, then release the brake pedal.
5. Repeat steps 3 and 4 until all the air in the fluid has been bled out.
6. Tighten the brake bleeder plug to 73 inch lbs. (8.3 Nm).
7. Repeat the above steps to bleed the air out of the brake line for each wheel.

Master Cylinder

If the master cylinder is reinstalled or if the reservoir becomes empty, bleed the air from the master cylinder. To prevent brake fluid from adhering, cover nearly painted surfaces with a shop rag or a piece of cloth.

1. Using a union nut wrench (10 mm), disconnect the 2 brake lines from the master cylinder, using a suitable brake line wrench.
2. Have an assistant slowly depress the brake pedal and hold it.
3. Cover the 2 outer holes with your fingers, and have your assistant release the brake pedal.
4. Repeat the previous 2 steps 3 or 4 times.
5. Using a union nut wrench (10 mm), connect the 2 the brake lines to the master cylinder and tighten to 11 ft. lbs. (15 Nm).

➡**Use a torque wrench with a fulcrum length of 250 mm (9.84 in.).**

➡**This torque value is effective when the union nut wrench is parallel to the torque wrench.**

BRAKES — ANTI-LOCK BRAKE SYSTEM (ABS)

WHEEL SPEED SENSORS

REMOVAL & INSTALLATION

See Figure 2.

1. Before servicing the vehicle, refer to the Precautions Section.
2. Remove the front wheel.
3. Remove the 3 screws and the front wheel opening extension pad
4. Remove the front fender liner.
5. Remove the front speed sensor, as follows:

a. Disconnect the speed sensor connector and clamp.

b. Remove the 2 bolts and separate the speed sensor harness from the body and shock absorber assembly.

c. Disengage the 2 claws from the steering knuckle.

d. Remove the bolt and the front speed sensor.

➡**Do not allow foreign matter to attach to the sensor tip.**

➡**Clean the installation hole and surface for the speed sensor every time the speed.**

To install:

➡**Do not twist the wire harness for the front speed sensor when installing the speed sensor.**

➡**The bolt B tightens the brake flexible hose and front speed sensor together. Make sure that the flexible hose is positioned over the front speed sensor.**

➡**Install the stopper firmly to the hole in the body.**

6. To install, reverse removal procedure.
7. Tighten the front speed sensor with the bolt to 71 inch lbs. (8 Nm).

➡**Do not allow foreign matter to attach to the sensor tip.**

8. Tighten the sensor harness brackets with the 2 bolts to the body and shock absorber assembly and 44 inch lbs. (5 Nm), and then to 14 ft. lbs. (19 Nm).
9. Install the front wheel and tighten the lug nuts to 76 ft. lbs. (103 Nm).
10. Check ABS speed sensor signal.

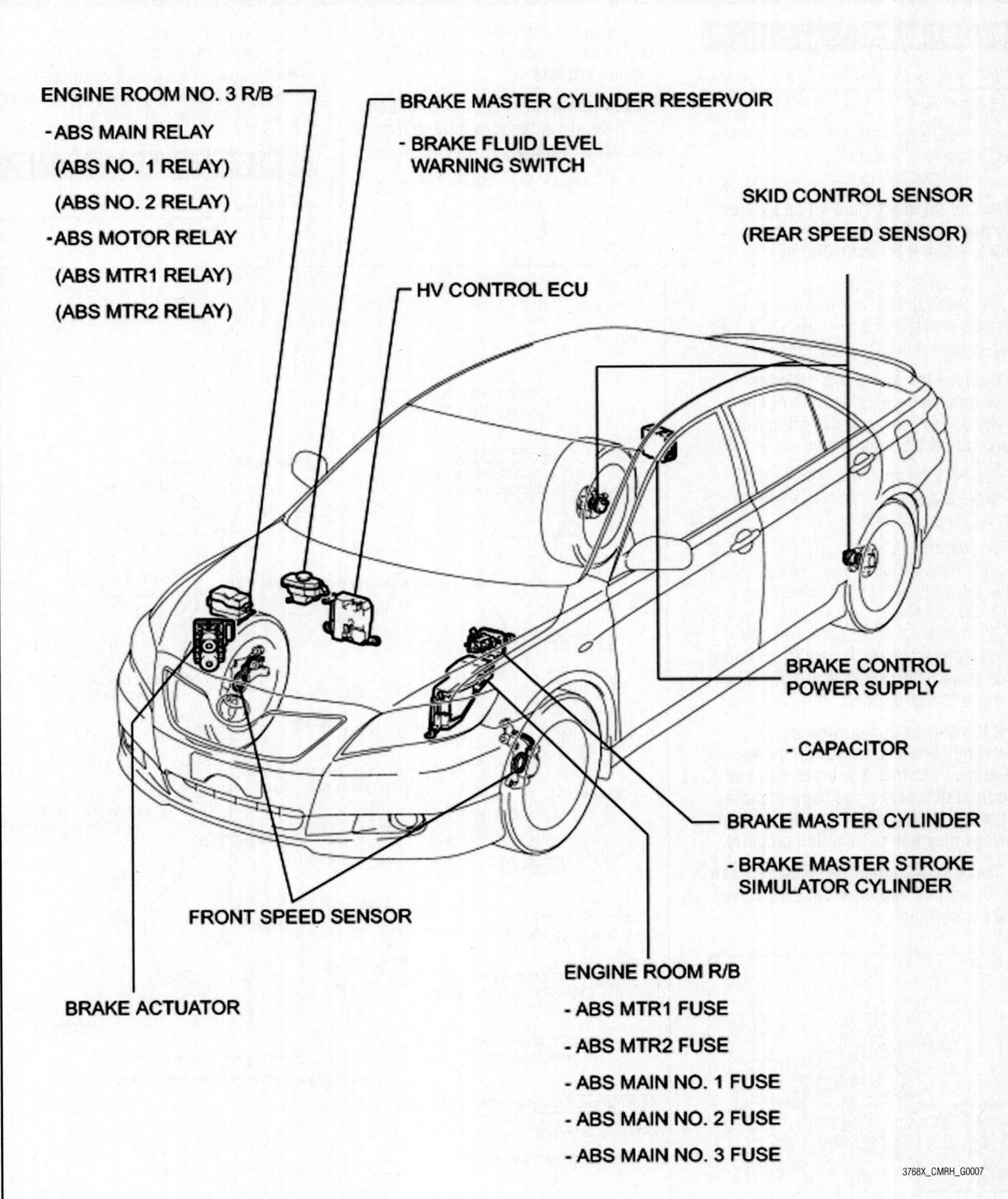

Fig. 2 Electronically controlled brake system parts location

BRAKE CALIPER

REMOVAL & INSTALLATION

See Figure 3.

1. Before servicing the vehicle, refer to the Precautions Section.
2. Remove the front wheel.

➡Do not let brake fluid sit on painted surfaces, as it will eat through the paint. Wash it off immediately.

3. Drain brake fluid.
4. Remove the union bolt and gasket from the disc brake cylinder assembly, then disconnect the flexible hose.

➡Remove the disc brake cylinder assembly while holding both of the brake pads or the anti-squeal springs may fall off the brake pads.

5. Hold the front disc brake cylinder slide pin and remove the 2 bolts and disc brake cylinder assembly.

To install:

6. Install the disc brake cylinder assembly with the 2 bolts and tighten to 25 ft. lbs. (34 Nm).
7. Check the installation of the anti-squeal springs. Visually check for any clearance between the brake pad and front disc brake pad support plates.

➡If the anti-squeal springs are installed correctly, there will be no clearance between the brake pad and the front disc brake pad support plates. If there is a clearance, the anti-squeal springs may not be installed properly.

➡Check all 4 contact surfaces between the brake pad and the front disc brake pad support plates.

Fig. 3 Removing the union bolt and gasket

8. Connect the flexible hose with the union bolt and a new gasket and tighten to 21 ft. lbs. (29 Nm).

➡Install the front brake flexible hose lock securely in the lock hole in the disc brake cylinder.

9. Fill reservoir with brake fluid.
10. Bleed master cylinder.
11. Bleed brake line.
12. Bleed brake actuator assembly.
13. Inspect for brake fluid leak.
14. Inspect brake fluid level in reservoir.
15. Install the front wheel and tighten the lug nuts to 76 ft. lbs. (103 Nm).

DISC BRAKE PADS

REMOVAL & INSTALLATION

See Figures 4 and 5.

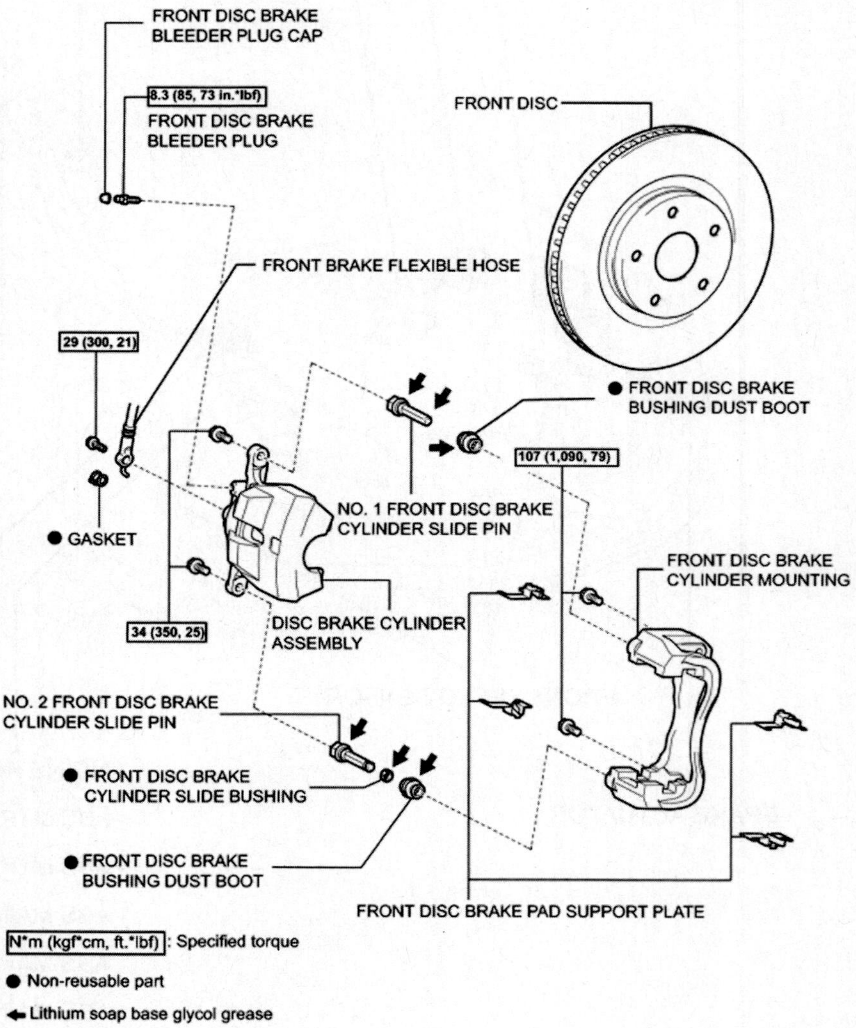

Fig. 4 Exploded view of the front brake components (1 of 2)

1. Before servicing the vehicle, refer to the Precautions Section.

2. Remove the 2 front disc brake cylinder slide pins (upper and lower) from the front disc brake cylinder mounting.

3. Remove brake cylinder.

4. Remove the 2 anti-squeal springs.

5. Remove the 2 brake pads from the front disc brake cylinder mounting.

To install:

6. Install the 2 brake pads with front anti-squeal shims to the front disc brake cylinder mounting.

➡When replacing worn pads, the front anti-squeal springs must be replaced at the same time.

➡Be sure to install the anti-squeal springs into the front disc brake pad installation holes as far as they will go.

7. Install the 2 front disc brake cylinder slide pins (upper and lower) from the front disc brake cylinder mounting.

8. Install the brake cylinder.

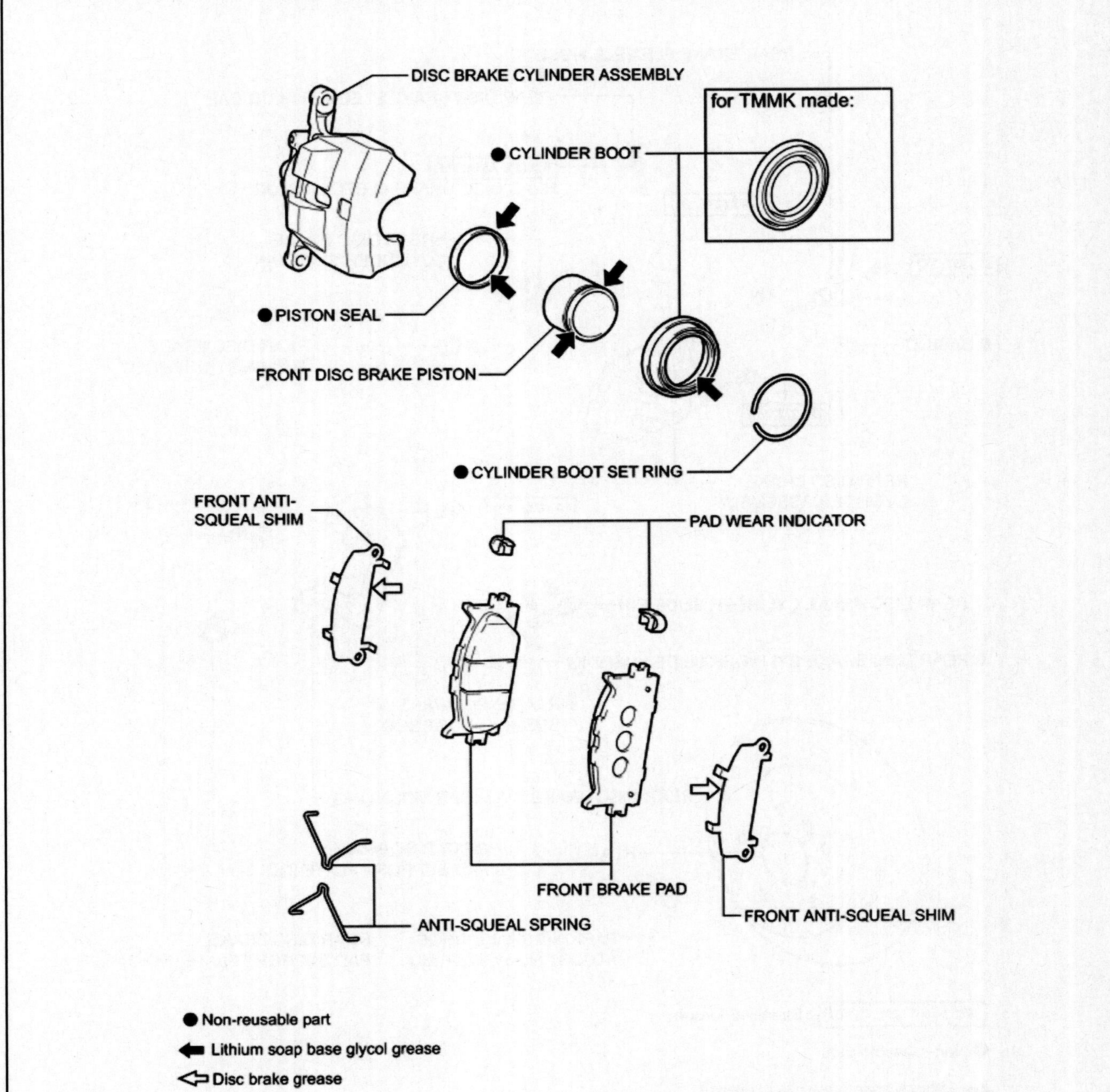

Fig. 5 Exploded view of the front brake components (2 of 2)

BRAKE CALIPER

REMOVAL & INSTALLATION

See Figures 6 and 7.

1. Before servicing the vehicle, refer to the Precautions Section.
2. Remove the rear wheel.
3. Drain brake fluid.

***** WARNING**

Do not let brake fluid sit on painted surfaces, as it will eat through the paint. Wash it off immediately.

4. Remove the union bolt and the gasket from the rear disc brake cylinder assembly, and then disconnect the rear brake flexible hose.
5. Hold the 2 rear disc brake cylinder slide pins and remove the 2 bolts and rear disc brake cylinder assembly.

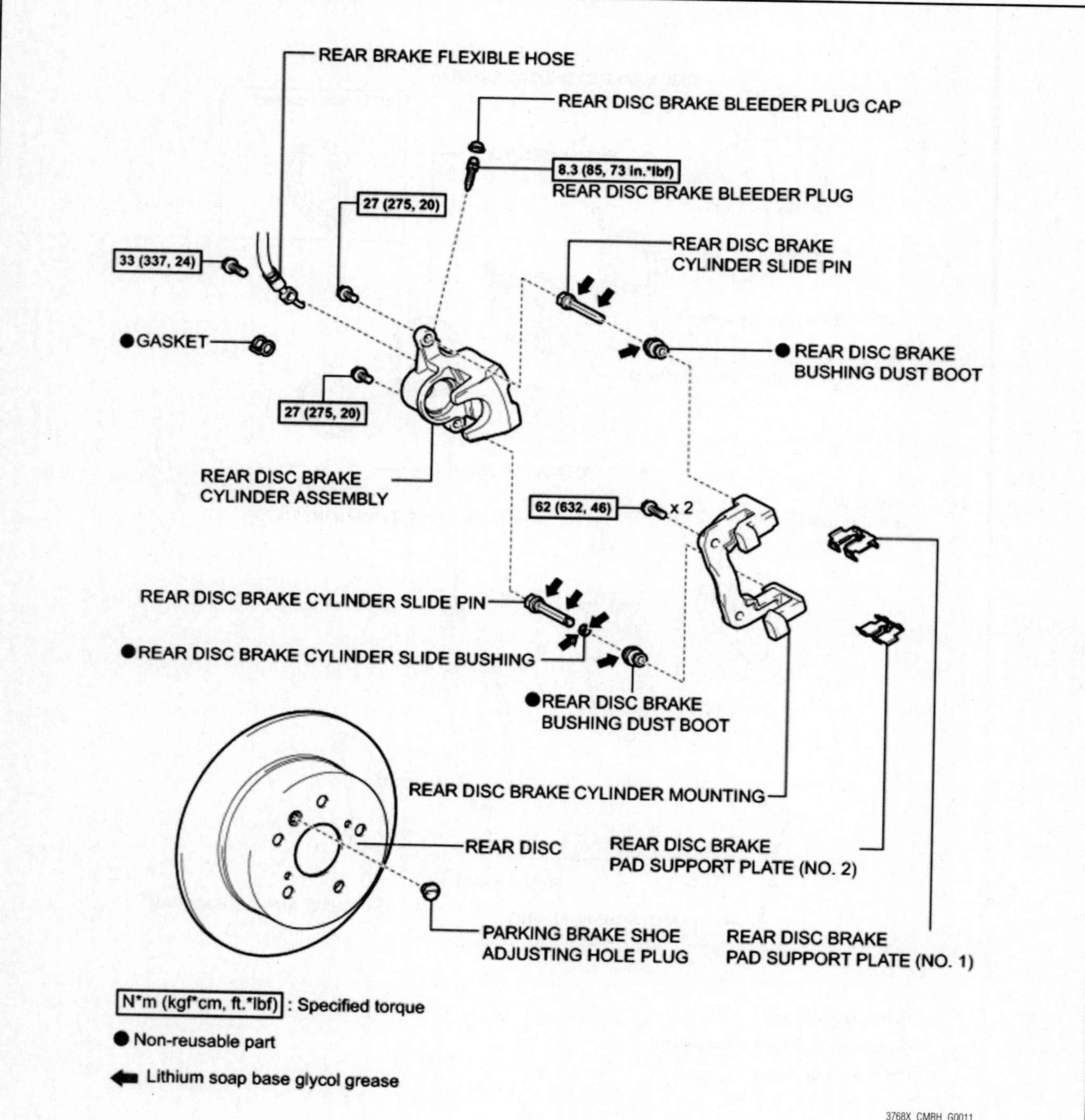

Fig. 6 Exploded view of the rear brake components (1 of 2)

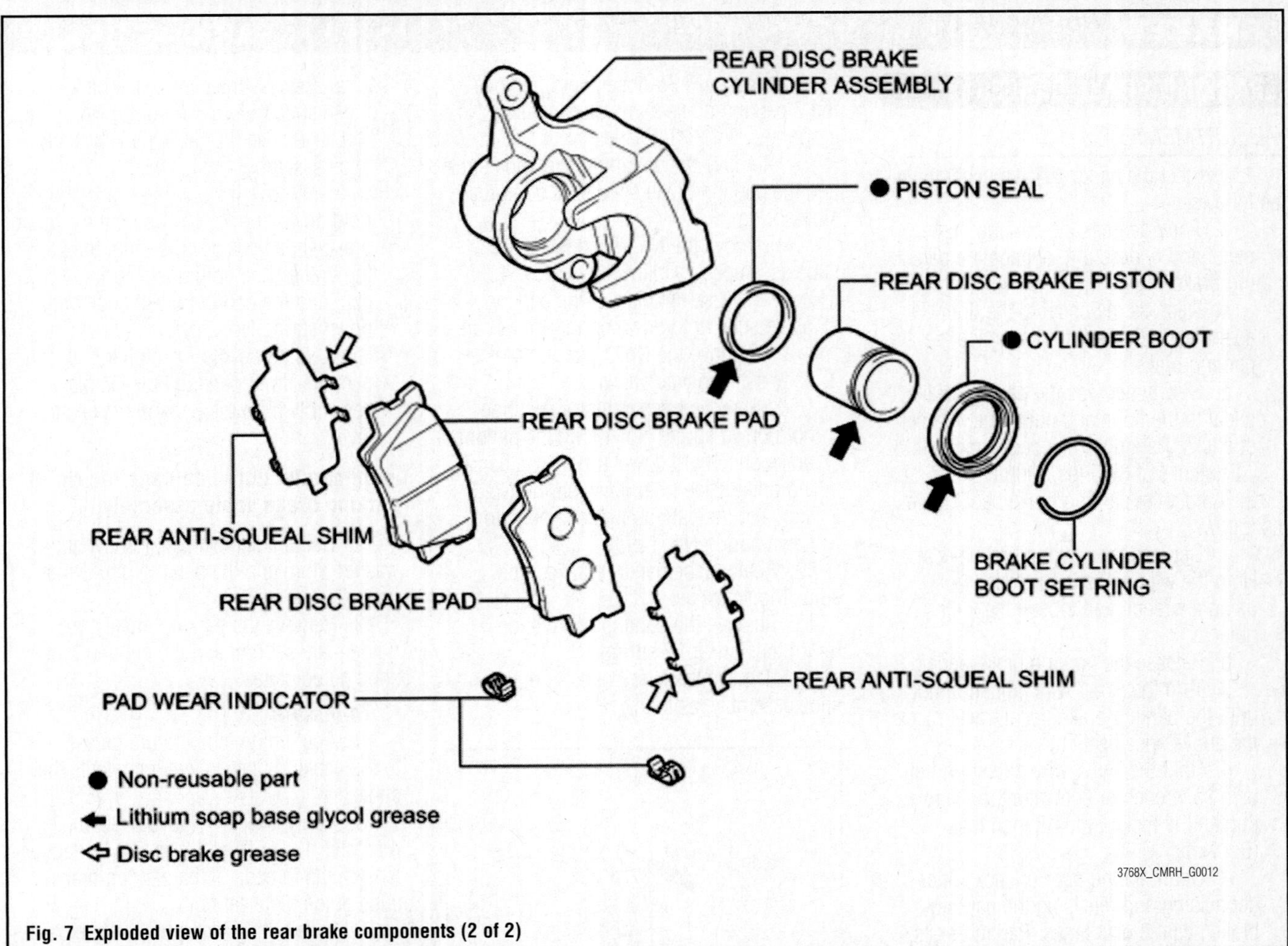

Fig. 7 Exploded view of the rear brake components (2 of 2)

To install:

6. Install the rear disc brake cylinder assembly with the 2 bolts and tighten to 20 ft. lbs. (77 Nm).
7. Connect the rear brake flexible hose with the union bolt and a new gasket and tighten to 24 ft. lbs. (33 Nm).
8. Fill reservoir with brake fluid.
9. Bleed brake line.
10. Inspect for brake fluid leak.
11. Inspect brake fluid level in reservoir.
12. Install the rear wheel and tighten the lug nuts to 76 ft. lbs. (103 Nm).

DISC BRAKE PADS

REMOVAL & INSTALLATION

See Figures 6 and 7.

1. Before servicing the vehicle, refer to the Precautions Section.
2. Remove the 2 rear disc brake cylinder slide pins (upper and lower) from the rear disc brake cylinder mounting.
3. Remove brake cylinder.
4. Remove the 2 brake pads with the rear anti-squeal shims.

To install:

5. Installation is the reverse of removal procedure.

PARKING BRAKE CABLES

ADJUSTMENT

1. Inspect the parking brake pedal travel, as follows:

a. Fully depress the parking brake pedal and release it to engage the parking brake.

b. Depress the pedal to the floor again, and release it to disengage the parking brake.

c. Slowly depress the parking brake pedal to the floor, and count the number of clicks. Parking brake pedal travel: 9 to 11 notches at 67.5 lbs. (300 N).

2. Adjust the parking brake pedal travel, as follows:

a. Depress the parking brake pedal. Hold the wire adjusting nut No. 1 using a wrench and loosen the lock nut.

b. Release the parking brake pedal.

c. Turn the No. 1 wire adjusting nut until the parking brake pedal travel meets the above specification.

d. Hold the No. 1 wire adjusting nut using a wrench or equivalent tool and tighten the lock nut to 48 inch lbs. (5.4 Nm).

e. Count the number of clicks after depressing and releasing the parking brake pedal 3 or 4 times. Parking brake pedal travel: 9 to 11 notches at 67.5 lbs. (300 N).

f. Check whether the parking brake drags or not.

g. When operating the parking brake pedal, check that the parking brake indicator light comes on.

PARKING BRAKE SHOES

REMOVAL & INSTALLATION

See Figures 8 through 10.

1. Remove the rear wheel.

2. Remove the 2 bolts and separate the rear disc brake caliper assembly. Do not disconnect the flexible hose from the disc brake caliper assembly.

3. Remove the parking brake shoe adjusting hole plug from the rear disc.

4. Release the parking brake and place the matchmarks on the rear disc and the axle hub.

5. Remove the rear disc.

➡If the disc cannot be removed easily, turn the shoe adjuster until the disc turns freely.

6. Using needle-nose pliers, remove the 2 parking brake shoe return tension No. 1 springs.

7. Remove the parking brake shoe strut and the parking brake shoe strut compression spring.

8. Remove the No. 1 parking brake shoe assembly, as follows:

a. Release the claw of the parking brake shoe hold down spring No. 2 cup.

b. Remove the No. 1 parking brake shoe assembly as shown.

c. Remove the parking brake shoe hold down spring No. 1 cup, the parking brake shoe hold down spring, the parking brake shoe hold down spring No. 2 cup, and the parking brake shoe hold down spring No. 1 pin.

9. Remove the parking brake shoe adjusting screw set.

10. Remove the parking brake shoe return tension No. 2 spring.

11. Remove the No. 2 parking brake shoe assembly, as follows:

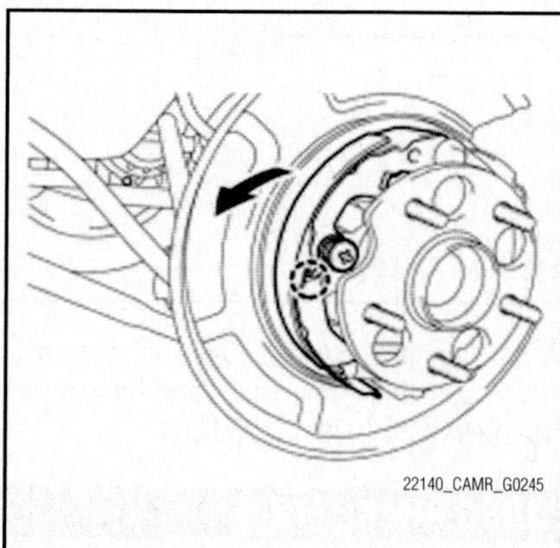

Fig. 8 Remove the No. 1 parking brake shoe assembly

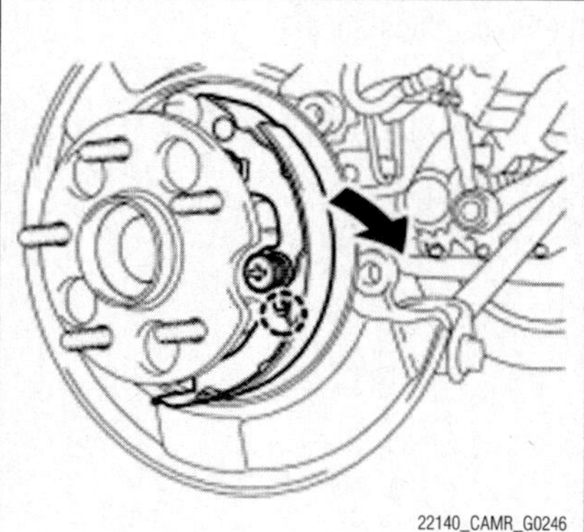

Fig. 9 Remove the No. 2 parking brake shoe assembly

a. Release the claw of the parking brake shoe hold down spring No. 2 cup.

b. Remove the No. 2 parking brake shoe assembly as shown.

c. Remove the parking brake shoe hold down spring No. 1 cup, the parking brake shoe hold down spring, the parking brake shoe hold down spring No. 2 cup, and the parking brake shoe hold down spring No. 2 pin.

d. Using needle-nose pliers, disconnect the No. 3 parking brake cable assembly from the parking brake shoe lever.

➡Be careful not to damage the No. 3 parking brake cable assembly.

12. Using a screwdriver, remove the C-washer, shim and the parking brake shoe lever.

13. Remove the parking brake shoe guide plate set bolt and the parking brake shoe guide plate.

To install:

14. Apply high temperature grease to the backing plate where it contacts the shoe.

15. Apply adhesive (Toyota Genuine Adhesive 1344, Three Bond 1344 or equivalent) to the threads of the parking brake shoe guide plate set bolt.

16. Install the parking brake shoe guide plate with the parking brake shoe guide plate set bolt and tighten to 13 ft. lbs. (18 Nm).

17. Install the parking brake shoe lever and shim to the No. 2 parking brake shoe assembly with a new C-washer.

18. Using a feeler gauge, measure the clearance between the No. 2 parking brake shoe assembly and parking brake shoe lever. Standard clearance: Less than 0.35 mm (0.014 in.).

19. If the clearance is not as specified, replace the shim with one of the correct size.

20. Install the No. 2 parking brake shoe assembly as follows:

a. Using needle-nose pliers, connect the No. 3 parking brake cable assembly to the parking brake shoe lever.

b. Install the No. 2 parking brake shoe assembly with the parking brake shoe hold down spring No. 2 pin, the parking brake shoe hold down spring No. 2 cup, the parking brake shoe hold down spring and the parking brake shoe hold down spring No. 1 cup.

c. Engage the claw of the parking brake shoe hold down spring No. 2 cup to the No. 2 parking brake shoe assembly.

21. Install the parking brake shoe adjusting screw set, as follows:

a. Apply high temperature grease to the parking brake shoe adjusting screw set as shown in the illustration.

b. Install the parking brake shoe return tension No. 2 spring to the No. 1 parking brake shoe assembly and the No. 2 parking brake shoe assembly.

c. Install the parking brake shoe adjusting screw set to the No. 1 parking brake shoe assembly and the No. 2 parking brake shoe assembly.

22. Install the No. 1 parking brake shoe assembly as follows:

a. Install the No. 1 parking brake shoe assembly with the parking brake shoe hold down spring No. 1 pin, parking brake shoe hold down spring No. 2 cup, parking brake shoe hold down spring and parking brake shoe hold down spring No. 1 cup.

b. Engage the claw of the parking brake shoe hold down spring No. 2 cup to the No. 1 parking brake shoe assembly.

23. Attach the parking brake shoe strut and the parking brake shoe strut compression spring to the No. 1 parking brake shoe assembly and No. 2 parking brake shoe assembly.

24. Using needle-nose pliers, install the 2 parking brake shoe return tension No. 1 springs. First install the front side spring and then the rear side spring.

25. Inspect parking brake installation and check that each part is installed properly.

➡There should be no oil or grease on the friction surfaces of the shoe linings and discs.

26. Install the rear disc.

27. Install the parking brake shoe adjusting hole plug.

28. Adjust parking brake shoe clearance.

29. Install the rear disc brake caliper assembly with the 2 bolts and tighten to 46 ft. lbs. (62 Nm).

30. Install the rear wheel.

31. Adjust the parking brake pedal travel.

32. Bed in parking brake shoes to discs, as follows:

a. Drive the vehicle at about 31 mph (50 km/h) on a safe, level and dry road.

b. Depress the parking brake pedal with 34 lbs. (150 N) of force.

33. Drive the vehicle about 0.25 miles (400 m) in this condition.

a. Repeat this procedure 3 times using 5-minute intervals between each procedure to prevent the parking brake assembly from overheating.

34. Remove the rear wheel.

35. Adjust parking brake shoe clearance.

36. For A/T vehicles, adjust the parking brake pedal travel.

37. For M/T vehicles, adjust the parking brake pedal travel.

38. Install the rear wheel and tighten the lug nuts to 76 ft. lbs. (103 Nm).

ADJUSTMENT

1. Adjust parking brake shoe clearance, as follows:

a. Temporarily install the hub nuts.

b. Remove the shoe adjusting hole plug, turn the adjuster and expand the shoes until the disc locks.

c. Contract the shoe adjuster until the disc rotates smoothly. Standard: returns 8 notches

d. Check that the disc has no brake drag.

e. Install the shoe adjusting hole plug.

Parking Brake Lever Travel Adjustment (M/T)

1. Pull up the parking brake lever. Hold the No. 1 wire adjusting nut using a wrench and loosen the lock nut.
2. Release the parking brake lever.
3. Turn the No. 1 wire adjusting nut until the parking brake lever travel meets the above specification.
4. Hold the No. 1 wire adjusting nut using a wrench or an equivalent tool and tighten the lock nut to 44 inch lbs. (5 Nm).
5. Count the number of clicks after depressing and releasing the parking brake lever 3 or 4 times.
6. Check whether the parking brake drags.
7. When operating the parking brake lever, check that the parking brake indicator light comes on.

Parking Brake Lever Travel Adjustment (A/T)

1. Depress the parking brake pedal. Hold the No. 1 wire adjusting nut using a wrench and loosen the lock nut.
2. Release the parking brake pedal.
3. Turn the No. 1 wire adjusting nut until the parking brake pedal travel meets the above specification.
4. Hold the No. 1 wire adjusting nut using a wrench or an equivalent tool and tighten the lock nut to 48 inch lbs. (5.4 Nm).
5. Count the number of clicks after depressing and releasing the parking brake pedal 3 or 4 times.
6. Check whether the parking brake drags.
7. When operating the parking brake pedal, check that the parking brake indicator light comes on.

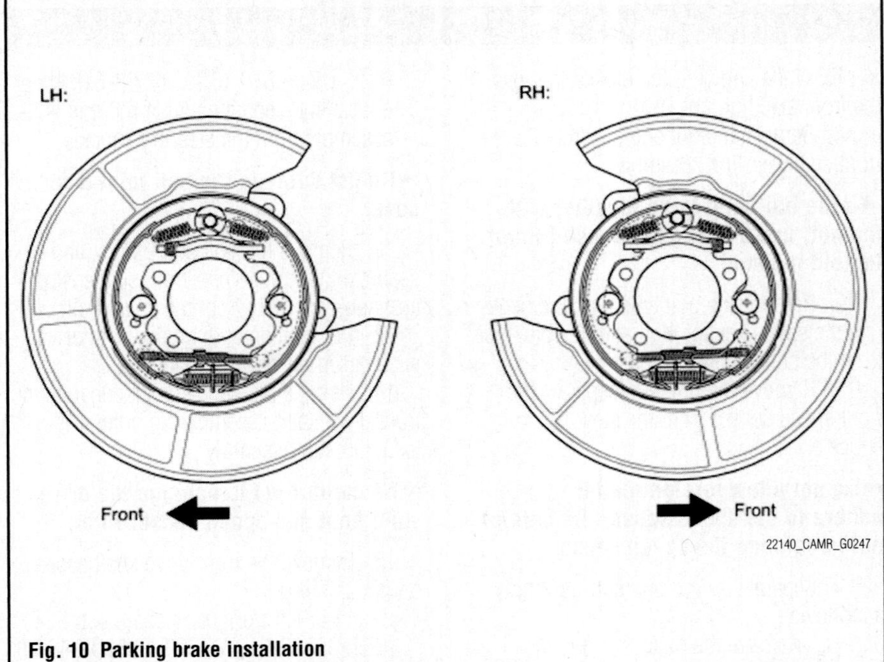

Fig. 10 Parking brake installation

CHASSIS ELECTRICAL

AIR BAG (SUPPLEMENTAL RESTRAINT SYSTEM)

GENERAL INFORMATION

This vehicle is equipped with a Supplemental Restraint System (SRS). It consists of a driver airbag, front passenger airbag, driver side knee airbag, front seat side airbag and curtain shield airbag. Failure to carry out service operations in the correct sequence could cause the SRS to unexpectedly deploy during servicing, possibly leading to a serious accident. Further, if a mistake is made in servicing the SRS, it is possible that the SRS may fail to operate when required. Before performing servicing (including removal or installation of parts, inspection or replacement), be sure to read the following carefully, then follow the correct procedures indicated in the repair manual.

SERVICE PRECAUTIONS

CAUTION

Disconnect and isolate the battery negative cable before beginning any airbag system component diagnosis, testing, removal, or installation procedures. Wait at least 90 seconds after the ignition switch is turned off and the negative (-) terminal cable is disconnected from the battery before starting the operation. The SRS is equipped with a backup power source, so if work is started within 90 seconds after disconnecting the negative (-) terminal cable from the battery, the SRS may be deployed. Failure to disable the airbag system may result in accidental airbag deployment, personal injury, or death.

DISARMING THE SYSTEM

To avoid personal injury when working on vehicles equipped with an air bag, the negative battery cable must be disconnected and at least 90 seconds must elapse before working on the system. Failure to do so may result in deployment of the air bag.

ARMING THE SYSTEM

To arm the system after service is finished, connect the negative battery cable.

CLOCKSPRING CENTERING

See Figures 11 and 12.

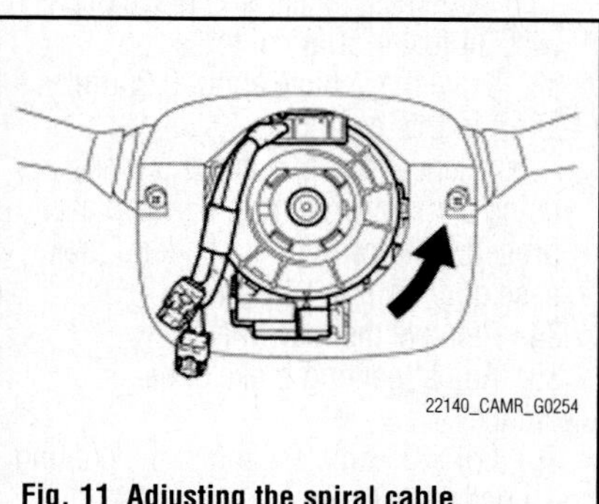

Fig. 11 Adjusting the spiral cable

1. Before servicing the vehicle, refer to the Precautions Section.
2. Check that the ignition switch is **OFF**.
3. Check that the battery negative (-) terminal is disconnected.

CAUTION

After removing the terminal, wait for at least 90 seconds before starting the operation.

4. Rotate the spiral cable counterclockwise slowly by hand until it feels firm.
5. Rotate the spiral cable clockwise approximately 2.5 turns to align the marks.

➡**Do not turn the spiral cable by the airbag wire harness.**

➡**The spiral cable will rotate approximately 2.5 turns to both the left and right from the center.**

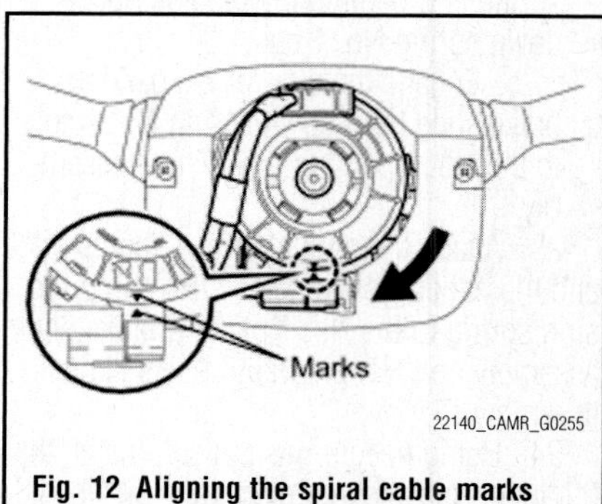

Fig. 12 Aligning the spiral cable marks

DRIVE TRAIN

FRONT HALFSHAFT

REMOVAL & INSTALLATION

See Figures 13 through 18.

1. Before servicing the vehicle, refer to the Precautions Section.
2. Remove the engine under cover.
3. Remove the drain plug and gasket, and then drain the automatic transaxle fluid.
4. Install a new gasket and drain plug and tighten to 36 ft. lbs. (49 Nm).
5. Remove front wheel.
6. Using SST (SST: 09930-00010) and hammer, release the staked part of the front axle hub nut.

➡**Loosen the staked part of the nut completely, otherwise the screw of the drive shaft may be damaged.**

7. While applying the brakes, remove the front axle hub nut.
8. Remove the nut and separate the front stabilizer link assembly.

➡**If the ball joint turns together with the nut, use a hexagon wrench (6mm) to hold the stud.**

9. Remove the bolt and clip, and separate the speed sensor wire and flexible hose from the shock absorber.
10. Remove the bolt and separate the front speed sensor from the steering knuckle.

➡**Do not allow foreign matter to adhere to the speed sensor. Be careful not to damage the speed sensor.**

11. Separate tie rod end sub-assembly, as follows:
    a. Remove the cotter pin and nut.
    b. Using SST (SST: 09628-62011) or equivalent, separate the tie rod end sub-assembly from the steering knuckle.

➡**Do not damage the ball joint dust cover.**

12. Remove the bolt and 2 nuts, and separate the lower No. 1 front suspension arm sub-assembly from the lower ball joint.
13. Put matchmarks on the front drive shaft assembly and the axle hub.
14. Using a plastic hammer, separate the front drive shaft assembly from the front axle hub sub-assembly.

➡**Be careful not to damage the drive shaft boot and speed sensor rotor.**

15. Remove the front drive shaft assembly(s), as follows:
    a. For left front drive shaft, use SST (SST: 09520-01010, SST: 09520-24010)

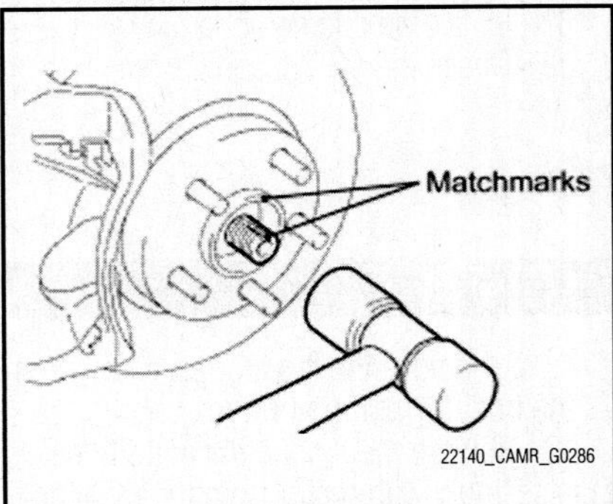

Fig. 13 Put matchmarks on the front drive shaft assembly and the axle hub

or equivalent, and remove the left front drive shaft assembly.

➡**Be careful not to damage the drive shaft dust cover, boot and oil seal. Be careful not to drop the drive shaft assembly.**

b. For right front drive shaft, use a screwdriver and remove the bearing bracket hole snap ring.

c. Remove the bolt and right front drive shaft assembly from the drive shaft bearing bracket.

➡**Do not damage the boot and oil seal.**

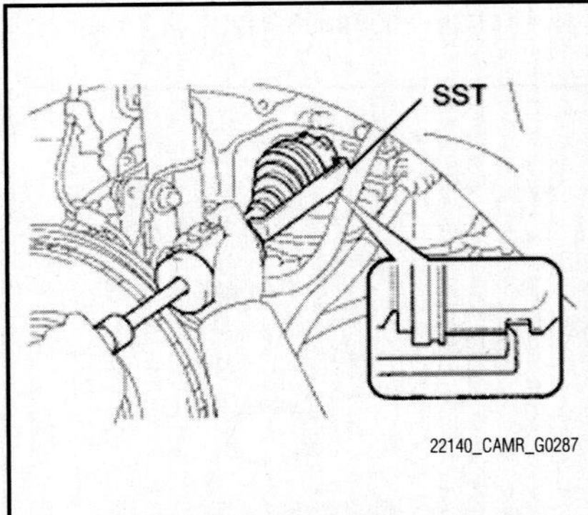

Fig. 14 Remove the left front drive shaft assembly

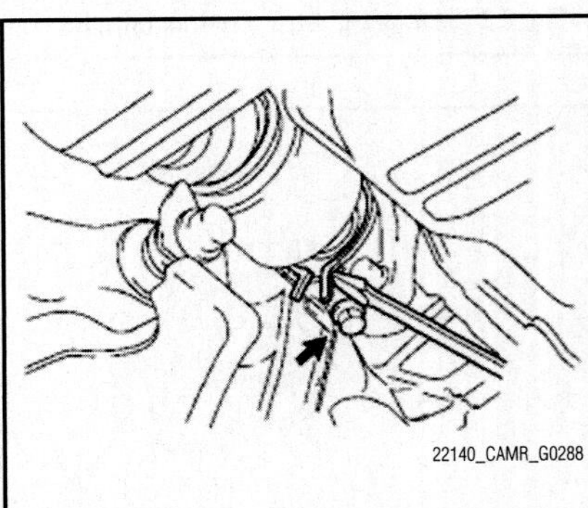

Fig. 15 Remove the right front drive shaft assembly

16. Fix front axle hub bearing. The hub bearing could be damaged if it is subjected to the vehicle's full weight, such as moving the vehicle with the drive shaft removed. If it is necessary to place the vehicle's weight on the hub bearing, first support it with SST (SST: 09608-16042).

To install:

17. Install the front drive shaft assembly(s), as follows:

a. Coat the spline of the inboard joint shaft assembly with automatic transaxle fluid.

b. For the left front drive shaft, align the shaft splines and install the drive shaft assembly with a brass bar and hammer.

➡**Set the shaft snap ring with the opening side facing down. Be careful not to damage the drive shaft dust cover, boot, and oil seal. Move the drive shaft assembly while keeping it level.**

c. For the right front drive shaft, install the drive shaft and use a screwdriver to install a new bearing bracket hole snap ring, and install a new bolt tightened to 24 ft. lbs. (32 Nm).

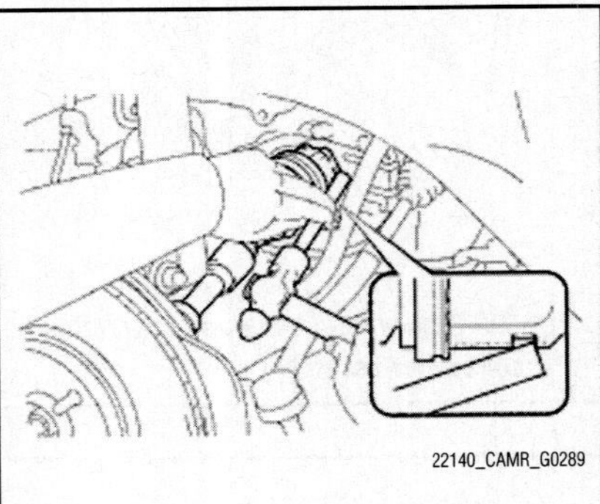

Fig. 16 Install the left front drive shaft assembly

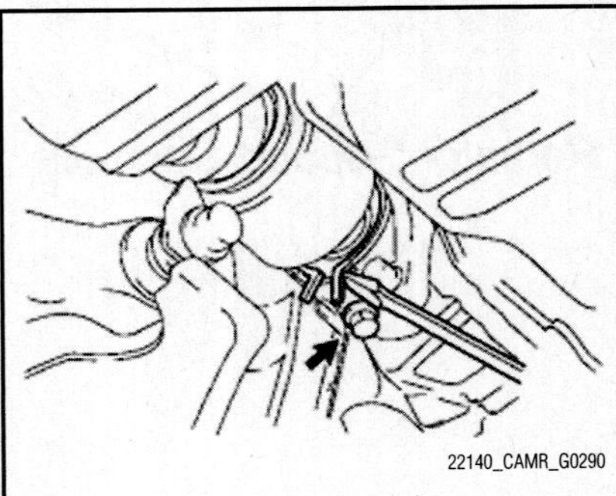

Fig. 17 Install the right front drive shaft assembly

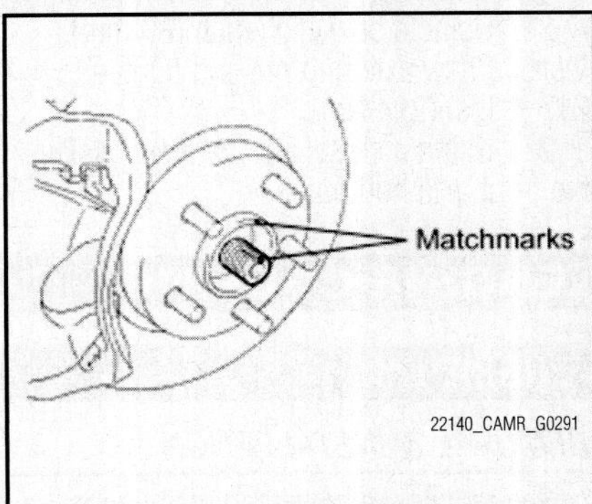

Fig. 18 Align the matchmarks and install the front drive shaft assembly

➡**Be careful not to damage the drive shaft dust cover, boot and oil seal.**

➡**Move the drive shaft assembly while keeping it level.**

18. Align the matchmarks and install the front drive shaft assembly to the front axle hub sub-assembly.

➡**Be careful not to damage the drive shaft boot and speed sensor rotor.**

19. Install the lower ball joint to the lower No. 1 front suspension arm sub-assembly with the bolt and 2 nuts and tighten to 55 ft. lbs. (75 Nm).

20. Install the tie rod end sub-assembly to the steering knuckle with the nut and tighten to 36 ft. lbs. (49 Nm).

21. Install a new cotter pin. If the holes for the cotter pin are not aligned, tighten the nut up to 60° further.

22. Install the front speed sensor to the steering knuckle with the bolt and tighten to 71 inch lbs. (8 Nm).

23. Install the flexible hose and the speed sensor to the shock absorber with the bolt and set the sensor clip on the knuckle and tighten to 14 ft. lbs. (19 Nm).

➡**Be careful not to damage the speed sensor. Do not allow foreign matter to adhere to the speed sensor. Do not twist the sensor wire when installing the speed sensor.**

24. Install the stabilizer link assembly with the nut and tighten to 55 ft. lbs. (74 Nm).

➡**If the ball joint turns together with the nut, use a hexagon wrench (6 mm) to hold the stud.**

25. Clean the threaded parts on the drive shaft and axle hub nut using a non-residue solvent.

➡**Be sure to perform this work for a new drive shaft. Keep the threaded parts free of oil and foreign objects.**

26. Using a socket wrench (30 mm), install a new axle hub nut and tighten to 217 ft. lbs. (294 Nm).
27. Using a chisel and hammer, stake the front axle hub nut.
28. Install front wheel.
29. Add automatic transaxle fluid.
30. Inspect automatic transaxle fluid.
31. Inspect and adjust front wheel alignment.
32. Install the engine under cover.
33. Check the ABS speed sensor signal.

ENGINE COOLING

ENGINE FAN

REMOVAL & INSTALLATION

See radiator removal and installation.

RADIATOR

REMOVAL & INSTALLATION

See Figures 19 through 21.

1. Before servicing the vehicle, refer to the Precautions Section.
2. Drain engine coolant.
3. Remove both front wheel opening extension pads.
4. Remove both engine under cover.
   a. Remove air cleaner cap sub-assembly.
   b. Remove air cleaner inlet assembly.
5. Remove front bumper assembly.
6. Remove front bumper energy absorber.
7. Separate the radiator reserve tank hose from the radiator assembly.
8. Disconnect the inlet and outlet radiator hose from the radiator assembly.
9. Disconnect the oil from the radiator assembly.
10. Disconnect the oil cooler inlet hose and outlet hose from the radiator assembly.
11. Remove the upper radiator support, as follows:
    a. Disconnect the horn connector.
    b. Remove the 3 bolts and separate the hood lock assembly from the radiator support upper.
    c. Remove the clamp and separate the hood lock control cable from the radiator support upper.
    d. Remove the 5 bolts and radiator support upper.
12. Remove the radiator assembly, as follows:
    a. Remove the 3 clamps and 2 connectors.
    b. Remove the 4 bolts and separate the condenser assembly from the radiator assembly.
    c. Remove the radiator assembly from the body by pulling upwards.
13. Release the 3 snap fits and lift the fan assembly with motor from the radiator.
14. Remove the 2 radiator support cushions from the radiator assembly.
15. Remove the 2 radiator support lowers from the radiator assembly.

To install:

16. Installation is the reverse of removal.
17. After installation, inspect for coolant leak.

Fig. 19 Removing 5 bolts and upper radiator support

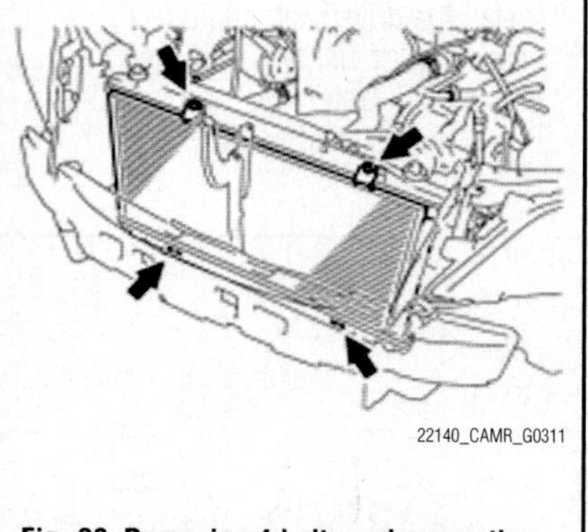

Fig. 20 Removing 4 bolts and separating the condenser assembly

Fig. 21 Remove the 2 radiator support cushions

THERMOSTAT

REMOVAL & INSTALLATION

See Figures 22 and 23.

1. Before servicing the vehicle, refer to the Precautions Section.
2. Remove both front wheel opening extension pads.
3. Remove both engine under cover.
4. Drain engine coolant.

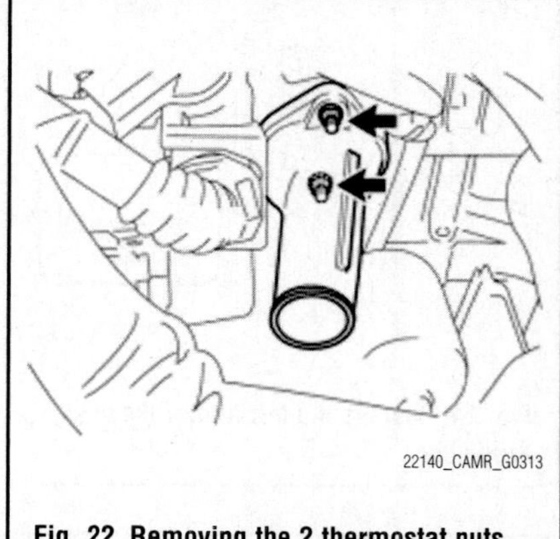

Fig. 22 Removing the 2 thermostat nuts

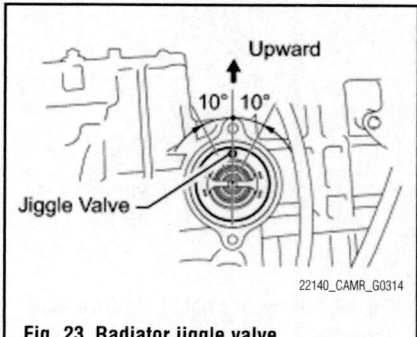

Fig. 23 Radiator jiggle valve

5. Separate the radiator hose outlet.
6. Remove the 2 nuts and disconnect the water inlet from the cylinder block.
7. Remove the thermostat and the gasket from the thermostat.

To install:

8. Install a new gasket to the thermostat.
9. Install a new gasket to the thermostat.

➡The jiggle valve may be set within 10°of either side of the prescribed position.

10. Install the thermostat with the jiggle valve facing up.
11. Install the water inlet and tighten to 7 ft. lbs. (10 Nm).
12. The remainder of installation is the reverse of removal.
13. After installation, inspect for coolant leak.

WATER PUMP

REMOVAL & INSTALLATION

See Figures 24 and 25.

1. Before servicing the vehicle, refer to the Precautions Section.
2. Disconnect the negative battery cable.
3. Remove both front wheel opening extension pads.
4. Remove both engine under covers.
5. Drain and recycle the engine coolant.
6. Remove RH front fender apron seal.
7. Remove RH No. 2 engine mounting stay.
8. Remove engine moving control rod sub-assembly.
9. Remove RH No. 2 engine mounting bracket

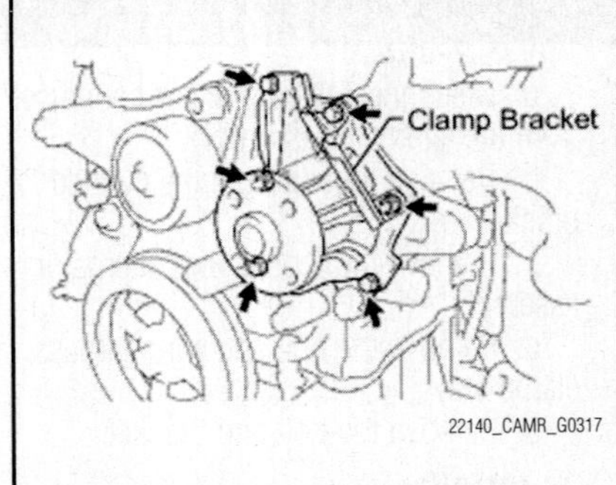

Fig. 24 Removing the 4 bolts, 2 nuts and clamp bracket

10. Remove the drive belt.
11. Remove generator assembly.
12. Using SST: 09960-10010, remove the 4 bolts and water pump pulley.
13. Remove water pump assembly by performing the following:
    a. Remove the clamp of the Crankshaft Position (CKP) sensor from the water pump.
    b. Disconnect the wire of the Crankshaft Position (CKP) sensor from the clamp bracket.
    c. Remove the 4 bolts, 2 nuts and clamp bracket.

➡Be careful not to damage the contact surfaces of the water pump and cylinder block.

   d. Using a screwdriver, pry between the water pump and cylinder block, and then remove the water pump. Tape the screwdriver tip before use.

To install:

14. Install water pump assembly.
    a. Remove any old seal packing material from the water pump assembly contact surface.

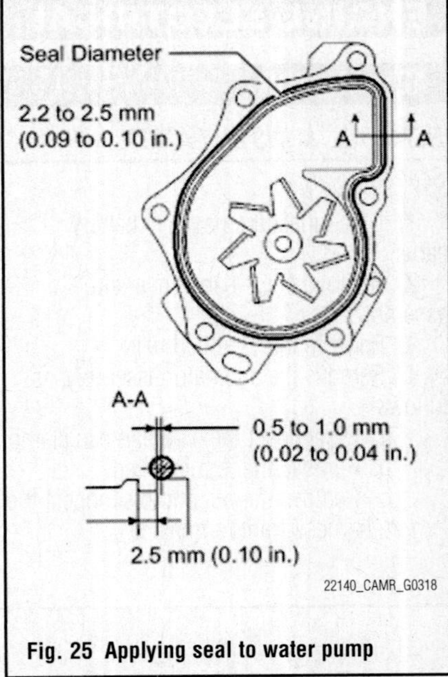

Fig. 25 Applying seal to water pump

➡Remove any oil from the contact surface. The parts must be set within 3 minutes after applying seal packing. Otherwise, the material must be removed and reapplied.

   b. Apply a continuous line of seal packing. Seal packing: Toyota Genuine Seal Packing Black, Three Bond 1207B or Equivalent. Standard seal diameter: 0.09 to 0.10 in. (2.2 to 2.5 mm).

15. Install the water pump and clamp bracket with the 4 bolts and 2 nuts and tighten to 80 inch lbs. (9 Nm).
16. To complete installation, reverse removal procedure.
17. Tighten the water pump bolts to 19 ft. lbs. (26 Nm).
18. Add engine coolant.
19. Inspect for coolant leak.

ALTERNATOR

REMOVAL & INSTALLATION

See Figure 26.

1. Disconnect the negative battery cable.
2. Remove the V-bank cover sub-assembly.
3. Remove the V-ribbed belt.
4. Remove the alternator assembly, as follows:
    a. Disconnect the wire harness clamp.
    b. Remove the terminal cap.
    c. Remove the nut and disconnect the wire harness from terminal B.
    d. Disconnect the alternator connector from the alternator assembly.
    e. Remove the nut from the cylinder block.
    f. Remove the 2 bolts and alternator assembly.
    g. Remove the bolt and wire harness clamp stay.
    h. Remove the bolt and bracket.

To install:

5. Install the alternator assembly, as follows:
    a. Install the bracket with the bolt and tighten to 15 ft. lbs. (20 Nm).
    b. Install the wire harness clamp stay and tighten to 74 inch lbs. (8.4 Nm).
    c. Install the alternator assembly with the 2 bolts and tighten to 32 ft. lbs. (43 Nm).
    d. Install the nut to the cylinder block and tighten to 15 ft. lbs. (20 Nm).
    e. Connect the alternator connector to the alternator assembly.
    f. Install the alternator wire with the nut and tighten to 87 inch lbs. (9.8 Nm).
    g. Install the terminal cap.
    h. Connect the wire harness clamp.
6. Install the V-ribbed belt.
7. Install the V-bank cover sub-assembly.
8. Connect the negative battery cable.
9. Perform initialization.

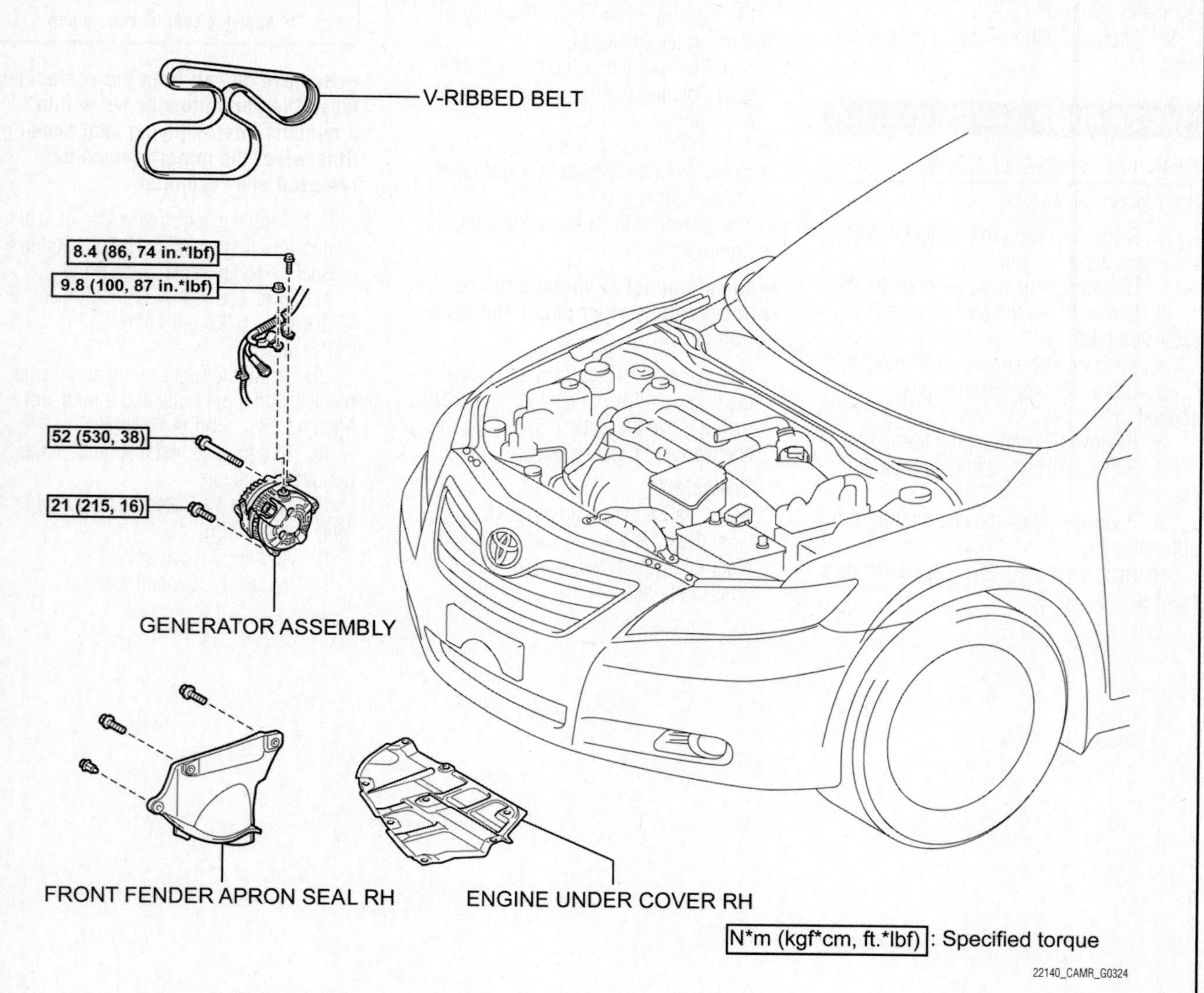

Fig. 26 Alternator (generator) system component locations

ENGINE ELECTRICAL

IGNITION COIL

REMOVAL & INSTALLATION

See Figure 27.

1. Before servicing the vehicle, refer to the Precautions Section.
2. Disconnect the negative battery cable.
3. Remove engine cover(s).

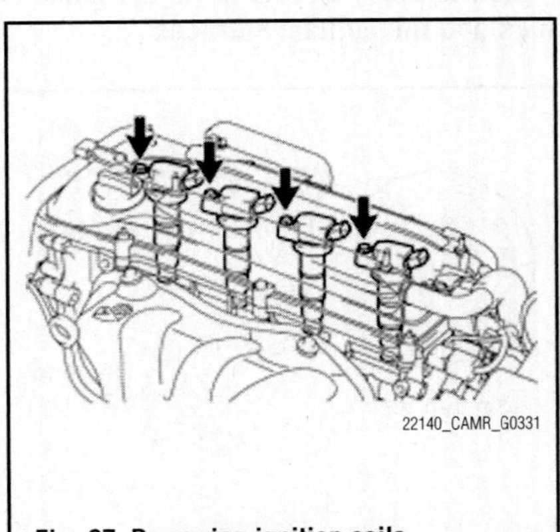

Fig. 27 Removing ignition coils

4. Disconnect the 4 ignition coil connectors. Remove the 4 bolts and 4 ignition coils.

To install:

5. To install, reverse removal procedure. Tighten ignition coil bolts to 80 inch lbs. (9 Nm).

IGNITION TIMING

INSPECTION

All engines are equipped with a Distributorless Ignition System (DIS). No timing adjustment is possible.

SPARK PLUGS

REMOVAL & INSTALLATION

See Figure 28.

1. Before servicing the vehicle, refer to the Precautions Section.
2. Remove the plastic engine cover.
3. Disconnect the 4 ignition coil connectors and remove the 4 bolts and ignition coils.
4. Using a 16 mm (0.63 in.) spark plug wrench, remove the 4 spark plugs.

To install:

5. Installation is the reverse of removal.
   a. Torque the ignition coils to 80 inch lbs. (9 Nm) and the spark plugs to 13 ft. lbs (18 Nm).

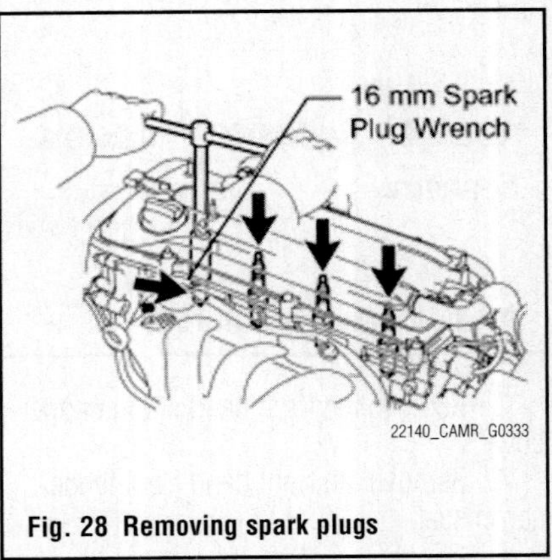

Fig. 28 Removing spark plugs

ENGINE MECHANICAL

ACCESSORY DRIVE BELTS

ACCESSORY BELT ROUTING

See Figure 29.

INSPECTION

See Figure 30.

Visually check the V-ribbed belt for excessive wear, frayed cords, etc. If any defect has been found, replace the V-ribbed belt.

- Cracks on the rib side of a belt are considered acceptable. If the belt has chunks missing from the ribs, it should be replaced
- A "new belt" is a belt which has been used for less than 5 minutes with the engine running
- A "used belt" is a belt which has been used for 5 minutes or more with the engine running

ADJUSTMENT

This vehicle is equipped with an auto-tensioner and cannot be adjusted.

REMOVAL & INSTALLATION

See Figure 29.

1. Before servicing the vehicle, refer to the Precautions Section.
2. Remove the right hand front wheel.

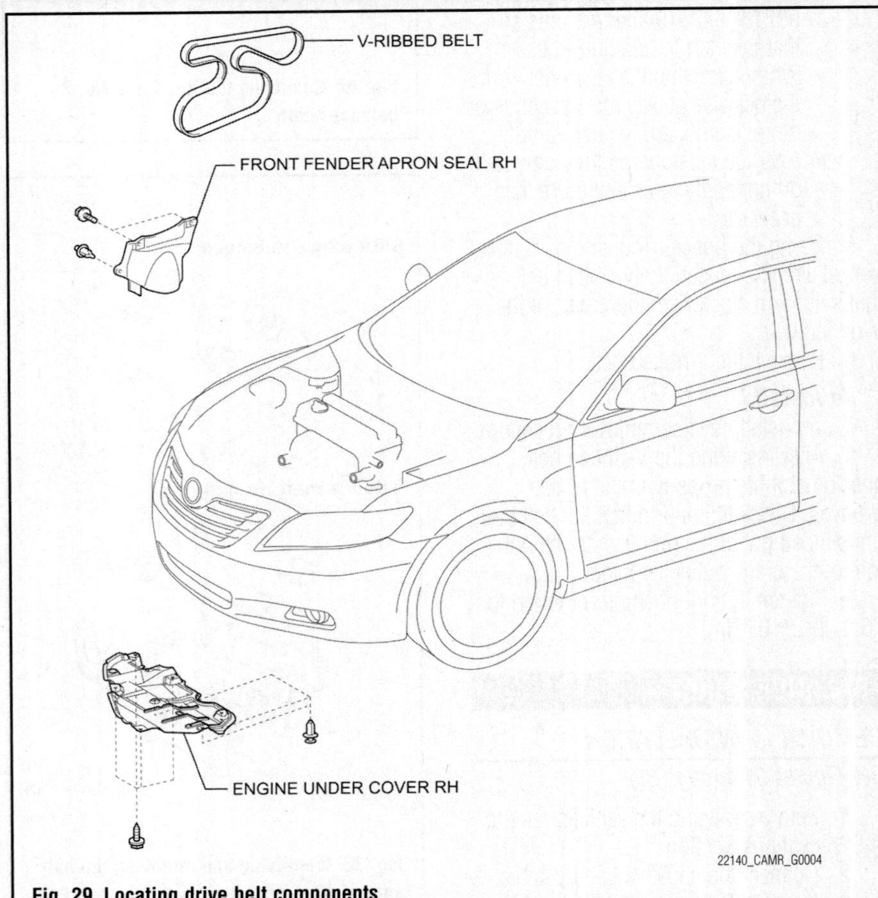

Fig. 29 Locating drive belt components

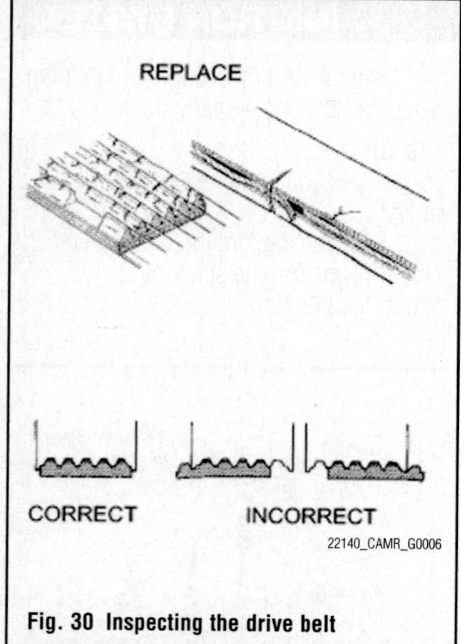

Fig. 30 Inspecting the drive belt

3. Remove the right hand engine under cover.

4. Remove the right hand front fender apron seal.

➡Before removing, take note of the following:

- Be sure to connect Special Tool: 09216—42010 and the tools so that they are in line during use
- When retracting the tensioner, turn it clockwise slowly for 3 seconds or more. Do not apply force rapidly
- After the tensioner is fully retracted, do not apply force any more than necessary

5. Using the Special Tool and a 19 mm socket wrench, loosen the v-ribbed belt tensioner arm clockwise, then remove the v-ribbed belt.

6. Remove the v-ribbed belt.

To install:

7. To install, reverse removal procedure.

8. After installing the V-ribbed belt, check that it fits properly in the ribbed grooves. Check to confirm that the belt has not slipped out of the grooves on the bottom of the crank pulley by hand.

9. Tighten the right hand front wheel to: 76 ft. lbs. (103 Nm).

BALANCE SHAFT

REMOVAL & INSTALLATION

See Figures 31 through 38.

1. Before servicing the vehicle, refer to the Precautions Section.

2. Connect the negative battery cable.

3. Drain the engine oil.

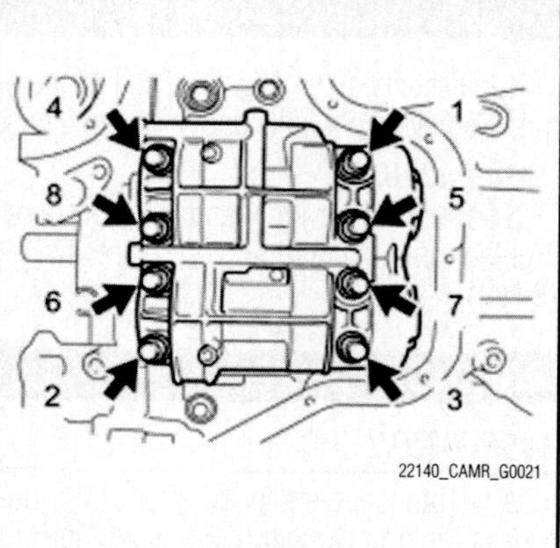

Fig. 31 Sub-assembly bolt removal sequence

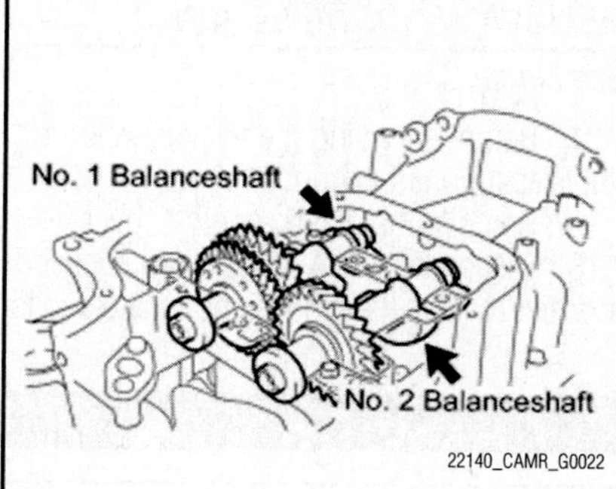

Fig. 32 Removing the No. 1 and No. 2 balance shafts

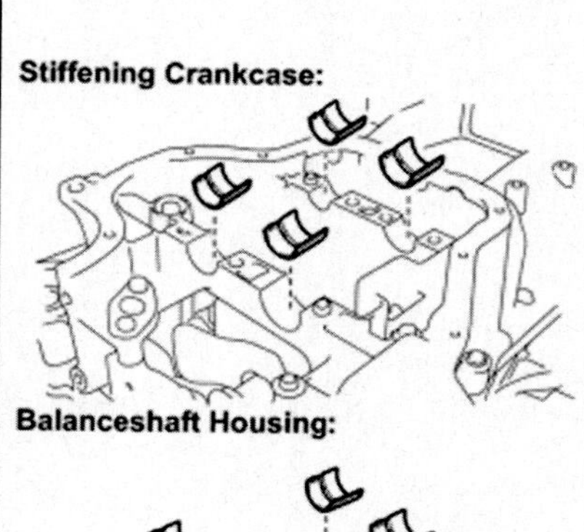

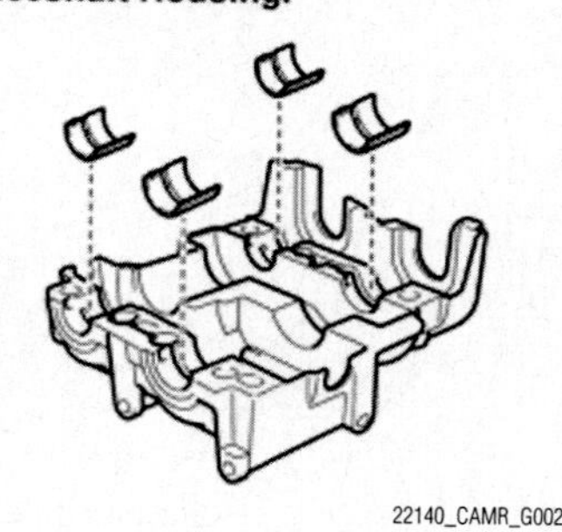
Fig. 33 Identifying and removing the balance shaft bearings

4. Remove the oil pump. Refer to Oil Pump below for removal procedure.

5. Remove the No. 1 and No. 2 balance shaft sub-assembly. Remove the eight bolts in sequence.

6. Remove the No. 1 and No. 2 balance shafts.

7. Remove the balance shaft bearings if necessary.

To install:

➡Do not apply engine oil to the bearings and the contact surfaces.

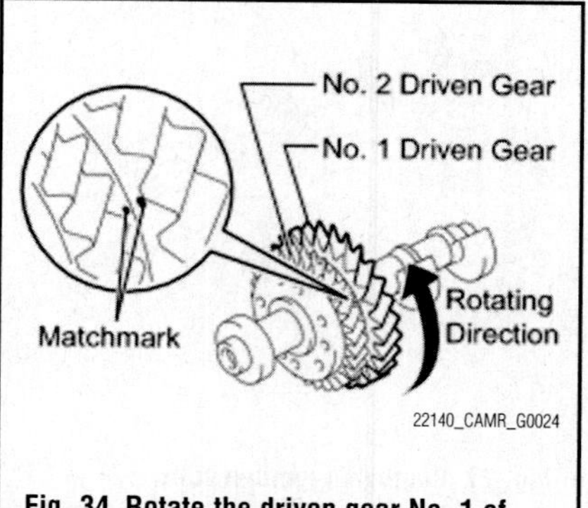

Fig. 34 Rotate the driven gear No. 1 of balance shaft No. 1

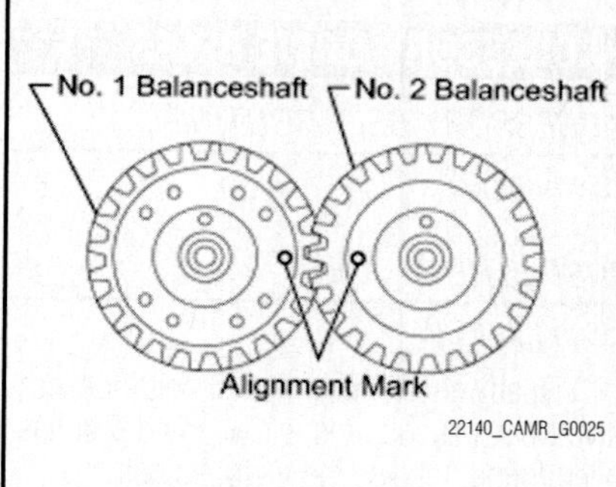

Fig. 35 Aligning marks of the No. 1 and No. 2 balance shafts

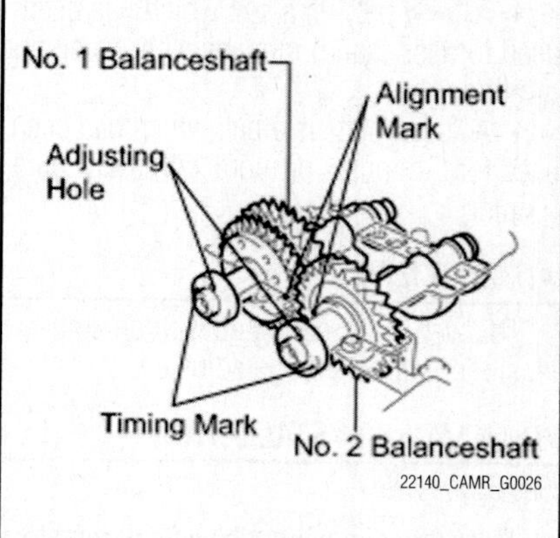

Fig. 36 Placing the No. 1 and No. 2 balance shafts on the crankcase

8. Install the bearings in the crankcase and balance shaft housing.

9. Apply a light coat of engine oil to the bearings.

➡Confirm that the match marks on driven gears No. 1 and No. 2 are matched.

10. Install No. 1 and No. 2 balance shaft sub-assembly. Rotate the driven gear No. 1 of balance shaft No. 1 in the rotating direction until it hits the stopper.

11. Align the alignment marks of the No. 1 and No. 2 balance shafts as shown.

12. Place the No. 1 and No. 2 balance shafts on the crankcase.

13. Apply a light coat of engine oil under the heads of the balance shaft housing bolts.

14. Install the balance shaft housing bolts. The balance shaft housing bolts should be tightened in 2 progressive steps as follows:

a. Tighten the eight balance shaft housing bolts in sequence to: 16 ft. lbs. (22 Nm).

b. Mark the front side of each balance shaft housing bolt head with paint.

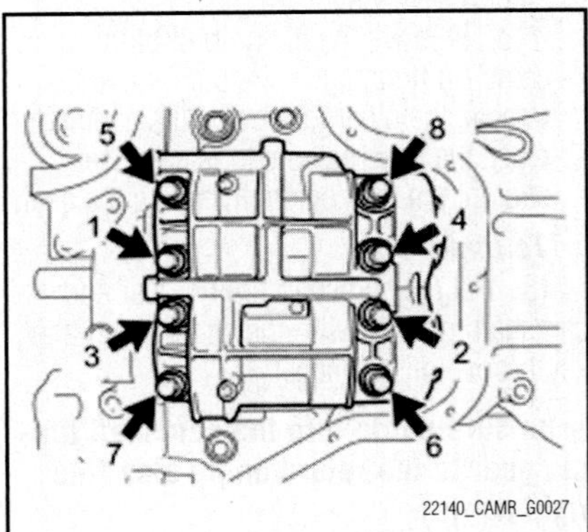

Fig. 37 Balance shaft housing bolt tightening sequence

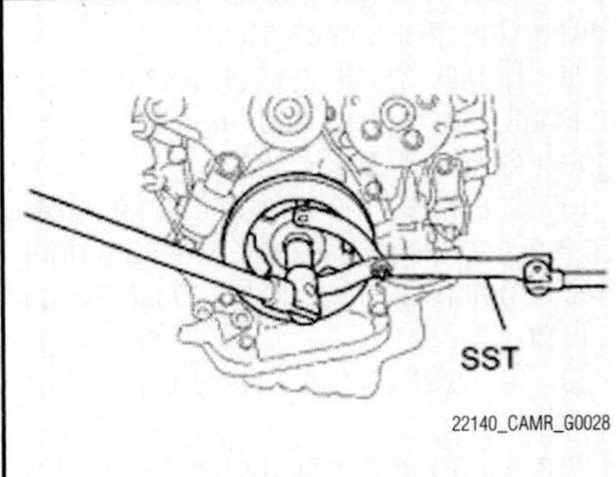

Fig. 38 Marking the bolt head and tightening procedure

Retighten the bolts by 90°. Check that the paint marks are now at a 90°angle to the front.

15. To complete installation, reverse remaining removal procedure.

16. Check the engine for leaks.

CAMSHAFT AND TIMING GEAR

CAMSHAFT BEARING REPLACEMENT

To remove and replace the camshaft bearings, refer to the camshaft removal procedure.

INSPECTION

➡Be careful not to damage the camshafts.

1. To inspect the No.1 camshaft, perform the following:

a. Clamp the camshaft in a vise, and confirm that the camshaft timing gear is locked.

b. Release the lock pin.

➡The 2 advance side paths are provided in the groove of the camshaft. Plug one of the paths with a rubber piece.

c. Cover the 4 oil paths of the cam journal with vinyl tape.

d. Break through the tape of the advance side path and the retard side path on the opposite side to the hole of the advance side path.

➡Cover the paths with a piece of cloth when applying pressure to keep oil from splashing.

e. Apply approximately 28 psi of air pressure to the two broken paths.

f. Check that the camshaft timing gear revolves in the advance direction when reducing the air pressure of the retard side path.

➡This operation releases the lock pin for the most retarded position.

➡Do not remove the air gun from the advance side path first. The gear may abruptly shift in the retard direction and break the lock pin.

g. When the camshaft timing gear reaches the most advanced position, remove the air gun from the retard side path and advance side path, in that order.

➡Do not use an air gun to check for smooth operation.

h. Rotate the camshaft timing gear within its movable range several times, but do not turn it to the most retarded position. Check that the gear rotates smoothly.

i. Check the lock in the most retarded position. Confirm that the camshaft timing gear is locked at the most retarded position.

j. To inspect the camshaft for runout, place the camshaft on V-blocks.

k. Using a dial indicator, measure the circle runout at the center journal. Maximum circle runout: 0.0012 inches (0.03 mm). If the circle runout is greater than the maximum, replace the camshaft.

l. Inspect the cam lobes by using a micrometer and measure the cam lobe height. Standard cam lobe height: 1.8624 to 1.8664 inches. (47.306 to 47.406 mm). Minimum cam lobe height: 1.8581 inches (47.196 mm).

m. If the cam lobe height is less than the minimum, replace the No.1 camshaft.

n. Inspect the camshaft journals by using a micrometer and measure the journal diameter. If the journal diameter is not as specified, check the oil clearance.

2. To inspect the No.2 Camshaft, perform the following:

a. Place the camshaft on V-blocks.

b. Using a dial indicator, measure the circle runout at the center journal. Maximum circle runout: 0.0012 inches. (0.03 mm). If the circle runout is greater than the maximum, replace the No. 2 camshaft.

c. Inspect the cam lobes by using a micrometer and measure the cam lobe height. Standard cam lobe height: 1.8104 to 1.8143 inches. (45.983 to 46.083 mm). Minimum cam lobe height: 1.8060 inches (45.873 mm).

d. If the cam lobe height is less than the minimum, replace the No.2 camshaft.

e. Inspect the camshaft journals by using a micrometer and measure the journal diameter. If the journal diameter is not as specified, check the oil clearance.

REMOVAL & INSTALLATION

See Figures 39 through 50.

⁂ CAUTION

All models are equipped with a Supplemental Restraint System (SRS), which uses an air bag. Whenever working near any of the SRS components, such as the impact sensors, the air bag module, steering column and instrument panel, disable the SRS.

1. Before servicing the vehicle, refer to the Precautions Section.
2. Disconnect the negative battery cable.
3. Loosen the lug nuts on the front right hand wheel.
4. Apply the parking brake, block the rear wheels, then raise and safely support the front of the vehicle securely on jackstands.
5. Remove the front right hand wheel.
6. Remove the left and right hand under cover.
7. Remove the No.1 engine cover sub-assembly and 2 nuts.
8. Remove the ignition coil assembly.
9. Remove the cylinder head cover sub-assembly
10. Set No.1 Cylinder to TDC/ Compression by performing the following:
- Turn the crankshaft pulley until its groove and the timing mark "0" of the timing chain cover are aligned
- Check that each timing mark of the camshaft timing gear and sprocket is aligned with each timing mark located on the No. 1 and No. 2 bearing caps as shown in the illustration. If not, turn the crankshaft by 1 revolution (360°) to align the timing marks

➡Do not turn the crankshaft without the chain tensioner.

11. Remove No.1 Chain tensioner assembly. Remove the 2 nuts, tensioner and gasket.
12. Remove the No. 2 camshaft by performing the following:
    a. While holding the camshaft with a wrench, loosen the camshaft timing set bolt.
    b. Using several steps, uniformly loosen and remove the 10 bearing cap bolts in the sequence shown.

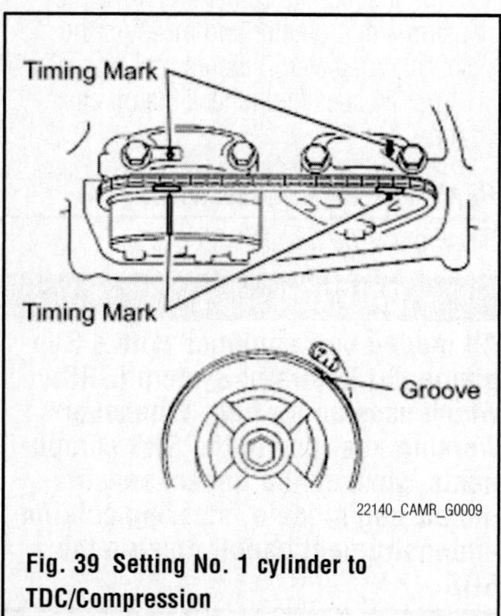

Fig. 39 Setting No. 1 cylinder to TDC/Compression

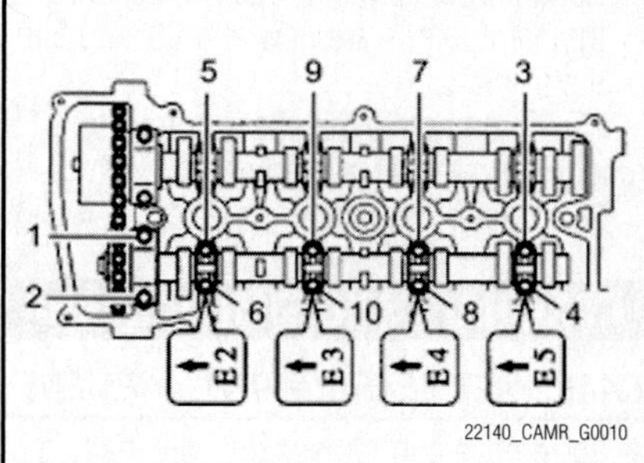

Fig. 40 No. 2 camshaft bearing cap bolt removal sequence

    c. While holding the No. 2 camshaft by hand, remove the camshaft timing sprocket set bolt.
    d. Remove the camshaft timing sprocket from the No. 2 camshaft with the timing chain wrapped on the sprocket.
    e. Remove the camshaft timing sprocket from the timing chain.

13. Remove the No. 1 camshaft by performing the following:
    a. Using several steps, uniformly loosen and remove the 10 bearing cap bolts in the sequence shown.
    b. Remove the 5 bearing caps.
    c. Remove the camshaft and camshaft timing gear while holding the timing chain by hand.

➡Be careful not to drop anything inside the timing chain cover.

    d. Tie the timing chain with a string.

14. Remove the camshaft timing gear assembly by performing the following:
    a. Clamp the camshaft in a vise, and make sure that the camshaft timing gear does not rotate.
    b. Cover all the oil ports except the advance side port shown in the illustration with vinyl tape

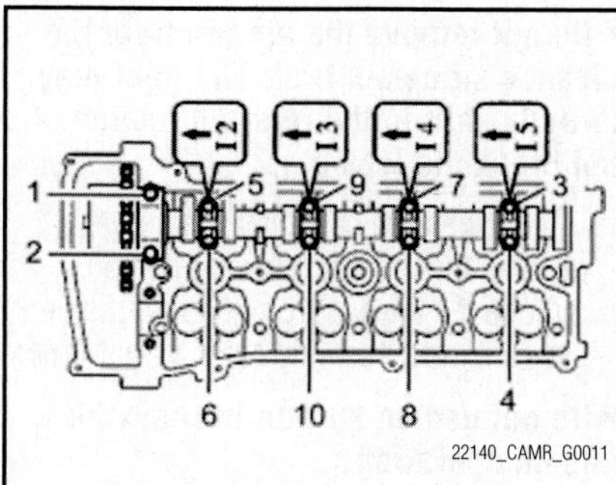

Fig. 41 No.1 camshaft bearing cap bolt removal sequence

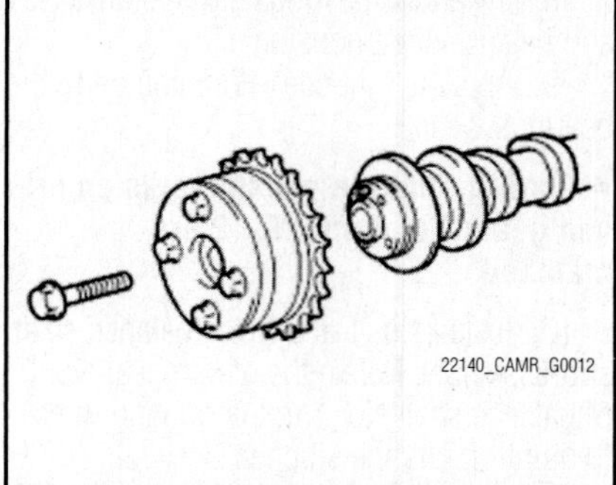

Fig. 42 Removing the flange bolt from the camshaft timing gear

➡Cover the paths with a shop rag or piece of cloth to avoid oil splashes.

➡Depending on the air pressure, the camshaft timing gear will turn to the advance angle side without applying force by hand. Also, if the pressure is difficult to apply because of air leakage from the port, the lock may be difficult to release.

    c. Apply air pressure of 14 psi to the oil path, then turn the camshaft timing gear in the advance direction (counterclockwise) by hand.
    d. Remove the flange bolt of the camshaft timing gear. Be sure not to remove the other four bolts. If planning to reuse the gear, be sure to release the straight pin lock before installing the gear.

To install:

15. Put the camshaft timing gear and camshaft together with the straight pin and key groove misaligned.

➡Be sure not to turn the camshaft timing gear to the retard angle side (the right angle).

16. Turn the camshaft timing gear as shown in the illustration while pushing it gently against the camshaft. Push further at the position where the pin fits into the groove.
17. Check that there is no clearance between the gear and camshaft.
18. Tighten the flange bolt with the camshaft timing gear fixed in place and tighten to 40 ft. lbs. (54 Nm).
19. Check that the camshaft timing gear can move to the retard angle side (the right direction) and is locked in the most retarded position.
20. To install the No.1 camshaft, perform the following:
    a. Apply a light coat of engine oil to the journal portion of the camshaft.
    b. Install the timing chain onto the camshaft timing gear with the paint mark

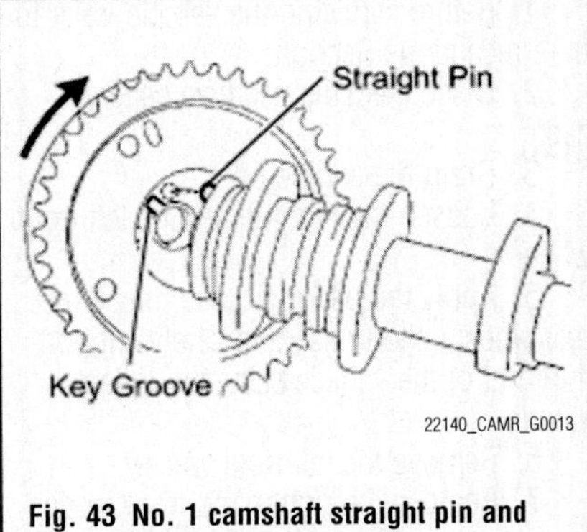

Fig. 43 No. 1 camshaft straight pin and key groove misaligned

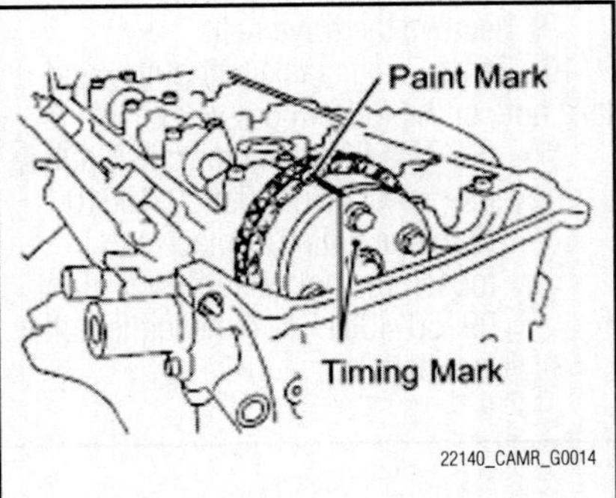

Fig. 44 Aligning No. 1 camshaft paint mark with the timing mark

aligned with the timing mark in the camshaft timing gear.

c. Examine the front marks and numbers, and check that the order is as shown in the illustration below. Then install the bearing caps into the cylinder head.

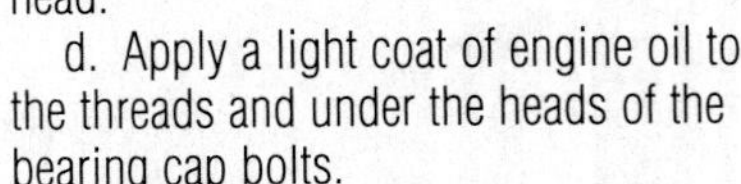

d. Apply a light coat of engine oil to the threads and under the heads of the bearing cap bolts.

e. Using several steps, uniformly tighten the 10 bearing cap bolts in the sequence shown. Tighten the following to specification:

- No. 1 Bearing cap: 22 ft. lbs. (30 Nm)
- No. 3 Bearing cap: 80 inch lbs. (9 Nm)

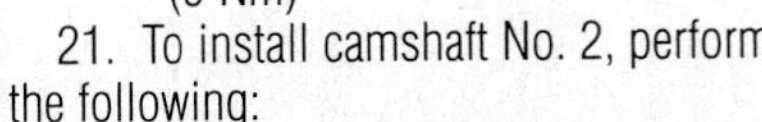

21. To install camshaft No. 2, perform the following:

a. Apply a light coat of engine oil to the journal portion of the No. 2 camshaft.

b. Put the No. 2 camshaft on the cylinder head with the paint mark of the chain aligned with the timing mark on the camshaft timing sprocket.

c. While holding the No. 2 camshaft by hand, temporarily tighten the camshaft timing sprocket set bolt.

d. Examine the front marks and numbers, and check that the order is as

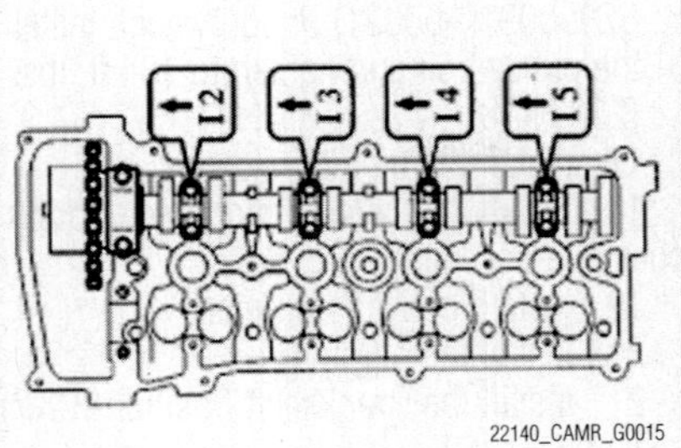

Fig. 45 Checking the No. 1 camshaft bearing cap front marks, numbers and order

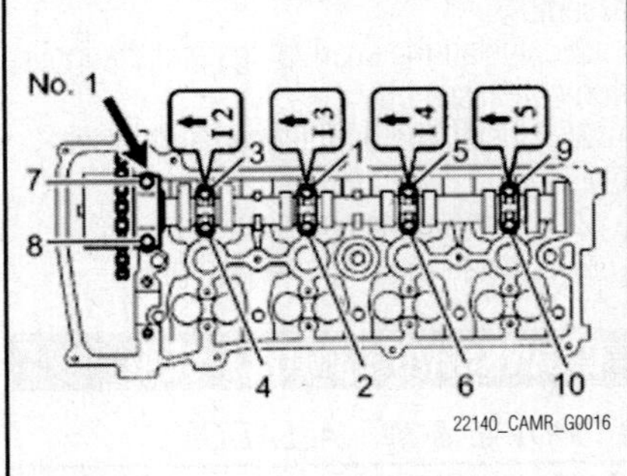

Fig. 46 No. 1 camshaft bearing cap tightening sequence

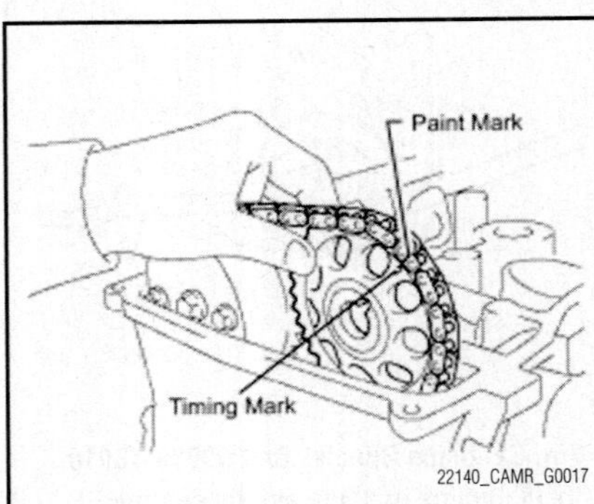

Fig. 47 Aligning No. 2 camshaft paint mark with the timing mark

shown in the illustration. Then install the bearing caps onto the cylinder head.

e. Apply a light coat of engine oil to the threads and under the heads of the bearing cap bolts.

f. Using several steps, uniformly tighten the 10 bearing cap bolts in the sequence shown. Tighten the following to specification:

- No. 1 Bearing cap: 22 ft. lbs. (30 Nm)
- No. 3 Bearing cap: 80 inch lbs. (9 Nm)

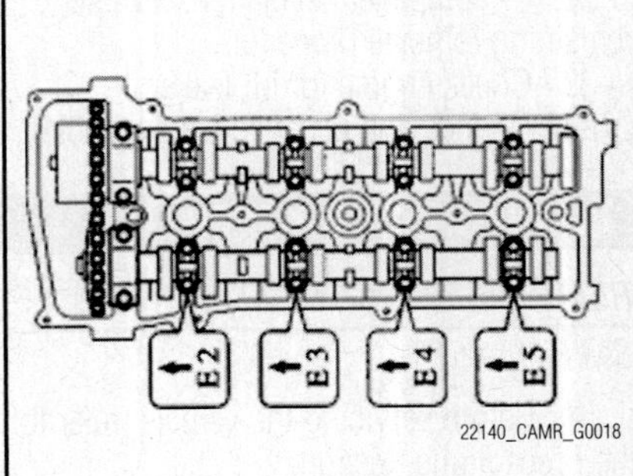

Fig. 48 Checking the No. 2 camshaft bearing cap front marks, numbers and order

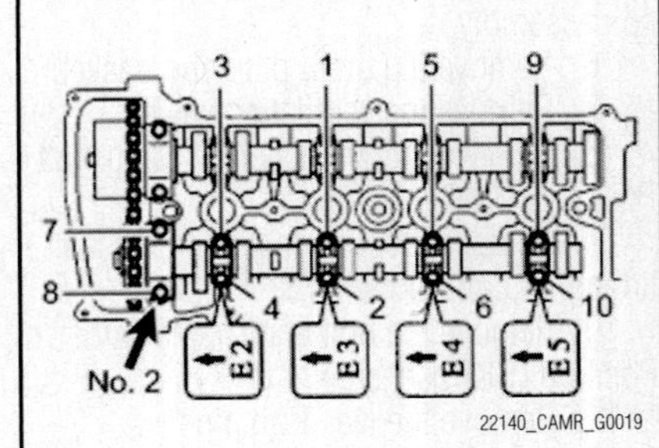

Fig. 49 No. 2 camshaft bearing cap tightening sequence

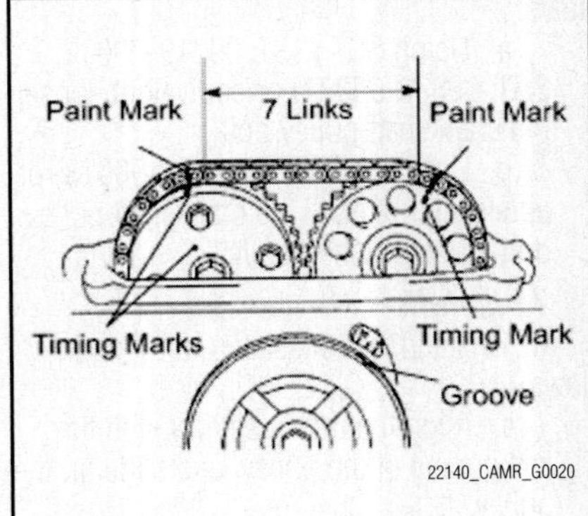

Fig. 50 Aligning paint marks on the chain with the timing marks on the camshaft timing gear and camshaft timing sprocket

g. While holding the camshaft with a wrench, tighten the camshaft timing sprocket set bolt and tighten to 40 ft. lbs. (54 Nm).

h. Check that the paint marks on the chain are aligned with the timing marks on the camshaft timing gear and camshaft timing sprocket. Also, check that the crankshaft pulley groove is aligned with the timing mark "0" of the timing chain cover.

22. To complete installation, reverse remaining removal procedure.
23. Check engine for oil leaks.
24. Connect the negative battery cable.

CRANKSHAFT DAMPER

REMOVAL & INSTALLATION

See Figure 51.

1. Before servicing the vehicle, refer to the Precautions Section.
2. Remove the engine assembly.
3. Install on engine stand.
4. Remove the oil filler cap and gasket.
5. Remove the spark plugs and ignition coil assembly.
6. Remove the drain plug and gasket.
7. Remove the ventilation valve.
8. Remove the 4 bolts and 4 Camshaft Position (CMP) sensors.
9. Remove the 4 bolts and 4 camshaft timing oil control valves.
10. Remove the bolt and Crankshaft Position (CKP) sensor.
11. Remove the No. 1 oil pipe.
12. Remove the oil pipe.
13. Remove the cylinder block water drain cock sub-assembly.
14. Remove the oil filter element.
15. Remove the crankshaft pulley, as follows:
    a. Using SST (SST: 09213-70011, SST: 09330-00021) or equivalent, loosen the crankshaft pulley bolt.
    b. Using SST (SST: 09950-50013) or equivalent, remove the crankshaft pulley bolt and crankshaft pulley.

To install:

16. Install the crankshaft pulley, as follows:
    a. Align the pulley set key with the key groove of the pulley, and slide on the pulley.
    b. Using SST (SST: 09213-70011, SST: 09330-00021) or equivalent, install the pulley bolt and tighten to 184 ft. lbs. (250 Nm).
17. Install the oil filter element.
18. Install the cylinder block water drain cock sub-assembly.
19. Install the No. 1 oil pipe.
20. Install the oil pipe.
21. Install the Crankshaft Position (CKP) sensor.
22. Install the camshaft timing oil control valve assembly.
23. Install the Camshaft Position (CMP) sensor.
24. Install the ventilation valve sub-assembly.
25. Install the spark plugs and the ignition coil assembly.
26. Install the oil filler cap sub-assembly.
27. Remove the engine stand.
28. Install the engine assembly.

CRANKSHAFT FRONT SEAL

REMOVAL & INSTALLATION

See Figures 52 through 56.

1. Before servicing the vehicle, refer to the Precautions Section.
2. Disconnect the negative battery cable.
3. Drain the engine oil
4. Loosen the lug nuts on the left front wheel.
5. Apply the parking brake, block the rear wheels, then raise and safely support the front of the vehicle securely on jack stands.
6. Remove the left front wheel.
7. Remove the right hand front fender apron seal.
8. Remove the left and right hand engine under cover.
9. Remove the drive belt.
10. Remove the crankshaft pulley and take note of the following:
    - For TMMK made crankshaft pulley, use Special Tool 09960-10010 to fix the pulley in place and to loosen the bolt. Use Special Tool 09950-40011 to remove the pulley and bolt.

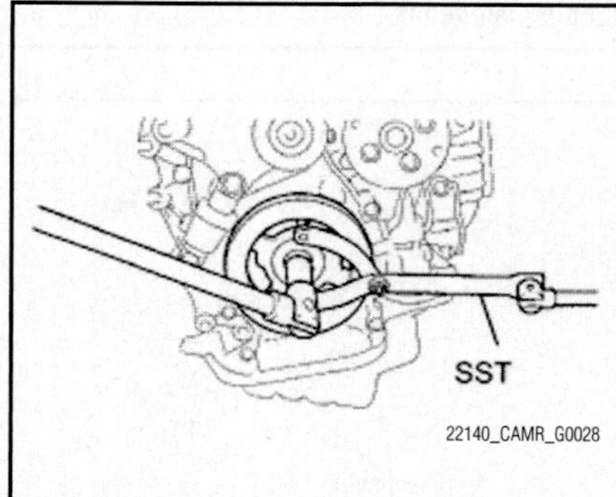

Fig. 52 Using Special Tool 09960-10010, fix the pulley in place and loosen the bolt—TMMK Pulley

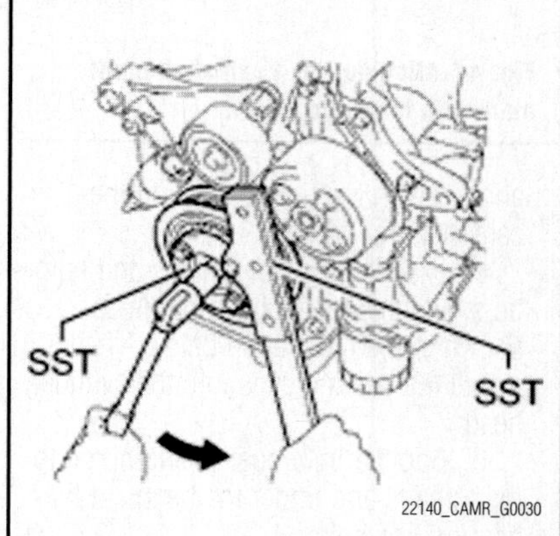

Fig. 54 Using Special Tools 09213-54015 and 09330-00021 to fix the pulley in place and to loosen the bolt—TMC Pulley

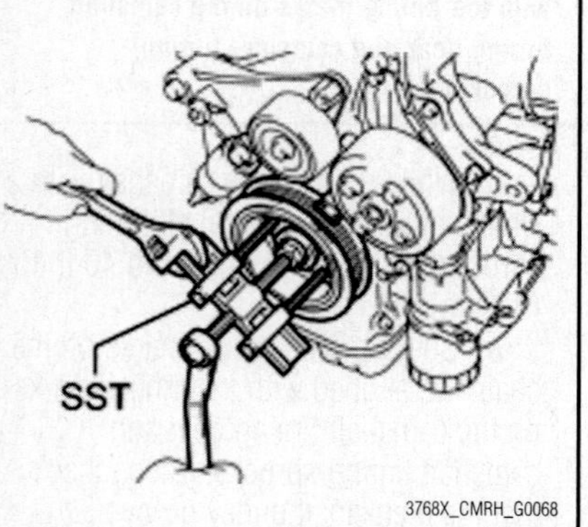

Fig. 51 Removing the crankshaft pulley

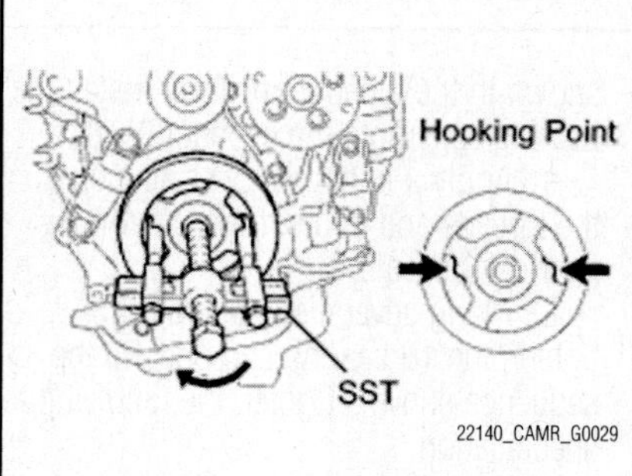

Fig. 53 Using Special Tool 09950-40011 to remove the pulley and bolt—TMMK Pulley

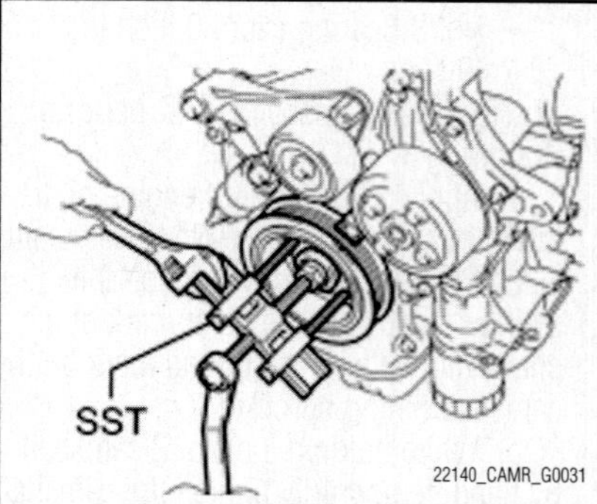

Fig. 55 Using Special Tools 09950-50013 and 09950-40011 to remove the pulley and bolt—TMC Pulley

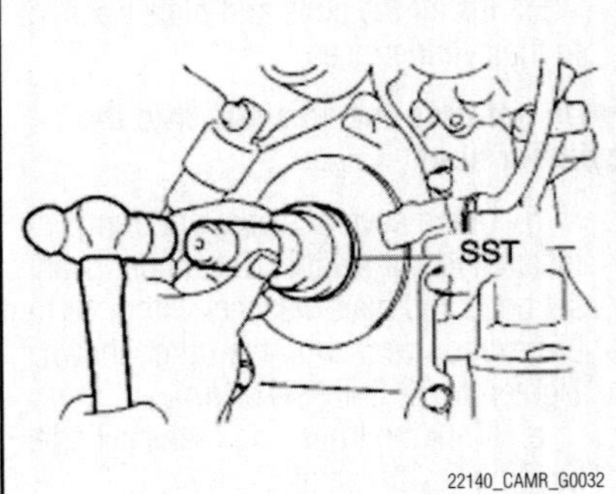

Fig. 56 Using Special Tool 09223-22010 and hammer to install new oil seal

- For TMC made crankshaft pulley, use Special Tools 09213-54015 and 09330-00021 to fix the pulley in place and to loosen the bolt. Use Special Tools 09950-50013 and 09950-40011 to remove the pulley and bolt.

11. Using a knife, cut off the oil seal lip. Using a screwdriver with the tip taped, pry out the oil seal.
12. After the removal, check the crankshaft for damage. If it is damaged, smooth the surface with 400-grit sandpaper.

To install:

13. Apply MP grease to a new oil seal lip. Keep the lip free from foreign matter.
14. Using Special Tool 09223-22010 and a hammer, tap in the oil seal until its surface is flush with the rear oil seal retainer edge.
15. For TMMK made pulleys, align the pulley set key with the key groove of the pulley. Using Special Tool 09960-10010, keep the pulley in place and tighten the bolt to: 125 ft. lbs. (170 Nm).
16. For TMC made pulleys, align the pulley set key with the key groove of the pulley. Using Special Tools 09213-54015 and 09330-00021, keep the pulley in place and tighten the bolt to 133 ft. lbs. (180 Nm).
17. To complete installation, reverse remaining removal procedure.
18. Check engine for oil leaks.

CYLINDER HEAD

REMOVAL & INSTALLATION

See Figures 57 through 67.

1. Discharge the fuel system pressure.
2. Remove the luggage trim service hole cover.
3. Disconnect the cable from the negative battery terminal.
4. Remove the engine under covers.
5. Remove the front fender apron seal RH.
6. Remove the No. 1 engine cover sub assembly.
7. Drain the engine coolant.
8. Drain the engine oil.
9. Remove the wiper arms and blade assemblies.
10. Remove the front fender to cowl side seals.
11. Remove the cowl top ventilator louver sub assembly.
12. Remove the windshield wiper motor and link assembly.
13. Separate the brake master cylinder reservoir sub assembly.
14. Remove the cowl top panel outer sub assembly.
15. Remove the air cleaner inlet assembly.
16. Remove the air cleaner cap sub assembly.
17. Remove the air cleaner case sub assembly.
18. Remove the throttle body assembly.
19. Disconnect the fuel tube sub assembly.
20. Remove the fuel delivery pipe with the injector.
21. Remove the intake manifold.
22. Remove the No. 1 intake manifold insulator.
23. Remove the front exhaust pipe assembly.
24. Remove the No. 2 engine mounting stay RH.
25. Remove the engine moving control rod sub assembly.
26. Remove the No. 2 engine mounting bracket RH.
27. Remove the V-ribbed belt tensioner cover sub assembly.
28. Remove the V-ribbed belt.
29. Remove the oil level gage sub assembly.
30. Remove the oil level gauge guide.
31. Remove the manifold stay.
32. Remove the No. 2 manifold stay.
33. Remove the exhaust manifold converter sub assembly.
34. Remove the chain sub assembly.
35. Remove the No. 2 camshaft.
    a. Using several steps, uniformly loosen and remove the 10 bearing cap bolts in the sequence shown.
    b. Remove the 5 bearing caps and No. 2 camshaft.
36. Remove the camshaft.

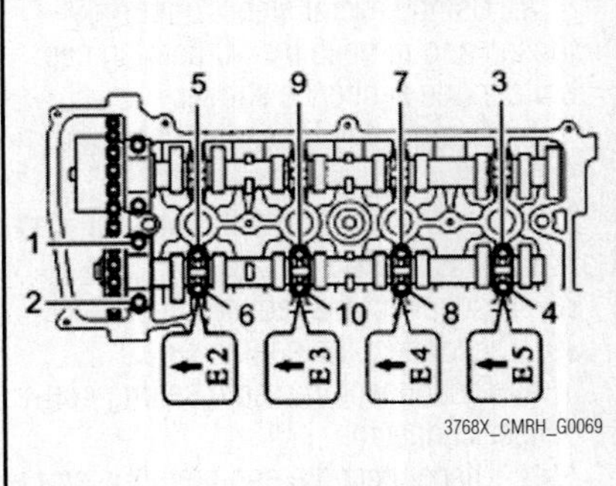

Fig. 57 Removing the bearing cap bolts in sequence

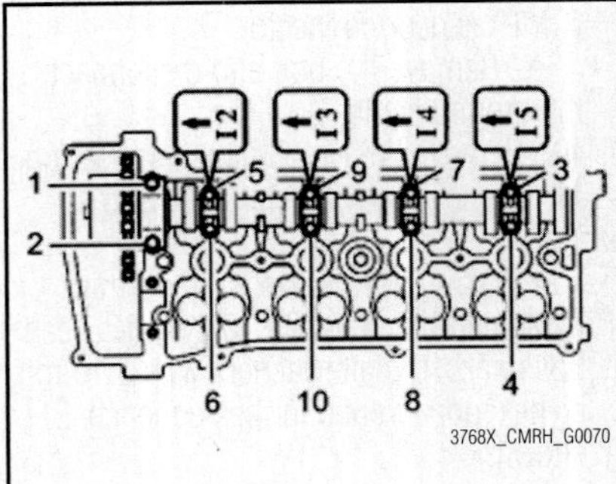

Fig. 58 Removing the 10 bearing cap bolts

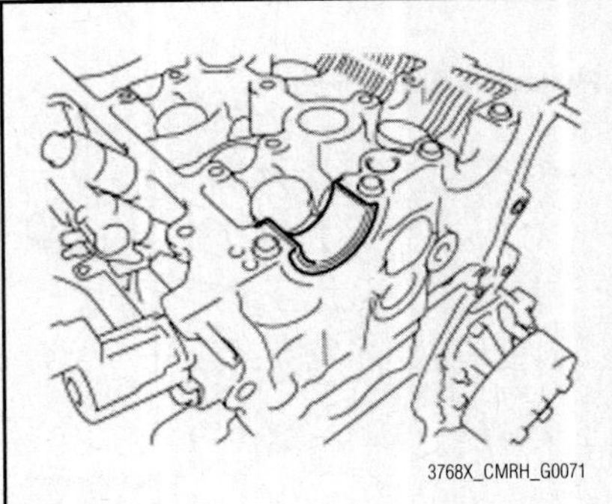

Fig. 59 Removing the No. 2 camshaft bearing

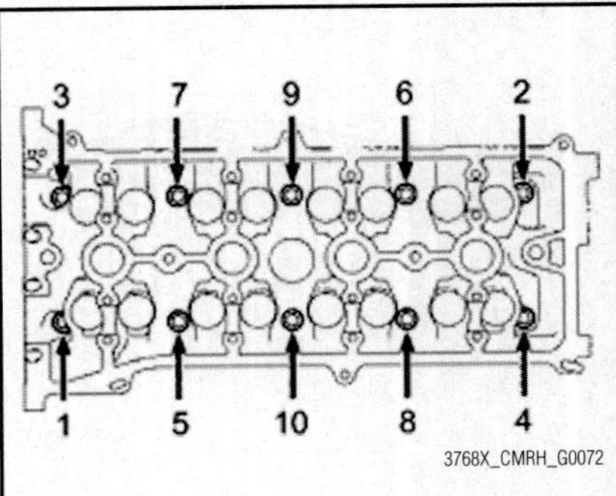

Fig. 60 Removing the 10 cylinder head bolts

a. Using several steps, uniformly loosen and remove the 10 bearing cap bolts in the sequence shown.

b. Remove the 5 bearing caps and camshaft.

37. Remove the camshaft timing oil control valve assembly.

38. Disconnect the radiator hose inlet.

39. Disconnect the engine wire.

a. Disconnect the radio setting condenser connector.

b. Disconnect the engine oil pressure switch connector.

c. Disconnect the engine coolant temperature sensor connector.

d. Disconnect the Camshaft Position (CMP) sensor connector.

e. Remove the bolt and disconnect the ground cable.

40. Remove the No. 2 camshaft bearing.

41. Remove the cylinder head sub assembly.

a. Using several steps, uniformly loosen and remove the 10 cylinder head bolts and 10 plate washers with a 10 mm bi-hexagon wrench in the sequence shown.

➡Head warpage or cracking could result from removing the bolts in the wrong order.

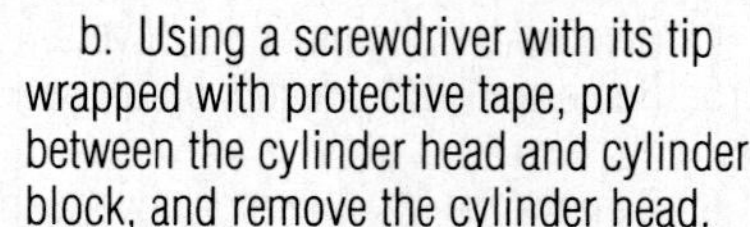

b. Using a screwdriver with its tip wrapped with protective tape, pry between the cylinder head and cylinder block, and remove the cylinder head.

➡Be careful not to damage the contact surfaces between the cylinder head and cylinder block.

42. Remove the cylinder head gasket.

To install:

43. Install the cylinder head gasket.

a. Place a new gasket on the cylinder block surface with the Lot No. stamp facing upward.

➡Remove any oil from the contact surfaces.

➡Make sure that the gasket is installed in the correct direction.

44. Install the cylinder head sub assembly.

a. Install the cylinder head on the cylinder block.

➡The cylinder head bolts are tightened in 2 progressive steps.

b. Apply a light coat of engine oil to the bolt threads and the area beneath the bolt heads that come in contact with the washers.

c. Install the bolts and plate washers to the cylinder head.

➡Do not drop the washers into the cylinder head.

d. Using several steps, uniformly install and tighten the 10 cylinder head set bolts and plate washers with a 10 mm bi-hexagon wrench in the order shown. Tighten to 52 ft. lbs. (70 Nm).

e. Mark the front side of the cylinder head bolts with paint.

f. Retighten the cylinder head bolts 90°in the sequence shown.

g. Check that the paint mark is now at a 90°angle from the front.

45. Connect the engine wire.

a. Connect the ground cable with the bolt. Tighten the bolt to 74 inch lbs. (8.4 Nm).

b. Connect the Camshaft Position (CMP) sensor connector.

c. Connect the engine coolant temperature sensor connector.

d. Connect the engine oil pressure switch connector.

e. Connect the radio setting condenser connector.

46. Connect the radiator hose inlet.

47. Install the No. 2 camshaft bearing.

Protective Tape

3768X_CMRH_G0073

Fig. 61 Removing the cylinder head

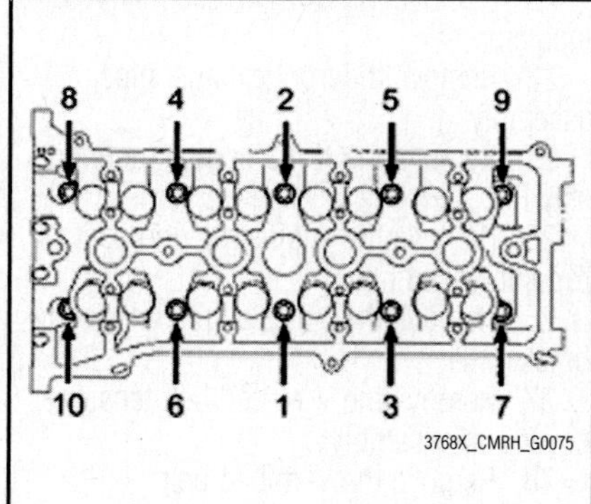

Fig. 63 Installing the cylinder head bolts

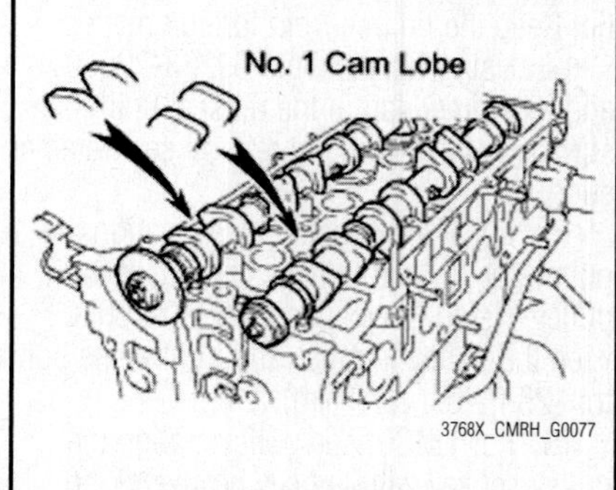

Fig. 65 Placing the camshafts on the cylinder head

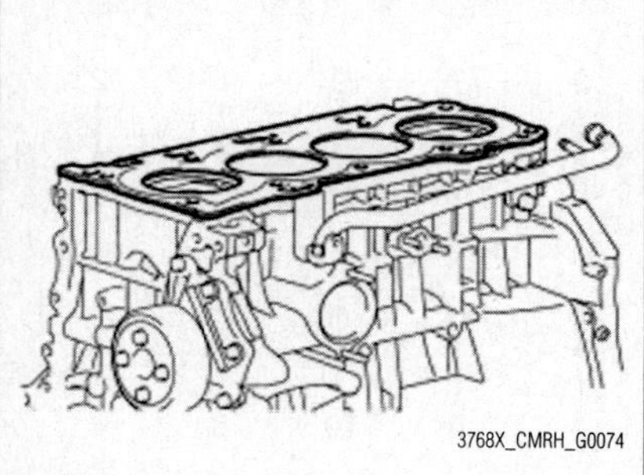

Fig. 62 Removing the cylinder head gasket

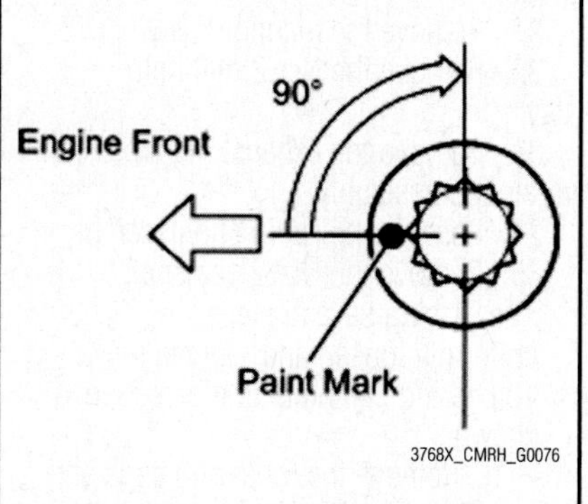

Fig. 64 Marking the cylinder head bolts

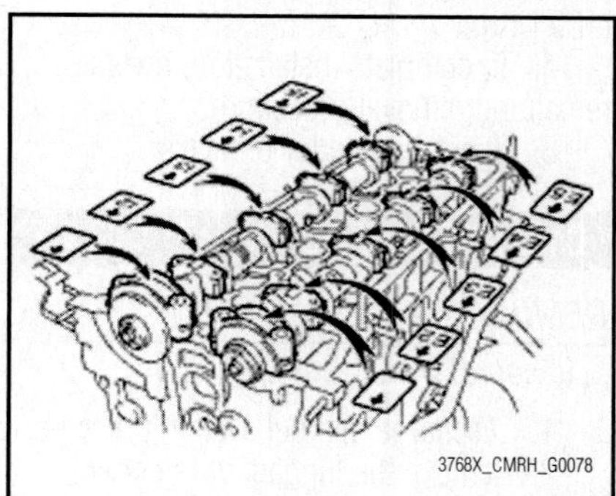

Fig. 66 Examining the front marks and numbers

a. Install the No. 2 camshaft bearing.

48. Install the camshaft timing oil control valve assembly.

49. Install the camshaft.

a. Apply a light coat of engine oil to the journal portion of the camshaft.

b. Place the 2 camshafts on the cylinder head with the No. 1 cam lobes facing the directions shown.

c. Examine the front marks and numbers, and check that the order is as shown in the illustration. Then install the bearing caps onto the cylinder head.

d. Apply a light coat of engine oil to the threads and under the heads of the bearing cap bolts.

e. Using several steps, uniformly tighten the 20 bearing cap bolts in the sequence shown. Tighten the No.1 and No. 2 bearing cap to 22 ft. lbs. (30 Nm). Tighten the No. 3 bearing cap to 80 inch lbs. (9 Nm).

50. Install the chain sub assembly.

51. Install the exhaust manifold converter sub assembly.

52. Install the No. 2 manifold stay.

53. Install the manifold stay.

54. Install the oil level gauge guide.

55. Install the oil level gauge sub assembly.

56. Install the V-ribbed belt.

57. Install the V-ribbed belt tensioner cover sub assembly.

58. Install the No. 2 engine mounting bracket RH.

59. Install the engine moving control rod sub assembly.

60. Install the No. 2 engine mounting stay RH.

61. Install the front exhaust pipe assembly.

62. Install the No. 1 intake manifold insulator.

63. Install the intake manifold.

64. Install the fuel delivery pipe with the injector.

65. Connect the fuel tube sub assembly.

66. Install the throttle body assembly.

67. Install the air cleaner case sub assembly.

68. Install the air cleaner cap sub assembly.

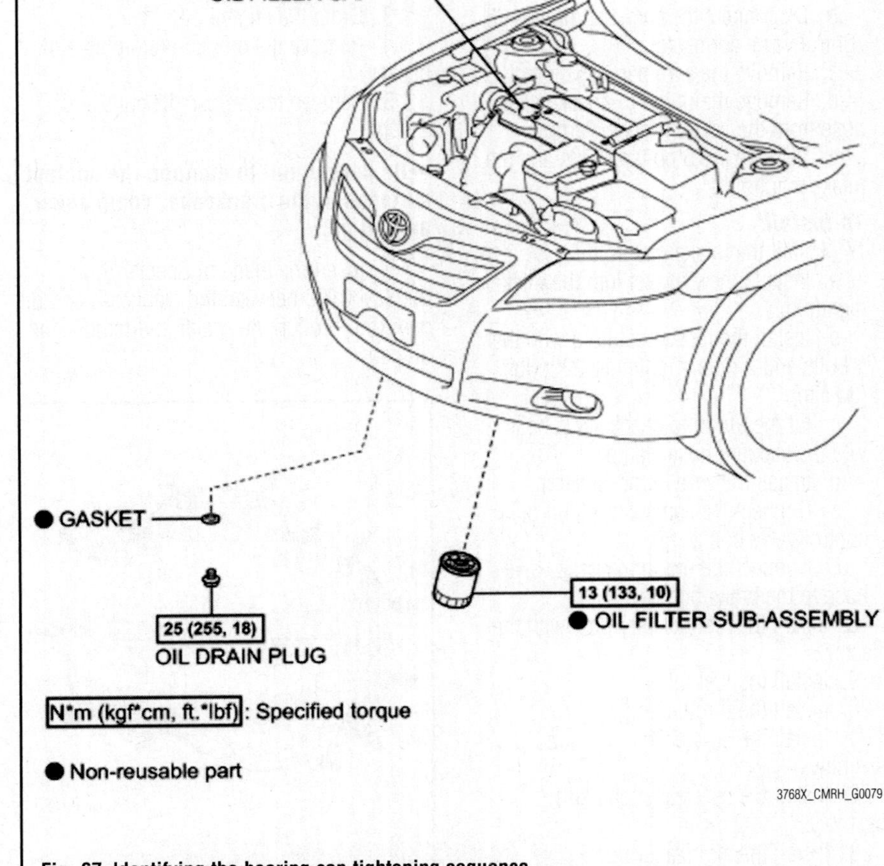

Fig. 67 Identifying the bearing cap tightening sequence

69. Install the air cleaner inlet assembly.

70. Install the cowl top panel outer sub assembly.

71. Install the brake master cylinder reservoir sub assembly.

72. Install the windshield wiper motor and link assembly.

73. Install the cowl top ventilator louver sub assembly.

74. Install the front fender to the cowl side seals.

75. Install the front wiper arm and blade assemblies.

76. Connect the cable to the negative battery terminal.

77. Install the luggage trim service hole cover.

78. Add engine oil.

79. Inspect for fuel leak.

80. Inspect for exhaust gas leak.

81. Inspect the ignition timing.

82. Inspect the idle speed.

83. Inspect the compression.

84. Inspect CO/HC.

85. Install the No. 1 engine cover sub assembly.

86. Install the front fender apron seal RH.

87. Install the engine under covers.

88. Install the front wheel RH.

89. Perform initialization.

FLEXPLATE

REMOVAL & INSTALLATION

See Figures 68 and 69.

1. Before servicing the vehicle, refer to the Precautions Section.

2. Disconnect the negative battery cable.

3. To gain access to the flywheel or flexplate, remove the transaxle.

a. For TMMK made flexplates: use Special Tool: 09960-10010 to hold the crankshaft.

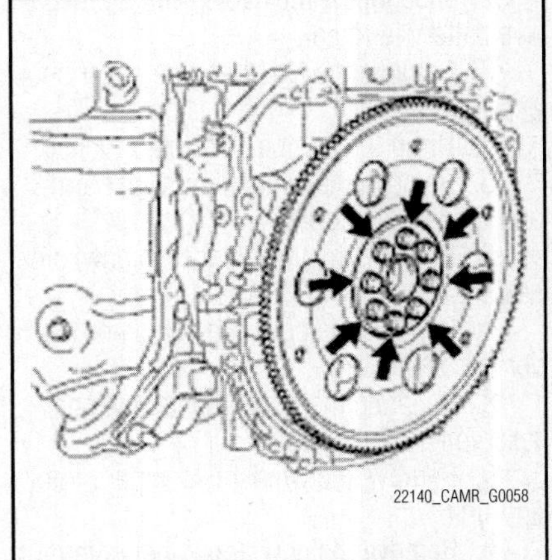

Fig. 68 Removing A/T flexplate bolts

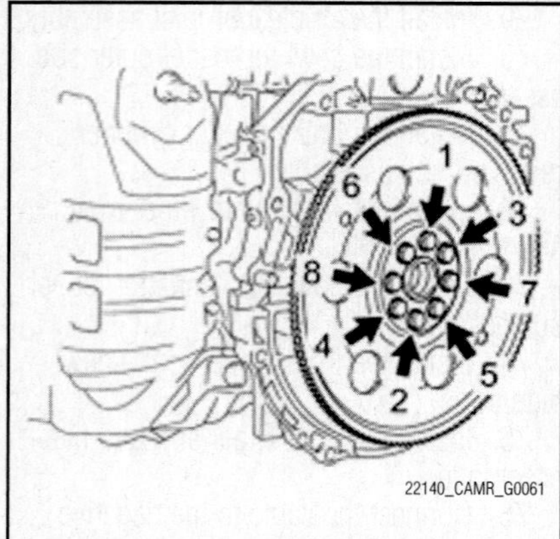

Fig. 69 A/T flywheel bolt tightening sequence

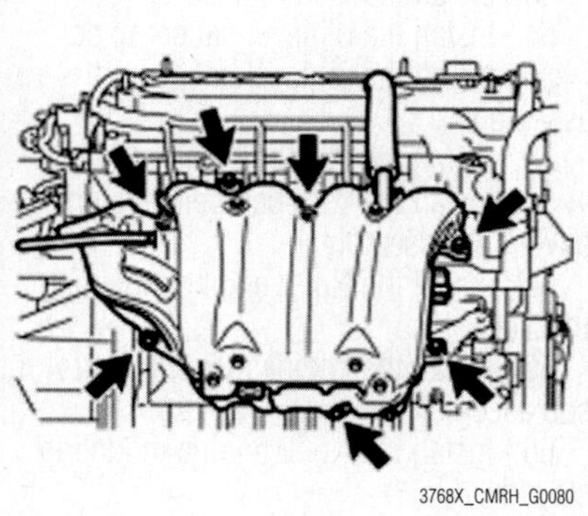

Fig. 70 Removing and installing the intake manifold

b. For TMC made flexplate: use Special Tools: 09213-54015and 09330-00021 to hold the crankshaft.

c. Remove the eight bolts, rear spacer, drive plate and front spacer.

To install:

4. For TMMK made flywheel: use Special Tool: 09960-10010 to hold the crankshaft.
5. For TMC made flywheels: use Special Tools: 09213-54015and 09330-00021 to hold the crankshaft.
6. Clean the bolt and the bolt hole.
7. Apply adhesive to 2 or 3 threads of the bolt end. Adhesive: Part No. 08833-00070, Three Bond or equivalent.
8. Install the front spacer, drive plate and rear spacer with the 8 bolts. Uniformly tighten the eight bolts in the sequence. Tighten 72 ft. lbs. (98 Nm).

INTAKE MANIFOLD

REMOVAL & INSTALLATION

See Figure 70.

1. Disconnect the fuel system pressure.
2. Disconnect the cable from the negative battery terminal.
3. Remove the No. 1 engine cover sub assembly.
4. Drain the engine coolant.
5. Remove the windshield wiper arm and blade assemblies.
6. Remove the front fender to cowl side seals.
7. Remove the cowl top ventilator louver sub assembly.
8. Separate the brake master cylinder reservoir sub assembly.
9. Remove the windshield wiper motor and link.
10. Remove the cowl top panel outer sub assembly.
    a. Remove the 4 bolts, 4 nuts and the cowl top panel outer sub assembly.
11. Remove the air cleaner cap sub assembly.
12. Remove the air cleaner case sub assembly.
13. Remove the throttle body.
14. Disconnect the fuel tube.
15. Remove the fuel delivery pipe with the injector.
16. Remove the intake manifold.
    a. Disconnect the union to check valve hose from the brake booster.
    b. Disconnect the camshaft timing oil control valve connector.
    c. Remove the wire harness clamp.
    d. Remove the union to check valve hose from the vacuum hose clamp.
    e. Remove the 5 bolts, 2 nuts and the intake manifold.

To install:

17. Install the intake manifold.
    a. Install a new gasket into the intake manifold.
    b. Install the intake manifold with the 5 bolts and 2 nuts. Tighten to 22 ft. lbs. (30 Nm).
    c. Fit the union to check valve hose into the vacuum hose clamp.
    d. Install the wire harness clamp.
    e. Connect the camshaft timing oil control valve connector.
    f. Connect the union to check valve hose to the brake booster.
18. Install the fuel delivery pipe with the injector.
19. Install the fuel tube.
20. Install the throttle body.
21. Install the air cleaner case sub assembly.
22. Install the air cleaner cap sub assembly.
23. Install the air cleaner inlet assembly.
24. Connect the cable to the negative battery terminal.
25. Add engine coolant.
26. Inspect for coolant leaks.
27. Inspect for fuel leaks.
28. Install the No. 1 engine cover sub assembly.
29. Install the cowl top panel outer sub assembly. Tighten the bolts to 44 inch lbs. (5 Nm). Tighten the nuts to 63 ft. lbs. (85 Nm).
30. Install the brake master cylinder reservoir sub assembly.
31. Install the windshield wiper motor and link.
32. Install the cowl top ventilator louver sub assembly.
33. Install the front fender to cowl side seals.
34. Install the windshield wiper arm and blade assemblies.
35. Perform initialization.

OIL PAN

REMOVAL & INSTALLATION

See Figures 71 and 72.

1. Before servicing the vehicle, refer to the Precautions Section.
2. Disconnect the negative battery cable.
3. Drain the engine oil.
4. Remove the oil pan drain plug and gasket.
5. Remove the oil pan 12 bolts and 2 nuts.

➡Be careful not to damage the contact surfaces of the crankcase, chain cover and oil pan.

6. Insert the blade of Special Tool: 09032-00100 between the crankcase and oil pan. Cut through the sealer and remove the oil pan.

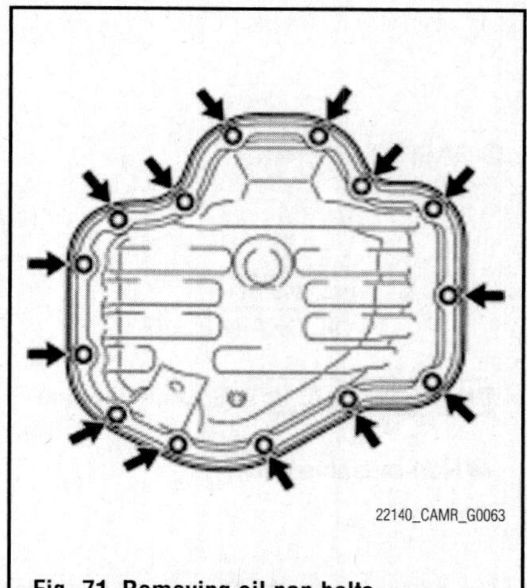

Fig. 71 Removing oil pan bolts

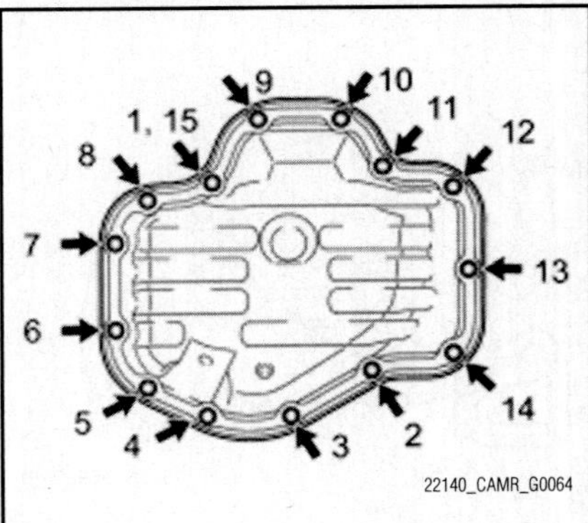

Fig. 72 Oil pan bolt and nut tightening sequence

To install:

7. Remove any old packing material and be careful not to drop any oil on the contact surfaces of the cylinder block and oil pan.
8. Apply a continuous bead of seal packing (diameter 0.118 to 0.157 inches (3.0 to 4.0 mm)). Use Toyota Genuine Seal Packing Block, Three Bond 1207B or Equivalent.

➡Remove any oil from the contact surfaces. Install the oil pan within 3 minutes after applying seal packing. Do not start the engine for at least 2 hours after installing.

9. Uniformly tighten the 12 bolts and 2 nuts in sequence. Tighten the bolts and nuts to 80 inch lbs. (9 Nm).
10. To complete installation, reverse remaining removal procedure.
11. Add oil and check for leaks.

OIL PUMP

REMOVAL & INSTALLATION

See Figure 73.

1. Before servicing the vehicle, refer to the Precautions Section.
2. Disconnect the negative battery cable.
3. Drain the engine oil.
4. Remove or disconnect the following:
   - Plastic engine cover and two nuts
   - Front right wheel
   - LH and RH engine under cover
   - RH front fender apron seal
   - Front exhaust pipe assembly
   - RH engine mount stay
   - Engine mount
   - RH engine mounting bracket
   - Drive belt.
   - Generator assembly
   - Vane pump
   - Ignition coil assembly
   - Ventilation hoses
   - Valve cover
5. Turn the crankshaft pulley until its groove and the timing mark "0" of the timing chain cover are aligned.
6. Check that each timing mark of the camshaft timing gear and sprocket is aligned with each timing mark located on the intake and exhaust bearing caps. If not, turn the crankshaft by 1 revolution (360°) to align the timing marks.
7. Remove or disconnect the following:
   - Crankshaft pulley
   - Crankshaft Position (CKP) sensor
   - Oil pan
   - Chain upper tensioner assembly
8. Install the No. 1 engine hanger (12281-28010) and No. 2 engine hanger (12282-28010) with the bolts (91512-61020) and tighten to: 28 ft. lbs. (38 Nm).
9. Remove or disconnect the following:
   - Drive belt tensioner
   - Engine mount insulator
   - RH engine mount bracket
   - Using a E10 Torx® socket, remove the stud bolt for the drive belt tensioner from the cylinder block
   - Timing chain cover twelve bolts and two nuts

➡Be careful not to damage the contact surfaces of the timing chain cover, cylinder block and cylinder head. Tape the screwdriver tip before use.

a. Timing chain cover by prying between the timing chain cover and cylinder head or cylinder block with a screwdriver.

➡Tape the screwdriver tip before use.

b. Using a screwdriver and a hammer, tap out the timing chain case oil seal.
c. Crankshaft Position (CKP) sensor plate.
d. Chain tensioner slipper.
e. Chain vibration damper.
f. Timing chain guide.
g. Upper chain assembly.
h. Crankshaft timing sprocket.
i. Lower chain assembly.
j. Three bolts, oil pump and gasket.

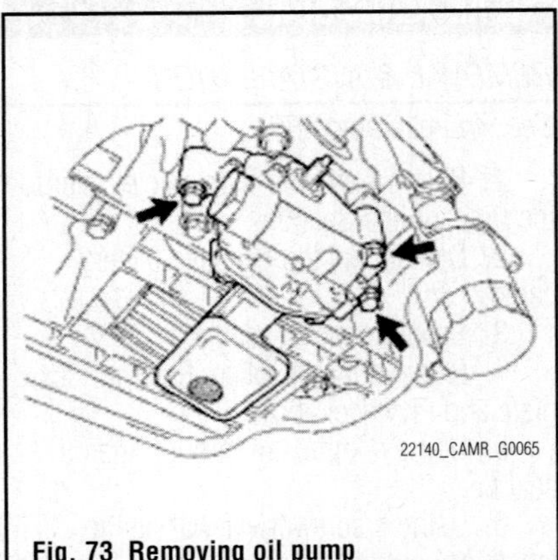

Fig. 73 Removing oil pump

To install:

10. Install a new gasket and the oil pump with the 3 bolts. Tighten the bolts 14 ft. lbs. (19 Nm).
11. To complete installation, reverse removal procedure. Refer to the appropriate sections to install components correctly.

INSPECTION

See Figures 74 through 77.

1. Check the oil jet located above the timing chain sprocket for damage or clogging. If necessary, repair the cylinder block.

➡If the valve does not fall smoothly, replace the relief valve. If necessary, replace the oil pump assembly.

2. Coat the relief valve with engine oil, and then check that the valve falls smoothly into the valve hole under its own weight.

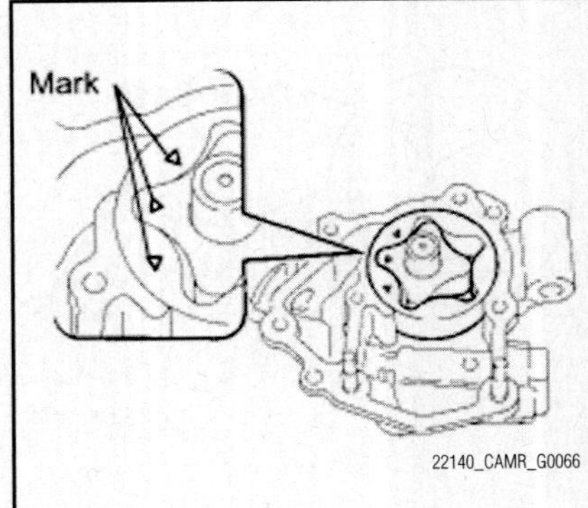

Fig. 74 Aligning the oil pump driven rotors and oil pump cover side marks

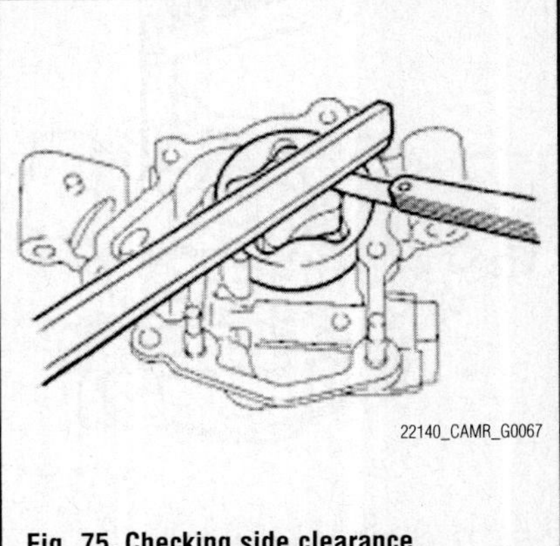

Fig. 75 Checking side clearance

3. Install the oil pump rotor. Coat the drive rotor and driven rotor with engine oil.
4. Place the drive and driven rotors into the oil pump with the marks facing the pump cover side.
5. To check the side clearance, perform the following:
 a. Using a feeler gauge and precision straightedge, measure the clearance between the rotors and precision straightedge.
 b. Standard clearance: 0.0012 to 0.0033 inches (0.030 to 0.085 mm).
 c. Maximum clearance: 0.0063 inches (0.16 mm).
 d. If the side clearance is greater than the maximum, replace the oil pump assembly.
6. To check the tip clearance, perform the following:
 a. Using a feeler gauge, measure the clearance between the drive and driven rotor tips.
 b. Standard clearance: 0.0031 to 0.0063 inches (0.080 to 0.160 mm).
 c. Maximum clearance: 0.0138 inches (0.35 mm).

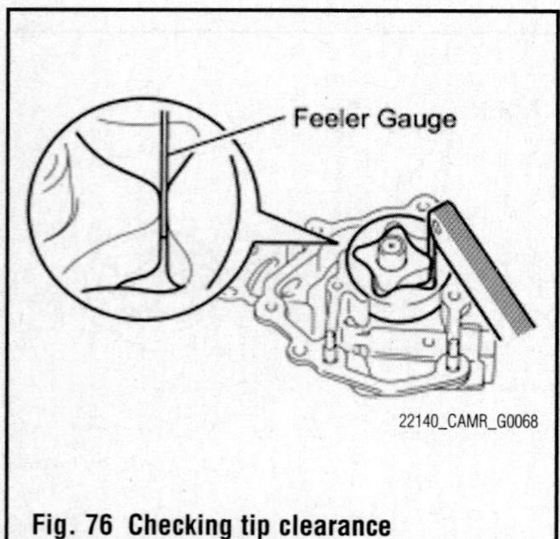

Fig. 76 Checking tip clearance

Fig. 77 Checking body clearance

 d. If the tip clearance is greater than the maximum, replace the oil pump assembly.
7. To check the body clearance, perform the following:
 a. Using a feeler gauge, measure the clearance between the driven rotor and pump body.
 b. Standard clearance: 0.0039 to 0.0067 inches (0.100 to 0.170 mm).
 c. Maximum clearance: .0128 inches. (0.325 mm).
 d. If the body clearance is greater than the maximum, replace the oil pump assembly.
8. Inspect the lower chain assembly and replace as necessary.
9. Inspect the oil pump drive sprocket and replace as necessary.
10. Inspect the chain tensioner plate and replace as necessary.

PISTON AND RING

POSITIONING

See Figure 78.

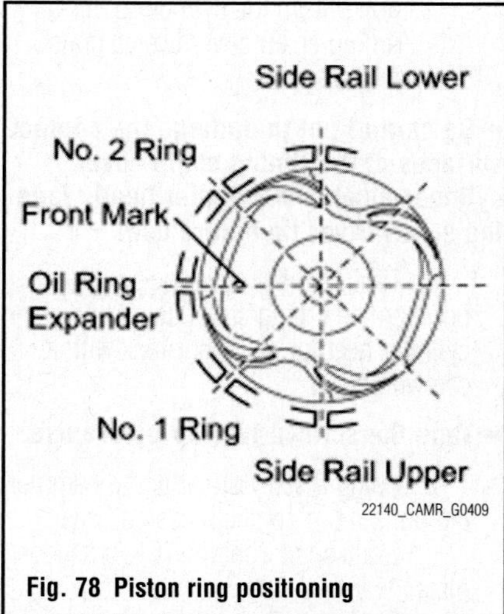

Fig. 78 Piston ring positioning

REAR MAIN SEAL

REMOVAL & INSTALLATION

See Figures 79 and 80.

1. Before servicing the vehicle, refer to the Precautions Section.
2. Disconnect the negative battery cable.
3. Separate the transaxle.
4. Remove the flexplate. Refer to Flexplate and Flywheel above.
5. Using a knife, cut through the oil seal lip.
6. Using a screwdriver with its tip taped, pry out the oil seal.

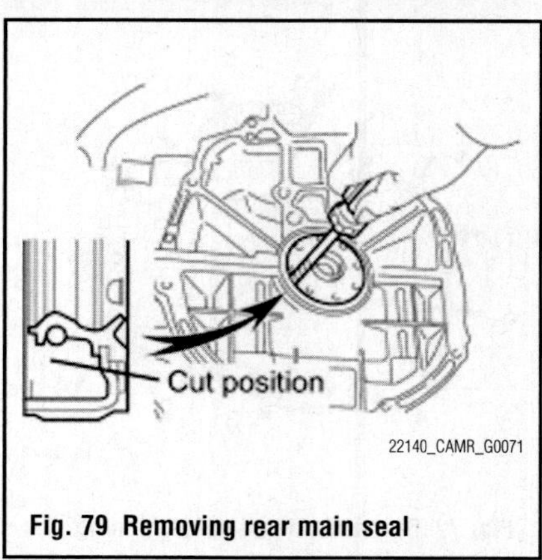

Fig. 79 Removing rear main seal

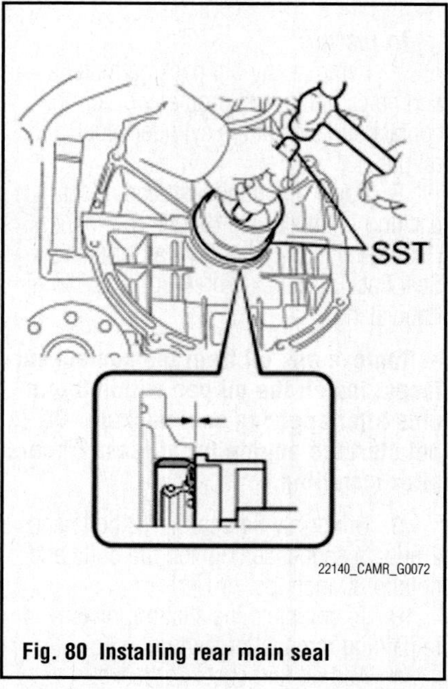

Fig. 80 Installing rear main seal

7. After the removal, check the crankshaft for damage. If it is damaged, smooth the surface with 400-grit sandpaper.

To install:

➡Keep the lip free from foreign matter.

8. Apply MP grease to a new oil seal lip.
9. Using Special Tools: 09223-15030, 09950-70010 and a hammer, tap in the oil seal until its surface is flush with the rear oil seal retainer edge. Wipe off extra grease from the crankshaft.
10. To complete installation, reverse remaining removal procedure.

TIMING CHAIN & SPROCKETS

REMOVAL & INSTALLATION

See Figures 81 through 103.

1. Before servicing the vehicle, refer to the Precautions Section.

Fig. 81 Removing the Crankshaft Position (CKP) sensor plate

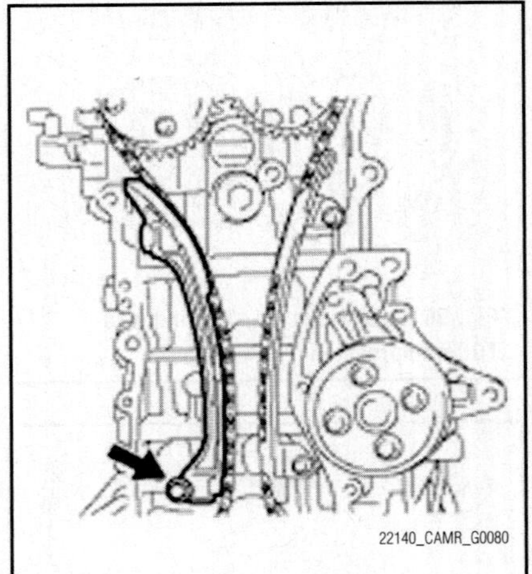

Fig. 82 Removing the bolt and chain tensioner slipper

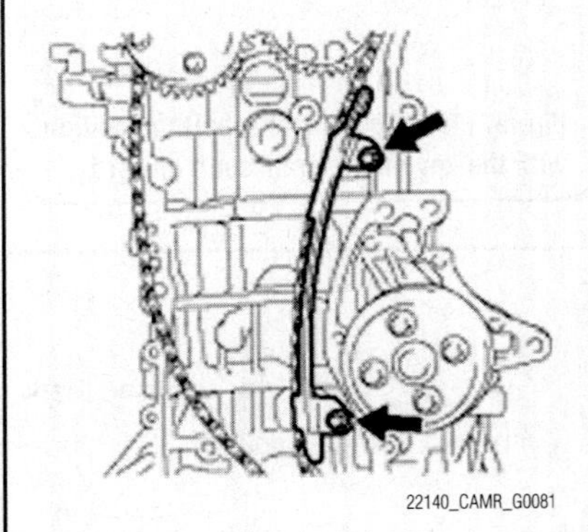

Fig. 83 Removing the two bolts and upper chain vibration damper

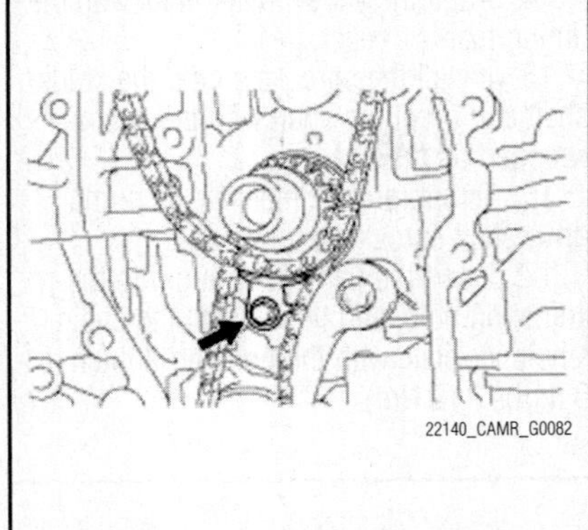

Fig. 84 Removing the bolt and timing chain guide

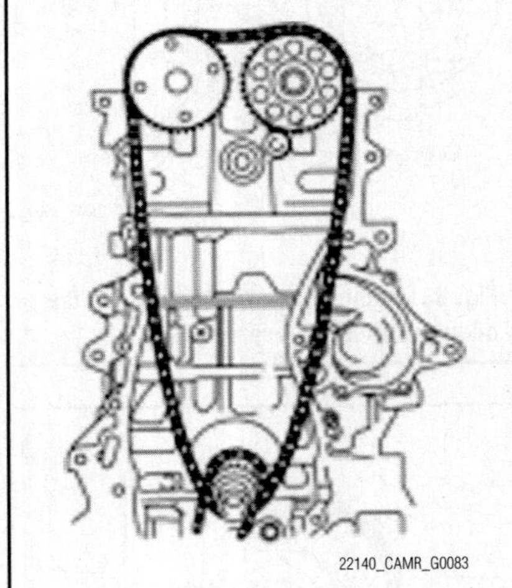

Fig. 85 Removing the upper chain assembly

Fig. 86 Removing the crankshaft timing sprocket

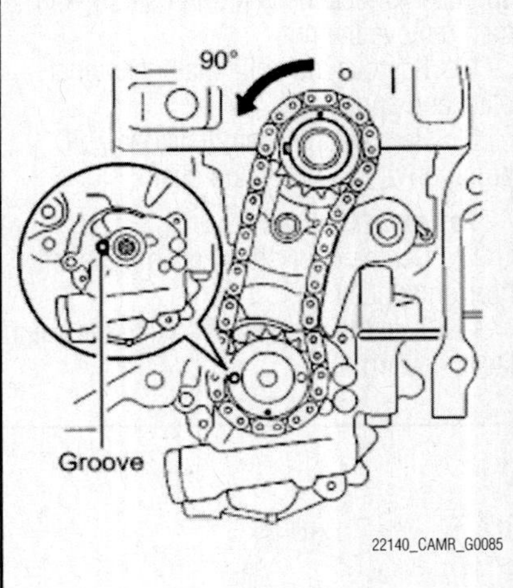

Fig. 87 Turning the crankshaft by 90°counterclockwise to align the adjusting hole

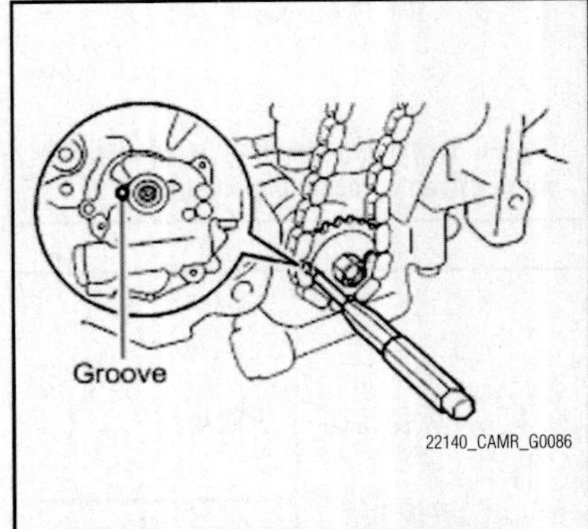

Fig. 88 Locking the gear in position

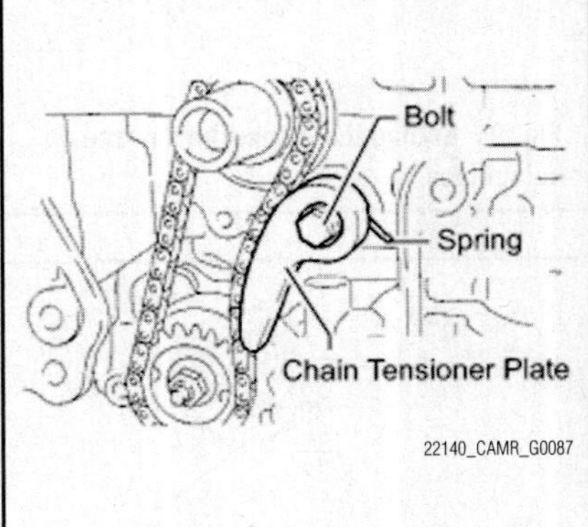

Fig. 89 Removing the bolt, chain tensioner plate and spring

2. Remove the Crankshaft Position (CKP) sensor plate.

3. Remove the bolt and chain tensioner slipper.

4. Remove the two bolts and chain vibration damper.

5. Remove the bolt and timing chain guide.

6. Remove the upper chain assembly.

7. Remove the crankshaft timing sprocket.

8. To remove the lower chain assembly. Turn the crankshaft by 90°counterclockwise to align the adjusting hole of the oil pump drive shaft sprocket with the groove of the oil pump.

9. Insert a 4 mm diameter bar into the adjusting hole of the oil pump drive shaft

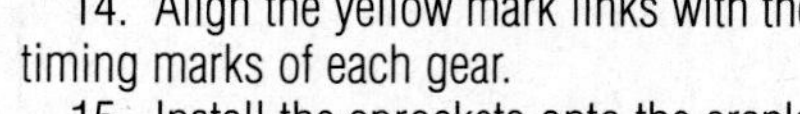

sprocket to lock the gear in position, and then remove the nut.

10. Remove the bolt, chain tensioner plate and spring.

11. Remove the chain tensioner, oil pump driven sprocket and chain.

To install:

12. Set the crankshaft key into the left horizontal position.

13. Turn the drive shaft so that the cutout faces upward.

14. Align the yellow mark links with the timing marks of each gear.

15. Install the sprockets onto the crankshaft and oil pump shaft with the chain wrapped on the gears.

16. Temporarily tighten the oil pump drive shaft sprocket with the nut.

17. Insert the damper spring into the adjusting hole, and then install the chain tensioner plate with the bolt and tighten to 9 ft. lbs. (12 Nm).

18. Align the adjusting hole of the oil pump drive shaft sprocket with the groove of the oil pump.

19. Insert a 4 mm diameter bar into the adjusting hole of the oil pump drive shaft gear to lock the gear in position, and then tighten the nut to 22 ft. lbs. (30 Nm).

20. Rotate the crankshaft clockwise by 90°, and align the crankshaft key to the top.

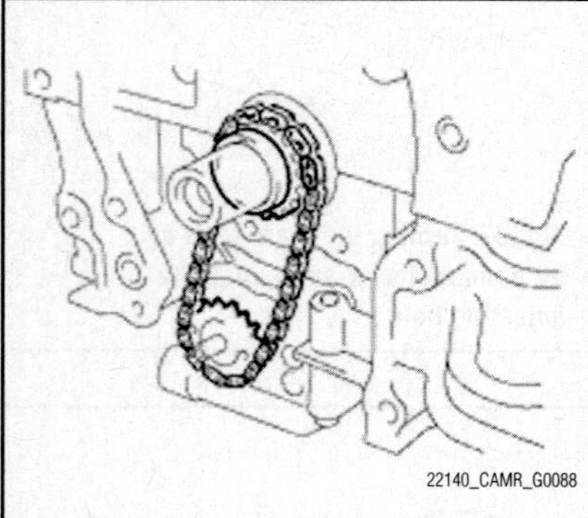

Fig. 90 Removing the chain tensioner, oil pump driven sprocket and chain

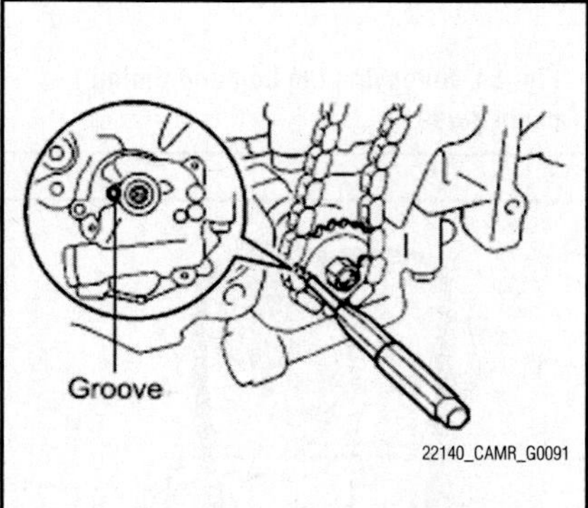

Fig. 93 Aligning the adjusting hole of the oil pump drive shaft sprocket

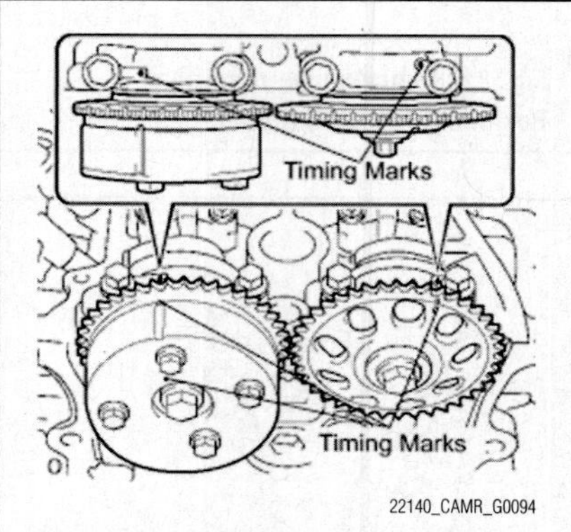

Fig. 96 Setting the No. 1 cylinder to TDC/compression

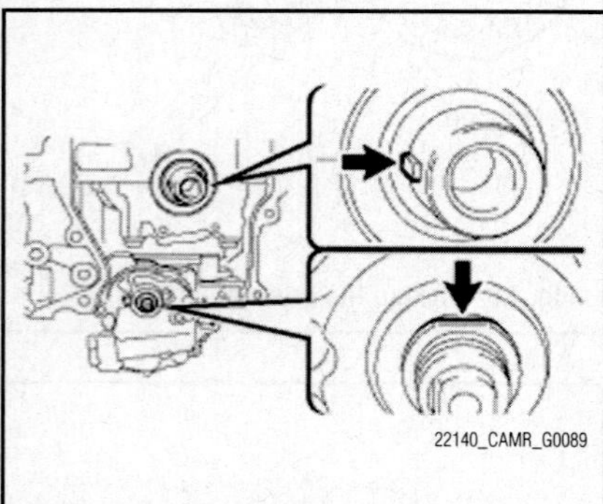

Fig. 91 Setting the crankshaft key and cutout face

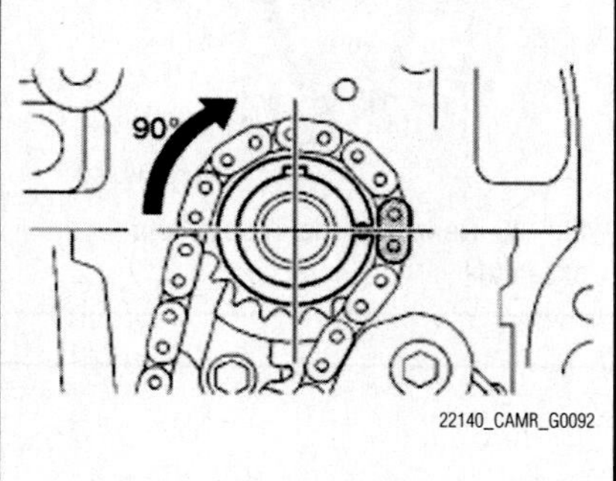

Fig. 94 Rotating the crankshaft clockwise by 90°

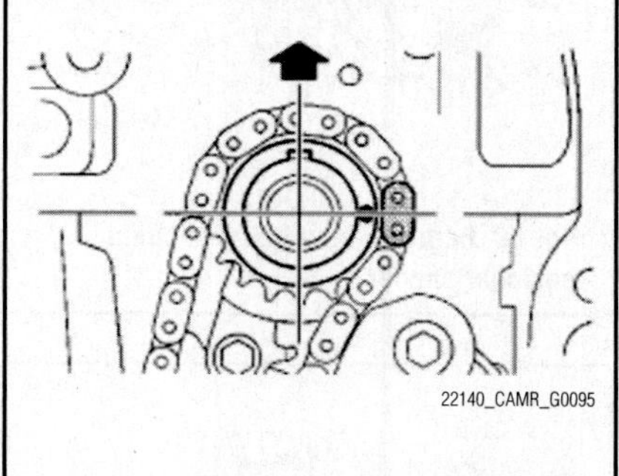

Fig. 97 Turning the crankshaft to position with the key on the crankshaft upward

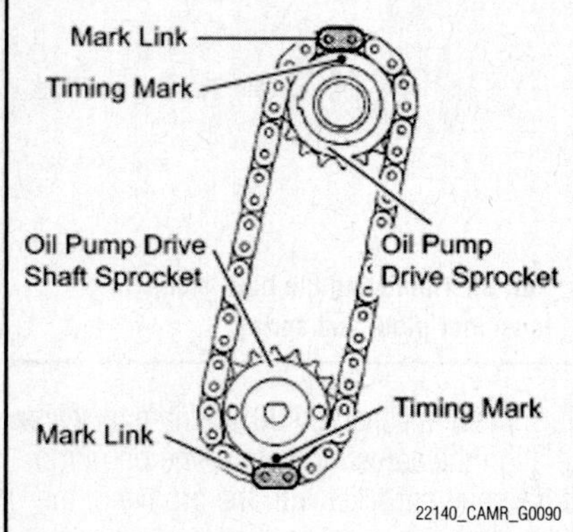

Fig. 92 Aligning the yellow mark links with the timing marks of each gear

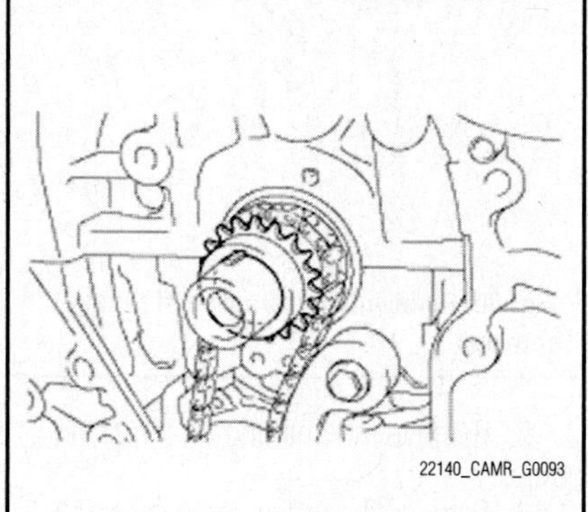

Fig. 95 Installing the crankshaft timing sprocket

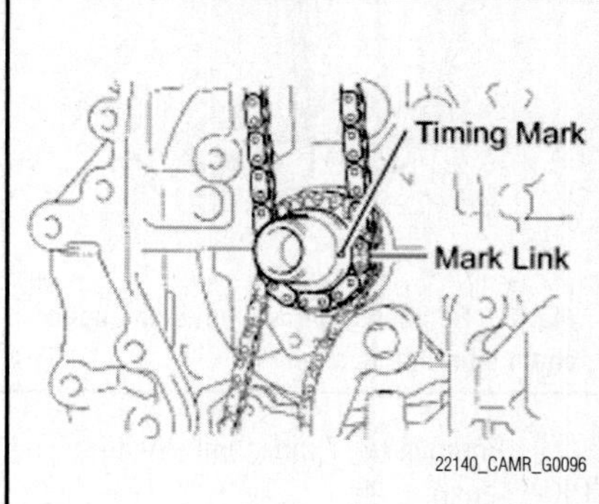

Fig. 98 Installing the chain onto the crankshaft timing sprocket with the gold or pink mark link

21. Install the crankshaft timing sprocket.
22. Install the chain vibration damper with the 2 bolts and tighten to 80 inch lbs. (9 Nm).
23. Set the No. 1 cylinder to TDC/compression.
24. Turn the camshafts with a wrench (using the hexagonal lobe) to align the timing marks of the camshaft timing gear with each timing mark located on the No. 1 and No. 2 bearing caps.
25. Using the crankshaft pulley bolt, turn the crankshaft to position with the key on the crankshaft upward.
26. Install the chain onto the crankshaft timing sprocket with the gold or pink mark link aligned with the timing mark on the crankshaft.
27. Using Special Tool: 09309-37010 and a hammer, tap in the crankshaft timing sprocket.
28. Align the gold or yellow link with each timing mark located on the camshaft timing gear and sprocket, then install the chain.
29. Install the chain tensioner slipper with the bolt and tighten to: 14 ft. lbs. (19 Nm).
30. Install the timing chain guide with the bolt and tighten to: 80 inch lbs. (9 Nm).
31. Install the crankshaft sensor plate with the "F" mark facing forward.

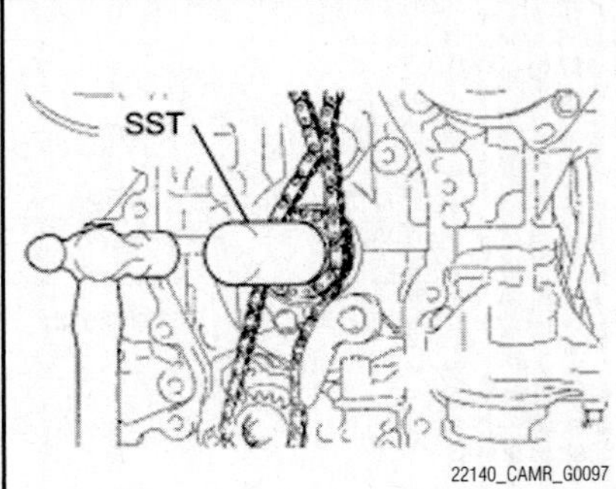
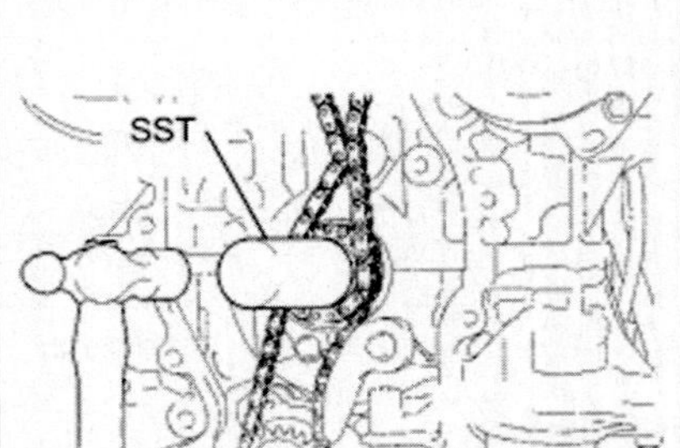

Fig. 99 Using Special Tool: 09309-37010 and a hammer to install the crankshaft timing sprocket

Fig. 100 Aligning the gold or yellow link with each timing mark located on the camshaft timing gear and sprocket

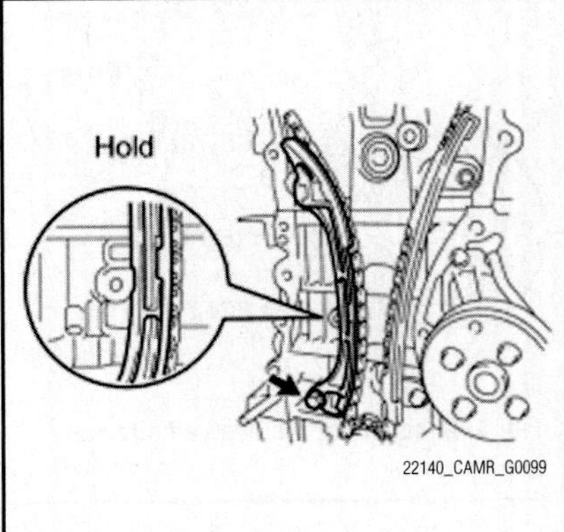

Fig. 101 Installing the chain tensioner slipper and bolt

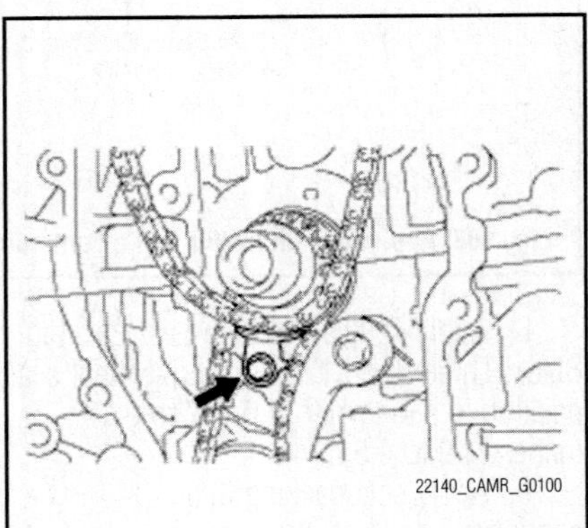

Fig. 102 Installing the timing chain guide and bolt

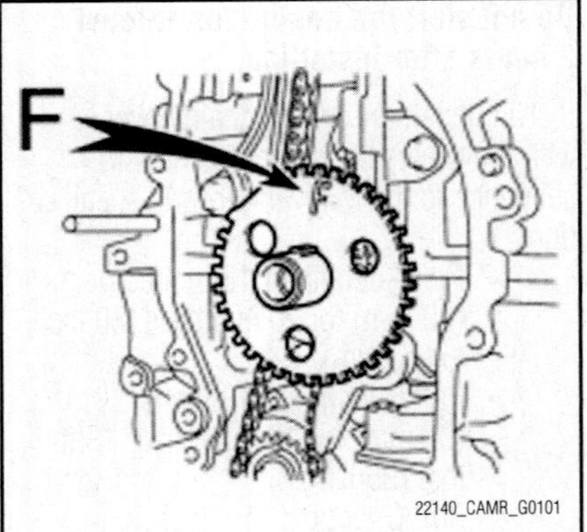

Fig. 103 Installing the crankshaft sensor plate with the "F" mark facing forward

TIMING CHAIN COVER & SEAL

REMOVAL & INSTALLATION

See Figures 104 through 109.

1. Before servicing the vehicle, refer to the Precautions Section.
2. Using a E10 Torx® socket, remove the stud bolt for the drive belt tensioner from the cylinder block.
3. Remove the 12 bolts and 2 nuts.

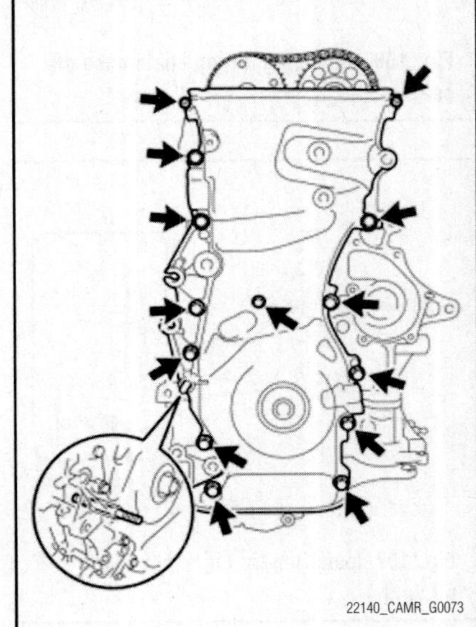

Fig. 104 Locating timing chain cover bolts and nuts

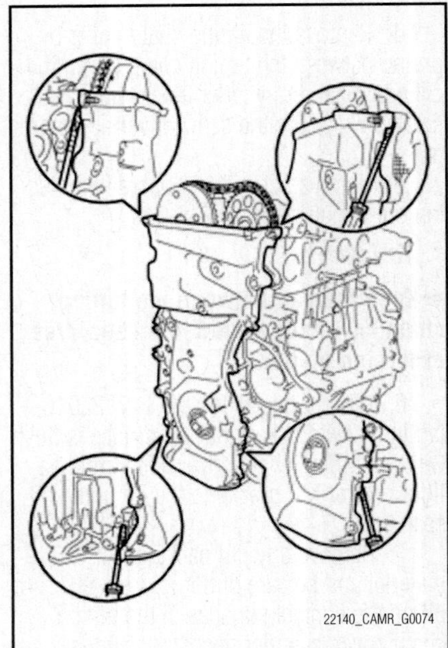

Fig. 105 Removing timing chain cover

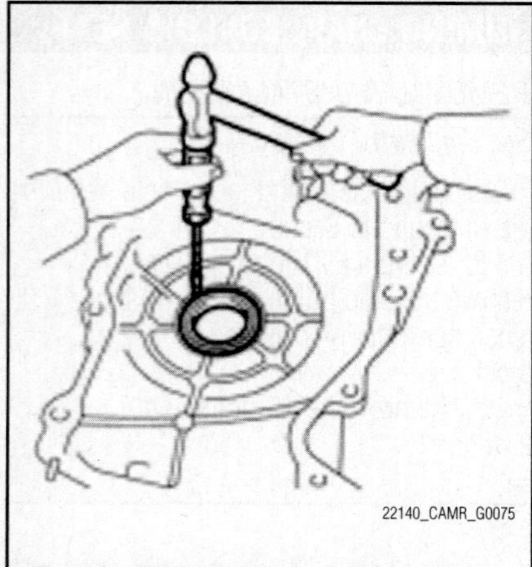

Fig. 106 Removing timing chain case oil seal

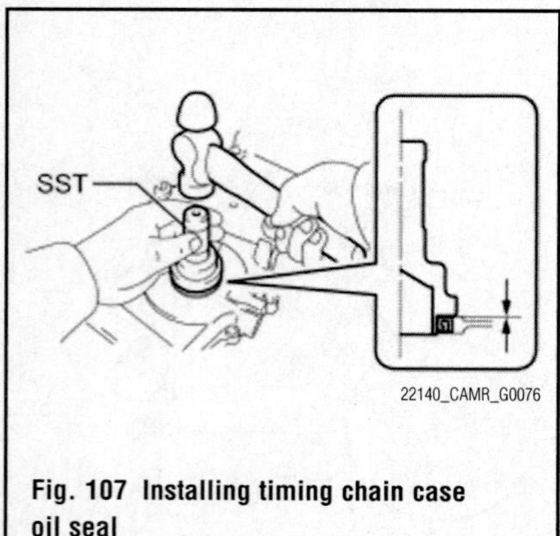

Fig. 107 Installing timing chain case oil seal

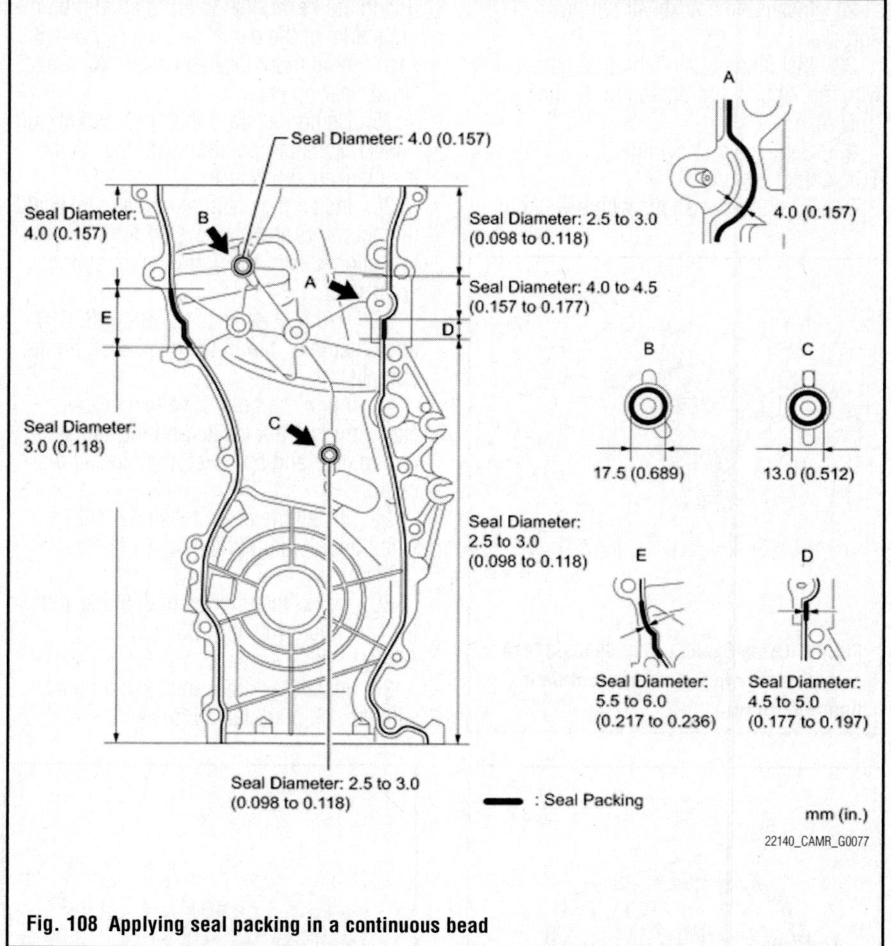

Fig. 108 Applying seal packing in a continuous bead

➡Be careful not to damage the contact surfaces of the timing chain cover, cylinder block and cylinder head.

4. Remove the timing chain cover by prying between the timing chain cover and cylinder head or cylinder block with a screwdriver. Tape the screwdriver tip before use.

5. Using a screwdriver and a hammer, tap out the oil seal.

To install:

➡Keep the gap between the timing chain cover edge and the oil seal free of foreign matter.

6. Using Special Tool: 09223-22010, tap in a new oil seal until its surface is flush with the timing chain cover edge. Apply a light coat of MP grease to the lip of the oil seal.

7. Remove any old packing (FIPG) material and be careful not to drop any oil on the contact surfaces of the timing chain cover, cylinder head and cylinder block.

8. Apply Toyota Genuine Seal Packing Black, Three Bond 1207B or Equivalent seal packing in a diameter of 0.157 to 0.177 inches (4.0 to 4.5 mm).

9. Apply seal packing in a continuous bead as shown in the illustration.

➡Remove any oil from the contact surface. Install the chain cover within 3 minutes after applying seal packing. Do not start the engine for at least 2 hours after installing.

10. Install the timing chain cover with the twelve bolts and two nuts in sequence to the following torque specification:

- Bolt A length: 1.18 inches (30 mm) for 10 mm head: 80 inch lbs. (9 Nm)
- Bolt B length: 1.18 inches (30 mm) for 12 mm head: 18 ft. lbs. (25 Nm)
- Bolt C length: 1.57 inches (40 mm) for 14 mm head: 41 ft. lbs. (55 Nm)
- Nut: 8 ft. lbs. (11 Nm)

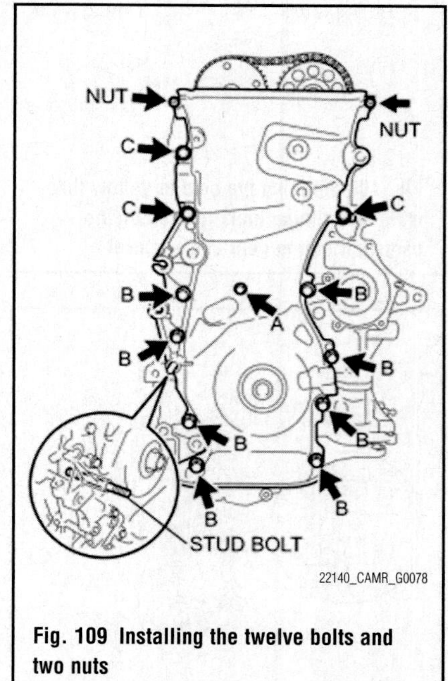

Fig. 109 Installing the twelve bolts and two nuts

11. Using a E10 Torx® socket, install the stud bolt to the drive belt tensioner and tighten to 16 ft. lbs. (22 Nm).

VALVE LASH

ADJUSTMENT

➡**Keep the lash adjuster free of dirt and foreign objects.**

➡**Only use clean engine oil.**

1. Place the lash adjuster into a container filled with engine oil.
2. Insert the SST's (SST: 09276-75010) tip into the lash adjuster's plunger and use the tip to press down on the check ball inside the plunger.
3. Squeeze the SST and lash adjuster together to move the plunger up and down 5 to 6 times.
4. Check the movement of the plunger and bleed the air. OK: Plunger moves up and down.

➡**When bleeding air from the high-pressure chamber, make sure that the tip of the SST is actually pressing the check ball as shown in the illustration. If the check ball is not pressed, air will not bleed.**

5. After bleeding the air, remove the SST. Then, try to press the plunger quickly and firmly with a finger. OK: Plunger is very difficult to move. If the result is not as specified, replace the lash adjuster.
6. Install the lash adjusters.

➡**Install the lash adjuster to the same place where it was removed from.**

ENGINE PERFORMANCE & EMISSION CONTROLS

ACCELERATOR PEDAL POSITION (APP) SENSOR

LOCATION

See Figure 110.

Refer to the accompanying illustration for sensor location.

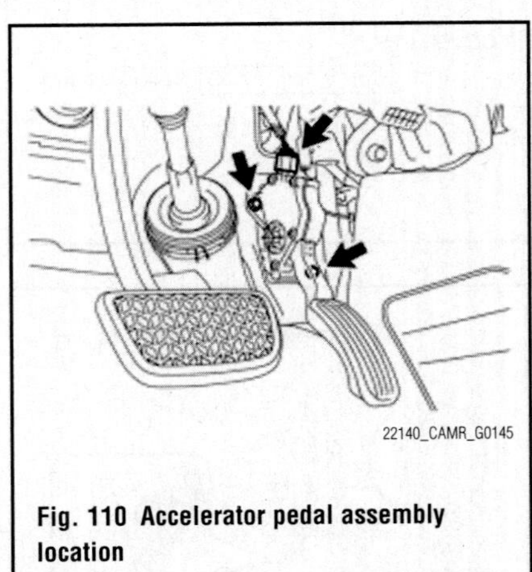

Fig. 110 Accelerator pedal assembly location

REMOVAL & INSTALLATION

See Figure 111.

1. Before servicing the vehicle, refer to the Precautions Section.
2. Remove left center floor carpet cover.
3. Disconnect the accelerator pedal connector.
4. Remove the 2 nuts and accelerator pedal assembly.

➡**Avoid physical shock to the accelerator pedal assembly.**

➡**Do not disassemble the accelerator pedal assembly.**

To install:

5. Installation is the reverse of the removal procedure. Tighten the accelerator pedal rod nuts to 43 inch lbs. (5.4 Nm).

CAMSHAFT POSITION (CMP) SENSOR

LOCATION

See Figure 111.

Refer to the accompanying illustration for sensor location.

REMOVAL & INSTALLATION

See Figure 112.

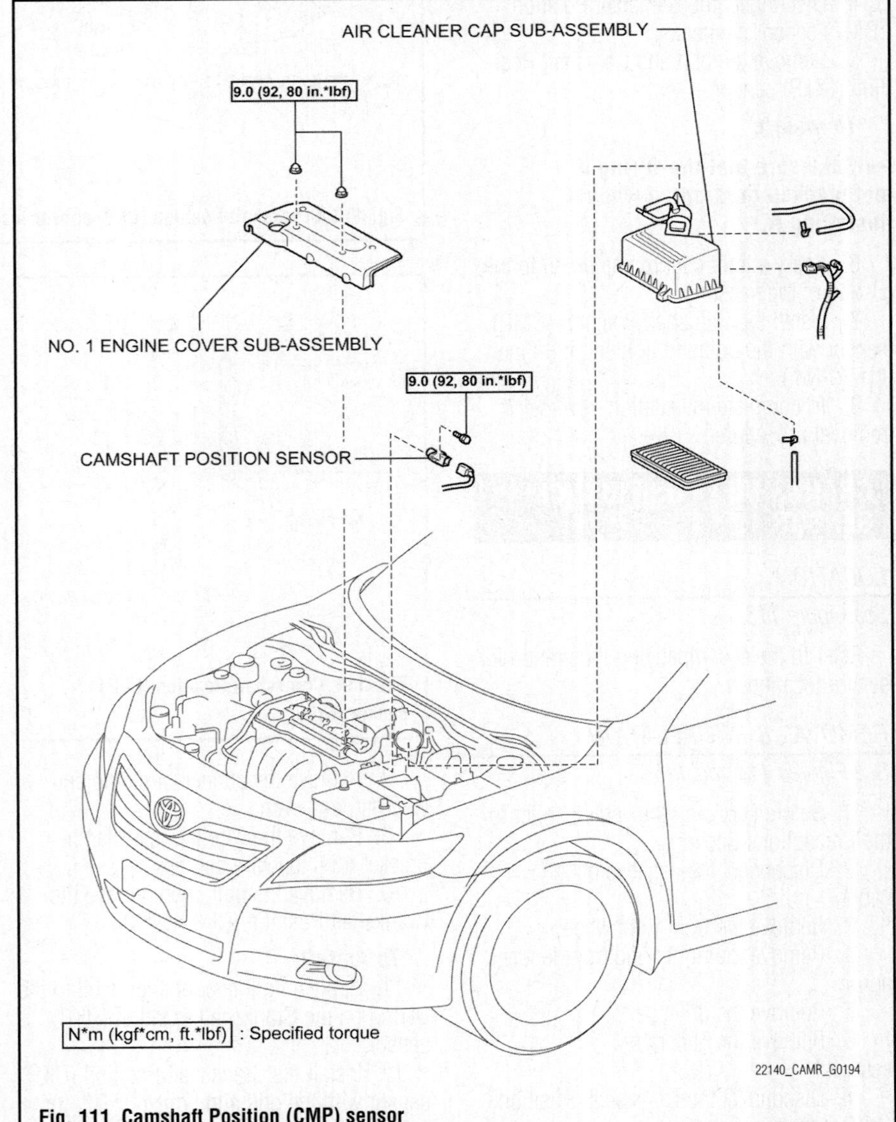

Fig. 111 Camshaft Position (CMP) sensor

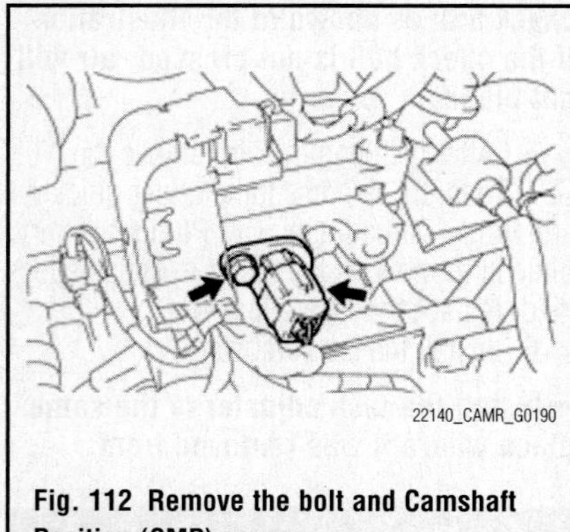

Fig. 112 Remove the bolt and Camshaft Position (CMP) sensor

1. Before servicing the vehicle, refer to the Precautions Section.
2. Remove the plastic engine cover.
3. Remove air cleaner cap sub-assembly.
4. Disconnect the Camshaft Position (CMP) sensor connector.
5. Remove the bolt and Camshaft Position (CMP) sensor.

To install:

➡Make sure that the O-ring is not cracked or jammed when installing it.

6. Apply a light coat of engine oil to the O-ring of the sensor.
7. Install the Camshaft Position (CMP) sensor with the bolt and tighten to: 80 inch lbs. (9 Nm).
8. To complete installation, reverse removal procedure.

CRANKSHAFT POSITION (CKP) SENSOR

LOCATION

See Figure 113.

Refer to the accompanying illustration for sensor location.

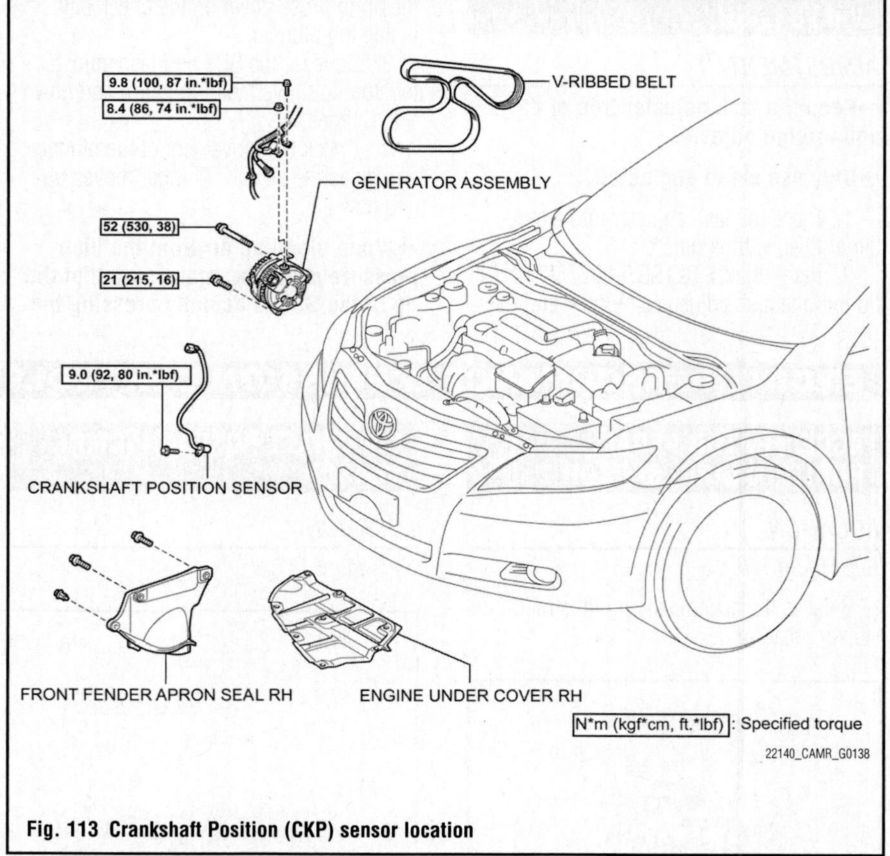

Fig. 113 Crankshaft Position (CKP) sensor location

REMOVAL & INSTALLATION

See Figures 114 and 115.

1. Before servicing the vehicle, refer to the Precautions Section.
2. Disconnect the negative battery cable.
3. Remove the front right wheel.
4. Remove the right hand front fender apron.
5. Remove the drive belt.
6. Remove the generator assembly.
7. Disconnect the Crankshaft Position (CKP) sensor connector.

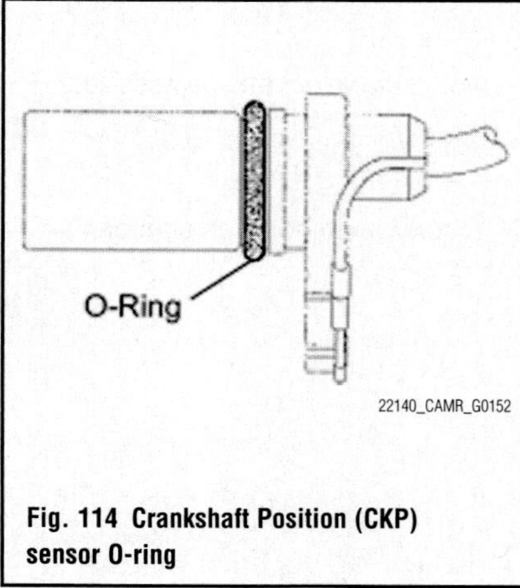

Fig. 114 Crankshaft Position (CKP) sensor O-ring

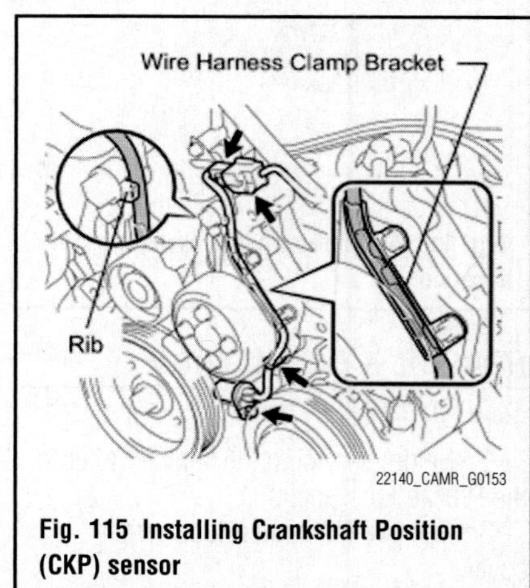

Fig. 115 Installing Crankshaft Position (CKP) sensor

8. Remove the connector clamp and wire harness clamp.
9. Remove the wire harness clamp bracket from the wire harness.
10. Remove the bolt, then remove the Crankshaft Position (CKP) sensor.

To install:

11. Apply a light coat of engine oil to the O-ring on the Crankshaft Position (CKP) sensor.
12. Install the Crankshaft Position (CKP) sensor with the bolt and tighten to 7ft. lbs. (10 Nm).
13. Connect the Crankshaft Position (CKP) sensor connector.
14. The remainder of installation is the reverse of the removal procedure.

ELECTRONIC CONTROL MODULE (ECM)

LOCATION

See Figure 116.

Refer to the accompanying illustration for ECM location.

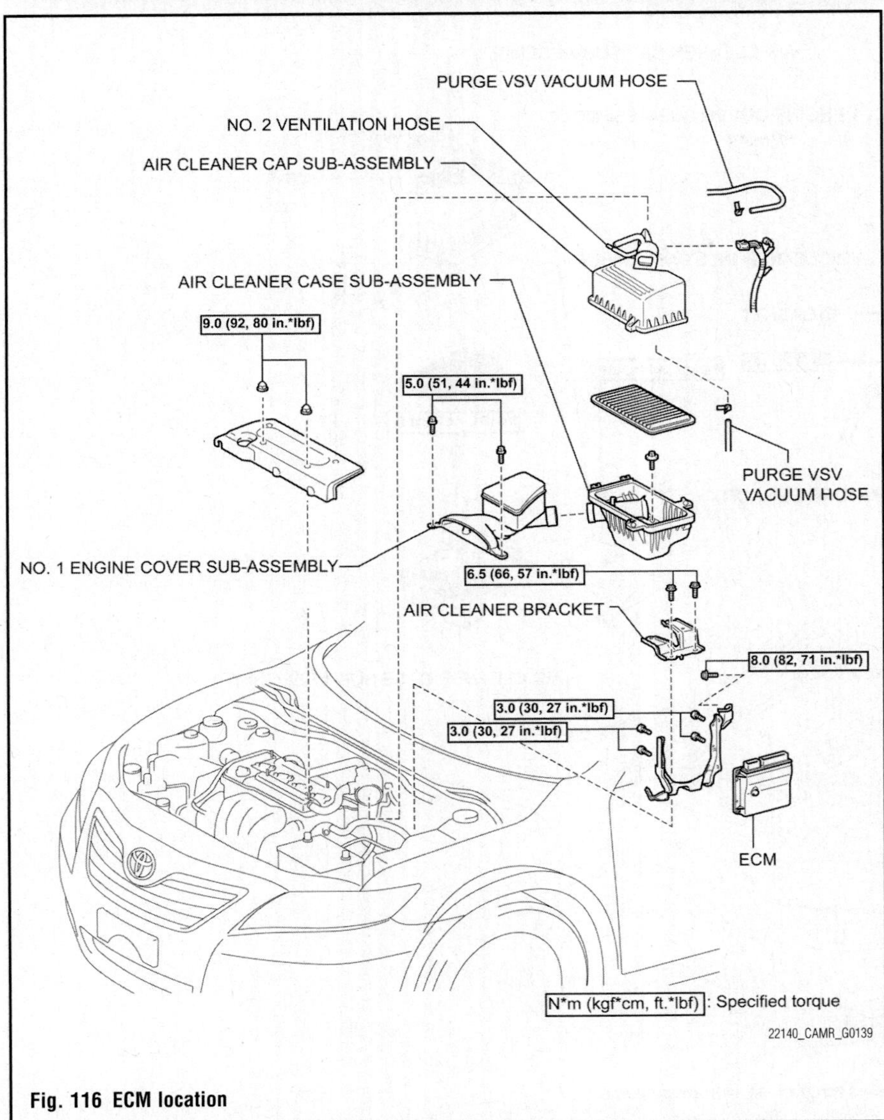

Fig. 116 ECM location

OPERATION

The ECM controls all engine computer related functions.

REMOVAL & INSTALLATION

See Figure 117.

1. Before servicing the vehicle, refer to the Precautions Section.
2. Disconnect the negative battery cable.
3. Remove the plastic engine cover.
4. Remove the air cleaner inlet assembly.
5. Remove the air cleaner cap sub-assembly.
6. Remove the air cleaner case sub-assembly.
7. Remove the two bolts and air cleaner bracket.
8. Disconnect the two ECM connectors by raising the two levers. While pushing the locks on the two levers, disconnect the two ECM connectors.

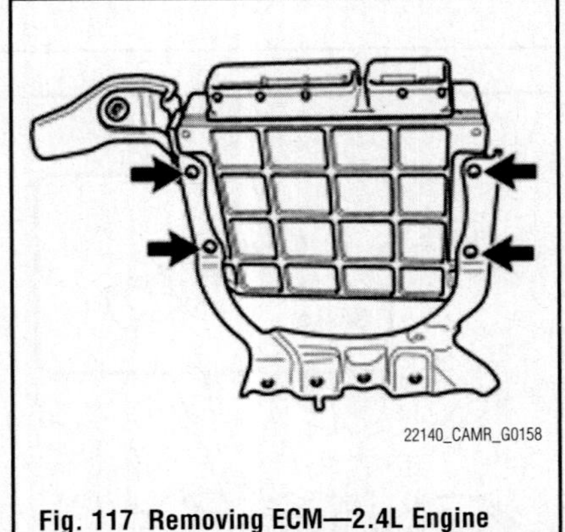

Fig. 117 Removing ECM—2.4L Engine

➡After disconnecting the connectors, make sure that dirt, water or other foreign matter does not contact the connections of the connectors.

a. Remove the ECM with the bracket and three bolts.

b. Remove the four screws and ECM brackets.

To install:

9. Install the bracket to the ECM with the 4 screws, and tighten to 27 inch lbs. (3 Nm).
10. Attach the ECM with the three nuts and tighten to 71 inch lbs. (8 Nm).

➡Make sure that dirt, water or other foreign matter does not contact the connections of the connectors.

11. Connect the two ECM connectors and lower the two levers.
12. Install the ECM to the body.
13. Install the two bolts and air cleaner bracket.
14. Install the air cleaner case sub-assembly.
15. Install the air cleaner cap sub-assembly.
16. Install the air cleaner inlet assembly.
17. Remove the plastic engine cover.
18. Connect the negative battery cable.
19. Register the immobilizer communication ID. If the ECM is replaced, register the ECM communication ID for the immobilizer system (refer to the Service Bulletin for registration).
20. Perform initialization. After replacing the ECM on vehicles with a dynamic laser cruise control system, it is necessary to initialize the ECM so that the ECM can recognize the dynamic laser cruise control system.
21. Be sure to perform the following procedure after replacing the ECM:
    a. Turn the ignition switch on (IG).
    b. Turn the cruise control main switch on.
    c. With the brake pedal depressed, push the cruise control main switch to RES/ACC 3 times within 3 seconds. Check that the buzzer sounds at this time.

➡Do not turn the headlight dimmer switch on at this time because the optical axis automatic adjustment mode has already started, which may lead to an incorrect optical axis setting. If the headlight dimmer switch is turned on by mistake, readjust the optical axis.

ENGINE COOLANT TEMPERATURE (ECT) SENSOR

LOCATION

See Figure 118.

Refer to the accompanying illustration for sensor location.

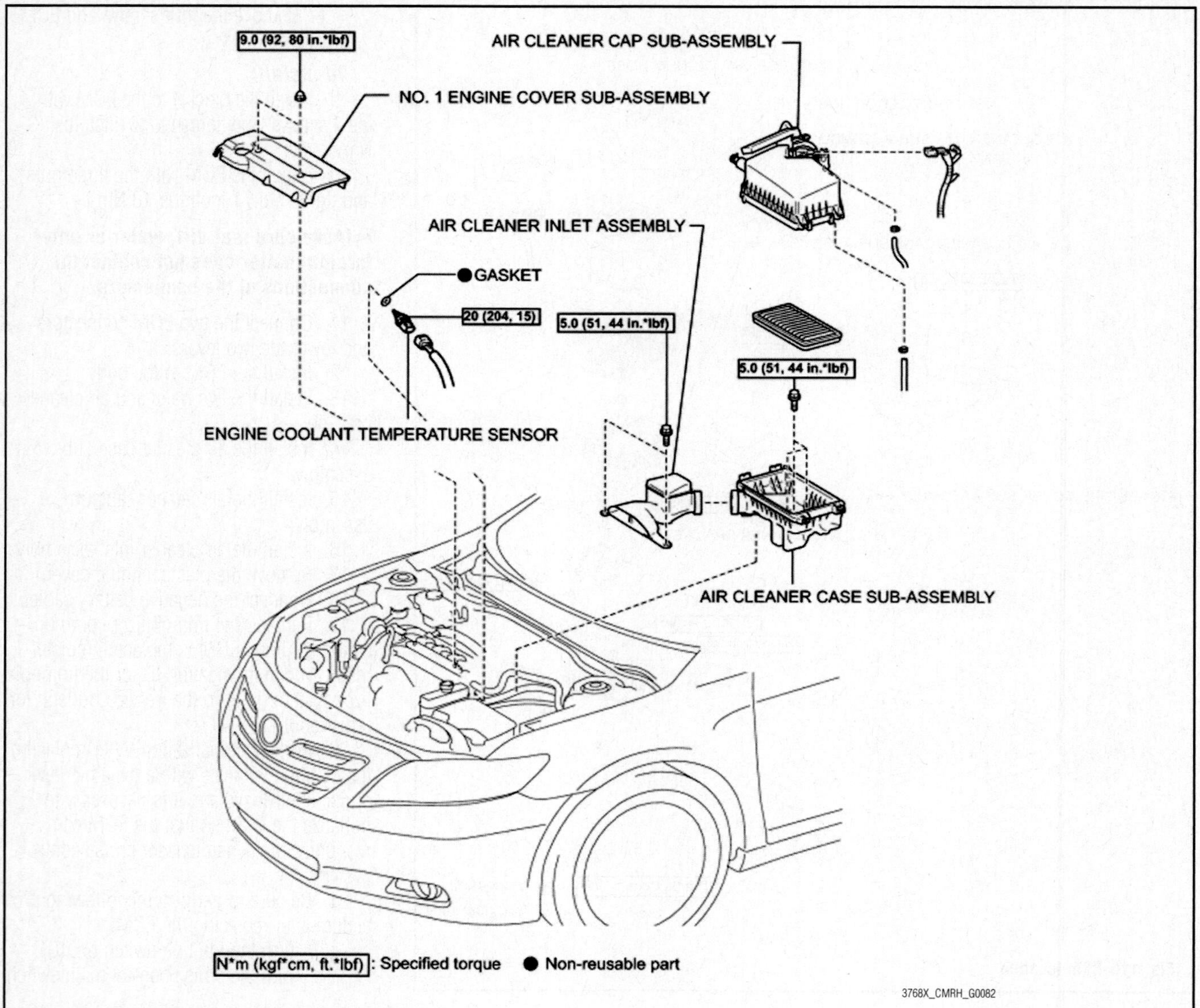

Fig. 118 Engine Coolant Temperature (ECT) sensor

REMOVAL & INSTALLATION

See Figure 119.

1. Drain the engine coolant.
2. Remove the No. 1 engine cover sub assembly.
3. Remove the air cleaner inlet assembly.
4. Remove the air cleaner cap sub assembly.
5. Remove the air cleaner case sub assembly.
6. Remove the engine coolant temperature sensor.
   a. Disconnect the engine coolant temperature sensor connector.
   b. Using the SST (09817-33190), remove the engine coolant temperature sensor and gasket.

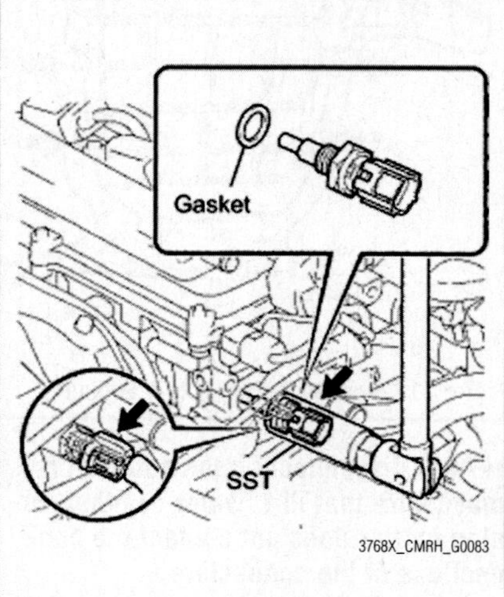

Fig. 119 Removing the ECT sensor

To install:

7. Install the ECT sensor.
   a. Install a new gasket onto the ECT.
   b. Using the SST (09817-33190), install the ECT. Tighten to 15 ft. lbs. (20 Nm).
8. Install the air cleaner case sub assembly.
9. Install the air cleaner cap sub assembly.
10. Install the air cleaner inlet assembly.
11. Add engine coolant.
12. Inspect for coolant leaks.
13. Install the No. 1 engine cover sub assembly.

EVAPORATIVE EMISSIONS (EVAP) CANISTER

LOCATION

See Figure 120.

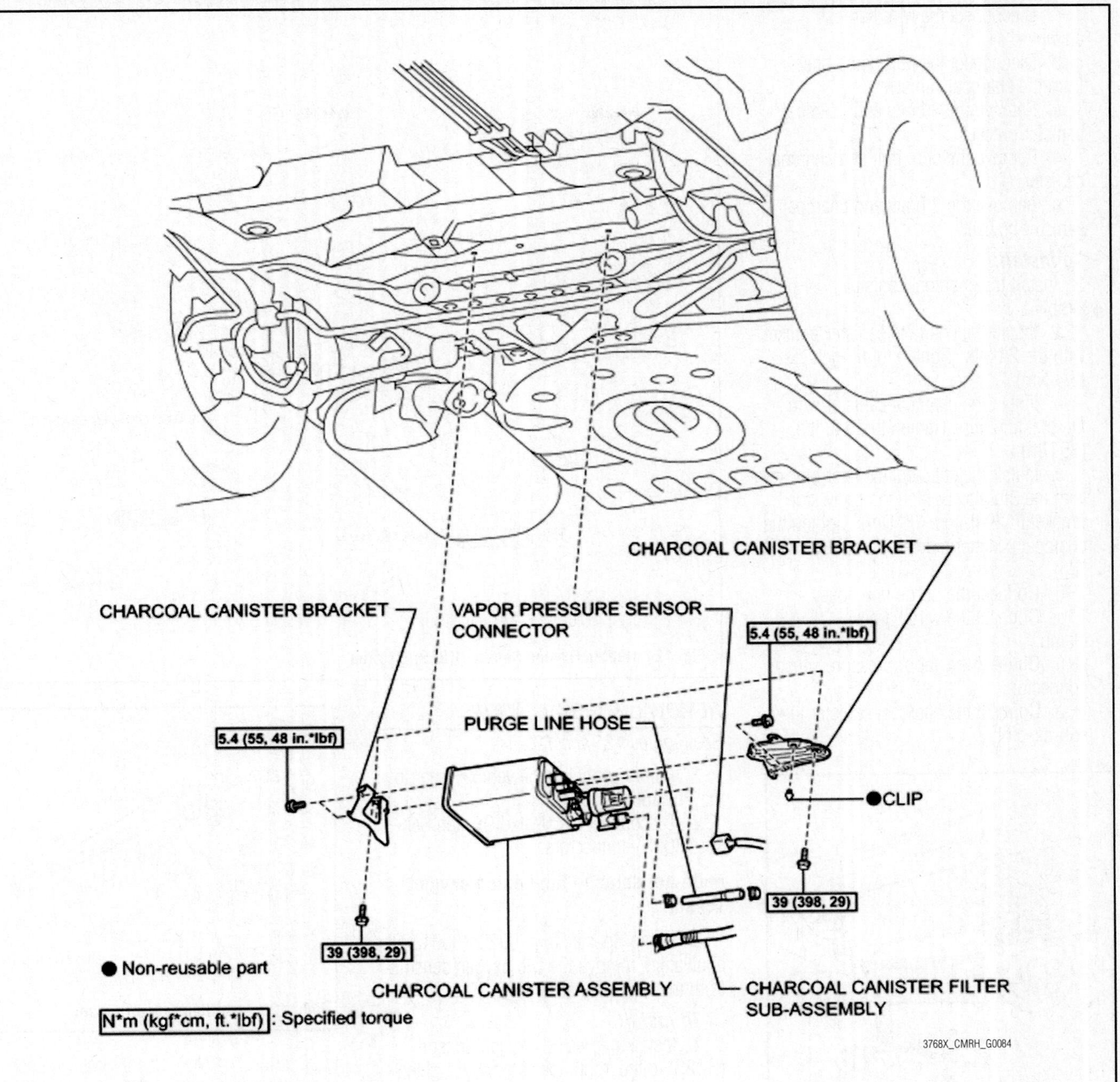

CHARCOAL CANISTER BRACKET
CHARCOAL CANISTER BRACKET
VAPOR PRESSURE SENSOR CONNECTOR
5.4 (55, 48 in.*lbf)
5.4 (55, 48 in.*lbf)
PURGE LINE HOSE
●CLIP
39 (398, 29)
39 (398, 29)
● Non-reusable part
CHARCOAL CANISTER ASSEMBLY
CHARCOAL CANISTER FILTER SUB-ASSEMBLY
N*m (kgf*cm, ft.*lbf) : Specified torque
3768X_CMRH_G0084

Fig. 120 Evaporative Emissions (EVAP) canister

Refer to the accompanying illustration for EVAP canister location.

REMOVAL & INSTALLATION

See Figures 121 through 123.

1. Remove the fuel tank.
2. Remove the charcoal canister assembly.
   a. Disconnect the charcoal canister filter sub assembly from the charcoal canister.
   b. Disconnect the vapor pressure sensor connector.

3768X_CMRH_G0085

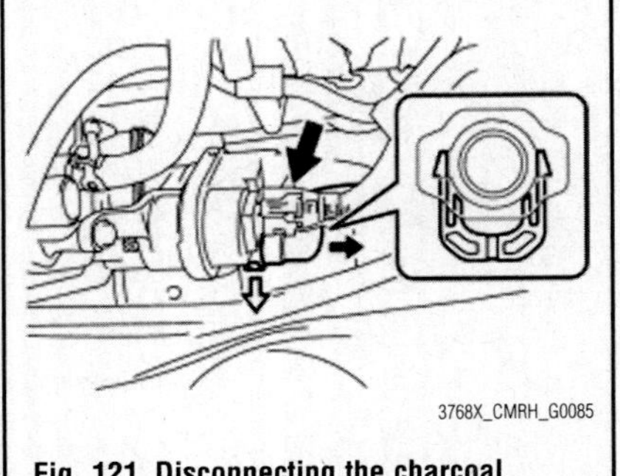

Fig. 121 Disconnecting the charcoal canister

3768X_CMRH_G0086

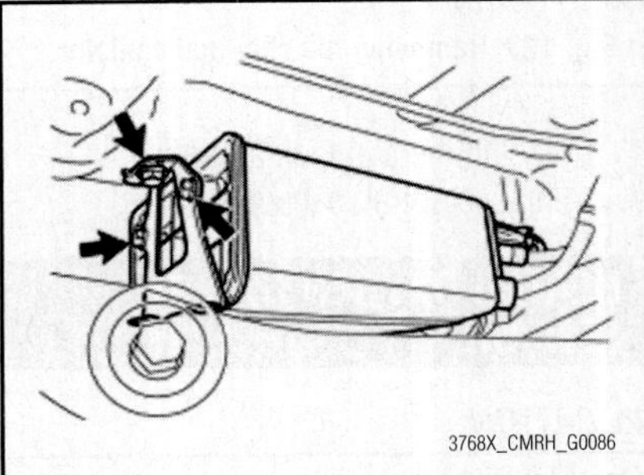

Fig. 122 Removing the charcoal canister bracket

c. Disconnect the wire harness clamp.

d. Disconnect the purge line hose from the charcoal canister.

e. Remove the 3 bolts and charcoal canister bracket.

f. Remove the clip, bolt and charcoal canister.

g. Remove the 2 bolts and charcoal canister bracket.

To install:

3. Install the charcoal canister assembly.

a. Install the charcoal canister bracket with the 2 bolts. Tighten to 48 inch lbs. (5.4 Nm).

b. Install the charcoal canister with the bolt and clip. Tighten to 29 ft. lbs. (39 Nm).

c. Install the charcoal canister bracket with the 3 bolts. Tighten the body side bracket to 29 ft. lbs. (39 Nm). Tighten the charcoal canister side to 48 inch lbs. (5.4 Nm).

d. Connect the purge line hose.

e. Connect the wire harness clamp.

f. Connect the vapor pressure sensor connector.

g. Connect the charcoal canister filter sub-assembly.

Fig. 123 Removing the charcoal canister

4. Install the fuel tank assembly.
5. Inspect for an exhaust leak.

HEATED OXYGEN (HO2S) SENSOR

LOCATION

See Figure 124.

Refer to the accompanying illustration for sensor location.

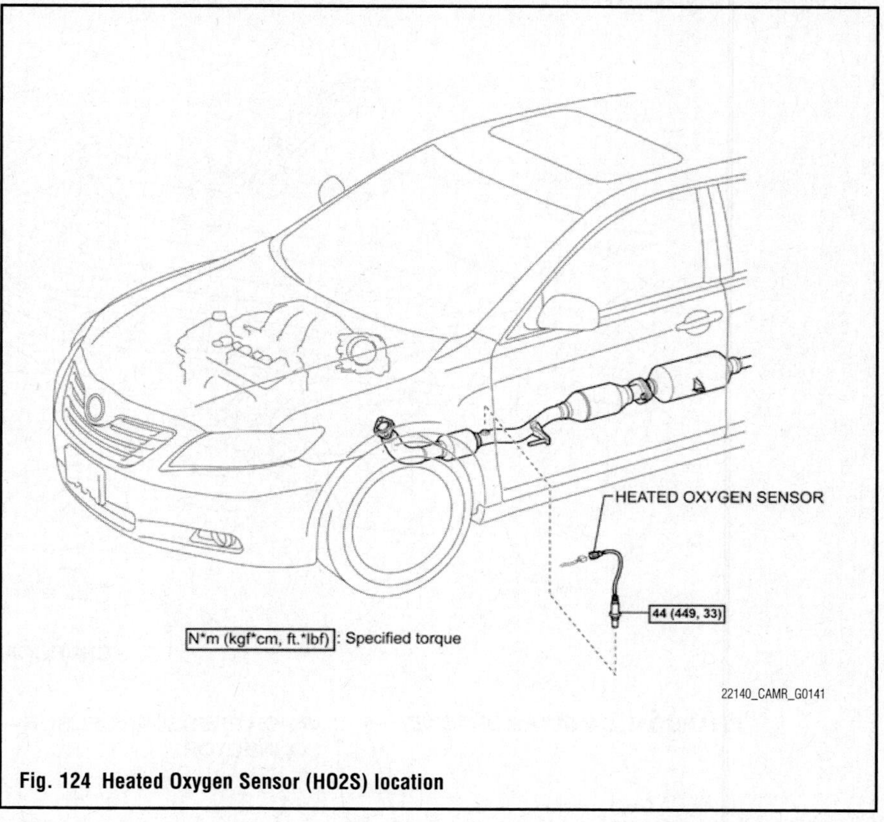

Fig. 124 Heated Oxygen Sensor (HO2S) location

REMOVAL & INSTALLATION

See Figures 125 and 126.

1. Before servicing the vehicle, refer to the Precautions Section.

2. Disconnect the Heated Oxygen Sensor (HO2S) connectors.

➡Do not damage the heated oxygen sensor.

3. Using Special Tool: 09224-00010 or equivalent, remove the two oxygen sensors from the front pipe assembly.

To install:

4. Install the Heated Oxygen Sensor (HO2S) to the front pipe assembly. Tighten to 32 ft. lbs. (44 Nm). Use a torque wrench with a fulcrum length of 300 mm (11.81 in.).

5. Connect the Heated Oxygen Sensor (HO2S) connector.

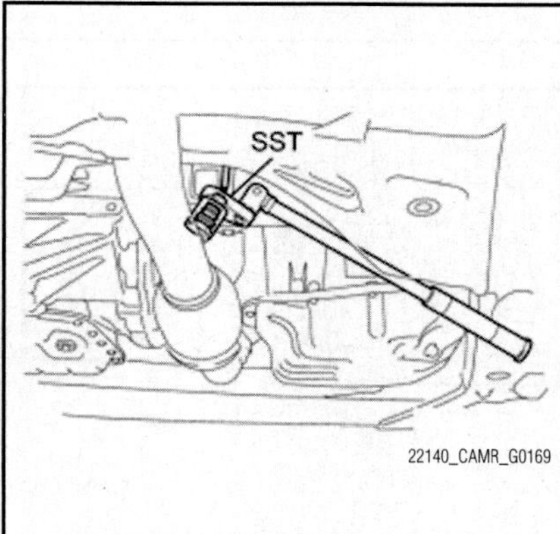

Fig. 125 Removing oxygen sensor

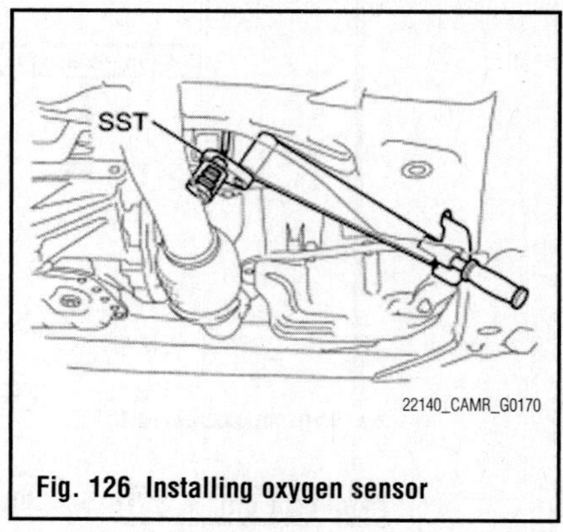

Fig. 126 Installing oxygen sensor

INTAKE AIR TEMPERATURE (IAT) SENSOR

REMOVAL & INSTALLATION

See Mass Air Flow Meter.

KNOCK SENSOR (KS)

LOCATION

See Figure 127.

Refer to the accompanying illustration for sensor location.

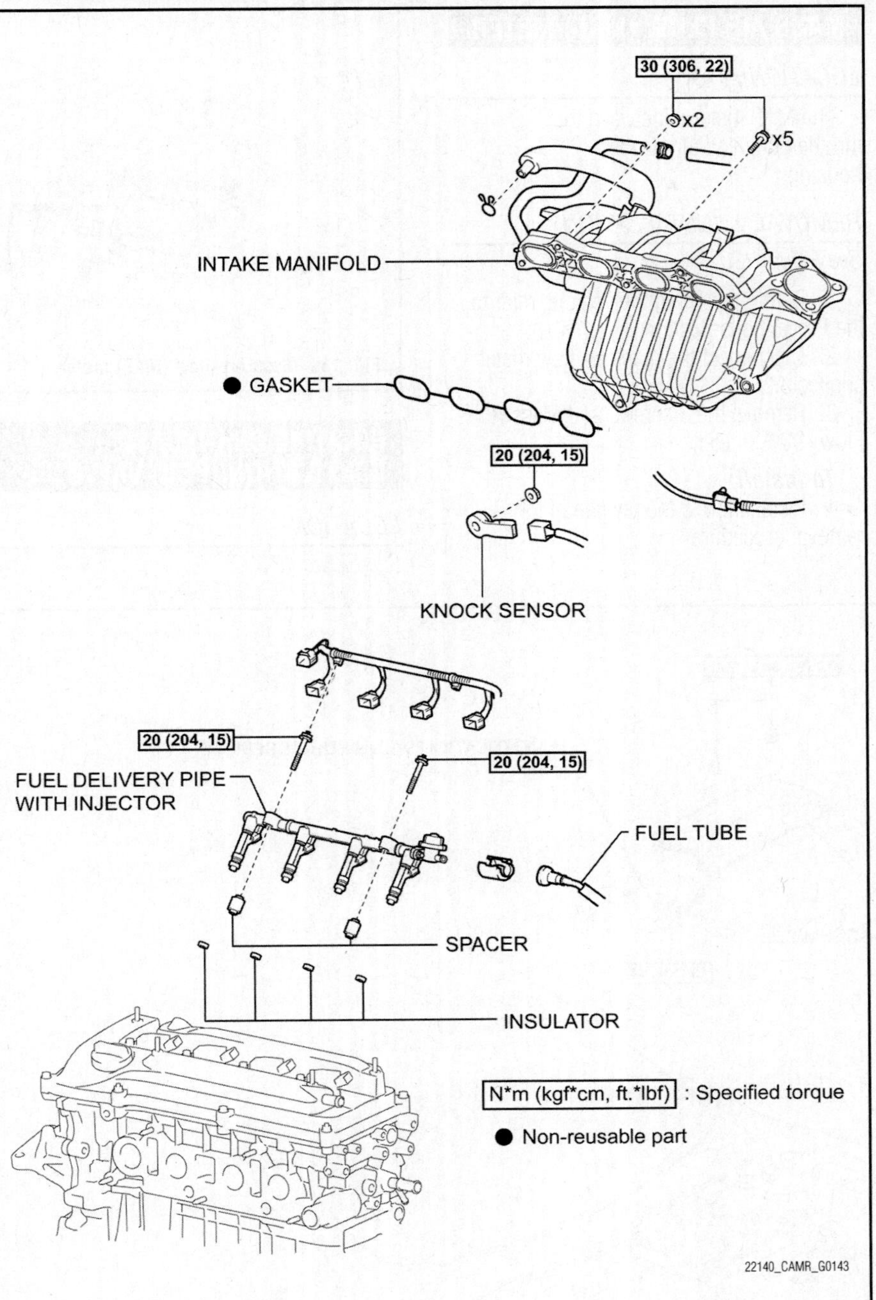

30 (306, 22)
x2
x5
INTAKE MANIFOLD
● GASKET
20 (204, 15)
KNOCK SENSOR
20 (204, 15)
FUEL DELIVERY PIPE WITH INJECTOR
20 (204, 15)
FUEL TUBE
SPACER
INSULATOR
N*m (kgf*cm, ft.*lbf) : Specified torque
● Non-reusable part
22140_CAMR_G0143

Fig. 127 Knock Sensor (KS) location

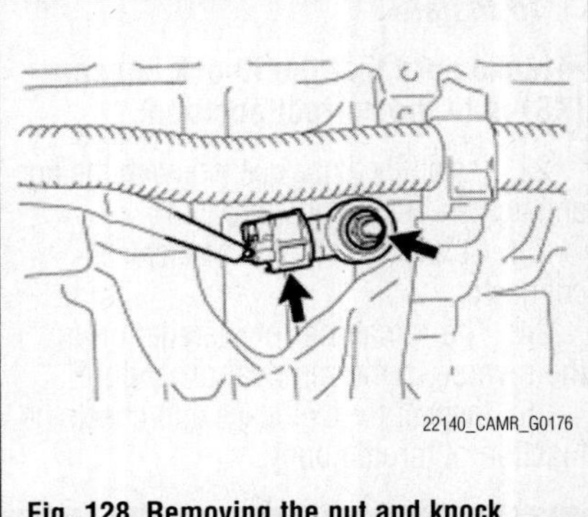

Fig. 128 Removing the nut and knock sensor—2.4L Engine

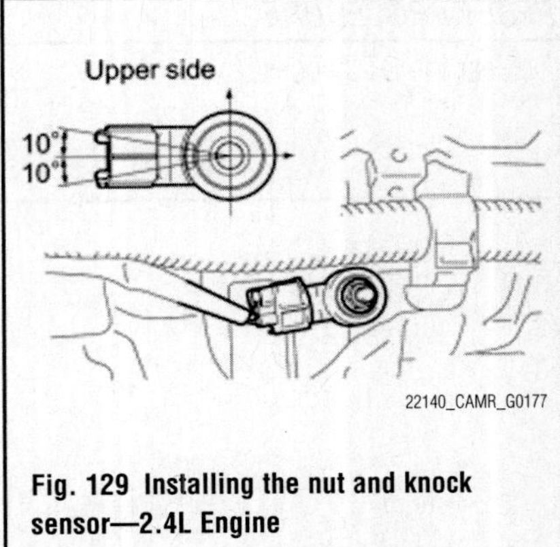

Fig. 129 Installing the nut and knock sensor—2.4L Engine

REMOVAL & INSTALLATION

See Figures 128 and 129.

CAUTION

Observe all applicable safety precautions when working around fuel. Whenever servicing the fuel system, always work in a well ventilated area. Do not allow fuel spray or vapors to come in contact with a spark or open flame. Keep a dry chemical fire extinguisher near the work area. Always keep fuel in a container specifically designed for fuel storage; also, always properly seal fuel containers to avoid the possibility of fire or explosion.

1. Before servicing the vehicle, refer to the Precautions Section.
2. Properly discharge the fuel system pressure.
3. Disconnect battery negative cable.
4. Remove plastic engine cover.
5. Drain and recycle the engine coolant.
6. Remove windshield wiper arms and blade assemblies.
7. Remove both front fender to cowl side seals.
8. Remove cowl top ventilator louver sub-assembly.
9. Remove windshield wiper motor and link.
10. Remove the four bolts, four nuts and cowl top panel outer sub-assembly.
11. Remove air cleaner cap sub-assembly.
12. Remove air cleaner case sub-assembly.
13. Remove throttle body.
14. Disconnect fuel tube.
15. Remove fuel delivery pipe with injector.
16. Disconnect the union to check valve hose from the brake booster.
17. Disconnect the camshaft timing oil control valve connector.
18. Remove the wire harness clamp.
19. Remove the union to check valve hose from the vacuum hose clamp.
20. Remove the 5 bolts, 2 nuts and intake manifold. Remove the gasket from the intake manifold.
21. Disconnect the Knock Sensor (KS) connector. Remove the nut and knock sensor.

To install:

➡Make sure that the Knock Sensor (KS) is in the correct position.

22. Install the knock sensor with the nut and tighten to 15 ft. lbs. (20 Nm).
23. Connect the knock sensor connector.
24. The remainder of installation is the reverse of the removal procedure.
25. Inspect for fuel leaks and check the function of throttle body.

MALFUNCTION INDICATOR LIGHT (MIL)

RESET PROCEDURE

Clear the DTC codes.

MASS AIR FLOW (MAF) METER

LOCATION

The MAF meter is between the throttle body and air cleaner housing.

REMOVAL & INSTALLATION

See Figure 130.

1. Before servicing the vehicle, refer to the Precautions Section.
2. Disconnect the mass air flow meter connector.
3. Remove the 2 screws and Mass Air Flow (MAF) meter.

To install:

4. Installation is the reverse of the removal procedure.

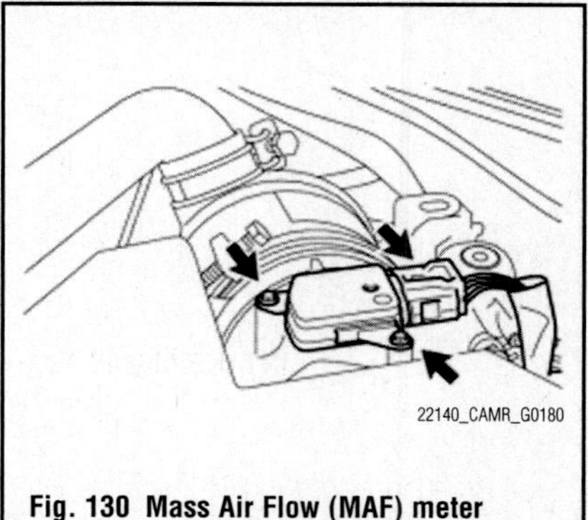

Fig. 130 Mass Air Flow (MAF) meter

POSITIVE CRANKCASE VENTILATION (PCV) VALVE

LOCATION

See Figure 131.

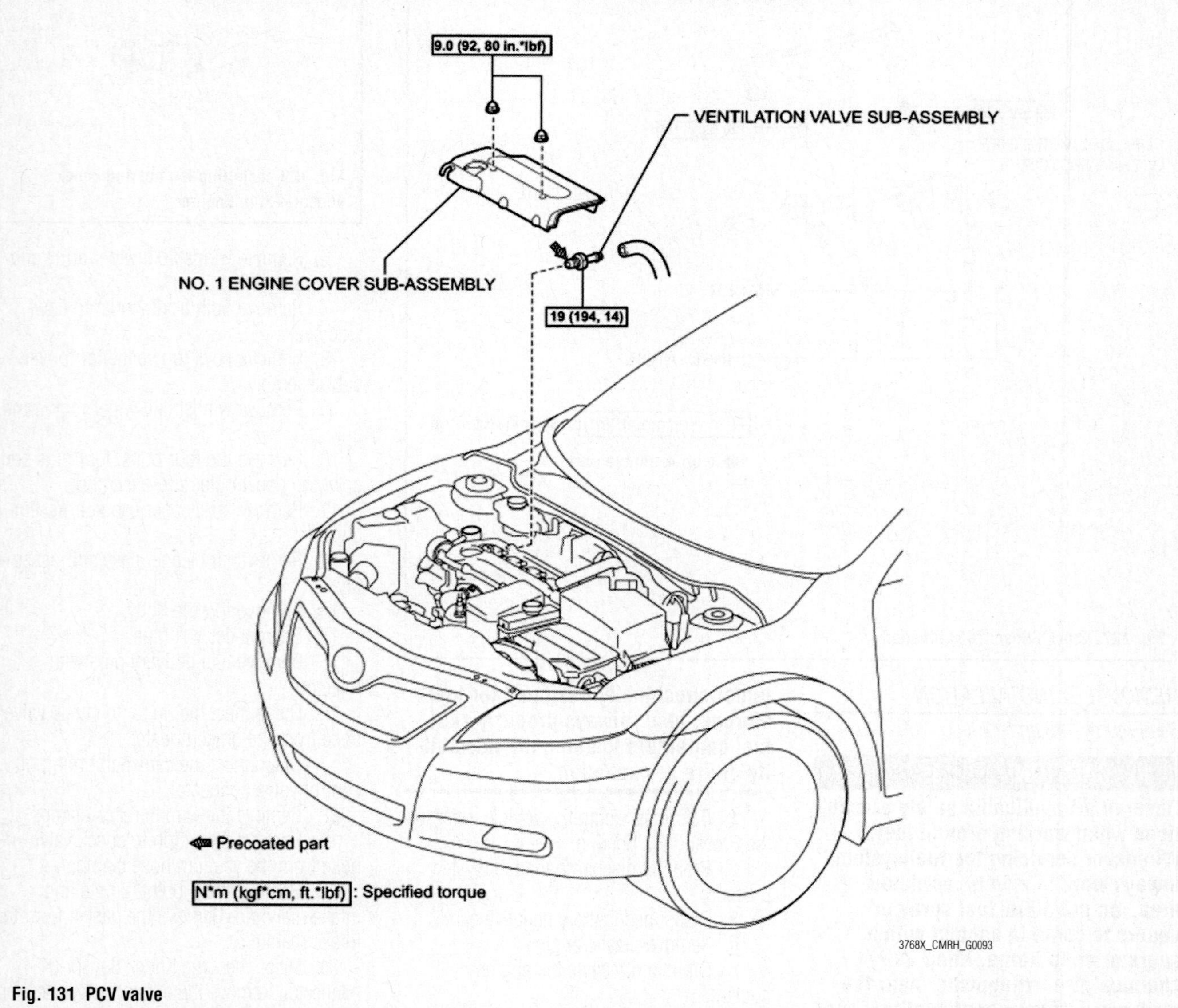

Fig. 131 PCV valve

Refer to the accompanying illustration for PCV valve location.

REMOVAL & INSTALLATION

See Figures 132 and 133.

Fig. 132 Disconnecting the ventilation hose from the ventilation valve sub assembly

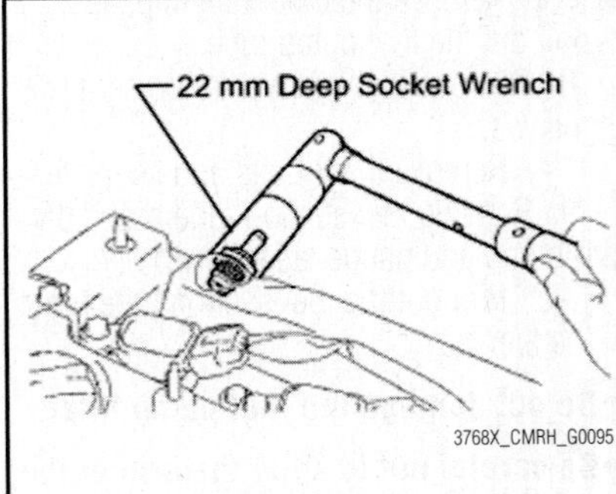

Fig. 133 Removing the ventilation valve sub assembly

1. Separate the brake master cylinder reservoir sub assembly.
2. Remove the No. 1 engine cover sub assembly.
3. Remove the ventilation valve sub assembly.
   a. Disconnect the ventilation hose from the ventilation valve sub assembly.
   b. Using a 22 mm deep socket wrench, remove the ventilation valve sub-assembly.

To install:

CAUTION

Do not start the engine within 1 hour after installation.

To install, reverse the removal procedure. Tighten the ventilation valve to 14 ft. lbs. (19 Nm).

THROTTLE POSITION SENSOR (TPS)

LOCATION

Refer to the Throttle Body removal and installation procedures.

REMOVAL & INSTALLATION

Refer to the Throttle Body removal and installation procedures.

VARIABLE CAMSHAFT TIMING OIL CONTROL SOLENOID

REMOVAL & INSTALLATION

See Figure 134.

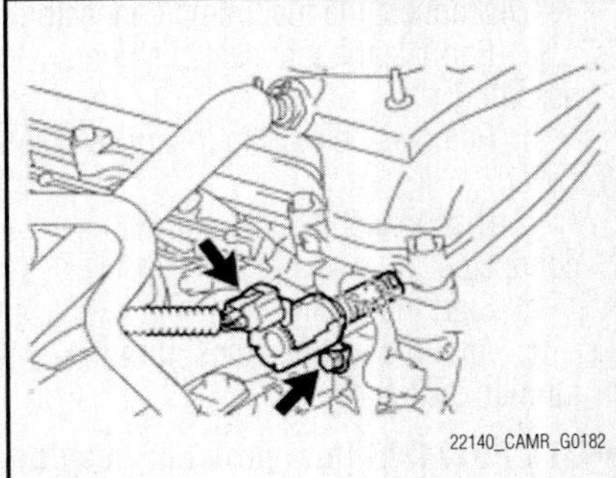

Fig. 134 Removing the camshaft timing oil control valve assembly

1. Before servicing the vehicle, refer to the Precautions Section.
2. Remove the plastic engine cover.
3. Remove the bolt and disconnect the vacuum hose clamp.
4. Remove the clip nut.
5. Disconnect the camshaft timing oil control valve assembly connector.
6. Remove the bolt and camshaft timing oil control valve assembly.

To install:

7. Apply a light coat of engine oil to an O-ring of the camshaft timing oil control valve assembly sensor.
8. Install the camshaft timing oil control valve assembly with the bolt and tighten to 80 inch lbs. (9 Nm).
9. To complete installation, reverse removal procedure.

FUEL — GASOLINE FUEL INJECTION SYSTEM

FUEL SYSTEM SERVICE PRECAUTIONS

Safety is the most important factor when performing not only fuel system maintenance, but any type of maintenance. Failure to conduct maintenance and repairs in a safe manner may result in serious personal injury or death. Work on a vehicle's fuel system components can be accomplished safely and effectively by adhering to the following rules and guidelines.

- To avoid the possibility of fire and personal injury, always disconnect the negative battery cable unless the repair or test procedure requires that battery voltage be applied.
- Always relieve the fuel system pressure prior to disconnecting any fuel system component (injector, fuel rail, pressure regulator, etc.) fitting or fuel line connection. Exercise extreme caution whenever relieving fuel system pressure to avoid exposing skin, face and eyes to fuel spray. Please be advised that fuel under pressure may penetrate the skin or any part of the body that it contacts.
- Always place a shop towel or cloth around the fitting or connection prior to loosening to absorb any excess fuel due to spillage. Ensure that all fuel spillage is quickly removed from engine surfaces. Ensure that all fuel-soaked cloths or towels are deposited into a flame-proof waste container with a lid.
- Always keep a dry chemical (Class B) fire extinguisher near the work area.
- Do not allow fuel spray or fuel vapors to come into contact with a spark or open flame.
- Always use a second wrench when loosening or tightening fuel line connection fittings. This will prevent unnecessary stress and torsion on fuel piping. Always follow the proper torque specifications.
- Always replace worn fuel fitting O-rings with new ones. Do not substitute fuel hose where rigid pipe is installed.

FUEL SYSTEM PRESSURE

RELIEVING

CAUTION

Perform the following procedures to prevent fuel from spilling out before removing any fuel system parts.

CAUTION

Pressure will still remain in the fuel line even after performing the following procedures. When disconnecting the fuel line, cover it with a shop rag or a piece of cloth to prevent fuel from spraying or coming out.

1. Disconnect the fuel pump connector:
   a. Remove the rear seat cushion assembly.
   b. Remove the rear floor service hole cover.
   c. Disconnect the fuel pump connector.
   d. Start the engine.
   e. After the engine stops, turn the ignition switch off.

➡**DTC P0171/25 (fuel problem) may be detected.**

   f. Crank the engine again. Check that the engine does not start.
   g. Remove the fuel tank cap to discharge pressure from the fuel tank.
   h. Disconnect the cable from the negative (-) battery terminal.
   i. Reconnect the fuel pump connector.
   j. Install the rear floor service hole cover.
   k. Install the rear seat.
2. Check that there are no fuel leaks from the fuel system after doing any maintenance or repairs.

FUEL FILTER

REMOVAL & INSTALLATION

See Figure 135.

1. Before servicing the vehicle, refer to the precautions section.
2. Remove the fuel pump from the vehicle.
3. Remove the fuel pump filter, as follows:

➡**Do not damage the fuel pump filter. Do not remove the suction filter.**

   a. Using a screwdriver, pry out the clips.
   b. Pull out the fuel pump filter from the fuel pump.

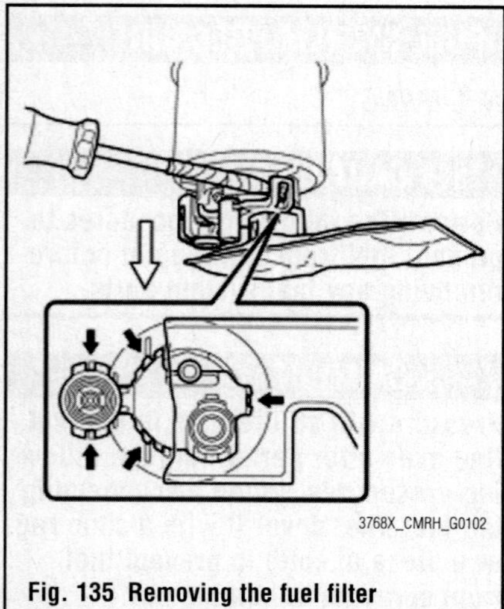

Fig. 135 Removing the fuel filter

To install:

4. Install the fuel pump filter with a new clip.
5. Install the fuel pump.

FUEL LEVEL SENDING UNIT

REMOVAL & INSTALLATION

1. Before servicing the vehicle, refer to the precautions section.
2. Remove the fuel pump from the vehicle.
3. Remove the fuel sender gauge assembly, as follows:
   a. Disconnect the fuel sender gauge connector.
   b. Unlock the fuel sender gauge, and slide it to remove.

To install:

4. Slide the fuel sender gauge to engage with the claw.
   a. Connect the fuel sender gauge connector.
5. Install the fuel pump.

FUEL PUMP

REMOVAL & INSTALLATION

See Figures 136 and 137.

1. Before servicing the vehicle, refer to the precautions section.
2. Discharge fuel system pressure.
3. Disconnect battery negative cable.
4. Remove the rear seat cushion assembly.
5. Remove the rear floor service hole cover.

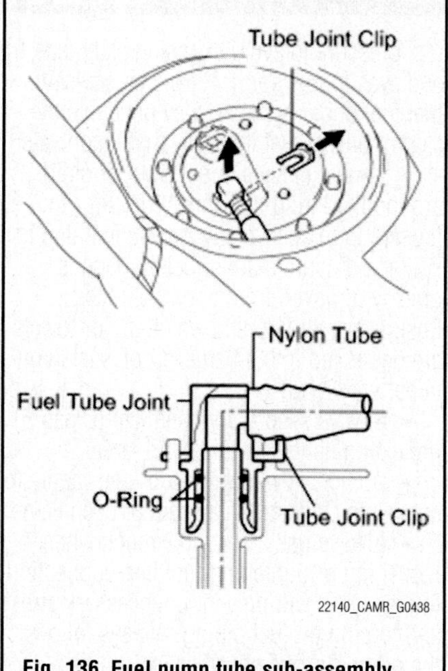

Fig. 136 Fuel pump tube sub-assembly

6. Disconnect the fuel pump connector.
7. Separate the fuel pump tube sub-assembly, as follows:

➡**Check if there is any dirt or mud around the connector before this operation and remove the dirt as necessary.**

➡**Be careful of mud because the quick connector has an O-ring which seals the pipe and connector that can be contaminated.**

➡**Do not use any tools in this operation.**

➡**Do not bend or twist the nylon tube. Cover the fuel tube joint with a plastic bag.**

➡**When the fuel tube joint and fuel suction plate are stuck, pinch the fuel tank tube between fingers, and turn it carefully to release it. Disconnect the fuel tank tube.**

   a. Remove the tube joint clip, and pull out the fuel pump tube.
8. Remove fuel tank vent tube set plate, as follows:
   a. Remove the 8 bolts and set plate.
9. Remove fuel suction tube assembly with pump and gauge, as follows:
   a. Pull out the fuel suction tube from the fuel tank.

➡**Do not damage the fuel pump filter.**

➡**Be careful not to bend the arm of the fuel sender gauge.**

   b. Remove the gasket from the fuel suction tube.

To install:

10. Install a new gasket to the fuel suction tube.
11. Install the fuel suction tube.

➡**Do not damage the fuel pump filter.**

➡**Be careful not to bend the arm of the fuel sender gauge.**

12. Install the fuel tank vent tube set plate, as follows:
    a. Align the mark of the set plate with the fuel suction tube.
    b. Install the set plate with the 8 bolts and tighten to 52 inch lbs. (5.9 Nm).
13. Install the fuel pump tube with the tube joint clip.

➡**Check that there are no scratches or foreign objects on the connecting part.**

➡**Check that the fuel tube joint is inserted securely.**

➡**Check that the tube joint clip is on the collar of the fuel tube joint.**

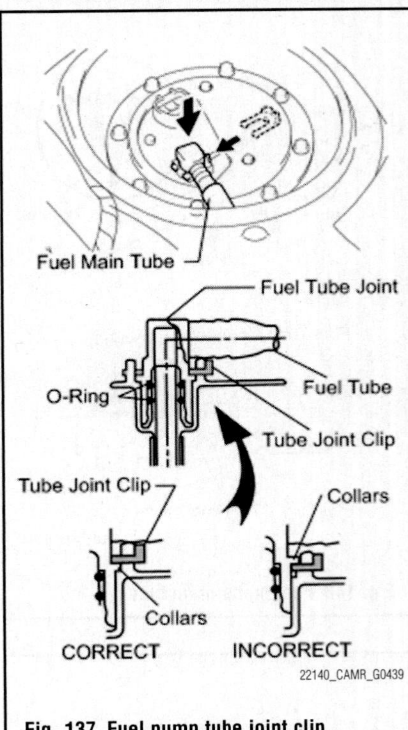

Fig. 137 Fuel pump tube joint clip

➡After installing the tube joint clip, check that the fuel tube joint is pulled off.

14. Connect battery negative cable.
15. Inspect for fuel leak.
16. Install the rear floor service hole cover.
17. Install the rear seat cushion assembly.

FUEL PRESSURE REGULATOR

REMOVAL & INSTALLATION

See Figure 138.

1. Before servicing the vehicle, refer to the precautions section.
2. Remove the fuel pump from the vehicle.
3. Using a screwdriver with its tip wrapped in protective tape, remove the fuel pressure regulator from the fuel filter.
4. Remove the 2 O-rings from the fuel pressure regulator.

To install:

5. Apply a light coat of spindle oil or gasoline to a new O-ring, and install it to the fuel pressure regulator.
6. Push in the fuel pressure regulator to the fuel tank fuel filter.
7. Install the fuel pump filter.
8. Install the fuel pump.

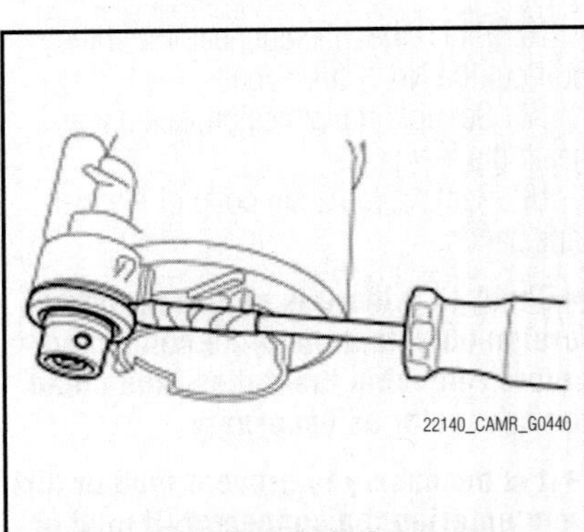

Fig. 138 Removing the fuel pressure regulator from the fuel filter—2.4L

FUEL RAIL AND INJECTOR

REMOVAL & INSTALLATION

See Figures 139 through 142.

1. Before servicing the vehicle, refer to the precautions section.
2. Discharge fuel system pressure.
3. Disconnect battery negative cable.
4. Remove air cleaner cap sub-assembly.
5. Remove No. 1 engine cover sub-assembly.

➡Check for foreign matter on the pipe and around the connector before disconnecting the quick connector. Clean the connector if necessary.

6. Disconnect fuel tube sub-assembly, as follows:
   a. Remove the No. 1 fuel pipe clamp.

⁂ CAUTION

Do not use any tools in this following procedure. Check for foreign matter on the sealing surface of the disconnected pipe. Clean it if necessary.

   b. If the connector and pipe are stuck, pinch the connector, and push and pull the pipe to disconnect them.
   c. Separate the fuel tube from the fuel hose clamp.

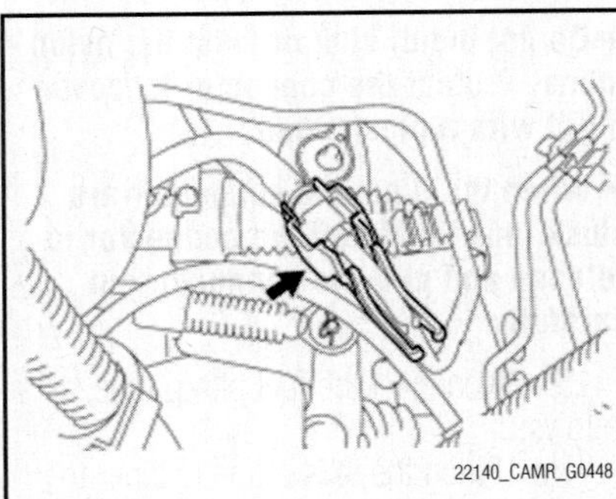

Fig. 139 Removing the No. 1 fuel pipe clamp

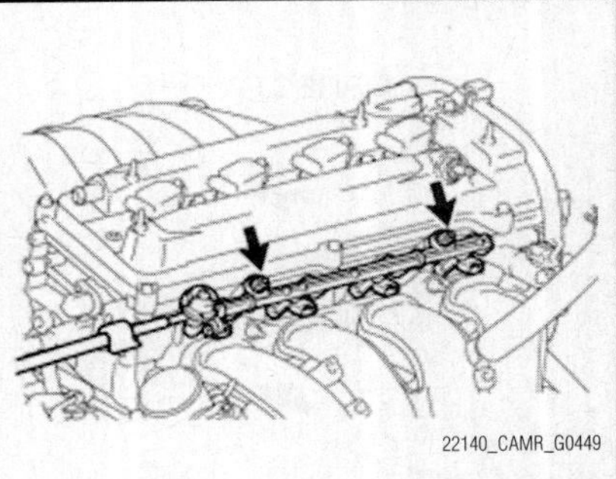

Fig. 140 Removing the fuel delivery pipe together with the 4 fuel injectors

7. Disconnect the No. 2 ventilation hose from the ventilation valve.
8. Remove fuel delivery pipe with injector as follows:
   a. Remove the 2 wire harness clamps.

⁂ CAUTION

Be careful not to drop the fuel injectors when removing the fuel delivery pipe.

   b. Remove the 2 bolts, then remove the fuel delivery pipe together with the 4 fuel injectors.
   c. Remove the 2 delivery pipe spacers from the cylinder head.
   d. Remove the 4 insulators from the cylinder head.
9. Pull out the 4 injectors from the delivery pipe.
10. Remove the 4 O-rings from the injectors.

To install:

11. Install the fuel injector assembly, as follows:
   a. Apply a light coat of spindle oil or gasoline to new O-rings, and install one to each injector.
   b. Apply a light coat of spindle oil or gasoline where the fuel delivery pipe contacts the O-ring.
   c. Apply a light coat of gasoline or spindle oil to the O-ring again, then install the right and left fuel injectors onto the fuel delivery pipe.

➡Make sure that the O-ring is not cracked or jammed before installing the injector.

   d. Check that the fuel injector rotates smoothly. If the fuel injector does not rotate, replace the O-ring.
12. Install fuel delivery pipe with injector as follows:
   a. Install 4 new insulators into the cylinder head.

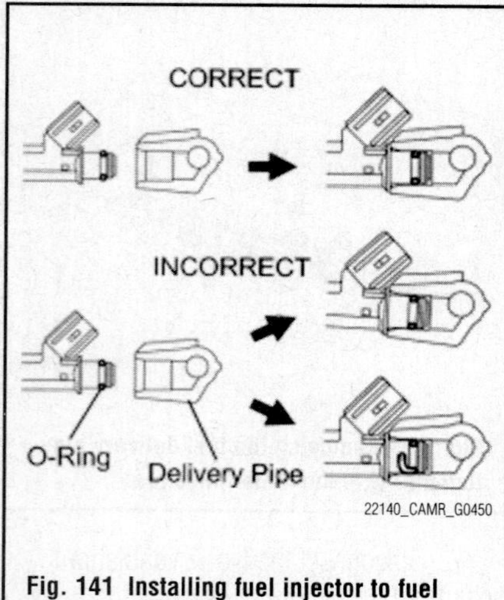

Fig. 141 Installing fuel injector to fuel tube

b. Install the 2 delivery pipe spacers onto the cylinder head.

c. Install the fuel delivery pipe together with the 4 fuel injectors, then temporarily tighten the 2 bolts.

➡Be careful not to drop the fuel injectors when installing the fuel delivery pipe.

d. Check that the fuel injector rotates smoothly. If the fuel injector does not rotate smoothly, replace the O-ring.

e. Tighten the 2 bolts to the specified torque and tighten to 15 ft. lbs. (20 Nm).

f. Connect the 4 fuel injector connectors.

g. Install the 2 wire harness clamps.

13. Connect the No. 2 ventilation hose to the ventilation valve.
14. Install the fuel tube to the fuel hose clamp.
15. Push the fuel tube connector until it makes a "click" sound.
16. Install the No. 1 fuel pipe clamp.
17. Install air cleaner cap sub-assembly.
18. Connect cable to negative battery terminal.

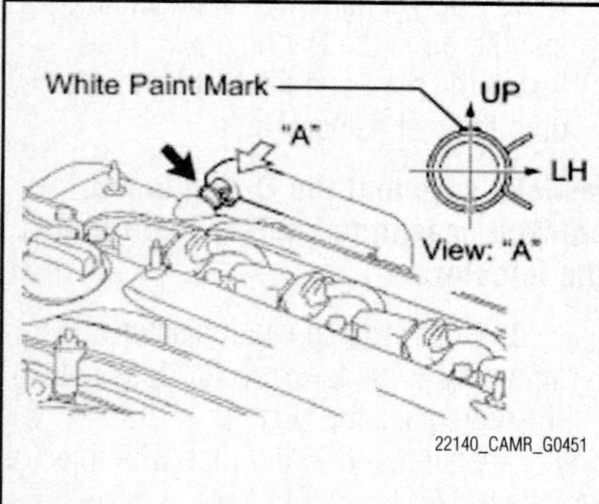

Fig. 142 Positioning paint mark and hose clamp

19. Check for fuel leaks.
20. Install No. 1 engine cover sub-assembly.

FUEL TANK

REMOVAL & INSTALLATION

See Figures 143 through 148.

1. Before servicing the vehicle, refer to the precautions section.
2. Discharge fuel system pressure.
3. Disconnect battery negative cable.
4. Remove rear seat cushion assembly.
5. Remove rear floor service hole cover.
6. Separate fuel pump tube sub-assembly.
7. Remove fuel tank vent tube set plate.
8. Remove fuel suction tube assembly with pump and gauge.
9. Drain fuel.
10. Remove center exhaust pipe assembly.
11. Disconnect no. 2 parking brake cable assembly.
12. Disconnect no. 3 parking brake cable assembly.
13. Remove rear stabilizer BAR No. 1 bracket.
14. Remove lower center fuel tank protector, as follows:

a. Remove the 4 bolts and the (Except SE grade).

b. Remove the 4 bolts and 2 clips (for SE Grade).

c. Remove the fuel tank protector (for SE Grade).

➡Check that there is no dirt or other foreign objects around the connector before removing fuel tubes, and clean the connector as necessary.

➡It is necessary to prevent mud or dirt from entering the connector. If mud or dirt gets in the connector, the O-rings may not seal properly.

➡Do not use any tools in these operations.

➡Do not bend, kink or twist the nylon tubes. Protect the connector by covering it with a plastic bag.

➡When the pipe and connector are stuck, push and pull the connector to release and pull the connector out carefully.

15. Disconnect the fuel pump tube, as follows:

a. Pinch the tabs of the retainer to remove the lock claws and pull it down as shown.

b. Pull out the fuel tank main tube.

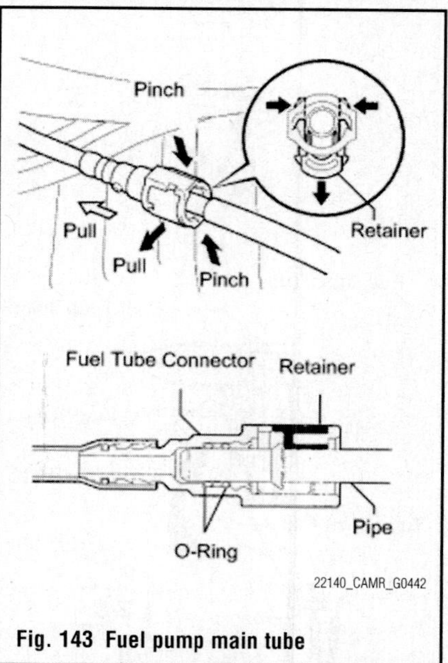

Fig. 143 Fuel pump main tube

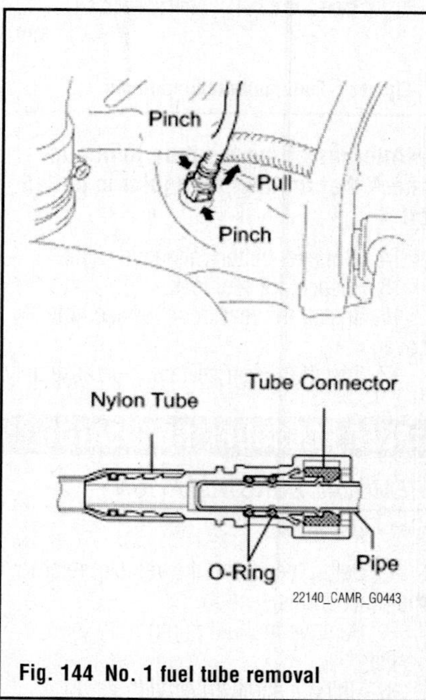

Fig. 144 No. 1 fuel tube removal

16. Pinch the tube connector and then pull out the No. 1 fuel tube.
17. Set up a transmission jack underneath the fuel tank.
18. Remove the 2 set bolts of the fuel tank bands.

➡Check that there is no dirt or other foreign objects around the connector before removing fuel tubes, and clean the connector as necessary.

➡It is necessary to prevent mud or dirt from entering the connector. If mud or dirt gets in the connector, the O-rings may not seal properly.

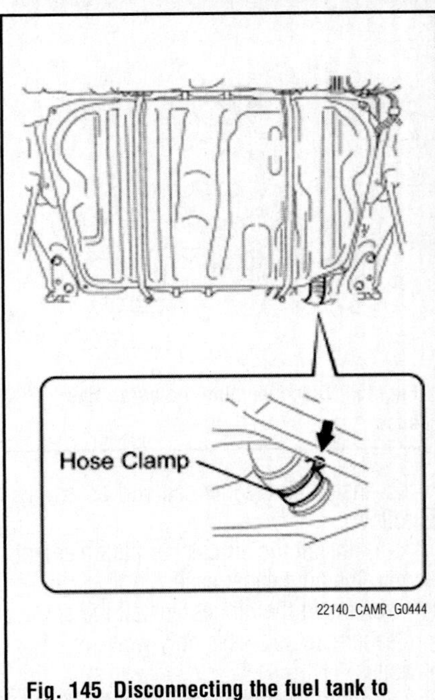

Fig. 145 Disconnecting the fuel tank to filter pipe hose—except PZEV

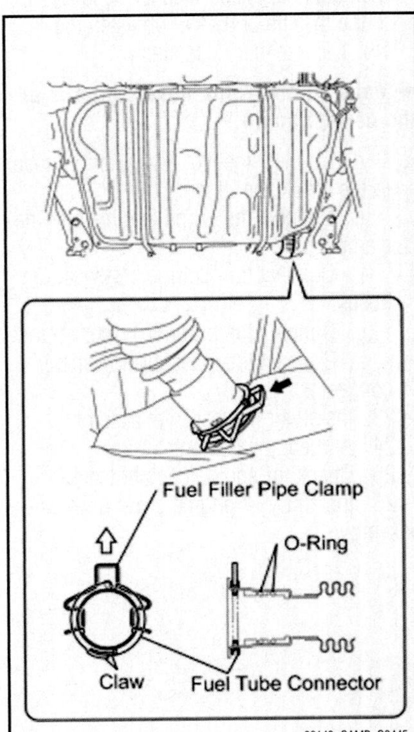

Fig. 146 Disconnecting the fuel tank to filter pipe hose—PZEV

➡**Do not use any tools in these operations.**

➡**Do not bend, kink or twist the nylon tubes. Protect the connector by covering it with a plastic bag.**

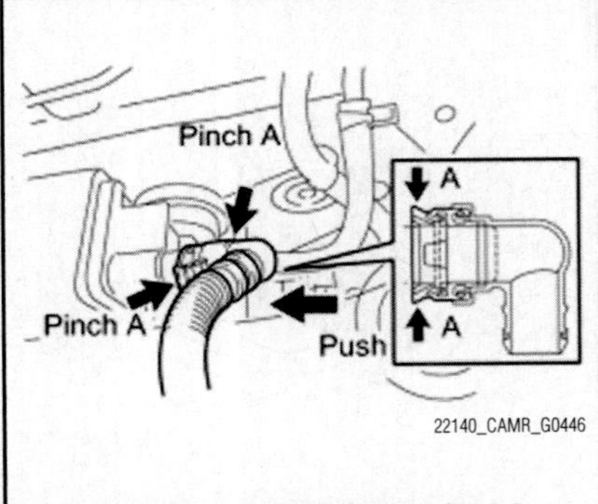

Fig. 147 Disconnect the fuel tank vent hose from the charcoal canister

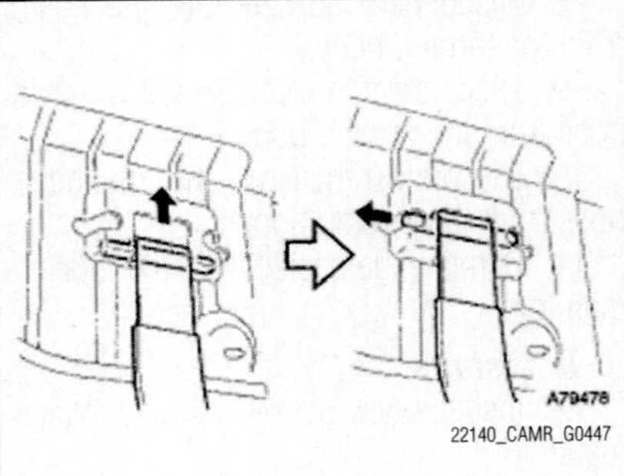

Fig. 148 Removing the 2 pins and 2 fuel tank bands

➡**When the pipe and connector are stuck, push and pull the connector to release and pull the connector out carefully.**

19. Remove the hose clamp and disconnect the fuel tank to filter pipe hose (except PZEV).
20. Remove the clamp and disconnect the fuel tank to filter pipe hose (for PZEV).
21. Slightly lower the transmission jack.
22. Disconnect the fuel tank vent hose from the charcoal canister, as follows:
    a. Push the connector deep into the charcoal canister to release the locking pin.
    b. Pinch portion A.
    c. Pull out the connector.
23. Remove the 2 pins and 2 fuel tank bands as shown in the illustration.
24. Remove the 4 clip nuts.

To install:

25. Install the 4 clip nuts.
26. Install the 2 fuel tank bands with the 2 pins.
27. Connect the fuel tank vent hose.
28. Connect the fuel tank inlet pipe with the fuel filter pipe clamp.
29. Tighten the 2 set bolts of the fuel tank bands to 29 ft. lbs. (39 Nm).
30. Connect the No. 1 fuel tube, as follows:
    a. Push the fuel tube connector into the pipe until the fuel tube connector makes a "click" sound.

➡**Check that there is no damage or foreign objects on the connected part.**

➡**After connecting, check if the fuel tube connector and the pipe are securely connected by trying to pull them apart.**

31. Connect the fuel pump tube, as follows:
    a. Push in the fuel pump tube connector to the pipe and push up the retainer so that the claws engage.

➡**Check that there is no damage or foreign objects on the connected part.**

➡**After connecting, check if the fuel tube connector and the pipe are securely connected by trying to pull them apart.**

32. Install the lower center fuel tank protector and tighten to 48 inch lbs. (5.4 Nm).
33. Install the No. 3 parking brake cable assembly with the bolt and nut and tighten to 53 inch lbs. (6 Nm), and 75 inch lbs. (8.5 Nm).
34. Install the No. 2 parking brake cable assembly with the bolt and nut and tighten to 53 inch lbs. (6 Nm), and 75 inch lbs. (8.5 Nm).
35. Install the center exhaust pipe assembly.
36. Install the fuel suction tube assembly with pump and gauge.
37. Install the fuel tank vent tube set plate.
38. Connect the fuel pump tube subassembly.
39. Add fuel.
40. Connect battery negative cable.
41. Inspect for fuel leak and exhaust gas leak.
42. Install the rear floor service hole cover.
43. Install the rear seat cushion assembly.

IDLE SPEED

ADJUSTMENT

Idle speed is maintained by the ECM. No adjustment is necessary or possible.

THROTTLE BODY

REMOVAL & INSTALLATION

See Figures 149 through 151.

1. Drain engine coolant.

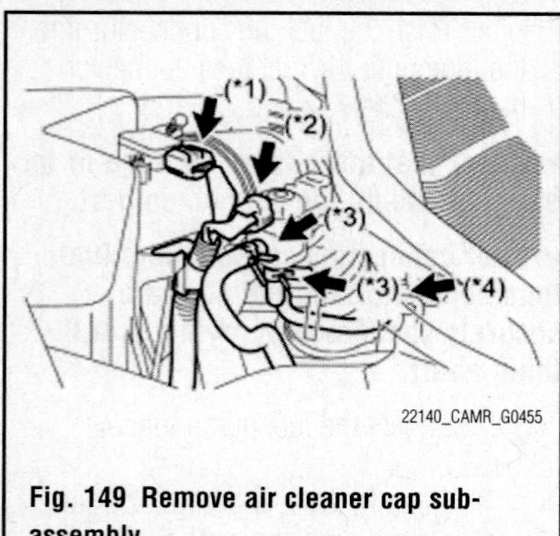

Fig. 149 Remove air cleaner cap sub-assembly

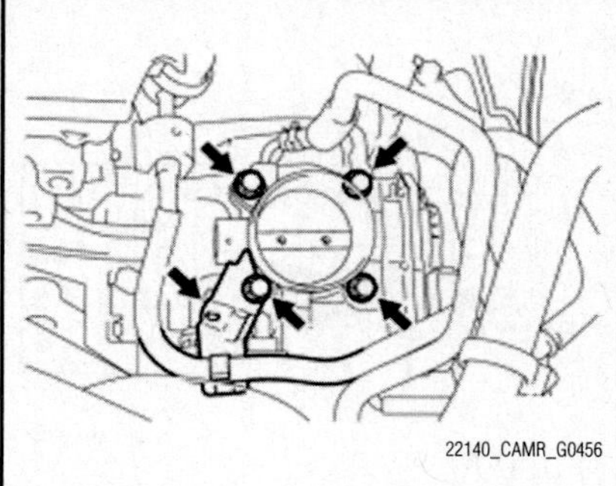

Fig. 150 Removing the 4 bolts and fuel pipe support

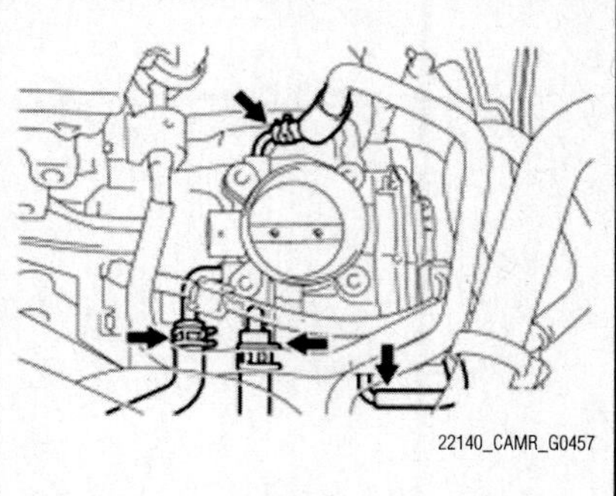

Fig. 151 Disconnecting the purge line hose

2. Remove No. 1 engine cover sub-assembly.

3. Remove air cleaner cap sub-assembly, as follows:

a. Disconnect the mass air flow meter connector.

b. Disconnect the purge VSV connector.

c. Disconnect the 2 purge VSV vacuum hoses.

d. Disconnect the purge line hose from the clamp.

e. Disconnect the No. 2 ventilation hose from the air cleaner hose.

f. Lock the No. 1 air cleaner hose clamp, and then disconnect the No. 1 air cleaner hose from the throttle body.

g. Remove the 2 bolts and air cleaner cap.

h. Remove the air cleaner filter element from the air cleaner case.

4. Remove air cleaner case sub-assembly.

5. Disconnect the throttle position sensor connector and wire harness clamp.

6. Remove the 4 bolts, and then remove the fuel pipe support and throttle body.

7. Disconnect the purge line hose from the throttle body.

8. Disconnect the water by-pass hose from the throttle body.

9. Disconnect the No. 2 water by-pass hose from the throttle body.

10. Disconnect the No. 1 throttle body hose from the throttle body.

11. Remove the gasket from the intake manifold.

To install:

12. Install a new gasket onto the intake manifold.

13. Connect the purge line hose to the throttle body.

14. Connect the water by-pass hose to the throttle body.

15. Connect the No. 2 water by-pass hose to the throttle body.

16. Connect the No. 1 throttle body hose to the throttle body.

17. Install the throttle body and fuel pipe clamp with the 4 bolts and tighten to 22 ft. lbs. (30 Nm).

18. Connect the fuel tube into the clamp.

19. Connect the throttle position sensor connector.

20. Connect the wire harness clamp.

21. Install air cleaner case sub-assembly.

22. Install air cleaner cap sub-assembly, as follows:

a. Install the air cleaner filter element onto the air cleaner case.

b. Insert the hinges. Install the air cleaner cap sub-assembly with the 2 bolts.

c. Align the matchmarks of the No. 1 air cleaner hose and throttle body, and then connect the air cleaner hose No. 1 to the throttle body and unfasten the No. 1 air cleaner hose clamp.

➡Make sure that the hose clamp is at the correct angle.

d. Connect the No. 2 ventilation hose to the air cleaner hose.

e. Connect the purge line hose to the clamp.

f. Connect the 2 purge VSV vacuum hoses.

g. Connect the purge VSV connector.

h. Connect the mass air flow meter connector.

23. Install air cleaner inlet assembly.

24. Add engine coolant.

25. Check for engine coolant leaks.

26. Install No. 1 engine cover sub-assembly.

HEATING & AIR CONDITIONING SYSTEM

BLOWER MOTOR

REMOVAL & INSTALLATION

See Figures 152 and 153.

1. Drain and recycle the engine coolant.
2. Disconnect the negative battery cable.
3. Remove instrument panel.
4. For TMC made:
   a. Disconnect the connector.
   b. Remove the 2 screws and blower assembly.
   c. Remove the 3 screws and blower with fan motor sub-assembly.
5. For TMMK made:
   a. Remove cooler expansion valve.
   b. Remove the connector and clamp, and disconnect the wire harness.
   c. Remove the 6 screws and then the blower assembly with the cooler evaporator sub-assembly.
   d. Remove the 3 screws and blower with fan motor sub-assembly.

To install:

6. To install, reverse removal procedure.

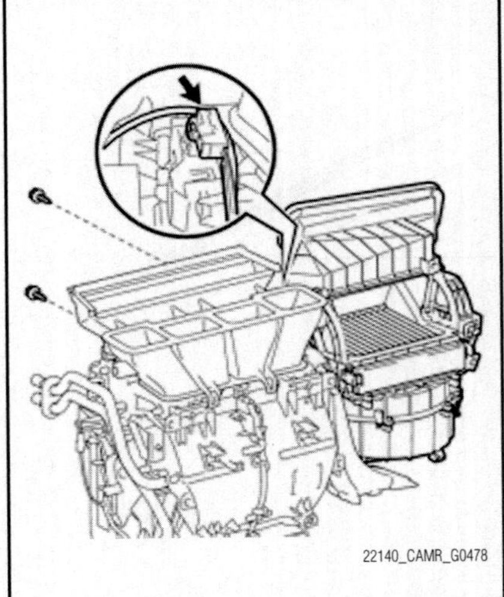

Fig. 152 Disconnecting the connector and removing the 2 screws—TMC

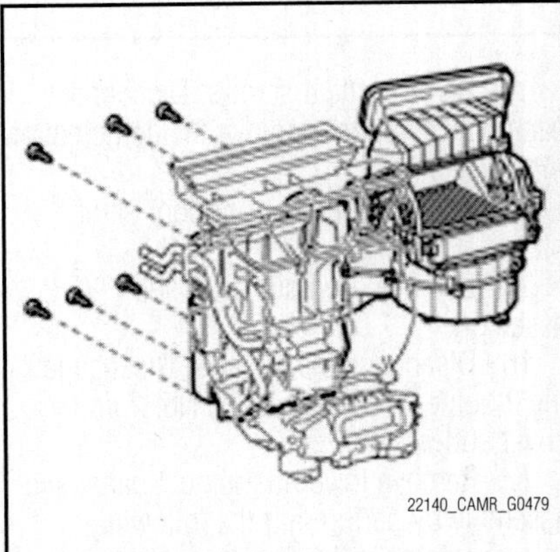

Fig. 153 Removing the 6 screws and blower assembly—TMMK

HEATER CORE

REMOVAL & INSTALLATION

See Figures 154 through 161.

CAUTION

Models are equipped with a Supplemental Restraint System (SRS), which uses an air bag. Whenever working near any of the SRS components, such as the impact sensors, the air bag module, steering column and instrument panel, disable the SRS.

1. Before servicing the vehicle, refer to the Precautions Section.

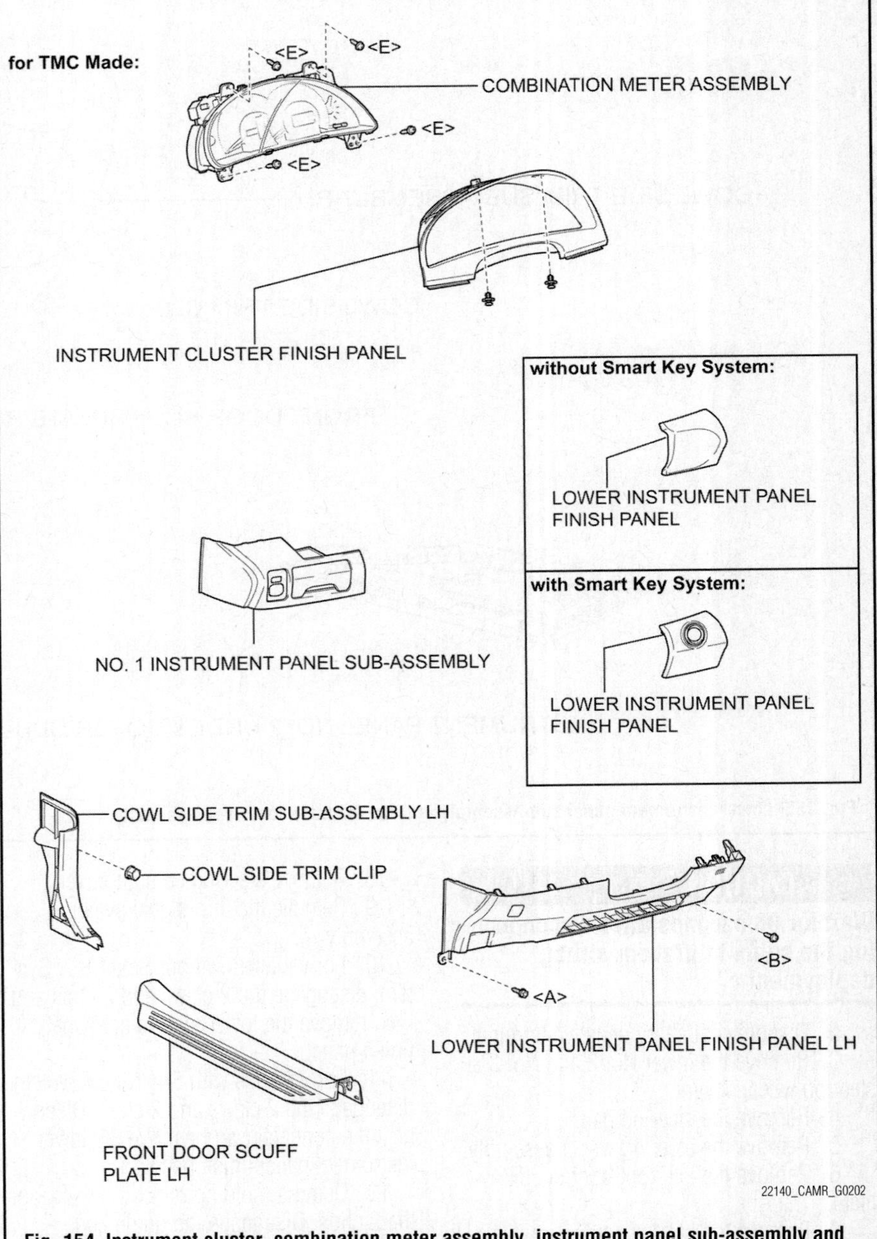

Fig. 154 Instrument cluster, combination meter assembly, instrument panel sub-assembly and lower trip sub-assembly

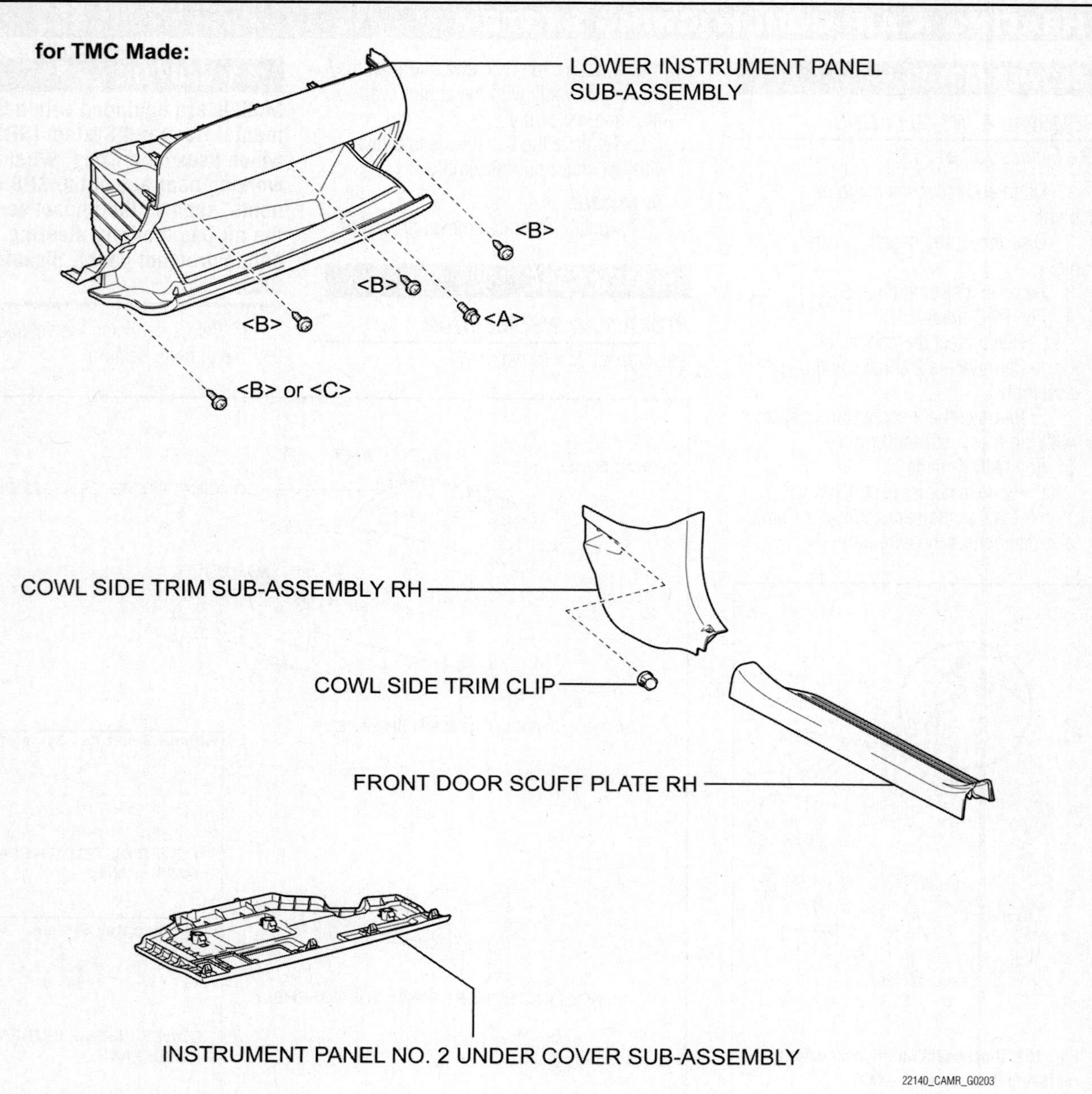

Fig. 155 Lower instrument panel sub-assembly and No. 2 under cover sub-assembly

CAUTION

Wait for 90 seconds after disconnecting the cable to prevent airbag deployment.

2. Disconnect battery negative terminal.
3. Remove the lower No. 2 and No. 3 steering wheel covers.
4. Remove the steering pad.
5. Remove the steering wheel assembly.
6. Remove the LH front door scuff plate.
7. Remove the LH cowl side trim sub-assembly.
8. Remove steering column cover.
9. Remove the turn signal switch assembly.
10. For vehicles without Smart Key System, disengage the 2 claws and 2 clips and then remove the lower instrument panel finish panel.
11. For vehicles with Smart Key System, disengage the 2 claws and 2 clips. Disconnect the connector and remove the lower instrument panel finish panel.
12. Using a molding remover, disengage the 2 clips. Disengage the guide and 4 claws, and then remove the No. 1 instrument cluster finish panel.
13. Remove the 4 screws. Disconnect each connector and remove the combination meter assembly.
14. Remove the RH front door scuff plate.
15. Remove cowl side trim sub-assembly.
16. Disengage the 4 claws. Disengage the 2 guides and remove the No. 2 under cover sub-assembly.
17. Remove lower instrument panel sub-assembly by performing the following:
    a. Remove the 4 screws.
    b. Disengage the 3 claws and the 3 clips.

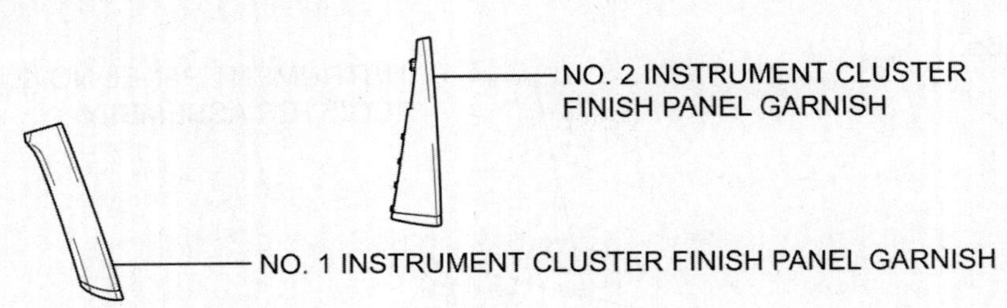

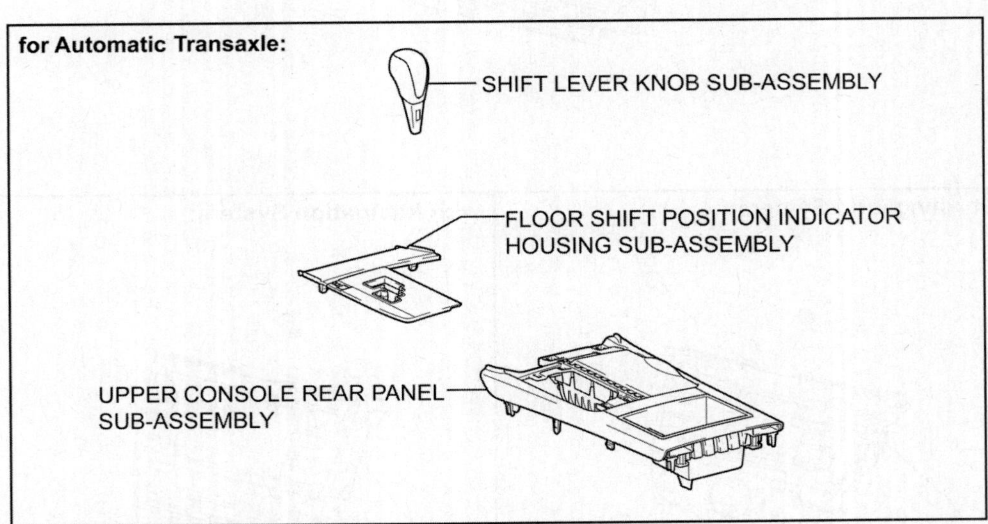

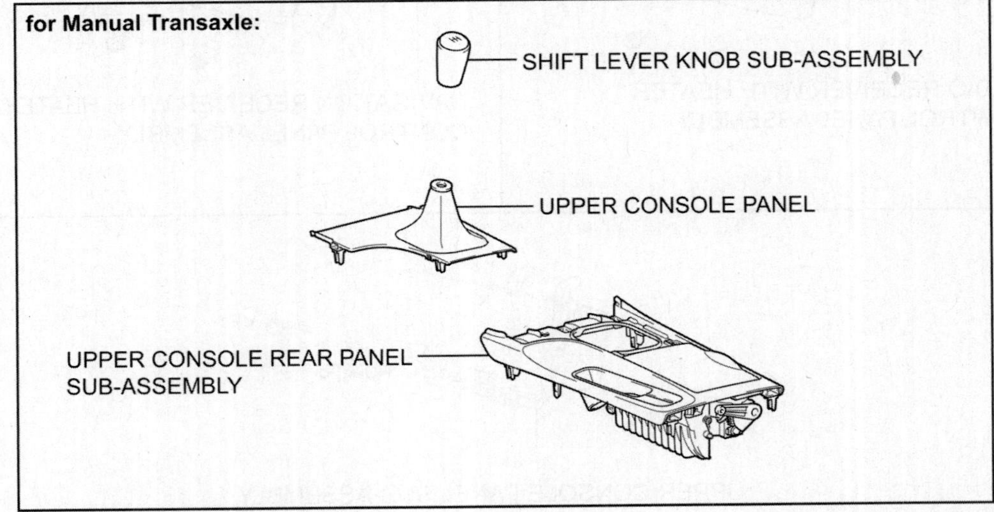

Fig. 156 Instrument cluster panel garnish, floor housing sub-assembly and upper console rear panel sub-assembly

c. Disconnect the connector and remove the lower instrument panel sub-assembly.

18. Turn the shift lever knob counterclockwise and remove the shift lever knob sub-assembly.

19. Disengage the 2 clips and remove the No. 1 instrument cluster finish panel garnish.

20. Disengage the 2 clips and remove the No. 2 instrument cluster finish panel garnish.

21. Disengage the 6 claws and the 3 clips, and then remove the floor shift position indicator housing sub-assembly. If equipped with Seat Heater System, disconnect each connector.

22. Disengage the 3 claws and the 5 clips. Disconnect the connector and remove the upper console rear panel sub-assembly.

23. Remove instrument panel no. 2 register assembly by performing the following:

a. Apply protective tape to the areas.

b. Using a moulding remover, disengage the 3 clips.

c. Using a moulding remover, disengage the 4 clips.

d. Disconnect the connector and remove the instrument panel No. 2 register assembly.

for TMC Made:

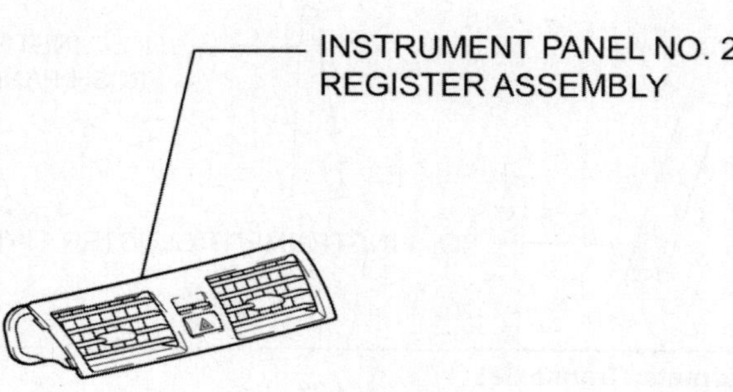

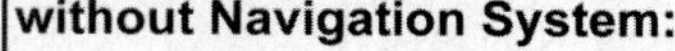

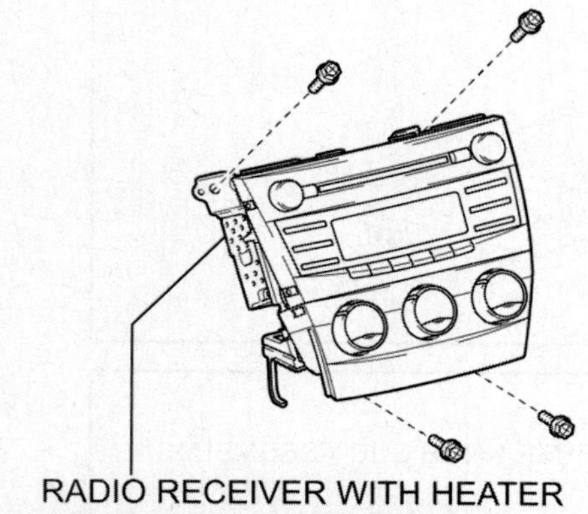

with Navigation System:

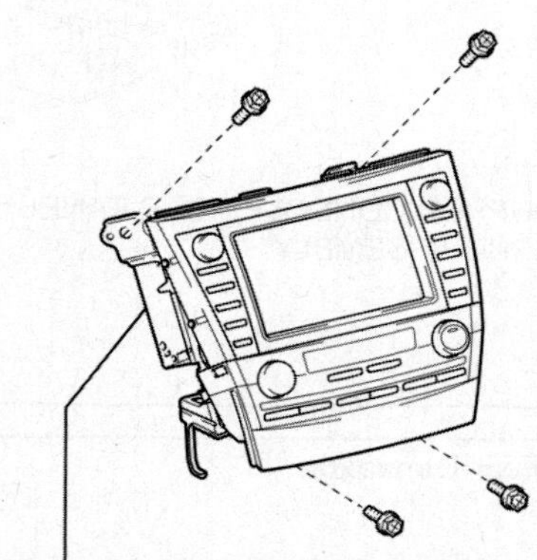

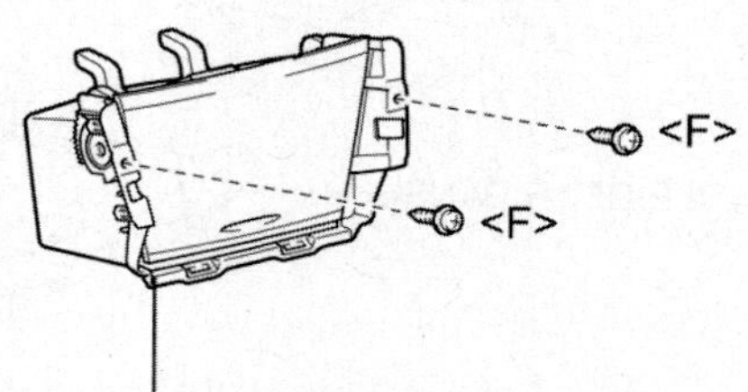

22140_CAMR_G0205

Fig. 157 Instrument panel no. 2 register assembly, control panel assembly and upper console sub-assembly

24. Remove radio receiver with heater control panel assembly.
25. Remove the console box pocket.
26. Remove the console box pocket.
27. Remove the console box assembly by performing the following:
    a. Remove the 2 screws.
    b. Disengage the clamp.
    c. Remove the 2 bolts and the console box assembly.
28. Remove both of the front console box inserts by performing the following:
    a. Remove the 3 screws.
    b. Disengage the clip and remove the front console box insert.
29. Remove the LH front pillar garnish.
30. Disengage the 4 clips and remove the instrument panel No. 1 register assembly.
31. Remove instrument panel no. 1 speaker panel sub-assembly by performing the following:
    a. Disengage the 6 claws and the 2 clips.
    b. Disengage the 2 guides and remove the instrument panel No. 1 speaker panel sub-assembly.
32. Remove RH front no. 2 speaker assembly.
33. Remove the RH front pillar garnish.

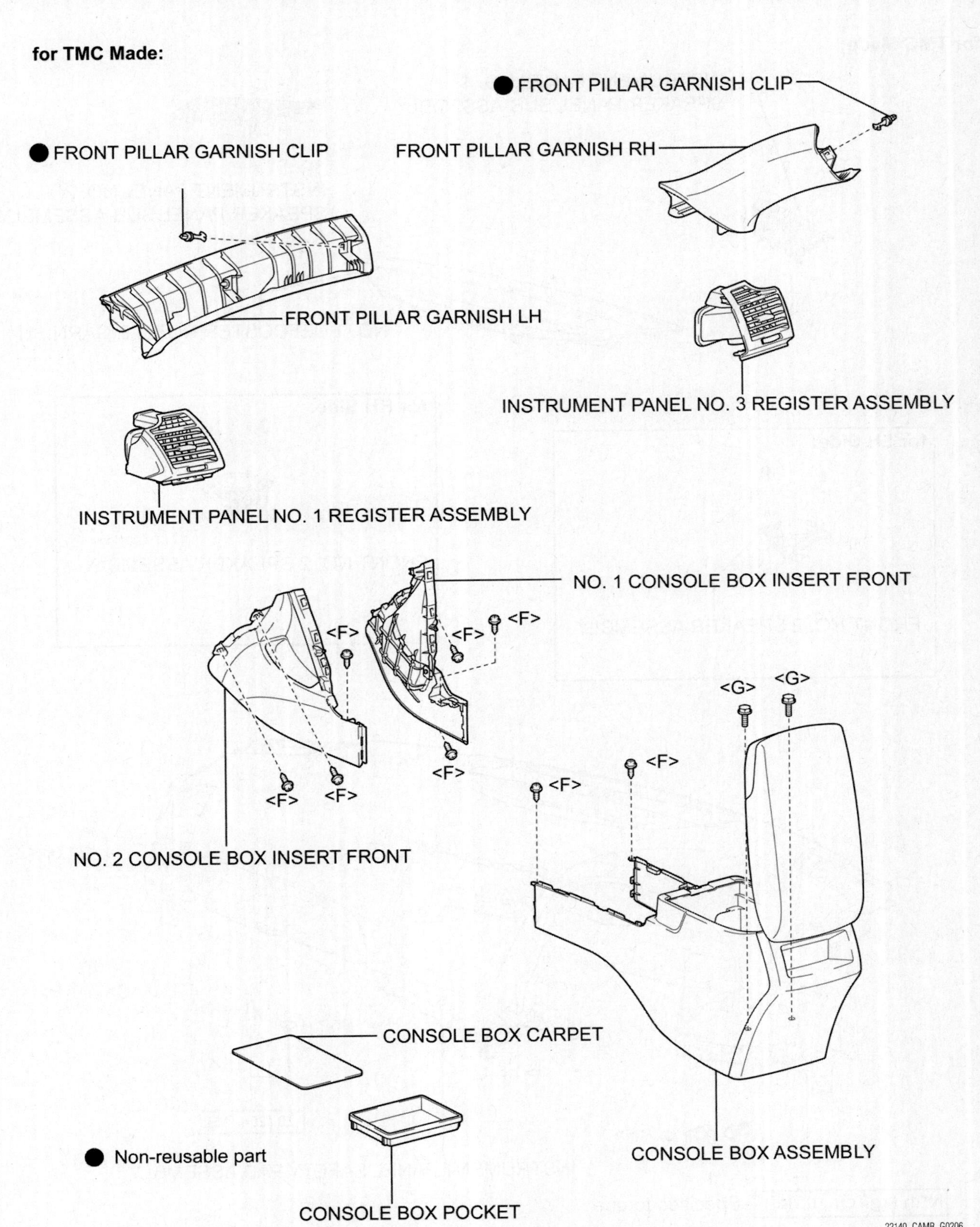

Fig. 158 Console box assembly, front console box inserts, LH front pillar garnish and instrument panel No. 1 and No. 3 register assembly

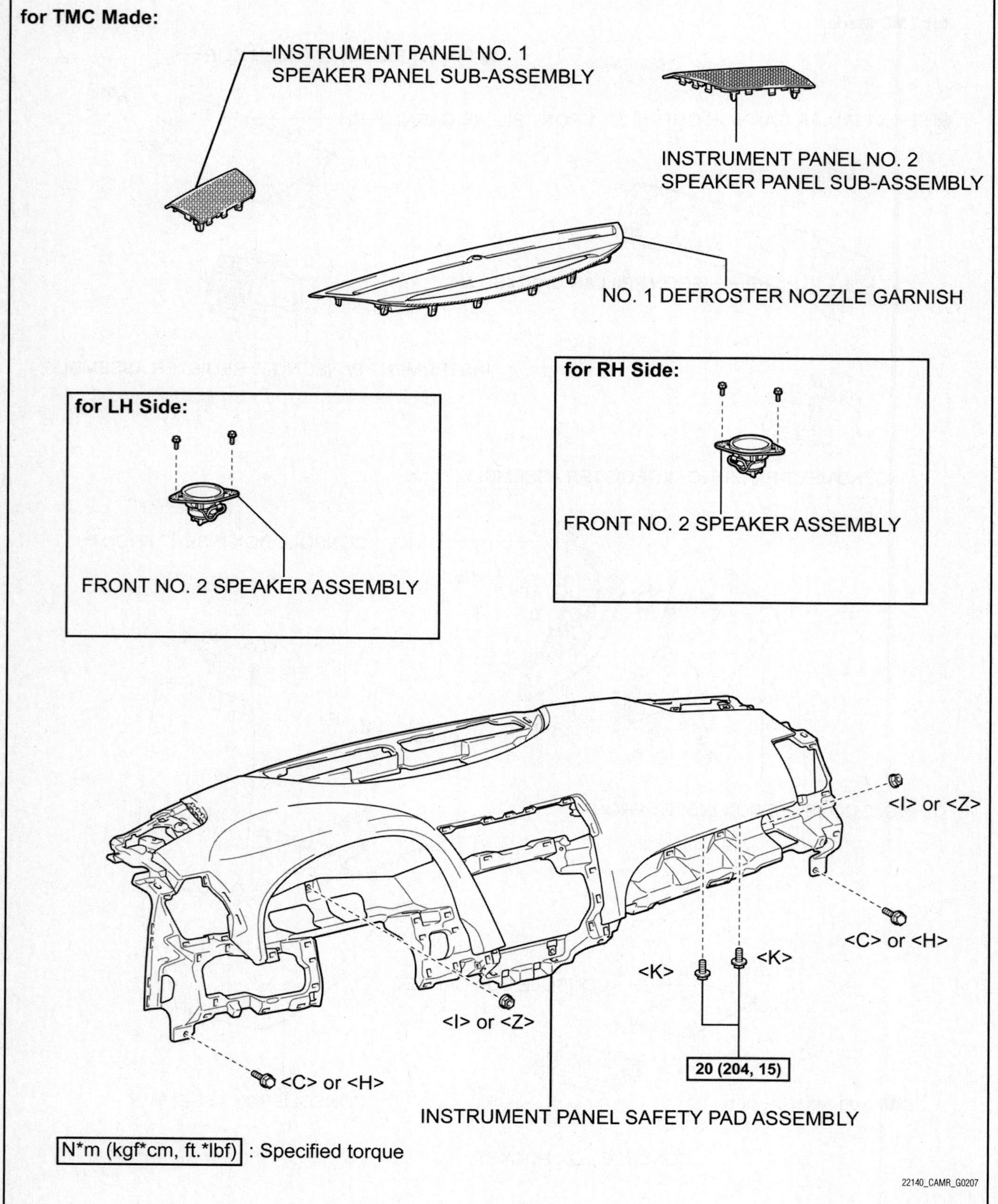

Fig. 159 Speakers, speaker panel sub-assembly and No. 1 defroster nozzle garnish

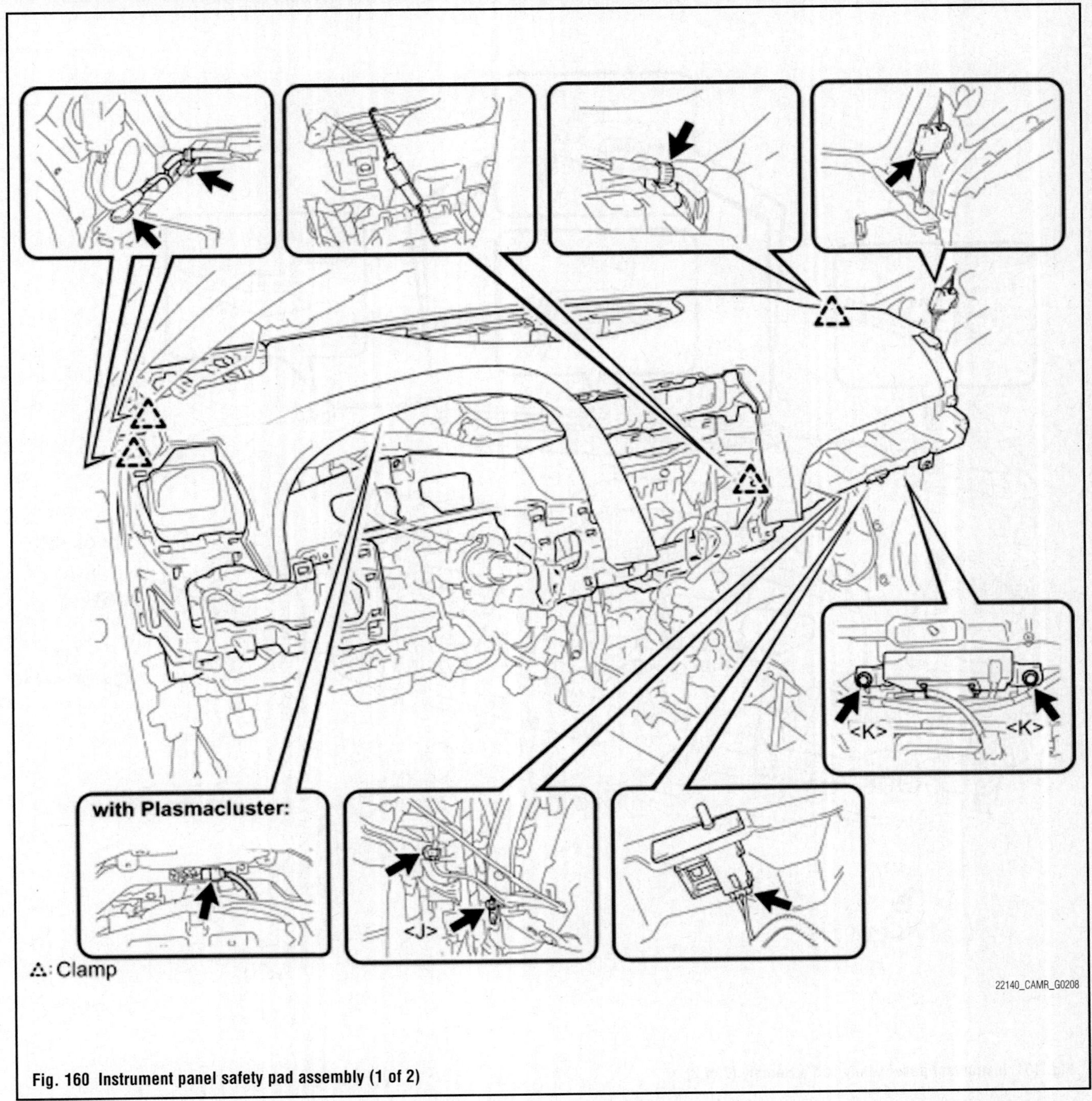

Fig. 160 Instrument panel safety pad assembly (1 of 2)

34. Disengage the 4 clips and remove the instrument panel No. 3 register assembly.

35. Remove instrument panel no. 2 speaker panel sub-assembly by performing the following:

a. Disengage the 6 claws and the 2 clips.

b. Disengage the 2 guides and remove the instrument panel No. 2 speaker panel sub-assembly.

36. Remove LH and RH front no. 2 speaker assemblies.

37. Remove no. 1 defroster nozzle garnish by performing the following:

a. Disengage the 8 clips and the 4 guides.

b. Disconnect each connector and remove the No. 1 defroster nozzle garnish.

38. Disconnect instrument panel wire assembly.

39. Remove instrument panel safety pad assembly by performing the following:

a. Disengage each clamp.

b. Disconnect each connector.

c. Remove the bolt (J).

CAUTION

Models are equipped with a Supplemental Restraint System (SRS), which uses an air bag. Whenever working near any of the SRS components, such as the impact sensors, the air bag module, steering column and instrument panel, disable the SRS.

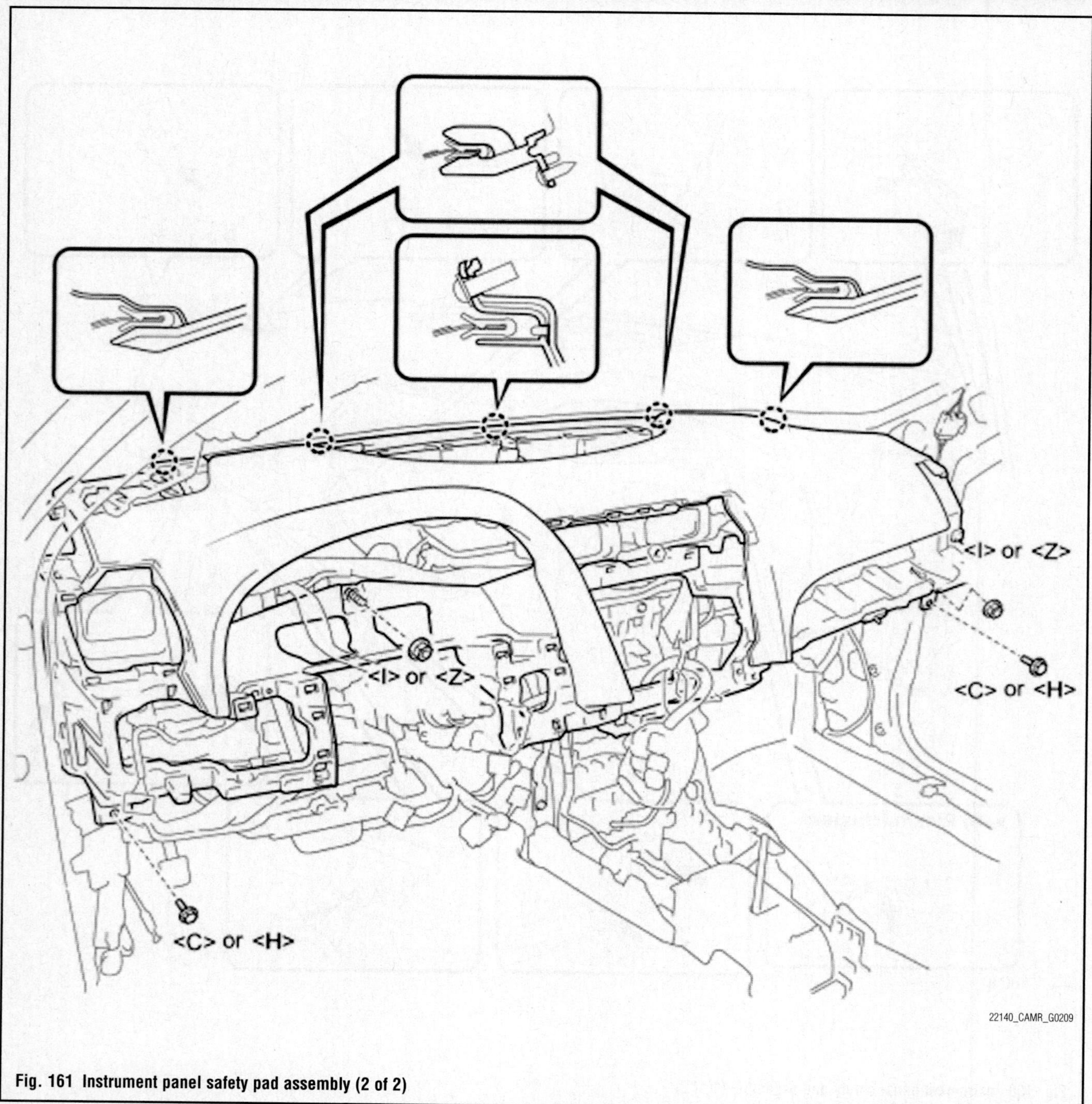

Fig. 161 Instrument panel safety pad assembly (2 of 2)

d. Remove the 2 passenger airbag bolts (K).

e. If equipped with Plasmacluster, disconnect the connector

f. Disconnect the connector

g. Remove the 2 bolts (C) or (H).

h. Disengage the 5 claws and remove the instrument panel safety pad assembly.

i. Disengage the claw and remove the 5 instrument panel stays.

40. Remove heater core as necessary.

To install:

41. Installation is the reverse of the removal procedure.
42. Perform initialization.
43. Inspect the steering pad.
44. Inspect the SRS warning light.

STEERING

POWER RACK & PINION STEERING GEAR

REMOVAL & INSTALLATION

See Figures 162 through 165.

CAUTION

Models are equipped with a Supplemental Restraint System (SRS), which uses an air bag. Whenever working near any of the SRS components, such as the impact sensors, the air bag module, steering column and instrument panel, disable the SRS.

➡Be sure to turn the front wheels straight ahead when removing and installing the power steering link assembly.

➡If disconnecting the steering sliding yoke and the pinion shaft of the power steering link assembly, be sure to put matchmarks before starting the operation.

1. Before servicing the vehicle, refer to the precautions section.
2. Place front wheels facing straight ahead.
3. Disconnect the negative battery cable.
4. Remove front wheels.
5. Separate steering sliding yoke, as follows:
   a. Secure the steering wheel with the seat belt in order to prevent rotation. This operation is useful to prevent damage to the spiral cable.
   b. Remove the bolt and slide the steering sliding yoke. Do not separate the steering sliding yoke from the power steering link assembly.
   c. Put matchmarks on the steering sliding yoke and the power steering link assembly.
   d. Separate the steering sliding yoke from the power steering link assembly.
6. Separate both tie rod assemblies, as follows:
   a. Remove the cotter pin and the nut.
   b. Using SST: 09628-00011, separate the tie rod assembly LH from the steering knuckle.
7. Remove engine assembly with transaxle.
8. Disconnect pressure feed tube assembly, as follows:
   a. Using SST: 09023-12701, disconnect the pressure feed tube assembly (return tube side) from the power steering link assembly.
   b. Using SST: 09023-12701, disconnect the pressure feed tube assembly (pressure feed tube side) from the power steering link assembly.
   c. Remove the 2 bolts and separate the pressure feed tube clamp.

➡Because the nut has its own stopper, do not turn the nut. Loosen the bolt with the nut fixed.

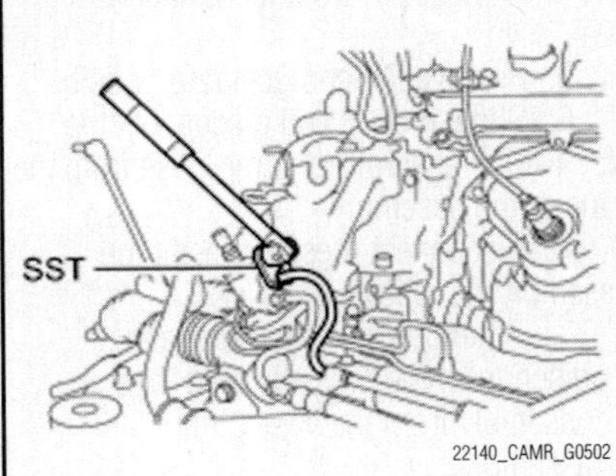

Fig. 162 Disconnect the pressure feed tube assembly—return tube side

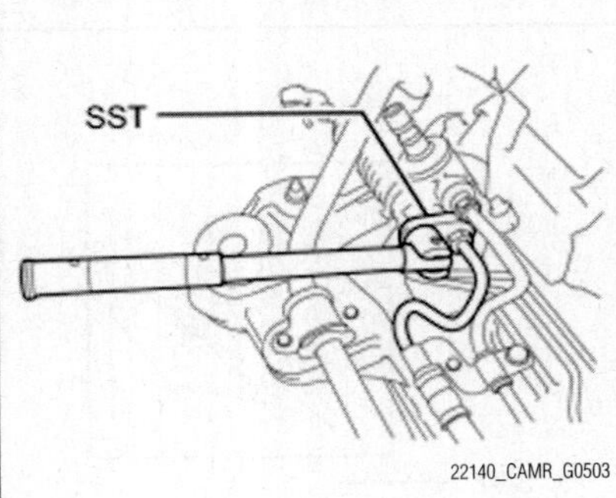

Fig. 163 Disconnect the pressure feed tube assembly—pressure feed tube side

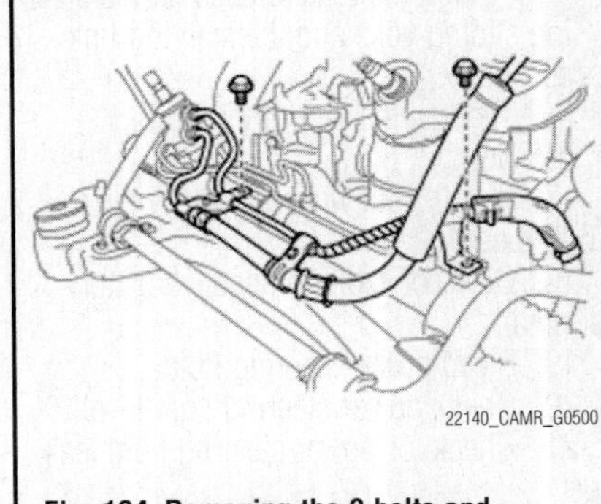

Fig. 164 Removing the 2 bolts and pressure feed tube clamp

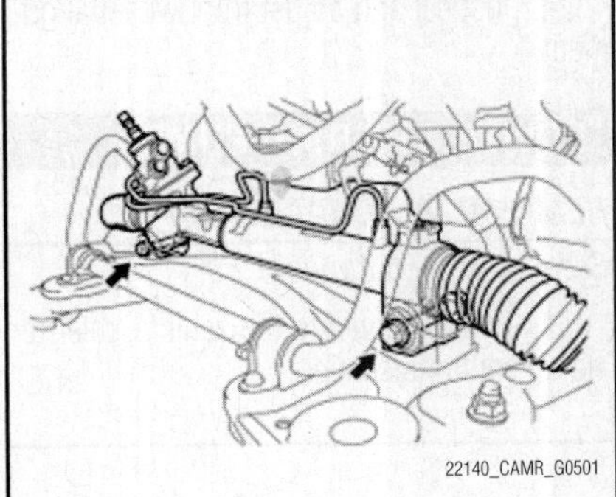

Fig. 165 Removing power steering link assembly

9. Remove the 2 bolts, 2 nuts, and the power steering link assembly.
10. If equipped, remove the power steering rack housing heat insulator from the power steering link assembly.

To install:

11. If equipped, install the power steering rack housing heat insulator to the power steering link assembly.
12. Install the power steering link assembly with the 2 bolts and 2 nuts. Tighten to 52 ft. lbs. (70 Nm).
13. Connect pressure feed tube assembly, as follows:
   a. Temporarily connect the pressure feed tube assembly to the power steering link assembly.
   b. Install the pressure feed tube assembly clamp with the 2 bolts and tighten to 87 inch lbs. (9.8 Nm).

➡Use a torque wrench with a fulcrum length of 11.81 inches (300 mm).

➡This torque value is effective when SST is parallel to the torque wrench.

   c. Using SST: 09023-12701, tighten the pressure feed tube assembly (pressure feed tube side) to 16 ft. lbs. (22 Nm).
   d. Using SST: 09023-12701, tighten the pressure feed tube assembly (return tube side) to 16 ft. lbs. (22 Nm).
14. Install engine assembly with transaxle.
15. Connect both of the tie rod assemblies to the steering knuckle with the nut. Tighten to 36 ft. lbs. (49 Nm). Install a new cotter pin. Further tighten the nut up to 60°if the holes for the cotter pin are not aligned.

16. Connect steering sliding yoke, as follows:

a. Align the matchmarks on the steering sliding yoke and the steering link assembly.

b. Install the bolt to 26 ft. lbs. (35 Nm).

17. Install front wheels and tighten to 76 ft. lbs. (103 Nm).

18. Connect cable to negative battery terminal.

19. Bleed power steering fluid.

20. Check power steering fluid level.

21. Check for power steering fluid leakage.

22. Check for exhaust gas leaks.

23. Place front wheels facing straight ahead.

24. Inspect and adjust front wheel alignment.

POWER STEERING PUMP

REMOVAL & INSTALLATION

See Figures 166 through 169.

1. Before servicing the vehicle, refer to the precautions section.

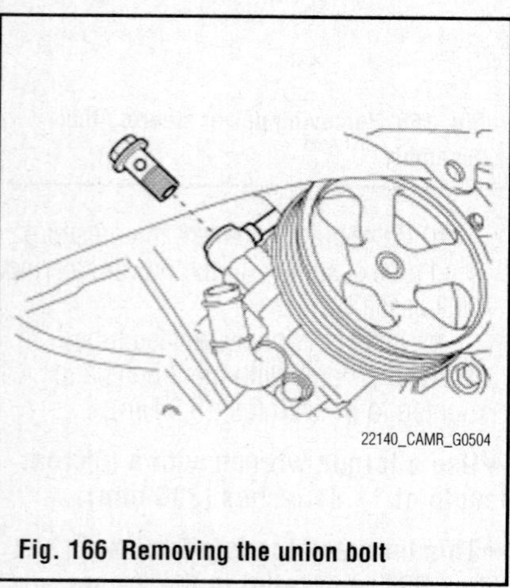

Fig. 166 Removing the union bolt

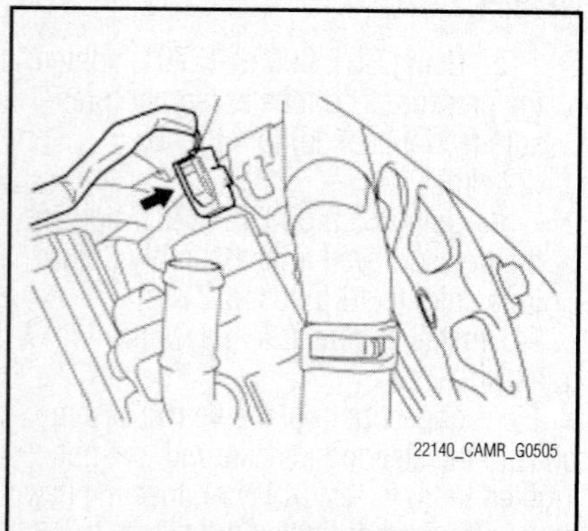

2. Drain power steering fluid.

3. Remove RH engine under cover.

4. Remove RH front fender apron seal.

5. Remove fan and generator v belt.

6. Slide the clip and disconnect the No. 1 fluid reservoir to pump hose from the vane pump assembly.

7. Disconnect pressure feed tube assembly, as follows:

a. Remove the union bolt and disconnect the pressure feed tube assembly from the vane pump assembly.

b. Remove the gasket from the pressure feed tube assembly.

8. Disconnect the power steering fluid pressure switch connector.

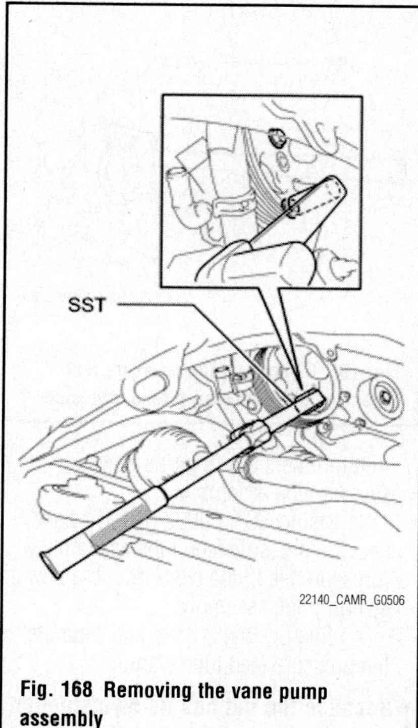

Fig. 168 Removing the vane pump assembly

Fig. 167 Disconnecting the power steering fluid pressure switch connector

Fig. 167 Disconnecting the power steering fluid pressure switch connector

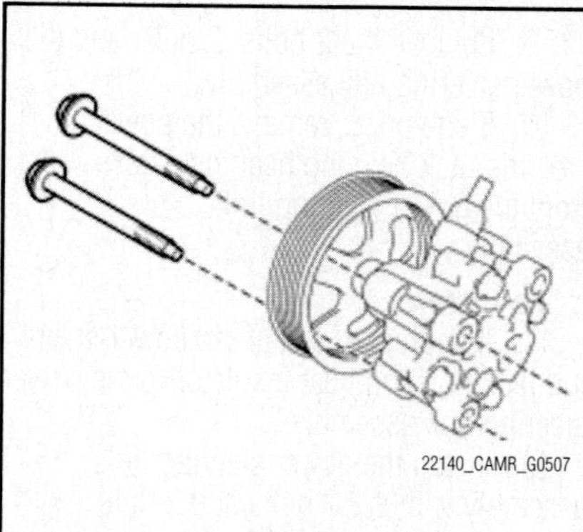

Fig. 169 Removing 2 bolts from the vane pump assembly

9. Remove vane pump assembly, as follows:

a. Using SST: 09249-63010, loosen the 2 bolts and remove the vane pump assembly.

b. Remove the 2 bolts from the vane pump assembly.

To install:

10. Install vane pump assembly, as follows:

a. Temporarily install the 2 bolts to the vane pump assembly.

b. Install the vane pump assembly.

➡Use a torque wrench with a fulcrum length of 11.81 inches (300 mm).

➡This torque value is effective when SST is parallel to the torque wrench.

c. Using SST: 09249-63010, tighten the 2 bolts to 32 ft. lbs. (43 Nm).

11. Connect the connector to the power steering fluid pressure switch.

12. Connect pressure feed tube assembly, as follows:

a. Install a new gasket to the pressure feed tube assembly.

➡Make sure that the stopper of the pressure feed tube assembly contacts the vane pump assembly securely as shown in the illustration.

b. Connect the pressure feed tube assembly to the vane pump assembly with the union bolt. Tighten to 37 ft. lbs. (50 Nm).

13. Connect No. 1 fluid reservoir to pump hose, as follows:

➡Connect the No. 1 oil reservoir to pump hose with the paint mark facing toward the rear of the vehicle.

➡Push the No. 1 oil reservoir to pump hose as far as it will go.

➡Install the clip at the position specified in the illustration.

a. Connect the No. 1 fluid reservoir to pump hose to the vane pump assembly with the clip.

14. To complete installation, reverse removal procedure.

BLEEDING

1. Before servicing the vehicle, refer to the precautions section.

2. Check the fluid level.

3. Jack up the front of the vehicle and support it with stands.

4. With the engine stopped, turn the wheel slowly from lock to lock several times.
5. Lower the vehicle.
6. Start the engine.
7. Run the engine at idle for a few minutes.
8. With the engine idling, turn the wheel left or right to the full lock position and keep it there for 2 to 3 seconds, then turn the wheel to the opposite full lock position and keep it there for 2 to 3 seconds.
9. Repeat the above steps several times.
10. Stop the engine.
11. Check for foaming or emulsification. If the system has to be bled twice because of foaming or emulsification, check for fluid leaks in the system.
12. Check the fluid level.

SUSPENSION

FRONT SUSPENSION

COIL SPRING

REMOVAL & INSTALLATION

See Figures 170 through 175.

1. Before servicing the vehicle, refer to the precautions section.
2. Remove the front shock absorber.
3. As shown in the illustration, secure the front shock absorber with coil spring in a vise using aluminum plates by clamping onto a double nutted bolt affixed to the bracket at the bottom of the absorber.

➡Do not use an impact wrench.

➡If the front coil spring is compressed at an angle, using 2 SST will make the work easier.

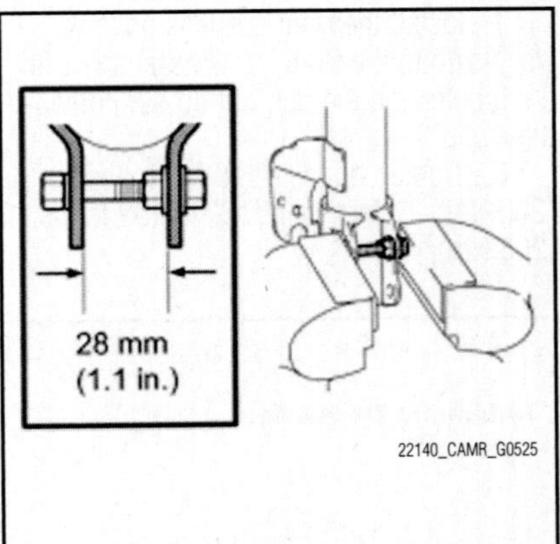

Fig. 170 Secure the front shock absorber

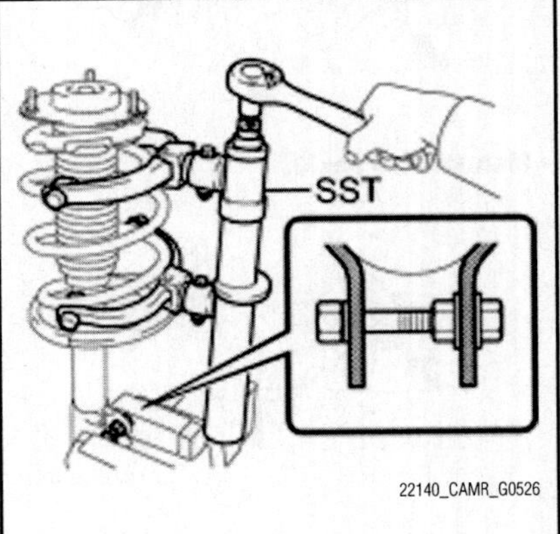

Fig. 171 Removing coil spring components

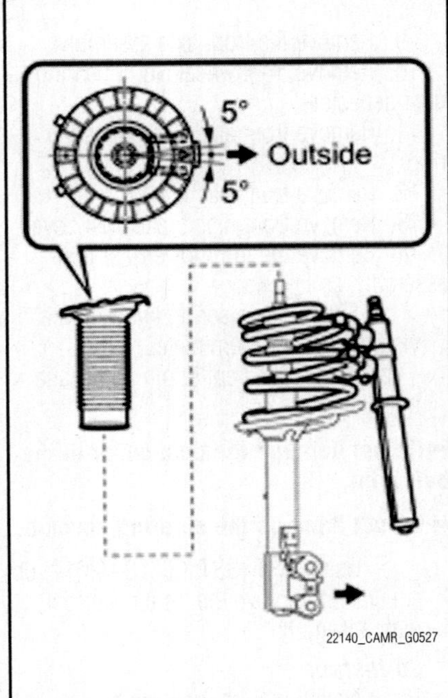

Fig. 172 Installing the front coil spring upper insulator

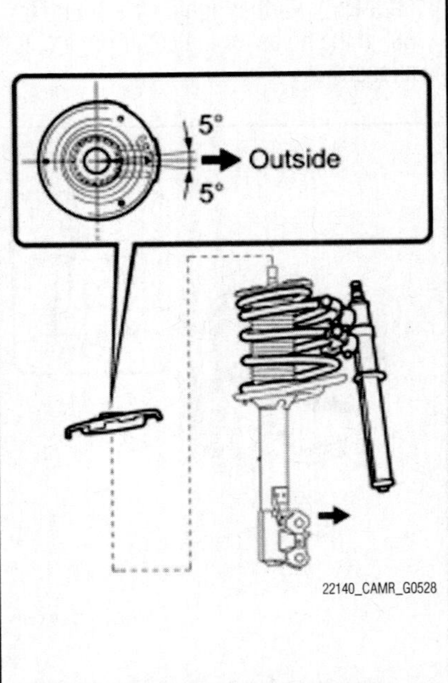

Fig. 173 Installing the front coil spring upper seat

4. Using SST: 09727-30021, compress the front coil spring.
5. Remove the front suspension support sub-assembly, front suspension support bearing, front coil spring upper seat, front coil spring upper insulator, front coil spring, front spring bumper, and front coil spring lower insulator from the front shock absorber.

To install:

6. Install front coil spring as follows:
   a. Install the front spring bumper to the piston rod.

➡Align the 2 protrusions of the front coil spring lower insulator and the 2 holes in the front shock absorber.

➡Do not use an impact wrench.

   b. Install the front coil spring lower insulator onto the front shock absorber.
   c. Using SST: 09727-30021, compress the front coil spring.

➡The smaller diameter end of the front coil spring must face upward.

➡Fit the lower end of the front coil spring into the gap of the insulator.

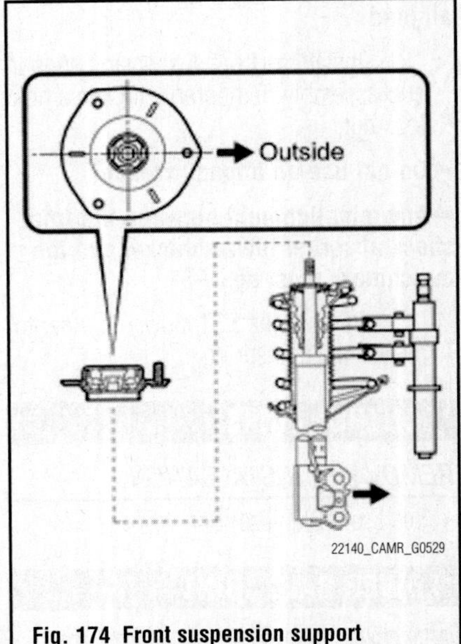

Fig. 174 Front suspension support sub-assembly

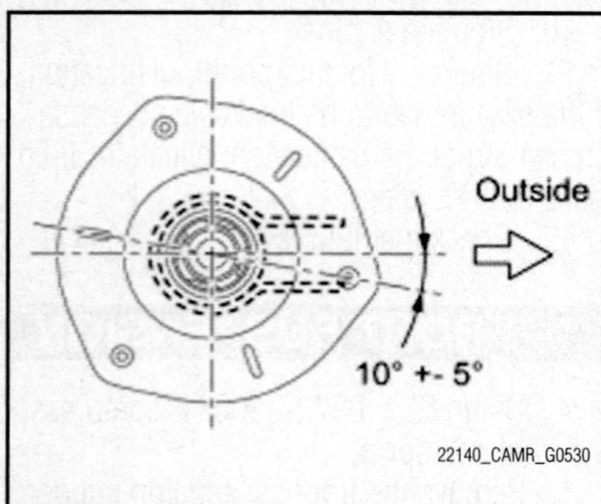

Fig. 175 Aligning the front shock absorber lower bracket and arrows

d. Install the front coil spring to the front shock absorber.

➡Any misalignment between the front shock absorber lower bracket and the matchmark must be +/-5°.

e. Install the front coil spring upper insulator as shown in the illustration.

➡Any misalignment between the front shock absorber lower bracket and the matchmark must be +/-5°.

7. Install the front coil spring upper seat with the mark facing to the outside of the vehicle.

➡If there is foreign matter inside the front suspension support bearing, replace it with a new one.

a. Install a new front suspension support bearing.

➡Check that the flats on the piston rod and the flats on the front suspension support sub-assembly are aligned.

b. Install the front suspension support sub-assembly. Temporarily tighten a new lock nut.

➡Do not use an impact wrench.

➡Any misalignment between the front shock absorber lower bracket and the matchmark must be +/-5°.

c. Remove the SST slowly in order to release the coil spring.

CONTROL LINKS

REMOVAL & INSTALLATION

See Front Stabilizer Bar.

LOWER BALL JOINT

REMOVAL & INSTALLATION

See Figure 176.

1. Before servicing the vehicle, refer to the precautions section.
2. Remove the front wheel.
3. Remove the front axle hub nut.
4. Separate the front speed sensor.
5. Separate the front disc the brake caliper assembly.
6. Remove front disc.
7. Separate the tie rod assembly.
8. Separate the No. 1 front lower suspension arm.
9. Remove the front axle assembly.
10. Remove front wheel No. 1 bearing dust deflector.
11. Remove front axle hub hole snap ring.
12. Remove front axle hub.
13. Remove front disc brake dust cover.
14. Remove the front lower ball joint assembly, as follows:
    a. Secure the steering knuckle in a vise using aluminum plates.
    b. Remove the cotter pin and castle nut.

➡Do not damage the dust cover of the ball joint.

➡Do not damage the steering knuckle.

c. Using SST (SST: 09628-62011) or equivalent, remove the front lower ball joint assembly.

To install:

15. Installation is the reverse of the removal procedure, noting the following:
    a. Install the front lower ball joint assembly to the steering knuckle with the castle nut and tighten to 91 ft. lbs. (123 Nm). Further tighten the nut up to 60° if the holes for the cotter pin are not aligned.

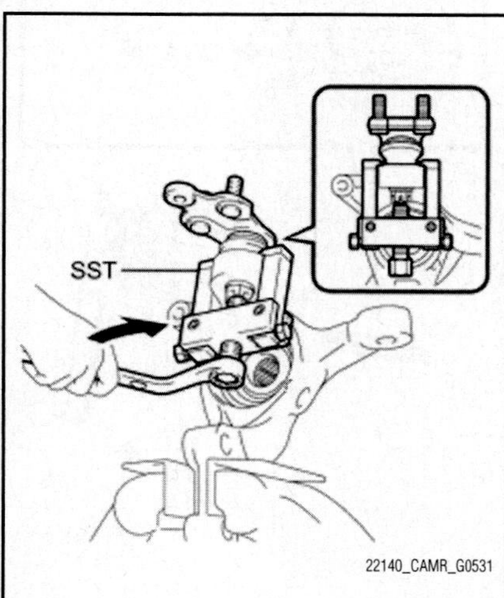

Fig. 176 Remove the front lower ball joint assembly

b. Inspect and adjust the front wheel alignment.

c. Inspect the ABS speed sensor signal.

LOWER CONTROL ARM

REMOVAL & INSTALLATION

See Figures 177 through 179.

1. Before servicing the vehicle, refer to the precautions section.
2. Remove the engine assembly with transaxle.

➡Use the same procedures for the RH side and the LH side. The procedures listed below are for the LH side.

3. Remove the 3 nuts and the engine mounting insulator.
4. Remove the 3 bolts and the nut on the front suspension lower No. 1 arm and remove it from the front frame assembly.
5. Remove the front lower arm bushing stopper.

To install:

6. Install the front lower arm bushing stopper.
7. Install the front suspension lower No. 1 arm to the front frame assembly with the 3 bolts and the nut, but do not tighten them yet.
8. Tighten bolts "A" to 148 ft. lbs. (200 Nm). Tighten bolts "B" to 152 ft. lbs. (206 Nm).

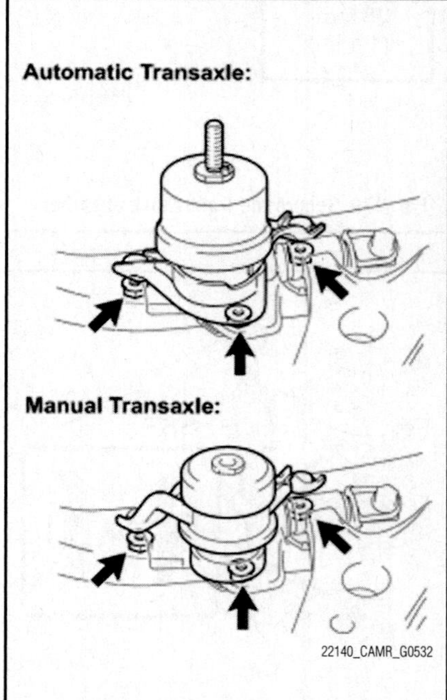

Fig. 177 Removing the 3 nuts and the engine mounting insulator

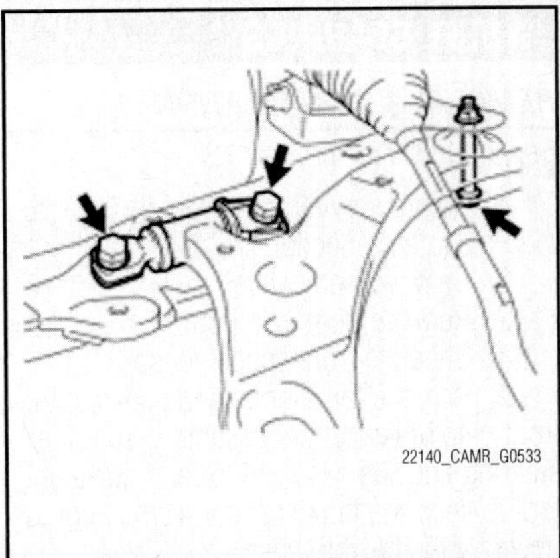

Fig. 178 Removing the 3 bolts and the nut on the front suspension lower No. 1 arm

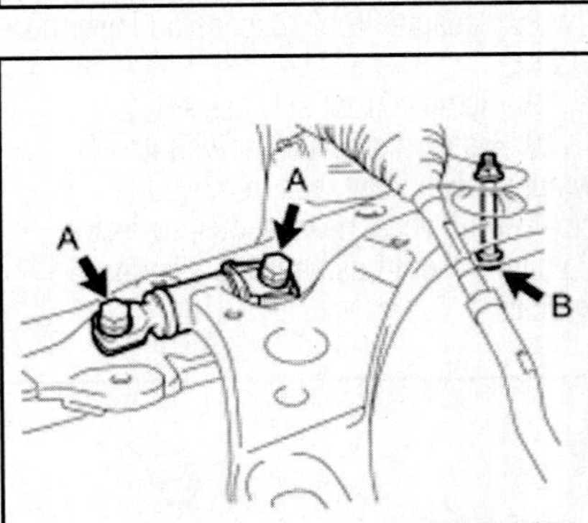

Fig. 179 Tightening the 3 bolts and the nut on the front suspension lower No. 1 arm

9. Install the engine mounting insulator with the 3 nuts and tighten to 64 ft. lbs. (87 Nm).
10. Install the engine assembly with transaxle.

SHOCK ABSORBERS

REMOVAL & INSTALLATION

See Figures 180 through 182.

➡Use the same procedures for the RH side and the LH side. The procedures listed below are for the LH side.

1. Before servicing the vehicle, refer to the precautions section.
2. Remove the front wheel.
3. Remove the nut and disconnect the front stabilizer link assembly from the front shock absorber assembly.
4. Remove front shock absorber with coil spring, as follows:
    a. Loosen the lock nut of the front shock absorber with coil spring.

➡Do not remove the lock nut.

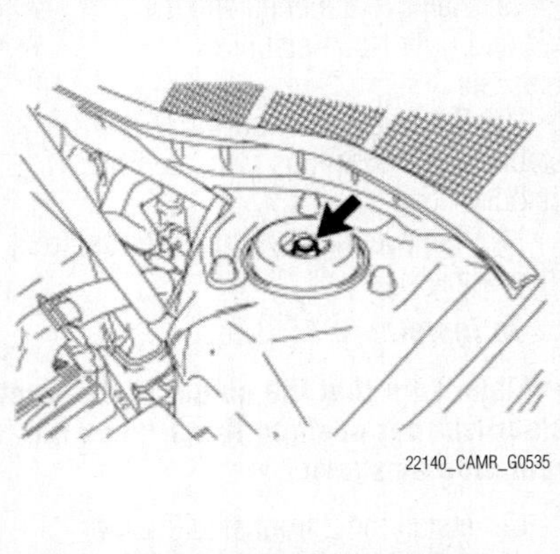

Fig. 180 Loosening the lock nut of the front shock absorber arm

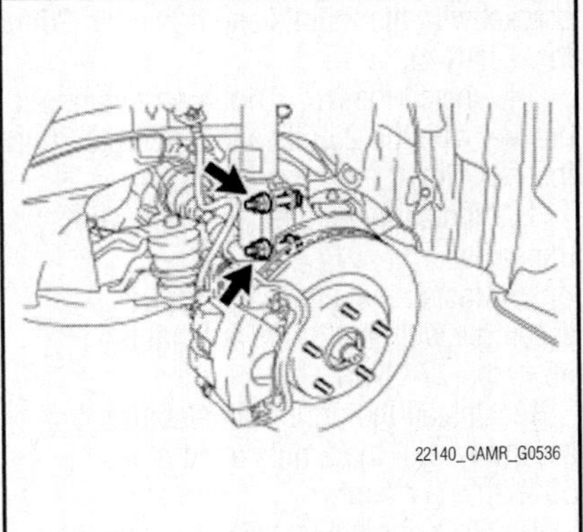

Fig. 181 Remove the 2 nuts on the lower side of the front shock absorber

➡Only loosen the nut when disassembling the front shock absorber with coil spring.

b. Remove the bolt and disconnect the front flexible hose and front speed sensor wire harness from the front shock absorber with coil spring.

➡Be sure to remove the front speed sensor from the front shock absorber with coil spring.

c. Remove the 2 nuts on the lower side of the front shock absorber with coil spring.

➡When removing the nuts, keep the bolts from rotating.

➡Keep the bolts inserted to secure the front axle assembly.

d. Remove the 3 nuts on the upper side of the front shock absorber with coil spring.

e. Lower the front axle assembly, and remove the 2 bolts on the lower side of the front shock absorber.

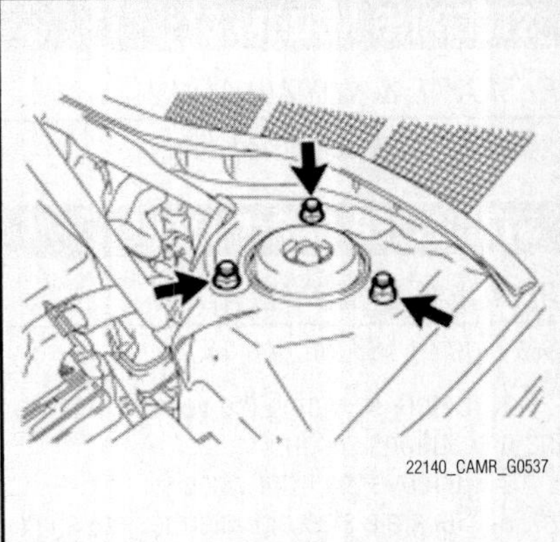

Fig. 182 Removing the 3 nuts on the upper side of the front shock absorber

➡Make sure that the front speed sensor is disconnected from the front shock absorber with coil spring.

f. Remove the front shock absorber with coil spring.

To install:

5. Install front shock absorber with coil spring, as follows:
    a. Install the front shock absorber with coil spring to the front axle assembly and insert the 2 bolts from the front side of the vehicle.
    b. Slowly jack up the vehicle using a wooden block and install the front shock absorber with coil spring (upper side) to the vehicle.
    c. Install the 3 nuts to the upper side of the front shock absorber with coil spring and tighten to 63 ft. lbs. (85 Nm).

➡When installing the nuts, keep the bolts from rotating.

d. Install the 2 nuts to the lower side of the front shock absorber with coil spring and tighten to 155 ft. lbs. (210 Nm).

e. Install the front flexible hose and front speed sensor wire harness with the bolt and tighten to 14 ft. lbs. (19 Nm).

f. Fully tighten the lock nut and tighten to 52 ft. lbs. (70 Nm).

➡If the ball joint turns together with the nut, use a hexagon wrench (6 mm) to hold the stud.

6. Install the front stabilizer link assembly with the nut and tighten to 55 ft. lbs. (74 Nm).
7. Install front wheel and tighten to 76 ft. lbs. (103 Nm).
8. Inspect and adjust front wheel alignment.

STEERING KNUCKLE

REMOVAL & INSTALLATION

See Wheel Hub and Bearing.

STABILIZER BAR & LINKS

REMOVAL & INSTALLATION

See Figures 183 through 185.

1. Before servicing the vehicle, refer to the precautions section.
2. Remove the front wheels.
3. Separate steering intermediate shaft assembly.
4. Separate tie rod end sub-assembly.

➡If the ball joint turns together with the nut, use a hexagon wrench (6 mm) to hold the stud.

5. Remove the 2 nuts and the front stabilizer link assembly
6. Remove the engine assembly with transaxle.
7. Remove the bolts and the left and right No. 1 stabilizer brackets.
8. Remove the engine assembly with transaxle.

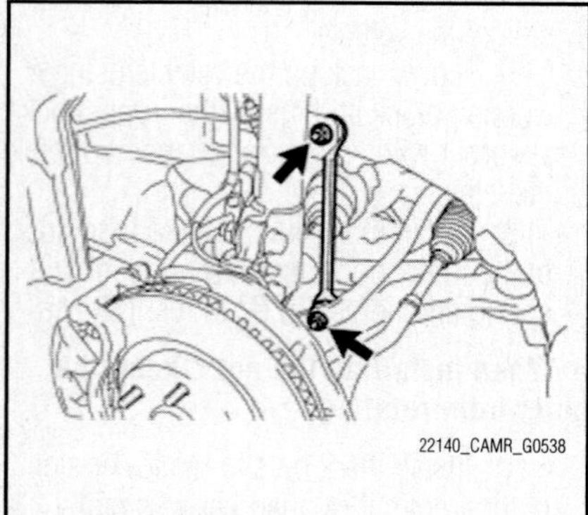

Fig. 183 Remove the 2 nuts and front stabilizer link assembly (left hand shown)

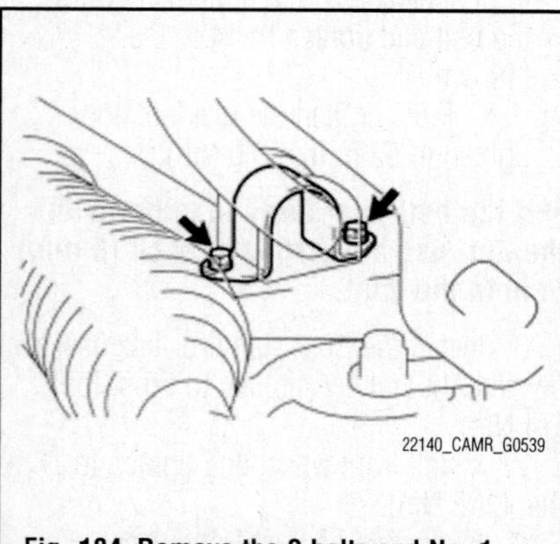

Fig. 184 Remove the 2 bolts and No. 1 stabilizer bracket (left hand shown)

9. Remove the bolts and the left and right No. 1 stabilizer brackets.
10. Remove the 2 front No. 1 stabilizer bar bushings from the front stabilizer bar.
11. Remove the front stabilizer bar from the vehicle.

To install:

➡Make sure that the cutout of the front stabilizer bar bushing No. 1 faces the rear side as shown.

12. Install the 2 front stabilizer bar bushings No. 1 to the outside of the bushing stopper on the front stabilizer bar.
13. Install the No. 1 left front stabilizer bracket with the 2 bolts and tighten to 20 ft. lbs. (27 Nm).
14. Install the No. 1 right front stabilizer bracket with the 2 bolts and tighten to 20 ft. lbs. (27 Nm).
15. Install the engine assembly with transaxle.
16. Install the left front stabilizer link assembly with the 2 nuts and tighten to 55 ft. lbs. (74 Nm).
17. Install the right front stabilizer link assembly with the 2 nuts and tighten to 55 ft. lbs. (74 Nm).
18. To complete installation, reverse removal procedure.
19. Inspect and adjust the front wheel alignment.

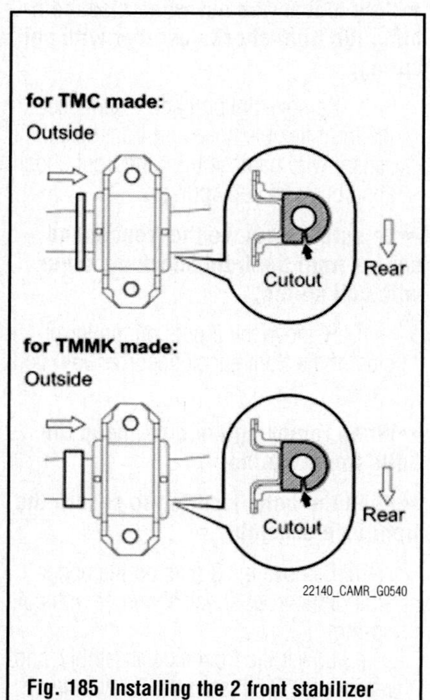

Fig. 185 Installing the 2 front stabilizer bar bushings

WHEEL BEARINGS

REMOVAL & INSTALLATION

See Figures 186 through 191.

1. Before servicing the vehicle, refer to the Precautions Section.
2. Remove front wheel.
3. Remove front axle hub nut.
4. Separate front speed sensor.
5. Remove the 2 bolts and separate the front disc brake caliper assembly from the steering knuckle. Use wire or an equivalent tool to keep the brake caliper from hanging down by the flexible hose.
6. Remove front disc.
7. Separate tie rod end sub-assembly.
8. Separate front suspension lower no. 1 arm.
9. Remove front axle assembly.
10. Using a screwdriver with its tip wrapped with vinyl tape, remove the No. 1 front wheel bearing dust deflector. Be careful not to damage the steering knuckle.

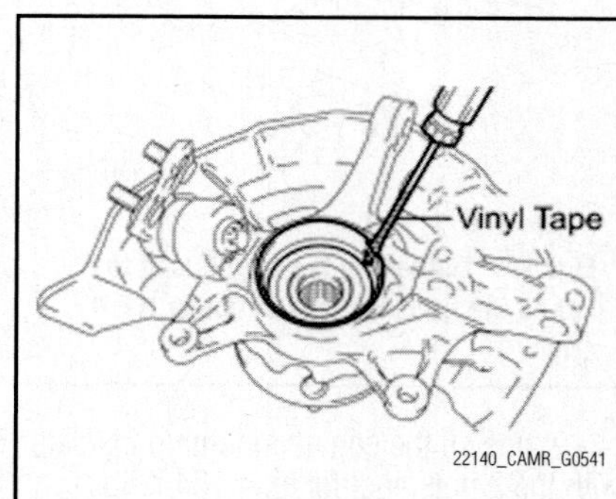

Fig. 186 Remove the No. 1 front wheel bearing dust deflector

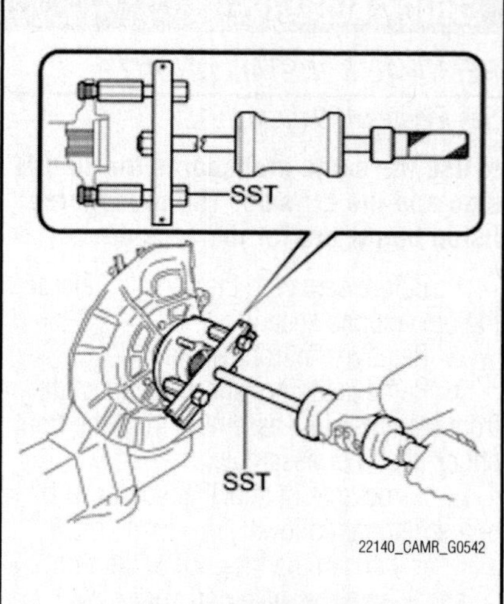

Fig. 187 Remove the front axle hub sub-assembly

11. Using snap ring pliers, remove the front axle hub hole snap ring.

12. Remove front axle hub sub-assembly by performing the following:

a. Hold the front axle assembly between aluminum plates in a vise.

➡**Do not overtighten the vise.**

b. Using SST 09520-00031, remove the front axle hub sub-assembly.

➡**Be careful not to drop the front axle hub sub-assembly.**

c. Using SST 09555-55010, SST: 09950-60010 and SST: 09950-70010 and a press, remove the bearing inner race (outside) from the front axle hub sub-assembly.

13. Remove the 4 bolts and disc brake dust cover from the steering knuckle.

14. Remove front lower ball joint assembly.

15. Remove front axle hub bearing by performing the following:

a. Place the bearing inner race (outside) on the front axle hub bearing.

b. Using SST 09527-17011, SST: 09950-60010 and a press, press the front axle hub bearing until it contacts the SST: 09950-70010.

c. Using SST: 09527-20011, SST: 09950-60010 to make the steering knuckle horizontal, fix it to the V-block.

16. Using SST: 09950-70010 and a press, remove the front axle hub bearing from the steering knuckle.

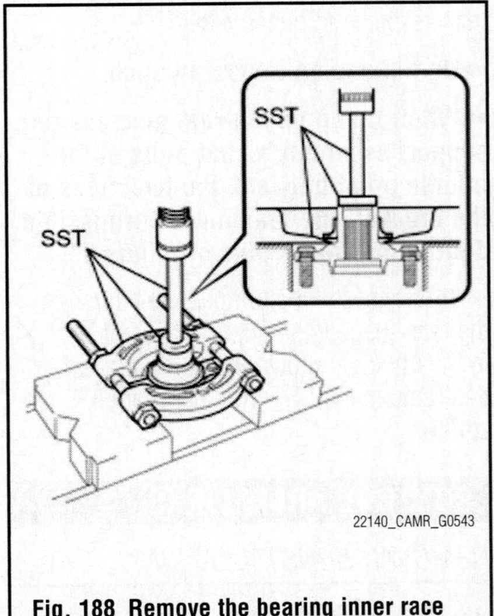

Fig. 188 Remove the bearing inner race (outside) from the front axle hub sub-assembly

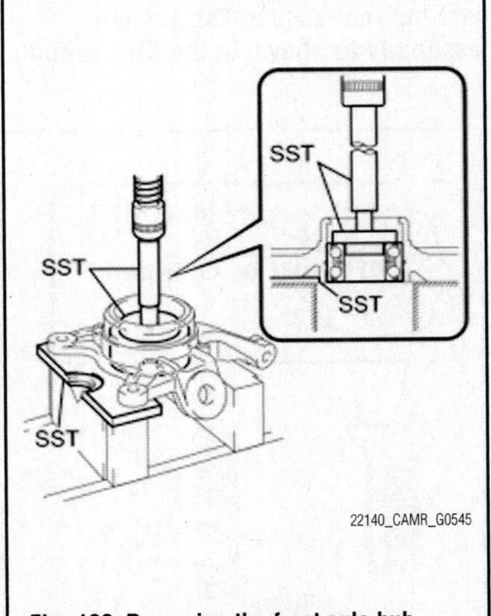

Fig. 190 Removing the front axle hub bearing

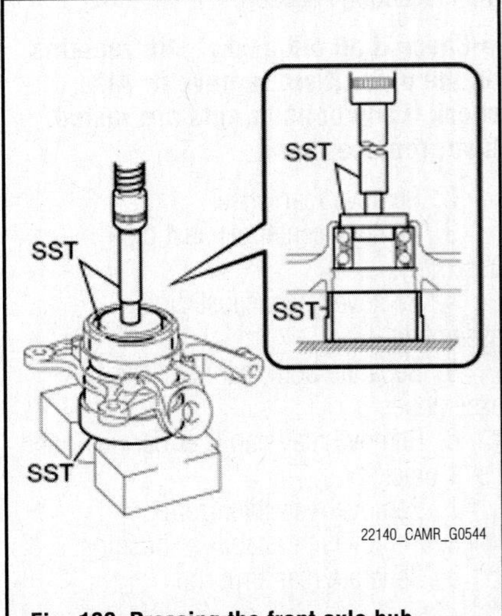

Fig. 189 Pressing the front axle hub bearing

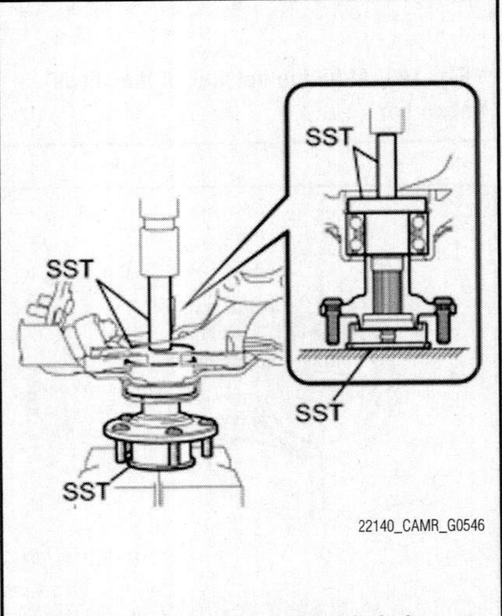

Fig. 191 Installing the front axle hub sub-assembly

To install:

17. Using SST's: 09950-60020, 09950-70010 and a press, install a new front axle hub bearing to the steering knuckle.

18. Install front lower ball joint assembly.

19. Install the disc brake dust cover to the steering knuckle with the 4 bolts and tighten to 73 inch lbs. (8.3 Nm).

20. Using SST's: 09608-32010, 09950-60020, 09950-70010 and a press, install the front axle hub sub-assembly.

21. Using snap ring pliers, install a new front axle hub hole snap ring.

➡**Align the hole for the speed sensor in the No. 1 front wheel bearing dust deflector with the steering knuckle.**

22. Using SST's: 09316-60011, 09608-32010 and a hammer, install a new No. 1 front wheel bearing dust deflector.

➡**Only when reusing the bolts and nuts, apply the small amount of engine oil to the screw part of the nuts.**

➡**Be careful not to damage the drive shaft boot or speed sensor rotor.**

23. Align the matchmarks and install the front drive shaft assembly to the front axle hub sub-assembly.

24. Install the steering knuckle with the front axle hub sub-assembly to the front shock absorber assembly with the 2 bolts and 2 nuts and tighten to 155 ft. lbs. (210 Nm).

25. Install the lower No. 1 front suspension arm sub-assembly.

26. Install the tie rod end sub-assembly.

27. Install the front disc.

28. Install the front disc brake caliper assembly with the 2 bolts to the steering knuckle and tighten to 79 ft. lbs. (107 Nm).

29. Clean the threaded parts on the drive shaft and axle hub nut using a non-residue solvent.

➡**Be sure to perform this work for a new drive shaft.**

➡**Keep the threaded parts free of oil and foreign objects.**

30. Using a 30 mm socket wrench, install the front axle hub nut and tighten to 217 ft. lbs. (294 Nm).

31. Remove the 2 bolts and separate the front disc brake caliper assembly from the steering knuckle.

32. Remove the front disc.

33. Inspect front axle hub bearing looseness.
34. Inspect front axle hub runout.
35. Install the front disc.
36. Install the front disc brake caliper assembly with the 2 bolts to the steering knuckle and tighten to 79 ft. lbs. (107 Nm).
37. Install the front speed sensor.
38. Using a chisel and hammer, stake the axle hub nut.
39. Install the front wheel.
40. Inspect and adjust front wheel alignment.
41. Check ABS speed sensor signal.

SUSPENSION

REAR SUSPENSION

COIL SPRING

REMOVAL & INSTALLATION

See Figures 192 through 194.

1. Before servicing the vehicle, refer to the precautions section.
2. Secure the rear shock absorber with coil spring in a vise using aluminum plates by closing the vise onto the double nutted bolt affixed to the bracket at the bottom of the absorber.

➡Do not use an impact wrench.

➡If the rear coil spring is compressed at an angle, using 2 SST will make the work easier.

3. Using SST: 09727-30021, compress the rear coil spring
4. Remove the nut, rear shock absorber collar and rear suspension support assembly.
5. Remove the rear coil spring, rear No. 1 spring bumper, and rear coil spring lower insulator.

To install:

6. Install the rear No. 1 spring bumper to the piston rod.
7. Install the rear coil spring lower insulator onto the rear shock absorber.

➡Do not use an impact wrench.

8. Using SST: 09727-30021, compress the rear coil spring.

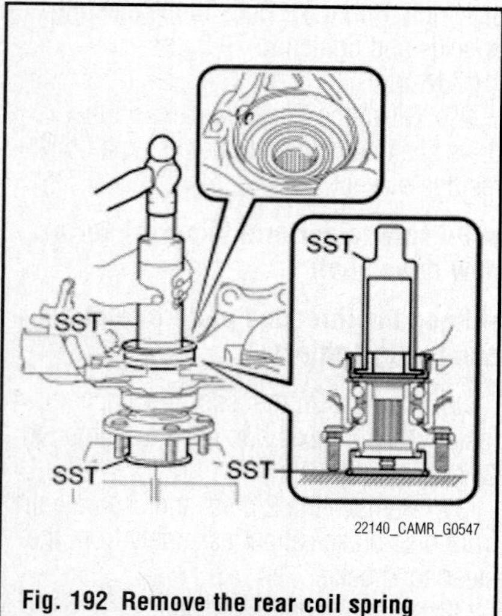

Fig. 192 Remove the rear coil spring

➡The smaller diameter end must face upward.

➡Fit the lower end of the rear coil spring into the gap of the lower seat.

➡If the front coil spring is compressed at an angle, using 2 SST will make the work easier.

9. Install the rear coil spring to the rear shock absorber.

➡Align the notches of the piston rod and the rear suspension support assembly as shown in the illustration before installing the rear suspension support assembly.

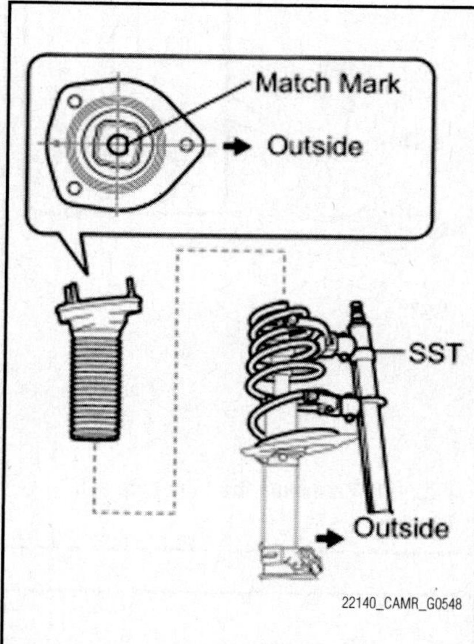

Fig. 193 Align the notches of the shock absorber

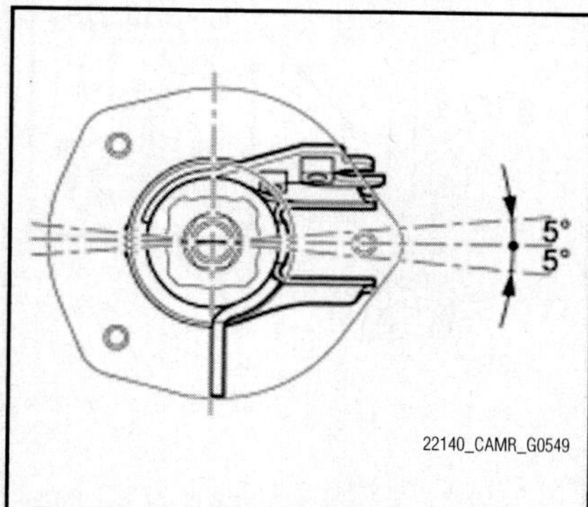

Fig. 194 Lining up the rear suspension support assembly's stud bolts

10. Install the rear suspension support assembly.
11. Align the notches of the shock absorber with the notch of the rear suspension support assembly so that the notches face the outside of the vehicle.
12. Install the rear shock absorber collar.
13. Loosely tighten a new lock nut to the rear suspension piston rod.

➡Do not use an impact wrench.

➡When lining up the rear suspension support assembly's stud bolts at the middle point between the two sides of the bracket, the maximum permissible degree of error is plus or minus 5°.

14. Release the spring while adjusting the rear suspension support assembly to the position shown in the illustration, and remove the SST from the rear coil spring.

CONTROL ARMS/LINKS

REMOVAL & INSTALLATION

No. 1 Suspension Arm

See Figures 195 through 200.

1. Before servicing the vehicle, refer to the precautions section.

➡Check if an old gasket still remains on the pipe. If so, remove it. Also, check if any bolts or nuts are rusted. If so, replace them.

2. Remove rear wheel.
3. Remove center exhaust pipe assembly.
4. Remove tail exhaust pipe assembly.
5. Separate both rear stabilizer link assemblies.
6. Remove rear stabilizer bar no. 2 and no. 1 bracket.
7. Remove rear stabilizer bar.
8. Remove rear stabilizer bushing.
9. Separate rear strut rod.

➡When removing the bolt, keep the nut from rotating.

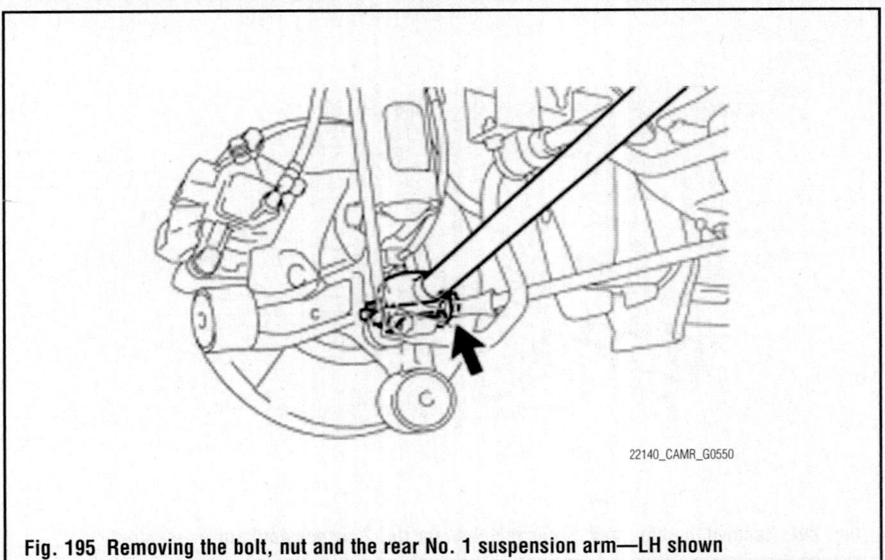

Fig. 195 Removing the bolt, nut and the rear No. 1 suspension arm—LH shown

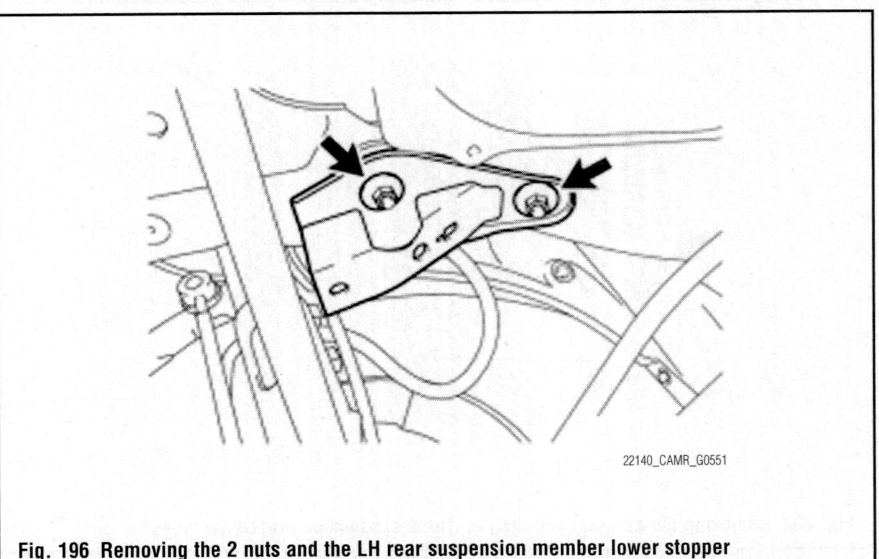

Fig. 196 Removing the 2 nuts and the LH rear suspension member lower stopper

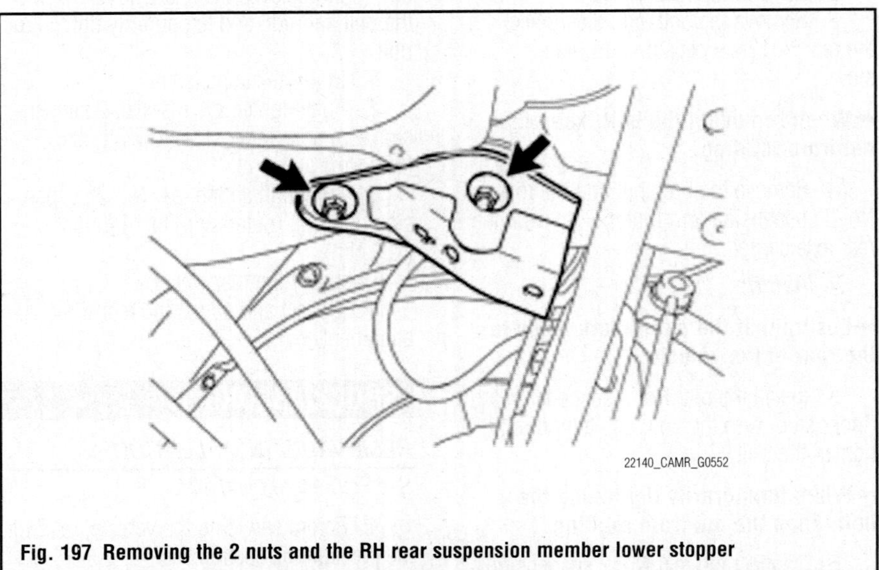

Fig. 197 Removing the 2 nuts and the RH rear suspension member lower stopper

10. Remove the bolt, nut and separate the rear suspension No. 2 arm (outer side) from the rear axle carrier.

➡When removing the bolt, keep the nut from rotating.

11. Remove the bolt, nut and the rear No. 1 suspension arm (outer side) from the rear axle carrier.
12. Remove the 2 nuts and the LH rear suspension member lower stopper.
13. Remove the 2 nuts and the RH rear suspension member lower stopper.
14. Support the rear suspension member with a jack.
15. Remove the 2 bolts, and the rear suspension member sub-assembly.
16. Remove the bolt and rear No. 1 suspension arm assembly.

To install:

17. Install the No. 1 rear suspension arm (inner side) with the bolt, and temporarily tighten the bolt.
18. Install the rear No. 1 suspension arm so that the bracket leans toward the front side of the vehicle.
19. Ensure that the paint mark faces the rear side of the vehicle.
20. Set the rear No.1 suspension arm in the position shown in the illustration, and fully tighten the bolt to 74 ft. lbs. (100 Nm).
21. Raise the rear suspension member with a jack. Install the rear suspension member with the 2 bolts and tighten to 41 ft. lbs. (56 Nm).
22. Install both the rear suspension member lower stoppers with the 2 nuts and tighten to:
    a. Nut A: 41 ft. lbs. (55 Nm).
    b. Nut B: 28 ft. lbs. (38 Nm).

➡Insert the bolt from the front of the vehicle and temporarily install the bolt.

23. Connect the rear No.1 suspension arm (outer side) to the rear axle carrier with the bolt and nut and temporarily tighten the bolt and nut. When temporarily tightening the bolt, keep the nut from rotating.

➡Insert the bolt from the inside of the vehicle and temporarily install the bolt.

24. Connect the strut rod assembly rear to the axle carrier with the bolt and nut and temporarily tighten the bolt. When temporarily tightening the bolt, keep the nut from rotating.
25. Jack up the rear axle carrier, placing a wooden block to avoid damage. Apply load to the suspension so that the installed bolt of the rear No. 1 suspension arm (inner

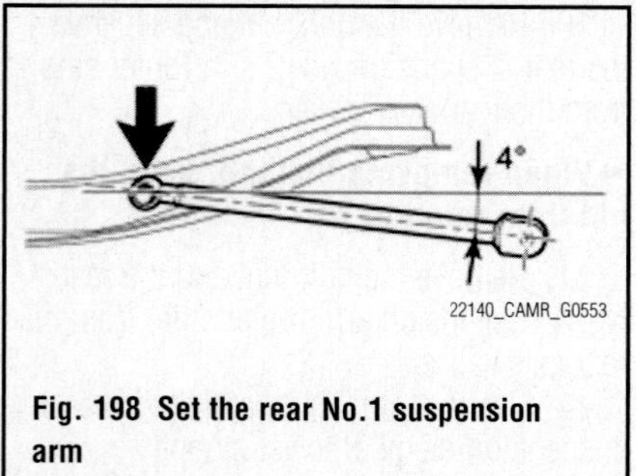

Fig. 198 Set the rear No.1 suspension arm

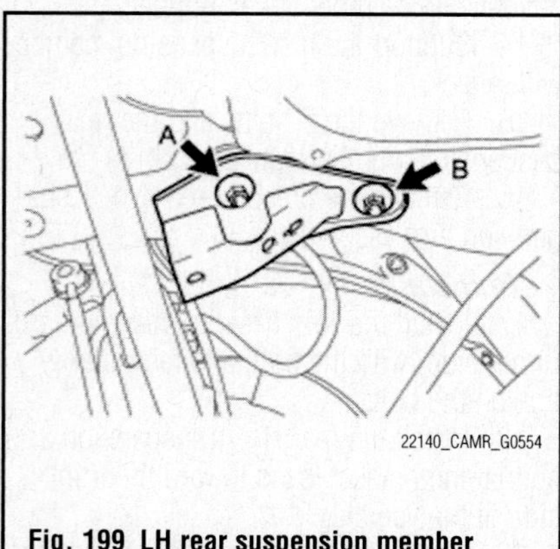

Fig. 199 LH rear suspension member lower stopper tightening sequence

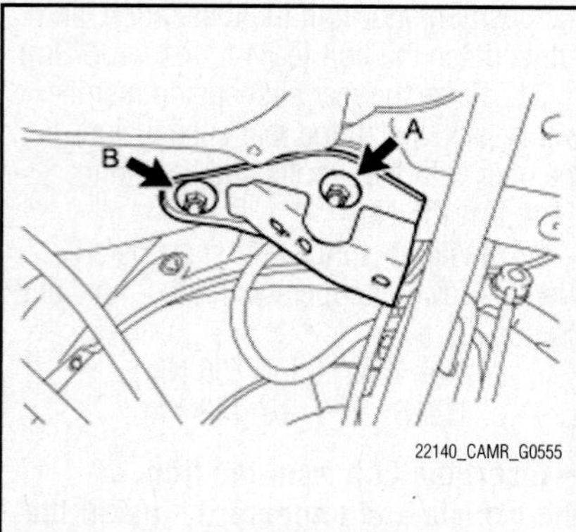

Fig. 200 LH rear suspension member lower stopper tightening sequence

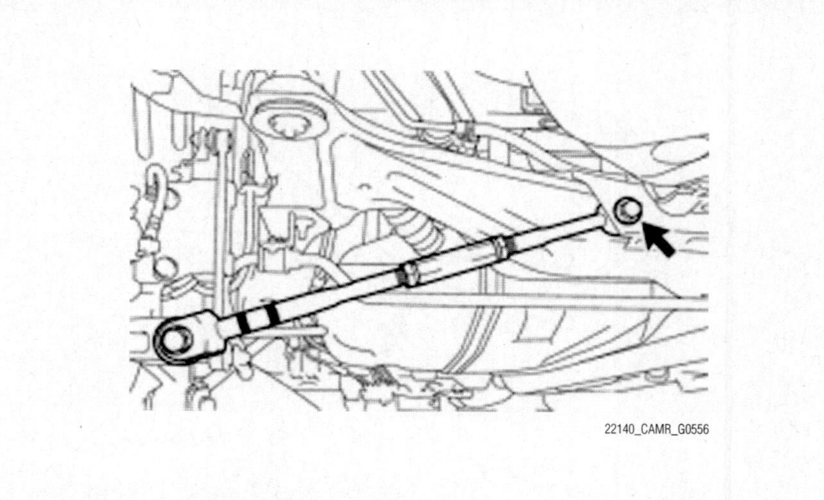

Fig. 201 Remove the bolt, and disconnect the rear No. 2 suspension arm (inner side)

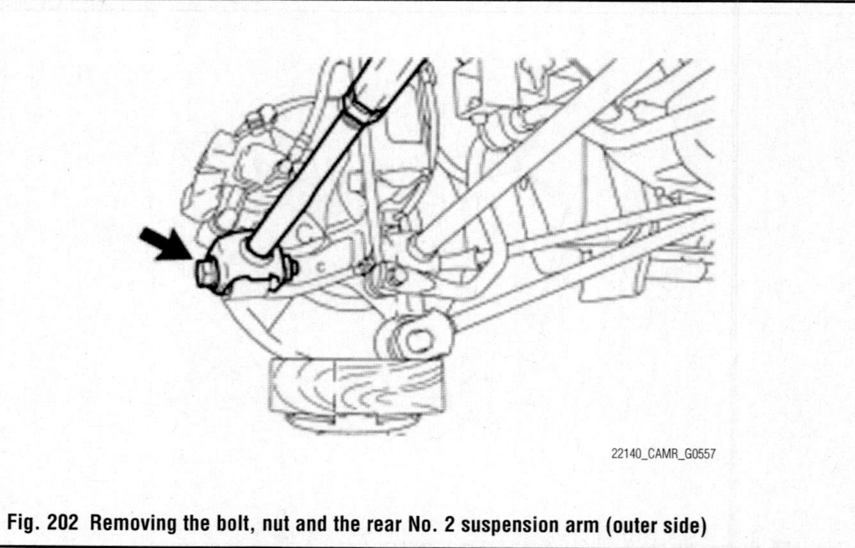

Fig. 202 Removing the bolt, nut and the rear No. 2 suspension arm (outer side)

side) is horizontally aligned with the center of the rear axle hub.

26. Fully tighten rear No. 1 suspension arm and tighten the bolt to 74 ft. lbs. (100 Nm).

27. Fully tighten rear No. 2 suspension arm and tighten the bolt to 74 ft. lbs. (100 Nm).

28. To complete installation, reverse removal procedure.

No. 2 Suspension Arm

See Figures 201 and 202.

1. Before servicing the vehicle, refer to the precautions section.
2. Remove the rear wheel.
3. Remove the bolt, and disconnect the rear No. 2 suspension arm (inner side).

➡When removing the bolt, keep the nut from rotating.

4. Remove the bolt, nut and the rear No. 2 suspension arm (outer side) from the rear axle carrier.

To install:

➡Ensure that the paint mark faces to the rear of the vehicle.

5. Install the rear No. 2 suspension arm (inner side) with the bolt, and temporarily tighten the bolt.

➡When temporarily tightening the bolt, keep the nut from rotating.

6. Connect the rear No. 2 suspension arm (outer side) to the rear axle carrier with the bolt and nut, and temporarily tighten the bolt.
7. Stabilize suspension.
8. Fully tighten the rear No. 2 suspension arm bolt (inner side) to 74 ft. lbs. (100 Nm).
9. Fully tighten the rear No. 2 suspension arm bolt (outer side) to 74 ft. lbs. (100 Nm).
10. Install the rear wheel.
11. Inspect and adjust the rear wheel alignment.

SHOCK ABSORBER

REMOVAL & INSTALLATION

See Figures 203 and 204.

1. Before servicing the vehicle, refer to the precautions section.

2. Remove the rear seat cushion assembly.
3. Remove rear seat headrest plate cover.
4. Remove rear seat headrest assembly.
5. Remove the rear seatback assembly.
6. Remove the rear wheel.
7. Separate LH rear stabilizer link assembly.
8. Remove the 2 bolts, and disconnect the rear brake flexible hose and rear speed sensor from the rear shock absorber with coil spring and rear axle carrier.
9. Remove the 4 claws and the rear suspension support No. 1 cover.
10. Remove the rear shock absorber with coil spring, as follows:

➡Do not remove the lock nut.

➡Only loosen the nut when disassembling the rear shock absorber with coil spring.

a. Loosen the lock nut of the rear shock absorber with coil spring.

➡When removing the nuts, keep the bolts from rotating.

➡Keep one bolt inserted to secure the hub and disc rotor.

b. Remove the 2 nuts and 2 bolts on the lower side of the rear shock absorber with coil spring.

c. Remove the 3 nuts on the upper side of the rear shock absorber with coil spring.

➡Make sure that the rear speed sensor is disconnected from the rear shock absorber with coil spring.

d. Lower the rear axle carrier, and remove the 2 bolts on the lower side of the rear shock absorber with coil spring.

To install:

11. Install the rear shock absorber with coil spring to the rear axle carrier assembly and insert the 2 bolts from the rear of the vehicle.
12. Slowly jack up the vehicle using a wooden block and install the rear shock absorber with coil spring (upper side) to the vehicle.
13. Install the 3 nuts to the upper side of the rear shock absorber with coil spring and tighten to 29 ft. lbs. (39 Nm).

➡When installing the nuts, keep the bolts from rotating.

14. Install the 2 nuts and 2 bolts to the lower side of the rear shock absorber with coil spring and tighten 133 ft. lbs. (180 Nm).
15. Fully tighten the lock nut to 41 ft. lbs. (55 Nm).
16. Connect rear speed sensor.
17. Install the LH rear stabilizer link assembly.
18. Engage the 4 claws and install the rear suspension support No. 1 cover.
19. To complete installation, reverse remaining removal.
20. Check abs speed sensor signal.
21. Inspect and adjust rear wheel alignment.

TESTING

1. Inspect the shock absorber.

a. Compress and extend the shock absorber rod 4 or more times.

➡A normal shock has no abnormal resistance or sound and operation resistance is normal.

➡If there is any abnormality, replace the rear shock absorber with a new one.

WHEEL BEARINGS

REMOVAL & INSTALLATION

See Figure 205.

➡Use the same procedures for the RH side and LH side.

➡The procedures listed below are for the LH side.

1. Before servicing the vehicle, refer to the precautions section.
2. Remove the rear wheel.
3. Separate the rear disc brake caliper assembly, as follows:

a. Remove the bolt and separate the flexible hose from the shock absorber.

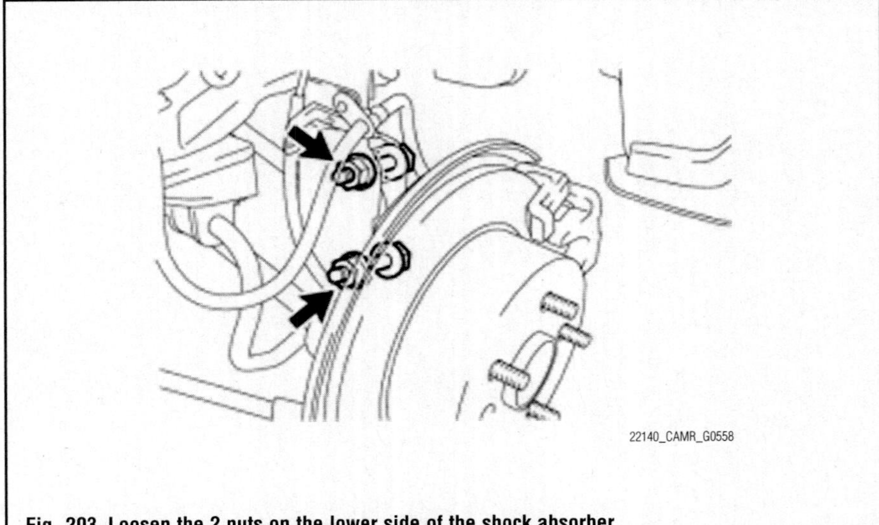

Fig. 203 Loosen the 2 nuts on the lower side of the shock absorber

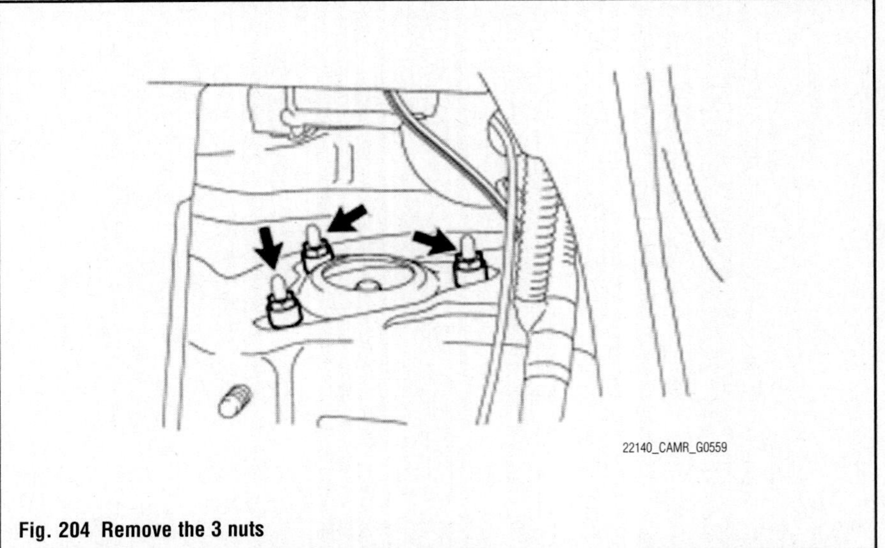

Fig. 204 Remove the 3 nuts

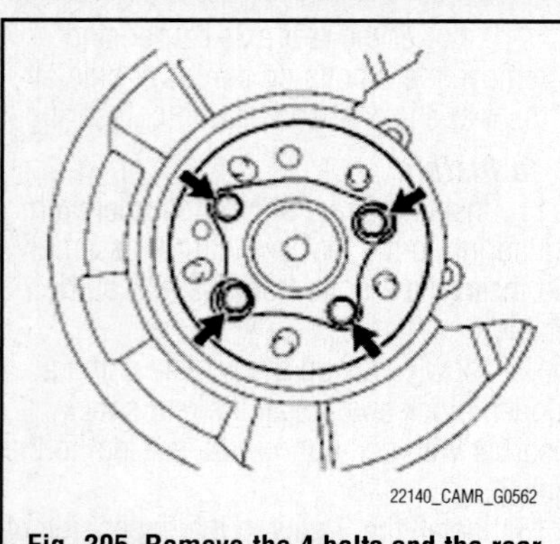

Fig. 205 Remove the 4 bolts and the rear axle hub and bearing assembly

b. Remove the 2 bolts and separate the rear disc brake caliper assembly.

4. Remove the rear disc.
5. Disconnect the skid control sensor connector.
6. Remove the 4 bolts and the rear axle hub and bearing assembly.

To install:

7. Install the hub and bearing assembly with the 4 bolts and tighten to 59 ft. lbs. (80 Nm).
8. Connect the skid control sensor connector. Do not twist the sensor wire.
9. Inspect rear axle hub bearing looseness.
10. Inspect rear axle hub runout.
11. Install the rear disc.
12. Install the rear disc brake caliper assembly, as follows:
    a. Install the rear disc brake caliper with the 2 bolts and tighten to 46 ft. lbs. (62 Nm).
    b. Install the flexible hose with the bolt and tighten to 14 ft. lbs. (19 Nm).
13. Install the rear wheel.
14. Inspect and adjust the rear wheel alignment.
15. Check ABS speed sensor signal.

TOYOTA

7

Corolla

SPECIFICATIONS AND MAINTENANCE CHARTS

ENGINE AND VEHICLE IDENTIFICATION

Engine							Model Year	
Code ①	Liters (cc)	Cu. In.	Cyl.	Fuel Sys.	Engine Type	Eng. Mfg.	Code ②	Year
2ZR-FE	1.8 (1794)	109.5	4	EFI	DOHC	Toyota	9	2009
2AZ-FE	2.4 (2400)	146	4	EFI	DOHC	Toyota	A	2010

EFI: Electronic Fuel Injection

DOHC: Double Overhead Camshaft

① 5th digit of VIN

② 10th digit of VIN

3768X_CORO_C0001

GENERAL ENGINE SPECIFICATIONS

Year	Model	Engine Displacement Liters (cc)	Engine Series (ID/VIN)	Fuel System	Net Horsepower @ rpm	Net Torque @ rpm (ft. lbs.)	Bore x Stroke (in.)	Com-pression Ratio	Oil Pressure @ rpm
2009	Corolla	1.8 (1794)	2ZR-FE	EFI	132@6000	128@4400	3.16x3.48	10.0:1	4.3
		2.4 (2400)	2AZ-FE	EFI	158@6000	162@4000	3.48x3.78	9.8	NA
2010	Corolla	1.8 (1794)	2ZR-FE	EFI	132@6000	128@4400	3.16x3.48	10.0:1	4.3
		2.4 (2400)	2AZ-FE	EFI	158@6000	162@4000	3.48x3.78	9.8	NA

NA: Not Available

EFI: Electronic Fuel Injection

3768X_CORO_C0002

ENGINE TUNE-UP SPECIFICATIONS

Year	Engine Displacement Liters	Engine ID/VIN	Spark Plug Gap (in.)	Ignition Timing (deg.)	Fuel Pump (psi)	Idle Speed (rpm)		Valve Clearance	
						MT	AT	In.	Ex.
2009	1.8	2ZR-FE	0.043	①	44-50	600-700	600-700	0.007-0.014	0.015-0.018
	2.4	2AZ-FE	0.043	②	44-50	600-700	600-700	NA	NA
2010	1.8	2ZR-FE	0.043	①	44-50	600-700	600-700	0.007-0.014	0.015-0.018
	2.4	2AZ-FE	0.043	②	44-50	600-700	600-700	NA	NA

Note: The Vehicle Emission Control Information label often reflects specification changes made during production.

NA: Not Available

BTDC: Before Top Dead Center

① With Techstream (ODB-II scanner or equivalent): BTDC 5-15 at idle
Without Techstream: BTDC 8-12 at idle. Connect terminals 13 (TC) and 4 (CG)

② BTDC 8-12 with terminal TC and CG of DLC3 connected

3768X_CORO_C0003

CAPACITIES

Year	Model	Engine Displacement Liters	Engine ID/VIN	Engine Oil with Filter	Transmission (pts.) 5-Spd	Transmission (pts.) Auto.	Drive Axle Front (pts.)	Drive Axle Rear (pts.)	Fuel Tank (gal.)	Cooling System (qts.)
2009	Corolla	1.8	2ZR-FE	4.4	4.0	5.2	NA	NE	13.2	5.8
		2.4	2AZ-FE	4.0	4.0	7.4	NA	NE	13.2	6.0
2010	Corolla	1.8	2ZR-FE	4.4	4.0	5.2	NA	NE	13.2	5.8
		2.4	2AZ-FE	4.0	4.0	7.4	NA	NE	13.2	6.0

Note: All capacities are approximate. Add fluid gradually and check to be sure a proper fluid level is obtained. Auto trans is a drain and fill capacity.

NE: Not Equipped

NA: Not Available

3768X_CORO_C0004

FLUID SPECIFICATIONS

Year	Model	Engine Displ. Liters	Engine Oil	Man. Trans.	Auto. Trans.	Drive Axle	Power Steering Fluid	Brake Master Cylinder
2009	Corolla	1.8	5-30W	SAE 75W	ATF Fluid	NA	ATF Fluid	DOT 3
		2.4	5-30W	SAE 75W	ATF Fluid	NA	ATF Fluid	DOT 3
2010	Corolla	1.8	5-30W	SAE 75W	ATF Fluid	NA	ATF Fluid	DOT 3
		2.4	5-30W	SAE 75W	ATF Fluid	NA	ATF Fluid	DOT 3

NA: Not Available

DOT: Department Of Transpotation

3768X_CORO_C0005

VALVE SPECIFICATIONS

Year	Engine Displacemen Liters	Engine ID/VIN	Seat Angle (deg.)	Face Angle (deg.)	Spring Test Pressure (lbs. @ in.)	Spring Installed Height (in.)	Stem-to-Guide Clearance (in.) Intake	Stem-to-Guide Clearance (in.) Exhaust	Stem Diameter (in.) Intake	Stem Diameter (in.) Exhaust
2009	1.8	2ZR-FE	45	44.5	35.7-39.5@ 1.323	1.323	0.0010-0.0024	0.0012-0.0025	0.2154-0.2159	0.2152-0.2158
	2.4	2AZ-FE	45	44.5	35.7-39.5@ 1.323	1.323	0.0010-0.0024	0.0012-0.0025	0.2154-0.2159	0.2152-0.2158
2010	1.8	2ZR-FE	45	44.5	35.7-39.5@ 1.323	1.323	0.0010-0.0024	0.0012-0.0025	0.2154-0.2159	0.2152-0.2158
	2.4	2AZ-FE	45	44.5	35.7-39.5@ 1.323	1.323	0.0010-0.0024	0.0012-0.0025	0.2154-0.2159	0.2152-0.2158

3768X_CORO_C0007

CAMSHAFT AND BEARING SPECIFICATIONS

All measurements are given in inches.

Year	Engine Displacement Liters	Engine VIN	Journal Diameter	Brg. Oil Clearance	Shaft End-play	Runout	Journal Bore	Lobe Lift Intake	Lobe Lift Exhaust
2009	1.8	2ZR-FE	①	0.0012-0.0025	NA	0.0016	NA	1.685-1.6890	1.745-1.7490
	2.4	2AZ-FE	②	③	NA	0.00118	NA	1.865-1.8660	1.813-1.8170
2010	1.8	2ZR-FE	①	0.0012-0.0025	NA	0.0016	NA	1.685-1.6890	1.745-1.7490
	2.4	2AZ-FE	②	③	NA	0.00118	NA	1.865-1.8660	1.813-1.8170

NA: Not Available

① No. 1 journal diameter: 1.356
Other: 0.903 to 0.904

② No. 1 journal diamter: 1.416
Other: 0.903 to 0.904

③ Intake: 0.00276 (Max.)
Exhaust: 0.00394 (Max.)

3768X_CORO_C0006

CRANKSHAFT AND CONNECTING ROD SPECIFICATIONS

All measurements are given in inches.

Year	Engine Displacement Liters	Engine ID/VIN	Crankshaft Main Brg. Journal Dia.	Crankshaft Main Brg. Oil Clearance	Crankshaft Shaft End-play	Crankshaft Thrust on No.	Connecting Rod Journal Diameter	Connecting Rod Oil Clearance	Connecting Rod Side Clearance
2009	1.8	2ZR-FE	1.8893-1.8898	0.0006-0.0015	0.0008-0.0087	3	1.7320-1.7323	0.0012-0.0024	0.0063-0.0135
	2.4	2AZ-FE	2.164-2.165	0.00236	NA	3	0.283-0.2870	0.00248	0.0063-0.0143
2010	1.8	2ZR-FE	1.8893-1.8898	0.0006-0.0015	0.0008-0.0087	3	1.7320-1.7323	0.0012-0.0024	0.0063-0.0135
	2.4	2AZ-FE	2.164-2.165	0.00236	NA	3	0.283-0.2870	0.00248	0.0063-0.0143

NA: Not Available

3768X_CORO_C0010

PISTON AND RING SPECIFICATIONS

All measurements are given in inches.

Year	Engine Displacement Liters	Engine ID/VIN	Piston Clearance	Ring Gap Top Compression	Ring Gap Bottom Compression	Ring Gap Oil Control	Ring Side Clearance Top Compression	Ring Side Clearance Bottom Compression	Ring Side Clearance Oil Control
2009	1.8	2ZR-FE	0.0011-0.0020	0.0079-0.0118	0.0118-0.0197	0.0039-0.0157	0.0008-0.0028	0.0008-0.0024	0.0008-0.0026
	2.4	2AZ-FE	0.00082-0.0017	0.0094-0.0122	0.0130-0.0169	0.0039-0.0118	0.0007-0.0027	0.0007-0.0023	0.0007-0.0027
2010	1.8	2ZR-FE	0.0011-0.0020	0.0079-0.0118	0.0118-0.0197	0.0039-0.0157	0.0008-0.0028	0.0008-0.0024	0.0008-0.0026
	2.4	2AZ-FE	0.00082-0.0017	0.0094-0.0122	0.0130-0.0169	0.0039-0.0118	0.0007-0.0027	0.0007-0.0023	0.0007-0.0027

3768X_CORO_C0009

TORQUE SPECIFICATIONS

All readings in ft. lbs.

Year	Engine Displacement Liters	Engine ID/VIN	Cylinder Head Bolts	Main Bearing Bolts	Rod Bearing Bolts	Crankshaft Damper Bolts	Flywheel Bolts	Manifold Intake	Manifold Exhaust	Spark Plugs	Oil Pan Drain Plug
2009	1.8	2ZR-FE	①	②	③	102	65	21	16	15	27
	2.4	2AZ-FE	④	NA	⑤	NA	⑥	22	27	14	27
2010	1.8	2ZR-FE	①	②	③	102	65	21	16	15	27
	2.4	2AZ-FE	④	NA	⑤	NA	⑥	22	27	14	27

NA: Not Available

① Step 1: 36 ft. lbs.
Step 2: 90 degree turn
Step 3: 45 degree turn

② Inner 12 point bolts:
Step 1: 16 ft. lbs.
Step 2: 32 ft. lbs.
Step 3: 45 degree turn
Step 4: 45 degree turn
Outer cap bolts: 13 ft. lbs.

③ Step 1: 15 ft. lbs.
Step 2: 90 degree turn

④ Step 1: 52 ft. lbs.
Step 2: 90 degree turn

⑤ Step 1: 18 ft. lbs.
Step 2: 90 degree turn

⑥ A/T: 72 ft. lbs.
M/T: 96 ft. lbs.

3768X_CORO_C0008

WHEEL ALIGNMENT

Year	Model		Caster Range (+/-Deg.)	Caster Preferred Setting (Deg.)	Camber Range (+/-Deg.)	Camber Preferred Setting (Deg.)	Toe-in (Deg.)	Steering Axis Inclination (Deg.)
2009	Corolla	F	0.75	+2.50	0.75	-0.35	0+/-0.2	12.53+/-0.75
		R	—	—	0.5	-1.45	①	—
2010	Corolla	F	0.75	+2.50	0.75	-0.35	0+/-0.2	12.53+/-0.75
		R	—	—	0.5	-1.45	①	—

① For P185/65R15: 0.25+/-0.25 for P195/65R15: 0.26+/-0.26 for 195/55R16: 0.34+/-0.25

3768X_CORO_C0011

TIRE, WHEEL AND BALL JOINT SPECIFICATIONS

Year	Model	OEM Tires Standard	OEM Tires Optional	Tire Pressures (psi) Front	Tire Pressures (psi) Rear	Wheel Size	Ball Joint Inspection	Lug Nut (ft. lbs.)
2009	Corolla	P195/65R15 89S	None	30	30	NA	NA	76
		P195/65R15 91H	None	30	30	NA	NA	76
		P205/55R16 89H	None	32	32	NA	NA	76
		P215/45R17 87W	None	32	32	NA	NA	76
		P205/55R16 91V	None	32	32	NA	NA	76
2010	Corolla	P195/65R15 89S	None	30	30	NA	NA	76
		P195/65R15 91H	None	30	30	NA	NA	76
		P205/55R16 89H	None	32	32	NA	NA	76
		P215/45R17 87W	None	32	32	NA	NA	76
		P205/55R16 91V	None	32	32	NA	NA	76

PSI: Pounds Per Square Inch

NA: Not Available

3768X_CORO_C0012

BRAKE SPECIFICATIONS

All measurements in inches unless noted

Year	Model		Brake Disc Original Thickness	Brake Disc Minimum Thickness	Brake Disc Maximum Runout	Brake Drum Diameter Original Inside Diameter	Brake Drum Diameter Max. Wear Limit	Brake Drum Diameter Maximum Machine Diameter	Minimum Lining Thickness	Brake Caliper Bracket Bolts (ft. lbs.)	Brake Caliper Mounting Bolts (ft. lbs.)
2009	Corolla	F	0.866	0.748	0.0019	NE	NE	NE	0.039	NA	25
		R	0.35	0.30	0.00591	9.00	9.04	NA	0.039	NA	26
2010	Corolla	F	0.866	0.748	0.0019	NE	NE	NE	0.039	NA	25
		R	0.35	0.30	0.00591	9.00	9.04	NA	0.039	NA	26

NE: Not Equipment

NA: Not Available

3768X_CORO_C0013

SCHEDULED MAINTENANCE INTERVALS

TOYOTA—COROLLA

TO BE SERVICED	TYPE OF SERVICE	VEHICLE MILEAGE INTERVAL (x1000)													
		5	10	15	20	25	30	35	40	45	50	55	60	90	120
Engine oil & filter	R	✓	✓	✓	✓	✓	✓	✓	✓	✓	✓	✓	✓	✓	✓
Drive belts	S/I						✓						✓	✓	✓
Automatic transaxle fluid & filter	S/I						✓						✓	✓	✓
Brake line pipes & hoses	S/I	✓	✓	✓	✓	✓	✓	✓	✓	✓	✓	✓	✓	✓	✓
Brake linings & drums	S/I	✓	✓	✓	✓	✓	✓	✓	✓	✓	✓	✓	✓	✓	✓
Brake pads & discs (front & rear if equipped)	S/I	✓	✓	✓	✓	✓	✓	✓	✓	✓	✓	✓	✓	✓	✓
Cabin air filter	R				✓				✓				✓		✓
Differential oil	S/I						✓						✓	✓	✓
Drive shaft boots	S/I	✓	✓	✓	✓	✓	✓	✓	✓	✓	✓	✓	✓	✓	✓
Drive shaft bolt (tighten)	S/I	✓	✓	✓	✓	✓	✓	✓	✓	✓	✓	✓	✓	✓	✓
Engine coolant	S/I			✓			✓			✓			✓	✓	✓
Manual transaxle oil	S/I						✓						✓	✓	✓
Steering gear housing oil	S/I	✓	✓	✓	✓	✓	✓	✓	✓	✓	✓	✓	✓	✓	✓
Steering linkage	S/I	✓	✓	✓	✓	✓	✓	✓	✓	✓	✓	✓	✓	✓	✓
Air filter	R						✓						✓	✓	✓
Rotate tires	S/I	✓	✓	✓	✓	✓	✓	✓	✓	✓	✓	✓	✓	✓	✓
Spark plugs	R									✓				✓	
Fuel lines & connections	S/I						✓			✓			✓	✓	✓
Fuel tank cap gasket	R									✓				✓	
Charcoal canister	S/I									✓				✓	

R: Replace S/I: Service or Inspect

FREQUENT OPERATION MAINTENANCE (SEVERE SERVICE)

If a vehicle is operated under any of the following conditions it is considered severe service:

- Extremely dusty areas.

- 50% or more of the vehicle operation is in 32°C (90°F) or higher temperatures, or constant operation in temperatures below 0°C (32°F).

- Prolonged idling (vehicle operation in stop and go traffic).

- Frequent short running periods (engine does not warm to normal operating temperatures).

- Police, taxi, delivery usage or trailer towing usage.

Oil & oil filter: change every 5000 miles.

Bolts & nuts on chassis & body: tighten every 5000 miles.

Ball joints & dust covers: service or inspect every 5,000 miles.

Drive shaft boots: service or inspect every 12,000 miles.

Steering linkage: service or inspect every 12,000 miles.

Air filter: service or inspect every 5,000 miles.

Exhaust system: service or inspect every 15,000 miles.

Timing belt: replace every 60,000 miles.

3768X_CORO_C0014

BRAKES — INFORMATION AND PRECAUTIONS

ANTI-LOCK SYSTEMS

- Certain components within the ABS system are not intended to be serviced or repaired individually.
- Do not use rubber hoses or other parts not specifically specified for and ABS system. When using repair kits, replace all parts included in the kit. Partial or incorrect repair may lead to functional problems and require the replacement of components.
- Lubricate rubber parts with clean, fresh brake fluid to ease assembly. Do not use shop air to clean parts; damage to rubber components may result.
- Use only DOT 3 brake fluid from an unopened container.
- If any hydraulic component or line is removed or replaced, it may be necessary to bleed the entire system.
- A clean repair area is essential. Always clean the reservoir and cap thoroughly before removing the cap. The slightest amount of dirt in the fluid may plug an orifice and impair the system function. Perform repairs after components have been thoroughly cleaned; use only denatured alcohol to clean components. Do not allow ABS components to come into contact with any substance containing mineral oil; this includes used shop rags.
- The Anti-Lock control unit is a microprocessor similar to other computer units in the vehicle. Ensure that the ignition switch is **OFF** before removing or installing controller harnesses. Avoid static electricity discharge at or near the controller.
- If any arc welding is to be done on the vehicle, the control unit should be unplugged before welding operations begin.

DISC AND DRUM SYSTEMS

CAUTION

Dust and dirt accumulating on brake parts during normal use may contain asbestos fibers from production or aftermarket brake linings. Breathing excessive concentrations of asbestos fibers can cause serious bodily harm. Exercise care when servicing brake parts. Do not sand or grind brake lining unless equipment used is designed to contain the dust residue. Do not clean brake parts with compressed air or by dry brushing. Cleaning should be done by dampening the brake components with a fine mist of water, then wiping the brake components clean with a dampened cloth. Dispose of cloth and all residue containing asbestos fibers in an impermeable container with the appropriate label. Follow practices prescribed by the Occupational Safety and Health Administration (OSHA) and the Environmental Protection Agency (EPA) for the handling, processing, and disposing of dust or debris that may contain asbestos fibers.

BRAKES — BLEEDING THE BRAKE SYSTEM

BLEEDING PROCEDURE

BLEEDING PROCEDURE

➡After bleeding the air from the brake system, if the height or feel of the brake pedal cannot be obtained, perform air bleeding in the brake actuator assembly with the Scan Tool by following the procedures below.

1. Depress the brake pedal more than 20 times with the engine off.
2. Connect the Scan Tool to the DLC3, and turn the ignition switch to the ON position.
3. Select "AIR BLEEDING" on the Scan Tool.
4. Bleed the air out of the brake line as usual when "Step 1: Increase" appears on the Scan Tool display.

➡Bleed the air by following the steps displayed on the Scan Tool. Make sure that the brake fluid in the master cylinder reservoir tank does not become empty.

   a. Connect the vinyl tube to either one of the bleeder plugs.
   b. Depress the brake pedal several times, then loosen the bleeder plug connected to the vinyl tube with the pedal depressed.
   c. When fluid stops coming out, tighten the bleeder plug and release the brake pedal.
   d. Repeat this procedure until all air in the fluid is completely bled out.
   e. Tighten the bleeder plug completely.
   f. Repeat the above procedures for each wheel to bleed the air out of the brake line.

5. Bleed the air out of the suction line when "Step 2: Inhalation" appears on the Scan Tool display.

➡Bleed the air by following the steps displayed on the Scan Tool. Make sure that the brake fluid in the master cylinder reservoir tank does not become empty.

   a. Connect the vinyl tube to the bleeder plug at the right front wheel or the right rear wheel and loosen the bleeder plug.
   b. Operate the brake actuator assembly to bleed the air using the Scan Tool.

➡At this time, be sure to release the brake pedal.

➡This operation stops automatically after 4 seconds.

   c. Check if the operation has stopped by referring to the Scan Tool display.
   d. Repeat this until all air in the fluid is completely bled out.
   e. Tighten the bleeder plug.
   f. Repeat the above procedures for the other wheels to bleed the air out of the brake line.

6. Bleed the air out of the pressure reduction line when "Step 3: Decrease" appears on the Scan Tool display.

➡Bleed the air by following the steps displayed on the Scan Tool. Make sure that the brake fluid in the master cylinder reservoir tank does not become empty.

   a. Connect a vinyl tube to either one of the bleeder plugs.
   b. Loosen the bleeder plug.
   c. Using the Scan Tool, operate the brake actuator assembly, completely depress the brake pedal and keep it.

➡During this procedure, the pedal will feel heavy, but completely depress it so that the brake fluid comes out from the bleeder plug. Be sure to keep depressing the brake pedal. Do not depress and release the pedal repeatedly.

➡The operation stops automatically after 4 seconds. When performing this procedure continuously, set an interval of at least 20 seconds. When the operation is complete, the brake pedal goes down slightly. This is a normal phenomenon caused when the solenoid opens.

d. Tighten the bleeder plug, then release the brake pedal.

e. Repeat this until all the air in the fluid is completely bled out.

f. Tighten the bleeder plug.

g. Repeat the above procedures for the other wheels to bleed the air out of the brake line.

7. Bleed the air out of the brake line as usual again when "Step 4: Increase" appears on the Scan Tool display.

➡Bleed the air by following the steps displayed on the Scan Tool. Make sure that the brake fluid in the master cylinder reservoir tank does not become empty.

a. Connect the vinyl tube to either one of the bleeder plugs.

b. Depress the brake pedal several times, then loosen the bleeder plug connected to the vinyl tube with the pedal depressed.

c. When fluid stops coming out, tighten the bleeder plug, then release the brake pedal.

d. Repeat this until all the air in the fluid is completely bled out.

e. Tighten the bleeder plug.

f. Repeat the above procedures for the other wheels to bleed the air out of the brake line.

g. Make sure that the air bleeding is complete by referring to the Scan Tool display.

h. Check the fluid level and add fluid if necessary.

BLEEDING THE ABS SYSTEM

The following automated ABS bleed procedure is required when one of the following occur:

- Manual bleeding at the wheel cylinders does not achieve the desired pedal height or feel.
- BPMV replacement
- Extreme loss of brake fluid has occurred.
- Air ingestion is suspected.

1. If none of the above conditions apply, use standard bleed procedures.

2. The auto bleed procedure is used on BOSH 5.3 equipped vehicles. This procedure uses a scan tool to cycle the system solenoid valves and run the pump in order to purge the air from the secondary circuits. These secondary circuits are normally closed off, and are only opened during system initialization at vehicle start up and during ABS operation. The automated bleed procedure opens these secondary circuits and allows any air trapped inside the BPMV to flow out toward the wheel cylinders or calipers where it can be purged out of the system.

3. Inspect the battery for full charge, repair the battery and charging system, as necessary.

4. Connect a scan tool to the data link connector (DLC) and select current and history DTCs. Repair any DTCs prior to performing the ABS bleed procedure.

Inspect for visual damage and leaks and repair, as needed.

5. Raise and vehicle on a suitable support.

6. Turn the ignition switch to the **OFF-**position.

7. Remove all 4 tires.

8. Connect the pressure bleeding tool according to the manufacturer's instructions.

9. Turn the ignition switch to **RUN** position, engine off.

10. Connect a scan tool and establish communications with the ABS system.

11. Pressurize the bleeding tool to 30–35 psi (206–241 kPa).

12. Performing the Automated Bleed Procedure

➡The Auto Bleed Procedure may be terminated at any time during the process by pressing the EXIT button. No further Scan Tool prompts pertaining to the Auto Bleed procedure will be given. After exiting the bleed procedure, relieve bleed pressure and disconnect bleed equipment per manufacturer's instructions. Failure to properly relieve pressure may result in spilled brake fluid causing damage to components and painted surfaces.

13. With the pressure bleeding tool at 30–35 psi (206–241 kPa), and all bleeder screws in closed position, select Automated Bleed Procedure on the scan tool and follow the instructions.

14. The first part of the automated bleed procedure will cycle the pump and front release valves for one minute. After the cycling has stopped the scan tool will enter a cool down mode and display a 3 minute timer. The auto bleed will not continue until this time expired, and cannot be overridden.

15. During the next step, the scan tool will request the technician to open one of the bleeder screws. The scan tool will then cycle the respective release valve and pump motor for one minute.

16. The scan tool will repeat step 3 for the remaining bleeder screws.

17. With the bleeder tool still attached to the vehicle and maintaining 30–35 psi (206–241 kPa), the scan tool will instruct the technician to independently open each bleeder screw for approximately 20 seconds. This should allow any remaining air to be purged from the brake lines.

18. When the automated bleed procedure is completed the scan tool will display the appropriate message.

19. Install all 4 tires.

20. Remove pressure from the pressure bleeding tool and then disconnect the tool from the vehicle.

21. Depress the brake pedal to gage pedal height and feel. Repeat steps 1-8 until the pedal is acceptable.

22. Remove the scan tool from the DLC connector.

23. Lower the vehicle.

24. Inspect the brake fluid level in master cylinder.

25. Road test the vehicle while making sure the brake pedal remains high and firm.

26. If the vehicle is equipped with a traction control system (TCS), the scan tool will cycle both the ABS and the TCS solenoid valves. This bleed procedure is the same as above.

WHEEL SPEED SENSORS

REMOVAL & INSTALLATION

Front

See Figure 1.

1. The right side and the left side procedures are the same.
2. Remove the wheel.
3. Remove the fender liner.
4. Disengage the speed sensor wire harness clamp from the body.
5. Disconnect the speed sensor connector.
6. Remove the 2 clamp bolts holding the sensor harness from the body and shock absorber.
7. Remove the bolt and speed sensor.

CAUTION

Keep the tip of the sensor clean.

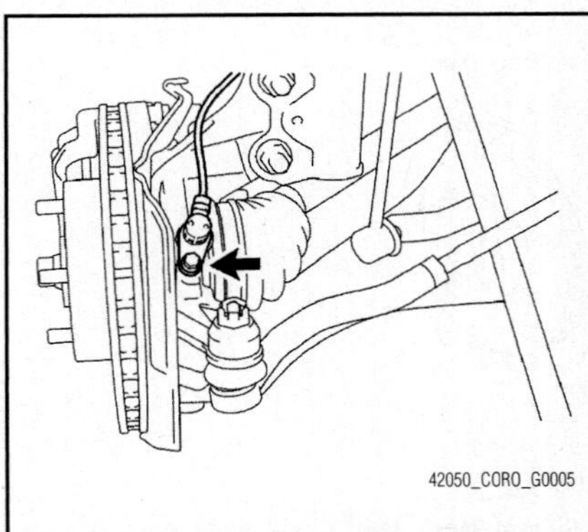

Fig. 1 Removal and installation of the speed sensor

To install:

8. Install the speed sensor with the bolt. Tighten to 71 inch lbs. (8 Nm) of torque.
9. Install the sensor harness clamp with the 2 bolts to the body and shock absorber.
10. Connect the speed sensor connector.
11. Engage the speed sensor wire harness clamp to the body.
12. Install the fender liner.
13. Install the wheel.

Rear

See Figures 2 and 3.

➡Use the same procedure for the RH side and LH side. The procedure listed below is for the LH side.

➡If the sensor rotor needs to be replaced, replace it together with the rear axle hub and bearing assembly with rear speed sensor.

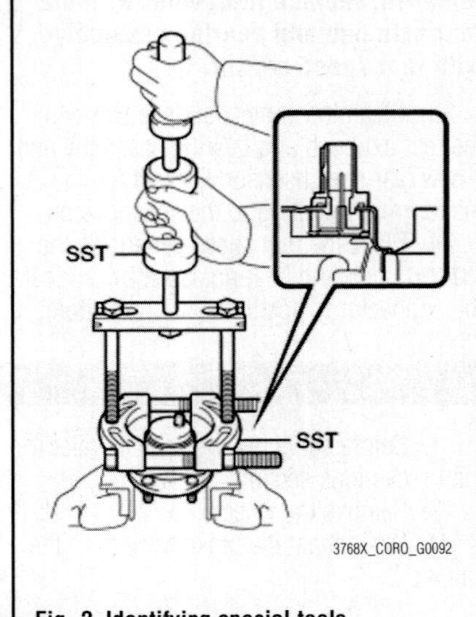

Fig. 2 Identifying special tools

➡Special tools needed for this procedure: SST: 09520-00031, 09521-00010, 09520-00040) (SST: 09521-00020)

1. Disconnect cable from negative battery terminal.
2. For vehicles equipped with disc brakes, remove lower LH and RH instrument panel finish panel LH.
3. For vehicles equipped with disc brakes, remove shift lever knob sub-assembly.
4. For vehicles equipped with disc brakes, remove center instrument cluster finish panel assembly.
5. For vehicles equipped with disc brakes, remove upper console panel sub-assembly.
6. Remove rear wheel.
7. Loosen parking brake cable (for disc type). Refer to Parking Brake in this section.
8. Disconnect rear speed sensor wire (for Drum Type). Using a screwdriver, disconnect the connector from the rear speed sensor. Be careful not to damage the rear speed sensor.
9. Disconnect rear speed sensor wire (for Disc Type). Using a screwdriver, disconnect the connector from the rear speed sensor. Be careful not to damage the rear speed sensor.
10. Remove parking brake lever protector.
11. Separate no. 3 parking brake cable assembly (for Disc Type). Refer to Parking Brake in this section.
12. Separate rear disc brake caliper assembly (for Disc Type). Refer to Brake Caliper under Rear Disc Brake in this section.
13. Remove rear disc (for disc type). Refer to Disc under Rear Disc Brake in this section.
14. Remove rear brake drum (for drum type). Refer to Brake Drum under Rear Drum Brake in this section.
15. Remove rear axle hub and bearing assembly with rear speed sensor (for drum type). Refer to Wheel Hub and Bearing Assembly.
16. Remove rear axle hub and bearing assembly with rear speed sensor (for disc type). Refer to Wheel Hub and Bearing Assembly.
17. Remove rear speed sensor:
    a. Install the 3 hub nuts and mount the rear axle hub and bearing assembly in a vise using aluminum plates.
    b. Replace the rear axle hub and bearing assembly if it is dropped or receives a strong shock.
    c. Using a pin punch and a hammer, drive out the 2 pins and remove the 2 attachments from SST.
    d. Using SST and 2 bolts (diameter: 12 mm, pitch: 1.5 mm), remove the rear speed sensor from the rear axle hub and bearing assembly.
18. Keep the rear speed sensor away from magnets.
19. Pull the rear speed sensor off straight, taking care not to allow it to contact with the rear speed sensor rotor.

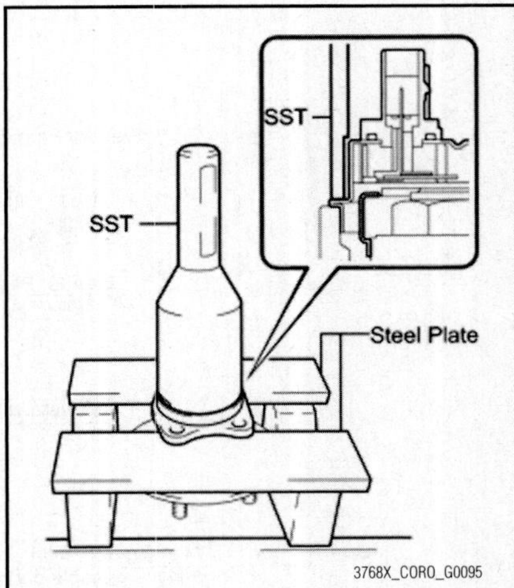

Fig. 3 Installing the new speed sensor to the rear axle hub and bearing assembly using SST: 09214-76011

20. If the rear speed sensor rotor is damaged or deformed, replace the rear axle hub and bearing assembly.

21. Do not scratch the contact surface between the rear axle hub and bearing assembly and the rear speed sensor.

22. Prevent foreign matter from attaching to the speed sensor rotor or tip.

➡If the sensor rotor needs to be replaced, replace it together with the rear axle hub and bearing assembly with rear speed sensor.

To install:

➡If the sensor rotor needs to be replaced, replace it together with the rear axle hub and bearing assembly with rear speed sensor.

23. Clean the contact surface between the rear axle hub and bearing assembly and a new rear speed sensor. Prevent foreign matter from attaching to the sensor rotor.

24. Place the rear speed sensor on the rear axle hub and bearing assembly so that the connector is positioned at the bottom.

25. Using SST: 09214-76011, steel plates and a press, install a new speed sensor to the rear axle hub and bearing assembly.

➡Keep the rear speed sensor away from magnets. Do not use a hammer to install the rear speed sensor. Check that there is no foreign matter such as iron chips on the detecting portion of the rear speed sensor. Slowly press in the rear speed sensor straight.

26. To complete the installation, reverse remaining removal procedure.

BRAKES

FRONT DISC BRAKES

BRAKE CALIPER

REMOVAL & INSTALLATION

See Figure 4.

1. Before servicing the vehicle, refer to the Precautions section.

2. Remove the wheels.

3. Disconnect the brake hose from the caliper.

4. Remove the bolts that attach the caliper to the torque plate. If applicable, hold the flats of the sliding pin with a wrench while loosening the caliper attaching bolts.

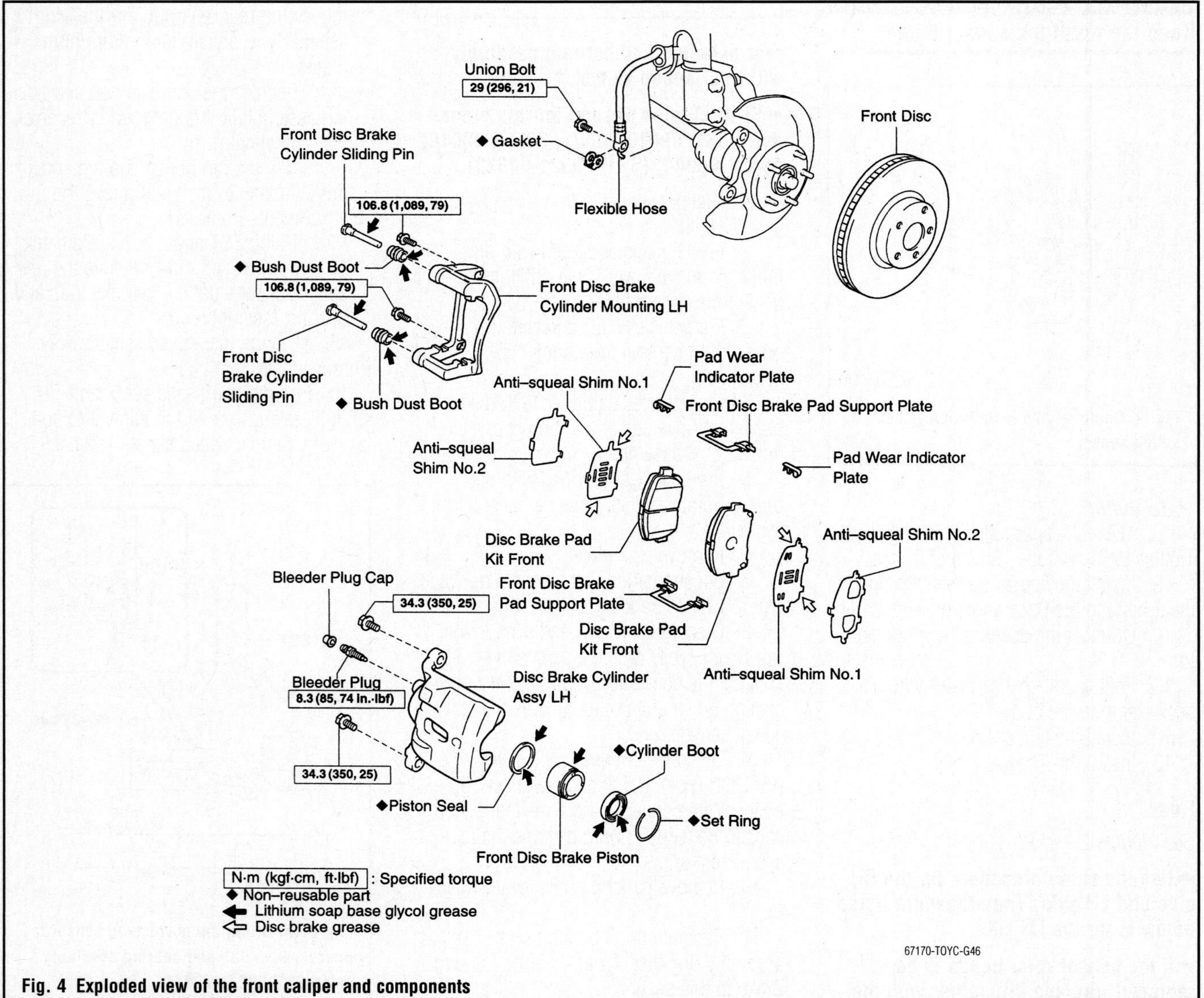

Fig. 4 Exploded view of the front caliper and components

5. Lift up and remove the caliper assembly.

To install:

6. Install the caliper and loosely install the bolts.
7. Hold the flats of the sliding pin with a wrench, then tighten the bolts. Tighten to 25 ft. lbs. (34 Nm).
8. Connect the brake hose to the caliper, using 2 new washers.
9. Fill the brake system to the proper level and bleed the brake system.
10. Add brake fluid to the reservoir to fill to the correct level.
11. Lower the vehicle to the ground.

DISC BRAKE PADS

REMOVAL & INSTALLATION

See Figure 4.

1. Before servicing the vehicle, refer to the Precautions section.
2. Remove the wheels.
3. Loosen and remove the caliper mounting bolts, then remove the caliper assembly, without disconnecting the brake line. Position it aside.
4. Slide out the old brake pads along with any anti-squeal shims, springs, pad wear indicators and pad support plates.

To install:

5. Install the pad support plates into the torque plate.
6. Install the pad wear indicators onto the pads. Be sure the arrow on the indicator plate is pointing in the direction of rotation.
7. Install the anti-squeal shims on the outside of each pad and then install the pad assemblies into the torque plate.
8. Compress the caliper piston into the bore.
9. Position the caliper back down over the pads.
10. Install and tighten the caliper mounting bolts.
11. Install the wheels. Check the brake fluid level.

BRAKES — REAR DISC BRAKES

BRAKE CALIPER

REMOVAL & INSTALLATION

See Figure 5.

1. Before servicing the vehicle, refer to the Precautions section.
2. Remove the wheels.
3. Disconnect the brake hose from the caliper.
4. Remove the bolts that attach the caliper to the torque plate. If applicable, hold the flats of the sliding pin with a wrench while loosening the caliper attaching bolts.

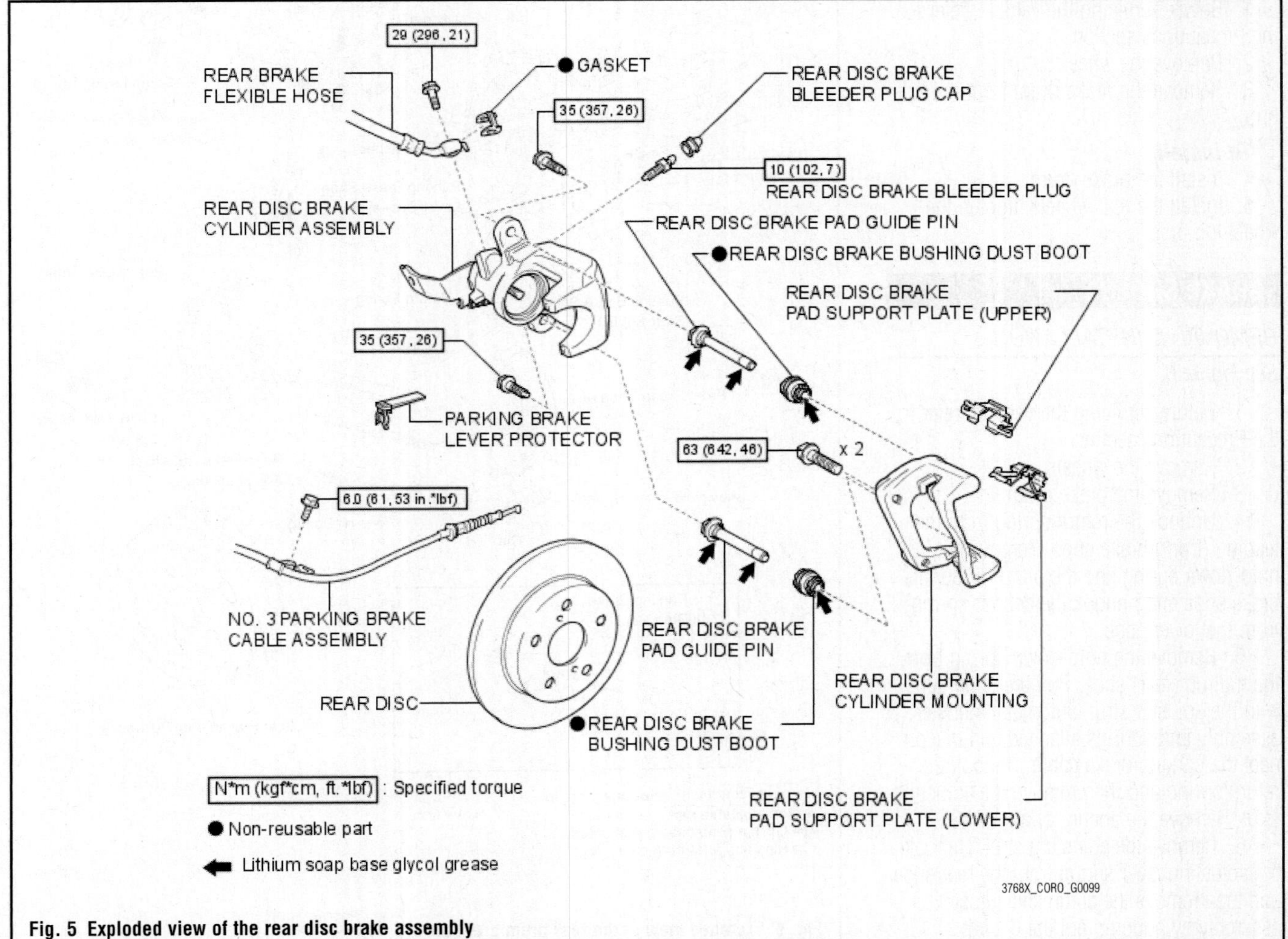

Fig. 5 Exploded view of the rear disc brake assembly

5. Lift up and remove the caliper assembly.

To install:

6. Install the caliper and loosely install the bolts.
7. Hold the flats of the sliding pin with a wrench, then tighten the bolts. Tighten the bolts on rear calipers to 34 ft. lbs. (46 Nm).
8. Connect the brake hose to the caliper, using 2 new washers.
9. Fill the brake system to the proper level and bleed the brake system.
10. Add brake fluid to the reservoir to fill to the correct level.
11. Lower the vehicle to the ground.

DISC BRAKE PADS

REMOVAL & INSTALLATION

See Figure 5.

1. Before servicing the vehicle, refer to the Precautions section.
2. Remove the wheels.
3. Loosen and remove the caliper mounting bolts, then remove the caliper assembly, without disconnecting the brake line. Position it aside.
4. Slide out the old brake pads along with any anti-squeal shims, springs, pad wear indicators and pad support plates.

To install:

5. Install the pad support plates into the torque plate.
6. Install the pad wear indicators onto the pads. Be sure the arrow on the indicator plate is pointing in the direction of rotation.
7. Install the anti-squeal shims on the outside of each pad and then install the pad assemblies into the torque plate.
8. Compress the caliper piston into the bore. For Corolla rear calipers, use tool SST 09719-14020, to rotate the piston clockwise while pressing it into the bore until it locks.
9. Position the caliper back down over the pads.
10. Install and tighten the caliper mounting bolts.
11. Install the wheels. Check the brake fluid level.

BRAKES — REAR DRUM BRAKES

BRAKE DRUM

REMOVAL & INSTALLATION

See Figure 6.

1. Before servicing the vehicle, refer to the Precautions section.
2. Remove the wheels.
3. Remove the brake drum from the axle hub.

To install:

4. Install the brake drum.
5. Install the rear wheels, tighten the wheel lug nuts.

BRAKE SHOES

REMOVAL & INSTALLATION

See Figure 6.

1. Before servicing the vehicle, refer to the Precautions section.
2. Remove the wheels.
3. Remove the brake drum.
4. Unhook the return spring from the leading (front) brake shoe. Remove the hold-down spring and the pin. Pull out the brake shoe and unhook the anchor spring from the lower edge.
5. Remove the hold-down spring from the trailing (rear) shoe. Pull the shoe out with the adjuster strut, automatic adjuster assembly and springs attached and disconnect the parking brake cable. Unhook the return spring and then remove the adjusting strut. Remove the anchor spring.
6. Remove the adjusting strut. Unhook the adjusting lever spring from the rear shoe and then remove the automatic adjuster assembly by popping out the C-clip.

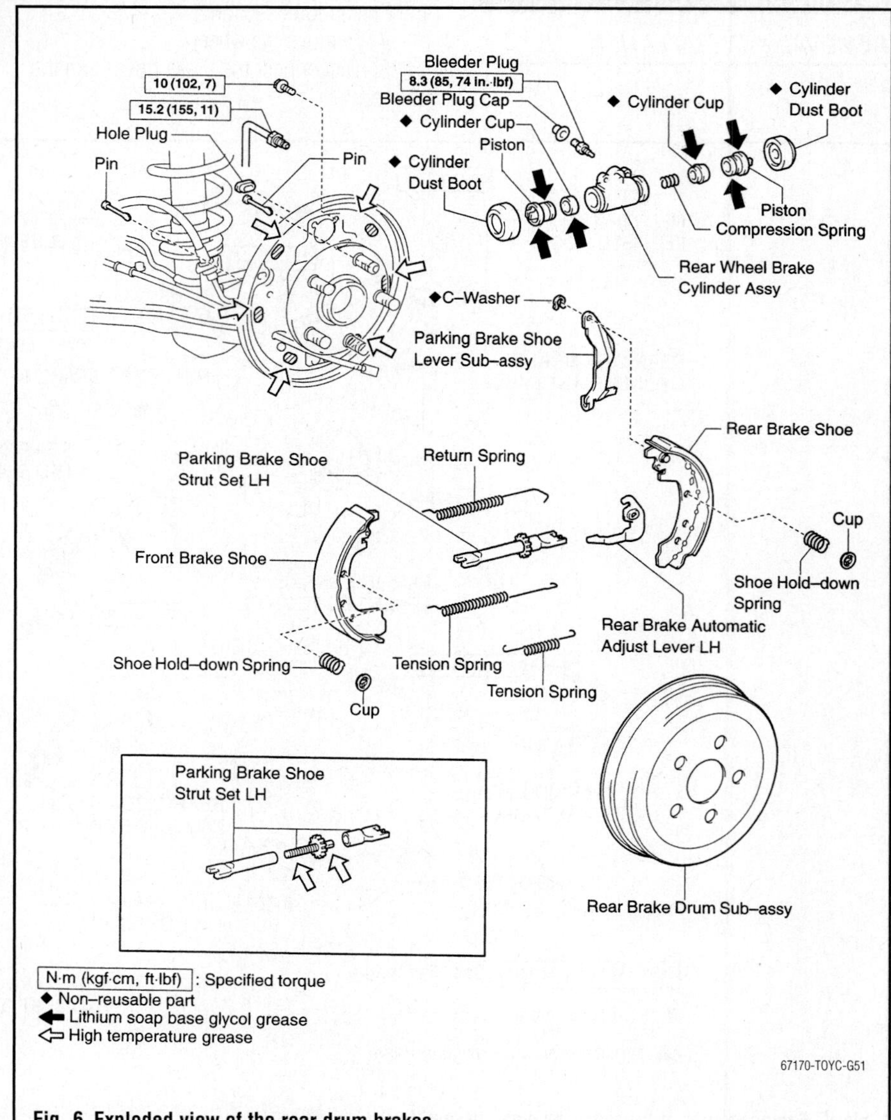

Fig. 6 Exploded view of the rear drum brakes

To install:

7. Mount the automatic adjuster assembly onto a new rear brake shoe. Make sure the C-clip fits properly. Connect the adjusting strut/return spring and then install the adjusting spring.
8. Connect the parking brake cable to the rear shoe and then position the shoe so the lower end rides in the anchor plate and the upper end is against the boot in the wheel cylinder. Install the pin and the hold-down spring.
9. Install the anchor spring between the front and rear shoes. Install the hold-down spring and pin.
10. Connect the return spring/adjusting strut between the 2 shoes so it rides freely.
11. Install the drum.
12. Install the wheel.

ADJUSTMENT

1. Temporarily install the hub nuts.
2. Remove the hole plug, and turn the adjuster and expand the shoe until the drum locks.
3. Using a screwdriver, back off the adjuster 8 notches.
4. Install the hole plug.

BRAKES

PARKING BRAKE

PARKING BRAKE CABLES

ADJUSTMENT

See Figure 7.

1. Remove the rear wheel.
2. Adjust the brake shoe clearance.
3. Install the rear wheel.
4. Pull the parking brake lever to the fully applied position, and count the number of clicks (6 to 9 is optimal).
5. Remove the rear console box sub-assembly.
6. Loosen the lock nut and turn the adjusting nut until the lever travel turns correct (6-9 Clicks).
7. Tighten the lock nut to 44 in. lbs. (5 Nm) of torque.
8. Install the rear console box assembly.

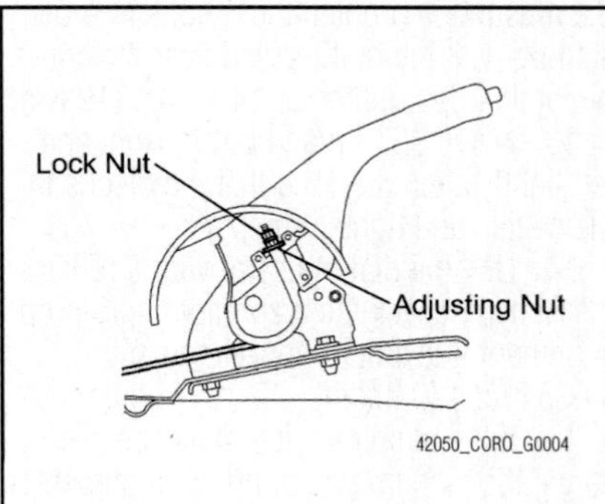

Fig. 7 Adjusting the parking brake cable

PARKING BRAKE SHOES

REMOVAL & INSTALLATION

Drum

1. Before servicing the vehicle, refer to the Precautions section.
2. Remove the wheels.
3. Remove the brake drum.
4. Unhook the return spring from the leading (front) brake shoe. Remove the hold-down spring and the pin. Pull out the brake shoe and unhook the anchor spring from the lower edge.
5. Remove the hold-down spring from the trailing (rear) shoe. Pull the shoe out with the adjuster strut, automatic adjuster assembly and springs attached and disconnect the parking brake cable. Unhook the return spring and then remove the adjusting strut. Remove the anchor spring.
6. Remove the adjusting strut. Unhook the adjusting lever spring from the rear shoe and then remove the automatic adjuster assembly by popping out the C-clip.

To install:

7. Mount the automatic adjuster assembly onto a new rear brake shoe. Make sure the C-clip fits properly. Connect the adjusting strut/return spring and then install the adjusting spring.
8. Connect the parking brake cable to the rear shoe and then position the shoe so the lower end rides in the anchor plate and the upper end is against the boot in the wheel cylinder. Install the pin and the hold-down spring.
9. Install the anchor spring between the front and rear shoes. Install the hold-down spring and pin.
10. Connect the return spring/adjusting strut between the 2 shoes so it rides freely.
11. Install the drum.
12. Install the wheel.

ADJUSTMENT

Inspect Parking Brake Lever Travel

1. Pull the parking brake lever firmly.
2. Release the parking brake lock, and return the parking brake lever to its off position.
3. Slowly pull the parking brake lever all the way up, and count the number of clicks.
4. Parking brake lever travel: 6 to 9 notches at 45 Lbs (200 N).

Adjust Parking Brake Lever Travel

1. Remove the upper console box assembly.
2. Completely release the parking brake lever.
3. Loosen the lock nut and the adjusting nut to completely release the parking brake cable.
4. Fully depress the brake lever 3 to 5 times with the engine stopped.
5. Turn the adjusting nut until the parking brake lever travel is corrected to within the specified range.
6. Parking brake lever travel: 6 to 9 notches at 45 Lbs (200 N).
7. Using a wrench or an equivalent tool, hold the adjusting nut and tighten the lock nut to 53 inch. lbs. (200 Nm).
8. Operate the parking brake lever 3 to 4 times, and check the parking brake lever travel.
9. Check whether the parking brake drags or not.
10. Install the upper console box assembly.

Inspect Rear Disc Brake Cylinder Operation Lever And Stopper Clearance (For Rear Disc Brake)

1. Release the parking brake lever and check that the clearance measurement between the rear disc brake cylinder operation lever and the stopper is within the specified range:
   a. Clearance: 0.0197 inch. (0.5 mm) or less
   b. If the clearance is not within the specified range, replace the rear disc brake caliper assembly.

CHASSIS ELECTRICAL

AIR BAG (SUPPLEMENTAL RESTRAINT SYSTEM)

GENERAL INFORMATION

This vehicle is equipped with a Supplemental Restraint System (SRS). It consists of a driver airbag, front passenger airbag, side airbag, curtain shield airbag and front seat belt pretensioner. Failure to carry out service operations in the correct sequence could cause the SRS to unexpectedly deploy during servicing, possibly leading to a serious accident. Further, if a mistake is made in servicing the SRS, it is possible that the SRS may fail to operate when required. Before performing servicing (including removal or installation of parts, inspection or replacement), be sure to read the following carefully, then follow the correct procedures indicated in the repair manual.

SERVICE PRECAUTIONS

✻✻ CAUTION

Wait at least 90 seconds after the ignition switch is turned off and the negative (-) terminal cable is disconnected from the battery before starting the operation. (The SRS is equipped with a backup power source, so that if work is started within 90 seconds after disconnecting the negative (-) terminal cable of the battery, the SRS may be deployed.)

DISARMING THE SYSTEM

To avoid personal injury when working on vehicles equipped with an air bag, the negative battery cable must be disconnected and at least 90 seconds must elapse before working on the system. Failure to do so may result in deployment of the air bag.

ARMING THE SYSTEM

After vehicle service is completed, reattach the battery cables (positive cable first!) to rearm the air bag system.

CLOCKSPRING CENTERING

1. Check that the ignition switch is off.
2. Check that the battery negative (-) terminal is disconnected.
3. Rotate the spiral cable counterclockwise slowly by hand until it feels firm.

➡Do not turn the spiral cable by the airbag wire harness.

4. Rotate the spiral cable clockwise approximately 2.5 turns to align the marks.

➡Do not turn the spiral cable by the airbag wire harness. The spiral cable will rotate approximately 2.5 turns to both the left and right from the center.

DRIVE TRAIN

CLUTCH DRIVEN DISC & PRESSURE PLATE

REMOVAL & INSTALLATION

See Figures 8 and 9.

1. Remove the manual transaxle assembly.
2. Remove the clutch release fork with the clutch release bearing from the manual transaxle.
3. Remove the clutch release fork boot from the manual transaxle.
4. Remove the release bearing and clip from the clutch release fork.
5. Remove the release fork support from the manual transaxle.
6. Put matchmarks on the clutch cover assembly and the flywheel sub-assembly.
7. Loosen each set bolt one turn at a time until the spring tension is released.
8. Remove the set bolts and pull off the clutch cover.

➡Do not drop the clutch disc.

9. Remove clutch disc assembly.

➡Keep the parts clutch disc lining, the pressure plate, and the surface of the flywheel sub-assembly free of oil and foreign matter.

To install:

10. Insert the alignment tool (SST: 09301-00110) into the clutch disc assembly, then insert them both into the flywheel sub-assembly.

➡Insert the clutch disc assembly in the correct direction.

11. Align the matchmark on the clutch cover assembly with the one on the flywheel sub-assembly.
12. Following the procedure shown in the illustration, tighten the 6 bolts in order, starting with the bolt located near the knock pin at the top. Tighten to 14 ft. lbs. (19 Nm).
13. Move SST up and down, right and left lightly after checking that the disc is in the center, and tighten the bolts.
14. Using a dial indicator with a roller instrument, check the diaphragm spring tip alignment. Maximum non-alignment: 0.0354 inch (0.9 mm)
15. If the alignment is not as specified, adjust the diaphragm spring tip alignment using SST: 09333-00013.
16. Install the release fork support onto the transaxle assembly. Tighten to 27 ft. lbs. (37 Nm).
17. Install the clutch release fork boot to the manual transaxle.
18. Apply release hub grease to the contact surfaces between the release fork and release bearing assembly, release fork and push rod, and release fork and fork support.
19. Install the release fork onto the release bearing assembly with the clip.
20. Apply clutch spline grease to the input shaft spline. Do not grease the smooth collar area of the spline.

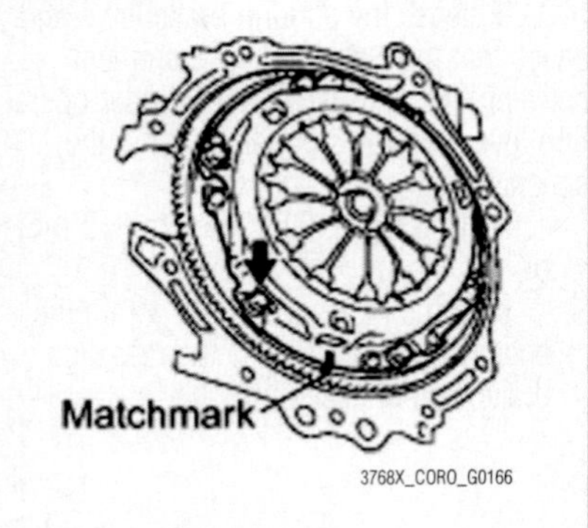

Fig. 8 Applying matchmark

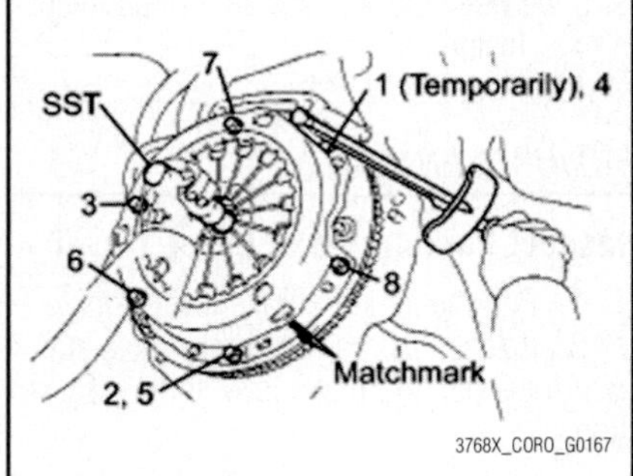

Fig. 9 Clutch assembly tightening sequence

21. Install the clutch release bearing with the release fork onto the transaxle assembly.

➡**After installation, move the fork back and forth to check that the release bearing slides smoothly.**

22. Install the manual transaxle.

HYDRAULIC SYSTEM BLEEDING

BLEEDING PROCEDURE

➡**If any maintenance on the clutch system was performed or the system is suspected of containing air, bleed the system. Use care; brake fluid will remove the paint from any surface. If the brake fluid spills onto any painted surface, wash it off immediately with soap and water.**

1. Before servicing the vehicle, refer to the Precautions section.
2. Fill the clutch reservoir with brake fluid. Check the reservoir level frequently and add fluid as needed.
3. Connect one end of a vinyl tube to the bleeder plug on the slave cylinder and submerge the other end into a clear container half-filled with brake fluid.
4. Slowly pump the clutch pedal several times.
5. Have an assistant hold the clutch pedal down and loosen the bleeder plug until fluid and/or air starts to run out of the bleeder plug. Close the bleeder plug while the pedal is held to the floor.

➡**Do not allow the pedal to rise backup while the bleeder is still open. If this happens, it will allow air to reenter the slave cylinder and cause the clutch system not to work properly.**

6. Repeat Steps 2 and 3 until all the air bubbles are removed from the system.
7. Tighten the bleeder plug when all the air is gone.
8. Refill the master cylinder to the proper level as required.
9. Check the system for leaks.

FRONT HALFSHAFT

REMOVAL & INSTALLATION

See Figures 10 and 11.

➡**The hub bearing could be damaged if subjected to the full weight of the vehicle, such as if the vehicle is moved without the halfshafts. If it is absolutely necessary to place the full vehicle weight on the hub bearing, first support the bearing with SST No. 09608–16042.**

➡**This procedure is for removing both LH and RH halfshafts.**

1. Remove the LH and RH engine under covers.
2. Drain transmission fluid.
3. Raise and support the vehicle.
4. Remove the front wheels.
5. Remove front axle shaft nut:
   a. Using SST: 09930-00010 and a hammer, release the staked part of the front axle shaft nut.
   b. Insert SST into the groove with the flat surface facing up.
   c. Do not damage the tip of SST using grinders.
   d. Completely unstake the staked part before removing the axle hub nut.
   e. Do not damage the threads of the drive shaft.
   f. Using a socket wrench (30 mm), remove the front axle shaft nut.
6. Separate the wheel speed sensors.
7. Separate the front stabilizer link assembly. Refer to Stabilizer in Front Suspension.
8. Separate the tie rod end sub-assembly. Refer to Tie Rod in Front Suspension.
9. Separate front lower suspension. Refer to Front Suspension.
10. Separate front axle assembly:
    a. Using a plastic hammer, tap the end of the drive shaft and disengage the fitting between the drive shaft and front axle assembly.
    b. If it is difficult to disengage the fitting, tap the end of the drive shaft with a brass bar and hammer.
    c. Push the front axle assembly out of the vehicle to remove the drive shaft from the front axle assembly.

➡**Do not push out the front axle further than necessary. Do not damage the outboard joint boot. Do not damage the speed sensor rotor.**

11. If equipped, remove the 2 bolts and the manual transmission case protector.
12. Using a halfshaft puller, remove the LH halfshaft assembly.
13. Remove front RH drive shaft assembly:
    a. For 1.8L Engines:
    b. Using a brass bar and hammer, remove the front drive shaft assembly.

➡**Do not damage the oil seal. Do not damage the inboard joint boot. Do not drop the drive shaft.**

    c. For 2.4L Engines:
    d. Remove the bearing bracket hole snap ring.
    e. Remove the bolt and front drive shaft assembly RH from the drive shaft bearing bracket.

➡**Do not damage the boot or oil seal.**

14. Using a screwdriver, remove the snap ring from the front LH and RH drive inboard joint assembly.

To install:

15. Install a new front drive shaft hole snap ring to the front drive inboard joint assembly.
16. Align the inboard joint splines, and using a brass bar and a hammer, install the front drive shaft assembly.

➡**Face the end gap of the front drive inboard joint hole snap ring downward. Do not damage the oil seal. Do not damage the inboard joint boot.**

➡**Confirm whether the drive shaft is securely driven in by checking the reaction force and sound.**

17. For 2.4L engines:

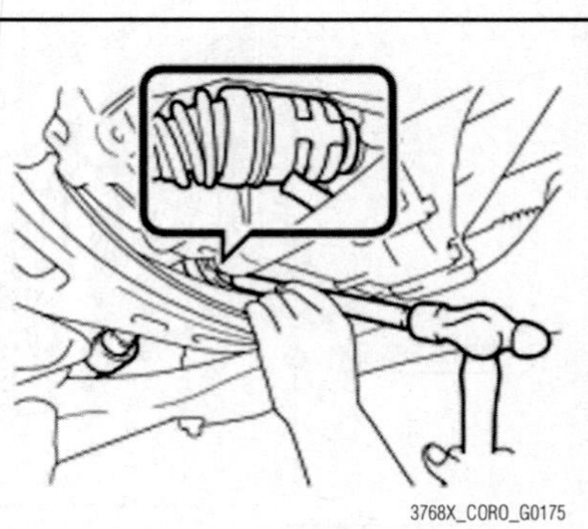

Fig. 10 Removing the halfshaft assembly—1.8L Engine

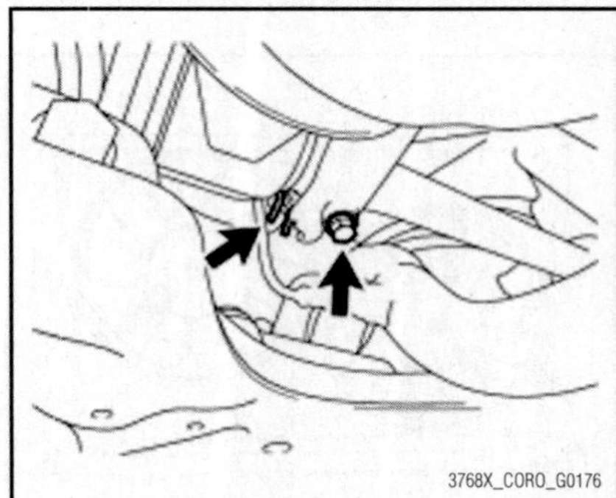

Fig. 11 Removing the bearing bracket hole snap ring and halfshaft assembly—2.4L Engine

a. Using a screwdriver, install a new bearing bracket hole snap ring.

➡**Do not damage the boot or oil seal. Move the drive shaft assembly while keeping it level.**

b. Install a new bolt. Tighten to 24 ft. lbs. (32 Nm).

18. If equipped, install the manual transmission case protector with the 2 bolts. Tighten to 13 ft. lbs. (18 Nm).

19. To complete installation, reverser remaining removal procedure.

20. Install front axle shaft nut:

a. Clean the threaded parts on the drive shaft and axle shaft nut using a non-residue solvent.

➡**Be sure to perform this work for a new drive shaft. Keep the threaded parts free of oil and foreign matter.**

b. Using a socket wrench (30 mm), install a new axle shaft nut. Tighten to 160 ft. lbs. (216 Nm).

c. Using a chisel and hammer, caulk the axle shaft nut.

21. Tighten wheel lug nuts to 76 ft. lbs. (103 Nm).

22. Add transmission fluid.

23. Inspect for fluid leaks.

24. Inspect and adjust the wheel alignment.

ENGINE COOLING

ENGINE FAN

REMOVAL & INSTALLATION

1.8L Engine

See Figure 12.

1. Disconnect the negative battery cable.
2. Remove the radiator assembly. Refer to Radiator in this section.
3. Remove the nut, then remove the fan.
4. To install, reverse removal procedure.

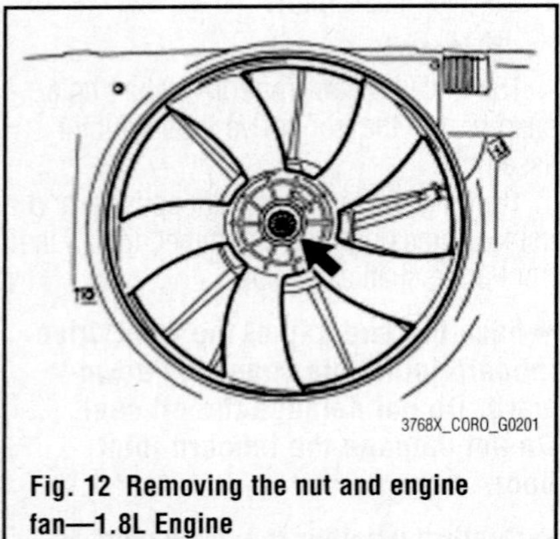

Fig. 12 Removing the nut and engine fan—1.8L Engine

2.4L Engine

See Figure 13.

1. Disconnect the negative battery cable.
2. Remove the radiator assembly. Refer to Radiator in this section.
3. Remove the 2 nuts and 2 fans.

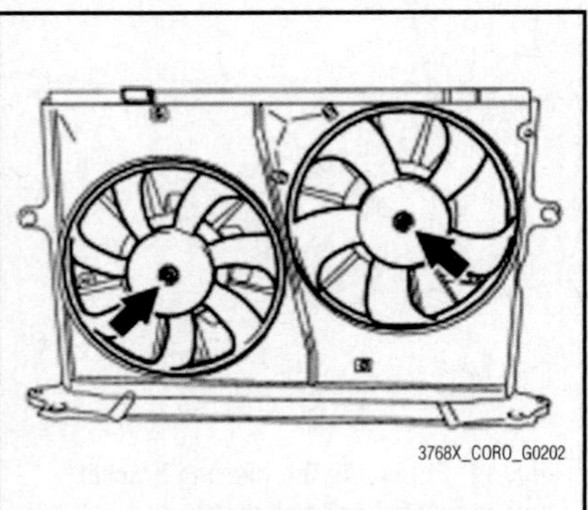

Fig. 13 Removing the nut and engine fan—2.4L Engine

RADIATOR

REMOVAL & INSTALLATION

1.8L Engine

See Figure 14.

1. Disconnect the negative battery cable.
2. Drain the coolant.
3. Remove battery.
4. Remove the 2 radiator grille protectors.
5. Remove front bumper assembly:

a. Remove the clip.

b. Using a screwdriver, turn the pin 90 degrees and remove the pin hold clip.

c. Put protective tape around the front bumper assembly.

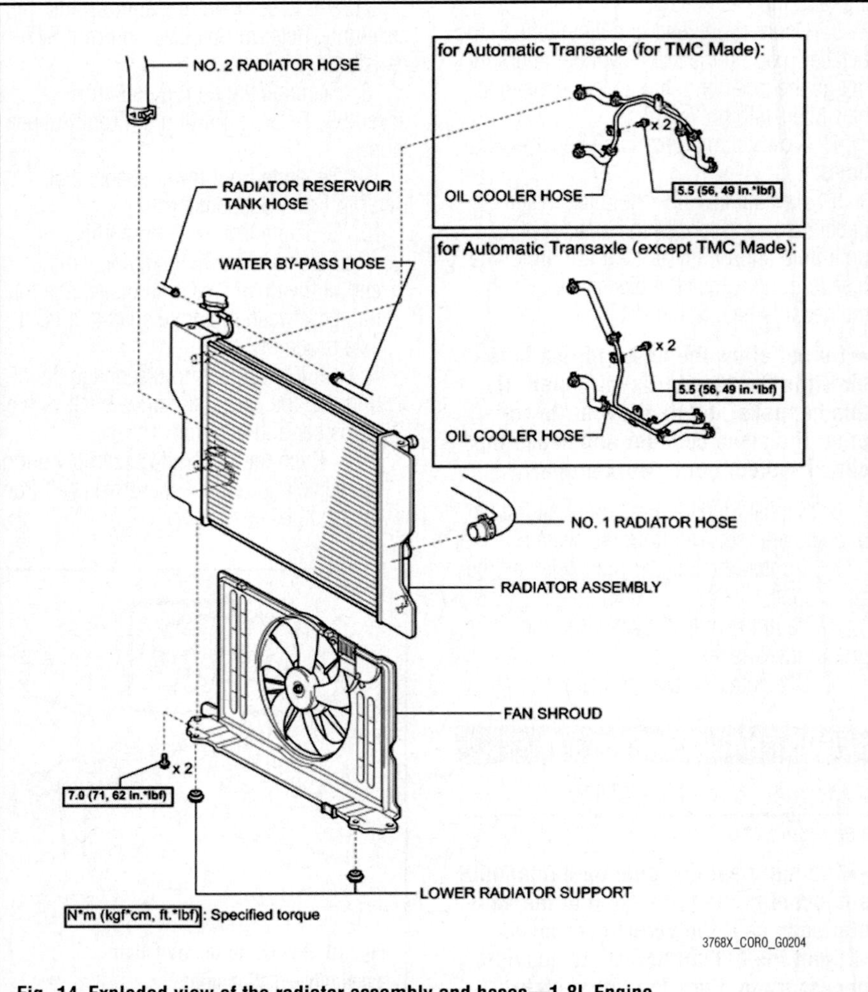

Fig. 14 Exploded view of the radiator assembly and hoses—1.8L Engine

d. Remove the 2 screws and 3 clips.
e. Remove the 4 screws.
f. Remove the 2 clips.
g. Disengage the 6 claws and remove the front bumper assembly.
h. Disconnect the fog light connector. (w/ Fog Light).

6. Disconnect the radiator reservoir tank hose from the radiator assembly.
7. Disconnect the both radiator hose from the radiator assembly.
8. Remove the 2 bolts and 2 upper radiator supports.
9. Remove the 2 support cushions from the 2 upper radiator supports.
10. Remove the 3 bolts from the hood lock assembly.
11. Disconnect the hood lock control cable and remove the hood lock assembly.
12. Separate the water by-pass hose from the 3 clamps.
13. Disconnect the water by-pass hose from the radiator assembly.
14. Remove the 2 bolts and hood lock support sub-assembly.
15. Disconnect the horn connector.
16. Remove the 4 bolts and upper radiator support sub-assembly.
17. Remove the 2 bolts, disengage the 2 claws, and remove the No. 2 fan shroud from the radiator assembly.
18. Disconnect the cooling fan ECU connector and wire harness clamp.
19. Remove the radiator assembly with the fan shroud.
20. Remove the 2 lower radiator supports.
21. For A/T equipped vehicles:
a. Disconnect the 2 oil cooler hoses from the radiator.
b. Remove the 2 bolts and oil cooler hose.
22. Remove the 2 bolts and fan shroud from the radiator assembly.
23. To install, reverse removal procedure.

➡Do not apply any excessive force to the cooler condenser assembly or pipe when installing the radiator assembly.

2.4L Engine

See Figure 15.

1. Disconnect the negative battery cable.
2. Drain the coolant.
3. Remove battery.
4. Remove the 2 radiator grille protectors.
5. Remove front bumper assembly:
a. Remove the clip.
b. Using a screwdriver, turn the pin 90 degrees and remove the pin hold clip.
c. Put protective tape around the front bumper assembly.
d. Remove the 2 screws and 3 clips.
e. Remove the 4 screws.
f. Remove the 2 clips.
g. Disengage the 6 claws and remove the front bumper assembly.
h. Disconnect the fog light connector. (w/ Fog Light).

6. Disconnect the radiator reserve tank hose from the radiator assembly.
7. Disconnect both radiator hoses from the radiator assembly.
8. Disconnect oil cooler hose (for automatic transaxle).
9. Remove the 2 bolts and 2 upper radiator supports.
10. Remove the 2 support cushions from the 2 upper radiator supports.
11. Remove the 3 bolts from the hood lock assembly.
12. Disconnect the hood lock control cable and remove the hood lock assembly.
13. Separate the water by-pass hose from the 2 clamps.
14. Disconnect the water by-pass hose from the radiator assembly.
15. Remove the 2 bolts and hood lock support sub-assembly.
16. Disconnect the horn connector.
17. Remove the 4 bolts and upper radiator support sub-assembly.
18. Remove the 2 bolts, disengage the 2 claws, and remove the No. 2 fan shroud from the radiator assembly.
19. Disconnect the 2 cooling fan motor connectors and wire harness clamp.

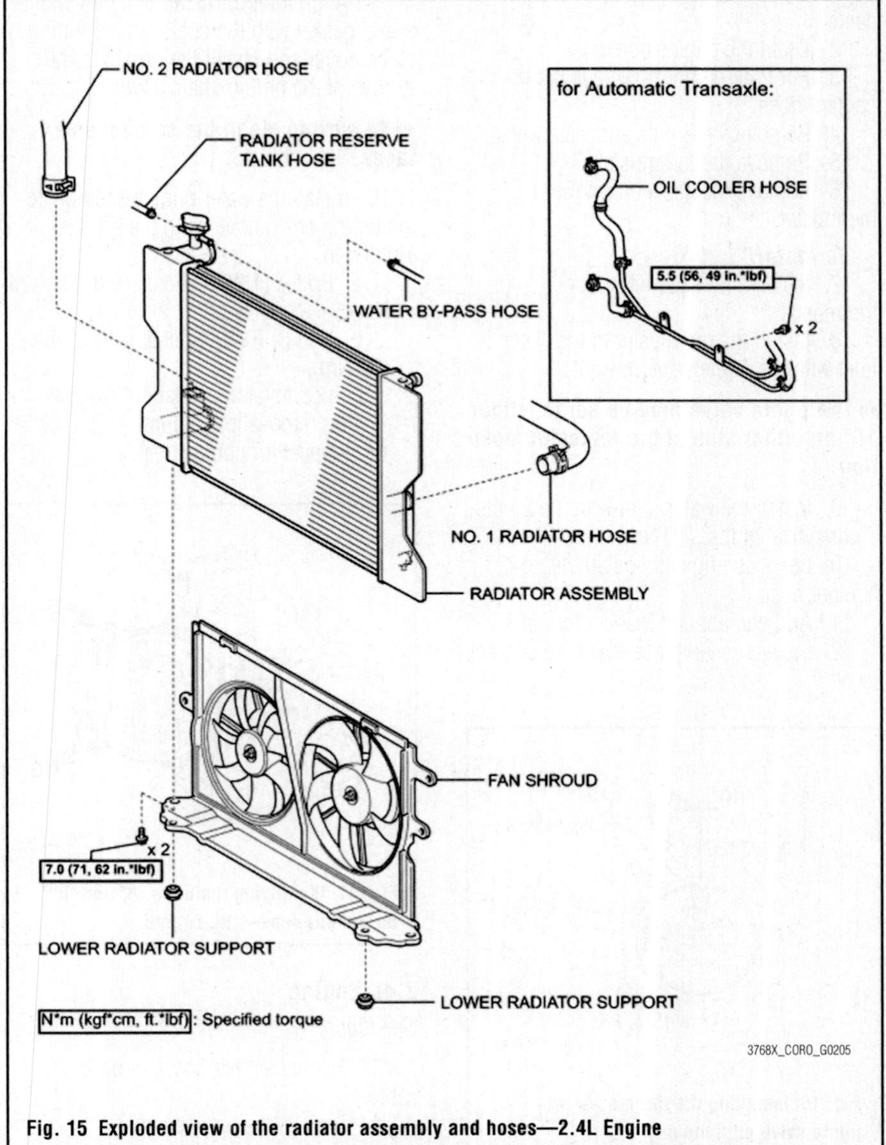

Fig. 15 Exploded view of the radiator assembly and hoses—2.4L Engine

20. Remove the radiator assembly with the fan shroud.

➡**Do not apply any excessive force to the cooler condenser assembly or pipe when removing the radiator assembly.**

21. Disconnect the 2 oil cooler hoses from the radiator.
22. Disconnect the clamp from the fan shroud.
23. Remove the 2 bolts and oil cooler hose.
24. Remove the 2 bolts and fan shroud from the radiator assembly.
25. To install, reverse removal procedure.

THERMOSTAT

REMOVAL & INSTALLATION

See Figure 16.

1. Disconnect the negative battery cable.
2. Drain the engine coolant.
3. For 2.4L engines, remove the no. 2 coolant hose.
4. Remove the 2 nuts and water inlet.
5. Remove the thermostat.
6. Remove the gasket from the thermostat.

To install:

7. Install a new gasket on the thermostat.
8. Install the thermostat to the water inlet with the jiggle valve upward.

➡**The jiggle valve may be set to within 10° on either side of the indicated position.**

9. Install the water inlet with the 2 nuts. Tighten to 7 ft. lbs. (10 Nm).
10. For 2.4L engines, install the no. 2 coolant hose
11. Add coolant and inspect for leaks.
12. Inspect reservoir tank engine coolant level.

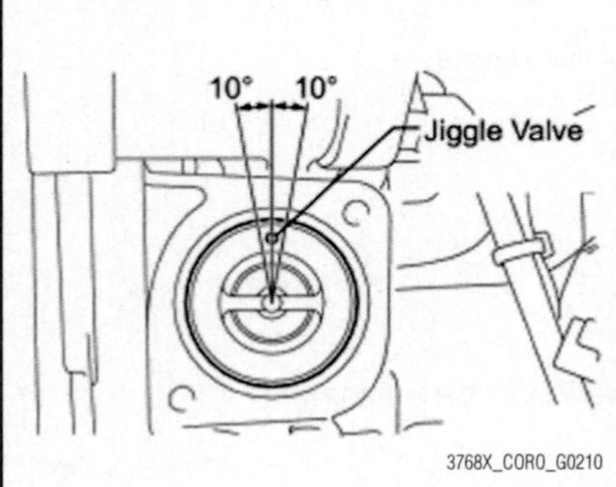

Fig. 16 Installing the thermostat and jiggle valve alignment

WATER PUMP

REMOVAL & INSTALLATION

1.8L Engine

See Figure 17.

1. Disconnect the negative battery cable.
2. Remove the engine cover.
3. Remove the RH engine under cover.
4. Drain the engine coolant.
5. Remove the accessory drive belt.
6. Remove the alternator. Refer to Alternator in Charging System under Engine Electrical.
7. Remove the 5 bolts and water pump assembly from the timing chain cover.
8. Remove the water pump gasket from the timing chain cover.

To install:

9. Align the protrusion of a new water pump gasket with the cutout in the timing chain cover and install the gasket to the groove of the timing chain cover.

➡**Be sure to clean the contact surfaces.**

10. Install the water pump assembly to the timing chain cover with the 5 bolts. Tighten to:
    a. Bolt A (1.38 inches): 19 ft. lbs. (26 Nm).
    b. Bolt B: (0.709 inches): 18 ft. lbs. (24 Nm).
11. To complete installation, reverse remaining removal procedure.
12. Inspect for coolant leaks.

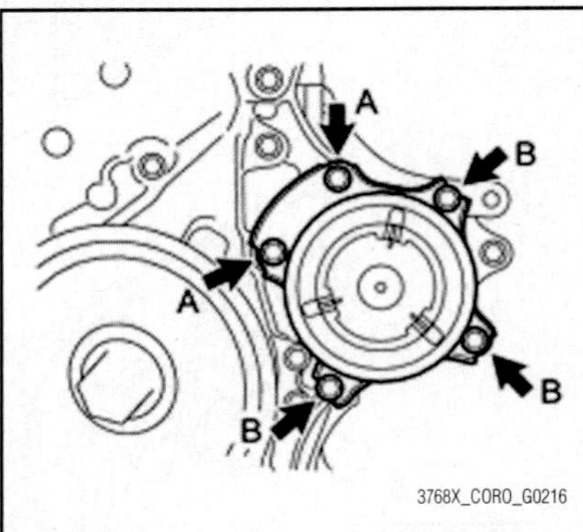

Fig. 17 Identifying the bolts "A" and "B" for installation—1.8L Engine

2.4L Engine

See Figure 18.

1. Disconnect the negative battery cable.
2. Remove the engine cover.
3. Remove the LH and RH engine under cover.
4. Drain the engine coolant.
5. Remove the accessory drive belt.
6. Remove the alternator. Refer to Alternator in Charging System under Engine Electrical.
7. Remove the RH engine mounting insulator sub-assembly. Refer to Engine under Engine Mechanical.
8. Using a pulley remover, remove the water pump pulley.
9. Remove the clamp of the crankshaft position sensor from the water pump.
10. Disconnect the wire of the crankshaft position sensor from the clamp bracket.
11. Remove the 4 bolts, 2 nuts and clamp bracket.

➡**Tape the screwdriver tip before use.**

12. Using a screwdriver, pry between the water pump and cylinder block, and then remove the water pump.

➡**Be careful not to damage the contact surfaces of the water pump and cylinder block.**

➡**Remove any oil from the contact surface. The parts must be set within 3 minutes after applying seal packing. Otherwise, the material must be removed and reapplied.**

13. To install, reverse the removal procedure and pay attention to the following:
    a. Apply a continuous line of seal packing. Use Toyota Genuine Seal Packing Black, Three Bond 1207B or equivalent.
    b. Tighten the water pump bolts to 80 inch lbs. (9 Nm).
    c. Tighten the water pump pulley to 19 ft. lbs. (26 Nm).

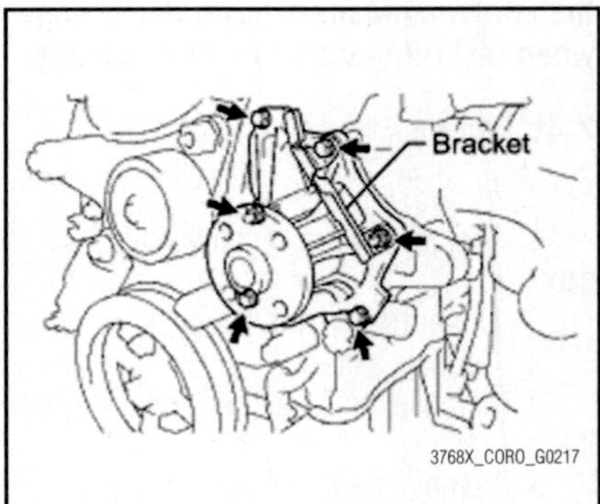

Fig. 18 Locating and identifying fasteners—2.4L Engine

ENGINE ELECTRICAL — CHARGING SYSTEM

ALTERNATOR

REMOVAL & INSTALLATION

1.8L Engine

See Figure 19.

1. Disconnect the negative battery cable.
2. Remove the RH engine under cover.
3. Remove the accessory drive belt.
4. Remove generator assembly:

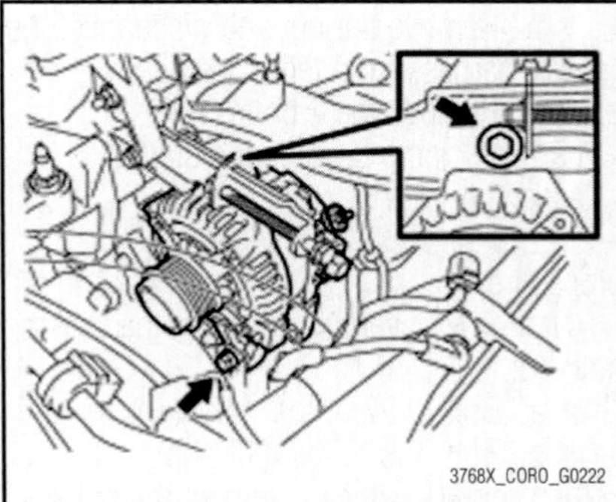

Fig. 19 Removing the alternator assembly—1.8L Engine

   a. Remove the terminal cap.
   b. Remove the nut and disconnect the wire harness from terminal B.
   c. Disconnect the connector and harness clamp.
   d. Remove the 2 bolts and generator assembly.
   e. Remove the bolt and wire harness clamp bracket.
5. To install, reverse removal procedure. Pay attention to the following:
   a. Alternator upper mounting bolt: Tighten to 32 ft. lbs. (43 Nm).
   b. Alternator lower mounting bolt: Tighten to 14 ft. lbs. (19 Nm).

2.4L Engine

See Figure 20.

1. Disconnect the negative battery cable.
2. Remove the RH engine under cover.
3. Remove the accessory drive belt.
4. Remove generator assembly:
   a. Disconnect the generator connector.
   b. Remove the nut and disconnect the wire harness from terminal B.
   c. Separate the 2 wire harness clamps.
   d. Remove the 2 bolts and generator assembly.
   e. Remove the bolt and wire harness clamp bracket.
5. To install, reverse removal procedure. Pay attention to the following:
   a. Alternator upper mounting bolt: Tighten to 16 ft. lbs. (21 Nm).
   b. Alternator lower mounting bolt: Tighten to 38 ft. lbs. (52 Nm).

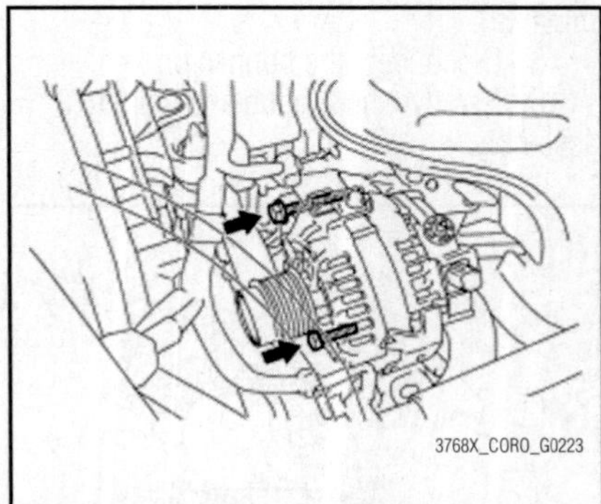

Fig. 20 Removing the alternator assembly—2.4L Engine

ENGINE ELECTRICAL — IGNITION SYSTEM

FIRING ORDER

See Figure 21.

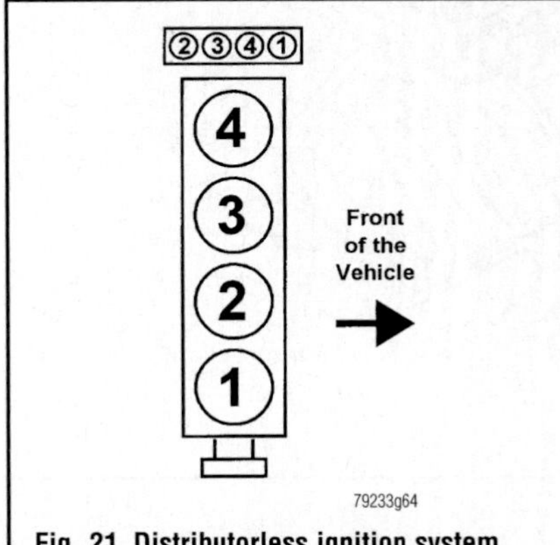

Fig. 21 Distributorless ignition system

IGNITION COIL

REMOVAL & INSTALLATION

See Figure 22.

1. Disconnect the battery's negative terminal.
2. Remove the two nuts and clips and remove the cylinder head cover.

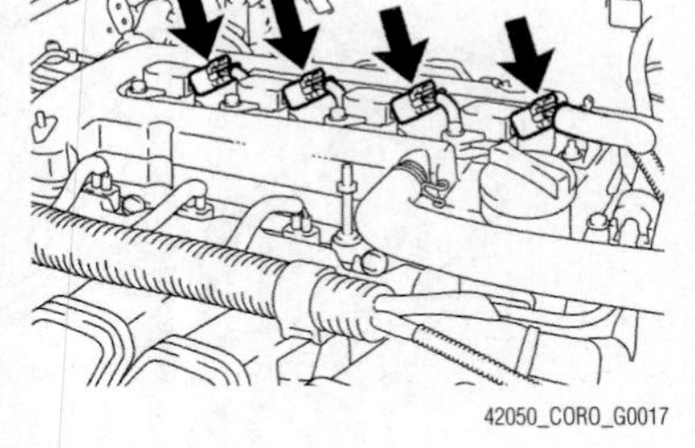

Fig. 22 Location of the ignition coils

3. Disconnect the four ignition coil connectors.
4. Remove the four bolts that retain the coils.

To install:

5. Install the coils and the four bolts.
6. Connect the coil connections.
7. Install the cylinder head cover.

IGNITION TIMING

ADJUSTMENT

The ignition timing is controlled by the Powertrain Control Module (PCM). No adjustment is necessary or possible.

SPARK PLUGS

REMOVAL & INSTALLATION

See Figure 23.

1. Disconnect the negative battery cable.
2. Remove the ignition coil pack. Refer to Ignition Coil Pack under Ignition System.
3. Remove the 4 spark plugs.
4. To install, use a 14 mm spark plug wrench, install the 4 spark plugs and tighten to 15 ft. lbs. (20 Nm).
5. To complete installation, reverse remaining removal procedure.

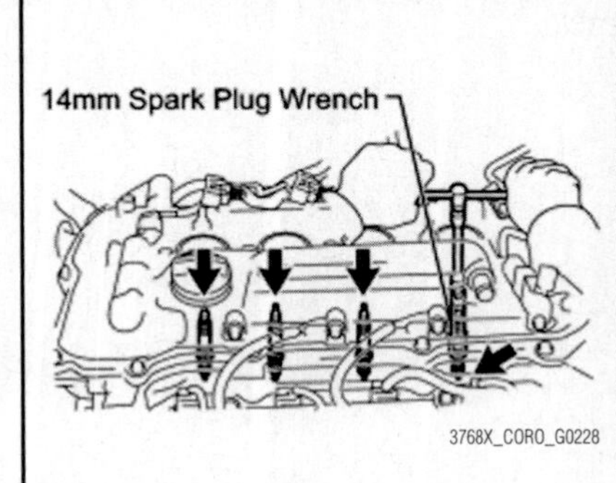

Fig. 23 Removing the spark plugs

STARTER

REMOVAL & INSTALLATION

1.8L Engine

See Figure 24.

1. Disconnect the negative battery cable.
2. Remove transmission oil filler tube sub-assembly (for automatic transaxle).
3. Remove the terminal cap.
4. Remove the nut and disconnect terminal 30.
5. Disconnect the connector.
6. Remove the 2 bolts and starter assembly.

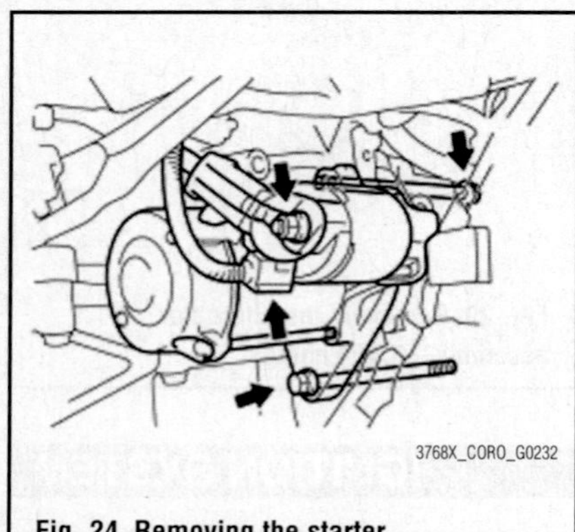

Fig. 24 Removing the starter assembly—1.8L Engine

To install:

7. Install the starter assembly with the 2 bolts and tighten to 27 ft. lbs. (37 Nm).
8. Connect the connector.
9. Connect terminal 30 with the nut.
10. Close the terminal cap.
11. To complete the installation, reverse remaining removal procedure.

2.4L Engine

See Figures 25 and 26.

1. Disconnect the negative battery cable.
2. Remove no. 1 engine cover sub-assembly.
3. Remove air cleaner cap sub-assembly with hose and air cleaner case.
4. Remove battery remove battery carrier.
5. Release the claw, and disconnect the wire harness.

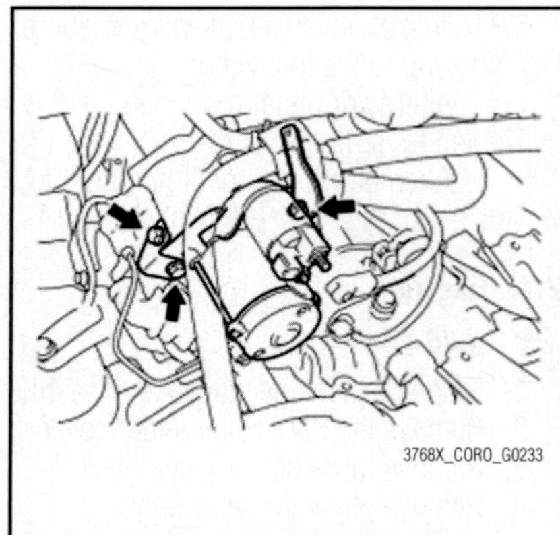

Fig. 25 Removing the starter—M/T vehicles

6. For M/T vehicles, perform the following:
   a. Disconnect the terminal 50 connector from the starter assembly.
   b. Remove the nut and disconnect the wire harness from terminal 30.
   c. Remove the 3 bolts, clutch accumulator bracket, wire harness clamp bracket and starter assembly.
7. For A/T vehicles:
   a. Disconnect the terminal 50 connector from the starter assembly.
   b. Remove the nut and disconnect the wire harness from terminal 30.
   c. Remove the 2 bolts, wire harness clamp bracket and starter assembly.
8. To install, reverse removal procedure.
9. For M/T vehicles, tighten the starter bolts to:
   a. Bolt A: 27 ft. lbs. (27 Nm).
   b. Bolt B: 9 ft. lbs. (12 Nm).
10. For A/T vehicles, tighten the bolts to 27 ft. lbs. (37 Nm).

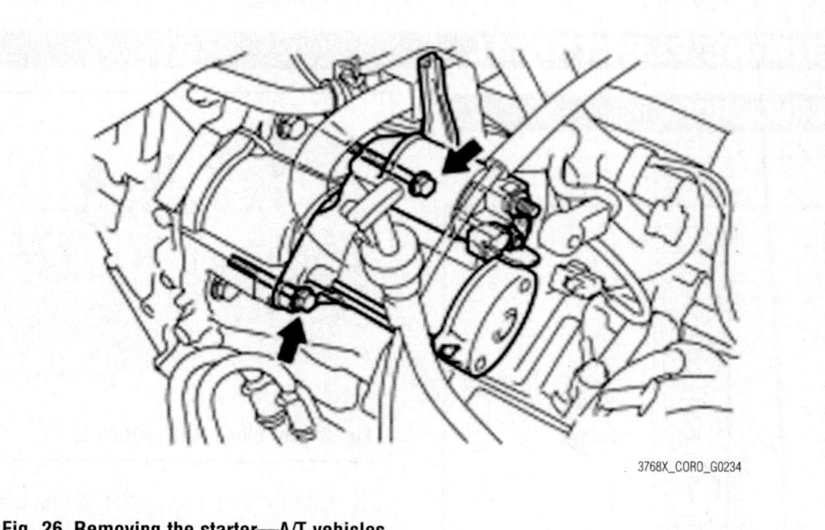

Fig. 26 Removing the starter—A/T vehicles

ENGINE MECHANICAL

ACCESSORY DRIVE BELTS

ACCESSORY BELT ROUTING

See Figures 27 and 28.

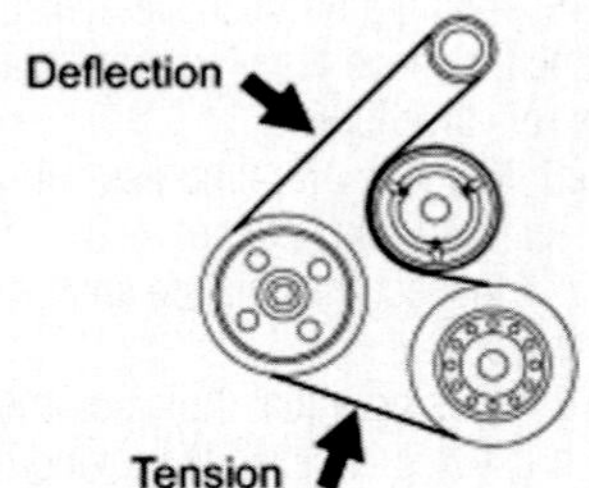

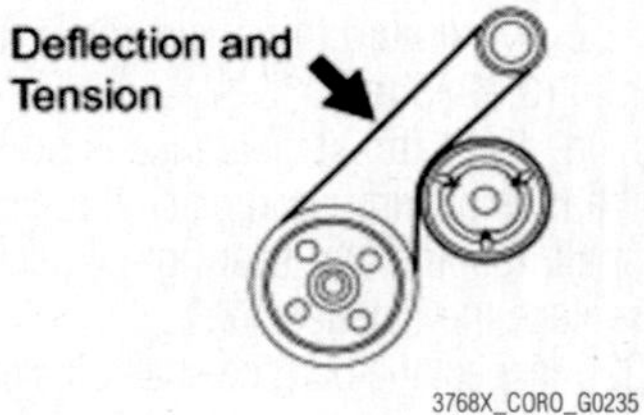

Fig. 27 Accessory belt routing—1.8L Engine

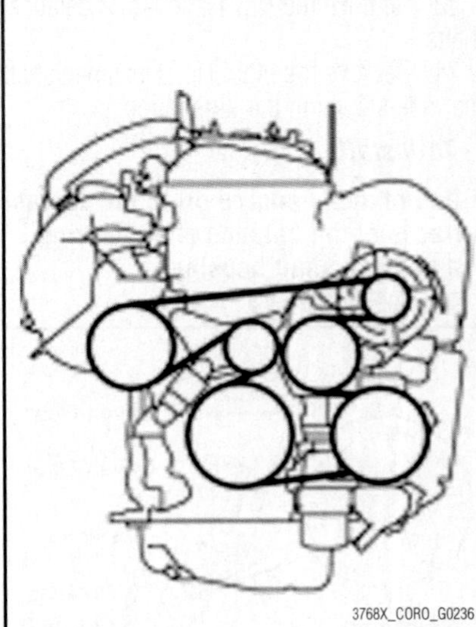

Fig. 28 Accessory belt routing—2.4L Engine

INSPECTION

Inspect the drive belt for signs of glazing or cracking. A glazed belt will be perfectly smooth from slippage, while a good belt will have a slight texture of fabric visible. Cracks will usually start at the inner edge of the belt and run outward. All worn or damaged drive belts should be replaced immediately.

ADJUSTMENT

Only the 1.8L engine belt requires adjustment. Refer to the removal procedure for adjustment procedure.

REMOVAL & INSTALLATION

1.8L Engine

See Figure 29.

➡Refer to the illustration for bolt identification.

1. Remove the engine cover.
2. Remove the RH engine under cover.
3. Loosen bolts A and B.
4. Loosen bolt C, then remove the V-ribbed belt.

➡Do not loosen bolt D.

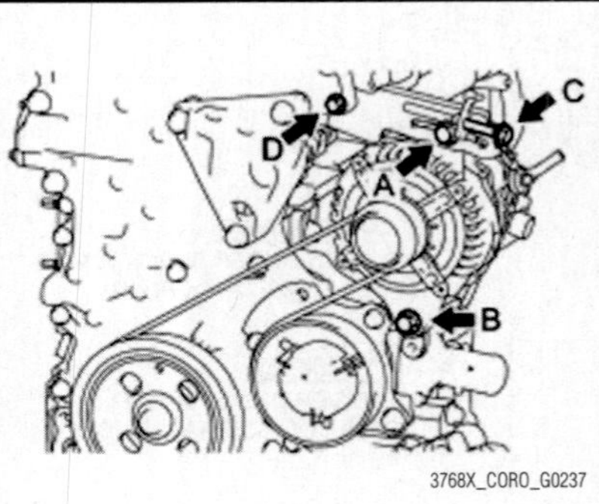

Fig. 29 Removing the accessory belt—1.8L Engine

To install:

5. Install the belt.
6. Turn bolt C to adjust the tension of the V-ribbed belt.
7. Tighten bolts:
   a. A: 14 ft. lbs. (19 Nm).
   b. B: 32 ft. lbs. (43 Nm).

➡Confirm that bolt D is not loosened.

8. Inspect the belt.
9. To complete installation, reverse remaining removal procedure.

2.4L Engine

See Figure 30.

➡Adjustment is not possible or necessary with the auto-tensioner.

1. Remove the right hand cover under the engine.
2. Turn the drive belt tensioner clockwise to relieve tension on the belt.
3. Remove the fan and generator V belt.
4. Return the tensioner to the unloaded position.

➡When retracting the tensioner, turn it clockwise slowly in 3 sec. or more. Be sure not to apply force rapidly. After the tensioner is retracted all the way, do not apply force any more than necessary.

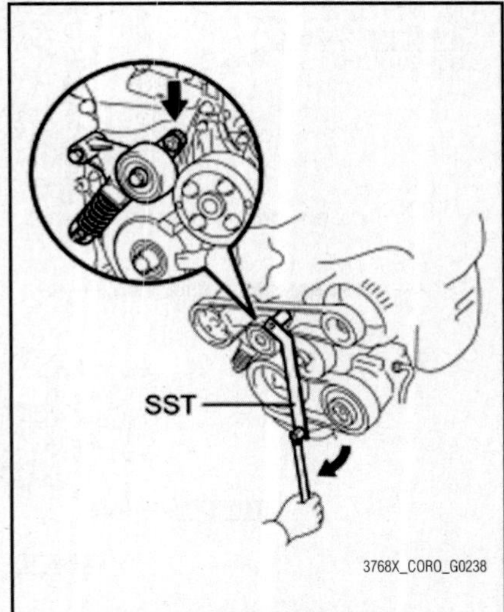

Fig. 30 Removing the serpentine belt—2.4L Engine

To install:

5. Turn the drive belt tensioner clockwise.
6. Install the belt.
7. Install the right hand cover under the engine.

BALANCE SHAFT

REMOVAL & INSTALLATION

2.4L Engine

See Figures 31 through 36.

1. Remove the valve cover.
2. Using a 22 mm deep socket wrench, remove the ventilation valve from the cylinder head cover.

3. Remove the spark plugs.
4. Remove the oil filter and union.
5. Remove the drive belt.
6. Remove the crankshaft position sensor. Refer to Crankshaft Position Sensor under Engine Performance & Emission Controls.
7. Remove the bolt and camshaft position sensor. Refer to Camshaft Position Sensor under Engine Performance & Emission Controls.
8. Set the no. 1 cylinder to TDC/compression.
9. Remove camshaft timing oil control valve assembly. Refer to Camshaft and Valve Lifters in this section.
10. Remove the no. 1 chain tensioner assembly. Refer to Timing Chain and Sprockets in this section.
11. Remove water pump assembly. Refer to Water Pump in Engine Cooling.
12. Remove the oil pan sub-assembly. Refer to Oil Pan in this section.
13. Remove the timing chain cover and assembly. Refer to Timing Chain Front Cover and Timing Chain Sprockets in this section.

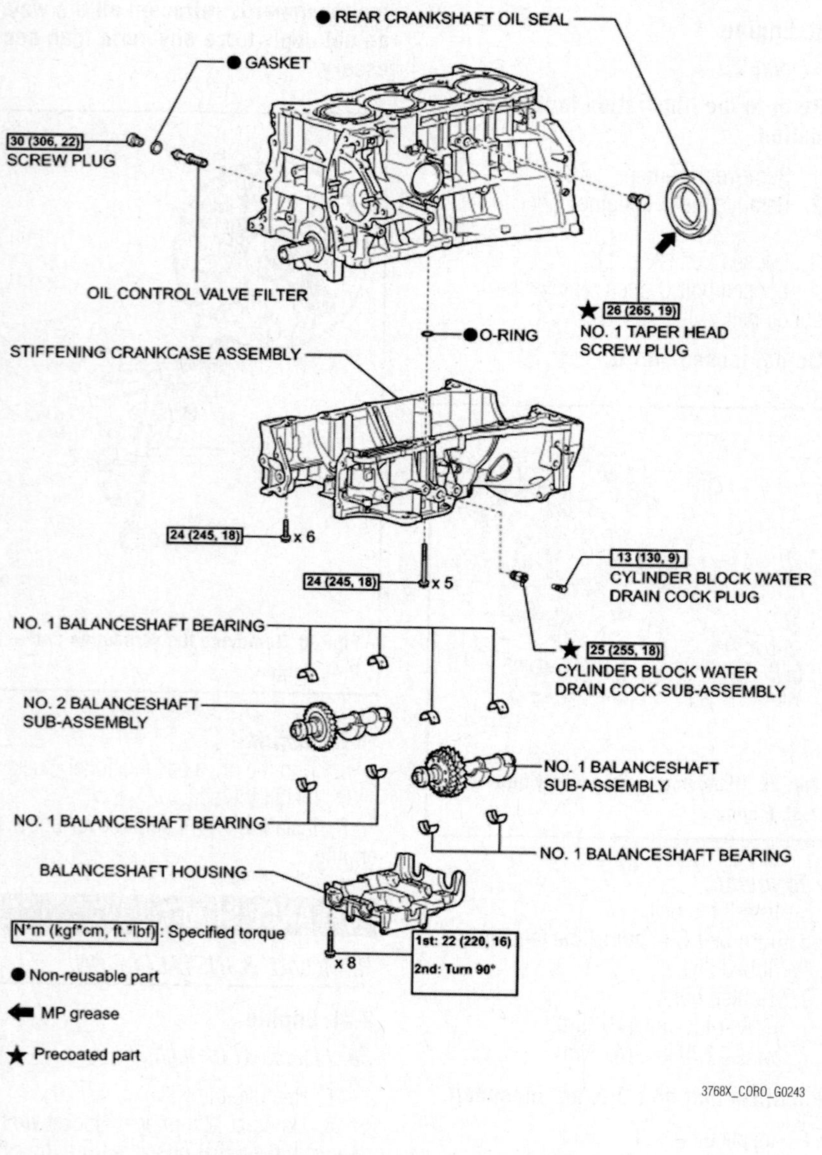

Fig. 31 Exploded view of the balance shaft and related components—2.4L Engine

14. Remove the front crankshaft oil seal. Refer to Crankshaft Front Seal in this section.
15. Remove the no. 1 crankshaft position sensor plate. Refer to Crankshaft in this section.
16. Remove crankshaft timing gear or sprocket. Refer to Crankshaft in this section.
17. Remove the camshaft assembly. Refer to Camshaft and Valve Lifters in this section.
18. Remove the cylinder head sub-assembly. Refer to Cylinder Head in this section.
19. Using a 8mm socket hexagon wrench, remove the plug, gasket and oil control valve filter.
20. Remove oil pump assembly. Refer to Oil Pump in this section.
21. Inspect the balance shaft thrust clearance:
   a. Using a dial indicator, measure the thrust clearance while moving the balance shaft back and forth.
   b. Standard thrust clearance: 0.05 to 0.09 mm (0.0020 to 0.0035 in.)
   c. Maximum thrust clearance: 0.09 mm (0.0035 in.)
   d. If the thrust clearance is greater than the maximum, replace the balance shaft housing and bearings. If necessary, replace the balance shaft.
22. Inspect the balance shaft oil clearance:
   a. Clean each bearing and journal.
   b. Check each bearing and journal for pitting and scratches.
   c. If a bearing or journal is damaged, replace the bearings. If necessary, replace the balance shaft.
23. Remove the No. 1 and No. 2 balance shafts.
24. Remove the eight No. 1 balance shaft bearings shown in the illustration.

To install:

➡Do not apply engine oil to the contact surfaces of the balance shaft bearing and balance shaft housing.

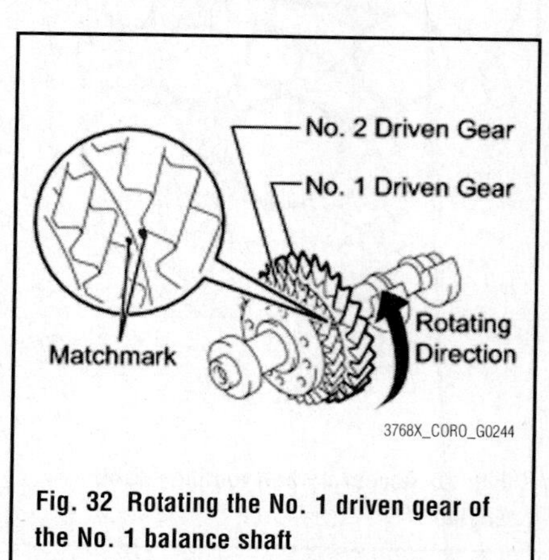

Fig. 32 Rotating the No. 1 driven gear of the No. 1 balance shaft

25. Align the bearing claw with the claw groove, and push in the 8 bearings.

26. Apply a light coat of the engine oil to the bearings.

27. Rotate the No. 1 driven gear of the No. 1 balance shaft in the rotating direction until it hits the stopper.

➡Confirm that the match marks on the No. 1 and No. 2 driven gears are matched.

28. Confirm that the alignment marks on the No. 1 and No. 2 balance shafts are aligned.

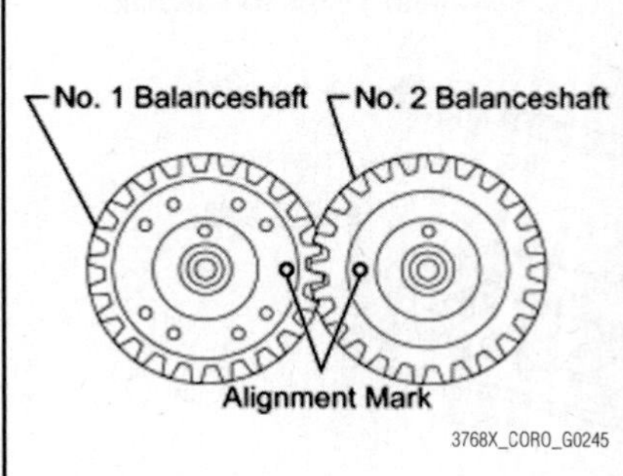

Fig. 33 Confirming that the alignment marks on the No. 1 and No. 2 balance shafts are aligned

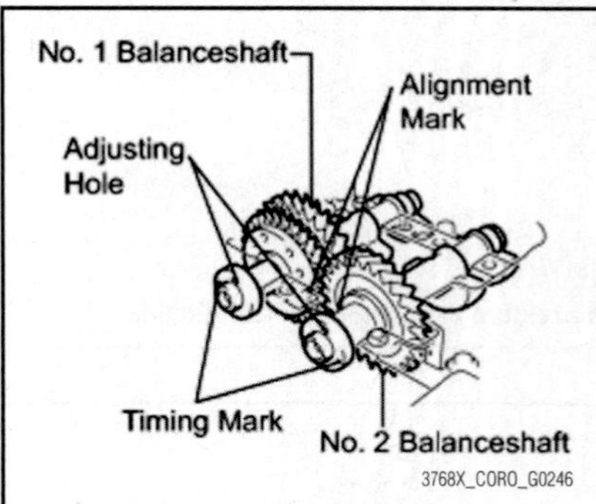

Fig. 34 Aligning the alignment marks on the No. 1 and No. 2 balance shafts

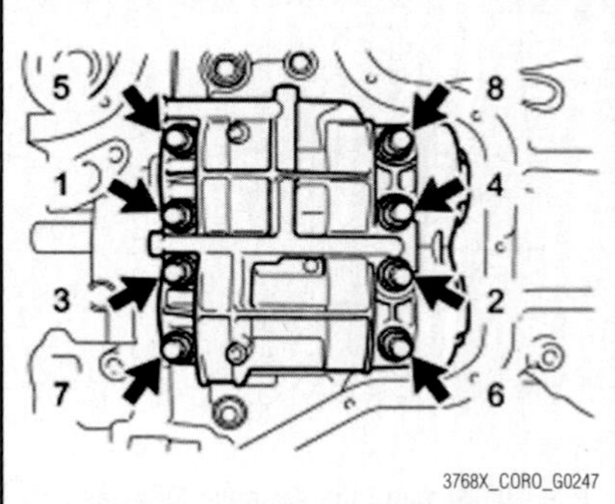

Fig. 35 Balance shaft bolt tightening sequence

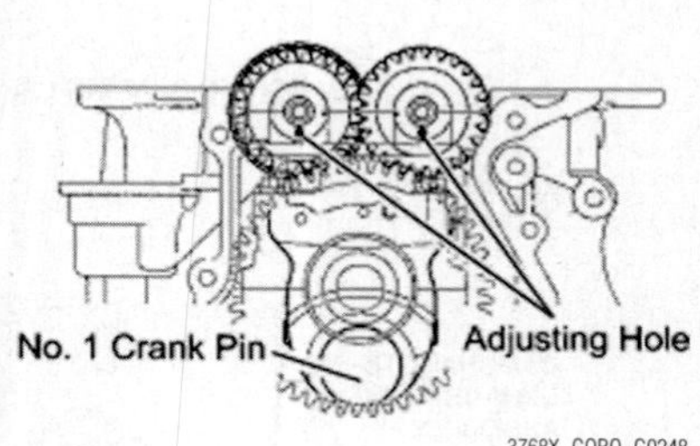

Fig. 36 Installing the No. 1 and No. 2 balance shafts and aligning the adjusting holes

29. Align the alignment marks on the No. 1 and No. 2 balance shafts.

30. Place the No. 1 and No. 2 balance shafts onto the crankcase.

31. Apply a light coat of engine oil to the threads and under the heads of the bolts.

32. Using several steps, uniformly install and tighten the 8 bolts in the sequence. Tighten to 16 ft. lbs. (22Nm). Mark the front of the bolts with paint for reference and tighten the bolts an extra 90°.

33. Check that the paint mark is now at a 90°angle to the front.

34. Place a new O-ring on the cylinder block.

35. With the No. 1 crank pin of the crankshaft placed at the TDC position, install the No. 1 and No. 2 balance shafts and align the adjusting holes.

36. To complete installation procedure, reverse remaining removal procedure.

CAMSHAFT AND VALVE LIFTERS

INSPECTION

1. To check the lock of the camshaft timing gear, clamp the camshaft in a vice and confirm that the camshaft timing gear is locked.

⁂ CAUTION

Be careful not to damage the camshaft.

2. To release the lock pin, cover the four oil paths on the cam journal with vinyl tape. The two advance side paths are provided in the groove of the camshaft. Plug one of the paths with a piece of rubber.

3. Puncture the tape over the advance side path and retard the side path on the opposite side of the groove.

4. Apply air pressure to the two broken paths (the advance side path and the retard side path).

⁂ WARNING

Cover the paths with a shop rag to avoid oil splashes.

5. Confirm that the camshaft timing gear revolves in the timing advance direction when weakening the air pressure of the timing retard path.

6. When the camshaft timing gear comes to the most advanced position, release the air pressure of the timing retard side path, then release that of the timing advance side path.

7. There should be a smooth revolution of the gear.

REMOVAL & INSTALLATION

1.8L Engine

See Figures 37 through 44.

1. Disconnect the negative battery cable.

2. Remove engine assembly with transaxle.

3. Install the engine stand.

4. Remove the intake manifold. Refer to Intake Manifold in this section.

5. Remove the fuel tube sub-assembly. Refer to Fuel System.

6. Remove the fuel delivery pipe sub-assembly. Refer to Fuel System.

7. Remove the fuel injector assembly. Refer to Fuel System.

8. Remove the ignition coil assembly.

9. Remove the oil level dipstick sub-assembly.

10. Remove the exhaust manifold. Refer to Exhaust Manifold in this section.

11. Remove the ventilation hose.

12. Disconnect the no. 3 water by-pass hose.

13. Remove the no. 1 water by-pass pipe.

14. Remove the water by-pass hose.

15. Remove the inlet water hose.

16. Remove the inlet water.

17. Remove the thermostat. Refer to Thermostat in Engine Cooling.

18. Remove the radio setting condenser.

19. Remove the cylinder head cover sub-assembly.

20. Turn the crankshaft so that the No. 1 piston is at TDC on the compression stroke. Check to see that the point marks on the camshaft sprockets are facing each other, if not, rotate the crankshaft 1 full revolution.

21. Remove the crankshaft pulley.

22. Remove the no. 1 chain tensioner assembly.

23. Remove the timing chain cover sub-assembly.
24. Remove the timing chain cover oil seal.
25. Remove the chain tensioner slipper.
26. Remove the no. 1 chain vibration damper.
27. Remove the chain sub-assembly.
28. Remove the no. 2 chain vibration damper.
29. Uniformly loosen and remove the 10 bearing cap bolts in the sequence shown in the illustration.
30. Uniformly loosen and remove the 15 bearing cap bolts in the sequence shown in the illustration.

➡Uniformly loosen the bolts while keeping the camshaft level.

31. Remove the 5 bearing caps.
32. Remove the camshaft.
33. Remove the No. 2 camshaft.
34. Remove the rocker arm. Refer to Rocker Arm in this section.
35. Remove the valve lash adjuster assembly. Refer to Valve Lash Adjuster in this section.
36. Remove the 2 No. 1 camshaft bearings.

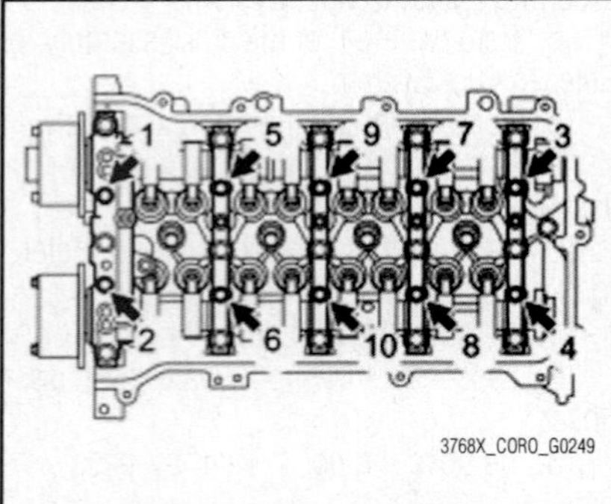

Fig. 37 Removing the 10 bearing cap bolts in sequence

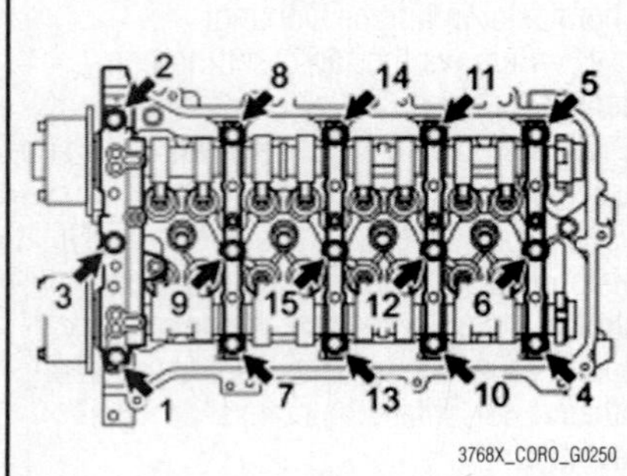

Fig. 38 Remove the 15 bearing cap bolts in sequence

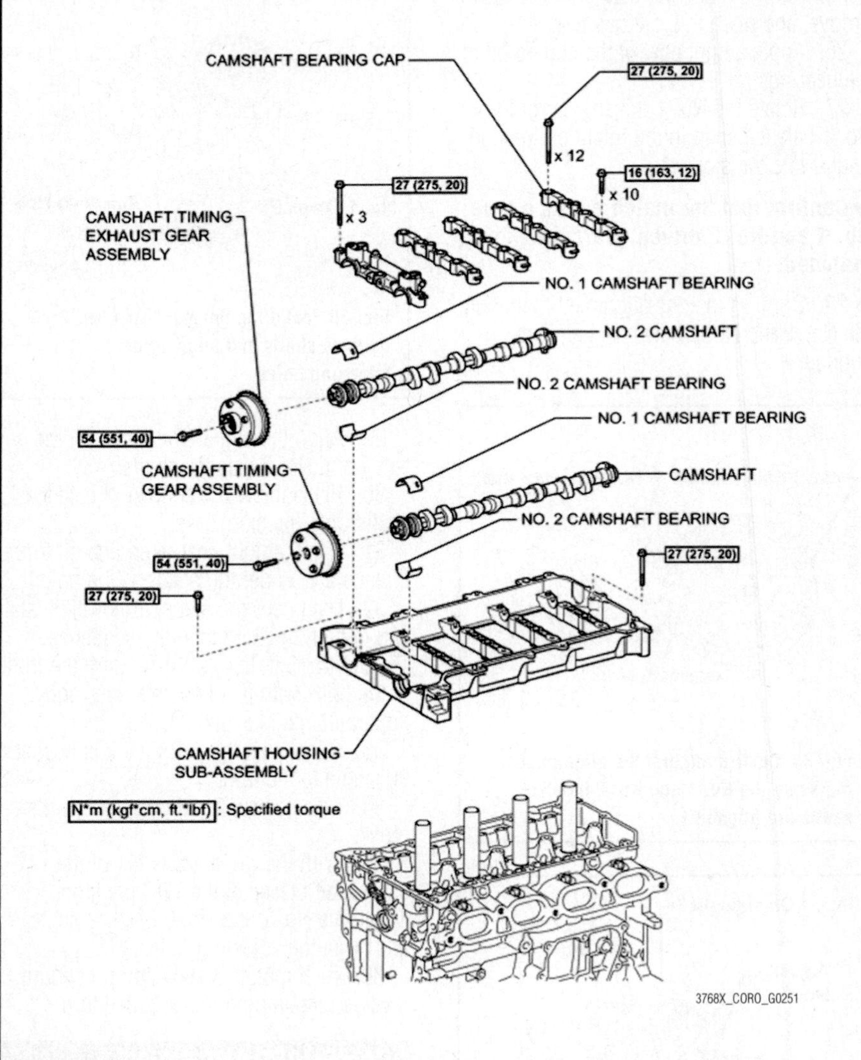

Fig. 39 Exploded view of the camshaft assembly and related components—1.8L Engine

37. Remove the 2 No. 2 camshaft bearings.
38. If necessary, remove camshaft housing sub-assembly:
    a. Remove the 2 bolts.
    b. Remove the camshaft housing by prying between the cylinder head and camshaft housing with a screwdriver.

➡Be careful not to damage the contact surfaces of the cylinder head and camshaft housing. Tape the screwdriver tip before use.

To install:

39. Remove the valve lash adjuster assembly. Refer to Valve Lash Adjuster in this section.
40. Install the rocker arm. Refer to Rocker Arm in this section.
41. Install the no. 1 camshaft bearing:
    a. Clean both surfaces of the bearings.

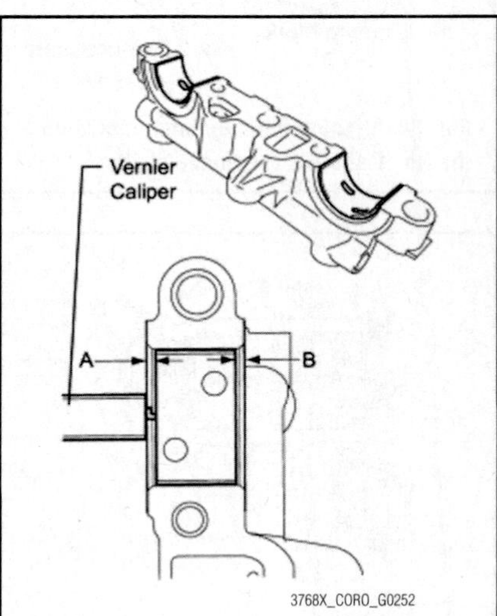

Fig. 40 Measure the distance between the bearing cap edge and the camshaft bearing edge—No. 1 camshaft

b. Install the 2 No. 1 camshaft bearings.

c. Using a vernier caliper, measure the distance between the bearing cap edge and the camshaft bearing edge. Dimension A-B: 0.7 mm (0.0276 in.) or less.

➡Position the bearings to the center of the bearing cap by measuring dimensions A.

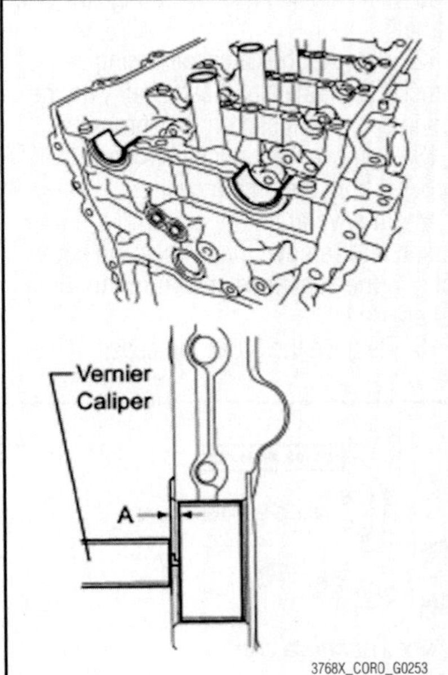

Fig. 41 Measure the distance between the bearing cap edge and the camshaft bearing edge—No. 2 camshaft

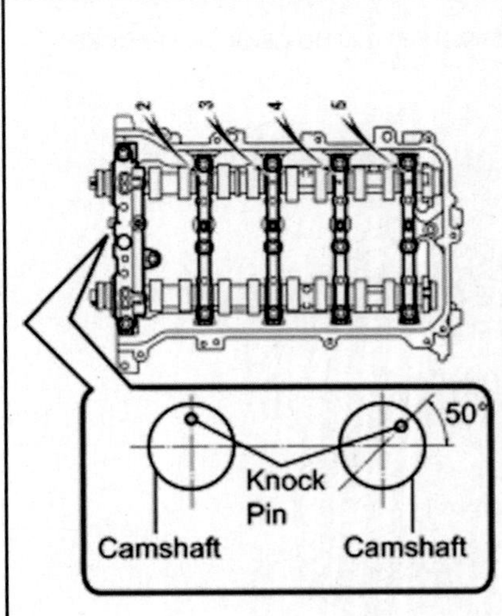

Fig. 42 Making sure of the marks and numbers on the camshaft bearing caps

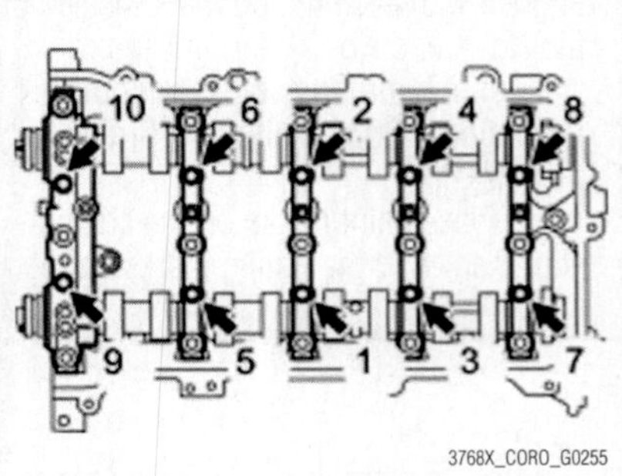

Fig. 43 Bearing cap 10 bolt tightening sequence

42. Install the no. 2 camshaft bearing:

a. Clean both surfaces of the bearings.

b. Install the 2 No. 2 camshaft bearings.

c. Using a vernier caliper, measure the distance between the bearing cap edge and the camshaft bearing edge. Dimension: 1.05 to 1.75 mm (0.0413 to 0.0689 in.)

➡Position the bearings to the center of the bearing cap by measuring dimensions A.

43. Install the no. 2 camshaft:

a. Clean the camshaft journals.

b. Apply a light coat of engine oil to the camshaft journals, camshaft housings and bearing caps.

c. Install the No. 2 camshaft to the camshaft housing.

44. Install the camshaft:

a. Apply a light coat of engine oil to the camshaft journals, camshaft housings and bearing caps.

b. Install the camshaft to the camshaft housing.

45. Apply engine oil to the camshaft journals, camshaft housings and bearing caps.

46. Make sure of the marks and numbers on the camshaft bearing caps and place them in each proper position and direction.

➡Make sure that the knock pin of the camshaft is positioned as shown in the illustration.

47. Tighten the 10 bolts in the order shown in the illustration.

48. If removed, install the camshaft housing sub-assembly:

a. Make sure that the valve rocker arm is installed correctly. Refer to Rocker Arms.

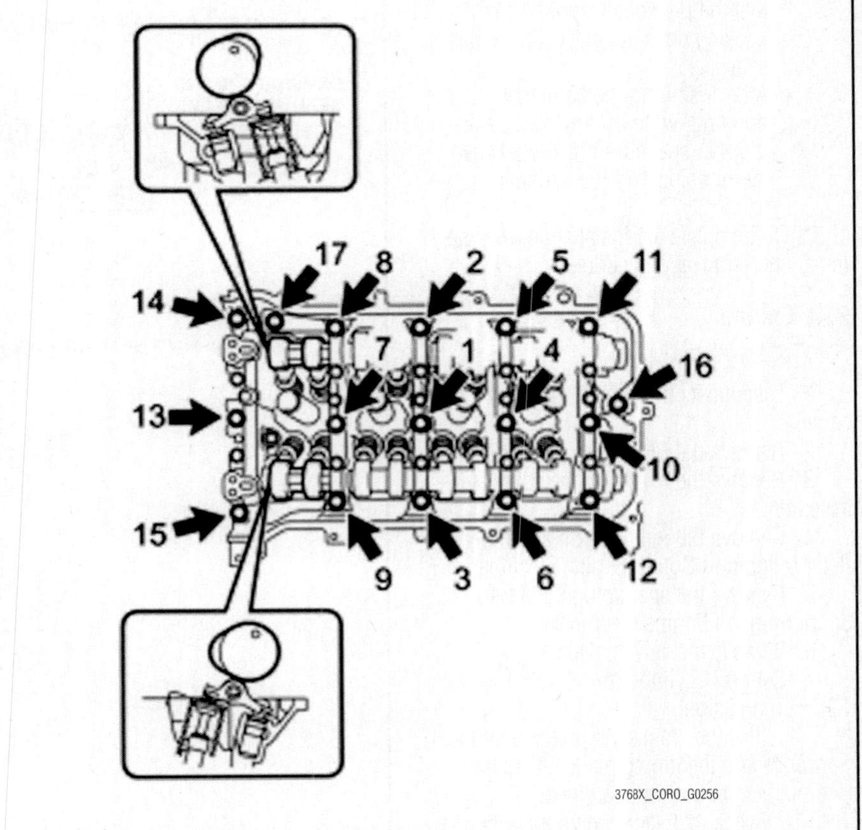

Fig. 44 Camshaft housing 17 bolt tightening sequence

b. Apply seal packing in a continuous bead. Use Toyota Genuine Seal Packing Black, Three Bond 1207B or equivalent.

➡**Remove any oil from the contact surface. Install the camshaft housing sub-assembly within 3 minutes and tighten the bolts within 15 minutes after applying seal packing. Do not start the engine for at least 2 hours after installing.**

49. Set the camshaft and No. 2 camshaft as shown in the illustration.
50. Install the camshaft housing and tighten the 17 bolts in the order shown in the illustration. Tighten to 20 ft. lbs. (27 Nm).

➡**After installing the camshaft housing, make sure that the cam lobes are positioned as shown in the illustration.**

- If any of the bolts are loosened during installation, remove the camshaft housing, clean the installation surfaces, and reapply seal packing.
- If the camshaft housing is removed because any of the bolts are loosened during installation, make sure that the previously applied seal packing does not enter any oil passages.
- After installing the camshaft housing, wipe off any seal packing that seeped out from between the housing and the cylinder head.

51. To complete the installation, reverse remaining removal procedure.

2.4L Engine

See Figures 45 through 51.

1. Disconnect the negative battery cable.
2. Remove the RH engine under cover.
3. Remove the No. 1 engine cover sub-assembly.
4. Remove the ignition coil assembly. Refer to Ignition Coil in Engine Electrical.
5. Remove the spark plugs. Refer to Spark Plugs in Engine Electrical.
6. Remove the valve cover.
7. Set no. 1 cylinder to TDC/compression:
   a. Turn the crankshaft pulley until the groove and the timing mark "0" on the timing chain cover are aligned.
   b. Check that each timing mark on the camshaft timing gear and sprocket is aligned with each timing mark located on the No. 1 and No. 2 bearing caps.
   c. If not, turn the crankshaft pulley 1 revolution (360°) to align the timing marks.
   d. Place paint marks on the chain in alignment with the timing marks on the camshaft timing gear and camshaft timing sprocket.
8. While holding the No. 2 camshaft with a wrench, loosen the No. 2 camshaft timing set bolt.
9. Using several steps, uniformly loosen and remove the 10 bearing cap bolts in the sequence shown in the illustration.
10. Remove the 5 bearing caps.
11. While holding the No. 2 camshaft by hand, remove the camshaft timing sprocket set bolt.
12. Remove the camshaft timing sprocket from the No. 2 camshaft with the timing chain wrapped on the sprocket.
13. Remove the camshaft timing sprocket from the timing chain.
14. In several steps, uniformly loosen and remove the 10 bearing caps bolts in the sequence shown in the illustration.
15. Remove the 5 bearing caps.

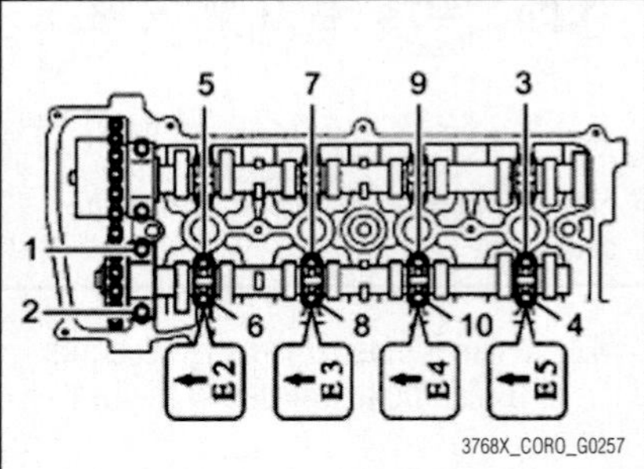

Fig. 45 10 bearing cap bolts removal sequence and identifying the 5 bearing caps

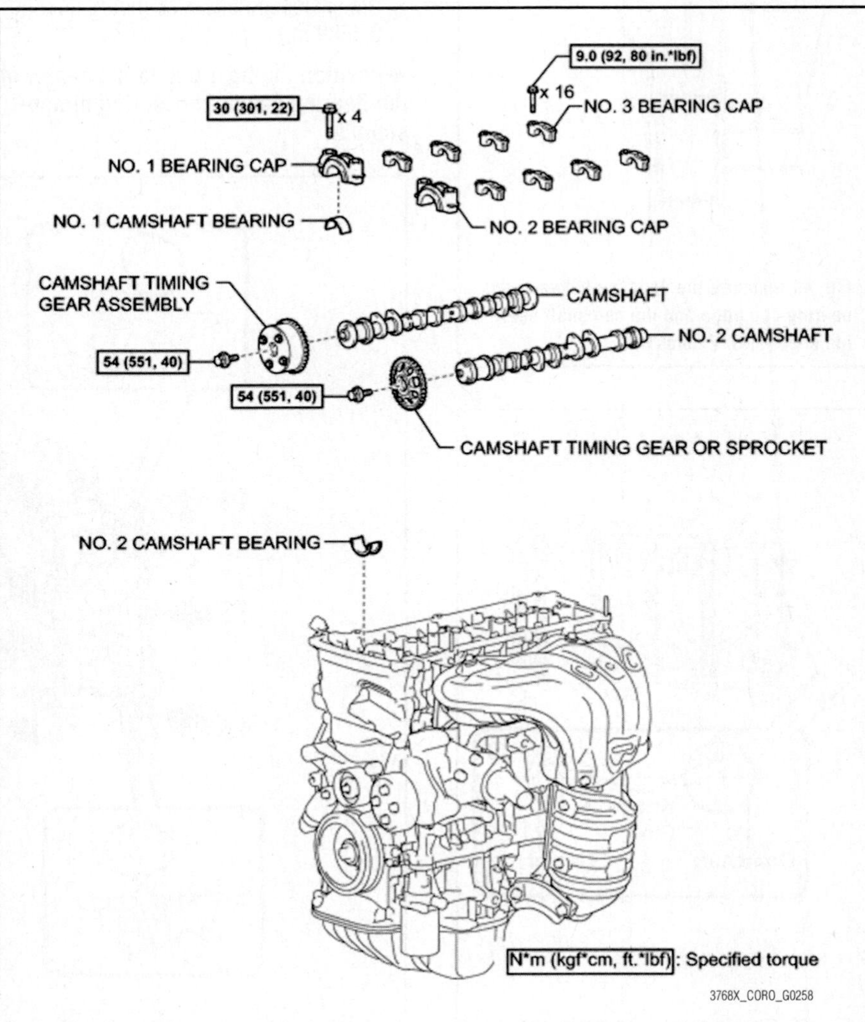

Fig. 46 Exploded view of the camshaft assembly and related components

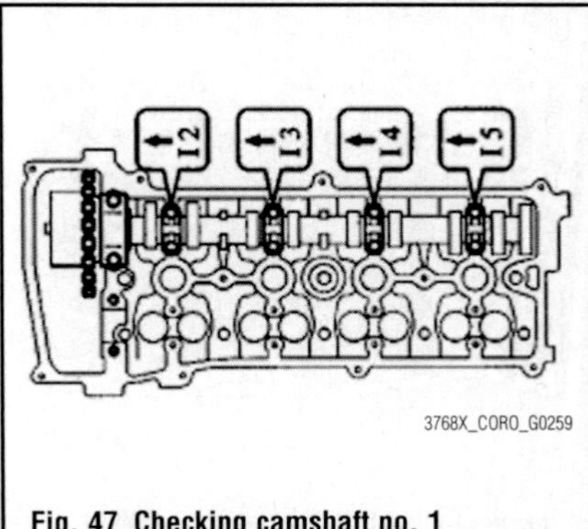

Fig. 47 Checking camshaft no. 1 bearing cap orientation

16. Remove the camshaft and camshaft timing gear assembly while holding the timing chain by hand.
17. Support the timing chain with a string to prevent it from slipping off the crankshaft sprocket.

➡Be careful not to drop anything inside the timing chain cover.

To install:

18. Apply a light coat of engine oil to the journal portion of the camshaft.
19. Install the timing chain onto the camshaft timing gear with the paint mark aligned with the timing mark on the camshaft timing gear.
20. Examine the front marks and numbers, and check that the order is as shown in the illustration. Then install the bearing caps into the cylinder head.
21. Apply a light coat of engine oil on the threads and under the heads of the bearing cap bolts.
22. Using several steps, uniformly tighten the 10 bearing cap bolts in the sequence shown in the illustration.
    a. No. 1 bearing: 22 ft. lbs. (30 Nm).
    b. No. 3 bearing: 80 inch lbs. (9 Nm).
23. Install the no. 2 camshaft:
    a. Apply a light coat of engine oil to the journal portion of the No. 2 camshaft.

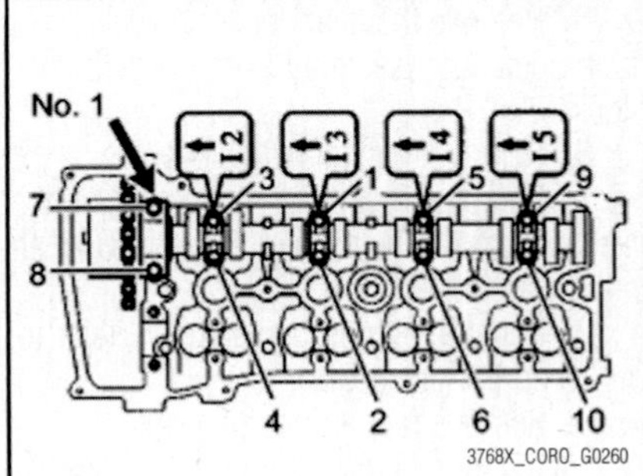

Fig. 48 10 bearing cap bolts tightening sequence

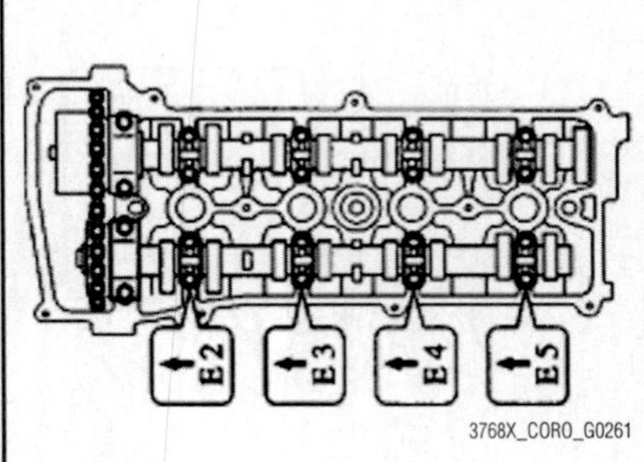

Fig. 49 No. 2 camshaft bearing cap orientation

b. Put the No. 2 camshaft on the cylinder head with the paint mark on the chain aligned with the timing mark on the camshaft timing sprocket.
c. While holding the No. 2 camshaft by hand, temporarily tighten the camshaft timing sprocket set bolt.
d. Examine the front marks and numbers, and check that the order is as shown in the illustration. Then install the bearing caps onto the cylinder head.
e. Apply a light coat of engine oil to the threads and under the heads of the bearing cap bolts.
f. Using several steps, uniformly tighten the 10 bearing cap bolts in the sequence shown in the illustration.
g. No. 2 bearing: 22 ft. lbs. (30 Nm).
h. No. 3 bearing: 80 inch lbs. (9 Nm).
i. While holding the camshaft with a wrench, tighten the camshaft timing sprocket set bolt. Tighten to 40 ft. lbs. (54 Nm.)

➡Be careful not to damage the valve lifter.

j. Check that the paint marks on the chain are aligned with the timing marks on the camshaft timing gear and camshaft timing sprocket. Also, check that the crankshaft pulley groove is aligned with the timing mark "0" on the timing mark chain cover.

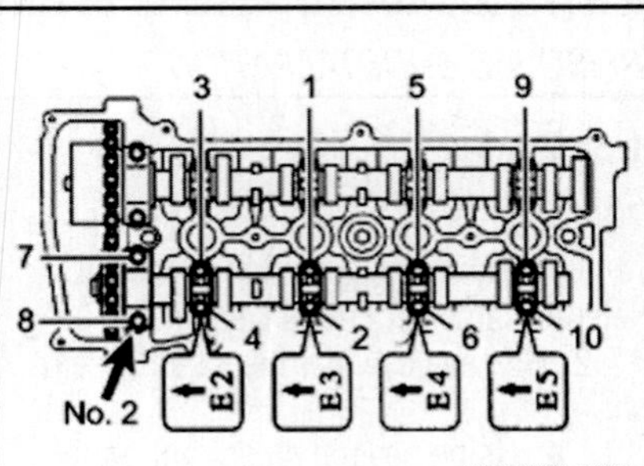

Fig. 50 No. 2 camshaft bearing cap Tightening sequence

24. Install the no. 1 chain tensioner assembly:
    a. Release the ratchet pawl, then fully push in the plunger and set the hook to the pin so that the plunger is in the position shown in the illustration.
    b. Install a new gasket and the chain tensioner with the 2 nuts. Tighten the nuts to 80 inch. lbs. (9 Nm).

➡When installing the chain tensioner, set the hook again if the hook releases the plunger.

25. Turn the crankshaft counterclockwise, then disconnect the hook from the pin.
26. Turn the crankshaft clockwise, then check that the plunger is extended.
27. Set no. 1 cylinder to TDC/compression:
    a. Turn the crankshaft pulley until the groove and the timing mark "0" on the timing chain cover are aligned.
    b. Check that each timing mark on the camshaft timing gear and sprocket is aligned with each timing mark located on the No. 1 and No. 2 bearing caps.
    c. If not, turn the crankshaft pulley 1 revolution (360°) to align the timing marks.
    d. Place paint marks on the chain in alignment with the timing marks on the camshaft timing gear and camshaft timing sprocket.
28. Check valve clearance.
29. Adjust valve clearance.
30. To complete installation, reverse remaining removal procedure.

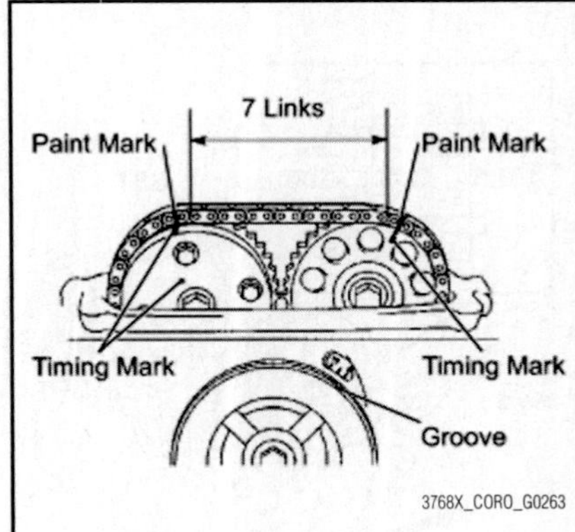

Fig. 51 Check that the paint marks on the chain are aligned with the timing marks on the camshaft timing gear and camshaft timing sprocket

CATALYTIC CONVERTER

REMOVAL & INSTALLATION

The catalytic converter is integrated with the exhaust pipe and cannot be removed separately. The catalytic converter must be replaced with the exhaust pipe as a unit. Refer to Exhaust Manifold in this section.

CRANKSHAFT DAMPER

REMOVAL & INSTALLATION

See Figures 52 and 53.

1. Disconnect the negative battery cable.
2. Remove the RH front wheel.
3. Remove the RH engine under cover.
4. Remove the engine cover.
5. Remove the accessory drive belt.
6. Using a pulley bolt remover (SST's 09213-58013 and 09330-00021), hold the pulley in place and loosen the pulley bolt.

➡Check the SST installation positions when installing them to prevent the SST fixing bolts from coming into contact with the timing chain cover sub-assembly.

7. Using a pulley puller (SST 09950-50013), remove the crankshaft pulley and pulley bolt.
8. If necessary, remove timing chain cover front oil seal:
    a. Using a knife, cut off the lip of the oil seal.
    b. Using a screwdriver with its tip wrapped with tape, pry out the oil seal.

➡After removing, check the crankshaft for damage. If damaged, smooth the surface with 400-grit sandpaper.

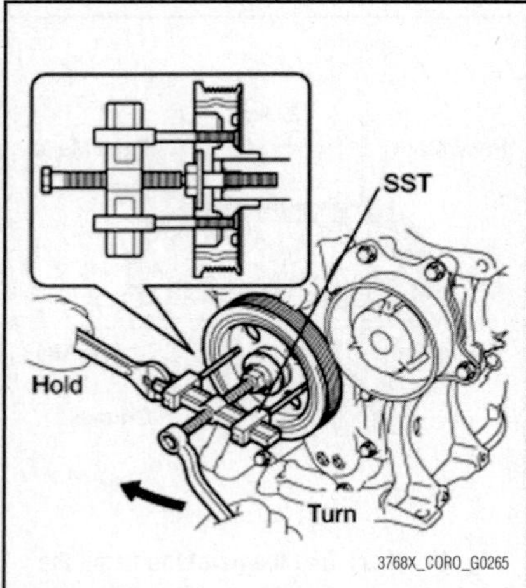

Fig. 52 Removing the crankshaft pulley and pulley bolt

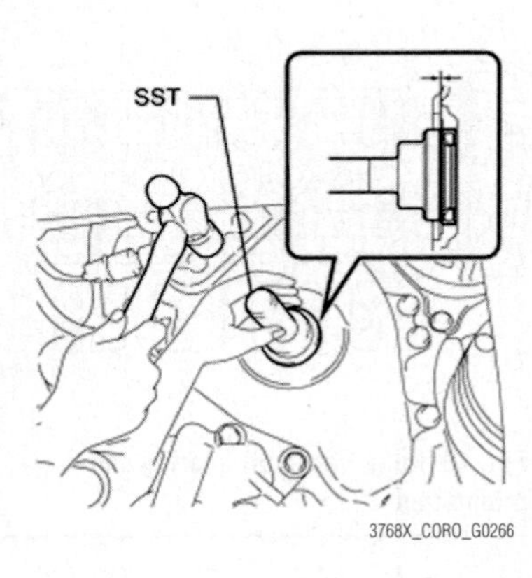

Fig. 53 Installing the crankshaft front seal

To install:

➡Keep the lip free of foreign matter.

9. Apply MP grease to the lip of a new oil seal.
10. Using a seal installer (SST 09223-22010) and a hammer, tap in the oil seal until its surface is flush with the rear oil seal retainer edge.

➡Wipe off extra grease from the crankshaft.

11. Align the pulley set key with the key groove of the pulley.
12. Using a pulley bolt remover (SST's 09213-58013 and 09330-00021), hold the pulley in place and tighten the pulley bolt to 140 ft. lbs. (190 Nm).
13. To complete installation, reverse remaining removal procedure.

CRANKSHAFT FRONT SEAL

REMOVAL & INSTALLATION

Refer to Crankshaft Damper to remove the crankshaft front seal.

CYLINDER HEAD

REMOVAL & INSTALLATION

1.8L Engine

See Figures 54 through 57.

1. Remove the camshaft assembly. Refer to Camshaft in this section.
2. Remove cylinder head sub-assembly:
    a. Using several steps, uniformly loosen and remove the 10 cylinder head bolts and 10 plate washers with a 10 mm bi-hexagon wrench in the sequence shown in the illustration.

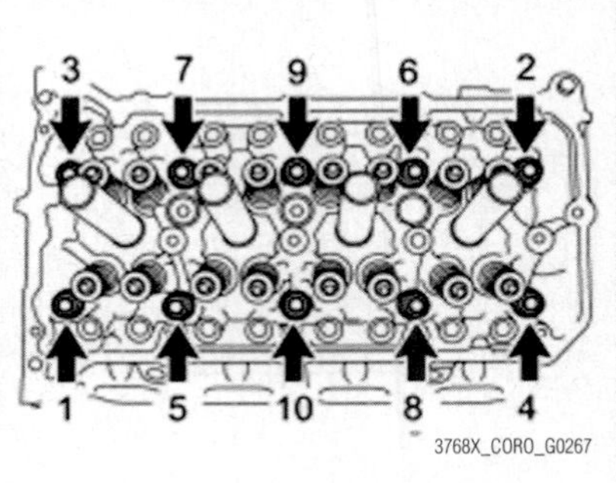

Fig. 54 Cylinder head sub-assembly removal sequence

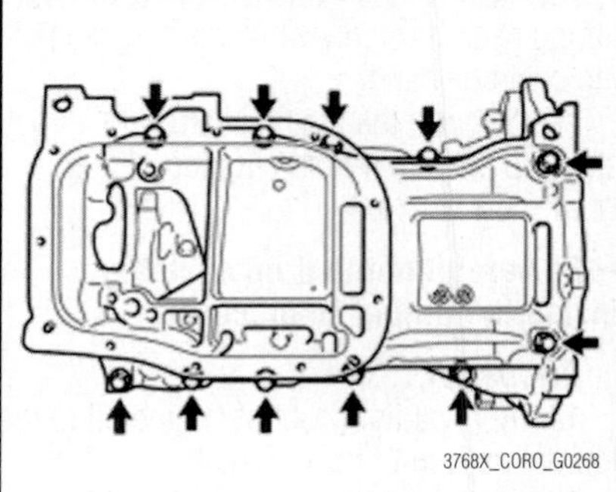

Fig. 55 Location of the 11 stiffening crankcase assembly bolts

➡Head warpage or cracking could result from removing the bolts in the wrong order.

    b. Using a screwdriver with its tip wrapped with tape, pry between the cylinder head and cylinder block, and remove the cylinder head.

➡Be careful not to damage the contact surfaces of the cylinder head and cylinder block.

3. Remove the cylinder head gasket.
4. Remove the water drain cock plug from the water drain cock sub-assembly.
5. Remove the cylinder block water drain cock sub-assembly from the cylinder block.
6. Remove the ventilation valve.
7. Remove the oil pan drain plug and gasket. Remove the oil pan. Refer to Oil Pan in this section.
8. Remove oil pump assembly. Refer to Oil Pump in this section.
9. Remove rear engine oil seal:
    a. Using a knife, cut off the oil seal lip.
    b. Using a screwdriver with its tip taped, pry out the oil seal.

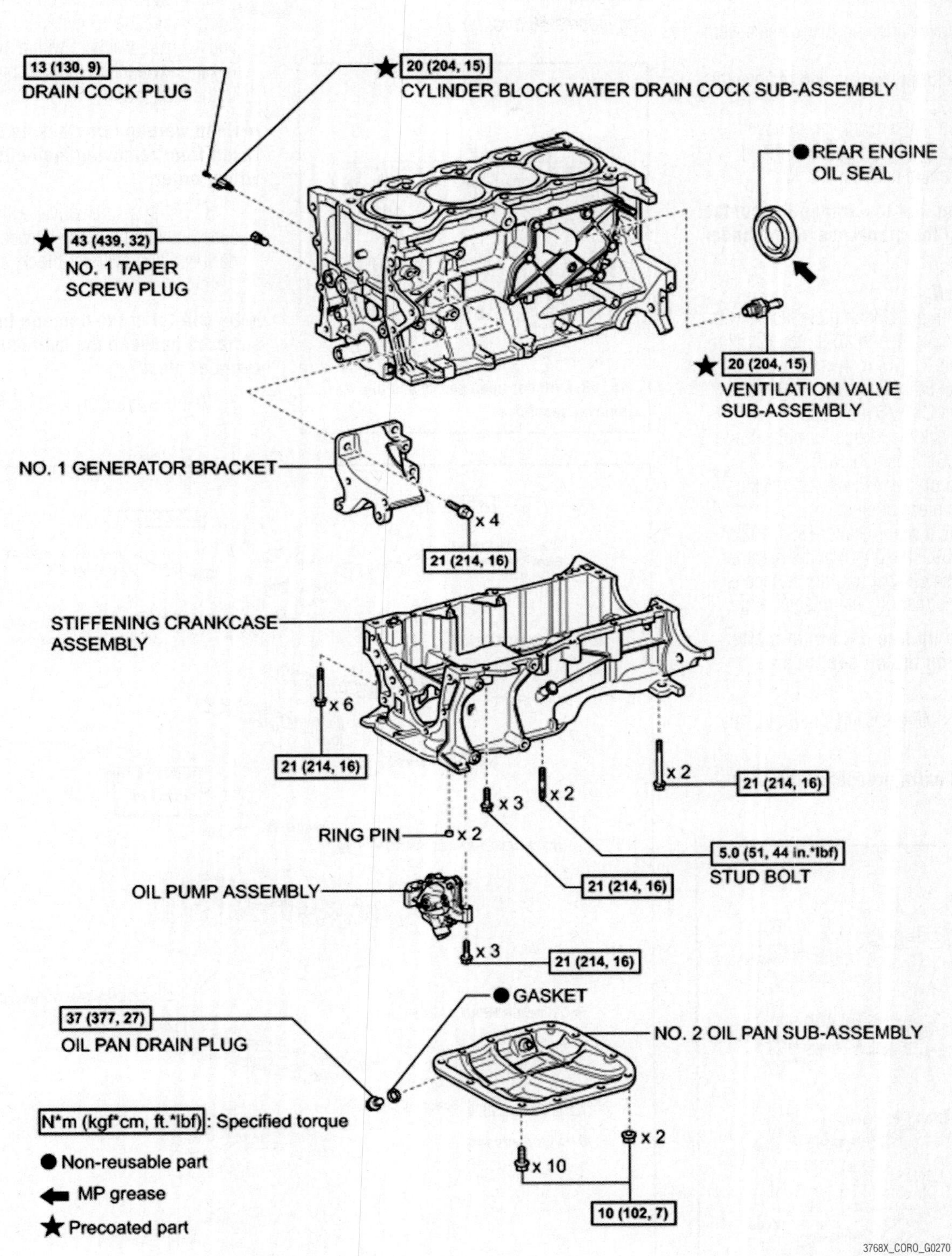

Fig. 56 Exploded view of the cylinder head and related components—1.8L Engine

➡**After removing the oil seal, check the crankshaft for damage. If it is damaged, smooth the surface with 400-grit sandpaper.**

10. Remove stiffening crankcase assembly:

a. Uniformly loosen and remove the 11 bolts.

b. Using a screwdriver, remove the crankcase by prying between the crankcase and cylinder block.

➡**Be careful not to damage the contact surfaces of the crankcase and cylinder block.**

To install:

11. Install the stiffening crankcase with the 11 bolts. Tighten to 16 ft. lbs. (21 Nm).

a. Bolt A: 5.43 inches.

b. Bolt B: 1.38 inches.

c. Bolt C: 2.76 inches.

12. Recheck the torque for bolts 1 and 2. Tighten to 16 ft. lbs. (21 Nm).

13. Wipe off any excess seal packing with a clean piece of cloth.

14. Using a seal installer (SST 09223-15030 and 09950-70010) and a hammer, evenly tap the oil seal until its surface is flush with the rear oil seal retainer edge.

➡**Keep the lip free of foreign matter. Do not tap on the oil seal at an angle.**

15. Apply MP grease to a new oil seal lip.

➡**Wipe off extra grease on the crankshaft.**

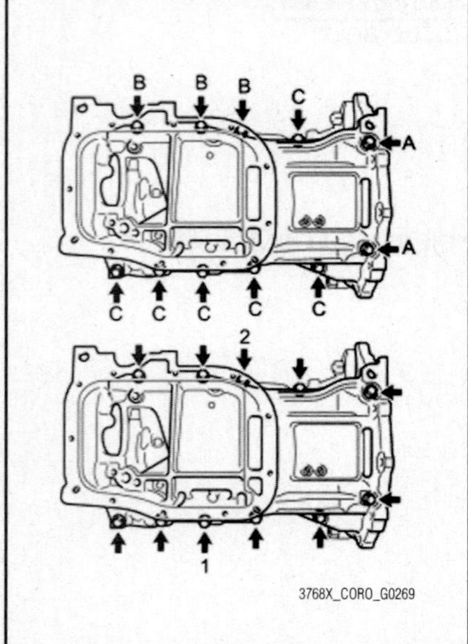

Fig. 57 Identifying and location of the 11 stiffening crankcase bolts

16. To complete the installation, reverse remaining removal procedure.

2.4L Engine

See Figures 58 through 60.

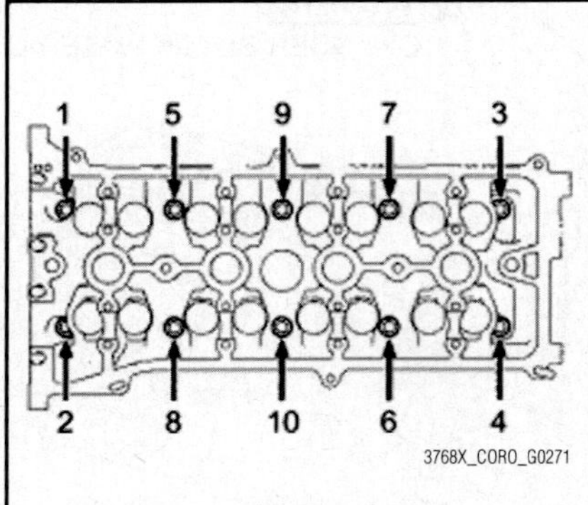

Fig. 58 Cylinder head sub-assembly removal sequence

1. Remove the camshaft assembly. Refer to Camshaft in this section.

2. Remove cylinder head sub-assembly:

a. In several steps, uniformly loosen and remove the 10 cylinder head bolts and 10 plate washers with a 10 mm bi-hexagon wrench in the sequence shown in the illustration.

➡**Head warpage or cracking could result from removing the bolts in the wrong order.**

b. Using a screwdriver with its tip wrapped with tape, pry between the cylinder head and cylinder block, and remove the cylinder head.

➡**Be careful not to damage the contact surfaces between the cylinder head and cylinder block.**

3. Remove cylinder head gasket.

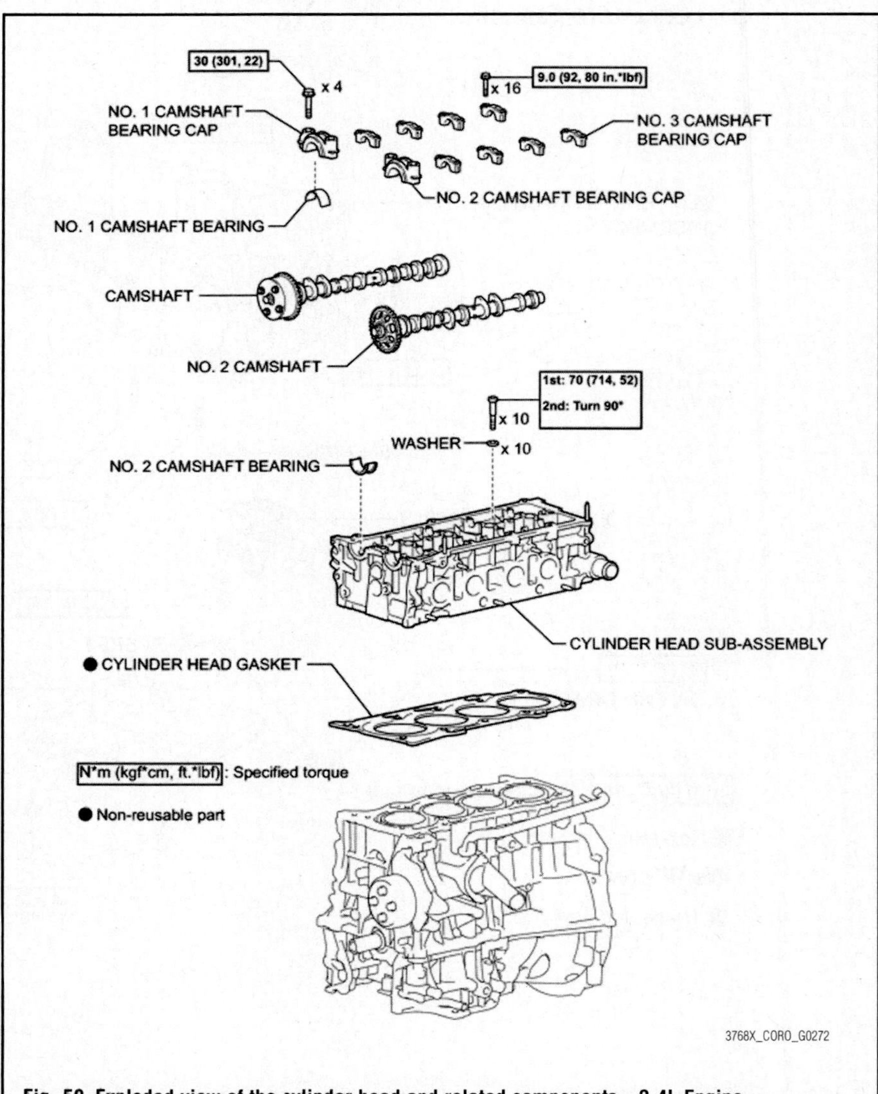

Fig. 59 Exploded view of the cylinder head and related components—2.4L Engine

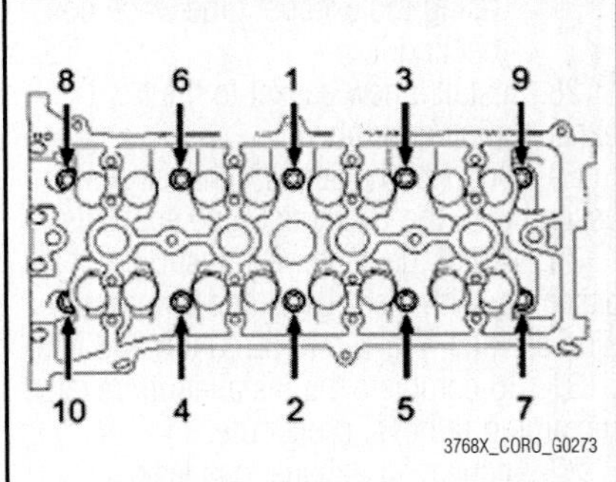

Fig. 60 Cylinder head tightening sequence

To install:

4. Place a new cylinder head gasket on the cylinder block surface with the Lot No. stamp facing upward.

➡Remove any oil from the contact surface. Be careful of the installation direction.

5. Place the cylinder head on the cylinder head gasket.

➡Place the cylinder head gently in order to avoid damaging the cylinder head gasket.

➡The cylinder head bolts are tightened in 2 successive steps.

6. Install the cylinder head bolts:
   a. Apply a light coat of engine oil to the threads and under the heads of the cylinder head set bolts.
   b. Using several steps, uniformly install and tighten the 10 cylinder head set bolts and plate washers with a 10 mm bi-hexagon wrench in the order shown in the illustration. Tighten to 52 ft. lbs. (70 Nm).
   c. Mark the front of the cylinder head bolts with paint.
   d. Further tighten the cylinder head bolts 90°.
   e. Check that the paint mark is now at a 90°angle to the front.
7. To complete the installation, reverse remaining removal procedure.

EXHAUST MANIFOLD

REMOVAL & INSTALLATION

1.8L Engine

See Figure 61.

1. Disconnect the negative battery cable.
2. Remove front wiper arm motor and assembly.

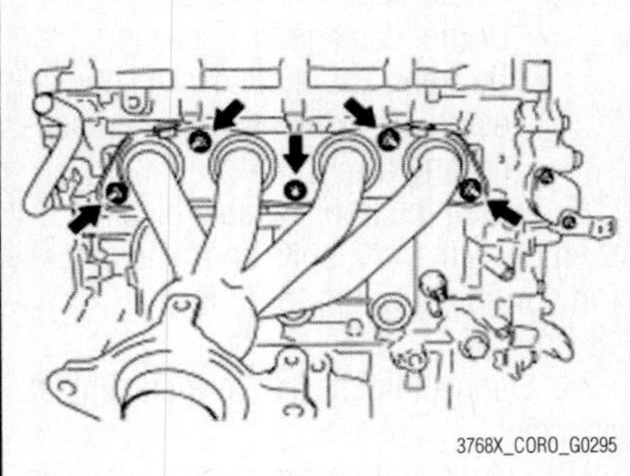

Fig. 61 Removing the 5 nuts and exhaust manifold

3. Remove the valve cover.
4. Remove air fuel ratio sensor.
5. Remove the 4 bolts and the exhaust manifold heat insulator.
6. Disconnect the heated oxygen sensor connector. Refer to Heated Oxygen Sensor in Engine Performance and Emission control.
7. Remove the 4 bolts and 4 compression springs.
8. Remove the front exhaust pipe assembly from the 2 exhaust pipe supports.
9. Remove the 3 bolts and manifold stay.
10. Remove the 5 nuts and exhaust manifold.
11. Remove the exhaust manifold gasket.
12. Remove the 3 bolts and exhaust manifold heat insulator.

To install:

13. Install the No. 2 exhaust manifold heat insulator with the 3 bolts. Tighten to 9 ft. lbs. (12 Nm).
14. Install a new exhaust manifold gasket.
15. Install the exhaust manifold with the 5 nuts. Tighten to 16 ft. lbs. (21 Nm).
16. Install the manifold stay with the 3 bolts. Tighten to 32 ft. lbs. (43 Nm).
17. Using a vernier caliper, measure the free length of the compression springs.
    a. Front springs: Minimum: 1.63 inches.
    b. Rear springs: Minimum: 1.52 inches.

➡If the free length is less than the minimum, replace the compression spring.

18. Fully insert new gaskets to the exhaust manifold and front exhaust pipe assembly.
19. Using a plastic hammer and wooden block, tap in the 2 new gaskets until their surface are flush with the exhaust manifold and front exhaust pipe assembly.

➡Be sure to install the gaskets in the correct direction.

- Do not reuse the gaskets.
- Do not damage the gaskets.
- Do not push in the gaskets by using the exhaust pipe when connecting it.

20. Connect the front exhaust pipe assembly to the 2 exhaust pipe supports.
21. Install the front exhaust pipe assembly with the 4 bolts and 4 compression springs. Tighten to 32 ft. lbs. (43 Nm).
22. Connect the heated oxygen sensor connector.
23. Install the outer exhaust manifold heat insulator with the 4 bolts. Tighten to 9 ft. lbs. (12 Nm).
24. To complete the installation, reverse remaining removal procedure.
25. Inspect for exhaust gas leak.

2.4L Engine

See Figures 62 & 63.

1. Disconnect the negative battery cable.
2. Remove the drive belt. Refer to Accessory Drive Belts in the section.
3. Remove the alternator. Refer to Alternator in Engine Electrical.
4. Remove the 4 bolts and 2 compression springs.
5. Remove the center exhaust pipe assembly from the 2 exhaust pipe supports.
6. Remove the 2 gaskets.
7. Remove front exhaust pipe assembly:
   a. Disconnect the heated oxygen sensor connector.
   b. Remove the 2 bolts, 2 compression springs and front exhaust pipe assembly.
   c. Remove the gasket from the exhaust manifold.
8. Remove the bolt, nut and LH manifold stay.

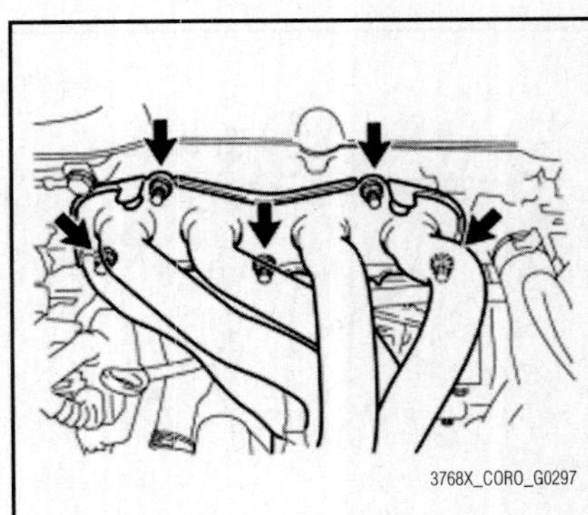

Fig. 62 Remove the 5 nuts and exhaust manifold converter sub-assembly

9. Remove the bolt, nut and RH manifold stay.

10. Remove the 4 bolts and upper exhaust manifold heat insulator.

11. Remove air fuel ratio sensor.

12. Remove the 5 nuts and exhaust manifold converter sub-assembly.

13. Remove the gasket.

14. Remove the 2 bolts and underside exhaust manifold heat insulator.

15. Remove the 4 bolts and catalytic converter insulator.

To install:

16. Install the 4 bolts and catalytic converter insulator. Tighten to 9 ft. lbs. (12 Nm).

17. Install the underside exhaust manifold heat insulator with the 2 bolts. Tighten to 9 ft. lbs. (12 Nm).

18. Install the exhaust manifold converter sub-assembly:

  a. Install a new gasket.

  b. Install the exhaust manifold converter sub-assembly with the 5 nuts in the order shown in the illustration. Tighten to 27 ft. lbs. (37 Nm).

19. Install the RH and LH manifold stay with the bolt and nut. Tighten to 33 ft. lbs. (44 Nm).

20. Install the air fuel ratio sensor.

21. Install the outer exhaust manifold heat insulator with the 4 bolts. 9 ft. lbs (12 Nm).

22. Using a vernier caliper, measure the free length of the compression springs.

  a. Front springs: Minimum: 1.63 inches.

  b. Rear springs: Minimum: 1.52 inches.

23. Remove the remains of exhaust manifold converter with wire brash.

24. Fully insert a new gasket to the exhaust manifold.

➡Be sure to install the gaskets in the correct direction.

- Do not reuse the gaskets.
- Do not damage the gaskets.
- Do not push in the gaskets by using the exhaust pipe when connecting it.

25. Install the front exhaust pipe assembly with the 2 bolts and 2 compression springs. Tighten to 32 ft. lbs. (43 Nm).

26. Connect the heated oxygen sensor connector.

27. Fully insert a new gasket to the center exhaust pipe assembly.

➡Be sure to install the gaskets in the correct direction.

- Do not reuse the gaskets.
- Do not damage the gaskets.
- Do not push in the gaskets by using the exhaust pipe when connecting it.

28. Install a new gasket to the front exhaust pipe assembly.

29. Connect the center exhaust pipe assembly to the 2 exhaust pipe supports.

30. Install the center exhaust pipe assembly with the 4 bolts and 2 compression springs. Tighten to 32 ft. lbs. (43 Nm).

31. To complete the installation, reverse remaining removal procedure.

32. Inspect for exhaust gas leak.

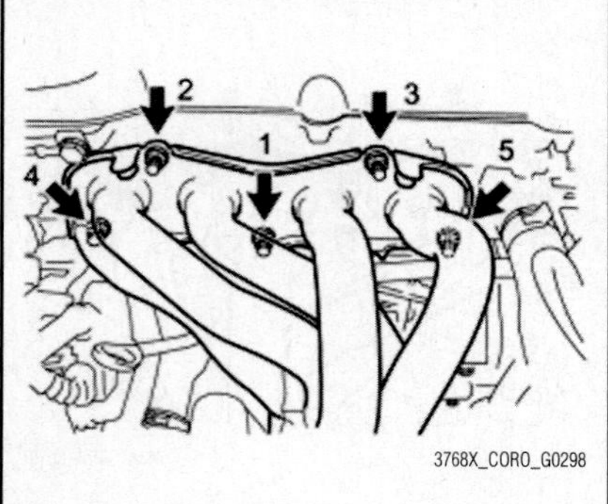

Fig. 63 Exhaust manifold tightening sequence

INTAKE MANIFOLD

REMOVAL & INSTALLATION

1.8L Engine

See Figure 64.

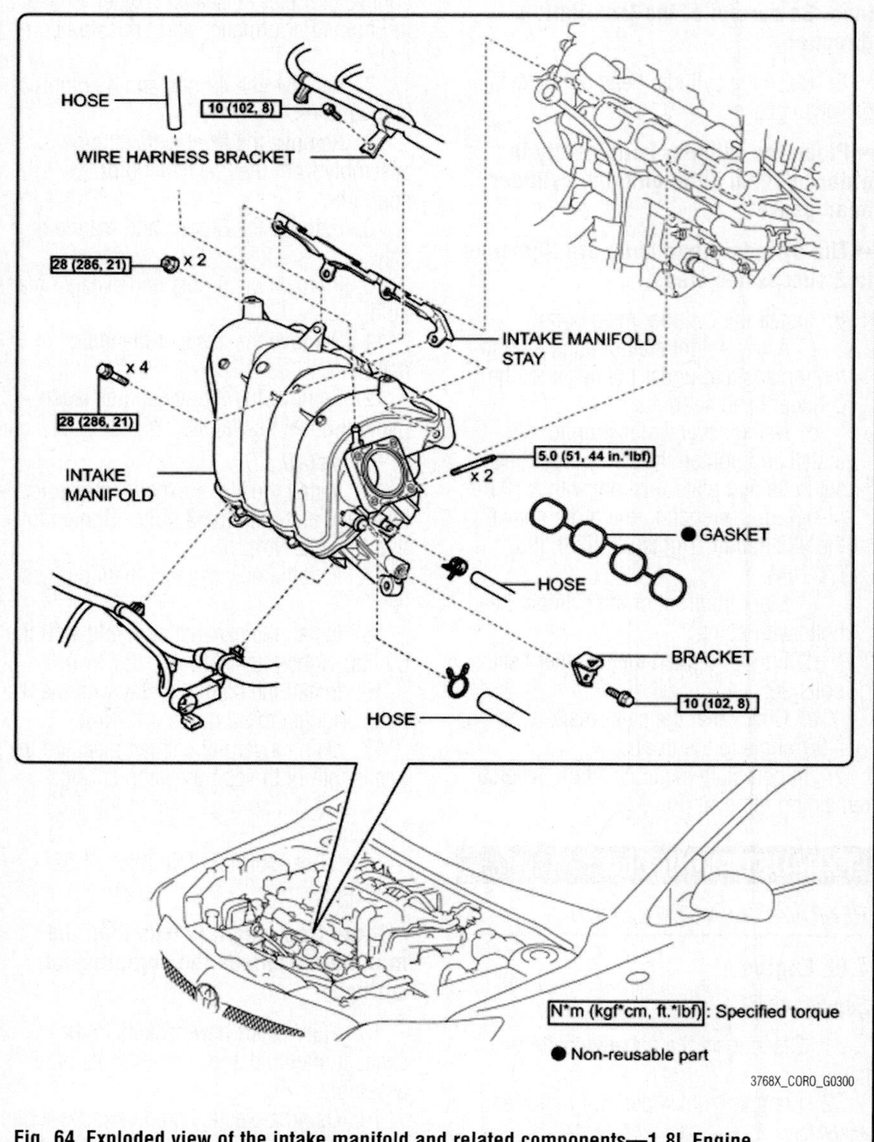

Fig. 64 Exploded view of the intake manifold and related components—1.8L Engine

1. Disconnect the negative battery cable.
2. Drain and recycle the engine coolant.
3. Remove the engine cover.
4. Remove the air cleaner assembly.
5. Remove the throttle body. Refer to Throttle Body in Fuel System.
6. Remove the intake manifold:
   a. Remove the bolt and disconnect the wire harness bracket.
   b. Disconnect the 3 hoses.
   c. Remove the 4 bolts, 2 nuts, intake manifold stay and intake manifold.
   d. Remove the gasket from the intake manifold.
   e. Remove the bolt and bracket.
   f. Using a TORX® socket E6, remove the 2 stud bolts from the intake manifold.

To install:

7. Using a TORX® socket E6, install the 2 stud bolts to the intake manifold. Tighten to 44 inch lbs. (5 Nm).
8. Install the bracket with the bolt. Tighten to 8 ft. lbs. (10 Nm).
9. Install a new gasket into the intake manifold.
10. Install the intake manifold and intake manifold stay with the 4 bolts and 2 nuts. Tighten to 21 ft. lbs. (28 Nm).
11. To complete the installation, reverse remaining removal procedure.

2.4L Engine

See Figures 65 and 66.

CAUTION

Observe all applicable safety precautions when working around fuel. Whenever servicing the fuel system, always work in a well ventilated area. Do not allow fuel spray or vapors to come in contact with a spark or open flame. Keep a dry chemical fire extinguisher near the work area. Always keep fuel in a container specifically designed for fuel storage; also, always properly seal fuel containers to avoid the possibility of fire or explosion.

1. Discharge fuel system pressure.
2. Disconnect the cable from negative battery terminal.
3. Remove the engine cover.
4. Drain and recycle the engine coolant.
5. Remove the windshield wiper motor and link assembly.
6. Remove suspension tower damper assembly (with front strut bar). Refer to Suspension Tower Damper Assembly in Front Suspension.

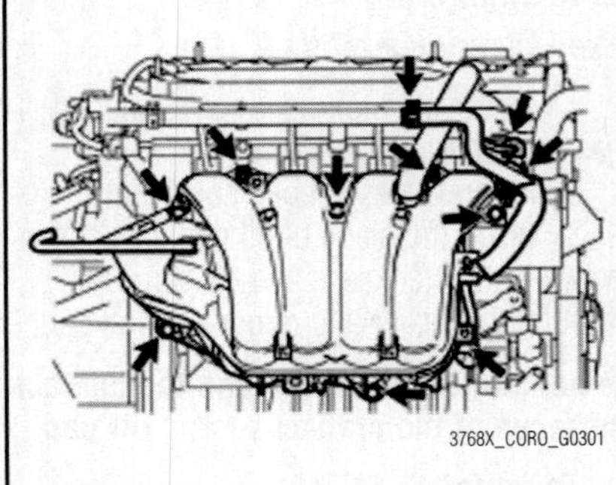

Fig. 65 Removing the 5 bolts, 2 nuts and intake manifold

7. Remove the air cleaner assembly.
8. Remove the throttle body. Refer to Throttle Body in Fuel System.
9. Disconnect the main fuel tube.
10. Disconnect the ventilation hose.
11. Remove fuel delivery pipe with injector. Refer to Fuel Rail and Injectors in Fuel System.
12. Remove intake manifold:
    a. Disconnect the union to connector tube hose from the No. 2 hose to hose tube.
    b. Disconnect the camshaft timing oil control valve connector.

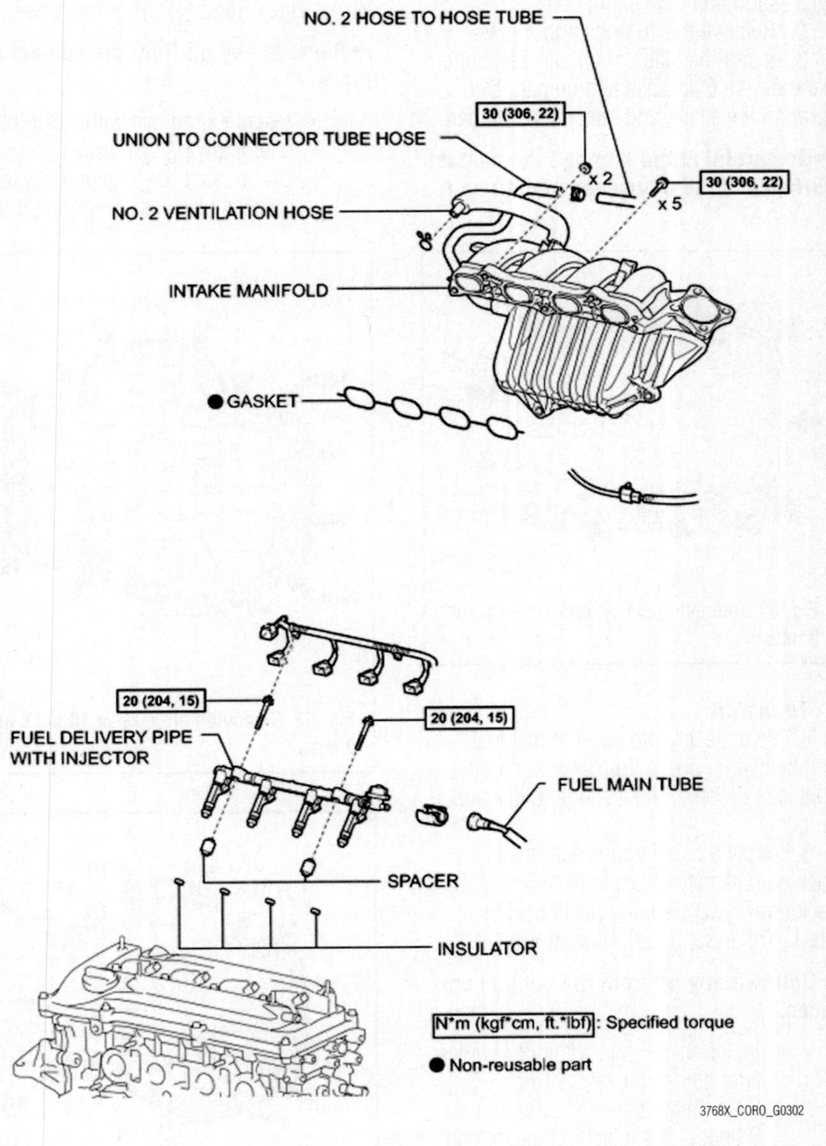

Fig. 66 Exploded view of the intake manifold and related components—2.4L Engine

c. Remove the wire harness clamp.

d. Remove the union to connector tube hose from the vacuum hose clamp.

e. Remove the 5 bolts, 2 nuts and intake manifold.

f. Remove the gasket from the intake manifold.

13. To install, reverse removal procedure. Refer to exploded view illustration for torque values.

OIL PAN

REMOVAL & INSTALLATION

1.8L Engine

See Figure 67.

1. Remove the oil pan drain plug and gasket and drain the engine oil.
2. Remove the 10 bolts and 2 nuts.
3. Insert the blade of oil pan seal cutter between the crankcase and oil pan. Cut through the sealer and remove the oil pan.

➡Be careful not to damage the contact surfaces of the crankcase and oil pan.

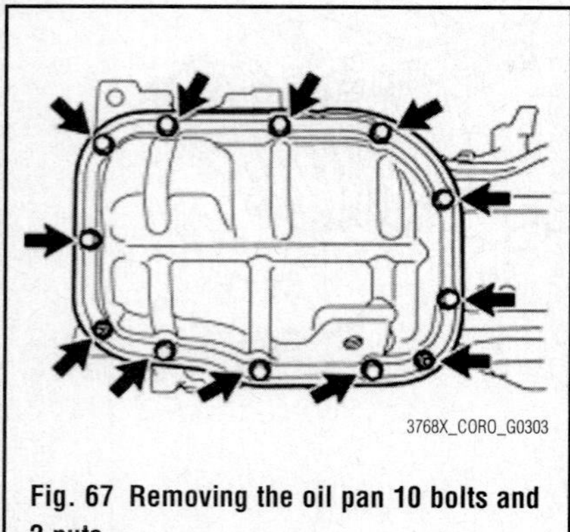

Fig. 67 Removing the oil pan 10 bolts and 2 nuts

To install:

4. Remove any old packing material and be careful not to drop any oil on the contact surfaces of the cylinder block and oil pan.
5. Apply a continuous bead of seal packing (Diameter 4.0 mm (0.157 in.)). Seal packing: Toyota Genuine Seal Packing Black, Three Bond 1207B or equivalent.

➡Remove any oil from the contact surfaces.

- Install the oil pan within 3 minutes after applying seal packing.
- Do not start the engine for at least 2 hours after installing the oil pan.

6. Install the oil pan with the 10 bolts and 2 nuts. Tighten to 7 ft. lbs. (10 Nm).

2.4L Engine

See Figures 68 and 69.

1. Remove the oil pan drain plug and gasket and drain the engine oil.
2. Remove the 12 bolts and 2 nuts.
3. Insert the blade of oil pan seal cutter between the crankcase and oil pan. Cut through the sealer and remove the oil pan.

➡Be careful not to damage the contact surfaces of the crankcase and oil pan.

To install:

4. Remove any old packing material and be careful not to drop any oil on the contact surfaces of the cylinder block and oil pan.
5. Apply a continuous bead of seal packing (Diameter 4.0 mm (0.157 in.)). Seal packing: Toyota Genuine Seal Packing Black, Three Bond 1207B or equivalent.

➡Remove any oil from the contact surfaces.

- Install the oil pan within 3 minutes after applying seal packing.
- Do not start the engine for at least 2 hours after installing the oil pan.

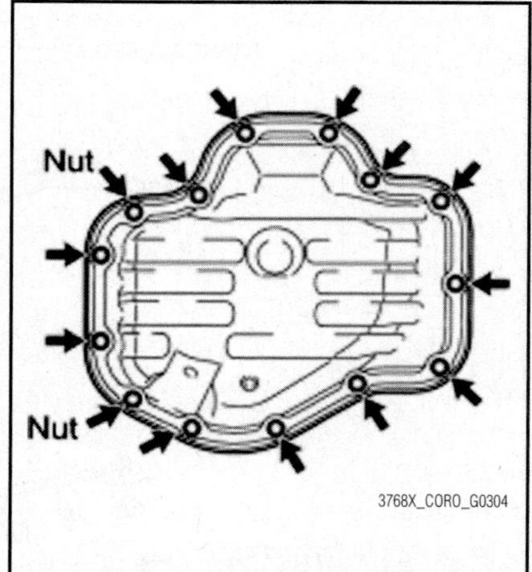

Fig. 68 Removing the oil pan 10 bolts and 2 nuts

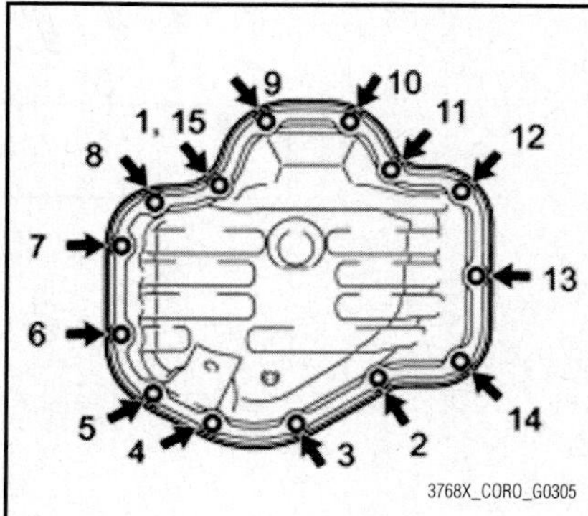

Fig. 69 Oil pan tightening sequence

6. Uniformly tighten the 12 bolts and 2 nuts in the sequence shown in the illustration. Tighten to 80 inch lbs. (9 Nm).

OIL PUMP

REMOVAL & INSTALLATION

1. Remove the engine assembly with the transaxle.
2. Remove the timing chain front cover. Refer to Timing Chain Front Cover in this section.
3. Remove the oil pan. Refer to Oil Pan in this section.
4. Remove the 3 bolts and oil pump.
5. For 2.4L engines, remove the gasket.
6. To install, reverse the removal procedure. Tighten the oil pump assembly bolts to 16 ft. lbs. (21 Nm).

PISTON AND RING

POSITIONING

See Figures 70 through 72.

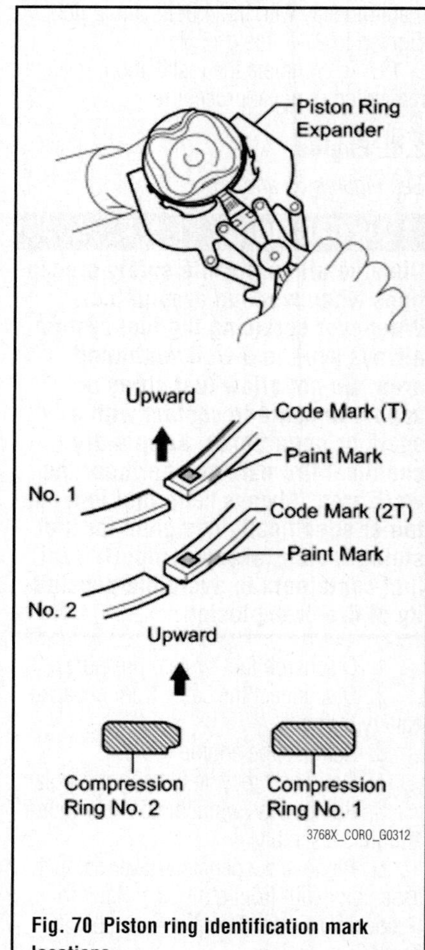

Fig. 70 Piston ring identification mark locations

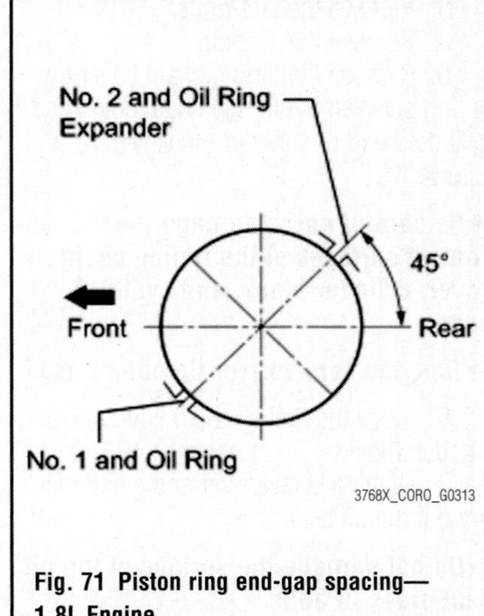

Fig. 71 Piston ring end-gap spacing—1.8L Engine

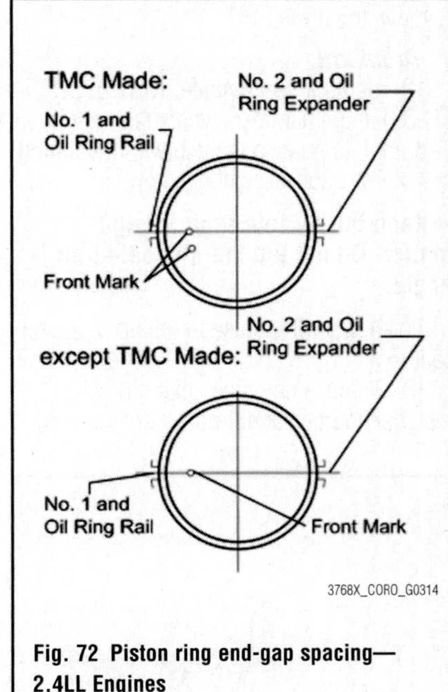

Fig. 72 Piston ring end-gap spacing—2.4LL Engines

REAR MAIN SEAL

REMOVAL & INSTALLATION

See Figures 73 and 74.

1. Remove the appropriate transaxle. Refer to Drivetrain.
2. Remove the flywheel or flexplate.
3. Remove rear engine oil seal:
   a. Using a knife, cut off the oil seal lip.
   b. Using a screwdriver with its tip taped, pry out the oil seal.

➡After removing the oil seal, check the crankshaft for damage. If it is damaged, smooth the surface with 400-grit sandpaper.

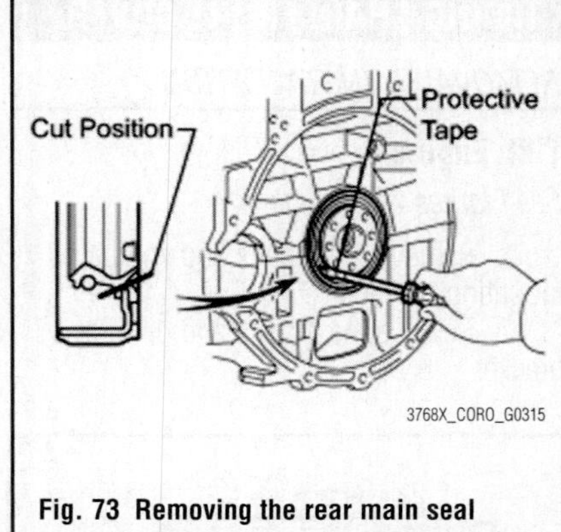

Fig. 73 Removing the rear main seal

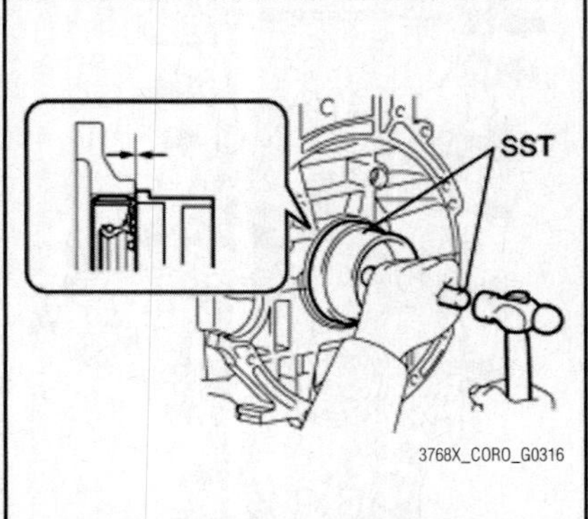

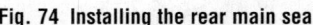

Fig. 74 Installing the rear main seal

To install:

4. Using a seal installer (SST 09223-15030 and 09950-70010) and a hammer, evenly tap the oil seal until its surface is flush with the rear oil seal retainer edge.

➡Keep the lip free of foreign matter. Do not tap on the oil seal at an angle.

5. Apply MP grease to a new oil seal lip.

➡Wipe off extra grease on the crankshaft.

6. To complete installation, reverse remaining removal procedure.

ROCKER ARMS

REMOVAL & INSTALLATION

1.8L Engine

See Figures 75 through 78.

1. Remove the camshaft assembly. Refer to Camshafts in this section.
2. Remove the 16 valve rocker arms.

➡Arrange the removed parts in the correct order.

3. Remove the 16 valve lash adjusters from the cylinder head.

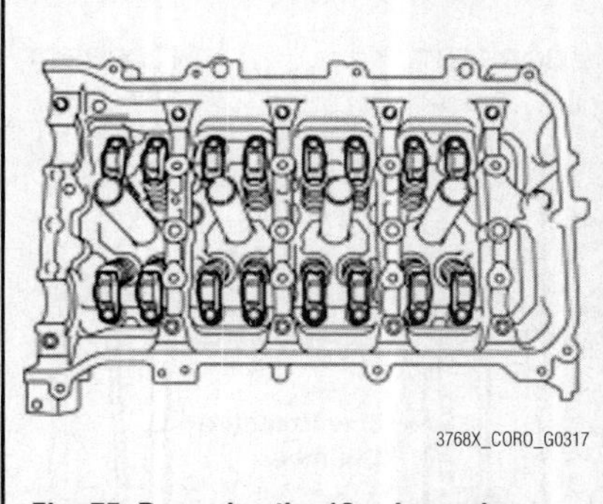

Fig. 75 Removing the 16 valve rocker arms

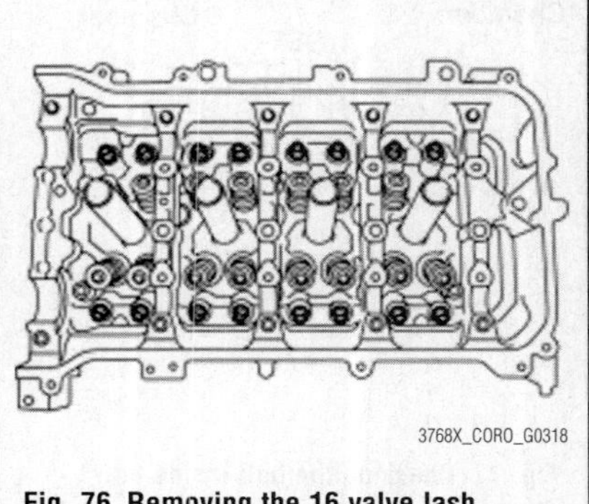

Fig. 76 Removing the 16 valve lash adjusters

➡Arrange the removed parts in the correct order.

To install:

➡Keep the lash adjuster free of dirt and foreign matter. Only use clean engine oil.

4. Place the lash adjuster into a container filled with engine oil.
5. Insert the tip of SST 09276-75010 into the lash adjuster plunger and use the tip to press down on the check ball inside the plunger.
6. Squeeze SST and the lash adjuster together to move the plunger up and down 5 to 6 times.
7. Check the movement of the plunger and bleed it. If the plunger moves up and down it is ok.

➡When bleeding air from the high-pressure chamber, make sure that the tip of SST is actually pressing the check ball as shown in the illustration. If the check ball is not pressed, the high-pressure chamber will not be bled.

8. After bleeding, remove SST. Then, try to press the plunger quickly and firmly by hand.

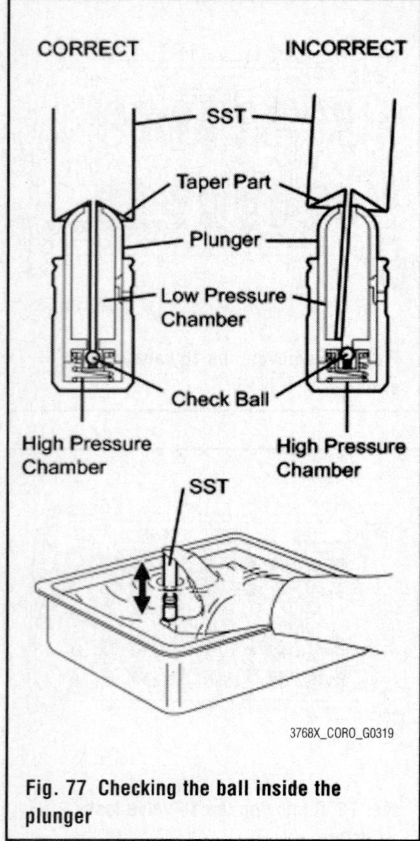

Fig. 77 Checking the ball inside the plunger

9. If the plunger is very difficult to move it is ok. If the result is not as specified, replace the lash adjuster.

10. Install the lash adjusters

➡Install the lash adjuster to the same place it was removed from.

11. Apply engine oil to the lash adjuster tip and valve stem cap end.

12. Make sure that the valve rocker arms are installed as shown in the illustration.

13. To complete installation, reverse remaining removal procedure.

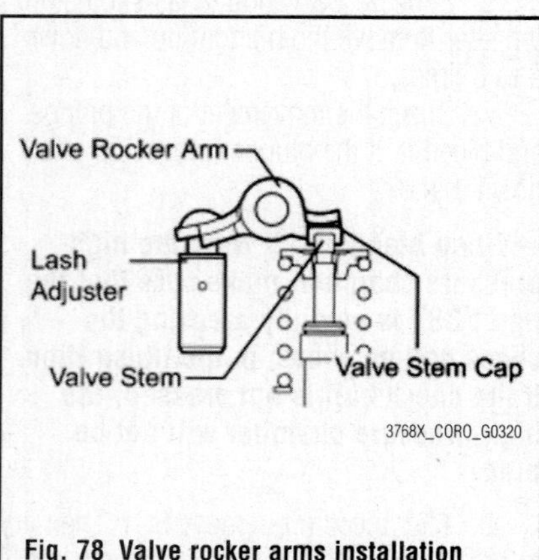

Fig. 78 Valve rocker arms installation orientation

TIMING CHAIN FRONT COVER

REMOVAL & INSTALLATION

1.8L Engine

See Figures 79 through 84.

1. Remove the 3 bolts and engine mounting bracket.

2. Remove the 4 bolts and oil filter bracket.

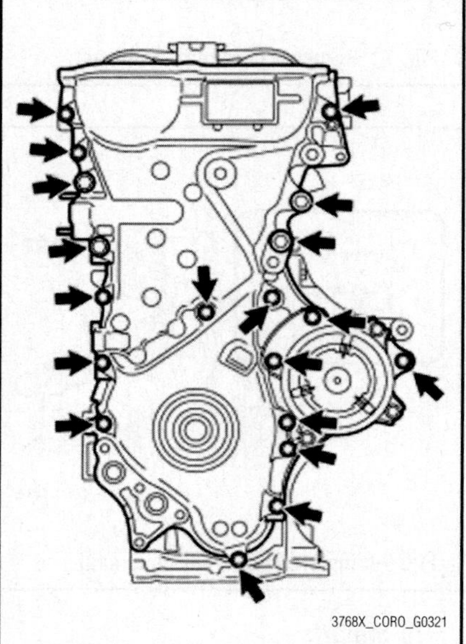

Fig. 79 Removing the 19 bolts and the timing chain front cover

3. Remove the 2 O-rings.

4. Remove the 19 bolts.

5. Remove the timing chain cover by prying between the timing chain cover and cylinder head or cylinder block with a screwdriver.

➡Be careful not to damage the contact surfaces of the timing chain cover, cylinder block, and cylinder head.

➡Tape the screwdriver tip before use.

6. Place the timing chain cover on wooden blocks.

7. Using a screwdriver and a hammer, pry out the oil seal.

➡Do not damage the surface of the oil seal press fit hole.

8. Remove the 3 O-rings.

9. Remove the 3 bolts and water pump. Remove the gasket.

To install:

10. Using a seal installer (SST 09223-15030) and a hammer, evenly tap the oil seal until its surface is flush with the timing chain cover sub-assembly edge.

➡Keep the lip free from foreign matter. Do not tap the oil seal at an angle.

11. Apply MP grease to the lip of the oil seal.

12. Remove any old packing (FIPG) material and be careful not to drop any oil

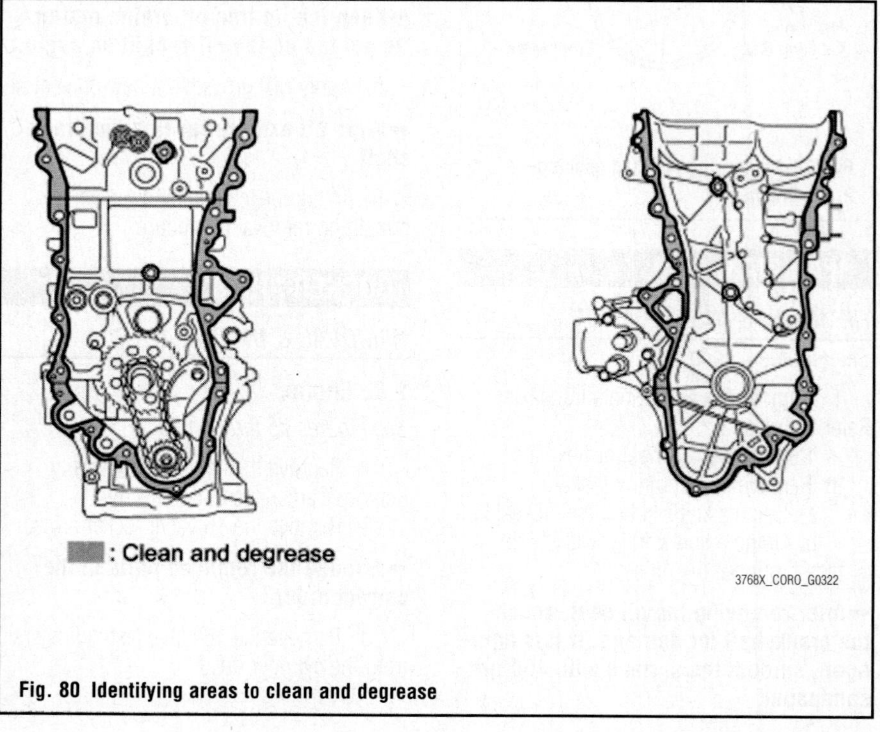

Fig. 80 Identifying areas to clean and degrease

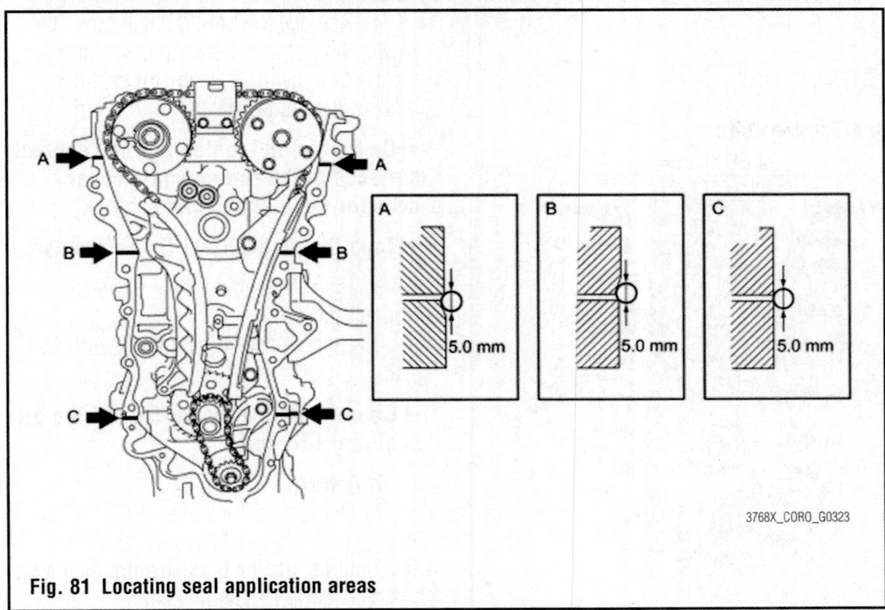

Fig. 81 Locating seal application areas

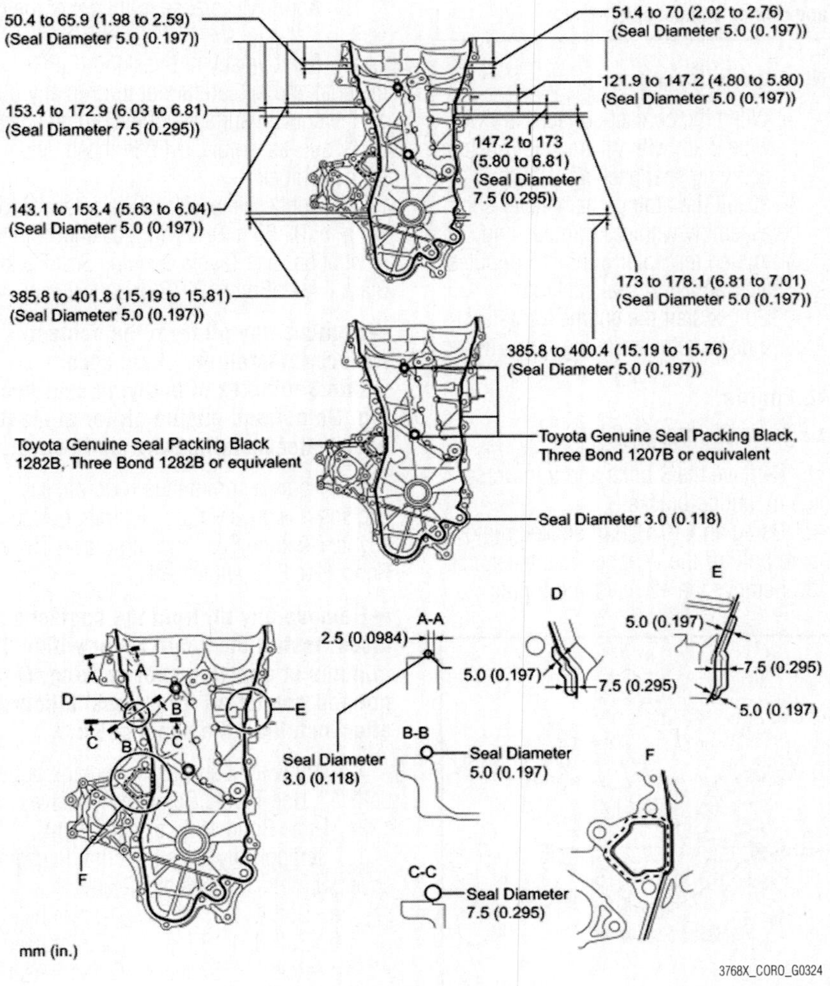

50.4 to 65.9 (1.98 to 2.59) (Seal Diameter 5.0 (0.197))
51.4 to 70 (2.02 to 2.76) (Seal Diameter 5.0 (0.197))
121.9 to 147.2 (4.80 to 5.80) (Seal Diameter 5.0 (0.197))
153.4 to 172.9 (6.03 to 6.81) (Seal Diameter 7.5 (0.295))
147.2 to 173 (5.80 to 6.81) (Seal Diameter 7.5 (0.295))
143.1 to 153.4 (5.63 to 6.04) (Seal Diameter 5.0 (0.197))
173 to 178.1 (6.81 to 7.01) (Seal Diameter 5.0 (0.197))
385.8 to 401.8 (15.19 to 15.81) (Seal Diameter 5.0 (0.197))
385.8 to 400.4 (15.19 to 15.76) (Seal Diameter 5.0 (0.197))
Toyota Genuine Seal Packing Black 1282B, Three Bond 1282B or equivalent
Toyota Genuine Seal Packing Black, Three Bond 1207B or equivalent
Seal Diameter 3.0 (0.118)
A-A
2.5 (0.0984)
5.0 (0.197)
D
7.5 (0.295)
E
5.0 (0.197)
7.5 (0.295)
5.0 (0.197)
B-B
Seal Diameter 3.0 (0.118)
Seal Diameter 5.0 (0.197)
F
C-C
Seal Diameter 7.5 (0.295)
mm (in.)
3768X_CORO_G0324

Fig. 82 Applying seal packing to the timing chain cover sub-assembly in a continuous line

on the contact surfaces of the timing chain cover sub-assembly, cylinder head, and cylinder block.

13. Install 3 new O-rings.

14. Apply seal packing as shown in the illustration. Seal packing: Toyota Genuine Seal Packing Black, Three Bond 1207B or equivalent.

➡Remove any oil from the contact surfaces. Install the chain cover within 3 minutes after applying seal packing. Do not start the engine for at least 2 hours after installing the timing chain cover sub-assembly. Apply seal packing to the timing chain cover sub-assembly in a continuous line as shown in the following illustration.

➡

- When the contact surfaces are wet, wipe them with oil-free cloth before applying seal packing.
- Install the timing chain cover sub-assembly within 3 minutes and tighten the bolts within 15 minutes after applying seal packing.
- Do not start the engine for at least 2 hours after installing.

15. For the continuous line area, use: Toyota Genuine Seal Packing Black, Three Bond 1207B or equivalent.

16. For the dashed line area, use: Toyota Genuine Seal Packing 1282B, Three Bond 1282B or equivalent.

17. Apply adhesive to the threads of the bolt E. Use Toyota Genuine Adhesive 1324, Three Bond 1324 or equivalent.

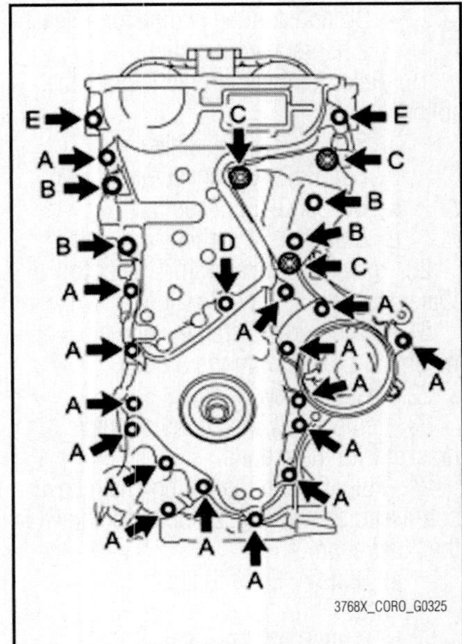

Fig. 83 Identifying the timing chain front cover bolts

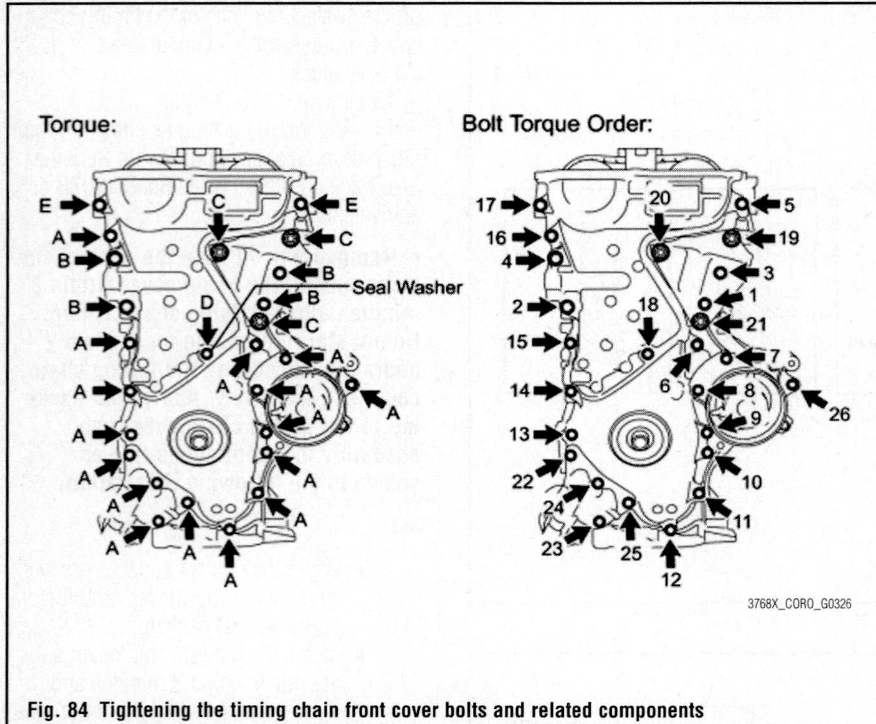

Fig. 84 Tightening the timing chain front cover bolts and related components

18. Temporarily install the timing chain cover sub-assembly with 19 bolts.

➡**Note the following:**

- When the contact surfaces are wet, wipe them with oil-free cloth before applying seal packing.
- Install the timing chain cover "sub-assembly within 3 minutes and tighten the bolts within 15 minutes after applying seal packing.
- Do not start the engine for at least 2 hours after installing.

19. Refer to illustration to identify the following:
    a. Bolt A, E: 1.38 inches.
    b. Bolt B: 2.16 inches.
    c. Bolt C: 3.15 inches.
    d. Bolt D: 1.57 inches
20. Install the water pump. Refer to Water Pump in Engine Cooling.
21. Temporarily install the engine mounting bracket with the 3 bolts.
22. Install 2 new O-rings.
23. Temporarily install the oil filter bracket with the 4 bolts.
24. Fully tighten the timing chain cover sub-assembly with the 26 bolts as shown in the illustration.
    a. Bolts A, E: 19 ft. lbs. (26 Nm).
    b. Bolts B, C: 37 ft. lbs. (51 Nm).
    c. Bolt D: 7 ft. lbs. (10 Nm).

➡**Note the following:**

- When the contact surfaces are wet, wipe them with oil-free cloth before applying seal packing.
- Install the timing chain cover sub-assembly within 3 minutes and tighten the bolts within 15 minutes after applying seal packing.
- Do not start the engine for at least 2 hours after installing.

2.4L Engine

See Figures 85 through 88.

1. Remove the 3 bolts and transverse engine mounting bracket.
2. Using an E10 TORX® socket, remove the stud bolt for the V-ribbed belt tensioner.
3. Remove the 12 bolts and 2 nuts.

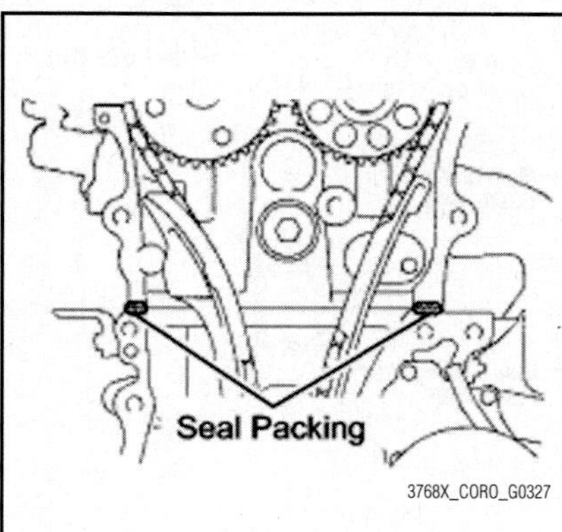

Fig. 85 Locating seal application areas

4. Remove the timing chain cover by prying the portions between the timing chain cover, cylinder head and cylinder block with a screwdriver.

➡**Be careful not to damage the contact surfaces of the timing chain cover, cylinder head and cylinder block.**

➡**Tape the screwdriver tip before use.**

5. Place the timing chain cover on wooden blocks.
6. Using a screwdriver, pry out the oil seal.

➡**Do not damage the surface of the oil seal press fit hole.**

To install:

7. Using a seal installer (SST 09223-22010) and a hammer, evenly tap the oil seal until its surface is flush with the timing chain cover sub-assembly edge.

➡**Keep the lip free from foreign matter. Do not tap the oil seal at an angle.**

8. Apply MP grease to the lip of the oil seal.
9. Remove any old packing (FIPG) material and be careful not to drop any oil on the contact surfaces of the timing chain cover sub-assembly, cylinder head, and cylinder block.
10. Apply seal packing (Diameter 4.0 to 4.5 mm (0.157 to 0.177 in.)) as shown in the illustration. Use Toyota Genuine Seal Packing Black, Three Bond 1207B or equivalent.

➡**Remove any oil from the contact surfaces. Install the chain cover within 3 minutes of applying seal packing. Do not add engine oil for at least 2 hours after installing the chain cover.**

11. Apply a continuous bead of seal packing as shown in the illustration. Use Toyota Genuine Seal Packing Black, Three Bond 1207B or equivalent.

➡**Remove any oil from the contact surfaces. Install the chain cover within 3 minutes of applying seal packing. Do not add engine oil for at least 2 hours after installing the chain cover.**

12. Apply adhesive to the threads of the bolt "A". Use Toyota Genuine Adhesive 1324, Three Bond 1324 or equivalent.
13. Temporarily install the timing chain cover with the 12 bolts and 2 nuts.:
    a. Bolt A: 1.18 inches (10 mm head)
    b. Bolt B: 1.18 inches (12 mm head
    c. Bolt C: 1.57 inches (14 mm head)
14. Temporarily install the engine transverse engine mounting bracket with the 3 bolts.

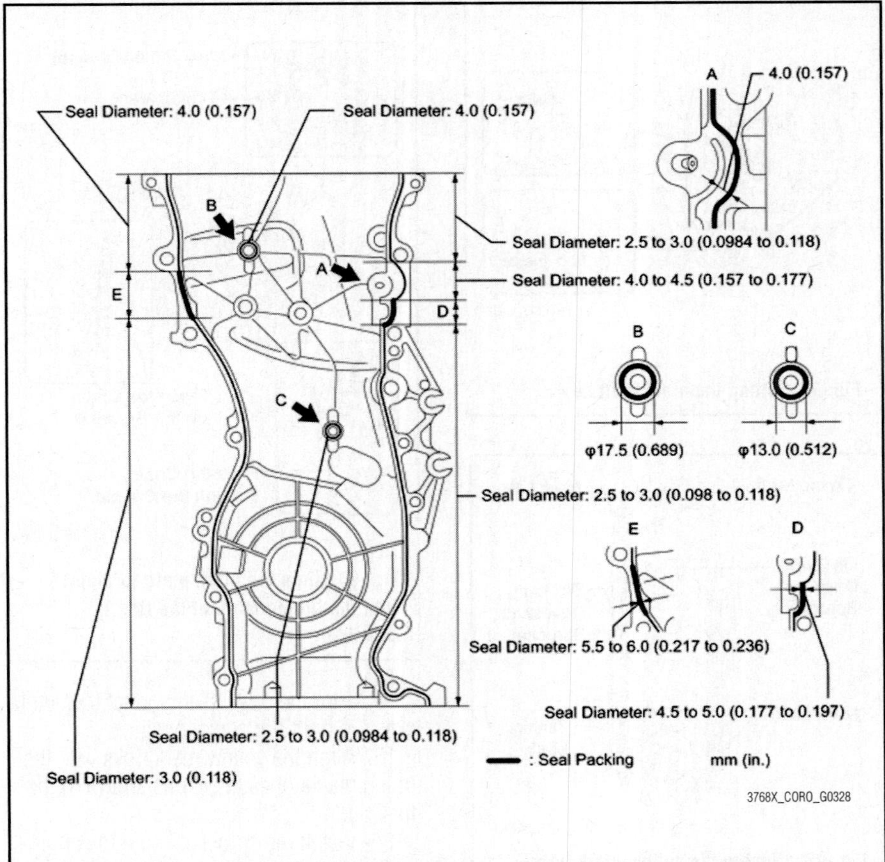

Fig. 86 Applying seal packing to the timing chain cover sub-assembly in a continuous line

15. Fully tighten the timing chain cover with the 15 bolts and 2 nuts as shown in the illustration.
    a. Bolt A: 80 inch lbs. (9 Nm).
    b. Bolt B: 18 ft. lbs. (25 Nm).
    c. Bolt C: 41 ft. lbs. (55 Nm).
    d. Nut: 8 ft. lbs. (11 Nm).

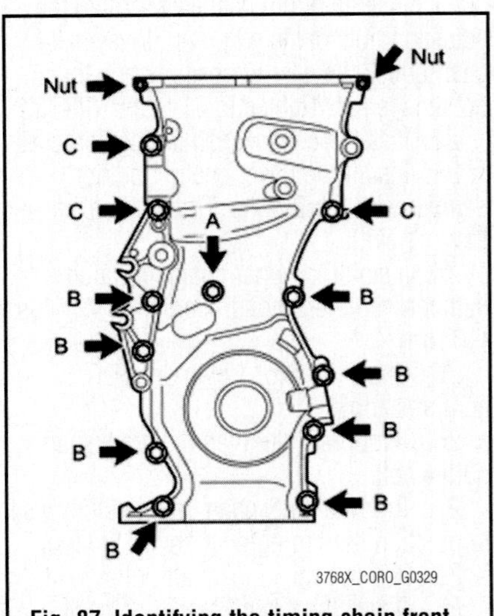

Fig. 87 Identifying the timing chain front cover bolts

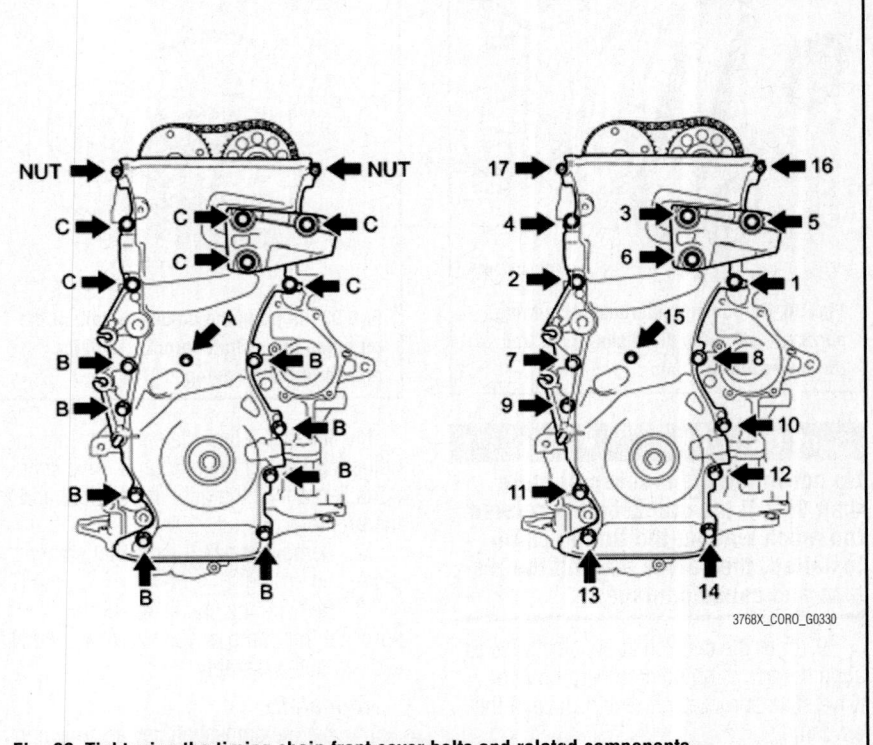

Fig. 88 Tightening the timing chain front cover bolts and related components

TIMING CHAIN & SPROCKETS

REMOVAL & INSTALLATION

1.8L Engine

See Figures 89 through 97.

➡Do not turn the crankshaft without the chain tensioner installed.

1. Remove the 2 nuts, bracket, tensioner and gasket.
2. Remove the chain tensioner slipper.
3. Remove the 2 bolts and LH chain vibration damper.
4. Hold the hexagonal portion of the camshaft with a wrench and turn the camshaft timing gear assembly counterclockwise to loosen the chain between the camshaft timing gears.
5. With the chain loosened, release the chain from the camshaft timing gear assembly and place it on the camshaft timing gear assembly.

➡Be sure to release the chain from the sprocket completely.

6. Turn the camshaft clockwise to return it to the original position and remove the chain.
7. Remove the 2 bolts and No. 2 chain vibration damper.
8. Remove the crankshaft timing sprocket.

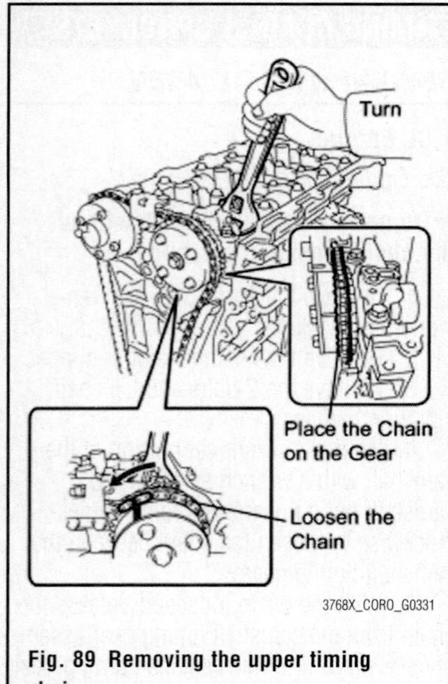

Fig. 89 Removing the upper timing chain

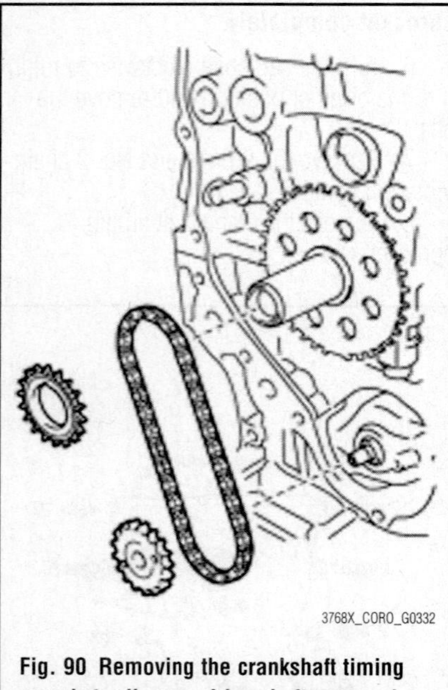

Fig. 90 Removing the crankshaft timing sprocket, oil pump drive shaft gear and the lower timing chain

⁂ CAUTION

Do not rotate the crankshaft more than 90°. If the crankshaft is rotated too much without the timing chain installed, the valves may hit the pistons and cause damage.

9. Turn the crankshaft 90°clockwise to align the adjusting hole of the oil pump drive shaft sprocket with the groove of the oil pump.
10. Remove the crank pulley bolt.

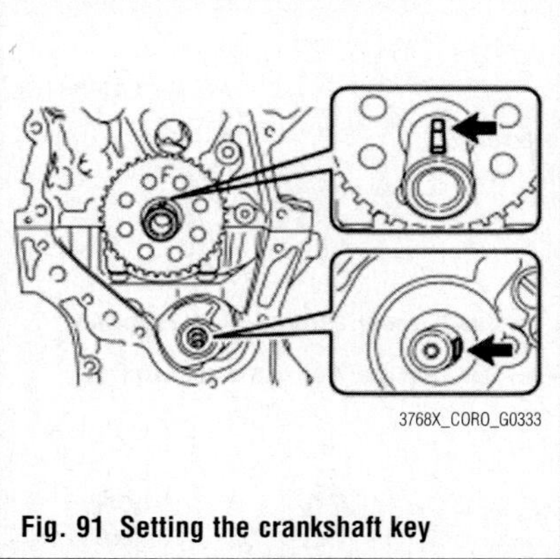

Fig. 91 Setting the crankshaft key

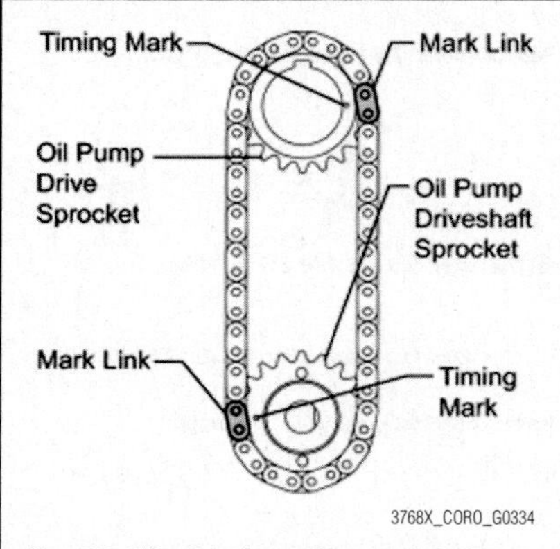

Fig. 92 Aligning the yellow mark links with the timing marks of each gear

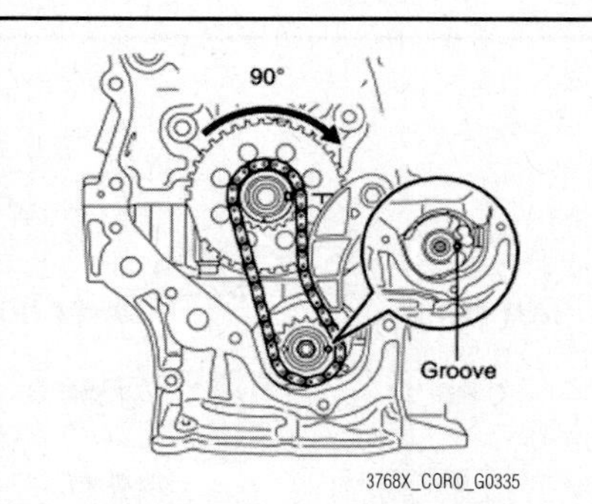

Fig. 93 Aligning the adjusting hole of the oil pump drive shaft sprocket with the groove of the oil pump

11. Insert a 3 mm diameter bar into the adjusting hole of the oil pump drive shaft sprocket to lock the gear in position, and then remove the nut.
12. Remove the bolt, chain tensioner plate, and spring.
13. Remove the crankshaft timing sprocket, oil pump drive shaft gear, and No. 2 chain sub-assembly.

To install:

14. Set the crankshaft key as shown in the illustration.

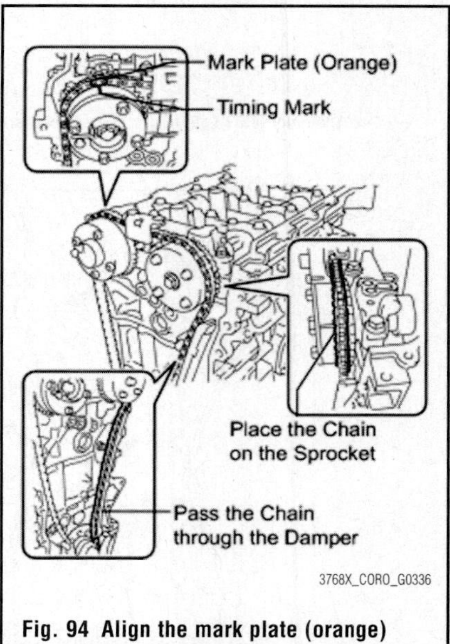

Fig. 94 Align the mark plate (orange) with the timing mark of the No. 2 camshaft

15. Turn the drive shaft so that the cutout faces the right horizontal position.
16. Align the yellow mark links with the timing marks of each gear as shown in the illustration.
17. Install the sprockets onto the crankshaft and oil pump shaft with the chain on the gears.
18. Temporarily tighten the oil pump drive shaft sprocket with the nut.
19. Insert the damper spring into the adjusting hole, and then install the chain tensioner plate with the bolt. Tighten to 7 ft. lbs. (10 Nm).
20. Align the adjusting hole of the oil pump drive shaft sprocket with the groove of the oil pump.
21. Insert a 3 mm diameter bar into the adjusting hole of the oil pump drive shaft gear to lock the gear in position, and then tighten the nut. Tighten to 21 ft. lbs. (28 Nm).
22. Install the crankshaft timing sprocket.
23. Install the lower chain vibration damper with the 2 bolts. Tighten to 16 ft. lbs. (21 Nm).
24. Install the upper chain vibration damper with the 2 bolts. Tighten to 7 ft. lbs. (10 Nm).
25. Check the No. 1 cylinder TDC/ compression.
26. Temporarily tighten the crankshaft pulley bolt.
27. Turn the crankshaft counterclockwise to position the timing gear key to the top.
28. Remove the crankshaft pulley bolt.
29. Check the timing marks on each camshaft timing gear.

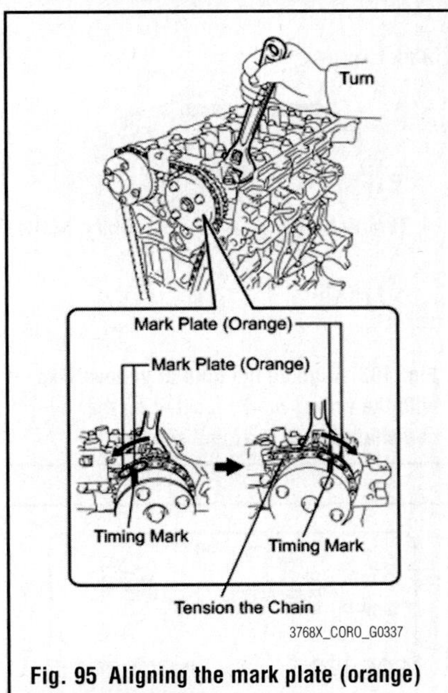

Fig. 95 Aligning the mark plate (orange) and timing mark

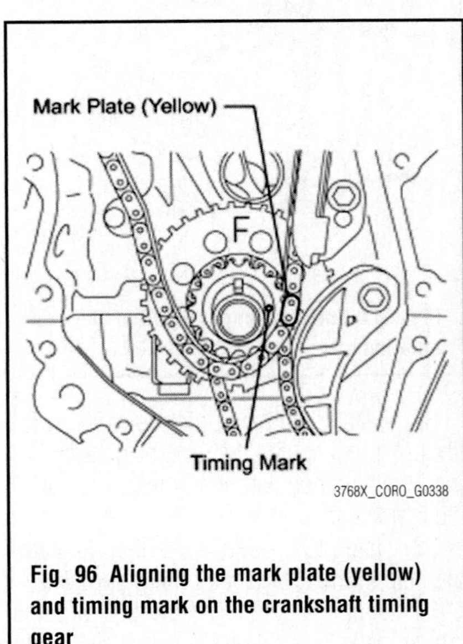

Fig. 96 Aligning the mark plate (yellow) and timing mark on the crankshaft timing gear

30. Align the mark plate (orange) with the timing mark of the No. 2 camshaft as shown in the illustration and install the chain.

➡Be sure to position the mark plate at the front of the engine. The mark plate on the camshaft side is colored orange. Do not pass the chain around the sprocket of the camshaft timing gear assembly. Only place it on the sprocket. Pass the chain through the No. 1 vibration damper.

31. Place the chain on the crankshaft without passing it around the shaft.
32. Hold the hexagonal portion of the camshaft with a wrench and turn the camshaft timing gear assembly counterclockwise to align the mark plate (orange) and timing mark.

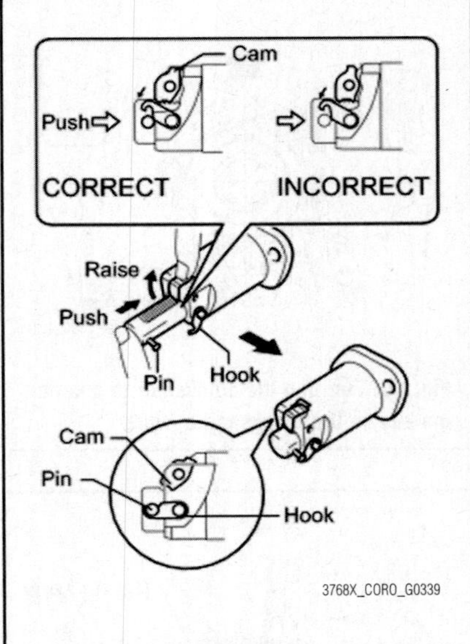

Fig. 97 Correct plunger position and orientation.

➡Be sure to position the mark plate at the front of the engine. The mark plate on the camshaft side is colored orange.

33. Hold the hexagonal portion of the camshaft with a wrench and turn the camshaft timing gear assembly clockwise.

➡To tension the chain, slowly turn the camshaft timing gear assembly clockwise to prevent the chain from being misaligned.

34. Align the mark plate (yellow) and timing mark and install the chain to the crankshaft timing gear.

➡The mark plate on the crankshaft side is colored yellow.

35. Recheck each timing mark at TDC/compression.
36. Install the chain tensioner slipper.
37. Install the upper chain tensioner assembly:
    a. Release the ratchet pawl, then fully push in the plunger and engage the hook to the pin so that the plunger is in the position.

➡Make sure that the cam engages the first tooth of the plunger to allow the hook to pass over the pin.

    b. Install a new gasket, bracket and No. 1 chain tensioner with the 2 nuts.

➡If the hook releases the plunger while the chain tensioner is being installed, engage the hook again.

    c. Turn the crankshaft counterclockwise, then disconnect the hook from plunger knock pin.
    d. Turn the crankshaft clockwise, then check that the plunger is extended.

38. To complete installation, reverse remaining removal procedure.

2.4L Engine

See Figures 98 through 104.

1. Turn the crankshaft pulley until the groove and the timing mark "0" on the timing chain cover are aligned.
2. Check that each timing mark on the camshaft timing gear and sprocket is aligned with the timing marks located on the No. 1 and No. 2 bearing caps.
3. Remove the crankshaft pulley.

➡Do not turn the crankshaft without the chain tensioner installed.

➡Do not lift the engine more than necessary.

4. Remove the 2 nuts, bracket, tensioner and gasket.
5. Lift the engine upward.
6. Remove the bolt, nut and V-ribbed belt tensioner assembly.
7. Remove the crankshaft sensor.
8. Remove the timing chain and sprocket cover.
9. Remove the No. 1 crankshaft position sensor plate.
10. Remove the bolt and timing chain guide.
11. Remove the bolt and chain tensioner slipper.

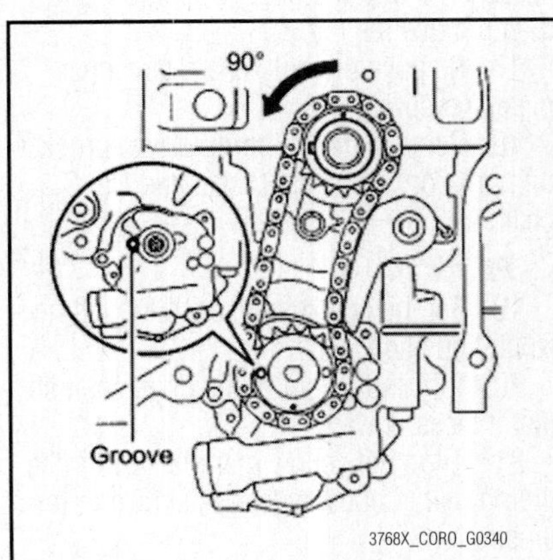

Fig. 98 Align the adjusting hole on the oil pump drive shaft sprocket with the groove on the oil pump

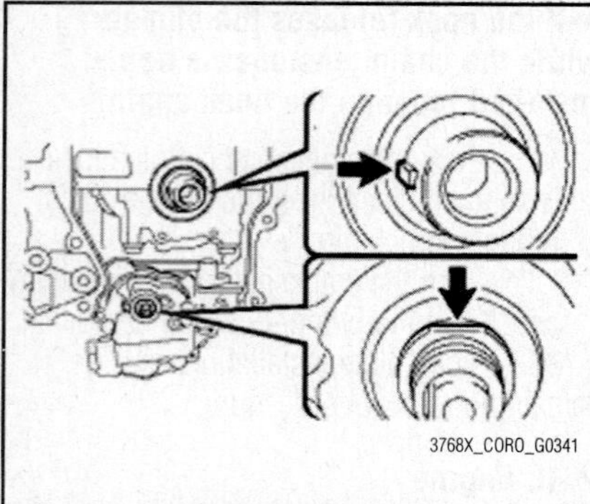

Fig. 99 Setting the crankshaft key in the left horizontal position

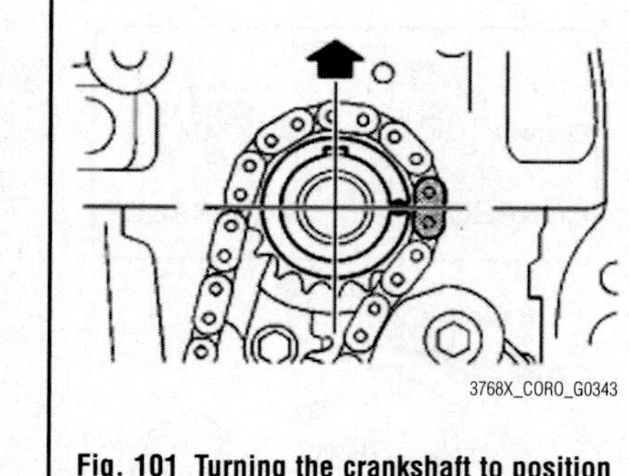

Fig. 101 Turning the crankshaft to position the key on the crankshaft upward

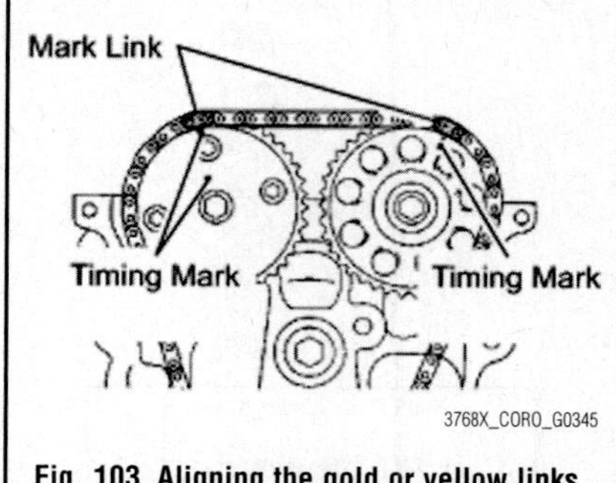

Fig. 103 Aligning the gold or yellow links with the timing marks located on the camshaft timing gear and sprocket

Fig. 100 Aligning the yellow mark links with the timing marks of each gear

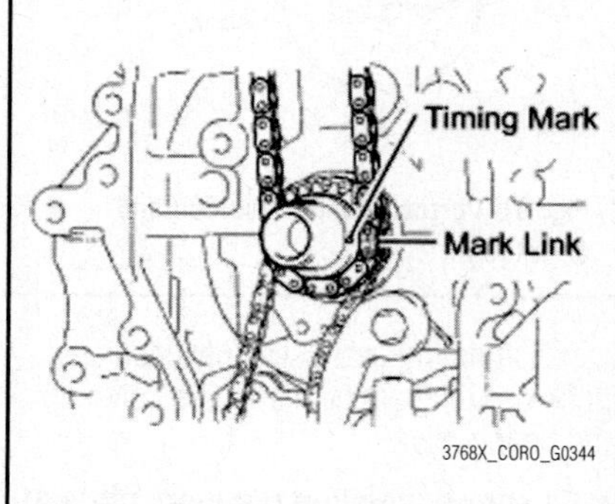

Fig. 102 Install the chain onto the crankshaft timing sprocket

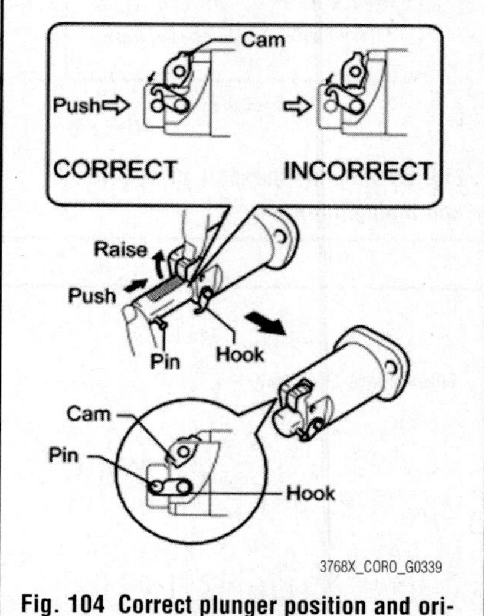

Fig. 104 Correct plunger position and orientation.

12. Remove the 2 bolts and No. 1 chain vibration damper.

13. Remove the chain sub-assembly.

14. Remove the crankshaft timing gear or sprocket from the crankshaft.

15. Turn the crankshaft 90° counterclockwise to align the adjusting hole on the oil pump drive shaft sprocket with the groove on the oil pump.

16. Insert a 4 mm diameter bar into the adjusting hole of the oil pump drive shaft sprocket to lock the gear in position, and then remove the nut.

17. Remove the bolt, chain tensioner plate and spring.

18. Remove the oil pump drive sprocket, oil pump drive shaft sprocket and No. 2 chain.

To install:

19. Set the crankshaft key in the left horizontal position.

20. Turn the cutout of the drive shaft so that it faces upward.

21. Align the yellow mark links with the timing marks of each gear as shown in the illustration.

22. Install the sprockets onto the crankshaft and oil pump shaft with the chain wrapped on the gears.

23. Temporarily tighten the oil pump drive shaft sprocket with the nut.

24. Insert the damper spring into the adjusting hole, and then install the chain tensioner plate with the bolt. Tighten to 9 ft. lbs. (12 Nm).

25. Align the adjusting hole on the oil pump drive shaft sprocket with the groove on the oil pump.

26. Insert a 4 mm diameter bar into the adjusting hole on the oil pump drive shaft gear to lock the gear in position, and then tighten the nut. Tighten to 22 ft. lbs. (30 Nm).

27. Install the crankshaft timing gear or sprocket to the crankshaft.

28. Install the No. 1 chain vibration damper with the 2 bolts. Tighten to 80 inch lbs. (9 Nm).

29. Set the No. 1 cylinder to TDC/compression.

30. Turn the camshafts with a wrench (using the hexagonal lobe) to align the timing marks on the camshaft timing gear with the timing marks located on the No. 1 and No. 2 bearing caps

31. Using the crankshaft pulley bolt, turn the crankshaft to position the key on the crankshaft upward.

32. Install the chain onto the crankshaft timing sprocket with the gold or orange mark link aligned with the timing mark on the crankshaft.

33. Using SST 09309-37010 and a hammer, tap in the crankshaft timing sprocket.

34. Align the gold or yellow links with the timing marks located on the camshaft timing gear and sprocket, then install the chain.

35. Install the chain tensioner slipper with the bolt. Tighten to 14 ft. lbs. (19 Nm).

36. Install the timing chain guide with the bolt. Tighten to 80 inch lbs. (9 Nm).

37. Install the sensor plate with the "F" mark facing forward.

38. Install the timing chain cover.

39. Release the ratchet pawl, then fully push in the plunger and set the hook to the pin so that the plunger is in the position.

40. Install a new gasket and the chain tensioner with the 2 nuts. Tighten to 80 inch lbs. (9 Nm).

41. To complete installation, reverse remaining removal procedure.

ENGINE PERFORMANCE & EMISSION CONTROLS

ACCELERATOR PEDAL POSITION (APP) SENSOR

LOCATION

See Figure 105.

Refer to the accompanying illustration for sensor location.

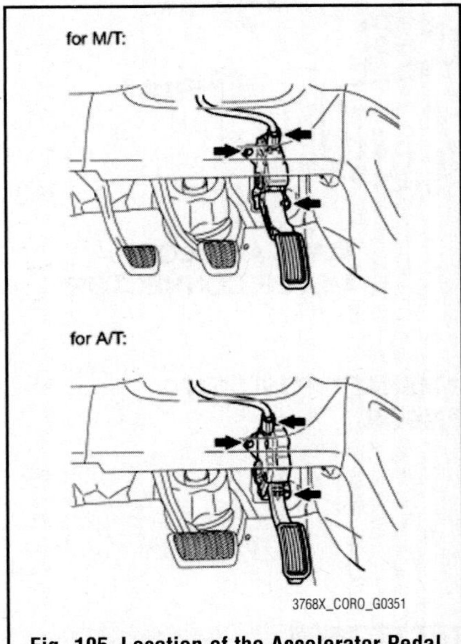

Fig. 105 Location of the Accelerator Pedal Position (APP) sensor

REMOVAL & INSTALLATION

1. Disconnect the negative battery cable.
2. Disconnect the Accelerator Pedal Position (APP) sensor connector.
3. Remove the 2 bolts and accelerator pedal..
4. To install, reverse removal procedure.

CAMSHAFT POSITION (CMP) SENSOR

LOCATION

See Figures 106 and 107.

Refer to the accompanying illustrations for sensor location.

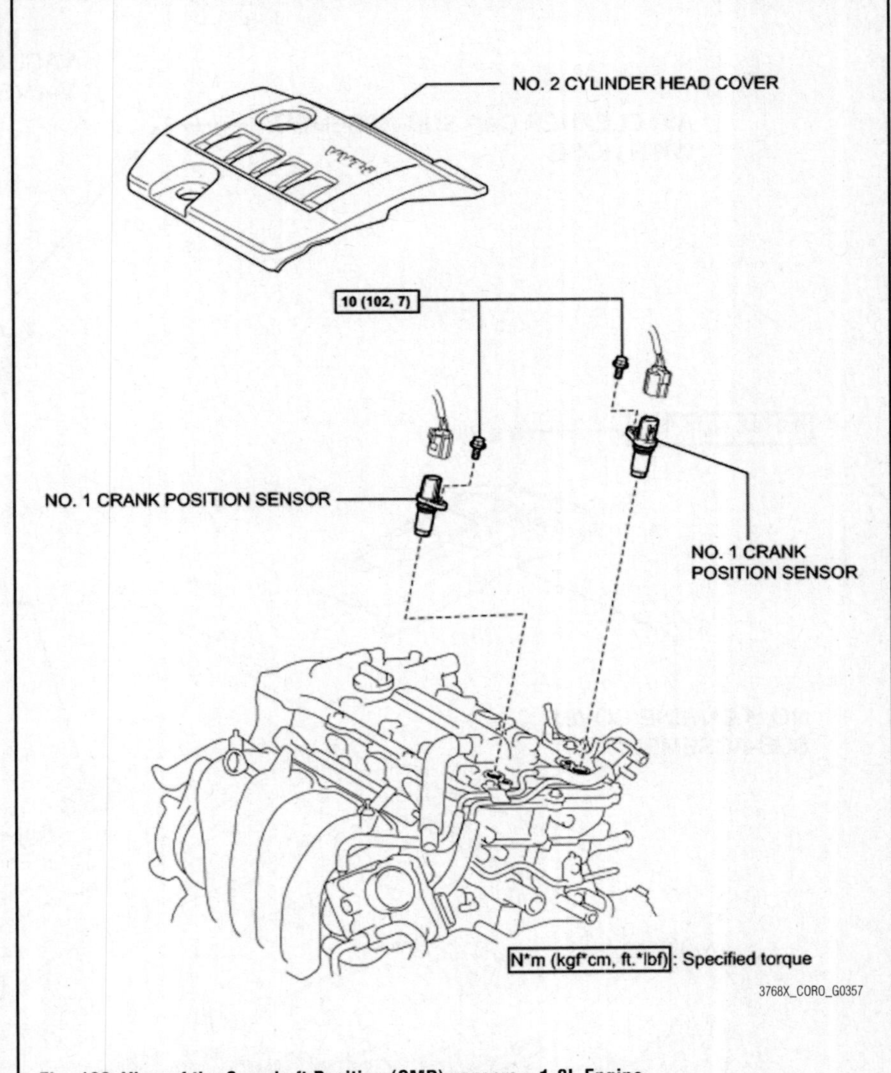

Fig. 106 View of the Camshaft Position (CMP) sensors—1.8L Engine

REMOVAL & INSTALLATION

1.8L Engine

➡**Although the part name refers to the No. 1 crank position sensors, this procedure is for the camshaft position sensors.**

1. Disconnect the negative battery cable.
2. Exhaust camshaft side:
   a. Disconnect the duty vacuum switching valve connector and 3 engine wire harness clamps.
   b. Disconnect the No. 1 crank position sensor connector.
   c. Remove the bolt and No. 1 crank position sensor.
3. Intake camshaft side:
   a. Disconnect the No. 1 crank position sensor connector.
   b. Remove the bolt and No. 1 crank position sensor.
4. To install, reverse removal procedure.
5. Apply a light coat of engine oil to the O-rings on the No. 1 crank position sensors.
6. Tighten the CMP sensor bolts to 7 ft. lbs. (10 Nm).
7. Inspect for oil leaks.

2.4L Engine

1. Disconnect the negative battery cable.
2. Air cleaner cap sub-assembly with hose.
3. Disconnect the camshaft position sensor connector.
4. Remove the bolt and the camshaft position sensor.
5. To install, reverse removal procedure.
6. Apply a light coat of engine oil to the O-ring of the sensor.
7. Tighten the camshaft position sensor with the bolt to 80 inch lbs. (9 Nm).

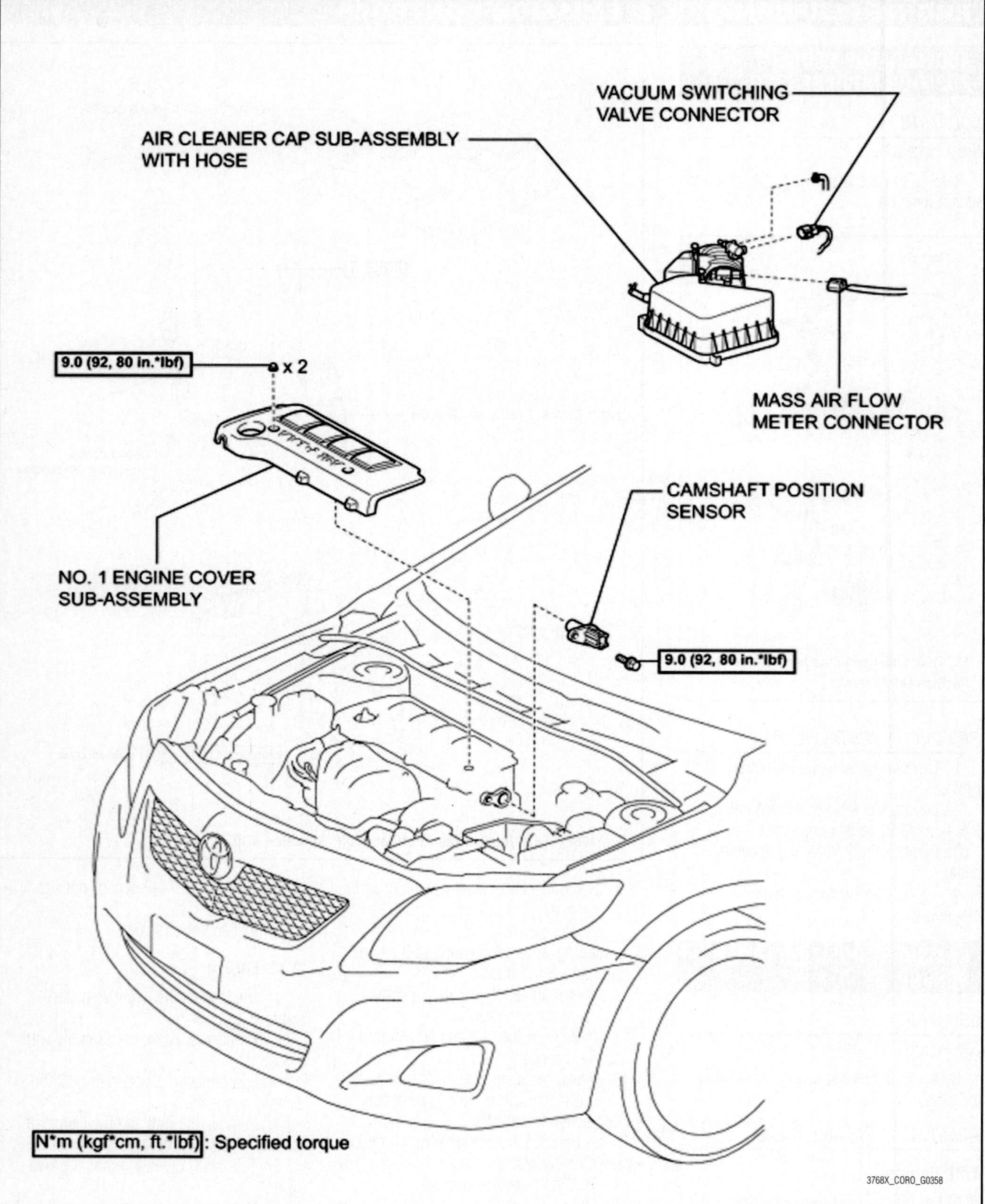

Fig. 107 View of the Camshaft Position (CMP) sensors—2.4L Engine

CRANKSHAFT POSITION (CKP) SENSOR

LOCATION

See Figures 108 and 109.

Refer to the accompanying illustrations for sensor location.

REMOVAL & INSTALLATION

1.8L Engine

1. Disconnect the negative battery cable.
2. Remove the RH engine under cover.
3. Disconnect the crank position sensor connector.
4. Remove the bolt and crank position sensor.
5. To install, apply a light coat of engine oil to the O-ring on the crank position sensor.
6. Install the crank position sensor with the bolt. Tighten to 7 ft. lbs. (10 Nm).

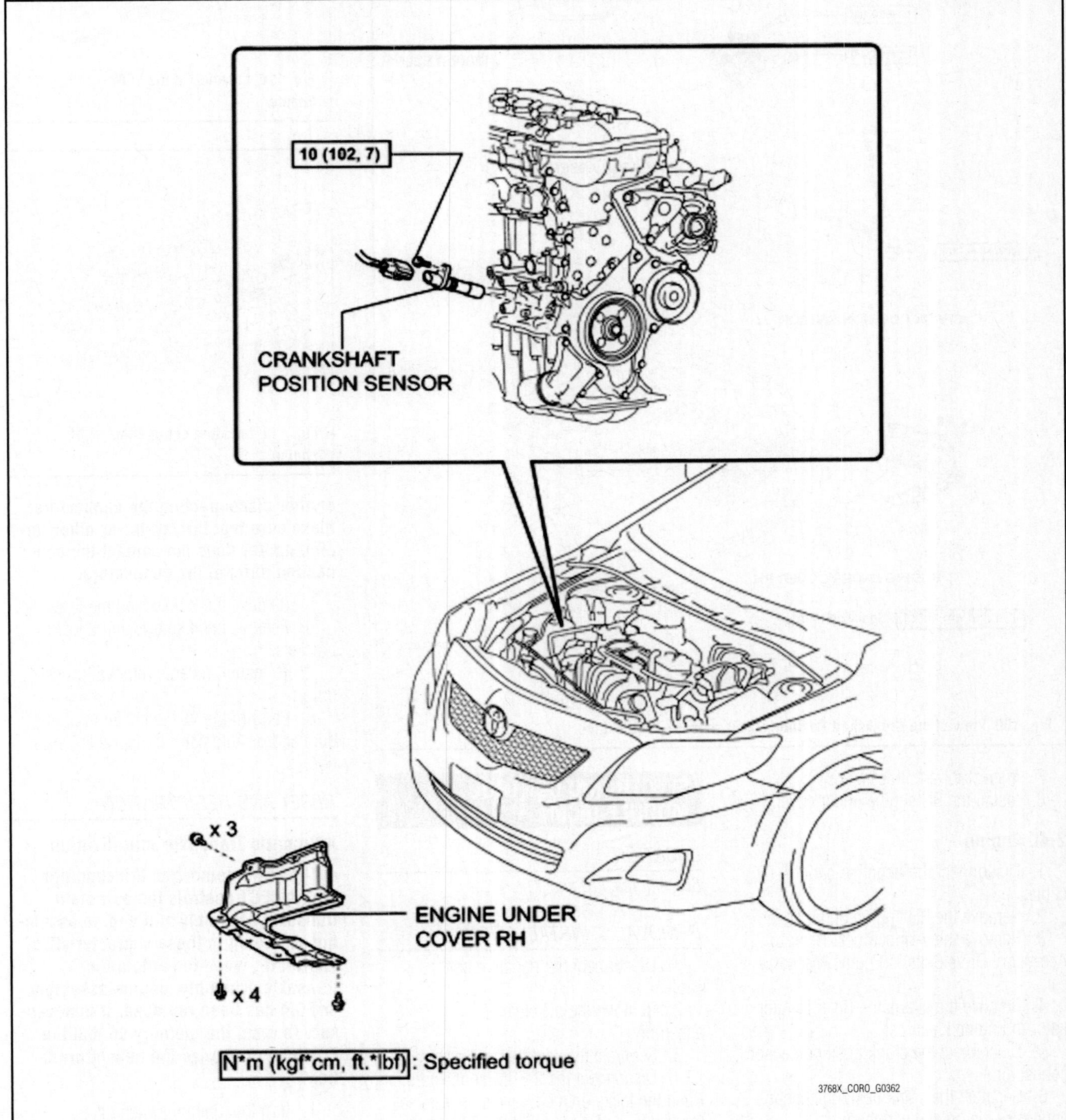

Fig. 108 View of the Crankshaft Position (CKP) sensor—1.8L Engine

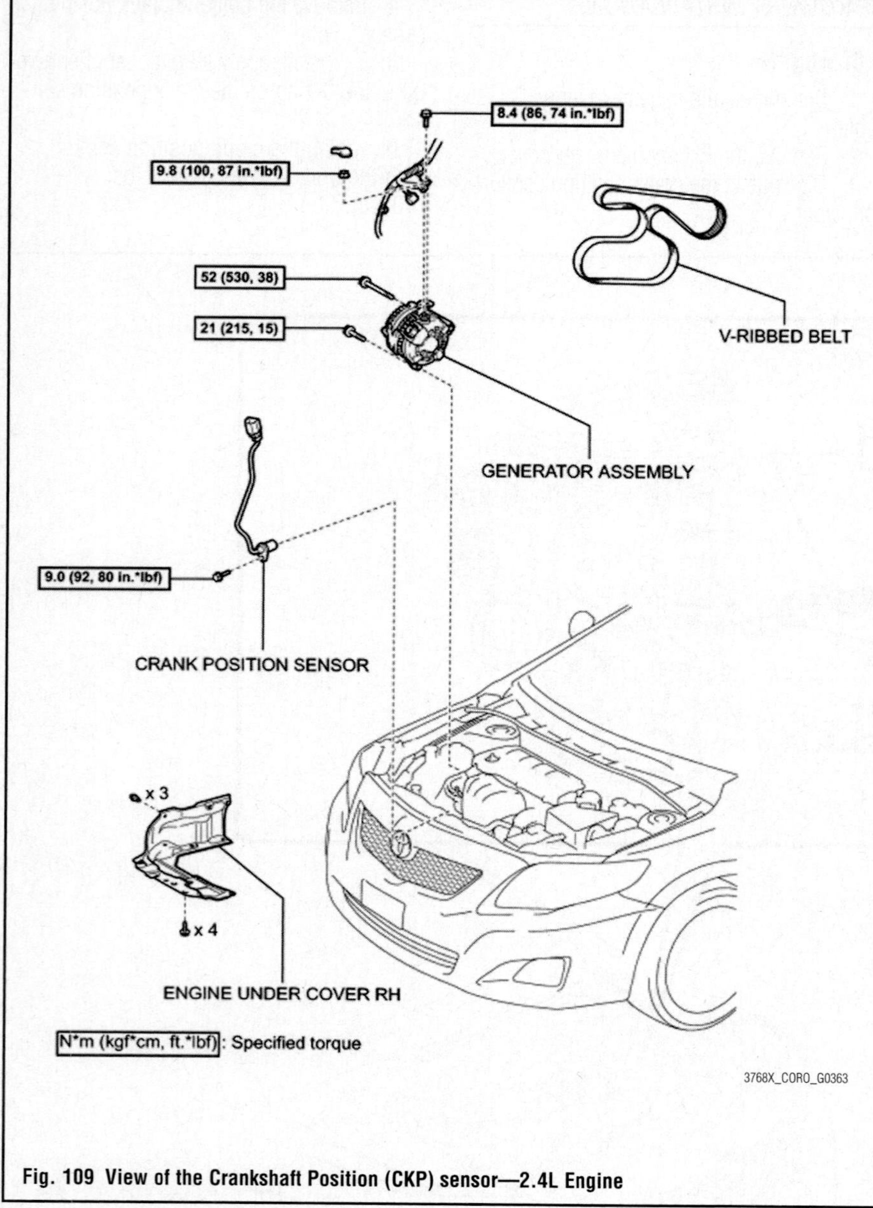

Fig. 109 View of the Crankshaft Position (CKP) sensor—2.4L Engine

7. Inspect for an oil leak.
8. Install the RH engine under cover.

2.4L Engine

1. Disconnect the negative battery cable.
2. Remove the RH engine under cover.
3. Remove the v-ribbed belt. Refer to Accessory Drive Belts in Engine Mechanical.
4. Remove the alternator. Refer to Alternator in Engine Electrical.
5. Disconnect the crank position sensor connector.
6. Separate the crank position sensor connector clamp and wire harness.
7. Remove the bolt and crank position sensor.

ELECTRONIC CONTROL MODULE (ECM)

LOCATION

See Figures 110 and 111.

REMOVAL & INSTALLATION

1. Disconnect the negative battery cable.
2. Remove the air cleaner assembly.
3. Separate the wire harness clamp.
4. Disconnect the 2 ECM connectors. Push the locks on the 2 levers, then raise the levers, and disconnect the 2 ECM connectors.

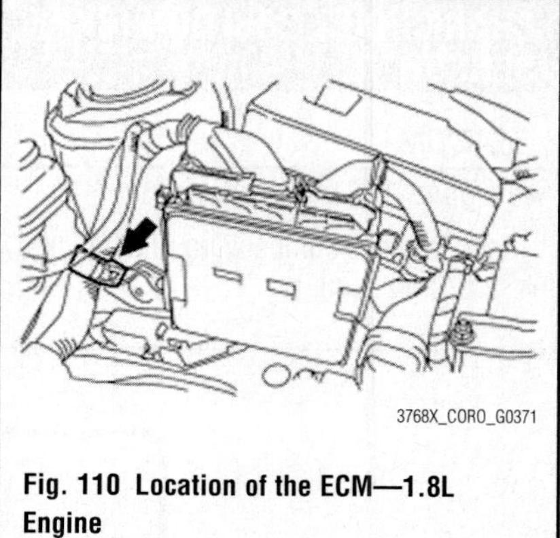

Fig. 110 Location of the ECM—1.8L Engine

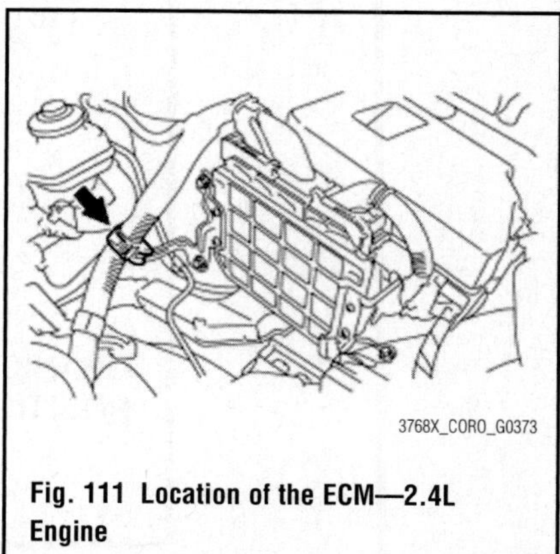

Fig. 111 Location of the ECM—2.4L Engine

➡After disconnecting the connectors, make sure that dirt, water or other foreign matter does not contact the connecting parts of the connectors.

5. Remove the 3 bolts and the ECM.
6. Remove the 4 screws and 2 ECM brackets.
7. To install, reverse removal procedure.
8. Perform the VIN registration procedure and/or Automatic Transaxle Initialization.

RESET AND REGISTRATION

Automatic Transaxle Initialization

➡The ECM memorizes the condition that the ECT controls the automatic transaxle assembly and engine assembly according to those characteristics. Therefore, when the automatic transaxle assembly, engine assembly, or ECM has been replaced, it is necessary to reset the memory so that the ECM can memorize the new information.

1. Turn the ignition switch off.
2. Connect the Techstream to the DLC3.

3. Turn the ignition switch to ON and the Techstream main switch on.

4. Enter the following menus: Powertrain / Engine and ECT / Utility / Reset Memory. Then, press "Next".

CAUTION

After performing the reset memory, be sure to perform the road test.

5. The ECM is learned by performing the road test.

VIN Registration

➡The Vehicle Identification Number (VIN) must be input into a replacement ECM.

➡The VIN is a 17-digit alphanumeric vehicle identification number. The Techstream is required to register the VIN.

➡This registration section consists of two parts: Read VIN and Write VIN.

1. Read VIN: This process allows the VIN stored in the ECM to be read in order to confirm that the two VINs, provided with the vehicle and stored in the vehicle's ECM, are the same.

2. Write VIN: This process allows the VIN to be input into the ECM. If the ECM is changed, or the ECM VIN and vehicle VIN do not match, the VIN can be registered, or overwritten in the ECM by following this procedure.

3. Read the VIN:
    a. Confirm the vehicle VIN.
    b. Connect the Techstream to the DLC3.
    c. Turn the ignition switch to ON.
    d. Turn the Techstream on.
    e. Enter the following menus: Powertrain / Engine and ECT / Utility / VIN / VIN Read.

4. Write the VIN:
    a. Confirm the vehicle VIN.
    b. Connect the Techstream to the DLC3.
    c. Turn the ignition switch to ON.
    d. Turn the Techstream on.
    e. Enter the following menus: Powertrain / Engine and ECT / Utility / VIN / VIN Write.

ROAD TEST

1. Perform the test at the ATF (Automatic Transmission Fluid) temperature 50 to 80 °C (122 to 176 °F) in the normal operation.

a. D position test: Shift into the D position and fully depress the accelerator pedal and check the following points.

- Check up-shift operation. Check that 1 → 2, 2 → 3, 3 → 4 and 4 → 5th up-shifts take place, and that the shift points conform to the automatic shift schedule.

➡5th Gear Up-shift Prohibition Control: Engine coolant temperature is 55°C (131°F) or less and vehicle speed is at 70km/h (43 mph) or less. ATF temperature is -2°C (28°F) or less.

➡5th and 4th Gear Lock-up Prohibition Control: Brake pedal is depressed. Accelerator pedal is released. Engine coolant temperature is 60°C (140°F) or less.

- Check for shift shock and slip. Check for shock and slip at the 1 → 2, 2 → 3, 3 → 4, and 4→ 5th up-shifts.
- Check for abnormal noise and vibration. Check for abnormal noise and vibration when up-shifting from 1 → 2, 2 → 3, 3 → 4, and 4→ 5 while driving with the shift lever in the D position, and also check while driving in the lock-up condition. The check for the cause of abnormal noise and vibration must be done thoroughly as it could also be due to loss of balance in the differential, torque converter clutch, etc.
- Check kick-down operation. Check vehicle speeds when the 2nd to 1st, 3rd to 2nd, 4th to 3rd, and 5th to 4th kick-downs take place while driving with the shift lever in the D position. Confirm that each speed is within the applicable vehicle speed range indicated in the automatic shift schedule.
- Check for abnormal shock and slip at kick-down.
- Check the lock-up mechanism: Drive in the D position (5th gear), at a steady speed (lock-up ON). Lightly depress the accelerator pedal and check that the engine speed does not change abruptly.

➡There is no lock-up function in the 1st, 2nd and 3rd gears. If there is a big jump in engine speed, there is no lock-up.

b. S position test: Shift to the S position, depress the accelerator pedal and check the following points:

- Check shift operation.
- While driving in the D position and 5th gear, shift into the S position and back to the D position. Check that the gear change 5 → 4down-shift and 4 → 5up-shift can be performed.
- With the shift lever in the S position (while the vehicle is stopped), shift into the "+" position to check that the shift position on the combination meter changes as follows: 1 → 2, 2 → 3, 3 → 4, and 4 → 5.
- While driving in the 4(S) position and 4th gear (at a vehicle speed of approximately 40 to 50 km/h (25 to 31mph)), shift into the "-" position and check if the 3rd gear down-shift occurs and the engine brake performs properly.
- While driving in the 3(S) position and 3rd gear (at a vehicle speed of approximately 30 to 40 km/h (19 to 25 mph)), shift into the "-" position and check if the 2nd gear down-shift occurs and the engine brake performs properly.
- While driving in the 2(S) position and 2nd gear (at a vehicle speed of approximately 20 to 30 km/h (12 to 19 mph)), shift into the "-" position and check if the 1st gear down-shift occurs and the engine brake performs properly.

➡Manual shift (S position) is prohibited under either of the following conditions: Down-shifting may cause engine overrun. The driver continuously down-shifts. (Down-shifting to 1st gear may not be performed.)

c. R position test: Shift into the R position, lightly depress the accelerator pedal, and check that the vehicle moves backward without any abnormal noise or vibration.

CAUTION

Before conducting this test ensure that the test area is free from people and obstruction.

d. P position test: Stop the vehicle on a grade (more than 5°) and after shifting into the P position, release the parking brake. Then, check that the parking lock pawl holds the vehicle in place.

e. Uphill/downhill control function:

- Check that the gear does not up-shift to the 4th or 5th gear while the vehicle is driving uphill.
- Check that the gear automatically down-shifts from the 5th to 4th or from the 4th to 3rd gear when brake is applied while the vehicle is driving downhill.

ENGINE COOLANT TEMPERATURE (ECT) SENSOR

LOCATION

See Figures 112 and 113.

Refer to the accompanying illustrations for sensor location.

REMOVAL & INSTALLATION

1.8L Engine

1. Disconnect the negative battery cable.
2. Drain engine coolant.
3. Remove the cylinder head cover.
4. Remove air cleaner cap sub-assembly.
5. Disconnect the Engine Coolant Temperature (ECT) sensor connector.
6. Remove the ECT sensor.
7. To install, reverse removal procedure. Tighten the ECT sensor 14 ft. lbs. (20 Nm).

2.4L Engine

1. Disconnect the negative battery cable.
2. Drain engine coolant.
3. Remove the cylinder head cover.
4. Remove air cleaner cap sub-assembly with hose.
5. Remove air cleaner case sub-assembly.
6. Disconnect the Engine Coolant Temperature (ECT) sensor connector.
7. Remove the Engine Coolant Temperature (ECT) sensor and gasket.
8. To install, reverse removal procedure. Use a new gasket and tighten the ECT sensor 14 ft. lbs. (20 Nm).

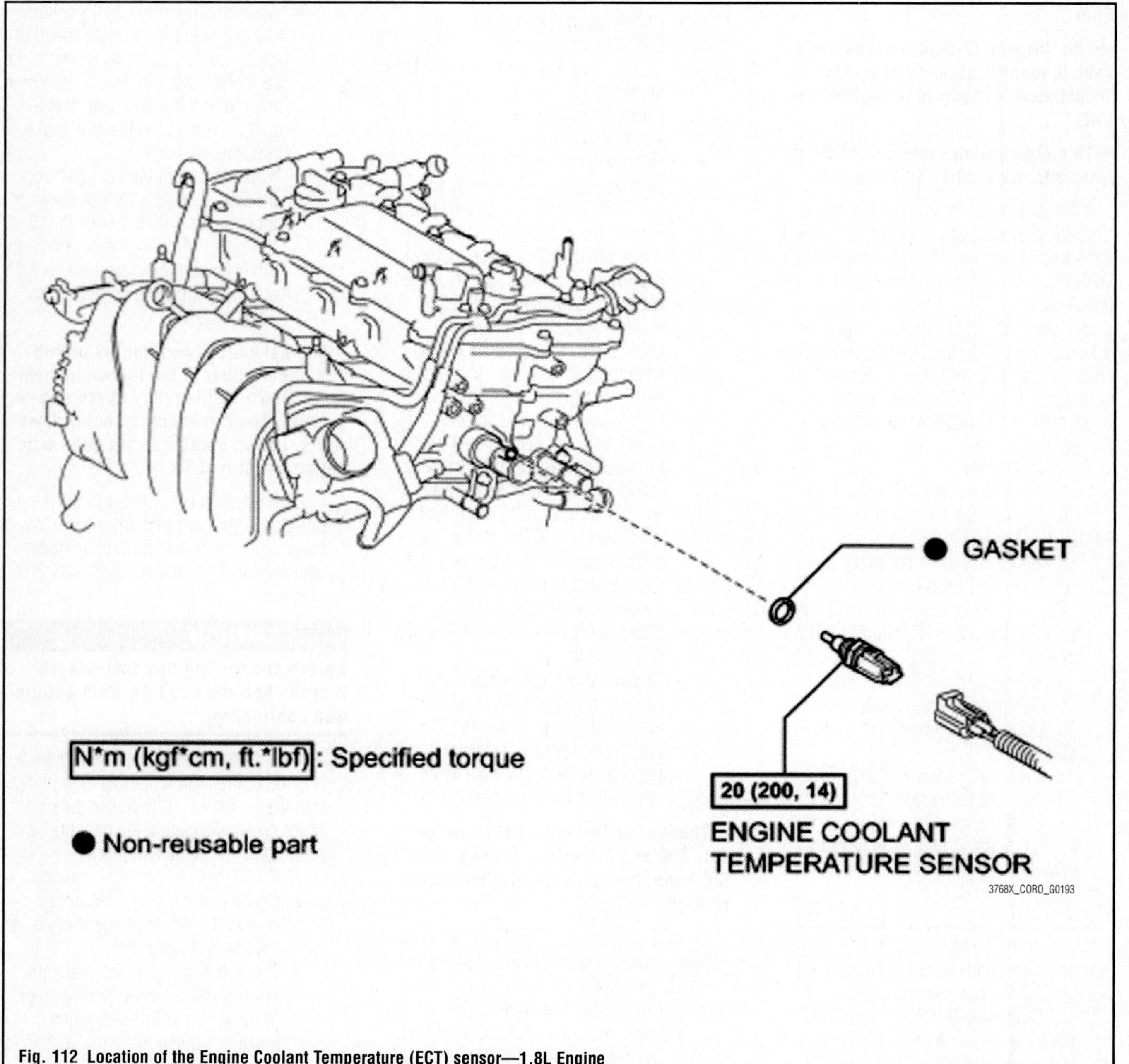

Fig. 112 Location of the Engine Coolant Temperature (ECT) sensor—1.8L Engine

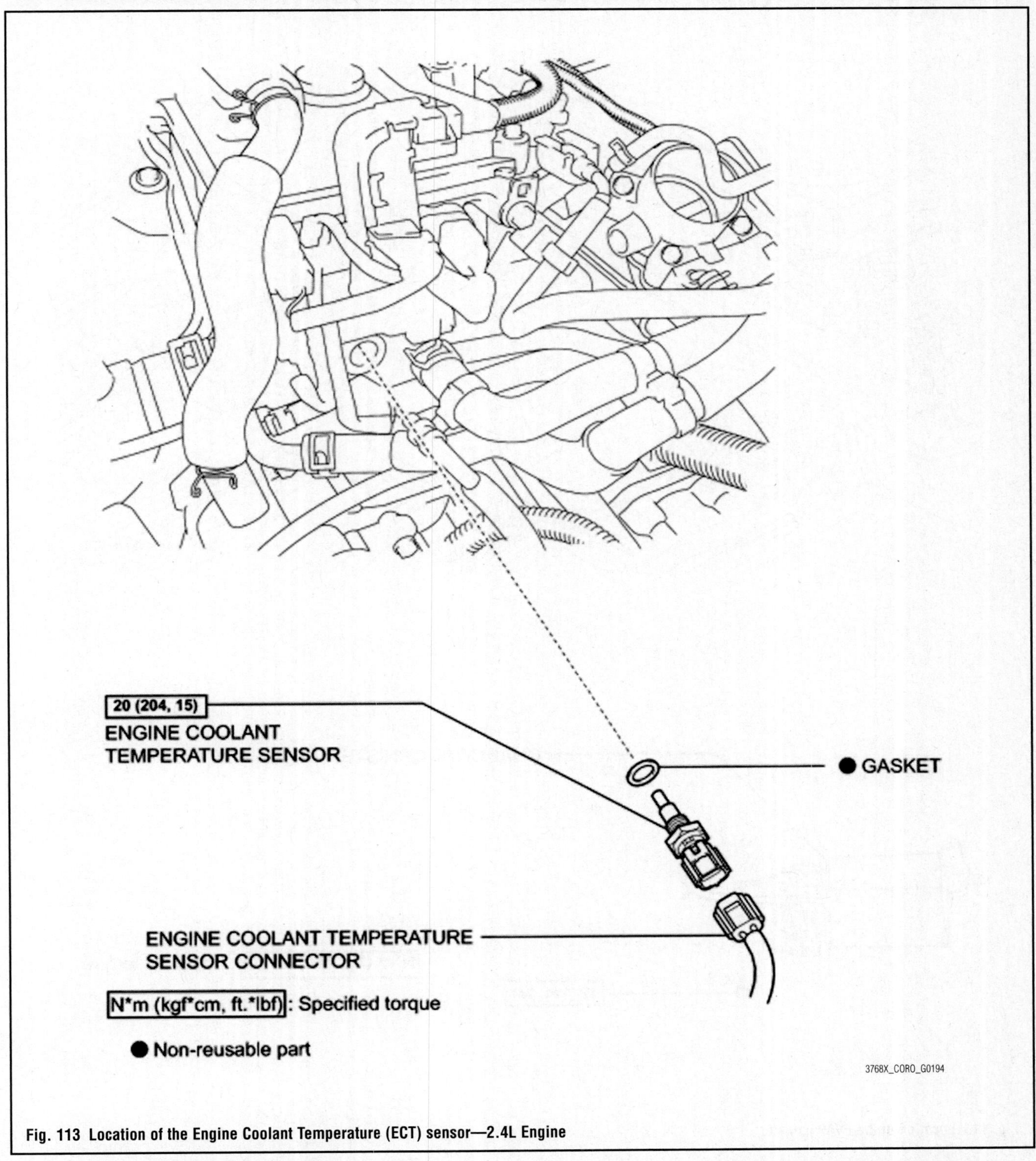

Fig. 113 Location of the Engine Coolant Temperature (ECT) sensor—2.4L Engine

EVAPORATIVE EMISSIONS (EVAP) CANISTER

LOCATION

See Figure 114.

Refer to accompanying illustration for EVAP canister location.

REMOVAL & INSTALLATION

**** CAUTION**

Observe all applicable safety precautions when working around fuel. Whenever servicing the fuel system, always work in a well ventilated area. Do not allow fuel spray or vapors to come in contact with a spark or open flame. Keep a dry chemical fire extinguisher near the work area. Always keep fuel in a container specifically designed for fuel storage; also, always properly seal fuel containers to avoid the possibility of fire or explosion.

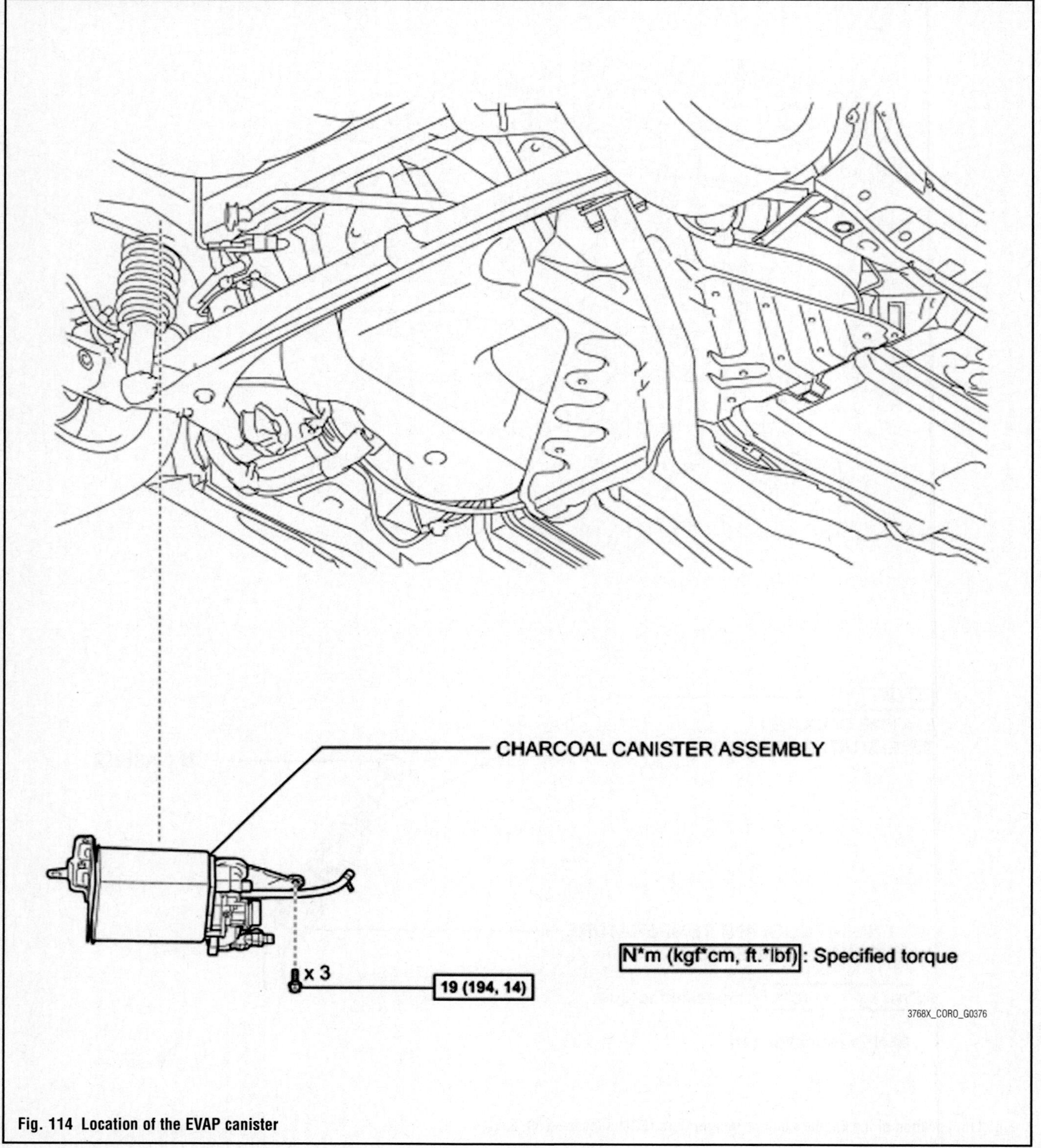

Fig. 114 Location of the EVAP canister

1. Disconnect the fuel tank vent hose from the charcoal canister assembly.
2. Disconnect the vapor pressure sensor connector.
3. Disconnect the wire harness clamp.
4. Disconnect the purge line hose from the charcoal canister assembly.
5. Disconnect the charcoal canister filter sub-assembly from the charcoal canister assembly.
6. Remove the 3 bolts and charcoal canister.
7. To install, reverse removal procedure.
8. Tighten the 3 bolts to 14 ft. lbs. (19 Nm).

HEATED OXYGEN SENSOR (HO2S)

LOCATION

See Figures 115 and 116.

Refer to the accompanying illustrations for sensor location.

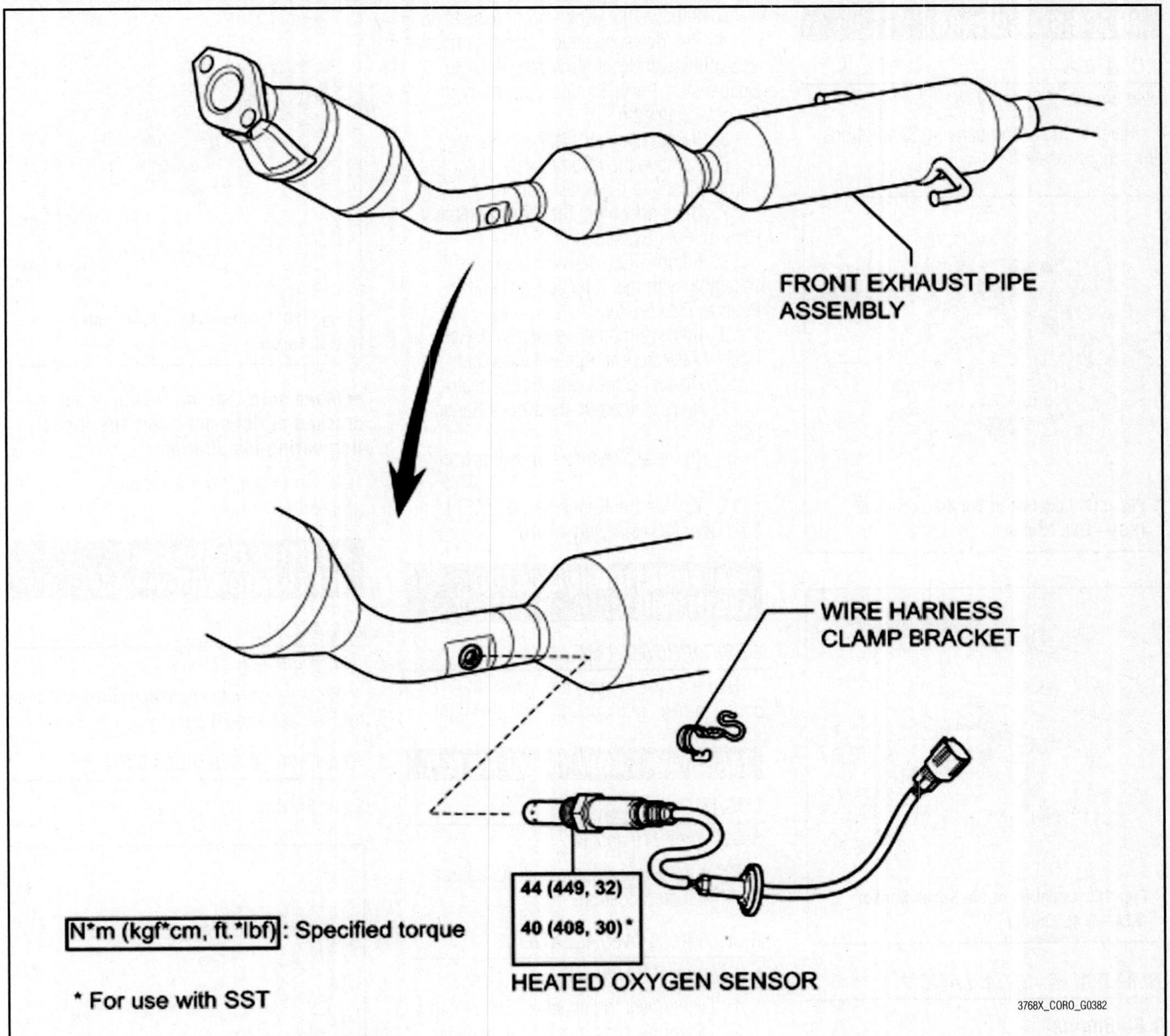

Fig. 115 Location of the HO2S sensor—1.8L Engine

REMOVAL & INSTALLATION

1. Disconnect the negative battery cable.
2. Disconnect the heated oxygen sensor connector.
3. Remove the grommet and pull the sensor connector out of the cabin through the floor panel.
4. Remove the wire harness clamp bracket and disconnect the wire harness clamp.
5. Using a wrench or SST: 09224-00010, remove the Heated Oxygen Sensor (HO2S).
6. To install, reverse removal procedure.
7. Tighten to 32 ft. lbs. (44 Nm).

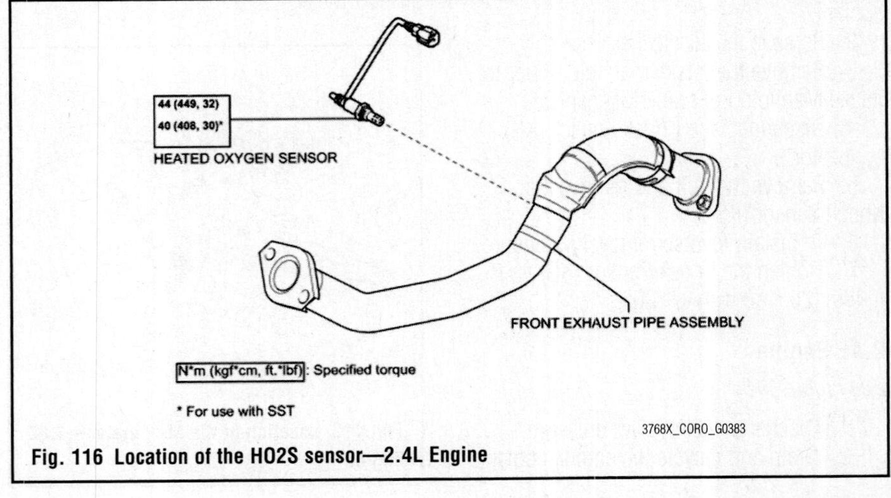

Fig. 116 Location of the HO2S sensor—2.4L Engine

KNOCK SENSOR (KS)

LOCATION

See Figures 117 and 118.

Refer to the accompanying illustrations for sensor location.

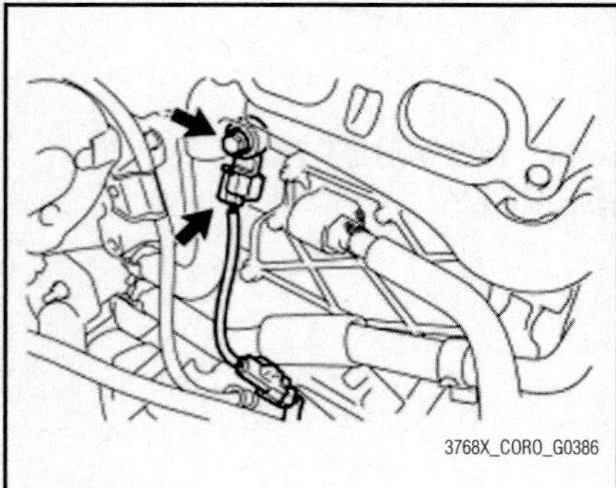

Fig. 117 Location of the Knock Sensor (KS)—1.8L Engine

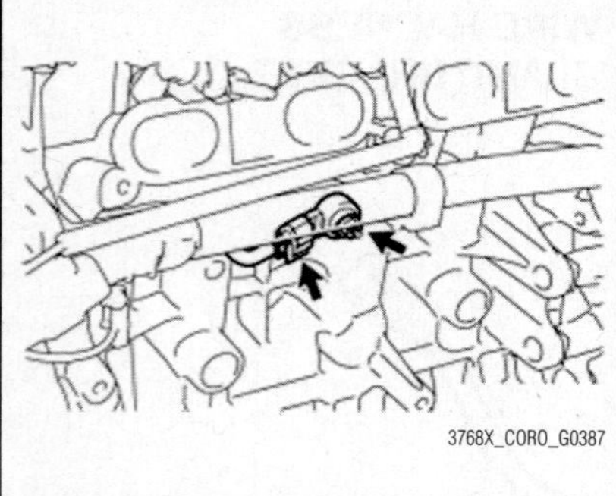

Fig. 118 Location of the Knock Sensor (KS)—2.4L Engine

REMOVAL & INSTALLATION

1.8L Engine

See Figure 117.

1. Drain and recycle the engine coolant.
2. Remove the air cleaner assembly.
3. Remove the intake manifold. Refer to Intake Manifold in Engine Mechanical.
4. Disconnect the Knock Sensor (KS) connector.
5. Remove the bolt and remove the Knock Sensor (KS).
6. To install, reverse removal procedure.
7. Tighten the Knock Sensor (KS) to 15 ft. lbs. (20 Nm) with +/- 20°.

2.4L Engine

See Figure 118.

1. Discharge fuel system pressure.
2. Drain and recycle the engine coolant.
3. Remove the wiper motor assembly.
4. Remove suspension tower damper assembly (with front strut bar). Refer to Suspension Tower Damper Assembly in Front Suspension.
5. Remove the air cleaner assembly.
6. Remove the throttle body. Refer to Throttle Body in the Fuel Section.
7. Disconnect fuel tube sub-assembly. Refer to Fuel System.
8. Remove fuel delivery pipe sub-assembly with fuel tube sub-assembly. Refer to Fuel System.
9. Remove the intake manifold. Refer to Intake Manifold in Engine Mechanical.
10. Disconnect the sensor connector.
11. Remove the nut and Knock Sensor (KS).
12. To install, reverse removal procedure.
13. Tighten the Knock Sensor (KS) to 15 ft. lbs. (20 Nm) with +/- 10°.

MALFUNCTION INDICATOR LIGHT (MIL)

RESET PROCEDURE

Using a ODB II scan tool, clear all DTC codes to reset malfunction indicator light.

MASS AIR FLOW (MAF) METER

LOCATION

See Figures 119 and 120.

Refer to the accompanying illustrations for MAF meter location.

REMOVAL & INSTALLATION

See Figures 119 and 120.

1. Disconnect the negative battery cable.
2. Disconnect the mass air flow meter connector.
3. Remove the 2 screws and the mass air flow meter.

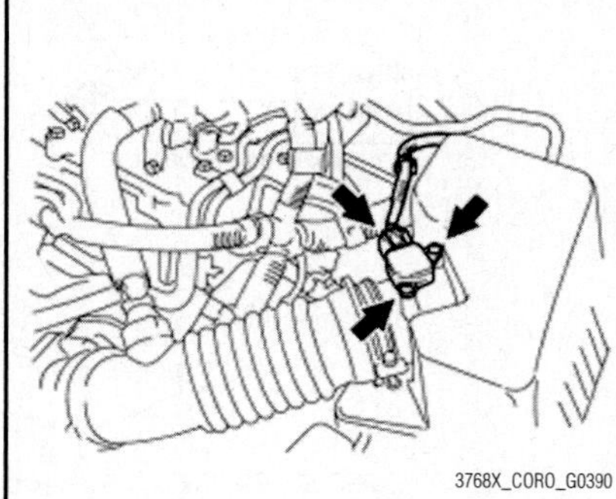

Fig. 119 Location of the MAF meter—1.8L Engine

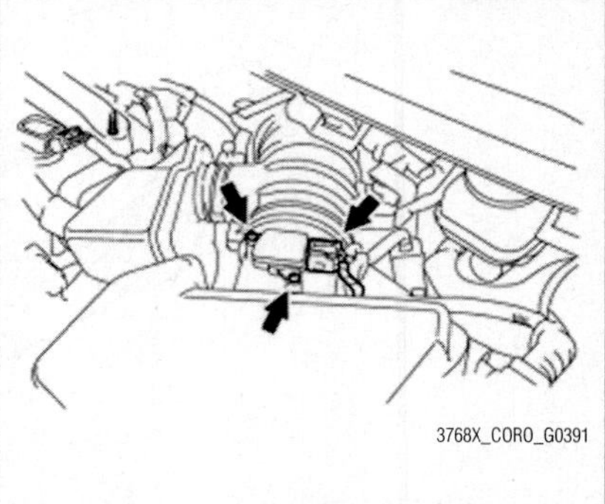

Fig. 120 Location of the MAF meter—2.4L Engine

➡Make sure that the O-ring is not cracked or does not jump out of position during installation.

4. To install, reverse removal procedure.

POSITIVE CRANKCASE VENTILATION (PCV) VALVE

LOCATION

See Figure 121.

Refer to the accompanying illustrations for PCV valve location.

REMOVAL & INSTALLATION

1.8L Engine

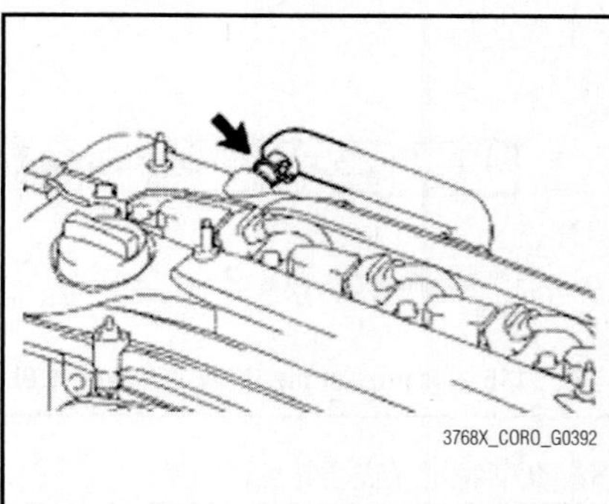

Fig. 121 Location of the PCV valve—2.4L Engine

1. Remove the intake manifold. Refer to Intake Manifold in Engine Mechanical.
2. Using a ball joint lock nut wrench (22 mm), remove the ventilation valve sub-assembly.

To install:

3. To install, reverse the removal procedure.
4. Apply adhesive to 2 or 3 threads of the ventilation valve sub-assembly.

5. Using a ball joint lock nut wrench (22 mm), install the ventilation valve sub-assembly. Tighten to:
    a. If tightening without a ball joint lock nut wrench: 15 ft. lbs. (20 Nm).
    b. If tightening with a ball joint lock nut wrench: 7 ft. lbs. (10 Nm).

2.4L Engine

1. Disconnect the ventilation hose from the ventilation valve sub-assembly.
2. Using a 22 mm deep socket wrench, remove the ventilation valve sub-assembly.

To install:

3. To install, reverse the removal procedure.
4. Apply adhesive to 2 or 3 threads of the ventilation valve. Use Toyota genuine adhesive 1324, three bond 1324 or equivalent.
5. Using a 22 mm deep socket wrench, install the ventilation valve and tighten to 14 ft. lbs. (19 Nm).
6. Inspect for an oil leak.

VARIABLE CAMSHAFT OIL CONTROL VALVE

LOCATION

See Figures 122 through 124.

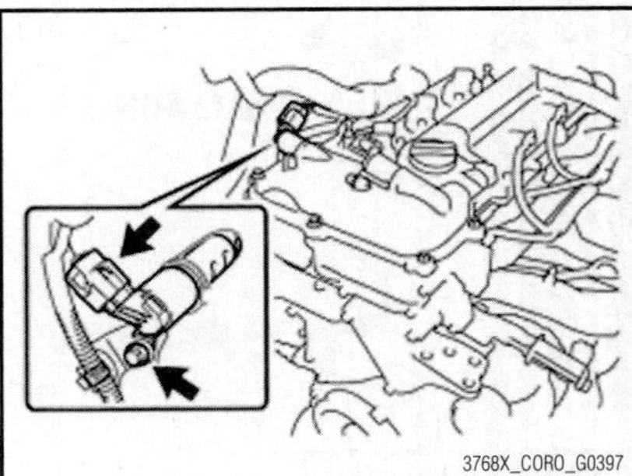

Fig. 122 Location of the camshaft oil control valve—Exhaust camshaft side—1.8L Engine

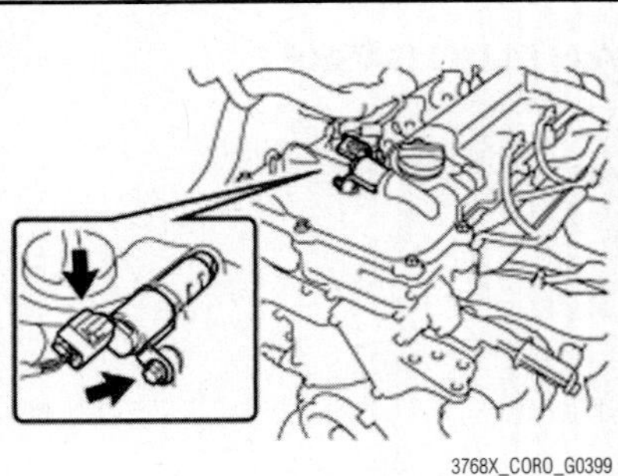

Fig. 123 Location of the camshaft oil control valve—Intake camshaft side—1.8L Engine

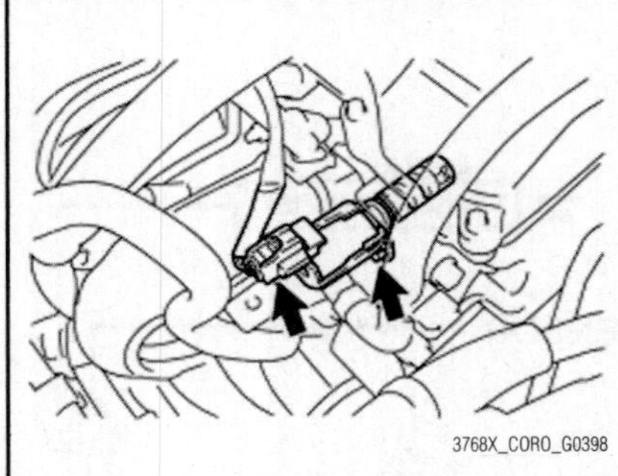

Fig. 124 Location of the camshaft oil control valve—2.4L Engine

Refer to the accompanying illustrations for valve location.

REMOVAL & INSTALLATION

1.8L Engine

See Figure 125.

1. Disconnect the negative battery cable.
2. Exhaust camshaft side:
    a. Remove the bolt and the wire harness bracket.
    b. Disconnect the camshaft timing oil control valve assembly connector.
    c. Remove the bolt and the camshaft timing oil control valve assembly and separate the wire harness bracket.
    d. Remove the O-ring from the camshaft timing oil control valve assembly.
3. Intake camshaft side:
    a. Disconnect the camshaft timing oil control valve assembly connector.
    b. Remove the bolt and remove the camshaft timing oil control valve assembly.
    c. Remove the O-ring from the camshaft timing oil control valve assembly.
4. To install, reverse removal procedure.

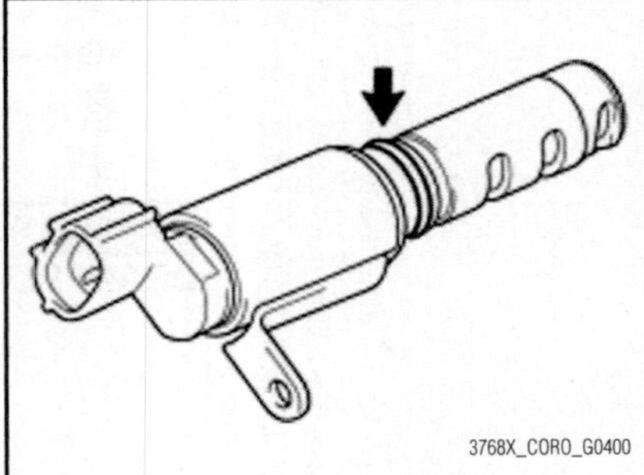

Fig. 125 Removing the O-ring from the camshaft timing oil control valve assembly

5. Apply a light coat of engine oil to a new O-ring and install it onto the camshaft timing oil control valve assembly

➡Do not twist the O-ring.

6. Tighten harness bracket with the bolt to 7 ft. lbs. (10 Nm).

2.4L Engine

See Figure 125.

1. Disconnect the negative battery cable.
2. Remove the bolt and the vacuum hose clamp.
3. Disconnect the camshaft timing oil control valve connector.
4. Remove the bolt and the camshaft timing oil control valve assembly.
5. Remove the O-ring from the camshaft timing oil control valve assembly.

VEHICLE SPEED SENSOR (VSS)

LOCATION

See Figures 126 and 127.

Refer to the accompanying illustrations for sensor location.

REMOVAL & INSTALLATION

U250E Automatic Transaxle

See Figures 128 and 129.

1. Disconnect the negative battery cable.
2. Remove the 2 bolts and the 2 speed sensors from the transaxle assembly.
3. Remove the 2 O-rings from the 2 speed sensors.

To install:

4. Coat 2 new O-rings with ATF, and install them to the 2 speed sensors.
5. Apply adhesive to the bolts threads. Toyota Genuine Adhesive 1344, Three Bond 1344 or Equivalent.
6. Install the 2 speed sensors to the transaxle case with the 2 bolts:
    a. Bolt A: 78 inch lbs. (9 Nm).
    b. Bolt B: 8 ft. lbs. (11 Nm).

U341E Automatic Transaxle

1. Disconnect the negative battery cable.
2. Disconnect the electrical connector
3. Remove the bolt and speed sensor from the transaxle case.
4. To install:
    a. Coat a new O-ring with ATF, and install it to the speed sensor.
    b. Install the speed sensor to the transaxle case with the bolt. Tighten to 48 inch. lbs. (5.4 Nm).

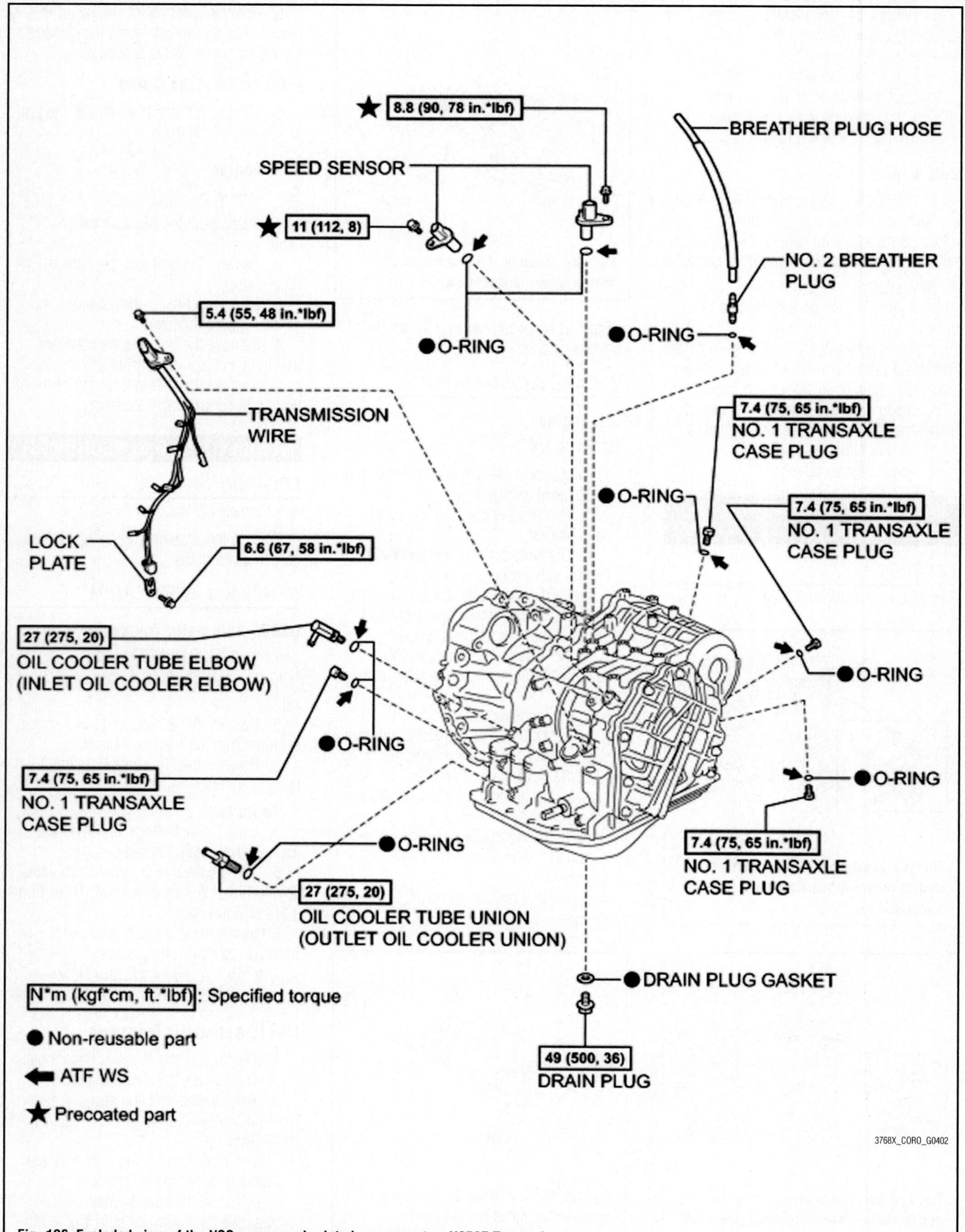

Fig. 126 Exploded view of the VSS sensor and related components—U250E Transaxle

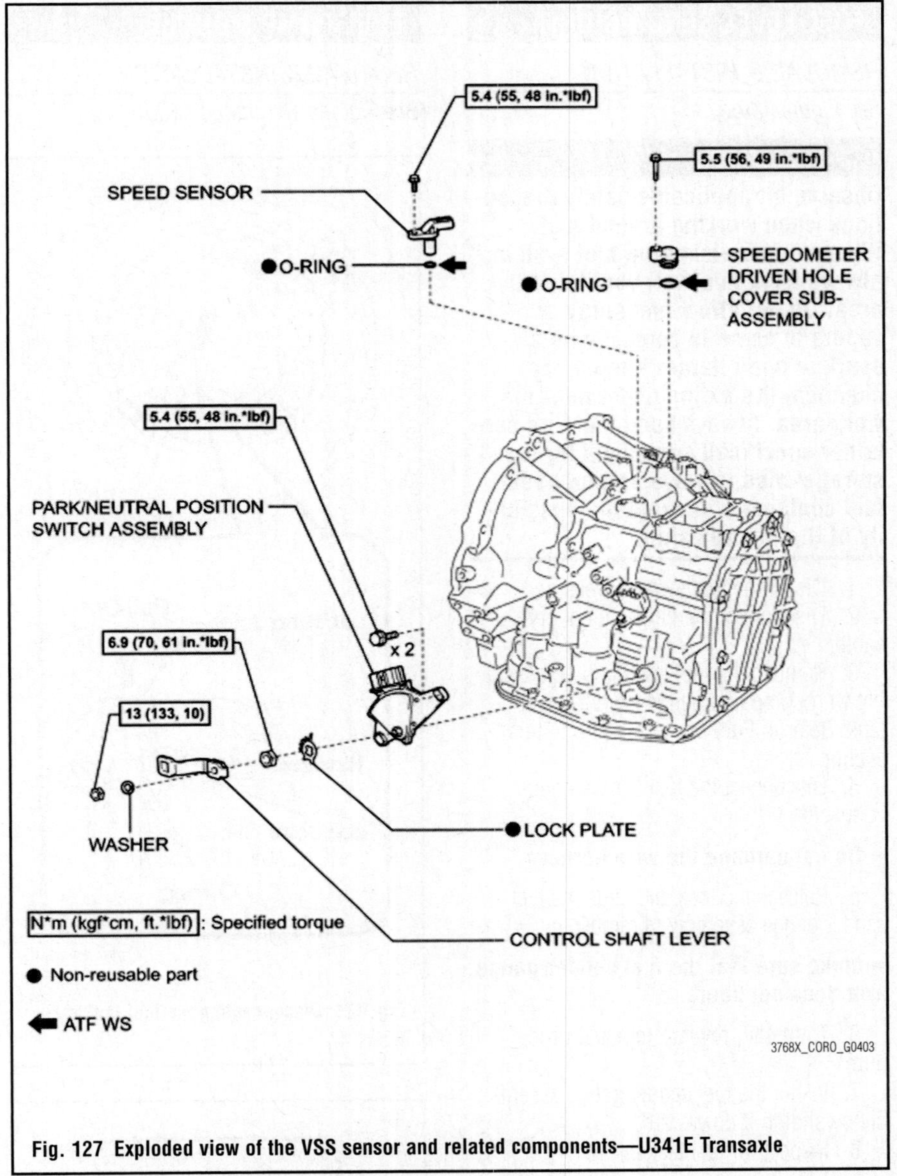

Fig. 127 Exploded view of the VSS sensor and related components—U341E Transaxle

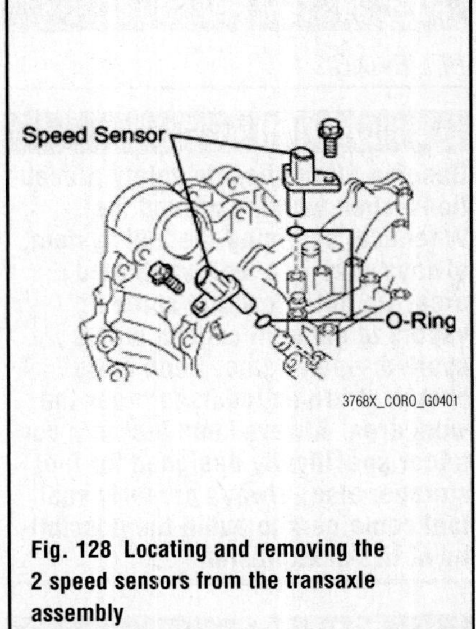

Fig. 128 Locating and removing the 2 speed sensors from the transaxle assembly

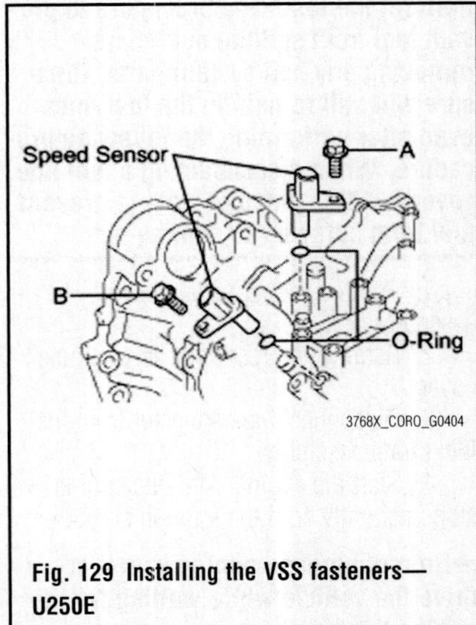

Fig. 129 Installing the VSS fasteners—U250E

FUEL

GASOLINE FUEL INJECTION SYSTEM

FUEL SYSTEM SERVICE PRECAUTIONS

Safety is the most important factor when performing not only fuel system maintenance but any type of maintenance. Failure to conduct maintenance and repairs in a safe manner may result in serious personal injury or death. Maintenance and testing of the vehicle's fuel system components can be accomplished safely and effectively by adhering to the following rules and guidelines.

- To avoid the possibility of fire and personal injury, always disconnect the negative battery cable unless the repair or test procedure requires that battery voltage be applied.
- Always relieve the fuel system pressure prior to disconnecting any fuel system component (injector, fuel rail, pressure regulator, etc.), fitting or fuel line connection. Exercise extreme caution whenever relieving fuel system pressure to avoid exposing skin, face and eyes to fuel spray. Please be advised that fuel under pressure may penetrate the skin or any part of the body that it contacts.
- Always place a shop towel or cloth around the fitting or connection prior to loosening to absorb any excess fuel due to spillage. Ensure that all fuel spillage (should it occur) is quickly removed from engine surfaces. Ensure that all fuel soaked cloths or towels are deposited into a suitable waste container.
- Always keep a dry chemical (Class B) fire extinguisher near the work area.
- Do not allow fuel spray or fuel vapors to come into contact with a spark or open flame.
- Always use a back-up wrench when loosening and tightening fuel line connection fittings. This will prevent unnecessary stress and torsion to fuel line piping.
- Always replace worn fuel fitting O-rings with new Do not substitute fuel hose or equivalent where fuel pipe is installed.

Before servicing the vehicle, make sure to also refer to the precautions in the beginning of this section as well.

FUEL SYSTEM PRESSURE

RELIEVING

CAUTION

Observe all applicable safety precautions when working around fuel. Whenever servicing the fuel system, always work in a well ventilated area. Do not allow fuel spray or vapors to come in contact with a spark or open flame. Keep a dry chemical fire extinguisher near the work area. Always keep fuel in a container specifically designed for fuel storage; also, always properly seal fuel containers to avoid the possibility of fire or explosion.

CAUTION

Perform the following procedure to prevent fuel from spilling out before removing any fuel system parts. Pressure will still remain in the fuel lines even after performing the following procedure. When disconnecting a fuel line, cover it with a piece of cloth to prevent fuel from spraying or coming out.

1. Remove the rear seat cushion assembly.
2. Remove the rear floor service hole cover.
3. Disconnect the connector from the fuel pump assembly.
4. Start the engine. After the engine stops naturally, turn the ignition switch off.

➡Do not increase engine speed or drive the vehicle while waiting for the engine to stop naturally.

➡DTC P0171/25 (system too lean) may be set.

5. Crank the engine again and make sure that the engine does not start.
6. Remove the fuel tank cap and discharge the pressure from the fuel tank.
7. Disconnect the cable from the negative (-) battery terminal.
8. Connect the connector of the fuel pump assembly.
9. Install the rear floor service hole cover.
10. Install the rear seat cushion assembly.

FUEL FILTER

REMOVAL & INSTALLATION

The fuel filter is integrated in the fuel pump and must be replaced as a unit. Refer to Fuel Pump in this section.

FUEL LEVEL SENDING UNIT

REMOVAL & INSTALLATION

See Figure 130.

CAUTION

Observe all applicable safety precautions when working around fuel. Whenever servicing the fuel system, always work in a well ventilated area. Do not allow fuel spray or vapors to come in contact with a spark or open flame. Keep a dry chemical fire extinguisher near the work area. Always keep fuel in a container specifically designed for fuel storage; also, always properly seal fuel containers to avoid the possibility of fire or explosion.

1. Discharge fuel system pressure.
2. Disconnect the negative battery cable.
3. Remove the fuel suction tube assembly with pump and gauge from the fuel tank. Refer to Fuel Pump Module in this section.
4. Disconnect the fuel sender gauge connector.

➡Do not damage the wire harness.

5. Release the lock and slide the fuel sender gauge assembly to remove it.

➡Make sure that the fuel sender gauge arm does not bend.

6. To install, reverse removal procedure.
7. Install the fuel sender gauge assembly by sliding it downward.
8. Inspect for fuel leaks.

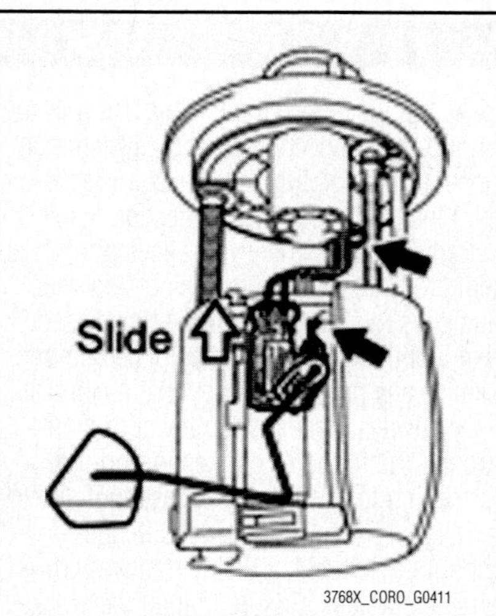

Fig. 130 Removing the fuel sending unit

FUEL PUMP MODULE

REMOVAL & INSTALLATION

See Figures 131 through 136.

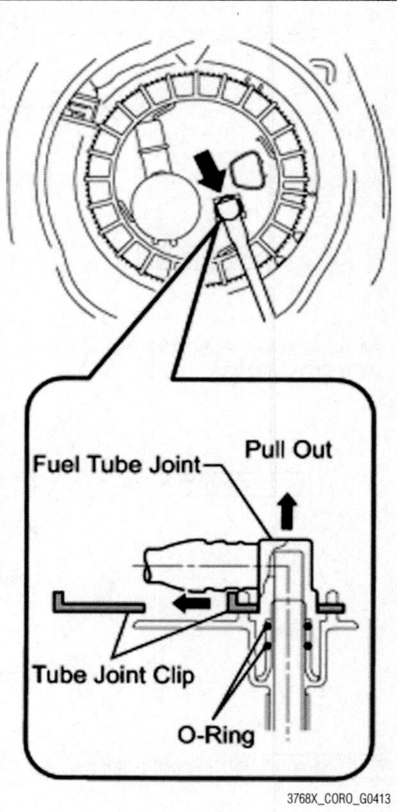

Fig. 131 Disconnecting the fuel tank main tube

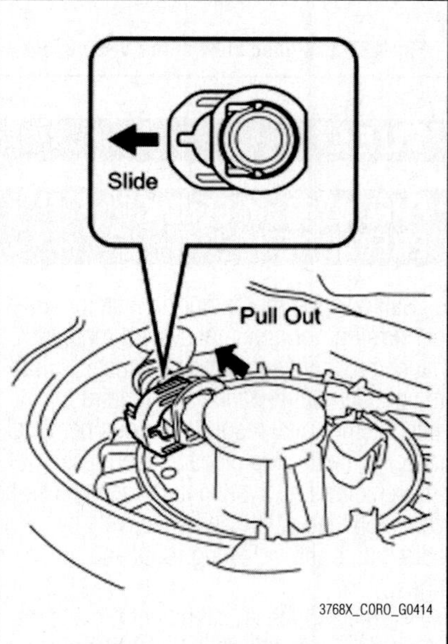

Fig. 132 Disconnecting the fuel tank vent hose

CAUTION

Observe all applicable safety precautions when working around fuel. Whenever servicing the fuel system, always work in a well ventilated area. Do not allow fuel spray or vapors to come in contact with a spark or open flame. Keep a dry chemical fire extinguisher near the work area. Always keep fuel in a container specifically designed for fuel storage; also, always properly seal fuel containers to avoid the possibility of fire or explosion.

1. Remove the rear seat cushion assembly.
2. Remove the rear floor service hole cover.
3. Disconnect the connector from the fuel suction tube assembly.
4. Properly relieve the fuel system pressure.
5. Disconnect the negative battery cable.
6. Disconnect the fuel tank main tube:
   a. Remove the tube joint clip, and pull the fuel tube joint out of the plug of the fuel suction plate.

➡Check that there is no dirt or other foreign objects around the fuel tube joint before disconnecting it. Clean the joint if necessary.

- It is necessary to prevent mud or dirt from entering the joint. If mud or dirt gets in the joint, the O-rings may not seal properly.

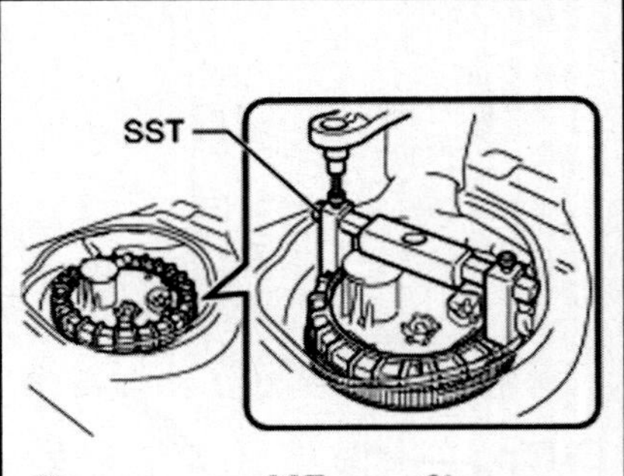

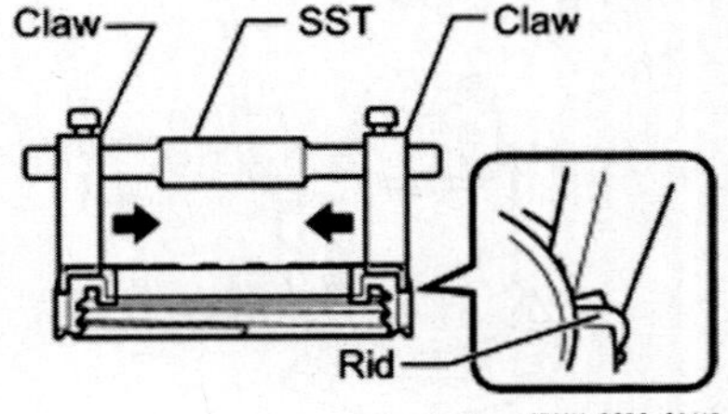

Fig. 133 Installing the special tools to the fuel pump gauge retainer

- Only disconnect the joint by hand.
- Do not bend, kink or twist the nylon tubes.
- Protect the joint by covering it with a plastic bag.

7. Disconnect the fuel tank vent hose:
   a. Pinch the retainer and pull the fuel tank vent connector out of the fuel tank to disconnect the fuel tank vent hose from the fuel suction plate.

➡Check that there is no dirt or other foreign objects around the fuel tube joint before disconnecting it. Clean the joint if necessary.

- It is necessary to prevent mud or dirt from entering the joint. If mud or dirt gets in the joint, the O-rings may not seal properly.
- Only disconnect the joint by hand.
- Do not bend, kink or twist the nylon tubes.
- Protect the joint by covering it with a plastic bag.
- If the pipe and connector are stuck, carefully try wiggling or pushing and pulling on the connector to release it. Pull the connector off the pipe carefully.

8. Remove the fuel pump gauge retainer:
   a. Using a 6 mm socket hexagon wrench, set SST 09808-14020 to the fuel pump gauge retainer.

➡Engage the SST claws securely with the fuel pump gauge retainer ribs to secure SST. Install SST while pressing the SST claws toward the fuel pump gauge retainer (towards the center of SST).

   b. Using SST 09808-14020, loosen the fuel pump gauge retainer.

➡Do not use any tools other than specified in this operation. Damage to the fuel pump gauge retainer or the fuel tank. Loosen the retainer by turning it counterclockwise while holding SST down. Do not allow the claw of the tank suction tube support to slip out of its groove on the fuel tank.

➡The ribs on the fuel pump gauge retainer can be fitted into the tips of SST.

9. Remove the fuel pump gauge retainer while holding the fuel suction tube assembly by hand.

➡Make sure the fuel level sending unit arm does not bend.

10. Remove the fuel pump module from the fuel tank.
11. Remove the gasket from the fuel tank.

To install:

12. Install a new gasket onto the fuel tank.
13. Set the fuel pump module in to the fuel tank.

➡Make sure the fuel level sending unit arm does not bend.

14. Install the fuel pump gauge retainer:
    a. Align the protrusion of the fuel suction tube assembly with the notch of the fuel tank.
    b. While holding the fuel suction tube assembly by hand, align the marks of a new fuel pump gauge retainer and fuel tank as shown in the illustration, then install the fuel pump gauge retainer.

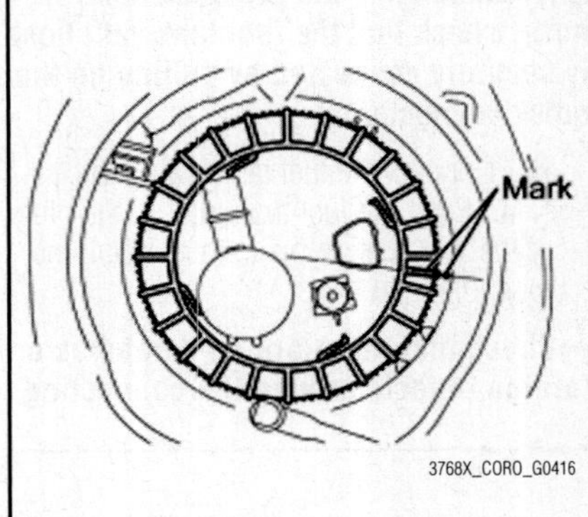

Fig. 134 Aligning the marks of a new fuel pump gauge retainer

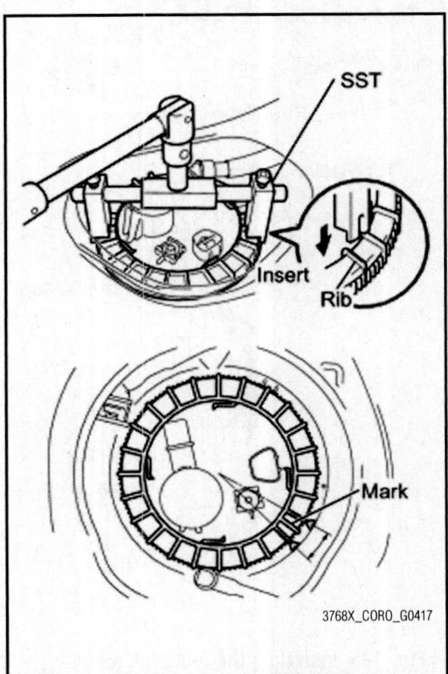

Fig. 135 Identifying installation range

c. Using SST and your hand, tighten the fuel pump gauge retainer 2 revolutions so that the mark of the retainer comes within the range shown in the illustration.

➡**Use the SST. Do not use any other tools such as a screwdriver.**

➡**Insert the notch of SST into the rib of the fuel pump gauge retainer.**

15. Connect the fuel tank vent hose:
   a. Align the fuel tank vent connector with the pipe, then push in the fuel tank vent connector until the retainer makes a "click" sound to connect the fuel tank vent hose to the fuel suction plate.

➡**Check that there are no scratches or foreign objects around the connecting surfaces of the fuel tank vent connector and pipe before performing this work. After connecting the fuel tank vent hose, check that the fuel tank vent hose is securely connected by pulling on the quick connector.**

16. Connect the fuel tank main tube:
   a. Push the fuel tube joint in the plug of the fuel suction plate, then install the tube joint clip.

➡**Check that there are no scratches or foreign objects around the connecting surfaces of the fuel tube joint and plug before performing this work. Check that the fuel tube joint is securely inserted to the end. Check that the tube joint clip is on the collar of the fuel tube joint. After installing the tube joint clip, check that the fuel tank main tube cannot be pulled out.**

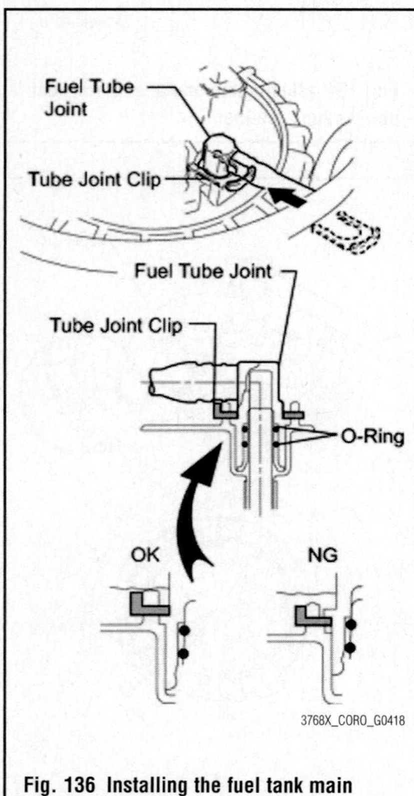

Fig. 136 Installing the fuel tank main tube

FUEL PUMP

REMOVAL & INSTALLATION

See Figures 137 and 138.

✲✲ CAUTION

Observe all applicable safety precautions when working around fuel. Whenever servicing the fuel system, always work in a well ventilated area. Do not allow fuel spray or vapors to come in contact with a spark or open flame. Keep a dry chemical fire extinguisher near the work area. Always keep fuel in a container specifically designed for fuel storage; also, always properly seal fuel containers to avoid the possibility of fire or explosion.

1. Remove the fuel pump module. Refer to Fuel Pump Module in this section.
2. Release the claw and disconnect the fuel pump filter hose.
3. Remove the E-ring and separate the 2 claws, then remove the fuel pump sub-tank.

➡**Do not separate the tube indicated in the illustration.**

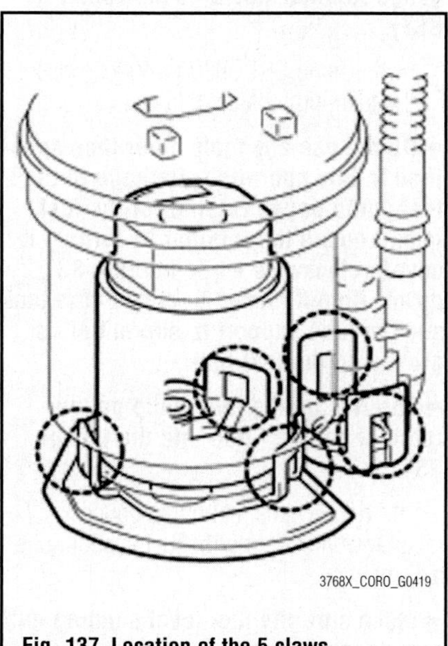

Fig. 137 Location of the 5 claws

4. Disconnect the fuel pump harness connector. Do not damage the wire harness.
5. Using a screwdriver with its tip wrapped in protective tape, disengage the 2 claws and remove the No. 1 fuel suction support.
6. Using a screwdriver with its tip wrapped in protective tape, disengage the 5 claws, and remove the suction filter and fuel pump from the fuel filter.

➡**Do not damage the fuel pump filter or fuel filter. Do not remove the suction filter. Do not use either the fuel pump or the suction filter if the suction filter is removed from the fuel pump.**

7. Disconnect the fuel pump connector.
8. Remove the O-ring from the fuel pump.

➡**Do not disassemble the fuel pump and the fuel filter because they are non-reusable parts.**

To install:

9. Apply a light coat of gasoline or spindle oil to a new O-ring, then install it into the fuel pump.
10. Connect the fuel pump harness connector.
11. Apply a light coat of gasoline or spindle oil to the O-ring of the fuel pump again.
12. Engage the 5 claws, and install the fuel pump filter onto the fuel pump with fuel filter.

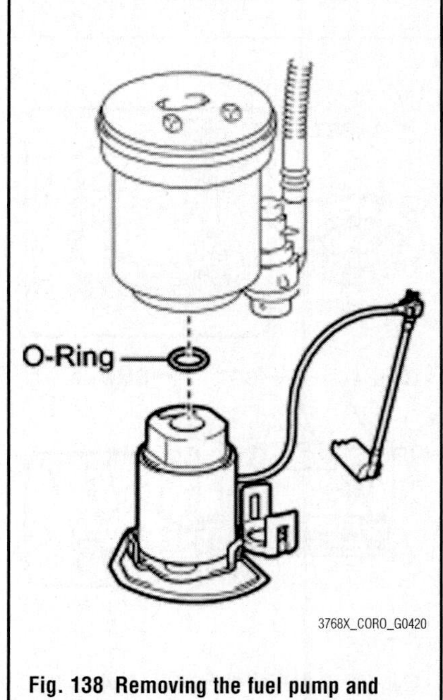

Fig. 138 Removing the fuel pump and o-ring

➡Make sure that the O-ring is not cracked or does not jump out of position during installation.

13. Engage the 2 claws of the No. 1 fuel suction support.
14. Connect the fuel pump harness connector.
15. Engage the 2 claws and install a new E-ring and fuel pump sub-tank.
16. Align the groove of the fuel pump filter hose with the cutout of the fuel sub-tank and install the hose.
17. To complete installation reverse remaining removal procedure.

FUEL PRESSURE REGULATOR

REMOVAL & INSTALLATION

See Figures 139 and 140.

✻✻ CAUTION

Observe all applicable safety precautions when working around fuel. Whenever servicing the fuel system, always work in a well ventilated area. Do not allow fuel spray or vapors to come in contact with a spark or open flame. Keep a dry chemical fire extinguisher near the work area. Always keep fuel in a container specifically designed for fuel storage; also, always properly seal fuel containers to avoid the possibility of fire or explosion.

1. Properly relieve the fuel system pressure.
2. Disconnect the negative battery cable.
3. Remove the fuel pump module. Refer to Fuel Pump Module in this section.
4. Remove the fuel pump. Refer to Fuel Pump in this section.

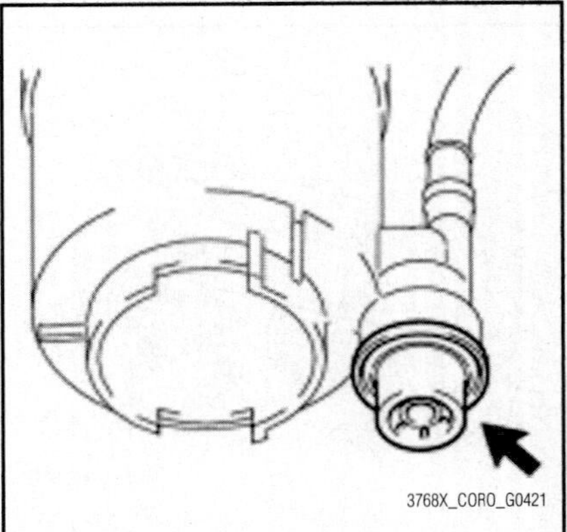

Fig. 139 Identifying and locating the fuel pressure regulator

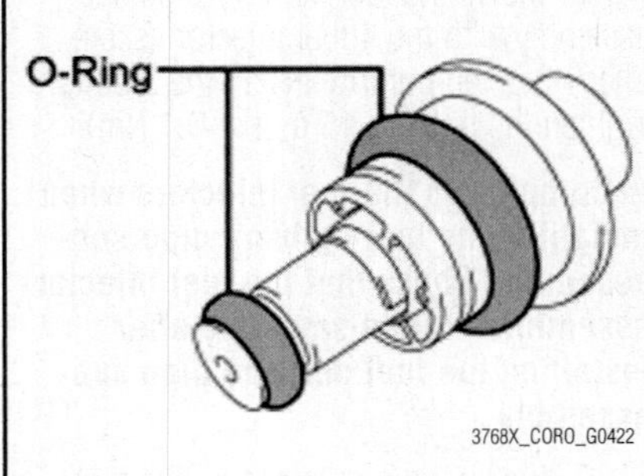

Fig. 140 Removing the 2 O-rings from the fuel pressure regulator assembly

5. Using a screwdriver with its tip wrapped in protective tape, remove the fuel pressure regulator assembly.

➡Slowly pull out the fuel pressure regulator assembly because the O-ring is firmly installed between the regulator and the fuel filter.

6. Remove the 2 O-rings from the fuel pressure regulator assembly.
7. To install, reverse removal procedure.
8. Apply gasoline to 2 new O-rings and then install them to the fuel pressure regulator assembly.

FUEL RAIL AND INJECTOR

REMOVAL & INSTALLATION

1.8L Engine

See Figures 141 through 144.

✻✻ CAUTION

Observe all applicable safety precautions when working around fuel. Whenever servicing the fuel system, always work in a well ventilated area. Do not allow fuel spray or vapors to come in contact with a spark or open flame. Keep a dry chemical fire extinguisher near the work area. Always keep fuel in a container specifically designed for fuel storage; also, always properly seal fuel containers to avoid the possibility of fire or explosion.

Fig. 141 Removing the No. 2 fuel pipe clamp (Type A)

1. Properly relieve the fuel system pressure.
2. Disconnect the negative battery cable.
3. Disconnect the No. 2 ventilation hose.
4. Remove the 2 bolts and disconnect the ground wires.
5. Disconnect the 4 fuel injector assembly connectors.
6. Disconnect the 2 wire harness clamps.

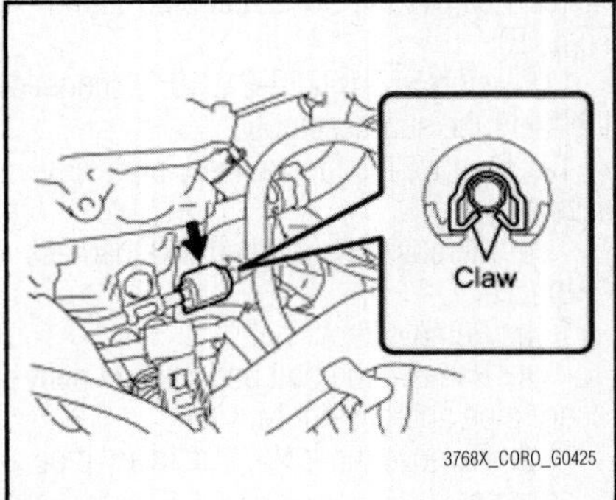

Fig. 142 Removing the No. 2 fuel pipe clamp (Type B)

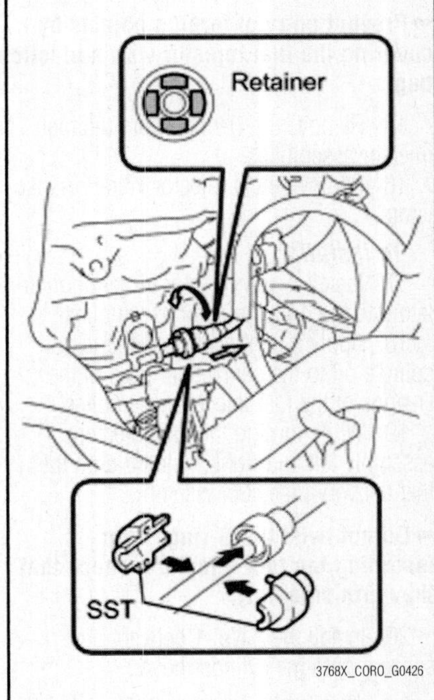

Fig. 143 Using SST 09268-21010, disconnect the fuel tube sub-assembly

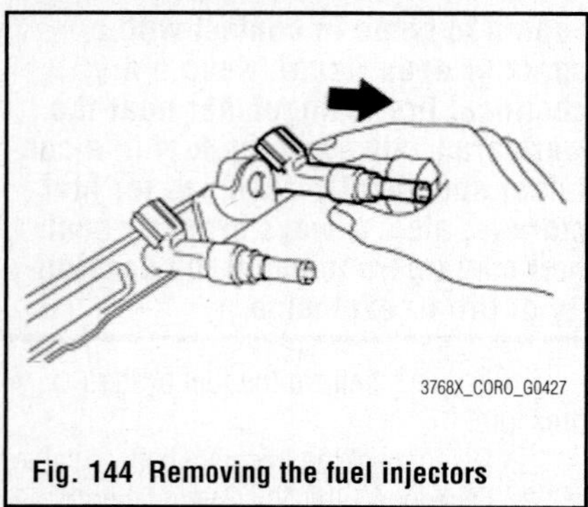

Fig. 144 Removing the fuel injectors

7. Disconnect the 4 wire harness clamps.
8. Remove the 2 bolts and 2 wire harness brackets.
9. Remove the No. 2 fuel pipe clamp (Type A).
10. Remove the No. 2 fuel pipe clamp (Type B).
11. Using SST 09268-21010, disconnect the fuel tube sub-assembly.
12. Remove the fuel delivery pipe sub-assembly:
    a. Remove the bolt and wire harness bracket.
    b. Remove the 2 bolts.
    c. Remove the bolt and the fuel delivery pipe sub-assembly.
    d. Remove the 2 No. 1 delivery pipe spacers.
13. Pull the 4 fuel injector assemblies out of the fuel delivery pipe sub-assembly.
14. For reinstallation, attach a tag or label to the injector shaft.

➡Prevent entry of foreign objects by covering the fuel injector with a plastic bag.

15. Remove the O-rings from the fuel injector assemblies.
16. Remove the 4 injector vibration insulators.

To install:

17. Install 4 new injector vibration insulators to the 4 fuel injector assemblies.
18. Apply a light coat of gasoline or spindle oil to the contact surfaces of the O-rings of the fuel injector assemblies.
19. While turning the fuel injector assembly left and right, install it onto the fuel delivery pipe sub-assembly.

➡Do not twist the O-ring. After installing the fuel injectors, check that they turn smoothly.

20. Install the 2 No. 1 delivery pipe spacers onto the cylinder head.

➡Install the No. 1 delivery pipe spacers in the correct direction.

21. Install the fuel delivery pipe sub-assembly with the 4 fuel injector assemblies, then temporarily install the 2 bolts. Tighten the bolts to 15 ft. lbs. (21 Nm).

➡Do not drop the fuel injectors when installing the fuel delivery pipe sub-assembly. Check that the fuel injector assemblies rotate smoothly after installing the fuel delivery pipe sub-assembly.

22. Install the bolt to secure the fuel delivery pipe sub-assembly. Tighten the bolt to 15 ft. lbs. (21 Nm).
23. Install the wire harness bracket with the bolt. Tighten to 44 inch lbs. (5 Nm).
24. Insert the fuel tube sub-assembly connector into the fuel delivery pipe until a "click" sound can be heard.

➡Check that there are no scratches or foreign matter around the contact surfaces of the fuel tube connector and pipe before performing this work. After connecting the fuel tube, check that the fuel tube connector and pipe are securely connected by pulling on them.

25. Install a new No. 2 fuel pipe clamp (Type B).
26. Install a new No. 2 fuel pump clamp (Type A).
27. Install the 2 wire harness brackets with the 2 bolts. Tighten to 10 ft. lbs. (13 Nm).
28. Connect the 4 wire harness clamps.
29. Connect the 4 fuel injector assembly connectors.
30. Connect the 2 wire harness clamps.
31. Connect the ground wires with the 2 bolts.
32. Install the air cleaner assembly.
33. Connect the No. 2 ventilation hose.
34. Connect cable to negative battery terminal.
35. Inspect for fuel leaks.

2.4L Engine

See Figures 145 through 149.

⁂ CAUTION

Observe all applicable safety precautions when working around fuel. Whenever servicing the fuel system, always work in a well ventilated area. Do not allow fuel spray or vapors to come in contact with a spark or open flame. Keep a dry chemical fire extinguisher near the work area. Always keep fuel in a container specifically designed for fuel storage; also, always properly seal fuel containers to avoid the possibility of fire or explosion.

1. Properly relieve the fuel system pressure.
2. Disconnect the negative battery cable.
3. Remove the air cleaner assembly.

➡Do not forcibly bend, kink or twist the fuel main tube.

4. Remove the fuel tube from the fuel hose clamp.
5. Remove the fuel pipe clamp.
6. Wipe off any dirt on the fuel tube connector.
7. Hold the fuel tube connector, and then install SST 09268-21010.
8. Turn SST to align the retainer inside the fuel tube connector with the chamfered part of SST.

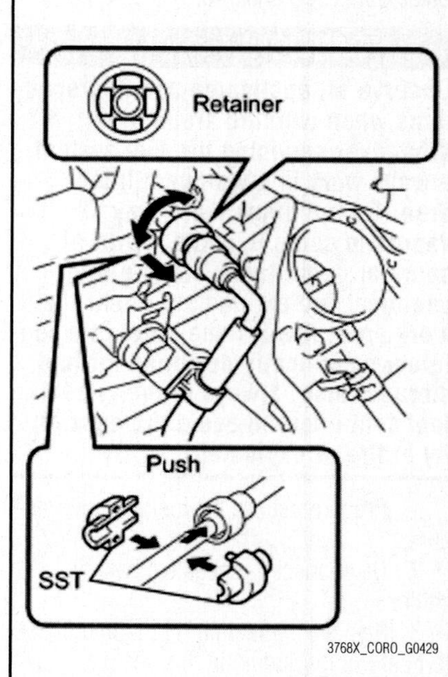

Fig. 145 Installing SST: 09268-21010 to the fuel tube connector

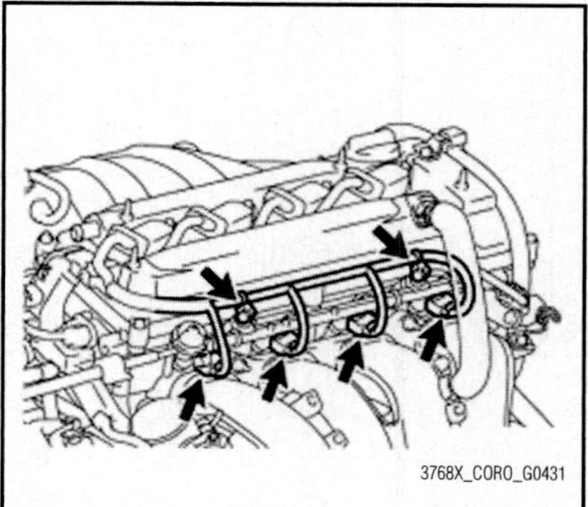

Fig. 146 Disconnecting the 4 fuel injector connectors

9. Insert SST into the fuel tube and hold it. Then push the fuel tube connector toward SST.

10. Mount the retainer of the fuel tube connector onto the chamfered part of SST.

11. Slide SST and fuel tube connector together towards the fuel tube until they make a "click" sound, and then disconnect the fuel tube.

12. Drain the fuel remaining inside the fuel tube.

13. Cover the fuel tube and fuel pipe with a plastic bag to protect the disconnected part.

14. Disconnect the No. 2 ventilation hose from the ventilation valve.

15. Remove the 2 wire harness clamps.

16. Disconnect the 4 fuel injector connectors.

17. Remove the 2 bolts, then remove the fuel delivery pipe together with the 4 fuel injectors.

➡**Be careful not to drop the fuel injectors when removing the fuel delivery pipe.**

18. Remove the 2 delivery pipe spacers from the cylinder head.

19. Remove the 4 insulators from the cylinder head.

20. Pull the 4 fuel injectors out of the fuel delivery pipe.

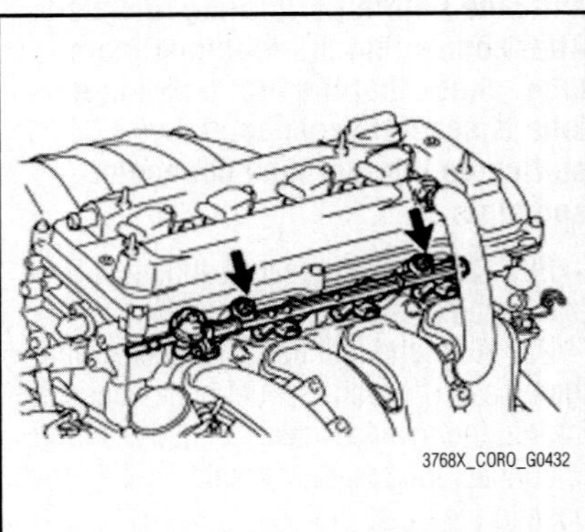

Fig. 147 Removing the 2 bolts, then remove the fuel delivery pipe together with the 4 fuel injectors

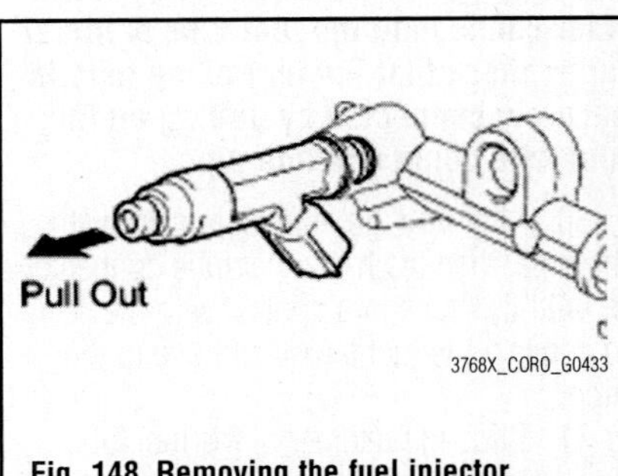

Fig. 148 Removing the fuel injector

21. Remove the 4 O-rings from the 4 fuel injectors.

To install:

22. Apply a light coat of gasoline or spindle oil to new O-rings, then install them to the fuel injectors.

23. Apply a light coat of gasoline or spindle oil to the part of the fuel delivery pipe which comes into contact with the O-ring of the fuel injector.

24. Apply a light coat of gasoline or spindle oil to the O-ring again, then install the right and left fuel injectors onto the fuel delivery pipe.

➡**Make sure that the O-rings are not cracked or does not jump out of position during installation.**

25. Check that the fuel injectors rotate smoothly. If the fuel injector does not rotate, replace the O-ring.

26. Install 4 new insulators to the cylinder head.

27. Install the 2 delivery pipe spacers onto the cylinder head.

28. Install the fuel delivery pipe together with the 4 fuel injectors, then temporarily tighten the 2 bolts.

➡**Be careful not to drop the fuel injectors when installing the fuel delivery pipe.**

29. Check that the fuel injector rotates smoothly.

30. If the fuel injector does not rotate smoothly, replace the O-ring.

31. Tighten the 2 fuel rail bolts to 15 ft. lbs. (20 Nm).

32. Connect the 4 fuel injector connectors.

33. Install the 2 wire harness clamps.

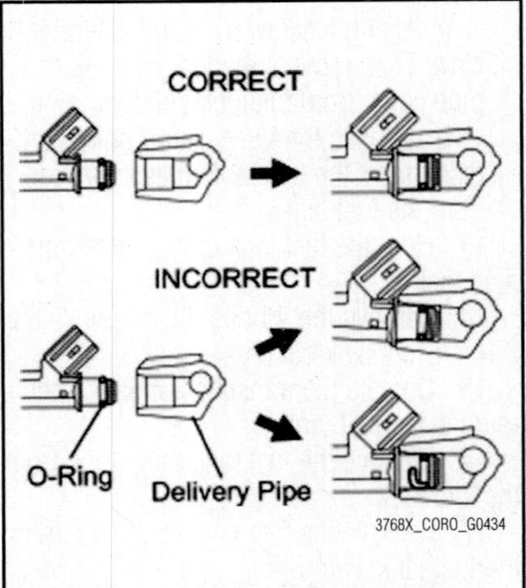

Fig. 149 Correct fuel injector installation position

34. Connect the ventilation hose to the ventilation valve.

➡**Make sure that the paint mark and hose clamp are at the correct angle when installing the hose.**

35. Connect the fuel main tube.

36. Push the fuel tube connector until it makes a "click" sound.

37. Install the fuel pipe clamp.

38. Install the fuel tube to the fuel hose clamp.

39. Install the air cleaner assembly.

40. Connect cable to negative battery terminal.

41. Inspect for fuel leaks.

FUEL TANK

REMOVAL & INSTALLATION

See Figures 150 and 151.

⁂ CAUTION

Observe all applicable safety precautions when working around fuel. Whenever servicing the fuel system, always work in a well ventilated area. Do not allow fuel spray or vapors to come in contact with a spark or open flame. Keep a dry chemical fire extinguisher near the work area. Always keep fuel in a container specifically designed for fuel storage; also, always properly seal fuel containers to avoid the possibility of fire or explosion.

1. Properly relieve the fuel system pressure.

2. Disconnect the negative battery cable.

3. Remove the fuel pump module. Refer to Fuel Pump Module in this section.

4. Drain fuel.

5. Remove front exhaust pipe assembly. Refer to Exhaust Manifold in Engine Mechanical.

6. Remove the 4 bolts and the No. 1 fuel tank protector.

7. Remove the 4 bolts, and separate the parking brake cables.

8. Disconnect the fuel tank vent hose.

9. Pull the fuel tank vent hose out of the pipe.

➡**Check that there is no dirt or other foreign objects around the connector before disconnecting it. Clean the connector if necessary.**

- It is necessary to prevent mud or dirt from entering the connector. If mud or dirt gets in the connector, the O-rings may not seal properly.

- Only disconnect the quick connector by hand.
- Do not bend, kink or twist the nylon tubes.
- Protect the connector by covering it with a plastic bag.

10. Disconnect breather tube fuel hose:

a. Pinch the retainer of the fuel tube connector, then pull the fuel tube connector out of the pipe.

➡Check that there is no dirt or other foreign objects around the connector before disconnecting it. Clean the connector if necessary.

- It is necessary to prevent mud or dirt from entering the connector. If mud or dirt gets in the connector, the O-rings may not seal properly.
- Only disconnect the quick connector by hand.
- Do not bend, kink or twist the nylon tubes.
- Protect the connector by covering it with a plastic bag.
- If the pipe and connector are stuck, carefully try wiggling or pushing and pulling on the connector to release it. Pull the connector off the pipe carefully.

b. Separate the fuel breather tube fuel hose.

11. Disconnect the fuel tank main tube sub-assembly:

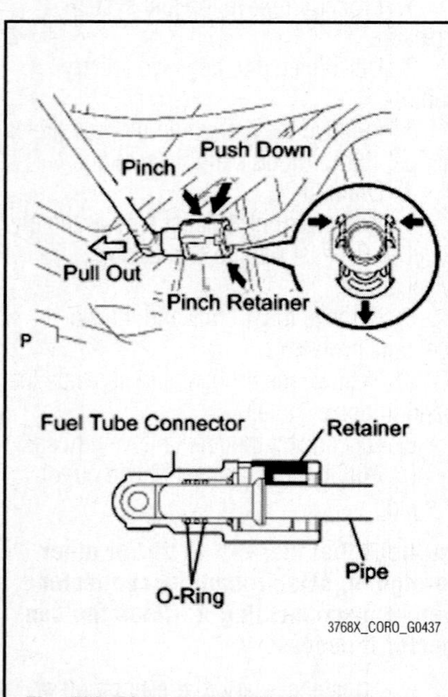

Fig. 150 Disconnecting the fuel tank main tube sub-assembly

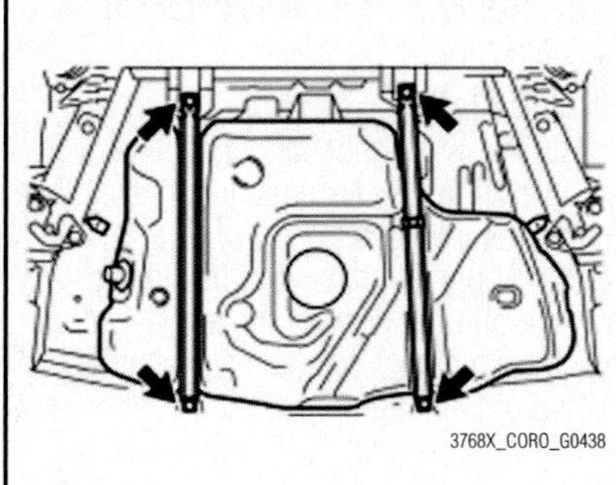

Fig. 151 Removing the 4 bolts, then remove the 2 No. 1 fuel tank bands

a. Pinch the tabs of the retainer of the fuel tube connector to disengage the lock claws and push it down as shown in the illustration.

b. Pull the fuel tank main tube out of the pipe.

➡Check that there is no dirt or other foreign objects around the connector before disconnecting it. Clean the connector if necessary.

- It is necessary to prevent mud or dirt from entering the connector. If mud or dirt gets in the connector, the O-rings may not seal properly.
- Only disconnect the quick connector by hand.
- Do not bend, kink or twist the nylon tubes.
- Protect the connector by covering it with a plastic bag.
- If the pipe and connector are stuck, carefully try wiggling or pushing and pulling on the connector to release it. Pull the connector off the pipe carefully.

12. Disconnect fuel tank filler pipe sub-assembly:

a. Using a screwdriver, unfasten the claw. Then remove the fuel tank filler pipe cover from the fuel tank filler pipe.

b. Loosen the hose clamp bolt, then disconnect the fuel tank filler pipe hose from the fuel tank.

13. Hold the fuel tank using a transmission jack.

14. Remove the 4 bolts, then remove the 2 No. 1 fuel tank bands.

15. Operate the transmission jack, then remove the fuel tank.

16. Remove the fuel tank main tube from the fuel tank.

17. Remove the fuel tank vent hose from the fuel tank clamp.

18. Remove the fuel tank cushions from the fuel tank.

To install:

19. Install new fuel tank cushions onto the fuel tank.

20. Install the fuel tank vent hose onto the fuel tank clamp.

21. Install the fuel tank main tube onto the fuel tank.

22. Set the fuel tank on a transmission jack.

23. Operate the transmission jack, then install the fuel tank into the vehicle.

24. Install the 2 No. 1 fuel tank bands with the 4 bolts. Tighten to 29 ft. lbs. (39 Nm).

25. Connect the fuel tank filler pipe to the fuel tank.

➡Make sure that the hose clamp is facing in the correct direction when installing.

26. Engage the claw, then install the fuel tank filler pipe cover onto the fuel tank filler pipe.

27. Align the fuel tube connector with the pipe, then push the fuel tube connector in until it comes into contact with the seat to connect the fuel tank main tube to the pipe, then push the retainer up until the claws lock.

➡Check that there are no scratches or foreign objects around the connecting surfaces of the fuel tube connector and pipe before performing this work. After connecting the fuel tank main tube, check that the fuel tank main tube is securely connected by pulling on the fuel tube connector and pipe.

28. Connect the breather tube fuel hose to the clamp.

29. Align the fuel hose connector with the pipe, then push the fuel hose connector in until the retainer makes a "click" sound to connect the fuel tank breather tube fuel hose to the pipe.

➡Check that there are no scratches or foreign objects around the connecting surfaces of the fuel tube connector and pipe before performing this work. After connecting the fuel tank breather tube, check that the fuel pump tube is securely connected by pulling on the fuel tube connector and pipe.

30. Align the fuel tube connector with the pipe, then push the fuel tube connector in until it comes into contact with the seat to connect the fuel tank vent hose to the pipe.

31. Slide the retainer of the fuel tube connector to lock the claws

➡Check that there are no scratches or foreign objects around the connecting surfaces of the fuel tube connector and pipe before performing this work. After connecting the fuel tank breather tube, check that the fuel pump tube is securely connected by pulling on the fuel tube connector and pipe.

32. Install the fuel tank protector sub-assembly with the 4 bolts. Tighten to 49 inch lbs. (5.5 Nm).
33. Install the parking cables with the 4 bolts. Tighten to 49 inch lbs. (5.5 Nm).
34. To complete installation, reverse remaining removal procedure.
35. Inspect for fuel leaks.
36. Inspect for exhaust gas leak.

IDLE SPEED

ADJUSTMENT

Adjustment is not available or necessary.

THROTTLE BODY

REMOVAL & INSTALLATION

1.8L Engine

See Figure 152.

1. Drain and recycle the engine coolant.
2. Disconnect the negative battery cable.
3. Disconnect the mass air flow meter connector and the 2 wire harness clamps.
4. Remove the air cleaner assembly.
5. Disconnect the throttle body connector and the 2 water by-pass hoses.
6. Remove the 2 bolts and 2 nuts, and remove the throttle body assembly.
7. Remove the gasket from the intake manifold.

To install:

8. Install a new gasket onto the intake manifold.
9. Install the throttle body assembly with the 2 bolts and 2 nuts. Tighten to 7 ft. lbs. (10 Nm).
10. Connect the 2 water by-pass hoses and the throttle body connector.
11. To complete installation, reverse remaining removal procedure.
12. Perform the initialization procedure.

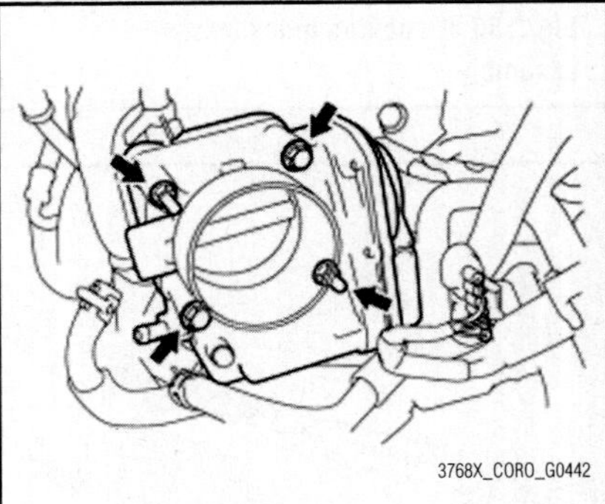

Fig. 152 Removing the 2 bolts and 2 nuts, and removing the throttle body assembly

2.4L Engine

See Figure 153.

1. Drain and recycle the engine coolant.
2. Disconnect the negative battery cable.
3. Disconnect the mass air flow meter connector and the 2 wire harness clamps.
4. Remove the air cleaner assembly.
5. Disconnect the 2 water by-pass hoses.
6. Disconnect the throttle body assembly connector.
7. Disconnect the throttle body hose.
8. Remove the 4 bolts and throttle body assembly.
9. Remove the gasket from the intake manifold.

To install:

10. Install a new gasket onto the intake manifold.
11. Install the throttle body assembly with the 4 bolts. Tighten to 22 ft. lbs. (30 Nm).
12. Connect the 2 water by-pass hoses and the throttle body connector.
13. To complete installation, reverse remaining removal procedure.
14. Perform the initialization procedure.

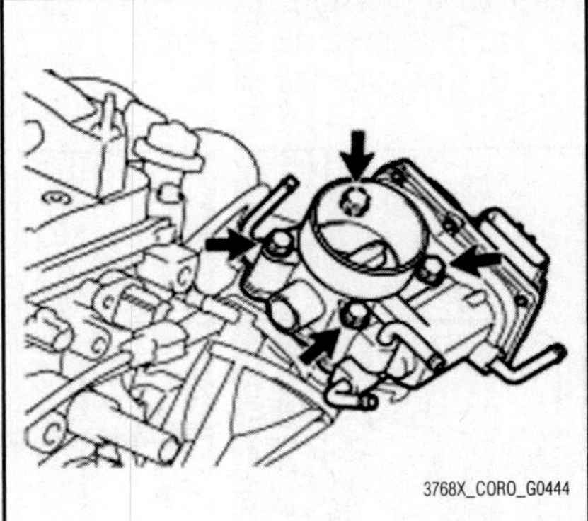

Fig. 153 Removing the 4 bolts and throttle body assembly

INITIALIZATION

➡Be sure to perform this procedure after reassembling the throttle body assembly, removing and reinstalling any throttle body component or replacing the ECM.

1. Disconnect the cable from the negative (-) battery terminal. Wait at least 60 seconds and reconnect the cable.
2. Turn the ignition switch to ON without operating the accelerator pedal.

➡If the accelerator pedal is operated, perform the above steps again.

3. Connect the Techstream to the DLC3 and clear the DTC's.
4. Start the engine and check that the MIL is not illuminated and that the idle speed is within the specified range when the air conditioning is switched off after the engine is warmed up.
   a. Standard:
      - Condition: A/C switched off
      - Engine idle speed: 600 to 700 rpm

➡Be sure to perform this step with all accessories off. Make sure that the shift lever is in N or P.

5. Enter the following menus: Powertrain/ Engine and ECT/ Data List/ Throttle Pos. Sensor Output. Fully depress the accelerator pedal and check that the value is 60% or more.
6. Perform a road test and confirm that there are no abnormalities. Refer to Drive Train.

HEATING & AIR CONDITIONING SYSTEM

BLOWER MOTOR

REMOVAL & INSTALLATION

See Figures 154 through 156.

1. Disconnect the negative battery cable.

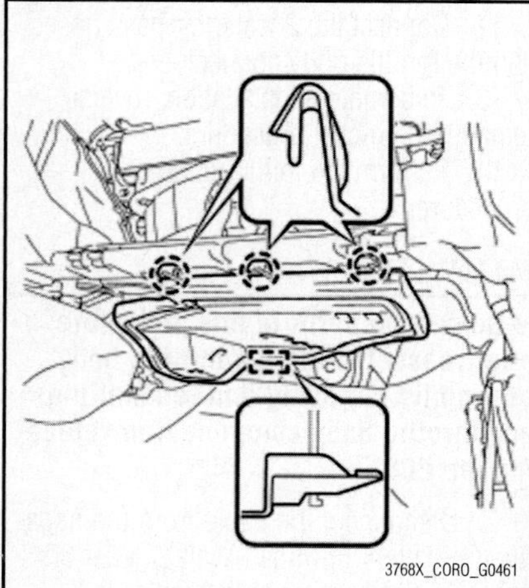

Fig. 154 Removing the No. 2 instrument panel under cover sub-assembly

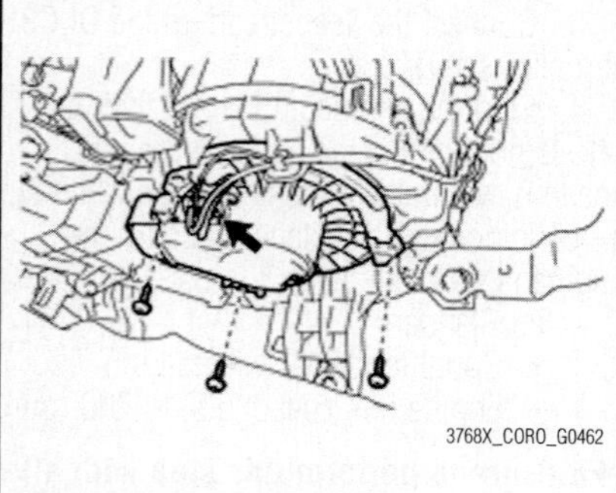

Fig. 155 Removing blower motor sub-assembly (without PTC heater)

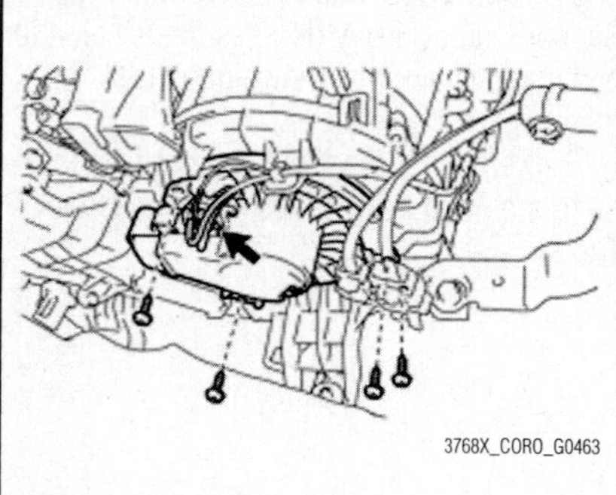

Fig. 156 Remove blower motor sub-assembly (with PTC heater)

2. Remove the no. 2 instrument panel under cover sub-assembly (w/ instrument panel under cover)
   a. Disengage the 3 claws.
   b. Disengage the guide and remove the No. 2 instrument panel under cover sub-assembly.
3. Remove blower motor sub-assembly (without PTC heater):
   a. Disconnect the connector.
   b. Remove the 3 screws and the blower motor sub-assembly.
4. Remove blower motor sub-assembly (with PTC heater):
   a. Remove the quick heater connector screw.
   b. Disconnect the connector.
   c. Remove the 4 screws and the blower motor sub-assembly.
5. To install, reverse removal procedure.

HEATER CORE

REMOVAL & INSTALLATION

See Figures 157 through 162.

1. Remove Heater and A/C Unit, as outlined in this section.
2. Remove immobilizer code ECU (with smart key system):
   a. Disconnect the connector.
   b. Disengage the clamp.
   c. Remove the bolt and the immobilizer code ECU.
3. Remove transponder key ECU (without smart key system):
   a. Disconnect the connector.
   b. Disengage the clamp.

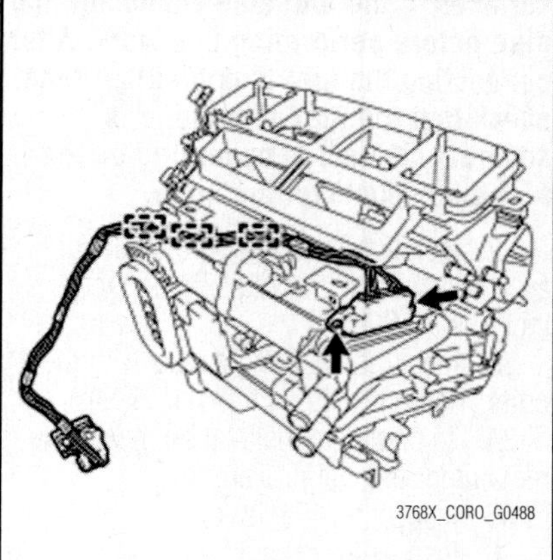

Fig. 158 Disengaging the clamps and removing the 2 quick heater screws

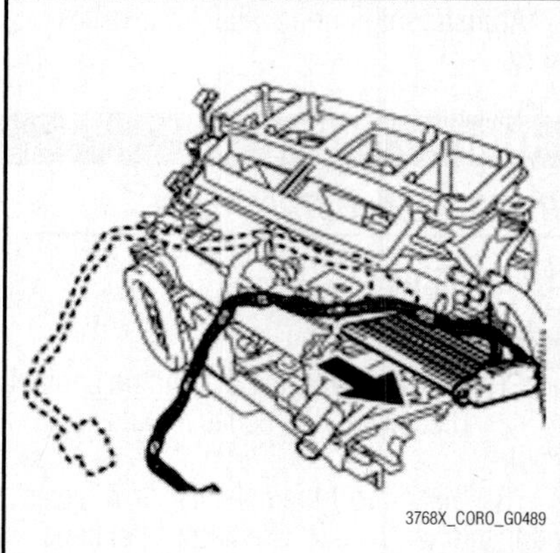

Fig. 159 Removing quick heater assembly

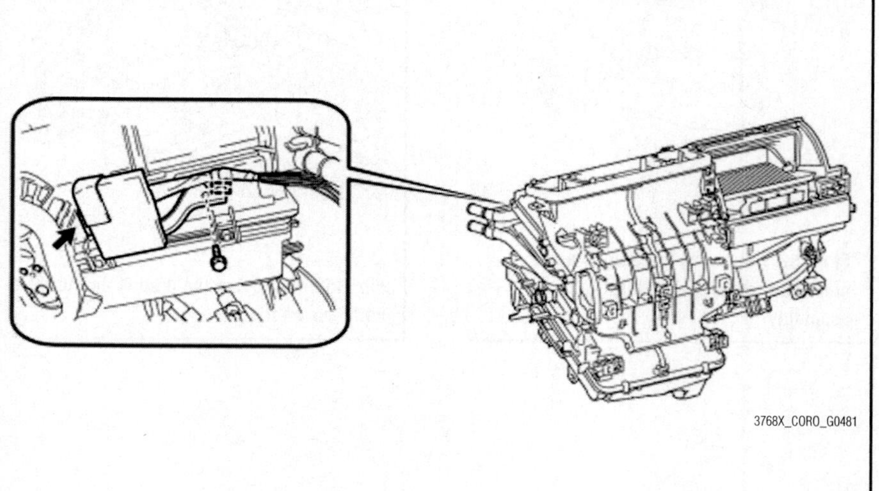

Fig. 157 Removing the immobilizer code ECU/ transponder key ECU

c. Remove the bolt and the transponder key ECU.

4. Remove the air outlet control servo motor (for automatic air conditioning system).

5. Remove the air mix control servo motor (for TMC made with automatic air conditioning system).

6. Remove the no. 2 heater control cable sub-assembly (for manual air conditioning system).

7. Remove the air mix damper control cable sub-assembly (for manual air conditioning system).

8. Remove the quick heater assembly (with PTC heater):

a. Disengage each clamp.

b. Remove the 2 screws.

c. Remove the quick heater assembly as shown in the illustration.

9. Remove the air conditioning harness assembly (for automatic air conditioning system):

a. Disconnect the connector.

b. Disengage each clamp and remove the air conditioning harness.

10. Disengage the 2 claws and remove the RH console mounting bracket.

11. Remove the screw and the console LH mounting bracket.

12. Remove air conditioning amplifier assembly:

a. Remove the screw.

b. Disengage the claw and remove the air conditioning amplifier assembly.

13. Remove the air conditioning duct sub-assembly (for automatic air conditioning system).

14. Remove the drain cooler hose from the air conditioning radiator assembly.

15. Remove heater radiator unit sub-assembly:

a. Remove the screw and the clamp.

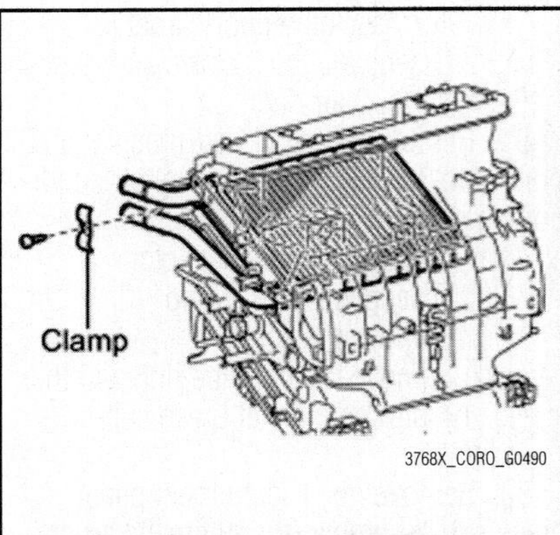

Fig. 160 Removing the heater radiator unit sub-assembly from the air conditioning unit assembly

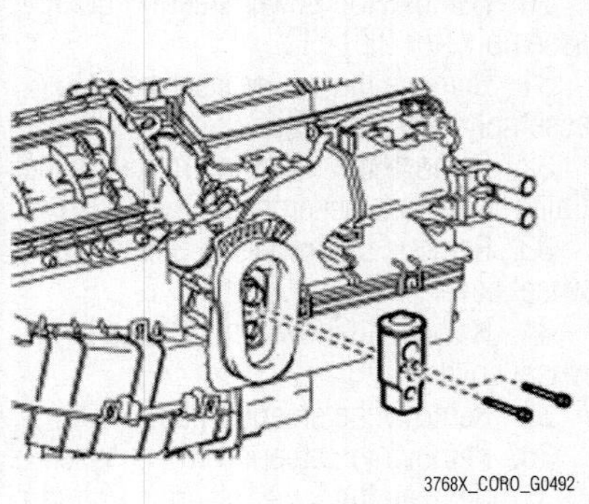

Fig. 161 Remove the 2 hexagon bolts and cooler expansion valve

b. Remove the heater radiator unit sub-assembly from the air conditioning unit assembly.

16. Using a 4 mm hexagon wrench, remove the 2 hexagon bolts and cooler expansion valve.

17. Remove no. 1 cooler evaporator sub-assembly:

a. Remove the 4 screws.

b. Disengage the 4 claws and remove the lower heater case.

c. Disengage the clamp, and remove the No. 1 cooler evaporator sub-assembly together with the No. 1 cooler thermistor.

d. Remove the 2 O-rings.

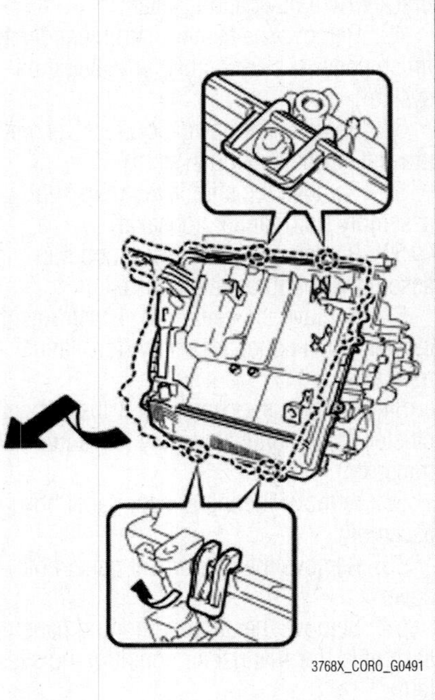

Fig. 162 Disengaging the 4 claws and remove the lower heater case.

To install:

18. Sufficiently apply compressor oil to 2 new O-rings and the fitting surfaces. Install the 2 O-rings to the No. 1 cooler evaporator sub-assembly. Compressor oil: ND-OIL 8 or equivalent.

➡Keep the O-rings and O-ring fitting surfaces clean from dirt or foreign matter.

19. Install the No. 1 cooler evaporator sub-assembly together with the No. 1 cooler thermistor as a set.

20. Engage the clamp.

21. Engage the 4 claws

22. Install the lower heater case with the 4 screws.

23. Using a 4 mm hexagon wrench, install the cooler expansion valve with the 2 hexagon bolts. Tighten to 31 inch lbs. (4 Nm).

24. To complete installation, reverse remaining removal procedure.

HEATER AND A/C UNIT

REMOVAL & INSTALLATION

See Figure 163.

✲✲ CAUTION

These models are equipped with a Supplemental Restraint System (SRS), which uses an air bag. Whenever working near any of the SRS components, such as the impact sensors, the air bag module, steering column and instrument panel, disable the SRS, as described in Chassis Electrical.

1. Align the front wheels facing straight ahead.

2. Disconnect the negative battery cable.

✲✲ CAUTION

Wait at least 90 seconds after disconnecting the cable from the negative (-) battery terminal to disable the SRS system.

➡Before disconnecting the cable, set the air conditioning control switch to DEF-MODE. (for Automatic Air Conditioning System).

3. Drain and recycle the engine coolant.

4. Remove the wiper motor and assembly.

5. Remove the bolt and slide the hook connector.

6. Disconnect the suction pipe sub-assembly.

➡Seal the openings of the disconnected parts using vinyl tape to prevent entry of moisture and foreign matter.

7. Remove the O-ring from the suction hose sub-assembly.
8. Disconnect the air conditioning tube assembly.
9. Remove the O-ring from the air conditioning tube assembly.

➡Seal the openings of the disconnected parts using vinyl tape to prevent entry of moisture and foreign matter.

10. Using pliers, grip the claws of the clip and slide the clip to disconnect the heater outlet water hose.

➡Do not apply excessive force to the heater outlet water hose. Prepare a drain pan or a piece of cloth in case the coolant leaks.

11. Disconnect heater water hose inlet.

➡Disconnection procedure for the heater inlet water hose is the same as that for the heater outlet water hose.

12. Remove the lower LH instrument panel finish panel.
13. Remove the lower RH instrument panel finish panel.
14. Remove the LH instrument panel finish panel end.
15. Remove the RH instrument panel finish panel end .
16. Remove the center instrument panel register assembly.
17. Remove the instrument cluster finish panel assembly.
18. Remove the combination meter assembly.
19. Remove the LH front pillar garnish (w/o curtain shield airbag).
20. Remove the LH front pillar garnish (w/ curtain shield airbag).
21. Remove the RH front pillar garnish (w/o curtain shield airbag).
22. Remove the RH front pillar garnish (w/ curtain shield airbag).
23. Remove the lower instrument panel finish panel assembly.
24. Disconnect the LH front door opening trim weather strip.
25. Remove the glove compartment door assembly.
26. Remove the no. 1 instrument panel box door sub-assembly.
27. Disconnect the RH front door opening trim weather strip.
28. Disconnect the instrument panel wire assembly
29. Remove the upper instrument panel sub-assembly.
30. Remove the power steering ECU assembly (for 2ZR-FE).
31. Remove the power steering ECU assembly (for 2AZ-FE).
32. Remove the lower instrument panel finish panel sub-assembly.
33. Remove the lower no. 3 steering wheel cover.
34. Remove the lower no. 2 steering wheel cover.
35. Remove the steering pad.
36. Remove the steering wheel assembly.
37. Remove the lower steering column cover.
38. Remove the upper steering column cover.
39. Remove the turn signal switch assembly with spiral cable sub-assembly.
40. Remove the column hole cover silencer sheet.
41. Separate the no. 2 steering intermediate shaft assembly.
42. Remove the instrument panel sub reinforcement.
43. Remove the no. 3 air duct sub-assembly.
44. Remove the transponder key amplifier (w/ smart key system).
45. Remove the stop light switch assembly.
46. Remove the steering post assembly (for 2ZR-FE).
47. Remove the steering post assembly (for 2AZ-FE).
48. Remove the radio receiver with bracket (w/o navigation system).
49. Remove the center instrument cluster finish panel sub-assembly (w/ navigation system).
50. Remove the navigation receiver with bracket (w/ navigation system).
51. Remove the shift lever knob sub-assembly (for manual transaxle).
52. Remove the shift lever knob sub-assembly (for automatic transaxle).
53. Remove the center no. 1 instrument cluster finish panel assembly (for manual transaxle).
54. Remove the center no. 1 instrument cluster finish panel assembly (for automatic transaxle).
55. Remove the instrument panel box assembly.
56. Remove the instrument panel hole cover.
57. Remove the air conditioning panel assembly (for manual air conditioning system).
58. Remove the air conditioning control assembly (for automatic air conditioning system).
59. Remove the LH front door scuff plate.
60. Remove the LH cowl side trim board.
61. Remove the front no. 1 console box insert
62. Remove the no. 1 switch hole base (w/o smart key system).
63. Remove the no. 1 switch hole base (w/ smart key system).
64. Remove the RH front door scuff plate.
65. Remove the RH cowl side trim board.
66. Remove the no. 2 instrument panel under cover sub-assembly (w/ instrument panel under cover).
67. Remove the front no. 2 console box insert.
68. Remove the upper console panel sub-assembly.
69. Remove the console box carpet.
70. Remove the console box assembly (for manual transaxle).
71. Remove the console box assembly (for automatic transaxle).
72. Remove the no. 2 antenna cord sub-assembly.
73. Remove the lower instrument panel sub-assembly.
74. Remove the smart key ECU assembly (w/ smart key system).
75. Remove the no. 1 air duct sub-assembly.
76. Remove the 2 nuts and the No. 1 air duct sub-assembly.
77. Disengage the 6 claws and remove the lower defroster nozzle assembly.
78. Remove the 2 bolts and the center instrument panel to cowl brace.
79. Remove the rear no. 2 air duct (w/ rear air duct):
    a. Remove the clip.
    b. Turn back the floor carpet.
    c. Disengage the 2 claws and remove the rear No. 2 air duct.
80. Remove the rear no. 3 air duct (w/ rear air duct):
    a. Remove the clip.
    b. Turn back the floor carpet.
    c. Disengage the 2 claws and remove the rear No. 3 air duct.
81. Remove the no. 1 instrument panel brace sub-assembly (for manual air conditioning system):
    a. Disconnect the connector.
    b. Disengage each clamp.
    c. Remove the screw.
    d. Remove the bolt, the nut, and the No. 1 instrument panel brace sub-assembly.
82. Remove no. 1 instrument panel brace sub-assembly (for automatic air conditioning system):
    a. Disengage each clamp.
    b. Remove the screw.

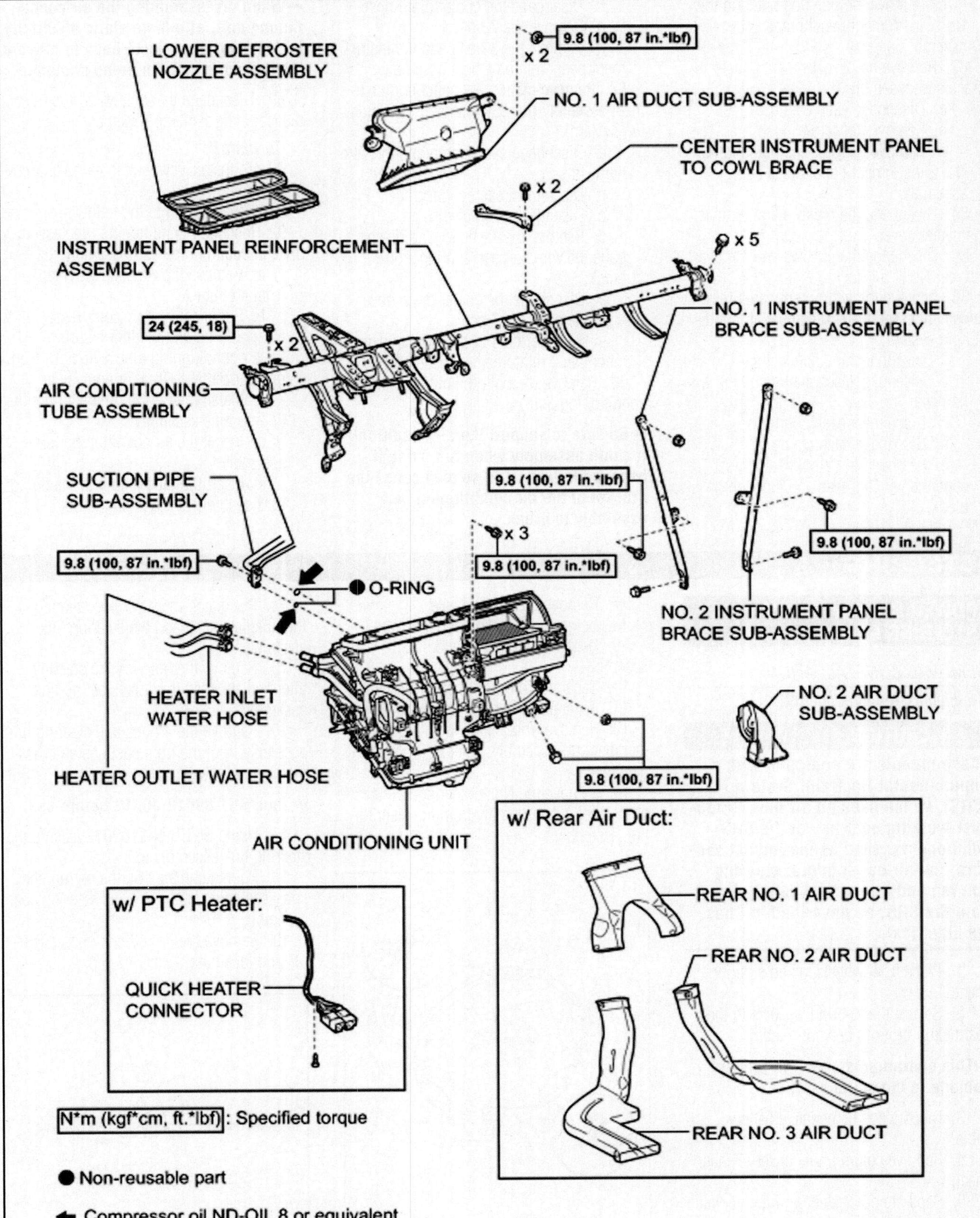

Fig. 163 Exploded view of the heater, A/C unit and related components

c. Remove the bolt, the nut, and the No. 1 instrument panel brace sub-assembly.

83. Remove no. 2 instrument panel brace sub-assembly:

a. Disengage each clamp.

b. Remove the screw.

c. Remove the bolt, the nut, and the No. 2 instrument panel brace sub-assembly.

84. Disengage the 4 claws and remove the heater cover.

85. Disengage the 4 claws and remove the rear No. 1 air duct.

86. Remove the instrument panel reinforcement assembly (for manual air conditioning system):

a. Disconnect the connector.

b. Disengage each clamp and the wire harness.

c. Disconnect each connector.

d. Disengage each clamp.

e. Remove the 12 bolts, and disengage the wire harness and junction block.

f. Disengage the cooler drain hose.

g. Remove the 7 bolts.

h. Remove the 3 bolts and the instrument panel reinforcement assembly.

87. Remove instrument panel reinforcement assembly (for automatic air conditioning system):

a. Disengage each clamp and the wire harness.

b. Disconnect each connector.

c. Disengage each clamp.

d. Remove the 12 bolts, and disengage the wire harness and junction block.

e. Disengage the cooler drain hose.

f. Remove the 7 bolts.

g. Remove the 3 bolts and the instrument panel reinforcement assembly.

88. Remove the bolt, the nut, and the air conditioning unit.

➡Be sure to support the air conditioning unit assembly when removing it because failure to do so may cause the bracket of the air conditioning unit assembly to break.

➡When disassembling the air conditioning unit, eliminate static electricity by touching the vehicle body to prevent the components from being damaged.

89. Disengage the 2 claws and remove the No. 2 air duct sub-assembly.

To install:

90. To install, reverse the removal procedure.

91. Pay attention to the following:

92. Install air conditioning tube assembly and suction pipe sub-assembly:

a. Remove the attached vinyl tape from the tube.

b. Sufficiently apply compressor oil to a new O-ring and the fitting surface of the air conditioning tube assembly. Compressor oil: ND-OIL 8 or equivalent.

c. Install the O-ring on the air conditioning tube assembly.

d. Install the air conditioning tube assembly.

e. Insert the pipe joint into the fitting hole securely and tighten the bolt.

STEERING

POWER RACK & PINION STEERING GEAR

REMOVAL & INSTALLATION

See Figures 164 through 168.

⁂ CAUTION

These models are equipped with a Supplemental Restraint System (SRS), which uses an air bag. Whenever working near any of the SRS components, such as the impact sensors, the air bag module, steering column and instrument panel, disable the SRS, as described in Chassis Electrical.

1. Place front wheels facing straight ahead.

2. Secure the steering wheel with the seat belt in order to prevent rotation.

➡This operation is useful to prevent damage to the spiral cable.

3. Disconnect the negative battery cable.

4. Remove column hole cover silencer sheet.

5. Separate no. 2 steering intermediate shaft assembly.

6. Remove clips A and the No. 1 steering column hole cover sub-assembly and disengage clip B from the body.

7. Remove the front wheels.

8. Remove the engine under covers.

9. Separate the tie rod end sub-assembly.

10. Separate front stabilizer link assembly.

11. Separate front lower suspension arms.

12. Remove front suspension cross-member sub-assembly. Refer to Front Suspension.

13. Remove the No. 1 steering column hole cover sub-assembly from the steering link assembly.

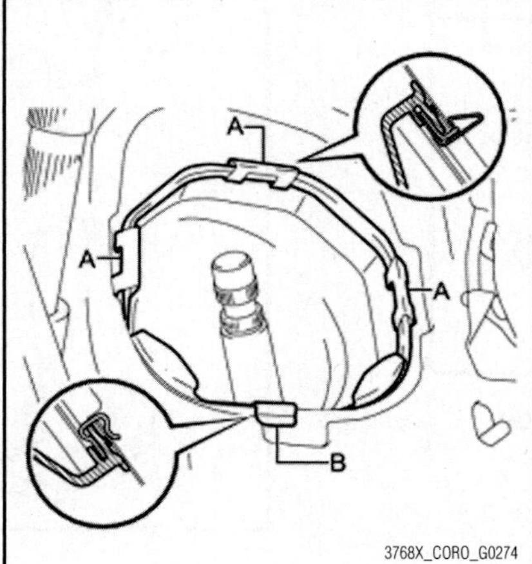

Fig. 164 Remove clips A and the No. 1 steering column hole cover sub-assembly

14. Put match marks on the steering intermediate shaft and the steering link assembly.

15. Remove the bolt and the steering intermediate shaft from the steering link assembly.

16. Remove the 4 bolts and steering link assembly from the front suspension cross-member sub-assembly.

➡Tape SST 09612-00012 before use.

17. Using SST 09612-00012, secure the steering link assembly in a vise.

18. Put matchmarks on the tie rod end sub-assembly LH and RH and steering gear assembly.

19. Remove the tie rod end sub-assembly and lock nuts.

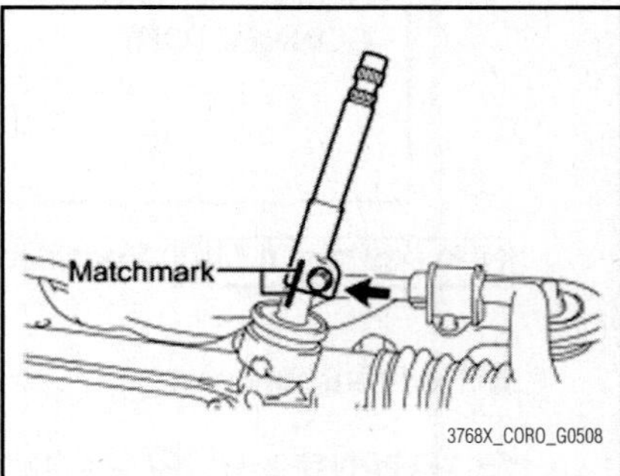

Fig. 165 Putting match marks on the steering intermediate shaft and the steering link assembly

To install:

20. Install the lock nut and the tie rod end sub-assembly to the steering gear assembly until the match marks are aligned.

➡After adjusting the toe-in, tighten the lock nut.

21. Install the steering link assembly to the front suspension crossmember sub-assembly with the 4 bolts:
    a. Temporarily tighten the bolts in order of (A), (B), (C), and then (D).

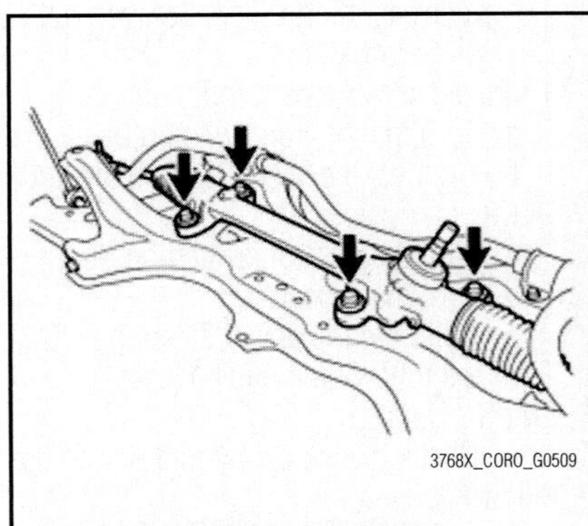

Fig. 166 Remove the 4 bolts and steering link assembly

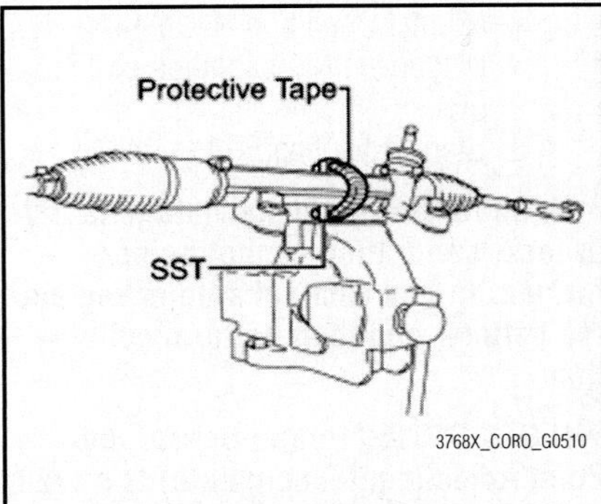

Fig. 167 Installing SST 09612-00012 to the steering link assembly

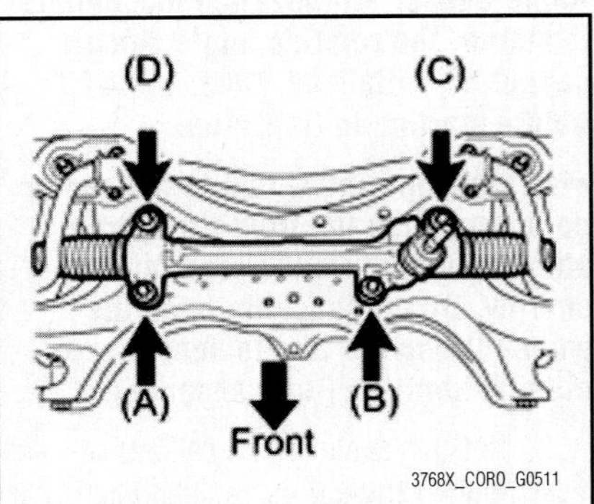

Fig. 168 Steering link assembly bolt tightening sequence

    b. Tighten the bolts in order of (A), (B), (C), and then (D).
    c. Tighten the bolts to 43 ft. lbs. (58 Nm).

22. Align the matchmarks and install the steering intermediate shaft to the steering link assembly.
23. Install the bolt. Tighten to 26 ft. lbs. (35 Nm).
24. To complete installation, reverse remaining removal procedure.

POWER STEERING ECU

REMOVAL & INSTALLATION

1.8L Engine

See Figures 169 and 170.

1. Place front wheels facing straight ahead.
2. Disconnect the negative battery cable.
3. Remove the upper instrument panel.
4. Separate the wire harness clamp from the power steering ECU assembly.

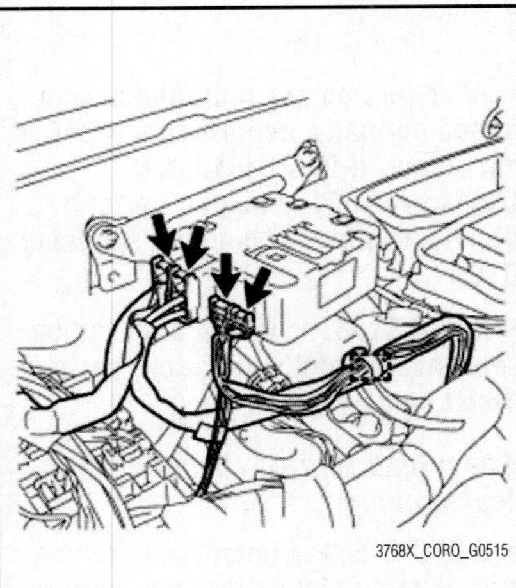

Fig. 169 Disconnecting the 4 connectors from the power steering ECU assembly

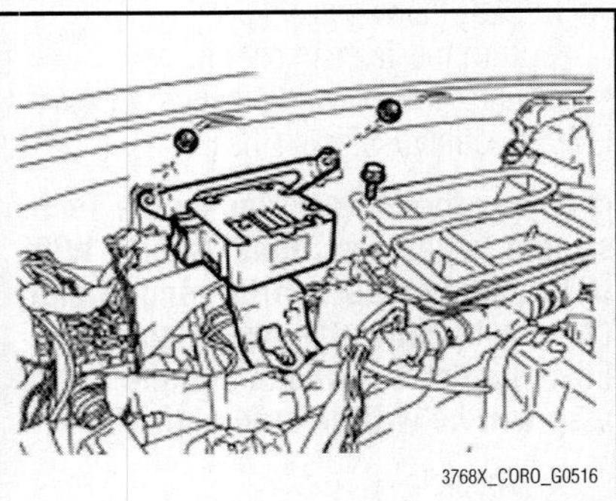

Fig. 170 Removing the bolt, 2 nuts, and the power steering ECU assembly

5. Disconnect the 4 connectors from the power steering ECU assembly.
6. Remove the bolt, 2 nuts, and the power steering ECU assembly.

To install:

7. Install the power steering ECU assembly with the bolt and 2 nuts. Tighten to 71 inch lbs. (8 Nm).
8. Connect the 4 connectors to the power steering ECU assembly.
9. Install the wire harness clamp to the power steering ECU assembly.
10. To complete the installation, reverse remaining removal procedure.
11. Perform the Assist Map Writing procedure.
12. Calibrate torque sensor zero point.

2.4L Engine

See Figures 171 through 173.

1. Place front wheels facing straight ahead.
2. Disconnect the negative battery cable.
3. Remove the upper instrument panel.
4. Separate the wire harness clamp from the power steering ECU assembly.
5. Disconnect the connector from the power steering ECU assembly.

➡As shown in the illustration, pull out the lock of the lock lever and turn the lock lever to disconnect the connector.

6. Disconnect the 3 connectors from the power steering ECU assembly.

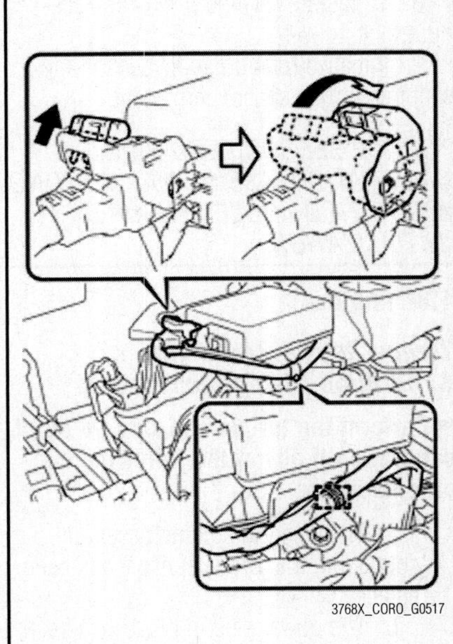

Fig. 171 Separating the wire harness clamp

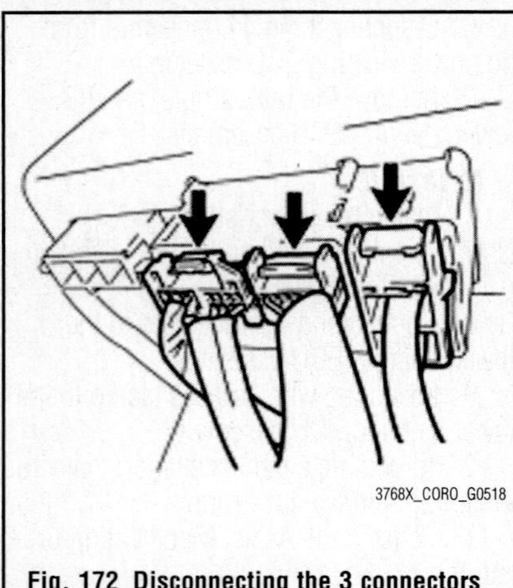

Fig. 172 Disconnecting the 3 connectors

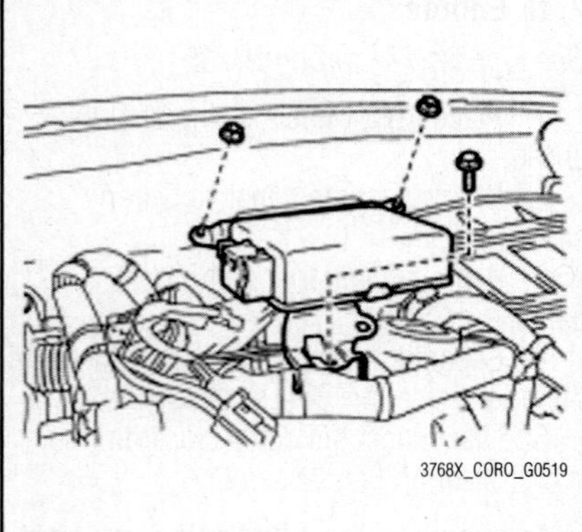

Fig. 173 Removing the bolt, 2 nuts, and the power steering ECU assembly

7. Remove the bolt, 2 nuts, and the power steering ECU assembly.
8. To install, reverse removal procedure.
9. Initialize rotation angle sensor and calibrate torque sensor zero point.

TORQUE SENSOR ZERO POINT CALIBRATION, ASSIST MAP WRITING & ROTATION ANGLE SENSOR INITIALIZATION

1.8L Engine

Torque Sensor Zero Point Calibration (Using Techstream)

➡Perform the torque sensor zero point calibration if any of the following conditions occur:

a. The steering column assembly (containing the torque sensor) has been replaced.
b. The power steering ECU has been replaced.
c. There is a difference in steering effort between turning right and left.

➡When torque sensor zero point calibration is performed, the assist map is written automatically at the same time.

➡If DTC C1516 (Torque Sensor Zero Point Adjustment Incomplete) is stored, the torque sensor zero point cannot be calibrated. Clear the DTC before starting calibration.

➡Do not touch the steering wheel during the torque sensor zero point calibration. Perform the calibration only when the vehicle is stopped.

1. Perform the torque sensor zero point calibration:
   a. Set the steering wheel to the center point and align the front wheels straight ahead.
   b. Turn the ignition switch off.
   c. Connect the Techstream to the DLC3.
   d. Start the engine.
   e. Turn the Techstream on.
   f. Calibrate the power steering ECU. Enter the following menus: Chassis / EMPS / Utility / Torque Sensor Adjustment.
2. Check for DTCs

➡After zero point calibration is completed normally, confirm that a DTC is not output. If DTC C1515, C1516, C1534 or C1581 is output, perform troubleshooting for the corresponding DTC.

➡If DTC C1581 is output after torque sensor zero point calibration, perform assist map writing.

Assist Map Writing (Using Techstream)

➡If DTC C1581 is output after torque sensor zero point calibration, perform assist map writing.

1. Turn the ignition switch off.
2. Connect the Techstream to the DLC3.
3. Start the engine.
4. Turn the Techstream on.
5. Enter the following menus: Chassis / EMPS / Utility / Signal Check.

➡Follow the instructions on the Techstream to perform Signal Check. With DTC C1581 output, performing Signal Check will cause the power steering ECU to enter Test Mode and the assist map will be written automatically.

6. Check for DTCs.
7. After writing the assist map, if DTC C1581 is output, perform the troubleshooting procedure for DTC C1581.

2.4L Engine

Rotation Angle Sensor Initialization and Torque Sensor Zero Point Calibration

➡Clear the rotation angle sensor calibration value, initialize the rotation angle sensor, and calibrate the torque sensor zero point if any of the following has occurred:

- The power steering ECU has been replaced.
- The steering column assembly has been replaced.
- Steering effort differs between left and right.

1. Inspection before calibration:
   a. Turn the ignition switch off.
   b. Connect the Techstream to the DLC3.
   c. Turn the ignition switch on (IG).
   d. Turn the Techstream on.
   e. Calibrate the power steering ECU. Enter the following menus: Chassis / EMPS / Data List.
   f. Check the values by referring to the following:
      - Tester Display: IG Power Supply
      - Measurement Item/Range: ECU power source voltage/Min.: 0 V. Max.: 20.1531 V.
      - Normal Condition: 11 to 14 V
      - Diagnostic Note: Ignition switch on (IG)
   g. Standard voltage: 11 to 14 V.

➡If the IG power supply voltage is 9 V or less, calibration cannot be performed. In this case, charge or replace the battery, and then perform calibration.

➡If DTC C1516 (Torque Sensor Zero Point Adjustment Incomplete) is stored, the torque sensor zero point cannot be calibrated. Clear the DTC before starting calibration. If DTC C1526 (Rotation Angle Sensor Initialization Incomplete) is stored, the rotation angle sensor cannot be initialized. Clear the DTC before starting initialization.

➡Set the steering wheel to the center point and align the front tires straight ahead. Do not turn the steering wheel sharply. Do not touch the steering wheel during the torque sensor zero point calibration (for 3 seconds).

2. Rotation angle sensor calibration value clear, rotation angle sensor initialization, and torque sensor zero point calibration:
   a. Turn the ignition switch off.

b. Connect the Techstream to the DLC3.
c. Turn the ignition switch on (IG).
d. Turn the Techstream on.
e. Enter the following menus: Chassis / EMPS / Utility / Torque Sensor Adjustment.

POWER STEERING PUMP

REMOVAL & INSTALLATION

➡The 2009–10 Corolla models do not use a power steering pump. The Corolla utilizes a Power Steering Motor that works in conjunction with a power steering ECU. The power steering motor is integrated in to the steering column assembly and must be replaced as a unit.

SUSPENSION

FRONT SUSPENSION

COIL SPRING

REMOVAL & INSTALLATION

Refer to MacPherson Strut.

CONTROL LINKS

REMOVAL & INSTALLATION

See Figure 174.

1. Remove front wheel.

➡If the ball joint turns together with the nut, use a hexagon wrench (6 mm) to hold the stud bolt.

2. Remove the 2 nuts and stabilizer bar link.
3. Installation is reverse of removal.
4. Tighten the link assembly nuts to 55 ft. lbs. (74 Nm).

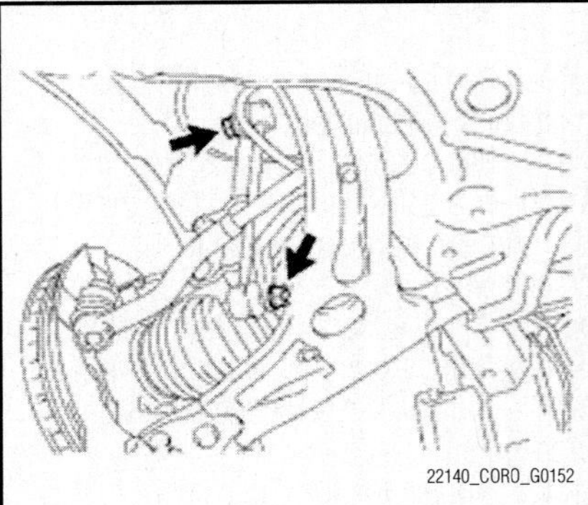

Fig. 174 Removing front stabilizer link assembly

LOWER BALL JOINT

REMOVAL & INSTALLATION

See Figure 175.

1. Before servicing the vehicle, refer to the Precautions section.
2. Remove or disconnect the following:
   - Negative battery cable. On vehicles equipped with an air bag, wait at least 90 seconds before proceeding
   - Front wheels
   - Cotter pin from the bearing locknut cap, then remove the cap
3. Depress the brake pedal and loosen the axle nut
4. Remove or disconnect the following:
   - Brake caliper attaching hardware, position the caliper aside with the hydraulic line still attached and suspend it with a wire
   - ABS speed sensor, if equipped
   - Rotor
5. Loosen the 2 nuts holding the strut to the steering knuckle assembly. Do not remove at this time.
6. Remove or disconnect the following:
   - Cotter pin and nut from the tie rod end. Using a tie rod end removal tool, separate the tie rod end from the steering knuckle
   - Steering knuckle from the strut assembly
   - Axle nut and grasp the hub and knuckle assembly. With a plastic hammer tap the axle shaft to remove knuckle and hub

➡Cover the halfshaft boot with a shop rag to protect it from any damage.

7. Clamp the steering knuckle in a vise and remove the dust deflector. Remove the nut holding the steering knuckle to the ball joint. Press the ball joint out of the steering knuckle.
8. Remove the ball joint from the arm.

To install:

9. Install the Lower ball joint to the lower arm. Tighten the fasteners to: 66 ft. lbs. (89 Nm).

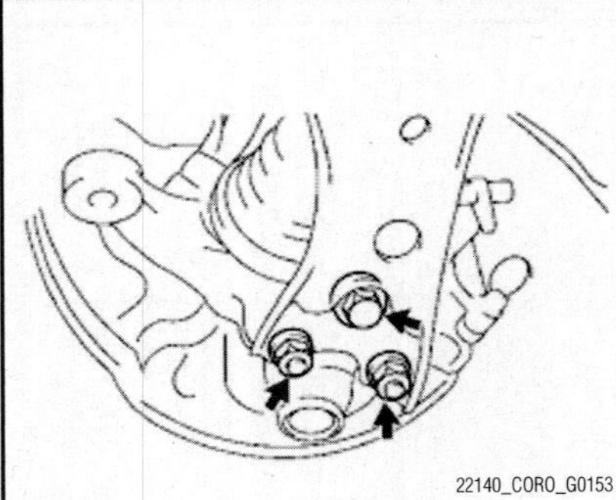

Fig. 175 Removing lower ball joint/ control arm

10. Install the ball joint to the steering knuckle. Tighten the ball joint-to-steering knuckle nut to: 76 ft. lbs. (103 Nm).
11. Install or connect the following:
   - New cotter pin. Drive the deflector shield onto the knuckle
   - Knuckle and hub assembly to the axle and temporarily tighten the axle nut
   - Knuckle assembly to the lower strut bracket. Temporarily insert the mounting bolts from the rear and install the nuts
   - Tie rod end to the knuckle
12. Tighten the bolts on the lower side of the strut assembly.
13. Install or connect the following:
   - ABS speed sensor
   - Brake disc and the caliper
14. Tighten the axle nut.
15. Connect the negative battery cable.
16. Check and adjust the alignment, if needed.

LOWER CONTROL ARM

REMOVAL & INSTALLATION

See Figures 175 through 179.

1. Remove front wheel.
2. Disconnect front stabilizer link assembly.
3. Separate front suspension arm sub assembly lower left and right:
   a. Remove the bolt and 2 nuts, and separate the front suspension arm sub-assembly lower No. 1 LH from the lower ball joint assembly front LH.
4. Separate rack and pinion power steering gear assembly:
   a. Remove the 4 bolts, separate the rack & pinion power steering gear assembly.

➡Loosen the bolt since the nut cannot be rotated.

   b. Suspend the rack & pinion power steering gear assembly.
5. Suspended engine assembly:
6. Separate front suspension cross member subassembly:
   a. Remove the 3 bolts and 3 nuts, disconnect the transverse engine

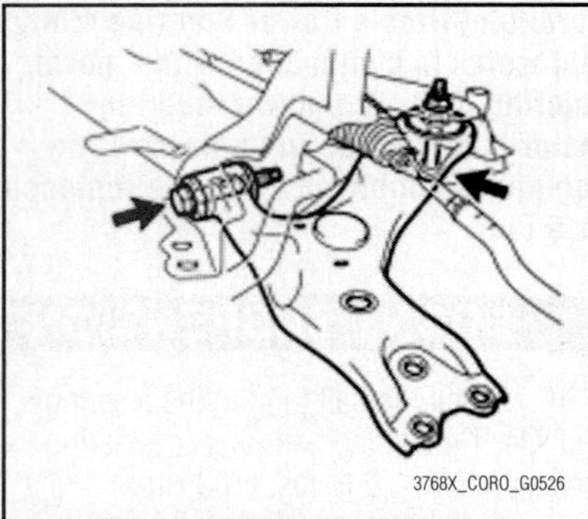

Fig. 176 Remove the 2 bolts, nut and front lower control arm

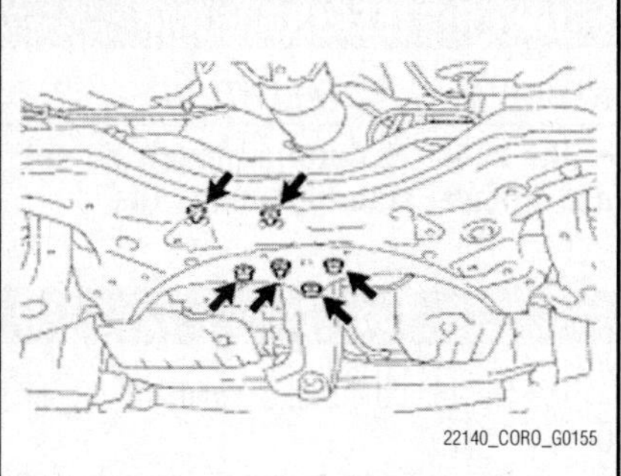

Fig. 178 Installing engine mounting member subassembly

mounting insulator and engine mounting member sub-assembly center from the front suspension cross member sub-assembly.

b. Remove the 4 bolts.

c. Lower the transmission jack, remove the front suspension cross member sub-assembly.

7. Remove front suspension arm sub assembly lower:

a. Remove the 2 bolts, nut and front suspension arm sub-assembly lower No. 1 LH from the front suspension cross member sub-assembly.

To install:

8. Temporarily tighten front suspension arm subassembly lower.

a. Install the front suspension arm sub-assembly lower, temporarily tighten the 2 bolts and nut.

9. Install front suspension cross member subassembly:

a. Lift the front suspension cross member sub-assembly up with a transmission jack.

b. Insert service tool to the base hole of the RH side cross member and RH side of the vehicle.

c. Tighten the bolts temporarily.

d. Insert service tool to the base hole of the LH side of cross member and LH side of the vehicle.

e. Tighten the bolts temporarily.

f. Then tighten the bolt A and B by the specified torque A: 116 ft. lbs. (157 Nm) B: 83 ft. lbs. (113 Nm)

10. Connect the transverse engine mounting insulator and engine mounting member sub-assembly center to the front suspension cross member sub-assembly.

11. Install the 3 bolts and 3 nuts. Tighten to 38 ft. lbs. (52 Nm)

12. Install rack and pinion power steering gear assembly:

a. Install the rack & pinion power steering gear assembly with the 4 bolts. Tighten to 66 ft. lbs. (89 Nm)

13. Install front suspension arm sub assembly lowers:

a. Install the front suspension arm sub-assembly lowers with the 2 nuts and bolt to the lower ball joint assembly front.

14. Install front stabilizer link assemblies.

15. Stabilizes suspension:

a. Install the front wheel and jack down the vehicle.

b. Bounce the vehicle up and down several times to stabilize the suspension.

16. Fully tighten front suspension arm subassembly lower:

a. Fully tighten the 2 bolts and nut.

17. Inspect and adjust front wheel alignment.

➡Tighten the bolt since the nut cannot be rotated.

STABILIZER BAR (SWAY BAR) & LINKS

REMOVAL & INSTALLATION

See Figure 180.

1. Remove the column hole cover silencer sheet.
2. Separate no. 2 steering intermediate shaft assembly.
3. Separate no. 1 steering column hole cover sub-assembly.
4. Remove front wheels.
5. Separate tie rod end sub-assembly.
6. Remove the front stabilizer link assembly.
7. Separate front lower suspension arm.
8. Remove front suspension cross-member sub-assembly.
9. Remove the 4 bolts, 2 No. 1 front stabilizer brackets and front stabilizer bar from the front suspension crossmember.
10. Remove the 2 No. 1 front stabilizer bar bushings from the front stabilizer bar.
11. To install, reverse the removal procedure.
12. Tighten the four bolts to 14 ft. lbs. (19 Nm).
13. Inspect and adjust front wheel alignment.

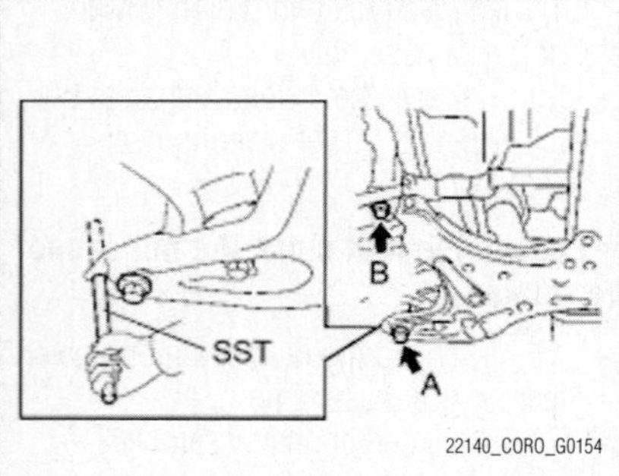

Fig. 177 Installing front suspension arm lower

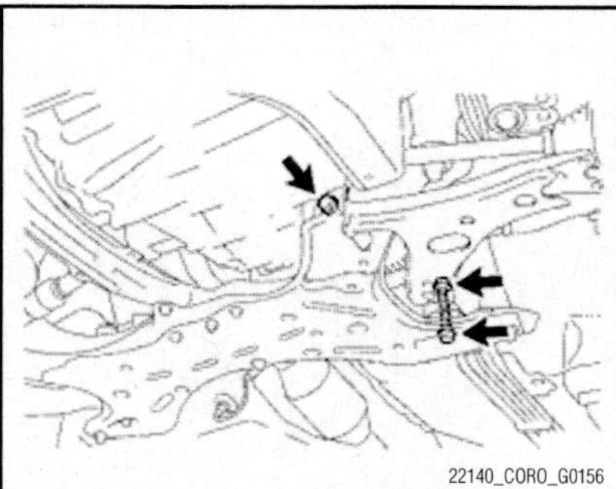

Fig. 179 Tightening front suspension arm lower

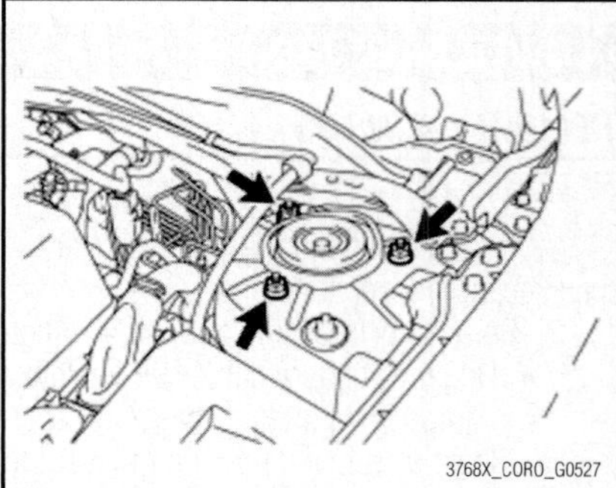

Fig. 180 View of the stabilizer bar, steering knuckle & strut

STEERING KNUCKLE

REMOVAL & INSTALLATION

See Figure 181.

1. Remove front wheel.
2. Remove front axle hub nut.
3. Separate front speed sensor.
4. Separate front disc brake caliper assembly.
5. Remove front disc.
6. Separate tie rod end sub-assembly.
7. Separate front lower suspension arm.
8. Remove front axle assembly.
9. Remove front lower ball joint
10. Remove front axle hub hole snap ring.
11. Remove front axle hub sub-assembly.
12. Remove front disc brake dust cover.
13. Place the front axle hub bearing inner race (outside) on the front axle hub bearing.
14. Using SST's 09223-15020, 09387-02010, 09950-60010, 09950-70010 and a press, remove the front axle hub bearing from the steering knuckle.

To install:

15. Using SST's and a press, install a new front axle hub bearing to the steering knuckle.
16. To compete the installation, reverse the removal procedure.
17. Check the wheel alignment and for speed sensor signal.

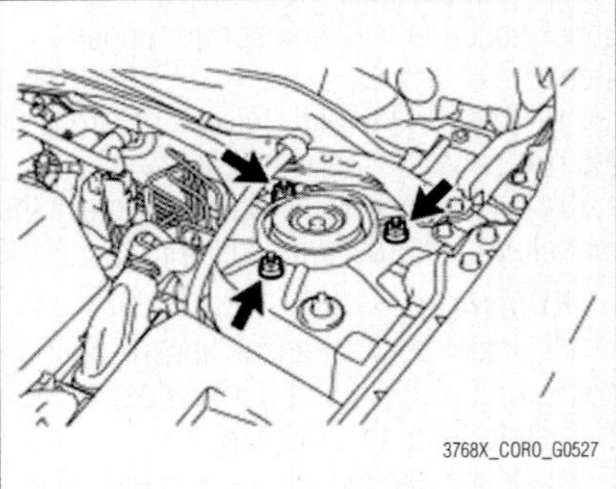

Fig. 181 Identifying the special tools and removing the steering knuckle

STRUT

REMOVAL & INSTALLATION

See Figure 180.

1. Before servicing the vehicle, refer to the Precautions section.
2. Remove front wheel.
3. Remove front wiper arm head cap.
4. Remove front LH and RH wiper arm and blade assembly.
5. Remove hood to cowl top seal.
6. Remove center no.1 cowl top ventilator louver.
7. Remove LH cowl top ventilator louver.
8. Remove windshield wiper motor and link assembly
9. Remove outer cowl top panel.
10. Separate front stabilizer link assembly.
11. Remove the bolt and separate the front flexible hose and the front speed sensor from the front shock absorber with coil spring.

➡Be sure to separate the front speed sensor from the front shock absorber with coil spring completely.

12. Remove the front suspension support dust cover.
13. Loosen the front support to front shock absorber nut of the front shock absorber.

➡Do not remove the front support to front shock absorber nut. Loosen the nut only when the front shock absorber with coil spring needs to be disassembled.

14. Support the front axle using a jack and wooden block.
15. Remove the 2 bolts and 2 nuts, and separate the front shock absorber with coil spring (lower side) from the steering knuckle.
16. Remove the 3 nuts and front shock absorber with coil spring.

➡Make sure that the front speed sensor is completely separated from the front shock absorber with coil spring.

To install:

17. Install or connect the following:
    - Nuts holding the strut to the strut tower. Nuts to 29 ft. lbs. (39 Nm)
    - Steering knuckle to the strut lower bracket
18. Insert the 2 bolts from the rear side and tighten the strut-to-steering knuckle arm bolts. Tighten as follows: 113 ft. lbs. (153 Nm).
19. Install or connect the following:
    - Brake line to the steering knuckle
    - If equipped with ABS, secure the wiring harness
    - Wheel
    - Negative battery cable
20. Check and adjust the alignment, if needed.

OVERHAUL

See Figure 182.

1. To disassemble the strut:
    a. Install a bolt and 2 nuts to the bracket at the lower portion of the strut shell and secure it in a vise.
    b. Compress the coil spring.
    c. Remove the dust cover and hold the spring seat so that it will not turn. Remove the nut on the top of the strut.
    d. Remove the suspension support, bearing, dust seal, spring seat, spring, insulators and bumper.
    e. Remove front coil spring lower insulator and front shock absorber

To install:

2. To assemble the strut:
    a. Install the spring bumper to piston.
    b. Using a spring compressor, compress the spring.
    c. Install the coil spring to the strut. Fit the lower end of the coil spring into the gap of the lower seat.
    d. Install the spring seat with the insulator.
    e. Install the dust seal on the spring seat.
    f. Install the suspension support and tighten 35 ft. lbs. (47 Nm). After the nut has been tighten, release the compressor tool tension.
    g. Pack multipurpose grease into the suspension support. Install the dust cover.

➡Do not use an impact wrench to tighten the nut. Also, check that the bearing fits into the recess in the suspension support.

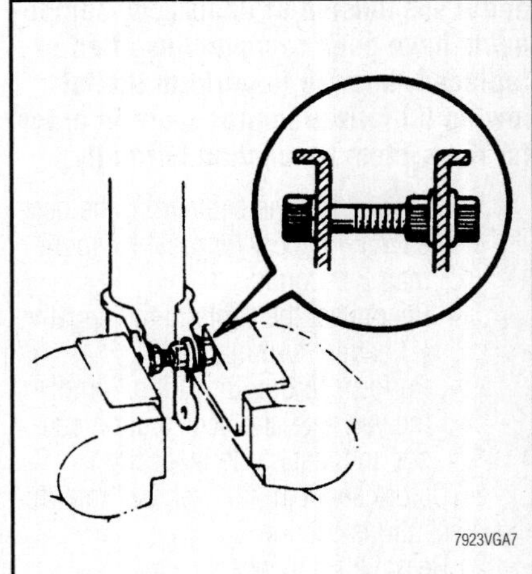

Fig. 182 Proper method of supporting the strut in a vise

WHEEL HUB & BEARING (SEALED UNIT)

REMOVAL & INSTALLATION

See Figures 180, 183 and 184.

1. Remove front wheel.
2. Remove front axle hub nut. Refer to Front Brakes.
3. Separate front speed sensor. Refer to Brakes.
4. Separate front disc brake caliper assembly. Refer to Front Brakes.
5. Remove front disc. Refer to Front Brakes.
6. Separate tie rod end sub-assembly.
7. Separate front lower suspension arm.
8. Put match marks on the front drive shaft assembly and the front axle hub sub-assembly.
9. Using a plastic hammer, separate the front drive shaft assembly from the front axle assembly.

➡Be careful not to damage the drive shaft boot or speed sensor rotor.

10. Remove the 2 bolts, 2 nuts, and front axle assembly.
11. Remove the lower ball joint.
12. Using snap ring pliers, remove the front axle hub hole snap ring.

➡Do not over tighten the vise.

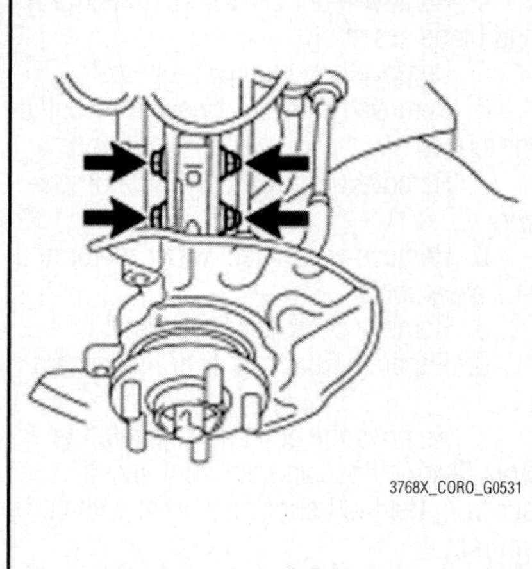

Fig. 183 Removing the 2 bolts, 2 nuts and front axle assembly

13. Remove front axle hub sub-assembly:
    a. Using a hub puller (SST 09520-00031), remove the front axle hub sub-assembly.

➡Apply a small amount of grease to the threads and tip of SST (09953-04020) before use.

    b. Using SST 09950-40011 and 09950-60010, remove the bearing inner race (outside) from the front axle hub sub-assembly.

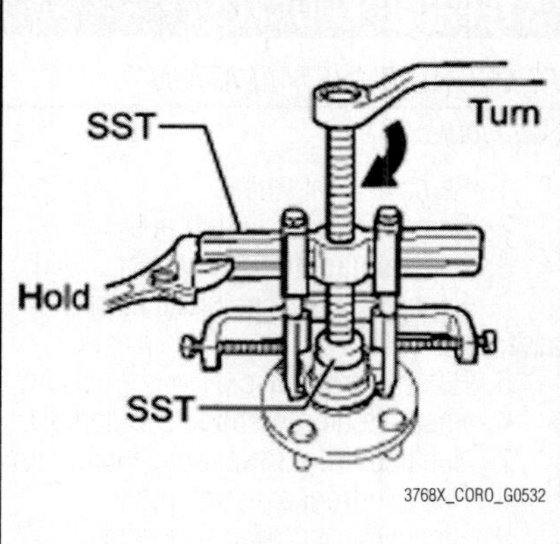

Fig. 184 Identifying the special tools and removing the bearing inner race from the front axle hub sub-assembly

14. Remove the 3 bolts and the front disc brake dust cover from the steering knuckle.
15. Remove front axle hub bearing:
    a. Place the front axle hub bearing inner race (outside) on the front axle hub bearing.
    b. Using SST and a press, remove the front axle hub bearing from the steering knuckle.
16. Using SST's 09223-15020, 09387-02010, 09950-60010, 09950-70010 and a press, remove the front axle hub bearing from the steering knuckle.

SUSPENSION

REAR SUSPENSION

MACPHERSON STRUTS

REMOVAL & INSTALLATION

See Figures 185.

➡For vehicles equipped with VSC, if the wheel alignment has been adjusted, and if suspension or underbody components have been removed/installed or replaced, be sure to perform the following initialization procedure in order for the system to function normally:

- Disconnect the cable from the negative (-) battery terminal for more than 2 seconds.
- Reconnect the cable to the negative (-) battery terminal.
- Perform zero point calibration of the yaw rate and acceleration sensor and test mode inspection.

1. Before servicing the vehicle, refer to the Precautions section.
2. Remove spare wheel cover.
3. Remove rear floor finish plate:
    a. Using a clip remover, remove the 2 clips.
    b. Disengage the 8 claws and remove the rear floor finish plate.
4. Remove the 4 clips and inner luggage compartment trim cover.
5. Remove rear wheel.
6. Support the rear axle beam assembly using a jack and wooden block.
7. Remove the nut and rear shock absorber cushion retainer.

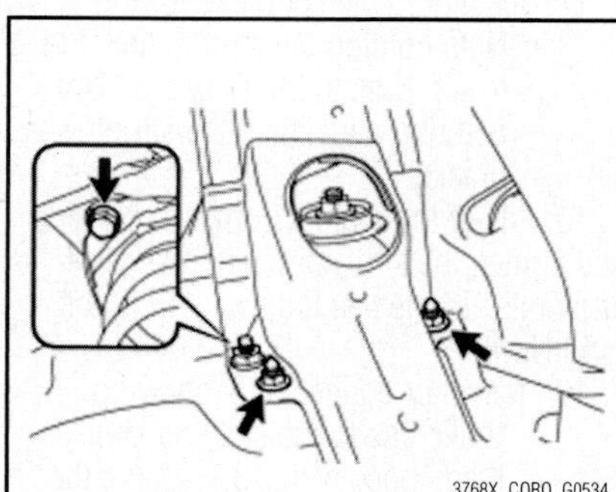

Fig. 185 Removing the 2 nuts from the rear shock absorber with coil spring (upper side)

8. Remove the 2 nuts from the rear shock absorber with coil spring (upper side).
9. Remove the bolt (lower side) from the rear shock absorber with coil spring.
10. Slowly lower the jack and remove the rear shock absorber with coil spring.

To install:

11. Install the rear shock absorber with coil spring with the bolt (lower side). Tighten to 59 ft. lbs. (80 Nm).
12. Install the rear shock absorber with coil spring with the 2 nuts (upper side). Tighten to 59 ft. lbs. (80 Nm).
13. Support the rear axle beam using a jack and wooden block.
14. Temporarily install the rear shock absorber with coil spring (lower side) and rear shock absorber cushion retainer with the nut.
15. Install the rear wheel.
16. Lower the vehicle.
17. Bounce the vehicle up and down several times to stabilize the suspension.
18. Jack up the rear axle beam, placing a wooden block underneath to avoid damage.

Apply load to the suspension so that the rear shock absorber with coil spring is positioned as shown in the illustration.

a. Length A: 8.11 inches

b. If the rear shock absorber with coil spring cannot be positioned as shown in the illustration even when the rear axle beam is jacked up, apply additional load to the vehicle such as by having a person sit in the rear seat.

CAUTION

Do not jack up the rear axle beam too high as the vehicle may fall.

19. Fully tighten the nut on the rear shock absorber with coil spring (lower side). Tighten to 59 ft. lbs. (80 Nm).

➡The final torque must be applied under standard vehicle height conditions.

20. To complete the installation, reverse the removal procedure.

21. Perform yaw rate sensor zero point calibration.

22. Check steering angle sensor zero point calibration:

a. Drive the vehicle straight ahead at 35 km/h (22 mph) or more for at least 5 seconds.

b. Check that the center position of the steering wheel is correctly set while driving straight ahead.

➡If front wheel alignment and steering position are adjusted as a result of an off-centered position of the steering wheel, acquire yaw rate and acceleration sensor zero point again after the adjustments are completed.

c. If the center position of the steering wheel is correctly set, reconfirm the DTC.

d. If the center position of the steering wheel is not correctly set, adjust the front wheel alignment or steering position.

OVERHAUL

1. Place the strut assembly in a pipe vise or strut vise.

➡Do not attempt to clamp the strut assembly in a flat jaw vise as this will result in damage to the strut tube.

2. Compress the spring until the upper suspension support is free of any spring tension. Do not over-compress the spring.

3. Hold the upper support, then remove the nut on the end of the shock piston rod.

4. Remove the support, coil spring, insulator, and bumper.

5. Inspect the strut as follows:

a. Check the shock absorber by moving the piston shaft through its full range of travel. It should move smoothly and evenly throughout its entire travel without any trace of binding or notching.

b. Use a small straightedge to check the piston shaft for any bending or deformation.

c. Inspect the spring for any sign of deterioration or cracking. The waterproof coating on the coils should be intact to prevent rusting.

YAW RATE SENSOR ZERO POINT CALIBRATION (WITH VSC)

See Figure 187.

➡While obtaining the zero point, keep the vehicle stationary and do not vibrate, tilt, move, or shake it. (Do not start the engine.) Be sure to perform this procedure on a level surface (with an inclination of less than 1 degree).

1. Clear the zero point calibration data:

a. Turn the ignition switch off.

b. Check that the steering wheel is centered.

c. Check that the shift lever is in P (for automatic transaxle model) or the parking brake is applied (for manual transaxle model).

d. Turn the ignition switch to ON.

e. The warning and indicator light come on for 3 seconds to indicate that the initial check is completed.

f. Connect and disconnect terminals TS and CG of the DLC3 (ODB II connector) 4 times or more within 8 seconds.

g. Check that the VSC OFF indicator light comes on.

➡If the ignition switch is turned to ON for more than 15 seconds with the shift lever in P (for automatic transaxle model) or the parking brake applied (for manual transaxle model) after zero point of the yaw rate and acceleration sensor has been cleared, only the zero point of the yaw rate sensor will be stored. If the vehicle is driven under these conditions, the skid control ECU will store the zero point calibration for the acceleration sensor as not being completed. The skid control ECU will then also indicate this as a malfunction of the VSC system using the indicator lights.

2. Perform zero point calibration of the yaw rate and acceleration sensor:

a. Turn the ignition switch off.

b. Check that the steering wheel is centered.

c. Check that the shift lever is in P (for automatic transaxle model) or the parking brake is applied (for manual transaxle model).

➡DTCs C1210/36 and C1336/39 will be recorded if the shift lever is not in P (for automatic transaxle model) or the parking brake is not applied (for manual transaxle model)

d. Connect terminals TS and CG of the DLC3 (ODB II connector).

e. Turn the ignition switch to ON.

f. Keep the vehicle stationary on a level surface for 2 seconds or more.

g. Check that the VSC OFF indicator light comes on for several seconds and then blinks in Test Mode.

➡Note the following:

- The slip indicator light remains on during Test Mode because traction control operation is prohibited (The slip indicator light goes off when the VSC OFF switch is on).
- If the VSC OFF indicator light does not blink, perform zero point calibration again.
- The zero point calibration is performed only once after the system enters Test Mode.

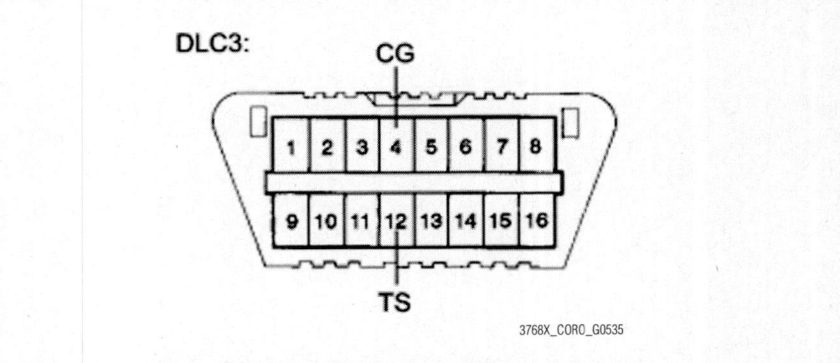

Fig. 187 Identifying the TS and CG terminals on the DLC3 (ODB II connector)

- Calibration cannot be performed again until the stored data is cleared.

h. Turn the ignition switch off and disconnect SST from the DLC3 (ODB II connector).

STABILIZER BAR

REMOVAL & INSTALLATION

See Figure 188.

1. Remove the rear axle beam damper.
2. Remove the 2 bolts, 2 nuts and rear stabilizer bar.

➡Be sure to loosen the nuts.

3. To install, reverse removal procedure.

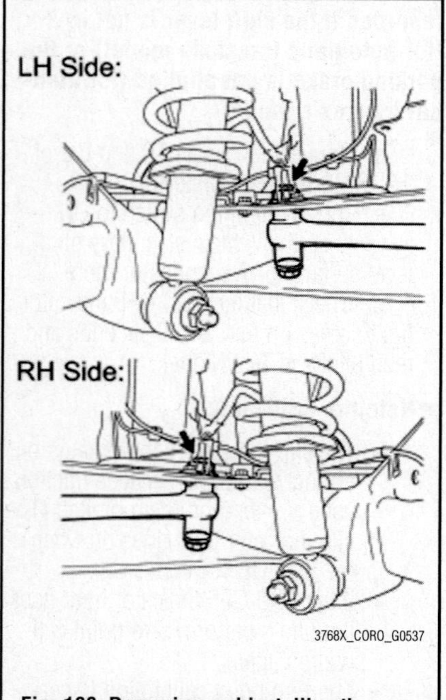

Fig. 188 Removing and installing the rear stabilizer bar

4. Check that the identification mark of the rear stabilizer bar is positioned on the right side of the vehicle.
5. Install the rear stabilizer bar with the 2 bolts and 2 nuts. Tighten to 184 ft. lbs. (250 Nm).

➡Be sure to tighten the nuts. If reusing the bolts, insert them from the upper side of the vehicle.

WHEEL HUB & BEARING

REMOVAL & INSTALLATION

See Figures 189 and 190.

1. Disconnect cable from negative battery terminal.
2. Remove lower instrument panel finish panel.
3. Remove shift lever knob sub-assembly.
4. Remove center instrument cluster finish panel assembly.
5. Remove upper console panel sub-assembly.
6. Loosen parking brake cable (for rear disc brake).

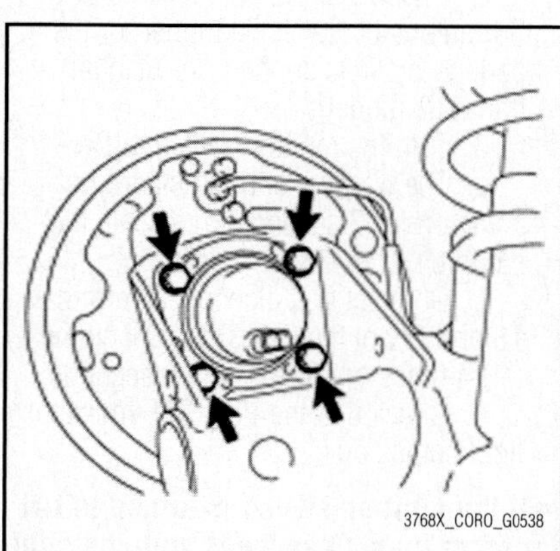

Fig. 189 Removing the 4 bolts and the rear axle hub and bearing assembly—Drum brakes

7. Remove rear wheel.
8. Disconnect rear speed sensor wire (for rear drum brake).
9. Disconnect rear speed sensor wire (for rear disc brake).
10. Remove parking brake lever protector (for rear disc brake).
11. Separate no. 3 parking brake cable assembly (for rear disc brake).
12. Separate rear disc brake caliper assembly (for rear disc brake).
13. Remove rear disc (for rear disc brake).
14. Remove rear brake drum (for rear drum brake).
15. Remove the 4 bolts and the rear axle hub and bearing assembly.

To install:

16. Install the rear axle hub and bearing assembly with the 4 bolts. Tighten to 74 ft. lbs. (100 Nm).
17. To complete installation, reverse remaining removal procedure.
18. Inspect rear wheel alignment.
19. Check for speed sensor signal.

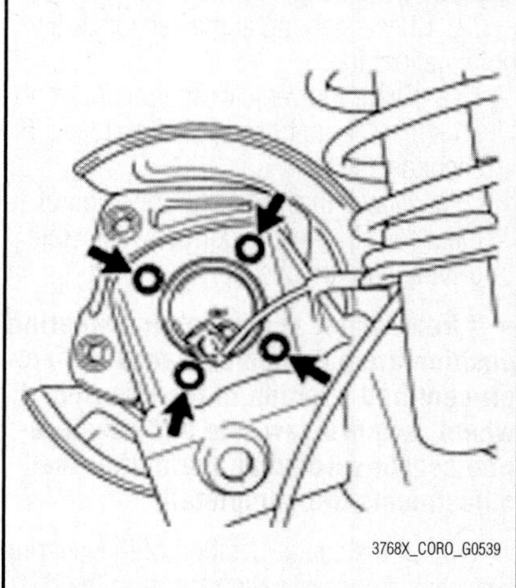

Fig. 190 Removing the 4 bolts and the rear axle hub and bearing assembly—Disc brakes

TOYOTA

FJ Cruiser

8

SPECIFICATIONS AND MAINTENANCE CHARTS

ENGINE AND VEHICLE IDENTIFICATION

Engine							Model Year	
Code/ID ①	Liters (cc)	Cu. In.	Cyl.	Fuel Sys.	Engine Type	Eng. Mfg.	Code ②	Year
1GR-FE/U	4.0 (3956)	241	6	SFI	DOHC	Toyota	9	2009
							A	2010

MFI: Multi-port Fuel Injection

DOHC: Double Overhead Camshaft

① 1GR-FE engine: stamped on the right side of the engine block. Engine ID is the fifth character of the VIN number.

② 10th digit of the VIN number

3768X_FJCR_C0001

GENERAL ENGINE SPECIFICATIONS

Year	Model	Engine Displacement Liters	Engine Series Code/ID	Net Horsepower @ rpm	Net Torque @ rpm (ft. lbs.)	Bore x Stroke (in.)	Compression Ratio	Oil Pressure @ rpm
2009	FJ Cruiser	4.0	1GR-FE/U	239@5200	278@3800	3.70x3.74	10.0:1	43-85@3000
2010	FJ Cruiser	4.0	1GR-FE/U	260@5600	271@4400	3.70x3.74	10.4:1	43-85@3000

3768X_FJCR_C0002

ENGINE TUNE-UP SPECIFICATIONS

Year	Engine Displacement Liters	Engine Code/ID	Spark Plug Gap (in.)	Ignition Timing (deg.)	Fuel Pump (psi)	Idle Speed (rpm)		Valve Clearance	
						MT	AT	Intake	Exhaust
2009	4.0	1GR-FE/U	0.039-0.043	7-24B ①	41-42	650-750	650-750	0.006-0.010	0.011-0.015
2010	4.0	1GR-FE/U	0.039-0.043	7-24B ①	41-42	650-750	650-750	NA	NA

NOTE: The Vehicle Emission Control Information label often reflects specification changes made during production.

The label figures must be used if they differ from those in this chart.

B: Before top dead center idle

① With terminals TC and CG of the DLC3 (ODB II connector) connected

3768X_FJCR_C0003

CAPACITIES

Year	Model	Engine Displacement Liters	Engine Code/ID	Engine Oil with Filter (qts.)	Transmission (pts.) 6-Spd	Transmission (pts.) Auto.	Transfer Case (pts.)	Drive Axle Front (pts.)	Drive Axle Rear (pts.)	Fuel Tank (gal.)	Cooling System (qts.)
2009	FJ Cruiser	4.0	1GR-FE/U	5.5	3.8	6.0	①	②	6.4 ③	21.1	⑤
2010	FJ Cruiser	4.0	1GR-FE/U	5.5	3.8	6.0	①	④	6.4 ③	21.1	⑤

① MT: 3.0 pts
AT: 2.2 pts
② Full Time 4WD: 3.0 pts
Part Time 4WD: 3.2 pts
③ W/ Locking Differential: 6.2 pts
④ MT: 3.0 pts
AT: 3.4 pts
⑤ MT: 9.9
AT: 10.4

3768X_FJCR_C0004

FLUID SPECIFICATIONS

Year	Model	Engine Displ. Liters	Engine Oil	Man. Trans.	Auto. Trans.	Drive Axle Front	Drive Axle Rear	Transfer Case	Power Steering Fluid	Brake Master Cylinder	Cooling System
2009	FJ Cruiser	4.0	5W-30	75W-90	Dexron III	①	80W-90	75W-90	Dexron II or III	DOT 3	S-LLC
2010	FJ Cruiser	4.0	5W-30	75W-90	Dexron III	②	80W-90	75W-90	Dexron II or III	DOT 3	S-LLC

DOT: Department Of Transportation
S-LLC: Toyota Super Long Life Coolant
① Full Time 4WD: 80W-90
Part Time 4WD: 75W-85
② MT: 80W-90
AT: 75W-85

3768X_FJCR_C0014

VALVE SPECIFICATIONS

Year	Engine Displacement Liters	Engine Code/ID	Seat Angle (deg.)	Face Angle (deg.)	Spring Test Pressure (lbs. @ in.)	Spring Installed Height (in.)	Stem-to-Guide Clearance (in.) Intake	Stem-to-Guide Clearance (in.) Exhaust	Stem Diameter (in.) Intake	Stem Diameter (in.) Exhaust
2009	4.0	1GR-FE/U	NA	44.5	41.9-46.3@ 1.311	1.882	0.0010-0.0024	0.0012-0.0026	0.2154-0.2159	0.2152-0.2158
2010	4.0	1GR-FE/U	NA	44.5	41.9-46.3@ 1.311	1.910	0.0010-0.0024	0.0012-0.0026	0.2154-0.2160	0.2152-0.2160

NA: Not Available

3768X_FJCR_C0005

CAMSHAFT SPECIFICATIONS

All measurements in inches unless noted

Year	Engine Displacement Liters	Engine Code/ID	Journal Dia.	Brg. Oil Clearance	Shaft End-play	Circle Runout	Lobe Height	
							Intake	Exhaust
2009	4.0	1GR-FE/U	①	②	0.0160-0.0350	0.0024	1.7389-1.7428	1.7551-1.7591
2010	4.0	1GR-FE/U	③	④	NA	0.00157	1.728-1.7320	1.743-1.7470

NA: Not Available

① No. 1: 1.4162-1.4167
All others: 0.9039-0.9045

② No. 1: 0.0016-0.0031
All others: 0.0010-0.0024

③ No. 1: 1.4152-1.4157
All others: 1.0221-1.0226

④ No. 1: 0.00157-0.00311
All others: 0.000984-0.0224

3768X_FJCR_C0006

CRANKSHAFT AND CONNECTING ROD SPECIFICATIONS

All measurements are given in inches.

Year	Engine Displacement Liters	Engine Code/ID	Crankshaft				Connecting Rod		
			Main Brg. Journal Dia.	Main Brg. Oil Clearance	Shaft End-play	Thrust on No.	Journal Diameter	Oil Clearance	Side Clearance
2009	4.0	1GR-FE/U	2.8342-2.8346	0.0007-0.0012	0.0016-0.0094	NA	2.2044-2.2047	0.0010-0.0018	0.0059-0.0118
2010	4.0	1GR-FE/U	2.8342-2.8346	0.0007-0.0012	0.0016-0.0094	NA	2.2044-2.2047	0.0010-0.0018	0.0059-0.0118

NA: Not Available

3768X_FJCR_C0007

PISTON AND RING SPECIFICATIONS

All measurements are given in inches.

Year	Engine Displacement Liters	Engine Code/ID	Piston Clearance	Ring Gap			Ring Side Clearance		
				Top Compression	Bottom Compression	Oil Control	Top Compression	Bottom Compression	Oil Control
2009	4.0	1GR-FE/U	0.0031-0.0040	0.0118-0.0157	0.0157-0.0197	0.0039-0.0157	0.0008-0.0028	0.0008-0.0024	0.0028-0.0060
2010	4.0	1GR-FE/U	0.0035-0.0020	0.008-0.0142	0.0138-0.0193	0.0039-0.0173	0.0007-0.0027	0.0007-0.0023	0.0027-0.0059

3768X_FJCR_C0008

TORQUE SPECIFICATIONS

All readings in ft. lbs.

Year	Engine Displacement Liters	Engine Code/ID	Cylinder Head Bolts	Main Bearing Bolts	Rod Bearing Bolts	Crankshaft Damper Bolts	Flywheel Bolts	Manifold Intake	Manifold Exhaust	Spark Plugs	Oil Pan Drain Plug
2009	4.0	1GR-FE/U	①	②	③	185	61	19	22	15	30
2010	4.0	1GR-FE/U	①	②	③	185	61	19	22	15	30

① Right side: 27 ft. lbs. plus 180 degrees
Left side (recessed head): 27 ft. lbs. plus 180 degrees
Left side (0.55 inch head): 22 ft. lbs.

② 12 pointed head: 45 ft. lbs. plus 90 degrees
12mm head: 18 ft. lbs.

③ Step 1: 18 ft. lbs.
Step 2: Plus 90 degrees

3768X_FJCR_C0009

WHEEL ALIGNMENT

Year	Model	Caster Range (+/-Deg.)	Caster Preferred Setting (Deg.)	Camber Range (+/-Deg.)	Camber Preferred Setting (Deg.)	Toe-in (in.)	Steering Axis Inclination (Deg.)
2009	2WD	0.50	3.57	0.50	-0.57	0.04+/-0.08	12.92+/-0.50
	4WD	0.50	2.82	0.50	0.15	0.04+/-0.08	12.35+/-0.50
2010	2WD	0.50	3.57	0.50	-0.57	0.04+/-0.08	12.92+/-0.50
	4WD	0.50	2.82	0.50	0.15	0.04+/-0.08	12.35+/-0.50

NOTE: All alignment figures based on the following nominal ride heights:
2WD Front: 4.56 in.
2WD Rear: 3.20 in.
4WD Front: 3.43 in.
4WD Rear: 2.43 in.

3768X_FJCR_C0010

TIRE, WHEEL AND BALL JOINT SPECIFICATIONS

Year	Model	OEM Tires		Tire Pressures (psi)		Wheel	Ball Joint	Lug Nut Torque
		Standard	Optional	Front	Rear	Size	Inspection	(ft. lbs.)
2009	FJ Cruiser	P265/70R17	P265/75R16	32	32	NA	①	82
2010	FJ Cruiser	P265/70R17	P265/75R16	32	32	NA	①	82

① Lower ball joint excessive play, all models: 0.020 inch

3768X_FJCR_C0011

BRAKE SPECIFICATIONS

All measurements in inches unless noted

Year		Brake Disc			Brake Drum Diameter		Minimum Lining Thickness		Brake Caliper	
		Original Thickness	Minimum Thickness	Maximum Runout	Original Inside Diameter	Maximum Machine Diameter	Front	Rear	Bracket Bolts (ft. lbs.)	Mounting Bolts (ft. lbs.)
2009	F	1.102	1.024	0.0020	NE	NE	0.039	NA	NA	91
	R	0.709	0.630	0.0079	NE	NE	NA	0.039	NA	65
2010	F	1.102	1.024	0.0020	NE	NE	0.039	NA	NA	91
	R	0.709	0.630	0.0079	NE	NE	NA	0.039	NA	65

NE: Not Equipped
NA: Not Available

3768X_FJCR_C0012

SCHEDULED MAINTENANCE INTERVALS

FJ CRUISER

TO BE SERVICED	TYPE OF SERVICE	VEHICLE MILEAGE INTERVAL (x1000)												
		5	10	15	20	25	30	35	40	45	50	55	60	65
Automatic transmission fluid	S/I						✓						✓	
Ball joints & dust covers	S/I			✓			✓			✓			✓	
Brake line pipes & hoses	S/I			✓			✓			✓			✓	
Brake pads & discs/linings & drums (front & rear)	S/I	✓	✓	✓	✓	✓	✓	✓	✓	✓	✓	✓	✓	✓
Cabin air filter	R						✓						✓	
Drive belts	S/I	At 60,000 miles, then every 15,000 miles thereafter												
Driveshaft boots (4WD)	S/I			✓			✓			✓			✓	
Engine air filter	R						✓						✓	
Engine coolant	S/I			✓			✓			✓			✓	
Engine coolant	R	At 100,000 miles, then every 50,000 miles												
Engine oil & filter	R	✓	✓	✓	✓	✓	✓	✓	✓	✓	✓	✓	✓	✓
Exhaust pipes & mountings	S/I			✓			✓			✓			✓	
Front differential oil (4WD)	S/I			✓			✓			✓			✓	
Fuel tank cap gasket	S/I						✓						✓	
Fuel lines & connections	S/I						✓						✓	
Manual transmission oil	S/I						✓						✓	
Propeller shaft (4WD)	L			✓			✓			✓			✓	
Propeller shaft bolt	S/I			✓			✓			✓			✓	
Radiator, condenser and/or intercooler	S/I			✓			✓			✓			✓	
Rear differential oil	S/I			✓			✓			✓			✓	
Rotate tires	S/I	✓	✓	✓	✓	✓	✓	✓	✓	✓	✓	✓	✓	✓
Spark plugs	R						✓						✓	
Steering gear box	S/I			✓			✓			✓			✓	
Steering linkage & boots	S/I			✓			✓			✓			✓	
Transfer Case (4WD)	R						✓						✓	

FREQUENT OPERATION MAINTENANCE (SEVERE SERVICE)

Driving on dirt roads or dusty roads:

Air filter: Inspect every 15,000 miles

Ball joints and dust covers: Inspect every 5,000 miles

Drive shaft boots (4WD): Inspect every 5,000 miles

Engine air filter: Inspect every 5,000 miles

Lubricate propeller shaft (4WD): Lubricate every 5,000 miles

Nuts and bolts on chassis: Tighten every 5,000 miles

Propeller shaft bolt: Re-torque every 5,000 miles

Steering linkage and boots: Inspect every 5,000 miles

Driving while towing:

Automatic transmission fluid: Replace every 60,000 miles

Front differential oil (4WD): Replace every 15,000 miles

Manual transmission oil (4WD): Replace every 30,000 miles

Nuts and bolts on chassis: Tighten every 5,000 miles

Propeller shaft (4WD): Lubricate every 5,000 miles

Propeller shaft bolt: Re-torque every 5,000 miles

Rear differential oil : Replace every 15,000 miles

Transfer case oil (4WD): Replace every 60,000 miles

3768X_FJCR_C0013

BRAKES — INFORMATION AND PRECAUTIONS

ANTI-LOCK SYSTEMS

• Certain components within the ABS system are not intended to be serviced or repaired individually.

• Do not use rubber hoses or other parts not specifically specified for and ABS system. When using repair kits, replace all parts included in the kit. Partial or incorrect repair may lead to functional problems and require the replacement of components.

• Lubricate rubber parts with clean, fresh brake fluid to ease assembly. Do not use shop air to clean parts; damage to rubber components may result.

• Use only DOT 3 brake fluid from an unopened container.

• If any hydraulic component or line is removed or replaced, it may be necessary to bleed the entire system.

• A clean repair area is essential. Always clean the reservoir and cap thoroughly before removing the cap. The slightest amount of dirt in the fluid may plug an orifice and impair the system function. Perform repairs after components have been thoroughly cleaned; use only denatured alcohol to clean components. Do not allow ABS components to come into contact with any substance containing mineral oil; this includes used shop rags.

• The Anti-Lock control unit is a microprocessor similar to other computer units in the vehicle. Ensure that the ignition switch is **OFF** before removing or installing controller harnesses. Avoid static electricity discharge at or near the controller.

• If any arc welding is to be done on the vehicle, the control unit should be unplugged before welding operations begin.

DISC AND DRUM SYSTEMS

**** CAUTION**

Dust and dirt accumulating on brake parts during normal use may contain asbestos fibers from production or aftermarket brake linings. Breathing excessive concentrations of asbestos fibers can cause serious bodily harm. Exercise care when servicing brake parts. Do not sand or grind brake lining unless equipment used is designed to contain the dust residue. Do not clean brake parts with compressed air or by dry brushing. Cleaning should be done by dampening the brake components with a fine mist of water, then wiping the brake components clean with a dampened cloth. Dispose of cloth and all residue containing asbestos fibers in an impermeable container with the appropriate label. Follow practices prescribed by the Occupational Safety and Health Administration (OSHA) and the Environmental Protection Agency (EPA) for the handling, processing, and disposing of dust or debris that may contain asbestos fibers.

BRAKES — BLEEDING THE BRAKE SYSTEM

BLEEDING PROCEDURE

BLEEDING & FILLING PROCEDURE

See Figure 1.

1. Before servicing the vehicle, refer to the precautions.

**** WARNING**

Depressing the brake pedal with the reservoir cap removed will cause the fluid to spray. When bleeding, maintain the amount of fluid in the reservoir between the Min. and Max. lines.

2. Fill reservoir with brake fluid.
3. Bleed brake booster with accumulator pump assembly.
4. If the brake master cylinder is disassembled, the brake line is disconnected from the brake master cylinder or if the reservoir becomes empty, bleed the brake master cylinder.
   a. Turn the ignition switch to **ON**, and wait until the pump motor has stopped. Pump operating sound can be heard.
   b. Turn the ignition switch to **OFF**, and depress the brake pedal more than 20 times.
   c. When pressure in the accumulator is released, the reaction force becomes lighter and the stroke becomes longer.
   d. Repeat the first two steps 5 times.
   e. Turn the ignition switch to **ON**, and check that the pump stops after approximately 8 to 14 seconds.

**** WARNING**

If the pump does not stop, repeat the procedure again.

5. Bleed brake line.
   a. Turn the ignition switch to **ON**, and wait until the pump motor has stopped. Pump operating sound can be heard.
   b. For the front brake line, connect the vinyl tube to the brake caliper. Depress the brake pedal several times, then loosen the bleeder plug with the pedal held down.

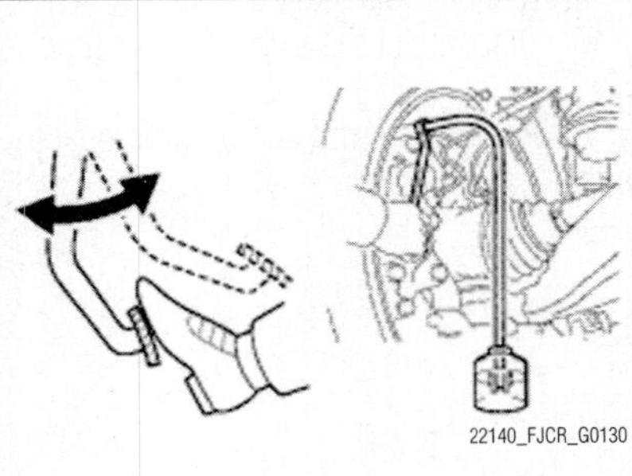

Fig. 1 Connect the vinyl tube to the brake caliper

   c. At the point when the fluid stops coming out, tighten the bleeder plug, then release the brake pedal.
   d. Repeat the steps until all the air in the fluid has been bled out. Tighten the bleeder plug to 8 ft. lbs. (11 Nm).
   e. Repeat the above procedures to bleed the other brake line.
6. For the rear brake line, connect the vinyl tube to the brake caliper. Depress the brake pedal, hold it, and then loosen the bleeder plug. Brake fluid is pumped out automatically.
   a. Loosen the bleeder plug and release the air. Keep the brake fluid in the reservoir tank above the **MIN** line during the above procedures.
   b. When the air is completely bled out of the brake fluid through the bleeder plug, tighten the bleeder plug. Tighten the bleeder plug to 8 ft. lbs. (11 Nm).
   c. Repeat the above procedures to bleed the other brake line.
7. Bleed master cylinder solenoid.
8. If the brake master cylinder is disassembled, the brake line is disconnected from the brake master cylinder or if the reservoir becomes empty, bleed the brake master cylinder.
   a. Connect the scan tool to the diagnostic link connector.
   b. Turn the ignition switch to **ON**.

c. Select **ACTIVE TEST** mode on the scan tool.

d. Connect the vinyl tube to the rear brake caliper.

e. Loosen the bleeder plug.

f. Select **SRMF** to drive the solenoids and bleed air from the rear brake caliper.

⁂ WARNING

Do not depress the brake pedal. Keep the brake fluid in the reservoir tank above the MIN line during the above procedures. Brake fluid is sent through the pump.

➡To protect the solenoids, the Techstream turns OFF automatically 2 seconds after every solenoid has been turned ON.

g. Repeat steps until all the air in the brake fluid is bled out.

h. When the air is completely bled out of the brake fluid through the bleeder plug, tighten the bleeder plug to 8 ft. lbs. (11 Nm).

i. Repeat the above procedures to bleed the other brake line.

j. Turn the ignition switch to **OFF**. Turn the ignition switch to **ON**.

k. Clear DTC.

9. Check fluid level in reservoir

a. Turn the ignition switch to **OFF**, and depress the brake pedal more than 20 times (until the pedal reaction feels light and pedal stroke becomes longer), and adjust the fluid level to the MAX level.

b. When the ignition switch is turned to **ON**, brake fluid is sent to the accumulator and the fluid level decreases by approximately 5 mm from the level when the ignition switch is **OFF** (normal).

BRAKES — ANTI-LOCK BRAKE SYSTEM (ABS)

WHEEL SPEED SENSORS

REMOVAL & INSTALLATION

Front

See Figure 2.

1. Before servicing the vehicle, refer to the precautions.
2. Disconnect the negative battery cable.
3. Raise and safely support the vehicle securely on jackstands.
4. Remove front wheel
5. Disconnect the skid control sensor wire.
6. Using a 5 mm hexagon wrench, remove the bolt and the front speed sensor.

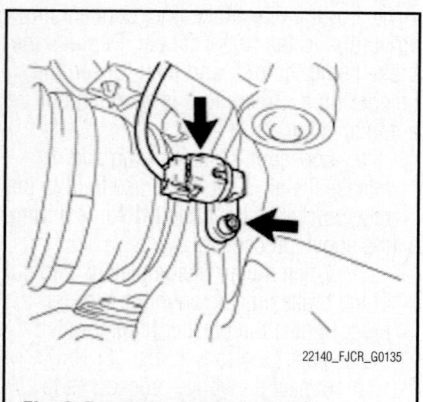

Fig. 2 Front wheel speed sensor

⁂ WARNING

Do not let any foreign matter attach to the sensor tip or enters the sensor bore.

To install:

⁂ WARNING

Make sure that the sensor tip is clean.

7. Install the wheel speed sensor and using a 5 mm hexagon wrench, tighten the bolt to 73 inch lbs. (98 Nm).
8. Connect the skid control sensor wire.
9. Install front wheel and lower the vehicle.
10. Connect the negative battery cable.
11. Using a scan tool, check VSC sensor signal.

Rear

See Figure 3.

1. Before servicing the vehicle, refer to the precautions.
2. Disconnect the negative battery cable.
3. Raise and safely support the vehicle securely on jackstands.
4. Remove rear wheel
5. Disconnect the skid control sensor wire.
6. Remove the nut and the rear speed sensor.

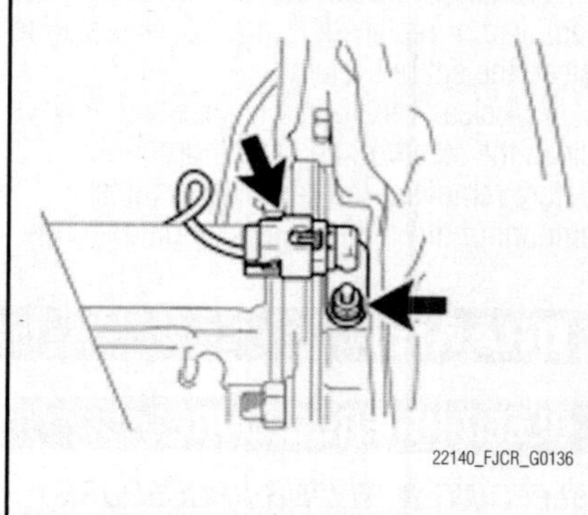

Fig. 3 Rear wheel speed sensor

⁂ WARNING

Do not let any foreign matter attach to the sensor tip or enters the sensor bore.

To install:

⁂ WARNING

Make sure that the sensor tip is clean.

7. Install the wheel speed sensor and tighten the bolt to 73 inch lbs. (98 Nm).
8. Connect the skid control sensor wire.
9. Install rear wheel and lower the vehicle.
10. Connect the negative battery cable.
11. Using a scan tool, check VSC sensor signal.

BRAKE CALIPER

REMOVAL & INSTALLATION

See Figures 4 and 5.

1. Before servicing the vehicle, refer to the precautions.
2. Raise and safely support the vehicle securely on jackstands.
3. Remove the front wheels.
4. Drain brake fluid.
5. Remove front disc brake anti rattle with hole pin.
   a. Remove the 2 pin hold clips, then remove the 2 hole pins from the disc brake caliper.
   b. Remove the anti rattle spring from the disc brake pad.
6. Remove the disc brake pads with anti squeal shims from the disc brake caliper.
7. Remove the anti squeal shims from each of the disc brake pads.
8. Using a union nut wrench, separate the brake tube from the disc brake caliper.

➡Use a container to collect the brake fluid as it drains out.

9. Remove the disc brake caliper.

To install:

10. Install the disc brake caliper and tighten the two bolts to 91 ft. lbs. (123 Nm).
11. Install the brake tube onto the disc brake caliper and tighten to 10 ft. lbs. (14 Nm).
12. If necessary, replace the anti squeal shim when replacing the brake pad.
13. Apply disc brake grease to both sides of each No. 1 shim.
14. Install the No. 1 and No. 2 anti squeal shims onto each brake pad.
15. Install the disc brake pads with anti squeal shims onto the disc brake caliper.

**** WARNING**

There should be no oil or grease on the friction surfaces of the disc brake pads or the front disc.

16. Install the anti rattle spring and hole pins onto the disc brake caliper.

➡The anti rattle spring is installed onto the lower hole pin.

17. Install the pin hold clip with its handle oriented in the center of the vehicle.
18. Fill reservoir with brake fluid.
19. Bleed the brakes.
20. Install the front wheel and lower the vehicle.

Fig. 4 Disc brake pads with anti squeal shims

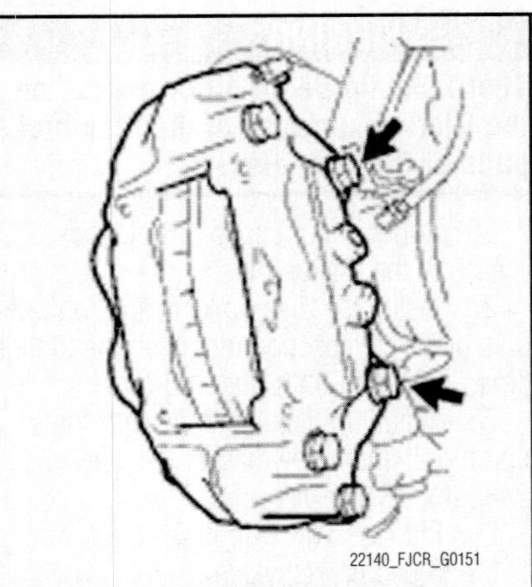

Fig. 5 Disc brake caliper retaining bolts

DISC BRAKE PADS

REMOVAL & INSTALLATION

See Figure 4.

1. Before servicing the vehicle, refer to the precautions.
2. Raise and safely support the vehicle securely on jackstands.
3. Remove the front wheels.
4. Drain brake fluid.
5. Remove front disc brake anti rattle with hole pin.
   a. Remove the 2 pin hold clips, then remove the 2 hole pins from the disc brake caliper.
   b. Remove the anti rattle spring from the disc brake pad.
6. Remove the disc brake pads with anti squeal shims from the disc brake caliper.
7. Remove the anti squeal shims from each of the disc brake pads.

To install:

8. If necessary, replace the anti squeal shim when replacing the brake pad.
9. Apply disc brake grease to both sides of each No. 1 shim.
10. Install the No. 1 and No. 2 anti squeal shims onto each brake pad.
11. Install the disc brake pads with anti squeal shims onto the disc brake caliper.

**** WARNING**

There should be no oil or grease on the friction surfaces of the disc brake pads or the front disc.

12. Install the anti rattle spring and hole pins onto the disc brake caliper.

➡The anti rattle spring is installed onto the lower hole pin.

13. Install the pin hold clip with its handle oriented in the center of the vehicle.
14. Fill reservoir with brake fluid.
15. Bleed the brakes.
16. Install the front wheel and lower the vehicle.

BRAKE CALIPER

REMOVAL & INSTALLATION

See Figure 6.

1. Before servicing the vehicle, refer to the precautions.
2. Raise and safely support the vehicle securely on jackstands.
3. Remove the rear wheels.
4. Drain brake fluid.
5. Using a union nut wrench, separate the brake tube from the disc brake caliper.

➡Use a container to collect the brake fluid as it drains out.

6. Remove the disc brake caliper by remove the 2 slide pins.
7. Remove the 2 disc brake pads with anti squeal shim from the disc brake caliper mounting.
8. Remove the 2 anti squeal shims from each brake.
9. Remove the indicator plate from the inner side of the brake.
10. Remove the 4 pad support plates from the disc brake caliper mounting.
11. Remove the disc brake caliper mounting.
12. Remove the slide bushing from the disc brake caliper mounting.
13. Remove the dust boot from the disc brake caliper mounting.

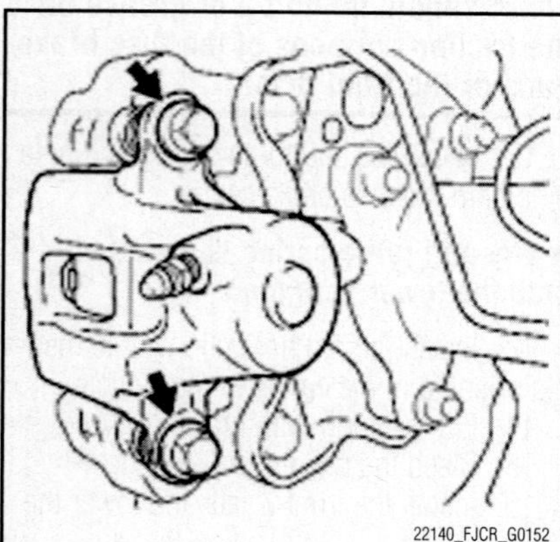

Fig. 6 Disc brake caliper retaining bolts

14. Remove the hole plug from the disc brake caliper mounting.

To install:

15. Install a new hole plug onto the disc brake caliper mounting.
16. Apply lithium soap base glycol grease to a new dust boot.
17. Install the dust boot onto the disc brake caliper mounting.
18. Apply lithium soap base glycol grease to a new slide bushing.
19. Install the slide bushing onto the disc brake caliper mounting.
20. Install the disc brake caliper mounting with the 2 bolts and 2 washers. Tighten to 78 ft. lbs. (105 Nm).
21. Install the 4 pad support plates onto the disc brake caliper mounting.
22. Install the indicator plate onto the inner side brake pad.

➡Install the indicator plate facing downward.

23. If necessary, replace the anti squeal shim when replacing the brake pad.
24. Install the 2 anti squeal shims onto the brake pads.
25. Install the 2 disc brake pads with anti squeal shims onto the disc brake caliper mounting.

⁂ WARNING

There should be no oil or grease on the friction surfaces of the disc brake pads or the rear disc.

26. Apply lithium soap base glycol grease to the 2 slide pins.
27. Install the disc brake caliper onto the disc brake caliper mounting with the 2 slide pins. Tighten to 65 ft. lbs. (88 Nm).
28. Install the flexible hose with the union bolt and a new gasket. Tighten to 23 ft. lbs. (31Nm).
29. Fill reservoir with brake fluid.
30. Bleed brake system.
31. Check fluid level in reservoir.
32. Check for brake fluid leakage.
33. Install rear wheel.

DISC BRAKE PADS

REMOVAL & INSTALLATION

See Figure 4.

1. Before servicing the vehicle, refer to the precautions.
2. Raise and safely support the vehicle securely on jackstands.
3. Remove the front wheels.
4. Drain brake fluid.
5. Remove front disc brake anti rattle with hole pin.
    a. Remove the 2 pin hold clips, then remove the 2 hole pins from the disc brake caliper.
    b. Remove the anti rattle spring from the disc brake pad.
6. Remove the disc brake pads with anti squeal shims from the disc brake caliper.
7. Remove the anti squeal shims from each of the disc brake pads.

To install:

8. If necessary, replace the anti squeal shim when replacing the brake pad.
9. Apply disc brake grease to both sides of each No. 1 shim.
10. Install the No. 1 and No. 2 anti squeal shims onto each brake pad.
11. Install the disc brake pads with anti squeal shims onto the disc brake caliper.

⁂ WARNING

There should be no oil or grease on the friction surfaces of the disc brake pads or the front disc.

12. Install the anti rattle spring and hole pins onto the disc brake caliper.

➡The anti rattle spring is installed onto the lower hole pin.

13. Install the pin hold clip with its handle oriented in the center of the vehicle.
14. Fill reservoir with brake fluid.
15. Bleed the brakes.
16. Install the front wheel and lower the vehicle.

PARKING BRAKE CABLES

ADJUSTMENT

See Figure 7.

1. Before servicing the vehicle, refer to the precautions.
2. Inspect the parking brake lever travel.
   a. Slowly pull the parking brake lever to the fully applied position, counting the number of clicks.
   b. The parking brake lever should travel 5–7 clicks when pulled at 45 lbs. of force.
3. Turn the adjusting nut until the parking brake lever travel is corrected to within the specified range.
4. Operate the parking brake lever 3 to 4 times, and check the parking brake lever travel.
5. Check whether the parking brake drags or not.
6. When operating the parking brake lever, check that the brake warning light illuminates. The brake warning light always illuminates at the first click.

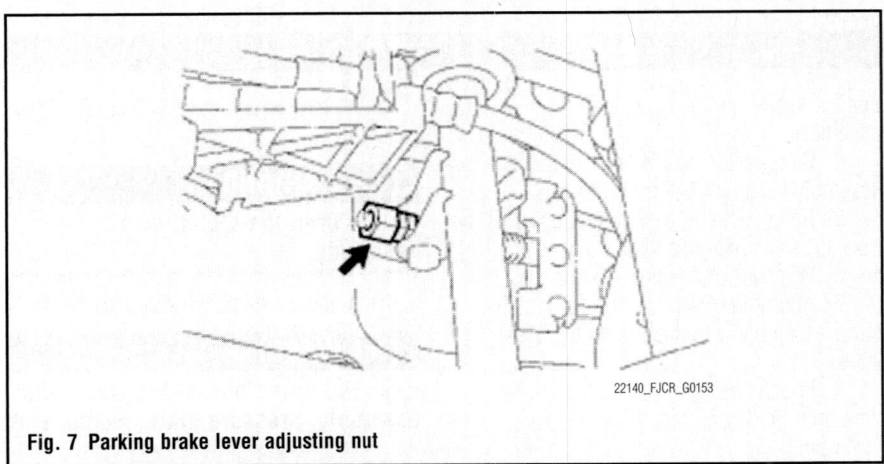

Fig. 7 Parking brake lever adjusting nut

PARKING BRAKE SHOES

REMOVAL & INSTALLATION

See Figure 8.

1. Before servicing the vehicle, refer to the precautions.
2. Raise and safely support the vehicle securely on jackstands.
3. Remove rear wheel.
4. Remove the disc brake caliper assembly and suspend it out of the way.
5. Release the parking brake lever.
6. Place matchmarks on the disc and axle hub and remove the disc.

➡If the disc cannot be removed easily, turn the shoe adjuster until the wheel turns freely.

To install:

7. Align the matchmarks of the disc and axle hub and install the disc.
8. Temporarily install 2 hub nuts.
9. Remove the hole plug, and turn the adjuster to expand the shoe until the disc locks.
10. Contract the shoe adjuster until the disc rotates smoothly. Then return 8 notches.
11. Install the hole plug.
12. Install the disc brake caliper.
13. Install rear wheel and lower the vehicle.
14. Inspect parking brake lever travel.
15. Adjust parking brake lever travel.

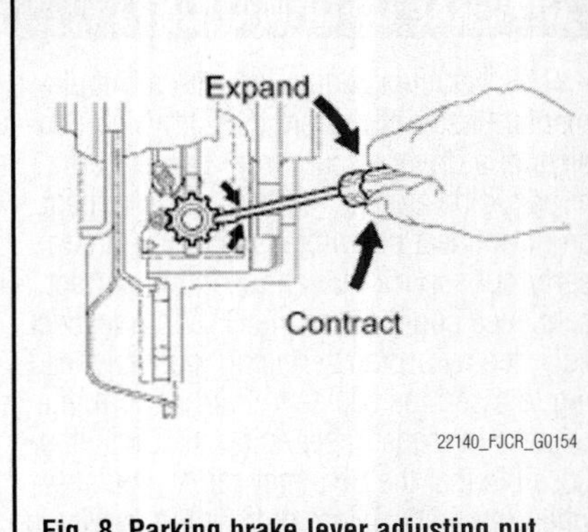

Fig. 8 Parking brake lever adjusting nut

ADJUSTMENT

See Figure 8.

1. Before servicing the vehicle, refer to the precautions.
2. Raise and safely support the vehicle securely on jackstands.
3. Remove rear wheel.
4. Temporarily install 2 hub nuts.
5. Remove the hole plug, and turn the adjuster to expand the shoe until the disc locks.
6. Contract the shoe adjuster until the disc rotates smoothly. Then return 8 notches.
7. Install rear wheel and lower the vehicle.
8. Inspect parking brake lever travel.
9. Adjust parking brake lever travel.

GENERAL INFORMATION

This vehicle is equipped with a Supplemental Restraint System (SRS), which consists of a driver airbag, front passenger airbag, side airbags, curtain shield airbags and front seat belt pretensioner. Failure to carry out service operations in the correct sequence could cause the SRS to unexpectedly deploy during servicing, possibly leading to a serious accident. Furthermore, if a mistake is made in servicing the SRS, it is possible that the SRS may fail to operate when required. Before performing servicing (including removal or installation of parts, inspection or replacement), be sure to read the following carefully, then follow the correct procedures as indicated in the repair manual.

SERVICE PRECAUTIONS

**** CAUTION**

Wait at least 90 seconds after the ignition switch is turned off and the negative (-) terminal cable is disconnected from the battery before starting the operation. The SRS is equipped with a back-up power source, so if work is started within 90 seconds of disconnecting the negative (-) terminal cable of the battery, the SRS may be deployed.

DISARMING THE SYSTEM

Wait at least 90 seconds after the ignition switch is turned off and the negative (-) terminal cable is disconnected from the battery before starting the operation. The SRS is equipped with a back-up power source, so if work is started within 90 seconds of disconnecting the negative (-) terminal cable of the battery, the SRS may be deployed.

ARMING THE SYSTEM

Reconnect the negative battery cable and perform an SRS warning light check.

CLOCKSPRING CENTERING

See Figure 9.

1. Check that the ignition switch is turned to **OFF**.
2. Rotate the spiral cable counterclockwise slowly by hand until it feels firm.

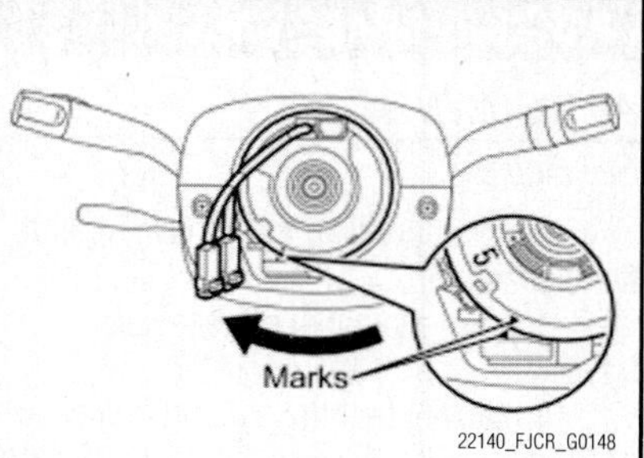

Fig. 9 Rotate the spiral cable counterclockwise slowly by hand until it feels firm, then rotate the spiral cable clockwise approximately 2.5 turns to align the marks

**** WARNING**

Do not turn the spiral cable with the airbag wire harness or connector.

3. Rotate the spiral cable clockwise approximately 2.5 turns to align the marks.
4. The spiral cable will rotate approximately 2.5 turns both ways from the center.

DRIVE TRAIN

CLUTCH DRIVEN DISC & PRESSURE PLATE

REMOVAL & INSTALLATION

See Figures 10 through 16.

1. Before servicing the vehicle, refer to the precautions.
2. Remove manual transmission assembly.
3. Remove the clutch release fork together with the clutch release bearing from the transmission assembly.
4. Remove the clutch release bearing assembly from the clutch release fork.
5. Remove the release fork support from the transaxle assembly.
6. Remove release bearing hub clip.
7. Align the matchmark on the clutch cover assembly with the one on the flywheel.
8. Loosen each set bolt one turn at a time until the spring tension is released.
9. Remove the 6 bolts and clutch cover assembly.

**** WARNING**

Do not drop the clutch disc assembly.

10. Remove the clutch disc assembly.

**** WARNING**

Keep the lining part of the clutch disc assembly, pressure plate and surface

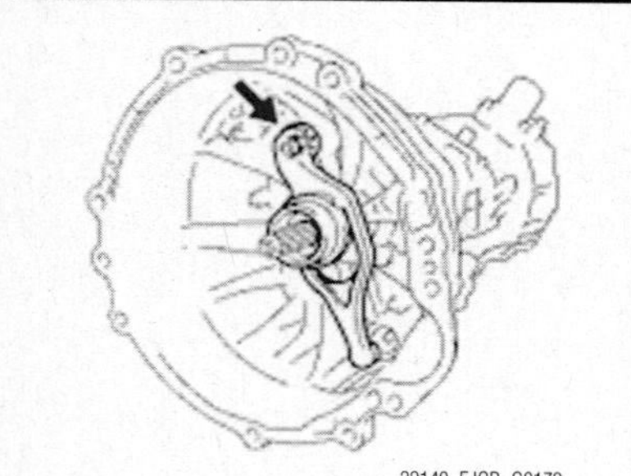

Fig. 10 Remove the clutch release fork together with the clutch release bearing from the transmission assembly

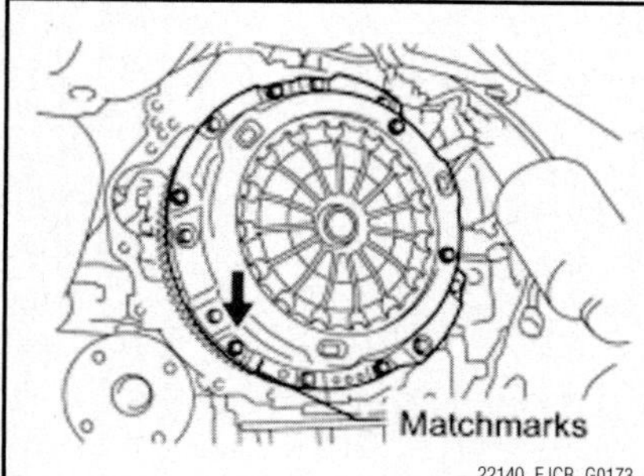

Fig. 11 Align the matchmark on the clutch cover assembly with the one on the flywheel

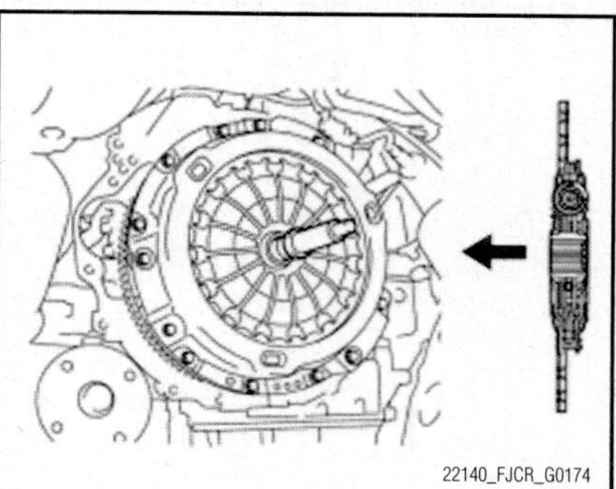

Fig. 12 Insert the clutch disc assembly in the correct orientation

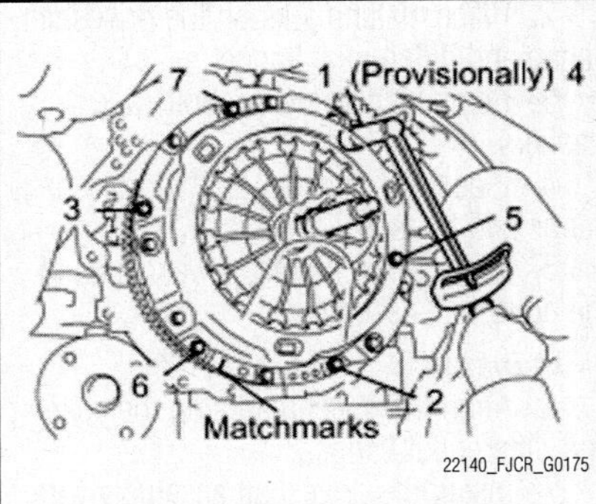

Fig. 13 Clutch disc assembly tightening sequence

of the flywheel free from oil and foreign matter.

To install:

11. Insert SST: 09301-00220 into the clutch disc assembly, then insert them into the flywheel.

➡Insert the clutch disc assembly in the correct orientation.

12. Align the matchmarks on the clutch cover assembly and flywheel. Move SST up and down, right and left lightly, after checking that the disc is in the center.
13. Following the pattern shown in the illustration, tighten the 6 bolts, in the order starting with the bolt located near the knock pin on the top. Following the order in the illustration, uniformly tighten the bolts to 14 ft. lbs. (19 Nm).
14. Using a dial indicator with roller instrument, check the diaphragm spring tip alignment. Runout should not exceed 0.020 in. (0.5 mm).
15. If the alignment is not as specified, adjust the diaphragm spring tip alignment using SST: 09333-00013.
16. Install the release fork support onto the transaxle assembly and tighten to 35 ft. lbs. (47 Nm).
17. Install release bearing hub clip.
18. Apply Toyota Genuine Release Hub Grease or equivalent to the release fork and release bearing assembly contact surfaces, release fork and push rod contact surface and release fork pivot point.
19. Install the release fork onto the release bearing assembly.
20. Apply Toyota Genuine Clutch Spline Grease or equivalent to the input shaft spline.

WARNING

Do not apply grease to portion A shown in the illustration.

21. Install the bearing onto the release fork, and then install them onto the transaxle assembly.
22. After installation, move the fork forward and backward to check that the release bearing slides smoothly.
23. Install manual transmission assembly.

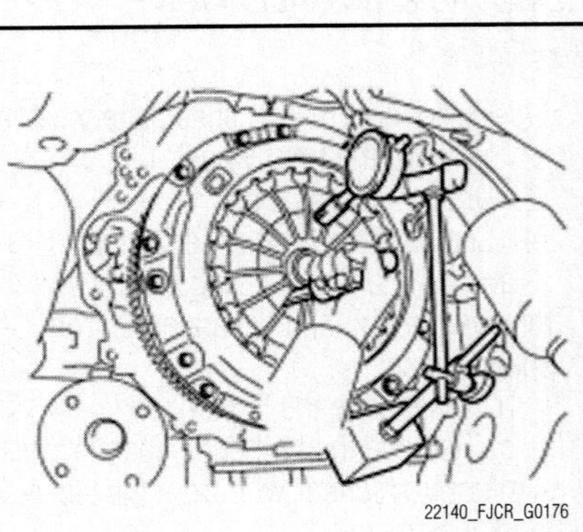

Fig. 14 Using a dial indicator with roller instrument, check the diaphragm spring tip alignment. Runout should not exceed 0.020 in. (0.5 mm)

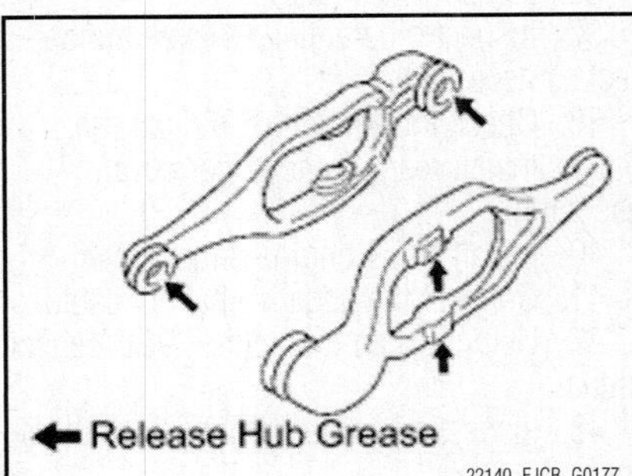

Fig. 15 Apply Toyota Genuine Release Hub Grease or equivalent to the release fork and release bearing assembly contact surfaces

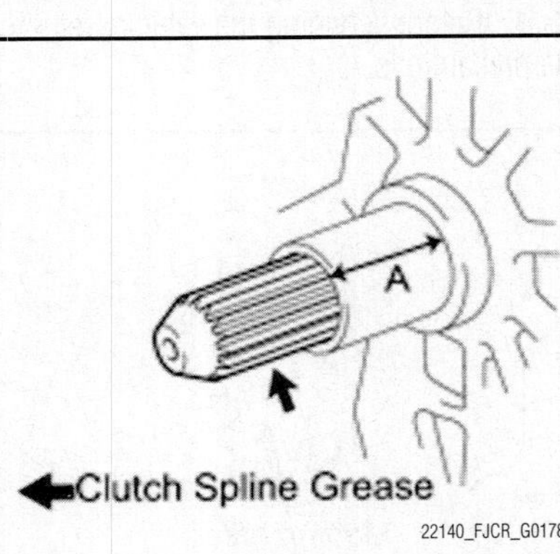

Fig. 16 Apply Toyota Genuine Clutch Spline Grease or equivalent to the input shaft spline. Do not apply grease to portion A shown in the illustration

HYDRAULIC SYSTEM BLEEDING

BLEEDING PROCEDURE

➡If any maintenance on the clutch system was performed or the system is suspected of containing air, bleed the system. Use care; brake fluid will remove the paint from any surface. If the brake fluid spills onto any painted surface, wash it off immediately with soap and water.

1. Before servicing the vehicle, refer to the precautions section.
2. Fill the clutch reservoir with brake fluid. Check the reservoir level frequently and add fluid as needed.
3. Connect one end of a vinyl tube to the bleeder plug on the slave cylinder and submerge the other end into a clear container half-filled with brake fluid.
4. Slowly pump the clutch pedal several times.
5. Have an assistant hold the clutch pedal down and loosen the bleeder plug until fluid and/or air starts to run out of the bleeder plug. Close the bleeder plug while the pedal is held to the floor.

➡Do not allow the pedal to rise backup while the bleeder is still open. If this happens, it will allow air to reenter the slave cylinder and cause the clutch system not to work properly.

6. Repeat Steps 2 and 3 until all the air bubbles are removed from the system.
7. Tighten the bleeder plug when all the air is gone.
8. Refill the master cylinder to the proper level as required.
9. Check the system for leaks.

TRANSFER CASE ASSEMBLY

REMOVAL & INSTALLATION

See Figure 17.

1. Before servicing the vehicle, refer to the precautions.
2. Disconnect the negative battery cable.
3. Drain transfer oil.
4. Remove lower transfer case protector.
5. Remove the transmission assembly.
6. Remove transfer case assembly.

To install:

7. Install the transfer case onto the transmission. Tighten the bolts to 17 ft. lbs. (24 Nm).

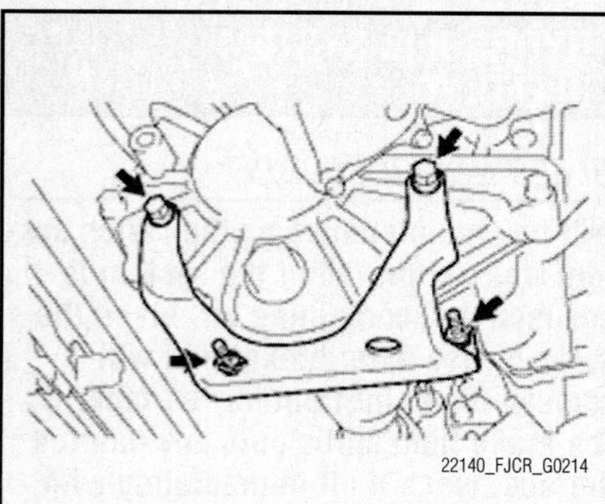

Fig. 17 Lower transfer case protector mounting bolts

8. Install the transmission.
9. Install the lower transfer case protector with the 4 bolts. Tighten the bolts to 13 ft. lbs. (18 Nm).
10. Inspect and adjust the transfer case oil.
11. Connect the negative battery cable.
12. Using a scan tool, perform the reset memory procedure to initialize the transmission.
13. Check for transfer oil leakage.

FRONT DIFFERENTIAL CARRIER

REMOVAL & INSTALLATION

1. Before servicing the vehicle, refer to the precautions.
2. Disconnect the negative battery cable.
3. Remove the front wheel.
4. Remove the No.1 engine under cover.
5. Remove the rear engine under cover assembly.
6. Drain the differential oil.
7. Remove the front halfshaft assembly.
8. Disconnect the front speed sensors.
9. Remove front axle hub grease cap.
10. Separate tie rod end.
11. Separate front lower ball joint attachment.
12. Remove front axle hub nut.
13. Remove front drive shaft assembly.
14. Remove the bolt and disconnect the differential breather tube bracket.
15. Support the differential with a jack.
16. Remove the No. 1 differential mounting nut.
17. Remove the 2 mounting bolts and 2 nuts.
18. Lower the jack and remove the front differential.
19. Remove the 3 bolts and front No. 1 differential support.
20. Remove the 2 bolts and front No. 2 differential support.
21. Remove the 2 bolts and front No. 3 differential support.

To install:

22. Install the front No. 3 differential support with the 2 bolts. Tighten to 80 ft. lbs. (108 Nm).
23. Install the front No. 2 differential support with the 2 bolts. Tighten to 118 ft. lbs. (160 Nm).
24. Install the front No. 1 differential support with the 3 bolts. Tighten to 137 ft. lbs. (186 Nm).
25. Support the front differential with a jack.
26. Install the 2 front mounting bolts and 2 nuts. Tighten to 101 ft. lbs. (137 Nm).
27. Install front differential mounting nut No. 1. Tighten to 64 ft. lbs. (87 Nm).
28. Install the front differential breather tube bracket with the bolt. Tighten to 10 ft. lbs. (13 Nm).
29. Lower the jack.
30. Install front halfshaft assembly.
31. Attach front lower ball joint
32. Install tie rod end.
33. Install front axle hub nut.
34. Install front axle hub grease cap.
35. Install front speed sensor.
36. Install front wheel.
37. Inspect differential oil level and correct as necessary.
38. Check for differential oil leakage.
39. Install rear engine under cover assembly.
40. Install No. 1 engine under cover.
41. Connect the negative battery cable.
42. Using a scan tool, check VSC sensor signal.
43. Inspect and adjust front wheel alignment

FRONT DRIVESHAFT

REMOVAL & INSTALLATION

See Figure 18.

1. Before servicing the vehicle, refer to the precautions.

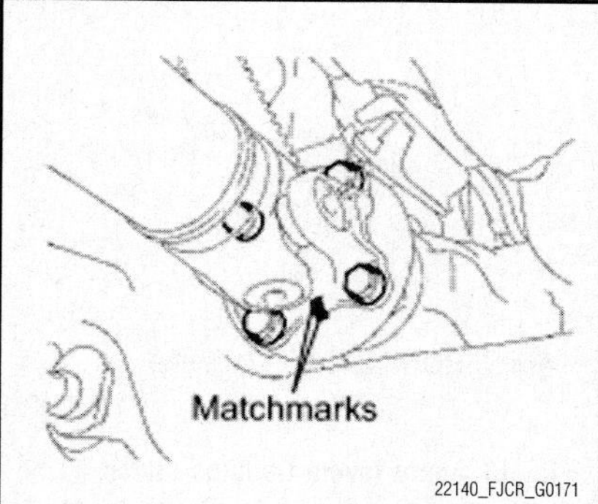

Fig. 18 Place matchmarks on the driveshaft flange for installation reference

2. Place matchmarks on the driveshaft flange and differential flange.
3. Remove the 4 nuts, 4bolts and 4 washers.
4. Place matchmarks on the driveshaft flange and transfer flange.
5. Remove the 4 nuts, 4 washers and the driveshaft assembly.

To install:

6. Align the matchmarks on the yoke and differential flange.
7. Install the driveshaft assembly with the 4 bolts, 4 nuts and 4 washers. Tighten to 65 ft. lbs. (88 Nm).
8. Align the matchmarks on the yoke and transfer flange.
9. Install the driveshaft assembly with the 4 nuts and 4 washers. Tighten to 65 ft. lbs. (88 Nm).
10. Before servicing the vehicle, refer to the precautions.
11. Place matchmarks on the driveshaft flange and differential flange.
12. Remove the 4 nuts, 4bolts and 4 washers.
13. Place matchmarks on the driveshaft flange and transfer flange.
14. Remove the 4 nuts, 4 washers and the driveshaft assembly.

To install:

15. Align the matchmarks on the yoke and differential flange.
16. Install the driveshaft assembly with the 4 bolts, 4 nuts and 4 washers. Tighten to 65 ft. lbs. (88 Nm).
17. Align the matchmarks on the yoke and transfer flange.
18. Install the driveshaft assembly with the 4 nuts and 4 washers. Tighten to 65 ft. lbs. (88 Nm).

FRONT HALFSHAFT

REMOVAL & INSTALLATION

See Figure 19.

1. Disconnect the negative battery cable.
2. Remove the front wheel.
3. Drain differential oil.
4. Separate the front speed sensor:
    a. Using a 5 mm hexagon wrench, remove the bolt and separate the front speed sensor.
    b. Disengage the 3 clamps.
    c. Remove the bolt and separate the skid control sensor wire from the steering knuckle.
5. Using a screwdriver and a hammer, remove the front axle hub grease cap.
6. Remove the cotter pin and adjusting lock cap.

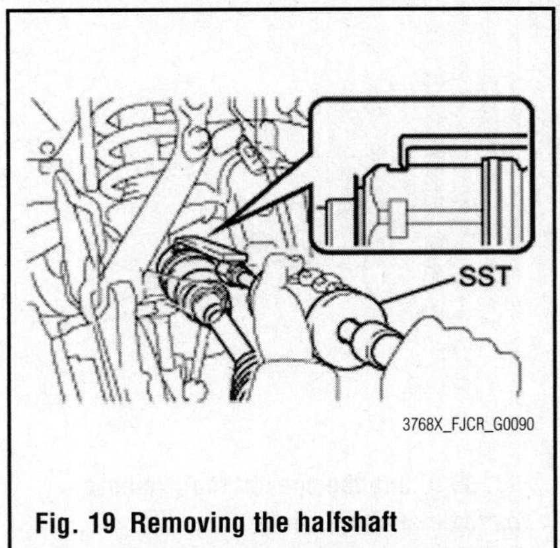

Fig. 19 Removing the halfshaft

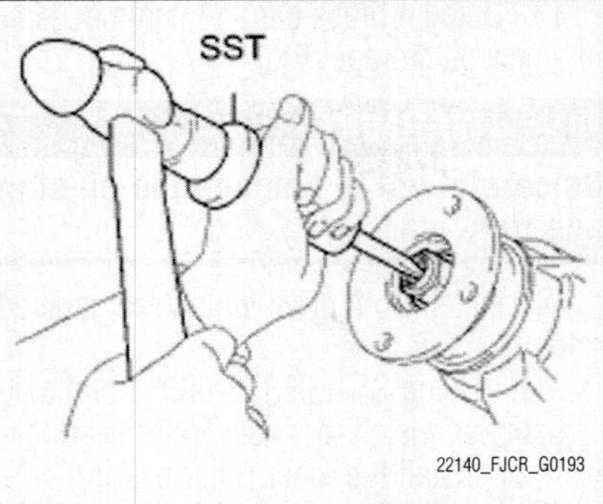

Fig. 20 Using the special tool and a hammer, loosen the staked part of the nut

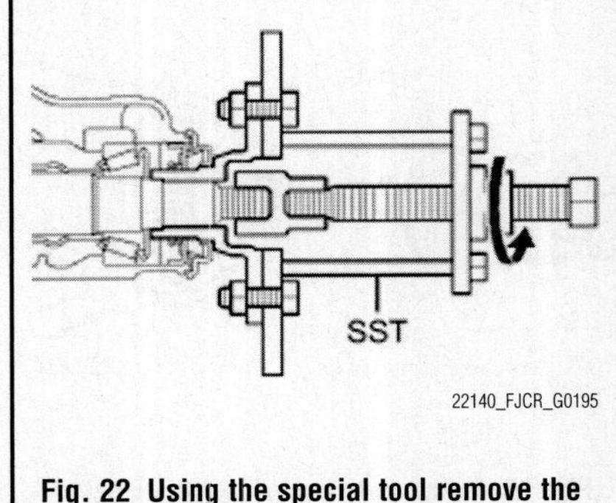

Fig. 22 Using the special tool remove the companion flange

7. Remove the front axle hub nut.
8. Separate tie rod end sub-assembly. Refer to Tie Rod End in Steering.
9. Separate front lower ball joint attachment. Refer to Lower Ball Joint in Front Suspension.
10. Using SST 09520-01010 and 09520-24010, remove the halfshaft.

To install:

11. Coat the spline of the inboard joint shaft with ATF.
12. Align the shaft splines and install the drive shaft with a brass bar and hammer.

➡Set the snap ring with the opening side facing downward. Do not damage the oil seal.

➡Whether the inboard joint shaft is in contact with the pinion shaft or not can be confirmed from the sound or feeling when driving it.

13. To complete installation, reverse remaining removal procedure.
14. Inspect differential oil.
15. Check VSC sensor signal.
16. Check for differential oil leakage.
17. Inspect and adjust the front wheel alignment.

FRONT PINION SEAL

REMOVAL & INSTALLATION

See Figures 20 through 29.

1. Before servicing the vehicle, refer to the precautions.
2. Remove No. 1 engine under cover.
3. Remove the 4 bolts, then remove the rear engine under cover.
4. Remove front halfshaft assembly.
5. Remove front drive pinion companion flange nut.
   a. Using SST: 09930-00010 and a hammer, loosen the staked part of the nut.

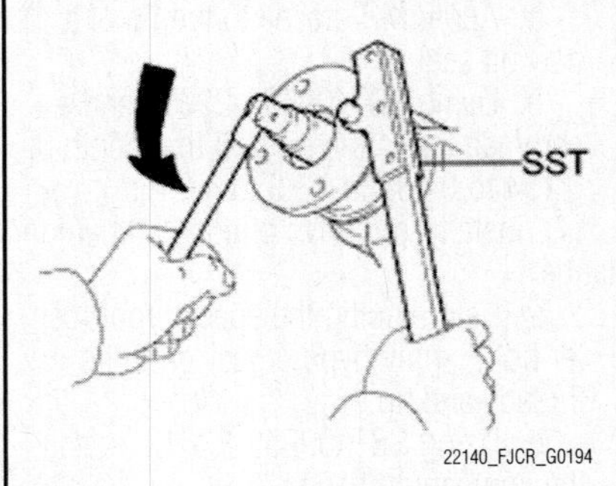

Fig. 21 Using the special tool hold the companion flange and remove the nut

   b. Using SST: 09330-00021 to hold the companion flange, remove the nut.
6. Remove front drive pinion companion flange.
   a. Before using the special tool (center bolt), apply hypoid gear oil to its threads and tip.
   b. Using SST: 09950-30012, remove the companion flange.
7. Remove front differential carrier oil seal.
   a. Using SST: 09308-10010, remove the oil seal.
8. Remove front differential drive pinion oil slinger.
9. Remove front drive pinion rear tapered roller bearing.
   a. Using SST: 09556-22010, remove the inner roller bearing.
   b. Using SST: 09308-00010, tap out the outer roller bearing.
10. Using a screwdriver and hammer, tap out the oil storage ring.
11. Remove front differential drive pinion bearing spacer.

To install:

12. Install front differential drive pinion bearing spacer.

➡Install the spacer in the correct direction.

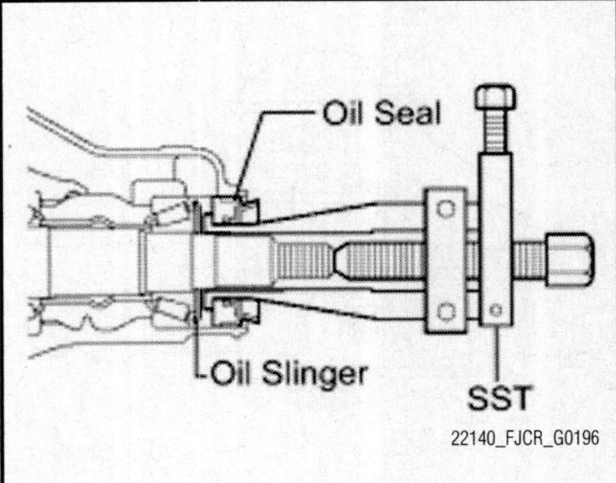

Fig. 23 Using the special tool remove the oil seal and oil slinger

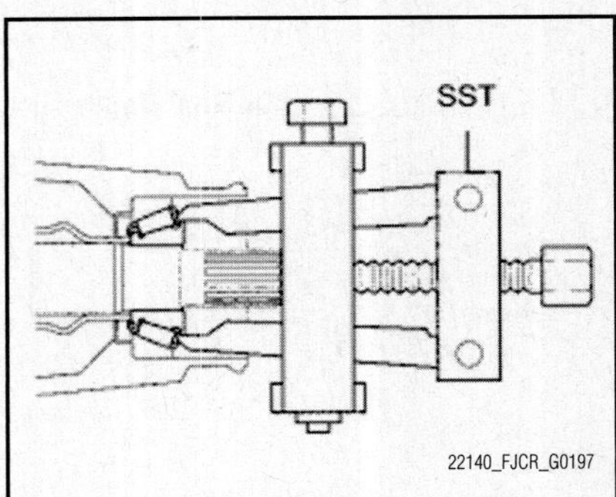

Fig. 24 Using the special tool remove the inner roller bearing

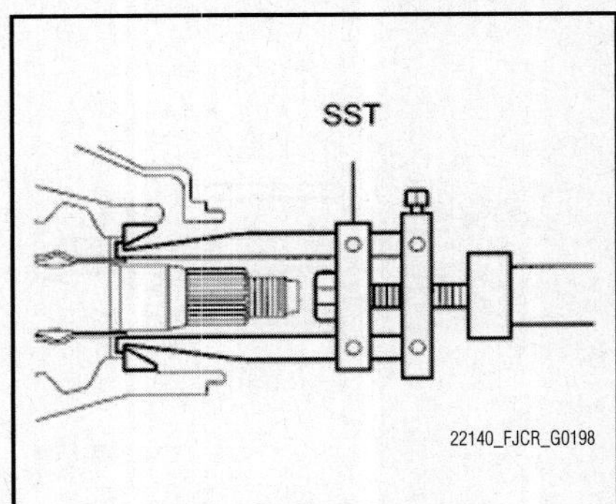

Fig. 25 Using the special tool remove the outer roller bearing

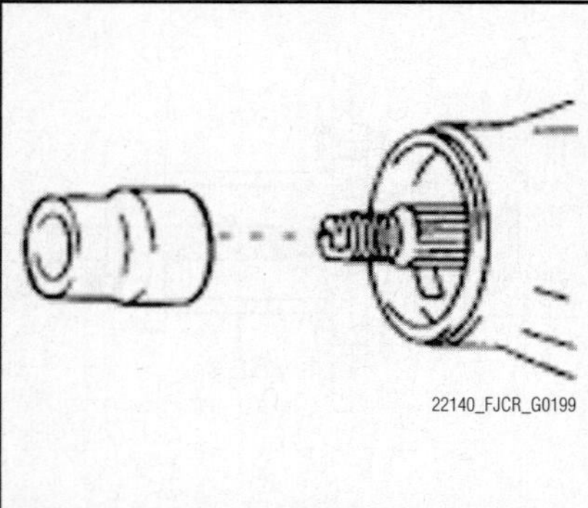

Fig. 26 Install the spacer in the correct direction

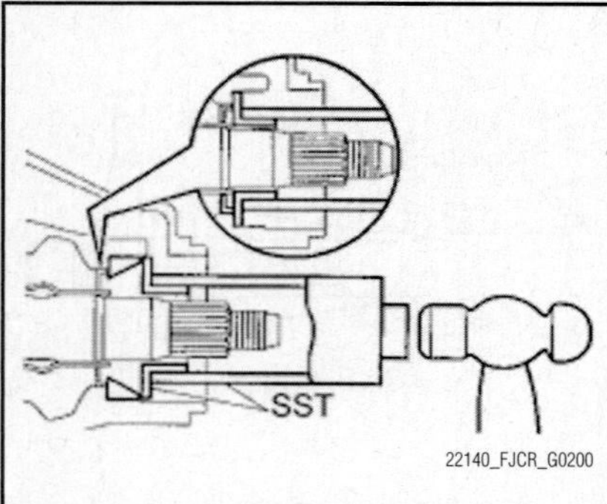

Fig. 27 Using the special tool and a hammer, install the outer roller bearing

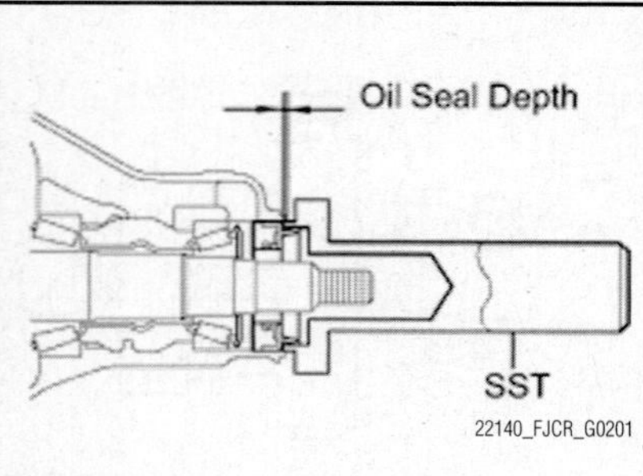

Fig. 28 Using the special tool and a hammer, tap in the oil seal to a depth of 0.154 to 0.188 in. (3.9 to 4.8 mm)

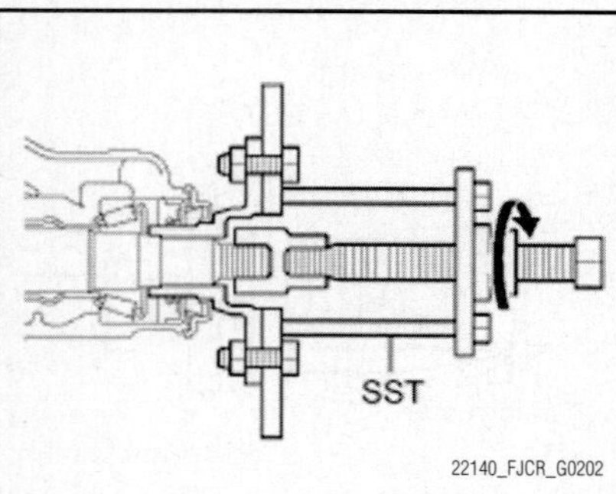

Fig. 29 Using the special tool install the companion flange

13. Using a brass bar and hammer, tap in a new oil storage ring.

⁂ WARNING

Be careful not to damage the oil storage ring.

14. Install front drive pinion rear tapered roller bearing.
    a. Using SST: 09316-60011 and a hammer, install the outer roller bearing.
    b. Install the inner roller bearing.
15. Install front differential drive pinion oil slinger.
16. Install front differential carrier oil seal.
    a. Apply MP grease to the lip of a new oil seal.
    b. Using SST: 09554-22010 and a hammer, tap in the oil seal to a depth of 0.154 to 0.188 in. (3.9 to 4.8 mm).
17. Install front drive pinion companion flange.
    a. Before using the special tool (center bolt), apply hypoid gear oil to its threads and tip.
    b. Using SST: 09950-30012, install the companion flange.
    c. Using SST: 09330-00021 to hold the companion flange, install the nut. Tighten to 273 ft. lbs. (370 Nm).
18. Using a torque wench, measure the differential drive pinion preload.
    a. Preload for a new bearing should be 8.7–13.9 inch lbs. (0.98–1.57 Nm).
    b. Preload for a used bearing should be 4.3–6.9 inch lbs. (0.49–0.78 Nm).
    c. If the result is not as specified, adjust the preload.
19. Using a chisel and hammer, stake the nut.
20. Install front halfshaft assembly.
21. Add differential oil.
22. Check for differential oil leakage.
23. Install rear engine under cover assembly.
24. Install No. 1 engine under cover.

REAR AXLE SHAFT, BEARING & SEAL

REMOVAL & INSTALLATION

See Figures 30 and 31.

1. Before servicing the vehicle, refer to the precautions.
2. Disconnect the negative battery cable.
3. Raise and safely support the vehicle securely on jackstands.
4. Remove rear wheel.
5. Drain brake fluid.

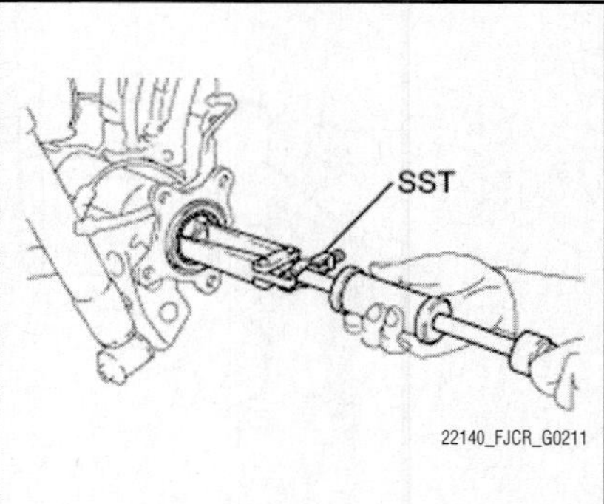

Fig. 30 Using the special tool, remove rear axle shaft oil seal

6. Remove rear disc brake caliper assembly and hang out of the way.
7. Remove rear disc.
8. Remove parking brake shoe return tension spring.
9. Remove parking brake shoe strut.
10. Remove parking brake shoe.
11. Remove the 2 bolts and separate the parking brake cable from the parking brake plate.
12. Remove rear speed sensor.
13. Remove rear axle shaft.
    a. Remove the 4 nuts and rear axle shaft with parking brake plate.
    b. Remove the O-ring.
14. Using SST: 09308-00010, remove rear axle shaft oil seal.

To install:

15. Using SST: 09950-60020, install rear axle shaft oil seal.
    a. Install the O-ring.
    b. Install the 4 nuts and rear axle shaft with parking brake plate. Tighten to 89 ft. lbs. (120 Nm).
16. Install rear axle shaft.
17. Inspect the axle shaft backlash.
    a. Using a dial indicator, check the backlash near the center of the axle shaft. Maximum backlash is 0.0020 in. (0.05 mm).

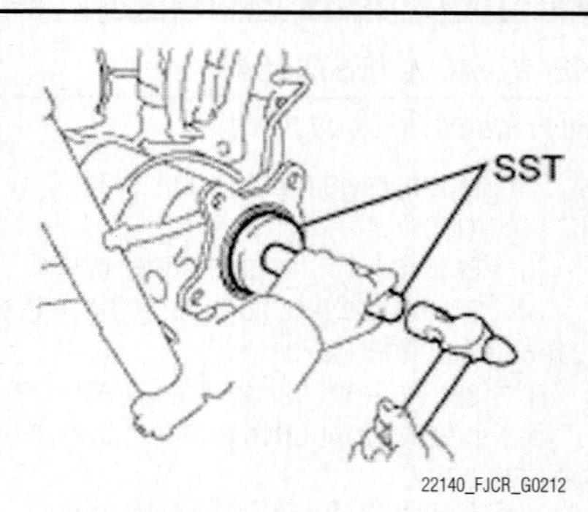

Fig. 31 Using the special tool, install rear axle shaft oil seal

b. If the backlash is greater than the maximum, replace the bearing.
18. Inspect the axle shaft runout.
a. Using a dial indicator, check the runout of the surface of the axle shaft. Maximum runout is 0.0020 in. (0.05 mm).
b. If the runout is greater than the maximum, replace the bearing.
19. Install rear speed sensor. Tighten to 71 inch lbs. (8 Nm).
20. Install the 2 bolts and separate the parking brake cable from the parking brake plate.
21. Install parking brake shoe.
22. Install parking brake shoe strut.
23. Install parking brake shoe return tension spring.
24. Install rear disc.
25. Install rear disc brake caliper assembly and hang out of the way.
26. Drain brake fluid.
27. Install rear wheel and lower the vehicle.
28. Inspect differential oil.
29. Check for differential oil leakage.
30. Check and adjust fluid level in reservoir.
31. Check for brake fluid leakage.
32. Inspect and adjust parking brake lever travel.
33. Using a scan tool, check VSC sensor signal.

REAR DIFFERENTIAL HOUSING

REMOVAL & INSTALLATION

See Figures 32 through 34.

1. Before servicing the vehicle, refer to the precautions.
2. Disconnect the negative battery cable.
3. Raise and safely support the vehicle securely on jackstands.
4. Remove rear wheel.
5. Drain brake fluid.
6. Remove rear disc brake caliper assembly and hang out of the way.
7. Remove rear disc.
8. Remove parking brake shoe return tension spring.
9. Remove parking brake shoe strut.
10. Remove parking brake shoe.
11. Remove the 2 bolts and separate the parking brake cable from the parking brake plate.
12. Remove rear speed sensor.
13. Remove rear axle shaft.
14. Remove rear driveshaft assembly.
15. Remove rear differential carrier assembly:
a. For vehicles with differential lock: Remove the 11 nuts and the differential carrier.

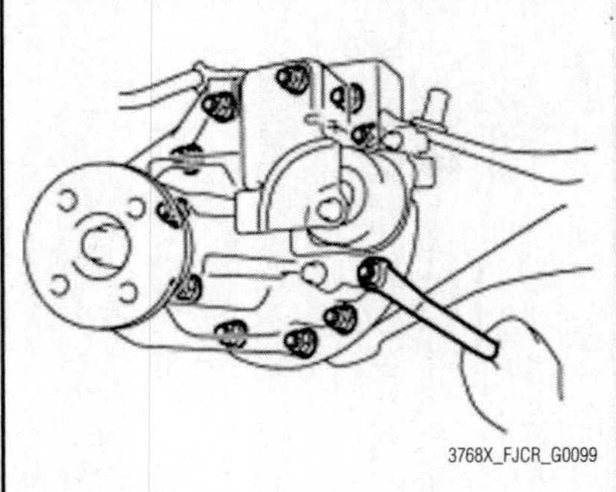

Fig. 32 Remove the 11 nuts and the differential carrier—with differential lock

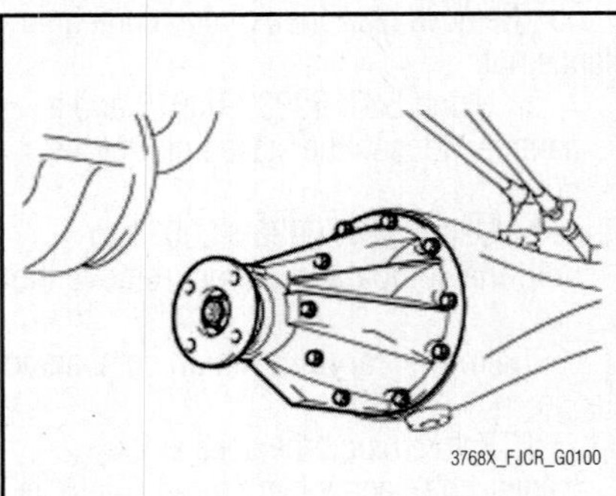

Fig. 33 Remove the 10 nuts and the differential carrier—without differential lock

b. For vehicles without differential lock: Remove the 10 nuts and the differential carrier.

⁂ WARNING

Be careful not to damage the contact surface.

16. Remove rear differential carrier gasket.

To install:

17. Remove any dust and oil from the differential carrier assembly and the contact surfaces of the axle housing.

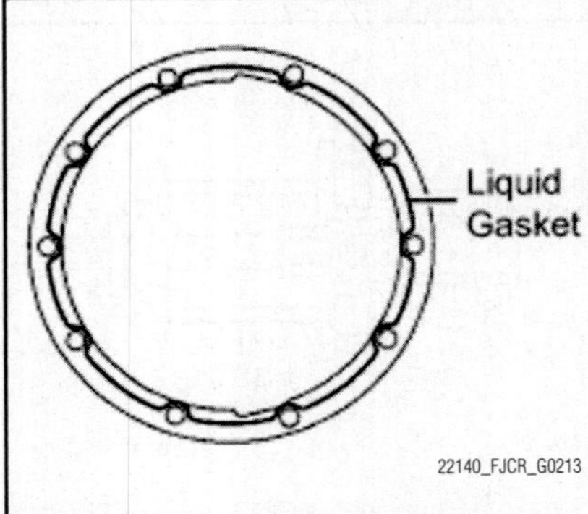

Fig. 34 Apply liquid gasket to both sides of a new gasket

18. Apply liquid gasket to both sides of a new gasket.

➡Do not apply the liquid gasket to the stud bolt.

19. Install a new gasket and the differential carrier assembly with the 10 and/or 11 nuts and 10 and/or 11 washers. Tighten to 38 ft. lbs. (52 Nm).
20. Install rear axle shaft.
21. Inspect the axle shaft backlash.
a. Using a dial indicator, check the backlash near the center of the axle shaft. Maximum backlash is 0.0020 in. (0.05 mm).
b. If the backlash is greater than the maximum, replace the bearing.
22. Inspect the axle shaft runout.
a. Using a dial indicator, check the runout of the surface of the axle shaft. Maximum runout is 0.0020 in. (0.05 mm).
b. If the runout is greater than the maximum, replace the bearing.
23. Install rear speed sensor. Tighten to 71 inch lbs. (8 Nm).
24. Install the 2 bolts and separate the parking brake cable from the parking brake plate.
25. Install parking brake shoe.
26. Install parking brake shoe strut.
27. Install parking brake shoe return tension spring.
28. Install rear disc.
29. Install rear disc brake caliper assembly and hang out of the way.
30. Drain brake fluid.
31. Install rear wheel and lower the vehicle.
32. Inspect differential oil.
33. Check for differential oil leakage.
34. Check and adjust fluid level in reservoir.
35. Check for brake fluid leakage.
36. Inspect and adjust parking brake lever travel.
37. Using a scan tool, check VSC sensor signal.

REAR DRIVESHAFT

REMOVAL & INSTALLATION

See Figure 35.

1. Before servicing the vehicle, refer to the precautions.
2. Place matchmarks on the driveshaft flange and differential flange.
3. Remove the 4 nuts, 4bolts and 4 washers.
4. Place matchmarks on the driveshaft flange and transfer flange.

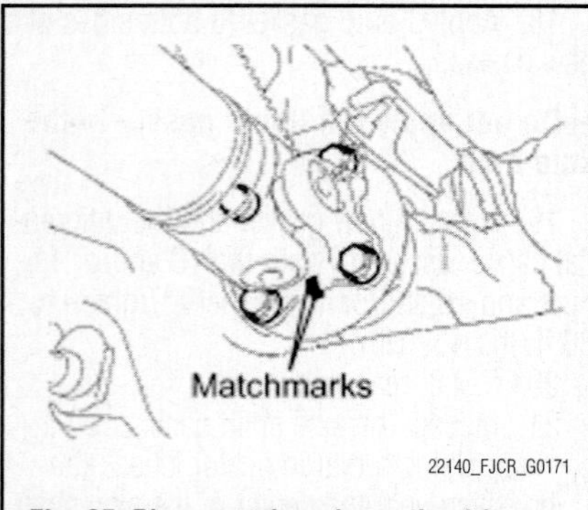

Fig. 35 Place matchmarks on the driveshaft flange for installation reference

5. Remove the 4 nuts, 4 washers and the driveshaft assembly.

To install:

6. Align the matchmarks on the yoke and differential flange.
7. Install the driveshaft assembly with the 4 bolts, 4 nuts and 4 washers. Tighten to 65 ft. lbs. (88 Nm).
8. Align the matchmarks on the yoke and transfer flange.
9. Install the driveshaft assembly with the 4 nuts and 4 washers. Tighten to 65 ft. lbs. (88 Nm).
10. Before servicing the vehicle, refer to the precautions.
11. Place matchmarks on the driveshaft flange and differential flange.
12. Remove the 4 nuts, 4bolts and 4 washers.
13. Place matchmarks on the driveshaft flange and transfer flange.
14. Remove the 4 nuts, 4 washers and the driveshaft assembly.

To install:

15. Align the matchmarks on the yoke and differential flange.
16. Install the driveshaft assembly with the 4 bolts, 4 nuts and 4 washers. Tighten to 65 ft. lbs. (88 Nm).
17. Align the matchmarks on the yoke and transfer flange.
18. Install the driveshaft assembly with the 4 nuts and 4 washers. Tighten to 65 ft. lbs. (88 Nm).

REAR PINION SEAL

REMOVAL & INSTALLATION

See Figures 36 through 45.

1. Before servicing the vehicle, refer to the precautions.
2. Remove No. 1 engine under cover.
3. Remove the 4 bolts, then remove the rear engine under cover.
4. Remove rear halfshaft assembly.

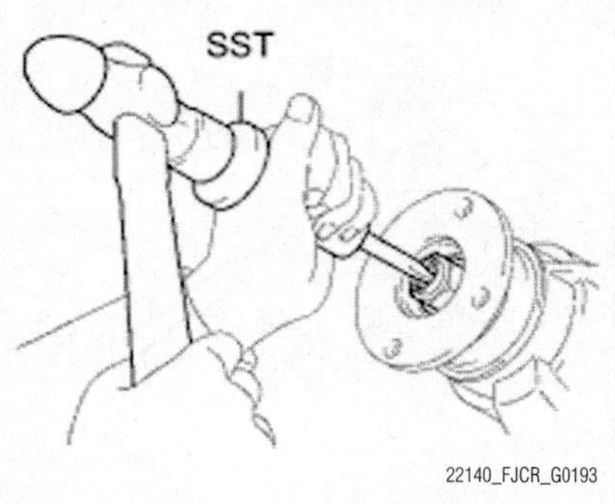

Fig. 36 Using the special tool and a hammer, loosen the staked part of the nut

5. Remove rear drive pinion companion flange nut.
    a. Using SST: 09930-00010 and a hammer, loosen the staked part of the nut.
    b. Using SST: 09330-00021 to hold the companion flange, remove the nut.
6. Remove rear drive pinion companion flange.
    a. Before using the special tool (center bolt), apply hypoid gear oil to its threads and tip.
    b. Using SST: 09950-30012, remove the companion flange.

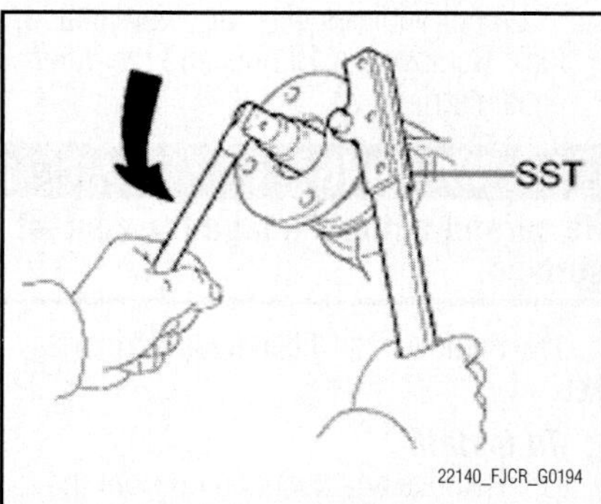

Fig. 37 Using the special tool hold the companion flange and remove the nut

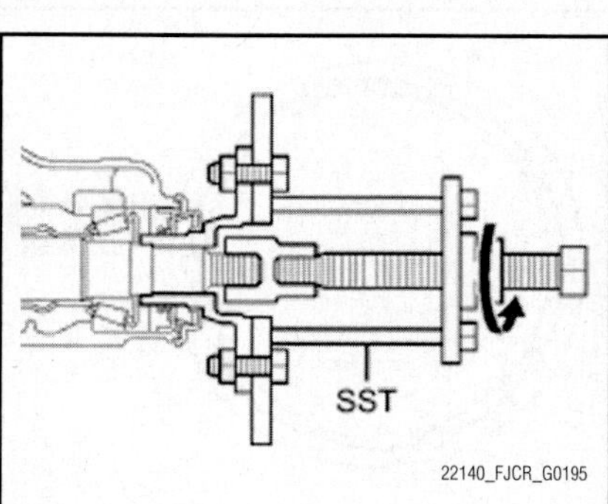

Fig. 38 Using the special tool remove the companion flange

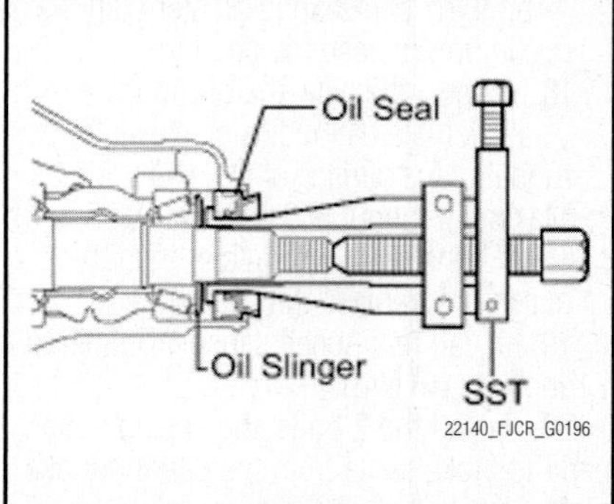

Fig. 39 Using the special tool remove the oil seal and oil slinger

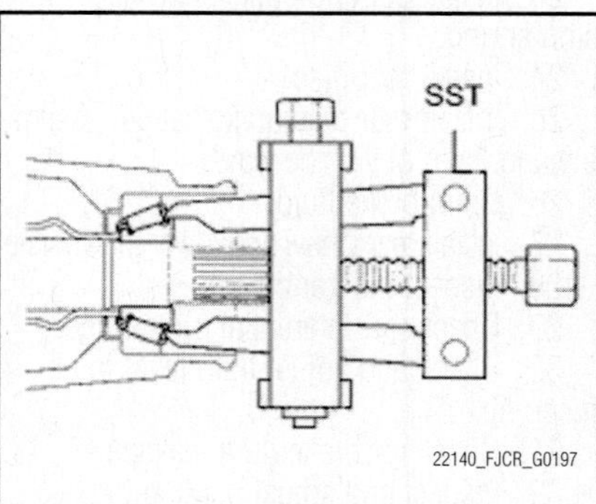

Fig. 40 Using the special tool remove the inner roller bearing

7. Remove rear differential carrier oil seal.
    a. Using SST: 09308-10010, remove the oil seal.
8. Remove rear differential drive pinion oil slinger.
9. Remove rear drive pinion rear tapered roller bearing.
    a. Using SST: 09556-22010, remove the inner roller bearing.
    b. Using SST: 09308-00010, tap out the outer roller bearing.
10. Using a screwdriver and hammer, tap out the oil storage ring.
11. Remove rear differential drive pinion bearing spacer.

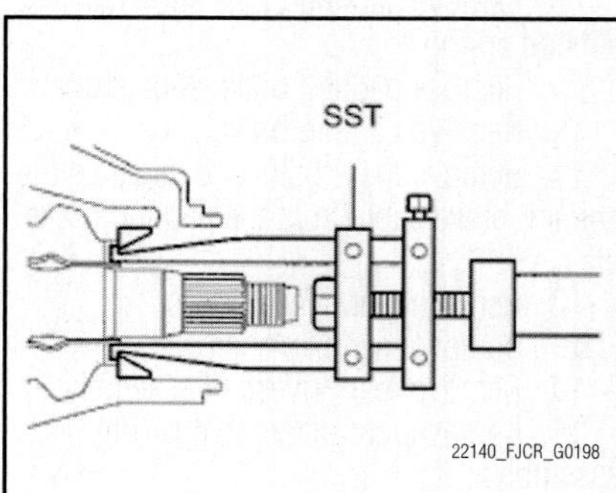

Fig. 41 Using the special tool remove the outer roller bearing

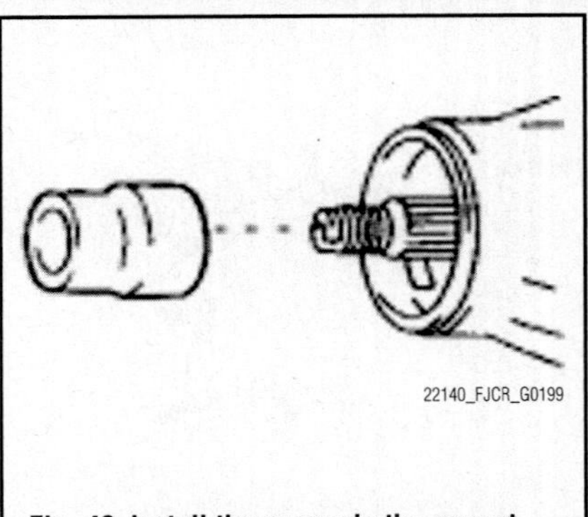

Fig. 42 Install the spacer in the correct direction

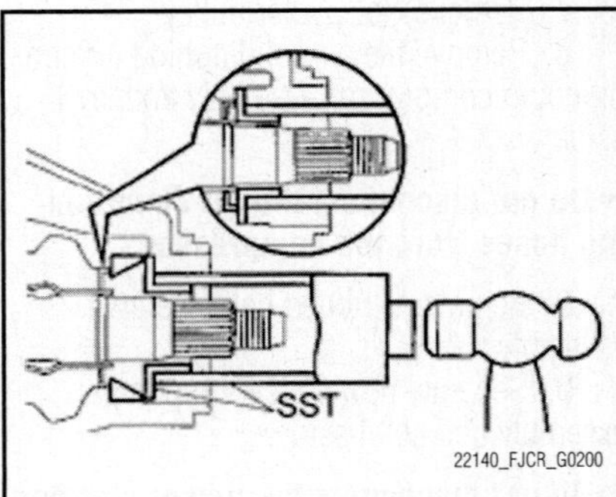

Fig. 43 Using the special tool and a hammer, install the outer roller bearing

To install:

12. Install rear differential drive pinion bearing spacer.

➡Install the spacer in the correct direction.

13. Using a brass bar and hammer, tap in a new oil storage ring.

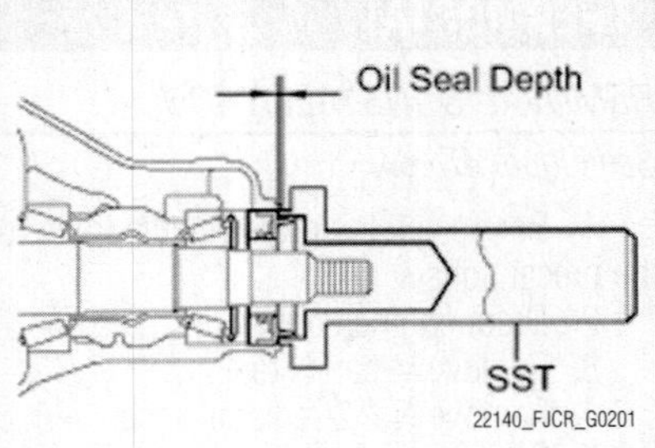

Fig. 44 Using the special tool and a hammer, tap in the oil seal to a depth of 0.154 to 0.188 in. (3.9 to 4.8 mm)

WARNING

Be careful not to damage the oil storage ring.

14. Install rear drive pinion rear tapered roller bearing.
   a. Using SST: 09316-60011 and a hammer, install the outer roller bearing.
   b. Install the inner roller bearing.
15. Install rear differential drive pinion oil slinger.
16. Install rear differential carrier oil seal.
   a. Apply MP grease to the lip of a new oil seal.
   b. Using SST: 09554-22010 and a hammer, tap in the oil seal to a depth of 0.154 to 0.188 in. (3.9 to 4.8 mm).
17. Install rear drive pinion companion flange.
   a. Before using the special tool (center bolt), apply hypoid gear oil to its threads and tip.

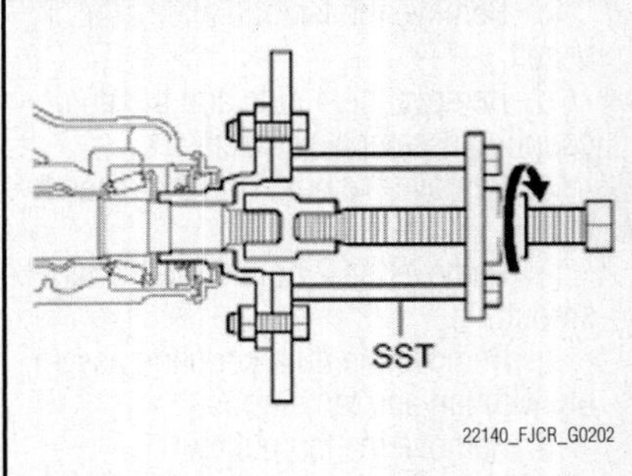

Fig. 45 Using the special tool install the companion flange

   b. Using SST: 09950-30012, install the companion flange.
   c. Using SST: 09330-00021 to hold the companion flange, install the nut. Tighten to 273 ft. lbs. (370 Nm).
18. Using a torque wench, measure the differential drive pinion preload.
   a. Preload for a new bearing should be 8.7–13.9 inch lbs. (0.98–1.57 Nm).
   b. Preload for a used bearing should be 4.3–6.9 inch lbs. (0.49–0.78 Nm).
   c. If the result is not as specified, adjust the preload.
19. Using a chisel and hammer, stake the nut.
20. Install rear halfshaft assembly.
21. Add differential oil.
22. Check for differential oil leakage.
23. Install rear engine under cover assembly.
24. Install No. 1 engine under cover.

ENGINE COOLING

ENGINE FAN

REMOVAL & INSTALLATION

See Figure 46.

1. Before servicing the vehicle, refer to the precautions.
2. Remove V-bank cover.
3. Remove no.1 engine under cover.
4. Remove radiator support seal upper.
5. Remove fan shroud.
6. Remove the 4 nuts and the fan from the fluid coupling.

To install:

7. Install the 4 nuts and the fan from the fluid coupling. Tighten to 80 inch lbs. (9 Nm).
8. Install fan shroud.
9. Install radiator support seal upper.
10. Install no.1 engine under cover.
11. Install V-bank cover.

RADIATOR

REMOVAL & INSTALLATION

1. Before servicing the vehicle, refer to the precautions.
2. Remove V-bank cover.
3. Remove No. 1 engine under cover.
4. Drain engine coolant.
5. Disengage the 7 clips, then remove the radiator support seal.
6. Remove fan shroud.
   a. Disconnect the hose from the radiator.
   b. Loosen the 4 nuts from the fan pulley.

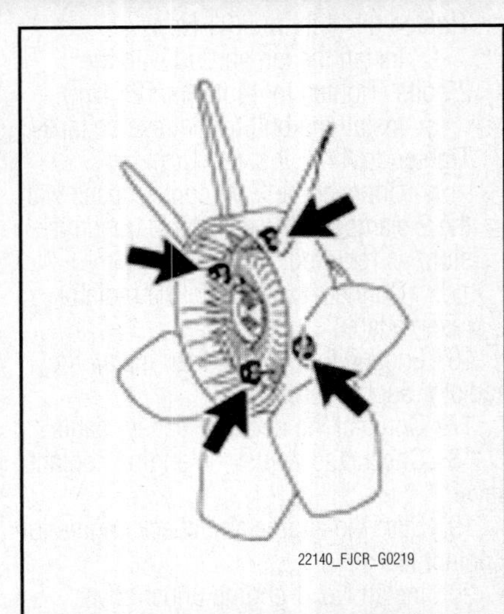

Fig. 46 Fan coupling fastener locations

c. Remove the fan and alternator V-belt.
d. Remove the 4 nuts and the fluid coupling assembly with fan.
e. Remove the bolt from the reserve tank.
f. Remove the 2 bolts from the fan shroud.
g. Remove the fluid coupling assembly with fan and fan shroud.
h. Remove the fan pulley.
7. Remove radiator grille.
8. Remove radiator assembly.
a. Disconnect the 2 radiator hoses.
b. Disconnect the 2 oil cooler hoses for the automatic transmission, as required.
c. Remove the 4 bolts, then remove the radiator.
9. Remove radiator hoses.
10. Remove inlet oil cooler hose for automatic transmission, as required.

To install:

11. Install inlet oil cooler hose for automatic transmission, as required.
12. Install the radiator hoses.
13. Install the radiator with the 4 bolts. Tighten to 44 ft. lbs. (21 Nm).
a. Connect the 2 oil cooler hoses automatic transmission, as required.
b. Install the 2 radiator hoses.
14. Install radiator grille.
15. Install fan shroud.
a. Temporarily install the fan shroud together with the fluid coupling assembly with fan.
b. Temporarily install the fluid coupling assembly with fan with the 4 nuts.
c. Install the fan and alternator V-belt.
d. Install the fan pulley with the 4 nuts. Tighten to 44 ft. lbs. (21 Nm).
e. Install the bolt to the reserve tank. Tighten to 44 ft. lbs. (21 Nm).
f. Install the fan shroud with the 2 bolts. Tighten to 44 ft. lbs. (21 Nm).
g. Install the bolt to the reserve tank. Tighten to 44 ft. lbs. (21 Nm).
h. Connect the 2 oil cooler hoses with the 2 clamps to the automatic transmission, as required.
i. Connect the hose to the radiator reserve tank.
16. Engage the 7 clips, then install the radiator support seal.
17. Connect the negative battery cable.
18. Check and adjust the engine coolant level.
19. Start the engine and check engine for coolant leaks.
20. Install No. 1 engine under cover.
21. Install V-bank cover

THERMOSTAT

REMOVAL & INSTALLATION

See Figure 47.

1. Before servicing the vehicle, refer to the precautions.
2. Drain engine coolant.
3. Remove V-bank cover.
4. Remove No. 2 radiator hose.
5. Remove the 3 nuts, then remove the water inlet with thermostat and gasket.

To install:

6. Install a new gasket onto the water inlet with thermostat.
7. Install the water inlet with thermostat with the 3 nuts. Tighten to 80 inch lbs. (9 Nm)
8. Connect No. 2 radiator hose.
9. Connect the negative battery cable.
10. Add engine coolant.
11. Check for engine coolant leakage.
12. Install V-bank cover.

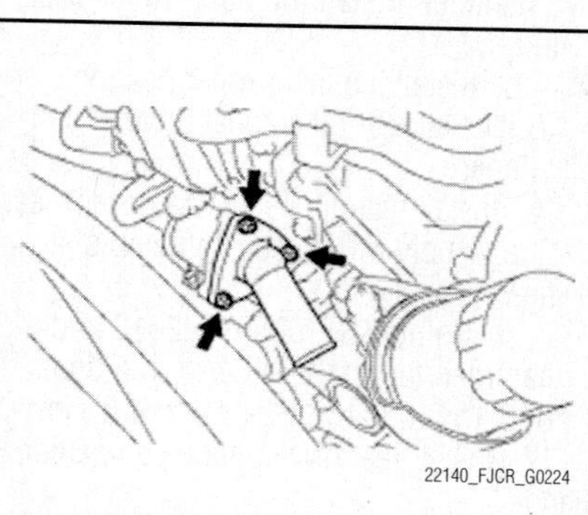

Fig. 47 Remove the 3 nuts, then remove the water inlet

WATER PUMP

REMOVAL & INSTALLATION

See Figures 48 and 49.

1. Before servicing the vehicle, refer to the precautions.
2. Remove fan.
3. Remove V-bank cover.
4. Remove air cleaner assembly.
5. Remove alternator assembly.
6. Remove water inlet.
a. Disconnect the 2 radiator hoses.
b. Disconnect the 5 water by-pass hoses.
c. Remove the 5 bolts and water inlet.
d. Remove the O-ring from the water outlet pipe.
e. Remove the gasket from the water pump.

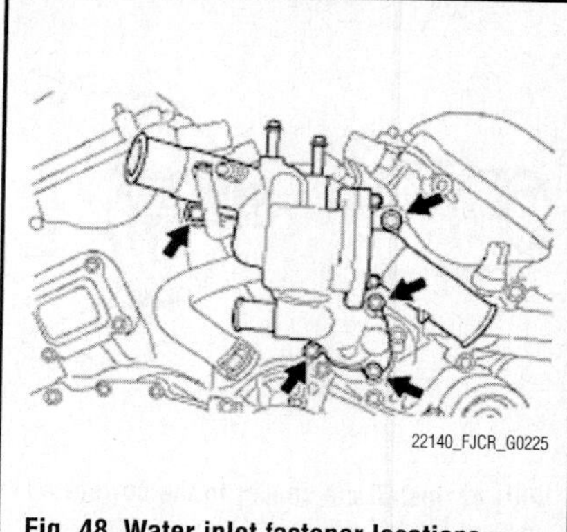

Fig. 48 Water inlet fastener locations

7. Remove No. 2 idler pulley.
8. Remove the air conditioning air conditioning compressor assembly and lay it aside.

➡Do not disconnect the air conditioning hoses from the compressor.

9. Remove V-ribbed belt tensioner assembly.
10. Separate power steering pump assembly and set it aside.

➡Do not disconnect the power steering hoses.

11. Remove the 17 bolts, then remove the water pump and gasket.

To install:

12. Install a new gasket and the water pump with the 17 bolts. Tighten bolts **A** to 80 inch lbs. (9 Nm) and bolts **B** to 17 ft. lbs. (23 Nm).
13. Install power steering pump assembly and set it aside.

➡Do not disconnect the power steering hoses.

14. Install V-ribbed belt tensioner assembly.
15. Install air conditioning compressor assembly.
16. Install No. 2 idler pulley.

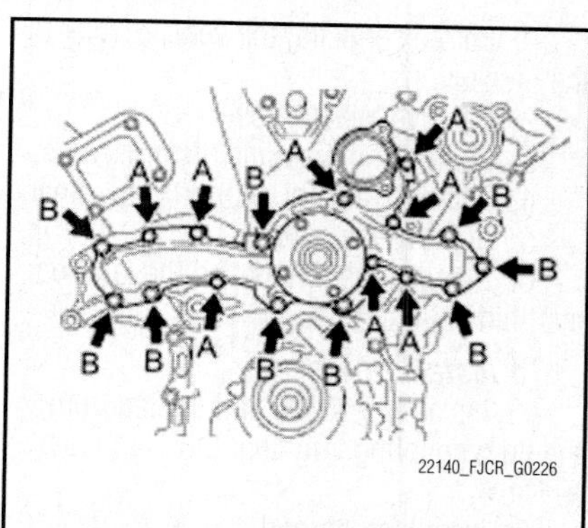

Fig. 49 Water pump bolt tightening torque

17. Install water inlet.
    a. Install a new O-ring onto the water outlet pipe.
    b. Install a new gasket onto the water pump.
    c. Apply soapy water to the O-ring.
    d. Install the water inlet with the 5 bolts. Tighten bolt to 80 inch lbs. (9 Nm).
    e. Connect the 5 water by-pass hoses.
    f. Connect the 2 radiator hoses.
18. Install alternator assembly.
19. Install air cleaner assembly.
20. Install fan.
21. Connect the negative battery cable.
22. Install V-bank cover.

ENGINE ELECTRICAL — CHARGING SYSTEM

ALTERNATOR

REMOVAL & INSTALLATION

See Figure 50.

1. Before servicing the vehicle, refer to the precautions.
2. Remove battery.
3. Remove V-bank cover.
4. Remove No. 1 engine under cover.
5. Remove fan and alternator V-belt.
6. Disconnect the wire harness.
7. Remove the bolt and wire harness bracket.
8. Disconnect the connector from the alternator assembly.
9. Remove the terminal cap and nut.
10. Disconnect the wire harness from terminal B.
11. Remove the 2 bolts, then separate the wire harness clamp bracket from the alternator assembly.

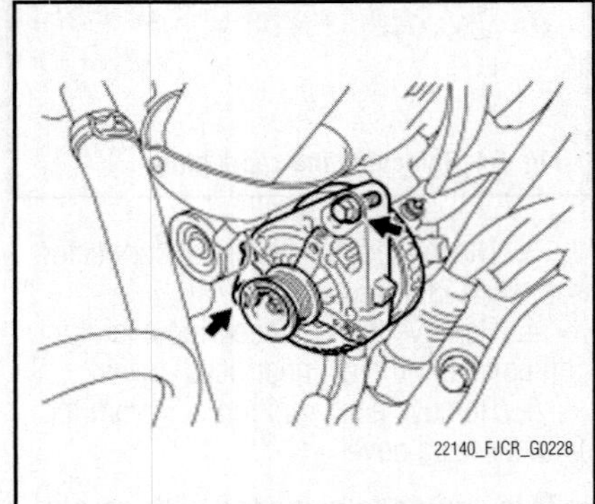

Fig. 50 Alternator mounting bolts

12. Remove the 2 bolts, then remove the alternator assembly.

To install:

13. Install the alternator assembly with the 2 bolts. Tighten to 32 ft. lbs. (43 Nm).
14. Install the wire harness clamp bracket with the 2 bolts. Tighten to 71 inch lbs. (8 Nm).
15. Connect the wire harness.
16. Connect the wire harness to terminal B and install the nut. Tighten to 7 ft. lbs. (10 Nm).
17. Connect the connector to the alternator assembly.
18. Install the wire harness bracket with the bolt. Tighten to 51 inch lbs. (6 Nm).
19. Install fan and alternator V-belt.
20. Install No. 1 engine under cover.
21. Install V-bank cover.
22. Install battery

ENGINE ELECTRICAL — IGNITION SYSTEM

FIRING ORDER

See Figure 51.

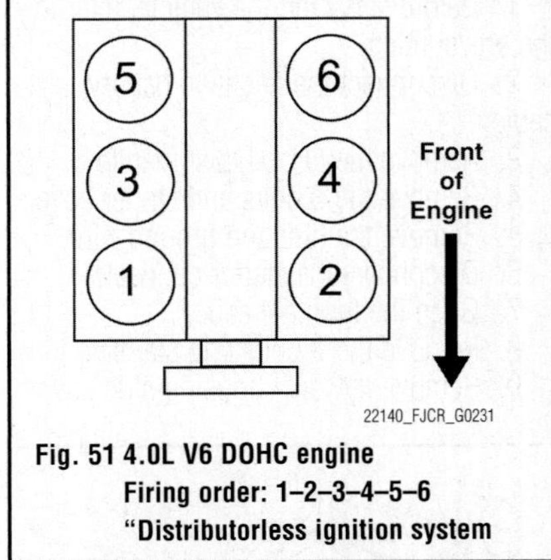

Fig. 51 4.0L V6 DOHC engine Firing order: 1–2–3–4–5–6 "Distributorless ignition system

IGNITION COIL

REMOVAL & INSTALLATION

2009 Vehicles

See Figure 52.

1. Before servicing the vehicle, refer to the precautions.
2. Disconnect the negative battery cable.
3. Remove V-bank cover.
4. Remove air cleaner assembly.
5. Remove oil baffle plate.
6. Remove No. 1 surge tank bracket.
7. Disconnect the ventilation hose from the intake air surge tank.
8. Disconnect the 3 connectors, remove the 3 bolts, then remove the 3 ignition coils.

To install:.

9. Install the 3 ignition coils with the 3 bolts. Tighten the bolts to 7 ft. lbs. (10 Nm).

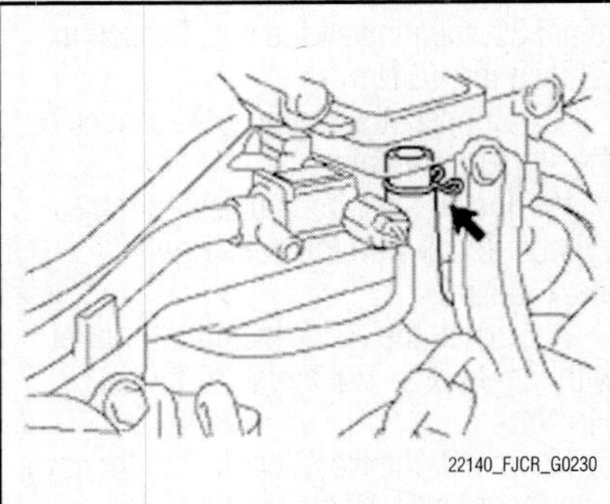

Fig. 52 Ignition coils mounted on the cylinder head

10. Connect the 3 ignition coil connectors.
11. Connect the ventilation hose to the intake air surge tank.
12. Install No. 1 surge tank bracket.
13. Install oil baffle plate.
14. Install air cleaner assembly.
15. Connect the negative battery cable.
16. Inspect ignition timing.
17. Inspect engine idling speed.
18. Install V-bank cover.

2010 Vehicles

See Figure 53.

1. Before servicing the vehicle, refer to the precautions.
2. Disconnect the negative battery cable.
3. Remove V-bank cover.
4. Remove air cleaner assembly.
5. Disconnect the 6 ignition coil connectors.
6. Remove the 6 bolts and 6 ignition coils.
7. To install, reverse removal procedure.
8. Tighten the bolts to 7 ft. lbs. (10 Nm).

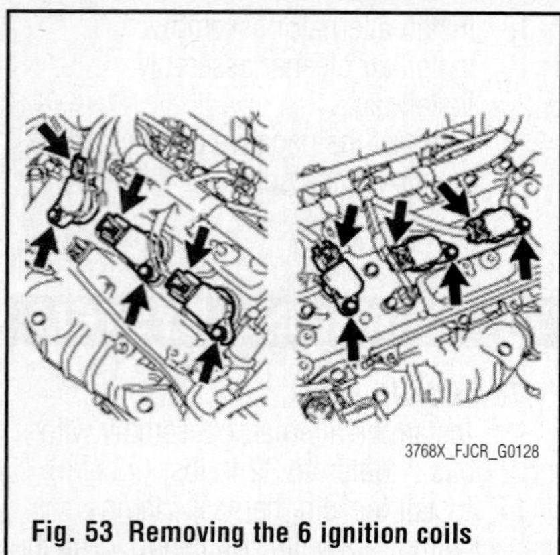

Fig. 53 Removing the 6 ignition coils

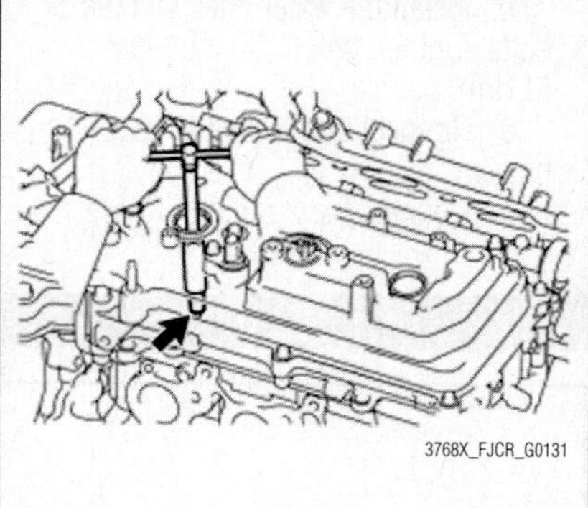

Fig. 54 Removing the spark plug

IGNITION TIMING

ADJUSTMENT

The ignition timing is controlled by the Electronic Control Module (ECM). No adjustment is possible.

SPARK PLUGS

REMOVAL & INSTALLATION

See Figure 54.

1. Before servicing the vehicle, refer to the precautions.
2. Remove the V-bank cover.
3. Disconnect the electrical connector from each ignition coil.
4. Remove the bolt securing each ignition coil from the cylinder head cover.
5. Remove each ignition coil from the cylinder head cover.

➡This engine is equipped with an aluminum cylinder head. Allow the engine to cool before removing spark plugs. Removing the spark plugs from an engine at operating temperature may damage the spark plug threads in the cylinder head. Also be sure to clean any dirt or debris from around spark plug holes prior to removing spark plugs.

6. Remove the spark plugs from the cylinder head.
7. Inspect the spark plugs for electrode wear, carbon deposits and insulator damage.

To install:

⁂ WARNING

Do not touch the tip of the spark plug. Do not damage the iridium surface of the electrode when gapping the plug.

➡Do not adjust the gap on used spark plugs. Replace the spark plug if the gap is greater than specification.

8. Apply a small quantity of anti-seize compound to the plug threads, and screw the plugs into the cylinder head finger-tight.
9. Tighten the spark plugs to 15 ft. lbs. (20 Nm).
10. Install each ignition coil to the cylinder head cover.
11. Secure the ignition coils using the bolts. Tighten the bolts to 80 inch lbs (9 Nm).
12. Connect each ignition coil electrical connector.
13. Install the V-bank cover.

ENGINE ELECTRICAL

STARTING SYSTEM

STARTER

REMOVAL & INSTALLATION

2009 Vehicles

See Figure 55.

1. Before servicing the vehicle, refer to the precautions.
2. Disconnect the negative battery cable.
3. Remove the 4 bolts, then remove the No. 2 engine under cover assembly.
4. Remove the 3 bolts, then remove the No. 2 manifold bracket.

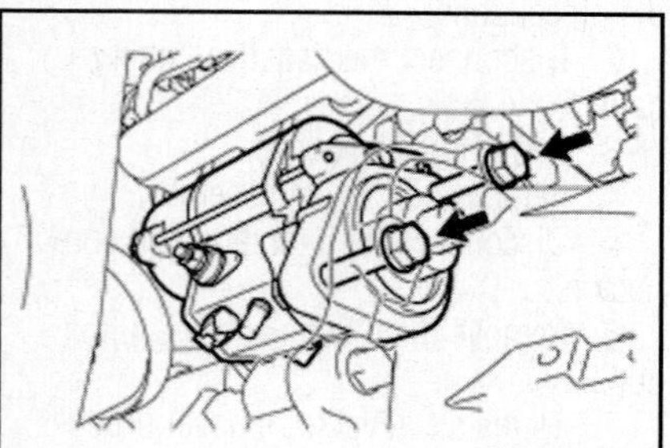

Fig. 55 Removing the starter assembly—2009 vehicles

5. Remove the bolt, then separate the earth wire harness from the cylinder block.
6. Remove starter assembly.
   a. Disconnect the terminal 50 connector from the starter assembly.
   b. Remove the nut, then disconnect the wire harness from the terminal 30.
   c. Remove the 2 bolts, then remove the starter assembly.

To install:

7. Install the starter assembly with 2 bolts. Tighten to 27 ft. lbs. (37 Nm).
8. Connect the wire harness to the terminal 30, then install the nut. Tighten to 52 inch lbs. (6 Nm).
9. Connect the terminal 50 connector to the starter assembly.
10. Install the wire harness onto the cylinder block with the bolt. Tighten to 10 ft. lbs. (13 Nm).
11. Install the No. 2 manifold bracket with the 3 bolts. Tighten to 30 ft. lbs. (40 Nm).
12. Install the No. 2 engine under cover with the 4 bolts. Tighten to 21 ft. lbs. (29 Nm).
13. Connect the negative battery cable.

2010 Vehicles

See Figure 56.

1. Before servicing the vehicle, refer to the precautions.
2. Disconnect the negative battery cable.
3. Remove the LH exhaust manifold.
4. Remove the 3 bolts and starter cover.
5. Remove the bolt and ground wire.
6. Disconnect the starter connector.
7. Open the terminal cap.
8. Remove the 2 bolts and starter.
9. Remove flywheel housing side cover.

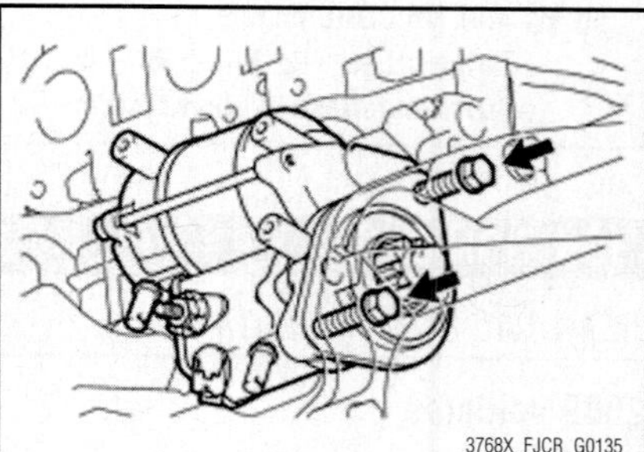

Fig. 56 Removing the starter assembly—2010 vehicles

ENGINE MECHANICAL

ACCESSORY DRIVE BELTS

ACCESSORY BELT ROUTING

See Figure 57.

INSPECTION

Inspect the drive belt for signs of glazing or cracking. A glazed belt will be perfectly smooth from slippage, while a good belt will have a slight texture of fabric visible. Cracks will usually start at the inner edge of the belt and run outward. All worn or damaged drive belts should be replaced immediately.

ADJUSTMENT

The belt tension is maintained by an automatic tensioner. No adjustment is possible.

REMOVAL & INSTALLATION

See Figure 58.

1. Before servicing the vehicle, refer to the precautions.
2. Remove the 4 bolts, then remove the No. 1 engine under cover.
3. While releasing the belt tension by turning the belt tensioner counterclockwise, remove the V-ribbed belt from the belt tensioner.

To install:

4. Check that nothing gets caught in the tensioner by turning it clockwise and counterclockwise. If a malfunction exists, replace the tensioner.
5. While turning the belt tensioner counterclockwise, align the holes as shown, and then insert a bar of 0.24 in. (6 mm) into the holes to fix the belt tensioner.
6. Install the V-ribbed belt.

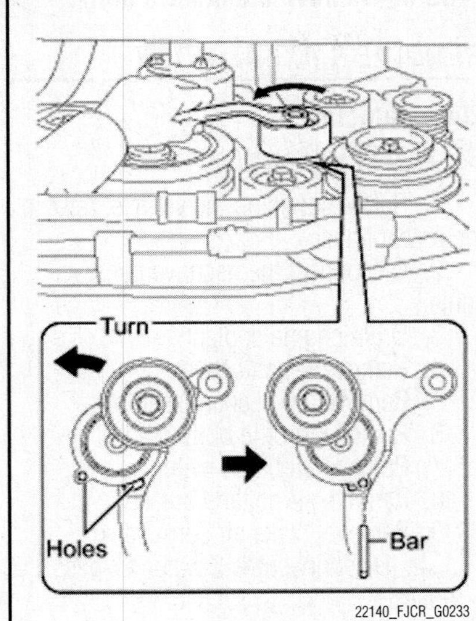

Fig. 58 While turning the belt tensioner counterclockwise, align the holes as shown, and then insert a bar of 0.24 in. (6 mm) into the holes to fix the belt tensioner

7. While turning the belt tensioner counterclockwise, remove the bar.
8. If it is hard to install the V-ribbed belt, perform the following procedure:
   a. Put the V-ribbed belt on all parts except the P/S pump, as shown in the illustration.
   b. While releasing the belt tension by turning the belt tensioner counterclockwise, put the V-ribbed belt on the P/S pump.
9. Install the No. 1 engine under cover with the 4 bolts. Tighten to 21 ft. lbs. (29 Nm).

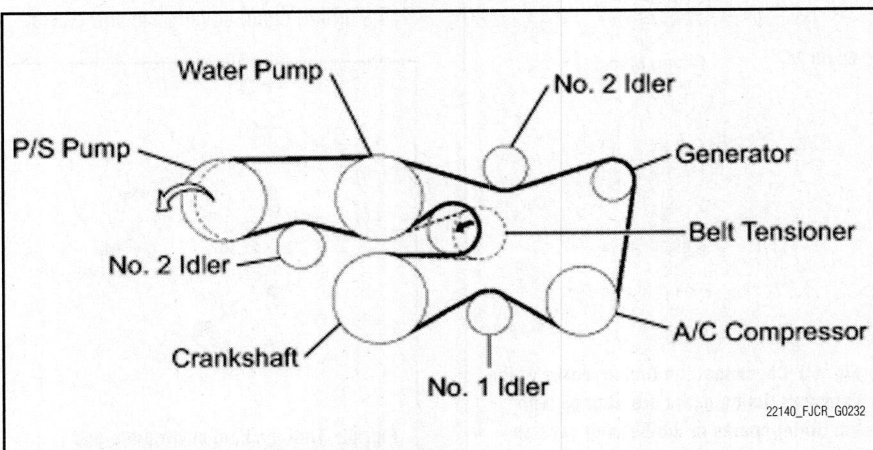

Fig. 57 Accessory belt routing—4.0L DOHC engine

CAMSHAFT AND VALVE LIFTERS

INSPECTION

See Figure 59.

1. Inspect the camshaft for runout.
   a. Place the camshaft on V-blocks.
   b. Using a dial indicator, measure the circle runout at the center journal. Maximum runout is0.0024 in. (0.06 mm).
   c. If the circle runout is greater than the maximum, replace the camshaft.
2. Inspect the cam lobes.
   a. Using a micrometer, measure the cam lobe height. Minimum cam lobe height is 1.7330 in. (44.018 mm) for the intake and 1.7492 in. (44.430 mm) for the exhaust lobes.
   b. If the cam lobe height is less than the minimum, replace the camshaft.
3. Using a micrometer, measure the journal diameter. No. 1 journal diameter is 1.4162 to 1.4167 in. (35.971 to 35.985 mm) and all other journal diameters are 0.9039 to 0.9045 in. (22.959 to 22.975 mm)
   a. If the journal diameter is not as specified, check the oil clearance.
4. Inspect camshaft timing gear assembly.
   a. Fix the intake camshaft in a vise.

✱✱ WARNING

Be careful not to damage the camshaft.

   b. Align the knock pin hole in the camshaft timing gear assembly with the knock pin of the camshaft, and install the camshaft timing gear assembly with the bolt. Tighten to 74 ft. lbs. (100 Nm).
   c. Confirm that the camshaft timing gear assembly is locked.
   d. Release that lock pin.
   e. Cover the 4 oil paths of the cam journal with masking tape as shown in the illustration.
   f. One of the 2 grooves on the cam journal is for the retard side path (upper) and the other is for the advance side path (lower). Each groove has 2 oil paths. Plug one of the oil paths for each groove with a piece of rubber before wrapping the cam journal with tape.
   g. Puncture the tape covering the advance oil path and retard oil path on the opposite side of the groove as shown in the illustration.
   h. Apply compressed air at 29 PSI (200 kPa) into the 2 broken paths (the

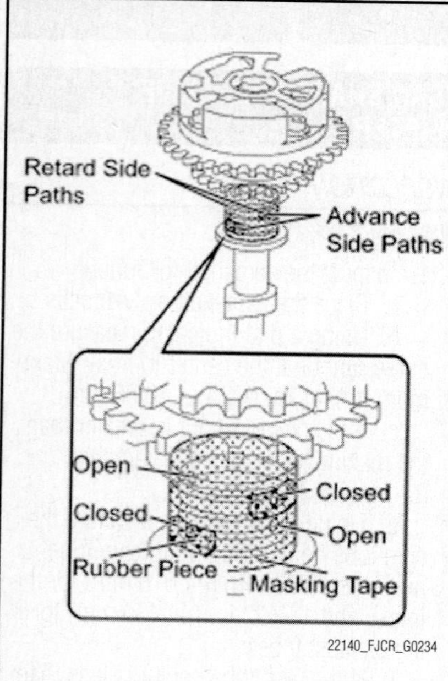

Fig. 59 Cover the 4 oil paths of the cam journal with masking tape as shown

advance side path and the retard side path).

i. Cover the paths with a shop rag or piece of cloth to avoid oil splashes.

j. Confirm that the camshaft timing gear assembly rotates in the timing advance direction when reducing the compressed air on the timing retard path.

k. When the lock pin is released, the camshaft timing gear rotates in the advance direction.

l. When the camshaft timing gear comes to the most advanced position, release the compressed air on the timing retard side path, and release that on the timing advance side path.

m. Camshaft timing assembly gear occasionally shifts to the retard side abruptly if the air compression on the advanced side path is released first. This often causes breakage of the lock pin.

n. Check the smooth revolution.

o. Rotate the camshaft timing gear several times within the movable range except for the most retarded position and check it rotates smoothly. The gear should moves smoothly in a range of about 31°.

p. Be sure to perform this check by hand, instead of using compressed air.

q. Check that the lock is in the most retarded position.

r. Confirm that the camshaft timing gear assembly is locked in the most retarded position.

s. Remove the set bolt, then remove the camshaft timing gear assembly.

➡**Do not remove the other 3 bolts.**

REMOVAL & INSTALLATION

2009 Vehicles

See Figures 60 through 79.

1. Before servicing the vehicle, refer to the precautions.
2. Disconnect the negative battery cable.
3. Drain engine coolant.
4. Remove V-bank cover.
5. Remove air cleaner assembly.
6. Remove throttle body bracket.
7. Remove oil baffle plate.
8. Remove surge tank bracket.
9. Remove intake air surge tank.

a. Disconnect the 2 water by-pass hoses.

b. Disconnect the fuel vapor feed hose.

c. Disconnect the ventilation hose.

d. Disconnect the 2 VSV connectors.

e. Disconnect the throttle body with motor connector.

f. Separate the 3 wire harness clamps and the hose clamp.

g. Remove the 2 nuts.

h. Using a socket hexagon wrench 8, remove the 4 bolts, intake air surge tank and the gasket.

10. Remove ignition coil assembly.
11. Remove the 10 bolts, 3 seal washers, 2 nuts, cylinder head cover and gasket.

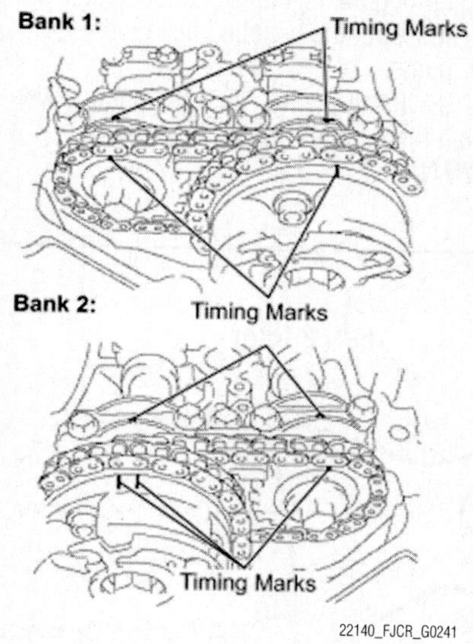

Bank 1: Timing Marks / Bank 2: Timing Marks / Timing Marks / 22140_FJCR_G0241

Fig. 60 Check that the timing marks of the camshaft timing gears are aligned with the timing marks of the bearing caps as shown

12. Set No. 1 cylinder to TDC on the compression stroke.

a. Turn the crankshaft pulley until its groove and the "0" timing mark of the timing chain cover are aligned.

b. Check that the timing marks of the camshaft timing gears are aligned with the timing marks of the bearing caps as shown in the illustration.

13. If not, turn the crankshaft 1 complete revolution (360°) and align the timing marks above.

a. Place paint marks on the No. 1 chain links corresponding to the timing marks of the camshaft timing gears.

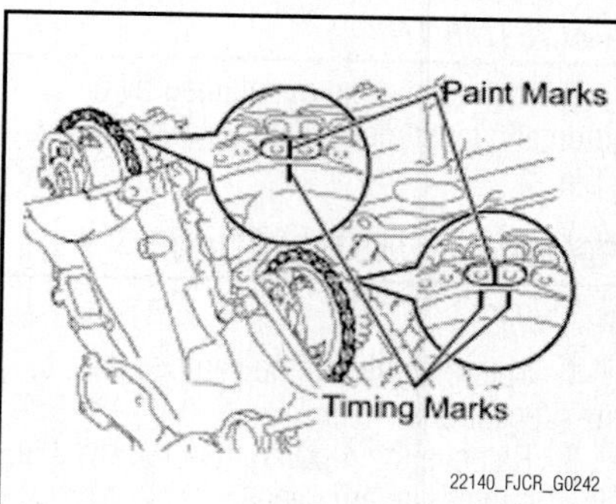

Fig. 61 Place paint marks on the No. 1 chain links corresponding to the timing marks of the camshaft timing gears

14. Remove No. 1 chain tensioner assembly.

⁂ WARNING

Never rotate the crankshaft with the chain tensioner removed. When rotating the camshaft with the timing chain removed, rotate the crankshaft counterclockwise 40°from the TDC first.

a. Remove the 4 bolts, then remove the timing chain cover plate and gasket.

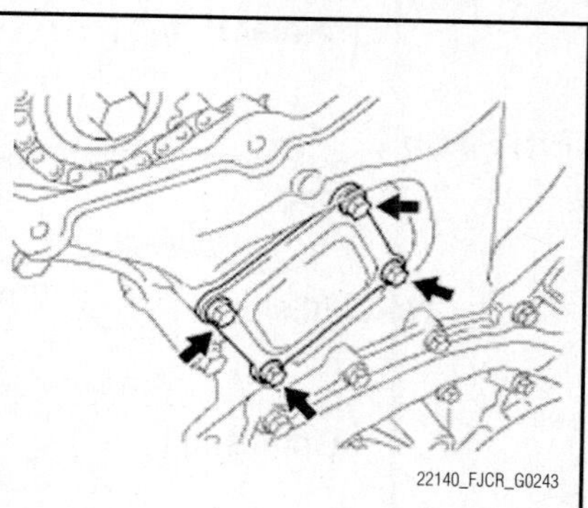

Fig. 62 Timing chain cover plate and gasket

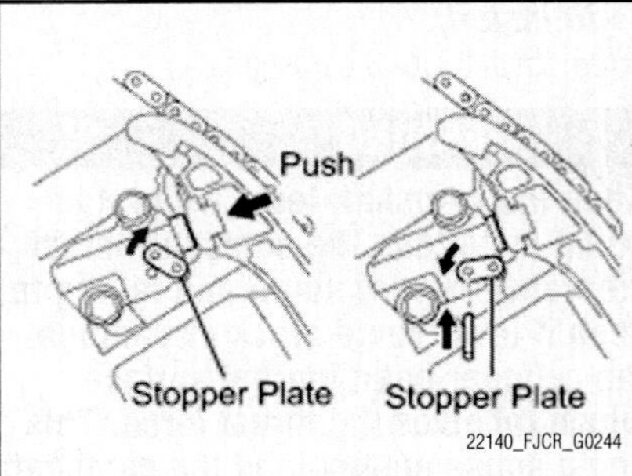

Fig. 63 While turning the stopper plate of the tensioner upward, push in the plunger of the chain tensioner as shown

b. While turning the stopper plate of the tensioner upward, push in the plunger of the chain tensioner as shown in the illustration.

c. While turning the stopper plate of the tensioner downward, insert a pin of 0.138 in. (3.5 mm) diameter into the holes in the stopper plate and tensioner to fix the stopper plate.

d. Remove the 2 bolts, then remove the chain tensioner.

15. Remove No. 2 camshaft.

WARNING

Keep the camshaft level while it is being removed. The camshaft thrust clearance is very small and failing to keep it level could crack or damage the cylinder head journal surface, which receives the thrust force. This could subsequently lead the camshaft to seize or break.

16. Perform the following steps to avoid damaging the camshaft during removal.

a. While raising the chain tensioner No. 2, insert a pin of 0.138 in. (3.5 mm) diameter into the hole to hold it.

b. Hold the hexagonal portion of the No. 2 camshaft with a wrench, and remove the camshaft timing gear set bolt.

WARNING

Be careful not to damage the cylinder head or valve lifter with the wrench.

c. Separate the camshaft timing gear from the No. 2 camshaft.

d. Rotate the camshaft counterclockwise using the wrench so that the cam lobes of No. 1 cylinder face upward as shown in the illustration.

e. Using several steps, uniformly loosen and remove the 8 bearing cap bolts in the sequence shown in the illustration.

f. Remove the 4 bearing caps and No. 2 camshaft.

17. Remove the No. 2 chain tensioner bolt, then remove the No. 2 chain tensioner and camshaft timing gear.

18. Remove camshaft.

WARNING

Keep the camshaft level while it is being removed. The camshaft thrust clearance is very small and failing to keep it level could crack or damage the cylinder head journal surface, which receives the thrust force. This could subsequently lead the camshaft to seize or break.

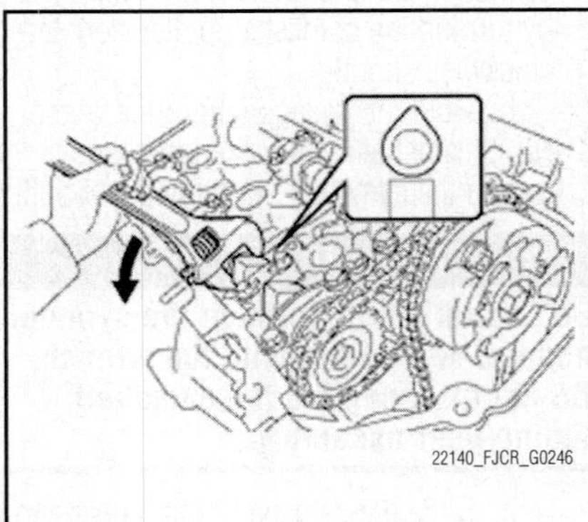

Fig. 65 Cam lobes facing upward

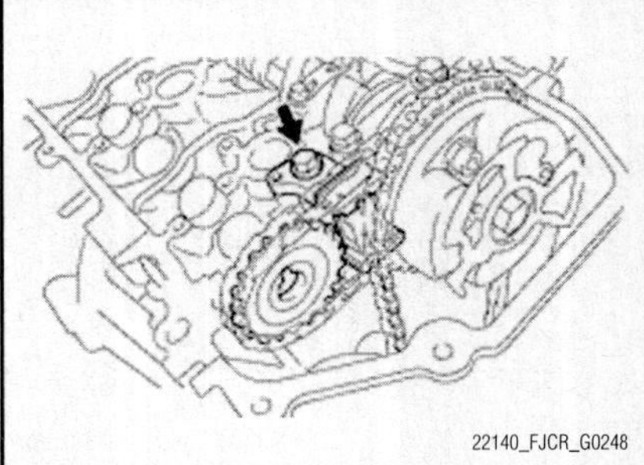

Fig. 67 Remove the No. 2 chain tensioner bolt, then remove the No. 2 chain tensioner and camshaft timing gear

19. Perform the following steps to avoid damaging the camshaft during removal.

a. Hold the hexagonal portion of the No. 1 camshaft with a wrench, and loosen the camshaft timing gear set bolt.

WARNING

Be careful not to damage the cylinder head or valve lifter with the wrench.

➡Do not disassemble the camshaft timing gear assembly.

b. Slide the camshaft timing gear and separate the No. 1 chain from the camshaft timing gear.

c. Rotate the No. 1 camshaft counterclockwise using the wrench so that the cam lobes of No. 1 cylinder face downward as shown in the illustration.

d. Using several steps, loosen and remove the 8 bearing cap bolts in the sequence shown in the illustration.

e. Remove the 4 bearing caps.

f. Remove the camshaft timing gear set bolt with the No. 1 camshaft lifted up,

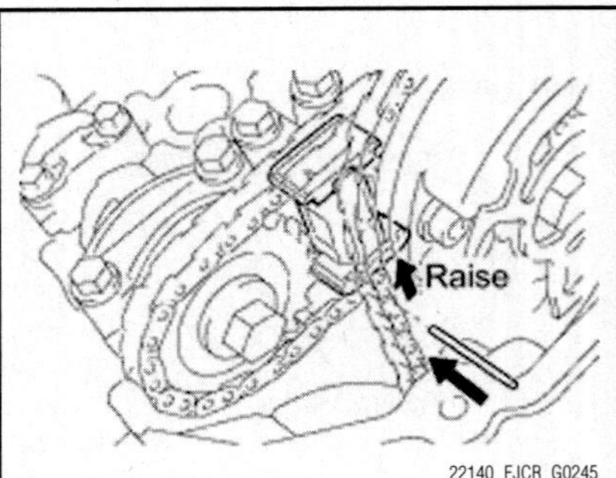

Fig. 64 While raising the chain tensioner No. 2, insert a pin of 0.138 in. (3.5 mm) diameter into the hole to hold it

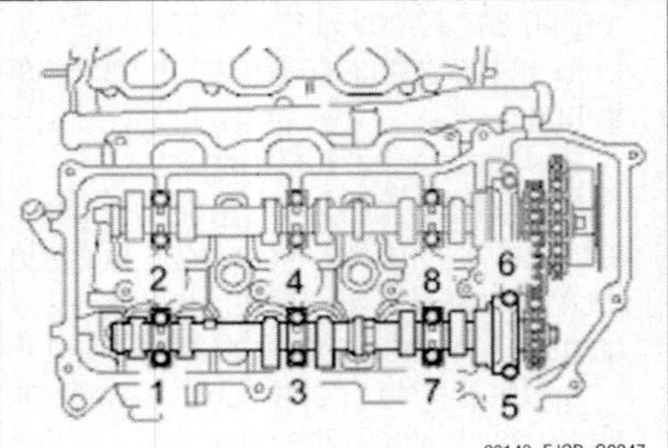

Fig. 66 Using several steps, uniformly loosen and remove the 8 bearing cap bolts in the sequence shown—No. 2 and 4 camshafts

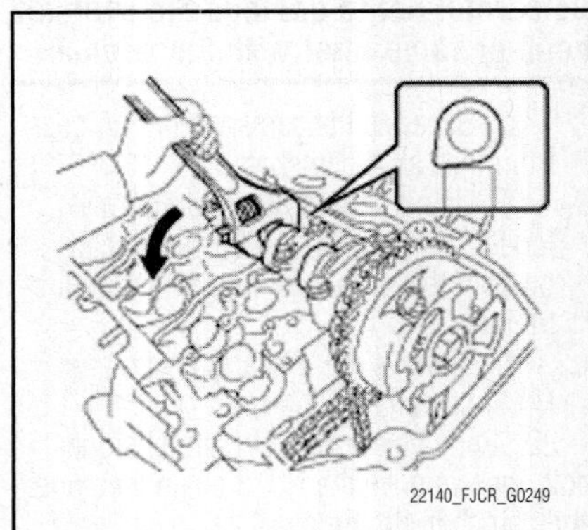

Fig. 68 Cam lobes facing downward

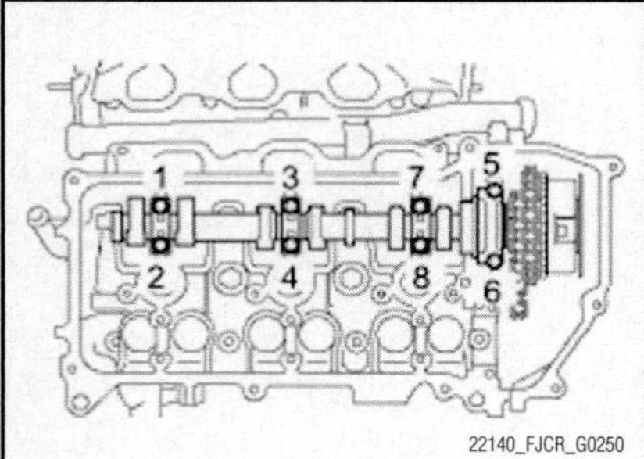

Fig. 69 Using several steps, loosen and remove the 8 bearing cap bolts in the sequence shown—No. 1 and 3 camshafts

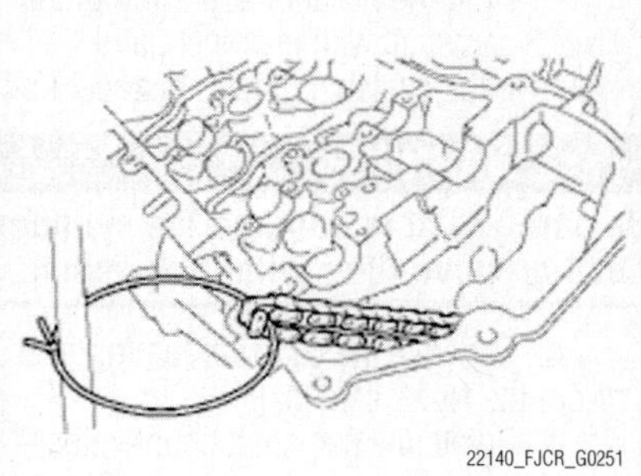

Fig. 70 Tie the No. 1 chain with a piece of string as shown

then remove the No. 1 camshaft and camshaft timing gear with No. 2 chain.

g. Tie the No. 1 chain with a piece of string as shown in the illustration.

⁂ WARNING

Be careful not to drop anything inside the timing chain cover.

20. Remove No. 4 camshaft.

⁂ WARNING

Keep the camshaft level while it is being removed. The camshaft thrust clearance is very small and failing to keep it level could crack or damage the cylinder head journal surface, which receives the thrust force. This could subsequently lead the camshaft to seize or break.

21. Perform the following steps to avoid damaging the camshaft during removal.

a. While pushing down the No. 3 chain tensioner, insert a pin of 0.138 in. (3.5 mm) diameter into the hole to hold it.

b. Hold the hexagonal portion of the No. 4 camshaft with a wrench, and remove the camshaft timing gear set bolt.

⁂ WARNING

Be careful not to damage the cylinder head or valve lifter with the wrench.

c. Separate the camshaft timing gear from the No. 4 camshaft.

d. Using several steps, uniformly loosen and remove the 8 bearing cap bolts in the sequence shown in the illustration.

e. Remove the 4 bearing caps and No. 4 camshaft.

22. Remove the No. 3 chain tensioner bolt, then remove the No. 3 chain tensioner and camshaft timing gear.

23. Remove No. 3 camshaft.

⁂ WARNING

Keep the camshaft level while it is being removed. The camshaft thrust clearance is very small and failing to keep it level could crack or damage the cylinder head journal surface, which receives the thrust force. This could subsequently lead the camshaft to seize or break.

24. Perform the following steps to avoid damaging the camshaft during removal.

a. Release the chain tension between the camshaft timing gear on the left cylinder head and crankshaft timing gear by turning the crankshaft pulley counterclockwise slightly.

b. Hold the hexagonal portion of the No. 3 camshaft with a wrench, then loosen the camshaft timing gear set bolt.

⁂ WARNING

Be careful not to damage the cylinder head or valve lifter with the wrench. Do not disassemble the camshaft timing gear assembly.

c. Slide the camshaft timing gear and separate the No. 1 chain from the camshaft timing gear.

d. Using several steps, uniformly loosen and remove the 8 bearing cap bolts in the sequence shown in the illustration.

e. Remove the 4 bearing caps.

f. Remove the camshaft timing gear set bolt with the No. 3 camshaft lifted up, then remove the No. 3 camshaft and camshaft timing gear with No. 2 chain.

g. Tie the No. 1 chain with a piece of string as shown in the illustration.

⁂ WARNING

Be careful not to drop anything inside the timing chain cover.

To install:.

25. Install No. 3 camshaft.

⁂ WARNING

Keep the camshaft level while it is being installed. The camshaft thrust clearance is very small and failing to keep it level could crack or damage the cylinder head journal surface, which receives the thrust force. This could subsequently lead the camshaft to seize or break.

26. Perform the following steps to avoid damaging the camshaft during removal.

a. Align the yellow mark link with the timing mark (2 dot marks) of the camshaft timing gear as shown in the illustration.

b. Apply new engine oil to the thrust portion and journal of the camshafts.

c. Temporarily put the No. 1 chain on the No. 2 chain of the camshaft timing gear.

d. Align the knock pin hole in the camshaft timing gear with the knock pin of the No. 3 camshaft, and insert the No. 3 camshaft into the camshaft timing gear.

e. Temporarily install the camshaft timing gear set bolt.

f. Set the No. 3 camshaft onto the left cylinder head with the cam lobes of the No. 2 cylinder facing downward as shown in the illustration.

g. Install the 4 bearing caps in the proper locations as shown.

h. Apply a light coat of engine oil to the threads and under the heads of the bearing cap bolts.

i. Using several steps, uniformly install and tighten the 8 bearing cap bolts in the sequence shown in the illustration. Tighten the 0.39 in. (10 mm) head bearing cap bolts to 80 inch lbs. (9 Nm) and the 0.47 in. (12 mm) head bearing cap bolts to 18 ft. lbs. (24 Nm).

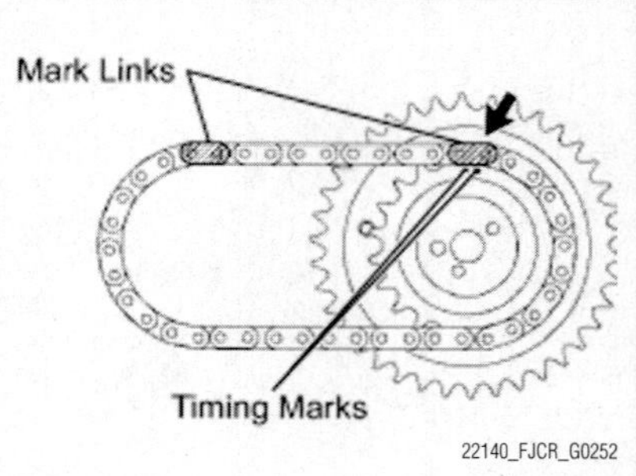

Fig. 71 Align the yellow mark link with the timing mark (2 dot marks) of the camshaft timing gear as shown

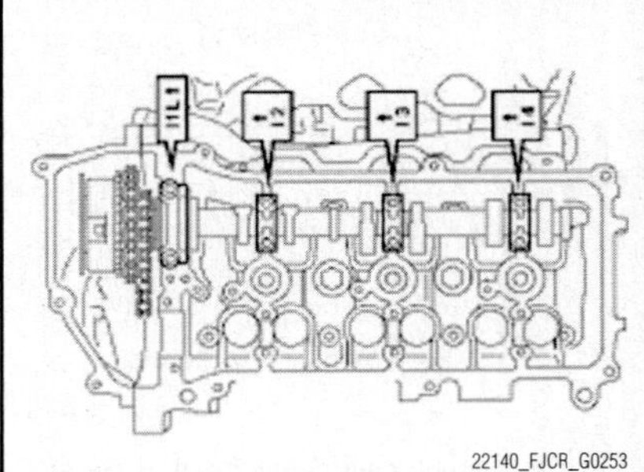

Fig. 72 Install the 4 bearing caps in the proper locations as shown

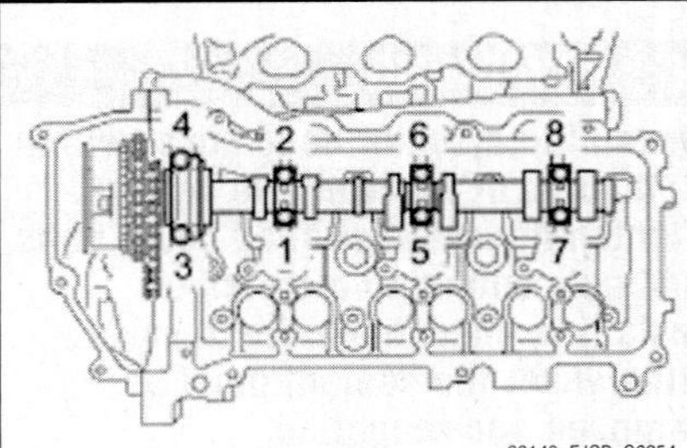

Fig. 73 Using several steps, uniformly install and tighten the 8 bearing cap bolts in the sequence shown—No. 1 and 3 camshafts

j. Align the paint mark of the No. 1 chain with the timing marks of the camshaft timing gear.

k. Hold the hexagonal portion of the No. 3 camshaft with a wrench, and tighten the camshaft timing gear set bolt. Tighten to 74 ft. lbs. (100 Nm).

27. Install No. 3 chain tensioner assembly.

a. While pushing in the tensioner, insert a pin of 0.138 in. (3.5 mm) diameter into the hole to hold it.

b. Temporarily install the camshaft timing gear and No. 3 chain tensioner and align the yellow mark links with the timing marks (1 dot mark and 2 dot marks) of the camshaft timing gears.

c. Tighten the No. 3 chain tensioner bolt. Tighten to 14 ft. lbs. (19 Nm).

28. Install No. 4 camshaft.

WARNING

Keep the camshaft level while it is being installed. The camshaft thrust clearance is very small and failing to keep it level could crack or damage the cylinder head journal surface, which receives the thrust force. This could subsequently lead the camshaft to seize or break.

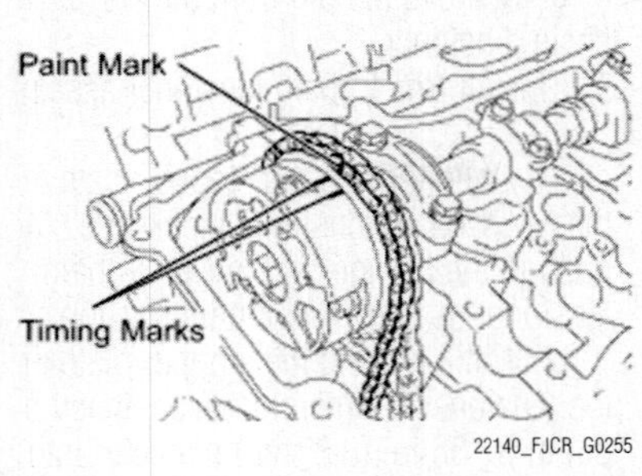

Fig. 74 Temporarily install the camshaft timing gear and No. 3 chain tensioner and align the yellow mark links with the timing marks (1 dot mark and 2 dot marks) of the camshaft timing gears

29. Perform the following steps to avoid damaging the camshaft during removal.

a. Align the knock pin hole in the camshaft timing gear with the knock pin of the No. 4 camshaft, and insert the No. 4 camshaft into the camshaft timing gear.

b. Temporarily install the camshaft timing gear set bolt.

c. Install the 4 bearing caps in the proper locations as shown.

d. Apply a light coat of engine oil to the threads of the bearing cap bolts.

e. Using several steps, uniformly install and tighten the 8 bearing cap bolts in the sequence shown in the illustration. Tighten the 0.39 in. (10 mm) head bearing cap bolts to 80 inch lbs. (9 Nm) and the 0.47 in. (12 mm) head bearing cap bolts to 18 ft. lbs. (24 Nm).

f. Hold the hexagonal portion of the No. 4 camshaft with a wrench, and tighten the camshaft timing gear set bolt. Tighten to 74 ft. lbs. (100 Nm).

g. Remove the pin from the No. 3 chain tensioner.

h. Release the chain tension between the camshaft timing gear on the right cylinder head and crankshaft timing gear by turning the crankshaft pulley clockwise slightly.

30. Install camshaft.

WARNING

Keep the camshaft level while it is being installed. The camshaft thrust clearance is very small and failing to keep it level could crack or damage the cylinder head journal surface, which receives the thrust force. This could subsequently lead the camshaft to seize or break.

31. Perform the following steps to avoid damaging the camshaft during removal.

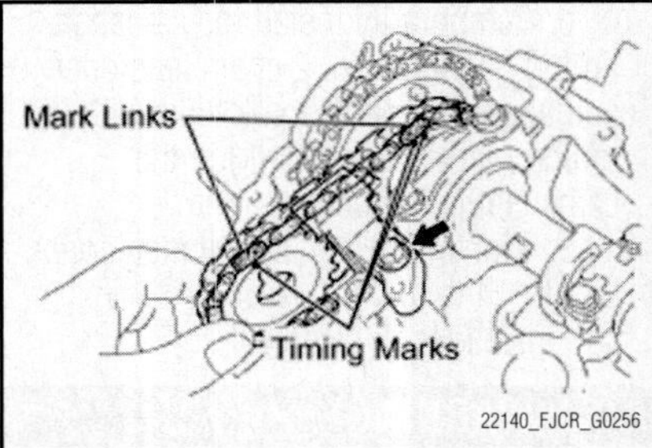

Fig. 75 Using several steps, uniformly install and tighten the 8 bearing cap bolts in the sequence shown—No. 2 and 4 camshafts

a. Align the yellow mark link with the timing mark (1 dot mark) of the camshaft timing gear as shown in the illustration.

b. Apply new engine oil to the thrust portion and journal of the camshafts.

c. Temporarily install the No. 1 chain onto the No. 2 chain of the camshaft timing gear.

d. Align the knock pin hole in the camshaft timing gear with the knock pin of the No. 1 camshaft, and insert the No. 1 camshaft into the camshaft timing gear.

e. Temporarily install the camshaft timing gear set bolt.

f. Install the No. 1 camshaft onto the right cylinder head with the cam lobes of the No. 1 cylinder facing downward as shown in the illustration.

g. Install the 4 bearing caps in the proper locations as shown.

h. Apply a light coat of engine oil to the threads and under the heads of the bearing cap bolts.

i. Using several steps, uniformly install and tighten the 8 bearing cap bolts in the sequence shown in the illustration. Tighten the 0.39 in. (10 mm) head bearing cap bolts to 80 inch lbs. (9 Nm) and the 0.47 in. (12 mm) head bearing cap bolts to 18 ft. lbs. (24 Nm).

j. Rotate the No. 1 camshaft clockwise using a wrench so that the timing mark of the camshaft timing gear is aligned with the timing mark of the camshaft bearing cap.

k. Align the paint mark of the No. 1 chain with the timing mark of the camshaft timing gear.

l. Hold the hexagonal portion of the No. 1 camshaft with a wrench, and tighten the camshaft timing gear set bolt. Tighten to 74 ft. lbs. (100 Nm).

32. Install No. 2 chain tensioner assembly.

a. While pushing in the tensioner, insert a pin of 0.138 in. (3.5 mm) diameter into the hole to hold it.

b. Temporarily install the camshaft timing gear and No. 2 chain tensioner and align the yellow mark links with the timing marks (1 dot mark) of the camshaft timing gears.

c. Tighten the No. 2 chain tensioner bolt to 14 ft. lbs. (19 Nm).

33. Install No. 2 camshaft.

WARNING

Keep the camshaft level while it is being installed. The camshaft thrust clearance is very small and failing to keep it level could crack or damage the cylinder head journal surface, which receives the thrust force. This could subsequently lead the camshaft to seize or break.

34. Perform the following steps to avoid damaging the camshaft during removal.

a. Install the No. 2 camshaft onto the right cylinder head with the cam lobes of No. 1 cylinder facing upward as shown in the illustration.

b. Install the 4 bearing caps in the proper locations as shown.

c. Apply a light coat of engine oil to the threads and under the heads of the bearing cap bolts.

d. Using several steps, uniformly install and tighten the 8 bearing cap bolts in the sequence shown in the illustration. Tighten the 0.39 in. (10 mm) head bearing cap bolts to 80 inch lbs. (9 Nm) and the 0.47 in. (12 mm) head bearing cap bolts to 18 ft. lbs. (24 Nm).

e. Rotate the No. 2 camshaft clockwise using a wrench so that the knock pin of the No. 2 camshaft is aligned with the knock pin hole in the camshaft timing gear.

f. Hold the hexagonal portion of the No. 2 camshaft with a wrench, and install the camshaft timing gear set bolt. Tighten to 74 ft. lbs. (100 Nm).

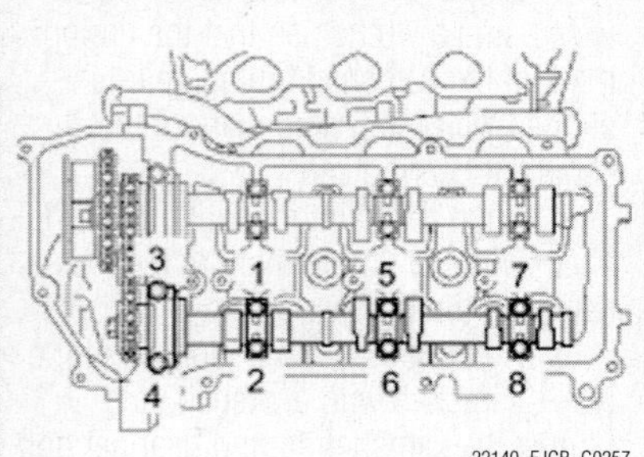

Fig. 76 Using several steps, uniformly install and tighten the 8 bearing cap bolts in the sequence shown—No. 2 and 4 camshafts

g. Remove the pin from the No. 2 chain tensioner.

35. Install No. 1 chain tensioner assembly.

a. While turning the stopper plate of the No. 1 chain tensioner clockwise, push in the plunger of the No. 1 chain tensioner as shown in the illustration.

b. While turning the stopper plate of the tensioner counterclockwise, insert a pin of 0.138 in. (3.5 mm) diameter into the holes in the stopper plate and No. 1 chain tensioner to fix the stopper plate.

c. Install the No. 1 chain tensioner with the 2 bolts. Tighten to 7 ft. lbs. (10 Nm).

d. Remove the bar from the No. 1 chain tensioner.

e. Install a new gasket and the timing chain cover plate with the 4 bolts. Tighten to 80 inch lbs. (9 Nm).

f. Turn the crankshaft pulley 2 complete revolutions slowly until its groove and the "0" timing mark of the timing chain cover are aligned.

g. Check that the timing marks of the camshaft timing gears are aligned with the timing marks of the bearing cap as shown in the illustration.

36. Set No. 1 cylinder to TDC on the compression stroke.

37. Inspect and adjust valve clearance.

38. Install cylinder head cover.

a. Remove any old sealant.

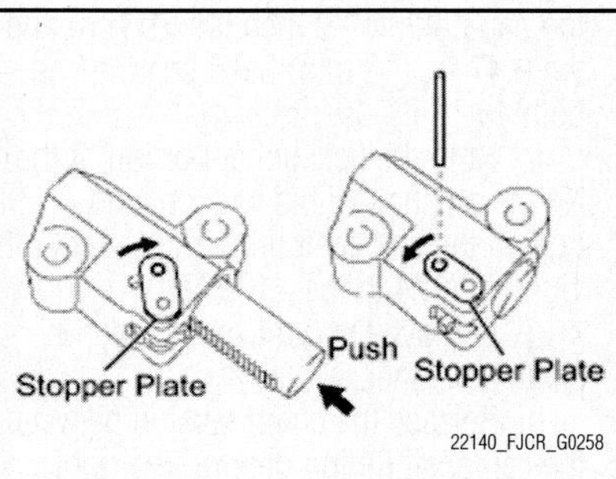

Fig. 77 While turning the stopper plate of the No. 1 chain tensioner clockwise, push in the plunger of the No. 1 chain tensioner as shown

WARNING

Do not drop any oil on the contact surfaces of the cylinder head, timing chain cover and cylinder head cover.

b. Apply a continuous bead of Three Bond 1207B or the equivalent sealant 0.08–0.12 in. (2–3 mm) diameter to the cylinder head and timing chain cover as shown in the illustration.

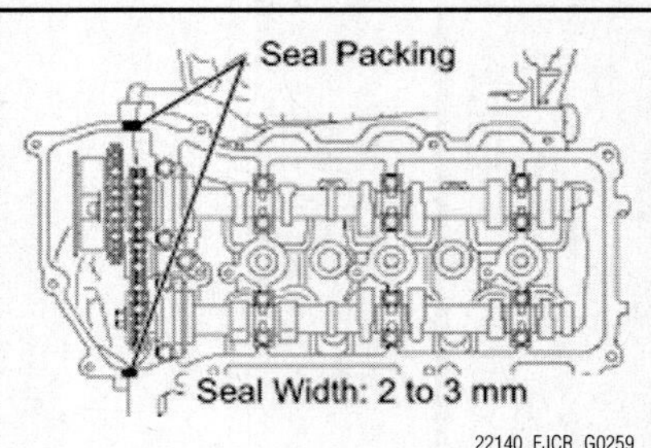

Fig. 78 Apply a continuous bead of Three Bond 1207B or the equivalent sealant 0.08–0.12 in. (2–3 mm) diameter to the cylinder head and timing chain cover as shown

WARNING

Install the cylinder head cover within 3 minutes of applying the sealant. Tighten the cylinder head cover bolts and nuts within 15 minutes of installing the cylinder head cover. Otherwise, the sealant must be removed and reapplied.

c. Install the seal washers onto the bolts.

d. Install the cylinder head cover with the 10 bolts and 2 nuts. Tighten the bolts and nuts uniformly in several steps. Tighten bolts **A** to 7 ft. lbs. (10 Nm), bolts **B** to 80 inch lbs. (9 Nm) and the nuts to 80 inch lbs. (9 Nm).

➡Bolt A are 0.98 in. (25 mm) and bolt B 2.36 in. (60 mm).

39. Install ignition coil assembly.
40. Install intake air surge tank.
41. Install No. 2 surge tank bracket.
42. Install No. 1 surge tank bracket.
43. Install oil baffle plate.
44. Install throttle body bracket.

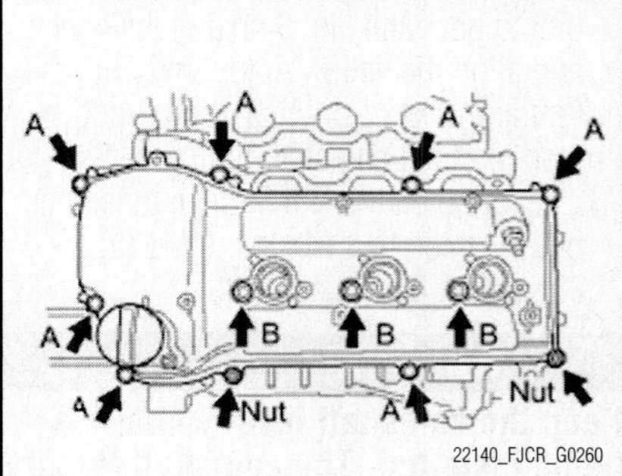

Fig. 79 Tighten the cylinder head cover bolts and nuts uniformly in several steps. Tighten bolts A to 7 ft. lbs. (10 Nm), bolts B to 80 inch lbs. (9 Nm) and the nuts to 80 inch lbs. (9 Nm)

45. Install air cleaner assembly.
46. Connect the negative battery cable.
47. Add engine coolant.
48. Check for engine coolant leakage.
49. Check for engine oil leakage.
50. Inspect ignition timing.
51. Install V-bank cover.

2010 Vehicles

See Figures 80 through 108.

1. Remove no. 1 engine under cover sub-assembly.
2. Drain engine oil.
3. Drain engine coolant.
4. Remove v-bank cover.
5. Remove air cleaner assembly.
6. Remove intake manifold.
7. Remove fan shroud.
8. Remove ignition coil assembly.
9. Separate vane pump assembly.
10. Remove generator assembly.
11. Remove engine oil level dipstick guide.
12. Remove no. 1 oil pipe.
13. Remove no. 2 oil pipe.
14. Remove rear cylinder head cover.
15. Disconnect fuel pipe sub-assembly.
16. Remove the valve covers.
17. Remove the 4 bolts, timing chain cover plate and gasket.
18. Turn the crankshaft pulley and align the notch with the "0" timing mark of the timing chain cover.

➡If the marks are not aligned, turn the crankshaft again to align the marks.

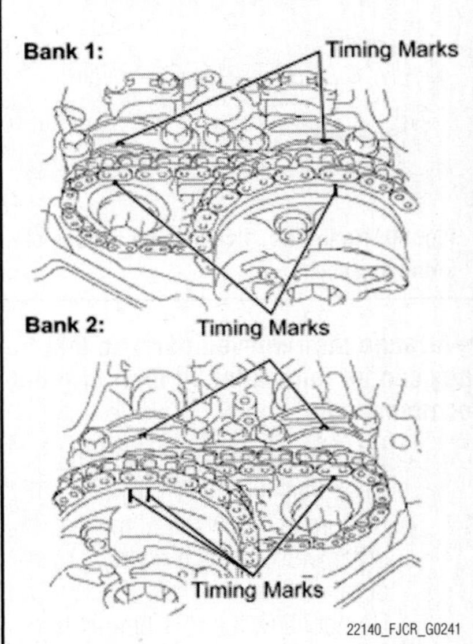

Fig. 80 Check that the timing marks of the camshaft timing gears are aligned with the timing marks of the bearing caps as shown

19. Check that the timing marks of the camshaft timing gears are aligned with the timing marks of the bearing cap.

➡Be sure to place the paint marks on 2 links of the chain and on the sprockets of the camshaft timing gears at the locations of the timing marks of the camshaft timing gears.

20. Place paint marks on the timing marks and sprockets of each camshaft timing gear and on the links of the No. 1 chain.

➡This prevents the valves and pistons from interfering with each other.

21. Turn the crankshaft approximately 30° counterclockwise so that there is some slack in the chain.
22. Align the hole in the lever of the tensioner with the hole in the tensioner body as shown in the illustration, and then insert a pin with a diameter of 0.0500 inches. (1.27 mm) into the hole.
23. Turn the crankshaft clockwise, and align the notch with the timing mark "0" of the timing chain cover.
24. Remove the 2 bolts and chain tensioner.

➡Do not drop the No. 1 chain tensioner assembly or bolts into the timing chain cover.

25. Disconnect the chain sub-assembly (for bank 1):

➡When turning the crankshaft, engine oil may spray out of the oil holes.

a. Turn the crankshaft clockwise until it is in the position shown in the illustration so that there is some slack in the chain between the banks.

CAUTION

As the camshafts turn suddenly, do not touch the camshafts or camshaft timing gears.

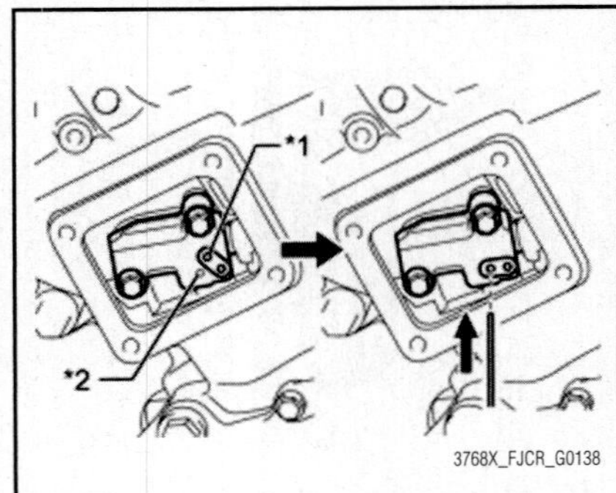

Fig. 81 Lever hole (1) and tensioner hole (2)

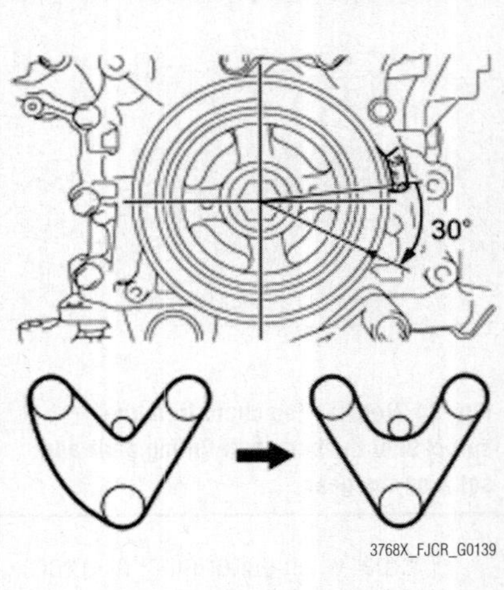

Fig. 82 Turn the crankshaft clockwise until it is in the position shown to release tension on the chain

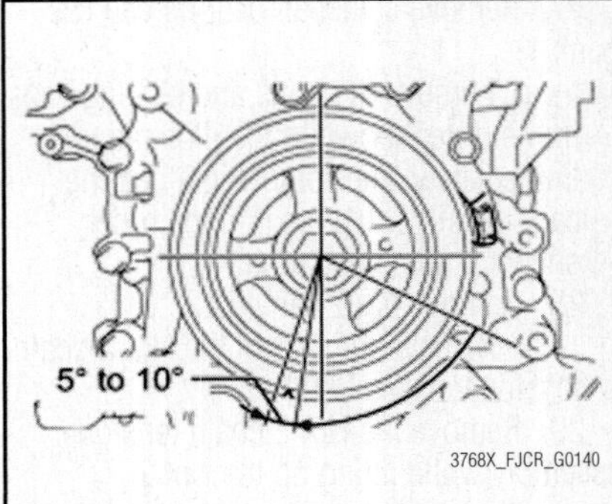

Fig. 83 Turn the crankshaft clockwise until it is in the position shown in the illustration so that the chain can be removed easily

b. Turn the crankshaft clockwise until it is in the position shown in the illustration so that the chain can be removed easily.

CAUTION

As the camshaft may turn suddenly and pinch your fingers when the chain is removed, pinch the chain and lift it upward to remove it from the sprocket.

26. Remove the chain from the sprocket of the camshaft timing gear and set it on the gear.

➡Arrange the removed parts so that they can be reinstalled in their original locations:

- Do not install the bearing caps when installing VVT bolt kit.
- Be sure to follow the numerical

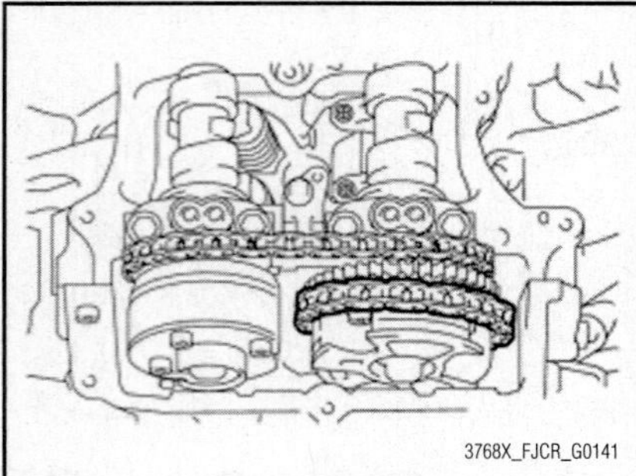

Fig. 84 Remove the chain from the sprocket of the camshaft timing gear and set it on the gear

order when performing this procedure.

- Do not allow the VVT bolt kit to contact the camshaft.
- Do not drop the VVT bolt kit into the cylinder head.

27. Remove camshaft bearing cap (for bank 1):

a. Remove the bolts and bearing caps in the order shown in the illustration. Immediately after removing a bearing cap, install VVT bolt kit in the order shown in the illustration.

28. Remove no. 2 camshaft:

a. Remove the bolt of the No. 2 chain tensioner assembly.

29. Remove the No. 2 chain tensioner assembly while lifting up the No. 2 camshaft.

30. While lifting up the No. 2 camshaft,

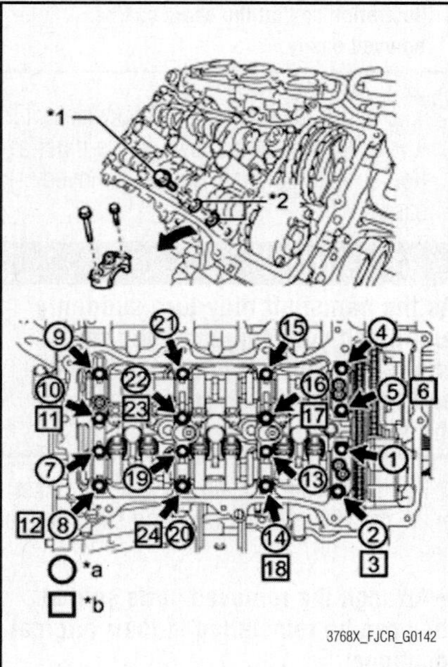

Fig. 85 Bank 1 bearing cap removal sequence—VVT bolt kit (1), part removal (a), VVT bolt kit installation (b)

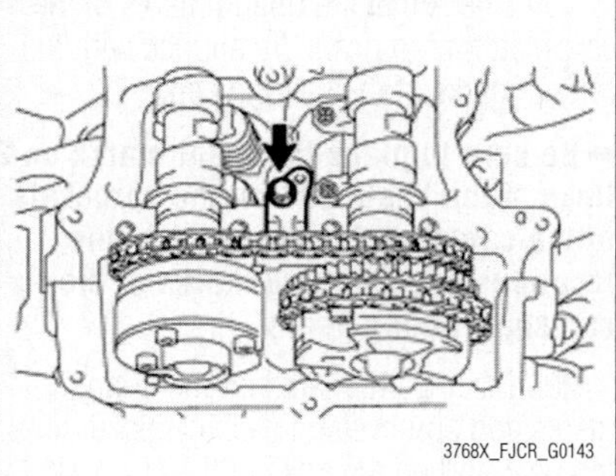

Fig. 86 Remove the bolt of the No. 2 chain tensioner assembly

pass it through the No. 2 chain and pull it out towards the front of the vehicle to remove it.

31. Remove camshaft:

a. Lift up the rear of the camshaft so that it is at an angle.

b. Remove the chain from the camshaft timing gear and pull out the camshaft and No. 2 chain towards the rear of the vehicle to remove them.

➡Do not drop the chain into the gap between the engine and cover.

32. Suspend the chain with a string or equivalent.

33. Disconnect chain sub-assembly (for bank 2):

a. Turn the crankshaft counterclockwise, and align the notch with the "0" timing mark of the timing chain cover.

b. Remove the chain from the sprocket of the camshaft timing gear and set it on the gear.

CAUTION

As the camshaft may turn suddenly and pinch your fingers when the chain is removed, pinch the chain and lift it upward to remove it from the sprocket.

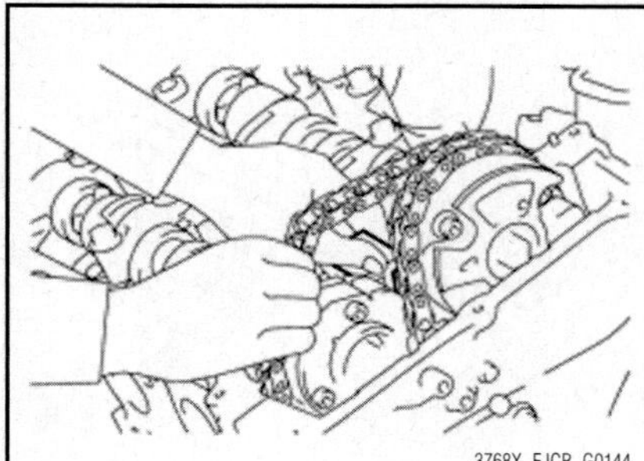

Fig. 87 Remove the No. 2 chain tensioner assembly while lifting up the No. 2 camshaft

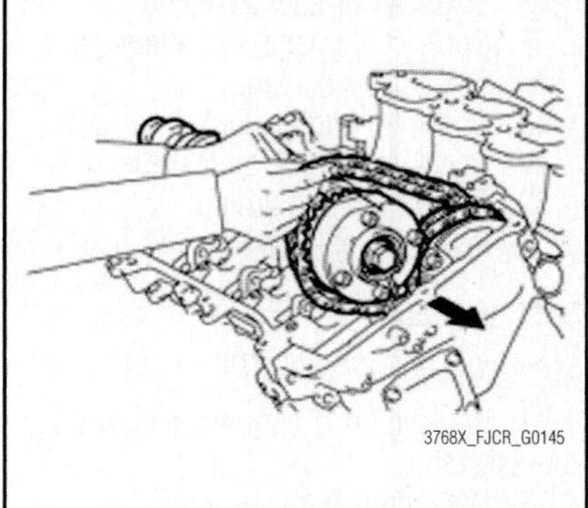

Fig. 88 Removing the No. 2 camshaft

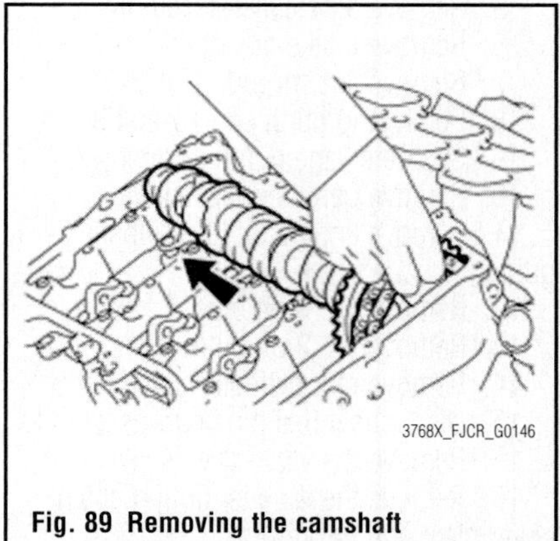

Fig. 89 Removing the camshaft

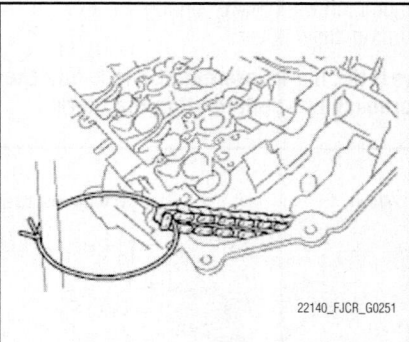

Fig. 90 Tie the No. 1 chain with a piece of string as shown

➡Arrange the removed parts so that they can be reinstalled in their original locations.

- Do not install the bearing caps when installing VVT bolt kit.
- Be sure to follow the numerical order when performing this procedure.
- Do not allow the VVT bolt kit to contact the camshaft.
- Do not drop the VVT bolt kit into the cylinder head.

34. Remove camshaft bearing cap (for bank 2):

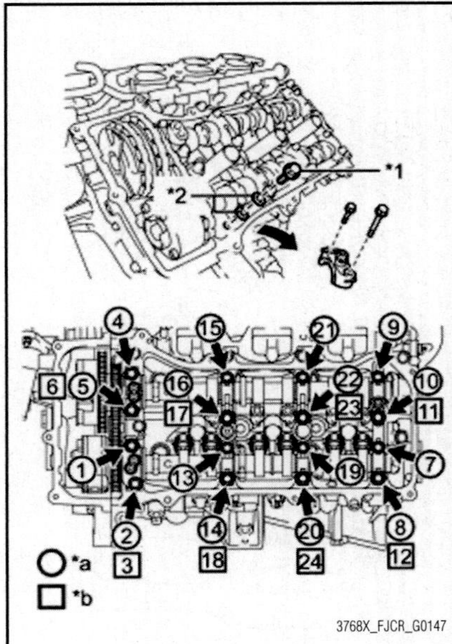

Fig. 91 Bank 2 bearing cap removal sequence—VVT bolt kit (1), part removal (a), VVT bolt kit installation (b)

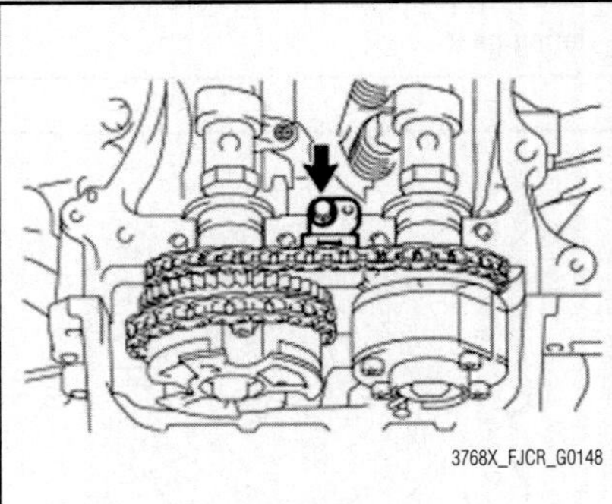

Fig. 92 Remove the bolt of the No. 3 chain tensioner assembly

a. Remove the bolts and bearing caps in the order shown in the illustration. Immediately after removing a bearing cap, install VVT bolt kit in the order shown in the illustration.

35. Remove no. 4 camshaft sub-assembly:

a. Remove the bolt of the No. 3 chain tensioner assembly.

36. Remove the No. 3 chain tensioner assembly while lifting up the No. 4 camshaft.

37. While lifting up the No. 4 camshaft, pass it through the No. 2 chain and pull it out towards the front of the vehicle to remove it.

38. Remove no. 3 camshaft sub-assembly:

a. Lift up the rear of the camshaft so that it is at an angle.

b. Remove the chain from the

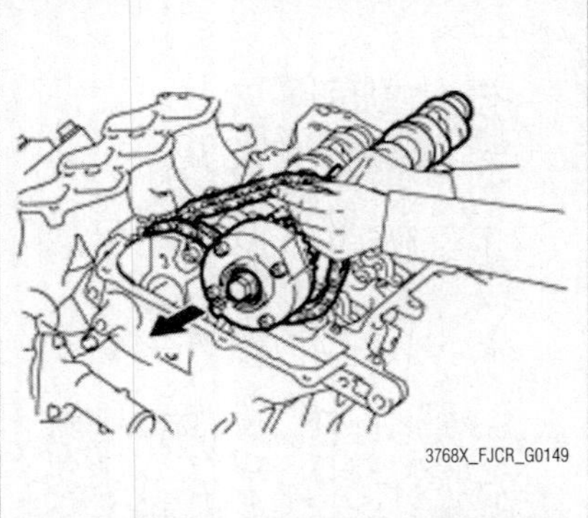

Fig. 93 Removing the No. 4 camshaft

camshaft timing gear and pull out the No. 3 camshaft and No. 2 chain towards the rear of the vehicle to remove them.

➡Do not drop the chain into the gap between the engine and cover.

c. Suspend the chain with a string or equivalent.

➡Do not remove the other 3 bolts. If planning to reuse the camshaft timing gear, be sure to release the straight pin lock before installing the camshaft timing gear.

To install:

39. Install no. 3 camshaft sub-assembly:

a. Check that the notch is aligned with the "0" timing mark of the timing chain cover.

b. Align the mark plate (yellow) with the timing mark of the camshaft timing gear as shown in the illustration and install the No. 2 chain to the camshaft timing gear.

c. Make sure that the No. 1 valve rocker arm sub-assembly is installed as shown in the illustration.

d. Install the chain to the No. 3 camshaft, and then install the camshaft to the LH camshaft housing.

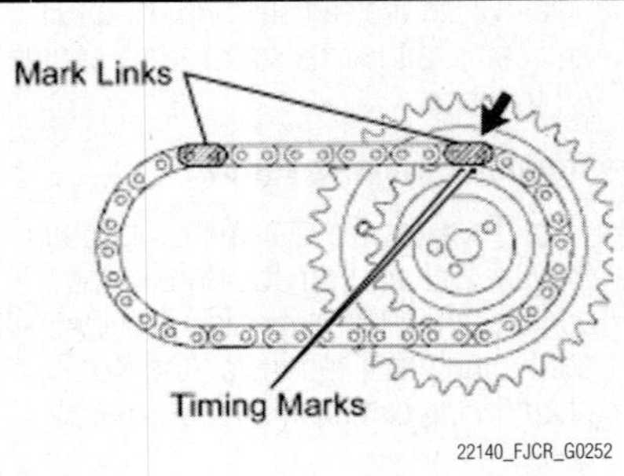

Fig. 94 Align the yellow mark link with the timing mark (2 dot marks) of the camshaft timing gear as shown

➡Place the chain on the camshaft timing gear but do not engage the teeth of the sprocket and the chain. Install the camshaft so that the timing mark is facing upward.

40. Install no. 4 camshaft sub-assembly:

a. Clean the camshaft housing LH and camshaft journals and apply engine oil to them.

b. Pass the No. 4 camshaft through the No. 2 chain from the front of the vehicle, align the mark plate (yellow) with the timing mark and install the No. 2 chain to the camshaft timing exhaust gear.

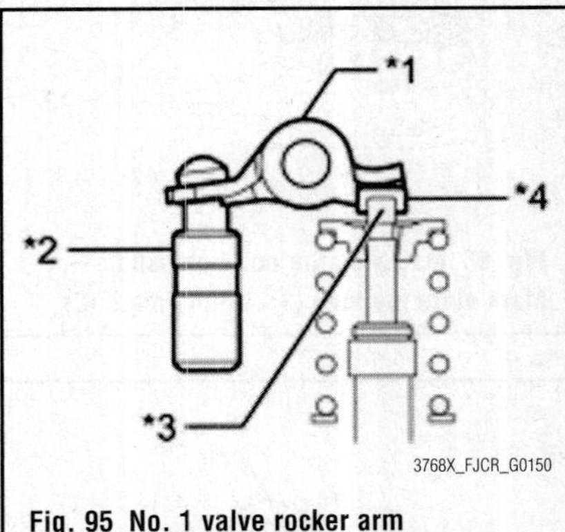

Fig. 95 No. 1 valve rocker arm sub-assembly installation orientation

c. While lifting up the No. 4 camshaft, pass the No. 3 chain tensioner assembly through the No. 2 chain and set it in place.

d. Install the No. 4 camshaft to the camshaft housing LH, and then install the No. 3 chain tensioner assembly with the bolt. Tighten to 15 ft. lbs. (21 Nm).

41. Install camshaft bearing cap (for bank 2):

a. Clean the camshaft bearing caps and apply engine oil to them.

b. Make sure that the No. 1 valve rocker arm sub-assembly is installed as shown in the illustration.

➡Be sure to follow the numerical order when performing this procedure. Do not drop the VVT bolt kit into the cylinder head.

c. Check the marks and numbers on the camshaft bearing caps, and then remove VVT bolt kit in the order shown in the illustration. Immediately after removing VVT bolt kit in the location for a bearing cap, install the bearing cap with the bolts in the order shown in the illustration.

- Bolt A: 21 ft. lbs. (28 Nm).
- Bolt B: 12 ft. lbs. (16 Nm).

d. Check the torque of each bolt again.

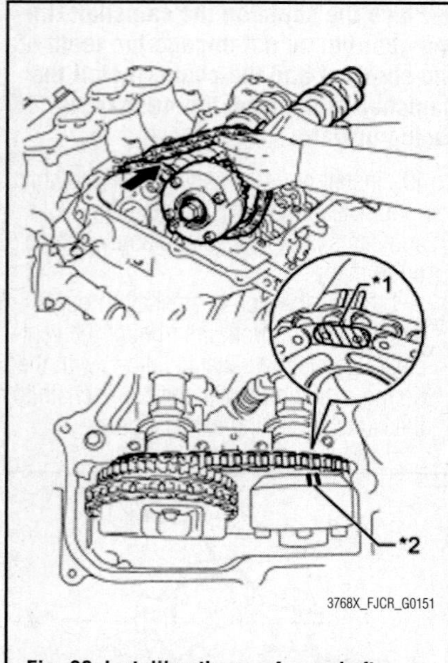

Fig. 96 Installing the no. 4 camshaft—Mark plate (yellow) (1), timing mark (2)

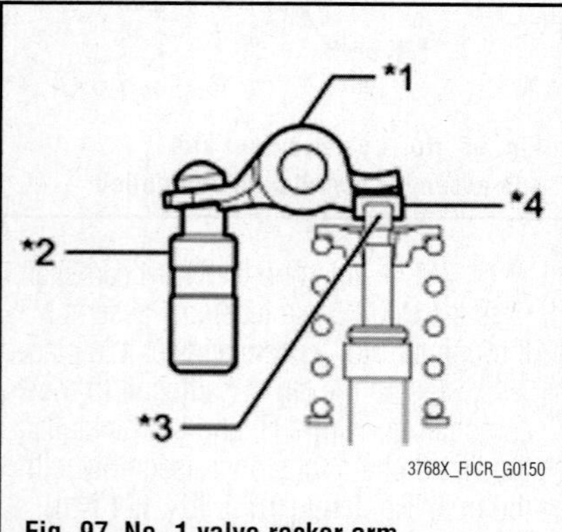

Fig. 97 No. 1 valve rocker arm sub-assembly installation orientation

➡If the paint marks are not aligned, align them by turning the camshaft slightly.

42. Connect chain sub-assembly (for bank 2):

a. Align the paint marks on the camshaft timing gear and No. 1 chain and install the No. 1 chain to the camshaft timing gear.

43. Install camshaft:

a. Turn the crankshaft clockwise until it is in the position shown in the illustration so that the chain can be removed easily.

➡When turning the crankshaft, engine oil may spray out of the oil holes.

b. Align the mark plate (yellow) with the timing mark of the camshaft timing gear as shown in the illustration and install the No. 2 chain to the camshaft timing gear.

➡The mark plate is yellow.

c. Clean the camshaft housing RH and camshaft journals and apply engine oil to them.

d. Make sure that the No. 1 valve rocker arm sub-assembly is installed as shown in the illustration.

e. Install the chain to the camshaft, and then install the camshaft to the RH camshaft housing.

44. Install no. 2 camshaft:

a. Clean the camshaft housing RH and camshaft journals and apply engine oil to them.

➡The mark plate is yellow.

b. Pass the No. 2 camshaft through the No. 2 chain from the front of the vehicle, align the mark plate (yellow) with the timing mark and install the No. 2 chain to the camshaft timing exhaust gear.

c. While lifting up the No. 2 camshaft, pass the No. 2 chain tensioner assembly through the No. 2 chain and set it in place.

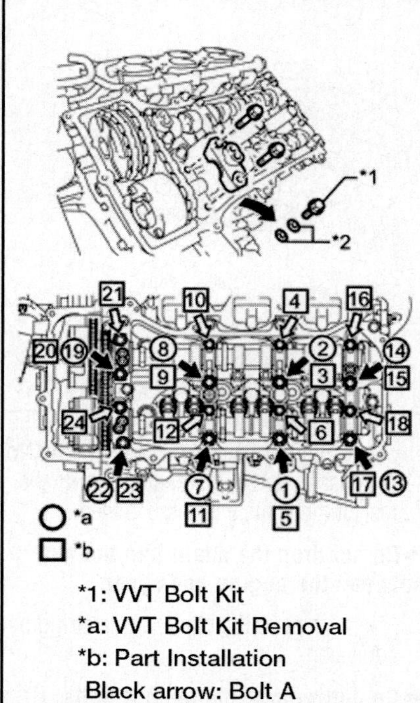

Fig. 98 Install the bank 2 bearing cap with the bolts in the order shown in the illustration

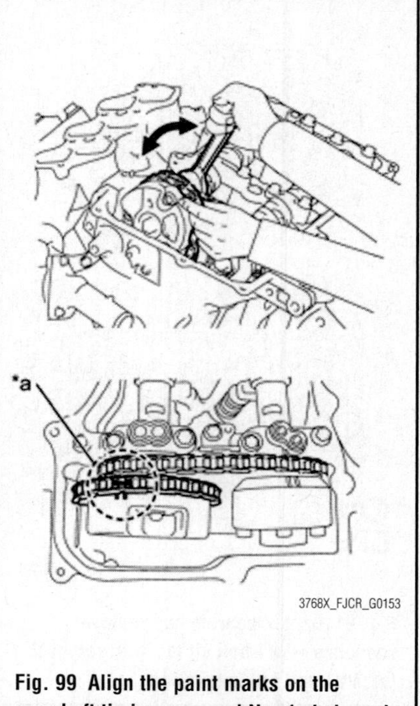

Fig. 99 Align the paint marks on the camshaft timing gear and No. 1 chain and install the No. 1 chain to the camshaft timing gear

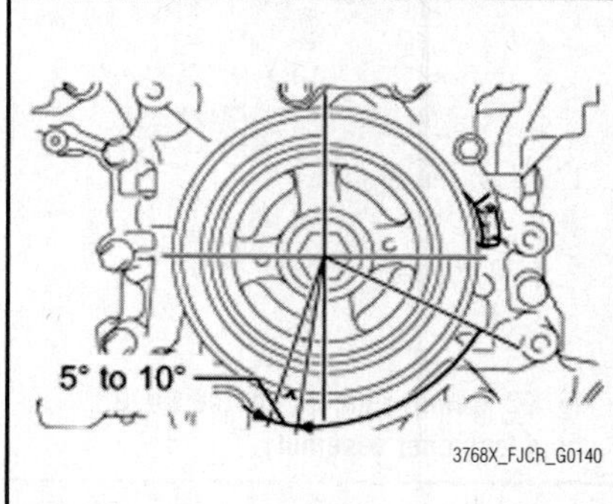

Fig. 100 Turn the crankshaft clockwise until it is in the position shown in the illustration so that the chain can be removed easily

Fig. 101 Align the mark plate (yellow) with the timing mark of the camshaft timing gear as shown

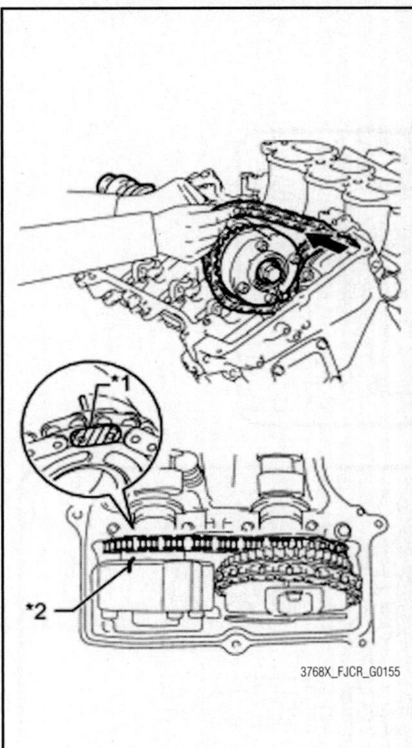

Fig. 102 Installing the No. 2 camshaft—Mark plate (yellow) (1), Timing mark (2)

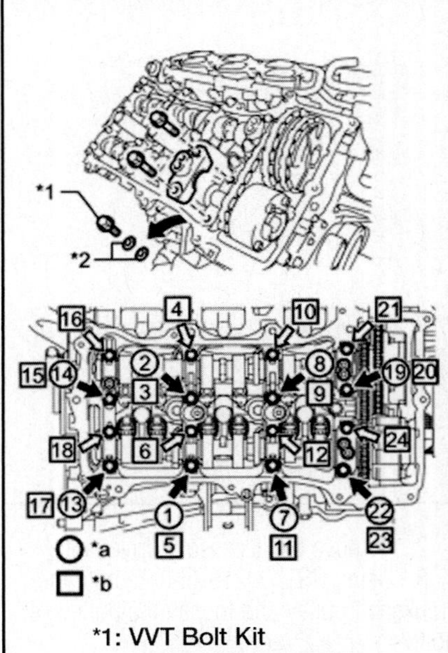

Fig. 103 Install the bank 1 bearing cap with the bolts in the order shown in the illustration

d. Install the No. 2 camshaft to the camshaft housing RH, and then install the No. 2 chain tensioner assembly with the bolt. Tighten to 15 ft. lbs. (21 Nm).

45. Install camshaft bearing cap (for bank 1):

a. Clean the camshaft bearing caps and apply engine oil to them.

b. Make sure that the No. 1 valve rocker arm sub-assembly is installed as shown in the illustration.

➡Be sure to follow the numerical order when performing this procedure. Do not drop the VVT bolt kit into the cylinder head.

c. Check the marks and numbers on the camshaft bearing caps, and then remove VVT bolt kit in the order shown in the illustration. Immediately after removing VVT bolt kit in the location for a bearing cap, install the bearing cap with the bolts in the order shown in the illustration.

- Bolt A: 21 ft. lbs. (28 Nm).
- Bolt B: 12 ft. lbs. (16 Nm).

d. Check the torque of each bolt again.

➡If the paint marks are not aligned, align them by turning the camshaft slightly.

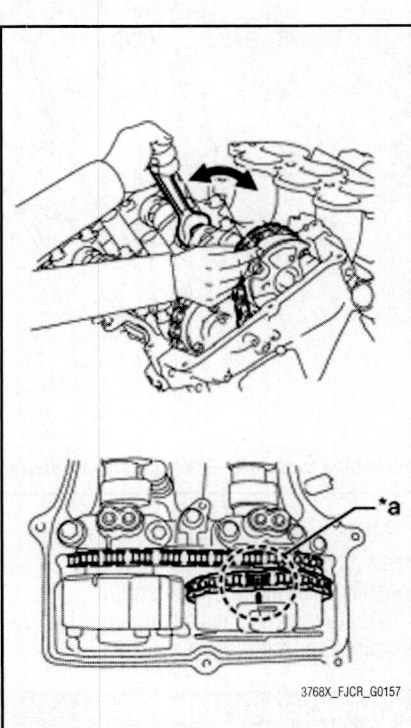

Fig. 104 Align the paint marks on the camshaft timing gear and No. 1 chain and install the No. 1 chain to the camshaft timing gear

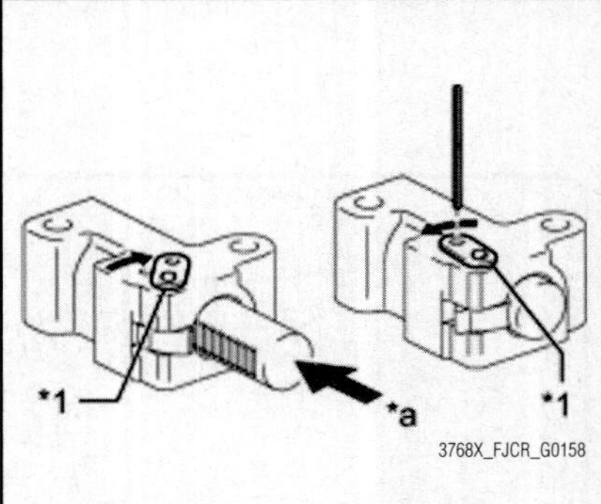

Fig. 105 While turning the stopper plate (1) of the tensioner clockwise, push in the plunger (a) of the tensioner as shown

46. Connect chain sub-assembly (for bank 1):

a. Align the paint marks on the camshaft timing gear and No. 1 chain and install the No. 1 chain to the camshaft timing gear.

47. Install no. 1 chain tensioner assembly:

a. Turn the crankshaft counterclockwise 30° past the "0" timing mark, and then turn it clockwise to align the notch with the "0" timing mark.

b. Turn the crankshaft slightly to eliminate the slack in the chain.

➡Make sure there is some slack in the chain around the area where the chain tensioner is installed.

c. While turning the stopper plate of the tensioner clockwise, push in the plunger of the tensioner as shown in the illustration.

d. While turning the stopper plate of the tensioner counterclockwise, insert a pin of _1.27 mm (0.0500 in.) into the holes on the stopper plate and tensioner to fix the stopper plate in place.

e. Install the chain tensioner with the 2 bolts. Tighten to 7 ft. lbs. (10 Nm).

f. Remove the pin from the No. 1 chain tensioner.

48. Inspect valve timing:

➡Check each timing mark from a viewpoint directly in-line with the center of the camshaft and the timing mark on each camshaft timing gear. If the timing marks are checked from any other viewpoint, the valve timing may appear misaligned.

a. Check the camshaft timing marks.

b. Check that each camshaft timing mark is positioned as shown in the illustration.

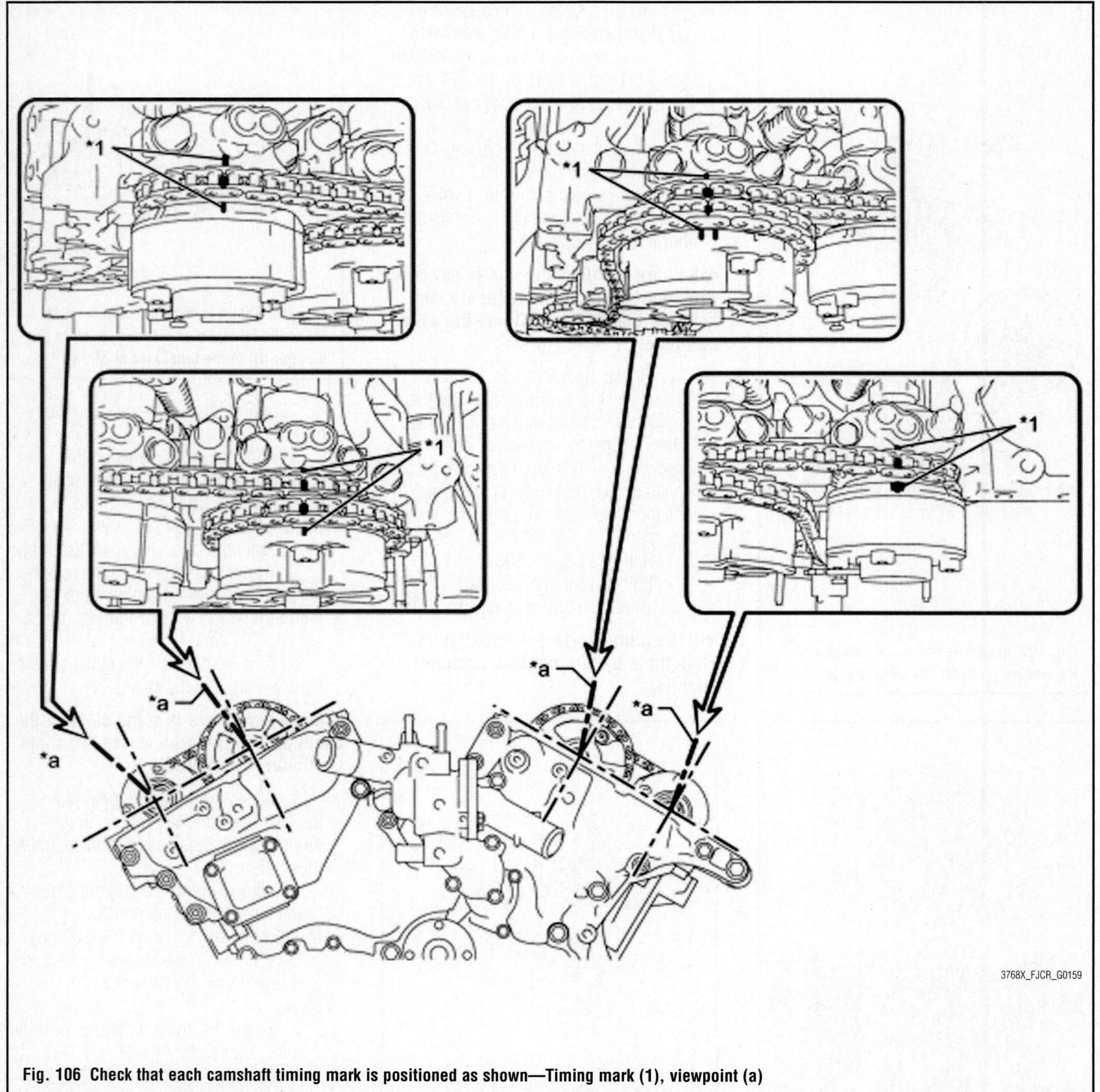

Fig. 106 Check that each camshaft timing mark is positioned as shown—Timing mark (1), viewpoint (a)

➡For the intake camshaft: Be sure to check mark A at the point when marks B, C, and D are positioned in line. If the marks are checked from any other viewpoint, they cannot be checked correctly.

c. If the valve timing is misaligned, reinstall the timing chain.

d. Turn the crankshaft 2 revolutions, set the No. 1 cylinder to TDC/compression and check the timing marks again.

49. Install a new gasket and the timing chain cover plate with the 4 bolts.

50. Before installing the cylinder head cover, pour engine oil into the locations shown in the illustration.

51. To complete installation, reverse remaining removal procedure.

CRANKSHAFT DAMPER

REMOVAL & INSTALLATION

See Figures 109 and 110.

1. Before servicing the vehicle, refer to the precautions.

2. Remove the accessory drive belt.

3. Using SST: 09213-54015, hold the crankshaft pulley and loosen the pulley set bolt.

4. Using the pulley set bolt and SST: 09950-50013, remove the crankshaft pulley.

To install:

5. Install the crankshaft pulley.

6. Using SST: 09213-54015, install the pulley set bolt. Tighten to184 ft. lbs. (250 Nm).

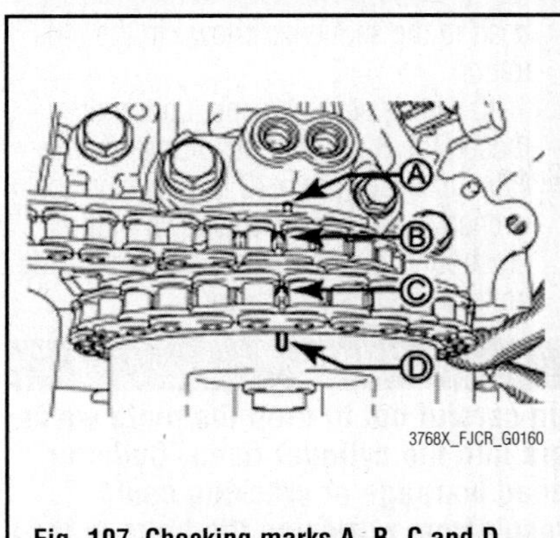

Fig. 107 Checking marks A, B, C and D

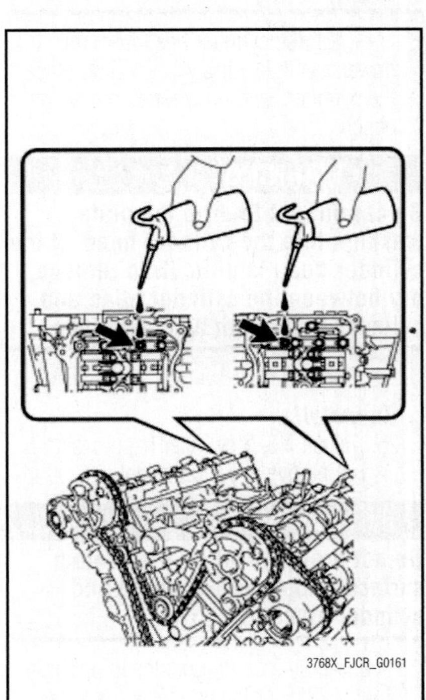

Fig. 108 Oil lubrication points

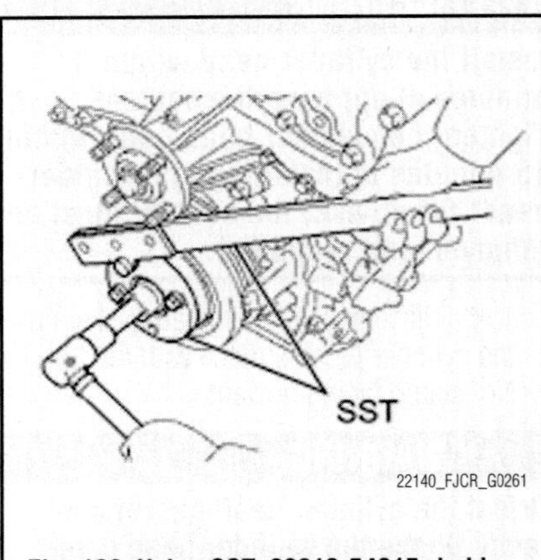

Fig. 109 Using SST: 09213-54015, hold the crankshaft pulley and loosen the pulley set bolt

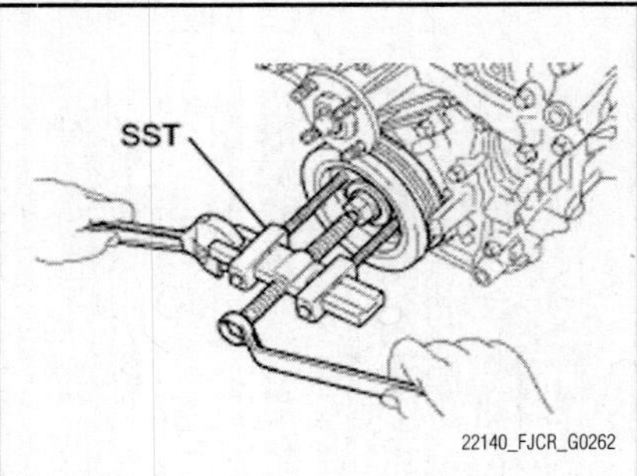

Fig. 110 Using the pulley set bolt and SST: 09950-50013, remove the crankshaft pulley

CRANKSHAFT FRONT SEAL

REMOVAL & INSTALLATION

See Figures 111 and 112.

1. Disconnect the negative battery cable.
2. Remove the fan shroud.
3. Remove oil filter bracket. Refer to Oil Filter in this section.
4. Remove the crankshaft damper.

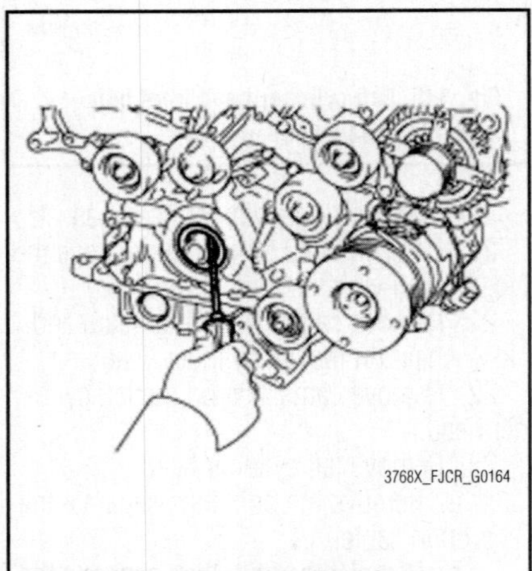

Fig. 111 Prying out the oil seal

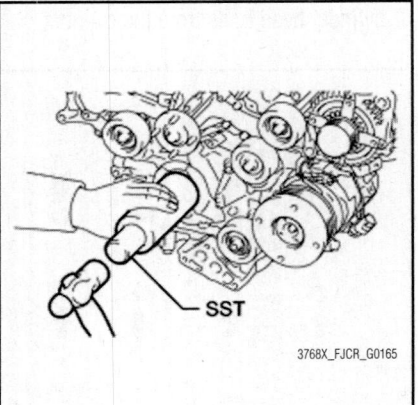

Fig. 112 Using SST: 09226-10010 and a hammer, tap in the oil seal

➡Do not damage the surface of the oil seal press fit hole and crankshaft.

➡Tape the screwdriver tip before use.

5. Using a screwdriver, pry out the oil seal.

To install:

6. Apply MP grease to the lip of a new oil seal.
7. Using SST: 09226-10010 and a hammer, tap in the oil seal until its surface is flush with the timing chain cover edge.

➡Keep the lip free from foreign matter. Do not tap the oil seal at an angle.

8. To complete installation, reverse remaining removal procedure.
9. Inspect for engine oil leak.
10. Inspect for coolant leak.

CYLINDER HEAD

REMOVAL & INSTALLATION

See Figures 113 through 121.

1. Before servicing the vehicle, refer to the precautions.
2. Discharge fuel system pressure.
3. Remove timing chain.
4. Remove the 2 bolts and the No. 1 cool air inlet.
5. On 4WD vehicles, remove exhaust pipe stopper bracket.
   a. Remove the 2 bolts, then remove the exhaust pipe stopper bracket.
6. Remove No. 2 front exhaust pipe assembly.
   a. Disconnect the oxygen sensor connector.
   b. Remove the 2 bolts and 2 nuts.
   c. Disengage the support and remove the front exhaust pipe and 2 gaskets.
7. Remove front exhaust pipe assembly.
   a. Disconnect the oxygen sensor connector.
   b. Remove the 2 bolts, 2 springs and 2 nuts, then separate the front exhaust pipe from the exhaust manifold.
8. Remove the 3 bolts and manifold bracket.
9. Remove exhaust manifolds.
   a. Disconnect the oxygen sensor connector.
   b. Remove the 6 nuts and exhaust manifold.
   c. Remove the gasket.
10. Disconnect fuel pipes.
11. Remove intake manifold.
12. Disconnect heater water inlet hose.

13. Remove water by-pass joint.
    a. Disconnect the engine coolant temperature sensor connector.
    b. Remove the 2 bolts and 4 nuts, then remove the water by-pass joint RR and 2 gaskets.
    c. Remove the O-ring from the water outlet hose.
14. Remove camshaft timing gears and No. 2 chain on the right cylinder head.
15. Remove the bolt, then remove the No. 2 chain tensioner.
16. Remove camshafts on the right cylinder head.
17. Remove No. 2 camshaft bearing.
18. Remove right cylinder head.
    a. Remove the bolt and separate the ground cable.
    b. Using several steps, loosen the 8 cylinder head bolts on the cylinder head uniformly with a 10 mm bi-hexagon wrench in the sequence shown in the illustration. Remove the 8 cylinder head bolts and 8 plate washers.

WARNING

Be careful not to drop the plate washers into the cylinder head. Cylinder head warpage or cracking could result from removing the bolts in the wrong order.

    c. Lift the cylinder head from the dowels on the cylinder block, and place the cylinder head on wooden blocks on a bench.

WARNING

Be careful not to drop the plate washers into the cylinder head. If the cylinder head is difficult to lift off, pry between the cylinder head and cylinder block with a screwdriver.

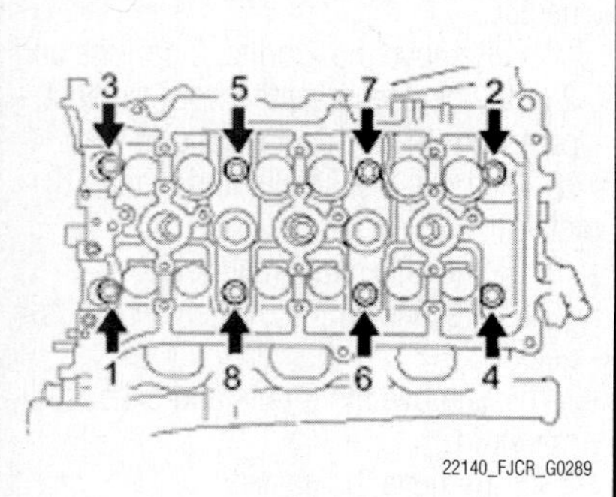

Fig. 113 Right cylinder head bolt loosening sequence—2009 vehicles

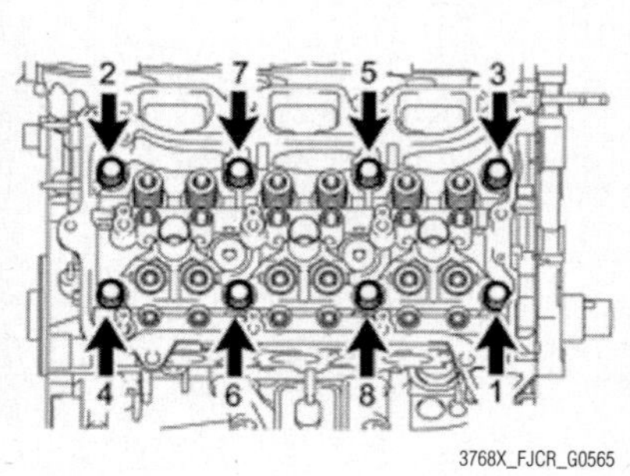

Fig. 114 Right cylinder head bolt loosening sequence—2010 vehicles

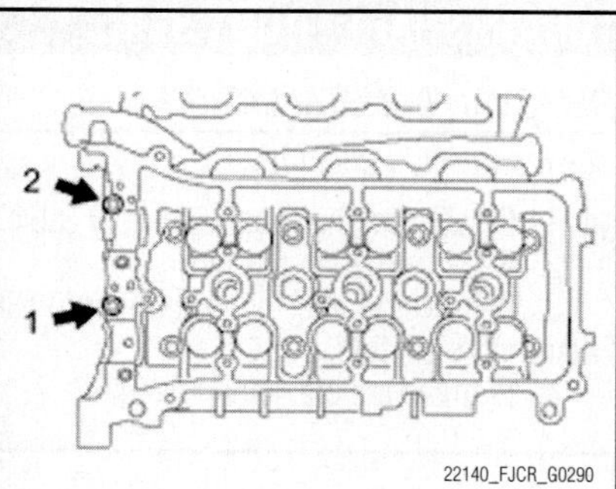

Fig. 115 Left cylinder head front bolt loosening sequence

19. Remove right cylinder head gasket.
20. Remove the 2 bolts, then remove the No. 1 chain vibration damper.
21. Remove camshaft timing gears and No. 2 chain on the left cylinder head.
22. Remove camshafts on the left cylinder head.
23. Remove left cylinder head.
    a. Remove the bolt, then separate the ground cable.
    b. Remove the bolt, then separate the oxygen connector bracket.
    c. Using several steps, remove the 2 cylinder head bolts from the cylinder head in the sequence shown in the illustration.
    d. Using several steps, uniformly loosen the 8 cylinder head bolts on the cylinder head with a 10 mm bi-hexagon wrench in the sequence shown in the illustration. Remove the 8 cylinder head bolts and 8 plate washers.

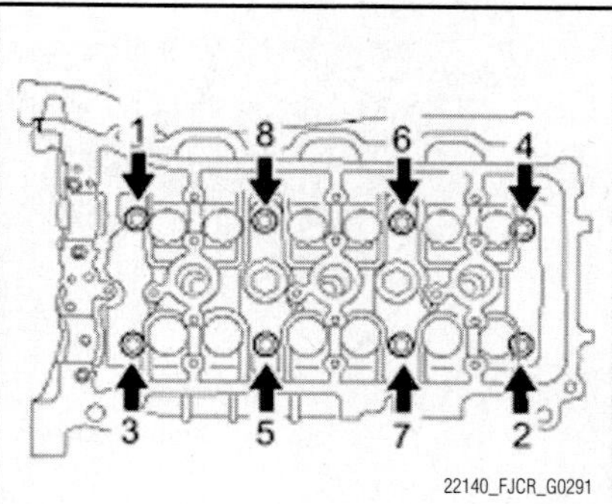

Fig. 116 Left cylinder head bolt loosening sequence

WARNING

Be careful not to drop the plate washers into the cylinder head. Cylinder head warpage or cracking could result from removing the bolts in the wrong order.

    e. Lift the cylinder head from the dowels on the cylinder block, and place the cylinder head on wooden blocks on a bench.

WARNING

Be careful not to drop the plate washers into the cylinder head. If the cylinder head is difficult to remove, pry between the cylinder head and cylinder block with a screwdriver.

24. Remove right cylinder head gasket.

To install:.

25. Install No. 2 cylinder head gasket.
    a. Remove any old sealant.

WARNING

Do not drop any oil on the contact surface of the cylinder head and cylinder block.

    b. Apply a continuous bead of Three Bond 1207B or the equivalent sealant 0.098–0.118 in. (2–3 mm) diameter to a new cylinder head gasket as shown in the illustration.

WARNING

Install the cylinder head within 3 minutes of applying the sealant. Tighten the cylinder head bolts within 15 minutes of installing the cylinder head. Otherwise, the sealant must be removed and reapplied.

    c. Place the cylinder head gasket on the cylinder block surface with the Lot No. stamp facing upward.

WARNING

Orient the cylinder head gasket correctly. Place the cylinder head carefully in order not to damage the gasket with the bottom part of the head.

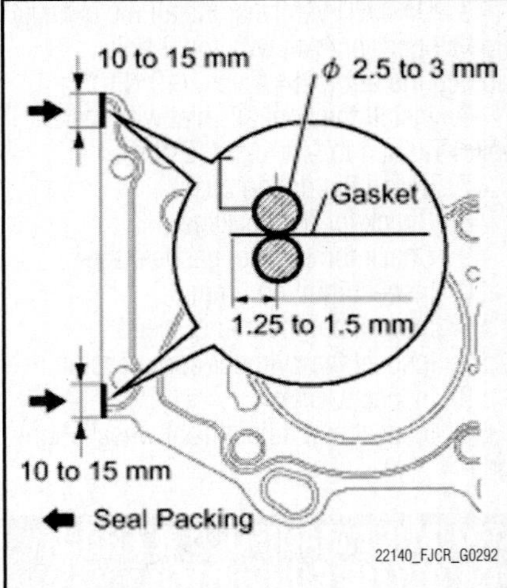

Fig. 117 Apply a continuous bead of Three Bond 1207B or the equivalent sealant 0.098–.118 in. (2–3 mm) diameter

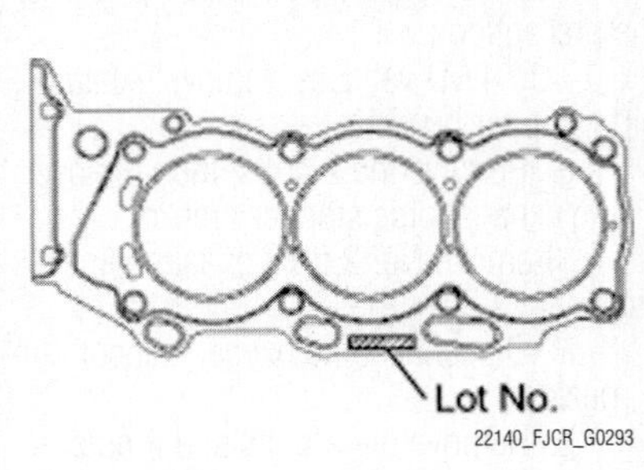

Fig. 118 Apply a continuous bead sealant in the pattern shown—2009 Vehicle shown, 2010 similar

26. Install the left cylinder head.
    a. Place the cylinder head on the cylinder head gasket.
    b. Install the 8 cylinder head bolts.

➡The cylinder head bolts are tightened in 2 successive steps. If any cylinder head bolts are broken or deformed, replace them.

c. Apply a light coat of engine oil to the threads of the cylinder head bolts.
d. Install the plate washer onto the cylinder head bolt.
e. Using several steps, uniformly tighten each bolt with a 10 mm bi-hexagon wrench in the sequence shown in the illustration. Tighten to 27 ft. lbs. (36 Nm).

⁂ WARNING

If any cylinder head bolts do not meet the torque specification, replace them.

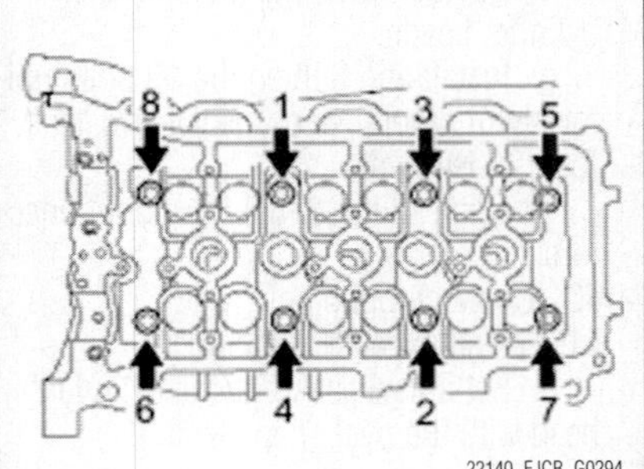

Fig. 119 Using several steps, uniformly tighten each bolt with a 10 mm bi-hexagon wrench in the sequence shown—left cylinder head

⁂ WARNING

Do not drop the washers into the cylinder head.

f. Mark the front side of each cylinder head bolt with paint.
g. Retighten the cylinder head bolts an additional 180°.
h. Check that the painted marks are now at 180°from the engine front.
i. Install the front 2 cylinder head bolts.
j. Apply a light coat of engine oil to the threads of the cylinder head bolts.
k. Using several steps, uniformly install and tighten the 2 cylinder head bolts in the sequence shown in the illustration. Tighten to 22 ft. lbs. (30 Nm).
l. Install the ground cable with the bolt. Tighten to 71 ft. lbs. (96 Nm).
m. Install the oxygen sensor connector bracket with bolt. Tighten to 14 ft. lbs. (19 Nm).

27. Install camshafts on the left cylinder head.
28. Install No. 3 chain tensioner assembly.

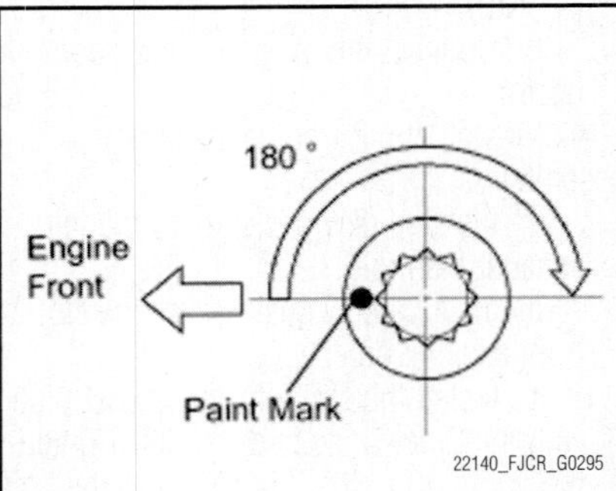

Fig. 120 Check that the painted marks are now at 180°from the engine front

29. Install camshaft timing gears and No. 2 chain on the left cylinder head.
30. Install No. 1 chain vibration damper.
    a. Install the chain vibration damper with the 2 bolts. Tighten to 14 ft. lbs. (19 Nm).
31. Install the right cylinder head gasket.
    a. Remove any old sealant.

⁂ WARNING

Do not drop any oil on the contact surfaces of the cylinder head and cylinder block.

b. Apply a continuous bead of Three Bond 1207B or the equivalent sealant 0.098–.118 in. (2–3 mm) diameter to a new cylinder head gasket as shown in the illustration.

⁂ WARNING

Install the cylinder head within 3 minutes of applying the sealant. Tighten the cylinder head bolts within 15 minutes of installing the cylinder head. Otherwise, the sealant must be removed and reapplied.

c. Place the cylinder head gasket on the cylinder block surface with the Lot No. stamp facing upward.

⁂ WARNING

Orient the cylinder head gasket correctly. Place the cylinder head carefully in order not to damage the gasket.

32. Install the right cylinder head.
    a. Place the cylinder head on the cylinder head gasket.
    b. Install the 8 cylinder head bolts.

➡The cylinder head bolts are tightened in 2 successive steps. If any cylinder head bolts are broken or deformed, replace them.

c. Apply a light coat of engine oil to the threads of the cylinder head bolts.
d. Install the plate washer onto the cylinder head bolt.
e. Using several steps, uniformly tighten each bolt with a 10 mm bi-hexagon wrench in the sequence shown in the illustration. Tighten to 27 ft. lbs. (36 Nm).

⁂ WARNING

If any cylinder head bolts do not meet the torque specification, replace them.

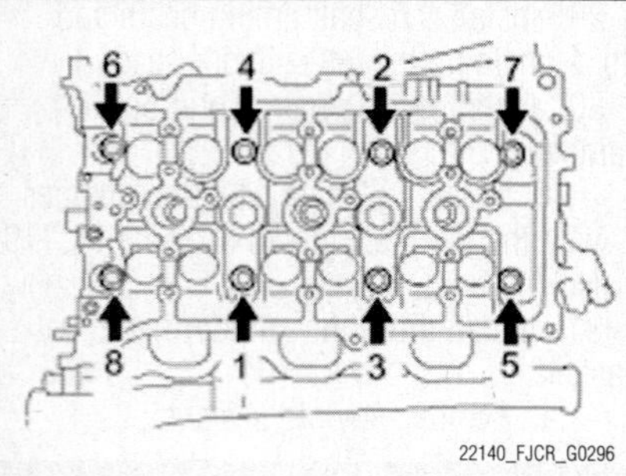

Fig. 121 Using several steps, uniformly tighten each bolt with a 10 mm bi-hexagon wrench in the sequence shown—right cylinder head

WARNING

Do not drop the washers into the cylinder head.

f. Mark the front side of each cylinder head bolt with paint.

g. Retighten the cylinder head bolts an additional 180°.

h. Check that the painted marks are now at 180°from the engine front.

i. Install the ground cable with the bolt. Tighten to 71 ft. lbs. (96 Nm).

33. Install No. 2 camshaft bearing.

a. Install the No. 2 camshaft bearing onto the cylinder head.

WARNING

Clean the installation planes of the back side of the bearing and cylinder head and keep them free of oil.

34. Install camshafts on the right cylinder head.

35. Install No. 2 chain tensioner assembly.

36. Install camshaft timing gears and No. 2 chain on the right cylinder head.

37. Install water by-pass joint.

a. Install a new O-ring onto the water outlet pipe.

b. Apply soapy water to the O-ring.

c. Install 2 new gaskets and water by-pass joint rear with the 2 bolts and 4 nuts. Tighten to 80 inch lbs. (9 Nm).

d. Connect the engine coolant temperature sensor connector.

38. Connect heater water inlet hose.

39. Install intake manifold.

a. Set a new gasket on each cylinder head.

➡Align the port holes of the gasket and cylinder head. Orient the gasket correctly.

b. Set the intake manifold on the cylinder heads.

c. Install and tighten the 10 bolts uniformly in several steps. Tighten to 19 ft. lbs. (26 Nm).

d. Connect the 6 fuel injector connectors.

40. Connect fuel pipes.

41. Install exhaust manifolds.

a. Set a new gasket to the cylinder head with the oval shape facing forward.

➡Orient the new gasket correctly.

b. Install the exhaust manifold with the 6 nuts. Tighten to 22 ft. lbs. (30 Nm).

c. Connect the oxygen sensor connector.

42. Install the manifold bracket with the 3 bolts. Tighten to 30 ft. lbs. (40 Nm).

43. Install front exhaust pipe assembly.

a. Check the free length.

b. Using Vernier calipers, measure the free length of the compression spring. Minimum length should be 0.594 in. (5mm). If the free length is less than the minimum, replace the compression spring.

c. Install a new gasket onto the right exhaust manifold.

d. Install a new gasket onto the exhaust front pipe.

e. Using a wooden block and plastic-faced hammer, tap in the new gasket until it is flush with the exhaust front pipe.

➡Make sure that the gasket is in the correct direction when installing. Do not reuse the removed gasket.

f. Install the exhaust front pipe with new 2 nuts. Tighten to 40 ft. lbs. (54 Nm).

g. Install the exhaust center pipe onto the front exhaust pipe with the 2 bolts and 2 springs. Tighten to 32 ft. lbs. (43 Nm).

h. Connect the oxygen sensor connector.

44. Install No. 2 front exhaust pipe assembly.

a. Install a new gasket onto the left exhaust manifold.

b. Install a new gasket onto the No. 2 front exhaust pipe.

c. Install the No. 2 front exhaust pipe with the 2 new nuts and 2 bolts. Tighten the nut to 40 ft. lbs. (54 Nm) and the bolt to 35 ft. lbs. (48 Nm).

d. Connect the oxygen sensor connector.

45. On 4WD vehicles, install the exhaust pipe stopper bracket with the 2 bolts. Tighten the nut to 14 ft. lbs. (19 Nm).

46. Install the cool air inlet with the 2 bolts. Tighten to 9 ft. lbs. (12 Nm).

47. Install the timing chain.

48. Check for fuel leakage.

49. Check for exhaust gas leakage.

50. Inspect ignition timing.

51. Inspect engine idling speed.

52. Inspect the cylinder compression.

53. Inspect CO/HC.

54. Inspect and adjust front wheel alignment

EXHAUST MANIFOLD & FRONT CATALYTIC CONVERTER

REMOVAL & INSTALLATION

2009 Vehicles

See Figures 122 and 123.

1. Before servicing the vehicle, refer to the precautions.

2. On 4WD vehicles, remove exhaust pipe stopper bracket.

a. Remove the 2 bolts, then remove the exhaust pipe stopper bracket.

3. Remove No. 2 front exhaust pipe assembly.

a. Disconnect the oxygen sensor connector.

b. Remove the 2 bolts and 2 nuts.

c. Disengage the support and remove the front exhaust pipe and 2 gaskets.

4. Remove front exhaust pipe assembly.

a. Disconnect the oxygen sensor connector.

b. Remove the 2 bolts, 2 springs and 2 nuts, then separate the front exhaust pipe from the exhaust manifold.

5. Remove the 3 bolts and manifold bracket.

6. Remove exhaust manifolds.

a. Disconnect the oxygen sensor connector.

b. Remove the 6 nuts and exhaust manifold.

c. Remove the gasket.

To install:

7. Install left exhaust manifold.

a. Set a new gasket to the left cylinder head with the oval shape facing the correct direction.

➡On the left cylinder head, orient the oval shape to the rear. On the right cylinder head, orient the oval shape to the front.

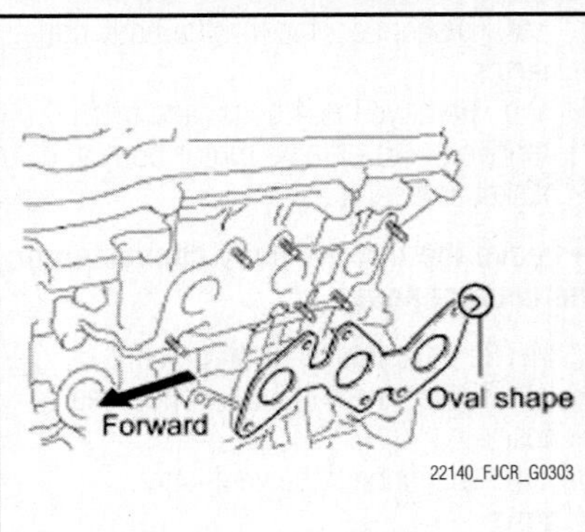

Fig. 122 Using Vernier calipers, measure the free length of the compression spring

b. Install the exhaust manifold with the 6 nuts. Tighten to 22 ft. lbs. (30 Nm).

c. Connect the oxygen sensor connector.

8. Install the No. 2 manifold bracket with the 3 bolts. Tighten to 30 ft. lbs. (40 Nm).

9. Install right exhaust manifold.

a. Set a new gasket to the right cylinder head with the oval shape facing forward.

➡**Orient the new gasket correctly.**

b. Install the exhaust manifold with the 6 nuts. Tighten to 22 ft. lbs. (30 Nm).

c. Connect the oxygen sensor connector.

10. Install the manifold bracket with the 3 bolts. Tighten to 30 ft. lbs. (40 Nm).

11. Install front exhaust pipe assembly.

a. Check the free length.

b. Using Vernier calipers, measure the free length of the compression spring. Minimum length should be 0.594 in. (5mm). If the free length is less than the minimum, replace the compression spring.

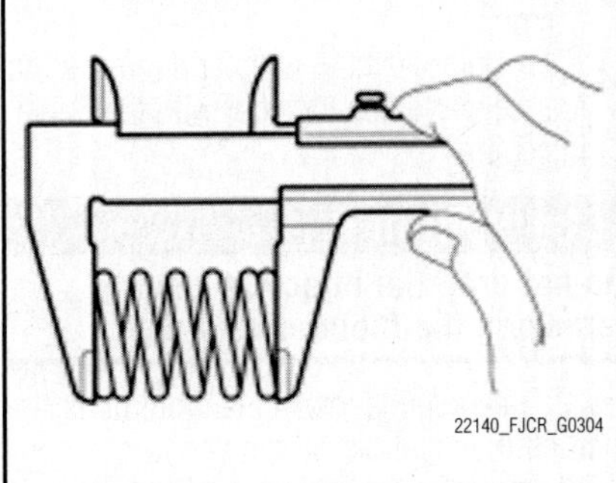

Fig. 123 Using Vernier calipers, measure the free length of the compression spring

c. Install a new gasket onto the exhaust manifold.

d. Install a new gasket onto the exhaust front pipe.

e. Using a wooden block and plastic-faced hammer, tap in the new gasket until it is flush with the exhaust front pipe.

➡**Make sure that the gasket is in the correct direction when installing. Do not reuse the removed gasket.**

f. Install the exhaust front pipe with new 2 nuts. Tighten to 40 ft. lbs. (54 Nm).

g. Install the exhaust center pipe onto the front exhaust pipe with the 2 bolts and 2 springs. Tighten to 32 ft. lbs. (43 Nm).

h. Connect the oxygen sensor connector.

12. Install No. 2 front exhaust pipe assembly.

a. Install a new gasket onto the left exhaust manifold.

b. Install a new gasket onto the No. 2 front exhaust pipe.

c. Install the No. 2 front exhaust pipe with the 2 new nuts and 2 bolts. Tighten the nut to 40 ft. lbs. (54 Nm) and the bolt to 35 ft. lbs. (48 Nm).

d. Connect the oxygen sensor connector.

13. On 4WD vehicles, install the exhaust pipe stopper bracket with the 2 bolts. Tighten the nut to 14 ft. lbs. (19 Nm).

14. Check for exhaust gas leakage.

15. Inspect CO/HC.

2010 Vehicles

See Figures 124 through 127.

1. Remove the v-bank cover.
2. Remove the air cleaner assembly.
3. Remove the 5 clips and LH fender apron seal.
4. Remove the 4 clips and RH fender apron seal.
5. Remove the 5 clips and LH frame seal.
6. Remove the 5 clips and RH fender apron seal.
7. Remove the 2 bolts and the exhaust pipe stopper bracket.
8. Remove the exhaust pipe assembly.
9. Remove the 3 bolts and manifold stay.
10. Remove the 3 bolts and heat insulator.
11. Remove RH exhaust manifold:

a. Disconnect the air fuel ratio sensor connector.

b. Remove the 6 nuts, manifold and gasket.

12. Remove air fuel ratio sensor (for Bank 1 Sensor 1).
13. Remove the 3 bolts and No. 2 manifold stay.
14. Remove the 3 bolts and heat insulator.
15. Remove LH exhaust manifold:

a. Disconnect the air fuel ratio sensor connector.

b. Remove the 6 nuts, manifold and gasket.

16. Remove air fuel ratio sensor (for Bank 2 Sensor 1).

To install:

17. Install air fuel ratio sensor (for Bank 2 Sensor 1).
18. Install the LH exhaust manifold:

➡**Be careful of the installation direction.**

a. Install a new gasket onto the cylinder head.

b. Temporarily install the manifold with the 6 nuts.

c. Tighten the 6 nuts in the sequence shown in the illustration. Tighten to 15 ft. lbs. (21 Nm).

d. Connect the air fuel ratio sensor connector.

Fig. 124 Install a new LH gasket onto the cylinder head

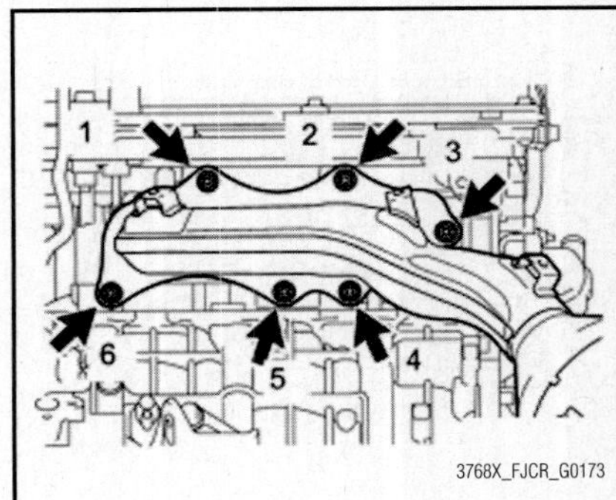

Fig. 125 LH exhaust manifold tightening sequence

19. Install no. 2 exhaust manifold heat insulator. Tighten the three bolts to 10 ft. lbs. (13 Nm).

20. Install the manifold stay with the 3 bolts. Tighten to 30 ft. lbs. (13 Nm).

21. Install air fuel ratio sensor (for Bank 1 Sensor 1).

22. Install the RH exhaust manifold:

➡Be careful of the installation direction.

a. Install a new gasket onto the cylinder head.

b. Temporarily install the manifold with the 6 nuts.

c. Tighten the 6 nuts in the sequence shown in the illustration. Tighten to 15 ft. lbs. (21 Nm).

d. Connect the air fuel ratio sensor connector.

23. Install no. 1 exhaust manifold heat insulator. Tighten the three bolts to 10 ft. lbs. (13 Nm).

24. Install the no. 2 manifold stay with the 3 bolts. Tighten to 30 ft. lbs. (13 Nm).

25. To complete the installation, reverse remaining removal procedure.

26. Install the exhaust pipe stopper bracket with the 2 bolts. Tighten to 14 ft. lbs. (19 Nm).

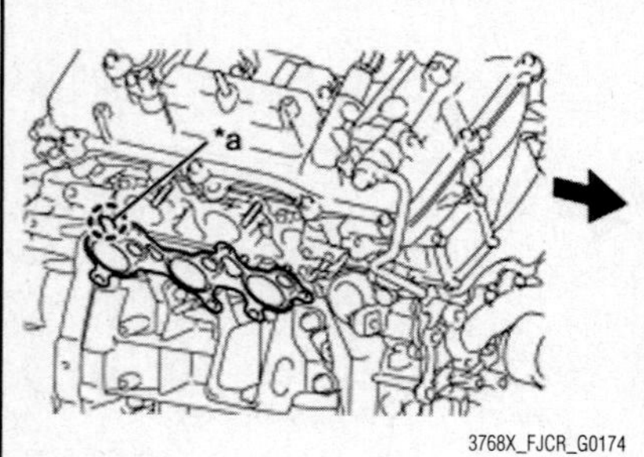

Fig. 126 Install a new RH gasket onto the cylinder head

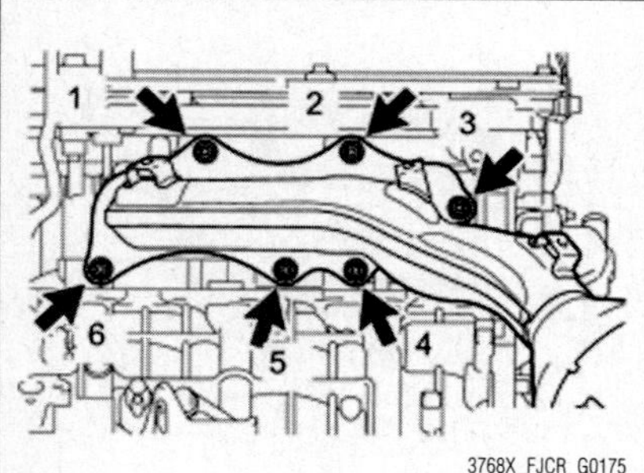

Fig. 127 RH exhaust manifold tightening sequence

FLYWHEEL

REMOVAL & INSTALLATION

See Figure 128.

1. Before servicing the vehicle, refer to the precautions.
2. Remove the transmission assembly.
3. Using SST: 09213-54015, hold the crankshaft.
4. Remove the 8 bolts, then remove the flywheel and 2 spacers.

To install:

5. Apply Three Bond 1324 or the equivalent adhesive to the 2 or 3 end threads of the bolts.
6. Install the flywheel and 2 spacers onto the crankshaft.
7. Install the flywheel with the 6 bolts and tighten in the order shown in the illustration to 61 ft. lbs. (83 Nm.
8. Install the transmission assembly.

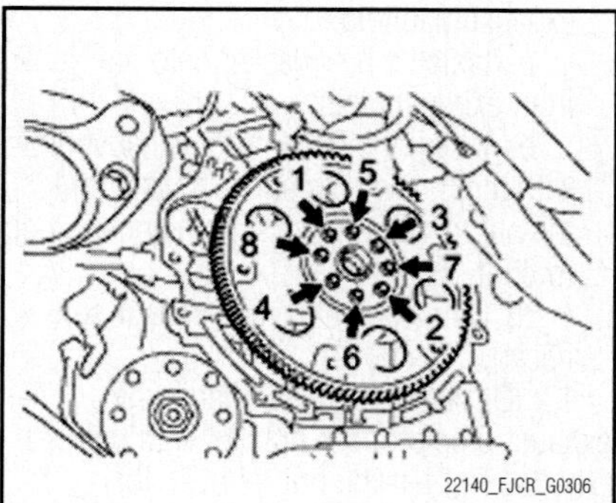

Fig. 128 Install the flywheel with the 8 bolts and tighten in the order shown

INTAKE MANIFOLD

REMOVAL & INSTALLATION

2009 Vehicles

See Figures 129 and 130.

1. Before servicing the vehicle, refer to the precautions.
2. Remove air cleaner assembly.

a. Remove the engine V-bank cover by removing the two acorn nuts.

b. Disconnect the ventilation hose No. 2.

c. Disconnect the vacuum hose.

d. Disconnect the mass air flow meter connector. Remove the 2 wire harness clamps.

e. Loosen the 2 hose clamps. Remove the 2 bolts, then remove the air cleaner.

3. Remove throttle body with motor body assembly.

a. Disconnect the throttle body connector.

b. Remove the 4 bolts, and then remove the throttle w/ motor body and gasket and set it aside

➡Leave the throttle body connected to the coolant hoses.

4. Remove intake air surge tank.

a. Disconnect the fuel vapor feed hose.

b. Disconnect the ventilation hose.

c. Disconnect the 2 VSV connectors.

d. Remove the 3 wire harness clamps and hose clamps.

➡Removing the No. 1 VSV at the rear passenger side of the surge tank makes it easier to get to the clamps.

e. Remove the 3 upper bolts which are used to secure the 2 surge tank stays and throttle body bracket.

f. Loosen the 2 bolts that hold the brackets to the side of the cylinder head and remove the brackets.

✱✱ WARNING

Before proceeding, make sure there is no dirt or debris on or around the base of the surge tank. If there is, you must remove it so that it will not enter the engine when the surge tank is removed.

g. Using an 8mm Allen socket, remove the 4 bolts and the 2 nuts with a 12mm socket. Remove the surge tank and gasket.

5. Disconnect fuel pipes.

a. Remove the fuel pipe clamp.

b. Pinch the tube connector, and then pull the fuel pipe out of the delivery pipe as shown in the illustration.

6. Remove fuel delivery pipe assembly.

a. Disconnect the 6 fuel injector connectors.

b. Remove the 6 bolts and remove the fuel delivery pipe together with the 6 fuel injectors.

✱✱ WARNING

Do not drop the injectors when removing the fuel delivery pipe.

7. Remove the intake manifold bolts, then lift the manifold off the engine.
8. Remove the intake manifold gaskets.

To install:

9. Install new manifold gaskets.

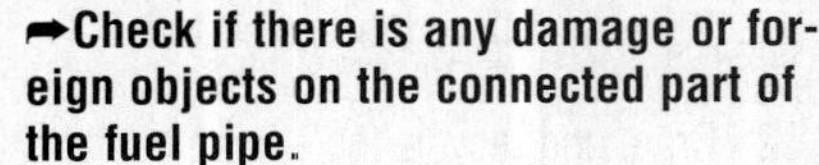

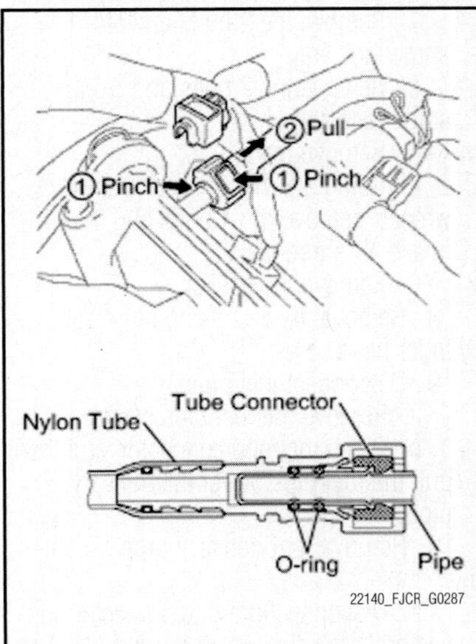

Fig. 129 Pinch the tube connector, and then pull the fuel pipe out of the delivery pipe as shown

10. Install the intake manifold bolts and tighten to 19 ft. lbs. (26 Nm).
11. Install fuel delivery pipe assembly.
   a. Place the fuel delivery pipe together with the 6 fuel injectors on the intake manifold.
   b. Temporarily install the 6 bolts, which are used to hold the fuel delivery pipe, onto the intake manifold.
   c. Check that the fuel injectors rotate smoothly. If the fuel injectors do not rotate smoothly, replace the O-ring.
   d. Position each fuel injector connector facing outward.
   e. Tighten the 6 fuel delivery pipe attaching bolts to 11 ft. lbs (15 Nm).
   f. Connect the 6 fuel injector connectors.
12. Connect fuel pipes.
   a. Push the tube connector into the pipe until the tube connector makes a "click" sound.

➡Check if there is any damage or foreign objects on the connected part of the fuel pipe.

   b. After connecting, check that the pipe and connector are securely connected by pulling them.
   c. Install the fuel pipe clamp.
13. Install intake air surge tank.
   a. Install the surge tank using a new gasket. Using an 8mm Allen socket, install the 4 bolts and the 2 nuts with a 12mm socket.
   b. Install the brackets and tighten the 2 bolts that hold the brackets to the side of the cylinder head.
   c. Install the 3 upper bolts which are used to secure the 2 surge tank stays and throttle body bracket.
   d. Install No. 1 VSV at the rear passenger side of the surge tank if removed.
   e. Install the 3 wire harness clamps and hose clamps.
   f. Connect the 2 VSV connectors.
   g. Connect the ventilation hose.
   h. Connect the fuel vapor feed hose.
14. Install throttle body with motor body assembly.
   a. Install the throttle body with motor using a new gasket Install the 4 bolts, and then and set it aside
   b. Connect the throttle body connector.
15. Install air cleaner assembly.
   a. Install the air cleaner and 2 bolts. Tighten the 2 hose clamps.
   b. Connect the mass air flow meter connector. Install the 2 wire harness clamps.
   c. Connect the vacuum hose.
   d. Connect the ventilation hose No. 2.
16. Install the engine V-bank cover.

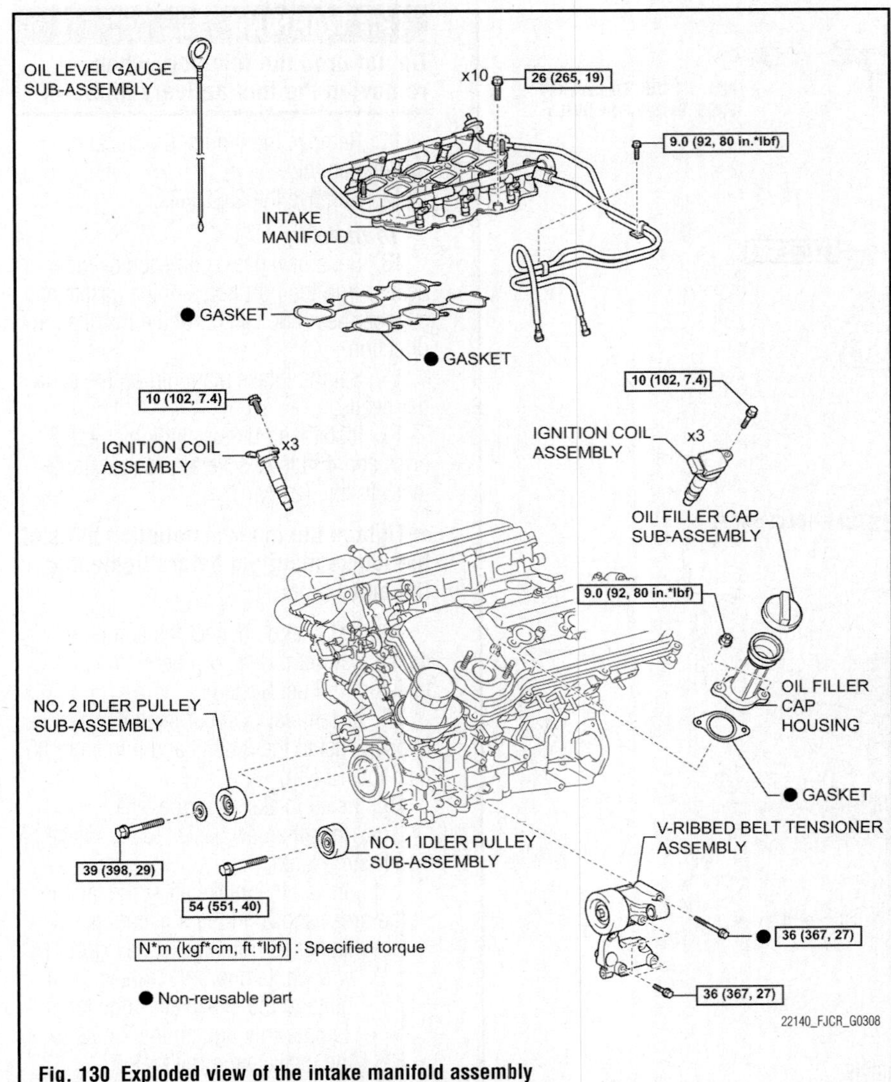

Fig. 130 Exploded view of the intake manifold assembly

2010 Vehicles

See Figures 131 and 132.

⁂ CAUTION

Observe all applicable safety precautions when working around fuel. Whenever servicing the fuel system, always work in a well ventilated area. Do not allow fuel spray or vapors to come in contact with a spark or open flame. Keep a dry chemical fire extinguisher near the work area. Always keep fuel in a container specifically designed for fuel storage; also, always properly seal fuel containers to avoid the possibility of fire or explosion.

1. Properly relieve the fuel system pressure.
2. Drain and recycle the engine coolant.
3. Raise the front of the V-bank cover to disengage the 2 pins.
4. Remove the 2 V-bank cover hooks from the bracket.
5. Remove the V-bank cover.
6. Remove the air cleaner assembly.
7. Remove intake air surge tank:
   a. Disconnect the No. 4 water bypass hose.
   b. Disconnect the No. 5 water bypass hose.
   c. Disconnect the throttle body connector.
   d. Disconnect the No. 1 fuel vapor feed hose.
   e. Disconnect the No. 1 vacuum switching valve connector.
   f. Disconnect the No. 1 ventilation hose.
   g. Disconnect the 2 heater water hose clamps.
   h. Remove the 2 bolts and throttle body bracket.
   i. Using a clip remover, disengage the wire harness clamp.
   j. Remove the 2 bolts and No. 1 surge tank stay.
   k. Remove the 2 bolts and No. 2 surge tank stay.
   l. Remove the 2 nuts.
   m. Using a 8 mm socket hexagon wrench, remove the 4 bolts and the intake air surge tank.
   n. Remove the gasket.
8. Remove the 3 bolts and the rear cylinder head cover.
9. Disconnect fuel pipes.
   a. Remove the fuel pipe clamp.
   b. Pinch the tube connector, and then pull the fuel pipe out of the delivery pipe.
10. Remove fuel delivery pipe assembly.
   a. Disconnect the 6 fuel injector connectors.
   b. Remove the 6 bolts and remove the fuel delivery pipe together with the 6 fuel injectors.

WARNING

Do not drop the injectors when removing the fuel delivery pipe.

11. Remove the 4 nuts, 6 bolts and intake manifold.
12. Remove the 2 gaskets.

To install:

13. Set a new gasket on each cylinder head. Align the port holes of the gasket and cylinder head. Be careful of the installation direction.
14. Set the intake manifold on the cylinder heads.
15. Install and uniformly tighten the 6 bolts and 4 nuts in several passes. Tighten to 15 ft. lbs. (21 Nm).

➡Tighten the inner installation bolts of the intake manifold before tightening the outer bolts.

16. Install and connect the fuel pipe delivery system. Refer to Fuel Rail and Injectors in Fuel System.
17. Install rear cylinder head cover. Loosely install the 3 bolts and tighten to 80 inch lbs. (9 Nm).
18. Install intake air surge tank:
   a. Install a new gasket to the intake air surge tank.
   b. Install the intake air surge tank in the order shown in the illustration.
      - Using a 8 mm socket hexagon wrench, temporarily tighten the 4 bolts to the intake air surge tank.
      - Temporarily tighten the 2 nuts to the intake air surge tank.

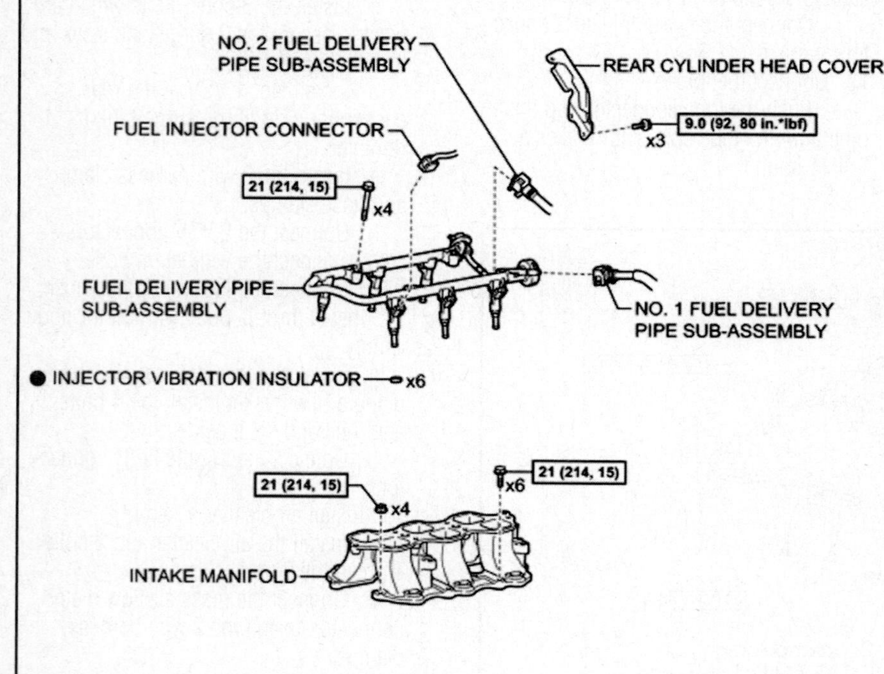

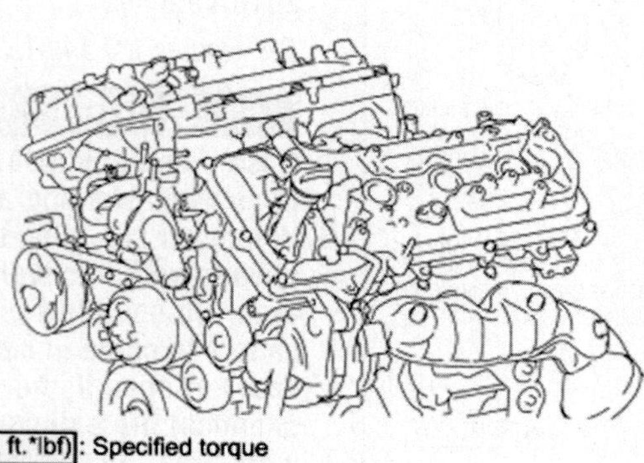

Fig. 131 Exploded view of the intake manifold assembly

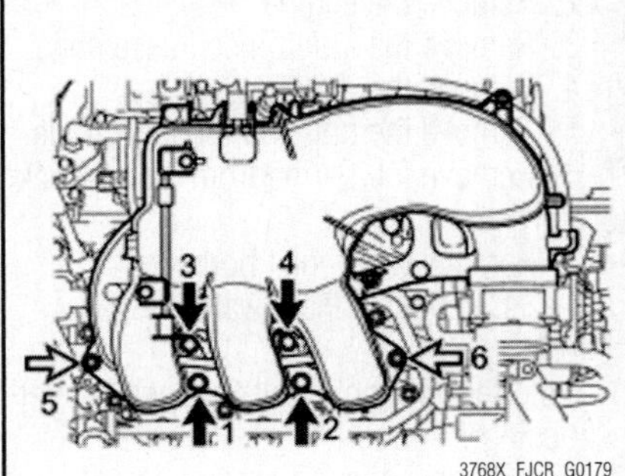

Fig. 132 Install the intake air surge tank in the order shown

- Tighten the 4 bolts and 2 nuts to 21 ft. lbs. (28 Nm).

19. To complete installation, reverse remaining removal procedure.
20. Refer to the exploded view illustration for torque values.
21. Add engine coolant. Inspect for coolant leak .
22. Inspect for fuel leak.
23. Engage the 2 V-bank cover hooks to the bracket.
24. Align the 2 V-bank cover grommets with the 2 pins and press down on the V-bank cover to engage the pins.

OIL PAN

REMOVAL & INSTALLATION

See Figures 133 through 136.

1. Before servicing the vehicle, refer to the precautions.
2. Drain engine oil.
3. Remove No. 2 oil pan.
   a. Remove the 10 bolts and 2 nuts.
   b. Insert the blade of SST: 09032-00100 between the oil pan and No. 2 oil pan, cut off applied sealer and remove the No. 2 oil pan.

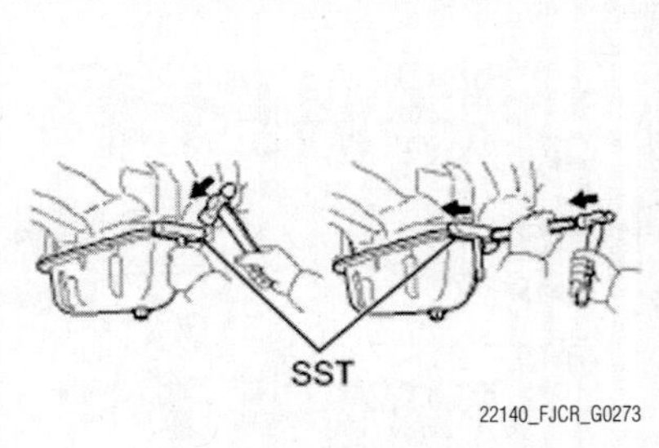

Fig. 133 Insert the blade of SST: 09032-00100 between the oil pan and No. 2 oil pan, cut off applied sealer

**** WARNING**

Be careful not to damage the contact surfaces of the oil pan and No. 2 oil pan. Be careful not to damage the No. 2 oil pan flange.

4. Remove the 2 nuts, then remove the oil strainer and gasket.
5. Remove oil pan.

To install:

6. Install oil pan.
   a. Remove any old sealant.

**** WARNING**

Do not drop any oil on the contact surfaces of the cylinder block, rear oil seal and oil pan.

   b. Install a new O-ring onto the oil pump.
   c. Apply a continuous bead of Three Bond or the equivalent sealant 0.12–0.16 in. (3–4 mm) diameter to the oil pan as shown in the illustration.

**** WARNING**

Install the oil pan within 3 minutes of applying the sealant. Tighten the oil pan bolts and nuts within 15 minutes of installing the oil pan. Otherwise, the sealant must be removed and reapplied.

   d. Install the oil pan with the 17 bolts and 2 nuts, and tighten the bolts and nuts uniformly in several steps. Tighten the 0.39 in. (10 mm) head oil pan bolt to 7 ft. lbs. (10 Nm), the 0.47 in. (12 mm) head oil pan bolt to 16 ft. lbs. (21 Nm) and the nuts to 16 ft. lbs. (21 Nm).

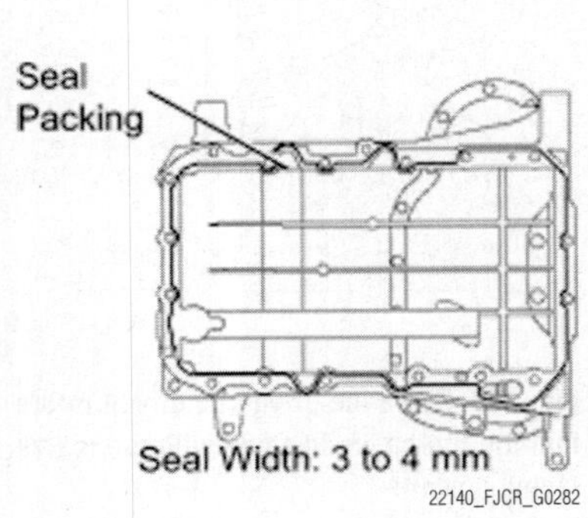

Fig. 134 Apply a continuous bead of Three Bond or the equivalent sealant 0.12–0.16 in. (3–4 mm) diameter to the oil pan as shown

**** WARNING**

Install bolts in their proper locations. Bolt lengths are as follows: A is 0.98 in. (25 mm) B 1.77 in. (45 mm) and C 0.55 in. (14 mm).

To install:

   e. Remove any old sealant.

**** WARNING**

Do not drop any oil on the contact surfaces of the oil pan and No. 2 oil pan.

   f. Apply a continuous bead of Three Bond or the equivalent sealant 0.12–0.16 in. (3–4 mm) diameter to the oil pan as shown in the illustration.

**** WARNING**

Install the oil pan within 3 minutes of applying the sealant. Tighten the oil pan bolts and nuts within 15 minutes of installing the oil pan. Otherwise, the sealant must be removed and reapplied.

   g. Install the oil pan with the 17 bolts and 2 nuts, and tighten the bolts and

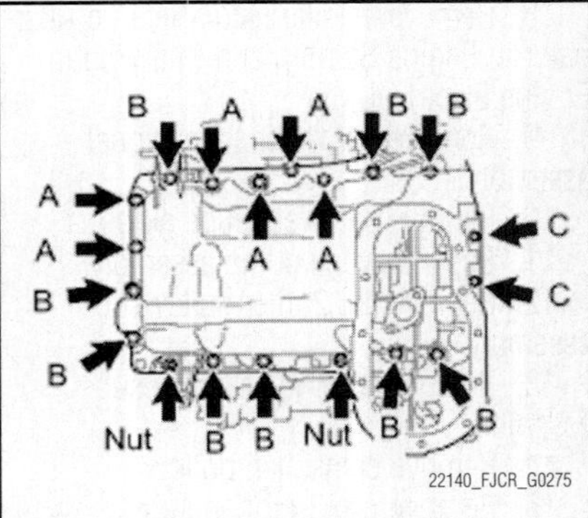

Fig. 135 Oil pan fastener locations

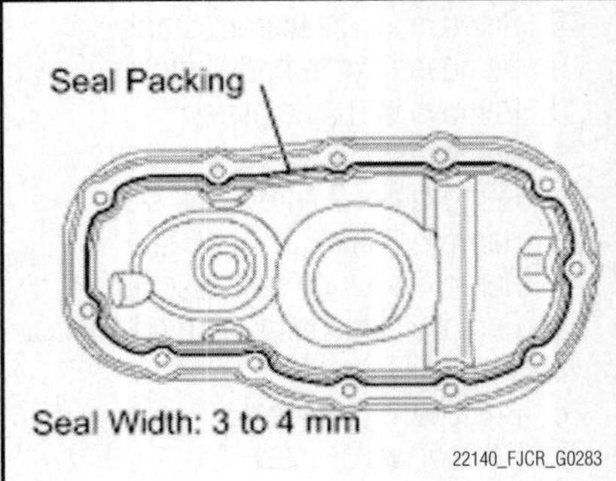

Fig. 136 Apply a continuous bead of Three Bond or the equivalent sealant 0.12–0.16 in. (3–4 mm) diameter to the oil pan

nuts uniformly in several steps. Tighten the bolts to 80 inch lbs. (9 Nm) and the nuts to 7 ft. lbs. (10 Nm).

7. Add engine oil.

8. Add power steering fluid and bleed the system, as necessary.

9. On 4WD vehicles, inspect and adjust the differential oil.

10. Check for engine oil leakage.

OIL PUMP

REMOVAL & INSTALLATION

2009 Vehicles

See Figure 137.

1. Before servicing the vehicle, refer to the precautions.
2. Remove the battery.
3. Dain the engine oil.
4. Remove power steering link assembly. Refer to Front Suspension.
5. Remove front differential carrier assembly (for 4WD). Refer to Drive Train.
6. Remove the cooling fan. Refer to Cooling Fan in Engine Cooling.
7. Remove the A/C compressor. Refer to Compressor in Heating, ventilation & air conditioning.
8. Remove the alternator. Refer to Alternator in Engine Starting and Charging in Engine Electrical.
9. Remove the v-ribbed tensioner assembly.
10. Remove oil level gauge guide.
11. Separate vane pump assembly.
12. Remove no. 2 idler pulley sub-assembly.
13. Remove no. 1 idler pulley sub-assembly.
14. Remove crankshaft pulley.
15. Remove no. 2 oil pan sub-assembly.
16. Remove oil strainer sub-assembly.
17. Remove oil pan sub-assembly
18. Remove air cleaner assembly.
19. Remove throttle body bracket.
20. Remove oil baffle plate.
21. Remove no. 1 surge tank stay.
22. Remove no. 2 surge tank stay.
23. Remove intake air surge tank.
24. Remove ignition coil assembly.
25. Remove camshaft timing oil control valve assembly.
26. Remove VVT sensor.
27. Remove water inlet.
28. Remove cylinder head cover sub-assembly.
29. Remove cylinder head cover sub-assembly LH.
30. Remove the 3 bolts, 2 nuts, oil filter bracket and gasket.
31. Remove timing chain sub-assembly. Refer to Timing Cover and Seal in this section.
32. Remove the rear main seal. Refer to Rear Main Seal in this section.
33. Remove oil pump cover.
    a. Remove the 3 bolts, then remove the oil pipe.
    b. Remove the 2 O-rings.
    c. Remove the 7 bolts, then remove the oil pump cover, drive rotor and driven rotor.
34. Remove oil pump relief valve.
    a. Remove the relief valve plug.
    b. Remove the relief valve spring and relief valve.

To install:

35. Install oil pump relief valve.
    a. Coat the relief valve with engine oil and insert the relief valve and spring into the valve hole.
    b. Install the relief valve plug and tighten to 36 ft. lbs. (49 Nm).
36. Install oil pump cover.
    a. Apply fresh engine oil to the drive and driven rotors.
    b. Place the drive and driven rotors into the timing chain cover with the marks facing the oil pump cover side.
    c. Install the oil pump cover with the 7 bolts. Tighten to 80 inch lbs. (9 Nm).
    d. Install a new O-ring onto the oil pump cover.
    e. Install a new O-ring onto the oil pipe.
    f. Install the oil pipe with the 3 bolts. Tighten to 80 inch lbs. (9 Nm).
37. To complete installation, reverse remaining removal procedure.

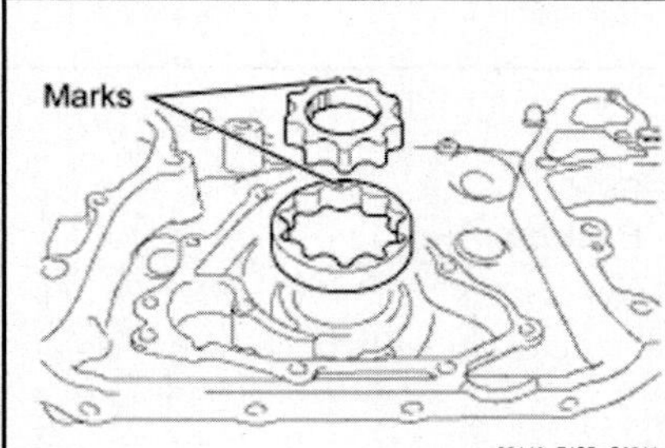

Fig. 137 Place the drive and driven rotors into the timing chain cover with the marks facing upward

2010 Vehicles

See Figures 138 through 143.

1. Before servicing the vehicle, refer to the precautions.
2. Remove the engine.
3. Remove the ignition coil assembly.
4. Remove the dipstick.
5. Remove the bolt and dipstick guide.
6. Remove the O-ring from the dipstick guide.
7. Remove water inlet housing:
    a. Disconnect the 3 water by-pass hoses.
    b. Remove the 5 bolts and water inlet housing.
    c. Remove the O-ring and gasket from the water outlet pipe and water pump.
    d. Remove the 3 water by-pass hoses.
8. Remove no. 1 and no. 2 idler pulley sub-assembly.
9. Remove the 5 bolts and V-ribbed belt tensioner.
10. Remove the 2 nuts, bolt, oil filter bracket and gasket.
11. Remove crankshaft pulley.
12. Remove no. 1 oil pipe:
    a. Remove the 2 oil pipe unions, gaskets and No. 1 oil pipe.
    b. Remove the oil control valve filter LH and gaskets.

➡ **Do not touch the mesh when removing the oil control valve filter.**

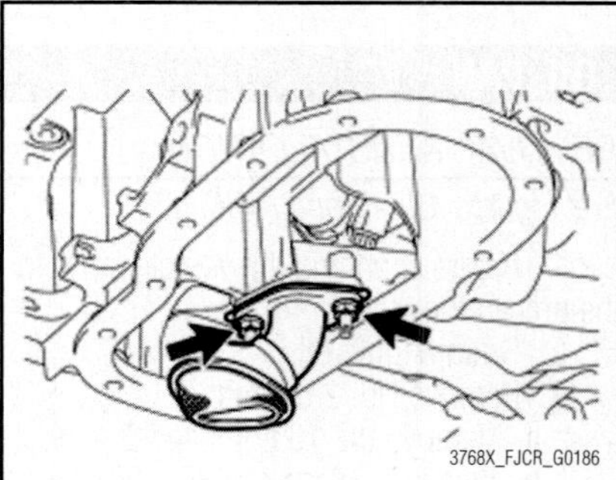

Fig. 138 Remove the 2 nuts, oil strainer and gasket

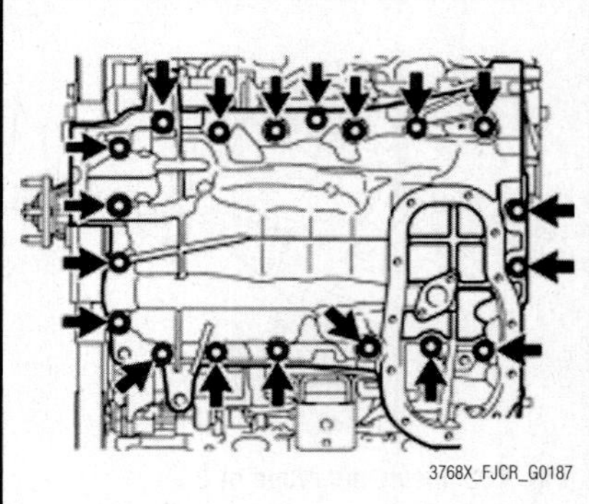

Fig. 139 Remove the 17 bolts and 2 nuts

13. Remove no. 2 oil pipe:
   a. Remove the 2 oil pipe unions, gaskets and No. 2 oil pipe.
   b. Remove the oil control valve filter RH and gaskets.

➡ Do not touch the mesh when removing the oil control valve filter.

14. Remove rear cylinder head cover.
15. Remove the 2 bolts and disconnect the fuel pipe.
16. Remove the LH and RH valve covers.

➡Make sure the removed parts are returned to the same places they were removed from.

17. Remove the oil pan.
18. Remove the 2 nuts, oil strainer and gasket.
19. Remove oil pan sub-assembly:
   a. Remove the 17 bolts and 2 nuts.

➡Make sure the removed parts are returned to the same places they were removed from.

   b. Using a screwdriver, remove the oil pan by prying between the oil pan and cylinder block as shown in the illustration.

➡Be careful not to damage the contact surfaces of the cylinder block or oil pan.

➡Tape the screwdriver tip before use.

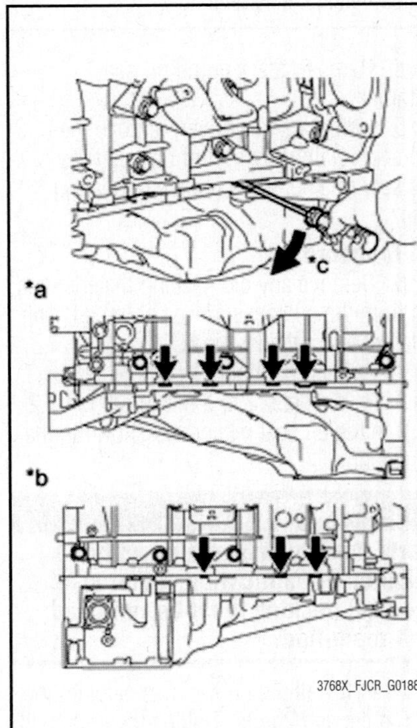

Fig. 140 Remove the oil pan by prying at the indicated locations between the oil pan and cylinder block as shown

Fig. 141 Remove the 8 bolts

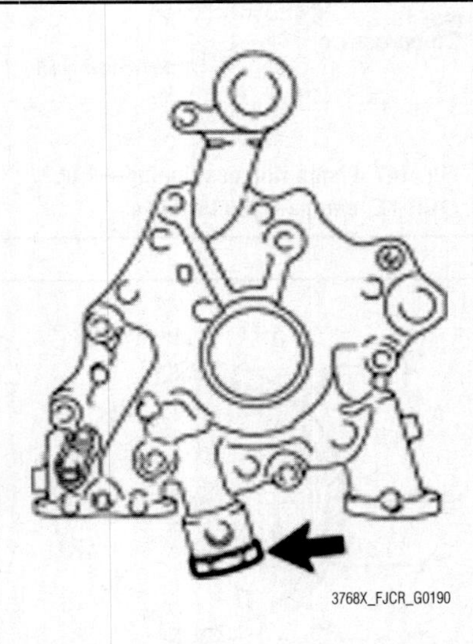

Fig. 142 Remove oil pump relief valve

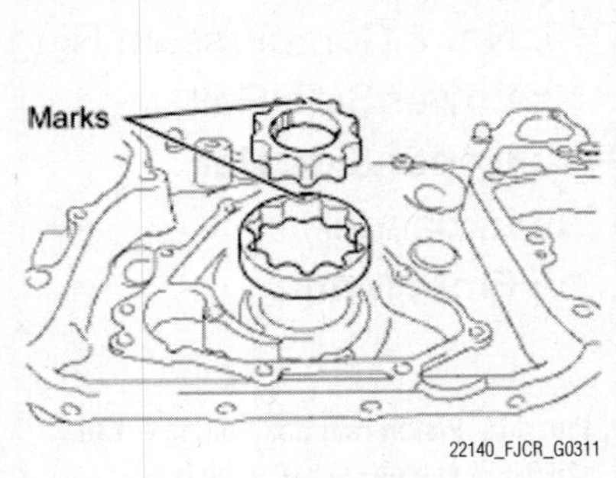

Fig. 143 Place the drive and driven rotors into the timing chain cover with the marks facing the oil pump cover side

   c. Remove the 3 O-rings from the timing chain cover.
   d. Remove the 3 O-rings from the timing chain cover.
20. Remove timing chain cover sub-assembly.
21. Remove front crankshaft oil seal.
22. Remove the 8 bolts, screw, oil pump cover, drive rotor and driven rotor.
23. Remove oil pump relief valve:
   a. Remove the relief valve plug.
   b. Remove the spring and relief valve.

To install:

24. Install oil pump relief valve
   a. Coat the relief valve with engine oil and insert the relief valve and spring into the valve hole.
   b. Install the relief valve plug. Tighten to 36 ft. lbs. (49 Nm).
25. Install oil pump cover:
   a. Apply fresh engine oil to the drive and driven rotors.
   b. Place the drive and driven rotors into the timing chain cover with the matchmarks facing the oil pump cover side.
26. Install the oil pump cover with the 8 bolts and screw:
   a. Screw: 7 ft. lbs. (10 Nm).
   b. Bolts: 80 inch. lbs. (9 Nm).
27. To complete installation, reverse remaining removal procedure.

INSPECTION

See Figures 143 through 146.

1. Inspect oil pump relief valve.
   a. Coat the valve with engine oil, then check that it falls smoothly into the valve hole by its own weight. If not, replace the relief valve. If necessary, replace the oil pump assembly.
2. Inspect oil pump rotor set.
   a. Place the drive and driven rotors into the timing chain cover with the marks facing upward.
   b. Check the rotor tip clearance.
   c. Using a feeler gauge, measure the clearance between the drive and driven rotor tips.
   d. Standard clearance is 0.0024–0.0063 in. (0.06–0.16 mm). Maximum tip clearance is 0.0063 in. (0.16 mm).
   e. If the clearance is greater than the maximum, replace the drive and driven rotors together.
3. Check the rotor side clearance.
   a. Using a feeler gauge and precision straight edge, measure the clearance between the rotors and precision straight edge.
   b. Standard clearance is 0.0012–0.0035 in. (0.03–0.09 mm). Maximum side clearance is 0.0035 in. (0.09 mm).
   c. If the clearance is greater than the maximum, replace the drive and driven

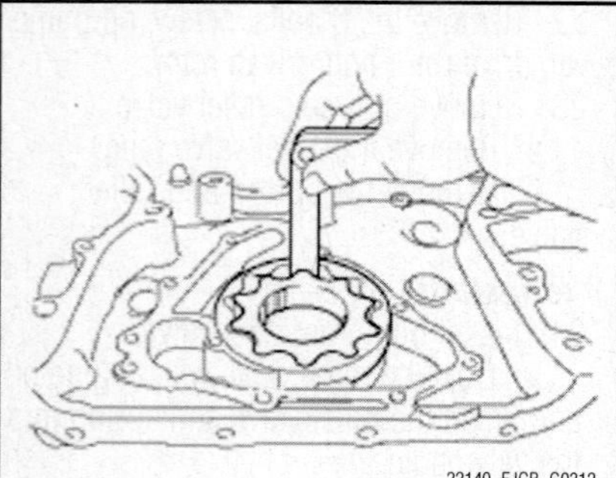
Fig. 144 Using a feeler gauge, measure the clearance between the drive and driven rotor tips

rotors. If necessary, replace the timing chain cover assembly.

4. Check the rotor body clearance.
   a. Using a feeler gauge, measure the clearance between the driven rotor and body.
   b. Standard clearance is 0.0098–0.0128 in. (0.250–0.325 mm). Maximum body clearance is 0.0128 in. (0.325 mm)
5. If the clearance is greater than the maximum, replace the drive and driven

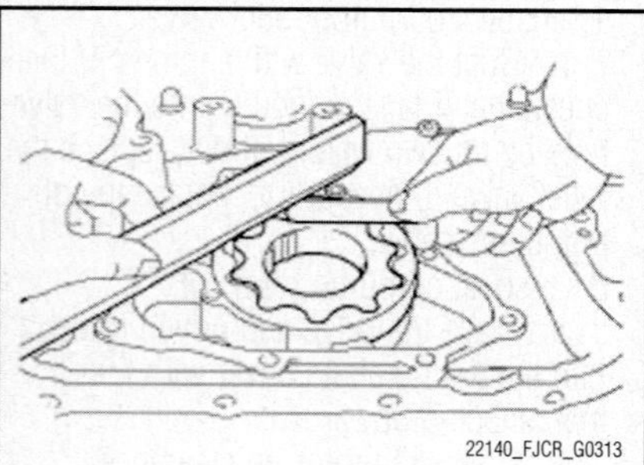
Fig. 145 Using a feeler gauge and precision straight edge, measure the clearance between the rotors and precision straight edge

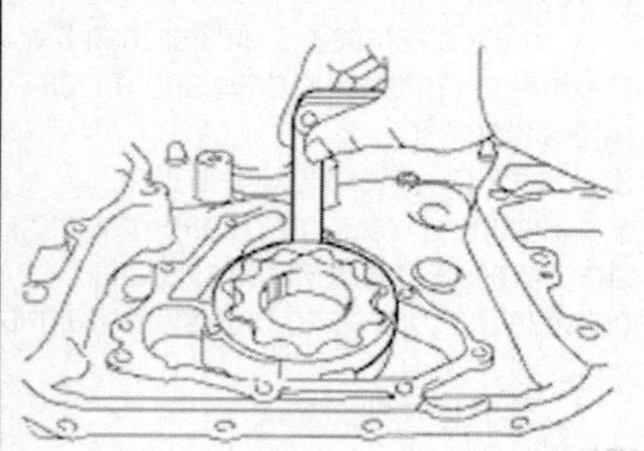
Fig. 146 Using a feeler gauge, measure the clearance between the driven rotor and body

rotors. If necessary, replace the timing chain cover assembly.

PISTON AND RING

POSITIONING

See Figures 147 and 148.

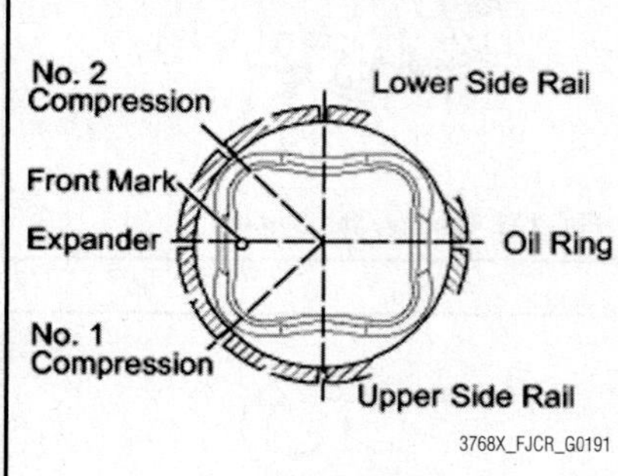

Fig. 147 Piston ring positioning—4.0L (1GR-FE) engine—2009 vehicles

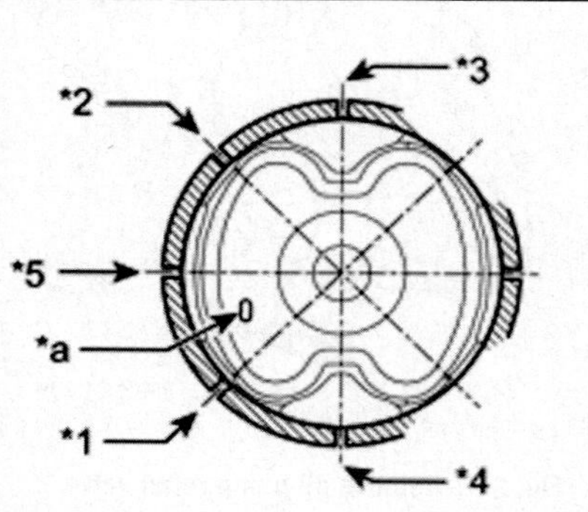

Fig. 148 Piston ring positioning—4.0L (1GR-FE) engine—2010 vehicles

REAR MAIN SEAL

REMOVAL & INSTALLATION

2009 Vehicles

See Figures 149 and 150.

1. Before servicing the vehicle, refer to the precautions.
2. Remove the flywheel/driveplate.
3. Remove the 5 bolts and 2 nuts.

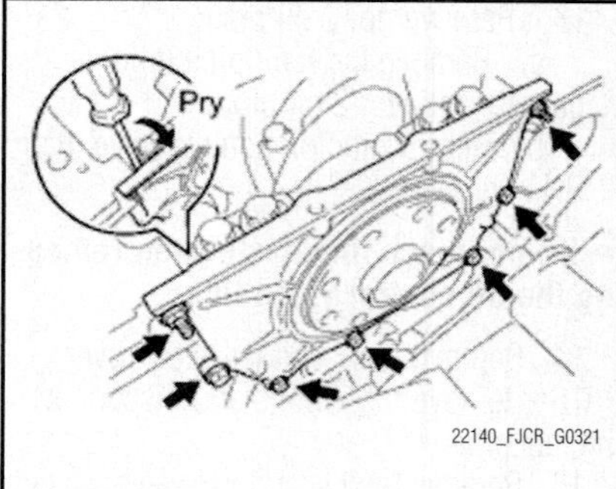

Fig. 149 Rear engine oil seal retainer attaching bolts

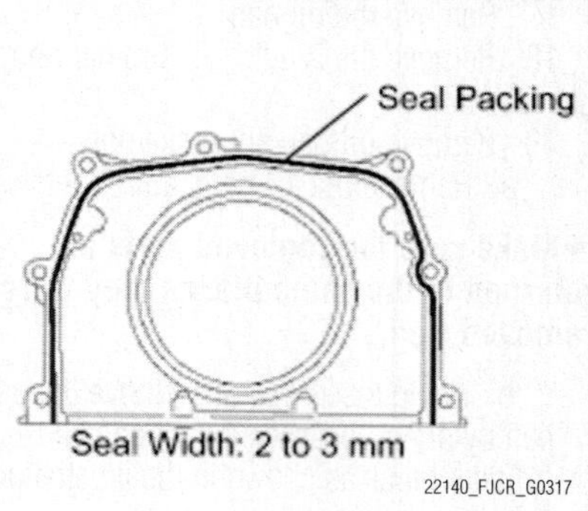

Fig. 150 Apply a continuous bead of Three Bond 1207B or the equivalent sealant 0.08 to 0.12 in. (2 to 3 mm) diameter to the oil seal retainer as shown

4. Remove rear engine oil seal retainer.
5. Using a screwdriver, remove the oil seal retainer by prying between the oil seal retainer and crankshaft bearing cap.

To install:

6. Remove any old packing material and oil from the contact surfaces of the oil seal retainer and cylinder block.
7. Apply a continuous bead of seal packing (diameter 2 to 3 mm (0.08 to 0.12 in.)) to the oil seal retainer as shown in the illustration.

✻✻ WARNING

Parts must be assembled within 3 minutes of application. Otherwise, the seal packing must be removed and reapplied.

   a. Install the oil seal retainer with the 5 bolts and 2 nuts. Tighten the nuts to 80 inch lbs. (9 Nm) and the bolts to 7 ft. lbs. (10 Nm).
8. Install the flywheel/driveplate.

2010 Vehicles

See Figures 151 and 152.

1. Remove the transmission.
2. For M/T, remove the clutch cover, disc assembly and flywheel.
3. For A/T, remove the flexplate.
4. Remove rear crankshaft oil seal:
   a. Using a knife, cut off the lip of the oil seal.
   b. Using a screwdriver, pry out the oil seal.

➡ Do not damage the surface of the oil seal press fit hole and crankshaft. Tape the screwdriver tip before use.

To install:

5. Install rear crankshaft oil seal:
   a. Apply MP grease to the lip of a new oil seal.
   b. Using SST 09223-78010 and a hammer, tap in the oil seal until its surface is flush with the rear oil seal retainer edge.
6. To complete installation, reverse remaining removal procedure.

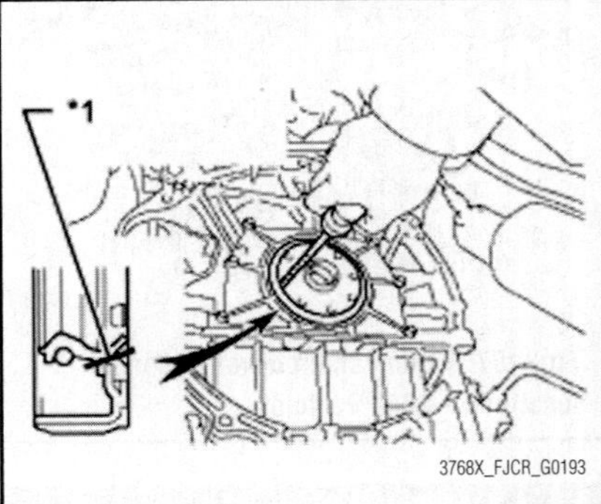

Fig. 151 Identifying the cut position—Using a screwdriver, pry out the oil seal

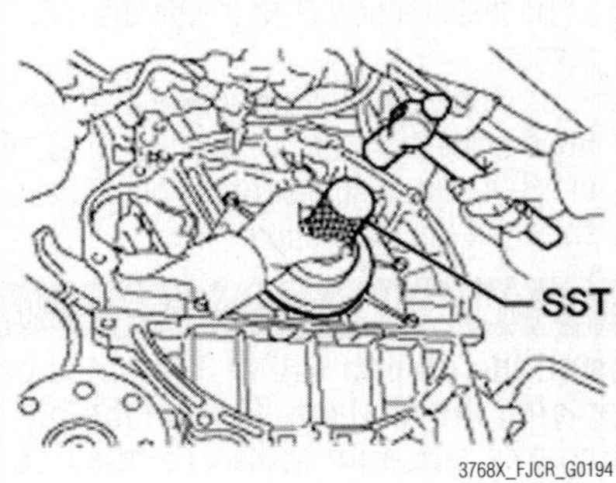

Fig. 152 Using SST 09223-78010 and a hammer, tap in the oil seal until its surface is flush with the rear oil seal retainer edge

TIMING CHAIN FRONT COVER

REMOVAL & INSTALLATION

See Figures 153 through 160.

1. Before servicing the vehicle, refer to the precautions.
2. Remove battery.
3. Drain engine coolant.
4. Drain engine oil.
5. Remove power steering rack assembly.
6. On 4WD vehicles, remove front differential carrier assembly.
7. Remove fan.
8. Remove alternator assembly.
9. Remove the air conditioning compressor and set it aside.

➡Do not disconnect the air conditioning lines.

10. Remove the 5 bolts, then remove the V-ribbed belt tensioner.
11. Remove the oil level gauge.
12. Remove the power steering pump assembly and set it aside.

➡Do not disconnect the power steering lines.

13. Remove crankshaft pulley.
14. Remove No. 2 oil pan.
    a. Remove the 10 bolts and 2 nuts.
    b. Insert the blade of SST: 09032-00100 between the oil pan and No. 2 oil pan, cut off applied sealer and remove the No. 2 oil pan.

⁂ WARNING

Be careful not to damage the contact surfaces of the oil pan and No. 2 oil pan. Be careful not to damage the No. 2 oil pan flange.

15. Remove the 2 nuts, then remove the oil strainer and gasket.
16. Remove oil pan.
    a. Remove the 4 housing bolts.
    b. Remove the flywheel housing under cover.
    c. Remove the 17 bolts and 2 nuts.
    d. Using a screwdriver, remove the oil pan by prying between the oil pan and cylinder block.

⁂ WARNING

Be careful not to damage the contact surfaces of the cylinder block and oil pan.

    e. Remove the O-ring from the oil pump.
17. Remove air cleaner assembly.

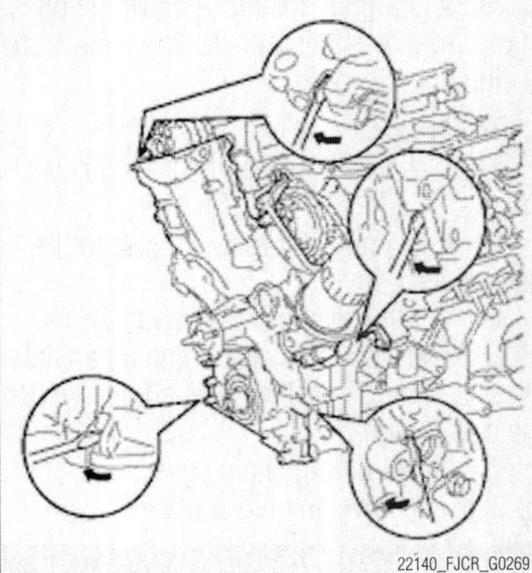

Fig. 153 Remove the timing chain cover by prying between the timing chain cover, cylinder head or cylinder block with a screwdriver

18. Remove throttle body bracket.
19. Remove oil baffle plate.
20. Remove surge tank bracket.
21. Remove intake air surge tank.
22. Remove ignition coil assembly.
23. Remove camshaft timing oil control valve assembly.
    a. Disconnect the 2 connectors.
    b. Remove the 2 bolts, then remove the 2 camshaft timing oil control valves.
24. Remove VVT sensor.
    a. For the right cylinder head sensor, disconnect the VVT sensor connector and remove the sensor bolt. Then remove the sensor.
    b. For the left cylinder head sensor, disconnect the No. 4 water by-pass hose and No. 5 water by-pass hose. Disconnect the VVT sensor connector and remove the sensor bolt. Then remove the sensor.
25. Remove the water inlet.
26. Remove the valve covers and gaskets.
27. Remove timing chain cover.
    a. Remove the 24 bolts and 2 nuts.
    b. Remove the timing chain cover by prying between the timing chain cover, cylinder head or cylinder block with a screwdriver.

⁂ WARNING

Be careful not to damage the contact surfaces of the timing chain cover, cylinder block and cylinder head.

    c. Remove the O-ring from the left cylinder head.
28. Pry out the seal from the cover with a flat-bladed tool.

29. It is a good idea to remove the oil pump from the timing cover and replace the O-ring.

To install:

30. Clean and inspect the timing cover area.
31. Apply multi-purpose grease to the new oil seal lip.
32. Tap the seal into place with SST: 09226-10010 or equivalent, and a hammer. Do this until the seal surface is flush with the cover edge.
33. Install timing chain cover.
    a. Remove any old sealant.

⁂ WARNING

Do not drop any oil on the contact surfaces of the timing chain cover, cylinder head and cylinder block.

b. Install a new O-ring onto the left cylinder head as shown in the illustration.

c. Apply a continuous bead of Three Bond 1207B or the equivalent sealant 0.12–0.16 in. (3–4 mm) diameter to the 4 locations shown in the illustration.

d. Keep the seal surface between the cylinder block and the cylinder head shown in the illustration free of oil before installing the chain cover.

e. Apply a continuous bead of Three Bond or the equivalent sealant 0.12–0.16 in. (3–4 mm) diameter to the timing chain cover as shown in the illustration. Use Three Bond 1282B for the water pump part and Three Bond 1207B for the remainder.

⁂ WARNING

Install the timing chain cover within 3 minutes of applying the sealant. The timing chain cover bolts and nuts must be tightened within 15 minutes of the installation. Otherwise the sealant must be removed and reapplied.

Do not apply sealant to portion A shown in the illustration.

f. Align the key way of the oil pump drive rotor with the rectangular portion of the crankshaft timing gear, and slide the timing chain cover into place.

g. Install the timing chain cover with the 24 bolts and 2 nuts. Tighten the bolts and nuts to 17 ft. lbs. (23 Nm) uniformly in several steps.

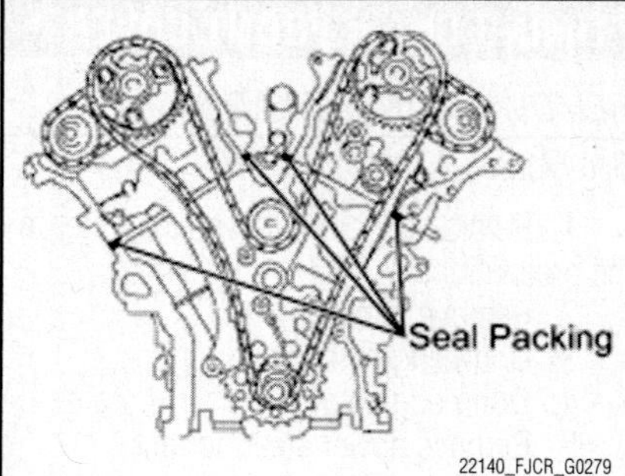

Fig. 154 Apply a continuous bead of Three Bond 1207B or the equivalent sealant 0.12–0.16 in. (3–4 mm) diameter to the 4 locations shown

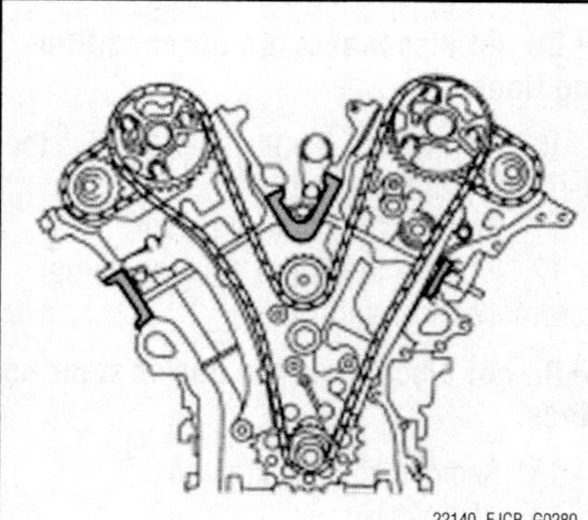

Fig. 155 Keep the seal surface between the cylinder block and the cylinder head shown in the illustration free of oil before installing the chain cover

➡Pay attention not to wrap the chain and slipper over the timing chain cover seal line.

⁂ WARNING

Place bolts in their proper positions. Bolt A length is 0.98 in. (25 mm) and bolt B length is 2.17 in. (55 mm).

34. Install cylinder head covers.
35. Install water inlet.
36. Install VVT sensor. Tighten to 71 inch lbs. (8 Nm).
37. Install camshaft timing oil control valve assembly. Tighten to 80 inch lbs. (9 Nm).
38. Install ignition coil assembly.
39. Install intake air surge tank.
40. Install surge tank brackets.
41. Install oil baffle plate.
42. Install throttle body bracket.
43. Install air cleaner assembly.
44. Install oil pan.
    a. Remove any old sealant.

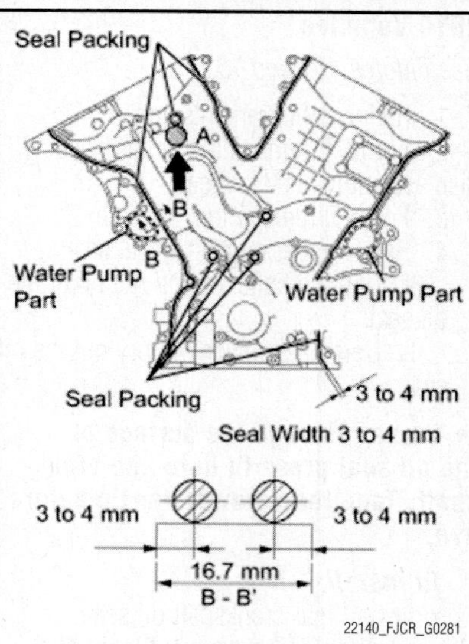

Fig. 156 Apply a continuous bead of Three Bond or the equivalent sealant 0.12–0.16 in. (3–4 mm) diameter to the timing chain cover as shown

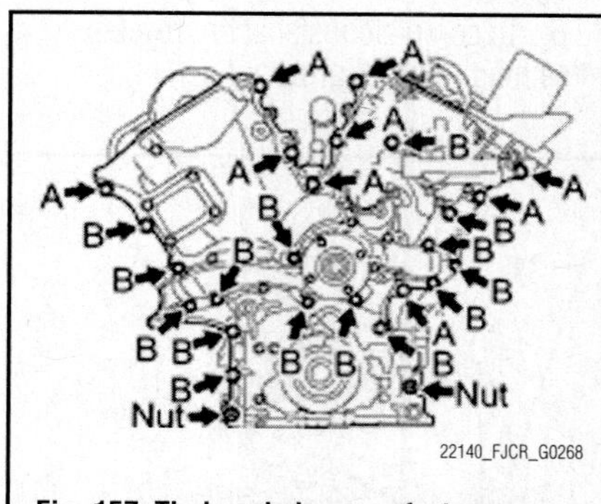

Fig. 157 Timing chain cover fastener locations—2009 vehicles

⁂ WARNING

Do not drop any oil on the contact surfaces of the cylinder block, rear oil seal and oil pan.

b. Install a new O-ring onto the oil pump.

c. Apply a continuous bead of Three Bond or the equivalent sealant 0.12–0.16 in. (3–4 mm) diameter to the oil pan as shown in the illustration.

⁂ WARNING

Install the oil pan within 3 minutes of applying the sealant. Tighten the oil pan bolts and nuts within 15 minutes of installing the oil pan. Otherwise, the sealant must be removed and reapplied.

d. Install the oil pan with the 17 bolts and 2 nuts, and tighten the bolts

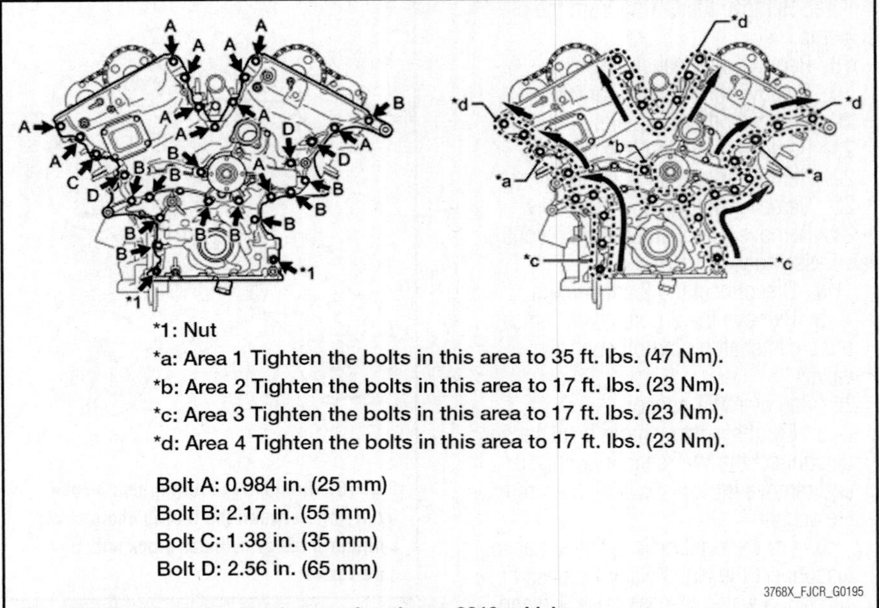

Fig. 158 Timing chain cover fastener locations—2010 vehicles

and nuts uniformly in several steps. Tighten the 0.39 in. (10 mm) head oil pan bolt to 7 ft. lbs. (10 Nm), the 0.47 in. (12 mm) head oil pan bolt to 16 ft. lbs. (21 Nm) and the nuts to 16 ft. lbs. (21 Nm).

✲✲ WARNING

Install bolts in their proper locations. Bolt lengths are as follows: A is 0.98 in. (25 mm) B 1.77 in. (45 mm) and C 0.55 in. (14 mm).

e. Install the 4 housing bolts. Tighten to 27 ft. lbs. (37 Nm).

f. Install the flywheel housing under cover.

45. Install oil strainer.

a. Install a new gasket and the oil strainer with the 2 nuts. Tighten to 80 inch lbs. (9 Nm).

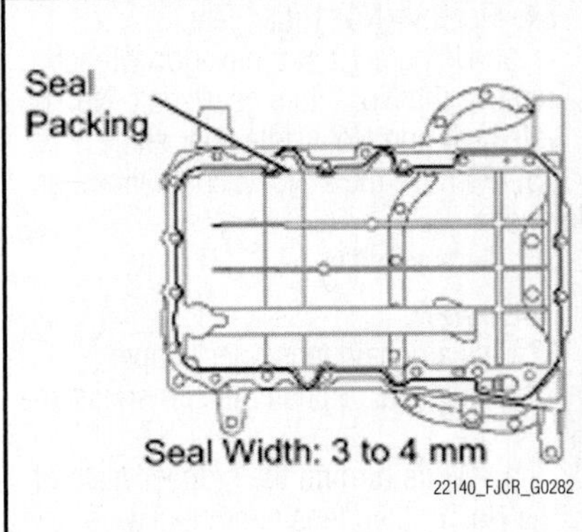

Fig. 159 Apply a continuous bead of Three Bond or the equivalent sealant 0.12–0.16 in. (3–4 mm) diameter to the oil pan as shown

46. Install No. 2 oil pan.

a. Remove any old sealant.

✲✲ WARNING

Do not drop any oil on the contact surfaces of the oil pan and No. 2 oil pan.

b. Apply a continuous bead of Three Bond or the equivalent sealant 0.12–0.16 in. (3–4 mm) diameter to the oil pan.

✲✲ WARNING

Install the oil pan within 3 minutes of applying the sealant. Tighten the oil pan bolts and nuts within 15 minutes of installing the oil pan. Otherwise, the sealant must be removed and reapplied.

c. Install the oil pan with the 17 bolts and 2 nuts, and tighten the bolts and nuts uniformly in several steps. Tighten the bolts to 80 inch lbs. (9 Nm) and the nuts to 7 ft. lbs. (10 Nm).

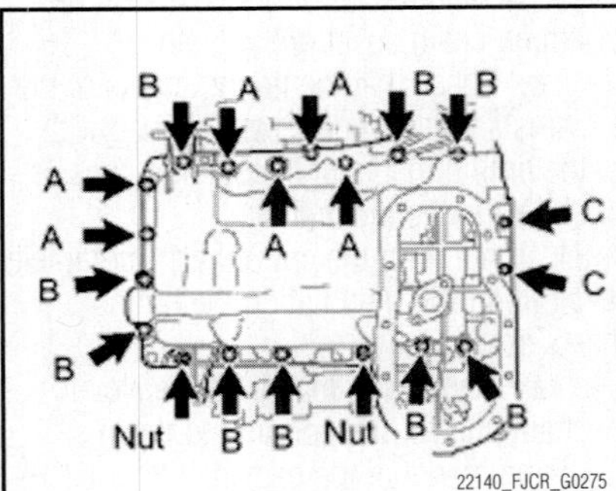

Fig. 160 Oil pan fastener locations

47. Install crankshaft pulley.

48. Install No. 1 idler pulley.

a. Install the idler pulley with the bolt. Tighten to 29 ft. lbs. (39 Nm).

➡DOUBLE is marked on the No. 1 idler pulley to distinguish it from the No. 2 idler pulley.

49. Install No. 2 idler pulley.

a. Install the 2 No. 2 idler pulleys with the 2 bolts. Tighten to 40 ft. lbs. (54 Nm).

50. Install power steering pump assembly and set it aside.

➡Do not disconnect the power steering hoses.

a. Install the power steering pump with the 2 bolts. Tighten to 32 ft. lbs. (43 Nm).

✲✲ WARNING

Do not hit the pulley with other parts when installing the power steering pump.

b. Connect the power steering pressure switch connector.

51. Install oil level gauge guide.

a. Install a new O-ring onto the oil level gauge guide.

b. Apply a light coat of engine oil to the O-ring.

c. Push the oil level gauge guide end into the guide hole in the oil pan.

d. Install the oil level gauge guide with the bolt. Tighten to 80 inch lbs. (9 Nm).

e. Install the oil level gauge guide.

52. Install V-ribbed belt tensioner assembly.

53. Install air conditioner compressor assembly.

54. Install alternator assembly.

55. Install fan.

56. On 4WD vehicles, install front differential carrier assembly.

57. Install power steering link assembly.

58. Install battery.

59. Add engine coolant.

60. Add engine oil.

61. Add power steering fluid and bleed the system, as necessary.

62. On 4WD vehicles, inspect and adjust the differential oil.

63. Check for engine coolant leakage.

64. Check for engine oil leakage.

65. Check for power steering fluid leakage.

66. Check for differential oil leakage.
67. Inspect and adjust front wheel alignment.

TIMING CHAIN & SPROCKETS

REMOVAL & INSTALLATION

See Figures 161 through 165.

1. Before servicing the vehicle, refer to the precautions.
2. Remove battery.
3. Drain engine coolant.
4. Drain engine oil.
5. Remove power steering rack assembly
6. On 4WD vehicles, remove front differential carrier assembly.
7. Remove fan.
8. Remove alternator assembly.
9. Remove the air conditioning compressor and set it aside.

➡Do not disconnect the air conditioning lines.

10. Remove the 5 bolts, then remove the V-ribbed belt tensioner.
11. Remove the oil level gauge.
12. Remove the power steering pump assembly and set it aside.

➡Do not disconnect the power steering lines.

13. Remove idler pulleys.
14. Remove crankshaft pulley.
15. Remove No. 2 oil pan.
    a. Remove the 10 bolts and 2 nuts.
    b. Insert the blade of SST: 09032-00100 between the oil pan and No. 2 oil pan, cut off applied sealer and remove the No. 2 oil pan.

⁂ WARNING

Be careful not to damage the contact surfaces of the oil pan and No. 2 oil pan. Be careful not to damage the No. 2 oil pan flange.

16. Remove the 2 nuts, then remove the oil strainer and gasket.
17. Remove oil pan.
    a. Remove the 4 housing bolts.
    b. Remove the flywheel housing under cover.
    c. Remove the 17 bolts and 2 nuts.
    d. Using a screwdriver, remove the oil pan by prying between the oil pan and cylinder block.

⁂ WARNING

Be careful not to damage the contact surfaces of the cylinder block and oil pan.

    e. Remove the O-ring from the oil pump.
18. Remove air cleaner assembly.
19. Remove throttle body bracket.
20. Remove oil baffle plate.
21. Remove surge tank bracket.
22. Remove intake air surge tank.
23. Remove ignition coil assembly.
24. Remove camshaft timing oil control valve assembly.
    a. Disconnect the 2 connectors.
    b. Remove the 2 bolts, then remove the 2 camshaft timing oil control valves.
25. Remove VVT sensor.
    a. For the right cylinder head sensor, disconnect the VVT sensor connector and remove the sensor bolt. Then remove the sensor.
    b. For the right cylinder head sensor, disconnect the No. 4 water by-pass hose and No. 5 water by-pass hose. Disconnect the VVT sensor connector and remove the sensor bolt. Then remove the sensor.
26. Remove the water inlet.
27. Remove the 10 bolts, 3 seal washers, 2 nuts, cylinder head cover and gasket.
28. Remove timing chain cover.
    a. Remove the 24 bolts and 2 nuts.
    b. Remove the timing chain cover by prying between the timing chain cover, cylinder head or cylinder block with a screwdriver.

⁂ WARNING

Be careful not to damage the contact surfaces of the timing chain cover, cylinder block and cylinder head.

    c. Remove the O-ring from the left cylinder head.
29. Remove timing gear case or timing chain case oil seal.
30. Set No. 1 cylinder to TDC on the compression stroke.
    a. Turn the crankshaft pulley until its groove and the "0" timing mark of the timing chain cover are aligned.
    b. Check that the timing marks of the camshaft timing gears are aligned with the timing marks of the bearing caps as shown in the illustration.
31. If not, turn the crankshaft 1 complete revolution (360°) and align the timing marks above.
    a. Place paint marks on the No. 1 chain links corresponding to the timing marks of the camshaft timing gears.
32. Remove No. 1 chain tensioner assembly.

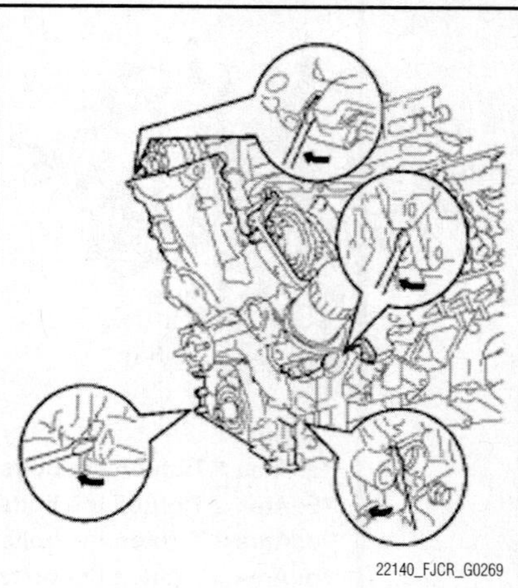

Fig. 161 Remove the timing chain cover by prying between the timing chain cover, cylinder head or cylinder block with a screwdriver

⁂ WARNING

Never rotate the crankshaft with the chain tensioner removed. When rotating the camshaft with the timing chain removed, rotate the crankshaft counterclockwise 40° from the TDC first.

    a. Remove the 4 bolts, then remove the timing chain cover plate and gasket.
    b. While turning the stopper plate of the tensioner upward, push in the plunger of the chain tensioner.
    c. While turning the stopper plate of the tensioner downward, insert a pin of 0.138 in. (3.5 mm) diameter into the holes in the stopper plate and tensioner to fix the stopper plate.
    d. Remove the 2 bolts, then remove the chain tensioner.
33. Remove chain tensioner slipper.
34. Remove No. 1 idle gear.
    a. Using a 10 mm hexagon wrench, remove the No. 2 idle gear shaft, No. 1 idle gear and No. 1 idle gear shaft.
35. Remove the 2 No. 2 chain vibration dampers.
36. Remove chain.

To install:

37. Install chain tensioner slipper.
38. Install No. 1 chain tensioner assembly.
    a. While turning the stopper plate of the No. 1 chain tensioner clockwise, push in the plunger of the No. 1 chain tensioner.
    b. While turning the stopper plate of the tensioner counterclockwise, insert a pin of 0.138 in. (3.5 mm) diameter into

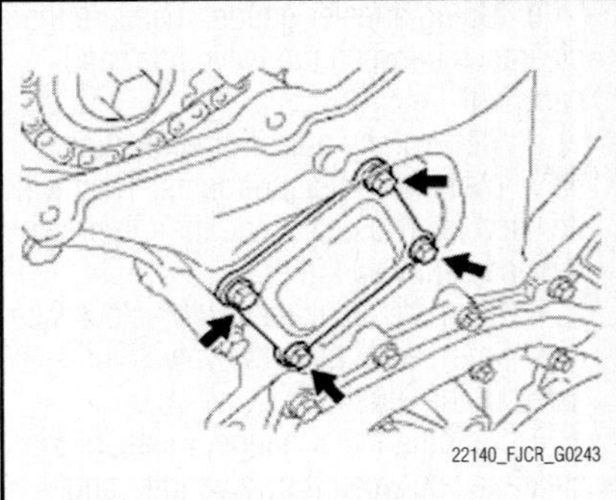

Fig. 162 Timing chain cover plate and gasket

the holes in the stopper plate and No. 1 chain tensioner to fix the stopper plate.

c. Install the No. 1 chain tensioner with the 2 bolts. Tighten to 7 ft. lbs. (10 Nm).

39. Install the timing chain.

a. Set the No. 1 cylinder to TDC on the compression stroke.

b. Align the timing marks of the camshaft timing gears and bearing caps.

c. Using the crankshaft pulley set bolt, turn the crankshaft to align the crankshaft set key with the timing line of the cylinder block.

d. Align the yellow mark link with the timing mark of the crankshaft timing link.

e. Align the orange mark links with the timing marks of the camshaft timing gears, and install the chain.

40. Install the 2 No. 2 chain vibration dampers.

41. Install the idle gear.

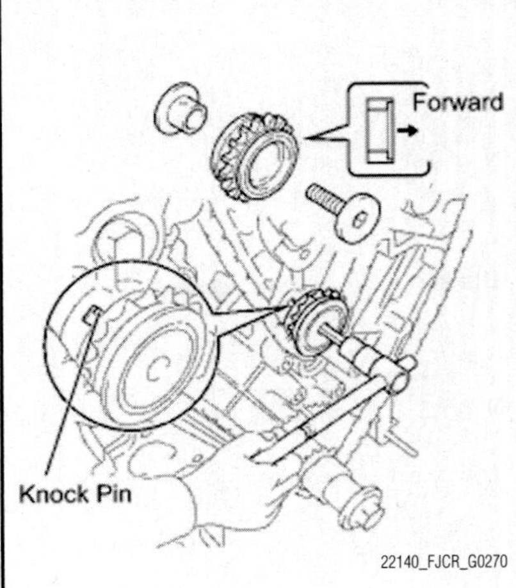

Fig. 163 Using a 10 mm hexagon wrench, remove the No. 2 idle gear shaft, No. 1 idle gear and No. 1 idle gear shaft

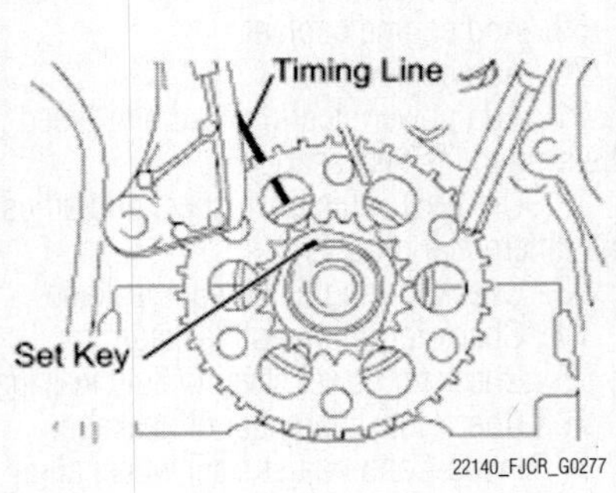

Fig. 164 Using the crankshaft pulley set bolt, turn the crankshaft to align the crankshaft set key with the timing line of the cylinder block

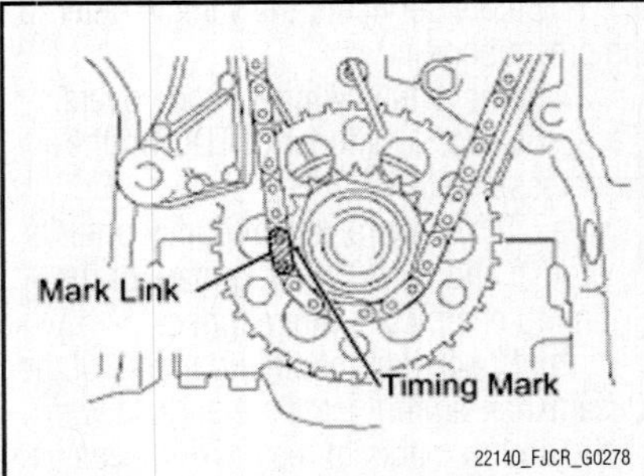

Fig. 165 Align the yellow mark link with the timing mark of the crankshaft timing link

a. Apply a light coat of engine oil to rotating surface of the No. 1 idle gear shaft.

b. Temporarily install the No. 1 idle gear shaft together with the No. 2 idle gear shaft with the knock pin of the No. 1 idle gear shaft and the knock pin groove of the cylinder block are aligned.

➡Orient the idle gear shafts correctly.

c. Using a 10 mm hexagon wrench, tighten the No. 2 idle gear shaft to 44 ft. lbs. (60 Nm).

d. Remove the bar from the chain tensioner.

42. Install the timing gear case oil seal.

43. Install timing chain cover.

44. Install cylinder head covers.

45. Install water inlet.

46. Install VVT sensor. Tighten to 71 inch lbs. (8 Nm).

47. Install camshaft timing oil control valve assembly. Tighten to 80 inch lbs. (9 Nm).

48. Install ignition coil assembly.

49. Install intake air surge tank.

50. Install surge tank brackets.

51. Install oil baffle plate.

52. Install throttle body bracket.

53. Install air cleaner assembly.

54. Install oil pan.

a. Remove any old sealant.

⁂ WARNING

Do not drop any oil on the contact surfaces of the cylinder block, rear oil seal and oil pan.

b. Install a new O-ring onto the oil pump.

c. Apply a continuous bead of Three Bond or the equivalent sealant 0.12–0.16 in. (3–4 mm) diameter to the oil pan.

⁂ WARNING

Install the oil pan within 3 minutes of applying the sealant. Tighten the oil pan bolts and nuts within 15 minutes of installing the oil pan. Otherwise, the sealant must be removed and reapplied.

d. Install the oil pan with the 17 bolts and 2 nuts, and tighten the bolts and nuts uniformly in several steps. Tighten the 0.39 in. (10 mm) head oil pan bolt to 7 ft. lbs. (10 Nm), the 0.47 in. (12 mm) head oil pan bolt to 16 ft. lbs. (21 Nm) and the nuts to 16 ft. lbs. (21 Nm).

⁂ WARNING

Install bolts in their proper locations. Bolt lengths are as follows: A is 0.98 in. (25 mm) B 1.77 in. (45 mm) and C 0.55 in. (14 mm). Refer to the Oil Pan procedure for bolt locations.

e. Install the 4 housing bolts. Tighten to 27 ft. lbs. (37 Nm).

f. Install the flywheel housing under cover.

55. Install oil strainer.

a. Install a new gasket and the oil strainer with the 2 nuts. Tighten to 80 inch lbs. (9 Nm).

56. Install No. 2 oil pan.

a. Remove any old sealant.

⁂ WARNING

Do not drop any oil on the contact surfaces of the oil pan and No. 2 oil pan.

b. Apply a continuous bead of Three Bond or the equivalent sealant 0.12–0.16 in. (3–4 mm) diameter to the oil pan.

⁂ WARNING

Install the oil pan within 3 minutes of applying the sealant. Tighten the oil pan bolts and nuts within 15 minutes of installing the oil pan. Otherwise, the sealant must be removed and reapplied.

c. Install the oil pan with the 17 bolts and 2 nuts, and tighten the bolts and nuts uniformly in several steps. Tighten the bolts to 80 inch lbs. (9 Nm) and the nuts to 7 ft. lbs. (10 Nm).

57. Install crankshaft pulley.
58. Install No. 1idler pulley.

a. Install the idler pulley with the bolt. Tighten to 29 ft. lbs. (39 Nm).

➡DOUBLE is marked on the No. 1 idler pulley to distinguish it from the No. 2 idler pulley.

59. Install No. 2idler pulley.

a. Install the 2 No. 2 idler pulleys with the 2 bolts. Tighten to 40 ft. lbs. (54 Nm).

60. Install power steering pump assembly and set it aside.

➡Do not disconnect the power steering hoses.

a. Install the power steering pump with the 2 bolts. Tighten to 32 ft. lbs. (43 Nm).

⁂ WARNING

Do not hit the pulley with other parts when installing the power steering pump.

b. Connect the power steering pressure switch connector.

61. Install oil level gauge guide.

a. Install a new O-ring onto the oil level gauge guide.

b. Apply a light coat of engine oil to the O-ring.

c. Push the oil level gauge guide end into the guide hole in the oil pan.

d. Install the oil level gauge guide with the bolt. Tighten to 80 inch lbs. (9 Nm).

e. Install the oil level gauge guide.

62. Install V-ribbed belt tensioner assembly.
63. Install air conditioner compressor assembly.
64. Install alternator assembly.
65. Install fan.
66. On 4WD vehicles, install front differential carrier assembly.
67. Install power steering link assembly.
68. Install battery.
69. Add engine coolant.
70. Add engine oil.
71. Add power steering fluid and bleed the system, as necessary.
72. On 4WD vehicles, inspect and adjust the differential oil.
73. Check for engine coolant leakage.
74. Check for engine oil leakage.
75. Check for power steering fluid leakage.
76. Check for differential oil leakage.
77. Inspect and adjust front wheel alignment.

VALVE LASH

ADJUSTMENT

See Figures 166 and 167.

1. Before servicing the vehicle, refer to the precautions.
2. Remove the cylinder head covers.
3. Set No. 1 cylinder to TDC on the compression stroke.

a. Turn the crankshaft pulley until its groove and the "0" timing mark of the timing chain cover are aligned.

b. Check that the timing marks of the camshaft timing gears are aligned with the timing marks of the bearing caps as shown in the illustration.

4. If not, turn the crankshaft 1 complete revolution (360°) and align the timing marks above.
5. Inspect valve clearance.

a. Check the valves indicated in the illustration.

b. Using a feeler gauge, measure the clearance between the valve lifter and camshaft.

c. Record any out-of-specification valve clearance measurements. They will be used later to determine the required replacement valve lifter.

d. Turn the crankshaft 240° clockwise (first time), and check the valves indicated in the illustration.

e. Using a feeler gauge, measure the clearance between the valve lifter and camshaft.

f. Record any out-of-specification valve clearance measurements. They will be used later to determine the required replacement valve lifter.

g. Turn the crankshaft 240° clockwise, and check the valves indicated in the illustration.

h. Using a feeler gauge, measure the clearance between the valve lifter and camshaft.

i. Record any out-of-specification valve clearance measurements. They will be used later to determine the required replacement valve lifter.

6. Adjust valve clearance.

a. Set the No. 1 cylinder to TDC/compression.

b. Turn the crankshaft pulley until its groove and the "0" timing mark of the timing chain cover are aligned.

c. Check that the timing marks of the camshaft timing gears are aligned with the timing marks of the bearing caps as shown in the illustration.

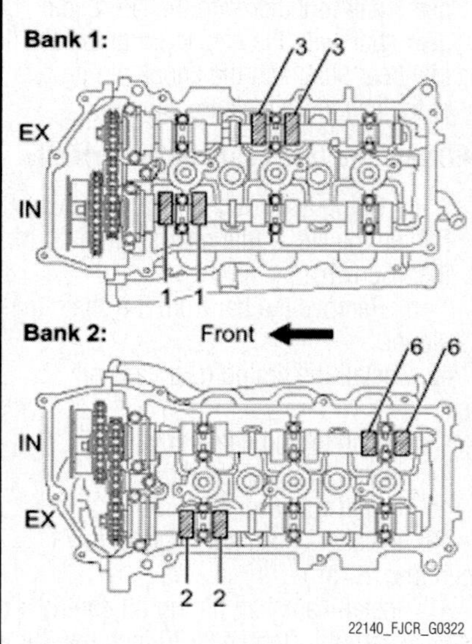

Fig. 166 With the No. 1 cylinder to TDC on the compression stroke check the valves indicated

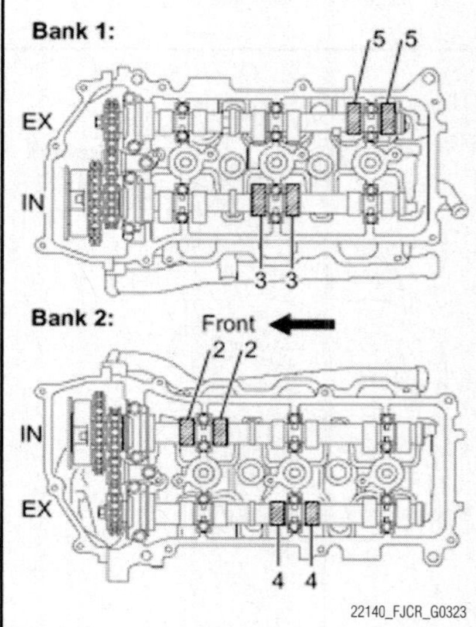

Fig. 167 After rotating the crankshaft 240° clockwise (first time) check the valves indicated

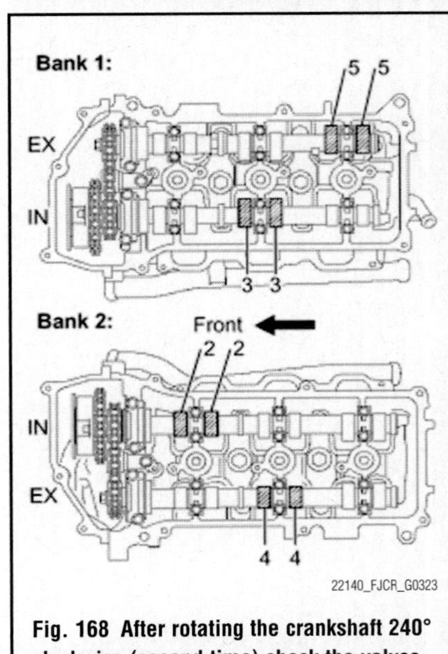

Fig. 168 After rotating the crankshaft 240° clockwise (second time) check the valves indicated

d. If not, turn the crankshaft 1 complete revolution (360°) and align the timing marks as above.

e. Place paint marks on the No. 1 chain links corresponding to the timing marks of the camshaft timing gears.

7. Remove the No. 1 chain tensioner assembly.
8. Remove the No. 2 camshaft.
9. Remove the No. 2 chain tensioner assembly.
10. Remove the camshaft.
11. Remove the No. 4 camshaft.
12. Remove the No. 3 chain tensioner assembly.
13. Remove the No. 3 camshaft.
14. Remove the valve lifters.

To install:

15. Determine the replacement valve lifter size according to the following formulas:

a. Using a micrometer, measure the thickness of the removed lifter.

b. Calculate the thickness of a new lifter so that the valve clearance comes within the specified value.

c. For the intake side: Thickness of new lifter = Thickness of removed lifter + Measured valve clearance - 0.008 in. (0.20 mm).

d. For the exhaust side: Thickness of new lifter = Thickness of removed lifter + Measured valve clearance - 0.013 in. (0.34 mm).

e. Select a new lifter with a thickness as close as possible to the calculated value.

➡Lifters are available in 35 sizes in increments of 0.0008 in. (0.020 mm), from 0.1992 in.–0.2260 in. (5.740–5.060mm).

16. Install the No. 3 camshaft.
17. Install the No. 3 chain tensioner assembly.
18. Install the No. 4 camshaft.
19. Install the camshaft.
20. Install the No. 2 chain tensioner assembly.
21. Install the No. 2 camshaft.
22. Install the No. 1 chain tensioner assembly.
23. Check that the timing marks of the camshaft timing gears are aligned with the timing marks of the bearing cap as shown in the illustration.
24. Install the cylinder head covers.

ENGINE PERFORMANCE & EMISSION CONTROLS

ACCELERATOR PEDAL POSITION (APP) SENSOR

LOCATION

See Figure 169.

Refer to the accompanying illustration for sensor location.

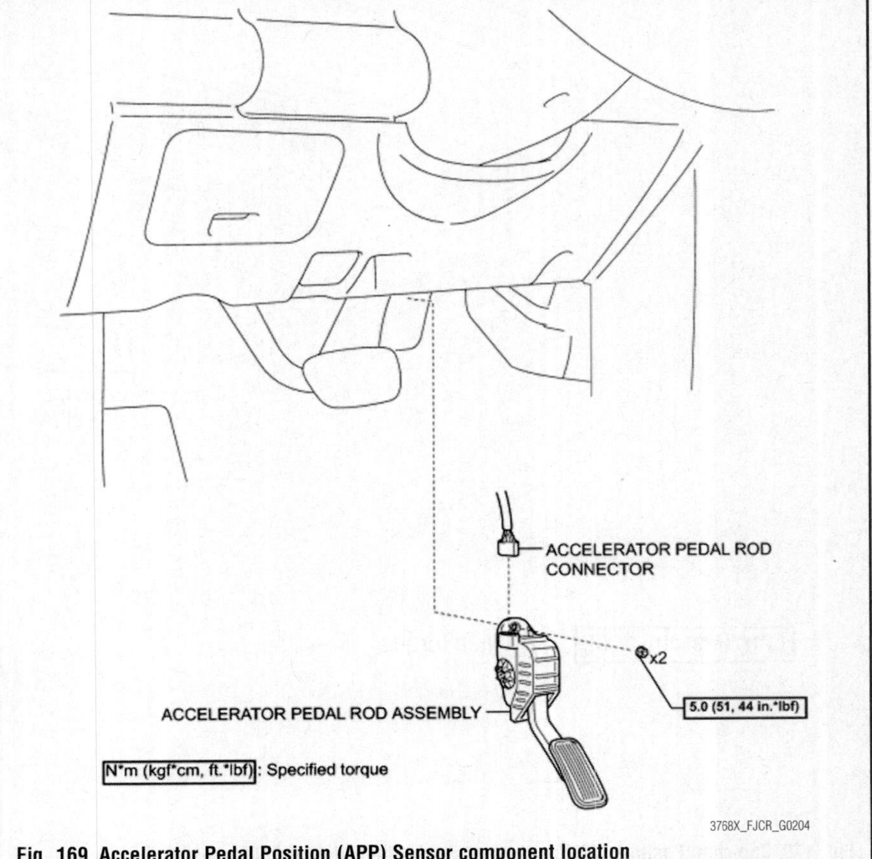

Fig. 169 Accelerator Pedal Position (APP) Sensor component location

REMOVAL & INSTALLATION

1. Before servicing the vehicle, refer to the precautions section.
2. Disconnect the negative battery cable.
3. Disconnect the sensor connector.
4. Remove the sensor attaching bolts, then remove the accelerator pedal.

To install:

5. Install the accelerator pedal. Tighten the pedal attaching bolts to 44 inch lbs. (5 Nm).
6. Connect the sensor connector.
7. Connect the negative battery cable.

CAMSHAFT POSITION (CMP) SENSOR (VVT)

LOCATION

See Figures 170 and 171.

Refer to the accompanying illustrations for sensor location.

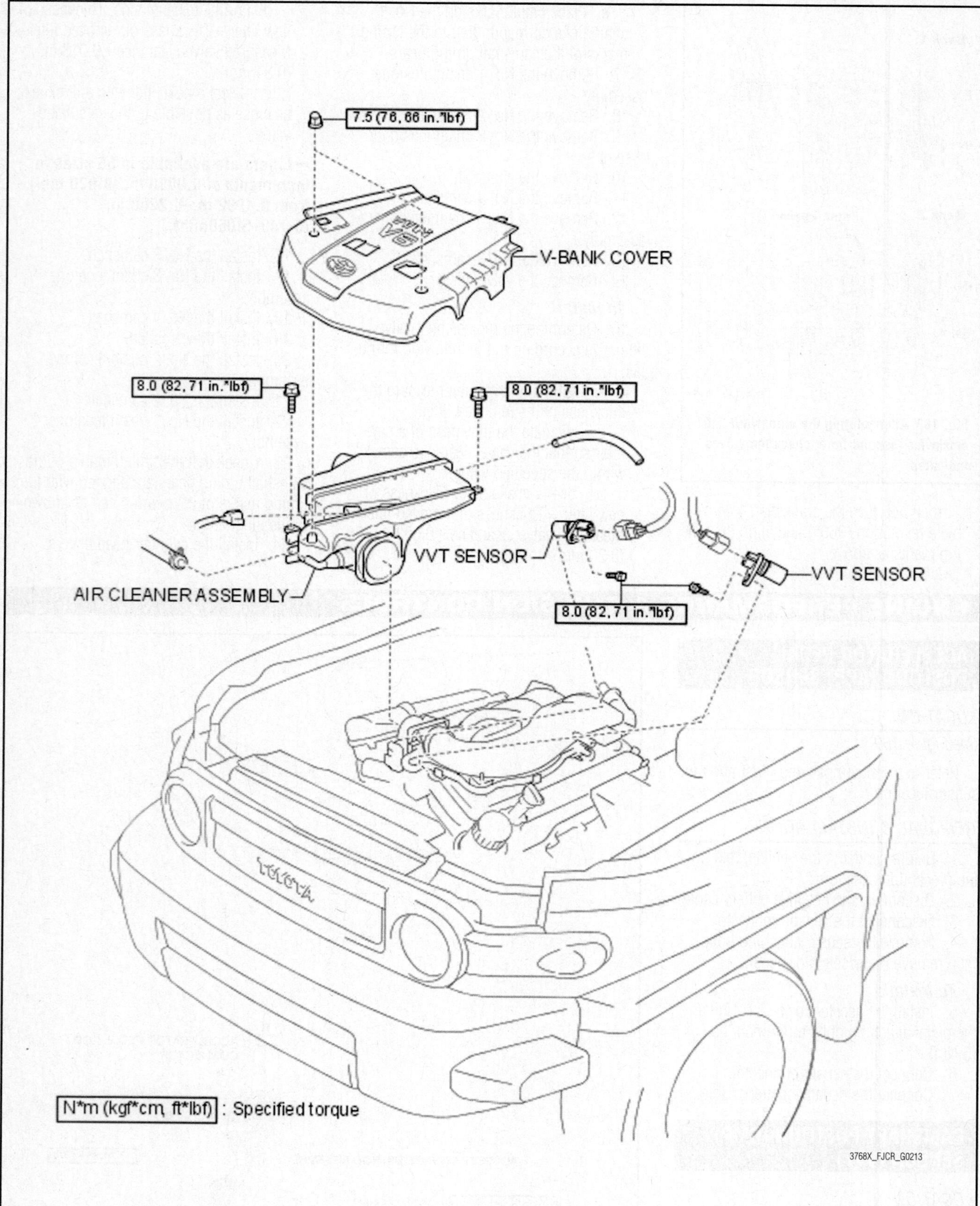

Fig. 170 Camshaft Position (CMP) sensor component locations—2009 vehicles

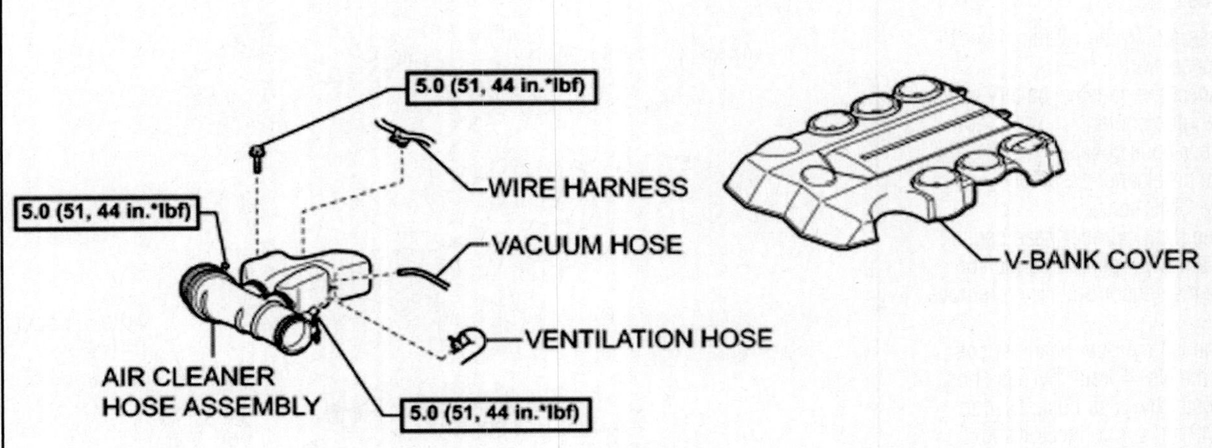

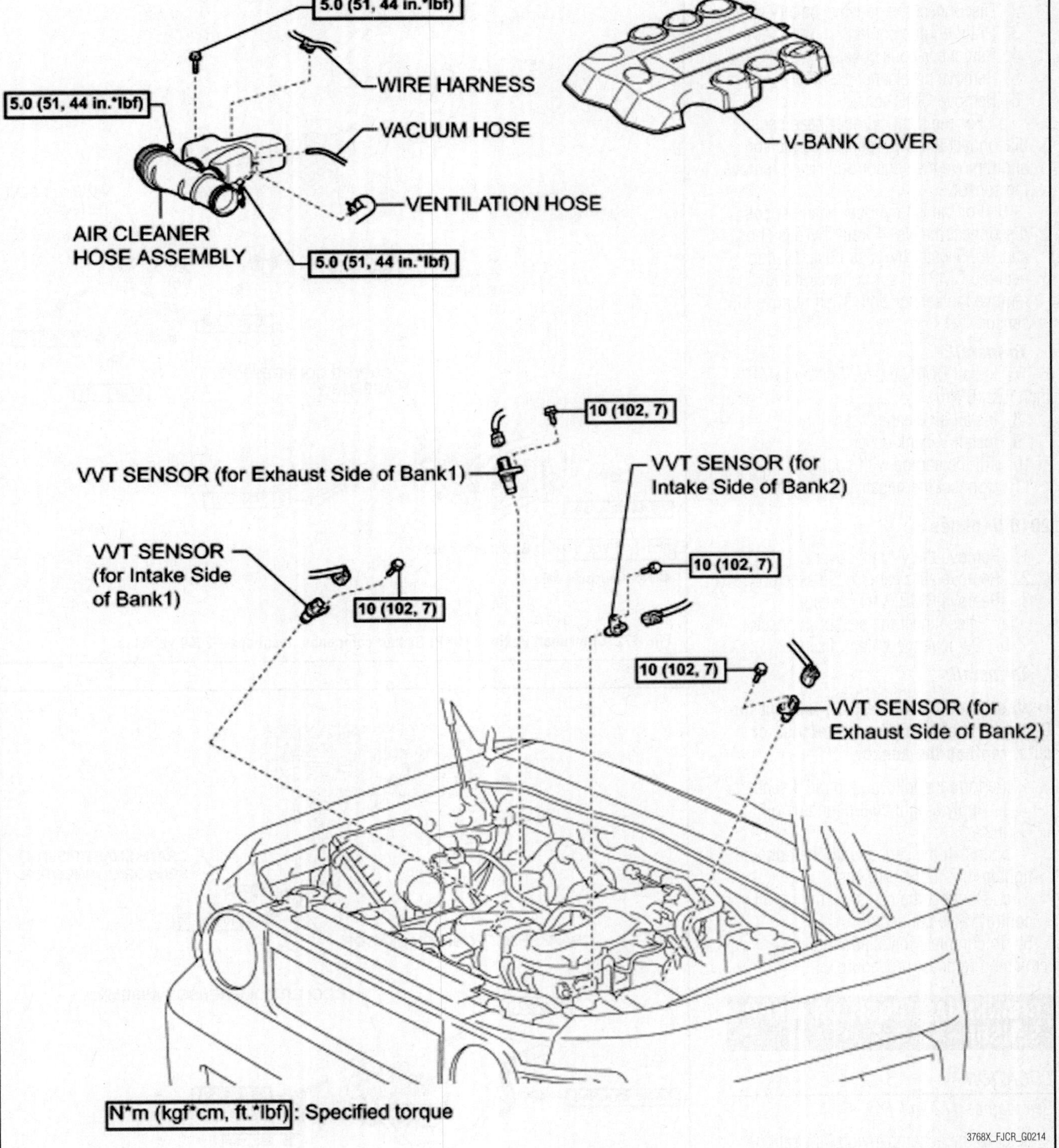

Fig. 171 Camshaft Position (CMP) sensor component locations—2010 vehicles

REMOVAL & INSTALLATION

2009 Vehicles

1. Before servicing the vehicle, refer to the precautions section.
2. Disconnect the negative battery cable.
3. Drain engine coolant.
4. Remove v-bank cover.
5. Remove air cleaner assembly.
6. Remove CMP sensor.
   a. For the right cylinder head sensor, disconnect the CMP sensor connector and remove the sensor bolt. Then remove the sensor.
   b. For the left cylinder head sensor, disconnect the No. 4 water by-pass hose and No. 5 water by-pass hose. Disconnect the CMP sensor connector and remove the sensor bolt. Then remove the sensor.

To install:

7. Install CMP sensor. Tighten to 71 inch lbs. (8 Nm).
8. Install air cleaner assembly.
9. Install V-bank cover.
10. Fill the engine with coolant.
11. Connect the negative battery cable.

2010 Vehicles

1. Remove the v-bank cover.
2. Remove air cleaner hose assembly.
3. Remove CMP (VVT) sensor:
   a. Disconnect the sensor connector.
   b. Remove the bolt and sensor.

To install:

➡When reusing the sensor, inspect the O-ring. If the O-ring has scratches or cuts, replace the sensor.

4. Perform the following for all 4 sensors:
   a. Apply a light coat of engine oil to O-ring.
   b. Install the sensor with the bolt and tighten to 7 ft. lbs. (10 Nm).
   c. Connect the camshaft timing oil control valve connector.
5. To complete installation, reverse remaining removal procedure.

CRANKSHAFT POSITION (CKP) SENSOR

LOCATION

See Figures 172 and 173.

Refer to the accompanying illustrations for sensor location.

REMOVAL & INSTALLATION

1. Before servicing the vehicle, refer to the precautions section.

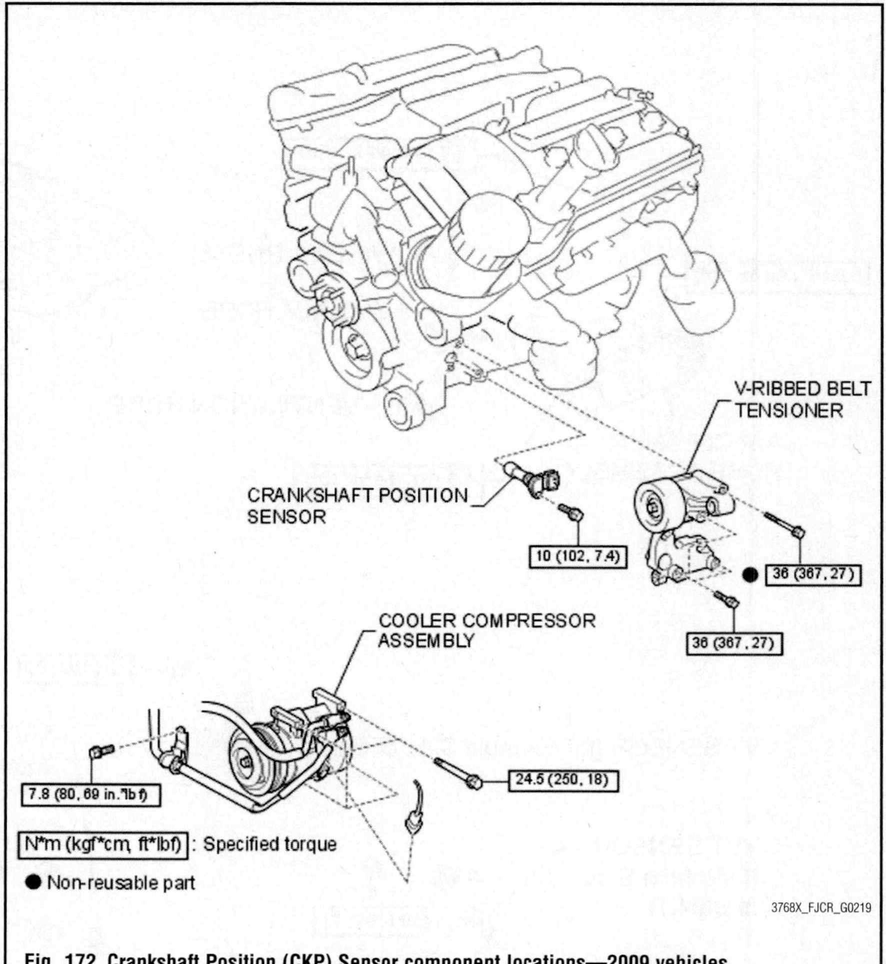

Fig. 172 Crankshaft Position (CKP) Sensor component locations—2009 vehicles

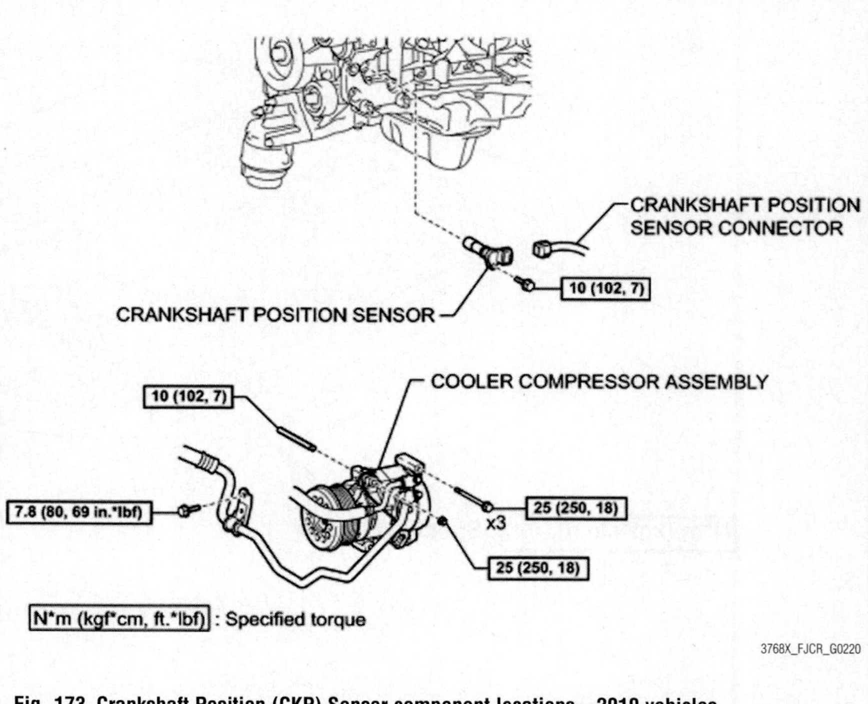

Fig. 173 Crankshaft Position (CKP) Sensor component locations—2010 vehicles

2. For 2009 vehicles, remove the fan.
3. For 2009 vehicles, remove the alternator.
4. Remove the air conditioning compressor and set it aside.

➡Do not disconnect the air conditioning hoses from the compressor.

5. Disconnect the sensor electrical connector.
6. Remove the attaching bolt and sensor.

To install:

7. Install the sensor into cylinder head. Tighten the sensor to 80 inch lbs. (9 Nm).
8. Connect the sensor electrical connector.
9. Install the air conditioning compressor.
10. Install the alternator.
11. Install the fan.

ELECTRONIC CONTROL MODULE (ECM)

LOCATION

See Figures 174 and 175.

Refer to the accompanying illustrations for ECM location.

REMOVAL & INSTALLATION

2009 Vehicles

1. Disconnect the negative battery cable.
2. Remove glove compartment door assembly.
3. On 4WD automatic transmission vehicles, remove four wheel drive control ECU.
   a. Disconnect the connector.
   b. Remove the screw and four wheel drive control ECU.
4. Remove ECM.
   a. Disconnect the 5 connectors.
   b. Remove the 3 nuts, then remove the ECM.

To install:

5. Install the ECM with the 3 nuts. Tighten to 71 inch lbs. (8 Nm).
   a. Connect the 5 connectors.
6. On 4WD automatic transmission vehicles, install four wheel drive control ECU.
   a. Install the four wheel drive control ECU with the bolt. Tighten to 71 inch lbs. (8 Nm).
   b. Connect the connector.
7. Install glove compartment door assembly.
8. Connect the negative battery cable.
9. Connect a scan tool and perform the reset memory automatic transmission initialization.
10. Connect a scan tool and perform the registration (VIN registration) when replacing the ECM.
11. Using a scan tool, set up the function of the ATF (Automatic Transmission Fluid) temperature warning lamp.

2010 Vehicles

1. Disconnect the negative battery cable.
2. Remove glove compartment door assembly.
3. On 4WD automatic transmission vehicles, remove four wheel drive control ECU.
   a. Disconnect the connector.
   b. Remove the bolt and four wheel drive control ECU.
4. Remove ECM.
   a. Disconnect the 5 connectors.
   b. Remove the 3 nuts, then remove the ECM.

To install:

5. Install the ECM with the 3 nuts. Tighten to 49 inch lbs. (5.5 Nm).
   a. Connect the 6 connectors.
6. On 4WD automatic transmission vehicles, install four wheel drive control ECU.
   a. Install the four wheel drive control ECU with the bolt. Tighten to 71 inch lbs. (8 Nm).
   b. Connect the connector.
7. Install glove compartment door assembly.
8. Connect the negative battery cable.

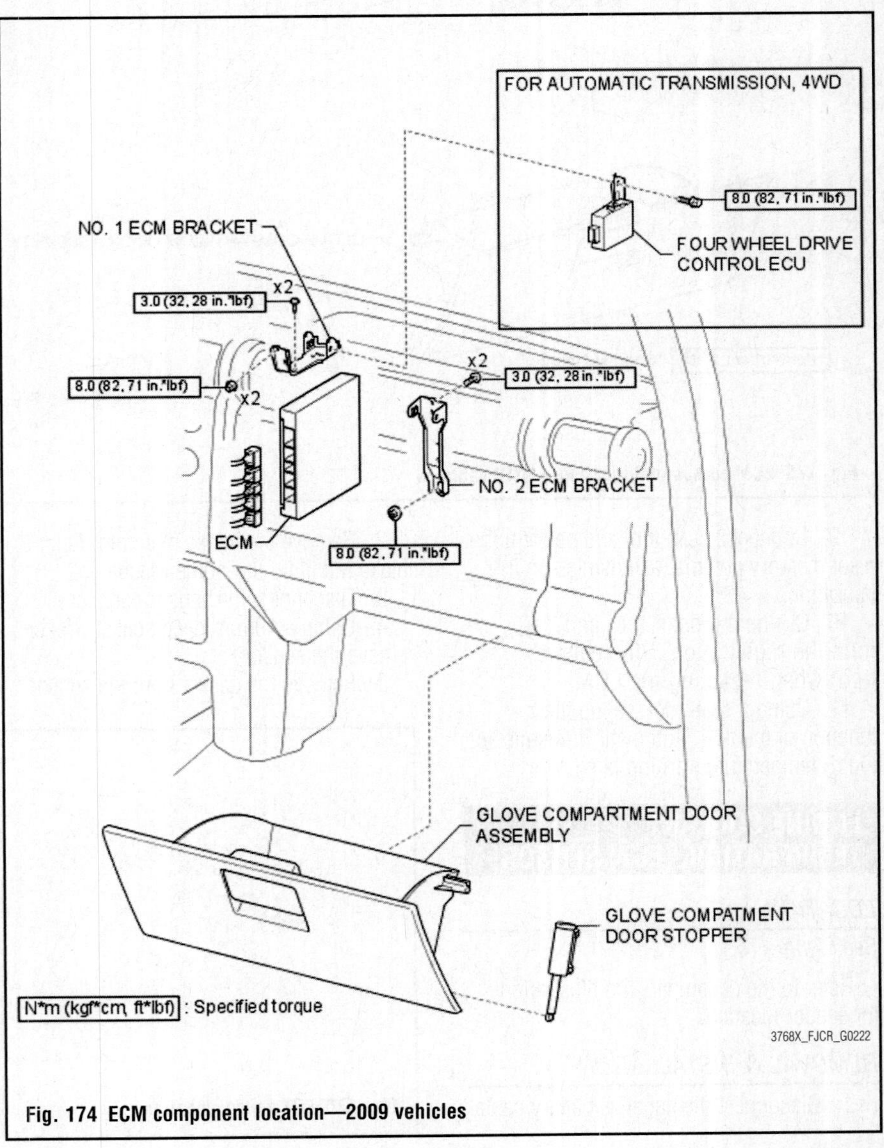

Fig. 174 ECM component location—2009 vehicles

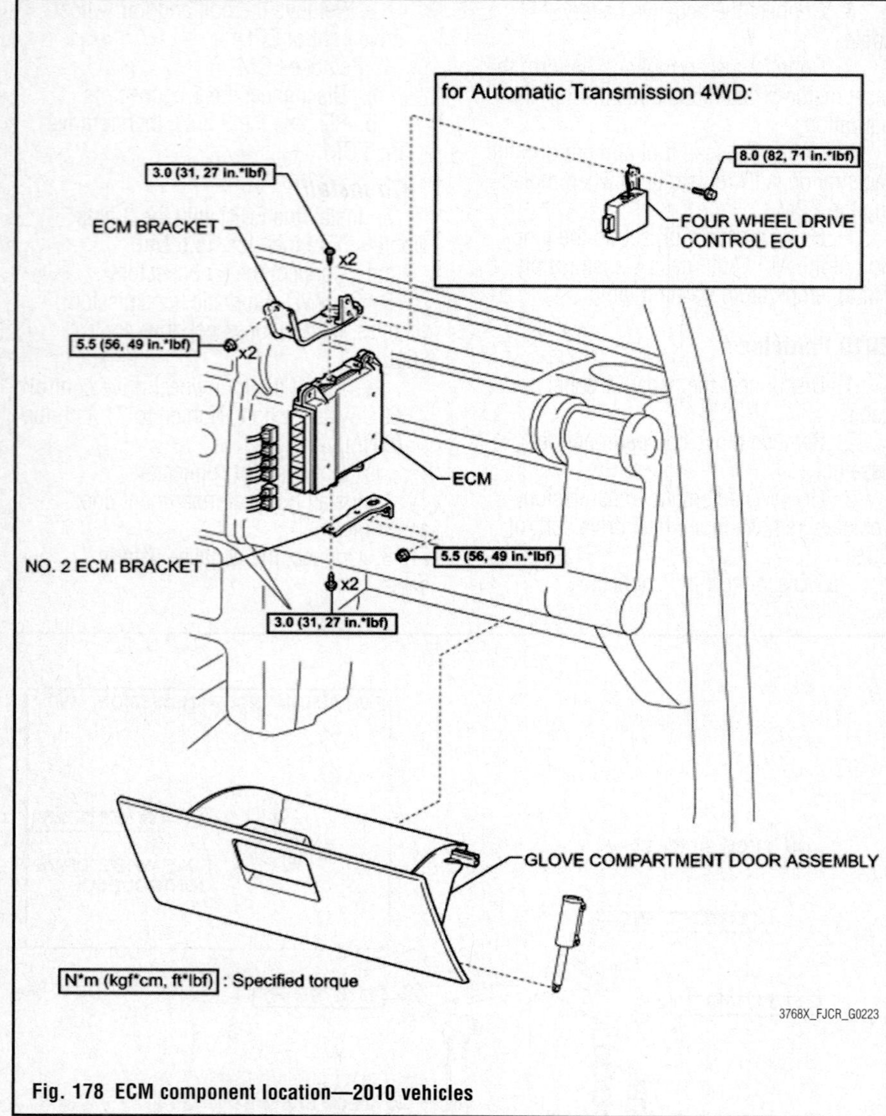

Fig. 178 ECM component location—2010 vehicles

9. Connect a scan tool and perform the reset memory automatic transmission initialization.
10. Connect a scan tool and perform the registration (VIN registration) when replacing the ECM.
11. Using a scan tool, set up the function of the ATF (Automatic Transmission Fluid) temperature warning lamp.

ENGINE COOLANT TEMPERATURE (ECT) SENSOR

LOCATION

See Figure 176.

Refer to the accompanying illustrations for sensor location.

REMOVAL & INSTALLATION

1. Disconnect the negative battery cable.
2. Remove the intake manifold. Refer to Intake Manifold in Engine Mechanical.
3. Disconnect the sensor connector.
4. Using a 19 mm deep socket wrench, remove the sensor.
5. Remove the gasket from the sensor.

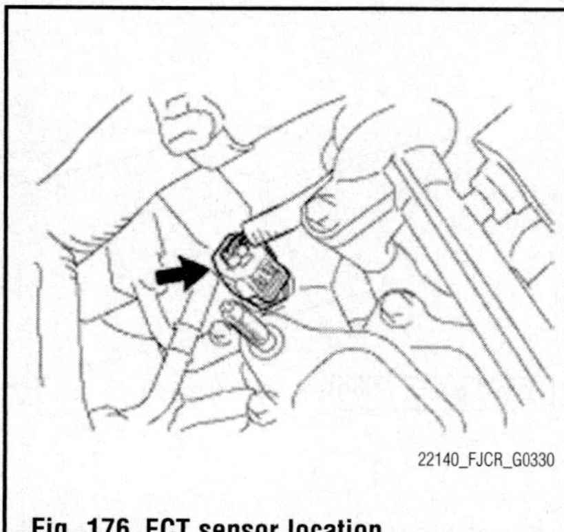

Fig. 176 ECT sensor location

EVAPORATIVE EMISSIONS (EVAP) CANISTER

LOCATION

See Figure 177.

Refer to the accompanying illustrations for EVAP canister location.

REMOVAL & INSTALLATION

See Figures 178 through 180.

CAUTION

Observe all applicable safety precautions when working around fuel. Whenever servicing the fuel system, always work in a well ventilated area. Do not allow fuel spray or vapors to come in contact with a spark or open flame. Keep a dry chemical fire extinguisher near the work area. Always keep fuel in a container specifically designed for fuel storage; also, always properly seal fuel containers to avoid the possibility of fire or explosion.

➡Note the following:

- Remove any dirt and foreign objects from the fuel tank vent hose connector before performing the work.
- Do not allow any scratches or foreign objects on the parts when disconnecting, as the fuel tank vent hose connector has the O-ring that seals the pipe.
- Perform the work by hand. Do not use any tools.
- Do not forcibly bend, twist or turn the nylon tube.
- Protect the disconnected part by covering it with a vinyl bag after disconnecting the fuel tank vent hose.
- If the fuel vent connector and pipe are stuck, push and pull them to release.

1. Properly relieve the fuel system pressure.
2. Remove charcoal canister assembly:
   a. Disconnect the fuel tank breather tube from the fuel tank.
   b. Disconnect the No. 2 canister outlet hose sub-assembly.
3. Disconnect the connector.
4. Disconnect the EVAP hose.
5. Remove the 3 bolts, then remove the charcoal canister assembly.

To install:

6. Install the canister with the 3 bolts. Tighten to 15 ft. lbs. (20 Nm).

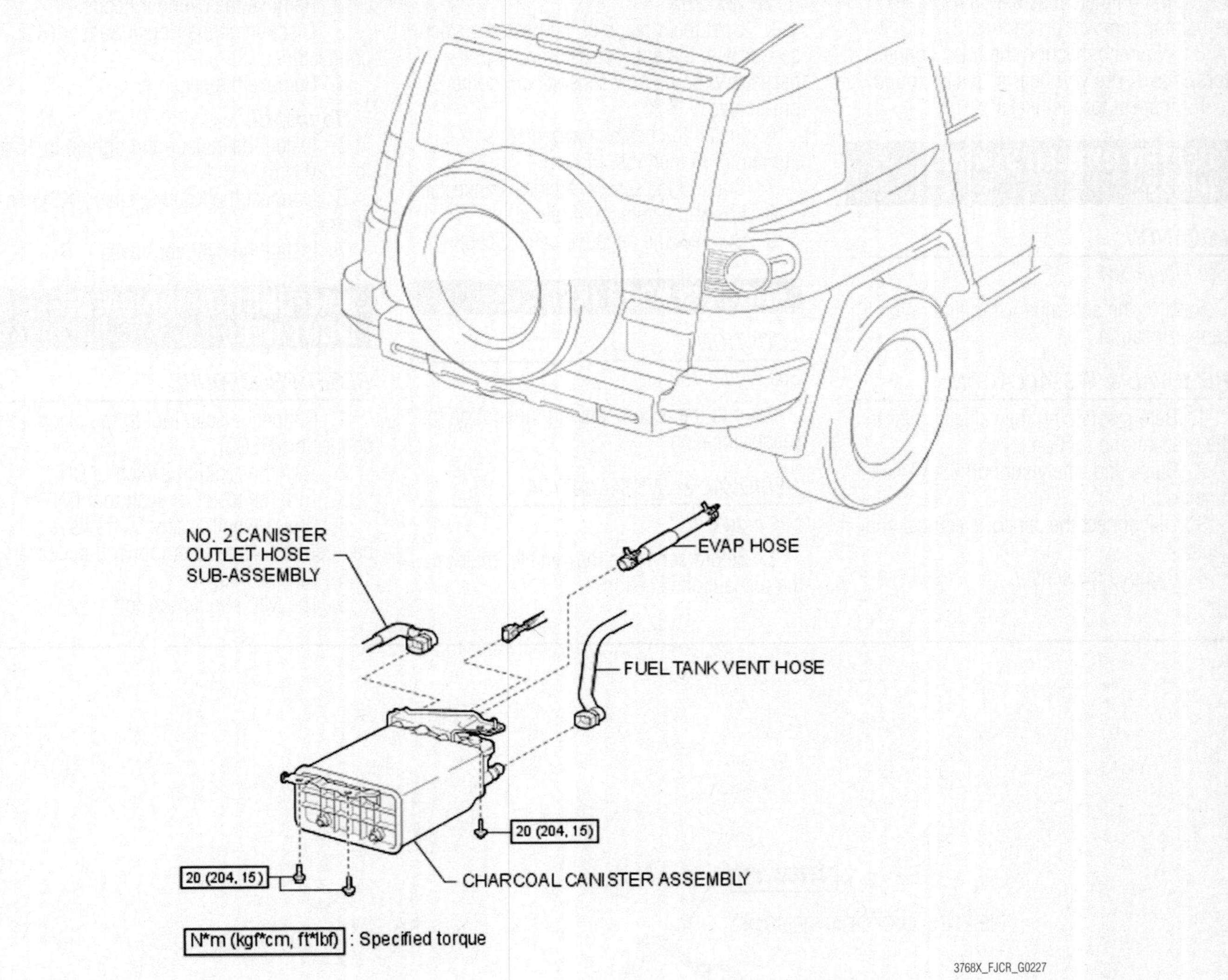

Fig. 180 EVAP canister component location

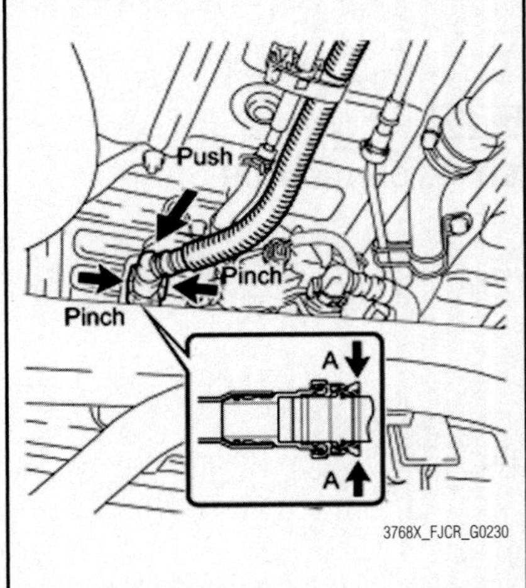

Fig. 178 Disconnect the fuel tank breather tube—Push the connector deep inside, pinch portion A as shown, pull out the connector

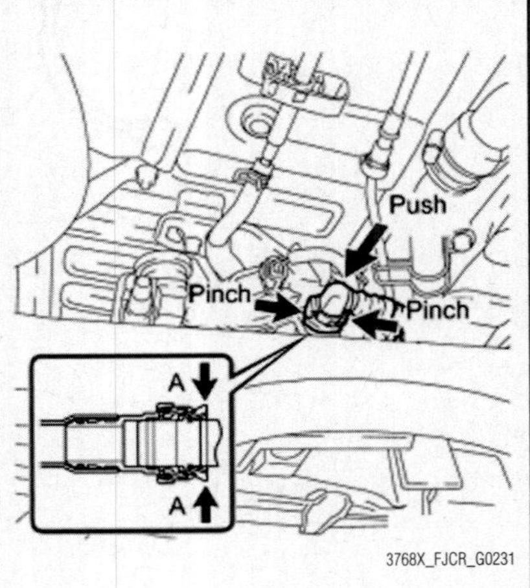

Fig. 179 Disconnect the No. 2 canister outlet hose—Push the connector deep inside, pinch portion A as shown, pull out the connector

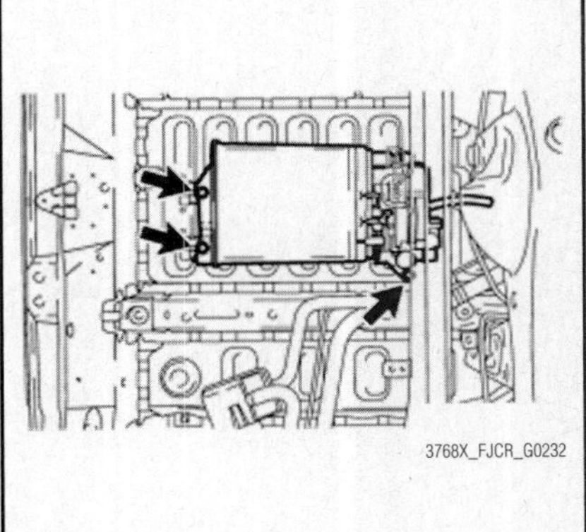

Fig. 180 Remove the 3 bolts, then remove the charcoal canister assembly

7. To complete installation, reverse remaining removal procedure.

8. When connecting the hose connectors, make sure you hear a "click" sound.

9. Inspect for fuel leaks.

HEATED OXYGEN SENSOR (HO2S)

LOCATION

See Figure 181.

Refer to the accompanying illustration for sensor location.

REMOVAL & INSTALLATION

1. Before servicing the vehicle, refer to the precautions section.
2. Raise and safely support the vehicle, if necessary.
3. Disconnect the sensor electrical connector.
4. Remove the sensor.

To install:

5. Coat the threads of the heated oxygen sensor with the anti-seize compound, if necessary. Some new sensors will come pre-coated.
6. Install the heated oxygen sensor tighten to 35 inch lbs. (4 Nm).
7. Connect the sensor electrical connector.
8. Lower the vehicle, if raised.
9. Connect the negative battery cable.

KNOCK SENSOR (KS)

LOCATION

See Figure 182.

Refer to the accompanying illustration for sensor location.

REMOVAL & INSTALLATION

See Figure 183.

1. Before servicing the vehicle, refer to the precautions section.
2. Remove the cylinder heads.
3. Disconnect the Knock Sensor (KS) connectors.
4. Remove the sensors.

To install:

5. Install the sensor and tighten to 15 ft. lbs. (20 Nm).
6. Connect the Knock Sensor (KS) connector.
7. Install the cylinder heads.

MALFUNCTION INDICATOR LIGHT (MIL)

RESET PROCEDURE

1. Connect a scan tool to the diagnostic connector (DLC3).
2. Turn the ignition switch to ON.
3. Turn the tester or scan tool ON
4. Check whether any DTCs have been stored. Note them down if necessary.
5. Clear DTCs.
6. The MIL should turn off.

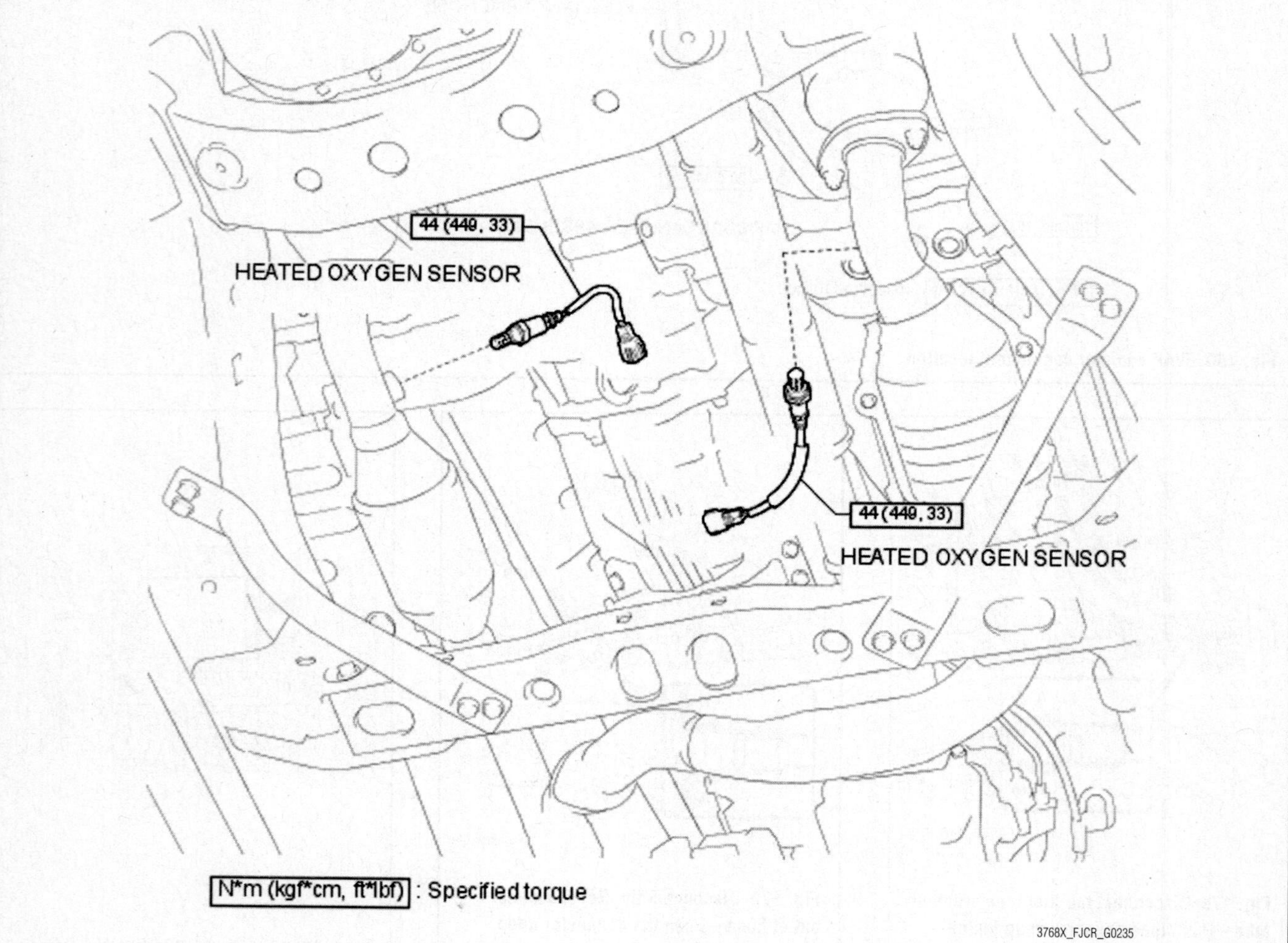

Fig. 181 Heated oxygen sensor component location

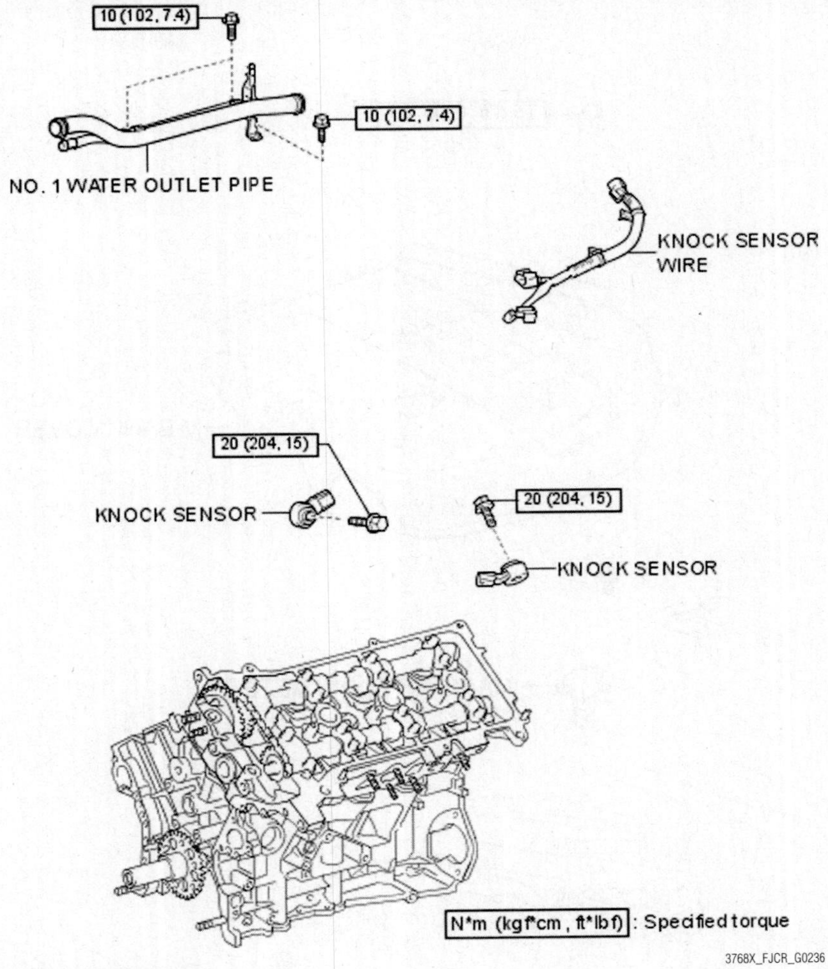

Fig. 182 Knock Sensor (KS) component location

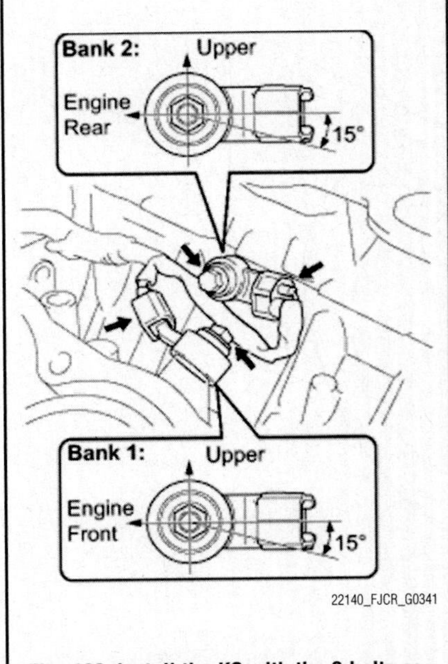

Fig. 183 Install the KS with the 2 bolts as shown

MASS AIR FLOW (MAF) METER

LOCATION

See Figures 184 and 185.

Refer to the accompanying illustrations for MAF meter location.

REMOVAL & INSTALLATION

1. Before servicing the vehicle, refer to the precautions section.
2. Disconnect the negative battery cable.
3. Remove V-bank cover.
4. Remove Mass Air Flow (MAF) meter.
   a. Disconnect the connector.
   b. Remove the 2 screws, then remove the MAF meter.

To install:

5. Install MAF meter.
6. Connect the connector.
7. Install V-bank cover.
8. Connect the negative battery cable.

POSITIVE CRANKCASE VENTILATION (PCV) VALVE

LOCATION

See Figures 186 and 187.

Refer to the accompanying illustrations for PCV valve location.

REMOVAL & INSTALLATION

1. On 2009 vehicles:
   a. Disconnect the ventilation hose.
   b. Remove the ventilation valve.
2. On 2010 vehicles:
   a. Disconnect the heater hose clamp.
   b. Disconnect the ventilation hose from the ventilation valve.
   c. Remove the ventilation valve.

To install:

3. To install, reverse removal procedure.
4. Apply adhesive to 2 or 3 threads of the ventilation valve. Use Toyota Genuine Adhesive 1324, Three Bond 1324 or equivalent
5. Tighten the valve to 20 ft. lbs. (27 Nm).

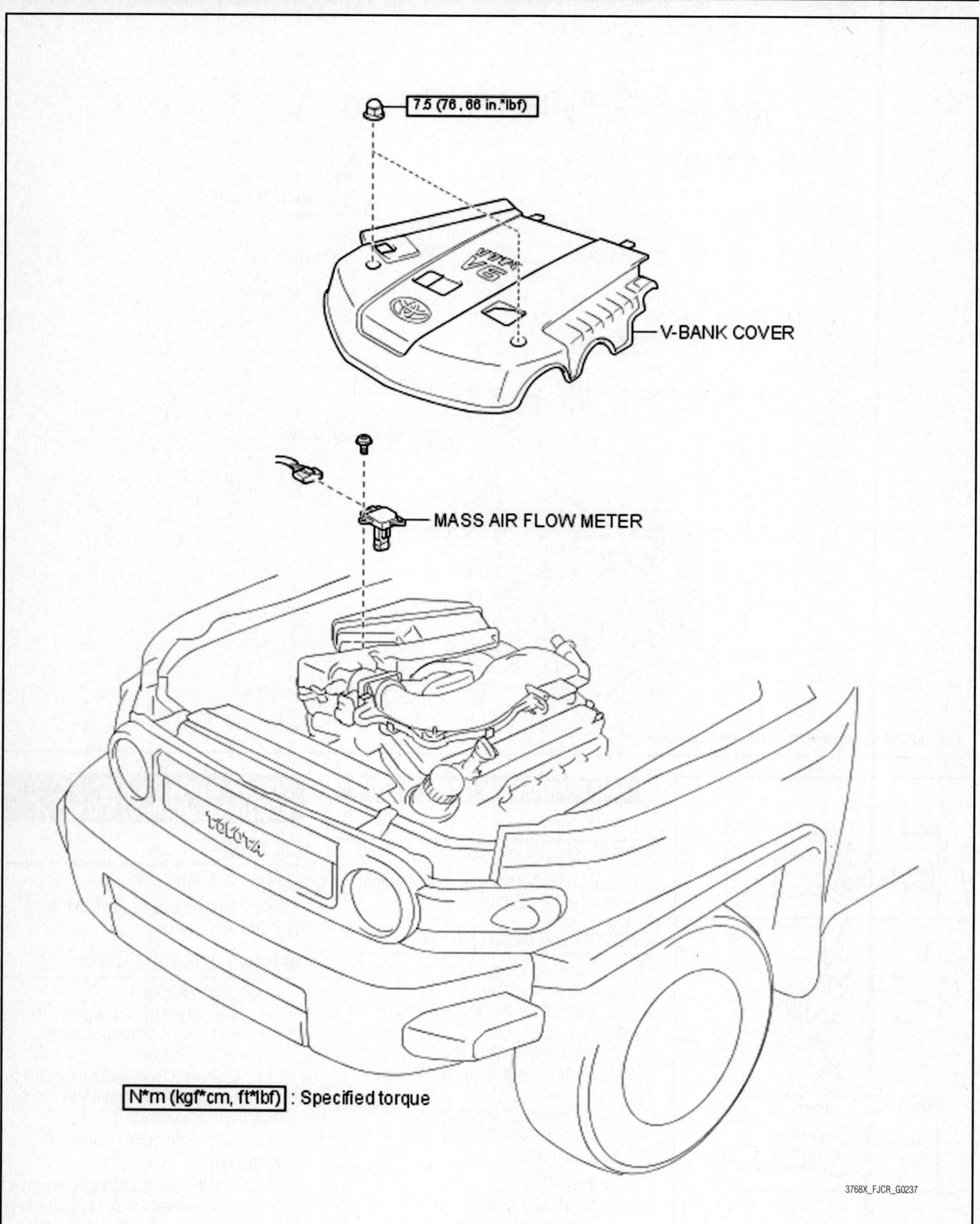

Fig. 184 MAF meter component location—2009 vehicles

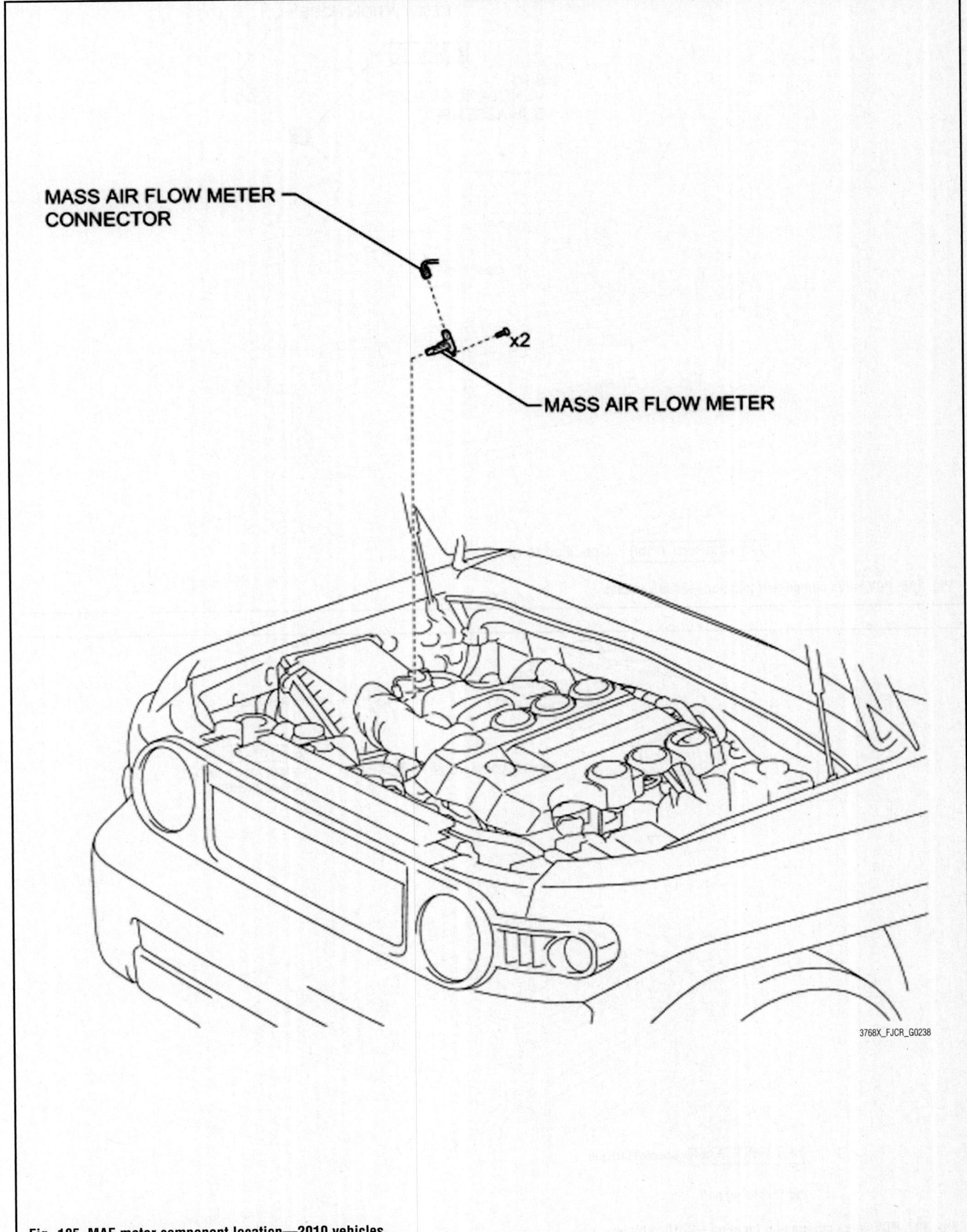

Fig. 185 MAF meter component location—2010 vehicles

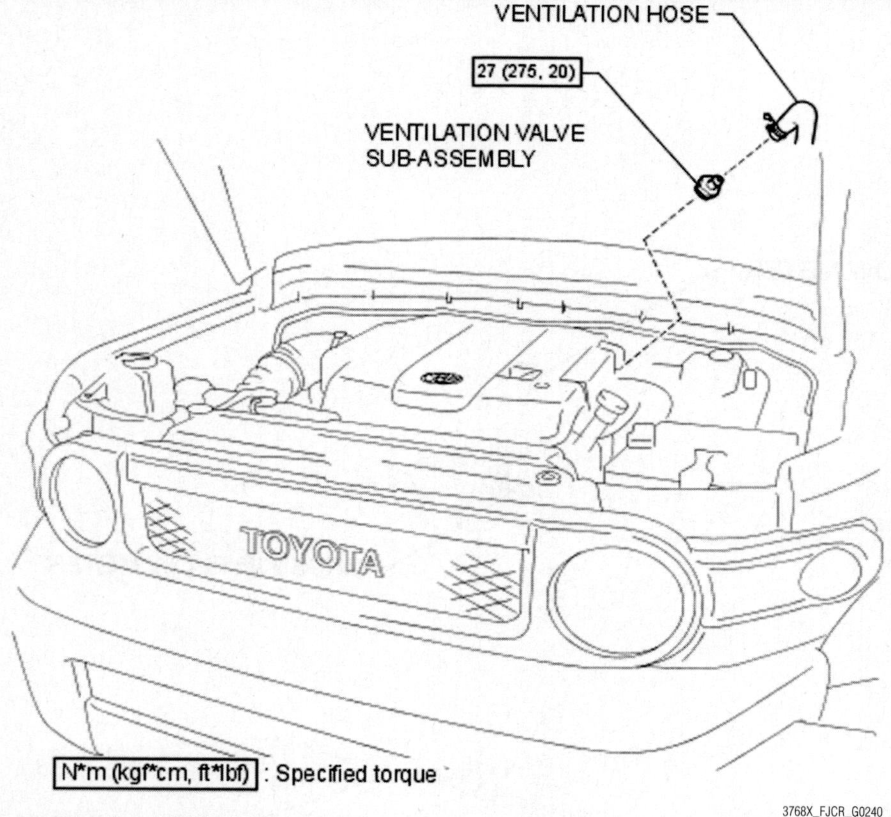

Fig. 186 PCV valve component location—2009 vehicles

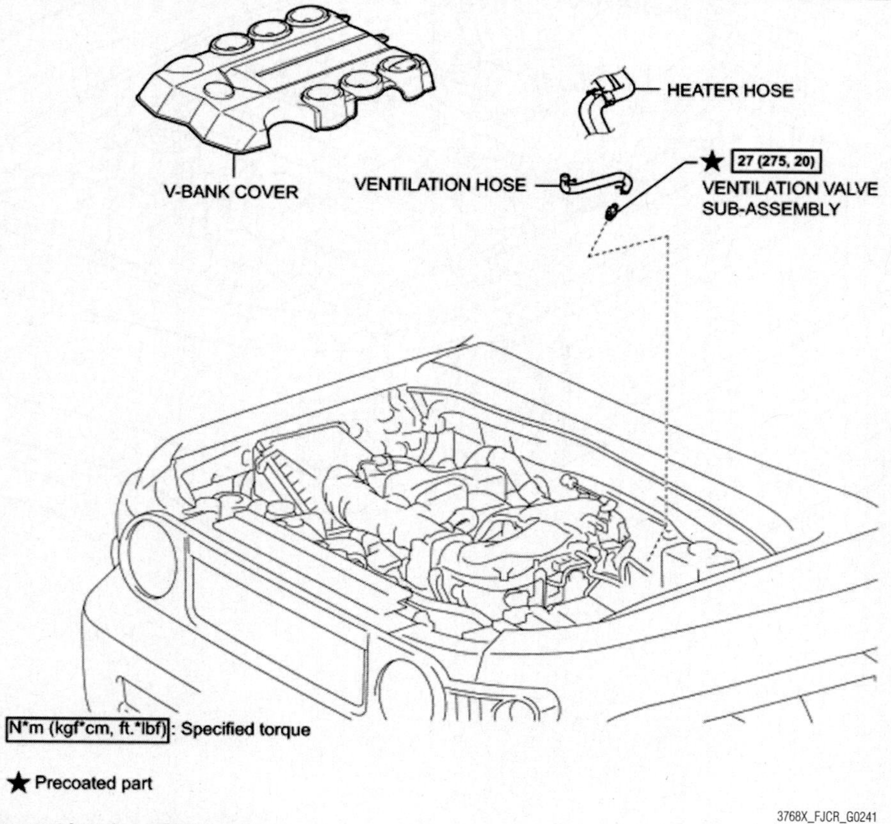

Fig. 187 PCV valve component location—2010 vehicles

VEHICLE SPEED SENSOR (VSS)

LOCATION

See Figures 188 and 189.

Refer to the accompanying illustrations for sensor location.

REMOVAL & INSTALLATION

1. Disconnect the negative battery cable.
2. Disconnect the 2 sensor connectors.
3. Remove the 2 bolts and transmission revolution sensors.
4. Remove the O-ring from each sensor.
5. To install, reverse removal procedure.
6. Coat 2 new O-rings with Toyota Genuine ATF WS or equivalent and install one onto each transmission revolution sensor.
7. Tighten the bolts to 48 inch lbs. (5.4 Nm).

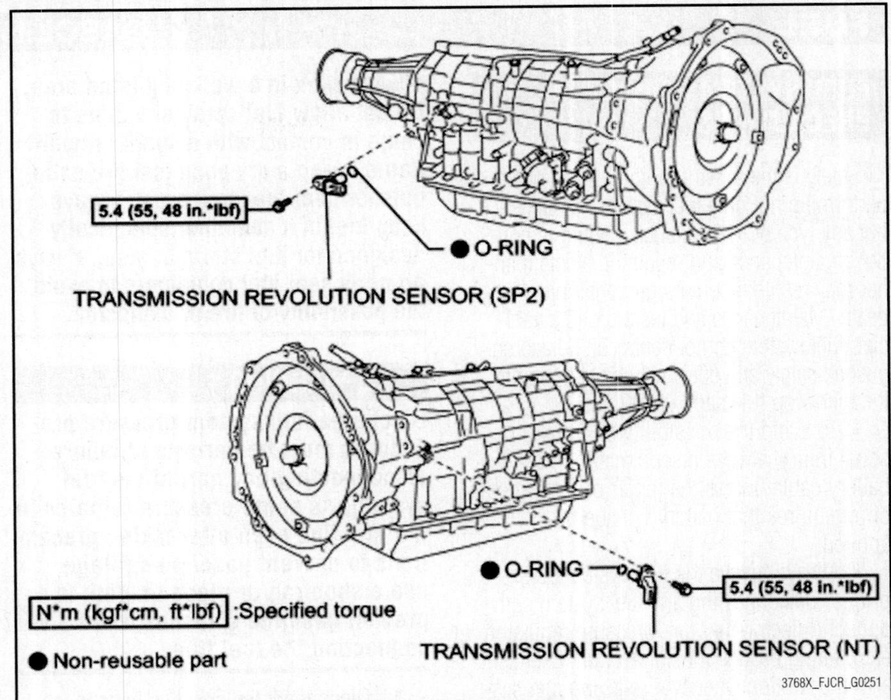

Fig. 188 Vehicle Speed Sensor (VSS) location—A750E A/T transmission

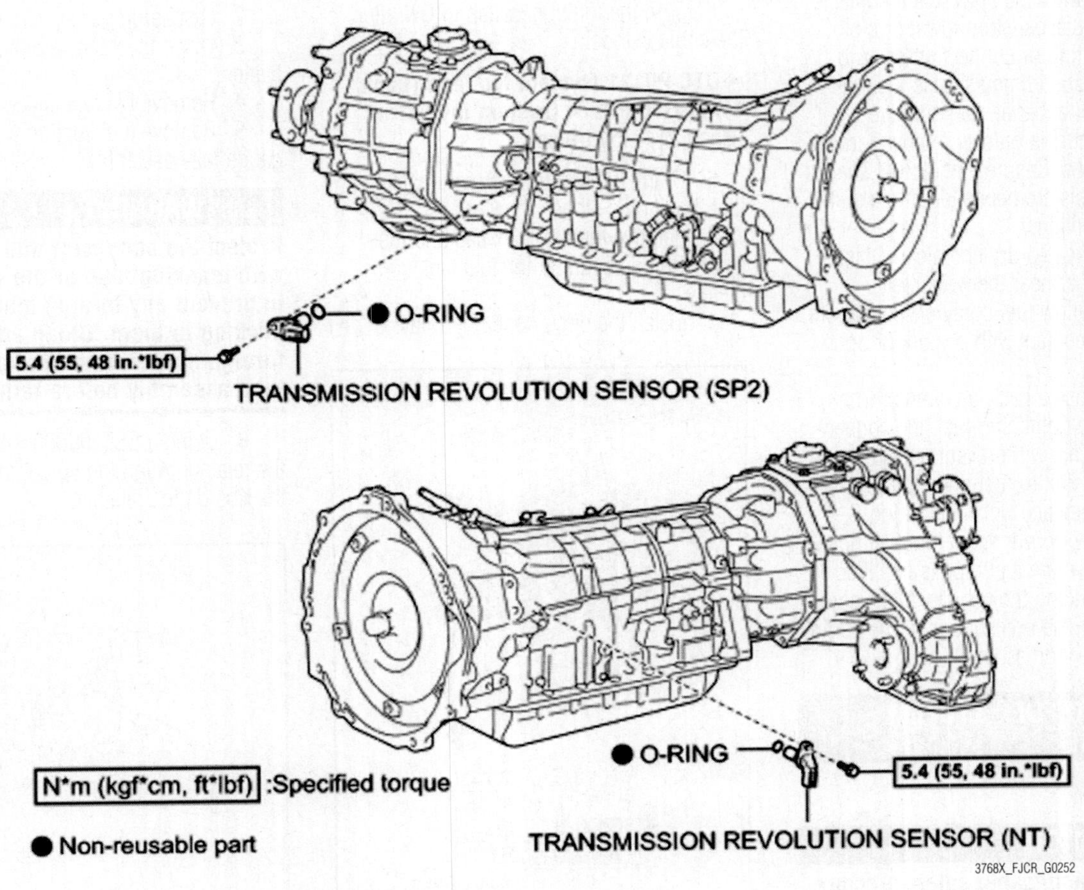

Fig. 189 Vehicle Speed Sensor (VSS) location—A750EF A/T transmission

FUEL SYSTEM SERVICE PRECAUTIONS

Safety is the most important factor when performing not only fuel system maintenance but any type of maintenance. Failure to conduct maintenance and repairs in a safe manner may result in serious personal injury or death. Maintenance and testing of the vehicle's fuel system components can be accomplished safely and effectively by adhering to the following rules and guidelines.

- To avoid the possibility of fire and personal injury, always disconnect the negative battery cable unless the repair or test procedure requires that battery voltage be applied.
- Always relieve the fuel system pressure prior to disconnecting any fuel system component (injector, fuel rail, pressure regulator, etc.), fitting or fuel line connection. Exercise extreme caution whenever relieving fuel system pressure to avoid exposing skin, face and eyes to fuel spray. Please be advised that fuel under pressure may penetrate the skin or any part of the body that it contacts.
- Always place a shop towel or cloth around the fitting or connection prior to loosening to absorb any excess fuel due to spillage. Ensure that all fuel spillage (should it occur) is quickly removed from engine surfaces. Ensure that all fuel soaked cloths or towels are deposited into a suitable waste container.
- Always keep a dry chemical (Class B) fire extinguisher near the work area.
- Do not allow fuel spray or fuel vapors to come into contact with a spark or open flame.
- Always use a back-up wrench when loosening and tightening fuel line connection fittings. This will prevent unnecessary stress and torsion to fuel line piping.
- Always replace worn fuel fitting O-rings with new Do not substitute fuel hose or equivalent where fuel pipe is installed.

Before servicing the vehicle, make sure to also refer to the precautions in the beginning of this section as well.

RELIEVING FUEL SYSTEM PRESSURE

See Figure 190.

** CAUTION

Observe all applicable safety precautions when working around fuel. Whenever servicing the fuel system, always work in a well ventilated area. Do not allow fuel spray or vapors to come in contact with a spark or open flame. Keep a dry chemical fire extinguisher near the work area. Always keep fuel in a container specifically designed for fuel storage; also, always properly seal fuel containers to avoid the possibility of fire or explosion.

** WARNING

Discharge fuel system pressure procedures must be performed before disconnecting any part of the fuel system. As some pressure remains in the fuel line even after taking precautions to prevent gasoline spillage, use a shop rag or piece of cloth to prevent gasoline splashes when disconnecting the fuel line.

1. Disconnect the negative battery cable.
2. Remove the circuit opening relay.
3. Connect the negative battery cable.
4. Start the engine.
5. Turn the ignition switch to **ON** after the engine stops.

➡DTC P0171 (system too lean (Bank 1)) or DTC P0174 (system too learn (Bank 2)) may be set.

6. Crank the engine again and check that the engine stops.
7. Remove the fuel tank cap and discharge the pressure in the fuel tank completely.
8. Install the circuit opening relay.

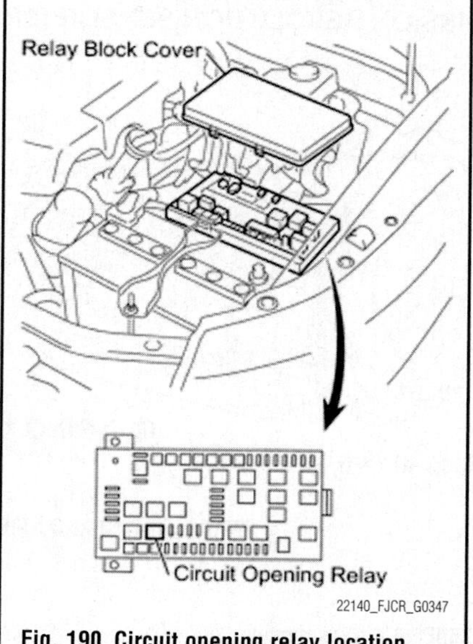

Fig. 190 Circuit opening relay location

FUEL FILTER

REMOVAL & INSTALLATION

See Figures 191 through 194.

** CAUTION

Observe all applicable safety precautions when working around fuel. Whenever servicing the fuel system, always work in a well ventilated area. Do not allow fuel spray or vapors to come in contact with a spark or open flame. Keep a dry chemical fire extinguisher near the work area. Always keep fuel in a container specifically designed for fuel storage; also, always properly seal fuel containers to avoid the possibility of fire or explosion.

➡The fuel filter is a integral part of the fuel pump unit and is not normally serviced.

1. Before servicing the vehicle, refer to the precautions.
2. Discharge fuel system pressure.
3. Disconnect the negative battery cable.
4. Remove the fuel tank.
5. Remove fuel suction with pump and gauge tube assembly.

** WARNING

Protect the connector and tube joint with masking tape or the equivalent to prevent any foreign matter from sticking to them. Clean any dirt and foreign matter from the fuel suction tube assembly before removing.

6. Using a SST: 09808-14020, loosen the retainer. Align the tips of the SST with the ribs on the retainer.

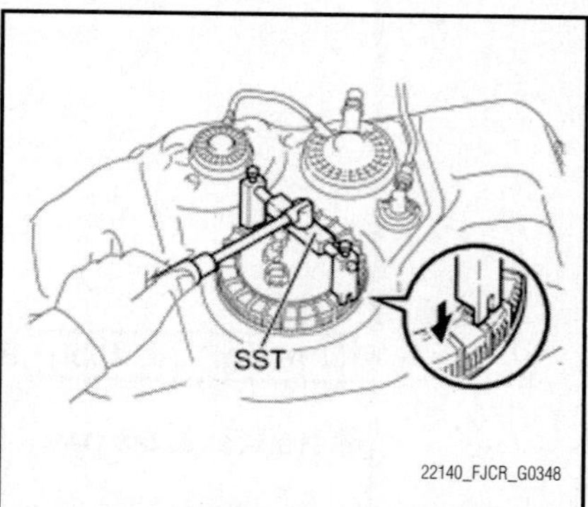

Fig. 191 Using a SST: 09808-14020, loosen the retainer

7. Remove the retainer.
8. Pull the fuel pump assembly out of the fuel tank.

⁂ WARNING

Do not bend the arm of the sender gauge.

9. Remove the gasket from the fuel tank.
10. Remove fuel sender gauge assembly.
    a. Disconnect the connector.
    b. Disengage the claw and remove the sender gauge by sliding it upward.
11. Remove No. 1 fuel sub-tank.
    a. Disengage the 5 claws and remove the fuel pump tank.
    b. Separate the connector and disengage the clamp.
12. Remove fuel pump assembly.
    a. Disengage the clamp, then disconnect the connector.
    b. Disengage the 5 claws and separate the fuel pump from the fuel pump case.
    c. Disconnect the connector from the fuel pump.
13. Remove the fuel filter from the fuel pump.

To install:

14. Install the fuel filter onto the fuel pump.
15. Install fuel pump assembly.
    a. Connect the fuel filter to the fuel pump.
    b. Engage the 5 claw fittings and install the fuel pump onto the fuel pump case.
    c. Engage the clamp, then connect the connector.

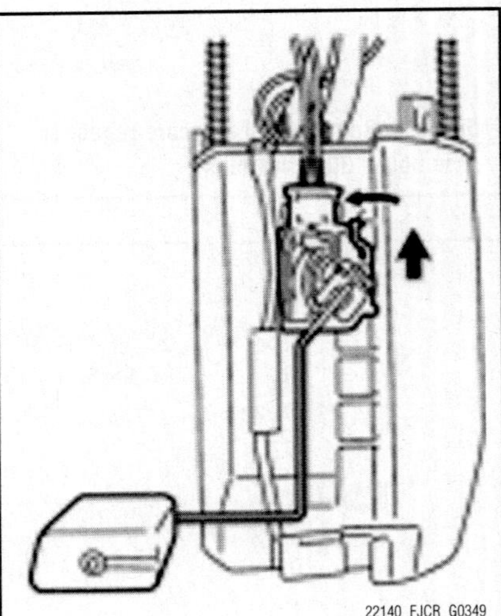

Fig. 192 Disengage the claw and remove the sender gauge by sliding it in the direction shown

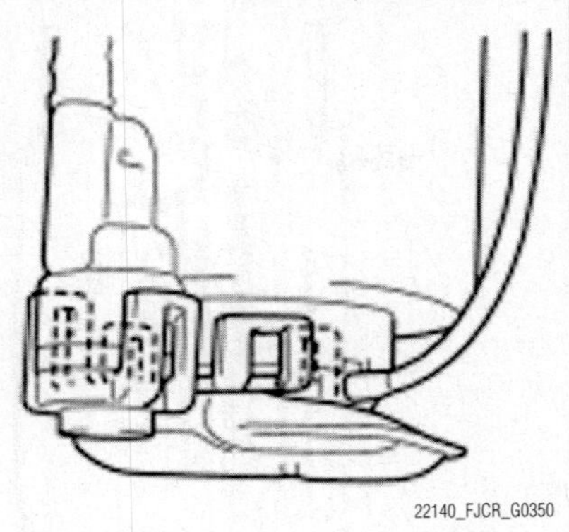

Fig. 193 Disengage the 5 claws and separate the fuel pump from the fuel pump case

16. Install No. 1 fuel sub-tank.
    a. Install the connector and engage the clamp.
    b. Engage the 5 claws and install the fuel pump tank.
17. Install fuel sender gauge assembly.
    a. Engage the claw and install the sender gauge by sliding it in the direction as shown in the illustration.
    b. Connect the connector.
18. Install the fuel tank.
19. Connect the negative battery cable.

FUEL LEVEL SENDING UNIT

REMOVAL & INSTALLATION

See Figure 195.

⁂ CAUTION

Observe all applicable safety precautions when working around fuel. Whenever servicing the fuel system, always work in a well ventilated area. Do not allow fuel spray or vapors to come in contact with a spark or open flame. Keep a dry chemical fire extinguisher near the work area. Always keep fuel in a container specifically designed for fuel storage; also, always properly seal fuel containers to avoid the possibility of fire or explosion.

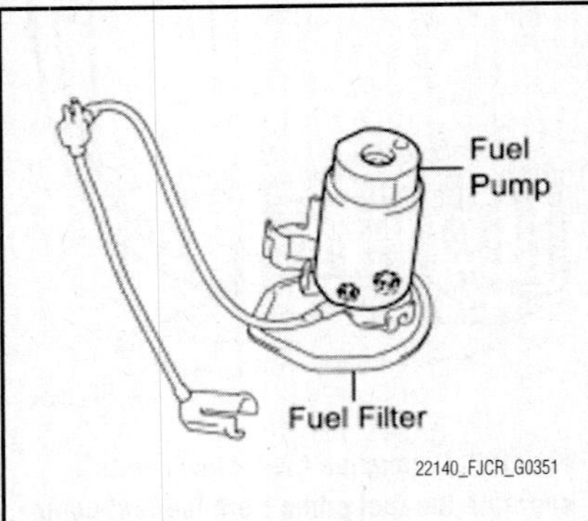

Fig. 194 Fuel filter location

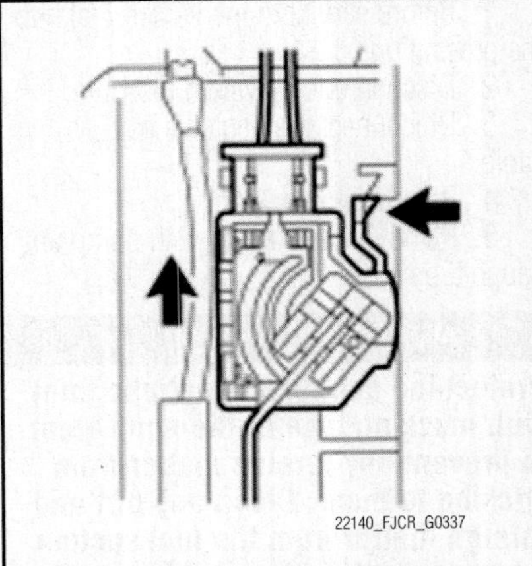

Fig. 195 Unlock the fuel sender gauge, and slide and remove it

1. Before servicing the vehicle, refer to the precautions.
2. Discharge fuel system pressure.
3. Disconnect the negative battery cable.
4. Remove the fuel tank.
5. Disconnect the fuel sender gauge connector.
6. Unlock the fuel sender gauge, and slide and remove it.

To install:

7. Install the fuel sender gauge onto the fuel suction with pump and gauge tube.
8. Connect the fuel sender gauge connector.
9. Install the fuel tank.
10. Connect the negative battery cable.

FUEL PUMP

REMOVAL & INSTALLATION

See Figures 196 through 198.

⁂ CAUTION

Observe all applicable safety precautions when working around fuel. Whenever servicing the fuel system, always work in a well ventilated area. Do not allow fuel spray or vapors to come in contact with a spark or open flame. Keep a dry chemical fire extinguisher near the work area. Always keep fuel in a container specifically designed for fuel storage; also, always properly seal fuel containers to avoid the possibility of fire or explosion.

1. Before servicing the vehicle, refer to the precautions.
2. Discharge fuel system pressure.
3. Disconnect the negative battery cable.
4. Remove the fuel tank.
5. Remove fuel suction with pump and gauge tube assembly.

⁂ WARNING

Protect the connector and tube joint with masking tape or the equivalent to prevent any foreign matter from sticking to them. Clean any dirt and foreign matter from the fuel suction tube assembly before removing.

6. Using a SST: 09808-14020, loosen the retainer. Align the tips of the SST with the ribs on the retainer.
7. Remove the retainer.
8. Pull the fuel pump assembly out of the fuel tank.

⁂ WARNING

Do not bend the arm of the sender gauge.

9. Remove the gasket from the fuel tank.
10. Remove fuel sender gauge assembly.
    a. Disconnect the connector.
    b. Disengage the claw and remove the sender gauge by sliding it upward.
11. Remove No. 1 fuel sub-tank.
    a. Disengage the 5 claws and remove the fuel pump tank.
    b. Separate the connector and disengage the clamp.
12. Remove fuel pump assembly.
    a. Disengage the clamp, then disconnect the connector.
    b. Disengage the 5 claws and separate the fuel pump from the fuel pump case.
    c. Disconnect the connector from the fuel pump.

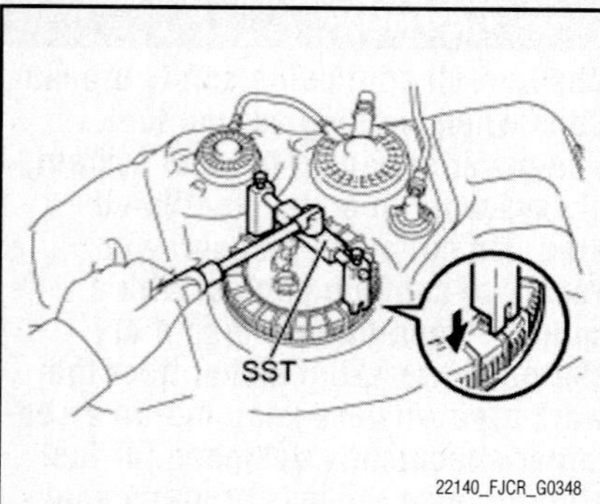

Fig. 196 Using a SST: 09808-14020, loosen the retainer

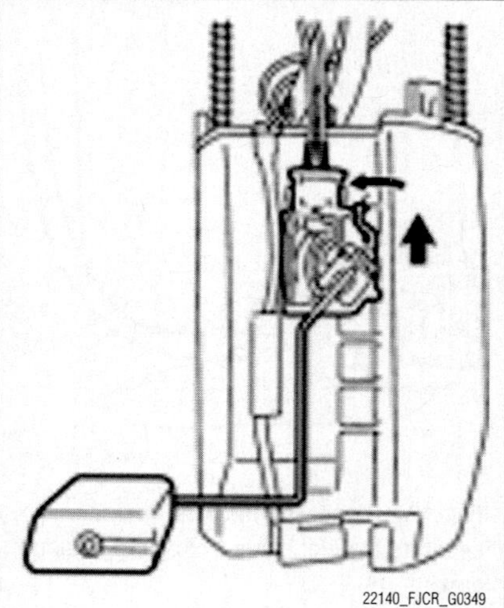

Fig. 197 Disengage the claw and remove the sender gauge by sliding it in the direction shown

To install:

13. Install fuel pump assembly.
    a. Connect the fuel filter to the fuel pump.
    b. Engage the 5 claw fittings and install the fuel pump onto the fuel pump case.
    c. Engage the clamp, then connect the connector.
14. Install No. 1 fuel sub-tank.
    a. Install the connector and engage the clamp.
    b. Engage the 5 claws and install the fuel pump tank.
15. Install fuel sender gauge assembly.
    a. Engage the claw and install the sender gauge by sliding it in the direction as shown in the illustration.
    b. Connect the connector.

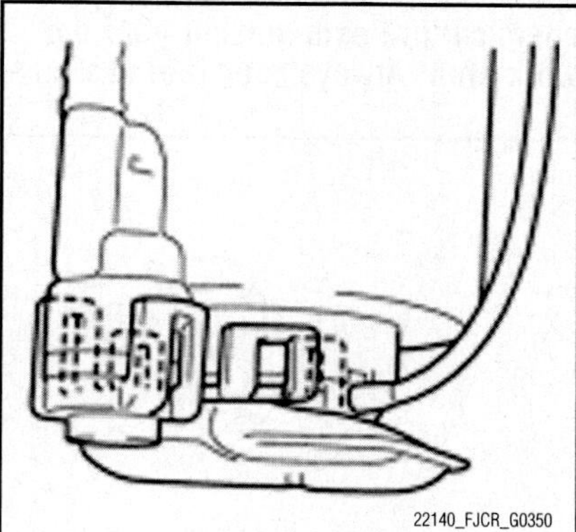

Fig. 198 Disengage the 5 claws and separate the fuel pump from the fuel pump case

16. Install the fuel tank.
17. Connect the negative battery cable.

FUEL PRESSURE REGULATOR

REMOVAL & INSTALLATION

See Figures 199 and 200.

⁂ CAUTION

Observe all applicable safety precautions when working around fuel. Whenever servicing the fuel system, always work in a well ventilated area. Do not allow fuel spray or vapors to come in contact with a spark or open flame. Keep a dry chemical fire extinguisher near the work area. Always keep fuel in a container specifically designed for fuel storage; also, always properly seal fuel containers to avoid the possibility of fire or explosion.

1. Before servicing the vehicle, refer to the precautions.
2. Discharge fuel system pressure.
3. Disconnect the negative battery cable.
4. Remove v-bank cover.
5. For 2009 vehicles, remove air cleaner assembly.

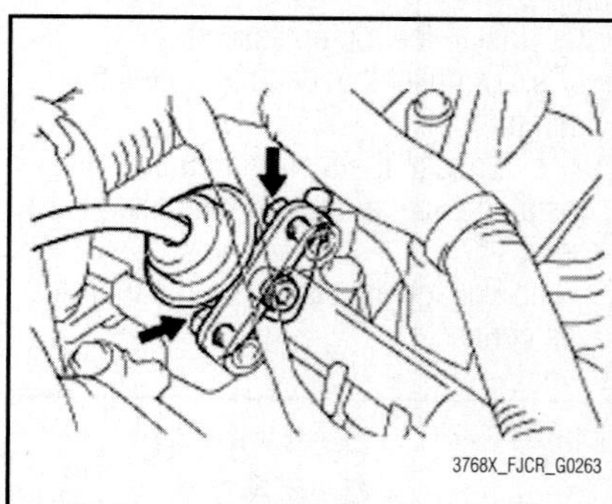

Fig. 199 Remove fuel pressure regulator assembly—2009 vehicles

Fig. 200 Remove fuel pressure regulator assembly—2010 vehicles

6. For 2010 vehicles, remove the intake air surge tank.

7. Disconnect no. 2 fuel pipe sub-assembly.

8. Remove fuel pressure regulator assembly.

a. Remove the vacuum hose.

b. Remove the 2 bolts, then remove the fuel pressure regulator.

c. If necessary, remove the O-ring from the fuel pressure regulator.

To install:

9. Apply a light coat of spindle oil or gasoline to a new O-ring and install it onto the fuel pressure regulator.

10. To complete installation, reverse remaining removal procedure.

11. Check for fluid leaks.

FUEL RAIL AND INJECTOR

REMOVAL & INSTALLATION

2009 Vehicles

See Figures 201 and 202.

CAUTION

Observe all applicable safety precautions when working around fuel. Whenever servicing the fuel system, always work in a well ventilated area. Do not allow fuel spray or vapors to come in contact with a spark or open flame. Keep a dry chemical fire extinguisher near the work area. Always keep fuel in a container specifically designed for fuel storage; also, always properly seal fuel containers to avoid the possibility of fire or explosion.

1. Before servicing the vehicle, refer to the precautions.

2. Remove air cleaner assembly.

a. Remove the engine V-bank cover by removing the two acorn nuts.

b. Disconnect the ventilation hose No. 2.

c. Disconnect the vacuum hose.

d. Disconnect the mass air flow meter connector. Remove the 2 wire harness clamps.

e. Loosen the 2 hose clamps. Remove the 2 bolts, then remove the air cleaner.

3. Remove throttle body with motor body assembly.

a. Disconnect the throttle body connector.

b. Remove the 4 bolts, and then remove the throttle w/ motor body and gasket and set it aside

➡Leave the throttle body connected to the coolant hoses.

4. Remove intake air surge tank.

a. Disconnect the fuel vapor feed hose.

b. Disconnect the ventilation hose.

c. Disconnect the 2 VSV connectors.

d. Remove the 3 wire harness clamps and hose clamps.

➡Removing the No. 1 VSV at the rear passenger side of the surge tank makes it easier to get to the clamps.

e. Remove the 3 upper bolts which are used to secure the 2 surge tank stays and throttle body bracket.

f. Loosen the 2 bolts that hold the brackets to the side of the cylinder head and remove the brackets.

WARNING

Before proceeding, make sure there is no dirt or debris on or around the base of the surge tank. If there is, you must remove it so that it will not enter the engine when the surge tank is removed.

g. Using an 8mm Allen socket, remove the 4 bolts and the 2 nuts with a 12mm socket. Remove the surge tank and gasket.

5. Disconnect fuel pipes.

a. Remove the fuel pipe clamp.

b. Pinch the tube connector, and then pull the fuel pipe out of the delivery pipe as shown in the illustration.

6. Remove fuel delivery pipe assembly.

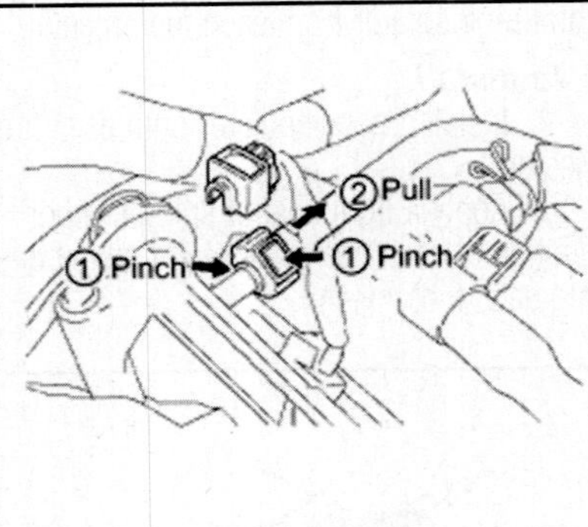

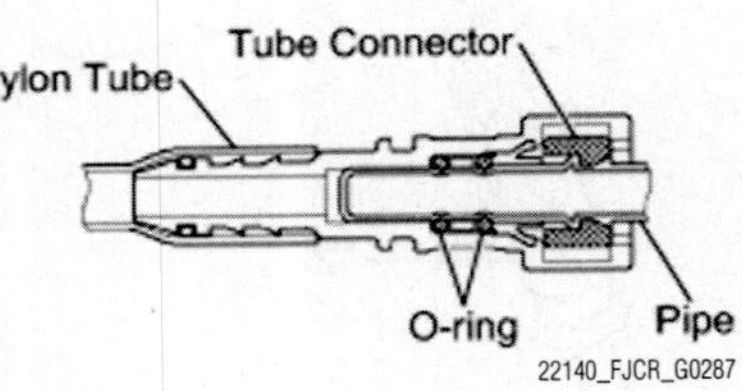

Fig. 201 Pinch the tube connector, and then pull the fuel pipe out of the delivery pipe as shown

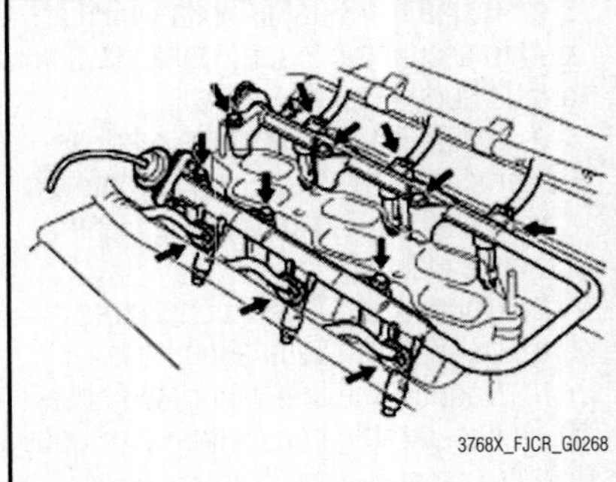

Fig. 202 Locating the 6 fuel injector connectors

a. Disconnect the 6 fuel injector connectors.

b. Remove the 6 bolts and remove the fuel delivery pipe together with the 6 fuel injectors.

WARNING

Do not drop the injectors when removing the fuel delivery pipe.

To install:

7. Install fuel delivery pipe assembly.

a. Place the fuel delivery pipe together with the 6 fuel injectors on the intake manifold.

b. Temporarily install the 6 bolts, which are used to hold the fuel delivery pipe, onto the intake manifold.

c. Check that the fuel injectors rotate smoothly. If the fuel injectors do not rotate smoothly, replace the O-ring.

d. Position each fuel injector connector facing outward.

e. Tighten the 6 fuel delivery pipe attaching bolts to 11 ft. lbs (15 Nm).

f. Connect the 6 fuel injector connectors.

8. Connect fuel pipes.

a. Push the tube connector into the pipe until the tube connector makes a "click" sound.

➡Check if there is any damage or foreign objects on the connected part of the fuel pipe.

b. After connecting, check that the pipe and connector are securely connected by pulling them.

c. Install the fuel pipe clamp.

9. Install intake air surge tank.

a. Install the surge tank using a new gasket. Using an 8mm Allen socket, install the 4 bolts and the 2 nuts with a 12mm socket.

b. Install the brackets and tighten the 2 bolts that hold the brackets to the side of the cylinder head.

c. Install the 3 upper bolts which are used to secure the 2 surge tank stays and throttle body bracket.

d. Install No. 1 VSV at the rear passenger side of the surge tank if removed.

e. Install the 3 wire harness clamps and hose clamps.

f. Connect the 2 VSV connectors.

g. Connect the ventilation hose.

h. Connect the fuel vapor feed hose.

10. Install throttle body with motor body assembly.

a. Install the throttle body with motor using a new gasket Install the 4 bolts, and then and set it aside

b. Connect the throttle body connector.

11. Install air cleaner assembly.

a. Install the air cleaner and 2 bolts. Tighten the 2 hose clamps.

b. Connect the mass air flow meter connector. Install the 2 wire harness clamps.

c. Connect the vacuum hose.

d. Connect the ventilation hose No. 2.

12. Install the engine V-bank cover.

2010 Vehicles

See Figures 203 through 205.

CAUTION

Observe all applicable safety precautions when working around fuel. Whenever servicing the fuel system, always work in a well ventilated area. Do not allow fuel spray or vapors to come in contact with a spark or open flame. Keep a dry chemical fire extinguisher near the work area. Always keep fuel in a container specifically designed for fuel storage; also, always properly seal fuel containers to avoid the possibility of fire or explosion.

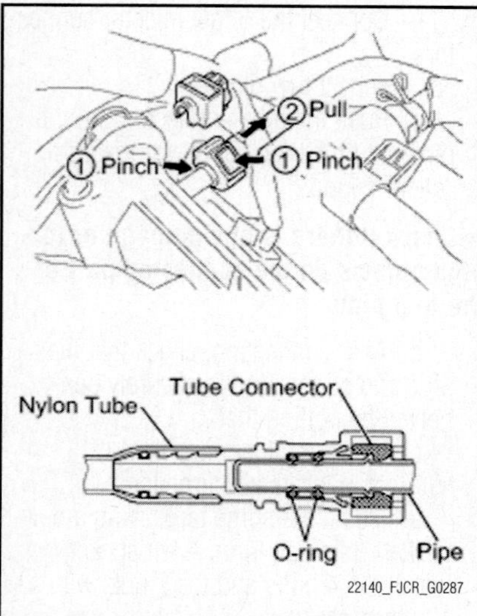

Fig. 203 Pinch the tube connector, and then pull the fuel pipe out of the delivery pipe as shown

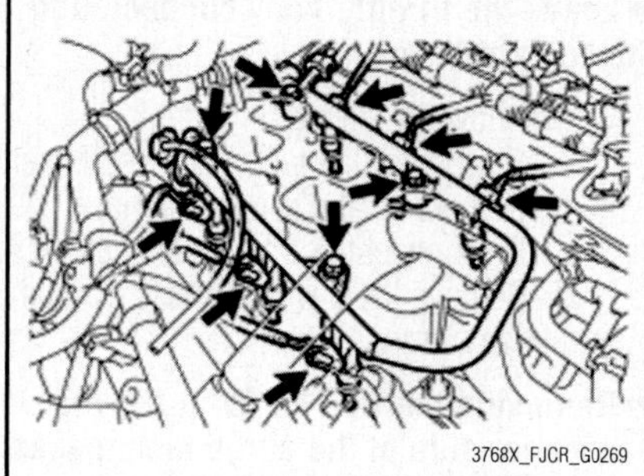

Fig. 204 Identifying the 4 bolts and fuel delivery pipe together with the 6 fuel injectors

1. Properly relieve the fuel system pressure.

2. Remove the intake air surge tank.

3. Remove the 3 bolts and rear cylinder head cover.

4. Disconnect fuel pipes.

a. Remove the fuel pipe clamp.

b. Pinch the tube connector, and then pull the fuel pipe out of the delivery pipe as shown in the illustration.

5. Disconnect the 6 fuel injector connectors.

6. Remove the 4 bolts and fuel delivery pipe together with the 6 fuel injectors.

7. Pull the 6 fuel injectors from the fuel delivery pipe.

8. Remove the O-ring and injector vibration insulator from each fuel injector.

To install:

9. Install a new insulator onto each fuel injector.

10. Apply a light coat of spindle oil or gasoline to each new O-ring and install one onto each fuel injector.

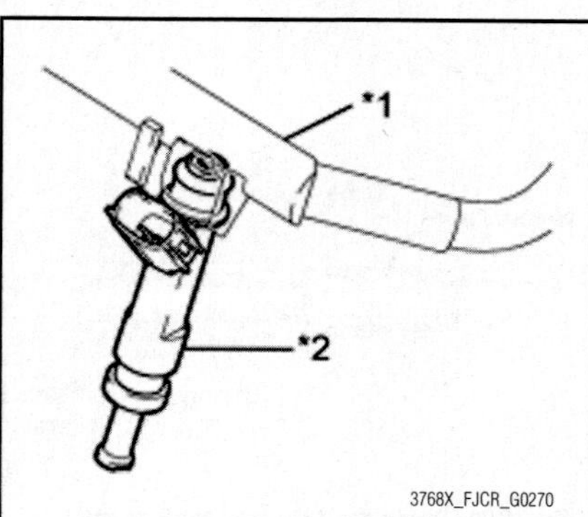

Fig. 205 Pulling the fuel injector from the fuel delivery pipe

11. Install the 6 injectors by twisting side to side and then pushing in.

12. While turning each fuel injector left and right, install it onto the fuel delivery pipe.

13. Position the fuel injector connectors facing outward.

14. Place the fuel delivery pipe together with the 6 fuel injectors on the intake manifold.

15. Temporarily install the 6 bolts, which are used to hold the fuel delivery pipe, to the intake manifold.

16. Check that the fuel injectors rotate smoothly.

➡If the fuel injectors do not rotate smoothly, replace the O-ring of any injector that does not rotate smoothly.

17. Position the fuel injector connectors facing outward.

18. Tighten the 6 bolts, which are used to hold the fuel delivery pipe, to the intake manifold. Tighten to 15 ft. lbs. (21 Nm).

19. Connect the 6 fuel injector connectors.

20. To complete the installation, reverse remaining removal procedure.

21. Inspect for fuel leaks.

FUEL TANK

REMOVAL & INSTALLATION

See Figures 206 and 207.

CAUTION

Observe all applicable safety precautions when working around fuel. Whenever servicing the fuel system, always work in a well ventilated area. Do not allow fuel spray or vapors to come in contact with a spark or open flame. Keep a dry chemical fire extinguisher near the work area. Always keep fuel in a container specifically designed for fuel storage; also, always properly seal fuel containers to avoid the possibility of fire or explosion.

1. Before servicing the vehicle, refer to the precautions.

2. Discharge fuel system pressure.

3. Disconnect the negative battery cable.

4. Remove left front door scuff plate.

5. Remove left rear door scuff plate.

6. Remove left rear seat cushion assembly.

7. Remove rear floor service hole cover.

a. Pull back the floor carpet.

b. Remove the left center floor silencer pad.

c. Remove the 3 screws from the service hole cover.

d. Disconnect the connector.

8. Remove fuel tank to filler pipe hose.

a. Loosen the clamp bolt, and disconnect the fuel tank to filler pipe hose.

9. Remove fuel tank breather tube.

a. Pinch the retainer to disengage the lock claws and pull out the No.1 fuel tank breather tube.

⁂ WARNING

Check that there is no dirt or mud around the quick connector before performing this work, because the quick connector has an O-ring which seals the pipe and the connector. Clean the connector if necessary.

⁂ WARNING

Do not use any tools in this work. Do not bend or twist the nylon tube. To protect the tube, cover it with a vinyl bag after disconnecting it. When the connector and the pipe are stuck, turn the retainer carefully to free and then disconnect the fuel tank tube.

10. Remove No. 1 fuel tank protector.

a. Remove the 2 nuts and the fuel tank protector bracket. (for automatic transmission)

b. Remove the nut and the fuel tank protector bracket. (for manual transmission)

c. Remove the 4 nuts and the fuel tank protector.

11. Disconnect fuel tank main tube and fuel tank return tube.

a. Pinch the retainer to disengage the lock claws and pull out the 2 fuel tank tubes.

12. Disconnect fuel tank vent hose.

a. Disconnect the fuel tank breather tube from the fuel tank.

b. Push the connector deep inside.

c. Pinch portion A, as shown in the illustration.

d. Pull out the connector.

13. Remove fuel tank assembly.

a. Hold the fuel tank using a mission jack.

b. Remove the 2 fuel tank bands.

c. Remove the 2 bolts.

d. Remove the 2 clips and 2 pins, then remove the 2 fuel tank bands.

14. Remove No. 3 fuel tank protector.

a. Remove the 2 bolts and disengage the 3 claws, then remove the No. 3 fuel tank protector.

15. Remove fuel tank cushion.

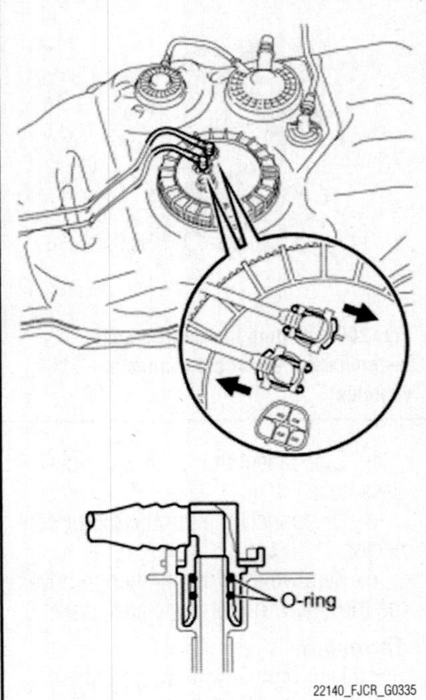

Fig. 206 Remove fuel tank main tube and fuel tank return tube connector at the fuel sender

16. Remove fuel tank main tube and fuel tank return tube.

a. Remove the 2 joint clips, and pull out the 2 fuel tank tubes.

➡After disconnecting, cover the fuel tube joint with a vinyl bag.

⁂ WARNING

When the fuel tube joint and fuel suction plate are stuck, turn the fuel tank main tube carefully to free and then disconnect it. Likewise, disconnect the fuel tank return tube.

17. Remove fuel suction with pump and gauge tube assembly.

18. Drain fuel.

19. Remove fuel tank breather tube.

a. Pinch the retainer to disengage the lock claws and pull out the fuel tank breather tube.

b. Loosen the clamp bolt, and disconnect the fuel tank to filler pipe hose.

To install:.

20. Install fuel tank to filler pipe hose.

a. Connect a new fuel tank to filler pipe hose, as shown in the illustration, with the clamp. Install fuel tank breather tube.

b. Push the tube connector into the pipe until the connector makes a "click" sound and install the retainer.

➡Check if there is any damage or foreign matter on the connected part of the pipe. After connecting, check if the pipe and the connector are securely connected by pulling them.

21. Install fuel suction with pump and gauge tube assembly.

22. Install fuel tank main tube and fuel tank return tube.

a. Install the fuel tank main tube and return tube with the 2 joint clips.

➡Check that there are no scratches or foreign matter around the connected part of the fuel tube joint and plug before performing this work. Check that the fuel tube joint is securely inserted into the end. Check that the tube joint clips are on the collar of the fuel tube joint. After installing the tube joint clip, check that the fuel tank main tube and return tube can be pulled out.

23. Install 4 new fuel tank cushions the fuel tank.

24. Install No. 3 fuel tank protector.

a. Install the No. 3 fuel tank protector with the 2 bolts and 3 claws. Tighten to 44 inch lbs. (5.0 Nm).

25. Install fuel tank assembly.

a. Set a transmission jack to the fuel tank.

b. Install the fuel tank and 2 fuel tank bands with the 2 clips, 2 pins and the 2 bolts. Tighten to 30 ft. lbs. (40 Nm).

26. Install fuel tank vent hose.

a. Connect the fuel tank vent hose.

b. Align the fuel tank vent hose connector with the pipe, then push in the fuel tank vent connector until the retainer makes a "click" sound to connect the fuel tank vent hose to the charcoal canister.

27. Connect fuel tank main tube and fuel tank return tube.

a. Push the tube connector into the pipe until the connector makes "click" sound.

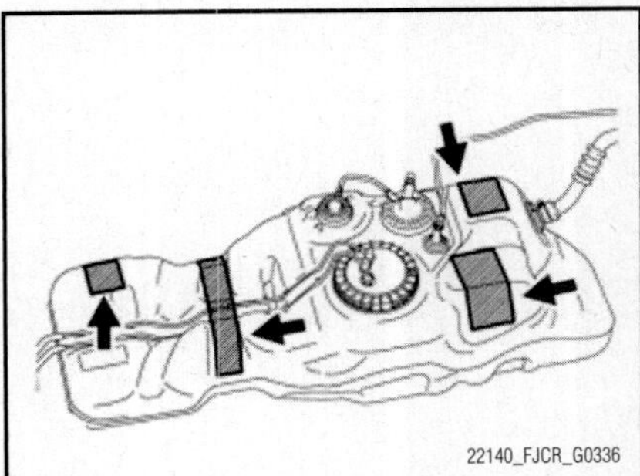

Fig. 207 Fuel tank cushion locations

b. Push the tube connector into the pipe until the connector makes "click" sound.

c. Connect the fuel tank to filler pipe hose, as shown in the illustration, with the clamp.

28. Install No. 1 fuel tank protector.

a. Install the fuel tank protector with the 4 nuts. Tighten to 15 ft. lbs. (20 Nm).

b. Install the fuel tank protector bracket with the 2 nuts. Tighten to 49 inch lbs. (6 Nm).

29. Install rear floor service hole cover.

a. Connect the connector.

b. Install the 3 screws from the center floor.

c. Install the left center floor silencer pad.

d. Install the floor carpet.

30. Install left rear door scuff plate.

a. Engage the 6 claws and install the rear door scuff plate.

31. Install left front door scuff plate.

32. Install rear seat cushion assembly. Tighten to 27 ft. lbs. (37 Nm).

33. Connect the negative battery cable.

34. Check for fuel leaks.

IDLE SPEED

ADJUSTMENT

The idle speed is controlled by the Electronic Control Module (ECM). No adjustment is possible.

THROTTLE BODY

REMOVAL & INSTALLATION

See Figures 208 and 209.

1. Disconnect the negative battery cable.
2. Drain engine coolant.
3. Remove the 2 nuts, then remove the V-bank cover.
4. Remove air cleaner assembly.
5. Remove throttle with motor body assembly.

a. Disconnect the No. 5 water by-pass hose.

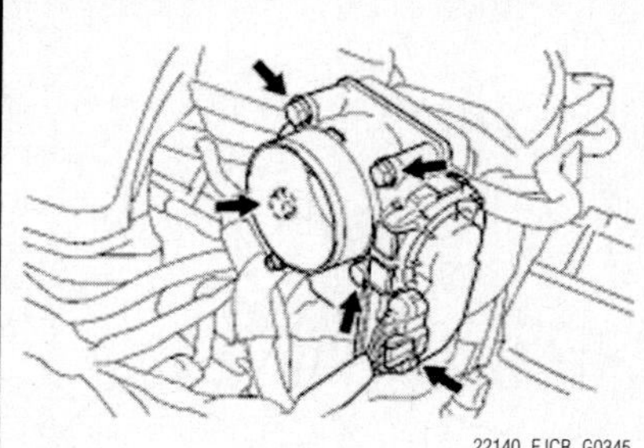

Fig. 208 Locating the throttle body fasteners and electrical connector—2009 vehicles

b. Disconnect the No. 4 water by-pass hose.

c. Disconnect the throttle motor connector.

d. Remove the 4 bolts, then remove the throttle w/ motor body and gasket.

To install:

6. Install throttle with motor body assembly.

a. Install a new gasket and the throttle with motor body with the 4 bolts. Tighten to 9 ft. lbs. (11 Nm).

b. Connect the throttle motor connector.

c. Connect the No. 4 water by-pass hose.

d. Connect the No. 5 water by-pass hose.

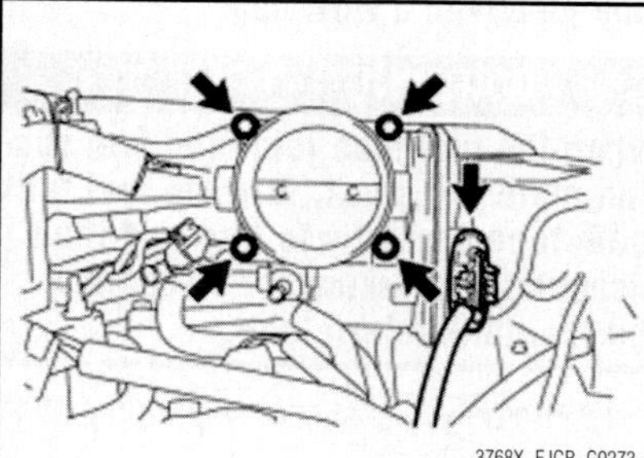

Fig. 209 Locating the throttle body fasteners and electrical connector—2010 vehicles

7. Install air cleaner assembly.
8. Connect the negative battery cable.
9. Add engine coolant.
10. Check for engine coolant leak.
11. Install v-bank cover.

INITIALIZATION

➡Be sure to perform this procedure after reassembling the throttle body or removing and reinstalling any throttle body component. Perform the following procedure after replacing the ECM, throttle body or any throttle body components. The following procedure should also be performed if the throttle body is cleaned. Be sure to perform this procedure after replacing the ECM and reconnecting the battery cable.

1. Disconnect the cable from the negative (-) battery terminal. Wait at least 60 seconds and reconnect the cable.
2. Turn the ignition switch to ON without operating the accelerator pedal.

➡If the accelerator pedal is operated, perform the above steps again.

3. Connect the Techstream to the DLC3 and clear the DTC's.
4. Start the engine and check that the MIL is not illuminated and that the idle speed is within the specified range when the A/C is switched off after the engine is warmed up.

a. Standard:

- Condition: A/C switched off
- Engine idle speed: 580 to 680 rpm

➡Be sure to perform this step with all accessories off. Make sure that the shift lever is in neutral.

5. Enter the following menus: Powertrain / Engine and ECT / Data List / All Data / Throttle Sensor Position. Fully depress the accelerator pedal and check that the value is 60% or more.
6. Perform a road test and confirm that there are no abnormalities.

HEATING & AIR CONDITIONING SYSTEM

AIR CONDITIONING UNIT

REMOVAL & INSTALLATION

See Figures 210 through 213.

1. Before servicing the vehicle, refer to the precautions.
2. Disconnect the negative battery cable.

➡Wait for at least 90 seconds after disconnecting the cable to prevent the airbag from working.

3. Discharge refrigerant from refrigeration system.
4. Drain engine coolant.
5. Remove roof antenna pole.
6. Remove windshield wiper arm cover.
7. Remove front wiper arm and blade assembly.
8. Remove left front fender side panel upper.
9. Remove antenna ornament.
10. Remove front fender side panel upper.
11. Remove cowl top ventilator louver assembly.
12. Remove cowl top ventilator louver.
13. Disconnect cooler refrigerant suction pipe **A** using SST: 09870-00015.
    a. Check the directions of the piping clamp and SST by referring to the illustration on the caution label.
    b. Push down SST and release the clamp lock.
    c. Do not deform the tube when pushing SST.
    d. Pull SST slightly, push the release lever, and then remove the piping clamp with SST.
    e. Disconnect the suction pipe.

⁂ WARNING

Do not use any tools when disconnecting the pipe.

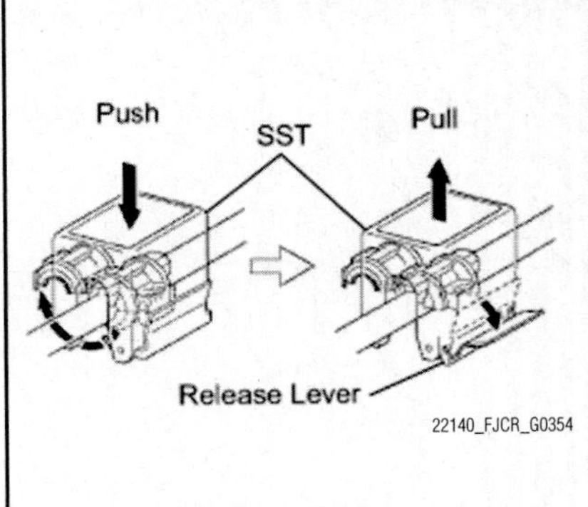

Fig. 210 Disconnect cooler refrigerant suction pipe

14. Seal the openings of the disconnected parts using vinyl tape to prevent moisture and foreign matter from entering.
15. Disconnect cooler refrigerant liquid pipe **A**. The disconnection procedure of the liquid pipe is the same as for the suction pipe.
16. Disconnect heater inlet water hose.
    a. Using pliers, grip the claw of the clip, slide the clip and disconnect the heater water inlet hose from the heater unit.
17. Using pliers, grip the claw of the clip, slide the clip and disconnect the heater water outlet hose from the heater unit.
18. Position front wheels facing straight ahead.
19. Remove lower steering wheel covers.
20. Remove steering pad.
21. Remove steering wheel assembly.
22. Remove lower steering column cover.
23. Remove upper steering column cover.
24. Remove combination switch assembly.
25. Remove front door scuff plates.
26. Remove front floor footrest.
27. Remove footrest clip.
28. Remove cowl side trim boards.
29. Remove front door opening trim weatherstrips.
30. Remove assist grip plug.
31. Remove assist grip assembly.
32. Remove front pillar garnishes.
33. Remove instrument panel garnishes.
34. Remove integration control and panel assembly.
35. Remove radio receiver assembly.
36. Remove parking brake hole cover.
37. On manual transmission vehicles, remove shift lever knob.
38. On 4WD vehicles, remove shift lever knob.
39. Remove console upper rear panel.
40. Remove box bottom mat.
41. Remove front console box.
42. Remove console upper panel no. 1 garnish.
43. Remove instrument lower cover.
44. Remove no 1 instrument panel register assembly.
45. Separate hood lock control lever.
46. Remove instrument panel finish plate.
47. Remove lower instrument panel finish panel.
48. Remove instrument cluster finish panel.
49. Remove combination meter assembly.
50. Remove glove compartment door assembly.
51. Remove instrument panel lower finish panel.
52. Remove no. 2 instrument panel register assembly.
53. Remove instrument panel speaker panel.
54. Remove front no 2 speaker assembly.
55. Remove assist grip retainers.
56. Disconnect passenger airbag connector.
57. Remove instrument panel.
58. Remove left instrument panel finish panel end.
59. Remove no. 1 heater to register duct.
60. Remove no. 2 heater to register duct.
61. Remove rear no. 1 air duct.
62. Remove rear no. 2 air duct.
63. Remove no. 1 air duct.
64. Remove no. 2 air duct.
65. Remove no. 1 instrument panel brace mounting bracket.
    a. On the left side remove the bolt and nut and remove the instrument panel brace mounting bracket.
    b. On the right side remove the bolt and nut and remove the instrument panel brace mounting bracket.
66. Remove ECM.
67. Remove steering column hole cover.
68. Remove the bolt and separate the steering shaft thrust stopper from the steering intermediate shaft assembly.
    a. Mark matchmarks on the steering column assembly and steering intermediate shaft.
    b. Pull the intermediate shaft assembly and steering shaft thrust stopper out of the steering column assembly.
69. Remove steering column assembly.
70. Remove the bolt and remove the instrument panel side bracket.
71. Separate main body ECU.
    a. Remove the 2 nuts and remove the driver side junction block.
72. Remove the cooler unit drain hose.
73. Remove instrument panel reinforcement.
    a. Remove the 3 bolts and 5 nuts and disconnect the wire harness.
    b. Disconnect the connectors.
    c. Disengage the clamps.
    d. Remove the 5 bolts and the 2 nuts of the air conditioning unit.
    e. Remove the 2 caps and the 7 bolts of the reinforcement.
    f. Disengage the reinforcement hook of the air conditioning unit, and remove the reinforcement.

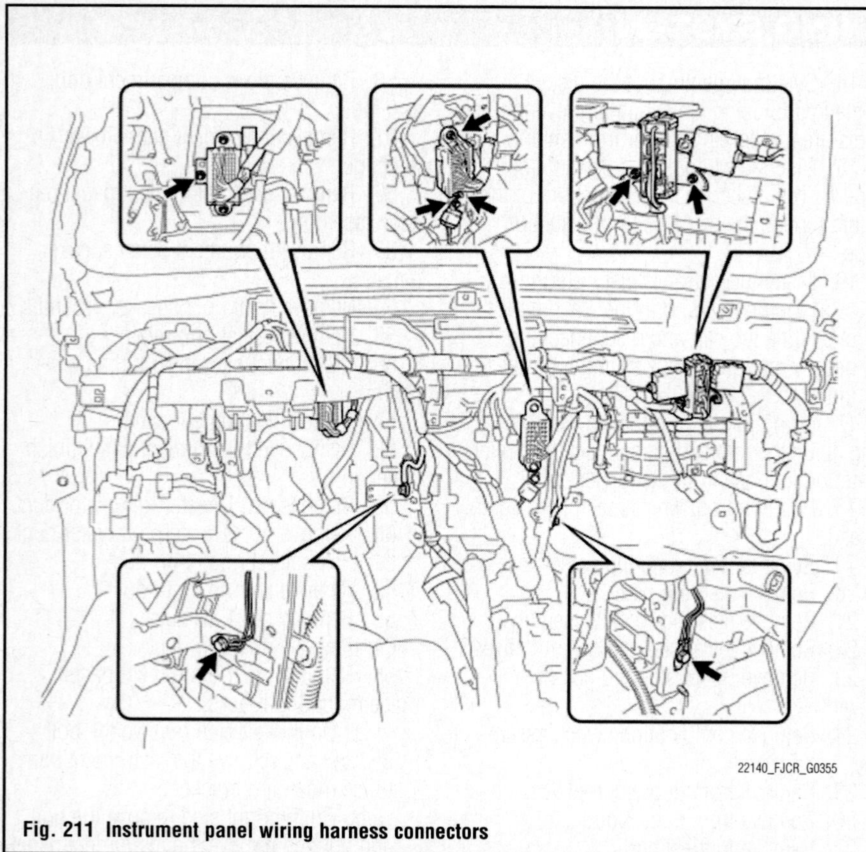

Fig. 211 Instrument panel wiring harness connectors

g. Remove the air conditioning unit.

74. Remove air conditioning unit assembly.
 a. Remove the 2 screws.
 b. Remove the air conditioning unit as shown in the illustration.

75. Remove heater to register duct assembly.
 a. Disengage the 4 claws and remove the heater to register duct.

76. Remove lower defroster nozzle assembly.
 a. Disengage the 4 claws and remove the lower defroster nozzle.

To install:

77. Install lower defroster nozzle assembly.

78. Install heater to register duct assembly.

79. Install air conditioning unit assembly.
 a. Install the air conditioning unit with the 2 screws as shown in the illustration.

80. Install instrument panel reinforcement.
 a. Temporarily install the air conditioning unit assembly.
 b. Insert the bracket hook into the holes of the reinforcement bracket, and temporarily install the reinforcement.

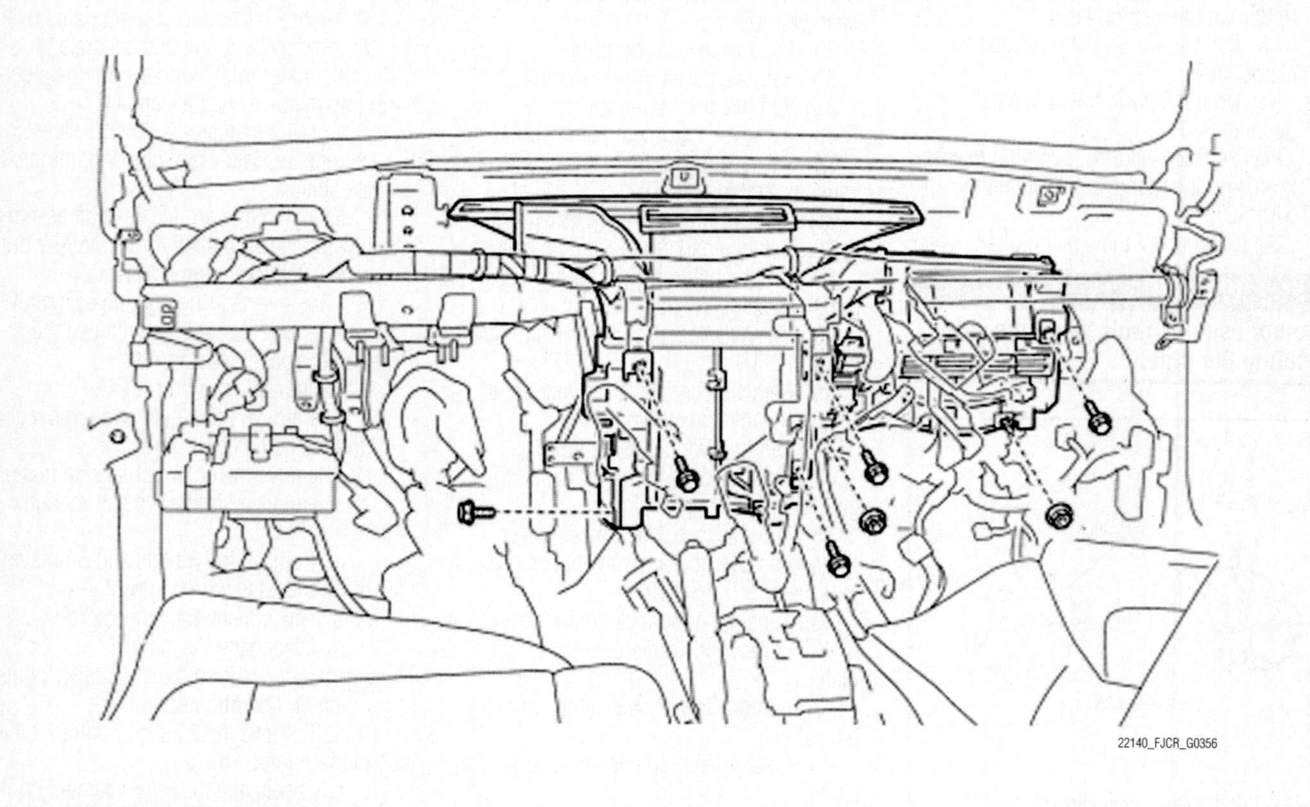

22140_FJCR_G0356

Fig. 212 Air conditioning unit fasteners

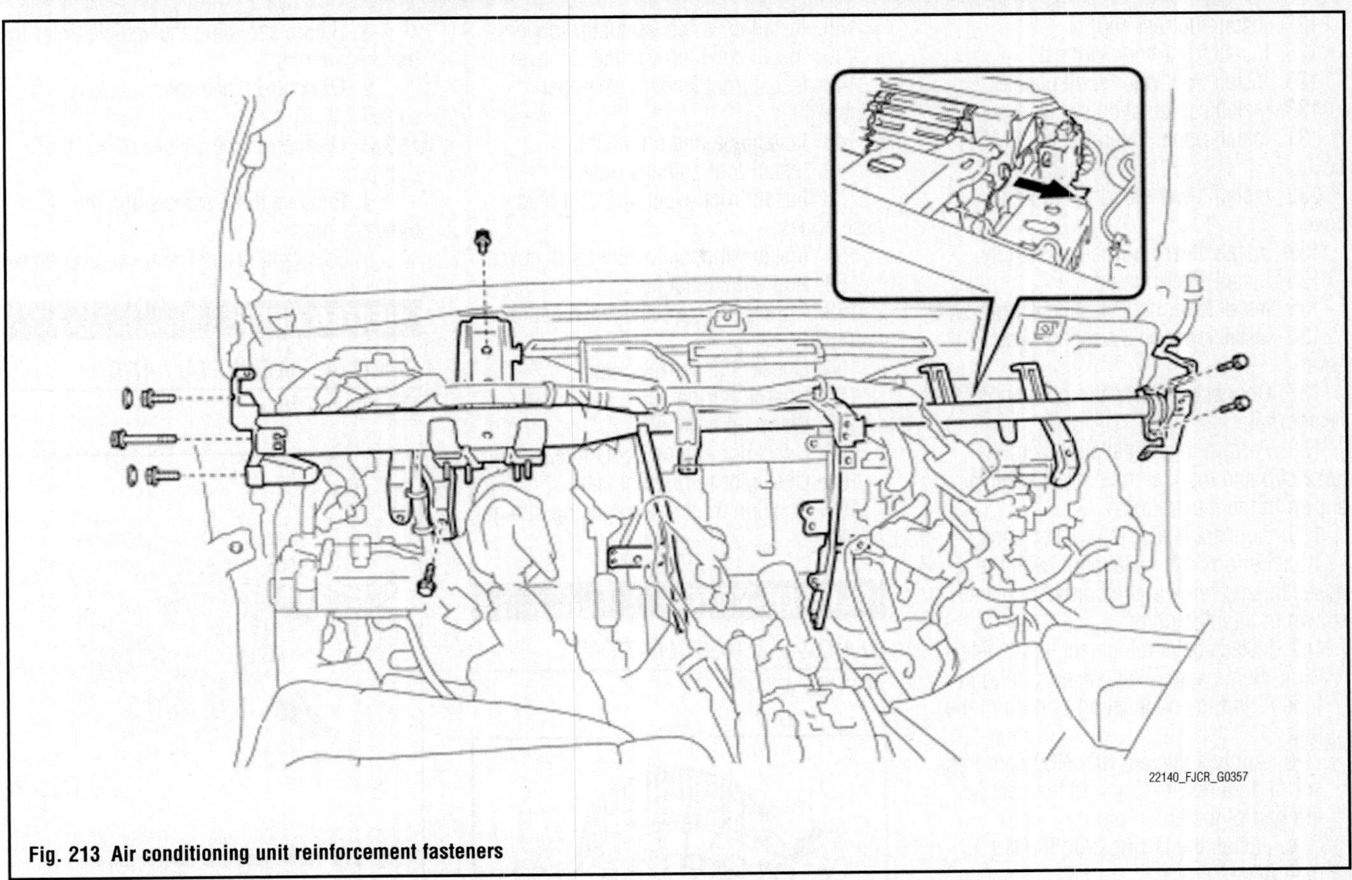

Fig. 213 Air conditioning unit reinforcement fasteners

c. Install the instrument panel reinforcement with the 7 bolts.

d. Install the 2 caps.

e. Install the 5 bolts. Tighten the bolts in the sequence order shown in the illustration to install the air conditioner unit assembly. Tighten to 87 inch lbs. (98 Nm).

f. Install the 2 nuts. Tighten to 48 inch lbs. (54 Nm).

g. Engage the clamps.

h. Connect the connectors.

i. Connect the wire harness with the 3 bolts and 5 nuts. Tighten to 65 inch lbs. (73 Nm).

81. Install cooler unit drain hose.

a.Install the cooler unit drain hose.

82. Install the main body ECU with the 2 nuts. Tighten to 74 inch lbs. (84 Nm).

83. Install the instrument panel side bracket with the bolt. Tighten to 74 inch lbs. 84 Nm).

84. Install steering column assembly.

85. Install steering intermediate shaft assembly.

a. Align the matchmarks on the steering column assembly and the steering intermediate shaft assembly.

b. Install the steering intermediate shaft assembly and thrust stopper onto the steering column assembly with the bolt. Tighten to 27 ft. lbs. (36 Nm).

86. Install steering column hole cover.

87. Install ECM.

88. Install no. 1 instrument panel brace mounting bracket.

a. On the left side, install the instrument panel brace mounting bracket with the bolt and nut.

b. On the right side, install the instrument panel brace mounting bracket with the bolt and nut.

89. Install no. 1 air duct.

90. Install no. 2 air duct.

91. Install rear no. 1 air duct.

92. Install rear no. 2 air duct.

93. Install no. 1 heater to register duct.

94. Install no. 2 heater to register duct.

95. Install instrument panel.

96. Connect passenger airbag connector.

97. Install left instrument panel finish panel end.

98. Install assist grip retainers.

99. Install front no. 2 speaker assembly.

100. Install no. 2 instrument panel speaker panels.

101. Install no. 2 instrument panel register assembly.

102. Install right instrument panel lower finish panel.

103. Install glove compartment door assembly.

104. Install combination meter assembly.

105. Install instrument cluster finish panel.

106. Install lower instrument panel.

107. Install left lower instrument panel finish panel.

108. Install instrument panel finish plate.

109. Connect hood lock control lever.

110. Install no 1 instrument panel register assembly.

111. Install instrument lower cover.

112. Install console upper panel no 1 garnish.

113. Install front console box.

114. Install box bottom mat.

115. Install console upper rear panel.

116. On manual transmission vehicles, install shift lever knob.

117. On 4WD vehicles, install shift lever knob.

118. Install parking brake hole cover.

119. Install radio receiver assembly.

120. Install integration control and panel assembly.

121. Install instrument panel garnishes.

122. Install front pillar garnishes.

123. Install assist grip assembly.

124. Install assist grip plug.

125. Install front door opening trim weatherstrips.

126. Install cowl side trim boards.

127. Install footrest clip.
128. Install front floor footrest.
129. Install front door scuff plates.
130. Install combination switch assembly.
131. Install upper steering column cover.
132. Install lower steering column cover.
133. Install steering wheel assembly.
134. Install steering pad.
135. Install lower no 2 steering wheel cover.
136. Install lower no 3 steering wheel cover.
137. Connect heater water outlet hose to heater unit.
138. Perform the installation with the hose clip and mark at the correct angle as shown in the illustration.
139. Connect heater inlet water hose.
140. Perform the installation with the hose clip and mark at the correct angle as shown in the illustration.
141. Install cooler refrigerant liquid pipe a.
   a. Remove the vinyl tape from liquid tube **A** and the connecting portion of the unit.
   b. Apply sufficient ND-OIL8 compressor oil to a new O-ring and the connecting part of the liquid pipe.
   c. Install the O-ring onto the liquid pipe **A**.
   d. Install the liquid and piping clamp.
142. After connection, check the claw fitting of the piping clamp.
143. Install cooler refrigerant suction pipe **A**.
144. Connection procedure of the suction pipe is the same as for the liquid pipe.
145. Install cowl top ventilator louvers.
146. Install cowl top ventilator louver.
147. Install front fender side upper panels.
148. Install antenna ornament.
149. Install roof antenna pole.
150. Install front wiper arm and blade assemblies.
151. Install windshield wiper arm cover.
152. Add engine coolant.
153. Connect cable to negative battery terminal.
154. Check SRS warning light.
155. Charge refrigerant.
156. Warm up engine.
157. Check for engine coolant leak.
158. Check for refrigerant leak.
159. Position front wheels facing straight ahead.

BLOWER MOTOR

REMOVAL & INSTALLATION

See Figure 214.

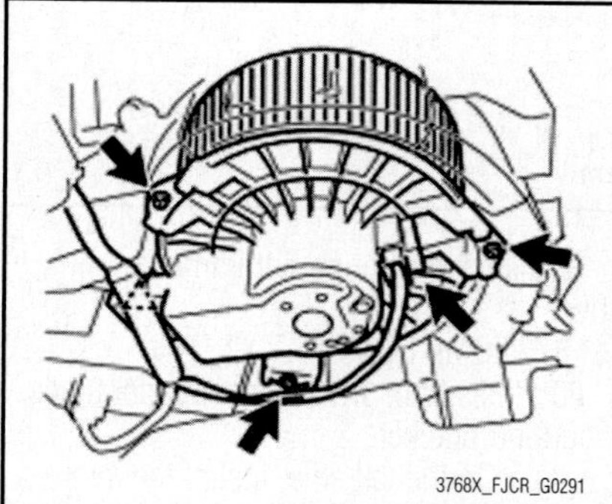

Fig. 214 Remove the blower motor as shown

1. Before servicing the vehicle, refer to the precautions.
2. Disconnect the negative battery cable.
3. Disconnect the connector and the clamp.
4. Remove the 3 screws and the blower motor.
5. To install, reverse removal procedure.

HEATER CORE

REMOVAL AND INSTALLATION

See Figure 215.

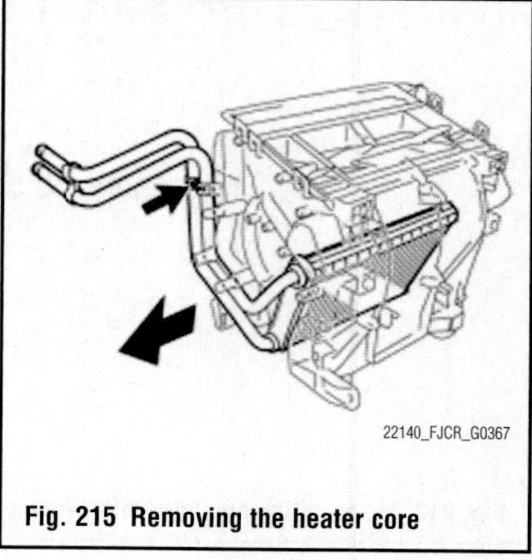

Fig. 215 Removing the heater core

1. Before servicing the vehicle, refer to the precautions.
2. Remove the air conditioning unit.
3. Remove the screw and clamp.
4. Remove the heater radiator unit from the heater case.
5. Installation is the reverse of the removal procedure.

STEERING

POWER RACK & PINION STEERING GEAR

REMOVAL & INSTALLATION

2WD Models

See Figures 216 through 218.

1. Before servicing the vehicle, refer to the precautions section.
2. Disconnect the negative battery cable.
3. Place front wheels facing straight ahead.
4. Remove no.1 engine under cover.
5. Remove fan and generator V-belt.
6. Remove front wheels.
7. Remove no.2 steering intermediate shaft.

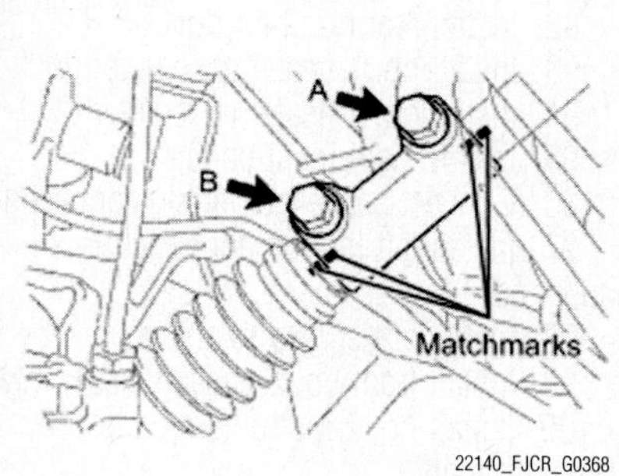

Fig. 216 Place matchmarks on the steering sliding yoke, No. 2 steering intermediate shaft and steering intermediate shaft

   a. Fix the steering wheel with the seat belt in order to prevent it from rotating. This operation is effective for preventing any damage to the spiral cable.
   b. Place matchmarks on the steering sliding yoke, No. 2 steering intermediate shaft and steering intermediate shaft.
   c. Remove bolts from the steering sliding yoke.
   d. Slide the steering sliding yoke up and separate it from the No. 2 steering intermediate shaft.
   e. Pull down the steering sliding yoke from the steering intermediate shaft to remove it.
   f. Place matchmarks on the No. 2 steering intermediate shaft and power steering link.

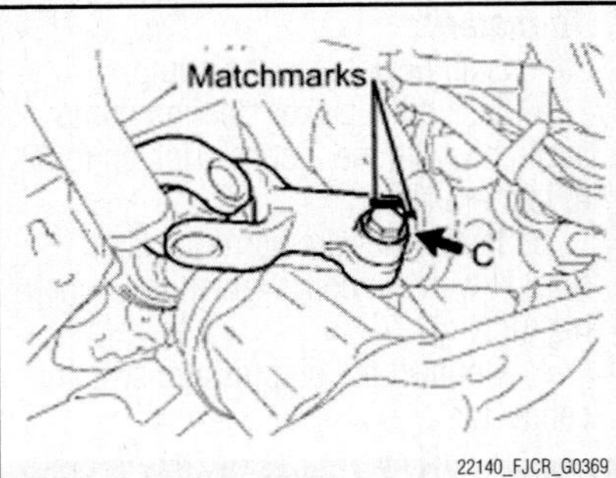

Fig. 217 Place matchmarks on the No. 2 steering intermediate shaft and power steering link

g. Remove bolt C from the No. 2 steering intermediate shaft.

h. Slide the No. 2 steering intermediate shaft up and remove it from the power steering link.

8. Separate tie rod end.

a. Remove the cotter pin and nut.

b. Using SST: 09628-62011, separate the tie rod end from the steering knuckle arm.

9. Separate pressure feed tube assembly.

a. Remove the 2 bolts and separate the tube support brackets.

b. Loosen the flare nut and separate the pressure feed tube.

c. Disengage the clip and disconnect the return hose.

10. Remove the 4 bolts and remove the air conditioning compressor and magnetic clutch.

11. Remove power steering link.

a. Remove the 2 bolts and 2 nuts.

WARNING

The nut has a detent, so never turn the nut. Be sure to turn only the bolt.

b. Tilt the transmission and remove the power steering link.

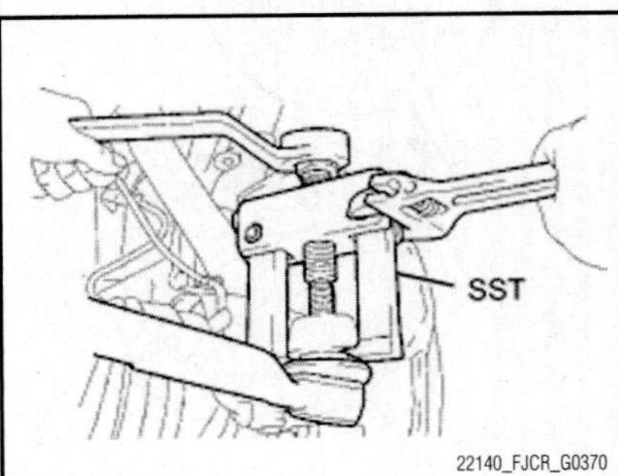

Fig. 218 Using SST: 09628-62011, separate the tie rod end from the steering knuckle arm

To install:.

12. Install the power steering link with the 2 bolts and 2 nuts. Tighten to 74 ft. lbs. (100 Nm).

WARNING

Never turn the nut since it has a detent. Be sure to turn only the bolt.

13. Install pressure feed tube assembly.

a. Connect the return hose with the clip.

b. Tighten the flare nut to 33 ft. lbs. (44 Nm) and connect the pressure feed tube.

c. Install the tube support brackets with the 2 bolts. Tighten to 21 ft. lbs. (28 Nm).

14. Install the air conditioning compressor and magnetic clutch with the 4 bolts.

15. Install tie rod end.

a. Install the tie rod end onto the steering knuckle arm.

b. Install the nut. Tighten to 67 ft. lbs. (91 Nm).

c. Install a new cotter pin.

16. Install No. 2 steering intermediate shaft.

a. Align the matchmarks on the No. 2 steering intermediate shaft and power steering link.

b. Install the No. 2 steering intermediate shaft onto the power steering link. Tighten to 27 ft. lbs. (36 Nm).

c. Align the matchmarks on the steering intermediate shaft and steering sliding yoke.

d. Install the steering sliding yoke onto the steering intermediate shaft and slide it upward.

e. Align the matchmarks on the steering sliding yoke and No. 2 steering intermediate shaft.

f. Install the steering sliding yoke. Tighten to 27 ft. lbs. (36 Nm).

17. Install fan and generator V-belt.

18. Install front wheels.

19. Place front wheels facing straight ahead.

20. Inspect steering wheel center point.

21. Connect the negative battery cable.

22. Add power steering fluid.

23. Bleed power steering fluid.

24. Check for power steering fluid leakage.

25. Inspect and adjust front wheel alignment.

26. Install No. 1 engine under cover

4WD Models

See Figures 216 through 218.

1. Before servicing the vehicle, refer to the precautions section.

2. Disconnect the negative battery cable.

3. Place front wheels facing straight ahead.

4. Remove no.1 engine under cover.

5. Remove fan and generator V-belt.

6. Remove front wheels.

7. Remove no. 1 engine under cover.

8. Remove rear engine under cover assembly.

9. Remove no. 2 exhaust front pipe assembly.

10. Remove front exhaust pipe assembly.

11. Remove driveshaft heat insulator.

12. Remove front halfshaft assembly.

13. Remove driveshaft assembly.

14. Remove front suspension member brackets.

15. Remove no. 3 frame crossmember.

16. Remove No. 2 steering intermediate shaft.

a. Fix the steering wheel with the seat belt in order to prevent it from rotating. This operation is effective for preventing any damage to the spiral cable.

b. Place matchmarks on the steering sliding yoke, No. 2 steering intermediate shaft and steering intermediate shaft.

c. Remove bolts from the steering sliding yoke.

d. Slide the steering sliding yoke up and separate it from the No. 2 steering intermediate shaft.

e. Pull down the steering sliding yoke from the steering intermediate shaft to remove it.

f. Place matchmarks on the No. 2 steering intermediate shaft and power steering link.

g. Remove bolt C from the No. 2 steering intermediate shaft.

h. Slide the No. 2 steering intermediate shaft up and remove it from the power steering link.

17. Separate tie rod end.

a. Remove the cotter pin and nut.

b. Using SST: 09628-62011, separate the tie rod end from the steering knuckle arm.

18. Separate pressure feed tube assembly.

a. Remove the 2 bolts and separate the tube support brackets.

b. Loosen the flare nut and separate the pressure feed tube.

c. Disengage the clip and disconnect the return hose.

19. Remove the 4 bolts and remove the air conditioning compressor and magnetic clutch.

20. Remove power steering link.

a. Remove the 2 bolts and 2 nuts.

**** WARNING**

The nut has a detent, so never turn the nut. Be sure to turn only the bolt.

b. Tilt the transmission and remove the power steering link.

To install:.

21. Install the power steering link with the 2 bolts and 2 nuts. Tighten to 74 ft. lbs. (100 Nm).

**** WARNING**

Never turn the nut since it has a detent. Be sure to turn only the bolt.

22. Install pressure feed tube assembly.
   a. Connect the return hose with the clip.
   b. Tighten the flare nut to 33 ft. lbs. (44 Nm) and connect the pressure feed tube.
   c. Install the tube support brackets with the 2 bolts. Tighten to 21 ft. lbs. (28 Nm).
23. Install the air conditioning compressor and magnetic clutch with the 4 bolts.
24. Install tie rod end.
   a. Install the tie rod end onto the steering knuckle arm.
   b. Install the nut. Tighten to 67 ft. lbs. (91 Nm).
   c. Install a new cotter pin.
25. Install No. 2 steering intermediate shaft.
   a. Align the matchmarks on the No. 2 steering intermediate shaft and power steering link.
   b. Install the No. 2 steering intermediate shaft onto the power steering link. Tighten to 27 ft. lbs. (36 Nm).
   c. Align the matchmarks on the steering intermediate shaft and steering sliding yoke.
   d. Install the steering sliding yoke onto the steering intermediate shaft and slide it upward.
   e. Align the matchmarks on the steering sliding yoke and No. 2 steering intermediate shaft.
   f. Install the steering sliding yoke. Tighten to 27 ft. lbs. (36 Nm).
26. Install no. 3 frame crossmember.
27. Install front suspension member brackets.
28. Install driveshaft assembly.
29. Install front halfshaft assembly.
30. Install driveshaft heat insulator.
31. Install front exhaust pipe assembly.
32. Install no. 2 exhaust front pipe assembly.
33. Install rear engine under cover assembly.
34. Install no. 1 engine under cover.
35. Install fan and generator V-belt.
36. Install front wheels.
37. Place front wheels facing straight ahead.
38. Inspect steering wheel center point.
39. Connect the negative battery cable.
40. Add power steering fluid.
41. Bleed power steering fluid.
42. Check for power steering fluid leakage.
43. Inspect and adjust front wheel alignment.
44. Install No. 1 engine under cover.

POWER STEERING PUMP

REMOVAL & INSTALLATION

2009 Vehicles

See Figure 219.

1. Before servicing the vehicle, refer to the precautions section.
2. Disconnect the negative battery cable.
3. Remove No. 1 engine under cover.
4. Remove fan and generator v belt.
5. Drain power steering fluid.
6. Disengage the clip and disconnect the No. 1 oil reservoir to pump hose.
7. Disconnect pressure feed tube assembly.
   a. Remove the union bolt, then disconnect the pressure feed tube.
   b. Remove the gasket from the pressure feed tube.
8. Remove power steering pump.
   a. Disconnect the oil pressure switch connector.
   b. Remove the bolt and the wire harness clamp bracket.
   c. Rotate the pulley on the power steering pump to access the mounting bolts
   d. Remove the 2 bolts and the power steering pump assembly.

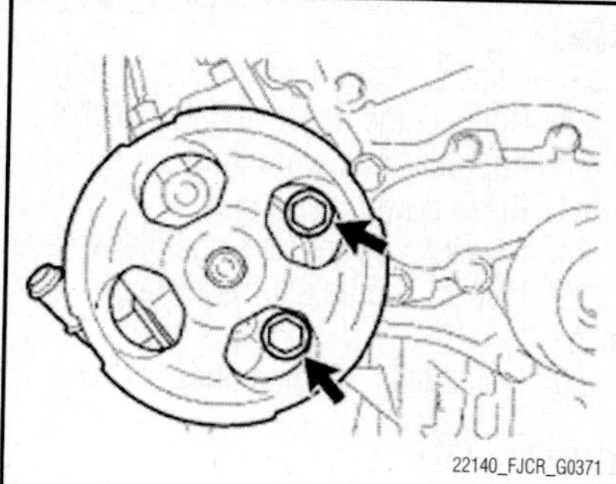

Fig. 219 Rotate the pulley on the power steering pump to access the mounting bolts

To install:.

9. Install power steering pump.
   a. Install the power steering pump assembly with the 2 bolts. Tighten to 32 ft. lbs. (43 Nm).
   b. Install the wire harness clamp bracket with the bolt. Tighten to 71 inch lbs. (8 Nm).
   c. Connect the oil pressure switch connector.

**** WARNING**

Make sure that no oil adheres to the connector.

10. Connect pressure feed tube assembly.
   a. Install a new gasket onto the pressure feed tube.
   b. Install the pressure feed tube with the union bolt. Tighten to 38 ft. lbs. (51 Nm).
11. Connect the No. 1 oil reservoir to pump hose with the clip.
12. Install fan and generator v belt.
13. Connect the negative battery cable.
14. Add power steering fluid.
15. Bleed power steering fluid.
16. Check for power steering fluid leakage.
17. Install No. 1 engine under cover.

2010 Vehicles

See Figures 220 through 222.

1. Remove the RH front wheel.
2. Remove RH front fender apron seal.
3. Remove fan and alternator v belt.
4. Drain power steering fluid.

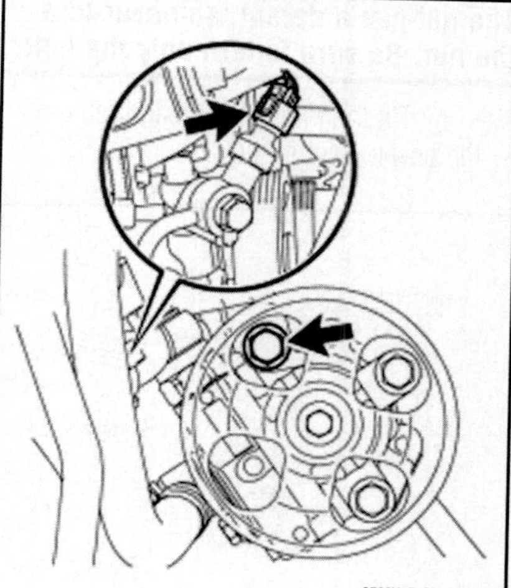

Fig. 220 Disconnect the oil pressure switch connector and remove the bolt and the wire harness clamp bracket

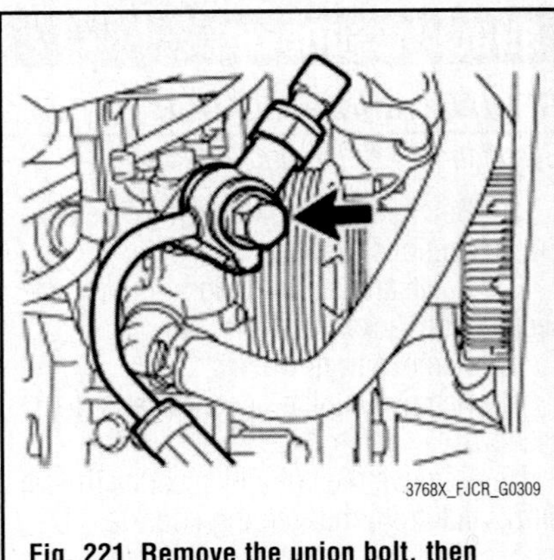

Fig. 221 Remove the union bolt, then separate the pressure feed tube

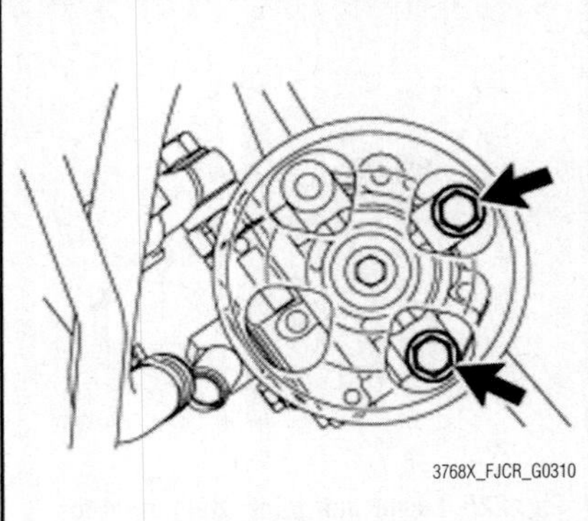

Fig. 222 Remove the 2 bolts and the power steering pump

5. Separate pressure feed tube assembly:
   a. Disconnect the oil pressure switch connector.
   b. Remove the bolt and the wire harness clamp bracket.
   c. Remove the union bolt, then separate the pressure feed tube.
   d. Remove the gasket from the pressure feed tube.
6. Disengage the clip and separate the No. 1 oil reservoir to pump hose.
7. Remove the 2 bolts and the power steering pump.

To install:

8. Install the power steering pump assembly with the 2 bolts. Tighten to 32 ft. lbs (43 Nm).
9. Connect the No. 1 oil reservoir to pump hose with the clip.
10. Install pressure feed tube assembly:
   a. Install a new gasket onto the pressure feed tube.
   b. Install the pressure feed tube with the union bolt. Tighten to 37 ft. lbs. (50 Nm).
   c. Install the wire harness clamp bracket with the bolt. Tighten to 32 ft. lbs. (43 Nm).

➡ Make sure that no oil adheres to the connector.

   d. Connect the oil pressure switch connector.
11. To complete installation, reverse remaining removal procedure.
12. Add power steering fluid.
13. Bleed power steering fluid.
14. Check for power steering fluid leakage.

BLEEDING

1. Check the fluid level.
2. Raise and safely support the vehicle securely on jackstands.
3. Turn the steering wheel.
4. With the engine stopped, turn the wheel slowly from lock to lock several times.
5. Lower the vehicle.
6. Start the engine.
7. Run the engine at idle for a few minutes.
8. Turn the steering wheel.
9. With the engine idling, turn the wheel to the left or right full lock position and keep it there for 2 to 3 seconds. Then turn the wheel to the opposite full lock position and keep it there for 2 to 3 seconds.
10. Repeat this step several times.
11. Stop the engine.
12. Check for foaming or emulsification.

⁂ WARNING

If the system has to be bled twice because of foaming or emulsification, check for fluid leakage in the system.

13. Check the fluid level.

SUSPENSION

FRONT SUSPENSION

COIL SPRING

REMOVAL & INSTALLATION

See Figure 223.

1. Before servicing the vehicle, refer to the precautions section.
2. Remove the shock absorber assembly.
3. Remove front support to front shock absorber nut
4. Using a coil spring compressor, compress the coil spring.
5. While holding the shock absorber rod, remove the nut.

⁂ WARNING

Do not use an impact wrench. It will damage the shock absorber rod.

6. Remove front shock absorber cushion retainer.
7. Remove front shock absorber no. 1 cushion.
8. Remove front suspension support.
9. Remove front shock absorber cushion retainer.
10. Remove front coil spring.

To install:

11. Using a coil spring compressor, compress the coil spring.
12. Install the coil spring onto the shock absorber.

➡Fit the lower end of the coil spring into the gap of the spring lower seat.

13. Install front coil spring.
14. Install front shock absorber cushion retainer.
15. Install front suspension support.
16. Install front shock absorber no. 1 cushion.
17. Install front shock absorber cushion retainer.
18. Install front support to front shock absorber nut.
19. Align the suspension support and the absorber bushing as shown in the illustration.
20. Fit and tighten a new lock nut. Tighten to 18 ft. lbs. (25 Nm).
21. Release the coil spring while checking the position of the suspension support.

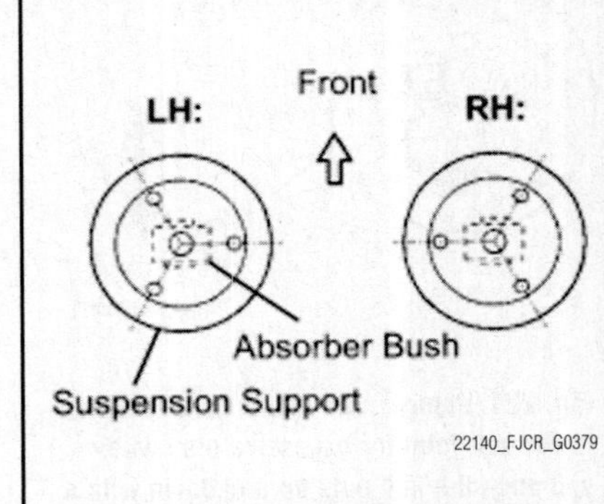

Fig. 223 Align the suspension support and the absorber bushing as shown

LOWER BALL JOINT

REMOVAL & INSTALLATION

The lower ball joint is an integral part of the lower control arm and is not serviced separately. See Front Suspension, Lower Control Arm, Removal and Installation.

LOWER CONTROL ARM

REMOVAL & INSTALLATION

See Figures 224 through 226.

1. Before servicing the vehicle, refer to the precautions section.
2. Remove front wheel.
3. Inspect front lower suspension arm.
    a. Install the hub nuts onto the disc.
    b. Using a dial indicator, check the lower ball joint for excessive play when you push the hub nuts up and down with a force of 66 lbs. (294 N). Maximum play is 0.020 in. (0.5 mm).
    c. If it is not within the specification, replace the lower arm.
4. Separate front shock absorber with coil spring.
    a. Remove the bolt, nut and washer.
    b. Separate the front shock absorber with coil spring from the lower arm.
5. Remove front lower suspension arm.
    a. Remove the 2 bolts and separate the lower ball joint attachment from the steering knuckle.
    b. Place matchmarks on the No. 2 camber adjust cam and toe adjust cam.
    c. Remove the nut, No. 2 camber adjust cam, No. 1 camber adjust cam, bolt, toe adjust cam, toe adjust plate and lower arm.
    d. Remove the cotter pin and the nut.
    e. Using SST: 09628-00011, remove the lower ball joint attachment.

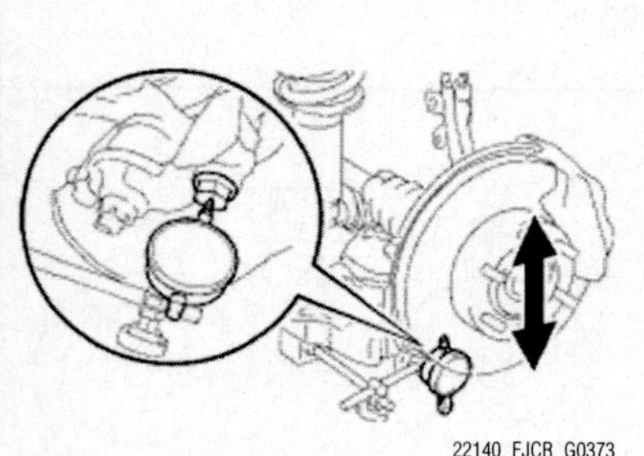

Fig. 224 Using a dial indicator, check the lower ball joint for excessive play when you push the hub nuts up and down with a force of 66 lbs. (294 N). Maximum play is 0.020 in. (0.5 mm)

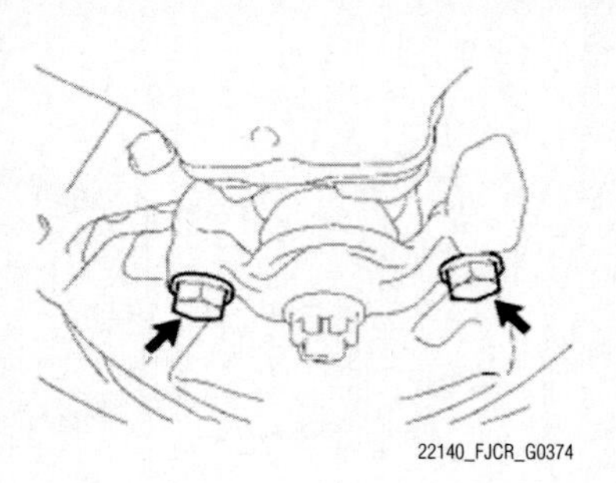

Fig. 225 Lower ball joint attachment to steering knuckle mounting bolts

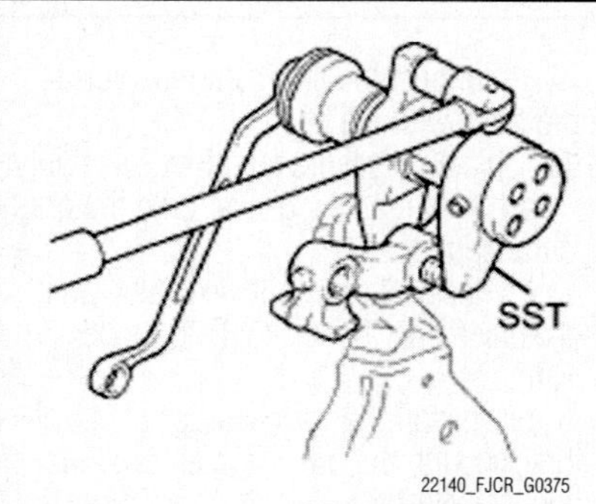

Fig. 226 Using SST: 09628-00011, remove the lower ball joint attachment

To install:.

6. Temporarily tighten front lower suspension arm.
    a. Align the matchmarks on the No. 2 camber adjust cam and toe adjust cam.
    b. Temporarily tighten the bolt and the nut.
    c. Install the lower ball joint attachment, a new nut and a new cotter pin. Tighten to 103 ft. lbs. (140 Nm).
    d. Install the lower ball joint attachment with the 2 bolts. Tighten to 118 ft. lbs. (160 Nm).
7. Temporarily tighten front shock absorber with coil spring.
    a. Install the front shock absorber with coil spring, bolt and washer, and temporarily tighten the nut.
8. Install front wheel and lower the vehicle.
9. Bounce the vehicle up and down several times to stabilize the suspension.
10. Fully tighten front lower suspension arm to 100 ft. lbs. (135 Nm).
11. Fully tighten front shock absorber with coil spring to100 ft. lbs. (135 Nm).
12. Inspect and adjust front wheel alignment.

SHOCK ABSORBERS

REMOVAL & INSTALLATION

See Figures 227 through 229.

1. Before servicing the vehicle, refer to the precautions section.
2. Raise and safely support the vehicle securely on jackstands.
3. Remove front wheels.
4. Remove engine under cover sub-assembly.
5. Remove the nut and separate the stabilizer link from the steering knuckle.

➡If the ball joint turns together with the nut, use a 6 mm hexagon wrench to hold the stud.

6. Remove front stabilizer bar.
7. Separate tie rod end sub-assembly.
8. Remove front shock absorber with coil spring.
    a. Remove the bolt, nut and washer.
    b. Remove the 3 nuts on the upper side of the front shock absorber with coil spring.
    c. Remove the front shock absorber with coil spring.

To install:

9. Temporarily tighten front shock absorber with coil spring.

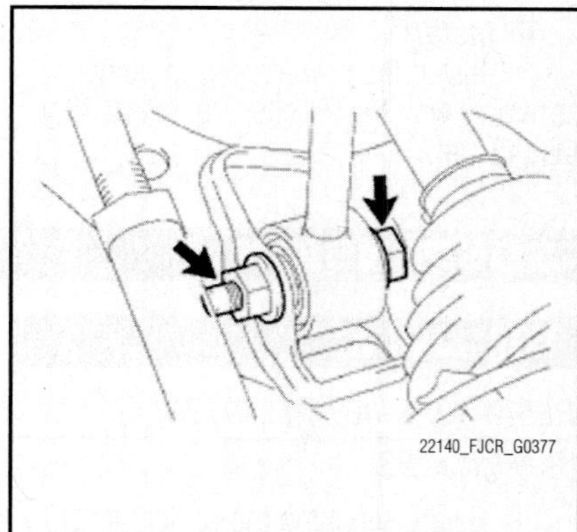

Fig. 227 Lower shock mounting bolts

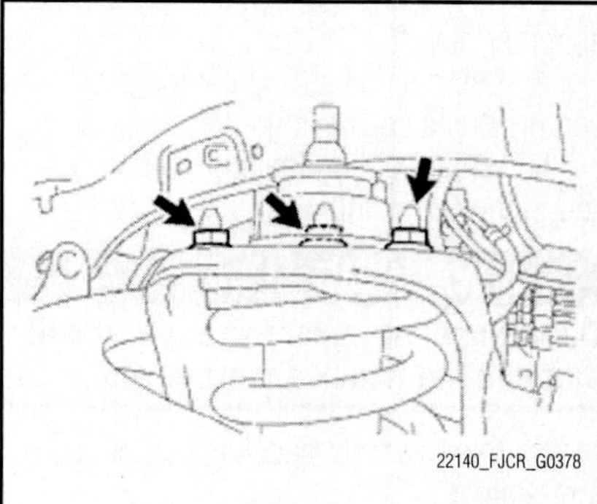

Fig. 228 Upper shock mounting nuts

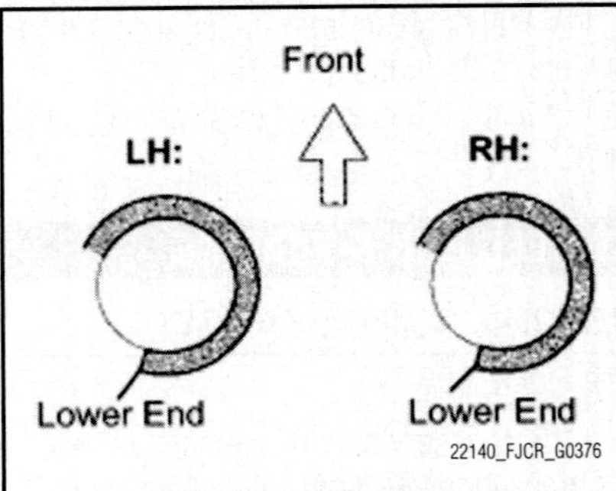

Fig. 229 Install the coil spring onto the body with the lower end of the coil spring facing the rear side of the vehicle

a. Install the coil spring onto the body with the lower end of the coil spring facing the rear side of the vehicle.

b. Install the 3 nuts onto the upper side of the front shock absorber with coil spring. Tighten to 47 ft. lbs. (64 Nm).

c. Temporarily tighten the bolt, nut and washer as shown in the illustration.

10. Install tie rod end sub-assembly.
11. Install front stabilizer bar.
12. Install front stabilizer link assembly.

a. Install the stabilizer link onto the steering knuckle with the nut. Tighten to 52 ft. lbs. (70 Nm).

13. Install engine under cover sub-assembly. Tighten to 21 ft. lbs. (29 Nm).
14. Install front wheels and lower the vehicle.
15. Bounce the vehicle up and down several times to stabilize the suspension.
16. Fully tighten the nut. Tighten to 100 ft. lbs. (135 Nm).
17. Inspect and adjust front wheel alignment.

STEERING KNUCKLE

REMOVAL & INSTALLATION

1. Before servicing the vehicle, refer to the precautions section.
2. Disconnect the negative battery cable.
3. Raise and safely support the vehicle securely on jackstands.
4. Remove front wheel.
5. Remove front disc brake caliper assembly and suspend it out of the way.
6. Remove front rotor.
7. Separate front speed sensor.
8. On 4WD vehicles, remove front axle hub grease cap.
9. On 4WD vehicles, remove front axle hub nut.
10. Separate front stabilizer link assembly.
11. Separate tie rod end sub-assembly.
12. Separate front lower ball joint attachment.
13. Separate front upper suspension arm.

a. Support the lower arm with a jack.

b. Remove the clip and nut.

c. Using SST: 09628-62011, separate the upper ball joint from the steering knuckle.

14. Remove front steering knuckle.

a. On 4WD vehicles, use a plastic hammer, separate the front axle hub from the front drive shaft.

b. Remove the front steering knuckle.

To install:

15. Install front steering knuckle.
16. Install front upper suspension arm.

a. Install a new nut and a new clip. Tighten to 81 ft. lbs. (110 Nm).

17. Install front lower ball joint attachment.
18. Install tie rod end sub-assembly.
19. Install front stabilizer link assembly.
20. On 4WD vehicles, install front axle hub nut.
21. Inspect front axle hub bearing.
22. On 4WD vehicles, install front axle hub grease cap.
23. Install front speed sensor.
24. Install front disc.
25. Install front disc brake caliper assembly.

a. Install the front disc brake caliper assembly with the 2 bolts. Tighten to 91 ft. lbs. (123 Nm).

26. Connect the negative battery cable.
27. Check fluid level in reservoir.
28. Check VSC sensor signal.
29. Inspect and adjust front wheel alignment.

STABILIZER BAR & LINKS

REMOVAL & INSTALLATION

See Figures 230 through 232.

1. Before servicing the vehicle, refer to the precautions section.
2. Raise and safely support the vehicle securely on jackstands.
3. Remove front wheels.
4. Remove engine under cover sub-assembly.
5. Remove the 2 nuts and stabilizer link.

➡If the ball joint turns together with the nut, use a 6 mm hexagon wrench to hold the stud.

6. Remove the 4 bolts and 2 stabilizer brackets and remove the stabilizer bar.

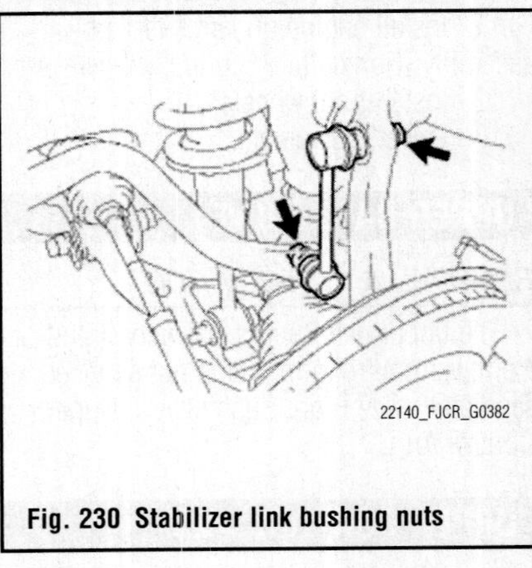

Fig. 230 Stabilizer link bushing nuts

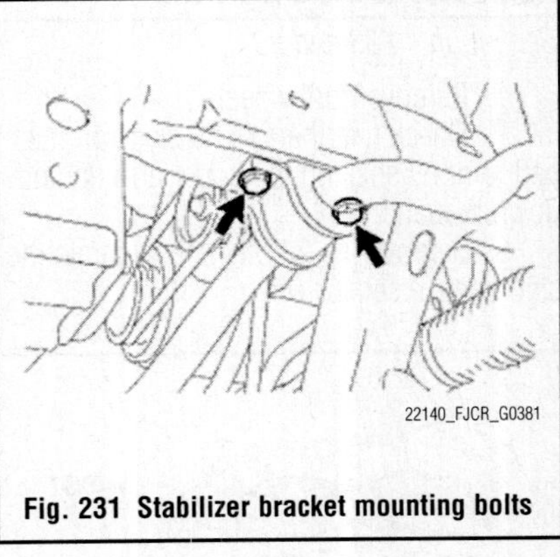

Fig. 231 Stabilizer bracket mounting bolts

7. Remove the 2 stabilizer bar bushings from the stabilizer bar.

To install:

8. Install the 2 stabilizer bar bushings onto the stabilizer bar.

a. Install the bushing onto the outer side of the bushing stopper on the stabilizer bar.

b. Install the bushing so that the protrusion faces inner side of the vehicle.

9. Install the stabilizer bar and 2 stabilizer brackets with the 4 bolts. Tighten to 30 ft. lbs. (40 Nm).
10. Install the stabilizer link with the 2 nuts. Tighten to 52 ft. lbs. (70 Nm).

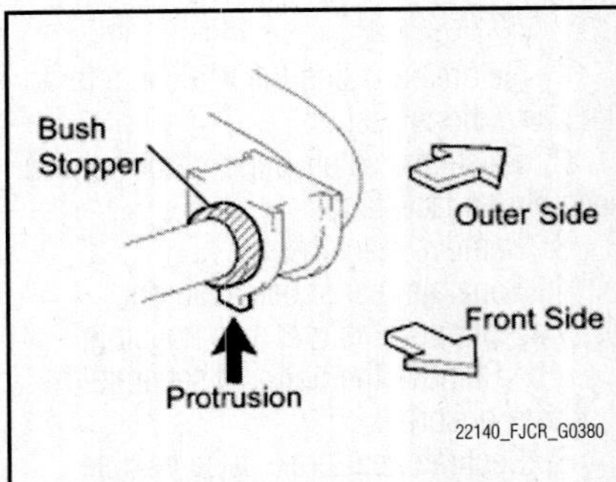

Fig. 232 Install the bushing onto the outer side of the bushing stopper and so the protrusion faces inner side of the vehicle

11. Install engine under cover sub-assembly. Tighten to 21 ft. lbs. (29 Nm).
12. Install front wheels.
13. Lower the vehicle.

UPPER BALL JOINT

REMOVAL & INSTALLATION

The upper ball joint is an integral part of the lower control arm and is not serviced separately. See Front Suspension, Upper Control Arm.

UPPER CONTROL ARM

REMOVAL & INSTALLATION

See Figures 233 and 234.

1. Remove front wheel.
2. Check that there is no slack on the ball joint by shaking the upper arm up and down by hand.
3. Remove the 2 bolts and separate the skid control sensor wire.

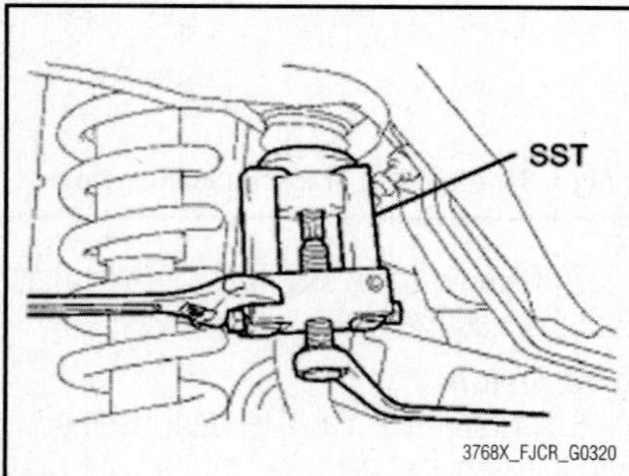

Fig. 233 Using SST: 09628-62011, separate the upper ball joint from the steering knuckle

Fig. 234 Remove the upper arm

4. Remove front upper suspension arm:
   a. Support the lower arm with a jack.
   b. Remove the clip and nut.
   c. Using SST: 09628-62011, separate the upper ball joint from the steering knuckle.
   d. Remove the bolt and separate the bracket.
   e. Remove the bolt, 2 washers and nut.
   f. Remove the upper arm.

To install:

5. Install the upper arm and temporarily tighten the bolt, 2 washers and nut.
6. Install the bracket with the bolt.
7. Install a new nut and a new clip. Tighten the nut to 81 ft. lbs. (110 Nm).
8. Install the skid control sensor wire with the 2 bolts.
9. Install front wheel.
10. Stabilize the suspension:
    a. Jack down the vehicle.
    b. Bounce the vehicle up and down several times to stabilize the suspension.
11. Fully tighten front upper suspension arm nut to 85 ft. lbs. (115 Nm).
12. Inspect and adjust front wheel alignment.

WHEEL HUB & BEARING

REMOVAL & INSTALLATION

See Figure 235.

1. Before servicing the vehicle, refer to the precautions section.
2. Remove the front steering knuckle.
3. Remove the 4 bolts, wheel hub and dust cover from the steering knuckle.
4. Remove the O-ring from the wheel hub.

To install:

5. Apply MP grease to a new O-ring.
6. Install the new O-ring onto the axle hub.
7. Install the dust cover and axle hub onto the steering knuckle with the 4 bolts. Tighten to 59 ft. lbs. (80 Nm).
8. Install the front steering knuckle.

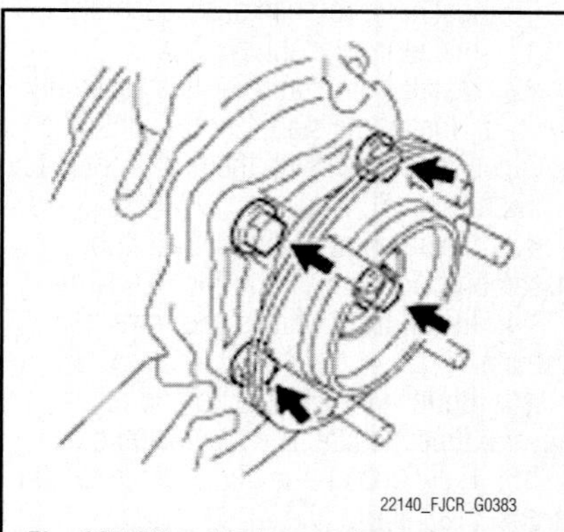

Fig. 235 Wheel hub mounting bolts

SUSPENSION

REAR SUSPENSION

COIL SPRING

REMOVAL & INSTALLATION

See Figure 236.

1. Before servicing the vehicle, refer to the precautions section.
2. Raise and safely support the vehicle securely on jackstands.
3. Remove rear wheel.
4. Separate rear shock absorber.
   a. Support the rear axle housing.
   b. Remove the bolt and separate the shock absorber.
5. Separate rear brake tube flexible hose.
   a. Using a union nut wrench, separate the 2 brake tubes.

➡Use a container to catch the brake fluid.

   b. Remove the 2 clips and disconnect the 2 flexible hoses.
6. Remove rear coil spring.
   a. Start to lower the rear axle housing.

⁂ WARNING

Do not snap the brake line or the parking brake cable.

   b. While lowering the rear axle housing, remove the coil spring.

To install:.

7. Install rear coil spring.
   a. Install the coil spring to the rear axle housing.

➡Fit the lower end of the coil spring into the gap of the spring lower seat.

8. Temporarily tighten rear shock absorber.
   a. Install the shock absorber and temporarily tighten the bolt.
9. Lower the vehicle.
10. Bounce the vehicle up and down several times to stabilize the suspension.
11. Fully tighten rear shock absorber to 72 ft. lbs. (98 Nm).
12. Install rear brake tube flexible hose.
    a. Install the 2 flexible hoses with 2 new clips.
    b. Install the 2 brake tubes onto the flexible hose. Tighten to 11 ft. lbs. (15 Nm).

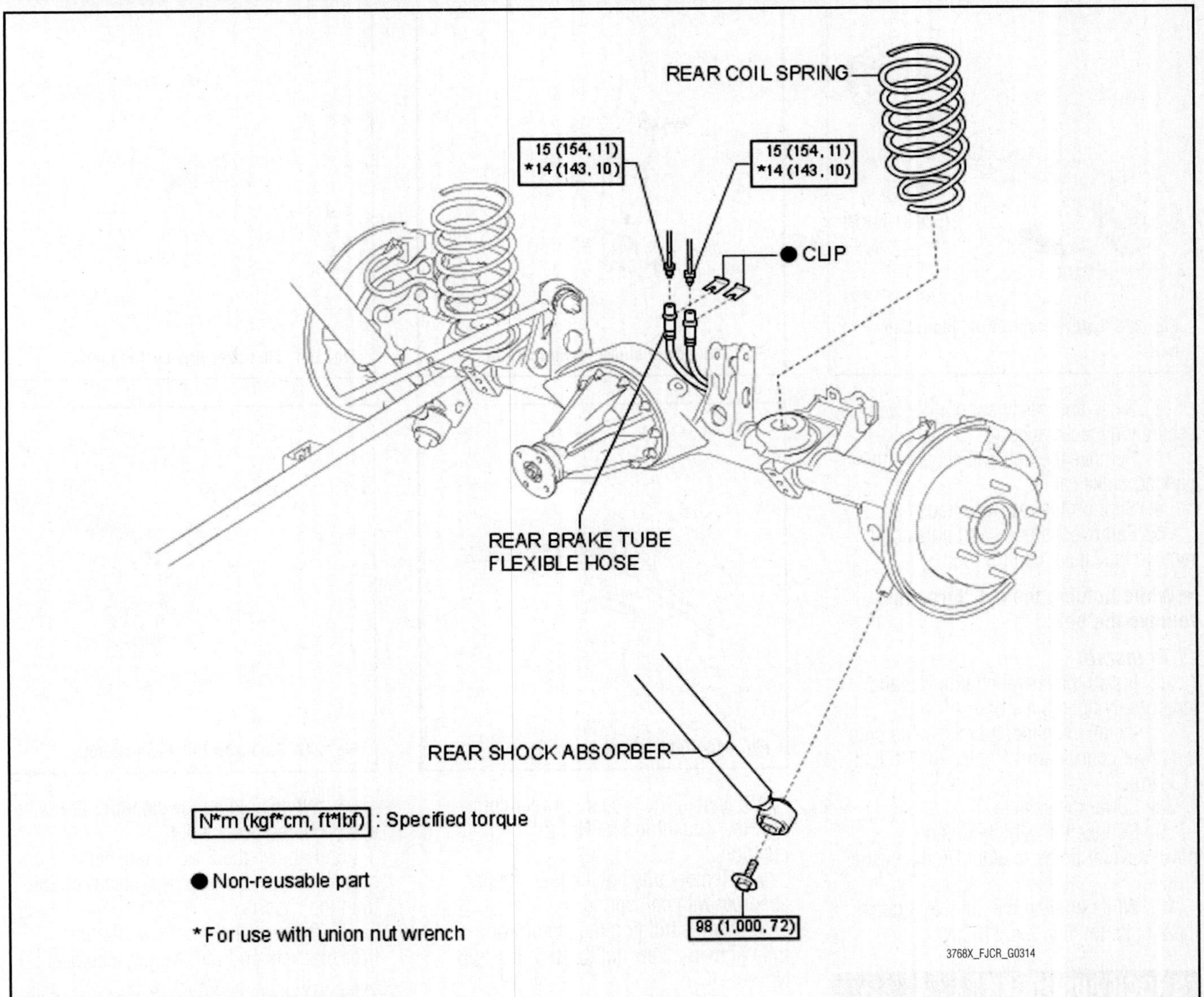

Fig. 236 Rear coil spring component locations

13. Fill reservoir with brake fluid.
14. Bleed brake line.
15. Check fluid level in reservoir.
16. Check for brake fluid leakage.
17. Install rear wheel.

CONTROL ARMS/LINKS

REMOVAL & INSTALLATION

Lateral Control Rod

See Figure 237.

1. Before servicing the vehicle, refer to the precautions section.
2. Raise and safely support the vehicle securely on jackstands.
3. Support the rear axle housing.
4. Remove the bolts and nuts and remove the lateral control rod.

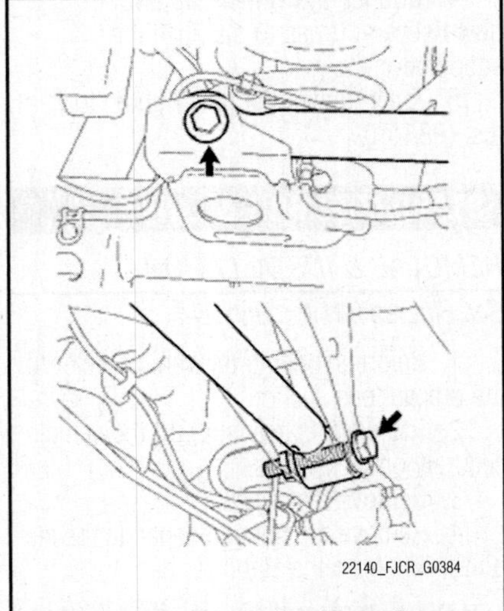

Fig. 237 Lateral control rod mounting bolts

➡While holding the nut, turn and remove the bolt.

To install:

5. Install the lateral control rod and temporarily tighten the bolt.
6. Temporarily tighten the bolt and nut.
7. Lower the vehicle.
8. Bounce the vehicle up and down several times to stabilize the suspension.
9. While holding the nut fully tighten the 2 bolts to 96 ft. lbs. (130 Nm).

Lower Control Arm

See Figure 238.

1. Before servicing the vehicle, refer to the precautions section.

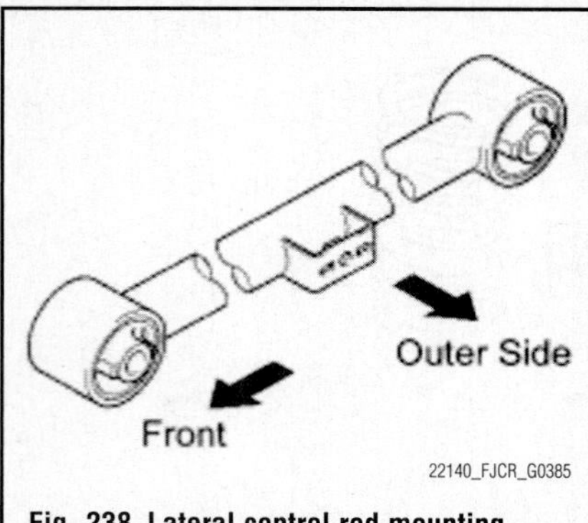

Fig. 238 Lateral control rod mounting bolts

2. Raise and safely support the vehicle securely on jackstands.
3. Remove the bolt and separate the parking brake cable.
4. Support the rear axle housing.
5. Remove the bolts and nuts and remove the lateral control rod.

➡While holding the nut, turn and remove the bolt.

To install:

6. Install the lateral control rod and temporarily tighten the bolt.
7. Install the parking brake cable onto the lower control arm. Tighten to 9 ft. lbs. (13 Nm).
8. Lower the vehicle.
9. Bounce the vehicle up and down several times to stabilize the suspension.
10. While holding the nut fully tighten the 2 bolts to 96 ft. lbs. (130 Nm).

SHOCK ABSORBER

REMOVAL & INSTALLATION

See Figures 239 and 240.

1. Before servicing the vehicle, refer to the precautions section.
2. Raise and safely support the vehicle securely on jackstands.
3. Remove rear wheels.
4. Support the rear axle housing.
5. Remove the bolt and separate the bottom of the shock absorber.
6. Remove the nut while keeping the piston rod from rotating.
7. Remove the 3 cushion retainers, the No. 1 cushion, the No. 2 cushion.
8. Remove the rear shock absorber.

To install:

9. Install the rear shock absorber.

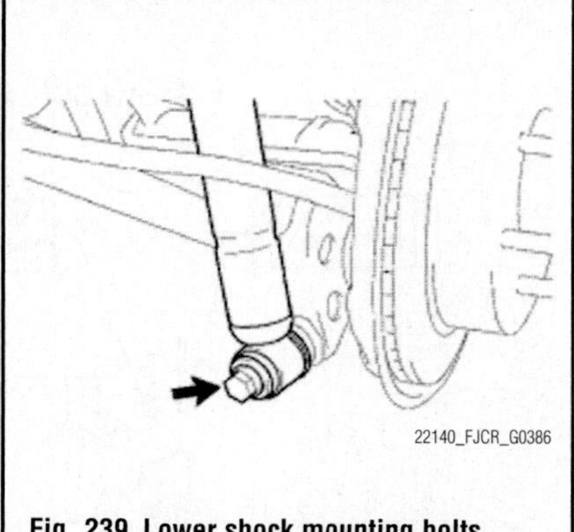

Fig. 239 Lower shock mounting bolts

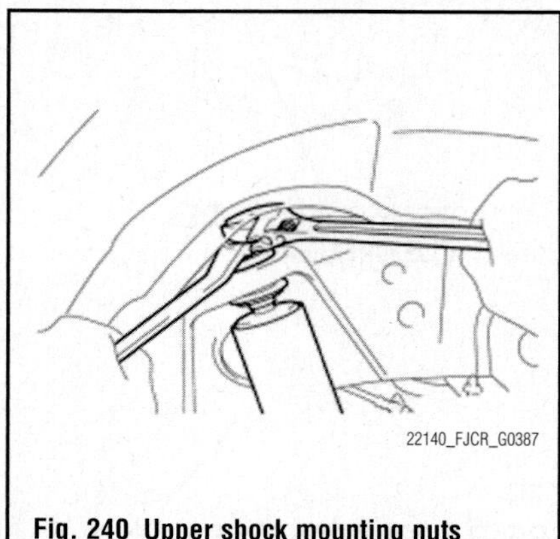

Fig. 240 Upper shock mounting nuts

10. Install the 3 cushion retainers, the No. 1 cushion, the No. 2 cushion.
11. Temporarily tighten rear shock absorber with coil spring.
12. While holding the piston rod, fully tighten a new nut to 18 ft. lbs. (25 Nm).
13. Install rear wheels and lower the vehicle.
14. Bounce the vehicle up and down several times to stabilize the suspension.
15. Fully tighten the nut. Tighten to 72 ft. lbs. (98 Nm).

STABILIZER BAR & LINKS

REMOVAL & INSTALLATION

See Figures 241 through 244.

1. Before servicing the vehicle, refer to the precautions section.
2. Raise and safely support the vehicle securely on jackstands.
3. Remove rear wheels.
4. Remove the nut and separate the stabilizer link from the stabilizer bar.

➡If the ball joint turns together with the nut, use a 6 mm hexagon wrench to hold the stud.

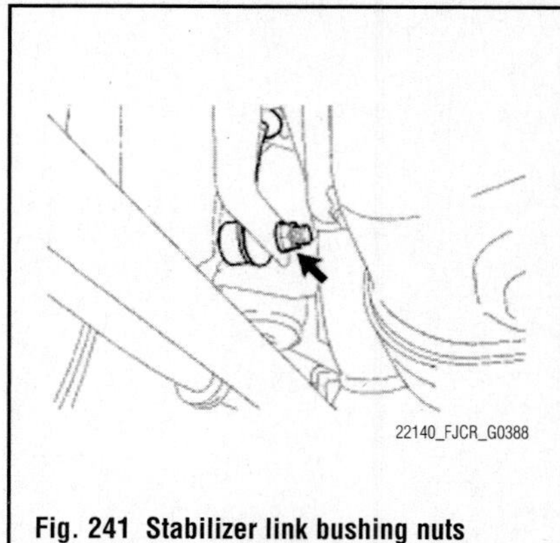

Fig. 241 Stabilizer link bushing nuts

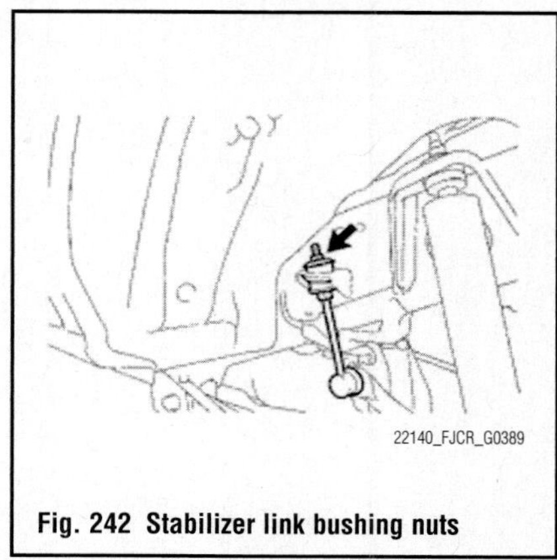

Fig. 242 Stabilizer link bushing nuts

5. While holding the stabilizer link with a wrench, remove the nut.
6. Remove the 2 No. 1 retainers, 2 cushions, No. 2 retainer and stabilizer link from the chassis.
7. Remove the 4 bolts and stabilizer brackets and remove the stabilizer bar.
8. Remove the 2 stabilizer bar bushings from the stabilizer bar.

To install:

9. Install the 2 stabilizer bar bushings onto the stabilizer bar.
10. Install the stabilizer bush onto the

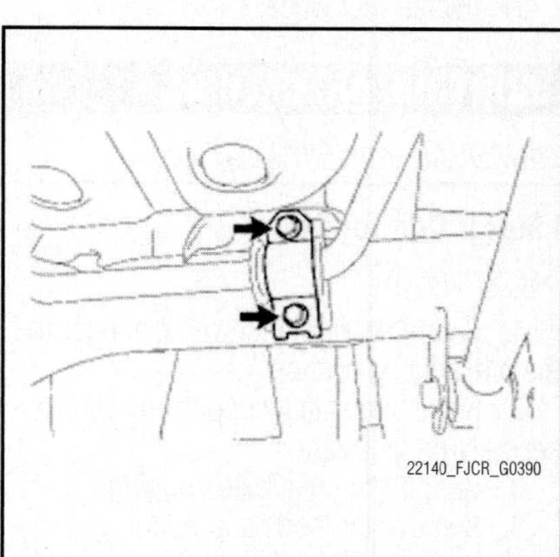

Fig. 243 Stabilizer bracket mounting bolts

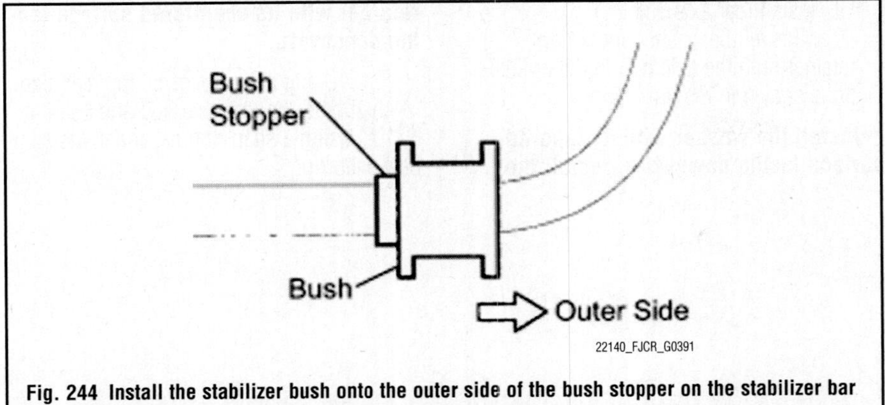

Fig. 244 Install the stabilizer bush onto the outer side of the bush stopper on the stabilizer bar

outer side of the bush stopper on the stabilizer bar.

11. Install the stabilizer bar and 2 stabilizer brackets with the 4 bolts. Tighten to 22 ft. lbs. (30 Nm).
12. Install the stabilizer link with the nuts. Tighten to 52 ft. lbs. (70 Nm).
13. Install the 2 No. 1 retainers, 2 cushions and No. 2 retainer.
14. While holding the stabilizer link with a spanner, install a new nut and tighten to 11 ft. lbs. (15 Nm).
15. Install rear wheels.
16. Lower the vehicle.

WHEEL BEARINGS

REMOVAL & INSTALLATION

See Figures 245 and 246.

1. Remove rear axle shaft from the housing.
2. Using a snap expander, remove the snap ring.
3. Using a press, remove the rear axle shaft.
    a. Remove the rear axle bearing inner retainer from the axle hub.
    b. Remove the rear axle shaft washer from the axle hub.
    c. Grind the rear axle bearing inner race surface using a grinder, then remove it with a chisel.
    d. Remove the rear axle shaft oil seal from the rear axle shaft.
4. Remove rear axle hub and bearing assembly.
    a. Attach the 4 nuts to the housing bolts.
    b. Using a hammer, remove the 4 housing bolts and rear axle hub and bearing assembly.

⁂ WARNING

Do not reuse the nuts previously removed from the vehicle.

5. Remove brake drum oil deflector.
6. Remove the deflector and deflector gasket to the rear axle shaft.

To install:

7. Install a new deflector gasket and deflector onto the rear axle shaft.
    a. Align the 2 notches.
    b. Install the washer and nut onto a new hub bolt, as shown in the illustration.
    c. Install the hub bolt by tightening the nut.

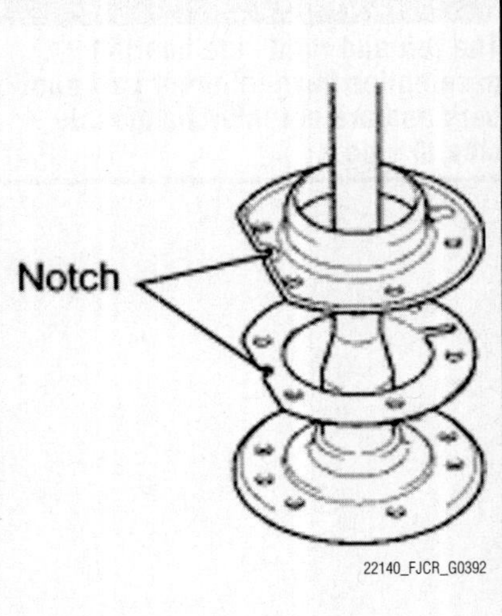

Fig. 245 Install a new deflector gasket and deflector onto the rear axle shaft and align the 2 notches

8. Install rear axle hub and bearing assembly.
    a. Position the parking brake plate on a new rear axle hub and bearing assembly and install the 4 housing bolts using 2 socket wrenches and a press.

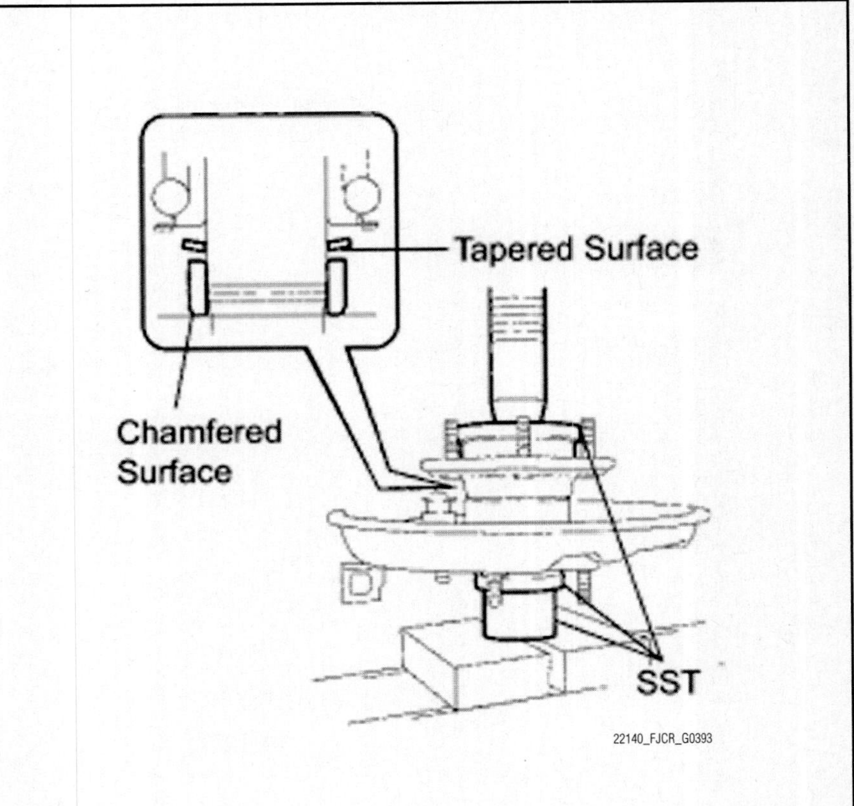

Fig. 246 Install the washer and a new retainer onto the axle hub in the orientations shown

WARNING

The left and right side bearing assemblies have different part numbers and are not interchangeable side to side.

9. Install rear axle shaft.
   a. Install the washer and a new retainer onto the axle hub in the orientations shown in the illustration.

➡Install the washer with its tapered surface facing downward. Install the retainer with its chamfered surface facing downward.

   b. Using a press, install the rear axle shaft onto the rear axle hub and bearing.
10. Using a snap ring expander, install a new snap ring.

TOYOTA

Highlander

9

SPECIFICATIONS AND MAINTENANCE CHARTS

ENGINE AND VEHICLE IDENTIFICATION

Engine						
Code	Liters (cc)	Cu. In.	Cyl.	Fuel Sys.	Engine Type	Eng. Mfg.
1AR-FE	2.7 (2672)	163	4	SFI	DOHC	Toyota
2GR-FE	3.5 (3456)	210	6	SFI	DOHC	Toyota

Model Year	
Code ①	Year
9	2009
A	2010

SFI: Sequential Fuel Injection

DOHC: Double Overhead Camshaft

① 10th digit of the VIN

3768X_HIGH_C0001

GENERAL ENGINE SPECIFICATIONS

Year	Model	Engine Displacement Liters	Engine Series ID	Net Horsepower @ rpm	Net Torque @ rpm (ft. lbs.)	Bore x Stroke (in.)	Com-pression Ratio	Oil Pressure @ rpm
2009	Highlander	2.7	1AR-FE	187@5800	186@4100	3.54x4.13	10:01	38@4000
		3.5	2GR-FE	270@6200	248@4700	3.70x3.27	10.8:1	55@3000
2010	Highlander	2.7	1AR-FE	187@5800	186@4100	3.54x4.13	10:01	38@4000
		3.5	2GR-FE	270@6200	248@4700	3.70x3.27	10.8:1	55@3000

3768X_HIGH_C0002

ENGINE TUNE-UP SPECIFICATIONS

Year	Engine Displacement Liters	Engine ID	Spark Plug Gap (in.)	Ignition Timing (deg.)*	Fuel Pump (psi)	Idle Speed (rpm)	Valve Clearance	
							Intake	Exhaust
2009	2.7	1AR-FE	0.043	8-12B	44-50	600-700	NA	NA
	3.5	2GR-FE	0.043	8-12B	44-50	600-700	NA	NA
2010	2.7	1AR-FE	0.043	8-12B	44-50	600-700	NA	NA
	3.5	2GR-FE	0.043	8-12B	44-50	600-700	NA	NA

NA: Not Available

NOTE: The Vehicle Emission Control Information label often reflects specification changes made during production.

The label figures must be used if they differ from those in this chart.

B: Before top dead center

* With terminals TC and CG connected to DLC3 (ODB II connector)

3768X_HIGH_C0003

CAPACITIES

Year	Model	Engine Displacement Liters	Engine ID	Engine Oil with Filter (qts.)	Transmission (pts.) 5-Spd	Transmission (pts.) Auto.*	Transfer Case (pts.)	Drive Axle Front (pts.)	Drive Axle Rear (pts.)	Fuel Tank (gal.)	Cooling System (qts.)
2009	Highlander	2.7	1AR-FE	4.6	NA	3.7	2.0	NE	2.0	19.1	①
		3.5	2GR-FE	6.4	NA	3.7	2.0	NE	2.0	19.1	①
2010	Highlander	2.7	1AR-FE	4.6	NA	3.7	2.0	NE	2.0	19.1	①
		3.5	2GR-FE	6.4	NA	3.7	2.0	NE	2.0	19.1	①

NA: Not Available

NE: Not Equipped

*After draining, add the following amounts, then, fill to the cold full line.

① Non-towing pkg. 7.3 qt. w/o rear air, 9.6 qts. w/ rear air
Towing pkg. 8 qt. w/o rear air, 10.4 qts. w/ rear air

3768X_HIGH_C0004

FLUID SPECIFICATIONS

Year	Model	Engine Displ. Liters	Engine Oil	Auto. Trans.	Drive Axle Rear ①	Transfer Case	Power Steering Fluid ②	Brake Master Cylinder	Cooling System
2009	Highlander	2.7	5W-20	ATF WS	80W-90	80W-90	NE	DOT 3	SLLC ③
		3.5	5W-30	ATF WS	80W-90	80W-90	NE	DOT 3	SLLC ③
2010	Highlander	2.7	5W-20	ATF WS	80W-90	80W-90	NE	DOT 3	SLLC ③
		3.5	5W-30	ATF WS	80W-90	80W-90	NE	DOT 3	SLLC ③

NE: Not Equipped

DOT: Department Of Transpotation

① Oil grade: Hypoid gear oil API GL-5

② The Highlander is not equipped with a power steering pump. It utilizes a power steering motor.

③ Toyota Super Long Life Coolant

3768X_HIGH_C0005

VALVE SPECIFICATIONS

Year	Engine Displacement Liters	Engine ID	Seat Angle (deg.)	Face Angle (deg.)	Spring Test Pressure (lbs. @ in.)	Spring Installed Height (in.)	Stem-to-Guide Clearance (in.) Intake	Stem-to-Guide Clearance (in.) Exhaust	Stem Diameter (in.) Intake	Stem Diameter (in.) Exhaust
2009	2.7	1AR-FE	45	44.5	NA	NA	0.0010-0.0024	0.0012-0.0026	0.2154-0.2159	0.2152-0.2157
	3.5	2GR-FE	45	44.5	NA	NA	0.0010-0.0024	0.0012-0.0026	0.2154-0.2159	0.2151-0.2157
2010	2.7	1AR-FE	45	44.5	NA	NA	0.0010-0.0024	0.0012-0.0026	0.2154-0.2159	0.2152-0.2157
	3.5	2GR-FE	45	44.5	NA	NA	0.0010-0.0024	0.0012-0.0026	0.2154-0.2159	0.2151-0.2157

NA: Information not available

3768X_HIGH_C0006

CAMSHAFT AND BEARING SPECIFICATIONS

All measurements are given in inches.

Year	Engine Displacement Liters	Engine VIN	Journal Diameter	Brg. Oil Clearance	Shaft End-play	Runout	Journal Bore	Lobe Lift Intake	Lobe Lift Exhaust
2009	2.7	1AR-FE	①	②	NA	0.0012	NA	1.73870-1.74429	1.73795-1.74354
	3.5	2GR-FE	③	④	NA	0.0016	NA	1.7447-1.7487	1.7426-1.7465
2010	2.7	1AR-FE	①	②	NA	0.0012	NA	1.73870-1.74429	1.73795-1.74354
	3.5	2GR-FE	③	④	NA	0.0016	NA	1.7447-1.7487	1.7426-1.7465

NA: Not Available

① No. 1 journal: 1.35626 to 1.53689 inches
Other journals: 0.90390 to 090453 inches

② Intake No. 1: 0.00137 to 0.00283 inches
Exhaust No. 1 journal: 0.00193 to 0.00339 inches
Other journals: 0.000984 to 0.00244 inches

③ No. 1 journal: 1.4152 to 1.4157 inches
Other journals: 1.0220 to 1.0226 inches

④ No. 1 journal: 0.0016 to 0.0031 inches
Other journals: 0.00098 to 0.0024 inches

3768X_HIGH_C0007

CRANKSHAFT AND CONNECTING ROD SPECIFICATIONS

All measurements are given in inches.

Year	Engine Displacement Liters	Engine ID	Crankshaft Main Brg. Journal Dia.	Crankshaft Main Brg. Oil Clearance	Crankshaft Shaft End-play	Crankshaft Thrust on No.	Connecting Rod Journal Diameter	Connecting Rod Oil Clearance	Connecting Rod Side Clearance
2009	2.7	1AR-FE	2.16531-2.16535	0.0006-0.0015	0.0016-0.0094	2	1.8894-1.8898	0.0011-0.0024	NA
	3.5	2GR-FE	2.4011-2.4016	0.0010-0.0019	0.0016-0.0095	2	2.0863-2.0866	0.0018-0.0026	0.0059-0.0157
2010	2.7	1AR-FE	2.16531-2.16535	0.0006-0.0015	0.0016-0.0094	2	1.8894-1.8898	0.0011-0.0024	NA
	3.5	2GR-FE	2.4011-2.4016	0.0010-0.0019	0.0016-0.0095	2	2.0863-2.0866	0.0018-0.0026	0.0059-0.0157

NA: Not Available

3768X_HIGH_C0008

PISTON AND RING SPECIFICATIONS

All measurements are given in inches.

Year	Engine Displ. Liters	Engine ID	Piston Clearance	Ring Gap Top Comp.	Ring Gap Bottom Comp.	Ring Gap Oil Control	Ring Side Clearance Top Comp.	Ring Side Clearance Bottom Comp.	Ring Side Clearance Oil Control
2009	2.7	1AR-FE	0.0003-0.0013	0.0087-0.0106	0.0146-0.0165	0.0039-0.0079	0.0008-0.0028	0.0008-0.0024	0.000787-0.00276
	3.5	2GR-FE	0.0018-0.0020	0.0098-0.0138	0.0197-0.0236	0.0039-0.0157	0.0008-0.0028	0.0008-0.0024	0.0028-0.0059
2010	2.7	1AR-FE	0.0003-0.0013	0.0087-0.0106	0.0146-0.0165	0.0039-0.0079	0.0008-0.0028	0.0008-0.0024	0.000787-0.00276
	3.5	2GR-FE	0.0018-0.0020	0.0098-0.0138	0.0197-0.0236	0.0039-0.0157	0.0008-0.0028	0.0008-0.0024	0.0028-0.0059

3768X_HIGH_C0009

TORQUE SPECIFICATIONS

All readings in ft. lbs.

Year	Engine Displacement Liters	Engine ID	Cylinder Head Bolts	Main Bearing Bolts	Rod Bearing Bolts	Crankshaft Damper Bolts	Flywheel Bolts	Manifold Intake	Manifold Exhaust	Spark Plugs	Oil Pan Drain Plug
2009	2.7	1AR-FE	①	29	②	192	72	15	26	18	30
	3.5	2GR-FE	③	④	⑤	184	61	15	15	18	33
2010	2.7	1AR-FE	①	29	②	192	72	15	26	18	30
	3.5	2GR-FE	③	④	⑤	184	61	15	15	18	33

① Step 1: 27
Step 2: 27
Step 3: Plus 90 degrees
Step 4: Plus 90 degrees

② Step 1: 15
Step 2: 30
Step 3: Plus 90 degrees

③ Step 1: 10mm bolts to 27 ft. lbs.
Step 2: 10mm point cap bolts plus 90 degrees
Step 3: 10mm point cap bolts plus 90 degrees
Step 4: Front bolts to 22 ft. lbs.

④ Step 1: 16 cap bolts to 45 ft. lbs.
Step 2: 16 cap bolts plus 90 degrees
Step 3: 8 side bolts to 38 ft. lbs.

⑤ Step 1: 18 ft. lbs.
Step 2: Plus 90 degrees

3768X_HIGH_C0010

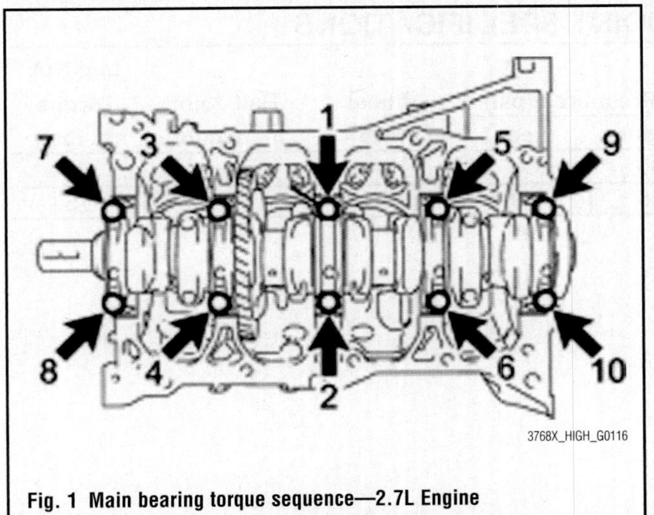

Fig. 1 Main bearing torque sequence—2.7L Engine

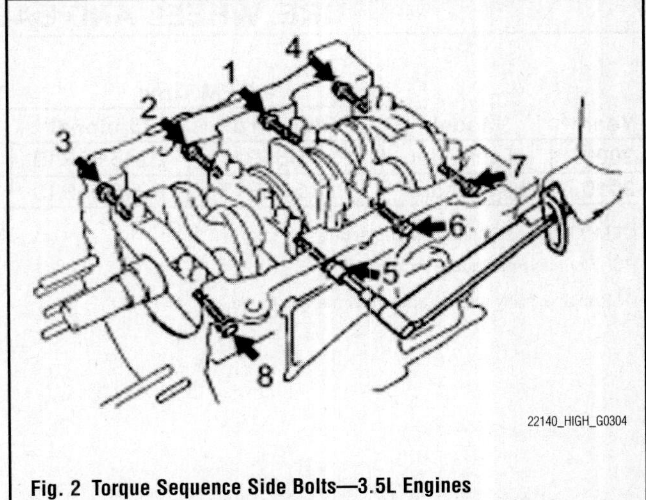

Fig. 2 Torque Sequence Side Bolts—3.5L Engines

WHEEL ALIGNMENT

Year	Model		Caster Range (+/-Deg.)	Caster Preferred Setting (Deg.)	Camber Range (+/-Deg.)	Camber Preferred Setting (Deg.)	Toe-in (in.)	Steering Axis Inclination (Deg.)
2009	Highlander	Front	0.75	+2.62	0.75	-0.63	0+/-0.08	11.02+/-0.75
		2WD R	NA	NA	0.75	-1.00	0.12+/-0.08	NA
		4WD R	NA	NA	0.75	-0.60	0.12+/-0.08	NA
2010	Highlander	Front	0.75	+2.62	0.75	-0.63	0+/-0.08	11.02+/-0.75
		2WD R	NA	NA	0.75	-1.00	0.12+/-0.08	NA
		4WD R	NA	NA	0.75	-0.60	0.12+/-0.08	NA

NA: Not Available

3768X_HIGH_C0011

TIRE, WHEEL AND BALL JOINT SPECIFICATIONS

Year	Model	OEM Tires		Tire Pressures (psi)		Wheel	Ball Joint	Lug Nut Torque
		Standard	Optional	Front	Rear	Size	Inspection	(ft. lbs.)
2009	Highlander	P245/65R17	P245/55R19	30	30	7.5-J	①	76
2010	Highlander	P245/65R17	P245/55R19	30	30	7.5-J	①	76

OEM: Original Equipment Manufacturer

PSI: Pounds Per Square Inch

① Replace if any measurable movement is found.

3768X_HIGH_C0012

BRAKE SPECIFICATIONS

All measurements in inches unless noted

Year	Model		Brake Disc Original Thickness	Brake Disc Minimum Thickness	Brake Disc Maximum Runout	Minimum Lining Thickness	Brake Caliper Bracket Bolts (ft. lbs.)	Brake Caliper Mounting Bolts (ft. lbs.)
2009	Highlander	F	1.102	1.024	0.0020	0.039	76	25
		R	0.394	0.335	0.0059	0.039	57	25
2010	Highlander	F	1.102	1.024	0.0020	0.039	76	25
		R	0.394	0.335	0.0059	0.039	56	25

F: Front

R: Rear

3768X_HIGH_C0013

SCHEDULED MAINTENANCE INTERVALS

TOYOTA—HIGHLANDER

TO BE SERVICED	TYPE OF SERVICE	VEHICLE MILEAGE INTERVAL (x1000)												
		7.5	15	22.5	30	37.5	45	52.5	60	67.5	75	82.5	90	97.5
Engine oil & filter	R	✓	✓	✓	✓	✓	✓	✓	✓	✓	✓	✓	✓	✓
Automatic transmission fluid	S/I		✓		✓		✓		✓		✓		✓	
Ball joints & dust covers	S/I		✓		✓		✓		✓		✓		✓	
Bolts & nuts on chassis & body	S/I		✓		✓		✓		✓		✓		✓	
Brake linings & drums	S/I		✓		✓		✓		✓		✓		✓	
Brake line pipes & hoses	S/I		✓		✓		✓		✓		✓		✓	
Brake pads & discs (front & rear)	S/I		✓		✓		✓		✓		✓		✓	
Propeller shaft grease	S/I		✓		✓		✓		✓		✓		✓	
Steering knuckle & chassis grease	S/I		✓		✓		✓		✓		✓		✓	
Steering linkage	S/I		✓		✓		✓		✓		✓		✓	
Air cleaner filter	R				✓				✓				✓	
Spark plugs ①	R				✓				✓				✓	
Drive belts	S/I				✓				✓				✓	
Exhaust pipes & mountings	S/I				✓				✓				✓	
Fuel lines & connections	S/I				✓				✓				✓	
Engine coolant	R						✓				✓			
Charcoal canister	R								✓					
Fuel tank cap gasket	R								✓					
Heated oxygen sensors (except Calif.) ②	R													

R: Replace S/I: Service or Inspect

① Platinum plugs are replaced at 100,000 mile intervals

② Heated oxygen sensors (except Calif.): replace every 80,000 miles.

FREQUENT OPERATION MAINTENANCE (SEVERE SERVICE)

If a vehicle is operated under any of the following conditions it is considered severe service:

- Extremely dusty areas.

- 50% or more of the vehicle operation is in 32°C (90°F) or higher temperatures, or constant temperatures below 0°C (32°F).

- Prolonged idling (vehicle operation in stop and go traffic).

- Frequent short running periods (engine does not warm to normal operating temperatures).

- Police, taxi, delivery usage or trailer towing usage.

Air cleaner filter: service or inspect every 3750 miles

Engine oil & filter: replace every 3750 miles.

Ball joints & dust covers: service or inspect every 7500 miles.

Bolts & nuts on chassis & body: service or inspect every 7500 miles.

Brake pads & discs (front & rear): service or inspect every 7500 miles.

Steering knuckle & chassis grease: service or inspect every 7500 miles.

Steering linkage: service or inspect every 7500 miles.

Exhaust pipes & mountings: service or inspect every 15,000 miles.

3768X_HIGH_C0014

BRAKES — INFORMATION AND PRECAUTIONS

ANTI-LOCK SYSTEMS

- Certain components within the ABS system are not intended to be serviced or repaired individually.
- Do not use rubber hoses or other parts not specifically specified for and ABS system. When using repair kits, replace all parts included in the kit. Partial or incorrect repair may lead to functional problems and require the replacement of components.
- Lubricate rubber parts with clean, fresh brake fluid to ease assembly. Do not use shop air to clean parts; damage to rubber components may result.
- Use only DOT 3 brake fluid from an unopened container.
- If any hydraulic component or line is removed or replaced, it may be necessary to bleed the entire system.
- A clean repair area is essential. Always clean the reservoir and cap thoroughly before removing the cap. The slightest amount of dirt in the fluid may plug an orifice and impair the system function. Perform repairs after components have been thoroughly cleaned; use only denatured alcohol to clean components. Do not allow ABS components to come into contact with any substance containing mineral oil; this includes used shop rags.
- The Anti-Lock control unit is a microprocessor similar to other computer units in the vehicle. Ensure that the ignition switch is **OFF** before removing or installing controller harnesses. Avoid static electricity discharge at or near the controller.
- If any arc welding is to be done on the vehicle, the control unit should be unplugged before welding operations begin.

DISC AND DRUM SYSTEMS

CAUTION

Dust and dirt accumulating on brake parts during normal use may contain asbestos fibers from production or aftermarket brake linings. Breathing excessive concentrations of asbestos fibers can cause serious bodily harm. Exercise care when servicing brake parts. Do not sand or grind brake lining unless equipment used is designed to contain the dust residue. Do not clean brake parts with compressed air or by dry brushing. Cleaning should be done by dampening the brake components with a fine mist of water, then wiping the brake components clean with a dampened cloth. Dispose of cloth and all residue containing asbestos fibers in an impermeable container with the appropriate label. Follow practices prescribed by the Occupational Safety and Health Administration (OSHA) and the Environmental Protection Agency (EPA) for the handling, processing, and disposing of dust or debris that may contain asbestos fibers.

BRAKES — BLEEDING THE BRAKE SYSTEM

BLEEDING PROCEDURE

BLEEDING & FLUID FILL PROCEDURE

When any work is done on the brake system that includes disconnecting fluid lines, or if air in the brake lines is suspected, bleed the air from the system.

WARNING

Do not let brake fluid remain on painted surfaces - it will eat away the paint if left on too long. Wash it off immediately.

Before proceeding, fill the brake fluid reservoir with brake fluid: SAE J1703 or FMVSS no. 116 DOT3

Bleeding the Master Cylinder

See Figure 3.

If the master cylinder has been disassembled or if the reservoir becomes empty, bleed the air from the master cylinder.

1. Remove the air cleaner assembly with hose.
2. Disconnect the brake lines from the master cylinder, using 12mm union nut wrench, or a suitable brake line wrench.
3. Have an assistant slowly depress the brake pedal and hold it.
4. Cover the outer holes with your fingers, and have your assistant release the brake pedal.
5. Repeat steps 3 and 4 several times.
6. Connect the brake lines. Tighten as follows:
    a. Use a torque wrench with a fulcrum length of 250 mm (9.84 inches) and tighten to 14 ft. lbs. (19 Nm). This torque value is effective when the union nut wrench is parallel to the torque wrench.
7. Install the air cleaner assembly with hose.

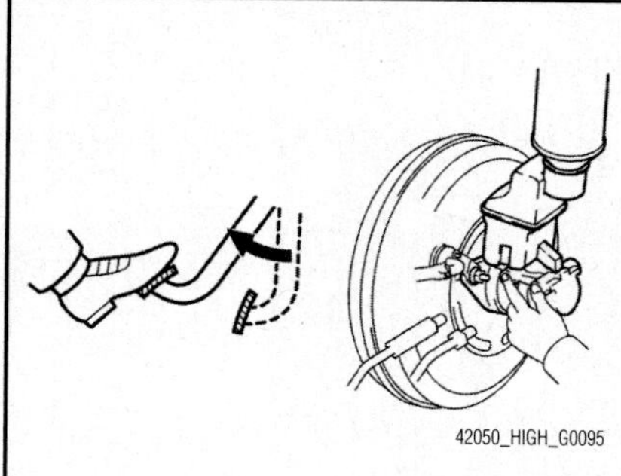

Fig. 3 Use your fingers to cover the outer holes, then release the brake pedal

Bleeding the Brake Lines

See Figure 4.

1. Raise and safely support the vehicle.
2. Connect a piece of vinyl tubing to the brake caliper.
3. Have an assistant depress the brake pedal several times, then loosen the bleeder plug while the pedal is depressed.
4. When fluid stops coming out, tighten the bleeder plug, then release the brake pedal.
5. Repeat steps 2 and 3 until all the air in the fluid has been bled out.
6. Tighten the brake bleeder plug to 73 inch lbs. (8.3 Nm).

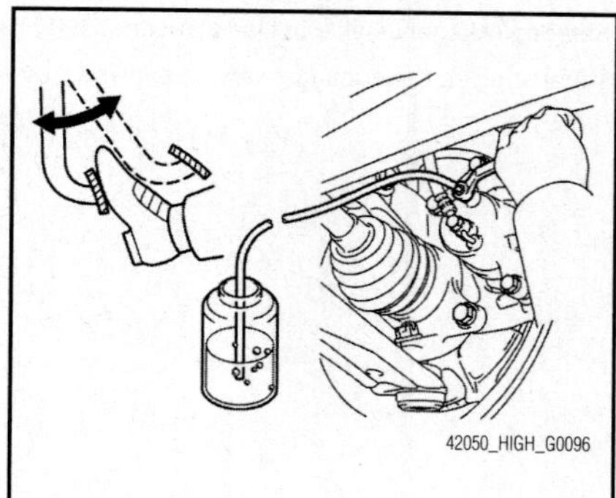

Fig. 4 Bleeding the brake lines at each wheel

7. Repeat the above steps to bleed the air out of the brake line for each wheel.

Bleeding the Actuator Assembly

See Figure 5.

➡After bleeding the air from the brake system, if the height or feel of the brake pedal cannot be obtained, perform air bleeding in the brake actuator assembly with a hand-held tester by following the procedures below.

1. Depress the brake pedal more than 20 times with the engine off.
2. Connect the hand-held tester to the DLC3, then turn the ignition switch to the **ON** position, but do NOT start the engine.
3. Select "AIR BLEEDING" on the hand-held tester.

➡Refer to the hand-held tester operator's manual for more details.

4. Bleed the air out of the regular brake line when "Step 1: Increase" appears on the hand-held tester display, as follows:

➡Bleed the air by following the steps displayed on the hand-held tester. Make sure that the brake fluid in the master cylinder reservoir tank does not become empty.

a. Connect the vinyl tube to either one of the bleeder plugs.
b. Have an assistant depress the brake pedal several times, then loosen the bleeder plug connected to the vinyl tube with the pedal depressed.
c. When fluid stops coming out, tighten the bleeder plug and release the brake pedal.
d. Repeat the previous 2 steps until all the air in the fluid is completely bled out.
e. Tighten the bleeder plug completely to 73 inch lbs. (8.3 Nm).
f. Repeat the above procedures for each wheel to bleed the air out of the brake line.

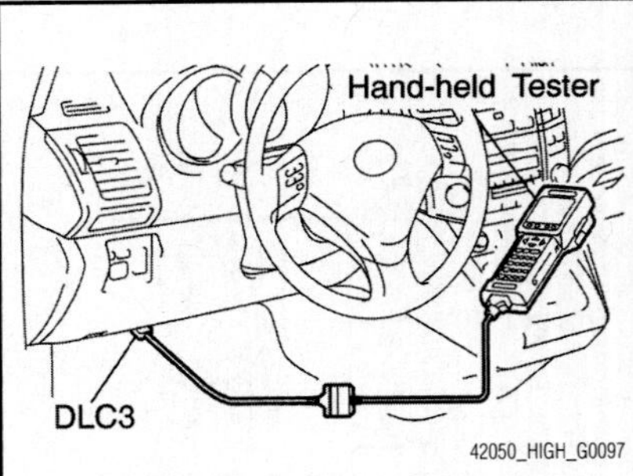

Fig. 5 Connect a suitable hand-held tester to the DLC3 to bleed the actuator assembly

5. Bleed the air out of the suction line when "Step 2: Inhalation" appears on the hand-held tester display, as follows:

➡Bleed the air by following the steps displayed on the hand-held tester. Make sure that the brake fluid in the master cylinder reservoir tank does not become empty.

a. Connect the vinyl tube to the bleeder plug at the right front wheel or the right rear wheel and loosen the bleeder plug.
b. Operate the brake actuator assembly to bleed the air using the hand-held tester.

➡This operation stops automatically after 4 seconds. At this time, be sure to release the brake pedal.

c. Check if the operation has stopped by referring to the hand-held tester display.
d. Repeat the previous 2 steps until all air in the fluid is completely bled out.
e. Tighten the bleeder plug completely to 73 inch lbs. (8.3 Nm).
f. Repeat the above procedures to bleed the air out of the brake line for each wheel.

6. Bleed the air out of the pressure reduction line when "Step 3: Decrease" appears on the hand-held tester display, as follows:

➡Bleed the air by following the steps displayed on the hand-held tester. Make sure that the brake fluid in the master cylinder reservoir tank does not become empty.

a. Connect a vinyl tube to either one of the bleeder plugs.
b. Loosen the bleeder plug.
c. Using the hand-held tester, operate the brake actuator assembly, completely depress the brake pedal and keep it depressed.

➡The operation stops automatically after 4 seconds. When performing this procedure continuously, set an interval of at least 20 seconds. When the operation is complete, the brake pedal goes down slightly. This is a normal phenomenon caused when the solenoid opens. During this procedure, the pedal will feel heavy, but completely depress it so that the brake fluid comes out from the bleeder plug. Be sure to keep depressing the brake pedal. Do not depress and release the pedal repeatedly.

d. Tighten the bleeder plug, then release the brake pedal.
e. Repeat the previous 3 steps until all the air in the fluid is completely bled out.
f. Tighten the bleeder plug completely to 73 inch lbs. (8.3 Nm).
g. Repeat the above procedures for each wheel to bleed the air out of the brake line.

7. Bleed the air out of the regular brake line again when "Step 4: Increase" appears on the hand-held tester display, as follows:

➡Bleed the air by following the steps displayed on the hand-held tester. Make sure that the brake fluid in the master cylinder reservoir tank does not become empty.

a. Connect the vinyl tube to either one of the bleeder plugs.
b. Depress the brake pedal several times, then loosen the bleeder plug connected to the vinyl tube with the pedal depressed.

8. When fluid stops coming out, tighten the bleeder plug, then release the brake pedal.
a. Repeat the previous 2 steps until all the air in the fluid is completely bled out.
b. Tighten the bleeder plug completely to 73 inch lbs. (8.3 Nm).
c. Repeat the above procedures for each wheel to bleed the air out of the brake line.

9. Check the fluid level and add fluid if necessary. Use SAE J1703 or FMVSS No. 116 DOT3 Brake fluid.

WHEEL SPEED SENSORS

REMOVAL & INSTALLATION

Front

See Figures 6 and 7.

➡If the sensor rotor needs to be replaced, replace it together with the front drive outboard joint shaft assembly.

1. Disconnect the negative battery cable.
2. Raise and safely support the vehicle.
3. Remove the front wheel and tire assembly.
4. Remove No. 1 engine under cover.
5. Remove the front fender molding sub-assembly, as follows:
   a. Remove the clip.
   b. Using a 4 mm hexagon wrench, remove the screw.
   c. Peel off the front fender side protector and disengage the 3 clips, and then remove the front fender molding sub-assembly.
   d. Remove the pad from the front fender molding sub-assembly.
   e. Remove the 2 No. 4 clips from the front fender molding sub-assembly.
   f. Remove the front fender side protector from the front fender molding sub-assembly.
6. Remove the front wheel opening extension pad.
7. Remove the front fender liner, as follows:
   a. Remove the screw.
   b. Using a screwdriver, turn the pin 90 degrees and remove the pin hold clip.
   c. Using a 4 mm hexagon wrench, remove the 2 screws.
   d. Remove the 2 grommets.

➡The grommets need to be replaced with new ones because they will break when they are removed.

   e. Remove the 5 clips.
   f. For left hand side, remove 8 screws and front fender liner.
   g. For right hand side, remove 7 screws and front fender liner.
8. Disconnect the front speed sensor connector and remove the 2 clamps.
9. Remove the bolt and No. 2 sensor clamp from the body.

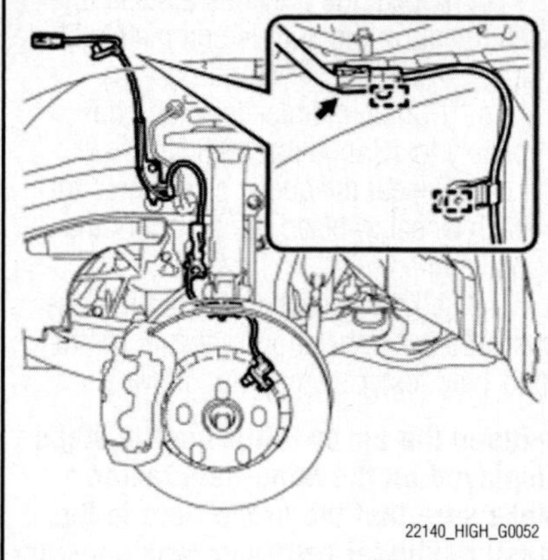

Fig. 6 Front speed sensor connector and clamps

10. Remove the bolt, No. 1 sensor clamp, and flexible hose together from the shock absorber assembly.
11. Remove the bolt, resin clamp, and the front speed sensor.

WARNING

Do not allow any foreign matter to come in contact with the tip of the sensor.

➡Clean the installation hole and the contact surface for the front speed sensor every time it is removed.

To install:

12. Install the resin clamp and the front speed sensor with bolt. Tighten to 71 inch lbs. (8 Nm).

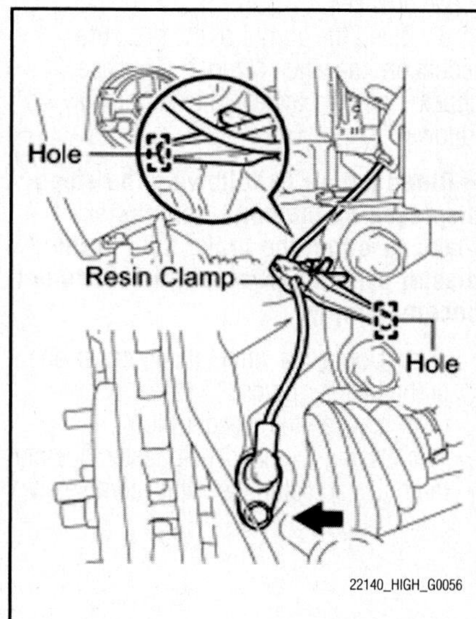

Fig. 7 Front speed sensor, bolt, and resin clamp removal

WARNING

Do not allow any foreign matter to come in contact with the tip of the sensor.

➡Firmly insert the front speed sensor body into the knuckle before tightening the bolt.

➡After installing the front speed sensor to the knuckle, make sure that there is no clearance between the front speed sensor stay and knuckle. Also make sure that no foreign matter is stuck between the parts.

➡Before installing the clamp, firmly insert the points of the clamp into the installation holes.

13. Temporarily install the No. 1 sensor clamp.

➡Be sure to insert the No. 1 sensor clamp claw into the stopper hole while installing the No. 1 sensor clamp.

14. Install the front flexible hose and the No. 1 sensor clamp together to the shock absorber with the bolt. Tighten to 14 ft. lbs. (19 Nm).

➡Do not twist the wire harness for the front speed sensor when installing it.

➡A bolt tightens the brake flexible hose and front speed sensor together. Make sure that the flexible hose is positioned over the front speed sensor.

15. Install the No. 2 sensor clamp to the body with the bolt. Tighten to 44 inch lbs. (5 Nm).

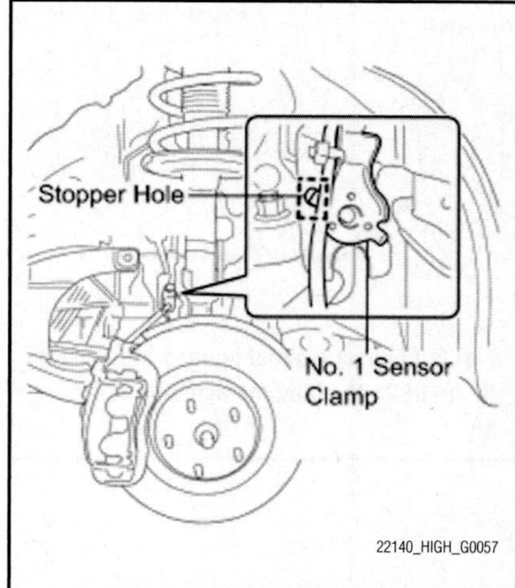

Fig. 8 Temporarily installing No. 1 sensor clamp

16. Install the 2 clamps and connect the front speed sensor connector.
17. Install the front fender liner.
18. Install the front wheel opening extension pad.
19. Install the front fender molding sub-assembly.
20. Install the No. 1 engine under cover.
21. Install the front wheel and tire assembly and tighten the lug nuts finger-tight.
22. Lower the vehicle, then final tighten the lug nuts to 76 ft. lbs. (103 Nm).
23. Connect the negative battery cable.
24. Check for speed sensor signal.

Rear–2WD Models

See Figures 9 through 12.

➡Use the same procedures for the Left side and Right side. The following procedure is for the Left side.

➡If the sensor rotor needs to be replaced, replace it together with the rear axle hub and bearing assembly with rear speed sensor.

1. Disconnect the negative battery cable.
2. Raise and safely support the vehicle.
3. Remove the rear tire and wheel assembly.
4. Using a screwdriver, disconnect the connector from the rear speed sensor.

✲✲ WARNING

Be careful not to damage the rear speed sensor.

5. Remove the two bolts and separate the rear disc brake caliper assembly.

➡Use wire or an equivalent tool to keep the brake caliper from hanging down by the flexible hose.

6. Put matchmarks on the rear disc and the axle hub.
7. Release the parking brake and remove the rear disc.

➡If the disc cannot be removed easily, turn and press firmly the shoe adjuster until the wheel comes free.

8. Remove the 4 bolts and the rear axle hub and bearing assembly, along with the wheel speed sensor.

➡Use wire or an equivalent tool to keep the parking brake assembly from hanging down by the parking brake cable assembly.

9. Install the hub nuts and mount the rear axle hub and bearing assembly in a vise using aluminum plates.

✲✲ WARNING

Replace the rear axle hub and bearing assembly if it is dropped or receives a strong shock.

10. Using appropriate tool, remove the rear speed sensor from the rear axle hub and bearing assembly.

To install:

11. Clean the contact surface between the rear axle hub and bearing assembly and a new rear speed sensor.

➡Do not allow foreign matter to attach to the sensor rotor.

12. Place the rear speed sensor on the axle hub so that the connector is positioned as shown in the illustration.
13. Using appropriate tool, steel plates, V-blocks and press, install a new rear speed sensor to the rear axle hub and bearing assembly.

✲✲ WARNING

Keep the rear speed sensor away from magnets.

✲✲ WARNING

Do not use a hammer to install the rear speed sensor.

✲✲ WARNING

Check that there is no foreign matter such as iron chips on the detecting portion of the rear speed sensor.

➡Slowly press the rear speed sensor in straight.

14. Install the parking brake assembly and the rear axle hub and bearing assembly with the 4 bolts, and tighten to 55 ft. lbs. (75 Nm).

➡Do not twist the No. 3 parking brake cable assembly when installing it.

15. Using a dial indicator, check for looseness near the center of the axle hub. Maximum looseness: 0 inches (0 mm). If the looseness exceeds the maximum, replace the rear axle hub and bearing assembly.

➡Ensure that the dial indicator is set perpendicular to the measurement surface.

16. Using a dial indicator, check for runout on the surface of the axle hub outside the hub bolt. Maximum runout: 0.0031 inches (0.08 mm). If the runout exceeds the maximum, replace the rear axle hub and bearing assembly.

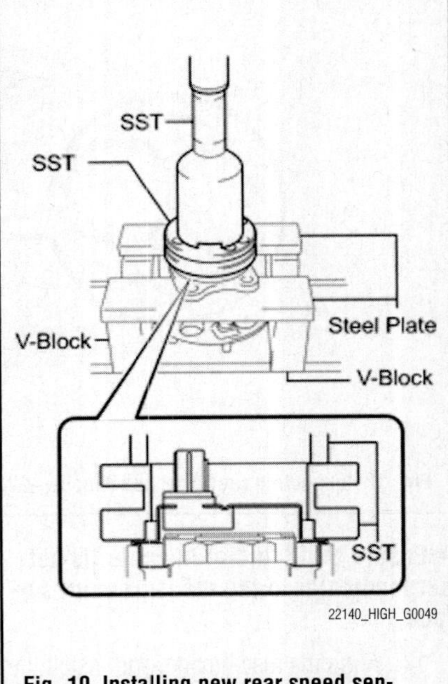

Fig. 10 Installing new rear speed sensor—2WD models

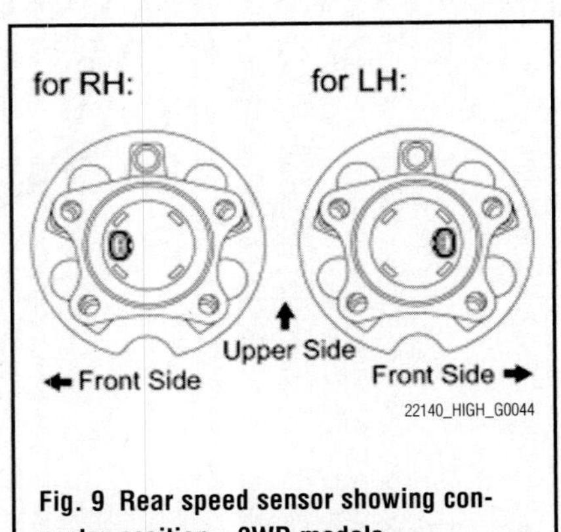

Fig. 9 Rear speed sensor showing connector position—2WD models

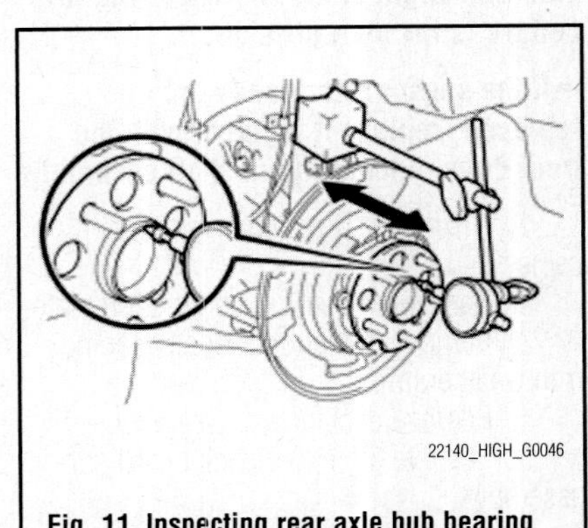

Fig. 11 Inspecting rear axle hub bearing looseness—2WD models

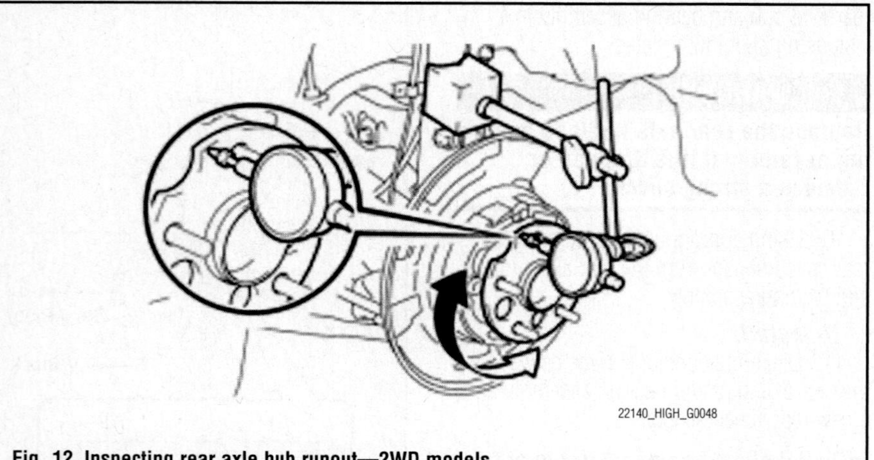

Fig. 12 Inspecting rear axle hub runout—2WD models

➡Ensure that the dial indicator is set perpendicular to the measurement surface.

17. Align the matchmarks and install the rear disc.

➡When replacing the rear disc with a new one, select the installation position where the rear disc has minimal runout.

18. Install the rear disc brake caliper assembly with the 2 bolts, and tighten to 57 ft. lbs. (78 Nm).
19. Connect the connector to the rear speed sensor.
20. Install the rear wheel and tire assembly and tighten the lug nuts finger-tight.
21. Lower the vehicle, then final tighten the lug nuts to 76 ft. lbs. (103 Nm).
22. Connect the negative battery cable.
23. Inspect and adjust rear wheel alignment.
24. Check for speed sensor signal.

Rear–4WD Models

See Figure 13.

➡Use the same procedures for the Left side and Right side. The following procedure is for the Left side.

➡If the sensor rotor needs to be replaced, replace it together with the front drive outboard joint shaft assembly.

1. Disconnect the negative battery cable.
2. Remove the left rear door scuff plate.
3. Remove the left rear door opening trim weatherstrip.
4. Remove deck board assembly.
5. Remove the No. 3 deck board sub-assembly.
6. Remove the No. 2 deck board sub-assembly.
7. Remove the tonneau cover assembly, as applicable.
8. Remove the rear No. 1 floor board, as applicable.
9. Remove both rear seat side covers, as applicable.
10. Remove both deck side trim boxes.
11. Remove jack carrier support.
12. Remove jack carrier cushion.
13. Remove jack assembly.
14. Remove jack carrier assembly.
15. Remove deck floor board assembly, as applicable.
16. Remove rear mat.
17. Remove rear deck floor box, as applicable.
18. Remove the rear No. 2 seat inner belt assembly, as applicable.
19. Disconnect both rear seat lap type belt assemblies, as applicable.
20. Remove the rear No. 2 seat assembly, as applicable.
21. Remove rear floor finish plate.
22. Remove left deck side trim cover and trim.
23. Remove left side trim cover or power outlet socket bezel, as applicable.
24. Remove left rear combination light service cover.
25. Remove rear power point socket assembly.
26. Remove rear power outlet socket cover.
27. Remove rear deck trim cover or left reclining remote control bezel, as applicable.
28. Remove left rope hook assembly.
29. Remove No. 2 deck side trim hook.
30. Remove left front deck side trim cover.
31. Remove left rear No. 1 seat outer belt assembly.
32. Remove left deck trim side panel assembly.

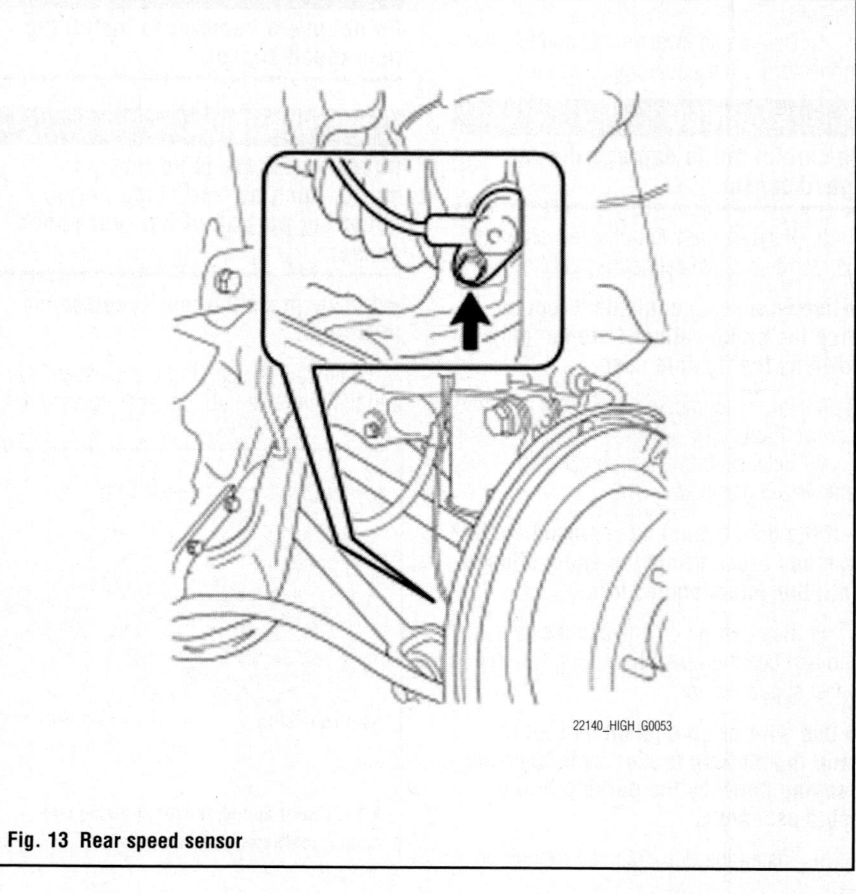

Fig. 13 Rear speed sensor

33. Raise and safely support the vehicle.
34. Remove rear wheel and tire assembly.
35. Disconnect the rear speed sensor connector.
36. Disconnect the grommet of the rear speed sensor wire from the hole of the wheel house.
37. Remove the 2 bolts, No. 1 clamp and No. 2 clamp from the body and absorber.
38. Remove the bolt and rear speed sensor body from the carrier.

WARNING

Do not allow any foreign matter to come in contact with the sensor tip or installation hole.

To install:

➡Use the same procedures for the Left side and Right side. The following procedure is for the Left side.

➡If the sensor rotor needs to be replaced, replace it together with the outboard joint shaft assembly.

39. Install the rear speed sensor with the bolt, and tighten to 71 inch lbs. (8 Nm).

WARNING

Do not allow any foreign matter to come in contact with the sensor tip or installation hole.

40. Install the No. 1 clamp and No. 2 clamp with the 2 bolts, and tighten to 44 inch lbs. (5 Nm).

WARNING

Do not twist the rear speed sensor wire when installing the clamp.

41. Insert the connector and grommet to the inside of the vehicle through the passage hole in the wheel well.

➡Make sure the grommet's band clamp remains on the outside of the vehicle.

42. Hold the grommet and pull it from the inside of the vehicle to the outside of the vehicle. Then fix it in place so that it is not tilted.

WARNING

When pulling out the grommet, do not grip the sensor wire.

43. Connect the rear speed sensor connector.
44. Install the rear wheel and tire assembly, and tighten the lug nuts finger-tight.
45. Lower the vehicle, then final tighten the lug nuts to 76 ft. lbs. (103 Nm).
46. Install left deck trim side panel assembly.
47. Connect left rear no. 1 seat outer belt assembly.
48. Install left front deck side trim cover.
49. Install no. 2 deck side trim hook.
50. Install left rope hook assembly.
51. Install rear deck trim cover, or left reclining remote control bezel, as applicable.
52. Install rear power outlet socket cover.
53. Install rear power point socket assembly.
54. Install left rear combination light service cover.
55. Install left side trim cover, or power outlet socket bezel, as applicable.
56. Install left deck side trim and trim cover.
57. Install rear floor finish plate.
58. Install rear no. 2 seat assembly, as applicable.
59. Connect rear seat lap type belt assemblies, both left and right, as applicable.
60. Install rear no. 2 seat inner belt assembly, as applicable.
61. Install rear deck floor box, as applicable.
62. Install deck floor board assembly, as applicable.
63. Install rear mat.
64. Install deck side trim boxes, both left and right.
65. Install rear seat side covers, both left and right, as applicable.
66. Install jack carrier assembly.
67. Install jack assembly.
68. Install jack carrier cushion.
69. Install jack carrier support.
70. Install rear no. 1 floor board, as applicable.
71. Install tonneau cover assembly, as applicable.
72. Install no. 2 deck board sub-assembly.
73. Install no. 3 deck board sub-assembly.
74. Install deck board assembly.
75. Install left rear door opening trim weatherstrip.
76. Install left rear door scuff plate.
77. Connect the negative battery cable.
78. Check for speed sensor signal.

BRAKES

FRONT DISC BRAKES

BRAKE CALIPER

REMOVAL & INSTALLATION

See Figure 14.

1. Disconnect the brake line from the caliper and plug it.
2. Hold the caliper slide pins and remove the mounting bolts.
3. Lift off the caliper.
4. Remove the pads and anti-squeal shims.
5. Remove the wear indicator from the inner pad.
6. Installation is the reverse of removal. Grease the caliper slides and bolts with lithium grease or equivalent. Apply disc brake grease to the anti-squeal shims. Torque the caliper bolts to 25 ft. lbs. (34 Nm); the brake line union bolt to 21 ft. lbs. (29 Nm).

DISC BRAKE PADS

REMOVAL & INSTALLATION

See Figure 14.

1. Hold the sliding pin and remove the lower bolt.
2. Lift the caliper up and secure it.
3. Remove the pads, 4 shims and wear indicator plate. Remove the 2 pad support plates.

➡The support plates can be reused, provided they have sufficient rebound, are not deformed or cracked, show no signs of wear and are cleaned of all rust and debris.

To install:

➡Always use new shims and wear indicators, even when re-installing the original pads.

4. Install a wear indicator plate on the inner pad.
5. Apply disc brake grease to both sides of the inner anti-squeal shims and install the shims.
6. Install the inner pad with the wear indicator plate facing upwards.
7. Install the outer pad.
8. Install the caliper. Torque the bolt to 25 ft. lbs. (34 Nm).
9. Install the wheel and tire assembly and carefully lower the vehicle.

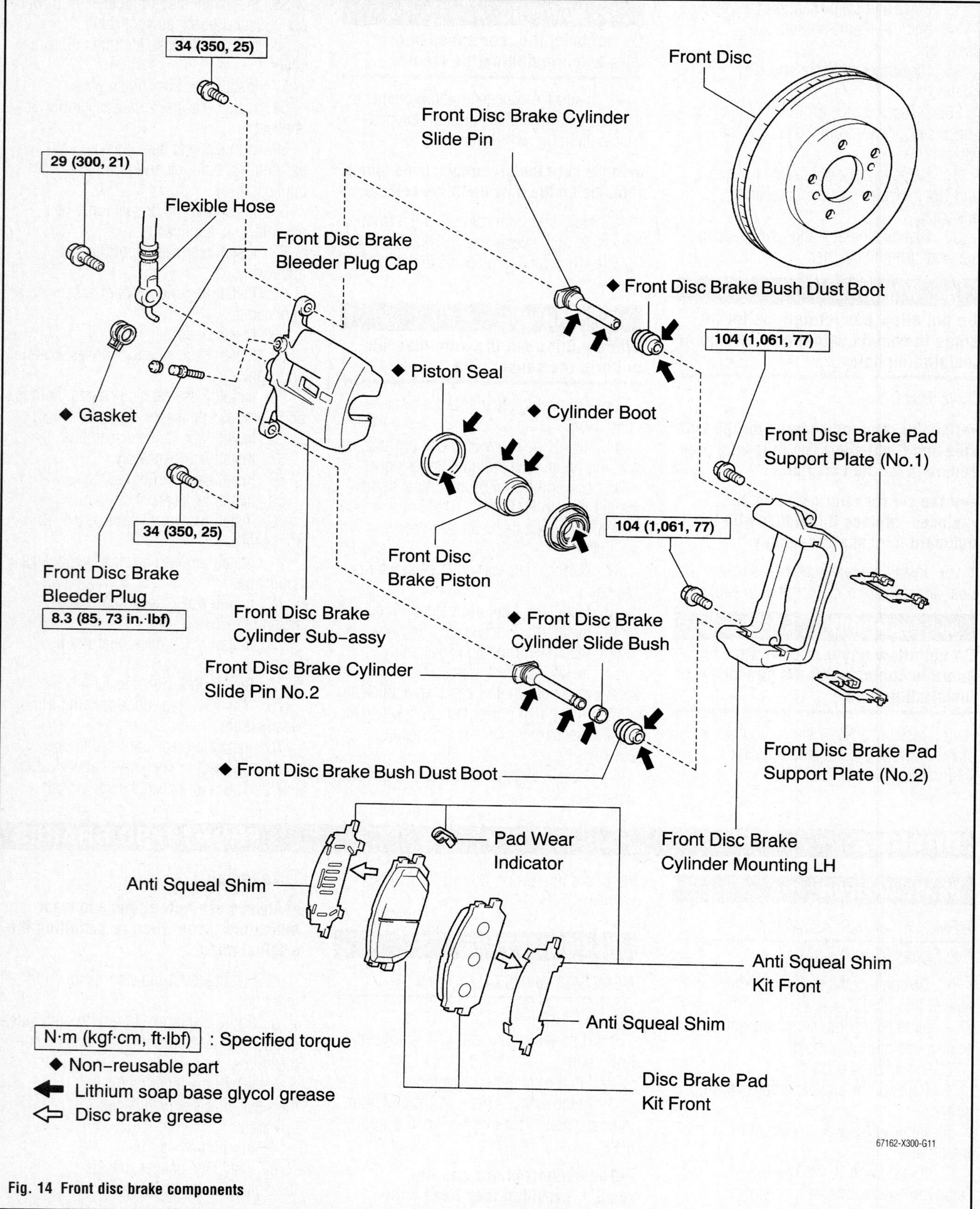
34 (350, 25)
Front Disc
Front Disc Brake Cylinder
Slide Pin
29 (300, 21)
Flexible Hose
Front Disc Brake
Bleeder Plug Cap
◆ Front Disc Brake Bush Dust Boot
104 (1,061, 77)
◆ Piston Seal
◆ Cylinder Boot
◆ Gasket
Front Disc Brake Pad
Support Plate (No.1)
104 (1,061, 77)
34 (350, 25)
Front Disc
Brake Piston
Front Disc Brake
Bleeder Plug
8.3 (85, 73 in.·lbf)
Front Disc Brake
Cylinder Sub-assy
◆ Front Disc Brake
Cylinder Slide Bush
Front Disc Brake Cylinder
Slide Pin No.2
◆ Front Disc Brake Bush Dust Boot
Front Disc Brake Pad
Support Plate (No.2)
Pad Wear
Indicator
Front Disc Brake
Cylinder Mounting LH
Anti Squeal Shim
Anti Squeal Shim
Kit Front
Anti Squeal Shim
N·m (kgf·cm, ft·lbf) : Specified torque
◆ Non-reusable part
Lithium soap base glycol grease
Disc brake grease
Disc Brake Pad
Kit Front
67162-X300-G11

Fig. 14 Front disc brake components

BRAKE CALIPER

REMOVAL & INSTALLATION

See Figure 15.

1. Disconnect the brake line from the caliper and plug it.
2. Remove the caliper mounting bolts.
3. Lift off the caliper.
4. Remove the pads and anti-squeal shims.
5. Remove the wear indicators from each pad.
6. Installation is the reverse of removal. Grease the caliper slides and bolts with lithium grease or equivalent. Apply disc brake grease to the anti-squeal shims. Torque the caliper bolts to 25 ft. lbs. (34 Nm); the brake line union bolt to 24 ft. lbs. (33 Nm).

DISC BRAKE PADS

REMOVAL & INSTALLATION

See Figure 15.

1. Disconnect the brake line from the caliper and plug it.

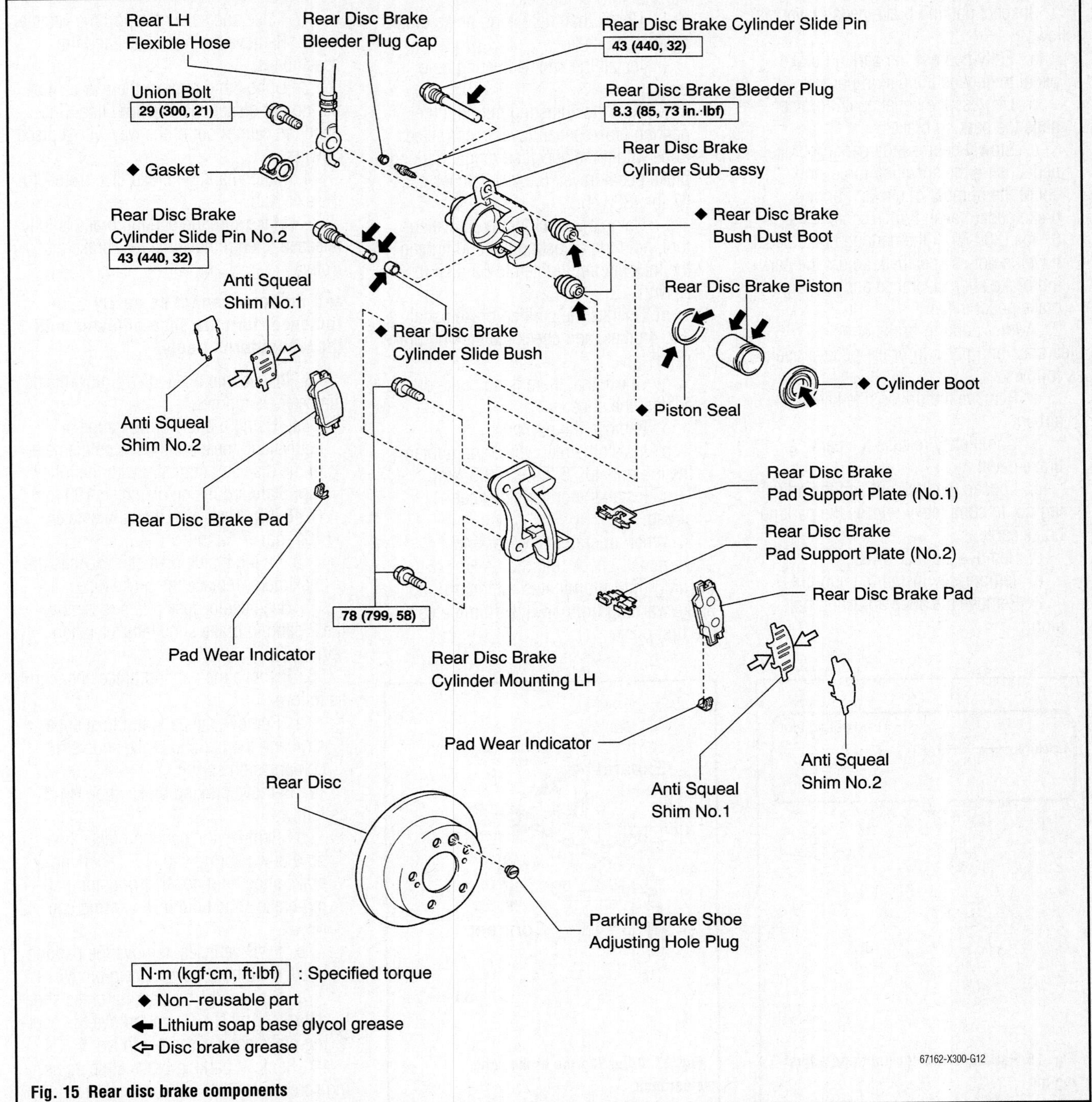

Fig. 15 Rear disc brake components

2. Remove the caliper mounting bolts.
3. Lift off the caliper.
4. Remove the pads and anti-squeal shims.
5. Remove the wear indicators from each pad.
6. Installation is the reverse of removal. Grease the caliper slides and bolts with lithium grease or equivalent. Apply disc brake grease to the anti-squeal shims. Torque the caliper bolts to 25 ft. lbs. (34 Nm); the brake line union bolt to 24 ft. lbs. (33 Nm).

BRAKES — PARKING BRAKE

PARKING BRAKE CABLES

ADJUSTMENT

See Figures 16 and 17.

1. Inspect parking brake pedal travel, as follows:
   a. Fully depress the parking brake pedal to engage the parking brake.
   b. Depress the pedal again to disengage the parking brake.
   c. Slowly depress the parking brake pedal using the specified force, and count the number of clicks. Parking brake pedal travel: 8 to 10 notches at 67 lbs (300 N). If the parking brake pedal travel is not as specified, adjust the parking brake shoe clearance and parking brake pedal travel.
2. Adjust parking brake shoe clearance and parking brake pedal travel, as follows:
   a. Remove the driver side knee airbag.
   b. Completely release the parking brake pedal.
   c. Loosen the lock nut and the adjusting nut to completely release the parking brake cable.
   d. Remove the rear wheel.
   e. Temporarily install the hub nuts.
   f. Remove the shoe adjusting hole plug.
   g. Turn the shoe adjuster and expand the shoe until the disc locks.
   h. Turn and contract the shoe adjuster until the disc can rotate smoothly. Standard: Return 8 notches.
   i. Check that there is no brake drag against the shoe.
   j. Install the shoe adjusting hole plug.
   k. Turn the adjusting nut until the parking brake pedal travel is corrected to be within the specified range. Parking brake pedal travel: 8 to 10 notches at 67 lbs (300 N).
   l. Using a wrench or an equivalent tool, hold the adjusting nut and tighten the lock nut and tighten to 62 inch lbs. (7 Nm).
   m. Operate the parking brake pedal 3 to 4 times, and check the parking brake pedal travel.
   n. Check that there is no brake drag against the shoe.
   o. Remove the hub nuts.
   p. Install the rear wheel and tighten the lug nuts to 76 ft. lbs. (103 Nm).
   q. Install the driver side knee airbag.
3. When operating the parking brake pedal, check that the brake warning light illuminates. Standard: the brake warning light always illuminates at the first click.

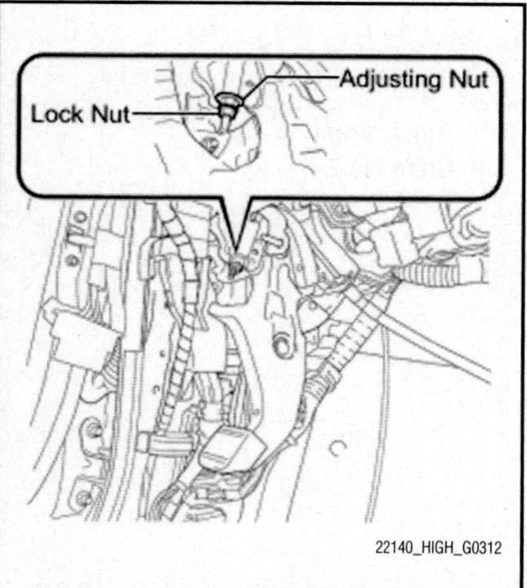

Fig. 16 Parking brake lock nut and adjusting nut

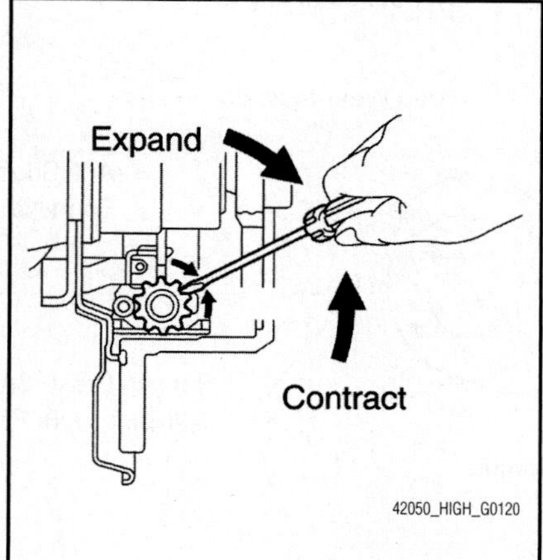

Fig. 17 Adjusting the brake shoe clearance

PARKING BRAKE SHOES

REMOVAL & INSTALLATION

See Figures 18 through 24.

1. Raise and safely support the vehicle.
2. Remove the rear wheel and tire assemblies.
3. Unbolt and remove the rear caliper, but do not disconnect the fluid line. Suspend the caliper out of the way with a piece of wire.
4. Matchmark the brake disc (rotor) to the axle hub.
5. Make sure the parking brake is fully released, then remove the rear brake disc (rotor).

➡If the rotor cannot be easily removed, turn the shoe adjuster until the wheel turns freely.

6. Inspect the brake disc (rotor) inside diameter, as follows:
   a. Using a brake drum gauge or equivalent, measure the inside diameter of the disc and compare with the following: Standard inside diameter: 190 mm (7.48 in.). Maximum inside diameter: 191 mm (7.52 in.)
   b. If the inside diameter exceeds the maximum, replace the brake disc.
7. Use needle-nose pliers to remove the 2 parking brake shoe return tension springs.
8. Remove the parking brake shoe strut, as follows:
   a. Remove the parking brake shoe strut and the parking brake shoe strut compression spring.
9. Remove parking brake shoe No. 1, as follows:
   a. Remove the parking brake shoe hold down spring cup No. 1, parking brake shoe hold down spring and parking brake shoe hold down spring cup No. 2.
   b. FWD vehicles, remove the parking brake shoe hold down spring pin No. 1.
   c. Disconnect the parking brake shoe return spring No. 2 and remove the parking brake shoe assembly lh No. 1.
10. Remove parking brake shoe adjusting screw set:

a. Remove the parking brake shoe adjusting screw set.

b. Remove the parking brake shoe return tension spring No. 2.

11. Remove parking brake shoe assembly No. 2:

a. Remove the parking brake shoe hold down spring cup No. 1, parking brake shoe hold down spring, parking brake shoe hold down spring cup No. 2 and parking brake shoe hold down spring pin No. 2.

b. Remove the parking brake shoe assembly lh No. 2.

c. Using needle-nose pliers, disconnect the parking brake cable No. 3 from the parking brake cable shoe lever.

WARNING

Be careful not to damage parking brake cable No. 3.

12. On 4WD models, separate the rear speed sensor.

13. On 4WD models, remove the rear axle shaft nut.

14. On 4WD models, remove rear axle hub & bearing assembly

15. On 4WD models, remove parking brake shoe hold down spring pin.

16. Remove parking brake shoe type C-washer, as follows:

a. Using a screwdriver, remove the c-washer.

b. Remove the shim and parking brake shoe lever from the parking brake shoe No. 2.

17. Inspect parking brake shoe lining thickness:

a. Using a ruler, measure the thickness of the shoe lining. Standard thickness is 2.5 mm (0.098 in.) and minimum thickness is 1.0 mm (0.039 in.). If the lining thickness is less than or equal to the minimum, or If there is severe or uneven wear, replace the brake shoe.

18. Inspect brake disc and parking brake shoe lining for proper contact

a. Apply chalk to the inside surface of the disc, then grind down the brake shoe lining to fit disc.

b. If the contact between the brake disc and the shoe lining is improper, repair it using a brake shoe grinder or replace the brake shoe assembly.

To install:

19. Install the parking brake shoe type C-washer, as follows

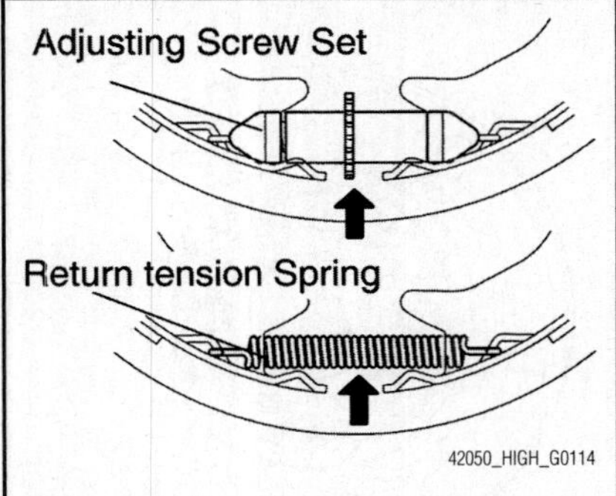

Fig. 18 Remove the parking brake shoe adjusting screw set and shoe return spring

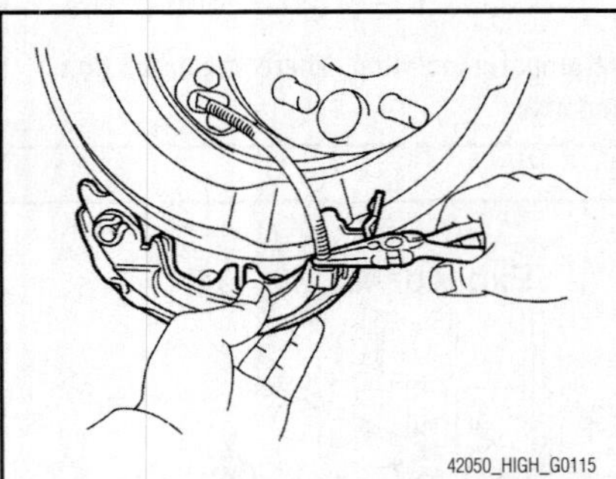

Fig. 19 Using needle-nose pliers, disconnect the parking brake cable No. 3 from the parking brake cable shoe lever

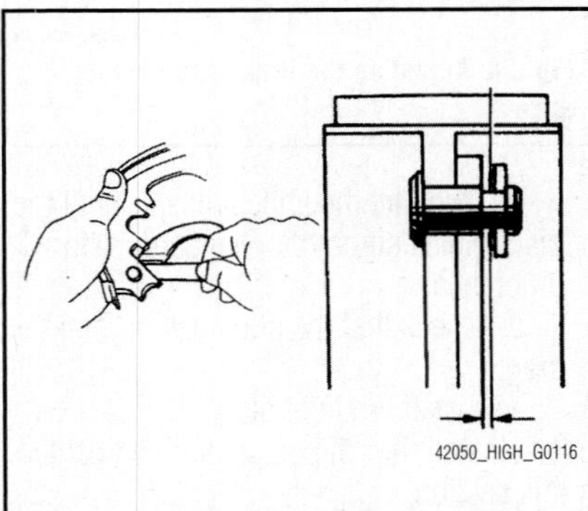

Fig. 20 Installing the parking brake shoe type C-washer

a. Using a feeler gauge, measure the clearance. Standard clearance: less than 0.35 mm (0.014 in.). If the clearance is not within the specifications, replace the shim with one of the correct size. The shim sizes: 0.3 mm (0.012 in.), 0.9 mm (0.035 in.) or 0.6 mm (0.024 in.).

b. Using pliers, install the parking brake shoe lever and the shim with a new C-washer.

20. Apply high temperature grease to the shaded parts shown in the illustration of the backing plate which make contact with the shoe.

21. On 4WD models, perform the following:

a. Install the parking brake shoe hold down spring pin.

b. Install the rear axle hub & bearing.

c. Install the rear axle shaft nut.

d. Install the rear speed sensor.

22. Install parking brake shoe No. 2, as follows:

a. Using needle-nose pliers, connect the parking brake cable No. 3 to the parking brake cable shoe lever.

➡Be careful not to damage the parking brake cable No. 3.

b. Install the parking brake shoe No. 2 with the parking brake shoe hold down spring, parking brake shoe hold down spring cup No. 1, parking brake shoe hold down spring cup No. 2 and parking brake shoe hold down spring pin No. 2.

23. Install the parking brake shoe adjusting screw set, as follows:

a. Apply high temperature grease to the parking brake shoe adjusting bolt and piece.

b. Attach the parking brake shoe return tension spring No. 2 to the parking brake shoe No. 1 and parking brake shoe assembly No. 2.

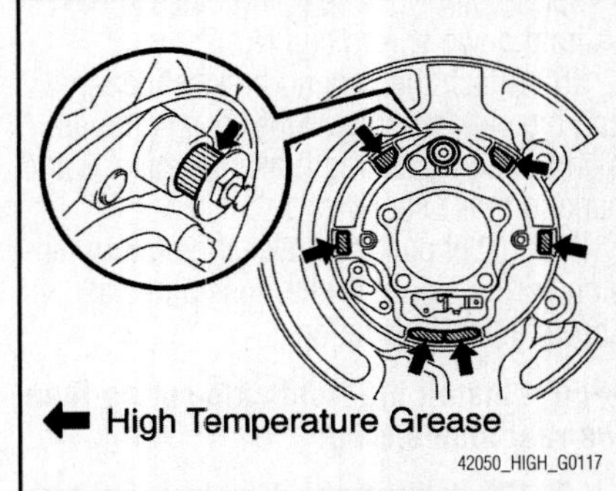

Fig. 21 Apply high temperature grease to the shaded parts of the backing plate which make contact with the shoe

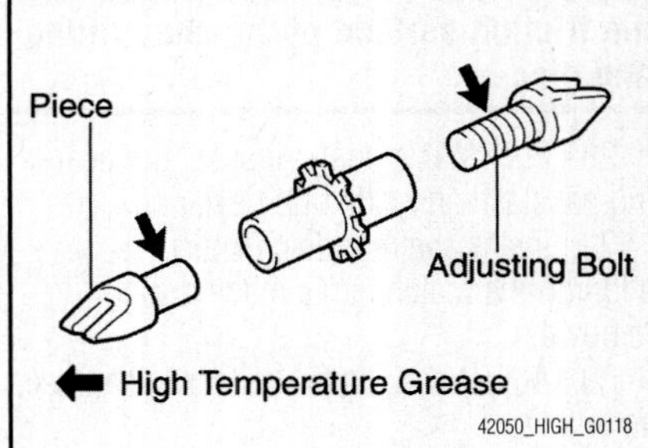

Fig. 22 Apply high temperature grease to the parking brake shoe adjusting bolt and piece

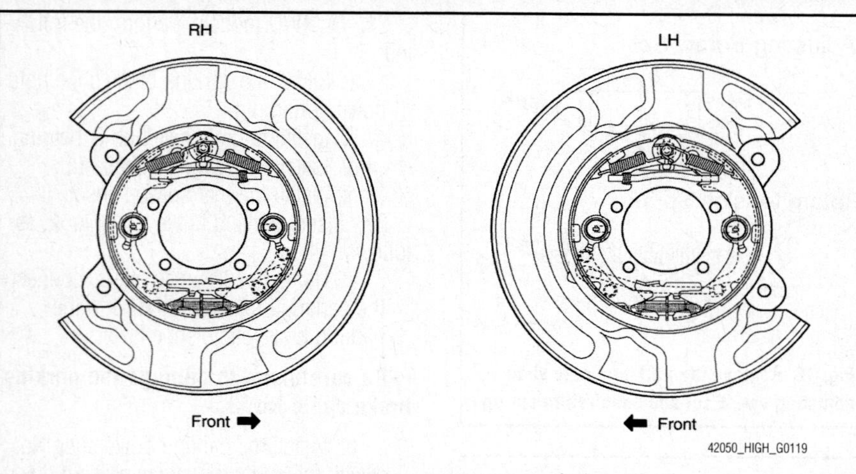

Fig. 23 Check that the parking brake components are properly installed. There should be no oil or grease on the friction surface of the shoe lining and disc

c. Attach the parking brake shoe adjusting screw set to the parking brake shoe No. 1 and parking brake shoe No. 2.

24. Install parking brake shoe No. 1:

a. For FWD models, install the parking brake shoe hold down spring pin No. 1.

b. Install the parking brake shoe No. 1 with the parking brake shoe hold down spring, parking brake shoe hold down spring cup No. 2, parking brake shoe hold down spring cup No. 2.

25. Attach the parking brake shoe strut and the parking brake shoe strut compression spring to parking brake shoe No. 2 and parking brake shoe No. 1.

26. Install parking brake shoe return tension spring using needle-nose pliers as shown in the illustration.

➡First install the front side spring then the rear side spring.

27. Check that the parking brake components are properly installed.

⁂ WARNING

There should be no oil or grease on the friction surface of the shoe lining and disc.

28. For 4WD models, inspect the bearing backlash and axle hub deviation.

29. Install the rear disc (rotor), aligning the matchmarks made during removal.

30. Adjust parking brake shoe clearance, as follows:

a. Temporarily install the hub nuts.

b. Remove the hole plug, turn the adjuster and expand the shoes until the disc locks.

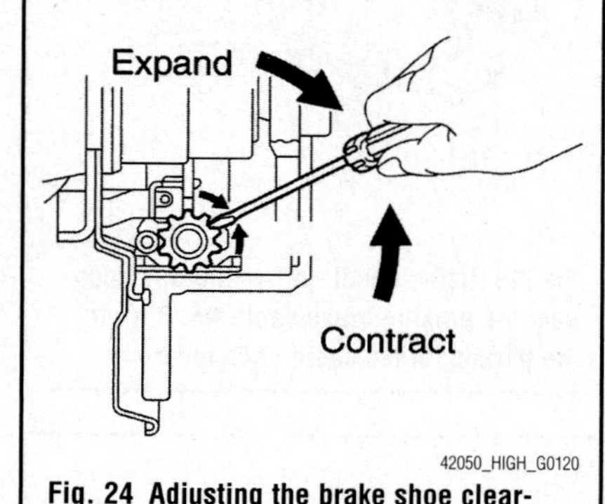

Fig. 24 Adjusting the brake shoe clearance

c. Contract the shoe adjuster until the disc rotates smoothly. Standard : return 8 notches

d. Check that the shoe has no brake drag.

e. Install the hole plug.

31. Install the caliper, as outlined earlier in this section.

32. Install the rear wheel and tire assembly and tighten the lug nuts to 76 ft. lbs. (103 Nm).

33. Inspect and adjust the parking brake pedal travel, as outlined in this section.

34. For 4WD models, check the ABS speed sensor signal.

ADJUSTMENT

See Figures 24 and 25.

1. Raise and safely support the vehicle.

2. Remove the rear wheel and tire assemblies.

3. Adjust parking brake shoe clearance, as follows:

a. Temporarily install the hub nuts.

b. Remove the hole plug, turn the adjuster and expand the shoes until the disc locks.

c. Contract the shoe adjuster until the disc rotates smoothly. Standard : return 8 notches

d. Check that the shoe has no brake drag.

e. Install the hole plug.

4. Install the rear wheel and tire assembly and tighten the lug nuts to 76 ft. lbs. (103 Nm).

5. Inspect the parking brake pedal travel, as follows:

a. Firmly step on the parking brake pedal.

b. Release the parking brake.

c. Once more, slowly depress the parking brake pedal all the way, and count the number of clicks. The parking brake pedal should travel 8 to 10 clicks at 67 lbs. (300 N).

6. If necessary, adjust parking brake pedal travel, as follows:

a. Remove the lower instrument panel finish panel sub-assembly.

b. Remove the lower instrument panel insert sub- assembly.

c. Depress the parking brake pedal 5 clicks to make room for the procedure, and loosen the lock nut with fixing adjusting nut by wrench.

d. Release the parking brake pedal to the original position.

e. Turn the parking brake wire adjusting nut until the parking brake pedal travel is correct.

f. Use a wrench to hold the parking brake adjusting nut, then tighten the lock nut to 62 inch lbs. (7 Nm).

g. Count the number of clicks after depressing and canceling the parking brake pedal 3 to 4 times.

h. Check whether the parking brake drags or not.

i. When operating the parking brake pedal, check that the parking brake pedal indicator light is lit.

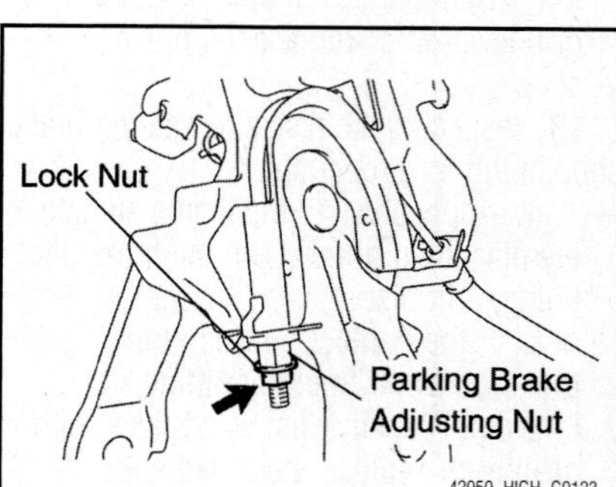

Fig. 25 Use a wrench to hold the parking brake adjusting nut secure while tightening the lock nut

CHASSIS ELECTRICAL

AIR BAG (SUPPLEMENTAL RESTRAINT SYSTEM)

GENERAL INFORMATION

These vehicles are equipped with an air bag system. The system must be disarmed before performing service on, or around, system components, the steering column, instrument panel components, wiring and sensors. Failure to follow the safety precautions and the disarming procedure could result in accidental air bag deployment, possible injury and unnecessary system repairs.

SERVICE PRECAUTIONS

**** CAUTION**

Disconnect and isolate the battery negative cable before beginning any airbag system component diagnosis, testing, removal, or installation procedures. Allow system capacitor to discharge for two minutes before beginning any component service. This will disable the airbag system. Failure to disable the airbag system may result in accidental airbag deployment, personal injury, or death.

DISARMING THE SYSTEM

To avoid personal injury when working on vehicles equipped with an air bag, the negative battery cable must be disconnected and at least 90 seconds must elapse before working on the system. Failure to do so may result in deployment of the air bag.

ARMING THE SYSTEM

To arm the system after service is finished, connect the negative battery cable.

CLOCKSPRING CENTERING

See Figure 26.

1. Check that the front wheels are facing straight ahead.
2. Set the turn signal switch to the neutral position.

**** WARNING**

If it is not in the neutral position, the turn signal switch pin may be snapped.

3. Check that the battery negative (-) cable is disconnected.

**** CAUTION**

Wait for at least 90 seconds after disconnecting the cable to prevent airbag deployment.

4. Rotate the spiral cable (clockspring) counterclockwise slowly by hand until it stops.

**** WARNING**

Do not turn the spiral cable using the airbag wire harness.

5. Rotate the spiral cable clockwise approximately 2.5 turns to align the marks.

➡The spiral cable will rotate approximately 2.5 turns to both the left and right from the center.

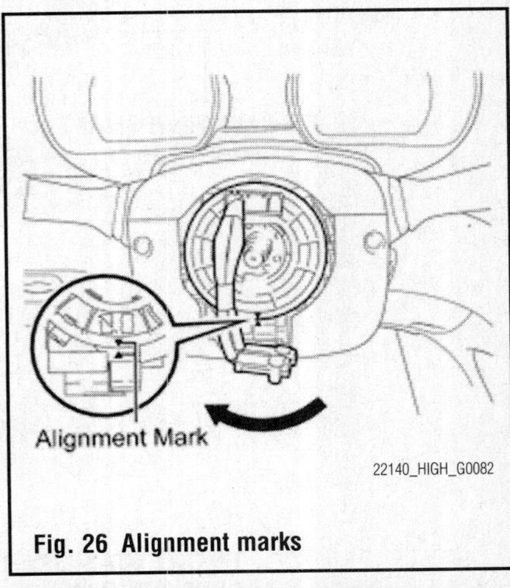

Fig. 26 Alignment marks

DRIVE TRAIN

TRANSFER CASE ASSEMBLY

REMOVAL & INSTALLATION

See Figures 27 and 28.

1. Remove the engine/transaxle assembly.
2. Drain the transaxle.
3. Separate the engine and transaxle.

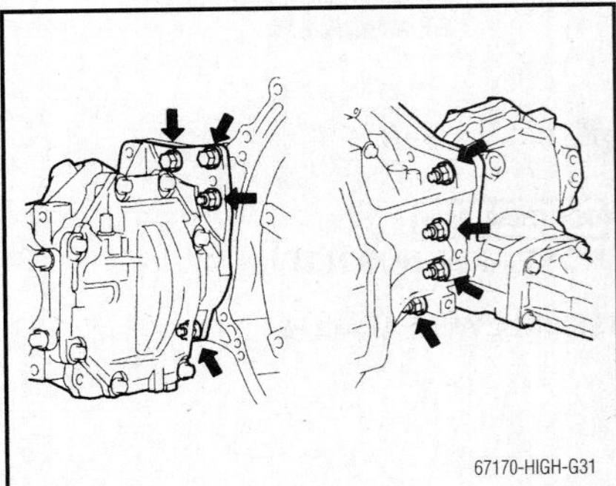

Fig. 27 Transfer case fastener locations—Highlander

4. Remove the 2 bolts and 6 nuts and separate the transfer case from the transaxle. It will be necessary to break it loose with a plastic mallet.

➡Keep the transfer case level during removal. Don't grasp the oil seals.

To install:

5. Clean all grease from the mating surfaces.
6. Apply a continuous 1.2mm diameter bead of silicone gasket material to the transaxle and transfer case as shown.
7. Join the transfer case to the transaxle within 10 minutes of gasket material application. If not, remove the material and start again.
8. Torque the nuts and bolts to 51 ft. lbs. (69 Nm).
9. The remainder of installation is the reverse of removal. Observe the following torques:
   - Engine mount bracket: 47 ft. lbs. (64 Nm)
   - Stiffener plate: 25 ft. lbs. (34 Nm)
   - Drain plug: 36 ft. lbs. (49 Nm)

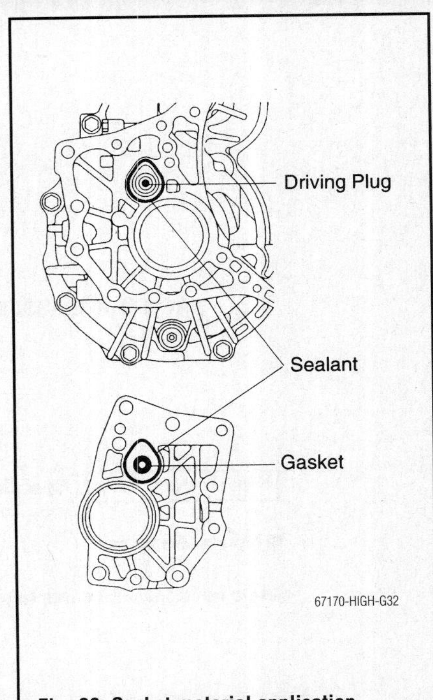

Fig. 28 Gasket material application—Highlander

FRONT HALFSHAFT

REMOVAL & INSTALLATION

2.7L Engines

See Figure 29.

1. Before servicing the vehicle, refer to the precautions section.
2. Remove or disconnect the following:
   - Front wheels
   - Fender apron seal
   - Transaxle fluid
   - Transfer case oil (4WD)
   - Hub nut
   - Stabilizer bar link
   - Speed sensor
   - Tie rod end
   - Lower arm from the ball joint
3. Slide the halfshaft from the hub, then, carefully, pry the shaft from the transaxle.
4. Installation is the reverse of removal. Torque the hub nut to 217 ft. lbs. (294 Nm).

3.5L Engines

See Figure 30.

1. Before servicing the vehicle, refer to the precautions section.
2. Remove or disconnect the following:
   - Front wheels
   - Fender apron seal
   - Transaxle fluid
   - Transfer case oil (4WD)
   - Hub nut
   - Stabilizer bar link

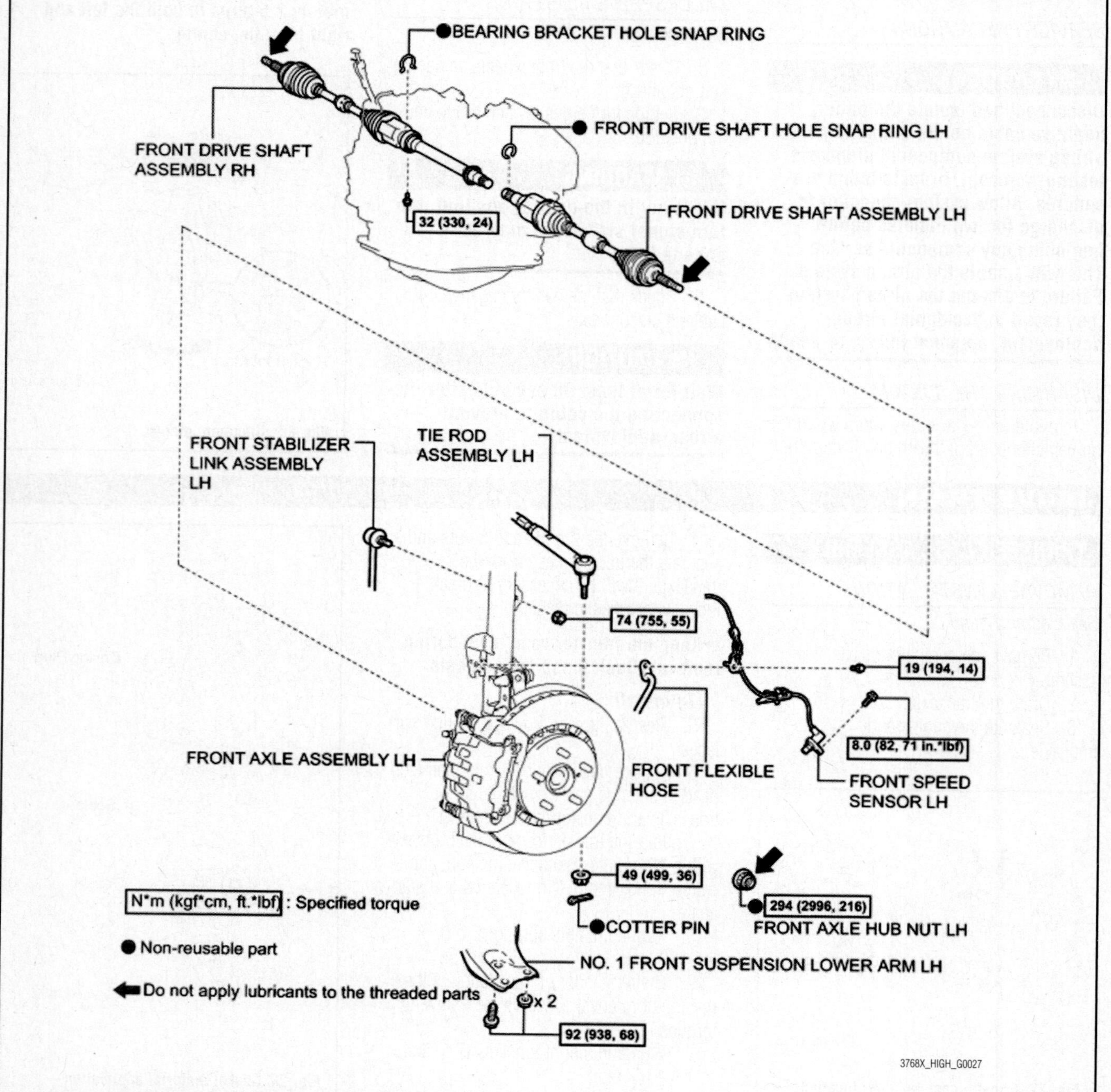

Fig. 29 Front halfshaft and related parts—2.7L engine

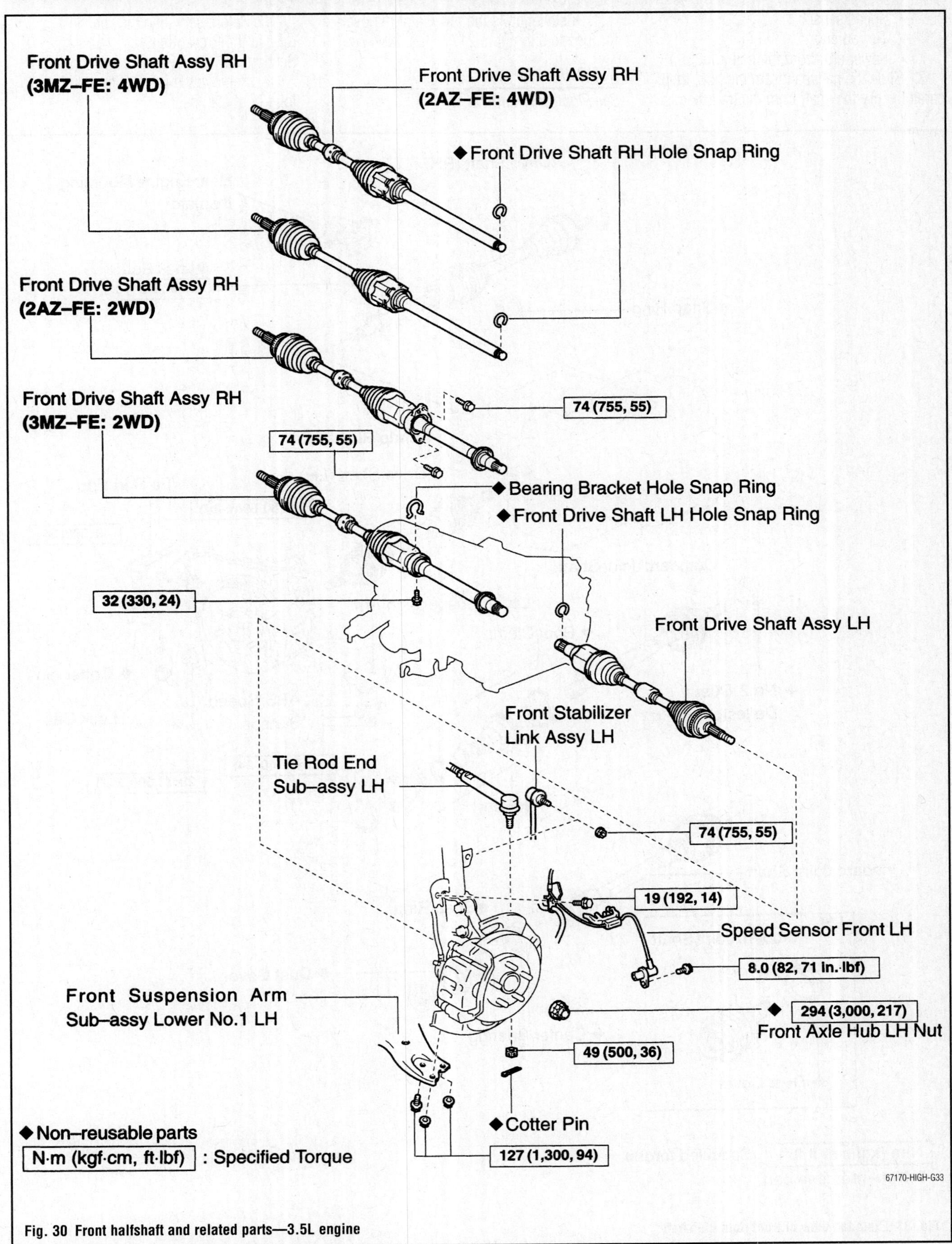
Front Drive Shaft Assy RH
(3MZ–FE: 4WD)
Front Drive Shaft Assy RH
(2AZ–FE: 4WD)
◆Front Drive Shaft RH Hole Snap Ring
Front Drive Shaft Assy RH
(2AZ–FE: 2WD)
Front Drive Shaft Assy RH
(3MZ–FE: 2WD)
74 (755, 55)
74 (755, 55)
◆Bearing Bracket Hole Snap Ring
◆Front Drive Shaft LH Hole Snap Ring
32 (330, 24)
Front Drive Shaft Assy LH
Front Stabilizer
Link Assy LH
Tie Rod End
Sub–assy LH
74 (755, 55)
19 (192, 14)
Speed Sensor Front LH
8.0 (82, 71 in.·lbf)
Front Suspension Arm
Sub–assy Lower No.1 LH
◆ 294 (3,000, 217)
Front Axle Hub LH Nut
49 (500, 36)
◆Cotter Pin
◆Non–reusable parts
N·m (kgf·cm, ft·lbf) : Specified Torque
127 (1,300, 94)
67170-HIGH-G33

Fig. 30 Front halfshaft and related parts—3.5L engine

- Speed sensor
- Tie rod end
- Lower arm from the ball joint

3. Slide the halfshaft from the hub, then, carefully, pry the shaft from the transaxle.

4. Installation is the reverse of removal. Torque the hub nut to 217 ft. lbs. (294 Nm).

OVERHAUL

See Figures 31 and 32.

1. Before servicing the vehicle, refer to the Precautions Section.

2. Remove the inboard and outboard joint boot clamps.

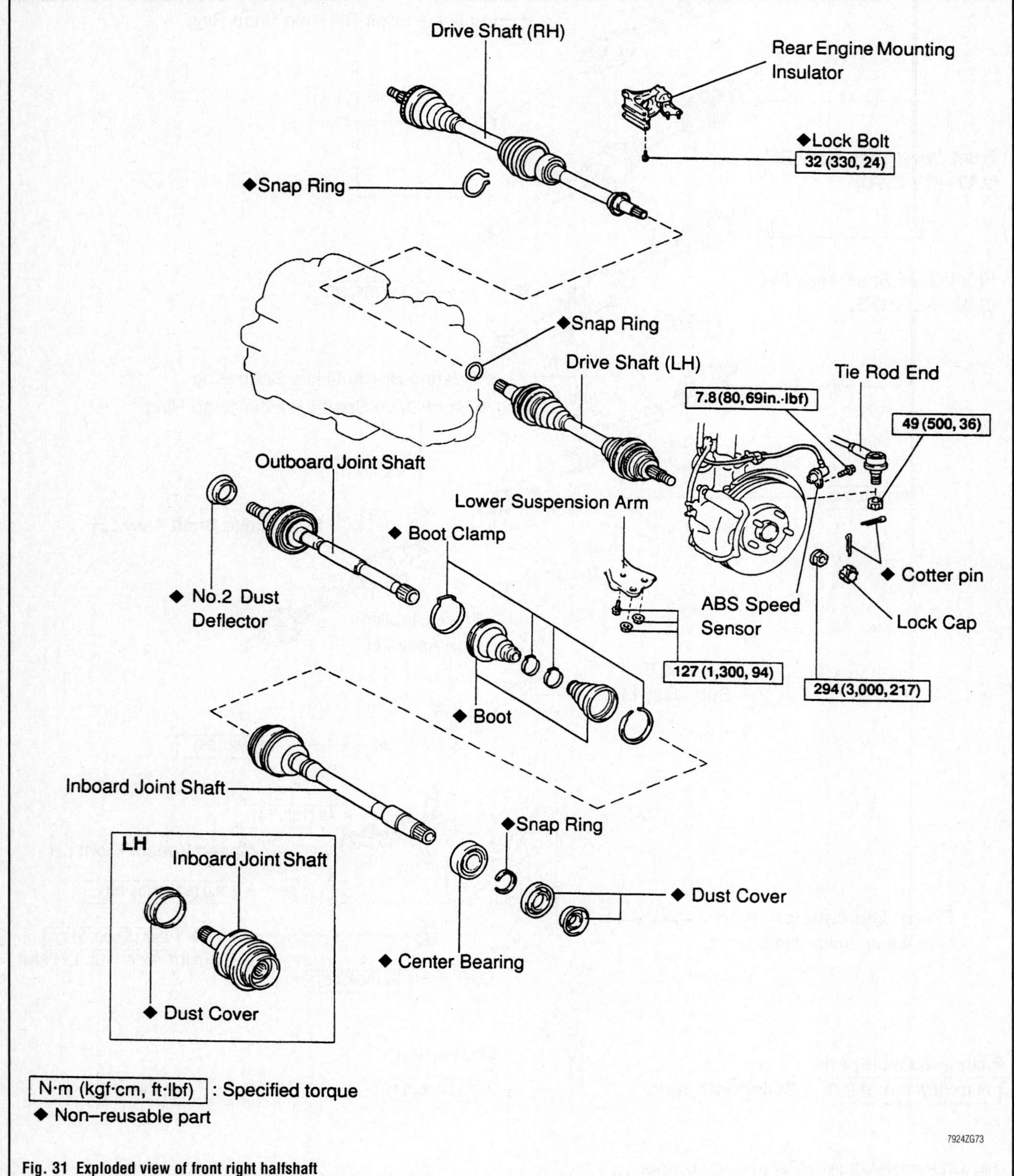

Fig. 31 Exploded view of front right halfshaft

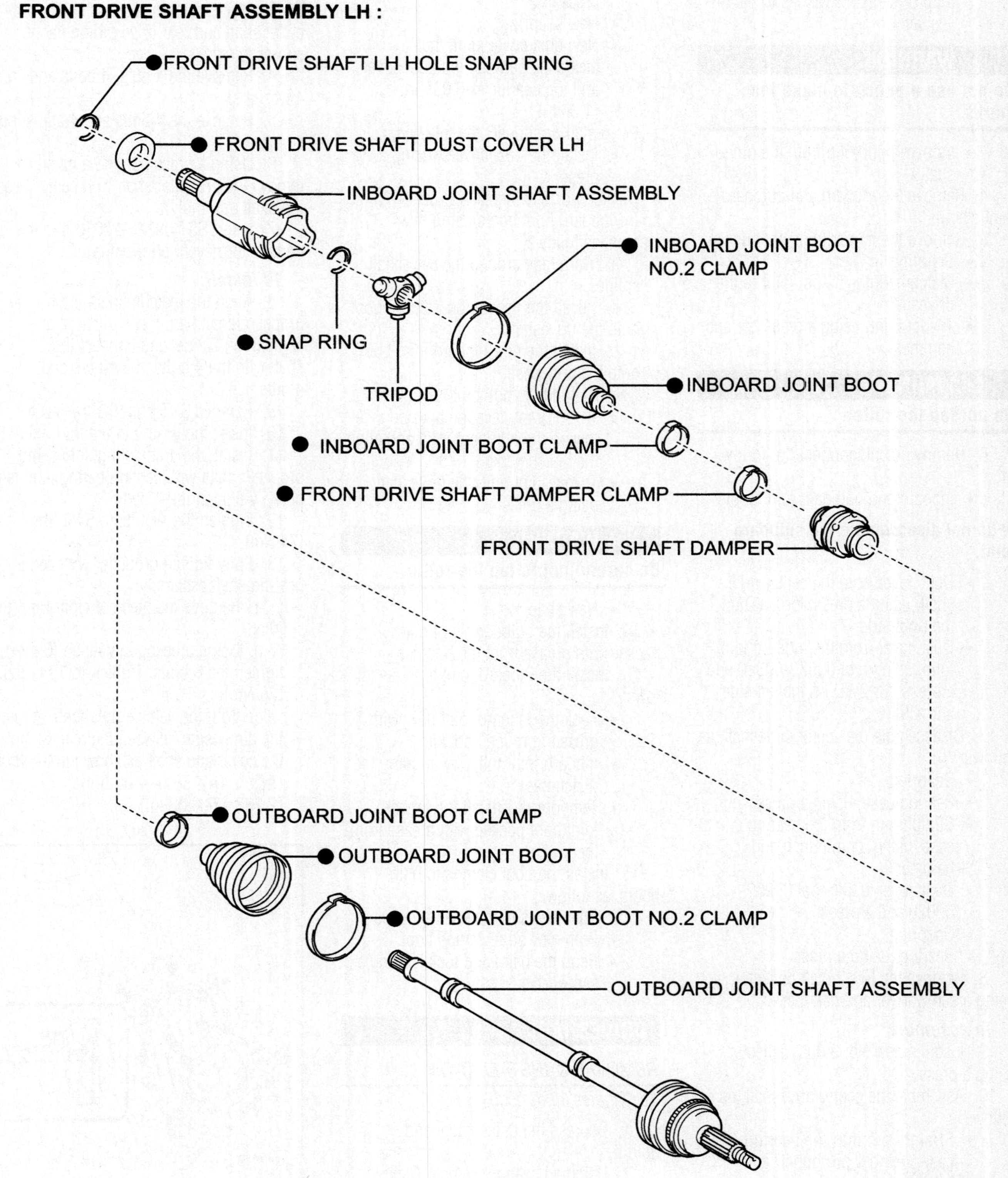

FRONT DRIVE SHAFT ASSEMBLY LH :

● FRONT DRIVE SHAFT LH HOLE SNAP RING
● FRONT DRIVE SHAFT DUST COVER LH
INBOARD JOINT SHAFT ASSEMBLY
● INBOARD JOINT BOOT NO.2 CLAMP
● SNAP RING
TRIPOD
● INBOARD JOINT BOOT
● INBOARD JOINT BOOT CLAMP
● FRONT DRIVE SHAFT DAMPER CLAMP
FRONT DRIVE SHAFT DAMPER
● OUTBOARD JOINT BOOT CLAMP
● OUTBOARD JOINT BOOT
● OUTBOARD JOINT BOOT NO.2 CLAMP
OUTBOARD JOINT SHAFT ASSEMBLY

09490_RX400H_G0036

Fig. 32 Exploded view of front left halfshaft

3. Disassemble the inboard joint tulip, as follows:
   - Matchmark the tri-pot, inboard joint tulip or center driveshaft to the driveshaft

⁂ WARNING

Do not use a punch to make the marks.

   - Inboard joint tulip from the driveshaft

4. Remove the inboard and outboard joint clamps.
5. Remove the tri-pot joint, as follows:
   - Snapring
   - Matchmark the tri-pot joint to the driveshaft
   - Tri-pot joint, using a brass bar and hammer

⁂ WARNING

Do not tap the roller.

6. Remove or disconnect the following:
   - Inboard and outboard joint boots

➡Do not disassemble the outboard joint.

   - Dust cover from the center driveshaft, using a press, for 2WD on the right side
   - Dust cover from the inboard joint tulip, using tool 09950-00020 and a press, for 2WD on the left side and 4WD

7. Disassemble the center driveshaft, as follows:
   - Snapring
   - Bearing case, using a press
   - Straight pin from the bearing case, using a pin punch and hammer
   - Dust cover, using tool 09950-00020 and a press
   - Snapring
   - Bearing, using a press

8. Remove the No. 2 dust deflector, using a screwdriver and hammer.

To assemble:

9. Install a new No. 2 dust deflector, using a press.
10. Assemble the center driveshaft, as follows:
    - Straight pin into the bearing case, using a pin punch and hammer
    - New bearing, using tools 09959-60010, 09950-70010 and a press
    - New snapring
    - Bearing with the bearing case assembly to the center driveshaft, using tool 09726-40010 and a press
    - New snapring
    - New dust cover, until the clearance between the dust cover and the bearing is 0.039 in. (1.0mm)
11. Install or connect the following:
    - Install the dust covers, using a press.
12. Temporarily install new outboard/inboard joint boots using new clamps, as follows:
    a. Warp tape around the driveshaft splines.
    b. Install the new outboard joint boot onto the driveshaft.
    c. Install the new inboard joint boot onto the driveshaft.
13. Install the tri-pot joint, as follows:
    - Tri-pot joint, face the beveled side toward the outboard joint and align the matchmarks
    - Tri-pot joint onto the driveshaft, using a press

⁂ WARNING

Be careful not to tap the roller.

    - New snapring
14. Install the outboard joint boot packed with grease from the boot kit.
15. Install the inboard joint tulip, as follows:
    - Pack the inboard joint boot with grease from the boot kit
    - Inboard joint tulip, by aligning the matchmarks
    - Temporarily, install the inboard joint boot packed with grease from the kit
16. Install the boot clamps to both boots, as follows:
    - Both boots to the shaft grooves
    - Both new boot clamps boot
    - Bend the band and lock it using a screwdriver

DRIVESHAFT

REMOVAL & INSTALLATION

See Figures 33 through 38.

1. Depress the brake pedal and hold it.
2. Using a hexagon wrench (6 mm), loosen the cross groove joint set bolts 1/2 turn.

➡Put a piece of cloth or equivalent into the inside of the universal joint cover so that the boot does not touch the inside of the universal joint cover.

3. Place matchmarks on the rear propeller shaft and rear drive pinion flange sub-assembly.
4. Remove the 4 nuts, 4 bolts and 4 washers.
5. Remove the 4 bolts and 4 adjusting shims.
6. Using a brass bar and a hammer, remove the propeller shaft with center bearing shaft assembly.
7. Insert SST 09325-20010 into the transfer to prevent oil leakage.

To install:

8. Align the matchmarks on the propeller shaft flange and differential companion flange, and connect the shaft with the 4 bolts, 4 washers and 4 nuts.
9. Remove SST from the transaxle.
10. Insert the yoke into the transaxle.
11. Install the 4 adjusting shims and propeller shaft with center bearing, and temporarily tighten the 4 bolts.
12. Tighten the 4 bolts to 54 ft. lbs. (74 Nm).
13. Fully tighten propeller with center bearing shaft assembly:
    a. Remove the piece of cloth from the joint.
    b. Using a hexagon wrench (6 mm), tighten the 6 bolts. Tighten to 19 ft. lbs. (26 Nm).
    c. With the vehicle unloaded, adjust the dimension between the rear side of the cover and shaft as shown in the illustration. (A): 58.0 +/- 0.5 mm (2.283 +/- 0.02 in.)

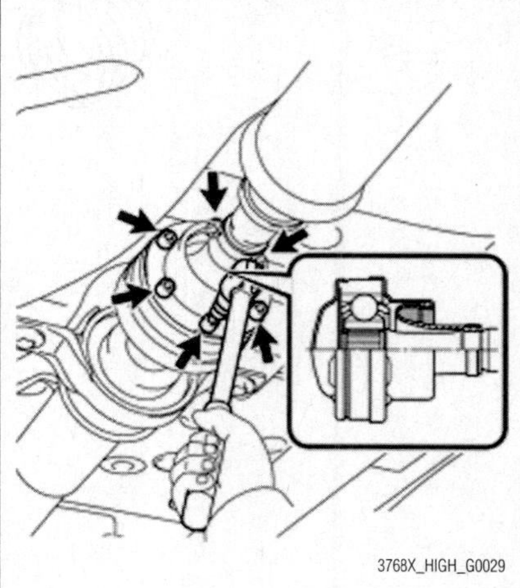

Fig. 33 Locating the cross groove joint set bolts

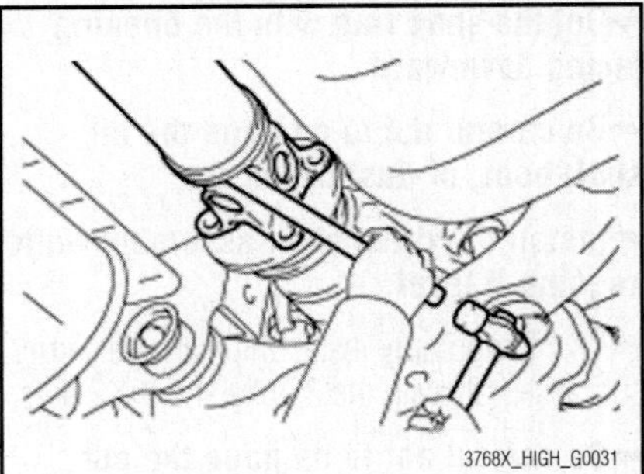

Fig. 34 Using a brass bar and a hammer, remove the propeller shaft with center bearing shaft assembly

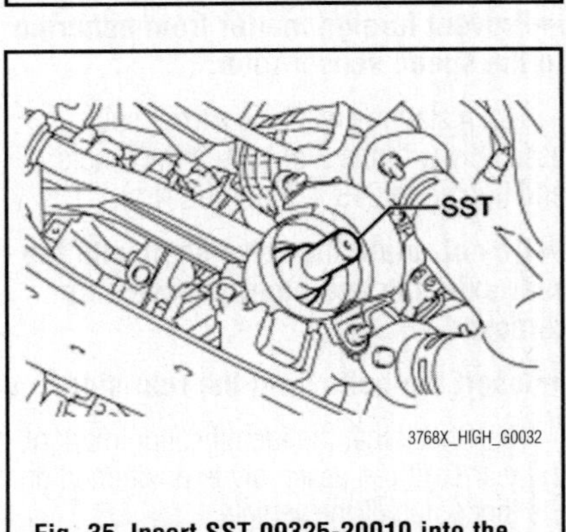

Fig. 35 Insert SST 09325-20010 into the transfer to prevent oil leakage

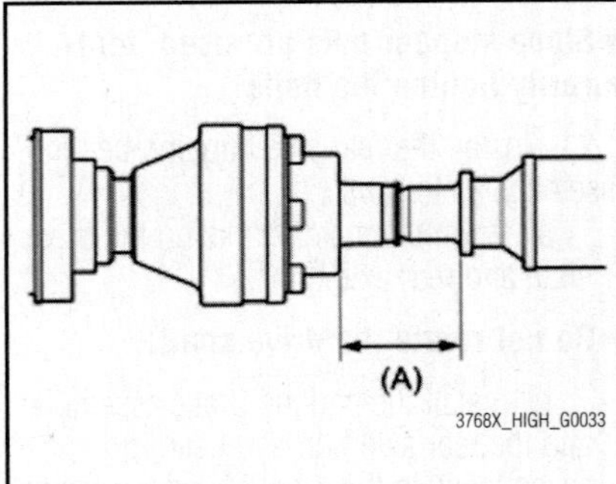

Fig. 36 Adjust the dimension between the rear side of the cover and shaft

d. Under the same condition as above, adjust the front and rear dimensions between the edge surface of the center support bearing and the edge surface of the cushion respectively as shown, and then tighten the bolts. (B): 12.5 +/- 1.0 mm (0.492 +/- 0.039 in.). Tighten to 27 ft. lbs. (37 Nm).

e. Check that the center line of the bracket is at the right angle in the shaft axial direction.

f. If any vibration or noise occurs, perform joint angle check as follows and replace the adjusting shim with a proper one.

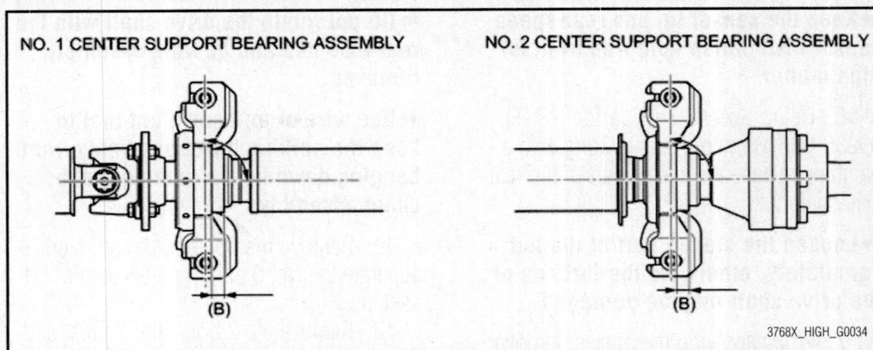

Fig. 37 Adjust the front and rear dimensions between the edge surface of the center support bearing and the edge surface of the cushion

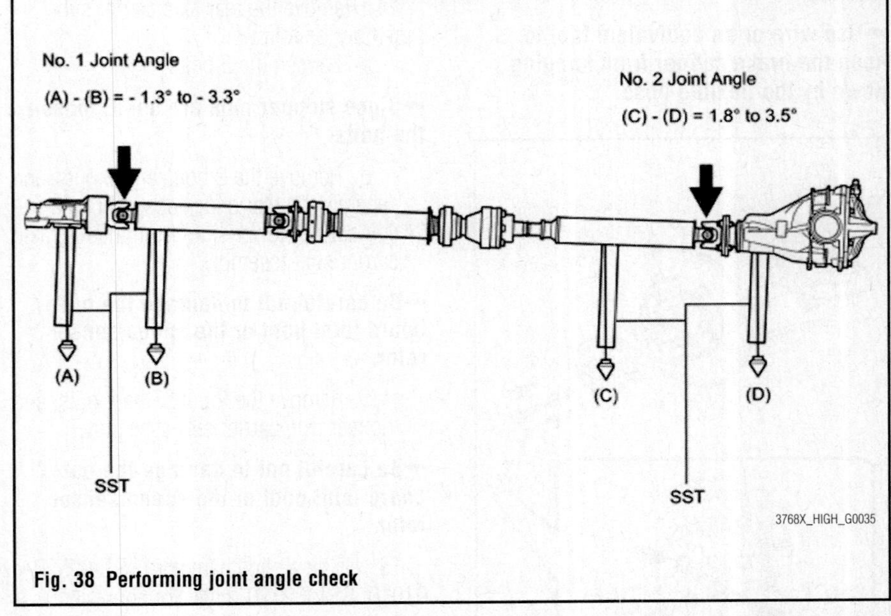

Fig. 38 Performing joint angle check

- Turn the propeller shaft several times by hand to stabilize the center support bearings.
- Using a jack, raise and lower the differential to stabilize the differential mounting cushion.
- Remove the transfer dynamic damper.
- Using SST 09370-50010, measure the transfer installation angle (A) and front propeller shaft installation angle (B). No. 1 joint angle: (A) - (B) = -1.3° to -3.3°
- Using SST 09370-50010, measure the rear propeller shaft installation angle (C) and rear differential shaft installation angle (D). No. 2 joint angle: (C) - (D) = 1.8° to 3.5°

➡If the measured angle is not within the specification, adjust it with the center support bearing adjusting shim.

14. Install the transfer dynamic damper. Tighten to 19 ft. lbs. (26 Nm).
15. Inspect and adjust the transfer oil.

REAR HALFSHAFT

REMOVAL & INSTALLATION

See Figures 39 through 42.

1. Raise and support the vehicle.
2. Remove the rear wheel.
3. Using a hexagon wrench (10 mm), remove the rear differential filler plug and rear differential filler plug gasket.
4. Using a hexagon wrench (10 mm), remove the rear differential drain plug and rear differential drain plug gasket, and drain the oil.
5. Remove the bolt and separate the rear speed sensor from the rear axle carrier sub-assembly.

➡Keep the sensor tip and rear speed sensor installation hole free from foreign matter.

6. Using Special Service Tool (SST) 09930-00010 or equivalent, along with a hammer, release the staked part of the rear axle shaft nut.

➡Loosen the staked part of the nut completely, otherwise the threads of the drive shaft may be damaged.

7. While applying the brakes, remove the rear axle shaft nut.
8. Remove the 2 caliper bracket bolts and separate the rear disc brake caliper assembly.

➡Use wire or an equivalent tool to keep the brake caliper from hanging down by the flexible hose.

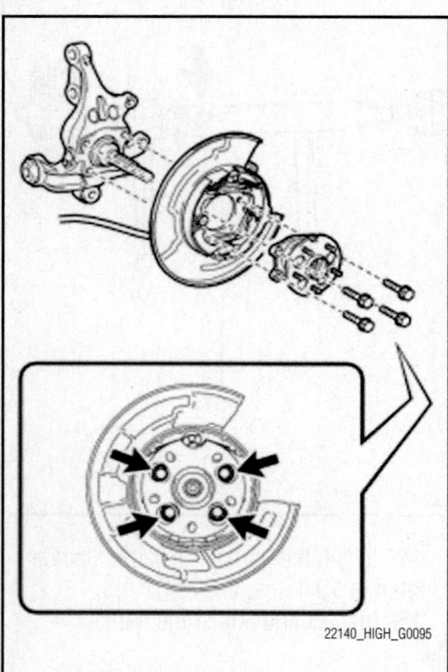

Fig. 39 Rear axle hub and bearing assembly and bolts

9. Put matchmarks on the rear disc and the axle hub.
10. Release the parking brake and remove the rear disc.

➡If the disc cannot be removed easily, turn and press firmly the shoe adjuster until the wheel comes free.

11. Remove rear axle hub and bearing assembly, as follows:
    a. Put matchmarks on the drive shaft and axle hub.

➡Do not punch the marks.

    b. Remove the 4 bolts and the rear axle hub and bearing assembly.

➡Do not rotate the drive shaft with the rear axle hub and bearing assembly removed.

➡Use wire or an equivalent tool to keep the parking brake assembly from hanging down by the parking brake cable assembly.

12. Remove the bolt and the nut, and separate the No. 3 parking brake cable assembly.
13. Remove the 2 bolts, the 2 nuts, and the rear strut rod assembly.

➡Since stopper nuts are used, loosen the bolts.

14. Remove the rear axle carrier sub-assembly, as follows:
    a. Loosen the 2 bolts.

➡Since stopper nuts are used, loosen the bolts.

    b. Remove the 2 bolts and 2 nuts, and separate the rear shock absorber with coil spring (lower side) from the rear axle carrier sub-assembly.

➡Be careful not to damage the outboard joint boot or the speed sensor rotor.

    c. Remove the 2 bolts, the 2 nuts, and the rear axle carrier sub-assembly.

➡Be careful not to damage the outboard joint boot or the speed sensor rotor.

15. Using a slide hammer (SST: 09520-01010, SST: 09520-24010 or equivalent), remove the rear drive shaft assembly (halfshaft) as shown in the illustration.

➡Remove the rear drive shaft assembly while keeping it level.

To install:

16. Align the shaft splines and install the rear drive shaft assembly (halfshaft) with a brass bar and hammer.

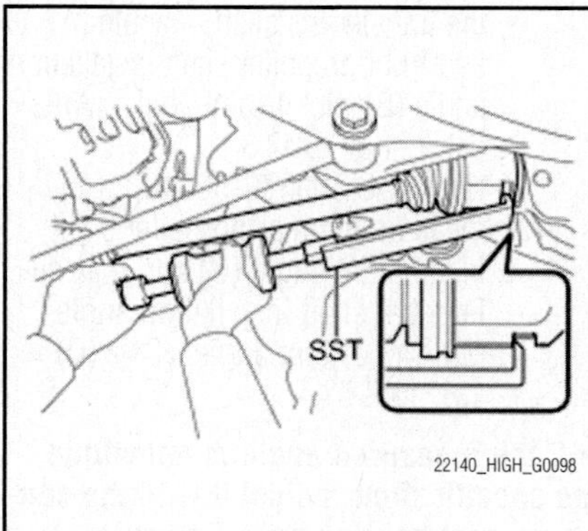

Fig. 40 Remove rear drive shaft assembly

➡Set the snap ring with the opening facing downward.

➡Be careful not to damage the oil seal, boot, or dust cover.

➡Install the drive shaft assembly while keeping it level.

17. Temporarily install the rear axle carrier sub-assembly with the 2 bolts and the 2 nuts.

➡Be careful not to damage the outboard joint boot.

➡Be careful not to damage the speed sensor rotor.

➡Prevent foreign matter from adhering to the speed sensor rotor.

18. Install the rear axle carrier sub-assembly with the 2 bolts and the 2 nuts, and tighten to 213 ft. lbs. (290 Nm).

➡Do not rotate the drive shaft with the rear axle hub and bearing assembly removed.

➡Insert the bolts from the rear side.

19. Check that the identification mark of the rear strut rod assembly is positioned on the inner side of the vehicle.
20. Temporarily install the rear strut rod assembly with the 2 bolts and the 2 nuts.

➡Since stopper nuts are used, temporarily tighten the bolts.

21. Install the rear axle hub and bearing assembly, as follows:
    a. Align the matchmarks on the drive shaft and rear axle hub.

➡Do not rotate the drive shaft.

    b. Install the parking brake assembly and the rear axle hub and bearing assembly with the 4 bolts, and tighten to 55 ft. lbs. (75 Nm).
22. Align the matchmarks and install the rear disc.

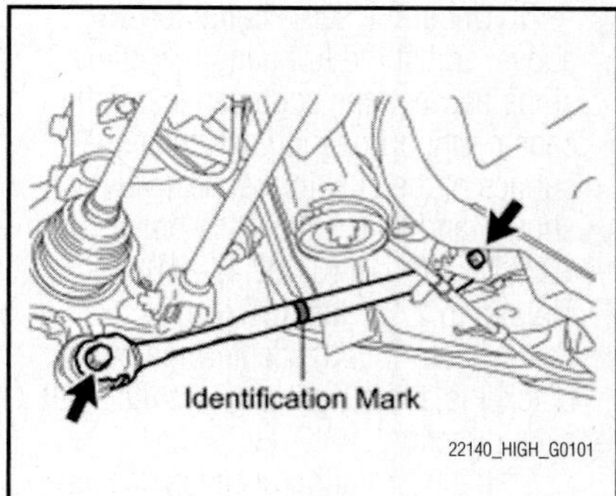

Fig. 41 Rear strut rod assembly identification mark—4WD models

➡**When replacing the rear disc with a new one, select the installation position where the rear disc has minimal runout.**

23. Install the rear disc brake caliper assembly with the 2 bolts, and tighten to 57 ft. lbs. (78 Nm).
24. Install the rear speed sensor to the rear axle carrier sub-assembly with the bolt, and tighten to 71 inch lbs. (8 Nm).

➡**Keep the rear speed sensor tip and sensor installation hole free from foreign matter.**

➡**Do not twist the rear speed sensor wire when installing.**

25. Jack up the rear axle carrier sub-assembly, placing a wooden block underneath to avoid damage. Apply load to the suspension so that the rear drive shaft assembly is positioned horizontally.

CAUTION

Do not jack up the rear axle carrier sub-assembly too high as the vehicle may fall.

➡**Do not bend the brake dust cover.**

➡**If the rear drive shaft assembly cannot be positioned horizontally as shown in the illustration even when the rear axle carrier sub-assembly is jacked up, apply additional load to the vehicle such as by having a person sit in the rear seat.**

➡**Use the same procedures for the RH side and LH side.**

26. Fully tighten the rear No. 1 and 2 suspension arm assemblies with the bolts and nuts, and tighten to 82 ft. lbs. (112 Nm).

➡**Since a stopper nut is used, tighten the bolt.**

➡**The final torque must be applied under standard vehicle height conditions.**

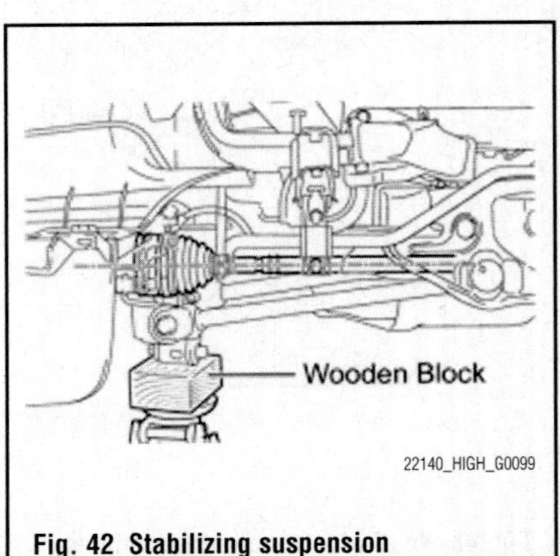

Fig. 42 Stabilizing suspension

27. Complete the installation of the rear strut rod assembly, tightening the bolts to 59 ft. lbs. (80 Nm).

➡**Since a stopper nut is used, fully tighten the bolt.**

➡**The final torque must be applied under standard vehicle height conditions.**

28. Install the No. 3 parking brake cable assembly with the bolt and the nut, and tighten to 29 ft. lbs. (39 Nm) and 53 inch lbs. (6 Nm).

➡**Do not twist the No. 3 parking brake cable assembly when installing it.**

29. Install the rear axle shaft nut, as follows:
   a. Clean the threaded parts on the drive shaft and axle hub nut using a non-residue solvent.

➡**Be sure to perform this work for a new drive shaft.**

➡**Keep the threaded parts free of oil and foreign objects.**

   b. Install a new rear axle shaft nut, and tighten to 216 ft. lbs. (294 Nm).
   c. Using a chisel and hammer, stake the rear axle shaft nut.
30. Install the rear differential drain plug, as follows:
   a. Using a hexagon wrench (10 mm), install the filler plug with a new gasket, and tighten to 36 ft. lbs. (49 Nm).
31. Fill the rear differential carrier assembly with hypoid gear oil.
32. Using a hexagon wrench (10 mm), install the rear differential filler plug with a new gasket, and tighten to 36 ft. lbs. (49 Nm).
33. Install the front wheel and tire assembly and tighten the lug nuts finger-tight.
34. Lower the vehicle, then final tighten the lug nuts to 76 ft. lbs. (103 Nm).
35. Inspect and adjust rear wheel alignment.
36. Check ABS speed sensor signal.

REAR PINION SEAL

REMOVAL & INSTALLATION

See Figure 43.

1. Before servicing the vehicle, refer to the precautions section.

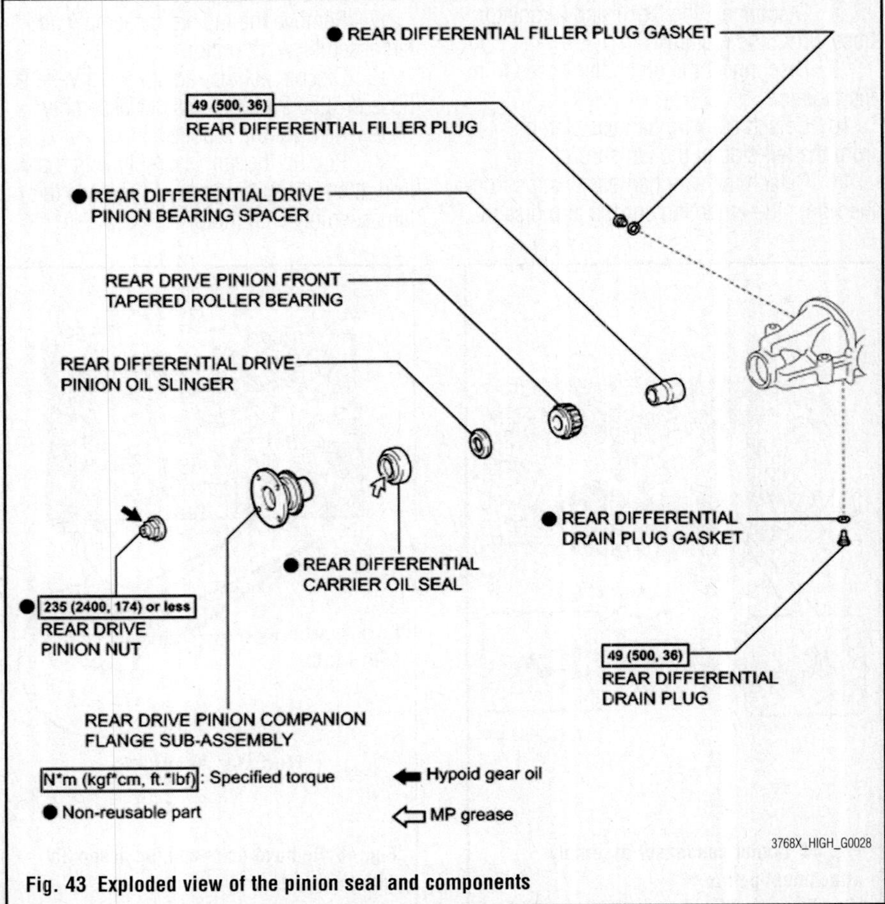

Fig. 43 Exploded view of the pinion seal and components

2. Drain the differential oil.
3. Remove or disconnect the following:
   - Exhaust pipe
   - Driveshaft by matchmarking it
   - Companion flange nut, by loosen the staked portion
   - Companion flange, using a screw-type extractor
   - Oil seal, using an extractor
   - Slinger
   - Front bearing
   - Spacer

To install:

4. Install or connect the following
   - New spacer
   - Bearing
   - Slinger
   - New seal

➡Seal installation depth: 2.0mm +/- 0.3mm

   - Companion flange
   - New nut. Coat the threads with clean differential oil. Torque the nut to 80 ft. lbs. (108 Nm).

5. The remainder of installation is the reverse of removal.

ENGINE COOLING

ENGINE FAN

REMOVAL & INSTALLATION

See Figures 44 through 46.

1. Remove the engine under cover assembly.
2. Remove the No. 1 engine under cover.
3. Drain engine coolant.
4. Remove the V-bank cover sub-assembly.
5. Remove the cool air intake duct seal.
6. Remove the battery and battery tray.
7. Remove the No. 1 and 2 air cleaner inlets.
8. Disconnect the No. 1 and 2 radiator hoses from the radiator.
9. Disconnect the oil cooler hoses from the radiator.
10. Detach the wire harness clamps from the left side of the fan shroud.
11. Detach the wire harness clamps from the right side of the fan shroud and disconnect the cooling fan Electronic Control Unit (ECU) connector.
12. Remove the radiator grill, as follows:
    a. Put protective tape around the radiator grill.
    b. Remove the 2 bolts and 4 clips.
    c. Disengage the 6 claws and 4 guides, and remove the radiator grill.
13. Remove the hood lock assembly.
14. Disconnect the low pitched and high pitched horn connectors.
15. Detach the hood lock control cable clamp and remove the 6 bolts and upper radiator support sub-assembly.
16. Remove the 4 bolts and move the cooler condenser assembly to remove the radiator assembly and fan assembly with motor.
17. Remove the radiator assembly and fan assembly with motor.
18. Disconnect the radiator reserve tank hose or pipe from the radiator assembly.
19. Remove the 3 bolts.
20. Pull up the fan assembly with motor from the radiator assembly to remove the fan assembly with motor.

To install:

21. Installation is the reverse of removal, noting the following:
    a. When installing the fan assembly with motor, tighten the 3 bolts to 69 inch lbs. (7.8 Nm).
    b. When installing the cooler condenser assembly, tighten the 4 bolts to 53 inch lbs. (6 Nm).
    c. When installing the upper radiator support sub-assembly, tighten the 6 bolts to 87 inch lbs. (9.8 Nm).
    d. When installing the hood lock assembly, tighten the 3 bolts to 71 inch lbs. (8 Nm).
    e. When installing the No. 1 air cleaner inlet, tighten bolt A to 62 inch lbs. (7 Nm), and bolt B to 44 inch lbs. (5 Nm).
    f. When installing the No. 2 air cleaner inlet, tighten the 2 bolts to 62 inch lbs. (7 Nm).
    g. Add engine coolant. Refer to radiator installation procedure for detailed instructions.

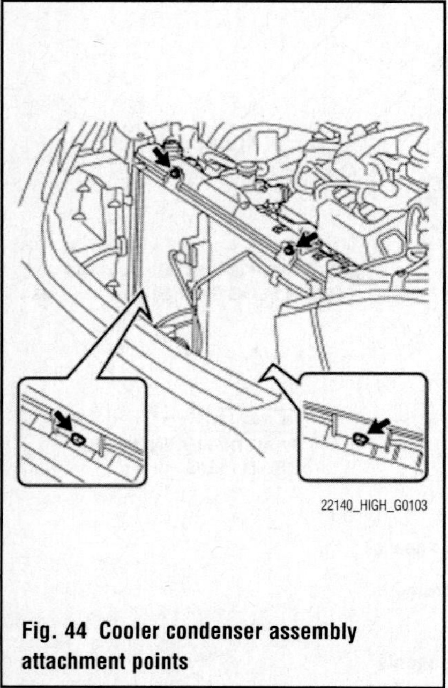

Fig. 44 Cooler condenser assembly attachment points

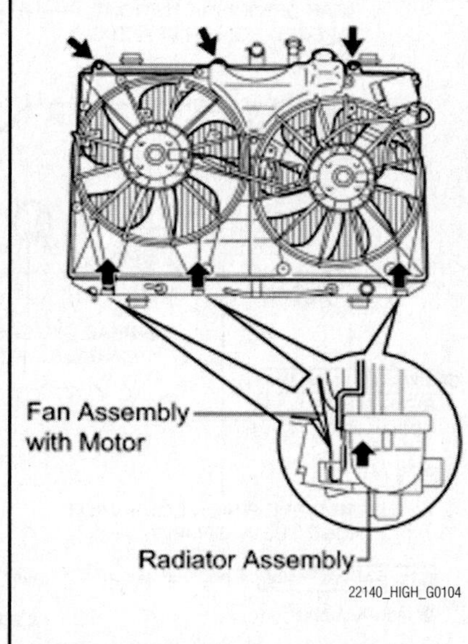

Fig. 45 Remove bolts and fan assembly with motor

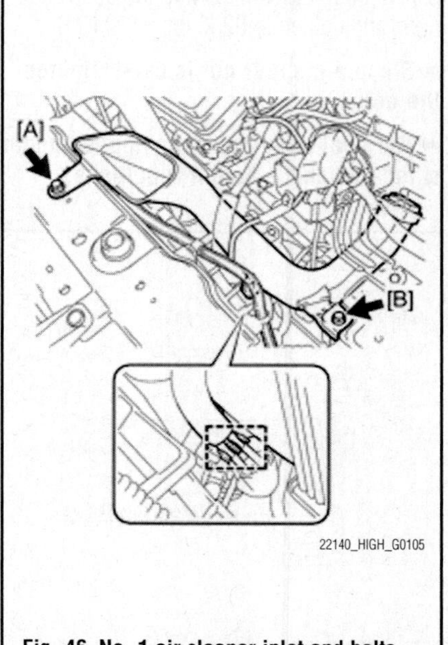

Fig. 46 No. 1 air cleaner inlet and bolts

h. Inspect for coolant leak. Refer to radiator installation procedure for detailed instructions.

i. Check the automatic transaxle fluid level.

j. Inspect for oil leaks.

RADIATOR

REMOVAL & INSTALLATION

See Figures 44 through 49.

1. Remove the engine under cover assembly.
2. Remove the No. 1 engine under cover.
3. Drain engine coolant.
4. Remove the V-bank cover sub-assembly.
5. Remove the cool air intake duct seal.
6. Remove the battery and battery tray.
7. Remove the No. 1 and 2 air cleaner inlets.
8. Disconnect the No. 1 and 2 radiator hoses from the radiator.
9. Disconnect the oil cooler hoses from the radiator.

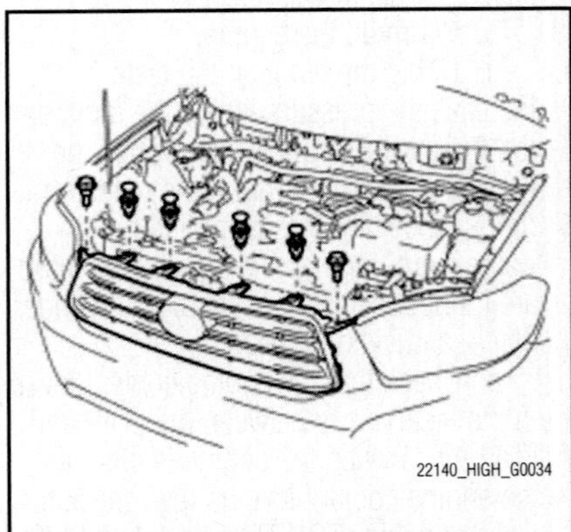

Fig. 47 Radiator grill, bolts, and clips

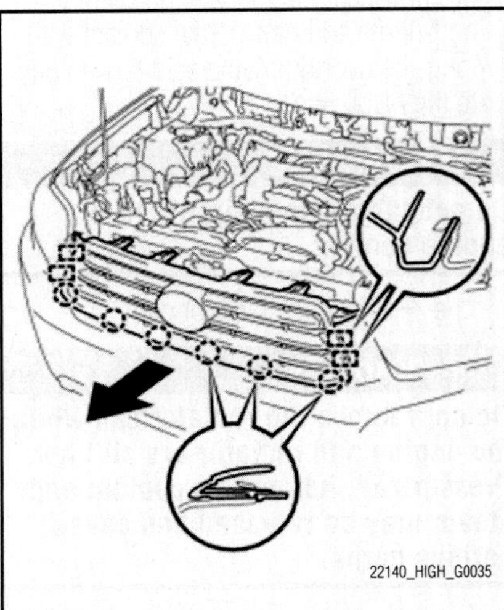

Fig. 48 Radiator grill, claws, and guides

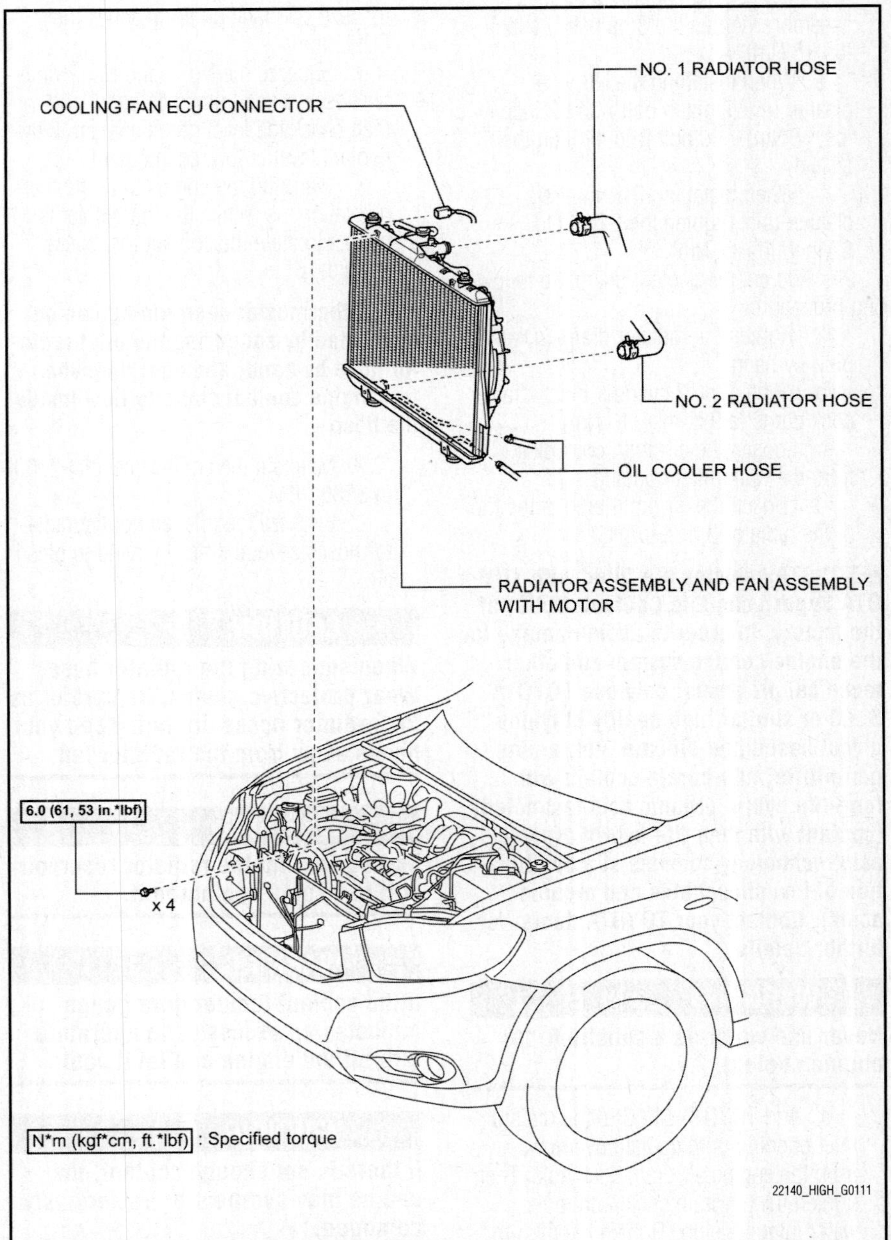

10. Detach the wire harness clamps from the left side of the fan shroud.
11. Detach the wire harness clamps from the right side of the fan shroud and disconnect the cooling fan Electronic Control Unit (ECU) connector.
12. Remove the radiator grill, as follows:
    a. Put protective tape around the radiator grill.
    b. Remove the 2 bolts and 4 clips.
    c. Disengage the 6 claws and 4 guides, and remove the radiator grill.
13. Remove the hood lock assembly.
14. Disconnect the low pitched and high pitched horn connectors.
15. Detach the hood lock control cable clamp and remove the 6 bolts and upper radiator support sub-assembly.
16. Remove the 4 bolts and move the cooler condenser assembly to remove the radiator assembly and fan assembly with motor.
17. Remove the radiator assembly and fan assembly with motor.
18. Disconnect the radiator reserve tank hose or pipe from the radiator assembly.
19. Remove the 3 bolts.
20. Pull up the fan assembly with motor from the radiator assembly to remove the fan assembly with motor.
21. Remove the 2 radiator support cushions.

Fig. 49 Radiator and fan assembly components

22. Remove the 2 lower radiator supports.

To install:

23. The installation procedure is the reverse of removal, with fluid levels being checked or fluids added, as noted. Use the following tightening specifications:
   a. When installing the fan assembly with motor, tighten the 3 bolts to 69 inch lbs. (7.8 Nm).
   b. When installing the cooler condenser assembly, tighten the 4 bolts to 53 inch lbs. (6 Nm).
   c. When installing the upper radiator support sub-assembly, tighten the 6 bolts to 87 inch lbs. (9.8 Nm).
   d. When installing the hood lock assembly, tighten the 3 bolts to 71 inch lbs. (8 Nm).
   e. When installing the No. 1 air cleaner inlet, tighten bolt A to 62 inch lbs. (7 Nm), and bolt B to 44 inch lbs. (5 Nm).
   f. When installing the No. 2 air cleaner inlet, tighten the 2 bolts to 62 inch lbs. (7 Nm).

24. Add engine coolant using the following procedure:
   a. Tighten the radiator drain cock plug by hand.
   b. Tighten the 2 cylinder block drain cock plugs to 9 ft. lbs. (13 Nm).
   c. Loosen the air drain cock plug from the water inlet housing.
   d. Loosen the air drain plug at the top of the radiator 3 or 4 turns.

➡TOYOTA vehicles are filled with TOYOTA Super Long Life Coolant (SLLC) at the factory. In order to avoid damage to the engine cooling system and other technical problems, only use TOYOTA SLLC or similar high quality ethylene glycol based non-silicate, non-amine, non-nitrite, non-borate coolant with long-life hybrid organic acid technology (coolant with long-life hybrid organic acid technology consists of a combination of low phosphates and organic acids). Contact your TOYOTA dealer for further details.

⁂ WARNING

Never use water as a substitute for engine coolant.

   e. Add TOYOTA SLLC to the radiator inlet opening until coolant overflows from the engine air drain cock hole. Then tighten the air drain cock plug to the water inlet housing. Tighten the plug to 9 ft. lbs. (13 Nm).
   f. Continue to add TOYOTA Super Long Life Coolant (SLLC) to the radiator inlet opening until coolant overflows from the radiator air drain hole. Then close the air drain plug at the top of the radiator and tighten to 13 inch lbs. (1.5 Nm).

➡If the coolant level at the radiator inlet opening drops after squeezing the No. 1 and No. 2 radiator hoses, add coolant.

   g. Slowly fill the radiator with TOYOTA Super Long Life Coolant (SLLC). For quantity information, refer to the capacities section.
   h. Slowly pour coolant into the radiator reservoir tank until it reaches the FULL line.
   i. Squeeze the No. 1 and No. 2 radiator hoses several times by hand, and then check the level of the coolant. If the coolant level is low, add coolant.
   j. Warm up the engine until the thermostat opens. While the thermostat is open, circulate the coolant for several minutes.

➡The thermostat open timing can be confirmed by squeezing the inlet radiator hose by hand, and checking when the engine coolant starts to flow inside the hose.

   k. Maintain the engine speed at 2500 to 3000 RPM.
   l. Squeeze the inlet and outlet radiator hoses several times by hand to bleed air.

⁂ CAUTION

When squeezing the radiator hoses: Wear protective gloves. Be careful as the radiator hoses are hot. Keep your hands away from the radiator fan.

⁂ WARNING

Make sure that the radiator reservoir still has some coolant in it.

⁂ WARNING

If the coolant temperature gauge indicates an excessive temperature, turn off the engine and let it cool.

⁂ WARNING

If there is not enough coolant, the engine may overheat or be seriously damaged.

⁂ WARNING

If the radiator reservoir does not have enough coolant, perform the following: 1) stop the engine, 2) wait until the coolant has cooled down, and 3) add coolant until the reservoir is filled to the FULL line.

   m. Stop the engine and wait until the engine coolant cools down.
   n. Add engine coolant to the FULL line on the radiator reservoir.

25. Inspect for coolant leak, using the following procedure:

⁂ CAUTION

Do not remove the radiator cap while the engine and radiator are still hot. Pressurized, hot engine coolant and steam may be released and cause serious burns.

➡Before performing each inspection, turn the A/C switch off.

   a. Fill the radiator with coolant and attach a radiator cap tester.
   b. Warm up the engine.
   c. Using the radiator cap tester, increase the pressure inside the radiator to 17 psi (118 kPa) and check that the pressure does not drop. If the pressure drops, check the hoses, radiator and water pump for leaks. If no external leaks are found, check the heater core, cylinder block, and cylinder head.
   d. Check that the engine coolant level in the reservoir is between the LOW and FULL lines when the engine is cold. If the engine coolant level is low, check for leaks and add "TOYOTA Super Long Life Coolant" or similar high quality ethylene glycol based non-silicate, non-amine, non-nitrite and non-borate coolant with long-life hybrid organic acid technology to the FULL line.

⁂ WARNING

Do not substitute plain water for engine coolant.

   e. Remove the radiator cap.

⁂ CAUTION

Do not remove the radiator cap while the engine and radiator are still hot. Pressurized, hot engine coolant and steam may be released and cause serious burns.

   f. Check if there are any excessive deposits of rust or scales around the radiator cap and radiator filler hole. Also, the coolant should be free of oil. If excessively dirty, clean the coolant passage and replace the coolant.

g. Install the radiator cap.

26. Check the automatic transaxle fluid level.

27. Inspect for oil leaks.

THERMOSTAT

REMOVAL & INSTALLATION

2.7L Engine

See Figures 50 and 51.

1. Disconnect the negative battery cable.
2. Raise and safely support the vehicle.
3. Remove the engine under cover assembly.
4. Disconnect the negative battery cable.
5. Remove v-ribbed belt.
6. Remove alternator assembly.
7. Disconnect no. 2 radiator hose.
8. Remove the 2 nuts and water inlet.
9. Remove the thermostat.
10. Remove the gasket from the thermostat.

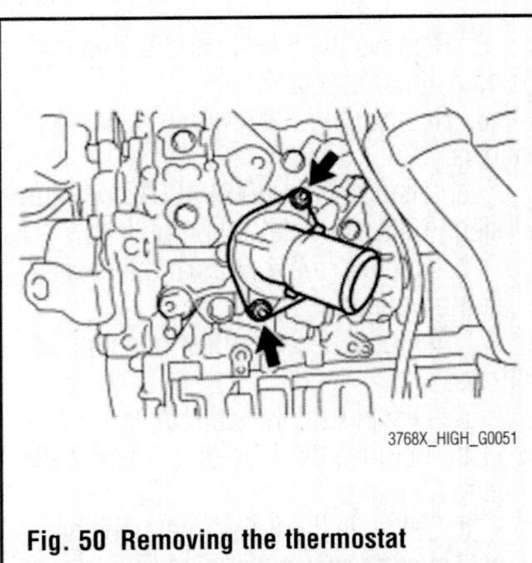

Fig. 50 Removing the thermostat

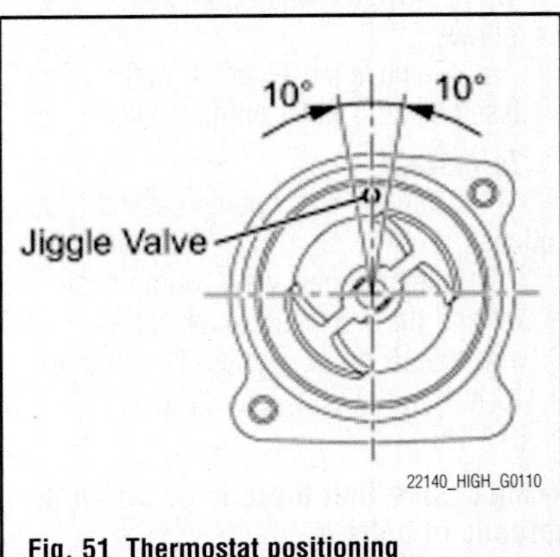

Fig. 51 Thermostat positioning

To install:

11. Install a new gasket to the thermostat. Install the thermostat with the jiggle valve facing upward.

➡The jiggle valve may be set to within 10° on either side of the prescribed position.

12. Install the water inlet with the 2 nuts. Tighten to 7 ft. lbs. (10 Nm).
13. To complete installation, reverse remaining removal procedure.

3.5L Engine

See Figures 52 through 56.

1. Remove the V-bank cover sub-assembly.
2. Raise and safely support the vehicle.
3. Remove the engine under cover assembly.
4. Remove the No. 1 engine under cover.
5. Remove the right front wheel.
6. Remove right front fender molding sub-assembly.
7. Remove the right front fender liner.
8. Remove the 2 bolts, clip and right front fender apron seal.
9. Drain the engine coolant.
10. Using a Special Service Tool (SST: 09961-00950), release the V-ribbed belt tension by turning the V-ribbed belt tensioner assembly counterclockwise, and remove the V-ribbed belt from the V-ribbed belt tensioner assembly.
11. While turning the V-ribbed belt tensioner assembly counterclockwise, align with its holes, and then insert the 5 mm bi-hexagon wrench into the holes to fix the V-ribbed belt tensioner assembly.
12. Disconnect the No. 2 radiator hose from the engine.

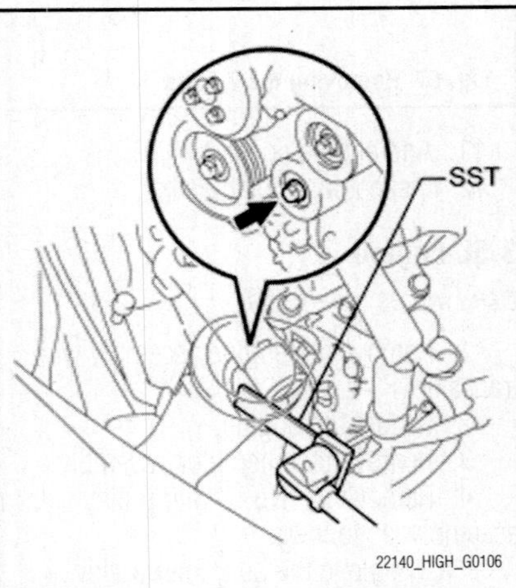

Fig. 52 Releasing tension on V-ribbed belt—3.5L engine

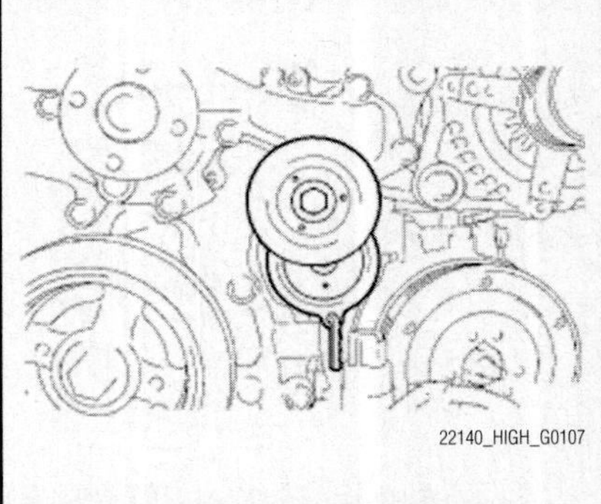

Fig. 53 V-ribbed belt tensioner assembly—3.5L engine

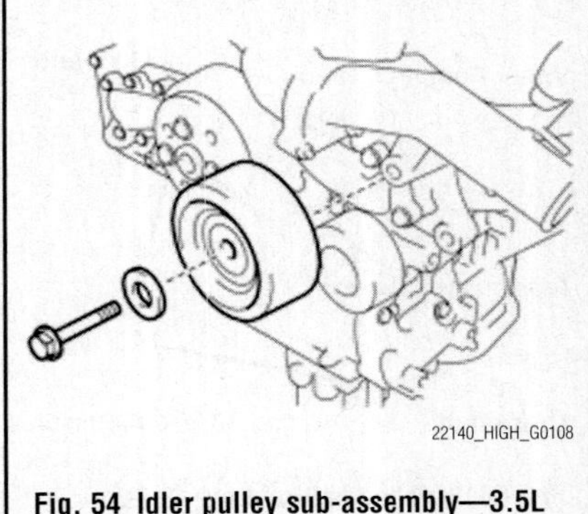

Fig. 54 Idler pulley sub-assembly—3.5L engine

13. Remove the bolt, idler pulley cover plate, and idler pulley sub-assembly.
14. Remove the 2 nuts and the water inlet.
15. Remove the thermostat from the water inlet housing.
16. Remove the gasket from the thermostat.

To install:

17. Install a new gasket to the thermostat.
18. Install the thermostat with the jiggle valve facing up.

➡The jiggle valve may be set within 10° on either side of the prescribed position.

19. Install the water inlet with the 2 nuts, and tighten to 7 ft. lbs. (10 Nm).
20. Install the idler pulley cover plate and idler pulley sub-assembly with the bolt, and tighten to 32 ft. lbs. (43 Nm).
21. Connect the No. 2 radiator hose to the engine.
22. Using SST: 09961-00950, turn the V-ribbed belt tensioner assembly counterclockwise and remove the bar.
23. If it is difficult to install the V-ribbed belt, perform the following procedure:
    a. Put the V-ribbed belt on every pulley except the tensioner pulley as shown in the illustration.

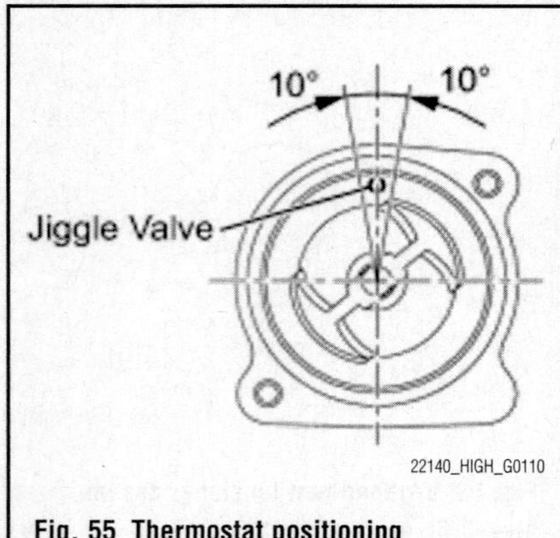

Fig. 55 Thermostat positioning

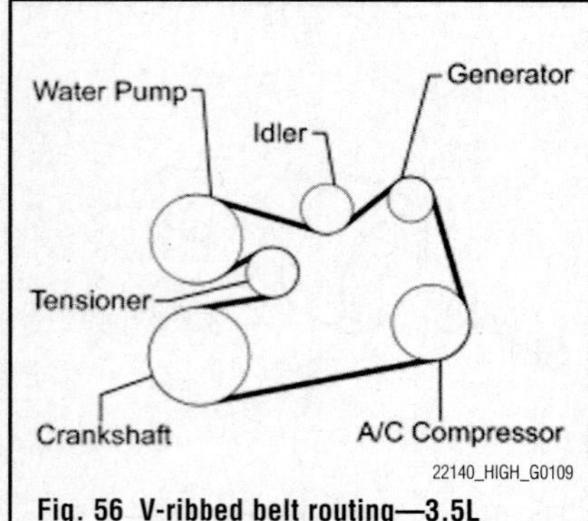

Fig. 56 V-ribbed belt routing—3.5L engine

b. Release the V-ribbed belt tension by turning the V-ribbed belt tensioner assembly counterclockwise, and put the V-ribbed belt on the V-ribbed tensioner assembly pulley.

➡Put the backside of the V-ribbed belt on the V-ribbed belt tensioner assembly pulley and the No. 2 idler pulley sub-assembly.

➡Check that the V-ribbed belt is properly set to each pulley.

c. After installing the V-ribbed belt, check that it fits properly in the ribbed grooves. Confirm that the belt has not slipped out of the grooves on the bottom of the crankshaft pulley by hand.

24. Install the right front fender apron seal with the 2 bolts and clip.
25. Install the right front fender liner.
26. Install the right front fender molding sub-assembly.
27. Install the right front wheel. Tighten the lug nuts to 76 ft. lbs. (103 Nm).
28. Add engine coolant. Refer to radiator installation procedure for detailed instructions.
29. Inspect for coolant leak. Refer to radiator installation procedure for detailed instructions.
30. Install the No. 1 engine under cover.
31. Install the engine under cover assembly.
32. Install the V-bank cover sub-assembly.

WATER PUMP

REMOVAL & INSTALLATION

2.4L Engine

See Figure 57.

1. Disconnect the negative battery cable.
2. Remove no. 1 engine under cover.
3. Drain and recycle the engine coolant.
4. Remove v-ribbed belt.
5. Remove alternator assembly.
6. Remove v-ribbed belt tensioner assembly.
7. Remove the 7 bolts, water pump and water pump gasket.

To install:

8. Install a new gasket and the water pump with the 7 bolts. Tighten to 15 ft. lbs. (21 Nm).
9. To complete installation, reverse remaining removal procedure.
10. When disconnecting the cable, some systems need to be initialized after the cable is reconnected.

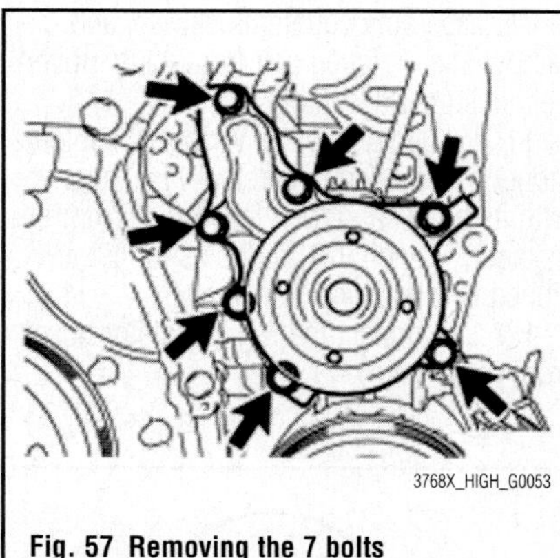

Fig. 57 Removing the 7 bolts

11. Add engine coolant.
12. Inspect for coolant leak.

3.5L Engine

See Figures 58 and 59.

1. Remove the engine assembly with transaxle.
2. Secure the engine.
3. Remove the alternator assembly.
4. Remove the No. 2 idler pulley sub-assembly, as follows:
   a. Remove the bolt, idler pulley cover plate, and idler pulley sub-assembly.

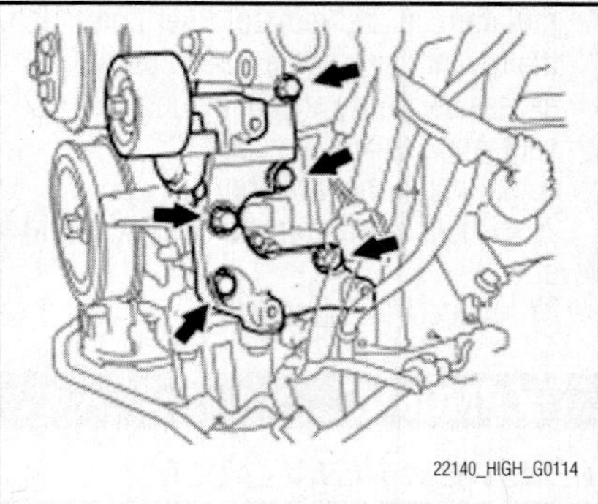

Fig. 58 V-ribbed belt tensioner assembly and bolts

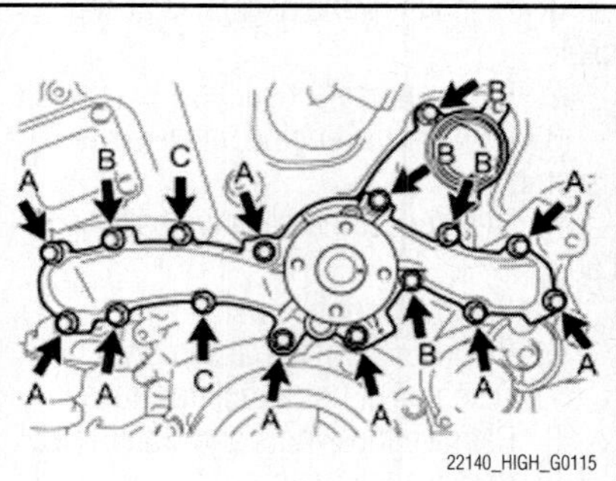

Fig. 59 Water pump assembly, gasket, and bolts

5. Remove the 5 bolts and V-ribbed belt tensioner assembly.
6. Remove the water pump pulley, as follows:
   a. Using SST: 09960-10010 or equivalent, hold the water pump pulley.
   b. Remove the 4 bolts and the water pump pulley.
7. Remove the water inlet housing, as follows:
   a. Disconnect the water hose.
   b. Remove the 2 bolts, nut and water inlet housing.
   c. Remove the water inlet housing gasket and water outlet pipe O-ring.
8. Remove the water pump assembly, as follows:
   a. Remove the 16 bolts, water pump assembly and water pump gasket.

To install:

9. Install the water pump assembly, as follows:
   a. Install a new water pump gasket and the water pump assembly with the 16 bolts. Tighten the bolts to 16 ft. lbs. (21 Nm) and 81 inch lbs. (9.1 Nm).

➡Make sure that there is no oil on the threads of bolts A.

➡Be sure to replace 2 bolts C with new ones or reuse them after applying adhesive (Toyota Genuine Adhesive 1344, Three Bond 1344 or equivalent.)

10. Install the water inlet housing, as follows:
   a. Install a new water inlet housing gasket and water outlet pipe O-ring.
   b. Install the water inlet housing with the 2 bolts and nut, and tighten to 7 ft. lbs. (10 Nm).

➡Be careful not to allow the O-ring to get caught between the parts.

   c. Connect the water hose.
11. Install the water pump pulley, as follows:
   a. Temporarily install the water pump pulley with the 4 bolts.
   b. Using SST: 09960-10010 or equivalent, hold the water pump pulley.
   c. Tighten the 4 bolts to 16 ft. lbs. (21 Nm).
12. Install the V-ribbed belt tensioner assembly with the 5 bolts, and tighten to 32 ft. lbs. (43 Nm).
13. Install the idler pulley cover plate and No. 2 idler pulley sub-assembly with the bolt, and tighten to 32 ft. lbs. (43 Nm).
14. Install the alternator assembly.
15. Install the engine hangers.
16. Remove the engine stand.
17. Install the engine assembly with transaxle.

ENGINE ELECTRICAL

CHARGING SYSTEM

ALTERNATOR

REMOVAL & INSTALLATION

2.7L Engine

See Figures 60 and 61.

1. Disconnect the negative battery cable.
2. Remove the v-belt.
3. Disconnect the alternator connector.
4. Remove the terminal cap.
5. Remove the nut and disconnect the alternator wire.

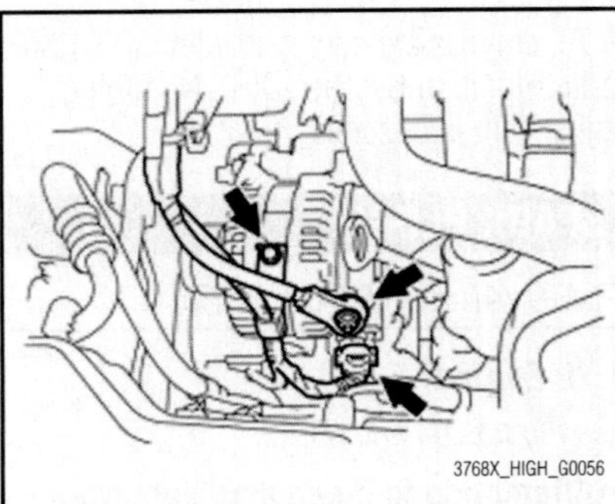

Fig. 60 Disconnect the alternator connector—2.7L Engine

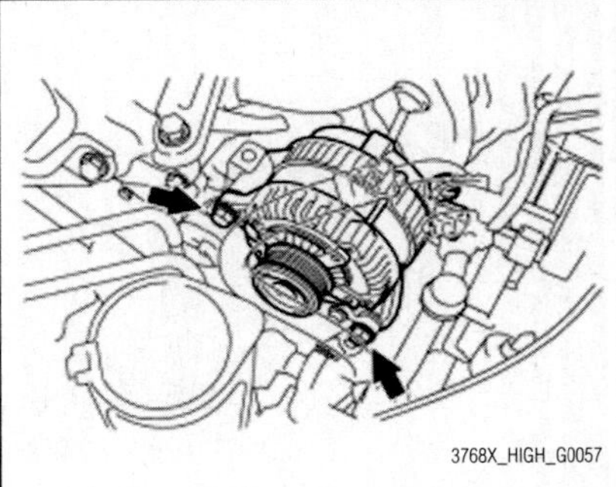

Fig. 61 Remove the 2 bolts and alternator—2.7L Engine

6. Remove the bolt and wire harness clamp bracket.
7. Remove the 2 bolts and alternator.
8. To install, reverse removal procedure.
9. Tighten the alternator bolts to 38 ft. lbs. (52 Nm).

3.5L Engine

See Figure 62.

1. Raise and safely support the vehicle.
2. Remove the front wheel.
3. Remove the engine under cover assembly.
4. Remove the No. 1 engine under cover.
5. Remove the right front fender molding sub-assembly.
6. Remove the right front fender l iner.
7. Remove the right front fender apron seal.
8. Drain engine coolant.
9. Remove the V-bank cover sub-assembly
10. Remove cool air intake duct seal.
11. Remove the battery.
12. Remove the No. 1 and 2 air cleaner inlets, as follows:
   a. Disconnect the 2 vacuum switching valve clamps.
   b. Disconnect the 2 vacuum hoses.
   c. Remove the 2 bolts and No. 2 air cleaner inlet.
   d. Disconnect the 2 vacuum hoses, and remove the 2 bolts and No. 1 air cleaner inlet.
13. Disconnect the No. 1 and 2 radiator hoses.
14. Disconnect the oil cooler hoses.
15. Detach the wire harness clamps from both sides of the fan shroud and disconnect the cooling fan ECU connector.
16. Remove the radiator grill. Refer to radiator removal procedure for detailed instructions.
17. Remove the hood lock assembly.

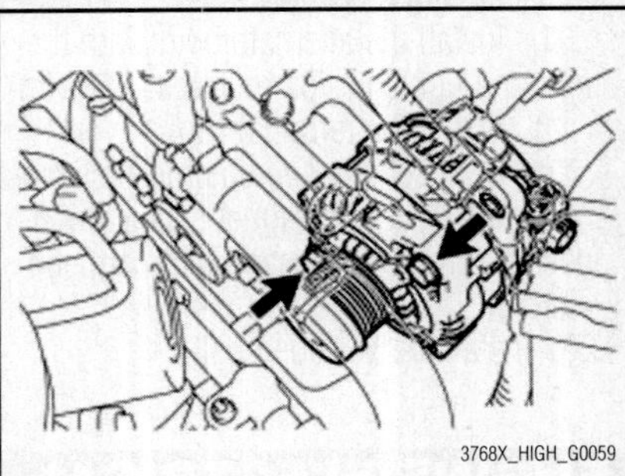

Fig. 62 Remove the 2 bolts and alternator—3.5L Engine

18. Disconnect the low pitched horn and high pitched horn connectors.
19. Detach the hood lock control cable clamp and remove the 6 bolts and upper radiator support sub-assembly.
20. Remove the 4 bolts and move the cooler condenser assembly.
21. Remove the radiator assembly and fan assembly with motor.
22. Remove the bolt and the No. 2 oil level dipstick guide.
23. Remove the V-ribbed belt.
24. Remove the alternator assembly, as follows:
   a. Remove the terminal cap.
   b. Remove the nut and disconnect the wire harness from terminal B.
   c. Disconnect the alternator connector from the alternator assembly.
   d. Disconnect the connector from the compressor and magnetic clutch.
   e. Disconnect the 3 wire harness clamps.
   f. Remove the 2 bolts, and then disconnect the bracket.
   g. Remove the 2 bolts and the alternator assembly.
   h. Disconnect the wire harness clamp, and then remove the alternator bracket.
   i. Remove the bolt and the wire harness clamp stay.

To install:

25. Install the alternator assembly, as follows:
  a. Install the wire harness clamp stay with the bolt, and tighten to 15 ft. lbs. (20 Nm).
  b. Connect the alternator bracket with the wire harness clamp.
  c. Install alternator assembly with the 2 bolts, and tighten to 32 ft. lbs. (43 Nm).
  d. Temporally install the 2 bolts., then fully tighten the 2 bolts to 15 ft. lbs. (20 Nm).
  e. Connect the alternator connector to the alternator assembly.
  f. Install the alternator wire with the nut, and tighten to 87 inch lbs. (9.8 Nm).
  g. Install the terminal cap.
  h. Connect the 3 wire harness clamps.
  i. Connect the magnetic clutch connector to the compressor and magnetic clutch.
26. Install the V-ribbed belt.
27. Install the No. 2 oil level dipstick guide, as follows:
  a. Install a new O-ring to the No. 2 oil level dipstick guide.
  b. Apply a light coat of engine oil to the O-ring.
  c. Push in the No. 2 oil level dipstick guide end into the No. 1 oil level dipstick guide.
  d. Install the No. 2 oil level dipstick guide with the bolt, and tighten to 15 ft. lbs. (20 Nm).
28. Install the radiator assembly and fan assembly with motor.
29. Install the cooler condenser assembly.
30. Install the upper radiator support sub-assembly.
31. Install the hood lock assembly.
32. Install the radiator grill.
33. Connect the cooling fan ECU connector.
34. Connect the oil cooler hose.
35. Connect the No. 1 and 2 radiator hose.
36. Install the No. 1 and 2 air cleaner inlet.
37. Install the battery.
38. Install the cool air intake duct seal.
39. Install the right front fender apron seal.
40. Install the right front fender liner.
41. Install the right front fender molding sub-assembly.
42. Add engine coolant. Refer to radiator installation procedure for detailed instructions.
43. Inspect for coolant leak. Refer to radiator installation procedure for detailed instructions.
44. Inspect automatic transaxle fluid.
45. Inspect for oil leaks.
46. Install the No. 1 engine under cover.
47. Install the engine under cover assembly.
48. Install the right front wheel. Tighten the lug nuts to 76 ft. lbs. (103 Nm).
49. Lower the vehicle.
50. Install the V-bank cover sub-assembly.

ENGINE ELECTRICAL

FIRING ORDER

Firing order for 2.7L engine: 1–3–4–2
Firing order for 3.5L engine: 1–2–3–4–5–6

IGNITION COIL

REMOVAL & INSTALLATION

Refer to Spark Plug procedures.

IGNITION TIMING

INSPECTION

See Figure 63.

1. Warm up the engine.
2. Using SST: 09843-18040, connect terminals 13 (TC) and 4 (CG) of the DLC3.

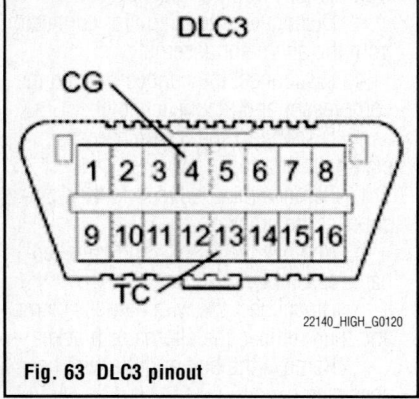

Fig. 63 DLC3 pinout

➡**Confirm the terminal numbers before connecting them. Connecting the wrong terminals can damage the engine.**

➡**Turn off all electrical systems before connecting the terminals.**

➡**Perform this inspection after the cooling fan motor is turned off.**

3. Remove the V-bank cover sub-assembly.
4. Pull out the red lead wire harness.
5. Connect the tester terminal of the timing light to the red lead wire as shown in the illustration.

➡**Use a timing light which detects the No. 1 cylinder ignition signal.**

6. Check the ignition timing at idle. Standard ignition timing: 8 to 12° BTDC at idle.

➡**When checking the ignition timing, the transmission should be in the neutral position.**

➡**Run the engine at 1000 to 1300 RPM for 5 seconds, and then check that the engine RPM returns to idle speed.**

7. Disconnect terminals 13 (TC) and 4 (CG) of the DLC3.
8. Check the ignition timing at idle. Standard ignition timing: 7 to 24° BTDC at idle.
9. Confirm that the ignition timing advances immediately when the engine RPM is increased.
10. Remove the timing light from the engine.

ADJUSTMENT

All engines are equipped with a Distributorless Ignition System (DIS). No timing adjustment is possible.

IGNITION SYSTEM

SPARK PLUGS

REMOVAL & INSTALLATION

2.7L Engine

See Figures 64 and 65.

➡**Attempting to disengage both front and rear clips at the same time may cause the No. 1 engine cover sub-assembly to break.**

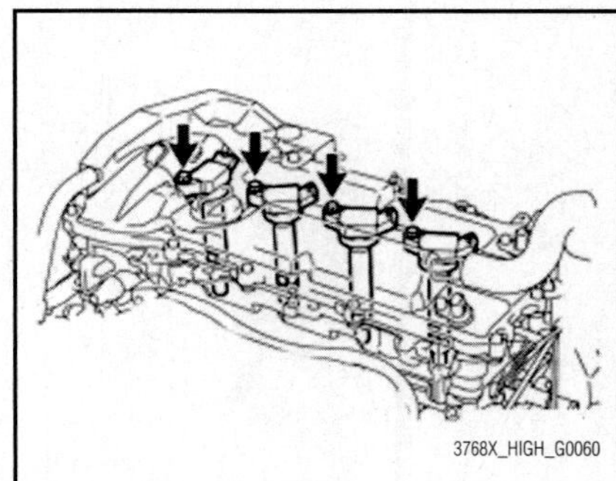

Fig. 64 Remove the 4 bolts and 4 ignition coil assemblies

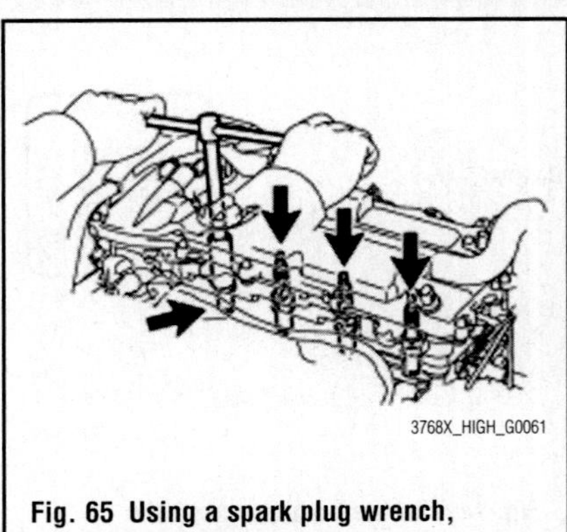

Fig. 65 Using a spark plug wrench, remove the 4 spark plugs

1. Disconnect the negative battery cable.
2. Lift the rear of the No. 1 engine cover sub-assembly to detach the cover from the 2 pins, and then lift the front of the No. 1 engine cover sub-assembly to detach the cover from the pin and remove the No. 1 engine cover sub-assembly.
3. Disconnect the 4 ignition coil assembly connectors.
4. Remove the 4 bolts and 4 ignition coil assemblies.
5. Using a spark plug wrench, remove the 4 spark plugs.
6. To install, reverse removal procedure.

3.5L Engine

See Figures 66 through 75.

1. Remove the engine under cover assembly.
2. Remove the No. 1 engine under cover.
3. Drain the engine coolant.
4. Remove the V-bank cover sub-assembly.
5. Remove both the front wiper arms and blade assemblies.
6. Remove the cowl top ventilator louver sub-assembly.
7. Remove the windshield wiper motor and link assembly.

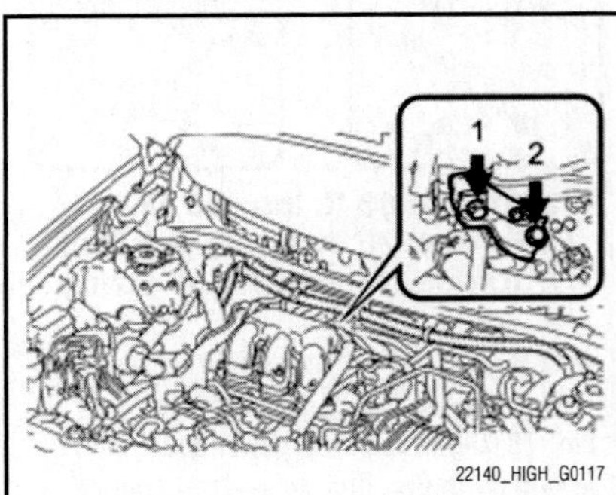

Fig. 66 Throttle body bracket bolts removal sequence—3.5L engine

Fig. 67 No. 1 surge tank stay bolt removal sequence

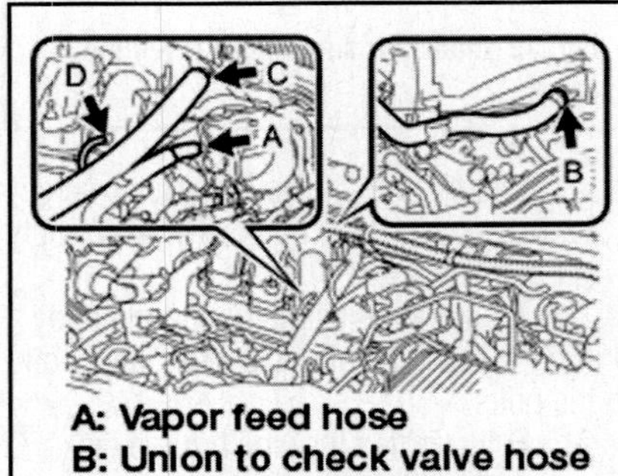

Fig. 68 Intake air surge tank assembly hoses

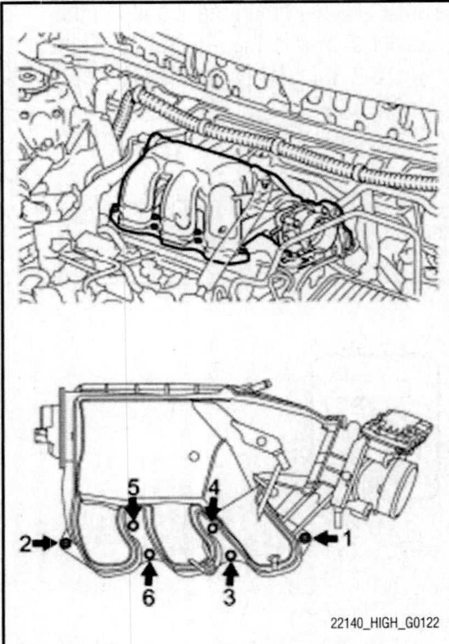

Fig. 69 Intake air surge tank assembly, bolts, and nuts removal sequence

8. Remove the outer cowl top panel sub-assembly.
9. Disconnect the engine room main wire.
10. Remove the 2 bolts in the order shown in the illustration and remove the throttle body bracket.
11. Remove the 2 bolts in the order shown in the illustration and remove the No. 1 surge tank stay.
12. Remove the air cleaner cap sub-assembly.
13. Remove the intake air surge tank assembly, as follows:
    a. Disconnect the 4 hoses.
    b. Disconnect the throttle body connector and clamp.
    c. Disconnect the connector.
    d. Disconnect the 2 water by-pass hoses from the throttle body.
    e. Remove the 4 bolts and 2 nuts in the order shown in the illustration.

➡Use a 5 mm hexagon socket wrench to remove the 4 bolts.

    f. Remove the gasket from the intake air surge tank.
14. Remove the ignition coil assembly, as follows:
    a. Remove the 4 bolts.
    b. Remove the nut.
    c. Disconnect the 2 harness clamps.
    d. Disconnect the 6 ignition coil connectors.
    e. Remove the 6 bolts and 6 ignition coils
15. Remove the 6 spark plugs.

To install:

16. Install the 6 spark plugs and tighten to 13 ft. lbs. (18 Nm).
17. Install the ignition coil assembly as follows:
    a. Install the 6 ignition coils with the 6 bolts and tighten to 7 ft. lbs. (10 Nm).
    b. Connect the 6 ignition coil connectors.
    c. Install the 4 bolts and tighten to 73 inch lbs. (8.3 Nm).
    d. Install the nut and tighten to 73 inch lbs. (8.3 Nm).
    e. Install the 2 clamps.
18. Install the intake air surge tank assembly as follows:
    a. Install the surge tank with the 4 bolts and 2 nuts in the order shown in the illustration, and tighten to 12 ft. lbs. (16 Nm) and 13 ft. lbs. (18 Nm), using a 5 mm hexagon socket wrench. **DO NOT** apply oil to the bolts.
    b. Connect the 2 water by-pass hoses to the throttle with motor body assembly.
    c. Connect the connector.
    d. Install the clamp and connect the throttle with motor body assembly connector.

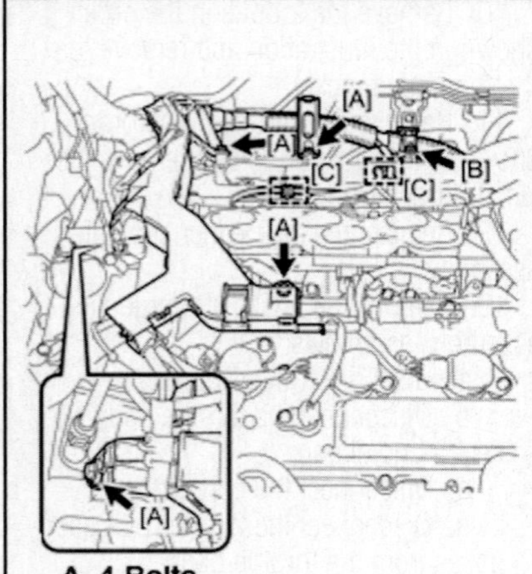

Fig. 70 Ignition coil assembly harness clamps

e. Connect the 4 hoses.

19. Temporarily install the No. 1 surge tank stay as follows:

a. Temporarily install the intake air surge tank assembly with 3 new gaskets on the intake manifold.

➡Do not allow the gaskets to slip out of place during installation.

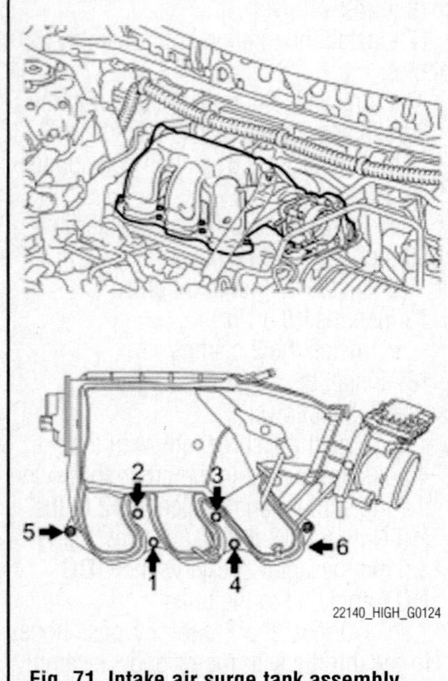

Fig. 71 Intake air surge tank assembly installation sequence

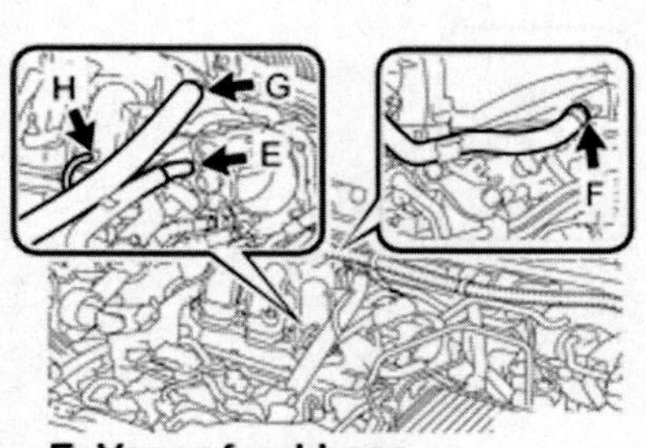

Fig. 72 Intake air surge tank assembly hoses

b. Temporarily install the No. 1 surge tank stay with the 2 bolts. **DO NOT** apply oil to the bolts.

20. Temporarily install the throttle body bracket with the 2 bolts. **DO NOT** apply oil to the bolts.

21. Fully tighten the No. 1 surge tank stay, as follows:

a. Fully tighten the 2 bolts in the order shown in the illustration. Tighten to 15 ft. lbs. (21 Nm). **DO NOT** apply oil to the bolts.

22. Fully tighten the throttle body bracket, as follows:

a. Fully tighten the 2 bolts in the order shown in the illustration. Tighten to 15 ft. lbs. (21 Nm). **DO NOT** apply oil to the bolts.

23. Connect the engine room main wire.

24. Install the air cleaner cap sub-assembly.

Fig. 73 No. 1 surge tank stay bolt tightening sequence

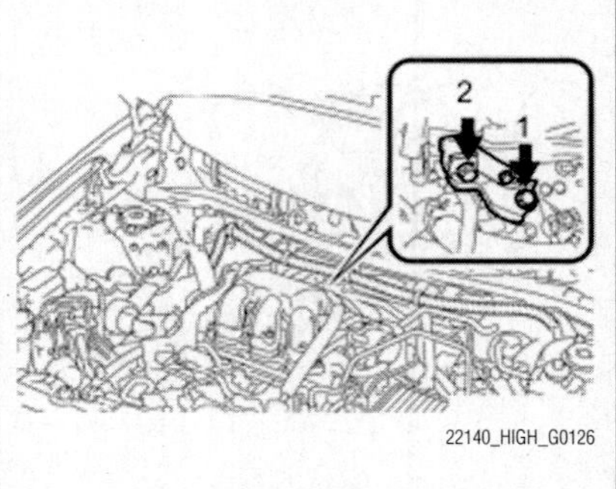

Fig. 74 Throttle body bracket bolts tightening sequence—3.5L engine

25. Install the outer cowl top panel sub-assembly, as follows:

a. Install the outer cowl top panel sub-assembly with the 8 bolts and 6 nuts.

b. Engage the 4 clamps.

26. Install the windshield wiper motor and link assembly.

27. Install the cowl top ventilator louver sub-assembly.

28. Install both front wiper arm and blade assemblies.

29. Install the V-bank cover sub-assembly.

30. Add engine coolant.

31. Inspect for engine coolant leak.

32. Install the No. 1 engine under cover.

33. Install the engine under cover assembly.

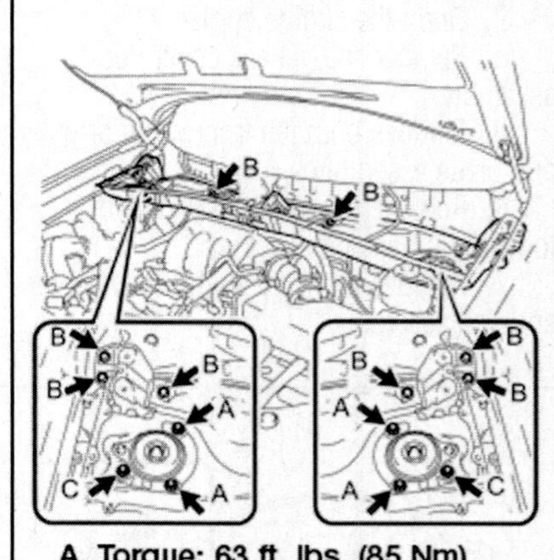

Fig. 75 Outer cowl top panel sub-assembly, bolts, and nuts—3.5L engine

ENGINE ELECTRICAL

STARTER

REMOVAL & INSTALLATION

2.7L Engine

See Figure 76.

1. Disconnect the negative battery cable.
2. Remove cool air intake duct seal.
3. Remove no. 1 engine cover sub-assembly.

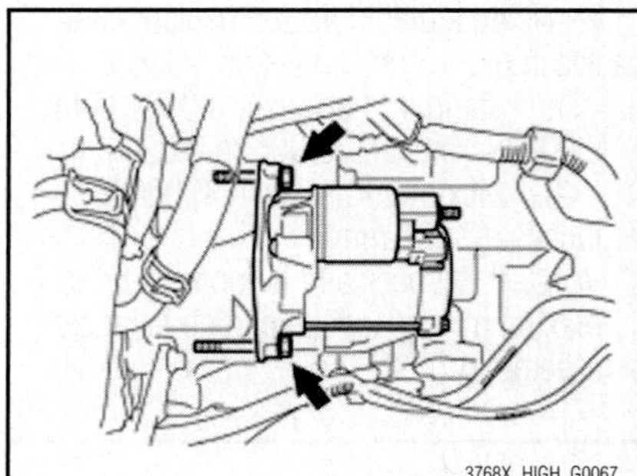

Fig. 76 Remove the 2 bolts and starter—2.7L Engine

4. Remove battery.
5. Remove air cleaner assembly.
6. Disconnect the starter connector.
7. Open the terminal cap, remove the nut and disconnect the starter wire.
8. Remove the 2 bolts and starter.
9. To install, reverse removal procedure.
10. Tighten the starter bolts to 27 ft. lbs. (37 Nm).

3.5L Engine

See Figure 77.

1. Remove the cool air intake duct seal.
2. Remove the battery.
3. Remove the No. 1 and 2 air cleaner inlets.
4. Remove the air cleaner cap sub-assembly.
5. Remove the air cleaner case sub-assembly.
6. Remove the 2 bolts and the air cleaner bracket.
7. Disconnect the starter connector.
8. Open the terminal cap, remove the nut and disconnect the starter wire.
9. Remove the 2 bolts and starter.

To install:

10. Installation is the reverse of removal procedure, noting the following:
    a. After installing the starter assembly with the 2 bolts, tighten to 27 ft. lbs. (37 Nm).
    b. After connecting the starter wire with the nut, tighten to 87 inch lbs. (9.8 Nm).
    c. After installing the bracket with the 2 bolts, tighten to 9 ft. lbs. (12 Nm).

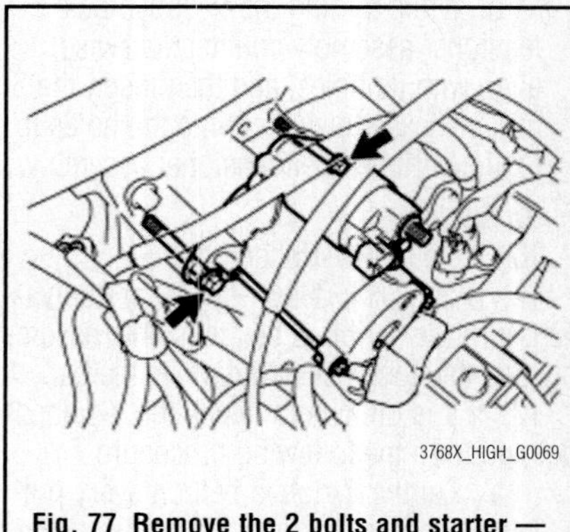

Fig. 77 Remove the 2 bolts and starter —3.5L Engine

ENGINE MECHANICAL

ACCESSORY DRIVE BELTS

ACCESSORY BELT ROUTING

See Figures 78 and 79.

INSPECTION

Inspect the drive belt for signs of glazing or cracking. A glazed belt will be perfectly smooth from slippage, while a good belt will have a slight texture of fabric visible. Cracks will usually start at the inner edge of the belt and run outward. All worn or damaged drive belts should be replaced immediately.

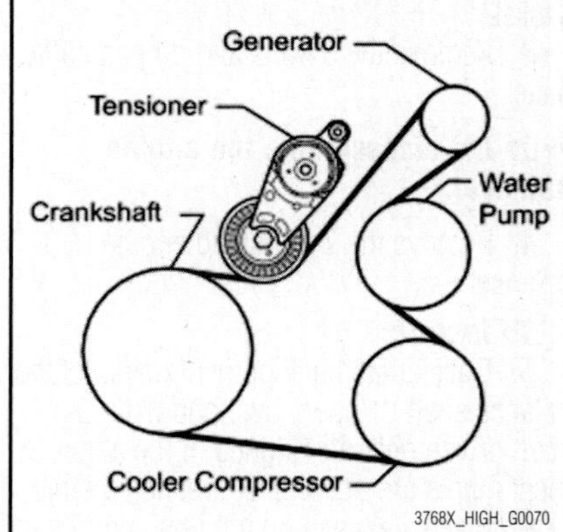

Fig. 78 Accessory drive belt routing —2.7L engine

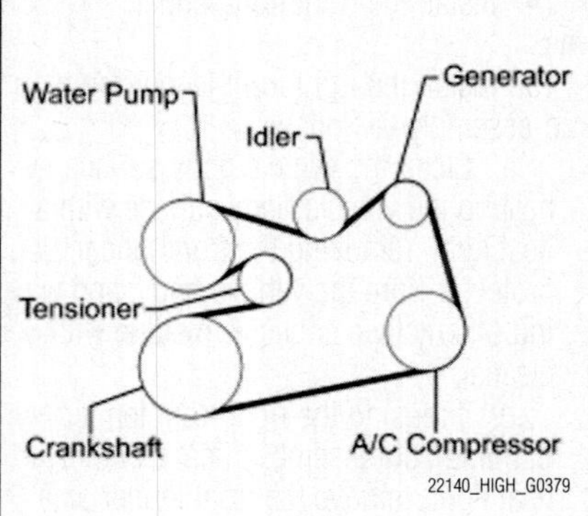

Fig. 79 Accessory drive belt routing —3.5L engine

ADJUSTMENT

Belt tension is maintained by an automatic tensioner. No adjustment is necessary of possible.

REMOVAL & INSTALLATION

See Figures 80 and 81.

1. Raise and safely support the vehicle.
2. Remove the right front wheel.
3. Remove the engine under cover assembly.
4. Remove the No. 1 engine under cover.
5. Remove the right front fender molding sub-assembly, as follows:
    a. Remove the clip.
    b. Using a 4 mm hexagon wrench, remove the screw.
    c. Peel off the front fender side protector and disengage the 3 clips, and then remove the right front fender molding sub-assembly.
    d. Remove the pad from the right front fender moulding sub-assembly.
    e. Remove the 2 clips No. 4 from the right front fender molding sub-assembly.
    f. Remove the front fender side protector from the right front fender molding sub-assembly.
6. Remove the right front fender liner.
7. Remove the right front fender apron seal.

8. For 2.7L engines, remove the V-ribbed belt as follows: Attach a wrench to the hexagonal portion of the belt tensioner as shown in the illustration, rotate the belt tensioner clockwise, and remove the V-ribbed belt.

9. For 3.5L engines, remove the V-ribbed belt as follows:

a. Using SST: 09961-00950 or equivalent, release the V-ribbed belt tension by turning the V-ribbed belt tensioner assembly counterclockwise, and remove the V-ribbed belt from the V-ribbed belt tensioner assembly.

b. While turning the V-ribbed belt tensioner assembly counterclockwise, align with its holes, and then insert the 5 mm bi-hexagon wrench into the holes to fix the V-ribbed belt tensioner assembly.

To install:

10. Install the V-ribbed belt.

11. Using SST: 09961-00950 or equivalent, turn the V-ribbed belt tensioner assembly counterclockwise and remove the bar.

12. If it is difficult to install the V-ribbed belt, perform the following procedure:

a. Put the V-ribbed belt on every pulley except the tensioner pulley as shown in the illustration.

b. Release the V-ribbed belt tension by turning the V-ribbed belt tensioner assembly counterclockwise, and put the V-ribbed belt on the V-ribbed tensioner assembly pulley.

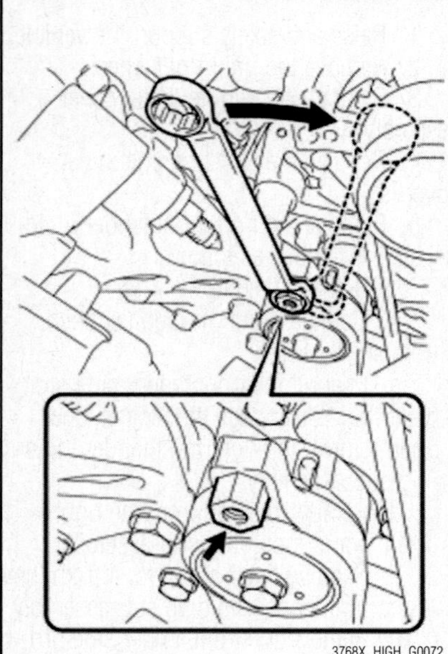

Fig. 80 Attach a wrench to the hexagonal portion of the belt tensioner rotate the belt tensioner clockwise, and remove the V-ribbed belt—2.7L Engines

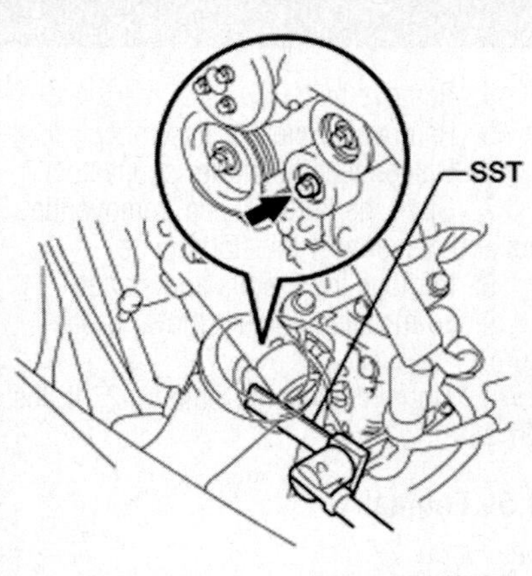

Fig. 81 Using SST: 09961-00950 or equivalent, remove the V-ribbed belt—3.5L Engines

➡Put the backside of the V-ribbed belt on the V-ribbed belt tensioner assembly pulley and No. 2 idler pulley sub-assembly.

➡Check that the V-ribbed belt is properly set to each pulley.

c. After installing the V-ribbed belt, check that it fits properly in the ribbed grooves. Confirm that the belt has not slipped out of the grooves on the bottom of the crankshaft pulley by hand.

13. Install the right front fender apron seal.

14. Install the right front fender liner.

15. Install the right front fender molding sub-assembly, as follows:

a. Clean the vehicle body surface by heating the vehicle body surface with a heat light, removing the front fender side protector from the vehicle body, and wiping off any tape adhesive residue with cleaner.

b. If reusing the right front fender molding sub-assembly, heat it with a heat light, remove the front fender side protector, wipe off any tape adhesive residue with cleaner, and install a new front fender side protector to the front fender molding sub-assembly.

c. Using a heat light, heat the vehicle body and the front fender moulding sub-assembly.

d. Remove the release paper from the front fender molding sub-assembly.

➡After removing the release paper, keep the exposed adhesive free from foreign matter.

e. Engage the 3 clips and install the right front fender molding sub-assembly.

f. Using a 4 mm hexagon wrench, install the screw.

g. Install the clip.

16. Install the no. 1 engine under cover.

17. Install the engine under cover assembly.

18. Install the right front wheel. Tighten the lug nuts to 76 ft. lbs. (103 Nm).

19. Lower the vehicle.

BALANCE SHAFT

INSPECTION

See Figure 82.

1. Using a dial indicator, measure the backlash of the crankshaft and balance shaft:

a. Standard backlash: 0.000197 to 0.000787 inches. (0.005 to 0.020 mm)

b. Maximum backlash: 0.000787 inches. (0.020 mm)

c. If the backlash is more than the maximum, replace the engine balancer assembly.

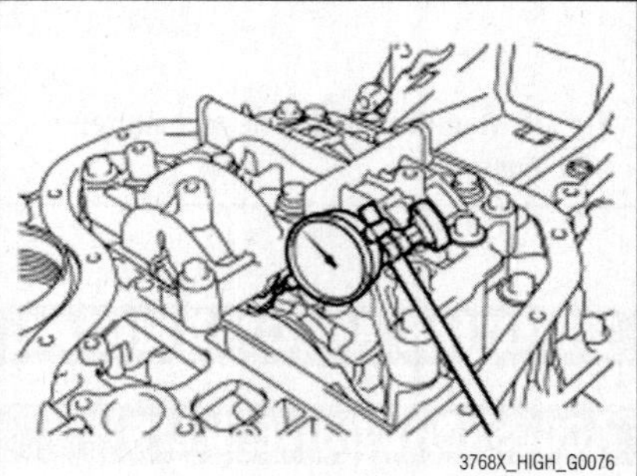

Fig. 82 Using a dial indicator, measure the backlash of the crankshaft and balance shaft

REMOVAL & INSTALLATION

See Figures 83 through 86.

1. Remove the oil pan.

2. Remove the 3 bolts, oil strainer and gasket.

3. Remove the 5 bolts and oil pan baffle plate.

➡Do not disassemble the engine balancer.

4. Remove the 7 bolts and engine balancer.

To install:

5. Check that the alignment marks of the balance shaft damper cover and balance shaft driven gear are aligned. If the alignment marks are not aligned, realign them.

6. Place a wrench on the rear cutout part of the No. 2 balance shaft and hold the shaft in place.

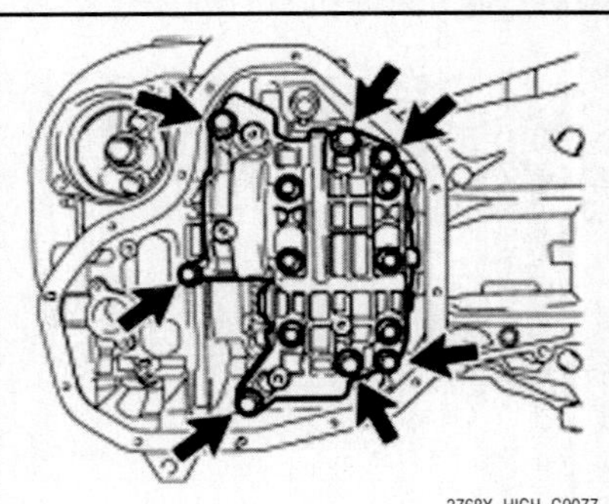
3768X_HIGH_G0077

Fig. 83 Remove the 7 bolts and engine balancer

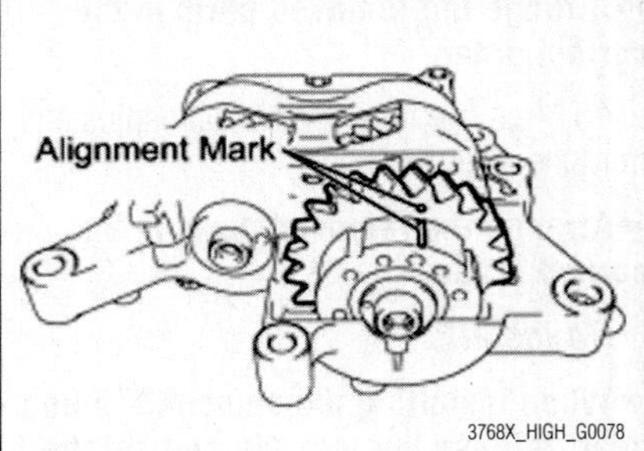

3768X_HIGH_G0078

Fig. 84 Check that the alignment marks of the balance shaft damper cover and balance shaft driven gear are aligned

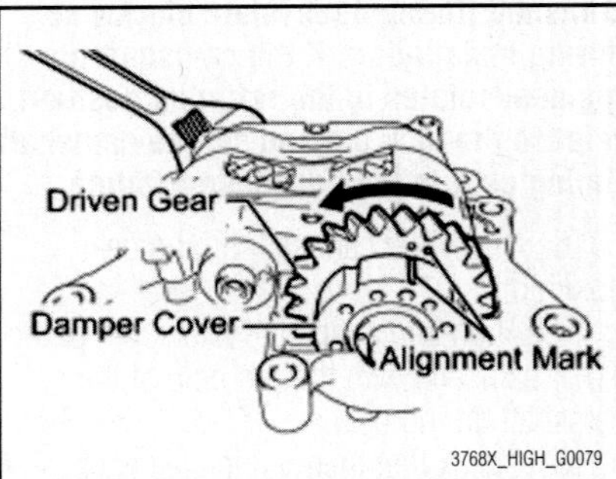

3768X_HIGH_G0079

Fig. 85 Rotate the balance shaft driven gear of the No. 1 balance shaft counterclockwise

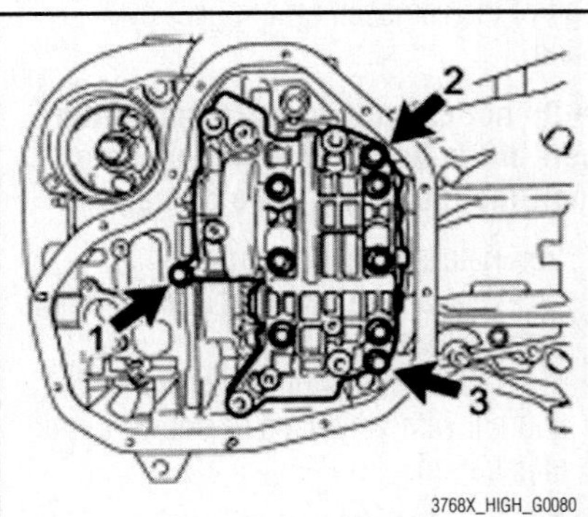

3768X_HIGH_G0080

Fig. 86 Tighten the 3 bolts in the sequence shown

7. Rotate the balance shaft driven gear of the No. 1 balance shaft counterclockwise to align the alignment mark of the balance shaft driven gear with the alignment mark of the balance shaft damper cover.

8. Install the engine balancer to the stiffening crankcase with the 3 bolts, and tighten the bolts in the sequence shown in the illustration. Tighten 18 ft. lbs. (24 Nm).

9. To complete installation, reverse remaining removal procedure.

CAMSHAFT AND VALVE LIFTERS

INSPECTION

See Figures 87 through 90.

1. To inspect the camshaft, place the camshaft on V-blocks.
   a. Using a dial indicator, measure the circle runout at the center journal.
      - 2.7L Maximum runout 0.0018 inches (0.03 mm).
      - 3.5L Maximum runout 0.0016 inches (0.04 mm).

➡Check the oil clearance after replacing the camshaft.

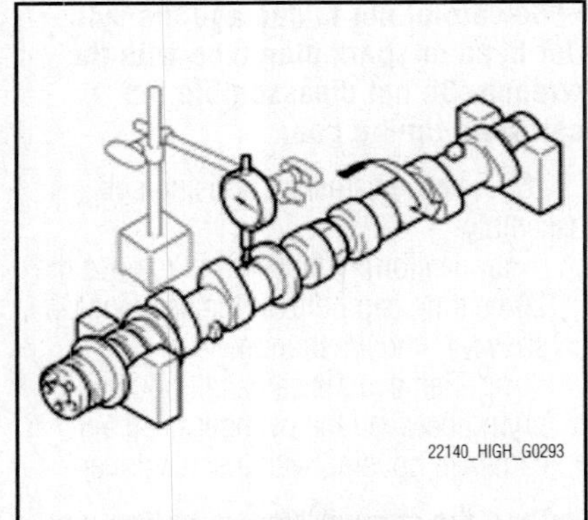
22140_HIGH_G0293

Fig. 87 Measuring Camshaft Runout

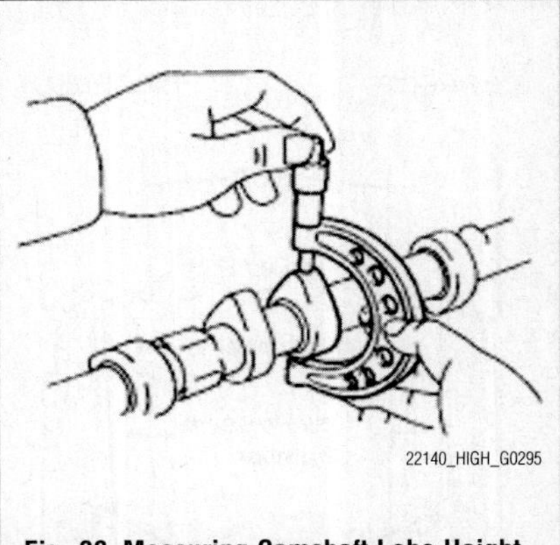
22140_HIGH_G0295

Fig. 88 Measuring Camshaft Lobe Height

2. Using a micrometer, measure the cam lobe height.
   a. For 2.7L Standard lobe height is 1.73870 to 1.74429 in. (44.163 to 44.305 mm) Minimum lobe height is 1.73279 in. (44.013 mm)
3. Using a micrometer, measure the journal diameter.
   a. For 2.7L standard journal diameter should be No. 1 Journal 1.35626 to 1.35689 in (34.449 to 34.465.
   b. If the journal diameter is not as specified, check the oil clearance.
4. For 3.5L the valve lash adjuster assembly can be inspected.
5. To inspect the lifters, place the valve lash adjuster assembly into a container filled with engine oil.
   a. Insert Service Tool tip into the valve lash adjuster assembly's plunger and use the tip to press down on the check ball inside the plunger.
   b. Squeeze Service Tool and valve lash adjuster assembly together to move the plunger up and down 5 to 6 times.
   c. Check the movement of the plunger and bleed the air. If plunger moves up and down freely, valve lash adjuster is OK.

Item	Specification
No. 1 journal	35.946 to 35.960 mm (1.4152 to 1.4157 in.)
Other journals	25.959 to 25.975 mm (1.0220 to 1.0226 in.

22140_HIGH_G0298

Fig. 89 Standard Journal Diameter—3.5L engine

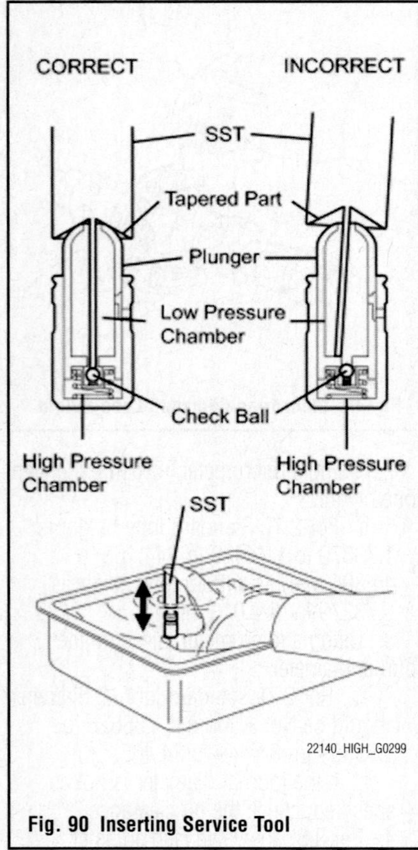

Fig. 90 Inserting Service Tool

➡When bleeding air from the high-pressure chamber, make sure that the tip of SST is actually pressing the check ball as shown in the illustration. If the check ball is not pressed, air will not bleed.

d. After bleeding the air, remove Service Tool. Then try to quickly and firmly press the plunger with by hand. If plunger is difficult to move, valve lash adjuster is OK.

e. If the result is not as specified, replace the valve lash adjuster assembly.

REMOVAL & INSTALLATION

2.7L Engine

See Figures 91 through 106.

1. Before servicing the vehicle, refer to the precautions section.
2. Disconnect the negative battery cable.
3. Drain the engine coolant.
4. Remove or disconnect the following:
   - Right front wheel
   - Right fender splash shield
   - Right fender apron seal
   - No. 1 engine undercover
   - Coil pack
   - Cylinder head cover

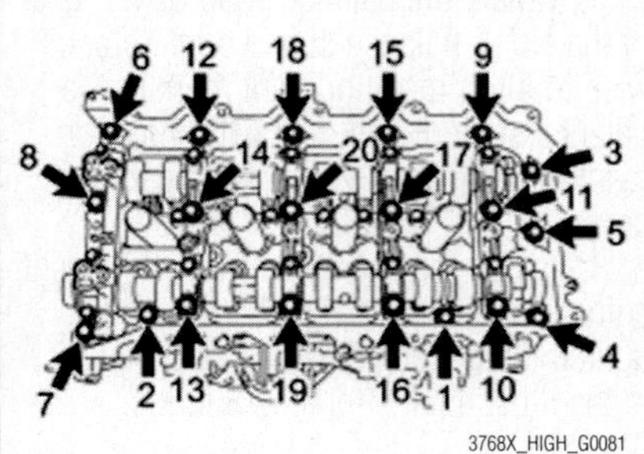

Fig. 91 Uniformly loosen and remove the 20 bearing cap bolts in the sequence shown

5. Remove the timing chain cover and timing chain assembly.
6. Remove camshaft timing gear assembly. Hold the hexagonal portion of the camshaft with a wrench and remove the bolt and camshaft timing gear.

➡Be careful not to damage the cylinder head or spark plug tube with the wrench. Do not disassemble the camshaft timing gear.

7. Remove camshaft timing exhaust gear assembly. Hold the hexagonal portion of the camshaft with a wrench and remove the bolt and camshaft timing gear.

➡Be careful not to damage the cylinder head or spark plug tube with the wrench. Do not disassemble the camshaft timing gear.

8. Remove camshaft housing sub-assembly:
   a. Uniformly loosen and remove the 20 bearing cap bolts in the sequence shown in the illustration.
   b. Remove the camshaft housing by prying between the cylinder head and camshaft housing with a screwdriver.

➡Tape the screwdriver tip before use. Be careful not to damage the contact surfaces of the cylinder head and camshaft housing.

9. Remove camshaft bearing cap:
   a. Remove the 11 bearing cap bolts in the sequence shown in the illustration.
   b. Remove the 5 bearing caps.

➡Arrange the removed parts in the correct order.

10. Remove the oil control valve filter from the No. 1 camshaft bearing cap.
11. Remove the No. 1 camshaft bearing.
12. Remove the No. 1 and No. 2 camshafts.
13. Remove the No. 2 camshaft bearing.
14. Remove the 16 valve rocker arms from the cylinder head.

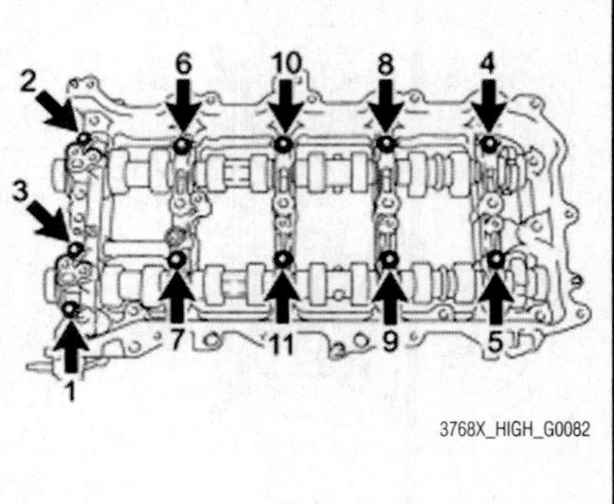

Fig. 92 Remove the 11 bearing cap bolts in the sequence shown

➡Arrange the removed parts in the correct order.

15. Remove the 16 valve lash adjusters from the cylinder head.

➡Arrange the removed parts in the correct order.

To install:

➡When installing the camshaft timing gear, release the lock pin and set the camshaft timing gear to the advanced position before installation.

➡If the camshaft timing gear is set to the advanced position, do not let the camshaft timing gear rotate clockwise during installation. If the camshaft timing gear rotates to the retarded position, release the lock pin and set the camshaft timing gear to the advanced position.

16. Check the camshaft timing gear position.
17. Align and attach the knock pin of the No. 1 camshaft with the pin hole of the camshaft timing gear.
18. Check that there is no clearance between the camshaft timing gear and camshaft flange.
19. Hold the camshaft in place by hand, and then install the installation bolt of the camshaft timing gear by hand.

➡Do not use any tools to install the bolt. If the bolt is installed using a tool, the lock pin will be damaged.

20. Release the lock pin:
    a. Clean the camshaft journal with a non-residue solvent.
    b. Cover the 4 oil paths of the cam journal with vinyl tape as shown in the illustration.

➡There are 4 oil paths in the grooves of the camshaft. Plug 3 of the paths with pieces of rubber.

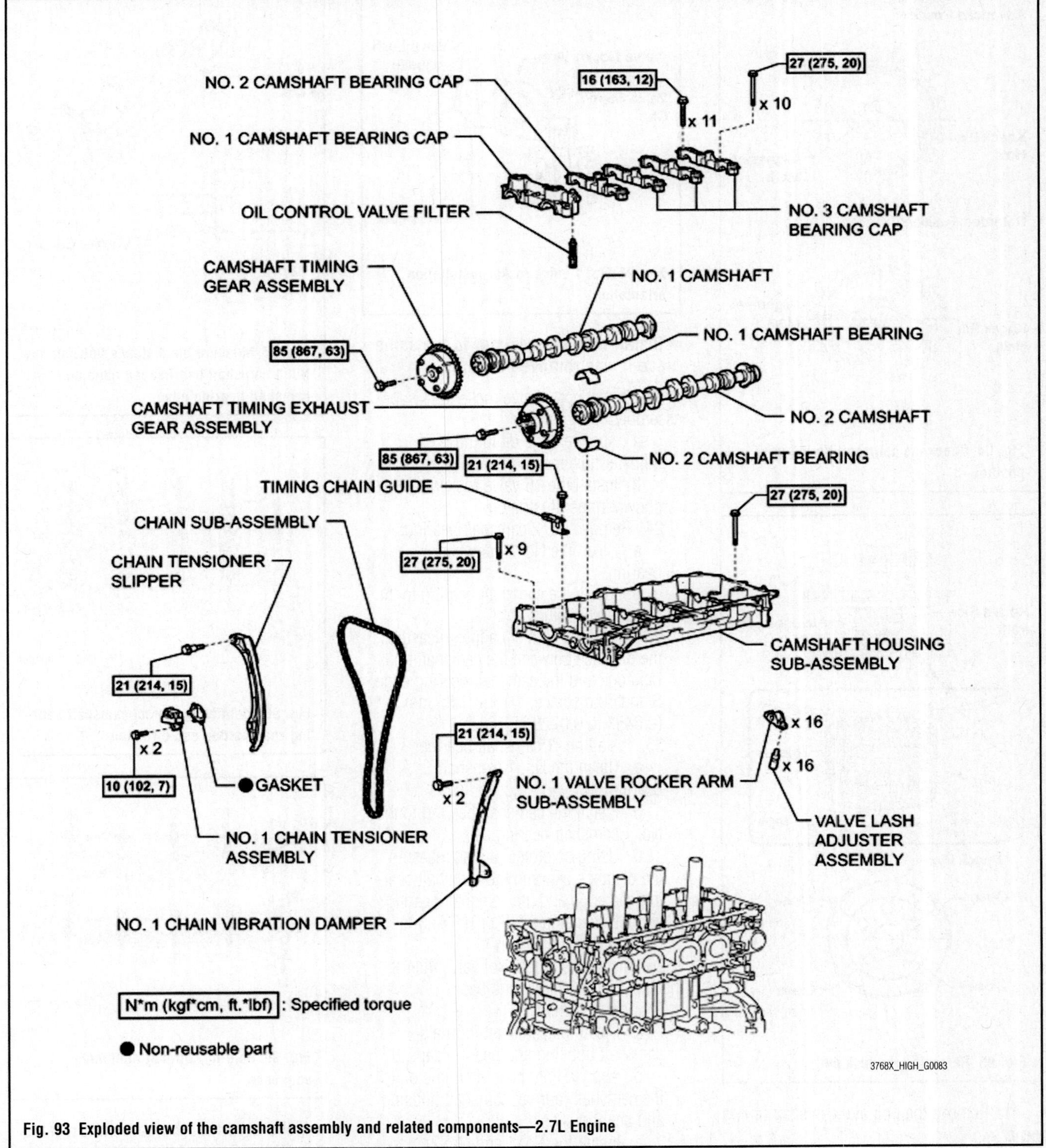

Fig. 93 Exploded view of the camshaft assembly and related components—2.7L Engine

c. Open a hole at port A shown in the illustration.

d. While applying approximately 200 kPa (2.0 kgf/ cm2, 29 psi) of air pressure to the oil path, forcibly turn the camshaft timing gear assembly in the advance direction (counterclockwise).

CAUTION

Cover the paths with a piece of cloth when applying pressure to keep oil from splashing.

➡Do not allow the camshaft timing gear assembly to lock. If it locks, release the lock pin again.

➡The camshaft timing gear assembly may be turned in the advance direction without applying any force. If enough air pressure cannot be applied because of air leakage from the port, releasing the lock pin may be difficult.

e. Remove the vinyl tape and rubber pieces from the camshaft.

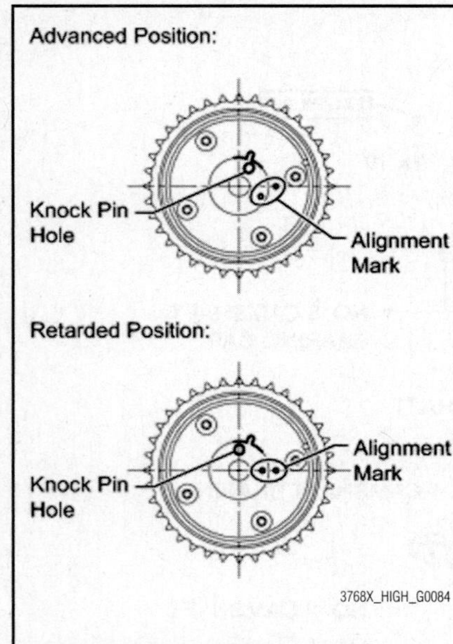

Fig. 94 Check the camshaft timing gear position

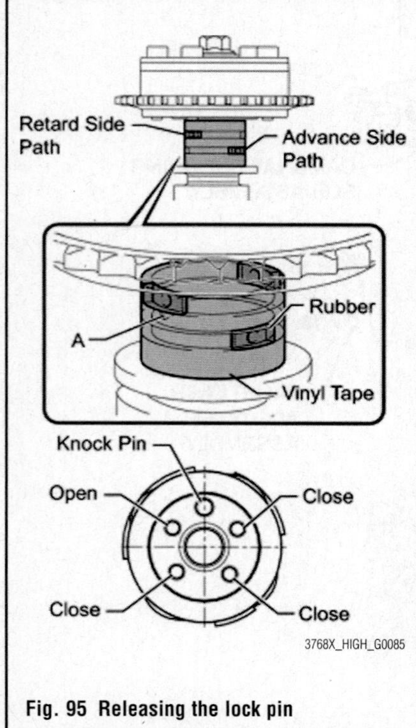

Fig. 95 Releasing the lock pin

21. Remove the bolt and camshaft timing gear.

➡Do not allow the camshaft timing gear assembly to lock. If it locks, release the lock pin again.

22. Install valve lash adjuster assembly:
   a. Inspect the valve lash adjuster before installing it.
   b. Install the 16 lash adjusters to the cylinder head.

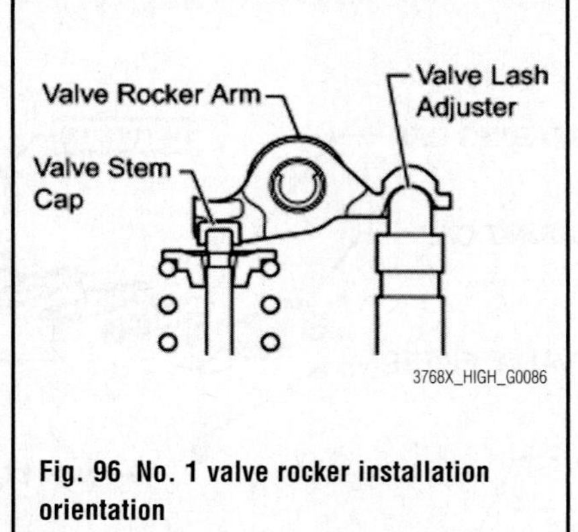

Fig. 96 No. 1 valve rocker installation orientation

➡Install the lash adjuster to the same place it was removed from.

23. Install no. 1 valve rocker arm sub-assembly:
   a. Apply engine oil to the lash adjuster tips and valve stem caps.
   b. Install the 16 valve rocker arms as shown in the illustration.
24. Remove no. 2 camshaft bearing:
   a. Clean the No. 2 camshaft bearing.
   b. Install the camshaft bearing to the camshaft housing.
   c. Using a vernier caliper, measure the distance between the camshaft housing edge and the camshaft bearing edge. Standard distance: 1.15 to 1.85 mm (0.0453 to 0.0728 in.)
25. Install no. 1 camshaft bearing:
   a. Clean the No. 1 camshaft bearing.
   b. Install the camshaft bearing to the No. 1 camshaft bearing cap.
   c. Using a vernier caliper, measure the distance between the camshaft bearing cap edge and the camshaft bearing edge. Dimension A - B or B - A: 0 to 0.7 mm (0 to 0.0276 in.).
26. Install the oil control valve filter to the No. 1 camshaft bearing cap.
27. Install camshaft:
   a. Clean the camshaft journals, camshaft housing and bearing caps.
   b. Apply a light coat of engine oil to the camshaft journal, camshaft housing and bearing caps.
   c. Install the No. 1 and No. 2 camshafts to the camshaft housing.
28. Install camshaft bearing cap:
   a. Confirm the marks and numbers on the camshaft bearing caps and place them in their proper positions and directions.
   b. Install the 11 bolts in the order shown in the illustration. Tighten to 12 ft. lbs. (16 Nm).

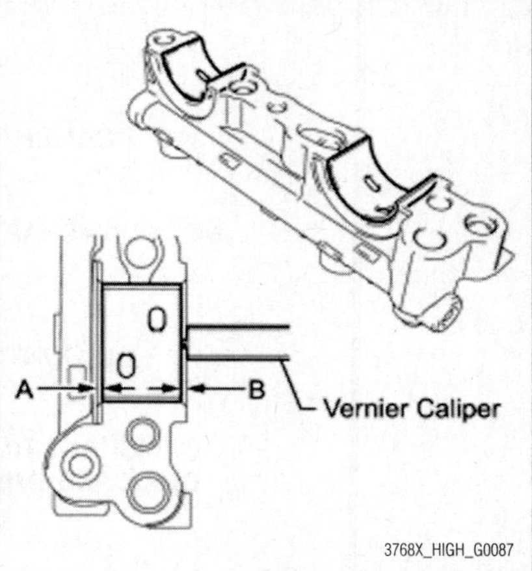

Fig. 97 Measure the distance between the No. 1 camshaft bearing cap edge and the camshaft bearing edge

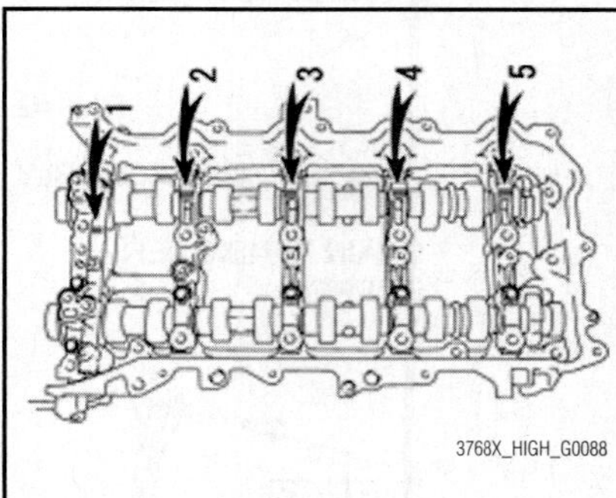

Fig. 98 Identifying correct camshaft bearing cap position and direction

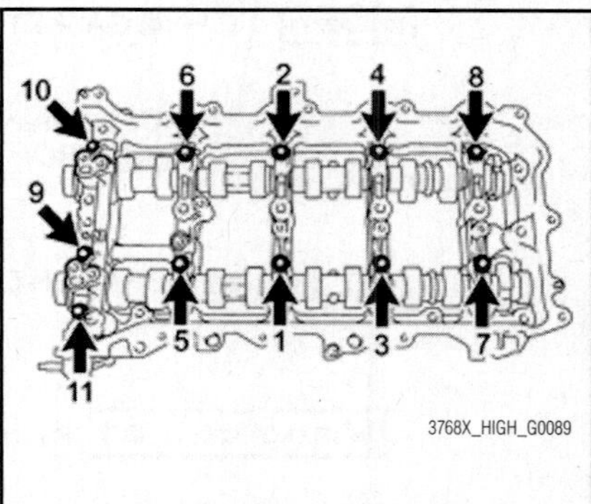

Fig. 99 Bearing cap bolt tightening sequence

➡Make sure that the camshaft rotates smoothly after installing the bearing caps.

29. Install camshaft housing sub-assembly:
   a. Check that the valve rocker arms are installed as shown in the illustration.
   b. Apply seal packing in a continuous line as shown in the illustration. Seal packing: Toyota Genuine Seal Packing

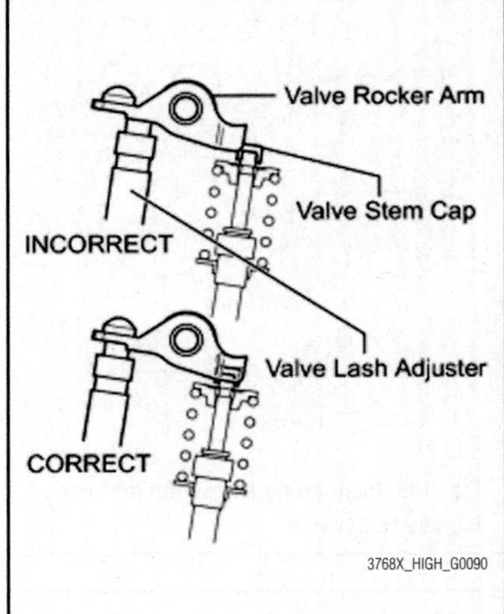

Fig. 100 Valve rocker arm installation orientation

Black, Three Bond 1207B or equivalent. Standard seal diameter: 3.0 to 4.0 mm (0.118 to 0.157 in.)

➡**Remove any oil from the contact surface. Install the camshaft housing within 3 minutes and tighten the bolts within 10 minutes after applying seal packing.**

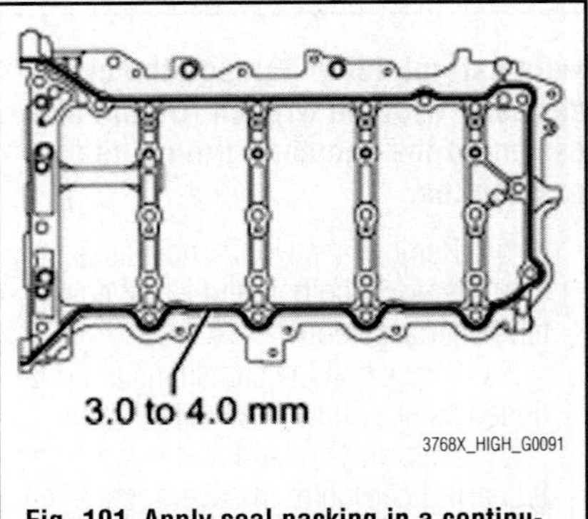

Fig. 101 Apply seal packing in a continuous line as shown

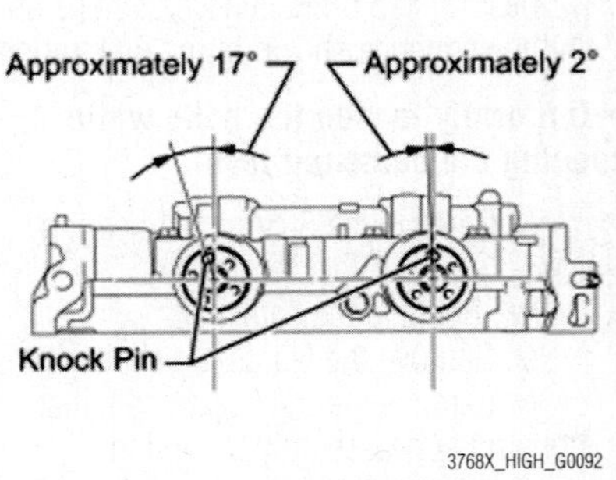

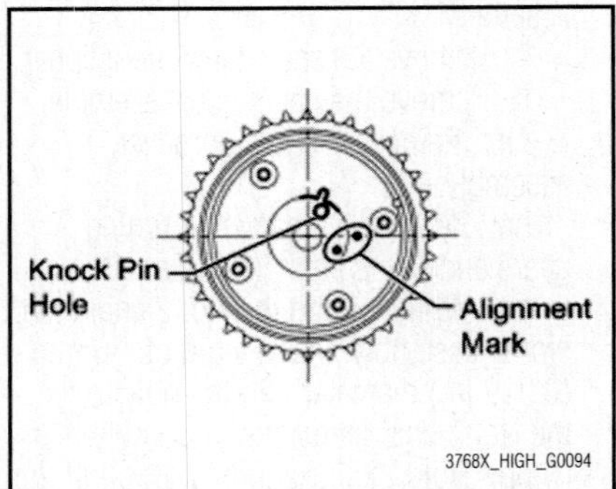

Fig. 104 Check the camshaft timing gear position

c. Position the knock pin of the No. 1 and No. 2 camshafts as shown in the illustration.

d. Install the camshaft housing, and then install the 20 bolts in the order shown in the illustration. Tighten to 20 ft. lbs. (27 Nm).

➡**Do not apply oil for at least 4 hours after the installation. Do not start the engine for at least 4 hours after the installation. Thoroughly wipe clean any seal packing.**

30. Install camshaft timing gear assembly:

a. Check the camshaft timing gear position. If the camshaft timing gear is not set to the advanced position, release the lock pin and reset the camshaft timing gear (Refer to the "SET CAMSHAFT TIMING GEAR ASSEMBLY" procedures").

b. Align and attach the knock pin of the No. 1 camshaft with the pin hole of the camshaft timing gear.

c. Check that there is no clearance between the camshaft timing gear and camshaft flange.

d. Using a wrench to hold the hexagonal portion of the No. 1 camshaft, install the bolt. Tighten to 63 ft. lbs. (85 Nm).

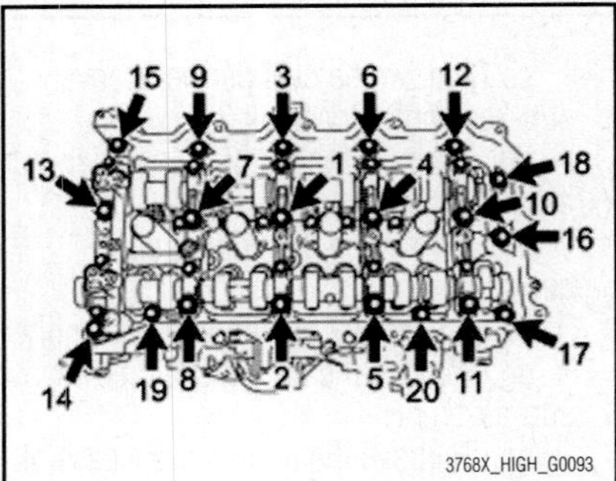

Fig. 103 Camshaft housing tightening sequence

Approximately 17° Approximately 2°

Knock Pin

3768X_HIGH_G0092

Fig. 102 Position the knock pin of the No. 1 and No. 2 camshafts as shown

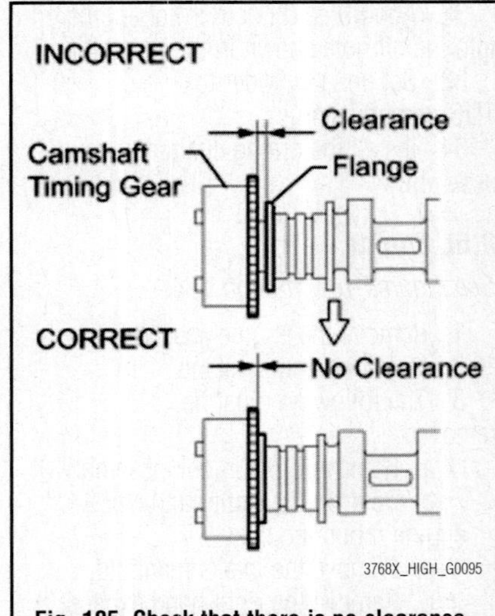

Fig. 105 Check that there is no clearance between the camshaft timing gear and camshaft flange

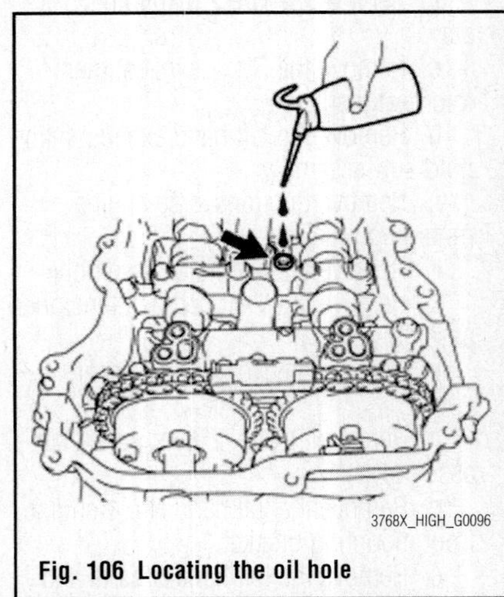

Fig. 106 Locating the oil hole

➡**Be careful not to damage the cylinder head or spark plug tube with the wrench. Do not disassemble the camshaft timing gear.**

31. Install camshaft timing exhaust gear assembly:

a. Align and attach the knock pin of the No. 2 camshaft with the pin hole of the camshaft timing exhaust gear.

b. Check that there is no clearance between the camshaft timing exhaust gear and camshaft flange.

c. Using a wrench to hold the hexagonal portion of the No. 2 camshaft, install the bolt. Tighten to 63 ft. lbs. (85 Nm).

➡**Oil must be added if the lash adjusters were removed. Make sure that the low pressure chamber and oil paths of the lash adjusters are full of engine oil.**

32. Add 50 cc (3.1 cu. in) of engine oil into the oil hole shown in the illustration.
33. Set no. 1 cylinder to TDC/compression.
34. Install the timing chain and cover assembly.

3.5L Engine

See Figures 107 through 124.

1. Remove the engine assembly.
2. Install on engine stand.
3. The following must be removed:
   a. Remove ignition coil assembly.
   b. Remove the right hand No. 2 engine mounting stay.
   c. Remove the intake manifold.
   d. Remove the right hand exhaust manifold sub-assembly.
   e. Remove the No. 2 engine oil level dipstick guide.
   f. Remove the No. 2 manifold stay.
   g. Remove the No. 2 exhaust manifold heat insulator.
   h. Remove the left hand exhaust manifold sub-assembly.
   i. Remove the transverse engine mounting bracket.
   j. Remove the alternator assembly.
   k. Remove the V-ribbed belt tensioner assembly.
   l. Remove the No. 2 timing gear cover.
   m. Remove the No. 2 idler pulley sub-assembly.
   n. Remove the left hand No. 1 engine front mounting bracket.
   o. Remove the left hand 6 bolts and No. 1 front engine mounting bracket.
   p. Remove the radio setting condenser.
   q. Remove the No. 1 vacuum switching valve.
   r. Remove the knock control sensor wire.
   s. Remove the knock control sensor.
   t. Remove the crankshaft position sensor.
   u. Remove the No. 1 oil pipe.
   v. Remove the oil pipe.
   w. Remove the crankshaft pulley.
   x. Remove the oil cooler assembly, if necessary.
   y. Remove the No. 1 oil cooler bracket, if necessary.
   z. Remove the water inlet housing.
   aa. Remove the water outlet.
   bb. Remove the cylinder head cover sub-assembly (for Bank 1).

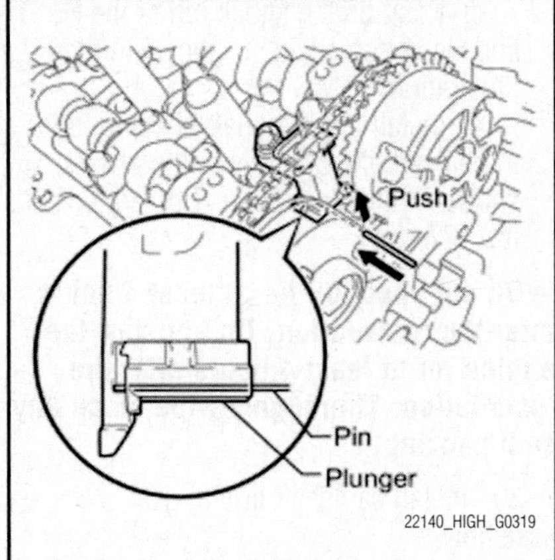

Fig. 107 Pinning tensioner

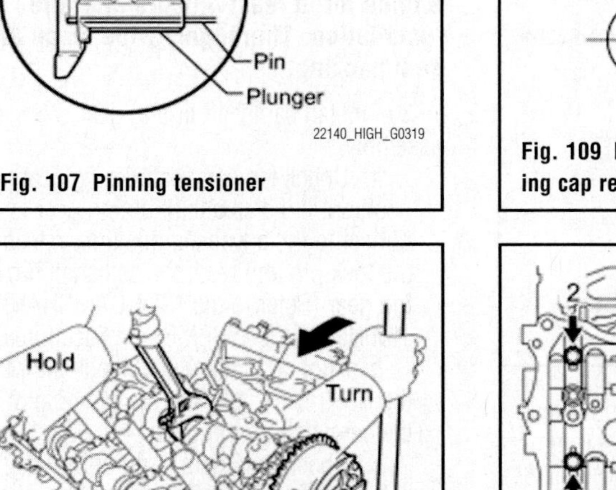

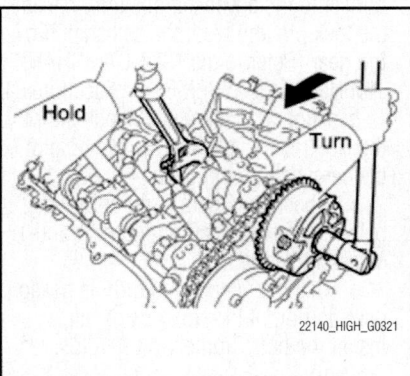

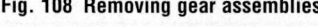

Fig. 108 Removing gear assemblies

   cc. Remove the cylinder head cover sub-assembly (for Bank 2).
   dd. Remove the No. 2 oil pan sub-assembly.
   ee. Remove the oil strainer sub-assembly.
   ff. Remove the oil pan sub-assembly.
   gg. Remove the timing chain cover sub-assembly
   hh. Remove the timing chain case oil seal.
   ii. Set the No. 1 cylinder to TDC/compression.
   jj. Remove the No. 1 chain tensioner assembly.
   kk. Remove the chain tensioner slipper.
   ll. Remove the chain sub-assembly.
   mm. Remove the idle sprocket assembly.
   nn. Remove the camshaft timing gears and No. 2 chain (for Bank 1).
   oo. While raising the No. 2 chain tensioner assembly, insert a pin of 1.0 mm (0.039 in.) diameter into the hole to fix the No. 2 chain tensioner assembly.
   pp. Hold the hexagonal portion of the camshaft with a wrench, and remove the 2 bolts and 2 camshaft timing gear assemblies.

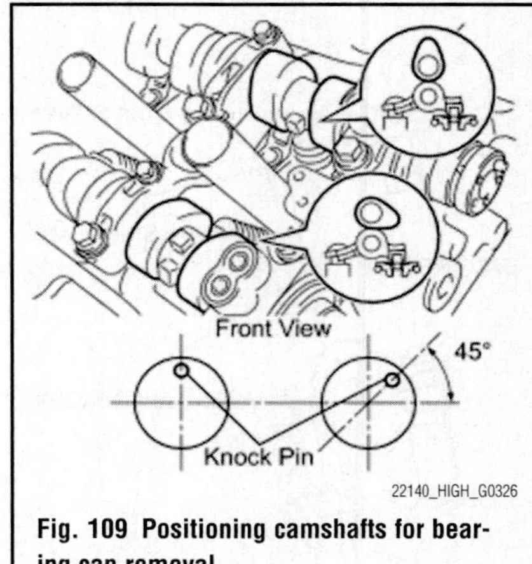

Fig. 109 Positioning camshafts for bearing cap removal

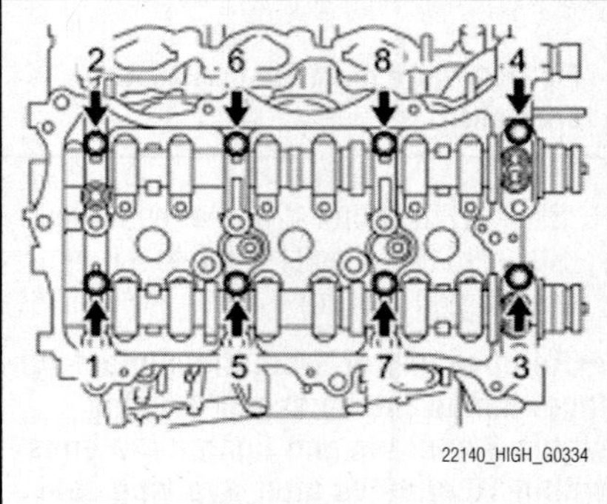

Fig. 110 Camshafts bearing cap removal sequence

➡**Be careful not to damage the cylinder head with the wrench. Do not disassemble the camshaft timing gear assemblies.**

   qq. Remove the No. 2 chain assembly.
   rr. Remove the bolt and No. 2 chain tensioner assembly.
   ss. Check that the camshafts are positioned as shown in the illustration.
   tt. Uniformly loosen and remove the 8 bearing cap bolts in several steps and in the sequence shown in the illustration.
   uu. Uniformly loosen and remove the 12 bearing cap bolts in several steps and in the sequence shown in the illustration.

➡**Uniformly loosen the bolts while keeping the camshaft level.**

   vv. Remove the 5 camshaft bearing caps.
   ww. Remove the camshaft.
   xx. Remove the No. 2 camshaft.
   yy. Remove the right hand camshaft housing sub-assembly by prying between the cylinder head and the right hand camshaft housing sub-assembly with a screwdriver.

➡Be careful not to damage the contact surfaces of the cylinder head and the right hand camshaft housing sub-assembly.

To install:

4. Install the following:
   a. Apply engine oil to the camshaft journals, camshaft housing sub-assembly RH and camshaft bearing caps.
   b. Install the camshaft and No. 2 camshaft to the right hand camshaft housing sub-assembly.
   c. Make sure of the marks and numbers on the camshaft bearing caps and place them in each proper position and direction.

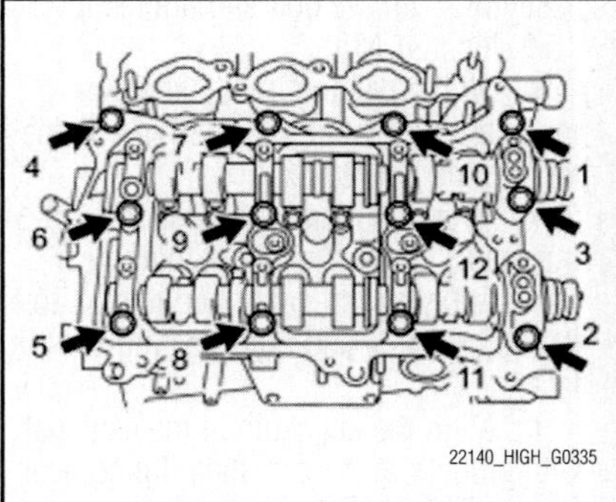

Fig. 111 Positioning camshafts for bearing cap removal

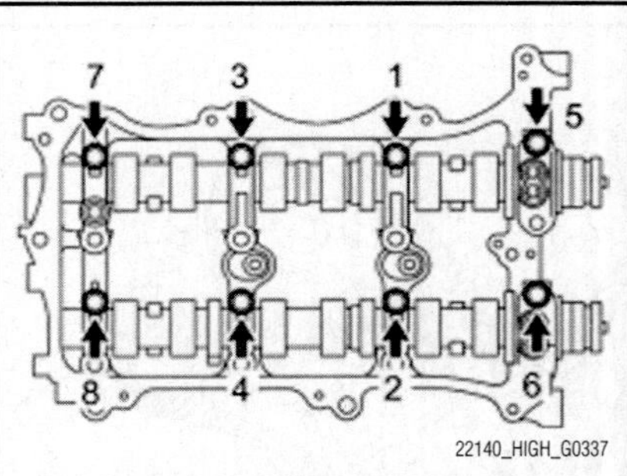

Fig. 112 Camshafts bearing cap tightening sequence

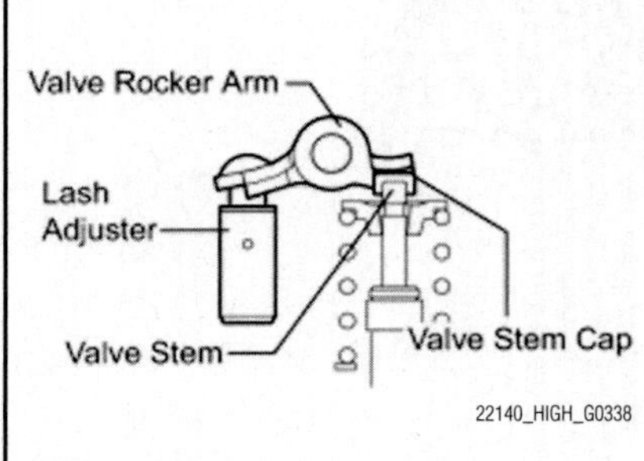

Fig. 113 Valve rocker arm sub-assembly positioning

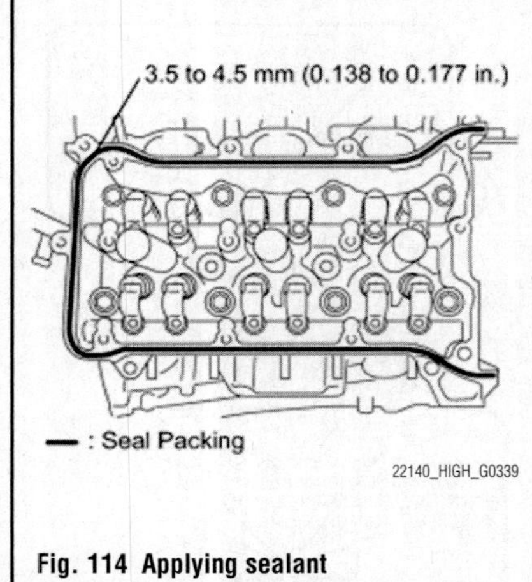

Fig. 114 Applying sealant

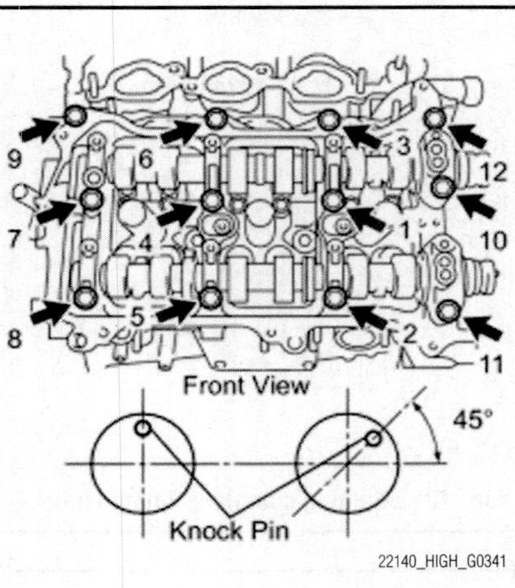

Fig. 115 Camshaft sub-assembly tightening sequence

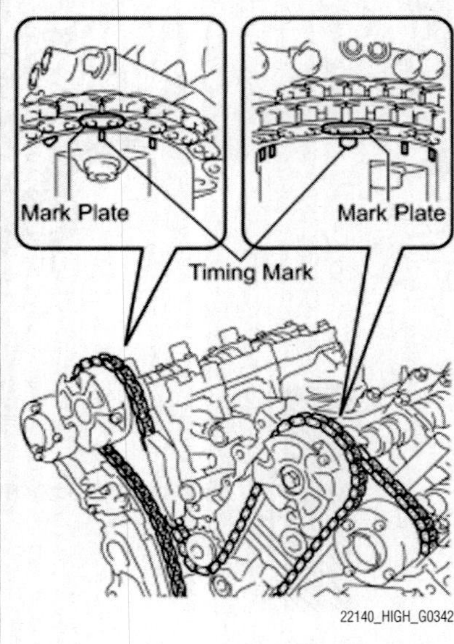

Fig. 116 Aligning timing chain sub-assembly

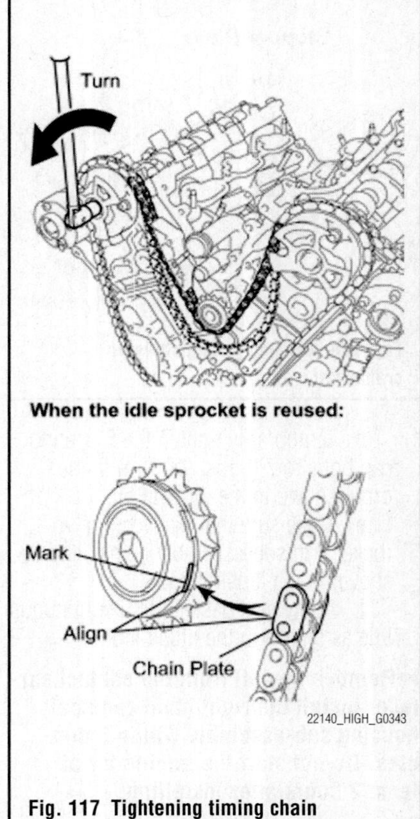

Fig. 117 Tightening timing chain

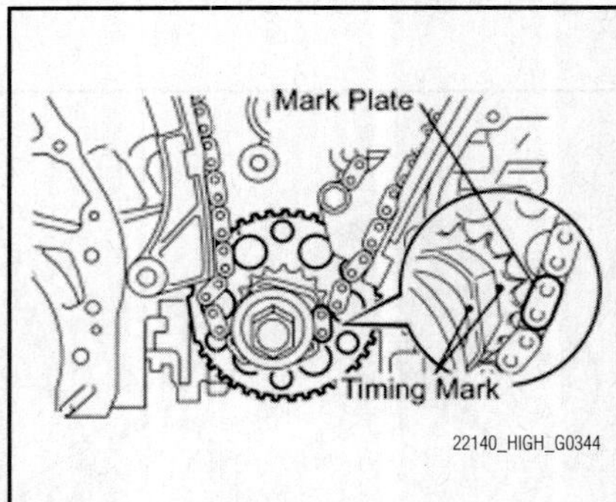

Fig. 118 Installing timing chain on crankshaft

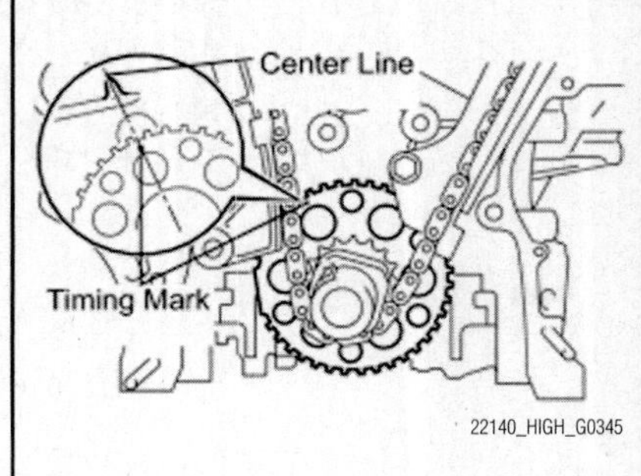

Fig. 119 Aligning timing chain on crankshaft

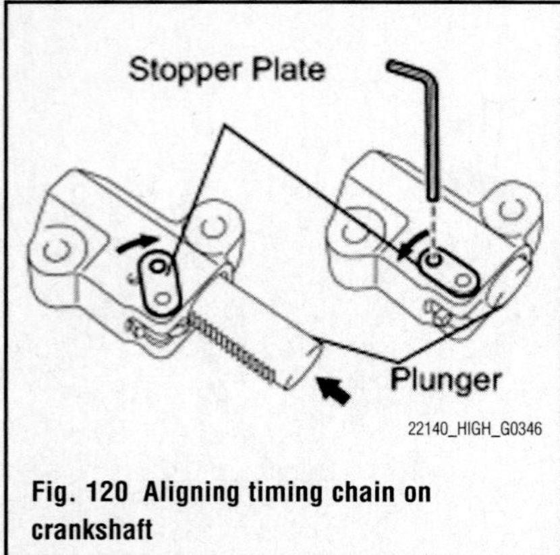

Fig. 120 Aligning timing chain on crankshaft

d. Temporarily tighten the 8 bearing cap bolts to 7 ft. lbs. (10 Nm) in the order shown in the illustration.

e. Make sure that the No. 1 valve rocker arm sub-assembly is installed as shown in the illustration.

f. Apply seal packing in a continuous line as shown in the illustration.

➡Remove any oil from the contact surface. Install the right hand camshaft housing sub-assembly within 3 minutes. Do not start the engine for at least 2 hours after installing.

g. Install the right hand camshaft housing sub-assembly and tighten the

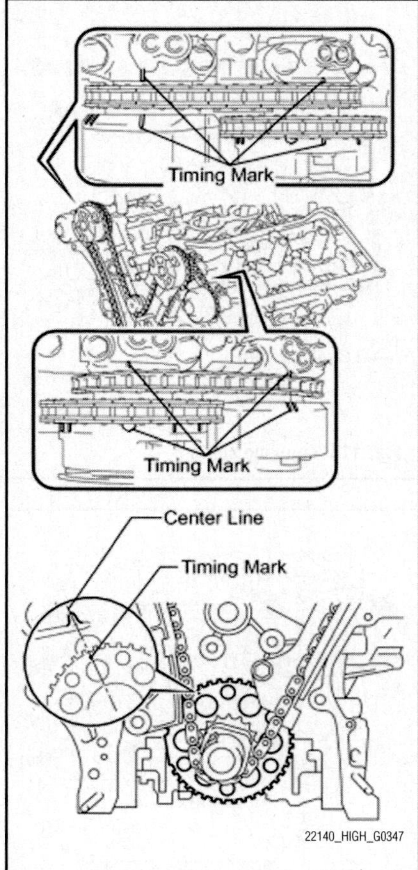

Fig. 121 Aligning complete timing chain

12 bolts to 21 ft. lbs. (28 Nm) in the order shown in the illustration.

➡When installing the right hand camshaft housing, it is necessary to correctly position the camshafts as shown in the removal illustration. If the camshaft housing sub-assembly is removed because any of the bolts are loosened during installation, make sure that the previously applied seal packing does not enter any oil passages.

h. Complete the tightening of the 8 bolts to 12 ft. lbs. (16 Nm) in the order shown above.

i. Install the No. 2 chain tensioner assembly with the bolt and tighten to 15 ft. lbs. (21 Nm).

j. While pushing in the tensioner, insert a pin of 1.0 mm (0.039 in.) diameter into the hole to fix it.

k. Align the mark plate with the timing marks of the camshaft timing gear.

l. Apply a light coat of engine oil to the bolt threads and bolts seating surface.

m. Align the knockpin of the camshaft with pinhole of the camshaft timing gear assembly. Install the camshaft timing gear assembly and camshaft timing exhaust timing gear assembly and

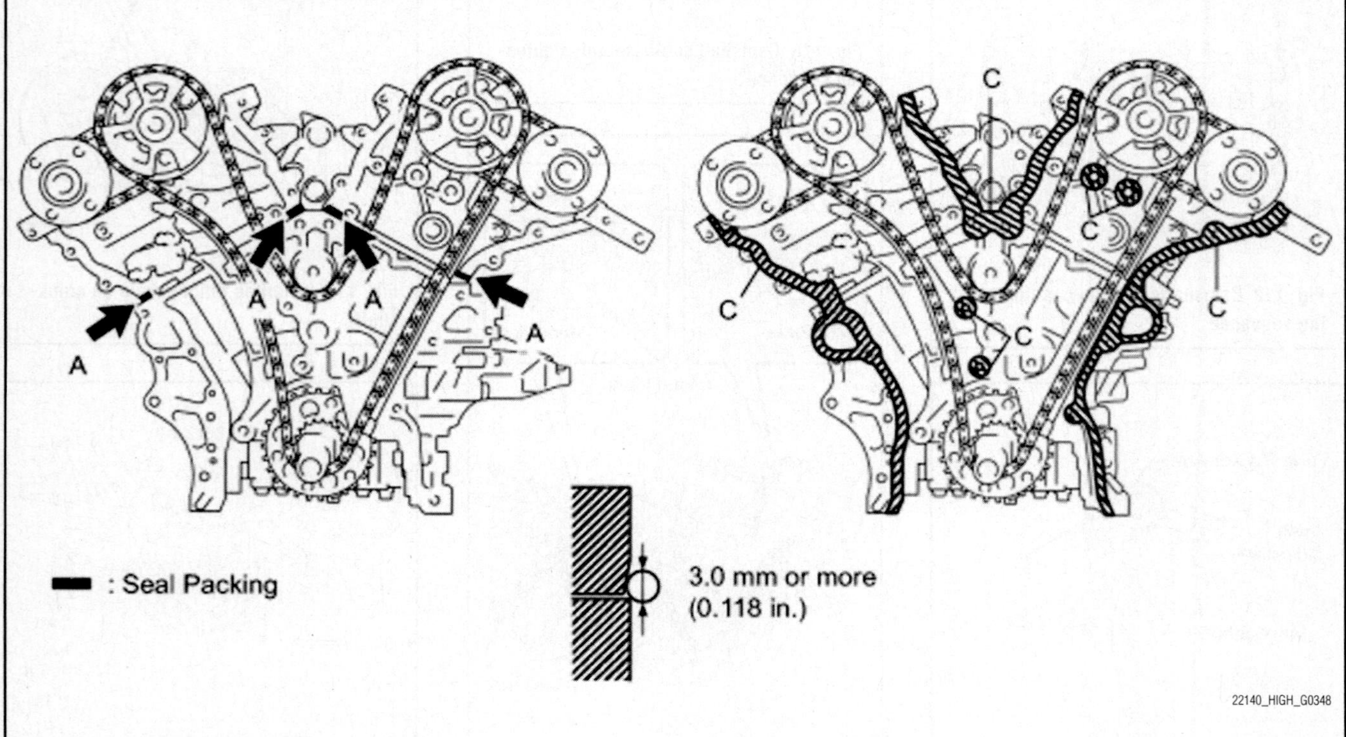

Fig. 122 Sealant application area

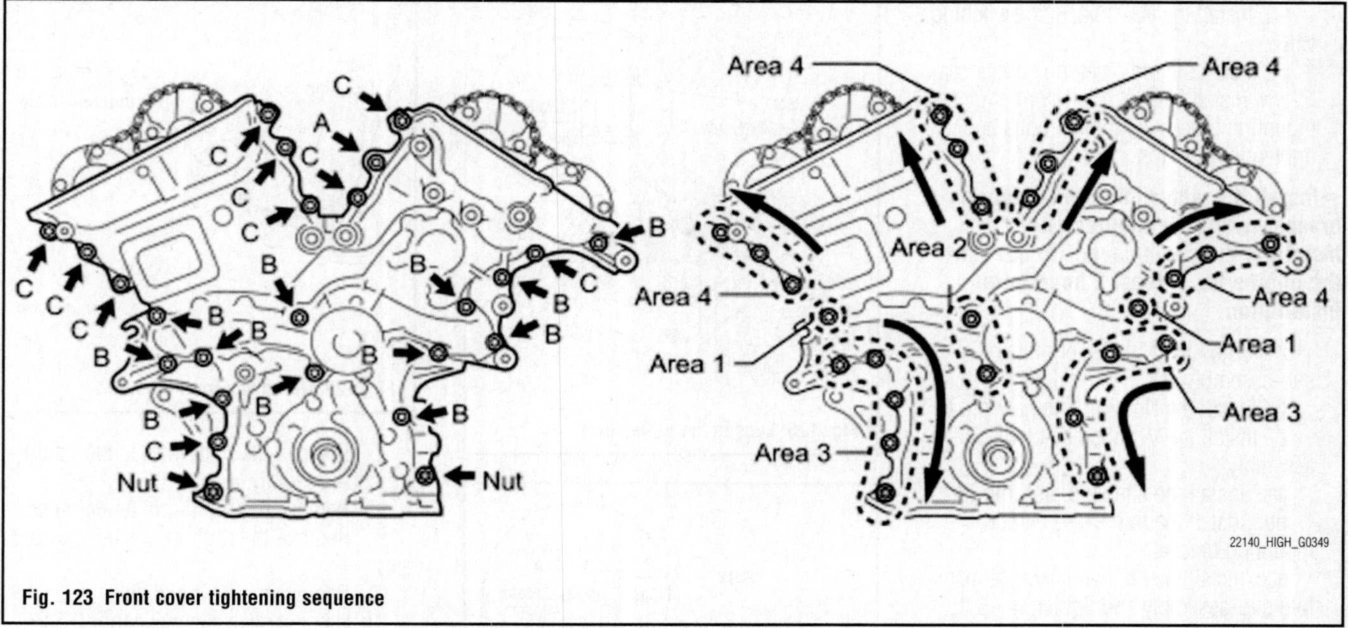

Fig. 123 Front cover tightening sequence

camshaft timing exhaust gear with the No. 2 chain sub-assembly installed.

n. Hold the hexagonal portion of the camshaft with the wrench and tighten the two bolts and camshaft timing gear assemblies to 74 ft. lbs. (100 Nm).

o. Remove the pan from the No 2 chain tensioner assembly.

p. Install idle sprocket assembly and tighten to 44 ft. lbs. (60 Nm).

q. Install chain sub-assembly.

r. Align the mark plate and timing marks as shown in the illustration and install the chain.

➡The camshaft mark plates are orange.

s. Do not pass the chain over the crankshaft, just temporarily place it on the crankshaft.

t. Turn the camshaft timing gear assembly on bank 1 counterclockwise to tighten the chain between the banks.

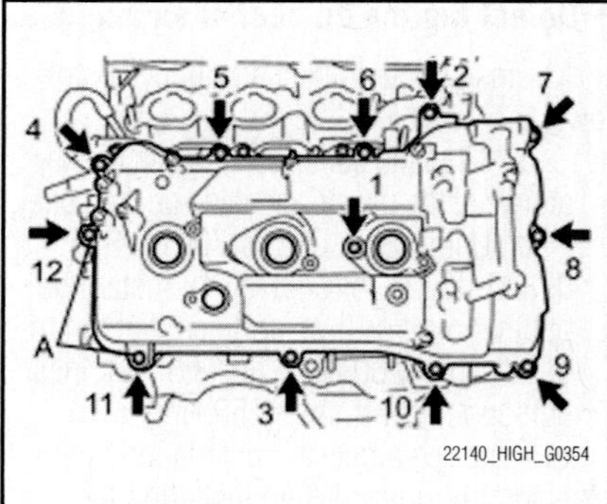

Fig. 124 Cylinder head cover tightening sequence

➡When the idle sprocket assembly is reused, align the timing chain plate with the mark where the plate has been in order to tighten the chain between the banks.

u. Align the mark plate and timing marks as shown in the illustration and install the chain onto the crankshaft timing sprocket. The crankshaft to mark plate is yellow.

v. Turn the crankshaft clockwise to set it to the right-hand block bore more centerline. (TDC Compression).

w. Install chain tensioner slipper.

x. Move the stopper plate upward to release the lock, and push the plunger deep into the tensioner.

y. Move the stopper plate downward to set the loss, and insert a hexagon wrench into the hole of the stopper plate.

z. Install No. 1 chain tensioner assembly and tighten bolts to 7 ft. lbs. (10 Nm).

aa. Remove the hexagon wrench from the No. 1 chain tensioner assembly. Check that the each timing mark is aligned with the crankshaft at TDC compression.

bb. Remove the pulley set bolt.

cc. Install timing chain case oil seal.

dd. Install sealant to timing chain cover sub-assembly.

ee. Install new O ring gasket on cylinder block.

ff. Align the oil pump's drive rotor spline and the crankshaft as shown in the illustration. Install the spline and chain cover to the crankshaft.

gg. Temporarily tighten the timing chain cover with the 23 bolts and 2 nuts.

- Tighten bolts in area 1 and 2: 15 ft. lbs. (21 Nm).
- Tighten bolt in area 3: 15 ft. lbs. (21 Nm).

➡First tighten the upper bolts and nuts, followed by the lower bolts and nuts as shown.

- Tighten bolt in area 4: 32 ft. lbs. (43 Nm)
- Tighten bolt in area 4: 15 ft. lbs. (21 Nm)

hh. Install oil pan sub-assembly and tighten 16 bolts and 2 nuts to 7 ft. lbs. (10 Nm) and 15 ft lbs. (21 Nm).

ii. Install oil strainers sub-assembly and tighten bolts and nuts to 7 ft. lbs. (10 Nm).

jj. Install No. 2 oil pan sub-assembly and tighten 16 bolts and 2 nuts to 7 ft. lbs. (10 Nm).

kk. Install cylinder head cover sub-assemblies and tighten to 7 ft. lbs. (10 Nm) and 15 ft. lbs. (21 Nm).

ll. Install the water inlet housing.

mm. Install No. 1 oil cooler bracket (w/ oil cooler).

nn. Install the oil cooler assembly (w/ oil cooler).

oo. Install the crankshaft pulley.

pp. Install the oil pipe.

qq. Install the No. 1 oil pipe.

rr. Install the crankshaft position sensor.

ss. Install the knock control sensor.

tt. Install the knock control sensor wire.

uu. Install the No. 1 vacuum switching valve.

vv. Install the radio setting condenser.

ww. Install the No. 1 left front engine mounting bracket with the 6 bolts and tighten to 40 ft. lbs. (54 Nm).

➡**Install the water inlet and mounting bracket within 15 minutes after installing the chain cover. Do not start the engine for at least 2 hours after installation.**

xx. Install the No. 2 idler pulley sub-assembly.

yy. Install the No. 2 timing gear cover

zz. Install the V-ribbed belt tensioner assembly.

aaa. Install the alternator assembly.

bbb. Install the transverse engine mounting bracket.

ccc. Install the left hand exhaust manifold sub-assembly and tighten to 15 ft. lbs. (21 Nm).

ddd. Install the No. 2 exhaust manifold heat insulator.

eee. Install the No. 2 manifold stay.

fff. Install the No. 2 engine oil level dipstick guide.

ggg. Install the right hand exhaust manifold sub-assembly and tighten to 15 ft. lbs. (21 Nm).

hhh. Install the intake manifold and tighten the 6 bolts and 4 nuts uniformly in several steps to 15 ft. lbs. (21 Nm).

iii. Install the right hand No. 2 engine mounting stay.

jjj. Install the ignition coil assembly.

kkk. Remove the engine stand.

lll. Install the engine assembly.

CRANKSHAFT DAMPER

REMOVAL & INSTALLATION

See Figures 125 and 126.

1. Raise and support the vehicle.
2. Remove the right front wheel.
3. Remove the engine under cover assembly.
4. Remove the No. 1 engine under cover.
5. Remove the right front fender molding sub-assembly.
6. Remove the right front fender liner.
7. Remove the right front fender apron seal.
8. Remove the V-ribbed belt.
9. Using a special service tool (SST: 09213-70011, SST: 09330-00021 or equivalent), loosen the crankshaft pulley bolt.
10. Using SST: 09950-50013 or equivalent, remove the crankshaft pulley bolt and crankshaft pulley.

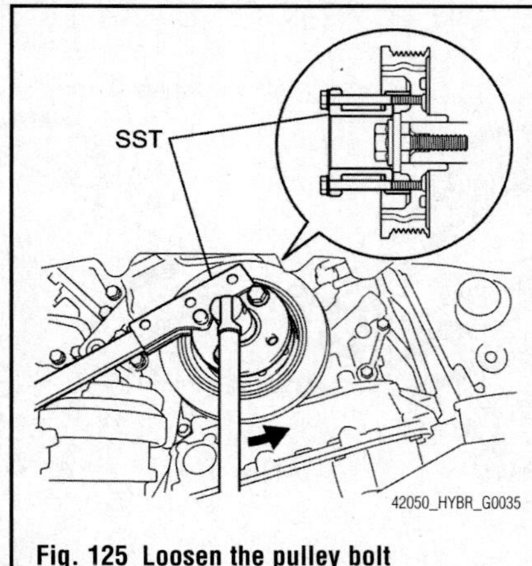

Fig. 125 Loosen the pulley bolt

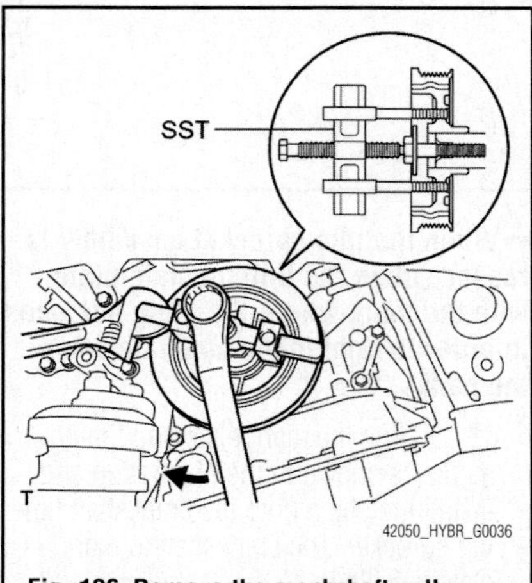

Fig. 126 Remove the crankshaft pulley

To install:

11. Install the crankshaft pulley, as follows:
    a. Align the pulley set key with the key groove of the pulley, and slide on the pulley.
    b. Using SST: 09213-70011, SST: 09330-00021, or equivalent, install the pulley bolt. For 2.7L engines, tighten to 192 ft. lbs. (260 Nm). For 3.5L engines, tighten to 184 ft. lbs. (250 Nm).
12. The remainder of installation is the reverse of removal. When installing the wheel, tighten the lug nuts to 76 ft. lbs. (103 Nm).

CRANKSHAFT FRONT SEAL

REMOVAL & INSTALLATION

See Figures 127 and 128.

1. Raise and support the vehicle.
2. Remove the right front wheel.
3. Remove the engine under cover assembly.
4. Remove the No. 1 engine under cover.

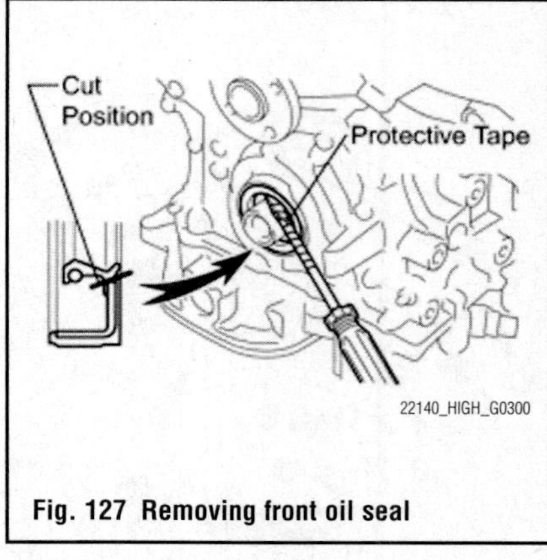

Fig. 127 Removing front oil seal

5. Remove the right front fender molding sub-assembly.
6. Remove the right front fender liner.
7. Remove the right front fender apron seal.
8. Remove the V-ribbed belt.
9. Using a special service tool (SST: 09213-70011, SST: 09330-00021 or equivalent), loosen the crankshaft pulley bolt.
10. Using SST: 09950-50013 or equivalent, remove the crankshaft pulley bolt and crankshaft pulley.
11. Using a screwdriver, pry out the timing chain case oil seal.

➡**Tape the screwdriver tip before use.**

➡**After the removal, check the crankshaft for damage. If it is damaged, smooth the surface with 400-grit sandpaper.**

To install:

12. Install timing chain case oil seal, as follows:
    a. Apply MP grease to a new oil seal lip.
    b. Using a Special Service Tool (SST: 09223-22010, SST: 09506-35010 or equivalent) and a hammer, tap in the oil seal until its surface is flush with the timing chain cover edge.

➡**Keep the lip free of foreign matter.**

➡**Do not tap the oil seal at an angle.**

13. Install the crankshaft pulley, as follows:
    a. Align the pulley set key with the key groove of the pulley, and slide on the pulley.
    b. Using SST: 09213-70011, SST: 09330-00021, or equivalent, install the pulley bolt. For 2.7L engines, tighten to 192 ft. lbs. (260 Nm). For 3.5L engines, tighten to 184 ft. lbs. (250 Nm).
14. The remainder of installation is the reverse of removal. When installing the wheel, tighten the lug nuts to 76 ft. lbs. (103 Nm).

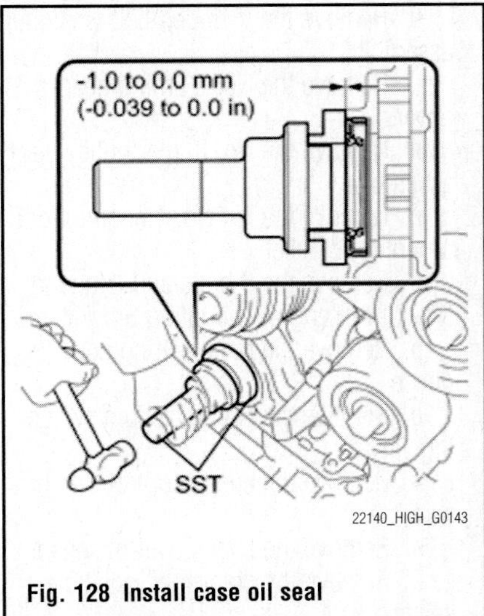

Fig. 128 Install case oil seal

CYLINDER HEAD

REMOVAL & INSTALLATION

2.7L Engine

See Figures 129 through 132.

1. Remove the engine and transaxle.
2. Remove the exhaust manifold assembly.
3. Remove throttle body assembly.
4. Remove vacuum switching valve assembly.
5. Remove no. 1 vacuum switching valve assembly.
6. Remove fuel delivery pipe sub-assembly.
7. Disconnect no. 2 ventilation hose.
8. Remove intake manifold.
9. Remove the camshafts.
10. Remove the 16 valve stem caps from the cylinder head.

➡Arrange the removed parts in the correct order.

11. Remove cylinder head sub-assembly:

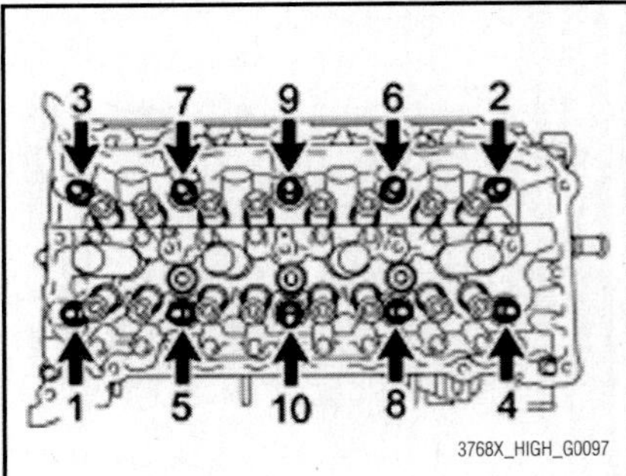

Fig. 129 Loosen the 10 bolts in the sequence shown

a. Using a 10 mm bi-hexagon wrench, uniformly loosen the 10 bolts in the sequence shown in the illustration. Remove the 10 cylinder head bolts and plate washers.

➡Be sure to keep the removed parts separate for each installation position.

➡Be careful not to drop washers into the cylinder head. Head warpage or cracking could result from removing bolts in an incorrect order.

b. Remove the cylinder head.

12. Remove the cylinder head gasket from the cylinder block.

To install:

13. Inspect cylinder head bolt:

a. Using a vernier caliper, measure the diameter of the threads at the measurement point.

b. Standard diameter: 10.85 to 11.00 mm (0.427 to 0.433 in.)

c. Minimum diameter: 10.6 mm (0.417 in.)

d. Measurement point (distance from the seat): 106 mm (4.17 in.)

➡If the diameter is less than the minimum, replace the cylinder head bolt. Failure to do so may lead to engine damage. If there is any thread deformation, replace the cylinder head bolt with a new one.

14. Inspect cylinder head sub-assembly:

a. Using a precision straightedge and feeler gauge, measure the warpage of the contact surfaces where the cylinder head contacts the cylinder block and manifold.

- Cylinder head lower side: 0.05 mm (0.00197 in.)
- Intake manifold side: 0.10 mm (0.00394 in.)
- Exhaust manifold side: 0.10 mm (0.00394 in.)

b. If the warpage is more than the maximum, replace the cylinder head.

c. Using a dye penetrant, check the intake ports, exhaust ports and cylinder surface for cracks. If cracked, replace the cylinder head.

➡Remove any oil from the contact surface. Install the cylinder head gasket within 3 minutes and tighten the bolts within 15 minutes after applying seal packing. Pay attention to the installation direction.

15. Install cylinder head gasket:

a. Clean the cylinder block and cylinder head with solvent.

b. Apply a continuous line of seal packing to a new cylinder head gasket as shown in the illustration. Seal packing: Toyota Genuine Seal Packing Black, Three Bond 1207B or equivalent. Standard seal dimension: 3.0 to 7.0 mm (0.118 to 0.276 in.) wide and 3.0 mm (0.118 in.) thick.

➡Apply at least 20 mm (0.787 in.) of seal packing from the inside edge of the protrusion of the cylinder block.

16. Install cylinder head sub-assembly:

a. The cylinder head bolts are tightened in 4 progressive steps.

b. Place the cylinder head on the cylinder block.

➡Ensure that no oil is on the mounting surface of the cylinder head. Place the cylinder head on the cylinder block gently in order not to damage the gasket with the bottom part of the head.

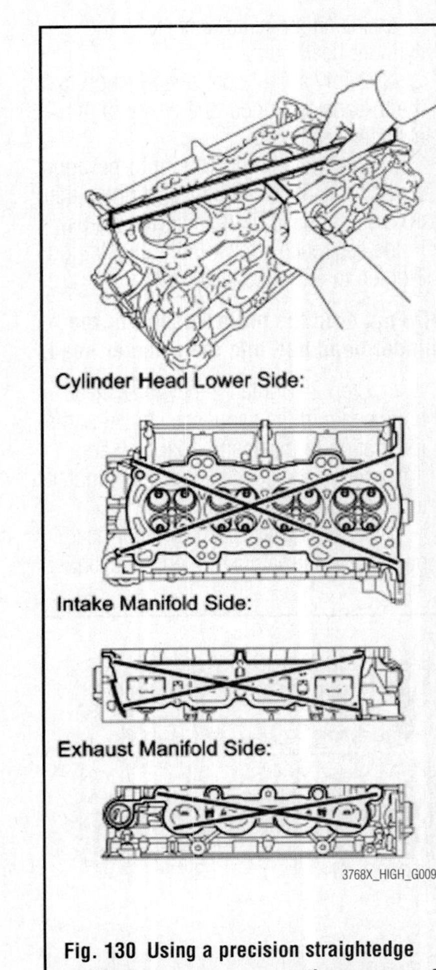

Fig. 130 Using a precision straightedge and feeler gauge, measure the warpage of the contact surfaces

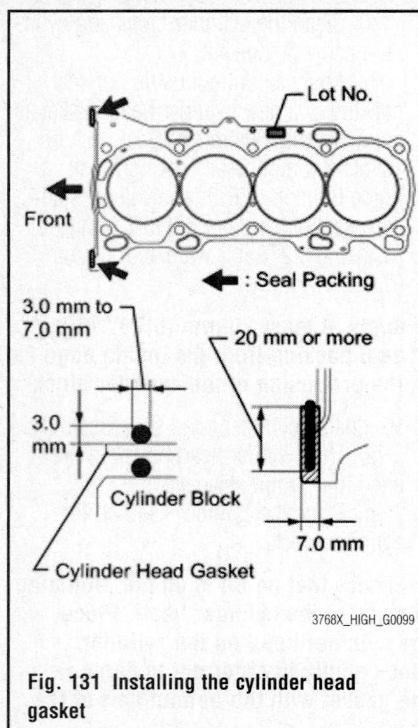

Fig. 131 Installing the cylinder head gasket

c. Install the plate washers to the cylinder head bolts.

d. Apply a light coat of engine oil to the threads and under the heads of the cylinder head bolts.

e. Step 1: Using a 10 mm bi-hexagon wrench, install and uniformly tighten the 10 cylinder head bolts in several steps, in the sequence shown in the illustration. Tighten to 27 ft. lbs. (36 Nm).

➡Do not drop the plate washer for the cylinder head bolt into the cylinder head.

f. Step 2: Tighten the cylinder head bolts again in the sequence shown in the illustration to make sure that they are tightened to the specified torque. Tighten to 27 ft. lbs. (36 Nm).

g. Step 3: Mark each cylinder head bolt head with paint. Tighten the cylinder head bolts 90° in the sequence shown in step 1.

h. Step 4: Tighten the cylinder head bolts another 90° in the sequence shown in step 1. Check that the painted marks are now facing rearward.

➡Do not apply oil for at least 4 hours after the installation. Do not start the engine for at least 4 hours after the installation. After the installation, if the seal packing has seeped out, wipe it off.

17. Apply a light coat of engine oil to the valve stem ends. Install the 16 valve stem caps to the cylinder head.

➡Do not drop the valve stem caps into the cylinder head.

18. To complete installation, reverse remaining removal procedure.

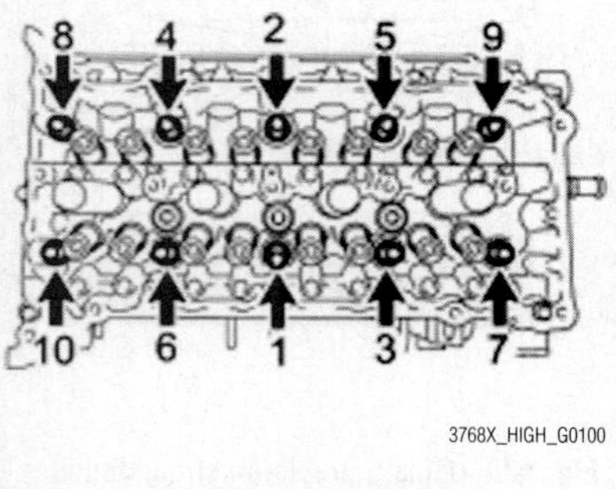

3.5L Engine

See Figures 133 through 154.

1. Remove the engine assembly.
2. Install on engine stand.
3. The following must be removed:
   a. Remove the ignition coil assembly.
   b. Remove right hand No. 2 engine mounting stay.
   c. Remove the intake manifold.
   d. Remove the right hand exhaust manifold sub-assembly.
   e. Remove the No. 2 engine oil level dipstick guide.
   f. Remove the No. 2 manifold stay.
   g. Remove the No. 2 exhaust manifold heat insulator.
   h. Remove the left exhaust manifold sub-assembly.
   i. Remove the transverse engine mounting bracket.
   j. Remove the alternator assembly.
   k. Remove the V-ribbed belt tensioner assembly.
   l. Remove the No. 2 timing gear cover.
   m. Remove the No. 2 idler pulley sub-assembly.
   n. Remove the left No. 1 engine front mounting bracket.
   o. Remove the 6 bolts and left hand No. 1 front engine mounting bracket.
   p. Remove the radio setting condenser.
   q. Remove the No. 1 vacuum switching valve.
   r. Remove the knock control sensor wire.
   s. Remove the knock control sensor.
   t. Remove the crankshaft position sensor.
   u. Remove the No. 1 oil pipe.
   v. Remove the oil pipe.
   w. Remove the crankshaft pulley.
   x. Remove the oil cooler assembly, if necessary.
   y. Remove the No. 1 oil cooler bracket, if necessary.
   z. Remove the water inlet housing.
   aa. Remove the water outlet.
   bb. Remove the cylinder head cover sub-assembly (for bank 1).
   cc. Remove the cylinder head cover sub-assembly (for bank 2).
   dd. Remove the No. 2 oil pan sub-assembly.
   ee. Remove the oil strainer sub-assembly.
   ff. Remove the oil pan sub-assembly.
   gg. Remove the timing chain cover sub-assembly.
   hh. Remove the timing chain case oil seal.
   ii. Set the No. 1 cylinder to TDC/compression.
   jj. Remove the No. 1 chain tensioner assembly.
   kk. Remove the chain tensioner slipper.

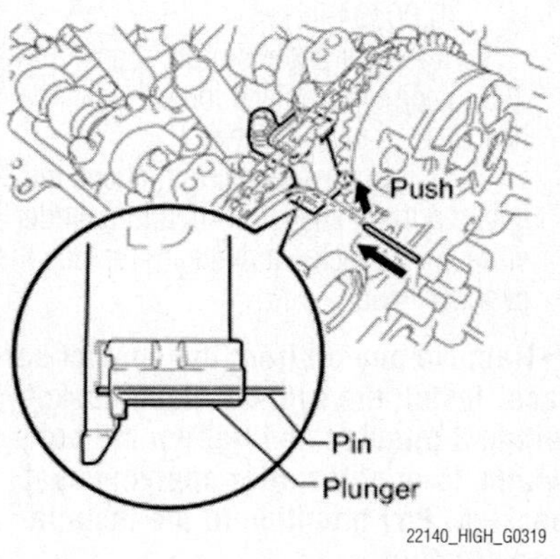

Fig. 132 Cylinder head tightening sequence

Fig. 133 Pinning tensioner

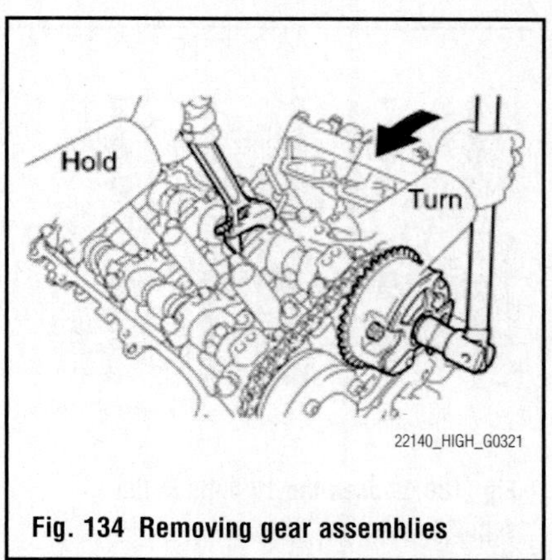

Fig. 134 Removing gear assemblies

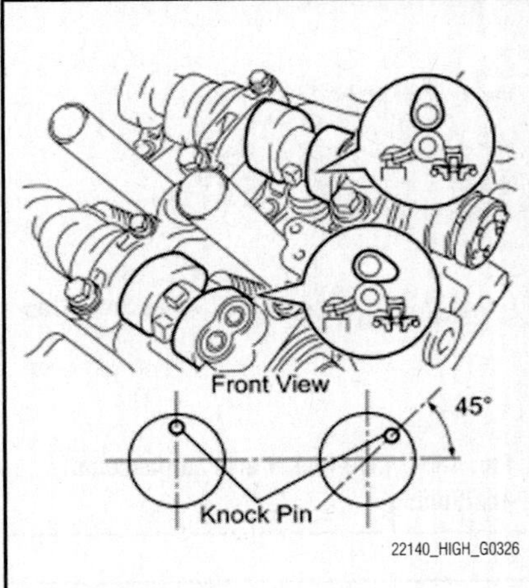

Fig. 135 Positioning camshafts for bearing cap removal

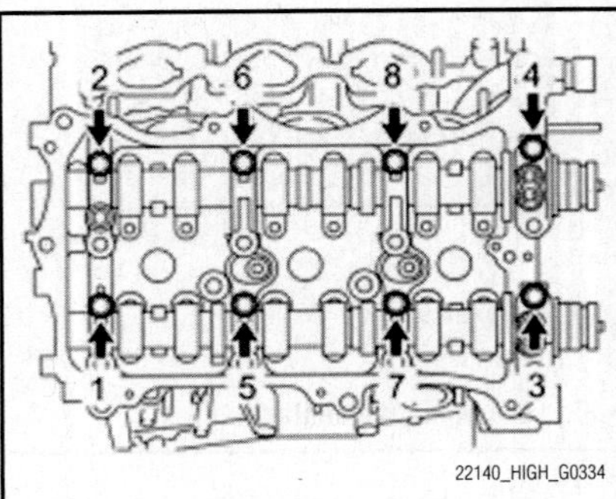

Fig. 136 Camshafts bearing cap removal sequence

ll. Remove the chain sub-assembly.

mm. Remove the idle sprocket assembly.

nn. Remove the camshaft timing gears and No. 2 chain (for bank 1).

oo. While raising the No. 2 chain tensioner assembly, insert a pin of 1.0 mm (0.039 in.) diameter into the hole to fix the No. 2 chain tensioner assembly.

pp. Hold the hexagonal portion of the camshaft with a wrench, and remove the 2 bolts and 2 camshaft timing gear assemblies.

➡Be careful not to damage the cylinder head with the wrench. Do not disassemble the camshaft timing gear assemblies.

qq. Remove the No. 2 chain assembly.

rr. Remove the bolt and No. 2 chain tensioner assembly.

ss. Check that the camshafts are positioned as shown in the illustration.

tt. Uniformly loosen and remove the 8 bearing cap bolts in several steps and in the sequence shown in the illustration.

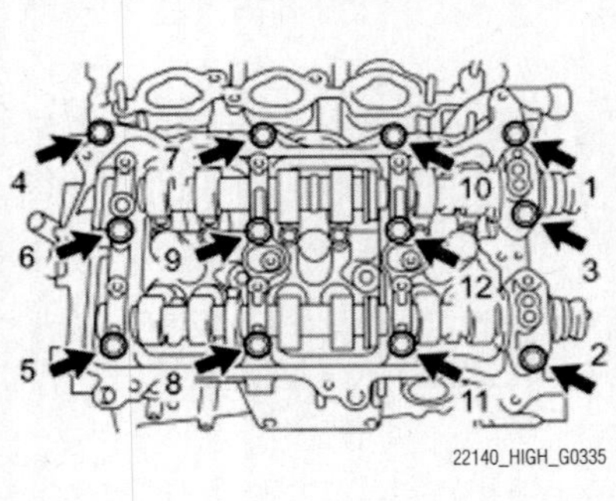

Fig. 137 Camshaft bearing cap loosening sequence

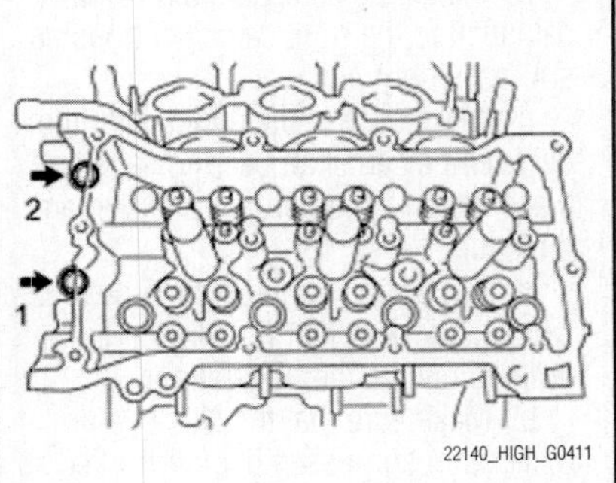

Fig. 138 Loosening cylinder head bolts LH shown

uu. Uniformly loosen and remove the 12 bearing cap bolts in several steps and in the sequence shown in the illustration.

➡Uniformly loosen the bolts while keeping the camshaft level.

vv. Remove the 5 camshaft bearing caps.

ww. Remove the camshaft.

xx. Remove the No. 2 camshaft.

yy. Remove the right camshaft housing sub-assembly by prying between the cylinder head and right camshaft housing sub-assembly with a screwdriver.

➡Be careful not to damage the contact surfaces of the cylinder head and the right camshaft housing sub-assembly.

4. Remove the 24 valve lash adjuster assemblies from the cylinder head.

➡Arrange the removed parts in the correct order.

5. Uniformly loosen and remove the 2 cylinder head set bolts in several steps and in the sequence shown in the illustration.

➡Be careful not to drop washers into the cylinder head. Cylinder head warpage or cracking could result from removing bolts in an incorrect order. Be sure to keep separate the removed parts for each installation position.

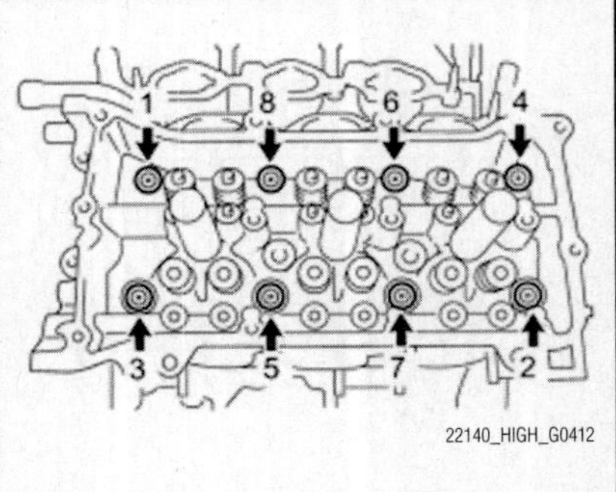

Fig. 139 Cylinder head bolt loosening sequence LH shown

6. Using a 10 mm bi-hexagon wrench, uniformly loosen the 8 bolts in the sequence shown in the illustration. Remove the 8 cylinder head bolts and plate washers.

7. Remove the cylinder head sub-assembly.

8. Remove the No. 2 cylinder head gasket.

To install:

9. Place the No. 2 cylinder head gasket on the cylinder block surface with the Lot No. stamp upward.

➡Gently lower the cylinder head in order not to damage the gasket with the bottom part of the head.

10. Place the cylinder head on the cylinder block.

➡Be careful not to allow oil to adhere to the bottom part of the cylinder head.

11. Apply a light coat of engine oil to the threads and under the heads of the cylinder head bolts.

➡The cylinder head bolts are tightened in 3 progressive steps.

a. Step 1: Using a 10 mm bi-hexagon wrench, install and uniformly tighten the 8 cylinder head bolts with the plate washers to 27 ft. lbs. (36 Nm) in several steps in the sequence shown in the illustration.

b. Step 2: Mark the cylinder head bolt head with paint. Tighten the cylinder head bolts another 90°.

c. Step 3: Tighten the cylinder head bolts an additional 90°.

12. Tighten the 2 bolts to 22 ft. lbs. (30 Nm) in the order shown in the illustration.

13. Install the 12 valve stem caps.

➡Keep the lash adjuster free of dirt and foreign objects. Only use clean engine oil.

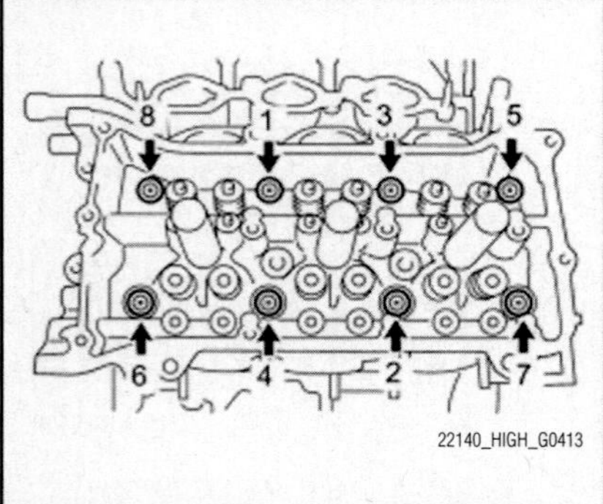

Fig. 140 Cylinder head bolt tightening sequence LH shown

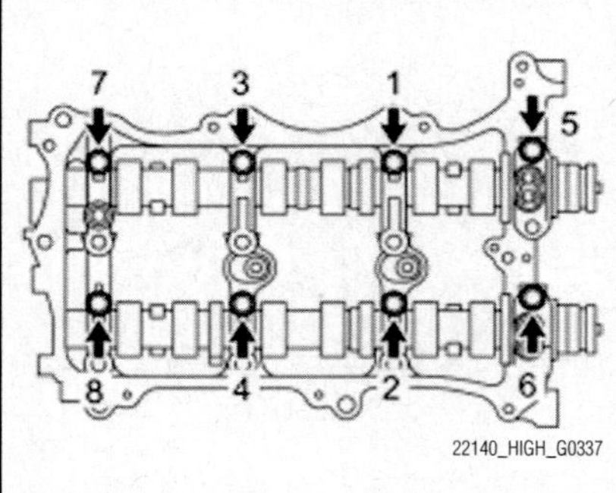

Fig. 142 Camshafts bearing cap tightening sequence

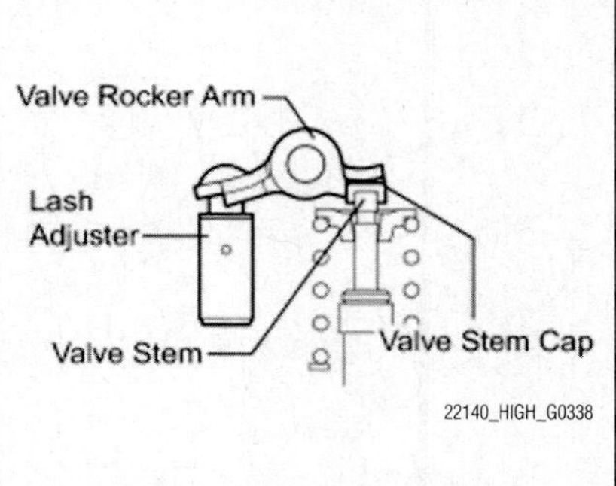

Fig. 143 Valve rocker arm sub-assembly positioning

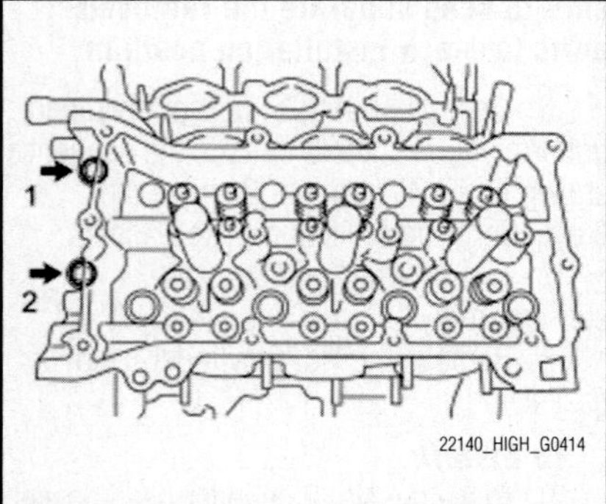

Fig. 141 Tightening cylinder head bolts LH shown

14. Place the lash adjuster into a container filled with engine oil.

a. Insert Service Tool's tip into the lash adjuster's plunger and use the tip to press down on the check ball inside the plunger.

b. Squeeze Service Tool and lash adjuster together to move the plunger up and down 5 to 6 times.

c. Check the movement of the plunger and bleed the air. Confirm that the plunger moves up and down freely.

d. When bleeding air from the high-pressure chamber, make sure that the tip of SST is actually pressing the check ball as shown in the illustration. If the check ball is not pressed, air will not bleed.

e. After bleeding the air, remove Service Tool. Then, try to press the plunger quickly and firmly with by hand. Confirm that the plunger is very difficult to move.

f. If the results are not as specified, replace the defective lash adjuster.

g. Install the lash adjusters.

➡Install the lash adjuster to the same place where it was removed from.

15. Install the following:

a. Apply engine oil to the camshaft journals, right camshaft housing sub-assembly, and camshaft bearing caps.

b. Install the camshaft and No. 2 camshaft to the right camshaft housing sub-assembly.

c. Make sure of the marks and numbers on the camshaft bearing caps and place them in each proper position and direction.

d. Temporarily tighten the 8 bearing cap bolts to 7 ft. lbs. (10 Nm) in the order shown in the illustration.

e. Make sure that the No. 1 valve rocker arm sub-assembly is installed as shown in the illustration.

f. Apply seal packing in a continuous line as shown in the illustration.

➡Remove any oil from the contact surface. Install the camshaft housing sub-assembly RH within 3 minutes. Do not start the engine for at least 2 hours after installing.

g. Install the right camshaft housing sub-assembly and tighten the 12 bolts to 21 ft. lbs. (28 Nm) in the order shown in the illustration.

➡When installing the camshaft housing RH, it is necessary to correctly position the camshafts as shown in the removal illustration. If the camshaft housing sub-assembly is removed because any of the bolts are loosened during installation, make sure that the previously applied seal packing does not enter any oil passages.

h. Complete the tightening of the 8 bolts to 12 ft. lbs. (16 Nm) in the order shown above.

i. Install the No. 2 chain tensioner assembly with the bolt and tighten to 15 ft. lbs. (21 Nm).

j. While pushing in the tensioner, insert a pin of 1.0 mm (0.039 in.) diameter into the hole to fix it.

k. Align the mark plate with the timing marks of the camshaft timing gear.

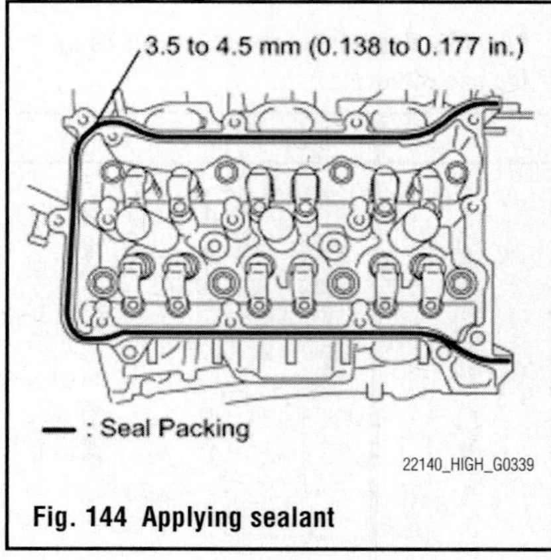

Fig. 144 Applying sealant

l. Apply a light coat of engine oil to the bolt threads and bolts seating surface.

m. Align the knockpin of the camshaft with pinhole of the camshaft timing gear assembly. Install the camshaft timing gear assembly and camshaft timing exhaust timing gear assembly and camshaft timing exhaust gear with the No. 2 chain sub-assembly installed.

n. Hold the hexagonal portion of the camshaft with the wrench and tighten the two bolts and camshaft timing gear assemblies to 74 ft. lbs. (100 Nm).

o. Remove the pan from the No 2 chain tensioner assembly.

p. Install idle sprocket assembly and tighten to 44 ft. lbs. (60 Nm).

q. Install chain sub-assembly.

r. Align the mark plate and timing marks as shown in the illustration and install the chain.

➡The camshaft mark plates are orange.

s. Do not pass the chain over the crankshaft, just temporarily place it on the crankshaft.

t. Turn the camshaft timing gear assembly on bank 1 counterclockwise to tighten the chain between the banks.

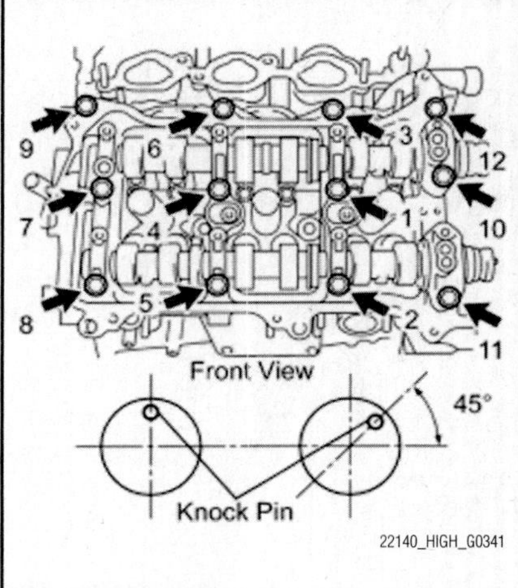

Fig. 145 Camshaft sub-assembly tightening sequence

➡When the idle sprocket assembly is reused, align the timing chain plate with the mark where the plate has been in order to tighten the chain between the banks.

u. Align the mark plate and timing marks as shown in the illustration and install the chain onto the crankshaft timing sprocket. The crankshaft to mark plate is yellow.

v. Turn the crankshaft clockwise to set it to the right-hand block bore more centerline. (TDC Compression)

w. Install chain tensioner slipper.

x. Move the stopper plate upward to release the lock, and push the plunger deep into the tensioner.

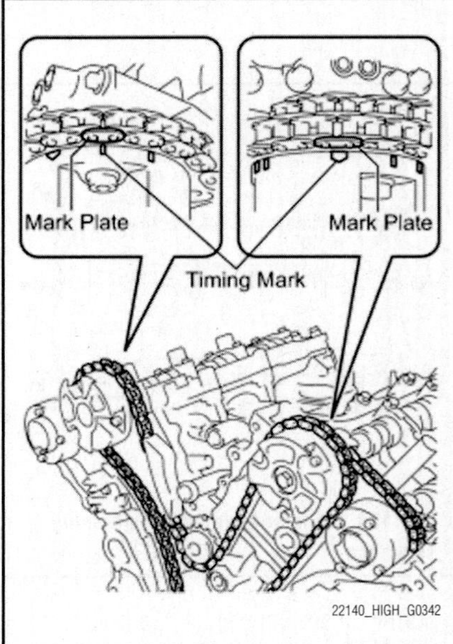

Fig. 146 Aligning timing chain sub-assembly

y. Move the stopper plate downward to set the loss, and insert a hexagon wrench into the hole of the stopper plate.

z. Install No. 1 chain tensioner assembly and tighten bolts to 7 ft. lbs. (10 Nm).

aa. Remove the hexagon wrench from the No. 1 chain tensioner assembly. Check that the each timing mark is aligned with the crankshaft at TDC compression.

bb. Remove the pulley set bolt.

cc. Install timing chain case oil seal.

dd. Install sealant to timing chain cover sub-assembly.

ee. Install new O ring gasket on cylinder block.

ff. Align the oil pump's drive rotor spline and the crankshaft as shown in the illustration. Install the spline and chain cover to the crankshaft.

gg. Temporarily tighten the timing chain cover with the 23 bolts and 2 nuts.

- Tighten bolts in area 1 and 2 15 ft. lbs. (21 Nm)
- Tighten bolt in area 3 15 ft. lbs. (21 Nm)

➡First tighten the upper bolts and nuts followed by the lower bolts and nuts as shown.

- Tighten bolt in area 4 32 ft. lbs. (43 Nm)
- Tighten bolt in area 4 15 ft. lbs. (21 Nm)

hh. Install oil pan subassembly and tighten 16 bolts and 2 nuts to 7 ft. lbs. (10 Nm) and 15 ft lbs. (21 Nm).

ii. Install oil strainers sub-assembly and tighten bolts and nuts to 7 ft. lbs. (10 Nm).

jj. Install No. 2 oil pan sub-assembly and tighten 16 bolts and 2 nuts to 7 ft. lbs. (10 Nm).

kk. Install cylinder head cover sub-assemblies and tighten to 7 ft. lbs. (10 Nm) and 15 ft. lbs. (21 Nm).

ll. Install water inlet housing

mm. Install the No. 1 oil cooler bracket (w/ oil cooler).

nn. Install the oil cooler assembly (w/ oil cooler).

oo. Install the crankshaft pulley.

pp. Install the oil pipe.

qq. Install the No. 1 oil pipe.

rr. Install the crankshaft position sensor.

ss. Install the knock control sensor.

tt. Install the knock control sensor wire.

uu. Install the No. 1 vacuum switching valve.

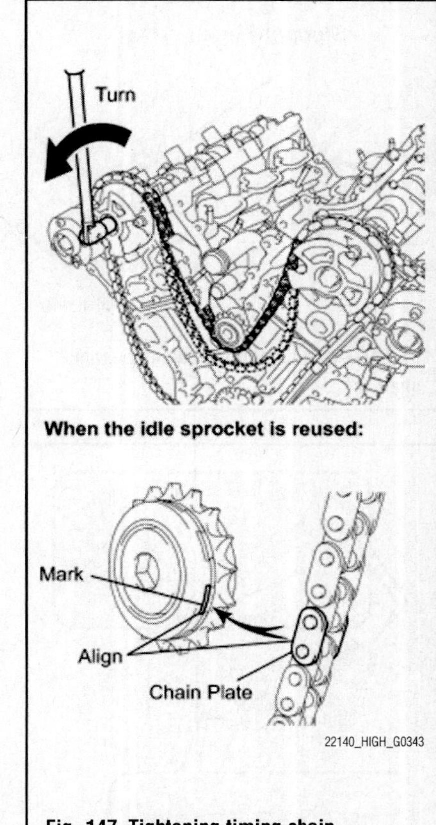

Fig. 147 Tightening timing chain

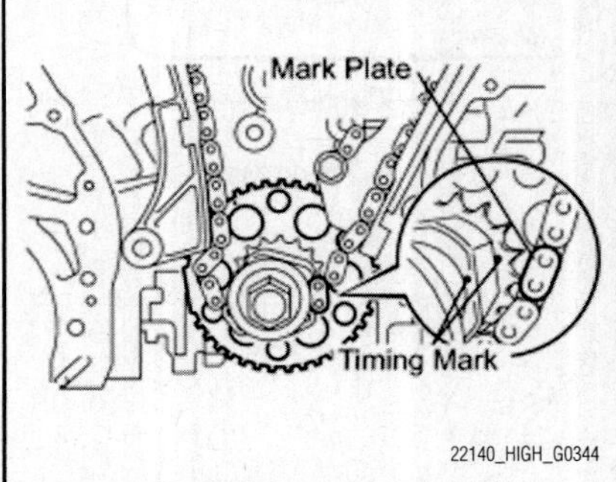

Fig. 148 Installing timing chain on crankshaft

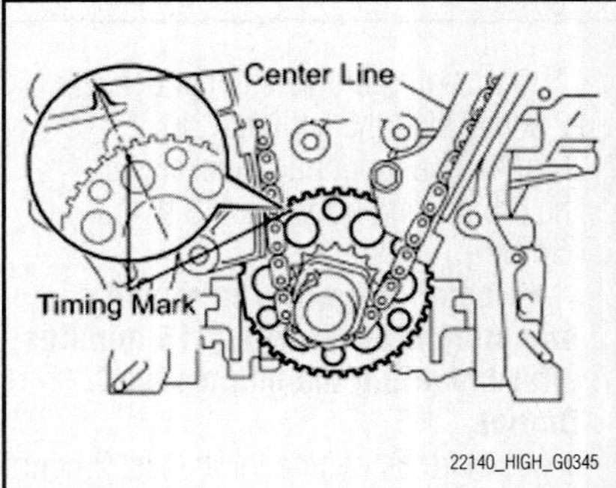

Fig. 149 Aligning timing chain on crankshaft

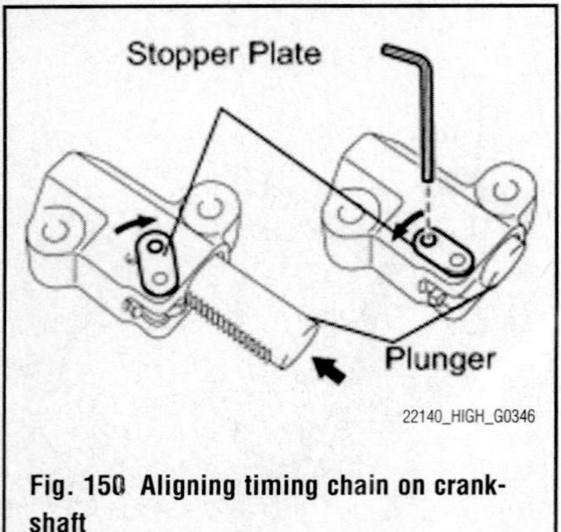

Fig. 150 Aligning timing chain on crankshaft

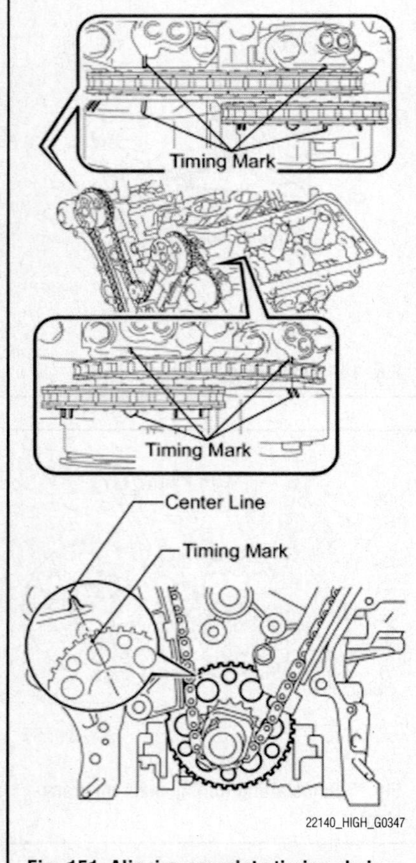

Fig. 151 Aligning complete timing chain

vv. Install radio setting condenser.

ww. Install the left hand No. 1 front engine mounting bracket with the 6 bolts and tighten to 40 ft. lbs. (54 Nm).

➡Install the water inlet and mounting bracket within 15 minutes after installing the chain cover. Do not

xx. start the engine for at least 2 hours after installation.

yy. Install the No. 2 idler pulley sub-assembly.

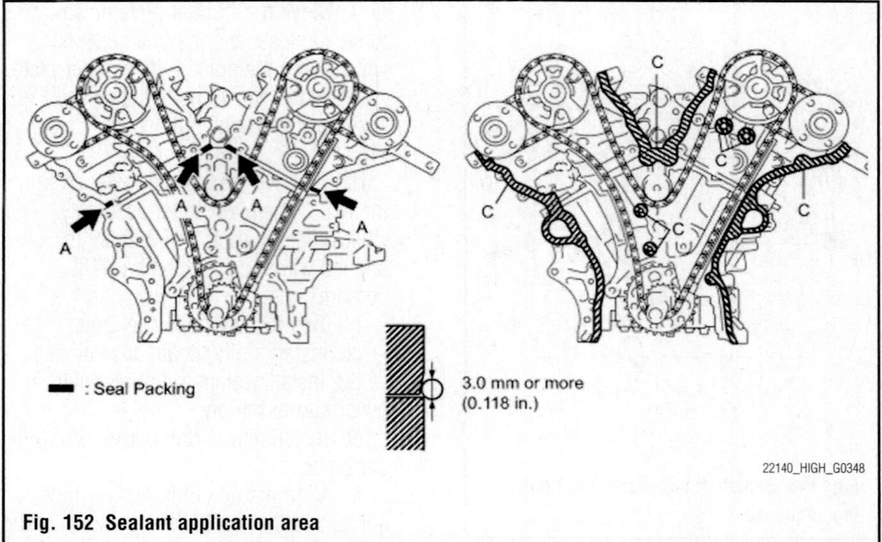

Fig. 152 Sealant application area

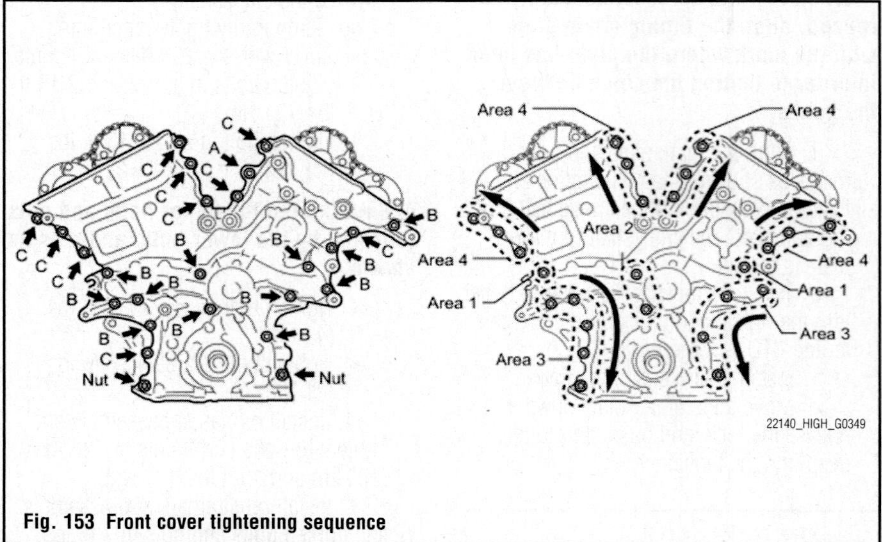

Fig. 153 Front cover tightening sequence

zz. Install the No. 2 timing gear cover.

53. Install the V-ribbed belt tensioner assembly.

54. Install the alternator assembly.

55. Install the transverse engine mounting bracket.

56. Install left hand exhaust manifold sub-assembly and tighten to 15 ft. lbs. (21 Nm).

57. Install the No. 2 exhaust manifold heat insulator.

58. Install the No. 2 manifold stay.

59. Install the No. 2 engine oil level dipstick guide.

60. Install right hand exhaust manifold sub-assembly and tighten to 15 ft. lbs. (21 Nm).

61. Install the intake manifold and tighten the 6 bolts and 4 nuts uniformly in several steps to 15 ft. lbs. (21 Nm).

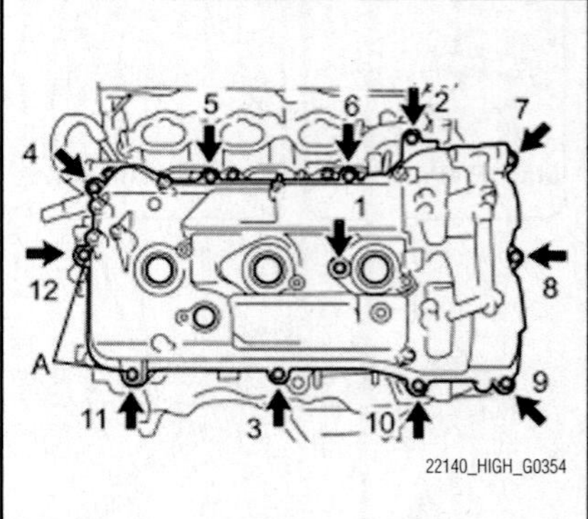

Fig. 154 Cylinder head cover tightening sequence

62. Install the right hand No. 2 engine mounting stay.

63. Install the ignition coil assembly.
64. Remove the engine stand.
65. Install the engine assembly.

EXHAUST MANIFOLD & CATALYTIC CONVERTER

REMOVAL & INSTALLATION

2.7L Engine

See Figures 155 through 157.

1. Before servicing the vehicle, refer to the precautions section.
2. Remove under no. 1 engine cover.
3. Remove front floor cover LH:
   a. Disconnect the 3 clips.
   b. Remove the 2 screws, 4 bolts and front floor cover LH.
4. Remove front exhaust pipe assembly:
   a. Disconnect the clamp and connector.
   b. Remove the 4 bolts, 2 compression springs and front exhaust pipe.
   c. Remove the gasket from the front exhaust pipe.
   d. Remove the gasket from the exhaust manifold assembly.
5. Remove the bolt, nut and stay.
6. Remove no. 2 manifold stay:
   a. Disconnect the connector.
   b. Remove the bolt and wire harness clamp.
   c. Remove the bolt, nut and No. 2 manifold stay.
7. Remove exhaust manifold assembly:
   a. Remove the bolt and plate.
   b. Remove the 5 nuts and exhaust manifold.
   c. Remove the exhaust manifold gasket.

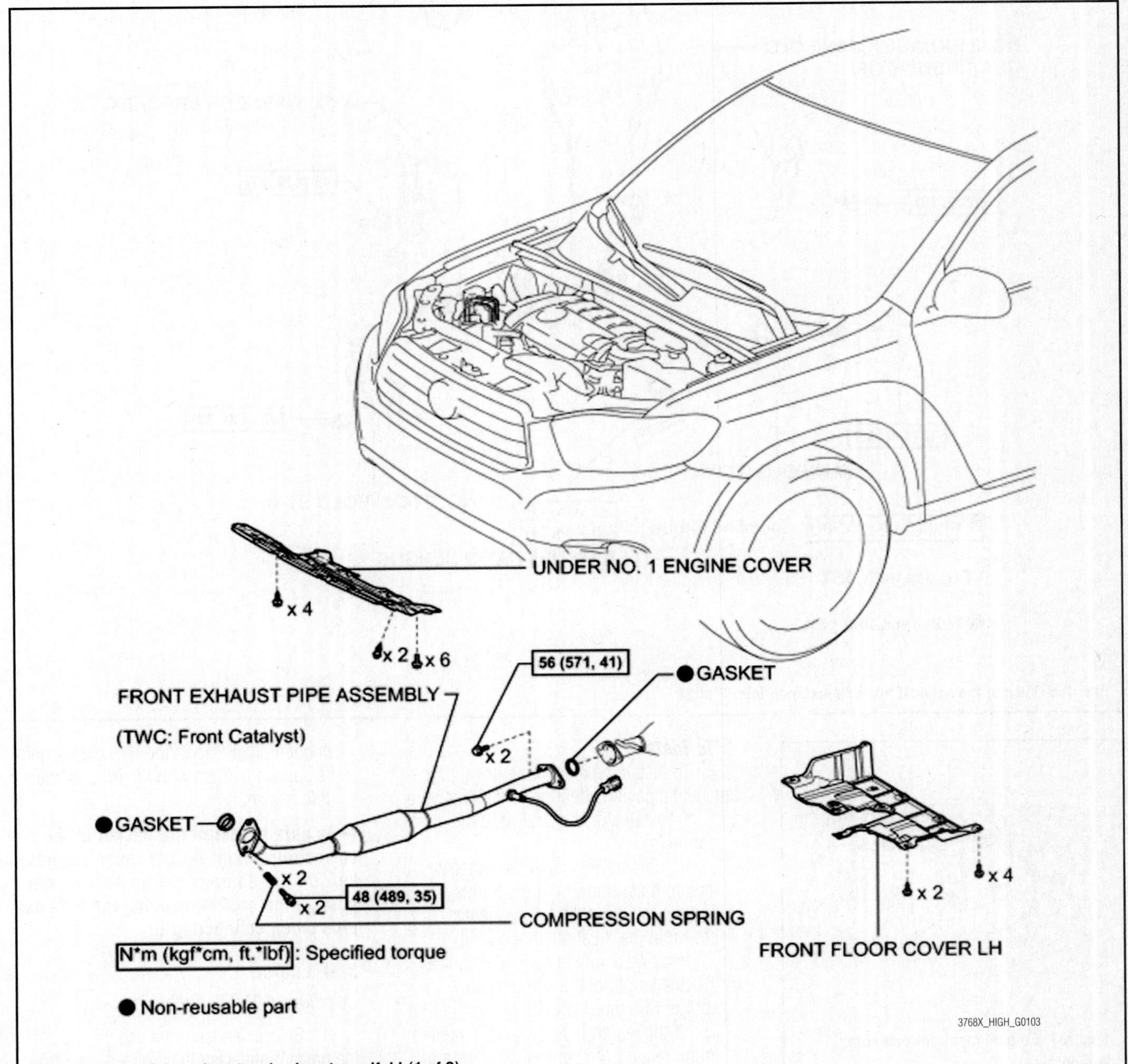

Fig. 155 View of the exhaust and exhaust manifold (1 of 2)

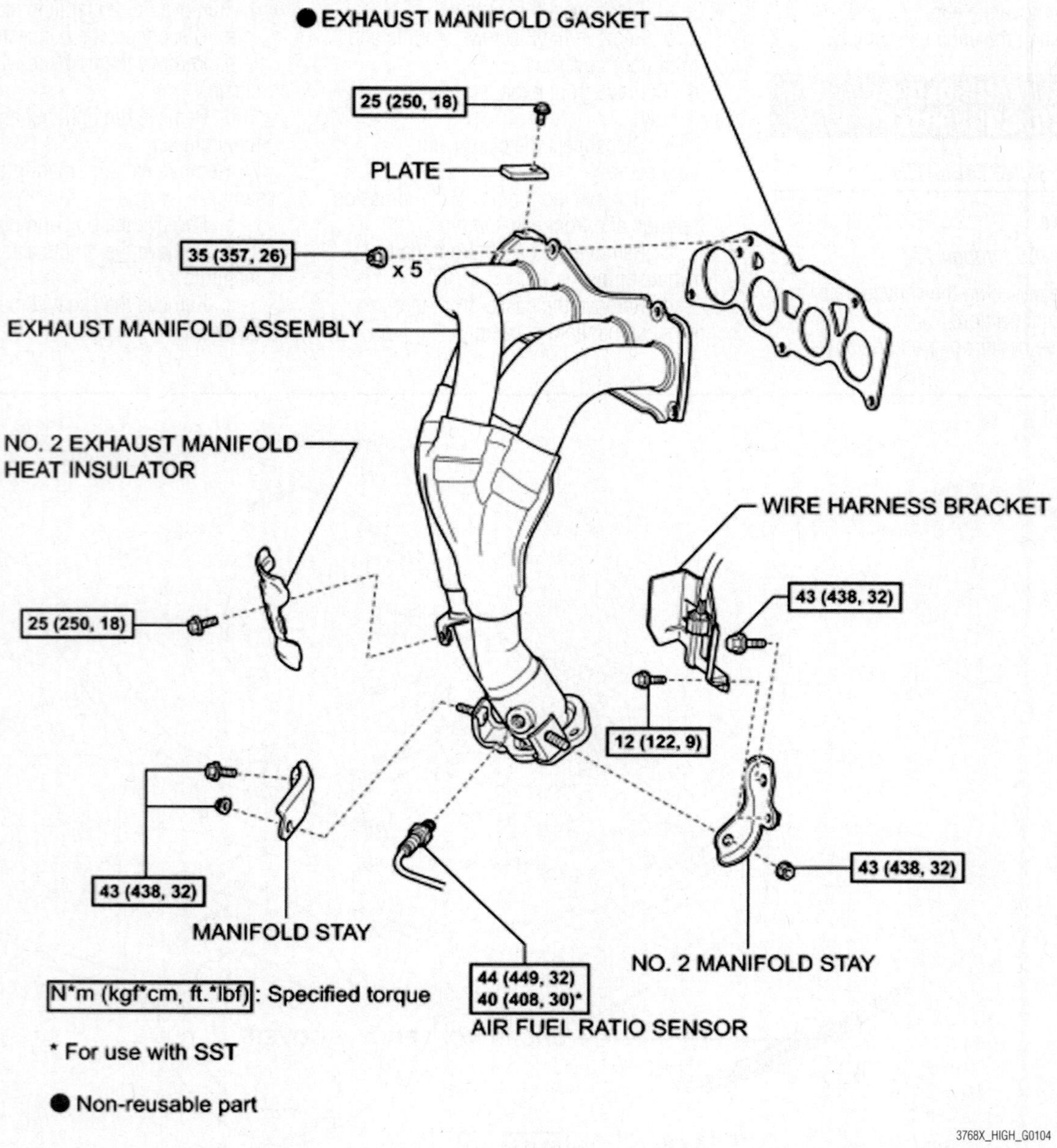

● EXHAUST MANIFOLD GASKET
25 (250, 18)
PLATE
35 (357, 26) x 5
EXHAUST MANIFOLD ASSEMBLY
NO. 2 EXHAUST MANIFOLD HEAT INSULATOR
WIRE HARNESS BRACKET
25 (250, 18)
43 (438, 32)
12 (122, 9)
43 (438, 32)
43 (438, 32)
MANIFOLD STAY
NO. 2 MANIFOLD STAY
N*m (kgf*cm, ft.*lbf) : Specified torque
44 (449, 32)
40 (408, 30)*
AIR FUEL RATIO SENSOR
* For use with SST
● Non-reusable part
3768X_HIGH_G0104

Fig. 156 View of the exhaust and exhaust manifold (2 of 2)

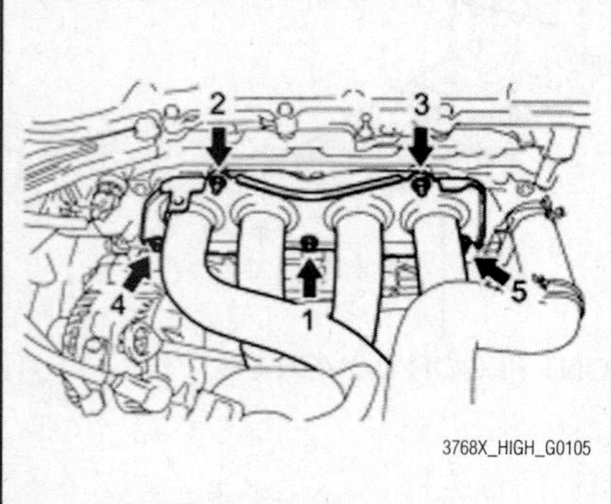

Fig. 157 Exhaust manifold tightening sequence

To install:

8. To install, reverse the removal procedure while paying attention to the following:
   a. Install a new gasket onto the cylinder head.
   b. Temporarily install the exhaust manifold assembly with the 5 nuts. Tighten the 5 nuts in the sequence shown in the illustration.
   c. Using a vernier caliper, measure the free length of the compression spring. Minimum length: 41.5 mm (1.64 in.). If the length is less than the minimum, replace the compression spring.
   d. Using a plastic hammer and wooden block, tap in the new gasket until its surface is flush with the exhaust manifold assembly.

➡Be sure to install the gasket in the correct direction. Do not reuse the gasket. Do not damage the gasket. Do not push in the gasket by using the exhaust pipe when connecting it.

3.5L Engine

See Figures 158 through 161.

1. Before servicing the vehicle, refer to the precautions section.
2. Remove the right front wheel.

3. Remove the V-bank cover sub-assembly.
4. Remove the engine under cover assembly.
5. Remove the No. 1 and 2 engine under covers.
6. Drain engine coolant.
7. Disconnect the No. 1 radiator hose.
8. Remove the radiator reserve tank assembly, as follows:
    a. Disconnect the hose.
    b. Remove the 2 bolts and the radiator reserve tank assembly.
9. Remove the No. 2 oil level dipstick guide.
10. Remove the air fuel ratio sensor (for Bank 2 Sensor 1).
11. Remove the 3 bolts and No. 2 exhaust manifold heat insulator.
12. For 4WD vehicles, remove the propeller with center bearing shaft assembly.
13. Remove the tail exhaust pipe assembly.
14. Remove the center exhaust pipe assembly.
15. Remove the front No. 3 exhaust pipe sub-assembly.
16. Remove the front exhaust pipe assembly.
17. Remove the bolt, nut and No. 2 manifold stay.
18. Remove the 6 nuts and left hand exhaust manifold sub-assembly.
19. Remove the gasket.
20. Remove the bolt, nut and manifold stay.
21. Remove the right hand exhaust manifold sub-assembly, as follows:
    a. Disconnect the air fuel ratio sensor (for bank 1 sensor 1) connector and remove the clamp.
    b. Remove the 6 nuts and the right hand exhaust manifold sub-assembly.
    c. Remove the gasket.
22. Remove the air fuel ratio sensor (for Bank 1 Sensor 1).

To install:

23. Install the air fuel ratio sensor (for Bank 1 Sensor 1).
24. Install the right hand exhaust manifold sub-assembly, as follows:
    a. Install a new gasket.
    b. Install the right hand exhaust manifold sub-assembly by tightening the 6 nuts in the order shown to 15 ft. lbs (21 Nm).
    c. Connect the air fuel ratio sensor (for Bank 1 Sensor 1) connector and install the clamp.
25. Install the manifold stay with the bolt and nut and tighten to 25 ft. lbs (34 Nm), 26 ft. lbs (35 Nm).

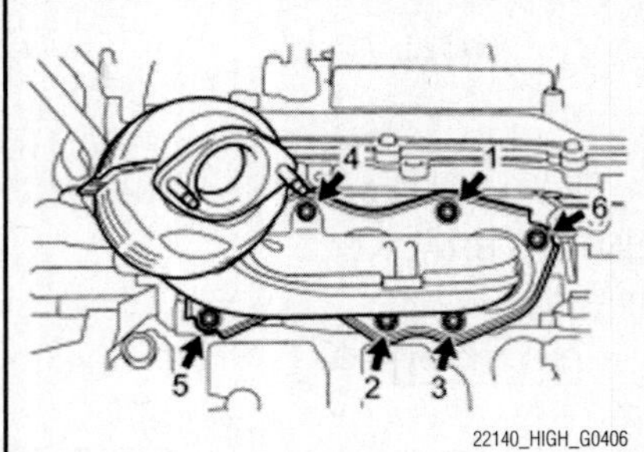

Fig. 158 Right exhaust manifold sub-assembly tightening sequence—3.5L engine

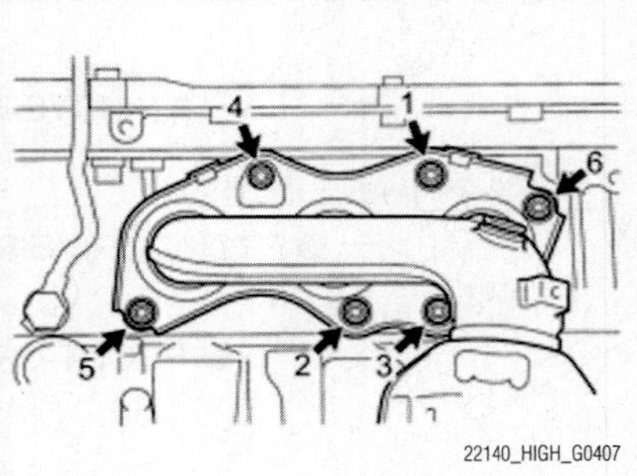

Fig. 159 Left exhaust manifold sub-assembly tightening sequence—3.5L engine

26. Install the left hand exhaust manifold sub-assembly, as follows:
    a. Install a new gasket.
    b. Install the left hand exhaust manifold sub-assembly by tightening the 6 nuts in the order shown to 15 ft. lbs (21 Nm).
27. Install the No. 2 manifold stay by tightening the bolt and nut in the order shown to 25 ft. lbs (34 Nm).
28. Install the front exhaust pipe assembly.
29. Install the front No. 3 exhaust pipe sub-assembly.
30. Install the center exhaust pipe assembly.
31. Install the tail exhaust pipe assembly.
32. For 4WD vehicles, temporarily tighten the propeller with center bearing shaft assembly.
33. For 4WD vehicles, fully tighten the propeller with center bearing shaft assembly.
34. Install the No. 2 exhaust manifold heat insulator by tightening the 3 bolts in the order shown to 75 inch lbs (8.5 Nm).
35. Install the air fuel ratio sensor (for Bank 2 Sensor 1).
36. Install the No. 2 oil level dipstick guide.

Fig. 160 No. 2 manifold stay tightening sequence—3.5L engine

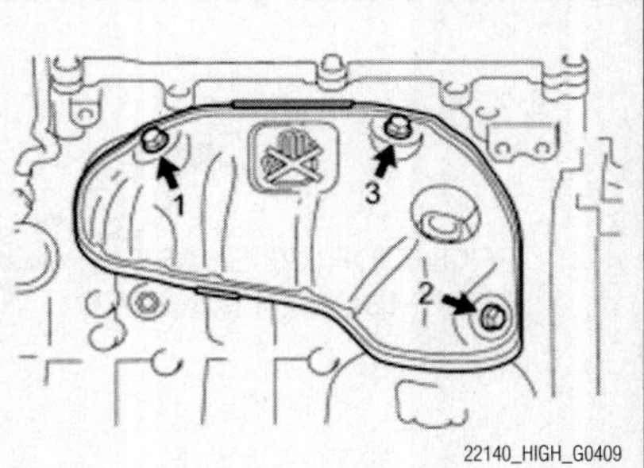

Fig. 161 No. 2 exhaust manifold heat insulator tightening sequence—3.5L engine

37. Install the radiator reserve tank assembly with the 2 bolts and tighten to 48 inch lbs (5.4 Nm).
38. Connect the hose.
39. Connect the No. 1 radiator hose.
40. Add engine coolant.
41. Inspect for coolant leak.
42. Inspect for gas leak, and repair as necessary.
43. For 4WD vehicles, inspect and adjust transfer oil.
44. Install the No. 1 and 2 engine under covers.
45. Install the engine under cover assembly.
46. Install the V-bank cover sub-assembly.
47. Install the right front wheel.

FLEXPLATE

REMOVAL & INSTALLATION

See Figures 162 and 163.

1. Remove automatic transaxle assembly
2. Hold the crankshaft and remove the 8 bolts, front spacer, drive plate and rear spacer.

To install:

3. Installation is reverse of removal.

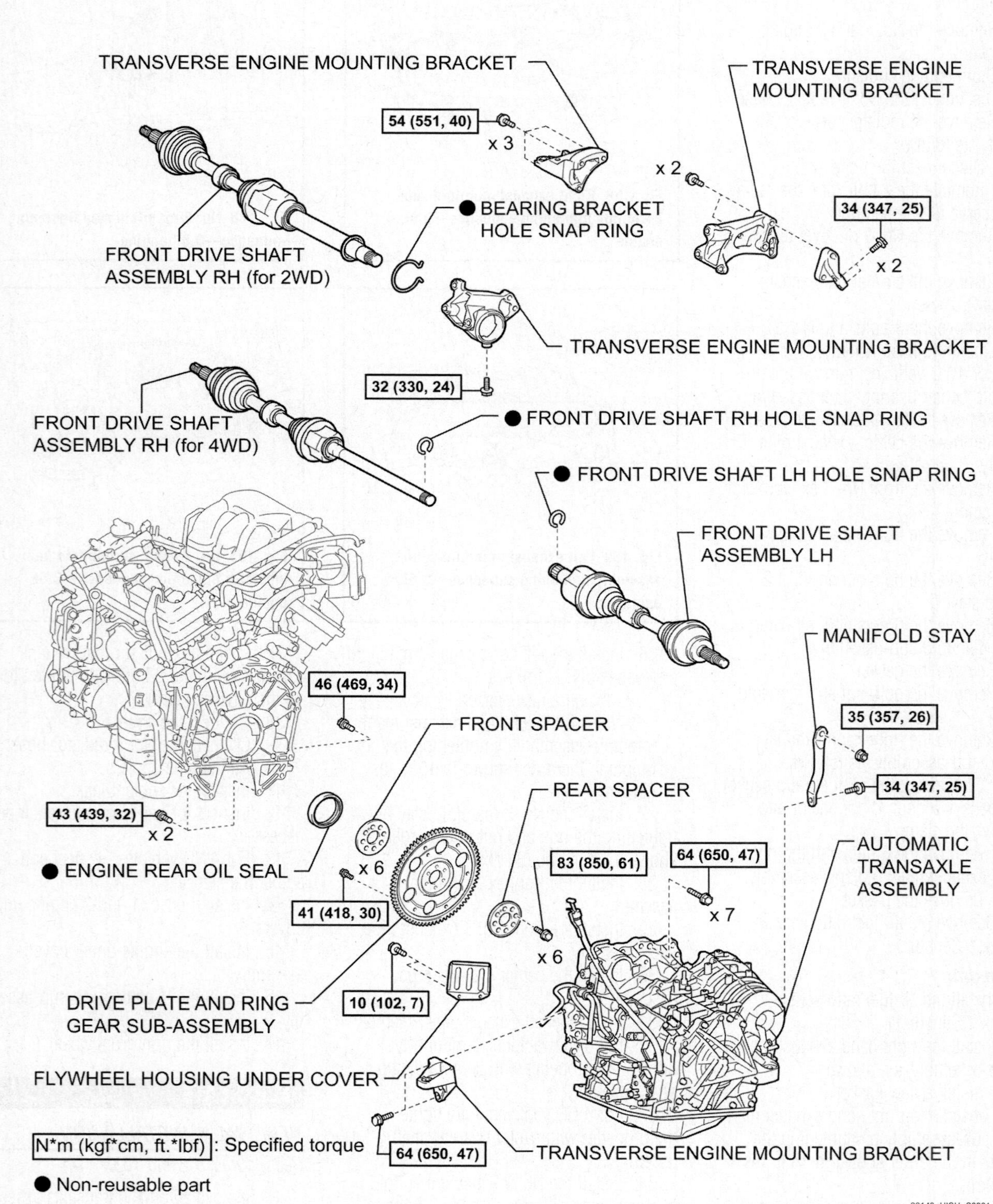

Fig. 162 Flywheel and related components

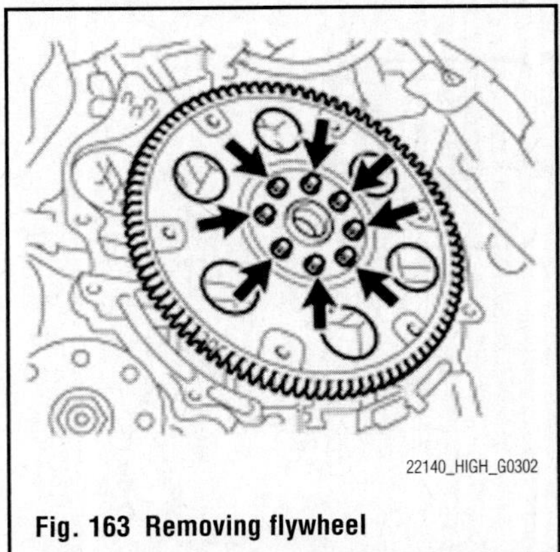

Fig. 163 Removing flywheel

INTAKE MANIFOLD

REMOVAL & INSTALLATION

2.7L Engine

See Figures 164 through 167.

CAUTION

Observe all applicable safety precautions when working around fuel. Whenever servicing the fuel system, always work in a well ventilated area. Do not allow fuel spray or vapors to come in contact with a spark or open flame. Keep a dry chemical fire extinguisher near the work area. Always keep fuel in a container specifically designed for fuel storage; also, always properly seal fuel containers to avoid the possibility of fire or explosion.

1. Properly relieve the fuel system pressure.
2. Disconnect the negative battery cable.
3. Remove the throttle body.
4. Disconnect the inlet heater water hose.
5. Disconnect the outlet heater water hose.
6. Remove vacuum switching valve assembly (for ACIS):
   a. Disconnect the 2 vacuum hoses, 2 union to connector tube hoses, clamp and connector.
   b. Remove the bolt and vacuum switching valve assembly (for ACIS).
7. Disconnect the No. 2 ventilation hose from the intake manifold.
8. Disconnect the union to connector tube hose from the intake manifold.
9. Remove no. 1 vacuum switching valve assembly:
   a. Slide the 2 clips and disconnect the 2 vacuum hoses and connector.
   b. Remove the bolt and No. 1 vacuum switching valve assembly.

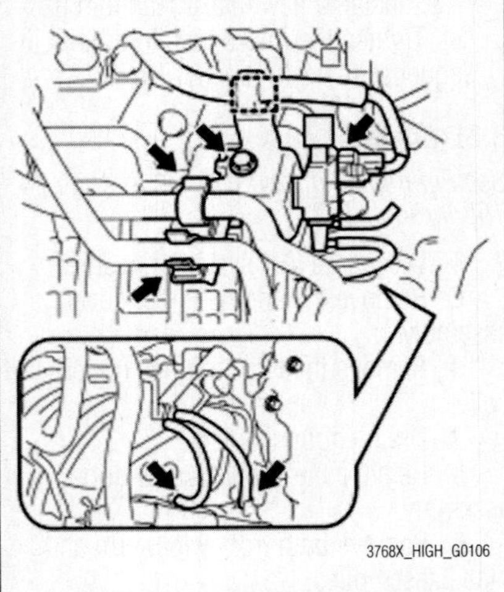

Fig. 164 Remove vacuum switching valve assembly (for ACIS)

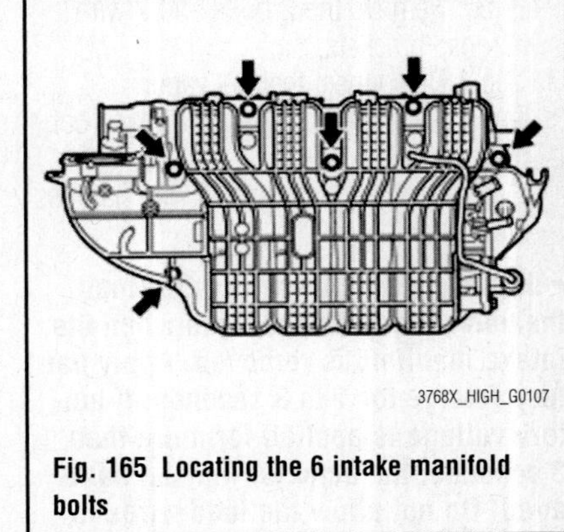

Fig. 165 Locating the 6 intake manifold bolts

10. Remove the fuel rail and injectors.
11. Remove intake manifold:
    a. Disconnect the 4 clamps and wire harness.

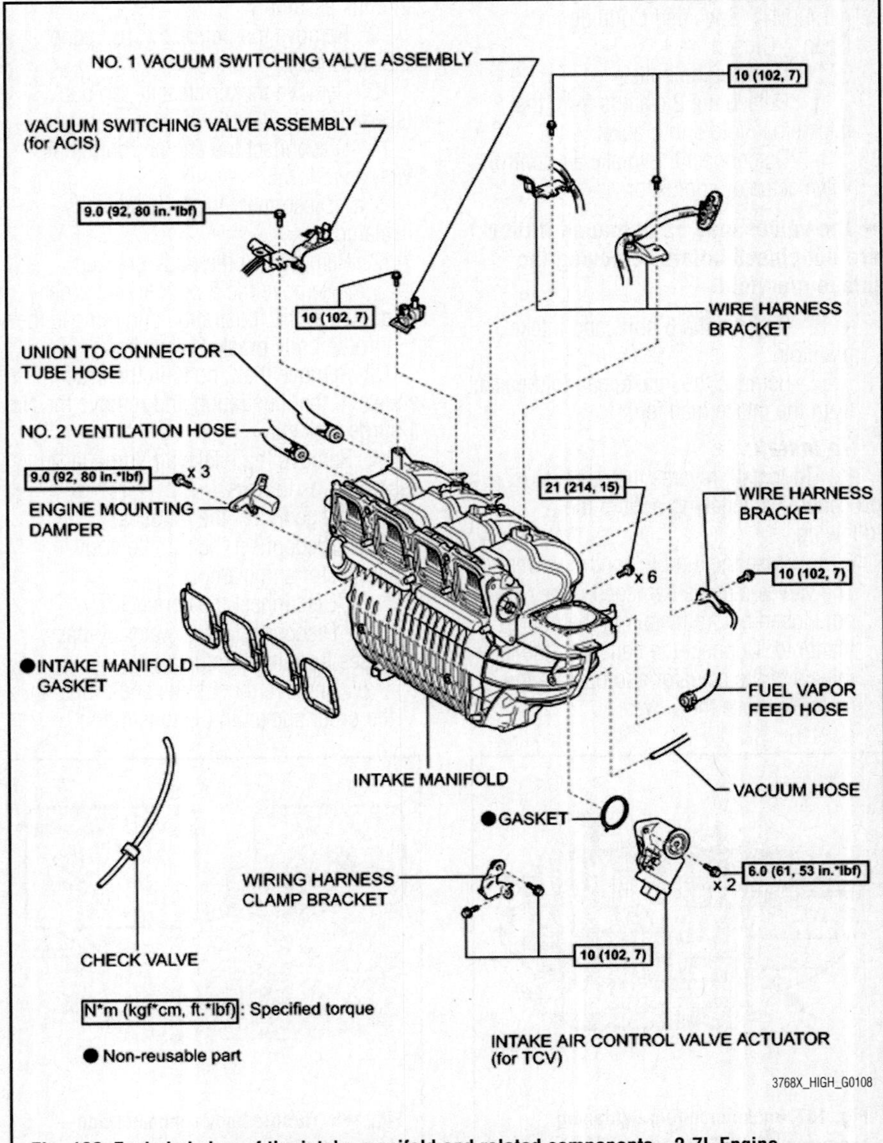

Fig. 166 Exploded view of the intake manifold and related components—2.7L Engine

b. Remove the 2 bolts and 2 wire harness brackets.

c. Disconnect the fuel vapor feed hose, vacuum hose, clamp and connector.

d. Remove the bolt and wire harness bracket.

➡If this procedure is not performed, the valves may be damaged when the intake manifold is removed. Apply battery voltage for 1 to 3 seconds. If battery voltage is applied for more than 3 seconds, the actuator may be damaged. Do not allow the lead wires to contact the other terminals.

e. Apply battery voltage to the terminals of the connector to close the tumble control valves. Standard: Positive battery voltage applied to terminal 8 (M-), and negative battery voltage applied to terminal 4 (M+). Specified Condition: Open _ Closed.

f. Remove the bolt.

g. Detach the 2 clamps from the intake manifold and bracket.

h. Disconnect the intake air control valve actuator connector.

➡The valves may be damaged if they are not closed before removing the intake manifold.

i. Remove the 6 bolts and intake manifold.

j. Remove the intake manifold gasket from the intake manifold.

To install:

12. To install, reverse the removal procedure while taking note of the following:

a. Close the tumble control valves. The valves may be damaged if they are not closed before installing the intake manifold. Connect the battery to the terminals of the actuator and operate the motor to close the valves.

b. Install a new intake manifold gasket. Tighten the intake manifold bolts in sequence to 15 ft. lbs. (21 Nm).

3.5L Engine

See Figures 66 through 69, 71 through 74, 168 through 172.

1. Discharge fuel system pressure.
2. Remove the engine under cover assembly.
3. Remove the No. 1 engine under cover.
4. Drain engine coolant.
5. Remove the V-bank cover sub-assembly.
6. Remove both front wiper arm and blade assemblies.
7. Remove the cowl top ventilator louver sub-assembly.
8. Remove the windshield wiper motor and link assembly.
9. Remove the outer cowl top panel sub-assembly.
10. Remove the air cleaner cap sub-assembly.
11. Disconnect the engine room main wire, as follows:

a. Disconnect the 5 harness clamps.

12. Remove throttle body bracket:

a. Remove the 2 bolts in the order shown in the illustration and remove the throttle body bracket.

13. Remove the 2 bolts in the order shown in the illustration and remove the No. 1 surge tank stay.
14. Remove the intake air surge tank assembly, as follows:

a. Disconnect the 4 hoses.

b. Disconnect the throttle body connector and clamp.

c. Disconnect the connector.

d. Disconnect the 2 water by-pass hoses from the throttle body.

e. Remove the 4 bolts and 2 nuts in the order shown in the illustration.

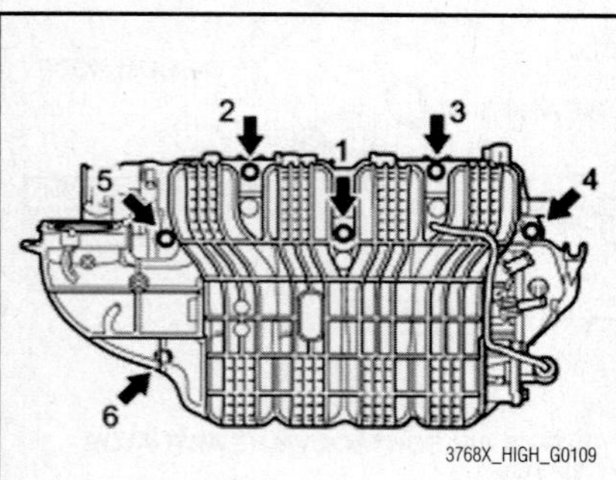

Fig. 167 Intake manifold tightening sequence—2.7L Engines

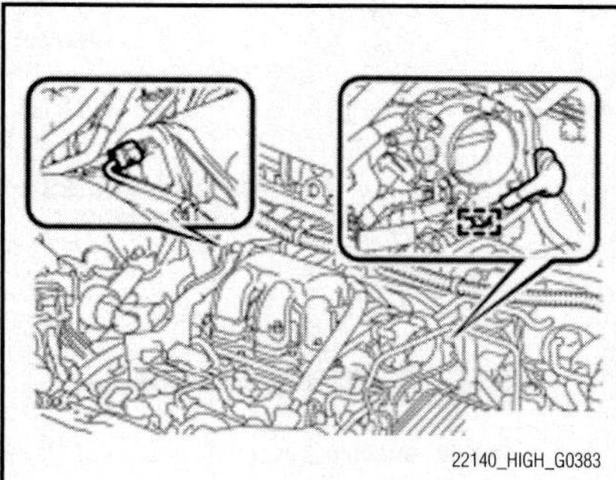

Fig. 168 Throttle body connector and clamp—3.5L engine

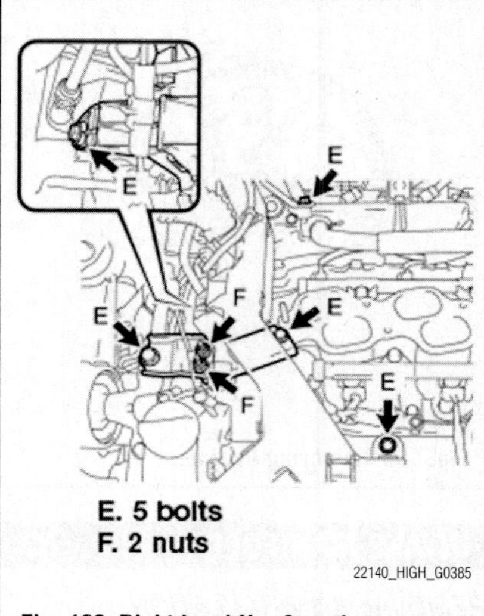

Fig. 169 Right hand No. 2 engine mounting stay bolt and nut removal sequence—3.5L engine

➡Use a 5 mm socket hexagon wrench to remove the 4 bolts

f. Remove the gasket from the intake air surge tank.

15. Remove the right hand No. 2 engine mounting stay, as follows:

a. Remove the 5 bolts (E).

b. Remove the 2 nuts (F).

c. Remove the right hand No. 2 engine mounting stay.

16. Disconnect the fuel main tube, as follows:

a. Remove the No. 2 fuel pipe clamp.

b. Pinch the tube connector and pull out the fuel pipe.

➡Check that there is no dirt or other foreign objects around the connector before disconnecting it. Clean the connector as necessary.

➡It is necessary to prevent dirt or foreign objects from entering the quick connector. If dirt or foreign objects enter the connector, the O-rings may not seal properly.

➡Only disconnect the quick connector by hand.

➡Do not bend, kink or twist the nylon tubes. Protect the connector by covering it with a plastic bag.

➡If the pipe and the connector are stuck, carefully try wiggling or pushing and pulling on the connector to release it. Pull the connector off the pipe carefully.

17. Remove the fuel delivery pipe sub-assembly, as follows:

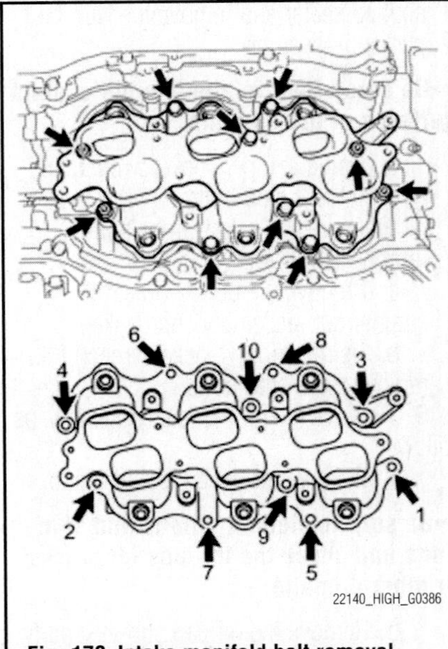

Fig. 170 Intake manifold bolt removal sequence—3.5L engine

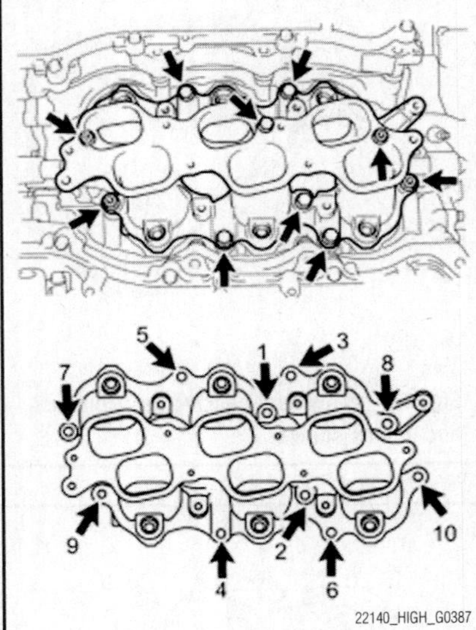

Fig. 171 Intake manifold bolt installation sequence—3.5L engine

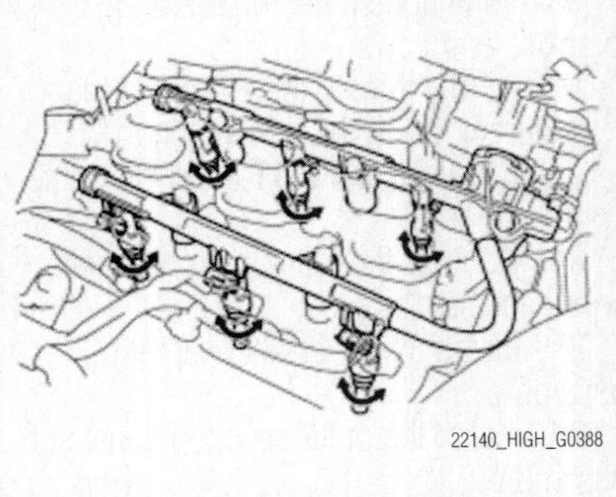

Fig. 172 Place fuel delivery pipe on intake manifold—3.5L engine

a. Disconnect the 6 fuel injector connectors.

18. Remove the 5 bolts and fuel delivery pipe sub-assembly together with the 6 fuel injectors.

➡Be careful not to drop the fuel injectors when removing the fuel delivery pipe sub-assembly.

a. Remove the 6 injector vibration insulators from the intake manifold.

19. Remove the intake manifold, as follows:

a. Remove the 6 bolts and 4 nuts in the order shown in the illustration and remove the intake manifold.

20. Remove the 2 No. 1 intake manifold to head gaskets.

To install:

21. Install the intake manifold, as follows:

a. Set 2 new gaskets on each cylinder head.

➡Align the port holes of the gaskets and cylinder head.

➡Make sure that the gaskets are installed in the correct direction.

b. Set the intake manifold on the cylinder heads.

c. Install the intake manifold with the 6 bolts and 4 nuts in the order shown in the illustration and tighten to 15 ft. lbs. (21 Nm). **DO NOT** apply oil to the bolts.

22. Install the right hand No. 2 engine mounting stay.

23. Install the bolt (A), and tighten to 15 ft. lbs. (21 Nm). **DO NOT** apply oil to the bolt.

24. Install the 2 nuts (B), and tighten to 17 ft. lbs. (23 Nm).

25. Install the bolt (C), and tighten to 28 ft. lbs. (38 Nm). **DO NOT** apply oil to the bolt.

26. Install the 3 bolts (D), and tighten to 73 inch lbs. (8.3 Nm). **DO NOT** apply oil to the bolts.

27. Install the fuel delivery pipe sub-assembly, as follows:

a. Install 6 new insulators to the intake manifold.

b. Place the fuel delivery pipe which has the 6 fuel injectors installed to it in position on the intake manifold.

➡Be careful not to drop the fuel injectors when installing the fuel delivery pipe.

c. Temporarily install the 5 bolts which are used to hold the fuel delivery pipe to the intake manifold. **DO NOT** apply oil to the bolts.

➡After installing the fuel injectors, check that they turn smoothly. If not, reinstall the injectors with new O-rings.

d. Tighten the 5 bolts which are used to hold the fuel delivery pipe to the intake manifold to 15 ft. lbs. (21 Nm). **DO NOT** apply oil to the bolts.

e. Connect the 6 fuel injector connectors.

28. Connect the fuel main tube, as follows:

a. Push in the tube connector onto the pipe until the tube connector clicks.

➡Before connecting the tube, make sure that it is not damaged. Make sure that there is no dirt present on the connecting surfaces.

➡After connecting, check that the fuel tube connector and the pipe are securely connected by pulling on them.

b. Install the No. 2 fuel pipe clamp.

29. Temporarily install the No. 1 surge tank stay, as follows:

a. Temporarily install the intake air surge tank assembly with 3 new gaskets on the intake manifold.

➡Do not allow the gaskets to slip out of place during installation.

b. Temporarily install the No. 1 surge tank stay with the 2 bolts. **DO NOT** apply oil to the bolts.

30. Temporarily install the throttle body bracket with the 2 bolts. **DO NOT** apply oil to the bolts.

31. Install the intake air surge tank assembly, as follows:

a. Install the surge tank with the 4 bolts and 2 nuts in the order shown in the illustration and tighten to 12 ft. lbs. (16 Nm) and 13 ft. lbs. (18 Nm). **DO NOT** apply oil to the bolts.

➡Use a 5 mm hexagon socket wrench to tighten the 4 bolts.

b. Connect the 2 water by-pass hoses to the throttle with motor body assembly.

c. Connect the connector.

d. Install the clamp and connect the throttle with motor body assembly connector.

e. Connect the 4 hoses.

32. Fully tighten the No. 1 surge tank stay, as follows:

a. Fully tighten the 2 bolts in the order shown in the illustration to 15 ft. lbs. (21 Nm). **DO NOT** apply oil to the bolts.

33. Fully tighten the throttle body bracket, as follows:
  a. Fully tighten the 2 bolts in the order shown in the illustration to 15 ft. lbs. (21 Nm). **DO NOT** apply oil to the bolts.
34. Connect the 5 engine room main wire harness clamps.
35. Install the air cleaner cap sub-assembly.
36. Install the outer cowl top panel sub-assembly.
37. Install the windshield wiper motor and link assembly.
38. Install the cowl top ventilator louver sub-assembly.
39. Install the both front wiper arm and blade assemblies.
40. Add engine coolant.
41. Inspect for engine coolant leak.
42. Inspect for fuel leak.
43. Install the V-bank cover sub-assembly.
44. Install the No. 1 engine under cover.
45. Install the engine under cover assembly.

OIL PAN

REMOVAL & INSTALLATION

2.7L Engine

See Figures 173 through 175.

1. Drain the oil in to a suitable container.
2. Remove the 11 bolts and 2 nuts.
3. Insert the blade of an oil pan seal cutter between the oil pan and stiffening crankcase, cut off the applied sealer and remove the oil pan.

➡Be careful not to damage the stiffening crankcase contact surface of the oil pan. Be careful not to damage the stiffening crankcase flange.

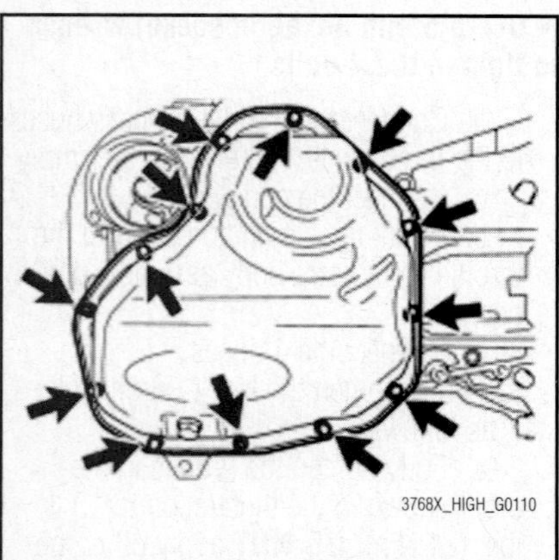

Fig. 173 Remove the 11 bolts and 2 nuts

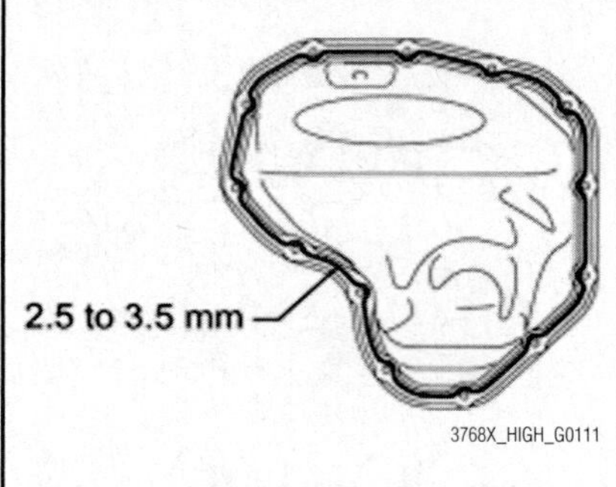

Fig. 174 Apply seal packing in a continuous line as shown

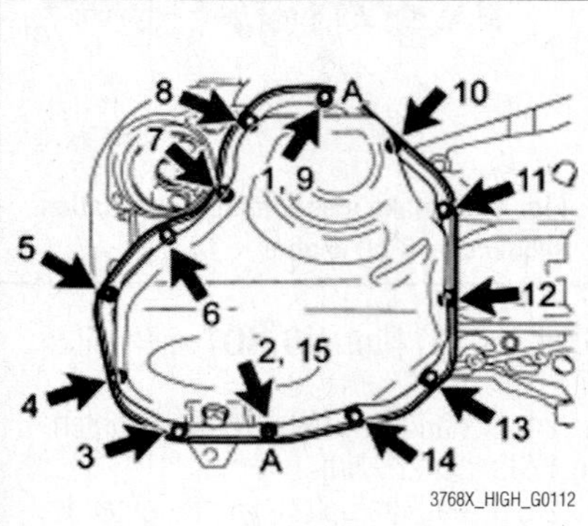

Fig. 175 Oil pan tightening sequence

To install:

4. Apply seal packing in a continuous line as shown in the illustration. Seal packing: Toyota Genuine Seal Packing Black, Three Bond 1207B or equivalent. Standard seal diameter: 2.5 to 3.5 mm (0.0984 to 0.138 in.)

➡Remove any oil from the contact surface. Install the oil pan within 3 minutes and tighten the bolts and nuts within 10 minutes after applying seal packing. Do not apply oil for at least 4 hours after the installation.

5. Install the oil pan by tightening the 11 bolts and 2 nuts in several steps, in the sequence shown in the illustration. Tighten to 7 ft. lbs. (10 Nm). Bolt "A" and nut "A" should be tightened twice.

3.5L Engine

See Figures 176 through 180.

1. Before servicing the vehicle, refer to the precautions section.
2. Drain the engine oil.
3. Remove the No. 2 oil pan sub-assembly, as follows:
  a. Remove the 16 bolts and 2 nuts.
  b. Insert the blade of oil pan seal cutter between the oil pans. Cut through the applied sealer and remove the No. 2 oil pan sub-assembly.

➡Be careful not to damage the contact surfaces of the oil pans.

  c. Using a Torx® socket wrench E6, remove the 2 stud bolts.
4. Remove the oil strainer sub-assembly, as follows:
  a. Remove the bolt, 2 nuts, oil strainer sub-assembly and gasket.
  b. Using a Torx® socket wrench E6, remove the 2 stud bolts.
5. Remove the oil pan sub-assembly, as follows:
  a. Remove the 16 bolts and 2 nuts.

➡Be sure to clean the bolts and stud bolts and check the threads for cracks or other damage.

  b. Remove the oil pan sub-assembly by prying between the oil pan sub-assembly and cylinder block sub-assembly with a screwdriver.

➡Be careful not to damage the contact surfaces of the cylinder block and oil pans.

➡Tape the screwdriver tip before use.

  c. Remove the 2 O-rings.
  d. Using a Torx® socket wrench E8, remove the 2 stud bolts.

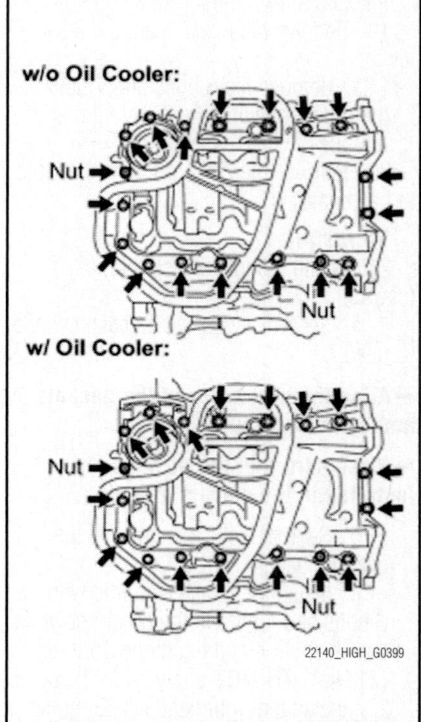

Fig. 176 Oil pan sub-assembly bolts and nuts—3.5L engine

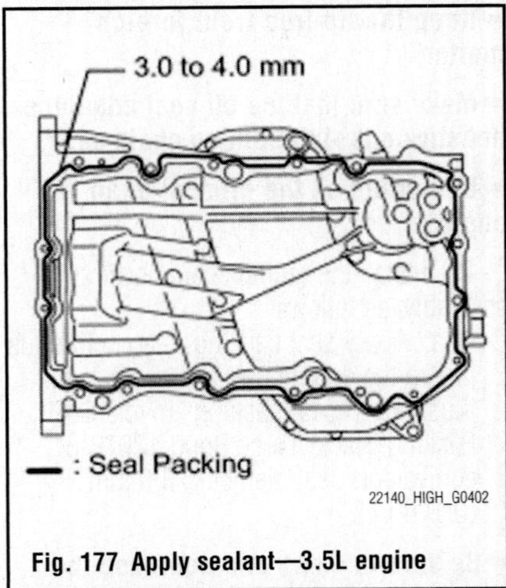

Fig. 177 Apply sealant—3.5L engine

To install:

6. Install the oil pan sub-assembly, as follows:

a. When replacing a stud bolt, install it by using an E8 Torx® socket wrench. Tighten to 7 ft. lbs. (10 Nm).

b. Install 2 new O-rings.

c. Apply seal packing in a continuous line as shown in the illustration. Seal packing: Toyota Genuine Seal Packing Black, Three Bond 1207B or equivalent. Seal diameter: 3.0 to 4.0 mm (0.118 to .0156 inches).

➡Remove any oil from the contact surface.

➡Install the oil pan within 3 minutes after applying seal packing.

➡Do not start the engine for at least 2 hours after installing.

d. Install the oil pan with the 16 bolts and 2 nuts and tighten to 7 ft. lbs. (10 Nm), 15 ft. lbs. (21 Nm).

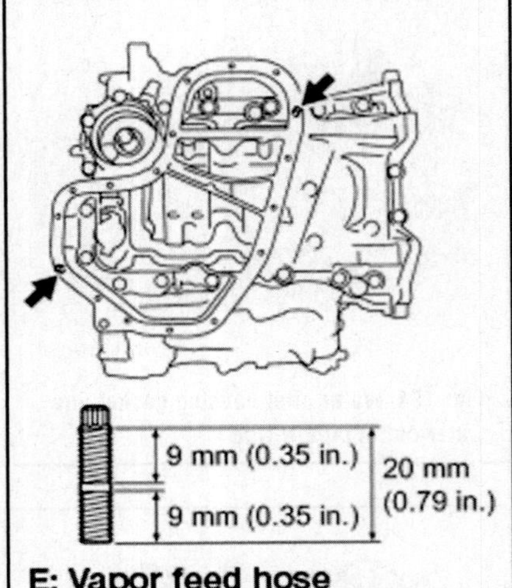

Fig. 179 No. 2 oil pan sub-assembly—3.5L engine

7. Install the oil strainer sub-assembly, as follows:

a. Using an E6 Torx® socket, install the stud bolts as shown in the illustration and tighten to 35 inch lbs. (4 Nm).

b. Install a new gasket and the oil strainer sub-assembly with the bolt and 2 nuts and tighten to 7 ft. lbs. (10 Nm).

8. Install the No. 2 oil pan sub-assembly, as follows:

a. Using an E6 Torx® socket, install the stud bolts as shown in the illustration and tighten to 35 inch lbs. (4 Nm).

b. Apply seal packing in a continuous line as shown in the illustration. Seal packing: Toyota Genuine Seal Packing Black, Three Bond 1207B or equivalent. Seal diameter: 3.0 to 4.0 mm (0.118 to .0156 inches).

➡Remove any oil from the contact surface.

➡Install the No. 2 oil pan sub-assembly within 3 minutes after applying seal packing.

➡Do not start the engine for at least 2 hours after installing.

c. Install the No. 2 oil pan sub-assembly with the 16 bolts and 2 nuts and tighten to 7 ft. lbs. (10 Nm).

9. Install a new oil pan drain plug gasket and the oil pan drain plug and tighten to 30 ft. lbs. (40 Nm).

OIL PUMP

REMOVAL & INSTALLATION

2.7L Engine

The oil pump is integrated in to the timing chain cover. Refer to Timing Chain Cover in this section.

3.5L Engine

See Figures 181 through 190.

1. Remove the engine assembly with transaxle.
2. Secure the engine.
3. Remove the No. 1 oil pipe, as follows:

a. Remove the 2 oil pipe unions, gaskets and No. 1 oil pipe.

b. Remove the left oil control valve filter and gaskets.

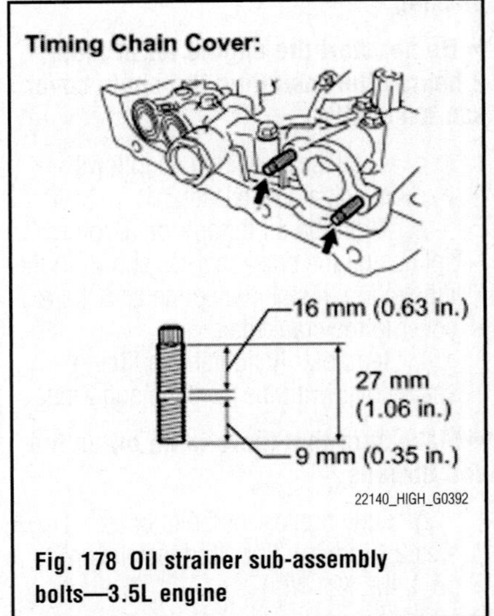

Fig. 178 Oil strainer sub-assembly bolts—3.5L engine

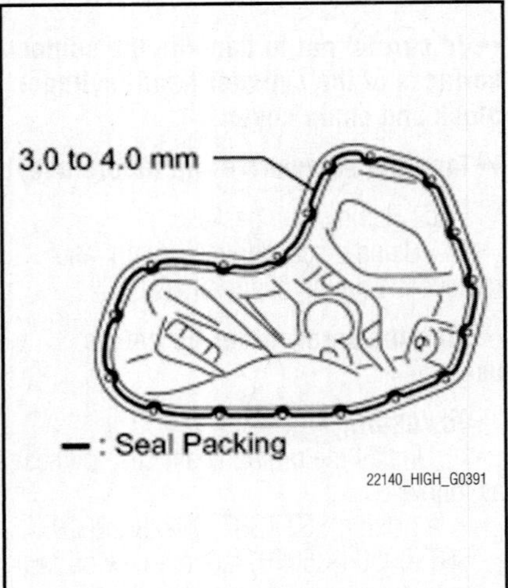

Fig. 180 Apply sealant—3.5L engine

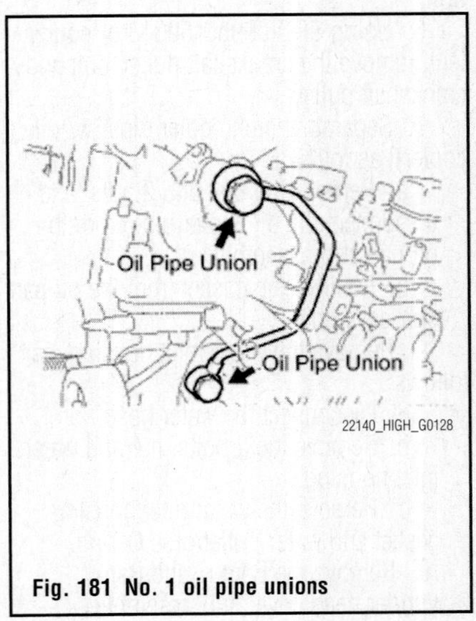

Fig. 181 No. 1 oil pipe unions

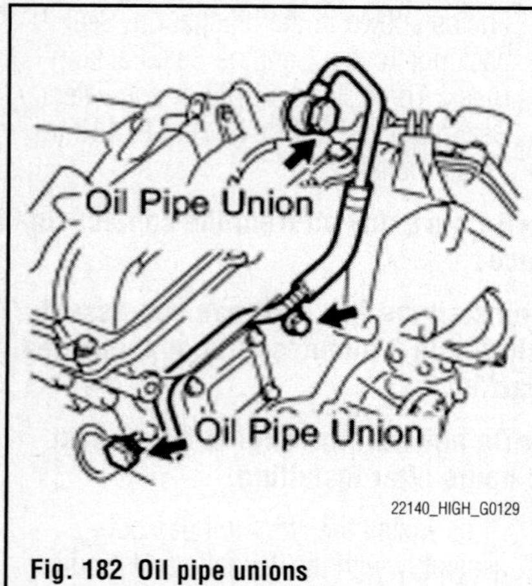

Fig. 182 Oil pipe unions

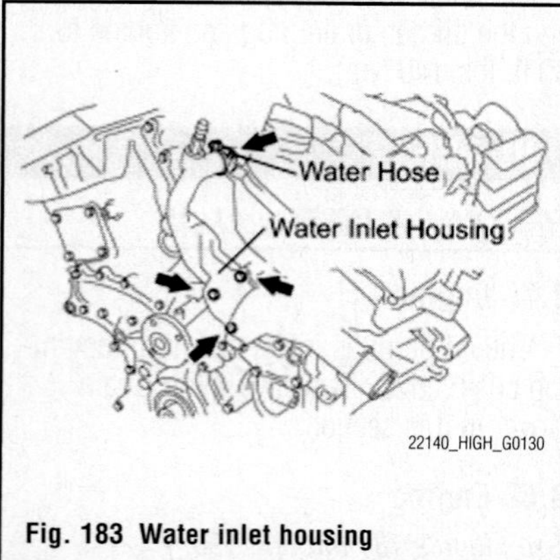

Fig. 183 Water inlet housing

4. Remove the oil pipe, as follows:
   a. Remove the bolt.
   b. Remove the 2 oil pipe unions and oil pipe.
   c. Remove the right oil control valve filter and gaskets.
5. Using a special service tool (SST: 09213-70011, SST: 09213-70011 or equivalent), loosen the crankshaft pulley bolt.
6. Using SST: 09950-50013 or equivalent, remove the crankshaft pulley bolt and crankshaft pulley.
7. Separate the oil cooler pipe (w/ oil cooler), as follows:
   a. Remove the bolt and 2 nuts, and disconnect the oil cooler pipe from the oil pan sub-assembly.
   b. Remove the gasket from the oil pan sub-assembly.
8. Remove the water inlet housing, as follows:
   a. Disconnect the water hose.
   b. Remove the 2 bolts, nut and water inlet housing.
   c. Remove the water inlet housing gasket and water outlet pipe O-ring.
9. Remove the Bank 1 and Bank 2 cylinder head cover sub-assemblies.

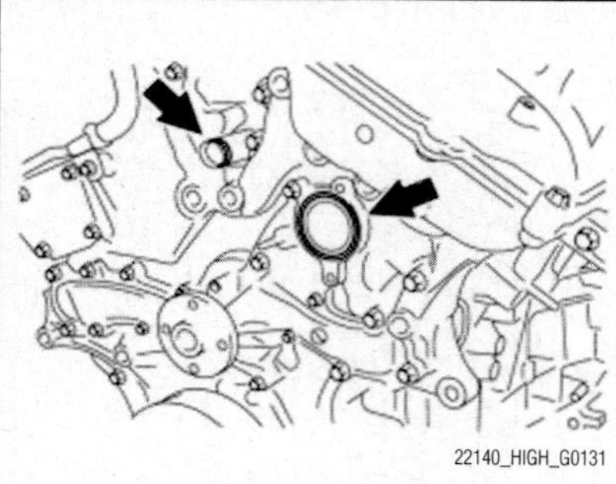

Fig. 184 Water inlet housing gasket and water outlet pipe O-ring

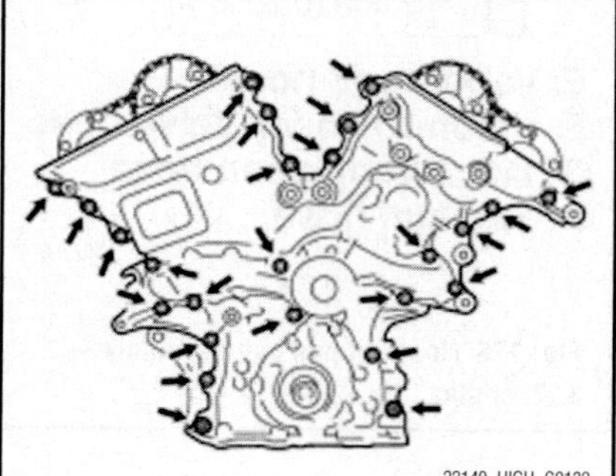

Fig. 185 Timing chain cover sub-assembly bolts and nuts

10. Remove the No. 2 oil pan sub-assembly.
11. Remove the oil strainer sub-assembly.
12. Remove the timing chain cover sub-assembly, as follows:
    a. Remove the 23 bolts and 2 nuts as shown in the illustration.
    b. Remove the timing chain cover by prying between the timing chain cover and cylinder head or cylinder block with a screwdriver.

➡Be careful not to damage the contact surfaces of the cylinder head, cylinder block and chain cover.

➡Tape the screwdriver tip before use.

    c. Remove the gasket.
13. Using a screwdriver, pry out and remove the timing chain oil seal.

➡Tape the screwdriver tip before use.

To install:

14. Install the timing chain case oil seal, as follows:
    a. Using SST (SST: 09223-22010, SST: 09506-35010) tap in a new oil seal until its surface is flush with the timing chain case edge.

➡Keep the lip free from foreign matter.

➡Make sure that the oil seal edge does not stick out of the timing chain case.

➡Do not tap on the oil seal at an angle.

15. Install the timing chain cover sub-assembly, as follows:
    a. Apply seal packing in a continuous line to the engine unit as shown in the illustration. Seal packing: Toyota Seal Packing Black, Three Bond 1207B or equivalent. Seal diameter: 3.0 mm (0.118 in.).

➡Be sure to clean and degrease the contact surfaces, especially the surfaces indicated by C in the illustration.

➡If the contact surfaces are wet, wipe them with an oil-free cloth before applying seal packing.

➡Install the chain cover sub-assembly within 3 minutes after applying seal packing.

➡Do not start the engine for at least 2 hours after installing the chain cover sub-assembly.

    b. Apply seal packing in a continuous line to the timing chain cover as shown in the illustration. Seal packing: Toyota Seal Packing Black, Three Bond 1207B, Three Bond 1282B or equivalent.

➡If the contact surfaces are wet, wipe them with an oil-free cloth before applying seal packing.

➡Install the chain cover sub-assembly within 3 minutes and tighten the bolts within 15 minutes after applying seal packing.

➡Do not start the engine for at least 2 hours after installing the chain cover sub-assembly.

    c. Apply seal packing as follows:
    d. Install a new gasket.
    e. Align the oil pump's drive rotor spline and the crankshaft as shown in the illustration. Install the spline and chain cover to the crankshaft.
    f. Temporarily tighten the timing chain cover with the 23 bolts and 2 nuts.

➡Make sure that there is no oil on the bolt threads.

    g. Fully tighten the bolts in area 1 and area 2 (from top to bottom as shown) to 15 ft. lbs. (21 Nm).

C
C
C
C
C
■ : Seal Packing
3.0 mm or more
22140_HIGH_G0389

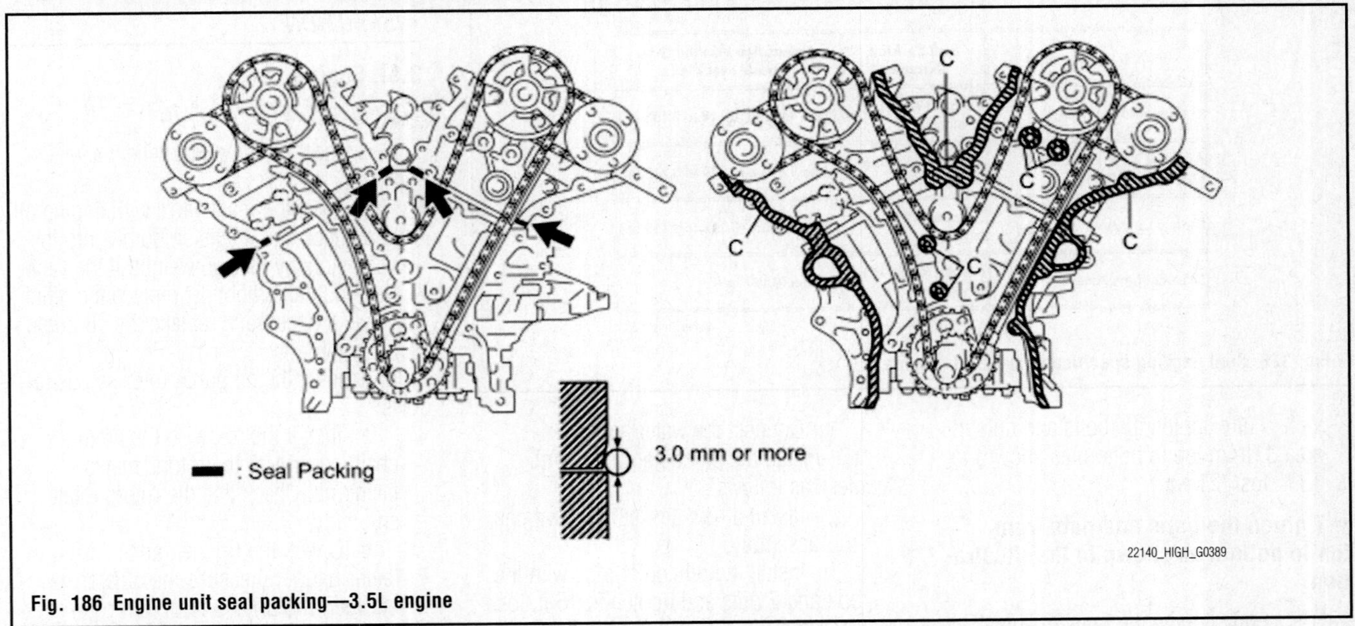

Fig. 186 Engine unit seal packing—3.5L engine

Be sure to apply seal packing
C
C
20 mm
20 mm
A
A
B
B
Be sure to apply seal packing
5.0 mm
A - A
B - B
3.0 to 4.0 mm
2.0 to 3.0 mm
1.0 to 2.0 mm
C - C
Dashed line area
(Seal packing: Toyota Genuine Seal Packing Black, Three Bond 1207B or equivalent)
Continuous line area
(Seal packing: Toyota Genuine Seal Packing Black, Three Bond 1207B or equivalent)
Alternate long and short dashed line area
(Seal packing: Toyota Genuine Seal Packing 1282B, Three Bond 1282B or equivalent)
Diagonal line area
(Seal packing: Toyota Genuine Seal Packing Black, Three Bond 1207B or equivalent)
22140_HIGH_G0384

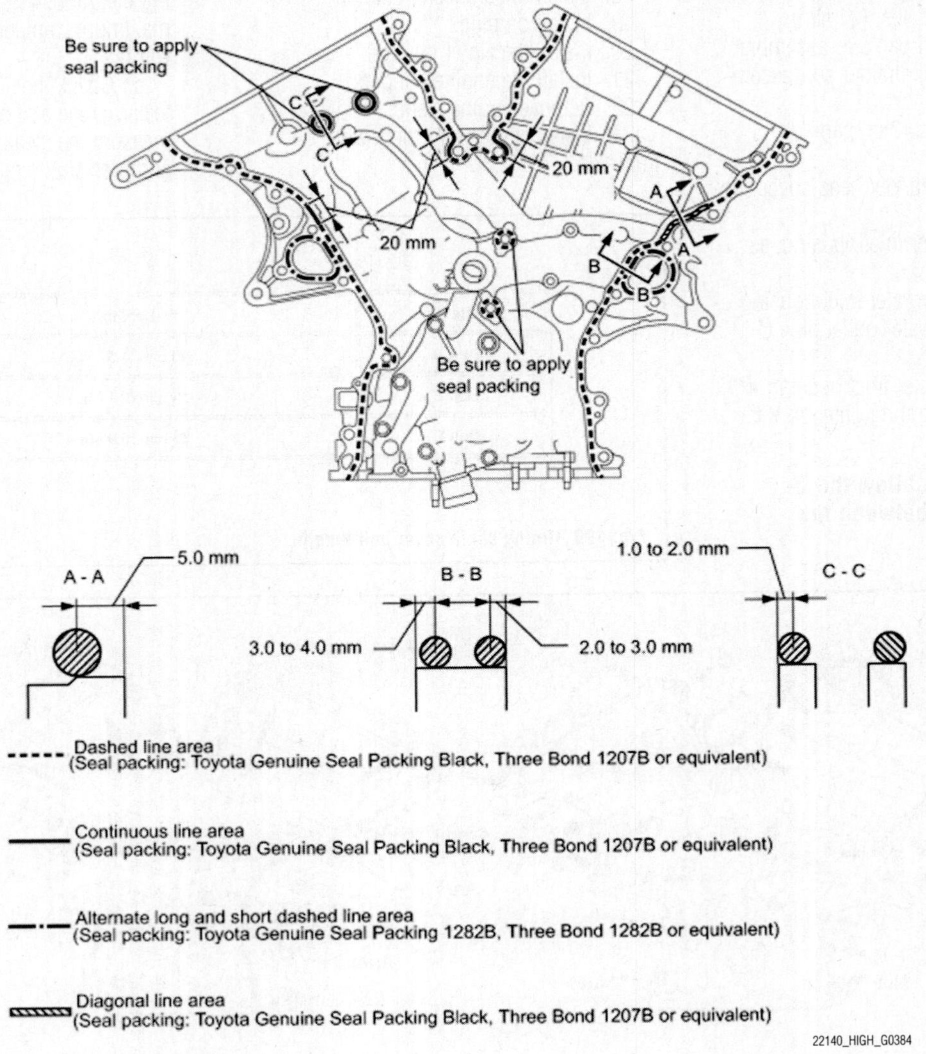

Fig. 187 Timing chain cover seal packing—3.5L engine

Area	Seal Packing Diameter	Application Position from Inside Seal Line
Continuous Line Area	4.5 mm (0.177 in.) or more	3.0 to 4.0 mm (0.118 to 0.158 in.)
Alternate Long and Short Dashed Line Area	3.5 mm (0.138 in.) or more	2.0 to 3.0 mm (0.079 to 0.118 in.)
Dashed Line Area	3.5 mm (0.138 in.) or more	3.0 to 4.0 mm (0.118 to 0.158 in.)
Diagonal Line Area	6.0 mm (0.236 in.) or more	5.0 mm (0.197 in.)

22140_HIGH_G0380

Fig. 188 Seal packing specifications—3.5L engine

h. Fully tighten the bolts and nuts in area 3 (from top to bottom as shown) to 15 ft. lbs. (21 Nm).

➡Tighten the bolts and nuts from top to bottom as shown in the illustration.

i. Fully tighten the bolts in area 4 (from bottom to top as shown) to 32 ft. lbs. (43 Nm), 15 ft. lbs. (21 Nm).

j. Install the oil pan sub-assembly.

k. Install the oil strainer sub-assembly.

l. Install the No. 2 oil pan sub-assembly.

m. Install the cylinder head cover sub-assembly.

16. Install the water inlet housing, as follows:

a. Install a new water inlet housing No. 1 gasket and water outlet pipe O-ring.

b. Install the water inlet housing with the 2 bolts and nut and tighten to 7 ft. lbs. (10 Nm).

➡Be careful not to allow the O-ring to get caught between the parts.

c. Connect the water hose.

17. Install the oil cooler pipe (w/oil cooler), as follows:

a. Install a new gasket to the oil pan sub-assembly.

b. Install the oil cooler pipe with the bolt and 2 nuts and tighten to 15 ft. lbs. (21 Nm).

18. Install the crankshaft pulley.

19. Install oil pipe.

20. Install the No. 1 oil pipe.

21. Install the engine hangers.

22. Remove engine stand.

23. Install engine assembly with transaxle.

INSPECTION

3.5L Engine

See Figures 191 through 197.

1. Inspect the oil pump relief valve, as follows:

a. Coat the relief valve with engine oil and check that it falls smoothly into the valve hole by its own weight. If the valve does not fall smoothly, replace the relief valve. If necessary, replace the oil pump assembly.

2. Inspect the oil pump rotor set, as follows:

a. Install the rotors to the timing chain cover with the rotors' marks facing up. Check that the rotors rotate smoothly.

b. Check the tip clearance: using a feeler gauge, measure the clearance between the drive and driven rotor tips, as shown in the illustration. If the clearance is greater than the maximum, replace the drive and driven rotors.

c. Check the side clearance: using a feeler gauge and precision straightedge, measure the clearance between the rotors and precision straightedge, as shown in

Item	Length
Bolt A	40 mm (1.57 in.)
Bolt B	55 mm (2.17 in.)
Bolt C	25 mm (0.98 in.)

22140_HIGH_G0133

Fig. 190 Timing chain cover bolt length

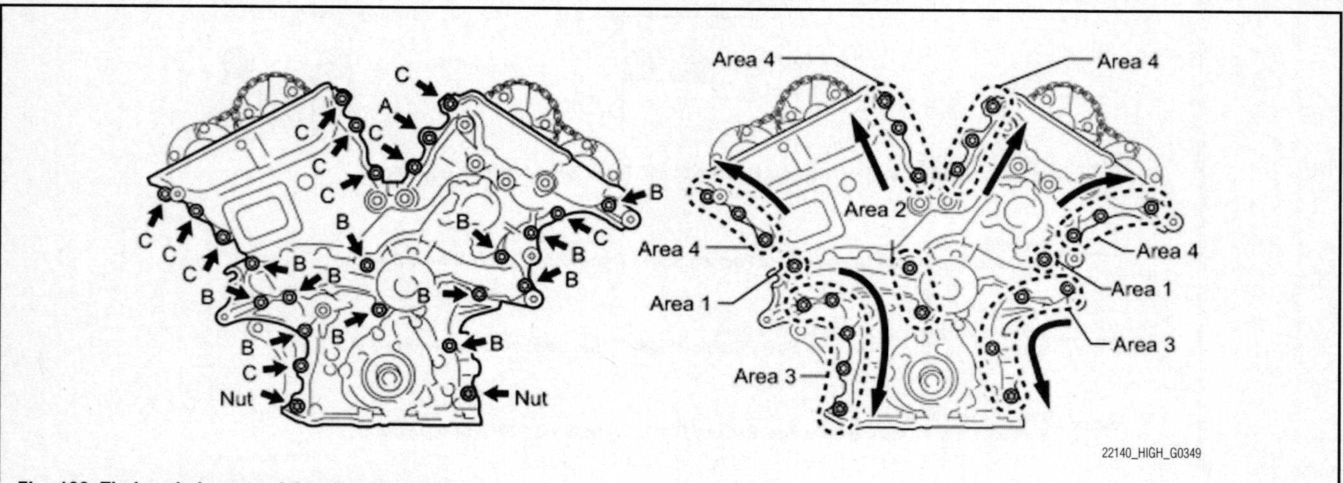

Fig. 189 Timing chain cover tightening sequence

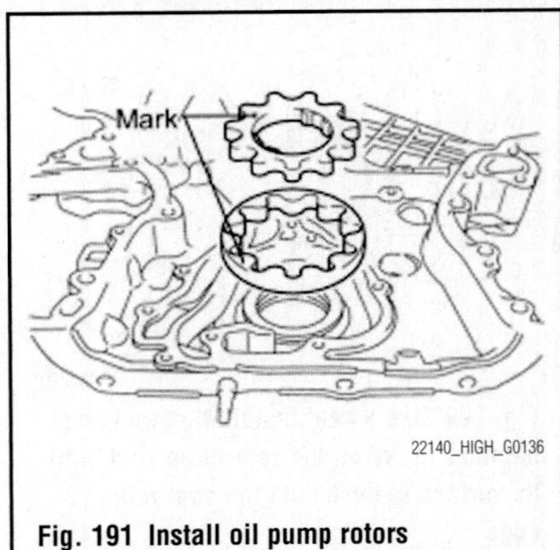

Fig. 191 Install oil pump rotors

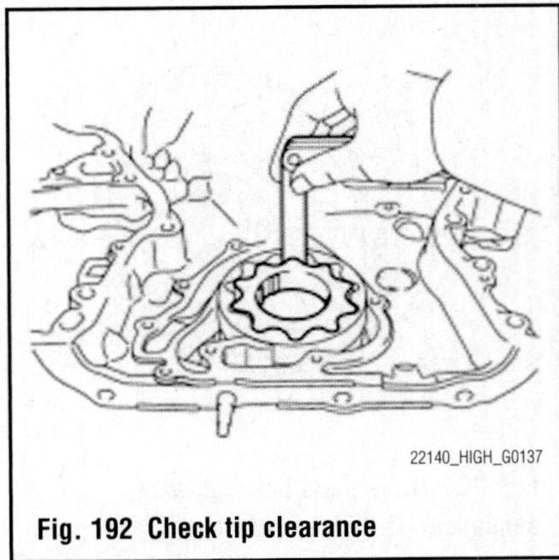

Fig. 192 Check tip clearance

Standard	Maximum
0.060 to 0.160 mm (0.0024 to 0.0063 in.)	0.160 mm (0.0063 in.)

22140_HIGH_G0138

Fig. 193 Tip clearance

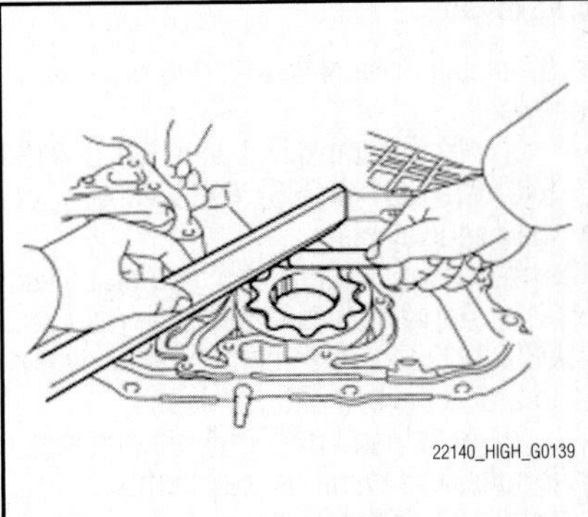

Fig. 194 Check side clearance

the illustration. If the side clearance is greater than the maximum, replace the timing chain cover sub-assembly.

d. Check the body clearance: using a feeler gauge, measure the clearance between the timing chain cover and driven rotor, as shown in the illustration. If the body clearance is greater than the maximum, replace the timing chain cover sub-assembly.

. Standard	Maximum
0.030 to 0.090 mm (0.0012 to 0.0035 in.)	0.090 mm (0.0035 in.)

22140_HIGH_G0140

Fig. 195 Side clearance

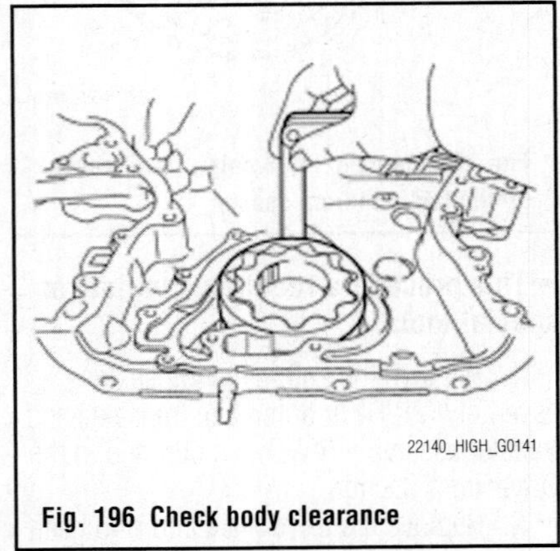

Fig. 196 Check body clearance

Standard	Maximum
0.250 to 0.325 mm (0.0098 to 0.0128 in.)	0.325 mm (0.0128 in.)

22140_HIGH_G0142

Fig. 197 Body clearance

PISTON AND RING

POSITIONING

See Figures 198 and 199.

REAR MAIN SEAL

REMOVAL & INSTALLATION

2.7L Engine

See Figures 200 and 201.

1. Remove the engine and transaxle.
2. Remove the automatic transaxle.
3. Remove the flexplate.

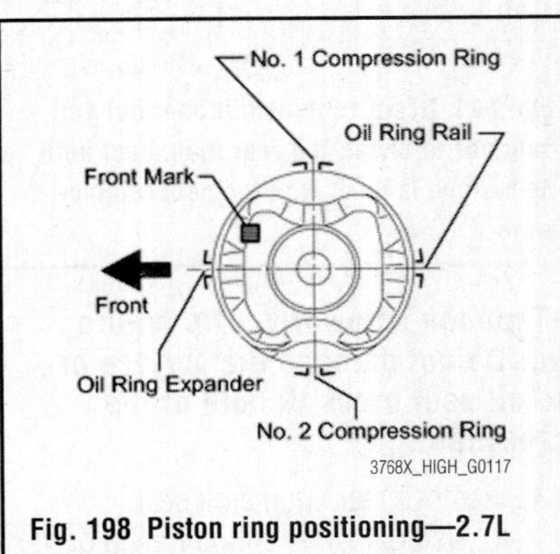

Fig. 198 Piston ring positioning—2.7L engine

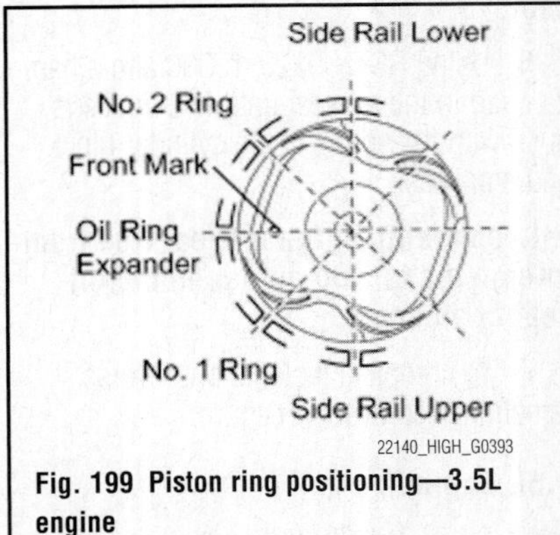

Fig. 199 Piston ring positioning—3.5L engine

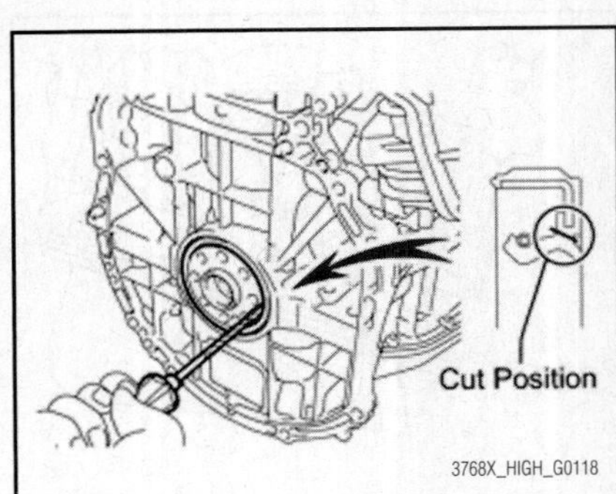

Fig. 200 Using a screwdriver, pry out the oil seal

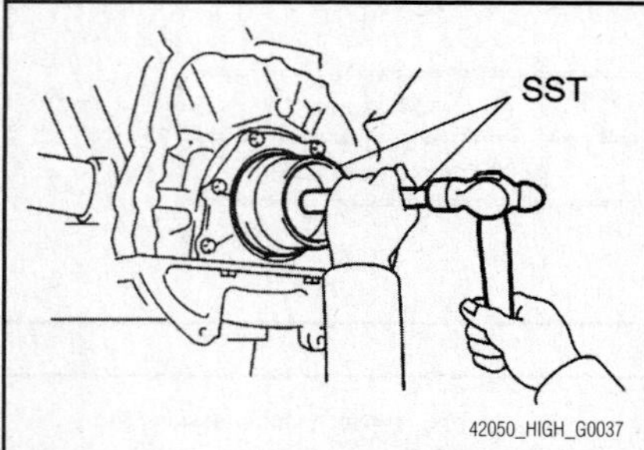

Fig. 201 Use a seal installation tool and hammer to install the rear main seal until its surface is flush with the seal retainer edge

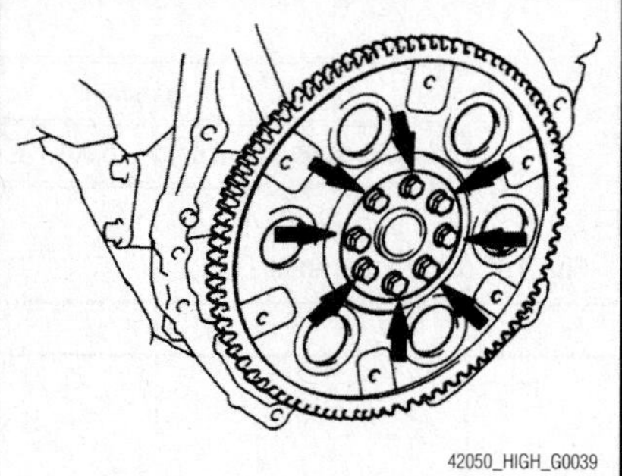

Fig. 203 Remove the 8 bolts, rear spacer, drive plate and front spacer

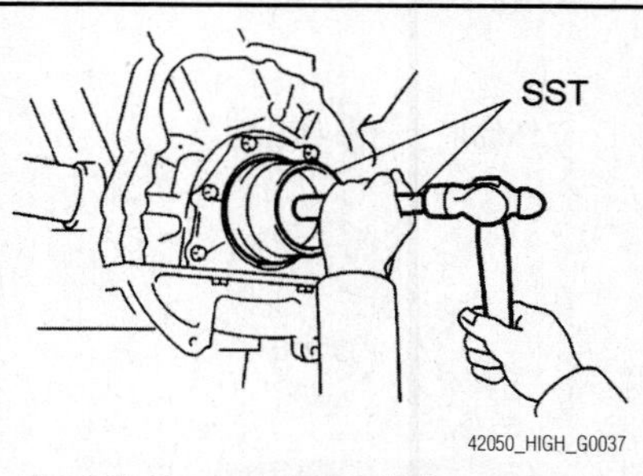

Fig. 205 Use a seal installation tool and hammer to install the rear main seal until its surface is flush with the seal retainer edge

➡Tape the screwdriver tip before use. Do not damage the surface of the oil seal press fit hole or the crankshaft.

4. Remove rear engine oil seal:
   a. Using a knife, cut off the lip of the oil seal.
   b. Using a screwdriver, pry out the oil seal.

To install:

5. Apply MP grease to the lip of a new oil seal.

➡Do not allow foreign matter to contact the lip of the oil seal. Do not allow MP grease to contact the dust seal.

6. Using SST: 09223-15030 and a hammer, tap in the oil seal until its surface is flush with the edges of the cylinder block and crankcase.

➡Keep the lip of the oil seal free from foreign matter. Do not tap in the oil seal at an angle.

7. To complete installation, reverse remaining removal procedure.

3.5L Engines

See Figures 202 through 206.

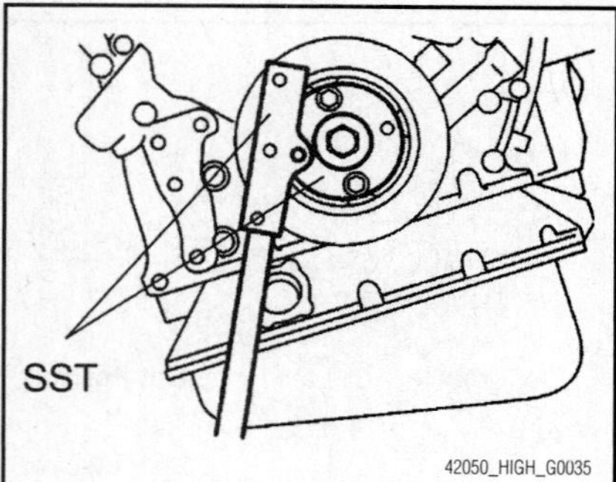

Fig. 202 Hold the crankshaft in place with the special tools as shown

➡This procedure requires a variety of special tools.

1. Remove the automatic transaxle assembly (2WD) or automatic transaxle and transfer assembly (4WD), as outlined in the Drive Train Section.
2. Remove the drive plate and ring gear, as follows:
   a. Secure the crankshaft with Special Service Tool (SST) 09213-54015 (91651-60855), 09330-00021 or their equivalents.
   b. Remove the 8 bolts, rear spacer, drive plate and front spacer.
3. Remove the rear main seal, as follows:
   a. Using a knife, carefully cut off the oil seal lip.
   b. Use a suitable prytool, or taped screwdriver, pry out the oil seal.

⁂ WARNING

After removing the seal, make sure the crankshaft is not damaged or scratched. If it is, you can mend it with a fine grit (No. 400) sandpaper.

To install:

4. Install engine rear oil seal, as follows:
   a. Apply suitable grease to a new oil seal lip. Make sure to keep the lip away from foreign materials to avoid picking up contamination or debris.

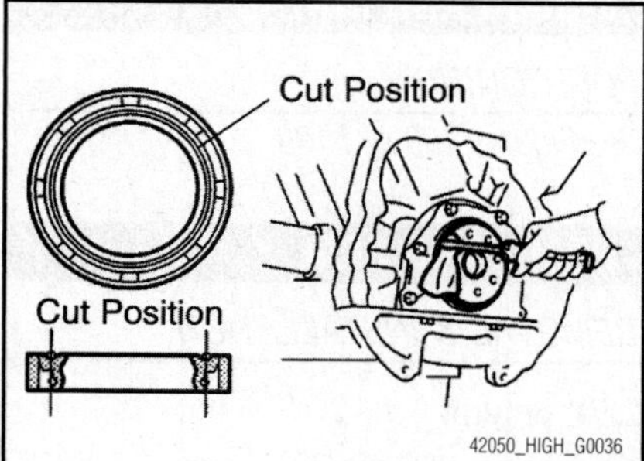

Fig. 204 If using a screwdriver wrapped in tape to remove the rear main seal, be very carefully not to damage the crankshaft

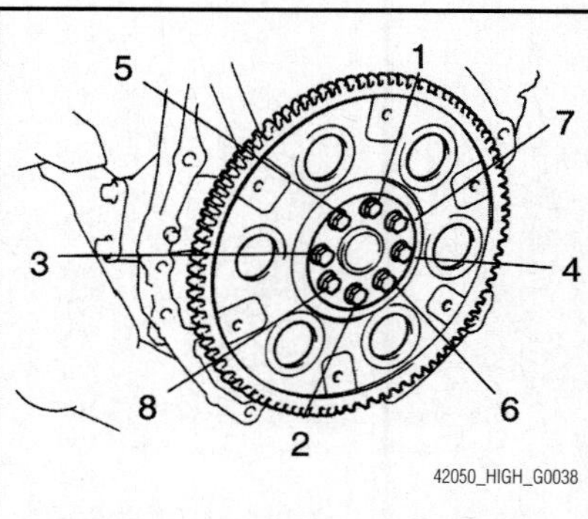

Fig. 206 Drive plate bolt tightening sequence—3.3L engines

   b. Using SST 09223-15030, 09950-70010 (09951-07100), or their equivalent, and a hammer, tap in the oil seal until its surface is flush with the rear oil seal retainer edge. Tap the seal in squarely to make sure it seats properly

➡Wipe the extra grease off of the crankshaft.

5. Install drive plate and ring gear, as follows:
   a. Fix the crankshaft with SST 09213-54015 (91651-60855), 09330-00021, or their equivalents.
   b. Clean the bolts and the bolt holes.
   c. Apply a suitable adhesive (part no. 08833-00070, three bond or equivalent) to 2 or 3 threads of the bolt end.
   d. Install and uniformly tighten the 8 bolts, in several passes in the sequence shown in the accompanying illustration to a final torque of 61 ft. lbs. (83 Nm).

⁂ WARNING

Do not start the engine for AT LEAST one hour after installing the seal!

6. Install the automatic transaxle assembly (2WD) or automatic transaxle and transfer assembly (4WD), as outlined in the Drive Train Section.

TIMING CHAIN FRONT COVER

REMOVAL & INSTALLATION

Refer to the Timing Chain & Sprocket procedure.

TIMING CHAIN & SPROCKETS

REMOVAL & INSTALLATION

2.7L Engine

See Figures 207 through 222.

➡Do not remove the oil pump or oil pump relief valve from the timing chain cover sub-assembly.

1. Remove the engine and transaxle.
2. Remove engine wire.
3. Remove ignition coil assembly.
4. Remove the valve cover.
5. Remove crankshaft position sensor.
6. Remove crankshaft pulley.
7. Remove the 5 bolts and engine mounting bracket RH.
8. Remove v-ribbed belt tensioner assembly.
9. Remove timing chain cover sub-assembly:
    a. Remove the 17 bolts and 2 nuts.
    b. Remove the timing chain cover by prying between the timing chain cover and cylinder head, camshaft housing, cylinder block and stiffening crankcase with a screwdriver as shown in the illustration.
    c. Remove the 3 gaskets from the stiffening crankcase.
10. Remove the 4 bolts, timing chain cover plate and gasket.
11. Set no. 1 cylinder to TDC/compression:
    a. Temporarily install the crankshaft pulley bolt.
    b. Rotate the crankshaft clockwise so that the timing marks on the crankshaft timing gear and camshaft timing gears are as shown in the illustration.
    c. If the timing marks do not align, rotate the crankshaft clockwise again and align the timing marks.
    d. Remove the crankshaft pulley bolt.
12. Remove the bolt and timing chain guide.
13. Remove no. 1 chain tensioner assembly:
    a. Allow the plunger to extend slightly, and then rotate the stopper plate

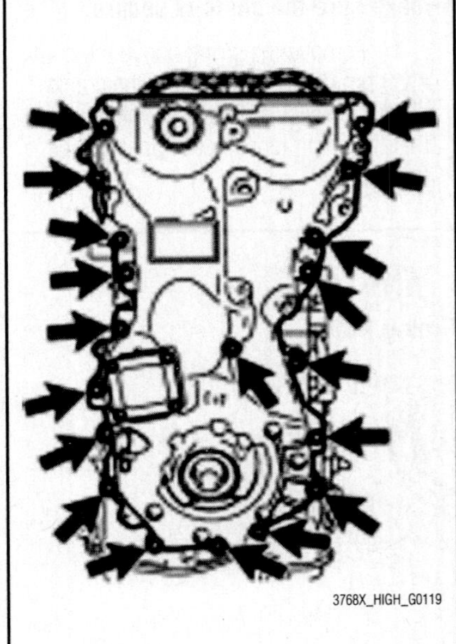

Fig. 207 Remove the 17 bolts and 2 nuts

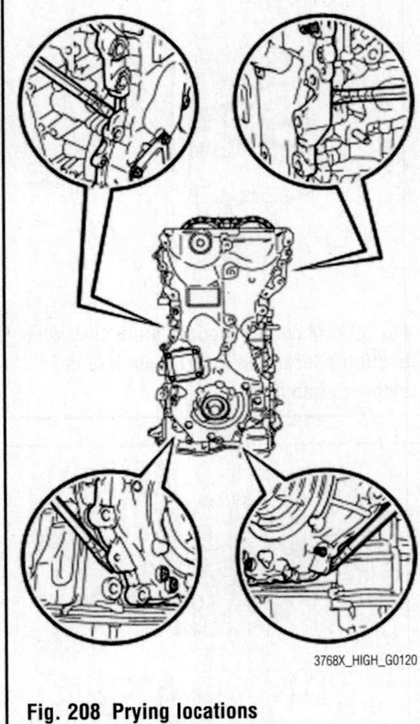

Fig. 208 Prying locations

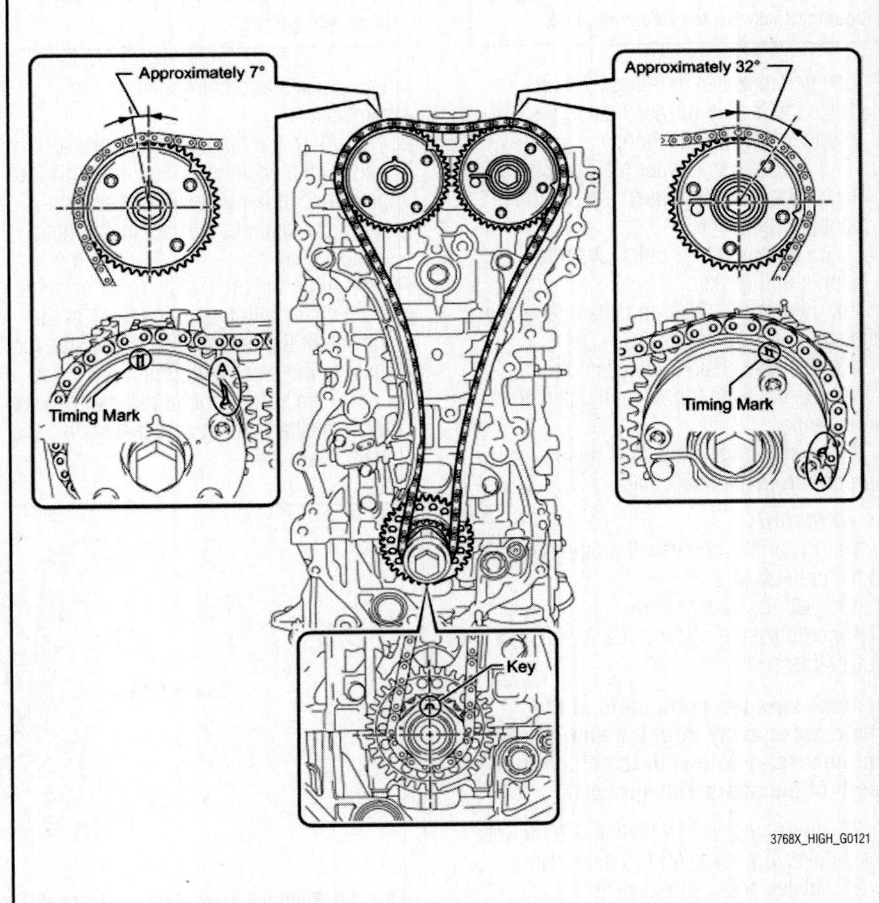

Fig. 209 Set no. 1 cylinder to TDC/compression—"A" is not a timing mark.

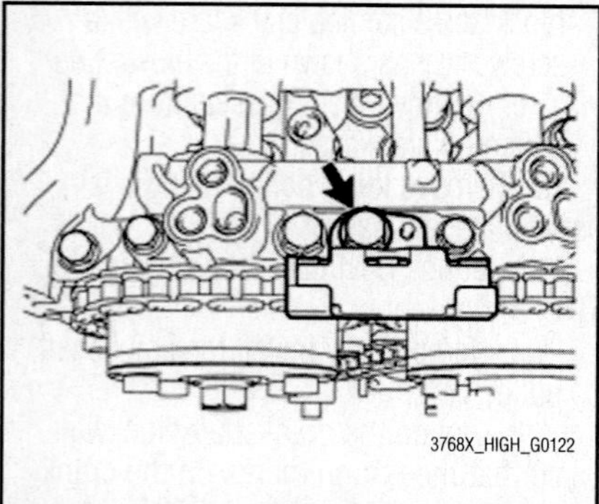

Fig. 210 Remove the bolt and timing chain guide

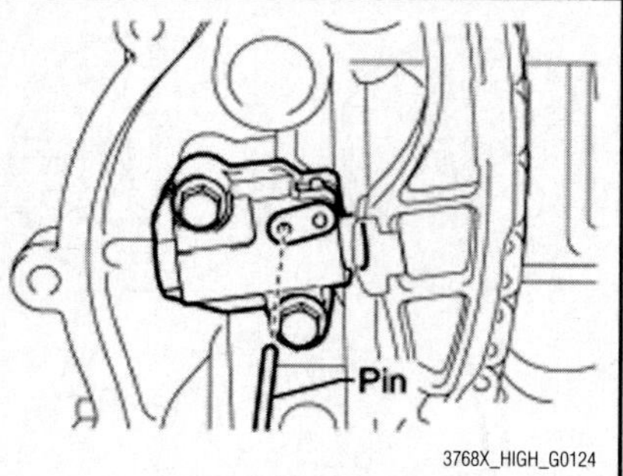

Fig. 212 Move the stopper plate clockwise to set the lock, and insert a pin into the stopper plate hole

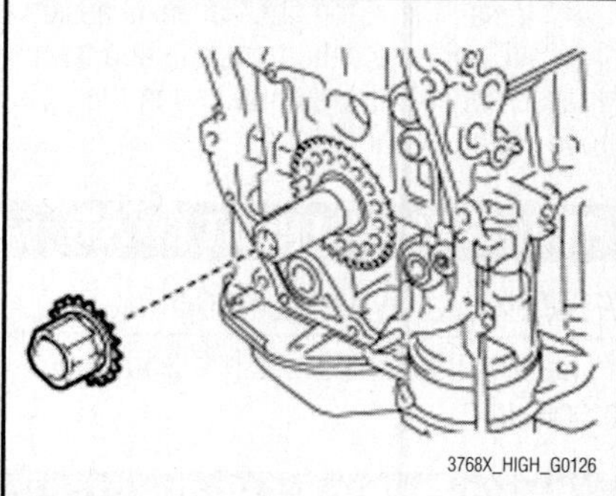

Fig. 214 Remove the crankshaft timing sprocket from the crankshaft

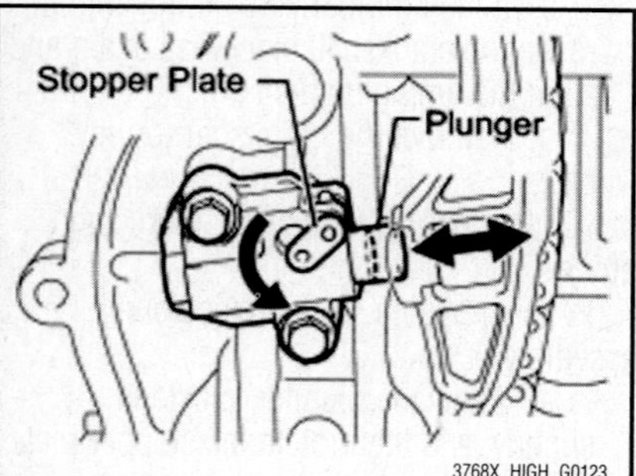

Fig. 211 Allow the plunger to extend slightly, and then rotate the stopper plate counterclockwise to release the lock

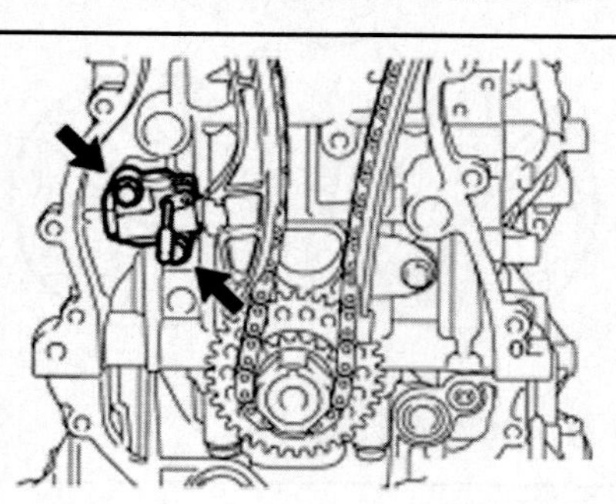

Fig. 213 Remove the 2 bolts, chain tensioner and gasket

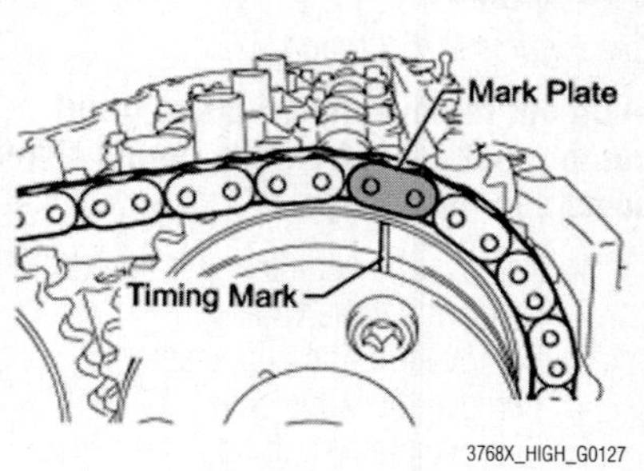

Fig. 215 Align the mark plate (yellow or gold) of the chain with the timing mark of the camshaft timing exhaust gear

counterclockwise to release the lock. Once the lock is released, push the plunger into the tensioner.

b. Move the stopper plate clockwise to set the lock, and insert a pin into the stopper plate hole.

c. Remove the 2 bolts, chain tensioner and gasket.

14. Remove the bolt and chain tensioner slipper.

15. Remove chain sub-assembly.

16. Remove the 2 bolts and chain vibration damper.

17. Remove the crankshaft timing sprocket from the crankshaft.

To install:

18. Install the crankshaft timing sprocket to the crankshaft.

19. Set no. 1 cylinder to TDC/compression. Refer to Camshaft and Valve Lifters.

➡Make sure the mark plate of the chain faces away from the engine. It is not necessary to install the chain to the teeth of the gears and sprocket.

20. Install the chain vibration damper with the 2 bolts. Tighten to 15 ft. lbs. (21 Nm).

21. Install chain sub-assembly:

a. Place the chain onto the camshaft

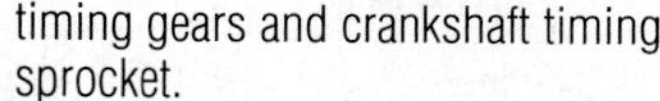

timing gears and crankshaft timing sprocket.

b. Align the mark plate (yellow or gold) of the chain with the timing mark of the camshaft timing exhaust gear and install the chain to the camshaft timing exhaust gear.

c. Align the mark plate (pink or gold) of the chain with the timing mark of the crankshaft timing sprocket and install the chain to the crankshaft timing sprocket.

d. Tie a string around the chain above the crankshaft timing sprocket so that the chain is secure.

e. Using the hexagonal portion of the intake camshaft, rotate the intake camshaft counterclockwise with a wrench, align the timing mark of the camshaft timing gear with the mark plate (yellow or gold) of the chain and install the chain to the camshaft timing gear.

➡Make sure the chain is secure.

f. Remove the string above the crankshaft timing sprocket, rotate the crankshaft clockwise, and loosen the chain so that the chain tensioner slipper can be installed.

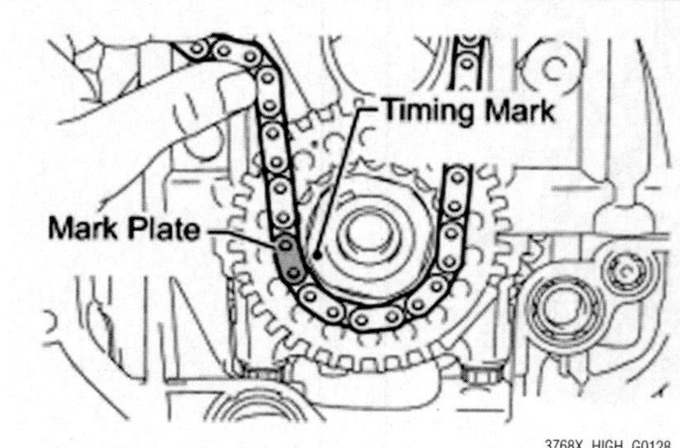

Fig. 216 Align the mark plate (pink or gold) of the chain with the timing mark of the crankshaft timing sprocket

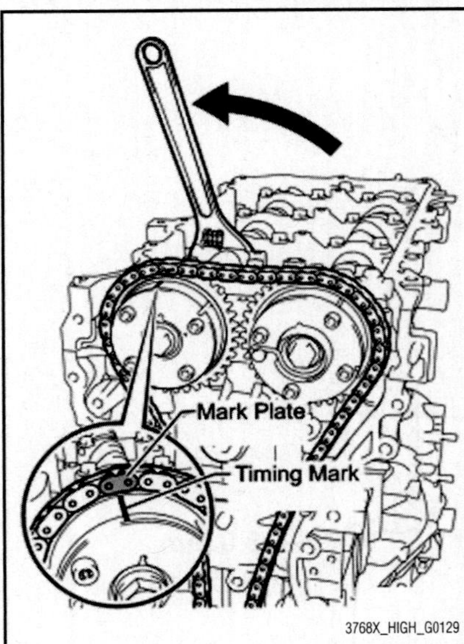

Fig. 217 Rotate the intake camshaft counterclockwise with a wrench

22. Install the chain tensioner slipper with the bolt. Tighten to 15 ft. lbs. (21 Nm).
23. Install a new gasket and the chain tensioner with the 2 bolts. Tighten to 7 ft. lbs. (10 Nm). Remove the pin from the stopper plate.
24. Install the timing chain guide with the bolt Tighten to 15 ft. lbs. (21 Nm).
25. Check no. 1 cylinder TDC/compression:
    a. Temporarily install the crankshaft pulley bolt.
    b. Rotate the crankshaft clockwise, and check that the timing marks on the crankshaft timing sprocket and camshaft timing gears are as shown in the illustration.
    c. Remove the crankshaft pulley bolt.
26. Install timing chain cover sub-assembly:
    a. Apply a light coat of engine oil to 3 new gaskets.

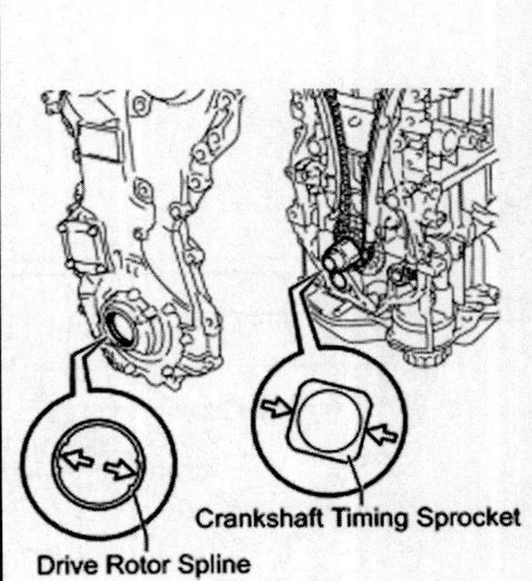

Fig. 219 Align the drive rotor spline and the crankshaft timing sprocket

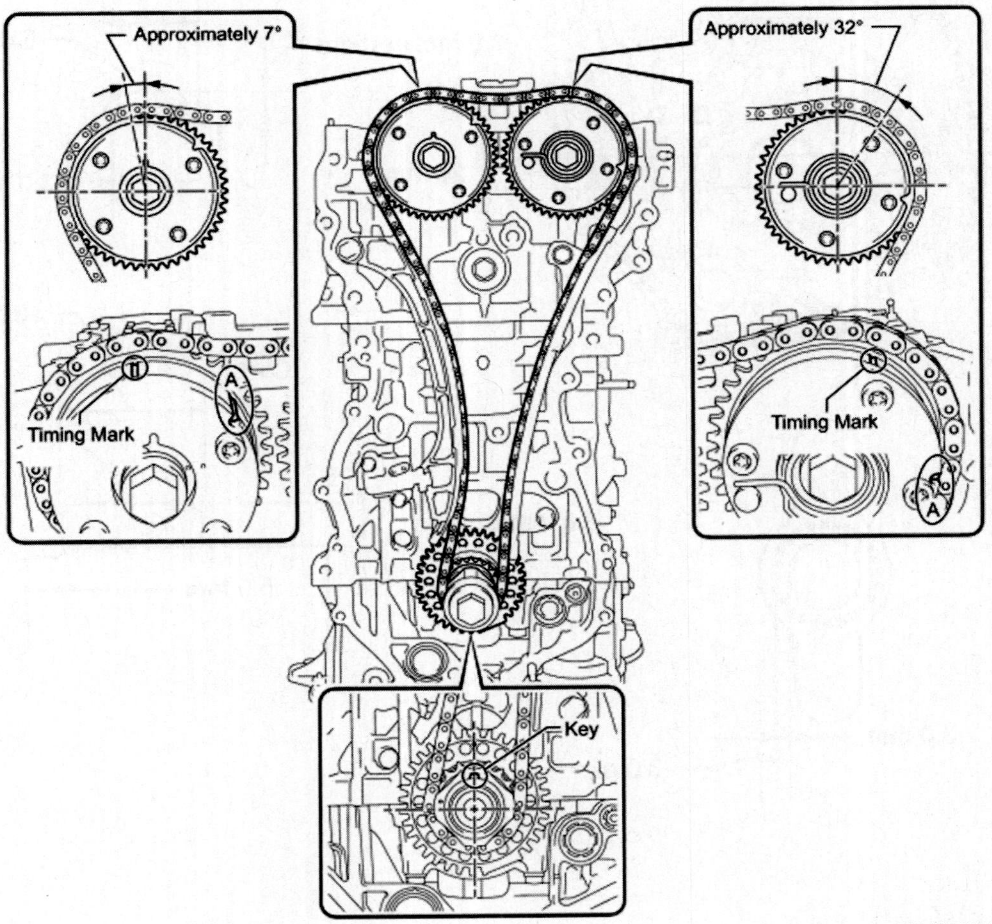

Fig. 218 Set no. 1 cylinder to TDC/compression—"A" is not a timing mark.

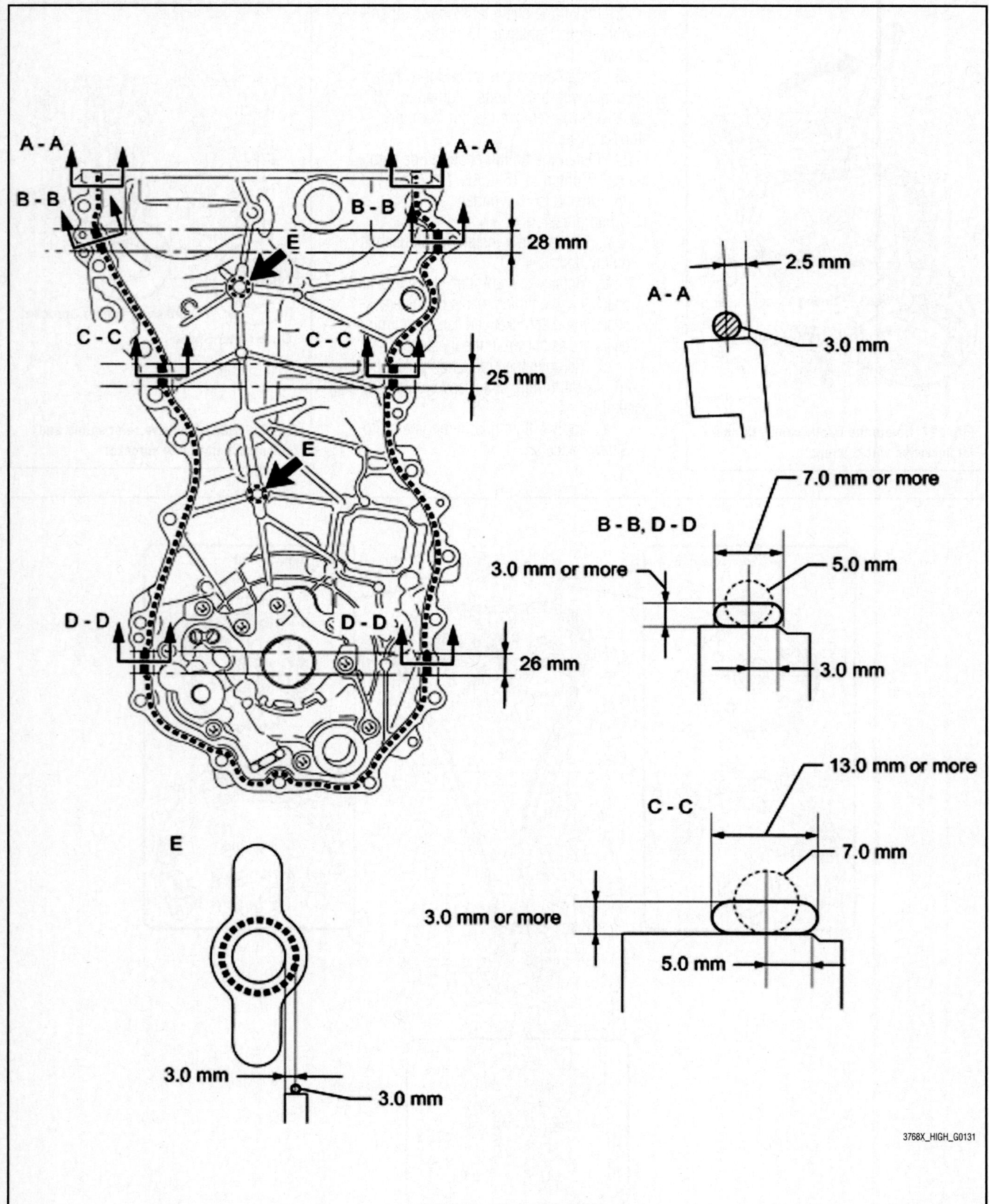

Fig. 220 Apply seal packing in a line to the timing chain cover as shown

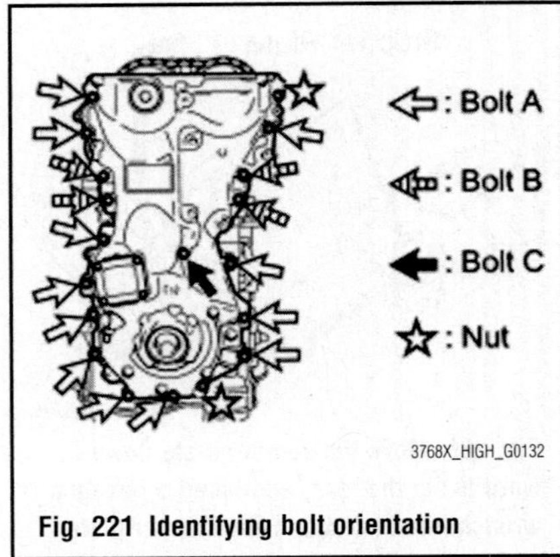

Fig. 221 Identifying bolt orientation

b. Install the 3 gaskets to the stiffening crankcase.

c. Align the drive rotor spline and the crankshaft timing sprocket as shown in the illustration.

d. Apply seal packing in a line to the timing chain cover as shown in the following illustration. Seal packing: Toyota Genuine Seal Packing Black, Three Bond 1207B or equivalent.

➡When the contact surfaces are wet, clean the surfaces with non-residue solvent before applying seal packing. Install the timing chain cover within 3 minutes and tighten the bolts within 10 minutes after applying seal packing. After applying seal packing to the timing chain cover, install the engine mounting bracket within 10 minutes. Do not apply oil for at least 4 hours after the installation. Do not start the engine for at least 4 hours after the installation.

e. Temporarily install the timing chain cover with the 17 bolts and 2 nuts. Bolt lengths:

- Bolt A: Length: 30 mm (1.18 in.). Thread Diameter: 8 mm (0.315 in.)

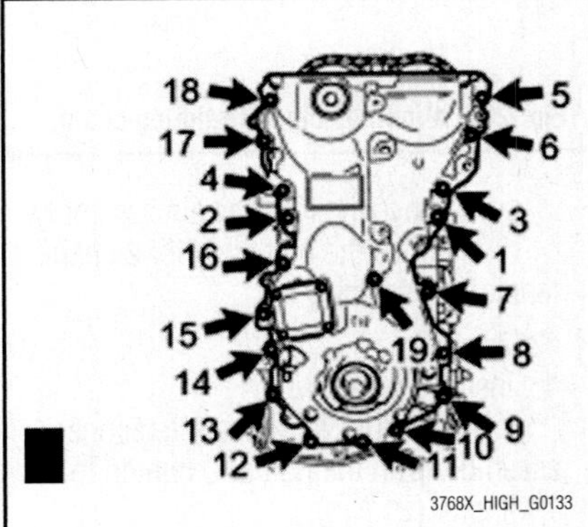

Fig. 222 Timing cover bolt and nut tightening sequence

- Bolt B: Length: 35 mm (1.38 in.). Thread Diameter: 10 mm (0.394 in.)
- Bolt C: Length: 45 mm (1.77 in.). Thread Diameter: 8 mm (0.315 in.)

f. Tighten the 17 bolts and 2 nuts in several steps, in the sequence shown in the illustration.

27. Install the engine mounting bracket, and install the 5 bolts in the order shown in the illustration:

a. For bolt 1, 2 and 3, tighten to 41 ft. lbs. (55 Nm).

b. For bolt 4 and 5, tighten to 15 ft. lbs. (21 Nm).

➡After applying seal packing to the timing chain cover, install the engine mounting bracket within 10 minutes.

3.5L Engine

See Figures 223 through 234.

1. Remove the engine assembly.
2. Install on engine stand.
3. The following must be removed:

a. Remove ignition coil assembly.

b. Remove the right hand No. 2 engine mounting stay.

c. Remove the intake manifold.

d. Remove the right hand exhaust manifold sub-assembly.

e. Remove the No. 2 engine oil level dipstick guide.

f. Remove the No. 2 manifold stay.

g. Remove the No. 2 exhaust manifold heat insulator.

h. Remove the left hand exhaust manifold sub-assembly.

i. Remove the transverse engine mounting bracket.

j. Remove the alternator assembly.

k. Remove the V-ribbed belt tensioner assembly.

l. Remove the No. 2 timing gear cover.

m. Remove the No. 2 idler pulley sub-assembly.

n. Remove the left hand No. 1 engine front mounting bracket.

o. Remove the left hand 6 bolts and No. 1 front engine mounting bracket.

p. Remove the radio setting condenser.

q. Remove the No. 1 vacuum switching valve.

r. Remove the knock control sensor wire.

s. Remove the knock control sensor.

t. Remove the crankshaft position sensor.

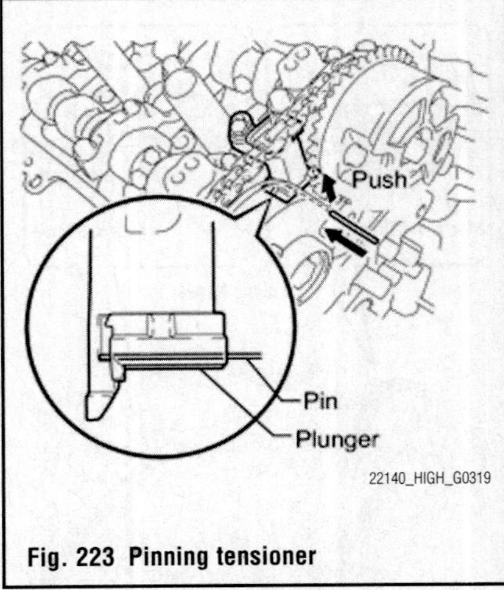

Fig. 223 Pinning tensioner

u. Remove the No. 1 oil pipe.

v. Remove the oil pipe.

w. Remove the crankshaft pulley.

x. Remove the oil cooler assembly, if necessary.

y. Remove the No. 1 oil cooler bracket, if necessary.

z. Remove the water inlet housing.

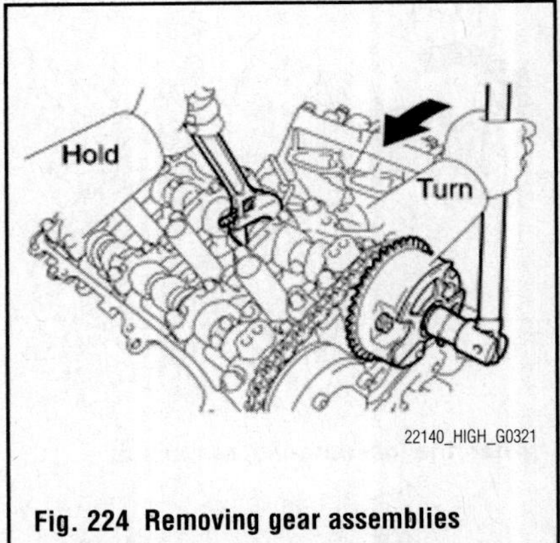

Fig. 224 Removing gear assemblies

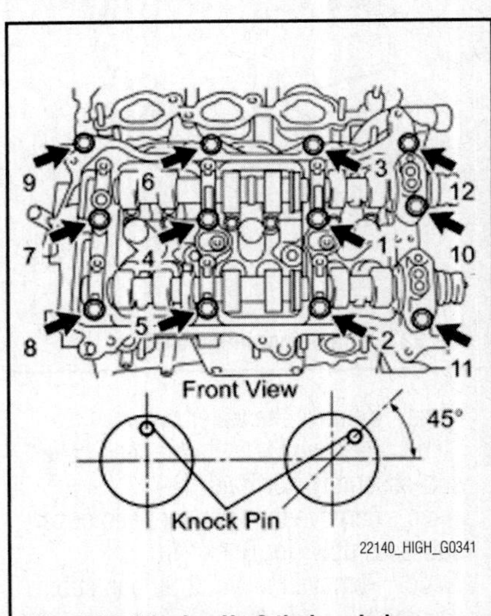

Fig. 225 Aligning No 2 timing chain

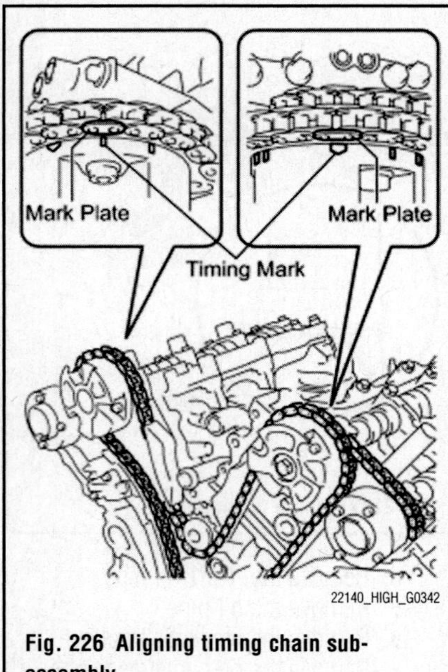

Fig. 226 Aligning timing chain sub-assembly

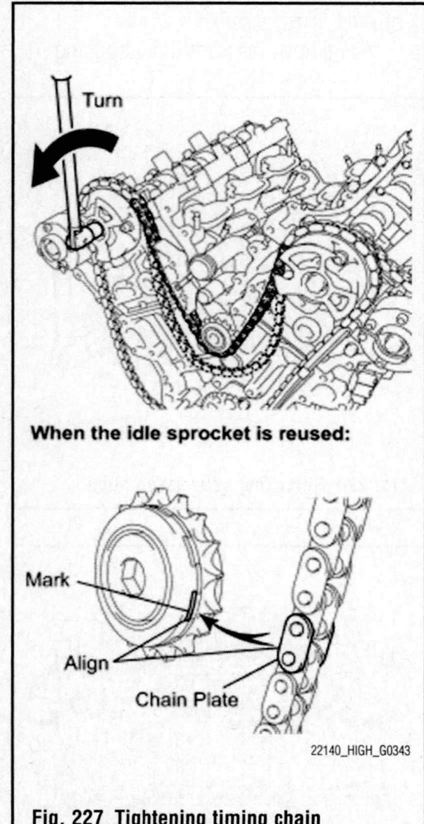

Fig. 227 Tightening timing chain

aa. Remove the water outlet.

bb. Remove the cylinder head cover sub-assembly (for Bank 1).

cc. Remove the cylinder head cover sub-assembly (for Bank 2).

dd. Remove the No. 2 oil pan sub-assembly.

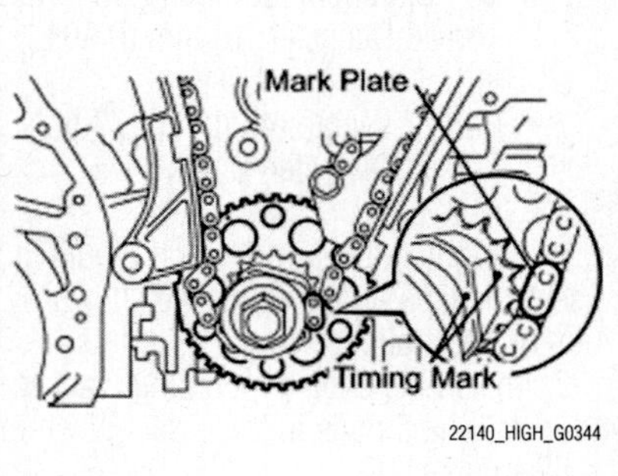

Fig. 228 Installing timing chain on crankshaft

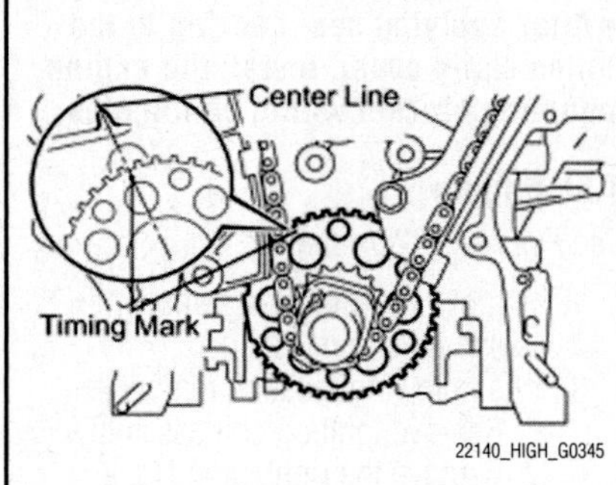

Fig. 229 Aligning timing chain on crankshaft

ee. Remove the oil strainer sub-assembly.

ff. Remove the oil pan sub-assembly.

gg. Remove the timing chain cover sub-assembly

hh. Remove the timing chain case oil seal.

ii. Set the No. 1 cylinder to TDC/compression.

jj. Remove the No. 1 chain tensioner assembly.

kk. Remove the chain tensioner slipper.

ll. Remove the chain sub-assembly.

mm. Remove the idle sprocket assembly.

nn. Remove the camshaft timing gears and No. 2 chain (for Bank 1).

oo. While raising the No. 2 chain tensioner assembly, insert a pin of 1.0 mm (0.039 in.) diameter into the hole to fix the No. 2 chain tensioner assembly.

pp. Hold the hexagonal portion of the camshaft with a wrench, and remove the 2 bolts and 2 camshaft timing gear assemblies.

➡Be careful not to damage the cylinder head with the wrench. Do not disassemble the camshaft timing gear assemblies.

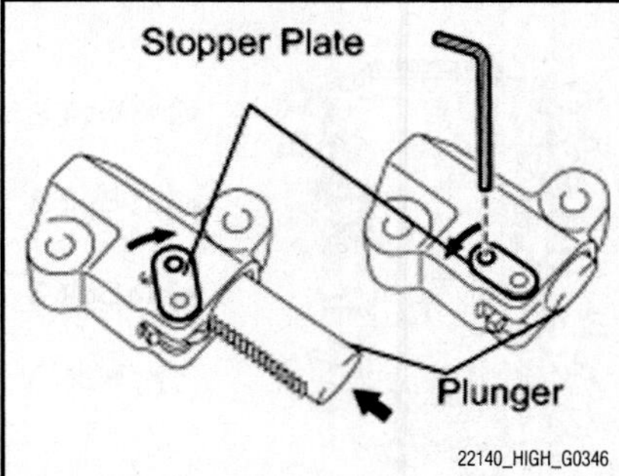

Fig. 230 Move the stopper plate downward to set the loss, and insert a hexagon wrench into the hole of the stopper plate

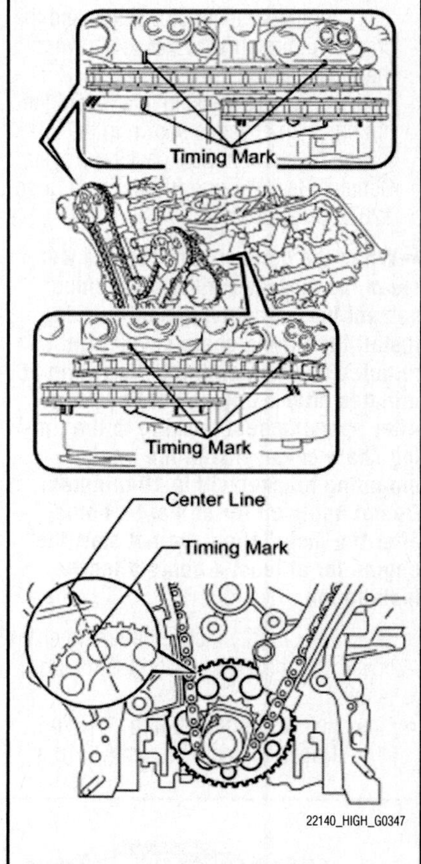

Fig. 231 Aligning complete timing chain

qq.Remove the No. 2 chain assembly.

rr. Remove the bolt and No. 2 chain tensioner assembly.

To install:

4. Install the following:

a. Install the No. 2 chain tensioner assembly with the bolt and tighten to 15 ft. lbs. (21 Nm).

b. While pushing in the tensioner, insert a pin of 1.0 mm (0.039 in.) diameter into the hole to fix it.

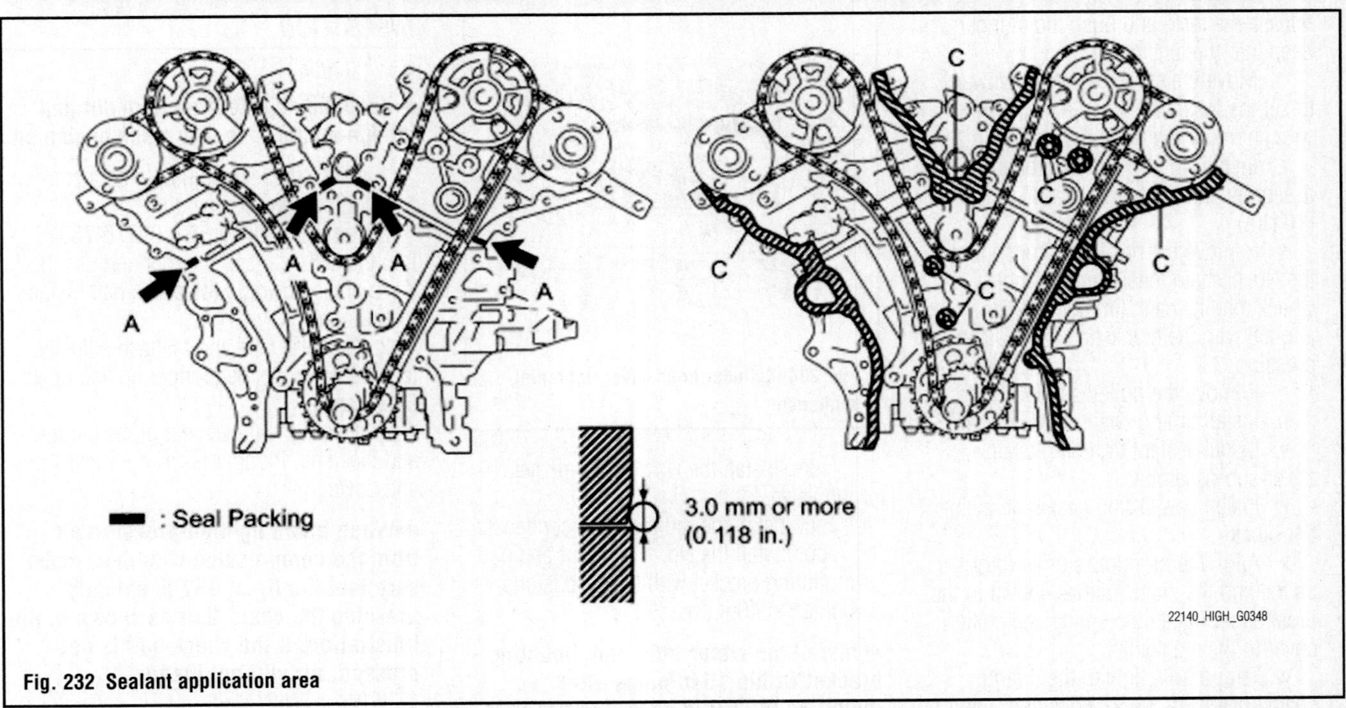

Fig. 232 Sealant application area

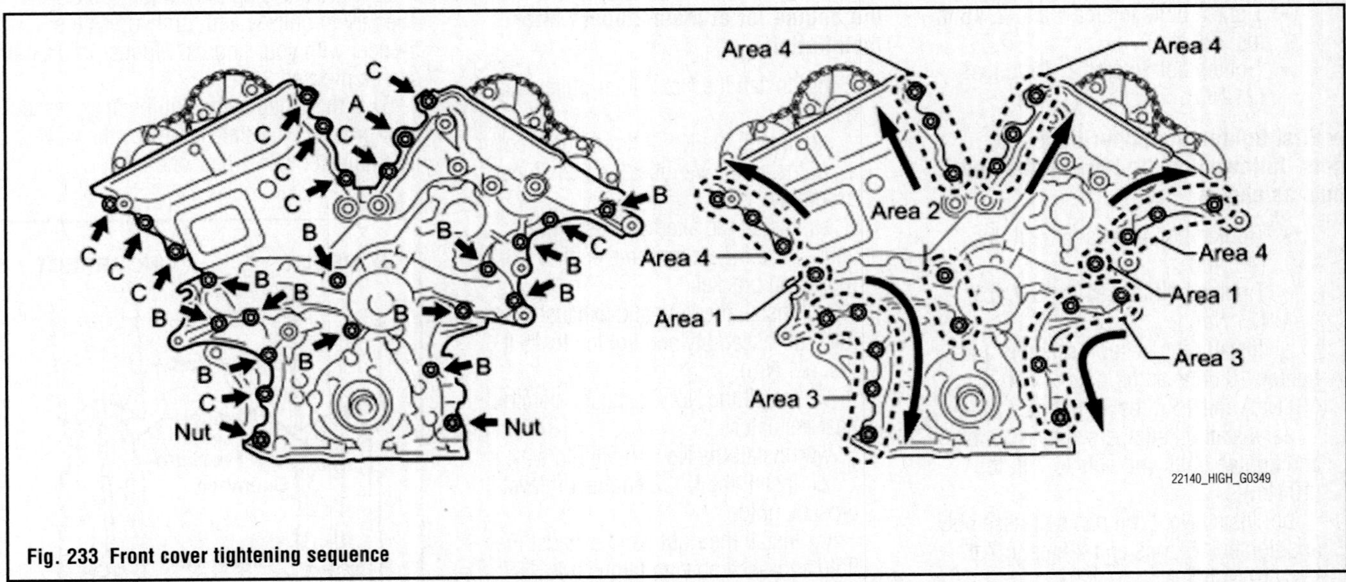

Fig. 233 Front cover tightening sequence

c. Align the mark plate with the timing marks of the camshaft timing gear as shown.

d. Apply a light coat of engine oil to the bolt threads and bolts seating surface.

e. Align the knockpin of the camshaft with pinhole of the camshaft timing gear assembly. Install the camshaft timing gear assembly and camshaft timing exhaust timing gear assembly and camshaft timing exhaust gear with the No. 2 chain sub-assembly installed.

f. Hold the hexagonal portion of the camshaft with the wrench and tighten the two bolts and camshaft timing gear assemblies to 74 ft. lbs. (100 Nm).

g. Remove the pan from the No 2 chain tensioner assembly.

h. Install idle sprocket assembly and tighten to 44 ft. lbs. (60 Nm).

i. Install chain sub-assembly.

j. Align the mark plate and timing marks as shown in the illustration and install the chain.

➡The camshaft mark plates are orange.

k. Do not pass the chain over the crankshaft, just temporarily place it on the crankshaft.

l. Turn the camshaft timing gear assembly on bank 1 counterclockwise to tighten the chain between the banks.

➡When the idle sprocket assembly is reused, align the timing chain plate with the mark where the plate has been in order to tighten the chain between the banks.

m. Align the mark plate and timing marks as shown in the illustration and install the chain onto the crankshaft timing sprocket. The crankshaft to mark plate is yellow.

n. Turn the crankshaft clockwise to set it to the right-hand block bore more centerline. (TDC Compression).

o. Install chain tensioner slipper.

p. Move the stopper plate upward to

release the lock, and push the plunger deep into the tensioner.

q. Move the stopper plate downward to set the loss, and insert a hexagon wrench into the hole of the stopper plate.

r. Install No. 1 chain tensioner assembly and tighten bolts to 7 ft. lbs. (10 Nm).

s. Remove the hexagon wrench from the No. 1 chain tensioner assembly. Check that the each timing mark is aligned with the crankshaft at TDC compression.

t. Remove the pulley set bolt.

u. Install timing chain case oil seal.

v. Install sealant to timing chain cover sub-assembly.

w. Install new O ring gasket on cylinder block.

x. Align the oil pump's drive rotor spline and the crankshaft as shown in the illustration. Install the spline and chain cover to the crankshaft.

y. Temporarily tighten the timing chain cover with the 23 bolts and 2 nuts.

- Tighten bolts in area 1 and 2: 15 ft. lbs. (21 Nm).
- Tighten bolt in area 3: 15 ft. lbs. (21 Nm).

➡First tighten the upper bolts and nuts, followed by the lower bolts and nuts as shown.

- Tighten bolt in area 4: 32 ft. lbs. (43 Nm)
- Tighten bolt in area 4: 15 ft. lbs. (21 Nm)

z. Install oil pan sub-assembly and tighten 16 bolts and 2 nuts to 7 ft. lbs. (10 Nm) and 15 ft lbs. (21 Nm).

aa. Install oil strainers sub-assembly and tighten bolts and nuts to 7 ft. lbs. (10 Nm).

bb. Install No. 2 oil pan sub-assembly and tighten 16 bolts and 2 nuts to 7 ft. lbs. (10 Nm).

cc. Install cylinder head cover sub-assemblies and tighten to 7 ft. lbs. (10 Nm) and 15 ft. lbs. (21 Nm).

dd. Install the water inlet housing.

ee. Install No. 1 oil cooler bracket (w/ oil cooler).

ff. Install the oil cooler assembly (w/ oil cooler).

gg. Install the crankshaft pulley.

hh. Install the oil pipe.

ii. Install the No. 1 oil pipe.

jj. Install the crankshaft position sensor.

kk. Install the knock control sensor.

ll. Install the knock control sensor wire.

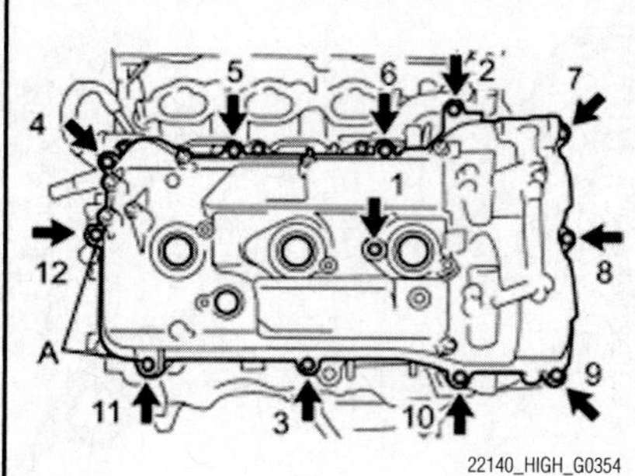

Fig. 234 Cylinder head cover tightening sequence

mm. Install the No. 1 vacuum switching valve.

nn. Install the radio setting condenser.

oo. Install the No. 1 left front engine mounting bracket with the 6 bolts and tighten to 40 ft. lbs. (54 Nm).

➡Install the water inlet and mounting bracket within 15 minutes after installing the chain cover. Do not start the engine for at least 2 hours after installation.

pp. Install the No. 2 idler pulley sub-assembly.

qq. Install the No. 2 timing gear cover

rr. Install the V-ribbed belt tensioner assembly.

ss. Install the alternator assembly.

tt. Install the transverse engine mounting bracket.

uu. Install the left hand exhaust manifold sub-assembly and tighten to 15 ft. lbs. (21 Nm).

vv. Install the No. 2 exhaust manifold heat insulator.

ww. Install the No. 2 manifold stay.

xx. Install the No. 2 engine oil level dipstick guide.

yy. Install the right hand exhaust manifold sub-assembly and tighten to 15 ft. lbs. (21 Nm).

zz. Install the intake manifold and tighten the 6 bolts and 4 nuts uniformly in several steps to 15 ft. lbs. (21 Nm).

53. Install the right hand No. 2 engine mounting stay.

54. Install the ignition coil assembly.

55. Remove the engine stand.

56. Install the engine assembly.

VALVE LASH

ADJUSTMENT

No adjustment is necessary on these engines.

INSPECTION

See Figure 235.

➡Keep the adjuster free from dirt and foreign matter. Use only clean engine oil.

1. Place the lash adjuster into a container full of new engine oil.
2. Insert the tip of SST: 09276-75010 into the lash adjuster plunger and use the tip to press down on the check ball inside the plunger.
3. Squeeze SST and the lash adjuster together to move the plunger up and down 5 to 6 times.
4. Check the movement of the plunger and bleed air. Plunger is OK if it moves up and down.

➡When bleeding high-pressure air from the compression chamber, make sure that the tip of SST is actually pressing the check ball as shown in the illustration. If the check ball is not pressed, air will not bleed.

5. After bleeding the air, remove SST. Then try to quickly and firmly press the plunger with your fingers. Plunger is OK if it can be pressed 3 times.
6. If the plunger can still be compressed after pressing it 3 times, replace the valve lash adjuster with a new one.

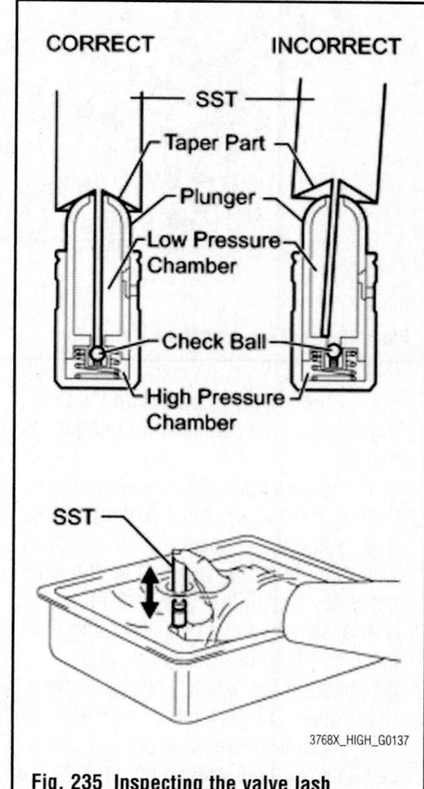

Fig. 235 Inspecting the valve lash adjuster assembly

ENGINE PERFORMANCE & EMISSION CONTROLS

ACCELERATOR PEDAL POSITION (APP) SENSOR

LOCATION

See Figure 236.

Refer to the accompanying illustration for sensor location.

REMOVAL & INSTALLATION

1. Disconnect the accelerator pedal assembly connector.
2. Remove the 2 nuts and accelerator pedal assembly.

To install:

3. Install the accelerator pedal assembly with the and tighten to 48 inch lbs. (5.4 Nm).

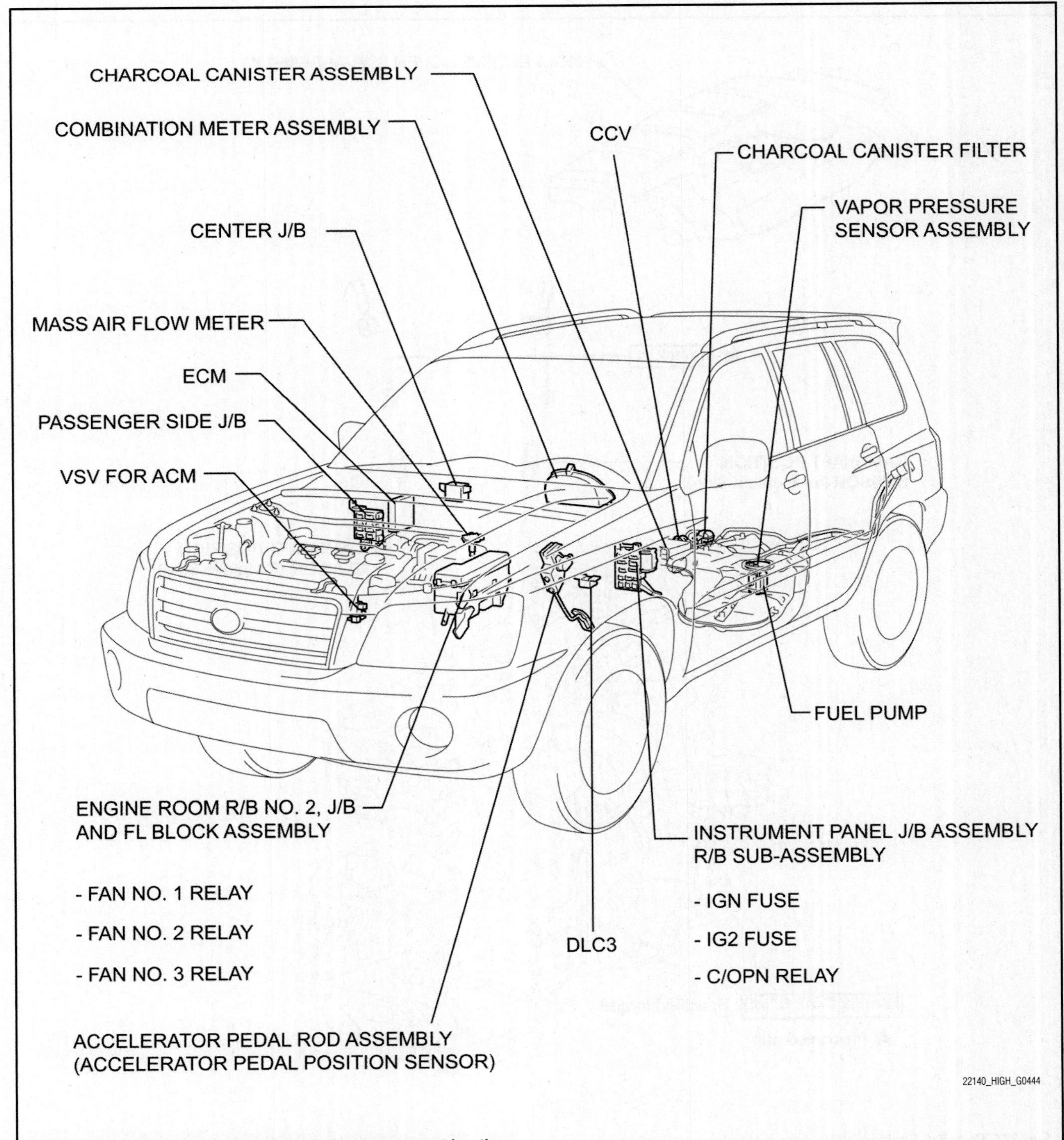

Fig. 236 Accelerator pedal assembly and related component location

4. Connect the accelerator pedal assembly connector.

CAMSHAFT POSITION (CMP) SENSOR

LOCATION

See Figures 237 and 238.

Refer to the accompanying illustrations for sensor location.

REMOVAL & INSTALLATION

2.7L Engine

1. Disconnect the negative battery cable.
2. Remove no. 1 engine cover sub-assembly.
3. Remove Camshaft Position (CMP) sensor (for exhaust side):
   a. Disconnect the sensor connector.
   b. Remove the bolt and sensor.
4. Remove Camshaft Position (CMP) sensor (for intake side):
   a. Disconnect the sensor connector.
   b. Remove the bolt and sensor.

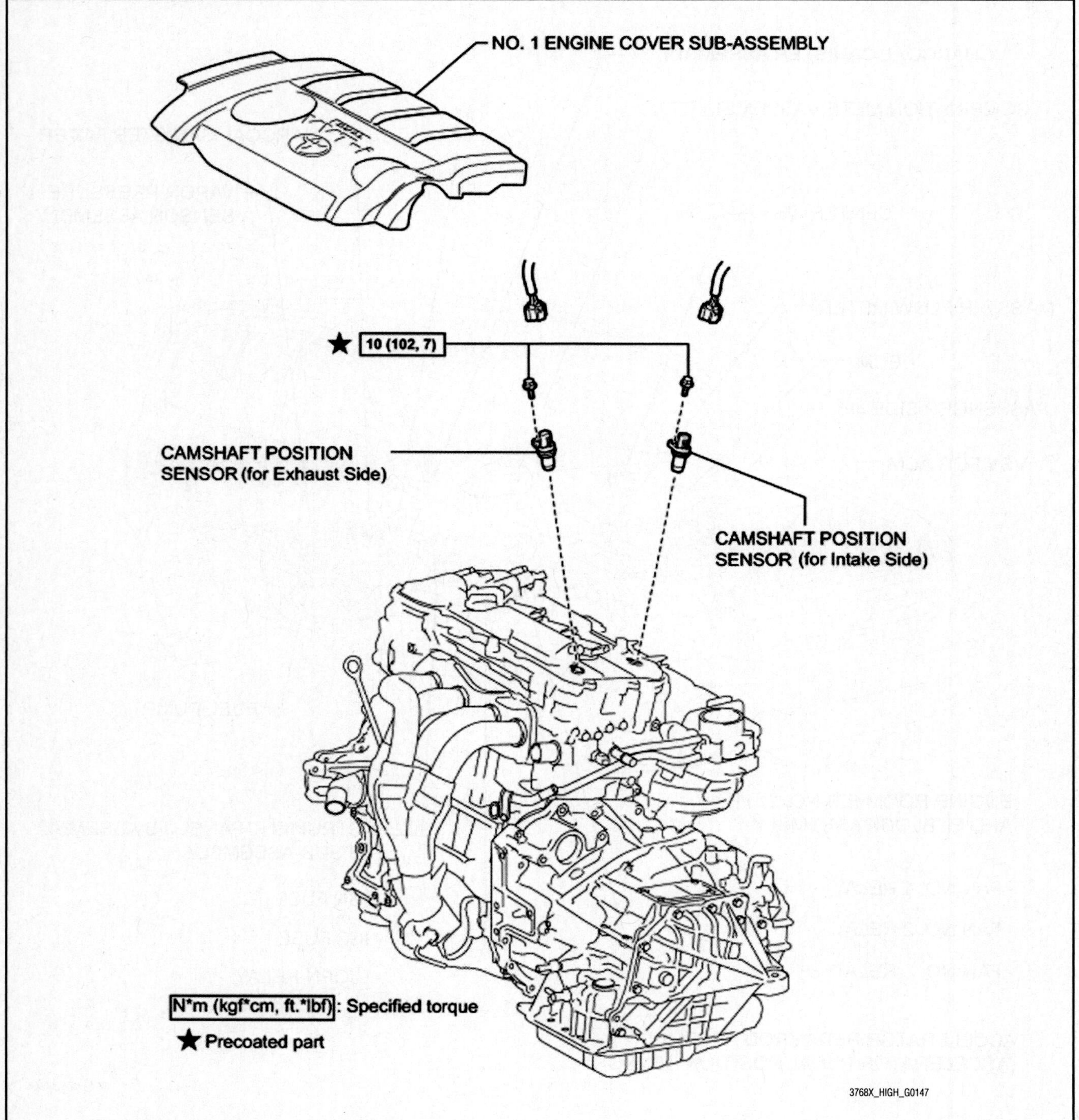

Fig. 237 Camshaft Position (CMP) sensor location—2.7L engine

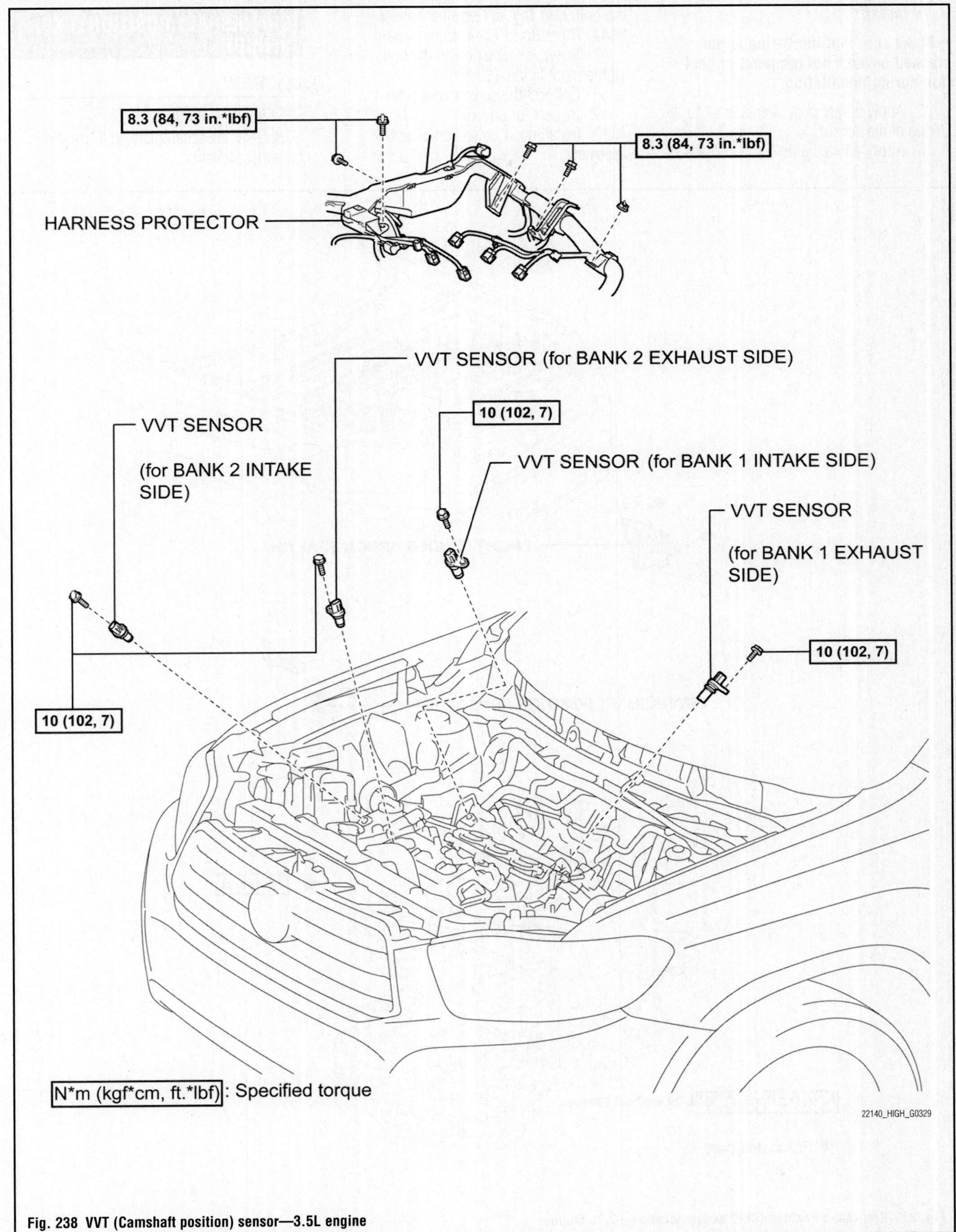

Fig. 238 VVT (Camshaft position) sensor—3.5L engine

To install:

➡Make sure that the O-ring is not cracked or does not jump out of position during installation.

5. Apply a light coat of engine oil to the O-ring of the sensor.
6. Apply adhesive to 2 or 3 threads of the bolt. Use Toyota Genuine Adhesive 1344, Three Bond 1344 or equivalent.
7. Install the sensor with the bolt. Tighten to 7 ft. lbs. (10 Nm).
8. Connect the sensor connector.
9. Inspect for oil leak.
10. Install no. 1 engine cover sub-assembly.

CRANKSHAFT POSITION (CKP) SENSOR

LOCATION

See Figures 239 and 240.

Refer to the accompanying illustrations for sensor location.

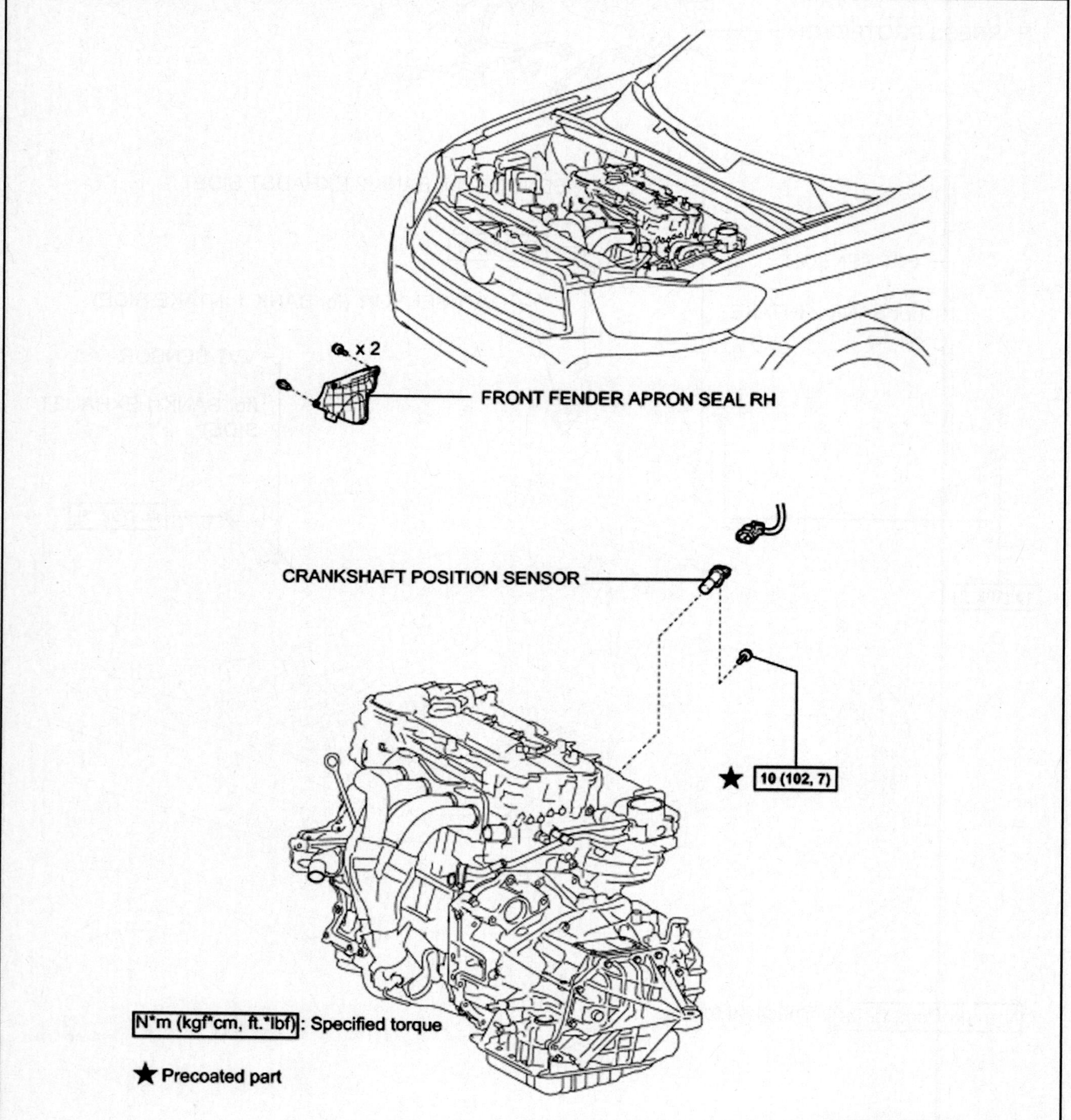

Fig. 239 Crankshaft Position (CKP) sensor location—2.7L Engine

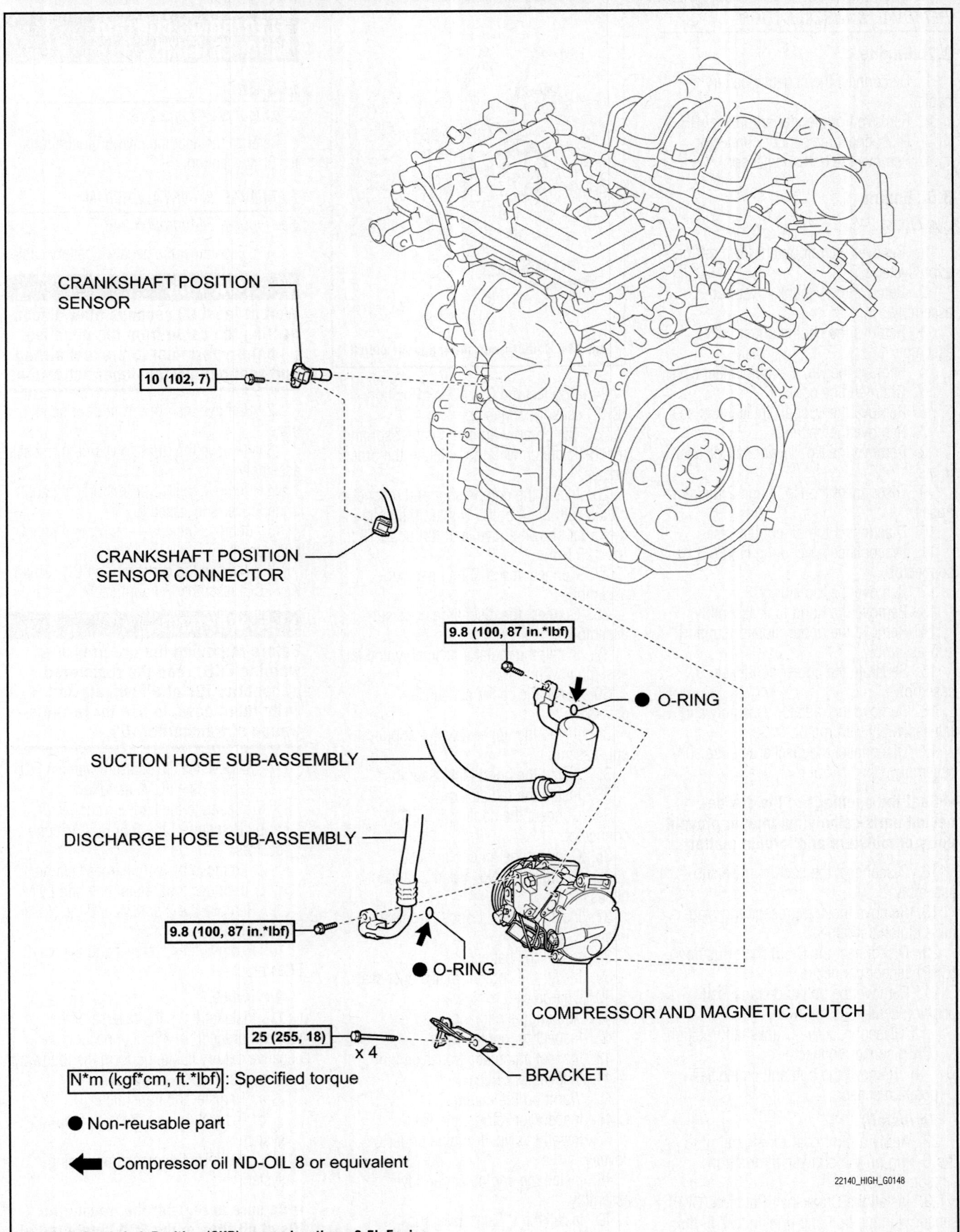

Fig. 240 Crankshaft Position (CKP) sensor location—3.5L Engine

REMOVAL & INSTALLATION

2.7L Engine

1. Disconnect the negative battery cable.
2. Remove front fender apron seal RH.
3. Disconnect the sensor connector.
4. Remove the bolt and sensor.

3.5L Engine

See Figure 242.

1. Recover the refrigerant from refrigeration system.
2. Remove the V-bank cover sub-assembly.
3. Remove the engine under cover assembly.
4. Remove the No. 1 engine under cover.
5. Drain engine coolant.
6. Remove the cool air intake duct seal.
7. Remove the battery.
8. Remove the No. 1 and 2 air cleaner inlets.
9. Disconnect the No. 1 and 2 radiator hoses.
10. Disconnect the oil cooler hose.
11. Disconnect the cooling fan ECU connector.
12. Remove the radiator grill.
13. Remove the hood lock assembly.
14. Remove the upper radiator support sub-assembly.
15. Separate the cooler condenser assembly.
16. Remove the radiator assembly and fan assembly with motor.
17. Disconnect the discharge hose sub-assembly.

➡Seal the openings of the disconnected parts using vinyl tape to prevent entry of moisture and foreign matter.

18. Disconnect the suction hose sub-assembly.
19. Remove the compressor and magnetic clutch as follows:
20. Disconnect the Crankshaft Position (CKP) sensor connector.
21. Remove the bolt and crankshaft position sensor, as follows:
    a. Disconnect the Crankshaft Position (CKP) sensor connector.
    b. Remove the bolt and crankshaft position sensor.

To install:

22. Apply a light coat of engine oil to the O-ring on the crankshaft position sensor.
23. Install the Crankshaft Position (CKP) sensor with the bolt and tighten to 7 ft. lbs. (10 Nm).

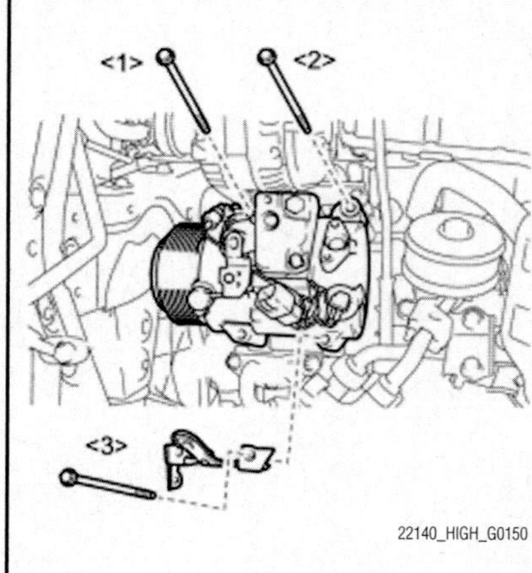

Fig. 241 Compressor and magnetic clutch

24. Connect the Crankshaft Position (CKP) sensor connector.
25. Temporarily tighten compressor and magnetic clutch with the bolts, in the order shown.
26. Install the compressor and magnetic clutch with the 4 bolts. Tighten the bolts in the order shown above, and torque to 18 ft. lbs. (25 Nm).
27. Connect the suction hose sub-assembly.
28. Connect the discharge hose sub-assembly.
29. Install the radiator assembly and fan assembly with motor.
30. Install the cooler condenser assembly.
31. Install the upper radiator support sub-assembly.
32. Install the hood lock assembly.
33. Install the radiator grill.
34. Connect the cooling fan ECU connector.
35. Connect the oil cooler hose.
36. Connect the No. 1 and 2 radiator hoses.
37. Install the No. 1 and 2 air cleaner inlets.
38. Install the battery.
39. Install the cool air intake duct seal.
40. Charge with refrigerant.
41. Add engine coolant.
42. Inspect for coolant leak.
43. Inspect automatic transaxle fluid.
44. Inspect for oil leak.
45. Warm up the engine.
46. Inspect for refrigerant leak.
47. Install the No. 1 engine under cover.
48. Install the engine under cover assembly.
49. Install the V-bank cover sub-assembly.

ELECTRONIC CONTROL MODULE (ECM)

LOCATION

See Figures 242 and 243.

Refer to the accompanying illustrations for ECM location.

REMOVAL & INSTALLATION

See Figures 244 through 246.

1. Disconnect the negative battery cable.

⁂ CAUTION

Wait at least 90 seconds after disconnecting the cable from the negative (-) battery terminal to prevent airbag and seat belt pretensioner activation.

2. Remove the right front door scuff plate.
3. Remove the right cowl side trim sub-assembly.
4. Remove the No. 2 instrument panel under cover sub-assembly.
5. Remove the lower instrument panel sub-assembly.
6. Remove the certification ECU (Smart Key ECU assembly), if applicable.

⁂ WARNING

Before removing the tire pressure warning ECU, read the registered transmitter IDs of all wheels and write them down to use for re-registration of transmitter IDs.

7. Remove the tire pressure warning ECU.
8. Remove the ECM, as follows:
    a. Separate the harness connector.
    b. Disconnect the 5 or 6 ECM connectors.
    c. Remove the wire harness clamp.
    d. Remove the 2 nuts, bolt and ECM.
9. Remove the 2 screws and the No. 1 ECM bracket.
10. Remove the 2 screws and the No. 2 ECM bracket.

To install:

11. Installation is the reverse of the removal procedure. After connecting the negative battery cable, perform the following procedures:
    a. Register the transmitter ID.
    b. Inspect the tire pressure warning system.
    c. Initialize tire pressure warning system.

➡Be sure to register the transmitter IDs of all tires in the ECU before initialization.

➡Be sure to inflate all tires to the proper inflation pressure before initialization.

d. Register immobilizer communication ID.

➡If the ECM is replaced, register the ECU communication ID for the immobilizer system.

e. Perform initialization.

➡If the ECM is replaced, perform RESET MEMORY (at initialization).

RESET

➡Perform the RESET MEMORY (AT initialization) when replacing the automatic transaxle assembly, engine assembly or ECM. The RESET MEMORY can be performed only with Techstream.

➡The ECM memorizes the condition that the ECT controls the automatic transaxle assembly and engine assembly according to those characteristics. Therefore, when the automatic transaxle assembly, engine assembly, or ECM has been replaced, it is necessary to reset the memory so that the ECM can memorize the new information.

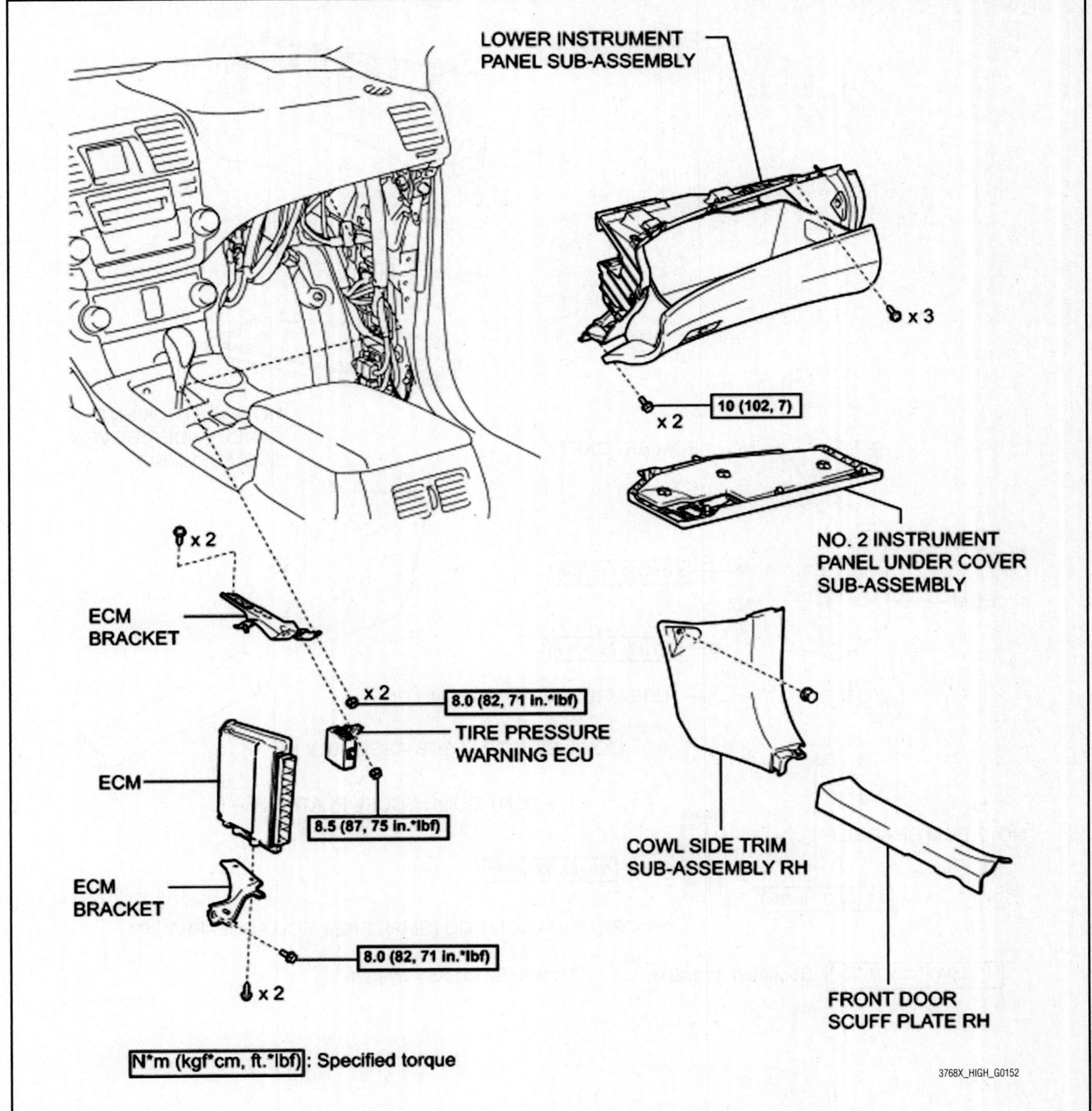

Fig. 242 ECM location—2.7L Engines

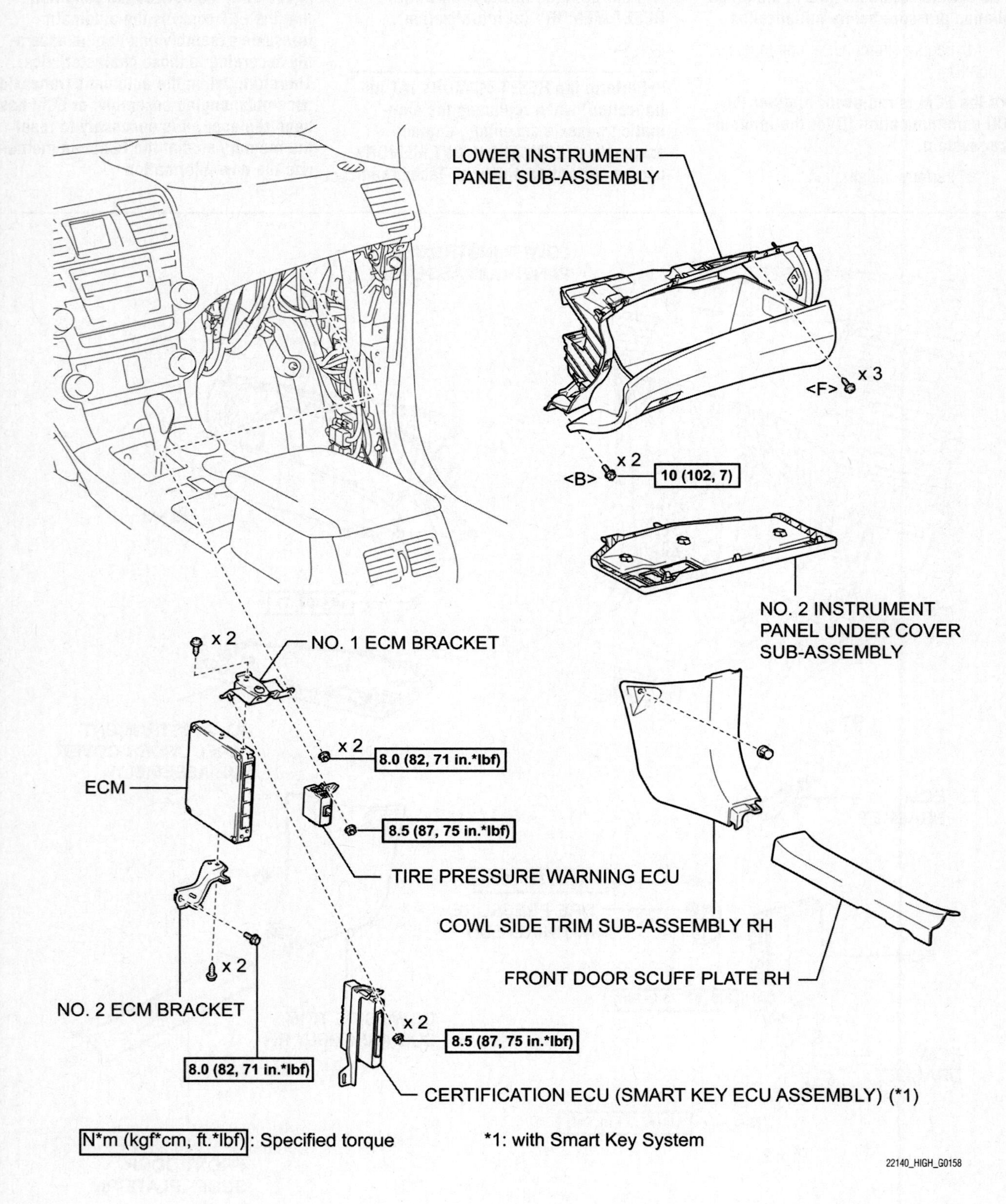

Fig. 243 ECM location—3.5L Engines

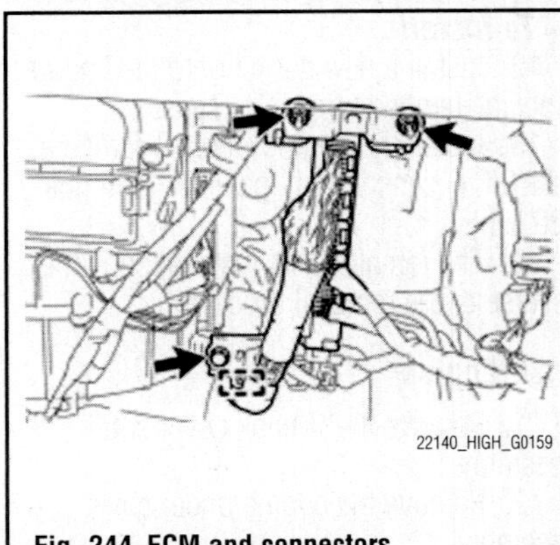
Fig. 244 ECM and connectors

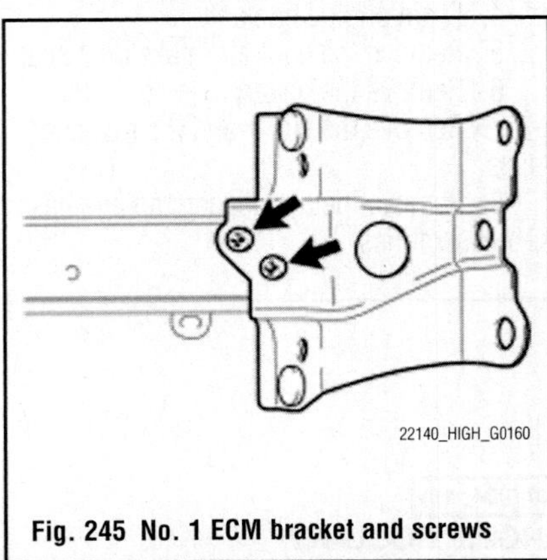
Fig. 245 No. 1 ECM bracket and screws

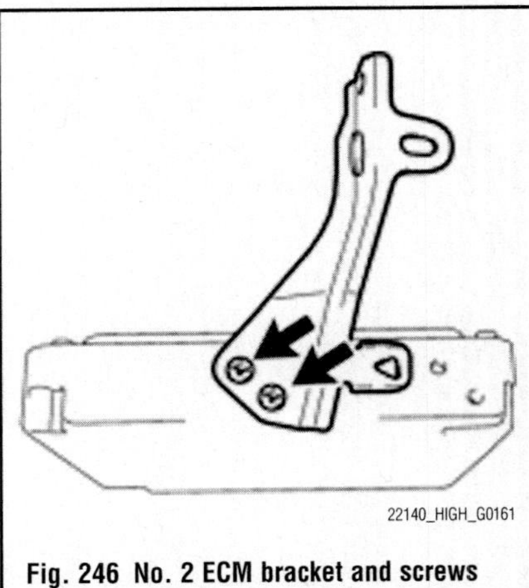
Fig. 246 No. 2 ECM bracket and screws

1. Reset procedure is as follows:
   a. Turn the ignition switch off.
   b. Connect Techstream to the DLC3.
   c. Turn the ignition switch on (IG) and push the Techstream main switch on.
   d. Enter the following menu: Powertrain / Engine and ECT / Utility / Reset Memory. Then, press "Next".
   e. Perform the reset memory procedure from the main menu.

⁂ CAUTION

After performing the RESET MEMORY, be sure to perform the ROAD TEST.

ROAD TEST

➡Perform the test at the ATF temperature 122 to 176°F (50 to 80°C) in the normal operation.

1. D position test: Shift into the D position and fully depress the accelerator pedal and check the following points:
   a. Check up-shift operation: Check that 1 → 2, 2 → 3, 3 → 4 and 4 → 5th up-shifts take place, and that the shift points conform to the automatic shift schedule.

➡5th Gear Up-shift Prohibition Control: Coolant temperature is 158°F (70°C) or less and vehicle speed is at 80 km/h (50 mph) or less. ATF temperature is 28°F (-2°C) or less.

➡4th Gear Up-shift Prohibition Control: Coolant temperature is 158°F (70°C) or less and vehicle speed is at 55 km/h (34 mph) or less.

➡5th and 4th Gear Lock-up Prohibition Control: Brake pedal is depressed. Accelerator pedal is released. Coolant temperature is 140°F (60°C) or less.

   b. Check for shift shock and slip. Check for shock and slip at the 1 → 2, 2 → 3, 3 → 4 and 4 → 5th up-shifts.
   c. Check for abnormal noise and vibration. Drive in the D position lock-up or 5th gear and check for abnormal noises and vibration.

➡The check for the cause of abnormal noise and vibration must be done very thoroughly as it could also be due to loss of balance in the differential, torque converter clutch, etc.

   d. Check kick-down operation. Check that the possible kick-down vehicle speed limits for 2nd to 1st, 3rd to 2nd, 4th to 3rd, 5th to 4th kick-downs conform to those indicated on the automatic shift schedule while driving through all gears with the shift lever in the D position.
   e. Check abnormal shock and slip at kick-down.
   f. Check the lock-up mechanism. Drive in D position (5th gear), at a steady speed (lock-up ON). Lightly depress the accelerator pedal and check that the engine speed does not change abruptly.
   - There is no lock-up in the 1st, 2nd, and 3rd gear.
   - 3rd lock-up operates while uphill control is active.
   - If there is a big jump in engine speed, there is no lock-up.

2. S position test:
   a. Shift to the S position, depress the accelerator pedal and check the following points. Check shift operation:
   - While driving in the D position and 5th gear, shift into the D position and S position and back to the D position. Check that the gear change 5 → 4 down-shift and 4 → 5 up-shift can be performed
   - With the shift lever in the S position (with the vehicle stopped), shift into the "+" position to check that the shift position on the combination meter changes as follows : 1 → 2, 2 → 3, 3 → 4 and 4 → 5
   - While driving in the 4(S) position and 4th gear (at a vehicle speed of about 40 to 50 km/h (25 to 31 mph)), shift into the "-" position and check if the 3rd gear down-shift occurs and the engine brake performs properly
   - While driving in the 3(S) position and 3rd gear (at a vehicle speed of about 30 to 40 km/h (19 to 25 mph)), shift into the "-" position and check if the 2nd gear down-shift occurs and the engine brake performs properly
   - While driving in the 2(S) position and 2nd gear (at a vehicle speed of about 20 to 30 km/h (12 to 19 mph)), shift into the "-" position and check if the 1st gear down-shift occurs and the engine brake performs properly

➡Manual shift (S position) prohibition control: Down-shifting causes engine overrun. Down-shifting is required continuously. (Down-shifting to 1st gear may not be performed.)

   b. R position test: Shift into the R position, lightly depress the accelerator pedal, and check that the vehicle moves backward without any abnormal noise or vibration.

⁂ CAUTION

Before conducting this test ensure that the test area is free from people and obstruction.

   c. P position test: Stop the vehicle on a grade (more the 5°) and after shifting into the P position, release the parking brake. Then, check that the parking lock pawl holds the vehicle in place.

d. Uphill/downhill control function test:

- Check that the gear does not up-shift to the 3rd to 4th or 4th to 5th gear while the vehicle is driving uphill.
- Check that the gear automatically down-shifts from the 5th to 4th or from the 4th to 3rd gear when brake is applied while the vehicle is driving downhill.

ENGINE COOLANT TEMPERATURE (ECT) SENSOR

LOCATION

See Figures 247 and 248.

Refer to the accompanying illustrations for sensor location.

REMOVAL & INSTALLATION

2.7L Engine

1. Remove windshield wiper motor and link.
2. Separate ejector tube.
3. Remove outer cowl top panel sub-assembly.
4. Remove no. 1 engine under cover.
5. Drain engine coolant.
6. Remove no. 1 engine cover sub-assembly.
7. Remove no. 1 vacuum switching valve.
8. Remove air cleaner cap sub-assembly.
9. Remove air cleaner case.
10. Disconnect the Engine Coolant Temperature (ECT) sensor connector.
11. Remove the Engine Coolant Temperature (ECT) sensor and gasket.

To install:

12. Install a new gasket onto the Engine Coolant Temperature (ECT) sensor.
13. Install the Engine Coolant Temperature (ECT) sensor and tighten to 15 ft. lbs. (20 Nm).
14. The remainder of installation is the reverse of the removal procedure.

3.5L Engine

1. Remove the V-bank cover sub-assembly.
2. Remove the engine under cover assembly.
3. Remove the No. 1 engine under cover.
4. Drain engine coolant.
5. Remove the cool air intake duct seal.
6. Remove the battery.
7. Remove the No. 1 and 2 air cleaner inlets.
8. Remove the air cleaner cap and case sub-assemblies.

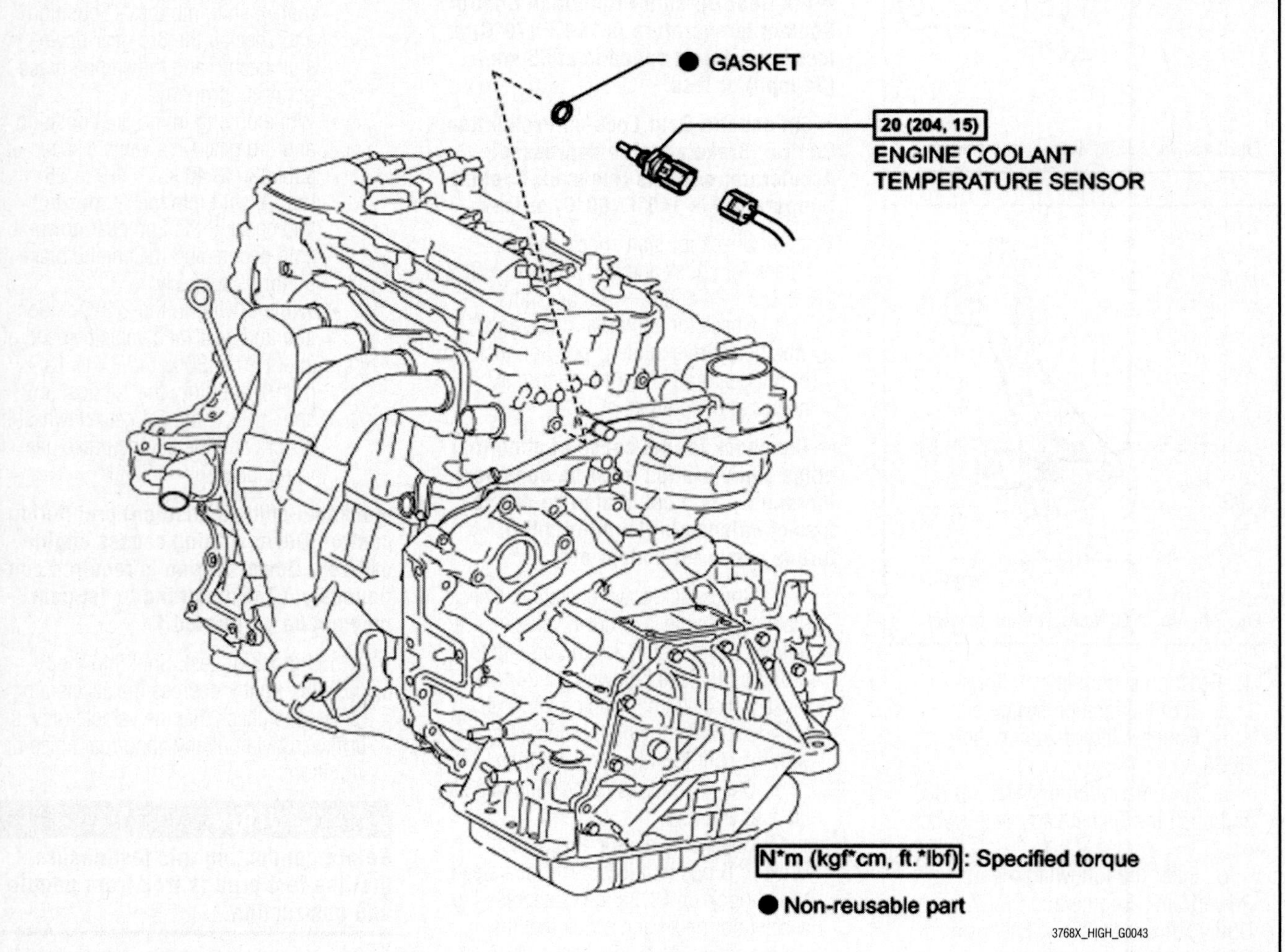

Fig. 247 Engine Coolant Temperature (ECT) sensor location—2.7L engine

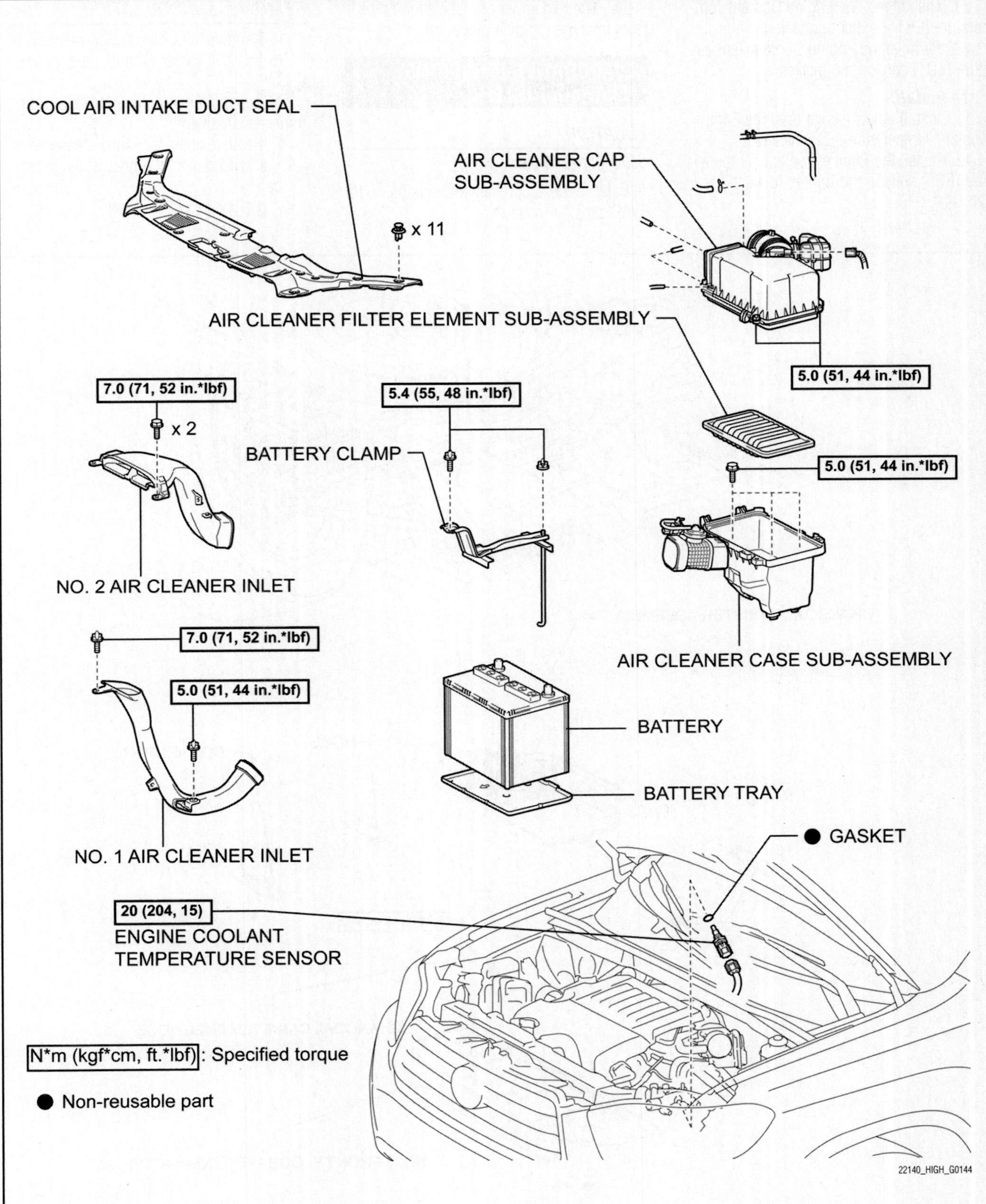

Fig. 248 Engine Coolant Temperature (ECT) sensor location—3.5L engine

9. Disconnect the Engine Coolant Temperature (ECT) sensor connector.

10. Remove the Engine Coolant Temperature (ECT) sensor and gasket.

To install:

11. Install a new gasket onto the Engine Coolant Temperature (ECT) sensor.

12. Install the Engine Coolant Temperature (ECT) sensor and tighten to 15 ft. lbs. (20 Nm).

13. The remainder of installation is the reverse of the removal procedure.

EVAPORATIVE EMISSIONS (EVAP) CANISTER

LOCATION

See Figure 249.

Refer to the accompanying illustration for EVAP canister location.

REMOVAL & INSTALLATION

1. Remove tail exhaust pipe assembly.
2. Remove center exhaust pipe assembly.
3. Remove the 4 bolts and No. 3 front floor heat insulator.
4. Remove charcoal canister assembly:
   a. Disconnect the hose and connector.
   b. Disconnect the tube.

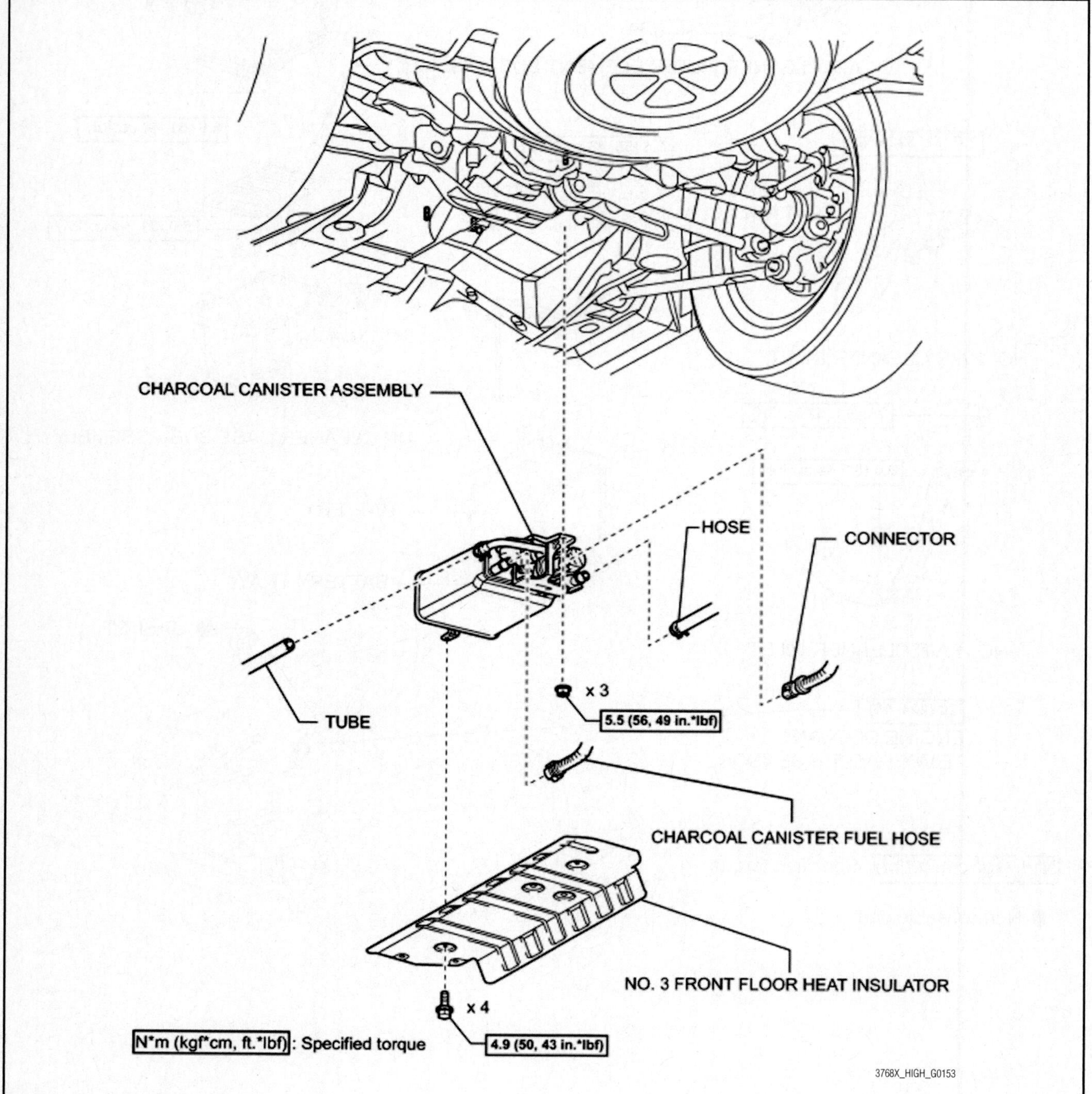

Fig. 249 EVAP canister location

c. Disconnect the charcoal canister fuel hose from the charcoal canister assembly. Pinch the retainer and pull out the quick connector with the quick connector pushed to the pipe side to disconnect the charcoal canister fuel hose from the charcoal canister assembly.

- Remove dirt or foreign objects on the quick connector before this work.
- Do not allow any scratches or foreign objects on the parts when disconnecting as the fuel hose connector has the O-ring that seals the pipe.
- Perform this work by hand. Do not use any tools.
- Do not forcibly bend, twist or turn the nylon tube. Protect the connected part by covering it with a vinyl bag after disconnecting the charcoal canister fuel hose.
- If the quick connector and pipe are stuck, push and pull them to release.

d. Remove the 3 nuts and charcoal canister assembly.

5. To install, reverse removal procedure.

HEATED OXYGEN SENSOR (HO2S)

LOCATION

See Figures 250 and 251.

Refer to the accompanying illustrations for sensor location.

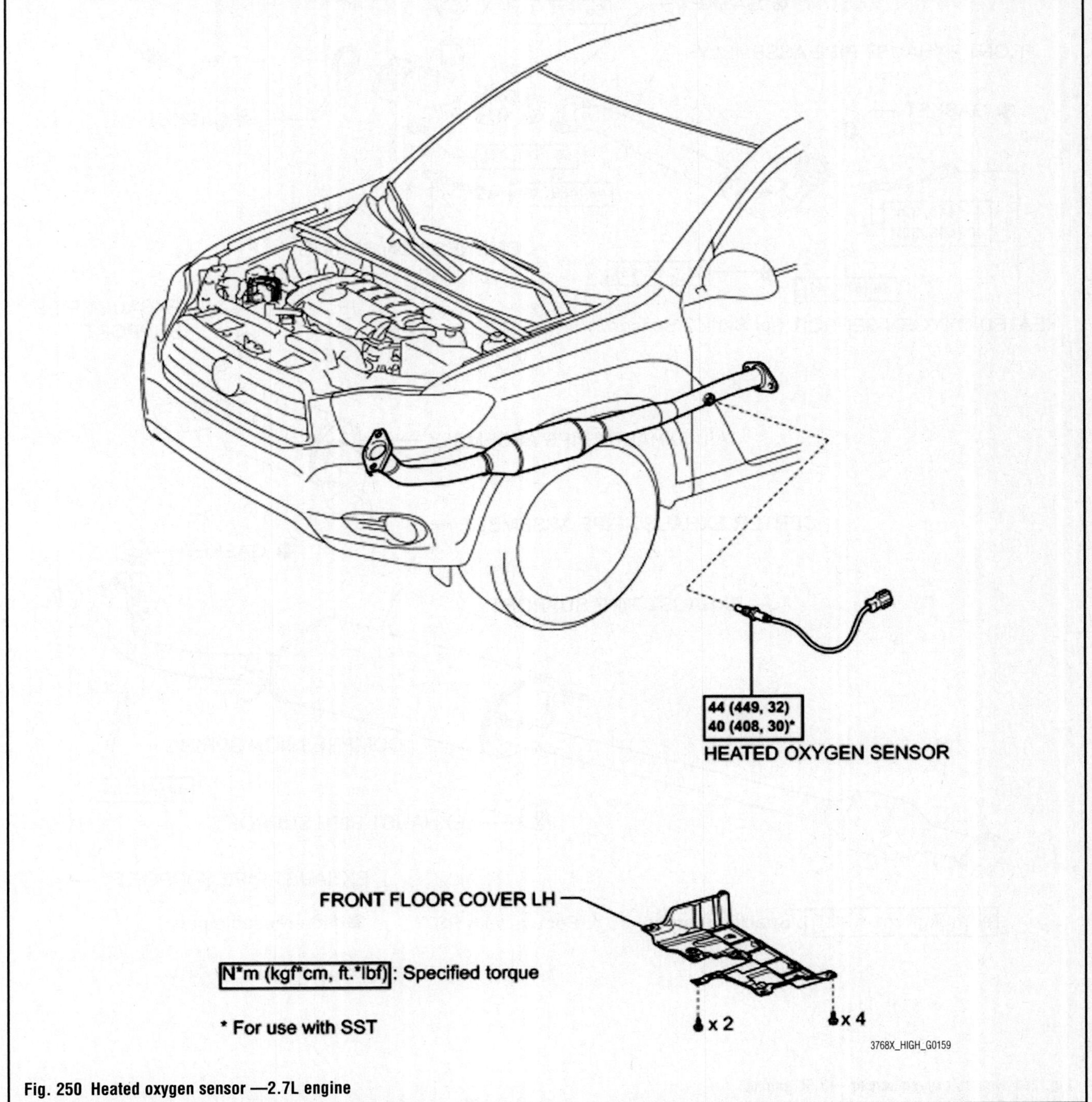

Fig. 250 Heated oxygen sensor —2.7L engine

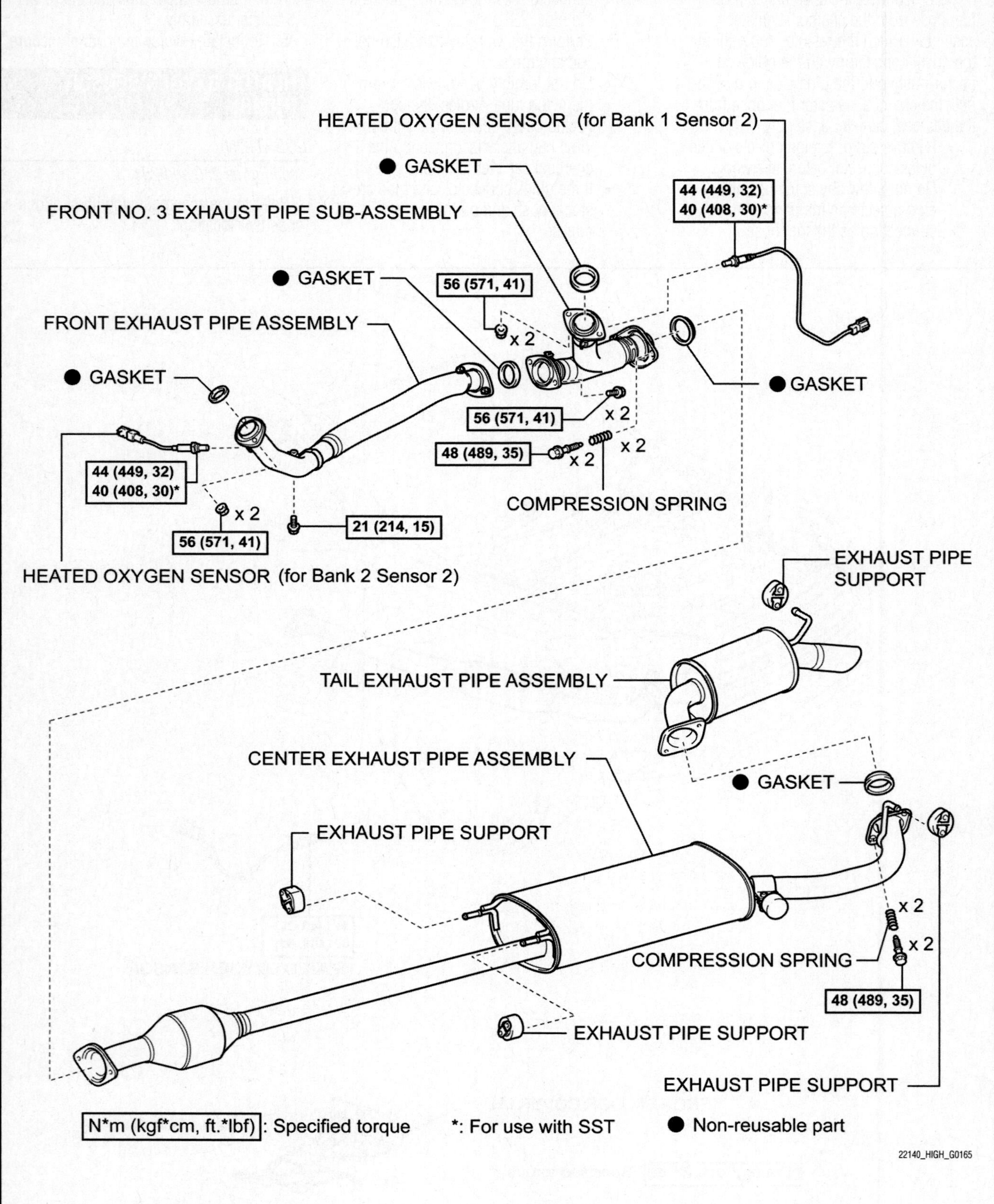

Fig. 251 Heated oxygen sensor —3.5L engine

REMOVAL & INSTALLATION

2.7L Engine

See Figure 252.

1. Remove front floor cover LH.
2. Disconnect the heated oxygen sensor connector.
3. Separate the 2 wire harness clamps.
4. Using SST: 09224-00010, remove the heated oxygen sensor from the front exhaust pipe.

To install:

➡When installing the heated oxygen sensors, use a torque wrench with a fulcrum length of 30 cm (11.81 inches), and make sure that SST and the wrench are connected in a straight line.

5. Using SST: 09224-00010 or equivalent, install the heated oxygen sensor to the front exhaust pipe sub-assembly, and tighten 30 ft. lbs. (40 Nm), 32 ft. lbs. (44 Nm).
6. To complete installation, reverse remaining removal procedure.

3.5L Engine

1. Remove the engine under cover assembly.
2. Remove the No. 1 and 2 engine under covers.
3. Remove the 3 clamps, 6 bolts and the left front floor cover.
4. For 4WD, remove the propeller shaft with center bearing shaft assembly.
5. Remove the tail exhaust pipe assembly.
6. Remove the center exhaust pipe assembly.
7. Remove the front No. 3 exhaust pipe sub-assembly.
8. Remove the front exhaust pipe assembly.

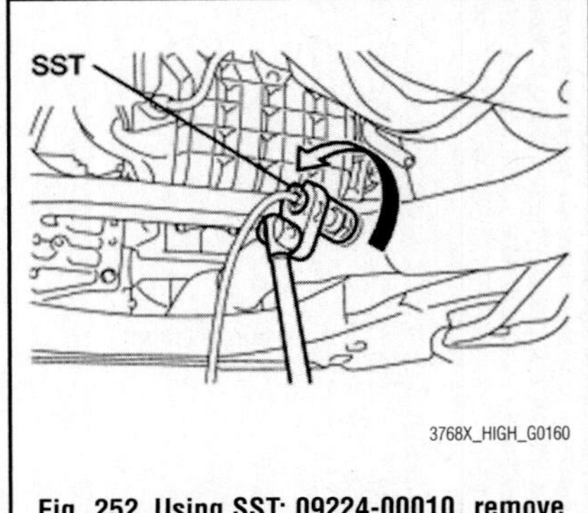

Fig. 252 Using SST: 09224-00010, remove the heated oxygen sensor

9. Using a Special Service Tool (SST: 09224-00010 or equivalent), remove the heated oxygen sensor (for Bank 1 Sensor 2) from the front No. 3 exhaust pipe sub-assembly.
10. Using SST: 09224-00010 or equivalent, remove the heated oxygen sensor (for Bank 2 Sensor 2) from the front exhaust pipe assembly.

To install:

➡When installing the heated oxygen sensors, use a torque wrench with a fulcrum length of 30 cm (11.81 inches), and make sure that SST and the wrench are connected in a straight line.

11. Using SST: 09224-00010 or equivalent, install the heated oxygen sensor (for Bank 1 Sensor 2) to the front No. 3 exhaust pipe sub-assembly, and tighten 30 ft. lbs. (40 Nm), 32 ft. lbs. (44 Nm).
12. Using SST: 09224-00010 or equivalent, install the heated oxygen sensor (for Bank 2 Sensor 2) to the front exhaust pipe assembly, and tighten 30 ft. lbs. (40 Nm), 32 ft. lbs. (44 Nm).
13. Install the front exhaust pipe assembly.
14. Install the front No. 3 exhaust pipe sub-assembly.
15. Install the center exhaust pipe assembly.
16. Install the tail exhaust pipe assembly.
17. For 4WD, temporarily tighten the propeller with center bearing shaft assembly.
18. For 4WD, fully tighten the propeller with center bearing shaft assembly.
19. Inspect for exhaust gas leak.
20. For 4WD, inspect and adjust transfer oil.
21. Install the 3 clamps, 6 bolts and the left front floor cover.
22. Install the No. 1 and 2 engine under covers.
23. Install the engine under cover assembly.

KNOCK SENSOR (KS)

LOCATION

See Figures 253 and 254.

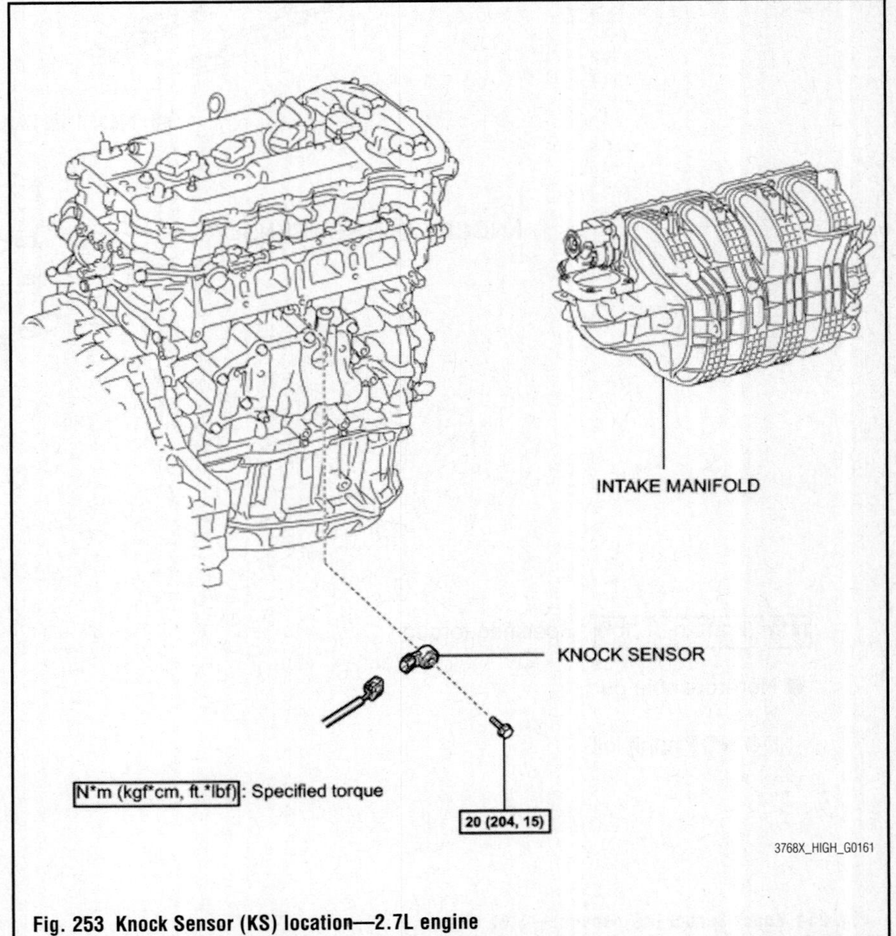

Fig. 253 Knock Sensor (KS) location—2.7L engine

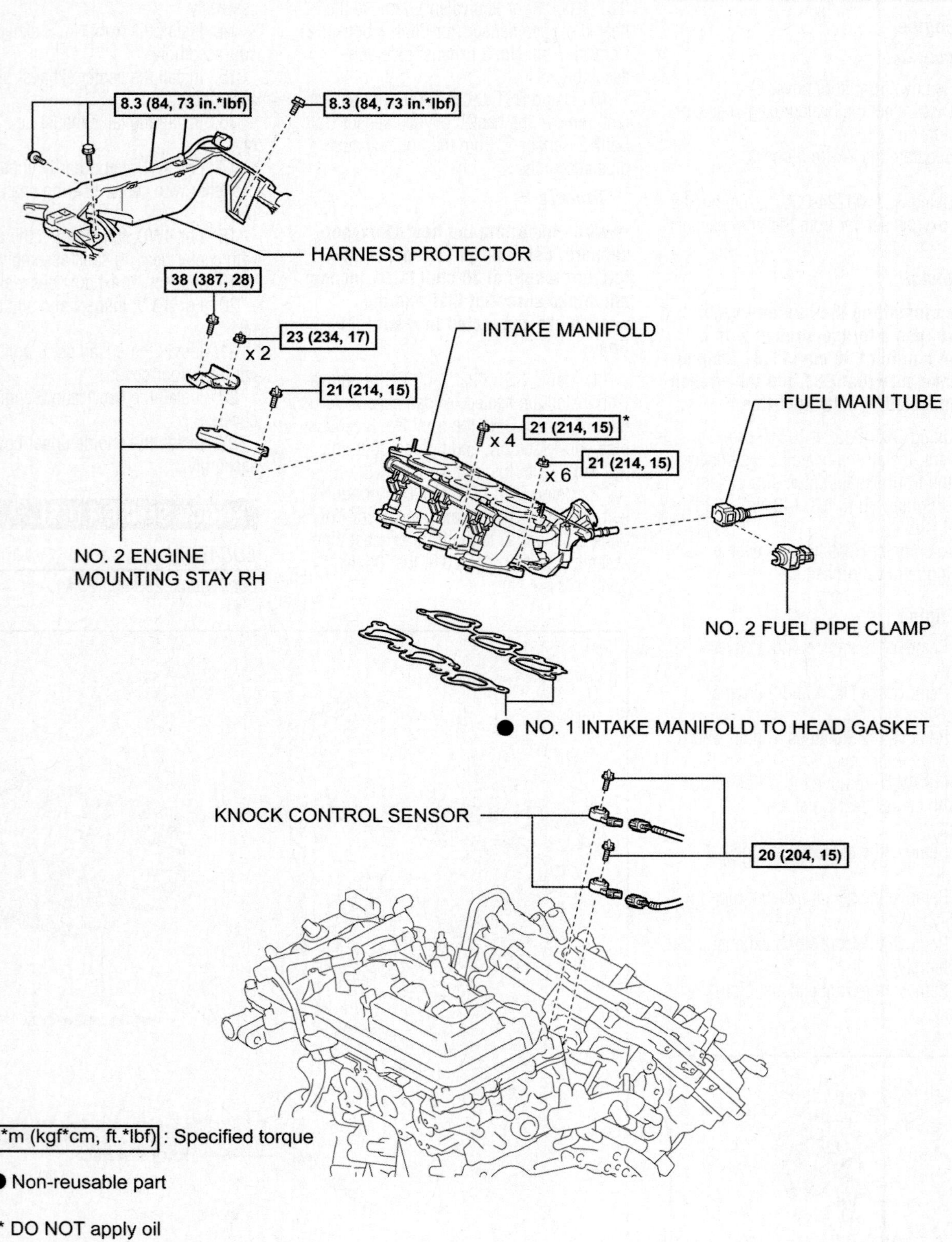

Fig. 254 Knock Sensor (KS) location—3.5L engine

REMOVAL & INSTALLATION

2.7L Engine

See Figure 255.

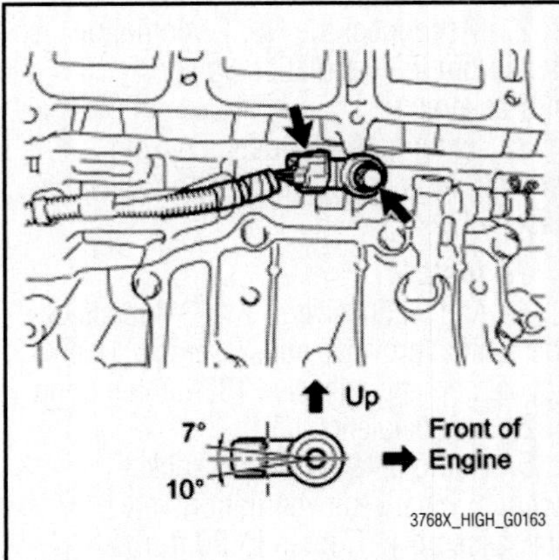

Fig. 255 Knock Sensor (KS) installation orientation

1. Disconnect the negative battery cable.
2. Remove the intake manifold.
3. Disconnect the sensor connector.
4. Remove the bolt and sensor.

To install:

➡The acceptable installation angle of the sensor is between 7° upward and 10° downward from the horizontal position.

5. Install the sensor with the bolt so that the sensor is angled as shown in the illustration. Tighten to 15 ft. lbs. (20 Nm).
6. Connect the sensor connector.
7. Install the intake manifold.

3.5L Engine

See Figure 256.

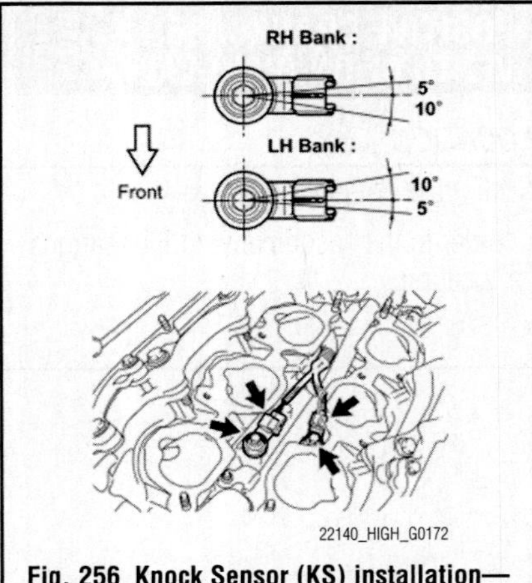

Fig. 256 Knock Sensor (KS) installation—3.5L engine

1. Remove the engine under cover assembly.
2. Remove the No. 1 engine under cover.
3. Drain engine coolant.
4. Discharge fuel system pressure.
5. Remove the front wiper arm and blade assemblies.
6. Remove the cowl top ventilator louver sub-assembly.
7. Remove the windshield wiper motor and link assembly.
8. Remove the outer cowl top panel sub-assembly.
9. Remove the V-bank cover sub-assembly.
10. Remove the air cleaner cap sub-assembly.
11. Disconnect the engine room main wire.
12. Remove the throttle body bracket.
13. Remove the No. 1 surge tank stay.
14. Remove the intake air surge tank assembly.
15. Remove the right No. 2 engine mounting stay.
16. Disconnect the fuel main tube.
17. Remove the intake manifold.
18. Disconnect the 2 knock control sensor connectors.
19. Remove the 2 bolts and then remove the 2 knock control sensors.

To install:

20. Install the 2 knock control sensors with the 2 bolts as shown in the illustration, and tighten to 15 ft. lbs. (20 Nm).
21. Connect the 2 knock control sensor connectors.
22. Install the intake manifold.
23. Install the right no. 2 engine mounting stay.
24. Connect the fuel main tube.
25. Temporarily install the No. 1 surge tank stay.
26. Temporarily install the throttle body bracket.
27. Install the intake air surge tank assembly.
28. Fully tighten the No. 1 surge tank stay.
29. Fully tighten the throttle body bracket.
30. Connect the engine room wire.
31. Install the air cleaner cap sub-assembly.
32. Add engine coolant.
33. Inspect for coolant leak.
34. Inspect for fuel leak.
35. Install the outer cowl top panel sub-assembly.
36. Install the windshield wiper motor and link assembly.
37. Install the cowl top ventilator louver sub-assembly.
38. Install the front wiper arm and blade assemblies.
39. Install the No. 1 engine under cover.
40. Install the engine under cover assembly.
41. Install the V-bank cover sub-assembly.

MALFUNCTION INDICATOR LIGHT (MIL)

RESET PROCEDURE

Using a ODB II scan tool, clear the DTC codes.

MASS AIR FLOW (MAF) METER

LOCATION

See Figure 257.

Refer to the accompanying illustration for sensor location.

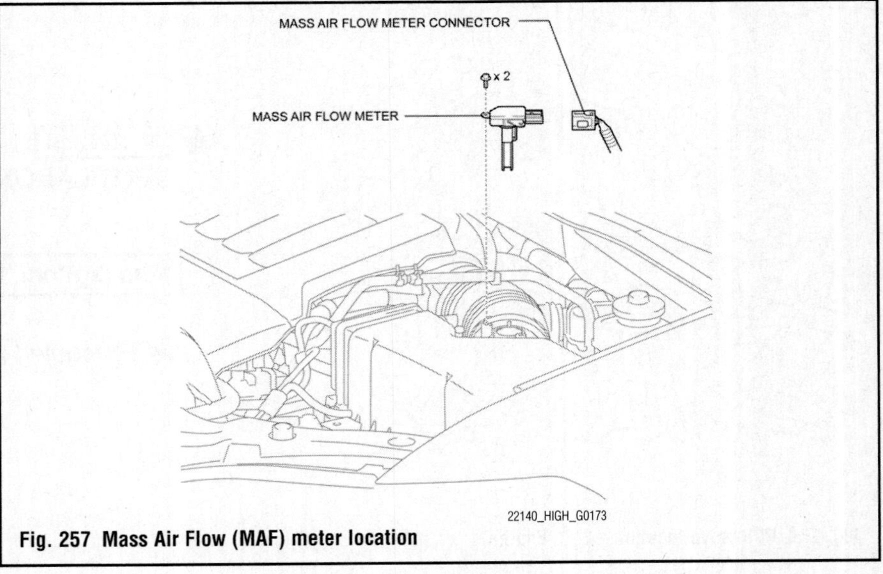

Fig. 257 Mass Air Flow (MAF) meter location

REMOVAL & INSTALLATION

See Figure 258.

1. Disconnect the Mass Air Flow (MAF) meter connector.
2. Remove the 2 screws and Mass Air Flow (MAF) meter.

To install:

3. Installation is the reverse of removal.

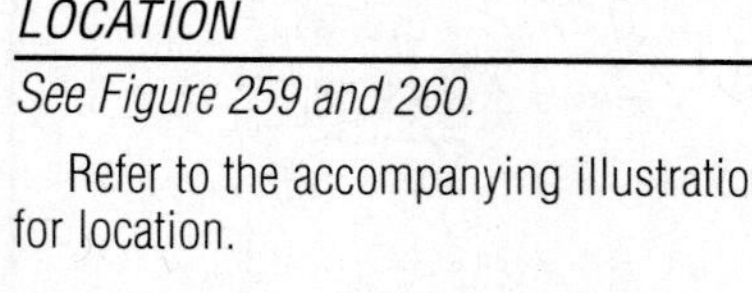

POSITIVE CRANKCASE VENTILATION (PCV) VALVE

LOCATION

See Figure 259 and 260.

Refer to the accompanying illustrations for location.

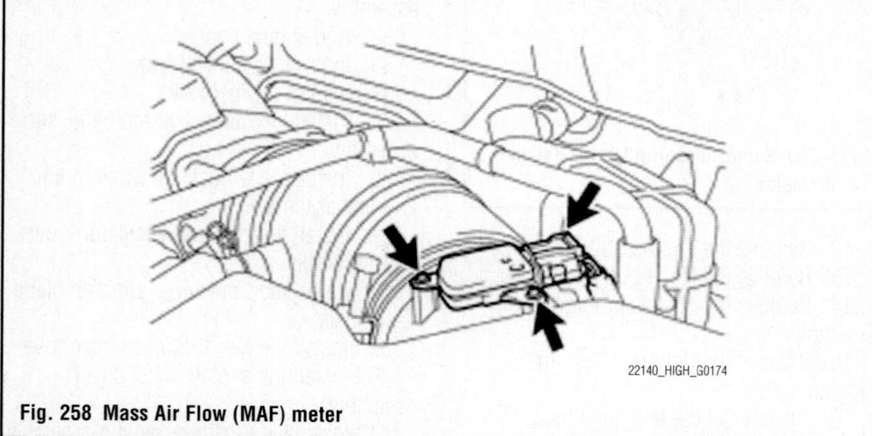

Fig. 258 Mass Air Flow (MAF) meter

REMOVAL & INSTALLATION

2.7L Engine

See Figure 261.

1. Remove the intake manifold.
2. Disconnect the No. 2 ventilation hose from the ventilation valve sub-assembly.
3. Using a 19 mm deep socket wrench, remove the ventilation valve sub-assembly.

To install:

4. Apply adhesive to 2 or 3 threads of the ventilation valve sub-assembly. Use Toyota genuine adhesive 1324, three bond 1324 or equivalent.
5. Using a 19 mm deep socket wrench, install the ventilation valve sub-assembly. Tighten to 20 ft. lbs. (27 Nm).
6. Connect the No. 2 ventilation hose to the ventilation valve sub-assembly.
7. Install the intake manifold.
8. Inspect for oil leak.

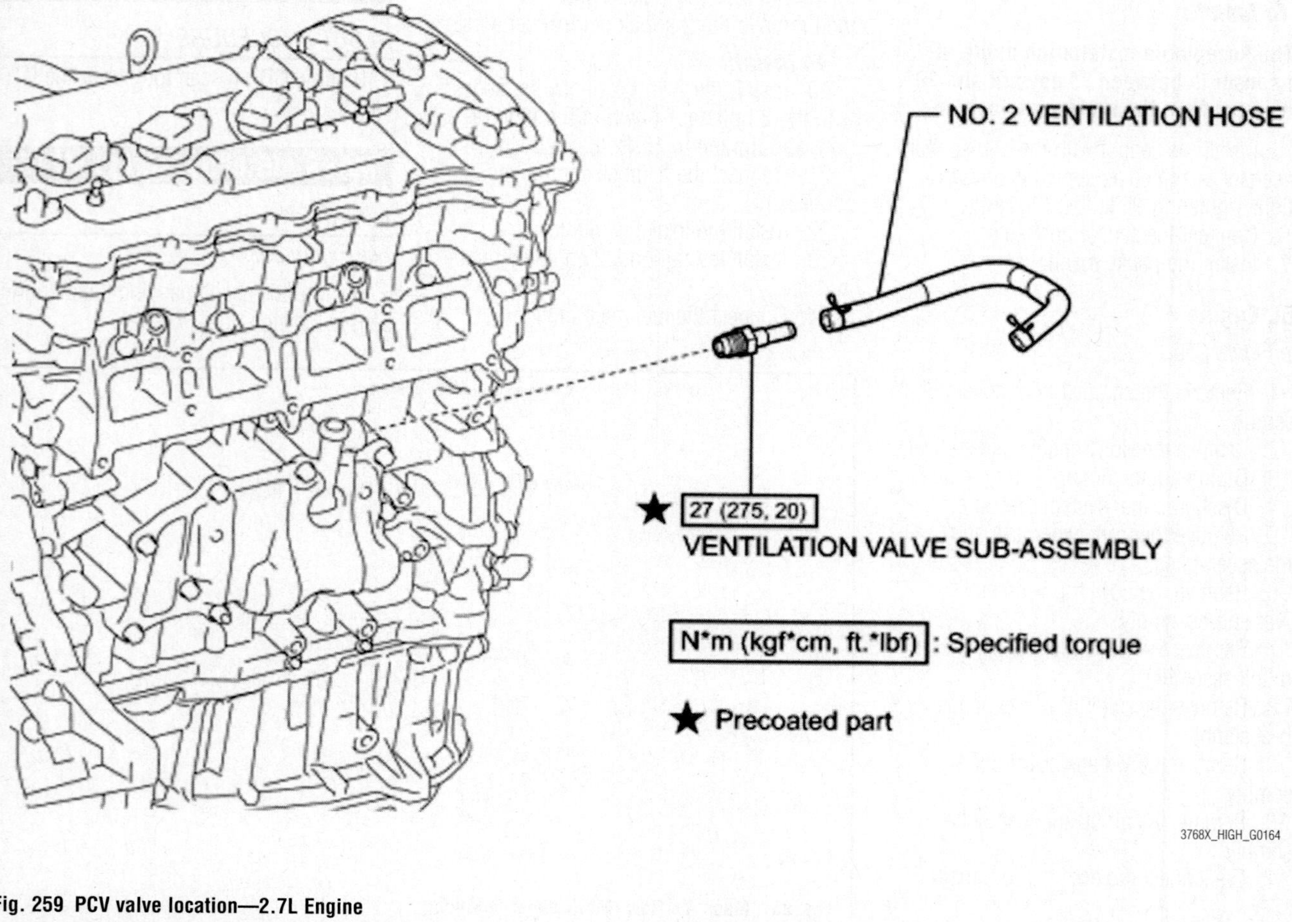

NO. 2 VENTILATION HOSE

★ 27 (275, 20)

VENTILATION VALVE SUB-ASSEMBLY

N*m (kgf*cm, ft.*lbf) : Specified torque

★ Precoated part

3768X_HIGH_G0164

Fig. 259 PCV valve location—2.7L Engine

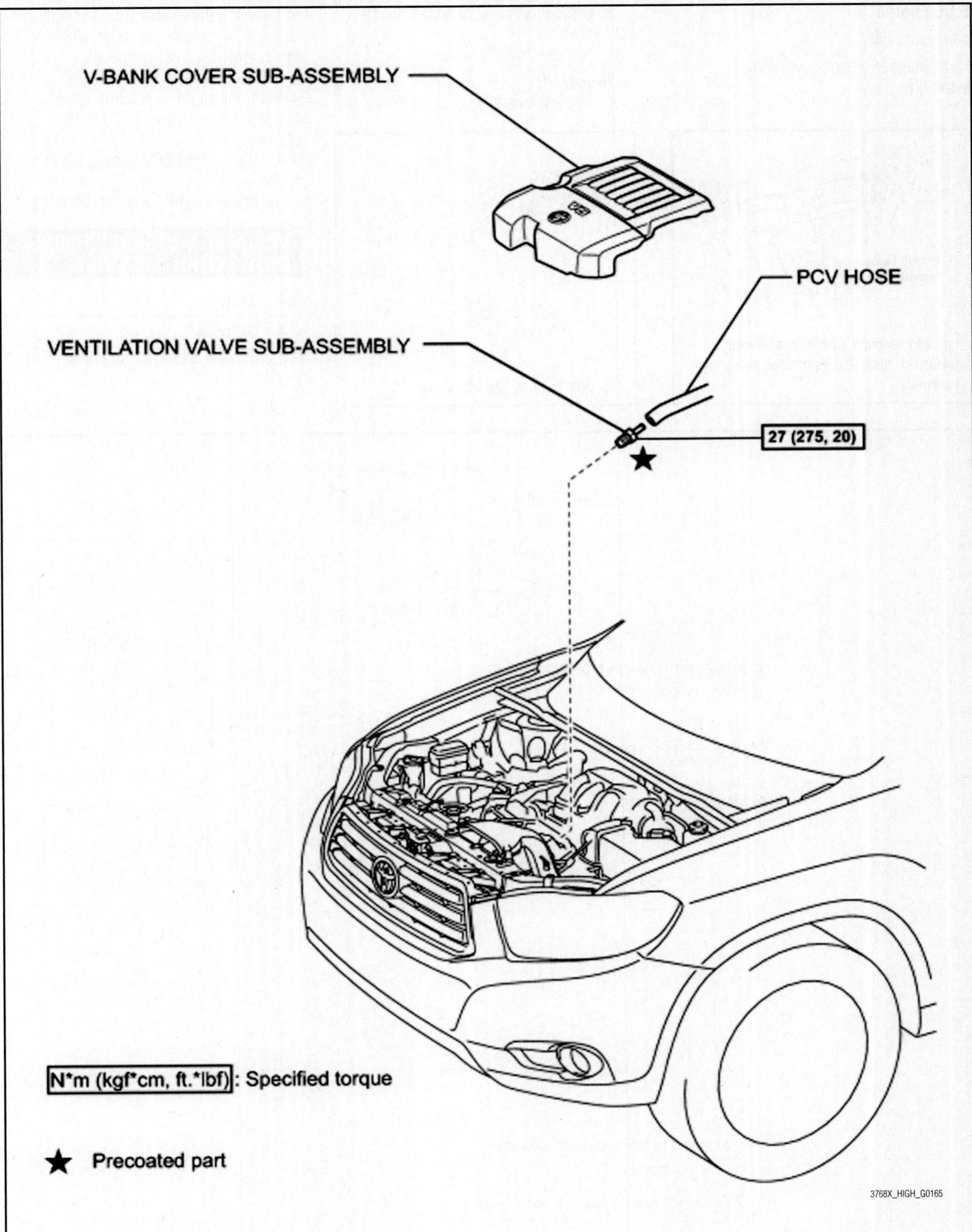

Fig. 260 PCV valve location—3.5L Engine

3.5L Engine

See Figure 262.

1. Remove v-bank cover sub-assembly.
2. Disconnect the PCV hose from the PCV valve.
3. Remove the PCV valve.

To install:

4. Apply adhesive to 2 or 3 threads of the ventilation valve sub-assembly. Use Toyota genuine adhesive 1324, three bond 1324 or equivalent.
5. Using a 19 mm deep socket wrench, install the ventilation valve sub-assembly. Tighten to 20 ft. lbs. (27 Nm).
6. Connect the PCV hose to the PCV valve.
7. Install v-bank cover sub-assembly

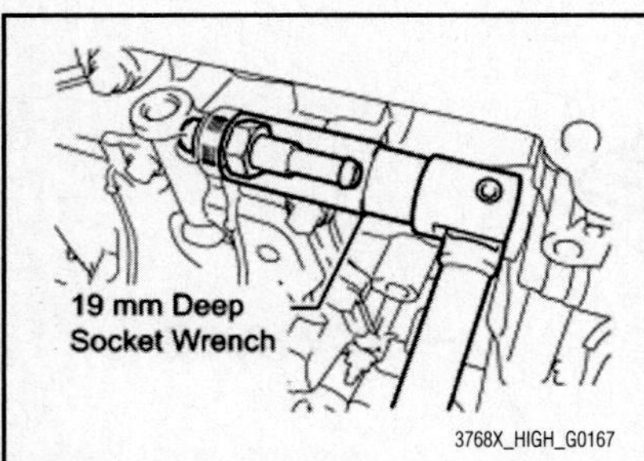

Fig. 261 Using a 19 mm deep socket wrench, remove the ventilation valve sub-assembly

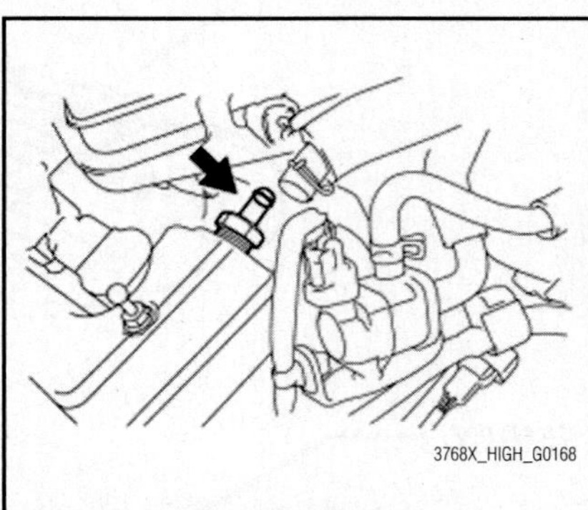

Fig. 262 Remove the PCV valve

VARIABLE CAMSHAFT TIMING OIL CONTROL SOLENOID

LOCATION

See Figure 263.

Refer to the accompanying illustration for location.

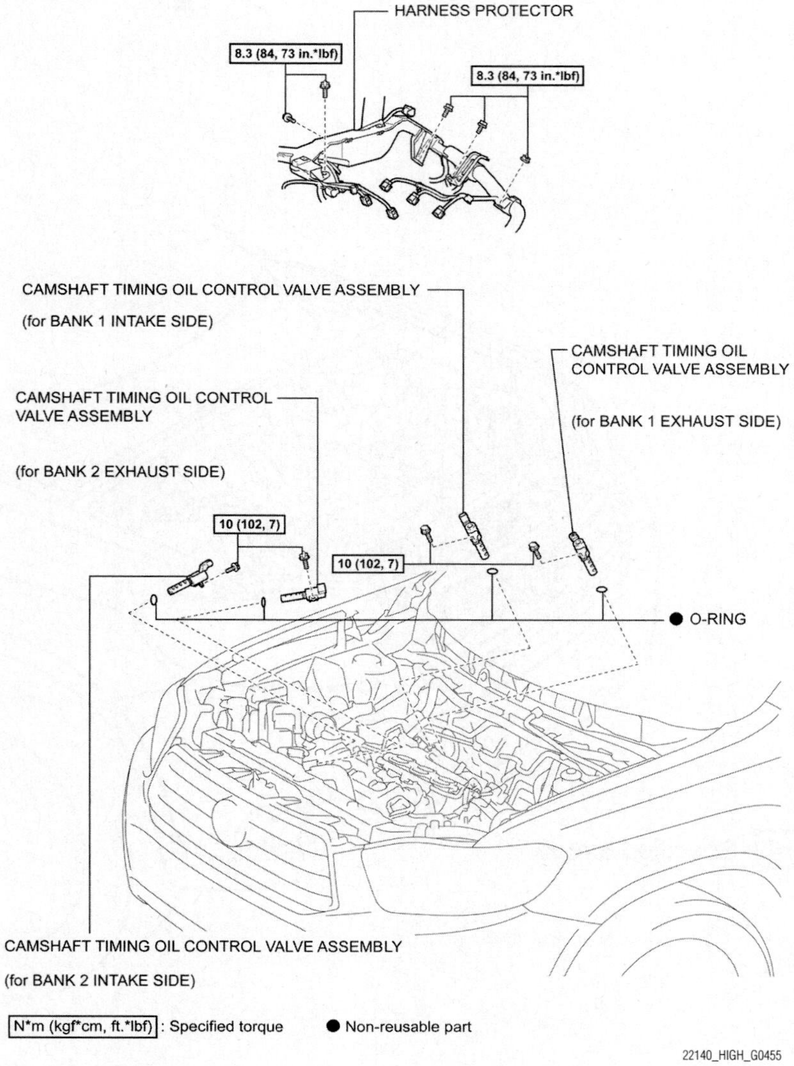

HARNESS PROTECTOR
8.3 (84, 73 in.*lbf)
8.3 (84, 73 in.*lbf)
CAMSHAFT TIMING OIL CONTROL VALVE ASSEMBLY
(for BANK 1 INTAKE SIDE)
CAMSHAFT TIMING OIL CONTROL VALVE ASSEMBLY
(for BANK 1 EXHAUST SIDE)
CAMSHAFT TIMING OIL CONTROL VALVE ASSEMBLY
(for BANK 2 EXHAUST SIDE)
10 (102, 7)
10 (102, 7)
● O-RING
CAMSHAFT TIMING OIL CONTROL VALVE ASSEMBLY
(for BANK 2 INTAKE SIDE)
N*m (kgf*cm, ft.*lbf) : Specified torque
● Non-reusable part
22140_HIGH_G0455

Fig. 263 Variable Camshaft Timing Oil Control Valve assembly

REMOVAL & INSTALLATION

3.5L Engine

See Figures 264 and 265.

1. Remove the engine under cover assembly.
2. Remove the No. 1 engine under cover.
3. Drain engine coolant.
4. Remove both front wiper arm and blade assemblies.
5. Remove the cowl top ventilator louver sub-assembly.
6. Remove the windshield wiper motor and link assembly.
7. Remove the outer cowl top panel sub-assembly.
8. Remove the V-bank cover sub-assembly.
9. Remove the air cleaner cap sub-assembly.
10. Disconnect the engine room main wire.
11. Remove the throttle body bracket.
12. Remove the No. 1 surge tank stay.
13. Remove the intake air surge tank assembly.
14. Separate the harness protector, as follows:
    a. Remove the 5 bolts and nut.
    b. Disconnect the 19 connectors.
    c. Disconnect the 2 clamps.
    d. Disconnect the 2 studs.
15. Remove the camshaft timing oil control valve assembly from Bank 1, intake and exhaust side, and Bank 2, intake and exhaust side, as follows:
    a. Remove the bolt and camshaft timing oil control valve assembly.
    b. Remove the O-ring from the camshaft timing oil control valve assembly.

To install:

16. Install the camshaft timing oil control valve assemblies to Bank 1, intake and exhaust side, and Bank 2, intake and exhaust side, as follows:
    a. Apply a coat of engine oil to a new O-ring and install it onto the camshaft timing oil control valve assembly.
    b. Install the camshaft timing oil control valve assembly with the bolt and tighten to 7 ft. lbs. (10 Nm).
17. Install the harness protector, as follows:
    a. Connect the 2 studs.
    b. Connect the 2 clamps.
    c. Connect the 19 connectors.
    d. Install the 5 bolts and nut and tighten to 73 inch lbs. (8.3 Nm).
18. Temporarily install the No. 1 surge tank stay.
19. Temporarily install the throttle body bracket.

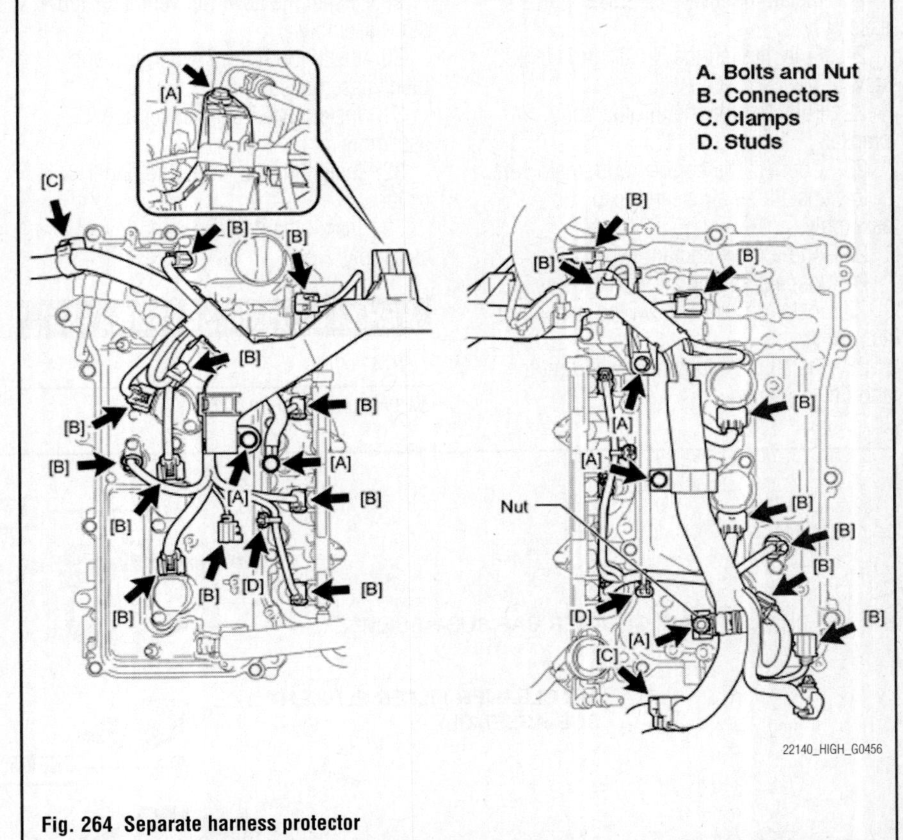

Fig. 264 Separate harness protector

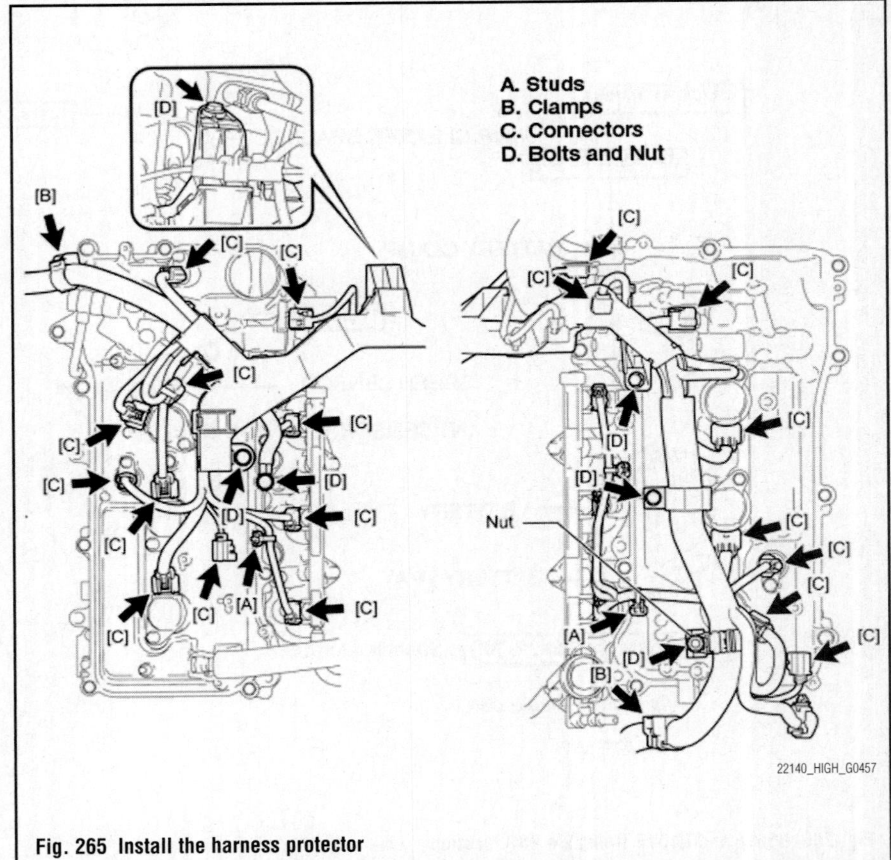

Fig. 265 Install the harness protector

20. Install the intake air surge tank assembly.
21. Fully tighten the No. 1 surge tank stay.
22. Fully tighten the throttle body bracket.
23. Connect the engine room main wire.
24. Install the air cleaner cap sub-assembly.
25. Add engine coolant.
26. Inspect for coolant leak.
27. Install the outer cowl top panel sub-assembly.
28. Install the windshield wiper motor and link assembly.
29. Install the cowl top ventilator louver sub-assembly.
30. Install both front wiper arm and blade assemblies.
31. Install the V-bank cover sub-assembly.
32. Install the No. 1 engine under cover.
33. Install the engine under cover assembly.

VEHICLE SPEED SENSOR (VSS)

LOCATION

See Figures 266 and 267.

Refer to the accompanying illustrations for sensor location.

REMOVAL & INSTALLATION

U151E and U151F Transaxles

CAUTION

Wait at least 90 seconds after disconnecting the cable from the negative (-) battery terminal to prevent airbag and seat belt pretensioner activation.

➡When disconnecting the cable, some systems need to be initialized after the cable is reconnected.

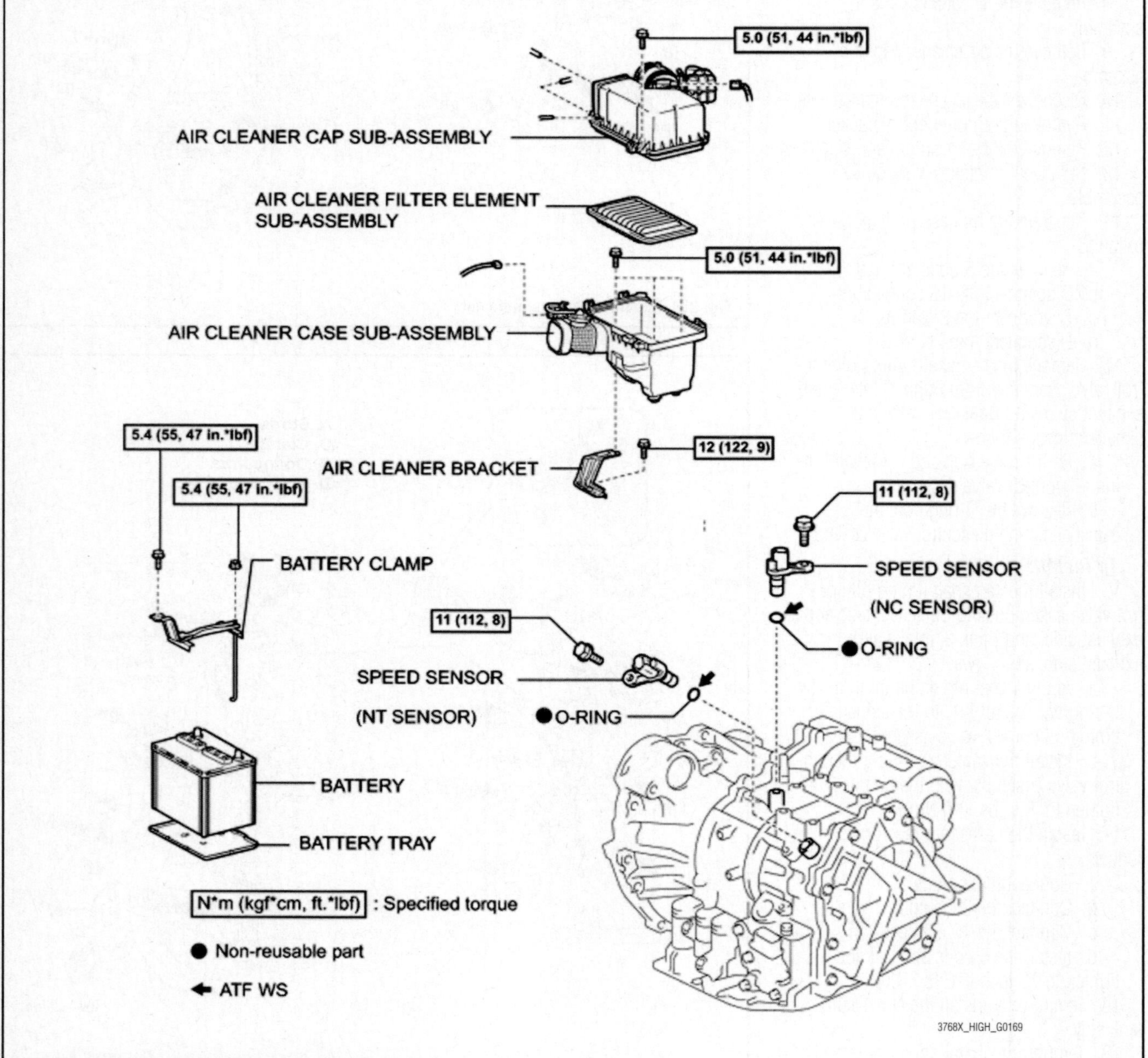

Fig. 266 U151E and U151F Transaxle VSS location

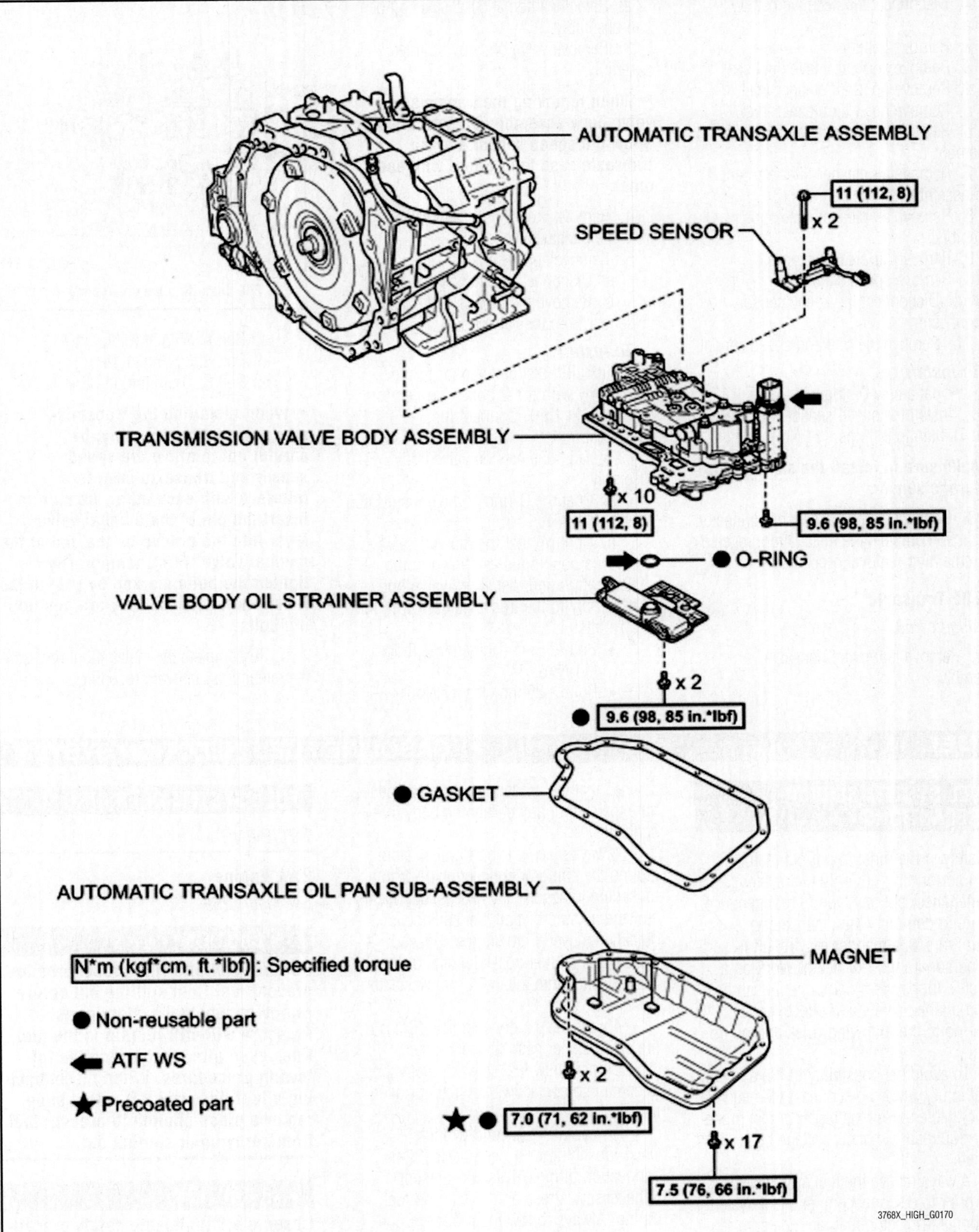

Fig. 267 U760E Transaxle VSS location

1. Disconnect the negative battery cable.
2. Remove battery.
3. Remove cool air intake duct seal.
4. Remove no. 2 air cleaner inlet.
5. Remove no. 1 air cleaner inlet.
6. Remove air cleaner cap sub-assembly.
7. Remove air cleaner filter element sub-assembly.
8. Remove air cleaner case sub-assembly.
9. Remove air cleaner bracket.
10. Remove speed sensors:
    a. Disconnect the speed sensor connector.
    b. Remove the bolt and speed sensor.

To install:

11. Coat a new O-ring with ATF.
12. Install the speed sensor with the bolt. Tighten to 8 ft. lbs. (11 Nm).

➡Make sure to install the same manufacturer's sensor.

13. Connect the speed sensor connector.
14. To complete the installation, reverse the remaining removal procedure.

U760E Transaxle

See Figure 268.

1. Remove automatic transaxle assembly.
2. Remove automatic transaxle oil pan sub-assembly
3. Remove valve body oil strainer assembly.

➡When removing the transmission valve body assembly, be careful not to allow the speed sensor and the transaxle case to interfere with each other.

4. Remove the 11 bolts and valve body from the transaxle.
5. Remove speed sensor:
    a. Disconnect the connector.
    b. Remove the 2 bolts and speed sensor from the valve body.

To install:

6. Install the speed sensor to the valve body with the 2 bolts. Tighten to 8 ft. lbs. (11 Nm). Connect the connector.
7. Install transmission valve body assembly:
    a. Coat the O-ring of the transmission wire with ATF.
    b. Confirm that the manual valve lever is positioned as shown in the illustration and install the valve body assembly to the transaxle case with the 11 bolts.
    - Bolt A, B, C: Tighten to 8 ft. lbs. (11 Nm).
    - Bolt D: 85 inch lbs. (7 Nm)

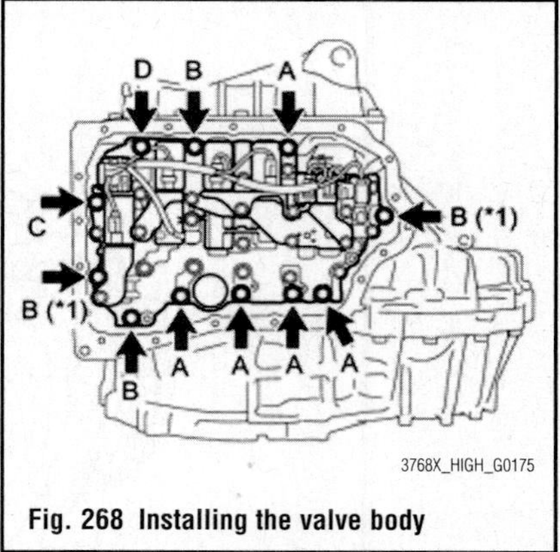

Fig. 268 Installing the valve body

    c. Bolt A: 25 mm (0.984 in.)
    d. Bolt B: 30 mm (1.18 in.)
    e. Bolt C, D: 35 mm (1.38 in.)

➡When installing the transmission valve body assembly, be careful not to allow the speed sensor and transaxle case to interfere with each other. Be sure to insert the pin of the manual valve lever into the groove on the end of the manual valve. First, temporarily tighten the bolts marked by (*1) in the illustration because they are positioning bolts.

8. To complete the installation, reverse the remaining removal procedure.

FUEL — GASOLINE FUEL INJECTION SYSTEM

FUEL SYSTEM SERVICE PRECAUTIONS

Safety is the most important factor when performing not only fuel system maintenance, but any type of maintenance. Failure to conduct maintenance and repairs in a safe manner may result in serious personal injury or death. Work on a vehicle's fuel system components can be accomplished safely and effectively by adhering to the following rules and guidelines.

- To avoid the possibility of fire and personal injury, always disconnect the negative battery cable unless the repair or test procedure requires that battery voltage be applied.
- Always relieve the fuel system pressure prior to disconnecting any fuel system component (injector, fuel rail, pressure regulator, etc.) fitting or fuel line connection. Exercise extreme caution whenever relieving fuel system pressure to avoid exposing skin, face and eyes to fuel spray. Please be advised that fuel under pressure may penetrate the skin or any part of the body that it contacts.
- Always place a shop towel or cloth around the fitting or connection prior to loosening to absorb any excess fuel due to spillage. Ensure that all fuel spillage is quickly removed from engine surfaces. Ensure that all fuel-soaked cloths or towels are deposited into a flame-proof waste container with a lid.
- Always keep a dry chemical (Class B) fire extinguisher near the work area.
- Do not allow fuel spray or fuel vapors to come into contact with a spark or open flame.
- Always use a second wrench when loosening or tightening fuel line connection fittings. This will prevent unnecessary stress and torsion on fuel piping. Always follow the proper torque specifications.
- Always replace worn fuel fitting O-rings with new ones. Do not substitute fuel hose where rigid pipe is installed.

FUEL SYSTEM PRESSURE

RELIEVING

2.7L Engine

See Figure 269.

✲✲ CAUTION

Perform the following procedures to prevent fuel from spilling out before removing any fuel system parts. Pressure will still remain in the fuel lines even after performing the following procedures. When disconnecting a fuel line, cover it with a shop rag or a piece of cloth to prevent fuel from spraying or coming out.

✲✲ CAUTION

Observe all applicable safety precautions when working around fuel. Whenever servicing the fuel system, always work in a well ventilated area. Do not allow fuel spray or

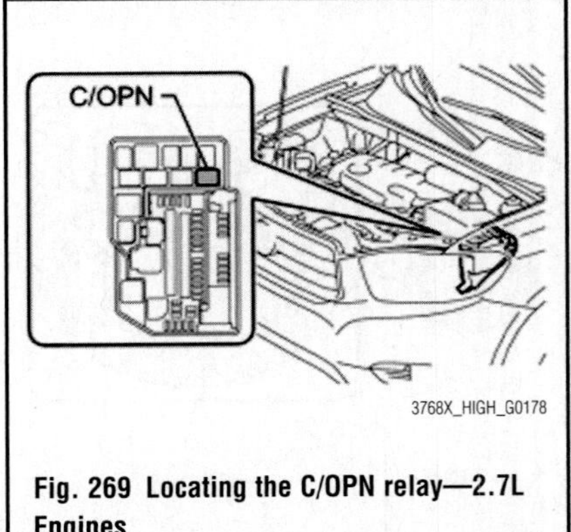

Fig. 269 Locating the C/OPN relay—2.7L Engines

vapors to come in contact with a spark or open flame. Keep a dry chemical fire extinguisher near the work area. Always keep fuel in a container specifically designed for fuel storage; also, always properly seal fuel containers to avoid the possibility of fire or explosion.

1. Separate the brake master cylinder reservoir assembly.
2. Remove the reservoir bracket.
3. Remove the No. 1 relay block cover.
4. Remove the C/OPN relay.
5. Start the engine.
6. After the engine stops, turn the ignition switch off.

➡DTC P0171/25 (fuel problem) may be detected.

7. Crank the engine again. Check that the engine does not start.
8. Remove the fuel tank cap to discharge pressure from the fuel tank.
9. Disconnect the cable from the negative (-) battery terminal.
10. Install the C/OPN relay.
11. Install the No. 1 relay block cover.
12. Install the reservoir bracket.
13. Install the brake master cylinder reservoir assembly.

3.5L Engines

See Figure 270.

⁂ CAUTION

Perform the following procedures to prevent fuel from spilling out before removing any fuel system parts. Pressure will still remain in the fuel lines even after performing the following procedures. When disconnecting a fuel line, cover it with a shop rag or a piece of cloth to prevent fuel from spraying or coming out.

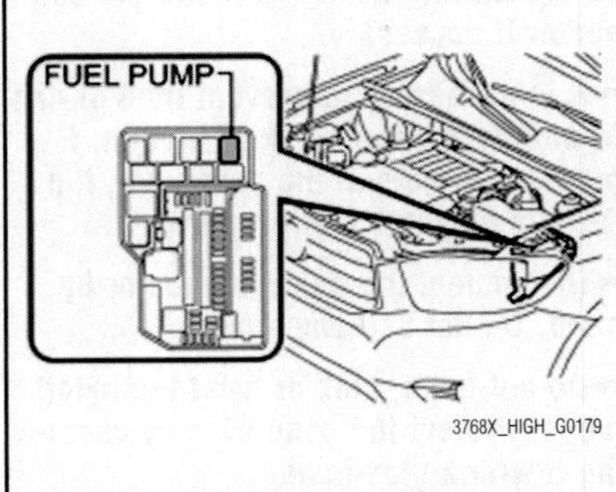

Fig. 270 Locating the fuel pump relay—3.5L Engines

⁂ CAUTION

Observe all applicable safety precautions when working around fuel. Whenever servicing the fuel system, always work in a well ventilated area. Do not allow fuel spray or vapors to come in contact with a spark or open flame. Keep a dry chemical fire extinguisher near the work area. Always keep fuel in a container specifically designed for fuel storage; also, always properly seal fuel containers to avoid the possibility of fire or explosion.

1. Remove the relay block cover.
2. Remove the FUEL PUMP relay.
3. Start the engine.
4. After the engine stops, turn the ignition switch off.

➡DTC P0171/25 (fuel problem) may be detected.

5. Crank the engine again. Check that the engine does not start.
6. Remove the fuel tank cap to discharge pressure from the fuel tank.
7. Disconnect the cable from the negative (-) battery terminal.
8. Install the FUEL PUMP relay.

FUEL FILTER

REMOVAL & INSTALLATION

The fuel filter is part of the fuel suction tube/fuel pump assembly and is located in the fuel tank. It is not a normally serviced item.

FUEL LEVEL SENDING UNIT

REMOVAL & INSTALLATION

See Figure 271.

1. Before servicing the vehicle, refer to the precautions section.

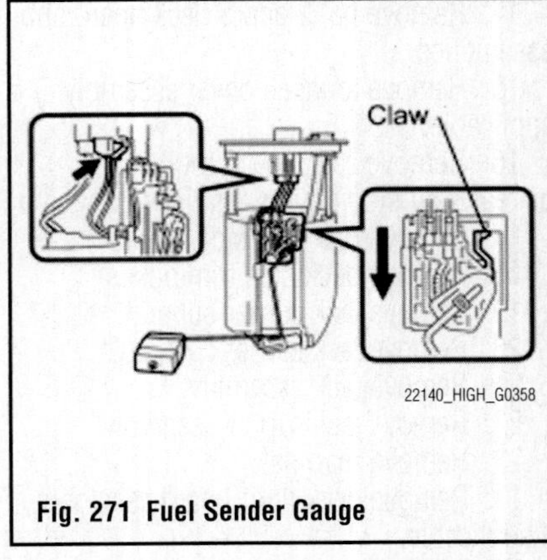

Fig. 271 Fuel Sender Gauge

2. Remove fuel suction tube assembly with pump and gauge. Refer to the Fuel Pump removal procedure for instructions.
3. Disconnect the connector and remove the Fuel Sender Gauge (Fuel Level Sending Unit) from the fuel suction tube.

To install:

4. Install the fuel sender gauge assembly by sliding the fuel sender gauge to fit the claw.
5. Install the fuel suction tube assembly with pump and gauge.
6. Connect the negative battery cable.
7. Inspect for fuel leak.
8. Disconnect the negative battery cable.
9. The remainder of installation is the reverse of the removal procedure. Refer to the Fuel Pump installation procedure for instructions.

FUEL PUMP MODULE

REMOVAL & INSTALLATION

See Figures 272 through 279.

1. Before servicing the vehicle, refer to the precautions section.
2. Discharge fuel system pressure.
3. Disconnect cable from negative battery terminal.
4. Remove the rear center seat assembly.
5. Remove the rear seat headrest assemblies.
6. Remove the seat track bracket covers.
7. Remove rear inner and outer track bracket covers.
8. Remove rear seat leg side covers.
9. Remove rear no. 1 seat lock cable assembly, if applicable.
10. Remove rear no. 1 seat assemblies.
11. Remove left rear door scuff plate.
12. Remove left rear door opening trim weatherstrip.
13. Remove deck board assembly.

14. Remove no. 2 and 3 deck board subassemblies.
15. Remove tonneau cover assembly, if applicable.
16. Remove rear no. 1 floor board, or rear seat side covers, as necessary.
17. Remove rear seat side covers.
18. Remove deck side trim boxes.
19. Remove jack carrier support.
20. Remove jack carrier cushion.
21. Remove jack assembly.
22. Remove jack carrier assembly.
23. Remove rear mat.
24. Remove deck floor board assembly, if applicable.
25. Remove rear deck floor box, if applicable.
26. Remove rear no. 2 seat inner belt assembly, if applicable.
27. Disconnect rear seat lap type belt assemblies.
28. Remove rear no. 2 seat assembly, if applicable.
29. Remove rear floor finish plate.
30. Remove deck side trim covers.
31. Remove left side trim cover, if applicable.
32. Remove power outlet socket bezel, if applicable.
33. Remove rear combination light service cover.
34. Remove rear power point socket assembly.
35. Remove rear power outlet socket cover
36. Remove rear deck trim cover, if applicable.
37. Remove left reclining remote control lever bezel, if applicable.
38. Remove left rope hook assembly.
39. Remove no. 2 deck side trim hook.
40. Remove left front deck side trim cover.
41. Disconnect left rear no. 1 seat outer belt assembly.
42. Remove left deck trim side panel assembly.
43. Remove the rear floor service hole cover, as follows:
    a. Lift the front floor carpet.
    b. Using the appropriate tool, remove the 3 clips and lift up the front floor carpet.
    c. Remove the rear floor service hole cover.
    d. Disconnect the fuel pump connector.
44. Remove the fuel suction tube assembly with pump and gauge, as follows:
    a. Remove the tube joint clip, and pull out the fuel pump tube.

➡Check that there is no dirt or other foreign objects around the connector before disconnecting it. Clean the connector if necessary.

➡It is necessary to prevent mud or dirt from entering the quick connector. If mud or dirt gets in the connector, the O-rings may not seal properly.

➡Disconnect the quick connector by hand. Do not use any tools.

➡Do not bend, kink or twist the nylon tubes. Protect the connector by covering it with a plastic bag.

➡If the pipe and connector are stuck, carefully try wiggling or pushing and pulling on the connector to release it. Pull the connector off carefully.

    b. Using a 6 mm socket hexagon wrench, install SST: 09808-14020 to the fuel pump gauge retainer.

➡Engage the SST claws securely with the fuel pump gauge retainer ribs to secure the SST.

➡Install the SST while pressing the SST claws toward the fuel pump gauge retainer (toward the center of SST).

    c. Using SST: 09808-14020, loosen the fuel pump gauge retainer.
    d. Remove the fuel pump gauge retainer.

WARNING

Do not use any tools other than specified in this operation. Damage to the fuel pump gauge retainer or the fuel tank may result.

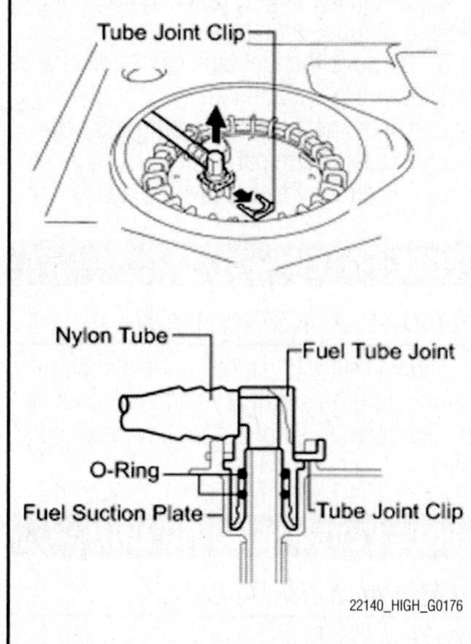

Fig. 272 Tube joint clip and fuel pump tube

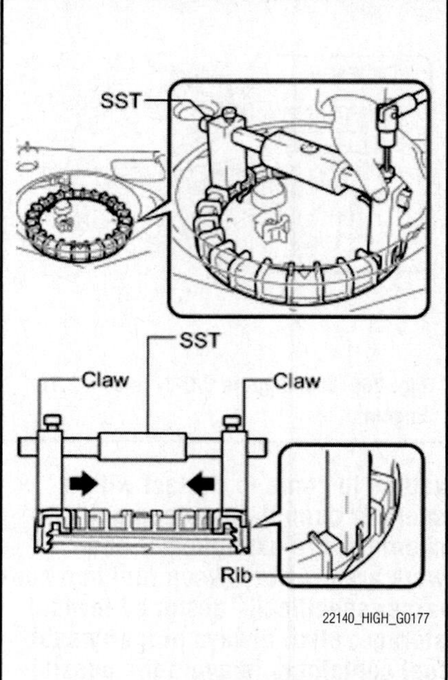
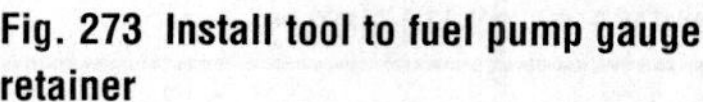

Fig. 273 Install tool to fuel pump gauge retainer

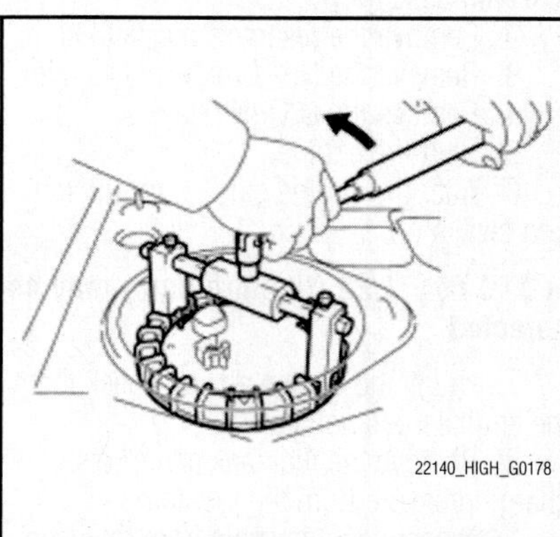

Fig. 274 Loosen fuel pump gauge retainer

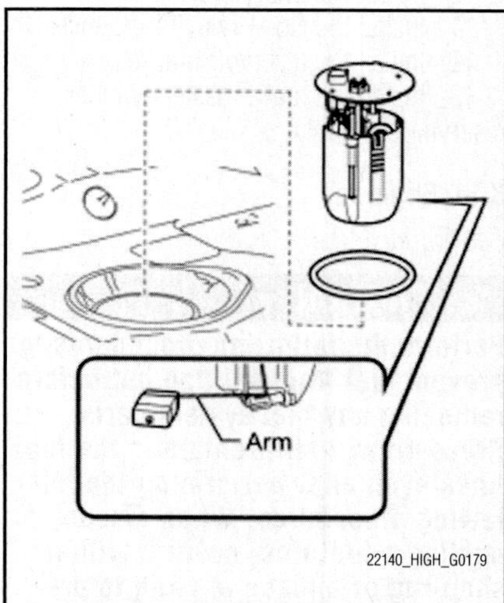

Fig. 275 Remove fuel pump gauge retainer

➡**Loosen the retainer by turning it counterclockwise while holding SST down. Do not allow the claw of the tank suction tube support to slip out of its groove on the fuel tank.**

➡**The ribs on the fuel pump gauge retainer can be fitted into the tips of SST.**

e. Remove the fuel suction tube with pump and gauge.

➡**Be careful not to bend the arm of the fuel sender gauge.**

f. Remove the gasket from the fuel tank.

To install:

45. Install the fuel suction tube assembly with pump and gauge, as follows:

a. Install a new gasket to the fuel tank.

b. Attach the fuel suction tube with pump and gauge to the fuel tank.

➡**Be careful not to bend the arm of the fuel sender gauge.**

c. Align the keyway of the fuel suction tube support with the key of the fuel suction tube with pump and gauge.

d. Align the triangle mark on a new fuel pump gauge retainer with the "S" mark on the fuel tank while pushing down the fuel suction tube with pump and gauge, and attach the fuel pump gauge retainer.

46. Using a 6 mm socket hexagon wrench, install SST: 09808-14020 to the fuel pump gauge retainer.

➡**Engage the SST claws securely with the fuel pump gauge retainer ribs to secure the SST.**

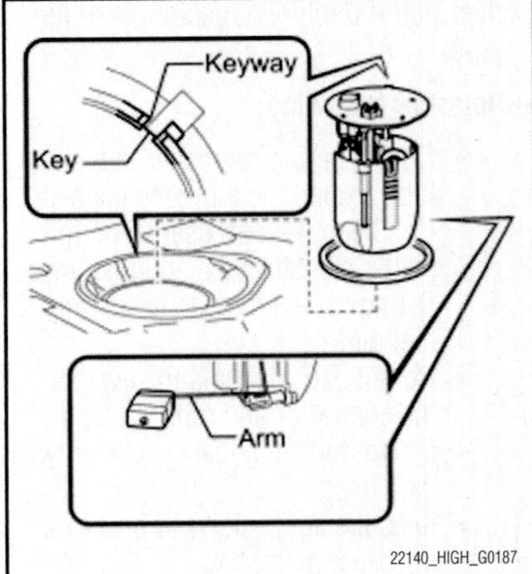

Fig. 276 Fuel suction tube assembly with pump and gauge

➡**Install the SST while pressing the SST claws toward the fuel pump gauge retainer (toward the center of SST).**

a. Rotate the fuel pump gauge retainer by hand, then tighten it one complete turn and another half turn using the SST: 09808-14020. The triangle mark on the fuel pump gauge retainer must be positioned between the "MIN" and "MAX" marks on the fuel tank.

WARNING

Do not use any tools other than specified in this operation. Damage to the fuel pump gauge retainer or the fuel tank may result.

➡**Fully tighten the retainer by turning it clockwise while holding the SST down. Do not allow the claw of the tank suction tube support to slip out of its groove on the fuel tank.**

➡**The ribs on the fuel pump gauge retainer can be fitted into the tips of SST.**

b. Install the fuel pump tube and the tube joint clip.

➡**Check that there are no scratches or foreign objects on the connecting part.**

➡**Check that the fuel tube joint is inserted securely.**

➡**Check that the tube joint clip is on the collar of the fuel tube joint.**

➡**After installing the tube joint clip, check that the fuel tube joint has not been pulled out of position.**

c. Connect the fuel pump connector.

47. Connect the negative battery cable.

48. Inspect for fuel leak.

49. Disconnect the negative battery cable.

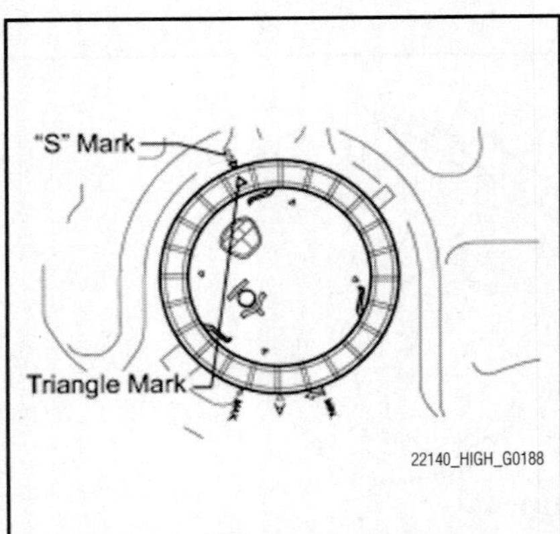

Fig. 277 Align and attach new fuel pump gauge retainer

50. Install new butyl tape to the rear floor service hole cover.

51. Install the rear floor service hole cover.

52. Install the front floor carpet with the 3 clips.

53. The remainder of installation is the reverse of the removal procedure.

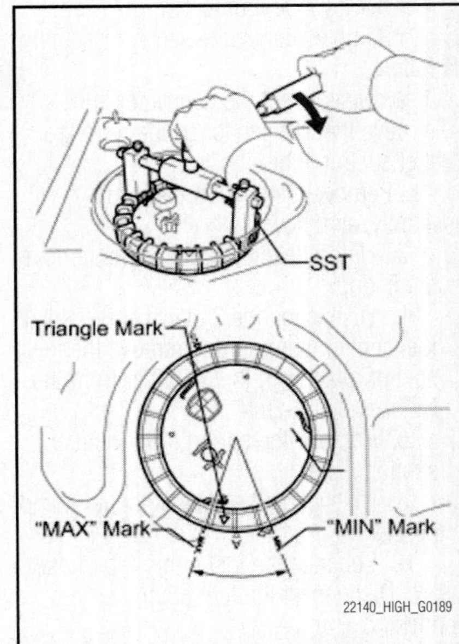

Fig. 278 Installing fuel pump gauge retainer

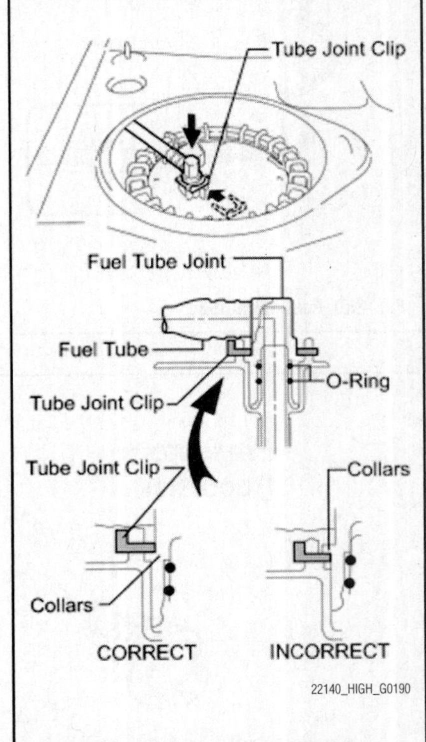

Fig. 279 Fuel pump tube and the tube joint clip

54. After installing all components and connecting the negative battery cable, inspect the SRS warning light..

FUEL PRESSURE REGULATOR

REMOVAL & INSTALLATION

See Figures 280 and 281.

1. Remove fuel pump.
2. Remove fuel gauge sender assembly, as follows:
   a. Disconnect the connector and remove the fuel sender gauge from the fuel suction tube.
3. Remove the suction plate sub-assembly, as follows:
   a. Using needle nozzle pliers, remove the E-ring.
   b. Disengage the 2 claws of the No. 1 fuel suction support and remove the fuel suction plate with the fuel filter from the No. 1 fuel sub-tank.
   c. Remove the spring from the fuel suction plate.
   d. Disengage the claw of the jet pump nozzle.
   e. Separate the fuel pump filter hose.
4. Disconnect the fuel pump harness connector.

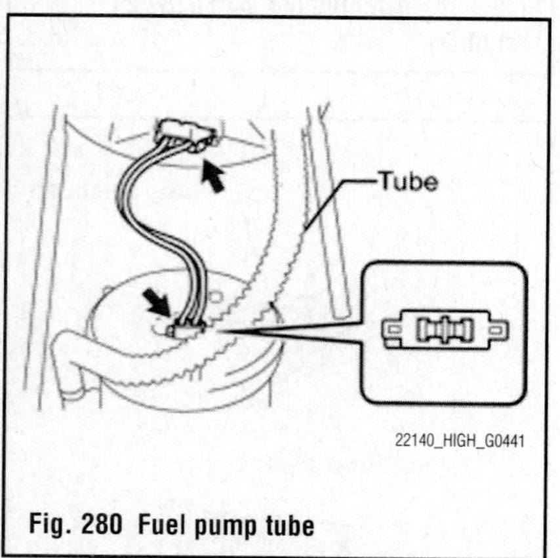

Fig. 280 Fuel pump tube

➡Do not separate the tube indicated in the illustration.

5. Using a screwdriver with its tip wrapped in protective tape, disengage the 5 claws on the filter and remove the fuel pump from the fuel filter.

➡Do not damage the fuel filter.

➡Do not remove the suction filter.

➡Do not use either the fuel pump or the suction filter if the suction filter is removed from the fuel pump.

6. Disconnect the fuel pump connector.
7. Remove the O-ring from the fuel filter.
8. Remove the fuel pressure regulator from the fuel filter.

To install:

9. Apply a light coat of gasoline or spindle oil to 2 new O-rings, then install them onto the fuel pressure regulator.
10. Install the fuel pressure regulator to the fuel filter.
11. Apply gasoline to a new O-ring and install it to the fuel filter.
12. Engage the 5 claws on the fuel filter and install the fuel pump with the pump filter.
13. Connect the fuel pump harness connector.

➡Do not separate the tube indicated in the illustration.

14. Install the fuel suction plate sub-assembly, as follows:
    a. Install the fuel pump filter tube while aligning it to the installation position of the No. 1 fuel sub-tank.
    b. Connect the jet pump nozzle.
    c. Make sure that the fuel tube passes under the protrusion of the fuel filter, and engage the claw of the No. 1 fuel suction support.
    d. Install the spring to the fuel suction plate shaft and install it to the No. 1 fuel sub-tank.
    e. Install a new E-ring.
15. Slide the fuel sender gauge to fit the claw.
16. Connect the fuel sender gauge connector.

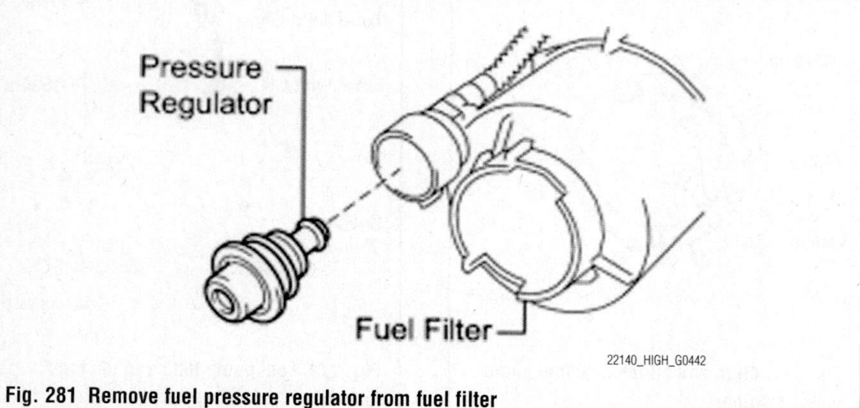

Fig. 281 Remove fuel pressure regulator from fuel filter

FUEL RAIL AND INJECTOR

REMOVAL & INSTALLATION

2.7L Engine

See Figures 282 through 285.

⁂ CAUTION

Observe all applicable safety precautions when working around fuel. Whenever servicing the fuel system, always work in a well ventilated area. Do not allow fuel spray or vapors to come in contact with a spark or open flame. Keep a dry chemical fire extinguisher near the work area. Always keep fuel in a container specifically designed for fuel storage; also, always properly seal fuel containers to avoid the possibility of fire or explosion.

1. Properly relieve the fuel system pressure.
2. Disconnect the negative battery cable.
3. Remove no. 1 engine cover sub-assembly.
4. Remove air cleaner cap sub-assembly.
5. Disconnect fuel tube sub-assembly:
   a. Remove the No. 1 fuel pipe clamp.
   b. Pinch the tube connector, and then pull the tube connector off of the pipe.

➡Note the following:

- Check for foreign matter in the fuel tube around the fuel tube connector. Clean it if necessary. Foreign matter can affect the ability of the O-ring to seal the connector and fuel pipe.
- Do not use any tools to separate the connector and pipe.
- Do not forcefully bend, kink or twist the hose.
- Keep the connector and pipe free from foreign matter.
- If the connector and pipe are stuck together, pinch the connector and turn it carefully to disconnect it.

- Put the connector in a plastic bag to prevent damage and contamination.

c. Remove the fuel tube sub-assembly from the fuel hose clamp.

6. Disconnect wire harness:

a. Disconnect the 4 fuel injector connectors.

b. Disconnect the 3 connectors.

c. Remove the 2 bolts and 2 wire harness brackets.

d. Detach the 2 clamps to disconnect the wire harness.

7. Remove vacuum switching valve assembly (for ACIS).

8. Remove fuel delivery pipe sub-assembly:

a. Remove the 2 bolts, and then remove the fuel delivery pipe together with the 4 fuel injectors.

➡Be careful not to drop the fuel injectors when removing the fuel delivery pipe.

b. Remove the 2 fuel delivery spacers from the cylinder head.

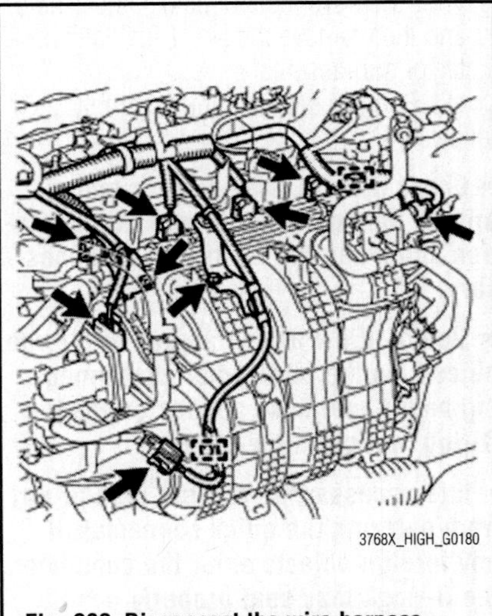

Fig. 282 Disconnect the wire harness

Fig. 283 Remove the 2 bolts, and then remove the fuel delivery pipe together with the 4 fuel injectors

Fig. 284 Remove the 2 fuel delivery spacers from the cylinder head

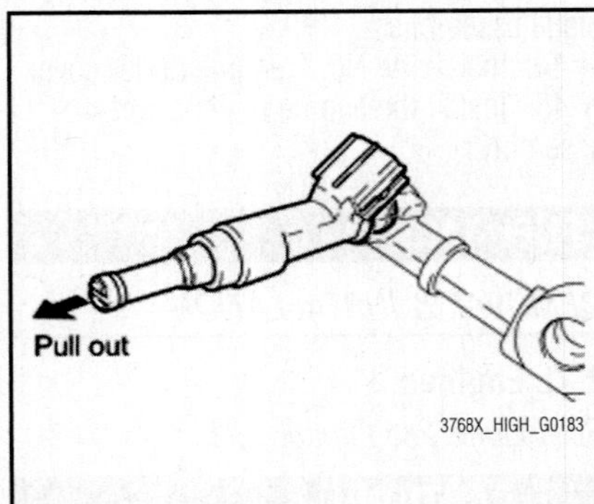

Fig. 285 Pull the 4 fuel injectors out of the fuel delivery pipe

c. Remove the 4 injector vibration insulators from the cylinder head.

9. Pull the 4 fuel injectors out of the fuel delivery pipe.

To install:

10. Apply a light coat of gasoline or spindle oil to new O-rings, and then install one onto each fuel injector.

11. Apply a light coat of gasoline or spindle oil to the part of the fuel delivery pipe which comes into contact with the O-ring of the fuel injector.

12. Apply a light coat of gasoline or spindle oil to the O-ring again, and then install the fuel injectors onto the fuel delivery pipe.

➡Make sure that the O-ring is not cracked or jammed when installing the injector.

13. Check that the fuel injector rotates smoothly. If the fuel injector does not rotate, replace the O-ring.

14. Install fuel delivery pipe sub-assembly:

a. Install 4 new injector vibration insulators to the cylinder head.

b. Install the 2 fuel delivery spacers onto the cylinder head.

➡Install the fuel delivery spacer so that the longer protrusion is on the cylinder head side.

c. Install the fuel delivery pipe together with the 4 fuel injectors to the cylinder head, and then temporarily install the 2 bolts.

d. Check that the fuel injector rotates smoothly. If the fuel injector does not rotate, replace the O-ring.

e. Tighten the 2 bolts to 15 ft. lbs. (21 Nm).

15. The remainder of installation is the reverse of the removal procedure.

3.5L Engine

See Figures 286 and 287.

1. Before servicing the vehicle, refer to the precautions section.
2. Relieve the fuel system pressure.
3. Disconnect the negative battery cable.
4. Remove the engine under cover assembly.
5. Remove the No. 1 engine under cover.
6. Drain the coolant.
7. Remove the front wiper arm and blade assemblies.
8. Remove the cowl top ventilator louver sub-assembly.
9. Remove the wiper motor and link assembly.
10. Remove the outer cowl top panel sub-assembly.
11. Remove the V-bank cover sub-assembly.
12. Remove the air cleaner cap sub-assembly.
13. Disconnect the engine room main wire.
14. Remove the throttle body bracket.
15. Remove the No. 1 surge tank stay.
16. Remove the intake air surge tank assembly.

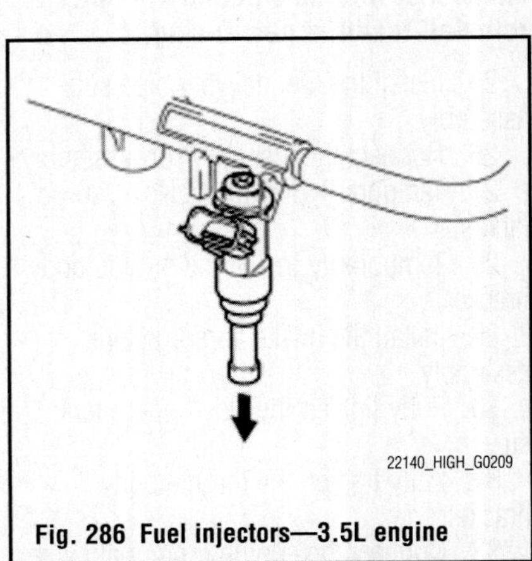

Fig. 286 Fuel injectors—3.5L engine

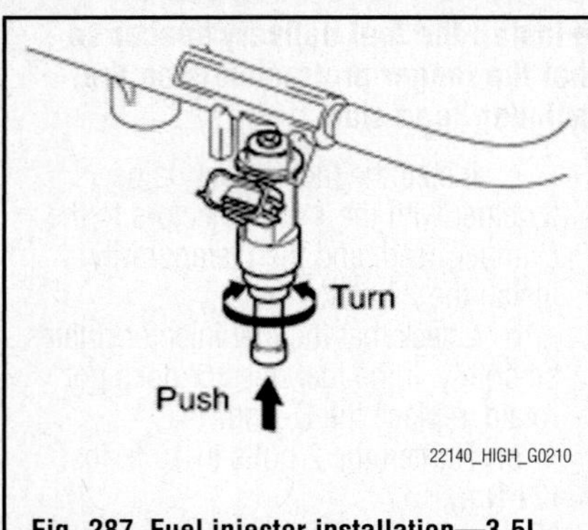

Fig. 287 Fuel injector installation—3.5L engine

17. Disconnect the fuel tube sub-assembly.
18. Remove the fuel delivery pipe sub-assembly.
19. Pull out the fuel injectors from the fuel delivery pipe.

➡If the injectors are to be reused, reinstall them to the same cylinder they came from.

20. Remove the 6 O-rings from the injectors.

To install:

21. Apply a light coat of spindle oil or gasoline to new O-rings, and install them to each injector.

➡The wound or the foreign body must not adhere in the ditch of O-ring.

22. Apply a light coat of spindle oil or gasoline where the fuel delivery pipe contacts the O-ring.
23. Push the fuel injector while turning it to install the injector in the fuel delivery pipe.
24. Position the fuel injector connector outward.

➡Be careful not to twist the O-ring.

➡After installing the fuel injector, check that it turns smoothly. If not, reinstall it with a new O-ring.

25. Install the fuel delivery pipe sub-assembly.
26. Connect the fuel tube sub-assembly.
27. Temporarily install the No. 1 surge tank stay.
28. Temporarily install the throttle body bracket.
29. Install the intake air surge tank assembly.
30. Fully tighten the No. 1 surge tank stay.
31. Fully tighten the throttle body bracket.
32. Connect the engine room main wire.
33. Install the air cleaner cap sub-assembly.
34. Connect the negative battery cable.
35. Inspect the SRS warning light.
36. Inspect for fuel leak.
37. Add engine coolant.
38. Inspect for engine coolant leak.
39. Install the V-bank cover sub-assembly.
40. Install the outer cowl top panel sub-assembly.
41. Install the windshield wiper motor and link assembly.
42. Install the cowl top ventilator louver sub-assembly.
43. Install both front wiper arm and blade assemblies.
44. Install the No. 1 engine under cover.
45. Install the engine under cover assembly.

FUEL TANK

REMOVAL & INSTALLATION

2.7L Engines

See Figures 288 through 298.

CAUTION

Observe all applicable safety precautions when working around fuel. Whenever servicing the fuel system, always work in a well ventilated area. Do not allow fuel spray or vapors to come in contact with a spark or open flame. Keep a dry chemical fire extinguisher near the work area. Always keep fuel in a container specifically designed for fuel storage; also, always properly seal fuel containers to avoid the possibility of fire or explosion.

1. Remove the fuel suction tube assembly with pump and gauge.
2. Drain fuel.

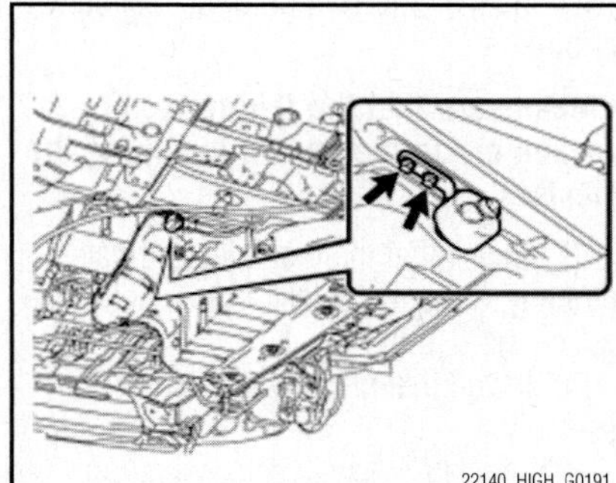

Fig. 288 No. 4 exhaust pipe support bracket

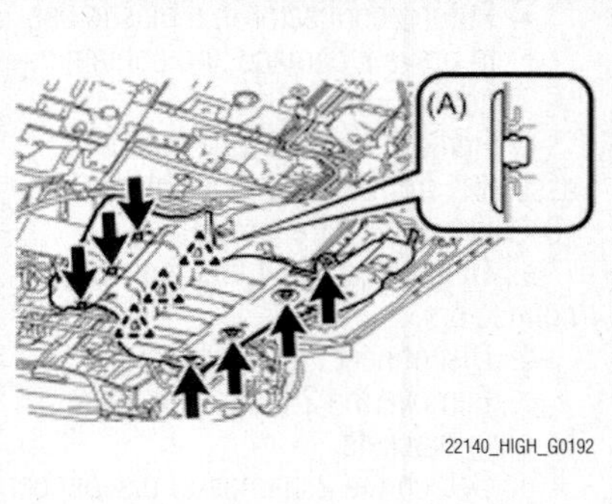

Fig. 289 No. 1 fuel tank protector sub-assembly, clips (A), and nuts

3. Remove center front floor cover.
4. Remove tail exhaust pipe assembly
5. Remove center exhaust pipe assembly.
6. Remove the No. 4 exhaust pipe support bracket, as follows:
   a. Remove the 2 bolts, and then remove the No. 4 exhaust pipe support bracket.
7. Remove the No. 1 fuel tank protector sub-assembly, as follows:
   a. Remove the 3 clips (A) and 7 nuts, and then remove the No. 1 fuel tank protector sub-assembly.
8. Remove the fuel tank assembly, as follows:

➡Check if there is any dirt or mud around the connector before this operation and clean the connector as necessary.

➡Do not allow any scratches or foreign objects on the parts when disconnecting as the fuel hose connector has the O-ring that seals the pipe.

➡It is necessary to prevent mud or dirt from entering the quick connector. If any foreign objects enter the connector, the O-rings may seal properly.

➡Perform this work by hand. Do not use any tools.

➡Do not forcibly bend, twist or turn the nylon tube.

➡Protect the connected part by covering it with a plastic bag after disconnecting the fuel tank vent hose.

➡If the connectors or pipe are stuck, push and pull them to release, and pull the connector out carefully.

   a. Disconnect the fuel pump tube by pinching the tab of the retainer to disengage the lock claws and pull it down as shown.

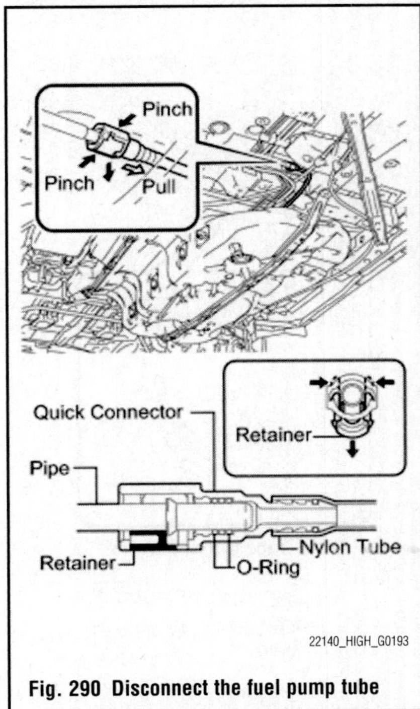

Fig. 290 Disconnect the fuel pump tube

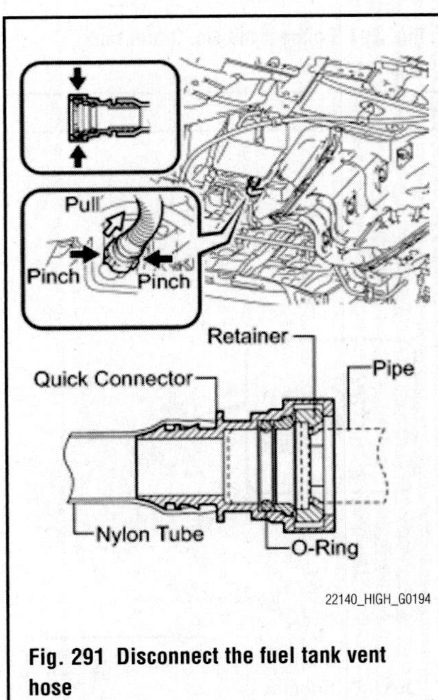

Fig. 291 Disconnect the fuel tank vent hose

b. Pull out the fuel pump tube.

c. Pinch the retainer and pull out the quick connector while pushing the quick connector against the pipe to disconnect the fuel tank vent hose from the charcoal canister assembly.

d. Pinch the tube connector and then pull out the No. 3 fuel tank breather tube.

e. Loosen the hose clamp bolt and disconnect the fuel tank to filler pipe hose.

f. Set a transmission jack under the fuel tank.

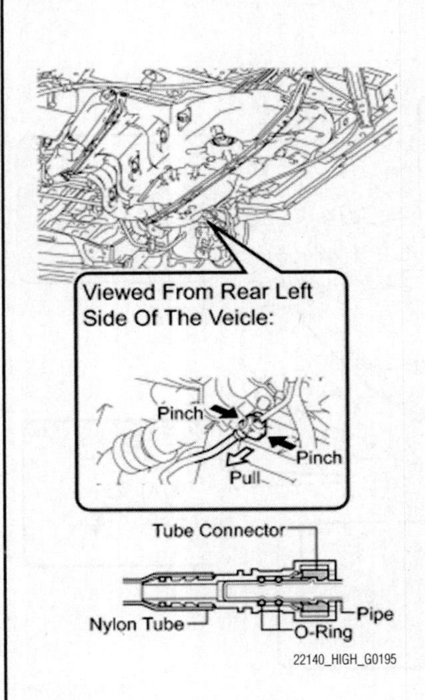

Fig. 292 Disconnect the No. 3 fuel tank breather tube

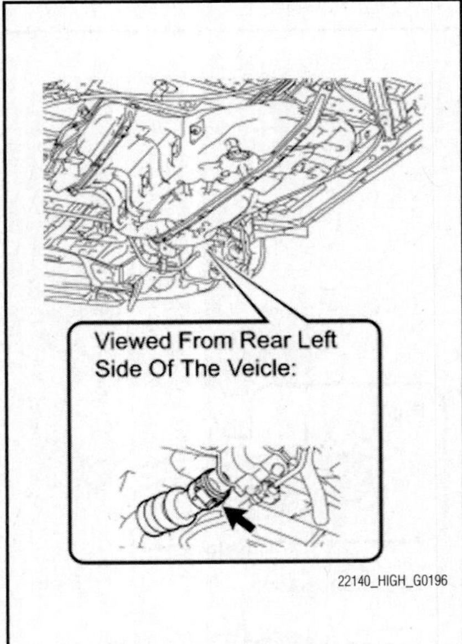

Fig. 293 Disconnect the fuel tank to filler pipe hose

g. Remove the 4 bolts, and then remove the 2 fuel tank bands.

h. Remove the 2 nuts.

i. Operate the transmission jack to remove the fuel tank.

To install:

9. Set the fuel tank onto a transmission jack.

10. Operating the transmission jack, install the fuel tank.

11. Tighten the 2 nuts to 14 ft. lbs. (20 Nm).

12. Install the 2 fuel tank bands with the 4 bolts and tighten to 29 ft. lbs. (39 Nm).

13. Align the matchmarks and install the fuel tank to filler pipe hose to the fuel tank.

14. Install the hose clamp within the range shown in the illustration.

15. Connect the fuel tank vent hose:

a. Align the quick connector with the pipe, then push in the quick connector until the retainer makes a click sound to connect the charcoal canister fuel hose to the charcoal canister assembly.

➡After connecting the charcoal canister fuel hose, check if the quick connector and the pipe are securely connected by pulling on them.

➡Check that there are no scratches or foreign objects around the connected part of the quick connector and pipe before this work.

16. Connect the No. 3 fuel tank breather tube:

a. Push the quick connector to the pipe until it makes a click sound.

➡After connecting, check if the quick connector and the pipe are securely connected by pulling on them.

➡Check if there is any damage or foreign objects on the connected part.

17. Connect the fuel pump tube:

a. Push the quick connector and push in the retainer to lock the claws.

➡After connecting, check if the quick connector and the pipe are securely connected by pulling on them.

➡Check if there is any damage or foreign objects on the connected part.

18. Install the No. 1 fuel tank protector sub-assembly with the 7 nuts and tighten to 49 inch lbs. (5.5 Nm).

19. Install 3 new clips.

20. Install the No. 3 front floor heat insulator.

21. Install the No. 4 exhaust pipe support bracket with the 2 bolts and tighten to 16 ft. lbs. (22 Nm).

22. Install center exhaust pipe assembly.

23. Install tail exhaust pipe assembly.

24. Install front center floor cover.

25. Install the fuel suction tube assembly with pump and gauge, as outlined in the fuel pump installation instructions.

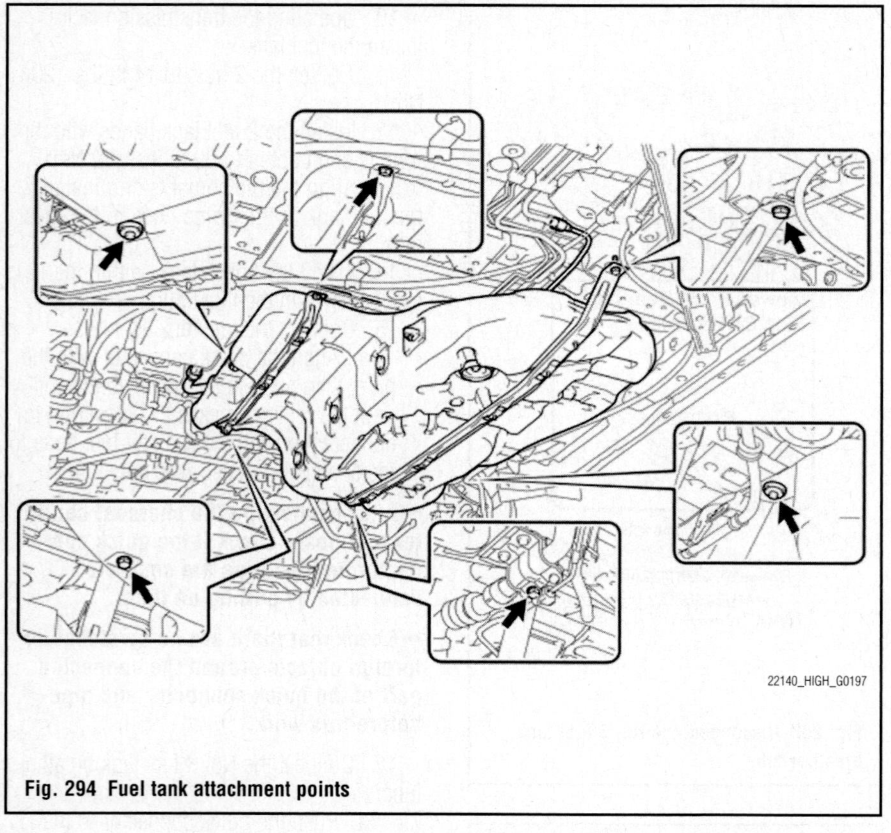

Fig. 294 Fuel tank attachment points

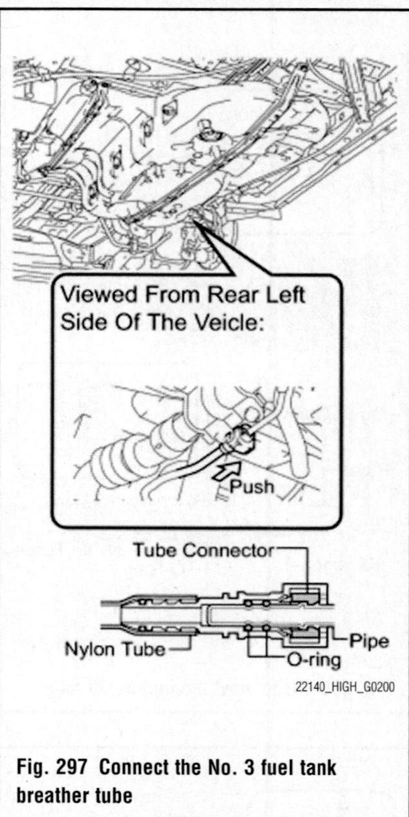

Fig. 297 Connect the No. 3 fuel tank breather tube

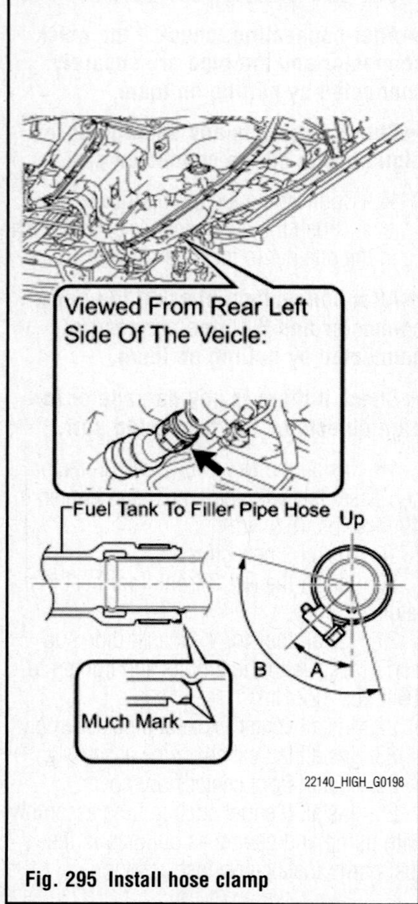

Fig. 295 Install hose clamp

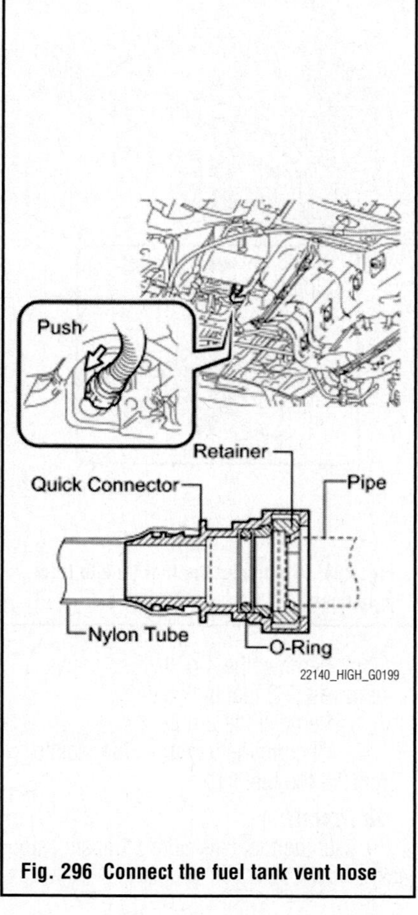

Fig. 296 Connect the fuel tank vent hose

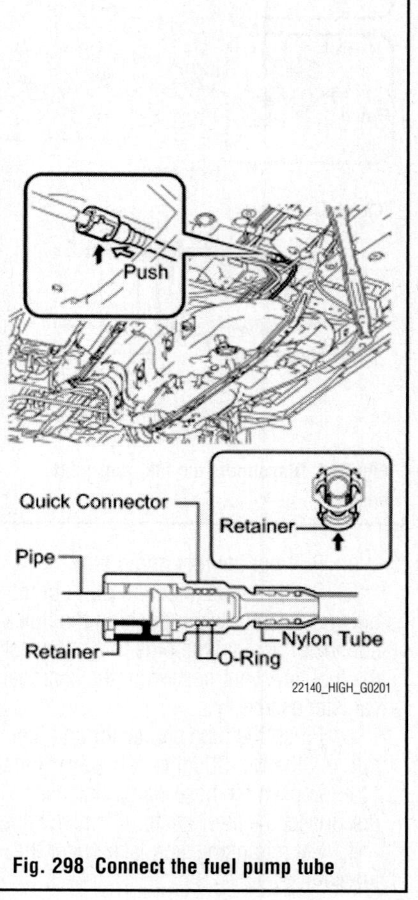

Fig. 298 Connect the fuel pump tube

26. After installing all components as outlined in the fuel pump installation instructions, connect battery negative cable and inspect SRS warning light.

27. Inspect for exhaust gas leak.

3.5L Engines

See Figures 288 through 292, 294 through 296 and 298.

1. Before servicing the vehicle, refer to the precautions section.
2. Discharge fuel system pressure.
3. Disconnect the negative battery cable.
4. Remove fuel suction tube assembly with pump and gauge, as outlined in the fuel pump removal instructions.
5. Drain fuel.
6. Remove the front center floor cover.
7. Remove the tail exhaust pipe assembly.
8. Remove the center exhaust pipe assembly.
9. Remove the No. 4 exhaust pipe support bracket, as follows:
    a. Remove the 2 bolts, and then remove the No. 4 exhaust pipe support bracket.
10. Remove the No. 3 front floor heat insulator.
11. For 4WD vehicles, remove the following:
    a. Remove the propeller with center bearing shaft assembly.
    b. Remove both rear wheels.
    c. Separate both rear speed sensors.
    d. Remove the rear axle shaft nuts.
    e. Remove both rear disc brake caliper assemblies.
    f. Remove both rear discs.
    g. Remove both rear axle hub and bearing assemblies.
    h. Remove the No. 2 and 3 parking brake cable assemblies.
    i. Remove both rear strut rod assemblies.
    j. Remove both rear axle carrier sub-assemblies.
    k. Remove both rear No. 2 suspension arm assemblies.
    l. Remove the right rear No. 1 suspension arm assembly.
    m. Remove the rear differential filler plug.
    n. Remove the rear differential drain plug.
    o. Remove the rear drive shaft assemblies and snap rings, left and right.
    p. Remove the rear suspension member.
12. Remove the No. 1 fuel tank protector sub-assembly, as follows:
    a. Remove the 3 clips (A) and 7 nuts, and then remove the No. 1 fuel tank protector sub-assembly.
13. Remove the fuel tank assembly, as follows:

➡Check if there is any dirt or mud around the connector before this operation and clean the connector as necessary.

➡Do not allow any scratches or foreign objects on the parts when disconnecting as the fuel hose connector has the O-ring that seals the pipe.

➡It is necessary to prevent mud or dirt from entering the quick connector. If any foreign objects enter the connector, the O-rings may seal properly.

➡Perform this work by hand. Do not use any tools.

➡Do not forcibly bend, twist or turn the nylon tube.

➡Protect the connected part by covering it with a plastic bag after disconnecting the fuel tank vent hose.

➡If the connectors or pipe are stuck, push and pull them to release, and pull the connector out carefully.

a. Disconnect the fuel pump tube by pinching the tab of the retainer to disengage the lock claws and pull it down as shown.
b. Pull out the fuel pump tube.
c. Pinch the retainer and pull out the quick connector while pushing the quick connector against the pipe to disconnect the fuel tank vent hose from the charcoal canister assembly.
d. Pinch the tube connector and then pull out the No. 3 fuel tank breather tube.
e. Loosen the hose clamp bolt and disconnect the fuel tank to filler pipe hose.
f. Set a transmission jack under the fuel tank.
g. Remove the 4 bolts, and then remove the 2 fuel tank bands.
h. Remove the 2 nuts.
i. Operate the transmission jack to remove the fuel tank.

To install:

14. Set the fuel tank onto a transmission jack.
15. Operating the transmission jack, install the fuel tank.
16. Tighten the 2 nuts to 14 ft. lbs. (20 Nm).
17. Install the 2 fuel tank bands with the 4 bolts and tighten to 29 ft. lbs. (39 Nm).
18. Align the matchmarks and install the fuel tank to filler pipe hose to the fuel tank.
19. Install the hose clamp within the range shown in the illustration.
20. Connect the fuel tank vent hose:
    a. Align the quick connector with the pipe, then push in the quick connector until the retainer makes a click sound to connect the charcoal canister fuel hose to the charcoal canister assembly.

➡After connecting the charcoal canister fuel hose, check if the quick connector and the pipe are securely connected by pulling on them.

➡Check that there are no scratches or foreign objects around the connected part of the quick connector and pipe before this work.

21. Connect the No. 3 fuel tank breather tube:
    a. Push the quick connector to the pipe until it makes a click sound.

➡After connecting, check if the quick connector and the pipe are securely connected by pulling on them.

➡Check if there is any damage or foreign objects on the connected part.

22. Connect the fuel pump tube:
    a. Push the quick connector and push in the retainer to lock the claws.

➡After connecting, check if the quick connector and the pipe are securely connected by pulling on them.

➡Check if there is any damage or foreign objects on the connected part.

23. Install the No. 1 fuel tank protector sub-assembly with the 7 nuts and tighten to 49 inch lbs. (5.5 Nm).
24. Install 3 new clips.
25. For 4WD vehicles, install the following:
    a. Temporarily install the rear suspension member.
    b. Fully tighten the rear suspension member.
    c. Install rear drive shaft assemblies and snap rings on left and right.
    d. Install the rear differential drain plug.
    e. Add differential oil.
    f. Inspect differential oil.
    g. Install the rear differential filler plug.
    h. Temporarily install the right rear No. 1 suspension arm assembly.
    i. Temporarily install both rear No. 2 suspension arm assemblies.

j. Install both rear axle carrier sub-assemblies.

k. Temporarily install both rear strut rod assemblies.

l. Install both rear axle hub and bearing assemblies.

m. Install both rear discs.

n. Install rear disc brake caliper assemblies.

o. Temporarily install both rear axle shaft nuts.

p. Install rear speed sensors.

q. Stabilize suspension.

r. Fully tighten both rear no. 1 suspension arm assemblies.

s. Fully tighten both rear no. 2 suspension arm assemblies.

t. Fully tighten rear strut rod assemblies.

u. Install No. 2 and 3 parking brake cable assemblies.

v. Separate both rear disc brake caliper assemblies.

w. Remove both rear discs.

x. Inspect rear axle hub bearing looseness, left and right.

y. Inspect rear axle hub runout, left and right.

z. Install both rear discs.

26. Install rear disc brake caliper assemblies.

a. Install rear axle shaft nuts.

b. Install rear wheels.

27. Install the No. 3 front floor heat insulator.

28. Install the No. 4 exhaust pipe support bracket with the 2 bolts and tighten to 16 ft. lbs. (22 Nm).

29. Install center exhaust pipe assembly.

30. Install tail exhaust pipe assembly.

31. For 4WD, temporarily tighten propeller with center bearing shaft assembly.

32. For 4WD, fully tighten propeller with center bearing shaft assembly.

33. Install front center floor cover.

34. Install the fuel suction tube assembly with pump and gauge, as outlined in the fuel pump installation instructions.

35. After installing all components as outlined in the fuel pump installation instructions, connect battery negative cable and inspect SRS warning light.

36. Inspect for exhaust gas leak.

37. For 4WD, inspect and adjust transfer oil.

38. For 4WD, inspect rear wheel alignment.

39. For 4WD, check for speed sensor signal.

IDLE SPEED

ADJUSTMENT

Idle speed is maintained by the Powertrain Control Module (PCM). No adjustment is necessary or possible.

THROTTLE BODY

REMOVAL & INSTALLATION

2.7L Engines

See Figure 299.

1. Remove windshield wiper motor and link.
2. Separate ejector tube.
3. Remove outer cowl top panel sub-assembly.
4. Remove cool air intake duct seal.
5. Remove no. 1 engine under cover.
6. Drain engine coolant.
7. Remove no. 1 engine cover sub-assembly.
8. Remove no. 1 vacuum switching valve.
9. Remove air cleaner cap sub-assembly.
10. Remove throttle body assembly:

a. Disconnect the throttle body assembly connector.

b. Disconnect the fuel tube from the clamp.

c. Disconnect the 2 water by-pass hoses from the throttle body assembly.

d. Remove the 4 bolts and the throttle body assembly with the fuel tube bracket.

e. Remove the bolt and fuel tube bracket.

f. Remove the gasket from the intake manifold.

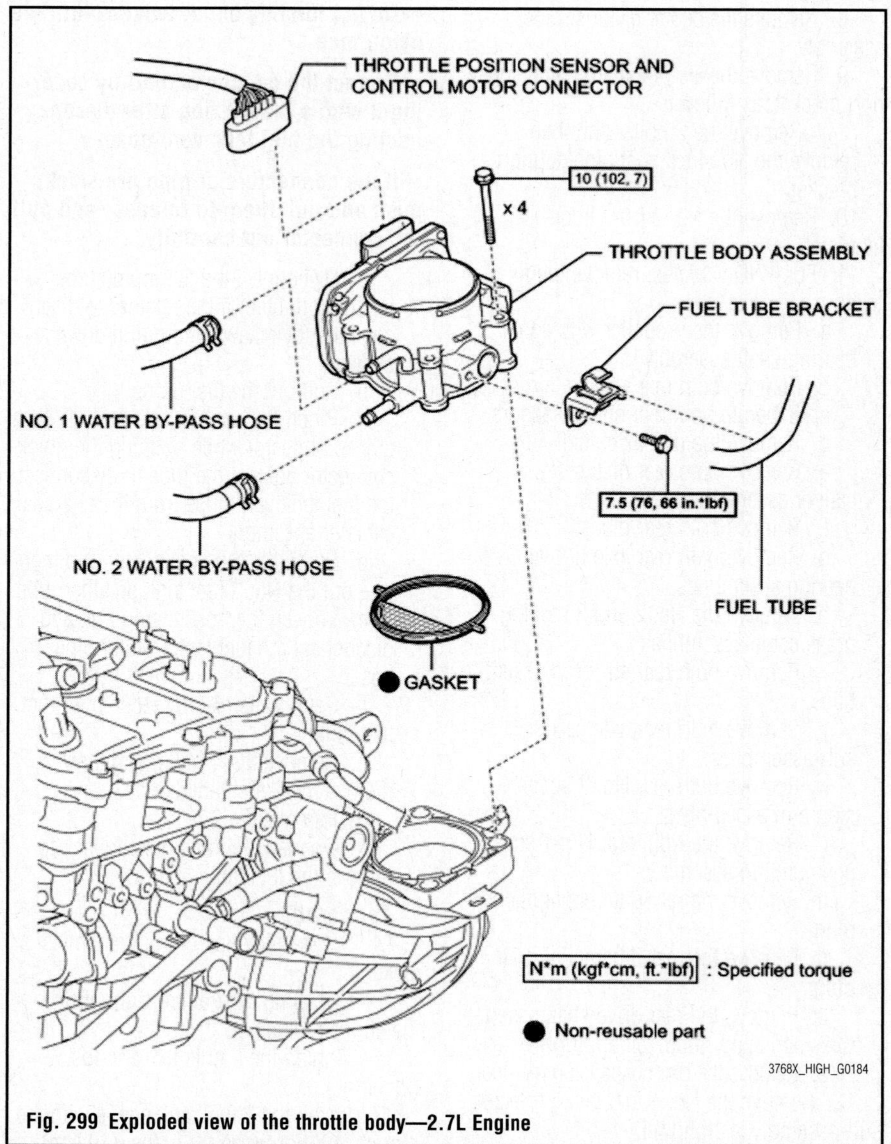

Fig. 299 Exploded view of the throttle body—2.7L Engine

To install:

11. Install a new gasket to the intake manifold.
12. Install the fuel tube bracket with the bolt.
13. Install the throttle body assembly with the 4 bolts. Tighten to 7 ft. lbs. (10 Nm).
14. Connect the 2 water by-pass hoses to the throttle body.
15. Connect the throttle body assembly connector.
16. Connect the fuel tube to the clamp.
17. The remainder of installation is the reverse of the removal procedure.

3.5L Engine

See Figure 300.

1. Remove the engine under cover assembly.
2. Remove the No. 1 engine under cover.
3. Drain engine coolant.
4. Remove both front wiper arm and blade assemblies.
5. Remove the cowl top ventilator louver sub-assembly.
6. Remove the windshield wiper motor and link assembly.
7. Remove the outer cowl top panel sub-assembly.
8. Remove the V-bank cover sub-assembly.
9. Remove the air cleaner cap sub-assembly.
10. Disconnect the throttle body connector and clamp.
11. Disconnect the 2 water by-pass hoses from the throttle body.
12. Remove the 4 bolts and throttle body.
13. Remove the throttle body gasket from the intake air surge tank.

To install:

14. Install a new throttle body gasket to the intake air surge tank.
15. Install the throttle body with the 4 bolts and tighten to 7 ft. lbs. (10 Nm).
16. Connect the 2 water by-pass hoses.

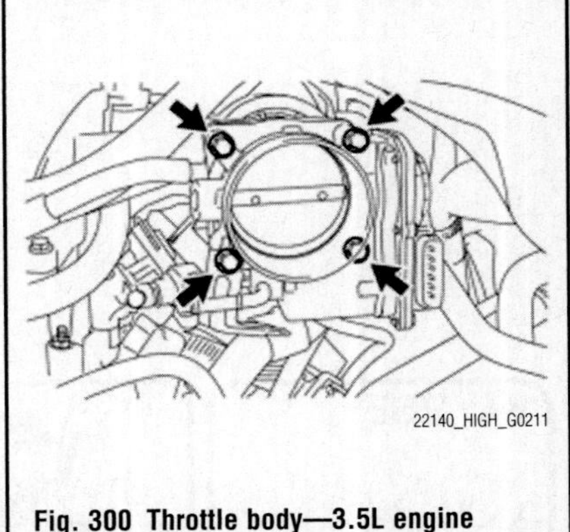

Fig. 300 Throttle body—3.5L engine

17. Connect the throttle body connector and clamp.
18. Install the air cleaner cap sub-assembly.
19. Add engine coolant.
20. Inspect for coolant leak.
21. The remainder of installation is the reverse of the removal procedure.

HEATING & AIR CONDITIONING SYSTEM

AIR CONDITIONING UNIT

REMOVAL & INSTALLATION

See Figures 301 through 306.

1. Recover refrigerant from refrigeration system.
2. Position front wheels straight ahead.
3. Remove both front wheels.

⁂ CAUTION

Wait for 90 seconds after disconnecting the cable to prevent the airbag from deploying.

4. Remove both front wiper arm and blade assemblies.
5. Remove the cowl top ventilator louver sub-assembly.
6. Remove the windshield wiper motor and link assembly.
7. Remove the cowl top outer panel sub-assembly.
8. Disconnect the heater inlet and outlet water hoses.
9. Disconnect the cooler refrigerant liquid pipe.
10. Disconnect the No. 1 cooler refrigerant suction pipe.
11. Remove the lower No. 2 and 3 steering wheel covers.
12. Remove the steering pad.
13. Remove the steering wheel assembly.
14. Remove the steering column cover.
15. Remove the turn signal switch assembly with spiral cable sub-assembly.
16. Remove the instrument cluster finish panel assembly.
17. Remove the combination meter assembly.
18. Remove the center instrument panel register assembly.
19. Remove the center instrument cluster finish panel assembly.
20. Remove the air conditioning control assembly.
21. Remove the radio receiver assembly with bracket.
22. Remove both front door scuff plates.
23. Remove both cowl side trim sub-assemblies.
24. Remove the lower instrument panel finish panel sub-assembly.
25. Remove the No. 2 instrument panel under cover sub-assembly.
26. Remove the lower instrument panel sub-assembly.
27. Remove the upper console panel sub-assembly.
28. Remove the No. 2 console box duct (w/o rear air conditioning system).
29. Remove the lower rear console box.
30. Remove the console box assembly.
31. Remove the front No. 1 No. 2 console box insert.
32. Remove the engine switch (W/ Smart Key System).
33. Remove the front pillar garnish.
34. Disconnect the front door opening trim weatherstrip.
35. Remove the No. 1 instrument panel speaker panel sub-assembly.
36. Remove the front No. 2 speaker assembly.
37. Disconnect the instrument panel wire assembly.
38. Remove the instrument panel safety pad assembly.
39. Remove the brake pedal return spring.
40. Remove the stop light switch assembly.
41. Separate the brake master cylinder push rod clevis.
42. Remove the brake pedal support sub-assembly.
43. Remove the driver side knee airbag assembly.
44. Remove the No. 1 air duct sub-assembly.
45. Separate the steering intermediate shaft sub-assembly.
46. Remove the steering column assembly.
47. Remove the certification ECU (smart key ECU assembly) (W/ smart key system).
48. Remove the air conditioning amplifier assembly.
49. Remove the rear No. 1 air duct.
50. Remove the rear No. 3 air duct.
51. Remove the No. 1 console box duct (w/o rear air conditioning system).

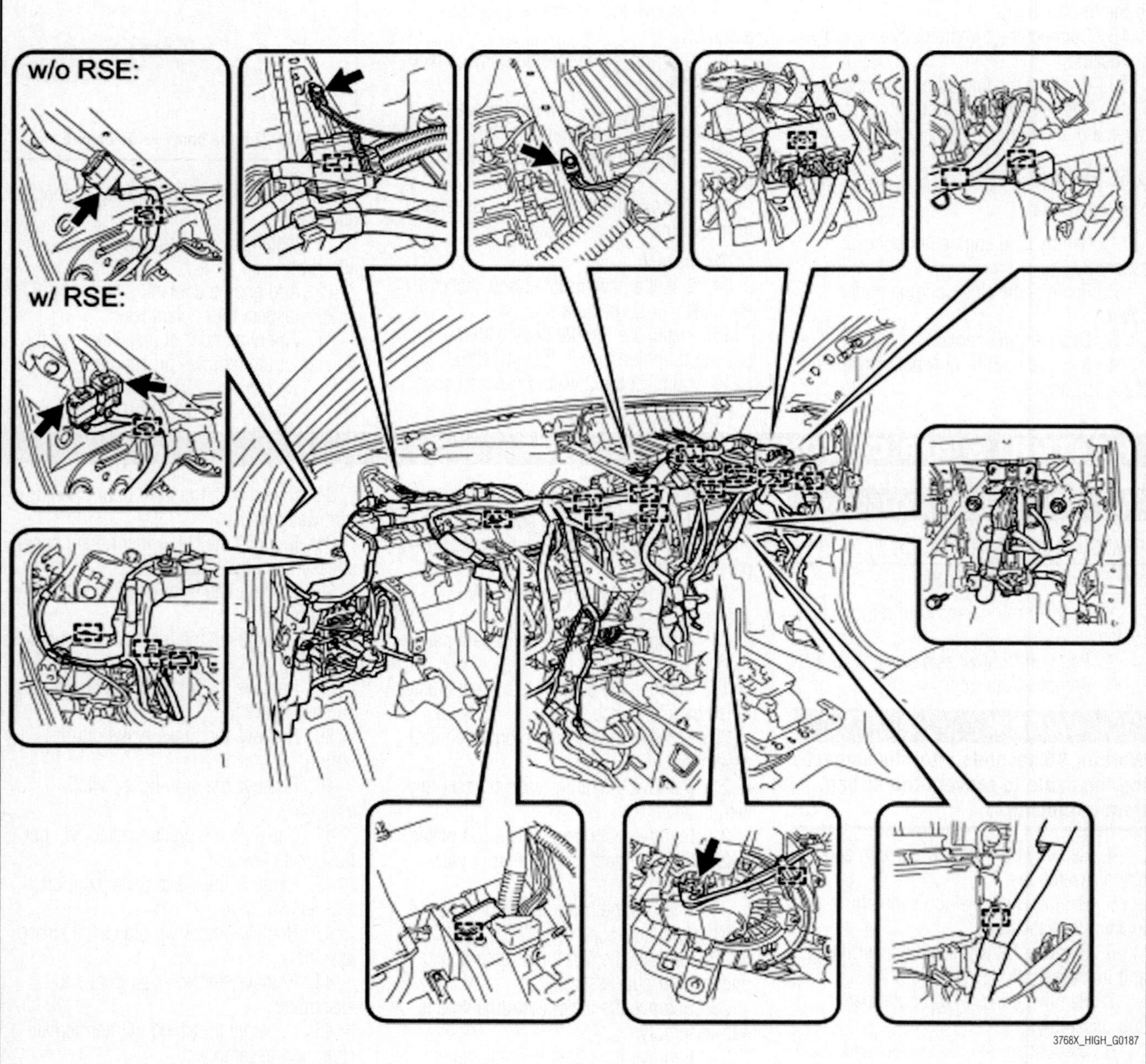

Fig. 301 Remove instrument panel reinforcement assembly with air conditioning unit—1 of 3

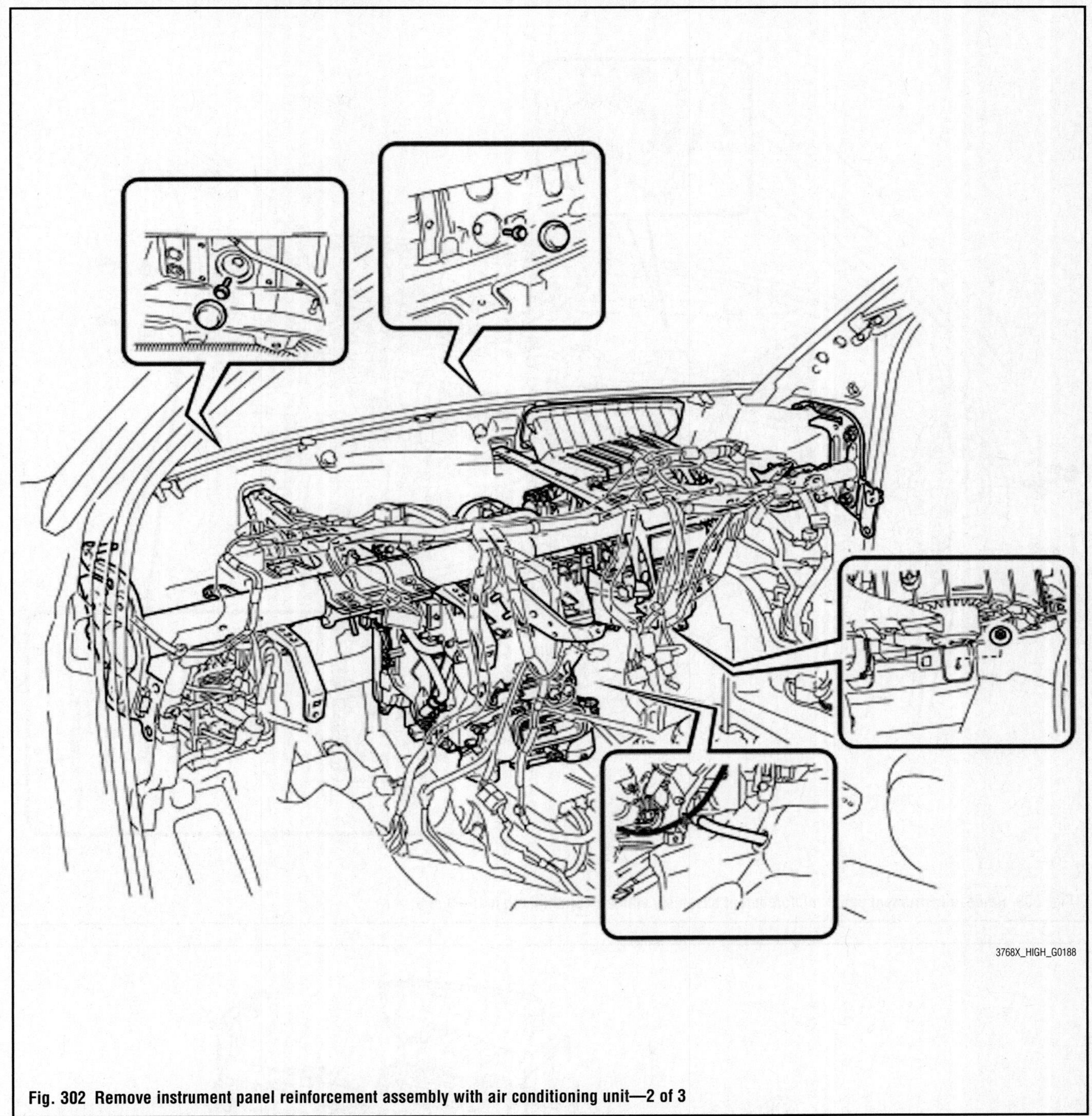

Fig. 302 Remove instrument panel reinforcement assembly with air conditioning unit—2 of 3

52. Remove the center heater to register duct.
53. Remove the No. 1 and 2 instrument panel brace sub-assemblies.
54. Remove instrument panel reinforcement assembly with air conditioning unit (w/o PTC Heater).
    a. Disengage each clamp.
    b. Remove the 2 bolts and disconnect the 2 earth wires.
    c. Disconnect each connector.
    d. Remove the 2 nuts and bolt.
    e. Disconnect the blower motor connector.
    f. Remove the 2 caps and the 2 bolts from the engine compartment side.
    g. Disconnect the cooler drain hose.
    h. Remove the nut.
    i. Using a TORX® socket wrench (T40), remove the 5 TORX® bolts.

➡The TORX® bolts on the passenger side can be removed with the collars for adjustment.

    j. Using a 12 mm hexagon wrench, remove the 2 collars and the instrument panel reinforcement assembly with air conditioning unit.
    k. Disconnect the connector.
55. Remove the 3 bolts and the air conditioning unit from the instrument panel reinforcement assembly.

To install:

56. To install, reverse the removal procedure. Refer to the illustrations for torque values.

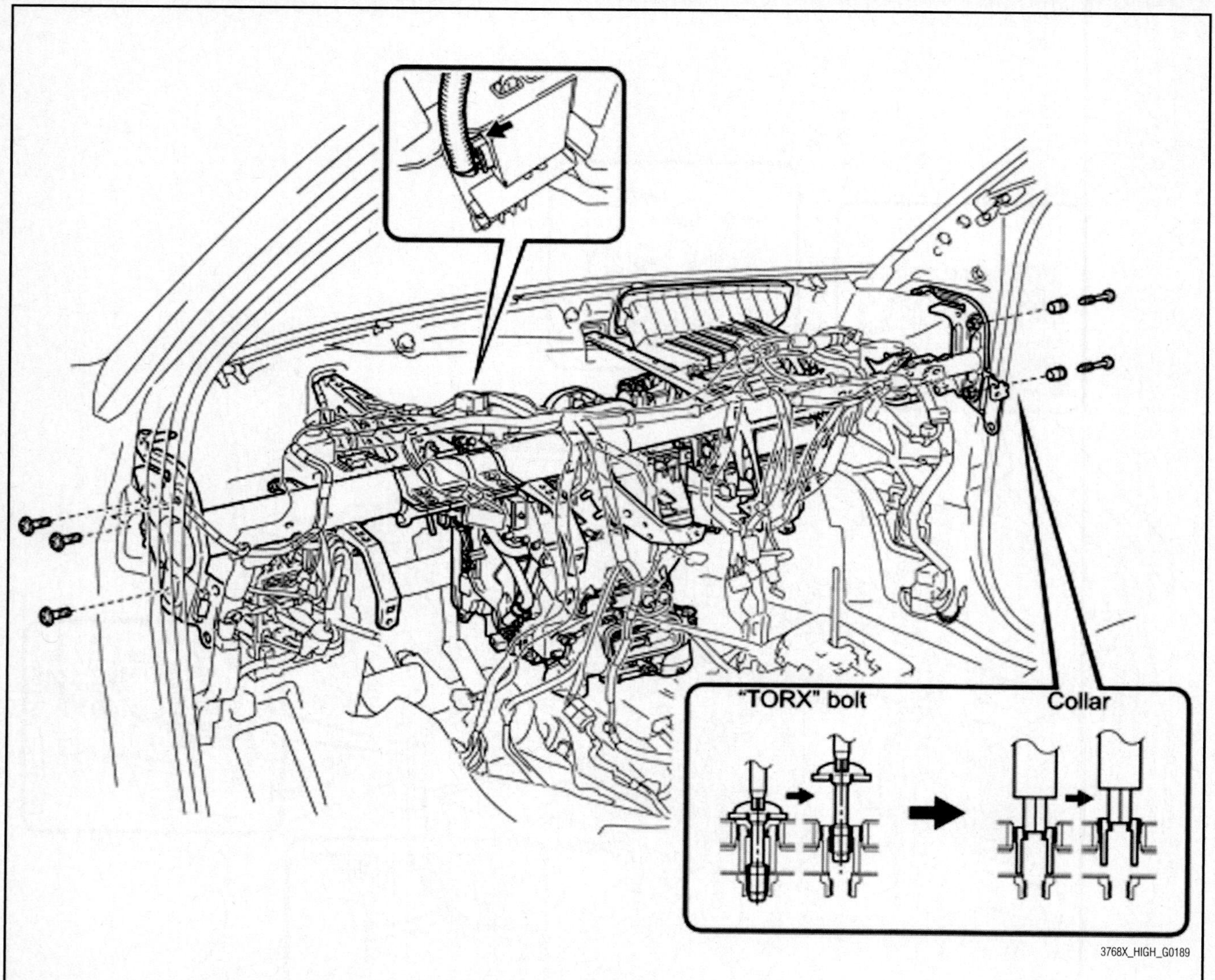

"TORX" bolt

Collar

3768X_HIGH_G0189

Fig. 303 Remove instrument panel reinforcement assembly with air conditioning unit—3 of 3

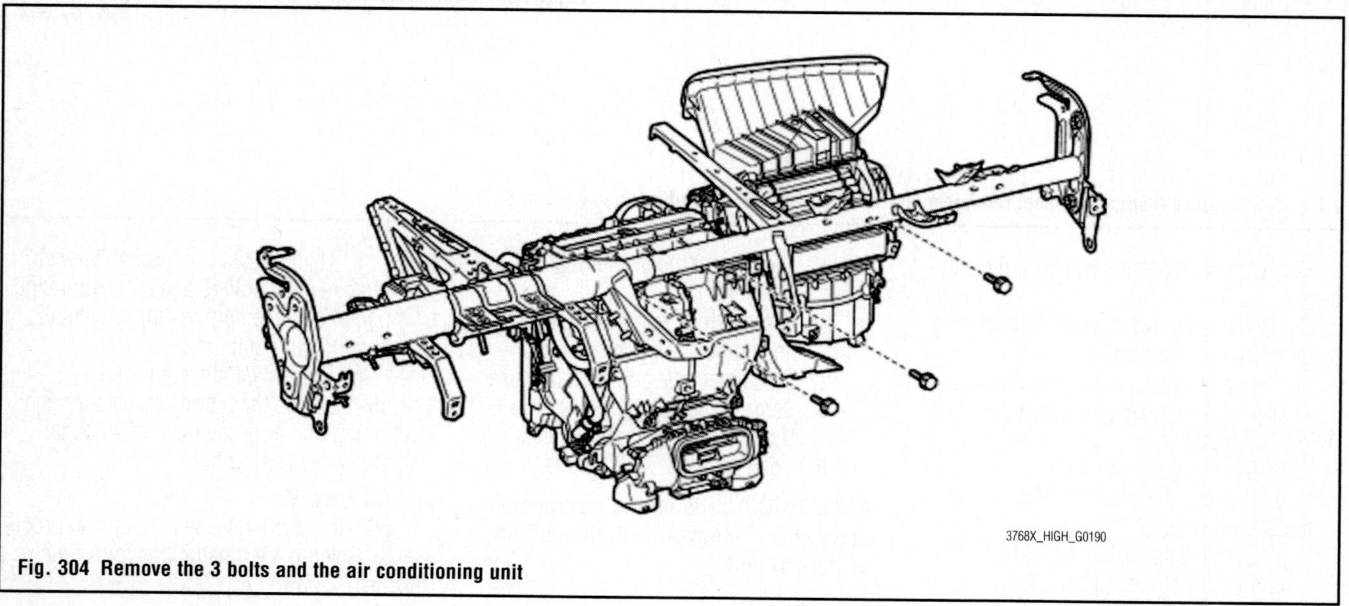

Fig. 304 Remove the 3 bolts and the air conditioning unit

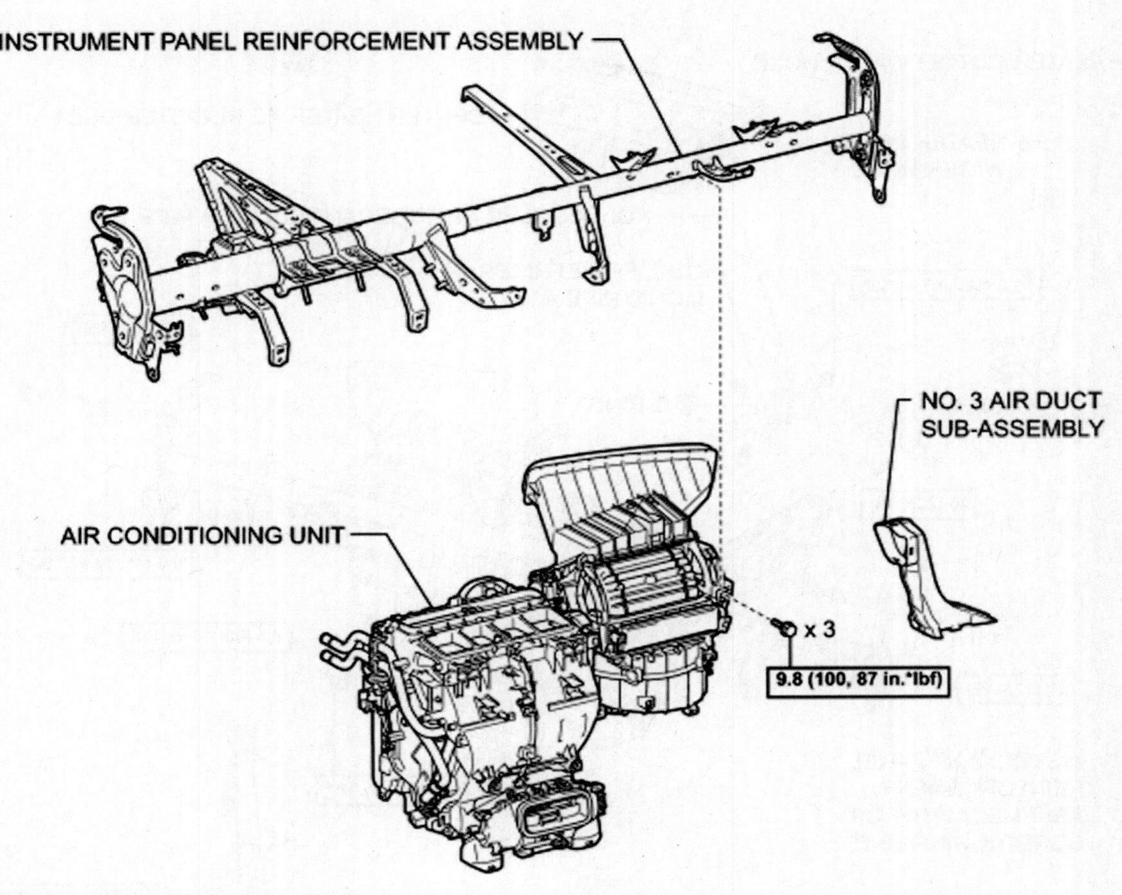

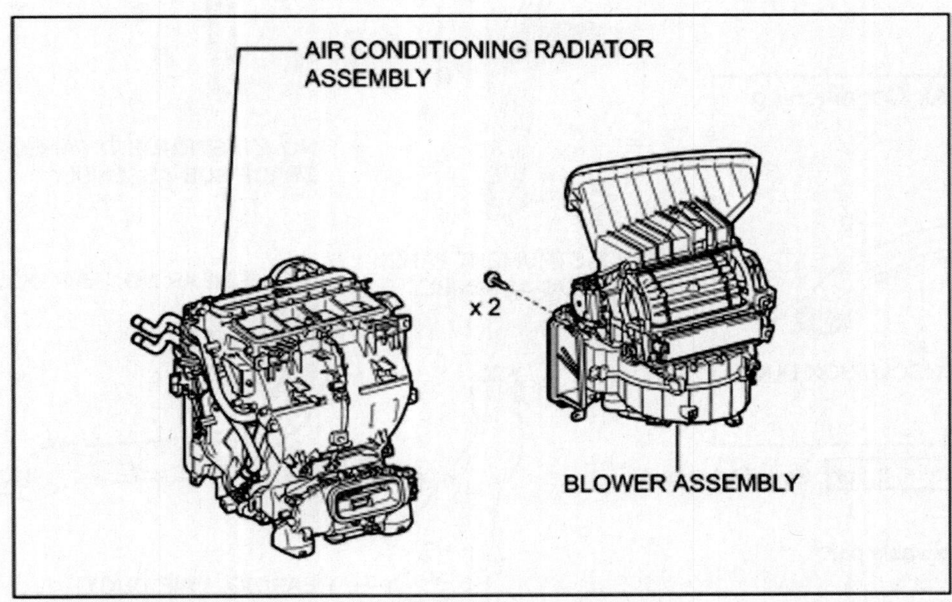

N*m (kgf*cm, ft.*lbf): Specified torque

3768X_HIGH_G0191

Fig. 305 Exploded view of the air conditioning unit and the instrument panel reinforcement assembly (1 Of 2)

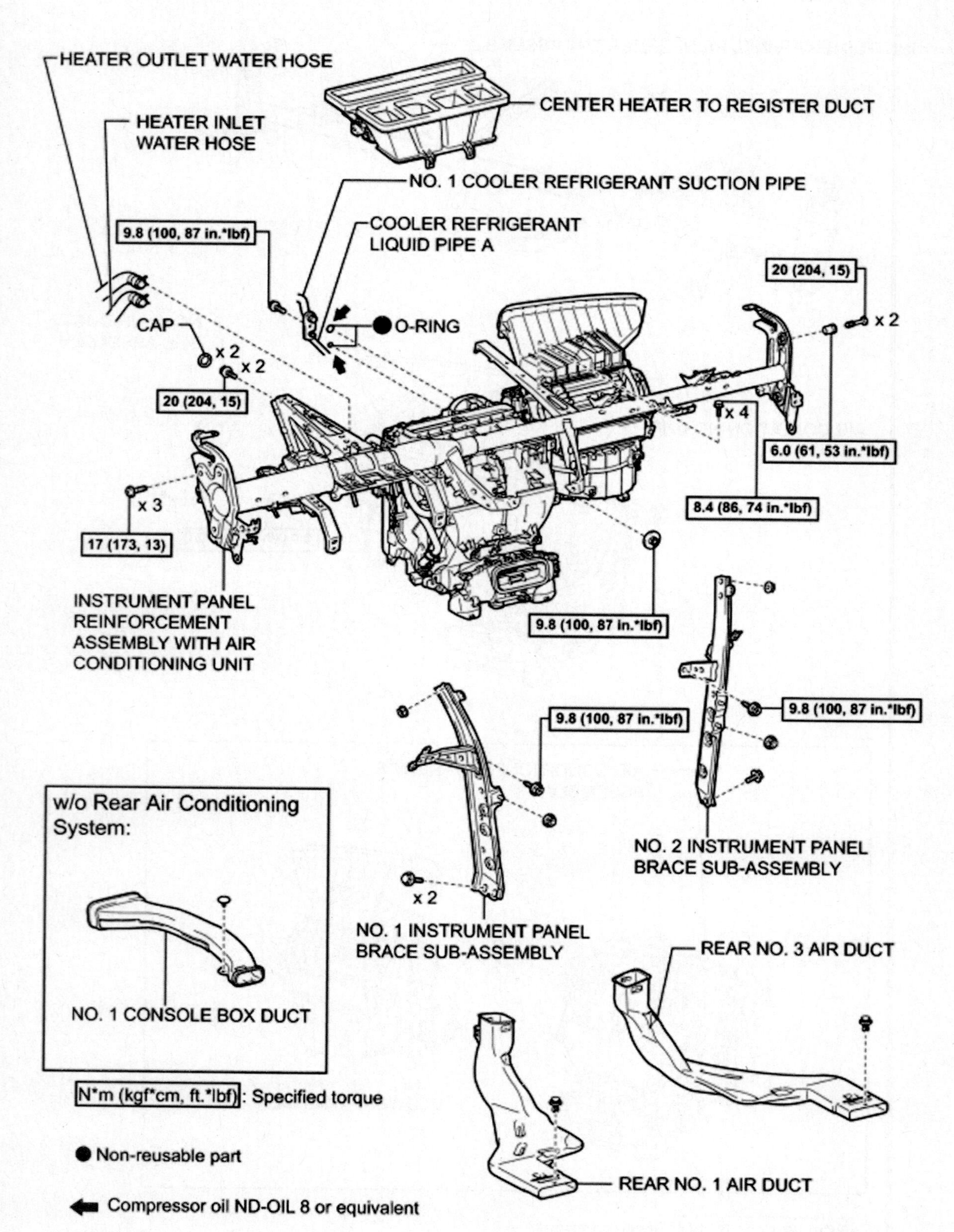

Fig. 306 Exploded view of the air conditioning unit and the instrument panel reinforcement assembly (2 Of 2)

BLOWER MOTOR

REMOVAL & INSTALLATION

See Figure 307.

1. Remove the No. 2 instrument panel under cover sub-assembly.
2. Remove front blower motor sub-assembly, as follows:
   a. Disconnect the connector.
   b. Remove the 3 screws and the front blower motor sub-assembly.

To install:

3. Installation is the reverse of the removal procedure.

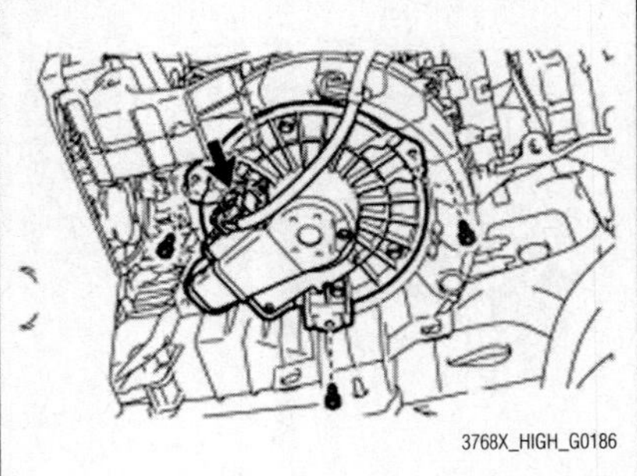

Fig. 307 Remove the 3 screws and the front blower motor sub-assembly

HEATER CORE

REMOVAL AND INSTALLATION

See Figures 308 and 309.

1. Remove the center instrument panel register assembly.
2. Remove the center instrument cluster finish panel assembly.
3. Remove the heater control and accessory assembly (for manual air conditioning system).
4. Remove the air conditioning control assembly (for automatic air conditioning system).
5. Remove the radio receiver assembly with bracket (w/o navigation system).
6. Remove the navigation receiver assembly with bracket (w/ navigation system).
7. Remove the right front door scuff plate.
8. Remove the right cowl side trim sub-assembly.
9. Remove the No. 2 instrument panel under cover sub-assembly.
10. Remove the lower instrument panel sub-assembly.
11. Remove the upper console panel sub-assembly.
12. Remove the lower rear console box.
13. Remove the console box assembly.
14. Remove the front No. 2 console box insert.
15. Disengage the 3 claws and remove the No. 3 air duct sub-assembly as shown in the illustration.
16. Remove quick heater assembly, as follows:
    a. Disconnect the 2 connectors.
    b. Remove the 3 screws and the quick heater assembly as shown in the illustration.

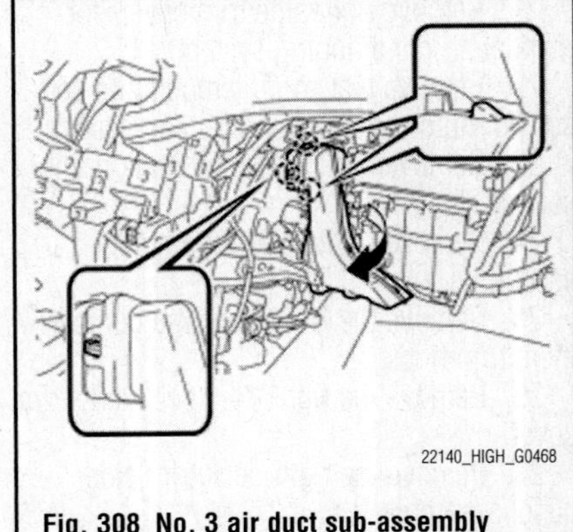

Fig. 308 No. 3 air duct sub-assembly

Fig. 309 Quick heater assembly and screws

To install:

17. Installation is the reverse of the removal procedure.

AUXILIARY HEATING & AIR CONDITIONING SYSTEM

BLOWER MOTOR

REMOVAL & INSTALLATION

See Figure 310.

1. Remove the right rear door scuff plate.
2. Remove the right rear door opening trim weatherstrip.
3. Remove the deck board assembly.
4. Remove the No. 2 and 3 deck board sub-assemblies.
5. Remove the tonneau cover assembly, as applicable.
6. Remove the rear seat side covers.
7. Remove the deck side trim boxes.
8. Remove the jack carrier support.
9. Remove the jack carrier cushion.
10. Remove jack assembly.
11. Remove the jack carrier assembly.
12. Remove rear mat.
13. Remove the deck floor board assembly.
14. Remove the rear No. 2 seat inner belt assembly.
15. Disconnect both rear seat lap type belt assemblies.
16. Remove the rear No. 2 seat assembly.
17. Remove the rear floor finish plate.
18. Remove the rear seat side garnish cap.
19. Remove right deck side trim cover.
20. Remove right deck side trim.

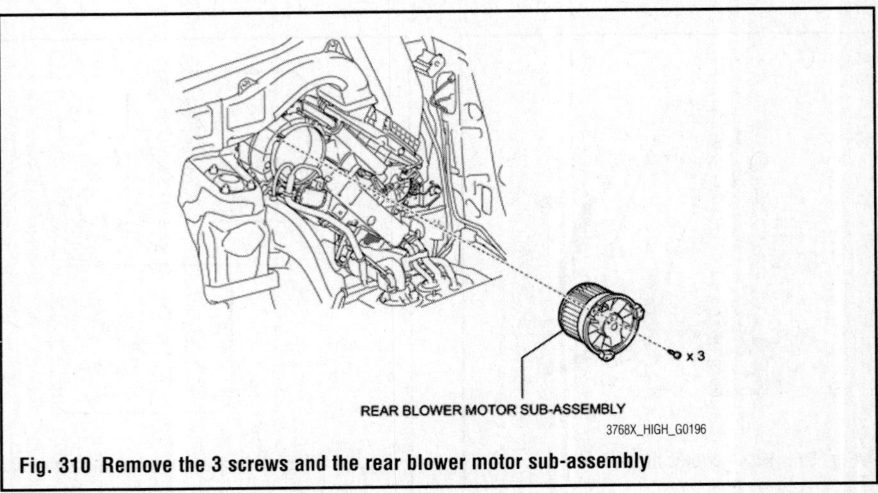

Fig. 310 Remove the 3 screws and the rear blower motor sub-assembly

21. Remove right side trim cover (for manual air conditioning system).
22. Remove rear room temperature sensor (for automatic air conditioning system).
23. Remove right rear combination light service cover.
24. Remove right hand rope hook assembly.
25. Remove the No. 1 luggage compartment trim hook
26. Remove the right front deck side trim cover.
27. Remove the right deck trim side panel assembly.
28. Remove the right roof side inner garnish assembly.
29. Remove the rear blower motor sub-assembly, as follows:
30. Disconnect the connector.
31. Remove the 3 screws and the rear blower motor sub-assembly.

To install:

32. Installation is the reverse of the removal procedure.

HEATER CORE

REMOVAL & INSTALLATION

See Figures 311 through 320.

➡Adjust the air outlet mode setting to FOOT.

1. Recover refrigerant from refrigeration system.
2. Disconnect the heater inlet and outlet water hoses.
3. Disconnect the cooler refrigerant liquid pipe C, as follows:
   a. Remove the bolt, and slide the hook connector.
   b. Disconnect the cooler refrigerant liquid pipe C.
   c. Remove the O-ring from the cooler refrigerant liquid pipe C.

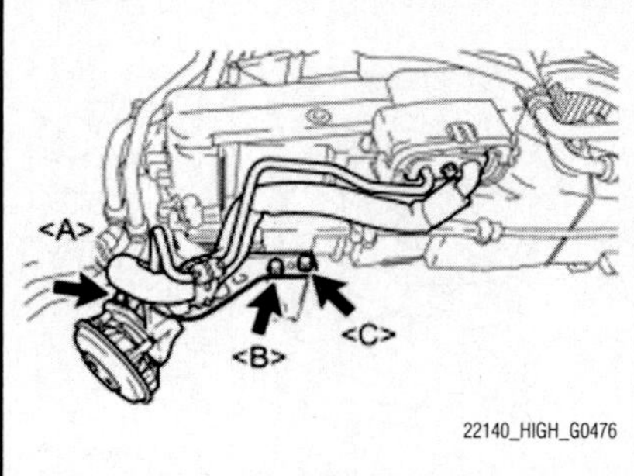

Fig. 312 Rear air conditioning tube and accessory assembly screw and bolt removal sequence

➡Seal the openings of the disconnected parts using vinyl tape to prevent entry of moisture and foreign matter.

4. Disconnect the rear cooler refrigerant suction hose.
5. Remove the O-ring from the rear cooler refrigerant suction hose.
6. Remove the right rear door scuff plate.
7. Remove the right rear door opening trim weatherstrip.
8. Remove the deck board assembly.
9. Remove the No. 2 and 3 deck board sub-assemblies.
10. Remove the tonneau cover assembly, as applicable.
11. Remove the rear seat side covers.
12. Remove the deck side trim boxes.
13. Remove the jack carrier support.
14. Remove the jack carrier cushion.
15. Remove jack assembly.
16. Remove the jack carrier assembly.
17. Remove rear mat.
18. Remove the deck floor board assembly.
19. Remove the rear No. 2 seat inner belt assembly.
20. Disconnect both rear seat lap type belt assemblies.
21. Remove the rear No. 2 seat assembly.

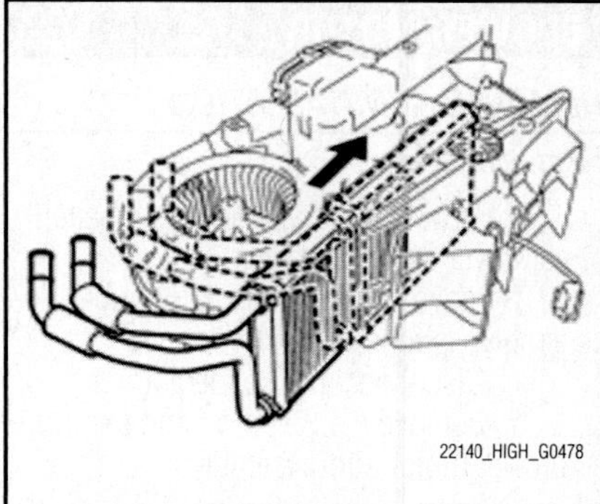

Fig. 313 Install heater radiator unit sub-assembly

22. Remove the rear floor finish plate.
23. Remove the rear seat side garnish cap.
24. Remove right deck side trim cover.
25. Remove right deck side trim.
26. Remove right side trim cover (for manual air conditioning system).
27. Remove rear room temperature sensor (for automatic air conditioning system).
28. Remove right rear combination light service cover.
29. Remove right hand rope hook assembly.
30. Remove the No. 1 luggage compartment trim hook
31. Remove the right front deck side trim cover.
32. Remove the right deck trim side panel assembly.
33. Remove the right roof side inner garnish assembly.
34. Disengage the 2 claws and remove the cooler plate.
35. Remove the No. 1 cooler air duct as shown in the illustration.
36. Remove the 2 clips and the rear No. 5 air duct.
37. Remove the rear cooling unit assembly, as follows:

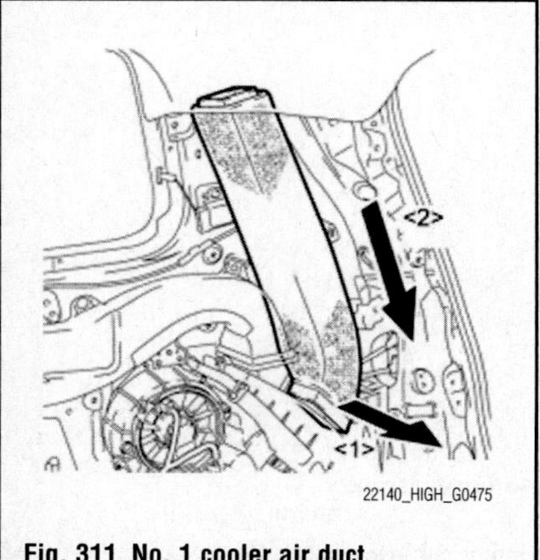

Fig. 311 No. 1 cooler air duct

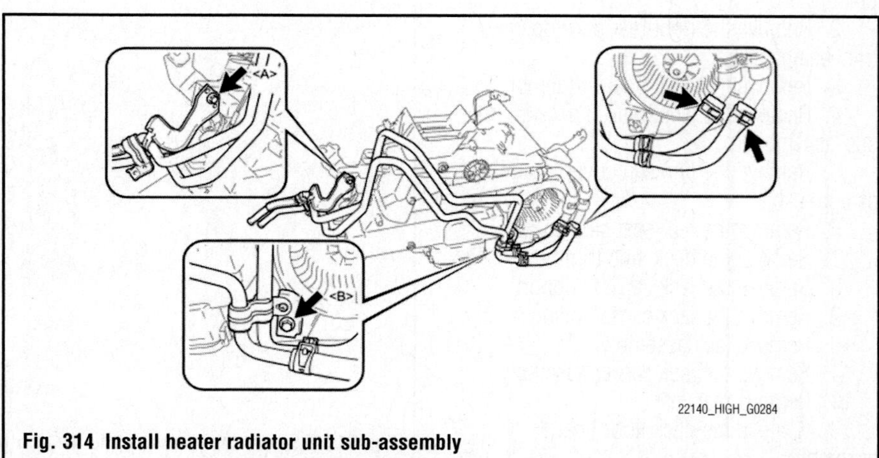

Fig. 314 Install heater radiator unit sub-assembly

a. Disengage each clamp.

b. Disconnect each connector.

c. Remove the 4 bolts and the rear cooling unit assembly.

38. Remove the drain cooler hose.

39. Remove the rear air mix control servo motor sub-assembly, as follows:

a. Disconnect the connector.

b. Remove the 2 screws and rear air mix control servo motor sub-assembly.

40. Remove the rear air outlet control servo motor sub-assembly, as follows:

a. Disconnect the connector.

b. Remove the 2 screws and rear air outlet control servo motor sub-assembly.

41. Remove the rear air conditioning tube and accessory assembly, as follows:

a. Remove the packing.

b. Remove the screws (C then B).

c. Remove the bolt (A).

d. Remove the bolt and slide the hook connector.

e. Remove the rear air conditioning tube and accessory assembly.

f. Remove the 2 O-rings from the rear air conditioning tube and accessory assembly.

42. Using pliers, grip the claws of the 2 clips and slide the clips to disconnect the heater water hose.

43. Remove the 2 screws and the heater water pipe and hose sub-assembly.

44. Remove the heater radiator unit sub-assembly, as follows:

a. Remove the screw and the heater clamp.

b. Remove the heater radiator unit sub-assembly from the rear cooling unit as shown in the illustration.

To install:

45. Install the heater radiator unit sub-assembly as shown in the illustration.

46. Install the clamp with the screw,

47. Install the heater water pipe and hose sub-assembly, as follows:

a. Install screw A, and tighten to 87 inch lbs. (9.8 Nm).

b. Install the heater water pipe and hose sub-assembly with the screw B and tighten to 87 inch lbs. (9.8 Nm).

c. Using pliers, grip the claws of the 2 clips and slide the clips to connect the heater water hose.

48. Install the rear air conditioning tube and accessory assembly, as follows:

a. Sufficiently apply compressor oil (ND-OIL 8 or equivalent) to 2 new O-rings and the fitting surfaces of the rear air conditioning tube and accessory assembly.

b. Install the 2 new O-rings on the rear air conditioning tube and accessory assembly.

c. Move the hook connector in the direction indicated by the arrow in the illustration.

d. Insert the pipe joint into the fitting hole securely and tighten the bolt to 87 inch lbs. (9.8 Nm).

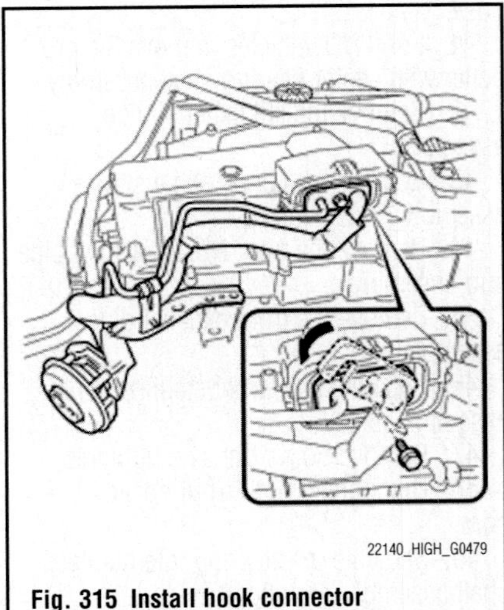

Fig. 315 Install hook connector

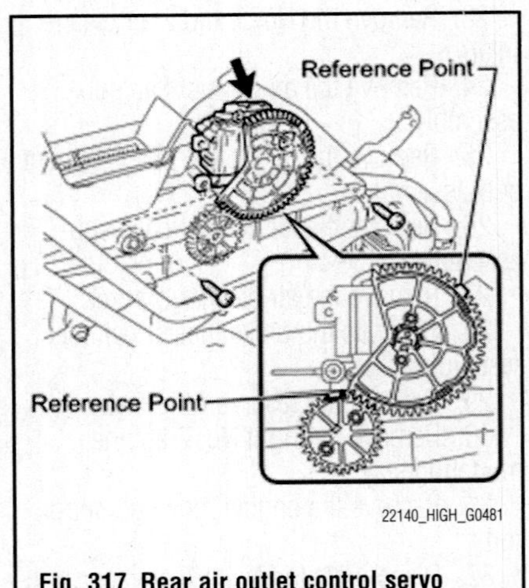

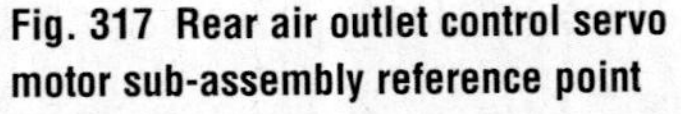

Fig. 317 Rear air outlet control servo motor sub-assembly reference point

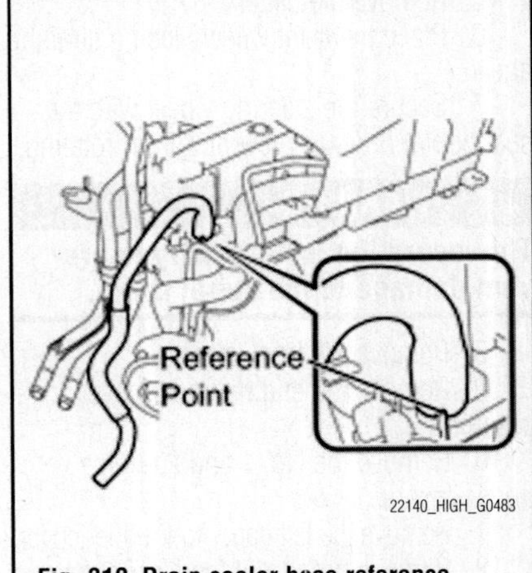

Fig. 319 Drain cooler hose reference point

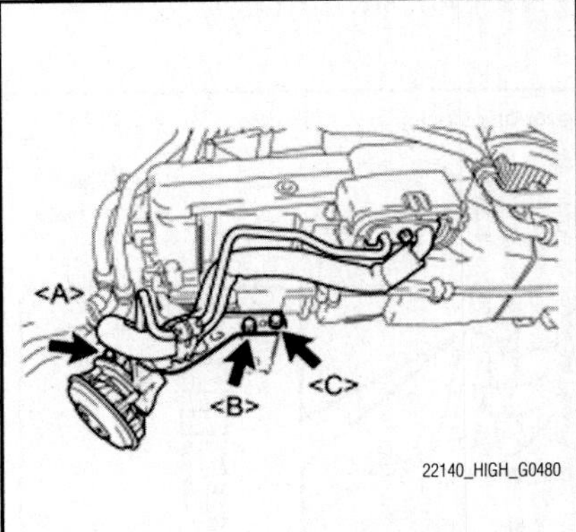

Fig. 316 Install rear air conditioning tube and accessory assembly

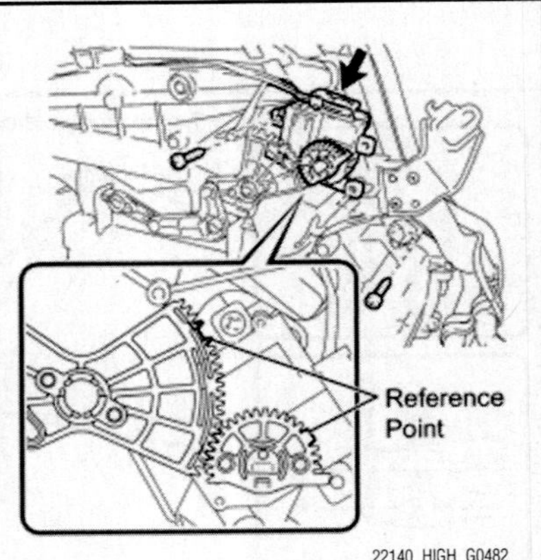

Fig. 318 Rear air mix control servo motor sub-assembly reference point

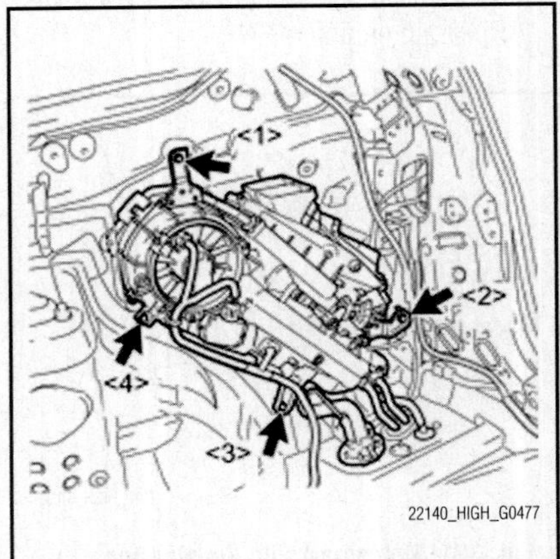

Fig. 320 Rear cooling unit assembly bolts tightening sequence

e. Install bolt A and tighten to 87 inch lbs. (9.8 Nm).

f. Install screw B and tighten to 87 inch lbs. (9.8 Nm).

g. Install the rear air conditioning tube and accessory assembly with screw C and tighten to 87 inch lbs. (9.8 Nm).

h. Install the packing.

49. Using the reference point, install the rear air outlet control servo motor sub-assembly with the 2 screws.

50. Connect the connector.

51. Using the reference point, install the rear air mix control servo motor sub-assembly with the 2 screws.

52. Connect the connector.

53. Using the reference point, install the drain cooler hose.

54. Install the rear cooling unit assembly with the 4 bolts and tighten in the order shown, to 87 inch lbs. (9.8 Nm).

55. Connect each connector.

56. Engage each clamp.

57. The remainder of installation is the reverse of the removal procedure.

STEERING

POWER RACK & PINION STEERING GEAR

REMOVAL & INSTALLATION

See Figures 321 through 323.

1. Before servicing the vehicle, refer to the precautions section.
2. Discharge fuel system pressure.
3. Recover refrigerant from refrigeration system.
4. Remove the cool air intake duct seal.
5. Remove the battery.
6. Place the front wheels facing straight ahead.
7. Secure the steering wheel with the seat belt in order to prevent it from rotating.

WARNING

This operation is necessary to prevent damage to the spiral cable.

8. Remove the front wheels.
9. Remove the engine under cover assembly.
10. Remove the No. 1 and 2 engine under covers.
11. Remove the left hand floor under cover.
12. Remove both front fender molding sub-assemblies.
13. Remove both front fender liners.
14. Remove both front fender apron seals.
15. Drain engine oil.
16. Drain engine coolant.
17. Drain automatic transaxle fluid.
18. Remove both front wiper arm and blade assemblies.
19. Remove the cowl top ventilator louver sub-assembly.
20. Remove the windshield wiper motor and link assembly.
21. Remove the outer cowl top panel sub-assembly.
22. Remove the V-bank cover sub-assembly.
23. Remove the No. 1 and 2 air cleaner inlets.
24. Remove the air cleaner cap sub-assembly.
25. Remove the air cleaner filter element sub-assembly.
26. Remove the air cleaner case sub-assembly.
27. Remove the air cleaner bracket.
28. Separate the brake master cylinder reservoir assembly.
29. Remove the reservoir bracket.
30. Remove the right No. 2 engine mounting stay.
31. Remove the engine moving control rod.
32. Disconnect the No. 1 fuel vapor feed hose.
33. Disconnect the No. 1 and 2 radiator hose.
34. Disconnect the heater water outlet hose B.
35. Disconnect the heater water inlet hose B.
36. Disconnect the fuel tube sub-assembly.
37. Disconnect the oil cooler inlet and outlet hoses.
38. Disconnect the transmission control cable assembly.
39. Disconnect engine wire.
40. Disconnect the union to check valve hose.
41. For 4WD vehicles, remove the propeller with center bearing shaft assembly.
42. Remove the tail exhaust pipe assembly.
43. Remove the center exhaust pipe assembly.
44. Remove the front No. 3 exhaust pipe sub-assembly.
45. Remove the front exhaust pipe assembly.
46. Separate both front stabilizer link assemblies.
47. Remove both front axle hub nuts.
48. Disconnect both front speed sensors.
49. Separate the steering intermediate shaft assembly, as follows:

a. Remove the bolt and slide the steering intermediate shaft assembly.

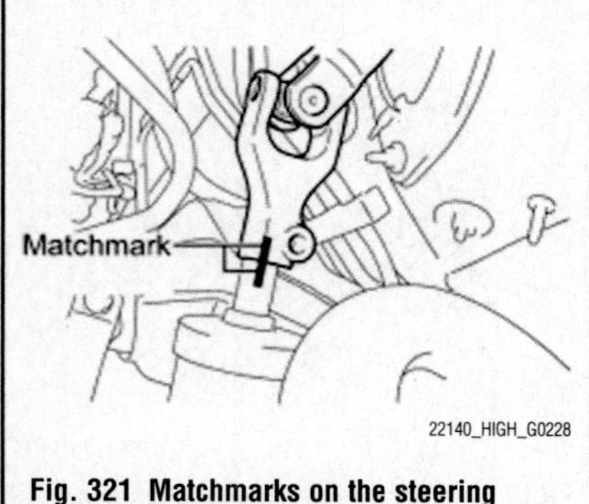

Fig. 321 Matchmarks on the steering intermediate shaft assembly and power steering link assembly

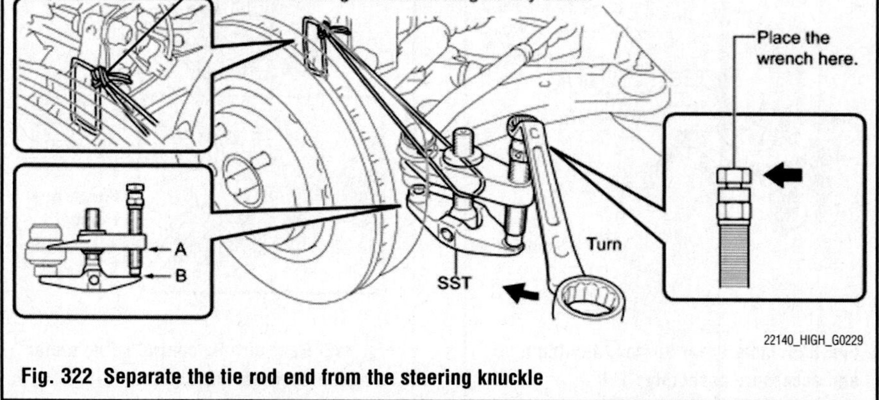

Fig. 322 Separate the tie rod end from the steering knuckle

➡Do not separate the steering intermediate shaft assembly from the power steering link assembly.

b. Put matchmarks on the steering intermediate shaft assembly and the power steering link assembly.

c. Separate the steering intermediate shaft assembly from the power steering link assembly.

50. Separate both tie rod assemblies:

a. Remove the cotter pin and the nut.

b. Install SST: 09960-20010 or equivalent to the tie rod end.

➡Make sure that the upper ends of the tie rod end and SST are aligned.

c. Using SST: 09960-20010, separate the tie rod end from the steering knuckle.

➡When securing SST to the steering knuckle, be sure to tighten the string of SST to prevent it from falling.

➡Install SST so that A and B are parallel.

➡Be sure to place the wrench on the part indicated in the illustration.

➡Do not damage the front disc brake dust cover.

➡Do not damage the ball joint dust cover.

➡Do not damage the steering knuckle.

51. Separate the No. 1 and 2 front suspension lower arms.
52. Separate both front axle assemblies.
53. Disconnect the discharge hose sub-assembly.
54. Disconnect the suction hose sub-assembly.
55. Remove the engine assembly with transaxle.
56. Remove both front No. 1 stabilizer brackets.
57. Remove the front stabilizer bar.
58. Remove the 2 bolts, 2 nuts, and power steering link assembly.

➡Because the nut has its own stopper, do not turn the nut. Loosen the bolt with the nut fixed.

59. Put matchmarks on the tie rod assemblies and the steering rack end sub-assemblies.
60. Loosen the lock nuts, and remove the tie rod assemblies and the lock nut.

To install:

61. Install the lock nuts and the tie rod assemblies to the steering rack end sub-assembly until the matchmarks are aligned.

➡After adjusting toe-in, torque the lock nut.

62. Install the power steering link assembly with the 2 bolts and 2 nuts and tighten to 51 ft. lbs. (70 Nm).

➡Make sure to tighten the bolts starting from the left side of the vehicle.

➡Because the nut has its own stopper, do not turn the nut. Tighten the bolt with the nut fixed.

63. Install the front stabilizer bar.
64. Install both front No. 1 stabilizer brackets.
65. Install the engine assembly with transaxle.
66. Reconnect the suction hose sub-assembly.
67. Reconnect the discharge hose sub-assembly.
68. Install both front axle assemblies.
69. Install both No. 1 front suspension lower arms.
70. Connect both tie rod assemblies and tighten to 36 ft. lbs. (49 Nm).
71. Install a new cotter pin.

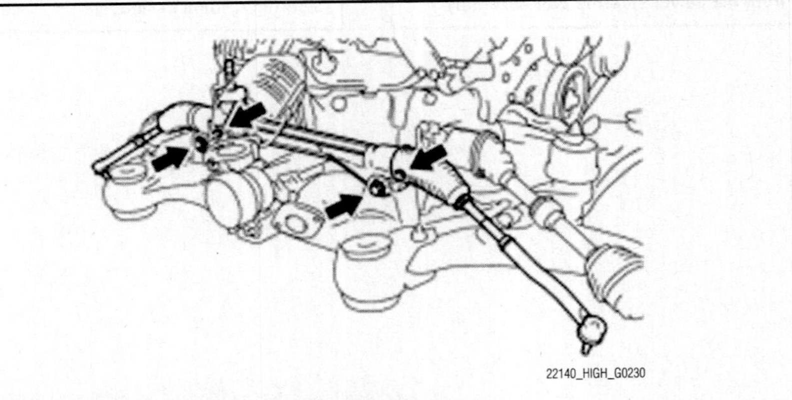

Fig. 323 Power steering link assembly bolts and nuts

➡Further tighten the nut up to 60° if the holes for the cotter pin are not aligned.

72. Align the matchmarks on the steering intermediate shaft assembly and the power steering link assembly.
73. Install the bolt and tighten to 26 ft. lbs. (35 Nm).
74. Install both front speed sensors.
75. Install both front axle hub nuts.
76. Install both front stabilizer link assemblies.
77. Install the front exhaust pipe assembly.
78. Install the front No. 3 exhaust pipe sub-assembly.
79. Install the center exhaust pipe assembly.
80. Install the tail exhaust pipe assembly.
81. For 4WD vehicles, temporarily tighten the propeller with center bearing shaft assembly.
82. For 4WD vehicles, fully tighten the propeller with center bearing shaft assembly.
83. Connect engine wire.
84. Connect the transmission control cable assembly.
85. Connect the fuel tube sub-assembly.
86. Connect the oil cooler inlet and outlet hoses.
87. Connect the heater water inlet hose B.
88. Connect the heater water outlet hose B.
89. Install the No. 1 and 2 radiator hoses.
90. Connect the union to check valve hose.
91. Connect the No. 1 fuel vapor feed hose.
92. Install the engine moving control rod.
93. Install the right No. 2 engine mounting stay.
94. Install the reservoir bracket.
95. Install the brake master cylinder reservoir assembly.
96. Install the air cleaner bracket.
97. Install battery.
98. Install the air cleaner case sub-assembly.
99. Install the air cleaner filter element sub-assembly.
100. Install the air cleaner cap sub-assembly.
101. Install the No. 1 and 2 air cleaner inlets.
102. Connect vacuum hoses.
103. Install the outer cowl top panel sub-assembly.
104. Install the windshield wiper motor and link assembly.
105.Install the cowl top ventilator louver sub-assembly.

106. Install both front wiper arm and blade assemblies.
107. Install front wheels. Tighten lug nuts to 76 ft. lbs. (103 Nm).
108. Add engine oil.
109. Add engine coolant.
110. Add automatic transaxle fluid.
111. Check automatic transaxle fluid.
112. Inspect for fuel leak.
113. Inspect for engine oil leak.
114. Inspect for coolant leak.
115. Inspect for exhaust gas leak.
116. Check shift lever position.
117. Place front wheels facing straight ahead.
118. Inspect and adjust front wheel alignment.
119. Check ignition timing.
120. Check engine idle speed.
121. Check CO/HC.
122. Check function of throttle body assembly.
123. Install both front fender apron seals.
124. Install both front fender liners.
125. Install both front fender molding sub-assemblies.
126. Install the left hand floor under cover.
127. Install the No. 1 and 2 engine under cover.
128. Install the engine under cover assembly.
129. Install the V-bank cover sub-assembly.
130. Check the ABS speed sensor signal.
131. Reset memory.

STEERING COLUMN AND POWER STEERING MOTOR

REMOVAL & INSTALLATION

See Figures 324 through 326.

1. Remove the steering linkage.
2. Remove steering column assembly:
   a. Disconnect the lower connector from the power steering ECU assembly.
   b. Disconnect the upper connector from the power steering ECU assembly.
   c. Disconnect the connectors and disengage the wire harness clamps from the steering column assembly.
   d. Remove the bolt, 2 nuts, and the steering column assembly.

To install:

➡If the bushings are missing or damaged, replace the steering column assembly with a new one.

3. Check that the 2 bushings are securely installed to the steering column assembly.
4. Install the steering column assembly with the bolt and 2 nuts. Tighten to 18 ft. lbs. (25 Nm).
5. To complete installation, reverse remaining removal procedure.

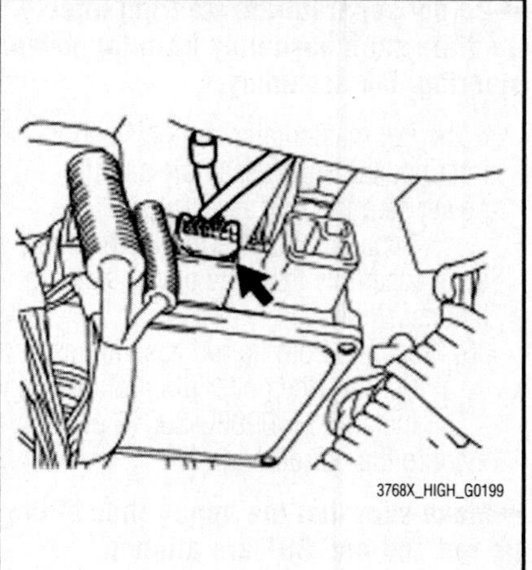

Fig. 325 Disconnect the upper connector from the power steering ECU assembly

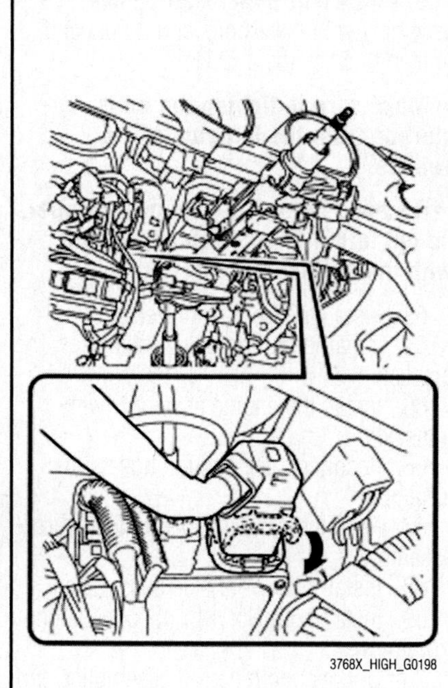

Fig. 324 Disconnect the lower connector from the power steering ECU assembly

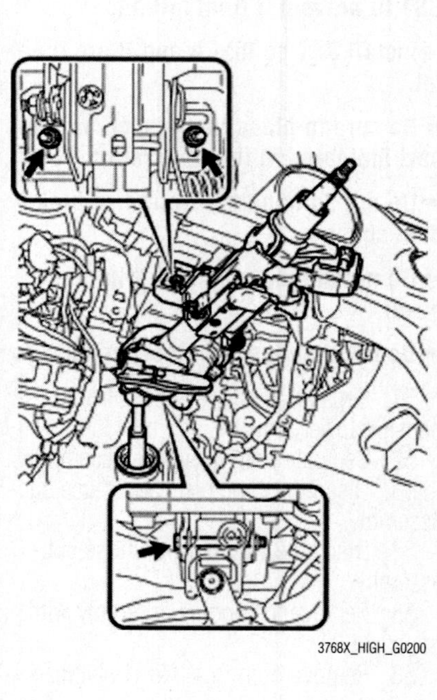

Fig. 326 Remove the bolt, 2 nuts, and the steering column assembly

CONTROL LINKS

REMOVAL & INSTALLATION

See Figures 327 and 328.

➡Perform the same procedure on each side.

1. Remove wheel and tire assembly.
2. For 2.7L engines, remove the two nuts and separate the front stabilizer link assembly.
3. For 3.5L engines, remove the nut and separate the front stabilizer link assembly.

➡If the ball joint turns together with the nut, use a hexagon wrench (6 mm) to hold the stud bolt.

To install:

4. Install the front stabilizer link assembly with the nut and tighten to 55 ft. lbs. (74 Nm).
5. Install the wheel and tire assembly.

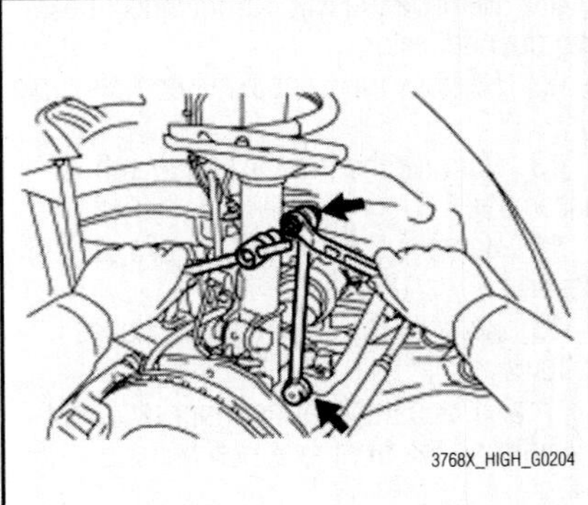

Fig. 327 Remove the two nuts and separate the front stabilizer link assembly—2.7L Engines

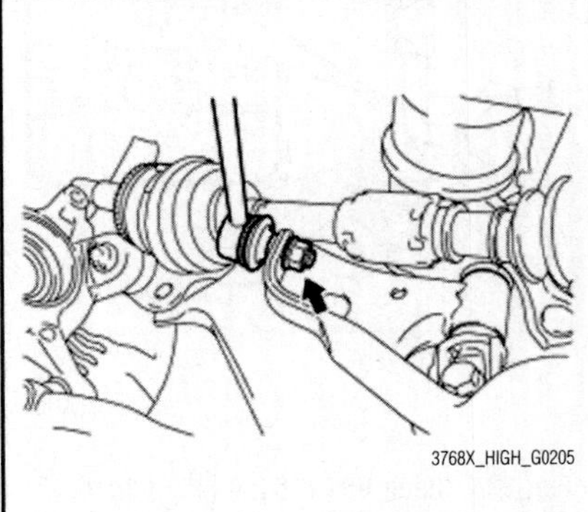

Fig. 328 Remove the nut and separate the front stabilizer link assembly—3.5L Engines

LOWER BALL JOINT

REMOVAL & INSTALLATION

See Figures 329 through 332.

1. Remove the front wheel.
2. Remove the front axle hub nut.
3. Remove the bolt and resin clamp, and separate the front speed sensor.

➡Be sure to completely separate the front speed sensor from the front shock absorber with coil spring.

➡Clean the installation hole and the surface for the speed sensor every time the speed sensor is removed.

➡Be careful not to damage the front speed sensor.

4. Put matchmarks on the front drive shaft assembly and the front axle hub sub-assembly.
5. Using a plastic hammer, separate the front drive shaft assembly from the front axle assembly.

➡Loosen the staked part of the front axle hub nut completely, otherwise the threads of the drive shaft may be damaged.

6. Remove the 2 bolts and separate the front disc brake caliper assembly.

➡Use wire or an equivalent tool to keep the brake caliper from hanging down by the flexible hose.

7. Remove the front disc.
8. Separate the tie rod assembly.
9. Remove the bolt, 2 nuts, and separate the front lower suspension arm from the lower ball joint.
10. Remove the 2 bolts, 2 nuts and front axle assembly.
11. Remove the front lower ball joint, as follows:

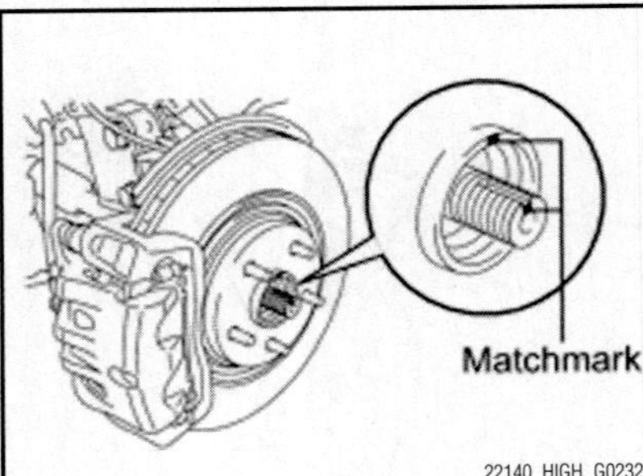

Fig. 329 Matchmark the front drive shaft assembly and the front axle hub sub-assembly

Fig. 330 Front lower suspension arm

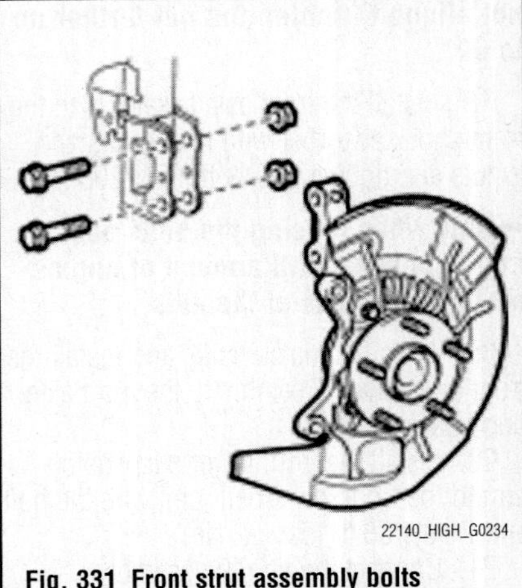

Fig. 331 Front strut assembly bolts

12. Secure the front axle assembly in a vise using aluminum plates.

➡When using a vise, do not overtighten it.

13. Remove the cotter pin and nut.
14. Install SST: 09960-20010 or equivalent to the front lower ball joint.
15. Using SST: 09960-20010 or equivalent, remove the front lower ball joint from the front axle assembly.

➡Install SST so that A and B are parallel.

➡Be sure to place a wrench on the part indicated in the illustration.

➡Do not damage the front lower ball joint dust cover.

To install:

16. Install the front lower ball joint to the steering knuckle with the nut and tighten to 91 ft. lbs. (123 Nm).

➡Prevent oil from adhering to the screw and tapered parts.

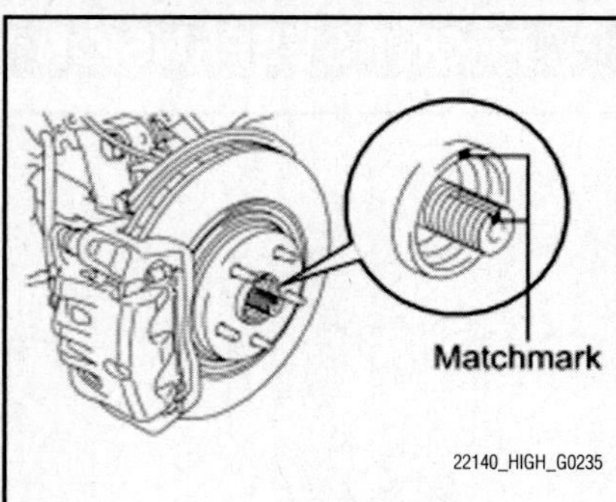

Fig. 332 Align the matchmarks and install the front drive shaft assembly

17. Install a new cotter pin.

➡If the holes for the cotter pin are not aligned, tighten the nut further up to 60°.

18. Install the front axle assembly to the front shock absorber with the 2 bolts and 2 nuts and tighten to 213 ft. lbs. (290 Nm).

➡Only when reusing the bolts and nuts, apply a small amount of engine oil to the threads of the nuts.

19. Align the matchmarks and install the front drive shaft assembly to the front axle hub sub-assembly.
20. Install the front lower suspension arm to the front lower ball joint with the bolt and 2 nuts 68 ft. lbs. (92 Nm).
21. Connect the tie rod assembly.
22. Install the front disc.
23. Install the front disc brake caliper assembly to the steering knuckle with the 2 bolts and tighten to 77 ft. lbs. (104 Nm).
24. Install the clamp and front speed sensor with the bolt and tighten to 71 inch lbs. (8 Nm).

➡Prevent foreign matter from attaching to the sensor tip.

➡Firmly insert the sensor body into the knuckle before tightening the bolt.

➡After installing the sensor to the knuckle, make sure that there is no clearance between the sensor stay and knuckle. Also make sure that no foreign matter is stuck between the parts.

➡To prevent interference between the sensor and magnetic rotor, do not rotate the sensor body during or after the insertion of the sensor body to the knuckle.

25. Install the front axle hub nut.
26. Check the ABS speed sensor signal

➡Check the ABS speed sensor signal.

27. Install front wheel and tighten lug nuts to 76 ft. lbs. (103 Nm).
28. Inspect and adjust front wheel alignment.

LOWER CONTROL ARM

REMOVAL & INSTALLATION

See Figure 333.

1. Remove the engine assembly with transaxle.
2. Remove the No. 1 front stabilizer brackets.
3. Remove the front stabilizer bar with front stabilizer link assembly.
4. Remove the power steering link assembly.
5. Install the engine hangers.
6. Separate the front frame assembly.
7. Remove the front lower suspension arm, as follows:
   a. Remove the 3 bolts, nut, and the front lower suspension arm from the front frame assembly.
   b. Remove the front lower arm bushing stopper from the front lower suspension arm.

To install:

8. Install the front lower suspension arm, as follows:
   a. Install the front lower arm bushing stopper to the front lower suspension arm.
   b. Install the front lower suspension arm to the front frame assembly with the 3 bolts and nut, but do not tighten them yet.
   c. Tighten the 3 bolts in numerical order shown. Tighten to 147 ft. lbs. (200 Nm), and 152 ft. lbs. (206 Nm).

➡Start installing the bolts from the front side of the vehicle.

9. Connect the front frame assembly.

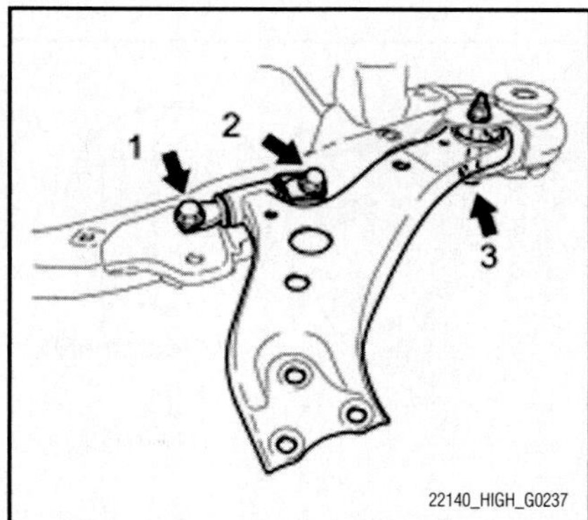

Fig. 333 Front lower control arm bolts tightening sequence

10. Remove the engine hangers.
11. Install the power steering link assembly.
12. Install the front stabilizer bar with front stabilizer link assembly.
13. Install the No. 1 front stabilizer brackets.
14. Install the engine assembly with transaxle.

STEERING KNUCKLE

REMOVAL & INSTALLATION

See Figure 334.

1. Remove the front wheel.
2. Remove the front axle hub nut.
3. Separate the front speed sensor.
4. Separate the front disc brake caliper assembly.
5. Remove the front disc.
6. Separate tie rod assembly.
7. Separate the front lower suspension arm.
8. Separate the front drive shaft assembly.
9. Remove the front axle assembly.
10. Remove the front lower ball joint.
11. Remove the No. 1 front wheel bearing dust deflector.
12. Remove the front axle hub hole snap ring.
13. Remove the front axle hub sub-assembly.
14. Remove the front disc brake dust cover.
15. Remove the steering knuckle, as follows:
   a. Place the bearing inner race (outside) on the front axle hub bearing.

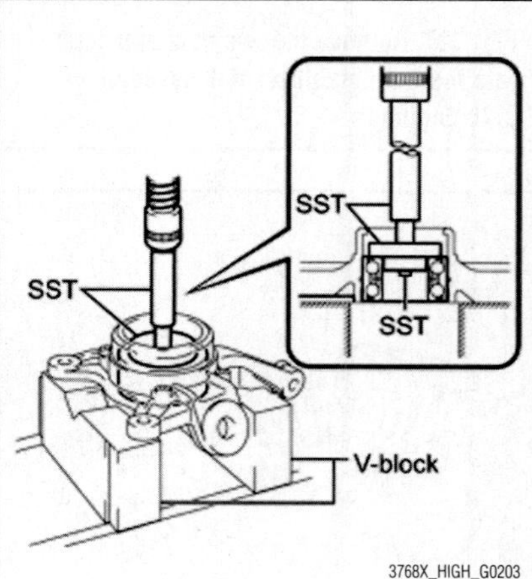

Fig. 334 Using SST (SST: 09950-60010, SST: 09950-70010, SST: 09950-60020, or equivalent),V-blocks and a press, remove the front axle hub bearing from the steering knuckle

b. Using SST (SST: 09950-60010, SST: 09950-70010, SST: 09950-60020, or equivalent),V-blocks and a press, remove the front axle hub bearing from the steering knuckle.

➡Keep the steering knuckle level.

To install:

16. Using SST (SST: 09950-70010, SST: 09950-60020, or equivalent), install a new front axle hub bearing to the steering knuckle.
17. Install the front disc brake dust cover.
18. Install the front axle hub sub-assembly.
19. Install the front axle hub hole snap ring.
20. Install the No. 1 front wheel bearing dust deflector.
21. Install the front lower ball joint.
22. Install the front axle assembly.
23. Install the front drive shaft assembly.
24. Install the front lower suspension arm.
25. Connect tie rod assembly.
26. Install the front disc.
27. Install the front disc brake caliper assembly.
28. Install the front axle hub nut.
29. Separate the front disc brake caliper assembly.
30. Remove the front disc.
31. Inspect the front axle bearing looseness.
32. Inspect front axle hub runout.
33. Install the front disc.
34. Install the front disc brake caliper assembly.
35. Install the front speed sensor.
36. Stake front axle hub nut.
37. Install the front wheel.
38. Inspect and adjust front wheel alignment.
39. Check for speed sensor signal.

STRUT

REMOVAL & INSTALLATION

See Figures 335 through 341.

1. Remove the front wheel.
2. Remove the front wiper arm and blade assemblies.
3. Loosen the front support to front shock absorber nut of the front shock absorber.

➡Do not remove the front support to front shock absorber nut.

➡Loosen the nut only when the front shock absorber with coil spring needs to be disassembled.

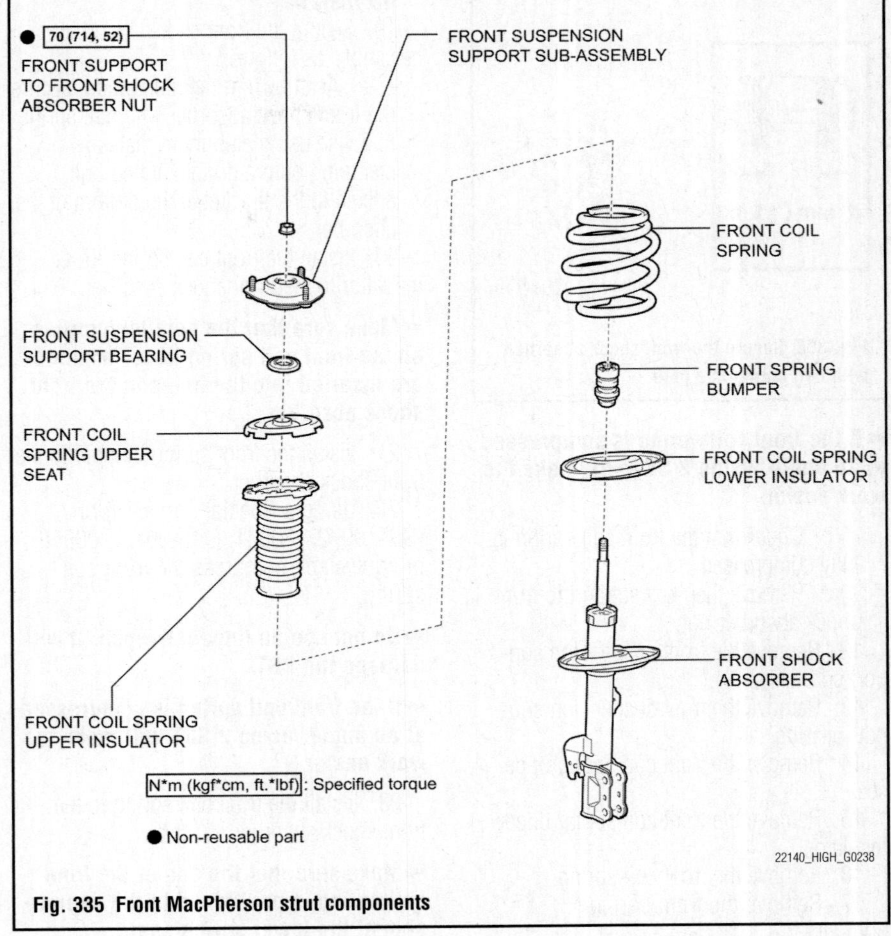

Fig. 335 Front MacPherson strut components

4. Remove the cowl top ventilator louver sub-assembly.
5. Remove the windshield wiper motor and link.
6. Remove the outer cowl top panel sub-assembly, as follows:
   a. Disengage the 4 clamps and separate the wiper wire harness from the outer cowl top panel sub-assembly.
   b. Remove the 8 bolts, 6 nuts, and the outer cowl top panel sub-assembly.
7. Remove the bolt and clamp, and separate the front speed sensor and front flexible hose.
8. Remove the nut and separate the front stabilizer link assembly from the front shock absorber.

➡If the ball joint turns together with the nut, use a hexagon wrench (6 mm) to hold the stud.

9. Remove the front shock absorber with coil spring, as follows:
   a. Support the front axle using a jack and wooden block.
   b. Remove the 2 bolts and 2 nuts, and separate the front shock absorber with coil spring (lower side) from the steering knuckle.

➡When removing the nuts, keep the bolts from rotating.

   c. Remove the nut and 2 spacers on the upper side of the front shock absorber with coil spring.

➡Make sure that the front speed sensor is completely separated from the front shock absorber with coil spring.

10. Secure the front shock absorber with coil spring, as follows:
    a. As shown in the illustration, secure the front shock absorber with coil spring in a vise using aluminum plates by clamping onto a double nutted bolt affixed to the bracket at the bottom of the absorber.
11. Remove the front support to front shock absorber nut, as follows:
    a. Using A Special Service Tool (SST: 09727-30021, SST: 09727-30021), compress the front coil spring.

➡Do not use an impact wrench. It will damage the SST.

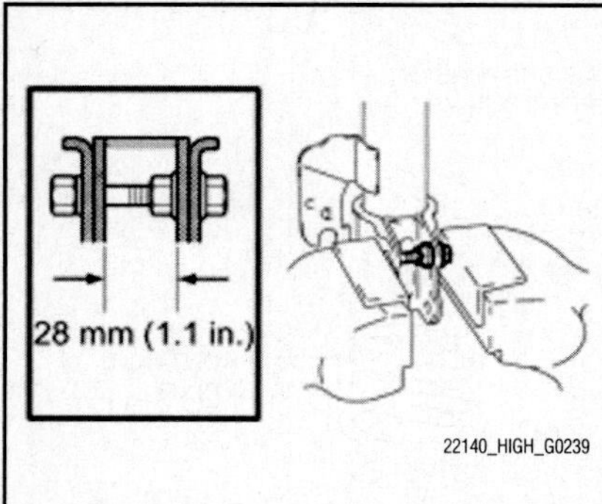

Fig. 336 Secure the front shock absorber with coil spring in a vise

➡If the front coil spring is compressed at an angle, using 2 SST will make the work easier.

b. Check that the front coil spring is fully compressed.

c. Remove the front support to front shock absorber nut.

12. Remove the front suspension support sub-assembly.
13. Remove the front suspension support bearing.
14. Remove the front coil spring upper seat.
15. Remove the front coil spring upper insulator.
16. Remove the front coil spring.
17. Remove the front spring bumper.
18. Remove the front coil spring lower insulator.

To install:

19. Secure the front shock absorber assembly, as follows:

a. As shown in the illustration, secure the front shock absorber with coil spring in a vise using aluminum plates by clamping onto a double nutted bolt affixed to the bracket at the bottom of the absorber.

20. Install the front coil spring lower insulator to the front shock absorber.

➡Make sure that the positioning pins on the front coil spring lower insulator are inserted into the holes in the front shock absorber.

21. Install the front spring bumper to the front shock absorber.
22. Using a Special Service Tool (SST: 09727-30021, SST: 09727-00050) or equivalent, compress the front coil spring.

➡Do not use an impact wrench. It will damage the SST.

➡If the front coil spring is compressed at an angle, using 2 SST will make the work easier.

23. Install the front coil spring to the front shock absorber.

➡Make sure that the end of the front coil spring is positioned in the depression of the lower spring seat.

24. Install the front coil spring upper insulator as shown in the illustration.

➡Any misalignment between the front shock absorber lower bracket and the alignment mark must be +/- 5°.

25. Install the front coil spring upper seat with the mark facing to the outside of the vehicle.

➡Any misalignment between the front shock absorber lower bracket and the alignment mark must be +/- 5°.

26. Install the front suspension support bearing as shown in the illustration.
27. Install the front suspension support sub-assembly as shown in the illustration.

➡Check that the slot on the piston rod and the slot on the front suspension support sub-assembly are aligned.

28. Temporarily tighten a new front support to front shock absorber nut.
29. Install the front shock absorber with coil spring (upper side) with the nut and 2 spacers and tighten to 63 ft. lbs. (85 Nm).
30. Install the front shock absorber with coil spring (lower side) to the steering knuckle and insert the 2 bolts and 2 nuts and tighten to 214 ft. lbs. (290 Nm).

➡When installing the nuts, keep the bolts from rotating.

31. Install the front stabilizer link assembly to the front shock absorber with the nut and tighten to 55 ft. lbs. (74 Nm).

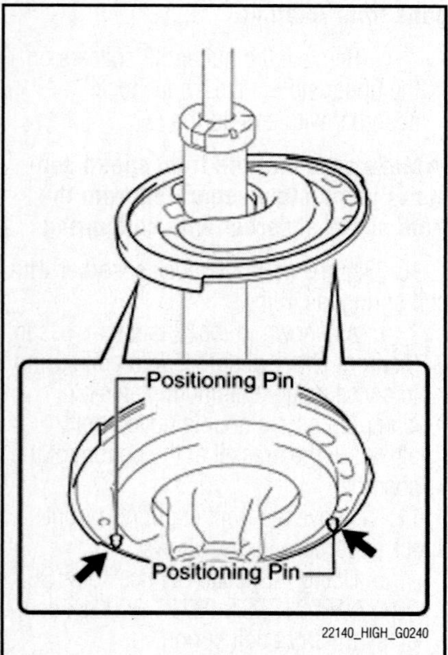

Fig. 337 Front coil spring lower insulator positioning pin

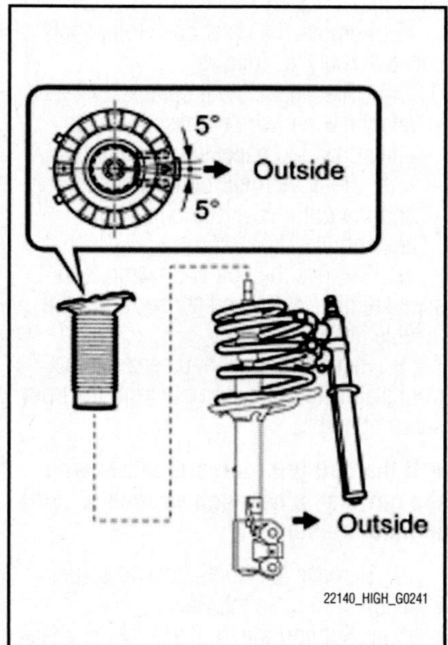

Fig. 338 Install the front coil spring upper insulator

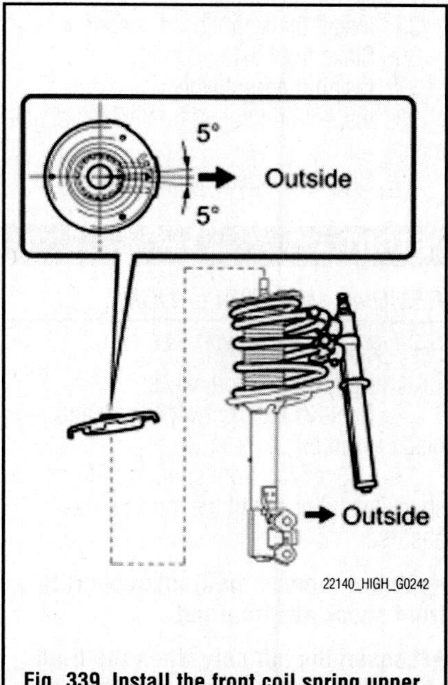

Fig. 339 Install the front coil spring upper seat

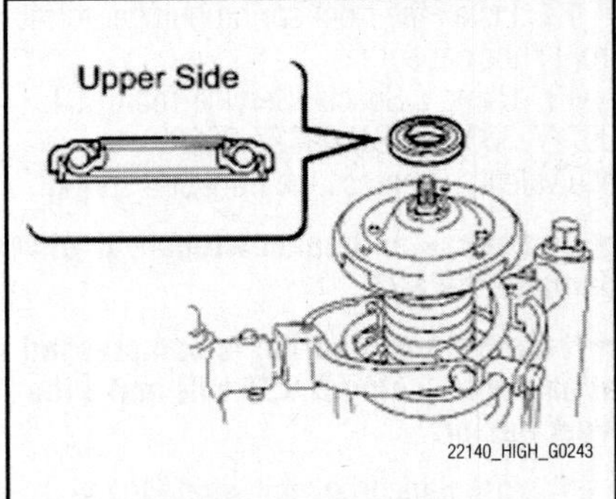

Fig. 340 Install the front suspension support bearing

➡**If the ball joint turns together with the nut, use a hexagon wrench (6 mm) to hold the stud bolt.**

32. Install the front speed sensor and front flexible hose with the bolt and tighten to 14 ft. lbs. (19 Nm).

➡**Do not twist the front speed sensor when installing it.**

33. Install the clamp.
34. Install the outer cowl top panel sub-assembly with the 8 bolts and 6 nuts and tighten to 63 ft. lbs. (85 Nm), 78 inch lbs. (8.8 Nm), 78 inch lbs. (8.8 Nm).
35. Engage the 4 clamps.
36. Fully tighten the front support to front shock absorber nut and tighten to 52 ft. lbs. (70 Nm).
37. Install the windshield wiper motor and link.
38. Install the cowl top ventilator louver sub-assembly.
39. Install the front wiper arm and blade assemblies.
40. Install the front wheel and tighten lug nuts to 76 ft. lbs. (103 Nm).
41. Inspect and adjust front wheel alignment.

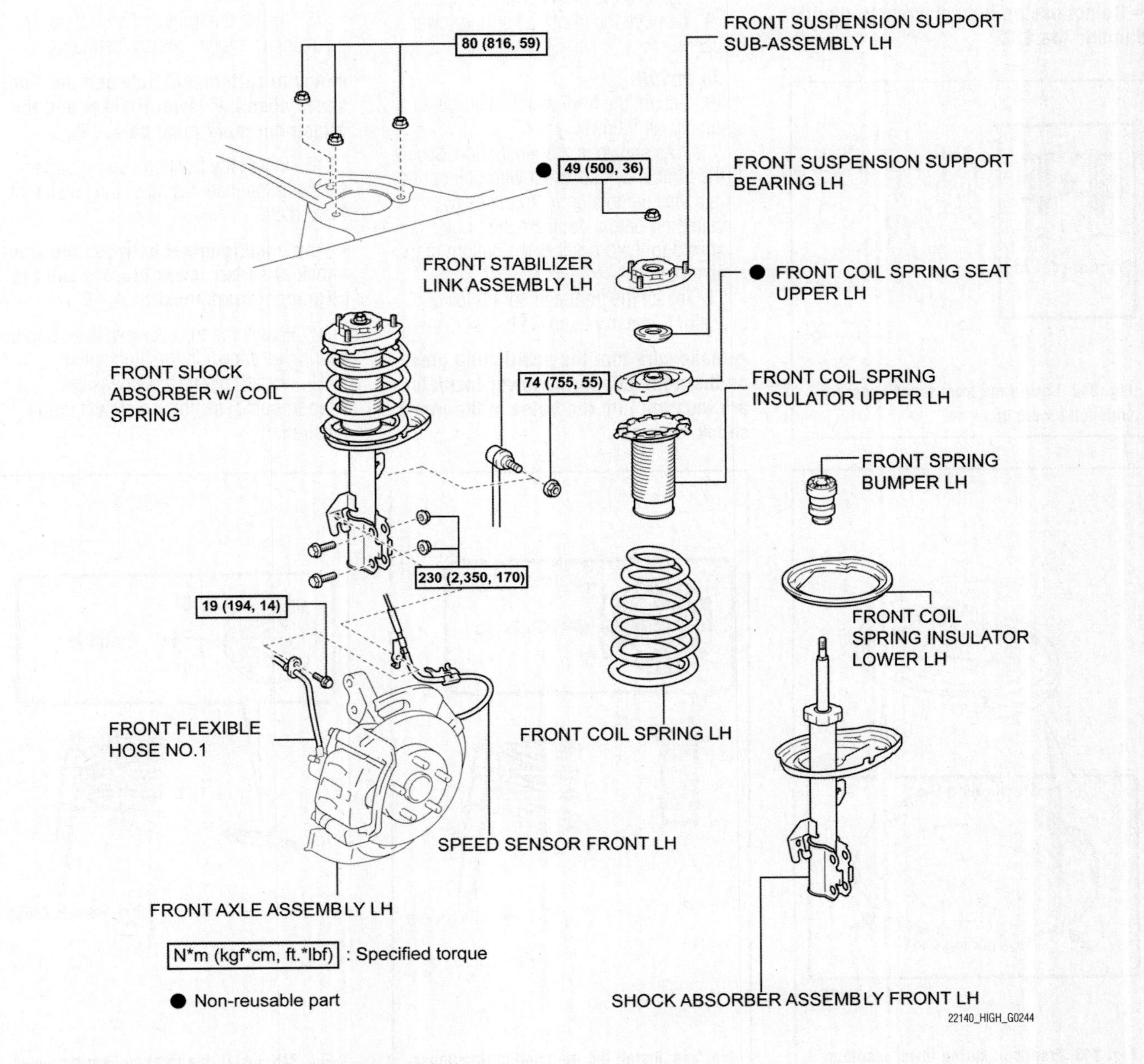

80 (816, 59)
FRONT SUSPENSION SUPPORT SUB-ASSEMBLY LH
49 (500, 36)
FRONT SUSPENSION SUPPORT BEARING LH
FRONT STABILIZER LINK ASSEMBLY LH
● FRONT COIL SPRING SEAT UPPER LH
FRONT SHOCK ABSORBER w/ COIL SPRING
74 (755, 55)
FRONT COIL SPRING INSULATOR UPPER LH
FRONT SPRING BUMPER LH
230 (2,350, 170)
19 (194, 14)
FRONT COIL SPRING INSULATOR LOWER LH
FRONT FLEXIBLE HOSE NO.1
FRONT COIL SPRING LH
SPEED SENSOR FRONT LH
FRONT AXLE ASSEMBLY LH
N*m (kgf*cm, ft.*lbf) : Specified torque
● Non-reusable part
SHOCK ABSORBER ASSEMBLY FRONT LH
22140_HIGH_G0244

Fig. 341 Install the front suspension support sub-assembly

OVERHAUL

See Figures 342 through 347.

1. Secure the front shock absorber with coil spring, as follows:
   a. As shown in the illustration, secure the front shock absorber with coil spring in a vise using aluminum plates by clamping onto a double nutted bolt affixed to the bracket at the bottom of the absorber.
2. Remove the front support to front shock absorber nut, as follows:
   a. Using A Special Service Tool (SST: 09727-30021, SST: 09727-30021), compress the front coil spring.

➡Do not use an impact wrench. It will damage the SST.

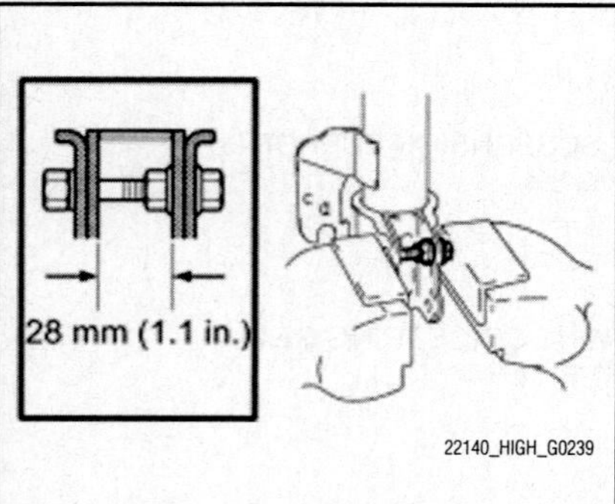

Fig. 342 Secure the front shock absorber with coil spring in a vise

➡If the front coil spring is compressed at an angle, using 2 SST will make the work easier.

   b. Check that the front coil spring is fully compressed.
   c. Remove the front support to front shock absorber nut.
3. Remove the front suspension support sub-assembly.
4. Remove the front suspension support bearing.
5. Remove the front coil spring upper seat.
6. Remove the front coil spring upper insulator.
7. Remove the front coil spring.
8. Remove the front spring bumper.
9. Remove the front coil spring lower insulator.

To install:

10. Secure the front shock absorber assembly, as follows:
    a. As shown in the illustration, secure the front shock absorber with coil spring in a vise using aluminum plates by clamping onto a double nutted bolt affixed to the bracket at the bottom of the absorber.
11. Install the front coil spring lower insulator to the front shock absorber.

➡Make sure that the positioning pins on the front coil spring lower insulator are inserted into the holes in the front shock absorber.

12. Install the front spring bumper to the front shock absorber.
13. Using a Special Service Tool (SST: 09727-30021, SST: 09727-00050) or equivalent, compress the front coil spring.

➡Do not use an impact wrench. It will damage the SST.

➡If the front coil spring is compressed at an angle, using 2 SST will make the work easier.

14. Install the front coil spring to the front shock absorber.

➡Make sure that the end of the front coil spring is positioned in the depression of the lower spring seat.

15. Install the front coil spring upper insulator as shown in the illustration.

➡Any misalignment between the front shock absorber lower bracket and the alignment mark must be +/- 5°.

16. Install the front coil spring upper seat with the mark facing to the outside of the vehicle.

➡Any misalignment between the front shock absorber lower bracket and the alignment mark must be +/- 5°.

17. Install the front suspension support bearing as shown in the illustration.
18. Install the front suspension support sub-assembly as shown in the illustration.

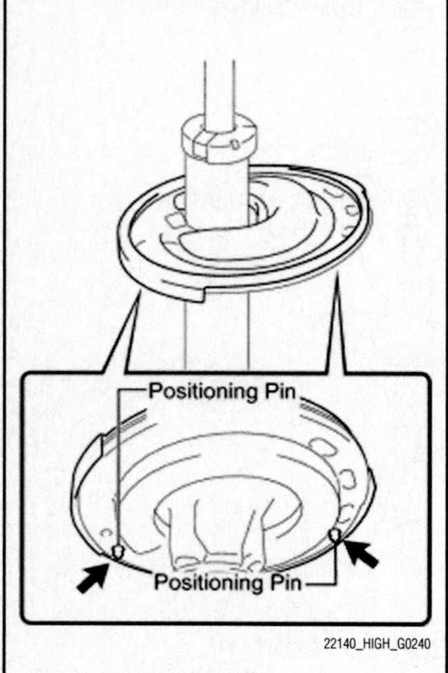

Fig. 343 Front coil spring lower insulator positioning pin

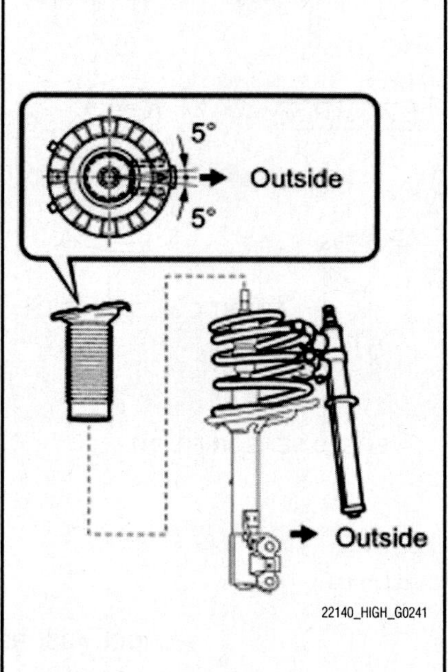

Fig. 344 Install the front coil spring upper insulator

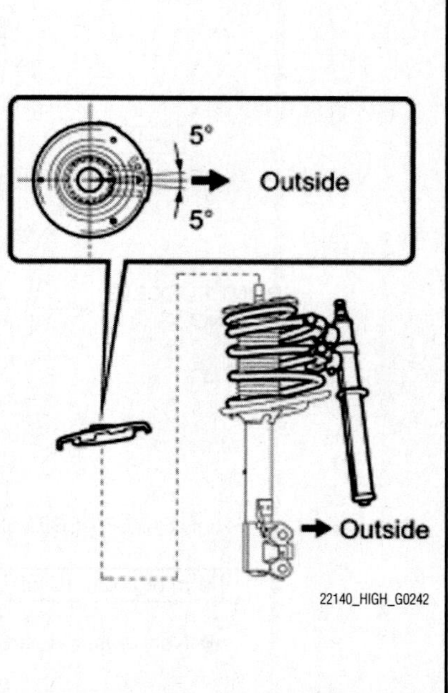

Fig. 345 Install the front coil spring upper seat

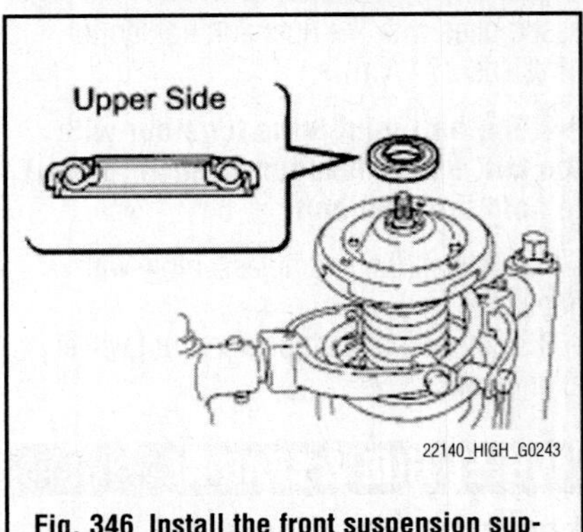

Fig. 346 Install the front suspension support bearing

➡Check that the slot on the piston rod and the slot on the front suspension support sub-assembly are aligned.

19. Temporarily tighten a new front support to front shock absorber nut.

STABILIZER BAR

REMOVAL & INSTALLATION

See Figure 348.

1. Remove the engine assembly with transaxle.
2. Remove the nuts and separate both front stabilizer link assemblies.

➡If the ball joint turns together with the nut, use a hexagon wrench (6 mm) to hold the stud.

3. Remove the 2 bolts and both No. 1 front stabilizer brackets from the front frame assembly.
4. Remove the front stabilizer bar.
5. Remove both No. 2 front stabilizer brackets from the front stabilizer bar bushing.
6. Remove the 2 No. 1 front stabilizer bar bushings from the front stabilizer bar.

To install:

7. Install the 2 No. 1 front stabilizer bar bushings to the front stabilizer bar as shown in the illustration.

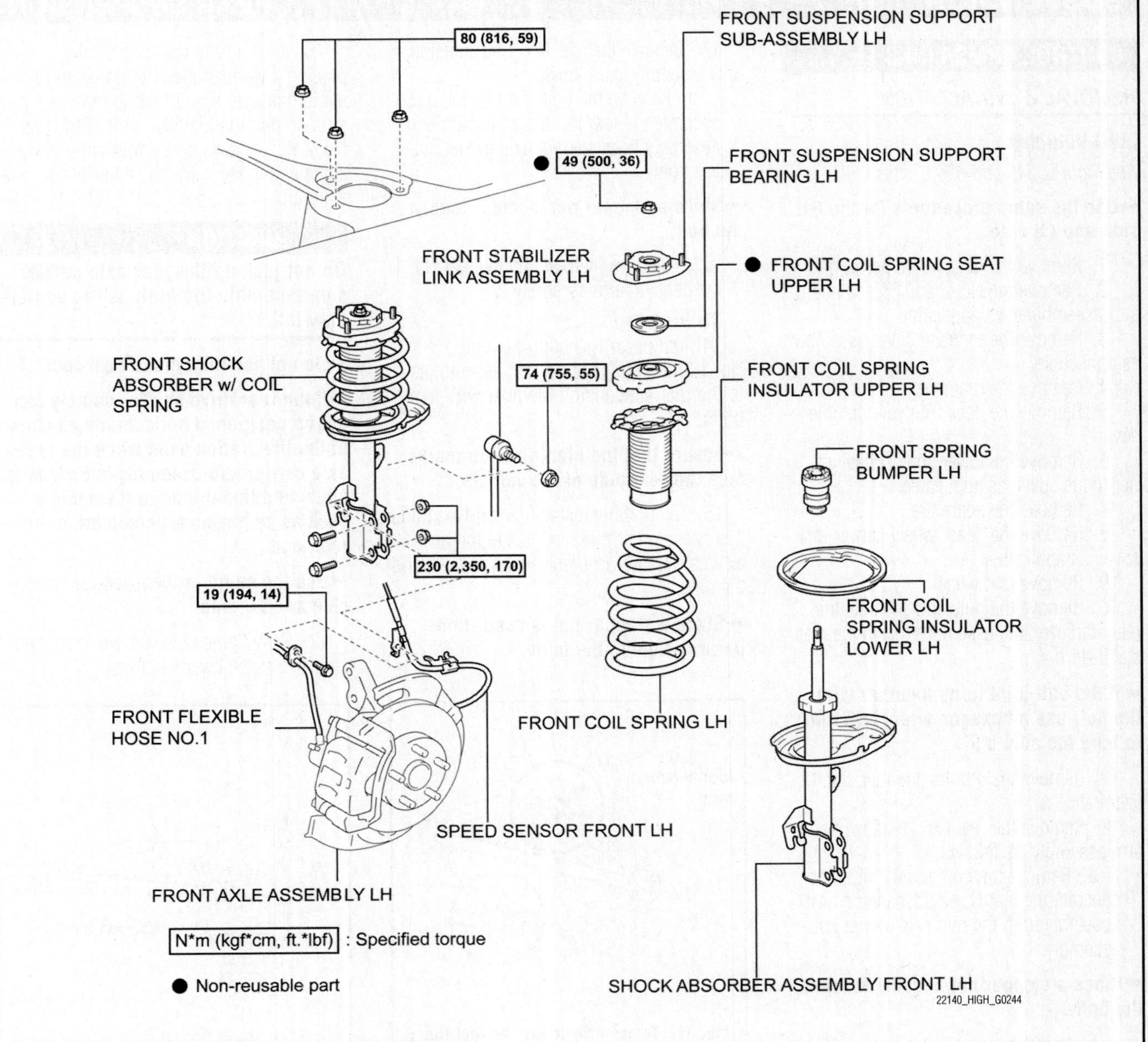

Fig. 347 Install the front suspension support sub-assembly

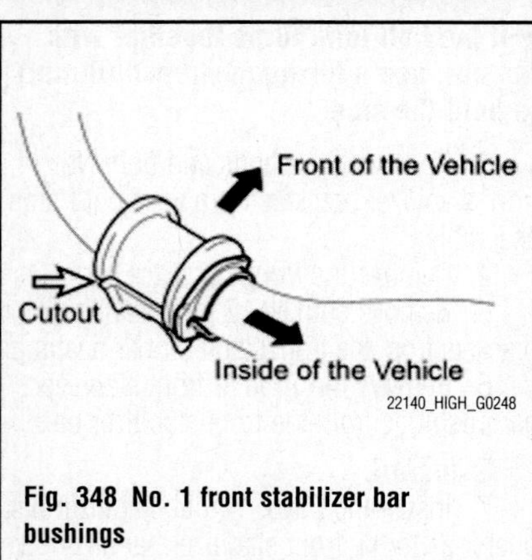

Fig. 348 No. 1 front stabilizer bar bushings

➡Install the No. 1 front stabilizer bar bushings so that the cutout faces the rear of the vehicle.

8. Install both No. 2 front stabilizer brackets to the No. 1 front stabilizer bar bushing.
9. Install the front stabilizer bar by inserting it from the right side of the vehicle.
10. Install both No. 1 front stabilizer brackets to the front frame assembly with the 2 bolts and tighten to 21 ft. lbs. (29 Nm).
11. Install both front stabilizer link assemblies with the nuts and tighten to 55 ft. lbs. (74 Nm).

➡If the ball joint turns together with the nut, use a hexagon wrench (6 mm) to hold the stud bolt.

12. Install the engine assembly with transaxle.
13. Inspect and adjust the front wheel alignment.

WHEEL HUB & BEARING

REMOVAL & INSTALLATION

See the Steering Knuckle procedure.

SUSPENSION

REAR SUSPENSION

CONTROL ARMS/LINKS

REMOVAL & INSTALLATION

2WD Vehicles

See Figures 349 and 350.

➡Use the same procedures for the RH side and LH side.

1. Remove the deck board assembly.
2. Remove the No. 2 and 3 deck board sub-assemblies, as applicable.
3. Remove the tonneau cover assembly, as applicable.
4. Remove the rear mat.
5. Remove the deck trim service hole cover.
6. Remove the lower spare wheel carrier hinge cover, as applicable.
7. Remove the spare tire.
8. Remove the spare wheel carrier lock cover, as applicable.
9. Remove rear wheel.
10. Remove the nuts and separate the rear stabilizer link assemblies from the rear stabilizer bar.

➡If the ball joint turns together with the nut, use a hexagon wrench (5 mm) to hold the stud bolt.

11. Remove the 4 bolts and rear stabilizer bar.
12. Remove the rear No. 2 suspension arm assembly, as follows:
    a. Remove the bolt and the nut, and separate the rear No. 2 suspension arm assembly from the rear axle carrier sub-assembly.

➡Since a stopper nut is used, loosen the bolt.

    b. Remove the bolt and the rear No. 2 suspension arm assembly.

13. Remove the rear No. 1 suspension arm assembly, as follows:
    a. Remove the bolt and the nut, and separate the rear No. 1 suspension arm assembly from the rear axle carrier sub-assembly.

➡Since a stopper nut is used, loosen the bolt.

    b. Remove the bolt and the rear No. 1 suspension arm assembly.

To install:

14. Temporarily install the rear No. 1 and 2 suspension arm assemblies to the rear suspension member with the bolts.

➡Ensure that the identification marks face the rear side of the vehicle.

15. Temporarily install the rear No. 1 and 2 suspension arm assemblies to the rear axle carrier sub-assembly with the bolts and the nuts.

➡Since a stopper nut is used, temporarily tighten the bolts.

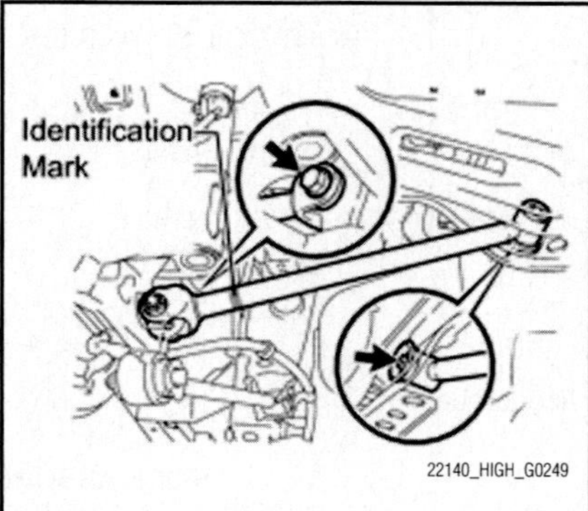

Fig. 349 Temporarily install the rear No. 1 and 2 suspension arm assemblies

16. Jack up the rear axle carrier, placing a wooden block underneath to avoid damage. Apply load to the suspension so that the installed bolt of the rear No. 1 suspension arm (inner side) is horizontally aligned with the center of the rear axle hub.

⁂ CAUTION

Do not jack up the rear axle carrier sub-assembly too high as the vehicle may fall.

➡Do not bend the brake dust cover.

➡If the rear drive shaft assembly cannot be positioned horizontally as shown in the illustration even when the rear axle carrier sub-assembly is jacked up, apply additional load to the vehicle such as by having a person sit in the rear seat.

➡Use the same procedures for the RH side and LH side.

17. Fully tighten the rear No. 1 suspension arm assembly, as follows:

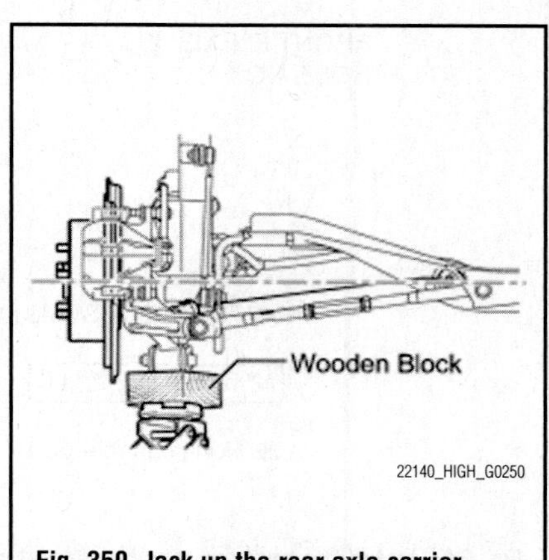

Fig. 350 Jack up the rear axle carrier

a. Using SST: 09961-00950 or equivalent and a socket wrench (19 mm), fully tighten the bolt to 88 ft. lbs. (120 Nm), 65 ft. lbs. (89 Nm).

➡Use a torque wrench with a fulcrum length of 425 mm (16.73 in.).

➡This torque value is effective when SST is parallel to the torque wrench.

➡The final torque must be applied under standard vehicle height conditions.

b. Fully tighten the bolt to 82 ft. lbs. (112 Nm).

➡Since a stopper nut is used, fully tighten the bolt.

➡The final torque must be applied under standard vehicle height conditions.

18. Fully tighten the rear No. 2 suspension arm assembly, as follows:

a. Fully tighten the bolts to 88 ft. lbs. (120 Nm), 82 ft. lbs. (112 Nm).

➡Since a stopper nut is used, fully tighten the bolt.

➡The final torque must be applied under standard vehicle height conditions.

19. Install the rear stabilizer bar with the 4 bolts and tighten to 14 ft. lbs. (19 Nm).

20. Install both rear stabilizer link assemblies to the rear stabilizer bar with the nuts and tighten to 29 ft. lbs. (39 Nm).

➡If the ball joint turns together with the nut, use a hexagon wrench (6 mm) to hold the stud bolt.

21. Install the wheels and tighten lug nuts to 76 ft. lbs. (103 Nm).

22. Inspect and adjust rear wheel alignment.

23. The remainder of installation is the reverse of the removal procedure.

4WD Vehicles

See Figures 351 through 355.

➡The removal procedures for the LH and RH sides are different.

➡When removing RH side components, it is not necessary to follow the steps with (for LH Side).

➡When removing LH side components, it is not necessary to follow the steps with (for RH side).

1. Remove the rear wheel (for RH Side).
2. Remove the rear wheels (for LH Side).
3. Remove the rear no. 2 suspension arm assembly LH (for LH Side), as follows:

a. Put matchmarks on the adjust cams and the rear suspension member sub-assembly.

b. Remove the bolt and the nut, and separate the rear No. 2 suspension arm assembly LH from the rear axle carrier sub-assembly LH.

➡Since a stopper nut is used, loosen the bolt.

c. Remove the nut, the No. 2 camber adjust cam, the rear suspension toe adjust cam sub-assembly, and the rear No. 2 suspension arm assembly LH.

➡When removing the nut, keep the rear suspension toe adjust cam sub-assembly from rotating.

4. Remove the tail exhaust pipe assembly.
5. Remove the center exhaust pipe assembly.
6. Remove the rear no. 2 suspension arm assembly RH (for RH Side).

➡Perform the same procedure as the rear No. 2 suspension arm assembly LH.

7. Remove the propeller with center bearing shaft assembly (for LH Side).
8. Remove the rear axle shaft nut LH (for LH Side).
9. Remove the rear axle shaft nut RH (for LH Side).
10. Separate the No. 3 parking brake cable assembly (for LH Side).

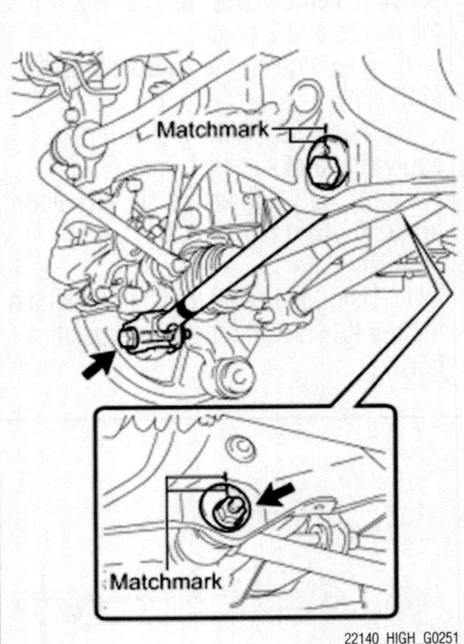

Fig. 351 Put matchmarks on the adjust cams and the rear suspension member sub-assembly

11. Separate the No. 2 parking brake cable assembly.
12. Remove the rear strut rod assembly LH (for LH Side).
13. Remove the rear strut rod assembly RH.
14. Remove the rear No. 1 suspension arm assembly RH (for RH Side). As follows:

a. Remove the bolt and the nut, and separate the rear No. 1 suspension arm assembly RH from the rear axle carrier sub-assembly RH.

b. Remove the bolt, the nut, and the rear No. 1 suspension arm assembly RH from the rear suspension member sub-assembly

➡Since stopper nuts are used, loosen the bolts.

15. Remove the rear speed sensor LH (for LH Side).
16. Remove the rear speed sensor RH (for LH Side).
17. Separate the rear disc brake caliper assembly LH (for LH Side).
18. Separate the rear disc brake caliper assembly RH (for LH Side).
19. Remove the rear disc (for LH Side), as follows:

a. Remove the rear disc from the rear axle hub and bearing assembly LH.

b. Remove the rear disc from the rear axle hub and bearing assembly RH.

➡Perform the same procedure as the LH side.

20. Remove the rear axle hub and bearing assembly LH (for LH Side).
21. Remove the rear axle hub and bearing assembly RH (for LH Side).
22. Remove the rear axle carrier sub-assembly LH (for LH Side), as follows:

a. Loosen the bolt.

➡Since a stopper nut is used, loosen the bolt.

b. Remove the 2 bolts and 2 nuts, and separate the rear shock absorber with coil spring from the rear axle carrier sub-assembly LH.

c. Remove the bolt, the nut, and the rear axle carrier sub-assembly LH.

➡Be careful not to damage the outboard joint boot.

➡Be careful not to damage the speed sensor rotor.

➡Use a rope or equivalent to hang the rear drive shaft assembly. Remove rear axle carrier sub-assembly RH (for LH Side), as follows:

d. Loosen the 2 bolts.

e. Remove the 2 bolts and 2 nuts, and separate the rear shock absorber with coil spring from the rear axle carrier sub-assembly RH.

f. Remove the 2 bolts, the 2 nuts, and the rear axle carrier sub-assembly RH.

➡Since stopper nuts are used, loosen the bolts.

➡Be careful not to damage the outboard joint boot.

➡Be careful not to damage the speed sensor rotor.

➡Use a rope or equivalent to hang the rear drive shaft assembly.

23. Remove the rear differential filler plug (for LH Side).
24. Remove the rear differential drain plug (for LH Side).
25. Remove the rear drive shaft assembly LH (for LH Side).
26. Remove the rear drive shaft snap ring LH (for LH Side).
27. Remove the rear drive shaft assembly RH (for LH Side).
28. Remove the rear drive shaft snap ring RH (for LH Side).
29. Remove the rear suspension member (for LH Side).
30. Remove the rear No. 1 suspension arm assembly LH (for LH Side), as follows:

a. Remove the bolt, the nut, and the rear No. 1 suspension arm assembly LH from the rear suspension member sub-assembly.

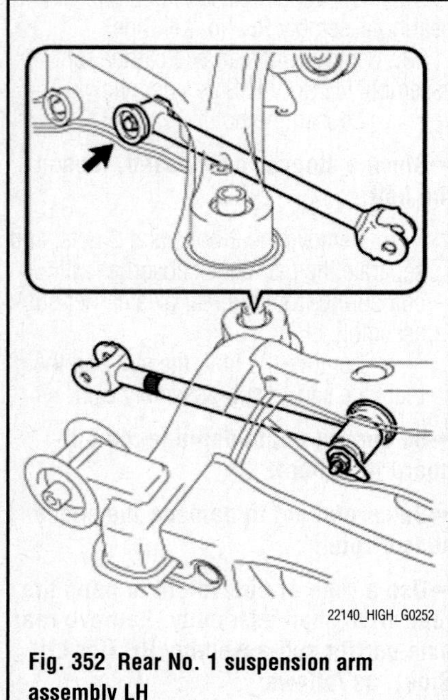

Fig. 352 Rear No. 1 suspension arm assembly LH

➡Since a stopper nut is used, loosen the bolt.

To install:

31. Temporarily install rear No. 1 suspension arm assembly LH (for LH Side), as follows:

a. Temporarily install the rear No. 1 suspension arm assembly LH to the rear suspension member sub-assembly with the bolt and the nut.

➡Ensure that the identification mark faces the rear side of the vehicle.

b. Set the rear No. 1 suspension arm LH in the position shown. Length A: 20 mm (0.787 in.)

c. Fully tighten the bolt to 59 ft. lbs. (80 Nm).

➡Since stopper nuts are used, temporarily tighten the bolts.

32. Install the rear suspension member (for LH Side), as follows:

a. Support the rear suspension member with a jack using a wooden block.

➡Use a properly sized wooden block to keep the jack and suspension member level.

➡Support the suspension member until retightening of the suspension member is complete.

b. Raise the rear suspension member with a jack.

c. Temporarily install the rear suspension member, 2 rear upper suspension member stoppers, and 2 rear lower suspension member stopper retainers with the 4 nuts and 2 bolts.

d. Fully tighten the 2 nuts 85 ft. lbs. (115 Nm).

e. Using SST: 09961-00950 or equivalent and a socket wrench (19 mm), fully tighten the nut (LH side) to 133 ft. lbs. (181 Nm), 98 ft. lbs. (134 Nm).

f. Using the same tools, fully tighten the nut (RH side) to the same specifications.

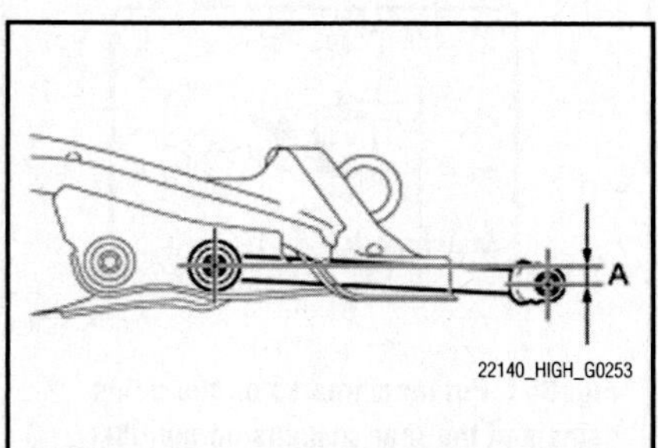

Fig. 353 Rear No. 1 suspension arm LH

➡Use a torque wrench with a fulcrum length of 425 mm (16.73 in.).

➡These torque values are effective when SST is parallel to the torque wrench .

33. Install the rear drive shaft snap ring LH (for LH Side).
34. Install the rear drive shaft assembly LH (for LH Side).
35. Install the rear drive shaft snap ring RH (for LH Side).
36. Install the rear drive shaft assembly RH (for LH Side).
37. Install the rear differential drain plug (for LH Side).
38. Add differential oil (for LH Side).
39. Inspect differential oil (for LH Side).
40. Install the rear differential filler plug (for LH Side).
41. Install the rear axle carrier sub-assembly LH (for LH Side), as follows:

a. Temporarily install the rear axle carrier sub-assembly LH with the bolt and the nut.

b. Install the rear axle carrier sub-assembly LH with the 2 bolts and the 2 nuts and tighten to 213 ft. lbs. (290 Nm).

➡Be careful not to damage the outboard joint boot.

➡Be careful not to damage the speed sensor rotor.

➡Prevent foreign matter from adhering to the speed sensor rotor.

➡Do not rotate the rear drive shaft assembly without the rear axle hub and bearing assembly installed.

➡Insert the bolts from the rear side.

42. Install the rear axle carrier sub-assembly RH (for LH Side), as follows:

a. Temporarily install the rear axle carrier sub-assembly LH with the 2 bolts and the 2 nuts.

b. Install the rear axle carrier sub-assembly LH with the 2 bolts and the 2 nuts and tighten to 213 ft. lbs. (290 Nm).

➡Be careful not to damage the outboard joint boot.

➡Be careful not to damage the speed sensor rotor.

➡Prevent foreign matter from adhering to the speed sensor rotor.

➡Do not rotate the rear drive shaft assembly without the rear axle hub and bearing assembly installed.

➡**Insert the bolts from the rear side.**

43. Temporarily install the rear No. 1 suspension arm assembly RH to the rear suspension member sub-assembly with the bolt and the nut.

➡**Ensure that the identification mark faces the rear side of the vehicle.**

44. Temporarily install the rear No. 1 suspension arm assembly RH to the rear axle carrier sub-assembly RH with the bolt and the nut.

➡**Since stopper nuts are used, temporarily tighten the bolts.**

45. Temporarily install the rear strut rod assembly LH (for LH Side).
46. Temporarily install the rear strut rod assembly RH.
47. Temporarily install the rear No. 2 suspension arm assembly LH (for LH Side).
   a. Temporarily install the rear No. 2 suspension arm assembly LH to the rear suspension member sub-assembly with the rear suspension toe adjust cam sub-assembly, the No. 2 camber adjust cam and the nut.

➡**Ensure that the identification mark faces the rear side of the vehicle.**

➡**When temporarily tightening the nut, keep the rear suspension toe adjust cam sub-assembly from rotating.**

   b. Temporarily install the rear No. 2 suspension arm assembly LH to the rear axle carrier sub-assembly LH with the bolt and the nut.

➡**Since a stopper nut is used, temporarily tighten the bolt.**

48. Temporarily install the rear No. 2 suspension arm assembly RH (for LH Side).

➡**Perform the same procedure as the rear No. 2 suspension arm assembly LH.**

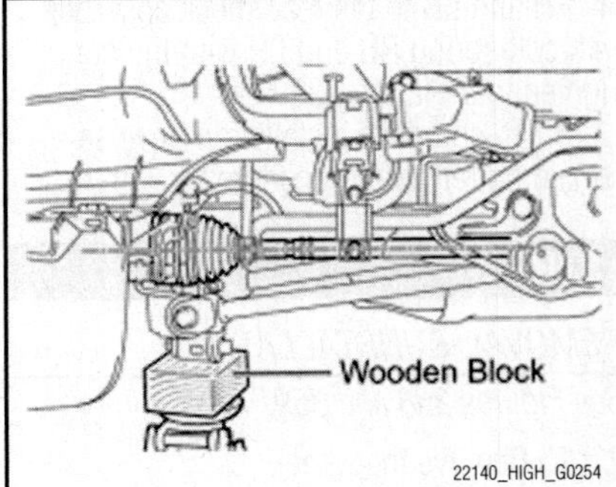

Fig. 354 Jack up the rear axle carrier sub-assembly

49. Install the rear axle hub and bearing assembly LH (for LH Side).
50. Install the rear axle hub and bearing assembly RH (for LH Side).
51. Install the rear disc to the rear axle hub and bearing assembly LH.
52. Install the rear disc to the rear axle hub and bearing assembly RH.

➡**Perform the same procedure as the LH side.**

53. Install the rear disc brake caliper assembly LH (for LH Side).
54. Install the rear disc brake caliper assembly RH (for LH Side).
55. Temporarily the install rear axle shaft nut LH (for LH Side).
56. Temporarily the install rear axle shaft nut RH (for LH Side).
57. Install the rear speed sensor LH (for LH Side).
58. Install the rear speed sensor RH (for LH Side).
59. Jack up the rear axle carrier sub-assembly, placing a wooden block underneath to avoid damage. Apply load to the suspension so that the rear drive shaft assembly is positioned horizontally.

CAUTION

Do not jack up the rear axle carrier sub-assembly too high as the vehicle may fall.

➡**Do not bend the brake dust cover.**

➡**If the rear drive shaft assembly cannot be positioned horizontally as shown in the illustration even when the rear axle carrier sub-assembly is jacked up, apply additional load to the vehicle such as by having a person sit in the rear seat.**

➡**Use the same procedures for the RH side and LH side.**

60. Fully tighten the rear No. 1 suspension arm assembly LH (for LH Side).
61. Fully tighten the rear No. 1 suspension arm assembly RH (for LH Side).

➡**Perform the same procedure as the rear No. 1 suspension arm assembly LH.**

62. Fully tighten the rear No. 1 suspension arm assembly RH (for RH Side).
63. Fully tighten the 2 bolts to 59 ft. lbs. (80 Nm), 82 ft. lbs. (112 Nm).

➡**Since a stopper nut is used, temporarily tighten the bolt.**

➡**The final torque must be applied under standard vehicle height conditions.**

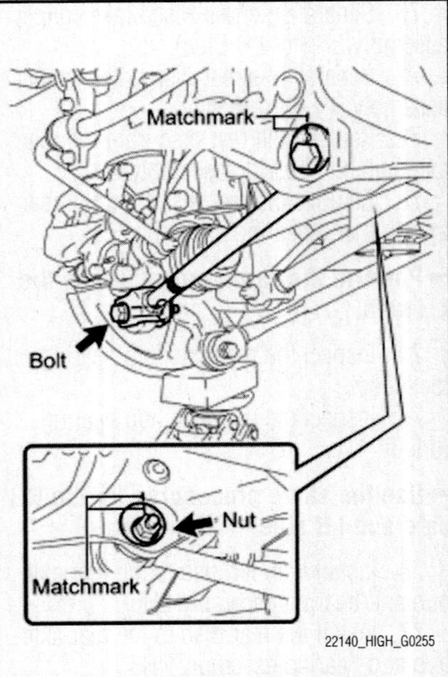

Fig. 355 Align the matchmarks on the adjust cams and rear suspension member sub-assembly

64. Fully tighten the rear No. 2 suspension arm assembly LH (for LH Side).
   a. Align the matchmarks on the adjust cams and rear suspension member sub-assembly.
   b. Fully tighten the nut to 74 ft. lbs. (100 Nm).

➡**The final torque must be applied under standard vehicle height conditions.**

➡**When fully tightening the nut, keep the rear suspension toe adjust cam sub-assembly from rotating.**

   c. Fully tighten the bolt to 82 ft. lbs. (112 Nm).

➡**Since a stopper nut is used, temporarily tighten the bolt.**

➡**The final torque must be applied under standard vehicle height conditions.**

65. Fully tighten the rear No. 2 suspension arm assembly RH (for RH Side).

➡**Perform the same procedure as the rear No. 2 suspension arm assembly LH.**

66. Fully tighten rear strut rod assembly LH (for LH Side).
67. Fully tighten rear strut rod assembly RH.
68. Install the No. 3 parking brake cable assembly (for LH Side).
69. Install the No. 2 parking brake cable assembly.

70. Separate the rear disc brake caliper assembly LH (for LH Side).
71. Separate the rear disc brake caliper assembly RH (for LH Side).
72. Remove the rear disc from the rear axle hub and bearing assembly LH.
73. Remove the rear disc from the rear axle hub and bearing assembly RH.

➡Perform the same procedure as the LH side.

74. Inspect the rear axle hub bearing looseness.
75. Inspect the rear axle hub bearing runout.

➡Use the same procedures for the RH side and LH side.

76. Install the rear disc to the rear axle hub and bearing assembly LH.
77. Install the rear disc to the rear axle hub and bearing assembly RH.

➡Perform the same procedure as the LH side.

78. Install the rear disc brake caliper assembly LH (for LH Side).
79. Install the rear disc brake caliper assembly RH (for LH Side).
80. Install the rear axle shaft nut LH (for LH Side).
81. Install the rear axle shaft nut RH (for LH Side).
82. Temporarily tighten the propeller with center bearing shaft assembly (for LH Side).
83. Fully tighten the propeller with center bearing shaft assembly (for LH Side).
84. Inspect and adjust transfer oil (for LH Side).
85. Install the center exhaust pipe assembly.
86. Install the tail exhaust pipe assembly.
87. Inspect for exhaust gas leak
88. Install the wheels and tighten lug nuts to 76 ft. lbs. (103 Nm).
89. Inspect and adjust rear wheel alignment.
90. Check for rear speed sensor signal (for LH Side).

STABILIZER BAR

REMOVAL & INSTALLATION

2WD Vehicles

See Figure 356.

1. Remove the nut and separate the rear stabilizer link assembly from the rear stabilizer bar.

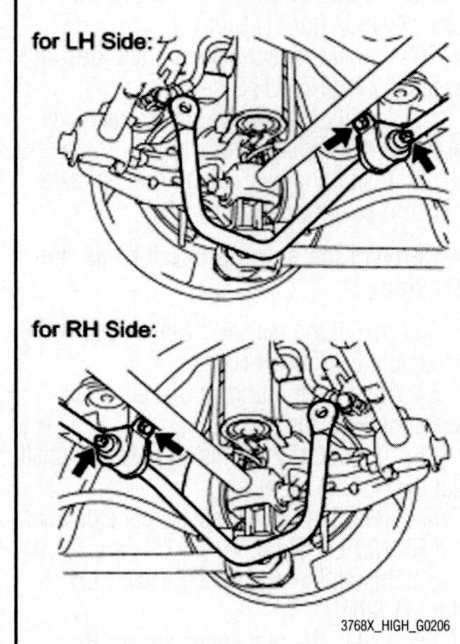

Fig. 356 Remove the 4 bolts and rear stabilizer bar—2WD vehicles

➡If the ball joint turns together with the nut, use a hexagon wrench (5 mm) to hold the stud bolt.

2. Remove the nut and the rear stabilizer link assembly from the rear shock absorber with coil spring.
3. Remove the 4 bolts and rear stabilizer bar.
4. Remove the rear No. 1 stabilizer bar bracket.
5. Remove the rear stabilizer bushing.

To install:

6. To install, reverse the removal procedure.
7. Tighten the rear stabilizer bar bolts to 14 ft. lbs. (19 Nm).
8. Tighten the rear stabilizer link assembly nut to 29 ft. lbs. (39 Nm).

4WD Vehicles

See Figure 357.

1. Remove deck board assembly.
2. Remove no. 3 deck board sub-assembly (w/ Tonneau Cover).
3. Remove no. 2 deck board sub-assembly (w/ Tonneau Cover).
4. Remove tonneau cover assembly (w/ Tonneau Cover).
5. Remove rear mat.
6. Remove deck trim service hole cover.
7. Remove lower spare wheel carrier hinge cover (w/ cover).
8. Remove spare tire.
9. Remove spare wheel carrier lock cover (w/ cover).

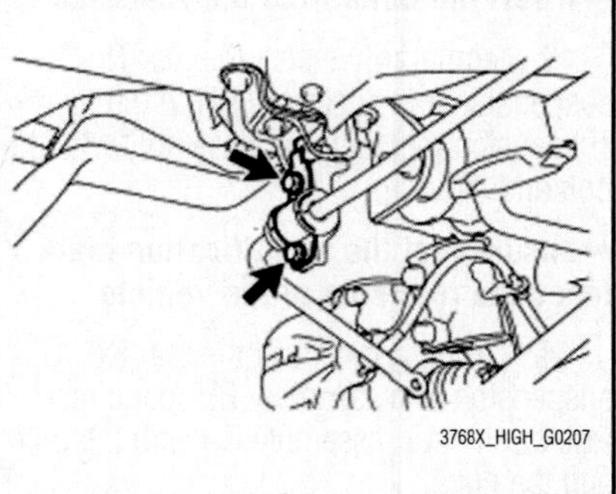

Fig. 357 Remove the 2 bolts and the rear stabilizer bar bracket

10. Remove rear wheels.
11. Remove tail exhaust pipe assembly.
12. Remove rear stabilizer link assembly.
13. Remove the 2 bolts and the rear stabilizer bar bracket.
14. Remove the rear stabilizer bushing.
15. Remove the bolt and the rear stabilizer bar bracket.

To install:

16. Temporarily install the rear stabilizer bar with the identification mark positioned on the left side of the vehicle.
17. Temporarily install the LH and RH rear stabilizer bar bracket (front side) with the bolt.
18. Loosely tighten the bolt so that the bracket can be moved by hand.
19. Install the rear stabilizer bushings to the rear stabilizer bar.

➡Make sure that the cutout of the rear stabilizer bushing is positioned towards the rear of the vehicle.

20. Install the rear stabilizer bar bracket LH and RH (rear side) with the 2 bolts. Tighten to 14 ft. lbs. (19 Nm).
21. Fully tighten the LH rear stabilizer bar brackets. Tighten to 40 ft. lbs. (54 Nm).
22. Fully tighten the RH rear stabilizer bar brackets. Tighten to 14 ft. lbs. (19 Nm).
23. Install the rear stabilizer link assembly RH and LH to the rear shock absorber with coil spring RH and LH with the nut. Tighten to 29 ft. lbs. (39 Nm).
24. To complete installation, reverse remaining removal procedure.

STRUT

REMOVAL & INSTALLATION

See Figures 358 and 359.

1. Remove the wheel.
2. Remove the deck side trim cover.
3. Remove the deck side trim.
4. For 4WD vehicles, remove the bolt

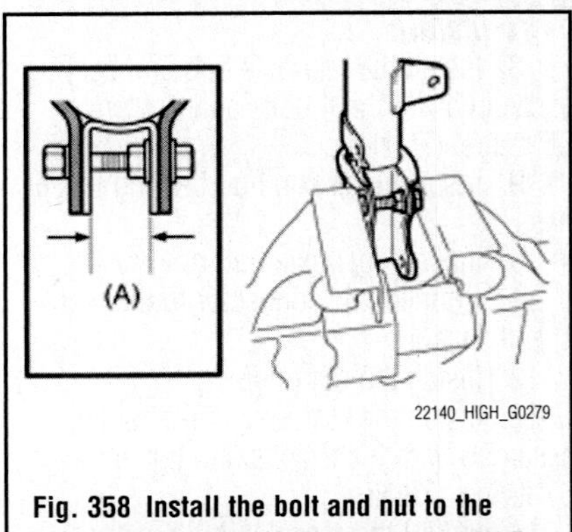

Fig. 358 Install the bolt and nut to the rear shock absorber with coil spring 2WD

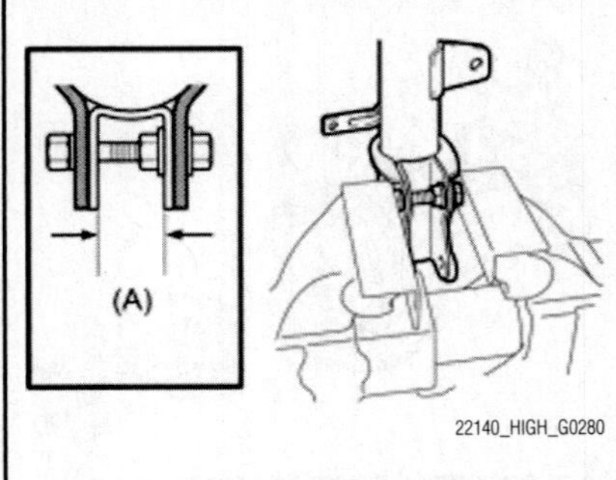

Fig. 359 Install the bolt and nut to the rear shock absorber with coil spring 4WD

and separate the rear flexible hose from the rear shock absorber with coil spring.

5. Remove the bolt and separate the rear speed sensor from the rear shock absorber with coil spring.

6. Remove the nut and separate the rear stabilizer link assembly from the rear shock absorber with coil spring.

➡If the ball joint turns together with the nut, use a hexagon wrench (5 mm) to hold the stud bolt.

7. Disengage the 4 claws and remove the rear No. 1 suspension support cover.

8. Using a jack and wooden block, support the rear axle carrier sub-assembly.

➡Do not deform the dust cover.

➡Support the rear axle carrier sub-assembly until reinstallation of the rear shock absorber with coil spring is complete.

9. Loosen the rear support to rear shock absorber nut.

➡Do not remove the rear support to rear shock absorber nut.

➡Loosen the nut only when the rear shock absorber with coil spring needs to be disassembled.

10. Remove the 2 bolts and 2 nuts, and separate the rear shock absorber with coil spring from the rear axle carrier sub-assembly,

➡When removing the nuts, keep the bolts from rotating.

11. Remove the 3 nuts and rear shock absorber with coil spring.

➡Make sure that the rear speed sensor and rear flexible hose are disconnected from the rear shock absorber with coil spring.

12. Remove the rear support to rear shock absorber nut, as follows:

a. Using SST: 09727-30021, compress the rear coil spring.

➡Do not use an impact wrench. It will damage SST.

b. Check that the front coil spring is fully compressed.

c. Hold the rear suspension support assembly and remove the rear support to rear shock absorber nut from the rear shock absorber assembly.

13. Remove the rear support to rear shock absorber collar from the rear shock absorber assembly.

14. Remove the rear suspension support assembly from the rear shock absorber assembly.

15. Remove the rear coil spring together with SST from the rear shock absorber assembly.

16. Remove the rear No. 1 spring bumper from the rear shock absorber assembly.

17. Remove the rear lower coil spring insulator from the rear shock absorber assembly.

To install:

18. Install the rear lower coil spring insulator onto the rear shock absorber assembly.

➡Fit the recessed part of the rear lower coil spring insulator into the recession on the shock absorber assembly.

19. Install the rear No. 1 spring bumper to the rear shock absorber assembly.

20. Temporarily install rear coil spring, as follows:

a. Using SST: 09727-30021 or equivalent, compress the rear coil spring.

➡Do not use an impact wrench. It will damage the SST.

b. Temporarily install the rear coil spring together with SST to the rear shock absorber assembly.

21. Install the rear suspension support assembly to the rear shock absorber assembly.

➡Align the cutout on the rear shock absorber assembly with the protrusion on the rear suspension support assembly by referring to the illustration.

22. Install the rear support to rear shock absorber collar to the rear shock absorber assembly.

23. Temporarily install the rear support to rear shock absorber nut to the rear shock absorber assembly.

24. Install the rear coil spring.

➡Do not use an impact wrench. It will damage the SST.

➡Make sure that the end of the rear coil spring is positioned in the depression of the rear lower coil spring insulator.

➡Ensure that the stud bolt is positioned 3.5° to the outside of the vehicle as shown in the illustration. The deviation should be within +-5°.

25. Install the rear shock absorber with coil spring with the 3 nuts and tighten to 43 ft. lbs. (58 Nm).

26. Install the rear shock absorber with coil spring with the 2 bolts and 2 nuts and tighten to 213 ft. lbs. (290 Nm).

➡When installing the nuts, keep the bolts from rotating.

27. Fully tighten the rear support to rear shock absorber nut 40 ft. lbs. (55 Nm).

28. Install the rear No. 1 suspension support cover.

29. Install the rear stabilizer link assembly to the rear shock absorber with coil spring with the nut and tighten to 29 ft. lbs. (39 Nm).

➡If the ball joint turns together with the nut, use a hexagon wrench (5 mm) to hold the stud bolt.

30. Install the rear speed sensor wire to the rear shock absorber with coil spring with the bolt and tighten to 44 inch lbs. (5 Nm).

➡Do not twist the rear speed sensor wire when installing it.

31. Install the rear flexible hose to the rear shock absorber with coil spring with the bolt and tighten to 14 ft. lbs. (19 Nm).

➡Do not twist the rear flexible hose when installing it.

32. Install deck side trim.
33. Install deck side trim cover.
34. Install the wheels and tighten lug nuts to 76 ft. lbs. (103 Nm).
35. Inspect and adjust rear wheel alignment.

WHEEL HUB & BEARING

REMOVAL & INSTALLATION

See Figure 360.

1. Disconnect the negative battery cable.
2. Remove the wheel.
3. Separate the rear flexible hose.
4. Remove the 2 bolts and separate the rear disc brake caliper assembly.
5. Remove the rear disc.
6. For 4WD vehicles: Using a screwdriver, disconnect the connector from the rear speed sensor.
7. Remove the 4 bolts and the rear axle hub & bearing assembly.

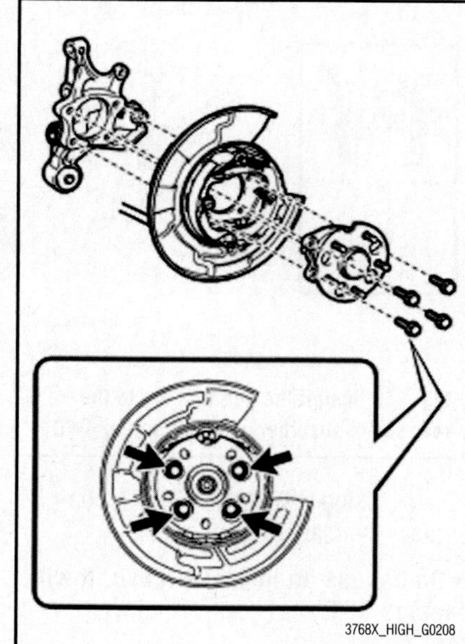

Fig. 360 Remove the 4 bolts and the rear axle hub & bearing assembly

To install:

8. Install the rear axle hub and bearing assembly with the 4 bolts and tighten to 55 ft. lbs. (75 Nm).
9. Inspect rear axle hub bearing looseness.
10. Inspect rear axle hub runout.
11. Connect the connector to the rear speed sensor.
12. Install the rear disc.
13. Install the rear disc brake caliper assembly with the 2 bolts and tighten to 57 ft. lbs. (78 Nm).
14. Install the rear flexible hose to the shock absorber with coil spring with the bolt and tighten to 14 ft. lbs. (19 Nm).
15. Install the wheel and tighten lug nuts to 76 ft. lbs. (103 Nm).
16. Connect the negative battery cable.
17. Inspect and adjust rear wheel alignment.
18. Check the ABS speed sensor signal.

TOYOTA

Highlander Hybrid

10

SPECIFICATIONS AND MAINTENANCE CHARTS

ENGINE AND VEHICLE IDENTIFICATION

Engine							Model Year	
Code ①	Liters (cc)	Cu. In.	Cyl.	Fuel Sys.	Engine Type	Eng. Mfg.	Code ②	Year
3MZ-FE	3.3 (3311)	202.1	6	SFI	DOHC	Toyota	9	2009
							A	2010

SFI: Sequential Fuel Injection

DOHC: Double Overhead Camshaft

① Stamped on the right side of the engine block

② 10th digit of the Vehicle Identification Number (VIN)

3768X_HIHY_C0001

GENERAL ENGINE SPECIFICATIONS

Year	Model	Engine Displacement Liters	Engine Series ID	Net Horsepower @ rpm	Net Torque @ rpm (ft. lbs.)	Bore x Stroke (in.)	Com-pression Ratio	Oil Pressure @ rpm
2009	Highlander Hybrid	3.3	3MZ-FE	268@5600	212@4400	3.62x3.27	10.8:1	36-78@3000
2010	Highlander Hybrid	3.3	3MZ-FE	268@5600	212@4400	3.62x3.27	10.8:1	36-78@3000

3768X_HIHY_C0002

ENGINE TUNE-UP SPECIFICATIONS

Year	Engine Displacement Liters	Engine ID	Spark Plug Gap (in.)	Ignition Timing (deg.)	Fuel Pump (psi)	Idle Speed (rpm)	Valve Clearance	
							Intake	Exhaust
2009	3.3	3MZ-FE	0.039-0.043	8-12B①	44-50	850-950	0.006-0.010	0.010-0.014
2010	3.3	3MZ-FE	0.039-0.043	8-12B①	44-50	850-950	0.006-0.010	0.010-0.014

NOTE: The Vehicle Emission Control Information label often reflects specification changes made during production.

The label figures must be used if they differ from those in this chart.

B: Before top dead center

① With terminals TC and CG of DLC3 connected

3768X_VENZ_C0003

CAPACITIES

Year	Model	Engine Displacement Liters	Engine ID	Engine Oil with Filter (qts.)	Transaxle (pts)	Rear Transaxle (pts.)	Rear Drive Axle (pts.)	Fuel Tank (gal.)	Cooling System (qts.)
2009	Highlander Hybrid	3.3	3MZ-FE	5.0	①	4.2	N/A	17.2	②
2010	Highlander Hybrid	3.3	3MZ-FE	5.0	①	4.2	N/A	17.2	②

N/A: Not Available

① With towing package: 8.8 pts.
Without towing package: 8.2 pts.

② With Rear Heater: 12.8 qts.
Without Rear Heater: 10.6 qts.

3768X_HIHY_C0005

FLUID SPECIFICATIONS

Year	Model	Engine Displacement Liters	Engine ID/VIN	Engine Oil	Auto. Trans. ①	Drive Axle ②	Power Steering Fluid	Brake Master Cylinder	Engine Coolant ③
2009	Highlander Hybrid	3.3	3MZ-FE	5W-30	ATF-WS	75W-90	NA	DOT 3	Toyota coolant
2010	Highlander Hybrid	3.3	3MZ-FE	5W-30	ATF-WS	75W-90	NA	DOT 3	Toyota coolant

DOT: Department Of Transpotation

NA: Not Available

① The use of genuine Toyota ATF-WS is recommended

② Synthetic GL-5 (75W-90) or equivalent

③The use of genuine Toyota engine coolant is recommended or similar
ethylene glycol based non-silicate, non-amine, non- nitrite, and non- borat coolant

3768X_HIHY_C0004

VALVE SPECIFICATIONS

All measurements are given in inches.

Year	Engine Displacement Liters	Engine ID	Seat Angle (deg.)	Face Angle (deg.)	Spring Test Pressure (lbs. @ in.)	Spring Installed Height (in.)	Stem-to-Guide Clearance (in.)		Stem Diameter (in.)	
							Intake	Exhaust	Intake	Exhaust
2009	3.3	3MZ-FE	45	40.5	41.9-46.3@ 1.331	1.331	0.0010-0.0024	0.0012-0.0026	0.2154-0.2159	0.2152 0.2157
2010	3.3	3MZ-FE	45	40.5	41.9-46.3@ 1.331	1.331	0.0010-0.0024	0.0012-0.0026	0.2154-0.2159	0.2152 0.2157

3768X_HIHY_C0006

CAMSHAFT AND BEARING SPECIFICATIONS CHART

All measurements are given in inches.

Year	Engine Displ. Liters	Engine ID/VIN	Journal Dia.	Brg. Oil Clearance	Shaft End-play	Runout	Journal Bore	Lobe Height Intake	Lobe Height Exhaust
2009	3.3	3MZ-FE	1.0614-1.0620	①	0.0016-0.0035	0.0024	NA	1.6981-1.7020	1.6933-1.6972
2010	3.3	3MZ-FE	1.0614-1.0620	①	0.0016-0.0035	0.0024	NA	1.6981-1.7020	1.6933-1.6972

NA: Not Available

① Intake Journals 4 and 5: 0.0010 - 0.0022 in.
All Others: 0.0010 - 0.0024 in.

3768X_HIHY_C0008

CRANKSHAFT AND CONNECTING ROD SPECIFICATIONS

All measurements are given in inches.

Year	Engine Displacement Liters	Engine ID	Crankshaft Main Brg. Journal Dia.	Crankshaft Main Brg. Oil Clearance	Crankshaft Shaft End-play	Crankshaft Thrust on No.	Connecting Rod Journal Diameter	Connecting Rod Oil Clearance	Connecting Rod Side Clearance
2009	3.3	3MZ-FE	2.4011-2.4016	①	0.0016-0.0094	2	2.0863-2.0866	0.0015-0.0026	0.0059-0.0118
2010	3.3	3MZ-FE	2.4011-2.4016	①	0.0016-0.0094	2	2.0863-2.0866	0.0015-0.0026	0.0059-0.0118

① Journals 1 and 4: 0.0006 - 0.0013 in.
Journals 2 and 3: 0.0010 - 0.0018 in.

3768X_HIHY_C0007

PISTON AND RING SPECIFICATIONS

All measurements are given in inches.

Year	Engine Displ. Liters	Engine ID	Piston Clearance	Ring Gap Top Comp.	Ring Gap Bottom Comp.	Ring Gap Oil Control	Ring Side Clearance Top Comp.	Ring Side Clearance Bottom Comp.	Ring Side Clearance Oil Control
2009	3.3	3MZ-FE	0.0013-0.0023	0.0118-0.0157	0.0197-0.0236	0.0059-0.0157	0.0012-0.0031	0.0008-0.0024	0.0012-0.0043
2010	3.3	3MZ-FE	0.0013-0.0023	0.0118-0.0157	0.0197-0.0236	0.0059-0.0157	0.0012-0.0031	0.0008-0.0024	0.0012-0.0043

3768X_HIHY_C0009

TORQUE SPECIFICATIONS

All readings in ft. lbs.

Year	Engine Displacement Liters	Engine ID	Cylinder Head Bolts	Main Bearing Bolts	Rod Bearing Bolts	Crankshaft Damper Bolts	Flywheel Bolts	Manifold Intake	Manifold Exhaust	Spark Plugs	Oil Pan Drain Plug
2009	3.3	3MZ-FE	①	②	③	162	61	11	36	18	33
2010	3.3	3MZ-FE	①	②	③	162	61	11	36	18	33

① Step 1: 12 point bolts to 40 ft. lbs.
Step 2: 12 point bolts plus 90 degrees
Step 3: Hex head recessed bolt to 14 ft. lbs.

② Step 1: 12 point cap bolts to 16 ft. lbs.
Step 2: 12 point cap bolts plus 90 degrees
Step 3: Hex head side bolts to 20 ft. lbs.

③ Step 1: 18 ft. lbs.
Step 2: Plus 90 degrees

3768X_HIHY_C0010

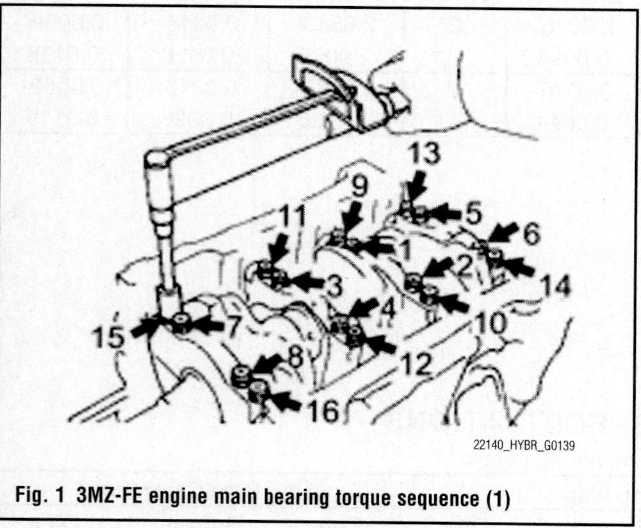

Fig. 1 3MZ-FE engine main bearing torque sequence (1)

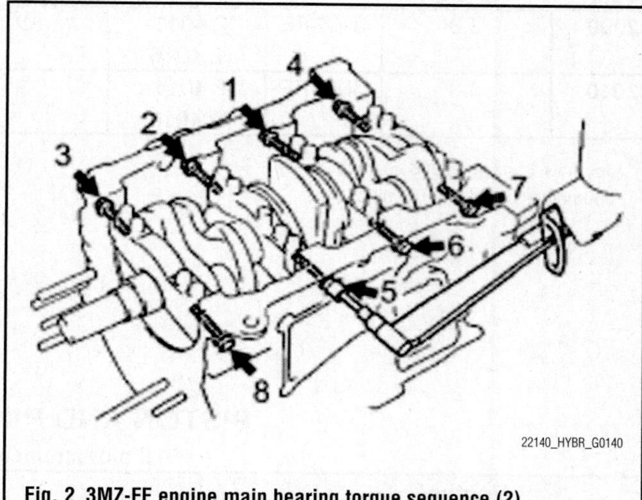

Fig. 2 3MZ-FE engine main bearing torque sequence (2)

WHEEL ALIGNMENT

Year	Model		Caster Range (+/-Deg.)	Caster Preferred Setting (Deg.)	Camber Range (+/-Deg.)	Camber Preferred Setting (Deg.)	Toe-in (in.)
2009	Highlander Hybrid	2WD F	0.75	+2.75	0.75	-0.58	0+/-0.08
		4WD F	0.75	+2.50	0.75	-0.58	0+/-0.08
		2WD R	NA	NA	0.75	-1.17	0.12+/-0.08
		4WD R	NA	NA	0.75	-0.67	0.12+/-0.08
2010	Highlander Hybrid	2WD F	0.75	+2.75	0.75	-0.58	0+/-0.08
		4WD F	0.75	+2.50	0.75	-0.58	0+/-0.08
		2WD R	NA	NA	0.75	-1.17	0.12+/-0.08
		4WD R	NA	NA	0.75	-0.67	0.12+/-0.08

NA: Not Available

F: Front

R: Rear

3768X_HIHY_C0011

TIRE, WHEEL AND BALL JOINT SPECIFICATIONS

Year	Model	OEM Tires Standard	OEM Tires Optional	Tire Pressures (psi) Front	Tire Pressures (psi) Rear	Wheel Size	Ball Joint Inspection	Lug Nut Torque (ft. lbs.)
2009	Highlander Hybrid	P225/65R17	NA	32	32	6.5-J	①	76
2010	Highlander Hybrid	P225/65R17	NA	32	32	6.5-J	①	76

NA: Not Available

OEM: Original Equipment Manufacturer

PSI: Pounds Per Square Inch

STD: Standard

OPT: Optional

① Replace if any measurable movement is found.

3768X_HIHY_C0012

BRAKE SPECIFICATIONS

All measurements in inches unless noted

Year	Model		Brake Disc Original Thickness	Brake Disc Minimum Thickness	Brake Disc Maximum Runout	Minimum Lining Thickness	Brake Caliper Bracket Bolts (ft. lbs.)	Brake Caliper Mounting Bolts (ft. lbs.)
2009	Highlander Hybrid	F	1.102	1.024	0.0020	0.039	78	25
		R	0.394	0.335	0.0059	0.039	56	32
2010	Highlander Hybrid	F	1.102	1.024	0.0020	0.039	78	25
		R	0.394	0.335	0.0059	0.039	56	32

F: Front

R: Rear

3768X_HIHY_C0013

SCHEDULED MAINTENANCE INTERVALS

TOYOTA Highlander Hybrid

TO BE SERVICED	TYPE OF SERVICE	VEHICLE MILEAGE INTERVAL (x1000)												
		5	10	15	20	25	30	35	40	45	50	55	60	65
Engine oil & filter	R	✓	✓	✓	✓	✓	✓	✓	✓	✓	✓	✓	✓	✓
Automatic transmission fluid	R												✓	
Ball joints & dust covers	S/I	✓	✓	✓	✓	✓	✓	✓	✓	✓	✓	✓	✓	✓
Bolts & nuts on chassis & body	S/I	✓	✓		✓	✓		✓	✓		✓	✓		✓
Brake line pipes & hoses	S/I			✓			✓			✓			✓	
Brake fluid	R						✓						✓	
Brake pads & discs (front & rear)	S/I	✓	✓	✓	✓	✓	✓	✓	✓	✓	✓	✓	✓	✓
Propeller shaft grease	S/I	✓	✓	✓	✓	✓	✓	✓	✓	✓	✓	✓	✓	✓
Steering knuckle & chassis grease	S/I	✓	✓	✓	✓	✓	✓	✓	✓	✓	✓	✓	✓	✓
Steering linkage	S/I	✓	✓	✓	✓	✓	✓	✓	✓	✓	✓	✓	✓	✓
Air cleaner filter	R						✓						✓	
Air conditioner filter	R						✓							
Spark plugs	R	Replace at 120,000 miles												
Exhaust pipes & mountings	S/I			✓			✓			✓			✓	
Fuel lines & connections	S/I												✓	
Engine coolant	R	Replace at 120,000 miles												
Timing belt	R	Replace at 90,000 miles												
Rear differential fluid	R												✓	
Rotate Tires	S/I	✓	✓	✓	✓	✓	✓	✓	✓	✓	✓	✓	✓	✓

R: Replace S/I: Service or Inspect

FREQUENT OPERATION MAINTENANCE (SEVERE SERVICE)

If a vehicle is operated under any of the following conditions it is considered severe service:

- Extremely dusty areas.

- 50% or more of the constant operation is in 32°C (90°F) or higher temperatures, or in temperatures below 0°C (32°F).

- Prolonged idling (vehicle operation in stop and go traffic).

- Frequent short running periods (engine does not warm to normal operating temperatures).

- Police, taxi, delivery usage or trailer towing usage.

Air cleaner filter: service or inspect every 3750 miles

Engine oil & filter: replace every 3750 miles.

Ball joints & dust covers: service or inspect every 7500 miles.

Bolts & nuts on chassis & body: service or inspect every 7500 miles.

Brake pads & discs (front & rear): service or inspect every 7500 miles.

Steering knuckle & chassis grease: service or inspect every 7500 miles.

Steering linkage: service or inspect every 7500 miles.

Exhaust pipes & mountings: service or inspect every 15,000 miles.

3768X_HIHY_C0014

BRAKES — INFORMATION AND PRECAUTIONS

ANTI-LOCK SYSTEMS

- Certain components within the ABS system are not intended to be serviced or repaired individually.
- Do not use rubber hoses or other parts not specifically specified for and ABS system. When using repair kits, replace all parts included in the kit. Partial or incorrect repair may lead to functional problems and require the replacement of components.
- Lubricate rubber parts with clean, fresh brake fluid to ease assembly. Do not use shop air to clean parts; damage to rubber components may result.
- Use only DOT 3 brake fluid from an unopened container.
- If any hydraulic component or line is removed or replaced, it may be necessary to bleed the entire system.
- A clean repair area is essential. Always clean the reservoir and cap thoroughly before removing the cap. The slightest amount of dirt in the fluid may plug an orifice and impair the system function. Perform repairs after components have been thoroughly cleaned; use only denatured alcohol to clean components. Do not allow ABS components to come into contact with any substance containing mineral oil; this includes used shop rags.
- The Anti-Lock control unit is a microprocessor similar to other computer units in the vehicle. Ensure that the ignition switch is **OFF** before removing or installing controller harnesses. Avoid static electricity discharge at or near the controller.
- If any arc welding is to be done on the vehicle, the control unit should be unplugged before welding operations begin.

DISC AND DRUM SYSTEMS

CAUTION

Dust and dirt accumulating on brake parts during normal use may contain asbestos fibers from production or aftermarket brake linings. Breathing excessive concentrations of asbestos fibers can cause serious bodily harm. Exercise care when servicing brake parts. Do not sand or grind brake lining unless equipment used is designed to contain the dust residue. Do not clean brake parts with compressed air or by dry brushing. Cleaning should be done by dampening the brake components with a fine mist of water, then wiping the brake components clean with a dampened cloth. Dispose of cloth and all residue containing asbestos fibers in an impermeable container with the appropriate label. Follow practices prescribed by the Occupational Safety and Health Administration (OSHA) and the Environmental Protection Agency (EPA) for the handling, processing, and disposing of dust or debris that may contain asbestos fibers.

BRAKES — BLEEDING THE BRAKE SYSTEM

BLEEDING PROCEDURE

BLEEDING PROCEDURE

See Figures 3 through 5.

➡This procedure requires specialized tools. Please read through the procedure and make sure you have access to the proper equipment before beginning the bleeding procedure.

CAUTION

Never bleed air from the brake hydraulic system without using the intelligent tester. failure to use the intelligent tester could cause serious injury or an accident.

Note the following before bleeding the brake system:

- Move the shift lever to the P position and apply the parking brake before bleeding.
- Add brake fluid carefully and check that the reservoir level remains between the min and max lines while bleeding the brakes.
- Do not stand the fluid can on the reservoir inlet when bleeding the brake actuator. doing so will cause brake fluid to overflow.
- The actuator pump motor and solenoid can be operated by the driver even if the ignition switch is off.
- If the pump motor operates while air still remains inside the brake actuator hose, air will enter the actuator, making it more difficult to bleed the brakes. If there is concern about air remaining in the actuator hose, remove the two motor relays (skid control relay no.2) until instructed to reinstall them.
- Although a buzzer may sound due to a decline in the accumulator pressure while bleeding, it is not necessary to stop bleeding.
- Daces indicating a malfunction in the motor relays (skid control relay no.2) or the pressure sensor are stored after bleeding. clear the Daces when instructed during or after bleeding.

1. Add SAE J1703 or FMVSS no. 116 DOT 3 brake fluid to the max line in the reservoir.

CAUTION

Add brake fluid carefully and check that the reservoir level remains between the min and max lines while bleeding the brakes. Do not stand the fluid can on the reservoir inlet when bleeding the brake actuator. Doing so will cause brake fluid to overflow.

2. Disable the brake control (ECB). When using the intelligent tester:

➡When using the intelligent tester , refer to the intelligent tester operator's manual for further details. Bleed the air by following the steps displayed on the intelligent tester.

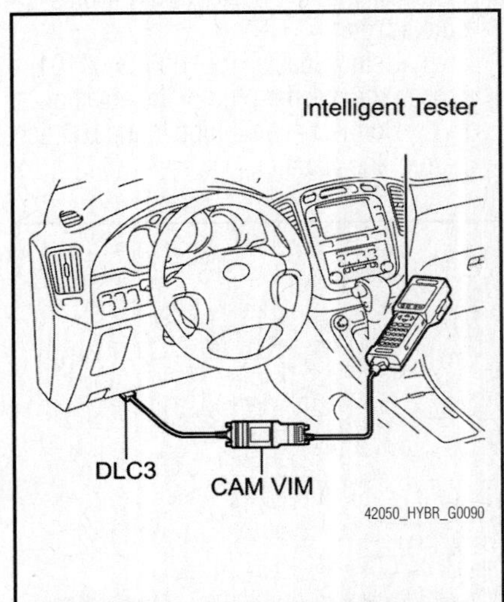

Fig. 3 Connect the intelligent tester to the DLC3 with the ignition switch OFF as shown in the illustration

a. Move the shift lever to the P position and apply the parking brake.

b. Connect the intelligent tester to the DLC3 with the ignition switch **OFF** as shown in the illustration.

c. Turn the ignition switch to the ON position and turn on the intelligent tester.

➡Do not start the engine.

d. Enter the following menus: DIAGNOSIS / OBD/ MOBD / ABS/TRAC/VSC / ECB UTILITY / ECB INVALID.

WARNING

If the pump motor operates while air remains inside the brake actuator hose, air will enter the actuator, and this will make bleeding the brakes more difficult.

e. When removing the ABS motor relay: Remove the 2 ABS motor relays with the ignition switch off in order to disable brake control.

WARNING

If the pump motor operates while air remains inside the brake actuator hose, air will enter the actuator, and this will make bleeding the brakes more difficult.

➡After the brake actuator assembly has been replaced, remove the ABS motor relay before bleeding the brakes.

3. **Bleed the brake actuator hose, as follows:**

a. Connect Special Tool 09992-00242, 0992-00350 or equivalent, to the reservoir with the brake reservoir pressure adapter.

b. Using Special Tool 09023-00101, loosen the bleeder plug of the actuator.

c. Connect a vinyl tube to the bleeder plug of the actuator.

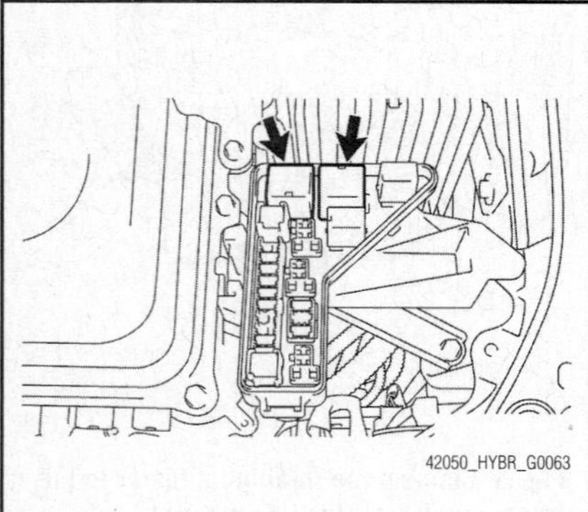

Fig. 4 Location of the 2 ABS motor relays

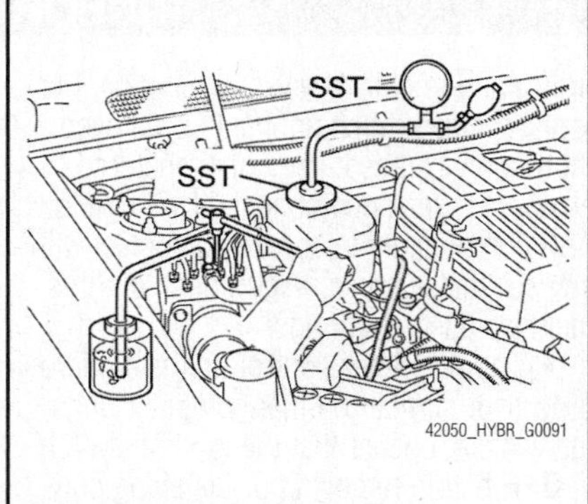

Fig. 5 View of the special tools

d. Use the SST to boost pressure in the reservoir. Standard pressure is 50 to 80 kPa (0.5 to 0.8 kgf/cm2, 7.3 to 11.6 psi)

e. Drain approximately 100 cc of fluid.

f. Tighten the bleeder plug and boost the pressure in the reservoir again (50 to 80 kPa (0.5 to 0.8 kgf/cm2)). Then, loosen the bleeder plug and bleed the brake actuator hose.

➡Repeat this procedure at least 5 times.

g. When air is completely bled out from the hose between the reservoir and the actuator, tighten the bleeder plug to 74 inch lbs. (8.3 Nm).

4. **Bleed the master cylinder, as follows:**

➡If the master cylinder has been disassembled or if the reservoir becomes empty, bleed the air from the master cylinder.

a. Enter the following menus: DIAGNOSIS / OBD/MOBD / ABS/TRAC/VSC / AIR BLEEDING.

b. Select "USUAL" if the front/rear brakes are removed, installed or disassembled.

c. Select "ACTUATOR" if the actuator is removed, installed or replaced.

d. Select "MASTER CYLINDER" if the brake master cylinder or the brake stroke simulator is removed, installed or replaced.

e. Disconnect the brake lines from the master cylinder.

f. Slowly depress and hold the brake pedal (Procedure A).

g. Cover the outer holes with fingers, and release the brake pedal (Procedure B).

h. Repeat procedure A and B 3 or 4 times.

i. Connect the brake lines to the master cylinder and tighten to 11 ft. lbs. (15 Nm).

5. **Bleed the front brake system, as follows:**

➡Air can be easily bled from the front brake system if air has been bled from the master cylinder when replacing the brake master cylinder assembly.

WARNING

If brake fluid leaks onto any painted surface of the vehicle, wash or otherwise remove it completely.

➡Bleed the air by following the steps displayed on the intelligent tester. (a) Depress the brake pedal several times and bleed the front brake system from the bleeder plugs on the front brake cylinder RH and LH.

➡Repeat the procedure until air is completely bled from the front brake system.

6. Tighten the bleeder plugs to 74 inch lbs. (8.3 Nm) after bleeding.

7. Cancel brake control (ECB) disable

a. Install the 2 motor relays (skid control relay No.2) if they have been removed.

b. Complete brake control prevention following the prompts on the tester screen. (If brake control has been prevented using the intelligent tester.)

8. Clear the DTC(s).

9. Bleed the rear brake system, as follows:

WARNING

Never bleed air from the brake hydraulic system without using the intelligent tester. Failure to use the intelligent tester could cause serious injury or an accident.

➡Bleed the air by following the steps displayed on the intelligent tester.

a. Connect the intelligent tester to the DLC3 with the ignition switch off.

b. Check that the parking brake is applied and turn the ignition switch to the **ON** position.

c. Enter the following menus: DIAGNOSIS / OBD/ MOBD / ABS/TRAC/VSC / ECB UTILITY / ECB INVALID.

d. With the brake pedal depressed, bleed the rear brake system from the bleeder plug on the rear disc brake cylinder LH while the pump motor and solenoid are operating.

WARNING

Keep the fluid inside the reservoir above the LOW level by replenishing.

➡Depress and hold the brake pedal. After the solenoid operates for approximately 30 seconds, release the brake pedal to stop the solenoid. Repeat the procedures until air is completely bled from the rear brake system. The ECB warning light comes on and the buzzer sounds while bleeding, but they do not indicate a malfunction.

e. Tighten the bleeder plug to 74 inch lbs. (8.3 Nm) after bleeding.

f. Enter the following menus: DIAGNOSIS / OBD/ MOBD / ABS/TRAC/ VSC / ECB UTILITY / ECB INVALID.

g. With the brake pedal depressed, bleed the rear brake system from the bleeder plug on the rear disc brake cylinder RH while the pump motor and solenoid are operating.

WARNING

Keep the fluid inside the reservoir above the LOW level by replenishing.

➡Depress and hold the brake pedal. After the solenoid operates for approximately 30 seconds, release the brake pedal to stop the solenoid. Repeat the procedures until air is completely bled from the rear brake system. The ECB warning light comes on and the buzzer sounds while bleeding, but they do not indicate a malfunction.

h. Tighten the bleeder plug to 74 inch lbs. (8.3 Nm) after bleeding.

10. Perform the accumulator zero down:

WARNING

Never bleed air from the brake hydraulic system without using the intelligent tester. Failure to use the intelligent tester could cause serious injury or an accident. Be sure to perform this procedure before replacement, removal, or installation of the actuator.

➡Perform accumulator zero down by following the steps displayed on the intelligent tester.

a. Connect the intelligent tester to the DLC3 with the ignition switch **OFF**.

b. Depressurize the accumulator:

- Check that the parking brake is applied and turn the ignition switch to the **ON** position.
- Enter the following menus: DIAGNOSIS / OBD/ MOBD / ABS/ TRAC/VSC / ECB UTILITY / ZERO DOWN.
- When the buzzer sounds, turn the ignition switch **OFF**.

c. Circulate the fluid in the accumulator.

d. Depressurize the accumulator 5 times.

➡Accumulator pressure is released and accumulated repeatedly, which circulates

the fluid inside the accumulator, when repeating accumulator zero down. The pump motor rotates and the accumulator is pressurized every time the ignition switch is turned from off to on.

11. Check the brake fluid level:

a. After performing accumulator zero down (accumulator depressurizing), return the fluid in the accumulator back to the reservoir and then adjust the fluid level in the master cylinder reservoir to the MAX level.

➡After performing accumulator zero down (accumulator depressurizing), fluid is built up in the accumulator by turning the ignition switch to the ON position and the fluid level of the reservoir lowers. If the fluid level is adjusted without performing accumulator zero down (accumulator depressurizing), fluid is sent from the accumulator to the reservoir. The fluid level may exceed the MAX level, but it is normal.

12. Clear the DTC(s).

13. When the brake actuator assembly is replaced, perform linear valve offset learning after bleeding is completed.

BRAKES — ANTI-LOCK BRAKE SYSTEM (ABS)

WHEEL SPEED SENSORS

REMOVAL & INSTALLATION

Front Sensors

See Figures 6 through 14.

➡If the sensor rotor needs to be replaced, replace it together with the front drive outboard joint shaft assembly.

1. Disconnect the negative battery cable.
2. Raise and safely support the vehicle.
3. Remove the front wheel and tire assembly.
4. Remove No. 1 engine under cover.
5. Remove the front fender molding sub-assembly, as follows:

a. Remove the clip.

b. Using a 4 mm hexagon wrench, remove the screw.

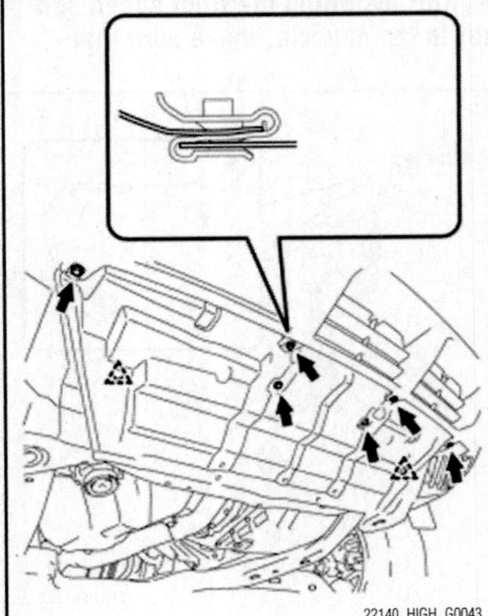

Fig. 6 No. 1 engine under cover, 6 bolts and 2 clips

c. Peel off the front fender side protector and disengage the 3 clips, and then remove the front fender molding sub-assembly.

d. Remove the pad from the front fender molding sub-assembly.

e. Remove the 2 No. 4 clips from the front fender molding sub-assembly.

f. Remove the front fender side protector from the front fender molding sub-assembly.

6. Remove the front wheel opening extension pad.
7. Remove the front fender liner, as follows:

a. Remove the screw.

b. Using a screwdriver, turn the pin 90 degrees and remove the pin hold clip.

c. Using a 4 mm hexagon wrench, remove the 2 screws.

d. Remove the 2 grommets.

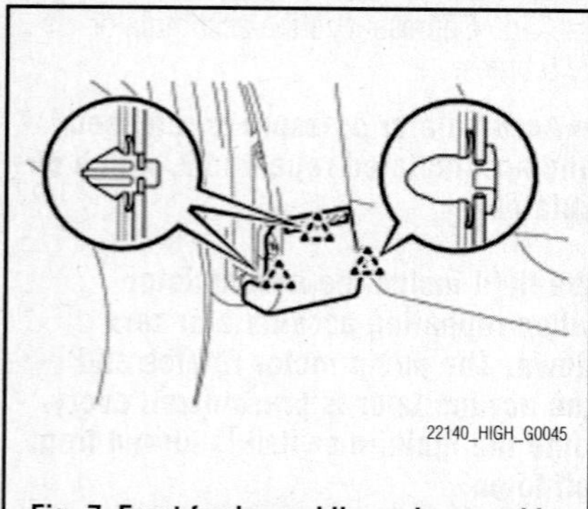

Fig. 7 Front fender molding sub-assembly and clips

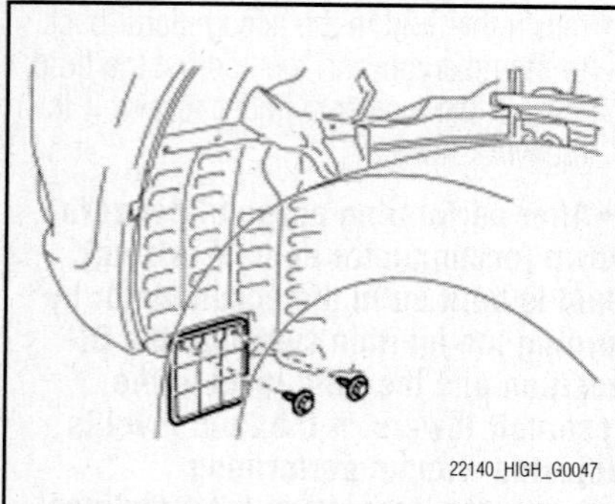

Fig. 8 Front wheel opening extension pad and screws

➡The grommets need to be replaced with new ones because they will break when they are removed.

e. Remove the 5 clips.

f. For left hand side, remove 8 screws and front fender liner.

g. For right hand side, remove 7 screws and front fender liner.

8. Disconnect the front speed sensor connector and remove the 2 clamps.

9. Remove the bolt and No. 2 sensor clamp from the body.

10. Remove the bolt, No. 1 sensor clamp, and flexible hose together from the shock absorber assembly.

11. Remove the bolt, resin clamp, and the front speed sensor.

⁂ WARNING

Do not allow any foreign matter to come in contact with the tip of the sensor.

➡Clean the installation hole and the contact surface for the front speed sensor every time it is removed.

To install:

12. Install the resin clamp and the front speed sensor with bolt. Tighten to 71 inch lbs. (8 Nm).

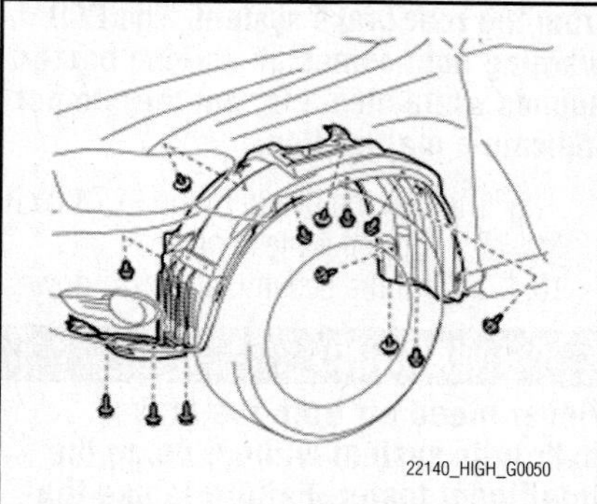

Fig. 9 Front fender liner removal—left hand side

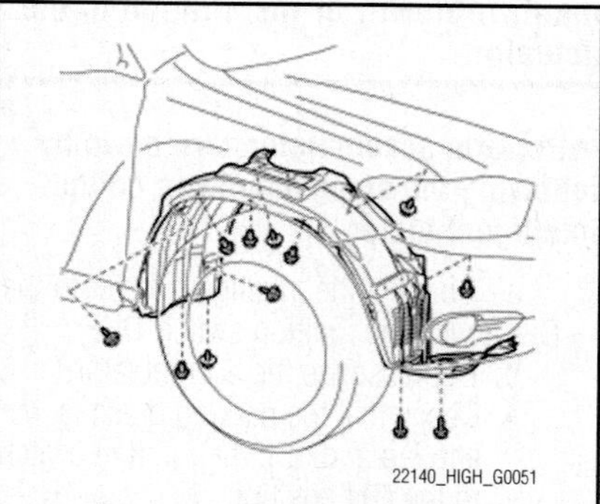

Fig. 10 Front fender liner removal—right hand side

⁂ WARNING

Do not allow any foreign matter to come in contact with the tip of the sensor.

➡Firmly insert the front speed sensor body into the knuckle before tightening the bolt.

➡After installing the front speed sensor to the knuckle, make sure that there is no clearance between the front speed sensor stay and knuckle. Also make sure that no foreign matter is stuck between the parts.

➡Before installing the clamp, firmly insert the points of the clamp into the installation holes.

13. Temporarily install the No. 1 sensor clamp.

➡Be sure to insert the No. 1 sensor clamp claw into the stopper hole while installing the No. 1 sensor clamp.

14. Install the front flexible hose and the No. 1 sensor clamp together to the shock absorber with the bolt. Tighten to 14 ft. lbs. (19 Nm).

➡Do not twist the wire harness for the front speed sensor when installing it.

➡A bolt tightens the brake flexible hose and front speed sensor together. Make sure that the flexible hose is positioned over the front speed sensor.

15. Install the No. 2 sensor clamp to the body with the bolt. Tighten to 44 inch lbs. (5 Nm).

16. Install the 2 clamps and connect the front speed sensor connector.

17. Install the front fender liner.

18. Install the front wheel opening extension pad.

19. Install the front fender molding sub-assembly.

20. Install the No. 1 engine under cover.

21. Install the front wheel and tire assembly and tighten the lug nuts finger-tight.

22. Lower the vehicle, then final tighten the lug nuts to 76 ft. lbs. (103 Nm).

23. Connect the negative battery cable.

24. Check for speed sensor signal.

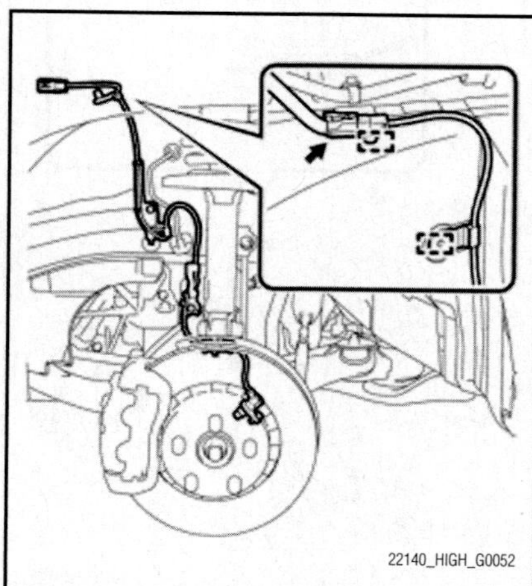

Fig. 11 Front speed sensor connector and clamps

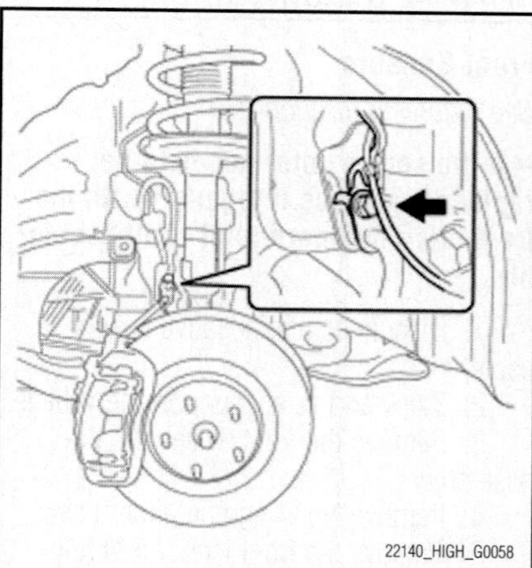

Fig. 12 Remove the bolt, No.1 sensor clamp, and flexible hose from shock absorber assembly

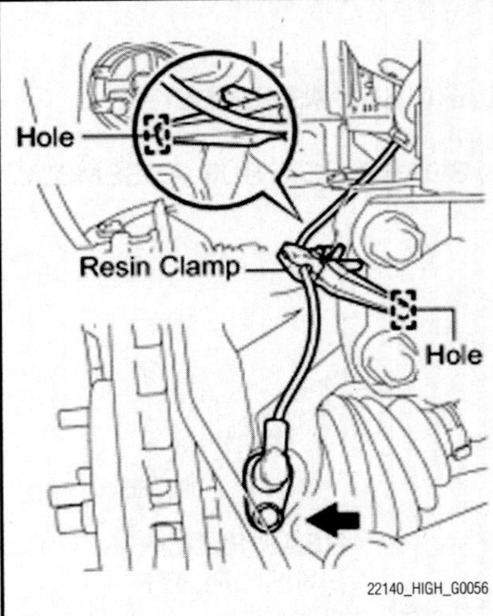

Fig. 13 Front speed sensor, bolt, and resin clamp removal

5. Remove the right and left side deck side trim boxes.
6. For models with the no. 2 rear seat, remove the rear seat track covers:
   a. Using a molding remover, disengage the 4 claws and remove the rear seat track cover. side.
7. Remove rear seat track bracket cover outer (w/ rear no. 2 seat):
   a. Disengage the 2 claws and clip and remove the rear seat track bracket cover outer.
8. Remove the rear door scuff plate.
9. For models with the no. 2 rear seat, remove the deck board no.1 check:
   a. Remove the 4 nuts and deck board no.1 check.
10. If equipped, remove rear no.2 seat assembly.

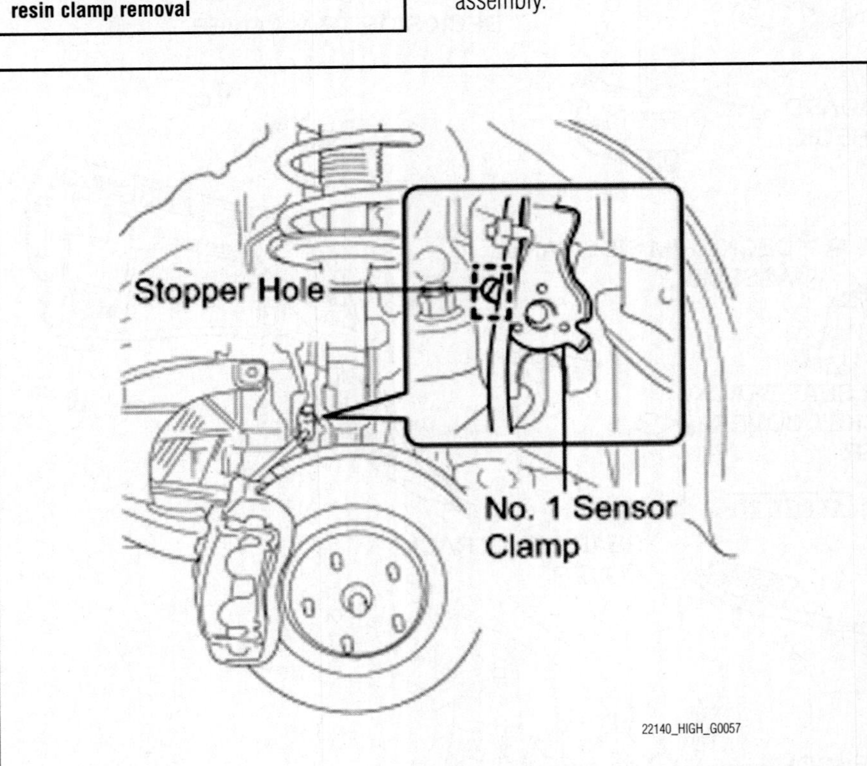

Fig. 14 Temporarily installing No. 1 sensor clamp

Rear (4WD Models Only)

See Figures 15 and 16.

1. Remove deck board sub-assembly.
2. Remove no. 2 deck board sub-assembly
   a. Remove no. 3 deck board sub-assembly
   b. Remove the rear deck floor box.
3. For models without the no. 2 rear seat, remove the front deck floor box front.
4. Remove rear floor finish plate
11. For models without the no. 2 rear seat, remove the deck side trim cover.
12. For models with the no. 2 rear seat, remove the left side deck side trim cover.
13. Remove the deck trim side panel assembly.
14. Raise and safely support the vehicle, then remove rear wheel and tire assembly.
15. Remove rear speed sensor, as follows:
   a. Disconnect the rear speed sensor connector.

⁂ WARNING

Be careful not to damage the wire harness when pulling out the harness.

   b. Pull out the sensor wire harness with the grommet.

➡Do not pull the sensor wire.

   c. Remove the 2 bolts that hold the sensor harness from the body and the shock absorber bracket.
   d. Remove the bolt and the rear speed sensor.

➡Prevent foreign matter from attaching to the sensor tip. Clean the installation hole and surface for the speed sensor every time the speed sensor is removed.

16. Inspect the rear speed sensor. If any of the following occurs, replace the rear speed sensor with a new one:
   a. The surface of the rear speed sensor is cracked, dented, or chipped off.
   b. The connector is scratched, cracked, or damaged.
   c. The rear speed sensor has been dropped.

To install:

17. Install rear speed sensor, as follows:
   a. Install the rear speed sensor with the bolt and tighten to 71 inch lbs. (8 Nm).

➡Prevent foreign matter from attaching to the sensor tip.

   b. Install the sensor harness clamps with the 2 bolts to the body and the shock absorber bracket. Tighten to 44 inch lbs. (5 Nm).

⁂ WARNING

Do not twist the sensor wire when installing the sensor.

   c. Push in the sensor wire harness with the grommet.

⁂ WARNING

When installing the grommet, do not pull the sensor wire. only pull the grommet.

   d. Connect the speed sensor rear connector.
18. Install rear wheel, then lower the vehicle. Tighten the lug nuts to 76 ft. lbs. (103 Nm).
19. Install the remaining components in the reverse of the removal procedure.

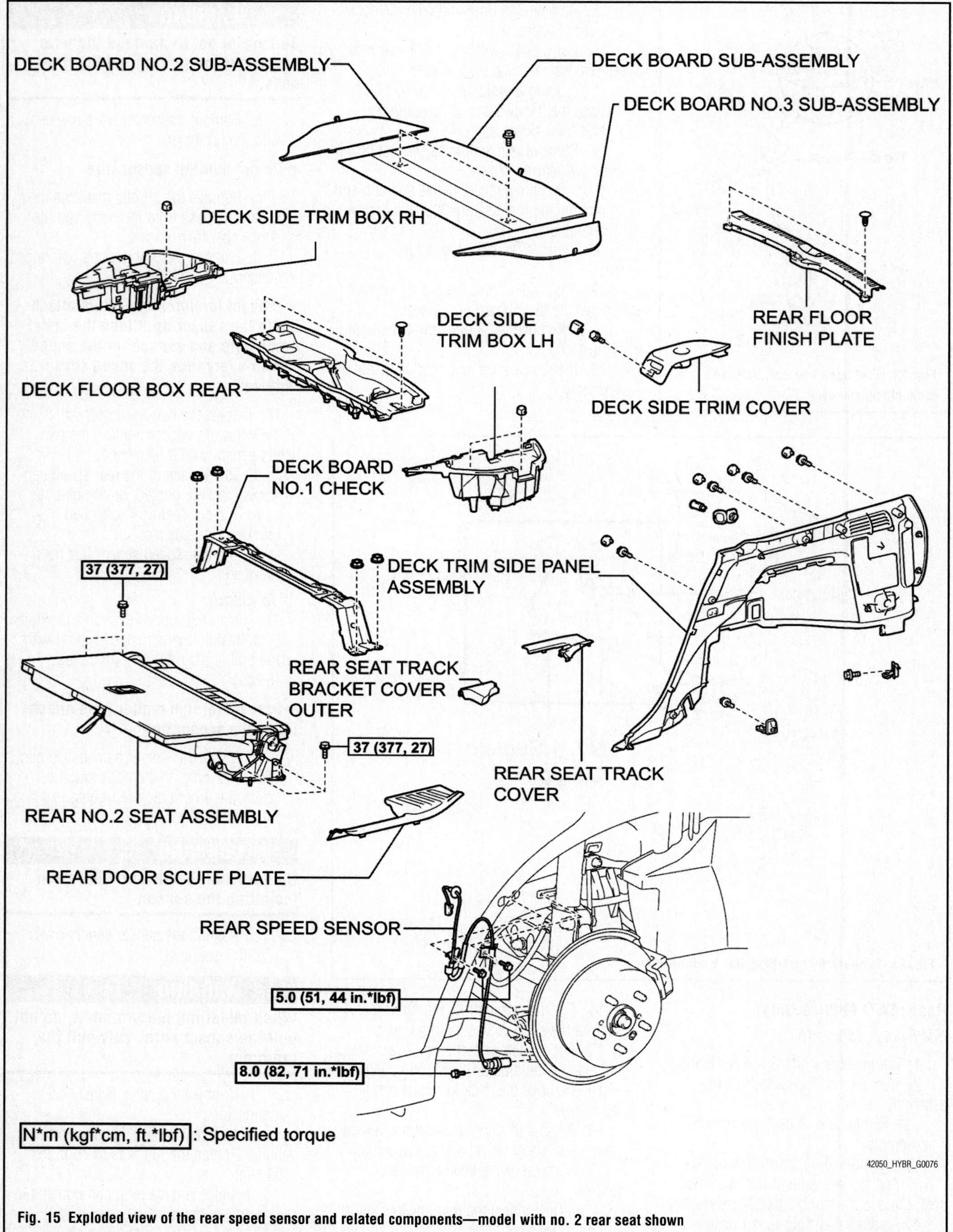

Fig. 15 Exploded view of the rear speed sensor and related components—model with no. 2 rear seat shown

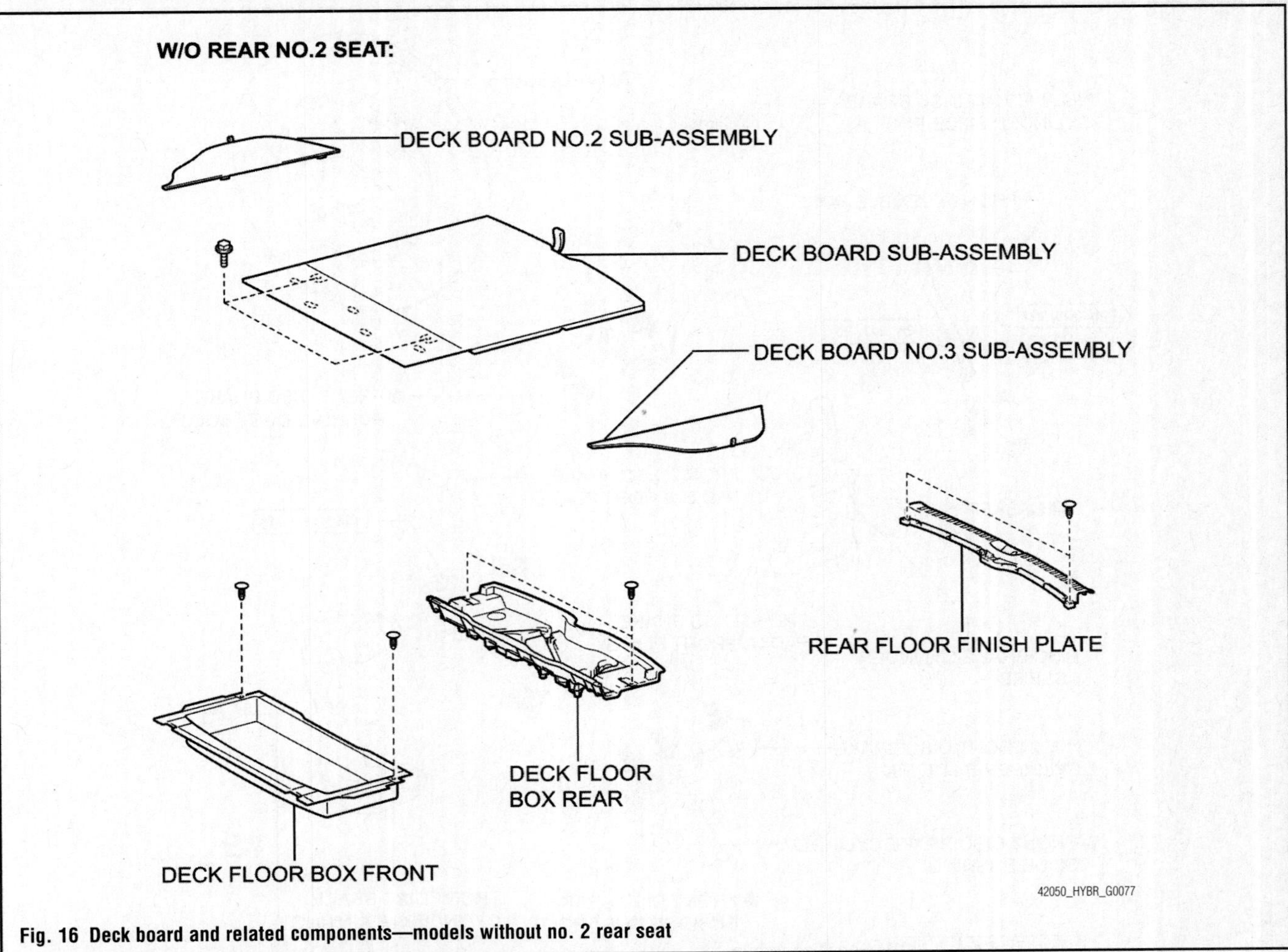

Fig. 16 Deck board and related components—models without no. 2 rear seat

BRAKES

FRONT DISC BRAKES

BRAKE CALIPER

REMOVAL & INSTALLATION

See Figure 17.

➡While the battery is connected, even if the power switch is off, the brake control system activates when the brake pedal is depressed or the door courtesy switch turns on. Therefore, during servicing of the brake system components, do not operate the brake pedal or open/close the doors while the battery is connected.

1. Carefully raise the vehicle.
2. Remove the front wheel.
3. Drain the brake fluid.
4. Remove the bolt and gasket, and disconnect the front flexible hose from the disc brake cylinder assembly.
5. Hold the front disc brake cylinder slide pins and remove the 2 bolts and disc brake cylinder assembly.

To install:

6. Hold the front disc brake cylinder slide pins and install the disc brake cylinder assembly with the 2 bolts. Tighten the bolts to 25 ft. lbs. 34 (Nm).
7. Connect the front flexible hose with the union bolt and a new gasket to the disc brake cylinder assembly. Tighten the banjo bolt to 20 ft. lbs. (30 Nm).

➡Install the front flexible hose lock securely in the lock hole in the disc brake cylinder assembly.

8. Fill the master cylinder reservoir with brake fluid.
9. Bleed the brake system.
10. Inspect for brake fluid leaks.
11. Perform the accumulator zero down procedure.
12. Inspect the brake fluid level.
13. Install the front wheel and tighten to 76 ft. lbs. (103 Nm).
14. Clear the Daces
15. Check for Daces. If any DTC is output, perform the troubleshooting for that DTC.

DISC BRAKE PADS

REMOVAL & INSTALLATION

See Figures 18 and 19.

1. Carefully raise the vehicle.
2. Remove the front wheel.
3. Hold the front disc brake cylinder slide pins and remove the 2 bolts and disc brake cylinder assembly.
4. Remove the 2 anti-squeal springs.
5. Remove the 2 brake pads from the front disc brake cylinder mounting.
6. Remove the 2 anti-squeal shims and the 2 pad wear indicators from the pads.
7. Remove the 4 front disc brake pad support plates from the front disc brake cylinder mounting.

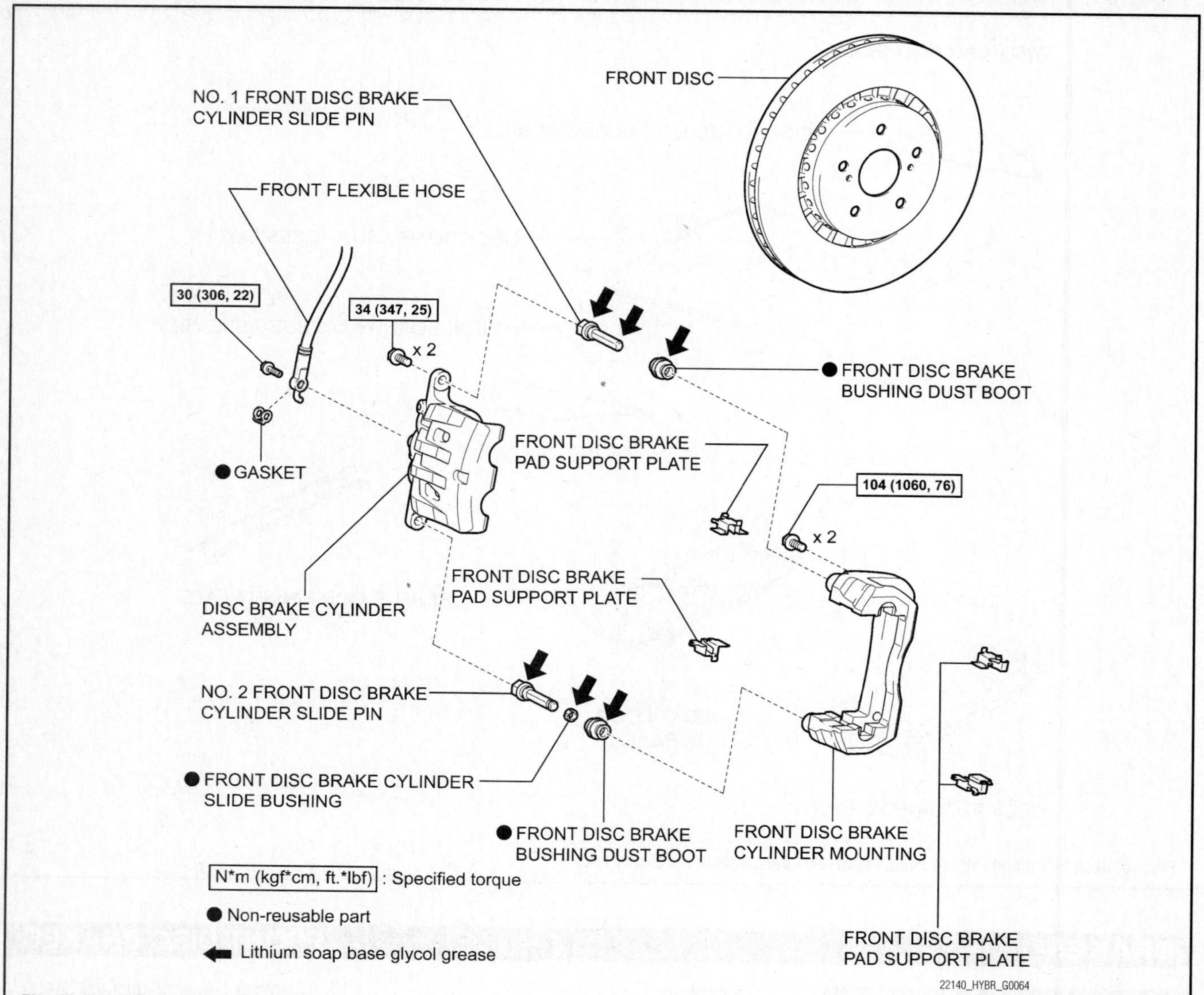

FRONT DISC

NO. 1 FRONT DISC BRAKE CYLINDER SLIDE PIN

FRONT FLEXIBLE HOSE

30 (306, 22)

34 (347, 25)

x 2

● FRONT DISC BRAKE BUSHING DUST BOOT

FRONT DISC BRAKE PAD SUPPORT PLATE

● GASKET

104 (1060, 76)

x 2

FRONT DISC BRAKE PAD SUPPORT PLATE

DISC BRAKE CYLINDER ASSEMBLY

NO. 2 FRONT DISC BRAKE CYLINDER SLIDE PIN

● FRONT DISC BRAKE CYLINDER SLIDE BUSHING

● FRONT DISC BRAKE BUSHING DUST BOOT

FRONT DISC BRAKE CYLINDER MOUNTING

N*m (kgf*cm, ft.*lbf) : Specified torque

● Non-reusable part

⬅ Lithium soap base glycol grease

FRONT DISC BRAKE PAD SUPPORT PLATE

22140_HYBR_G0064

Fig. 17 Front disc brake components

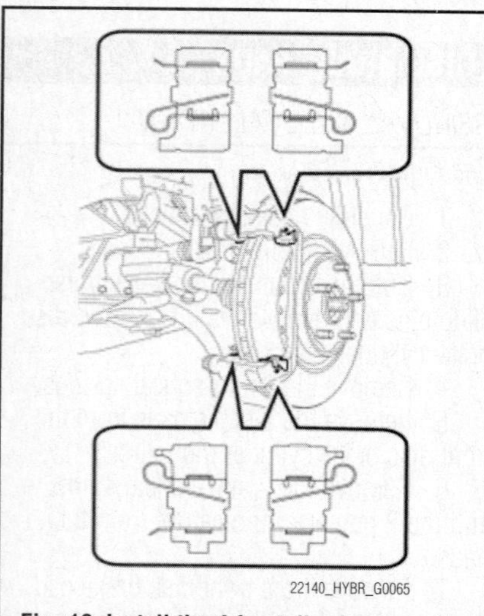

Fig. 18 Install the 4 front disc brake pad support plates as shown

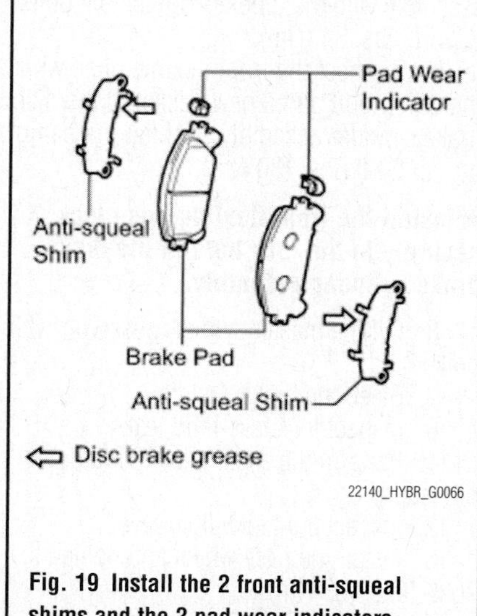

Fig. 19 Install the 2 front anti-squeal shims and the 2 pad wear indicators

To install:

8. Install the 4 front disc brake pad support plates to the front disc brake cylinder mounting as shown in the illustration.
9. Apply disc brake grease to the inside of the 2 front anti-squeal shims.
10. Install the 2 front anti-squeal shims and the 2 pad wear indicators to the pads.
11. Install the 2 brake pads with the front anti-squeal shims to the front disc brake cylinder mounting.
12. Install the 2 anti-squeal springs to the front disc brake pads.
13. Hold the front disc brake cylinder slide pins and install the disc brake cylinder assembly with the 2 bolts. Tighten the bolts to 25 ft. lbs. (34 Nm).
14. Tighten the wheel to 76 ft. lbs. (103 Nm).

BRAKE CALIPER

REMOVAL & INSTALLATION

See Figure 20.

1. Carefully raise the vehicle.
2. Remove the front wheel and tire assembly.
3. Disconnect the brake line from the caliper and plug it.
4. Remove the caliper mounting bolts.
5. Lift off the caliper.
6. Remove the pads and anti-squeal shims.
7. Remove the wear indicators from each pad.

To install:

8. Install the 2 rear brake pads with rear anti-squeal shims to the rear disc brake cylinder mounting.
9. Hold the rear disc brake cylinder slide pins and install the disc brake cylinder assembly with the 2 bolts. Tighten the mounting bolts to 25 ft. lbs. (34 Nm).
10. Install the brake hose to the caliper, tighten the banjo bolt to 24 ft. lbs. (33 Nm).
11. Tighten the wheel to 76 ft. lbs. (103 Nm).
12. Fill the master cylinder reservoir with brake fluid.
13. Bleed the brake system.
14. Inspect for brake fluid leaks.
15. Perform the accumulator zero down procedure.
16. Inspect the brake fluid level.
17. Install the front wheel and tighten to 76 ft. lbs. (103 Nm).
18. Clear the Daces

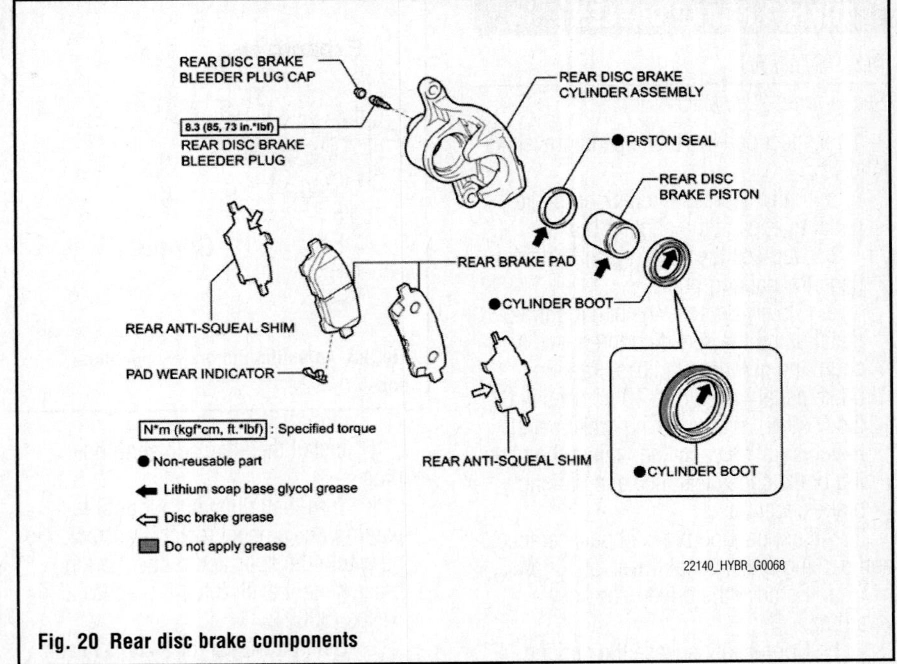

Fig. 20 Rear disc brake components

DISC BRAKE PADS

REMOVAL & INSTALLATION

See Figure 21.

1. Disconnect the brake line from the caliper and plug it.
2. Remove the caliper mounting bolts.
3. Lift off the caliper.
4. Remove the pads and anti-squeal shims.
5. Remove the wear indicators from each pad.
6. Installation is the reverse of removal. Grease the caliper slides and bolts with lithium grease or equivalent. Apply disc brake grease to the anti-squeal shims. Tighten the caliper bolts to 32 ft. lbs. (43 Nm); the brake line union bolt to 24 ft. lbs. (33 Nm).

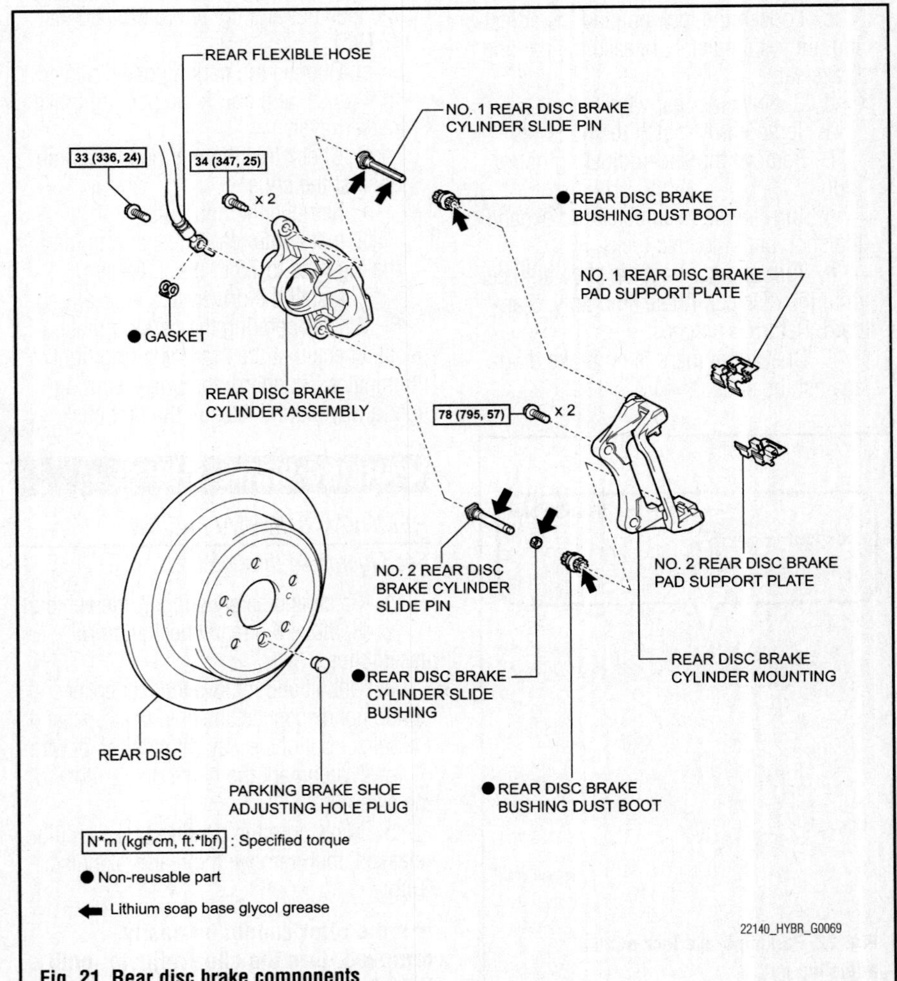

Fig. 21 Rear disc brake components

PARKING BRAKE CABLES

ADJUSTMENT

See Figures 22 and 23.

1. Inspect parking brake pedal travel, as follows:
   a. Fully depress the parking brake pedal to engage the parking brake.
   b. Depress the pedal again to disengage the parking brake.
   c. Slowly depress the parking brake pedal using the specified force, and count the number of clicks. Parking brake pedal travel: 8 to 10 notches at 67 lbs (300 N). If the parking brake pedal travel is not as specified, adjust the parking brake shoe clearance and parking brake pedal travel.
2. Adjust parking brake shoe clearance and parking brake pedal travel, as follows:
   a. Remove the driver side knee airbag.
   b. Completely release the parking brake pedal.
   c. Loosen the lock nut and the adjusting nut to completely release the parking brake cable.
   d. Remove the rear wheel.
   e. Temporarily install the hub nuts.
   f. Remove the shoe adjusting hole plug.
   g. Turn the shoe adjuster and expand the shoe until the disc locks.
   h. Turn and contract the shoe adjuster until the disc can rotate smoothly. Standard: Return 8 notches.
   i. Check that there is no brake drag against the shoe.

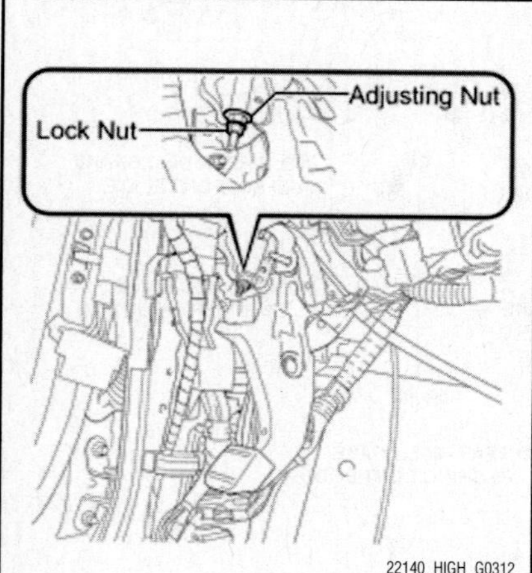

Fig. 22 Parking brake lock nut and adjusting nut

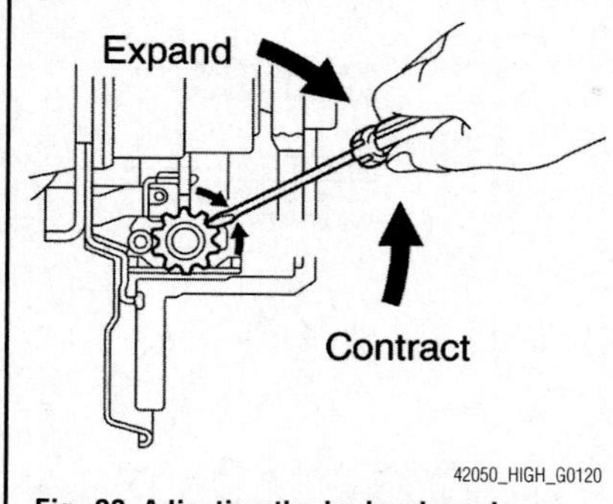

Fig. 23 Adjusting the brake shoe clearance

   j. Install the shoe adjusting hole plug.
   k. Turn the adjusting nut until the parking brake pedal travel is corrected to be within the specified range. Parking brake pedal travel: 8 to 10 notches at 67 lbs (300 N).
   l. Using a wrench or an equivalent tool, hold the adjusting nut and tighten the lock nut and tighten to 62 inch lbs. (7 Nm).
   m. Operate the parking brake pedal 3 to 4 times, and check the parking brake pedal travel.
   n. Check that there is no brake drag against the shoe.
   o. Remove the hub nuts.
   p. Install the rear wheel and tighten the lug nuts to 76 ft. lbs. (103 Nm).
   q. Install the driver side knee airbag.
3. When operating the parking brake pedal, check that the brake warning light illuminates. Standard: the brake warning light always illuminates at the first click.

PARKING BRAKE SHOES

REMOVAL & INSTALLATION

See Figures 24 through 34.

1. Raise and safely support the vehicle.
2. Remove the rear wheel and tire assemblies.
3. Unbolt and remove the rear caliper, but do not disconnect the fluid line. Suspend the caliper out of the way with a piece of wire.
4. Matchmark the brake disc (rotor) to the axle hub.
5. Make sure the parking brake is fully released, then remove the rear brake disc (rotor).

➡If the rotor cannot be easily removed, turn the shoe adjuster until the wheel turns freely.

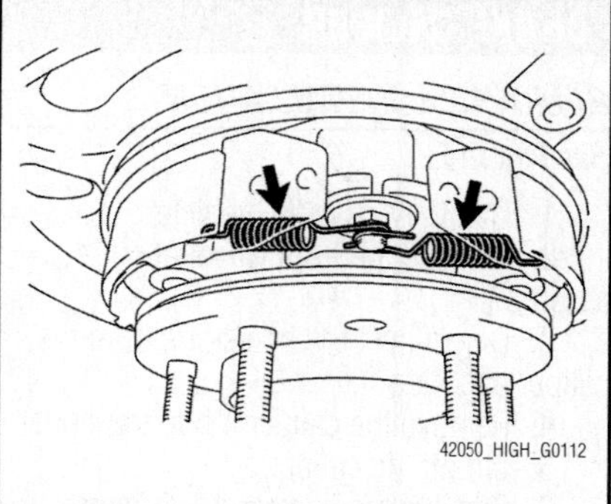

Fig. 24 Use a needle-nose pliers to remove the 2 return tension springs

6. Inspect the brake disc (rotor) inside diameter, as follows:
   a. Using a brake drum gauge or equivalent, measure the inside diameter of the disc and compare with the following: Standard inside diameter: 190 mm (7.48 in.). Maximum inside diameter: 191 mm (7.52 in.)
   b. If the inside diameter exceeds the maximum, replace the brake disc.
7. Use needle-nose pliers to remove the 2 parking brake shoe return tension springs.
8. Remove the parking brake shoe strut, as follows:
   a. Remove the parking brake shoe strut and the parking brake shoe strut compression spring.
9. Remove parking brake shoe no.1, as follows:
   a. Remove the parking brake shoe hold down spring cup No.1, parking brake shoe hold down spring and parking brake shoe hold down spring cup no.2.
   b. 2WD vehicles, remove the parking brake shoe hold down spring pin No.1.

Fig. 25 Remove the parking brake shoe hold-down spring cups

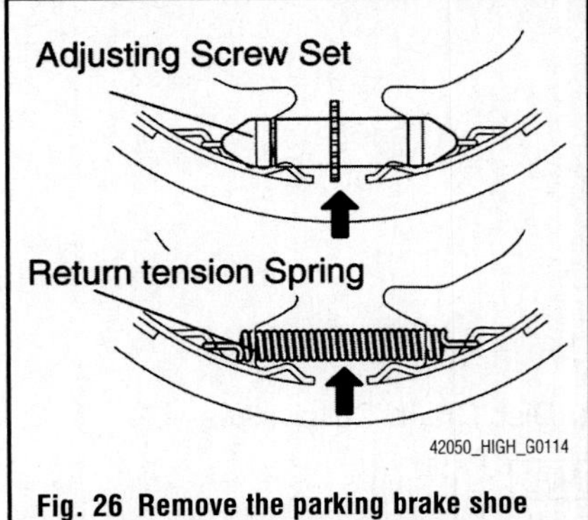

Fig. 26 Remove the parking brake shoe adjusting screw set and shoe return spring

c. Disconnect the parking brake shoe return spring no.2 and remove the parking brake shoe assembly lh no.1.

10. Remove parking brake shoe adjusting screw set:

a. Remove the parking brake shoe adjusting screw set.

b. Remove the parking brake shoe return tension spring No.2.

11. Remove parking brake shoe assembly no.2:

a. Remove the parking brake shoe hold down spring cup No.1, parking brake shoe hold down spring, parking brake shoe hold down spring cup no.2 and parking brake shoe hold down spring pin no.2.

b. Remove the parking brake shoe assembly lh no.2.

c. Using needle-nose pliers, disconnect the parking brake cable no.3 from the parking brake cable shoe lever.

WARNING

Be careful not to damage parking brake cable no.3.

12. On 4WD models, separate the rear speed sensor.

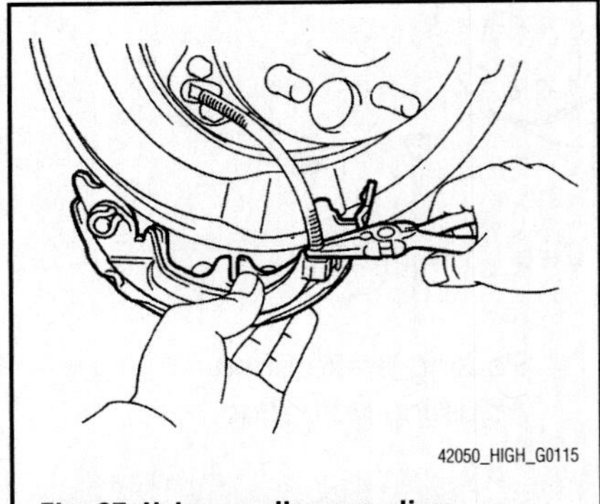

Fig. 27 Using needle-nose pliers, disconnect the parking brake cable no. 3 from the parking brake cable shoe lever

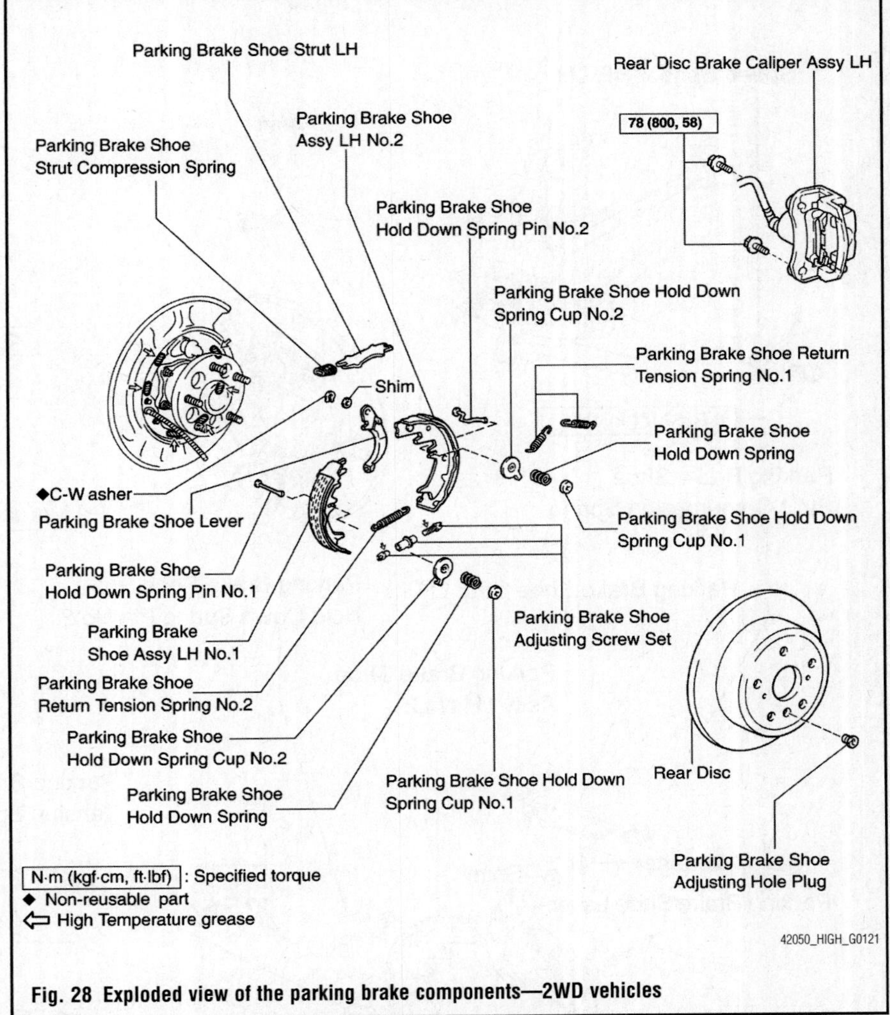

Fig. 28 Exploded view of the parking brake components—2WD vehicles

13. On 4WD models, remove the rear axle shaft nut.

14. On 4WD models, remove rear axle hub & bearing assembly

15. On 4WD models, remove parking brake shoe hold down spring pin.

16. Remove parking brake shoe type C-washer, as follows:

a. Using a screwdriver, remove the c-washer.

b. Remove the shim and parking brake shoe lever from the parking brake shoe no.2.

17. Inspect parking brake shoe lining thickness:

a. Using a ruler, measure the thickness of the shoe lining. Standard thickness is 2.5 mm (0.098 in.) and minimum thickness is 1.0 mm (0.039 in.). If the lining thickness is less than or equal to the minimum, or If there is severe or uneven wear, replace the brake shoe.

18. Inspect brake disc and parking brake shoe lining for proper contact

a. Apply chalk to the inside surface of the disc, then grind down the brake shoe lining to fit disc.

b. If the contact between the brake disc and the shoe lining is improper, repair it using a brake shoe grinder or replace the brake shoe assembly.

To install:

19. Install the parking brake shoe type C-washer, as follows

a. Using a feeler gauge, measure the clearance. Standard clearance: less than 0.35 mm (0.014 in.). If the clearance is not within the specifications, replace the shim with one of the correct size. The shim sizes: 0.3 mm (0.012 in.), 0.9 mm (0.035 in.) or 0.6 mm (0.024 in.).

b. Using pliers, install the parking brake shoe lever and the shim with a new C-washer.

20. Apply high temperature grease to the shaded parts shown in the illustration of the backing plate which make contact with the shoe.

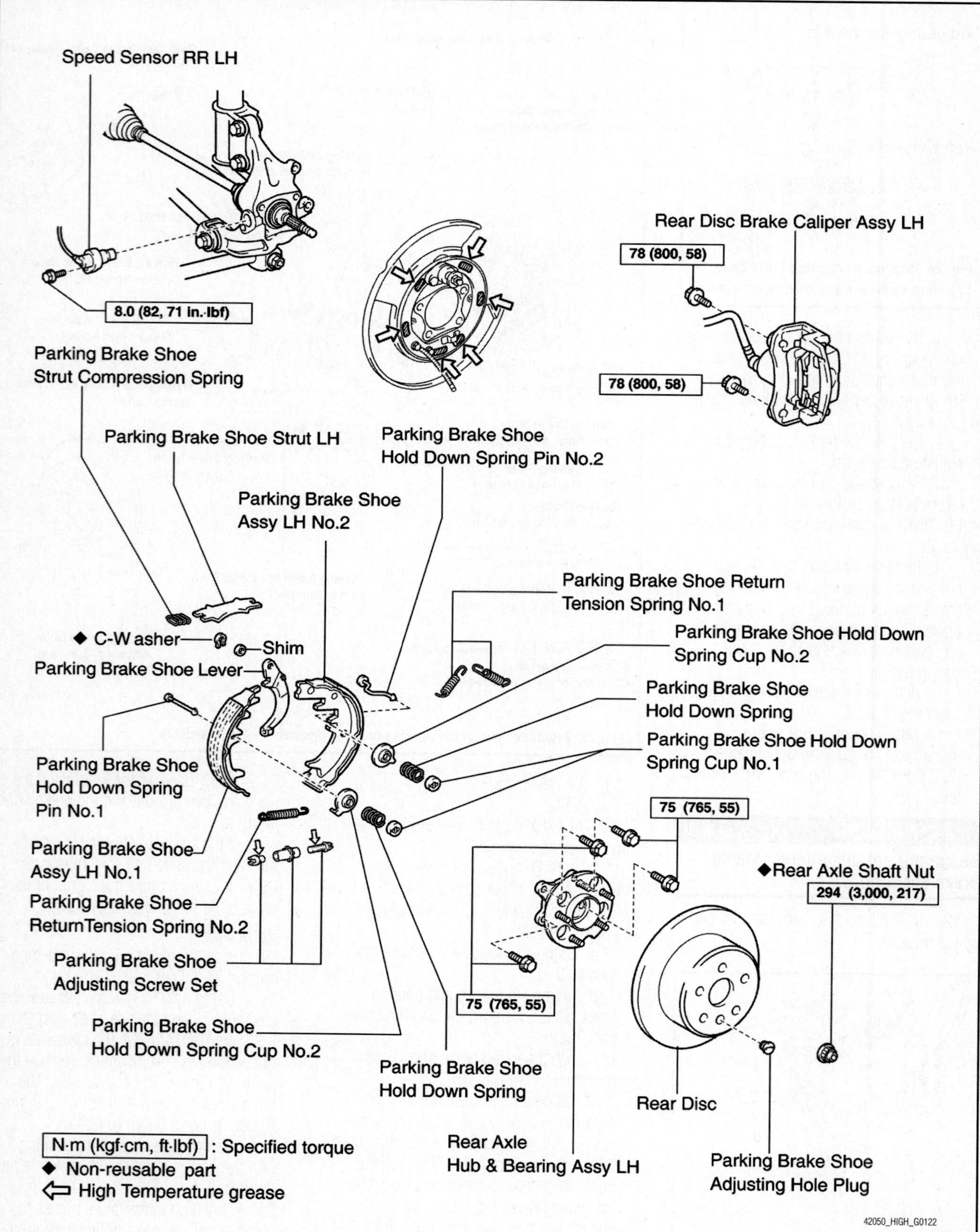

Fig. 29 Exploded view of the parking brake components—4WD vehicles

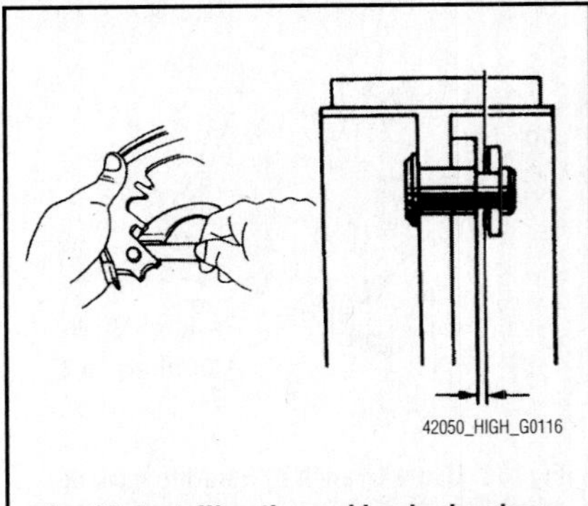

Fig. 30 Installing the parking brake shoe type C-washer

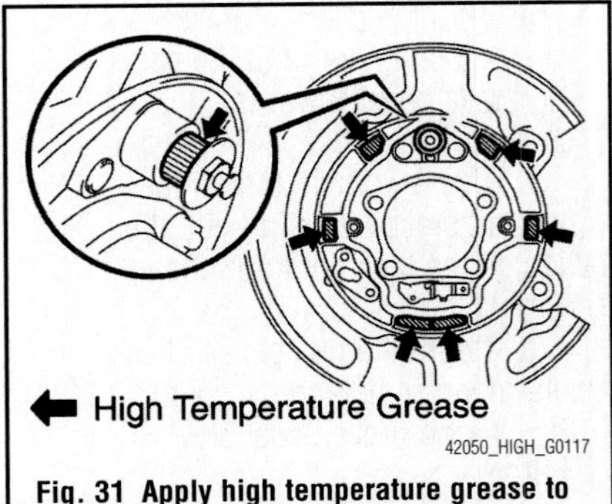

Fig. 31 Apply high temperature grease to the shaded parts of the backing plate which make contact with the shoe

21. On 4WD models, perform the following:
    a. Install the parking brake shoe hold down spring pin.
    b. Install the rear axle hub & bearing.
    c. Install the rear axle shaft nut.
    d. Install the rear speed sensor.
22. Install parking brake shoe no.2, as follows:
    a. Using needle-nose pliers, connect the parking brake cable no.3 to the parking brake cable shoe lever.

➡Be careful not to damage the parking brake cable no.3.

    b. Install the parking brake shoe no.2 with the parking brake shoe hold down spring, parking brake shoe hold down spring cup no.1, parking brake shoe hold down spring cup no.2 and parking brake shoe hold down spring pin no.2.
23. Install the parking brake shoe adjusting screw set, as follows:
    a. Apply high temperature grease to the parking brake shoe adjusting bolt and piece.
    b. Attach the parking brake shoe return tension spring no.2 to the parking brake shoe no.1 and parking brake shoe assembly no.2.

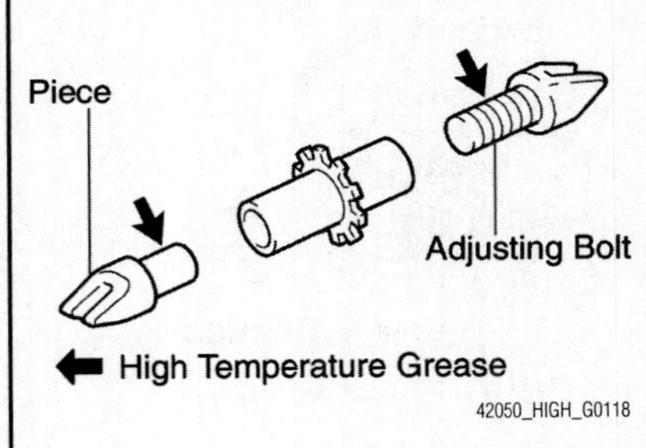

Fig. 32 Apply high temperature grease to the parking brake shoe adjusting bolt and piece

    c. Attach the parking brake shoe adjusting screw set to the parking brake shoe no.1 and parking brake shoe no.2.
24. Install parking brake shoe no.1:
    a. For 2WD models, install the parking brake shoe hold down spring pin no.1.
    b. Install the parking brake shoe no.1 with the parking brake shoe hold down spring, parking brake shoe hold down spring cup no.2, parking brake shoe hold down spring cup no.2.
25. attach the parking brake shoe strut and the parking brake shoe strut compression spring to parking brake shoe no.2 and parking brake shoe no.1.
26. Install parking brake shoe return tension spring using needle-nose pliers as shown in the illustration.

➡First install the front side spring then the rear side spring.

27. Check that the parking brake components are properly installed.

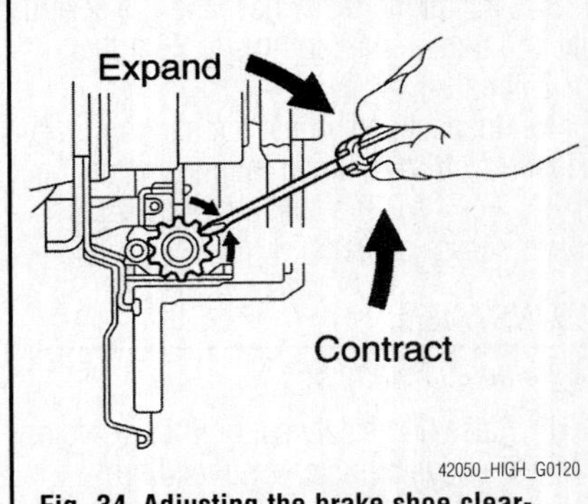

Fig. 34 Adjusting the brake shoe clearance

WARNING

There should be no oil or grease on the friction surface of the shoe lining and disc.

28. For 4WD models, inspect the bearing backlash and axle hub deviation.
29. Install the rear disc (rotor), aligning the matchmarks made during removal.
30. Adjust parking brake shoe clearance, as follows:
    a. Temporarily install the hub nuts.
    b. Remove the hole plug, turn the adjuster and expand the shoes until the disc locks.
    c. Contract the shoe adjuster until the disc rotates smoothly. Standard : return 8 notches
    d. Check that the shoe has no brake drag.
    e. Install the hole plug.
31. Install the caliper, as outlined earlier in this section.

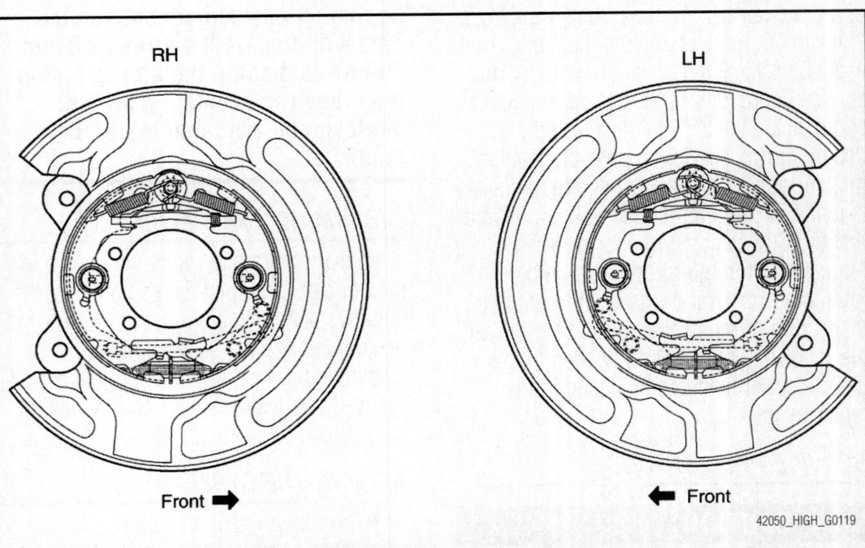

Fig. 33 Check that the parking brake components are properly installed. There should be no oil or grease on the friction surface of the shoe lining and disc

32. Install the rear wheel and tire assembly and tighten the lug nuts to 76 ft. lbs. (103 Nm).

33. Inspect and adjust the parking brake pedal travel, as outlined in this section.

34. For 4WD models, check the ABS speed sensor signal.

ADJUSTMENT

See Figures 35 and 36.

1. Raise and safely support the vehicle.
2. Remove the rear wheel and tire assemblies.
3. Adjust parking brake shoe clearance, as follows:
   a. Temporarily install the hub nuts.
   b. Remove the hole plug, turn the adjuster and expand the shoes until the disc locks.
   c. Contract the shoe adjuster until the disc rotates smoothly. Standard : return 8 notches
   d. Check that the shoe has no brake drag.
   e. Install the hole plug.
4. Install the rotor and caliper.
5. Install the rear wheel and tire assembly and tighten the lug nuts to 76 ft. lbs. (103 Nm).
6. Inspect the parking brake pedal travel, as follows:
   a. Firmly step on the parking brake pedal.

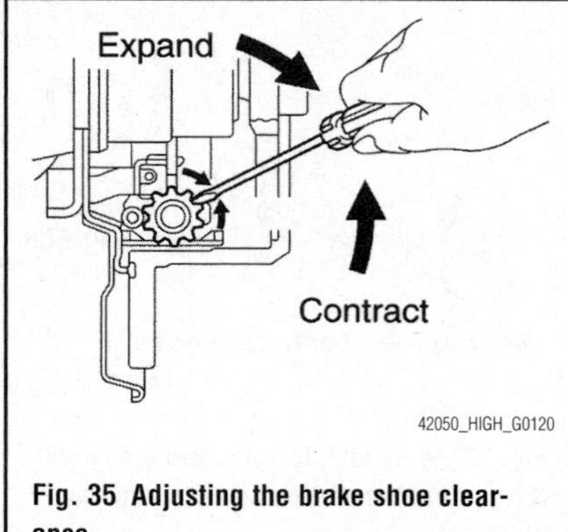

Fig. 35 Adjusting the brake shoe clearance

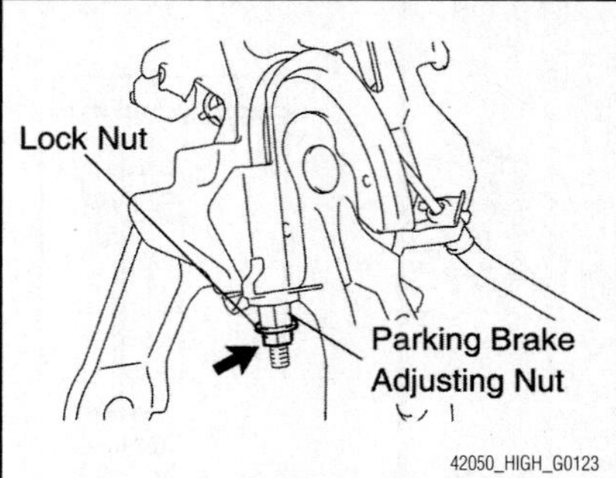

Fig. 36 Use a wrench to hold the parking brake adjusting nut secure while tightening the lock nut

   b. release the parking brake.
   c. Once more, slowly depress the parking brake pedal all the way, and count the number of clicks. The parking brake pedal should travel 5 to 7 clicks at 67 lbs. (300 N).
7. If necessary, adjust parking brake pedal travel, as follows:
   a. Remove the lower instrument panel finish panel sub-assembly.
   b. Remove the lower instrument panel insert sub- assembly.
   c. Depress the parking brake pedal 5 clicks to make room for the procedure, and loosen the lock nut with fixing adjusting nut by wrench.
   d. Release the parking brake pedal to the original position.
   e. Turn the parking brake wire adjusting nut until the parking brake pedal travel is correct.
   f. Use a wrench to hold the parking brake adjusting nut, then tighten the lock nut to 53 inch lbs. (6 Nm).
   g. Count the number of clicks after depressing and canceling the parking brake pedal 3 to 4 times.
   h. Check whether the parking brake drags or not.
   i. When operating the parking brake pedal, check that the parking brake pedal indicator light is lit.

CHASSIS ELECTRICAL

AIR BAG (SUPPLEMENTAL RESTRAINT SYSTEM)

GENERAL INFORMATION

This vehicle is equipped with a Supplemental Restraint System (SRS). It consists of a driver airbag, front passenger airbag, side airbag, curtain shield airbag, and front seat belt pretensioner. Failure to carry out service operations in the correct sequence could cause the SRS to unexpectedly deploy during servicing, possibly leading to a serious accident. Further, if a mistake is made in servicing the SRS, it is possible that the SRS may fail to operate when required. Before performing servicing (including removal or installation of parts, inspection or replacement), be sure to read the following items carefully, then follow the correct procedures indicated in the repair manual.

SERVICE PRECAUTIONS

⁂ CAUTION

Disconnect and isolate the battery negative cable before beginning any airbag system component diagnosis, testing, removal, or installation procedures. Allow system capacitor to discharge for two minutes before beginning any component service. This will disable the airbag system. Failure to disable the airbag system may result in accidental airbag deployment, personal injury, or death.

DISARMING THE SYSTEM

To avoid personal injury when working on vehicles equipped with an air bag, the negative battery cable must be disconnected and at least 90 seconds must elapse before working on the system. Failure to do so may result in deployment of the air bag.

ARMING THE SYSTEM

To arm the system after service is completed, connect the negative battery cable.

CLOCKSPRING CENTERING

See Figure 37.

1. Check that the front wheels are facing straight ahead.
2. Set the turn signal switch to the neutral position.

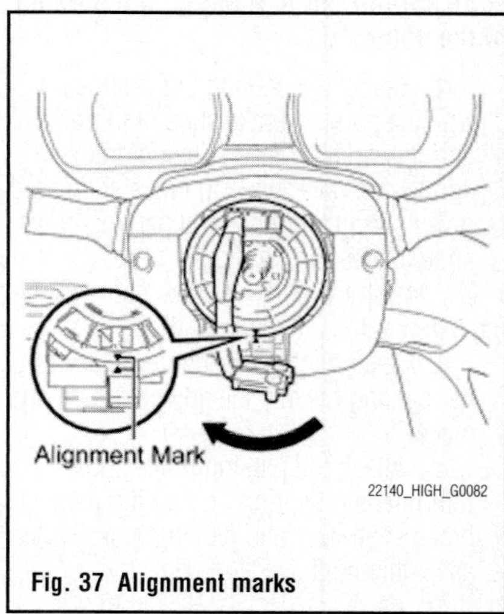

Fig. 37 Alignment marks

⁂ WARNING

If it is not in the neutral position, the turn signal switch pin may be snapped.

3. Check that the battery negative (-) cable is disconnected.

⁂ CAUTION

Wait for at least 90 seconds after disconnecting the cable to prevent airbag deployment.

4. Rotate the spiral cable (clockspring) counterclockwise slowly by hand until it stops.

⁂ WARNING

Do not turn the spiral cable using the airbag wire harness.

5. Rotate the spiral cable clockwise approximately 2.5 turns to align the marks.

➡The spiral cable will rotate approximately 2.5 turns to both the left and right from the center.

DRIVE TRAIN

FRONT HALFSHAFT

REMOVAL & INSTALLATION

See Figures 38 through 40.

1. Before servicing the vehicle, refer to the Precautions Section.
2. Remove or disconnect the following:
   - Engine under covers
   - Front wheels
   - Drain HV transaxle fluid
   - Cotter pin and hub nut
   - Front speed sensor
   - Brake caliper
   - Brake disc
   - Tie rod end, from the steering knuckle
   - Steering knuckle, from the lower control arm
   - Halfshaft from the axle hub, using a plastic hammer
   - Stabilizer link
3. Using Special Tool 095020-01010, remove the halfshaft from the transaxle.
4. For RH axle remove the bearing bracket hole snap ring from the drive shaft bearing bracket.

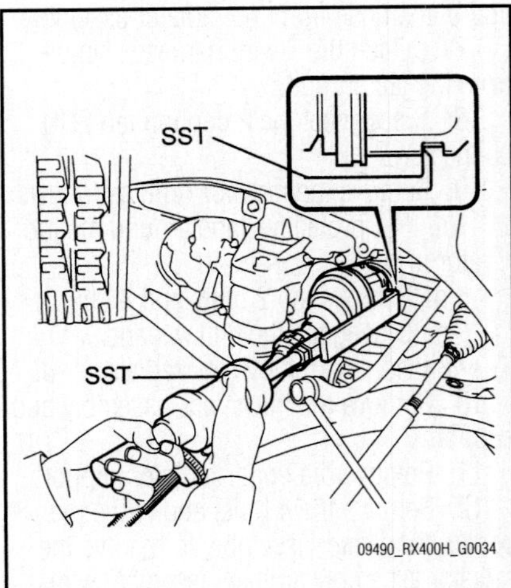

Fig. 38 Use the Special Tool to remove the halfshaft from the transaxle.

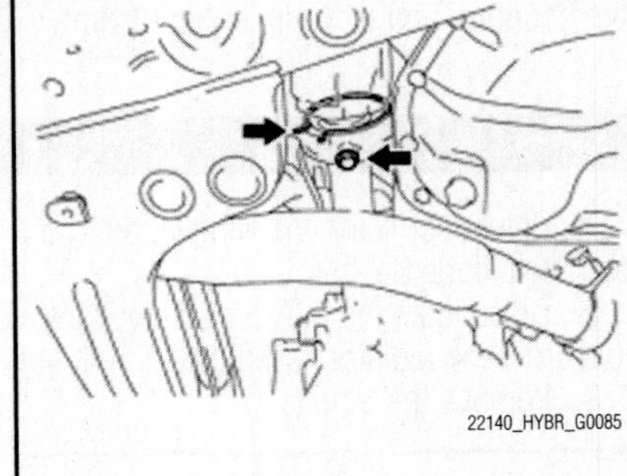

Fig. 39 Bearing bracket hole snap ring and bolt

5. Remove the bolt and front drive shaft assembly RH from the drive shaft bearing bracket.

To install:

6. Install a new halfshaft hole snapring.
7. Coat the splines of the inboard joint shaft assembly with ATF.
8. Align the shaft splines and install the halfshaft assembly with a brass drift and hammer.
9. For RH side axle install the bracket hole snap ring and bolt. Tighten the bolt to 24 ft. lbs. (32 Nm).
10. Tighten the axle shaft nut and tighten to 216 ft. lbs. (294 Nm).
11. The remainder of installation is the reverse order of removal.
12. Fill the HV transaxle with gear oil, install the engine under covers, check front end alignment and test drive.

➡If the cotter pin holes do not align, always correct by tightening the nut until the next hole aligns.

13. Install a new cotter pin.
14. Tighten the front wheels to 76 ft. lbs. (103 Nm).

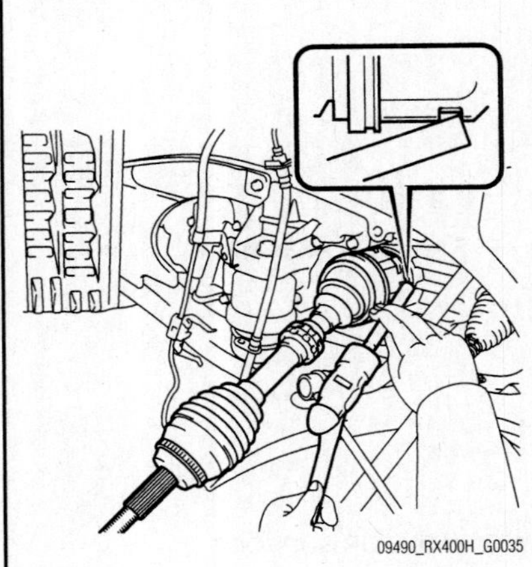

Fig. 40 Insert a brass drift into the groove to install the halfshaft.

REAR HALFSHAFT

REMOVAL & INSTALLATION

See Figure 41.

1. Before servicing the vehicle, refer to the Precautions Section.
2. Remove the rear wheel.
3. Disconnect and remove the speed sensor.
4. Remove the brake disc and brake caliper assembly.
5. Remove the axle shaft nut.
6. Separate the rear axle hub and bearing assembly.
7. Separate the parking brake cable.
8. Disconnect the strut rod.

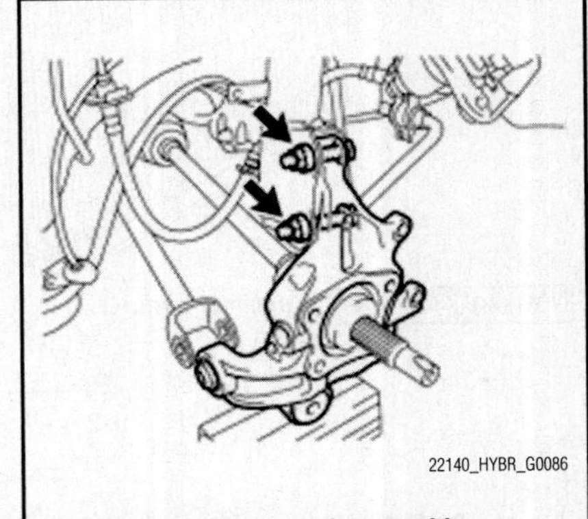

Fig. 41 Rear carrier sub assembly

9. Separate the rear suspension arms.
10. Remove the rear carrier sub assembly.
11. Put matchmarks on the rear drive shaft assembly and differential side gear shaft.
12. Remove the 4 nuts, washers and rear drive shaft assembly.

To install:

13. Align the matchmarks.
14. Install the rear drive shaft assembly with the 4 nuts and washers. Tighten to 41 ft. lbs. (56 Nm).
15. Install the rear carrier sub assembly. Tighten the 2 mounting strut bolts to 133 ft. lbs. (180 Nm).
16. Temporarily tighten the rear suspension arm assembly No.2 with the bolt and nut.
17. Temporarily tighten the rear suspension arm assembly No.1 with the bolt and nut.
18. Temporarily tighten the strut rod assembly rear with the bolt and nut.
19. Install the hub and bearing assembly with the 4 bolts and tighten to 55 ft. lbs. (75 Nm).
20. Install the rear brake disc.
21. Install the rear disc brake caliper assembly with the 2 bolts to the rear axle carrier. Tighten the bolts to 58 ft. lbs. (78 Nm).
22. Install the speed sensor and tighten the mounting bolt to 71 inch lbs. (8 Nm).
23. Using a socket wrench (30mm), install a new rear axle shaft nut. Tighten the axle nut to 217 ft. lbs. (294 Nm).
24. Using a chisel and hammer, stake the rear axle shaft nut.
25. Stabilize the suspension. Then tighten the control arms to 83 ft. lbs. (112 Nm). Tighten the strut rod to 133 ft. lbs. (180 Nm).
26. Install the parking brake cable assembly No.3 with the nut. Tighten the nut to 53 inch lbs. (6 Nm)
27. Install the rear wheel and tighten to 76 ft. lbs. (103 Nm)
28. Check rear wheel alignment.
29. Verify speeds sensor operation.

ENGINE COOLING

ENGINE FAN

REMOVAL & INSTALLATION

See Figure 42.

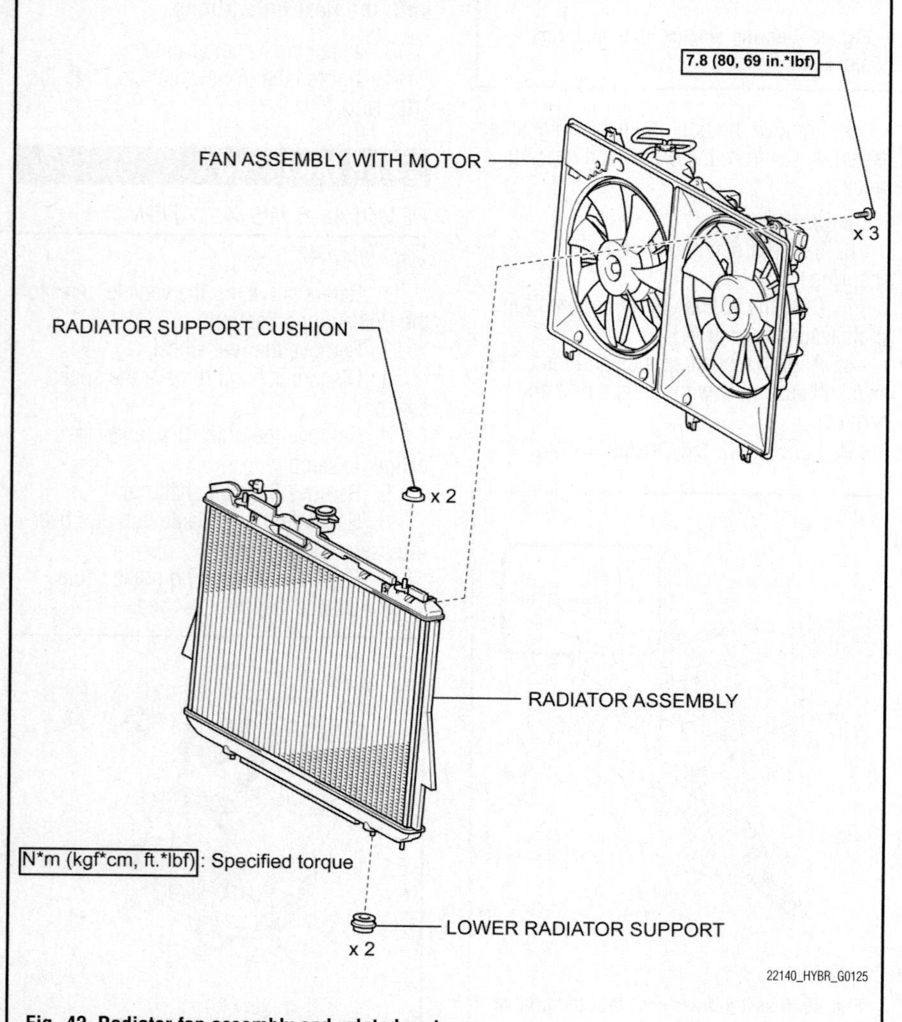

Fig. 42 Radiator fan assembly and related parts

1. Before servicing the vehicle, refer to the Precautions Section.
2. Disconnect the radiator reserve tank hose from the radiator assembly.
3. Remove the 3 bolts.
4. Pull up the fan assembly with motor from the radiator assembly to remove the fan assembly with motor.

To install:

5. Put in the fan assembly to the radiator assembly.
6. Install the fan assembly and tighten the 3 bolts to 69 inch lbs. (7.8 Nm).
7. Install the radiator reserve tank hose to the radiator assembly.

RADIATOR

REMOVAL & INSTALLATION

See Figures 43 through 46.

1. Before servicing the vehicle, refer to the Precautions Section.
2. Remove the engine under cover.
3. Drain the engine coolant.
4. Remove the cool air intake duct.
5. Remove the air cleaner cap assembly.
6. Disconnect the radiator inlet and outlet hose from the radiator assembly.
7. Detach the 7 wire harness clamps from the fan shroud.
8. Disconnect the 2 cooling fan ECU connectors.
9. Remove the radiator grill, as follows:
    a. Put protective tape around the radiator grill.
    b. Remove the 2 bolts and 4 clips.
    c. Disengage the 6 claws and 4 guides, and remove the radiator grill.
10. Remove the hood lock assembly and support.
11. Remove the upper radiator support.
12. Remove the 4 bolts and move the cooler condenser assembly to remove the radiator assembly and fan assembly with motor.
13. Separate the 3 water hose clamps from the fan shroud.

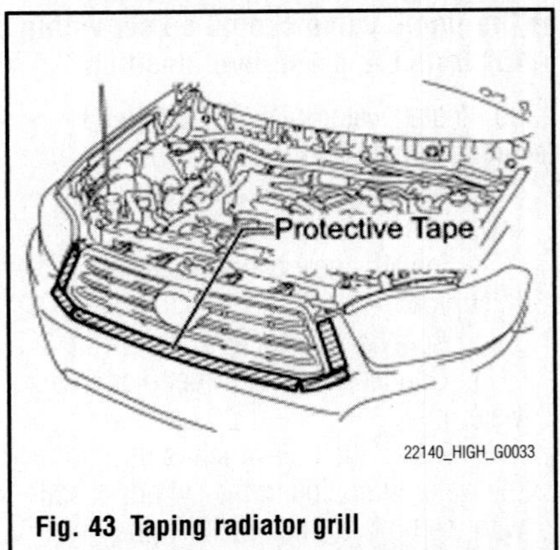

Fig. 43 Taping radiator grill

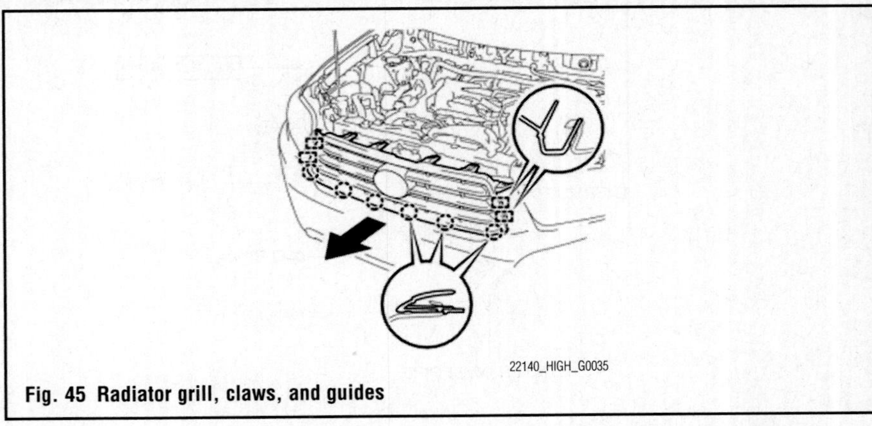

Fig. 45 Radiator grill, claws, and guides

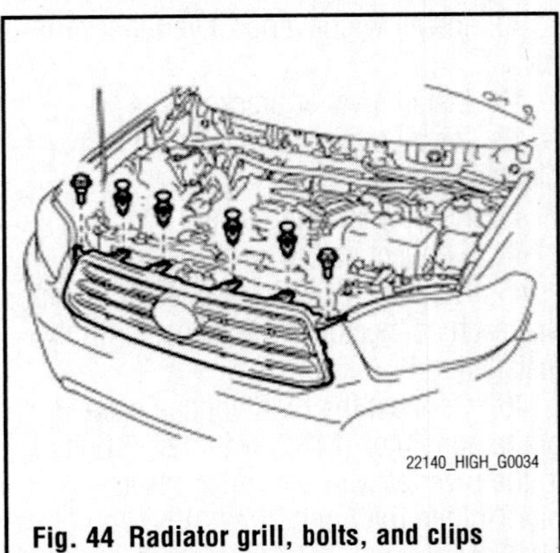

Fig. 44 Radiator grill, bolts, and clips

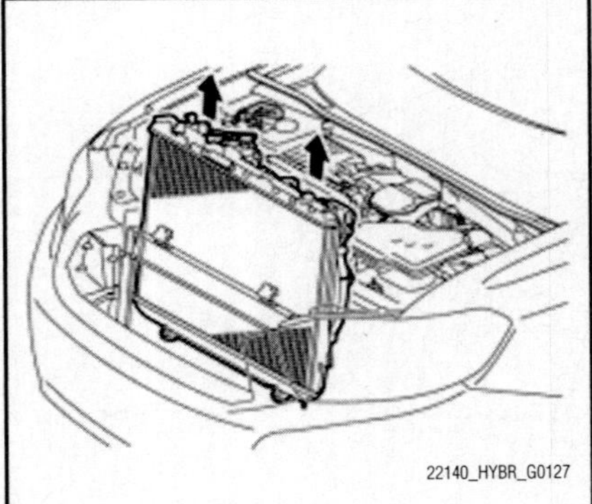

Fig. 46 Radiator removal

14. Remove the radiator assembly and fan assembly with motor.
15. Disconnect the radiator reserve tank hose from the radiator assembly. Remove the 3 bolts.
16. Pull up the fan assembly with motor from the radiator assembly to remove the fan assembly with motor.
17. Remove the upper and lower radiator support cushions.

To install:

18. Install the upper and lower radiator supports.
19. Install the fan assembly, and tighten bolts to 69 inch lbs. (7.8 Nm).
20. Connect the radiator reserve tank hose to the radiator assembly.
21. Install the radiator assembly and fan assembly.
22. Install the cooler condenser assembly, and tighten the 4 bolts to 53 inch lbs. (6 Nm).
23. Connect the 3 water hose clamps to the fan shroud.
24. Install the upper radiator support sub-assembly. Tighten the bolts to 62 inch lbs. (7 Nm).
25. Install the discharge hose bracket with the bolt and tighten to 87 inch lbs. (9.8 Nm).
26. Install the hood lock support sub-assembly and the bolts to 62 inch lbs. (7 Nm).
27. Install the hood lock assembly and tighten to 71 inch lbs. (8 Nm).
28. Install the radiator grill.
29. Reconnect the fan assembly wiring clamps and connectors.
30. Install and tighten the inlet and outlet radiator hoses.
31. Install the air cleaner cap assembly.
32. Install the intake cool air duct seal.
33. Replace the lost engine coolant. Check for coolant leaks.
34. Install the engine under cover.

THERMOSTAT

REMOVAL & INSTALLATION

See Figures 47 and 48.

1. Before servicing the vehicle, refer to the Precautions Section.

CAUTION

The HIGHLANDER HV has a hybrid system that operates at voltages up to 650 volts. Be sure to follow the instructions in this manual to handle the system correctly. Failure to do so may result in serious injury or electrocution. Engineer must undergo special training to be able to perform high-voltage system inspection and servicing.

CAUTION

All high-voltage wire harness connectors are colored orange. The HV battery and other high-voltage components have "High Voltage" caution labels. Do not carelessly touch these wires and components.

2. Before inspecting or servicing the high-voltage system, be sure to follow safety measures, such as wearing insulated gloves and removing the service plug to prevent electrocution. Carry the removed service plug in your pocket to prevent other technicians from reinstalling it while you are servicing the vehicle.
3. After removing the service plug, wait 5 minutes before touching any of the high-voltage connectors and terminals.
4. Disconnect the negative battery cable.
5. Remove the engine room left side cover.
6. Remove the engine room cover.
7. Drain the coolant for the inverter.

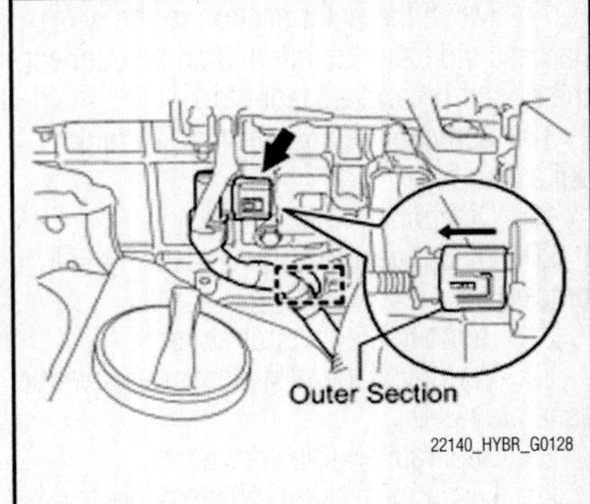

Fig. 47 No.1 sensor circuit breaker

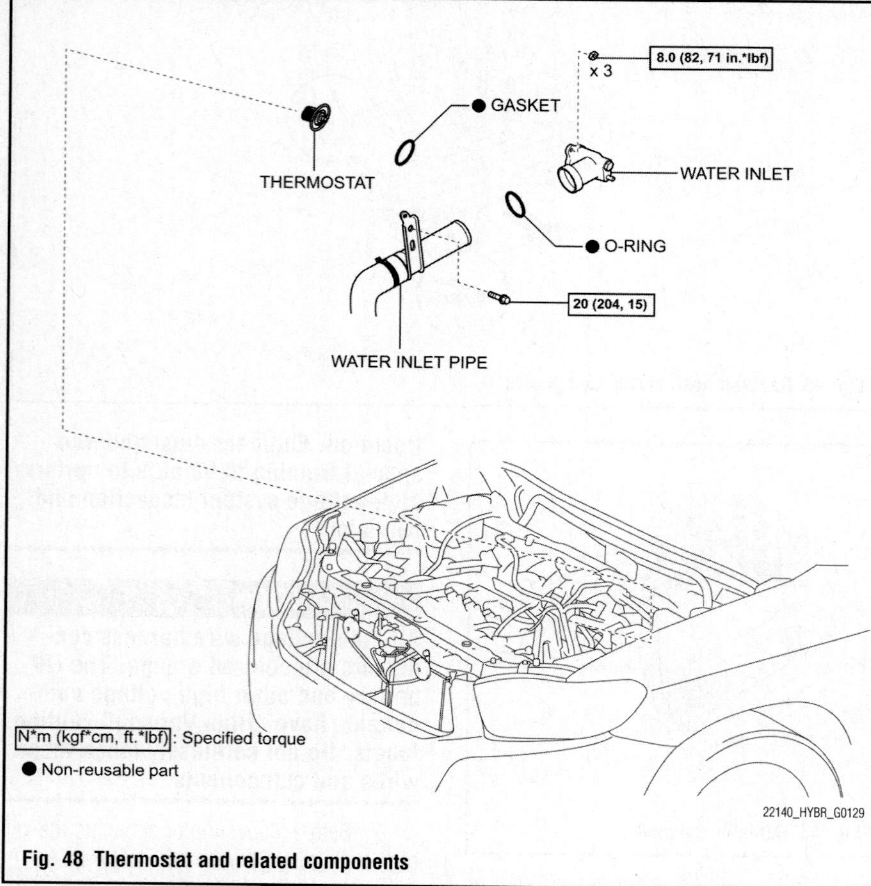

Fig. 48 Thermostat and related components

8. Remove the wiper arm assembly LH and RH.
9. Remove the cowl top ventilator top louver sub-assembly.
10. Remove the wiper motor and link assembly.
11. Remove the cowl top outer panel sub-assembly.
12. Remove the cool air intake duct seal.
13. Remove the air cleaner cap with inlet.
14. Remove the air cleaner with resonator.
15. Remove the bolt and inverter bracket No.5.
16. Remove bolt and ground cable terminal from power steering ECU assembly.
17. Move the outer section to the wire harness side as illustrated, then disconnect the circuit breaker sensor No.1.
18. Remove the inverter reserve tank subassembly.
19. Disconnect water hose.
20. Disconnect the power steering ECU bracket.
21. Remove the inverter cover.
22. Verify voltage of w/converter inverter assembly is 0 v.
23. Separate engine wire no.4.
24. Disconnect high voltage cable from the front motor.
25. Disconnect the MG ECU connector.
26. Disconnect the no.3 wire frame.
27. Install the inverter cover.
28. Separate engine room relay block assembly
29. Remove the inverter bracket no.4.
30. Remove w/converter inverter assembly. Since the inverter with converter assembly is very heavy, 2 people are needed to remove the inverter with converter assembly. When removing the inverter with converter assembly, do not damage the parts around it.
31. Remove the water inlet pipe, as follows:
32. Remove the bolt and the water inlet pipe.
33. Remove the o-ring from the water inlet pipe.
34. Disconnect the wire harness clamp.
35. Remove the 3 nuts and the water inlet.
36. Remove the thermostat and gasket.

To install:

37. Install a new gasket and thermostat.
38. Align the jiggle valve of the thermostat and water inlet, and insert the thermostat in the water inlet housing.

➡The jiggle valve should be set within +-15° from the prescribed position.

39. Install water inlet. Tighten the 3 retaining nuts to 71 inch lbs. (8 Nm).
    a. Install the water inlet pipe as follows:
    b. Install a new o-ring to the water inlet pipe.
    c. Apply soapy water to the o-ring.
    d. Connect the water inlet pipe to the water inlet.
    e. Install the bolt which is used to fix the water inlet pipe to the cylinder head with the bolt. Tighten to 15 ft. lbs. (20 Nm).
40. Install w/converter inverter assembly.
41. Install inverter bracket No.4.
42. Install engine compartment relay block assembly.
43. Remove inverter cover.
44. Connect no.3 wire frame.
45. Connect MG ECU connector. Check that each connector and terminal is firmly installed.
46. Connect the high voltage cable of the generator (MG1) with new 5 bolts to the inverter with converter assembly. Tighten the bolts to 7 ft. lbs. (10 Nm).
47. Connect engine wire No.4 and tighten the bolts to 48 inch lbs. (5.4 Nm).
48. Check high voltage cable connection.
49. Install inverter cover. Tighten the cover bolts to 7 ft. lbs. (10 Nm).
50. Install the power steering ECU bracket.
51. Connect the water hose.
52. Install inverter reserve tank sub-assembly.
53. Connect engine room wire No.2.
54. Connect circuit breaker sensor No.1.
55. Install the inverter bracket No.5.
56. Install air cleaner w/resonator.
57. Install air cleaner cap w/inlet.
58. Install air cleaner w/resonator.
59. Install air cleaner cap w/inlet.
60. Install the cool air intake duct seal.
61. Install the outer cowl top panel.
62. Install wiper motor and link assembly.
63. Install cowl top ventilator louver sub-assembly.
64. Install the LH and RH wiper arm assembly.
65. Install the service plug grip.
66. Connect the negative battery cable.
67. Refill Coolant for inverter and engine.
68. Check for leaks.

WATER PUMP

REMOVAL & INSTALLATION

See Figure 49.

1. Before servicing the vehicle, refer to the Precautions Section.
2. Drain the engine coolant.
3. Remove or disconnect the following:
   - Negative battery cable
   - Engine side covers
   - Right-hand front wheel
   - Engine splash shield
   - Right-hand front fender apron seal
   - Wiper arm and blade assembly
   - Top cowl ventilator louver assembly
   - Wiper motor and link assembly
   - Battery and battery tray
   - Air intake assembly
   - Brake master cylinder reservoir
   - Reservoir support bracket
   - Air cleaner support bracket
   - Engine moving control rod
   - Right-hand engine mount
   - Crankshaft pulley
   - Timing belt cover No.1 and 2
   - Timing belt guide
   - Timing belt
   - Timing belt idler sub-assembly No.1
   - Camshaft timing pulley
   - Timing belt cover No.3
   - Timing belt idler sub-assembly No.2
   - Water pump
4. Installation is the reverse order of removal. Tighten the water pump with a new gasket to 71 inch lbs. (8 Nm).
5. Refill the coolant to the correct level.
6. Start the engine and check for leaks.

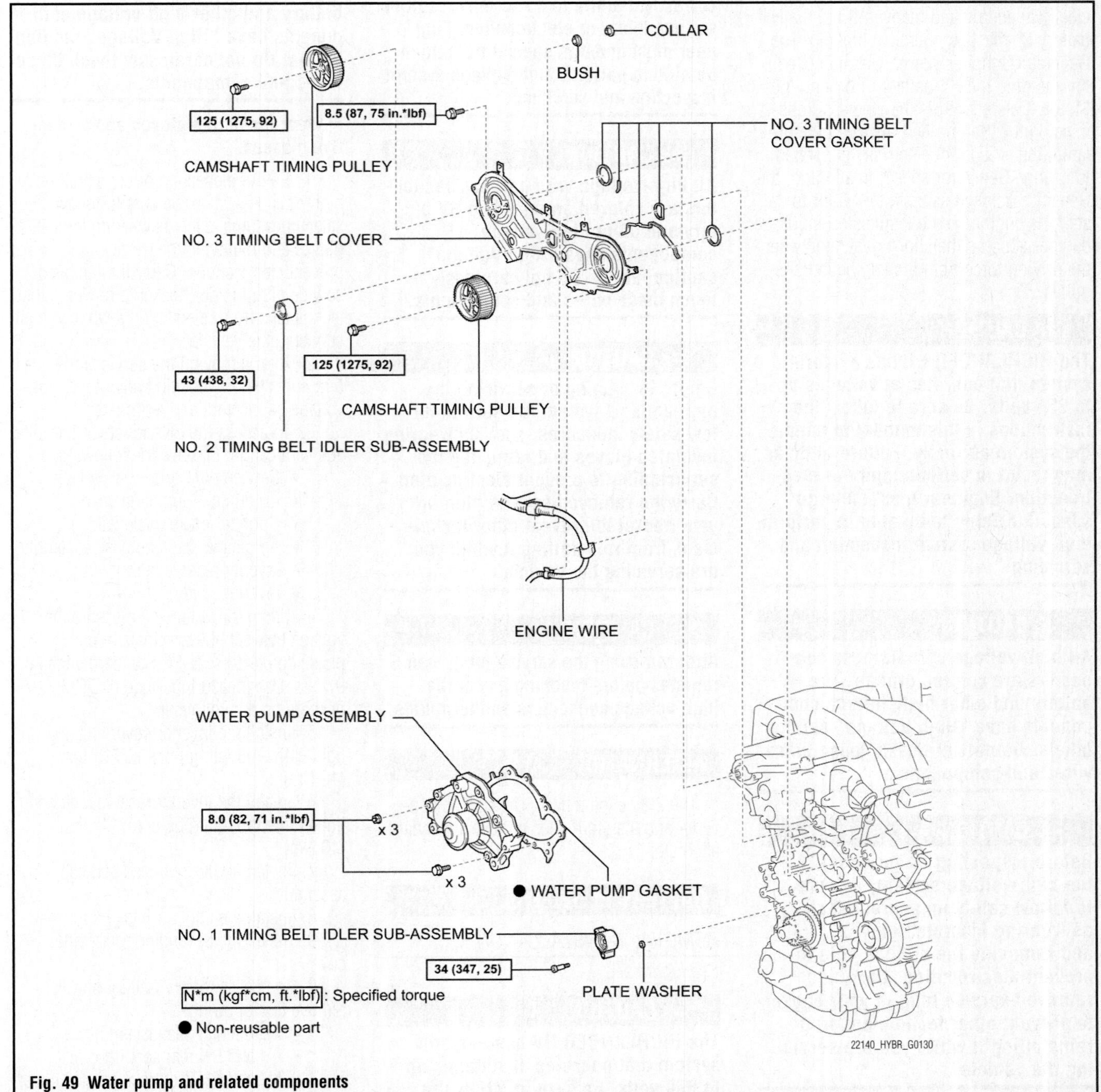

Fig. 49 Water pump and related components

PRECAUTIONS

The HV transaxle assembly consists of the planetary gear unit, Motor, and Generator. The gear unit uses the planetary gear to split engine output in accordance with a driving request while the vehicle is driven or the HV battery is charged. The Motor assists engine output while increasing vehicle driving force. The Motor also converts the energy, which is consumed in the form of heat during normal braking, into electrical energy and recover it into the HV battery to effect regenerative braking. The Generator supplies power, which is used for charging the HV battery or driving the Motor. It also controls the stepless transmission function of the transaxle by regulating the amount of electricity generated to change Generator speed. In addition, the Generator is used as a starter Motor to start the engine. The transmission input damper absorbs the shock generated when the driving force from the engine is transmitted.

CAUTION

The HIGHLANDER HV has a hybrid system that operates at voltages up to 650 volts. Be sure to follow the instructions in this manual to handle the system correctly. Failure to do so may result in serious injury or electrocution. Engineer must undergo special training to be able to perform high-voltage system inspection and servicing.

CAUTION

All high-voltage wire harness connectors are colored orange. The HV battery and other high-voltage components have "High Voltage" caution labels. Do not carelessly touch these wires and components.

CAUTION

Before inspecting or servicing the high-voltage system, be sure to follow safety measures, such as wearing insulated gloves and removing the service plug to prevent electrocution. Carry the removed service plug in your pocket to prevent other technicians from reinstalling it while you are servicing the vehicle.

CAUTION

After removing the service plug, wait 5 minutes before touching any of the high-voltage connectors and terminals.

CAUTION

The HIGHLANDER HV has a hybrid system that operates at voltages up to 650 volts. Be sure to follow the instructions in this manual to handle the system correctly. Failure to do so may result in serious injury or electrocution. Engineer must undergo special training to be able to perform high-voltage system inspection and servicing.

CAUTION

All high-voltage wire harness connectors are colored orange. The HV battery and other high-voltage components have "High Voltage" caution labels. Do not carelessly touch these wires and components.

CAUTION

Before inspecting or servicing the high-voltage system, be sure to follow safety measures, such as wearing insulated gloves and removing the service plug to prevent electrocution. Carry the removed service plug in your pocket to prevent other technicians from reinstalling it while you are servicing the vehicle.

CAUTION

After removing the service plug, wait 5 minutes before touching any of the high-voltage connectors and terminals.

ALTERNATOR

The 3.3L engine has a DC electric converter and therefore does not use a standard alternator.

BATTERY BLOWER

REMOVAL & INSTALLATION

See Figure 50.

CAUTION

The HIGHLANDER HV has a hybrid system that operates at voltages up to 650 volts. Be sure to follow the instructions in this manual to handle the system correctly. Failure to do so may result in serious injury or electrocution. Engineer must undergo special training to be able to perform high-voltage system inspection and servicing.

CAUTION

All high-voltage wire harness connectors are colored orange. The HV battery and other high-voltage components have "High Voltage" caution labels. Do not carelessly touch these wires and components.

➡Wear insulating gloves and protective glasses.

1. Before inspecting or servicing the high-voltage system, be sure to follow safety measures, such as wearing insulated gloves and removing the service plug to prevent electrocution. Carry the removed service plug in your pocket to prevent other technicians from reinstalling it while you are servicing the vehicle.
2. After removing the service plug, wait 5 minutes before touching any of the high-voltage connectors and terminals.
3. To access the inverter cover for voltage verification. Remove the following:
   - Service plug grip
   - LH engine room side cover
   - Cool air intake duct seal
   - Air cleaner cap and case assembly
   - Inverter reserve tank
   - Inverter cover
4. Using the voltmeter, measure the voltage between the terminals of the 2 phase connectors (N-P). Standard voltage: 0 volts. Use measuring range of DC 750 V or more on the voltmeter.
5. Install the inverter cover and tighten the mounting bolts to 7 ft. lbs. (10 Nm).
6. Install the inverter reserve tank and tighten the mounting bolts to 7 ft. lbs. (10 Nm).
7. Install air cleaner case and cap assembly.
8. Install the cool air intake duct seal.
9. Install the LH engine room side cover.
10. To access the HV battery blower remove the following:
    - Rear center seat assembly
    - LH and RH rear seat head rest
    - LH and RH seat track bracket cover

- LH and RH outer seat track bracket cover
- Rear seat leg side cover
- Disconnect the seat lock cables
- LH and RH seat assembly
- Deck board assembly No. 2 and No. 3
- Tonneau cover if equipped
- Rear No. 1 floor board LH (w/o rear No. 2 seat)
- LH and RH rear seat side cover (w/ rear No. 2 seat)
- Jack carrier support
- Jack carrier cushion
- Jack assembly and carrier assembly
- RH deck side trim box
- Deck floor board assembly (w/o rear No. 2 seat)
- Rear No. 2 seat belt assembly (w/ rear No. 2 seat)
- Rear seat belt lap assembly LH and RH (w/ rear No. 2 seat)
- Rear No. 2 seat assembly (w/ rear No. 2 seat)
- Rear floor finish plate
- Rear deck side trim and cover
- LH side trim cover (w/o rear seat entertainment system)
- Power outlet socket (w/ rear seat entertainment system)
- LH rear combination light service cover
- Power socket assembly and cover
- Rear deck trim cover (w/o rear seat entertainment system)
- Reclining remote control lever bezel LH (w/ remote folding function)
- LH rope hook assembly
- No.2 deck side trim hook
- LH front deck side trim cover
- Rear LH No. 1 outer seat belt assembly
- Deck trim side panel LH
- Quarter pillar garnish LH
- Roof side inner garnish assembly
- Rear seat side garnish cap
- Right side deck trim and cover
- Rear room temperature sensor (if equipped)
- Rear combination light service cover RH
- Rope hook assembly RH
- No. 1 luggage compartment trim hook
- RH front deck side trim cover
- RH rear No. 1 outer seat belt assembly
- RH deck trim side panel
- Air intake covers
- Turn back the front floor carpet assembly
- No. 1 HV battery tray
- Rear No.1. 2 and 3 floor boards
- Rear center seat inner belt assembly
- Battery cover lock striker
- Battery service hole cover.
- Battery cover sub-assembly

CAUTION

Be sure to wear insulated gloves and protective goggles.

11. Remove the battery cooling blower assembly as follows:
    a. Disconnect each battery cooling blower assembly connector and clamp.
    b. Remove the 9 nuts and 3 battery cooling blower assemblies.
12. Reverse the removal procedure and note the following:
    a. Install the 3 battery cooling blower assemblies with the 9 nuts. Tighten to 40 inch lbs. (4.5 Nm)
    b. Refer to HV battery removal for additional information.

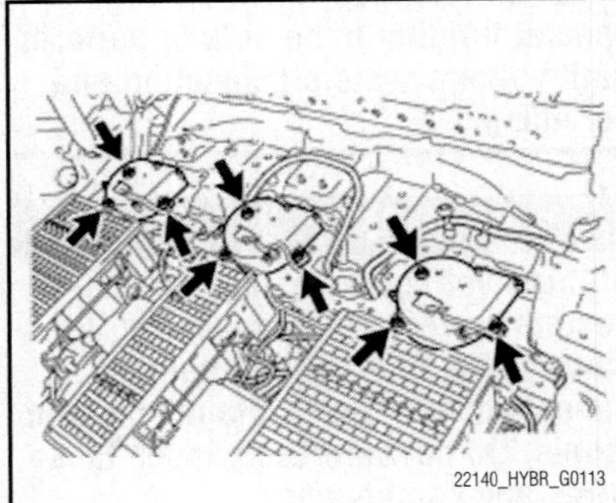

Fig. 50 Battery cooling blower assemblies

HV BATTERY FUSE

REMOVAL & INSTALLATION

See Figure 51.

CAUTION

The HIGHLANDER HV has a hybrid system that operates at voltages up to 650 volts. Be sure to follow the instructions in this manual to handle the system correctly. Failure to do so may result in serious injury or electrocution. Engineer must undergo special training to be able to perform high-voltage system inspection and servicing.

CAUTION

All high-voltage wire harness connectors are colored orange. The HV battery and other high-voltage components have "High Voltage" caution labels. Do not carelessly touch these wires and components.

1. Before inspecting or servicing the high-voltage system, be sure to follow safety measures, such as wearing insulated gloves and removing the service plug to prevent electrocution. Carry the removed service plug in your pocket to prevent other technicians from reinstalling it while you are servicing the vehicle.
2. After removing the service plug, wait 5 minutes before touching any of the high-voltage connectors and terminals.

➡Wear insulating gloves and protective glasses.

3. To access the inverter cover for voltage verification. Remove the following:
    - Service plug grip
    - LH engine room side cover
    - Cool air intake duct seal
    - Air cleaner cap and case assembly
    - Inverter reserve tank
    - Inverter cover
4. Using the voltmeter, measure the voltage between the terminals of the 2 phase connectors (N-P). Standard voltage: 0 volts. Use measuring range of DC 750 V or more on the voltmeter.
5. Install the inverter cover and tighten the mounting bolts to 7 ft. lbs. (10 Nm).
6. Install the inverter reserve tank and tighten the mounting bolts to 7 ft. lbs. (10 Nm).
7. Install air cleaner case and cap assembly.
8. Install the cool air intake duct seal.
9. Install the LH engine room side cover.
10. To access the HV battery fuse remove the following:
    - Rear center seat assembly
    - LH and RH rear seat head rest
    - LH and RH seat track bracket cover
    - LH and RH outer seat track bracket cover
    - Rear seat leg side cover
    - Disconnect the seat lock cables
    - LH and RH seat assembly
    - Deck board assembly No. 2 and No. 3
    - Tonneau cover if equipped
    - Rear No. 1 floor board LH (w/o rear No. 2 seat)
    - LH and RH rear seat side cover (w/ rear No. 2 seat)
    - Jack carrier support
    - Jack carrier cushion
    - Jack assembly and carrier assembly

- RH deck side trim box
- Deck floor board assembly (w/o rear No. 2 seat)
- Rear No. 2 seat belt assembly (w/ rear No. 2 seat)
- Rear seat belt lap assembly LH and RH (w/ rear No. 2 seat)
- Rear No. 2 seat assembly (w/ rear No. 2 seat)
- Rear floor finish plate
- Rear deck side trim and cover
- LH side trim cover (w/o rear seat entertainment system)
- Power outlet socket (w/ rear seat entertainment system)
- LH rear combination light service cover
- Power socket assembly and cover
- Rear deck trim cover (w/o rear seat entertainment system)
- Reclining remote control lever bezel LH (w/ remote folding function)
- LH rope hook assembly
- No.2 deck side trim hook
- LH front deck side trim cover
- Rear LH No. 1 outer seat belt assembly
- Deck trim side panel LH
- Quarter pillar garnish LH
- Roof side inner garnish assembly
- Rear seat side garnish cap
- Right side deck trim and cover
- Rear room temperature sensor (if equipped)
- Rear combination light service cover RH
- Rope hook assembly RH
- No. 1 luggage compartment trim hook
- RH front deck side trim cover
- RH rear No. 1 outer seat belt assembly
- RH deck trim side panel
- Air intake covers
- Turn back the front floor carpet assembly

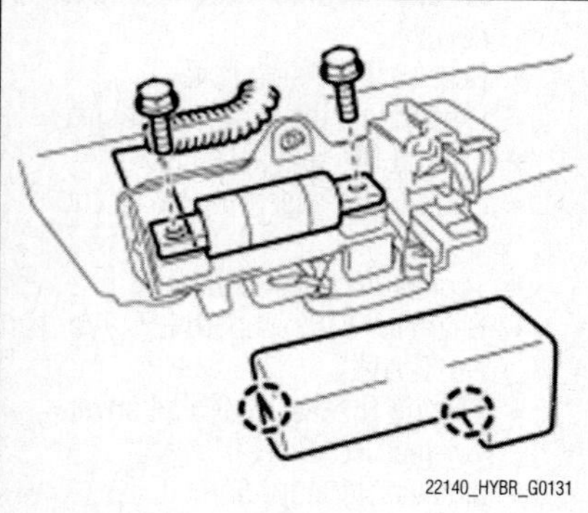

Fig. 51 Electric vehicle fuse

- Release the 2 claws and remove the fuse block cover.
- Remove the 2 bolts and electric vehicle fuse.

11. Reverse the removal procedure and note the following:
    a. Tighten the fuse mounting bolts to 48 inch lbs. (5.4 Nm).
    b. Refer to HV battery removal for additional information.

BATTERY SMART UNIT

REMOVAL & INSTALLATION

See Figures 52 and 53.

⁂ CAUTION

The HIGHLANDER HV has a hybrid system that operates at voltages up to 650 volts. Be sure to follow the instructions in this manual to handle the system correctly. Failure to do so may result in serious injury or electrocution. Engineer must undergo special training to be able to perform high-voltage system inspection and servicing.

⁂ CAUTION

All high-voltage wire harness connectors are colored orange. The HV battery and other high-voltage components have "High Voltage" caution labels. Do not carelessly touch these wires and components.

➡Wear insulating gloves and protective glasses.

1. Before inspecting or servicing the high-voltage system, be sure to follow safety measures, such as wearing insulated gloves and removing the service plug to prevent electrocution. Carry the removed service plug in your pocket to prevent other technicians from reinstalling it while you are servicing the vehicle.
2. After removing the service plug, wait 5 minutes before touching any of the high-voltage connectors and terminals.
3. To access the inverter cover for voltage verification. Remove the following:
    - Service plug grip
    - LH engine room side cover
    - Cool air intake duct seal
    - Air cleaner cap and case assembly
    - Inverter reserve tank
    - Inverter cover
4. Using the voltmeter, measure the voltage between the terminals of the 2 phase connectors (N-P). Standard voltage: 0 volts. Use measuring range of DC 750 V or more on the voltmeter.
5. Install the inverter cover and tighten the mounting bolts to 7 ft. lbs. (10 Nm).
6. Install the inverter reserve tank and tighten the mounting bolts to 7 ft. lbs. (10 Nm).
7. Install air cleaner case and cap assembly.
8. Install the cool air intake duct seal.
9. Install the LH engine room side cover.
10. To access the HV battery smart unit remove the following:
    - Rear center seat assembly
    - LH and RH rear seat head rest
    - LH and RH seat track bracket cover
    - LH and RH outer seat track bracket cover
    - Rear seat leg side cover
    - Disconnect the seat lock cables
    - LH and RH seat assembly
    - Deck board assembly No. 2 and No. 3
    - Tonneau cover if equipped
    - Rear No. 1 floor board LH (w/o rear No. 2 seat)
    - LH and RH rear seat side cover (w/ rear No. 2 seat)
    - Jack carrier support
    - Jack carrier cushion
    - Jack assembly and carrier assembly
    - RH deck side trim box
    - Deck floor board assembly (w/o rear No. 2 seat)
    - Rear No. 2 seat belt assembly (w/ rear No. 2 seat)
    - Rear seat belt lap assembly LH and RH (w/ rear No. 2 seat)
    - Rear No. 2 seat assembly (w/ rear No. 2 seat)
    - Rear floor finish plate
    - Rear deck side trim and cover
    - LH side trim cover (w/o rear seat entertainment system)
    - Power outlet socket (w/ rear seat entertainment system)
    - LH rear combination light service cover
    - Power socket assembly and cover
    - Rear deck trim cover (w/o rear seat entertainment system)
    - Reclining remote control lever bezel LH (w/ remote folding function)
    - LH rope hook assembly
    - No.2 deck side trim hook
    - LH front deck side trim cover
    - Rear LH No. 1 outer seat belt assembly
    - Deck trim side panel LH
    - Quarter pillar garnish LH

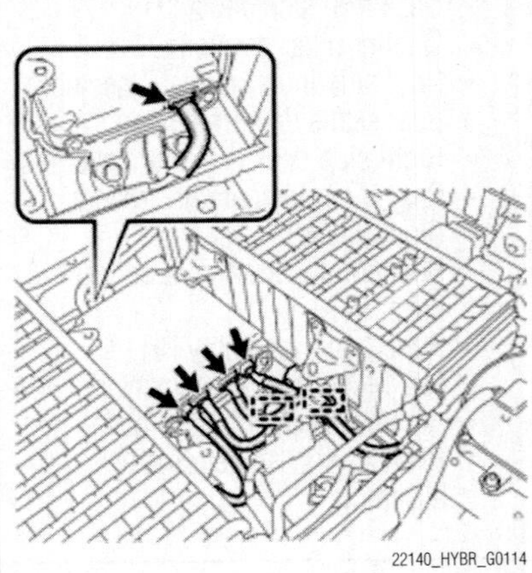
22140_HYBR_G0114

Fig. 52 Battery smart unit 5 connectors and 2 clamps

- Roof side inner garnish assembly
- Rear seat side garnish cap
- Right side deck trim and cover
- Rear room temperature sensor (if equipped)
- Rear combination light service cover RH
- Rope hook assembly RH
- No. 1 luggage compartment trim hook
- RH front deck side trim cover
- RH rear No. 1 outer seat belt assembly
- RH deck trim side panel
- Air intake covers
- Turn back the front floor carpet assembly
- No. 1 HV battery tray
- Rear No.1. 2 and 3 floor boards
- Rear center seat inner belt assembly
- Battery cover lock striker
- Battery service hole cover.
- Battery cover sub- assembly

11. Remove the battery smart unit as follows:

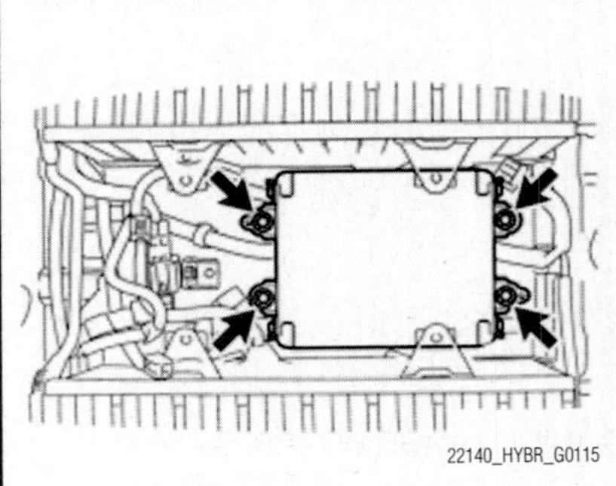
22140_HYBR_G0115

Fig. 53 Battery smart unit mounting bolts

a. Disconnect the 5 connectors and 2 clamps.

b. Remove the 4 bolts and battery smart unit.

12. Reverse the removal procedure and note the following:

a. Tighten the smart unit mounting bolts to 66 inch lbs. (7.5 Nm).

b. Refer to HV battery removal for additional information.

HV BATTERY

REMOVAL & INSTALLATION

See Figures 54 through 64.

CAUTION

The HIGHLANDER HV has a hybrid system that operates at voltages up to 650 volts. Be sure to follow the instructions in this manual to handle the system correctly. Failure to do so may result in serious injury or electrocution. Engineer must undergo special training to be able to perform high-voltage system inspection and servicing.

CAUTION

All high-voltage wire harness connectors are colored orange. The HV battery and other high-voltage components have "High Voltage" caution labels. Do not carelessly touch these wires and components.

1. Before inspecting or servicing the high-voltage system, be sure to follow safety measures, such as wearing insulated gloves and removing the service plug to prevent electrocution. Carry the removed service plug in your pocket to prevent other technicians from reinstalling it while you are servicing the vehicle.

2. After removing the service plug, wait 5 minutes before touching any of the high-voltage connectors and terminals.

➡Wear insulating gloves and protective glasses.

3. Before inspecting or servicing the high-voltage system, be sure to follow safety measures, such as wearing insulated gloves and removing the service plug to prevent electrocution. Carry the removed service plug in your pocket to prevent other technicians from reinstalling it while you are servicing the vehicle.

4. After removing the service plug, wait 5 minutes before touching any of the high-voltage connectors and terminals.

5. Check for Daces and confirm that P0AA6 (High voltage insulation is unusual) is not output before doing removal or installation inside the battery. If the DTC is output, perform troubleshooting first.

6. To access the inverter cover for voltage verification. Remove the following:

- Service plug grip
- LH engine room side cover
- Cool air intake duct seal
- Air cleaner cap and case assembly
- Inverter reserve tank
- Inverter cover

7. Using the voltmeter, measure the voltage between the terminals of the 2 phase connectors (N-P). Standard voltage: 0 volts. Use measuring range of DC 750 V or more on the voltmeter.

8. Install the inverter cover and tighten the mounting bolts to 7 ft. lbs. (10 Nm).

9. Install the inverter reserve tank and tighten the mounting bolts to 7 ft. lbs. (10 Nm).

10. Install air cleaner case and cap assembly.

11. Install the cool air intake duct seal.

12. Install the LH engine room side cover.

13. To access the HV battery remove the following:

- Rear center seat assembly
- LH and RH rear seat head rest
- LH and RH seat track bracket cover
- LH and RH outer seat track bracket cover
- Rear seat leg side cover
- Disconnect the seat lock cables
- LH and RH seat assembly
- Deck board assembly No. 2 and No. 3
- Tonneau cover if equipped
- Rear No. 1 floor board LH (w/o rear No. 2 seat)
- LH and RH rear seat side cover (w/ rear No. 2 seat)
- Jack carrier support
- Jack carrier cushion

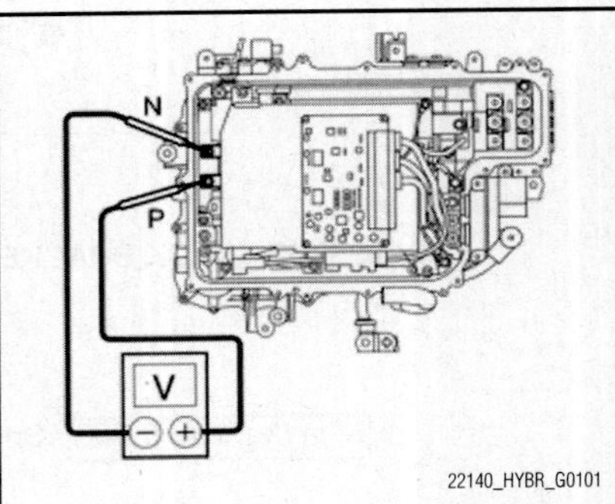

22140_HYBR_G0101

Fig. 54 Measure the voltage between the terminals of the 2 phase connectors

- Jack assembly and carrier assembly
- RH deck side trim box
- Deck floor board assembly (w/o rear No. 2 seat)
- Rear No. 2 seat belt assembly (w/ rear No. 2 seat)
- Rear seat belt lap assembly LH and RH (w/ rear No. 2 seat)
- Rear No. 2 seat assembly (w/ rear No. 2 seat)
- Rear floor finish plate
- Rear deck side trim and cover
- LH side trim cover (w/o rear seat entertainment system)
- Power outlet socket (w/ rear seat entertainment system)
- LH rear combination light service cover
- Power socket assembly and cover
- Rear deck trim cover (w/o rear seat entertainment system)
- Reclining remote control lever bezel LH (w/ remote folding function)
- LH rope hook assembly
- No.2 deck side trim hook
- LH front deck side trim cover
- Rear LH No. 1 outer seat belt assembly
- Deck trim side panel LH
- Quarter pillar garnish LH
- Roof side inner garnish assembly
- Rear seat side garnish cap
- Right side deck trim and cover
- Rear room temperature sensor (if equipped)
- Rear combination light service cover RH
- Rope hook assembly RH
- No. 1 luggage compartment trim hook
- RH front deck side trim cover
- RH rear No. 1 outer seat belt assembly

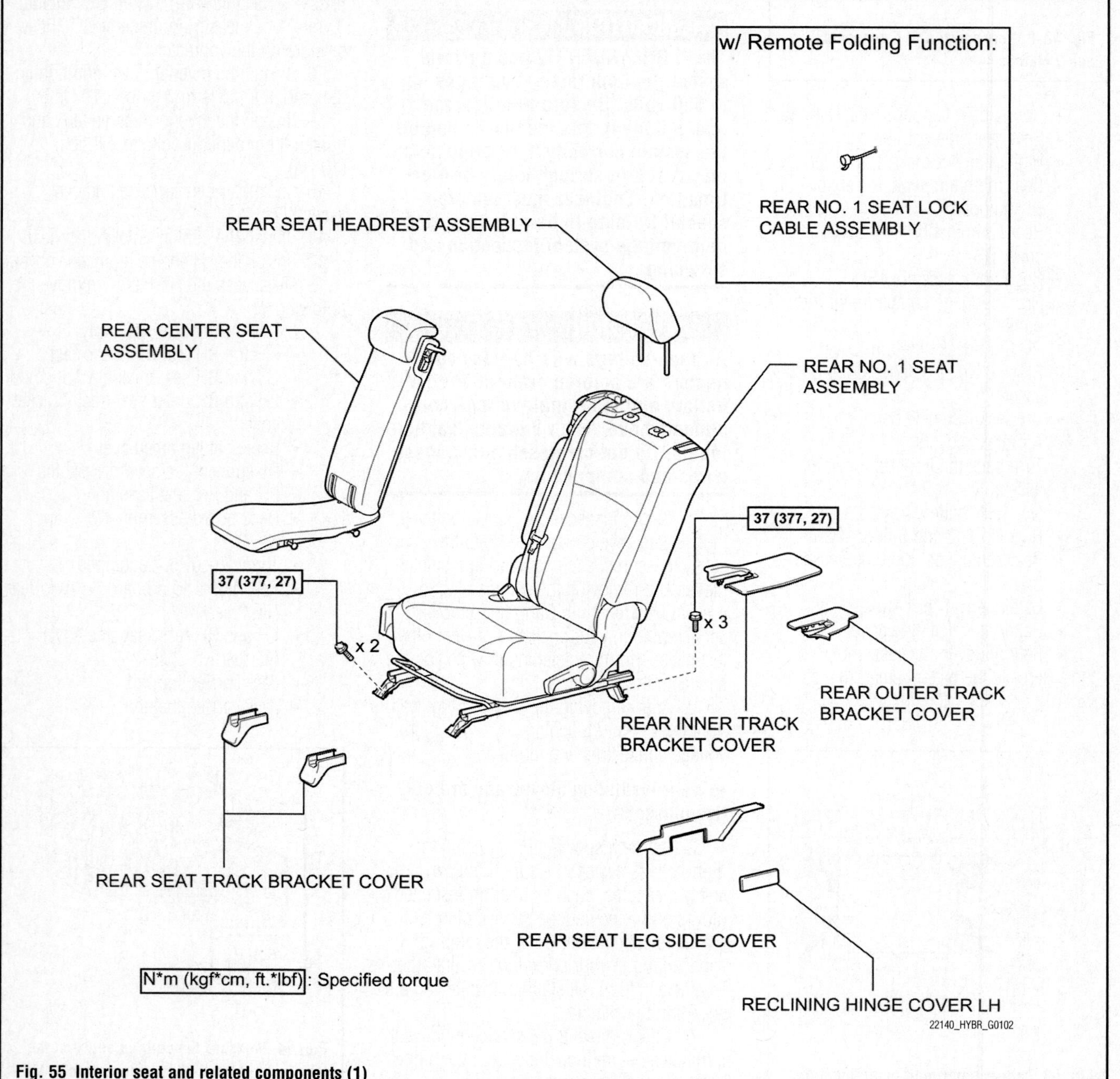

Fig. 55 Interior seat and related components (1)

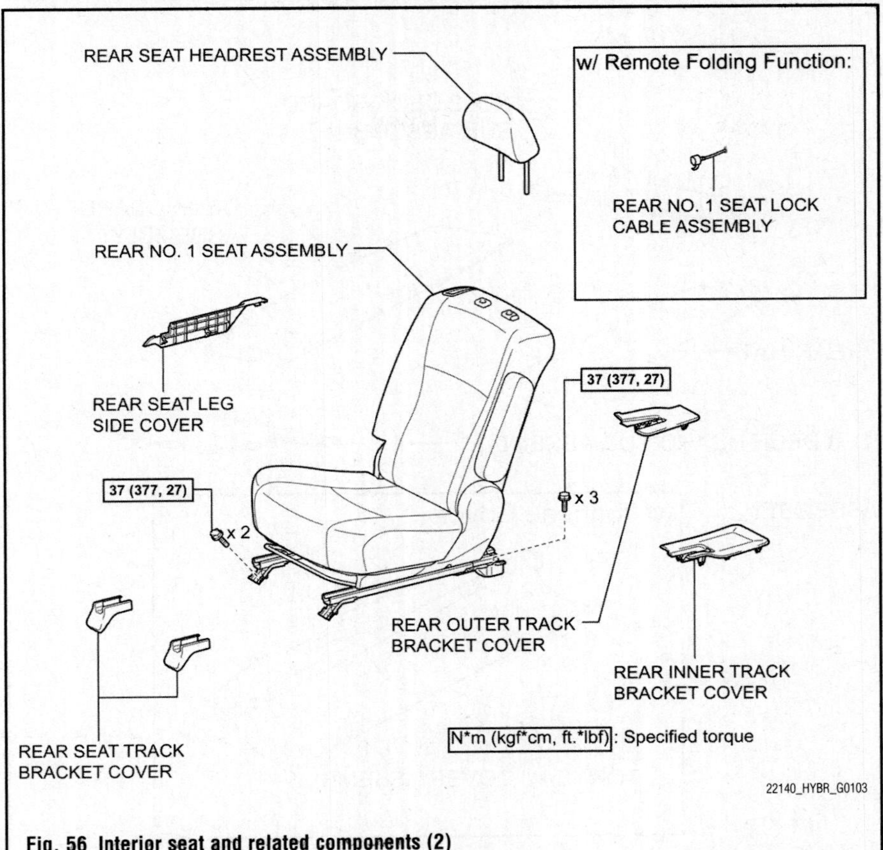

Fig. 56 Interior seat and related components (2)

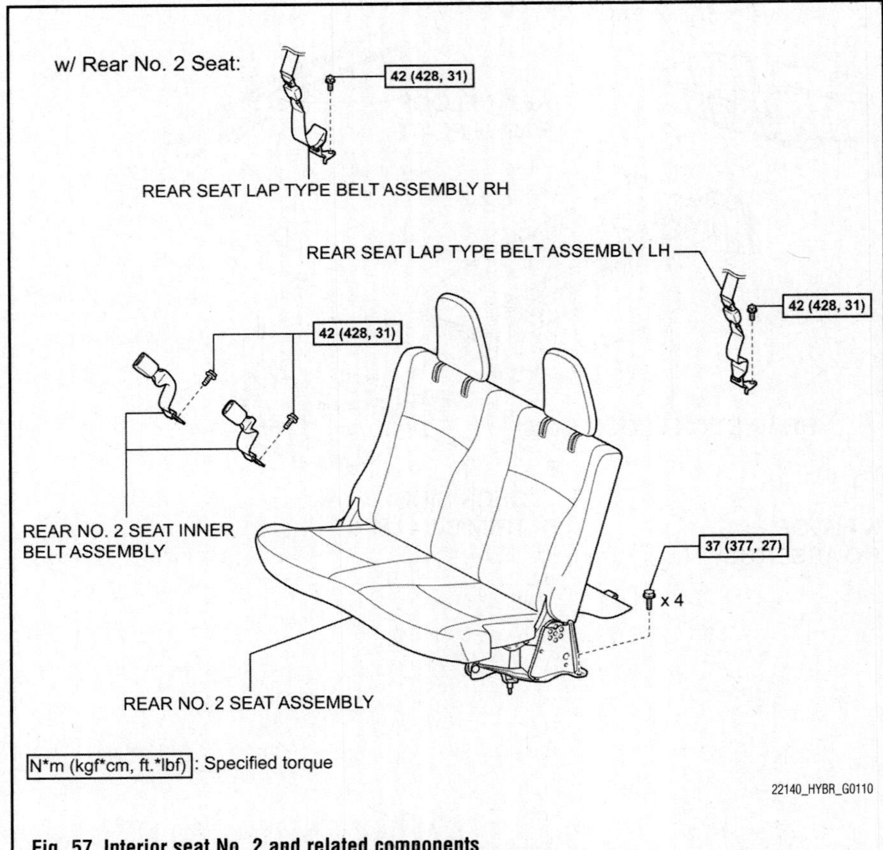

Fig. 57 Interior seat No. 2 and related components

- RH deck trim side panel
- Air intake covers
- Turn back the front floor carpet assembly
- No. 1 HV battery tray
- Rear No.1. 2 and 3 floor boards
- Rear center seat inner belt assembly
- Battery cover lock striker
- Battery service hole cover.
- Battery cover sub- assembly
- Battery smart unit

CAUTION

Be sure to wear insulated gloves and protective goggles.

14. Remove the HV battery as follows:
    a. Insulate the removed terminals with insulating tape
    b. Remove the battery room ventilation hose between the HV battery module LH and center HV battery module.
    c. Disconnect the wire harness clamp of the battery thermo sensor.
    d. Remove the nut and main battery cable from the HV battery module LH.
    e. Remove the nut, then disconnect the EV battery plug from the HV battery module LH.
    f. Remove the carry belts from the center HV battery module and install them to the HV battery module LH.
    g. Insert the ends of the carry belts into the installation holes. Pull the carry belts upward to securely install them.
    h. Remove the HV battery module LH.
    i. Remove the carry belts from the HV battery module LH.
    j. Remove the center HV battery module.
    k. Disconnect the 2 connectors from the HV relay assembly.
    l. Disconnect the 2 connectors and clamp.
    m. Remove the battery room ventilation hose between the center HV battery module and HV battery module RH.
    n. Install the carry belts to the center HV battery module.
    o. Remove the center HV battery module.
    p. Remove the HV battery module RH.
    q. Disconnect the 2 clamps.
    r. Remove the nut, then disconnect the main battery cable connector.
    s. Disconnect the clamp.
    t. Remove the battery room ventilation hose.
    u. Install the carry belts to the HV battery module RH.
    v. Remove the HV battery module RH.
15. Installation is in the reverse of the removal procedure

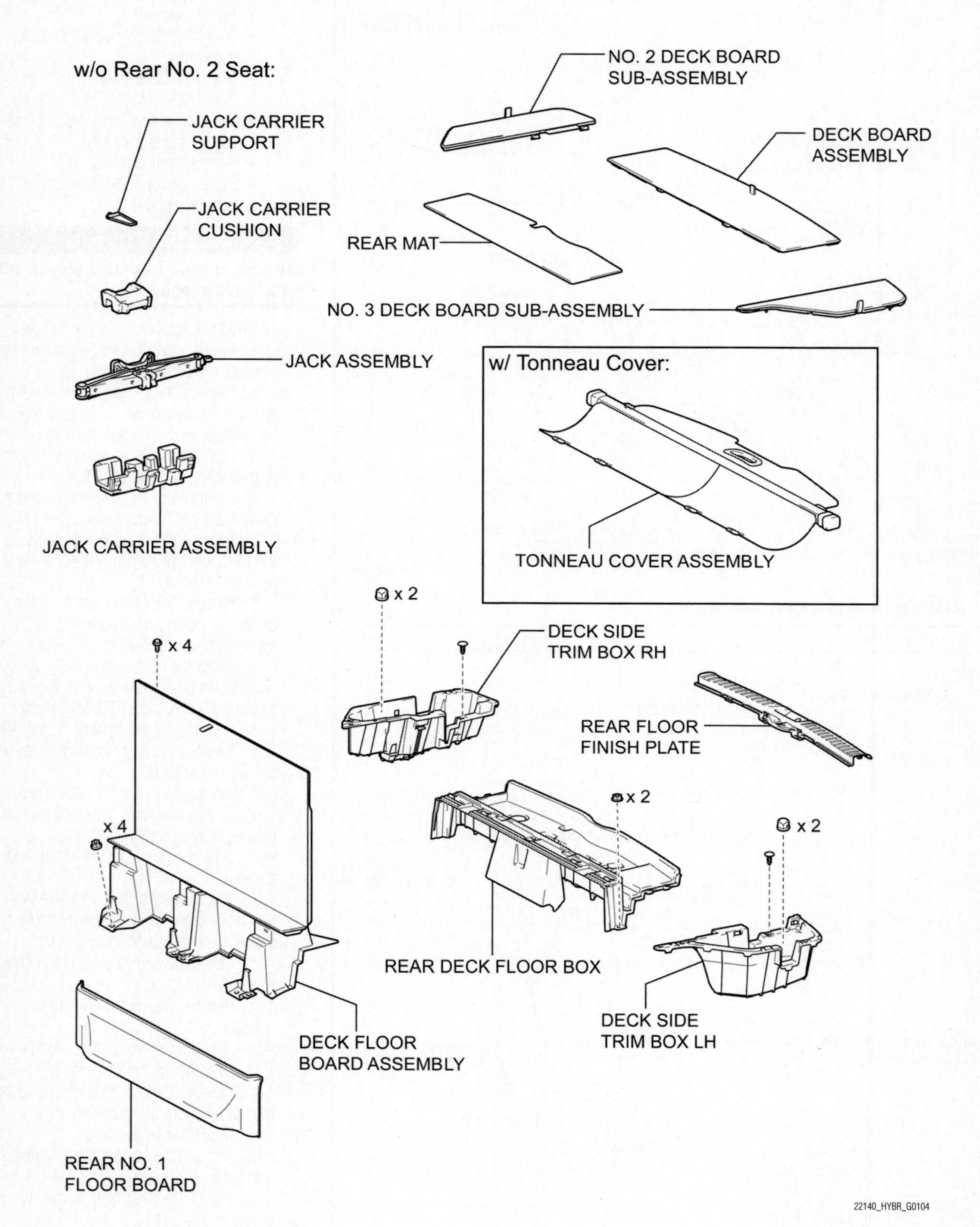

Fig. 58 Interior trim and related components (1)

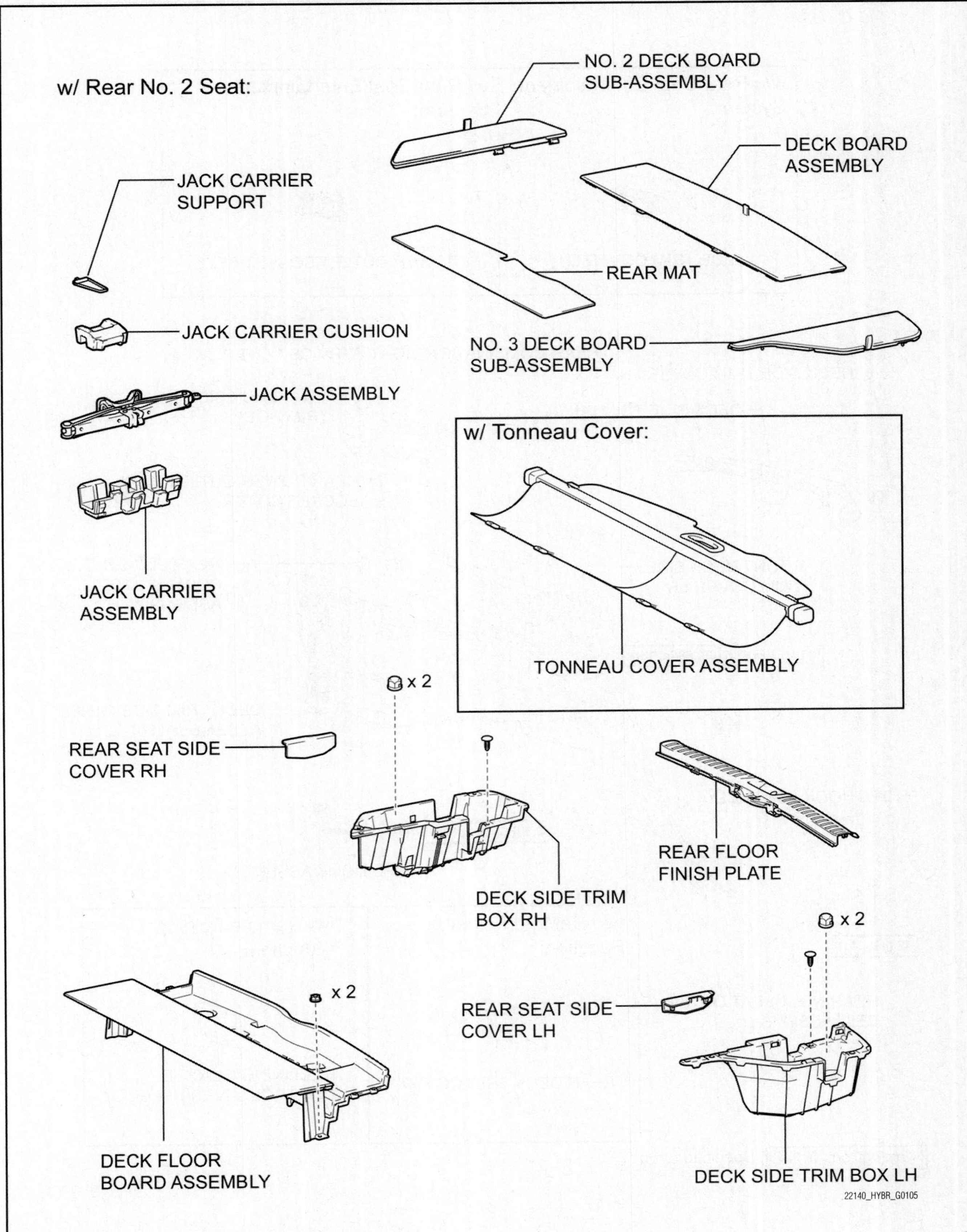

Fig. 59 Interior trim and related components (2)

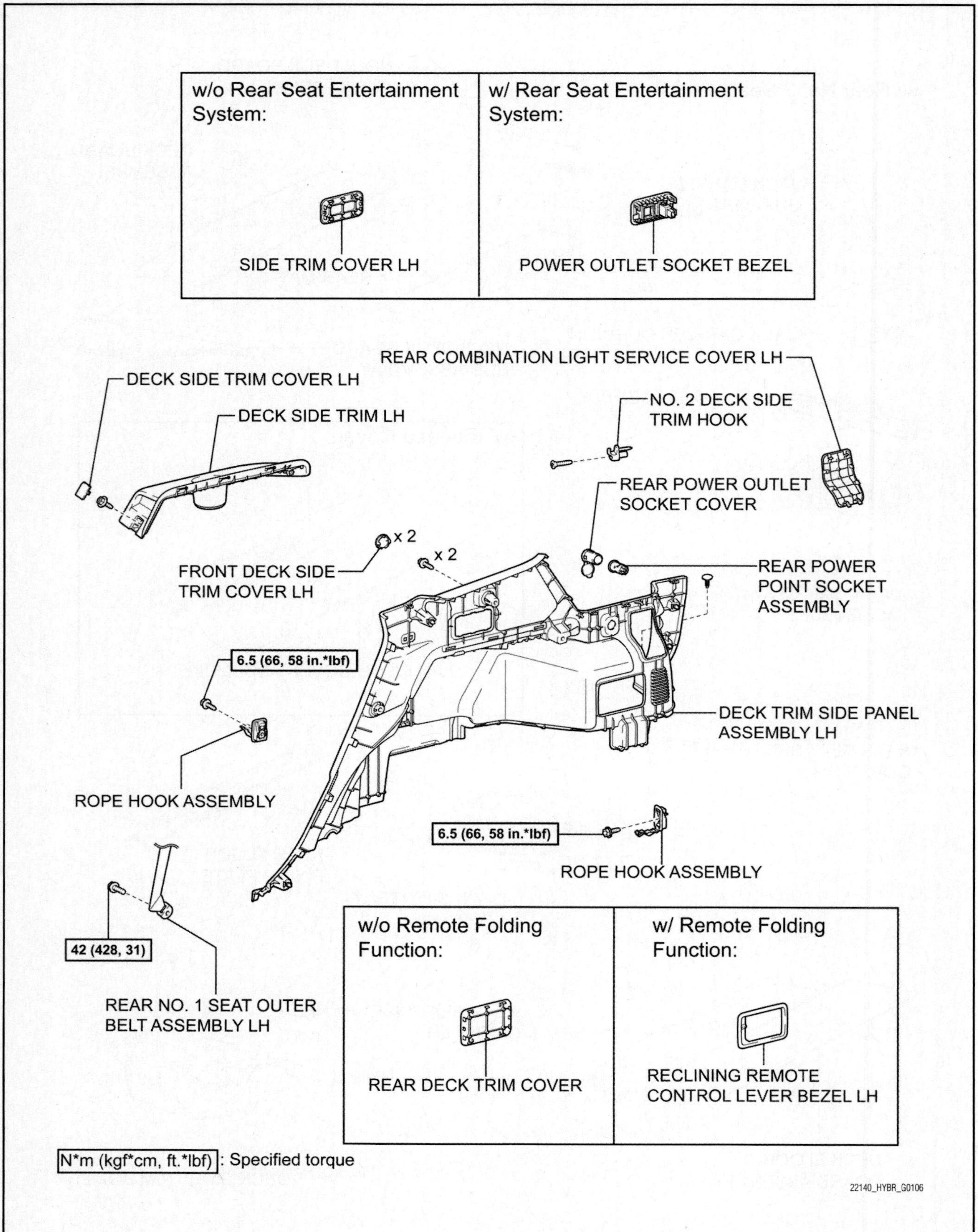

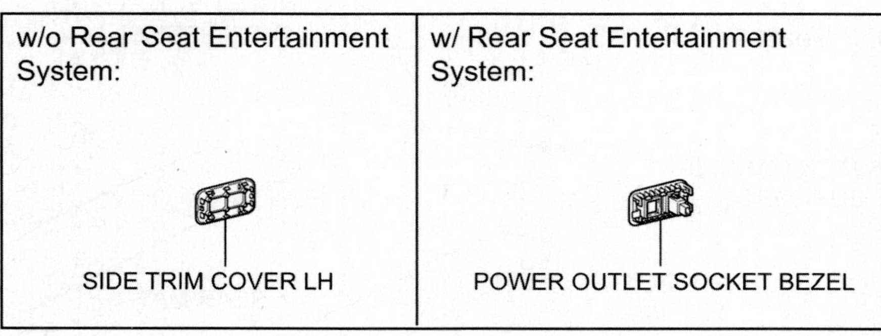

w/o Rear Seat Entertainment System:
SIDE TRIM COVER LH
w/ Rear Seat Entertainment System:
POWER OUTLET SOCKET BEZEL
REAR COMBINATION LIGHT SERVICE COVER LH
DECK SIDE TRIM COVER LH
DECK SIDE TRIM LH
NO. 2 DECK SIDE TRIM HOOK
REAR POWER OUTLET SOCKET COVER
x 2
x 2
FRONT DECK SIDE TRIM COVER LH
REAR POWER POINT SOCKET ASSEMBLY
6.5 (66, 58 in.*lbf)
DECK TRIM SIDE PANEL ASSEMBLY LH
ROPE HOOK ASSEMBLY
6.5 (66, 58 in.*lbf)
ROPE HOOK ASSEMBLY
42 (428, 31)
REAR NO. 1 SEAT OUTER BELT ASSEMBLY LH
w/o Remote Folding Function:
REAR DECK TRIM COVER
w/ Remote Folding Function:
RECLINING REMOTE CONTROL LEVER BEZEL LH
N*m (kgf*cm, ft.*lbf) : Specified torque
22140_HYBR_G0106

Fig. 60 Interior trim and related components (3)

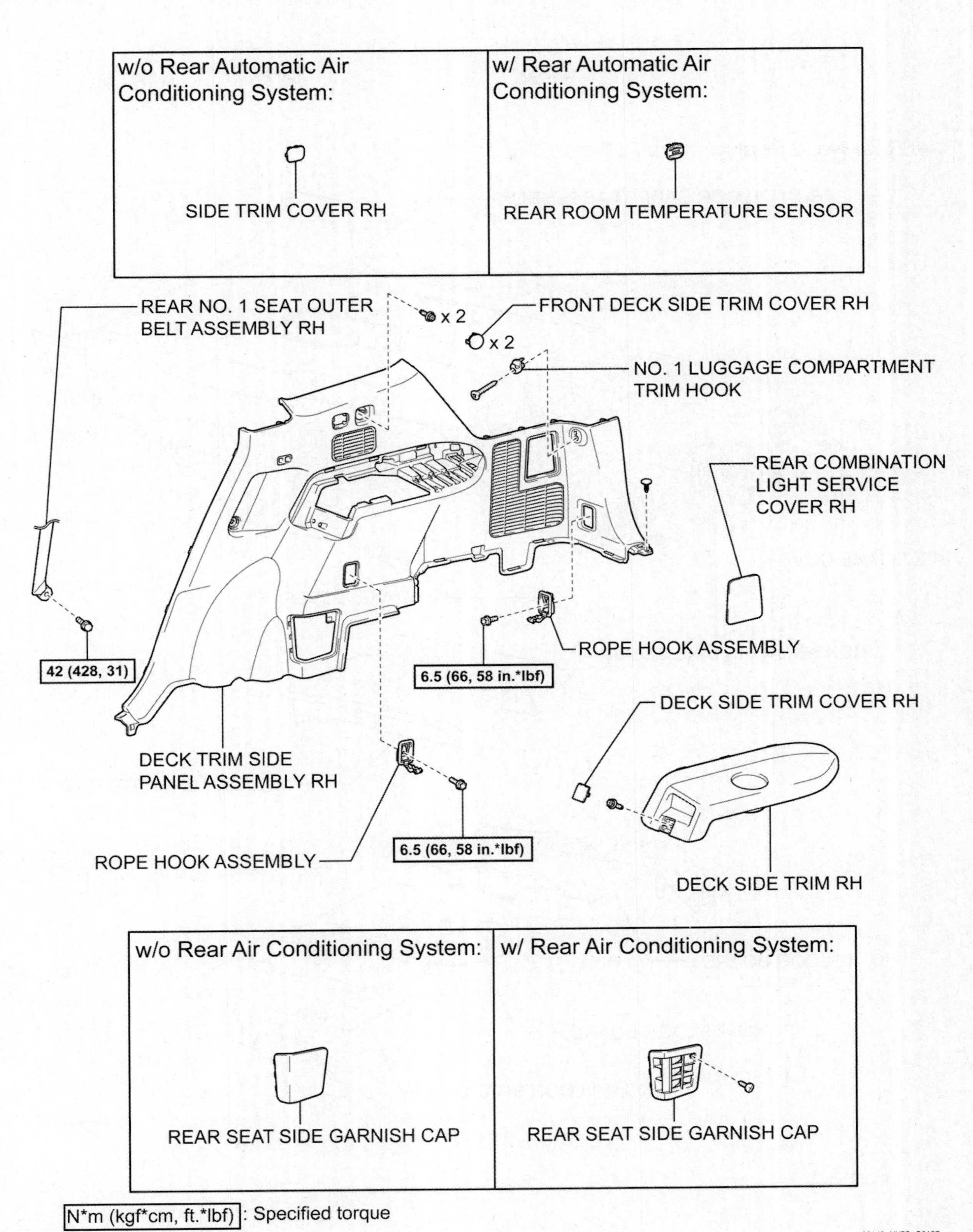

Fig. 61 Interior trim and related components (4)

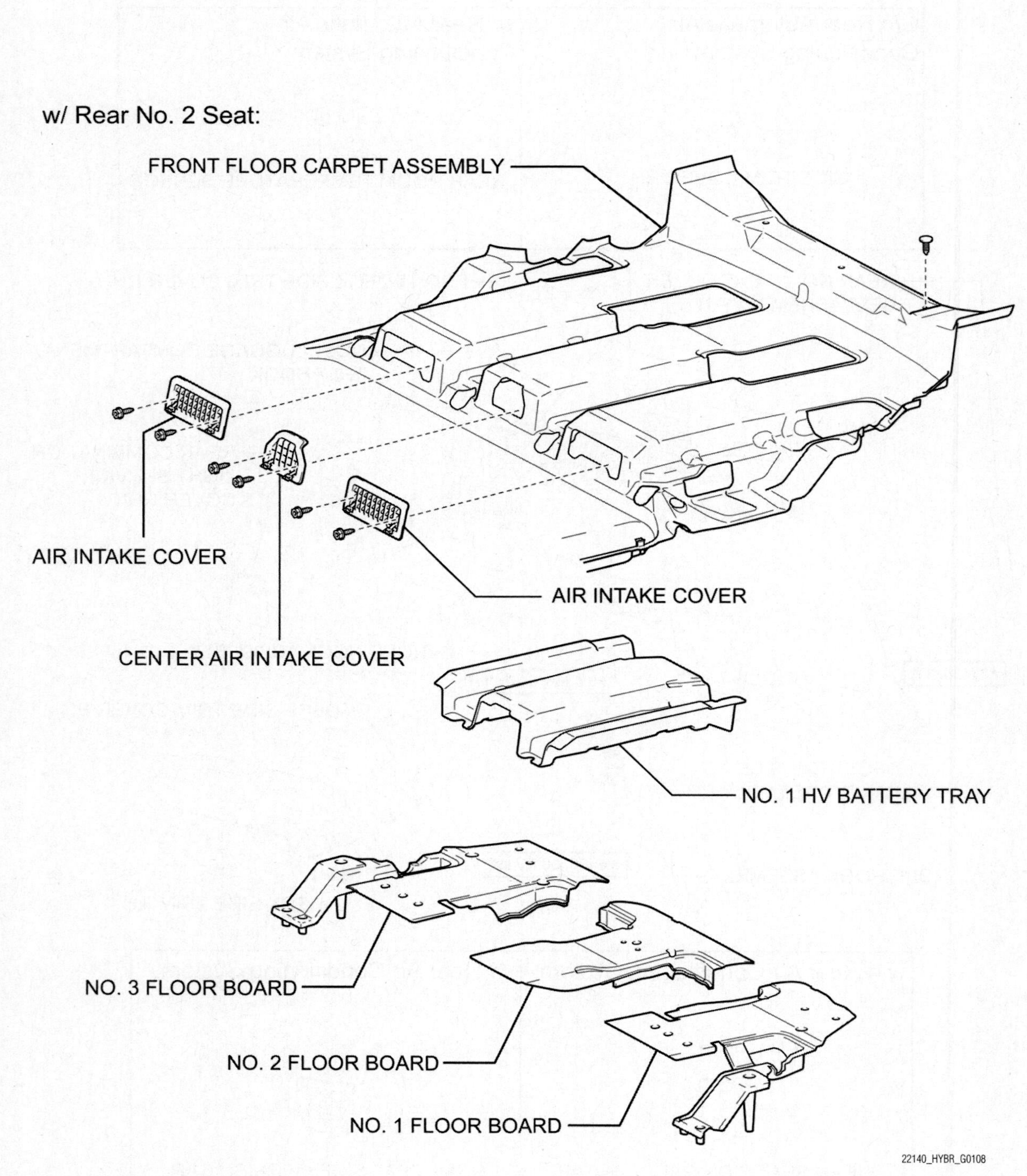

Fig. 62 Interior carpet and related parts

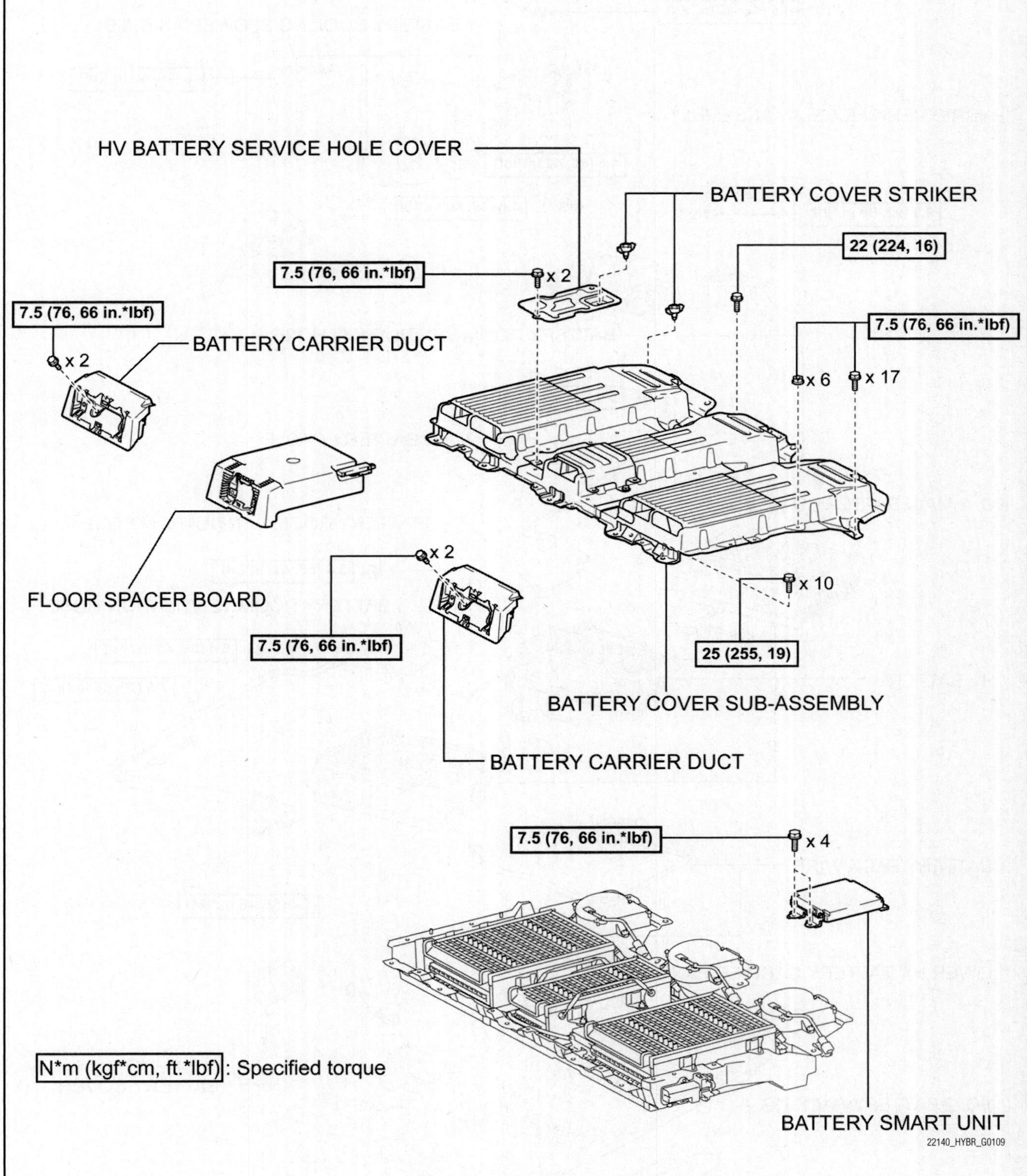

Fig. 63 Battery cover and related parts

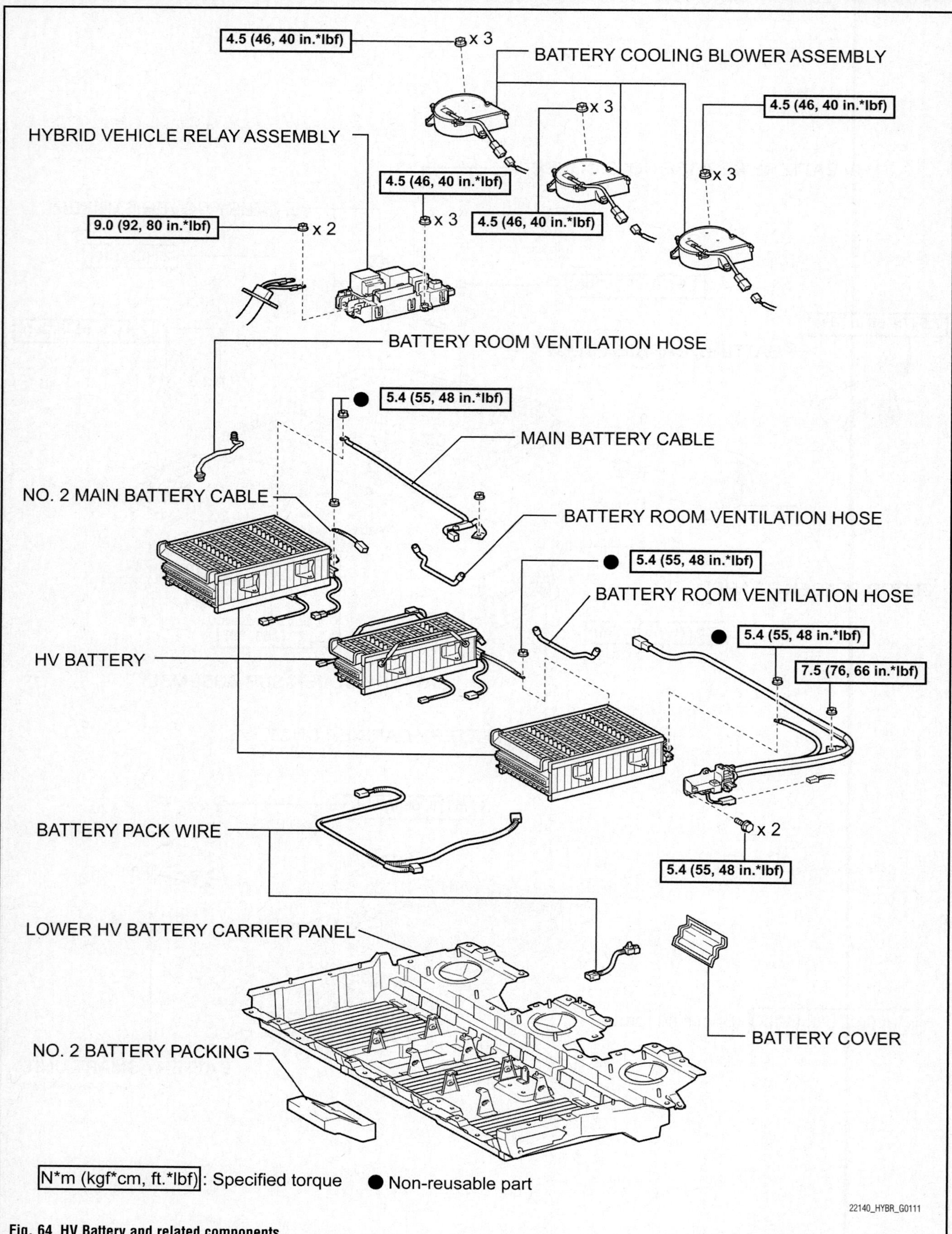

Fig. 64 HV Battery and related components

HV RELAY ASSEMBLY

REMOVAL & INSTALLATION

See Figure 65.

CAUTION

The HIGHLANDER HV has a hybrid system that operates at voltages up to 650 volts. Be sure to follow the instructions in this manual to handle the system correctly. Failure to do so may result in serious injury or electrocution. Engineer must undergo special training to be able to perform high-voltage system inspection and servicing.

CAUTION

All high-voltage wire harness connectors are colored orange. The HV battery and other high-voltage components have "High Voltage" caution labels. Do not carelessly touch these wires and components.

1. Before inspecting or servicing the high-voltage system, be sure to follow safety measures, such as wearing insulated gloves and removing the service plug to prevent electrocution. Carry the removed service plug in your pocket to prevent other technicians from reinstalling it while you are servicing the vehicle.
2. After removing the service plug, wait 5 minutes before touching any of the high-voltage connectors and terminals.
3. Refer to HV battery removal for interior removal.

CAUTION

Be sure to wear insulated gloves and protective goggles.

4. Remove the HV relay assembly as follows:
    a. Remove the 2 nuts, then disconnect the No. 3 frame wire from the hybrid vehicle relay assembly. b. Insulate the removed terminals with insulating tape.
    c. Disconnect the 2 main battery cable connectors from the hybrid vehicle relay assembly.
    d. Disconnect the 2 connectors from the hybrid vehicle relay assembly.
    e. Remove the 3 nuts and hybrid vehicle relay assembly.

To install:

  f. Install the hybrid vehicle relay assembly with the 3 nuts and tighten to 40 inch lbs. (4.5 Nm)

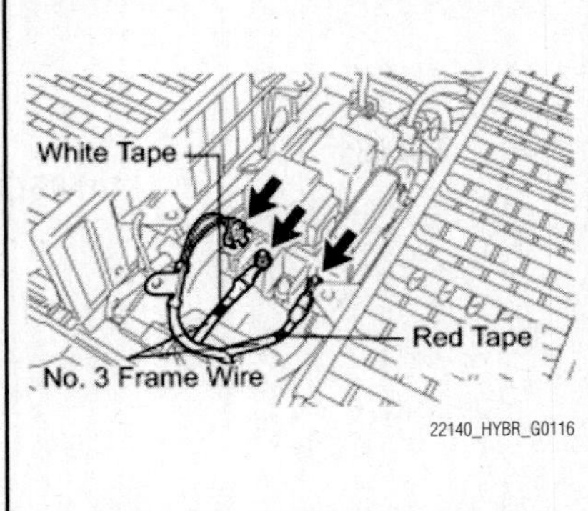

Fig. 65 HV relay No. 3 terminal wire view

  g. Connect the 2 connectors to the HV relay assembly.
  h. Connect the 2 main battery cable connectors to the HV relay assembly.
  i. Install the No. 3 frame wire to the hybrid vehicle relay with the 2 nuts.
  j. Be sure to connect the No. 3 frame wire to each correct terminal as shown in the illustration. Tighten the nuts to 80 inch lbs. (9 Nm).

5. Refer to HV battery for interior installation.

HYBRID TRANSAXLE ASSEMBLY

REMOVAL & INSTALLATION

See Figures 66 through 68.

CAUTION

The HIGHLANDER HV has a hybrid system that operates at voltages up to 650 volts. Be sure to follow the instructions in this manual to handle the system correctly. Failure to do so may result in serious injury or electrocution. Engineer must undergo special training to be able to perform high-voltage system inspection and servicing.

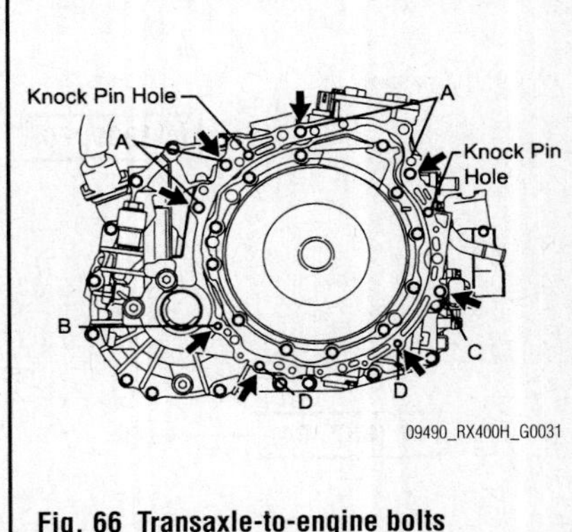

Fig. 66 Transaxle-to-engine bolts

CAUTION

All high-voltage wire harness connectors are colored orange. The HV battery and other high-voltage components have "High Voltage" caution labels. Do not carelessly touch these wires and components.

1. Before inspecting or servicing the high-voltage system, be sure to follow safety measures, such as wearing insulated gloves and removing the service plug to prevent electrocution. Carry the removed service plug in your pocket to prevent other technicians from reinstalling it while you are servicing the vehicle.
2. After removing the service plug, wait 5 minutes before touching any of the high-voltage connectors and terminals.
3. Before servicing the vehicle, refer to the Precautions Section.

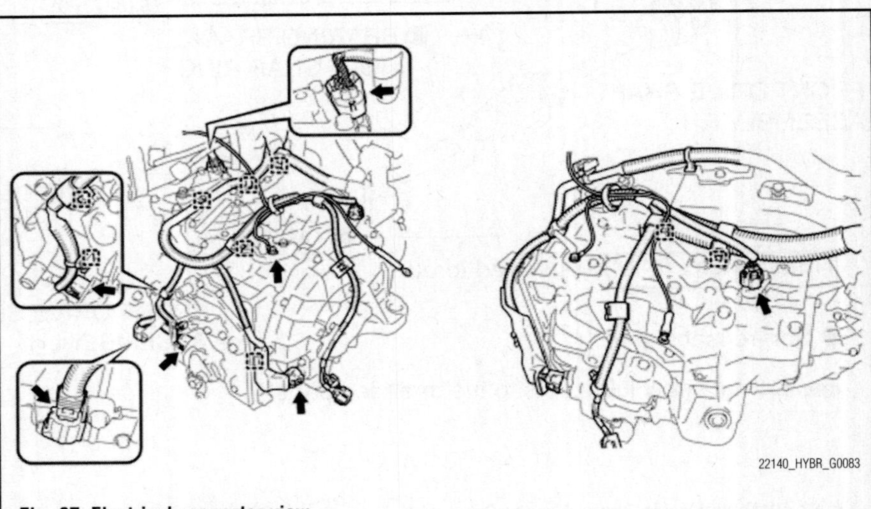

Fig. 67 Electrical connector view

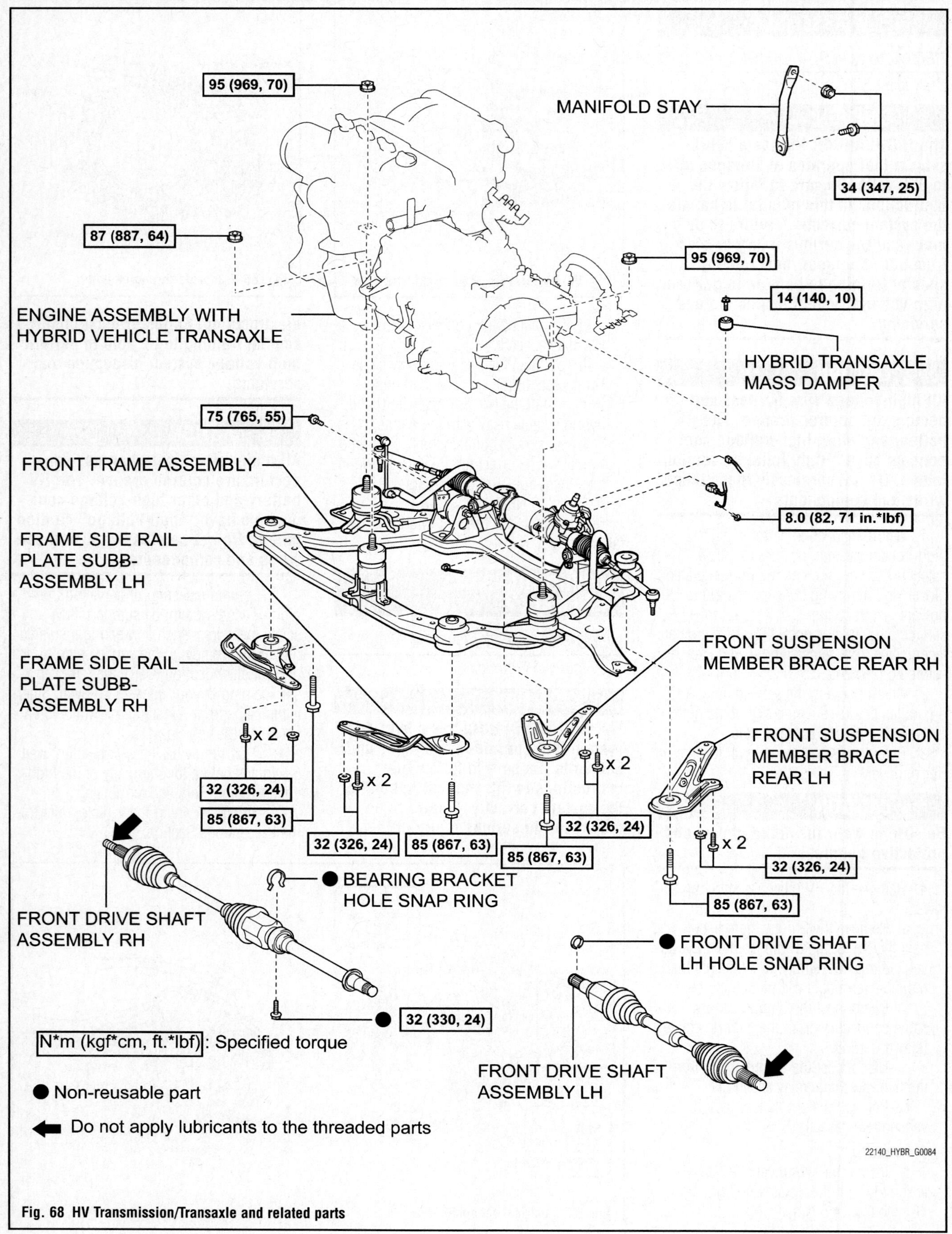

Fig. 68 HV Transmission/Transaxle and related parts

4. Remove or disconnect the following:
    - Engine/transaxle assembly.
    - Manifold stay
    - Transaxle damper
    - Front frame assembly
    - Halfshafts
    - Flywheel housing undercover
    - Engine wiring harnesses
    - Transaxle case cover
    - Coolant hose
    - Front engine mounting bracket
    - Transaxle oil cooler assembly
    - Transmission control cable bracket
5. Remove the 8 mounting bolts and separate the transaxle assembly from the vehicle.
6. Installation is the reverse of removal. Observe the following torques:
    - Transaxle-to-engine: Bolts A to 47 ft. lbs (64 Nm); Bolt B to 34 ft. lbs. (46 Nm); Bolts C to 47 ft. lbs. (64 Nm); Bolts D to 27 ft. lbs. (37 Nm)

➡**Do not reuse Bolt B.**

- Front engine mounting bracket: 47 ft. lbs. (64 Nm)
- Transaxle case cover: 74 inch lbs. (8.4 Nm)
- Undercover: 69 inch lbs. (8 Nm)

REAR TRACTION MOTOR

REMOVAL & INSTALLATION

See Figures 69 through 72.

CAUTION

The HIGHLANDER HV has a hybrid system that operates at voltages up to 650 volts. Be sure to follow the instructions in this manual to handle the system correctly. Failure to do so may result in serious injury or electrocution. Engineer must undergo special training to be able to perform high-voltage system inspection and servicing.

CAUTION

All high-voltage wire harness connectors are colored orange. The HV battery and other high-voltage components have "High Voltage" caution labels. Do not carelessly touch these wires and components.

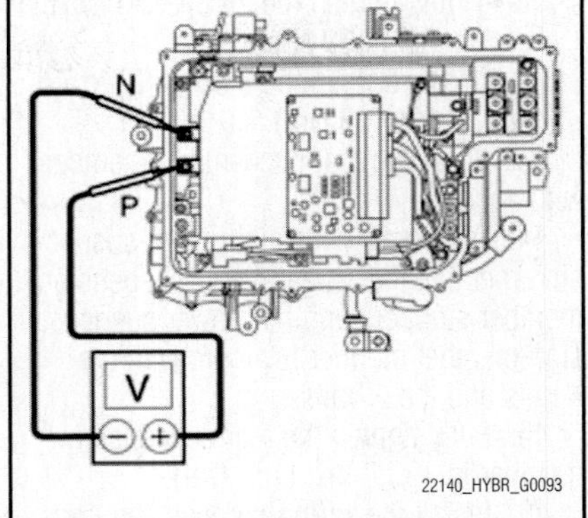
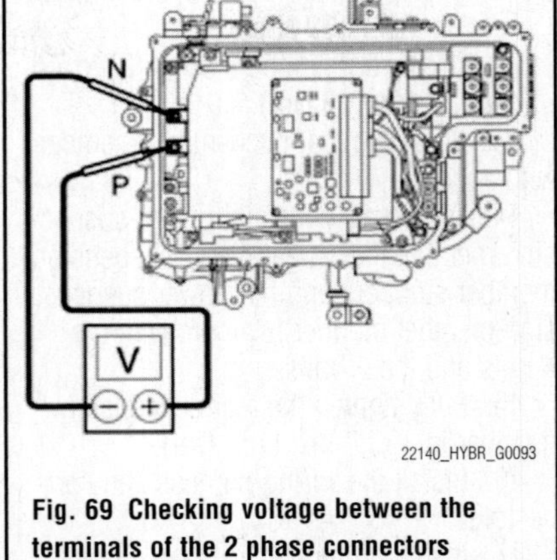

Fig. 69 Checking voltage between the terminals of the 2 phase connectors

1. Before inspecting or servicing the high-voltage system, be sure to follow safety measures, such as wearing insulated gloves and removing the service plug to prevent electrocution. Carry the removed service plug in your pocket to prevent other technicians from reinstalling it while you are servicing the vehicle.
2. After removing the service plug, wait 5 minutes before touching any of the high-voltage connectors and terminals.
3. Before servicing the vehicle, refer to the Precautions Section.
4. Disconnect the negative battery cable.
5. When disconnecting the cable, some systems need to be initialized after the cable is reconnected.
6. Remove the service plug grip, found underneath the Battery Service cover on the rear seat. Wait 5 minutes to discharge the high voltage capacitor.
7. Remove the LH and RH wiper arm assembly.
8. Remove the cowl top ventilator louver sub-assembly.
9. Remove the wiper motor and link assembly.
10. Remove the outer cowl top panel.
11. Remove the LH room side cover.
12. Remove the cool air intake duct seal.
13. Remove the air cleaner cap sub assembly.
14. Remove the air cleaner case sub-assembly.
15. Remove the inverter reserve tank sub assembly.
16. Remove the inverter cover.
17. Using the voltmeter, measure the voltage between the terminals of the 2 phase connectors (N-P). Standard voltage: 0 volts.
18. Install the inverter cover and tighten bolts to 7 ft. lbs. (10 Nm).
19. Install the inverter reserve tank sub-assembly with the 2 bolts. Tighten the bolts to 7 ft. lbs. (10 Nm).
20. Install the air cleaner case and cap assembly.
21. Install outer panel top cowl assembly.
22. Install wiper motor and link assembly.
23. Install the cowl top ventilator assembly.
24. Install the LH and RH wiper arm assembly.
25. Install the cool air duct seal.
26. Install the engine room side cover.
27. Drain the rear traction motor fluid.
28. Remove the rear wheels.
29. Remove exhaust pipe assembly.
30. Remove the RH and LH axle shaft nuts.
31. Remove the LH and RH strut rod assembly.
32. Remove the LH and RH suspension arm assembly. No.1 and No. 2.
33. Remove axle assembly for RH and LH side.
34. Remove the nuts, and separate the both parking brake cables.

Front of the Vehicle

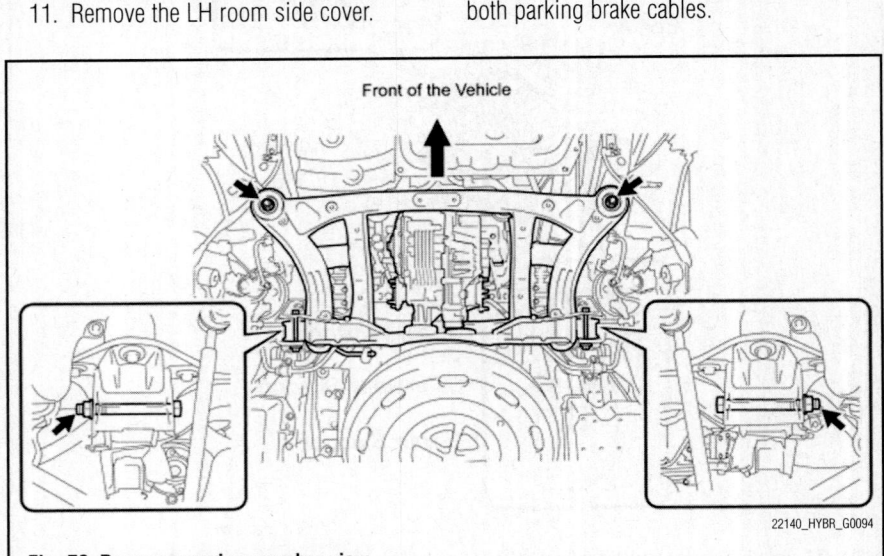

22140_HYBR_G0094

Fig. 70 Rear suspension member view

35. Remove the nut, and separate the ground cable. And all wiring harness clamps.

36. Wear insulated gloves. Remove the 2 nuts, and separate the No. 3 frame wire from the rear traction motor.

37. Support the rear suspension member with a jack.

38. Remove the rear suspension member as follows:

- Remove the 4 nuts, 2 bolts and 2 rear lower suspension member stopper retainers.
- Lower the rear suspension member.
- Remove the 2 rear upper suspension member stoppers.

39. Remove the 4 bolts and rear traction with transaxle motor assembly.

40. Remove the front differential support assembly.

To install:

41. Install the front differential support assembly to the rear traction with transaxle motor with new 2 bolts. Tighten the bolts to 59 ft. lbs. (80 Nm).

42. Install the rear traction with transaxle motor assembly as follows :

- Temporarily install the rear traction motor (front side) with the 2 lower stoppers, 2 upper supports, and 2 new bolts (A) as shown in the illustration.
- Temporarily install the rear traction motor (rear side) with the 2 new bolts (B).
- Fully tighten the 2 bolts (A) to 76 ft. lbs. (103 Nm).
- Fully tighten the 2 bolts (B) to 77 ft. lbs. (95 Nm).

43. Raise the rear suspension member with a jack.

44. Temporarily install the rear suspension member, the 2 rear upper suspension member stoppers and rear lower suspension member stopper retainers with the 4 nuts and the 2 bolts.

45. Fully tighten the rear suspension member to 133 ft lbs. (181 Nm).

46. Install the wiring harness and connectors.

47. Install both parking brake cables.

48. Install RH and LH axles.

49. Jack up the rear axle carrier, placing a wooden block to avoid damage.

50. Temporarily tighten the rear No. 2 suspension arm assembly LH with the bolt and nut.

51. Install the RH side by following the same procedures as for the LH side

52. Temporarily tighten the rear No. 1 suspension arm assembly LH with the bolt and nut.

53. Install the RH side by following the same procedures as for the LH side

54. Temporarily tighten the rear strut rod assembly with the bolt and nut.

55. Stabilize the suspension.

56. Fully tighten all suspension arm bolts to 82 ft. lbs. (112 Nm).

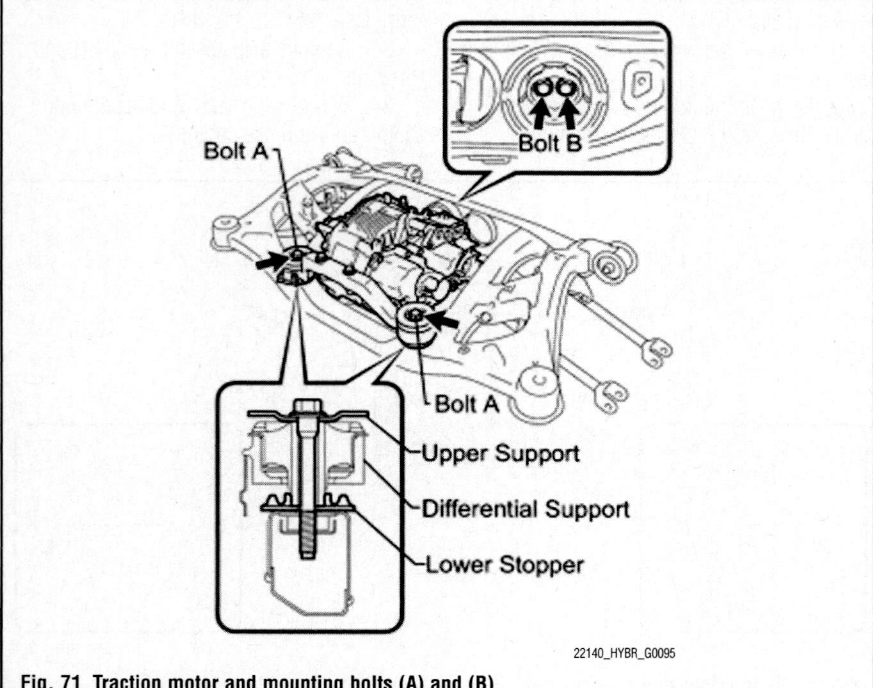

Fig. 71 Traction motor and mounting bolts (A) and (B)

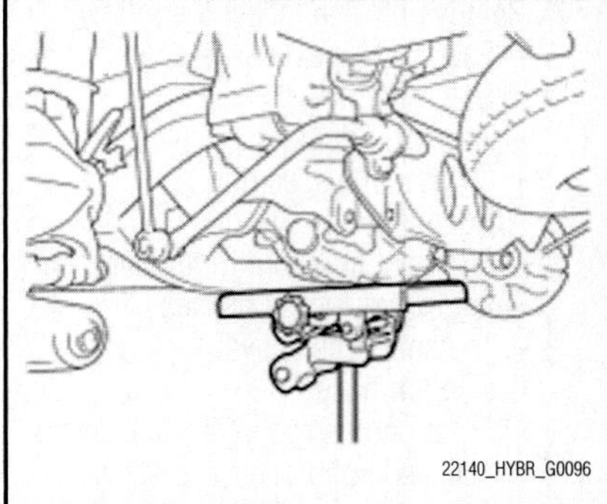

Fig. 72 Raise the rear suspension member with a jack

57. Fully tighten both strut rods to 59 ft. lbs. (80 Nm).

58. Install a new rear axle shaft nuts. Tighten both axle nuts to 216 ft. lbs. (294 Nm).

59. Install the rear wheels and tighten to 76 ft. lbs. (103 Nm).

60. Check and adjust rear wheel alignment.

61. Check speed sensor operation.

ELECTRIC WATER PUMP WITH MOTOR

REMOVAL & INSTALLATION

See Figures 73 and 74.

CAUTION

The HIGHLANDER HV has a hybrid system that operates at voltages up to 650 volts. Be sure to follow the instructions in this manual to handle the system correctly. Failure to do so may result in serious injury or electrocution. Engineer must undergo special training to be able to perform high-voltage system inspection and servicing.

CAUTION

All high-voltage wire harness connectors are colored orange. The HV battery and other high-voltage components have "High Voltage" caution labels. Do not carelessly touch these wires and components.

1. Before inspecting or servicing the high-voltage system, be sure to follow safety measures, such as wearing insulated gloves and removing the service plug to prevent electrocution. Carry the removed service plug in your pocket to prevent other technicians from reinstalling it while you are servicing the vehicle.

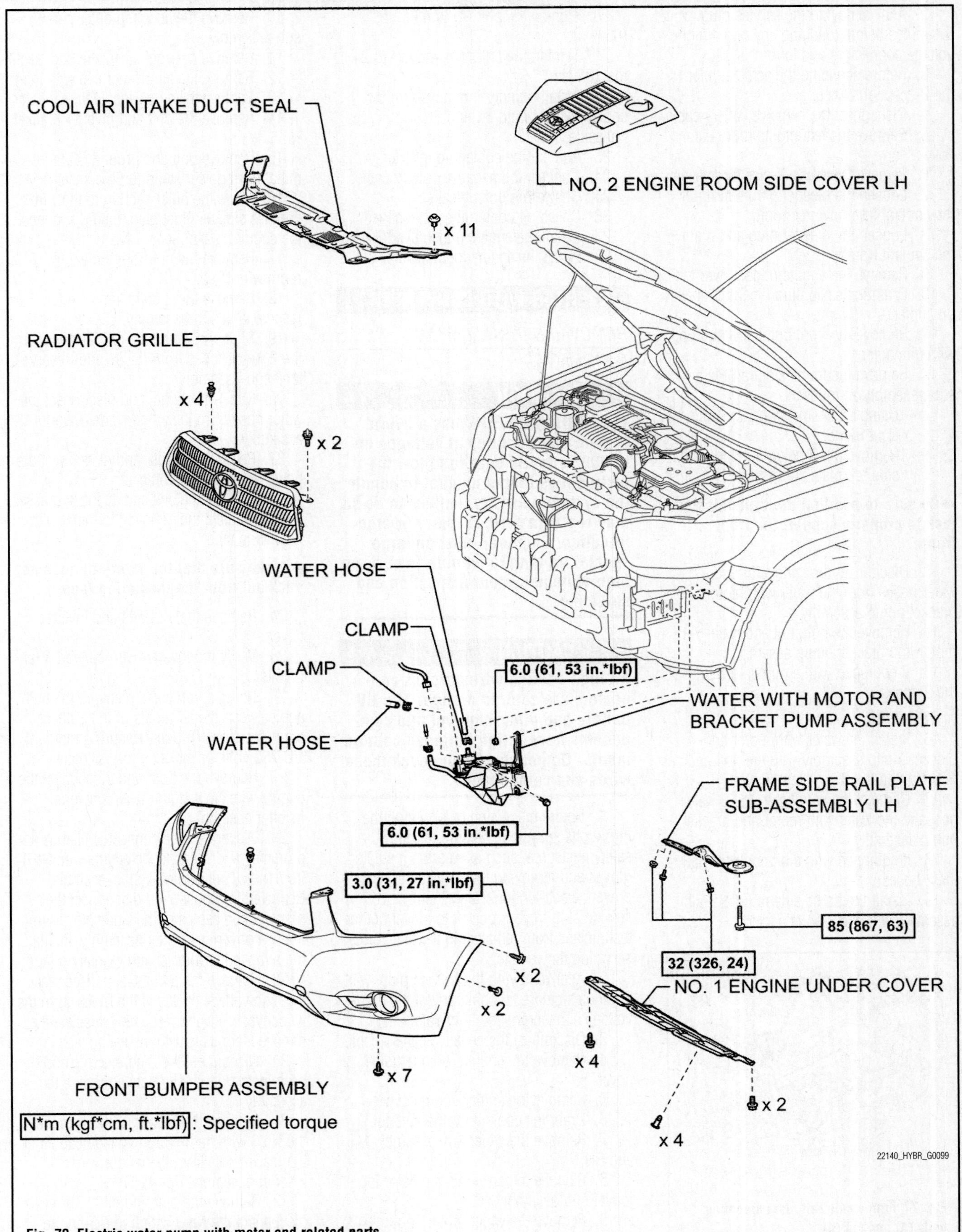

Fig. 73 Electric water pump with motor and related parts

2. After removing the service plug, wait 5 minutes before touching any of the high-voltage connectors and terminals.

3. Before servicing the vehicle, refer to the Precautions Section.

4. Disconnect the negative battery cable.

5. Remove the left engine room side cover.

6. Remove the transaxle side reserve tank.

7. Loosen the bleeder plug and drain the coolant from inverter cooler.

8. Loosen the bleeder plug and drain the coolant from inverter.

9. Remove the engine under cover.

10. Drain transaxle fluid if equipped with oil cooler.

11. Remove the front bumper if equipped with oil cooler.

12. Remove the frame side rail Plate sub-assembly as follows:

- Using a transmission jack, hold the front frame.
- Remove the 3 bolts, nut and frame side rail plate sub-assembly.

➡Be sure to position the transmission jack to properly support the front frame.

13. Disconnect the connector and 2 water hoses from the water with motor and bracket pump assembly.

14. Remove the bolt, nut and water with motor and bracket pump assembly.

15. If equipped with a oil cooler remove the 2 hoses.

To install:

16. Install the water with motor and bracket pump assembly with the bolt and nut. Tighten to 53 inch lbs. (6 Nm).

17. Connect the connector and 2 water hoses to the water with motor and bracket pump assembly.

18. If equipped with a oil cooler install the 2 hoses.

19. Install the frame side rail plate sub-assembly with the 3 bolts and nut.

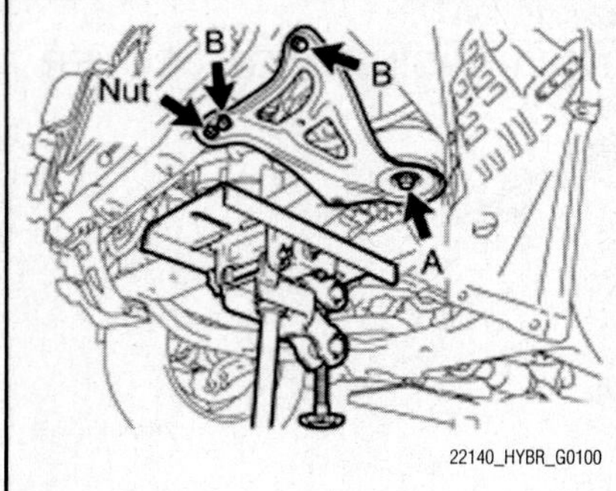

Fig. 74 Frame side rail plate mounting bolts (A) and nut (B)

20. Tighten the bolt (A) to 63 ft. lbs. (85 Nm).

21. Tighten the bolts (B) and nut to 24 ft. lbs. (32 Nm).

22. Install bumper if removed for oil cooler. Inspect fluid level for hybrid transaxle.

23. Add engine coolant to inverter.

24. Connect the negative battery cable.

25. Check for coolant leaks.

26. Check oil cooler lines if removed.

27. Install the engine undercover.

28. Install the engine room left side cover.

INVERTER WITH CONVERTER

REMOVAL & INSTALLATION

See Figure 75.

⁂ CAUTION

The HIGHLANDER HV has a hybrid system that operates at voltages up to 650 volts. Be sure to follow the instructions in this manual to handle the system correctly. Failure to do so may result in serious injury or electrocution. Engineer must undergo special training to be able to perform high-voltage system inspection and servicing.

⁂ CAUTION

All high-voltage wire harness connectors are colored orange. The HV battery and other high-voltage components have "High Voltage" caution labels. Do not carelessly touch these wires and components.

1. Before inspecting or servicing the high-voltage system, be sure to follow safety measures, such as wearing insulated gloves and removing the service plug to prevent electrocution. Carry the removed service plug in your pocket to prevent other technicians from reinstalling it while you are servicing the vehicle.

2. After removing the service plug, wait 5 minutes before touching any of the high-voltage connectors and terminals.

3. Disconnect the negative battery cable.

4. Remove the engine room left side cover.

5. Remove the engine room cover.

6. Drain the coolant for the inverter.

7. Remove the wiper arm assembly LH and RH.

8. Remove the cowl top ventilator top louver sub-assembly.

9. Remove the wiper motor and link assembly.

10. Remove the cowl top outer panel sub-assembly.

11. Remove the cool air intake duct seal.

12. Remove the air cleaner cap with inlet.

13. Remove the air cleaner with resonator.

14. Remove the bolt and inverter bracket No.5.

15. Remove bolt and ground cable terminal from power steering ECU assembly.

16. Move the outer section to the wire harness side as illustrated, then disconnect the circuit breaker sensor No.1.

17. Remove the nut from the engine room wire No.2.

18. Remove the 2 bolts and inverter reserve tank sub-assembly.

19. Slide the 2 clamps, and disconnect the 2 water hoses from the inverter reserve tank sub-assembly.

20. Slide the clamp, and disconnect the water hose from the w/ converter inverter assembly.

21. Remove the bolt, and disconnect the power steering ECU bracket.

22. Remove the bolt and interlock bracket.

23. Insulate the removed terminal with insulating tape.

➡Make sure that the terminal does not stick out from the insulating tape.

24. Remove the 12 bolts and inverter cover.

25. Verify the inverter with converter is 0 volts.

26. Using a voltmeter, measure the voltage between the terminals of the 2 phase connectors (N-P). Use measuring range of DC 750 volts or more on the voltmeter.

27. Remove the bolt, and disconnect the engine wire No.4 from the w/ converter inverter assembly.

28. Disconnect the connector, clamps and grommet, and separate the engine wire No.4 from the w/ converter inverter assembly. Remove the 5 bolts, and disconnect the high voltage cables of the Generator from the w/ converter inverter assembly. Insulate the removed terminals with insulating tape.

29. Remove the 5 bolts, and disconnect the high voltage cables of the Motor from the w/ converter inverter assembly. Insulate the removed terminals with insulating tape.

30. Disconnect the 2 connectors and grommets from the w/ converter inverter assembly.

31. Remove the 5 bolts, and disconnect the No.3 wire frame (high voltage cables of the rear motor) from the w/ converter inverter assembly. (for 4WD)

32. Remove the bolt and lift the lever to disconnect the No.3 wire frame from the w/ converter inverter assembly.

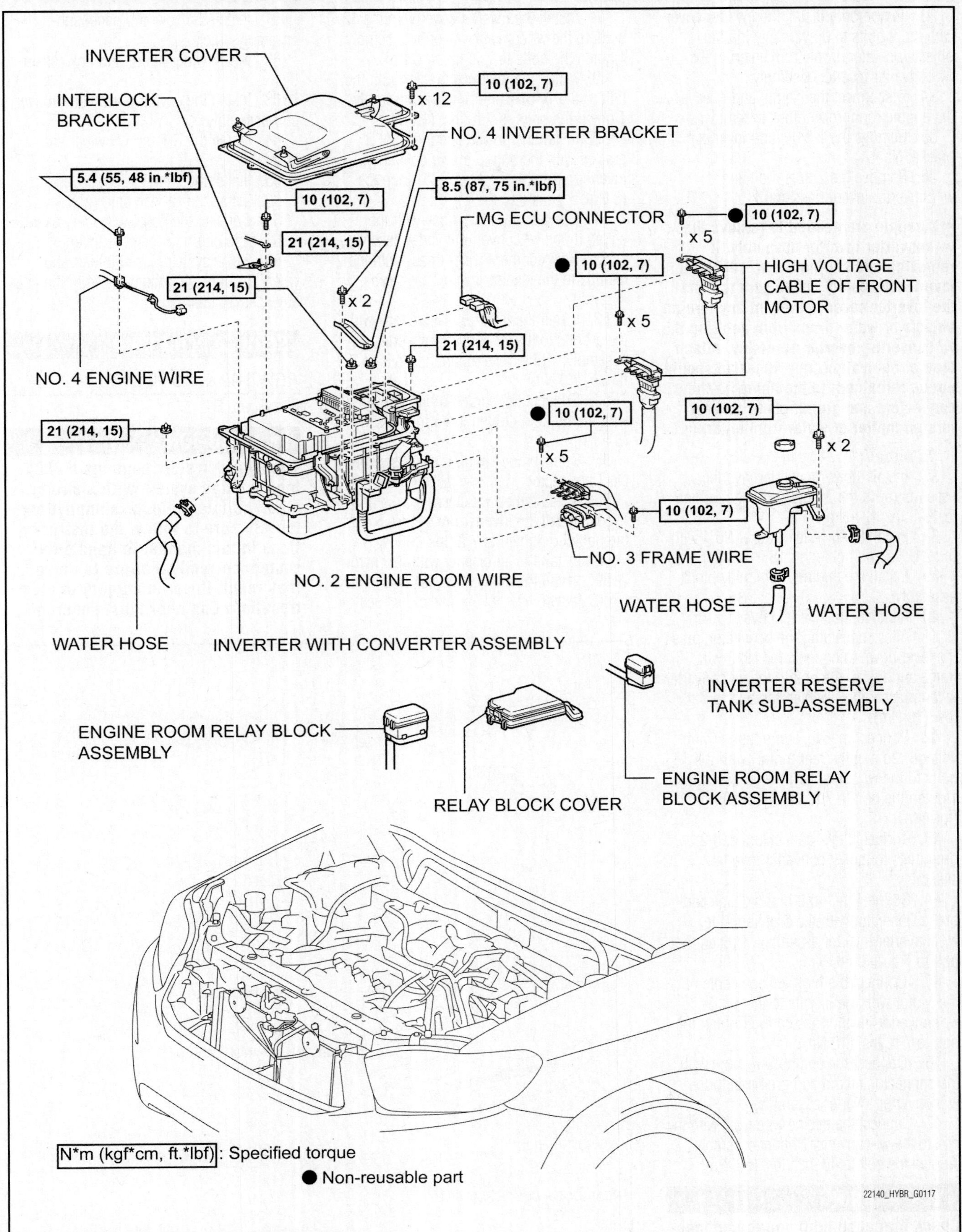

Fig. 75 Inverter converter and related components

33. Temporarily install the inverter cover with the 2 bolts to prevent any foreign objects or water drops from entering the w/ converter inverter assembly.

34. Disconnect the clamp, and release the engine room relay block assembly.

35. Remove the 2 bolts and inverter bracket No.4.

36. Remove the 2 nuts, bolt and w/ converter inverter assembly.

➡2 people are needed to remove the w/ converter inverter assembly. If removing and storing the w/ converter inverter assembly, make sure to install the inverter cover to prevent any foreign objects or water drops from entering the w/ converter inverter assembly. Attach tape or equivalent (any adhesive should not be remained) to the holes for the connectors and grommets to prevent foreign matter or water from entering.

To install:

37. Install the w/ converter inverter assembly with the 2 nuts and bolt. Tighten to 15 ft lbs. (21 Nm).

38. Install the inverter bracket No.4 with the 2 bolts and tighten to 15 ft lbs. (21 Nm).

39. Install the engine room relay block assembly.

40. Remove the inverter cover.

41. After connecting the connector, press the lever down to connect the No.3 wire frame and install the bolt to the w/ converter inverter assembly. Tighten the bolt to 7 ft. lbs. (10 Nm).

42. Connect the No.3 wire frame (high voltage cable of the rear motor) with the 5 bolts to the w/ converter inverter assembly. Tighten the bolt to 7 ft. lbs. (10 Nm). (for 4WD)

43. Connect the 2 connectors and 2 grommets to the w/ converter inverter assembly.

44. Connect the front high voltage cable of the Generator with the 5 bolts to the w/ converter inverter assembly. Tighten the bolt to 7 ft. lbs. (10 Nm).

45. Connect the high voltage cable of the Motor with the 5 bolts to the w/ converter inverter assembly. Tighten the bolt to 7 ft. lbs. (10 Nm).

46. Connect the engine wire No.4 with the connector, clamp and grommet to the w/ converter inverter assembly.

47. Connect the engine wire No.4 with the bolt to the w/ converter inverter assembly. Tighten the bolt to 48 inch lbs. (5.4 Nm).

⁂ WARNING

Check that each high voltage connector and terminal is firmly installed.

48. Install the inverter cover with the 12 bolts to the w/ converter inverter assembly. Tighten the bolts to 7 ft. lbs. (10 Nm).

49. Install the interlock bracket with the bolt to the w/ converter inverter assembly. Tighten the bolts to 7 ft. lbs. (10 Nm).

50. Install the power steering ECU bracket with the bolt to the w/ converter inverter assembly. Tighten to 71 inch lbs. (8 Nm).

51. Connect the water hose with the clamp to the w/ converter inverter assembly.

52. Connect the 2 water hoses with the 2 clamps to the inverter reserve tank sub-assembly.

53. Install the inverter reserve tank with the 2 bolts to the w/ converter inverter assembly. Tighten the mounting bolts to 7 ft. lbs. (10 Nm).

54. Connect the engine room wire No.2 with the nut and tighten to 75 inch lbs. (8.5 Nm).

55. Connect the circuit breaker sensor No.1 connector.

56. Install the power steering assembly.

57. Install the inverter bracket No.5 with the bolt and tighten to 7 ft. lbs. (10 Nm).

58. Install the air cleaner with resonator.

59. Install the air cleaner cap with inlet.

60. Install the cool air intake duct seal.

61. Install the cowl top panel sub-assembly outer.

62. Install wiper motor and link assembly.

63. Install the cowl top ventilator louver sub-assembly.

64. Install the RH and LH wiper arm assembly.

65. Install the engine room covers.

66. Install the service plug grip.

67. Connect the negative battery cable.

68. Add engine coolant to inverter.

69. Some systems need initialization when disconnecting the cable from the negative battery terminal.

SERVICE PLUG GRIP

REMOVAL & INSTALLATION

See Figure 76.

⁂ CAUTION

The hybrid system contains a 288V high-voltage system with a strong alkali solution of potassium hydroxide. Be sure to follow the instructions in this manual to handle the system correctly. Failure to do so may result in serious injury or electrocution. Engineer must undergo

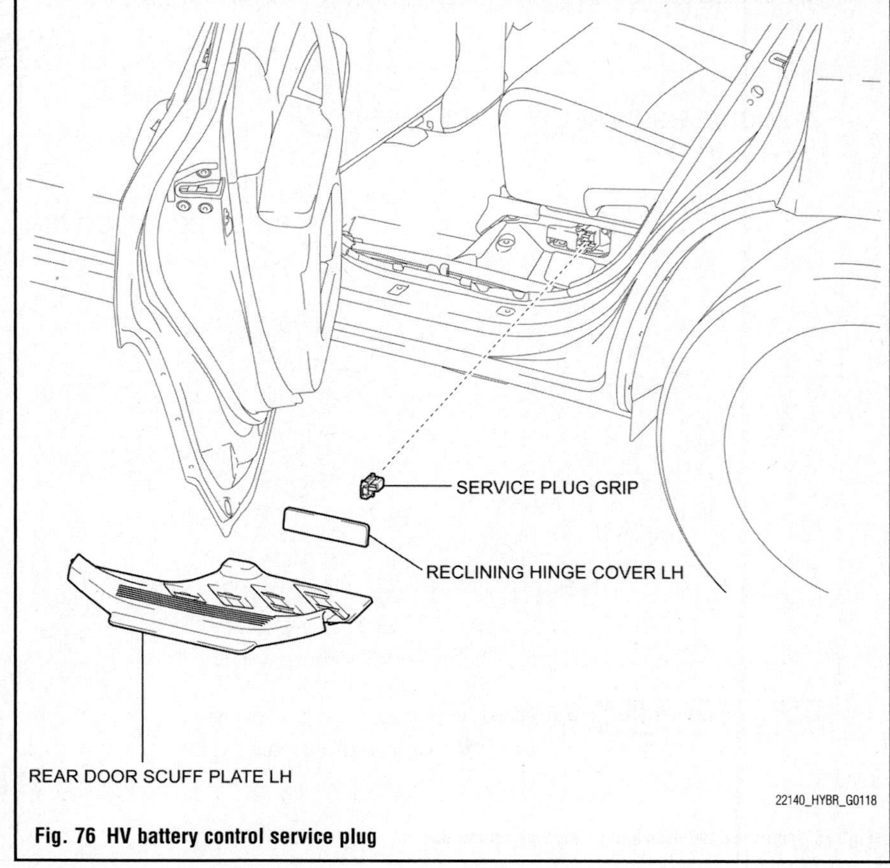

Fig. 76 HV battery control service plug

special training to be able to perform high-voltage system inspection and servicing.

⁂ CAUTION

All high-voltage wire harness connectors are colored orange. The HV battery and other high-voltage components have "High Voltage" caution labels. Do not carelessly touch these wires and components.

1. Check for Daces. Check for Daces and confirm that P0AA6 (High voltage insulation is unusual) is not output before doing removal or installation inside the battery. If the DTC is output, perform troubleshooting first.
2. Disconnect the negative battery cable.
3. Remove the 5 clips and door scuff plate LH.
4. Remove the 2 clips and reclining hinge cover.
5. Wear insulated glove and remove the service plug grip, after sliding up the lever of the service plug grip.

⁂ WARNING

After removing the service plug grip, do not operate the power switch as it may damage the hybrid vehicle control ECU.

⁂ CAUTION

Before connecting the service plug, check that no parts and tools remain and that the high voltage terminals and connectors are connected securely.

To install:

6. Wear insulated gloves, then insert the service plug.
7. Push down on the grip to lock.
8. Close the reclining hinge cover. Check that the 2 clips are securely connected to the battery carrier bracket (click sound).
9. Close the rear door scuff plate LH.
10. Connect the negative battery cable.
11. Some systems need initialization after reconnecting the cable to the negative battery terminal.

ENGINE ELECTRICAL — IGNITION SYSTEM

FIRING ORDER

See Figure 77.

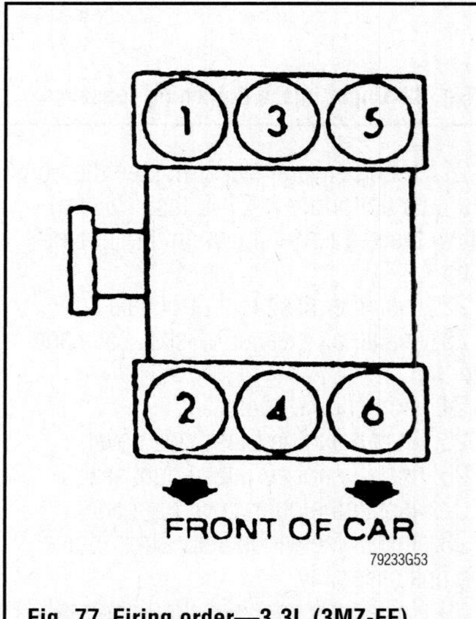

Fig. 77 Firing order—3.3L (3MZ-FE) Engine

IGNITION COIL

REMOVAL & INSTALLATION

See Figures 78 and 79.

1. Disconnect the negative battery cable.
2. Remove the RH and LH wiper arm assembly.
3. Remove the cowl and seal.
4. Remove the wiper motor and link assembly.
5. Remove the outer cowl top panel.
6. Remove the cool air intake duct seal.
7. Remove the LH engine side cover.
8. Drain the engine coolant.
9. Remove air cleaner cap and case assembly.
10. Remove the air cleaner bracket.
11. Remove the emission control valve set.
12. Remove the upper intake plenum.
13. Disconnect the 6 ignition coil connectors.
14. Remove the 6 bolts and 6 ignition coils.
15. Remove the 6 spark plugs.

To install:

16. Install the spark plug and tighten to 13 ft. lbs. (18 Nm).
17. Install the ignition coils and tighten retaining bolts to 71 inch lbs. (8 Nm).
18. Connect the ignition coil connectors.
19. Install intake plenum with a new gasket
20. Using a socket hexagon wrench 8 mm, install the upper intake plenum with the 4 bolts and 2 nuts.
21. Using several steps, tighten the bolts and nuts uniformly to 21 ft. lbs. (28 Nm) follow the sequence shown in the illustration.
22. Install emission control vale set.
23. Install air cleaner bracket, case and cap.
24. Add engine coolant.
25. Install engine room side cover.
26. Install cool air intake duct seal.
27. Install the outer cowl top panel.
28. Install the windshield wiper motor and link assembly.
29. Install the top cowl and hood seal.
30. Install the LH and RH wiper arm assembly.
31. Connect the negative battery cable.
32. Check for coolant leaks.

➡When disconnecting the cable, some systems need to be initialized after the cable is reconnected.

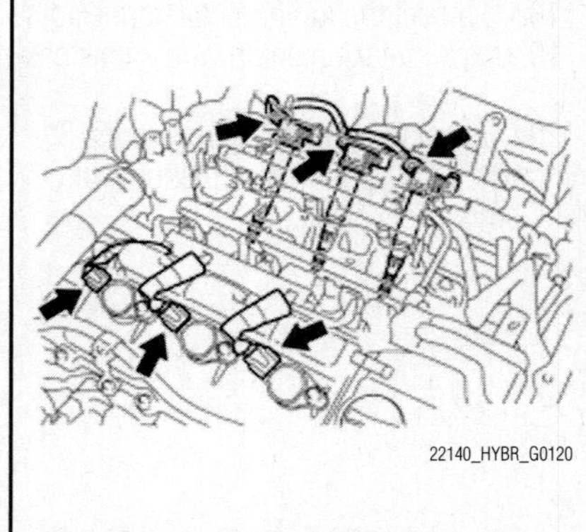

Fig. 78 Ignition coil and connector view

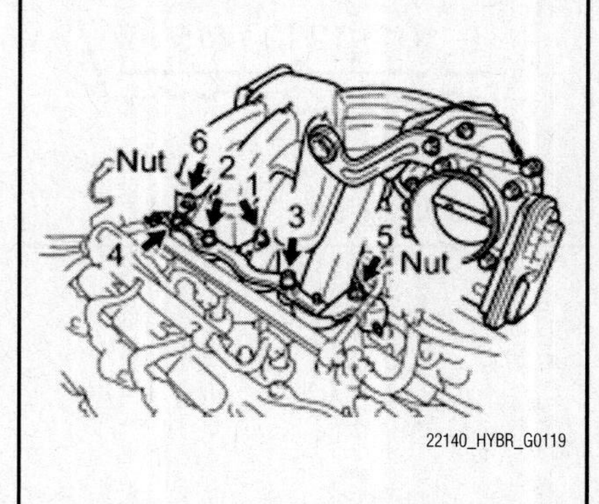

Fig. 79 Upper intake tightening sequence

IGNITION TIMING

INSPECTION

See Figure 80.

1. Warm up the engine.
2. Using SST: 09843-18040, connect terminals 13 (TC) and 4 (CG) of the DLC3.

➡Confirm the terminal numbers before connecting them. Connecting the wrong terminals can damage the engine.

➡Turn off all electrical systems before connecting the terminals.

➡Perform this inspection after the cooling fan motor is turned off.

3. Remove the V-bank cover sub-assembly.
4. Pull out the red lead wire harness.
5. Connect the tester terminal of the timing light to the red lead wire as shown in the illustration.

➡Use a timing light which detects the No. 1 cylinder ignition signal.

6. Check the ignition timing at idle. Standard ignition timing: 8 to 12° BTDC at idle.

➡When checking the ignition timing, the transmission should be in the neutral position.

➡Run the engine at 1000 to 1300 RPM for 5 seconds, and then check that the engine RPM returns to idle speed.

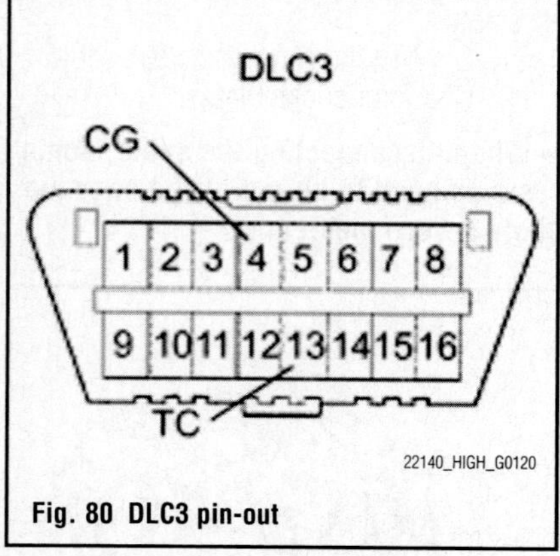

Fig. 80 DLC3 pin-out

7. Disconnect terminals 13 (TC) and 4 (CG) of the DLC3.
8. Check the ignition timing at idle. Standard ignition timing: 7 to 24° BTDC at idle.
9. Confirm that the ignition timing advances immediately when the engine RPM is increased.
10. Remove the timing light from the engine.

ADJUSTMENT

All engines are equipped with a Distributorless Ignition System (DIS). No timing adjustment is possible.

SPARK PLUGS & WIRES

REMOVAL & INSTALLATION

See Figures 81 and 82.

1. Disconnect the negative battery cable.
2. Remove the RH and LH wiper arm assembly.
3. Remove the cowl and seal.
4. Remove the wiper motor and link assembly.
5. Remove the outer cowl top panel.
6. Remove the cool air intake duct seal.
7. Remove the LH engine side cover.
8. Drain the engine coolant.
9. Remove air cleaner cap and case assembly.
10. Remove the air cleaner bracket.
11. Remove the emission control valve set.
12. Remove the upper intake plenum.
13. Disconnect the 6 ignition coil connectors.
14. Remove the 6 bolts and 6 ignition coils.
15. Remove the 6 spark plugs.

To install:

16. Install the spark plug and tighten to 13 ft. lbs. (18 Nm).
17. Install the ignition coils and tighten retaining bolts to 71 inch lbs. (8 Nm).
18. Connect the ignition coil connectors.
19. Install intake plenum with a new gasket
20. Using a socket hexagon wrench 8 mm, install the upper intake plenum with the 4 bolts and 2 nuts.

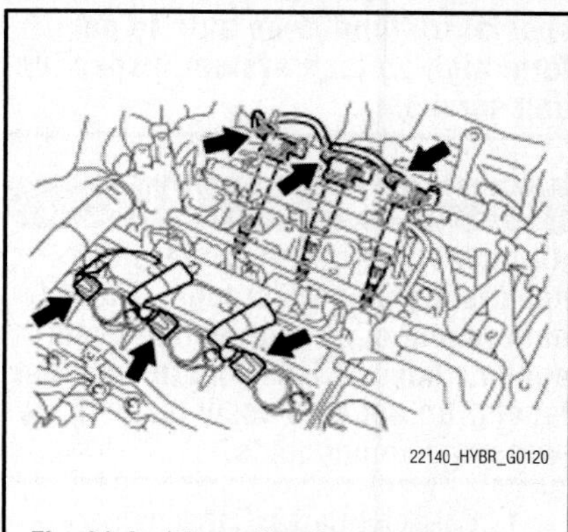

Fig. 81 Ignition coil and connector view

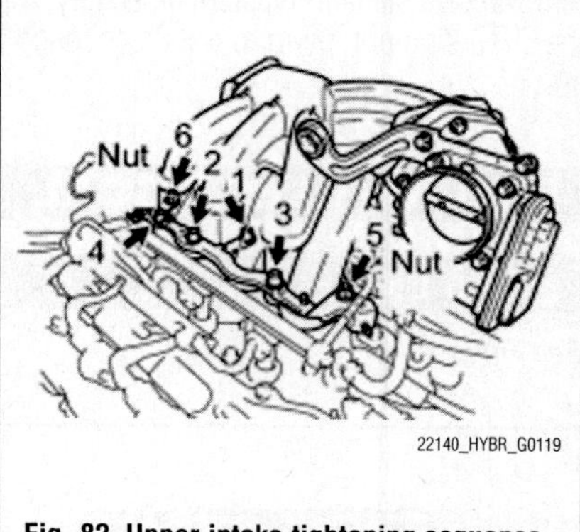

Fig. 82 Upper intake tightening sequence

21. Using several steps, tighten the bolts and nuts uniformly to 21 ft. lbs. (28 Nm) follow the sequence shown in the illustration.
22. Install emission control vale set.
23. Install air cleaner bracket, case and cap.
24. Add engine coolant.
25. Install engine room side cover.
26. Install cool air intake duct seal.
27. Install the outer cowl top panel.
28. Install the windshield wiper motor and link assembly.
29. Install the top cowl and hood seal.
30. Install the LH and RH wiper arm assembly.
31. Connect the negative battery cable.
32. Check for coolant leaks.

➡When disconnecting the cable, some systems need to be initialized after the cable is reconnected.

ENGINE ELECTRICAL

STARTER

The 3.3L engine has a DC electric converter and therefore does not use a standard starter.

ENGINE MECHANICAL

CAMSHAFT AND VALVE LIFTERS

INSPECTION

See Figures 83 through 86.

1. To inspect the camshaft, place the camshaft on V-blocks.
   a. Using a dial indicator, measure the circle runout at the center journal.
2. 3.3L Maximum runout 0.0024 inches (0.06 mm).
   a. If the runout is greater than the maximum, replace the camshaft.

➡Check the oil clearance after replacing the camshaft.

3. Using a micrometer, measure the cam lobe height.
4. Using a micrometer, measure the journal diameter.
   a. 3.3L standard journal diameter should be 1.0614 to 1.0620 in (26.959 to 26975 mm)
   b. If the journal diameter is not as specified, check the oil clearance.
5. To inspect the lifters, place the valve lash adjuster assembly into a container filled with engine oil.
   a. Insert Service Tool tip into the valve lash adjuster assembly's plunger and use the tip to press down on the check ball inside the plunger.
   b. Squeeze Service Tool and valve lash adjuster assembly together to move the plunger up and down 5 to 6 times.
   c. Check the movement of the plunger and bleed the air. If plunger moves up and down freely, valve lash adjuster is OK.

➡When bleeding air from the high-pressure chamber, make sure that the tip of SST is actually pressing the check ball as shown in the illustration. If the check ball is not pressed, air will not bleed.

   d. After bleeding the air, remove Service Tool. Then try to quickly and firmly press the plunger with by hand. If plunger is difficult to move, valve lash adjuster is OK.
   e. If the result is not as specified, replace the valve lash adjuster assembly.

REMOVAL & INSTALLATION

See Figures 87 through 102.

WARNING

The thrust clearance on both the intake and exhaust camshafts is very small; the camshafts must be kept level during removal. If the camshafts are removed without being kept level, the camshaft may be caught in the cylinder head, causing the head to break or the camshaft to seize.

1. Before servicing the vehicle, refer to the Precautions Section.

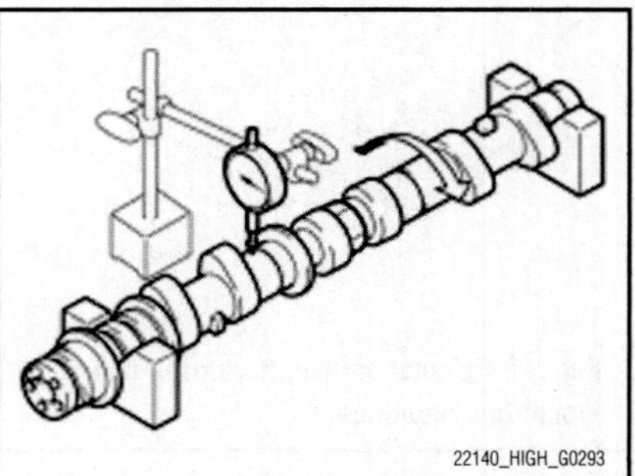

22140_HIGH_G0293

Fig. 83 Measuring Camshaft Runout

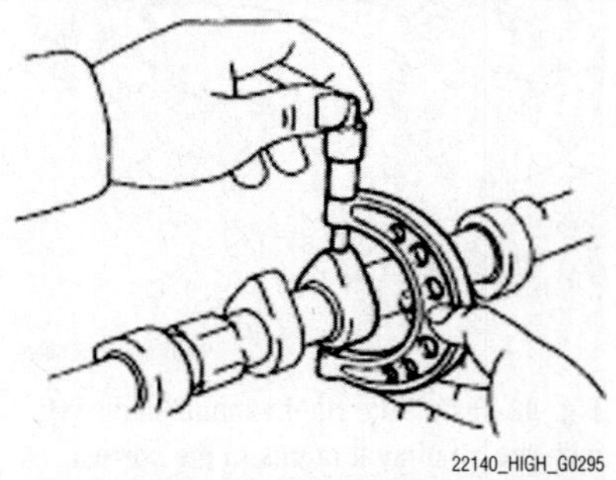

22140_HIGH_G0295

Fig. 84 Measuring Camshaft Lobe Height

Item	Specified Condition
Intake	43.132 to 43.232 mm (1.6981 to 1.7020 in.)
Exhaust	43.010 to 43.110 mm (1.6933 to 1.6972 in.)

22140_HIGH_G0317

Fig. 85 Standard Lobe Height—3.3L engine

Item	Specified Condition
Intake	42.98 mm (1.6921 in.)
Exhaust	42.86 mm (1.6874 in.)

22140_HIGH_G0318

Fig. 86 Minimum Lobe Height—3.3L engine

2. Relieve the fuel system pressure.
3. Drain the engine oil.
4. Drain the coolant from the engine radiator and hybrid transaxle.
5. Remove or disconnect the following:
- Negative battery cable
- Engine cover
- Right-hand front wheel
- Engine splash shields
- Right-hand front fender apron seal
- Wiper and blade assembly
- Top cowl ventilator louver assembly
- Wiper motor and link assembly
- Battery and battery tray
- Air intake assembly
- Emission control valve hoses
- Air intake surge tank
- Radiator intake hose
- Brake master cylinder reservoir and bracket
- Air cleaner bracket
- Engine moving control rod
- Right-hand engine No.2 mounting stay
- Ignition coil
- Valve covers
- Crankshaft pulley
- Timing belt No.1 and No.2 covers
- Right-hand engine mounting bracket
- No.2 timing belt guide
- Timing belt
- Timing belt idler

6. Using Special Tool 09960-10010, remove the camshaft timing pulleys.

➡Keep all valvetrain components in order for reassembly.

7. Disconnect the engine wiring harness clamps from the No.3 timing belt cover and remove the cover.
8. Remove the left camshafts as follows:
a. Align the timing marks (2-dot mark) of the camshaft drive and the driven gears by turning the camshaft with a wrench.
b. Secure the exhaust camshaft subgear to the main gear with a service bolt. A bolt 0.63–0.79 in. (16–20mm) long with a 6mm thread diameter and a 1mm pitch is recommended. Tighten bolt to 48 inch lbs. (5.4 Nm).

➡When removing the camshaft, make certain that the torsional spring force of the sub-gear has been eliminated by installing the service bolt.

c. Using several steps, loosen and remove the 10 bearing cap bolts uniformly in the sequence shown.
d. Remove the 5 bearing caps and the exhaust camshaft.
e. Using several steps, loosen and remove the 10 bearing cap bolts uniformly in the sequence shown.
f. Remove the 5 bearing caps and the intake camshaft.
g. Remove the oil seal from the intake camshaft.

9. Repeat the same process to remove the right-side camshafts, beginning with the intake camshaft.
10. Remove the valve lifter shims and hydraulic lifters. Identify each lifter and shim as it is removed so it can be reinstalled in the same position. If the lifters are to be reused, store them upside down in a sealed container.

To install:

11. Install the valve lifters into their original positions and install the shims. Check valve clearance and replace the shims as necessary.
12. When reinstalling, remember that the camshafts must be handled carefully and kept straight and level to avoid damage.
13. Install the right camshafts, as follows:
a. Apply new engine oil to the thrust portion and journal of the camshaft.
b. Position the exhaust camshaft on the head so that the alignment

Fig. 87 Removing the right-hand camshaft timing pulley, left-hand similar.

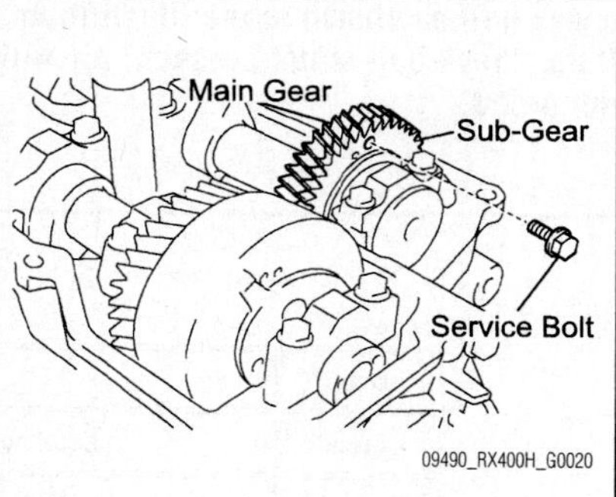

Fig. 89 Install a service bolt to secure the camshaft gears.

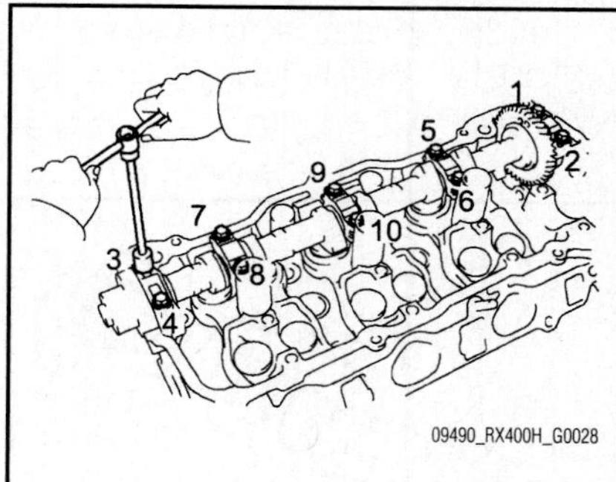

Fig. 91 Exhaust camshaft bearing cap loosening sequence

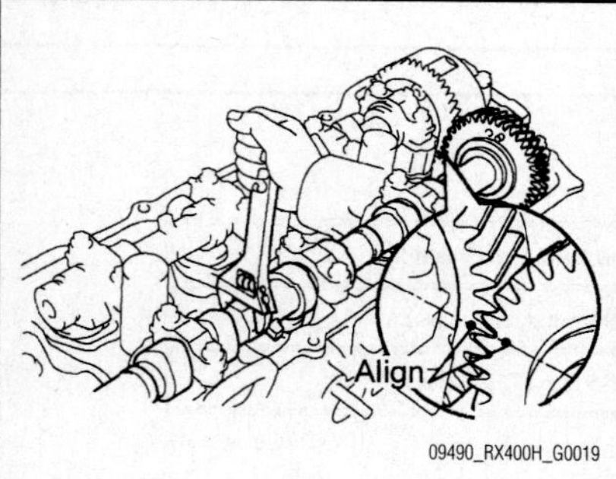

Fig. 88 Align the timing marks of the camshaft gears.

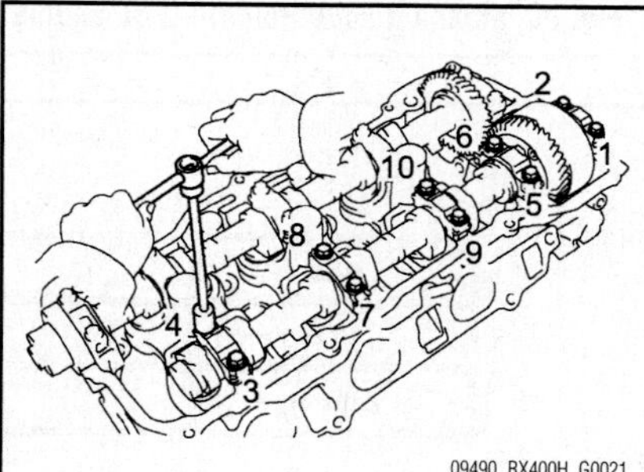

Fig. 90 Intake camshaft bearing cap loosening sequence

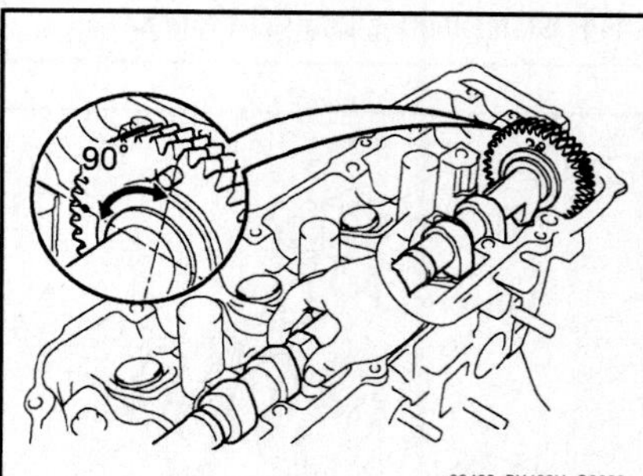

Fig. 92 Install the right exhaust camshaft with the alignment marks in the correct position.

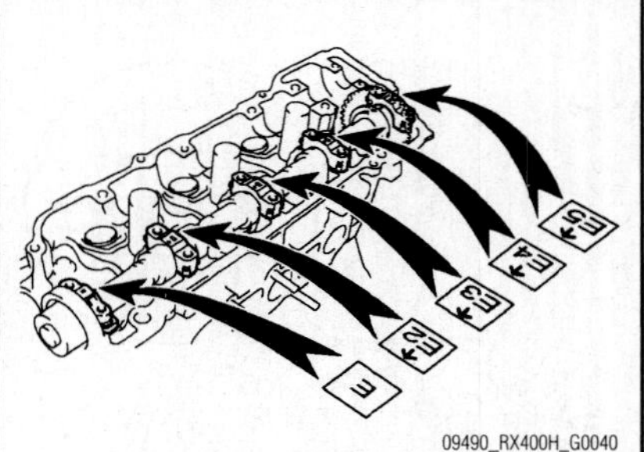

Fig. 93 Right exhaust camshaft bearing caps must be placed in their proper locations.

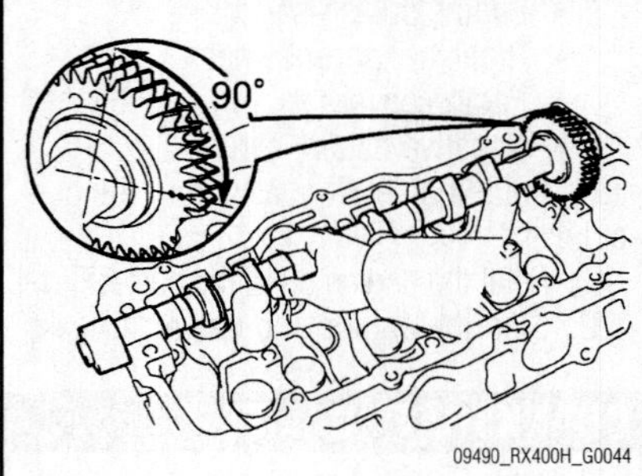

Fig. 97 Install the left exhaust camshaft with the alignment mark in the correct position.

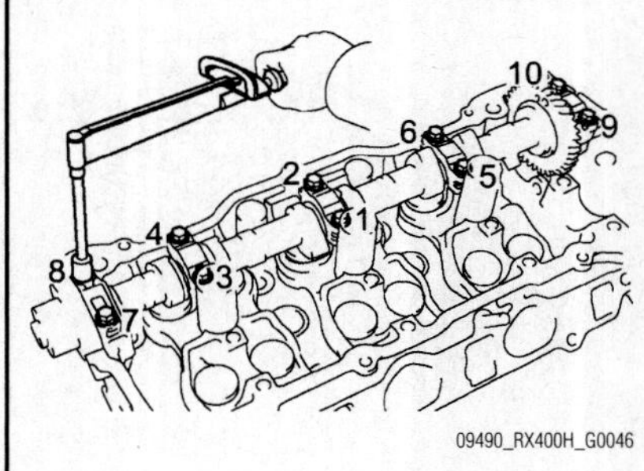

Fig. 99 Left exhaust camshaft bearing cap torque sequence

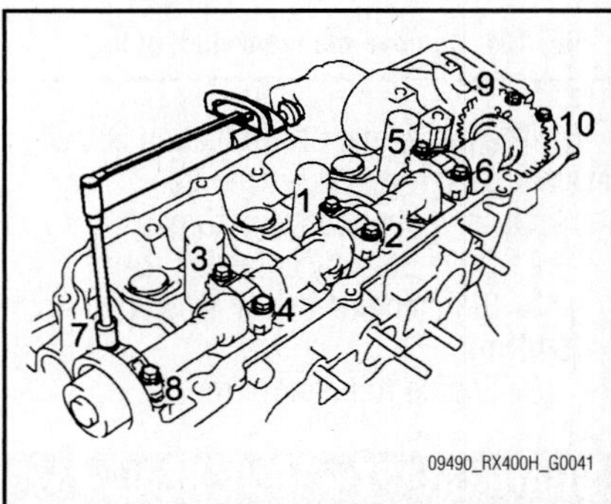

Fig. 94 Right exhaust camshaft bearing cap torque sequence

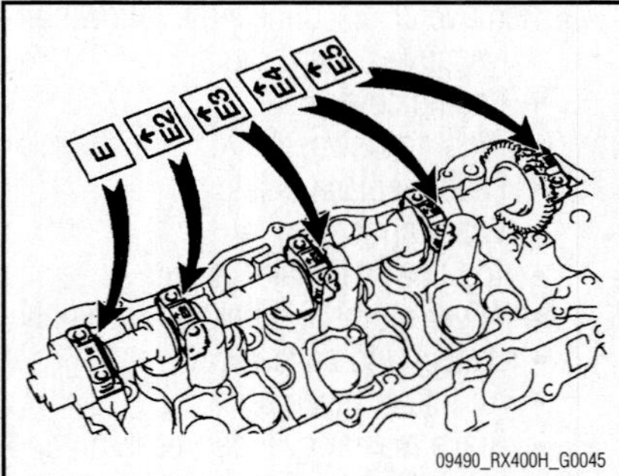

Fig. 98 Left exhaust camshaft bearing caps must be placed in their proper locations

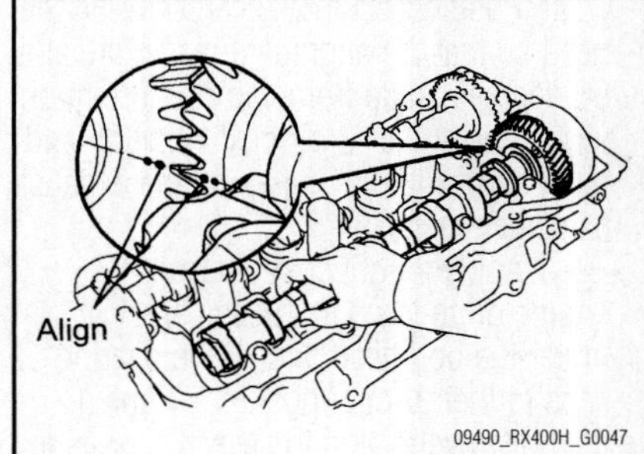

Fig. 100 Install the left intake camshaft with the alignment mark in the correct position.

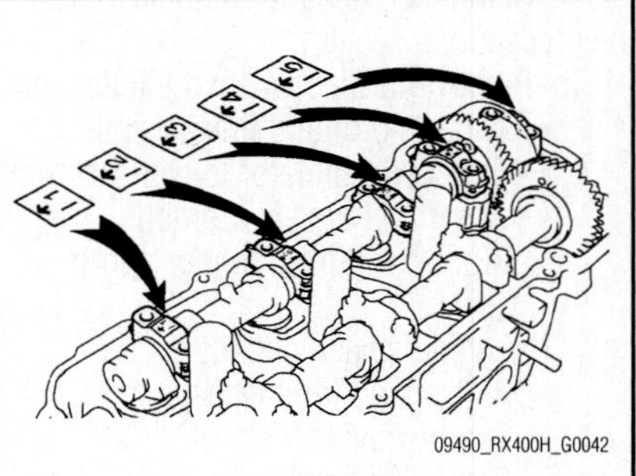

Fig. 95 Right intake camshaft bearing caps must be placed in their proper locations

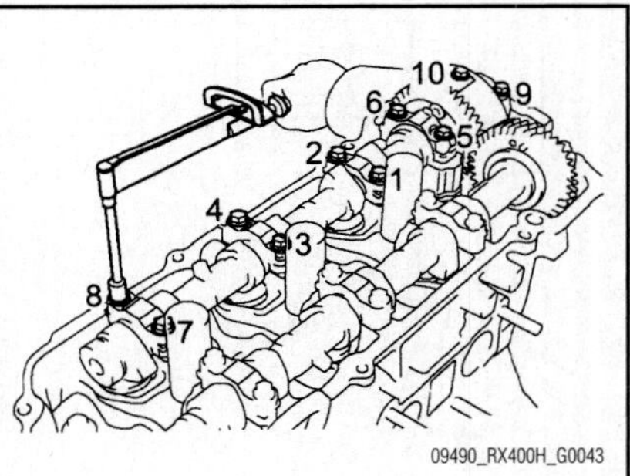

Fig. 96 Right intake camshaft bearing cap bolt tightening sequence

marks are at a 90 degrees angle from vertical.

c. Apply multi-purpose grease to the lip of a new oil seal.

d. Install the oil seal to the camshaft.

e. Apply sealant to the No. 1 bearing cap.

f. Apply a light coat of clean engine oil to the bolt threads and under the bolt head. Install the bearing caps to their proper position. Tighten the bolts evenly and in several passes to 12 ft. lbs. (16 Nm) in the proper sequence.

g. Position the intake camshaft on the head so that the alignment marks are at a 90 degrees angle from vertical. The mark should be at the "9 o'clock" position and must align with the marks on the other gear.

h. Apply a light coat of clean engine oil to the bolt threads and under the bolt head. Install the bearing caps to their proper position. Tighten the bolts evenly and in several passes to 12 ft. lbs. (16 Nm) in the proper sequence.

i. Remove the service bolt.

14. Install the left camshafts, as follows:

a. Apply new engine oil to the thrust portion and journal of the camshaft

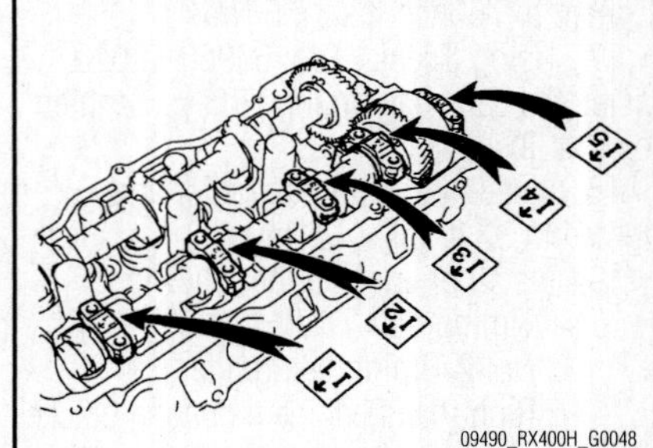

Fig. 101 Left exhaust camshaft bearing caps must be placed in their proper locations.

b. Position the exhaust camshaft on the head so that the alignment mark is at a 90 degrees angle from vertical. The mark should be at the "9 o'clock" position.

c. Apply multi-purpose grease to the oil seal lip and install the new oil seal to the camshaft.

d. Apply sealant to the No. 1 bearing cap.

e. Apply a light coat of clean engine oil to the bolt threads and under the bolt head. Install the bearing caps to their proper position. Tighten the bolts evenly and in several passes to 12 ft. lbs. (16 Nm) in the proper sequence.

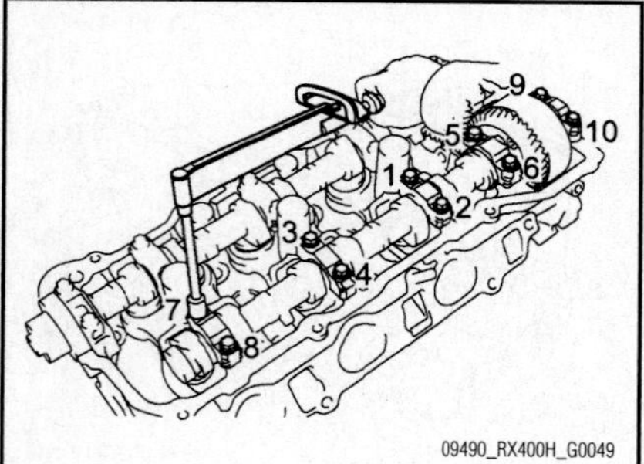

Fig. 102 Left exhaust bearing cap torque sequence

f. Position the intake camshaft on the head so that the alignment marks are at a 90 degrees angle from vertical. The mark should be at the "3 o'clock" position and must align with the marks on the exhaust camshaft gear.

g. Apply a light coat of clean engine oil to the bolt threads and under the bolt head. Install the bearing caps to their proper position. Tighten the bolts evenly and in several passes to 12 ft. lbs. (16 Nm) in the proper sequence.

h. Remove the service bolt.

15. Install or connect the following:

16. Install the timing belt cover No. 3. Tighten to 76 inch lbs. (8.5 Nm)

17. Using Special Tool 09960-10010, install the camshaft timing pulleys. Tighten to 92 ft. lbs. (125 Nm).

18. Install the timing belt idler and tighten to 32 ft. lbs. (43 Nm).

19. Install or connect the following:

- Timing belt
- No. 2 Timing belt guide
- Right-hand engine mounting bracket
- Timing belt covers Nos. 1 and 2
- Crankshaft pulley. Tighten to 162 ft. lbs. (220 Nm).
- Right-hand engine mounting stay No. 2
- Engine moving control rod
- Air cleaner bracket
- Brake master cylinder reservoir and bracket
- Valve covers
- Ignition coil
- Radiator intake hose
- Air intake surge tank
- Emission control valve hoses
- Air intake assembly
- Battery and battery tray
- Wiper motor and link assembly
- Top cowl ventilator louver assembly
- Wiper and blade assembly
- Right-hand front fender apron seal
- Engine splash shields
- Right-hand front wheel
- Engine covers
- Negative battery cable

20. Refill the cooling system to the correct level.

21. Refill the engine oil to the correct level.

22. Start the engine and check for leaks.

CRANKSHAFT DAMPER

REMOVAL & INSTALLATION

See Figures 103 and 104.

1. Before servicing the vehicle, refer to the precautions section.

2. Remove or disconnect the following:

- Negative battery cable
- Engine covers
- Right front wheel
- Fender splash shields
- Wiper arms
- Top cowl ventilator louver
- Wiper motor and linkage assembly
- Battery and battery tray
- Air intake assembly
- Brake master cylinder reservoir and bracket
- Air cleaner bracket
- Engine moving control rod
- Right-hand engine mounting stay No. 2

3. Use Special Tool 09213-54015 to hold the crankshaft pulley in order to loosen the pulley bolt.

4. Use Special Tool 09950-50013 to remove the crankshaft pulley.

To install:

5. Install the crankshaft pulley, as follows

- Align the keyway of the pulley with the key located on the crankshaft and slide the pulley into place
- Crankshaft pulley. Tighten the bolt to 162 ft. lbs. (219 Nm).

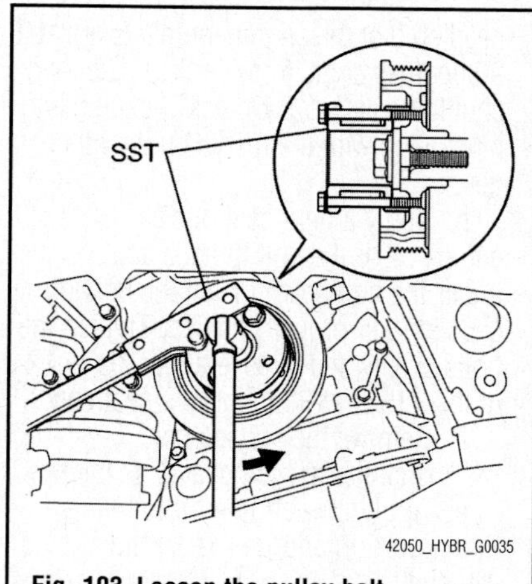

Fig. 103 Loosen the pulley bolt

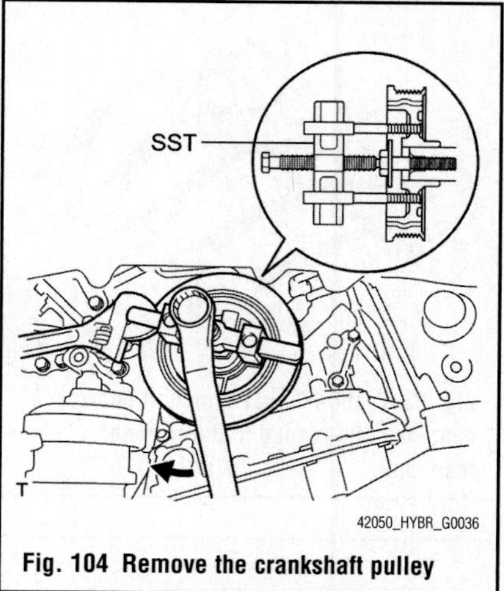

Fig. 104 Remove the crankshaft pulley

6. The remainder of installation is the reverse of the removal procedure.

7. Observe the following torques:

a. Drive belts by adjusting them

b. Right engine mount stay: 47 ft. lbs. (64 Nm).

c. Engine roll control rod.

CRANKSHAFT FRONT SEAL

REMOVAL & INSTALLATION

See Figures 105 and 106.

1. Before servicing the vehicle, refer to the precautions section.

2. Remove or disconnect the following:

- Wiper motor and link assembly
- Top cowl ventilator louver assembly
- Wiper and blade assembly
- Right-hand front fender apron seal
- Engine splash shields
- Right-hand front wheel
- Engine covers
- Battery and battery tray
- Air cleaner cap and case assembly

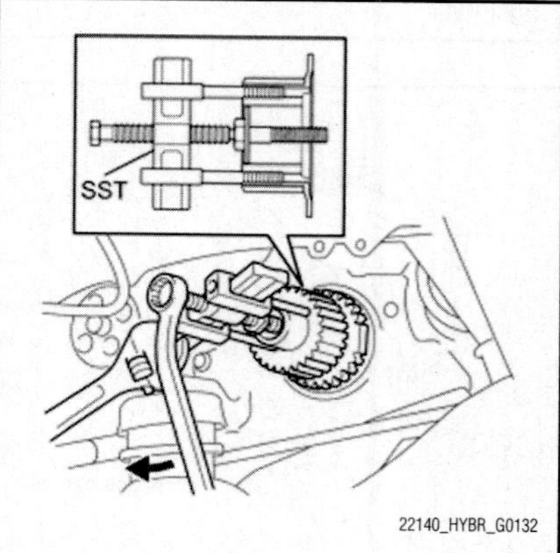

Fig. 105 Remove the timing crankshaft pulley

- Master cylinder Sub-assembly
- Air cleaner bracket
- Engine moving control rod
- RH engine mounting stay

3. Use Special Tool 09213-54015 to hold the crankshaft pulley in order to loosen the pulley bolt.
4. Use Special Tool 09950-50013 to remove the crankshaft pulley.
5. Remove the timing belt covers No.1 and No.2.
6. Remove the RH engine mounting bracket.
7. Remove the timing belt as follows:
   - Set No. 1 cylinder to TDC/compression.
   - Temporarily install the crankshaft pulley bolt and washer to the crankshaft.
   - Turn the crankshaft clockwise, and align the timing mark of the crankshaft timing pulley with the oil pump body.
   - Check that the timing marks of the camshaft timing pulleys and No. 3 timing belt cover are aligned.
   - If not, turn the crankshaft 1 revolution (360°).
8. Remove the bolt and the timing belt plate.
9. Install the pulley bolt to the crankshaft.
10. Using SST, remove the crankshaft timing pulley.
11. Using a knife, cut off the oil seal lip.
12. Using a screwdriver with its tip taped, pry out the oil seal.

To install:

13. Apply MP grease to a new oil seal lip.
14. Using SST and a hammer, tap in a new oil seal until its surface is flush with the oil pump edge.

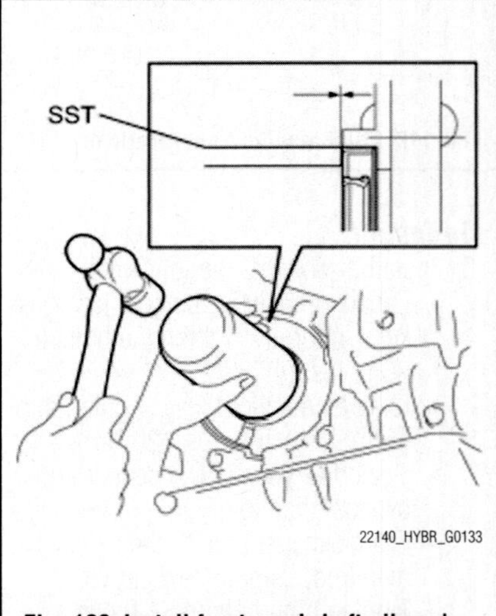

Fig. 106 Install front crankshaft oil seal

15. Reverse the removal procedure for installation and note the following:
   - Install new cover gaskets
   - Tighten timing belt covers to 74 inch lbs. (8 Nm).
   - Crankshaft pulley. Tighten the bolt to 162 ft. lbs. (219 Nm).
16. Start the vehicle and check for any leaks.
17. Recheck the ignition timing.

CYLINDER HEAD

REMOVAL & INSTALLATION

See Figures 107 through 111.

1. Before servicing the vehicle, refer to the Precautions Section.

CAUTION

The HIGHLANDER HV has a hybrid system that operates at voltages up to 650 volts. Be sure to follow the instructions in this manual to handle the system correctly. Failure to do so may result in serious injury or electrocution. Engineer must undergo special training to be able to perform high-voltage system inspection and servicing.

CAUTION

All high-voltage wire harness connectors are colored orange. The HV battery and other high-voltage components have "High Voltage" caution labels. Do not carelessly touch these wires and components.

2. Before inspecting or servicing the high-voltage system, be sure to follow safety measures, such as wearing insulated gloves and removing the service plug to prevent electrocution. Carry the removed service plug in your pocket to prevent other technicians from reinstalling it while you are servicing the vehicle.
3. After removing the service plug, wait 5 minutes before touching any of the high-voltage connectors and terminals.

➡Wear insulating gloves and protective glasses.

4. Relieve the fuel system pressure.
5. Drain the engine oil.
6. Drain the coolant from the engine radiator and hybrid transaxle.
7. Remove the service plug grip, found underneath the Battery Service cover on the rear seat. Wait 5 minutes to discharge the high voltage capacitor.
8. Remove or disconnect the following:
   - Negative battery cable
   - Engine cover
   - Right-hand front wheel
   - Engine splash shields
   - Right-hand front fender apron seal
   - Wiper and blade assembly
   - Top cowl ventilator louver assembly
   - Wiper motor and link assembly
   - Battery and battery tray
   - Air intake assembly
   - Converter with Inverter assembly
   - Emission control valve set
   - Intake air surge tank
   - Fuel supply hose
   - Heater inlet hose
   - Intake manifold
   - Radiator hoses
   - Water outlet from the cylinder heads
   - Brake master cylinder reservoir
   - Air cleaner bracket
   - Engine moving control rod
   - Right-hand No.2 engine mounting stay
   - Crankshaft pulley
   - Timing belt No.1 and No.2 covers
   - Right-hand engine mounting bracket
   - Timing belt guide No.2
   - Timing belt
   - Timing belt No.2 idler
   - Camshaft timing pulley
   - Timing belt No.3 cover
   - Front exhaust pipe assembly
   - Exhaust manifold heat insulator
   - Exhaust manifold stay
   - Right-hand exhaust manifold and gasket
   - Ignition coil
   - Right-hand cylinder head cover
   - Camshaft
   - VVT sensor connector
   - Camshaft timing oil control valve connecter

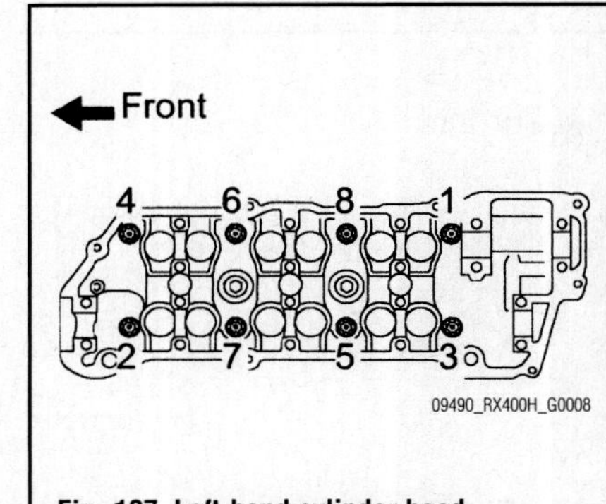

Fig. 107 Left-hand cylinder head loosening sequence

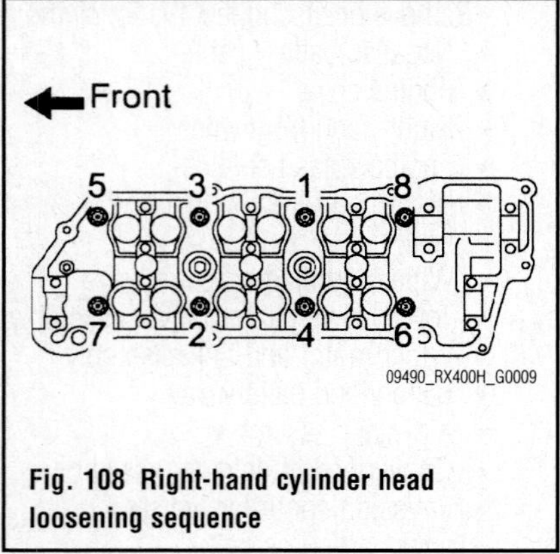

Fig. 108 Right-hand cylinder head loosening sequence

9. Loosen the right-hand cylinder head bolts in several steps in the sequence shown.
10. Remove the cylinder head bolts and plate washers.
11. Remove the right-hand cylinder head and gasket
12. Remove the manifold converter No.3 insulator.
13. Remove the exhaust manifold No.2 heat insulator.
14. Separate the cooling fan ECU and hang securely with mechanic's wire.
15. Remove or disconnect the following:

- Left-hand exhaust manifold
- Oil level gauge guide
- Water inlet pipe
- Left-hand cylinder head cover
- Camshaft
- Loosen the Right-hand cylinder head bolts in several steps in the sequence shown.

16. Remove the cylinder head bolts and plate washers.
17. Remove the Left-hand cylinder head and gasket.

To install:

18. Install the Left-hand cylinder head with a new gasket. Tighten the cylinder head bolts as follows:

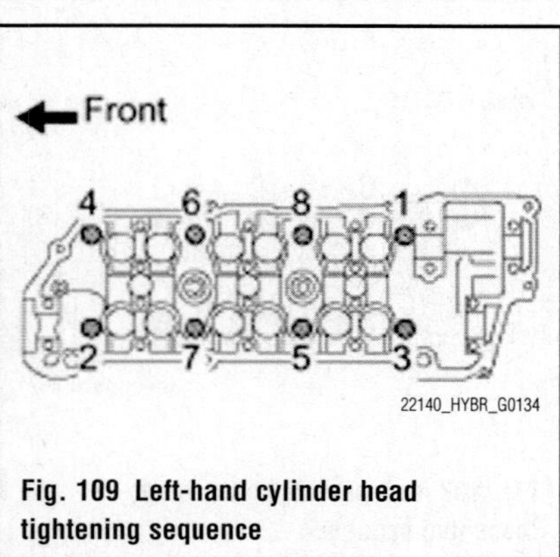

Fig. 109 Left-hand cylinder head tightening sequence

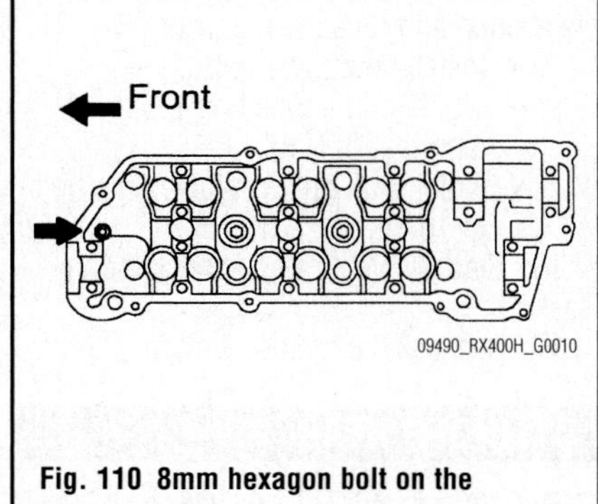

Fig. 110 8mm hexagon bolt on the cylinder head

- Step 1: Tighten the 8 cylinder head bolts to 40 ft. lbs. (54 Nm)
- Step 2: Tighten each bolt 90°
- Step 3: Tighten each bolt an additional 90°
- Step 4: Tighten the single 8mm hexagon bolt to 14 ft. lbs. (19 Nm)

19. Install the or connect the following:

- Camshaft
- Left-hand cylinder head cover. Tighten to 71 inch lbs. (8 Nm).
- Water inlet pipe
- Oil level gauge guide. Tighten to 71 inch lbs. (8 Nm).
- Cooling fan ECU
- Exhaust manifold No.2 heat insulator
- Manifold converter No.2 insulator

20. Install the Right-hand cylinder head with a new gasket. Tighten the cylinder head bolts as follows:

- Step 1: Tighten the 8 cylinder head bolts to 40 ft. lbs. (54 Nm)
- Step 2: Tighten each bolt 90°
- Step 3: Tighten each bolt an additional 90°
- Step 4: Tighten the single 8mm hexagon bolt to 14 ft. lbs. (19 Nm)

21. The remainder of installation is the reverse order of removal.
22. Refill the engine oil to the correct level.

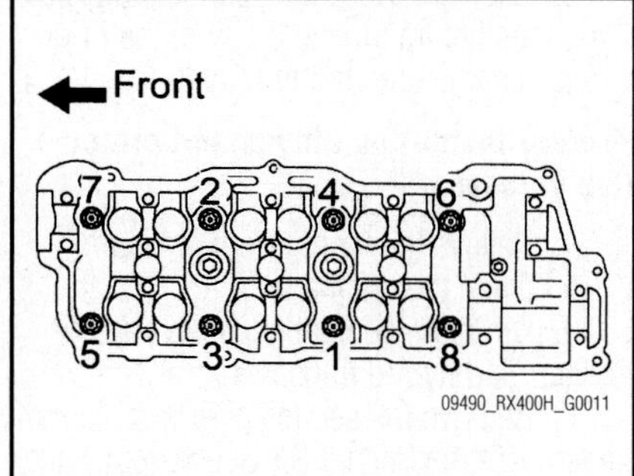

Fig. 111 Right-hand cylinder head tightening sequence

23. Refill the coolant to the engine radiator and hybrid transaxle to the correct level.
24. Replace the service plug grip.
25. Start the engine and check for leaks.

EXHAUST MANIFOLD

REMOVAL & INSTALLATION

Front

See Figure 112.

➡Removing the oil filter helps gain access to a lower bolt in the front exhaust manifold.

1. Before servicing the vehicle, refer to the Precautions Section.
2. Remove or disconnect the following:

- Negative battery cable
- Engine under covers
- Front exhaust pipe from the exhaust manifolds, by removing the nuts

➡Check for access to some of the manifold lower bolts, if so remove any possible.

- Heated Oxygen (HO_2) sensor
- Exhaust manifold stay, by removing the bolt and nut
- Remaining exhaust manifold nuts; then, separate the exhaust manifold from the engine

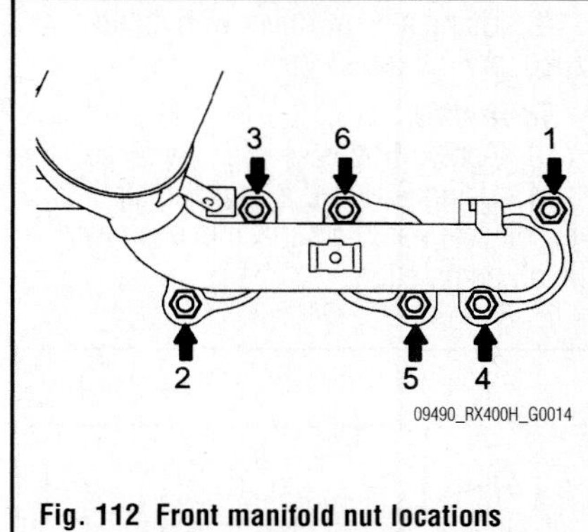

Fig. 112 Front manifold nut locations

To install:

3. Install or connect the following:

- Exhaust manifold, using a new gasket. Uniformly, tighten the bolts to 36 ft. lbs. (49 Nm).
- Exhaust manifold stay. Tighten the nut/bolt to 15 ft. lbs. (20 Nm).
- Heated Oxygen (HO_2) sensor to the exhaust manifold
- Front exhaust pipe to the exhaust manifold, using a new gasket. Tighten both nuts to 41 ft. lbs. (56 Nm).

- Engine under covers
- Negative battery cable

Rear

See Figures 113 and 114.

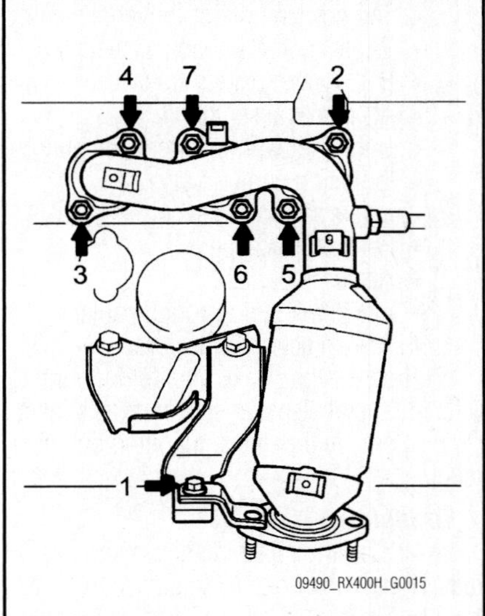

Fig. 113 Rear manifold nut locations

1. Before servicing the vehicle, refer to the Precautions Section.
2. Remove or disconnect the following:
   - Negative battery cable
   - Engine under covers
   - Front exhaust pipe from both exhaust manifolds, from below the engine
   - Exhaust Gas Recirculation (EGR) pipe from the rear exhaust manifold, by removing the 4 nuts
   - Heated Oxygen (HO_2) sensor wiring, from the right exhaust manifold
   - Exhaust manifold stay
   - 6 exhaust manifold nuts and the exhaust manifold

To install:

3. Install or connect the following:
   - Exhaust manifold to the engine, using a new gasket. Tighten the 6 nuts to 36 ft. lbs. (49 Nm).
   - Exhaust manifold stay. Tighten the nut/bolt to 25 ft. lbs. (34 Nm).
   - HO_2 sensor wiring to the exhaust manifold
   - EGR pipe to the exhaust manifold and the engine, using new gaskets. Tighten the 4 nuts to 108 inch lbs. (12 Nm).
   - Front exhaust pipe to the exhaust manifold, use a new gasket. Tighten both nuts to 41 ft. lbs. (56 Nm).
   - Engine under covers
   - Negative battery cable

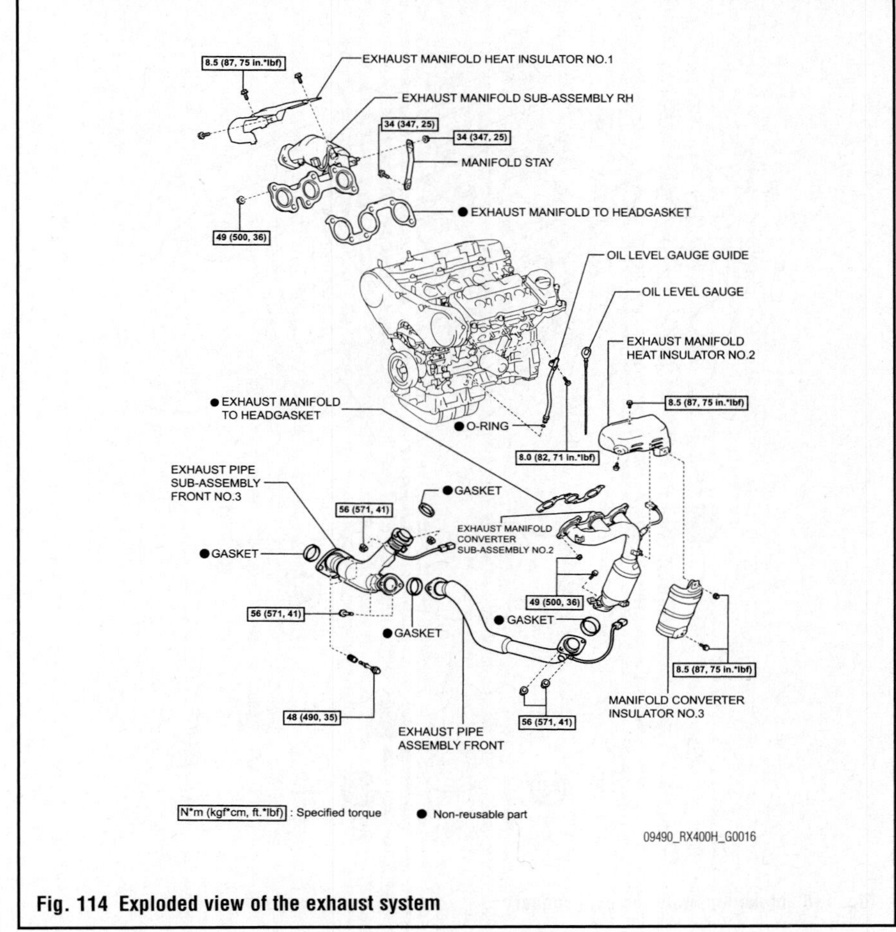

Fig. 114 Exploded view of the exhaust system

FLEXPLATE

REMOVAL & INSTALLATION

See Figures 115 through 117.

⁂ CAUTION

The HIGHLANDER HV has a hybrid system that operates at voltages up to 650 volts. Be sure to follow the instructions in this manual to handle the system correctly. Failure to do so may result in serious injury or electrocution. Engineer must undergo special training to be able to perform high-voltage system inspection and servicing.

⁂ CAUTION

All high-voltage wire harness connectors are colored orange. The HV battery and other high-voltage components have "High Voltage" caution labels. Do not carelessly touch these wires and components.

1. Before inspecting or servicing the high-voltage system, be sure to follow safety measures, such as wearing insulated gloves and removing the service plug to prevent electrocution. Carry the removed service plug in your pocket to prevent other technicians from reinstalling it while you are servicing the vehicle.
2. After removing the service plug, wait 5 minutes before touching any of the high-voltage connectors and terminals.

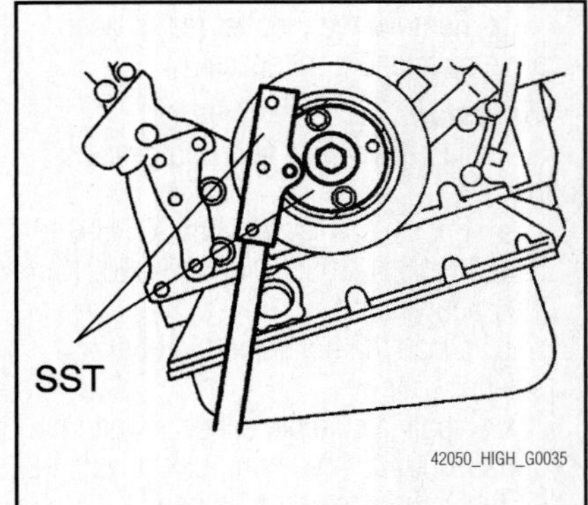

Fig. 115 Hold the crankshaft in place with the special tools as shown

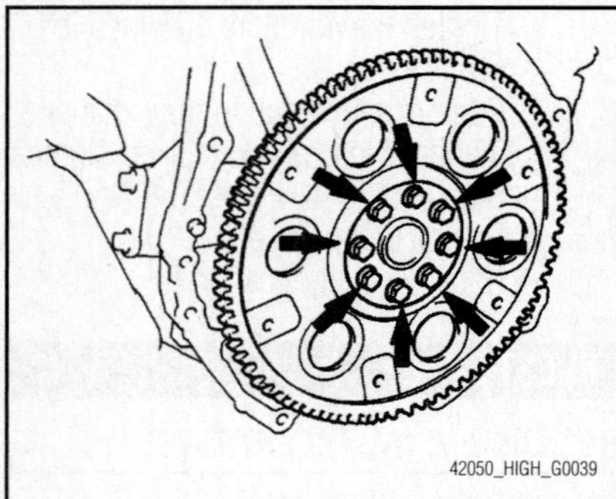

Fig. 116 Remove the 8 bolts, rear spacer, drive plate and front spacer

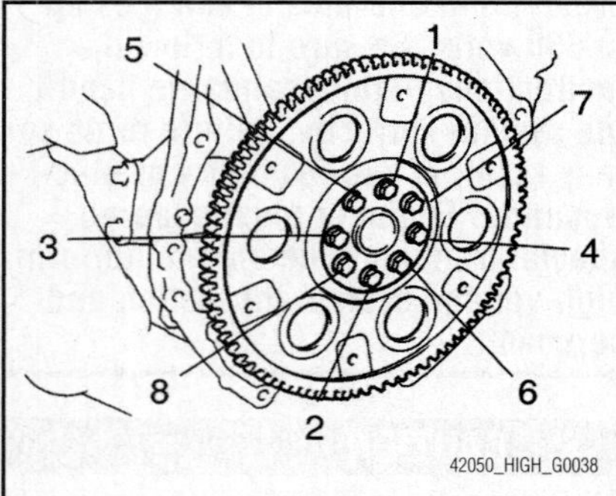

Fig. 117 Drive plate bolt tightening sequence

➡Wear insulating gloves and protective glasses.

➡This procedure requires a variety of special tools.

3. Remove the automatic transaxle assembly (2WD) or automatic transaxle and transfer assembly (4WD), as outlined in the Drive Train Section.
4. Remove the drive plate and ring gear, as follows:
   a. Secure the crankshaft with Special Service Tool (SST) 09213-54015 (91651-60855), 09330-00021 or their equivalents.
   b. Remove the 8 bolts, rear spacer, drive plate and front spacer.

To install:

5. Install drive plate and ring gear, as follows:
   a. Fix the crankshaft with SST 09213-54015 (91651-60855), 09330-00021, or their equivalents.
   b. Clean the bolts and the bolt holes.
   c. Apply a suitable adhesive (part no. 08833-00070, three bond or equivalent) to 2 or 3 threads of the bolt end.
   d. Install and uniformly tighten the 8 bolts, in several passes in the sequence shown in the accompanying illustration to a final torque of 61 ft. lbs. (83 Nm).

⁂ WARNING

Do not start the engine for AT LEAST one hour after installing the seal!

6. Install the automatic transaxle assembly (2WD) or automatic transaxle and transfer assembly (4WD), as outlined in the Drive Train Section.

INTAKE MANIFOLD

REMOVAL & INSTALLATION

See Figures 118 and 119.

1. Before servicing the vehicle, refer to the Precautions Section.
2. Relieve the fuel system pressure.
3. Drain the engine oil.
4. Drain the coolant from the engine radiator and hybrid transaxle.
5. Remove the service plug grip, found underneath the Battery Service cover on the rear seat. Wait 5 minutes to discharge the high voltage capacitor.
6. Remove or disconnect the following:
   - Negative battery cable
   - Engine cover
   - Right-hand front wheel
   - Engine splash shields
   - Right-hand front fender apron seal
   - Wiper and blade assembly
   - Top cowl ventilator louver assembly
   - Wiper motor and link assembly
   - Battery and battery tray
   - Air intake assembly
   - Converter with Inverter assembly
   - Emission control valve set
   - Intake air surge tank
   - Fuel supply hose
   - Heater inlet hose
   - Intake manifold ground cable
   - Fuel injector connectors
7. Loosen the intake manifold mounting bolts in several steps, in sequence as shown.
8. Remove the intake manifold and gaskets.

To install:

9. Install the intake manifold and gaskets. Tighten the bolts in sequence to 11 ft. lbs. (15 Nm).

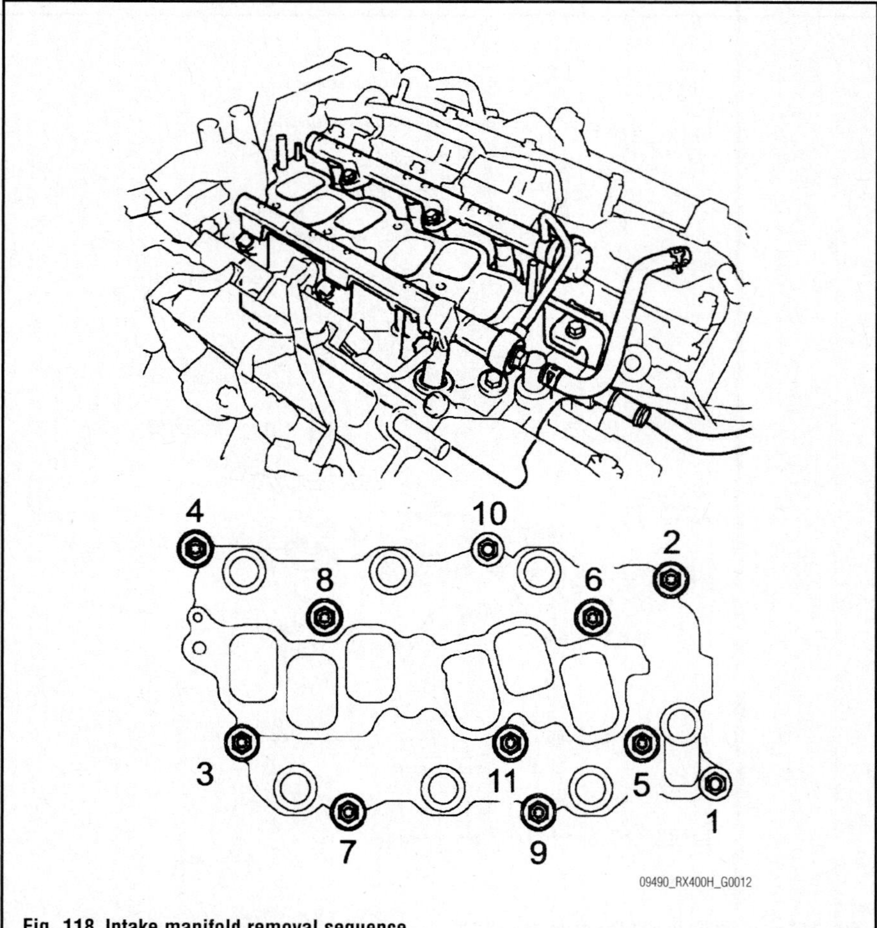

Fig. 118 Intake manifold removal sequence

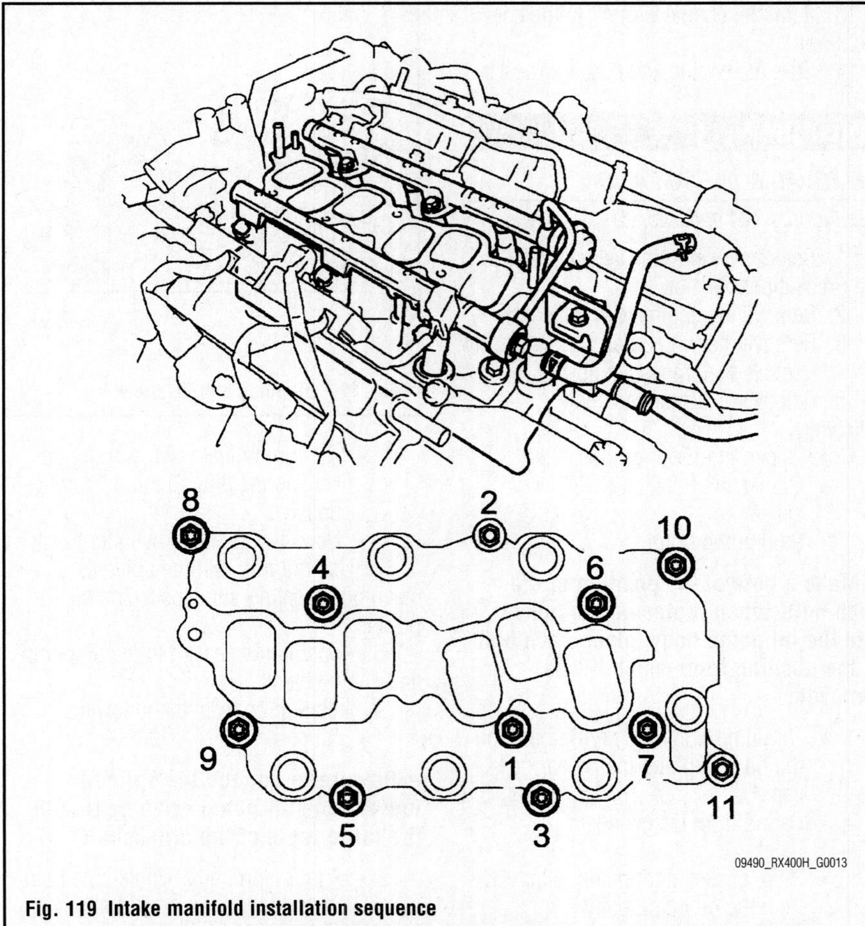

Fig. 119 Intake manifold installation sequence

10. Install or connect the following:
    - Fuel injector connectors
    - Intake manifold ground cable. Tighten to 11 ft. lbs. (15 Nm).
    - Heater inlet hose. Tighten to 74 inch lbs. (8.4 Nm).
    - Fuel supply hose
    - Intake air surge tank
    - Emission control valve set
    - Converter with Inverter assembly
    - Air intake assembly
    - Battery and battery tray
    - Wiper motor and link assembly
    - Top cowl ventilator louver assembly
    - Wiper and blade assembly
    - Right-hand front fender apron seal
    - Engine splash shields
    - Right-hand front wheel
    - Engine cover
    - Negative battery cable
11. Refill the engine oil to the correct level.
12. Refill the coolant to the engine radiator and hybrid transaxle to the correct level.
13. Replace the service plug grip.
14. Start the engine and check for leaks.

OIL PAN

REMOVAL & INSTALLATION

See Figures 120 through 123.

1. Before servicing the vehicle, refer to the Precautions Section.
2. Remove or disconnect the following:
    - Engine/transaxle assembly from the vehicle
    - Right-hand exhaust manifold
    - Transaxle mass damper
    - Front frame assembly
    - Halfshafts
    - Flywheel housing undercover
    - Front engine mounting bracket
    - Transaxle assembly from the engine
    - Transmission input damper assembly
    - Flywheel
3. Install the engine to a suitable engine stand.
4. Remove or disconnect the following:
    - Remaining exhaust manifold heat shields
    - Right-hand engine mounting bracket
    - Compressor mounting bracket
    - Crankshaft pulley
    - Timing belt
    - Timing belt idler
    - Crankshaft timing pulley
    - Oil level gauge assembly
5. Remove the lower oil pan as follows:
    a. Remove the mounting bolts and nuts
    b. Using Special Tool 09032-00100 or suitable seal cutter, cut the sealant between the upper and lower oil pans.
    c. Remove the lower oil pan.
6. Remove the oil strainer and gasket.
7. Remove the upper oil pan as follows:
    a. Uniformly loosen and remove the mounting bolts.
    b. Using a suitable pry tool, pry the upper oil pan from the cylinder block.

To install:

8. Remove any old sealant from the mating surface of the oil pans.
9. Install the upper oil pan as follows:

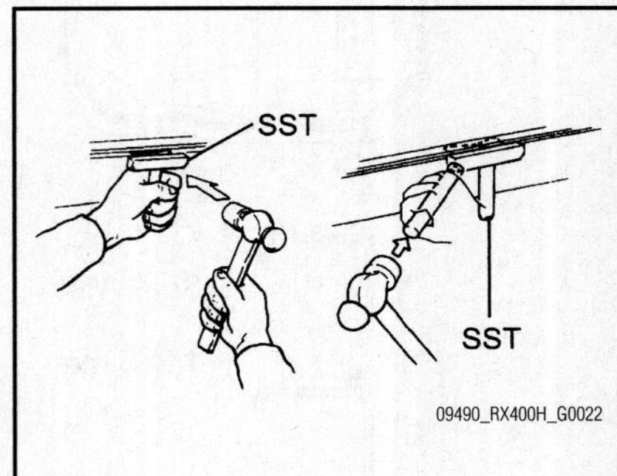

Fig. 120 Use a suitable tool to cut the sealant between the oil pans.

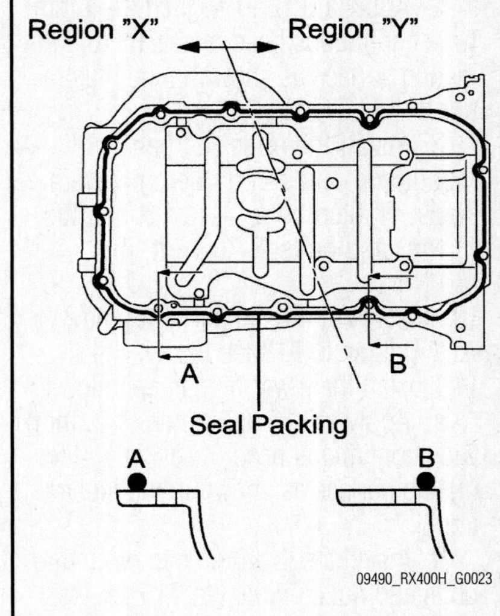

Fig. 121 Apply the sealant to the upper oil pan as shown.

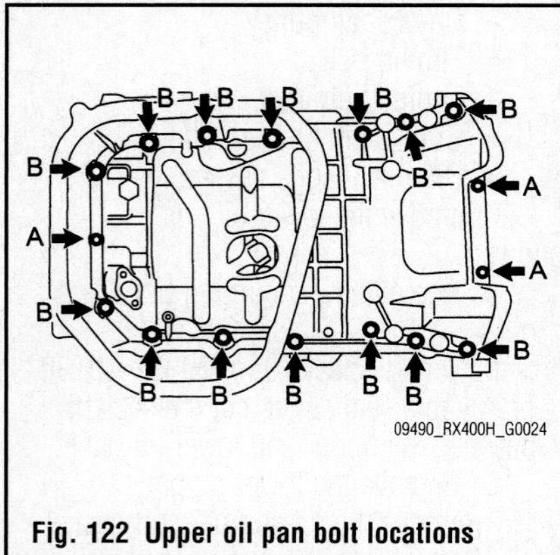

Fig. 122 Upper oil pan bolt locations

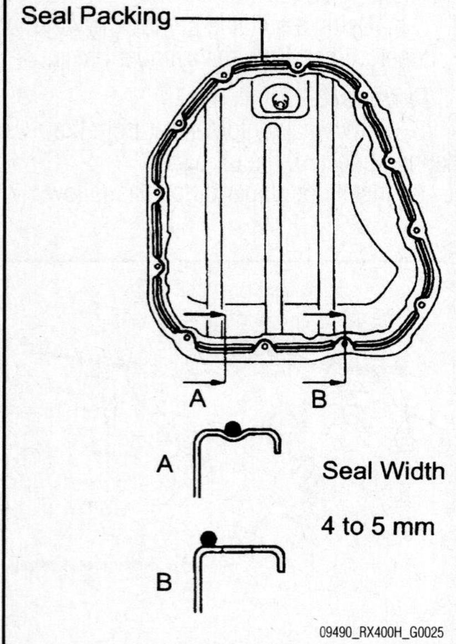

Fig. 123 Apply sealant to the lower oil pan as shown.

a. Apply a 0.12–0.16 inch (3–4 mm) wide continuous bead of sealant to the mating surface as shown in the illustration.

b. Install the upper oil pan mounting bolts and tighten in several steps. Tighten bolts 'A' to 71 inch lbs. (8 Nm) and bolts 'B' to 14 ft. lbs. (20 Nm).

10. Install the oil strainer with a new gasket. Tighten to 71 inch lbs. (8 Nm).

11. Install the lower oil pan as follows:

a. Apply a 0.16–0.20 inch (4–5 mm) wide continuous bead of sealant to the mating surface as shown in the illustration.

b. Install the lower oil pan mounting bolts and nuts. Tighten to 71 inch lbs. (8 Nm).

12. The remainder of the installation is the reverse order of removal.

13. Refill the engine with oil to the correct level.

14. Start the engine and check for leaks.

OIL PUMP

REMOVAL & INSTALLATION

See Figures 124 and 125.

1. Before servicing the vehicle, refer to the Precautions Section.
2. Remove the engine with transaxle.
3. Remove the timing belt.
4. Remove the crankshaft gear.
5. Remove or disconnect the following:
   - Upper and lower oil pans
   - Crankshaft Position (CKP) sensor
   - 9 oil pump bolts

➡Make a note of the position of the each bolt. When replacing the bolts into the oil pump body, place each bolt in the position from which it was removed.

   - Oil pump body, by prying between the oil pump and main bearing cap
   - O-ring from the cylinder block
   - Plug, gasket, spring and relief valve from the oil pump body
   - 9 screws, pump body cover, drive and driven rotors

To install:

6. Install or connect the following:
   - Driven rotors, drive, pump body cover, using the 9 screws

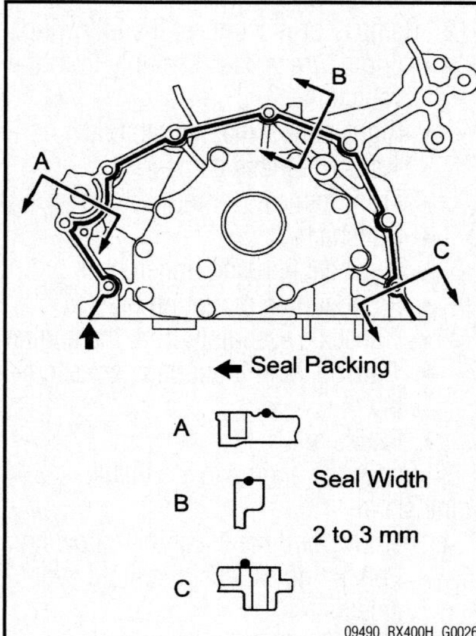

Fig. 124 Apply the sealant to the oil pump as shown

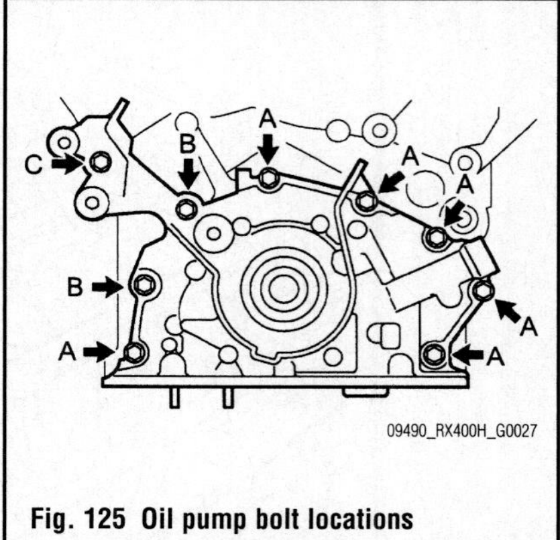

Fig. 125 Oil pump bolt locations

   - Oil pump relief valve, spring, gasket and the plug to the oil pump body
   - New O-ring on the cylinder block

7. Using a non residue solvent, clean both sealing surfaces to the oil pump.
8. Apply liquid sealant to the oil pump and engine block.
9. Install or connect the following:
   - Oil pump

➡Be sure to engage the splined teeth of the oil pump drive gear with the large teeth of the crankshaft.

   - 9 oil pump bolts. Tighten the bolts in several passes to 71 inch lbs. (8 Nm) for bolt 'A'; 14 ft. lbs. (20 Nm), for bolts 'B'; 32 ft. lbs. (43 Nm) for bolt 'C'
   - CKP sensor. Tighten the bolt to 71 inch lbs. (8 Nm).
   - Upper and lower oil pans
   - crankshaft gear
   - Timing belt, and covers
   - Engine with the transaxle.

10. Refill the engine with oil to the correct level.
11. Start the engine and inspect for leaks.
12. Recheck the engine oil level.

INSPECTION

Rotor Tip Clearance

See Figures 126 and 127.

1. Install the rotors to the oil pump body with the marks facing upward and matchmarks aligned. Check that the rotors revolves smoothly.
2. Using a feeler gauge, measure the rotor tip clearance between the drive and driven rotor tips.
3. The rotor tip clearance measurements are as follows:

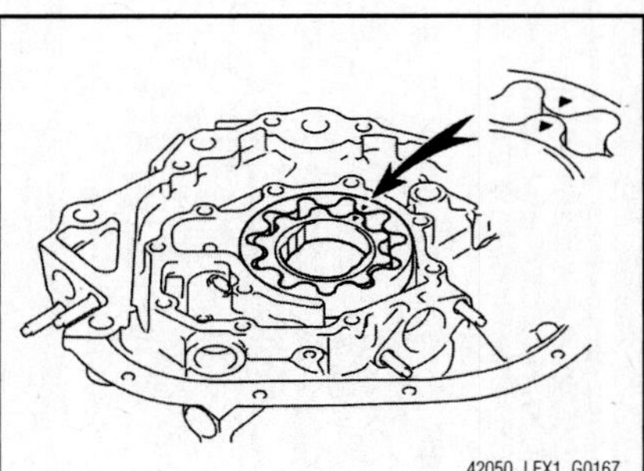

Fig. 126 Install the rotors to the oil pump body with the marks facing upward and matchmarks aligned.

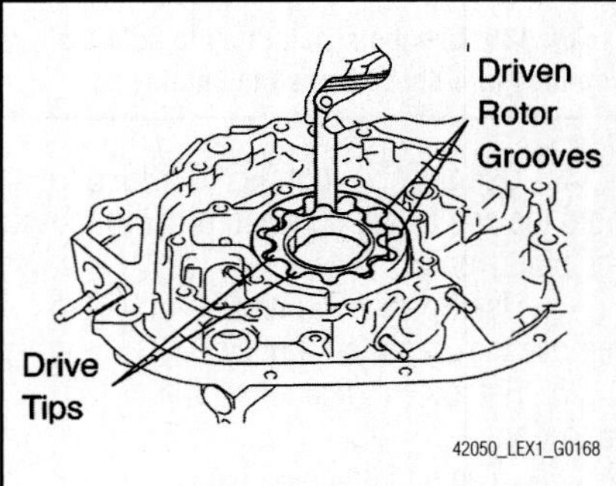

Fig. 127 Inspecting the oil pump rotor tip clearance.

- Standard tip clearance: 0.0024–0.0071 inch (0.06–0.18mm)
- Maximum tip clearance: 0.0138 inch. (0.35 mm)

4. If the tip clearance is greater than maximum, replace the rotors as a set.

Body Clearance

See Figure 128.

1. Install the rotors to the oil pump body with the marks facing upward and matchmarks aligned. Check that the rotors revolves smoothly.
2. Using a feeler gauge, measure the body clearance between the driven rotor and pump body.
3. The body clearance measurements are as follows:
   - Standard body clearance: 0.0098–0.0128 inch (0.250–0.325mm)
   - Maximum body clearance: 0.0118 inch. (0.30 mm)
4. If the body clearance is greater than maximum, replace the rotors as a set. If necessary, replace the oil pump assembly.

Rotor Side Clearance

See Figure 129.

1. Install the rotors to the oil pump body with the marks facing upward and matchmarks aligned. Check that the rotors revolves smoothly.
2. Using a feeler gauge and precision straight edge, measure the rotor side clearance between the rotors and precision straight edge.
3. The rotor side clearance measurements are as follows:
   a. 3.3L (3MZ-FE) engines
   - Standard side clearance: 0.0012–0.0035 inch. (0.03–0.09 mm)
   - Maximum side clearance: 0.0059 inch. (15 mm)
4. If the rotor side clearance is greater than maximum, replace the rotors as a set. If necessary, replace the oil pump assembly.

PISTON AND RING

POSITIONING

See Figures 130 through 132.

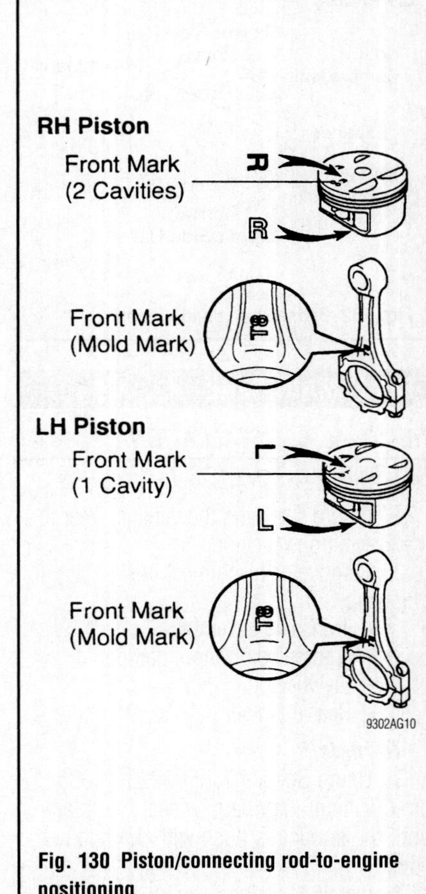

Fig. 130 Piston/connecting rod-to-engine positioning

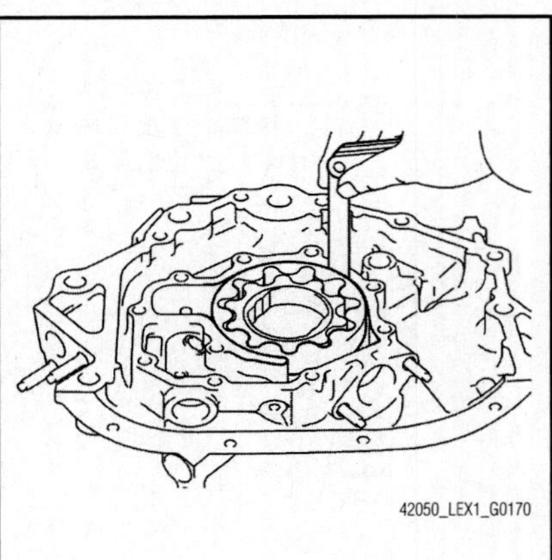

Fig. 128 Inspecting the oil pump rotor body clearance.

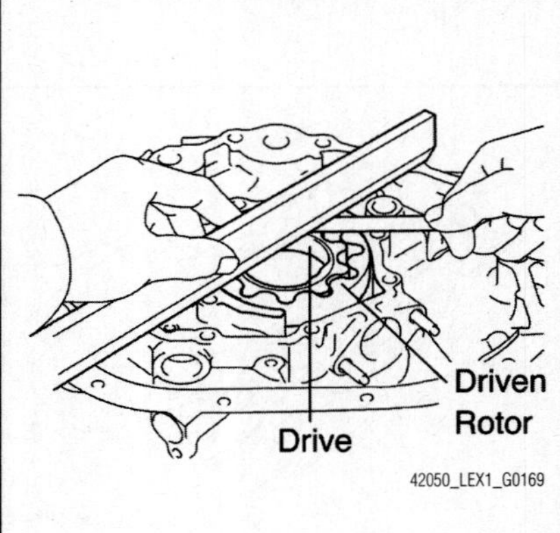

Fig. 129 Inspecting the oil pump rotor side clearance.

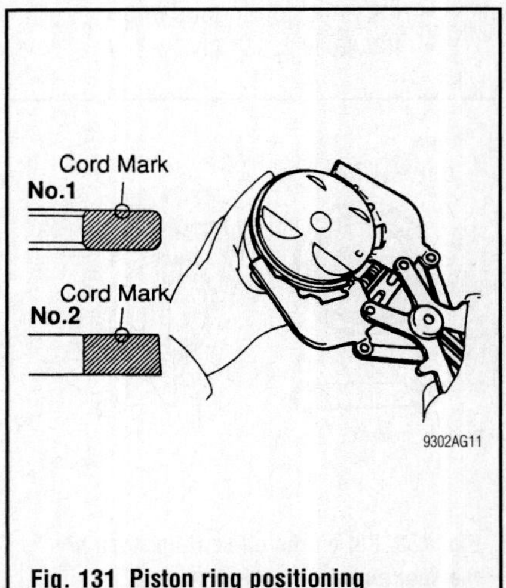

Fig. 131 Piston ring positioning

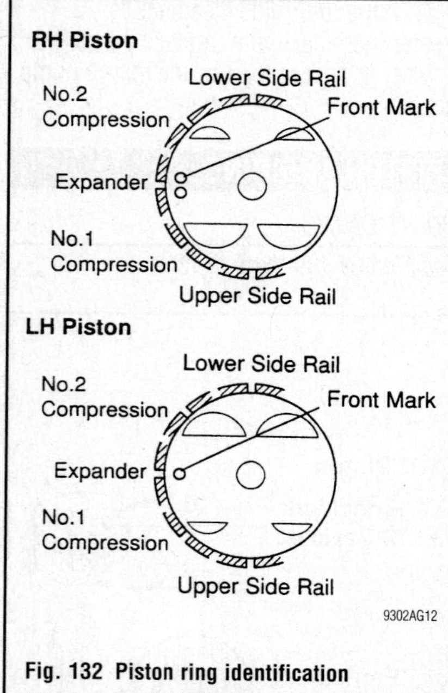

Fig. 132 Piston ring identification

REAR MAIN SEAL

REMOVAL & INSTALLATION

See Figures 133 and 134.

1. Before servicing the vehicle, refer to the Precautions Section.
2. Remove or disconnect the following:
   - Transaxle assembly
   - Transmission input damper
   - Flywheel
   - Rear main seal

To install:

3. Using Special Tool 09223-15030 or equivalent, tap the new seal into place until the surface is flush with the retainer edge.
4. Install or connect the following:
   - Flywheel
   - Transmission input damper
   - Transaxle assembly

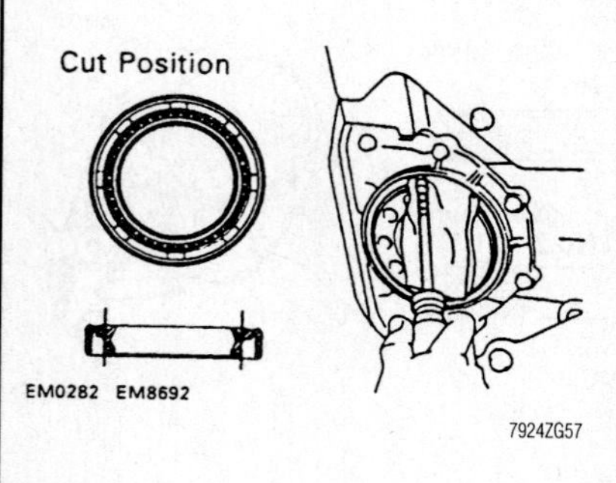

Fig. 133 Cut off the oil seal lip, then pry the seal out of the retaining plate

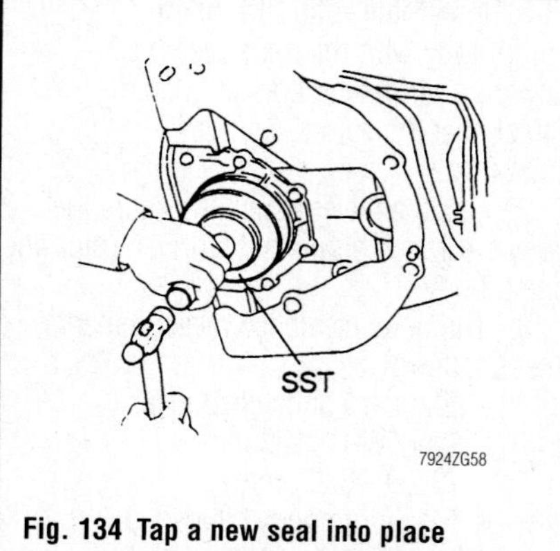

Fig. 134 Tap a new seal into place

TIMING BELT FRONT COVER

REMOVAL & INSTALLATION

Refer to Timing Belt & Sprockets.

TIMING BELT & SPROCKETS

REMOVAL & INSTALLATION

See Figures 135 through 147.

1. Before servicing the vehicle, refer to the Precautions Section.
2. Remove or disconnect the following:
   - Negative battery cable
   - Engine covers
   - Right front wheel
   - Fender splash shields
   - Wiper arms
   - Top cowl ventilator louver
   - Wiper motor and linkage assembly
   - Battery and battery tray
   - Air intake assembly
   - Brake master cylinder reservoir and bracket
   - Air cleaner bracket
   - Engine moving control rod
   - Right-hand engine mounting stay No. 2

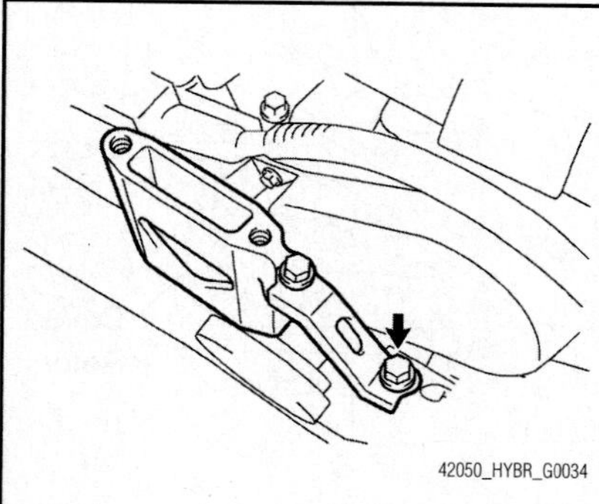

Fig. 135 Remove the right-hand engine mounting stay No. 2

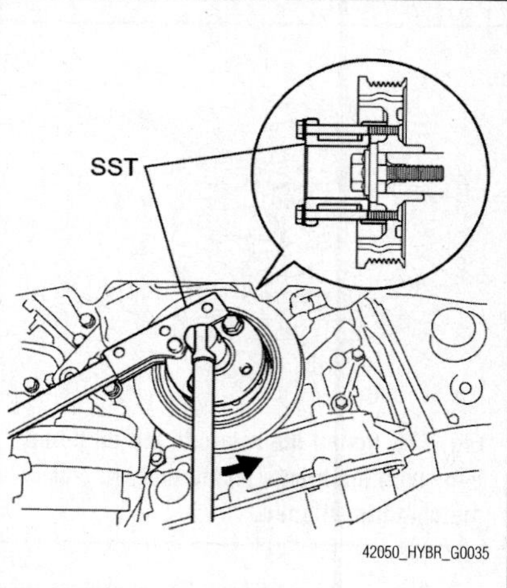

Fig. 136 Use the special tool to hold the pulley in order to loosen the pulley bolt

3. Use Special Tool 09213-54015 to hold the crankshaft pulley in order to loosen the pulley bolt.
4. Use Special Tool 09950-50013 to remove the crankshaft pulley.
5. Remove or disconnect the following:
   - Timing belt cover No. 1
   - Timing belt cover No. 2
   - Right-hand engine mounting bracket
   - Timing belt guide No. 2
6. Temporarily install the crank pulley bolt. Turn the crankshaft clockwise to align the timing mark on the crankshaft timing pulley with the notch in the oil pump body.
7. Check that the timing marks on the camshaft pulleys are aligned with the notches on the inner belt cover. If

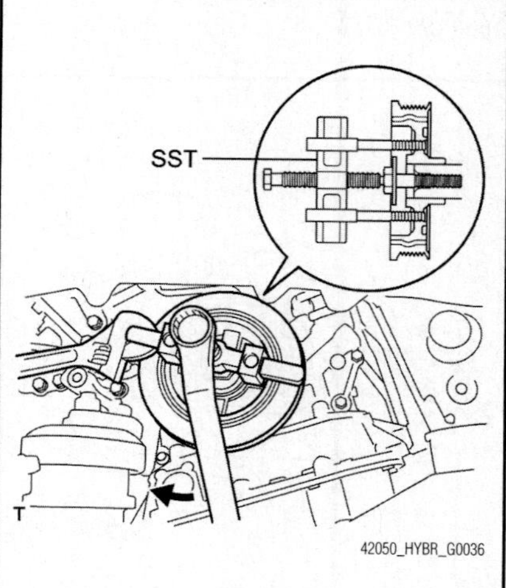

Fig. 137 Use the special tool to remove the crankshaft pulley

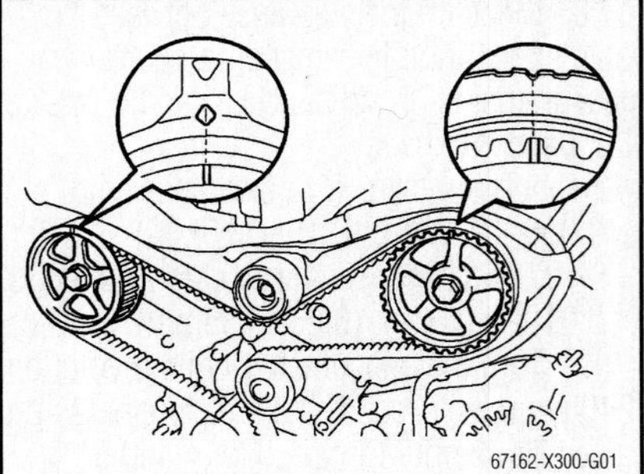

Fig. 138 Check that the timing marks on the camshaft pulleys are aligned with the notches on the inner belt cover

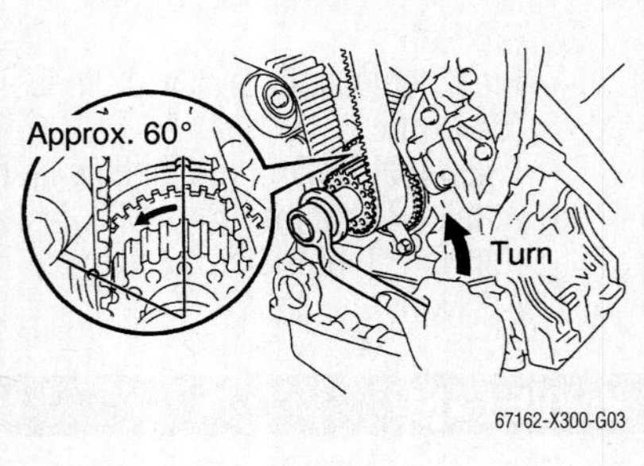

Fig. 141 Turn the crankshaft counterclockwise by 60 degrees

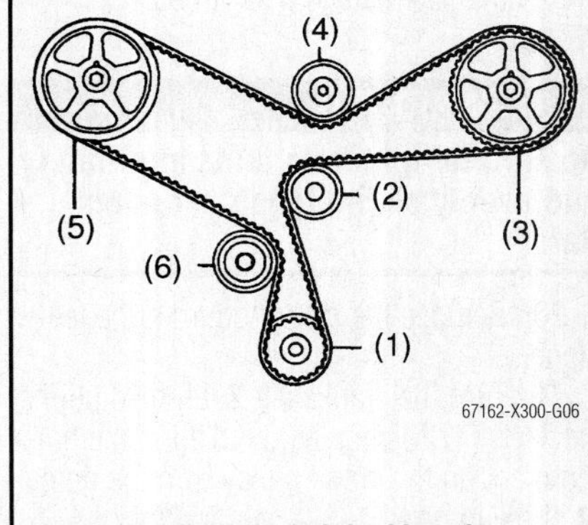

Fig. 144 Install the belt in this order

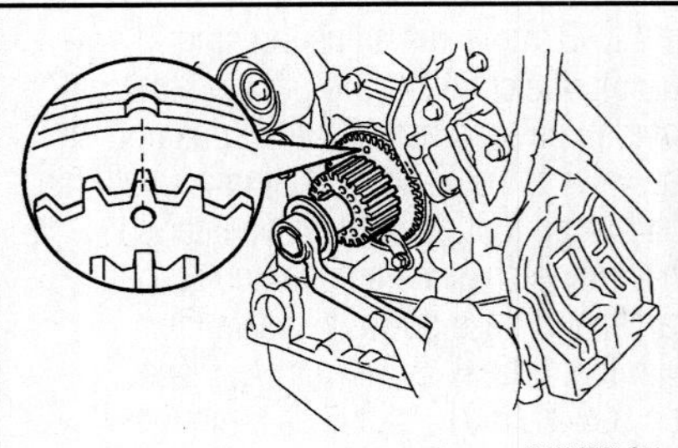

Fig. 139 Turn the crankshaft clockwise to align the timing mark on the crankshaft timing pulley with the notch in the oil pump body

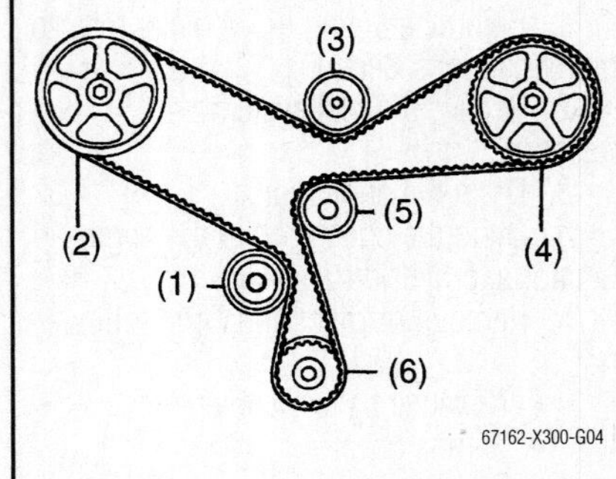

Fig. 142 Remove the belt from the pulleys in this order

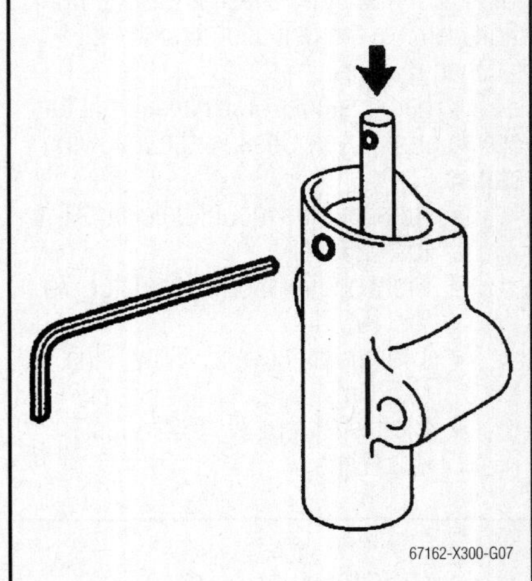

Fig. 145 Set the tensioner in a press and collapse the plunger. Do not apply more that 2,205 lbs (9.8 kn) of force. Insert a suitable metal rod through the holes to hold the plunger in position

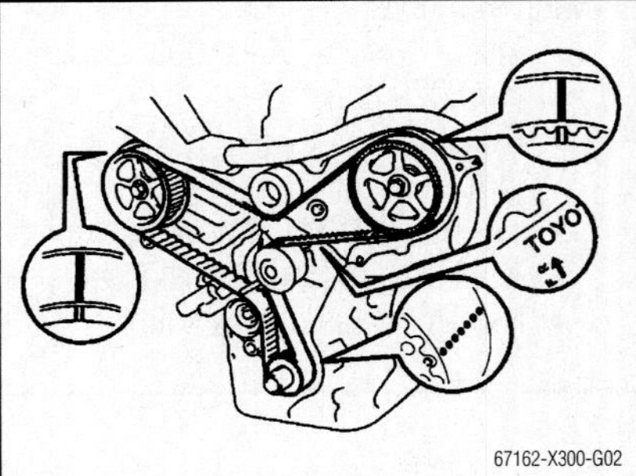

Fig. 140 If the timing belt is re-used, check that the 3 original installation marks are visible on the belt as shown

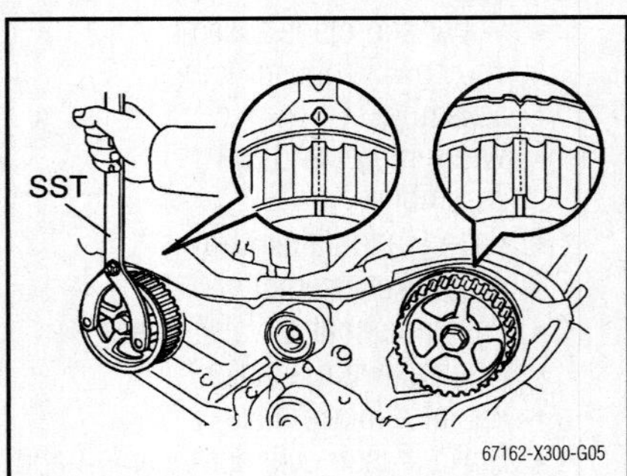

Fig. 143 Turn the camshaft pulleys back into alignment so the marks align with the notches on the inner cover

not, rotate the crankshaft 360 degrees clockwise.

➡If the timing belt is re-used, check that the 3 original installation marks are visible on the belt as shown. If not, paint three new marks on the belt.

8. Turn the crankshaft counterclockwise by 60 degrees. Make sure that the belt is still engaged.
9. Remove the timing belt tensioner.
10. Remove the belt from the pulleys in this order:
    - Lower idler pulley
    - Right camshaft pulley
    - Upper idler pulley
    - Left camshaft pulley
    - Water pump pulley
    - Crankshaft timing pulley
11. If the belt is being re-used, check it for wear or damage; don't twist it or turn it inside-out. If there is any doubt as to its condition, replace it.

To install:

12. Clean all the pulleys.
13. Turn the crankshaft another 60 degrees counterclockwise.
14. Turn the camshaft pulleys back into alignment so the marks align with the notches on the inner cover.
15. Turn the crankshaft back so that the timing mark aligns with the notch on the oil pump.
16. Align the installation marks on the belt with the timing marks on the pulleys.
17. Install the belt in this order:
    - Crankshaft
    - Water pump
    - Left camshaft
    - Upper idler
    - Right camshaft
    - Lower idler
18. Set the tensioner in a press and collapse the plunger. Do not apply more that 2,205 lbs (9.8 kn) of force. Insert a suitable metal rod through the holes to hold the plunger in position.
19. Install the tensioner and torque

the 2 bolts alternately to 20 ft. lbs. (27 Nm).

WARNING

Be sure to tighten to bolts alternately and evenly so the tensioner seats flat.

20. Remove the metal rod from the tensioner.
21. Turn the crankshaft 2 full revolutions clockwise (720 degrees), and align the timing mark on the crank pulley with the notch on the oil pump.
22. Check the timing marks on the camshaft pulleys for alignment with the notches on the inner cover. If they do not align, remove the belt and align the mismatched mark(s).
23. The remainder of installation is the reverse of removal. Observe the following torques:
    - Right engine mount bracket: 21 ft. lbs. (28 Nm)
    - Right engine mount insulator: 70 ft. lbs. (95 Nm)
    - Timing belt covers: 75 inch lbs. (8.5 Nm)
    - Crankshaft pulley: 162 ft. lbs. (220 Nm)
    - Alternator bracket: 21 ft. lbs. (28 Nm)
    - Right engine mount stay: 47 ft. lbs. (64 Nm)
    - Engine roll control rod: tighten first A, then B, and then C to 47 ft. lbs. (64 Nm). Torque D to 17 ft. lbs. (23 Nm)

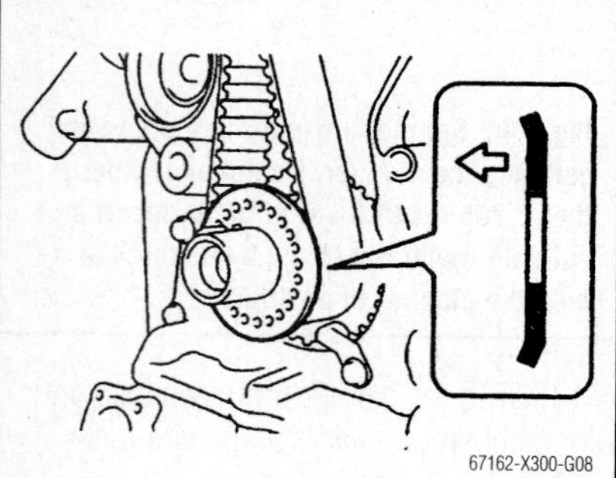

Fig. 146 Install the timing belt guide with the cupped side facing front

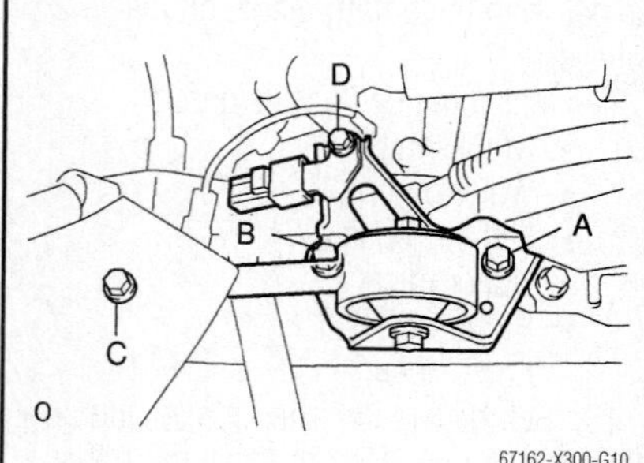

Fig. 147 Tighten the engine roll control rod bolts in this order

VALVE LASH

ADJUSTMENT

See Figures 148 through 150.

➡Adjust the valve clearance when the engine is cold.

1. Before servicing the vehicle, refer to the Precautions Section.
2. Relieve the fuel system pressure.
3. Drain the engine oil.
4. Drain the coolant from the engine radiator and hybrid transaxle.
5. Remove or disconnect the following:
    - Negative battery cable
    - Engine cover
    - Right-hand front wheel
    - Engine splash shields
    - Right-hand front fender apron seal
    - Wiper and blade assembly
    - Top cowl ventilator louver assembly
    - Wiper motor and link assembly
    - Battery and battery tray
    - Air intake assembly
    - Emission control valve hoses
    - Air intake surge tank
    - Radiator intake hose
    - Brake master cylinder reservoir and bracket
    - Air cleaner bracket
    - Engine moving control rod
    - Right-hand engine No.2 mounting stay
    - Ignition coil
    - Valve covers
6. Turn the crankshaft pulley and align its groove with the timing mark **0** of the No. 1 timing cover.
7. Check that the valve lifters on the No. 1 cylinder (intake and exhaust) are loose. If not, turn the crankshaft 1 complete revolution (360 degrees).

➡All measurements should be written down. These recorded measurements will need to be used in conjunction with a mathematical formula to determine the thickness of the replacement shims.

8. Measure the clearance between the valve lifters and the camshaft. Record the measurements on valves No. 1 and 6 intake; No. 2 and 3 exhaust.
    a. The intake valve clearance cold is 0.006–0.010 in. (0.15–0.25mm).
    b. The exhaust valve clearance cold is 0.010–0.014 in. (0.25–0.35mm).
9. Turn the crankshaft ⅔ of a revolution (240 degrees). Record the measurements on valves No. 2 and 3 intake; No. 4 and 5 exhaust.
10. Turn the crankshaft another ⅔ of a revolution (240 degrees). Record the measurements on valves No. 4 and 5 intake; No. 1 and 6 exhaust.
11. Remove the adjusting shim by turning the crankshaft to position the cam lobe of the camshaft in the up position on the valve to be adjusted. Using a small thin flat bladed tool, turn the valve lifter so that the notches are perpendicular to the camshaft. Press down the valve lifter with tool 09248-55010 part A. Place too 09248-55010 part B between the camshaft and the valve lifter; remove part A.
12. Remove the adjusting shim with a magnet and a small screwdriver.
13. Determine the replacement adjusting shim size by either using the charts or the following formulas:
    - Intake: N = T + (A - 0.008 in./0.020mm)
    - Exhaust: N = T + (A - 0.012 in./0.30mm)
    - T = Thickness of removed shim
    - A = Measured valve clearance
    - N = Thickness of new shim

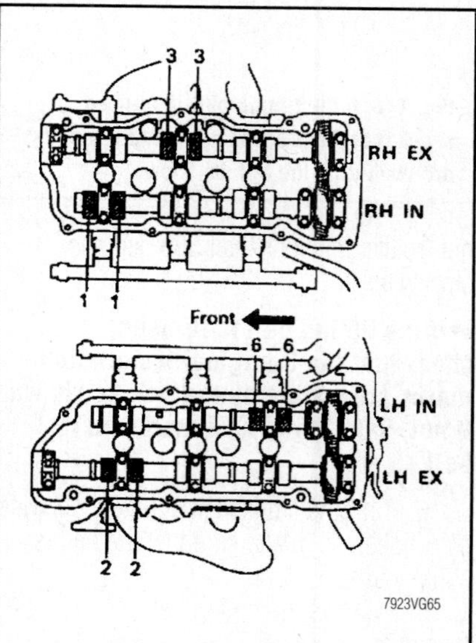

Fig. 148 Adjust these valves during the 1st step

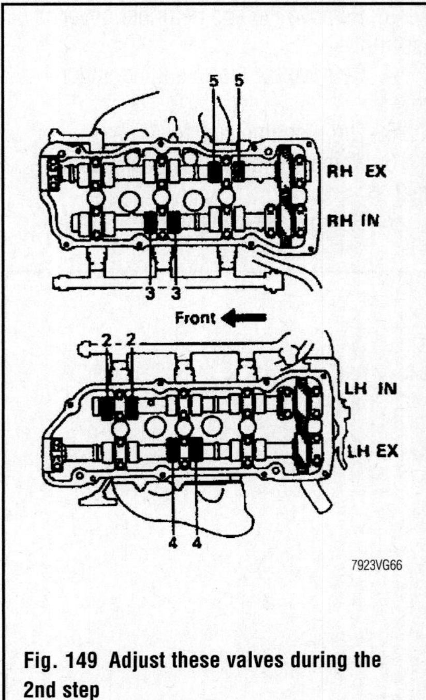

Fig. 149 Adjust these valves during the 2nd step

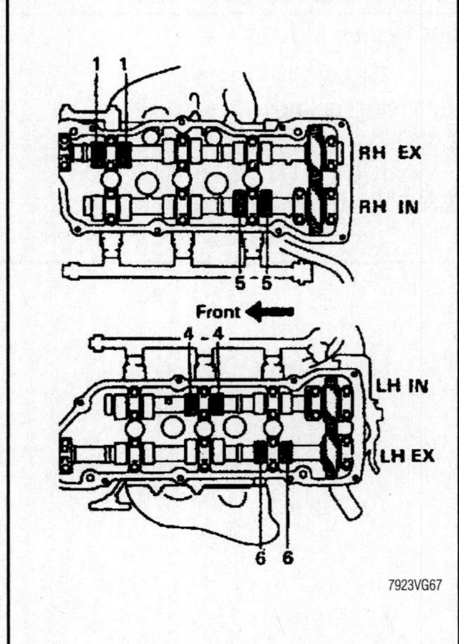

Fig. 150 Adjust these valves during the 3rd step

14. Select a new shim with a thickness as close as possible to the calculated value. Install the new replacement shim.

➡Shims are available in 17 sizes in increments of 0.0020 in. (0.050mm), from 0.0984 in. (2.500mm) to 0.1299 in. (3.300mm).

15. Recheck the valve clearance.
16. Install or connect the following:
    - Valve covers
    - Ignition coil
    - Radiator intake hose
    - Air intake surge tank
    - Emission control valve hoses
    - Air intake assembly
    - Battery and battery tray
    - Wiper motor and link assembly
    - Top cowl ventilator louver assembly
    - Wiper and blade assembly
    - Right-hand front fender apron seal
    - Engine splash shields
    - Right-hand front wheel
    - Engine covers
    - Negative battery cable
17. Refill the cooling system to the correct level.
18. Refill the engine oil to the correct level.
19. Start the engine and check for leaks.

ENGINE PERFORMANCE & EMISSION CONTROLS

ACCELERATOR PEDAL POSITION (APP) SENSOR

LOCATION

See Figure 151.

The Accelerator Pedal Position (APP) sensor is located inside the vehicle and is part of the accelerator pedal. It is referred to as the accelerator rod assembly.

REMOVAL & INSTALLATION

1. Disconnect the connector from the accelerator pedal rod assembly.
2. Remove the 2 nuts and accelerator pedal rod assembly.

To install:

3. Install the accelerator pedal rod assembly with the 2 nuts.
4. Tighten the mounting nuts to 43 inch lbs. (4.9 Nm).
5. Connect the connector to the accelerator pedal rod assembly.

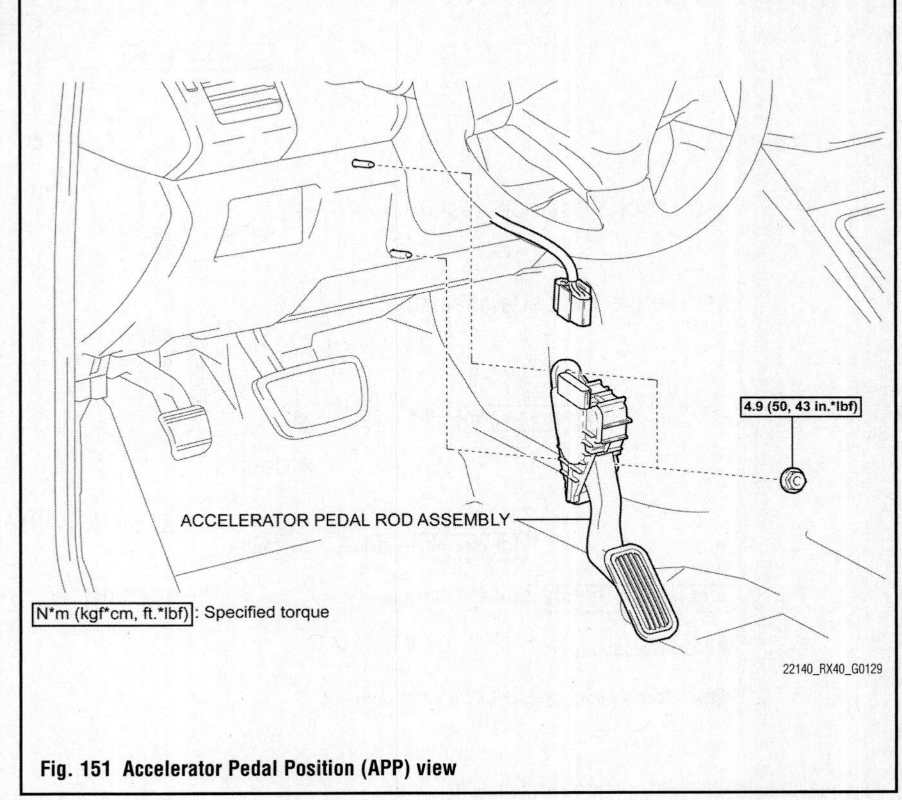

Fig. 151 Accelerator Pedal Position (APP) view

CRANKSHAFT POSITION (CKP) SENSOR

LOCATION

See Figure 152.

Refer to the accompanying illustration for sensor location.

REMOVAL & INSTALLATION

See Figures 153 and 154.

1. Recover the refrigerant from refrigeration system.
2. Remove the V-bank cover sub-assembly.
3. Remove the engine under cover assembly.
4. Remove the No. 1 engine under cover.
5. Drain engine coolant.
6. Remove the cool air intake duct seal.
7. Remove the battery.

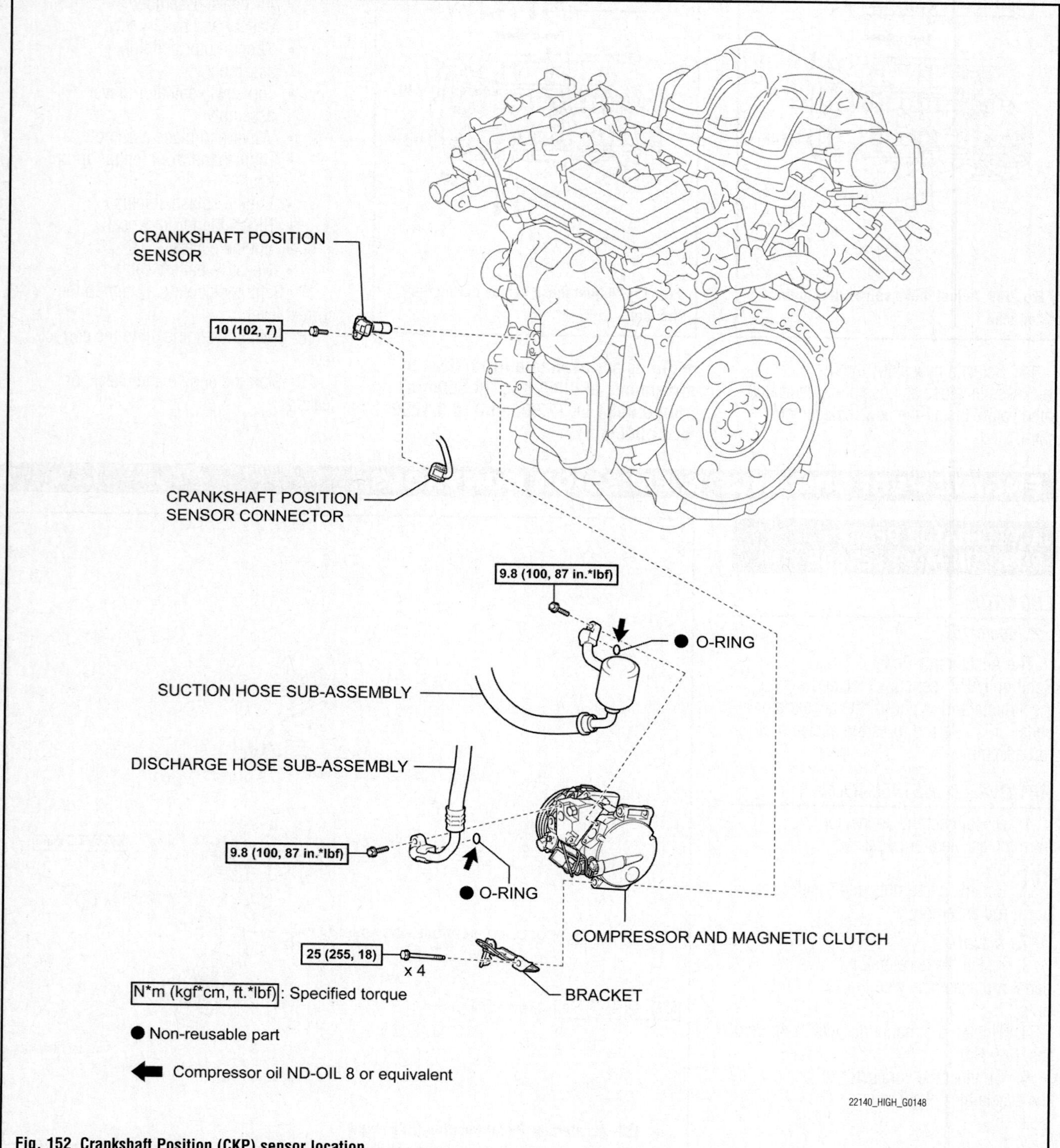

Fig. 152 Crankshaft Position (CKP) sensor location

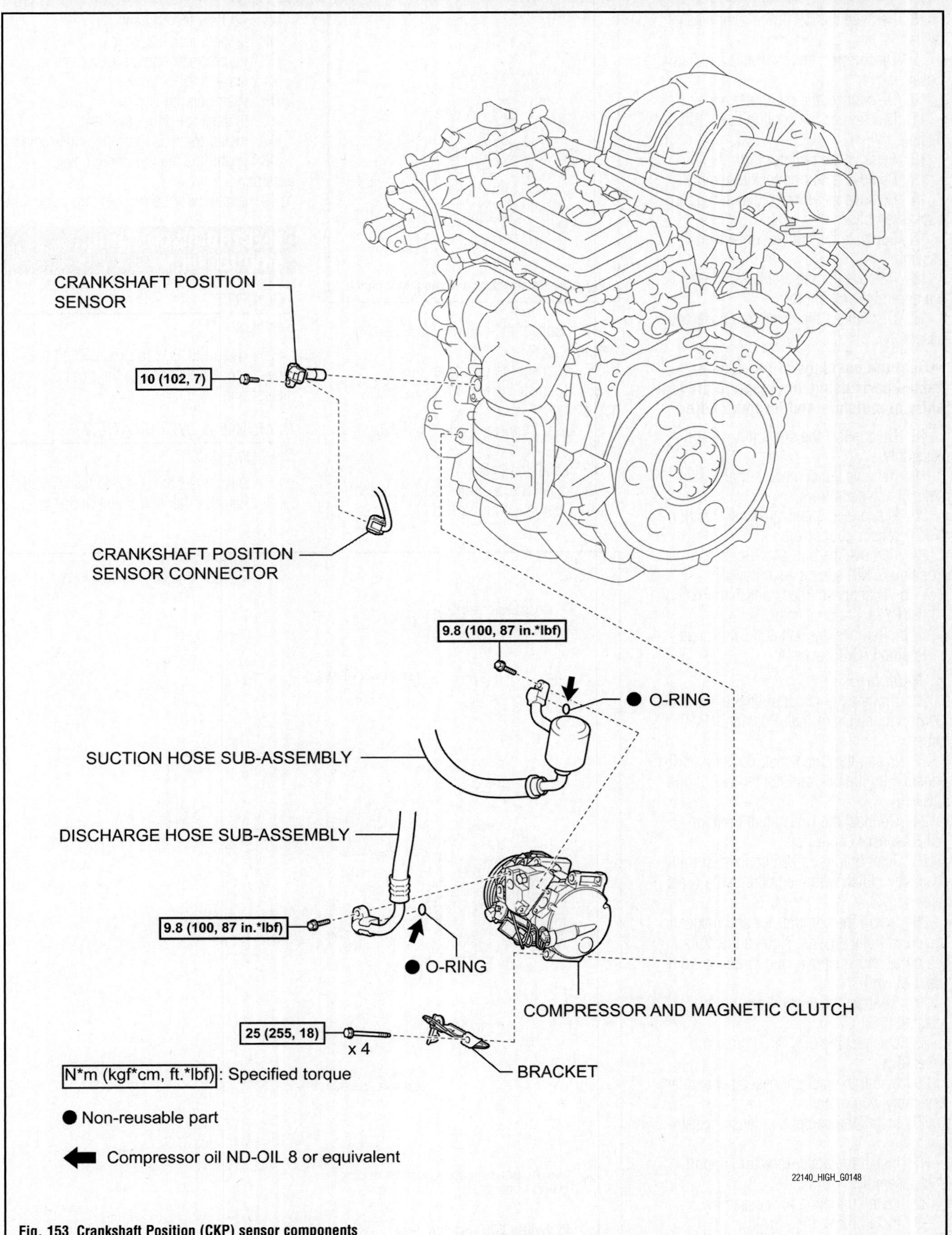

Fig. 153 Crankshaft Position (CKP) sensor components

8. Remove the No. 1 and 2 air cleaner inlets.
9. Disconnect the No. 1 and 2 radiator hoses.
10. Disconnect the oil cooler hose.
11. Disconnect the cooling fan ECU connector.
12. Remove the radiator grill.
13. Remove the hood lock assembly.
14. Remove the upper radiator support sub-assembly.
15. Separate the cooler condenser assembly.
16. Remove the radiator assembly and fan assembly with motor.
17. Disconnect the discharge hose sub-assembly.

➡**Seal the openings of the disconnected parts using vinyl tape to prevent entry of moisture and foreign matter.**

18. Disconnect the suction hose sub-assembly.
19. Remove the compressor and magnetic clutch as follows:
20. Disconnect the Crankshaft Position (CKP) sensor connector.
21. Remove the bolt and Crankshaft Position (CKP) sensor, as follows:
   a. Disconnect the Crankshaft Position (CKP) sensor connector.
   b. Remove the bolt and Crankshaft Position (CKP) sensor.

To install:

22. Apply a light coat of engine oil to the O-ring on the Crankshaft Position (CKP) sensor.
23. Install the Crankshaft Position (CKP) sensor with the bolt and tighten to 7 ft. lbs. (10 Nm).
24. Connect the Crankshaft Position (CKP) sensor connector.
25. Temporarily tighten compressor and magnetic clutch with the bolts, in the order shown.
26. Install the compressor and magnetic clutch with the 4 bolts. Tighten the bolts in the order shown above, and torque to 18 ft. lbs. (25 Nm).
27. Connect the suction hose sub-assembly.
28. Connect the discharge hose sub-assembly.
29. Install the radiator assembly and fan assembly with motor.
30. Install the cooler condenser assembly.
31. Install the upper radiator support sub-assembly.
32. Install the hood lock assembly.
33. Install the radiator grill.

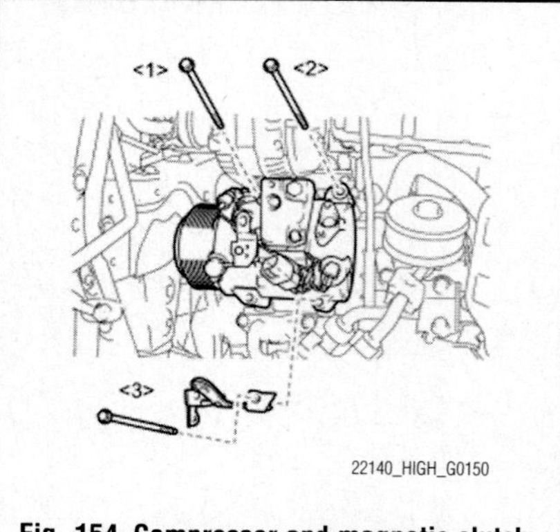

Fig. 154 Compressor and magnetic clutch

34. Connect the cooling fan ECU connector.
35. Connect the oil cooler hose.
36. Connect the No. 1 and 2 radiator hoses.
37. Install the No. 1 and 2 air cleaner inlets.
38. Install the battery.
39. Install the cool air intake duct seal.
40. Charge with refrigerant.
41. Add engine coolant.
42. Inspect for coolant leak.
43. Inspect automatic transaxle fluid.
44. Inspect for oil leak.
45. Warm up the engine.
46. Inspect for refrigerant leak.
47. Install the No. 1 engine under cover.
48. Install the engine under cover assembly.
49. Install the V-bank cover sub-assembly.

ELECTRONIC CONTROL MODULE (ECM)

LOCATION

See Figure 155.

The Hybrid ECU is located behind the glove compartment and to the right of the blower unit.

REMOVAL & INSTALLATION

See Figure 155.

1. Disconnect the negative battery cable.
2. Remove the instrument under cover.

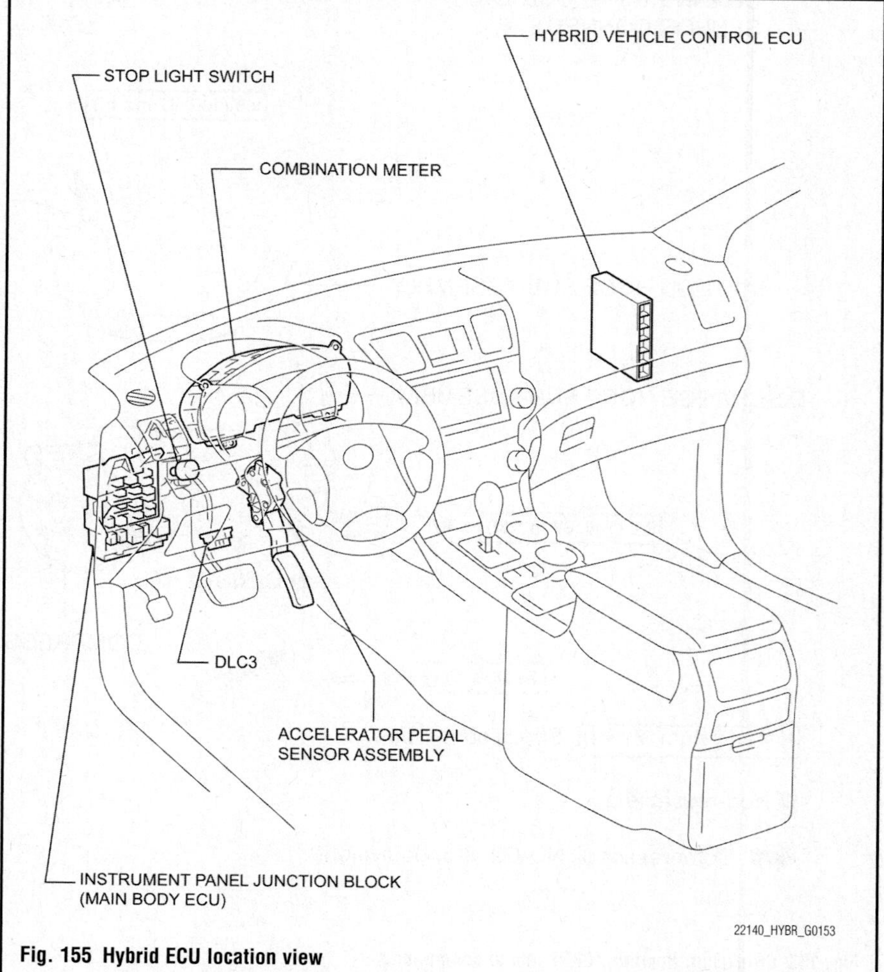

Fig. 155 Hybrid ECU location view

3. Remove the right door scuff plate.
4. Remove the cowl side trim.
5. Remove the glove compartment door assembly.
6. Disconnect the 3 wire harness clamps and 6 connectors from the hybrid vehicle control ECU
7. Remove the 2 nuts and hybrid vehicle control ECU.
8. Remove the ECU.
9. If ECU is to be changed remove the brackets.

To install:

10. Install brackets to the ECU if previously removed.
11. Install the wire harness clamp bracket with the 2 screws.
12. Install the hybrid vehicle control ECU with the 2 nuts. Tighten the mounting nuts to 49 inch lbs. (5.5 Nm).
13. Connect the 6 connectors to the hybrid vehicle control ECU.
14. Connect the 3 wire harness clamps.
15. Install the glove compartment door assembly.
16. Install the undercover.
17. Install the cowl side trim.
18. Install the scuff plate.
19. Connect the negative battery cable.

➡After replacing the hybrid vehicle control ECU on vehicles with a dynamic laser cruise control system, it is necessary to initialize the hybrid vehicle control ECU so that the ECU can recognize the dynamic laser cruise control system.

Be sure to perform the following procedures after replacing the ECU.

- Turn the ignition switch to the on position.
- Turn the cruise main switch on.
- With the brake pedal depressed, push the cruise control main switch to RES/ACC 3 times within 3 seconds. Check that the buzzer sounds at this time.

➡Do not turn the headlight dimmer switch on at this time because the optical axis automatic adjustment mode has already started, which may lead to an incorrect optical axis setting. If the headlight dimmer switch is turned on by mistake, readjust the optical axis.

20. Some systems need initialization when disconnecting the cable from the negative battery terminal.

ENGINE COOLANT TEMPERATURE (ECT) SENSOR

LOCATION

See Figure 156.

The Engine Coolant Temperature (ECT) sensor is located on the top front area of the engine.

REMOVAL & INSTALLATION

See Figure 156.

1. Remove the engine room covers.
2. Remove the RH and LH wiper arm assembly.
3. Remove the cowl and seal.
4. Remove the wiper motor and link assembly.
5. Remove the cool air intake duct seal.
6. Remove the air cleaner cap with inlet.
7. Remove the air cleaner case with resonator.
8. Air cleaner bracket.
9. Disconnect the ECT sensor connector.
10. Using a deep socket, remove the sensor from the top of the engine.

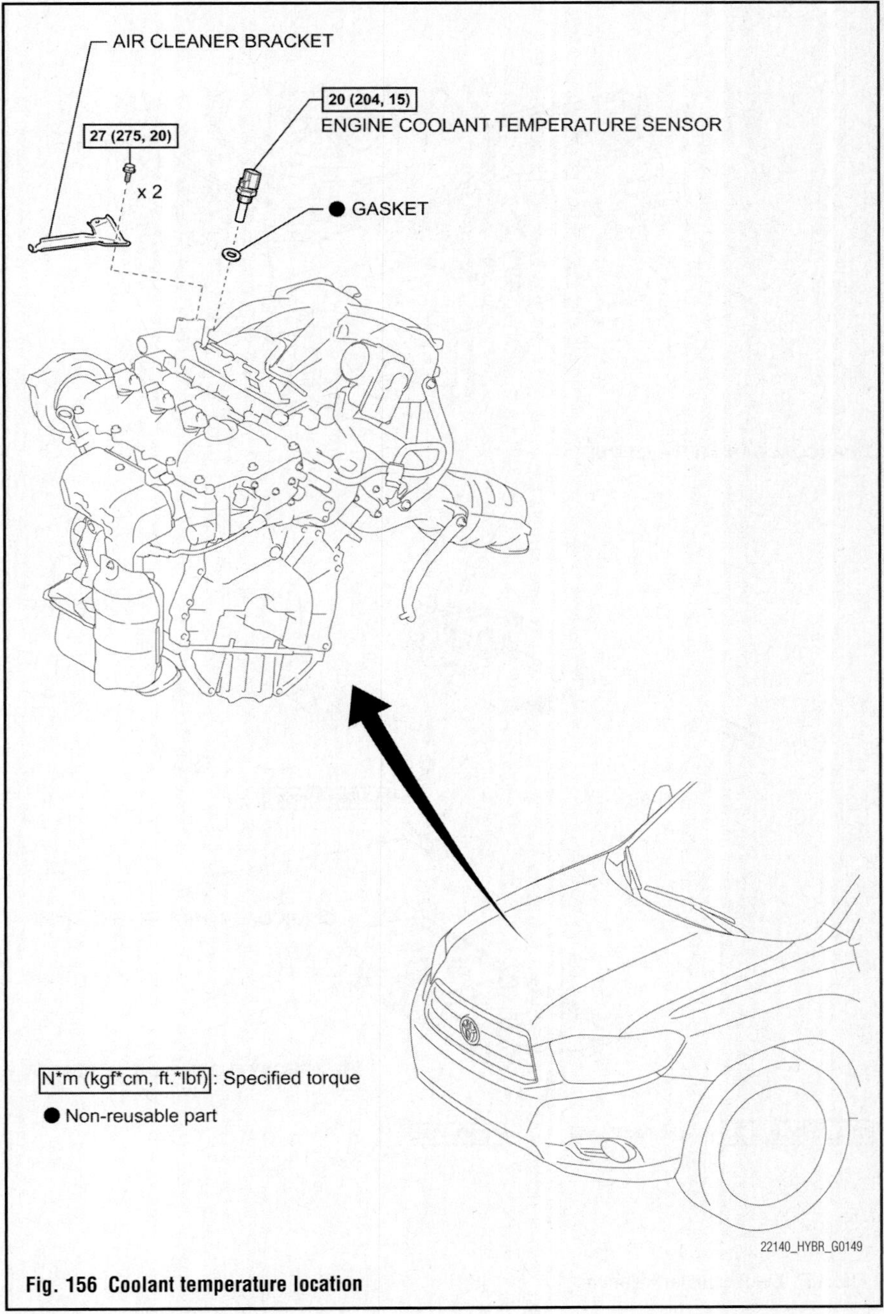

Fig. 156 Coolant temperature location

11. Remove the gasket from the sensor.
12. Installation is the reverse of the removal procedure.
13. Tighten the ECT sensor to 15 ft. lbs. (20 Nm).

EVAPORATIVE EMISSIONS (EVAP) CANISTER

LOCATION

See Figure 157.

Refer to the accompanying illustration for EVAP canister location.

REMOVAL & INSTALLATION

See Figure 158.

1. Remove tail exhaust pipe assembly.
2. Remove center exhaust pipe assembly.
3. Remove the 4 bolts and No. 3 front floor heat insulator.
4. Remove charcoal canister assembly:
    a. Disconnect the hose and connector.
    b. Disconnect the tube.
    c. Disconnect the charcoal canister fuel hose from the charcoal canister assembly. Pinch the retainer and pull out the quick connector with the quick connector pushed to the pipe side to disconnect the charcoal canister fuel hose from the charcoal canister assembly.

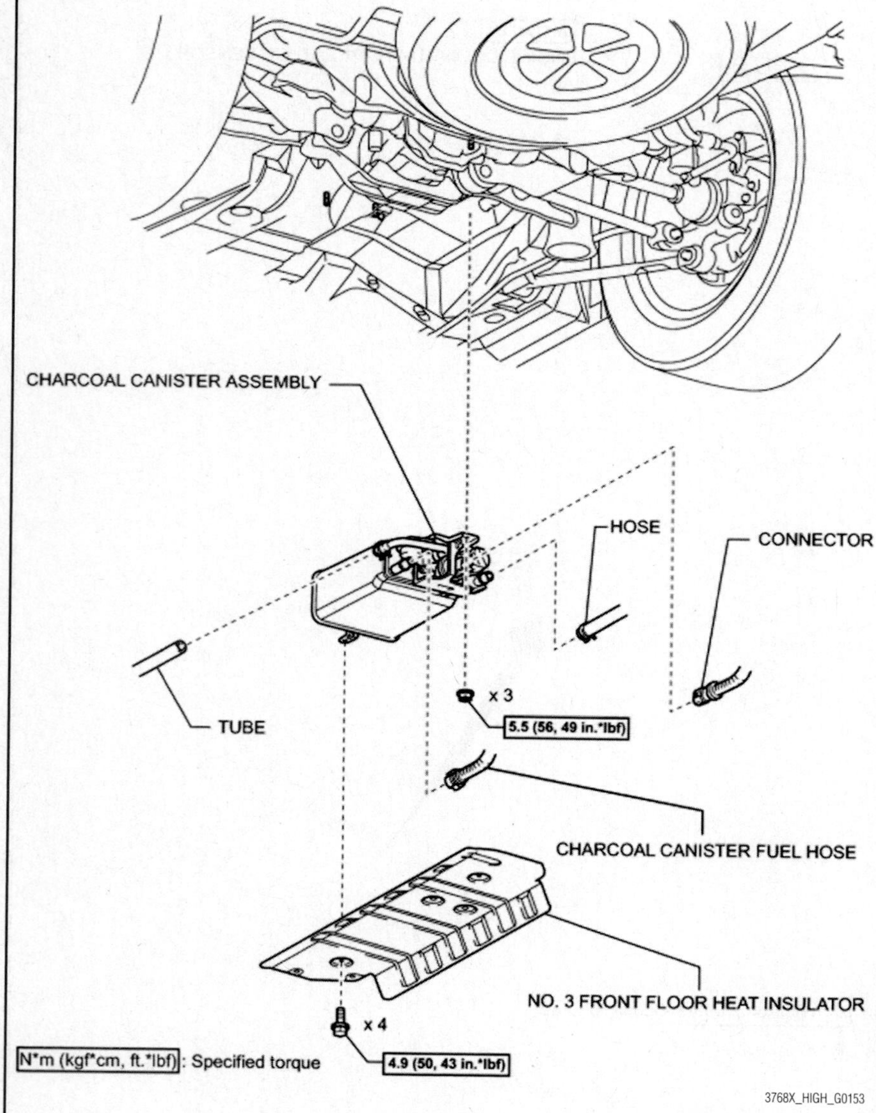

Fig. 157 EVAP canister location

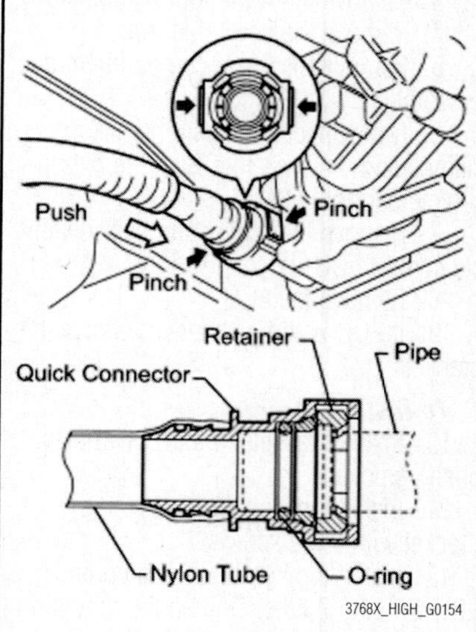

Fig. 158 Disconnect the charcoal canister fuel hose from the charcoal canister assembly

➡**Note the following:**

- Remove dirt or foreign objects on the quick connector before this work.
- Do not allow any scratches or foreign objects on the parts when disconnecting as the fuel hose connector has the O-ring that seals the pipe.
- Perform this work by hand. Do not use any tools.
- Do not forcibly bend, twist or turn the nylon tube. Protect the connected part by covering it with a vinyl bag after disconnecting the charcoal canister fuel hose.
- If the quick connector and pipe are stuck, push and pull them to release.

    d. Remove the 3 nuts and charcoal canister assembly.

5. To install, reverse removal procedure.

HEATED OXYGEN SENSOR (HO2S)

LOCATION

See Figure 159.

The Heated Oxygen Sensors (HO2S) are located below the converter. They are mounted in the front exhaust pipes.

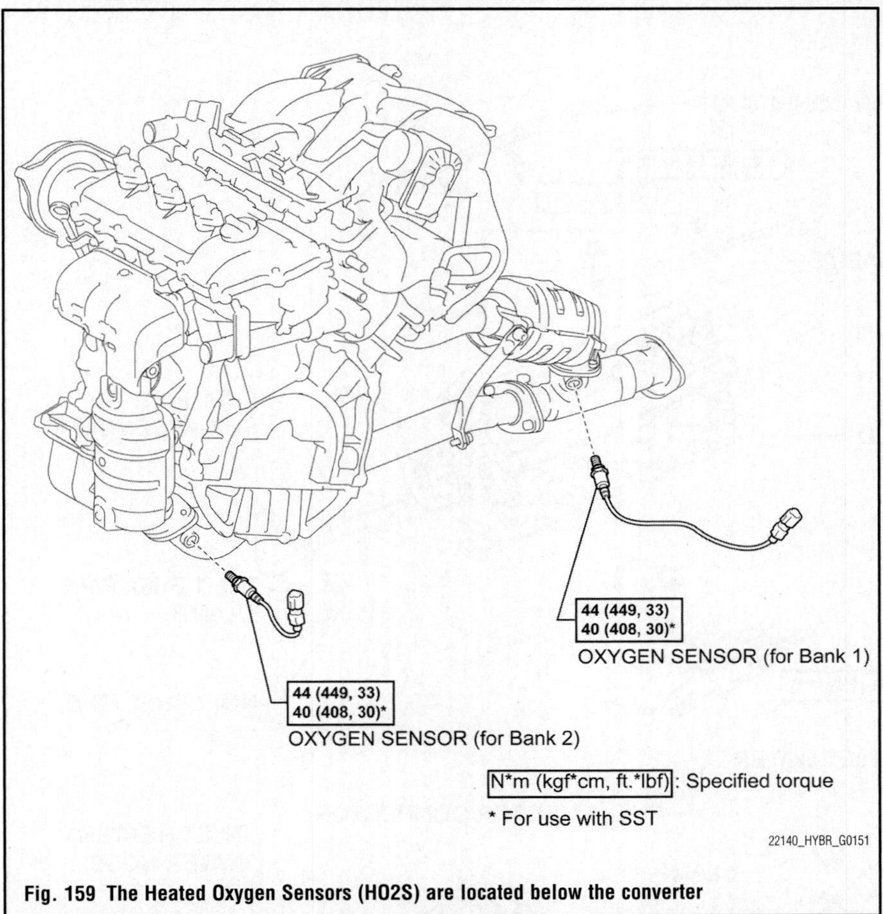

Fig. 159 The Heated Oxygen Sensors (HO2S) are located below the converter

REMOVAL & INSTALLATION

See Figures 160 and 161.

1. Bank 1:
   a. Disconnect the oxygen sensor connector.
   b. Using SST 09224-00010, remove the oxygen sensor from the front No. 3 exhaust pipe sub-assembly.
2. Bank 1:
   a. Using SST 09224-00010, remove the oxygen sensor from the front exhaust pipe sub-assembly.

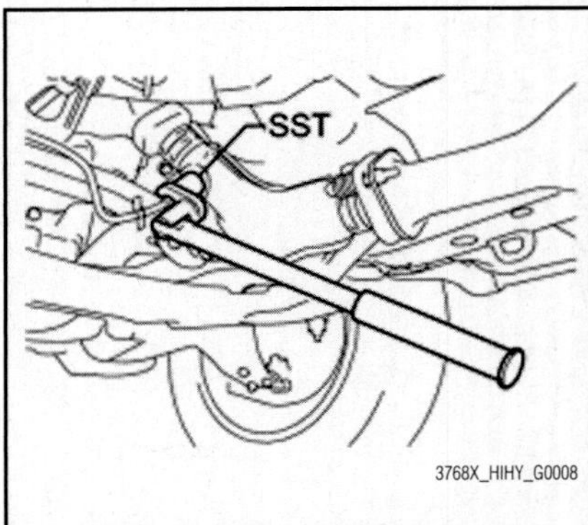

Fig. 160 Removing the oxygen sensor—Bank 1

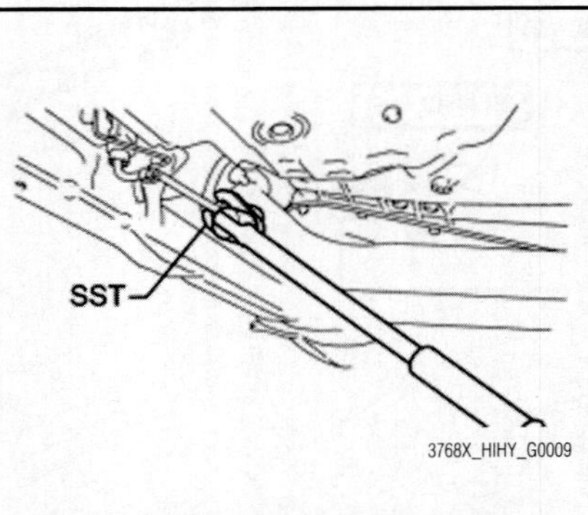

Fig. 161 Removing the oxygen sensor—Bank 2

To install:

3. Temporarily tighten the oxygen sensor to the exhaust pipe sub-assembly front.
4. Using SST, fully tighten the oxygen sensor to the exhaust pipe sub-assembly front.
5. Tighten the sensor to 33 ft. lbs. (44 Nm),
6. Connect the oxygen sensor connector.

KNOCK SENSOR (KS)

LOCATION

See Figure 162.

Refer to the accompanying illustration for sensor location.

REMOVAL & INSTALLATION

See Figures 163 through 166.

1. Discharge the fuel system.
2. Remove the engine room covers.
3. Disconnect the negative battery cable.
4. Drain the engine coolant.
5. Remove LH and RH wiper arm assembly.
6. Remove the cowl top ventilator louver.
7. Remove the wiper motor and link assembly.
8. Remove the cowl top panel.
9. Remove the cool air intake duct seal.
10. Remove the air cleaner assembly.
11. Remove the brake master cylinder reservoir.
12. Remove the air cleaner bracket.
13. Remove the air filter bracket.
14. Remove engine moving control rod.
15. Disconnect the VSV connector.
16. Remove the wire harness clamp.
17. Disconnect the fuel vapor feed hose No. 1.
18. Disconnect the fuel vapor feed hose No. 2.
19. Remove the 2 nuts, then remove the emission control valve set.
20. Disconnect the throttle motor connector.
21. Separate the water by-pass hose No. 2.
22. Separate the water by-pass hose No. 3.
23. Disconnect the ventilation hose.
24. Remove the 2 bolts, then remove the engine hanger No.1
25. Remove the 2 bolts, then remove the engine hanger No.1
26. Remove the 2 bolts, then remove the surge tank stay No. 1.
27. Remove the 2 bolts, then remove the surge tank stay No. 2.
28. Disconnect the ground cable connector.
29. Using a socket hexagon wrench 8 mm, remove the 4 bolts.
30. Remove the 2 nuts, then remove the emission control valve bracket and the intake air surge tank.
31. Remove the gasket from the intake air surge tank.

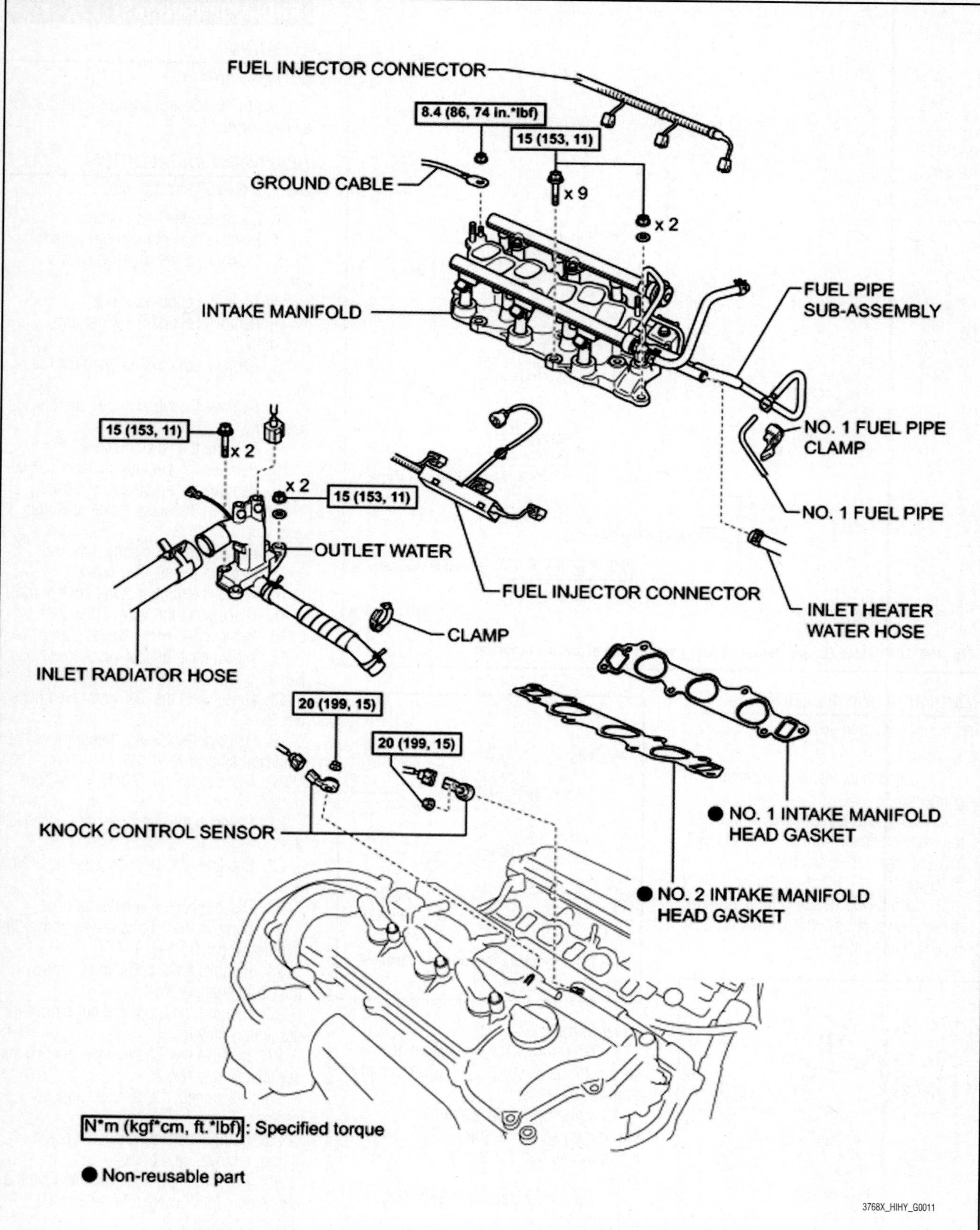

Fig. 162 Knock Sensor (KS) location

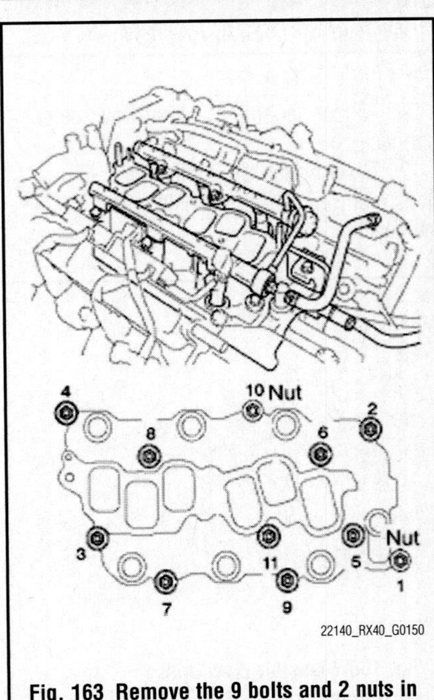

Fig. 163 Remove the 9 bolts and 2 nuts in the sequence shown

32. Remove the EFI fuel pipe clamp No.1.
33. Pinch the quick connector and then pull out the fuel pipe No. 1.
34. Disconnect the heater water inlet hose.
35. Disconnect the radiator hose inlet.
36. Remove the nut and ground cable.
37. Disconnect the 6 fuel injector connectors.
38. In order to remove the intake manifold, using several steps, remove the 9 bolts and 2 nuts in the sequence shown in the illustration.
39. Remove the water outlet.
40. Disconnect the 2 Knock Sensor (KS) connectors.
41. Remove the 2 nuts, and then remove the 2 knock sensors.

To install:

42. Install the 2 knock sensors so that it is horizontal as shown in the illustration. Then install the 2 bolts and tighten to 15 ft. lbs. (20 Nm).
43. Connect the 2 Knock Sensor (KS) connectors.
44. Install the intake manifold with the 9 bolts, 2 nuts and 2 washers. Using several steps, tighten the bolts and nuts uniformly in the removal sequence. Tighten to 11 ft. lbs. (1 Nm).
45. Connect the 6 fuel injector connectors.
46. Install the ground cable with the nut.
47. Connect the radiator hose inlet.
48. Connect the heater water inlet hose.
49. Align the quick connector with the pipe, then push in the quick connector until the retainer makes a "click" sound to connect the fuel hose to the fuel pipe.
50. Using a socket hexagon wrench 8 mm, install the intake manifold with the 4 bolts and 2 nuts . Using several steps, tighten the bolts and nuts uniformly in the sequence shown in the illustration. Tighten the nuts and bolts to 21 ft. lbs. (28 Nm).
51. Connect the ground cable connector.
52. Install the surge tank stay brackets and tighten to 14 ft. lbs. 20 (Nm).
53. Connect the ventilation hose.
54. Connect the fuel vapor feed hose.
55. Connect both of the water by-pass hoses.
56. Reconnect the throttle motor connector.
57. Install the emission control valve set with the 2 nuts and tighten to 80 inch lbs (8 Nm).
58. Reconnect both of the fuel vapor hose.
59. Connect the wire harness clamp.
60. Connect the VSV connector.
61. Install the engine moving control rod.
62. Install the reservoir bracket and tighten to 71 inch lbs. (9 Nm).
63. Install air filter assembly bracket and tighten to 14 ft. lbs. 20 (Nm).

FUEL INJECTOR CONNECTOR
8.4 (86, 74 in.*lbf)
15 (153, 11)
GROUND CABLE
FUEL PIPE SUB-ASSEMBLY
INTAKE MANIFOLD
15 (153, 11)
15 (153, 11)
EFI FUEL PIPE CLAMP NO.1
FUEL INJECTOR CONNECTOR
FUEL PIPE NO.1
WATER OUTLET
HEATER WATER INLET HOSE
CLAMP
15 (153, 11)
KNOCK CONTROL SENSOR
RADIATOR HOSE INLET
KNOCK CONTROL SENSOR
● INTAKE MANIFOLD TO HEAD GASKET NO.1
20 (199, 15)
● INTAKE MANIFOLD TO HEAD GASKET NO.2
N*m (kgf*cm, ft.*lbf) : Specified torque ● Non-reusable part
22140_RX40_G0152

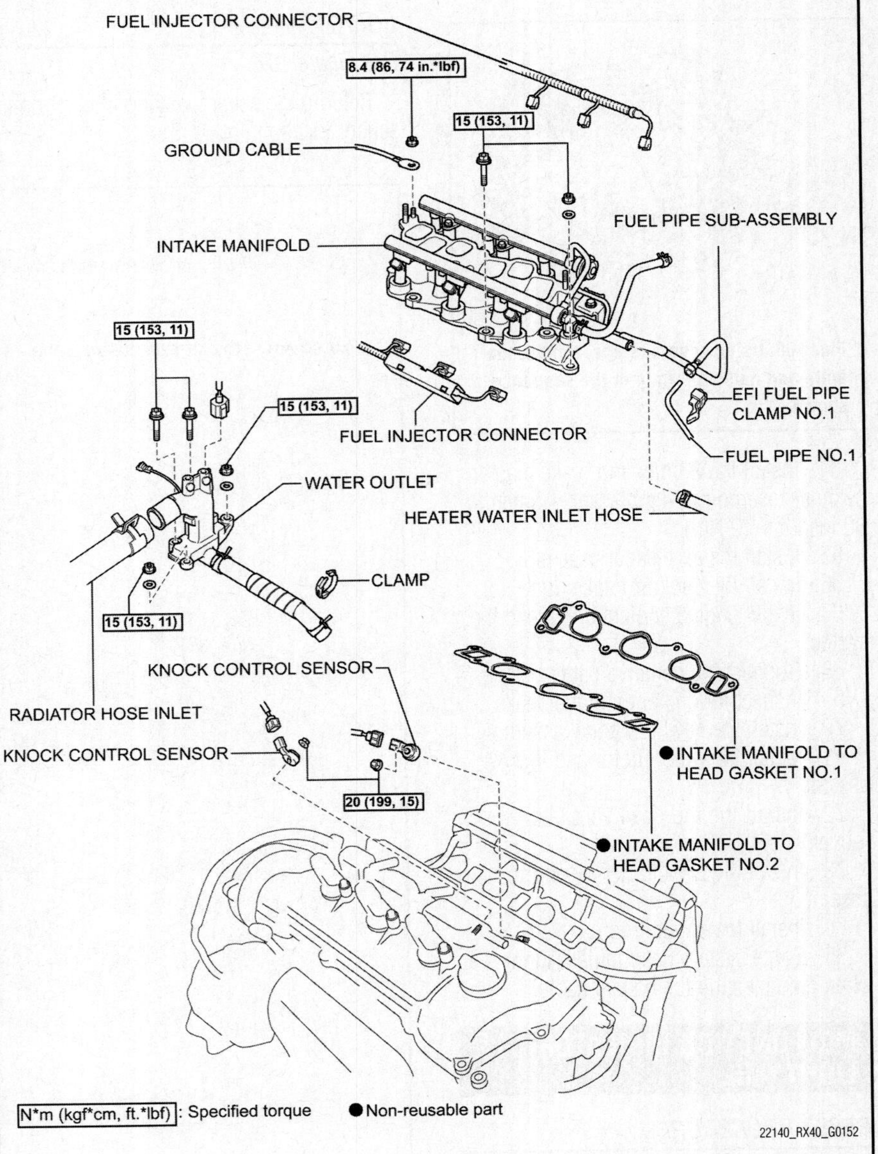

Fig. 164 Knock Sensor (KS) and related components

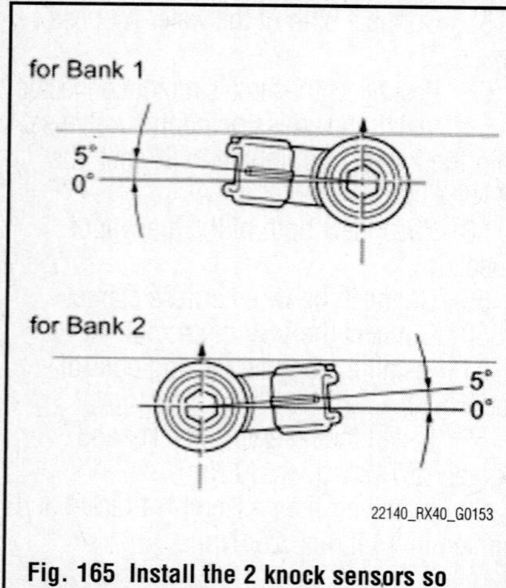

Fig. 165 Install the 2 knock sensors so that it is horizontal

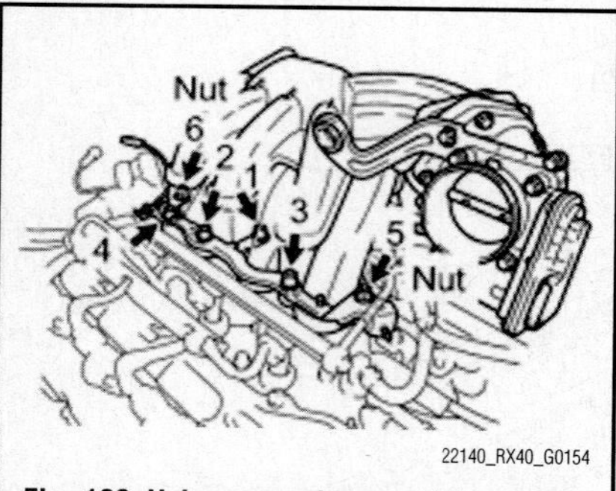

Fig. 166 Using several steps, tighten the bolts and nuts uniformly in the sequence shown

64. Install the 2 bolts and brake master cylinder reservoir to the bracket. Tighten to 80 inch lbs. (9 Nm).
65. Install the air cleaner assembly.
66. Install the cool air intake duct.
67. Install engine coolant and bleed the system.
68. Connect the negative battery cable.
69. Inspect for coolant and fuel leaks.
70. Install the cowl top panel assembly.
71. Install the wiper motor and link assembly.
72. Install the top cowl ventilator louver.
73. Install the LH and RH wiper arm assembly.
74. Install the engine room covers.
75. Some system need initialization when reconnecting the battery cable.

MALFUNCTION INDICATOR LIGHT (MIL)

RESET PROCEDURE

1. Clear DTC (Using the Techstream) as follows:
   a. Connect the Techstream to the DLC3.
   b. Turn the ignition switch ON.
   c. Enter the following menus: Powertrain / Engine and ECT / Trouble Codes.
   d. Press the YES button.
2. Clear DTC (Without using the Techstream). Perform either one of the following operations.
   a. Disconnect the negative (-) battery cable for more than 1 minute.
   b. Remove the EFI and ETCS fuses from the engine room No. 2 relay block located inside the engine compartment for more than 1 minute.

MASS AIR FLOW (MAF) SENSOR

LOCATION

See Figure 167.

Refer to the accompanying illustration for sensor location.

REMOVAL & INSTALLATION

See Figure 168.

1. Disconnect the mass air flow meter connector.
2. Disconnect the 2 wire harness clamps from the air cleaner assembly.

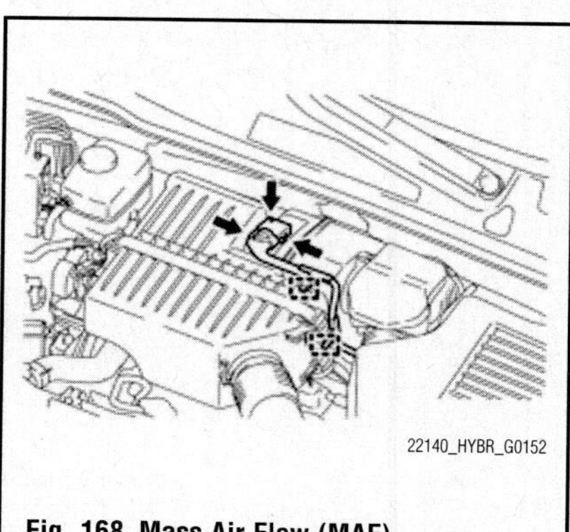

Fig. 168 Mass Air Flow (MAF) sensor/meter

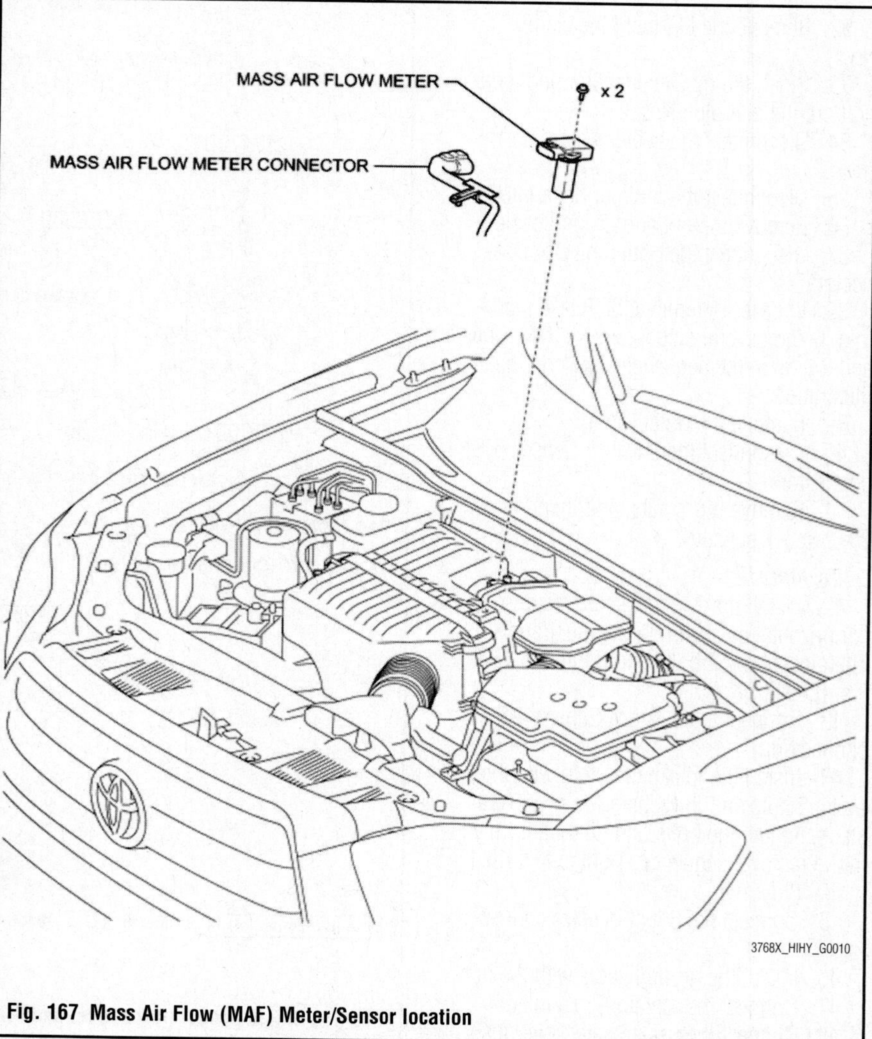

Fig. 167 Mass Air Flow (MAF) Meter/Sensor location

3. Remove the 2 screws and Mass Air Flow (MAF) meter.

To install:

4. Install the MAF meter with the 2 screws.
5. Connect the MAF meter connector.
6. Connect the 2 wire harness clamps to the air cleaner assembly.

POSITIVE CRANKCASE VENTILATION (PCV) VALVE

LOCATION

See Figure 169.

REMOVAL & INSTALLATION

See Figure 170.

1. Remove the intake manifold.
2. Remove the ventilation valve sub-assembly.

VARIABLE CAMSHAFT TIMING OIL CONTROL SOLENOID

LOCATION

See Figure 171.

The camshaft timing oil control valves are mounted in the rear of the cylinder heads.

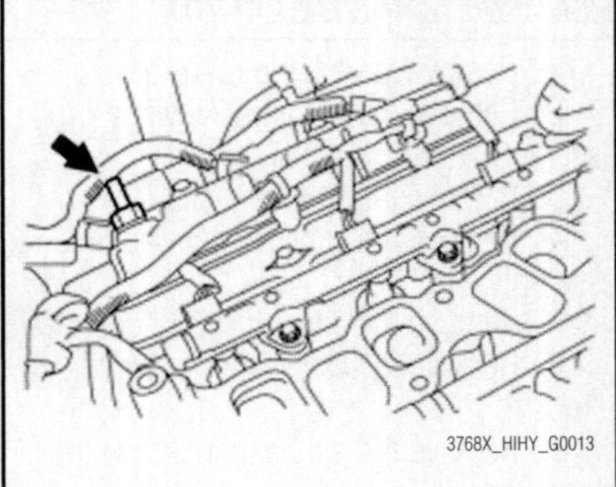

Fig. 170 Remove the ventilation valve sub-assembly

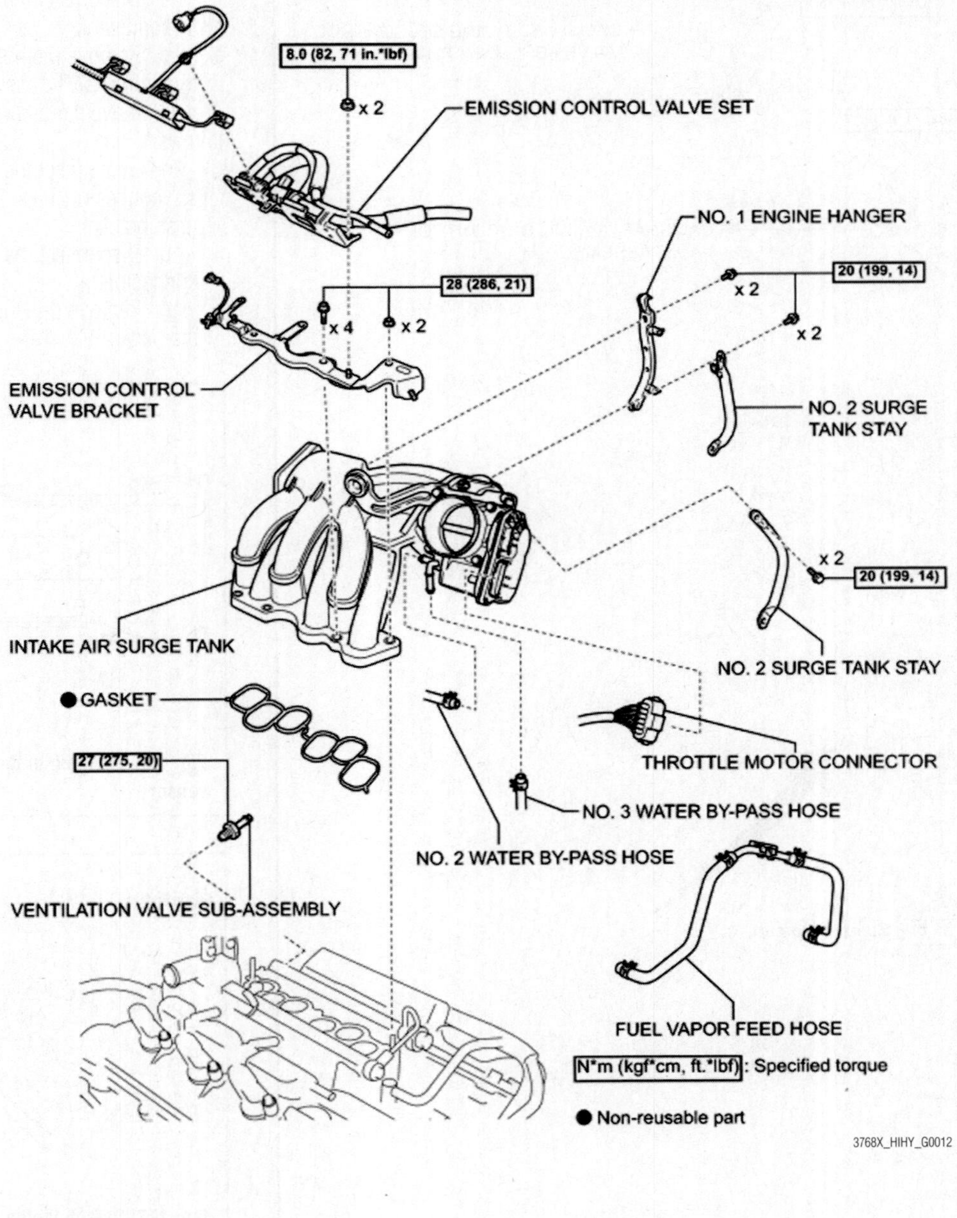

Fig. 169 Locating the PCV valve

REMOVAL & INSTALLATION

1. Remove the engine room covers.
2. Remove the RH and LH wiper arm assembly.
3. Remove the cowl and seal.
4. Remove the wiper motor and link assembly.
5. Remove the cool air intake duct seal.
6. Remove the air cleaner cap with inlet.
7. Remove the air cleaner case with resonator.
8. Disconnect the oil control valve connector.
9. Remove the bolt and oil control valve.

To install:

10. Apply a coat of engine oil to an O-ring of the sensor.
11. Install the camshaft timing oil control valve with the bolt.
12. Tighten the mounting bolt to 71 inch lbs. (8 Nm).
13. Connect the camshaft timing oil control valve connector.
14. Install the air cleaner case with resonator.
15. Install the air cleaner cap with inlet.
16. Install the cool air intake duct seal.
17. Install the wiper motor and link assembly.
18. Install the cowl and seal.
19. Install the RH and LH wiper arm assembly and engine room cover.

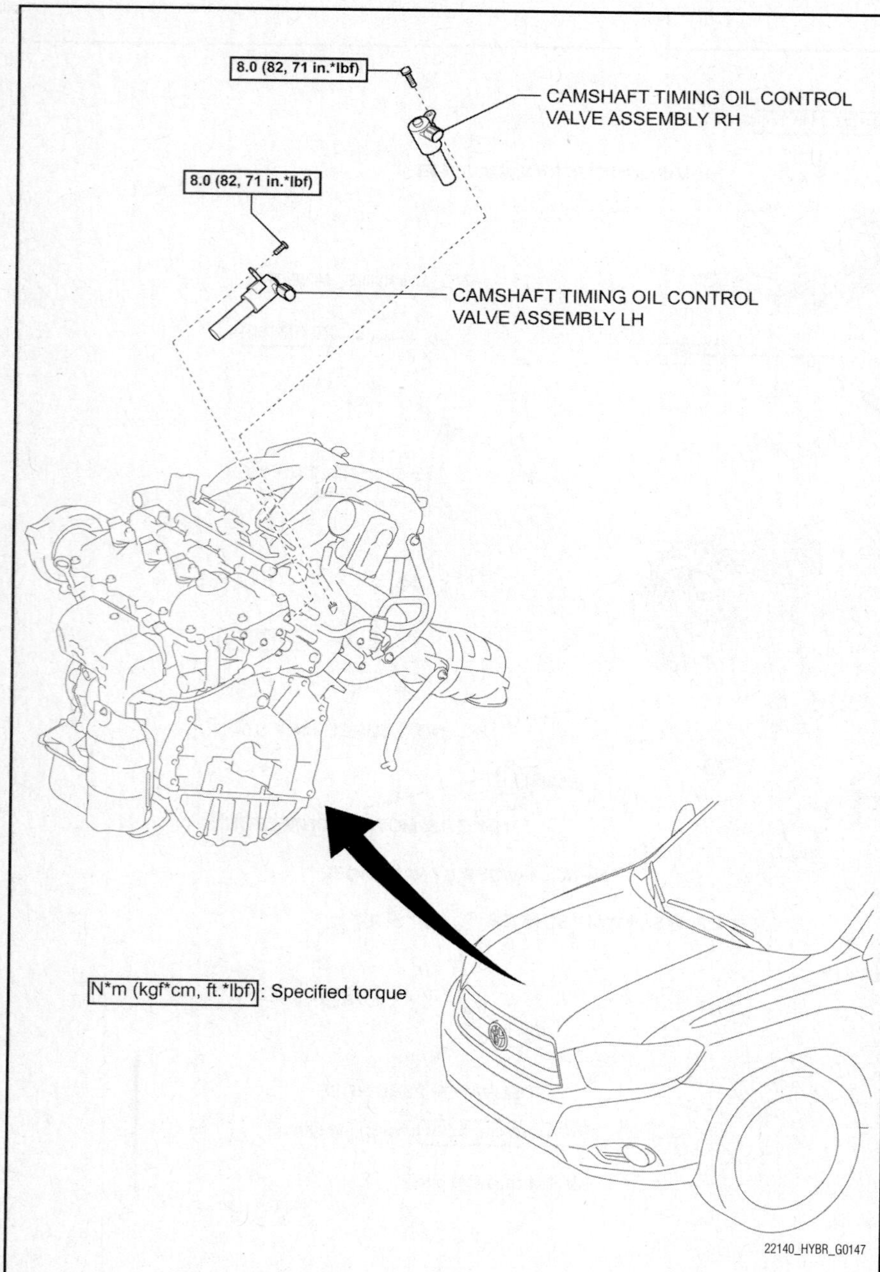

Fig. 171 Camshaft timing oil control valve location view

VARIABLE VALVE TIMING (VVT) SENSOR

REMOVAL & INSTALLATION

See Figures 172 through 176.

1. Disconnect the negative battery cable.
2. Remove air cleaner cap w/ inlet.
3. Remove the bank 1 VVT sensor, as follows:
   a. Disconnect the VVT sensor (bank 1) connector.
   b. Remove the bolt, and then remove the bank 1 VVT sensor.
4. Remove the bank 2 VVT sensor, as follows:
   a. Disconnect the heated oxygen sensor connector (bank 2) from the wire harness bracket.
   b. Disconnect the bank 2 VVT sensor connector.
   c. Remove the bolt, and then remove the bank 2 VVT sensor.

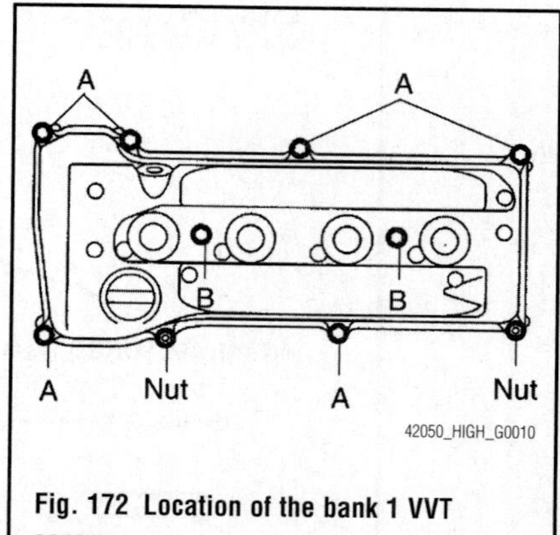

Fig. 172 Location of the bank 1 VVT sensor

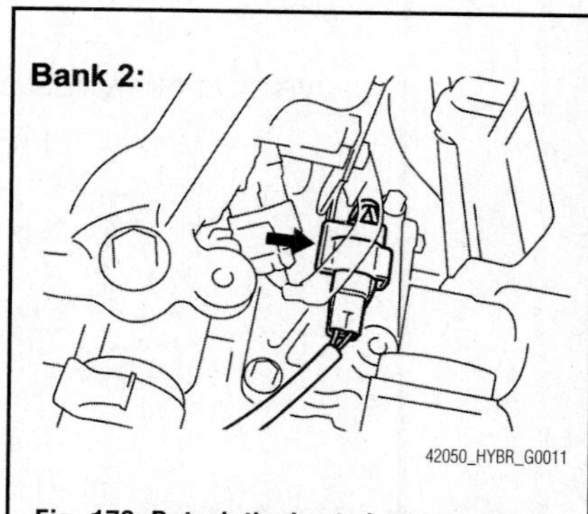

Fig. 173 Detach the heated oxygen sensor connector to access the bank 2 VVT sensor

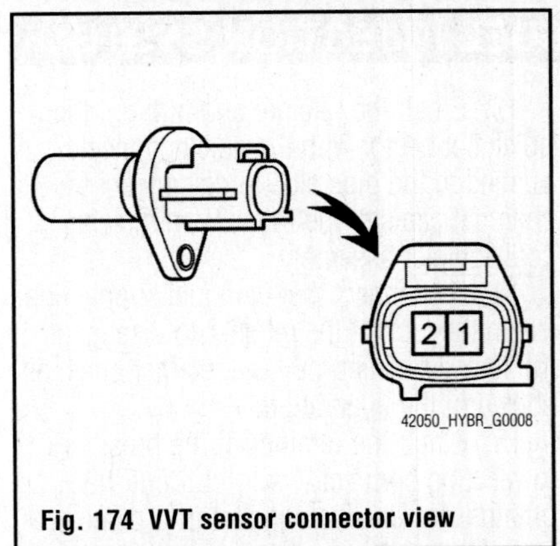

Fig. 174 VVT sensor connector view

Condition	Specified Condition
Cold	835 to 1,400 Ω
Hot	1,060 to 1,645 Ω

42050_HYBR_G0009

Fig. 175 VVT sensor resistance chart

5. Inspect the VVT sensor:

a. Using an ohmmeter, measure the resistance according to the value(s) in the accompanying chart. Standard resistance cold is 14–122°f (-10–50°c) and hot is 122–212°f (50–100°c. If the resistance is not within specifications, replace the sensor.

➡Cold and hot refer to the temperature of the coils themselves.

To install:

6. Install the bank 1 VVT sensor:

a. Apply a light coat of engine oil to the o-ring on the VVT sensor (bank 1).

b. Install the VVT sensor (bank 1) with the bolt an tighten to 71 inch lbs. (8 Nm).

c. Connect the VVT sensor (bank 1) connector.

7. Install the bank 2 VVT sensor:

a. Apply a light coat of engine oil to the o-ring on the VVT sensor (bank 2).

b. Install the VVT sensor (bank 2) with the bolt and tighten to 71 inch lbs. (8 Nm).

c. Connect the VVT sensor (bank 2) connector.

d. Install the heated oxygen sensor connector (bank 2) to the wire harness bracket.

8. Install air cleaner cap w/ inlet.

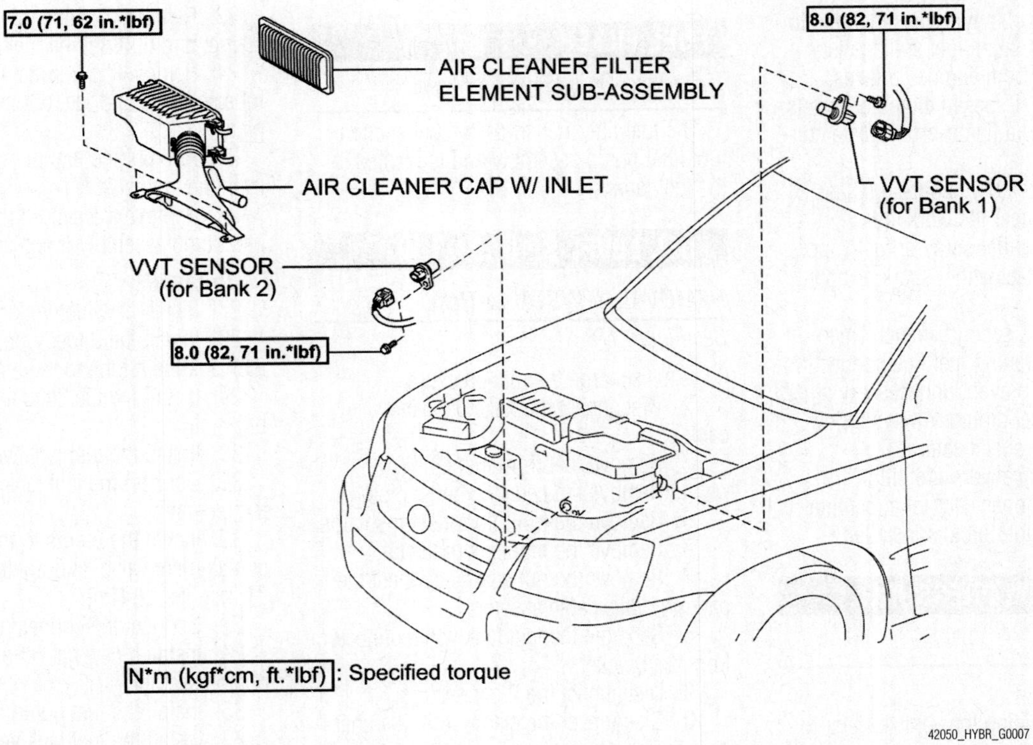

7.0 (71, 62 in.*lbf)
AIR CLEANER FILTER ELEMENT SUB-ASSEMBLY
8.0 (82, 71 in.*lbf)
AIR CLEANER CAP W/ INLET
VVT SENSOR (for Bank 1)
VVT SENSOR (for Bank 2)
8.0 (82, 71 in.*lbf)
N*m (kgf*cm, ft.*lbf) : Specified torque
42050_HYBR_G0007

Fig. 176 Exploded view of the VVT sensors and related components

FUEL SYSTEM SERVICE PRECAUTIONS

Safety is the most important factor when performing not only fuel system maintenance, but any type of maintenance. Failure to conduct maintenance and repairs in a safe manner may result in serious personal injury or death. Work on a vehicle's fuel system components can be accomplished safely and effectively by adhering to the following rules and guidelines.

- To avoid the possibility of fire and personal injury, always disconnect the negative battery cable unless the repair or test procedure requires that battery voltage be applied.
- Always relieve the fuel system pressure prior to disconnecting any fuel system component (injector, fuel rail, pressure regulator, etc.) fitting or fuel line connection. Exercise extreme caution whenever relieving fuel system pressure to avoid exposing skin, face and eyes to fuel spray. Please be advised that fuel under pressure may penetrate the skin or any part of the body that it contacts.
- Always place a shop towel or cloth around the fitting or connection prior to loosening to absorb any excess fuel due to spillage. Ensure that all fuel spillage is quickly removed from engine surfaces. Ensure that all fuel-soaked cloths or towels are deposited into a flame-proof waste container with a lid.
- Always keep a dry chemical (Class B) fire extinguisher near the work area.
- Do not allow fuel spray or fuel vapors to come into contact with a spark or open flame.
- Always use a second wrench when loosening or tightening fuel line connection fittings. This will prevent unnecessary stress and torsion on fuel piping. Always follow the proper torque specifications.
- Always replace worn fuel fitting O-rings with new ones. Do not substitute fuel hose where rigid pipe is installed.

FUEL SYSTEM PRESSURE

RELIEVING

See Figure 177.

1. Before servicing the vehicle, refer to the Precautions Section.
2. Disconnect the No. 3 relay block.
3. Remove the No. 2 junction block cover.
4. Remove the C/OPN RLY.
5. Put the vehicle in Inspection Mode and start the engine.

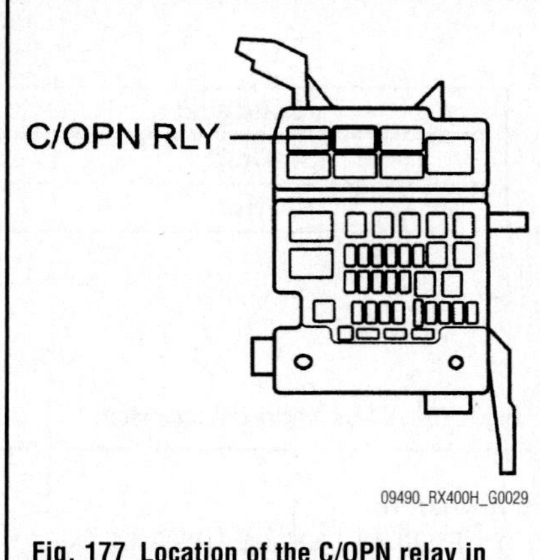

Fig. 177 Location of the C/OPN relay in the junction box.

6. Turn the ignition switch to **OFF** immediately after the engine comes to 'rough idle state'.

➡The hybrid system has a complicated process from an 'out of gas' to 'engine stall' condition. Therefore, 'rough idle' is regarded as 'stop'.

7. Disconnect the negative battery cable.
8. Reinstall the C/OPN RLY.

FUEL FILTER

REMOVAL & INSTALLATION

The fuel filter is part of the fuel suction tube/fuel pump assembly and is located in the fuel tank.

FUEL LEVEL SENDING UNIT

REMOVAL & INSTALLATION

See Figure 178.

1. Discharge the fuel pressure.
2. Disconnect the negative battery cable.
3. Remove fuel from fuel tank.
4. Remove the center exhaust pipe.
5. Remove the front floor heat insulator.
6. Remove the fuel tank protector.
7. Remove the nut and disconnect the parking brake cable assembly.
8. Separate the fuel tank wire connector from the bracket.
9. Disconnect the 2 connectors.
10. Disconnect the clamp from the charcoal canister protector.
11. Remove the 3 bolts and charcoal canister protector.
12. Disconnect the charcoal canister fuel hose from the charcoal canister assembly.
13. Pinch the retainer and pull out the quick connector with the quick connector pushed to the pipe side to disconnect the charcoal canister fuel hose from the charcoal canister assembly.
14. Disconnect the main fuel supply line. Pinch the tabs of the retainer to disengage the lock claws and pull the retainer down as shown in the illustration.
15. Pinch the retainer of the breather lower tube connector, and pull out the breather lower tube connector to disconnect the breather lower tube from the fuel tank.
16. Set a transmission jack to the fuel tank.
17. Remove the fuel filler pipe clamp and fuel tube connector from the fuel tank inlet pipe.
18. Remove the 4 bolts and the fuel tank bands.
19. Operate the transmission jack, and disconnect the fuel inlet pipe.
20. Operate the transmission jack, and remove the fuel tank.
21. Disconnect the fuel pump connector.
22. Disconnect the clamp and vapor pressure sensor connector.
23. Remove the tube joint clip and clamp, and pull out the fuel pump tube.
24. Remove the 8 bolts, fuel tank vent tube set plate and fuel pump connector plate.
25. Pull out the fuel pump assembly from the fuel tank.
26. Remove the sending unit from the fuel pump assembly if applicable.

To install:

27. Install new fuel tank cushions.
28. Install the 2 fuel tube clamps of the fuel tank main tube to the fuel tank.
29. Install the fuel tank main fuel assembly.
30. Install the fuel tank wire.
31. Connect the fuel tank vent hose to the fuel tank
32. Install the 2 nuts to the fuel vapor containment valve, and tighten the nuts to 71 inch lbs. (8 Nm).
33. Connect the connector and clamp.
34. Install a new gasket on the fuel pump assembly.
35. Install the fuel pump assembly.
36. Install the fuel tank vent tube set plate and fuel pump connector plate with the 8 bolts. Tighten the bolts to 53 inch lbs. (6 Nm).
37. Connect the fuel tank main tube with the tube joint clip and clamp.

38. Connect the fuel tank wire connector.
39. Connect the vapor pressure sensor connector and clamp.
40. Set the fuel tank on a transmission jack.
41. Operate the transmission jack, and install the fuel tank to the vehicle.
42. Operate the transmission jack, and connect the fuel tank inlet pipe with the tube connector and clamp.
43. Install the 2 fuel tank bands with the 4 bolts and tighten to 29 ft. lbs. (39 Nm).
44. Connect the breather lower tube. Push in the tube connector to the pipe until the tube connector makes a "click" sound.
45. Install the checker to the pipe.
46. Connect the fuel tank main supply line. Push in the fuel tube connector to the pipe and push up the retainer to engage the claws.
47. Align the quick connector with the pipe, then push in the quick connector until the retainer makes a "click" sound to connect the charcoal canister fuel hose to the charcoal canister assembly.
48. Connect the clamp to the charcoal canister protector.
49. Install the charcoal canister protector with the 3 bolts. Tighten the bolts to 48 inch lbs. (5.4 Nm).
50. Connect the fuel tank wire connectors, and install them to the bracket.
51. Install the parking brake cable assembly No.3 with the nut.
52. Install the fuel tank protector.
53. Install the front floor heat insulator and tighten the retaining nuts to 43 inch lbs. (4.9 Nm).
54. Install the center pipe exhaust assembly.
55. Connect the negative battery cable.
56. Add fuel and inspect for fuel leaks.
57. Inspect for exhaust leaks.
58. Some systems need initialization after reconnecting the cable to the negative battery terminal.

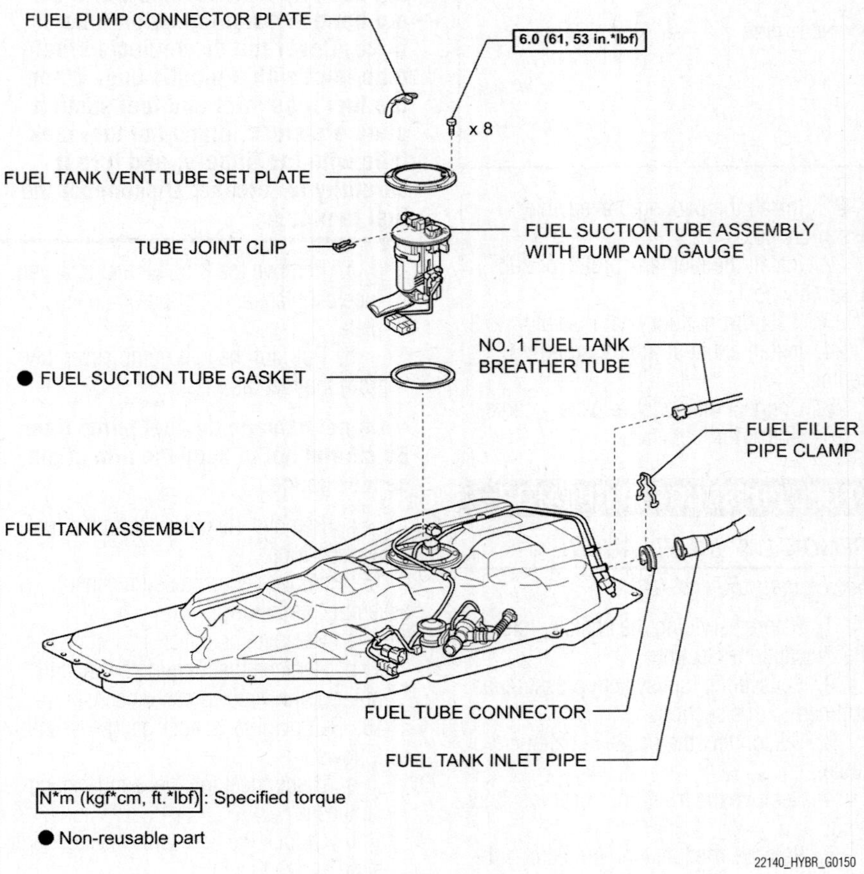

Fig. 178 Fuel level sending unit and related parts

FUEL PUMP

REMOVAL & INSTALLATION

See Figure 179.

1. Before servicing the vehicle, refer to the Precautions Section.
2. Relieve the fuel system pressure, as outlined in this section.
3. Disconnect the negative battery cable.
4. Remove the front floor heat insulator no.3.
5. Remove the fuel tank protector subassembly no.1.
6. Separate parking brake cable assembly no.3.
7. Disconnect the fuel tank wire.
8. Remove the charcoal canister protector.
9. Disconnect the charcoal canister fuel hose.
10. Disconnect the fuel tank main tube subassembly.
11. Disconnect the breather lower tube.
12. Remove the fuel tank assembly.
13. Remove the fuel pump and gauge with the suction tube, as follows:
    a. Disconnect the fuel pump connector.
    b. Disconnect the clamp and vapor pressure sensor connector.
14. Remove the tube joint clip and clamp, and pull out the fuel pump tube.

WARNING

Check that there is no dirt or foreign matter around the fuel tube joint and after cleaning off any excess dirt, proceed with the work. Be careful not to allow dirt or foreign matter to scratch or come into contact with the fuel tube connector and fuel suction plate that are sealed by o-rings. Do not use any tools for this work. Do not bend or twist the nylon tube by force. Cover the disconnected fuel tube joint with a plastic bag. When the fuel tube joint and fuel suction plate are stuck, pinch the fuel tank tube with the fingers, and turn it carefully to release. Disconnect the fuel tank tube.

a. Remove the 8 bolts, fuel tank vent tube set plate and fuel pump connector plate.
b. Pull out the fuel pump assembly from the fuel tank.

To install:

c. Install the fuel pump assembly.
d. Install the fuel tank vent tube set

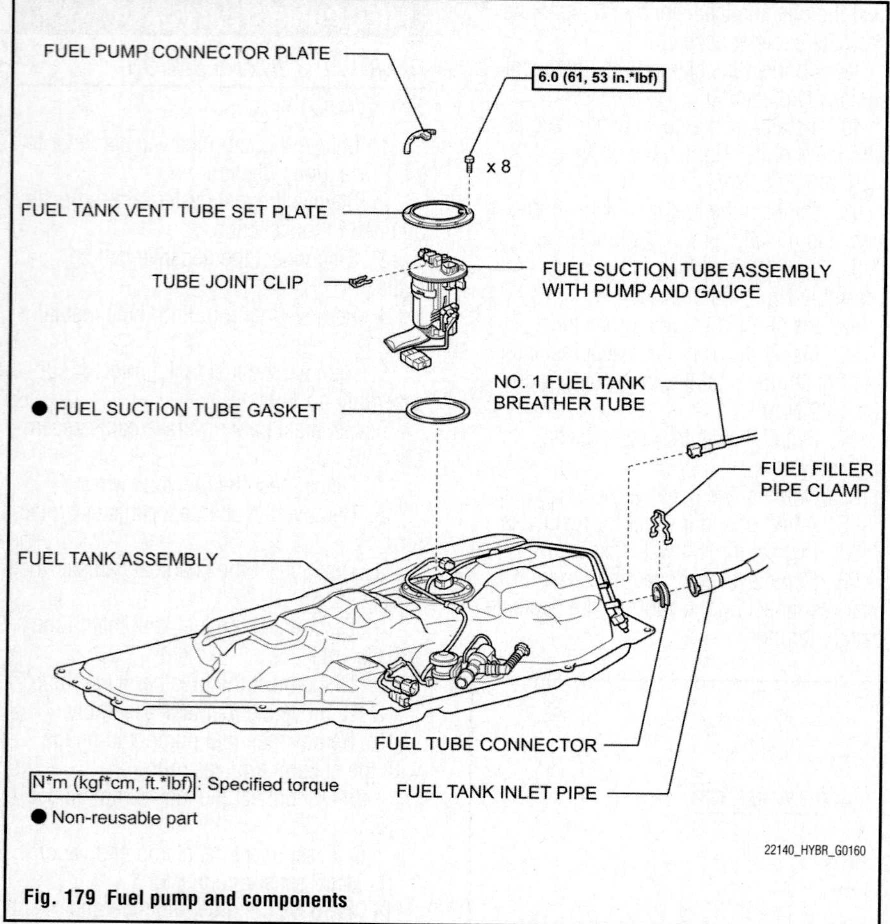

Fig. 179 Fuel pump and components

plate by aligning it with the cutout on the fuel pump assembly.

e. Install the fuel tank vent tube set plate and fuel pump connector plate with the 8 bolts. Tighten to 53 inch lbs. (6.0 Nm).

f. Connect the fuel tank main tube with the tube joint clip and clamp.

➡Check that there are no scratches or foreign objects on the connecting part. Check that the fuel tube joint is inserted securely. Check that the tube joint clip is on the collar of the fuel tube joint. after installing the tube joint clip, check that the fuel tube joint is not pulled off.

g. Connect the fuel tank wire connector.

h. Connect the vapor pressure sensor connector and clamp.

15. Install the fuel tank assembly.
16. Connect the breather lower tube,
17. Connect the fuel tank main tube sub-assembly.
18. Connect charcoal canister fuel hose.
19. Install charcoal canister protector.
20. Connect the fuel tank wire.
21. Install the parking brake cable assembly no.3.
22. Install the fuel tank protector sub-assembly no.1.
23. Install front floor heat insulator no.3.
24. Install exhaust pipe assembly center.
25. Connect the negative battery cable.
26. Check for fuel leaks.

FUEL PRESSURE REGULATOR

REMOVAL & INSTALLATION

See Figures 180 and 181.

1. Before servicing the vehicle, refer to the Precautions Section.
2. Relieve the fuel system pressure, as outlined in this section.
3. Disconnect the negative battery cable.
4. Remove the front floor heat insulator no.3.
5. Remove the fuel tank protector sub-assembly no.1.
6. Separate parking brake cable assembly no.3.
7. Disconnect the fuel tank wire.
8. Remove the charcoal canister protector.
9. Disconnect the charcoal canister fuel hose.
10. Disconnect the fuel tank main tube subassembly.
11. Disconnect the breather lower tube.
12. Remove the fuel tank assembly.
13. Remove the fuel pump and gauge with the suction tube, as follows:

a. Disconnect the fuel pump connector.

b. Disconnect the clamp and vapor pressure sensor connector.

14. Remove the tube joint clip and clamp, and pull out the fuel pump tube.

⁂ WARNING

Check that there is no dirt or foreign matter around the fuel tube joint and after cleaning off any excess dirt, proceed with the work. Be careful not to allow dirt or foreign matter to scratch or come into contact with the fuel tube connector and fuel suction plate that are sealed by o-rings. Do not use any tools for this work. Do not bend or twist the nylon tube by force. Cover the disconnected fuel tube joint with a plastic bag. When the fuel tube joint and fuel suction plate are stuck, pinch the fuel tank tube with the fingers, and turn it carefully to release. Disconnect the fuel tank tube.

a. Remove the 8 bolts, fuel tank vent tube set plate and fuel pump connector plate.

b. Pull out the fuel pump assembly from the fuel tank.

➡Do not damage the fuel pump filter. Be careful not to bend the arm of the sender gauge.

c. Remove the gasket from the fuel suction tube.

15. Remove vapor pressure sensor assembly, as follows:

a. Remove the cover.

b. Remove the tube joint clip, and pull out the vapor pressure sensor.

16. Remove fuel sender gauge assembly, as follows:

a. Disconnect the fuel sender gauge connector.

b. Unlock the fuel sender gauge, then slide and remove it.

17. Remove fuel suction plate sub-assembly:

a. Disconnect the fuel pump connector.

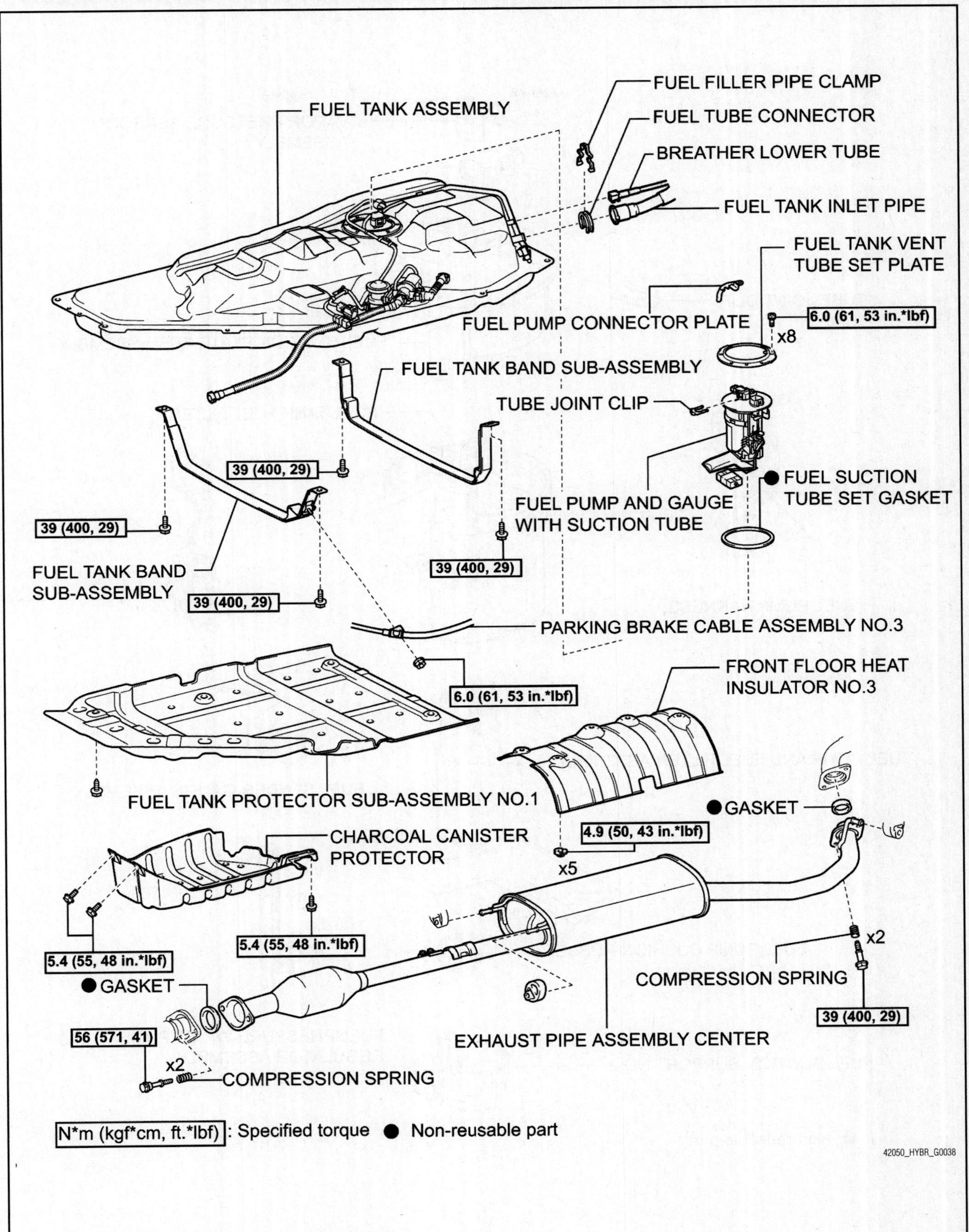

Fig. 180 Exploded view of the fuel pump and related components (1 of 2)

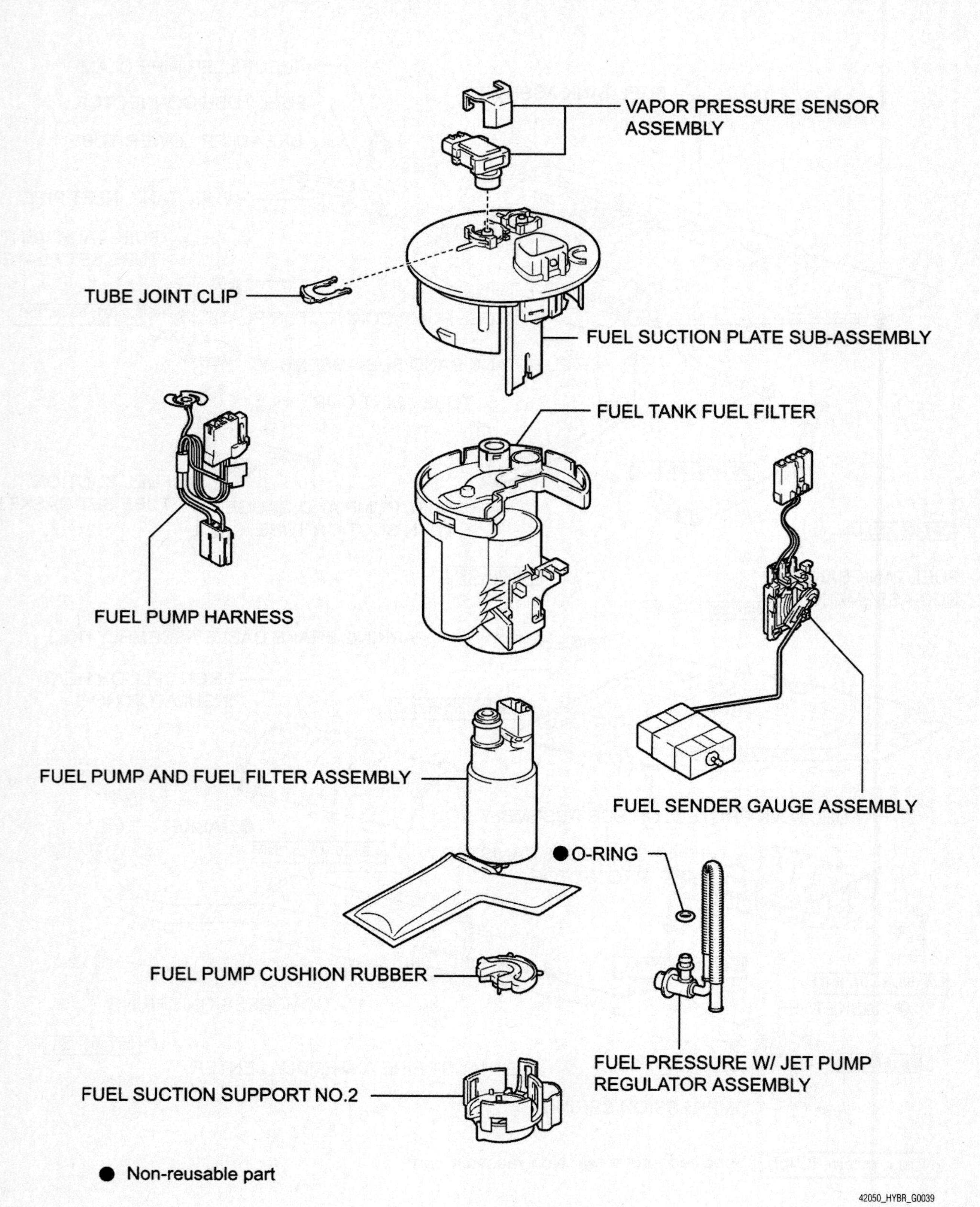

Fig. 181 Exploded view of the fuel pump and related components (2 of 2)

b. Using a screwdriver with its tip wrapped with tape, disengage the 4 snap claws from claw holes, and pull out the fuel suction plate.

➡Do not damage the fuel suction support.

18. Remove fuel pump harness, as follows:

a. Disconnect the connector.

b. Using a screwdriver, pry out the fuel pump harness.

19. Remove fuel suction support no.2, as follows:

a. Using a screwdriver with its tip wrapped with tape, disengage the 4 snap claws from the claw holes, and remove the fuel suction support.

WARNING

Do not damage the fuel suction support.

20. Remove fuel pump cushion rubber, as follows:

a. Remove the fuel pump cushion rubber from the fuel pump.

21. Remove fuel pump and fuel filter assembly, as follows:

a. Pull out the fuel pump from the fuel tank fuel filter.

22. Remove fuel pressure with jet pump regulator assembly:

a. Remove the fuel pressure with jet pump regulator assembly from the fuel tank fuel filter.

b. Remove the O-ring from the fuel pressure with jet pump regulator assembly.

To install:

23. Install fuel pressure w/ jet pump regulator assembly, as follows:

a. Apply a light coat of spindle oil or gasoline to a new o-ring, and install it.

b. Install the fuel pressure with jet pump regulator assembly to the fuel tank fuel filter.

24. Install fuel pump and fuel filter assembly:

a. Apply a light coat of spindle oil or gasoline to the o-ring of the fuel pump.

b. Push in the fuel pump to the fuel tank fuel filter.

25. Install fuel pump cushion rubber:

a. Install the fuel pump cushion rubber to the fuel pump.

26. Install the fuel suction support no.2.

27. Install fuel pump harness and attach the connector.

28. Install fuel suction plate sub-assembly, as follows:

a. Connect the fuel pump connector.

b. Install the fuel suction plate.

29. Install fuel sender gauge assembly:

a. Slide the fuel sender gauge to engage the claw.

b. Connect the fuel sender gauge connector.

30. Install vapor pressure sensor assembly:

a. Install the vapor pressure sensor assembly with the tube joint clip.

b. Install the cover.

31. Install fuel pump and gauge with suction tube:

a. Install a new gasket to the fuel pump assembly.

b. Install the fuel pump assembly.

c. Install the fuel tank vent tube set plate by aligning it with the cutout on the fuel pump assembly.

d. Install the fuel tank vent tube set plate and fuel pump connector plate with the 8 bolts. Tighten to 53 inch lbs. (6.0 Nm).

e. Connect the fuel tank main tube with the tube joint clip and clamp.

➡Check that there are no scratches or foreign objects on the connecting part. Check that the fuel tube joint is inserted securely. Check that the tube joint clip is on the collar of the fuel tube joint. after installing the tube joint clip, check that the fuel tube joint is not pulled off.

f. Connect the fuel tank wire connector.

g. Connect the vapor pressure sensor connector and clamp.

32. Install the fuel tank assembly.

33. Connect the breather lower tube,

34. Connect the fuel tank main tube sub-assembly.

35. Connect charcoal canister fuel hose.

36. Install charcoal canister protector.

37. Connect the fuel tank wire.

38. Install the parking brake cable assembly no.3.

39. Install the fuel tank protector sub-assembly no.1.

40. Install front floor heat insulator no.3.

41. Install exhaust pipe assembly center.

42. Connect the negative battery cable.

43. Check for fuel leaks.

FUEL RAIL AND INJECTOR

REMOVAL & INSTALLATION

See Figure 182.

1. Before servicing the vehicle, refer to the Precautions Section.
2. Relieve the fuel system pressure.
3. Drain the cooling system.
4. Remove or disconnect the following:
   - Negative battery cable
   - Engine cover
   - Wiper arm and blade assembly
   - Top cowl ventilator louver assembly
   - Wiper motor and linkage assembly
   - Air intake assembly
   - Emission control valve hoses
   - Air intake surge tank
5. Remove the mounting bolt and separate fuel hose No. 1
6. Remove the fuel pressure pulsation damper and gaskets.
7. Remove the fuel hose No. 2 union bolt and gaskets.
8. Disconnect the wiring at the injectors.
9. Remove the 4 bolts and each fuel rail with the injectors still attached.
10. Pull each injector from the fuel rail..

To install:

11. Install new O-rings on each injector. Apply a light coating of gasoline to the O-rings and mating points on the pipes.
12. Using a twisting motion, install the injectors on the pipes.

➡Be careful to avoid twisting the O-rings. After installation, check that the injectors turn smoothly. If not, use new O-rings.

13. Install the pipes and injectors.
14. Loosely install the bolts and make sure that the injectors still turn freely. If not, replace the O-rings.
15. Torque the bolts to 84 inch lbs. (10 Nm).
16. The remainder of installation is the reverse of removal. Observe the following torques:
    - Fuel hose No. 2 union bolt: 24 ft. lbs. (33 Nm)
    - Pulsation damper: 24 ft. lbs. (33 Nm)
    - Fuel hose No.1: 14 ft. lbs. (20 Nm)

FUEL TANK

REMOVAL & INSTALLATION

See Figure 183.

1. Before servicing the vehicle, refer to the Precautions Section.
2. Relieve the fuel system pressure, as outlined in this section.
3. Disconnect the negative battery cable.

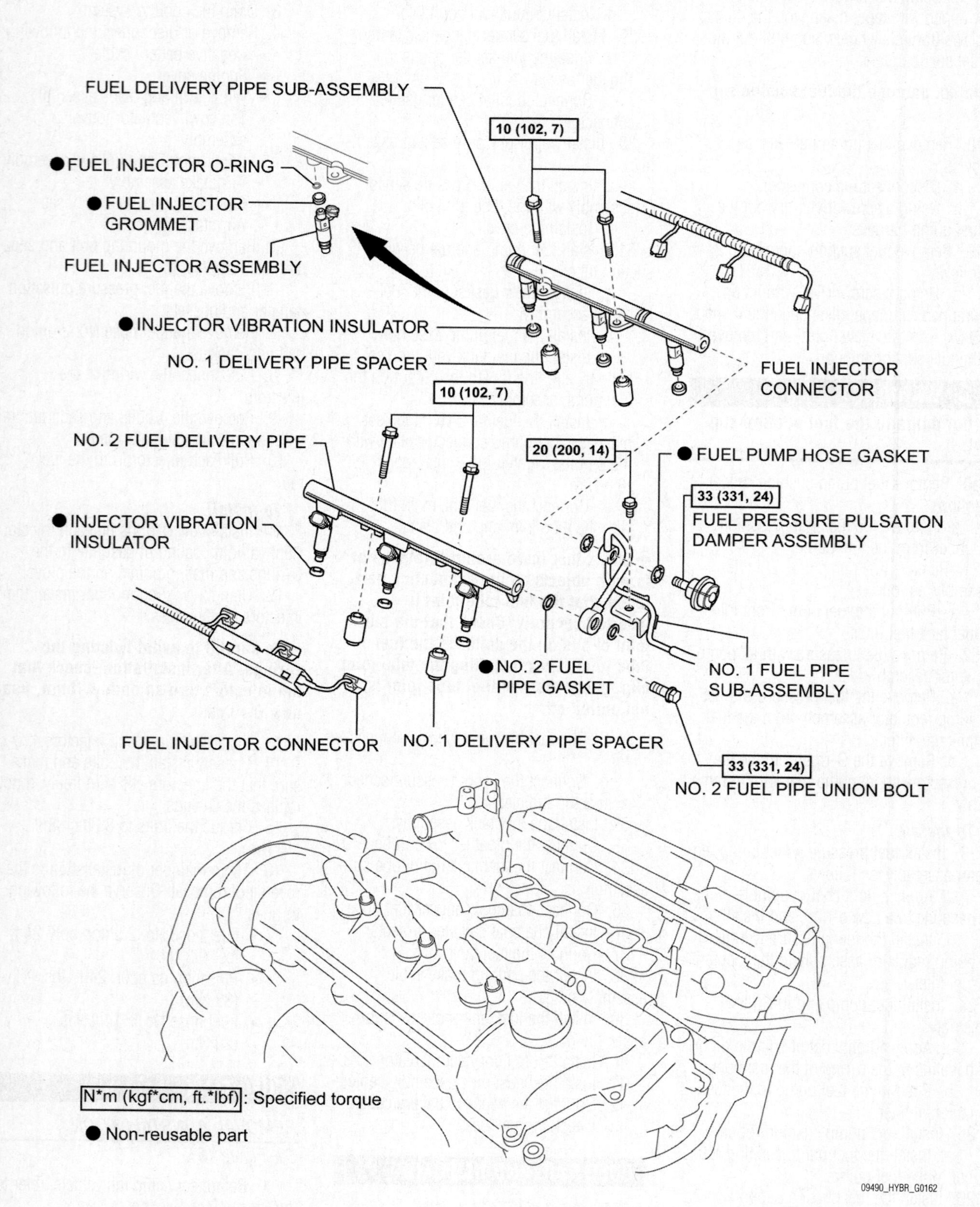

Fig. 182 Location of components for the fuel rail and injectors

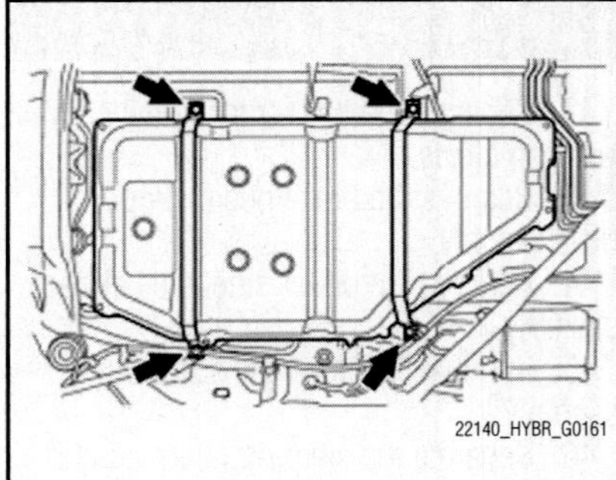

Fig. 183 Fuel tank straps and mounting bolts

4. Remove the front floor heat insulator no.3.
5. Remove the fuel tank protector sub-assembly no.1.
6. Separate parking brake cable assembly no.3.
7. Disconnect the fuel tank wire.
8. Remove the charcoal canister protector.
9. Disconnect the charcoal canister fuel hose.
10. Disconnect the fuel tank main tube subassembly.
11. Disconnect the breather lower tube.
12. Remove the fuel tank assembly.

To install:

13. Install the fuel tank assembly. Tighten the 4 mounting bolts to 29 ft. lbs. (39 Nm).
14. Connect the breather lower tube,
15. Connect the fuel tank main tube sub-assembly.
16. Connect charcoal canister fuel hose.
17. Install charcoal canister protector.
18. Connect the fuel tank wire.
19. Install the parking brake cable assembly no.3.
20. Install the fuel tank protector sub-assembly no.1.
21. Install front floor heat insulator no.3.
22. Install exhaust pipe assembly center.
23. Connect the negative battery cable.
24. Check for fuel leaks.

IDLE SPEED

ADJUSTMENT

Idle speed is maintained by the Engine Control Module (ECM). No adjustment is necessary or possible.

THROTTLE BODY

REMOVAL & INSTALLATION

See Figure 184.

1. Before servicing the vehicle, refer to the Precautions Section.
2. Relieve the fuel system pressure.
3. Drain the cooling system.
4. Remove or disconnect the following:
   - Negative battery cable
   - Engine cover
   - Wiper arm and blade assembly
   - Top cowl ventilator louver assembly

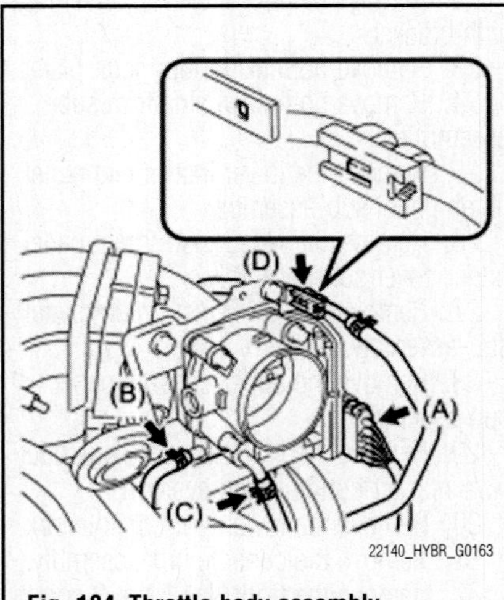

Fig. 184 Throttle body assembly

   - Wiper motor and linkage assembly
   - Cool air intake duct seal
   - Air cleaner cap and case assembly
5. Remove the throttle body as follows:
   a. Disconnect the throttle motor connector (A).
   b. Separate the No. 2 water by-pass hose (B).
   c. Separate the No. 3 water by-pass hose (C).
   d. Separate the fuel vapor feed hose (D).
   e. Remove the 4 bolts and the throttle body assembly.
   f. Remove the throttle body gasket from the intake air surge tank.

To install:

6. Install the throttle body assembly as follows:
7. Install a new throttle body gasket to the intake air surge tank.
   a. Tighten the 4 mounting bolts to 8 ft. lbs. (11 Nm).
   b. Connect the fuel vapor feed hose (A).
   c. Connect the No. 3 water by-pass hose (B).
   d. Connect the No. 2 water by-pass hose (C).
   e. Connect the throttle motor connector (D).
8. Install air cleaner case and cap assembly.
9. Add the engine coolant.
10. Inspect for any coolant leaks.
11. Connect the negative battery cable.
12. Install or connect the following:
    - Air cleaner cap and case assembly
    - Cool air intake duct seal
    - Wiper motor and linkage assembly
    - Top cowl ventilator louver assembly
    - Wiper arm and blade assembly
    - Engine cover
    - Negative battery cable

HEATING & AIR CONDITIONING SYSTEM

BLOWER MOTOR

REMOVAL & INSTALLATION

See Figure 185.

1. Before servicing the vehicle, refer to the Precautions Section.
2. Remove the No. 2 instrument panel under cover sub-assembly.
3. Remove front blower motor sub-assembly, as follows:
    a. Disconnect the connector.
    b. Remove the 3 screws and the front blower motor sub-assembly.

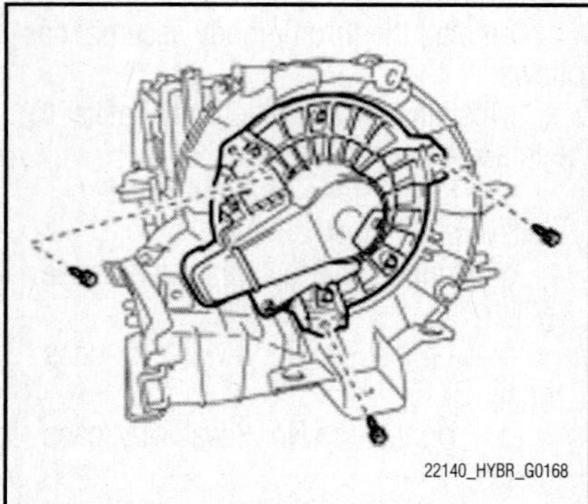

Fig. 185 Blower motor and retaining screws

To install:

4. Installation is the reverse of the removal procedure.

HEATER CORE

REMOVAL & INSTALLATION

See Figures 186 through 188.

1. Before servicing the vehicle, refer to the Precautions Section.
2. Recover refrigerant from refrigeration system.
3. Position front wheels straight ahead.
4. Remove both front wheels.

⁂ CAUTION

Wait for 90 seconds after disconnecting the cable to prevent the airbag from deploying.

5. Remove both front wiper arm and blade assemblies.
6. Remove the cowl top ventilator louver sub-assembly.
7. Remove the windshield wiper motor and link assembly.
8. Remove the cowl top outer panel sub-assembly.
9. Disconnect the heater inlet and outlet water hoses.
10. Disconnect the cooler refrigerant liquid pipe.
11. Disconnect the No. 1 cooler refrigerant suction pipe.
12. Remove the lower No. 2 and 3 steering wheel covers.
13. Remove the steering pad.
14. Remove the steering wheel assembly.
15. Remove the steering column cover.
16. Remove the turn signal switch assembly with spiral cable sub-assembly.
17. Remove the instrument cluster finish panel assembly.
18. Remove the combination meter assembly.
19. Remove the center instrument panel register assembly.
20. Remove the center instrument cluster finish panel assembly.
21. Remove the air conditioning control assembly.
22. Remove the radio receiver assembly with bracket.
23. Remove both front door scuff plates.
24. Remove both cowl side trim sub-assemblies.
25. Remove the lower instrument panel finish panel sub-assembly.
26. Remove the No. 2 instrument panel under cover sub-assembly.
27. Remove the lower instrument panel sub-assembly.
28. Remove the upper console panel sub-assembly.
29. Remove the No. 2 console box duct (w/o rear air conditioning system).
30. Remove the lower rear console box.
31. Remove the console box assembly.
32. Remove the front No. 1 No. 2 console box insert.
33. Remove the engine switch (W/ Smart Key System).
34. Remove the front pillar garnish.
35. Disconnect the front door opening trim weather strip.
36. Remove the No. 1 instrument panel speaker panel sub-assembly.
37. Remove the front No. 2 speaker assembly.
38. Disconnect the instrument panel wire assembly.
39. Remove the instrument panel safety pad assembly.
40. Remove the brake pedal return spring.
41. Remove the stop light switch assembly.
42. Separate the brake master cylinder push rod clevis.
43. Remove the brake pedal support sub-assembly.
44. Remove the driver side knee airbag assembly.
45. Remove the No. 1 air duct sub-assembly.
46. Separate the steering intermediate shaft sub-assembly.
47. Remove the steering column assembly.
48. Remove the certification ECU (smart key ECU assembly) (W/ smart key system).
49. Remove the air conditioning amplifier assembly.
50. Remove the rear No. 1 air duct.
51. Remove the rear No. 3 air duct.
52. Remove the No. 1 console box duct (w/o rear air conditioning system).
53. Remove the center heater to register duct.
54. Remove the No. 1 and 2 instrument panel brace sub-assemblies.
55. Remove instrument panel reinforcement assembly with air conditioning unit.
56. Remove the 2 caps and the 2 bolts from the engine compartment side.
57. Remove the heater core tube clamps.
58. Remove the heater core as shown in the illustration

To install:

59. Install heater core and tighten the tube brackets.
60. The remainder of installation is reverse of the removal procedure.
61. Check and note the following:
    a. Inspect the SRS warning light.
    b. Refill the engine coolant.
    c. Vacuum and recharge A/C system.
    d. Warm up the engine and check for leaks.

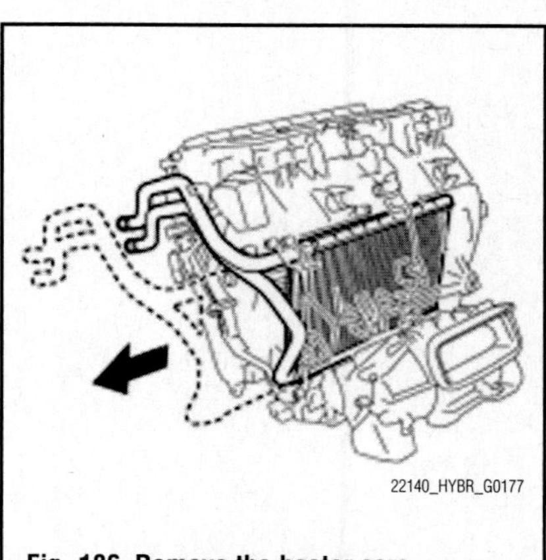

Fig. 186 Remove the heater core

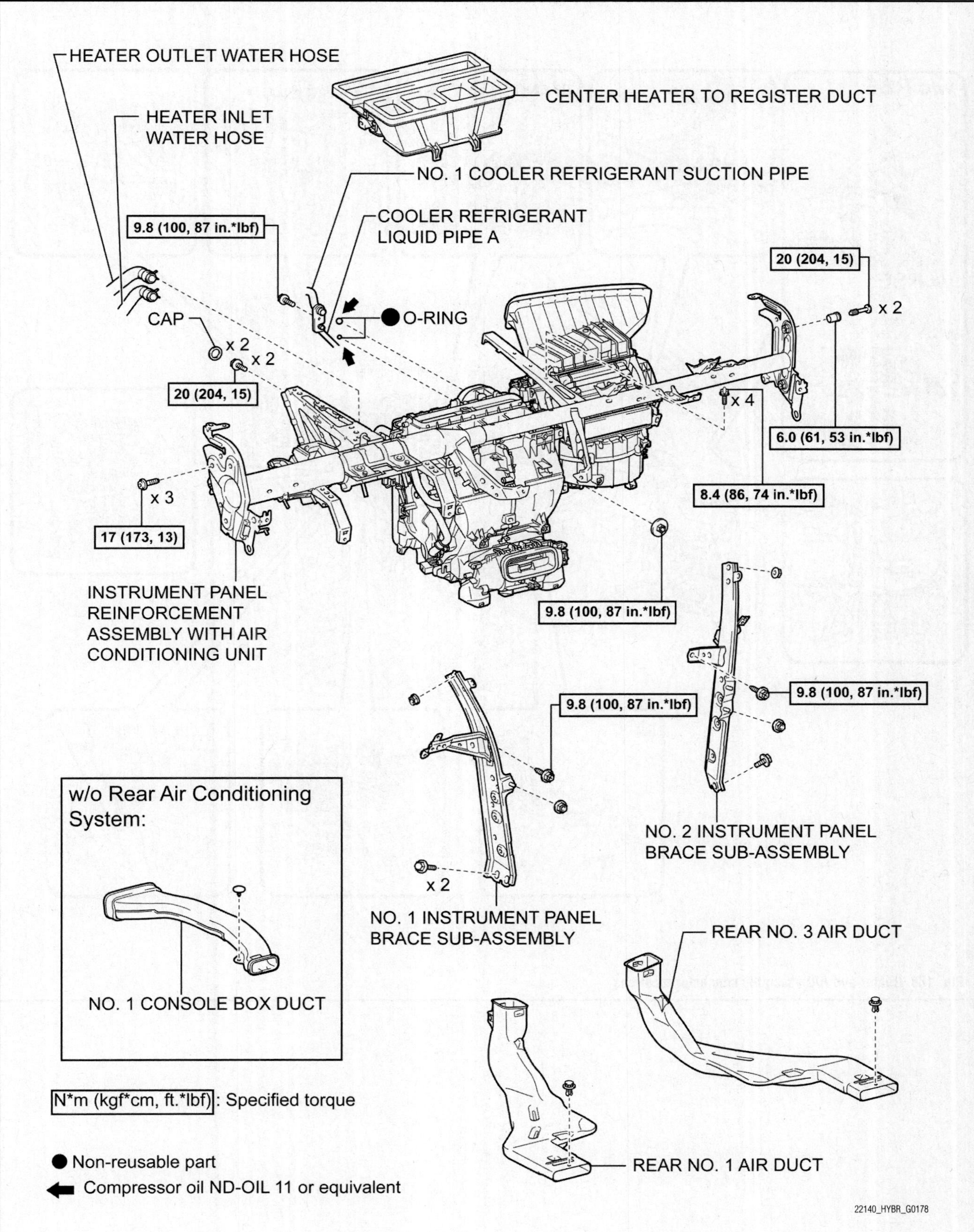

Fig. 187 Heater and A/C assembly, reinforcement assembly and related components

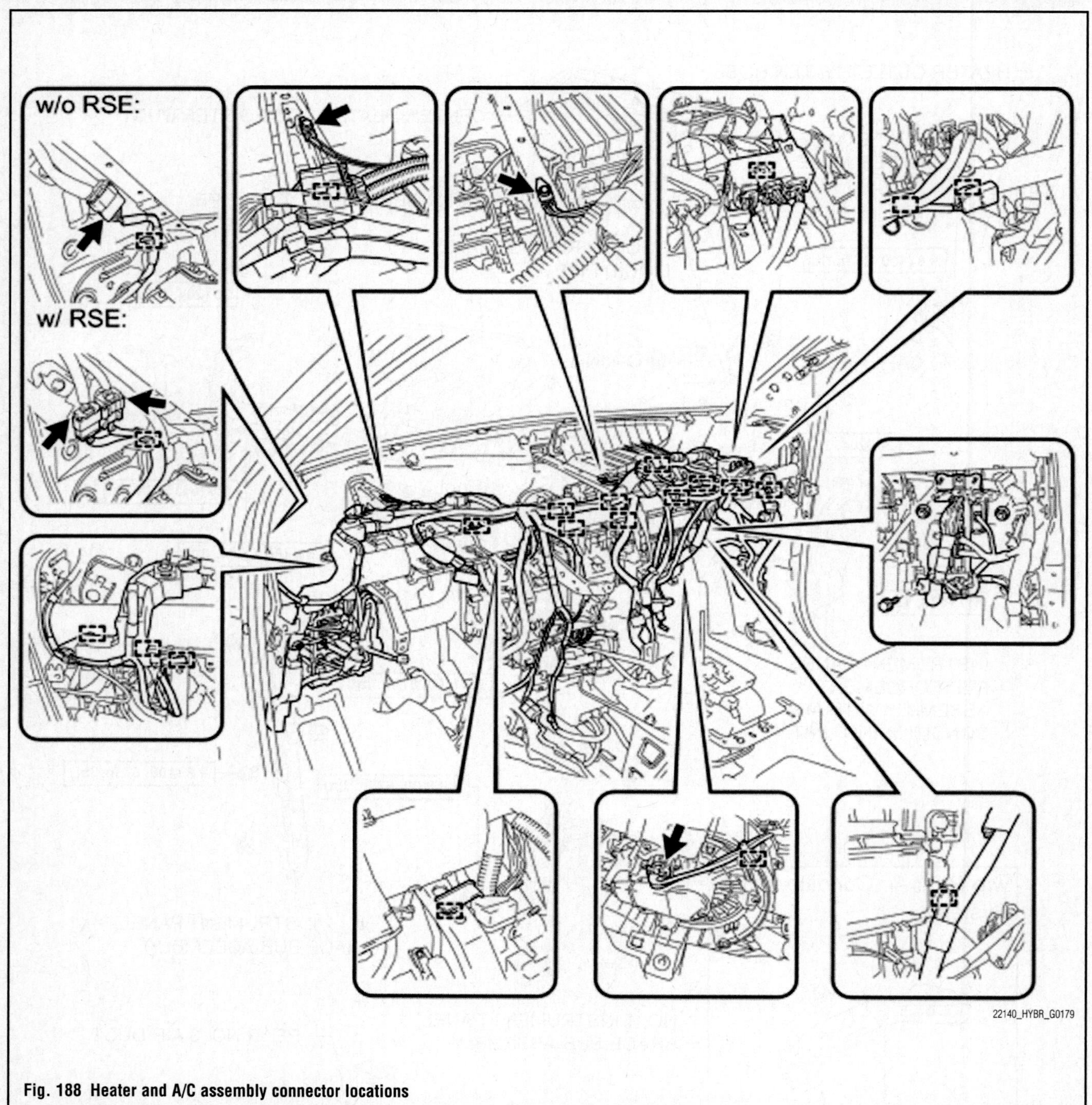

Fig. 188 Heater and A/C assembly connector locations

AUXILIARY HEATING & AIR CONDITIONING SYSTEM

BLOWER MOTOR

REMOVAL & INSTALLATION

See Figure 189.

1. Before servicing the vehicle, refer to the Precautions Section.
2. Remove the right rear door scuff plate.
3. Remove the right rear door opening trim weather strip.
4. Remove the deck board assembly.
5. Remove the No. 2 and 3 deck board sub-assemblies.
6. Remove the tonneau cover assembly, as applicable.
7. Remove the rear seat side covers.
8. Remove the deck side trim boxes.
9. Remove the jack carrier support.
10. Remove the jack carrier cushion.
11. Remove jack assembly.
12. Remove the jack carrier assembly.
13. Remove rear mat.
14. Remove the deck floor board assembly.
15. Remove the rear No. 2 seat inner belt assembly.
16. Disconnect both rear seat lap type belt assemblies.
17. Remove the rear No. 2 seat assembly.
18. Remove the rear floor finish plate.
19. Remove the rear seat side garnish cap.
20. Remove right deck side trim cover.
21. Remove right deck side trim.
22. Remove right side trim cover (for manual air conditioning system).
23. Remove rear room temperature sensor (for automatic air conditioning system).
24. Remove right rear combination light service cover.
25. Remove right hand rope hook assembly.
26. Remove the No. 1 luggage compartment trim hook
27. Remove the right front deck side trim cover.
28. Remove the right deck trim side panel assembly.
29. Remove the right roof side inner garnish assembly.
30. Remove the rear blower motor sub-assembly, as follows:
    a. Disconnect the connector.
    b. Remove the 3 screws and the rear blower motor sub-assembly.

To install:

31. Install the 3 screws and the rear blower motor sub-assembly
32. Reconnect the connector.
33. Install the right roof side inner garnish assembly.
34. Install the right deck trim side panel assembly.
35. Install the right front deck side trim cover.
36. Install the No. 1 luggage compartment trim hook
37. Install the right hand rope hook assembly.
38. Install the right rear combination light service cover.
39. Install rear room temperature sensor (for automatic air conditioning system).
40. Install right deck side trim.
41. Install right deck side trim cover.
42. Install the rear seat side garnish cap.
43. Install the rear floor finish plate.
44. Install the rear No. 2 seat assembly.
45. Install both rear seat lap type belt assemblies.
46. Install the rear No. 2 seat inner belt assembly.
47. Install the deck floor board assembly.
48. Install the rear mat.
49. Install the jack carrier cushion.
50. Install the jack carrier support.
51. Install the deck side trim boxes.
52. Install the rear seat side covers.
53. Install the tonneau cover assembly, as applicable.
54. Install the No. 2 and 3 deck board sub-assemblies.
55. Install the deck board assembly.
56. Install the right rear door opening trim weather strip.
57. Install the right rear door scuff plate.

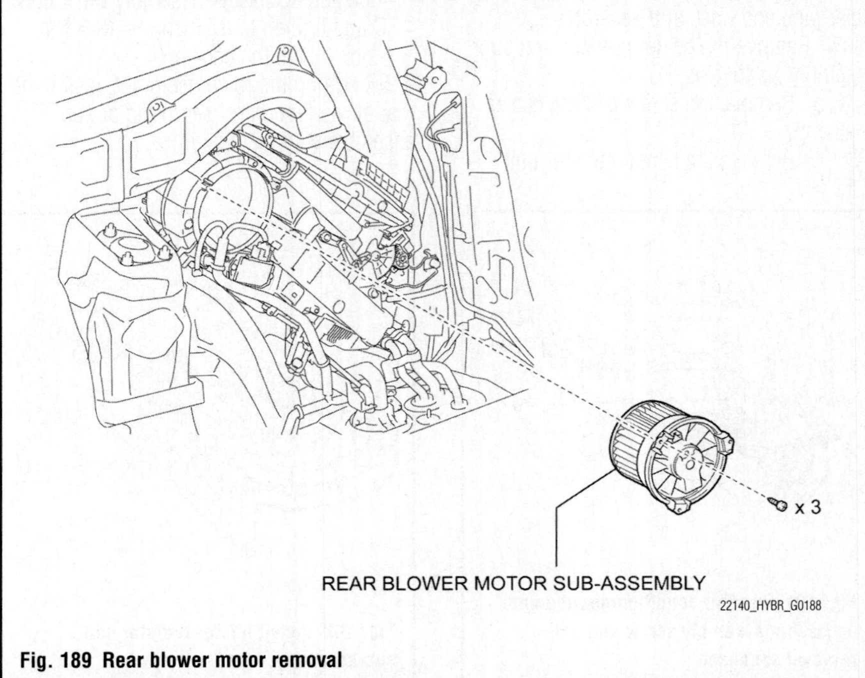

Fig. 189 Rear blower motor removal

HEATER CORE

REMOVAL & INSTALLATION

See Figures 190 through 199.

1. Before servicing the vehicle, refer to the Precautions Section.

➡Adjust the air outlet mode setting to FOOT.

2. Recover refrigerant from refrigeration system.
3. Disconnect the heater inlet and outlet water hoses.
4. Disconnect the cooler refrigerant liquid pipe C, as follows:
    a. Remove the bolt, and slide the hook connector.
    b. Disconnect the cooler refrigerant liquid pipe C.
    c. Remove the O-ring from the cooler refrigerant liquid pipe C.

➡Seal the openings of the disconnected parts using vinyl tape to prevent entry of moisture and foreign matter.

5. Disconnect the rear cooler refrigerant suction hose.
6. Remove the O-ring from the rear cooler refrigerant suction hose.
7. Remove the right rear door scuff plate.

8. Remove the right rear door opening trim weather strip.
9. Remove the deck board assembly.
10. Remove the No. 2 and 3 deck board sub-assemblies.
11. Remove the tonneau cover assembly, as applicable.
12. Remove the rear seat side covers.
13. Remove the deck side trim boxes.
14. Remove the jack carrier support.
15. Remove the jack carrier cushion.
16. Remove jack assembly.
17. Remove the jack carrier assembly.
18. Remove rear mat.
19. Remove the deck floor board assembly.
20. Remove the rear No. 2 seat inner belt assembly.
21. Disconnect both rear seat lap type belt assemblies.
22. Remove the rear No. 2 seat assembly.
23. Remove the rear floor finish plate.
24. Remove the rear seat side garnish cap.
25. Remove right deck side trim cover.
26. Remove right deck side trim.
27. Remove right side trim cover (for manual air conditioning system).
28. Remove rear room temperature sensor (for automatic air conditioning system).
29. Remove right rear combination light service cover.
30. Remove right hand rope hook assembly.
31. Remove the No. 1 luggage compartment trim hook
32. Remove the right front deck side trim cover.
33. Remove the right deck trim side panel assembly.
34. Remove the right roof side inner garnish assembly.
35. Disengage the 2 claws and remove the cooler plate.
36. Remove the No. 1 cooler air duct as shown in the illustration.
37. Remove the 2 clips and the rear No. 5 air duct.
38. Remove the rear cooling unit assembly, as follows:
    a. Disengage each clamp.
    b. Disconnect each connector.
    c. Remove the 4 bolts and the rear cooling unit assembly.
39. Remove the drain cooler hose.
40. Remove the rear air mix control servo motor sub-assembly, as follows:
    a. Disconnect the connector.
    b. Remove the 2 screws and rear air mix control servo motor sub-assembly.
41. Remove the rear air outlet control servo motor sub-assembly, as follows:
    a. Disconnect the connector.
    b. Remove the 2 screws and rear air outlet control servo motor sub-assembly.
42. Remove the rear air conditioning tube and accessory assembly, as follows:
    a. Remove the packing.
    b. Remove the screws (C then B).
    c. Remove the bolt (A).
    d. Remove the bolt and slide the hook connector.
    e. Remove the rear air conditioning tube and accessory assembly.
    f. Remove the 2 O-rings from the rear air conditioning tube and accessory assembly.
43. Using pliers, grip the claws of the 2 clips and slide the clips to disconnect the heater water hose.
44. Remove the 2 screws and the heater water pipe and hose sub-assembly.
45. Remove the heater radiator unit sub-assembly, as follows:
    a. Remove the screw and the heater clamp.
    b. Remove the heater radiator unit sub-assembly from the rear cooling unit as shown in the illustration.

To install:

46. Install the heater radiator unit sub-assembly as shown in the illustration.
47. Install the clamp with the screw,
48. Install the heater water pipe and hose sub-assembly, as follows:
    a. Install screw A, and tighten to 87 inch lbs. (9.8 Nm).
    b. Install the heater water pipe and hose sub-assembly with the screw B and tighten to 87 inch lbs. (9.8 Nm).
    c. Using pliers, grip the claws of the 2 clips and slide the clips to connect the heater water hose.
49. Install the rear air conditioning tube and accessory assembly, as follows:
    a. Sufficiently apply compressor oil (ND-OIL 8 or equivalent) to 2 new O-rings and the fitting surfaces of the rear air conditioning tube and accessory assembly.
    b. Install the 2 new O-rings on the rear air conditioning tube and accessory assembly.
    c. Move the hook connector in the direction indicated by the arrow in the illustration.
    d. Insert the pipe joint into the fitting hole securely and tighten the bolt to 87 inch lbs. (9.8 Nm).
    e. Install bolt A and tighten to 87 inch lbs. (9.8 Nm).
    f. Install screw B and tighten to 87 inch lbs. (9.8 Nm).
    g. Install the rear air conditioning tube and accessory assembly with screw C and tighten to 87 inch lbs. (9.8 Nm).
    h. Install the packing.
50. Using the reference point, install the rear air outlet control servo motor sub-assembly with the 2 screws.

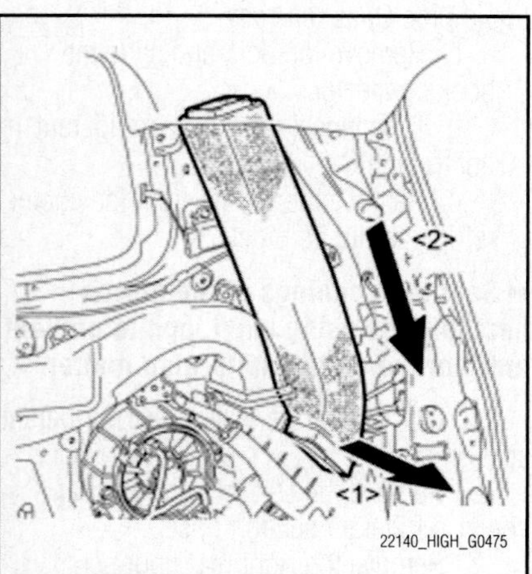

Fig. 190 No. 1 cooler air duct

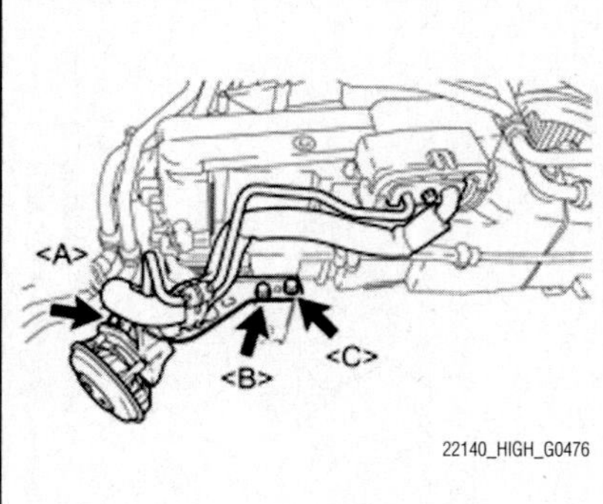

Fig. 191 Rear air conditioning tube and accessory assembly screw and bolt removal sequence

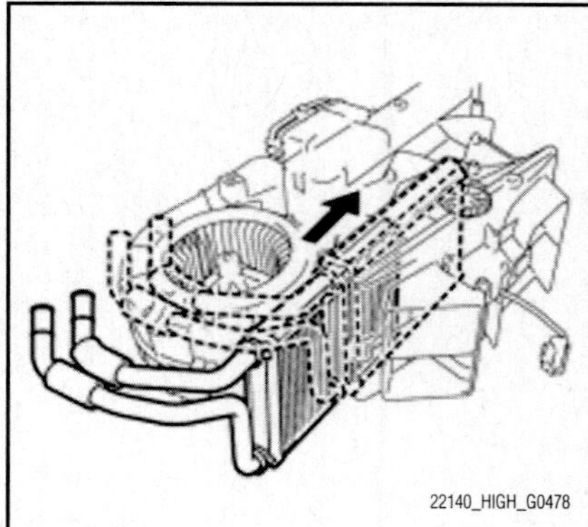

Fig. 192 Install heater radiator unit sub-assembly

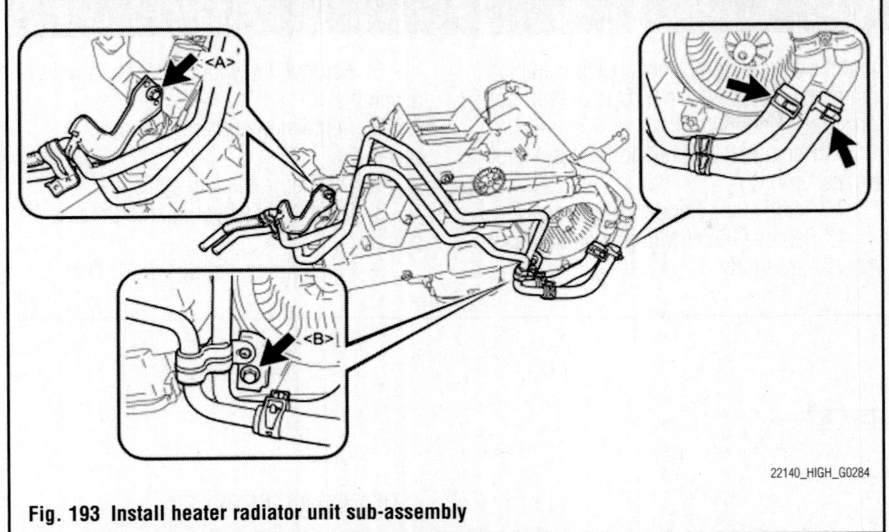

Fig. 193 Install heater radiator unit sub-assembly

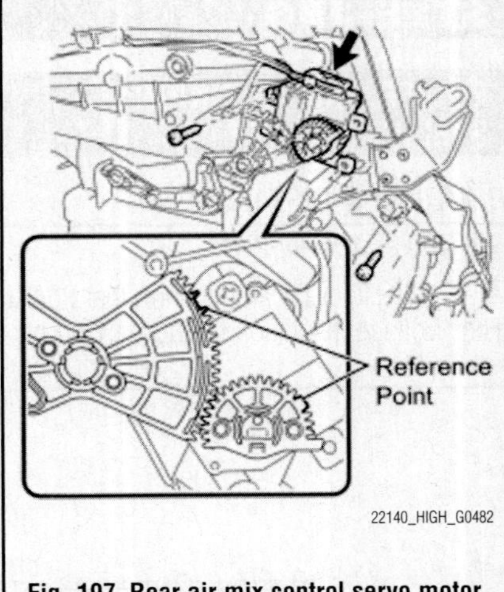

Fig. 197 Rear air mix control servo motor sub-assembly reference point

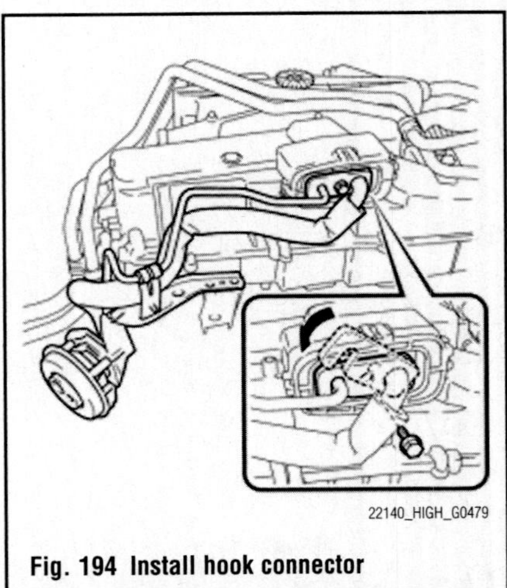

Fig. 194 Install hook connector

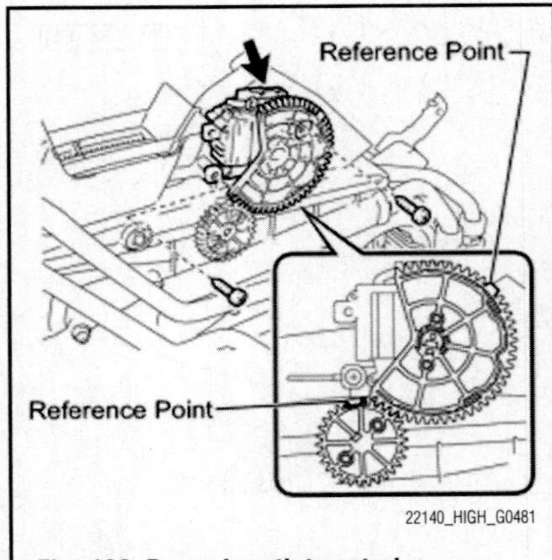

Fig. 196 Rear air outlet control servo motor sub-assembly reference point

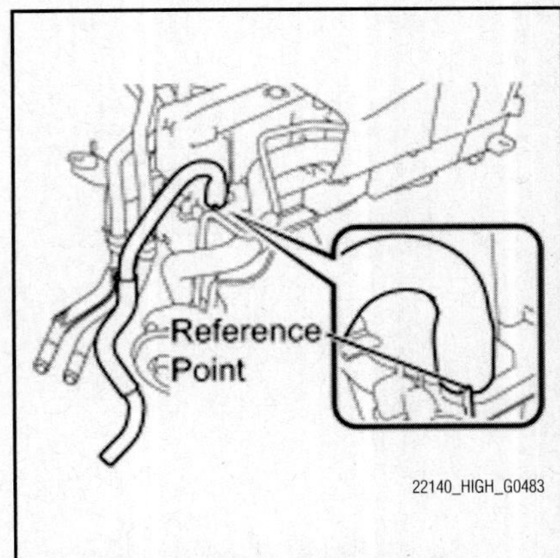

Fig. 198 Drain cooler hose reference point

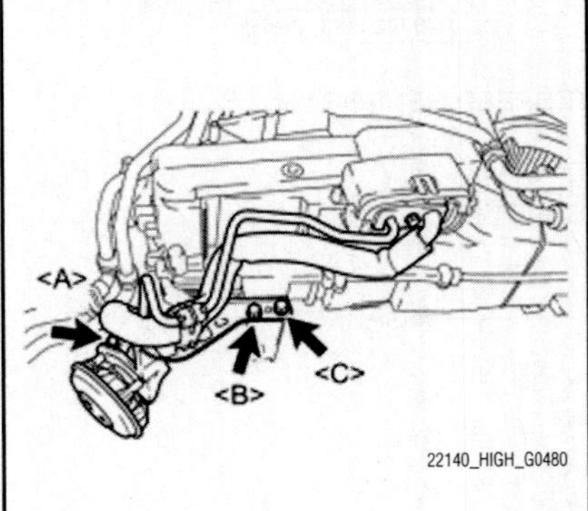

Fig. 195 Install rear air conditioning tube and accessory assembly

51. Connect the connector.
52. Using the reference point, install the rear air mix control servo motor sub-assembly with the 2 screws.
53. Connect the connector.
54. Using the reference point, install the drain cooler hose.
55. Install the rear cooling unit assembly with the 4 bolts and tighten in the order shown, to 87 inch lbs. (9.8 Nm).
56. Connect each connector.
57. Engage each clamp.
58. The remainder of installation is the reverse of the removal procedure.
59. Install engine coolant.
60. Vacuum and recharge A/C system.
61. Start the engine and check for leaks.

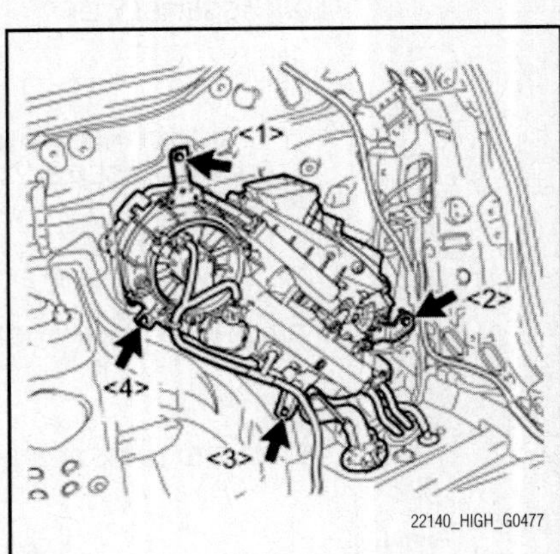

Fig. 199 Rear cooling unit assembly bolts tightening sequence

STEERING

POWER RACK & PINION STEERING GEAR

REMOVAL & INSTALLATION

See Figures 200 and 201.

1. Check for DT and confirm that P0AA6 (High voltage insulation is unusual) is not output before doing removal or installation inside the battery. If the DTC is output, perform troubleshooting first.
2. Remove the right and left front wiper arm assembly.
3. Remove the hood cowl top seal.
4. Remove the cowl top ventilator louver sub-assembly
5. Remove the wiper motor and link assembly.
6. Remove the outer cowl top panel.
7. Discharge the fuel system.
8. Place front wheel in straight ahead position.
9. Secure the steering wheel. This

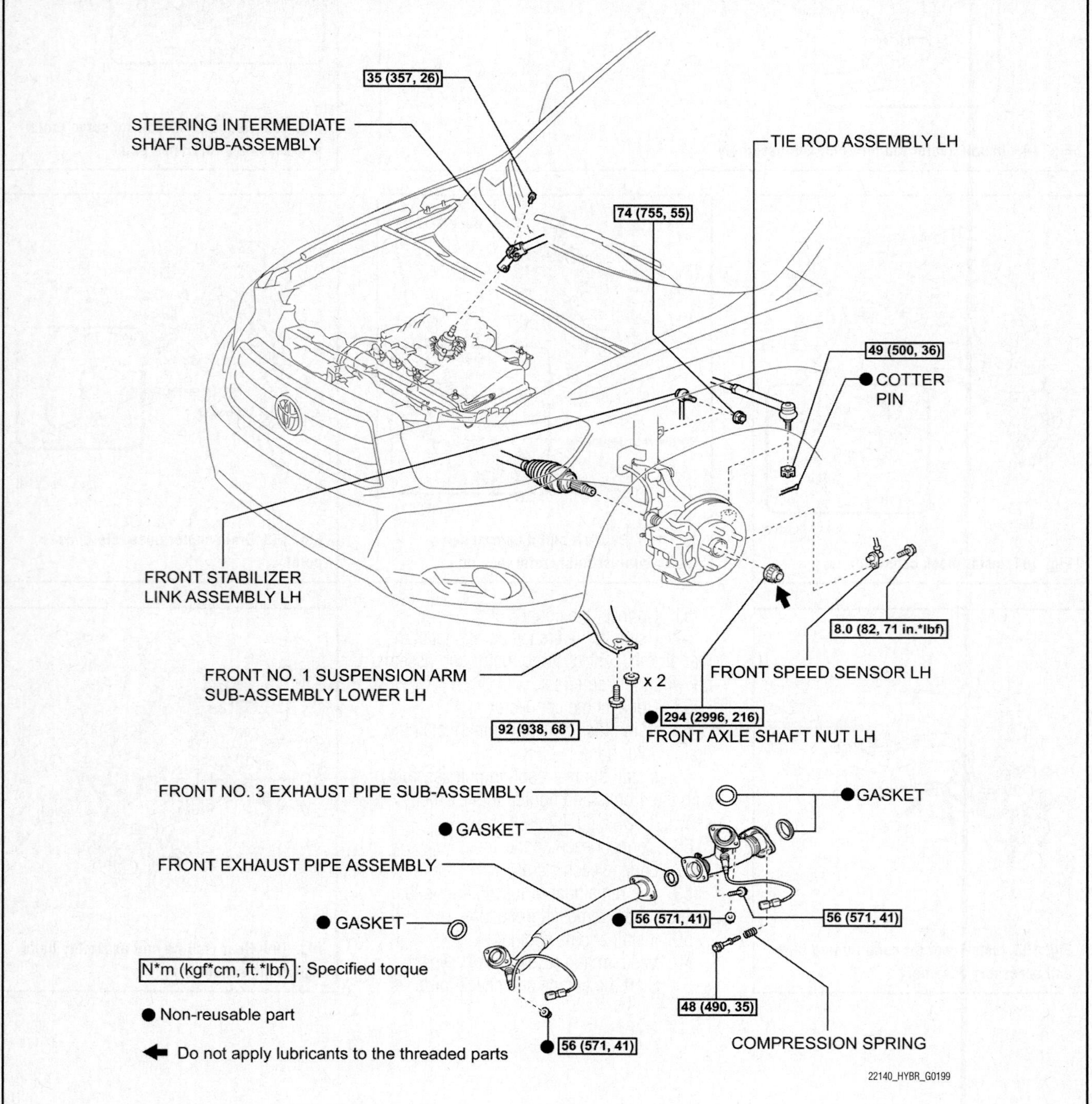

Fig. 200 Steering intermediate shaft and related components

operation is to prevent damage to the spiral cable.

10. Remove the service plug grip.
11. Remove the LH engine room side cover.
12. Remove the cool air intake duct seal.
13. Raise the vehicle and remove the front wheels.
14. Remove the No.1 engine cover.
15. Separate LH and RH fender liners.
16. Remove the front fender apron seals.
17. Remove the No.2 engine under cover.
18. Drain the coolant for the engine and inverter.
19. Drain the engine oil.
20. Drain the HV transaxle fluid.
21. Remove the primary batter, tray and carrier.
22. Remove the air cleaner cap and case assembly.
23. Move the outer section to the wire harness side, then disconnect the No. 1 circuit breaker sensor.
24. Disconnect No.3 frame wire.
25. Remove the inverter reserve tank assembly.
26. Slide the clamp, and disconnect the water hose from the inverter with converter assembly.
27. Remove the inverter cover.
28. Using the voltmeter, measure the voltage between the terminals of the 2 phase connectors (N-P). It should rear 0 volts.
29. Separate engine room block relay assembly.
30. Remove the No. 4 inverter bracket.
31. Remove the inverter with converter assembly.
32. Remove the brake master cylinder Reservoir.
33. Remove the air cleaner bracket.
34. Remove the No.2 engine stay.
35. Remove the engine moving control rod.
36. Remove the engine mounting control rod.
37. Remove the RH No.2 engine mounting bracket.
38. Separate compressor with motor assembly.
39. Remove the No.1 inverter bracket.
40. Separate transmission control cable assembly.
41. Disconnect the fuel vapor feed hose.
42. Disconnect the No.1 fuel pipe.
43. Disconnect heater hoses and radiator hoses.
44. Disconnect the oil cooler lines.
45. Disconnect the water hose.
46. Remove the RH front door scuff plate.
47. Remove the RH side cowl trim.
48. Remove the No.2 instrument under cover.
49. Remove the lower instrument panel.
50. Separate the engine wire.
51. Remove the exhaust pipes.
52. Remove the RH and LH stabilizer links.
53. Remove the front axle shaft nuts.
54. Remove the bolt and resin clamp, and separate the front speed sensors.
55. Separate the RH and LH tie rod ends from steering knuckle.
56. Remove the bolts and nuts, and separate the front lower suspension arms from the front lower ball joints.
57. Remove the LH and RH drive shafts.
58. Install matchmark and separate the steering intermediate shaft.
59. Remove the engine assembly with transaxle.
60. Remove the RH and LH stabilizer brackets.
61. Remove the stabilizer bar.
62. Remove the power steering link assembly.

To install:

63. Install the power steering link assembly with the 2 bolts and 2 nuts. Tighten to 51 ft. lbs. (70 Nm).
64. Install front stabilizer and brackets. Tighten bracket bolts to 21 ft. lbs. (29 Nm).
65. Install the front stabilizer links and tighten to 55 ft. lbs. (74 Nm).
66. Install engine assembly with transaxle.
67. Align the matchmarks on the steering intermediate shaft assembly and the power steering link assembly.
68. Install the bolt and tighten to 26 ft. lbs. (35 Nm).
69. Install front drive shafts and tighten the new nuts to 216 ft. lbs. (294 Nm).
70. Install both No.1 suspension arms and tighten mounting nuts and bolts to 68 ft. lbs. (92 Nm).
71. Connect the tie rod assembly LH and RH to the steering knuckle with the nut. Tighten to 36 ft. lbs. (49 Nm). Install new cotter pin.
72. Install the front speed sensors and tighten mounting bolt to 71 inch lbs. (8 Nm).

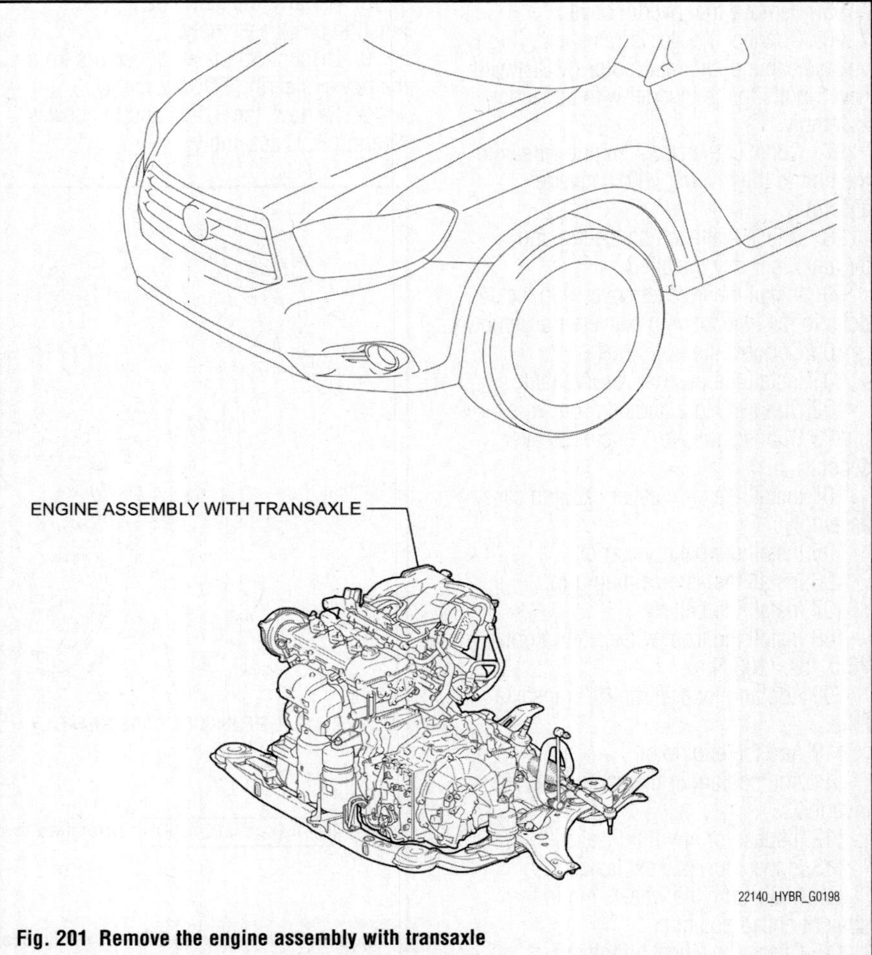

Fig. 201 Remove the engine assembly with transaxle

73. Install front stabilizer links. Tighten to 55 ft. lbs. (74 Nm).
74. Install the exhaust.
75. Connect the engine wire.
76. Install the lower instrument panel.
77. Install instrument panel under cover.
78. Install RH cowl side trim.
79. Install RH door scuff plate.
80. Connect the water hose.
81. Connect the oil cooler hoses.
82. Install the heater hoses.
83. Install the radiator hoses.
84. Install the No.1 fuel pipe.
85. Install transmission control cable.
86. Install No.1 inverter bracket.
87. Install compressor with motor assembly.
88. Install RH No.2 engine mounting bracket.
89. Install engine mounting control bracket.
90. Install the engine moving control rod.
91. Install the RH engine mounting stay.
92. Install the air cleaner bracket.
93. Install the brake master cylinder reservoir.
94. Install inverter with the converter assembly.
95. Remove the inverter cover.
96. Connect the No. 3 frame wire (high voltage cable of the rear motor (MGR) with new 5 bolts to the inverter with converter assembly.
97. Connect the No. 4 engine wire with the bolt to the inverter with converter assembly.
98. Check that each connector and terminal is firmly installed.
99. Install the inverter cover with the 12 bolts to the inverter with converter assembly.
100. Connect the water hose.
101. Install the inverter reserve tank.
102. Connect No.2 engine room wire,
103. Connect the No.1 circuit breaker sensor.
104. Install the air cleaner cap and case assembly.
105. Install the battery carrier.
106. Install the service plug grip.
107. Install the battery.
108. Install the front wheels and tighten to 76 ft. lbs. (103 Nm).
109. Add and inspect the HV transaxle fluid.
110. Add the engine oil.
111. Add coolant to the engine and the inverter.
112. Inspect for any fluid leaks.
113. Inspect for exhaust leaks.
114. Make sure the wheels are in the straight ahead position.
115. Check the wheel alignment.
116. Install the engine covers.
117. Install the front fender seals and liners.
118. Inspect the steering center point.
119. Check ignition timing and engine speed.
120. Check speed sensor operation.
121. Install the front wiper system in the reverse of the removal procedure.

POWER STEERING ECU

REMOVAL & INSTALLATION

See Figure 202.

1. Before servicing the vehicle, refer to the Precautions Section.
2. Place wheels in straight ahead position.
3. Disconnect the negative battery cable.
4. Remove the left front door scuff plate.
5. Remove the left side cowl trim.
6. Disconnect the hood lock control cable assembly and remove the lower instrument panel finish panel sub-assembly.
7. Remove the driver's side air bag assembly.
8. Remove the instrument panel junction block assembly.
9. Disconnect the 4 connectors from the power steering ECU assembly.
10. Remove the 3 nuts and the power steering ECU assembly.

To install:

11. Install the power steering ECU assembly with the 3 nuts, tighten the nuts to 10 ft. lbs. (14 Nm).
12. Check that the connector lever is at the fully unlocked position before installation.
13. Connect the 4 connectors to the power steering ECU assembly.
14. Connect the connectors to the back of the instrument panel junction block assembly.
15. Engage the wire harness clamp onto the instrument panel junction block assembly.
16. Install the instrument panel junction block assembly with the 3 nuts, tighten the nuts to 74 inch lbs. (8.4 Nm).
17. Connect the connectors to the instrument panel junction block assembly.
18. Engage the wire harness clamp onto the instrument panel junction block assembly.
19. Install the driver side knee airbag assembly with the 4 bolts. Tighten the bolts to 7 ft. lbs. (10 Nm).
20. Install the left cowl side trim.
21. Install the left front door scuff plate.
22. Connect the negative battery cable.

➡When disconnecting the cable, some systems need to be initialized after the cable is reconnected

23. Inspect the SRS warning light.

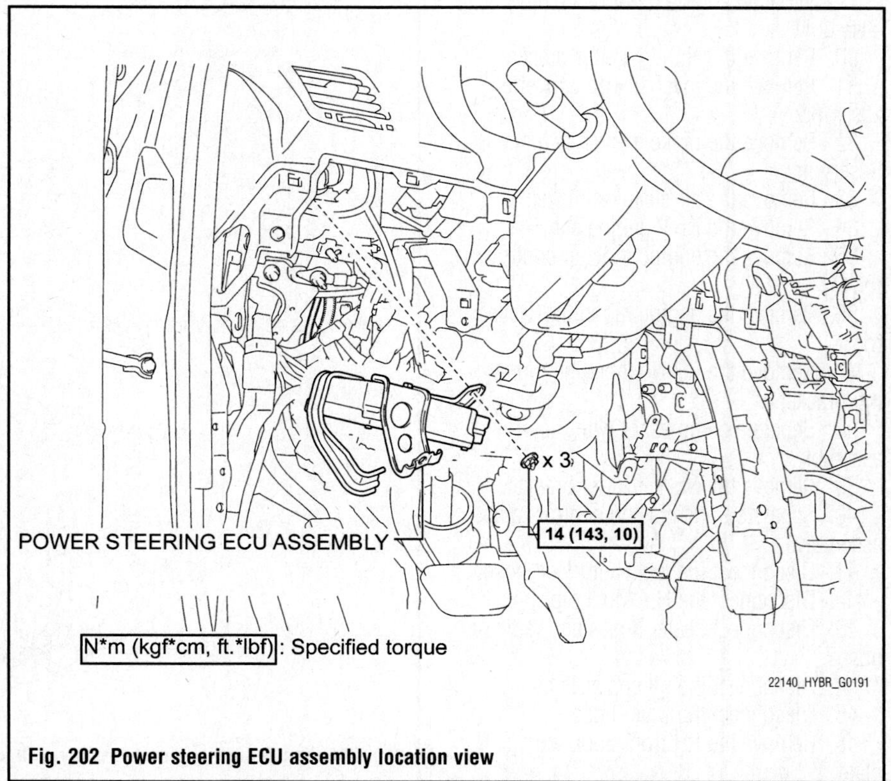

Fig. 202 Power steering ECU assembly location view

CONTROL LINKS

REMOVAL & INSTALLATION

See Figure 203.

➡**Perform the same procedure on each side.**

1. Remove wheel and tire assembly.
2. Remove the nut and separate the front stabilizer link assembly.

➡**If the ball joint turns together with the nut, use a hexagon wrench (6 mm) to hold the stud bolt.**

To install:

3. Install the front stabilizer link assembly with the nut and tighten to 55 ft. lbs. (74 Nm).
4. Install the wheel and tire assembly.

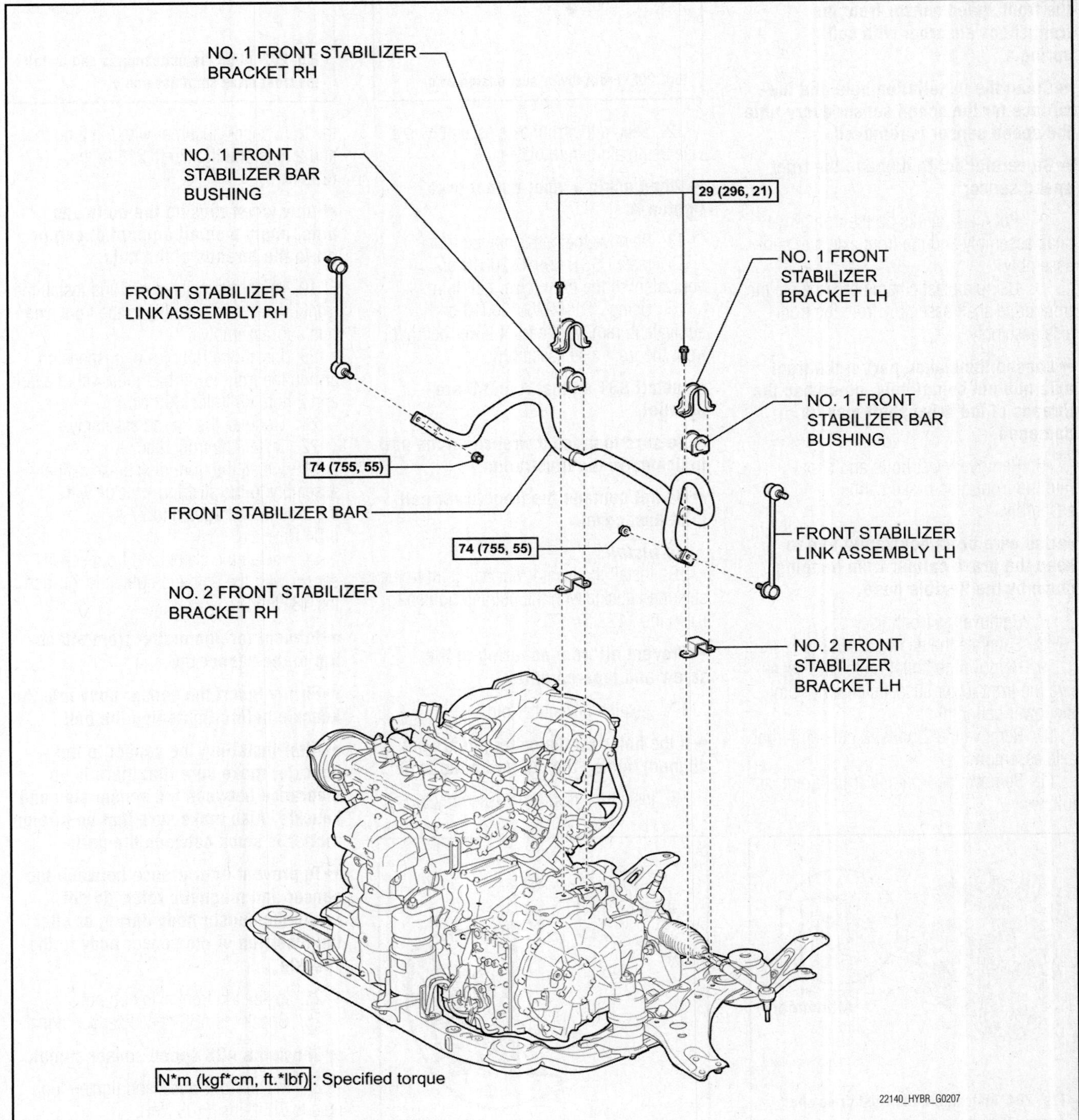

Fig. 203 Front stabilizer control links and related parts.

LOWER BALL JOINT

REMOVAL & INSTALLATION

See Figures 204 through 207.

1. Remove the front wheel.
2. Remove the front axle hub nut.
3. Remove the bolt and resin clamp, and separate the front speed sensor.

➡**Be sure to completely separate the front speed sensor from the front shock absorber with coil spring.**

➡**Clean the installation hole and the surface for the speed sensor every time the speed sensor is removed.**

➡**Be careful not to damage the front speed sensor.**

4. Put matchmarks on the front drive shaft assembly and the front axle hub sub-assembly.
5. Using a plastic hammer, separate the front drive shaft assembly from the front axle assembly.

➡**Loosen the staked part of the front axle hub nut completely, otherwise the threads of the drive shaft may be damaged.**

6. Remove the 2 bolts and separate the front disc brake caliper assembly.

➡**Use wire or an equivalent tool to keep the brake caliper from hanging down by the flexible hose.**

7. Remove the front disc.
8. Separate the tie rod assembly.
9. Remove the bolt, 2 nuts, and separate the front lower suspension arm from the lower ball joint.
10. Remove the 2 bolts, 2 nuts and front axle assembly.
11. Remove the front lower ball joint, as follows:

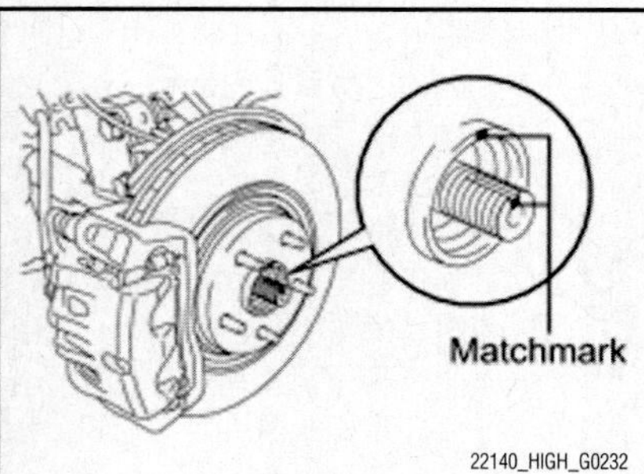

Fig. 204 Matchmark the front drive shaft assembly and the front axle hub sub-assembly

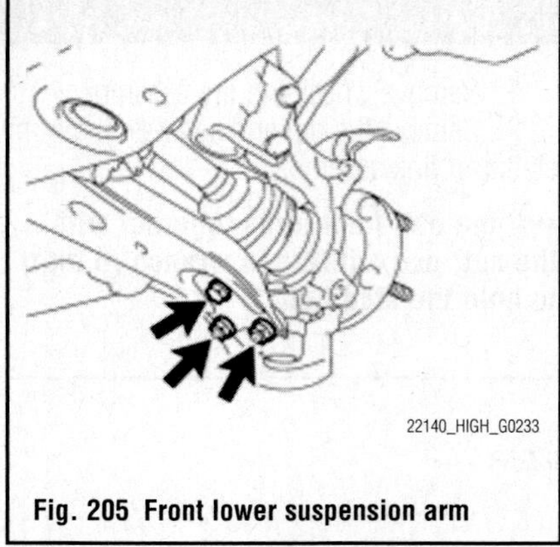

Fig. 205 Front lower suspension arm

12. Secure the front axle assembly in a vise using aluminum plates.

➡**When using a vise, do not over tighten it.**

13. Remove the cotter pin and nut.
14. Install SST: 09960-20010 or equivalent to the front lower ball joint.
15. Using SST: 09960-20010 or equivalent, remove the front lower ball joint from the front axle assembly.

➡**Install SST so that A and B are parallel.**

➡**Be sure to place a wrench on the part indicated in the illustration.**

➡**Do not damage the front lower ball joint dust cover.**

To install:

16. Install the front lower ball joint to the steering knuckle with the nut and tighten to 91 ft. lbs. (123 Nm).

➡**Prevent oil from adhering to the screw and tapered parts.**

17. Install a new cotter pin.

➡**If the holes for the cotter pin are not aligned, tighten the nut further up to 60°.**

18. Install the front axle assembly to the front shock absorber with the 2 bolts and 2 nuts and tighten to 213 ft. lbs. (290 Nm).

➡**Only when reusing the bolts and nuts, apply a small amount of engine oil to the threads of the nuts.**

19. Align the matchmarks and install the front drive shaft assembly to the front axle hub sub-assembly.
20. Install the front lower suspension arm to the front lower ball joint with the bolt and 2 nuts 68 ft. lbs. (92 Nm).
21. Connect the tie rod assembly.
22. Install the front disc.
23. Install the front disc brake caliper assembly to the steering knuckle with the 2 bolts and tighten to 77 ft. lbs. (104 Nm).
24. Install the clamp and front speed sensor with the bolt and tighten to 71 inch lbs. (8 Nm).

➡**Prevent foreign matter from attaching to the sensor tip.**

➡**Firmly insert the sensor body into the knuckle before tightening the bolt.**

➡**After installing the sensor to the knuckle, make sure that there is no clearance between the sensor stay and knuckle. Also make sure that no foreign matter is stuck between the parts.**

➡**To prevent interference between the sensor and magnetic rotor, do not rotate the sensor body during or after the insertion of the sensor body to the knuckle.**

25. Install the front axle hub nut.
26. Check the ABS speed sensor signal

➡**Check the ABS speed sensor signal.**

27. Install front wheel and tighten lug nuts to 76 ft. lbs. (103 Nm).
28. Inspect and adjust front wheel alignment.

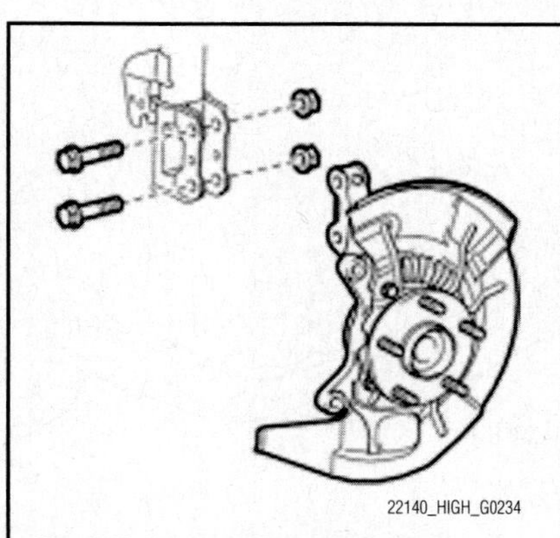

Fig. 206 Front strut assembly bolts

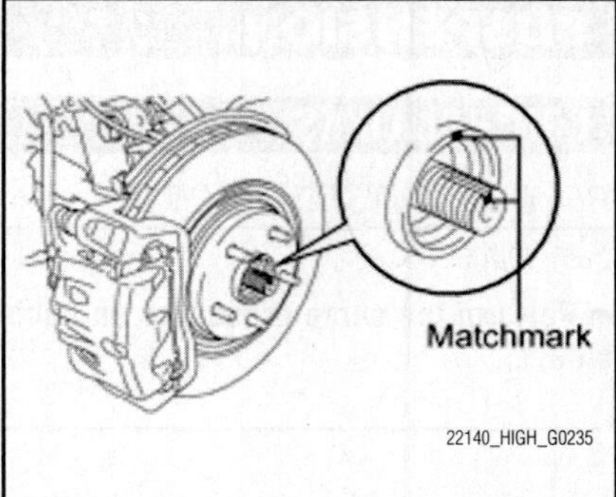

Fig. 207 Align the matchmarks and install the front drive shaft assembly

LOWER CONTROL ARM

REMOVAL AND & INSTALLATION

See Figure 208.

1. Remove the engine assembly with transaxle.
2. Remove the No. 1 front stabilizer brackets.
3. Remove the front stabilizer bar with front stabilizer link assembly.
4. Remove the power steering link assembly.
5. Install the engine hangers.
6. Separate the front frame assembly.
7. Remove the front lower suspension arm, as follows:
   a. Remove the 3 bolts, nut, and the front lower suspension arm from the front frame assembly.
   b. Remove the front lower arm bushing stopper from the front lower suspension arm.

To install:

8. Install the front lower suspension arm, as follows:
   a. Install the front lower arm bushing stopper to the front lower suspension arm.
   b. Install the front lower suspension arm to the front frame assembly with the 3 bolts and nut, but do not tighten them yet.
   c. Tighten the 3 bolts in numerical order shown. Tighten to 147 ft. lbs. (200 Nm), and 152 ft. lbs. (206 Nm).

➡Start installing the bolts from the front side of the vehicle.

9. Connect the front frame assembly.
10. Remove the engine hangers.
11. Install the power steering link assembly.
12. Install the front stabilizer bar with front stabilizer link assembly.
13. Install the No. 1 front stabilizer brackets.
14. Install the engine assembly with transaxle.

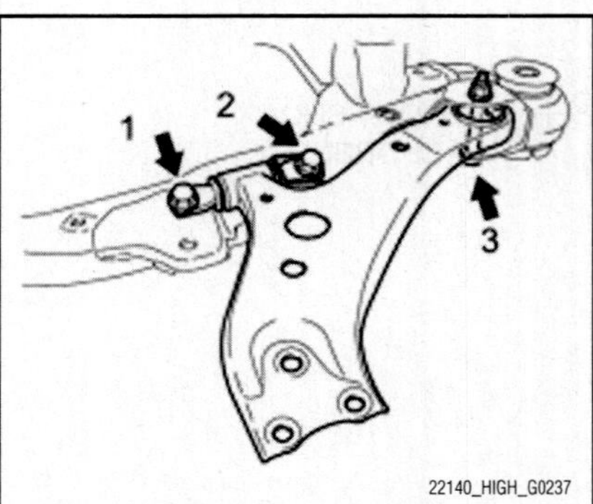

Fig. 208 Front lower control arm bolts tightening sequence

STEERING KNUCKLE

REMOVAL & INSTALLATION

See Figures 209 and 210.

1. Raise and support the vehicle.
2. Remove the tire and wheel assembly.
3. Remove the drive axle retaining nut.
4. Remove the speed sensor.
5. Remove the brake caliper assembly.
6. Remove the front brake rotor.
7. Disconnect outer tie rod end.
8. Remove the 2 bolts and 2 nuts, and separate the front shock absorber with coil spring (lower side) from the steering knuckle.
9. Remove the front lower ball joint from the steering knuckle.
10. Move the drive axle to the side.
11. Remove the lower arm to knuckle nuts and bolts.
12. Remove the steering knuckle.

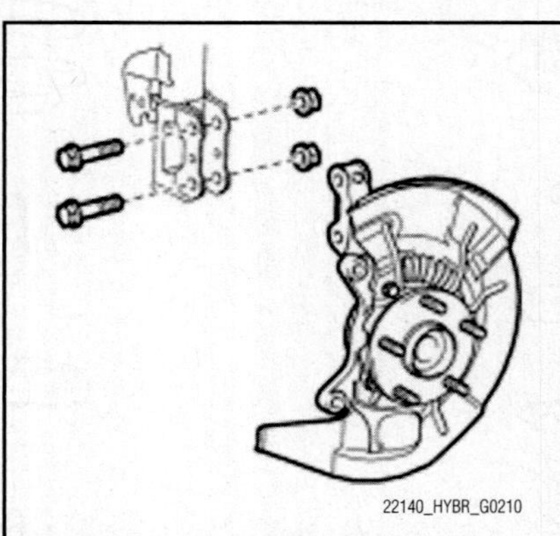

Fig. 209 Remove the 2 bolts and 2 nuts

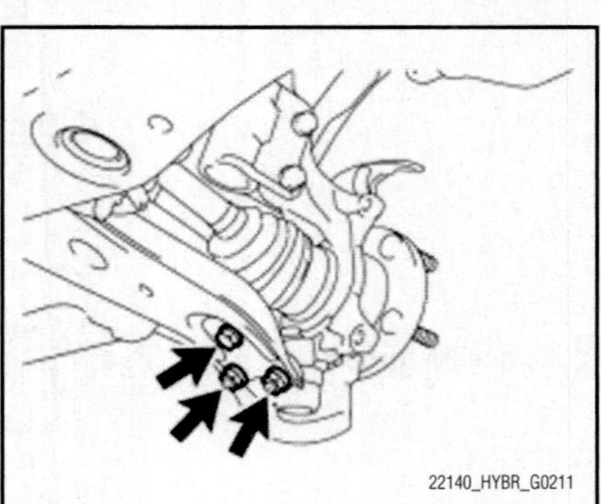

Fig. 210 Remove front lower suspension arm to the front lower ball joint

To install:

13. Install steering knuckle and drive axle.
14. Install ball joint retaining nut and tighten to 36 ft. lbs. (49 Nm). Install cotter pin.
15. Install the lower arm to knuckle nuts and bolts. Tighten to 68 ft. lbs. (92 Nm).
16. Install the 2 bolts and 2 nuts, and tighten the front strut with coil spring (lower side) to the steering knuckle. Tighten to 213 ft. lbs. (290 Nm).
17. Connect the outer tie rod end and tighten the castle nut to 36 ft. lbs. (49 Nm). Install a new cotter pin.
18. Install the front brake rotor.
19. Install front brake caliper assembly and tighten the mounting bolts to 77 ft. lbs. (104 Nm).
20. Install the speed sensor and tighten mounting bolt to 71 inch lbs. (8 Nm).
21. Install the new drive axle retaining nut and tighten to 216 ft. lbs. (294 Nm). Using a chisel and hammer, stake the front axle hub nut.
22. Install and tighten the front wheel assembly to 76 ft. lbs. (103 Nm).
23. Check and adjust wheel alignment.
24. Check speed sensor operation.

STRUT

REMOVAL & INSTALLATION

See Figures 211 through 218.

1. Remove the front wheel.
2. Remove the front wiper arm and blade assemblies.
3. Loosen the front support to front shock absorber nut of the front shock absorber.

➡Do not remove the front support to front shock absorber nut.

➡Loosen the nut only when the front shock absorber with coil spring needs to be disassembled.

4. Remove the cowl top ventilator louver sub-assembly.
5. Remove the windshield wiper motor and link.
6. Remove the outer cowl top panel sub-assembly, as follows:
   a. Disengage the 4 clamps and separate the wiper wire harness from the outer cowl top panel sub-assembly.
   b. Remove the 8 bolts, 6 nuts, and the outer cowl top panel sub-assembly.
7. Remove the bolt and clamp, and separate the front speed sensor and front flexible hose.
8. Remove the nut and separate the front stabilizer link assembly from the front shock absorber.

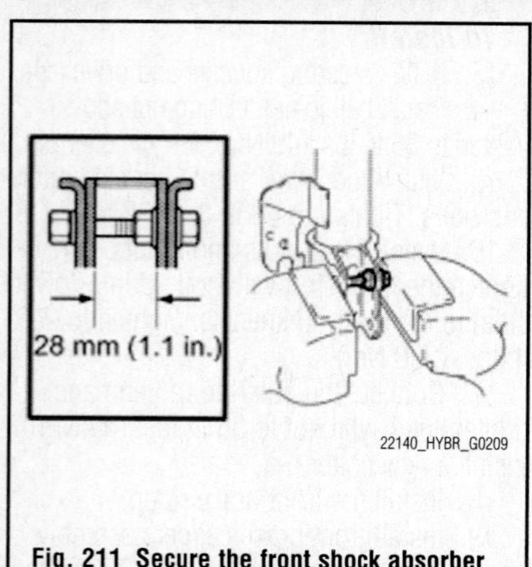

Fig. 211 Secure the front shock absorber with coil spring in a vise

➡If the ball joint turns together with the nut, use a hexagon wrench (6 mm) to hold the stud.

9. Remove the front shock absorber with coil spring, as follows:
   a. Support the front axle using a jack and wooden block.
   b. Remove the 2 bolts and 2 nuts, and separate the front shock absorber with coil spring (lower side) from the steering knuckle.'

➡When removing the nuts, keep the bolts from rotating.

   c. Remove the nut and 2 spacers on the upper side of the front shock absorber with coil spring.

➡Make sure that the front speed sensor is completely separated from the front shock absorber with coil spring.

10. Secure the front shock absorber with coil spring, as follows:
    a. As shown in the illustration, secure the front shock absorber with coil spring in a vise using aluminum plates by clamping onto a double nut bolt affixed to the bracket at the bottom of the absorber.
11. Remove the front support to front shock absorber nut, as follows:
    a. Using A Special Service Tool (SST: 09727-30021, SST: 09727-30021), compress the front coil spring.

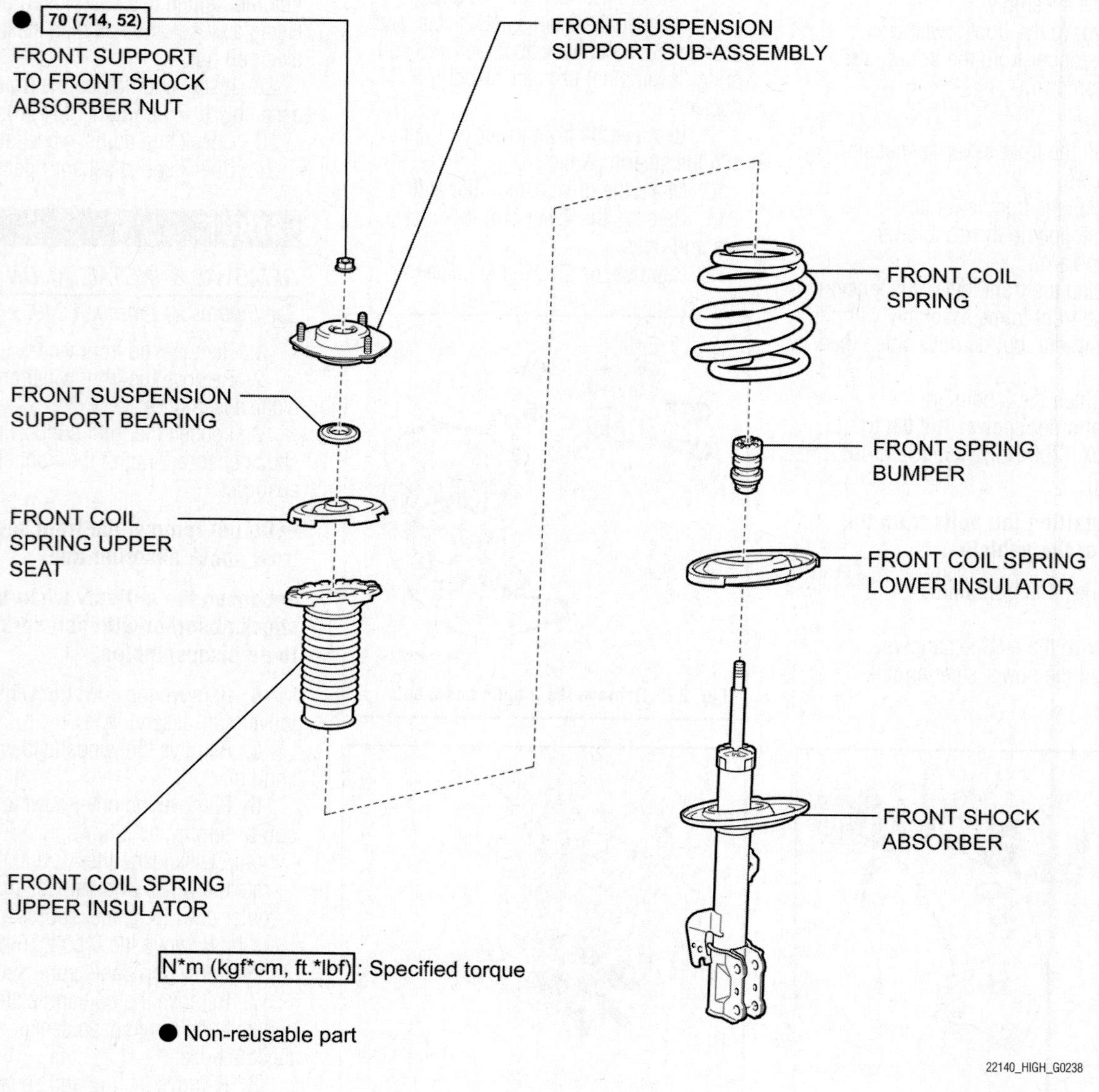

● 70 (714, 52)
FRONT SUPPORT TO FRONT SHOCK ABSORBER NUT
FRONT SUSPENSION SUPPORT SUB-ASSEMBLY
FRONT COIL SPRING
FRONT SUSPENSION SUPPORT BEARING
FRONT SPRING BUMPER
FRONT COIL SPRING UPPER SEAT
FRONT COIL SPRING LOWER INSULATOR
FRONT SHOCK ABSORBER
FRONT COIL SPRING UPPER INSULATOR
N*m (kgf*cm, ft.*lbf): Specified torque
● Non-reusable part
22140_HIGH_G0238

Fig. 212 Front MacPherson strut components

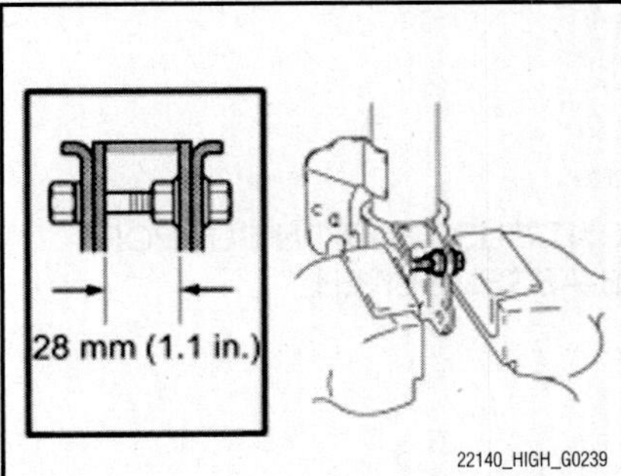

Fig. 213 Secure the front shock absorber with coil spring in a vise

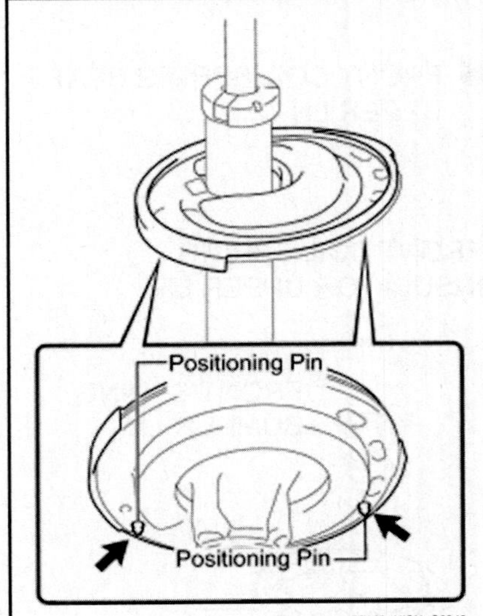

Fig. 214 Front coil spring lower insulator positioning pin

➡Do not use an impact wrench. It will damage the SST.

➡If the front coil spring is compressed at an angle, using 2 SST will make the work easier.

b. Check that the front coil spring is fully compressed.

c. Remove the front support to front shock absorber nut.

12. Remove the front suspension support sub-assembly.
13. Remove the front suspension support bearing.
14. Remove the front coil spring upper seat.
15. Remove the front coil spring upper insulator.
16. Remove the front coil spring.
17. Remove the front spring bumper.
18. Remove the front coil spring lower insulator.

To install:

19. Secure the front shock absorber assembly, as follows:

a. As shown in the illustration, secure the front shock absorber with coil spring in a vise using aluminum plates by clamping onto a double nut bolt affixed to the bracket at the bottom of the absorber.

20. Install the front coil spring lower insulator to the front shock absorber.

➡Make sure that the positioning pins on the front coil spring lower insulator are inserted into the holes in the front shock absorber.

21. Install the front spring bumper to the front shock absorber.

22. Using a Special Service Tool (SST: 09727-30021, SST: 09727-00050) or equivalent, compress the front coil spring.

➡Do not use an impact wrench. It will damage the SST.

➡If the front coil spring is compressed at an angle, using 2 SST will make the work easier.

23. Install the front coil spring to the front shock absorber.

➡Make sure that the end of the front coil spring is positioned in the depression of the lower spring seat.

24. Install the front coil spring upper insulator as shown in the illustration.

➡Any misalignment between the front shock absorber lower bracket and the alignment mark must be +/- 5°.

25. Install the front coil spring upper seat with the mark facing to the outside of the vehicle.

➡Any misalignment between the front shock absorber lower bracket and the alignment mark must be +/- 5°.

26. Install the front suspension support bearing as shown in the illustration.
27. Install the front suspension support sub-assembly as shown in the illustration.

➡Check that the slot on the piston rod and the slot on the front suspension support sub-assembly are aligned.

28. Temporarily tighten a new front support to front shock absorber nut.
29. Install the front shock absorber with coil spring (upper side) with the nut and 2 spacers and tighten to 63 ft. lbs. (85 Nm).
30. Install the front shock absorber with coil spring (lower side) to the steering knuckle and insert the 2 bolts and 2 nuts and tighten to 214 ft. lbs. (290 Nm).

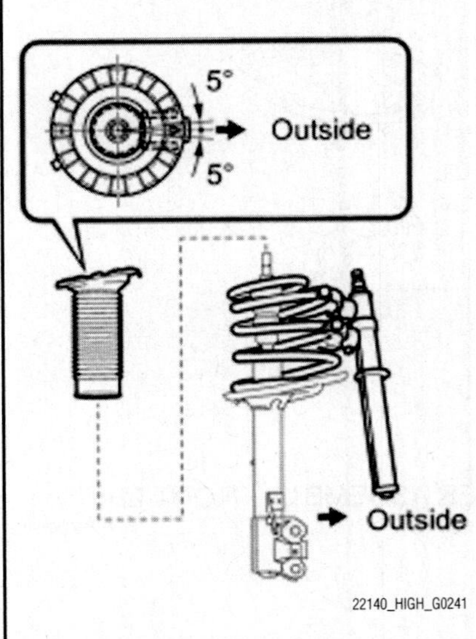

Fig. 215 Install the front coil spring upper insulator

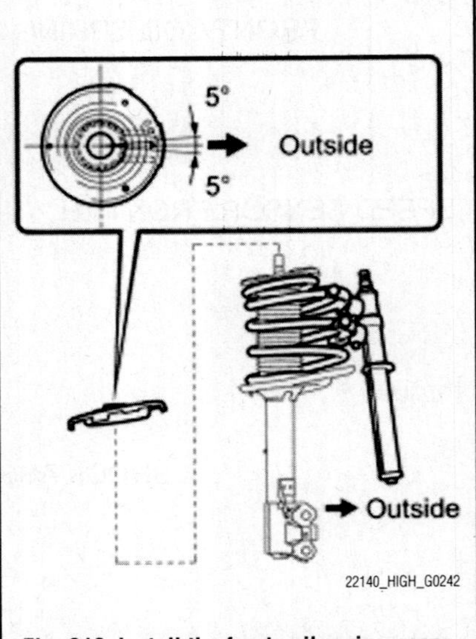

Fig. 216 Install the front coil spring upper seat

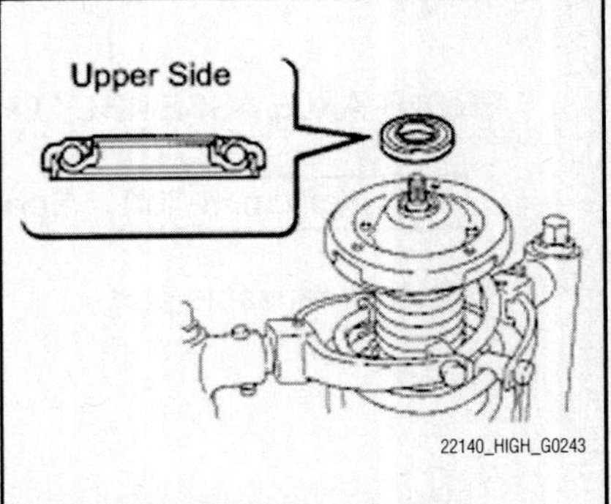

Fig. 217 Install the front suspension support bearing

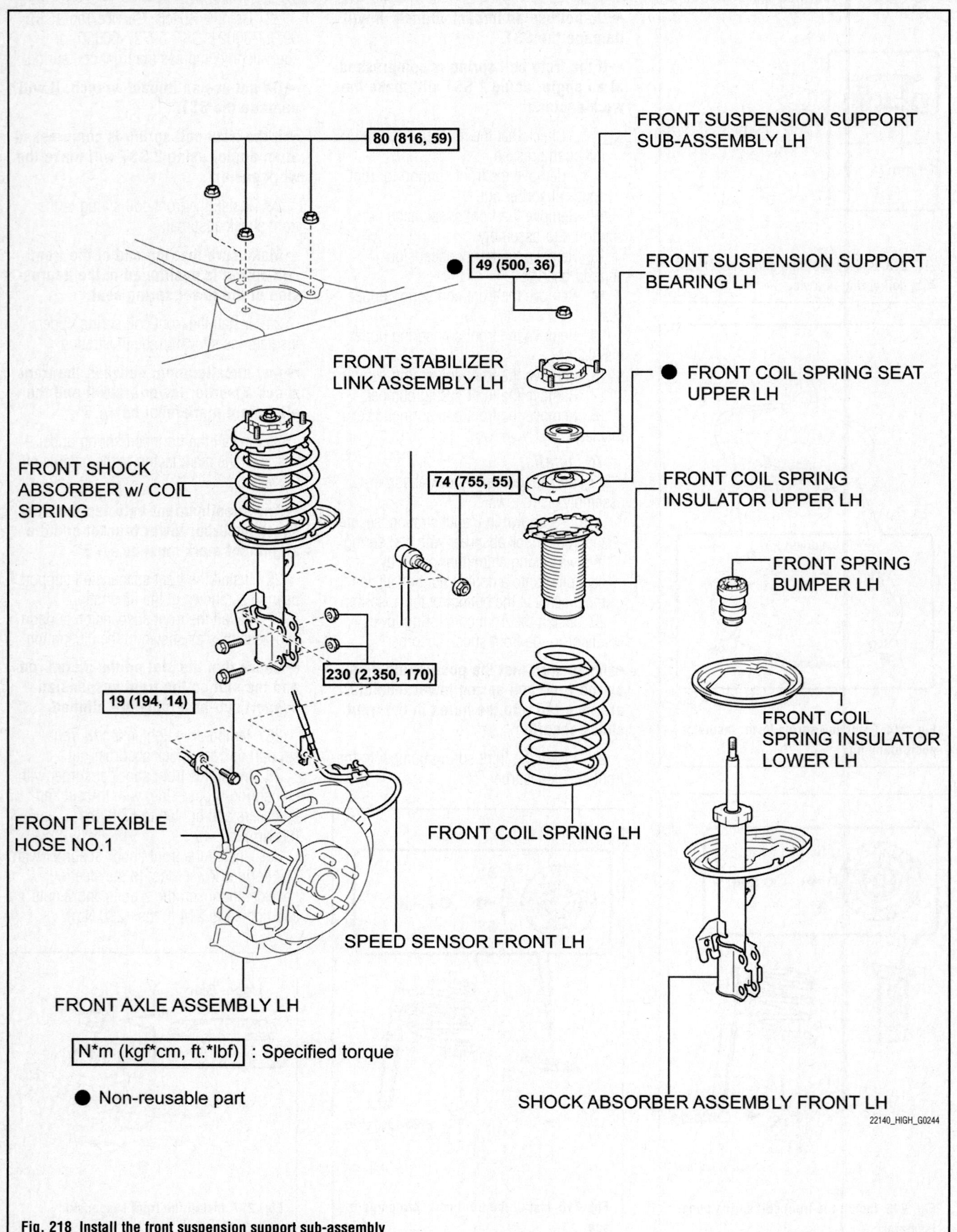

Fig. 218 Install the front suspension support sub-assembly

➡When installing the nuts, keep the bolts from rotating.

31. Install the front stabilizer link assembly to the front shock absorber with the nut and tighten to 55 ft. lbs. (74 Nm).

➡If the ball joint turns together with the nut, use a hexagon wrench (6 mm) to hold the stud bolt.

32. Install the front speed sensor and front flexible hose with the bolt and tighten to 14 ft. lbs. (19 Nm).

➡Do not twist the front speed sensor when installing it.

33. Install the clamp.
34. Install the outer cowl top panel subassembly with the 8 bolts and 6 nuts and tighten to 63 ft. lbs. (85 Nm), 78 inch lbs. (8.8 Nm), 78 inch lbs. (8.8 Nm).
35. Engage the 4 clamps.
36. Fully tighten the front support to front shock absorber nut and tighten to 52 ft. lbs. (70 Nm).
37. Install the windshield wiper motor and link.
38. Install the cowl top ventilator louver sub-assembly.
39. Install the front wiper arm and blade assemblies.
40. Install the front wheel and tighten lug nuts to 76 ft. lbs.

STABILIZER BAR

REMOVAL & INSTALLATION

See Figure 219.

1. Make sure the vehicle's front wheels are in the straight-ahead position.
2. Disconnect the negative battery cable.
3. Raise and safely support the vehicle.
4. Remove the remove front wheel and tire assemblies.
5. Remove the engine under cover.
6. Separate the steering intermediate shaft subassembly.
7. Separate the tie rod.
8. Remove the 2 nuts, then remove the front left stabilizer link assembly:

➡If the ball joint turns together with the nut, use a hexagon wrench (6 mm) to hold the stud.

9. Remove the 2 nuts, then remove the front right stabilizer link assembly:

➡If the ball joint turns together with the nut, use a hexagon wrench (6 mm) to hold the stud.

10. Remove the 2 bolts, then remove the left stabilizer bracket no. 1 from the front frame assembly.
11. Remove the 2 bolts, then remove the right stabilizer bracket no. 1 from the front frame assembly.
12. Remove the left and right no. 2 stabilizer brackets from the bushings.
13. Remove the 2 stabilizer bar bushings no.1 from the stabilizer bar.
14. Remove the front no.3 exhaust pipe subassembly
15. Remove the front exhaust pipe/
16. Remove the bolt, nut and the manifold stay.
17. Remove the power steering link.
18. Remove the front stabilizer bar from the right side of the vehicle.
19. Inspect the turning of the stabilizer link ball joint:
    a. secure the front stabilizer link assembly in a vise using aluminum plates.
    b. Install the nut to the front stabilizer link assembly stud.
    c. If using a torque wrench, turn the nut continuously at a rate of 3 to 5 seconds per turn and take the torque reading on the 5th turn. The turning torque should be 18 inch lbs. (2.0 Nm) or less.

➡If the turning torque is not within the specified range, replace the front stabilizer link assembly with a new one.

20. Inspect the dust cover:
    a. Check that the dust cover is not cracked and that there is no grease on it.

To install:

21. Install the front stabilizer bar by inserting it from the right side of the vehicle.
22. Install the power steering link.
23. Install the manifold stay with the bolt and nut. Tighten to 25 ft. lbs. (34 Nm).
24. Install the front exhaust pipe.

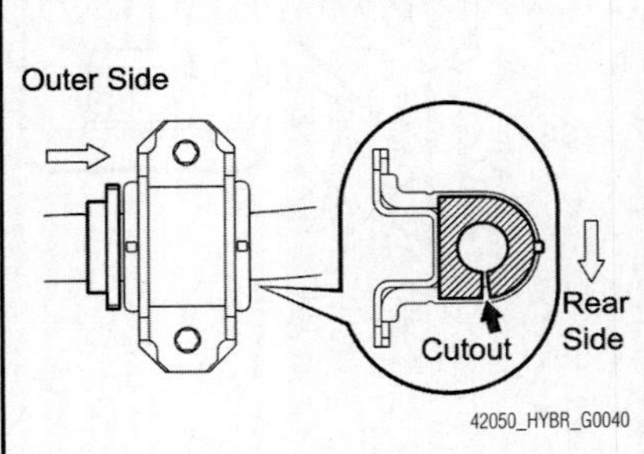

Fig. 219 Place the cutout of the front stabilizer bar bushing no.1 as facing the rear side as shown

25. Install the front no.3 exhaust pipe subassembly.
26. Install the 2 front stabilizer bar bushings no.1 to the outer side of the bushing stopper on the front stabilizer bar.

➡Place the cutout of the front stabilizer bar bushing no.1 as facing the rear side as shown in the illustration.

27. Install the right and left no.2 stabilizer brackets to the stabilizer bushings.
28. Install the right and left no.1 stabilizer bracket to the front frame assembly with the bolts. Tighten to 12 ft. lbs. (16 Nm).
29. Install the right and left stabilizer link assemblies with the nuts and tighten to 55 ft. lbs. (75 Nm).

➡If the ball joint turns together with the nut, use a hexagon (6 mm) wrench to hold the stud.

30. Connect the tie rod.
31. Connect steering intermediate shaft subassembly.
32. Make sure the front wheels are facing straight-ahead.
33. Install the engine under cover.
34. Install the front wheels.
35. Connect the negative battery cable.
36. Initialize the rotation angle sensor and calibrate torque sensor zero point.
37. Inspect and adjust the front wheel alignment.

WHEEL HUB & BEARING

REMOVAL & INSTALLATION

See Figures 220 through 226.

1. Before servicing the vehicle, refer to the Precautions Section.
2. Remove the front wheel.
3. Remove the front axle hub nut.
4. Separate the front speed sensor.
5. Separate the front disc brake caliper assembly.
6. Remove the front brake rotor.
7. Separate and remove the tie rod end.
8. Remove the bolt and 2 nuts, and separate the front suspension arm subassembly lower No.1 from the lower ball joint.
9. Using a plastic hammer, separate the drive shaft from the axle hub.
10. Remove the 2 bolts, nuts and steering knuckle.

⁂ WARNING

Be careful not to damage the boot and speed sensor rotor.

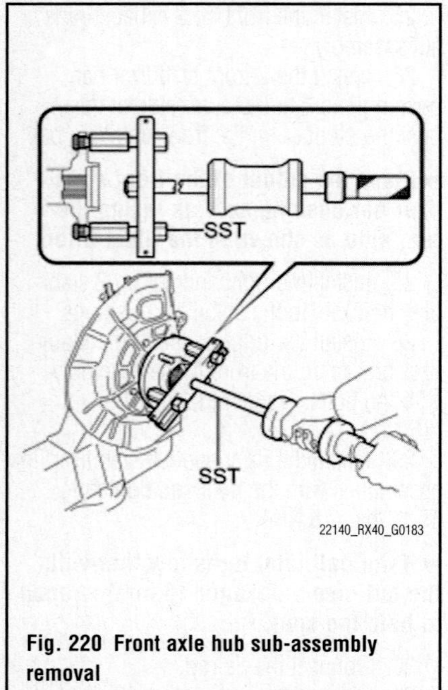

Fig. 220 Front axle hub sub-assembly removal

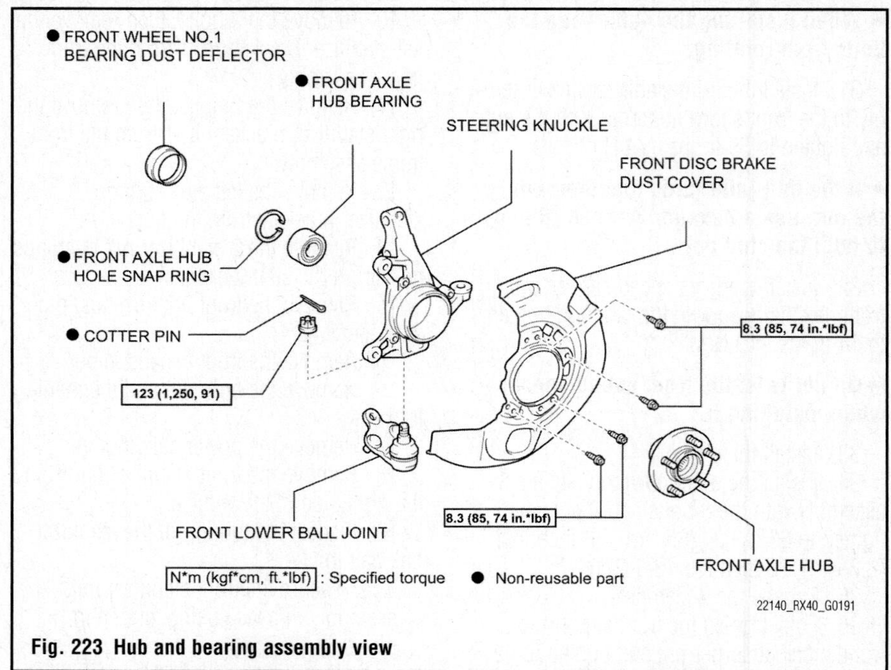

Fig. 223 Hub and bearing assembly view

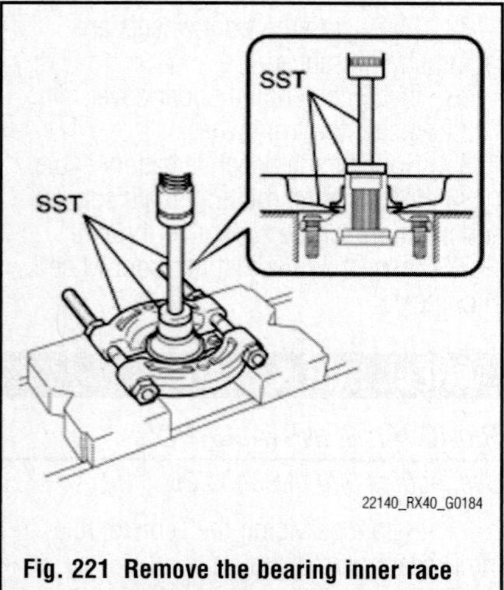

Fig. 221 Remove the bearing inner race

11. Using a screwdriver with its tip wrapped with vinyl tape, remove the bearing dust deflector No.1.
12. Using snap ring pliers, remove the front axle hub hole snap ring.
13. Hold the front axle assembly between aluminum plates in a vise.
14. Using SST 09520-00031, remove the front axle hub sub-assembly.
15. Using SST and a press, remove the bearing inner race (outside) from the front axle hub sub-assembly.
16. Remove the 4 bolts and disc brake dust cover from steering knuckle.
17. Remove the lower ball joint from the hub assembly.
18. Remove the front axle hub bearing as follows:
19. Place the bearing inner race (outside) on the front axle hub bearing.
20. Using SST and a press, press the front axle hub bearing until it contacts the SST.
21. Using SST to make the steering knuckle horizontal, fix it to the V-block, as shown in the illustration.
22. Using SST and a press, remove a front axle hub bearing to the steering knuckle.

To install:

23. Using SST and a press, install a new front axle hub bearing to the steering knuckle.
24. Install the lower ball joint into the steering knuckle and tighten castle nut to 91 ft. lbs. (93 Nm).

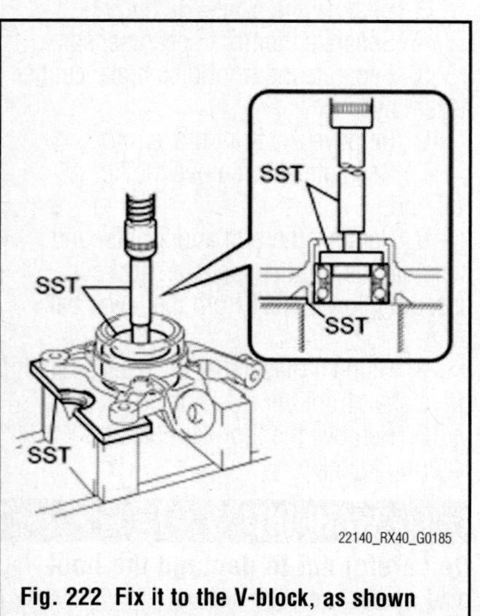

Fig. 222 Fix it to the V-block, as shown

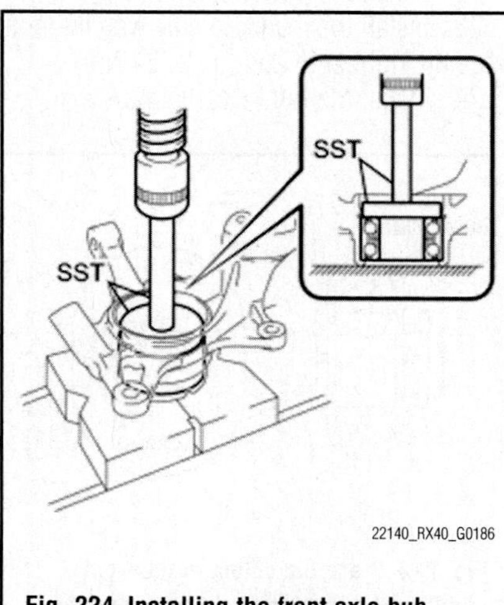

Fig. 224 Installing the front axle hub bearing to the steering knuckle

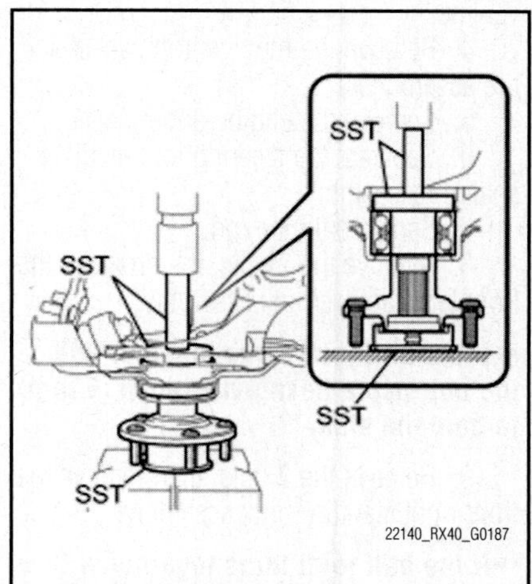

Fig. 225 Installing the front axle hub sub-assembly

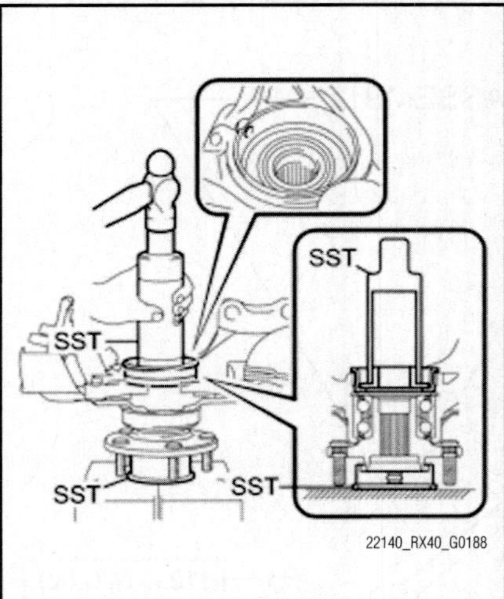

Fig. 226 Installing the bearing dust deflector No.1.

25. Install the disc brake dust cover to the steering knuckle with the 4 bolts. Tighten the bolts to 73 inch lbs. (8.3 Nm).
26. Using SST and a press, install the front axle hub sub-assembly.
27. Using snap ring pliers, install a new front axle hub hole snap ring.
28. Using SST and a hammer, install the bearing dust deflector No.1.

➡Align the hole for the speed sensor in the bearing dust deflector No.1 with the steering knuckle.

29. Install the front axle assembly to the front drive shaft assembly.
30. Install the front axle assembly to the front shock absorber assembly with the 2 bolts and nuts. Tighten the bolts and nuts to 217 ft. lbs. (290 Nm).
31. Install the lower ball joint to the front suspension arm sub-assembly lower with the bolt and 2 nuts. Tighten the nuts and bolts to 94 ft. lbs. (127 Nm).
32. Install the tie rod end to the steering knuckle with the nut. Tighten the nut to 36 ft. lbs. (49 Nm).
33. Install the front brake rotor.
34. Install the front brake caliper assembly.
35. Install the front axle hub nut and tighten to 217 ft. lbs. (294 Nm). Using a chisel and hammer, stake the axle hub nut.
36. Install the speed sensor to the steering knuckle with the bolt and tighten to 71 inch lbs. (8 Nm).
37. Install the front wheel and tighten to 76 ft. lbs. (103 Nm).
38. Adjust the front wheel alignment.
39. Check the speed sensor operation.

SUSPENSION

CONTROL ARMS/LINKS

REMOVAL & INSTALLATION

Rear Lower Arm

See Figure 227.

1. Remove the rear wheel.
2. Remove the rear exhaust.
3. Separate the No.3 parking brake cable.
4. Remove the 2 bolts, the 2 nuts, and the rear strut rod assembly.
5. Put matchmarks on the adjust cams and the rear suspension member sub-assembly.
6. Remove the bolt and nut, and separate the rear No. 2 suspension arm assembly from the rear axle carrier sub-assembly.
7. Remove the nut, the No. 2 camber adjust cam, the rear suspension toe adjust cam sub-assembly, and the rear No. 2 suspension arm assembly.
8. Remove the 2 bolts, the 2 nuts, and the rear No. 1 suspension arm assembly.

To install:

9. Temporarily install the rear No. 1 suspension arm assembly with the 2 bolts and the 2 nuts.
10. Ensure that the identification mark faces the rear side of the vehicle.
11. Temporarily install the rear No. 2 suspension arm assembly to the rear suspension member sub-assembly with the rear suspension toe adjust cam sub-assembly, the No. 2 camber adjust cam and the nut.
12. Temporarily install the rear No. 2 suspension arm assembly to the rear axle carrier sub-assembly with the bolt and the nut.
13. Temporarily install the rear strut rod assembly with the 2 bolts and the 2 nuts.
14. Stabilize the suspension.
15. Fully tighten the rear strut rod assembly to 59 ft. lbs. (80 Nm).
16. Install the No.3 parking brake cable.
17. Install the rear wheel and tighten to 76 ft. lbs. (103 Nm).

Rear Strut Rod Assembly

See Figure 228.

1. Remove the rear wheel.
2. Separate the No.3 parking brake cable.
3. Remove the 2 bolts, the 2 nuts, and the rear strut rod assembly.

To install:

4. Check that the identification mark of the rear strut rod assembly is positioned on the inner side of the vehicle.
5. Temporarily install the rear strut rod assembly with the 2 bolts and the 2 nuts.
6. Stabilize the suspension.
7. Fully tighten the rear strut rod assembly to 59 ft. lbs. (80 Nm).
8. Install the No.3 parking brake cable.
9. Install the rear wheel and tighten to 76 ft. lbs. (103 Nm).

STABILIZER BAR

REMOVAL & INSTALLATION

See Figures 229 and 230.

1. Remove the deck board assembly.
2. Remove the No.3 and No.2 deck board assembly. (w/ tonneau cover)
3. Remove the rear mat.
4. Remove the deck trim service hole cover.
5. Remove the spare tire.
6. Remove the rear wheels.
7. Remove the rear exhaust.
8. Remove the LH and RH rear stabilizer link assembly.
9. Remove both rear stabilizer bar brackets.
10. Remove stabilizer bar.

To install:

11. Temporarily install the rear stabilizer bar with the identification mark positioned on the left side of the vehicle.
12. Temporarily install the rear stabilizer bar bracket LH (front side) with the bolt. Loosely tighten the bolt so that the bracket can be moved by hand.
13. Temporarily install the rear stabilizer bar bracket RH (front side) with the 2 bolts. Loosely tighten the bolt so that the bracket can be moved by hand.
14. Install the rear stabilizer bushing (RH side) to the rear stabilizer bar.
15. Install the rear stabilizer bar bracket RH (rear side) with the 2 bolts.
16. Install the rear stabilizer bushing (LH side) to the rear stabilizer bar.
17. Install the rear stabilizer bar bracket LH (rear side) with the 2 bolts.
18. Tighten the stabilizer bracket bolts to 14 ft. lbs. (19 Nm).
19. Install the rear stabilizer link assembly LH and RH to the rear shock absorber. Tighten to 29 ft. lbs. (39 Nm).

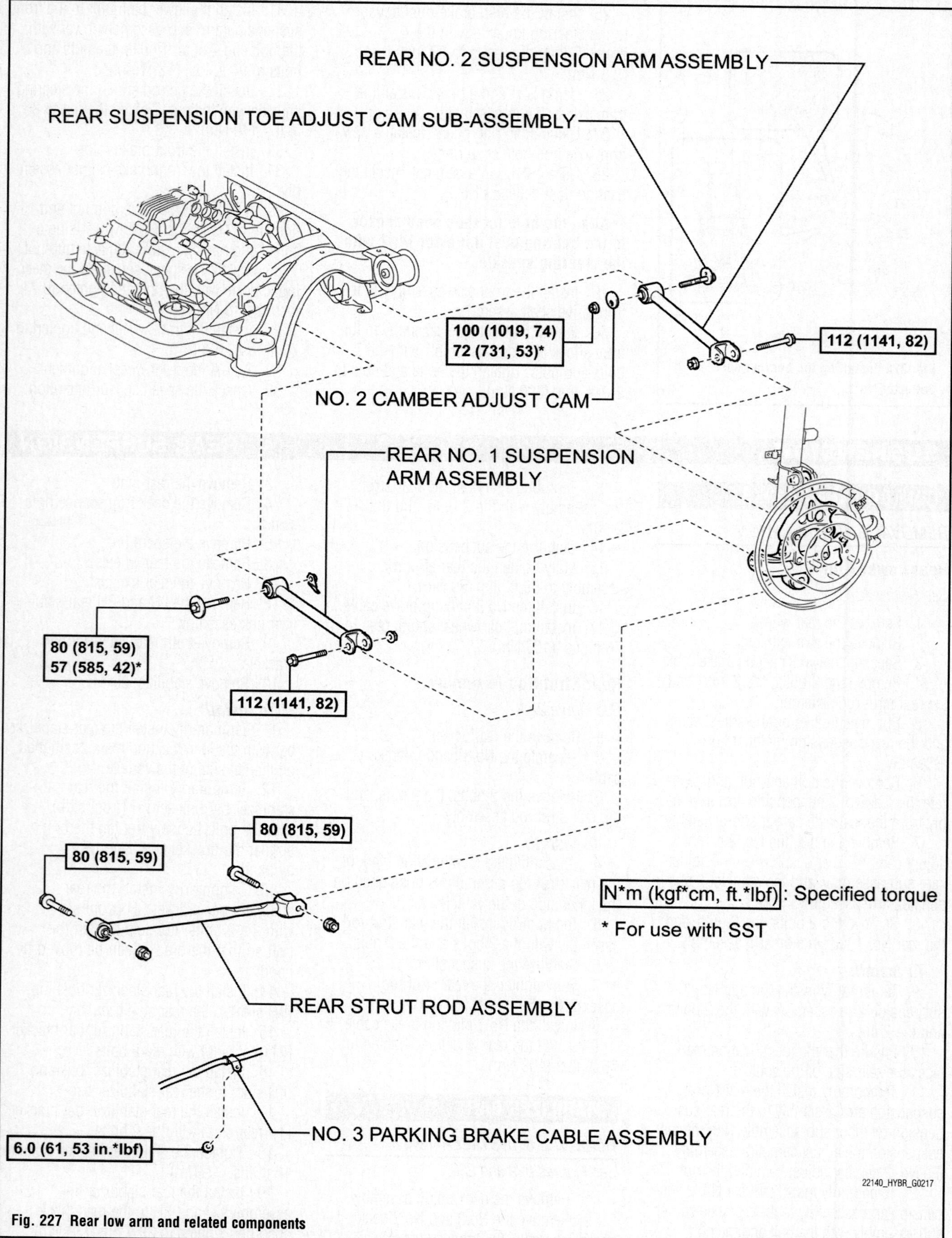

Fig. 227 Rear low arm and related components

22140_HYBR_G0218

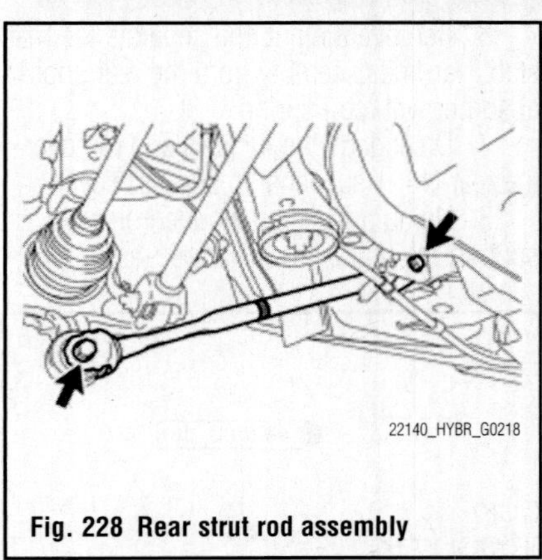

Fig. 228 Rear strut rod assembly

20. Install the rear exhaust assembly.
21. Inspect for exhaust leaks.
22. Install and tighten the rear wheels to 76 ft. lbs. (103 Nm).
23. Install the rear spare.
24. Install the rear mat.
25. Install the tonneau cover if equipped.
26. Install No.3 and No.2 deck board assembly. (if equipped with a tonneau cover)
27. Install the deck board assembly.

Front of the Vehicle
Identification Mark
22140_HYBR_G0222

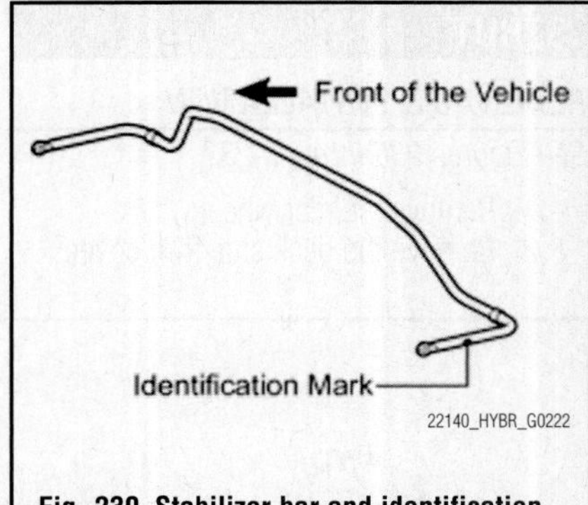

Fig. 230 Stabilizer bar and identification mark

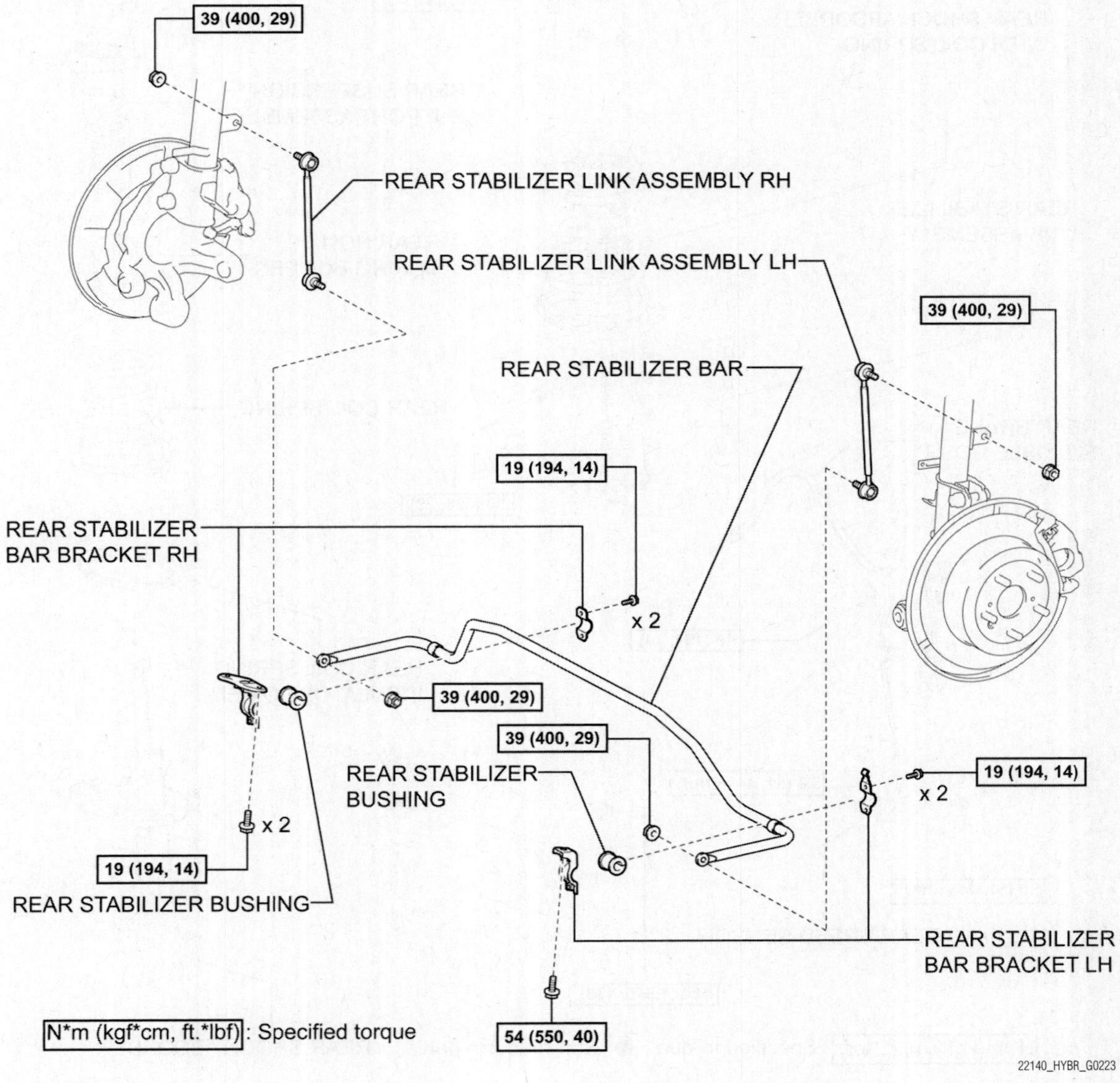

39 (400, 29)
REAR STABILIZER LINK ASSEMBLY RH
REAR STABILIZER LINK ASSEMBLY LH
39 (400, 29)
REAR STABILIZER BAR
19 (194, 14)
REAR STABILIZER BAR BRACKET RH
x 2
39 (400, 29)
39 (400, 29)
REAR STABILIZER BUSHING
19 (194, 14)
x 2
x 2
19 (194, 14)
REAR STABILIZER BUSHING
REAR STABILIZER BAR BRACKET LH
N*m (kgf*cm, ft.*lbf): Specified torque
54 (550, 40)
22140_HYBR_G0223

Fig. 229 Stabilizer bar and related components

STRUT

REMOVAL & INSTALLATION

See Figures 230 through 232.

1. Remove the rear wheel.
2. Remove the deck side trim cover.
3. Remove the deck side trim.
4. Remove the bolt and separate the rear flexible hose from the rear shock absorber with coil spring.
5. Remove the bolt and separate the rear speed sensor from the rear shock absorber with coil spring.
6. Remove the nut and separate the rear stabilizer link assembly from the rear shock absorber with coil spring.
7. Disengage the 4 claws and remove the rear No. 1 suspension support cover.
8. Using a jack and wooden block, support the rear axle carrier sub-assembly.

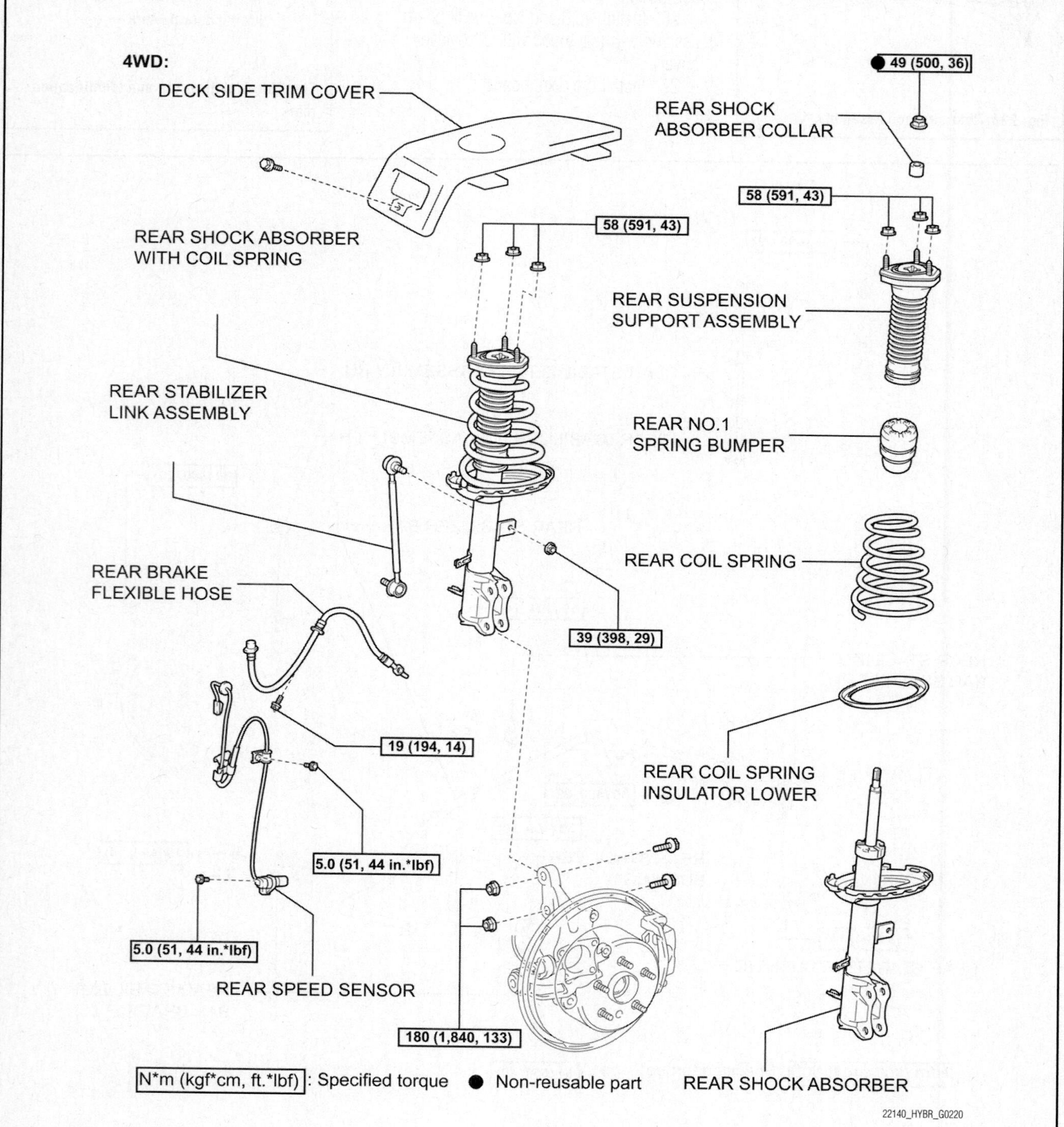

Fig. 231 Rear strut and related components

9. Loosen the rear support to rear shock absorber nut.

CAUTION

Do not remove the rear support to rear shock absorber nut.

10. Remove the 2 bolts and 2 nuts, and separate the rear shock absorber with coil spring from the rear axle carrier sub-assembly.

11. Remove the 3 nuts and rear shock absorber with coil spring.

12. Install the bolt and nut to the rear shock absorber with coil spring as shown in the illustration and secure the rear shock absorber with coil spring in a vise.

13. Using a spring compressor, compress the rear coil spring.

14. Check that the front coil spring is fully compressed.

15. Remove the rear support to rear shock absorber collar from the rear shock absorber assembly.

16. Remove the rear suspension support assembly from the rear shock absorber assembly.

17. Remove the rear coil spring together with SST from the rear shock absorber assembly.

18. Remove the rear No. 1 spring bumper from the rear shock absorber assembly.

19. Remove the rear lower coil spring insulator from the rear shock absorber assembly.

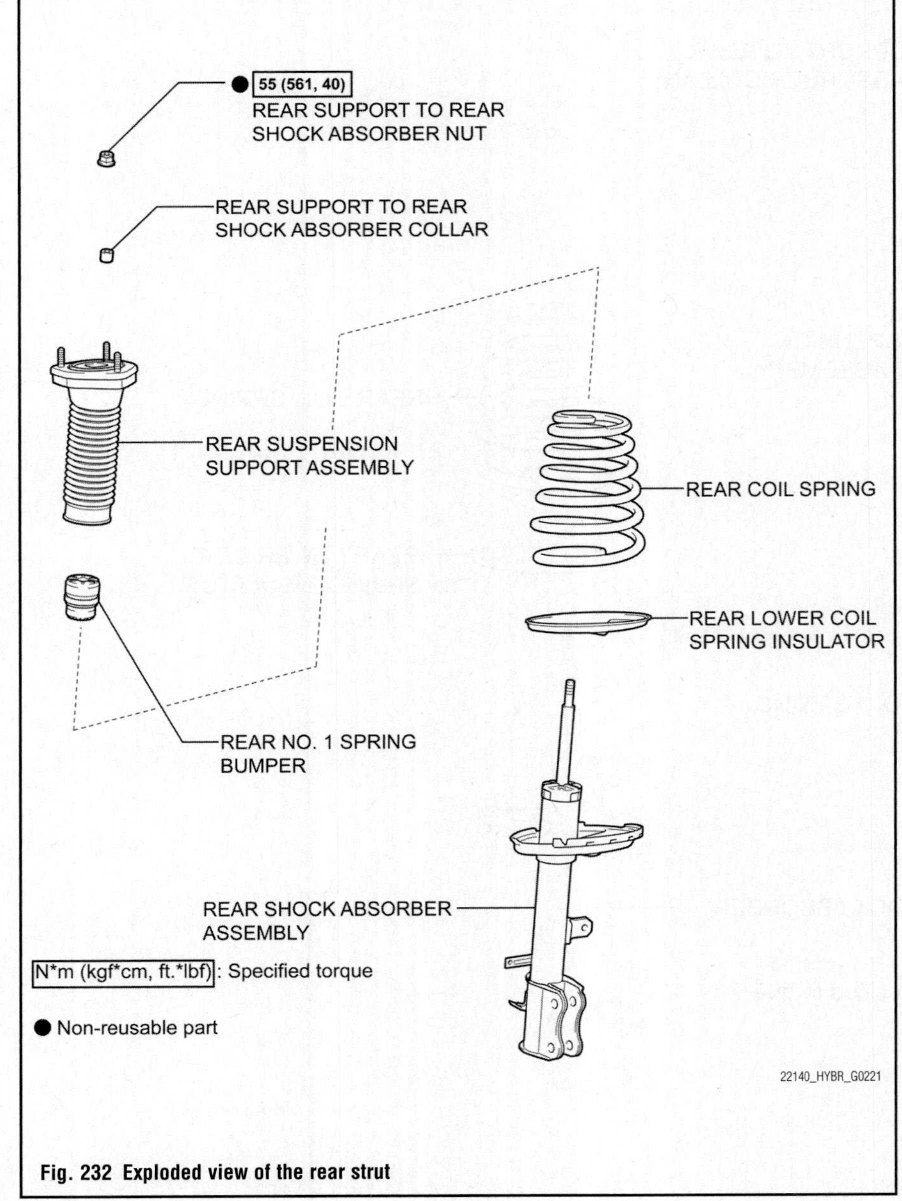

Fig. 232 Exploded view of the rear strut

To install:

20. Install the rear lower coil spring insulator onto the rear shock absorber assembly. Fit the recessed part of the rear lower coil spring insulator into the recession on the shock absorber assembly.

21. Install the rear No. 1 spring bumper to the rear shock absorber assembly.

22. Using spring, compress the rear coil spring.

23. Temporarily install the rear coil spring together with SST to the rear shock absorber assembly.

24. Install the rear suspension support assembly to the rear shock absorber assembly. Align the cutout on the rear shock absorber assembly with the protrusion on the rear suspension support assembly by referring to the illustration.

25. Install the rear support to rear shock absorber collar to the rear shock absorber assembly.

26. Temporarily install the rear support to rear shock absorber nut to the rear shock absorber assembly.

27. Install the rear coil spring. Ensure that the stud bolt is positioned 3.5° to the outside of the vehicle as shown in the illustration. The deviation should be within plus or minus 5°.

28. Install the rear shock absorber with coil spring with the 3 nuts. Tighten the upper mounting nuts to 43 ft. lbs. (58 Nm).

29. Install the rear shock absorber with coil spring with the 2 bolts and 2 nuts. Tighten to 213 ft. lbs. (290 Nm).

30. Fully tighten the rear support to rear shock absorber nut. Tighten the nut to 40 ft. lbs. (55 Nm).

31. Install the rear No. 1 suspension support cover.

32. Install the rear stabilizer link assembly to the rear shock absorber with coil spring with the nut.

33. Tighten the link assembly to 29 ft. lbs. (39 Nm).

34. Install the rear speed sensor to the rear shock absorber, and tighten the bolt to 54 inch lbs. (5 Nm).

35. Install the rear speed sensor to the rear shock absorber and tighten the bolt 14 ft. lbs. (19 Nm).

36. Install the deck side trim and cover.

37. Install the wheels and tighten lug nuts to 76 ft. lbs. (103 Nm).

38. Inspect and adjust rear wheel alignment.

39. Check the ABS speed sensor signal.

OVERHAUL

See Figure 233.

1. Before servicing the vehicle, refer to the Precautions Section.
2. Remove or disconnect the following:
   - Wheel

➡**If equipped, be careful not to damage the oil seal, driveshaft boot and/or speed sensor rotor when removing the steering knuckle.**

   - Shock absorber (strut assembly)

3. Install a nut/bolt to the bracket at the lower portion of the strut assembly and secure it in a vise.
4. Compress the coil spring with a spring compressor.

**** CAUTION**

The proper tools must be used for this procedure. The spring on the strut is under high pressure and can cause serious injury if not properly removed and installed.

5. Remove or disconnect the following:
   - Center retaining nut, by holding the spring seat
   - Support, dust seal, spring seat, insulator and spring from the strut assembly

To install:

6. Install the spring bumper and lower insulator to the strut assembly.
7. Compress the coil spring and fit the lower end of the spring into the spring seat gap.

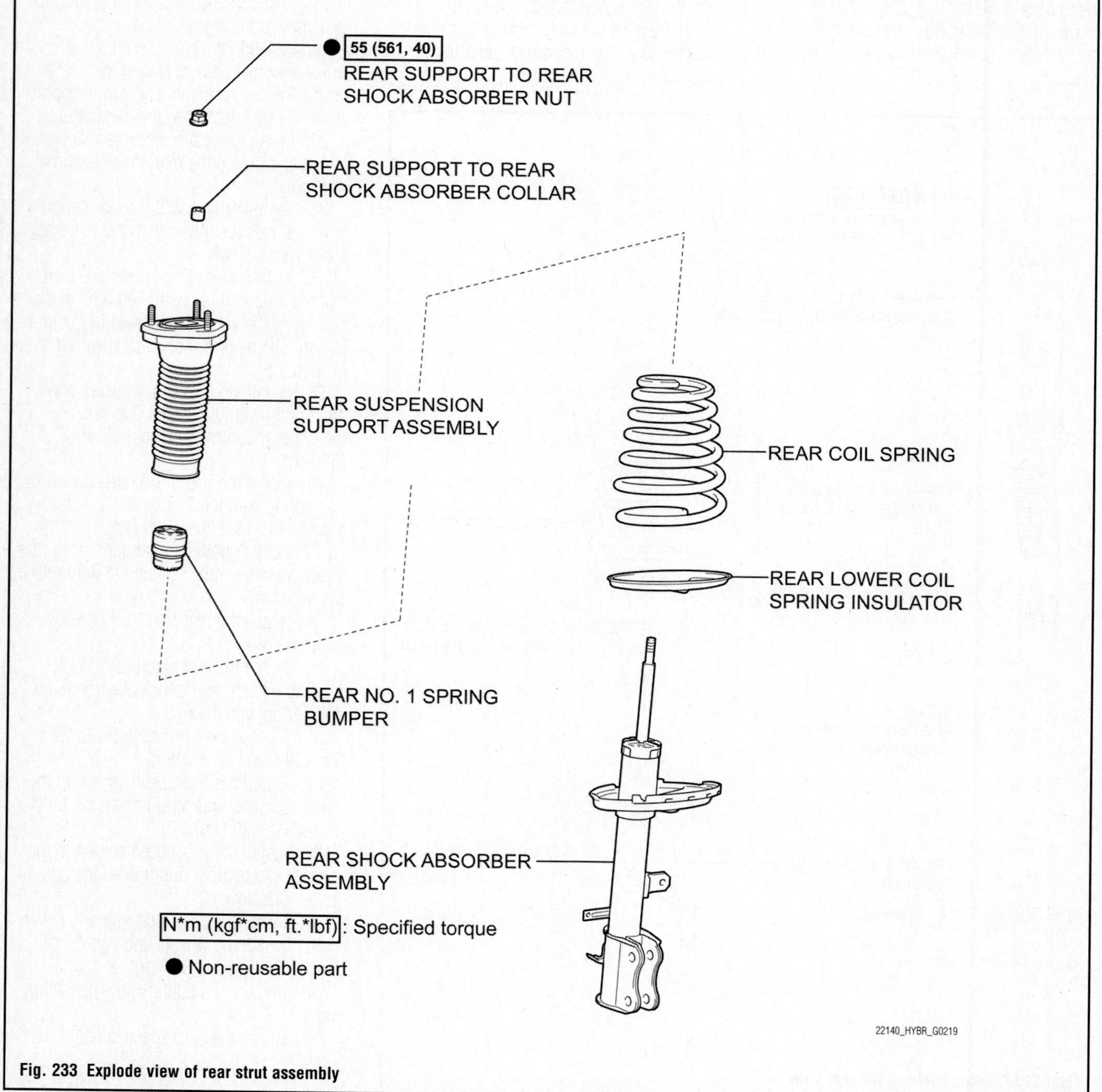

Fig. 233 Explode view of rear strut assembly

8. Install or connect the following:
- Upper insulator, spring seat, dust seal, support and spring seat. Tighten the new retaining nut to 36 ft. lbs. (49 Nm).
- Strut
- Wheel

9. If required, bleed the brake system and check for leaks.

10. Check and/or adjust the front wheel alignment.

WHEEL HUB & BEARING

REMOVAL & INSTALLATION

See Figure 234.

1. Disconnect the negative battery cable.
2. Remove the wheel.
3. Separate the rear flexible hose.
4. Remove the 2 bolts and separate the rear disc brake caliper assembly.
5. Remove the rear disc.
6. Using a screwdriver, disconnect the connector from the rear speed sensor.
7. For 4WD vehicles,

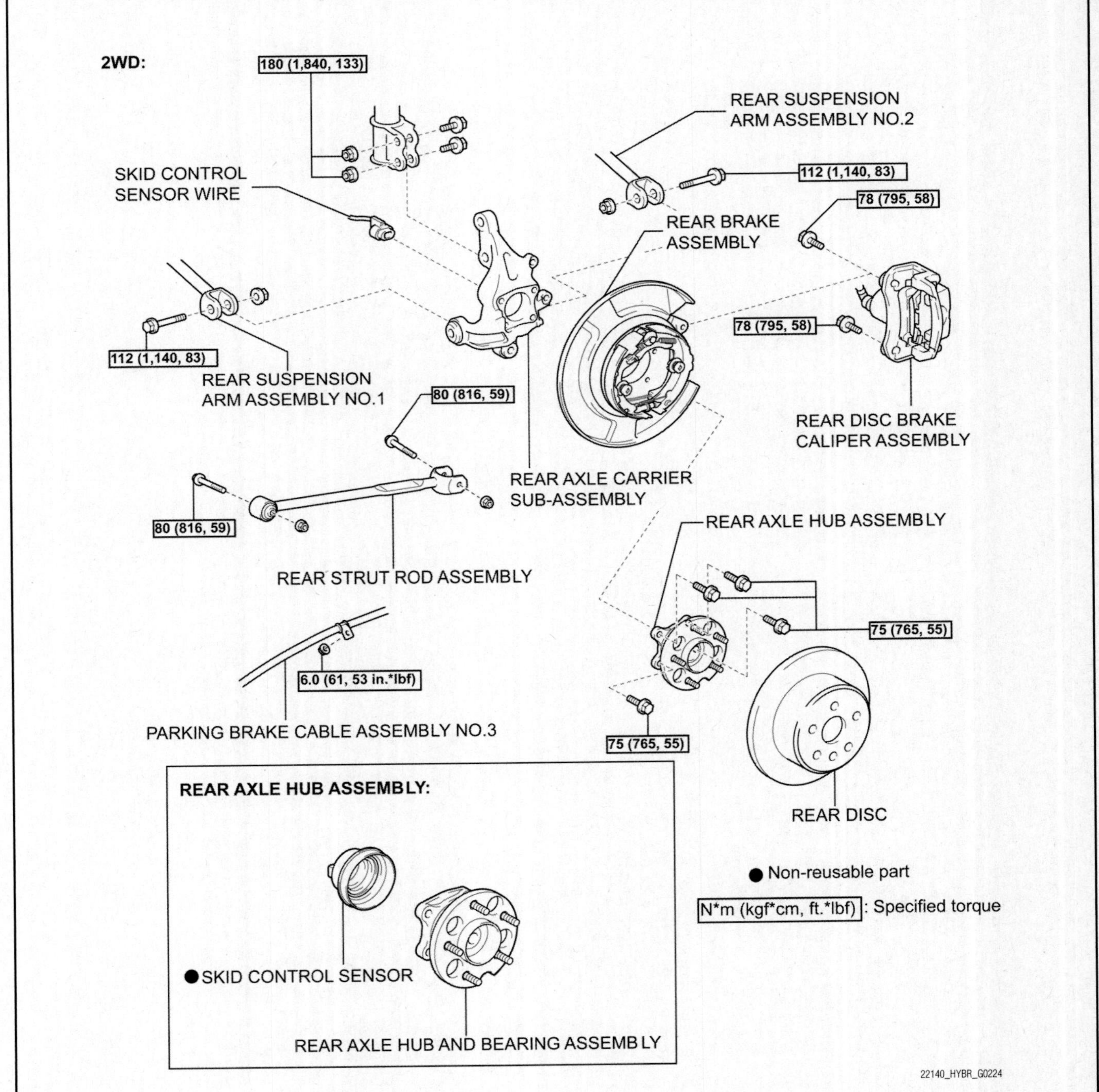

Fig. 234 Rear axle hub, bearing and component view

8. Remove the 4 bolts and the rear axle hub & bearing assembly.

To install:

9. Install the rear axle hub and bearing assembly with the 4 bolts and tighten to 55 ft. lbs. (75 Nm).
10. Inspect rear axle hub bearing looseness.
11. Inspect rear axle hub runout.
12. Connect the connector to the rear speed sensor.
13. Install the rear disc.
14. Install the rear disc brake caliper assembly with the 2 bolts and tighten to 57 ft. lbs. (78 Nm).
15. Install the rear flexible hose to the shock absorber with coil spring with the bolt and tighten to 14 ft. lbs. (19 Nm).
16. Install the wheel and tighten lug nuts to 76 ft. lbs. (103 Nm).
17. Connect the negative battery cable.
18. Inspect and adjust rear wheel alignment.
19. Check the ABS speed sensor signal.

TOYOTA

Land Cruiser

11

SPECIFICATIONS AND MAINTENANCE CHARTS

ENGINE AND VEHICLE IDENTIFICATION

Engine							Model Year	
Code ①	Liters (cc)	Cu. In.	Cyl.	Fuel Sys.	Engine Type	Eng. Mfg.	Code ②	Year
3UR-FE	5.7 (5700)	345.6	8	SFI	DOHC	Toyota	8	2008
							9	2009
							A	2010

SFI: Sequential Fuel Injection

DOHC: Double Overhead Camshaft

① Stamped on the left side of the engine block

3768X_LAND_C0001

GENERAL ENGINE SPECIFICATIONS

Year	Model	Engine Displacement Liters	Engine Series ID	Net Horsepower @ rpm	Net Torque @ rpm (ft. lbs.)	Bore x Stroke (in.)	Com-pression Ratio	Oil Pressure @ rpm
2008	Land Cruiser	5.7	3UR-FE	381@5600	401@3600	3.70x4.02	10.2:1	32 @ 2500
2009	Land Cruiser	5.7	3UR-FE	381@5600	401@3600	3.70x4.02	10.2:1	32 @ 2500
2010	Land Cruiser	5.7	3UR-FE	381@5600	401@3600	3.70x4.02	10.2:1	32 @ 2500

3768X_LAND_C0002

ENGINE TUNE-UP SPECIFICATIONS

Year	Engine Displacement Liters	Engine ID	Spark Plug Gap (in.)	Ignition Timing (deg.)*	Fuel Pump (psi)	Idle Speed (rpm)		Valve Clearance	
						MT	AT	Intake	Exhaust
2008	5.7	3UR-FE	0.043	8-12 B at idle	41-42	—	650-750	NA	NA
2009	5.7	3UR-FE	0.043	8-12 B at idle	41-42	—	650-750	NA	NA
2010	5.7	3UR-FE	0.043	8-12 B at idle	41-42	—	650-750	NA	NA

NOTE: The Vehicle Emission Control Information label often reflects specification changes made during production.
The label figures must be used if they differ from those in this chart.

B: Before top dead center

* With terminals TC and CG connected to DLC3 (Transmission in Neurtal and A/C switch OFF)

3768X_LAND_C0003

CAPACITIES

Year	Model	Engine Displacement Liters	Engine ID	Engine Oil with Filter (qts.)	Transmission (pts.) 5-Spd	Transmission (pts.) Auto.*	Transfer Case (pts.)	Drive Axle Front (pts.)	Drive Axle Rear (pts.)	Fuel Tank (gal.)	Cooling System (qts.)
2008	Land Cruiser	5.7	3UR-FE	7.4	—	10.6	3.1	3.9-5.2	8.8-8.9	24.6	16.3
2009	Land Cruiser	5.7	3UR-FE	7.4	—	10.6	3.1	3.9-5.2	8.8-8.9	24.6	16.3
2010	Land Cruiser	5.7	3UR-FE	7.4	—	10.6	3.1	3.9-5.2	8.8-8.9	24.6	16.3

*After draining, add the following amounts, then fill to the cold full line

① Without rear heater: 15.6

With rear heater: 16.2

3768X_LAND_C0004

FLUID SPECIFICATIONS

Year	Model	Engine Displacement Liters	Engine ID/VIN	Engine Oil	Auto. Trans.	Drive Axle	Power Steering Fluid	Brake Master Cylinder
2008	Land Cruiser	5.7	3UR-FE	5W-20	NA	—	ATF Dexron II Or III	DOT 3
2009	Land Cruiser	5.7	3UR-FE	5W-20	NA	—	ATF Dexron II Or III	DOT 3
2010	Land Cruiser	5.7	3UR-FE	5W-20	NA	—	ATF Dexron II Or III	DOT 3

DOT: Department Of Transpotation

NA: Not Available

3768X_LAND_C0013

VALVE SPECIFICATIONS

Year	Engine Displacement Liters	Engine ID	Seat Angle (deg.)	Face Angle (deg.)	Spring Test Pressure (lbs. @ in.)	Spring Installed Height (in.)	Stem-to-Guide Clearance (in.) Intake	Stem-to-Guide Clearance (in.) Exhaust	Stem Diameter (in.) Intake	Stem Diameter (in.) Exhaust
2008	5.7	3UR-FE	45	44.5	45.9-50.7@ 1.378	1.380	0.0010-0.0024	0.0012-0.0026	0.2154-0.2159	0.2152-0.2157
2009	5.7	3UR-FE	45	44.5	45.9-50.7@ 1.378	1.380	0.0010-0.0024	0.0012-0.0026	0.2154-0.2159	0.2152-0.2157
2010	5.7	3UR-FE	45	44.5	45.9-50.7@ 1.378	1.380	0.0010-0.0024	0.0012-0.0026	0.2154-0.2159	0.2152-0.2157

3768X_LAND_C0005

CAMSHAFT AND BEARING SPECIFICATIONS CHART

All measurements are given in inches.

Year	Engine Displ. Liters	Engine ID/VIN	Journal Dia.	Brg. Oil Clearance	Shaft End-play	Runout	Journal Bore	Lobe Height	
								Intake	Exhaust
2008	5.7	3UR-FE	①	②	NA	0.00157	NA	1.744-1.7500	1.740-1.7460
2009	5.7	3UR-FE	①	②	NA	0.00157	NA	1.744-1.7500	1.740-1.7460
2010	5.7	3UR-FE	①	②	NA	0.00157	NA	1.744-1.7500	1.740-1.7460

NA: Not Available

① No. 1 journal: 1.179-1.180

Other Journals: 1.022-1.023 in.

② No. 1 journal: 0.00118-0.00256 in.

Other journal: 0.000984-0.00244 in.

Other Maximum Journals: 0.00276 in.

3768X_LAND_C0014

CRANKSHAFT AND CONNECTING ROD SPECIFICATIONS

All measurements are given in inches.

Year	Engine Displacement Liters	Engine ID	Crankshaft				Connecting Rod		
			Main Brg. Journal Dia.	Main Brg. Clearance	Shaft End-play	Thrust on No.	Journal Diameter	Oil Clearance	Side Clearance
2008	5.7	3UR-FE	2.6373-2.6378	①	0.0008-0.0087	3	②	0.0010-0.0020	0.0059-0.0217
2009	5.7	3UR-FE	2.6373-2.6378	①	0.0008-0.0087	3	②	0.0010-0.0020	0.0059-0.0217
2010	5.7	3UR-FE	2.6373-2.6378	①	0.0008-0.0087	3	②	0.0010-0.0020	0.0059-0.0217

① No. 1 and No. 5: 0.0007-0.0012

All others: 0.0009-0.0015

② Mark 1: 2.32283-2.32307

Mark 2: 2.32307-2.32330

Mark 3: 2.32330-2.32354

Mark 2: 2.32354-2.32377

3768X_LAND_C0006

PISTON AND RING SPECIFICATIONS

All measurements are given in inches.

Year	Engine Displacement Liters	Engine ID	Piston Clearance	Ring Gap			Ring Side Clearance		
				Top Comp.	Bottom Comp.	Oil Control	Top Comp.	Bottom Comp.	Oil Control
2008	5.7	3UR-FE	0.0007-0.0028	0.0091-0.0130	0.0157-0.0197	0.0039-0.0157	0.00078-0.0028	0.00078-0.0024	0.0027-0.0057
2009	5.7	3UR-FE	0.0007-0.0028	0.0091-0.0130	0.0157-0.0197	0.0039-0.0157	0.00078-0.0028	0.00078-0.0024	0.0027-0.0057
2010	5.7	3UR-FE	0.0007-0.0028	0.0091-0.0130	0.0157-0.0197	0.0039-0.0157	0.00078-0.0028	0.00078-0.0024	0.0027-0.0057

3768X_LAND_C0007

TORQUE SPECIFICATIONS

All readings in ft. lbs.

Year	Engine Displacement Liters	Engine ID	Cylinder Head Bolts	Main Bearing Bolts	Rod Bearing Bolts	Crankshaft Damper Bolts	Flywheel Bolts	Manifold Intake	Manifold Exhaust	Spark Plugs	Oil Pan Drain Plug
2008	5.7	3UR-FE	①	②	③	NA	15	15	15	13	29
2009	5.7	3UR-FE	①	②	③	NA	15	15	15	13	29
2010	5.7	3UR-FE	①	②	③	NA	15	15	15	13	29

① Step 1: 27 ft. lbs.
Step 2: Plus 90 degrees
Step 3: Plus 90 degrees
For 12 mm head: 15 ft. lbs. (21 Nm)

② For inside position: 45 ft. lbs. (61 Nm)
For outside position: 20 ft. lbs.
Step 2: Plus 90 degrees
For cylinder block side position: 33 ft. lbs.

③ Step 1: 30 ft. lbs.
Step 2: Plus 90 degrees

3768X_LAND_C0008

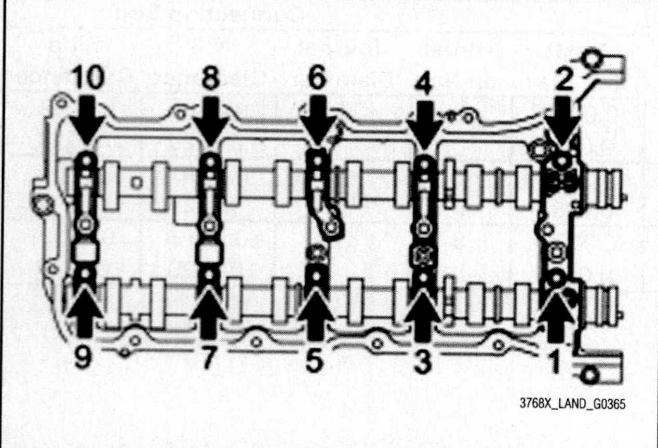

Fig. 1 Identifying the main bearing cap bolt tightening sequence RH

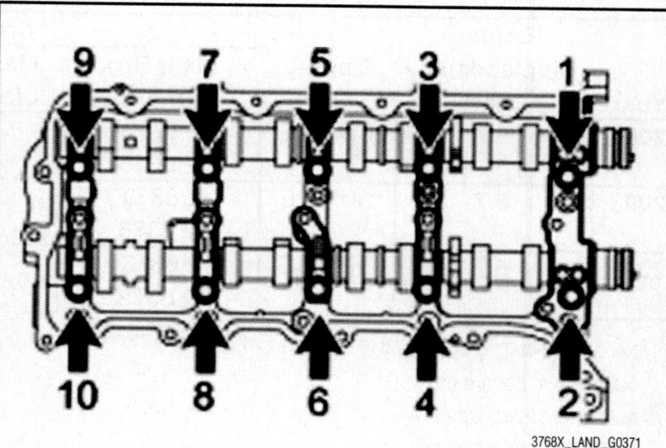

Fig. 2 Identifying the main bearing cap bolt tightening sequence LH

WHEEL ALIGNMENT

Year	Model	Caster Range (+/-Deg.)	Caster Preferred Setting (Deg.)	Camber Range (+/-Deg.)	Camber Preferred Setting (Deg.)	Toe-in (in.)	Steering Axis Inclination (Deg.)
2008	Land Cruiser	0.75	+2.90	0.75	+0.13	0.12+/-0.08	12.85+/-0.75
2009	Land Cruiser	0.75	+2.90	0.75	+0.13	0.12+/-0.08	12.85+/-0.75
2010	Land Cruiser	0.75	+2.90	0.75	+0.13	0.12+/-0.08	12.85+/-0.75

Note: All alignment specifications are based on nominal ride height and standard tires

3768X_LAND_C0009

TIRE, WHEEL AND BALL JOINT SPECIFICATIONS

Year	Model	OEM Tires		Tire Pressures (psi)		Wheel	Ball Joint	Lug Nut Torque
		Standard	Optional	Front	Rear	Size	Inspection	(ft. lbs.)
2008	Land Cruiser	P285/60R18	None	①	①	18x8J	②	97
2009	Land Cruiser	P285/60R18	None	①	①	18x8J	②	97
2010	Land Cruiser	P285/60R18	None	①	①	18x8J	②	97

OEM: Original Equipment Manufacturer

PSI: Pounds Per Square Inch

STD: Standard

OPT: Optional

NA: Not Available

① See placard on vehicle

② Upper arm ball joint turning torque: 9-39 inch lbs.
Lower arm ball joint turning torque: 27-89 inch lbs.
Stabilizer link ball joint turning torque: 4.4-30 inch lbs.

3768X_LAND_C0010

BRAKE SPECIFICATIONS

All measurements in inches unless noted

Year	Model		Brake Disc			Minimum	Brake Caliper	
			Original Thickness	Minimum Thickness	Maximum Runout	Lining Thickness	Bracket Bolts (ft. lbs.)	Mounting Bolts (ft. lbs.)
2008	Land Cruiser	F	1.260	1.150	0.0020	0.039	—	73
		R	0.709	0.630	0.0059	0.039	65	70
2009	Land Cruiser	F	1.260	1.150	0.0020	0.039	—	73
		R	0.709	0.630	0.0059	0.039	65	70
2010	Land Cruiser	F	1.260	1.150	0.0020	0.039	—	73
		R	0.709	0.630	0.0059	0.039	65	70

F: Front

R: Rear

3768X_LAND_C0011

SCHEDULED MAINTENANCE INTERVALS

Toyota - Land Cruiser

TO BE SERVICED	TYPE OF SERVICE	VEHICLE MILEAGE INTERVAL (x1000)																		
		5	10	15	20	25	30	35	40	45	50	55	60	65	70	75	80	85	90	95
Engine oil & filter	R	✓	✓	✓	✓	✓	✓	✓	✓	✓	✓	✓	✓	✓	✓	✓	✓	✓	✓	✓
Rotate tires	S/I	✓	✓	✓	✓	✓	✓	✓	✓	✓	✓	✓	✓	✓	✓	✓	✓	✓	✓	✓
Automatic transmission fluid & filter	S/I			✓			✓			✓			✓			✓			✓	
Ball joints & dust covers	S/I			✓			✓			✓			✓			✓			✓	
Bolts & nuts on chassis & body	S/I			✓			✓			✓			✓			✓			✓	
Brake line pipes & hoses	S/I			✓			✓			✓			✓			✓			✓	
Brake pads & discs	S/I	✓	✓	✓	✓	✓	✓	✓	✓	✓	✓	✓	✓	✓	✓	✓	✓	✓	✓	✓
Cabin air filter	S/I			✓			✓			✓			✓			✓			✓	
Propeller shaft grease	S/I			✓			✓			✓			✓			✓			✓	
bolt	S/I			✓			✓			✓			✓			✓			✓	
Steering knuckle & chassis grease	S/I			✓			✓			✓			✓			✓			✓	
Steering linkage	S/I			✓			✓			✓			✓			✓			✓	
Transfer and differential oil	S/I			✓			✓			✓			✓			✓			✓	
Air cleaner filter	R						✓						✓						✓	
Spark plugs ①	R						✓						✓						✓	
Drive belts	S/I						✓						✓						✓	
Exhaust pipes & mountings	S/I						✓						✓						✓	
Fuel lines & connections	S/I						✓						✓						✓	
Engine coolant	R															✓				
Charcoal canister	R						✓						✓						✓	
Fuel tank cap gasket	R						✓						✓						✓	
Heated oxygen sensors (exc. Cal.) ②	R																			

R: Replace S/I: Service or Inspect

① Platinum plugs, replace every 100,000 miles

② Heated oxygen sensors (except Calif.): replace every 80,000 miles.

FREQUENT OPERATION MAINTENANCE (SEVERE SERVICE)

If a vehicle is operated under any of the following conditions it is considered severe service:

- Extremely dusty areas.

- 50% or more of the vehicle operation is in 32°C (90°F) or higher temperatures, or constant operation in temperatures below 0°C (32°F).

- Prolonged idling (vehicle operation in stop and go traffic).

- Frequent short running periods (engine does not warm to normal operating temperatures).

- Police, taxi, delivery usage or trailer towing usage.

Air cleaner filter: service or inspect every 3750 miles

Engine oil & filter: replace every 3750 miles.

Ball joints & dust covers: service or inspect every 7500 miles.

Bolts & nuts on chassis & body: service or inspect every 7500 miles.

Brake pads & discs (front & rear): service or inspect every 7500 miles.

Steering knuckle & chassis grease: service or inspect every 7500 miles.

Steering linkage: service or inspect every 7500 miles.

Propeller shaft grease: service or inspect every 7500 miles.

Exhaust pipes & mountings: service or inspect every 15,000 miles.

3768X_LAND_C0012

BRAKES — INFORMATION AND PRECAUTIONS

ANTI-LOCK SYSTEMS

• Certain components within the ABS system are not intended to be serviced or repaired individually.

• Do not use rubber hoses or other parts not specifically specified for and ABS system. When using repair kits, replace all parts included in the kit. Partial or incorrect repair may lead to functional problems and require the replacement of components.

• Lubricate rubber parts with clean, fresh brake fluid to ease assembly. Do not use shop air to clean parts; damage to rubber components may result.

• Use only DOT 3 brake fluid from an unopened container.

• If any hydraulic component or line is removed or replaced, it may be necessary to bleed the entire system.

• A clean repair area is essential. Always clean the reservoir and cap thoroughly before removing the cap. The slightest amount of dirt in the fluid may plug an orifice and impair the system function. Perform repairs after components have been thoroughly cleaned; use only denatured alcohol to clean components. Do not allow ABS components to come into contact with any substance containing mineral oil; this includes used shop rags.

• The Anti-Lock control unit is a microprocessor similar to other computer units in the vehicle. Ensure that the ignition switch is **OFF** before removing or installing controller harnesses. Avoid static electricity discharge at or near the controller.

• If any arc welding is to be done on the vehicle, the control unit should be unplugged before welding operations begin.

DISC AND DRUM SYSTEMS

CAUTION

Dust and dirt accumulating on brake parts during normal use may contain asbestos fibers from production or aftermarket brake linings. Breathing excessive concentrations of asbestos fibers can cause serious bodily harm. Exercise care when servicing brake parts. Do not sand or grind brake lining unless equipment used is designed to contain the dust residue. Do not clean brake parts with compressed air or by dry brushing. Cleaning should be done by dampening the brake components with a fine mist of water, then wiping the brake components clean with a dampened cloth. Dispose of cloth and all residue containing asbestos fibers in an impermeable container with the appropriate label. Follow practices prescribed by the Occupational Safety and Health Administration (OSHA) and the Environmental Protection Agency (EPA) for the handling, processing, and disposing of dust or debris that may contain asbestos fibers.

BRAKES — BLEEDING THE BRAKE SYSTEM

BLEEDING PROCEDURE

➡If any work is done on the brake system or if air is suspected in the brake lines, bleed the air from the system. Bleeding Hydraulic Brake Booster is only possible with a Toyota proprietary scan system.

➡When bleeding, keep the amount of the fluid within the line of reservoir between Min. and Max. Do not let brake fluid remain on a painted surface. Wash it off immediately.

1. Before servicing the vehicle, refer to the precautions in the beginning of this section.
2. Check the fluid level in the reservoir after bleeding each wheel. Add DOT3 fluid, if necessary.
3. If the hydraulic brake booster was disassembled or if the reservoir becomes empty, bleeding Hydraulic Brake Booster is only possible with a Toyota proprietary scan system.
4. Bleeding the air from the hydraulic brake lines can be performed as follows:
   a. Turn the ignition switch OFF; depress the brake pedal 40 times or more.
   b. Turn the ignition switch to the ON position and start the brake booster pump. The pump stops after approximately 30 to 40 seconds.

➡If the pump does not operate as specified, repeat the above and recheck the operating time.

   c. Holding the brake pedal depressed, bleed the right and left rear brake cylinders.
   d. Turn the ignition switch ON, depress the brake pedal 20 times or more.
5. Bleeding Front Brake Lines:
   a. Turn the ignition switch to the ON position and wait until the pump motor has stopped.
   b. Connect the vinyl tube to the brake caliper.
   c. Depress the brake pedal several times, then loosen the bleeder plug with the pedal held down.
   d. At the point when the fluid stops coming out, tighten the bleeder plug, 8 ft. lbs. (11 Nm) then release the brake pedal.
   e. Repeat procedure until all the air in the fluid has been bled out.
   f. Repeat the above procedures to bleed the other brake line.
6. Bleeding Rear Brake Lines:
   a. Turn the ignition switch to the ON position and depress the brake pedal.
   b. Connect the vinyl tube to the brake caliper.
   c. Loosen the bleeder plug and release air.

➡Brake fluid is sent through the pump, so keep the brake pedal depressed until the air is completely bled out.

   d. When the air is completely bled out of the brake fluid through the bleeder plug, tighten the bleeder plug to 8 ft. lbs. (11 Nm) then release.
   e. Repeat the above procedures to bleed the other brake line.
7. Bleeding Hydraulic Brake Booster is only possible with a Toyota proprietary scan system.

WHEEL SPEED SENSORS

REMOVAL & INSTALLATION

Front

See Figures 3 through 6.

1. Disconnect the cable from the negative battery terminal.

CAUTION

Wait at least 90 seconds after disconnecting the cable from the negative (-) battery terminal to disable the SRS system.

2. Remove the front wheel.
3. Remove the front skid control sensor wire LH.
 a. Disconnect the connector from the front speed sensor.
 b. Remove the bolt and harness clamp.
 c. Remove the nut and harness clamp.
 d. Remove the 2 bolts and 2 harness clamps.
 e. Disconnect the connector and then remove the bolt, harness clamp and sensor wire.
4. Remove the front speed sensor LH.
 a. Using a 5 mm hexagon wrench, remove the bolt and speed sensor from the knuckles.

➡**Pull out the sensor while trying as much as possible not to rotate it.**

5. Remove the front skid control sensor wire RH.
 a. Disconnect the connector from the front speed sensor.

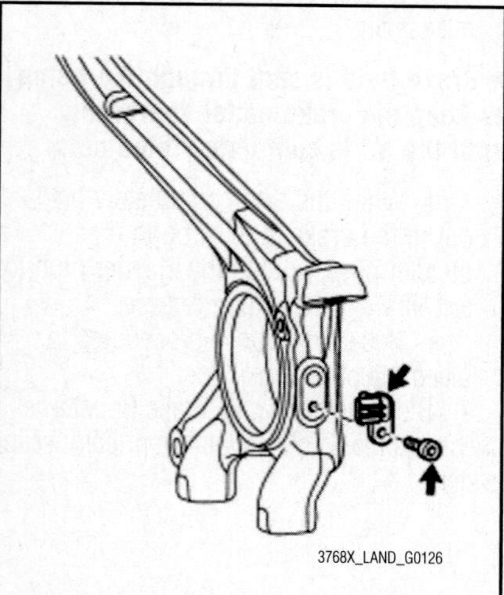

Fig. 3 Removing the LH speed sensor

 b. Remove the bolt and harness clamp.
 c. Remove the nut and harness clamp.
 d. Disconnect the connector.
 e. Detach the connector from the skid control sensor clamp.
 f. Remove the bolt and skid control sensor clamp.
 g. Remove the 2 bolts, 2 harness clamps and sensor wire.
6. Remove the front speed sensor RH.
 a. Using a 5 mm hexagon wrench, remove the bolt and speed sensor from the knuckle.

➡**Pull out the sensor while trying as much as possible not to rotate it.**

To install:

7. Install the front speed sensor LH.
 a. Using a 5 mm hexagon wrench, install the speed sensor with the bolt. Tighten to 73 inch lbs. (8.3 Nm).

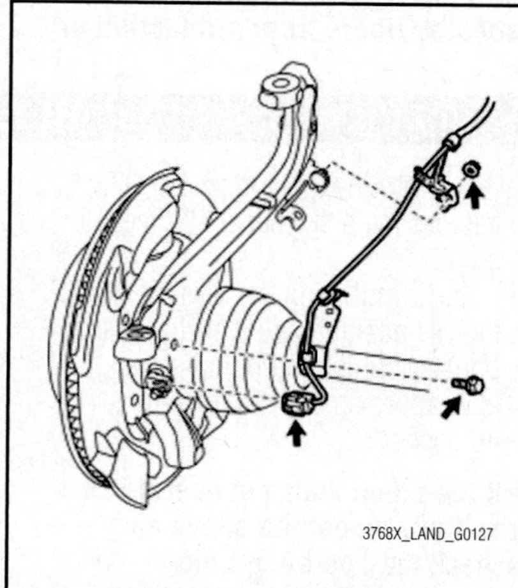

Fig. 4 Disconnecting the connector from the front speed sensor

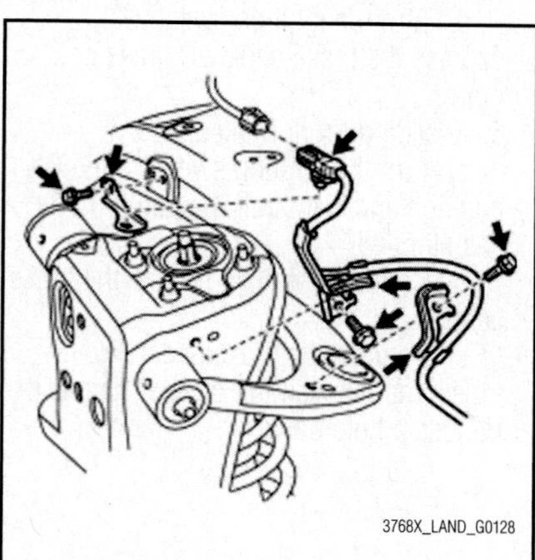

Fig. 5 Disconnecting the connector

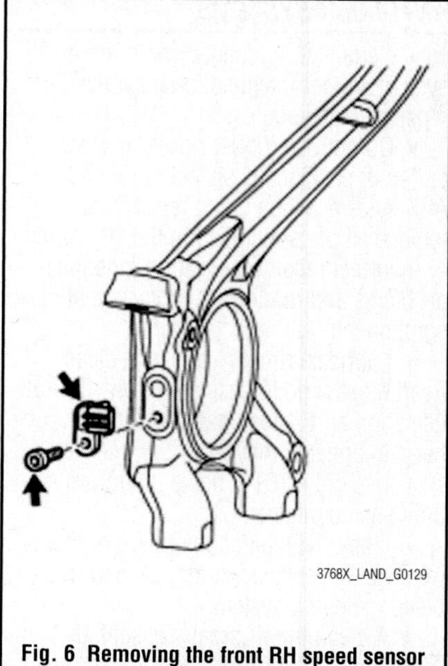

Fig. 6 Removing the front RH speed sensor

➡**Make sure there are no pieces of iron or other foreign matter attached to the sensor tip.**

➡**While inserting the speed sensor into the knuckle hole, do not strike or damage the sensor tip**

➡**After installing the speed sensor, make sure there is no clearance or foreign matter between the sensor stay part and the knuckle.**

➡**Make sure there is no foreign matter attached to the speed sensor rotor.**

8. Install the front skid control sensor wire LH.
 a. Connect the connector.
 b. Install the harness clamp with the bolt. Tighten to 9 ft. lbs. (13 Nm).

➡**When installing the clamp, do not twist the wire harness.**

➡**Make sure the clamp rotation stopper touches the installation position.**

 c. Install the 2 harness clamps with the 2 bolts. Tighten to 9 ft. lbs. (13 Nm).

➡**When installing the clamps, do not twist the wire harness.**

➡**Make sure the clamp rotation stopper touches the installation position.**

 d. Install the harness clamp with the nut. Tighten to 9 ft. lbs. (13 Nm).

➡**When installing the clamps, do not twist the wire harness.**

➡**Make sure the clamp rotation stopper touches the installation position.**

e. Install the harness clamp with the bolt. Tighten to9 ft. lbs. (13 Nm).

➡**Install the bracket so that the rotation stopper touches the knuckle.**

f. Connect the speed sensor connector.

➡**Securely connect the connector.**

9. Install the front RH speed sensor.
a. Using a 5 mm hexagon wrench, install the speed sensor with the bolt. Tighten to 73 inch lbs. (8.3 Nm).

➡**Make sure there are no pieces of iron or other foreign matter attached to the sensor tip.**

➡**While inserting the speed sensor into the knuckle hole, do not strike or damage the sensor tip**

➡**After installing the speed sensor, make sure there is no clearance or foreign matter between the sensor stay part and the knuckle.**

➡**Make sure there is no foreign matter attached to the speed sensor rotor.**

10. Install the front skid control sensor wire RH.
a. Install the skid control sensor clamp with the bolt (labeled A). Tighten to 44 inch lbs. (5 Nm).
b. Attach the connector to the skid control sensor clamp and then connect the connector.
c. Install the 2 harness clamps with the 2 bolts. Tighten to 9 ft. lbs. (13 Nm).

➡**When installing the clamps, do not twist the wire harness.**

➡**Make sure the clamp rotation stopper touches the installation position.**

d. Install the harness clamp with the nut. Tighten to 9 ft. lbs. (13 Nm).

➡**When installing the clamps, do not twist the wire harness.**

➡**Make sure the clamp rotation stopper touches the installation position.**

e. Install the harness clamp with the bolt. Tighten to 9 ft. lbs. (13 Nm).

➡**Install the bracket so that the rotation stopper touches the knuckle.**

f. Connect the speed sensor connector.

➡**Securely connect the connector.**

11. Install the front wheel. Tighten to 97 ft. lbs. (131 Nm).
12. Connect the cable to the negative battery terminal.
13. Check the speed sensor signal.

Rear

See Figures 7 through 11.

1. Disconnect the cable from the negative battery terminal.

CAUTION

Wait at least 90 seconds after disconnecting the cable from the negative (-) battery terminal to disable the SRS system.

2. Remove the rear wheel.
3. Remove the skid control sensor wire.
a. Disconnect the speed sensor LH connector.
b. Detach the 2 clamps and remove the bolt and sensor clamp.
c. Detach the 5 clamps.
4. Detach the 3 clamps.
a. Disconnect the speed sensor RH connector.
b. Remove the bolt.
c. Disconnect the connector, and remove the bolt and skid control sensor wire.

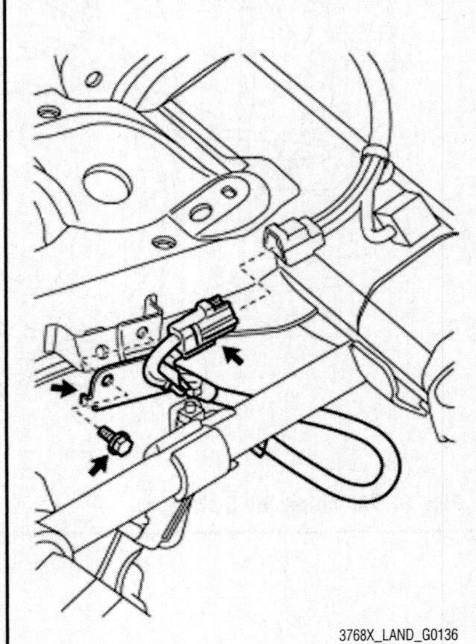

Fig. 7 Disconnecting the connector, and remove the bolt and skid control sensor wire

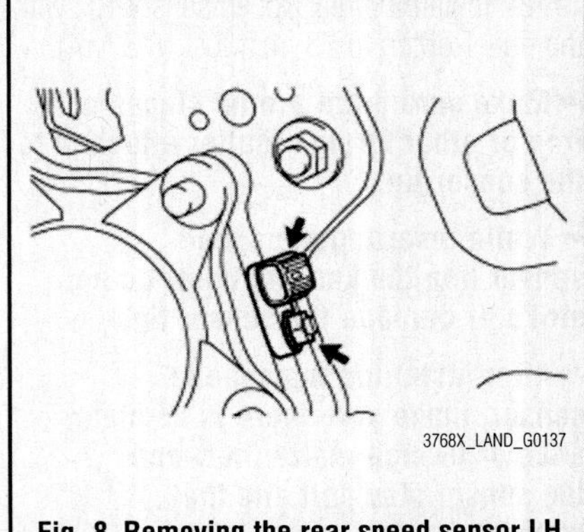

Fig. 8 Removing the rear speed sensor LH

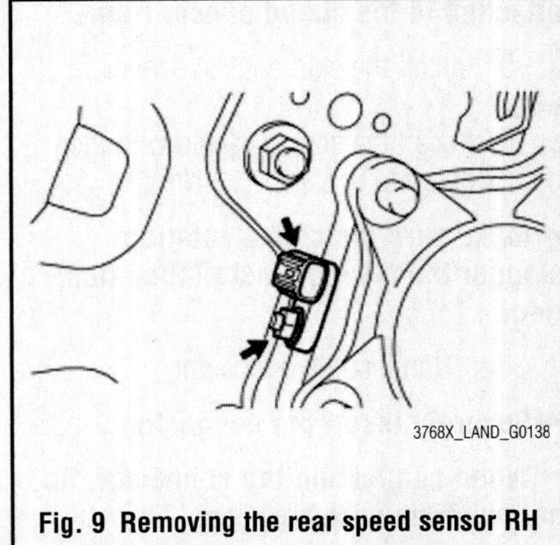

Fig. 9 Removing the rear speed sensor RH

5. Remove the rear speed sensor LH.
a. Remove the nut and speed sensor.

➡**Pull out the sensor while trying as much as possible not to rotate it.**

6. Remove the rear speed sensor RH.
a. Remove the nut and speed sensor.

➡**Pull out the sensor while trying as much as possible not to rotate it.**

To install:

7. Install the rear speed sensor LH with the nut. Tighten to 73 inch lbs. (8.3 Nm).

➡**Make sure there are no pieces of iron or other foreign matter attached to the sensor tip.**

➡**While inserting the speed sensor into the knuckle hole, do not strike or damage the sensor tip.**

➡**After installing the speed sensor, make sure there is no clearance or foreign matter between the sensor stay part and the knuckle.**

➡**Make sure there is no foreign matter attached to the speed sensor rotor.**

8. Install the rear speed sensor RH with the nut. Tighten to 73 inch lbs. (8.3 Nm).

➡Make sure there are no pieces of iron or other foreign matter attached to the sensor tip.

➡While inserting the speed sensor into the knuckle hole, do not strike or damage the sensor tip.

➡After installing the speed sensor, make sure there is no clearance or foreign matter between the sensor stay part and the knuckle.

➡Make sure there is no foreign matter attached to the speed sensor rotor.

9. Install the skid control sensor wire.

10. Install the sensor clamp with the bolt. Tighten to 9 ft. lbs. (13 Nm).

➡Make sure the clamp rotation stopper touches the installation position.

a. Connect the connector.

➡Securely insert the connector.

➡When connecting the connector, do not twist the wire harness.

11. Install the sensor clamp with the bolt. Tighten to 9 ft. lbs. (13 Nm).

➡Make sure the clamp rotation stopper touches the installation position.

a. Connect the connector.

➡Securely insert the connector.

➡When connecting the connector, do not twist the wire harness.

b. Connect the speed sensor RH connector.

➡Securely insert the connector.

➡When connecting the connector, do not twist the wire harness.

c. Attach the 3 clamps.

➡When attaching the clamps, do not twist the wire harness.

➡When attaching the clamps, securely insert them as shown in the illustration.

d. Attach the 5 clamps.

➡When attaching the clamps, do not twist the wire harness.

➡When attaching the clamps, securely insert them as shown in the illustration.

e. Attach the 2 clamps, and install the sensor clamp with the bolt. Tighten to 9 ft. lbs. (13 Nm).

f. Connect the speed sensor LH connector.

➡Securely insert the connector.

➡When connecting the connector, do not twist the wire harness.

12. Install the rear wheel. Tighten to 97 ft. lbs. (131 Nm).

13. Connect the cable to the negative battery terminal.

14. Check the speed sensor signal.

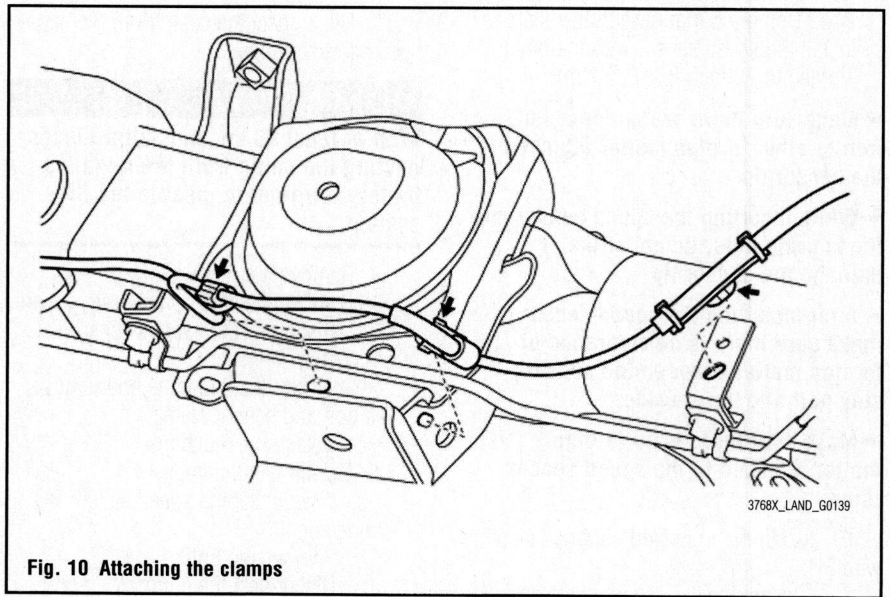

Fig. 10 Attaching the clamps

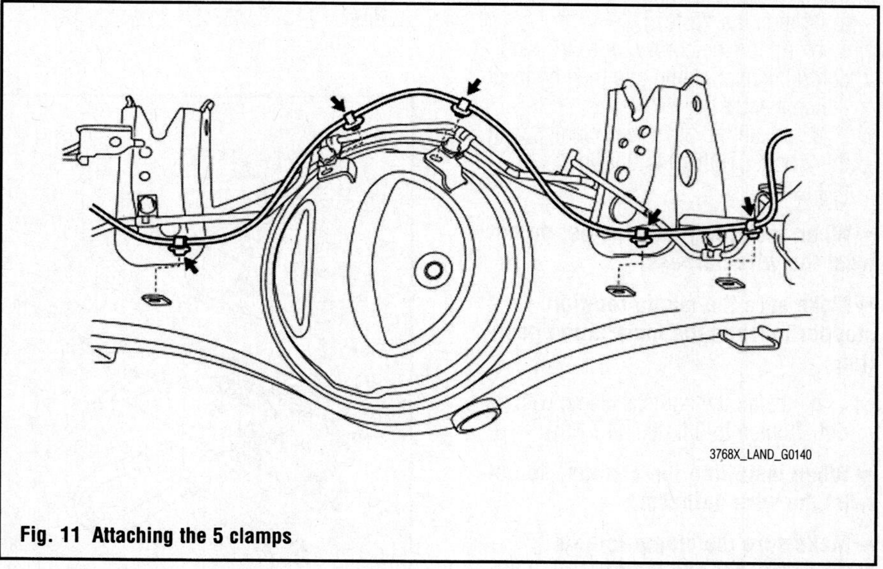

Fig. 11 Attaching the 5 clamps

BRAKE CALIPER

REMOVAL & INSTALLATION

See Figures 12 and 13.

➡Use the same procedures for the LH side and RH side.

➡The procedures listed below are for the LH side.

1. Remove the front wheel.
2. Drain the brake fluid.

➡Wash brake fluid off immediately if it is spilled on any painted surface.

3. Remove the brake pads.
4. Disconnect the front flexible hose.
   a. Remove the union bolt and gasket, and then disconnect the flexible hose.
5. Remove the 2 bolts and disc brake caliper from the knuckle.

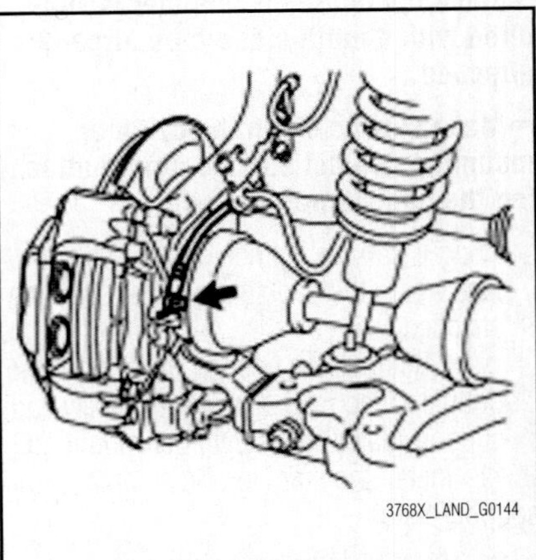

Fig. 12 Disconnecting the flexible hose

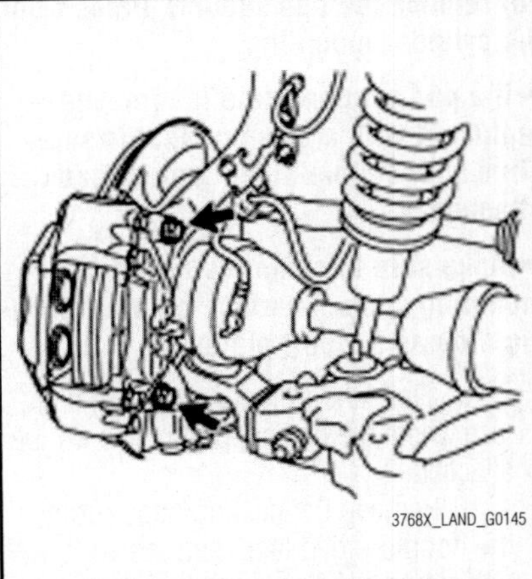

Fig. 13 Removing the disc brake caliper from the knuckle

To install:

6. Install the disc brake caliper assembly.
   a. Install the disc brake caliper with 2 new bolts. Tighten to 73 ft. lbs. (99 Nm).
7. Connect the front flexible hose.
   a. Install a new gasket and connect the flexible hose with the union bolt. Tighten to 22 ft. lbs. (30 Nm).
8. Fill the reservoir with brake fluid.
9. Bleed the brake line.
10. Check the brake fluid level in the reservoir.
11. Inspect for brake fluid leaks.
12. Install the front wheel. Tighten to 97 ft. lbs. (131 Nm).

DISC BRAKE PADS

REMOVAL & INSTALLATION

See Figures 14 through 16.

➡Use the same procedures for the LH side and RH side.

➡The procedures listed below are for the LH side.

1. Remove the front wheel.
2. Drain the brake fluid.

➡Wash brake fluid off immediately if it is spilled on any painted surface.

3. Remove the pin hold clip.

➡The pin hold clip can be reused if it has sufficient rebound; no deformation or wear; and has had all rust, dirt and foreign matter cleaned off.

4. Remove the 2 hold pins.
   a. Remove the front disc brake anti-rattle spring.

➡The anti-rattle spring can be reused if it has sufficient rebound; no deformation, cracks or wear; and has had all rust, dirt and foreign matter cleaned off.

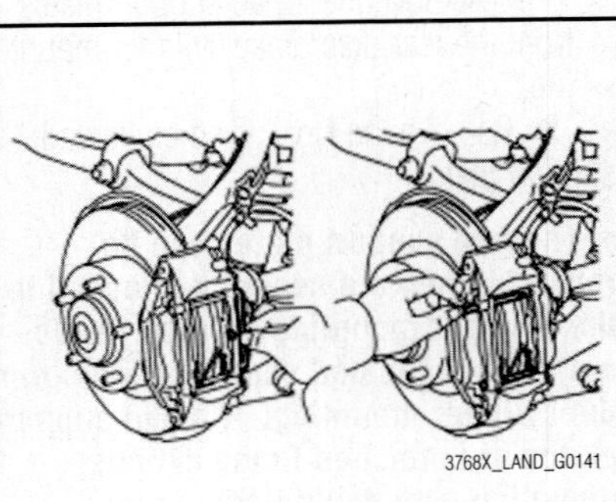

Fig. 14 Removing the pin hold clip

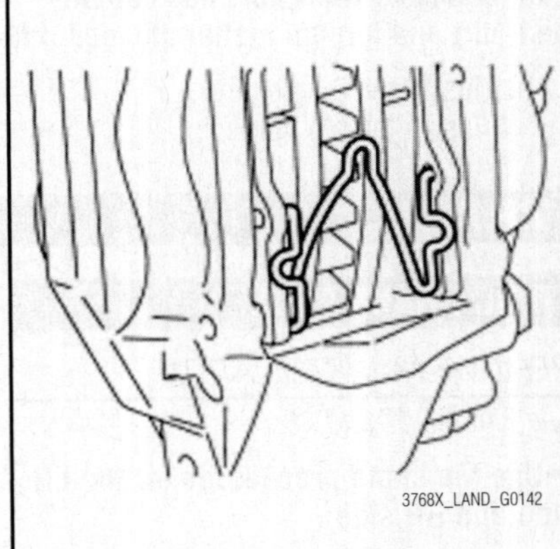

Fig. 15 Removing the front disc brake anti rattle spring

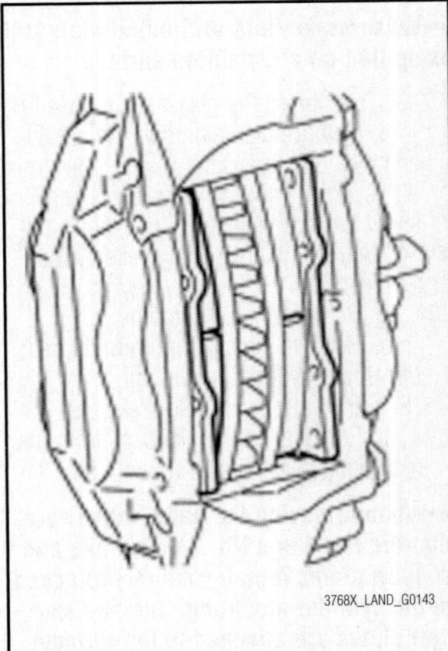

Fig. 16 Removing the 2 pads from the disc brake caliper

5. Remove the 2 pads from the disc brake caliper.
6. Remove the No. 1 anti squeal shims from the pads.

To install:

7. Install the No. 1 anti squeal shims to the pads.
8. Install the 2 pads to the disc brake caliper.
9. Install the front disc brake anti rattle spring between the 2 pads.

➡The anti-rattle spring can be reused if it has sufficient rebound; no deformation, cracks or wear; and has had all rust, dirt and foreign matter cleaned off.

10. Install the 2 hole pins.
11. Install the pin hold clip.

➡The pin hold clip can be reused if it has sufficient rebound; no deformation or wear; and has had all rust, dirt and foreign matter cleaned off.

12. Fill the reservoir with brake fluid.
13. Bleed the brake line.
14. Check the brake fluid level in the reservoir.
15. Inspect for brake fluid leaks.
16. Install the front wheel. Tighten to 97 ft. lbs. (131 Nm).

DISC BRAKE PADS

REMOVAL & INSTALLATION

See Figures 17 and 18.

➡Use the same procedures for the LH side and RH side.

➡The procedures listed below are for the LH side.

1. Remove the rear wheel.
2. Drain the brake fluid.

➡Wash brake fluid off immediately if it is spilled on any painted surface.

3. Disconnect the rear flexible hose LH.
   a. Remove the union bolt and gasket, and then disconnect the flexible hose from the rear disc brake cylinder. Use a container to catch brake fluid as it drains out.
4. Remove the rear disc brake cylinder assembly LH.
   a. Remove the 2 slide pins.
   b. Remove the cylinder from the rear disc brake cylinder mounting.
5. Remove the rear disc brake pad.
   a. Remove the 2 brake pads from the rear disc brake cylinder mounting.

➡When removing the pads, make sure that the No. 1 and No. 2 disc brake pad support plates remain securely attached to the cylinder mounting. The pad support plates are attached to the cylinder mounting with double-sided tape.

➡If a pad support plate is not securely attached to the cylinder mounting, replace it with a new one.

6. Remove the rear disc brake anti squeal shims from the disc brake pads.
7. Remove the rear disc brake pad wear indicator plate from the inner disc brake pad.
8. Remove the rear No. 1 disc brake pad support plate.

➡The pad support plates can be reused if they are securely attached to the cylinder mounting. Therefore, do not remove the pad support plates from the cylinder mounting. The pad support plates are attached to the cylinder mounting with double-sided tape.

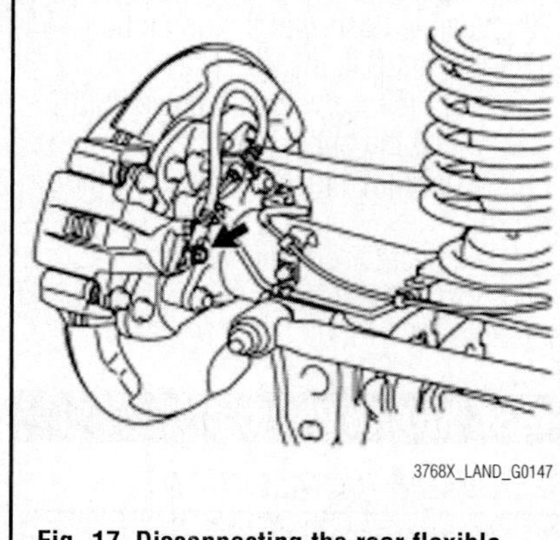

Fig. 17 Disconnecting the rear flexible hose LH

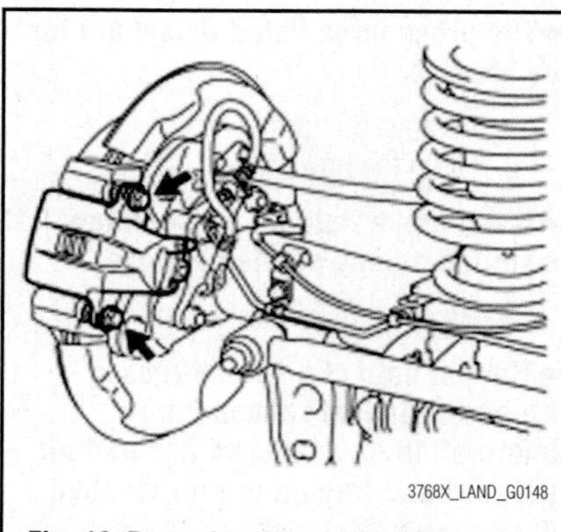

Fig. 18 Removing the rear disc brake cylinder assembly LH

➡If it is necessary to remove the pad support plates, replace them with new ones.

   a. Remove the 2 pad support plates from the rear disc brake cylinder mounting.
9. Remove the No. 2 disc brake pad support plate.

➡The pad support plates can be reused if they are securely attached to the cylinder mounting. Therefore, do not remove the pad support plates from the cylinder mounting. The pad support plates are attached to the cylinder mounting with double-sided tape.

➡If it is necessary to remove the pad support plates, replace them with new ones.

   a. Remove the 2 pad support plates from the rear disc brake cylinder mounting.

To install:

10. Install the rear No. 1 disc brake pad support plate.

➡The pad support plates can be reused if they are securely attached to the cylinder mounting. Therefore, do not remove the pad support plates from the cylinder mounting.

➡If a pad support plate is removed, replace it with a new one that is supplied with double-sided tape already attached.

➡Make sure to clean the cylinder mounting contact surface before attaching the pad support plates.

   a. Clean the cylinder mounting surface where the pad support plates will be attached.
   b. Remove the peeling paper from the double-sided tape, and install 2 new pad support plates to the cylinder mounting.
11. Install the rear No. 2 disc brake pad support plate.

➡The pad support plates can be reused if they are securely attached to the cylinder mounting. Therefore, do not remove the pad support plates from the cylinder mounting.

➡If a pad support plate is removed, replace it with a new one that is supplied with double-sided tape already attached.

➡Make sure to clean the cylinder mounting contact surface before attaching the pad support plates.

   a. Clean the cylinder mounting surface where the pad support plates will be attached.
   b. Remove the peeling paper from the double-sided tape, and install 2 new pad support plates to the cylinder mounting.

12. Install the pad wear indicator plate to the inner disc brake pad.

13. Install the rear disc brake anti squeal shims to the brake pads.

➡There should be no oil or grease on the friction surfaces of the brake pads and rear disc.

14. Install the rear disc brake pad.

➡When installing the pads, make sure that the No. 1 and No. 2 disc brake pad support plates are securely attached to the cylinder mounting. The pad support plates are attached to the cylinder mounting with double-sided tape.

➡If a pad support plate is not securely attached to the cylinder mounting, replace it with a new one.

a. Install the 2 disc brake pads to the disc brake cylinder mounting.

15. Install the rear disc brake cylinder assembly LH.

a. Apply lithium soap base glycol grease to the sliding part of the 2 cylinder slide pins.

b. Install the cylinder with the 2 cylinder slide pins. Tighten to 65 ft. lbs. (88 Nm).

16. Connect the rear flexible hose LH.

a. Install a new gasket and connect the flexible hose with the union bolt. Tighten to 22 ft. lbs. (30 Nm).

17. Fill the reservoir with brake fluid.

18. Bleed the brake line.

19. Check the brake fluid level in the reservoir.

20. Inspect for brake fluid leaks.

21. Adjust the parking brake lever travel.

22. Install the rear wheel. Tighten to 97 ft. lbs. (131 Nm).

BRAKES — PARKING BRAKE

PARKING BRAKE SHOES

REMOVAL & INSTALLATION

See Figures 19 through 23.

1. Remove the rear brake disc.
2. Remove the parking brake shoe return spring.
   a. Using the special tool (09703-30011), remove the return spring.
3. Remove the No. 1 parking brake shoe assembly LH.
   a. Using special tool (09718-00011), remove the shoe hold down spring cup, compression spring and shoe hold down spring pin.
   b. Disconnect the tension spring.
   c. Remove the No. 1 parking brake shoe and shoe adjuster screw set.
4. Remove the No. 2 parking brake shoe assembly LH.
   a. Using the special tool (09718-00011), remove the shoe hold down spring cup, compression spring and shoe hold down spring pin.

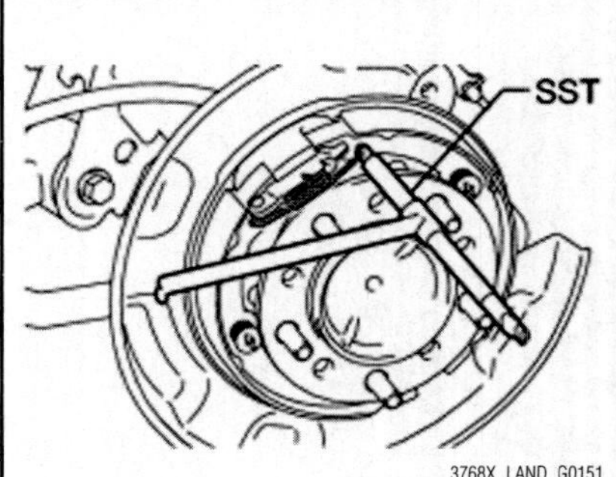

Fig. 19 Removing the parking brake shoe return spring

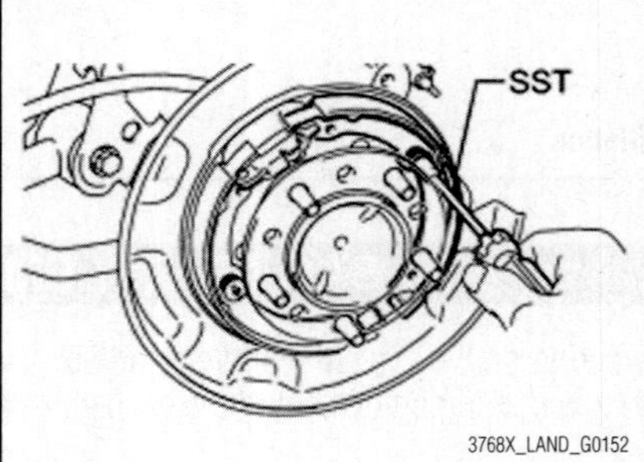

Fig. 20 Removing the shoe hold down spring cup, compression spring and shoe hold down spring pin

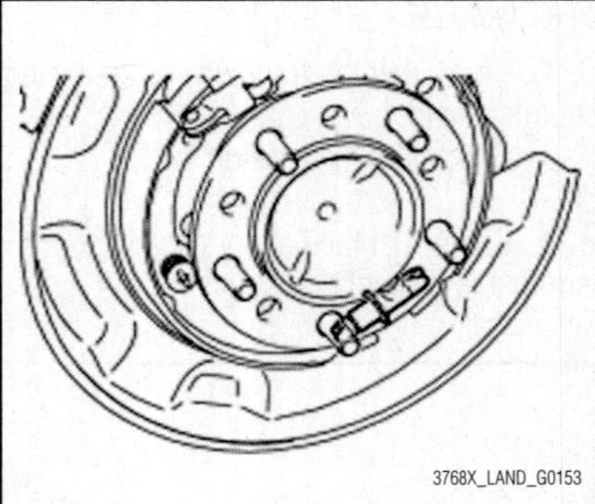

Fig. 21 Removing the No. 1 parking brake shoe and shoe adjuster screw set

   b. Remove the No. 2 parking brake shoe.

To install:

5. Install the No. 2 parking brake shoe assembly LH.
   a. Apply high temperature grease to the areas of the backing plate that contact the shoe.
   b. Using the special tool (09718-00011), install the No. 2 parking brake shoe with the shoe hold down spring pin, compression spring and shoe hold down spring cup.

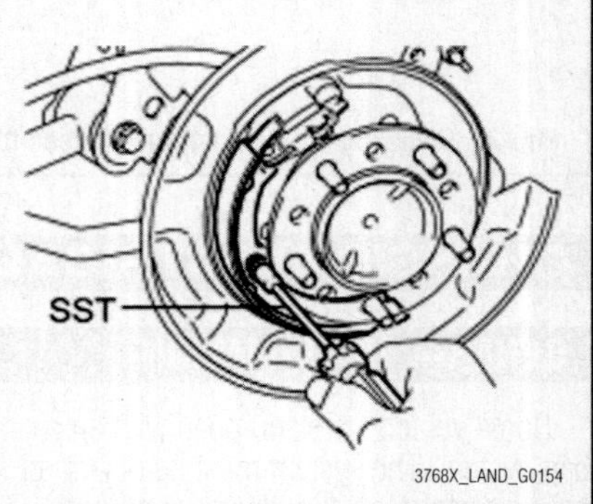

Fig. 22 Removing the No. 2 parking brake shoe assembly LH

6. Install the No. 1 parking brake show assembly LH.
   a. Apply high temperature grease to the thread and all joining areas of the parking brake shoe adjuster screw set.
   b. Set the No. 1 parking brake shoe and shoe adjuster screw set in place.
   c. Connect the tension spring.
   d. Using the special tool (09718-00011), install the No. 1 parking brake shoe with the shoe hold down spring pin, compression spring and shoe hold down spring cup.
7. Install the parking brake shoe return spring.
   a. Using the special tool (09703-30011), install the shoe return spring.
8. Check the parking brake installation.
   a. Check that each part is installed properly.

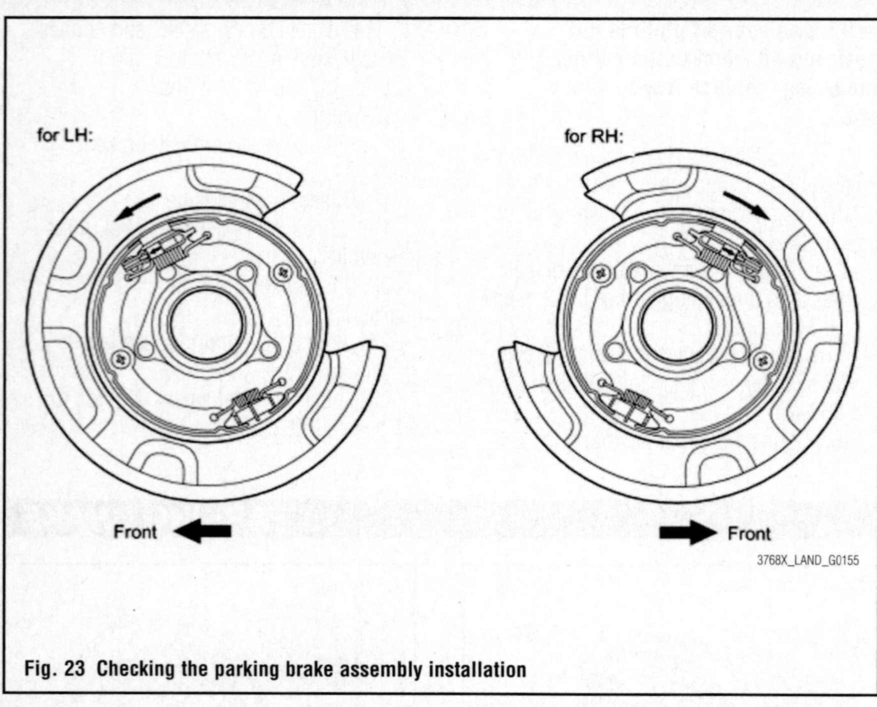

Fig. 23 Checking the parking brake assembly installation

9. Install the rear disc.
10. Check the parking brake lever travel.
11. Adjust the parking brake lever travel.
12. Install the rear wheel LH. Tighten to 97 ft. lbs. (131 Nm).
13. Drive the vehicle for approximately 0.25 miles (400 m) under the following conditions to settle the parking brake shoe and disc.
    a. The vehicle speed is approximately 31 mph (50 km/h) and the vehicle is on a safe, level and dry road.
    b. The parking brake lever is being pulled with a force of 33.7 ft. lbs. (150 N).
14. Repeat the procedure above 2 or 3 times.

➡Set a 5 minute interval between each procedure to prevent the brake assembly from overheating.

CHASSIS ELECTRICAL

AIR BAG (SUPPLEMENTAL RESTRAINT SYSTEM)

GENERAL INFORMATION

Some vehicles are equipped with an air bag system. The system must be disarmed before performing service on, or around, system components, the steering column, instrument panel components, wiring and sensors. Failure to follow the safety precautions and the disarming procedure could result in accidental air bag deployment, possible injury and unnecessary system repairs.

SERVICE PRECAUTIONS

⁂ CAUTION

Disconnect and isolate the battery negative cable before beginning any airbag system component diagnosis, testing, removal, or installation procedures. Allow system capacitor to discharge for two minutes before beginning any component service. This will disable the airbag system. Failure to disable the airbag system may result in accidental airbag deployment, personal injury, or death.

DISARMING THE SYSTEM

To avoid personal injury when working on vehicles equipped with an air bag, the negative battery cable must be disconnected and at least 90 seconds must elapse before working on the system. Failure to do so may result in deployment of the air bag.

ARMING THE SYSTEM

The system is self-arming when it is operating properly.

CLOCKSPRING CENTERING

See Figure 24.

1. Check that the front wheels are facing straight ahead.
2. Check that the ignition switch is off.
3. Check that the battery negative (-) terminal is disconnected

⁂ CAUTION

After removing the terminal, wait for at least 90 seconds before starting the operation.

4. Rotate the spiral cable clockwise slowly by hand until it feels firm.

➡Do not turn the spiral cable by the airbag wire harness

5. Rotate the spiral cable counterclockwise approximately 2.5 turns to align the marks

➡The spiral cable will rotate approximately 2.5 turns to both the left and right from the center

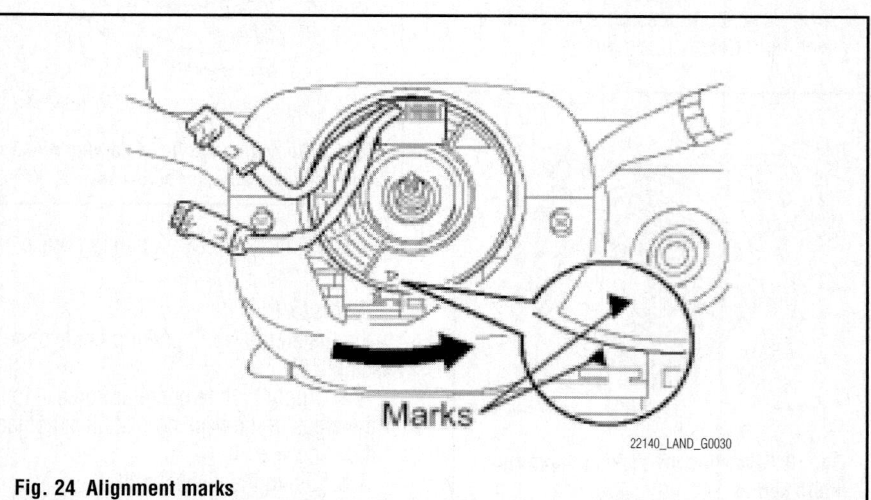

Fig. 24 Alignment marks

DRIVE TRAIN

TRANSFER CASE ASSEMBLY

REMOVAL & INSTALLATION

See Figure 25.

1. Drain the transfer oil.
2. Remove the automatic transmission with the transfer assembly.
3. Remove the transfer assembly.
   a. Remove the 8 transfer adaptor rear mounting bolts.
   b. Pull the transfer straight up and remove it from the transmission.

➡Take care not to damage the adaptor oil seal with the transfer input shaft spline.

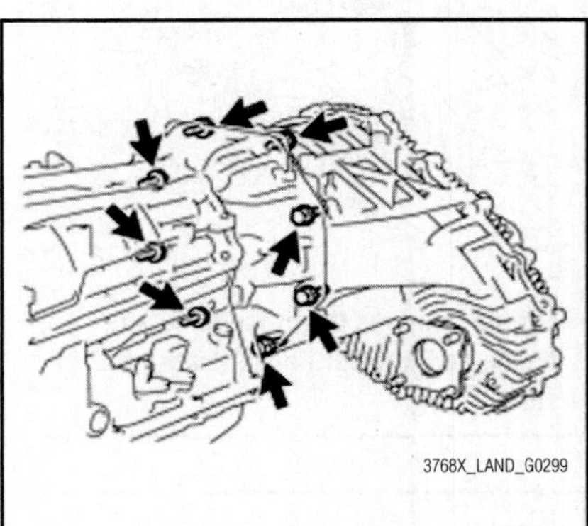

Fig. 25 Removing the transfer assembly

To install:

4. To install, reverse the removal procedure. Tighten the transfer bolts to 30 ft. lbs. (40 Nm). Check for transfer oil leaks.

FRONT DIFFERENTIAL CARRIER

REMOVAL & INSTALLATION

See Figures 26 through 29.

1. Remove the front propeller shaft assembly.
2. Remove the front drive shaft assembly LH.
3. Remove the front drive shaft assembly RH.
4. Remove the front suspension rebound stopper sub assembly LH.
   a. Remove the 3 bolts and suspension rebound stopper LH.
5. Remove the front suspension rebound stopper sub assembly RH.
   a. Remove the 3 bolts and suspension rebound stopper RH.

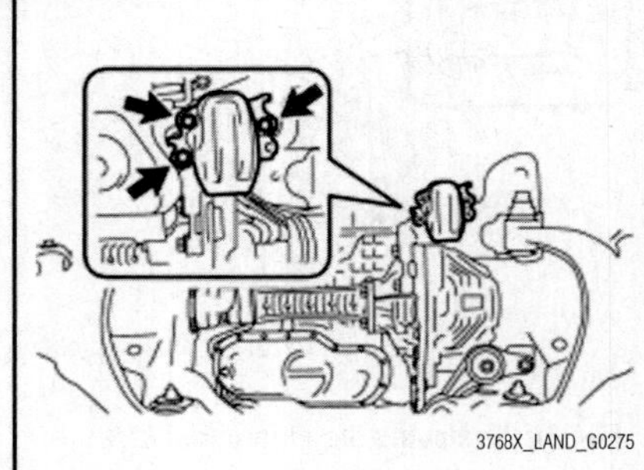

Fig. 26 Removing the front suspension rebound stopper sub assembly LH

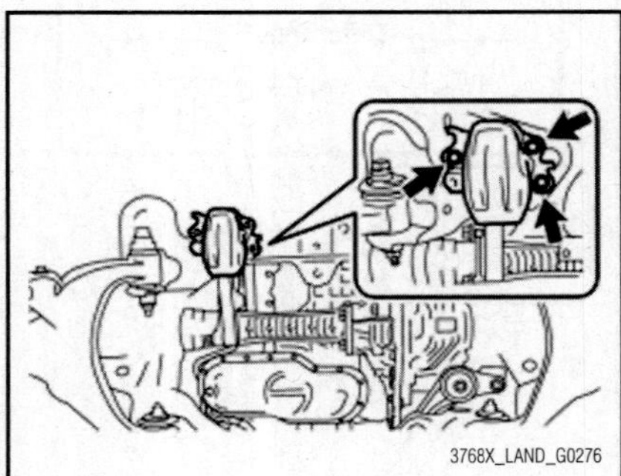

Fig. 27 Removing the front suspension rebound stopper sub assembly RH

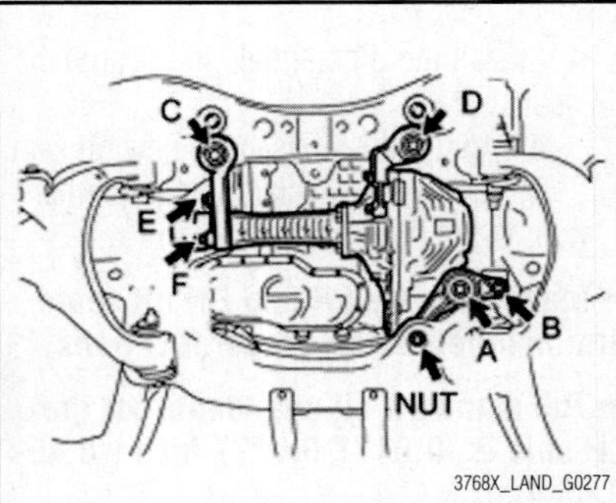

Fig. 28 Removing the front differential carrier

6. Remove the front differential carrier assembly.
   a. Disconnect the breather tube from the front differential union.
   b. Slowly lower the jack, and remove the front differential carrier.
   c. Remove bolts E and F, and the differential support assembly.

To install:

7. Install the front differential carrier assembly with bolts E and F. Tighten to 74 ft. lbs. (100 Nm).

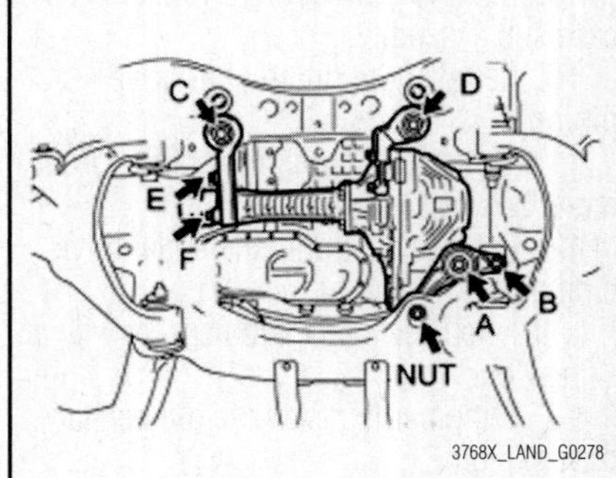

Fig. 29 Installing the front differential carrier assembly

   a. Using a jack, slowly raise the front differential carrier assembly to its installation position.
   b. Temporarily install the front differential support assembly with bolts A and B and the nut.
   c. Temporarily install the bolts C and D.
   d. Tighten bolt A. Tighten to 89 ft. lbs. (120 Nm).
   e. Tighten bolt B and the nut to 89 ft. lbs. (120 Nm).
   f. Tighten bolts C and D to 89 ft. lbs. (120 Nm).
8. Install the front suspension rebound stopper sub assembly LH with the 3 bolts. Tighten to 43 ft. lbs. (58 Nm).
9. Install the front suspension rebound stopper sub assembly RH with the 3 bolts. Tighten to 43 ft. lbs. (58 Nm).
10. Install the front drive shaft assembly LH and RH.
11. Install the front propeller shaft assembly.

FRONT DIFFERENTIAL CARRIER OIL SEAL

REPLACEMENT

1. Drain the differential oil.
2. Remove the front differential carrier assembly.
3. Remove the differential extension flange tube.
4. Remove the front drive pinion companion flange front nut.
5. Remove the front drive pinion companion flange front sub assembly.
6. Remove the front differential carrier oil seal using the special tool (09308-10010).
7. Remove the front differential drive pinion oil slinger.

8. Remove the front differential cross shaft bearing retainer.
9. Remove the front No. 1 differential case sub assembly.
10. Remove the differential drive pinion.
11. Remove the front drive pinion front radial ball bearing.
12. Replace the front differential drive pinion bearing spacer.
   a. Replace the bearing spacer with a new one.
13. Temporarily install the differential drive pinion.
14. Install the front No. 1 differential case sub assembly.
15. Install the front differential cross shaft bearing retainer.
16. Install the front drive pinion front radial ball bearing.
17. Install the front differential drive pinion oil slinger.
18. Install the front differential carrier oil seal.
   a. Using the special tool (09502-12010, 09316-60011 or 09316-00041) and a hammer, tap in a new oil seal. The standard depth is 0.221-0.260 inch (5.6-6.6 mm).
   b. Coat the lip of the oil seal with MP grease.
19. Install the front drive pinion companion flange sub assembly.
20. Inspect the differential drive pinion preload.
21. Inspect the total preload.
22. Stake the front drive pinion companion flange front nut.
23. Install the differential extension flange tube.
24. Install the front differential carrier assembly.
25. Add differential oil.

FRONT DIFFERENTIAL SIDE GEAR SHAFT OIL SEAL

REPLACEMENT

See Figures 30 and 31.

➡The procedure described below is for the LH side. Use the same procedure for both the RH and LH side, unless otherwise specified.

1. Drain the differential oil.
2. Remove the front drive shaft assembly LH.
3. Remove the differential side gear shaft oil seal.
   a. Using the special tool (09308-10010), tap out the 2 oil seal.

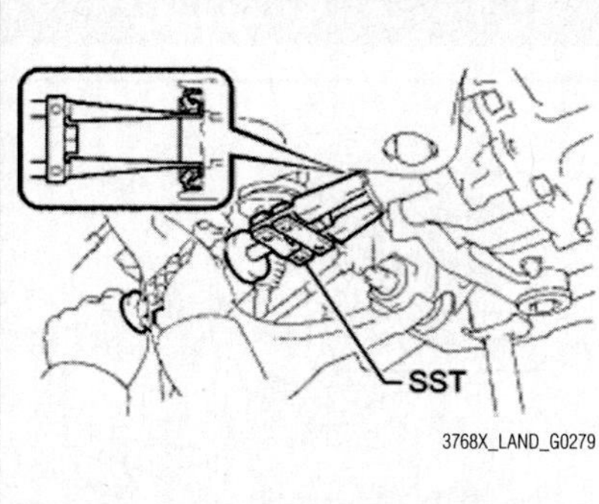

Fig. 30 Removing the differential side gear shaft oil seal

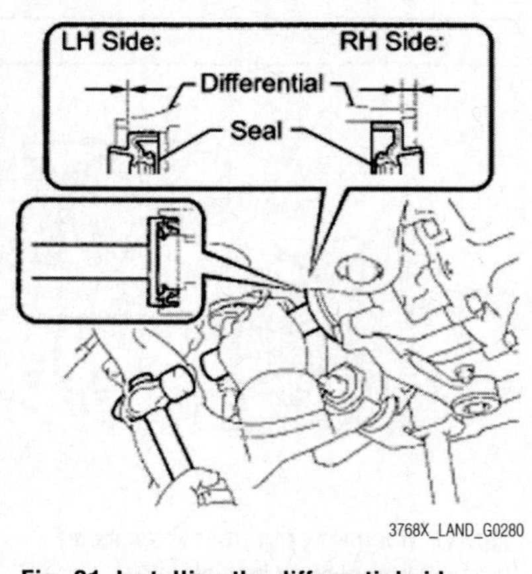

Fig. 31 Installing the differential side gear shaft oil seal

4. Install the differential side gear shaft oil seal.
   a. Apply MP grease to 2 new oil seal.
   b. Using the special tools and a hammer, tap in the 2 oil seal.

➡Make sure the LH and RH oil seals are installed in the proper locations.

➡The standard oil seal depth for the LH side is -0.0177-0.0177 inch (-0.45-0.45 mm).

➡The standard oil seal depth for the RH side is 0.189-0.228 inch (4.8-5.8 mm).

5. Install the front drive shaft assembly LH.
6. Add differential oil.

FRONT DRIVESHAFT

REMOVAL & INSTALLATION

See Figures 32 through 34.

1. Remove the front fender splash shield sub assembly LH.
2. Remove the front fender splash shield sub assembly RH.

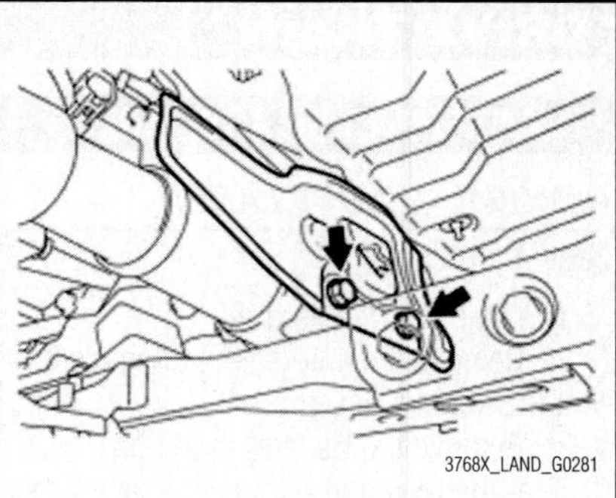

Fig. 32 Removing the propeller shaft heat insulator

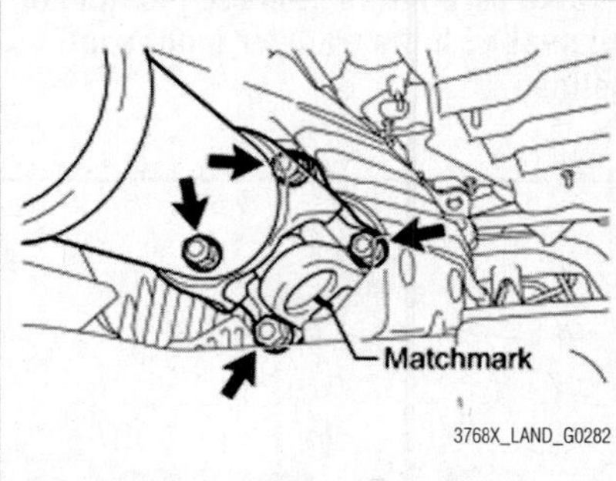

Fig. 33 Marking the propeller shaft flange and transfer flange (transfer side)

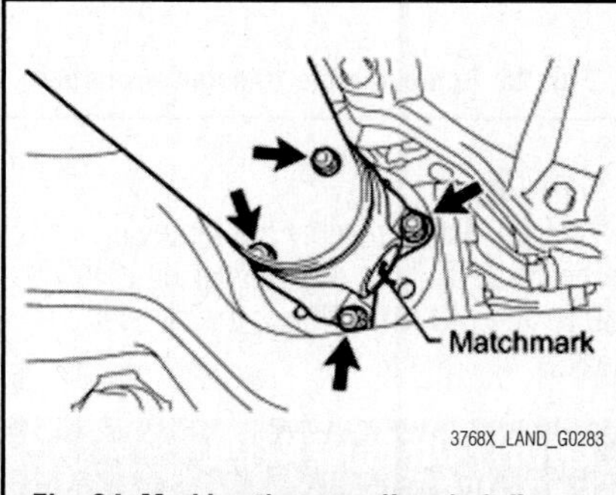

Fig. 34 Marking the propeller shaft flange and differential (differential side)

3. Remove the No. 1 engine under cover sub assembly.
4. Remove the No. 2 engine under cover.
5. Remove the propeller shaft heat insulator.
   a. Remove the 2 bolts and insulator.
6. Remove the front propeller shaft assembly.
   a. For the transfer side, place matchmarks on the propeller shaft flange and transfer flange. Remove the 4 nuts and 4 washers.

b. For the differential side, place matchmarks on the propeller shaft flange and differential. Remove the 4 nuts and 4 washers.

c. Remove the front propeller shaft.

To install:

7. Install the front propeller shaft assembly.

a. For the differential side, align the matchmarks on the yoke and differential flange. Connect the propeller shaft with the 4 nuts. Tighten to 65 ft. lbs. (88 Nm).

b. For the transfer side, align the matchmarks on the yoke and transfer flange. Install the propeller shaft with the 4 nuts. Tighten to 65 ft. lbs. (88 Nm).

8. Install the propeller shaft heat insulator with the 2 bolts. Tighten to 12 ft. lbs. (16 Nm).

9. Install the No. 2 engine under cover.

10. Install the No. 1 engine under cover sub assembly.

11. Install the front fender splash shield sub assembly LH.

12. Install the front fender splash shield sub assembly RH.

FRONT HALFSHAFT

REMOVAL & INSTALLATION

See Figure 35.

➡Use the same procedures for the RH and LH side.

➡The procedures listed below are for the LH side.

1. Remove the front fender splash shield sub assembly LH and RH.

2. Remove the No. 1 engine under cover sub assembly.

3. Remove the No. 2 engine under cover.

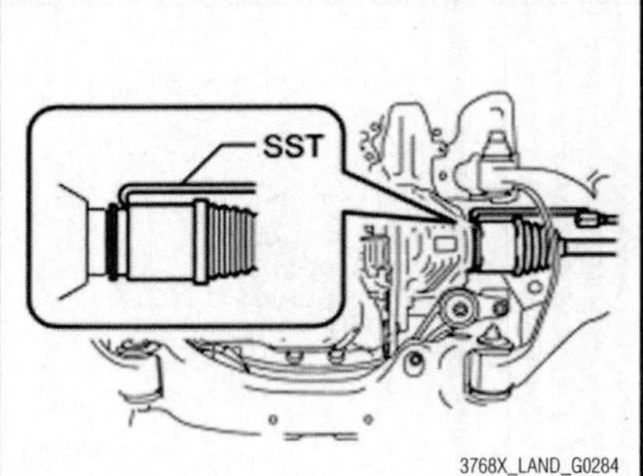

Fig. 35 Removing the front drive shaft assembly LH

4. Drain the differential oil.

5. Remove the front axle assembly LH.

6. Remove the front drive shaft assembly LH.

a. Using the special tool (09520-01010, 09520-024010 or 09520-32040), remove the drive shaft.

➡Be careful not to damage the dust cover and oil seal.

➡Move the drive shaft while keeping it level.

To install:

7. Install the front drive shaft assembly LH.

a. Coat the lip of the differential oil seal with MP grease.

➡Inspect the oil seal for wear or damage.

b. Coat the spline of the inboard joint shaft assembly with hypoid gear oil.

c. Align the shaft splines and install the drive shaft with a brass bar and hammer.

➡Set the snap ring with the opening side facing downward.

➡Be careful not to damage the oil seal, boot and dust cover.

➡If the outboard joint shaft is new, remove the anti-rust oil on the screw end with non-residue solvent.

➡Whether the inboard joint shaft is in contact with the pinion shaft or not can be confirmed from the sound or feeling when driving it.

8. Install the front axle assembly LH.

9. Add differential oil.

10. Check the speed sensor signal.

11. Inspect and adjust the front wheel alignment.

12. Adjust the headlight assembly.

13. Install the No. 2 engine under cover.

14. Install the No. 1 engine under cover sub assembly.

15. Install the front fender splash shield sub assembly LH.

16. Install the front fender splash shield sub assembly RH.

REAR AXLE SHAFT, BEARING & SEAL

REMOVAL & INSTALLATION

See Figures 36 through 38.

➡Use the same procedures for the LH and RH side.

➡The procedures listed below are for the LH side.

1. Remove the stabilizer control valve protector.

2. Open the stabilizer control with the accumulator housing shutter valve.

3. Disconnect the cable from the negative battery terminal.

4. Remove the rear wheel LH.

5. Drain the brake fluid.

6. Disconnect the rear brake flexible hose.

a. Disconnect the brake tube from the flexible hose with the special tool holding the flexible hose with a wrench.

Fig. 36 Disconnecting the rear brake flexible hose

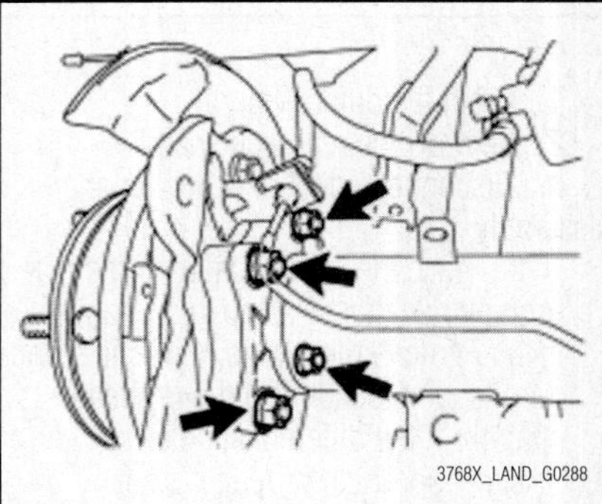

Fig. 37 Disconnecting the rear disc brake cylinder assembly LH

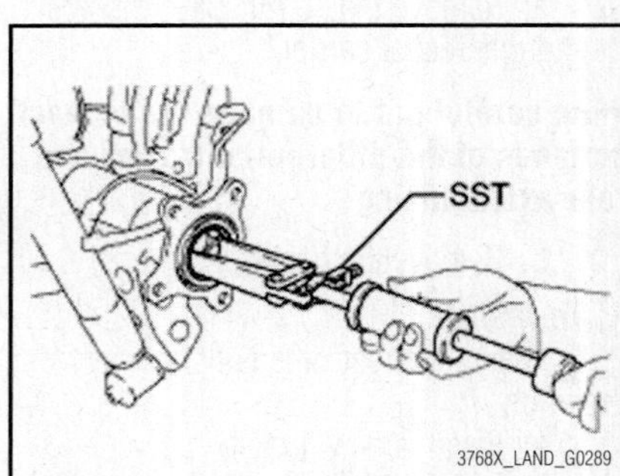

Fig. 38 Removing the rear axle shaft oil seal LH

➡Do not bend or damage the brake tube.

➡Do not allow any foreign matter such as dirt and dust to enter the brake tube from the connecting point.

b. Remove the clip.

7. Disconnect the rear disc brake cylinder assembly LH.

a. Remove the 2 bolts and disconnect the rear disc brake cylinder.

➡Do not twist or bend the flexible hose.

➡Do not disconnect the flexible hose from the disc brake cylinder.

b. Remove the o-ring.

8. Remove the rear axle shaft oil seal LH.

a. Using the special tool (09308-00010), tap out the oil seal.

To install:

9. To install, reverse the removal procedure. Tighten the axle shaft to 44 ft. lbs. (60 Nm). Tighten the rear disc brake cylinder LH to 70 ft. lbs. (95 Nm). Tighten the rear brake flexible hose to 11 ft lbs. (15 Nm). If tightening with a union nut wrench, tighten to 10 ft. lbs. (14 Nm).

REAR DIFFERENTIAL CARRIER ASSEMBLY

REMOVAL & INSTALLATION

See Figure 39.

1. Drain the differential oil.
2. Remove the rear axle shaft.
3. Disconnect the rear propeller shaft assembly.

a. Place the matchmarks on the rear propeller shaft and companion flange.

b. Remove the 4 nuts, 4 washers and 4 bolts, and disconnect the propeller shaft from the differential side.

c. Support the propeller shaft securely.

4. Remove the rear differential carrier sub assembly.

a. Remove the 10 nuts, 10 washers and differential carrier.

➡Be careful not to damage the contact surfaces of the differential carrier and rear axle housing.

b. Remove the gasket.

To install:

5. Install the rear differential carrier assembly.

a. Install a new gasket.

b. Install the differential carrier with the 10 washers and 10 nuts. Tighten to 53 ft. lbs. (72 Nm).

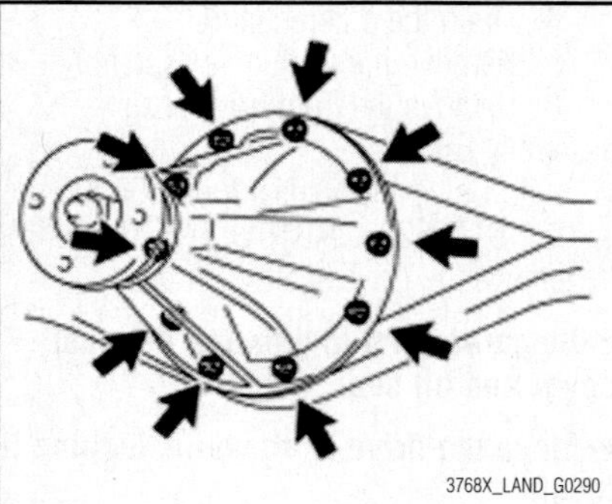

Fig. 39 Removing the rear differential carrier sub assembly

6. Connect the propeller shaft assembly.

a. Align the matchmarks on the propeller shaft flange and differential flange.

b. Connect the propeller shaft with the 4 bolts, 4 washers and 4 nuts. Tighten to 65 ft. lbs. (88 Nm).

7. Install the rear axle shaft.
8. Add differential oil.

REAR DIFFERENTIAL CARRIER OIL SEAL

REPLACEMENT

See Figures 40 through 45.

1. Drain the differential oil.
2. Disconnect the propeller shaft assembly.
3. Remove the drive pinion companion flange rear nut.

a. Using the special tool (09930-00010) and a hammer, loosen the staked part of the nut.

b. Using the special tool (09330-00021) to hold the companion flange, remove the nut.

4. Remove the rear drive pinion companion flange rear sub assembly using the special tool.
5. Remove the rear differential carrier oil seal using the special tool.
6. Remove the rear differential drive pinion oil slinger.
7. Remove the rear drive pinion front tapered roller bearing.

a. Using the special tool (09556-22010), remove the front bearing (inner race) from the drive pinion.

b. Using the special tool (09308-00010), remove the front bearing (outer race) from the drive pinion.

8. Replace the rear differential drive pinion bearing spacer.

a. Replace the bearing spacer with a new one.

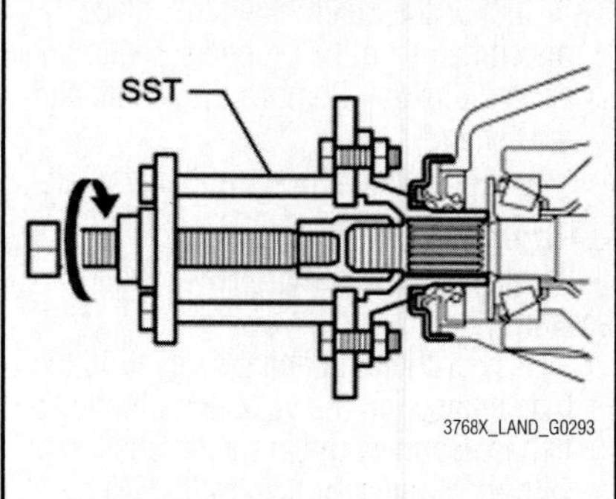

Fig. 40 Removing the rear drive pinion companion flange rear sub assembly

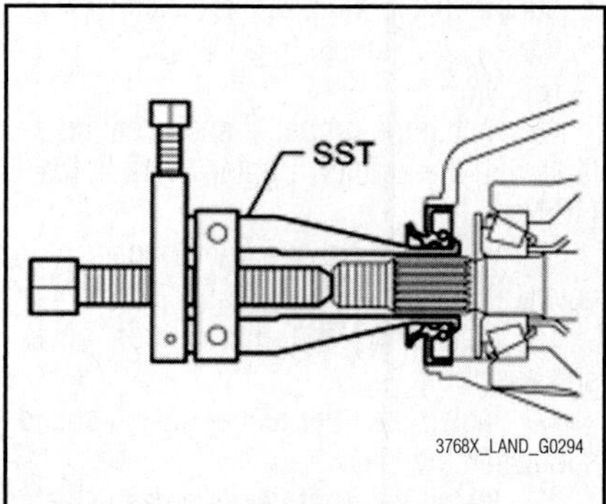

Fig. 41 Removing the rear differential carrier oil seal

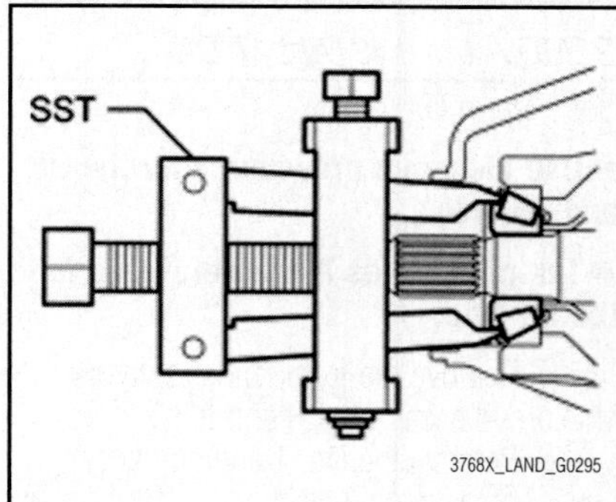

Fig. 42 Removing the front bearing (inner race) from the drive pinion

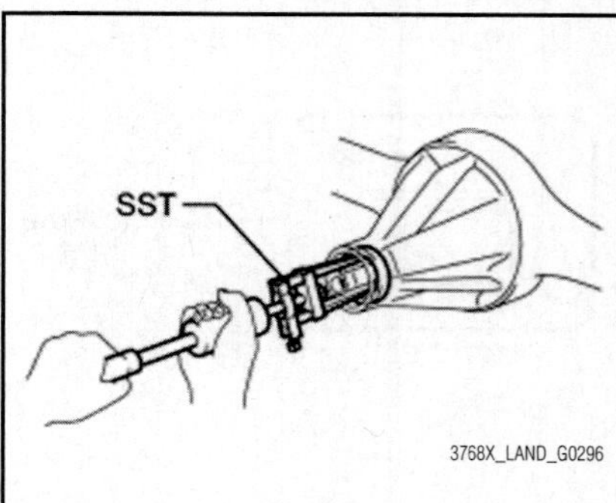

Fig. 43 Removing the front bearing (outer race) from the drive pinion

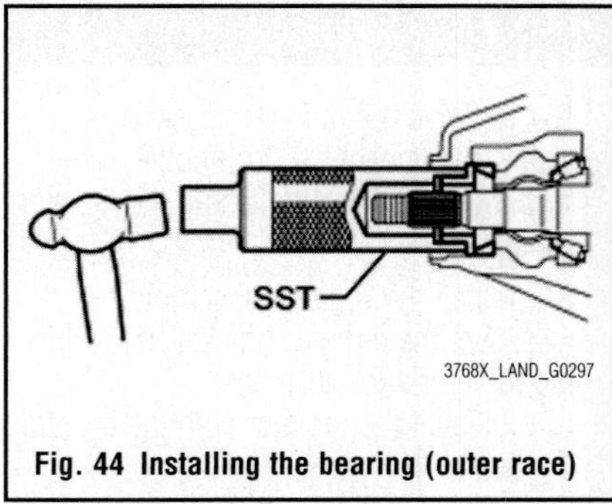

Fig. 44 Installing the bearing (outer race)

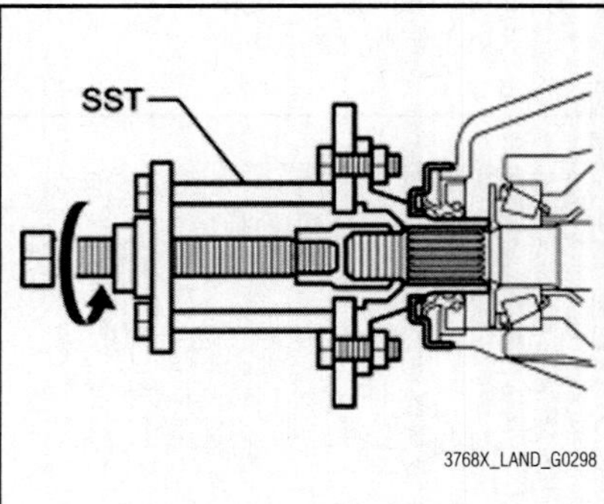

Fig. 45 Installing the rear drive pinion companion flange rear sub assembly

9. Install the rear drive pinion front tapered roller bearing.
   a. Using the special tool (09316-60011, 09316-00011 or 09316-00021) and a hammer, tap in the bearing (outer race).
10. Install the rear drive pinion companion flange rear sub assembly.
   a. Using the special tool (09951-03010, 09953-03010, 09954-03010, 09955-03030 or 09956-03040), install the companion flange on the drive pinion.
11. Adjust the drive pinion preload.
   a. Coat the threads of a new nut with hypoid gear oil.
   b. Using the special tool (09330-00021 or 09330-00030) to hold the companion flange, install the nut by tightening it until the standard preload is reached. Tighten to 325 ft. lbs. (441 Nm) or less.
   c. Using a torque wrench, measure the preload. The standard preload (at starting) for a new bearing is 10-14 inch lbs. (1.0-1.7 Nm). Standard preload (at starting) for reused bearings is 8-12 inch lbs. (0.9-1.4 Nm).

➡If the preload is not within specification, adjust the preload.

12. Stake the drive pinion companion flange rear nut.
   a. Using a chisel and hammer, stake the nut.
13. Connect the propeller shaft assembly.
14. Add differential oil.

REAR DRIVESHAFT

REMOVAL & INSTALLATION

See Figures 46 through 48.

1. Remove the transfer heat insulator.
   a. Remove the 4 bolts and insulator.
2. Remove the rear propeller shaft assembly.
   a. For the transfer side, place matchmarks on the propeller shaft flange and transfer flange.
   b. Remove the 4 nuts and 4 washers.
   c. For the differential side, place matchmarks on the propeller shaft flange and rear differential flange.
   d. Remove the 4 nuts, 4 bolts and 4 washers.
   e. Remove the rear propeller shaft.

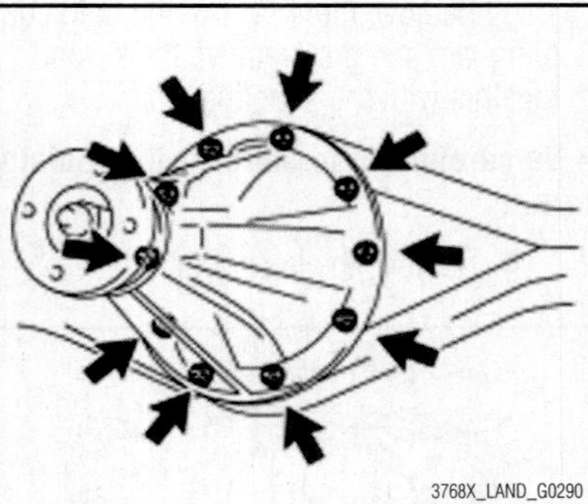

Fig. 46 Removing the transfer heat insulator

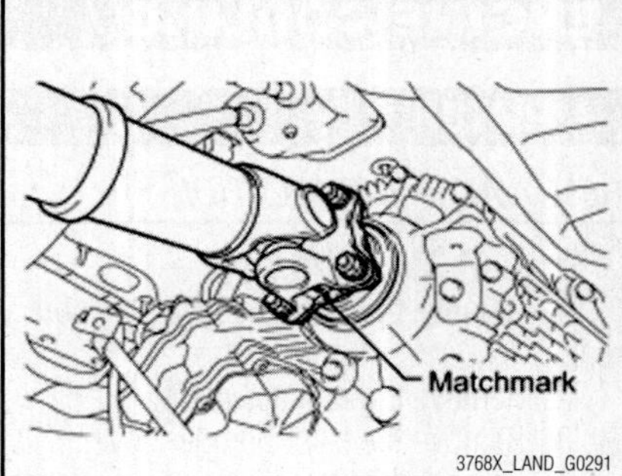

Fig. 47 Removing the rear propeller shaft assembly (transfer side)

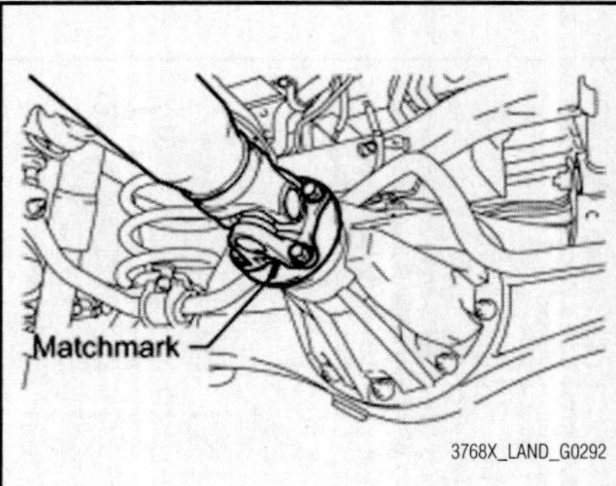

Fig. 48 Removing the rear propeller shaft assembly (differential side)

To install:

3. Install the rear propeller shaft.
   a. Align the matchmarks on the propeller shaft flange and transfer flange.
   b. Connect the propeller shaft with the 4 washers and 4 nuts. Tighten to 65 ft. lbs. (88 Nm).
   c. Align the matchmarks on the propeller shaft flange and differential flange.
   d. Install the propeller shaft with the 4 bolts, 4 washers and 4 nuts. Tighten to 65 ft. lbs. (88 Nm).
4. Install the transfer heat insulator.
   a. Install the insulator with the 4 bolts. Tighten to 22 ft. lbs. (30 Nm).

ENGINE COOLING

RADIATOR

REMOVAL & INSTALLATION

See Figures 49 through 55.

1. Remove the upper radiator support seal.
2. Remove the radiator grille.
3. Remove the front bumper cover.
4. Remove the transmission oil cooler air duct (with air cooled transmission oil cooler).
   a. Remove the 4 bolts and oil cooler air duct.
5. Disconnect the radiator side deflector RH (without air cooled transmission oil cooler).
   a. Using a clip remover, remove the 4 clips and disconnect the side deflector.
6. Disconnect the radiator side deflector LH.
   a. Using a clip remover, remove the 4 clips and disconnect the side deflector.
7. Remove the front fender splash shield sub assembly LH.
8. Remove the front fender splash shield sub assembly RH.
9. Remove the No. 1 engine under cover sub assembly.
10. Drain the engine coolant.
11. Remove the v-bank cover sub assembly.
12. Remove the air cleaner hose assembly.
13. Remove the No. 1 radiator hose.
14. Remove the No. 2 radiator hose.
15. Remove the fan shroud.
   a. Loosen the 4 nuts holding the fluid coupling fan.
   b. Remove the fan and generator v-belt.
   c. Disconnect the reservoir hose from the upper radiator tank.
   d. For vehicles with air cooled transmission oil coolers, detach the claw to open the flexible hose clamp.
   e. Remove the 2 bolts and disconnect the oil cooler tube from the fan shroud.
   f. Remove the 2 bolts holding the fan shroud.
   g. Remove the 4 nuts of the fluid coupling fan, and then remove the shroud together with the coupling fan.

➡Be careful not to damage the radiator core.

   h. Remove the fan pulley.
16. Remove the radiator assembly.
   a. Disconnect the 2 oil cooler hoses.
   b. Remove the 4 bolts and the radiator.

To install:

17. Install the radiator assembly.
   a. Set the radiator bracket hooks to the radiator support holes.
   b. Install the radiator with the 4 bolts. Tighten to 13 ft. lbs. (18 Nm).
   c. Connect the 2 oil cooler hoses.
18. Install the fan shroud.
   a. Install the fan pulley.

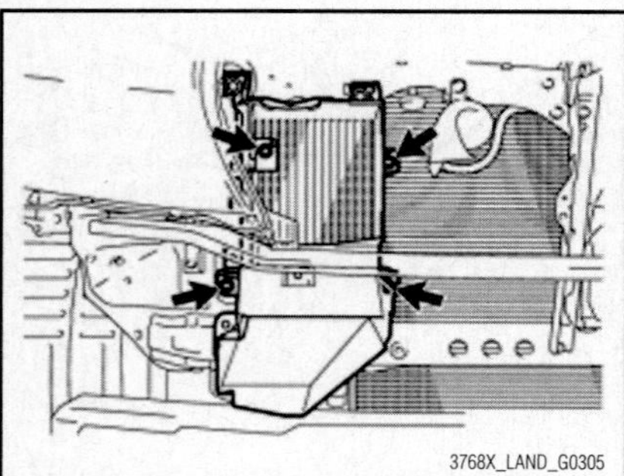

Fig. 49 Removing the transmission oil cooler air duct

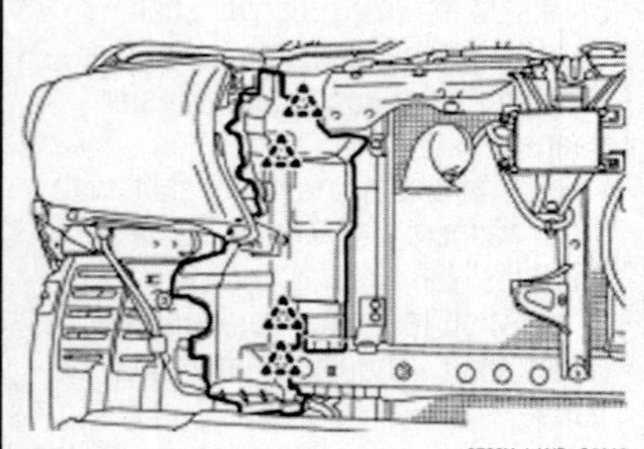

Fig. 50 Disconnecting the radiator side deflector RH

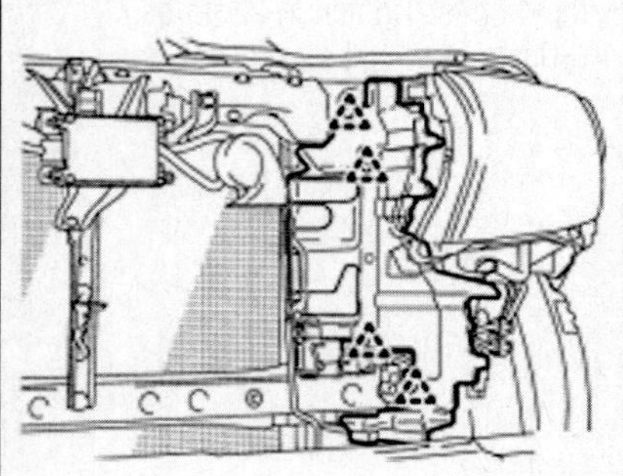

Fig. 51 Disconnecting the radiator side deflector LH

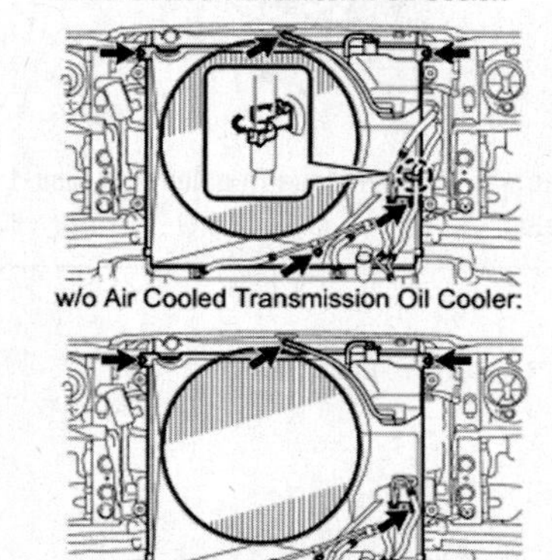

Fig. 52 Disconnecting the reservoir hose from the upper radiator tank

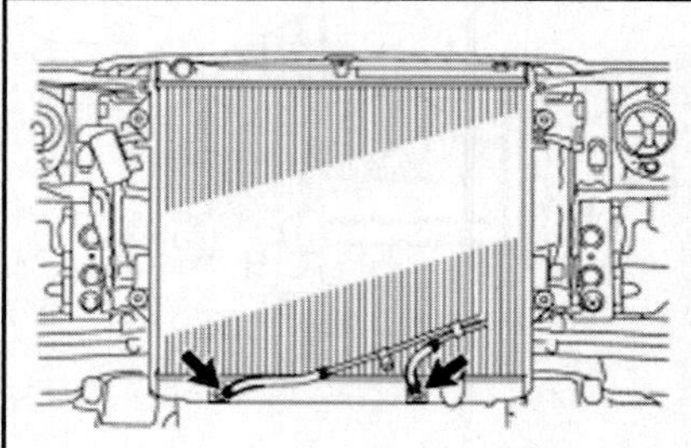

Fig. 53 Disconnecting the oil cooler hoses

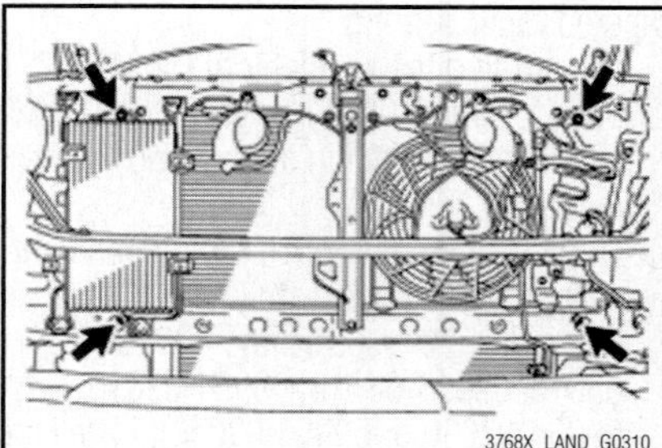

Fig. 54 Removing the radiator

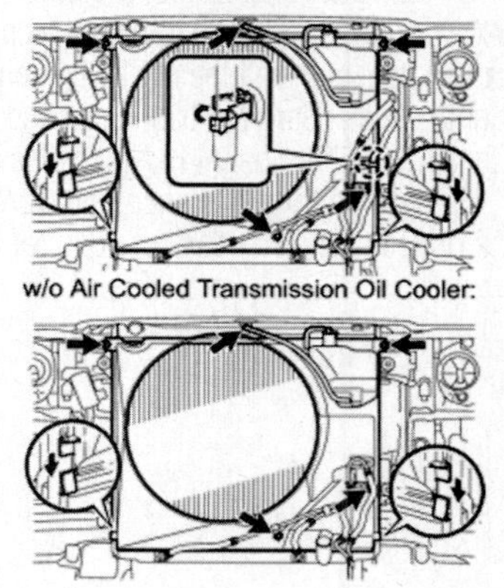

Fig. 55 Attaching the claws of the shroud to the radiator

b. Place the shroud together with the coupling fan between the radiator and engine.

➡**Be careful not to damage the radiator core.**

c. Temporarily install the fluid coupling fan to the fluid coupling bracket with the 4 nuts. Tighten the nuts as much as possible by hand.
d. Attach the claws of the shroud to the radiator.
e. Install the shroud with the 2 bolts. Tighten to 71 inch lbs. (8 Nm).
f. Connect the oil cooler tube to the fan shroud with the 2 bolts. Tighten to 44 inch lbs. (5 Nm).
g. Vehicles with air cooled transmission oil cooler, pass the hose through the flexible hose clamp and close the clamp.
h. Connect the reservoir hose to the upper radiator tank.
i. Install the fan and generator v-belt.
j. Tighten the 4 nuts of the fluid coupling fan. Tighten to 15 ft. lbs. (21 Nm).

19. Install the No. 2 radiator hose.
20. Install the No. 1 radiator hose.
21. Install the air cleaner hose assembly.
22. Install the v-bank cover sub assembly.
23. Add engine coolant.
24. Inspect for coolant leaks.
25. Install the No. 1 engine under cover sub assembly.
26. Install the front fender splash shield sub assembly LH.
27. Install the front fender splash shield sub assembly RH.
28. Connect the radiator side deflector LH.
    a. Connect the deflector with the 4 clips.
29. Connect the radiator side deflector RH (without air cooled transmission oil cooler).
    a. Connect the deflector with the 4 clips.
30. Install the transmission oil cooler air duct (with air cooled transmission oil cooler) with the 4 bolts. Tighten to 43 inch lbs. (4.9 Nm).
31. Install the front bumper cover.
32. Install the radiator grille.
33. Install the upper radiator support seal.

THERMOSTAT

REMOVAL & INSTALLATION

See Figures 56 through 59.

➡**If the thermostat is not installed, cooling efficiency decreases. Even if the engine tends to overheat, do not remove the thermostat.**

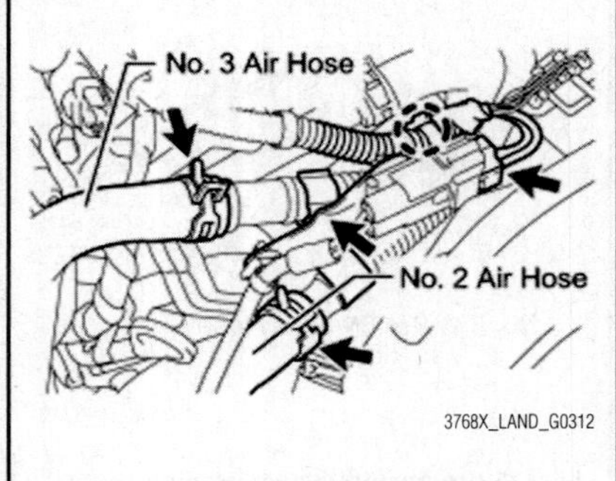

Fig. 56 Disconnecting the air hoses

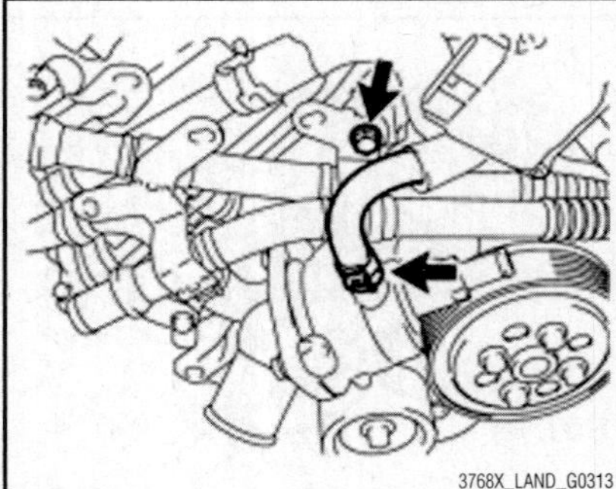

Fig. 57 Disconnecting the No. 5 water by pass hose

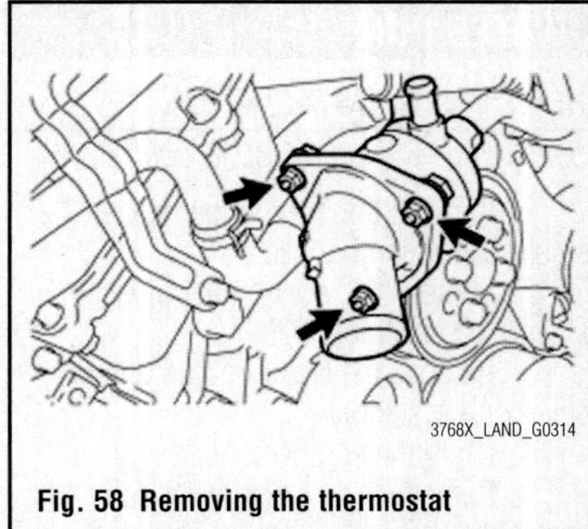

Fig. 58 Removing the thermostat

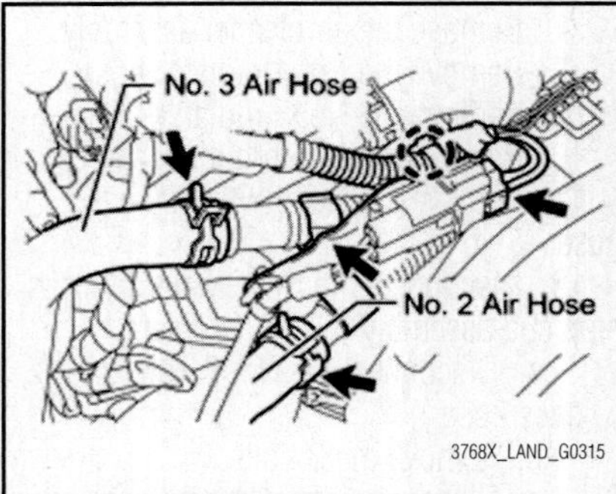

Fig. 59 Connecting the air pump connector clamp holder

1. Remove the front fender splash shield sub assembly LH.
2. Remove the front fender splash shield sub assembly RH.
3. Remove the No. 1 engine under cover sub assembly.
4. Drain the engine coolant.
5. Remove the v-bank cover sub assembly.
6. Remove the air cleaner hose assembly.
7. Remove the air cleaner assembly.
8. Remove the No. 1 radiator hose.
9. Remove the No. 2 radiator hose.
10. Remove the fan shroud.
11. Remove the water inlet sub assembly with the thermostat.
    a. Disconnect the No. 2 and No. 3 air hoses.
    b. Disconnect the air pump connector.
    c. Disconnect the air pump connector clamp holder.
    d. Using a clip remover, detach the wire harness clamp.
    e. Disconnect the No. 5 water by pass hose.
    f. Remove the air tube bracket bolt.
    g. Remove the 3 nuts, water inlet with the thermostat and gasket.

To install:

12. Install the water inlet sub assembly with the thermostat.
    a. Install a new gasket and the water inlet with the thermostat with the 3 nuts. Tighten to 7 ft. lbs. (10 Nm).
    b. Install the air tube bracket bolt. Tighten to 7 ft. lbs. (10 Nm).
    c. Connect the No. 5 water by pass hose.

➡**Install the hose so that the direction of the hose clamp is as indicated.**

    d. Connect the air pump connector clamp holder.
    e. Attach the wire harness clamp.
    f. Connect the air pump connector.
    g. Connect the No. 2 and No. 3 air hoses.
13. Install the fan shroud.
14. Install the No. 2 radiator hose.
15. Install the No. 1 radiator hose.
16. Install the air cleaner assembly.
17. Install the air cleaner hose assembly.
18. Add engine coolant.
19. Inspect for coolant leaks.
20. Install the v-bank cover sub assembly.
21. Install the No. 1 engine under cover sub assembly.
22. Install the front fender splash shield sub assembly LH.
23. Install the front fender splash shield sub assembly RH.

WATER PUMP

REMOVAL & INSTALLATION

See Figures 56 and 57, 60 through 67.

1. Remove the front fender splash shield sub assembly LH.
2. Remove the front fender splash shield sub assembly RH.
3. Remove the No. 1 engine under cover sub assembly.
4. Drain the engine coolant.
5. Remove the v-bank cover sub assembly.
6. Remove the air cleaner hose assembly.
7. Remove the air cleaner assembly.
8. Remove the No. 1 radiator hose.
9. Remove the No. 2 radiator hose.
10. Remove the fan shroud.
11. Remove the No. 1 water by pass hose.
12. Disconnect the No. 2 water by pass pipe sub assembly (with oil cooler).
    a. Disconnect the No. 6 water by-pass hose.
    b. Remove the 3 bolts.
    c. Disconnect the water by-pass pipe with water hose.
13. Remove the water inlet housing.
    a. Disconnect the No. 2 and No. 3 air hoses.
    b. Disconnect the air pump connector.
    c. Disconnect the air pump connector clamp holder.
    d. Using a clip remover, detach the wire harness clamp.
    e. Disconnect the No. 5 water by pass hose.
    f. Remove the air tube bracket bolt.
    g. Disconnect the No. 3 water by pass hose and remove the 3 bolts and water inlet housing.
    h. Remove the gasket from the water pump.
14. Remove the water pump pulley.
    a. Using the special tool (09960-10010, 09962-01000 or 09963-01000), hold the water pump pulley.
    b. Remove the 4 bolts and water pump pulley.
15. Remove the water pump assembly.
    a. Remove the 8 bolts, water pump and gasket.

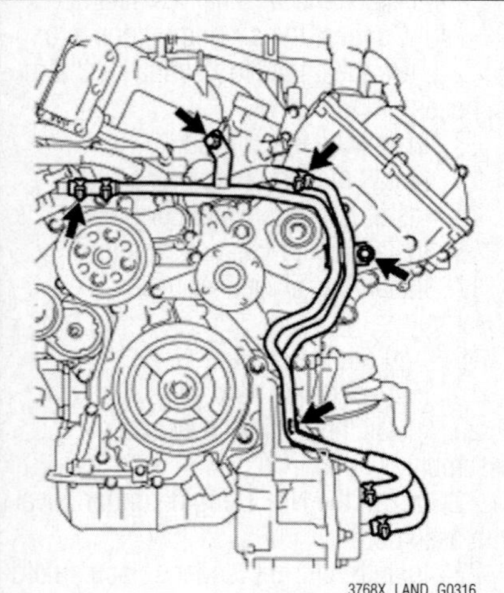

Fig. 60 Disconnecting the No. 2 water by pass pipe sub assembly (with oil cooler)

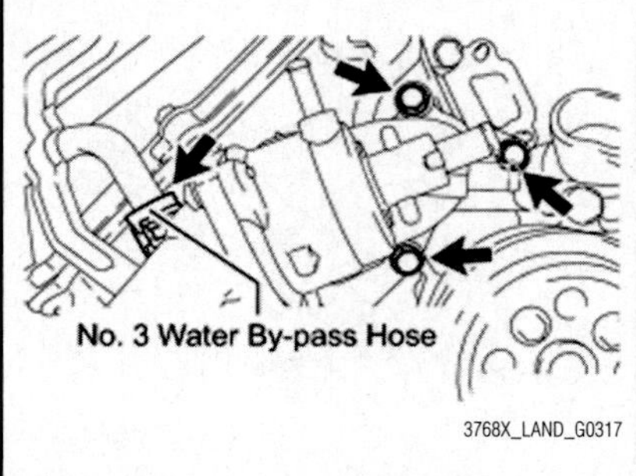

Fig. 61 Removing the water inlet housing

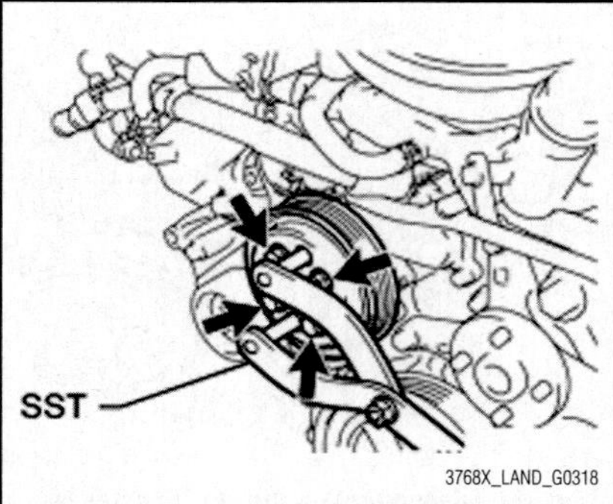

Fig. 62 Removing the water pump pulley

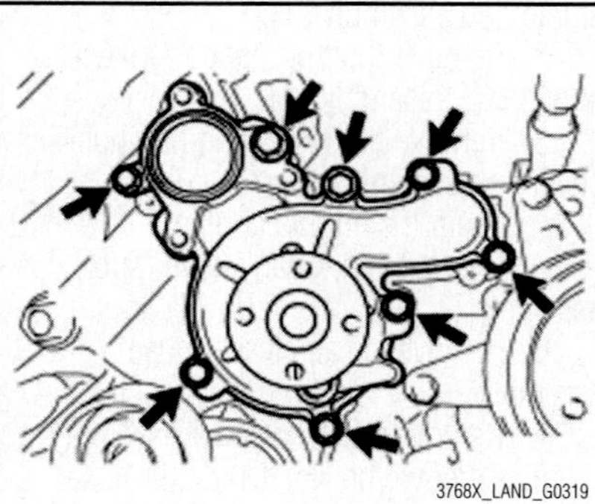

Fig. 63 Removing the water pump

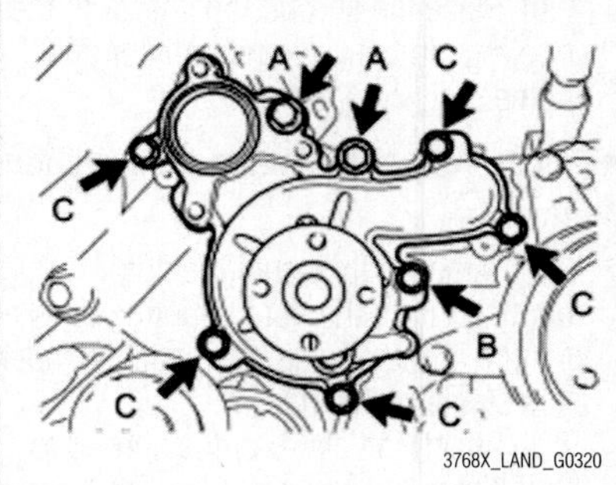

Fig. 64 Installing the water pump assembly

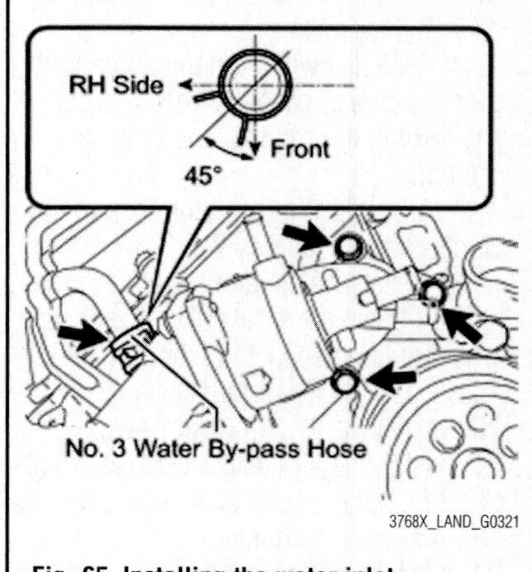

Fig. 65 Installing the water inlet housing

To install:

16. Install the water pump assembly.
    a. Install a new gasket and the water pump with the 8 bolts. Tighten bolt A to 35 ft. lbs. (47 Nm). Tighten bolt B to 17 ft. lbs. (23 Nm). Tighten bolt C to 15 ft. lbs. (20 Nm).
17. Install the water pump pulley.
    a. Temporarily install the pulley with the 4 bolts.
    b. Using the special tool, hold the pulley and tighten the 4 bolts to 15 ft. lbs. (21 Nm).
18. Install the water inlet housing.
    a. Install a new gasket to the water pump.
    b. Install the water inlet housing with the 3 bolts and connect the No. 3 water by pass hose. Tighten to 15 ft. lbs. (21 Nm).

➡Install the hose so that the direction of the hose clamp is as indicated.

    c. Install the air tube bracket bolt and tighten to 7 ft. lbs. (10 Nm).
    d. Connect the No. 5 water by pass hose.

➡Install the hose so that the direction of the hose clamp is as indicated.

e. Connect the air pump connector clamp holder.
f. Attach the wire harness clamp.
g. Connect the air pump connector.
h. Connect the No. 2 and No. 3 air hoses.
19. Connect the No. 2 water by pass pipe sub assembly (with oil cooler).
a. Connect the water by pass pipe with the water hose.

➡Install the hose so that the direction of the hose clamp is as indicated.

b. Install the 3 bolts and tighten to 7 ft. lbs. (10 Nm).
c. Connect the No. 6 water by pass hose.
20. Install the NO. 1 water by pass hose.
21. Install the fan shroud.
22. Install the No. 2 radiator hose.
23. Install the No. 1 radiator hose.
24. Install the air cleaner assembly.
25. Install the air cleaner hose assembly.
26. Add engine coolant.
27. Inspect for coolant leaks.
28. Install the v-bank cover sub assembly.
29. Install the No. 1 engine under cover sub assembly.
30. Install the front fender splash shield sub assembly LH.
31. Install the front fender splash shield sub assembly RH.

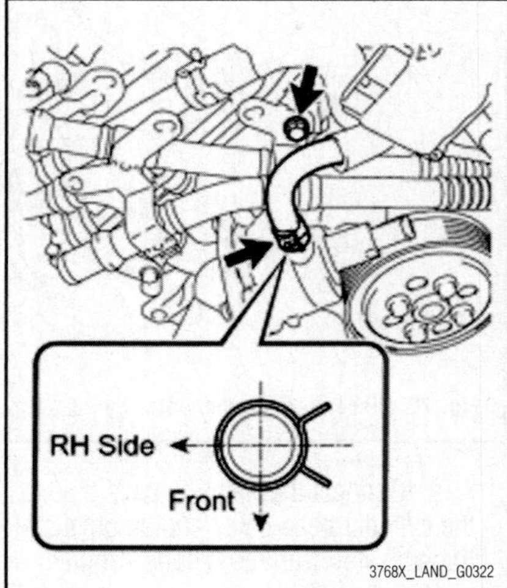

Fig. 66 Installing the air tube bracket bolt and connecting the No. 5 water by pass hose

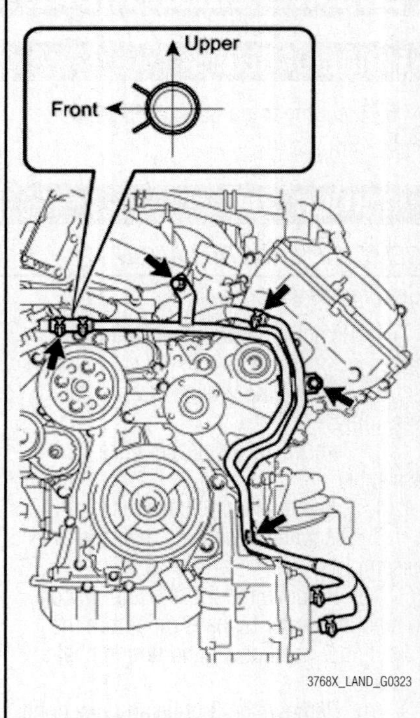

Fig. 67 Connecting the No. 2 water by pass pipe sub assembly (with oil cooler)

ENGINE ELECTRICAL

CHARGING SYSTEM

ALTERNATOR

REMOVAL & INSTALLATION

See Figures 68 and 69.

1.Disconnect the cable from the negative battery terminal.
2. Remove the v-bank cover sub assembly.
3. Remove the air cleaner hose assembly.
4. Remove the air cleaner assembly.
5. Remove the front fender splash shield sub assembly LH.
6. Remove the front fender splash shield sub assembly RH.
7. Remove the No. 1 engine under cover sub assembly.
8. Remove the fan and generator v-belt.
9. Remove the front fender apron seal front RH.
10. Disconnect the vane pump assembly.
11. Disconnect the oil cooler pipe assembly.
12. Disconnect the oil cooler pipe assembly.
13. Remove the generator.
a. Disconnect the generator connector.
b. Remove the terminal cap and nut, and disconnect the generator wire.
c. Remove the bolt and wire harness bracket from the generator.

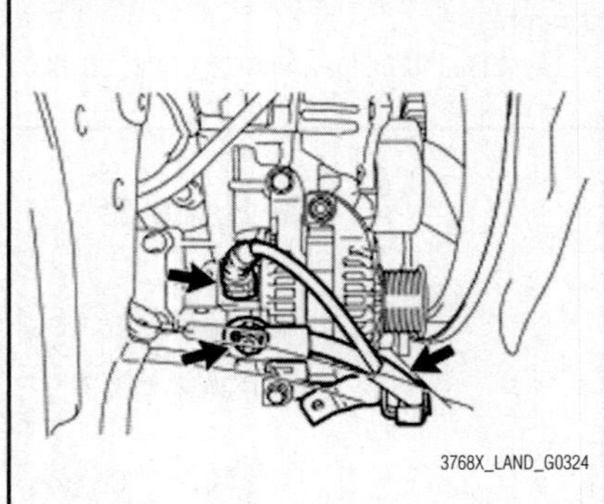

Fig. 68 Disconnecting the connector

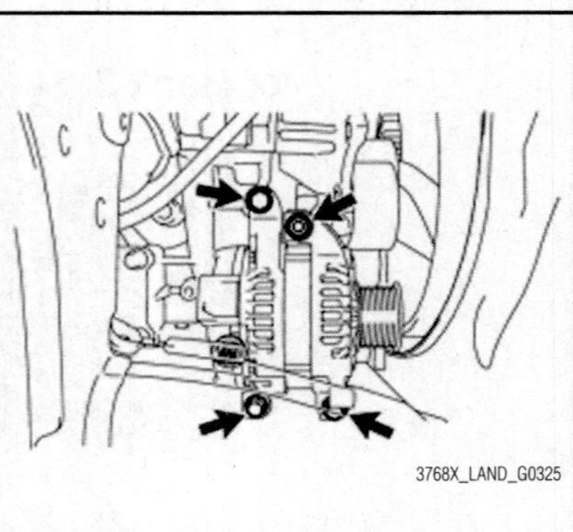

Fig. 69 Removing the generator

d. Remove the 3 bolts, nut and generator.

To install:

14. Install the generator assembly.
a. Install the generator with the 3 bolts and nut. Tighten to 32 ft. lbs. (43 Nm).
b. Connect the generator connector.
c. Connect the generator wire with the nut. Tighten to 9 ft. lbs. (12 Nm).
d. Install the terminal cap.
e. Install the harness bracket to the generator with the bolt. Tighten to 23 ft. lbs. (31 Nm).
15. Connect the oil cooler pipe assembly.
16. Connect the vane pump assembly.
17. Install the front fender apron seal front RH.
18. Install the fan and generator v-belt.
19. Install the No. 1 engine under cover sub assembly.
20. Install the front fender splash shield sub assembly LH.
21. Install the front fender splash shield sub assembly RH.
22. Install the air cleaner assembly.
23. Install the air cleaner hose assembly.
24. Install the v-bank cover sub assembly.
25. Connect the cable to the negative battery cable.

FIRING ORDERS

5.7L engine firing order: 1–8–4–3–6–5–7–2.

IGNITION COIL

REMOVAL & INSTALLATION

See Figures 71 through 73.

1. Remove the No. 2 engine under cover.
2. Remove the v-bank cover sub assembly.
3. Remove the air cleaner hose assembly.
4. Remove the air cleaner assembly.
5. Disconnect the water by pass pipe assembly.
   a. Remove the 2 bolts and disconnect the water by pass pipe from the cylinder head cover and timing chain cover.
   b. Remove the 3 bolts and disconnect the water by pass pipe from the cylinder head cover.
6. Remove the ignition coil assembly.
   a. Disconnect the 8 ignition coil connectors.
   b. Remove the 8 bolts and 8 ignition coils.
7. Remove the 8 spark plugs.

To install:

8. Install the spark plugs and tighten to 15 ft. lbs. (21 Nm).
9. Install the ignition coil assembly with the 8 bolts. Tighten to 7 ft. lbs. (10 Nm).
   a. Connect the 8 ignition coil connectors.
10. Connect the water by pass pipe assembly.

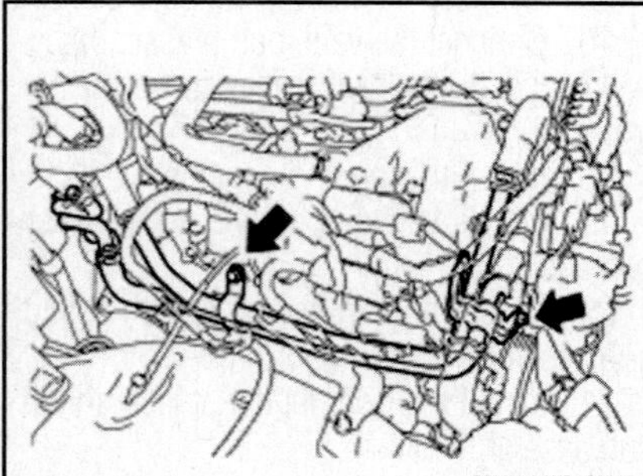

3768X_LAND_G0330

Fig. 71 Disconnecting the water by pass pipe from the cylinder head cover and timing chain cover

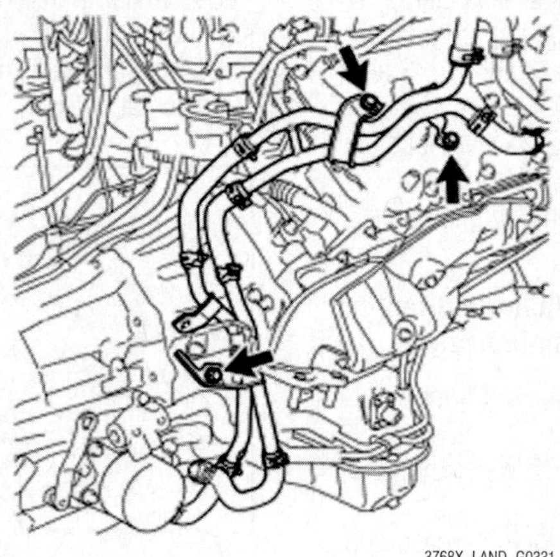

3768X_LAND_G0331

Fig. 72 Disconnecting the water by pass pipe from the cylinder head cover

   a. Connect the water by pass pipe to the cylinder head cover and automatic transmission with the 3 bolts. Tighten to 13 ft. lbs. (18 Nm).
   b. Connect the water by pass pipe to the cylinder head cover and timing chain cover with the 2 bolts. Tighten to 7 ft. lbs. (10 Nm).
11. Install the air cleaner assembly.
12. Install the air cleaner hose assembly.
13. Install the v-bank cover sub assembly.
14. Install the No. 2 engine under cover.

IGNITION TIMING

ADJUSTMENT

The ignition timing is controlled by the Powertrain Control Module (PCM). No adjustment is necessary or possible.

SPARK PLUGS

REMOVAL & INSTALLATION

Refer to IGNITION COIL.

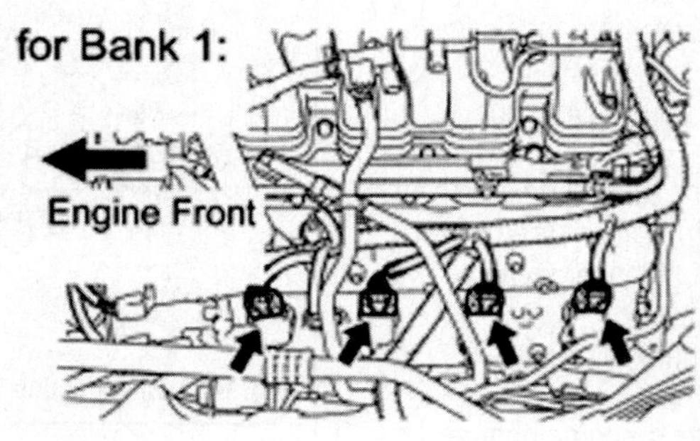

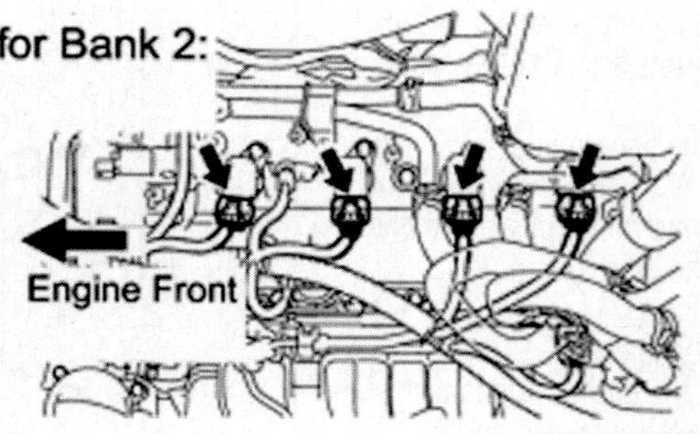

3768X_LAND_G0332

Fig. 73 Removing the ignition coil assembly

STARTER

REMOVAL & INSTALLATION

See Figures 74 through 77.

1. Disconnect the cable from the negative battery terminal.
2. Remove the front fender splash shield sub assembly LH.
3. Remove the front fender splash shield sub assembly RH.
4. Remove the No. 1 engine under cover sub assembly.
5. Remove the No. 2 engine under cover.
6. Remove the front fender apron seal front RH.
7. Remove the front fender apron seal rear RH.
8. Remove the engine oil level dipstick guide.
9. Remove the tailpipe assembly.
10. Remove the center exhaust pipe assembly.
11. Remove the front exhaust pipe assembly.
12. Remove the No. 1 manifold stay.
13. Remove the No. 1 exhaust manifold heat insulator.
14. Remove the exhaust manifold sub assembly RH.
15. Remove the 3 bolts and the starter cover.
16. Remove the starter assembly.
    a. Disconnect the starter connector.
    b. Remove the nut and disconnect the starter wire.
    c. Remove the 2 bolts and starter.
    d. Remove the flywheel housing side cover.

To install:

17. Install the starter assembly.
    a. Install the flywheel housing side cover.
    b. Install the starter with the 2 bolts. Tighten to 27 ft. lbs. (37 Nm).

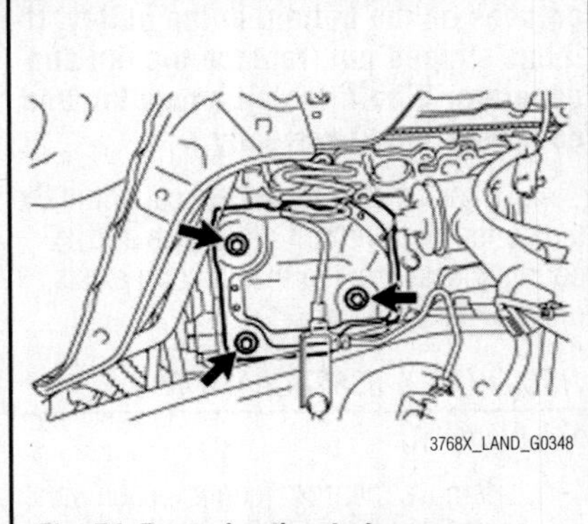

Fig. 74 Removing the starter cover

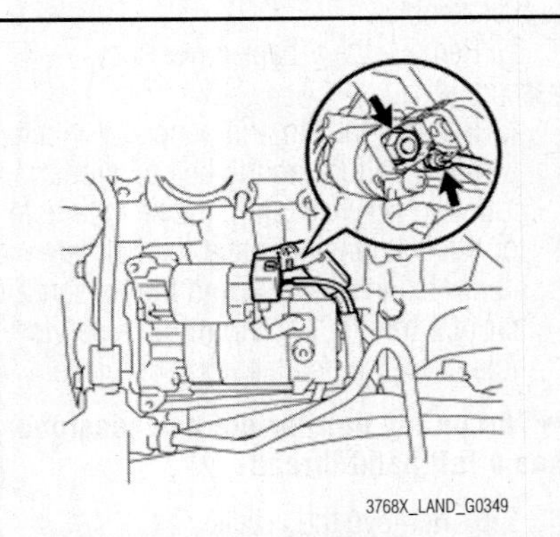

Fig. 75 Disconnecting the starter connector

    c. Install the starter wire with the nut. Tighten to 87 inch lbs. (9.8 Nm).
    d. Connect the starter connector.
18. Install the starter cover with the 3 bolts and tighten to 9 ft. lbs. (12 Nm).
19. Install the exhaust manifold sub assembly RH.
20. Install the No. 1 exhaust manifold heat insulator.
21. Install the No. 1 manifold stay.
22. Install the front exhaust pipe assembly.
23. Install the center exhaust pipe assembly.
24. Install the tailpipe assembly.
25. Install the engine oil level dipstick guide.

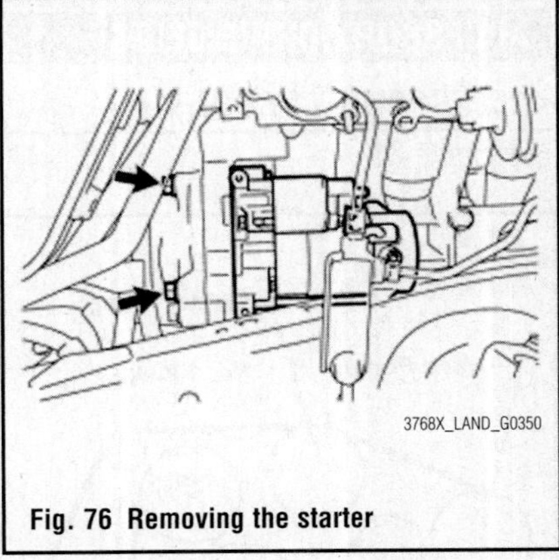

Fig. 76 Removing the starter

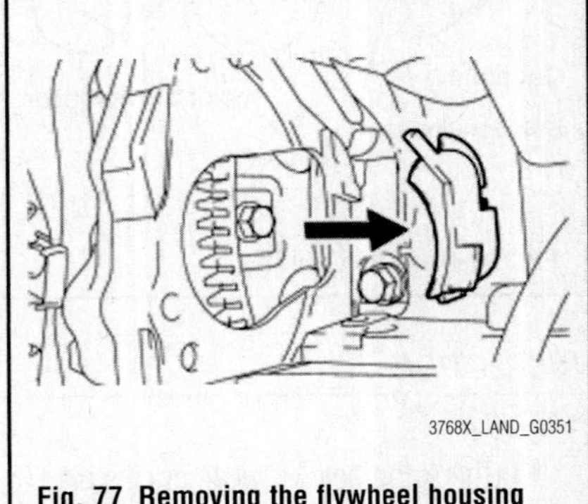

Fig. 77 Removing the flywheel housing side cover

26. Connect the cable to the negative battery terminal.
27. Inspect for exhaust gas leaks.
28. Install the No. 2 engine under cover.
29. Install the No. 1 engine under cover sub assembly.
30. Install the front fender apron seal rear RH.
31. Install the front fender apron seal front RH.
32. Install the front fender splash shield sub assembly RH.
33. Install the front fender splash shield sub assembly LH.

ENGINE MECHANICAL

ACCESSORY DRIVE BELTS

ACCESSORY BELT ROUTING

See Figure 78.

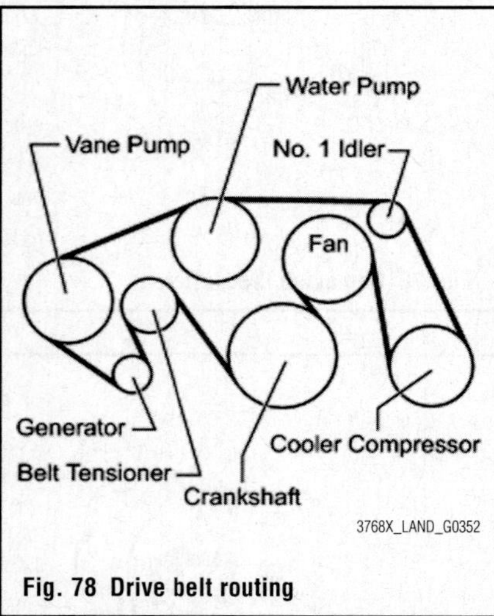

Fig. 78 Drive belt routing

INSPECTION

See Figure 79.

1. Check the belt for wear, cracks or other signs of damage.
2. If any of the following defects is found, replace the fan and generator V belt.
   - Belt is cracked
   - Belt is worn out to the extent that the cords are exposed
   - Belt has chunks missing from the ribs
3. Check that the belt fits properly in the ribbed grooves.

➡Check with your hand to confirm that the belt has not slipped out of the grooves on the bottom of the pulley. If it has slipped out, replace the fan and generator V belt. Install a new fan and generator V belt correctly.

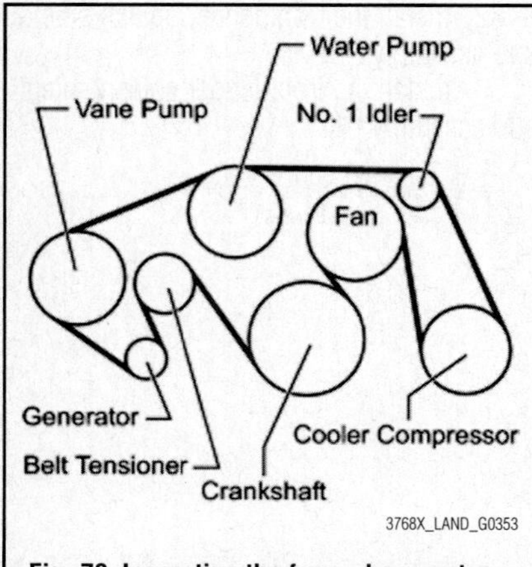

Fig. 79 Inspecting the fan and generator belt

4. Check that nothing gets caught in the tensioner by turning it clockwise and counterclockwise. If a malfunction exists, replace the tensioner.

REMOVAL & INSTALLATION

See Figure 80.

1. Remove the front fender splash shield sub assembly LH and RH.
2. Remove the No. 1 engine under cover sub assembly.
3. Remove the v-bank cover sub assembly.
4. Remove the fan and generator v belt.
   a. While turning the belt tensioner counterclockwise, align the service hole for the belt tensioner and the belt tensioner fixing position, and then insert a bar of 5 mm (0.197 in.) into the service hole to fix the belt tensioner in place.

➡The pulley bolt for the belt tensioner has a left hand thread.

   b. Remove the v-belt.

To install:

5. Install the fan and generator v-belt.
   a. Set the v-belt onto every part.
   b. While turning the belt tensioner counterclockwise, remove the bar.

➡Make sure that the v-belt is properly set to each pulley.

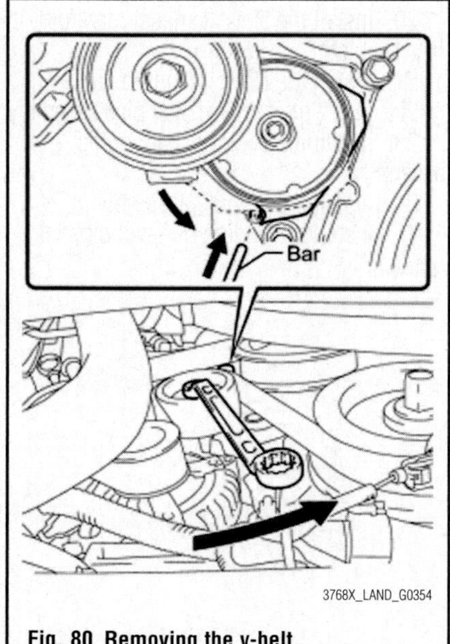

Fig. 80 Removing the v-belt

   c. After installing the belt, check that it fits properly in the ribbed grooves.

➡Make sure to check by hand that the belt has not slipped out of the grooves on the bottom of the pulley.

6. Install the v-bank cover sub assembly.
7. Install the No. 1 engine under cover sub assembly.
8. Install the front fender splash shield sub assembly LH and RH.

CAMSHAFT AND VALVE LIFTERS

REMOVAL & INSTALLATION

See Figures 81 through 100.

1. Discharge the fuel system pressure.
2. Disconnect the cable from the negative battery terminal.

✱✱ CAUTION

Wait at least 90 seconds after disconnecting the cable from the negative (-) battery terminal to disable to SRS system.

3. Remove the upper radiator support seal.
4. Remove the radiator grille.
5. Remove the front bumper cover.
6. Remove the transmission oil cooler air duct (with air cooled transmission oil cooler).
7. Disconnect the radiator side deflector (without air cooled transmission oil cooler).
8. Disconnect the radiator side deflector LH.
9. Remove the front fender main seal LH.
10. Remove the front fender main seal RH.
11. Remove the front wiper arm LH.
12. Remove the front wiper arm RH.
13. Remove the hood to cowl top seal.
14. Remove the cowl top ventilator louver sub assembly.
15. Remove the front fender splash shield sub assembly LH.
16. Remove the front fender splash shield sub assembly RH.
17. Remove the No. 1 engine under cover sub assembly.
18. Remove the No. 2 engine under cover.
19. Remove the front fender apron seal LH.

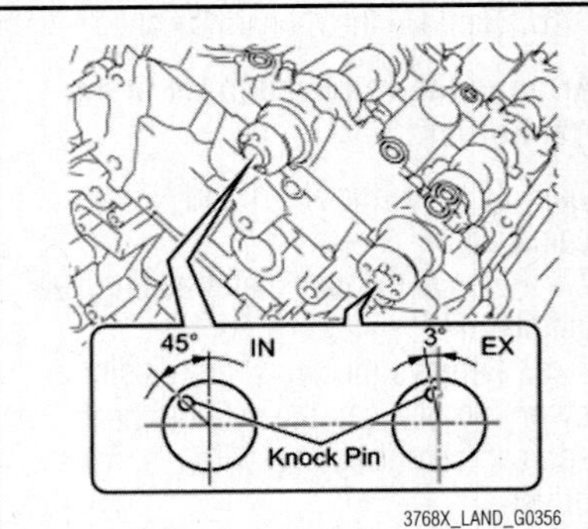

Fig. 81 Positioning the knock pin of the camshaft

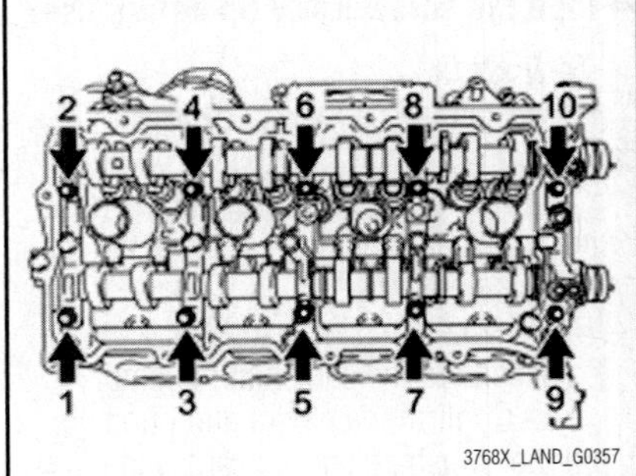

Fig. 82 Identifying the bearing cap bolt removal sequence (1 of 2)

20. Remove the front fender apron seal front RH.
21. Drain the engine oil.
22. Drain the engine coolant.
23. Remove the v-bank cover sub assembly.
24. Remove the air cleaner hose assembly.
25. Remove the air cleaner assembly.
26. Remove the No. 1 radiator hose.
27. Remove the No. 2 radiator hose.
28. Remove the fan shroud.
29. Remove the radiator assembly.
30. Disconnect the engine wire.
31. Disconnect the air pump hose and wire harness.
32. Disconnect the No. 2 water by pass pipe.
33. Disconnect the cooler compressor assembly.
34. Disconnect the No. 2 fuel tube sub assembly.
35. Remove the oil filter element.
36. Remove the engine oil level dipstick guide.
37. Remove the oil pressure sender gauge assembly.
38. Remove the No. 2 water by pass pipe sub assembly (with oil cooler).
39. Remove the No. 1 oil cooler bracket.

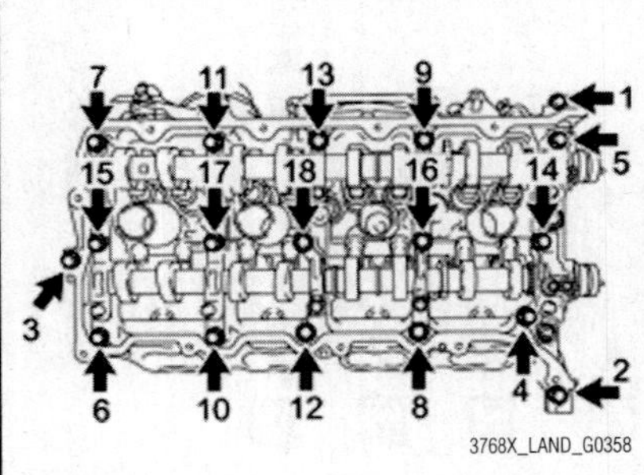

Fig. 83 Identifying the bearing cap bolt removal sequence (2 of 2)

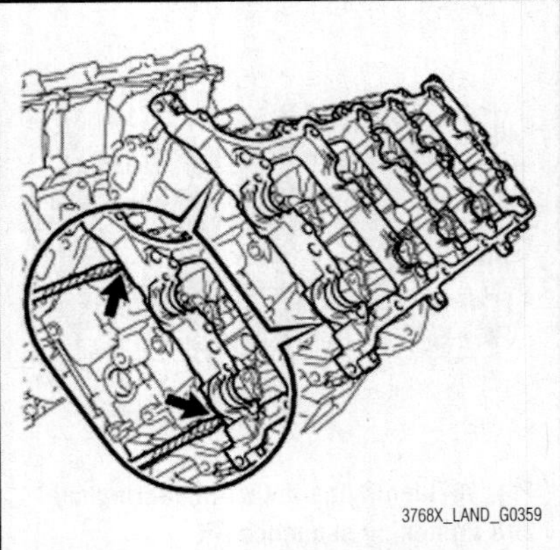

Fig. 84 Removing the camshaft housing sub assembly LH

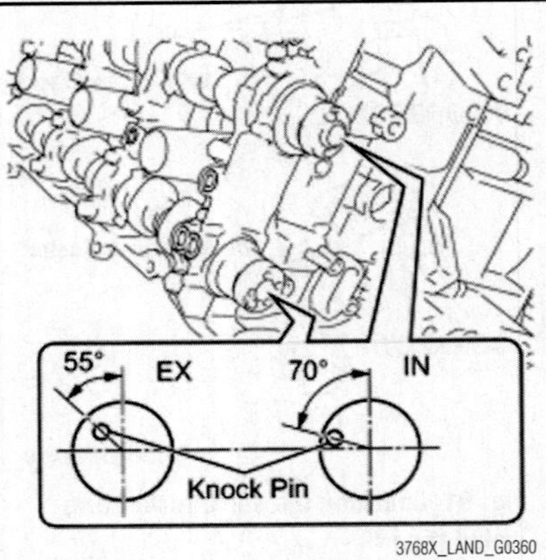

Fig. 85 Positioning the knock pin of the camshaft RH

40. Remove the intake manifold.
41. Disconnect the vane pump assembly.
42. Disconnect the oil cooler pipe assembly.
43. Remove the generator assembly.
44. Remove the No. 1 water by pass hose.
45. Remove the water by pass pipe sub assembly.
46. Remove the front water by pass joint.
47. Remove the No. 2 engine cover.
48. Remove the No. 1 engine cover.
49. Remove the air tube sub assembly.
50. Remove the water inlet housing.
51. Remove the water pump pulley.
52. Remove the No. 1 idler pulley sub assembly.
53. Remove the fluid coupling bracket.
54. Remove the v-ribbed belt tensioner assembly.
55. Remove the ignition coil assembly.
56. Remove the cylinder head cover sub assembly LH and RH.
57. Remove the spark plug tube gasket.
58. Remove the crankshaft pulley.
59. Disconnect the wire harness clamp bracket.
60. Remove the timing chain cover sub assembly.
61. Remove the water inlet pipe.
62. Set the No. 1 cylinder to TDC/compression.
63. Remove the No. 1 chain tensioner assembly LH.
64. Remove the No. 1 chain tensioner slipper LH.
65. Remove the No. 1 chain vibration damper LH.
66. Remove the No. 1 chain sub assembly LH.
67. Remove the No. 3 chain tensioner assembly.
68. Remove the No. 1 chain tensioner assembly RH.
69. Remove the No. 1 chain tensioner slipper RH.
70. Remove the No. 1 chain vibration damper RH.
71. Remove the No. 1 chain sub assembly RH.
72. Remove the No. 2 chain tensioner assembly.
73. Remove the crankshaft timing gear key.
74. Remove the camshaft bearing cap LH.
    a. Make sure that the knock pin of the camshaft is positioned as shown.
    b. Uniformly loosen and remove the 10 bearing cap bolts in the sequence shown in the illustration.
    c. Uniformly loosen and remove the 18 bearing cap bolts in the sequence shown.

➡Uniformly loosen the bolts while keeping the camshaft level.

    d. Remove the 6 bearing caps.

➡Arrange the removed parts in the correct order.

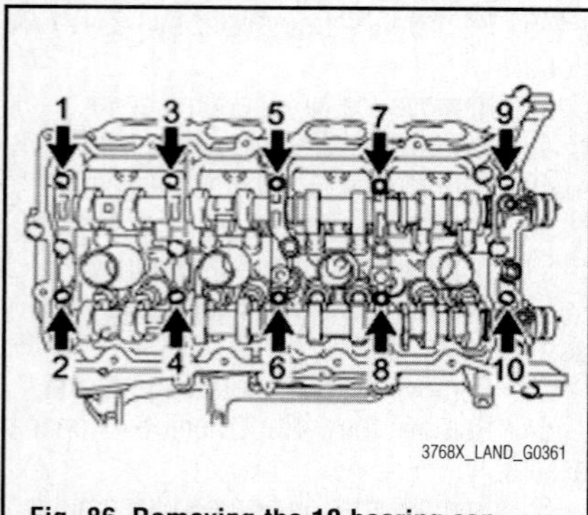

Fig. 86 Removing the 10 bearing cap bolts in sequence

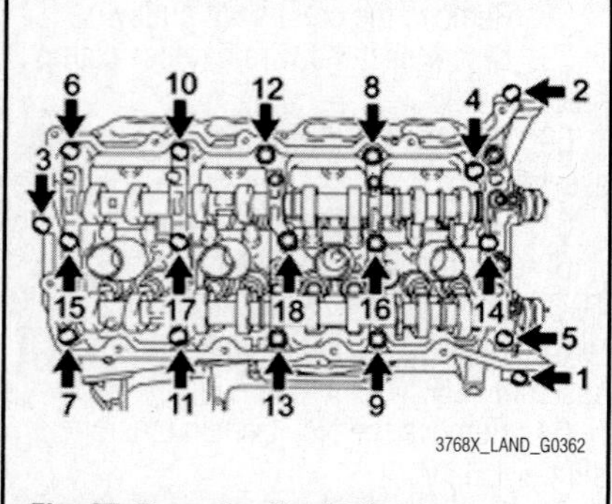

Fig. 87 Removing the 18 bearing cap bolts in sequence

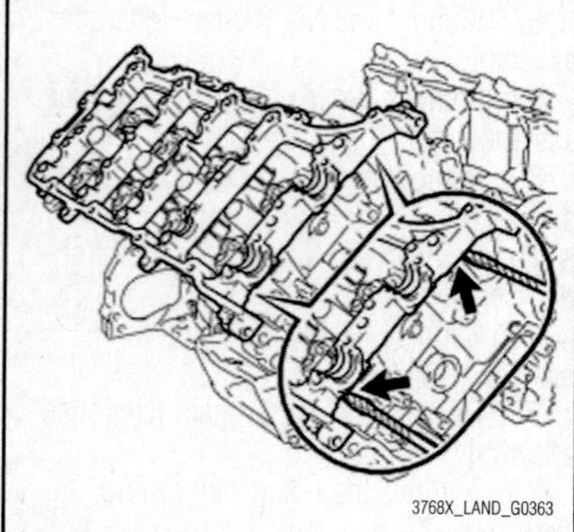

Fig. 88 Removing the camshaft housing sub assembly RH

e. Remove the No. 3 and No. 4 camshafts.

75. Remove the camshaft housing sub assembly LH.

a. Remove the camshaft housing by prying between the cylinder head and camshaft housing with a screwdriver.

➡**Be careful not to damage the contact surfaces of the cylinder head and camshaft housing.**

➡**Tape the screwdriver tip before use.**

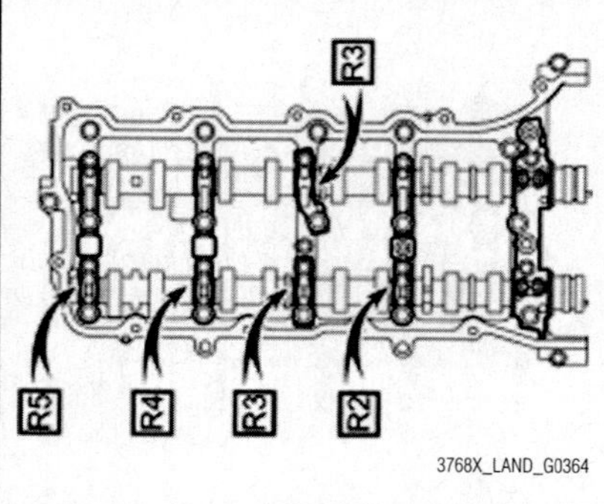

Fig. 89 Positioning the camshaft bearing caps

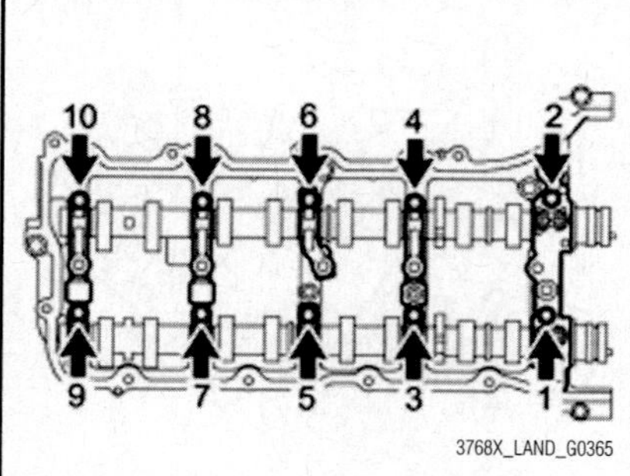

Fig. 90 Identifying the main bearing cap bolt tightening sequence RH

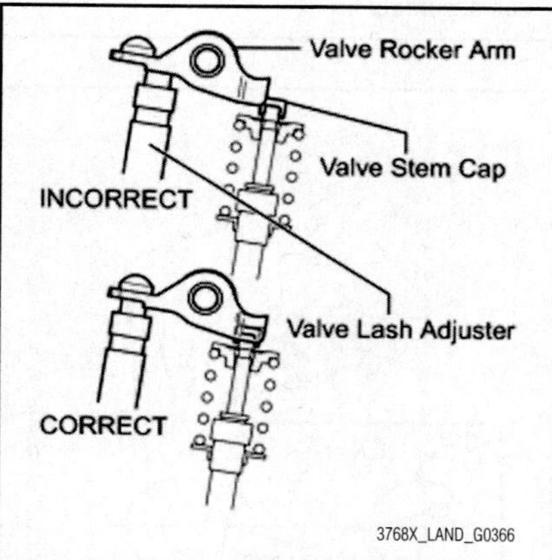

Fig. 91 Checking the valve rocker arm installation

76. Remove the camshaft bearing cap RH.

a. Make sure that the knock pin of the camshaft is positioned as shown.

b. Uniformly loosen and remove the 10 bearing cap bolts in the sequence shown.

c. Uniformly loosen and remove the 18 bearing cap bolts in the sequence shown.

➡**Uniformly loosen the bolts while keeping the camshaft level.**

d. Remove the 6 bearing caps.

➡**Arrange the removed parts in the correct order.**

e. Remove the No. 1 and No. 2 camshafts.

77. Remove the camshaft housing sub assembly RH.

a. Remove the camshaft housing by prying between the cylinder head and camshaft housing with a screwdriver.

➡**Be careful not to damage the contact surfaces of the cylinder head and camshaft housing.**

➡**Tape the screwdriver tip before use.**

To install:

78. Install the camshaft bearing cap RH.

a. Apply a light coat of engine oil to the camshaft journals, camshaft housings and bearing caps.

b. Install the No. 1 and No. 2 camshafts to the camshaft housing.

c. Confirm the marks and numbers on the camshaft bearing caps and place them in their proper positions and directions.

d. Temporarily install the 10 bolts in the order shown.

79. Install the camshaft housing sub assembly RH.

a. Make sure that the valve rocker arms are installed as shown.

b. Apply seal packing in a continuous line as shown with a seal diameter or 0.138-0.157 inch (3.5-4.0 mm).

➡**Remove any oil from the contact surface.**

➡**Install the camshaft housing within 3 minutes and tighten the bolts within 15 minutes after applying seal packing.**

c. Install the camshaft housing, and install the 18 bolts in the order shown. Tighten bolt A to 7 ft. lbs. (10 Nm). Tighten all bolts except A to 22 ft. lbs. (30 Nm).

➡**Do not start the engine for at least 2 hours after the installation.**

➡**Make sure that the knock pin of the camshaft is positioned as shown in the illustration before installing the camshaft housing.**

d. Tighten the 10 bolts in the order shown. Tighten to 12 ft. lbs. (16 Nm).

➡**Thoroughly wipe clean any seal packing.**

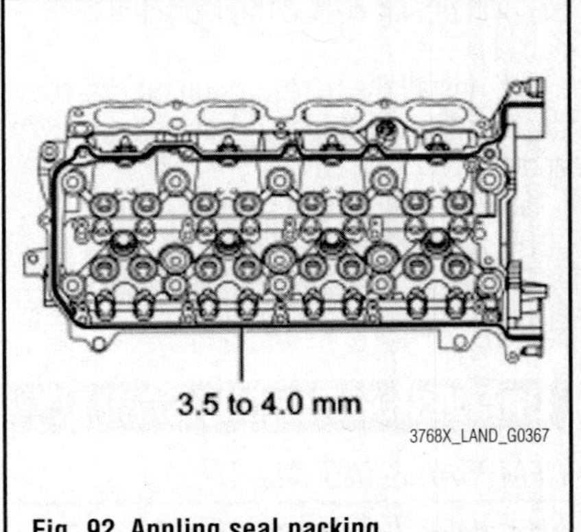

Fig. 92 Appling seal packing

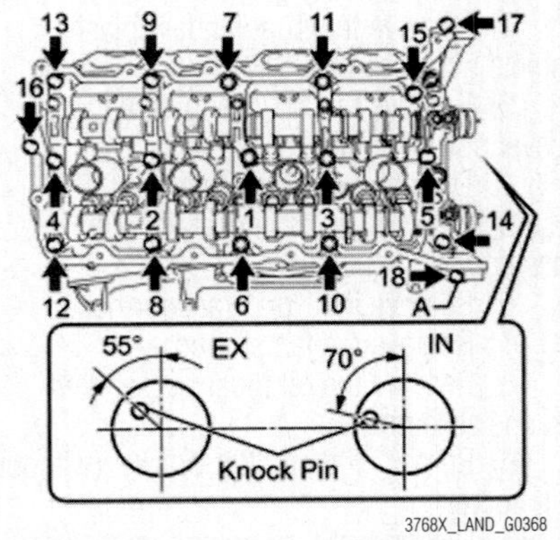

Fig. 93 Identifying the camshaft housing bolt tightening sequence RH

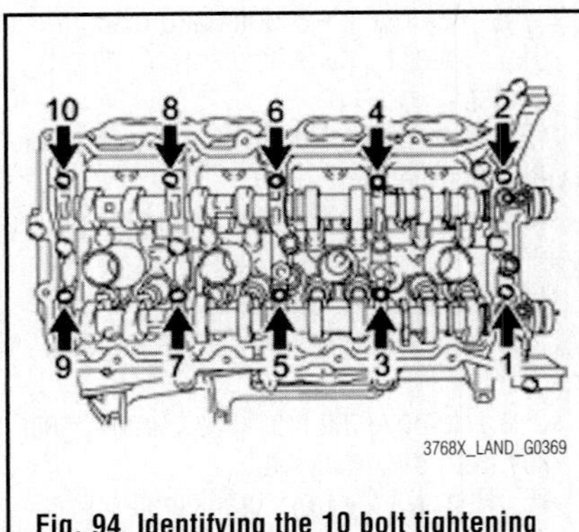

Fig. 94 Identifying the 10 bolt tightening sequence RH

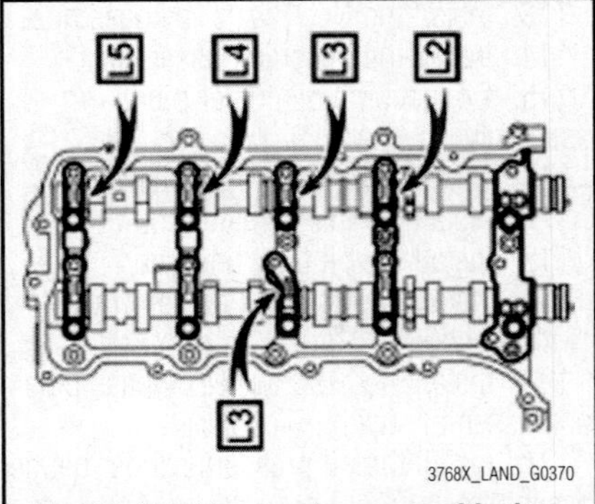

Fig. 95 Checking bearing cap positioning RH

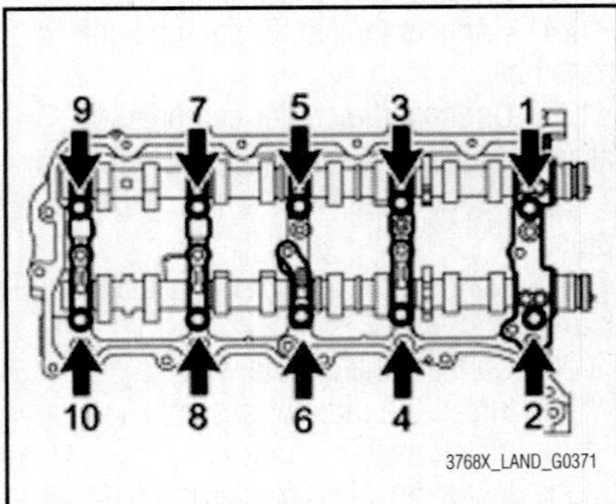

Fig. 96 Identifying the main bearing cap bolt tightening sequence LH

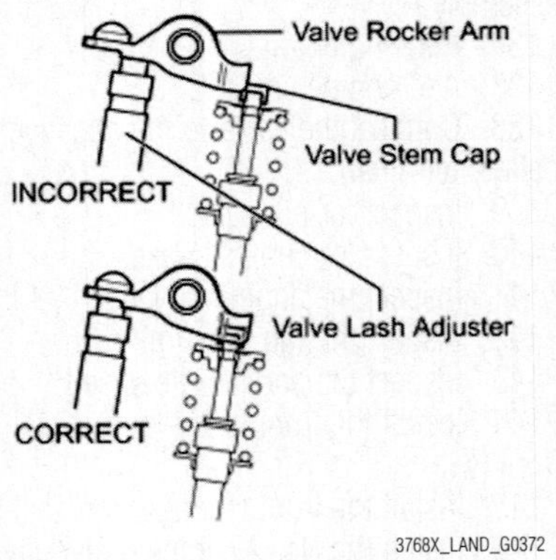

Fig. 97 Checking the valve rocker arm installation LH

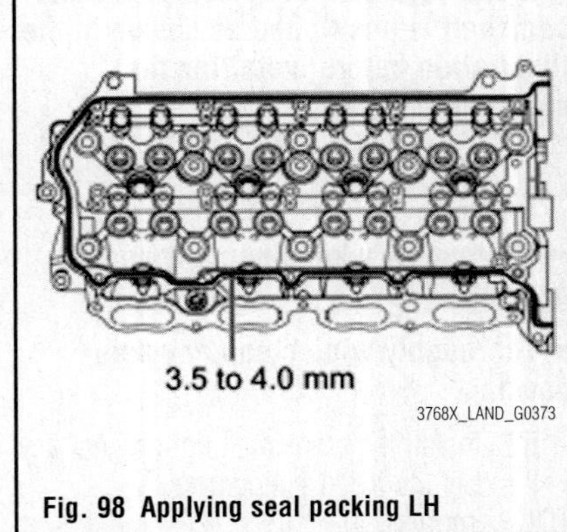

Fig. 98 Applying seal packing LH

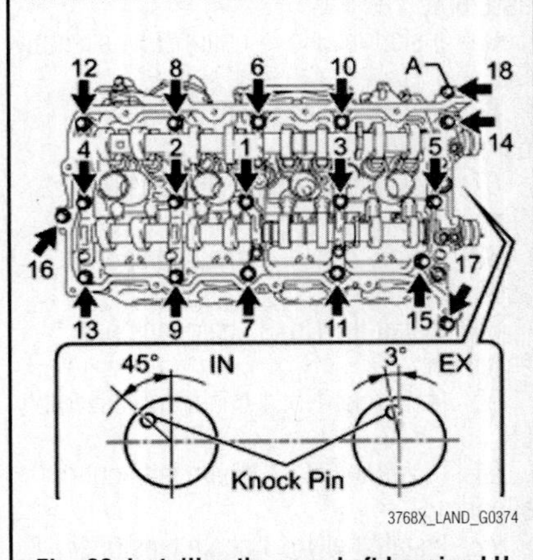

Fig. 99 Installing the camshaft housing LH

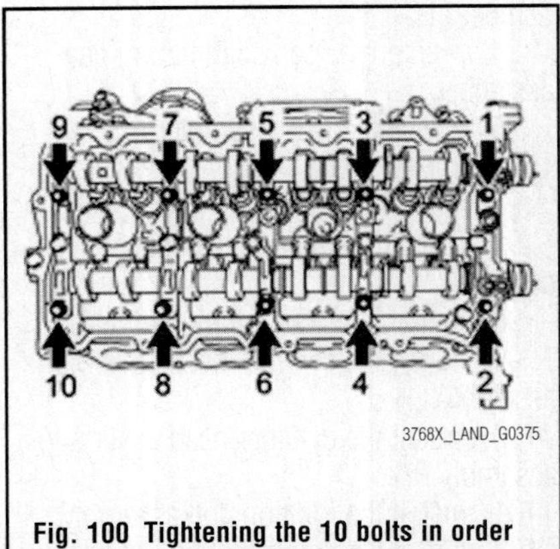

Fig. 100 Tightening the 10 bolts in order

80. Install the camshaft bearing cap LH.

➡Apply a light coat of engine oil to the camshaft journals, camshaft housings and bearing caps.

➡Install the No. 3 and No. 4 camshafts to the camshaft housing.

a. Confirm the marks and numbers on the camshaft bearing caps and place them in their proper positions and directions.

b. Temporarily install the 10 bolts in the order shown.

81. Install the camshaft housing sub assembly LH.

a. Make sure that the valve rocker arms are installed as shown.

b. Apply seal packing in a continuous line with a diameter of 0.138-0.157 inch (3.5-4.0 mm).

➡Remove any oil from the contact surface.

➡Install the camshaft housing within 3 minutes and tighten the bolts within 15 minutes after applying seal packing.

c. Install the camshaft housing, and install the 18 bolts in the order shown. Tighten bolt A to 7 ft. lbs. (10 Nm). Tighten all other bolts to 22 ft. lbs. (30 Nm).

➡Do not start the engine for at least 2 hours after the installation.

➡**Make sure that the knock pin of the camshaft is positioned as shown in the illustration before installing the camshaft housing.**

d. Tighten the 10 bolts in the order shown to 12 ft. lbs. (16 Nm).

➡**Thoroughly wipe clean any seal packing.**

➡**Thoroughly wipe clean any seal packing.**

82. Install the crankshaft timing gear key.
83. Set the No. 1 cylinder to TDC/compression.
84. Install the No. 2 chain tensioner assembly.
85. Install the No. 1 chain sub assembly RH.
86. Install the No. 1 chain vibration damper RH.
87. Install the No. 1 chain tensioner slipper RH.
88. Install the No. 1 chain tensioner assembly RH.
89. Install the No. 3 chain tensioner assembly.
90. Install the No. 1 chain sub assembly LH.
91. Install the No. 1 chain tensioner slipper LH.
92. Install the No. 1 chain tensioner assembly LH.
93. Install the No. 1 chain vibration damper LH.
94. Tighten the camshaft timing gear assembly.
95. Check the No. 1 cylinder to TCD/compression.
96. Install the water inlet pipe.
97. Install the timing chain cover sub assembly.
98. Install the spark plug tube gasket.
99. Install the cylinder head cover sub assembly LH.
100. Install the cylinder head cover sub assembly RH.
101. Install the ignition coil assembly.
102. Install the crankshaft timing gear key.
103. Install the crankshaft pulley.
104. Connect the wire harness clamp bracket.
105. Install the No. 1 idler pulley sub assembly.
106. Install the water pump pulley.
107. Install the water inlet housing.
108. Install the air tube sub assembly.
109. Install the No. 1 engine cover.
110. Install the No. 2 engine cover.
111. Install the front water by pass joint.
112. Install the water by pass pipe sub assembly.
113. Install the No. 1 water by pass hose.
114. Install the generator assembly.
115. Connect the oil cooler pipe assembly.
116. Connect the vane pump assembly.
117. Install the intake manifold.
118. Install the oil filter bracket.
119. Install the No. 1 oil cooler bracket (with oil cooler).
120. Install the No. 2 water by pass pipe sub assembly (with oil cooler).
121. Install the oil pressure sender gauge assembly.
122. Install the engine oil level dipstick guide.
123. Install the oil filter element.
124. Connect the No. 2 fuel tube sub assembly.
125. Connect the cooler compressor assembly.
126. Connect the No. 2 water by pass pipe.
127. Connect the air pump hose and wire harness.
128. Connect the engine wire.
129. Install the radiator assembly.
130. Install the fan shroud.
131. Install the No. 2 radiator hose.
132. Install the No. 1 radiator hose.
133. Install the air cleaner assembly.
134. Install the air cleaner hose assembly.
135. Install the v-bank cover sub assembly.
136. Add engine oil.
137. Add engine coolant.
138. Connect the cable to the negative battery terminal.
139. Inspect for oil leaks.
140. Inspect for coolant leaks.
141. Inspect the engine oil level.
142. Inspect the ignition timing.
143. Inspect the engine idle speed.
144. Install the front fender apron seal front RH.
145. Install the front fender apron seal LH.
146. Install the No. 2 engine under cover.
147. Install the No. 1 engine under cover sub assembly.
148. Install the front fender splash shield sub assembly LH.
149. Install the front fender splash shield sub assembly RH.
150. Install the cowl top ventilator louver sub assembly.
151. Install the hood to cowl top seal.
152. Install the front fender main seal LH.
153. Install the front fender main seal RH.
154. Install the front wiper arm LH and RH.
155. Connect the radiator side deflector LH.
156. Connect the radiator side deflector RH (without air cooled transmission oil cooler).
157. Install the transmission oil cooler air duct (with air cooled transmission oil cooler).
158. Install the front bumper cover.
159. Install the radiator grille.
160. Install the upper radiator support seal.

CRANKSHAFT FRONT SEAL

REMOVAL & INSTALLATION

See Figures 101 through 103.

1. Remove the front fender splash shield sub assembly LH.
2. Remove the front fender splash shield sub assembly RH.
3. Remove the No. 1 engine under cover sub assembly.
4. Drain the engine coolant.
5. Remove the v-bank cover sub assembly.
6. Remove the No. 1 radiator hose.
7. Remove the fan shroud.
8. Remove the oil pressure sender gauge assembly.
9. Remove the oil filter bracket (without oil cooler).
10. Disconnect the cooler compressor assembly (with the oil cooler).
11. Disconnect the No. 2 water by pass pipe sub assembly (with oil cooler).
    a. Remove the 3 bolts and disconnect the 2 water by pass hoses from the oil cooler.
12. Remove the No. 1 oil cooler bracket (with oil cooler).
13. Remove the oil filter bracket (with oil cooler).
14. Remove the crankshaft pulley.
15. Remove the crankshaft timing gear key.
    a. Remove the crankshaft timing gear key from the crankshaft.
16. Remove the front crankshaft oil seal.
    a. Using a screwdriver, pry out the oil seal.

➡**Do not damage the surface of the oil seal press fit hole and crankshaft.**

➡**Tape the screwdriver tip before use.**

To install:

17. Install the front crankshaft oil seal.
    a. Apply MP grease to the lip of a new oil seal.
    b. Using the special tools (09223-22010 or 09506-35010) and a hammer, tap in the oil seal to a depth between 0 and 0.0394 inch (0 and 1 mm) from the timing chain cover edge.

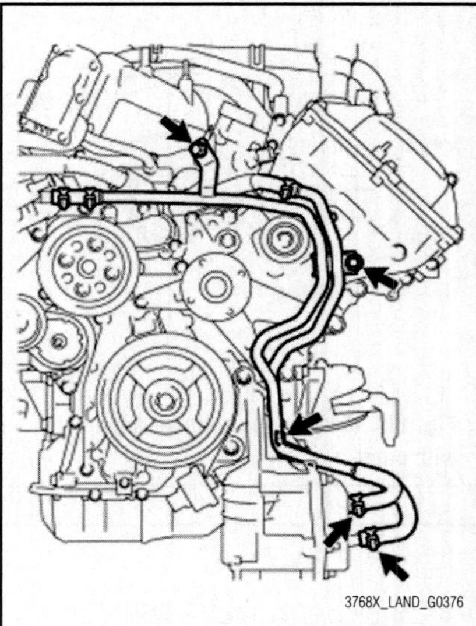

Fig. 101 Disconnecting the 2 water by pass hoses from the oil cooler

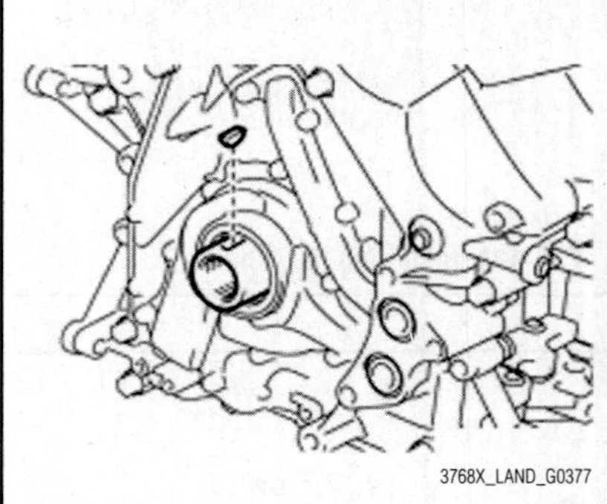

Fig. 102 Removing the crankshaft timing gear key

Fig. 103 Removing the front crankshaft oil seal

➡Keep the lip free from foreign matter.

➡Do not tap the oil seal at an angle.

18. Install the crankshaft timing gear key.
19. Install the crankshaft pulley.
20. Install the oil filter bracket (with oil cooler).
21. Install the No. 1 oil cooler bracket (with oil cooler).Install the No. 1 oil cooler bracket (with oil cooler).
22. Connect the No. 2 water by pass pipe sub assembly (with oil cooler).Connect the No. 2 water by pass pipe sub assembly (with oil cooler).
    a. Connect the 2 water by pass hoses to the oil cooler.
    b. Install the 3 by pass pipe bolts and tighten to 7 ft. lbs. (10 Nm).
23. Connect the cooler compressor assembly (with oil cooler).
24. Install the oil filter bracket (without oil cooler).
25. Install the oil pressure sender gauge assembly.
26. Install the fan shroud.
27. Install the No. 1 radiator hose.
28. Add engine coolant.
29. Inspect for coolant leaks.
30. Inspect for oil leaks.
31. Inspect the engine oil level.
32. Install the v-bank cover sub assembly.
33. Install the No. 1 engine under cover sub assembly.
34. Install the front fender splash shield sub assembly LH and RH.

CYLINDER HEAD

REMOVAL & INSTALLATION

See Figures 104 through 114.

1. Discharge the fuel system pressure.
2. Disconnect the cable from the negative battery terminal.

CAUTION

Wait at least 90 seconds after disconnecting the cable from the negative (-) battery terminal to disable the SRS system.

3. Remove the exhaust manifold sub assembly.
4. Remove the camshaft.
5. Remove the No. 1 valve rocker arm sub assembly.
6. Remove the valve lash adjuster assembly.
7. Remove the valve stem cap.
8. Remove the cylinder head sub assembly LH.
    a. Uniformly loosen and remove the 2 bolts in the sequence shown.
    b. Using a 10 mm bi-hexagon wrench, uniformly loosen the 10 cylinder head bolts in the sequence shown. Remove the 10linder head bolts and plate washers.

➡Be careful not to drop washers into the cylinder head.

➡Head warpage or cracking could result from removing bolts in an incorrect order.

➡Be sure to arrange the removed parts for each installation position separately.

    c. Remove the cylinder head and gasket.

9. Remove the cylinder head sub assembly RH.
    a. Uniformly loosen and remove the 2 bolts in the sequence shown.
    b. Using a 10 mm bi-hexagon wrench, uniformly loosen the 10 cylinder head bolts in the sequence shown in the illustration. Remove the 10 cylinder head bolts and plate washers.

➡Be careful not to drop washers into the cylinder head.

➡Head warpage or cracking could result from removing bolts in an incorrect order.

➡Be sure to arrange the removed parts for each installation position separately.

    c. Removing the cylinder head and gasket.

To install:

10. Inspect the cylinder head set bolt.
11. Inspect the cylinder head sub assembly.
12. Install the cylinder head sub assembly RH.
    a. Check the piston protrusions for each cylinder. Clean the cylinder block with solvent. Set the piston of the cylinder to be measured to slightly ATDS.
    b. Place the cylinder head gasket on the cylinder block surface with the front face of the Lot No. stamp upward.

➡Be careful of the installation direction.

➡Make sure that no oil is on the front end (indicated by the arrows) of the cylinder head gasket.

    c. Place the cylinder head on the cylinder block.

➡Ensure that no oil is on the mounting surface of the cylinder head.

➡Gently place the cylinder head in order not to damage the gasket with the bottom part of the head.

➡The cylinder head bolts are tightened in 3 progressive steps.

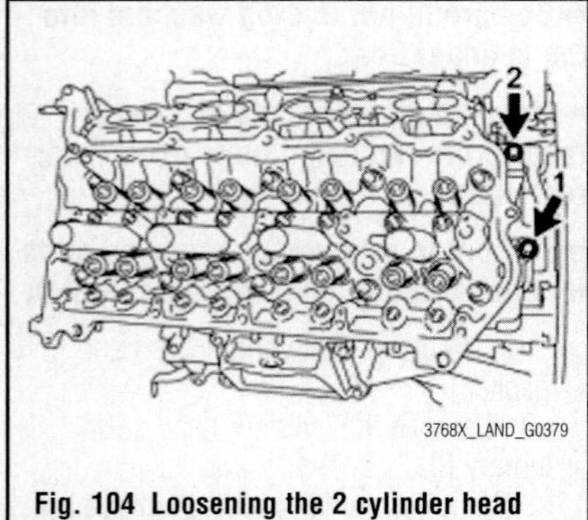

Fig. 104 Loosening the 2 cylinder head sub assembly bolts

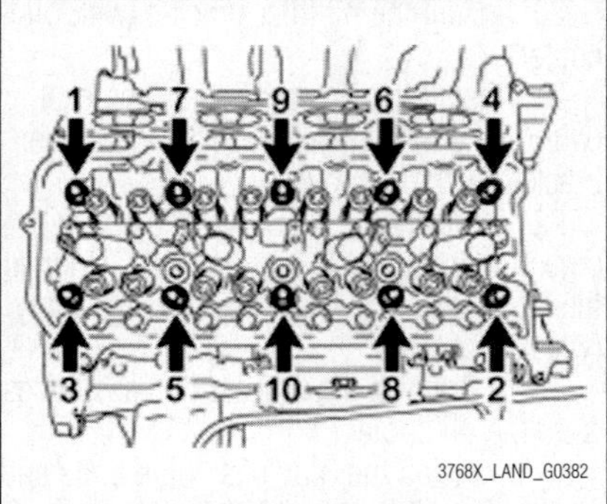

Fig. 107 Identifying the cylinder head bolt loosening sequence RH

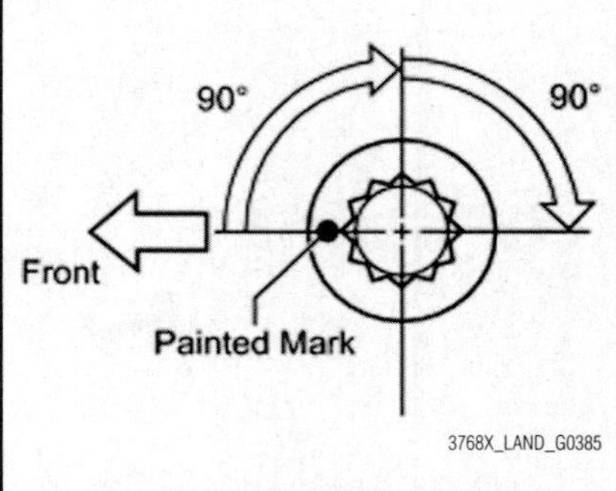

Fig. 110 Marking the cylinder head bolt with paint

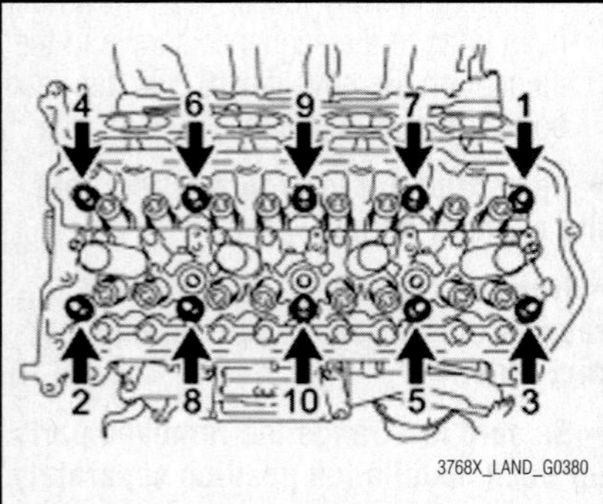

Fig. 105 Identifying the cylinder head bolt loosening sequence LH

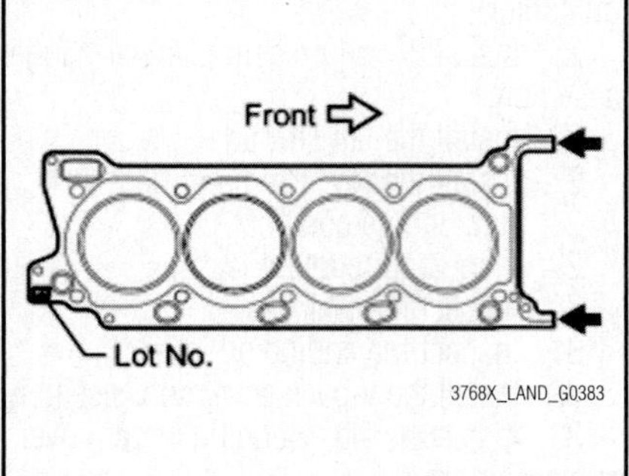

Fig. 108 Placing the cylinder head gasket on the cylinder block surface

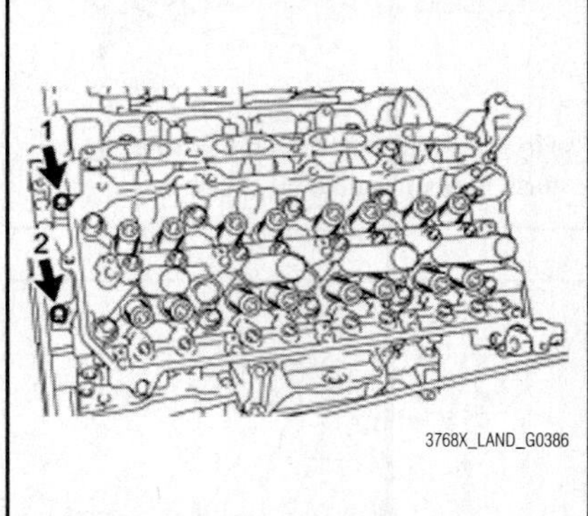

Fig. 111 Installing the 2 bolts

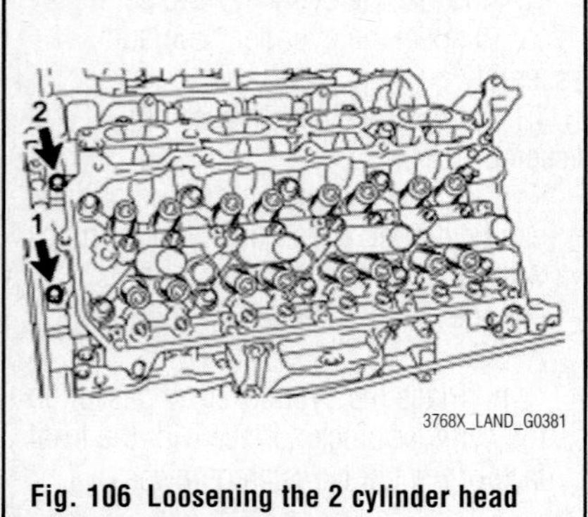

Fig. 106 Loosening the 2 cylinder head sub assembly bolts RH

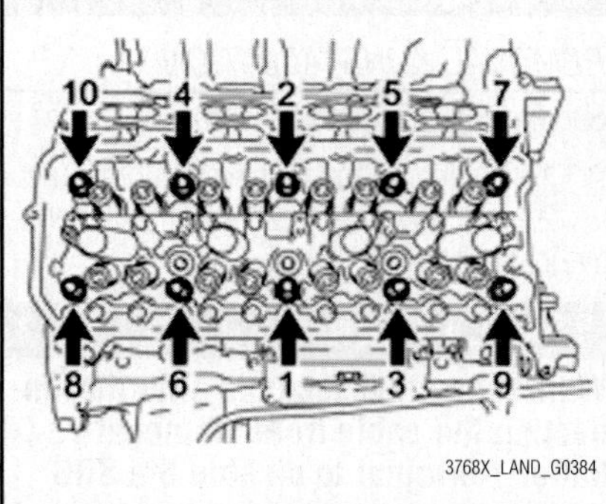

Fig. 109 Installing the RH cylinder head bolts in sequence

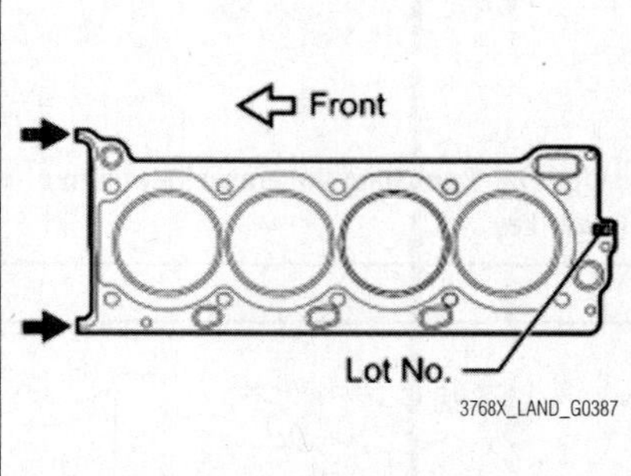

Fig. 112 Placing the cylinder head gasket on the cylinder block LH

d. Apply a light coat of engine oil to the threads and under the heads of the cylinder head bolts.

e. Step 1: Using a 10 mm bi-hexagon wrench, install and uniformly tighten the 10 cylinder head bolts with the plate washers in several steps in the sequence shown. Tighten to 27 ft. lbs. (36 Nm).

f. Step 2: Mark each cylinder head bolt with paint. Tighten the cylinder head bolts 90° in the same tightening sequence as the previous step.

g. Step 3: Tighten the cylinder head bolts another 90° in the sequence shown in step 1. Check that the paining marks are facing rearward.

h. Uniformly install the 2 bolts in the sequence shown.

13. Install the cylinder head sub assembly LH.

a. Check the piston protrusions for each cylinder.

b. Clean the cylinder block with solvent.

c. Set the piston of the cylinder to be measured to slightly ATDC.

d. Place the cylinder head gasket on the cylinder block surface with the front face of the Lot No. stamp upward.

➡Be careful of the installation direction.

➡Make sure that no oil is on the front end (indicated by the arrows) of the cylinder head gasket.

e. Place the cylinder head on the cylinder block.

➡Ensure that no oil is on the mounting surface of the cylinder head.

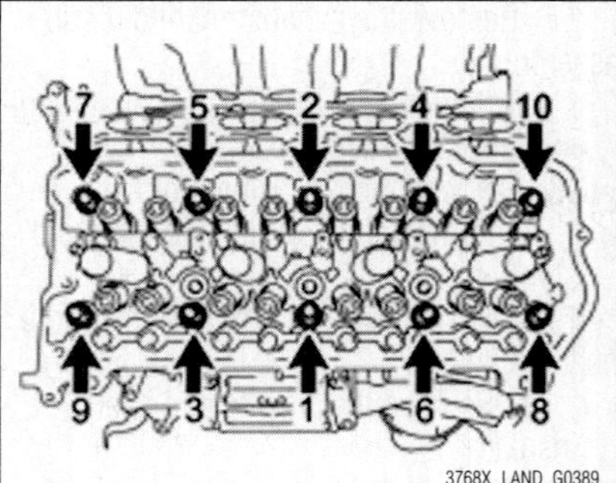

Fig. 113 Installing the LH cylinder head bolts in sequence

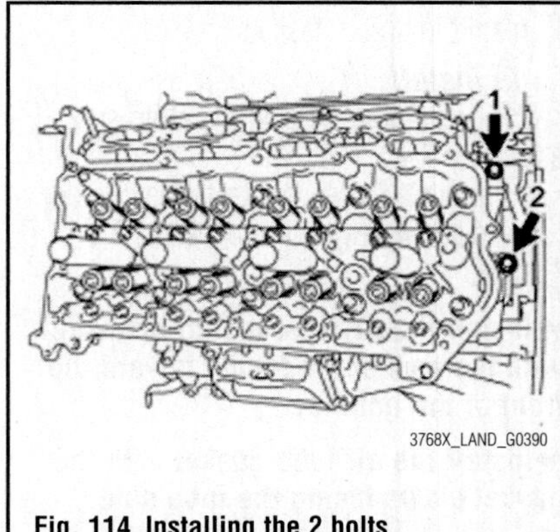

Fig. 114 Installing the 2 bolts

➡Gently place the cylinder head in order not to damage the gasket with the bottom part of the head.

➡The cylinder head bolts are tightened in 3 progressive steps.

f. Apply a light coat of engine oil to the threads and under the heads of the cylinder head bolts.

g. Step 1: Using a 10 mm bi-hexagon wrench, install and uniformly tighten the 10 cylinder head bolts with the plate washers in several steps in the sequence shown. Tighten to 27 ft. lbs. (36 Nm).

h. Step 2: Mark each cylinder head bolt with paint. Tighten the cylinder head bolts 90° in the same tightening sequence as the previous step.

i. Step 3: Tighten the cylinder head bolts another 90° in the sequence shown in step 1. Check that the paining marks are facing rearward.

j. Uniformly install the 2 bolts in the sequence shown.

14. Install the valve stem cap.
15. Install the valve lash adjuster assembly.
16. Install the No. 1 valve rocker arm sub assembly.
17. Install the camshaft.
18. Install the exhaust manifold sub assembly.
19. Connect the cable to the negative battery terminal.
20. Inspect for exhaust gas leaks.
21. Inspect the ignition timing.
22. Inspect the engine idle speed.

EXHAUST MANIFOLD

REMOVAL & INSTALLATION

See Figures 115 through 133

1. Remove the front fender splash shield sub assembly LH.
2. Remove the front fender splash shield sub assembly RH.
3. Remove the No. 1 engine under cover sub assembly.
4. Remove the No. 2 engine under cover.
5. Remove the front fender apron seal front RH.
   a. Using a clip remover, remove the 3 clips and fender apron seal.
6. Remove the front fender apron seal rear RH.
   a. Using a clip remover, remove the 4 clips and the fender apron seal.
7. Remove the front fender apron seal LH.
   a. Using a clip remover, remove the 3 clips and fender apron seal.
8. Remove the front fender apron seal rear LH.
   a. Using a clip remover, remove the 4 clips and fender apron seal.
9. Remove the engine oil level dipstick guide.
10. Remove the tailpipe assembly.
11. Remove the center exhaust pipe assembly.
12. Remove the front No. 2 exhaust pipe assembly.
13. Remove the front exhaust pipe assembly.
14. Remove the propeller shaft heat insulator.
    a. Remove the 2 bolts and heat insulator.
15. Remove the No. 2 manifold stay.
    a. Remove the 2 bolts and manifold stay.
16. Remove the No. 2 exhaust manifold heat insulator.
    a. Remove the 3 bolts and heat insulator.

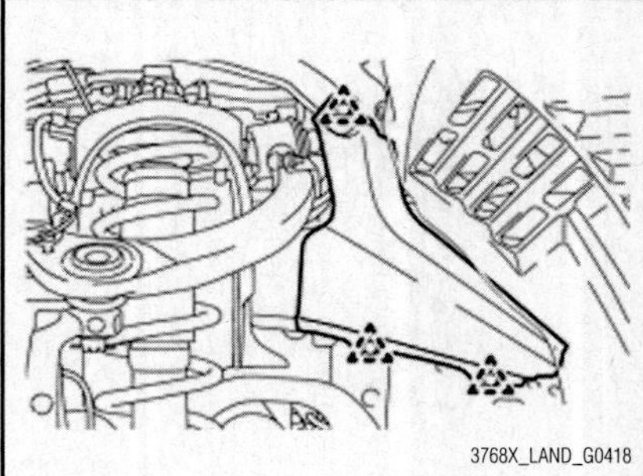

Fig. 115 Removing the front fender apron seal RH

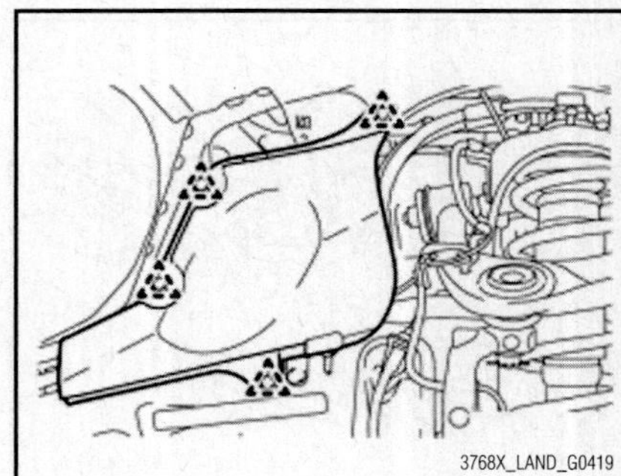

Fig. 116 Removing the front fender apron seal rear RH

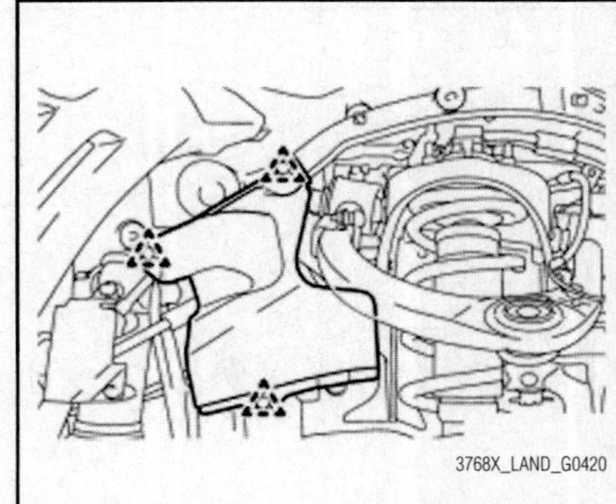

Fig. 117 Removing the front fender apron seal LH

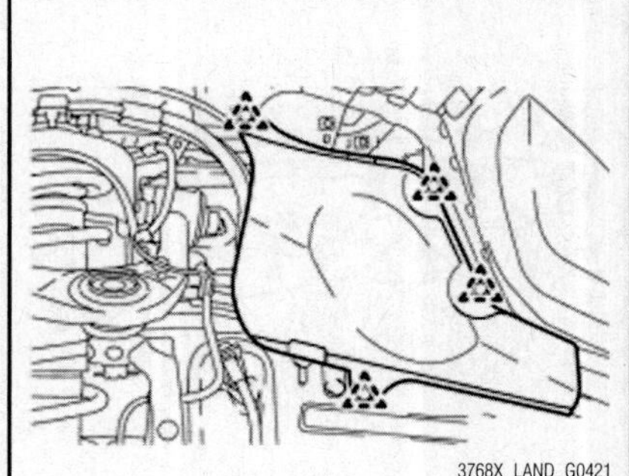

Fig. 118 Removing the front fender apron seal rear LH

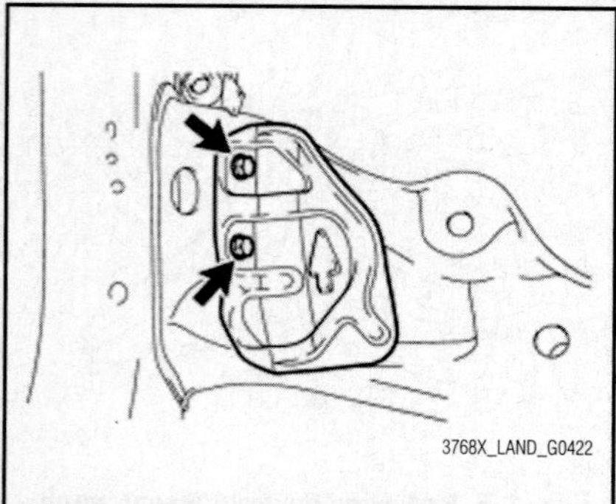

Fig. 119 Removing the propeller shaft heat insulator

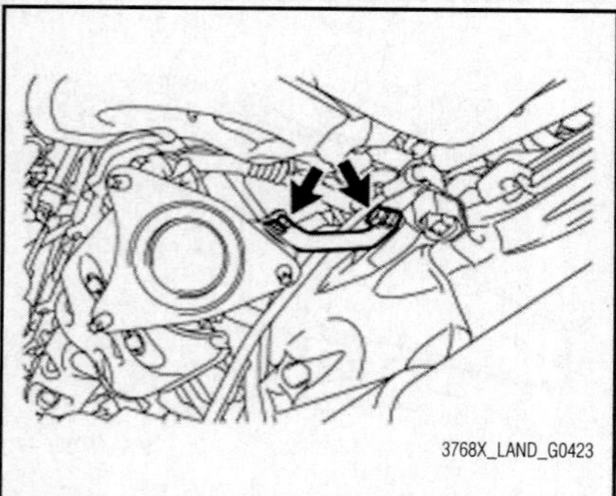

Fig. 120 Removing the No. 2 manifold stay

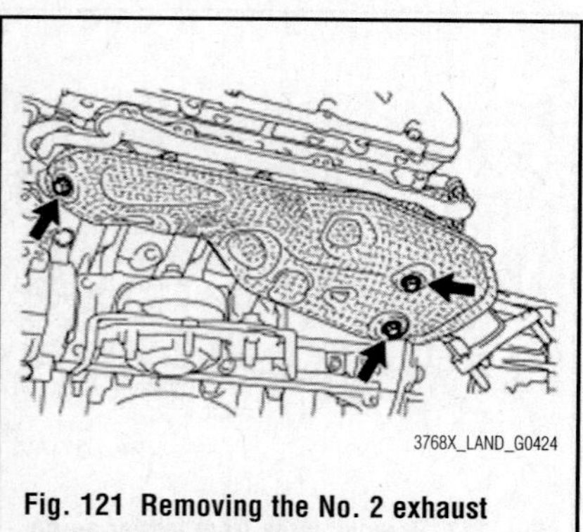

Fig. 121 Removing the No. 2 exhaust manifold heat insulator

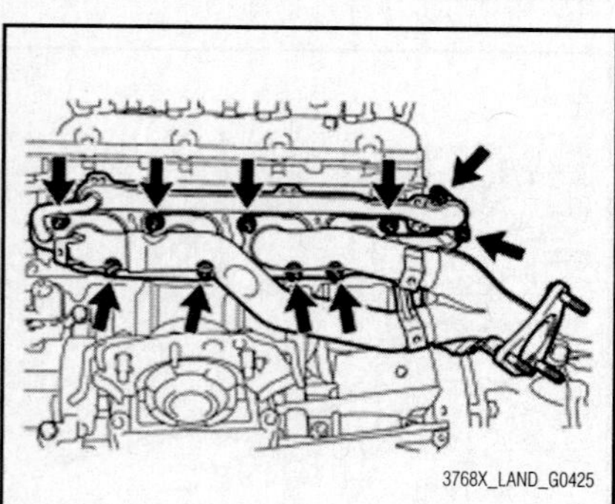

Fig. 122 Removing the exhaust manifold sub assembly LH

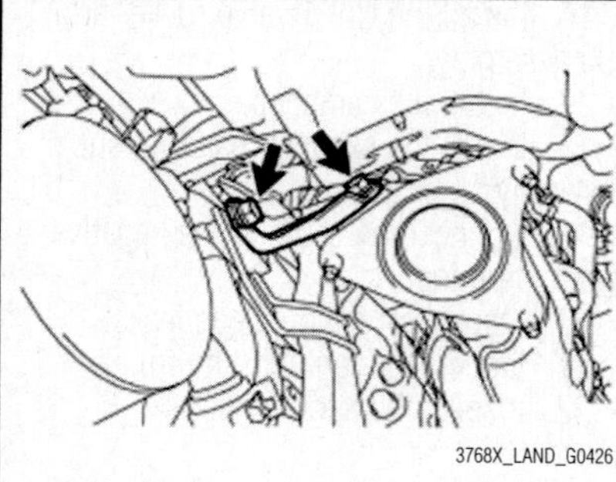

Fig. 123 Removing the No. 1 manifold stay

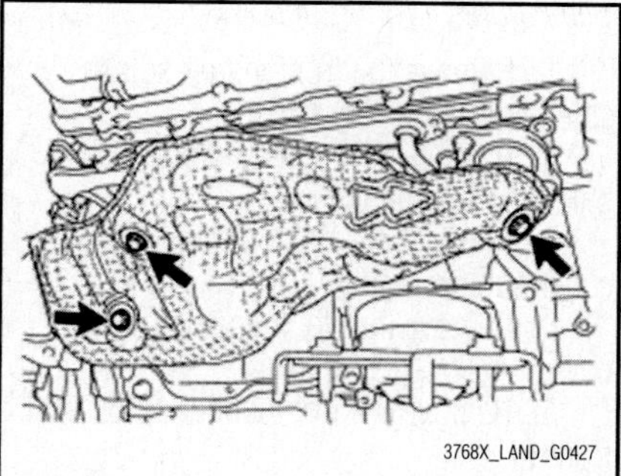

Fig. 124 Removing the No. 1 exhaust manifold heat insulator

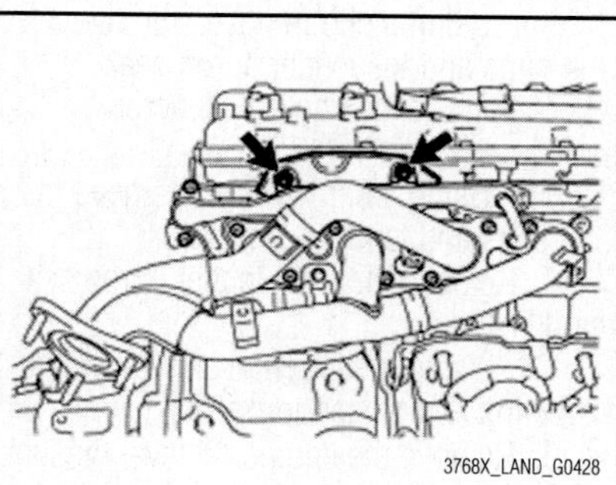

Fig. 125 Removing the 2 nuts and insulator from the exhaust manifold

Fig. 126 Removing the exhaust manifold

17. Remove the exhaust manifold sub assembly LH.
   a. Remove the 10 nuts, exhaust manifold and 2 gaskets.
18. Remove the No. 1 manifold stay.
   a. Remove the 2 bolts and manifold stay.
19. Remove the No. 1 exhaust manifold heat insulator.
   a. Remove the 3 bolts and heat insulator.
20. Remove the exhaust manifold sub assembly RH.
   a. Remove the 2 nuts and insulator from the exhaust manifold.
   b. Remove the 10 nuts, exhaust manifold and 2 gaskets.

To install:

21. Install the exhaust manifold sub assembly RH.
   a. Install a new gasket to the cylinder head and a new gasket to the No. 2 air tube.

➡Install the exhaust manifold gasket with the gasket tab facing toward the front of the engine.

➡Install the air tube gasket with the gasket claws facing the tube side.

   b. Temporarily install the exhaust manifold with the 2 nuts labeled A and 8 new nuts.
   c. Uniformly tighten the nuts that are not labeled A, and then tighten the 2 nuts labeled A. Tighten nut A to 7 ft. lbs. (10 Nm). Tighten all the other nuts to 15 ft. lbs. (21 Nm).
   d. Install the insulator to the exhaust manifold with the 2 nuts. Tighten to 7 ft. lbs. (10 Nm).
22. Install the No. 1 exhaust manifold heat insulator.
   a. Install the heat insulator with the 3 bolts and tighten to 7 ft. lbs. (10 Nm).
23. Install the No. 1 manifold stay.
   a. Temporarily install the manifold stay with the 2 bolts.
   b. Tighten the 2 bolts in the order shown to 30 ft. lbs. (40 Nm).
24. Install the exhaust manifold sub assembly LH.
   a. Install a new gasket to the cylinder head and a new gasket to the No. 3 air tube.

➡Install the exhaust manifold gasket with the gasket tab facing toward the rear of the engine.

➡Install the air tube gasket with the gasket claws facing the tube side.

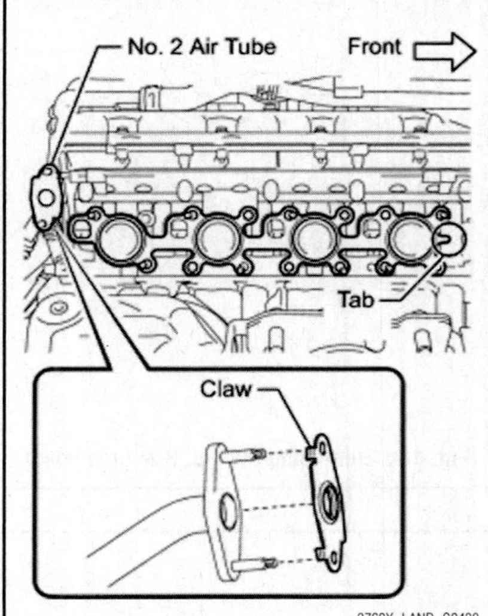

Fig. 127 Installing a new gasket to the cylinder head and No. 2 air tube

Fig. 128 Temporarily installing the exhaust manifold

b. Temporarily install the exhaust manifold with the 2 nuts labeled A and 8 new nuts.

c. Uniformly tighten the nuts that are not labeled A, and then tighten the 2 nuts labeled A. Tighten the A nuts to 7 ft. lbs. (10 Nm). Tighten all other nuts to 15 ft. lbs. (21 Nm).

25. Install the No. 2 exhaust manifold heat insulator.

a. Install the heat insulator with the 3 bolts to 7 ft. lbs. (10 Nm).

26. Install the No. 2 manifold stay.

a. Temporarily install the manifold stay with the 3 bolts.

b. Tighten the bolts in the order shown. Tighten to 30 ft. lbs. (40 Nm).

27. Install the propeller shaft heat insulator with the 2 bolts. Tighten to 12 ft. lbs. (16 Nm).

28. Install the front exhaust pipe assembly.

29. Install the front No. 2 exhaust pipe assembly.

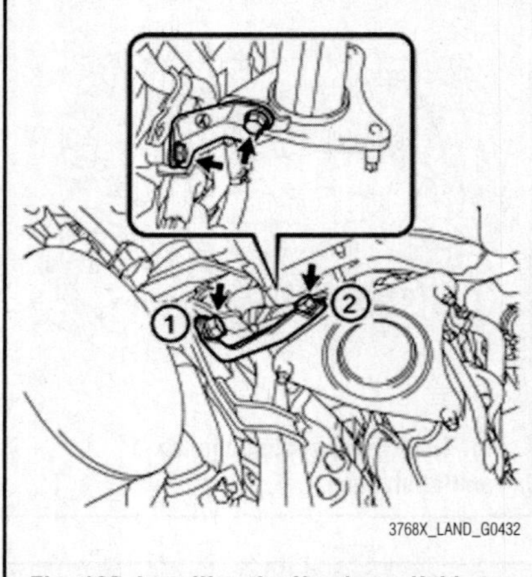

Fig. 129 Installing the No. 1 manifold stay

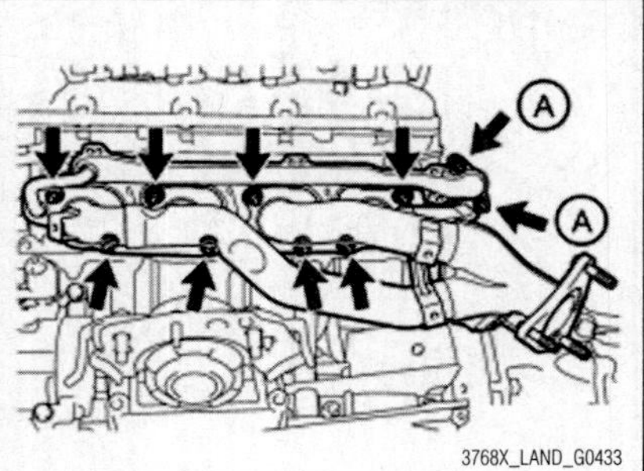

Fig. 130 Temporarily installing the exhaust manifold

30. Install the center exhaust pipe assembly.

31. Install the tailpipe assembly.

32. Install the engine oil level dipstick guide.

33. Inspect for exhaust gas leaks.

➡If gas is leaking, tighten the areas necessary to stop the leak. Replace any damaged parts as necessary.

34. Install the front fender apron seal LH.

a. Install the fender apron seal with the 3 clips.

35. Install the front fender apron seal rear LH with the 4 clips.

36. Install the front fender apron seal front RH with the 3 clips.

37. Install the front fender apron seal rear RH with the 4 clips.

38. Install the No. 1 engine under cover sub assembly.

39. Install the front fender splash shield sub assembly LH.

40. Install the front fender splash shield sub assembly RH.

41. Install the No. 2 engine under cover.

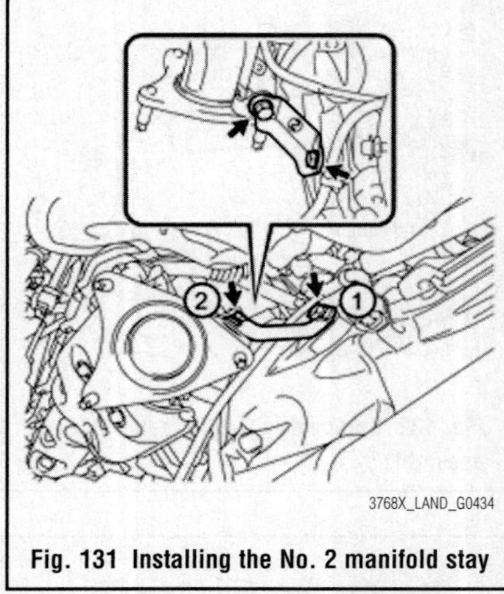

Fig. 131 Installing the No. 2 manifold stay

INTAKE MANIFOLD

REMOVAL & INSTALLATION

See Figures 132 through 148.

1. Disconnect the cable from the negative battery terminal.

2. Remove the cowl top ventilator louver sub assembly.

3. Remove the front fender splash shield sub assembly LH.

4. Remove the front fender splash shield sub assembly RH.

5. Remove the No. 1 engine under cover sub assembly.

6. Drain the engine coolant.

7. Remove the v-bank cover sub assembly.

8. Remove the air cleaner hose assembly.

a. Disconnect the vacuum hose and No. 2 ventilation hose.

b. Loosen the 2 hose clamps.

c. Remove the air cleaner hose.

9. Remove the intake manifold.

a. Disconnect the ventilation hose from the ventilation pipe of the cylinder head cover LH and RH.

b. Disconnect the 2 water by pass hoses.

c. Disconnect the throttle body connector.

d. Disconnect the No. 1 ventilation hose.

e. Disconnect the purge VSV connector.

f. Disconnect the purge line hose from the purge VSV.

g. Disconnect the vacuum switching valve connector (for ACIS).

h. Remove the No. 1 engine cover sub assembly.

i. Remove the No. 3 engine cover.

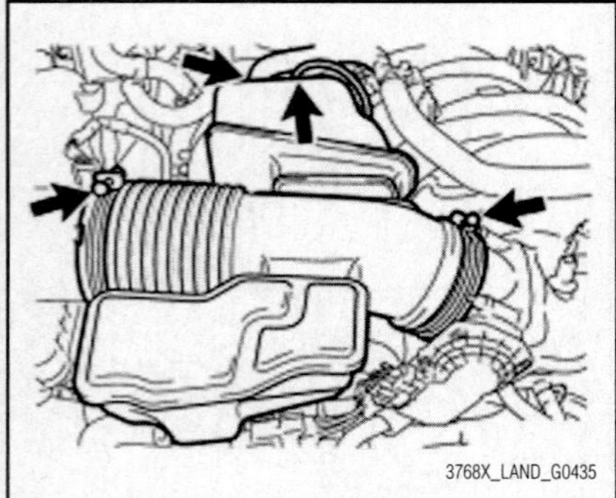

Fig. 132 Removing the air cleaner hose assembly

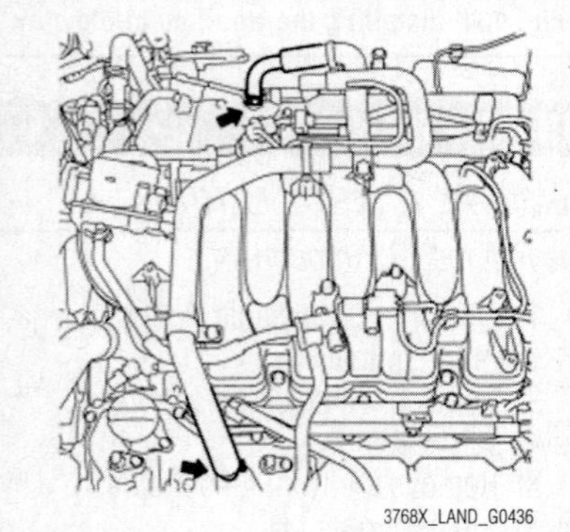

Fig. 133 Disconnecting the ventilation hose from the ventilation pipe of the cylinder head cover LH and RH

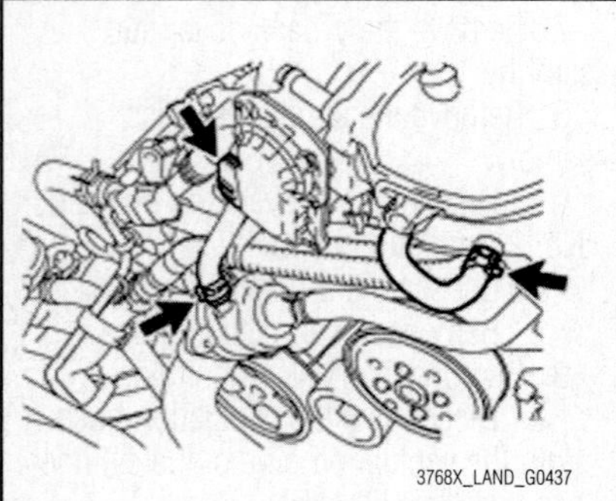
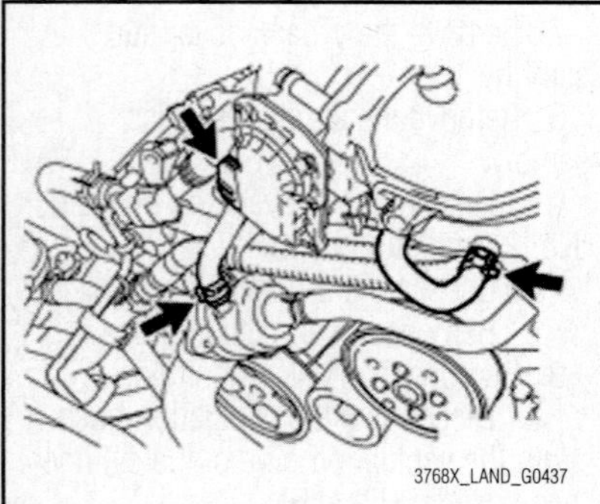

Fig. 134 Disconnecting the 2 water by pass hoses

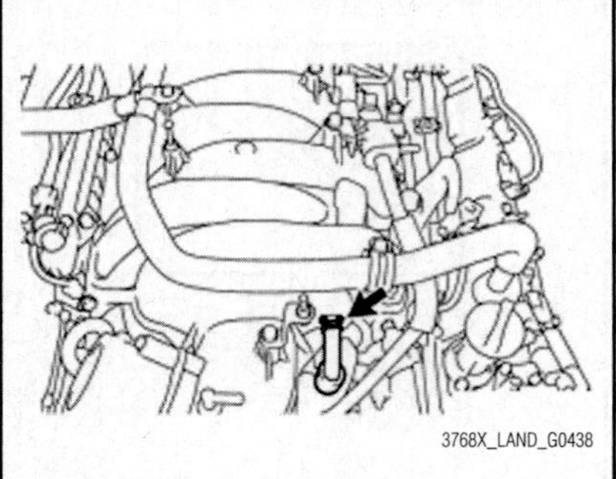

Fig. 135 Disconnecting the No. 1 ventilation hose

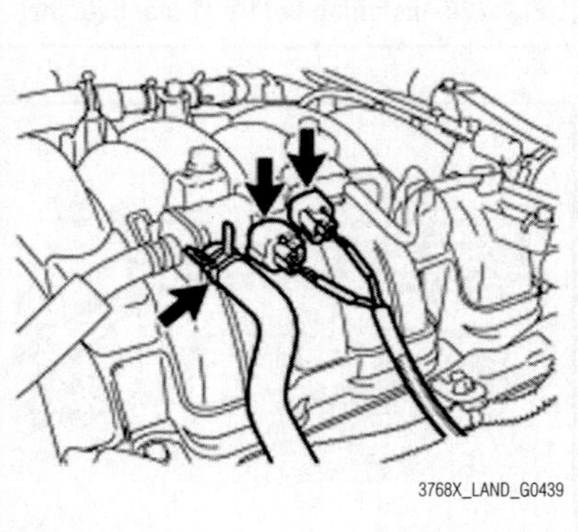

Fig. 136 Disconnecting the purge VSV connector

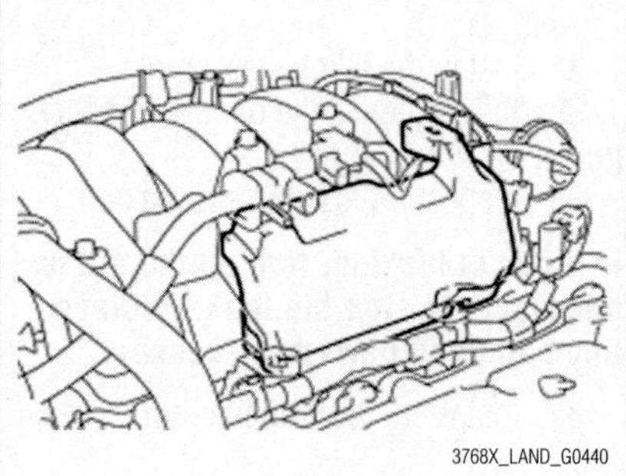

Fig. 137 Removing the No. 1 engine cover sub assembly

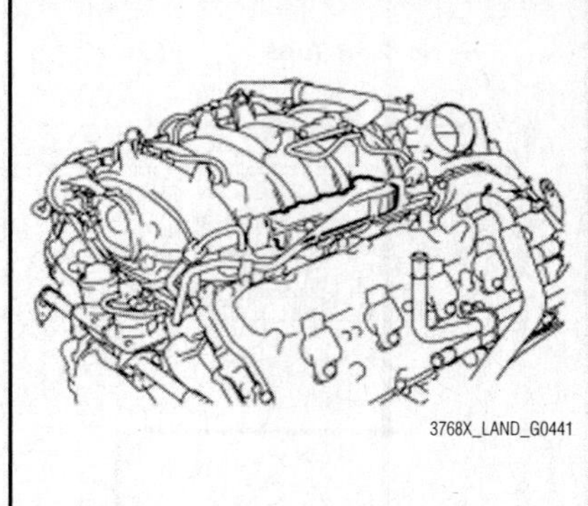

Fig. 138 Removing the No. 3 engine cover

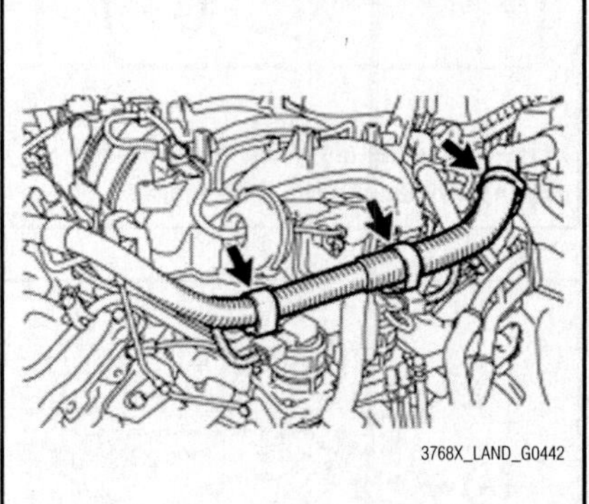

Fig. 139 Disconnecting the 3 wire clamps from the 3 wire brackets

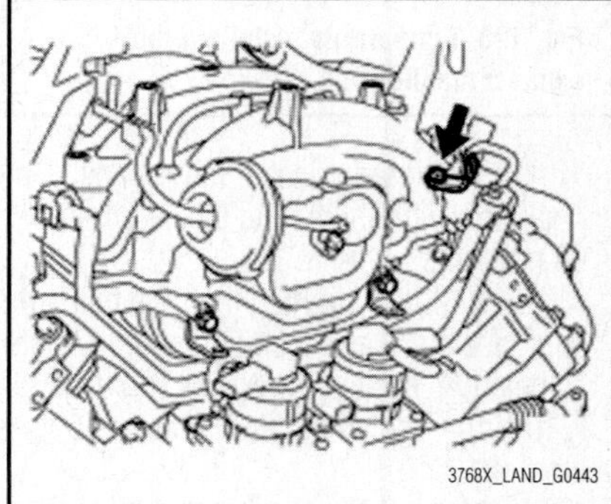

Fig. 140 Removing the bolt and wire bracket from the intake manifold

j. Disconnect the 3 wire clamps from the 3 wire brackets.

k. Remove the bolt and wire bracket from the intake manifold.

l. Remove the 2 nuts, 8 bolts, intake manifold and 2 gaskets.

10. Remove the ventilation hose assembly.

a. Remove the 2 bolts and ventilation hose from the intake manifold.

11. Remove the No. 1 v-bank cover bracket.

a. Remove the 2 bolts and the bracket.

12. Remove the v-bank cover bolt from the intake manifold.

13. Remove the No. 2 v-bank cover bracket sub assembly.

a. Remove the bolt and bracket.

14. Remove the throttle body assembly.

a. Remove the 4 bolts, throttle body and gasket.

15. Remove the vacuum switching valve assembly (for ACIS).

a. Disconnect the 2 vacuum hoses from the vacuum switching valve.

b. Remove the bolt and vacuum switching valve.

16. Remove the purge VSV.

a. Disconnect the purge line hose from the intake manifold.

b. Remove the bolt and purge VSV.

17. Remove the wire harness clamp bracket.

a. Remove the 2 bolts and 2 wire harness clamp brackets.

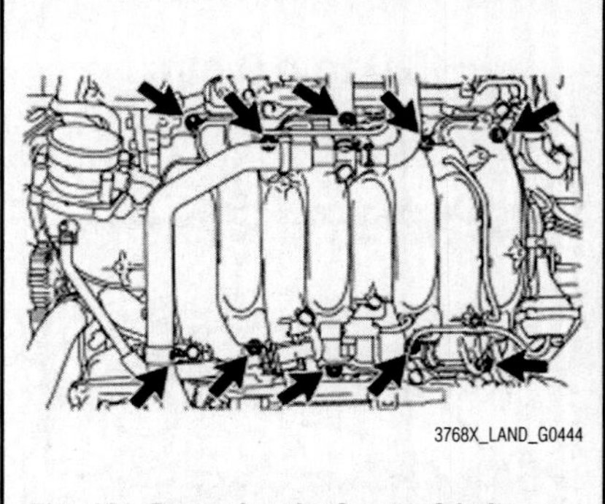

Fig. 141 Removing the 2 nuts, 8 bolts, intake manifold and 2 gaskets

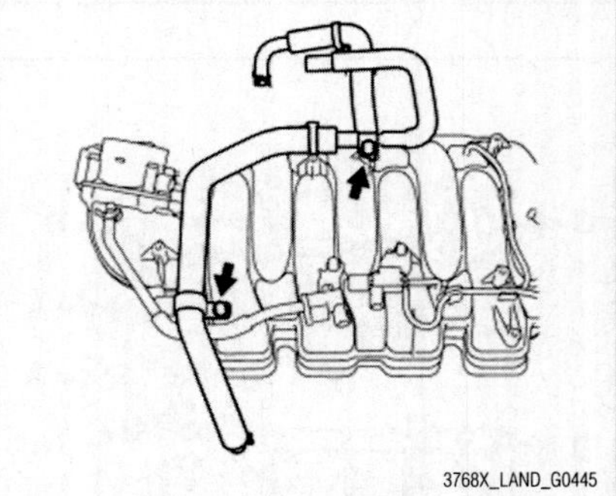

3768X_LAND_G0445

Fig. 142 Removing the 2 bolts and ventilation hose from the intake manifold

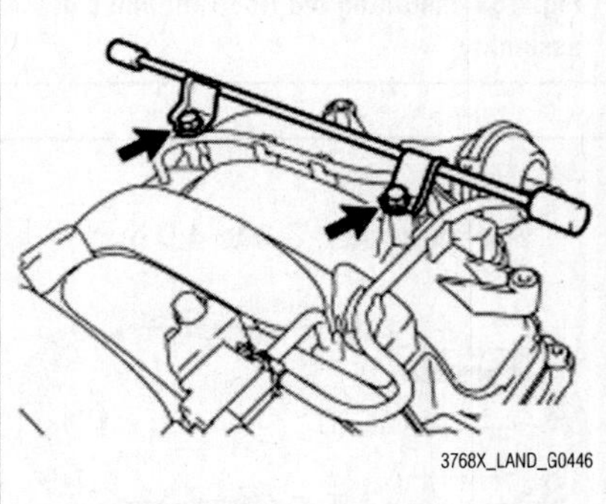

Fig. 143 Removing the No. 1 v-bank cover bracket

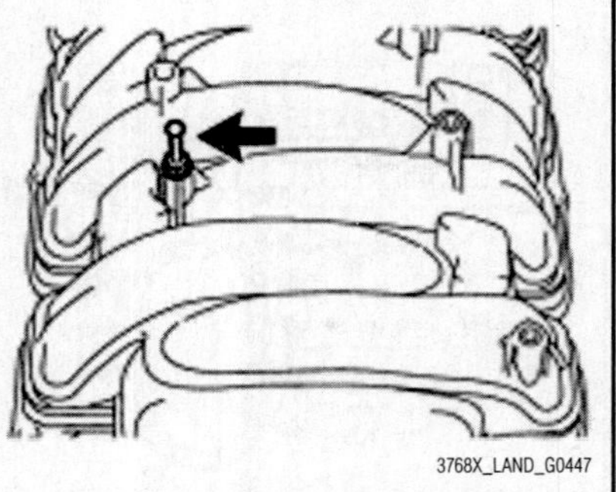

Fig. 144 Removing the v-bank cover bolt from the intake manifold

Fig. 145 Removing the No. 2 v-bank cover bracket sub assembly

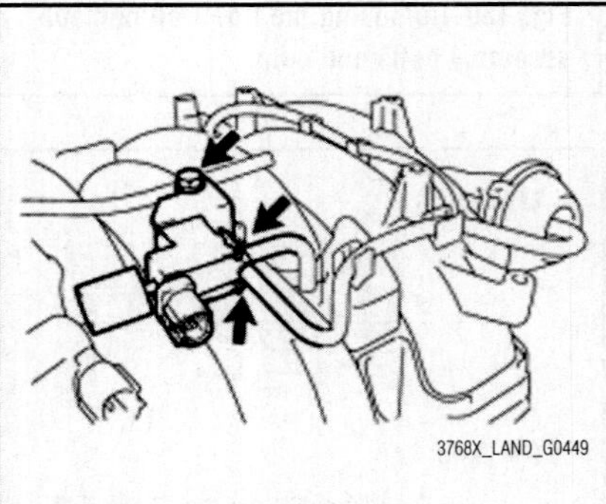

Fig. 146 Removing the vacuum switching valve assembly (for ACIS)

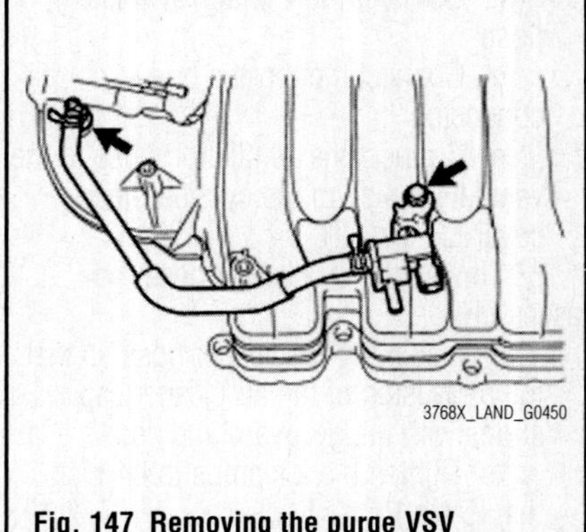

Fig. 147 Removing the purge VSV

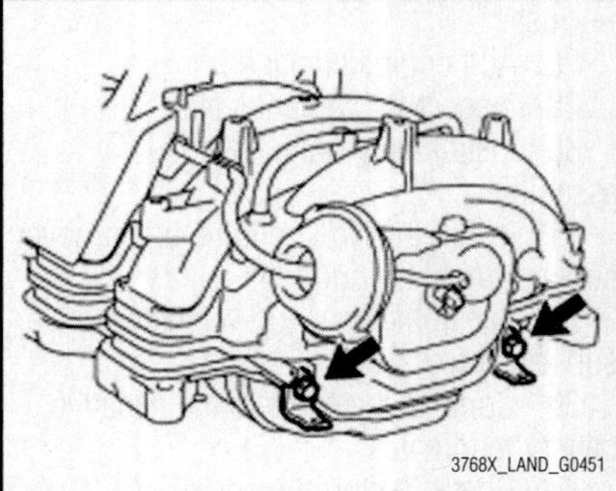

Fig. 148 Removing the wire harness clamp bracket

To install:

18. Install the wire harness clamp bracket with the 2 bolts. Tighten to 71 inch lbs. (8 Nm).
19. Install the purge VSV.
    a. Connect the purge line hose to the intake manifold.
    b. Install the purge VSV to the intake manifold with the bolt. Tighten to 15 ft. lbs. (21 Nm).
20. Install the vacuum switching valve assembly (for ACIS).
    a. Install the vacuum switching valve to the intake manifold with the bolt. Tighten to 80 inch lbs. (9 Nm).
    b. Connect the 2 vacuum hoses to the vacuum switching valve.
21. Install the throttle body assembly.
    a. Align the protrusion of a new gasket with the groove of the intake manifold, and install the gasket.
    b. Install the throttle body with the 4 nuts. Tighten to 7 ft. lbs. (10 Nm).
22. Install the No. 2 v-bank cover bracket sub assembly with the bolt. Tighten to 7 ft. lbs. (10 Nm).
23. Install the v-bank cover bolt to the intake manifold. Tighten to 7 ft. lbs. (10 Nm).
24. Install the No. 1 v-bank cover bracket with the 2 bolts. Tighten to 7 ft. lbs. (10 Nm).
25. Install the ventilation hose to the intake manifold with the 2 bolts. Tighten the A bolt to 15 ft. lbs. (21 Nm). Tighten the B bolt to 7 ft. lbs. (10 Nm).
26. Install the intake manifold.
    a. Place 2 new gaskets on the intake manifold.
    b. Place the intake manifold on the cylinder head.
    c. Install and uniformly tighten the 8 bolts and 2 nuts in several steps to 15 ft. lbs. (21 Nm).
    d. Install the wire bracket to the intake manifold with the bolt. Tighten to 71 inch lbs. (8 Nm).
    e. Connect the 3 wire clamps to the 3 wire brackets.
    f. Install the No. 3 engine cover.
    g. Install the No. 1 engine cover sub assembly.
    h. Connect the purge VSV connector.
    i. Connect the purge line hose to the purge VSV.
    j. Connect the vacuum switching valve connector (for ACIS).
    k. Connect the No. 1 ventilation hose.

l. Connect the 2 water by pass hoses.

m. Connect the throttle body connector.

n. Connect the ventilation hose to the ventilation pipe of the cylinder head cover LH and RH.

27. Install the air cleaner hose assembly.

a. Install the air cleaner hose so that the protrusion of the air cleaner cap aligns with the groove of the hose.

b. Tighten the 2 clamps to 44 inch lbs. (5 Nm).

c. Connect the vacuum hose.

d. Connect the No. 2 ventilation hose.

28. Add engine coolant.

29. Inspect for coolant leaks.

30. Install the v-bank cover sub assembly.

31. Install the No. 1 engine under cover sub assembly LH and RH.

32. Install the cowl top ventilator louver sub assembly.

33. Connect the cable to the negative battery terminal.

34. Check the throttle body assembly.

OIL PAN

REMOVAL & INSTALLATION

See Figures 149 through 154.

1. Remove the No. 2 oil pan sub assembly.

a. Vehicles with and oil level sensor, remove the engine oil level sensor.

b. Remove the 14 bolts and 2 nuts.

c. Insert the blade of oil pan seal cutter between the oil pans. Cut through the applied sealer and remove the No. 2 oil pan.

➡Be careful not to damage the contact surfaces of the oil pans

2. Remove the No. 1 oil pan sub assembly.

a. Remove the 14 bolts and 2 nuts.

➡Be sure to clean the bolts and stud bolts, and check the threads for cracks or other damage.

b. Remove the oil pan by prying between the oil pan and cylinder block with a screwdriver.

➡Be careful not to damage the contact surfaces of the cylinder block and oil pan.

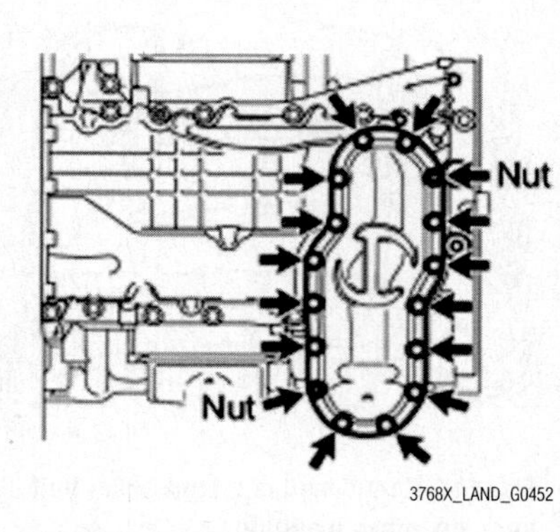

Fig. 149 Removing the No. 2 oil pan sub assembly 14 bolts and 2 nuts

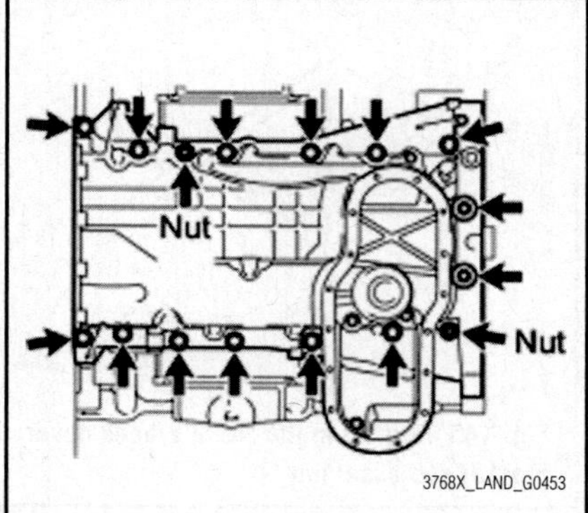

Fig. 150 Removing the No. 1 oil pan sub assembly bolts and nuts

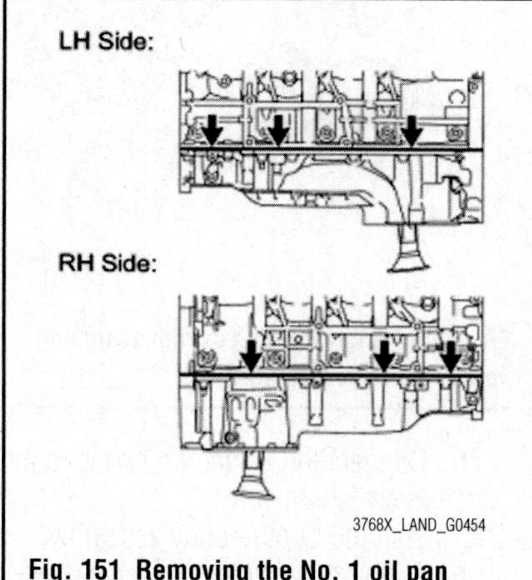

Fig. 151 Removing the No. 1 oil pan

➡Tape the screwdriver tip before use.

To install:

3. Install the No. 1 oil pan sub assembly.

a. Apply seal packing in a continuous line as shown.

➡Remove any oil from the contact surface.

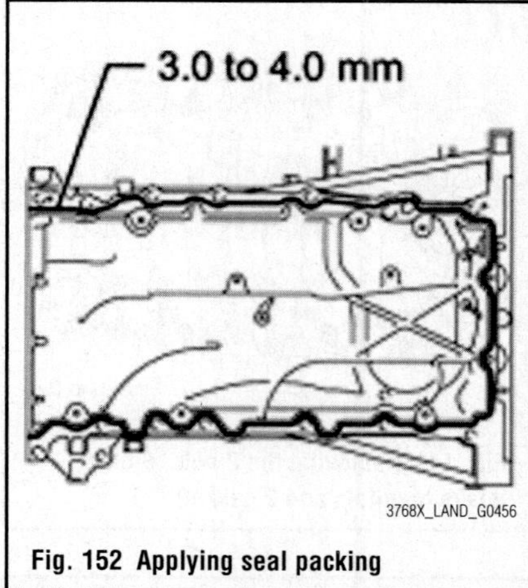

Fig. 152 Applying seal packing

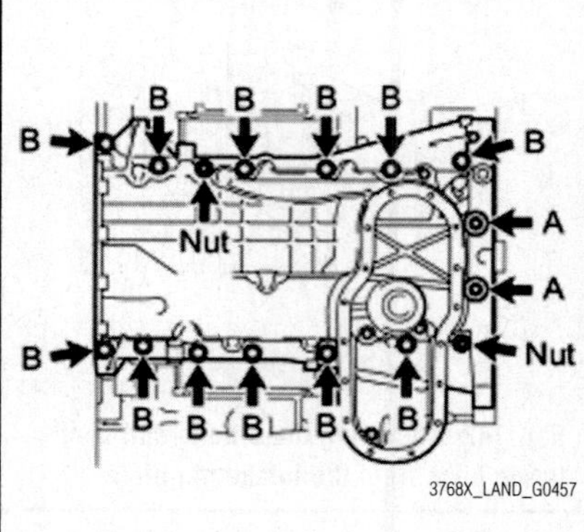

Fig. 153 Installing the No. 1 oil pan sub assembly

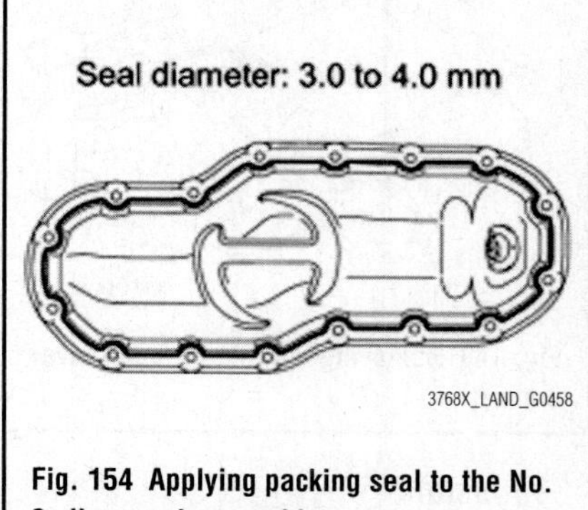

Fig. 154 Applying packing seal to the No. 2 oil pan sub assembly

➡Install the oil pan within 3 minutes and tighten the bolts and nuts within 15 minutes after applying seal packing.

b. Install the oil pan with the 14 bolts and 2 nuts. Tighten the A bolts to 7 ft. lbs. (10 Nm). Tighten the B bolts to 26 ft. lbs. (35 Nm). Tighten the nuts to 26 ft. lbs. (35 Nm).

➡Do not start the engine for at least 2 hours after installing.

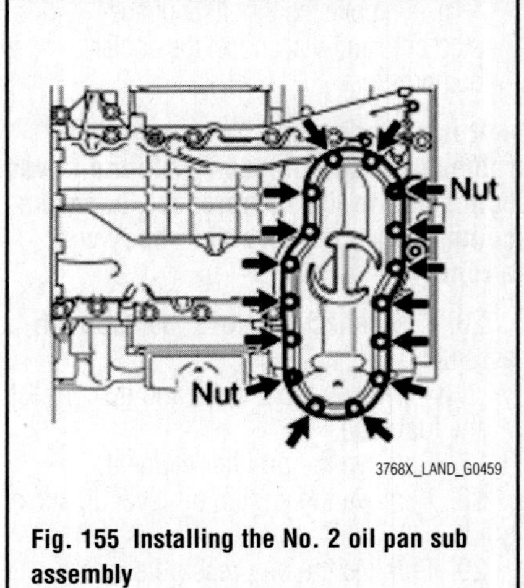

Fig. 155 Installing the No. 2 oil pan sub assembly

4. Install the No. 2 oil pan sub assembly.
   a. Apply seal packing in a continuous line as shown.

➡Remove any oil from the contact surface.

➡Install the oil pan within 3 minutes and tighten the bolts and nuts within 15 minutes after applying seal packing.

   b. Install the oil pan with the 14 bolts and 2 nuts. Tighten to 7 ft. lbs. (10 Nm).

➡Do not start the engine for at least 2 hours after installing.

OIL PRESSURE SENSOR/SWITCH

LOCATION

See Figure 156.

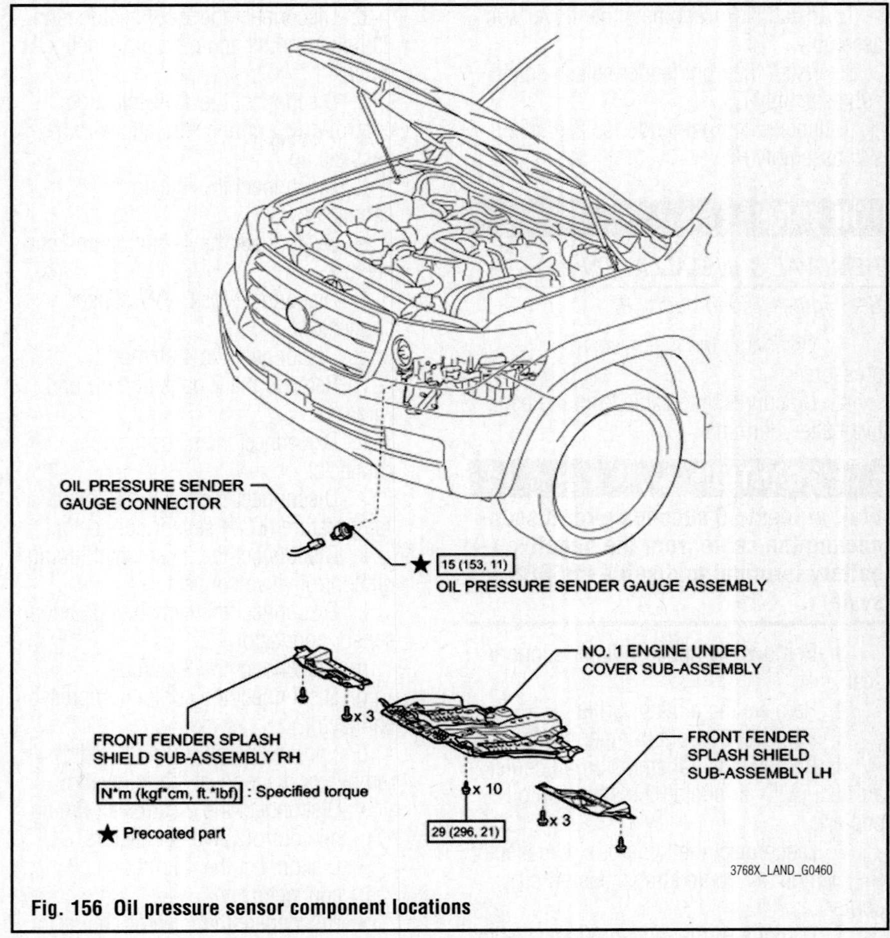

Fig. 156 Oil pressure sensor component locations

REMOVAL & INSTALLATION

See Figures 157 through 159.

1. Remove the front fender splash shield sub assembly LH.
2. Remove the front fender splash shield sub assembly RH.
3. Remove the No. 1 engine under cover sub assembly.
4. Remove the oil pressure sender gauge assembly.
   a. Disconnect the sender gauge connector.
   b. Remove the oil pressure sender gauge.

To install:

5. Install the oil pressure sender gauge assembly.

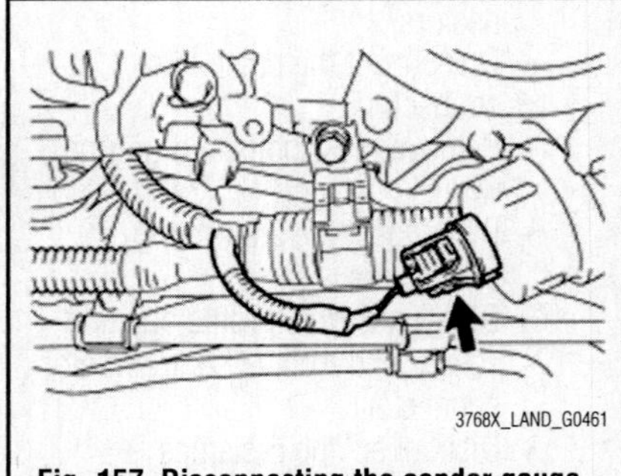

Fig. 157 Disconnecting the sender gauge connector

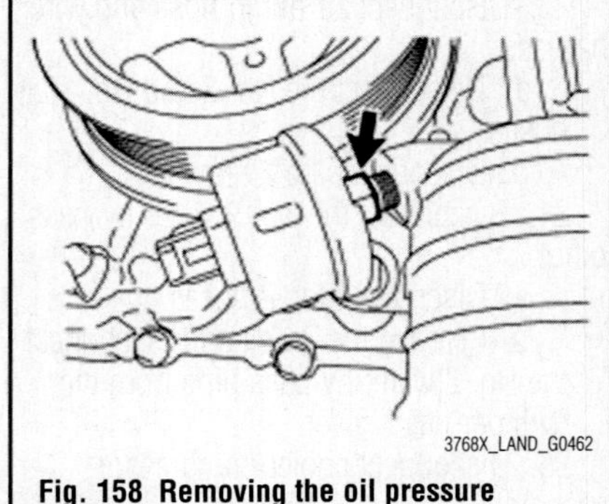

Fig. 158 Removing the oil pressure sender gauge

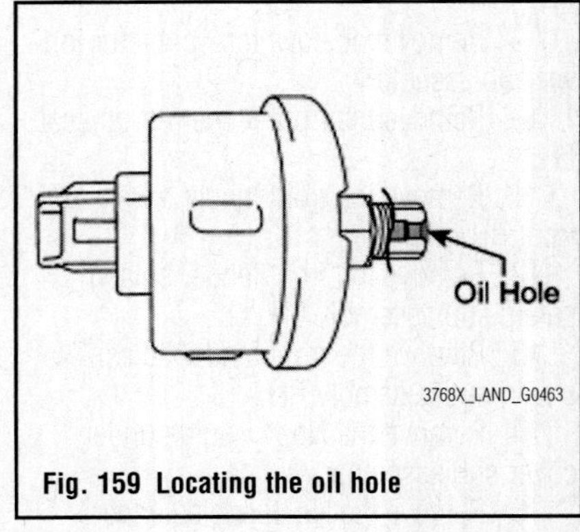

Fig. 159 Locating the oil hole

   a. Apply adhesive to 2 or 3 threads of the oil pressure sender gauge.

➡Do not allow adhesive to contact the oil hole.

   b. Install the oil pressure sender gauge and tighten to 11 ft. lbs. (15 Nm).

➡Do not start the engine within 1 hour after installation.

   c. Connect the sender gauge connector.

6. Inspect engine oil level.

7. Install No. 1 engine under cover sub assembly.
8. Install the front fender splash shield sub assembly LH.
9. Install the front fender splash shield sub assembly RH.

OIL PUMP

REMOVAL & INSTALLATION

See Figures 160 through 197.

1. Discharge the fuel system pressure.
2. Disconnect the cable from the negative battery terminal.

✻✻ CAUTION

Wait at least 90 seconds after disconnecting the cable from the negative (-) battery terminal to disable the SRS system.

3. Remove the upper radiator support seal.
4. Remove the radiator grille.
5. Remove the front bumper cover.
6. Remove the transmission oil cooler air duct (with air cooled transmission oil cooler).
7. Disconnect the radiator side deflector RH (without air cooled transmission oil cooler).
8. Disconnect the radiator side deflector LH.
9. Remove the cowl top ventilator louver sub assembly.
10. Remove the front fender apron seal LH.
11. Remove the front fender apron seal front RH.
12. Remove the front fender splash shield sub assembly LH.
13. Remove the front fender splash shield sub assembly RH.
14. Remove the No. 1 engine under cover sub assembly.
15. Remove the No. 2 engine under cover.
16. Drain engine oil.
17. Drain the engine coolant.
18. Remove the v-bank cover sub assembly.
19. Remove the air cleaner hose assembly.
20. Remove the air cleaner assembly.
21. Remove the radiator assembly.
22. Disconnect the engine wire.
   a. For the engine room LH side, remove the engine room relay block cover.
   b. Disconnect the 2 connectors and 2 clips from the engine room junction block.
   c. Disconnect the 4 air injection control driver connectors and wire harness clamp.
   d. Disconnect the injector connector.
   e. Disconnect the 4 ignition coil connectors.
   f. Disconnect the 2 VVT sensor connectors.
   g. Disconnect the 4 clamps.
   h. Remove the 2 bolts and ground wire.
   i. Disconnect the noise filter connector.
   j. Disconnect the Engine Coolant Temperature (ECT) sensor connector.
   k. Disconnect the 2 camshaft timing oil control valve connectors.
   l. Disconnect the camshaft position sensor connector.
   m. Disconnect the 3 clamps.
   n. Disconnect the cooler compressor connector.
   o. For the engine room RH side, remove or disconnect the following:
      - Disconnect the 2 camshaft timing oil control valve connectors
      - Disconnect the 4 ignition coil connectors
      - Disconnect the injector connector
      - Disconnect the 2 VVT sensor connectors
      - Disconnect the noise filter connector
      - Remove the 2 bolts and ground wire
      - Disconnect the 2 air pump connectors
      - Disconnect the throttle position sensor and throttle control motor connector
      - Disconnect the 5 clamps
   p. Disconnect the 2 clamps and power steering oil pressure switch connector.
23. Disconnect air pump hose and wire harness.
   a. Disconnect the No. 2 and No. 3 air hoses.
   b. Disconnect the 2 clamps.
24. Disconnect the No. 2 water by-pass pipe.
   a. Disconnect the 3 hoses.
   b. Remove the 3 bolt and disconnect the No. 2 water by pass pipe from the cylinder head cover.
25. Disconnect cooler compressor assembly.
   a. Remove the 3 bolts, nut and stud bolt, and disconnect the cooler compressor.

➡It is not necessary to completely remove the compressor. With the hoses connected to the compressor, hang the compressor on the vehicle body with a rope.

26. Disconnect the No. 2 fuel tube sub assembly.
   a. Remove the 2 bolts and disconnect the fuel tube.
27. Remove the oil filter element.
28. Remove the engine oil level dipstick guide.
29. Remove the oil pressure sender gauge assembly.
30. Remove the No. 2 water by pass pipe sub assembly (with oil cooler).

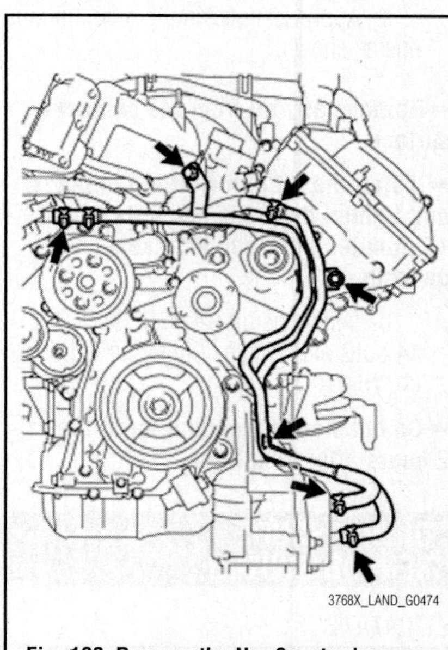

Fig. 160 Remove the No. 2 water by pass pipe sub assembly

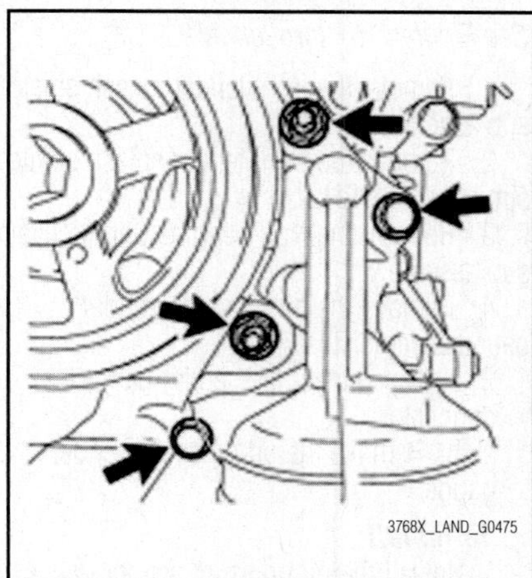

Fig. 161 Removing the oil filter bracket

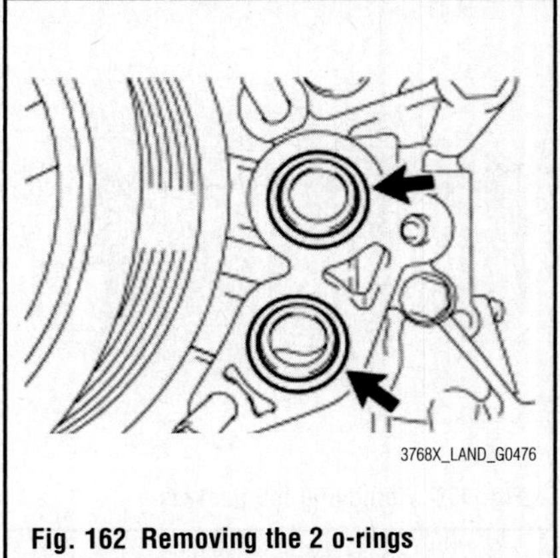

Fig. 162 Removing the 2 o-rings

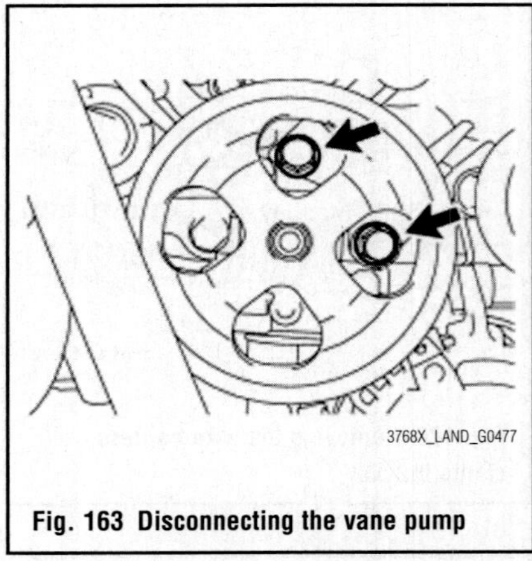

Fig. 163 Disconnecting the vane pump

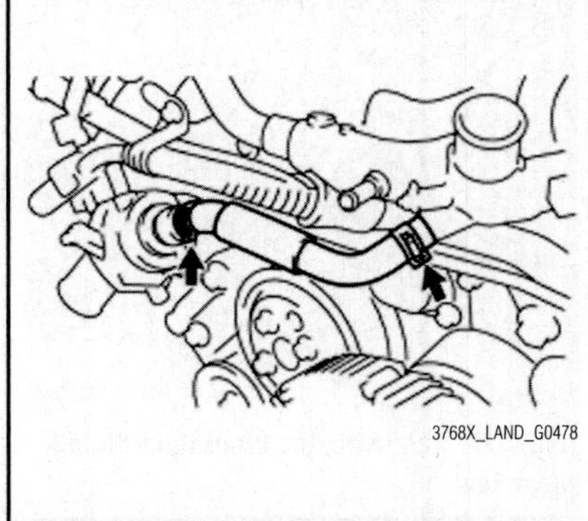

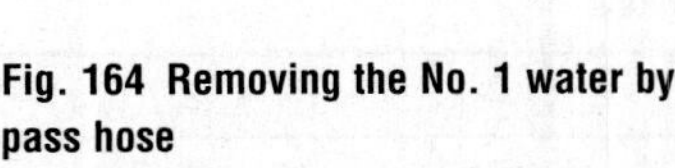
Fig. 164 Removing the No. 1 water by pass hose

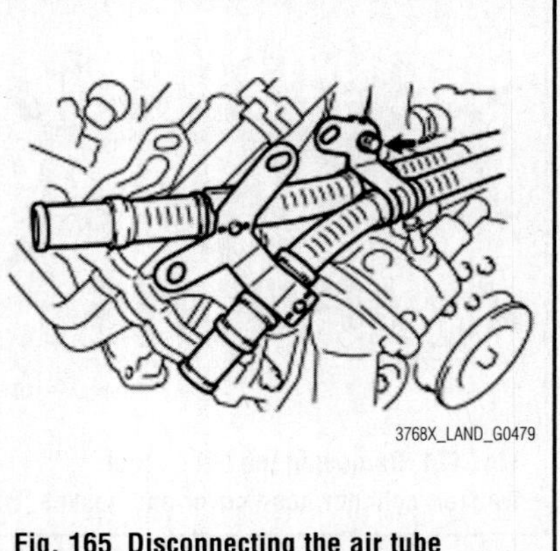

Fig. 165 Disconnecting the air tube

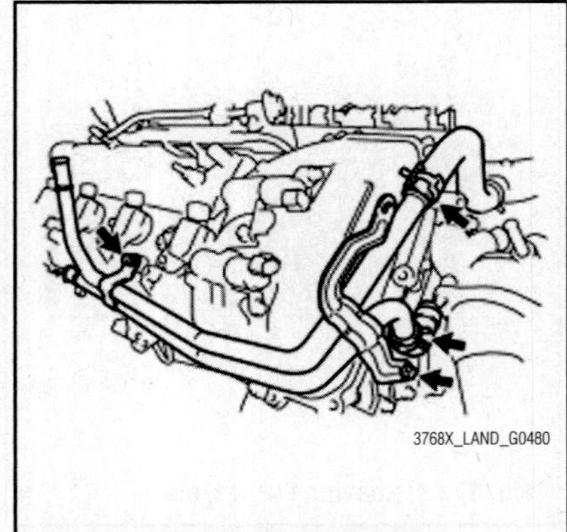

Fig. 166 Disconnecting the 2 hoses

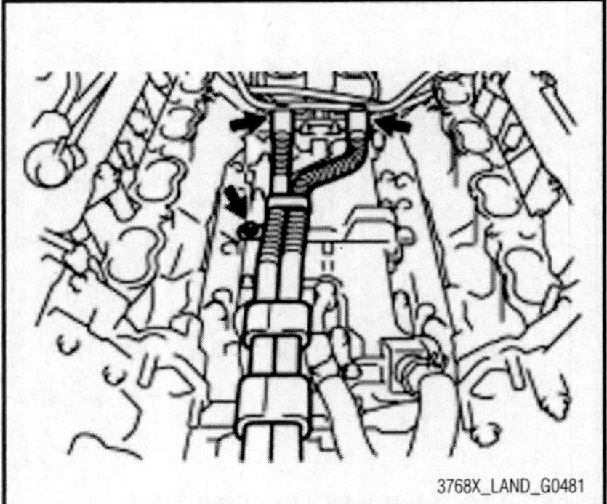

Fig. 167 Removing the air tube sub assembly

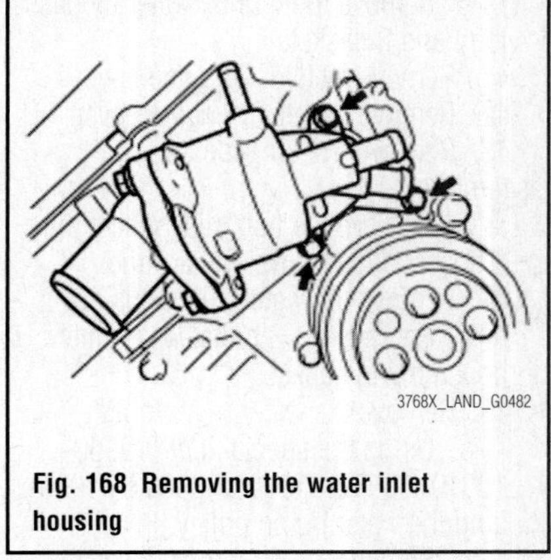

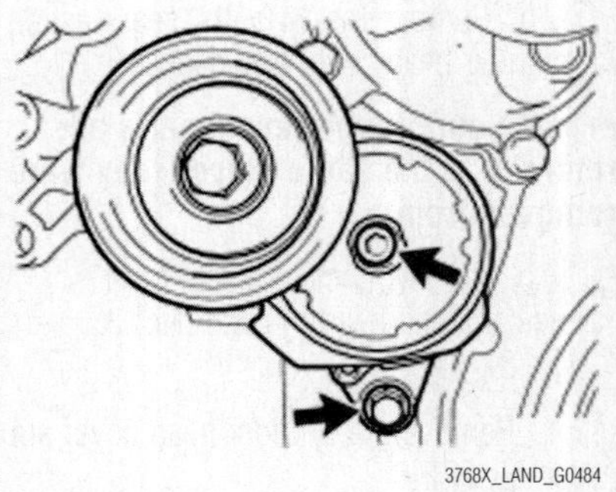

Fig. 168 Removing the water inlet housing

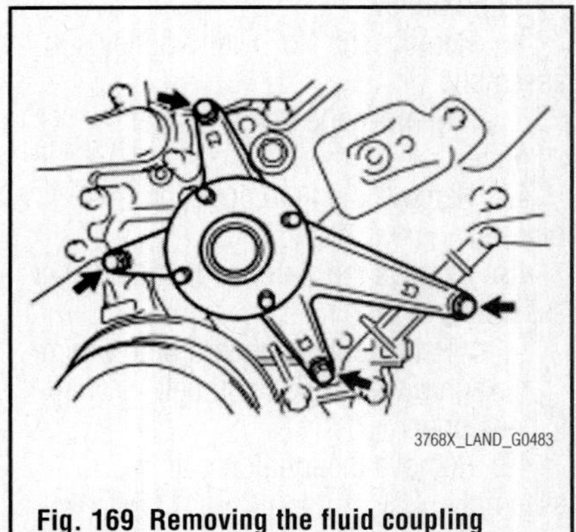

Fig. 169 Removing the fluid coupling bracket

3768X_LAND_G0484

Fig. 170 Removing the v-ribbed belt tensioner assembly

a. Remove the 3 bolts.

b. Disconnect the 4 hoses and remove the water by pass pipe.

31. Remove the No. 1 oil cooler bracket (with oil cooler).

32. Remove the oil filter bracket.

33. Vehicles without oil cooler perform the following.

a. Remove the 2 bolts, 2 nuts and oil filter bracket.

b. Remove the 2 o-rings.

34. Remove the intake manifold.

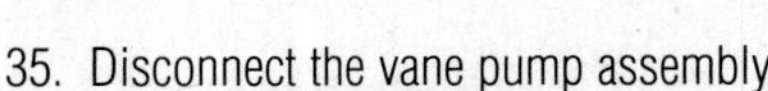

35. Disconnect the vane pump assembly.

a. Remove the 2 bolts and disconnect the vane pump.

36. Disconnect the oil cooler pipe assembly.

37. Remove the generator assembly.

38. Remove the No. 1 water by pass hose.

a. Remove the No. 1 water by pass hose by disconnecting the hose from the water inlet housing and front water by pass joint.

39. Remove the water by pass pipe sub assembly.

a. Remove the bolt and disconnect the air tube.

b. Disconnect the 2 hoses.

c. Remove the bolts and water by pass pipe.

40. Remove the front water by pass joint.

a. Disconnect the No. 2 water by pass hose from the water by pass joint.

b. Remove the 4 nuts, water by pass joint and 2 gaskets.

41. Remove the No. 2 engine cover.
42. Remove the No. 1 engine cover.
43. Remove the air tube sub assembly.
    a. Remove the bolt, disconnect the 2 hoses, and remove the air tube.
44. Remove the water inlet housing.
    a. Remove the 3 bolts, water inlet housing and gasket.
45. Remove the water pump pulley.
    a. Using the special tool (09960-10010, 09962-01000 or 09963-01000), hold the water pump pulley.
    b. Remove the 4 bolts and water pump pulley.
46. Remove the No. 1 idler pulley sub assembly.
    a. Remove the bolt and idler pulley.
47. Remove the fluid coupling bracket.
48. Remove the v-ribbed belt tensioner assembly.
    a. Remove the standard bolt, 6 mm hexagon wrench bolt and belt tensioner.
49. Remove the ignition coil assembly.
    a. Remove the 8 bolts and 8 ignition coils.
50. Remove the cylinder head cover sub assembly LH.
    a. Remove the 14 bolts, seal washer, cylinder head cover and gasket.

➡Make sure the removed parts are returned to the same places they were removed from.

    b. Remove the 5 gaskets from the camshaft bearing caps (No. 2, No. 3).

51. Remove the cylinder head cover sub assembly RH.
    a. Remove the bolt and noise filter.
    b. Remove the 14 bolts, seal washer, cylinder head cover and gasket.

➡Make sure the removed parts are returned to the same places they were removed from.

    c. Remove the 5 gaskets from the camshaft bearing caps (No. 1, No. 3).

52. Remove the spark plug tube gasket.
53. Remove the crankshaft pulley.
    a. Using the special tool (09213-70011 or 09330-00021), loosen the crankshaft pulley set bolt until 2 or 3 threads are engaged.

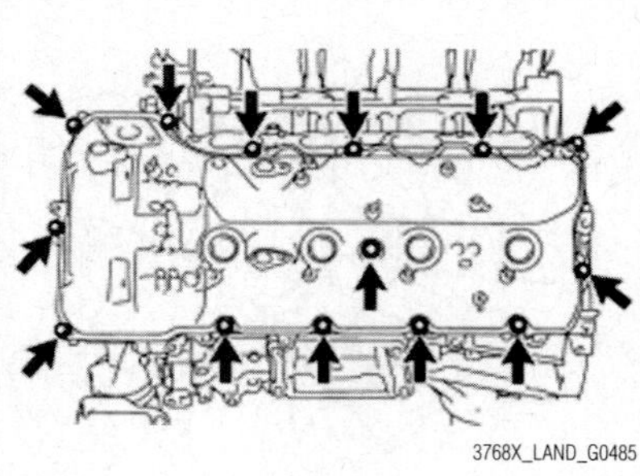

Fig. 171 Removing the bolts, seal washer, cylinder head cover and gasket

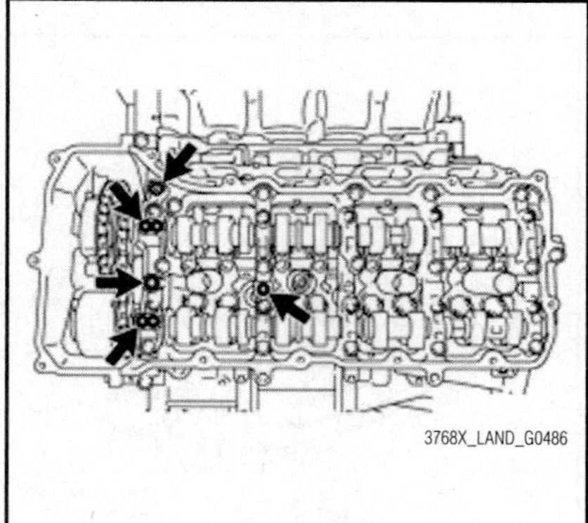

Fig. 172 Removing the gaskets

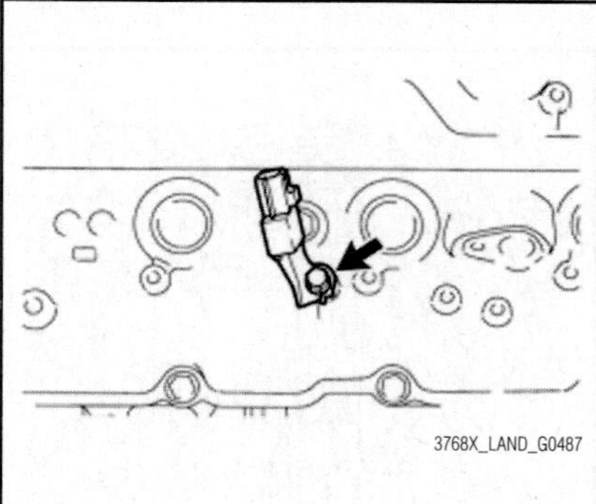

Fig. 173 Removing the noise filter

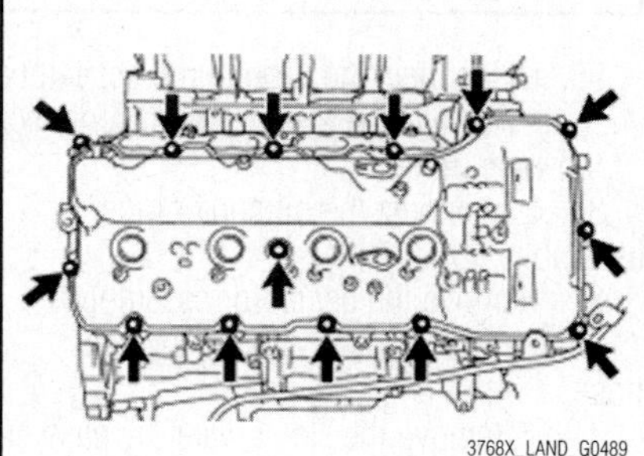

Fig. 174 Removing the seal washer, cylinder head cover and gasket

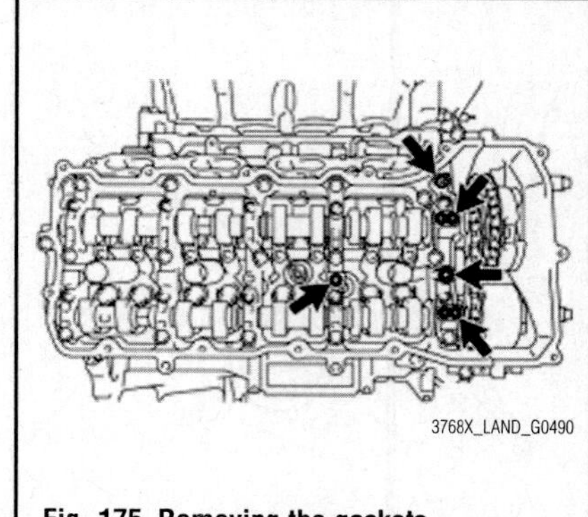

Fig. 175 Removing the gaskets

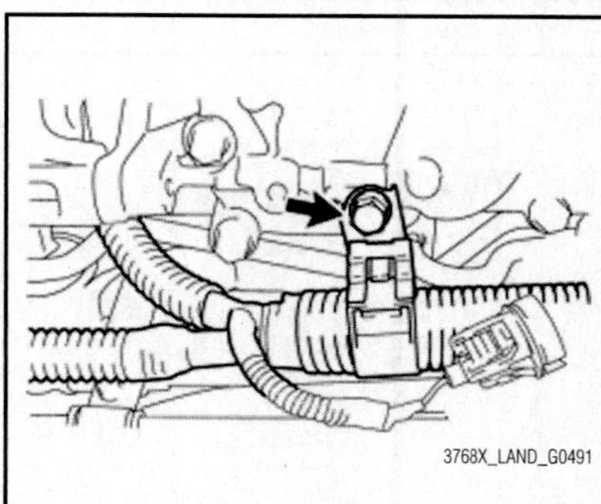

Fig. 176 Removing the wire harness clamp bracket

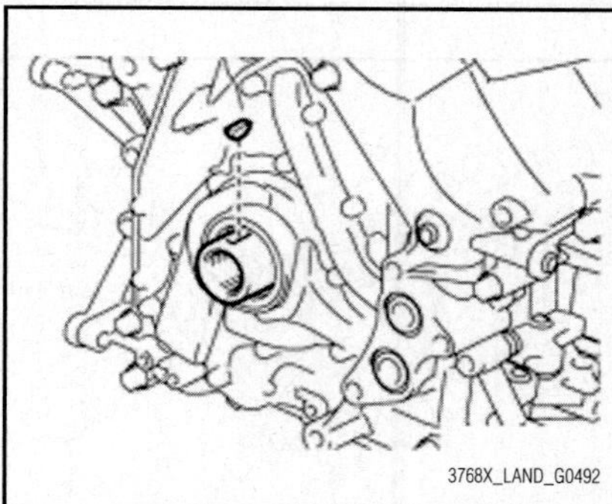

Fig. 177 Removing the crankshaft timing gear key

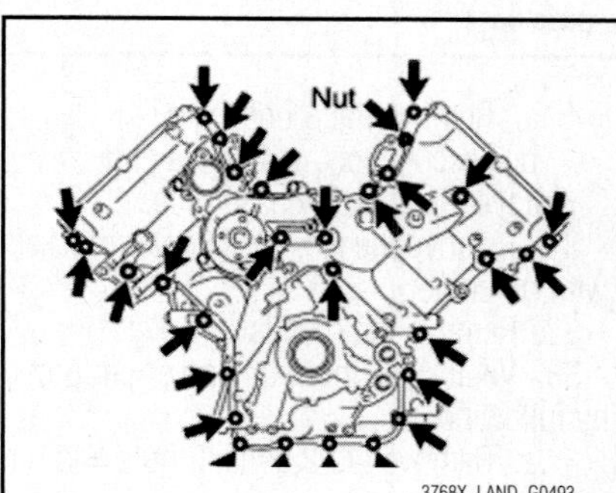

Fig. 178 Removing the timing chain cover sub assembly bolts and nut

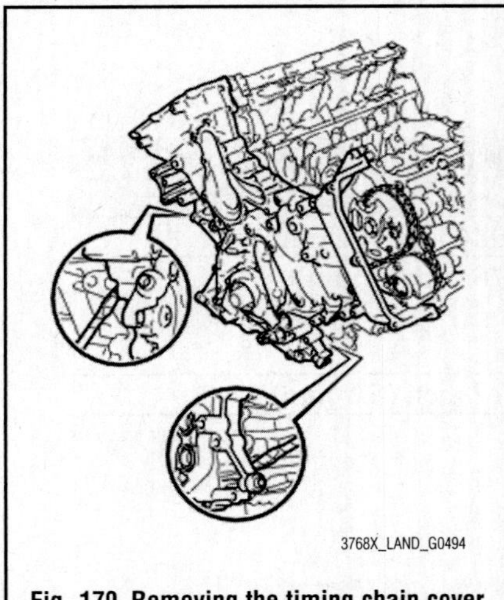

Fig. 179 Removing the timing chain cover

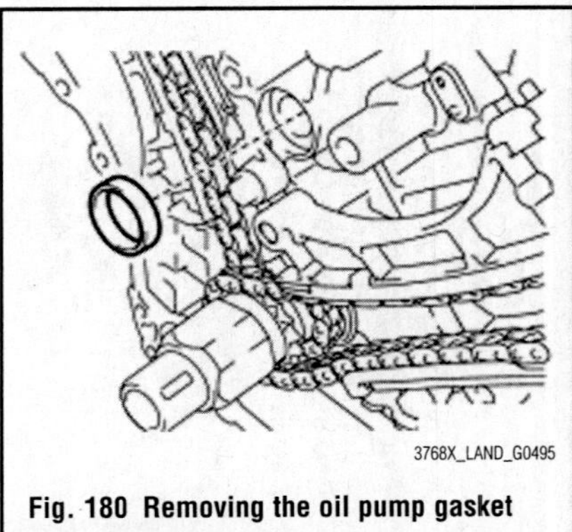

Fig. 180 Removing the oil pump gasket from the cylinder block

Fig. 181 Removing the O-ring from the oil pan

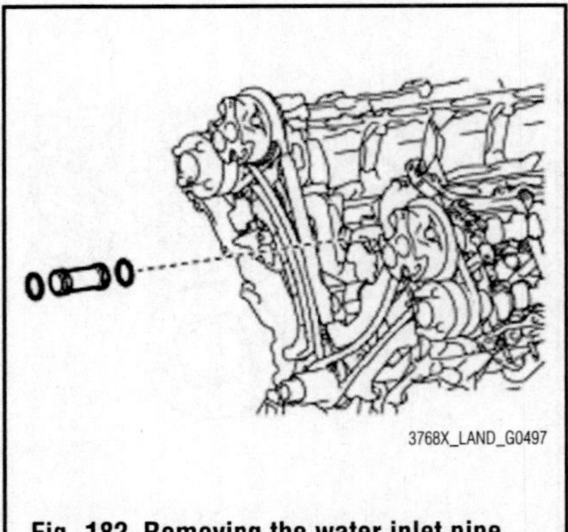

Fig. 182 Removing the water inlet pipe

b. Using the pulley set bolt and special tool (09950-50013 or 09951-05010), remove the crankshaft pulley.

54. Disconnect the wire harness clamp bracket.

a. Remove the bolt and disconnect the bracket.

55. Remove the crankshaft timing gear key.

a. Using a screwdriver, remove the timing gear key from the crankshaft.

56. Remove the timing chain cover sub assembly.

a. Remove the 28 bolts and nut shown.

b. Remove the timing chain cover by prying between the timing chain cover and cylinder head or cylinder block with a screwdriver as shown in the illustration.

➡**Be careful not to damage the contact surfaces of the cylinder head, cylinder block and chain cover.**

➡**Tape the screwdriver tip before use.**

c. Remove the oil pump gasket from the cylinder block.

d. Remove the O-ring from the oil pan.

57. Remove the water inlet pipe.

a. Remove the 2 o-rings from the water inlet pipe.

To install:

58. Install the water inlet pipe.

a. Apply soapy water to 2 new O-rings and install them to the inlet pipe.

b. Install the inlet pipe to the No. 1 heat exchanger cover.

59. Install the timing chain cover sub assembly.

a. Apply a light coat of engine oil to a new oil pump gasket.

b. Install the oil pump gasket.

c. Apply a light coat of engine oil to a new o-ring.

d. Install the o-ring.

e. Apply seal packing in a continuous line to the timing chain cover as shown.

➡**When the contact surfaces are wet, wipe them with an oil-free cloth before applying seal packing.**

➡**Install the chain cover within 3 minutes and tighten the bolts within 10 minutes after applying seal packing.**

➡**Do not start the engine for at least 2 hours after installing.**

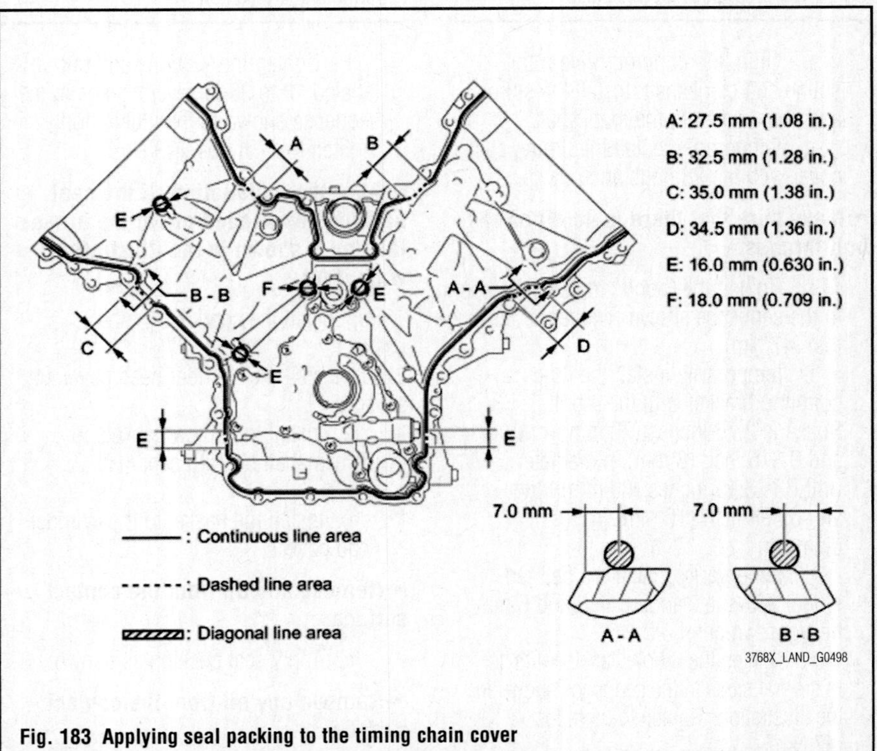

Fig. 183 Applying seal packing to the timing chain cover

Area	Seal Packing Diameter	Application Position From Inside Edge Of Cover
Continuous Line Area	0.118-0.157 inch (3-4 mm)	0.0984 inch (2.5 mm)
Dashed Line Area	0.252 inch (6.4 mm) or more, or within OK area shown in the illustration	0.276 inch (7 mm)
Diagonal Line Area	0.118-0.157 inch (3-4 mm)	0.217 inch (5.5 mm)

3768X_LAND_G0499

Fig. 184 Seal packing specifications

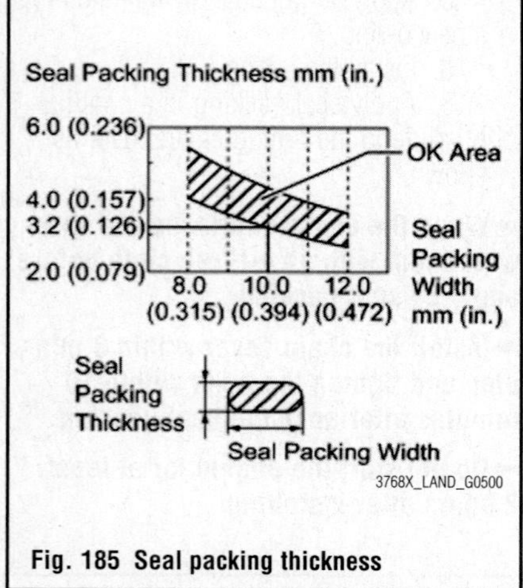

Fig. 185 Seal packing thickness

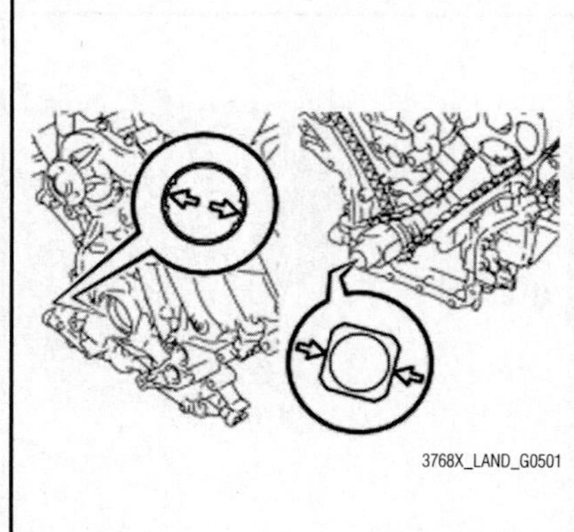

Fig. 186 Aligning the oil pump drive rotor spline and crankshaft

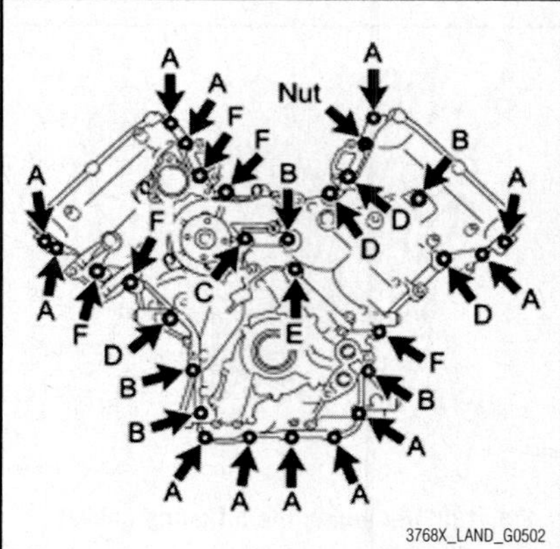

Fig. 187 Temporarily installing the timing chain cover

f. Align the oil pump drive rotor spline and crankshaft. Install the spline and chain cover to the crankshaft.

g. Temporarily install the timing chain cover with the 28 bolts and nut.

➡Make sure that there is no oil on the bolt threads.

h. Tighten the 3 bolts in several steps in the sequence shown. Tighten to 35 ft. lbs. (47 Nm).

i. Temporarily install the fluid coupling bracket with the 4 bolts. Bolt A is 2.76 inches (70 mm) long and 0.315 inch (8 mm) in diameter. Bolt B is 3.15 inches (80 mm) long and 0.394 inch (10 mm) in diameter.

j. Temporarily install the belt tensioner with the standard bolt and 6 mm hexagon wrench bolt.

k. Tighten the 8 bolts labeled 4 to 11 in several steps in the sequence shown in the illustration. Tighten to 35 ft. lbs. (47 Nm).

l. Tighten the 23 bolts and nut labeled 12 to 35 in several steps in the sequence shown in the illustration. Tighten to 17 ft. lbs. (23 Nm).

➡After the installation, if the seal packing has seeped out at the areas labeled A shown in the illustration, wipe it off.

60. Install the spark plug tube gasket.

61. Install the cylinder head cover sub assembly LH.

a. Install the 5 new gaskets to the camshaft bearing caps (No. 2, No. 3).

b. Install the gasket to the cylinder head cover.

➡Remove any oil from the contact surface.

c. Apply seal packing as shown.

➡Remove any oil from the contact surface.

➡Install the cylinder head cover within 3 minutes and tighten the bolts within 15 minutes after applying seal packing.

➡Do not start the engine for at least 2 hours after the installation.

d. Install the cylinder head cover and the seal washer with the 14 bolts in the order shown. Tighten Bolt A to 15 ft. lbs. (21 Nm). Tighten the remaining bolts to 9 ft. lbs. (12 Nm).

62. Install the cylinder head cover sub assembly RH.

a. Install 5 new gaskets to the camshaft bearing caps (No. 1, No. 3).

b. Install the gasket to the cylinder head cover.

➡Remove any oil from the contact surface.

c. Apply seal packing.

➡Remove any oil from the contact surface.

Item	Length	Thread Diameter
Bolt A	0.984 inch (25 mm)	0.315 inch (8 mm)
Bolt B	2.17 inch (55 mm)	0.315 inch (8 mm)
Bolt C	2.76 inch (70 mm)	0.315 inch (8 mm)
Bolt D	1.38 inch (35 mm)	0.394 inch (10 mm)
Bolt E	2.17 inch (55 mm)	0.394 inch (10 mm)
Bolt F	3.15 inch (80 mm)	0.394 inch (10 mm)

3768X_LAND_G0503

Fig. 188 Standard bolts

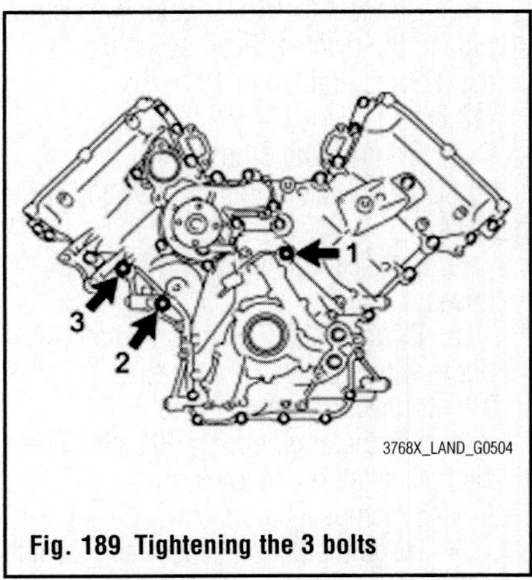

Fig. 189 Tightening the 3 bolts

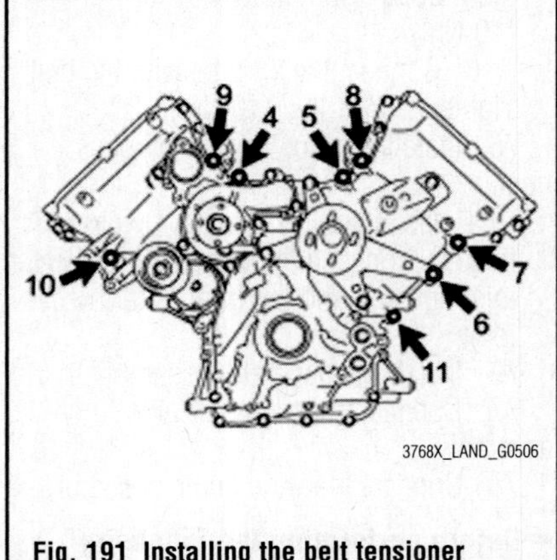

Fig. 191 Installing the belt tensioner

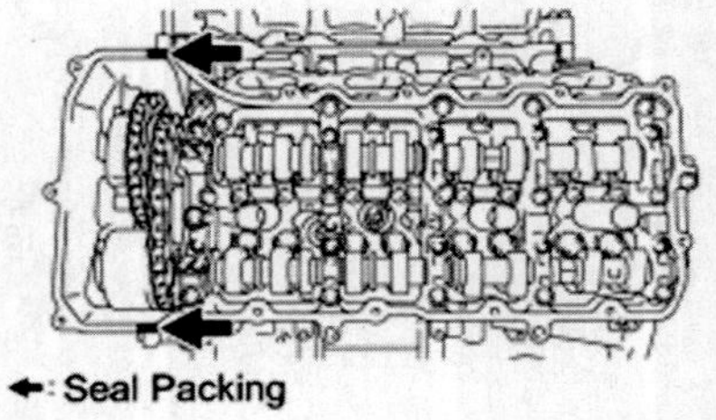

Seal Packing

3768X_LAND_G0508

Fig. 193 Applying seal packing

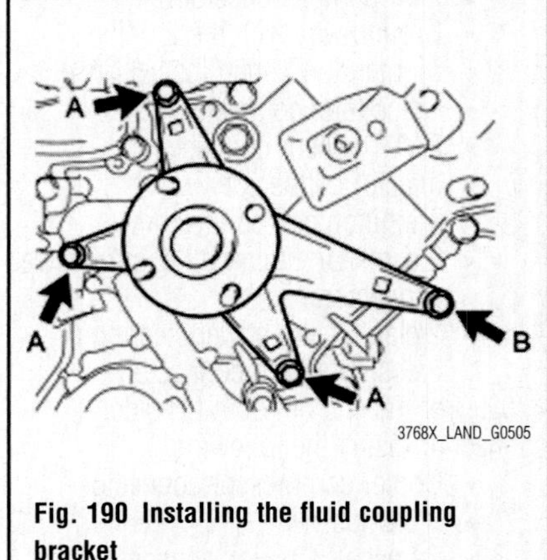

Fig. 190 Installing the fluid coupling bracket

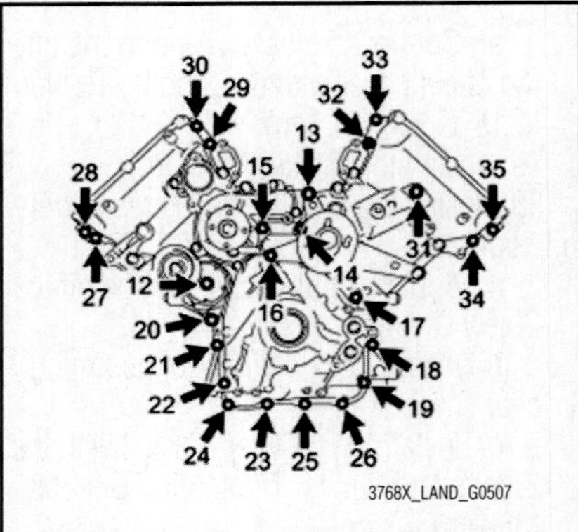

Fig. 192 Identifying the bolt tightening sequence

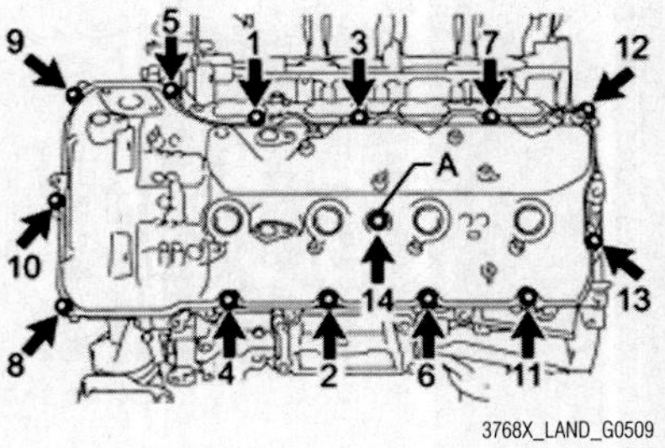

3768X_LAND_G0509

Fig. 194 Installing the cylinder head cover and seal washer

➡Install the cylinder head cover within 3 minutes and tighten the bolts within 15 minutes after applying seal packing.

➡Do not start the engine for at least 2 hours after the installation.

d. Install the cylinder head cover and the seal washer with the 14 bolts in the order shown. Tighten Bolt A to 15 ft. lbs. (21 Nm). Tighten the remaining bolts to 9 ft. lbs. (12 Nm).

e. Install the noise filter to the cylinder head cover with the bolt. Tighten to 62 inch lbs. (7 Nm).

63. Install the ignition coil assembly with the 8 bolts. Tighten to 7 ft. lbs. (10 Nm).

64. Install the crankshaft timing gear key to the crankshaft.

65. Connect the wire harness clamp bracket to the timing chain cover with the bolt. Tighten to 71 inch lbs. (8 Nm).

66. Install the crankshaft pulley.

a. Align the pulley set key with the groove of the pulley, and slide on the pulley.

b. Using the special tool (09213-70011 or 09330-00021), install the pulley set bolt. Tighten to 221 ft. lbs. (300 Nm).

67. Install the No. 1 idler pulley sub assembly with the bolt. Tighten to 32 ft. lbs. (43 Nm).

68. Install the water pump pulley.

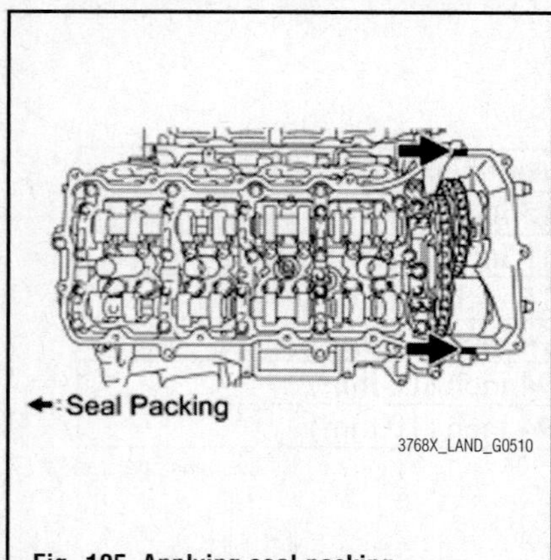

Fig. 195 Applying seal packing

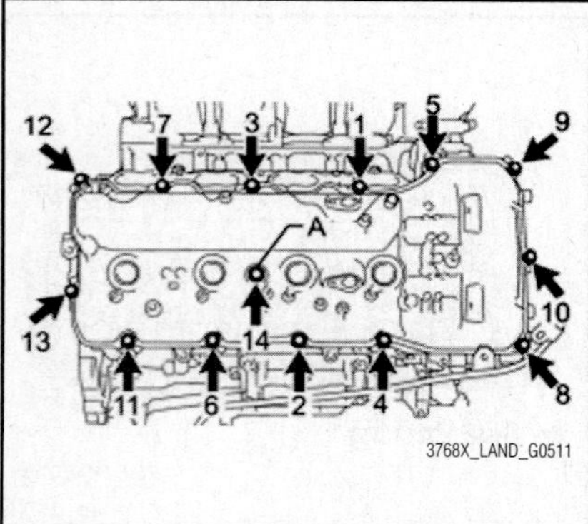

Fig. 196 Installing the cylinder head cover and seal washer

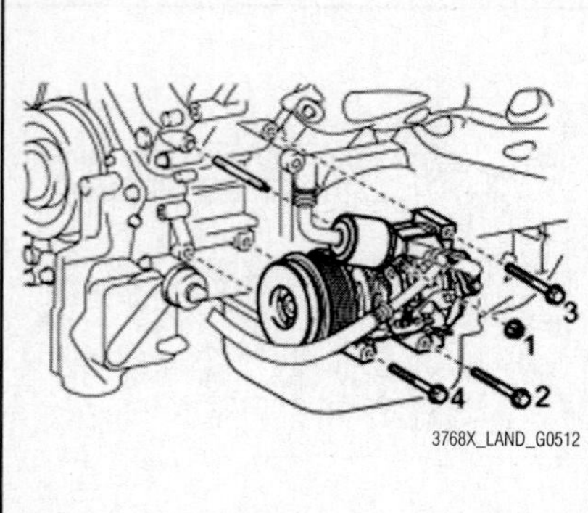

Fig. 197 Installing the cooler compressor assembly

a. Temporarily install the pulley with the 4 bolts.

b. Using the special tool (09960-10010, 09962-01000 or 09963-01000), hold the pulley and tighten the 4 bolts. Tighten to 15 ft. lbs. (21 Nm).

69. Install the water inlet housing.

a. Install a new gasket to the timing chain cover.

b. Install the water inlet with the 3 bolts. Tighten to 15 ft. lbs. (21 Nm).

70. Install the air tube sub assembly.

a. Connect the 2 hoses.

b. Install the air tube with the bolt. Tighten to 7 ft. lbs. (10 Nm).

71. Install the No. 1 engine cover.

72. Install the No. 2 engine cover.

73. Install the front water by pass joint.

a. Install the 2 new gaskets and the water by pass joint with the 4 nuts. Tighten to 15 ft. lbs. (21 Nm).

b. Connect the No. 2 water by pass hose to the water by pass joint.

74. Install the water by pass pipe sub assembly.

a. Connect the 2 hoses.

b. Install the water by pass pipe with the 2 bolts. Tighten to 7 ft. lbs. (10 Nm).

c. Connect the air tube with the bolt. Tighten to 7 ft. lbs. (10 Nm).

75. Install the No. 1 water by pass hose.

a. Install the No. 1 water by pass hose by connecting the hose to the water inlet housing and the front water by pass joint.

76. Install the generator assembly.

77. Connect the oil cooler pipe assembly.

78. Connect the vane pump assembly.

➡Before performing the following procedures, move the spacer until the vane pump can be installed.

a. Connect the vane pump to the timing chain cover with the 2 bolts. Tighten to 15 ft. lbs. (21 Nm).

79. Install the intake manifold.

80. Install the oil filter bracket (without oil cooler).

a. Apply a light coat of engine oil to 2 new O-rings.

b. Install the 2 o-rings to the timing chain cover.

c. Install the oil filter bracket with the 2 bolts and 2 nuts. Tighten to 26 ft. lbs. (35 Nm).

81. Install the oil filter bracket (with oil cooler).

82. Install the No. 1 oil cooler bracket (with oil cooler).

83. Install the No. 2 water by pass pipe sub assembly (with oil cooler).

a. Connect the 4 hoses.

b. Install the water by-pass pipe with the 3 bolts. Tighten to 7 ft. lbs. (10 Nm).

84. Install the oil pressure sender gauge assembly.

85. Install the engine oil level dipstick guide.

86. Install the oil filter element.

87. Connect the No. 2 fuel tube sub assembly.

a. Connect the fuel tube with the 2 bolts. Tighten to 7 ft. lbs. (10 Nm).

88. Connect the cooler compressor assembly.

a. Install the cooler compressor with the stud bolt. Tighten to 7 ft. lbs. (10 Nm).

b. Install the 3 bolts and nut. Tighten to 18 ft. lbs (25 Nm).

➡Tighten the bolts and nut in the order shown.

89. Connect the No. 2 water by pass pipe.

a. Connect the No. 2 water by pass pipe to the cylinder head cover with the 3 bolts. Tighten to 13 ft. lbs. (18 Nm).

b. Connect the 3 hoses.

90. Connect the air pump hose and wire harness.

a. Connect the No. 2 and No. 3 air hoses.

b. Connect the 2 clamps and air pump wire harness.

91. Connect the engine wire.

a. For the engine room RH side, connect or install the following:

- 5 clamps
- Throttle position sensor and throttle control motor connector
- 2 air pump connectors
- Ground wire with the 2 bolts (tighten to 71 inch lbs. (8 Nm))
- Noise filter connector
- 2 VVT sensor connectors
- Injector connector
- 4 ignition coil connectors
- 2 camshaft timing oil control valve connectors
- 2 clamps and power steering oil pressure switch connector.

92. For the engine room LH side, connect or install the following:

- Cooler compressor connector
- 3 clamps
- Camshaft position sensor connector
- 2 camshaft timing oil control valve connectors
- Engine Coolant Temperature (ECT) sensor connector
- 4 clamps
- Ground wire with the 2 bolts (tighten to 71 inch lbs. (8 Nm))
- Noise filter connector
- 2 VVT sensor connectors
- 4 ignition coil connectors
- Injector connector

- Wire harness clamp and 4 air injection control driver connectors
- 2 connectors and 2 clips to the engine room junction block
- Engine room relay block cover

93. Install the radiator assembly.
94. Install the air cleaner assembly.
95. Install the air cleaner hose assembly.
96. Install the v-bank cover sub assembly.
97. Add engine oil.
98. Add engine coolant.
99. Connect the cable to the negative battery terminal.
100. Inspect for oil leaks.
101. Inspect for coolant leaks.
102. Inspect the engine oil level.
103. Install the No. 2 engine under cover.
104. Install the No. 1 engine under cover sub assembly.
105. Install the front fender splash shield sub assembly LH.
106. Install the front fender splash shield sub assembly RH.
107. Install the front fender apron seal front RH.
108. Install the front fender apron seal LH.
109. Install the cowl top ventilator louver sub assembly.
110. Connect the radiator side deflector LH.
111. Connect the radiator side deflector RH (without air cooled transmission oil cooler).
112. Install the transmission oil cooler air duct (with air cooled transmission oil cooler).
113. Install the front bumper cover.
114. Install the radiator grille.
115. Install the upper radiator support seal.

REAR MAIN SEAL

REMOVAL & INSTALLATION

See Figure 198.

1. Disconnect the cable from the negative battery terminal.

CAUTION

Wait at least 90 seconds after disconnecting the cable from the negative (-) battery terminal to disable the SRS system.

2. Remove the automatic transmission assembly.

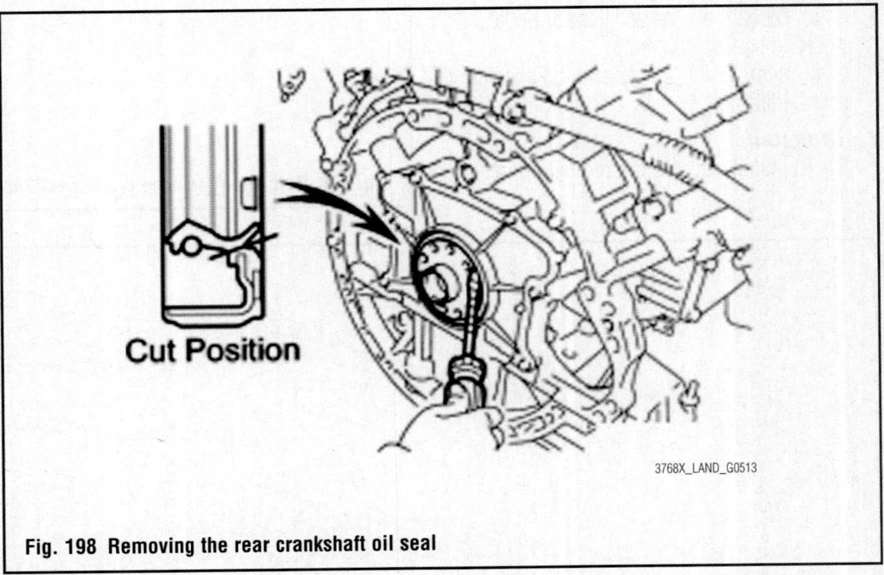

Fig. 198 Removing the rear crankshaft oil seal

3. Remove the drive plate and ring gear sub assembly.
4. Remove the rear crankshaft oil seal.
   a. Using a knife, cut off the lip of the oil seal.
   b. Using a screwdriver, pry out the oil seal.

➡Tape the screwdriver tip before use.

➡Do not damage the surface of the oil seal press fit hose and crankshaft.

To install:

5. To install, reverse the removal procedure.

TIMING CHAIN FRONT COVER

REMOVAL & INSTALLATION

Refer to OIL PUMP for removal and installation procedures.

TIMING CHAIN & SPROCKETS

REMOVAL & INSTALLATION

Refer to OIL PUMP for removal and installation procedures.

ENGINE PERFORMANCE & EMISSION CONTROLS

ACCELERATOR PEDAL POSITION (APP) SENSOR

LOCATION

See Figure 199.

Refer to the accompanying illustration for sensor location.

REMOVAL & INSTALLATION

See Figure 200.

1. Remove the NO. 1 instrument panel under cover sub assembly.
2. Remove the accelerator pedal assembly.
   a. Disconnect the accelerator pedal position sensor connector.
   b. Remove the 2 nuts and the accelerator pedal.

To install:

3. To install, reverse the removal procedure.

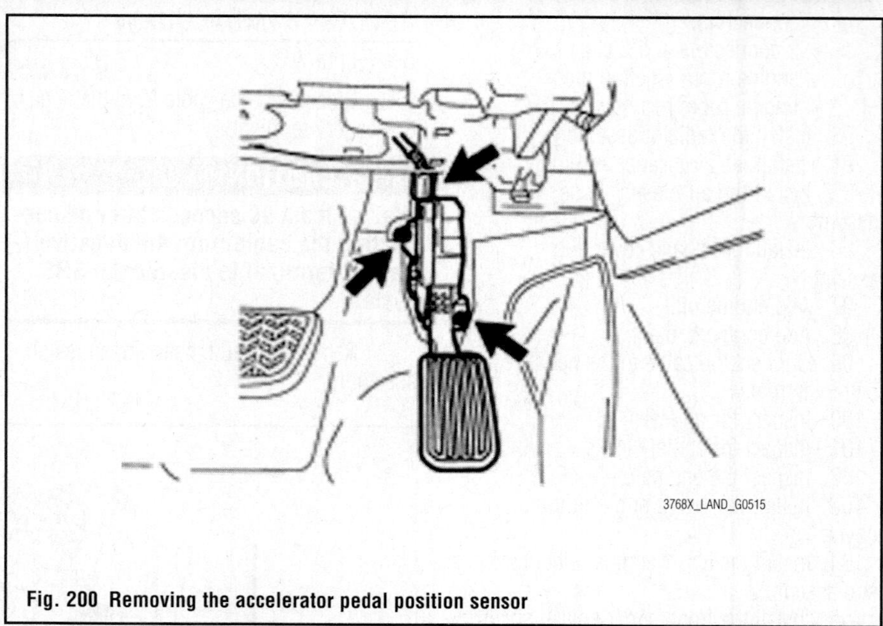

Fig. 200 Removing the accelerator pedal position sensor

ACCELERATOR PEDAL POSITION SENSOR CONNECTOR

x 2

NO. 1 INSTRUMENT PANEL UNDER COVER SUB-ASSEMBLY

x 2 5.4 (55, 48 in.*lbf)

N*m (kgf*cm, ft.*lbf) : Specified torque

ACCELERATOR PEDAL ASSEMBLY

3768X_LAND_G0514

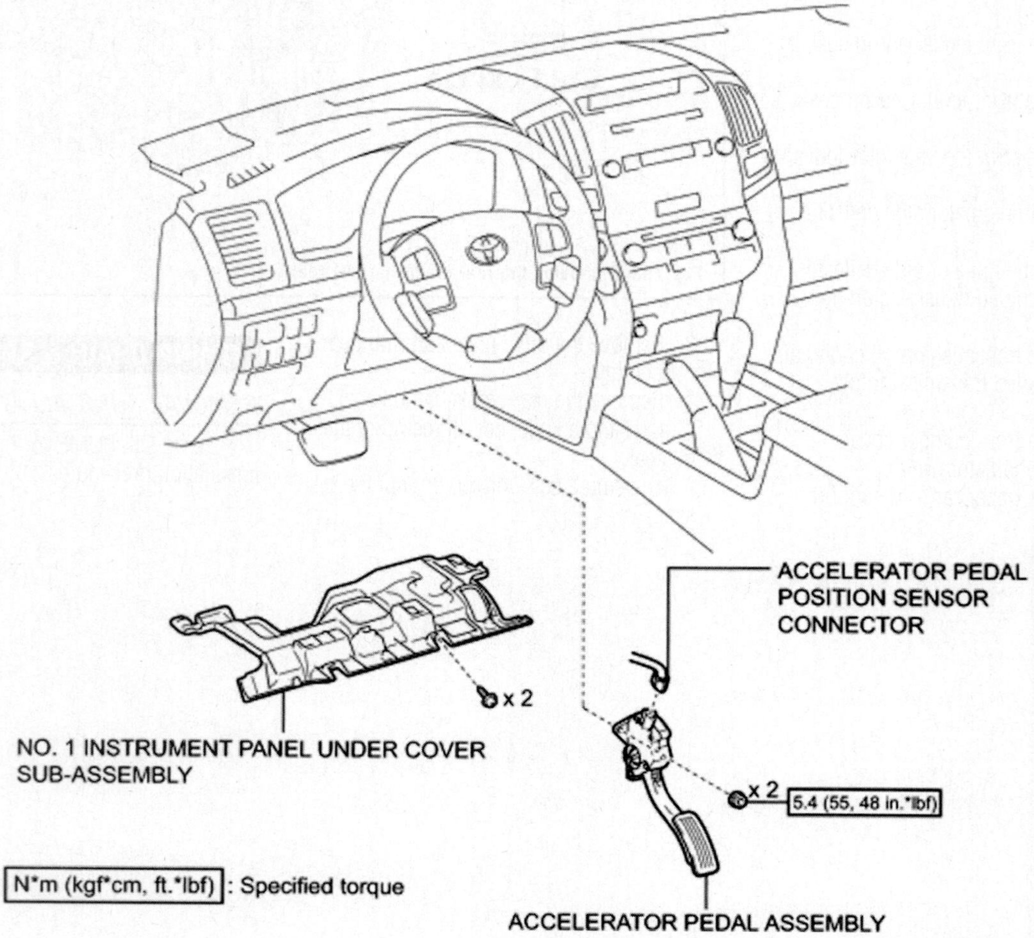

Fig. 199 Accelerator Pedal Position (APP) sensor and related components

CAMSHAFT POSITION (CMP) SENSOR

LOCATION

See Figure 201.

Refer to the accompanying illustration for sensor location.

REMOVAL & INSTALLATION

See Figures 202 and 203.

1. Remove the v-bank cover sub assembly.
2. Remove the air cleaner hose assembly.
3. Remove the camshaft position sensor.
    a. Disconnect the sensor connector.
    b. Remove the bolt and sensor.
4. Remove the VVT sensor.
    a. Disconnect the 4 sensor connectors.
    b. Remove the 4 bolts and 4 sensors.

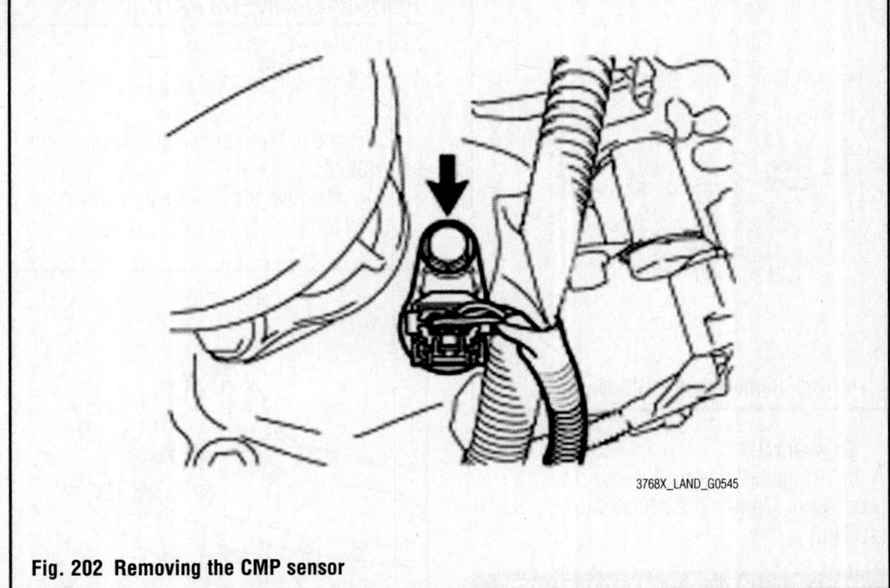

Fig. 202 Removing the CMP sensor

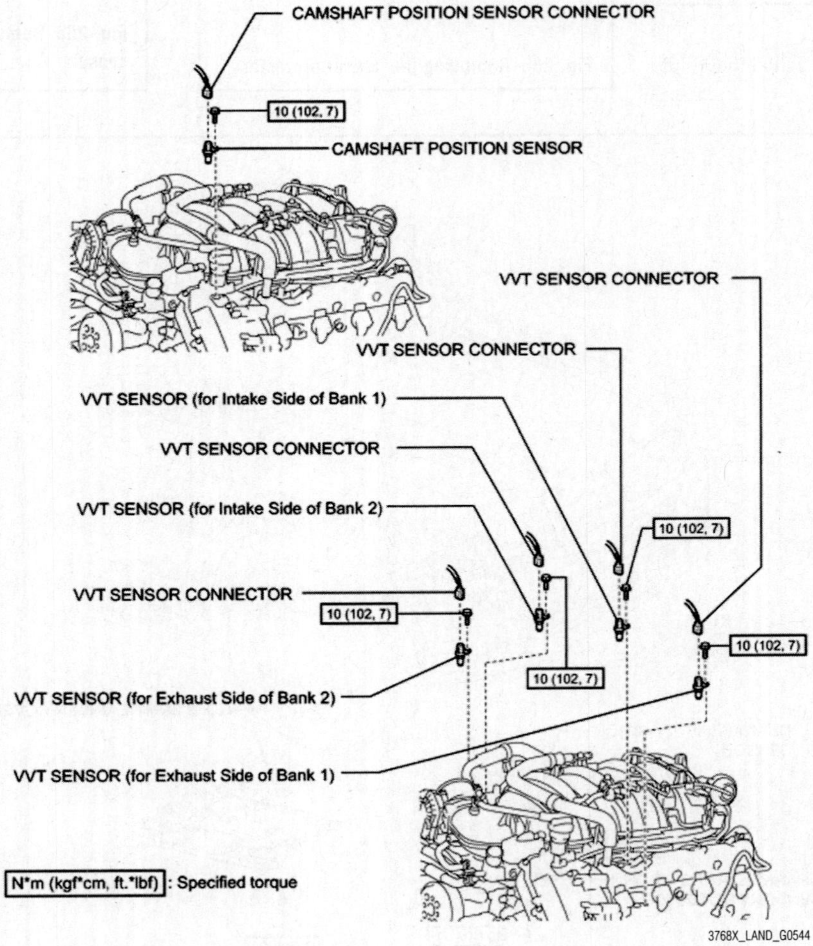

Fig. 201 Camshaft Position (CMP) sensor

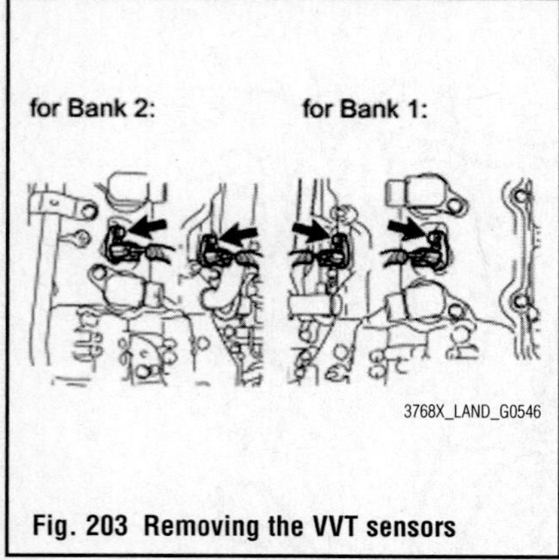

Fig. 203 Removing the VVT sensors

To install:

5. To install, reverse the removal procedure. Tighten the sensors to 7 ft. lbs. (10 Nm).

CRANKSHAFT POSITION (CKP) SENSOR

LOCATION

See Figure 204.

Refer to the accompanying illustration for sensor location.

REMOVAL & INSTALLATION

See Figures 205 and 206.

1. Remove the No. 2 engine under cover.
2. Remove the crankshaft position sensor protector.
   a. Remove the 2 bolts and sensor protector.

Fig. 205 Removing the sensor protector

3. Remove the crankshaft position sensor.
   a. Disconnect the sensor connector.
   b. Remove the bolt and sensor.

To install:

4. To install, reverse the removal procedure. Tighten the sensor and sensor protector to 7 ft. lbs. (10 Nm).

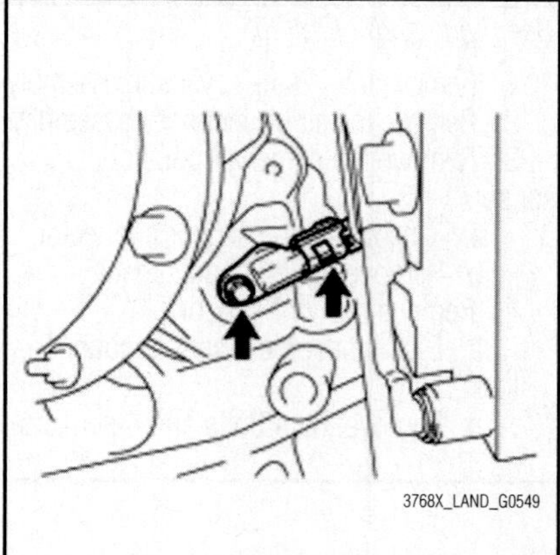

Fig. 206 Removing the crankshaft position sensor

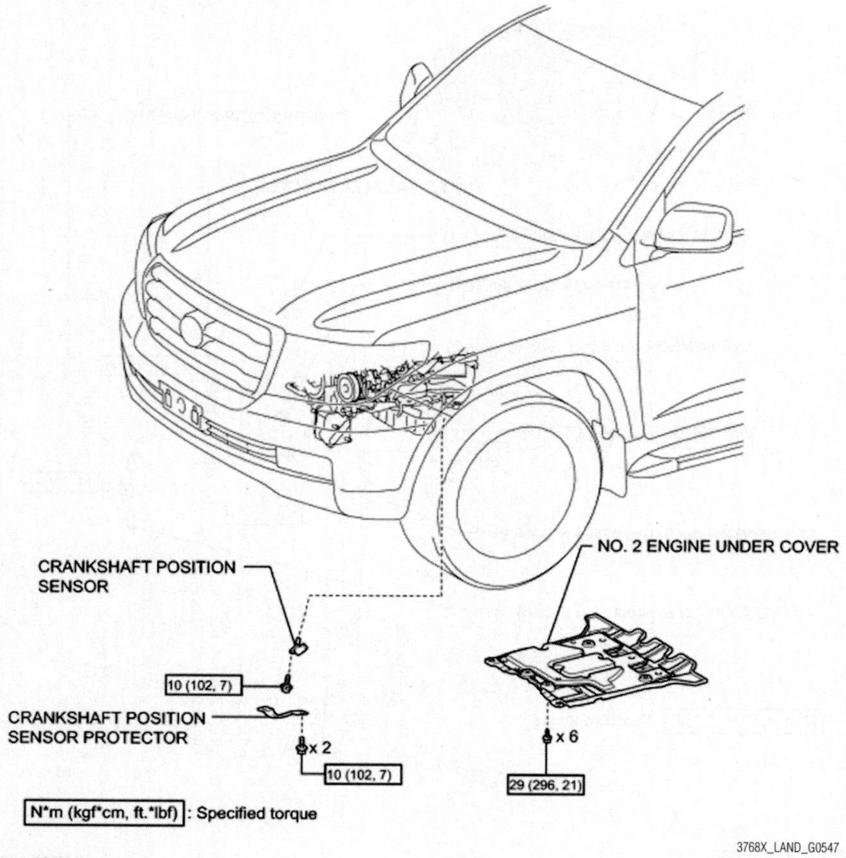

Fig. 204 Crankshaft Position (CKP) sensor

ELECTRONIC CONTROL MODULE (ECM)

LOCATION

See Figure 207.

Refer to the accompanying illustration for location.

REMOVAL & INSTALLATION

See Figures 208 through 210.

1. Disconnect the cable from the negative battery terminal.
2. Disconnect the connector holder block.
   a. Remove the 3 bolts and move the connector holder block so that the ECM can be removed in the next step.
3. Remove the ECM.
   a. Raise the 2 levers while pushing the locks on the 2 levers.

➡Make sure that the lock levers are raised 90° as shown in the illustration

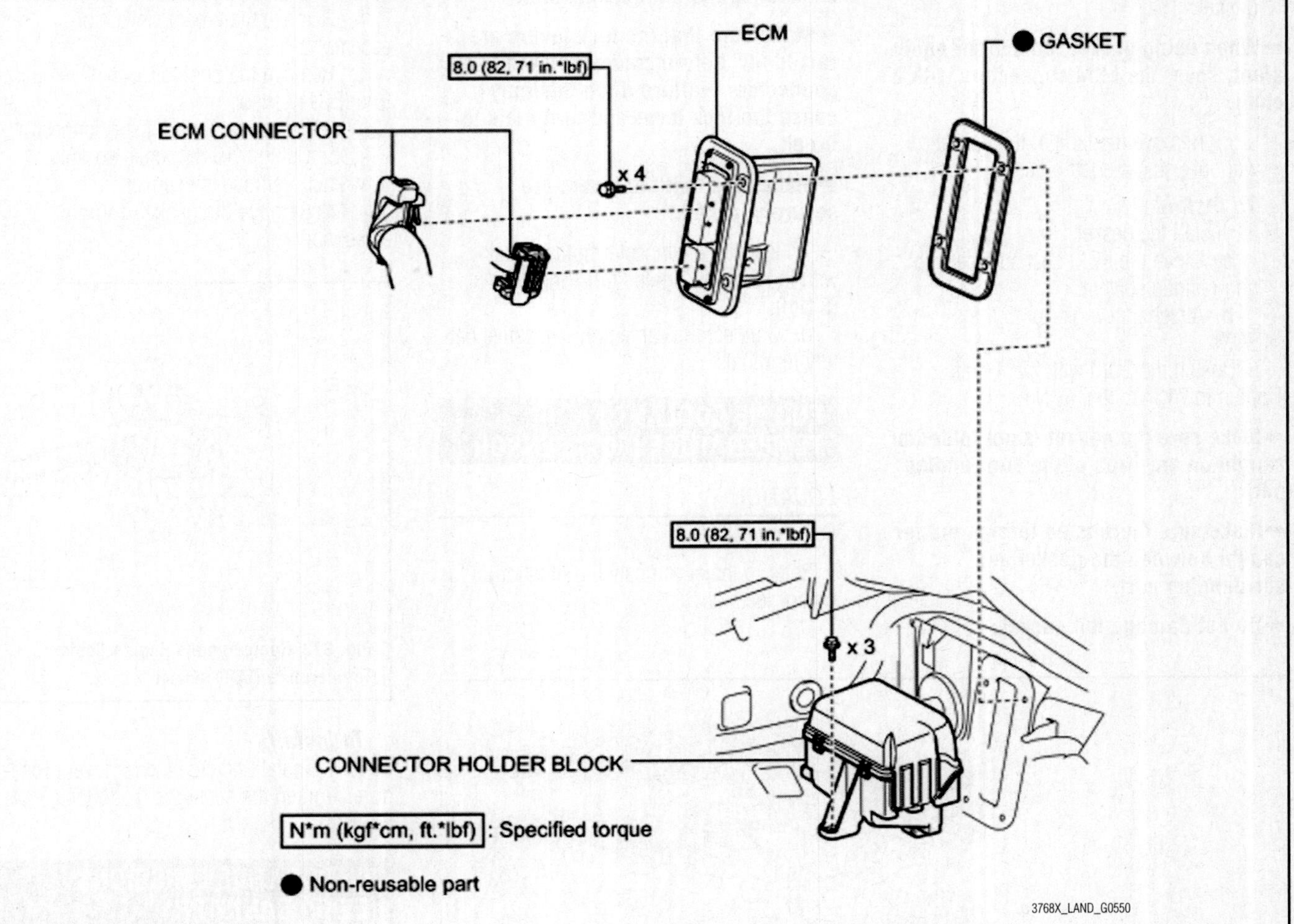

Fig. 207 ECM location

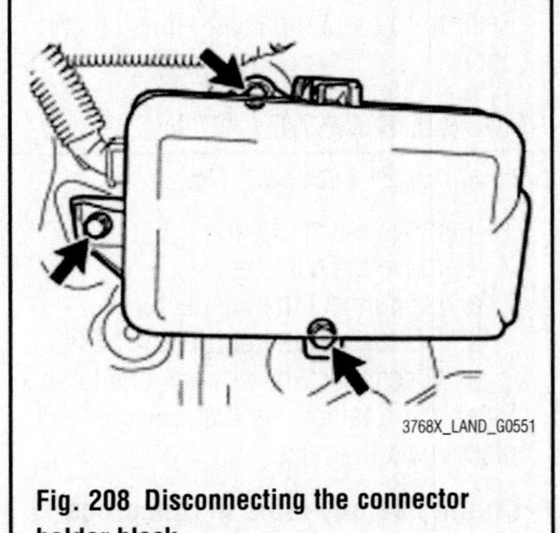

Fig. 208 Disconnecting the connector holder block

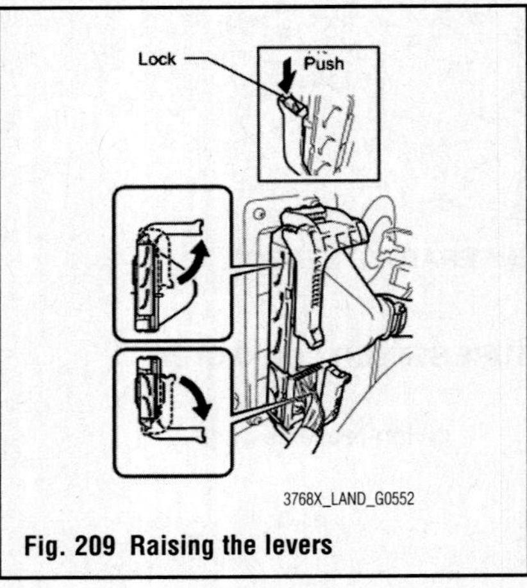

Fig. 209 Raising the levers

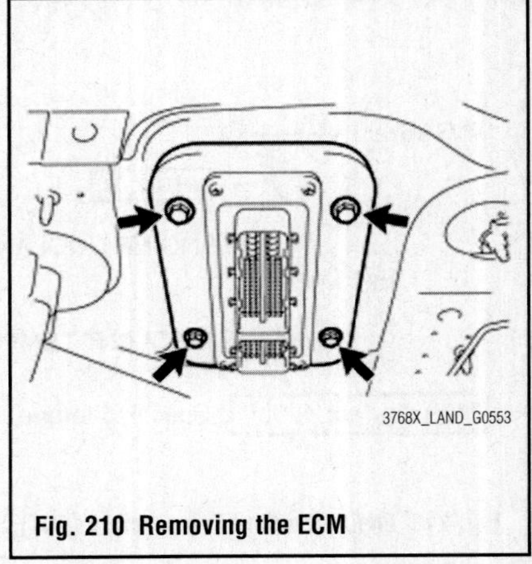

Fig. 210 Removing the ECM

before disconnecting the connectors. Failure to do this may cause the connectors to break.

b. Disconnect the 2 connectors.
c. Remove the 4 bolts and ECM.

4. Remove the gasket.
a. Peel off the gasket.
b. Spray gasket remover or equivalent on the remaining tape of the gasket.

➡When using gasket remover or equivalent, cover the ECM connectors with a cloth.

c. Remove the tape of the gasket without using bladed objects.

To install:

5. Install the gasket.
a. Clean the ECM seal surface with non residue solvent.
b. Attach a new gasket to the ECM.

6. Install the ECM with the 4 bolts. Tighten to 71 inch lbs. (8 Nm).

➡Make sure the gasket is not folded or caught on any area of the surrounding part.

➡Make sure there is no foreign matter caught between the gasket and surrounding part.

➡Do not damage the gasket.

a. Connect the 2 ECM connectors and push each lock lever down to lock the ECM connectors.

➡Be careful not to allow dirt, water, or other foreign matter to adhere to the connecting parts when connecting the connectors.

➡Do not twist the wire harness when connecting the ECM connectors.

➡Make sure that the lock levers are raised 90° before connecting the ECM connectors. Failure to do this may cause the lock lever and terminals to break.

➡Make sure that the levers are securely locked.

7. Install the connector holder block with the 3 bolts. Tighten to 71 inch lbs. (8 Nm).

8. Connect the cable to the negative battery terminal.

ENGINE COOLANT TEMPERATURE (ECT) SENSOR

LOCATION

See Figure 211.

Refer to the accompanying illustration for sensor location.

● GASKET
20 (200, 14)
ENGINE COOLANT TEMPERATURE SENSOR
ENGINE COOLANT TEMPERATURE SENSOR CONNECTOR
N*m (kgf*cm, ft.*lbf) : Specified torque
● Non-reusable part
3768X_LAND_G0554

Fig. 211 Engine Coolant Temperature (ECT) sensor

REMOVAL & INSTALLATION

See Figure 212.

1. Remove the front fender splash shield sub assembly RH.
2. Remove the front fender splash shield sub assembly LH.
3. Remove the No. 1 engine under cover sub assembly.
4. Drain the engine coolant.
5. Remove the v-bank cover sub assembly.
6. Remove the Engine Coolant Temperature (ECT) sensor.
a. Disconnect the sensor connector.
b. Using a 19 mm deep socket wrench, remove the sensor.
c. Remove the gasket from the sensor.

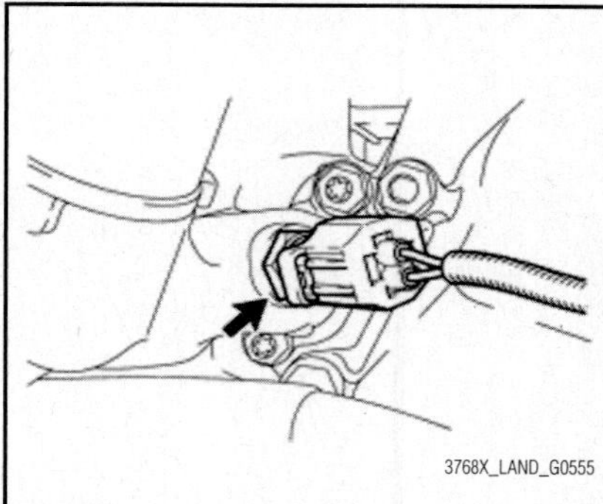

Fig. 212 Removing the Engine Coolant Temperature (ECT) sensor

To install:

7. To install, reverse the removal procedure. Tighten the sensor to 14 ft. lbs. (20 Nm).

EVAPORATIVE EMISSIONS (EVAP) CANISTER

LOCATION

See Figure 213.

Refer to the accompanying illustration for location.

REMOVAL & INSTALLATION

See Figures 214 through 216.

1. Remove the spare tire.
2. Remove the canister.
a. Disconnect the connector.
b. Disconnect the purge line hose.
c. Disconnect the air inlet line hose. Pinch the retainer and then raise it, and disconnect the hose.

➡Do not use any tools in this procedure.

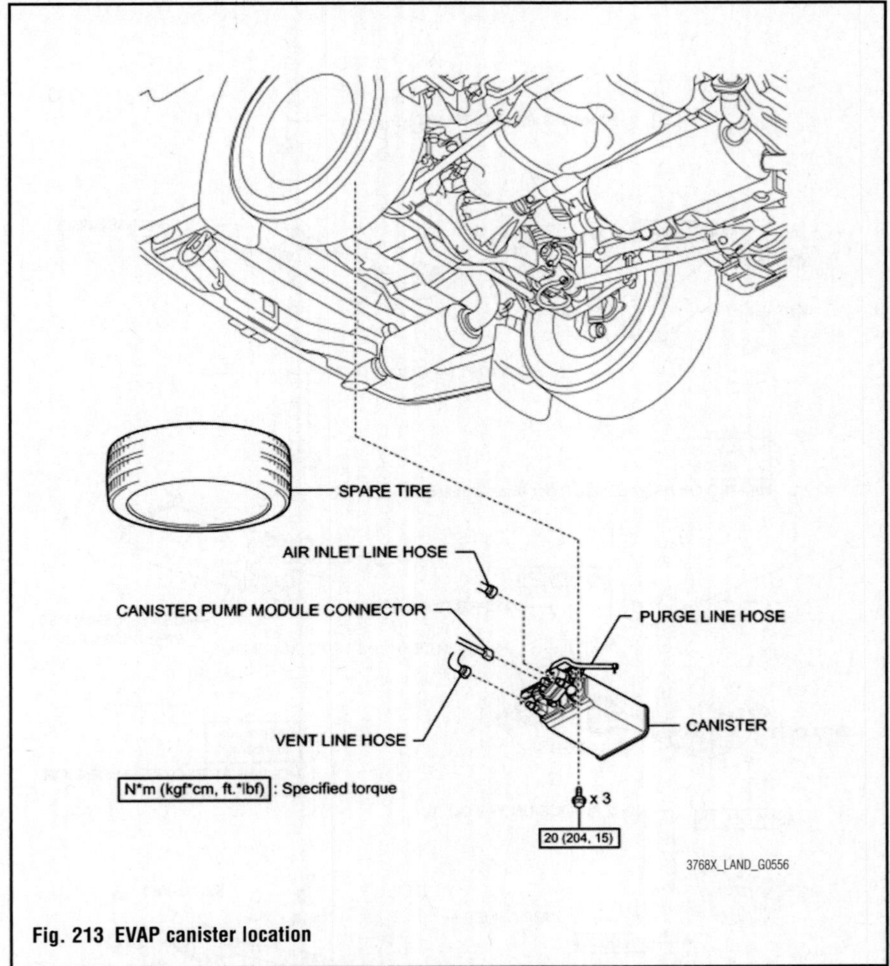

Fig. 213 EVAP canister location

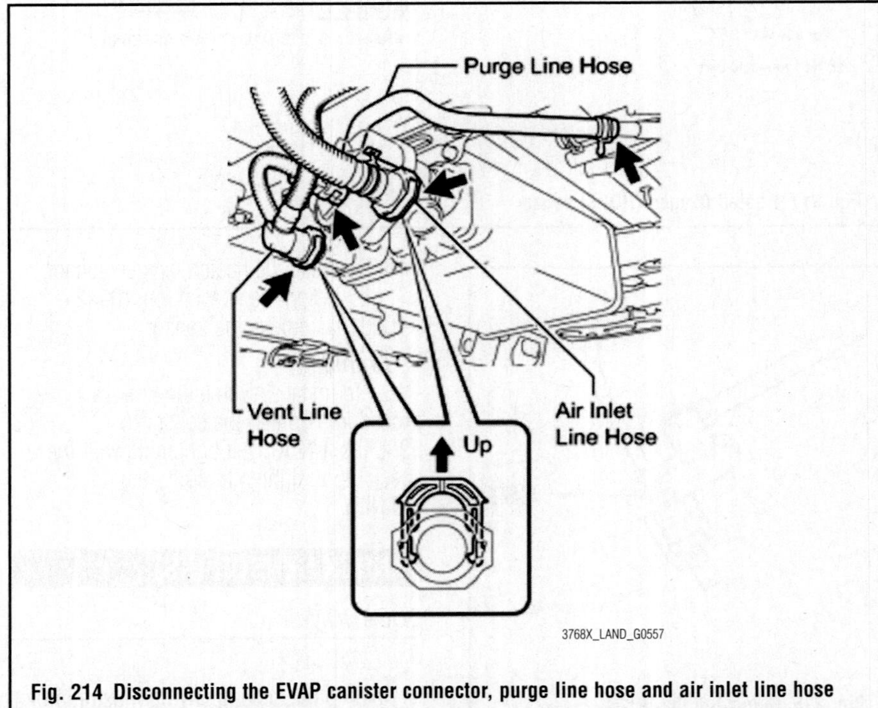

Fig. 214 Disconnecting the EVAP canister connector, purge line hose and air inlet line hose

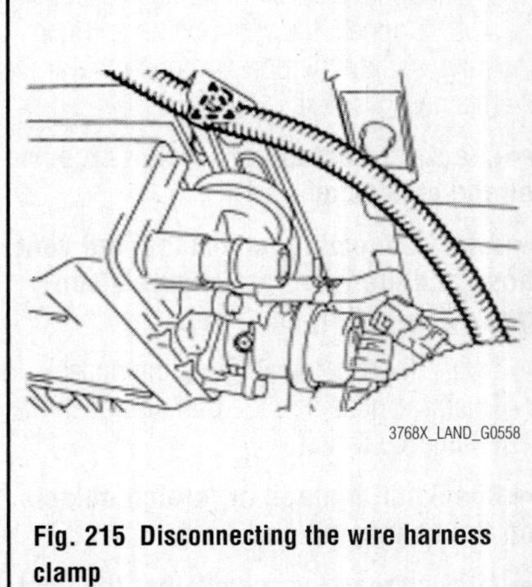

Fig. 215 Disconnecting the wire harness clamp

Fig. 216 Removing the canister

➡Check for any dirt and foreign matter contamination in the valve and around the connector. Clean if necessary. Foreign matter may damage the O-rings or cause leaks in the seal between the valve and connector.

d. Disconnect the vent line hose. Pinch the retainer and then raise it, and disconnect the hose.

➡Do not use any tools in this procedure.

➡Check for any dirt and foreign matter contamination in the valve and around the connector. Clean if necessary. Foreign matter may damage the O-rings or cause leaks in the seal between the valve and connector.

e. Disconnect the wire harness clamp.
f. Remove the 3 bolts and canister.

To install:

3. Install the canister with the 3 bolts. Tighten to 15 ft. lbs. (20 Nm).

a. Connect the wire harness clamp.

b. Connect the vent line hose. Push the hose into the port and push the retainer to lock it.

➡Check for damage or foreign objects on the connected part.

➡After connecting, check that the vent hose and the valve are securely connected by pulling on them.

c. Connect the air inlet line hose. Push the hose into the port and push the retainer to lock it.

➡Check for damage or foreign objects on the connected part.

➡After connecting, check that the vent hose and the valve are securely connected by pulling on them.

4. Connect the purge line hose.
5. Connect the connector.

HEATED OXYGEN SENSOR (HO2S)

LOCATION

See Figure 217.

Refer to the accompanying illustration for sensor location.

REMOVAL & INSTALLATION

See Figure 218.

⁂ CAUTION

The procedures should be performed by at least 2 people.

⁂ CAUTION

Wear protective gloves when removing the exhaust pipe.

⁂ CAUTION

The exhaust pipe is extremely hot immediately after the engine has stopped.

⁂ CAUTION

Confirm that the exhaust pipe has cooled down before removing it.

1. Remove the No. 2 engine under cover.
2. Remove the tailpipe assembly.
3. Remove the center exhaust pipe assembly.
4. Remove the front No. 2 exhaust pipe assembly.

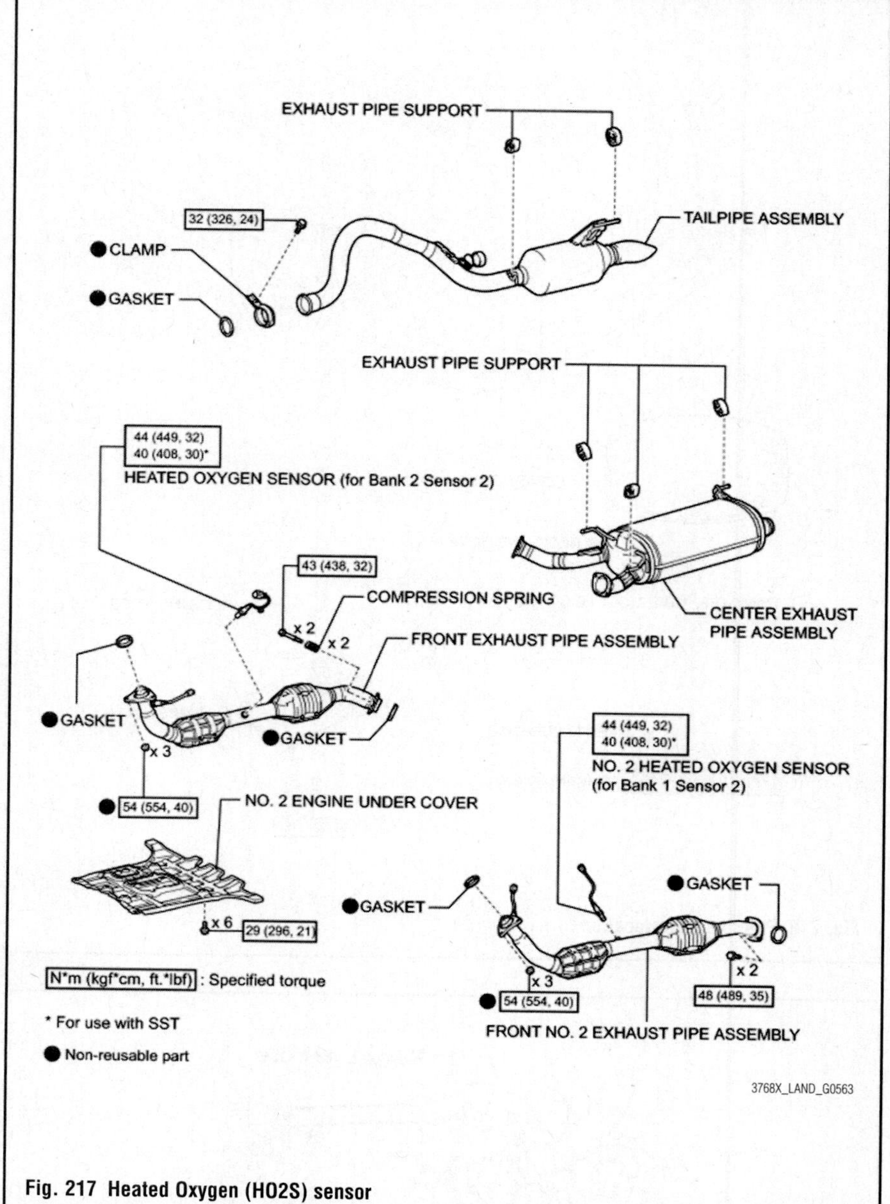

Fig. 217 Heated Oxygen (HO2S) sensor

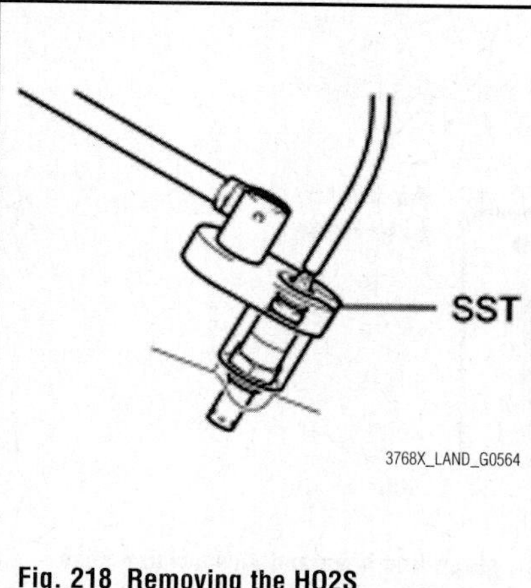

Fig. 218 Removing the HO2S

5. Remove the heated oxygen sensor.
   a. Using the special tool (09224-00010), remove the sensor.

To install:

6. To install, reverse the removal procedure. Tighten the sensor to 32 ft. lbs. (44 Nm). If tightening with the special tool, tighten to 30 ft. lbs. (40 Nm).

KNOCK SENSOR (KS)

LOCATION

See Figure 219.

Refer to the accompanying illustration for sensor location.

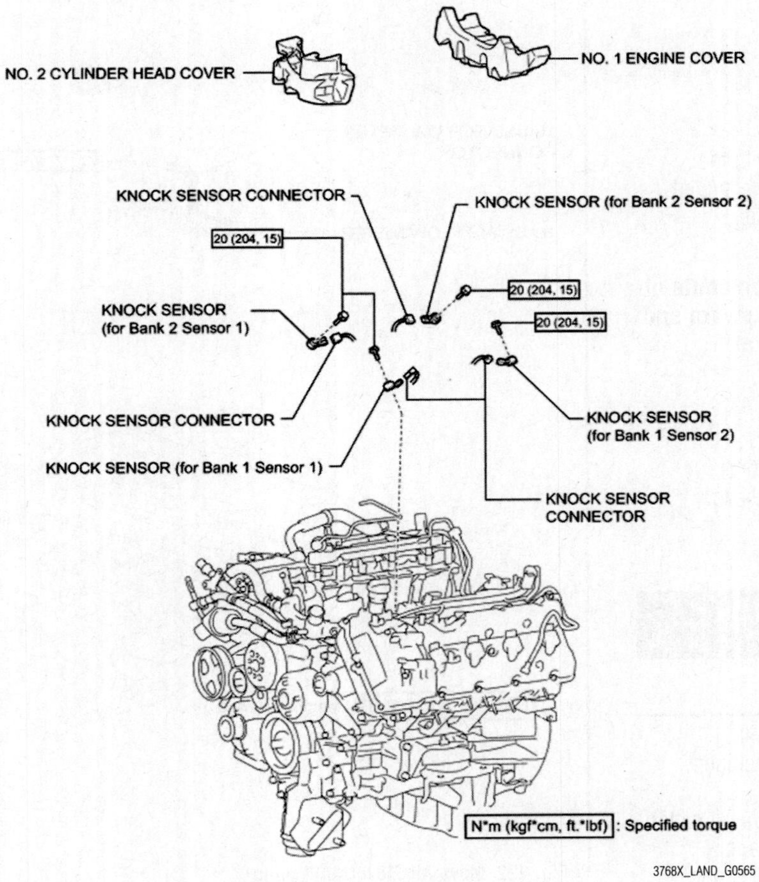

NO. 2 CYLINDER HEAD COVER
NO. 1 ENGINE COVER
KNOCK SENSOR CONNECTOR
KNOCK SENSOR (for Bank 2 Sensor 2)
20 (204, 15)
20 (204, 15)
20 (204, 15)
KNOCK SENSOR
(for Bank 2 Sensor 1)
KNOCK SENSOR CONNECTOR
KNOCK SENSOR
(for Bank 1 Sensor 2)
KNOCK SENSOR (for Bank 1 Sensor 1)
KNOCK SENSOR
CONNECTOR
N*m (kgf*cm, ft.*lbf) : Specified torque
3768X_LAND_G0565

Fig. 219 Knock Sensor (KS)

REMOVAL & INSTALLATION

See Figures 220 and 221.

1. Remove the intake manifold.
2. Remove the No. 2 cylinder head cover.
3. Remove the No. 1 engine cover.
4. Remove the Knock Sensor (KS):

3768X_LAND_G0566

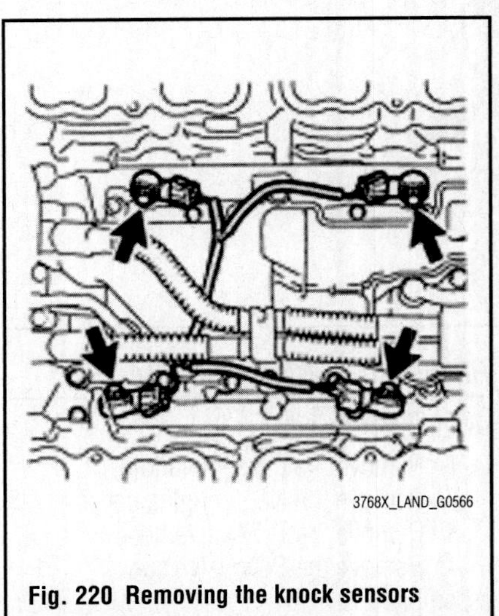

Fig. 220 Removing the knock sensors

A
C
Engine Front
B
D
A: Bank 2 Sensor 1
C: Bank 2 Sensor 2
Top
Top
Front
10°
10°
Front
10°
10°
B: Bank 1 Sensor 1
D: Bank 1 Sensor 2
Top
Top
10°
10°
Rear
Rear
10°
10°
3768X_LAND_G0567

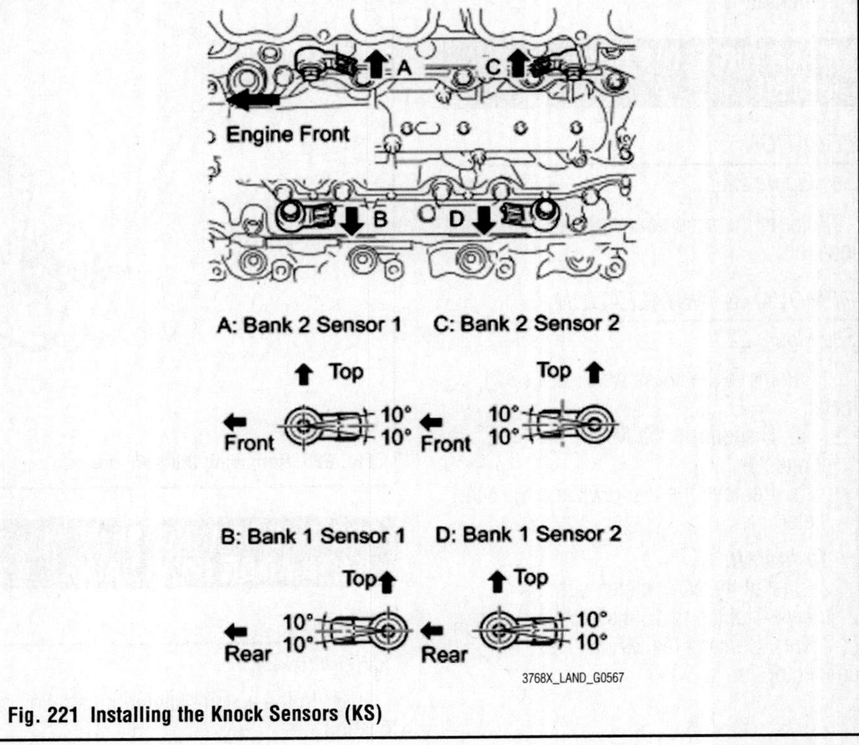

Fig. 221 Installing the Knock Sensors (KS)

a. Disconnect the 4 KS connectors.

b. Remove the 4 bolts and 4 KS.

To install:

5. Install the KS.

a. Install the 4 sensors with the 4 bolts so that the sensors are angled as shown. Tighten to 15 ft. lbs. (20 Nm).

➡The acceptable installation angle of the sensor is between 10° upward and downward from the horizontal position.

b. Connect the 4 sensor connectors.

6. Install the No. 1 engine cover.
7. Install the No. 2 cylinder head cover.
8. Install the intake manifold.

MALFUNCTION INDICATOR LIGHT (MIL)

RESET PROCEDURE

Clearing DTC codes resets MIL.

1. To clear codes without intelligent tester:

a. Disconnect the negative (-) battery cable for more than 1 minute, or Remove the EFI OR ECD NO. 1 and ETCS fuses from the engine room junction block located inside the engine compartment for more than 1 minute.

MASS AIR FLOW (MAF) METER

LOCATION

See Figure 222.

Refer to the accompanying illustration for location.

REMOVAL & INSTALLATION

See Figure 223.

1. Remove the Mass Air Flow (MAF) meter.

a. Disconnect the MAF connector.

b. Remove the 2 screws and the MAF meter.

To install:

2. Install the MAF meter with the 2 screws. Tighten to 15 inch lbs. (1.7 Nm). Connect the MAF connector.

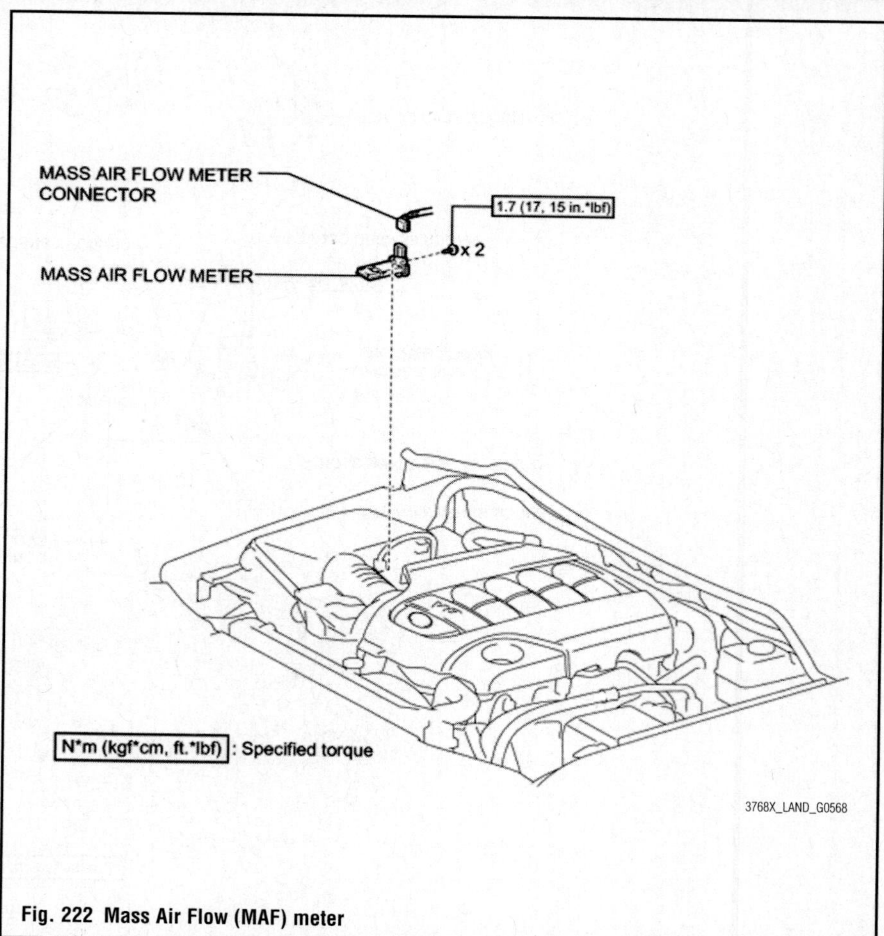

Fig. 222 Mass Air Flow (MAF) meter

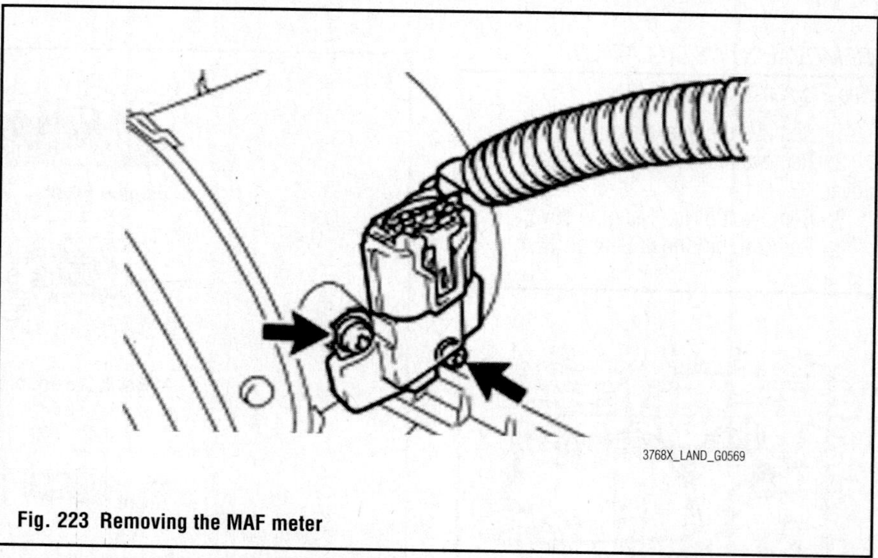

Fig. 223 Removing the MAF meter

POSITIVE CRANKCASE VENTILATION (PCV) VALVE

LOCATION

See Figure 224.

Refer to the accompanying illustration for location.

REMOVAL & INSTALLATION

See Figures 225 and 226.

1. Remove the intake manifold.
2. Remove the No. 2 engine cover.
3. Remove the PCV valve hose.
4. Remove the PCV valve sub assembly.

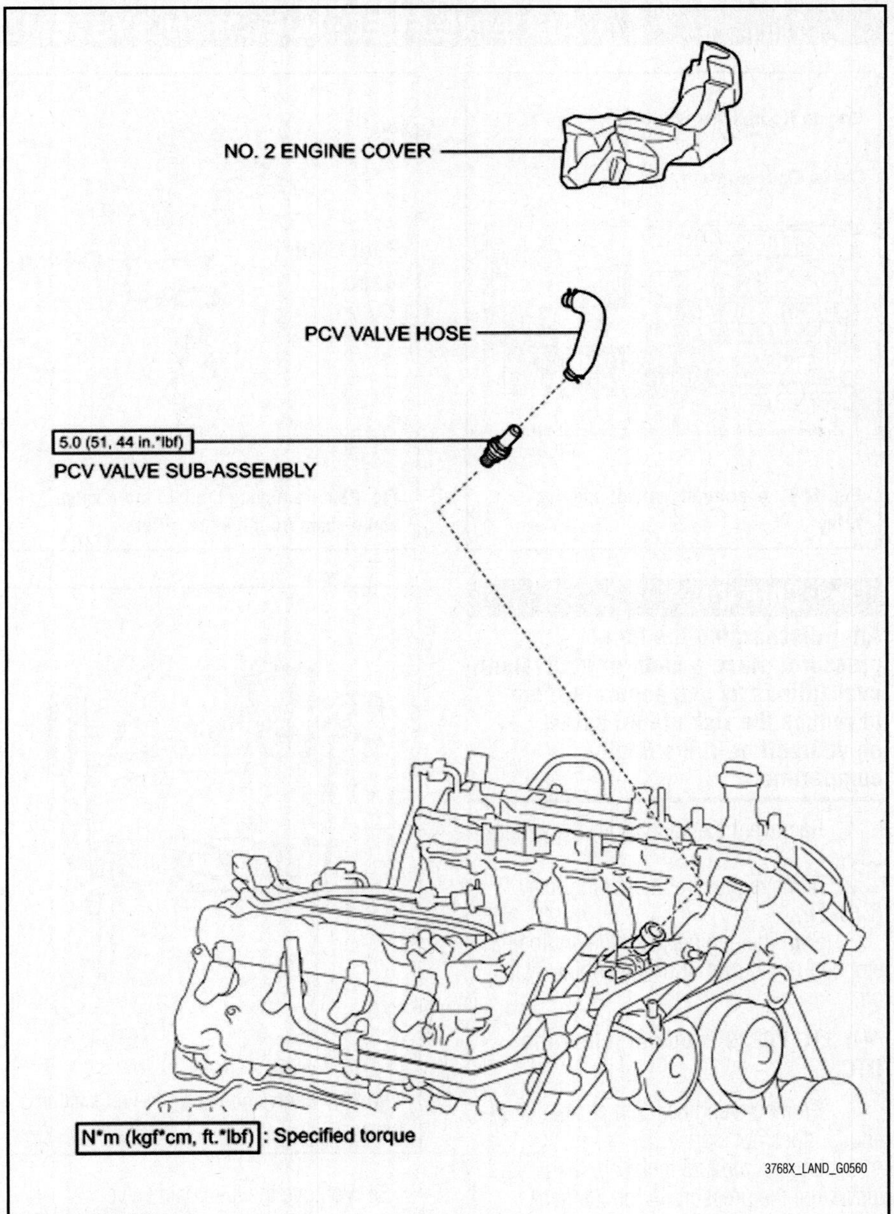

NO. 2 ENGINE COVER
PCV VALVE HOSE
5.0 (51, 44 in.*lbf)
PCV VALVE SUB-ASSEMBLY
N*m (kgf*cm, ft.*lbf) : Specified torque
3768X_LAND_G0560

Fig. 224 PCV valve

3768X_LAND_G0561

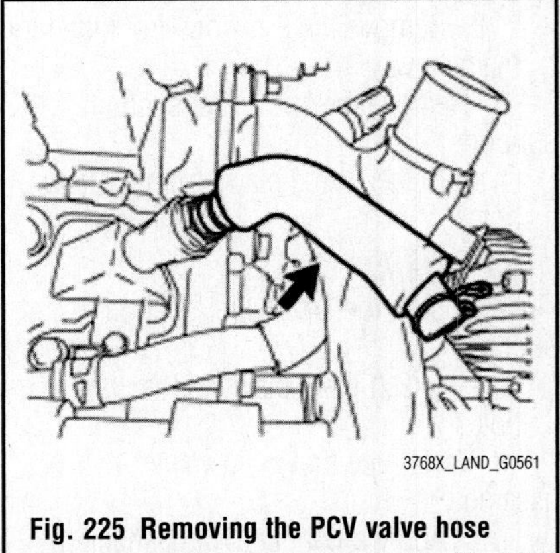

Fig. 225 Removing the PCV valve hose

3768X_LAND_G0562

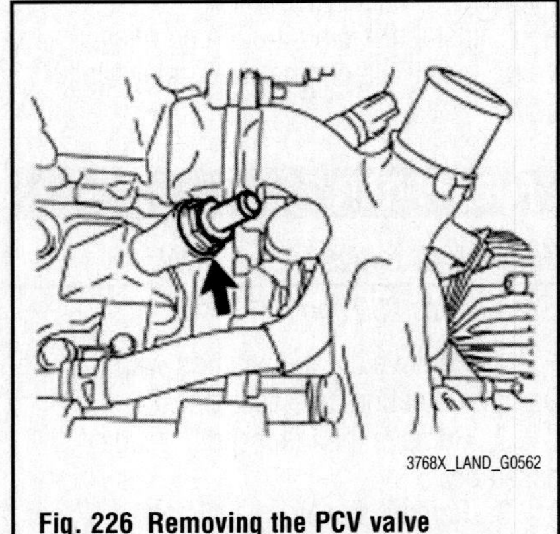

Fig. 226 Removing the PCV valve

To install:

5. Install the PCV valve sub assembly.
   a. Apply a light coat of engine oil to the o-ring of the valve.

➡When reusing the PCV valve inspect the o-ring. If the o-ring has scratches or cuts, replace the PCV valve.

   b. Install the PCM valve and tighten to 44 inch lbs. (5 Nm).
6. Install the PCV valve hose.
7. Install the No. 2 engine cover.
8. Install the intake manifold.

VEHICLE SPEED SENSOR (VSS)

REMOVAL & INSTALLATION

See Figures 227 and 228.

1. Remove the speed sensor NT.
   a. Disconnect the sensor connector.
   b. Remove the bolt and sensor.
   c. Remove the O-ring from the sensor.
2. Remove the speed sensor SP2.

(a) Disconnect the sensor connector.
(b) Remove the bolt and sensor.
(c) Remove the O-ring from the sensor.

To install:

3. To install, reverse the removal procedure. Tighten the speed sensors to 48 inch lbs. (5.4 Nm).

3768X_LAND_G0572

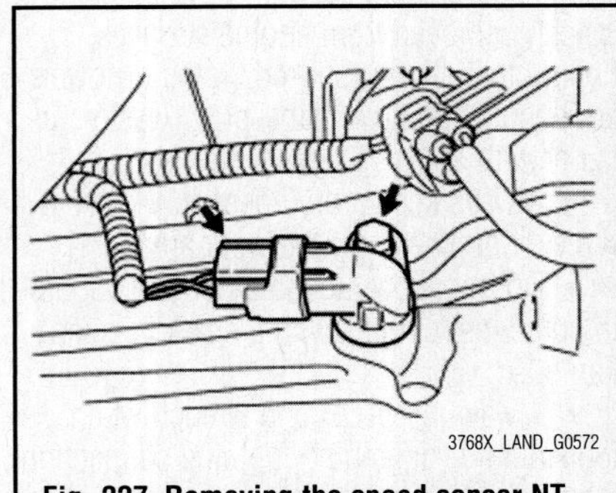

Fig. 227 Removing the speed sensor NT

3768X_LAND_G0573

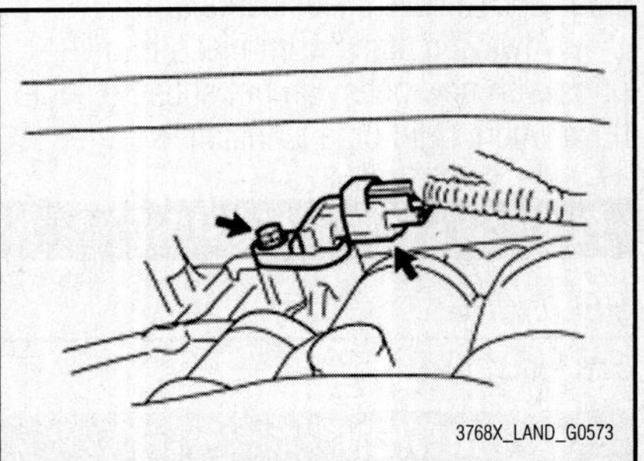

Fig. 228 Removing the speed sensor SP2

FUEL SYSTEM SERVICE PRECAUTIONS

Safety is the most important factor when performing not only fuel system maintenance, but any type of maintenance. Failure to conduct maintenance and repairs in a safe manner may result in serious personal injury or death. Work on a vehicle's fuel system components can be accomplished safely and effectively by adhering to the following rules and guidelines.

- To avoid the possibility of fire and personal injury, always disconnect the negative battery cable unless the repair or test procedure requires that battery voltage be applied.
- Always relieve the fuel system pressure prior to disconnecting any fuel system component (injector, fuel rail, pressure regulator, etc.) fitting or fuel line connection. Exercise extreme caution whenever relieving fuel system pressure to avoid exposing skin, face and eyes to fuel spray. Please be advised that fuel under pressure may penetrate the skin or any part of the body that it contacts.
- Always place a shop towel or cloth around the fitting or connection prior to loosening to absorb any excess fuel due to spillage. Ensure that all fuel spillage is quickly removed from engine surfaces. Ensure that all fuel-soaked cloths or towels are deposited into a flame-proof waste container with a lid.
- Always keep a dry chemical (Class B) fire extinguisher near the work area.
- Do not allow fuel spray or fuel vapors to come into contact with a spark or open flame.
- Always use a second wrench when loosening or tightening fuel line connection fittings. This will prevent unnecessary stress and torsion on fuel piping. Always follow the proper torque specifications.
- Always replace worn fuel fitting O-rings with new ones. Do not substitute fuel hose where rigid pipe is installed.

FUEL SYSTEM PRESSURE

RELIEVING

See Figure 229.

⁂ CAUTION

Do not disconnect any part of the fuel system until you have discharged the fuel system pressure.

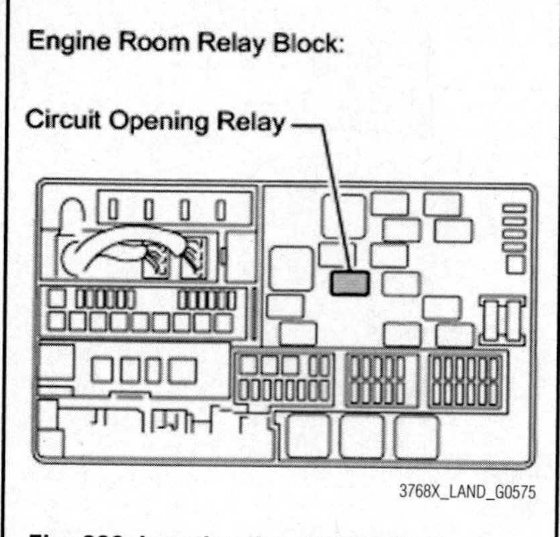

Fig. 229 Locating the circuit opening relay

⁂ CAUTION

After discharging the fuel pressure, place a cloth or equivalent over fittings as you separate them to reduce the risk of fuel spray on yourself or in the engine compartment.

1. Remove the engine room relay block cover.
2. Remove the circuit opening relay (C/OPN).
3. Start the engine. After the engine has stopped on its own, turn the engine switch off.

➡If DTC P0230 is output, clear the DTC.

4. Crank the engine, then check that the engine does not start.
5. Loosen the fuel tank cap, then discharge the pressure in the fuel tank completely.
6. Disconnect the cable from the negative (-) battery terminal.
7. Install the circuit opening relay.
8. Install the engine room relay block cover.

FUEL FILTER

REMOVAL & INSTALLATION

See Figures 230 and 231.

1. Remove the fuel suction with the pump and gauge tube assembly.
2. Remove the fuel sender gauge assembly.
3. Remove the No. 1 fuel sub tank.
4. Remove the fuel pump.

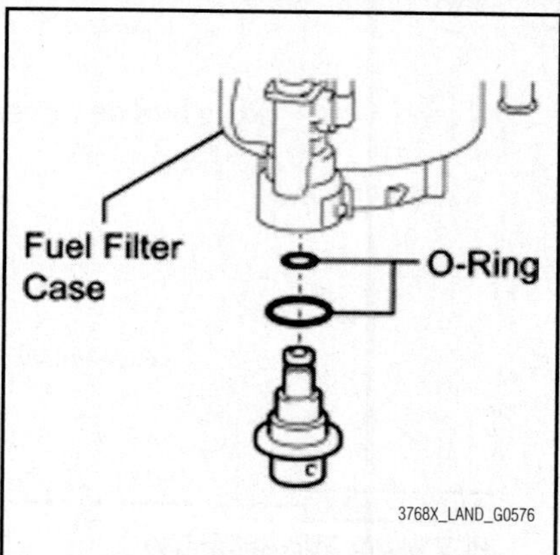

Fig. 230 Removing the fuel main valve and o-rings from the fuel filter

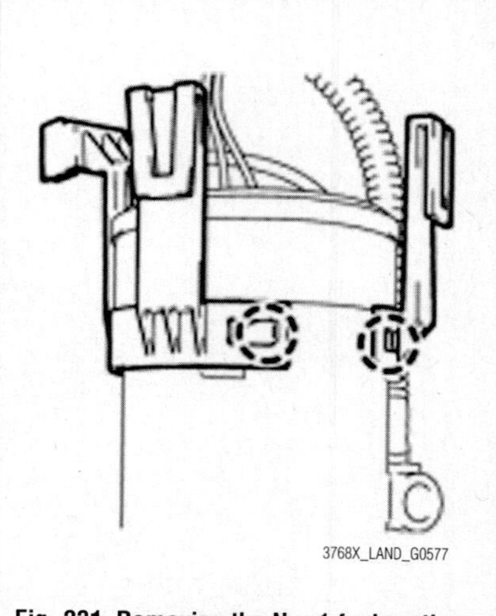

Fig. 231 Removing the No. 1 fuel suction support

5. Remove the fuel main valve assembly.
   a. Remove the fuel main valve from the fuel filter.
   b. Remove the 2 o-rings from the fuel main valve.
6. Remove the No. 1 fuel suction support.
   a. Detach the 2 claws from the claw holes.

To install:

7. Install the No. 1 fuel suction support.
   a. Attach the 2 claws to the claw holes.
8. Install the fuel main valve assembly.
   a. Apply a light coat of gasoline to 2 new o-rings, and install them onto the fuel main valve.

b. Install the fuel main valve to the fuel filter.

➡Make sure the o-rings are not cut or pinched during the installation.

9. Install the fuel pump.
10. Install the No. 1 fuel sub tank.
11. Install the fuel sender gauge assembly.
12. Install the fuel suction with the pump and gauge tube assembly.

FUEL PRESSURE REGULATOR

REMOVAL & INSTALLATION

See Figures 232 through 234.

1. Discharge the fuel system pressure.
2. Disconnect the cable from the negative battery terminal.
3. Remove the v-bank cover sub assembly.
4. Remove the air cleaner hose assembly.
5. Remove the fuel pressure regulator assembly.
    a. Disconnect the No. 2 fuel tube from the pressure regulator.

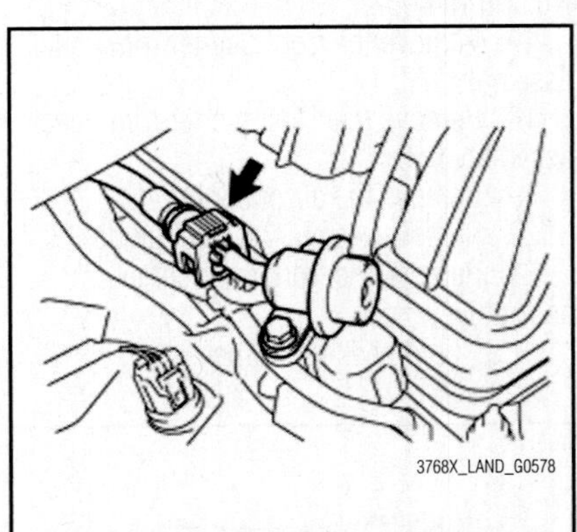

Fig. 232 Disconnecting the No. 2 fuel tube from the pressure regulator

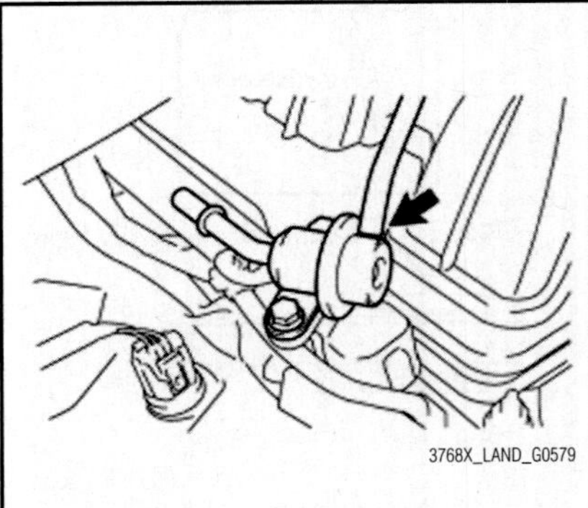

Fig. 233 Disconnecting the vacuum sensing hose from the pressure regulator

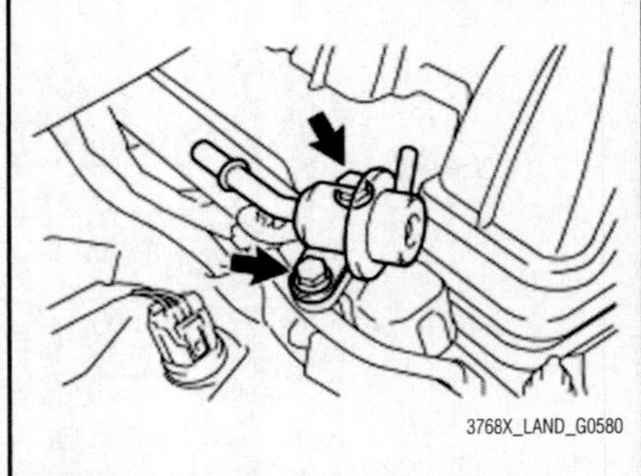

Fig. 234 Removing the 2 bolts, and pull out the pressure regulator

⁂ CAUTION

Put a cloth or equivalent under the pressure regulator.

6. Disconnect the vacuum sensing hose from the pressure regulator.
    a. Remove the 2 bolts, and pull out the pressure regulator.
    b. Remove the o-ring from the pressure regulator.

To install:

7. Install the fuel pressure regulator assembly.
    a. Apply a light coat of gasoline or spindle oil to a new O-ring, and install it to the pressure regulator.

➡Make sure that there are no scratches or foreign matter in or around the insertion hole of the delivery pipe.

➡When inserting the injector, be careful not to damage the O-ring.

    b. While turning the pressure regulator left and right, install it to the delivery pipe.
    c. Install the 2 bolts. Tighten to 7 ft. lbs. (10 Nm).
    d. Connect the vacuum sensing hose to the pressure regulator.
    e. Connect the No. 2 fuel tube to the pressure regulator.
8. Install the air cleaner hose assembly.
9. Install the v-bank cover sub assembly.
10. Connect the cable to the negative battery terminal.
11. Inspect for fuel leaks.

FUEL PUMP

REMOVAL & INSTALLATION

See Figure 235.

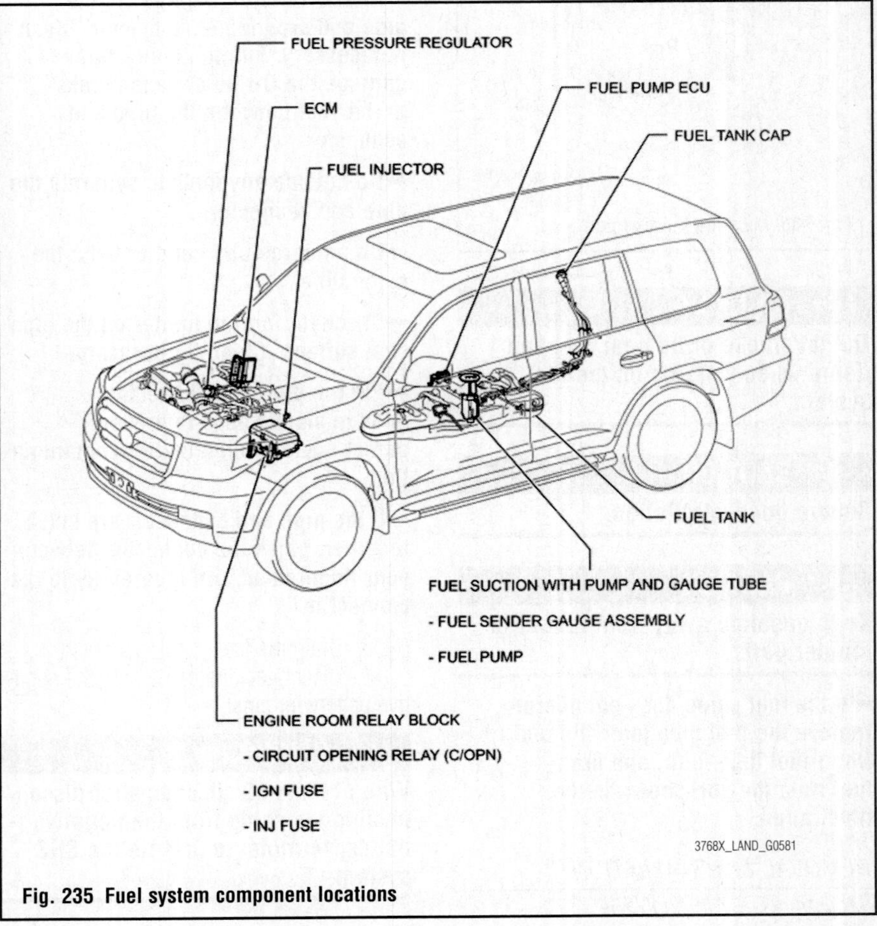

Fig. 235 Fuel system component locations

1. Remove the fuel suction with pump and gauge tube assembly.

To install:

2. To install, reverse the removal procedure.

FUEL TANK

DRAINING

See Figure 236.

1. Connect the cable to the negative (-) battery terminal.
2. Connect the Techstream to the DLC3.
3. Turn the engine switch on (IG).

➡Do not start the engine.

4. Turn the Techstream main switch on.
5. Enter the following menus: Powertrain/Engine and ECT/Active Test/Control the Fuel Pump/Speed.

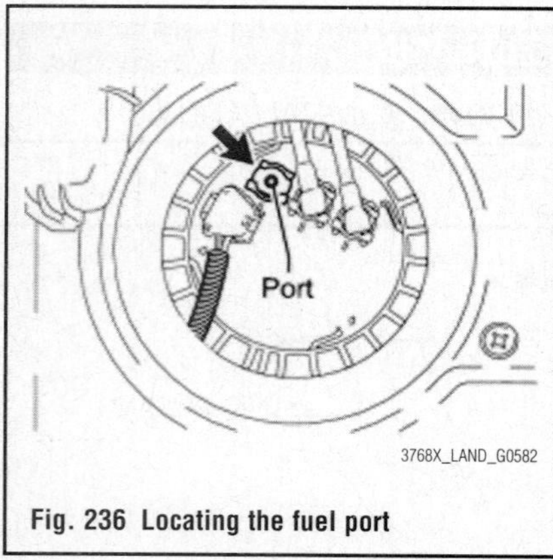

Fig. 236 Locating the fuel port

⁂ CAUTION

Do not smoke or be near an open flame when working on the fuel system.

⁂ CAUTION

Secure good ventilation.

⁂ CAUTION

Keep gasoline away from rubber or leather parts.

➡If the fuel pump does not operate, remove the fuel tube joint clip and No. 1 fuel tube joint, and drain fuel from the port shown in the illustration.

REMOVAL & INSTALLATION

See Figures 237 through 259.

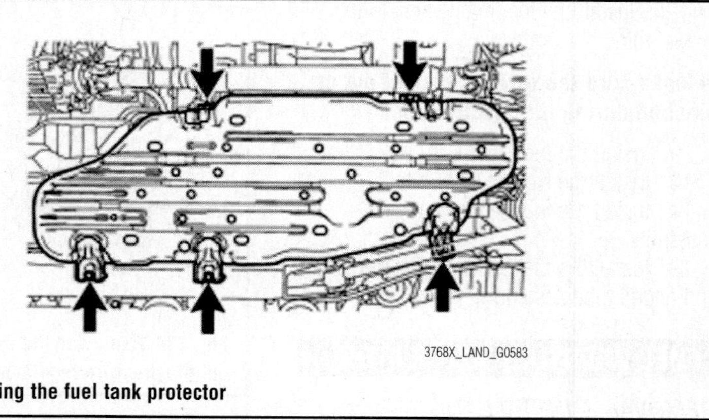

Fig. 237 Removing the fuel tank protector

1. Discharge the fuel system pressure.
2. Remove the fuel tank cap assembly.
3. Remove the No. 1 fuel tank protector sub assembly.
    a. Remove the 5 bolts and fuel tank protector.
4. Disconnect the fuel tank main tube sub assembly.
    a. Detach the fuel tube clamp.
    b. Disconnect the fuel tank main tube.

➡Check for foreign matter in the pipe and around the connector. Clean if necessary. Foreign matter may damage the O-ring or cause leaks in the seal between the pipe and connector.

➡Do not use any tools to separate the pipe and connector.

➡Do not forcefully bend or twist the nylon tube.

➡Check for foreign matter on the pipe seal surface. Clean if necessary.

➡Put the pipe and connector ends in plastic bags to prevent damage and foreign matter contamination.

➡If the pipe and connector are stuck together, pinch the connector between your fingers and turn it carefully to disconnect it.

5. Drain the fuel.
6. Disconnect the cable from the negative battery terminal.

⁂ CAUTION

Wait at least 90 seconds after disconnecting the cable from the negative (-) battery terminal to disable the SRS system.

7. Remove the rear No. 1 seat assembly LH.
8. Remove the rear No. 1 seat assembly RH.
9. Remove the rear No. 1 seat protector.
10. Remove the rear No. 2 seat protector.
11. Remove the rear step cover.
12. Remove the rear door scuff plate LH.
13. Remove the rear door scuff plate RH.
14. Remove the rear No. 2 seat assembly.
15. Remove the rear floor mat rear support plate.
16. Remove the luggage compartment No. 1 trim hook.
17. Remove the front quarter trim panel assembly LH.
18. Remove the front quarter trim panel assembly RH.
19. Remove the air duct plug.
20. Remove the rear air duct guide.
21. Remove the front floor carpet assembly.
    a. Fold back the carpet.

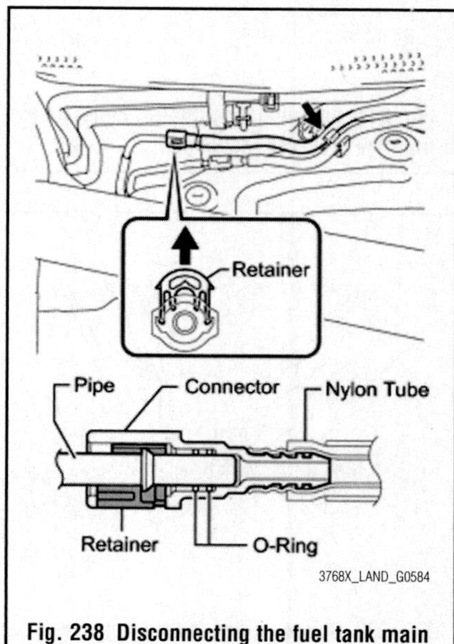

Fig. 238 Disconnecting the fuel tank main tube sub assembly

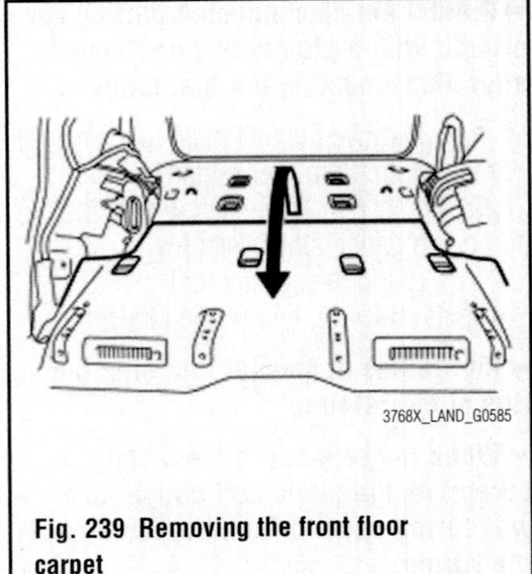

Fig. 239 Removing the front floor carpet

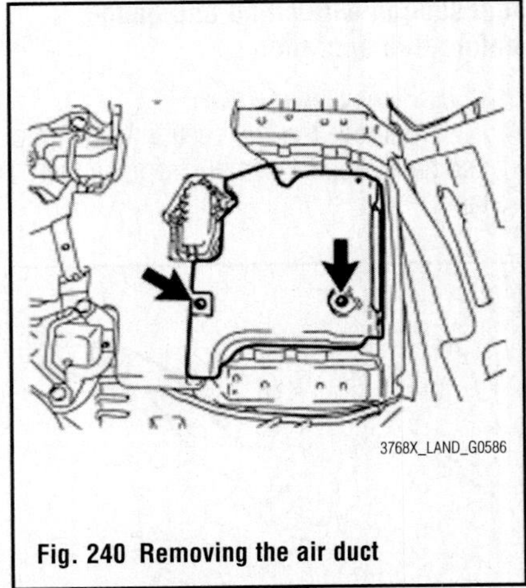
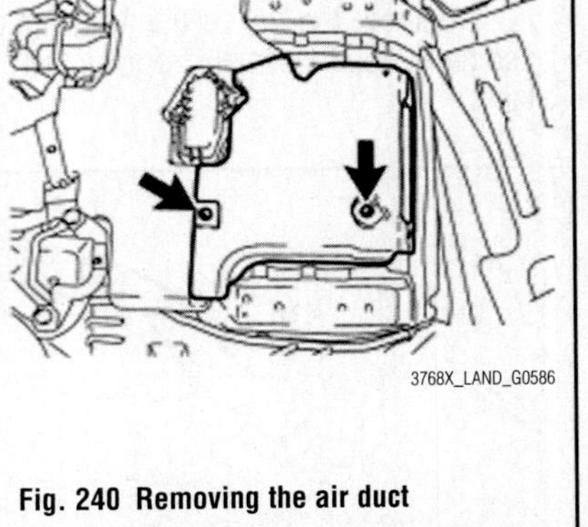

Fig. 240 Removing the air duct

➡Fold back the carpet until it is possible to remove the air duct.

22. Remove the rear floor No. 2 service whole cover.
    a. Remove the 2 screws and air duct.
    b. Remove the service hole cover.
    c. Disconnect the fuel pump and fuel sender gauge connector.
23. Disconnect the fuel tank return tube.
    a. Detach the fuel tube clamp.
    b. Disconnect the fuel tank return tube.

➡Check for foreign matter in the pipe and around the connector. Clean if necessary. Foreign matter may damage the O-ring or cause leaks in the seal between the pipe and connector.

➡Do not use any tools to separate the pipe and connector.

➡Do not forcefully bend or twist the nylon tube.

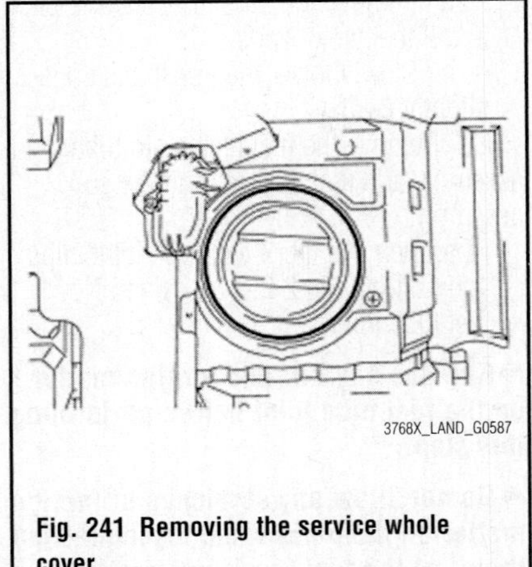

Fig. 241 Removing the service whole cover

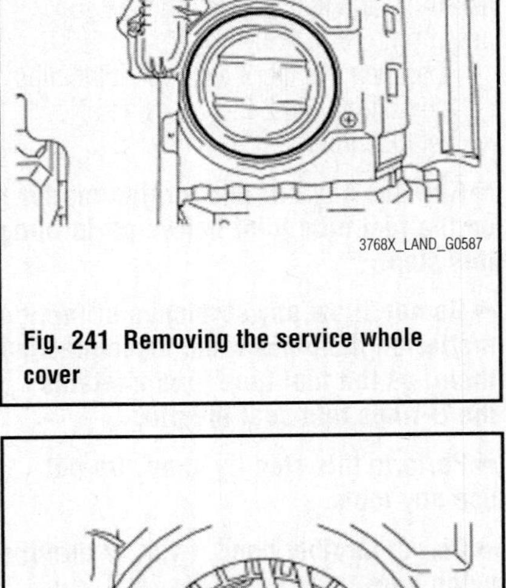
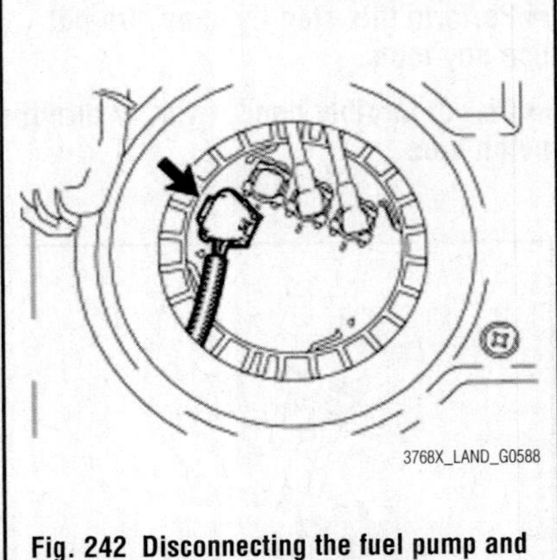

Fig. 242 Disconnecting the fuel pump and fuel sender gauge connector

➡Check for foreign matter on the pipe seal surface. Clean if necessary.

➡Put the pipe and connector ends in plastic bags to prevent damage and foreign matter contamination.

➡If the pipe and connector are stuck together, pinch the connector between your fingers and turn it carefully to disconnect it.

24. Disconnect the No. 2 fuel tank breather tube sub assembly.
    a. Detach the 2 fuel tube clamps.
    b. Disconnect the No. 2 fuel tank breather tube.

➡Do not use any tools in this procedure.

➡Check for any dirt and foreign matter contamination in the pipe and around the connector. Clean if necessary. Foreign matter may damage the O-rings or cause leaks in the seal between the pipe and connector.

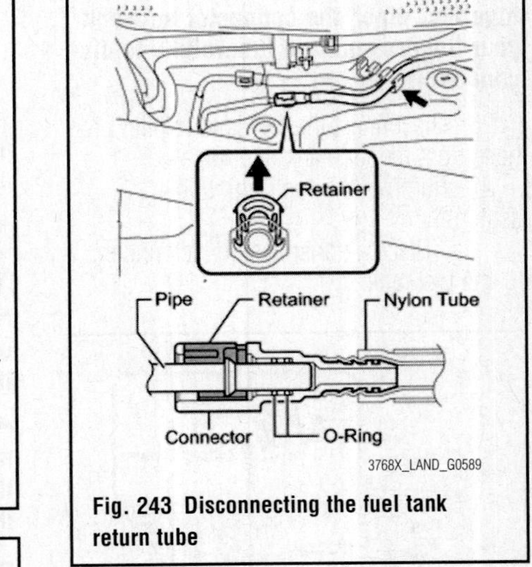

Fig. 243 Disconnecting the fuel tank return tube

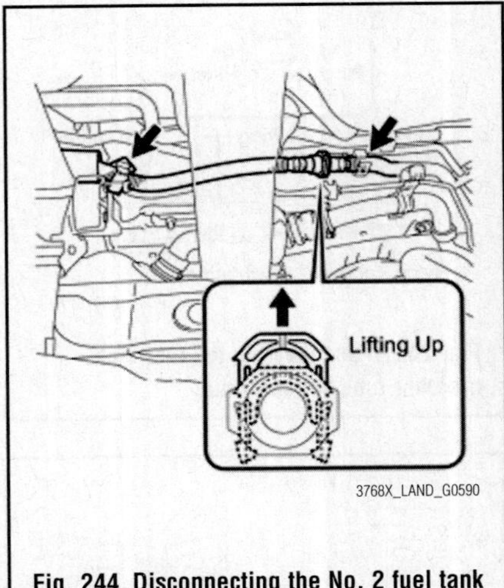

Fig. 244 Disconnecting the No. 2 fuel tank breather tube sub assembly

25. Disconnect the fuel tank breather tube sub assembly.
    a. Detach the fuel tube clamp.
    b. Disconnect the fuel tank breather tube.

➡Check for foreign matter in the pipe and around the connector. Clean if necessary. Foreign matter may damage the O-ring or cause leaks in the seal between the pipe and connector.

➡Do not use any tools to separate the pipe and connector.

➡Do not forcefully bend or twist the nylon tube.

➡Check for foreign matter on the pipe seal surface. Clean if necessary.

➡Put the pipe and connector ends in plastic bags to prevent damage and foreign matter contamination.

➡If the pipe and connector are stuck together, pinch the connector between your fingers and turn it carefully to disconnect it.

26. Disconnect the fuel tank to filler pipe hose from the fuel tank filler pipe.

27. Remove the fuel tank sub assembly.

a. Place a transmission jack under the fuel tank.

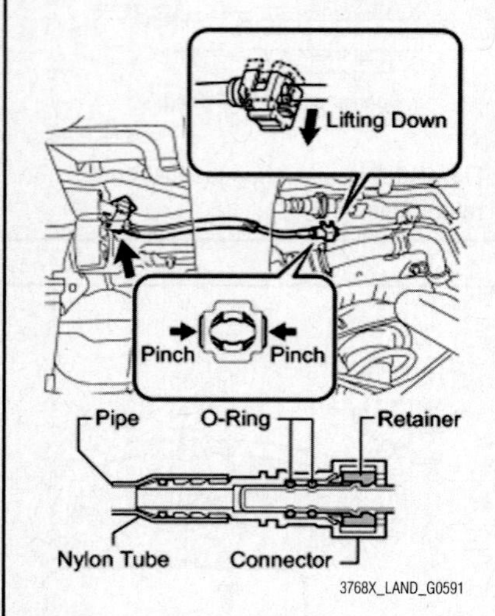

Fig. 245 Disconnecting the fuel tank breather tube sub assembly

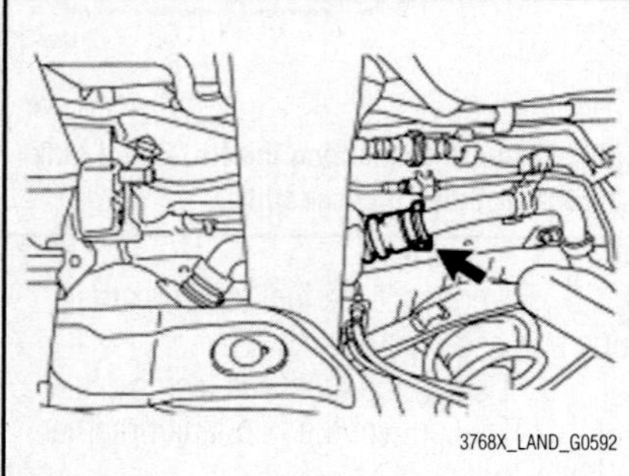

Fig. 246 Disconnecting the fuel tank to filler pipe hose

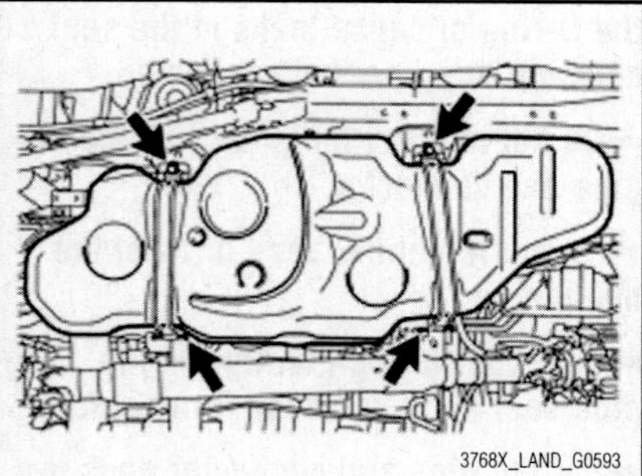

Fig. 247 Removing the fuel tank sub assembly

b. Remove the 2 bolts, 2 clips, 2 pins and 2 fuel tank bands.

c. Slowly lower the transmission jack slightly.

28. Remove the fuel tank main tube sub assembly and fuel tank return tube sub assembly.

a. Remove the 3 fuel tube joint clips and pull out the 2 fuel tubes and No. 1 fuel tube joint.

➡Remove any dirt and foreign matter on the fuel tube joint before performing this step.

➡Do not allow any scratches or foreign matter on the parts when disconnecting them, as the fuel tube joint contains the O-rings that seal the plug.

➡Perform this step by hand. Do not use any tools.

➡Do not forcibly bend, twist or turn the nylon tube.

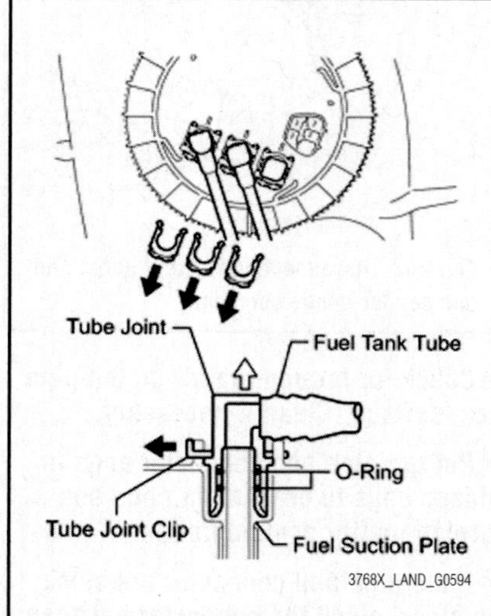

Fig. 248 Removing the 3 fuel tube joint clips and pulling out the 2 fuel tubes and No. 1 fuel tube joint

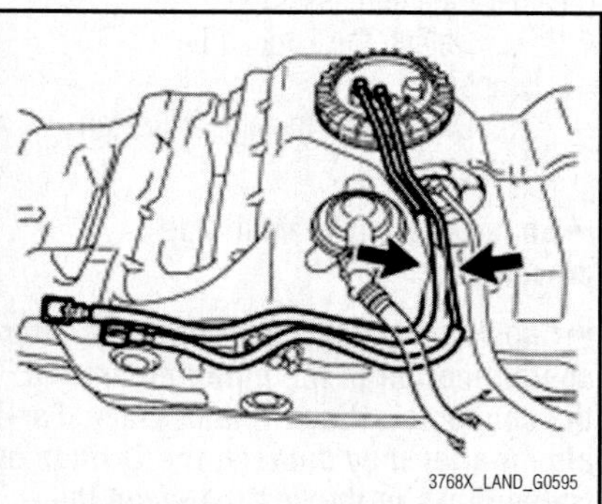

Fig. 249 Removing the fuel main tube and fuel return tube from the fuel tank

➡Protect the disconnected part by covering it with a plastic bag and tape after disconnecting the fuel tubes.

b. Remove the fuel main tube and fuel return tube from the fuel tank.

29. Remove the fuel suction with the pump and gauge tube assembly.

a. Using the special tool (09808-14020), loosen the retainer.

➡Fit the tips of special tool onto the ribs of the retainer.

➡When the retainer is loosened, be careful as the pump and gauge tube will spring upward from the force of the spring.

➡Remove foreign matter around the fuel suction with pump and gauge before this operation.

b. Remove the retainer.

c. Remove the fuel suction with pump and gauge tube assembly from the fuel tank.

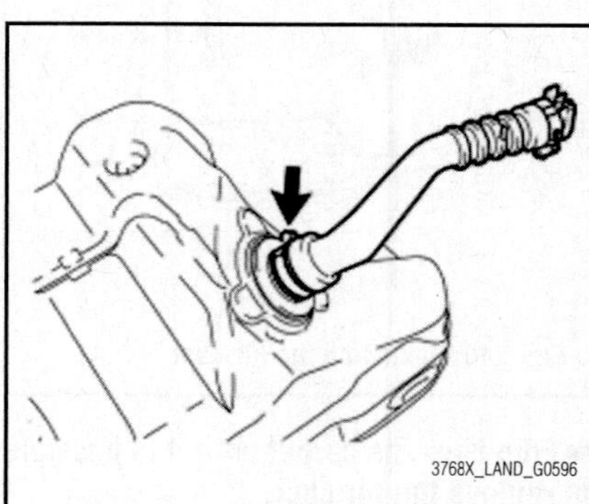

Fig. 250 Removing the fuel tank to filler pipe hose

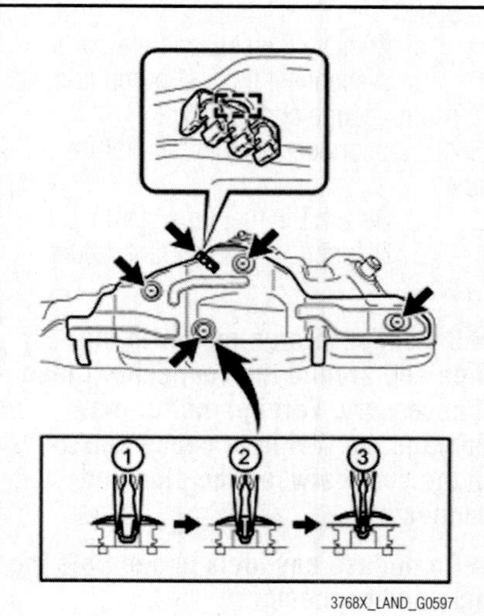

Fig. 251 Removing the No. 1 fuel tank heat insulator

➡Be careful not to bend the arm of the fuel sender gauge.

d. Remove the gasket from the fuel tank.

30. Remove the fuel tank to filler pipe hose.

a. Remove the fuel tank to filler pipe hose from the fuel tank.

31. Remove the No. 1 fuel tank heat insulator.

a. Remove the fuel tube clamp from the No. 1 fuel tank heat insulator.

b. Using needle nose pliers, remove the 4 clips and then remove the No. 1 fuel tank heat insulator.

To install:

32. Install the No. 1 fuel tank heat insulator with the 4 clips.

a. Install the fuel tube clamp to the No. 1 fuel tank heat insulator.

33. Install the fuel tank to the filler pipe hose.

a. Install the fuel tank to filler pipe hose to the fuel tank as shown in the illustration.

➡Align the fuel tank side mark with the hose side mark when installing the hose.

➡Tighten the hose clamp until the end of the hose clamp contacts the stopper as shown in the illustration.

34. Install the fuel suction with pump and gauge tube assembly.

a. Apply a light coat of gasoline or grease to a new gasket, and install it to the fuel tank.

b. Install the fuel suction with pump and gauge tube into the fuel tank.

➡Be careful not to bend the arm of the fuel sender gauge.

➡Align the protrusion of the fuel suction with pump and gauge tube with the groove of the fuel tank.

c. Put the retainer on the fuel tank. While holding the fuel suction with pump and gauge tube, tighten the retainer 1 complete turn by hand.

d. Using the special tool (09808-14020), tighten the retainer until the mark on the retainer is within range A on the fuel tank, as shown in the illustration.

➡Fit the tips of SST onto the ribs of the retainer.

35. Install the fuel tank main tube sub assembly and the fuel tank return tube sub assembly.

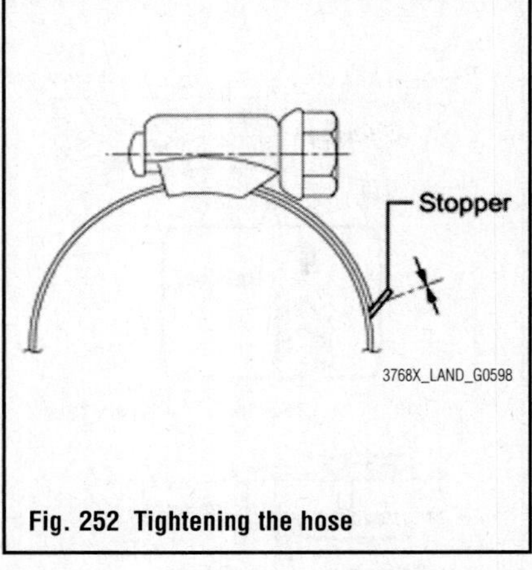

Fig. 252 Tightening the hose

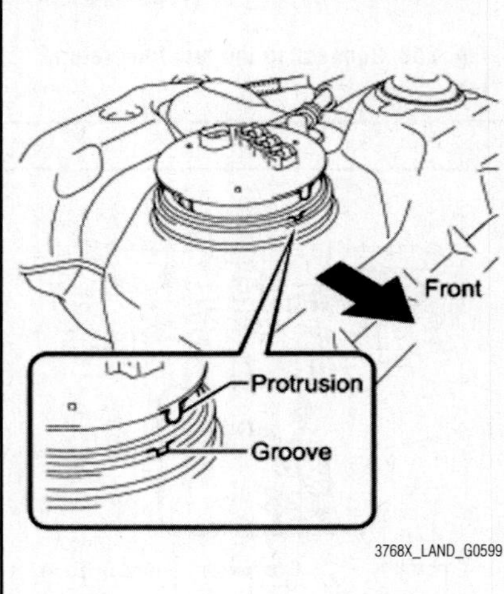

Fig. 253 Aligning the protrusion of the fuel suction with pump and gauge tube with the groove of the fuel tank

a. Install the 2 fuel tank tubes and No. 1 fuel tube joint with the 3 tube joint clips.

➡Check that there are no scratches or foreign objects on the connecting parts.

➡Check that the fuel tube joints are inserted securely.

➡Check that the tube joint clips are on the collars of the fuel tube joints.

➡After installing the tube joint clips, check that the fuel tube joints have not been pulled off.

b. Install the fuel tank main tube and fuel tank return tube to the fuel tank.

36. Install the fuel tank sub assembly.

a. Set the fuel tank on a transmission jack and raise the fuel tank.

b. Raise the transmission jack.

c. Install the 2 fuel tank bands with the 2 pins and 2 clips.

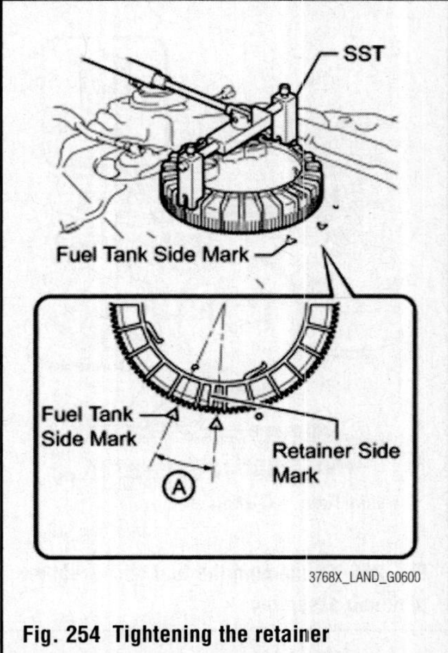

Fig. 254 Tightening the retainer

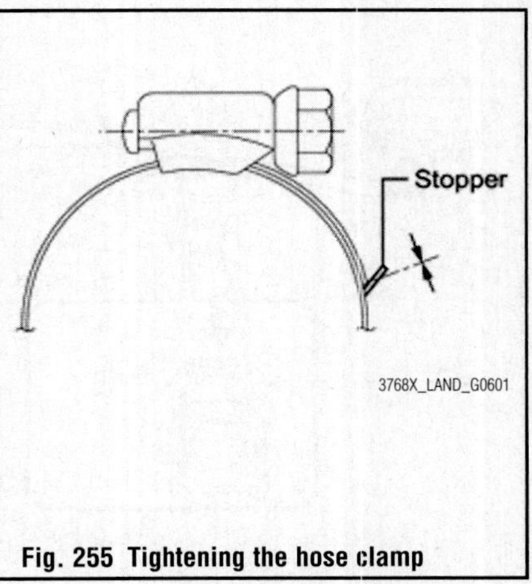

Fig. 255 Tightening the hose clamp

d. Connect the 2 fuel tank bands with the 2 bolts. Tighten to 30 ft. lbs. (40 Nm).

37. Connect the fuel tank to the filler pipe hose.

a. Connect the fuel tank to filler pipe hose to the fuel tank filler pipe as shown in the illustration.

➡Install the hose so that the distance between the fuel tank inlet pipe side mark and fuel tank to filler pipe hose side mark is 0 to 0.3 mm (0 to 0.0118 in.) as shown in the illustration.

➡Tighten the hose clamp until the end of the hose clamp contacts the stopper as shown in the illustration.

38. Connect the fuel tank breather tube sub assembly to the fuel tank filler pipe.

➡Push the parts together firmly until a "click" sound is heard.

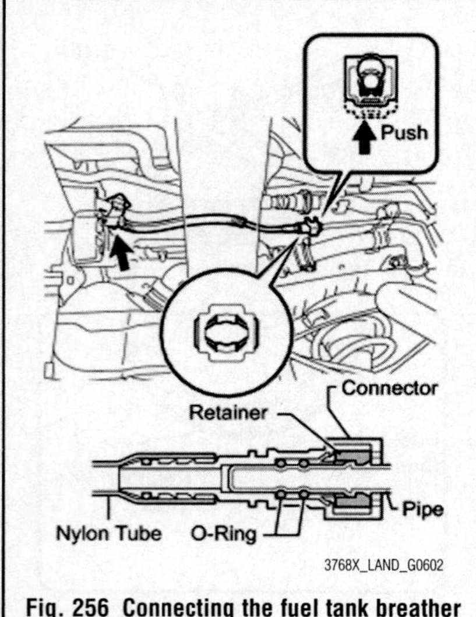

Fig. 256 Connecting the fuel tank breather tube sub assembly

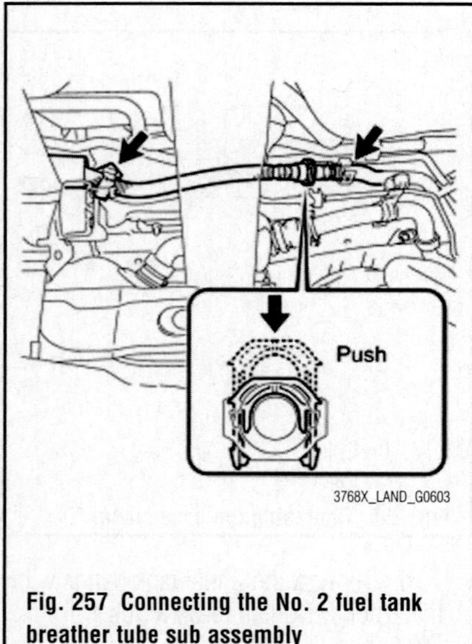

Fig. 257 Connecting the No. 2 fuel tank breather tube sub assembly

➡Before installing the tube connector to the pipe, check if there is any damage or foreign matter in the connector.

➡After the connection, check if the connectors and pipes are securely connected by trying to pull them apart.

a. Attach the fuel tube clamp.

39. Connect the No. 2 fuel tank breather tube sub assembly.

➡Before installing the hose, make sure that it is not damaged. Make sure that there is no foreign matter present on the connecting surfaces.

➡After connecting, check if the hose and the connector are securely connected by pulling on them.

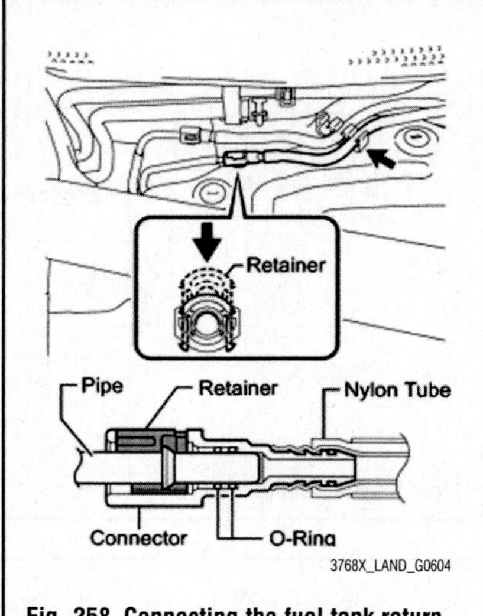

Fig. 258 Connecting the fuel tank return tube

Fig. 259 Connecting the fuel tank main tube sub assembly

a. Attach the 2 fuel tube clamps.

40. Connect the fuel tank return tube.

➡Push the parts together firmly until a "click" sound is heard.

➡Before installing the tube connector to the pipe, check if there is any damage or foreign matter in the connector.

➡After the connection, check if the connectors and pipes are securely connected by trying to pull them apart.

a. Attach the fuel tube clamp.

41. Connect the fuel tank main tube sub assembly.

➡Push the parts together firmly until a "click" sound is heard.

➡Before installing the tube connector to the pipe, check if there is any damage or foreign matter in the connector.

➡After the connection, check if the connectors and pipes are securely connected by trying to pull them apart.

a. Attach the fuel tube clamp.

42. Install the No. 1 fuel tank protector sub assembly with the 5 bolts. Tighten to 15 ft. lbs. (20 Nm).

43. Install the fuel tank cap assembly.

44. Install the rear floor No. 2 service hole cover.

a. Connect the fuel pump and fuel sender gauge connector.

b. Install the service hole cover with new butyl tape.

c. Install the air duct with the 2 screws.

45. Install the front floor carpet assembly.

46. Install the rear air duct guide.

47. Install the air duct plug.

48. Install the front quarter trim panel assembly RH.

49. Install the front quarter trim panel assembly LH.

50. Install the luggage compartment No. 1 trim hook.

51. Install the rear floor mat rear support plate.

52. Install the rear No. 2 seat assembly.

53. Install the rear door scuff plate RH.

54. Install the rear door scuff plate LH.

55. Install the rear step cover.

56. Install the rear No. 2 seat protector.

57. Install the rear No. 1 seat protector.

58. Install the rear No. 1 seat assembly RH.

59. Install the rear No. 1 seat assembly LH.

60. Connect the cable to the negative battery terminal.

61. Check the SRS warning light.

62. Add fuel.

63. Inspect for fuel leaks.

64. Inspect the rear No. 1 seat assembly RH and LH.

IDLE SPEED

ADJUSTMENT

Idle speed is maintained by the ECM. No adjustment is necessary or possible.

THROTTLE BODY

REMOVAL & INSTALLATION

See Figures 260 through 262.

1. Remove the front fender splash shield sub assembly LH.
2. Remove the front fender splash shield sub assembly RH.
3. Remove the No. 1 engine under cover sub assembly.
4. Drain the engine coolant.
5. Remove the v-bank cover sub assembly.
6. Remove the air cleaner hose assembly.

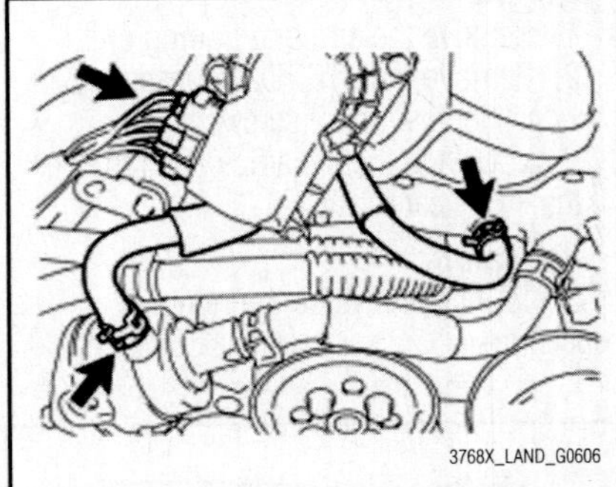

Fig. 260 Disconnecting the 2 water by pass hoses

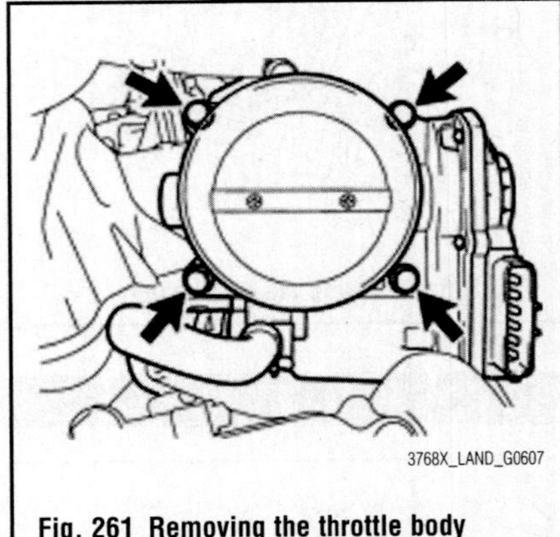

Fig. 261 Removing the throttle body

7. Remove the throttle body assembly.
   a. Disconnect the 2 water by pass hoses.
   b. Disconnect the throttle position sensor and control motor connector.
   c. Remove the 4 bolts, throttle body and gasket.
   d. Remove the 2 water by pass hoses from the throttle body.

To install:

8. Install the throttle body assembly.
   a. Install the 2 water by pass hoses to the throttle body.
   b. Install a new gasket to the intake air surge tank.

➡Align the protrusion of the gasket with the groove of the intake air surge tank.

   c. Install the throttle body with the 4 bolts. Tighten to 7 ft. lbs. (10 Nm).
   d. Connect the throttle position sensor and control motor connector.
   e. Connect the 2 water by pass hoses.
9. Add engine coolant.

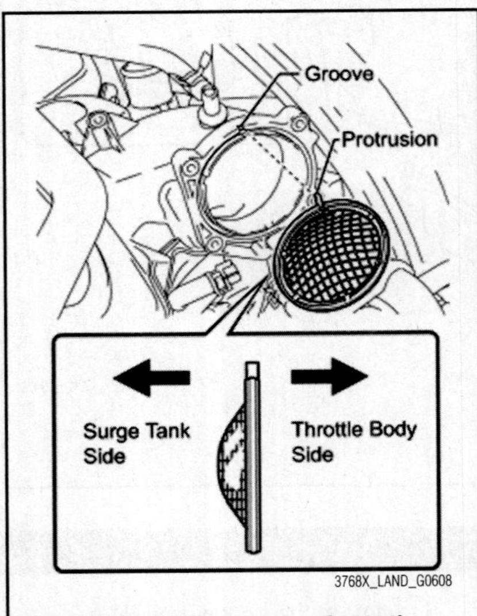

Fig. 262 Installing a new gasket to the intake air surge tank

10. Install the air cleaner hose assembly.
11. Inspect for coolant leaks.
12. Check the throttle body assembly.
13. Install the No. 1 engine under cover sub assembly.
14. Install the front fender splash shield sub assembly LH.
15. Install the front fender splash shield sub assembly RH.
16. Perform the initialization.

➡Perform the following procedure after replacing the ECM, throttle body assembly or any throttle body components. The following procedure should also be performed if the throttle body is cleaned.

   a. Disconnect the cable from the negative (-) battery terminal. Wait at least 60 seconds and reconnect the cable.
   b. Turn the ignition switch on (IG) without operating the accelerator pedal.

➡If the accelerator pedal is operated, perform the above steps again.

   c. Connect the Techstream to the DLC3 and clear the DTCs.
   d. Start the engine and check that the MIL is not illuminated and that the idle speed is within the specified range when the A/C is switched off after the engine is warmed up. With the A/C switched off the engine idle speed should be 650-750 rpm.

➡Be sure to perform this step with all accessories off.

➡Make sure that the shift lever is in N or P.

   e. Enter the following menus: Powertrain / Engine and ECT / Data List / Throttle Sensor Volt% Fully depress the accelerator pedal and check that the value is 60% or more.
   f. Perform a road test and confirm that there are no abnormalities.
17. Install the v-bank cover assembly.

HEATING, VENTILATION & AIR CONDITIONING

BLOWER MOTOR

REMOVAL & INSTALLATION

Front

See Figure 263.

1. Remove the air conditioning unit.
2. Remove the blower assembly.
   a. Remove the screw.
   b. Disconnect the connector and clamp.
   c. Detach the claw and remove the blower unit.

To install:

3. To install, reverse the removal procedure.

Rear

See Figure 264.

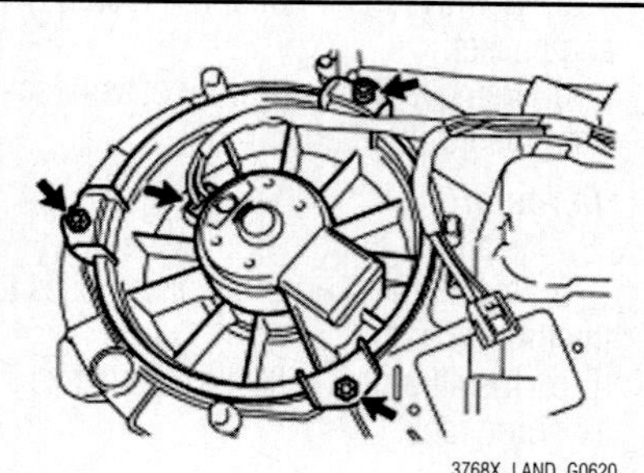

Fig. 264 Removing the rear blower motor

1. Remove the rear cooling unit assembly.
2. Remove the blower with the fan (rear) motor sub assembly.
   a. Disconnect the connector.
   b. Remove the 3 screws and the motor.

To install:

3. To install, reverse the removal procedure.

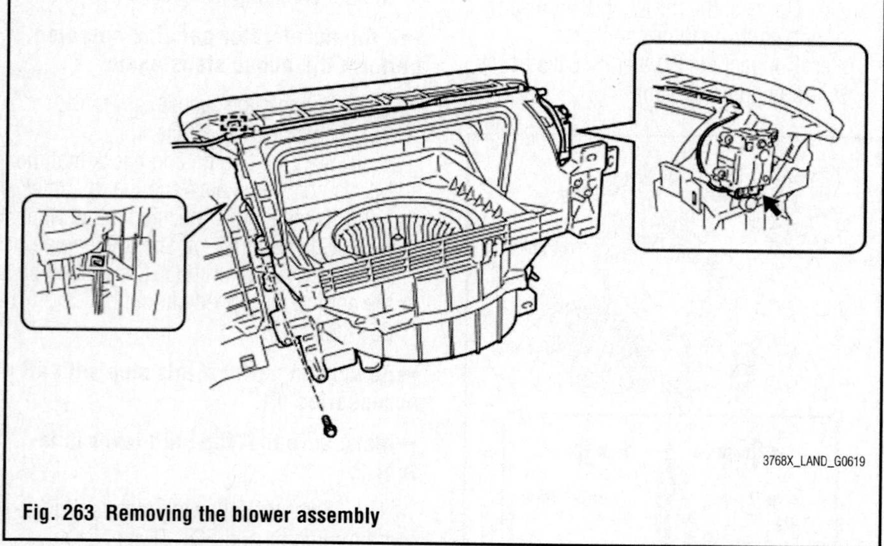

Fig. 263 Removing the blower assembly

HEATER UNIT

REMOVAL & INSTALLATION

See Figure 265.

1. Remove the air conditioning unit.
2. Remove the PTC heater assembly
   a. Remove the 2 screws.
   b. Detach the 3 clamps and remove the PTC heater.

To install:

3. To install, reverse the removal procedure.

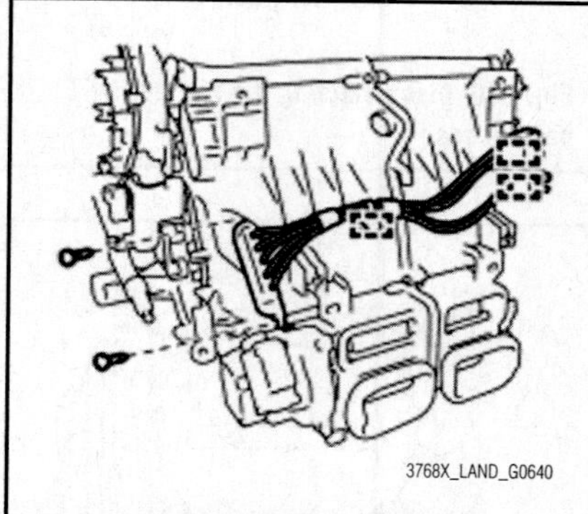

Fig. 265 Removing the PTC heater

STEERING

POWER STEERING (VANE) PUMP

REMOVAL & INSTALLATION

See Figures 266 through 268.

1. Disconnect the cable from the negative battery terminal.

⁂ CAUTION

Wait at least 90 seconds after disconnecting the cable from the negative (-) battery terminal to disable the SRS system.

➡When disconnecting the cable, some systems need to be initialized after the cable is reconnected.

2. Remove the front wheel RH.
3. Remove the front fender apron seal front RH.
4. Remove the v-bank cover sub assembly.
5. Remove the air cleaner hose assembly.
6. Remove the air cleaner assembly.
7. Remove the fan and generator v-belt.
8. Drain the power steering fluid.
9. Disconnect the suction hose.
   a. Slide the clip and disconnect the suction hose from the vane pump.
10. Disconnect the power steering oil pressure sensor connector.
   a. Detach the wire harness clamp and disconnect the connector.

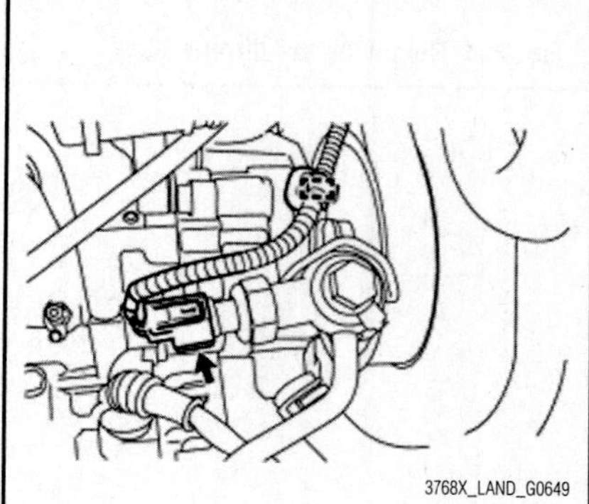

Fig. 266 Disconnecting the power steering oil pressure sensor connector

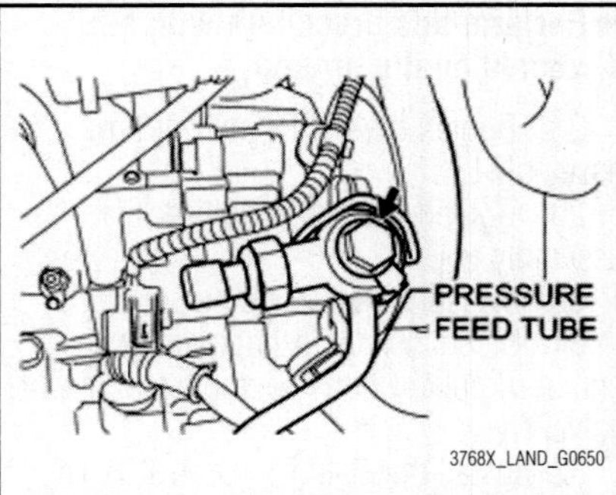

Fig. 267 Disconnecting the pressure feed tube

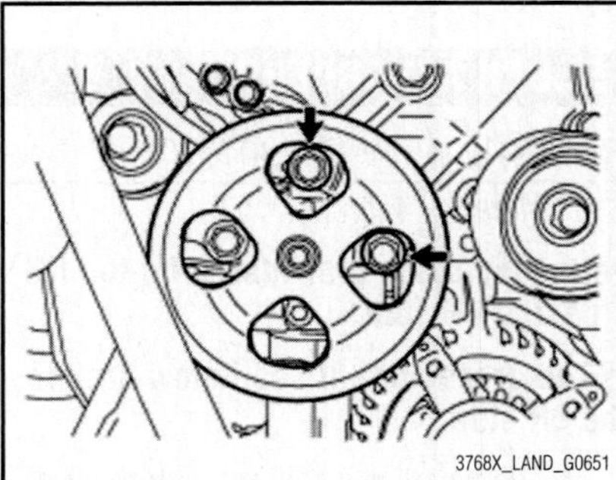

Fig. 268 Removing the vane pump assembly

11. Disconnect the pressure feed tube.
  a. Remove the gasket.
12. Remove the vane pump assembly.
  a. Remove the 2 bolts and vane pump.

To install:

13. Install the vane pump assembly. Tighten the 2 bolts to 15 ft. lbs. (21 Nm).
14. Connect the pressure feed tube. Tighten the union bolt to 37 ft. lbs. (50 Nm).
15. Connect the power steering oil pressure sensor connector. Attach the wire harness clamp.
16. Connect the suction hose to the vane pump assembly with the clip.
17. Install the fan and generator v-belt.
18. Install the air cleaner assembly.
19. Install the air cleaner hose assembly.
20. Install the v-bank cover sub assembly.
21. Install the front fender apron seal front RH.
22. Install the front wheel RH.
23. Add power steering fluid.
24. Bleed the power steering fluid.
25. Inspect for power steering fluid leaks.
26. Connect the cable to the negative battery terminal.

BLEEDING

1. Check the fluid level.
2. Jack up the front of the vehicle and support it with stands.
3. Turn the steering wheel.
  a. With the engine stopped, turn the wheel slowly from lock to lock several times.
4. Lower the vehicle.
5. Start the engine.
6. Idle the engine for a few minutes.
7. Turn the steering wheel.
  a. With the engine idling, turn the wheel left or right to the full lock position and keep it there for 2 to 3 seconds, then turn the wheel to the opposite full lock position and keep it there for 2 to 3 seconds.
  b. Repeat the previous step several times.
8. Stop the engine.
9. Check for foaming or emulsification. If the system has to be bled twice because of foaming or emulsification, check for fluid leaks in the system.
10. Check the fluid level.

SUSPENSION

FRONT SUSPENSION

LOWER CONTROL ARM

REMOVAL & INSTALLATION

See Figures 269 through 273.

➡Use the same procedures for the RH side and LH side.

➡The procedures listed below are for the LH side.

1. Remove the stabilizer control valve protector.
2. Open stabilizer control with accumulator housing shutter valve.
3. Remove the front wheel.
4. Remove the front fender splash shield sub assembly LH.
5. Remove the front fender splash shield sub assembly RH.
6. Remove the No. 1 engine under cover sub assembly.
7. Loosen the front No. 1 stabilizer bracket LH.
8. Loosen the front No. 1 stabilizer bracket RH.
9. Remove the front stabilizer link assembly LH.
10. Remove the front stabilizer link assembly RH.
11. Disconnect the front shock absorber with the coil spring LH.
  a. Remove the nut and bolt, and disconnect the shock absorber from the lower side.
12. Disconnect the front lower ball joint attachment LH.
  a. Remove the 2 bolts and disconnect the attachment from the steering knuckle.

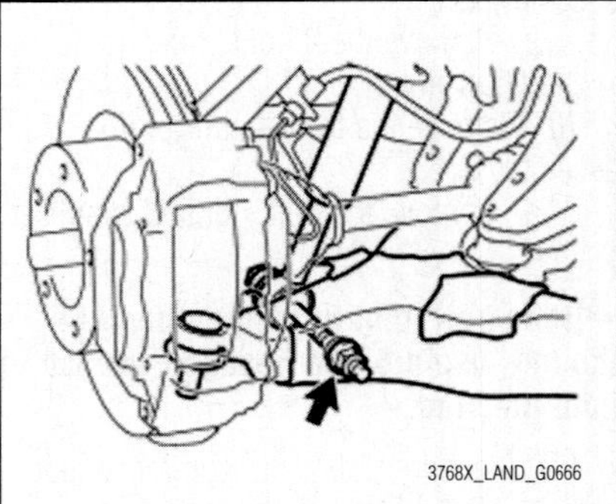

Fig. 269 Disconnecting the front shock absorber with the coil spring

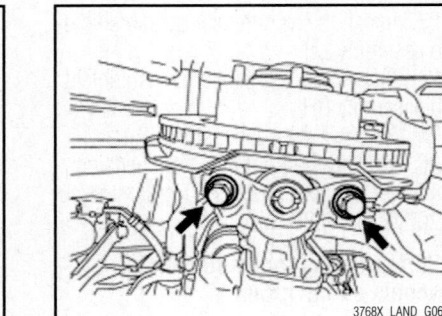

Fig. 270 Disconnecting the front lower ball joint attachment

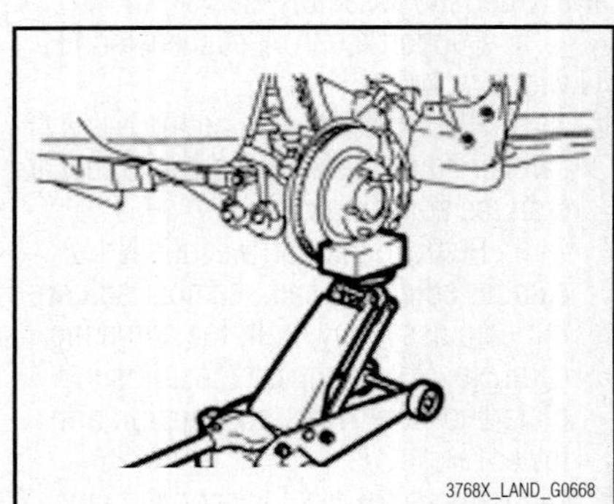

Fig. 271 Removing the front No. 1 suspension arm lower sub assembly

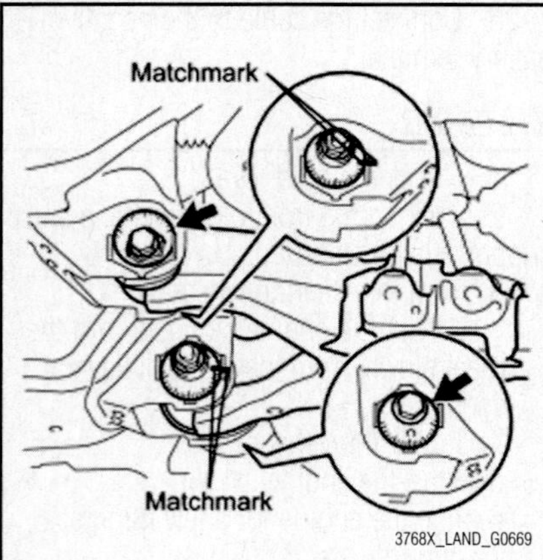

Fig. 272 Placing matchmarks on the No. 2 camber adjusting cam and No. 2 suspension toe adjusting plate

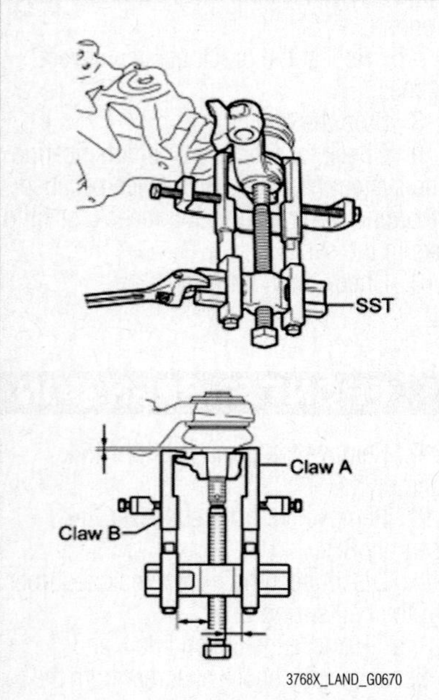

Fig. 273 Removing the lower ball joint attachment

13. Remove the front No. 1 suspension arm lower sub assembly LH.
    a. Support the front suspension LH with a jack.
    b. Place matchmarks on the No. 2 camber adjusting cam and No. 2 suspension toe adjusting plate.
    c. Remove the nut, washer, No. 2 camber adjusting cam, camber adjusting cam assembly, bolt, toe adjusting cam, No. 2 suspension toe adjusting plate and front No. 1 suspension arm lower LH.
14. Remove the front lower ball joint attachment LH.
    a. Remove the cotter pin and nut.
    b. Using the special tool (09950-40011), remove the lower ball joint attachment.

To install:

➡Use the same procedures for the RH side and LH side.

➡The procedures listed below are for the LH side.

15. Temporarily install the front No. 1 suspension arm lower sub assembly LH.
    a. Temporarily install the lower suspension arm, camber adjusting cam, No. 2 camber adjusting cam, No. 2 suspension toe adjusting plate, toe adjusting cam and washer with the bolt and nut.
    b. Align the matchmarks on the No. 2 camber adjusting cam and No. 2 suspension toe adjusting plate with the matchmarks on the vehicle body. Tighten the bolt and nut.

➡The bolt and nut will be tightened to the torque specification in the "Tighten Front No. 1 Suspension Arm Lower Sub-assembly LH" procedure.

16. Connect the front lower ball joint attachment LH.
    a. Connect the attachment to the steering knuckle with the 2 bolts. Tighten to 221 ft. lbs. (300 Nm).
17. Connect the front shock absorber with the coil spring LH.
    a. Connect the shock absorber with the bolt and nut.
18. Temporarily install the front stabilizer link assembly RH.
19. Temporarily install the front stabilizer link assembly LH.
20. Tighten the front No. 1 stabilizer bracket LH.
21. Tighten the front No. 1 stabilizer bracket RH.
22. Install the No. 1 engine under cover sub assembly.
23. Install the front fender splash shield sub assembly LH.
24. Install the front fender splash shield sub assembly RH.
25. Stabilize the suspension.
26. Tighten the front No. 1 suspension arm lower sub assembly LH. Tighten the nut to 207 ft. lbs. (280 Nm).

➡Perform this procedure with all 4 wheels on the ground.

27. Tighten the front shock absorber with the coil spring. Tighten the nut to 133 ft. lbs. (180 Nm).

➡Perform this procedure with all 4 wheels on the ground.

28. Tighten the front stabilizer link assembly LH.
29. Tighten the front stabilizer link assembly RH.
30. Measure the vehicle height.
31. Close the stabilizer control with and accumulator housing shutter valve.
32. Install the stabilizer control valve protector.
33. Inspect and adjust the front wheel alignment.
34. Adjust the headlight assembly.

SHOCK ABSORBERS

REMOVAL & INSTALLATION

See Figures 274 through 278.

➡Use the same procedures for the RH side and LH side.

➡The procedures listed below are for the LH side.

1. Remove the stabilizer control valve protector.
2. Open the stabilizer control with the accumulator housing shutter valve.
3. Remove the front wheel.
4. Remove the front fender splash shield sub assembly LH.
5. Remove the front fender splash shield sub assembly RH.
6. Remove the No. 1 engine under cover sub assembly.
7. Loosen the front No. 1 stabilizer bracket LH.
    a. Loosen the 2 bolts of the front stabilizer brackets.
8. Loosen the front No. 1 stabilizer bracket RH.
    a. Loosen the 2 bolts of the front stabilizer brackets.
9. Remove the front stabilizer link assembly LH.
    a. Remove the 2 bolts, nut and stabilizer link.
10. Remove the front stabilizer link assembly RH.
    a. Remove the bolt, nut and stabilizer link.

➡If the ball joint turns together with the nut, use a 6 mm hexagon wrench to hold the stud.

11. Disconnect the skid control sensor wire.
12. Disconnect the steering knuckle LH.

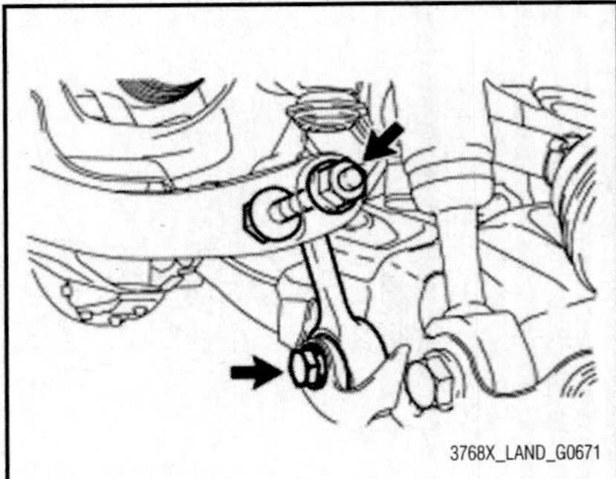

Fig. 274 Removing the front stabilizer link assembly LH

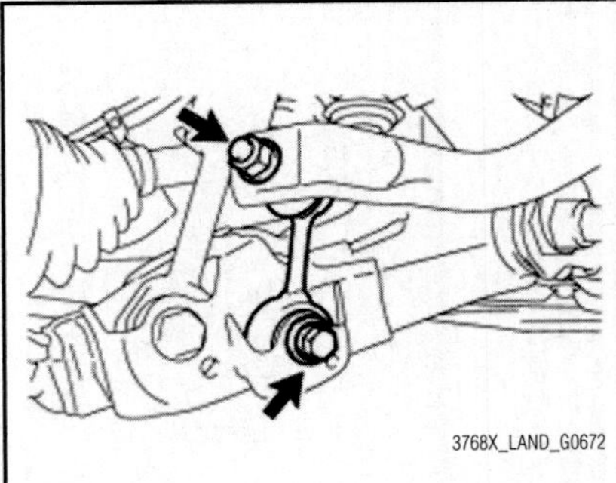

Fig. 275 Removing the front stabilizer link assembly RH

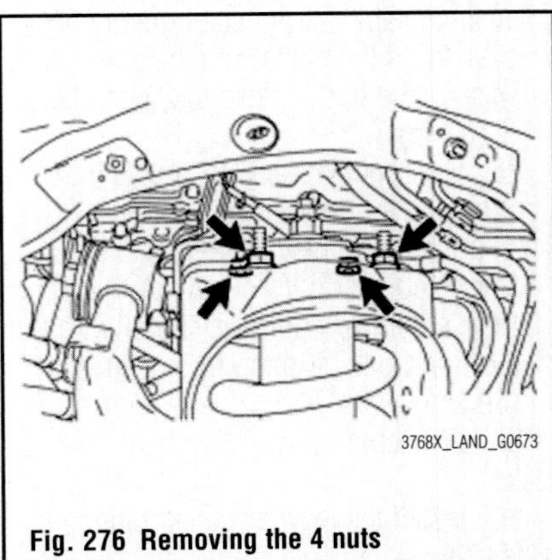

Fig. 276 Removing the 4 nuts

13. Remove the front shock absorber with the coil spring LH.

a. Remove the nut from the shock absorber lower side.

➡To prevent the shock absorber with the coil spring from falling, leave the bolt inserted.

b. Remove the 4 nuts.

c. Remove the bolt (lower side) and shock absorber with the coil spring.

To install:

14. Temporarily install the front shock absorber with the coil spring.

a. Temporarily install the upper side of the shock absorber to the chassis frame with the 4 nuts.

b. Temporarily install the lower side of the shock absorber to the lower suspension arm with the bolt and nut.

15. Connect the steering knuckle LH.

16. Connect the skid control sensor wire.

17. Temporarily install the front stabilizer link assembly RH.

a. Temporarily install the stabilizer link with the bolt.

b. Temporarily install the stabilizer link with the nut.

c. Tighten the nut. Tighten to 94 ft. lbs. (128 Nm).

18. Temporarily install the front stabilizer link assembly LH.

a. Temporarily install the stabilizer link to the lower arm with the bolt.

b. Temporarily install the stabilizer link to the stabilizer control arm with the bolt and nut.

➡If the front stabilizer control cylinder extends and it is difficult to temporarily install the stabilizer link to the stabilizer control arm, raise the stabilizer control arm with a jack and temporarily install the stabilizer link.

19. Tighten the front No. 1 stabilizer bracket LH.

a. Tighten the 2 bolts of the front stabilizer brackets. Tighten to 64 ft. lbs. (87 Nm).

➡Tighten the bolts in 3 steps, in the order shown in the illustration.

20. Tighten the front No. 1 stabilizer bracket RH.

a. Tighten the 2 bolts of the front stabilizer brackets. Tighten to 64 ft. lbs. (87 Nm).

➡Tighten the bolts in 3 steps, in the order shown in the illustration.

21. Install the No. 1 engine under cover sub assembly.

22. Install the front fender splash shield sub assembly LH and RH.

23. Stabilize the suspension.

a. Install the front wheels. Tighten to 97 ft. lbs. (131 Nm).

b. Lower the vehicle.

c. Press down on the vehicle several times to stabilize the suspension.

24. Tighten the front shock absorber with the coil spring.

a. Tighten the nut to 133 ft. lbs. (180 Nm).

➡Perform this procedure with all 4 wheel on the ground.

b. Tighten the 4 upper nuts in diametrically opposite pairs. Tighten to 33 ft. lbs. (45 Nm).

c. Check that the first nut that was tightened is at the torque specification.

25. Tighten the front stabilizer link assembly LH.

a. Tighten the bolt to 100 ft. lbs. (135 Nm).

b. Tighten the nut to 103 ft. lbs. (140 Nm).

➡Perform this procedure with all 4 wheels on the ground.

26. Tighten the front stabilizer link assembly RH.

a. Tighten the bolt to 100 ft. lbs. (135 Nm).

➡Perform this procedure with all 4 wheels on the ground.

27. Measure the vehicle height.

28. Close the stabilizer control with an accumulator housing shutter valve.

29. Install the stabilizer control valve protector.

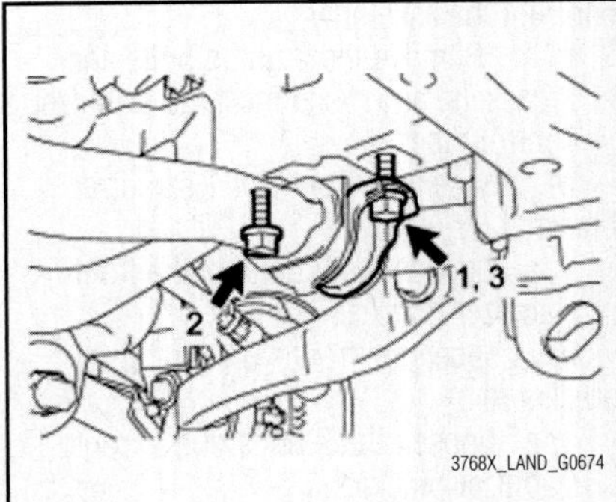

Fig. 277 Tightening the front No. 1 stabilizer bracket LH

Fig. 278 Tightening the nuts

STABILIZER BAR

REMOVAL & INSTALLATION

See Figures 279 through 294.

1. Remove the stabilize control valve protector.
2. Open the stabilizer control with the accumulator housing shutter valve.
3. Remove the front wheel.
4. Remove the tailpipe assembly.
5. Remove the exhaust center pipe assembly.
6. Discharge the suspension fluid pressure.
7. Remove the front fender splash shield sub assembly LH.
8. Remove the front fender splash shield sub assembly RH.
9. Remove the No. 1 engine under cover sub assembly.
10. Remove the front fender apron seal LH.
11. Remove the front fender apron seal rear LH.
12. Remove the stabilizer control tube protector.
    a. Remove the 2 bolts and tube protector.
13. Disconnect the front No. 1 stabilizer control tube assembly.
    a. Remove the 2 union bolts and 2 gaskets, and disconnect the stabilizer control tube.
14. Loosen the front No. 1 stabilizer bracket LH.
    a. Loosen the 2 bolts of the front stabilizer brackets.
15. Loosen the front No. 1 stabilizer bracket RH.
    a. Loosen the 2 bolts of the front stabilizer brackets.
16. Remove the front stabilizer link assembly LH.
    a. Remove the 2 bolts, nut and stabilizer link.
17. Remove the front stabilizer link assembly RH.
    a. Remove the bolt, nut and stabilizer link.

➡If the ball joint turns together with the nut, use a 6 mm hexagon wrench to hold the stud.

18. Remove the front No. 1 stabilizer bracket LH and RH.
    a. Remove the 2 bolts and stabilizer bracket from the front frame assembly.
19. Disconnect the front stabilizer control cylinder.

Fig. 279 Removing the stabilizer control tube protector

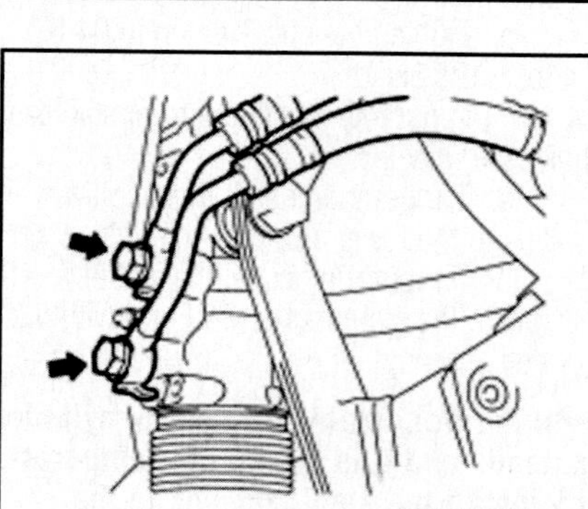

Fig. 280 Disconnecting the front No. 1 stabilizer control tube assembly

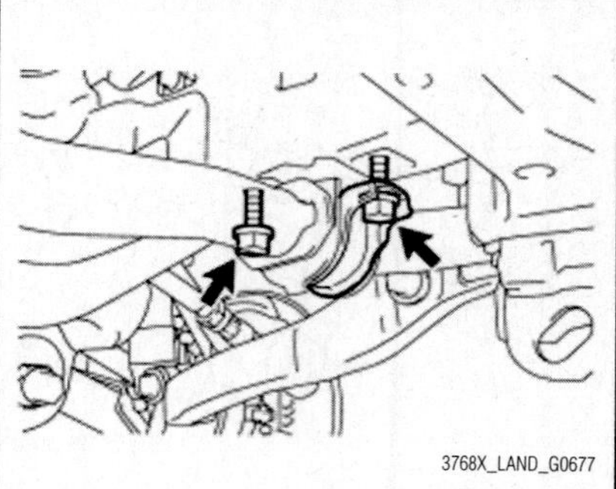

Fig. 281 Loosening the front No. 1 stabilizer bracket LH

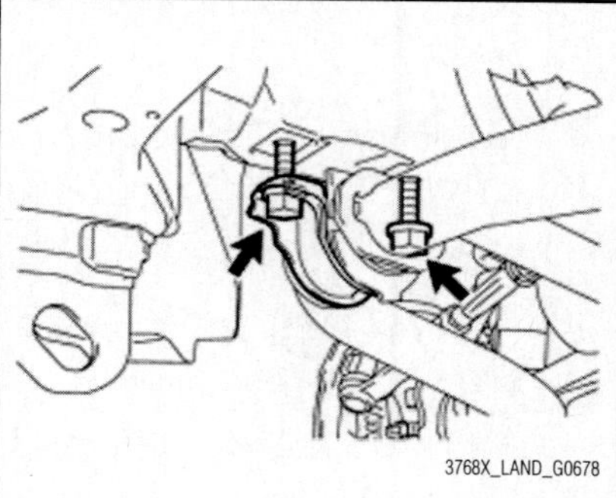

Fig. 282 Loosening the front No. 1 stabilizer bracket RH

    a. Remove the bolt, washer, nut, and stabilizer control cylinder from the frame assembly.
20. Remove the front stabilizer bar.
    a. Remove the bolt, nut and front stabilizer bar from the stabilizer control arm.
21. Remove the front No. 1 stabilizer bar bush.
    a. Remove the 2 stabilizer bar bushes from the front stabilizer bar.
22. Remove the front stabilizer control arm assembly.
    a. Remove the bolt, nut and front stabilizer control arm from the front stabilizer control cylinder.
23. Remove the front No. 1 suspension control bleeder plug.
    a. Remove the 2 front suspension control bleeder plugs from the front stabilizer control cylinder.

To install:

24. Install the front No. 1 suspension control bleeder plug.
    a. Install the 2 suspension control bleeder plugs to the front stabilizer control cylinder. Tighten to 7 ft. lbs. (10 Nm).
25. Install the front stabilizer control cylinder.
    a. Install the front stabilizer control cylinder to the frame with the bolt, washer and nut. Tighten to 203 ft. lbs. (275 Nm).
26. Connect the front No. 1 stabilizer control tube assembly.
    a. Install 2 new gaskets and the front stabilizer control cylinder to the front stabilizer control cylinder with the 2 union bolts. Tighten to 51 ft. lbs. (69 Nm).
27. Inspect for suspension fluid leaks.
28. Install the exhaust center pipe assembly.
29. Install the tailpipe assembly.
30. Install the stabilizer control tube protector with the 2 bolts. Tighten to 11 ft. lbs. (15 Nm).
31. Temporarily install the front stabilizer control arm assembly with the nut and bolt.
32. Bleed the air from the suspension fluid.
33. Install the front No. 1 stabilizer bar bush.
    a. Install the 2 bushes to the outer side of the bush stoppers on the front stabilizer bar as shown in the illustration.

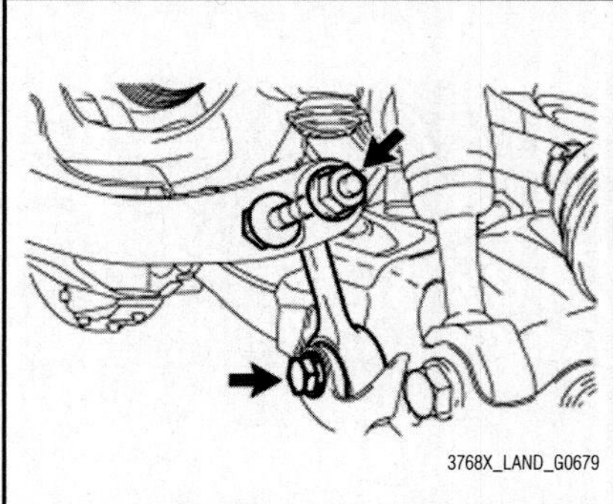

Fig. 283 Removing the front stabilizer link assembly LH

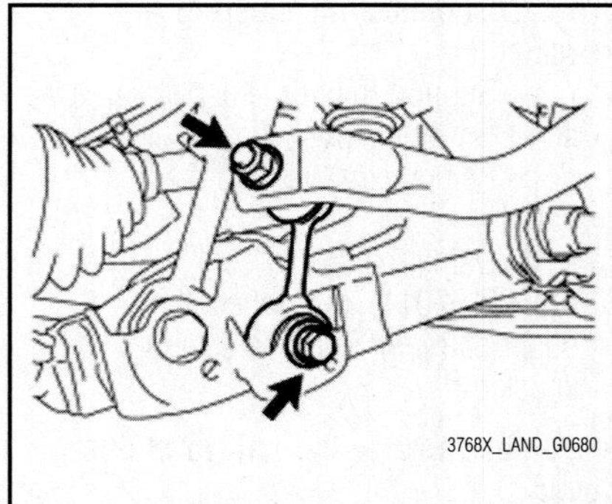

Fig. 284 Removing the front stabilizer link assembly RH

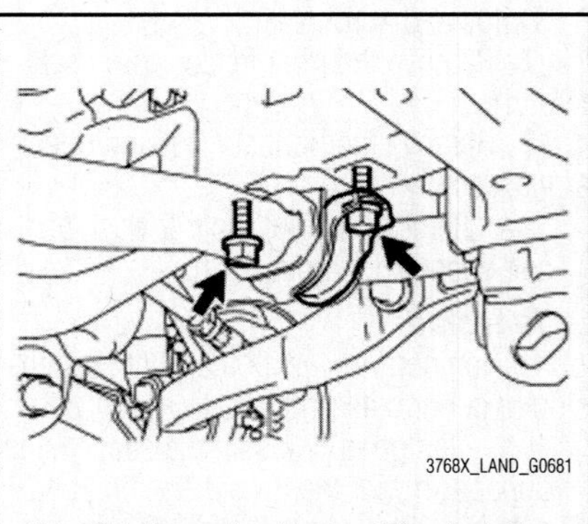

Fig. 285 Removing the front No. 1 stabilizer bracket LH

➡Be sure to install the front No. 1 stabilizer bar bushes so that the bump of each bush faces the inside of the vehicle.

34. Temporarily install the front stabilizer bar.

a. Set the stabilizer bracket so that the arrow mark is facing the front side of the vehicle.

b. Temporarily install the 2 stabilizer brackets and front stabilizer bar with the 4 bolts.

c. Temporarily install the front

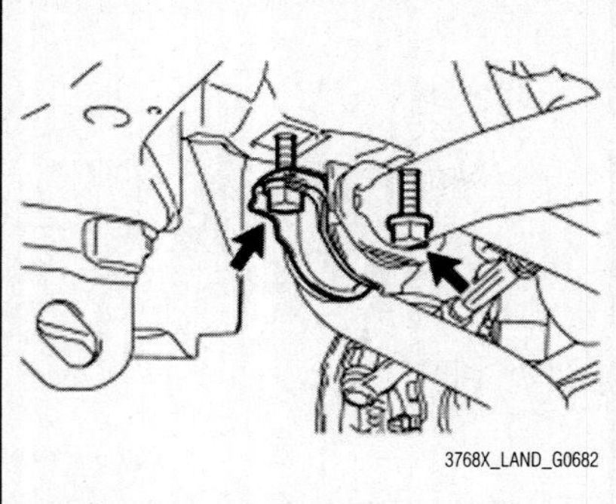

Fig. 286 Removing the front No. 1 stabilizer bracket RH

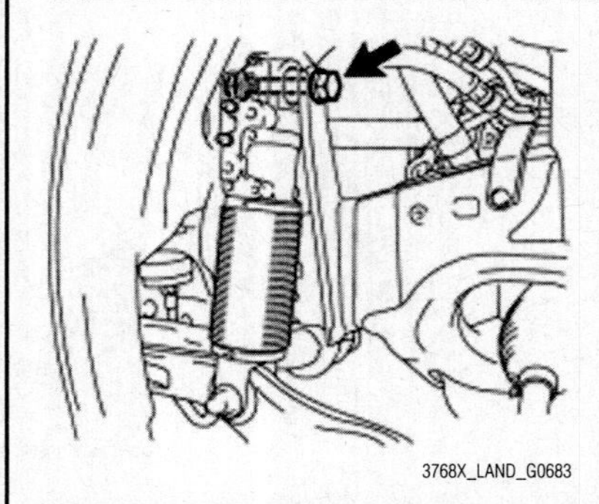

Fig. 287 Disconnecting the front stabilizer control cylinder

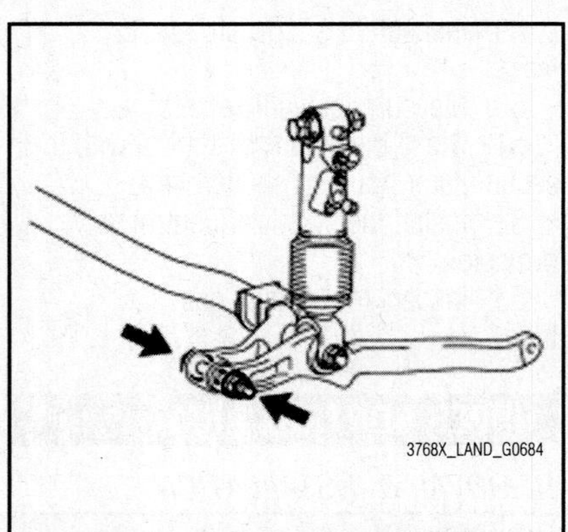

Fig. 288 Removing the front stabilizer bar

stabilizer bar to the stabilizer control arm with the bolt and nut.

35. Temporarily install the front stabilizer link assembly RH with the bolt and nut. Tighten to 94 ft. lbs. (128 Nm).

➡If the ball joint turns together with the nut, use a 6 mm hexagon wrench to hold the stud bolt.

36. Temporarily install the front stabilizer link assembly LH.

a. Temporarily install the stabilizer link to the lower arm with the bolt.

b. Temporarily install the stabilizer

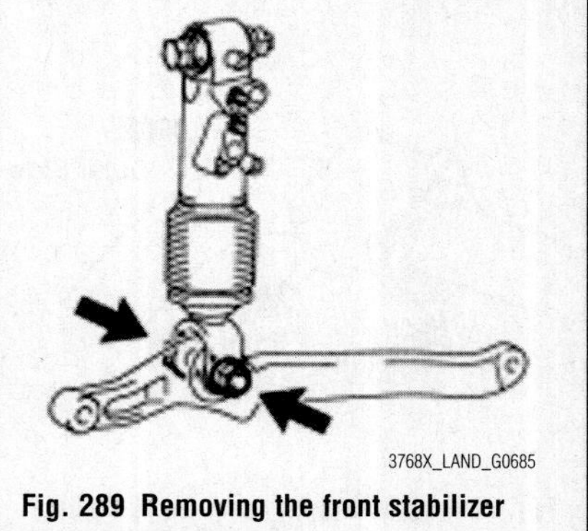

Fig. 289 Removing the front stabilizer control arm assembly

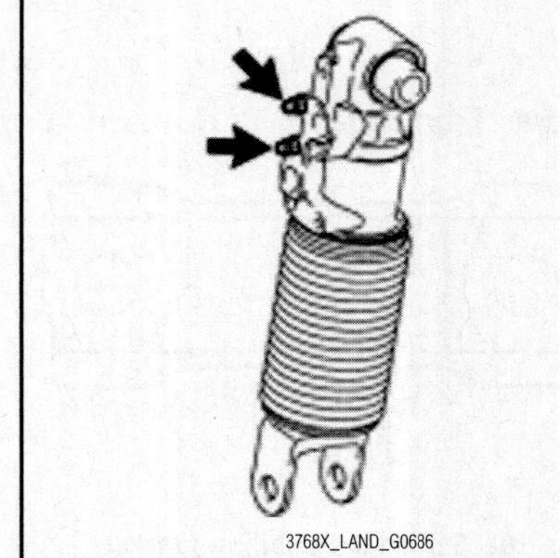

Fig. 290 Removing the front No. 1 suspension control bleeder plug

link to the stabilizer control arm with the bolt and nut.

➡If the front stabilizer control cylinder extends and it is difficult to temporarily install the stabilizer link to the stabilizer control arm, raise the stabilizer control arm with a jack and temporarily install the stabilizer link.

37. Tighten the front No. 1 stabilizer bracket LH with the 2 bolts. Tighten to 64 ft. lbs. (87 Nm).

➡Tighten the bolts in 3 steps, in the order shown in the illustration.

38. Install the front No. 1 stabilizer bracket RH with the 2 bolts. Tighten to 64 ft. lbs. (87 Nm).

➡Tighten the bolts in 3 steps, in the order shown in the illustration.

39. Install the No. 1 engine under cover sub assembly.

40. Install the front fender splash shield sub assembly LH.

41. Install the front fender splash shield sub assembly RH.

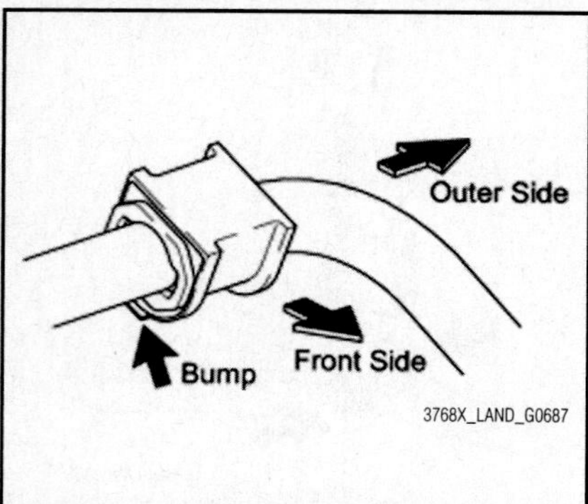

Fig. 291 Installing the front No. 1 stabilizer bar bush

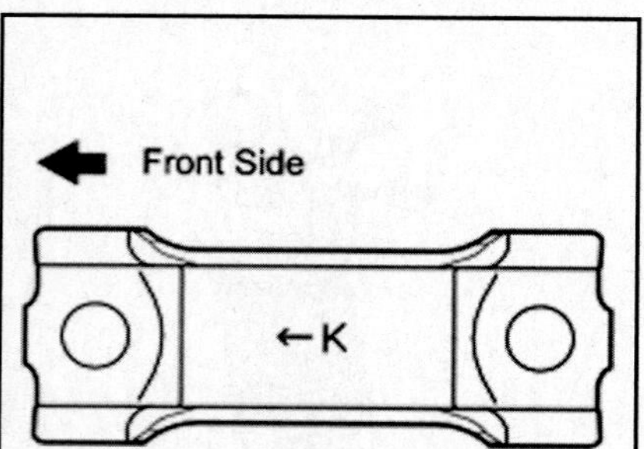

Fig. 292 Setting the stabilizer bracket

42. Stabilize the suspension.

43. Tighten the front stabilizer link assembly LH.

Perform this procedure with all 4 wheels on the ground.

a. Tighten the bolt to 100 ft. lbs. (135 Nm).

b. Tighten the nut to 103 ft. lbs. (140 Nm).

44. Tighten the front stabilizer link assembly RH.

a. Tighten the bolt to 100 ft. lbs. (135 Nm).

➡Perform this procedure with all 4 wheels on the ground.

45. Tighten the front stabilizer control arm assembly with the nut. Tighten to 140 ft. lbs. (190 Nm).

➡Perform this procedure with all 4 wheels on the ground.

46. Tighten the front stabilizer bar nut to 170 ft. lbs. (230 Nm).

➡Perform this procedure with all 4 wheels on the ground.

47. Install the front fender apron seal rear LH.

48. Install the front fender apron seal LH.

Fig. 293 Tightening the front No. 1 stabilizer bracket LH

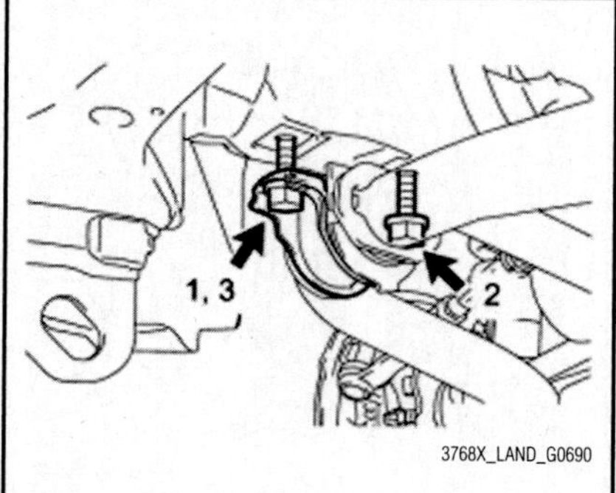

Fig. 294 Tightening the front No. 1 stabilizer bracket RH

49. Inspect for suspension fluid leaks.

50. Measure the vehicle height.

51. Close the stabilizer control with the accumulator housing shutter valve.

52. Install the stabilizer control valve protector.

53. Inspect for exhaust gas leaks.

UPPER CONTROL ARM

REMOVAL & INSTALLATION

See Figure 295.

➡Use the same procedures for the RH side and LH side.

➡The procedures listed below are for the LH side.

1. Remove the stabilizer control valve protector.

2. Open the stabilizer control with accumulator housing shutter valve.

3. Remove the front wheel.

4. Disconnect the skid control sensor wire.

a. Remove the bolt and nut, and disconnect the sensor wire from the steering knuckle and suspension upper arm.

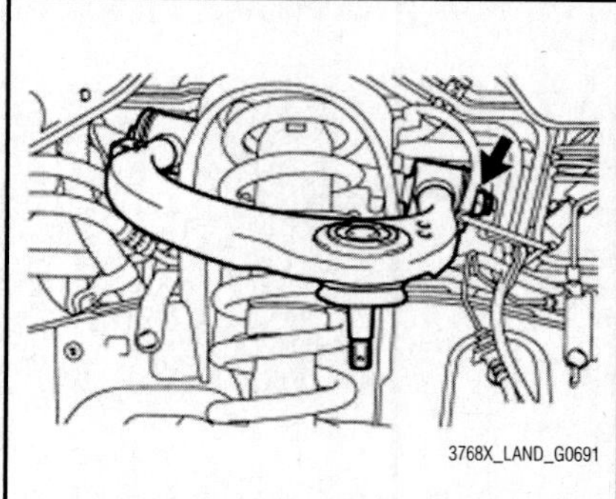

Fig. 295 Removing the front suspension upper arm assembly

5. Disconnect the steering knuckle LH.

a. Support the front suspension lower arm LH with a jack.

b. Remove the clip and the nut.

c. Using the special tool (09628-62011), disconnect the upper ball joint from the steering knuckle.

➡Do not damage the ball joint dust cover.

6. Remove the front fender apron seal LH.

a. Using the clip remover, remove the 3 clips and apron seal.

7. Remove the front fender apron seal rear LH.

8. Remove the front suspension upper arm assembly

a. Remove the nut, bolt, 2 washers and suspension upper arm.

To install:

9. Temporarily install the front suspension upper arm assembly LH.

a. Temporarily install the suspension upper arm with the 2 washers, bolt and nut.

➡ After stabilizing the suspension, tighten the nut.

10. Connect the steering knuckle LH.

a. Connect the steering knuckle to the suspension upper arm.

b. Install the nut and a new cotter pin. Tighten to 81 ft. lbs. (110 Nm).

➡ If the holes for the cotter pin are not aligned, tighten the nut further up to 60° .

11. Connect the skid control sensor wire to the steering knuckle and upper arm with the bolt and nut. Tighten to 9.6 ft. lbs. (13 Nm).

12. Stabilize the suspension.

13. Tighten the front suspension upper arm assembly LH. Tighten to 136 ft. lbs. (185 Nm).

14. Install the front fender apron seal rear LH with the 4 clips.

15. Install the front fender apron seal LH with the 3 clips.

16. Measure the vehicle height.

17. Close the stabilizer control with the accumulator housing shutter valve.

18. Install the stabilizer control valve protector.

19. Inspect and adjust the front wheel alignment.

20. Adjust the headlight assembly.

COIL SPRING

REMOVAL & INSTALLATION

See Figure 296.

➡Use the same procedures for the RH side and LH side.

➡The procedures listed below are for the LH side.

1. Remove the stabilizer control valve protector.
2. Open the stabilizer control with the accumulator housing shutter valve.
3. Remove the rear wheel.
4. Drain the brake fluid.
5. Disconnect the No. 3 parking brake cable assembly.
6. Disconnect the No. 2 parking brake cable assembly.
7. Disconnect the rear axle breather hose sub assembly.
8. Support rear axle housing assembly.
9. Remove the rear stabilizer end bracket.
10. Disconnect the rear stabilizer link assembly.
    a. Remove the nut and disconnect the stabilizer link assembly from the stabilizer bar.

➡Do not lose the ball Joint dust cover protector.

➡If the ball joint turns together with the nut, use a 6 mm hexagon wrench to hold the stud.

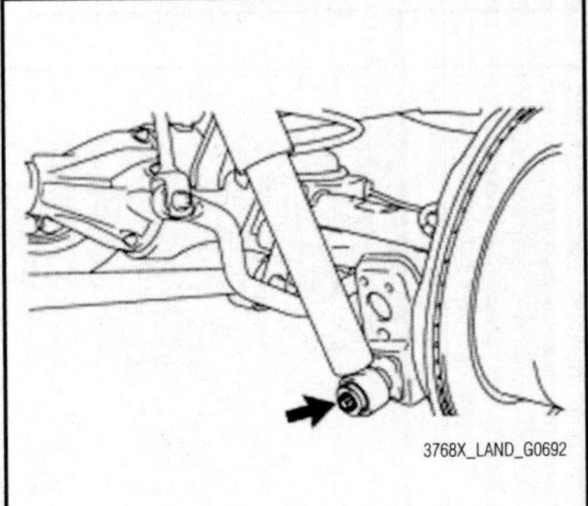

Fig. 296 Disconnecting the rear shock absorber assembly LH

11. Disconnect the rear lateral control rod assembly.
12. Disconnect the rear flexible hose.
13. Disconnect the rear shock absorber assembly LH.
    a. Remove the bolt on the lower side of the shock absorber.
    b. Disconnect the shock absorber from the axle housing.
14. Disconnect the rear shock absorber assembly RH.
    a. Remove the bolt on the lower side of the shock absorber.
    b. Disconnect the shock absorber from the axle housing.
15. Remove the rear coil spring LH.
    a. While lowering the axle housing, remove the coil spring.

➡Be careful not to snap the brake line.

16. Remove the hollow spring sub assembly.
17. Remove the hollow spring from the frame.

To install:

18. Install the hollow spring sub assembly.
19. Install the rear coil spring LH.

➡Before installing the coil spring, check that the coil spring end is in the correct position. If not, re-install it.

20. Temporarily install the rear shock absorber assembly LH.
    a. Temporarily install the lower side of the shock absorber with the bolt.
21. Temporarily install the rear shock absorber assembly RH.
    a. Temporarily install the lower side of the shock absorber with the bolt.
22. Temporarily install the rear lateral control rod assembly.
23. Connect the rear flexible hose.
24. Stabilize the suspension.
25. Tighten the rear shock absorber assembly LH. Tighten the bolt to 72 ft. lbs. (98 Nm).
26. Tighten the rear shock absorber assembly RH. Tighten the bolt to 72 ft. lbs. (98 Nm).
27. Tighten the rear lateral control rod assembly.
28. Install the rear stabilizer end bracket.
    a. Install the stabilizer end bracket to the vehicle with the 2 bolts. Tighten to 22 ft. lbs. (30 Nm).
29. Connect the rear stabilizer link assembly.
    a. Install the stabilizer link assembly to the ball joint dust cover protector.
    b. Connect the stabilizer link assembly to the stabilizer bar with the nut. Tighten to 74 ft. lbs. (100 Nm).

➡If the ball joint turns together with the nut, use a 6 mm hexagon wrench to hold the stud bolt.

➡If the rear stabilizer control cylinder extends and it is difficult to temporarily install the stabilizer link to the stabilizer bar, raise the stabilizer control arm with a jack and temporarily install the stabilizer link.

30. Connect the No. 3 parking brake cable assembly.
31. Connect the No. 2 parking brake cable assembly.
32. Connect the rear axle breather hose sub assembly.
33. Install the rear wheel. Tighten to 97 ft. lbs. (131 Nm).
34. Fill the reservoir with brake fluid.
35. Bleed the brake line/
36. Check the brake fluid level in the reservoir.
37. Inspect for brake fluid leaks.
38. Measure the vehicle height.
39. Close the stabilizer control with the accumulator housing shutter valve.
40. Install the stabilizer control valve protector.

CONTROL ARMS/LINKS

REMOVAL & INSTALLATION

See Figure 297.

Lower

➡Use the same procedures for the RH side and LH side.

➡The procedures listed below are for the LH side.

1. Remove the stabilizer control valve protector.
2. Open the stabilizer control with the accumulator housing shutter valve.
3. Remove the rear wheel.
4. Support the rear axle housing assembly.
5. Remove the lower control arm assembly LH.

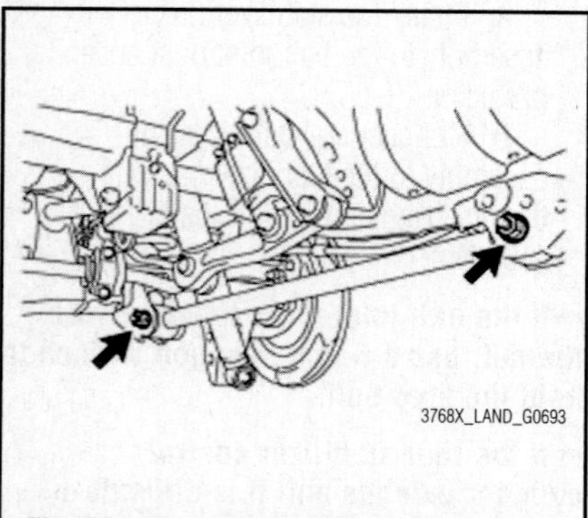

Fig. 297 Removing the lower control arm

a. Remove the 2 nuts, 2 washers, 2 bolts, and the lower control arm.

To install:

6. To install, reverse the removal procedure. Tighten the lower control arm to 111 ft. lbs. (150 Nm). Tighten the rear wheel to 97 ft. lbs. (131 Nm).

Upper

See Figures 298 and 299.

➡Use the same procedures for the RH side and LH side.

➡The procedures listed below are for the LH side.

1. Remove the stabilizer control valve protector.
2. Open the stabilizer control with the accumulator housing shutter valve.
3. Remove the rear wheel.
4. Disconnect the speed sensor wire harness.
   a. Remove the bolt and disconnect the speed sensor wire harness from the upper control arm.

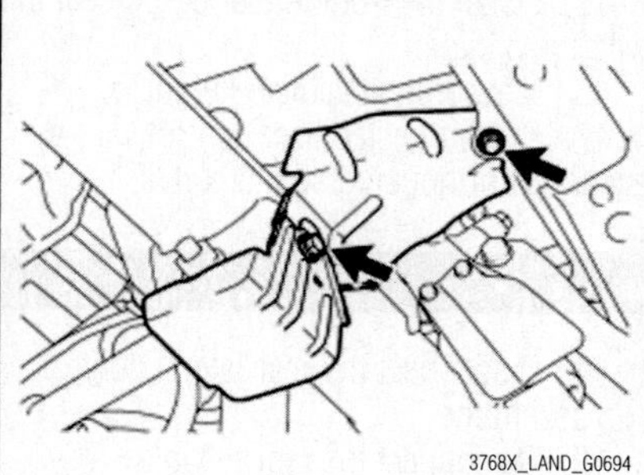

Fig. 298 Removing the rear stabilizer control tube insulator

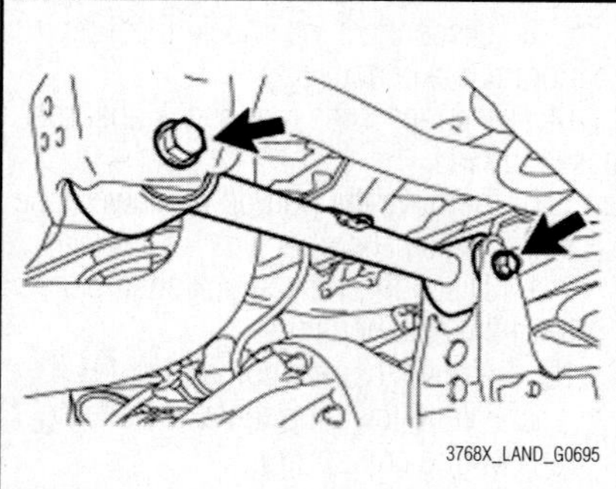

Fig. 299 Remove the rear upper control arm assembly LH

5. Remove the rear stabilizer control tube insulator.
   a. Remove the 2 bolts and rear stabilizer control tube insulator.
6. Support the rear axle housing assembly.
7. Remove the rear upper control arm assembly LH.
   a. Remove the 2 nuts, 2 washers and 2 bolts, and upper control arm.

To install:

8. Temporarily install the rear upper control arm assembly LH.
9. Stabilize the suspension.
   a. Install the rear wheels. Tighten to 97 ft. lbs. (131 Nm).
   b. Lower the vehicle.
   c. Press down on the vehicle several times to stabilize the suspension.
   d. Remove the rear wheels.
10. Tighten the rear upper control arm assembly LH.
    a. Tighten the 2 nuts to 111 ft. lbs. (150 Nm).
11. Install the rear stabilizer control tube insulator with the 2 bolts. Tighten to 13 ft. lbs. (18 Nm).
12. Connect the speed sensor wire harness with the bolt. Tighten to 9 ft. lbs. (13 Nm).
13. Install the rear wheel and tighten to 97 ft. lbs. (131 Nm).
14. Measure the vehicle height.
15. Close the stabilizer control with the accumulator housing shutter valve.
16. Install the stabilizer control valve protector.

SHOCK ABSORBER

REMOVAL & INSTALLATION

See Figures 300 through 304.

➡Use the same procedures for the RH side and LH side.

➡The procedures listed below are for the LH side.

1. Remove the stabilizer control valve protector.
2. Open the stabilizer control with the accumulator housing shutter valve.
3. Remove the rear wheel.
4. Disconnect the No. 3 parking brake cable assembly.
   a. Remove the bolt and disconnect the No. 3 parking brake cable.
5. Disconnect the No. 2 parking brake cable assembly.
   a. Remove the bolt and disconnect the No. 2 parking brake cable.

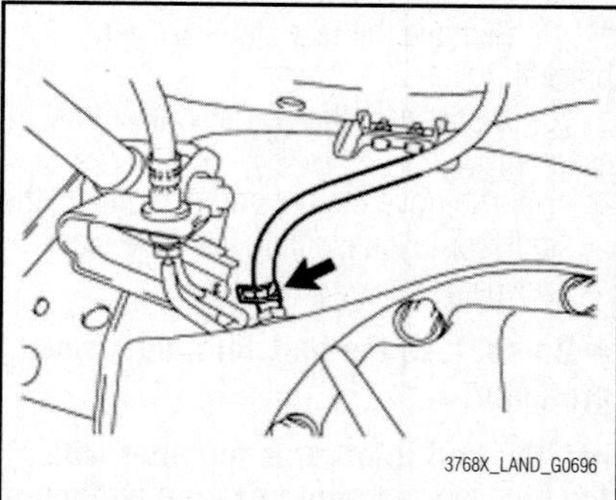

Fig. 300 Disconnect the rear axle breather hose sub assembly

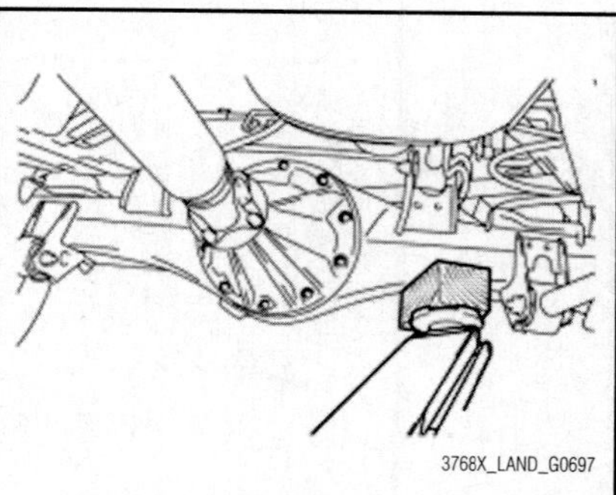

Fig. 301 Support the rear axle housing assembly

Fig. 302 Disconnecting the rear lateral control rod assembly

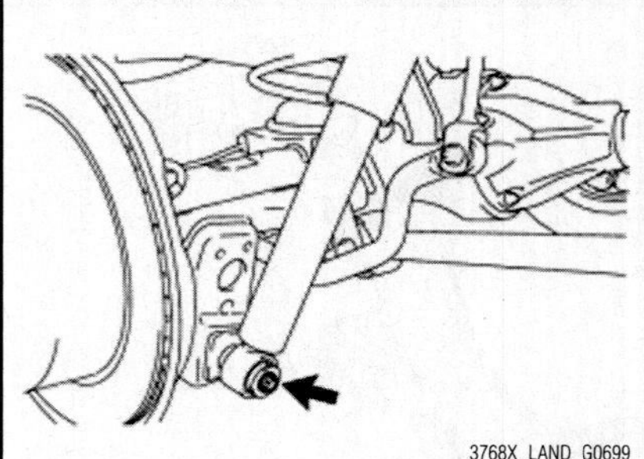

Fig. 303 Disconnecting the rear shock absorber assembly RH

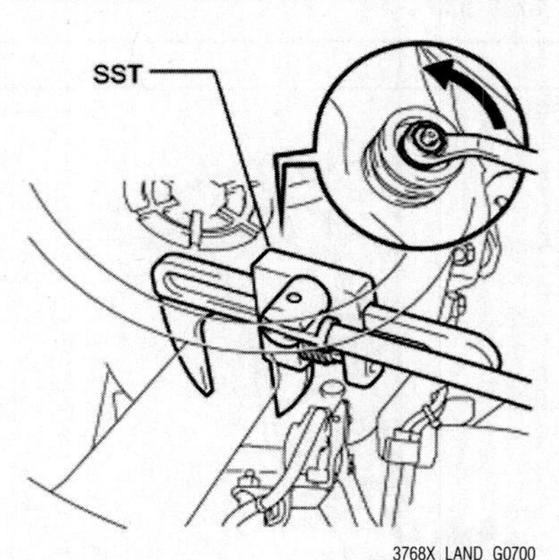

Fig. 304 Removing the rear shock absorber assembly LH

6. Disconnect the rear axle breather hose sub assembly.
   a. Disconnect the rear axle breather hose from the rear axle housing assembly.
7. Support the rear axle housing assembly.
   a. Support the rear axle housing with a jack using a wooden block to avoid damage.
8. Disconnect the rear lateral control rod assembly.
   a. Remove the bolt and nut, and disconnect the rear lateral control rod from the frame.
9. Disconnect the rear shock absorber assembly RH.
   a. Remove the bolt on the lower side of the shock absorber.
   b. Disconnect the shock absorber from the axle housing.
10. Remove the rear shock absorber assembly LH.
   a. Remove the bolt on the lower side of the shock absorber.
   b. Using the special tool (09922-10010), hold the rear shock absorber in place.
   c. Remove the nut, upper bracket and shock absorber assembly.
   d. Remove the lower bracket from the shock absorber.

To install:

11. Temporarily install the rear shock absorber assembly LH.
   a. Install the lower bracket to the shock absorber.
   b. Using special tool, hold the rear shock absorber in place.
   c. Temporarily install the rear shock absorber and upper bracket with the nut.
   d. Temporarily install the lower side of the shock absorber with the bolt.
12. Temporarily install the rear shock absorber assembly RH.
   a. Temporarily install the lower side of the shock absorber with the bolt.
13. Temporarily install the rear lateral control rod assembly.
   a. Temporarily install the lateral control rod with the nut and bolt.
14. Stabilize the suspension.
   a. Install the rear wheels. Tighten to 97 ft. lbs. (131 Nm).
   b. Lower the vehicle.
   c. Press down on the vehicle several times to stabilize the suspension.
   d. Remove the rear wheels.
15. Tighten the shock absorber assembly LH. Tighten the nut to 39 ft. lbs. (53 Nm). Tighten the bolt to 72 Nm (98 Nm).
16. Tighten the rear shock absorber assembly RH. Tighten the bolt and nut to 72 ft. lbs. (98 Nm).
17. Tighten the rear lateral control rod assembly nut to 111 ft. lbs. (150 Nm).
18. Connect the No. 3 parking brake cable assembly.
   a. Connect the No. 3 parking brake cable with the bolt to 9 ft. lbs. (13 Nm).
19. Connect the No. 2 parking brake cable.
   a. Connect the No. 2 parking brake cable with the bolt to 9 ft. lbs. (13 Nm).
20. Connect the rear axle breather hose sub assembly to the rear axle housing.
21. Install the rear wheel. Tighten to 97 ft. lbs. (131 Nm).
22. Measure the vehicle height.
23. Close stabilizer control with the accumulator housing shutter valve.
24. Install the stabilizer control valve protector.

TESTING

Compress and extend the shock absorber rod 4 times or more. Check that there is no abnormal resistance or unusual sound during the operation. If there is any abnormality, replace the shock absorber assembly with a new one.

STABILIZER BAR

REMOVAL & INSTALLATION

See Figures 305 through 314.

1. Remove stabilizer control valve protector.
2. Open stabilizer control with accumulator housing shutter valve.
3. Remove rear wheel.
4. Remove tailpipe assembly.
5. Remove exhaust center pipe assembly.
6. Discharge suspension fluid pressure.
7. Support rear axle housing assembly.
8. Remove rear stabilizer end bracket.
   a. Remove the 2 bolts and disconnect the stabilizer end bracket.
9. Remove the rear stabilizer bar sub assembly.
   a. Remove the nut and disconnect the stabilizer link assembly from the stabilizer bar.

➡Do not miss the ball joint dust cover protector.

➡If the ball joint turns together with the nut, use a 6 mm hexagon wrench to hold the stud.

   b. Remove the 4 bolts and stabilizer bar with the bushes and brackets.
10. Remove the stabilizer link sub assembly.
   a. Remove the nut, 3 retainers and 2 cushions, and disconnect the stabilizer link from the bracket.
   b. Remove the nut, bolt and stabilizer link.
11. Remove the rear stabilizer control tube insulator.
   a. Remove the 2 bolts and rear stabilizer control tube insulator.

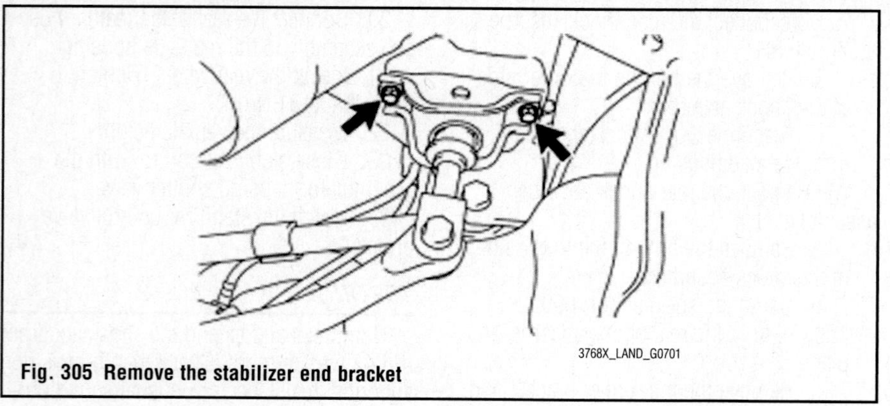

Fig. 305 Remove the stabilizer end bracket

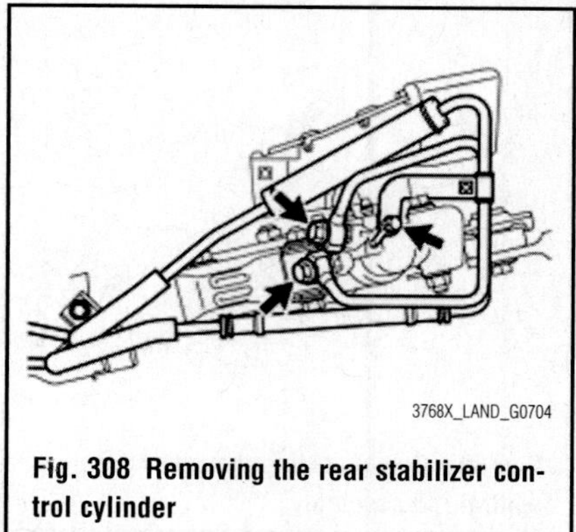

Fig. 308 Removing the rear stabilizer control cylinder

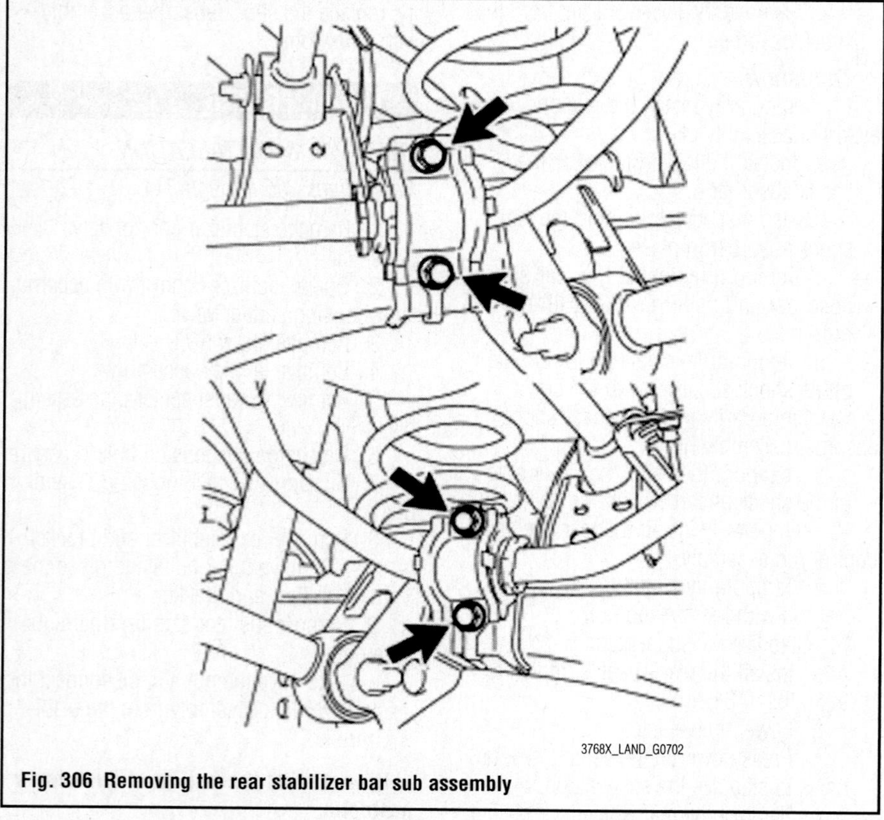
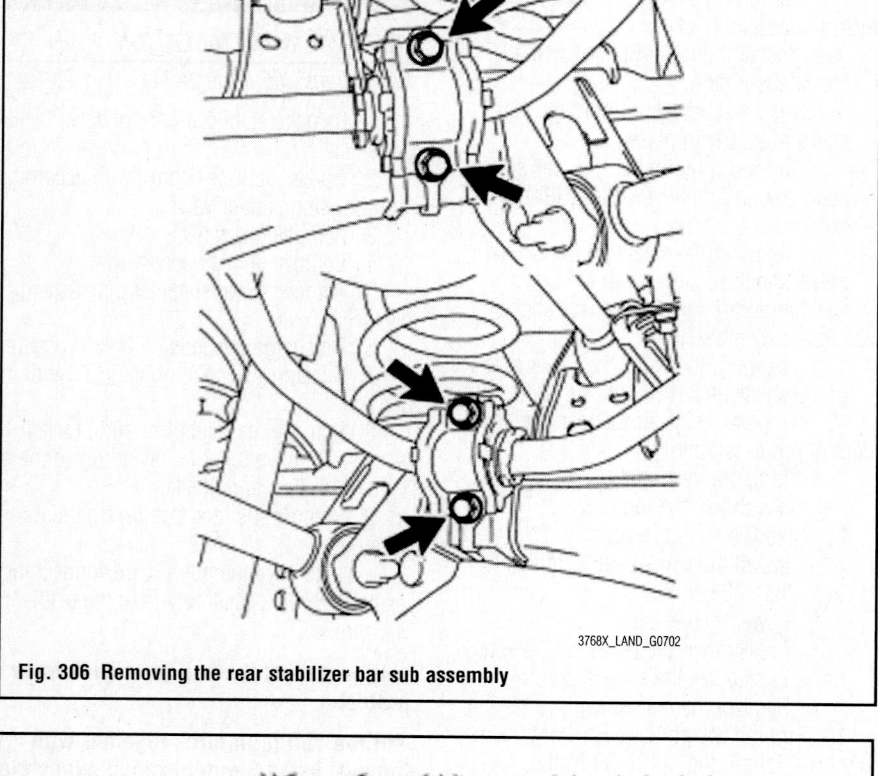

Fig. 306 Removing the rear stabilizer bar sub assembly

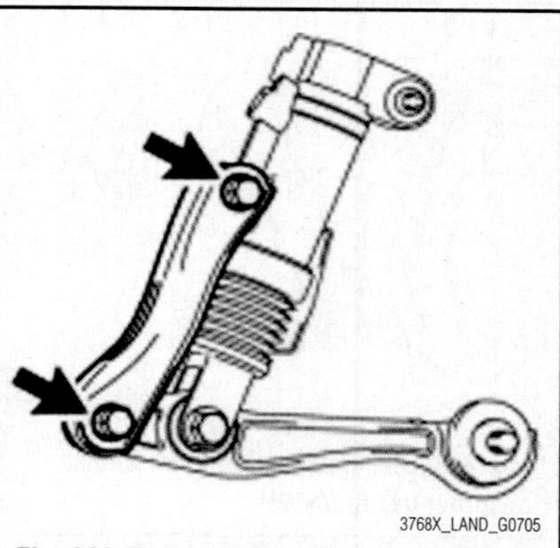

Fig. 309 Removing the rear No. 2 stabilizer link

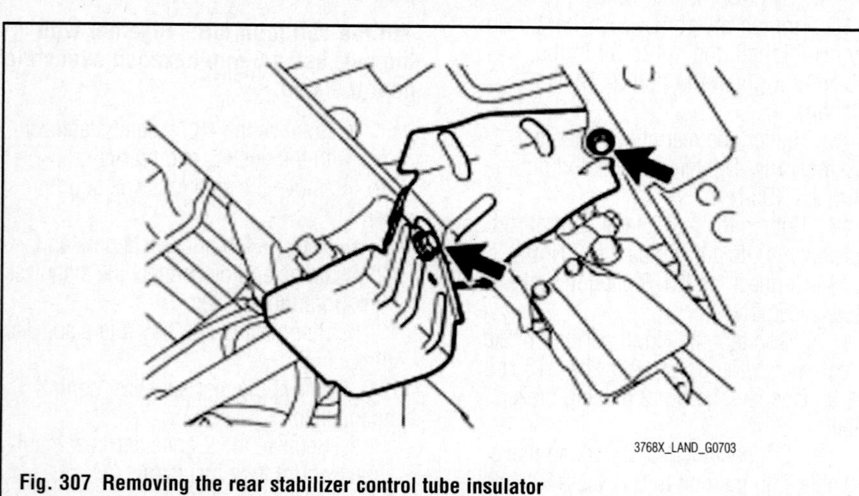

Fig. 307 Removing the rear stabilizer control tube insulator

Fig. 310 Removing the rear stabilizer link assembly

12. Remove the rear stabilizer control cylinder.
   a. Remove the 2 union bolts with the 2 gaskets and bolt, and disconnect the stabilizer control tube.
   b. Remove the bolt and stabilizer control cylinder.
13. Remove the rear No. 2 stabilizer link.
   a. Remove the 2 bolts, 2 nuts and No. 2 rear stabilizer link.

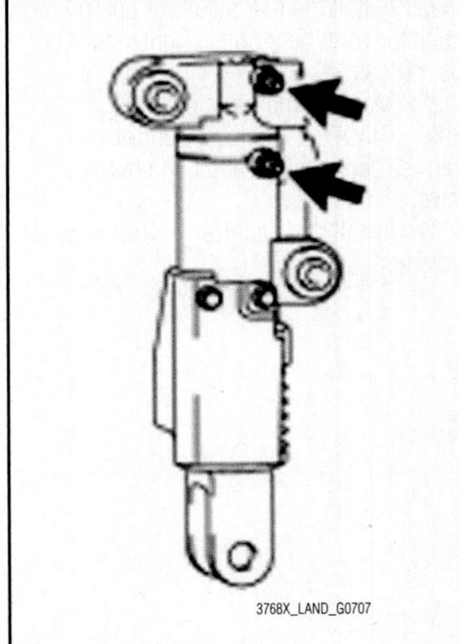

Fig. 311 Removing the rear suspension control bleeder plug

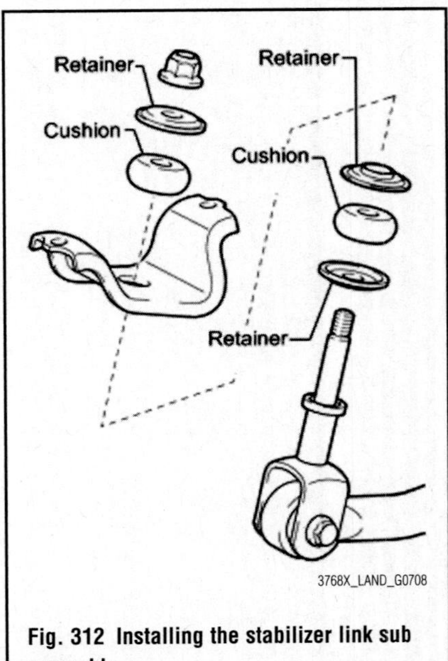

Fig. 312 Installing the stabilizer link sub assembly

14. Remove the rear stabilizer link assembly.
   a. Remove the bolt, nut and rear stabilizer link assembly.
15. Remove the rear suspension control bleeder plug.
   a. Remove the 2 rear suspension control bleeder plugs from the rear stabilizer control cylinder.

To install:

16. Install the rear suspension control bleeder plugs to the rear stabilizer control cylinder. Tighten to 7 ft. lbs. (10 Nm).

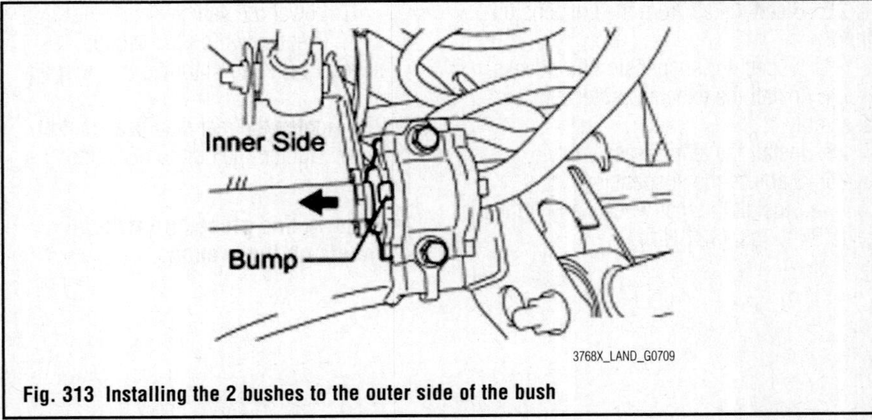

Fig. 313 Installing the 2 bushes to the outer side of the bush

17. Install the rear stabilizer link assembly with the bolt to 153 ft. lbs. (208 Nm).

➡Make sure the bolt is facing the correct direction.

18. Install the rear No. 2 stabilizer link with the 2 bolts and 2 nuts. Tighten to 72 ft. lbs. (97 Nm).

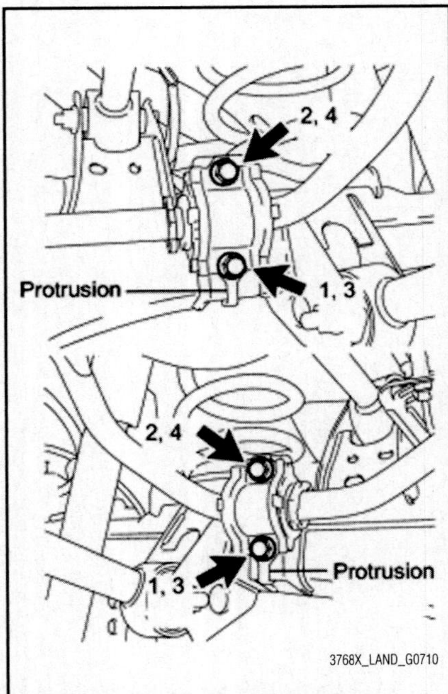

Fig. 314 Installing the stabilizer bar brackets

➡Make sure the bolt is facing the correct direction.

19. Temporarily install the rear stabilizer control cylinder to the frame with the bolt.
   a. Install the rear stabilizer control tube and 2 new gaskets to the rear stabilizer control cylinder with the 2 union bolts. Tighten to 51 ft. lbs. (69 Nm).
   b. Install the bolt and tighten to 21 ft. lbs. (29 Nm).
20. Install the stabilizer link sub assembly to the stabilizer bar with the bolt and nut. Tighten to 35 ft. lbs. (48 Nm).

➡Make sure the bolt is facing the correct direction.

   a. Make sure that all parts are facing as shown in the illustration and install the stabilizer end bracket, retainers and cushions with a new nut. Tighten to 22 ft. lbs. (30 Nm).
21. Install the rear stabilizer bar sub assembly.
   a. Install the 2 bushes to the outer side of the bush stoppers on the rear stabilizer bar as shown in the illustration.

➡Install the stabilizer bush so that the bump is on the inner side of the vehicle.

   b. Install the 2 stabilizer bar brackets to the vehicle with the 4 bolts. Tighten to 43 ft. lbs. (58 Nm).

➡Install the stabilizer bar bracket so that the protrusion is on the lower side.

➡Tighten the bolts in 4 steps, in the order shown in the illustration.

   c. Install the stabilizer end bracket to the vehicle with the 2 bolts. Tighten to 22 ft. lbs. (30 Nm).
   d. Install the stabilizer link assembly to the ball joint dust cover protector.
   e. Install the stabilizer bar to the stabilizer link assembly with the nut. Tighten to 74 ft. lbs. (100 Nm).

➡If the ball joint turns together with the nut, use a 6 mm hexagon wrench to hold the stud bolt.

➡If the rear stabilizer control cylinder extends and it is difficult to temporarily install the stabilizer link to the stabilizer bar, raise the stabilizer control arm with a jack and temporarily install the stabilizer link.

22. Bleed the air from the suspension fluid.
23. Inspect for suspension fluid leaks.
24. Install the exhaust center pipe assembly.
25. Install the tailpipe assembly.
26. Stabilize the suspension.
  a. Install the rear wheels and tighten to 97 ft. lbs. (131 Nm).
  b. Lower the vehicle.
  c. Press down on the vehicle several times to stabilize the suspension.
27. Tighten the rear stabilizer control cylinder. Tighten the bolt to 66 ft. lbs. (90 Nm).

➡Perform this procedure with all 4 wheels on the ground.

28. Install the rear stabilizer control tube insulator with the 2 bolts. Tighten to 13 ft. lbs. (18 Nm).
29. Measure the vehicle height.
30. Close the stabilizer control with the accumulator housing shutter valve.
31. Install the stabilizer control valve protector.

TOYOTA

12

Matrix

SPECIFICATIONS AND MAINTENANCE CHARTS

ENGINE AND VEHICLE IDENTIFICATION

Engine							Model Year	
Code ①	Liters (cc)	Cu. In.	Cyl.	Fuel Sys.	Engine Type	Eng. Mfg.	Code ②	Year
2ZR-FE	1.8 (1794)	109.5	4	EFI	DOHC	Toyota	9	2009
2AZ-FE	2.4 (2400)	109.5	4	EFI	DOHC	Toyota	A	2010

EFI: Electronic Fuel Injection

DOHC: Double Overhead Camshaft

① 5th digit of VIN

②10th digit of VIN

3768X_MATR_C0001

GENERAL ENGINE SPECIFICATIONS

Year	Model	Engine Displacement Liters (cc)	Engine Series (ID/VIN)	Net Horsepower @ rpm	Net Torque @ rpm (ft. lbs.)	Bore x Stroke (in.)	Com-pression Ratio	Oil Pressure @ rpm
2009	Matrix	1.8 (1794)	2ZR-FE	132@6000	128@4400	3.16x3.48	10.0:1	4.3
		2.4 (2400)	2AZ-FE	158@6000	162@4000	3.48x3.78	9.8	NA
2010	Matrix	1.8 (1794)	2ZR-FE	132@6000	128@4400	3.16x3.48	10.0:1	4.3
		2.4 (2400)	2AZ-FE	158@6000	162@4000	3.48x3.78	9.8	NA

NA: Not Available

EFI: Electronic Fuel Injection

3768X_MATR_C0002

ENGINE TUNE-UP SPECIFICATIONS

Year	Engine Displacement Liters	Engine ID/VIN	Spark Plug Gap (in.)	Ignition Timing (deg.)	Fuel Pump (psi)	Idle Speed (rpm) MT	Idle Speed (rpm) AT	Valve Clearance In.	Valve Clearance Ex.
2009	1.8	2ZR-FE	0.043	①	44-50	600-700	600-700	0.007-0.014	0.015-0.018
	2.4	2AZ-FE	0.043	②	44-50	600-700	600-700	NA	NA
2010	1.8	2ZR-FE	0.043	①	44-50	600-700	600-700	0.007-0.014	0.015-0.018
	2.4	2AZ-FE	0.043	②	44-50	600-700	600-700	NA	NA

Note: The Vehicle Emission Control Information label often reflects specification changes made during production.

NA: Not Available

BTDC: Before Top Dead Center

① With Techstream (ODB-II scanner or equivalent): BTDC 5-15 at idle
Without Techstream: BTDC 8-12 at idle. Connect terminals 13 (TC) and 4 (CG)

② BTDC 8-12 with terminal TC and CG of DLC3 connected

3768X_MATR_C0003

CAPACITIES

Year	Model	Engine Displacement Liters	Engine ID/VIN	Engine Oil with Filter	Transmission (pts.) 5-Spd	6-Spd	Auto.	Drive Axle Front (pts.)	Rear (pts.)	Fuel Tank (gal.)	Cooling System (qts.)
2009	Matrix	1.8	2ZR-FE	4.4	4.0	NE	16.0	NA	NE	13.2	6.0
		2.4	2AZ-FE	4.4	4.0	NE	6.0	NA	NE	13.2	6.0
2010	Matrix	1.8	2ZR-FE	4.4	4.0	NE	16.0	NA	NE	13.2	6.0
		2.4	2AZ-FE	4.4	4.0	NE	6.0	NA	NE	13.2	6.0

Note: All capacities are approximate. Add fluid gradually and check to be sure a proper fluid level is obtained. Auto trans is a drain and fill capacity.

NE: Not Equipped

NA: Not Available

3768X_MATR_C0004

FLUID SPECIFICATIONS

Year	Model	Engine Displacement Liters	Engine ID/VIN	Engine Oil	Auto. Trans.	Drive Axle	Power Steering Fluid	Brake Master Cylinder
2009	Matrix	1.8	2AZ-FE	5W-20	NA	—	ATF Dexron II Or III	DOT 3
		2.4	2AZ-FE	5W-20	NA	—	ATF Dexron II Or III	DOT 3
2010	Matrix	1.8	2AZ-FE	5W-20	NA	—	ATF Dexron II Or III	DOT 3
		2.4	2AZ-FE	5W-20	NA	—	ATF Dexron II Or III	DOT 3

DOT: Department Of Transpotation

NA: Not Available

3768X_MATR_C0014

VALVE SPECIFICATIONS

Year	Engine Displacement Liters	Engine ID/VIN	Seat Angle (deg.)	Face Angle (deg.)	Spring Test Pressure (lbs. @ in.)	Spring Installed Height (in.)	Stem-to-Guide Clearance (in.) Intake	Exhaust	Stem Diameter (in.) Intake	Exhaust
2009	1.8	2ZR-FE	45	44.5	35.7-39.5@ 1.323	1.323	0.0010-0.0024	0.0012-0.0025	0.2154-0.2159	0.2152-0.2158
	2.4	2AZ-FE	45	44.5	35.7-39.5@ 1.323	1.323	0.0010-0.0024	0.0012-0.0025	0.2154-0.2159	0.2152-0.2158
2010	1.8	2ZR-FE	45	44.5	35.7-39.5@ 1.323	1.323	0.0010-0.0024	0.0012-0.0025	0.2154-0.2159	0.2152-0.2158
	2.4	2AZ-FE	45	44.5	35.7-39.5@ 1.323	1.323	0.0010-0.0024	0.0012-0.0025	0.2154-0.2159	0.2152-0.2158

3768X_MATR_C0005

CAMSHAFT AND BEARING SPECIFICATIONS

All measurements are given in inches.

Year	Engine Displacement Liters	Engine VIN	Journal Diameter	Brg. Oil Clearance	Shaft End-play	Runout	Journal Bore	Lobe Lift Intake	Lobe Lift Exhaust
2009	1.8	2ZR-FE	①	0.0012-0.0025	NA	0.0016	NA	1.685-1.6890	1.745-1.7490
	2.4	2AZ-FE	②	③	NA	0.00118	NA	1.865-1.8660	1.813-1.8170
2010	1.8	2ZR-FE	①	0.0012-0.0025	NA	0.0016	NA	1.685-1.6890	1.745-1.7490
	2.4	2AZ-FE	②	③	NA	0.00118	NA	1.865-1.8660	1.813-1.8170

NA: Not Available

① No. 1 journal diameter: 1.356
Other: 0.903 to 0.904

② No. 1 journal diamter: 1.416
Other: 0.903 to 0.904

③ Intake: 0.00276 (Max.)
Exhaust: 0.00394 (Max.)

3768X_MATR_C0008

CRANKSHAFT AND CONNECTING ROD SPECIFICATIONS

All measurements are given in inches.

Year	Engine Displacement Liters	Engine ID/VIN	Crankshaft Main Brg. Journal Dia.	Crankshaft Main Brg. Oil Clearance	Crankshaft Shaft End-play	Crankshaft Thrust on No.	Connecting Rod Journal Diameter	Connecting Rod Oil Clearance	Connecting Rod Side Clearance
2009	1.8	2ZR-FE	1.8893-1.8898	0.0006-0.0015	0.0008-0.0087	3	1.7320-1.7323	0.0012-0.0024	0.0063-0.0135
	2.4	2AZ-FE	2.164-2.165	0.00236	NA	3	0.283-0.2870	0.00248	0.0063-0.0143
2010	1.8	2ZR-FE	1.8893-1.8898	0.0006-0.0015	0.0008-0.0087	3	1.7320-1.7323	0.0012-0.0024	0.0063-0.0135
	2.4	2AZ-FE	2.164-2.165	0.00236	NA	3	0.283-0.2870	0.00248	0.0063-0.0143

NA: Not Available

3768X_MATR_C0006

PISTON AND RING SPECIFICATIONS

All measurements are given in inches.

Year	Engine Disp. Liters	Engine ID/VIN	Piston Clearance	Ring Gap Top Compression	Ring Gap Bottom Compression	Ring Gap Oil Control	Ring Side Clearance Top Compression	Ring Side Clearance Bottom Compression	Ring Side Clearance Oil Control
2009	1.8	2ZR-FE	0.0011-0.0020	0.0079-0.0118	0.0118-0.0197	0.0039-0.0157	0.0008-0.0028	0.0008-0.0024	0.0008-0.0026
	2.4	2AZ-FE	0.00082-0.0017	0.0094-0.0122	0.0130-0.0169	0.0039-0.0118	0.0007-0.0027	0.0007-0.0023	0.0007-0.0027
2010	1.8	2ZR-FE	0.0011-0.0020	0.0079-0.0118	0.0118-0.0197	0.0039-0.0157	0.0008-0.0028	0.0008-0.0024	0.0008-0.0026
	2.4	2AZ-FE	0.00082-0.0017	0.0094-0.0122	0.0130-0.0169	0.0039-0.0118	0.0007-0.0027	0.0007-0.0023	0.0007-0.0027

3768X_MATR_C0007

TORQUE SPECIFICATIONS

All readings in ft. lbs.

Year	Engine Displacement Liters	Engine ID/VIN	Cylinder Head Bolts	Main Bearing Bolts	Rod Bearing Bolts	Crankshaft Damper Bolts	Flywheel Bolts	Manifold Intake	Manifold Exhaust	Spark Plugs	Oil Pan Drain Plug
2009	1.8	2ZR-FE	①	②	③	102	65	21	16	15	27
	2.4	2AZ-FE	④	NA	⑤	NA	⑥	22	27	14	27
2010	1.8	2ZR-FE	①	②	③	102	65	21	16	15	27
	2.4	2AZ-FE	④	NA	⑤	NA	⑥	22	27	14	27

NA: Not Available

① Step 1: 36 ft. lbs.
Step 2: 90 degree turn
Step 3: 45 degree turn

② Inner 12 point bolts:
Step 1: 16 ft. lbs.
Step 2: 32 ft. lbs.
Step 3: 45 degree turn
Step 4: 45 degree turn
Outer cap bolts: 13 ft. lbs.

③ Step 1: 15 ft. lbs.
Step 2: 90 degree turn

④ Step 1: 52 ft. lbs.
Step 2: 90 degree turn

⑤ Step 1: 18 ft. lbs.
Step 2: 90 degree turn

⑥ A/T: 72 ft. lbs.
M/T: 96 ft. lbs.

3768X_MATR_C0009

WHEEL ALIGNMENT

Year	Model		Caster Range (+/-Deg.)	Caster Preferred Setting (Deg.)	Camber Range (+/-Deg.)	Camber Preferred Setting (Deg.)	Toe-in (in.)
2009	Matrix - 2WD	F	0.75	①	0.75	-0.57	0+/-0.02
		R	—	—	③ or ⑥	④ or ⑦	⑤ or ⑧
	Matrix - 4WD	F	0.75	②	0.75	-0.58	0+/-0.02
		R	—	—	③ or ⑥	④ or ⑦	⑤ or ⑧
2010	Matrix - 2WD	F	0.75	①	0.75	-0.57	0+/-0.02
		R	—	—	③ or ⑥	④ or ⑦	⑤ or ⑧
	Matrix - 4WD	F	0.75	②	0.75	-0.58	0+/-0.02
		R	—	—	③ or ⑥	④ or ⑦	⑤ or ⑧

① 1.8L engine 205/55R16:+2.93
1.8L engine 215/45R17: +2.92
2.4L engine 205/55R16: +2.90
2.4L engine 215/45R17: +2.88
2.4L engine 215/45R18: +3.00

② 2.4L engine 205/55R16: +3.02
2.4L engine 215/45R17: +3.00

③ Torsion Beam Type Suspension: 0.5

④ Torsion Beam Type Suspension: -1.47

⑤ Torsion Beam Type Suspension: 0.114+/-0.106

⑥ Double Wishbone Type Suspension: 0.75

⑦ Double Wishbone Type Suspension: -1.05

⑧ Double Wishbone Type Suspension: 0.25+/- 0.25

3768X_MATR_C0010

TIRE, WHEEL AND BALL JOINT SPECIFICATIONS

Year	Model	OEM Tires Standard	OEM Tires Optional	Tire Pressures (psi) Front	Tire Pressures (psi) Rear	Wheel Size	Ball Joint Inspection	Lug Nuts (ft. lbs.)
2009	Matrix	205/55R16	—	②	31	6.5-JJ	9-26 in. ①	76
2010	Matrix	205/55R16	—	②	31	6.5-JJ	9-26 in. ①	76

OEM: Original Equipment Manufacturer

PSI: Pounds Per Square Inch

STD: Standard

OPT: Optional

① Torque required in inch lbs. to rotate ball joint when removed from the knuckle

② FWD: 32 PSI , AWD 36 PSI.

3768X_MATR_C0011

BRAKE SPECIFICATIONS

All measurements in inches unless noted

Year	Model		Brake Disc Original Thickness	Brake Disc Minimum Thickness	Brake Disc Maximum Runout	Minimum Lining Thickness	Brake Caliper Bracket Bolts (ft. lbs.)	Brake Caliper Mounting Bolts (ft. lbs.)
2009	Matrix	F	①	②	0.0020	0.039	79	25
		R	③	④	0.0059	0.039	43	20
2010	Matrix	F	①	②	0.0020	0.039	79	25
		R	③	④	0.0059	0.039	43	20

F: Front

R: Rear

① 1.8L engine: 0.866

2.4L engine 1.102

② 1.8L engine:0.748

2.4L engine: 0.984

③ Double Wishbone Type Suspension:0.394

Torsion Beam Type Suspension (1.8L): 0.354

Torsion Beam Type Suspension (2.4L): 0.394

④ Double Wishbone Type Suspension:0.334

Torsion Beam Type Suspension (1.8L): 0.295

Torsion Beam Type Suspension (2.4L): 0.335

3768X_MATR_C0012

SCHEDULED MAINTENANCE INTERVALS
TOYOTA MATRIX

TO BE SERVICED	TYPE OF SERVICE	VEHICLE MILEAGE INTERVAL (x1000)													
		5	10	15	20	25	30	35	40	45	50	55	60	90	120
Engine oil & filter	R	✓	✓	✓	✓	✓	✓	✓	✓	✓	✓	✓	✓	✓	✓
Drive belts	S/I						✓						✓	✓	✓
Automatic transaxle fluid & filter	S/I						✓						✓	✓	✓
Brake line pipes & hoses	S/I	✓	✓	✓	✓	✓	✓	✓	✓	✓	✓	✓	✓	✓	✓
Brake linings & drums	S/I	✓	✓	✓	✓	✓	✓	✓	✓	✓	✓	✓	✓	✓	✓
Brake pads & discs (front & rear if equipped)	S/I	✓	✓	✓	✓	✓	✓	✓	✓	✓	✓	✓	✓	✓	✓
Cabin air filter	R				✓				✓				✓		✓
Differential oil	S/I						✓						✓	✓	✓
Drive shaft boots	S/I	✓	✓	✓	✓	✓	✓	✓	✓	✓	✓	✓	✓	✓	✓
Drive shaft bolt (tighten)	S/I	✓	✓	✓	✓	✓	✓	✓	✓	✓	✓	✓	✓	✓	✓
Engine coolant	S/I			✓			✓			✓			✓	✓	✓
Manual transaxle oil	S/I						✓						✓	✓	✓
Steering gear housing oil	S/I	✓	✓	✓	✓	✓	✓	✓	✓	✓	✓	✓	✓	✓	✓
Steering linkage	S/I	✓	✓	✓	✓	✓	✓	✓	✓	✓	✓	✓	✓	✓	✓
Air filter	R						✓						✓	✓	✓
Rotate tires	S/I	✓	✓	✓	✓	✓	✓	✓	✓	✓	✓	✓	✓	✓	✓
Spark plugs	R									✓				✓	
Fuel lines & connections	S/I						✓			✓			✓	✓	✓
Fuel tank cap gasket	R									✓				✓	
Charcoal canister	S/I									✓				✓	

R: Replace S/I: Service or Inspect

FREQUENT OPERATION MAINTENANCE (SEVERE SERVICE)

If a vehicle is operated under any of the following conditions it is considered severe service:

- Extremely dusty areas.

- 50% or more of the vehicle operation is in 32°C (90°F) or higher temperatures, or constant operation in temperatures below 0°C (32°F).

- Prolonged idling (vehicle operation in stop and go traffic).

- Frequent short running periods (engine does not warm to normal operating temperatures).

- Police, taxi, delivery usage or trailer towing usage.

Oil & oil filter: change every 5000 miles.

Bolts & nuts on chassis & body: tighten every 5000 miles.

Ball joints & dust covers: service or inspect every 5,000 miles.

Drive shaft boots & except Supra): service or inspect every 12,000 miles.

Steering linkage: service or inspect every 12,000 miles.

Air filter: service or inspect every 5,000 miles.

Exhaust system: service or inspect every 15,000 miles.

Timing belt: replace every 60,000 miles.

3768X_MATR_C0013

BRAKES — INFORMATION AND PRECAUTIONS

ANTI-LOCK SYSTEMS

- Certain components within the ABS system are not intended to be serviced or repaired individually.
- Do not use rubber hoses or other parts not specifically specified for and ABS system. When using repair kits, replace all parts included in the kit. Partial or incorrect repair may lead to functional problems and require the replacement of components.
- Lubricate rubber parts with clean, fresh brake fluid to ease assembly. Do not use shop air to clean parts; damage to rubber components may result.
- Use only DOT 3 brake fluid from an unopened container.
- If any hydraulic component or line is removed or replaced, it may be necessary to bleed the entire system.
- A clean repair area is essential. Always clean the reservoir and cap thoroughly before removing the cap. The slightest amount of dirt in the fluid may plug an orifice and impair the system function. Perform repairs after components have been thoroughly cleaned; use only denatured alcohol to clean components. Do not allow ABS components to come into contact with any substance containing mineral oil; this includes used shop rags.
- The Anti-Lock control unit is a microprocessor similar to other computer units in the vehicle. Ensure that the ignition switch is **OFF** before removing or installing controller harnesses. Avoid static electricity discharge at or near the controller.
- If any arc welding is to be done on the vehicle, the control unit should be unplugged before welding operations begin.

DISC AND DRUM SYSTEMS

⁂ CAUTION

Dust and dirt accumulating on brake parts during normal use may contain asbestos fibers from production or aftermarket brake linings. Breathing excessive concentrations of asbestos fibers can cause serious bodily harm. Exercise care when servicing brake parts. Do not sand or grind brake lining unless equipment used is designed to contain the dust residue. Do not clean brake parts with compressed air or by dry brushing. Cleaning should be done by dampening the brake components with a fine mist of water, then wiping the brake components clean with a dampened cloth. Dispose of cloth and all residue containing asbestos fibers in an impermeable container with the appropriate label. Follow practices prescribed by the Occupational Safety and Health Administration (OSHA) and the Environmental Protection Agency (EPA) for the handling, processing, and disposing of dust or debris that may contain asbestos fibers.

BRAKES — BLEEDING THE BRAKE SYSTEM

BLEEDING PROCEDURE

See Figures 1 and 2.

➡If any work is performed on the brake system or if air in the brake lines is suspected, bleed the brake system.

➡Move the shift lever to P and apply the parking brake before bleeding the brakes.

➡Add brake fluid to keep the level between the MIN and MAX lines of the reservoir while bleeding the brakes.

➡If brake fluid leaks onto any painted surface, immediately wash it off.

1. Remove the center cowl top ventilator louver.
2. Fill the reservoir with brake fluid.
3. Bleed the brake master cylinder.

➡If the master cylinder is reinstalled or runs out of brake fluid, bleed the master cylinder.

➡To prevent brake fluid from damaging painted surface, cover any surrounding parts with a piece of cloth.

a. Using a union nut wrench (10 mm), disconnect the 2 brake lines from the master cylinder.

b. Slowly depress the brake pedal and hold it.

c. Cover the 2 outer holes with fingers, and release the brake pedal.

d. Repeat the previous 2 steps 3 or 4 times.

e. Using a union nut wrench (10 mm), connect the 2 brake lines to the master cylinder. If tightening with a union nut wrench tighten to 11 ft. lbs. (15 Nm). If tightening without a union nut wrench tighten to 10 ft. lbs. (14 Nm).

➡Use a torque wrench with a fulcrum length of 9.84 inch (250 mm).

➡This torque value is effective when the union nut wrench is parallel to the torque wrench.

4. Bleed the brake line.

➡Bleed the brake line of the wheel farthest from the master cylinder first.

➡Add brake fluid to keep the level between the MIN and MAX lines of the reservoir while bleeding the brakes.

a. Connect a vinyl tube to the bleeder plug.

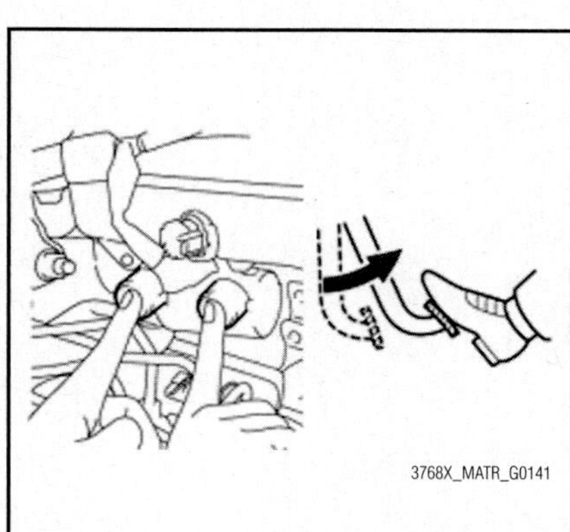

Fig. 1 Covering the 2 outer holes

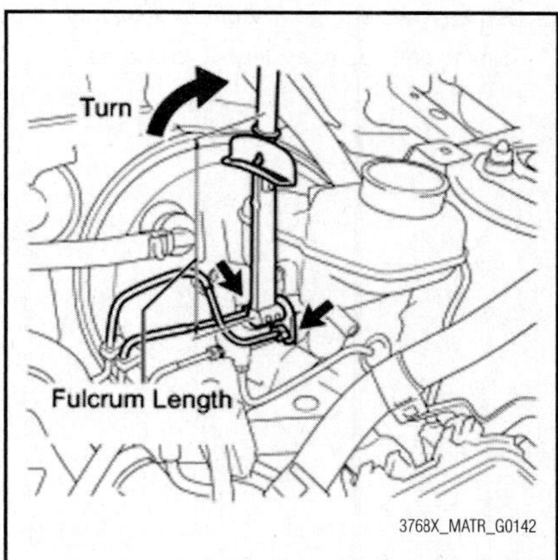

Fig. 2 Connecting the 2 brake lines to the master cylinder

b. Depress the brake pedal several times, and then loosen the bleeder plug with the pedal depressed.

c. When fluid stops coming out, tighten the bleeder plug, and then release the brake pedal.

d. Repeat the previous 2 steps until all the air in the fluid is completely bled out.

e. Tighten the bleeder plug completely. Tighten the front bleeder plug to 73 inch lbs. (8.3 Nm). Tighten the rear bleeder plug (double wishbone type suspension) to 73 inch lbs. (8.3 Nm). Tighten the rear bleeder plug (torsion beam type suspension) to 7 ft. lbs. (10 Nm).

f. Repeat the above procedure for each wheel to bleed the brake line.

5. Bleed the brake actuator (w/VSC).

➡After bleeding the brake system, if the specified height or feel of the brake pedal cannot be obtained, bleed the brake actuator assembly with the Techstream by following the procedure below.

a. Depress the brake pedal more than 20 times with the ignition switch off.

b. Connect the Techstream to the DLC3, and then turn the ignition switch to ON.

➡Do not start the engine.

c. Turn the Techstream on and select "Air Bleeding" on the screen.

➡Refer to the Techstream operator's manual for further details.

➡Bleed air by following the steps displayed on the Techstream.

d. Bleed air according to "Step 1: Increase Line" on the Techstream display.

➡Make sure that the master cylinder reservoir tank does not run out of brake fluid.

➡Add brake fluid to keep the level between the MIN and MAX lines of the reservoir while bleeding the brakes.

e. Connect a vinyl tube to either one of the bleeder plugs.

f. Depress the brake pedal several times, and then loosen the bleeder plug connected to the vinyl tube with the pedal depressed.

g. When fluid stops coming out, tighten the bleeder plug, and then release the brake pedal.

h. Repeat the previous 2 steps until all the air in the fluid is completely bled out.

i. Tighten the bleeder plug completely. Tighten the front bleeder plug to 73 inch lbs. (8.3 Nm). Tighten the rear bleeder plug (double wishbone type suspension) to 73 inch lbs. (8.3 Nm). Tighten the rear bleeder plug (torsion beam type suspension) to 7 ft. lbs. (10 Nm).

j. Repeat the above procedure for each wheel to bleed the brake line.

k. Bleed the suction line according to "Step 2: Inhalation Line" on the Techstream display.

➡Bleed the suction line by following the steps displayed on the Techstream.

➡Add brake fluid to keep the level between the MIN and MAX lines of the reservoir while bleeding the brakes.

l. Connect a vinyl tube to the bleeder plug at the right front wheel or the right rear wheel and loosen the bleeder plug.

m. Operate the brake actuator assembly to bleed air using the Techstream.

➡During this step, be sure to release the brake pedal.

➡The actuator operation stops automatically in 4 seconds.

n. Check that the actuator operation has stopped by referring to the Techstream display, and tighten the bleeder plug.

o. Repeat the previous 2 steps until all the air in the fluid is completely bled out.

p. Tighten the bleeder plug completely. Tighten the front bleeder plug to 73 inch lbs. (8.3 Nm). Tighten the rear bleeder plug (double wishbone type suspension) to 73 inch lbs. (8.3 Nm). Tighten the rear bleeder plug (torsion beam type suspension) to 7 ft. lbs. (10 Nm).

q. For the rest of the wheels, bleed air in the same way as stated in the above procedure.

r. Bleed the pressure reduction line according to "Step 3: Decrease Line" on the Techstream display.

➡Bleed the pressure reduction line by following the steps displayed on the Techstream.

➡Add brake fluid to keep the level between the MIN and MAX lines of the reservoir while bleeding the brakes.

s. Connect a vinyl tube to either one of the bleeder plugs.

t. Loosen the bleeder plug.

u. While keeping the brake pedal fully depressed, operate the brake actuator assembly using the Techstream.

➡The actuator operation stops automatically in 4 seconds. When performing this procedure continuously, an interval of at least 20 seconds is required.

➡After the operation is completed, the brake pedal goes down slightly. This is a normal phenomenon when the solenoid opens.

➡During this procedure, the pedal seems heavy, but completely depress it so that the brake fluid comes out from the bleeder plug.

➡Be sure to keep the brake pedal depressed. Never depress and release the pedal repeatedly.

v. Tighten the bleeder plug, and then release the brake pedal.

w. Repeat previous 3 steps until all the air in the fluid is completely bled out.

x. Tighten the bleeder plug completely. Tighten the front bleeder plug to 73 inch lbs. (8.3 Nm). Tighten the rear bleeder plug (double wishbone type suspension) to 73 inch lbs. (8.3 Nm). Tighten the rear bleeder plug (torsion beam type suspension) to 7 ft. lbs. (10 Nm).

y. Repeat the above procedure for the rest of the brakes to bleed the brake lines.

z. Bleed the brake lines again according to "Step 4: Increase Line" on the Techstream display.

➡Bleed air by following the steps displayed on the Techstream.

➡Add brake fluid to keep the level between the MIN and MAX lines of the reservoir while bleeding the brakes.

aa. Connect a vinyl tube to either one of the bleeder plugs.

bb. Depress the brake pedal several times, and then loosen the bleeder plug connected to the vinyl tube with the pedal depressed.

cc. When fluid stops coming out, tighten the bleeder plug, and then release the brake pedal.

dd. Repeat the previous 2 steps until all the air in the fluid is completely bled out.

ee. Tighten the bleeder plug completely. . Tighten the front bleeder plug to 73 inch lbs. (8.3 Nm). Tighten the rear bleeder plug (double wishbone type suspension) to 73 inch lbs. (8.3 Nm). Tighten the rear bleeder plug (torsion beam type suspension) to 7 ft. lbs. (10 Nm).

ff. Repeat the above procedure for each brake to bleed the brake lines.

gg. Finish "Air Bleeding" on the Techstream, and then turn the Techstream off.

hh. Disconnect the Techstream from the DLC3.

ii. Turn the ignition switch off.

6. Inspect for brake fluid leaks.
7. Inspect the fluid level.
8. Install the center cowl top ventilator louver.

WHEEL SPEED SENSORS

REMOVAL & INSTALLATION

Front

See Figure 3.

➡Use the same procedure for the RH side and LH side.

➡The procedure listed below is for the LH side.

➡If the sensor rotor needs to be replaced, replace it together with the front drive shaft assembly.

1. Disconnect the cable from the negative battery terminal.
2. Remove the front wheel.
3. Remove the front fender liner.
4. Remove the front speed sensor.

a. Disconnect the front speed sensor connector.

b. Remove the front speed sensor wire harness clamp from the body.

c. Remove the bolt and No. 2 sensor clamp from the body.

d. Remove the bolt and separate the brake flexible hose.

e. Remove the No. 1 sensor clamp from the shock absorber.

f. Remove the clamp.

g. Remove the bolt and front speed sensor.

3768X_MATR_G0131

Fig. 3 Removing the bolt and front speed sensor

➡Prevent foreign matter from attaching to the sensor tip.

➡Clean the installation hole and the contact surface for the speed sensor every time it is removed.

To install:

➡Use the same procedure for the RH side and LH side.

➡The procedure listed below is for the LH side.

➡If the sensor rotor needs to be replaced, replace it together with the front drive shaft assembly.

5. Install the front speed sensor with the bolt and tighten to 75 inch lbs. (8.5 Nm).

➡Prevent foreign matter from attaching to the sensor tip.

➡Do not file the hole or the contact surface because the gap between the magnet rotor and front speed sensor is important.

a. Install the clamp to the shock absorber.

b. Temporarily install the No. 1 sensor clamp to the shock absorber.

c. Install the brake flexible hose with the bolt. Tighten to 21 ft. lbs. (29 Nm).

➡Do not twist the wire harness for the front speed sensor when installing it.

➡Tighten the brake flexible hose and front speed sensor together with the bolt. Make sure that the front speed sensor is positioned over the flexible hose.

d. Install the No. 2 sensor clamp to the body with the bolt. Tighten to 71 inch lbs. (8 Nm).

e. Connect the speed sensor wire harness clamp.

f. Connect the speed sensor connector.

6. Install the front fender liner.
7. Install the front wheel. Tighten to 76 ft. lbs. (103 Nm).
8. Connect the cable to the negative battery terminal.
9. Check for speed sensor signal.

Rear

2WD Vehicles

See Figures 4 and 5.

➡Use the same procedure for the RH side and LH side.

➡The procedure listed below is for the LH side.

➡If the sensor rotor needs to be replaced, replace it together with the rear axle hub and bearing assembly with rear speed sensor.

1. Disconnect the cable from the negative battery terminal.
2. Remove the front upper console box (Torsion Beam Type suspension).
3. Loosen the parking brake cable (Torsion Beam Type suspension).
4. Disconnect the rear speed sensor wire (Torsion Beam Type suspension).

a. Using a screwdriver, disconnect the connector from the rear speed sensor.

➡Be careful not to damage the rear speed sensor.

5. Disconnect the rear speed sensor wire (Double Wishbone Type suspension).

a. Using a screwdriver, disconnect the connector from the rear speed sensor.

➡Be careful not to damage the rear speed sensor.

6. Remove the parking brake lever protector (Torsion Beam Type suspension).
7. Separate the No. 3 parking brake cable assembly (Torsion Beam Type suspension).
8. Separate the rear disc brake caliper assembly (Torsion Beam Type suspension).
9. Separate the rear disc brake caliper assembly (Double Wishbone Type suspension).
10. Remove the rear disc.
11. remove the rear axle hub and bearing assembly with the rear speed sensor.

12. Remove the rear speed sensor.
   a. Install 3 hub nuts and mount the rear axle hub and bearing assembly in a vise using aluminum plates.

➡Replace the rear axle hub and bearing assembly if it is dropped or receives a strong shock.

   b. Using a pin punch and hammer, drive out the 2 pins and remove the 2 attachments from special tool (09520-00031).
   c. Using special tool (09520-00031) and 2 bolts (diameter: 12 mm, pitch: 1.5 mm), remove the rear speed sensor from the rear axle hub and bearing assembly.

➡Keep the rear speed sensor away from magnets.

➡Pull the rear speed sensor off straight, taking care not to allow it to contact the rear speed sensor rotor.

➡If the rear speed sensor rotor is damaged or deformed, replace the rear axle hub and bearing assembly.

➡Do not scratch the contact surfaces between the rear axle hub and bearing assembly and the rear speed sensor.

➡Prevent foreign matter from attaching to the speed sensor rotor or tip.

➡If the sensor rotor needs to be replaced, replace it together with the rear axle hub and bearing assembly with rear speed sensor.

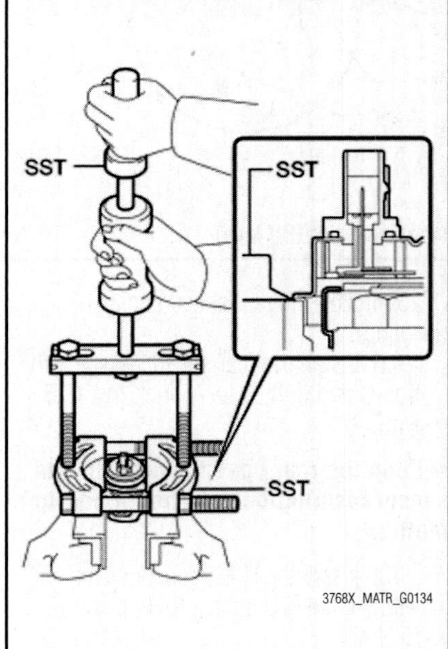

Fig. 4 Removing the rear speed sensor (2WD)

To install:

➡Use the same procedure for the RH side and LH side.

➡The procedure listed below is for the LH side.

➡If the sensor rotor needs to be replaced, replace it together with the rear axle hub and bearing assembly with rear speed sensor.

13. Install the rear speed sensor.
   a. Clean the contact surface between the rear axle hub and bearing assembly and a new rear speed sensor.

➡Prevent foreign matter from attaching to the sensor rotor.

   b. Place the rear speed sensor on the rear axle hub and bearing assembly so that the connector is positioned at the bottom when the sensor is installed on the vehicle.
   c. Using the special tool (09214-76011), steel plates and a press, install a new speed sensor to the rear axle hub and bearing assembly.

➡Keep the rear speed sensor away from magnets.

➡Do not use a hammer to install the rear speed sensor.

➡Check that there is no foreign matter such as iron chips on the detecting portion of the rear speed sensor.

➡Slowly press the rear speed sensor in straight.

14. Install the rear axle hub and bearing assembly with the rear speed sensor.
15. Inspect the rear axle hub bearing for looseness.
16. Inspect the rear axle hub runout.
17. Install the rear disc.
18. Install the rear disc brake caliper assembly.

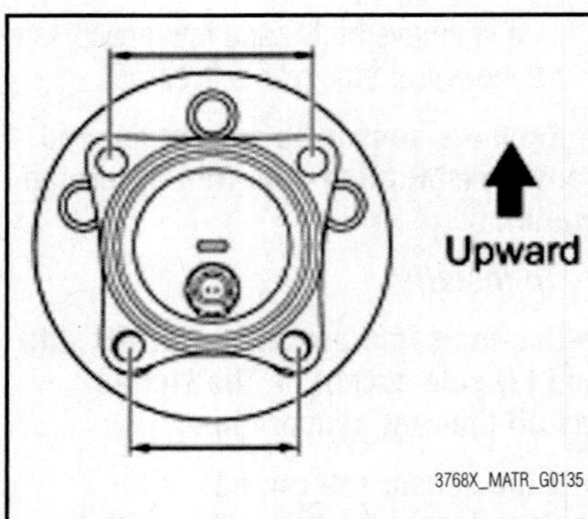

Fig. 5 Positioning the rear axle hub and bearing

19. Install the parking brake lever protector (Torsion Beam Type suspension).
20. Connect the rear speed sensor wire to the rear speed sensor.
21. Inspect the parking brake lever travel.
22. Adjust the parking brake lever travel (Torsion Beam Type suspension).
23. Inspect the rear disc brake caliper operation lever and stopper clearance (Torsion Beam Type suspension).
24. Adjust the parking brake shoe clearance and parking brake lever travel (Double Wishbone Type suspension).
25. Install the rear wheel and tighten to 76 ft. lbs. (103 Nm).
26. Connect the cable to the negative battery terminal.
27. Install the front upper console box (Torsion Beam Type suspension).
28. Inspect the brake warning light.
29. Check for speed sensor signal.
30. Inspect and adjust the rear wheel alignment (Double Wishbone Type suspension).
31. Inspect the rear wheel alignment.

4WD Vehicles

See Figures 6 and 7.

➡Use the same procedure for RH side and LH side except for the step to remove the rear sensor clip.

➡If the sensor rotor needs to be replaced, replace it together with the rear drive shaft assembly.

1. Disconnect the cable from the negative battery terminal.
2. Remove the rear seat cushion assembly.
3. Remove the rear seat headrest assembly (LH side).
4. Remove the rear seatback outer hinge cover (LH side).
5. remove the rear seatback inner hinge cover (LH side).
6. Remove the rear seatback assembly LH.
7. Remove the rear seat headrest assembly (RH side).
8. remove the rear seat center headrest assembly.
9. Disconnect the rear seat center outer belt assembly.
10. Remove the rear seatback outer hinge cover (RH side).
11. Remove the rear seatback inner hinge cover (RH side).
12. Remove the rear seatback assembly RH.

13. Remove the rear door scuff plate (LH side).
14. Remove the rear door opening trim weather-strips (LH side).
15. Remove the rear door scuff plate (RH).
16. Remove the rear door opening trim weather-strip (RH side).
17. Remove the rear seatback hinge sub assembly (LH side).
18. Remove the rear seat side garnish (LH side).
19. Remove the rear seatback hinge sub assembly (RH side).
20. Remove the rear seat side garnish (RH side).
21. Remove the rear wheel.
22. Remove the fuel tank filler pipe protector (LH side).
    a. Remove the 3 bolts and the fuel tank filler pipe protector.
23. Remove the rear speed sensor (LH side).
    a. Disconnect the rear speed sensor connector.
    b. Disconnect the grommet of the rear speed sensor wire from the hole in the wheel well.
    c. Remove bolt A and separate the No. 3 sensor clamp.
    d. Disconnect the sensor clip LH from the breather clamp.
    e. Remove bolt B and separate the No. 2 sensor clamp from the suspension member.
    f. Remove the nut and separate the No. 1 sensor clamp from the upper arm.
    g. Remove bolt C and the rear speed sensor body from the rear axle carrier.

➡Keep the sensor tip and rear speed sensor installation hole free of foreign matter.

24. Remove the rear speed sensor (RH side).
    a. Disconnect the rear speed sensor connector.
    b. Disconnect the grommet of the rear speed sensor wire from the hole in the wheel well.
    c. Remove bolt A and separate the No. 3 sensor clamp.
    d. Disconnect the sensor clip from the breather clamp.
    e. Remove bolt B and separate the No. 2 sensor clamp from the suspension member.
    f. Remove the nut and separate the No. 1 sensor clamp from the upper arm.

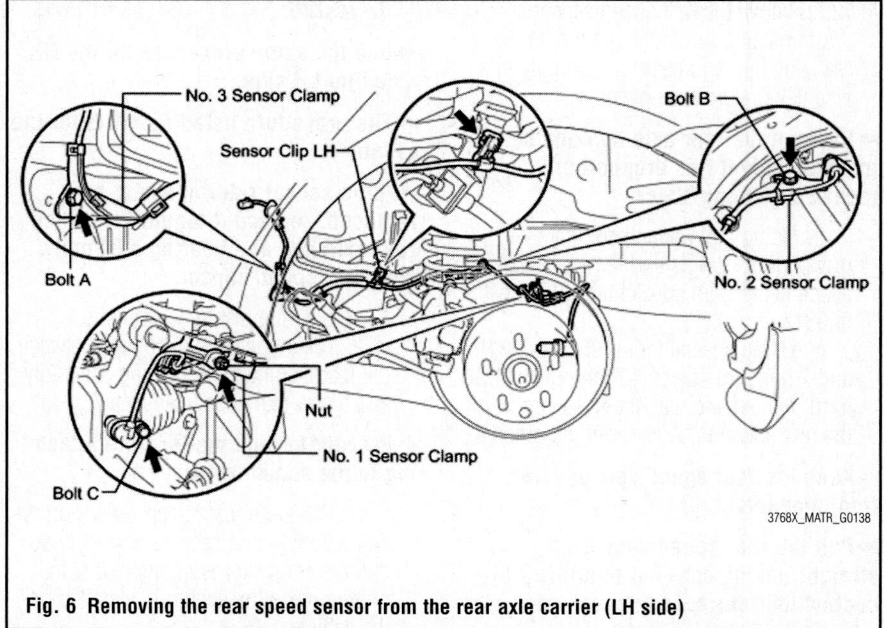

Fig. 6 Removing the rear speed sensor from the rear axle carrier (LH side)

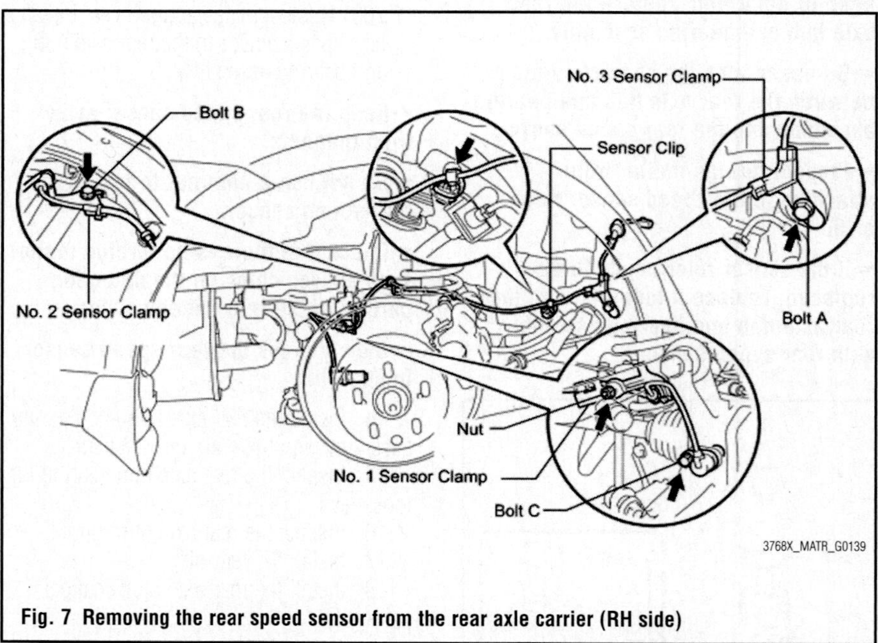

Fig. 7 Removing the rear speed sensor from the rear axle carrier (RH side)

    g. Remove bolt C and rear speed sensor body from the rear axle carrier.

➡Keep the sensor tip and rear speed sensor installation hole free of foreign matter.

To install:

➡Use the same procedure for RH side and LH side except for the step to install the rear sensor clip.

➡If the sensor rotor needs to be replaced, replace it together with the rear drive shaft assembly.

25. Install the rear speed sensor (LH side).
    a. Install the rear speed sensor with bolt C and tighten to 75 inch lbs. (8.5 Nm).

➡Keep the rear speed sensor tip and sensor installation hole free of foreign matter.

    b. Install the No. 1 sensor clamp with the nut and tighten to 44 inch lbs. (5 Nm).

➡Do not twist the rear speed sensor wire when installing the clamp.

c. Install the No. 2 sensor clamp with bolt B and tighten to 71 inch lbs. (8 Nm).

➡Do not twist the rear speed sensor wire when installing the clamp.

d. Install the sensor clip LH.

e. Install the No. 3 sensor clamp with bolt A and tighten to 71 inch lbs. (8 Nm).

➡Do not twist the rear speed sensor wire when installing the clamp.

f. Insert the connector and grommet to the inside of the vehicle through the passage hole in the wheel well.

➡Make sure that the grommet's band clamp remains on the outside of the vehicle.

g. Hold the grommet and pull it from the inside of the vehicle to the outside of the vehicle. Then hold in place so that it is not tilted.

➡When pulling out the grommet, do not pull on the sensor wire.

h. Connect the rear speed sensor connector.

26. Install the rear speed senor (RH side).

a. Install the rear speed sensor with bolt C and tighten to 75 inch lbs. (8.5 Nm).

➡Keep the rear speed sensor tip and sensor installation hole free of foreign matter.

b. Install the No. 1 sensor clamp with the nut and tighten to 44 inch lbs. (5 Nm).

➡Do not twist the rear speed sensor wire when installing the clamp.

c. Install the No. 2 sensor clamp with bolt B and tighten to 71 inch lbs. (8 Nm).

➡Do not twist the rear speed sensor wire when installing the clamp.

d. Install the sensor clip.

e. Install the No. 3 sensor clamp with bolt A and tighten to 71 inch lbs. (8 Nm).

➡Do not twist the rear speed sensor wire when installing the clamp.

f. Insert the connector and grommet to the inside of the vehicle through the passage hole in the wheel well.

➡Make sure that the grommet's band clamp remains on the outside of the vehicle.

g. Hold the grommet and pull it from the inside of the vehicle to the outside of the vehicle. Then hold it in place so that it is not tilted.

➡When pulling out the grommet, do not pull on the sensor wire.

h. Connect the rear speed sensor connector.

i. Install the fuel tank filler pipe protector (LH side) with the 3 screws.

27. Install the rear wheel and tighten to 76 ft. lbs. (103 Nm).

28. Install the rear seat side garnish LH and RH side.

29. Install the rear seatback hinge sub assembly LH and RH side.

30. Install the rear door opening trim weather-strip RH and LH side.

31. Install the rear door scuff plate RH and LH side.

32. Install the rear seatback assembly RH.

33. Install the rear seatback inner hinge cover (RH side).

34. Install the rear seatback outer hinge cover (RH side).

35. Connect the rear seat center outer belt assembly.

36. Install the rear seat center headrest assembly.

37. Install the rear seat headrest assembly (RH side).

38. Install the rear seatback assembly LH.

39. Install the rear seatback inner hinge cover (LH side).

40. Install the rear seatback outer hinge cover (LH side).

41. Install the rear seat headrest assembly (LH side).

42. Connect the cable to the negative battery terminal.

43. Check for the rear speed sensor signal.

BRAKES

FRONT DISC BRAKES

BRAKE CALIPER

REMOVAL & INSTALLATION

See Figures 8 through 14.

➡Use the same procedure for the LH side and RH side.

➡The following procedure listed is for the LH side.

1. Remove the front wheel.
2. Drain the brake fluid.

➡If brake fluid leaks onto any painted surface, immediately wash it off.

3. Disconnect the front flexible hose:

a. Remove the union bolt and gasket, and separate the front flexible hose from the disc brake caliper assembly.

4. Remove the front disc brake caliper assembly:

a. Hold the front disc brake caliper slide pin, and remove the 2 bolts and front disc brake caliper assembly.

5. Remove the front disc pad:

a. Remove the 2 disc brake pads from the disc brake caliper mounting.

6. Remove the front anti squeal shim:

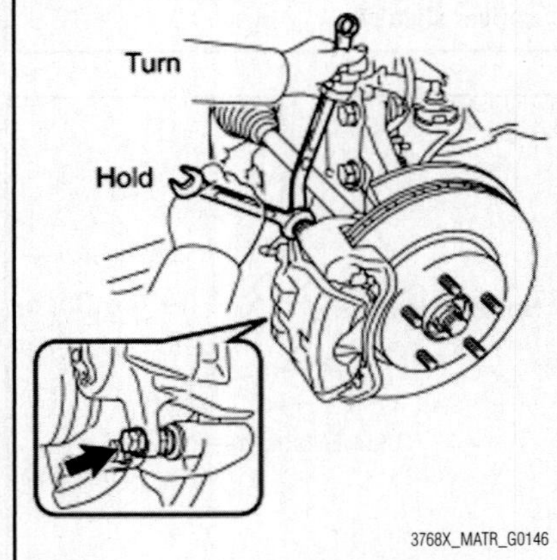

Fig. 8 Removing the front disc brake caliper assembly

a. Remove the 2 No. 1 anti-squeal shims and 2 No. 2 anti-squeal shims from the brake pads.

7. Remove the front disc brake pad support plate.

a. Remove the 2 No. 1 disc brake pad support plates and 2 No. 2 disc brake pad support plates from the front disc brake caliper mounting.

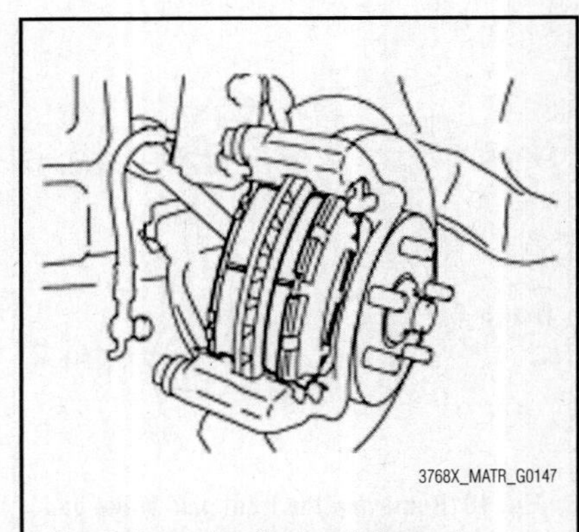

Fig. 9 Removing the front disc pad

➡Each front disc brake pad support plate has a different shape. Be sure to put an identification mark on each front disc brake pad support plate so that it can be reinstalled to its original position.

8. Removing the front disc brake caliper slide pin.
   a. Remove the No. 1 front disc brake caliper slide pin and No. 2 front disc brake caliper slide pin from the disc brake cylinder mounting.
9. Remove the front disc brake caliper slide bushing.
   a. Using a screwdriver, remove the front disc brake caliper slide bushing from the No. 2 front disc brake caliper slide pin.

➡Do not damage the No. 2 front disc brake caliper slide pin.

➡Tape the screwdriver tip before use.

10. Remove the front disc brake bushing dust boot.
   a. Remove the 2 front disc brake bushing dust boots from the disc brake cylinder mounting.
11. Remove the front disc brake caliper mounting.
   a. Remove the 2 bolts and front disc brake caliper mounting from the steering knuckle.
12. Installation is the reverse of the removal procedure.

DISC BRAKE PADS

REMOVAL & INSTALLATION

See Figures 8 through 10.

➡Use the same procedure for the LH side and RH side.

➡The following procedure listed is for the LH side.

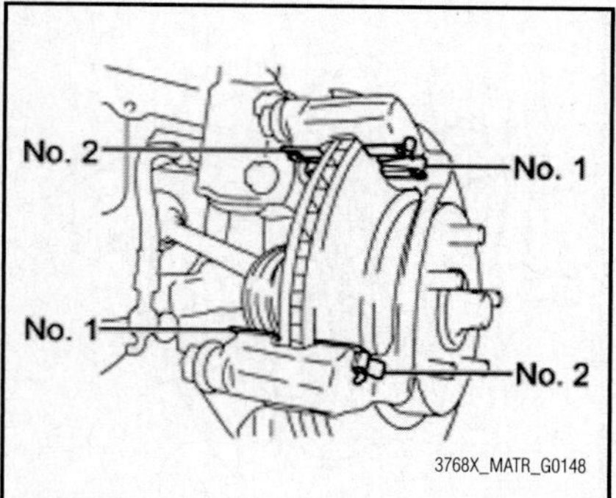

Fig. 10 Removing the front disc brake pad support plate

1. Remove the front wheel.
2. Drain the brake fluid.

➡If brake fluid leaks onto any painted surface, immediately wash it off.

3. Disconnect the front flexible hose.
   a. Remove the union bolt and gasket, and separate the front flexible hose from the disc brake cylinder assembly.
4. Remove the front disc brake caliper assembly.
   a. Hold the front disc brake caliper slide pin, and remove the 2 bolts and front disc brake caliper assembly.
5. Remove the front disc pad.
   a. Remove the 2 disc brake pads from the disc brake cylinder mounting.
6. Remove the front anti squeal shim.
   a. Remove the 2 No. 1 anti-squeal shims and 2 No. 2 anti-squeal shims from the brake pads.
7. Remove the front disc brake pad support plate.
   a. Remove the 2 No. 1 disc brake pad support plates and 2 No. 2 disc brake pad support plates from the front disc brake caliper mounting.

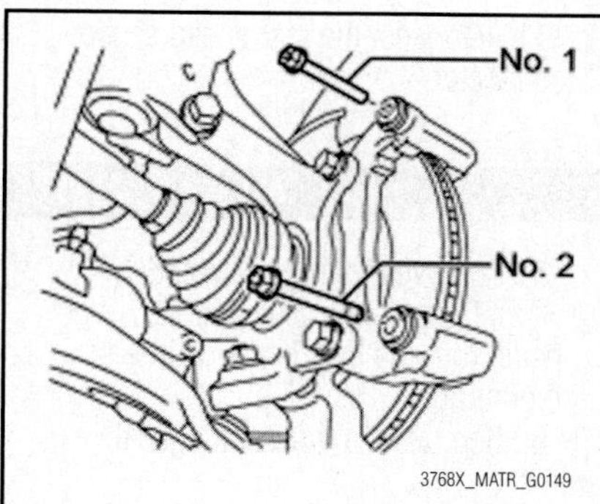

Fig. 11 Removing the front disc brake caliper slide pin

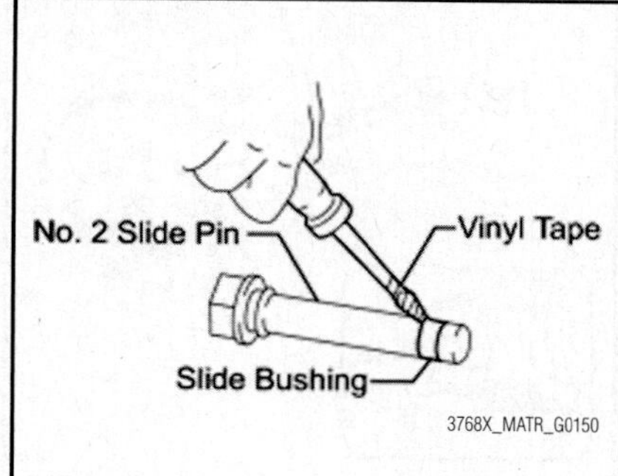

Fig. 12 Removing the front disc brake caliper slide bushing

➡Each front disc brake pad support plate has a different shape. Be sure to put an identification mark on each front disc brake pad support plate so that it can be reinstalled to its original position.

To install:

8. Install the front disc brake pad support plate.
   a. Install the 2 No. 1 front disc brake pad support plates and 2 No. 2 front disc brake pad support plates to the front disc brake caliper mounting.

➡Be sure to install each front disc brake pad support plate in the correct position and direction.

9. Install the front anti squeal shim (1.8L engine).
   a. Apply disc brake shim grease to both sides of each No. 1 anti-squeal shim as shown in the illustration.

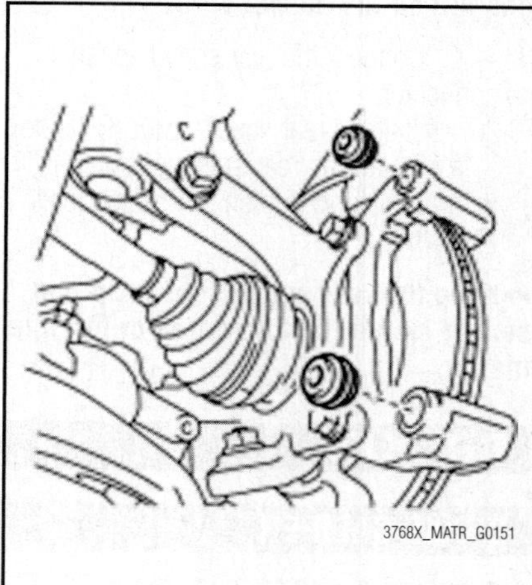

Fig. 13 Removing the front disc brake bushing dust boot

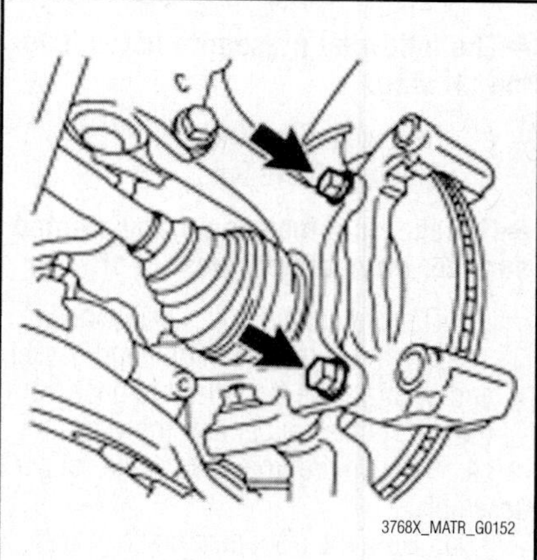

Fig. 14 Removing the front disc brake caliper mounting

b. Install the No. 1 anti-squeal shim and No. 2 anti-squeal shim to each brake pad.

➡**When replacing worn pads, the anti-squeal shims must be replaced together with the pads.**

➡**Install the shims in the correct positions and directions.**

➡**Apply disc brake grease to the area that contacts the anti-squeal shims.**

➡**Disc brake grease can seep out slightly from the area where the anti-squeal shim is installed.**

➡**Make sure that disc brake grease is not applied onto the lining surface.**

10. Install the front anti squeal shim (2.4 l engine).
 a. Apply disc brake grease to both sides of each No. 2 anti-squeal shim as shown in the illustration.
 b. Install the pad wear indicator, No. 1 anti squeal shim and No. 2 anti squeal shim to each brake pad.

➡**When replacing worn pads, the anti-squeal shims must be replaced together with the pads.**

➡**Install the shims in the correct positions and directions.**

➡**Apply disc brake grease to the area that contacts the anti-squeal shims.**

➡**Disc brake grease can seep out slightly from the area where the anti-squeal shim is installed.**

➡**Make sure that disc brake grease is not applied onto the lining surface.**

11. Install the front disc brake pad.
 a. Install the 2 front disc brake pads to the disc brake cylinder mounting.

➡**There should be no oil or grease on the friction surfaces of the disc brake pads or the front disc.**

12. Install the front disc brake caliper assembly.
 a. Hold the front disc brake caliper slide pin, and install the front disc brake caliper assembly to the front disc brake caliper mounting with the 2 bolts. Tighten to 25 ft. lbs. (34 Nm).
13. Connect the front flexible hose.
 a. Connect the front flexible hose to the front disc brake caliper assembly with the union bolt and a new gasket. Tighten to 21 ft. lbs. (29 Nm).

➡**Install the flexible hose lock securely into the lock hole in the disc brake cylinder.**

14. Fill the reservoir.
15. Bleed the brake line.
16. Inspect for fluid leaks.
17. Inspect the fluid level.
18. Install the front wheel. Tighten to 76 ft. lbs. (103 Nm).

BRAKES

REAR DISC BRAKES

BRAKE CALIPER

REMOVAL & INSTALLATION

Double Wishbone Type Suspension

See Figures 15 and 16.

➡**Use the same procedure for the LH side and RH side.**

➡**The following procedure listed is for the LH side.**

1. Remove the rear wheel.
2. Drain the brake fluid.

➡**If brake fluid leaks onto any painted surface, immediately wash it off.**

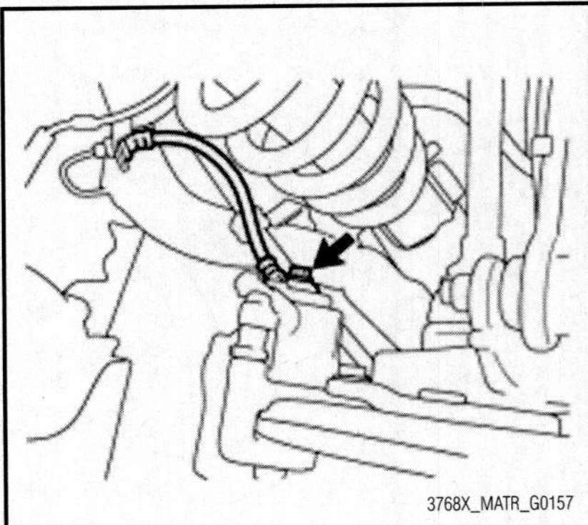

Fig. 15 Disconnecting the rear brake flexible hose

3. Disconnect the rear brake flexible hose.
 a. Remove the union bolt and the gasket from the rear disc brake caliper assembly, and then disconnect the rear brake flexible hose.
4. Remove the rear disc brake caliper assembly.
 a. Hold the 2 rear disc brake caliper slide pins and remove the 2 bolts and rear disc brake caliper assembly.

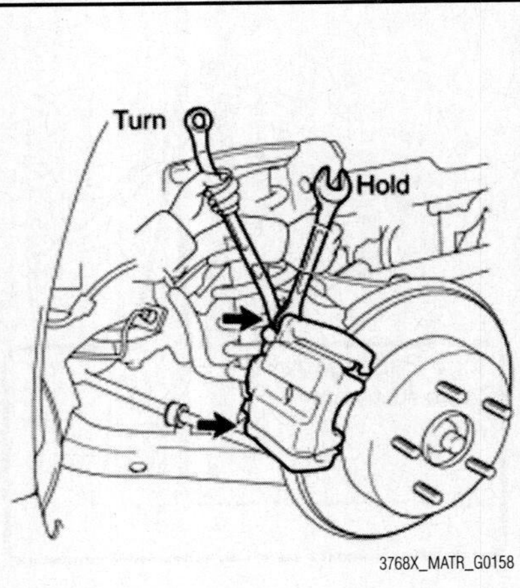

Fig. 16 Removing the rear disc brake caliper assembly

To install:

5. Install the rear brake cylinder assembly with the 2 bolts. Tighten to 20 ft. lbs. (27 Nm).
6. Connect the rear brake flexible hose with a union bolt and a new gasket. Tighten to 24 ft. lbs. (33 Nm).

Torsion Beam Type Suspension

See Figures 17 through 19.

➡**Use the same procedure for the LH side and RH side.**

➡**The following procedure listed is for the LH side.**

1. Remove the rear wheel.
2. Drain the brake fluid.

➡**If brake fluid leaks onto any painted surface, immediately wash it off.**

3. Remove the front upper console box.
4. Loosen the parking brake cable.
 a. Completely release the parking brake lever.
 b. Loosen the lock nut and the adjusting nut to completely release the parking brake cable.
5. Remove the parking brake lever protector.
6. Disconnect the No. 3 parking brake cable assembly.

a. Disengage the clamp and remove the bolt from the No. 3 parking brake cable assembly.

b. Separate the tip of the No. 3 parking brake cable assembly from the rear disc brake caliper assembly.

c. Separate the No. 3 parking brake cable assembly from the rear disc brake caliper assembly.

➡Insert an offset wrench (14 mm) at the base of the No. 3 parking brake cable assembly as shown in the illustration to disengage the clip. Pull out the No. 3 parking brake cable assembly from the rear disc brake caliper assembly.

7. Separate the rear brake flexible hose.

a. Remove the union bolt and gasket, and separate the rear brake flexible hose from the rear disc brake caliper assembly.

8. Remove the rear disc brake caliper assembly.

a. Hold the rear disc brake pad guide pin and remove the 2 bolts and rear disc brake caliper assembly.

To install:

9. Install the rear disc brake caliper assembly.

a. To compensate for pad wear when reusing the pad, use the special tool (09960-10010) to push and turn the piston (LH side: counterclockwise, RH side: clockwise) to the position where the protrusion on the pad lines up properly with the piston groove.

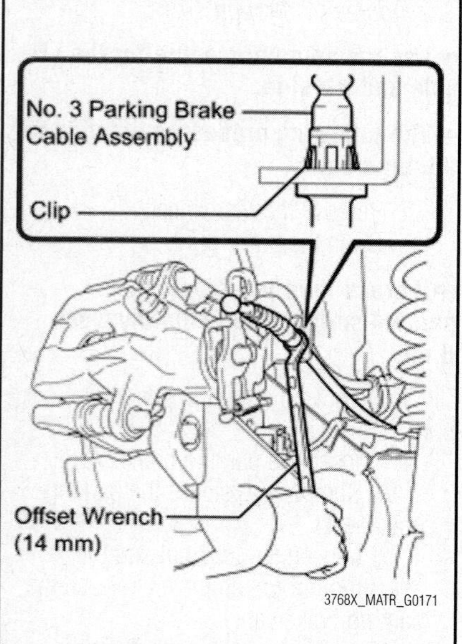

Fig. 17 Separating the No. 3 parking brake cable assembly from the rear disc brake caliper assembly

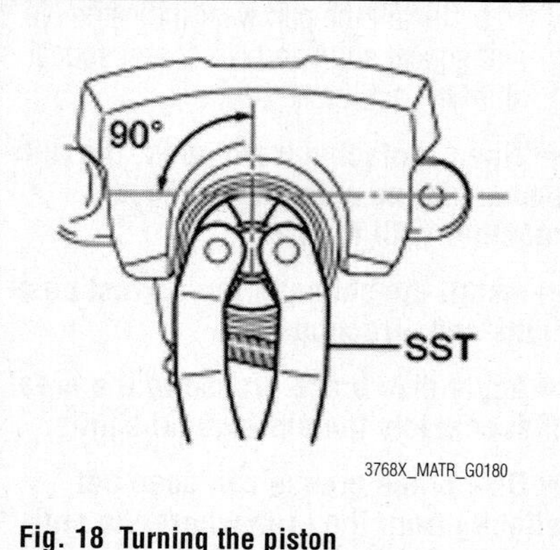

Fig. 18 Turning the piston

➡Place the disc between the 2 brake pads and determine the piston return value.

b. Hold the rear disc brake caliper slide pin, and install the rear disc brake caliper assembly to the rear disc brake caliper mounting with the 2 bolts. Tighten to 26 ft. lbs. (35 Nm).

10. Connect the rear brake flexible hose.

a. Connect the flexible hose to the rear disc brake caliper assembly with the union bolt and a new gasket. Tighten to 21 ft. lbs. (29 Nm).

➡Install the flexible hose lock securely into the lock hole in the disc brake cylinder.

11. Connect the No. 3 parking brake cable assembly.

a. Install the No. 3 parking brake cable assembly to the rear disc brake caliper assembly.

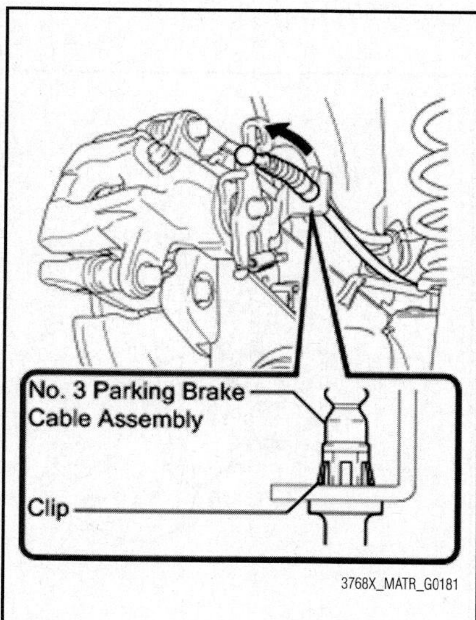

Fig. 19 Connecting the No. 3 parking brake cable assembly

➡Be sure to engage the No. 3 parking brake cable assembly clip onto the rear disc brake caliper assembly LH as shown in the illustration.

b. Connect the tip of the No. 3 parking brake cable assembly to the rear disc brake caliper assembly.

c. Install the bolt and tighten to 53 inch lbs. (6 Nm).

d. Engage the clamp to the No. 3 parking brake cable assembly.

12. Install the parking brake lever protector.

13. Fill the reservoir with brake fluid.

14. Bleed the brake line.

15. Inspect for brake fluid leaks.

16. Inspect the fluid level.

17. Adjust the parking brake lever travel.

18. Inspect the rear disc brake caliper operation lever and stopper clearance.

19. Install the front upper console box.

20. Install the rear wheel and tighten to 76 ft. lbs. (103 Nm).

DISC BRAKE PADS

REMOVAL & INSTALLATION

Double Wishbone Type Suspension

See Figures 20 through 23.

1. Remove the caliper.

2. Remove the rear brake pad.

a. Remove the 2 brake pads with the rear disc brake anti-squeal shims.

3. Remove the rear disc brake anti squeal shim and the pad wear indicator from each pad.

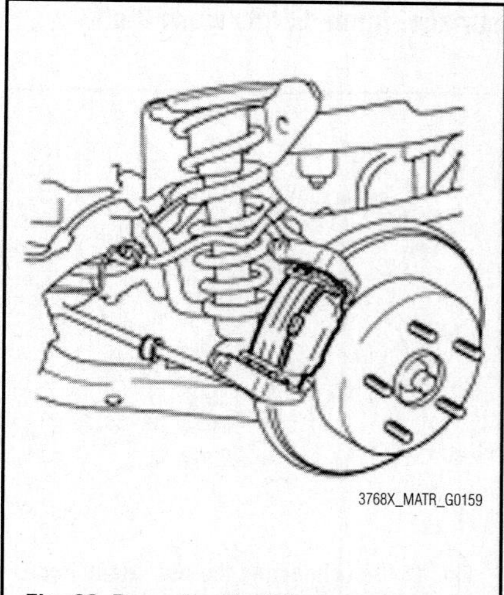

Fig. 20 Removing the 2 brake pads

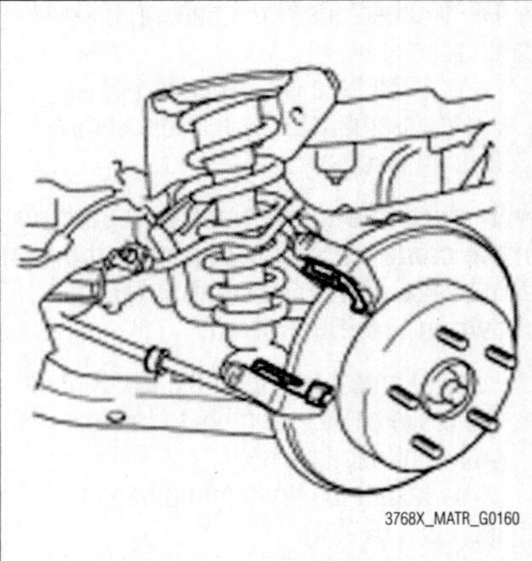

Fig. 21 Removing the rear disc brake pad support plate

4. Remove the rear disc brake pad support plate.
   a. Remove the rear disc brake pad support plates from the rear disc brake caliper mounting.
5. Remove the rear disc brake caliper mounting from the backing plate.

To install:

6. Install the rear disc brake pad support plate.
   a. Install the 2 rear disc brake pad support plates to the rear disc brake caliper mounting.

➡Be sure to install each rear disc brake pad support plate in the correct position and direction.

7. Install the rear disc brake anti squeal shim.
   a. Apply disc brake grease to the inside of the rear disc brake anti-squeal shims.

➡When replacing worn pads, the rear disc brake anti-squeal shims must be replaced together with the pads.

➡Apply disc brake grease to the area that contacts the rear disc brake anti-squeal shim.

➡Disc brake grease may seep out slightly from the areas where the rear disc brake anti-squeal shims are installed.

➡Make sure that disc brake grease is not applied onto the lining surface.

   b. Install the rear disc brake anti-squeal shim and the pad wear indicator to each brake pad.

➡Install the pad wear indicators and rear disc brake anti-squeal shims in the correct positions and directions.

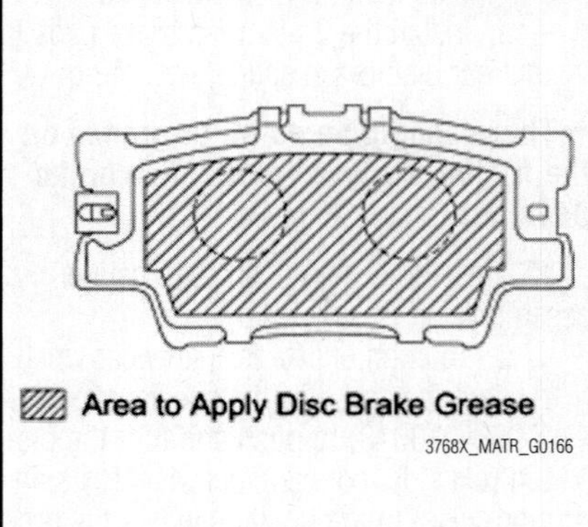

Fig. 22 Applying disc brake grease to the inside of the rear disc brake anti squeal shims

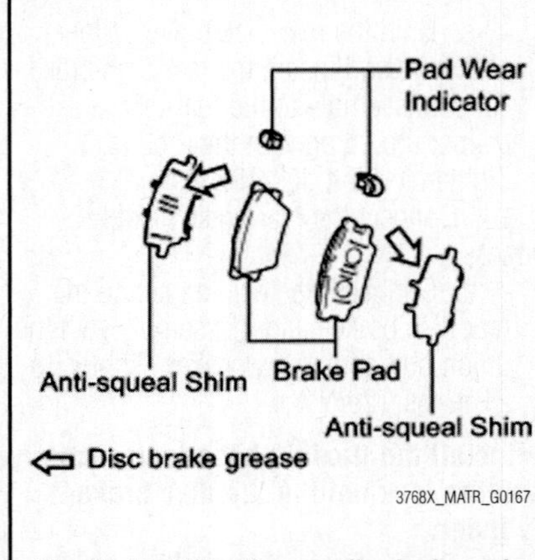

Fig. 23 Installing the rear disc brake anti squeal shim

8. Install the rear brake pad.
   a. Install the 2 brake pads with rear anti-squeal shims to the rear disc brake caliper mounting.
9. Install the rear disc brake caliper.

Torsion Beam Type Suspension

See Figures 17, 24 through 27.

➡Use the same procedure for the LH side and RH side.

➡The following procedure listed is for the LH side.

1. Remove the rear wheel.
2. Drain the brake fluid.

➡If brake fluid leaks onto any painted surface, immediately wash it off.

3. Remove the front upper console box.
4. Loosen the parking brake cable.
   a. Completely release the parking brake lever.
   b. Loosen the lock nut and the adjusting nut to completely release the parking brake cable.
5. Remove the parking brake lever protector.
6. Disconnect the No. 3 parking brake cable assembly.
   a. Disengage the clamp and remove the bolt from the No. 3 parking brake cable assembly.
   b. Separate the tip of the No. 3 parking brake cable assembly from the rear disc brake caliper assembly.
   c. Separate the No. 3 parking brake cable assembly from the rear disc brake caliper assembly.

➡Insert an offset wrench (14 mm) at the base of the No. 3 parking brake cable assembly as shown in the illustration to disengage the clip. Pull out the No. 3 parking brake cable assembly from the rear disc brake caliper assembly.

7. Separate the rear brake flexible hose.
   a. Remove the union bolt and gasket, and separate the rear brake flexible hose from the rear disc brake caliper assembly.
8. Remove the rear disc brake caliper assembly.
   a. Hold the rear disc brake pad guide pin and remove the 2 bolts and rear disc brake caliper assembly.
9. Remove the rear disc brake pad.
   a. Remove the 2 disc brake pads from the rear disc brake caliper mounting.
10. Remove the rear disc anti squeal shim.
   a. Remove the 2 anti-squeal shims from each brake pad.
11. Remove the rear disc brake pad support plate.
   a. Remove the rear disc brake pad support plate (upper) and rear disc brake pad support plate (lower) from the disc brake cylinder mounting.

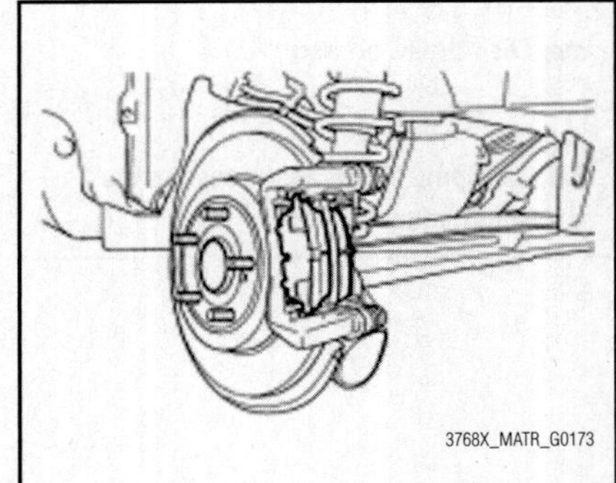

Fig. 24 Removing the 2 disc brake pads from the rear disc brake caliper mounting

➡Each rear disc brake pad support plate has a different shape. Be sure to put an identification mark on each rear disc brake pad support plate so that it can be installed to its original position.

To install:

12. Install the rear disc brake pad support plate.
    a. Install the 2 rear disc brake pad support plates to the rear disc brake caliper mounting.

➡Be sure to install each rear disc brake pad support plate in the correct position and direction.

13. Install the rear disc brake anti squeal shim.
    a. Apply disc brake grease to the 2 No. 1 anti-squeal shims.
    b. Install the No. 1 anti-squeal shim and No. 2 anti-squeal shim to each brake pad.

➡When replacing worn pads, the anti-squeal shims must be replaced together with the pads.

➡Apply disc brake grease to the area that contacts the anti-squeal shim.

➡Disc brake grease may seep out slightly from the areas where the anti-squeal shims are installed.

➡Make sure that disc brake grease is not applied onto the lining surface.

14. Install the rear disc brake pad.
    a. Install the 2 rear disc brake pads to the rear disc brake caliper mounting.

➡There should be no oil or grease on the friction surfaces of the disc brake pads or the rear disc.

15. Install the rear disc brake caliper assembly.
    a. To compensate for pad wear when reusing the pad, use the special tool (09960-10010) to push and turn the piston (LH side: counterclockwise, RH side: clockwise) to the position where the protrusion on the pad lines up properly with the piston groove.

➡Place the disc between the 2 brake pads and determine the piston return value.

    b. Hold the rear disc brake caliper slide pin, and install the rear disc brake caliper assembly to the rear disc brake caliper mounting with the 2 bolts. Tighten to 26 ft. lbs. (35 Nm).

16. Connect the rear brake flexible hose.
    a. Connect the flexible hose to the rear disc brake caliper assembly with the union bolt and a new gasket. Tighten to 21 ft. lbs. (29 Nm).

➡Install the flexible hose lock securely into the lock hole in the disc brake cylinder.

17. Connect the No. 3 parking brake cable assembly.
    a. Install the No. 3 parking brake cable assembly to the rear disc brake caliper assembly.

➡Be sure to engage the No. 3 parking brake cable assembly clip onto the rear disc brake caliper assembly LH as shown in the illustration.

    b. Connect the tip of the No. 3 parking brake cable assembly to the rear disc brake caliper assembly.
    c. Install the bolt and tighten to 53 inch lbs. (6 Nm).
    d. Engage the clamp to the No. 3 parking brake cable assembly.

18. Install the parking brake lever protector.
19. Fill the reservoir with brake fluid.
20. Bleed the brake line.
21. Inspect for brake fluid leaks.
22. Inspect the fluid level.
23. Adjust the parking brake lever travel.
24. Inspect the rear disc brake caliper operation lever and stopper clearance.
25. Install the front upper console box.
26. Install the rear wheel and tighten to 76 ft. lbs. (103 Nm).

Fig. 25 Apply disc brake grease to the 2 No. 1 anti-squeal shims

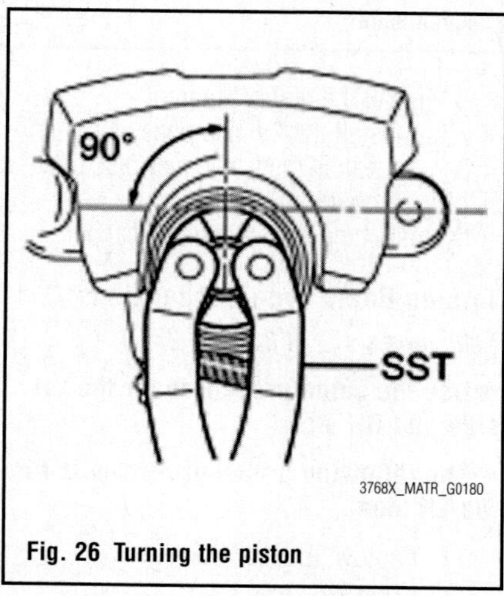

Fig. 26 Turning the piston

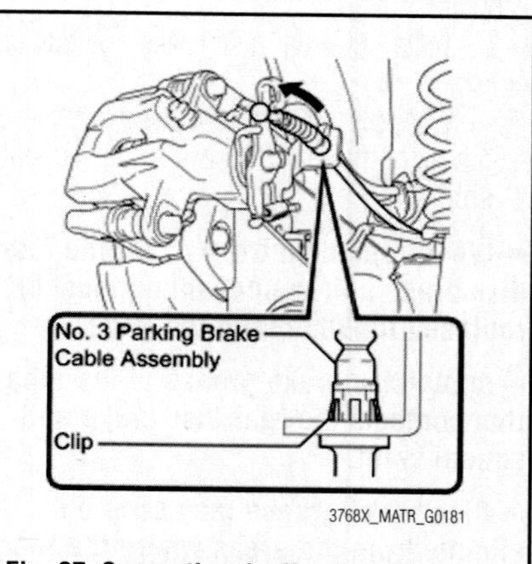

Fig. 27 Connecting the No. 3 parking brake cable assembly

PARKING BRAKE SHOES

REMOVAL & INSTALLATION

See Figures 28 through 48.

➡Perform the same procedure for the RH side and LH side.

➡The procedures listed below are for the LH side.

1. Remove the rear wheel.
2. Separate the rear disc brake caliper assembly.
3. Remove the rear disc.
4. Remove the No. 3 parking brake shoe return tension spring.
5. Remove the No. 2 parking brake shoe return tension spring.
6. Remove the No. 1 parking brake shoe hold down compression spring.
    a. Using the special tool (09718-00011), turn the No. 1 parking brake shoe hold down spring pin and remove the No. 1 parking brake shoe hold down compression spring as shown in the illustration.
7. Remove the parking brake shoe strut.
    a. Pull the No. 1 parking brake shoe assembly outward by hand.
    b. Remove the parking brake shoe strut.
8. Separate the No. 2 parking brake shoe hold down compression spring.

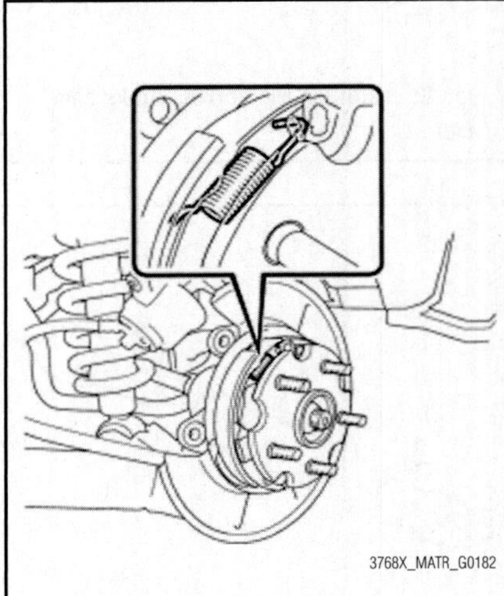

Fig. 28 Removing the No. 3 parking brake shoe return tension spring

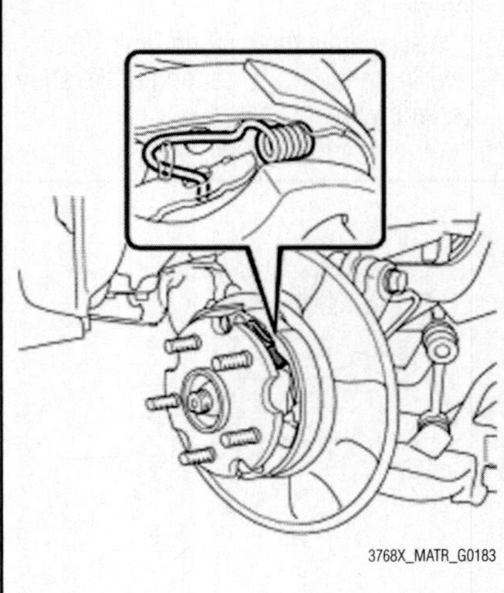

Fig. 29 Removing the No. 2 parking brake shoe return tension spring

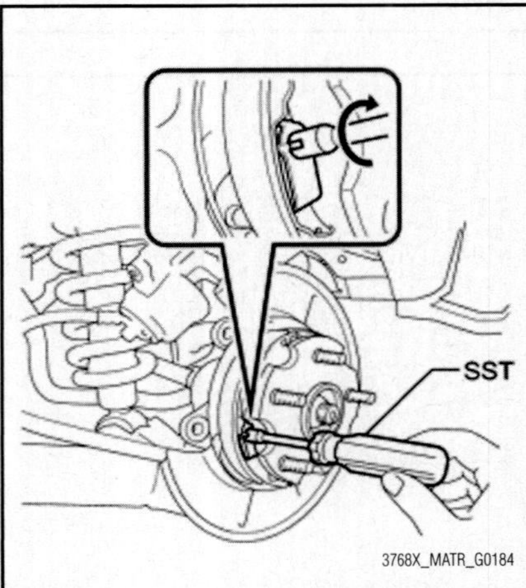

Fig. 30 Removing the No. 1 parking brake shoe hold down compression spring

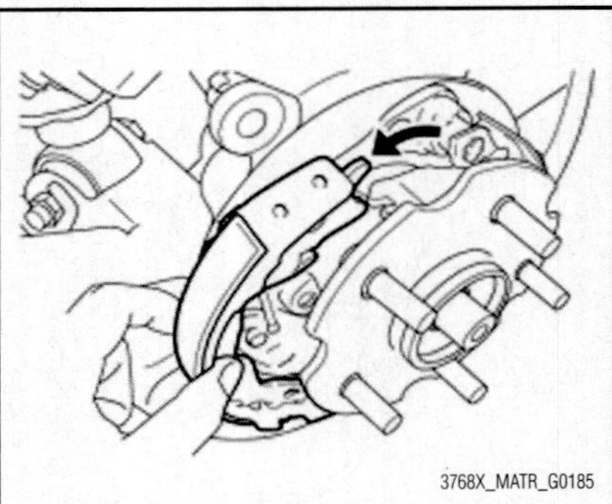

Fig. 31 Pulling out the No. 1 parking brake shoe assembly

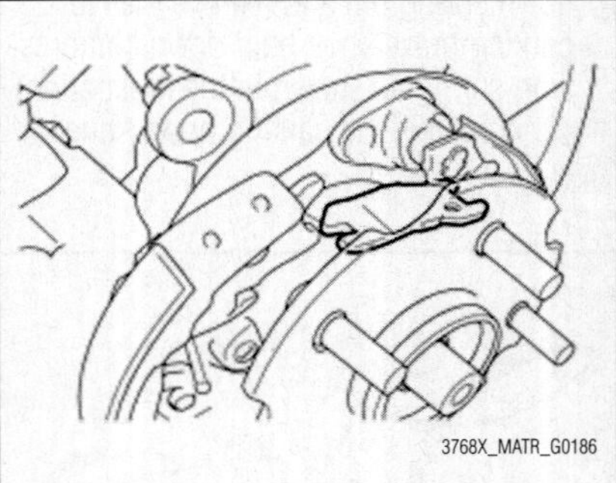

Fig. 32 Remove the parking brake shoe strut

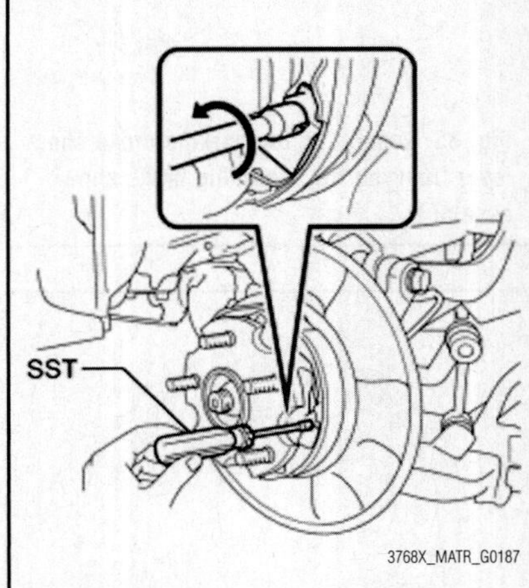

Fig. 33 Separating the No. 2 parking brake shoe hold down compression spring

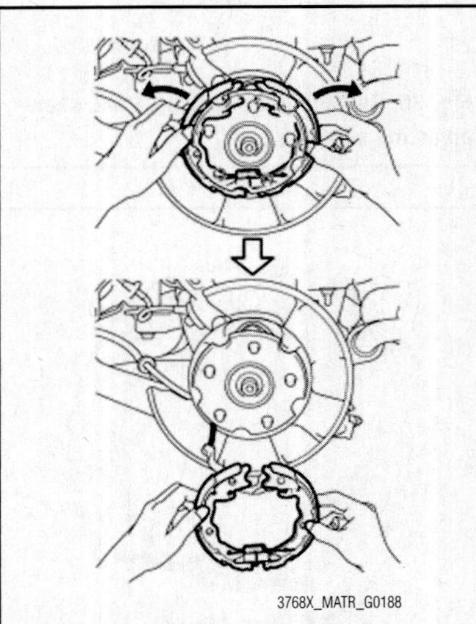

Fig. 34 Separating the parking brake shoe lever

a. Using special tool (09718-00011), turn the No. 2 parking brake shoe hold down spring pin and remove the No. 2 parking brake shoe hold down compression spring as shown in the illustration.

9. Separate the parking brake shoe lever.

a. Pull the No. 1 and No. 2 parking brake shoe assemblies outward by hand, then pull them out toward the lower of the vehicle.

b. Separate the parking brake shoe lever from the No. 2 parking brake shoe assembly.

10. Remove the parking brake shoe adjusting screw set.

11. Remove the No. 1 parking brake shoe return tension spring.

12. Remove the parking brake shoe lever.

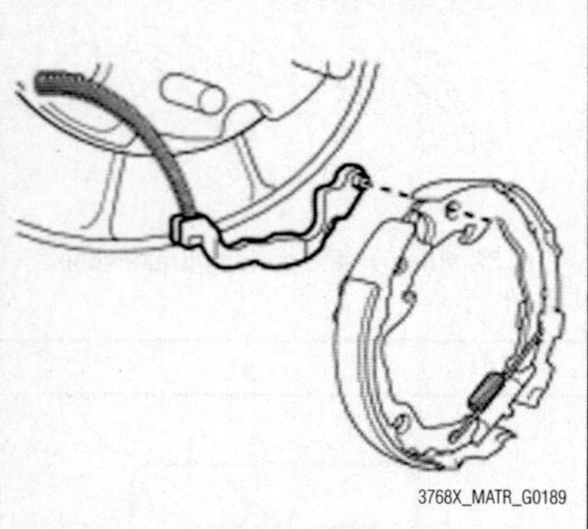

Fig. 35 Separating the parking brake shoe lever from the No. 2 parking brake shoe assembly

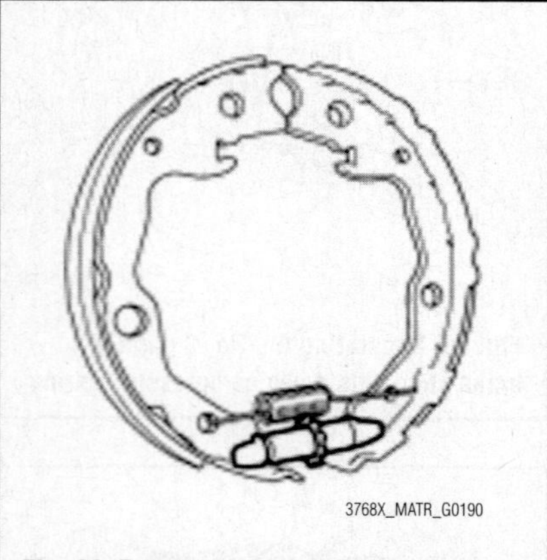

Fig. 36 Removing the parking brake shoe adjusting screw set

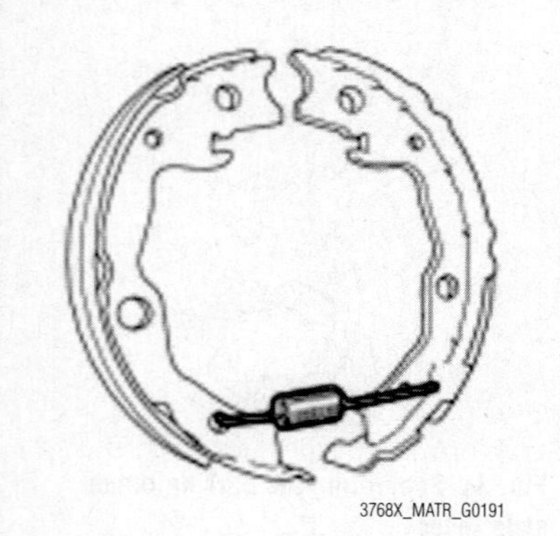

Fig. 37 Removing the No. 1 parking brake shoe return tension spring

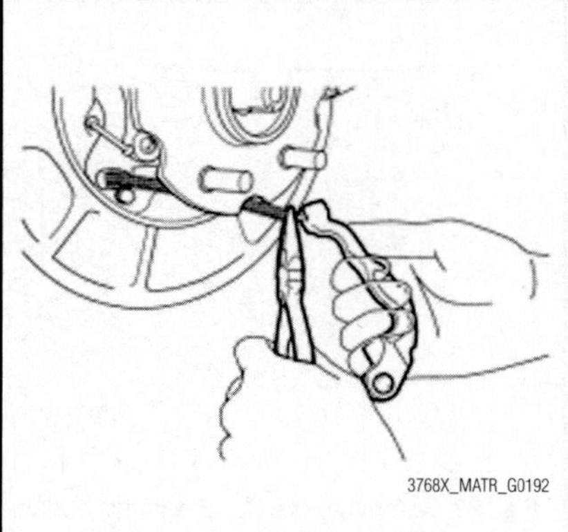

Fig. 38 Removing the parking brake show lever

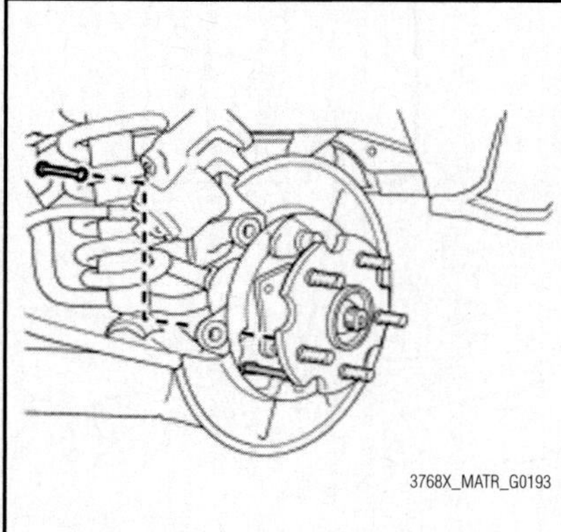

Fig. 39 Removing the No. 1 parking brake shoe hold down spring pin

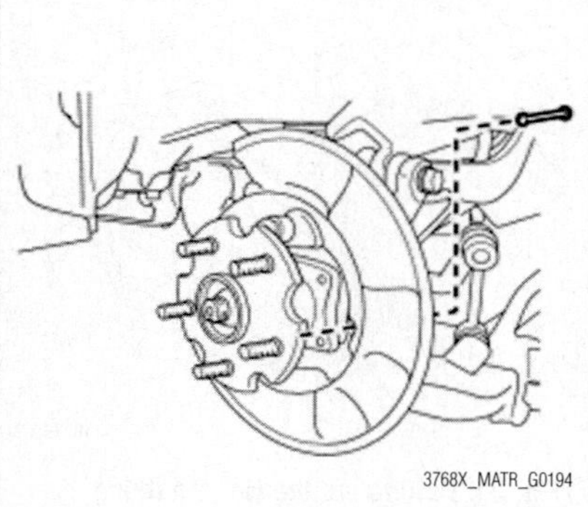

Fig. 40 Removing the No. 2 parking brake shoe hold down spring pin

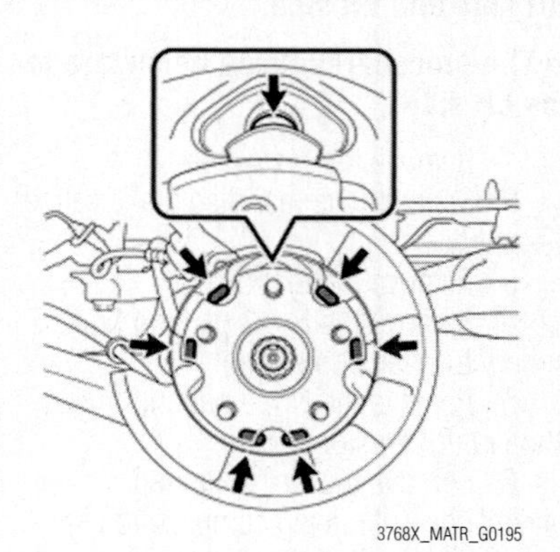

Fig. 41 Applying high temperature grease to the backing plate

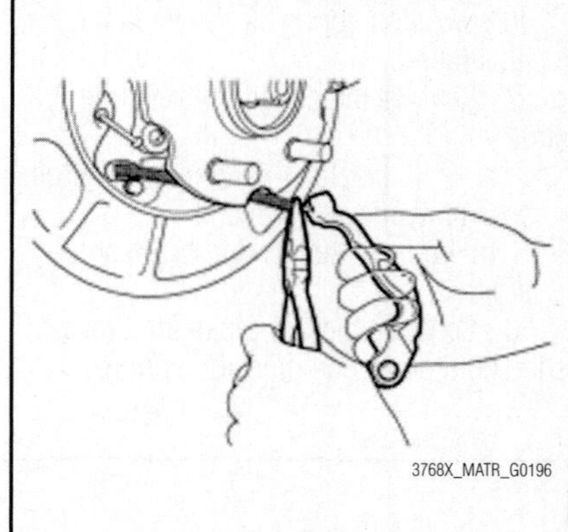

Fig. 42 Installing the parking brake shoe lever

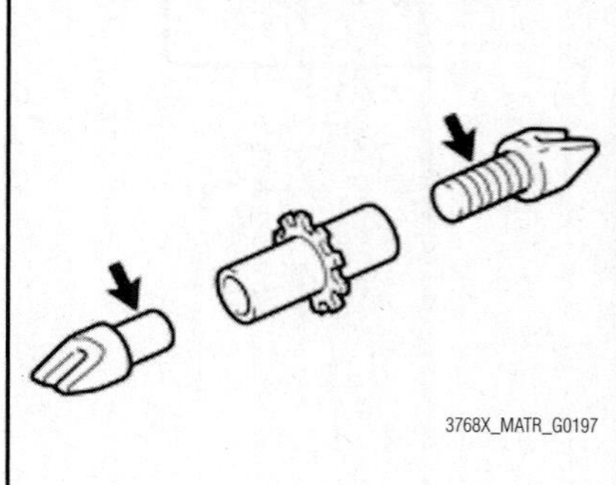

Fig. 43 Applying high temperature grease to the parking brake show adjusting screw set

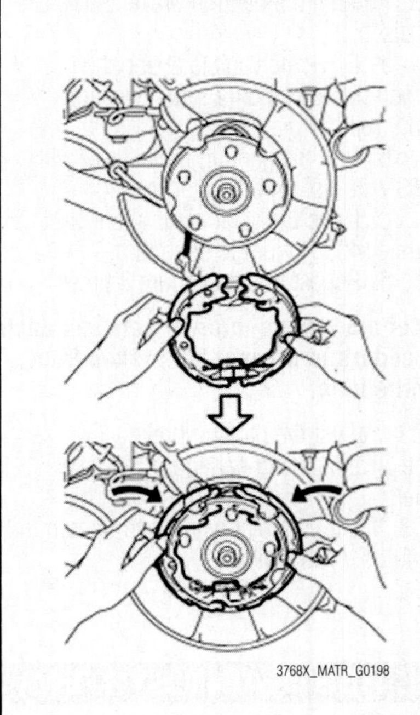

Fig. 44 Pulling the No. 1 and No. 2 parking brake shoe assembly outward

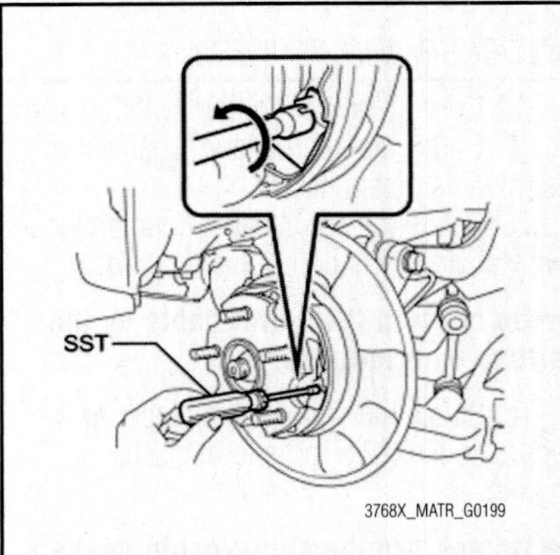

Fig. 45 Installing the No. 2 parking brake shoe hold down compression spring

Fig. 46 Positioning the No. 1 parking brake show assembly

a. Using needle-nose pliers, remove the parking brake shoe lever from the No. 3 parking brake cable assembly as shown in the illustration.

➡**Be careful not to damage the No. 3 parking brake cable assembly.**

13. Remove the No. 1 parking brake shoe hold down spring pin.

14. Remove the No. 2 parking brake shoe hold down spring pin.

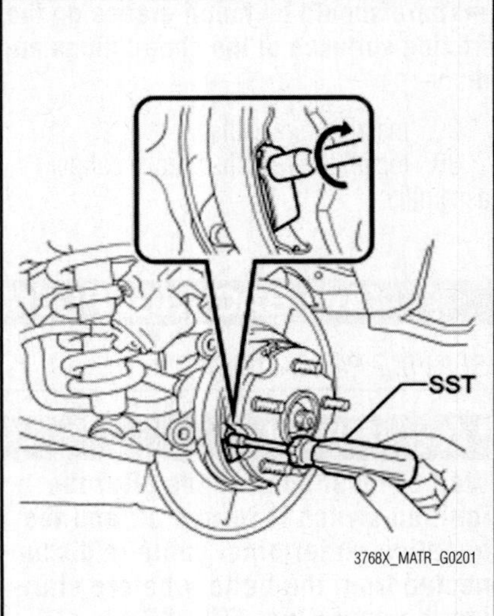

Fig. 47 Installing the No. 1 parking brake shoe hold down compression spring

To install:

➡**Before installation, apply high temperature grease to the areas indicated by the arrows.**

15. Apply high temperature grease to the backing plate.

16. Install the No. 1 parking brake shoe hold down spring pin.

17. Install the No. 2 parking brake shoe hold down spring pin.

18. Install the parking brake shoe lever.

19. Using needle-nose pliers, install the parking brake shoe lever to the No. 3 parking brake cable assembly as shown in the illustration.

➡**Be careful not to damage the No. 3 parking brake cable assembly.**

20. Install the No. 1 parking brake shoe return tension spring.

21. Install the parking brake shoe adjusting screw set.

a. Apply high temperature grease to the parking brake shoe adjusting screw set as shown in the illustration.

b. Install the parking brake shoe adjusting screw set.

22. Install the parking brake shoe lever.

a. Install the parking brake shoe lever to the No. 2 parking brake shoe assembly.

b. Pull the No. 1 and No. 2 parking brake shoe assemblies outward by hand and align them with the backing plates as shown in the illustration.

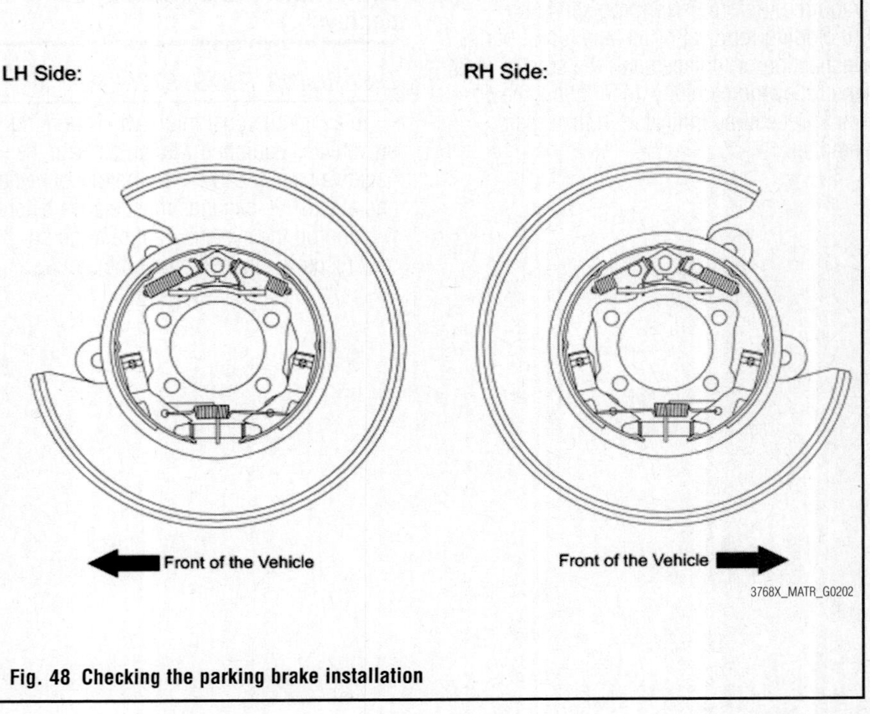

Fig. 48 Checking the parking brake installation

➡Do not allow the parking brake shoe lever to come off from the No. 2 parking brake shoe assembly.

23. Install the No. 2 parking brake shoe hold down compression spring.
    a. Using special tool (09718-00011), turn the No. 2 parking brake shoe hold down spring pin and install the No. 2 parking brake shoe hold down compression spring as shown in the illustration.
24. Install the parking brake shoe strut.
    a. Position the No. 1 parking brake shoe assembly correctly.
25. Install the No. 1 parking brake shoe hold down compression spring.
    a. Using special tool (09718-00011), turn the No. 1 parking brake shoe hold down spring pin and install the No. 1 parking brake shoe hold down compression spring as shown in the illustration.
26. Install the No. 2 parking brake shoe return tension spring.

➡Make sure to engage the tip of the No. 2 parking brake shoe return tension spring in the parking brake shoe strut.

27. Install the No. 3 parking brake shoe return tension spring.
28. Check the parking brake installation.

➡There should be no oil grease on the friction surfaces of the shoe linings and disc.

29. Install the rear disc.
30. Install the rear disc brake caliper assembly.
31. Bed in the parking brake shows to the discs.
    a. Drive the vehicle at about 31 mph (50 km/h) on a safe, level and dry road.
    b. Pull the parking brake lever with 33.7 ft. lbs. (150 N) of force.
    c. Drive the vehicle for about 0.25 mile (400 m) in this condition.
    d. Repeat this procedure 3 times.

➡Set a 5 minute interval between each procedure to prevent the brakes from overheating.

32. Adjust the parking brake shoe clearance and parking brake lever travel.
33. Install the rear wheel and tighten to 76 ft. lbs. (103 Nm).

CHASSIS ELECTRICAL

AIR BAG (SUPPLEMENTAL RESTRAINT SYSTEM)

GENERAL INFORMATION

This vehicle is equipped with a Supplemental Restraint System (SRS). It consists of a driver airbag, front passenger airbag, side airbag, curtain shield airbag and front seat belt pretensioner. Failure to carry out service operations in the correct sequence could cause the SRS to unexpectedly deploy during servicing, possibly leading to a serious accident. Further, if a mistake is made in servicing the SRS, it is possible that the SRS may fail to operate when required. Before performing servicing (including removal or installation of parts, inspection or replacement), be sure to read the following carefully, then follow the correct procedures indicated in the repair manual.

SERVICE PRECAUTIONS

⁂ CAUTION

Wait at least 90 seconds after the ignition switch is turned off and the negative (-) terminal cable is disconnected from the battery before starting the operation. (The SRS is equipped with a backup power source, so that if work is started within 90 seconds after disconnecting the negative (-) terminal cable of the battery, the SRS may be deployed.)

DISARMING THE SYSTEM

To avoid personal injury when working on vehicles equipped with an air bag, the negative battery cable must be disconnected and at least 90 seconds must elapse before working on the system. Failure to do so may result in deployment of the air bag.

ARMING THE SYSTEM

After vehicle service is completed, reattach the battery cables (positive cable first!) to rearm the air bag system.

CLOCKSPRING CENTERING

1. Check that the ignition switch is off.
2. Check that the battery negative (-) terminal is disconnected.
3. Rotate the spiral cable counterclockwise slowly by hand until it feels firm.

➡Do not turn the spiral cable by the airbag wire harness.

4. Rotate the spiral cable clockwise approximately 2.5 turns to align the marks.

➡Do not turn the spiral cable by the airbag wire harness. The spiral cable will rotate approximately 2.5 turns to both the left and right from the center.

DRIVE TRAIN

CLUTCH DRIVEN DISC & PRESSURE PLATE

REMOVAL & INSTALLATION

See Figures 49 and 50.

1. Remove the manual transaxle assembly.
2. Remove the clutch release fork with the clutch release bearing from the manual transaxle.
3. Remove the clutch release fork boot from the manual transaxle.
4. Remove the release bearing and clip from the clutch release fork.
5. Remove the release fork support from the manual transaxle.
6. Put matchmarks on the clutch cover assembly and the flywheel sub-assembly.
7. Loosen each set bolt one turn at a time until the spring tension is released.
8. Remove the set bolts and pull off the clutch cover.

➡ Do not drop the clutch disc.

9. Remove clutch disc assembly.

➡ Keep the parts clutch disc lining, the pressure plate, and the surface of the flywheel sub-assembly free of oil and foreign matter.

To install:

10. Insert the alignment tool (SST: 09301-00110) into the clutch disc assembly, then insert them both into the flywheel sub-assembly.

➡ Insert the clutch disc assembly in the correct direction.

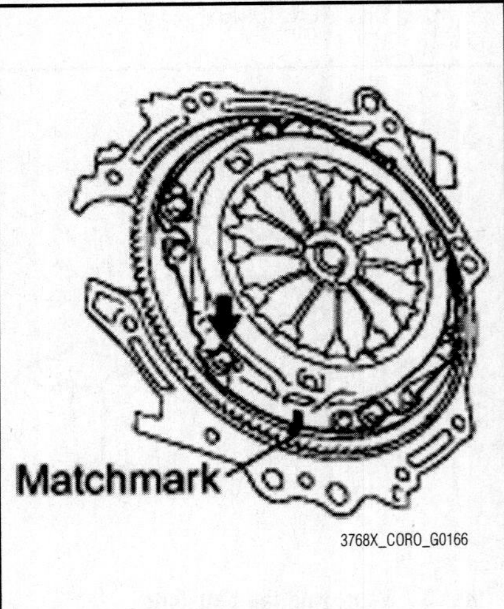

Fig. 49 Applying match mark

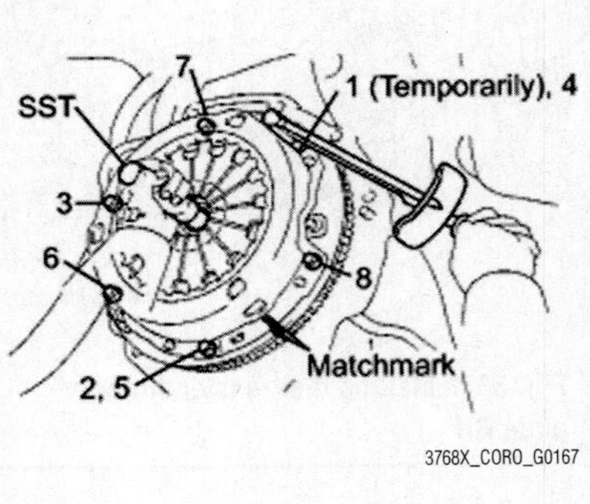

Fig. 50 Clutch assembly tightening sequence

11. Align the matchmark on the clutch cover assembly with the one on the flywheel sub-assembly.
12. Following the procedure shown in the illustration, tighten the 6 bolts in order, starting with the bolt located near the knock pin at the top. Tighten to 14 ft. lbs. (19 Nm).
13. Move SST up and down, right and left lightly after checking that the disc is in the center, and tighten the bolts.
14. Using a dial indicator with a roller instrument, check the diaphragm spring tip alignment. Maximum non-alignment: 0.0354 inch (0.9 mm)
15. If the alignment is not as specified, adjust the diaphragm spring tip alignment using SST: 09333-00013.
16. Install the release fork support onto the transaxle assembly. Tighten to 27 ft. lbs. (37 Nm).
17. Install the clutch release fork boot to the manual transaxle.
18. Apply release hub grease to the contact surfaces between the release fork and release bearing assembly, release fork and push rod, and release fork and fork support.
19. Install the release fork onto the release bearing assembly with the clip.
20. Apply clutch spline grease to the input shaft spline. Do not grease the smooth collar area of the spline.
21. Install the clutch release bearing with the release fork onto the transaxle assembly.

➡After installation, move the fork back and forth to check that the release bearing slides smoothly.

22. Install the manual transaxle.

HYDRAULIC SYSTEM BLEEDING

BLEEDING PROCEDURE

➡If any maintenance on the clutch system was performed or the system is suspected of containing air, bleed the system. Use care; brake fluid will remove the paint from any surface. If the brake fluid spills onto any painted surface, wash it off immediately with soap and water.

1. Before servicing the vehicle, refer to the Precautions section.
2. Fill the clutch reservoir with brake fluid. Check the reservoir level frequently and add fluid as needed.
3. Connect one end of a vinyl tube to the bleeder plug on the slave cylinder and submerge the other end into a clear container half-filled with brake fluid.
4. Slowly pump the clutch pedal several times.
5. Have an assistant hold the clutch pedal down and loosen the bleeder plug until fluid and/or air starts to run out of the bleeder plug. Close the bleeder plug while the pedal is held to the floor.

➡Do not allow the pedal to raise back-up while the bleeder is still open. If this happens, it will allow air to re-enter the slave cylinder and cause the clutch system not to work properly.

6. Repeat Steps 2 and 3 until all the air bubbles are removed from the system.
7. Tighten the bleeder plug when all the air is gone.
8. Refill the master cylinder to the proper level as required.
9. Check the system for leaks.

TRANSFER CASE ASSEMBLY

REMOVAL & INSTALLATION

See Figures 51 through 53.

1. Remove the engine assembly.
2. Remove the rear engine mounting insulator.
3. Install the engine hangers.
4. Remove the starter assembly.
5. Disconnect the wire harness.
6. Remove the rear engine mounting bracket.
7. Remove the rear drive plate and torque converter clutch setting bolt.

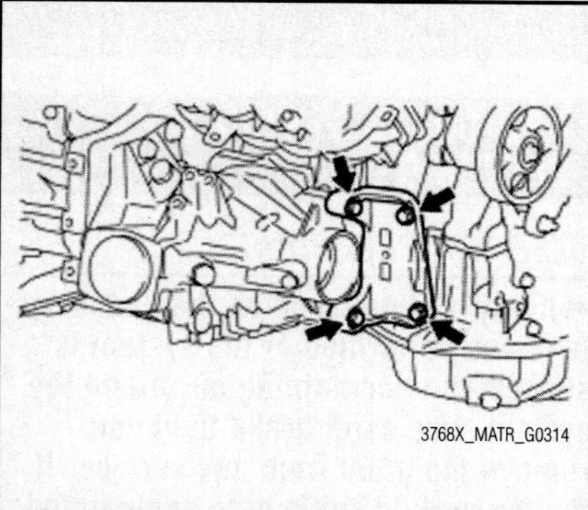

Fig. 51 Removing the transfer stiffener plate RH

8. Remove the transfer stiffener plate RH.
   a. Remove the 4 bolts and transfer stiffener plate RH.
9. Remove the automatic transaxle assembly.
10. Remove the transfer assembly.
   a. Remove the 2 bolts and 6 nuts.
   b. Using a plastic hammer, remove the transfer assembly from the transaxle assembly.

➡**Remove the transfer assembly from the transaxle assembly without tilting it.**

➡**During removal, do not hold the transfer assembly by the oil seals on either side of the assembly.**

To install:

11. Install the transfer assembly.
   a. Install the transfer assembly to the transaxle assembly with the 2 bolts and 6 nuts. Tighten to 51 ft. lbs. (69 Nm).

➡**Install the transfer assembly to the transaxle assembly horizontally.**

➡**Do not touch the transfer assembly oil seals during installation.**

12. Install the automatic transaxle assembly.

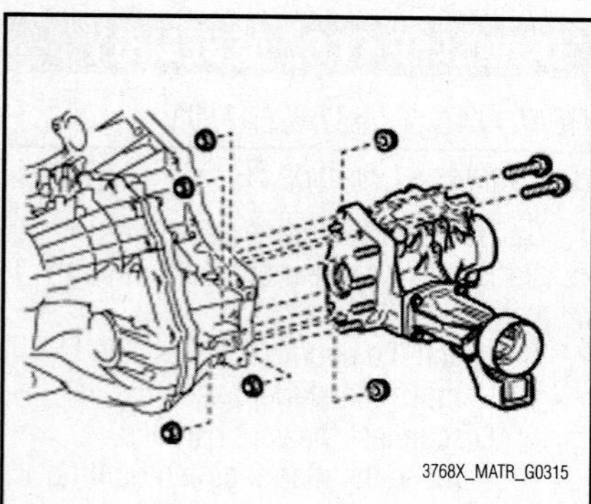

Fig. 52 Removing the transfer assembly

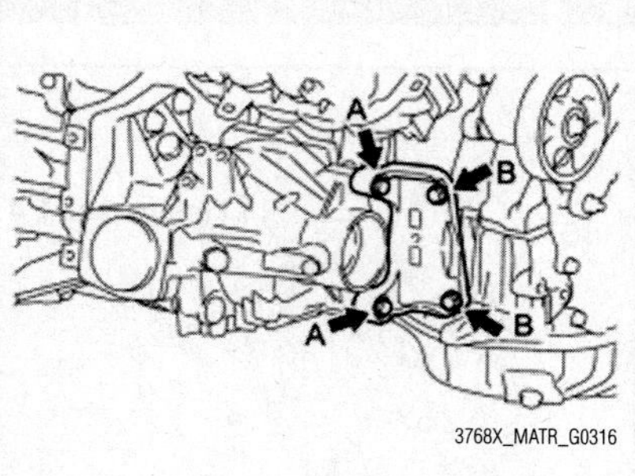

Fig. 53 Installing the transfer stiffener plate RH

13. Install the transfer stiffener plate RH with the 4 bolts. Tighten the bolts to 25 ft. lbs. (34 Nm). The A bolts are 1.7717 inch (45 mm) long. The B bolts are 1.1811 inch (30 mm) long.
14. Install the drive plate and the torque converter clutch setting bolt.
15. Install the rear engine mounting bracket.
16. Connect the wire harness.
17. Install the starter assembly.
18. Temporarily tighten the rear engine mounting insulator.
19. Install the engine assembly with the transaxle.

FRONT HALFSHAFT

REMOVAL & INSTALLATION

See Figures 54 and 55.

➡**The hub bearing could be damaged if subjected to the full weight of the vehicle, such as if the vehicle is moved without the halfshafts. If it is absolutely necessary to place the full vehicle weight on the hub bearing, first support the bearing with SST No. 09608–16042.**

➡**This procedure is for removing both LH and RH halfshafts.**

1. Remove the LH and RH engine under covers.
2. Drain transmission fluid.
3. Raise and support the vehicle.
4. Remove the front wheels.
5. Remove front axle shaft nut:
   a. Using SST: 09930-00010 and a hammer, release the staked part of the front axle shaft nut.
   b. Insert SST into the groove with the flat surface facing up.
   c. Do not damage the tip of SST using grinders.
   d. Completely unstake the staked part before removing the axle hub nut.
   e. Do not damage the threads of the drive shaft.
   f. Using a socket wrench (30 mm), remove the front axle shaft nut.
6. Separate the wheel speed sensors.
7. Separate the front stabilizer link assembly. Refer to Stabilizer in Front Suspension.
8. Separate the tie rod end sub-assembly.
9. Separate front lower suspension. Refer to Front Suspension.
10. Separate front axle assembly:
   a. Using a plastic hammer, tap the end of the drive shaft and disengage the fitting between the drive shaft and front axle assembly.
   b. If it is difficult to disengage the fitting, tap the end of the drive shaft with a brass bar and hammer.
   c. Push the front axle assembly out of the vehicle to remove the drive shaft from the front axle assembly.

➡**Do not push out the front axle further than necessary. Do not damage the outboard joint boot. Do not damage the speed sensor rotor.**

11. If equipped, remove the 2 bolts and the manual transmission case protector.
12. Using a halfshaft puller, remove the LH halfshaft assembly.
13. Remove front RH drive shaft assembly:
   a. For 1.8L Engines:
   b. Using a brass bar and hammer, remove the front drive shaft assembly.

➡**Do not damage the oil seal. Do not damage the inboard joint boot. Do not drop the drive shaft.**

   c. For 2.4L Engines:

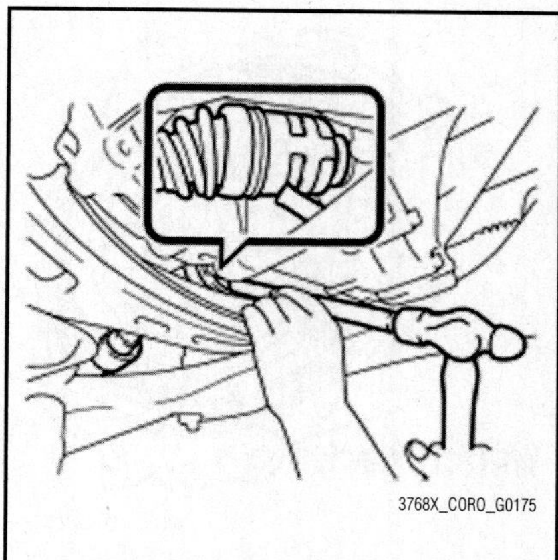

Fig. 54 Removing the halfshaft assembly—1.8L Engine

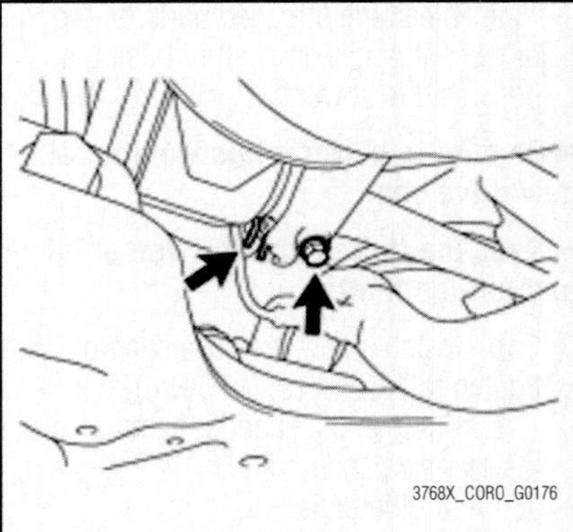
Fig. 55 Removing the bearing bracket hole snap ring and halfshaft assembly—2.4L Engine

d. Remove the bearing bracket hole snap ring.

e. Remove the bolt and front drive shaft assembly RH from the drive shaft bearing bracket.

➡Do not damage the boot or oil seal.

14. Using a screwdriver, remove the snap ring from the front LH and RH drive inboard joint assembly.

To install:

15. Install a new front drive shaft hole snap ring to the front drive inboard joint assembly.
16. Align the inboard joint splines, and using a brass bar and a hammer, install the front drive shaft assembly.

➡Face the end gap of the front drive inboard joint hole snap ring downward. Do not damage the oil seal. Do not damage the inboard joint boot.

➡Confirm whether the drive shaft is securely driven in by checking the reaction force and sound.

17. For 2.4L engines:

a. Using a screwdriver, install a new bearing bracket hole snap ring.

➡Do not damage the boot or oil seal. Move the drive shaft assembly while keeping it level.

b. Install a new bolt. Tighten to 24 ft. lbs. (32 Nm).

18. If equipped, install the manual transmission case protector with the 2 bolts. Tighten to 13 ft. lbs. (18 Nm).
19. To complete installation, reverser remaining removal procedure.
20. Install front axle shaft nut:

a. Clean the threaded parts on the drive shaft and axle shaft nut using a non-residue solvent.

➡Be sure to perform this work for a new drive shaft. Keep the threaded parts free of oil and foreign matter.

b. Using a socket wrench (30 mm), install a new axle shaft nut. Tighten to 160 ft. lbs. (216 Nm).

c. Using a chisel and hammer, caulk the axle shaft nut.

21. Tighten wheel lug nuts to 76 ft. lbs. (103 Nm).
22. Add transmission fluid.
23. Inspect for fluid leaks.
24. Inspect and adjust the wheel alignment.

REAR HALFSHAFT

REMOVAL & INSTALLATION

See Figures 56 and 57.

➡Use the same procedure for the RH side and LH side.

➡The procedure listed below is for the LH side.

1. Remove the rear wheel.
2. Remove the rear axle shaft nut.

a. Using the special tool (09930-00010) and a hammer, release the staked part of the rear axle shaft nut.

➡Loosen the staked part of the nut completely, otherwise the threads of the drive shaft may be damaged.

b. While applying the brakes, remove the rear axle shaft nut.

3. Remove the rear differential carrier cover plug.
4. Remove the rear differential drain plug.
5. Install the rear differential drain plug.
6. Separate the rear disc brake caliper assembly.
7. Remove the rear disc.
8. Remove the rear No. 3 parking brake shoe return tension spring.
9. Remove the No. 2 parking brake shoe return tension spring.
10. Remove the No. 1 parking brake shoe hold down compression spring.
11. Remove the parking brake shoe strut.
12. Separate the No. 2 parking brake shoe hold down compression spring.
13. Remove the parking brake shoe adjusting screw set.
14. Remove the No. 1 parking brake shoe return tension spring.
15. Remove the parking brake shoe lever.
16. Remove the No. 1 parking brake shoe hold down spring pin.
17. Remove the No. 2 parking brake shoe hold down spring pin.
18. Separate the rear stabilizer link assembly.
19. Separate the rear speed sensor.
20. Separate the No. 3 parking brake cable assembly.
21. Separate the upper control arm assembly.
22. Separate the rear No. 1 suspension arm assembly.
23. Remove the rear axle assembly.
24. Remove the rear drive shaft assembly.

a. Using a brass bar and a hammer, remove the rear drive shaft assembly.

➡Remove the rear drive shaft assembly while keeping it level.

25. Remove the rear drive shaft snap ring.

a. Using a screwdriver, remove the rear drive shaft snap ring.

To install:

26. Install the rear drive shaft snap ring.
27. Install the rear drive shaft assembly.

a. Align the shaft splines and install the rear drive shaft assembly with a brass bar and a hammer.

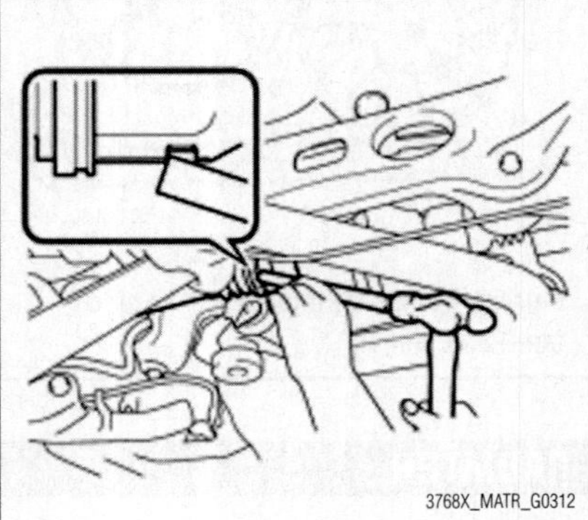
Fig. 56 Removing the rear drive shaft assembly

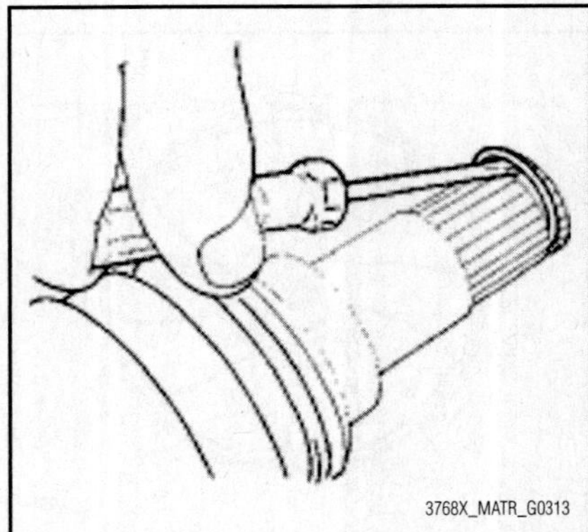
Fig. 57 Removing the rear drive shaft snap ring

➡**Set the snap ring with the opening facing downward.**

➡**Be careful not to damage the oil seal, boot, or dust cover.**

➡**Install the drive shaft assembly while keeping it level.**

28. Install the rear axle assembly.
29. Temporarily tighten the upper control arm assembly.
30. Install the rear axle assembly.
31. Temporarily tighten the upper control arm assembly.
32. Temporarily tighten the rear No. 1 suspension arm assembly.
33. Install the No. 3 parking brake cable assembly.
34. Install the rear stabilizer link assembly.
35. Install the rear speed sensor.
36. Apply high temperature grease.
37. Install the No. 1 parking brake shoe hold down spring pin.
38. Install the No. 2 parking brake shoe hold down spring pin.
39. Install the parking brake shoe lever.
40. Install the No. 1 parking brake shoe return tension spring.
41. Install the parking brake shoe adjusting screw set.
42. Install the parking brake shoe lever.
43. Install the No. 2 parking brake shoe hold down compression spring.
44. Install the parking brake shoe strut.
45. Install the No. 1 parking brake shoe hold down compression spring.
46. Install the No. 2 parking brake shoe return tension spring.
47. Install the No. 3 parking brake shoe return tension spring.
48. Check the parking brake installation.
49. Install the rear disc.
50. Install the rear disc brake caliper assembly.
51. Adjust the parking brake shoe clearance and parking brake pedal travel.
52. Install the rear axle shaft nut.
    a. Clean the threaded parts on the drive shaft and axle shaft nut using a non-residue solvent.

➡**Be sure to perform this work for a new drive shaft.**

➡**Keep the threaded parts free of oil and foreign matter.**

    b. Install a new rear axle shaft nut and tighten to 159 ft. lbs. (216 Nm).
    c. Using a chisel and a hammer, stake the rear axle shaft nut LH.
53. Add differential oil.
54. Inspect and adjust the differential oil.
55. Install the rear wheel. Tighten to 76 ft. lbs. (103 Nm).
56. Stabilize suspension.
57. Fully tighten the upper control arm assembly.
58. Fully tighten the rear No. 1 suspension arm assembly.
59. Inspect and adjust the rear wheel alignment.
60. Check the ABS speed sensor signal.

ENGINE COOLING

ENGINE FAN

REMOVAL & INSTALLATION

1.8L Engine

See Figure 58.

1. Disconnect the negative battery cable.
2. Remove the radiator assembly. Refer to Radiator in this section.
3. Remove the nut, then remove the fan.

To install:

To install, reverse removal procedure. Tighten the cooling fan motor screws to 35 inch lbs. (3.9 Nm).

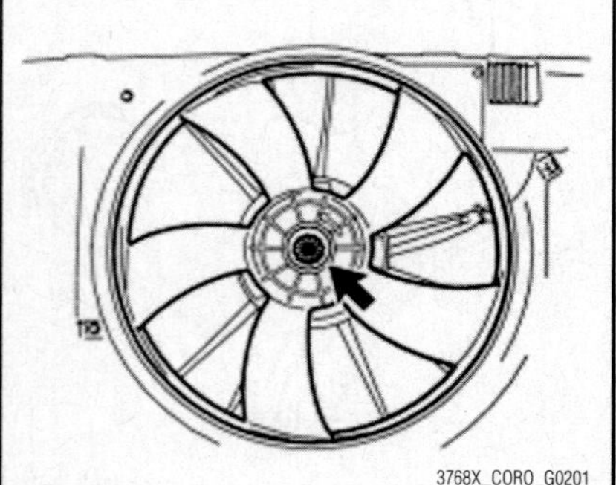

Fig. 58 Removing the nut and engine fan—1.8L engine

2.4L Engine

See Figure 59.

1. Disconnect the negative battery cable.
2. Remove the radiator assembly. Refer to Radiator in this section.
3. Remove the 2 nuts and 2 fans.

To install:

To install, reverse removal procedure. Tighten the cooling fan motor screws to 35 inch lbs. (3.9 Nm).

Fig. 59 Removing the nut and engine fan—2.4L engine

RADIATOR

REMOVAL & INSTALLATION

1.8L Engine

See Figure 60.

1. Disconnect the negative battery cable.
2. Drain the coolant.
3. Remove battery.
4. Remove the 2 radiator grille protectors.
5. Remove front bumper assembly:
    a. Remove the clip.
    b. Using a screwdriver, turn the pin 90 degrees and remove the pin hold clip.
    c. Put protective tape around the front bumper assembly.
    d. Remove the 2 screws and 3 clips.
    e. Remove the 4 screws.
    f. Remove the 2 clips.
    g. Disengage the 6 claws and remove the front bumper assembly.
    h. Disconnect the fog light connector. (w/ Fog Light).
6. Disconnect the radiator reservoir tank hose from the radiator assembly.
7. Disconnect the both radiator hose from the radiator assembly.
8. Remove the 2 bolts and 2 upper radiator supports.
9. Remove the 2 support cushions from the 2 upper radiator supports.
10. Remove the 3 bolts from the hood lock assembly.
11. Disconnect the hood lock control cable and remove the hood lock assembly.

12. Separate the water by-pass hose from the 3 clamps.
13. Disconnect the water by-pass hose from the radiator assembly.
14. Remove the 2 bolts and hood lock support sub-assembly.
15. Disconnect the horn connector.
16. Remove the 4 bolts and upper radiator support sub-assembly.
17. Remove the 2 bolts, disengage the 2 claws, and remove the No. 2 fan shroud from the radiator assembly.
18. Disconnect the cooling fan ECU connector and wire harness clamp.
19. Remove the radiator assembly with the fan shroud.
20. Remove the 2 lower radiator supports.
21. For A/T equipped vehicles:
a. Disconnect the 2 oil cooler hoses from the radiator.
b. Remove the 2 bolts and oil cooler hose.
22. Remove the 2 bolts and fan shroud from the radiator assembly.

To install:

To install, reverse removal procedure.

➡Do not apply any excessive force to the cooler condenser assembly or pipe when installing the radiator assembly.

2.4L Engine

See Figure 61.

1. Disconnect the negative battery cable.
2. Drain the coolant.

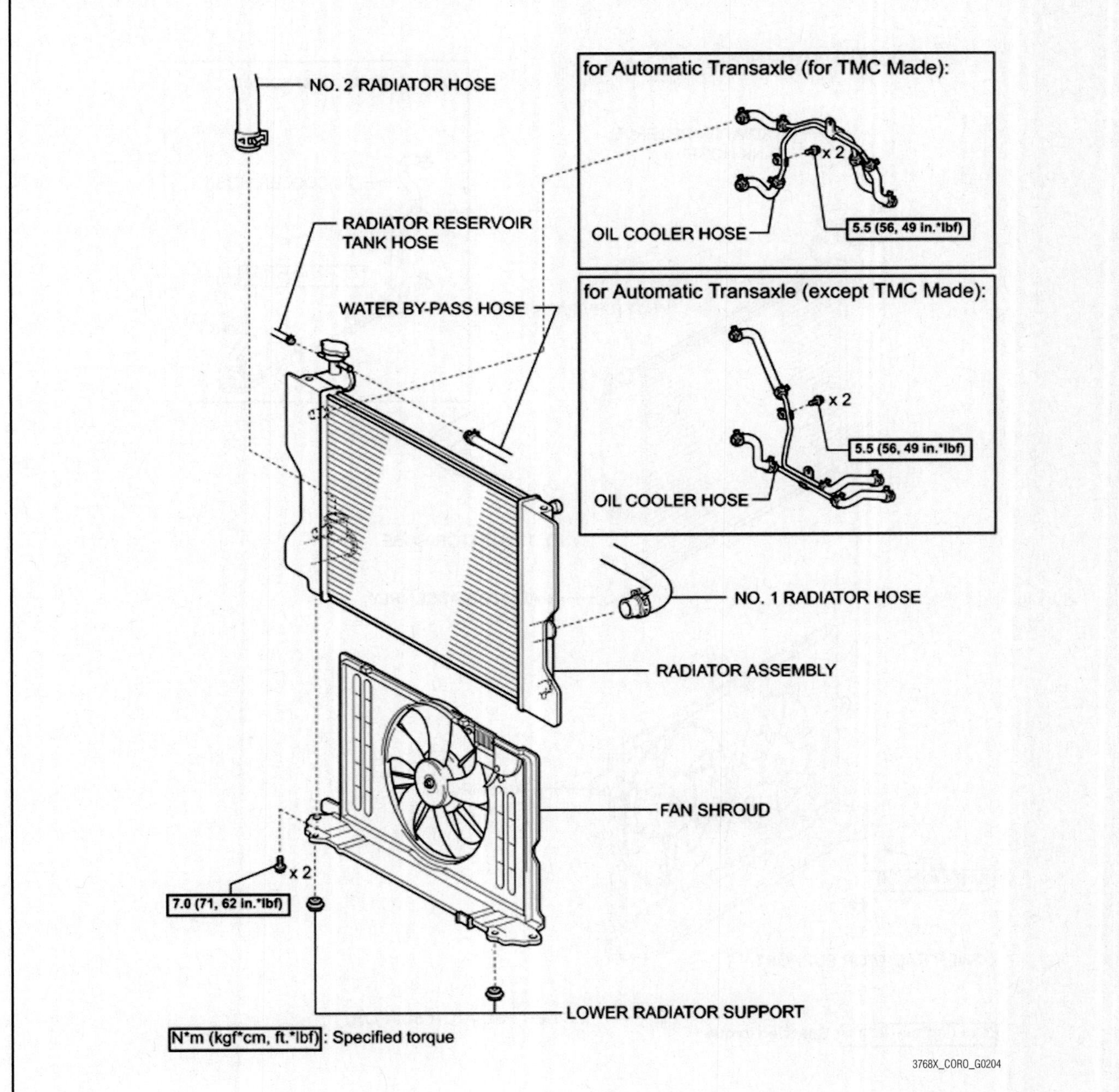

Fig. 60 Exploded view of the radiator assembly and hoses—1.8L Engine

3. Remove battery.
4. Remove the 2 radiator grille protectors.
5. Remove front bumper assembly:
   a. Remove the clip.
   b. Using a screwdriver, turn the pin 90 degrees and remove the pin hold clip.
   c. Put protective tape around the front bumper assembly.
   d. Remove the 2 screws and 3 clips.
   e. Remove the 4 screws.
   f. Remove the 2 clips.
   g. Disengage the 6 claws and remove the front bumper assembly.
   h. Disconnect the fog light connector. (w/ Fog Light).
6. Disconnect the radiator reserve tank hose from the radiator assembly.
7. Disconnect both radiator hoses from the radiator assembly.
8. Disconnect oil cooler hose (for automatic transaxle).
9. Remove the 2 bolts and 2 upper radiator supports.
10. Remove the 2 support cushions from the 2 upper radiator supports.
11. Remove the 3 bolts from the hood lock assembly.
12. Disconnect the hood lock control cable and remove the hood lock assembly.
13. Separate the water by-pass hose from the 2 clamps.
14. Disconnect the water by-pass hose from the radiator assembly.
15. Remove the 2 bolts and hood lock support sub-assembly.

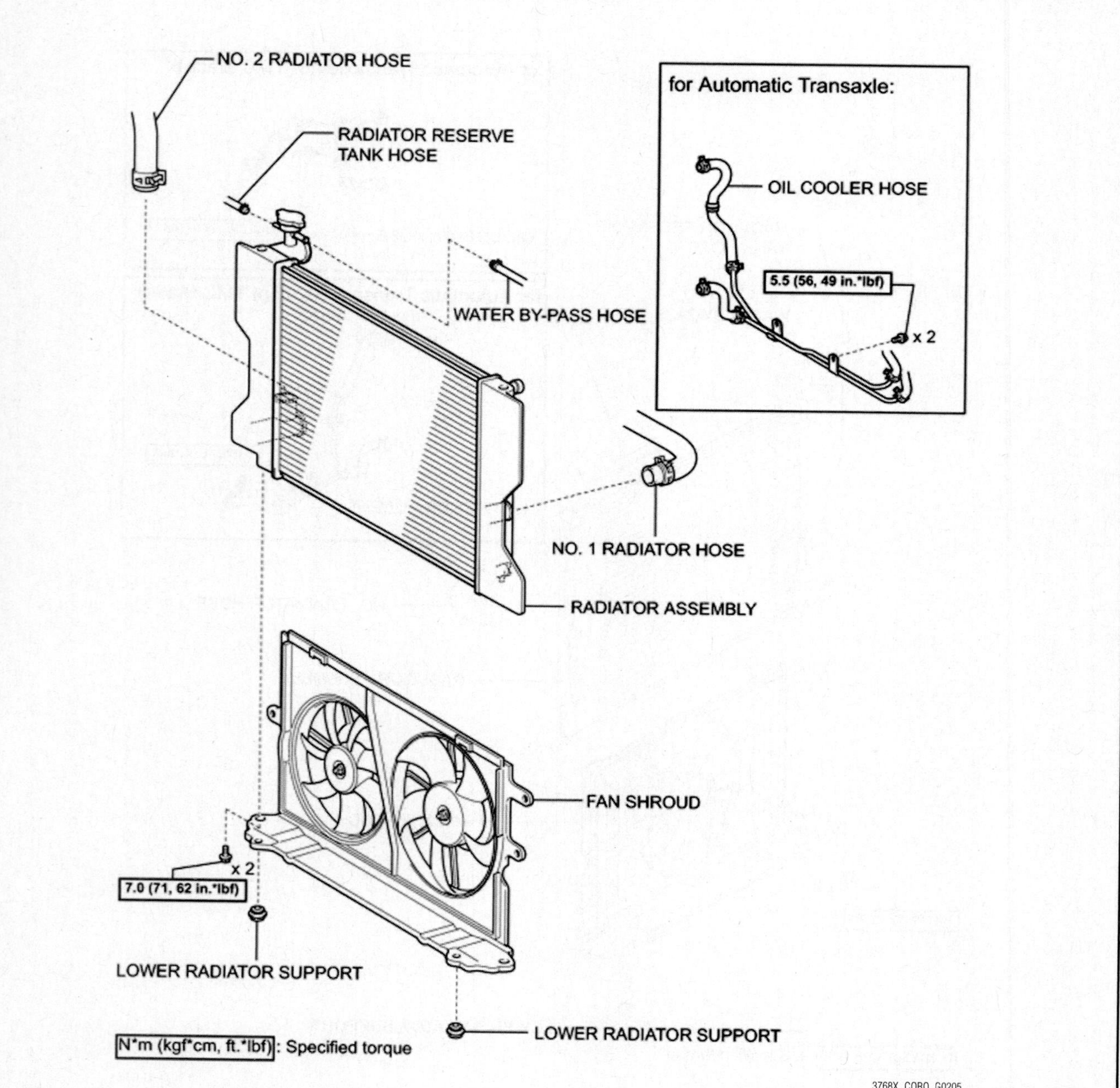

Fig. 61 Exploded view of the radiator assembly and hoses—2.4L Engine

16. Disconnect the horn connector.
17. Remove the 4 bolts and upper radiator support sub-assembly.
18. Remove the 2 bolts, disengage the 2 claws, and remove the No. 2 fan shroud from the radiator assembly.
19. Disconnect the 2 cooling fan motor connectors and wire harness clamp.
20. Remove the radiator assembly with the fan shroud.

➡Do not apply any excessive force to the cooler condenser assembly or pipe when removing the radiator assembly.

21. Disconnect the 2 oil cooler hoses from the radiator.
22. Disconnect the clamp from the fan shroud.
23. Remove the 2 bolts and oil cooler hose.
24. Remove the 2 bolts and fan shroud from the radiator assembly.

To install:

To install, reverse removal procedure.

THERMOSTAT

REMOVAL & INSTALLATION

See Figures 62 through 64.

1. Disconnect the negative battery cable.
2. Drain the engine coolant.
3. For 2.4L engines, remove the No. 2 coolant hose.
4. Remove the 2 nuts and water inlet.
5. Remove the thermostat.
6. Remove the gasket from the thermostat.

To install:

7. Install a new gasket on the thermostat.
8. Install the thermostat to the water inlet with the jiggle valve upward.

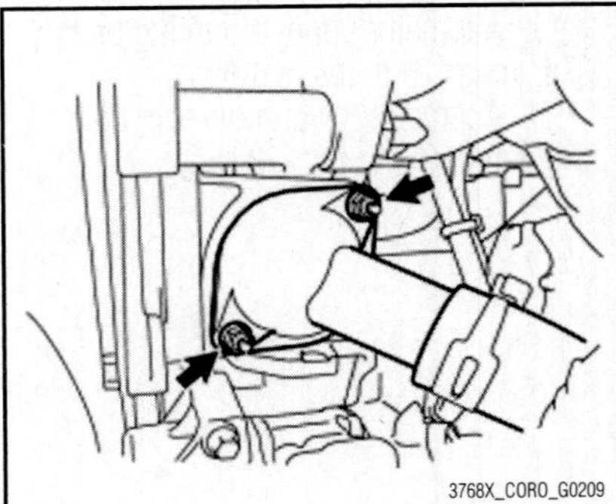

Fig. 62 Removing the 2 nuts and water inlet—1.8L Engine

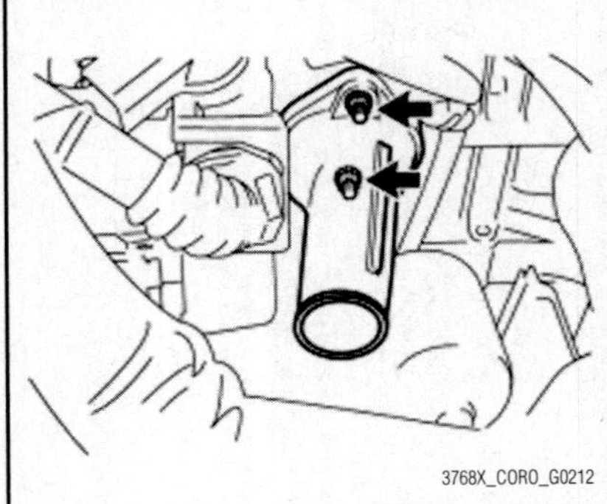

Fig. 63 Removing the 2 nuts and water inlet—2.4L Engine

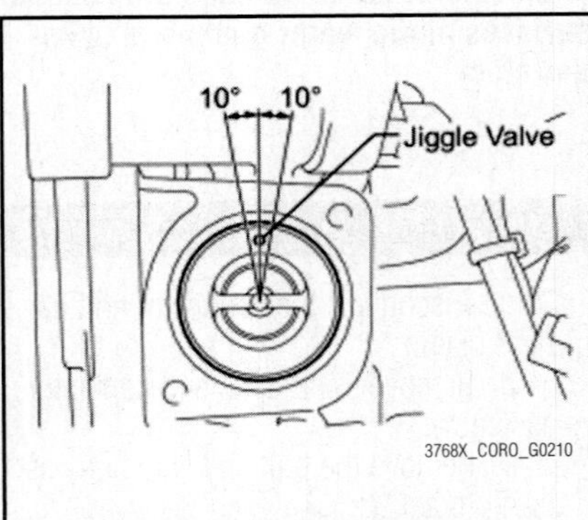

Fig. 64 Installing the thermostat and jiggle valve alignment

➡The jiggle valve may be set to within 10° on either side of the indicated position.

9. Install the water inlet with the 2 nuts. Tighten to 7 ft. lbs. (10 Nm).
10. For 2.4L engines, install the No. 2 coolant hose
11. Add coolant and inspect for leaks.
12. Inspect reservoir tank engine coolant level.

WATER PUMP

REMOVAL & INSTALLATION

1.8L Engine

See Figures 65 and 66.

1. Disconnect the negative battery cable.
2. Remove the engine cover.
3. Remove the RH engine under cover.
4. Drain the engine coolant.
5. Remove the accessory drive belt.
6. Remove the alternator. Refer to Alternator in Charging System under Engine Electrical.
7. Remove the 5 bolts and water pump assembly from the timing chain cover.

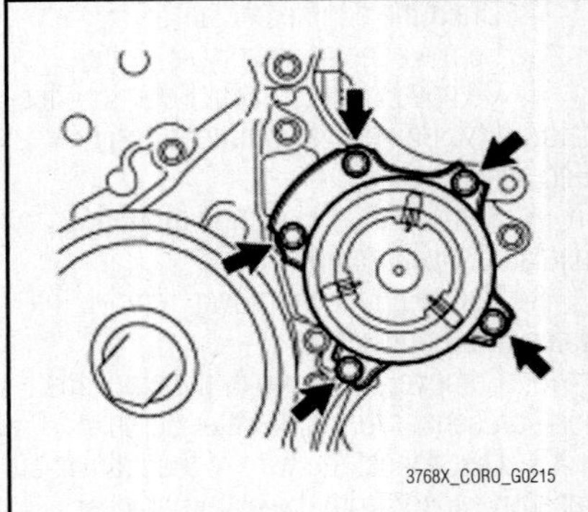

Fig. 65 Location of the 5 water pump bolts—1.8L engine

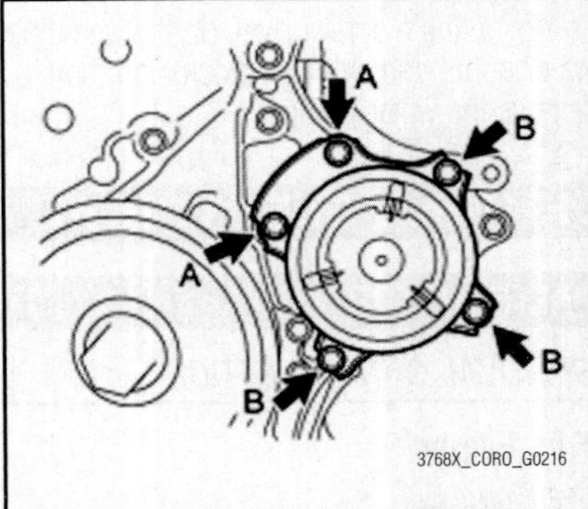

Fig. 66 Identifying the bolts "A" and "B" for installation—1.8L engine

8. Remove the water pump gasket from the timing chain cover.

To install:

9. Align the protrusion of a new water pump gasket with the cutout in the timing chain cover and install the gasket to the groove of the timing chain cover.

➡Be sure to clean the contact surfaces.

10. Install the water pump assembly to the timing chain cover with the 5 bolts. Tighten to:
    a. Bolt A (1.38 inches): 19 ft. lbs. (26 Nm).
    b. Bolt B: (0.709 inches): 18 ft. lbs. (24 Nm).
11. To complete installation, reverse remaining removal procedure.
12. Inspect for coolant leaks.

2.4L Engine

See Figure 67.

1. Disconnect the negative battery cable.
2. Remove the engine cover.
3. Remove the LH and RH engine under cover.

4. Drain the engine coolant.
5. Remove the accessory drive belt.
6. Remove the alternator. Refer to Alternator in Charging System under Engine Electrical.
7. Remove the RH engine mounting insulator sub-assembly.
8. Using a pulley remover, remove the water pump pulley.
9. Remove the clamp of the crankshaft position sensor from the water pump.
10. Disconnect the wire of the crankshaft position sensor from the clamp bracket.
11. Remove the 4 bolts, 2 nuts and clamp bracket.

➡**Tape the screwdriver tip before use.**

12. Using a screwdriver, pry between the water pump and cylinder block, and then remove the water pump.

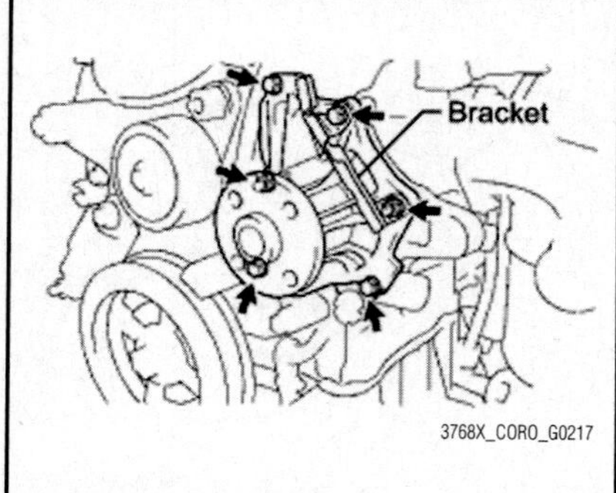

Fig. 67 Locating and identifying fasteners—2.4L engine

➡**Be careful not to damage the contact surfaces of the water pump and cylinder block.**

➡**Remove any oil from the contact surface. The parts must be set within 3 minutes after applying seal packing. Otherwise, the material must be removed and reapplied.**

To install:

13. To install, reverse the removal procedure and pay attention to the following:
   a. Apply a continuous line of seal packing. Use Toyota Genuine Seal Packing Black, Three Bond 1207B or equivalent.
   b. Tighten the water pump bolts to 80 inch lbs. (9 Nm).
   c. Tighten the water pump pulley to 19 ft. lbs. (26 Nm).

ENGINE ELECTRICAL

CHARGING SYSTEM

ALTERNATOR

REMOVAL & INSTALLATION

1.8L Engine

See Figure 68.

1. Disconnect the negative battery cable.
2. Remove the RH engine under cover.
3. Remove the accessory drive belt.
4. Remove generator assembly:
   a. Remove the terminal cap.
   b. Remove the nut and disconnect the wire harness from terminal B.

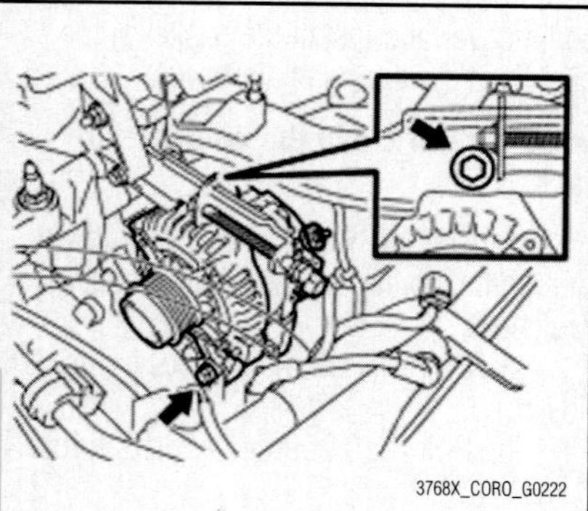

Fig. 68 Removing the alternator assembly—1.8L engine

   c. Disconnect the connector and harness clamp.
   d. Remove the 2 bolts and generator assembly.
   e. Remove the bolt and wire harness clamp bracket.

To install:

5. To install, reverse removal procedure. Pay attention to the following:
   a. Alternator upper mounting bolt: Tighten to 32 ft. lbs. (43 Nm).
   b. Alternator lower mounting bolt: Tighten to 14 ft. lbs. (19 Nm).

2.4L Engine

See Figure 69.

1. Disconnect the negative battery cable.
2. Remove the RH engine under cover.
3. Remove the accessory drive belt.
4. Remove generator assembly:
   a. Disconnect the generator connector.
   b. Remove the nut and disconnect the wire harness from terminal B.
   c. Separate the 2 wire harness clamps.

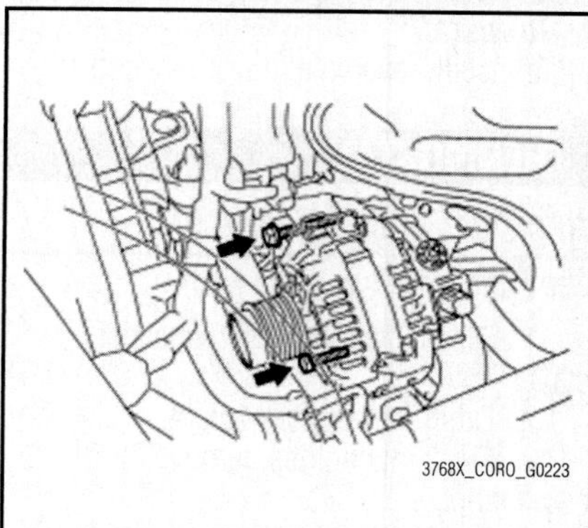

Fig. 69 Removing the alternator assembly—2.4L engine

   d. Remove the 2 bolts and generator assembly.
   e. Remove the bolt and wire harness clamp bracket.

To install:

5. To install, reverse removal procedure. Pay attention to the following:
   a. Alternator upper mounting bolt: Tighten to 16 ft. lbs. (21 Nm).
   b. Alternator lower mounting bolt: Tighten to 38 ft. lbs. (52 Nm).

FIRING ORDER

See Figure 70.

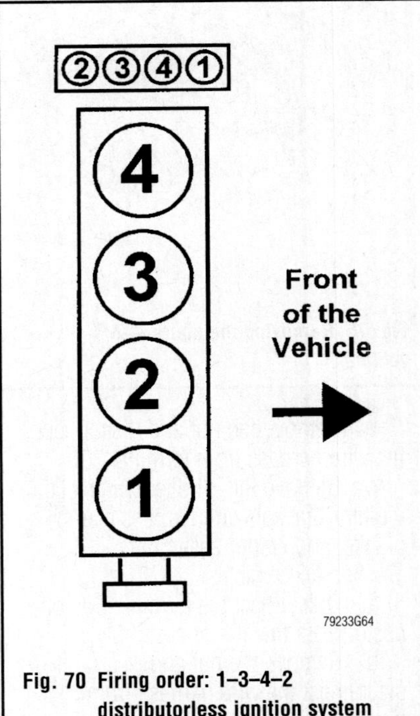

Fig. 70 Firing order: 1–3–4–2 distributorless ignition system

IGNITION COIL

REMOVAL & INSTALLATION

See Figure 71.

1. Disconnect the negative battery cable.
2. Remove the two nuts and clips and remove the cylinder head cover.
3. Disconnect the four ignition coil connectors.
4. Remove the four bolts that retain the coils.

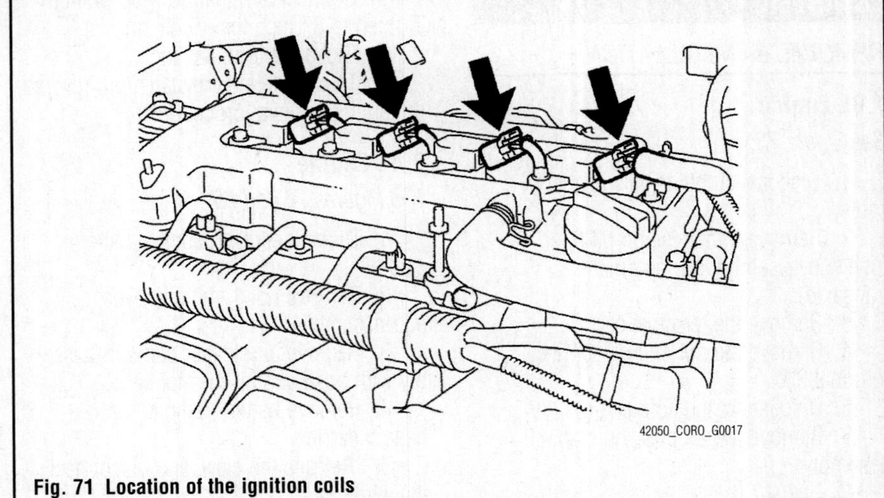

Fig. 71 Location of the ignition coils

To install:

5. Install the coils and the four bolts.
6. Connect the coil connections.
7. Install the cylinder head cover.

IGNITION TIMING

ADJUSTMENT

The ignition timing is controlled by the Powertrain Control Module (PCM). No adjustment is necessary or possible.

SPARK PLUGS

REMOVAL & INSTALLATION

See Figure 72.

1. Disconnect the negative battery cable.
2. Remove the ignition coil. Refer to Ignition Coil in this section.
3. Remove the 4 spark plugs.

To install:

4. To install, use a 14 mm spark plug wrench, install the 4 spark plugs and tighten to 15 ft. lbs. (20 Nm).
5. To complete installation, reverse remaining removal procedure.

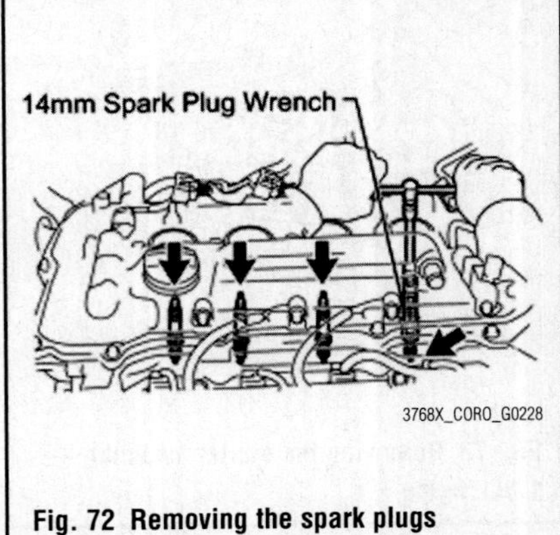

Fig. 72 Removing the spark plugs

STARTER

REMOVAL & INSTALLATION

1.8L Engine

See Figure 73.

1. Disconnect the negative battery cable.
2. Remove transmission oil filler tube sub-assembly (for automatic transaxle).
3. Remove the terminal cap.
4. Remove the nut and disconnect terminal 30.
5. Disconnect the connector.
6. Remove the 2 bolts and starter assembly.

To install:

7. Install the starter assembly with the 2 bolts and tighten to 27 ft. lbs. (37 Nm).
8. Connect the connector.

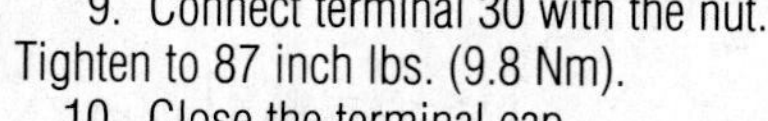

9. Connect terminal 30 with the nut. Tighten to 87 inch lbs. (9.8 Nm).
10. Close the terminal cap.
11. To complete the installation, reverse remaining removal procedure.

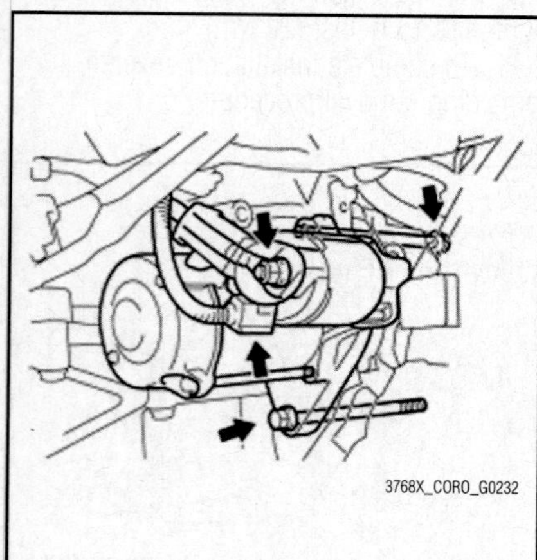

Fig. 73 Removing the starter assembly—1.8L Engine

2.4L Engine

See Figures 74 and 75.

1. Disconnect the negative battery cable.
2. Remove no. 1 engine cover sub-assembly.
3. Remove air cleaner cap sub-assembly with hose and air cleaner case.
4. Remove battery remove battery carrier.
5. Release the claw, and disconnect the wire harness.
6. For M/T vehicles, perform the following:
   a. Disconnect the terminal 50 connector from the starter assembly.

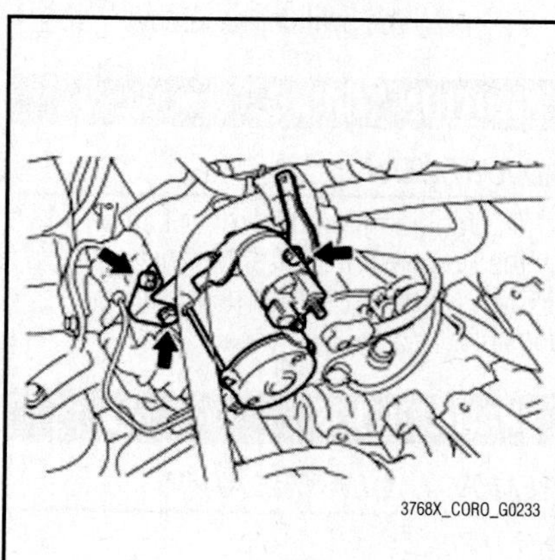

Fig. 74 Removing the starter—M/T vehicles

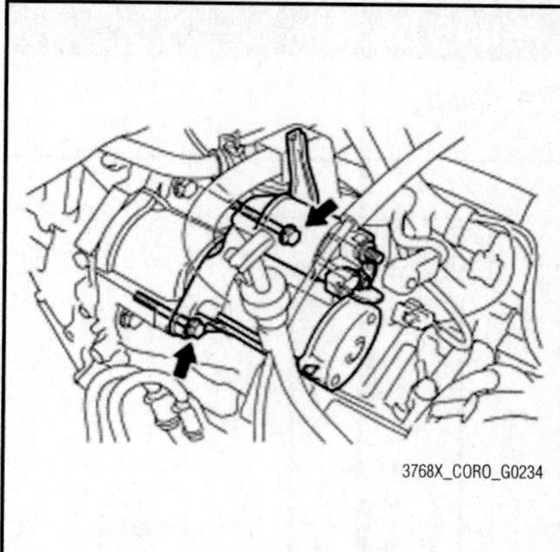

Fig. 75 Removing the starter—A/T vehicles

   b. Remove the nut and disconnect the wire harness from terminal 30.
   c. Remove the 3 bolts, clutch accumulator bracket, wire harness clamp bracket and starter assembly.
7. For A/T vehicles:
   a. Disconnect the terminal 50 connector from the starter assembly.
   b. Remove the nut and disconnect the wire harness from terminal 30.
   c. Remove the 2 bolts, wire harness clamp bracket and starter assembly.

To install:

8. To install, reverse removal procedure.
9. For M/T vehicles, tighten the starter bolts to:
   a. Bolt A: 27 ft. lbs. (37 Nm).
   b. Bolt B: 9 ft. lbs. (12 Nm).
10. For A/T vehicles, tighten the bolts to 27 ft. lbs. (37 Nm).

ENGINE MECHANICAL

ACCESSORY DRIVE BELTS

ACCESSORY BELT ROUTING

See Figures 76 and 77.

w/ Air Conditioning System:

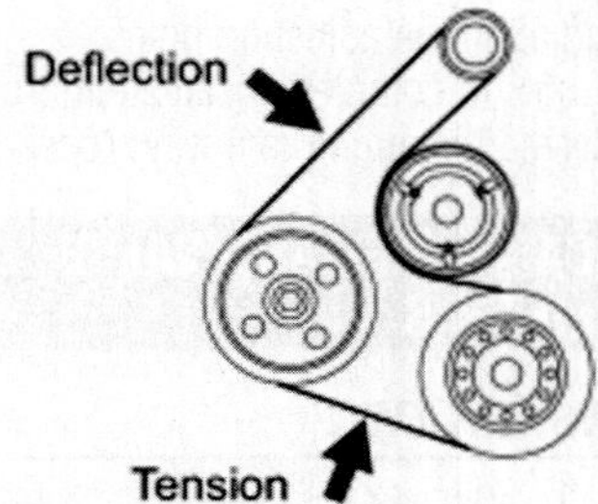

w/o Air Conditioning System:

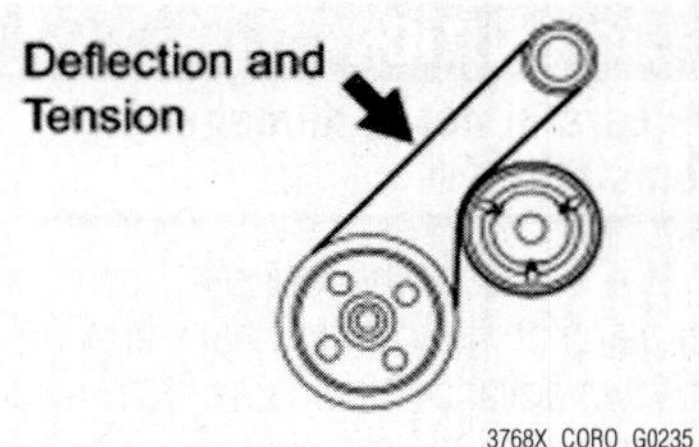

3768X_CORO_G0235

Fig. 76 Accessory belt routing—1.8L Engine

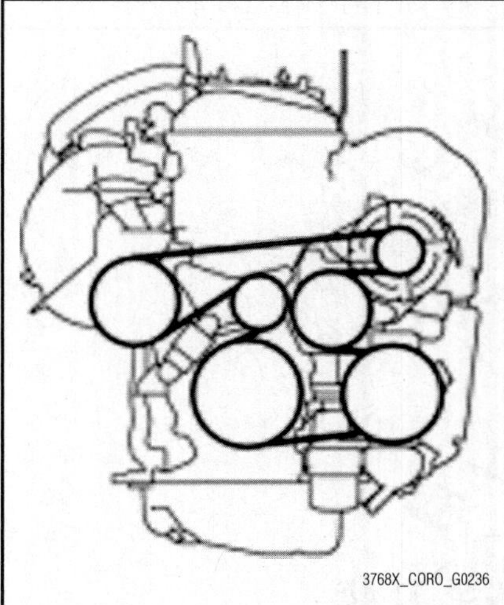

3768X_CORO_G0236

Fig. 77 Accessory belt routing—2.4L Engine

INSPECTION

Inspect the drive belt for signs of glazing or cracking. A glazed belt will be perfectly smooth from slippage, while a good belt will have a slight texture of fabric visible. Cracks will usually start at the inner edge of the belt and run outward. All worn or damaged drive belts should be replaced immediately.

ADJUSTMENT

Only the 1.8L engine belt requires adjustment. Refer to the removal procedure for adjustment procedure.

REMOVAL & INSTALLATION

1.8L Engine

See Figure 78.

➡Refer to the illustration for bolt identification.

1. Remove the engine cover.
2. Remove the RH engine under cover.
3. Loosen bolts A and B.
4. Loosen bolt C, then remove the V-ribbed belt.

➡Do not loosen bolt D.

To install:

5. Install the belt.
6. Turn bolt C to adjust the tension of the V-ribbed belt.
7. Tighten bolts:
   a. A: 14 ft. lbs. (19 Nm).
   b. B: 32 ft. lbs. (43 Nm).

➡Confirm that bolt D is not loosened.

8. Inspect the belt.
9. To complete installation, reverse remaining removal procedure.

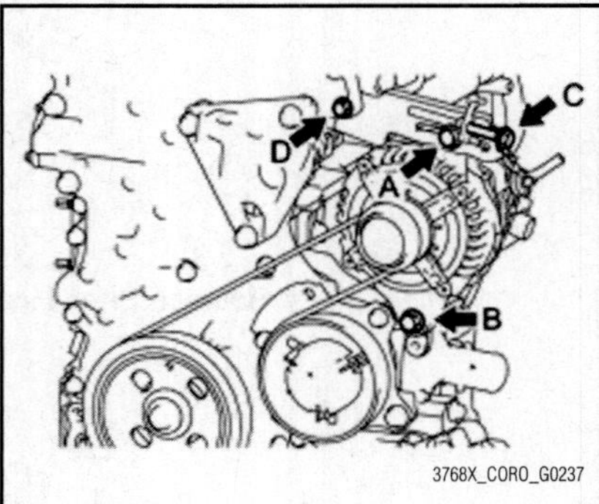

3768X_CORO_G0237

Fig. 78 Removing the accessory belt—1.8L engine

2.4L Engine

See Figure 79.

➡Adjustment is not possible or necessary with the auto-tensioner.

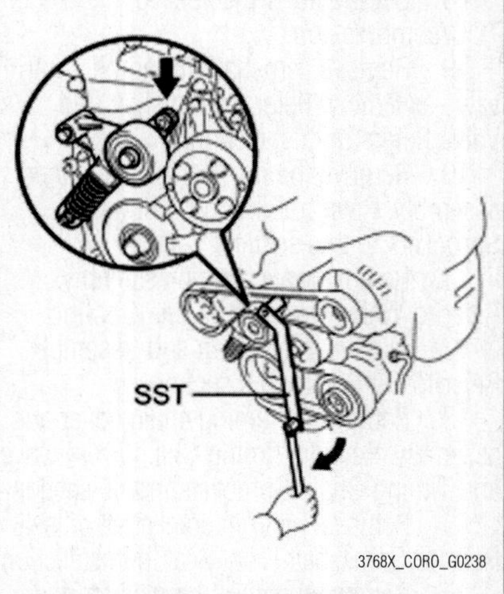

3768X_CORO_G0238

Fig. 79 Removing the serpentine belt—2.4L engine

1. Remove the right hand cover under the engine.
2. Turn the drive belt tensioner clockwise to relieve tension on the belt.
3. Remove the fan and generator V belt.
4. Return the tensioner to the unloaded position.

➡When retracting the tensioner, turn it clockwise slowly in 3 sec. or more. Be sure not to apply force rapidly. After the tensioner is retracted all the way, do not apply force any more than necessary.

To install:

5. Turn the drive belt tensioner clockwise.
6. Install the belt.
7. Install the right hand cover under the engine.

BALANCE SHAFT

REMOVAL & INSTALLATION

2.4L Engine

See Figures 80 through 87.

1. Remove the valve cover.
2. Using a 22 mm deep socket wrench, remove the ventilation valve from the cylinder head cover.
3. Remove the spark plugs.
4. Remove the oil filter and union.
5. Remove the drive belt.
6. Remove the crankshaft position sensor. Refer to Crankshaft Position Sensor under Engine Performance & Emission Controls.
7. Remove the bolt and camshaft position sensor. Refer to Camshaft Position

Sensor under Engine Performance & Emission Controls.

8. Set the no. 1 cylinder to TDC/compression.

9. Remove camshaft timing oil control valve assembly. Refer to Camshaft and Valve Lifters in this section.

10. Remove the no. 1 chain tensioner assembly. Refer to Timing Chain and Sprockets in this section.

11. Remove water pump assembly. Refer to Water Pump in Engine Cooling.

12. Remove the oil pan sub-assembly. Refer to Oil Pan in this section.

13. Remove the timing chain cover and assembly. Refer to Timing Chain Front Cover and Timing Chain Sprockets in this section.

14. Remove the front crankshaft oil seal. Refer to Crankshaft Front Seal in this section.

15. Remove the no. 1 crankshaft position sensor plate. Refer to Crankshaft in this section.

16. Remove crankshaft timing gear or sprocket. Refer to Crankshaft in this section.

17. Remove the camshaft assembly. Refer to Camshaft and Valve Lifters in this section.

18. Remove the cylinder head sub-assembly. Refer to Cylinder Head in this section.

19. Using an 8mm socket hexagon wrench, remove the plug, gasket and oil control valve filter.

20. Remove oil pump assembly. Refer to Oil Pump in this section.

21. Inspect the balance shaft thrust clearance:

a. Using a dial indicator, measure the thrust clearance while moving the balance shaft back and forth.

b. Standard thrust clearance: 0.05 to 0.09 mm (0.0020 to 0.0035 in.)

c. Maximum thrust clearance: 0.09 mm (0.0035 in.)

d. If the thrust clearance is greater than the maximum, replace the balance shaft housing and bearings. If necessary, replace the balance shaft.

22. Inspect the balance shaft oil clearance:

a. Clean each bearing and journal.

b. Check each bearing and journal for pitting and scratches.

c. If a bearing or journal is damaged, replace the bearings. If necessary, replace the balance shaft.

23. Remove the No. 1 and No. 2 balance shafts.

24. Remove the eight No. 1 balance shaft bearings shown in the illustration.

To install:

➡Do not apply engine oil to the contact surfaces of the balance shaft bearing and balance shaft housing.

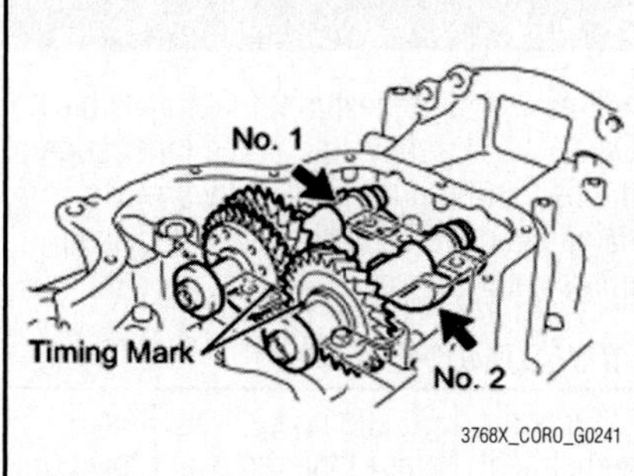

Fig. 80 Removing the No. 1 and No. 2 balance shafts

25. Align the bearing claw with the claw groove, and push in the 8 bearings.

26. Apply a light coat of the engine oil to the bearings.

27. Rotate the No. 1 driven gear of the No. 1 balance shaft in the rotating direction until it hits the stopper.

➡Confirm that the match marks on the No. 1 and No. 2 driven gears are matched.

28. Confirm that the alignment marks on the No. 1 and No. 2 balance shafts are aligned.

29. Align the alignment marks on the No. 1 and No. 2 balance shafts.

30. Place the No. 1 and No. 2 balance shafts onto the crankcase.

31. Apply a light coat of engine oil to the threads and under the heads of the bolts.

32. Using several steps, uniformly install and tighten the 8 bolts in the sequence. Tighten to 16 ft. lbs. (22Nm). Mark the front of the bolts with paint for reference and tighten the bolts an extra 90°.

33. Check that the paint mark is now at a 90° angle to the front.

34. Place a new O-ring on the cylinder block.

35. With the No. 1 crank pin of the crankshaft placed at the TDC position, install the No. 1 and No. 2 balance shafts and align the adjusting holes.

36. To complete installation procedure, reverse remaining removal procedure.

CAMSHAFT AND VALVE LIFTERS

INSPECTION

1. To check the lock of the camshaft timing gear, clamp the camshaft in a vice and confirm that the camshaft timing gear is locked.

CAUTION

Be careful not to damage the camshaft.

2. To release the lock pin, cover the four oil paths on the cam journal with vinyl tape. The two advance side paths are provided in the groove of the camshaft. Plug one of the paths with a piece of rubber.

3. Puncture the tape over the advance side path and retard the side path on the opposite side of the groove.

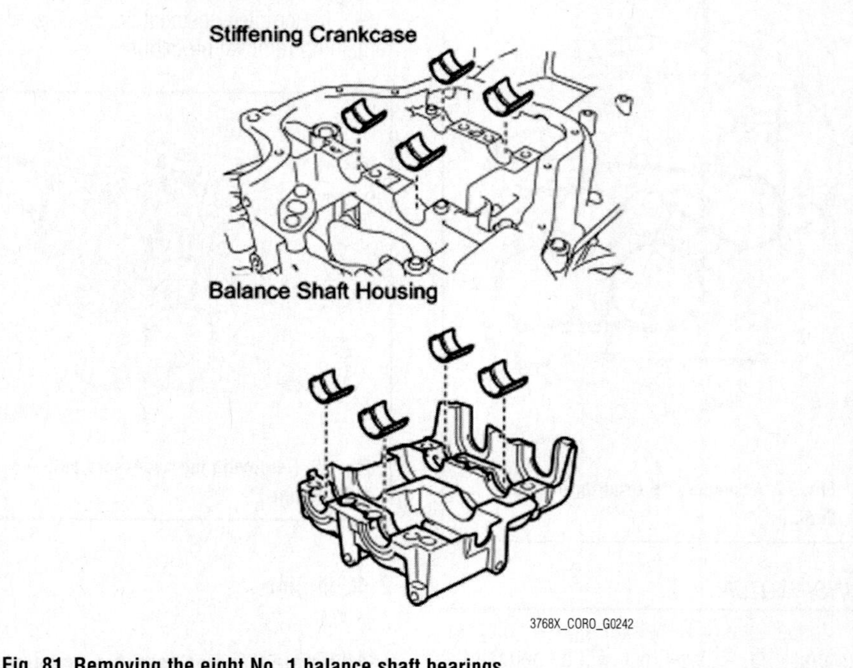

Fig. 81 Removing the eight No. 1 balance shaft bearings

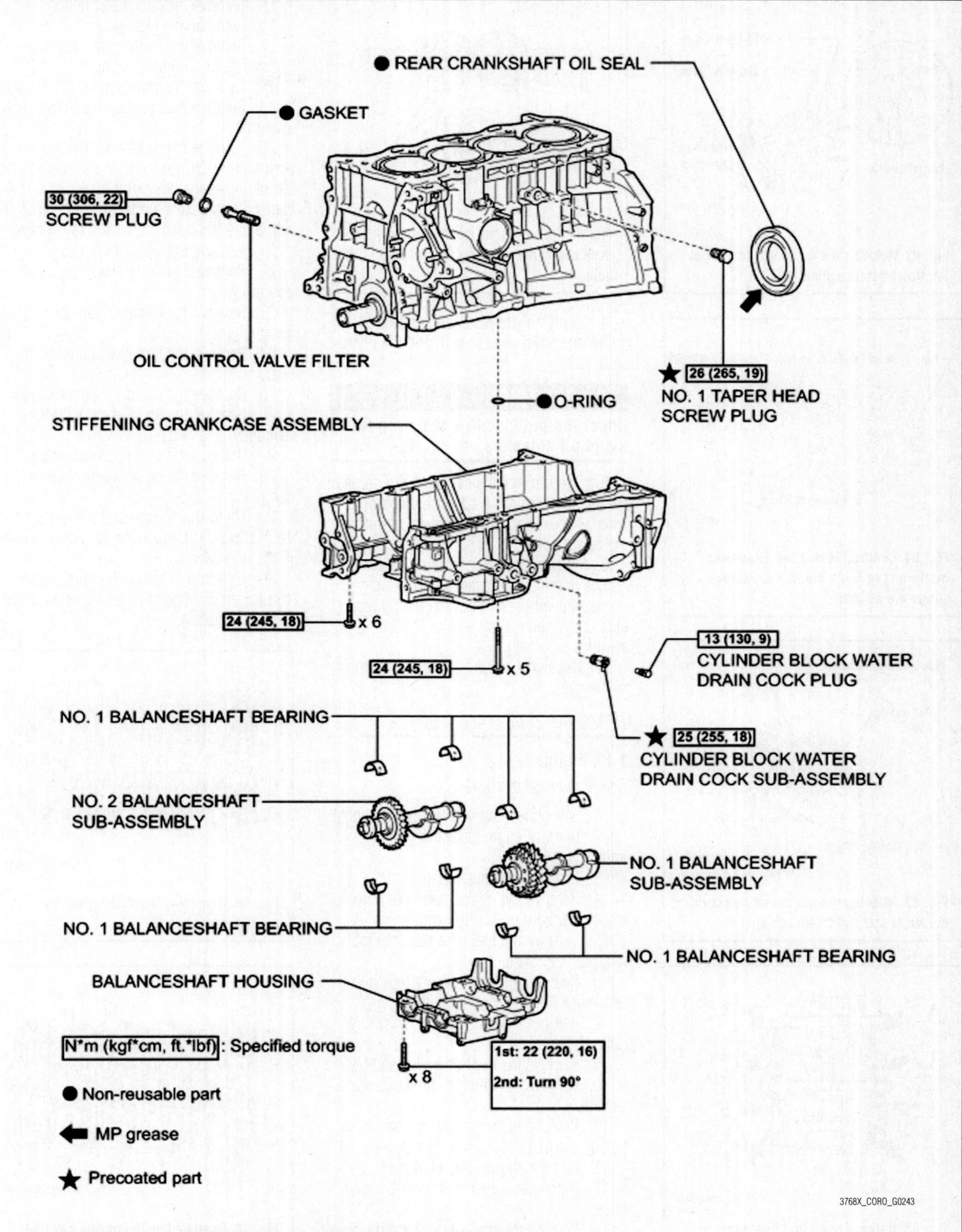

Fig. 82 Exploded view of the balance shaft and related components—2.4L Engine

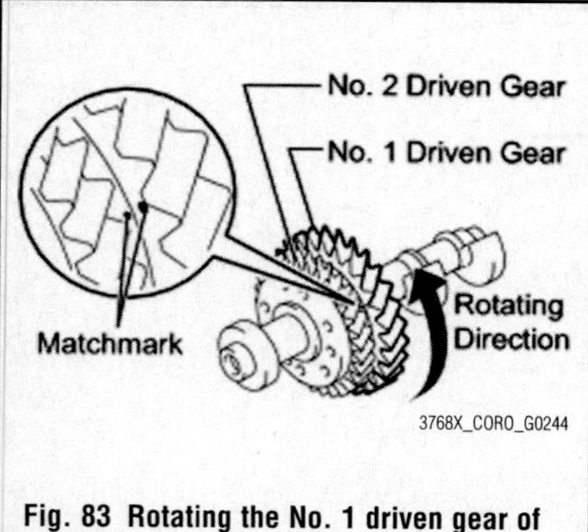

Fig. 83 Rotating the No. 1 driven gear of the No. 1 balance shaft

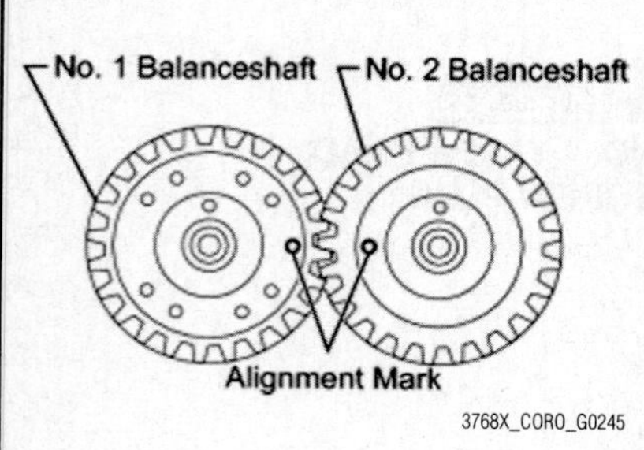

Fig. 84 Confirming that the alignment marks on the No. 1 and No. 2 balance shafts are aligned

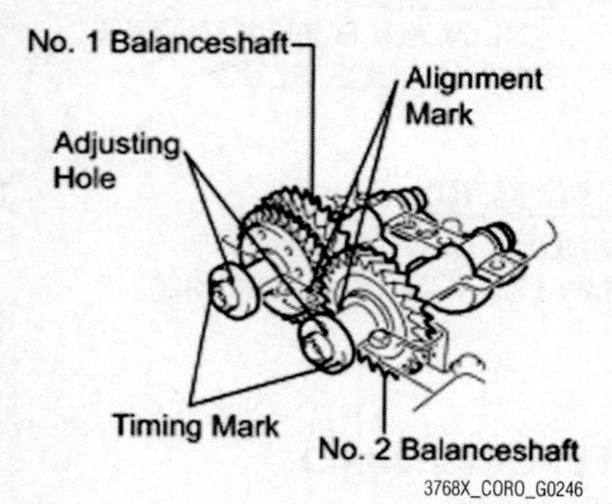

Fig. 85 Aligning the alignment marks on the No. 1 and No. 2 balance shafts

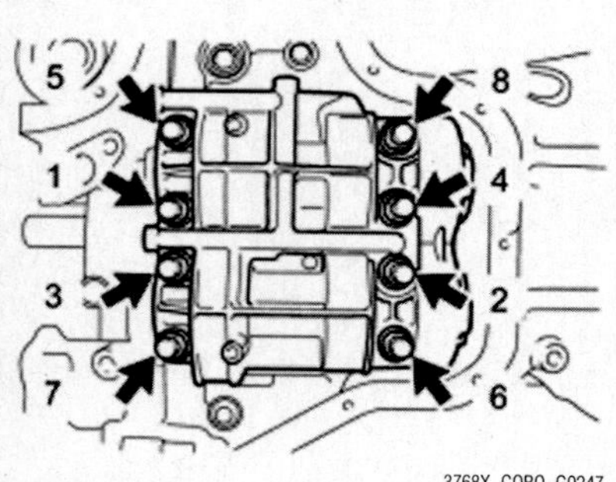

Fig. 86 Balance shaft bolt tightening sequence

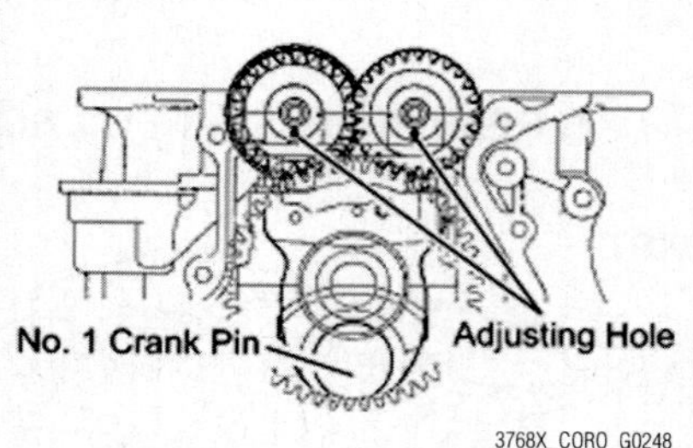

Fig. 87 Installing the No. 1 and No. 2 balance shafts and aligning the adjusting holes

4. Apply air pressure to the two broken paths (the advance side path and the retard side path).

WARNING

Cover the paths with a shop rag to avoid oil splashes.

5. Confirm that the camshaft timing gear revolves in the timing advance direction when weakening the air pressure of the timing retard path.
6. When the camshaft timing gear comes to the most advanced position, release the air pressure of the timing retard side path, then release that of the timing advance side path.
7. There should be a smooth revolution of the gear.

REMOVAL & INSTALLATION

1.8L Engine

See Figures 88 through 95.

1. Disconnect the negative battery cable.
2. Remove engine assembly with transaxle.
3. Install the engine stand.
4. Remove the intake manifold. Refer to Intake Manifold in this section.
5. Remove the fuel tube sub-assembly. Refer to Fuel System.
6. Remove the fuel delivery pipe sub-assembly. Refer to Fuel System.
7. Remove the fuel injector assembly. Refer to Fuel System.
8. Remove the ignition coil assembly.
9. Remove the oil level dipstick sub-assembly.
10. Remove the exhaust manifold. Refer to Exhaust Manifold in this section.
11. Remove the ventilation hose.
12. Disconnect the no. 3 water by-pass hose.
13. Remove the no. 1 water by-pass pipe.
14. Remove the water by-pass hose.
15. Remove the inlet water hose.
16. Remove the inlet water.
17. Remove the thermostat. Refer to Thermostat in Engine Cooling.
18. Remove the radio setting condenser.
19. Remove the cylinder head cover sub-assembly.
20. Turn the crankshaft so that the No. 1 piston is at TDC on the compression stroke. Check to see that the point marks on the camshaft sprockets are facing each other, if not, rotate the crankshaft 1 full revolution.
21. Remove the crankshaft pulley.
22. Remove the no. 1 chain tensioner assembly.
23. Remove the timing chain cover sub-assembly.
24. Remove the timing chain cover oil seal.
25. Remove the chain tensioner slipper.
26. Remove the no. 1 chain vibration damper.
27. Remove the chain sub-assembly.
28. Remove the no. 2 chain vibration damper.
29. Uniformly loosen and remove the 10 bearing cap bolts in the sequence shown in the illustration.
30. Uniformly loosen and remove the 15 bearing cap bolts in the sequence shown in the illustration.

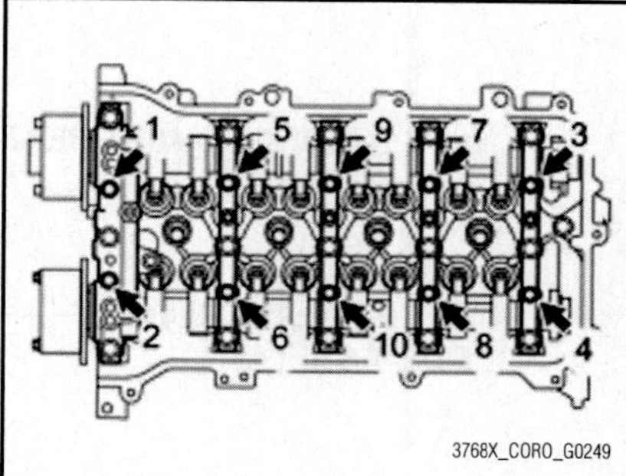

Fig. 88 Removing the 10 bearing cap bolts in sequence

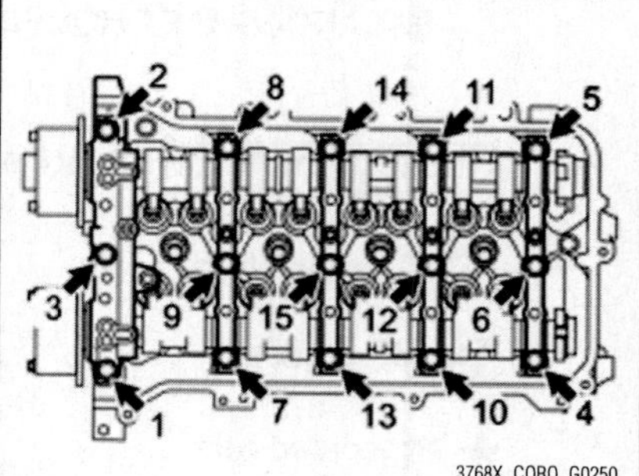

Fig. 89 Remove the 15 bearing cap bolts in sequence

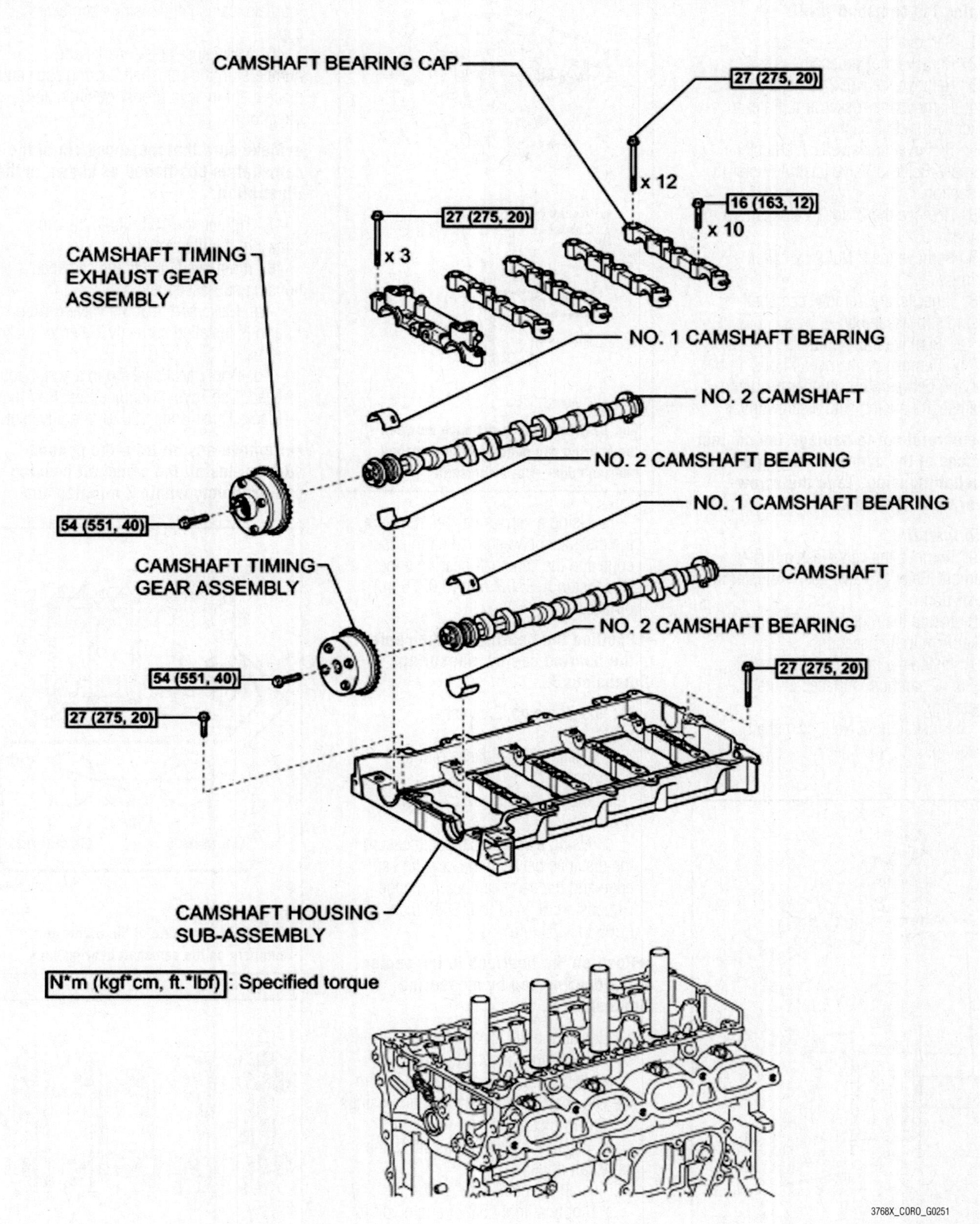

Fig. 90 Exploded view of the camshaft assembly and related components—1.8L engine

➡Uniformly loosen the bolts while keeping the camshaft level.

31. Remove the 5 bearing caps.
32. Remove the camshaft.
33. Remove the No. 2 camshaft.
34. Remove the rocker arm. Refer to Rocker Arm in this section.
35. Remove the valve lash adjuster assembly. Refer to Valve Lash Adjuster in this section.
36. Remove the 2 No. 1 camshaft bearings.
37. Remove the 2 No. 2 camshaft bearings.
38. If necessary, remove camshaft housing sub-assembly:
    a. Remove the 2 bolts.
    b. Remove the camshaft housing by prying between the cylinder head and camshaft housing with a screwdriver.

➡Be careful not to damage the contact surfaces of the cylinder head and camshaft housing. Tape the screw-driver tip before use.

To install:

39. Remove the valve lash adjuster assembly. Refer to Valve Lash Adjuster in this section.
40. Install the rocker arm. Refer to Rocker Arm in this section.
41. Install the no. 1 camshaft bearing:
    a. Clean both surfaces of the bearings.
    b. Install the 2 No. 1 camshaft bearings.

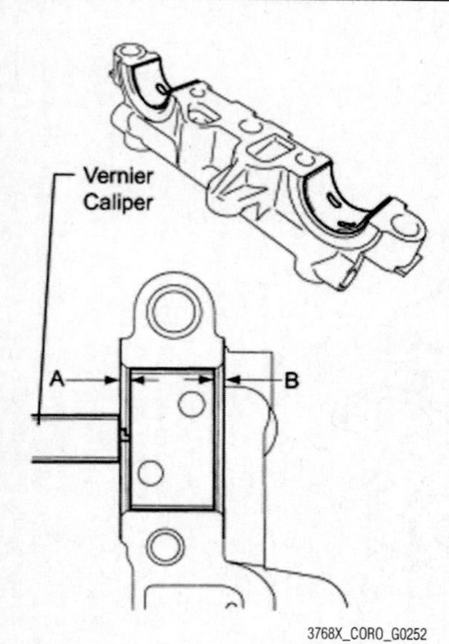

Fig. 91 Measure the distance between the bearing cap edge and the camshaft bearing edge—No. 1 camshaft

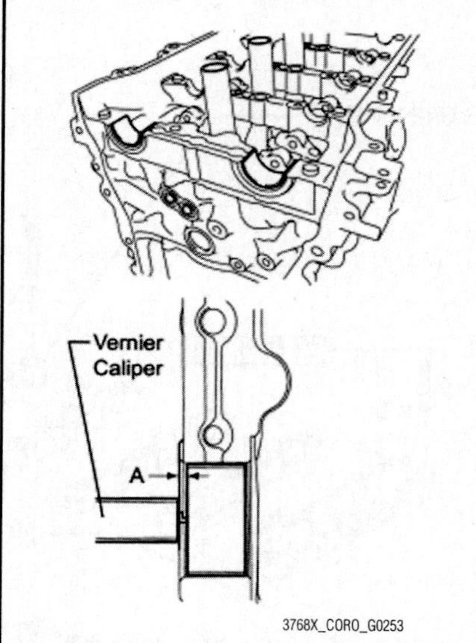

Fig. 92 Measure the distance between the bearing cap edge and the camshaft bearing edge—No. 2 camshaft

    c. Using a vernier caliper, measure the distance between the bearing cap edge and the camshaft bearing edge. Dimension A-B: 0.7 mm (0.0276 in.) or less.

➡Position the bearings to the center of the bearing cap by measuring dimensions A.

42. Install the no. 2 camshaft bearing:
    a. Clean both surfaces of the bearings.
    b. Install the 2 No. 2 camshaft bearings.
    c. Using a vernier caliper, measure the distance between the bearing cap edge and the camshaft bearing edge. Dimension: 0.0413 to 0.0689 inch (1.05 to 1.75 mm)

➡Position the bearings to the center of the bearing cap by measuring dimensions A.

43. Install the no. 2 camshaft:
    a. Clean the camshaft journals.
    b. Apply a light coat of engine oil to the camshaft journals, camshaft housings and bearing caps.
    c. Install the No. 2 camshaft to the camshaft housing.
44. Install the camshaft:
    a. Apply a light coat of engine oil to the camshaft journals, camshaft housings and bearing caps.
    b. Install the camshaft to the camshaft housing.
45. Apply engine oil to the camshaft journals, camshaft housings and bearing caps.
46. Make sure of the marks and numbers on the camshaft bearing caps and place them in each proper position and direction.

➡Make sure that the knock pin of the camshaft is positioned as shown in the illustration.

47. Tighten the 10 bolts in the order shown in the illustration.
48. If removed, install the camshaft housing sub-assembly:
    a. Make sure that the valve rocker arm is installed correctly. Refer to Rocker Arms.
    b. Apply seal packing in a continuous bead. Use Toyota Genuine Seal Packing Black, Three Bond 1207B or equivalent.

➡Remove any oil from the contact surface. Install the camshaft housing sub-assembly within 3 minutes and

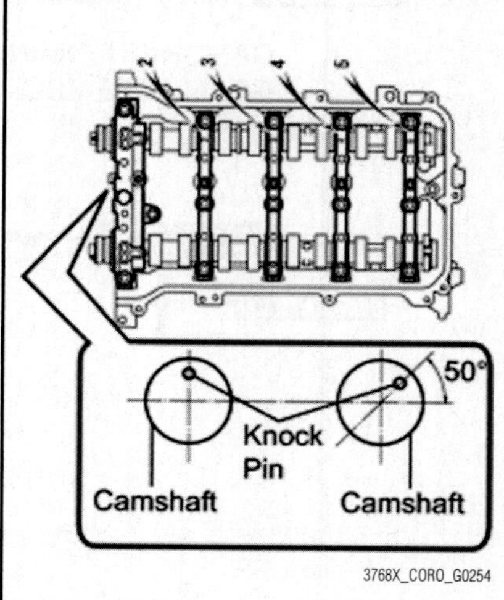

Fig. 93 Making sure of the marks and numbers on the camshaft bearing caps

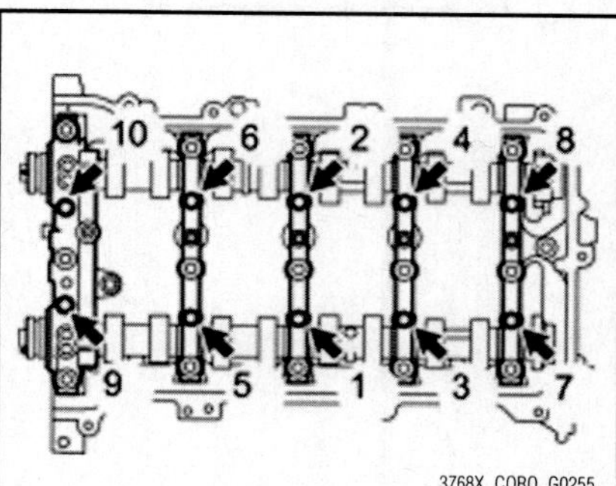

Fig. 94 Bearing cap 10 bolt tightening sequence

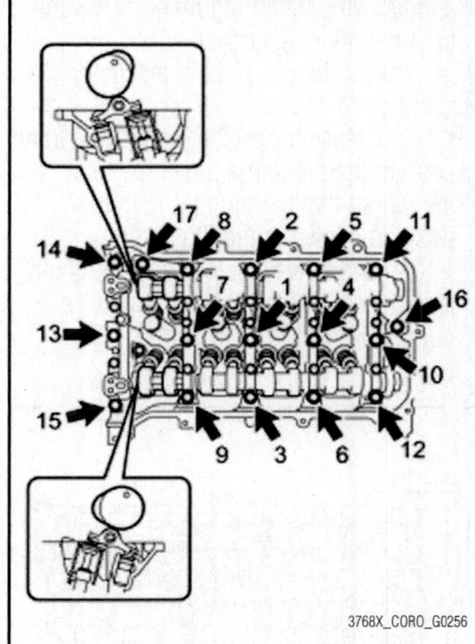

Fig. 95 Camshaft housing 17 bolt tightening sequence

tighten the bolts within 15 minutes after applying seal packing. Do not start the engine for at least 2 hours after installing.

49. Set the camshaft and No. 2 camshaft as shown in the illustration.

50. Install the camshaft housing and tighten the 17 bolts in the order shown in the illustration. Tighten to 20 ft. lbs. (27 Nm).

➡ **After installing the camshaft housing, make sure that the cam lobes are positioned as shown in the illustration.**

➡ **If any of the bolts are loosened during installation, remove the camshaft housing, clean the installation surfaces, and reapply seal packing.**

➡ **If the camshaft housing is removed because any of the bolts are loosened during installation, make sure that the previously applied seal packing does not enter any oil passages.**

➡ **After installing the camshaft housing, wipe off any seal packing that seeped out from between the housing and the cylinder head.**

51. To complete the installation, reverse remaining removal procedure.

2.4L Engine

See Figures 96 through 102.

1. Disconnect the negative battery cable.

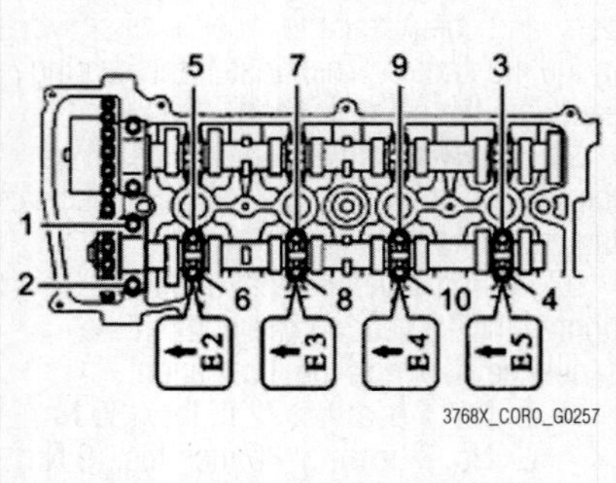

Fig. 96 10 bearing cap bolts removal sequence and identifying the 5 bearing caps

2. Remove the RH engine under cover.
3. Remove the No. 1 engine cover sub-assembly.
4. Remove the ignition coil assembly. Refer to Ignition Coil in Engine Electrical.
5. Remove the spark plugs. Refer to Spark Plugs in Engine Electrical.
6. Remove the valve cover.
7. Set no. 1 cylinder to TDC/compression:
   a. Turn the crankshaft pulley until the groove and the timing mark "0" on the timing chain cover are aligned.

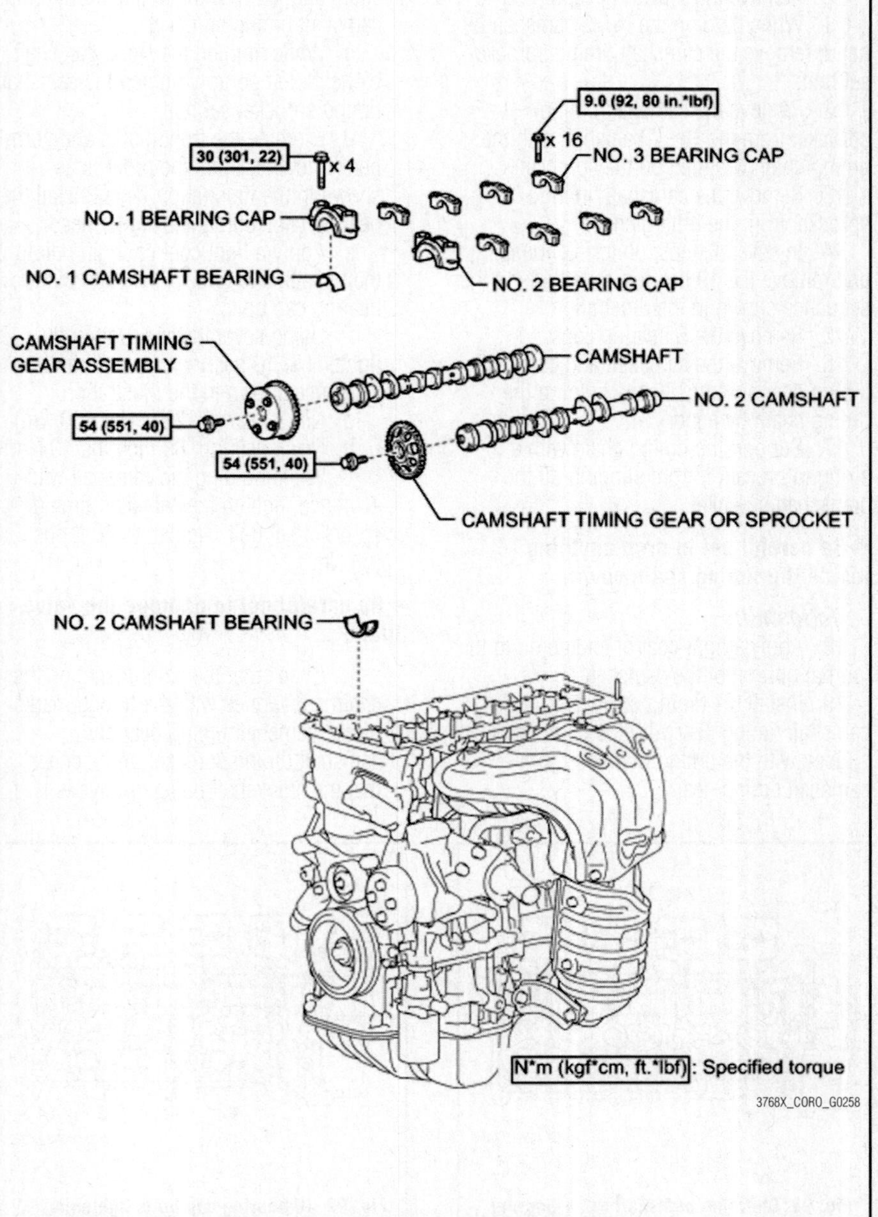

Fig. 97 Exploded view of the camshaft assembly and related components

b. Check that each timing mark on the camshaft timing gear and sprocket is aligned with each timing mark located on the No. 1 and No. 2 bearing caps.

c. If not, turn the crankshaft pulley 1 revolution (360°) to align the timing marks.

d. Place paint marks on the chain in alignment with the timing marks on the camshaft timing gear and camshaft timing sprocket.

8. While holding the No. 2 camshaft with a wrench, loosen the No. 2 camshaft timing set bolt.

9. Using several steps, uniformly loosen and remove the 10 bearing cap bolts in the sequence shown in the illustration.

10. Remove the 5 bearing caps.

11. While holding the No. 2 camshaft by hand, remove the camshaft timing sprocket set bolt.

12. Remove the camshaft timing sprocket from the No. 2 camshaft with the timing chain wrapped on the sprocket.

13. Remove the camshaft timing sprocket from the timing chain.

14. In several steps, uniformly loosen and remove the 10 bearing caps bolts in the sequence shown in the illustration.

15. Remove the 5 bearing caps.

16. Remove the camshaft and camshaft timing gear assembly while holding the timing chain by hand.

17. Support the timing chain with a string to prevent it from slipping off the crankshaft sprocket.

➡Be careful not to drop anything inside the timing chain cover.

To install:

18. Apply a light coat of engine oil to the journal portion of the camshaft.

19. Install the timing chain onto the camshaft timing gear with the paint mark aligned with the timing mark on the camshaft timing gear.

20. Examine the front marks and numbers, and check that the order is as shown in the illustration. Then install the bearing caps into the cylinder head.

21. Apply a light coat of engine oil on the threads and under the heads of the bearing cap bolts.

22. Using several steps, uniformly tighten the 10 bearing cap bolts in the sequence shown in the illustration.

a. No. 1 bearing: 22 ft. lbs. (30 Nm).

b. No. 3 bearing: 80 inch lbs. (9 Nm).

23. Install the no. 2 camshaft:

a. Apply a light coat of engine oil to the journal portion of the No. 2 camshaft.

b. Put the No. 2 camshaft on the cylinder head with the paint mark on the chain aligned with the timing mark on the camshaft timing sprocket.

c. While holding the No. 2 camshaft by hand, temporarily tighten the camshaft timing sprocket set bolt.

d. Examine the front marks and numbers, and check that the order is as shown in the illustration. Then install the bearing caps onto the cylinder head.

e. Apply a light coat of engine oil to the threads and under the heads of the bearing cap bolts.

f. Using several steps, uniformly tighten the 10 bearing cap bolts in the sequence shown in the illustration.

g. No. 2 bearing: 22 ft. lbs. (30 Nm).

h. No. 3 bearing: 80 inch lbs. (9 Nm).

i. While holding the camshaft with a wrench, tighten the camshaft timing sprocket set bolt. Tighten to 40 ft. lbs. (54 Nm.)

➡Be careful not to damage the valve lifter.

j. Check that the paint marks on the chain are aligned with the timing marks on the camshaft timing gear and camshaft timing sprocket. Also, check that the crankshaft pulley groove is aligned with the timing mark "0" on the timing mark chain cover.

24. Install the no. 1 chain tensioner assembly:

a. Release the ratchet pawl, then fully push in the plunger and set the hook to the pin so that the plunger is in the position shown in the illustration.

b. Install a new gasket and the chain tensioner with the 2 nuts. Tighten the nuts to 80 inch. lbs. (9 Nm).

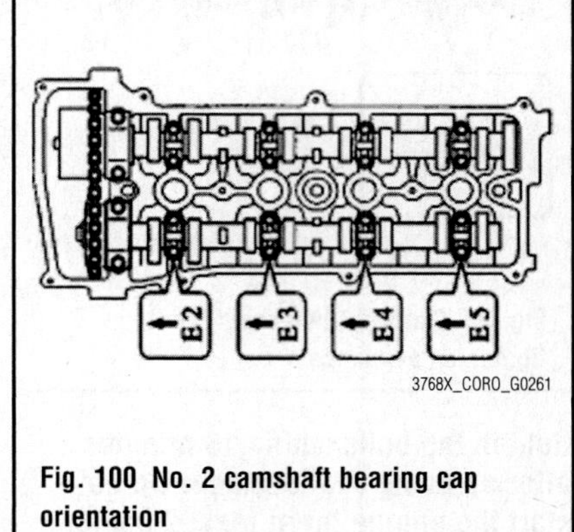

Fig. 100 No. 2 camshaft bearing cap orientation

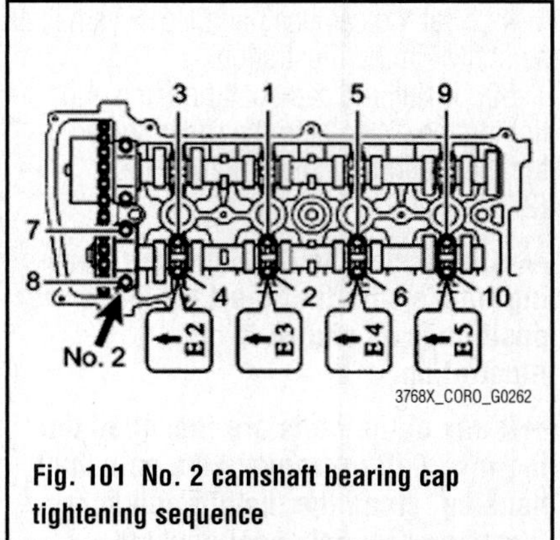

Fig. 101 No. 2 camshaft bearing cap tightening sequence

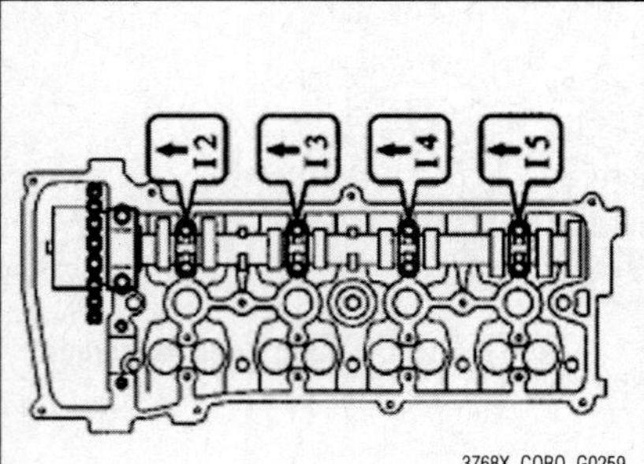

Fig. 98 Checking camshaft No. 1 bearing cap orientation

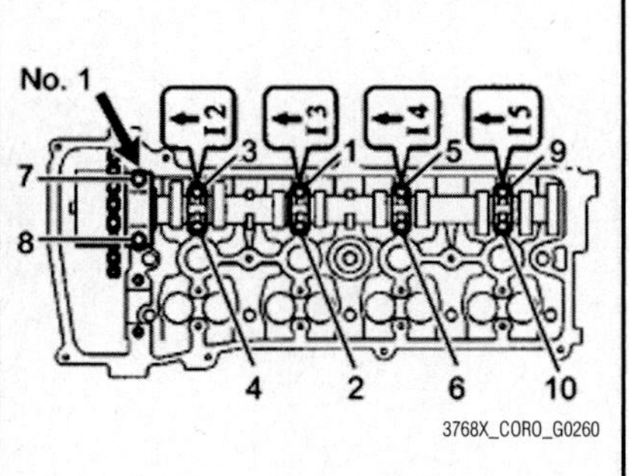

Fig. 99 10 bearing cap bolts tightening sequence

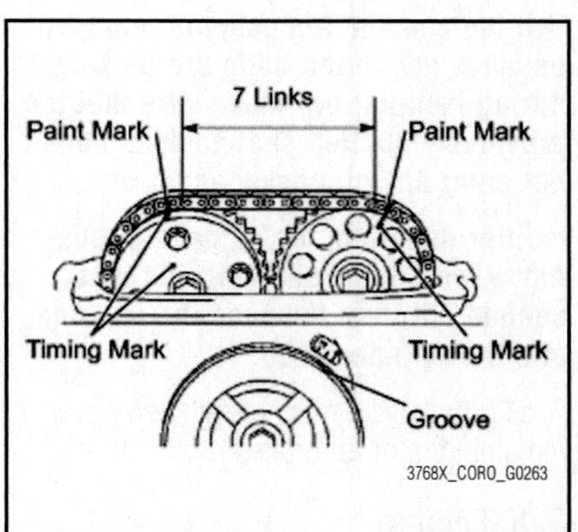

Fig. 102 Check that the paint marks on the chain are aligned with the timing marks on the camshaft timing gear and camshaft timing sprocket

➡When installing the chain tensioner, set the hook again if the hook releases the plunger.

25. Turn the crankshaft counterclockwise, then disconnect the hook from the pin.
26. Turn the crankshaft clockwise, then check that the plunger is extended.
27. Set no. 1 cylinder to TDC/compression:
    a. Turn the crankshaft pulley until the groove and the timing mark "0" on the timing chain cover are aligned.
    b. Check that each timing mark on the camshaft timing gear and sprocket is aligned with each timing mark located on the No. 1 and No. 2 bearing caps.
    c. If not, turn the crankshaft pulley 1 revolution (360°) to align the timing marks.
    d. Place paint marks on the chain in alignment with the timing marks on the camshaft timing gear and camshaft timing sprocket.
28. Check valve clearance.
29. Adjust valve clearance.
30. To complete installation, reverse remaining removal procedure.

CRANKSHAFT DAMPER

REMOVAL & INSTALLATION

See Figures 102 through 105.

1. Disconnect the negative battery cable.
2. Remove the RH front wheel.
3. Remove the RH engine under cover.
4. Remove the engine cover.
5. Remove the accessory drive belt.
6. Using a pulley bolt remover (SST's 09213-58013 and 09330-00021), hold the pulley in place and loosen the pulley bolt.

➡Check the SST installation positions when installing them to prevent the SST fixing bolts from coming into contact with the timing chain cover sub-assembly.

7. Using a pulley puller (SST 09950-50013), remove the crankshaft pulley and pulley bolt.
8. If necessary, remove timing chain cover front oil seal:
    a. Using a knife, cut off the lip of the oil seal.
    b. Using a screwdriver with its tip wrapped with tape, pry out the oil seal.

➡After removing, check the crankshaft for damage. If damaged, smooth the surface with 400-grit sandpaper.

To install:

➡Keep the lip free of foreign matter.

9. Apply MP grease to the lip of a new oil seal.
10. Using a seal installer (SST 09223-22010) and a hammer, tap in the oil seal until its surface is flush with the rear oil seal retainer edge.

➡Wipe off extra grease from the crankshaft.

11. Align the pulley set key with the key groove of the pulley.
12. Using a pulley bolt remover (SST's 09213-58013 and 09330-00021), hold the pulley in place and tighten the pulley bolt to 140 ft. lbs. (190 Nm).
13. To complete installation, reverse remaining removal procedure.

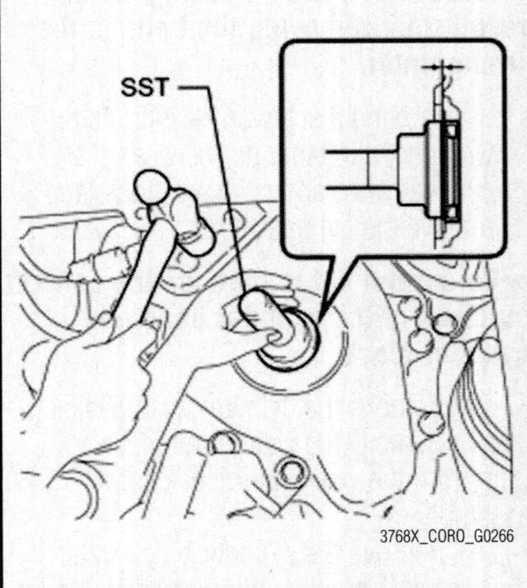

Fig. 105 Installing the crankshaft front seal

CRANKSHAFT FRONT SEAL

REMOVAL & INSTALLATION

Refer to Crankshaft Damper to remove the crankshaft front seal.

CYLINDER HEAD

REMOVAL & INSTALLATION

1.8L Engine

See Figures 106 through 109.

1. Remove the camshaft assembly. Refer to Camshaft in this section.
2. Remove cylinder head sub-assembly:
    a. Using several steps, uniformly loosen and remove the 10 cylinder head bolts and 10 plate washers with a 10 mm bi-hexagon wrench in the sequence shown in the illustration.

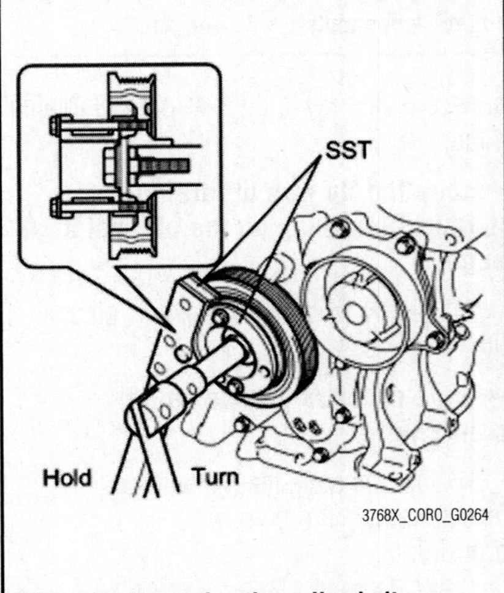

Fig. 103 Loosening the pulley bolt

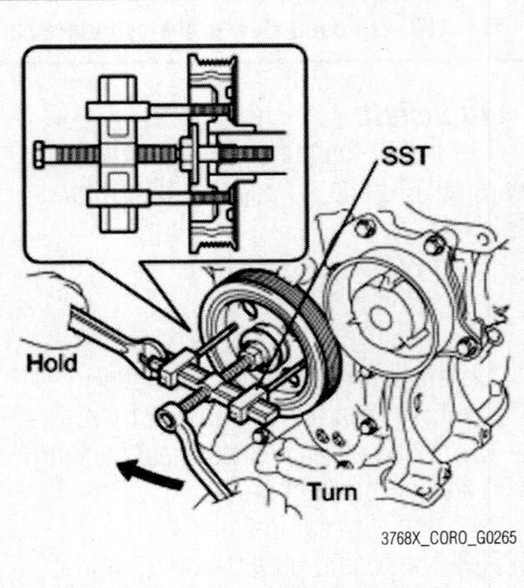

Fig. 104 Removing the crankshaft pulley and pulley bolt

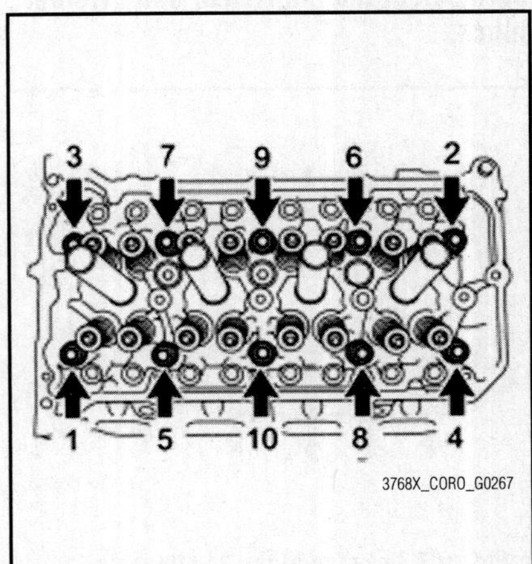

Fig. 106 Cylinder head sub-assembly removal sequence

➡Head warpage or cracking could result from removing the bolts in the wrong order.

b. Using a screwdriver with its tip wrapped with tape, pry between the cylinder head and cylinder block, and remove the cylinder head.

➡Be careful not to damage the contact surfaces of the cylinder head and cylinder block.

3. Remove the cylinder head gasket.
4. Remove the water drain cock plug from the water drain cock sub-assembly.
5. Remove the cylinder block water drain cock sub-assembly from the cylinder block.
6. Remove the ventilation valve.
7. Remove the oil pan drain plug and gasket. Remove the oil pan. Refer to Oil Pan in this section.
8. Remove oil pump assembly. Refer to Oil Pump in this section.
9. Remove rear engine oil seal:
   a. Using a knife, cut off the oil seal lip.
   b. Using a screwdriver with its tip taped, pry out the oil seal.

➡After removing the oil seal, check the crankshaft for damage. If it is damaged, smooth the surface with 400-grit sandpaper.

10. Remove stiffening crankcase assembly:
   a. Uniformly loosen and remove the 11 bolts.
   b. Using a screwdriver, remove the crankcase by prying between the crankcase and cylinder block.

➡Be careful not to damage the contact surfaces of the crankcase and cylinder block.

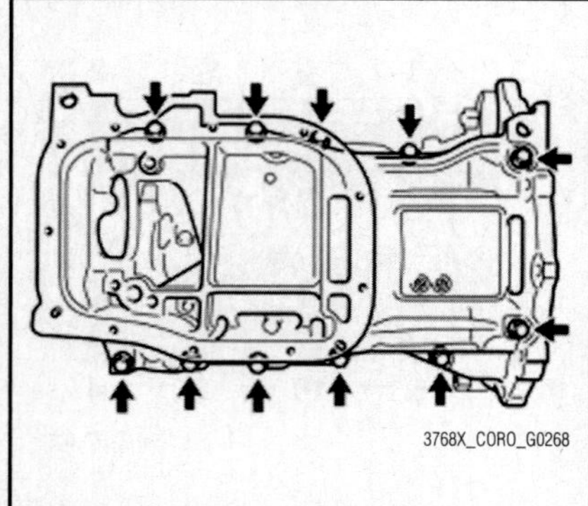

Fig. 107 Location of the 11 stiffening crankcase assembly bolts

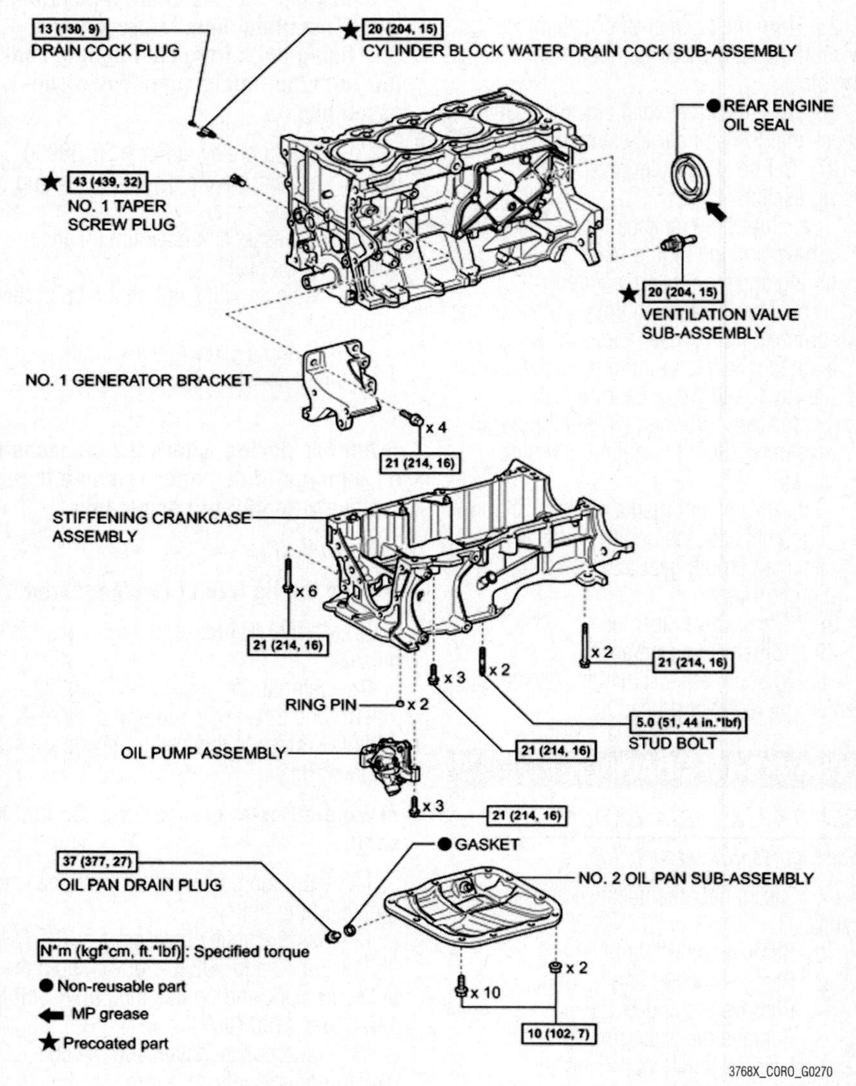

Fig. 108 Exploded view of the cylinder head and related components—1.8L engine

To install:

11. Install the stiffening crankcase with the 11 bolts. Tighten to 16 ft. lbs. (21 Nm).
   a. Bolt A: 5.43 inches.
   b. Bolt B: 1.38 inches.
   c. Bolt C: 2.76 inches.
12. Recheck the torque for bolts 1 and 2. Tighten to 16 ft. lbs. (21 Nm).
13. Wipe off any excess seal packing with a clean piece of cloth.
14. Using a seal installer (SST 09223-15030 and 09950-70010) and a hammer, evenly tap the oil seal until its surface is flush with the rear oil seal retainer edge.

➡Keep the lip free of foreign matter. Do not tap on the oil seal at an angle.

15. Apply MP grease to a new oil seal lip.

➡Wipe off extra grease on the crankshaft.

16. To complete the installation, reverse remaining removal procedure.

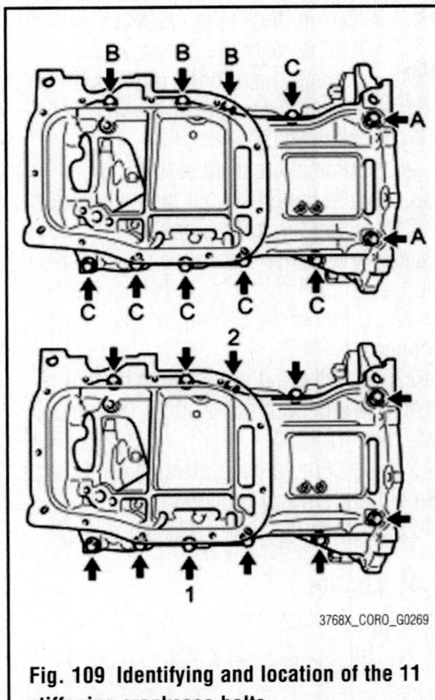

Fig. 109 Identifying and location of the 11 stiffening crankcase bolts

2.4L Engine

See Figures 110 through 112.

1. Remove the camshaft assembly. Refer to Camshaft in this section.
2. Remove cylinder head sub-assembly:
   a. In several steps, uniformly loosen and remove the 10 cylinder head bolts and 10 plate washers with a 10 mm bi-hexagon wrench in the sequence shown in the illustration.

➡Head warpage or cracking could result from removing the bolts in the wrong order.

   b. Using a screwdriver with its tip wrapped with tape, pry between the cylinder head and cylinder block, and remove the cylinder head.

➡Be careful not to damage the contact surfaces between the cylinder head and cylinder block.

3. Remove cylinder head gasket.

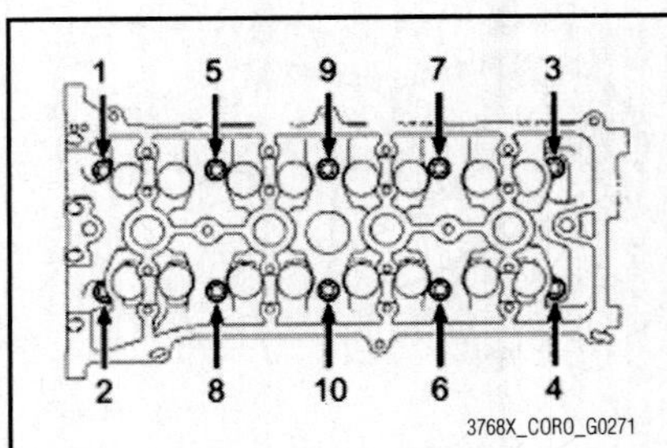

Fig. 110 Cylinder head sub-assembly removal sequence

To install:

4. Place a new cylinder head gasket on the cylinder block surface with the Lot No. stamp facing upward.

➡Remove any oil from the contact surface. Be careful of the installation direction.

5. Place the cylinder head on the cylinder head gasket.

➡Place the cylinder head gently in order to avoid damaging the cylinder head gasket.

➡The cylinder head bolts are tightened in 2 successive steps.

6. Install the cylinder head bolts:
   a. Apply a light coat of engine oil to the threads and under the heads of the cylinder head set bolts.
   b. Using several steps, uniformly install and tighten the 10 cylinder head set bolts and plate washers with a 10 mm bi-hexagon wrench in the order shown in the illustration. Tighten to 52 ft. lbs. (70 Nm).
   c. Mark the front of the cylinder head bolts with paint.

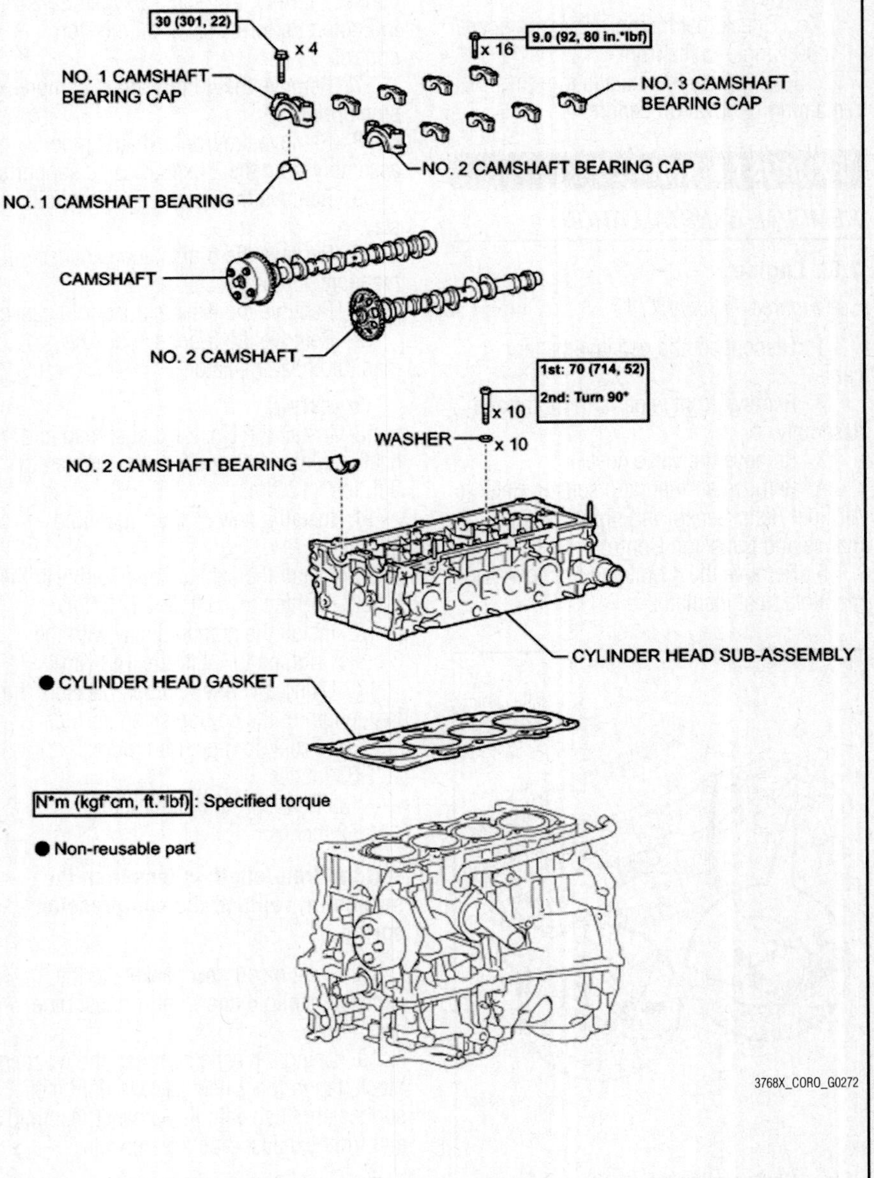

Fig. 111 Exploded view of the cylinder head and related components—2.4L engine

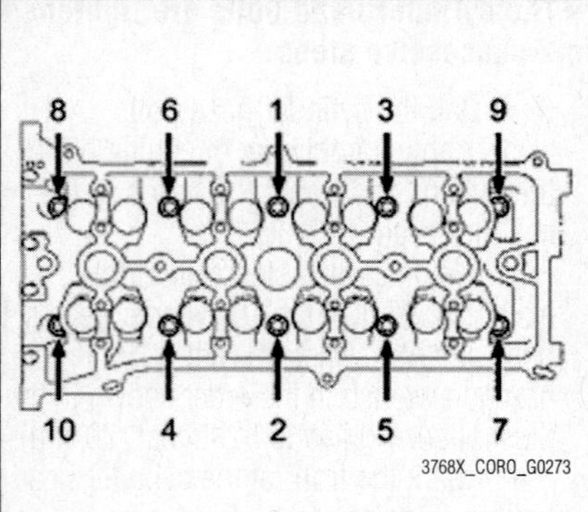

Fig. 112 Cylinder head tightening sequence

d. Further tighten the cylinder head bolts 90°.

e. Check that the paint mark is now at a 90° angle to the front.

7. To complete the installation, reverse remaining removal procedure.

EXHAUST MANIFOLD

REMOVAL & INSTALLATION

1.8L Engine

See Figures 113 and 114.

1. Disconnect the negative battery cable.
2. Remove front wiper arm motor and assembly.
3. Remove the valve cover.
4. Remove air fuel ratio sensor. Refer to Air Fuel Ratio Sensor in Engine Performance and Emission Controls.
5. Remove the 4 bolts and the exhaust manifold heat insulator.

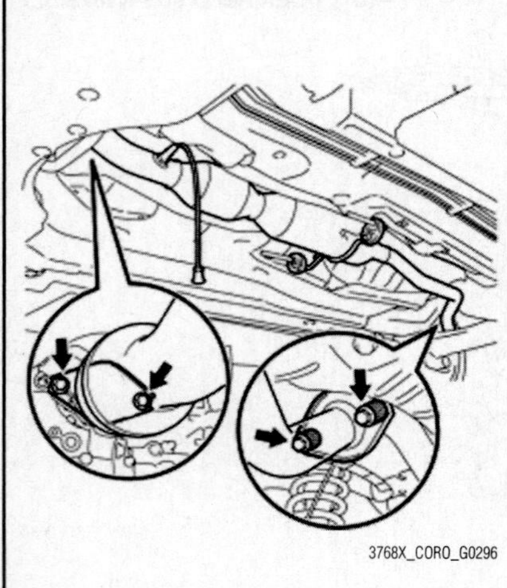

Fig. 113 Removing the 4 bolts and 4 compression springs

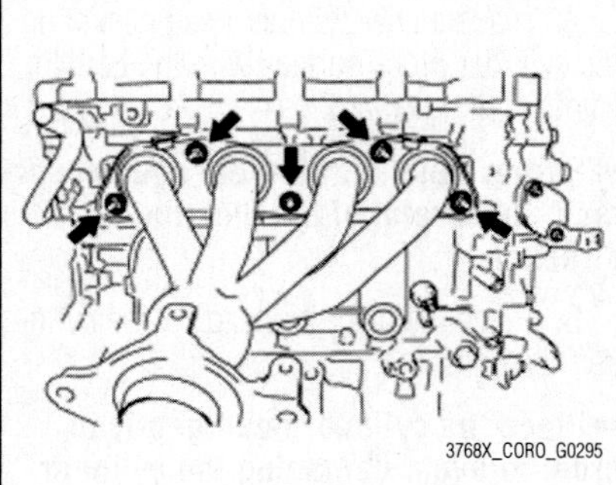

Fig. 114 Removing the 5 nuts and exhaust manifold

6. Disconnect the heated oxygen sensor connector. Refer to Heated Oxygen Sensor in Engine Performance and Emission control.
7. Remove the 4 bolts and 4 compression springs.
8. Remove the front exhaust pipe assembly from the 2 exhaust pipe supports.
9. Remove the 3 bolts and manifold stay.
10. Remove the 5 nuts and exhaust manifold.
11. Remove the exhaust manifold gasket.
12. Remove the 3 bolts and exhaust manifold heat insulator.

To install:

13. Install the No. 2 exhaust manifold heat insulator with the 3 bolts. Tighten to 9 ft. lbs. (12 Nm).
14. Install a new exhaust manifold gasket.
15. Install the exhaust manifold with the 5 nuts. Tighten to 16 ft. lbs. (21 Nm).
16. Install the manifold stay with the 3 bolts. Tighten to 32 ft. lbs. (43 Nm).
17. Using a vernier caliper, measure the free length of the compression springs.
    a. Front springs: Minimum: 1.63 inches.
    b. Rear springs: Minimum: 1.52 inches.

➡If the free length is less than the minimum, replace the compression spring.

18. Fully insert new gaskets to the exhaust manifold and front exhaust pipe assembly.
19. Using a plastic hammer and wooden block, tap in the 2 new gaskets until their surface are flush with the exhaust manifold and front exhaust pipe assembly.

➡Be sure to install the gaskets in the correct direction.

- Do not reuse the gaskets.
- Do not damage the gaskets.
- Do not push in the gaskets by using the exhaust pipe when connecting it.

20. Connect the front exhaust pipe assembly to the 2 exhaust pipe supports.
21. Install the front exhaust pipe assembly with the 4 bolts and 4 compression springs. Tighten to 32 ft. lbs. (43 Nm).
22. Connect the heated oxygen sensor connector.
23. Install the outer exhaust manifold heat insulator with the 4 bolts. Tighten to 9 ft. lbs. (12 Nm).
24. To complete the installation, reverse remaining removal procedure.
25. Inspect for exhaust gas leak.

2.4L Engine

See Figures 113, 115 and 116.

1. Disconnect the negative battery cable.
2. Remove the drive belt. Refer to Accessory Drive Belts in the section.
3. Remove the alternator. Refer to Alternator in Engine Electrical.
4. Remove the 4 bolts and 2 compression springs.
5. Remove the center exhaust pipe assembly from the 2 exhaust pipe supports.
6. Remove the 2 gaskets.
7. Remove front exhaust pipe assembly:
    a. Disconnect the heated oxygen sensor connector.
    b. Remove the 2 bolts, 2 compression springs and front exhaust pipe assembly.
    c. Remove the gasket from the exhaust manifold.
8. Remove the bolt, nut and LH manifold stay.
9. Remove the bolt, nut and RH manifold stay.
10. Remove the 4 bolts and upper exhaust manifold heat insulator.

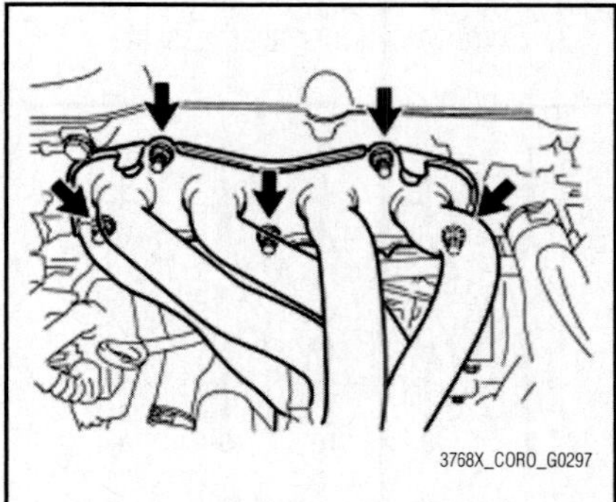

Fig. 115 Remove the 5 nuts and exhaust manifold converter sub-assembly

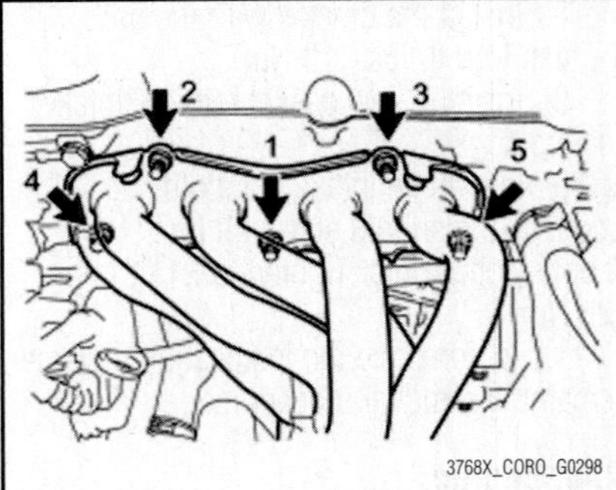

Fig. 116 Exhaust manifold tightening sequence

11. Remove air fuel ratio sensor.
12. Remove the 5 nuts and exhaust manifold converter sub-assembly.
13. Remove the gasket.
14. Remove the 2 bolts and underside exhaust manifold heat insulator.
15. Remove the 4 bolts and catalytic converter insulator.

To install:

16. Install the 4 bolts and catalytic converter insulator. Tighten to 9 ft. lbs. (12 Nm).
17. Install the underside exhaust manifold heat insulator with the 2 bolts. Tighten to 9 ft. lbs. (12 Nm).
18. Install the exhaust manifold converter sub-assembly:
    a. Install a new gasket.
    b. Install the exhaust manifold converter sub-assembly with the 5 nuts in the order shown in the illustration. Tighten to 27 ft. lbs. (37 Nm).
19. Install the RH and LH manifold stay with the bolt and nut. Tighten to 33 ft. lbs. (44 Nm).
20. Install the air fuel ratio sensor.
21. Install the outer exhaust manifold heat insulator with the 4 bolts. 9 ft. lbs (12 Nm).
22. Using a vernier caliper, measure the free length of the compression springs.
    a. Front springs: Minimum: 1.63 inches.
    b. Rear springs: Minimum: 1.52 inches.
23. Remove the remains of exhaust manifold converter with wire brash.
24. Fully insert a new gasket to the exhaust manifold.

➡Be sure to install the gaskets in the correct direction.

- Do not reuse the gaskets.
- Do not damage the gaskets.
- Do not push in the gaskets by using the exhaust pipe when connecting it.

25. Install the front exhaust pipe assembly with the 2 bolts and 2 compression springs. Tighten to 32 ft. lbs. (43 Nm).
26. Connect the heated oxygen sensor connector.
27. Fully insert a new gasket to the center exhaust pipe assembly.

➡Be sure to install the gaskets in the correct direction.

- Do not reuse the gaskets.
- Do not damage the gaskets.
- Do not push in the gaskets by using the exhaust pipe when connecting it.

28. Install a new gasket to the front exhaust pipe assembly.
29. Connect the center exhaust pipe assembly to the 2 exhaust pipe supports.
30. Install the center exhaust pipe assembly with the 4 bolts and 2 compression springs. Tighten to 32 ft. lbs. (43 Nm).
31. To complete the installation, reverse remaining removal procedure.
32. Inspect for exhaust gas leak.

INTAKE MANIFOLD

REMOVAL & INSTALLATION

1.8L Engine

See Figure 117.

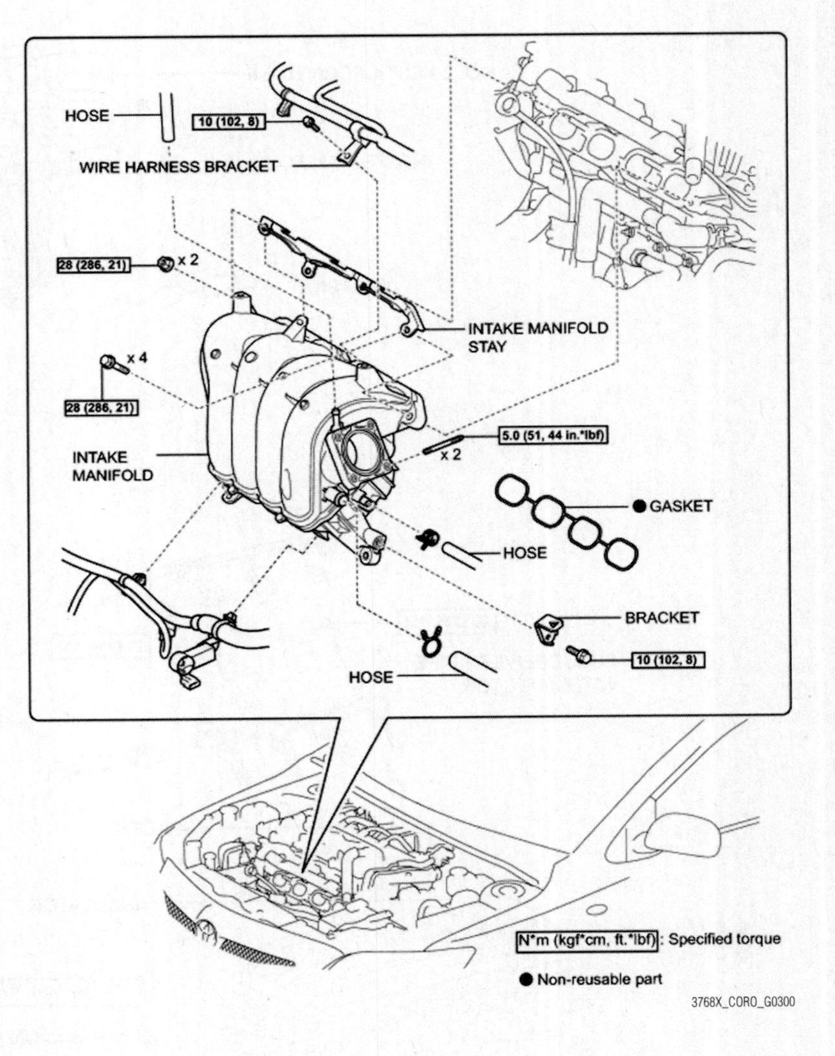

Fig. 117 Exploded view of the intake manifold and related components—1.8L engine

1. Disconnect the negative battery cable.
2. Drain and recycle the engine coolant.
3. Remove the engine cover.
4. Remove the air cleaner assembly.
5. Remove the throttle body. Refer to Throttle Body in the Fuel Section.
6. Remove the intake manifold:
    a. Remove the bolt and disconnect the wire harness bracket.
    b. Disconnect the 3 hoses.
    c. Remove the 4 bolts, 2 nuts, intake manifold stay and intake manifold.
    d. Remove the gasket from the intake manifold.
    e. Remove the bolt and bracket.
    f. Using a TORX® socket E6, remove the 2 stud bolts from the intake manifold.

To install:

7. Using a TORX® socket E6, install the 2 stud bolts to the intake manifold. Tighten to 44 inch lbs. (5 Nm).
8. Install the bracket with the bolt. Tighten to 8 ft. lbs. (10 Nm).
9. Install a new gasket into the intake manifold.
10. Install the intake manifold and intake manifold stay with the 4 bolts and 2 nuts. Tighten to 21 ft. lbs. (28 Nm).
11. To complete the installation, reverse remaining removal procedure.

2.4L Engine

See Figure 118.

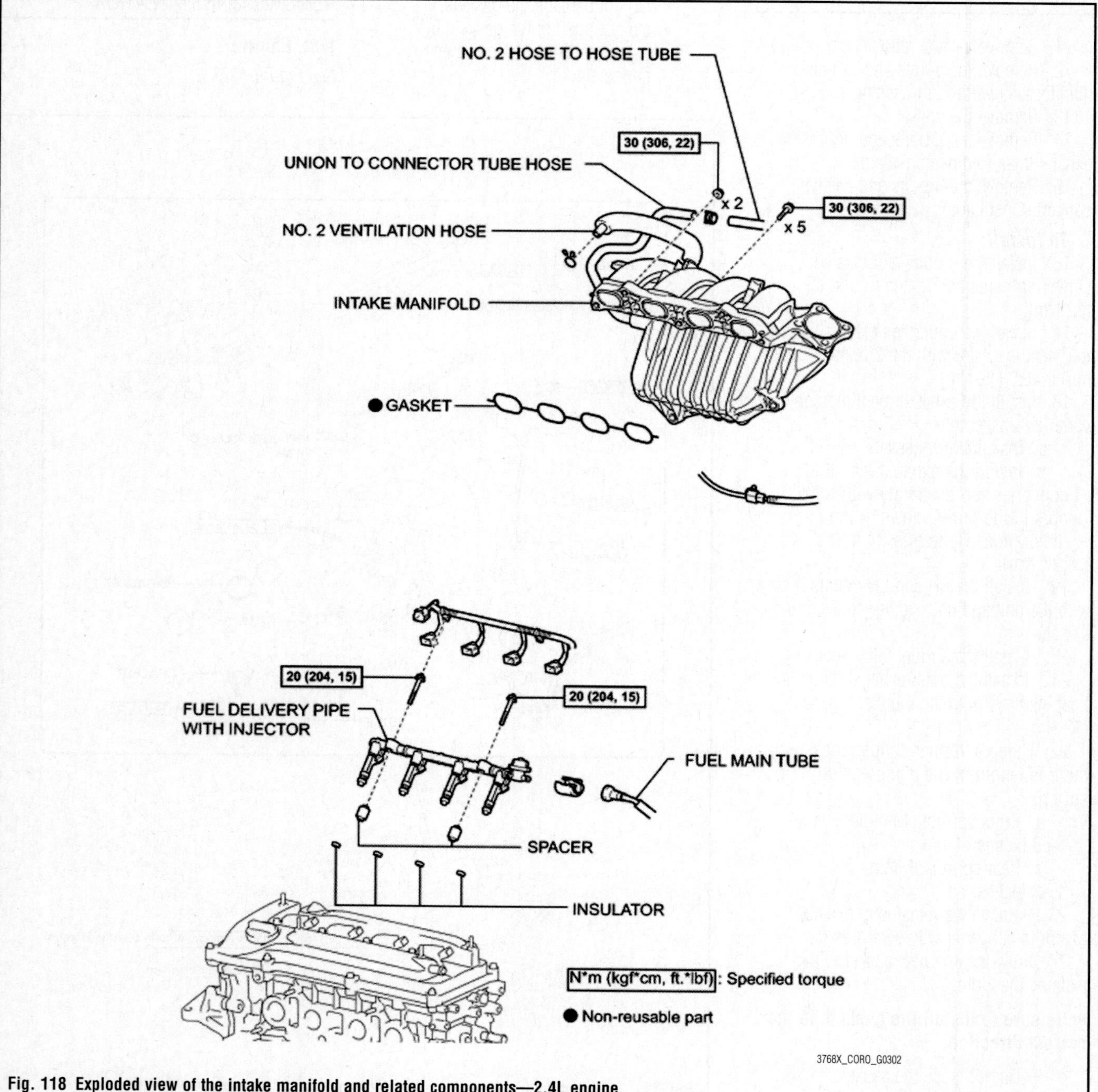

Fig. 118 Exploded view of the intake manifold and related components—2.4L engine

⁂ CAUTION

Observe all applicable safety precautions when working around fuel. Whenever servicing the fuel system, always work in a well ventilated area. Do not allow fuel spray or vapors to come in contact with a spark or open flame. Keep a dry chemical fire extinguisher near the work area. Always keep fuel in a container specifically designed for fuel storage; also, always properly seal fuel containers to avoid the possibility of fire or explosion.

1. Discharge fuel system pressure.
2. Disconnect the cable from negative battery terminal.
3. Remove the engine cover.
4. Drain and recycle the engine coolant.
5. Remove the windshield wiper motor and link assembly.
6. Remove suspension tower damper assembly (with front strut bar). Refer to Suspension Tower Damper Assembly in Front Suspension.
7. Remove the air cleaner assembly.
8. Remove the throttle body. Refer to Throttle Body in the Fuel Section.
9. Disconnect the main fuel tube.
10. Disconnect the ventilation hose.
11. Remove fuel delivery pipe with injector. Refer to Fuel Rail and Injectors in Fuel System.
12. Remove intake manifold:
    a. Disconnect the union to connector tube hose from the No. 2 hose to hose tube.
    b. Disconnect the camshaft timing oil control valve connector.
    c. Remove the wire harness clamp.
    d. Remove the union to connector tube hose from the vacuum hose clamp.
    e. Remove the 5 bolts, 2 nuts and intake manifold.
    f. Remove the gasket from the intake manifold.

To install:

13. To install, reverse the removal procedure. Refer to exploded view illustration for torque values.

OIL PAN

REMOVAL & INSTALLATION

1.8L Engine

See Figure 119.

1. Remove the oil pan drain plug and gasket and drain the engine oil.
2. Remove the 10 bolts and 2 nuts.
3. Insert the blade of oil pan seal cutter between the crankcase and oil pan. Cut through the sealer and remove the oil pan.

➡Be careful not to damage the contact surfaces of the crankcase and oil pan.

To install:

4. Remove any old packing material and be careful not to drop any oil on the contact surfaces of the cylinder block and oil pan.
5. Apply a continuous bead of seal packing (Diameter 4.0 mm (0.157 in.)). Seal packing: Toyota Genuine Seal Packing Black, Three Bond 1207B or equivalent.

➡Remove any oil from the contact surfaces.

- Install the oil pan within 3 minutes after applying seal packing.
- Do not start the engine for at least 2 hours after installing the oil pan.

6. Install the oil pan with the 10 bolts and 2 nuts. Tighten to 7 ft. lbs. (10 Nm).

2.4L Engine

See Figures 120 and 121.

1. Remove the oil pan drain plug and gasket and drain the engine oil.

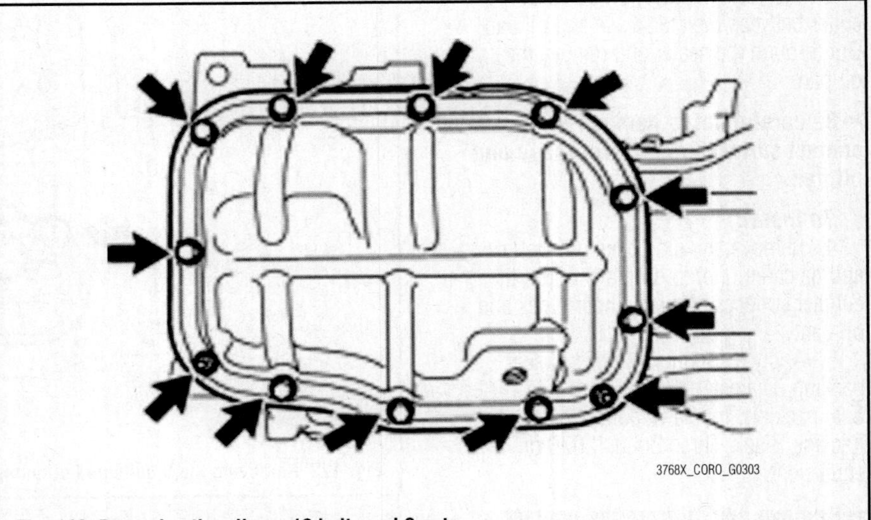

Fig. 119 Removing the oil pan 10 bolts and 2 nuts

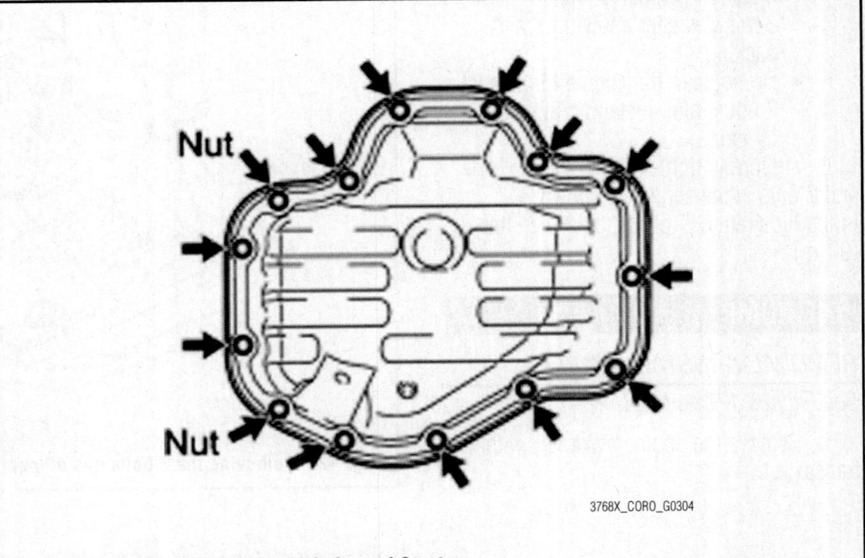

Fig. 120 Removing the oil pan 10 bolts and 2 nuts

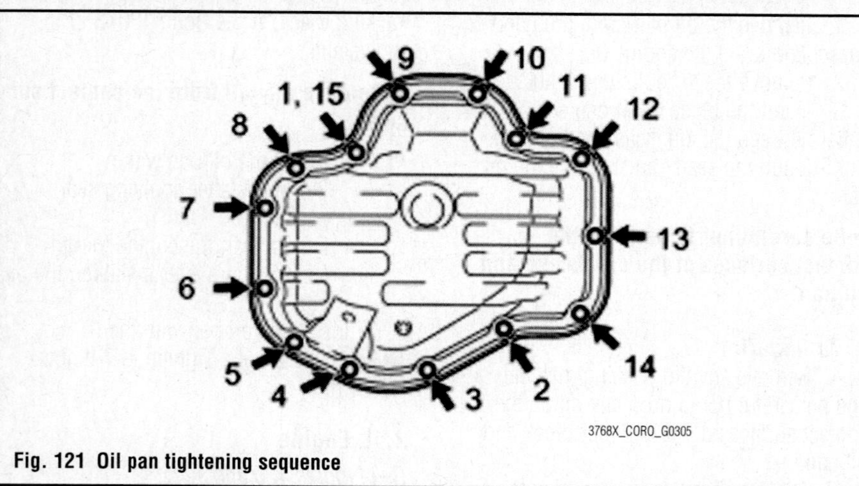

Fig. 121 Oil pan tightening sequence

2. Remove the 12 bolts and 2 nuts.

3. Insert the blade of oil pan seal cutter between the crankcase and oil pan. Cut through the sealer and remove the oil pan.

➡Be careful not to damage the contact surfaces of the crankcase and oil pan.

To install:

4. Remove any old packing material and be careful not to drop any oil on the contact surfaces of the cylinder block and oil pan.

5. Apply a continuous bead of seal packing (Diameter 4.0 mm (0.157 in.)). Seal packing: Toyota Genuine Seal Packing Black, Three Bond 1207B or equivalent.

➡Remove any oil from the contact surfaces.

- Install the oil pan within 3 minutes after applying seal packing.
- Do not start the engine for at least 2 hours after installing the oil pan.

6. Uniformly tighten the 12 bolts and 2 nuts in the sequence shown in the illustration. Tighten to 80 inch lbs. (9 Nm).

OIL PUMP

REMOVAL & INSTALLATION

See Figures 122 and 123.

1. Remove the engine assembly with the transaxle.

2. Remove the timing chain front cover. Refer to Timing Chain Front Cover in this section.

3. Remove the oil pan. Refer to Oil Pan in this section.

4. Remove the 3 bolts and oil pump.

5. For 2.4L engines, remove the gasket.

To install:

6. To install, reverse the removal procedure. Tighten the oil pump assembly bolts to 16 ft. lbs. (21 Nm).

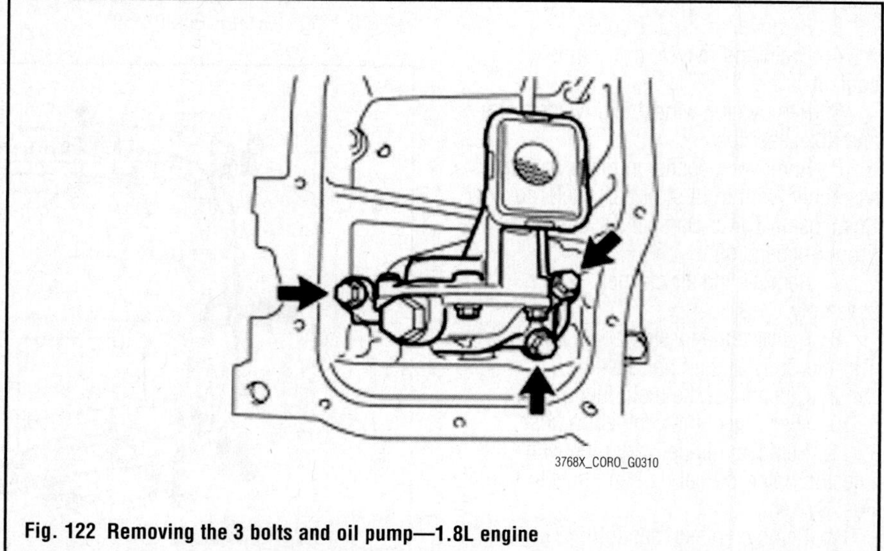

Fig. 122 Removing the 3 bolts and oil pump—1.8L engine

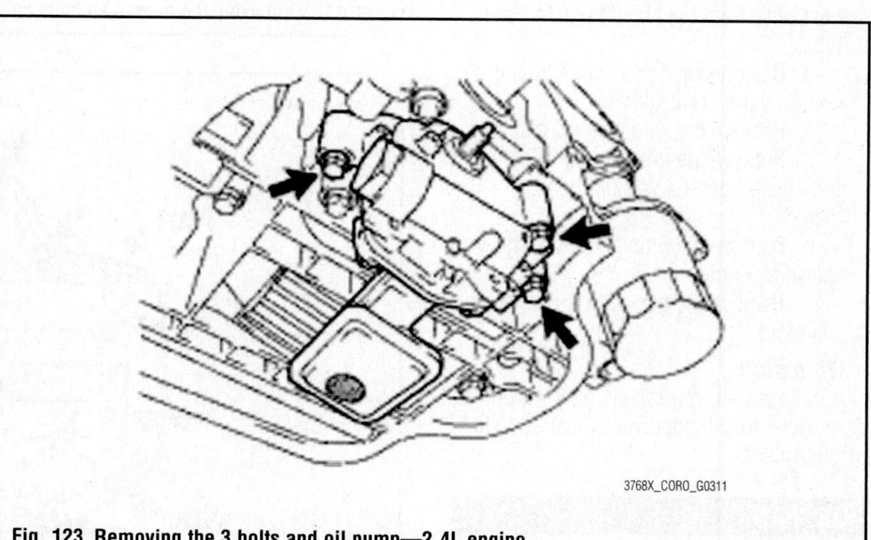

Fig. 123 Removing the 3 bolts and oil pump—2.4L engine

PISTON AND RING

POSITIONING

See Figures 124 through 126.

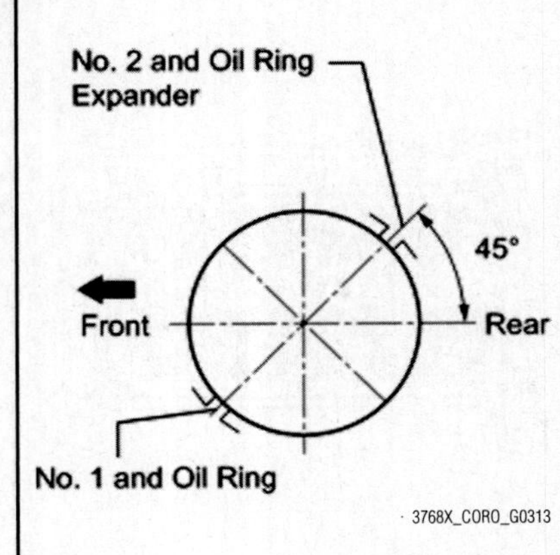

Fig. 125 Piston ring end-gap spacing—1.8L engine

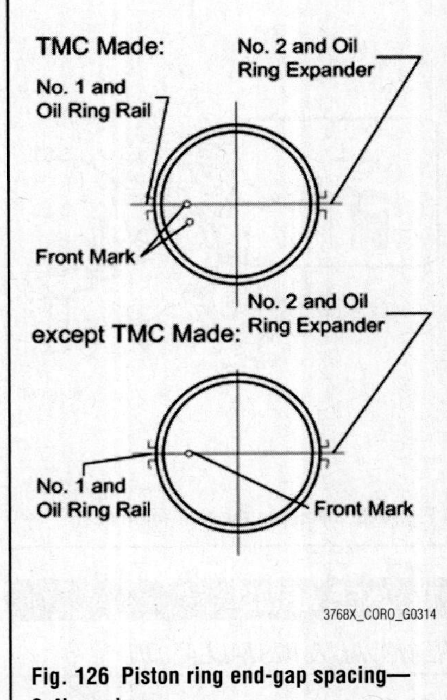

Fig. 126 Piston ring end-gap spacing—2.4L engines

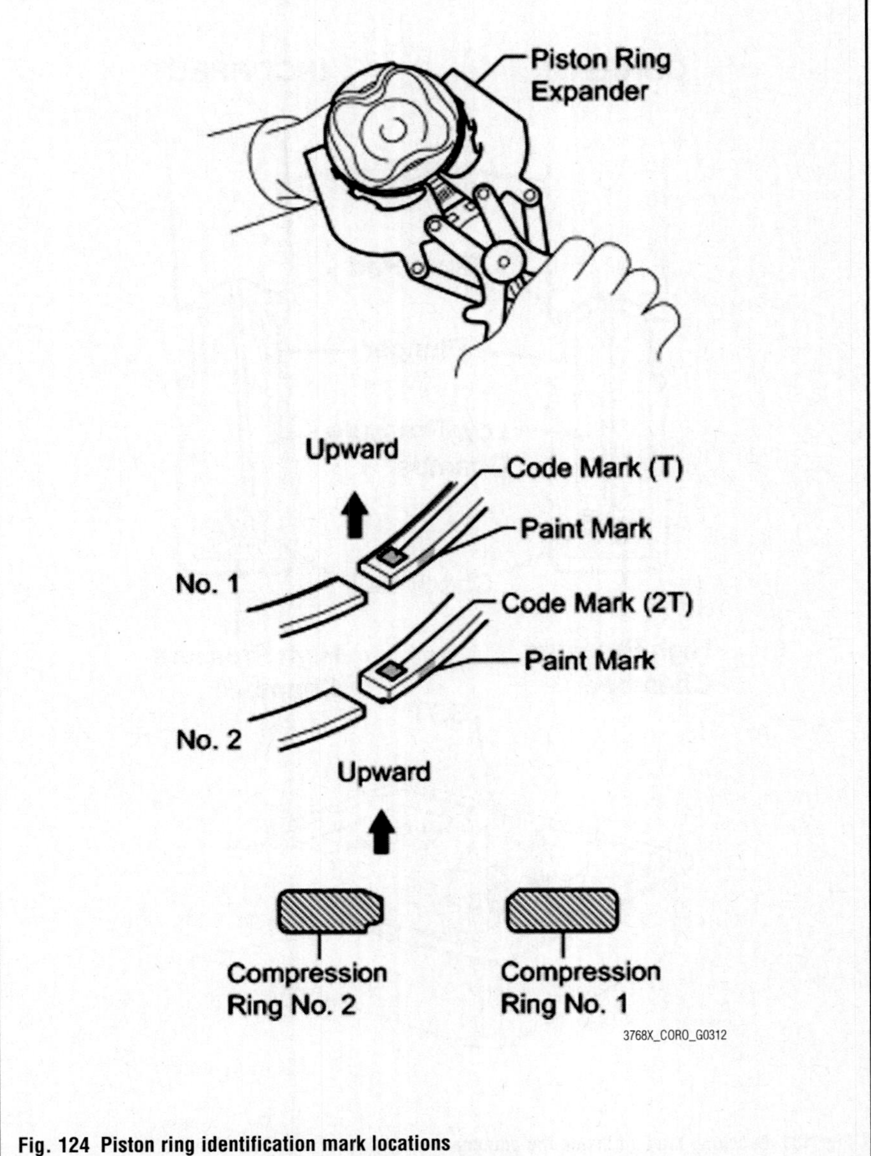

Fig. 124 Piston ring identification mark locations

REAR MAIN SEAL

REMOVAL & INSTALLATION

See Figures 127 and 128.

1. Remove the appropriate transaxle. Refer to Drivetrain.
2. Remove the flywheel or flexplate. Refer to Drivetrain.
3. Remove rear engine oil seal.
    a. Using a knife, cut off the oil seal lip.
    b. Using a screwdriver with its tip taped, pry out the oil seal.

➡After removing the oil seal, check the crankshaft for damage. If it is damaged, smooth the surface with 400-grit sandpaper.

To install:

4. Using a seal installer (SST 09223-15030 and 09950-70010) and a hammer, evenly tap the oil seal until its surface is flush with the rear oil seal retainer edge.

➡Keep the lip free of foreign matter. Do not tap on the oil seal at an angle.

5. Apply MP grease to a new oil seal lip.

➡Wipe off extra grease on the crankshaft.

6. To complete installation, reverse the remaining removal procedure.

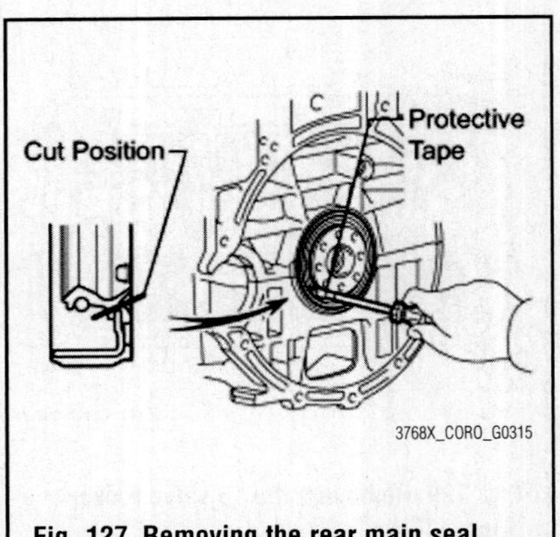

Fig. 127 Removing the rear main seal

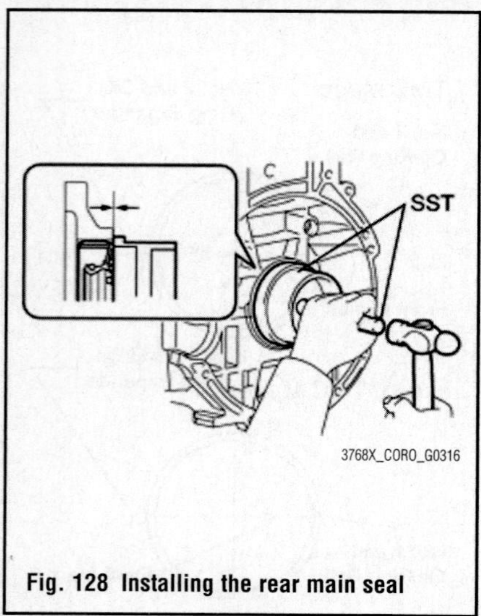

Fig. 128 Installing the rear main seal

ROCKER ARMS

REMOVAL & INSTALLATION

1.8L Engine

See Figures 129 through 132.

1. Remove the camshaft assembly. Refer to Camshafts in this section.
2. Remove the 16 valve rocker arms.

➡Arrange the removed parts in the correct order.

3. Remove the 16 valve lash adjusters from the cylinder head.

➡Arrange the removed parts in the correct order.

To install:

➡Keep the lash adjuster free of dirt and foreign matter. Only use clean engine oil.

4. Place the lash adjuster into a container filled with engine oil.

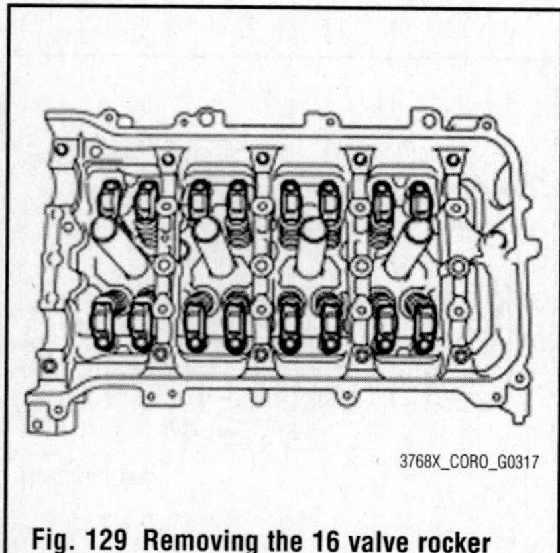

Fig. 129 Removing the 16 valve rocker arms

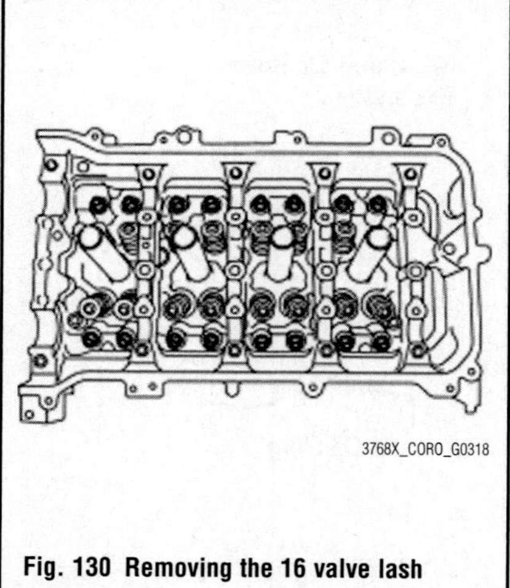

Fig. 130 Removing the 16 valve lash adjusters

5. Insert the tip of SST 09276-75010 into the lash adjuster plunger and use the tip to press down on the check ball inside the plunger.
6. Squeeze SST and the lash adjuster together to move the plunger up and down 5 to 6 times.
7. Check the movement of the plunger and bleed it. If the plunger moves up and down it is ok.

➡When bleeding air from the high-pressure chamber, make sure that the tip of SST is actually pressing the check ball as shown in the illustration. If the check ball is not pressed, the high-pressure chamber will not be bled.

8. After bleeding, remove SST. Then, try to press the plunger quickly and firmly by hand.

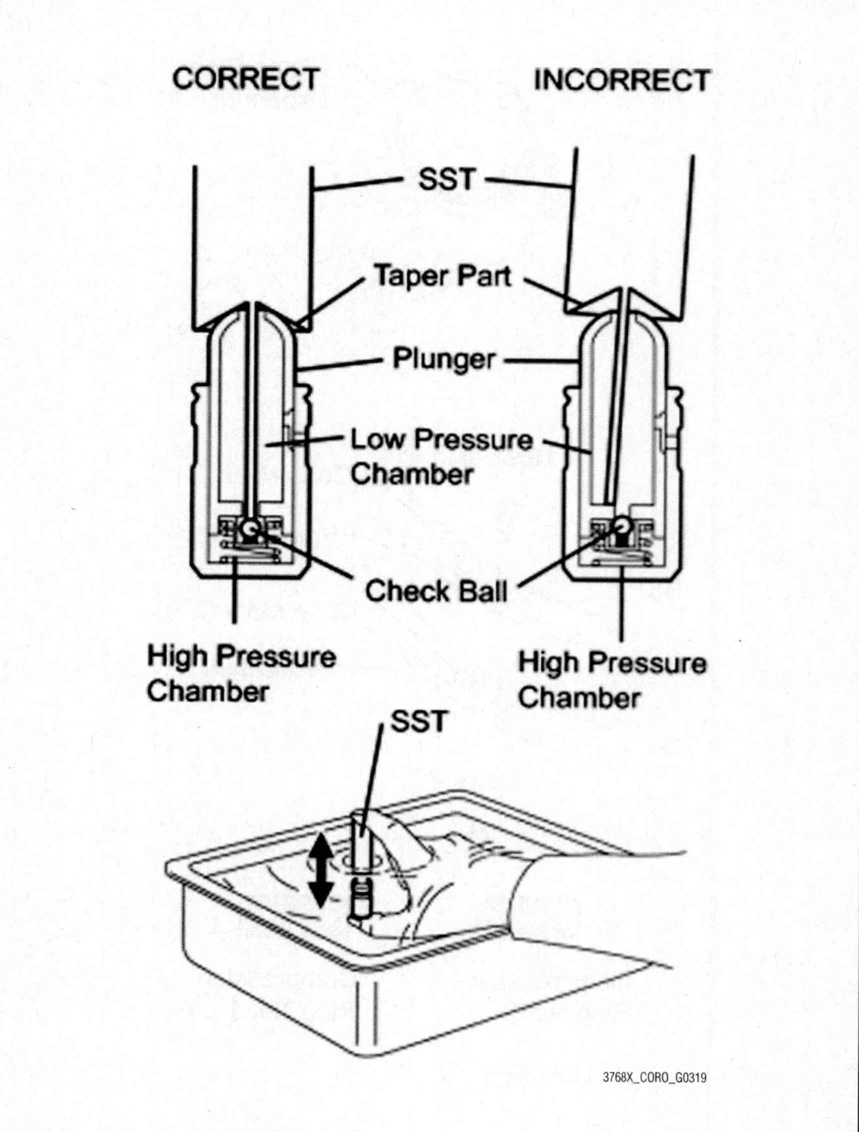

Fig. 131 Checking the ball inside the plunger

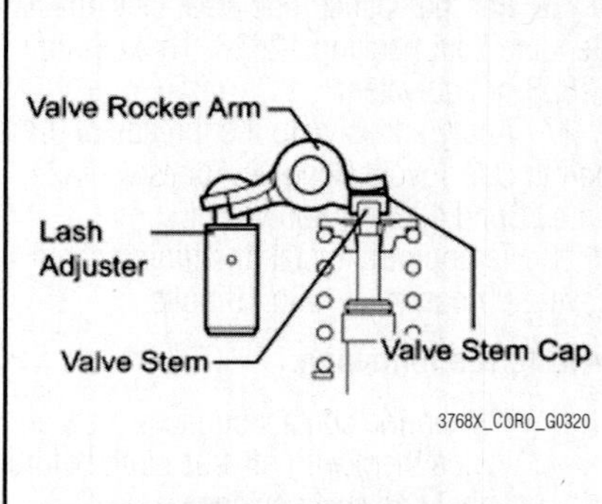

Fig. 132 Valve rocker arms installation orientation

9. If the plunger is very difficult to move it is ok. If the result is not as specified, replace the lash adjuster.
10. Install the lash adjusters

➡**Install the lash adjuster to the same place it was removed from.**

11. Apply engine oil to the lash adjuster tip and valve stem cap end.
12. Make sure that the valve rocker arms are installed as shown in the illustration.
13. To complete installation, reverse remaining removal procedure.

TIMING CHAIN FRONT COVER

REMOVAL & INSTALLATION

1.8L Engine

See Figures 133 through 138.

1. Remove the 3 bolts and engine mounting bracket.
2. Remove the 4 bolts and oil filter bracket.
3. Remove the 2 O-rings.
4. Remove the 19 bolts.
5. Remove the timing chain cover by prying between the timing chain cover and cylinder head or cylinder block with a screwdriver.

➡**Be careful not to damage the contact surfaces of the timing chain cover, cylinder block, and cylinder head.**

➡**Tape the screwdriver tip before use.**

6. Place the timing chain cover on wooden blocks.
7. Using a screwdriver and a hammer, pry out the oil seal.

➡**Do not damage the surface of the oil seal press fit hole.**

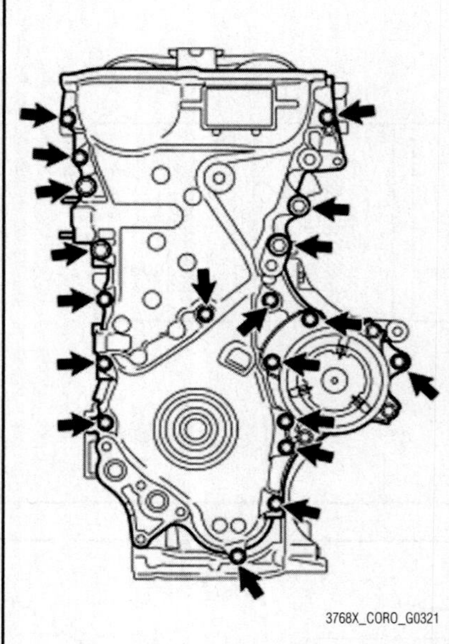

Fig. 133 Removing the 19 bolts and the timing chain front cover

8. Remove the 3 O-rings.
9. Remove the 3 bolts and water pump. Remove the gasket.

To install:

10. Using a seal installer (SST 09223-15030) and a hammer, evenly tap the oil seal until its surface is flush with the timing chain cover sub-assembly edge.

➡**Keep the lip free from foreign matter. Do not tap the oil seal at an angle.**

11. Apply MP grease to the lip of the oil seal.
12. Remove any old packing (FIPG) material and be careful not to drop any oil on the contact surfaces of the timing chain cover sub-assembly, cylinder head, and cylinder block.
13. Install 3 new O-rings.
14. Apply seal packing as shown in the illustration. Seal packing: Toyota Genuine Seal Packing Black, Three Bond 1207B or equivalent.

➡**Remove any oil from the contact surfaces. Install the chain cover within 3 minutes after applying seal packing. Do not start the engine for at least 2 hours after installing the timing chain cover sub-assembly.**

➡**Apply seal packing to the timing chain cover sub-assembly in a continuous line as shown in the following illustration.**

- When the contact surfaces are wet, wipe them with oil-free cloth before applying seal packing.
- Install the timing chain cover sub-assembly within 3 minutes and tighten the bolts within 15 minutes after applying seal packing.
- Do not start the engine for at least 2 hours after installing.

15. For the continuous line area, use: Toyota Genuine Seal Packing Black, Three Bond 1207B or equivalent.

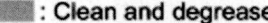

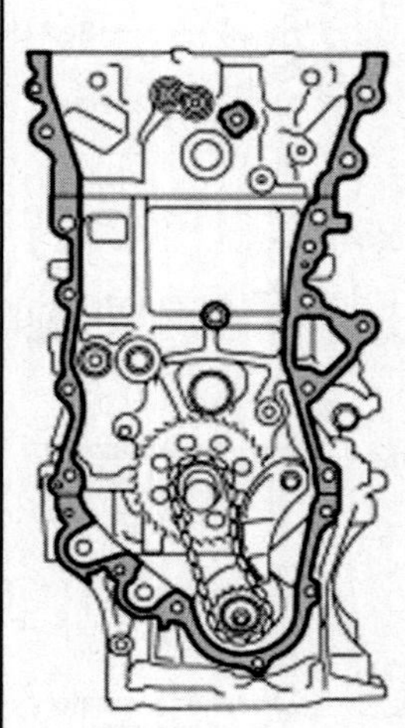
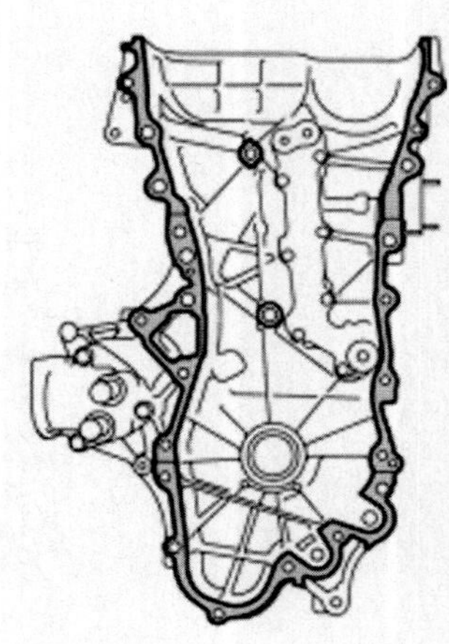

Fig. 134 Identifying areas to clean and degrease

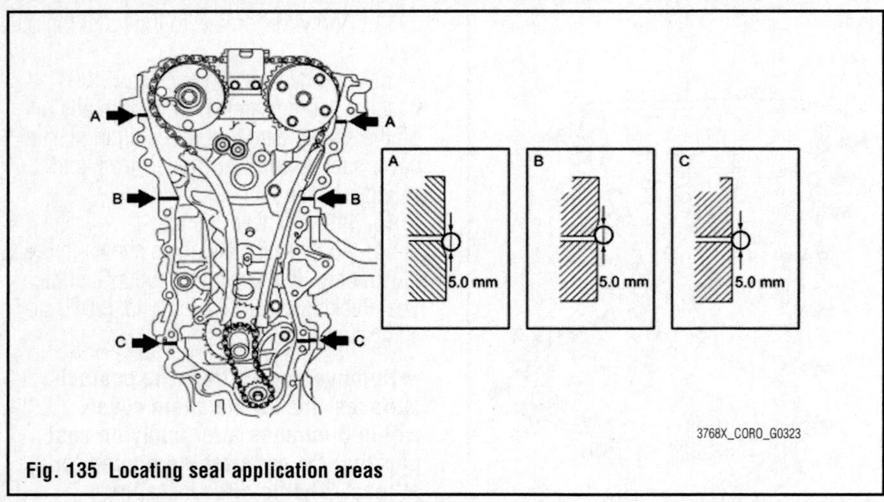

Fig. 135 Locating seal application areas

16. For the dashed line area, use: Toyota Genuine Seal Packing 1282B, Three Bond 1282B or equivalent.

17. Apply adhesive to the threads of the bolt E. Use Toyota Genuine Adhesive 1324, Three Bond 1324 or equivalent.

18. Temporarily install the timing chain cover sub-assembly with 19 bolts.

➡Note the following:

- When the contact surfaces are wet, wipe them with oil-free cloth before applying seal packing.
- Install the timing chain cover sub-assembly within 3 minutes and tighten the bolts within 15 minutes after applying seal packing.

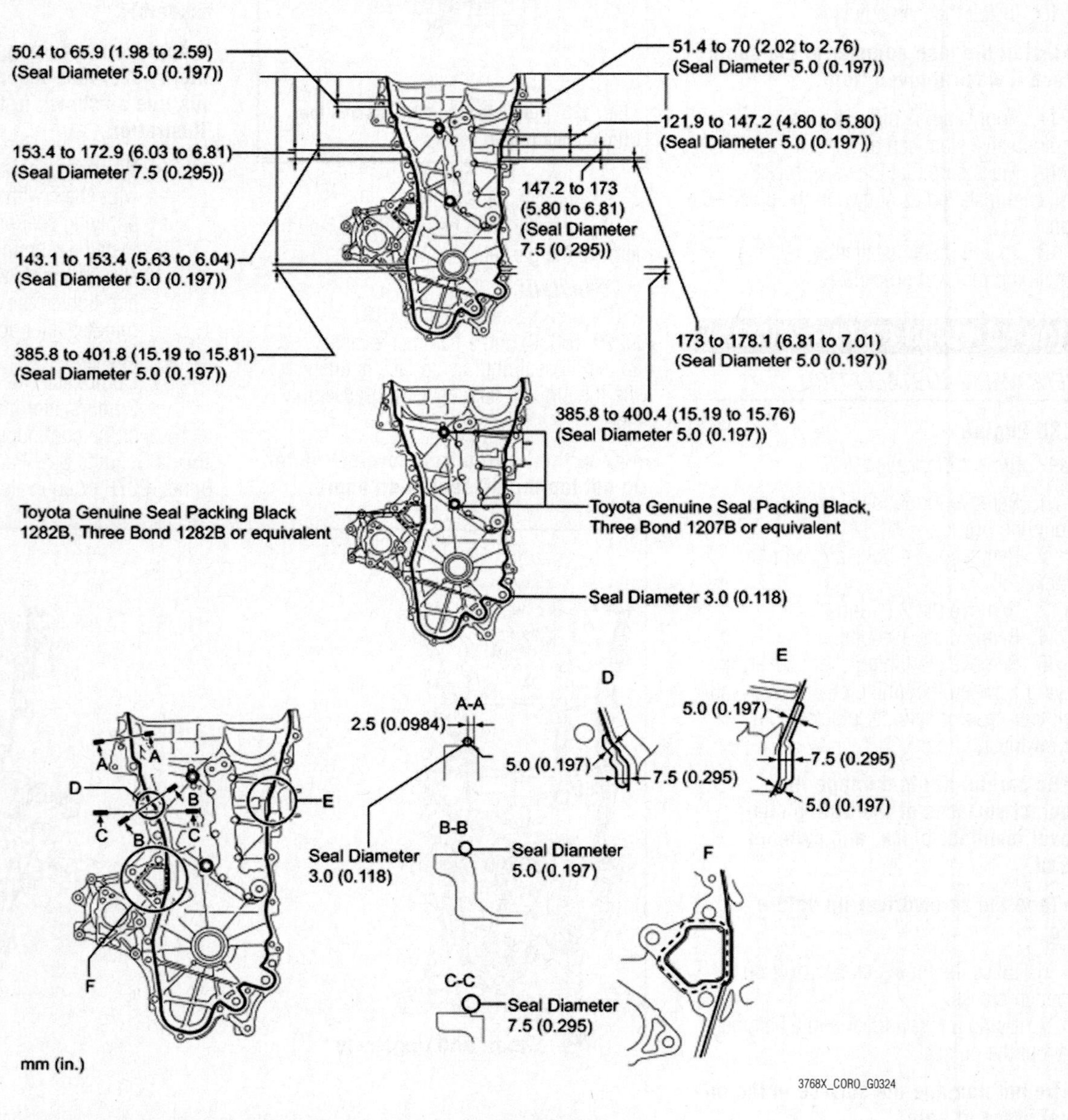

Fig. 136 Applying seal packing to the timing chain cover sub-assembly in a continuous line

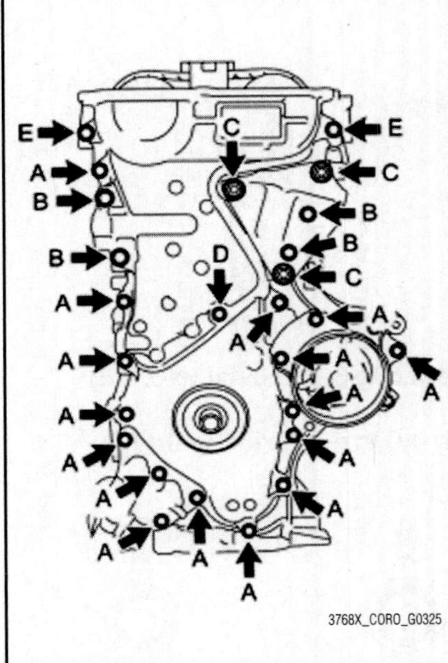

Fig. 137 **Identifying the timing chain front cover bolts**

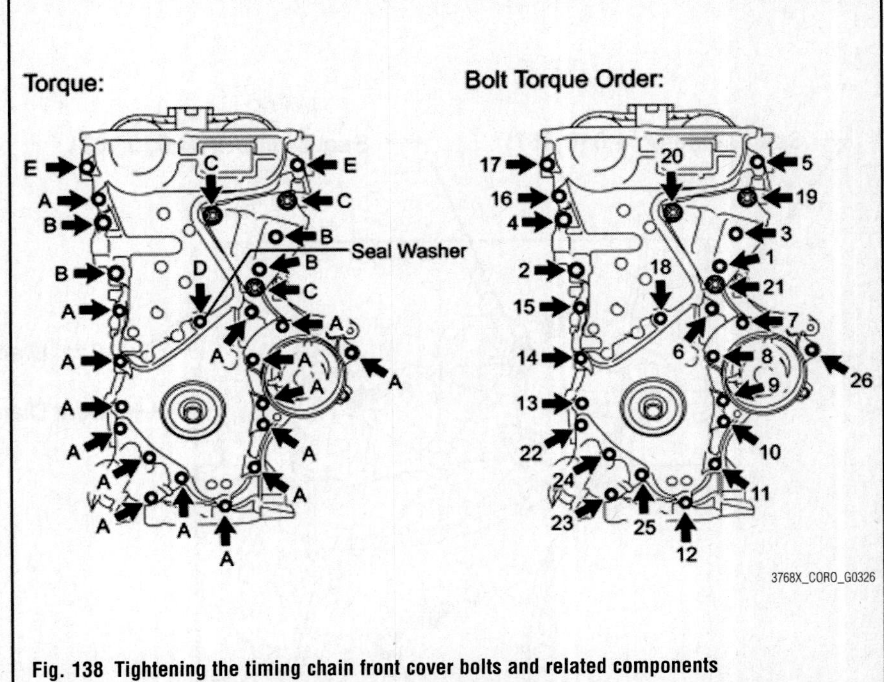

Fig. 138 **Tightening the timing chain front cover bolts and related components**

- Do not start the engine for at least 2 hours after installing.

19. Refer to illustration to identify the following:
    a. Bolt A, E: 1.38 inches.
    b. Bolt B: 2.16 inches.
    c. Bolt C: 3.15 inches.
    d. Bolt D: 1.57 inches
20. Install the water pump. Refer to Water Pump in Engine Cooling.
21. Temporarily install the engine mounting bracket with the 3 bolts.
22. Install 2 new O-rings.
23. Temporarily install the oil filter bracket with the 4 bolts.
24. Fully tighten the timing chain cover sub-assembly with the 26 bolts as shown in the illustration.
    a. Bolts A, E: 19 ft. lbs. (26 Nm).
    b. Bolts B, C: 37 ft. lbs. (51 Nm).
    c. Bolt D: 7 ft. lbs. (10 Nm).

➡Note the following:

- When the contact surfaces are wet, wipe them with oil-free cloth before applying seal packing.
- Install the timing chain cover sub-assembly within 3 minutes and tighten the bolts within 15 minutes after applying seal packing.
- Do not start the engine for at least 2 hours after installing.

2.4L Engine

See Figures 139 through 142.

1. Remove the 3 bolts and transverse engine mounting bracket.
2. Using an E10 TORX® socket, remove the stud bolt for the V-ribbed belt tensioner.
3. Remove the 12 bolts and 2 nuts.
4. Remove the timing chain cover by prying the portions between the timing chain cover, cylinder head and cylinder block with a screwdriver.

➡Be careful not to damage the contact surfaces of the timing chain cover, cylinder head and cylinder block.

➡Tape the screwdriver tip before use.

5. Place the timing chain cover on wooden blocks.
6. Using a screwdriver, pry out the oil seal.

➡Do not damage the surface of the oil seal press fit hole.

To install:

7. Using a seal installer (SST 09223-22010) and a hammer, evenly tap the oil seal until its surface is flush with the timing chain cover sub-assembly edge.

➡Keep the lip free from foreign matter. Do not tap the oil seal at an angle.

8. Apply MP grease to the lip of the oil seal.
9. Remove any old packing (FIPG) material and be careful not to drop any oil on the contact surfaces of the timing chain cover sub-assembly, cylinder head, and cylinder block.
10. Apply seal packing (Diameter 4.0 to 4.5 mm (0.157 to 0.177 in.)) as shown in the illustration. Use Toyota Genuine Seal Packing Black, Three Bond 1207B or equivalent.

➡Remove any oil from the contact surfaces. Install the chain cover within 3 minutes of applying seal packing. Do not add engine oil for at least 2 hours after installing the chain cover.

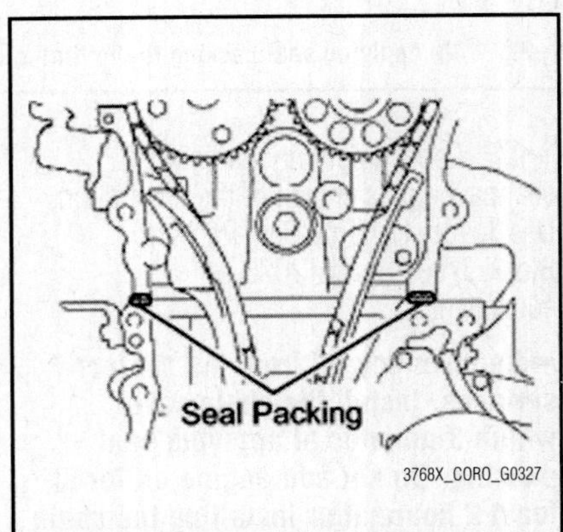

Fig. 139 **Locating seal application areas**

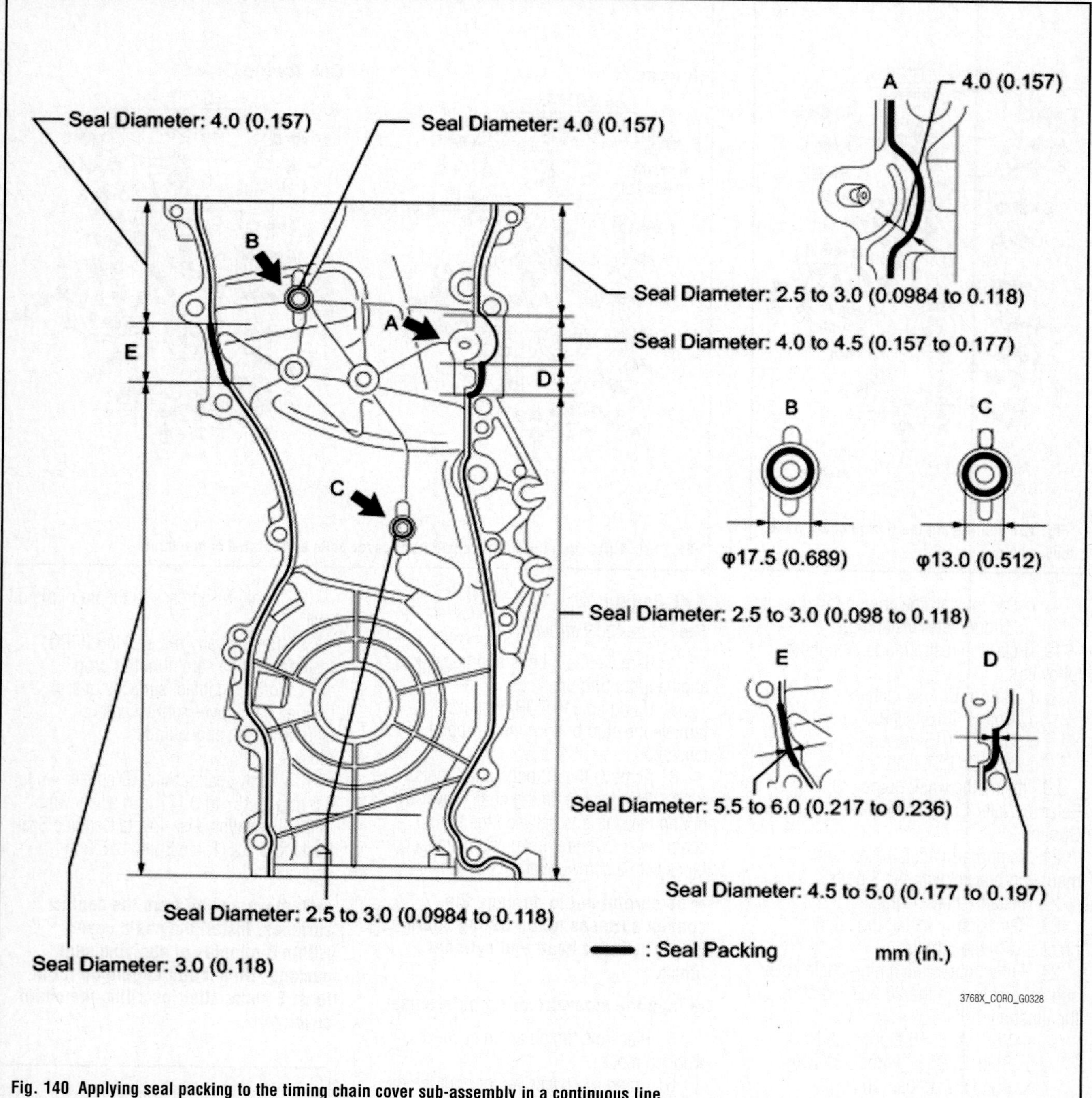

Fig. 140 Applying seal packing to the timing chain cover sub-assembly in a continuous line

11. Apply a continuous bead of seal packing as shown in the illustration. Use Toyota Genuine Seal Packing Black, Three Bond 1207B or equivalent.

➡Remove any oil from the contact surfaces. Install the chain cover within 3 minutes of applying seal packing. Do not add engine oil for at least 2 hours after installing the chain cover.

12. Apply adhesive to the threads of the bolt "A". Use Toyota Genuine Adhesive 1324, Three Bond 1324 or equivalent.

13. Temporarily install the timing chain cover with the 12 bolts and 2 nuts.:
   a. Bolt A: 1.18 inches (10 mm head)
   b. Bolt B: 1.18 inches (12 mm head
   c. Bolt C: 1.57 inches (14 mm head)

14. Temporarily install the engine transverse engine mounting bracket with the 3 bolts.

15. Fully tighten the timing chain cover with the 15 bolts and 2 nuts as shown in the illustration.
   a. Bolt A: 80 inch lbs. (9 Nm).
   b. Bolt B: 18 ft. lbs. (25 Nm).
   c. Bolt C: 41 ft. lbs. (55 Nm).
   d. Nut: 8 ft. lbs. (11 Nm).

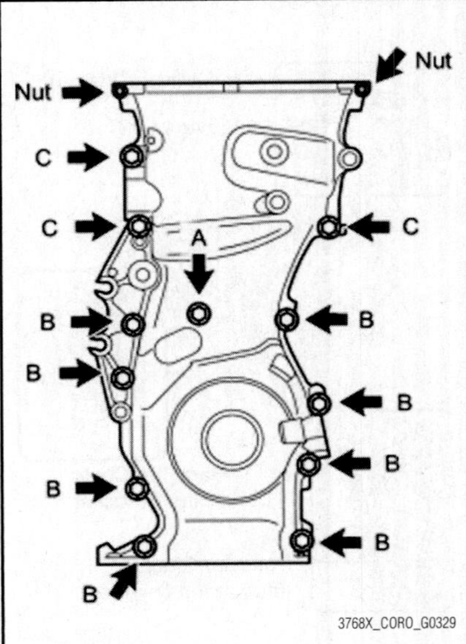

Fig. 141 Identifying the timing chain front cover bolts

3. Remove the 2 bolts and LH chain vibration damper.

4. Hold the hexagonal portion of the camshaft with a wrench and turn the camshaft timing gear assembly counterclockwise to loosen the chain between the camshaft timing gears.

5. With the chain loosened, release the chain from the camshaft timing gear assembly and place it on the camshaft timing gear assembly.

➡**Be sure to release the chain from the sprocket completely.**

6. Turn the camshaft clockwise to return it to the original position and remove the chain.

7. Remove the 2 bolts and No. 2 chain vibration damper.

8. Remove the crankshaft timing sprocket.

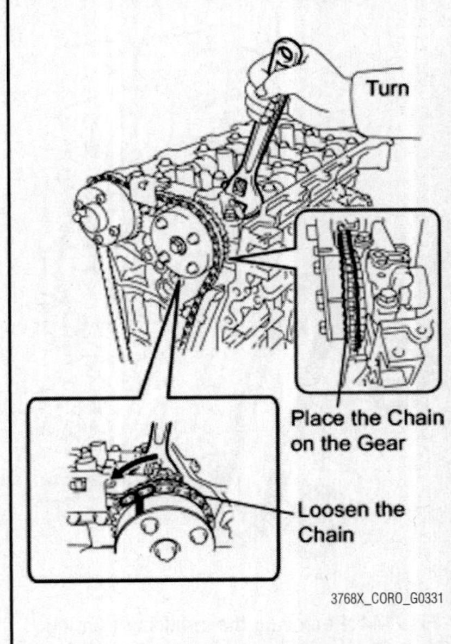

Fig. 143 Removing the upper timing chain

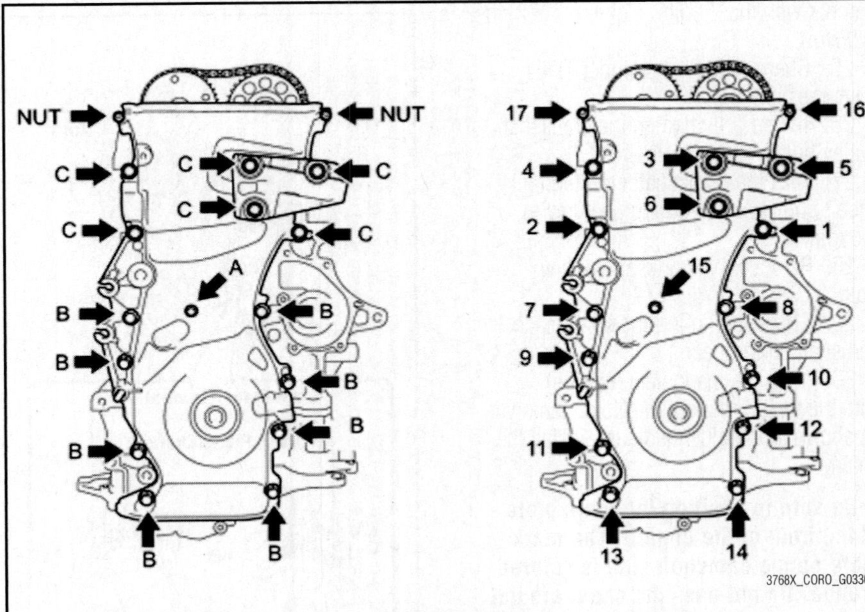

Fig. 142 Tightening the timing chain front cover bolts and related components

TIMING CHAIN & SPROCKETS

REMOVAL & INSTALLATION

1.8L Engine

See Figures 143 through 151.

➡**Do not turn the crankshaft without the chain tensioner installed.**

1. Remove the 2 nuts, bracket, tensioner and gasket.

2. Remove the chain tensioner slipper.

⁂ CAUTION

Do not rotate the crankshaft more than 90°. If the crankshaft is rotated too much without the timing chain installed, the valves may hit the pistons and cause damage.

9. Turn the crankshaft 90° clockwise to align the adjusting hole of the oil pump drive shaft sprocket with the groove of the oil pump.

10. Remove the crank pulley bolt.

11. Insert a 3 mm diameter bar into the adjusting hole of the oil pump drive shaft sprocket to lock the gear in position, and then remove the nut.

12. Remove the bolt, chain tensioner plate, and spring.

13. Remove the crankshaft timing sprocket, oil pump drive shaft gear, and No. 2 chain sub-assembly.

To install:

14. Set the crankshaft key as shown in the illustration.

15. Turn the drive shaft so that the cutout faces the right horizontal position.

16. Align the yellow mark links with the timing marks of each gear as shown in the illustration.

17. Install the sprockets onto the crankshaft and oil pump shaft with the chain on the gears.

18. Temporarily tighten the oil pump drive shaft sprocket with the nut.

19. Insert the damper spring into the adjusting hole, and then install the chain tensioner plate with the bolt. Tighten to 7 ft. lbs. (10 Nm).

20. Align the adjusting hole of the oil pump drive shaft sprocket with the groove of the oil pump.

21. Insert a 3 mm diameter bar into the adjusting hole of the oil pump drive shaft gear to lock the gear in position, and then tighten the nut. Tighten to 21 ft. lbs. (28 Nm).

22. Install the crankshaft timing sprocket.

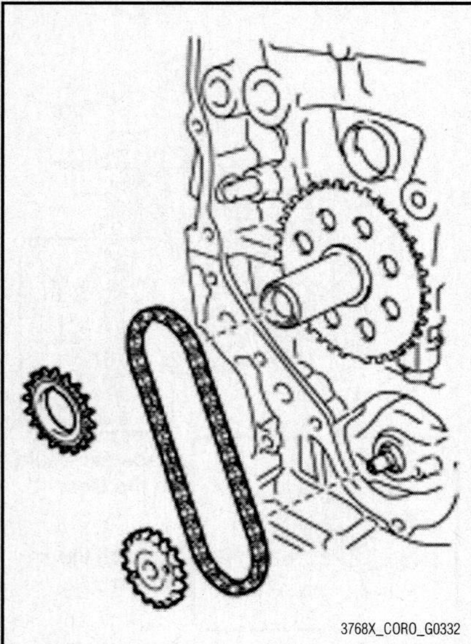

Fig. 144 Removing the crankshaft timing sprocket, oil pump drive shaft gear and the lower timing chain

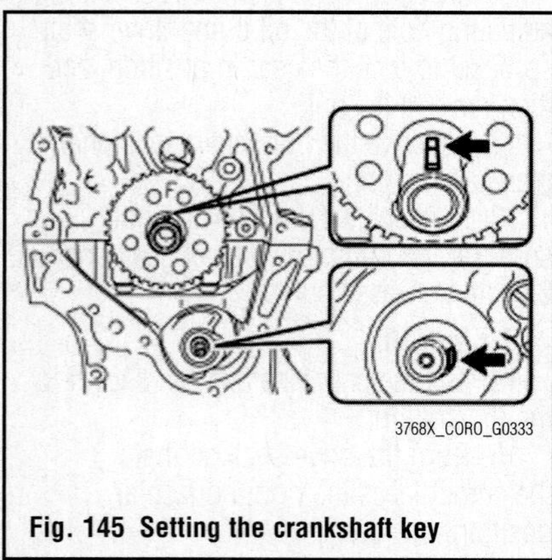

Fig. 145 Setting the crankshaft key

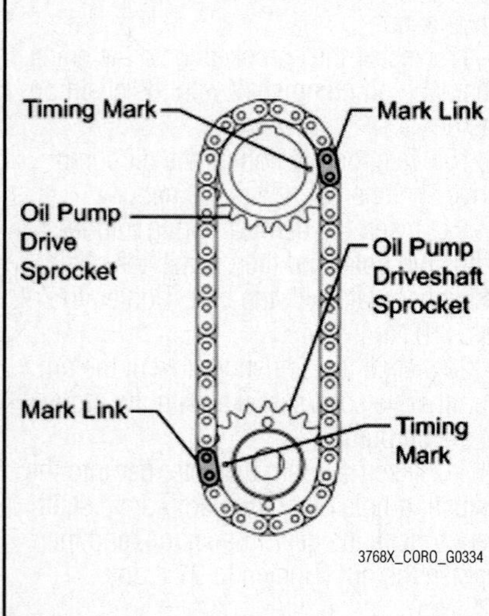

Fig. 146 Aligning the yellow mark links with the timing marks of each gear

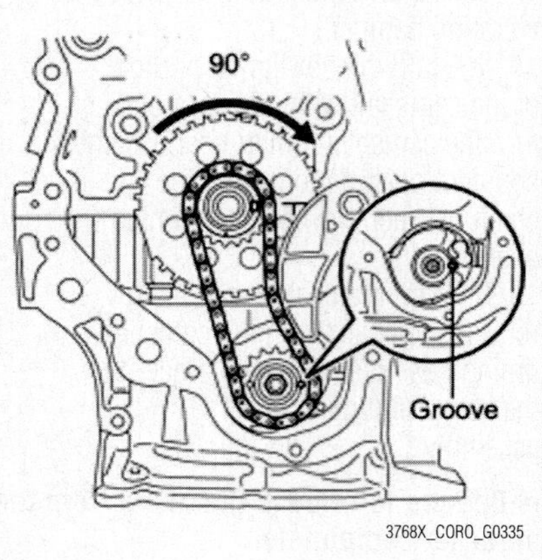

Fig. 147 Aligning the adjusting hole of the oil pump drive shaft sprocket with the groove of the oil pump

23. Install the lower chain vibration damper with the 2 bolts. Tighten to 16 ft. lbs. (21 Nm).
24. Install the upper chain vibration damper with the 2 bolts. Tighten to 7 ft. lbs. (10 Nm).
25. Check the No. 1 cylinder TDC/ compression.
26. Temporarily tighten the crankshaft pulley bolt.
27. Turn the crankshaft counterclockwise to position the timing gear key to the top.
28. Remove the crankshaft pulley bolt.
29. Check the timing marks on each camshaft timing gear.
30. Align the mark plate (orange) with the timing mark of the No. 2 camshaft as shown in the illustration and install the chain.

➡Be sure to position the mark plate at the front of the engine. The mark plate on the camshaft side is colored orange. Do not pass the chain around the sprocket of the camshaft timing gear assembly. Only place it on the sprocket. Pass the chain through the No. 1 vibration damper.

31. Place the chain on the crankshaft without passing it around the shaft.
32. Hold the hexagonal portion of the camshaft with a wrench and turn the camshaft timing gear assembly counterclockwise to align the mark plate (orange) and timing mark.

➡Be sure to position the mark plate at the front of the engine. The mark plate on the camshaft side is colored orange.

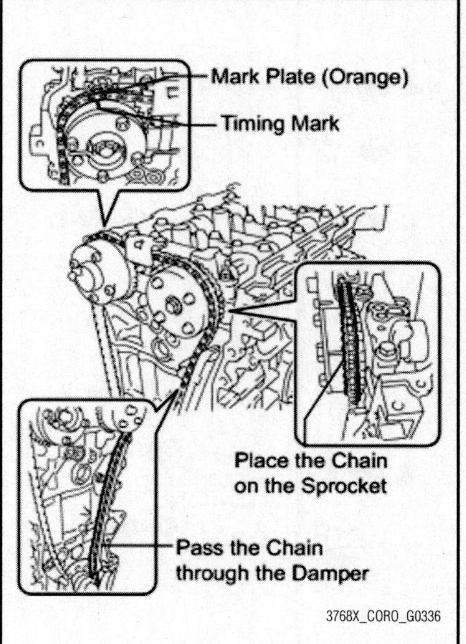

Fig. 148 Align the mark plate (orange) with the timing mark of the No. 2 camshaft

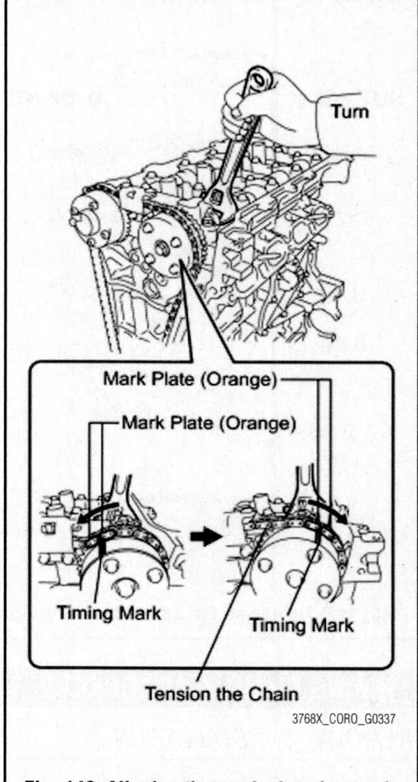

Fig. 149 Aligning the mark plate (orange) and timing mark

33. Hold the hexagonal portion of the camshaft with a wrench and turn the camshaft timing gear assembly clockwise.

➡To tension the chain, slowly turn the camshaft timing gear assembly

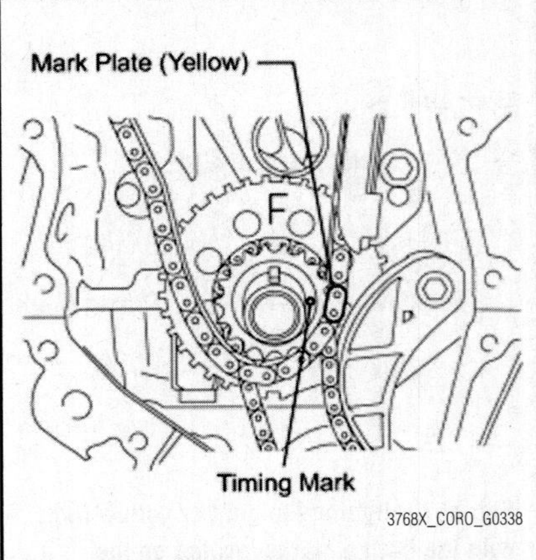

Fig. 150 Aligning the mark plate (yellow) and timing mark on the crankshaft timing gear

clockwise to prevent the chain from being misaligned.

34. Align the mark plate (yellow) and timing mark and install the chain to the crankshaft timing gear.

➡**The mark plate on the crankshaft side is colored yellow.**

35. Recheck each timing mark at TDC/ compression.
36. Install the chain tensioner slipper.
37. Install the upper chain tensioner assembly:
    a. Release the ratchet pawl, then fully push in the plunger and engage the hook to the pin so that the plunger is in the position.

➡**Make sure that the cam engages the first tooth of the plunger to allow the hook to pass over the pin.**

    b. Install a new gasket, bracket and No. 1 chain tensioner with the 2 nuts.

➡**If the hook releases the plunger while the chain tensioner is being installed, engage the hook again.**

    c. Turn the crankshaft counterclockwise, then disconnect the hook from plunger knock pin.
    d. Turn the crankshaft clockwise, then check that the plunger is extended.
38. To complete installation, reverse the remaining removal procedure.

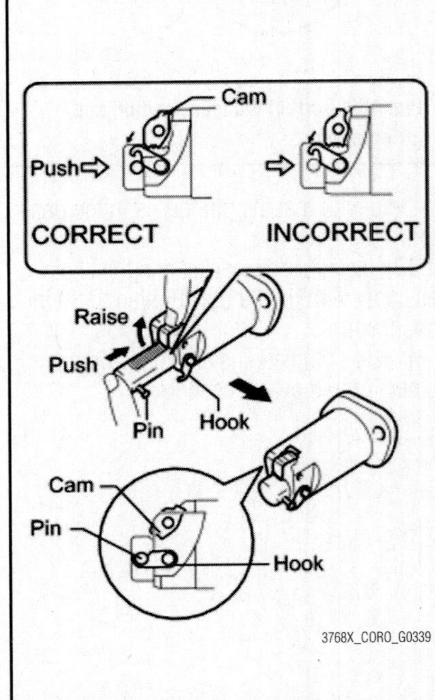

Fig. 151 Correct plunger position and orientation

2.4L Engine

See Figures 152 through 158.

1. Turn the crankshaft pulley until the groove and the timing mark "0" on the timing chain cover are aligned.
2. Check that each timing mark on the camshaft timing gear and sprocket is aligned with the timing marks located on the No. 1 and No. 2 bearing caps.
3. Remove the crankshaft pulley.

➡**Do not turn the crankshaft without the chain tensioner installed.**

➡**Do not lift the engine more than necessary.**

4. Remove the 2 nuts, bracket, tensioner and gasket.
5. Lift the engine upward.
6. Remove the bolt, nut and V-ribbed belt tensioner assembly.
7. Remove the crankshaft sensor.
8. Remove the timing chain and sprocket cover.
9. Remove the No. 1 crankshaft position sensor plate.
10. Remove the bolt and timing chain guide.
11. Remove the bolt and chain tensioner slipper.
12. Remove the 2 bolts and No. 1 chain vibration damper.
13. Remove the chain sub-assembly.
14. Remove the crankshaft timing gear or sprocket from the crankshaft.
15. Turn the crankshaft 90° counterclockwise to align the adjusting hole on the oil pump drive shaft sprocket with the groove on the oil pump.
16. Insert a 4 mm diameter bar into the adjusting hole of the oil pump drive shaft sprocket to lock the gear in position, and then remove the nut.
17. Remove the bolt, chain tensioner plate and spring.
18. Remove the oil pump drive sprocket, oil pump drive shaft sprocket and No. 2 chain.

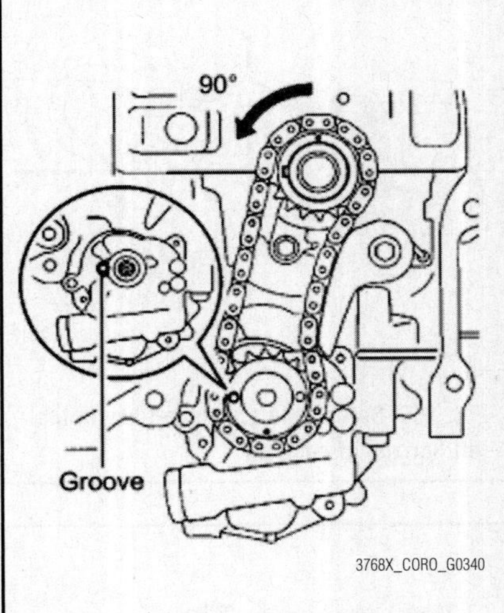

Fig. 152 Aligning the adjusting hole on the oil pump drive shaft sprocket with the groove on the oil pump

To install:

19. Set the crankshaft key in the left horizontal position.
20. Turn the cutout of the drive shaft so that it faces upward.
21. Align the yellow mark links with the timing marks of each gear as shown in the illustration.
22. Install the sprockets onto the crankshaft and oil pump shaft with the chain wrapped on the gears.
23. Temporarily tighten the oil pump drive shaft sprocket with the nut.
24. Insert the damper spring into the adjusting hole, and then install the chain tensioner plate with the bolt. Tighten to 9 ft. lbs. (12 Nm).
25. Align the adjusting hole on the oil pump drive shaft sprocket with the groove on the oil pump.
26. Insert a 4 mm diameter bar into the adjusting hole on the oil pump drive shaft gear to lock the gear in position, and then tighten the nut. Tighten to 22 ft. lbs. (30 Nm).
27. Install the crankshaft timing gear or sprocket to the crankshaft.
28. Install the No. 1 chain vibration damper with the 2 bolts. Tighten to 80 inch lbs. (9 Nm).

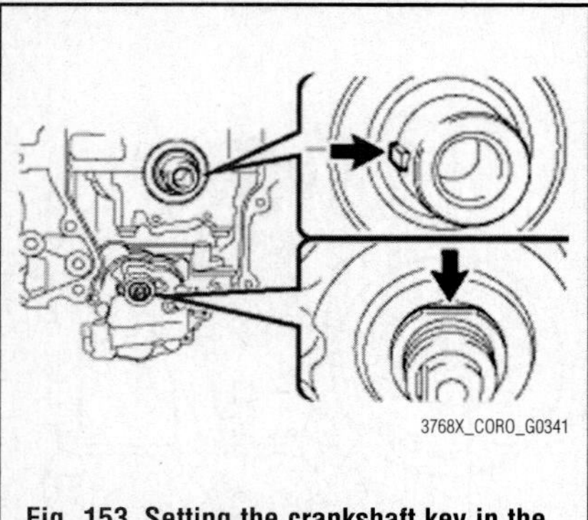

Fig. 153 Setting the crankshaft key in the left horizontal position

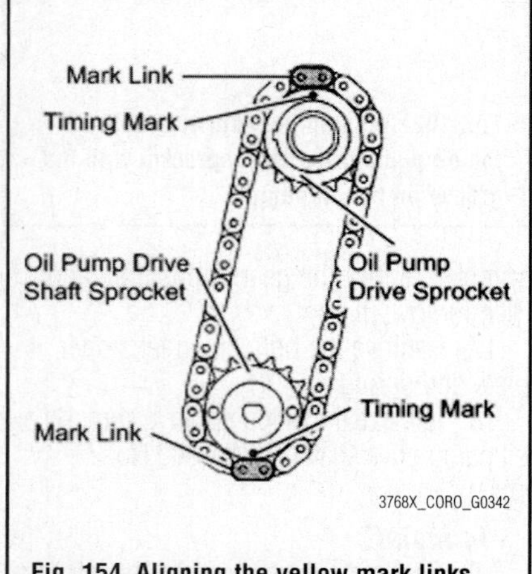

Fig. 154 Aligning the yellow mark links with the timing marks of each gear

29. Set the No. 1 cylinder to TDC/compression.

30. Turn the camshafts with a wrench (using the hexagonal lobe) to align the timing marks on the camshaft timing gear with the timing marks located on the No. 1 and No. 2 bearing caps

31. Using the crankshaft pulley bolt, turn the crankshaft to position the key on the crankshaft upward.

32. Install the chain onto the crankshaft timing sprocket with the gold or orange mark link aligned with the timing mark on the crankshaft.

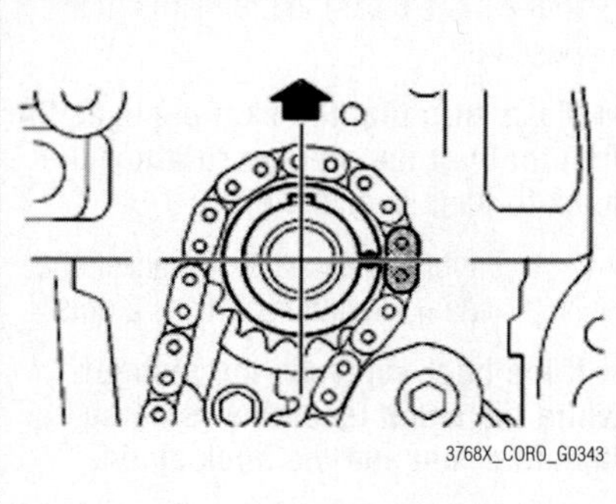

Fig. 155 Turning the crankshaft to position the key on the crankshaft upward

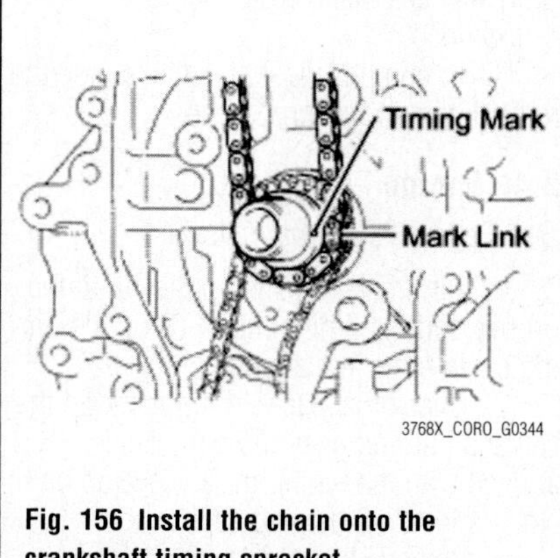

Fig. 156 Install the chain onto the crankshaft timing sprocket

33. Using SST 09309-37010 and a hammer, tap in the crankshaft timing sprocket.

34. Align the gold or yellow links with the timing marks located on the camshaft timing gear and sprocket, then install the chain.

35. Install the chain tensioner slipper with the bolt. Tighten to 14 ft. lbs. (19 Nm).

36. Install the timing chain guide with the bolt. Tighten to 80 inch lbs. (9 Nm).

37. Install the sensor plate with the "F" mark facing forward.

38. Install the timing chain cover.

39. Release the ratchet pawl, then fully push in the plunger and set the hook to the pin so that the plunger is in the position.

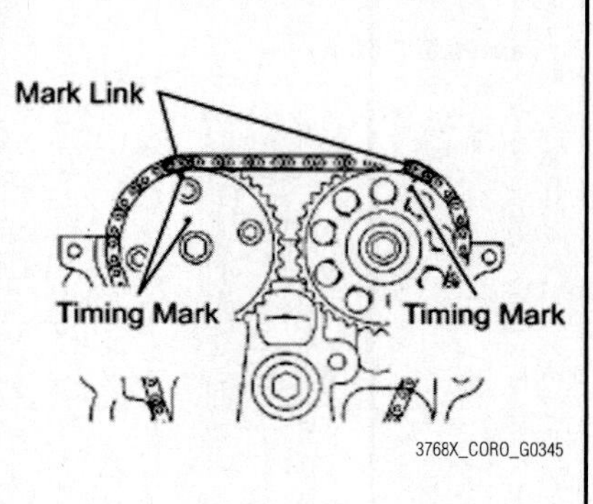

Fig. 157 Aligning the gold or yellow links with the timing marks located on the camshaft timing gear and sprocket

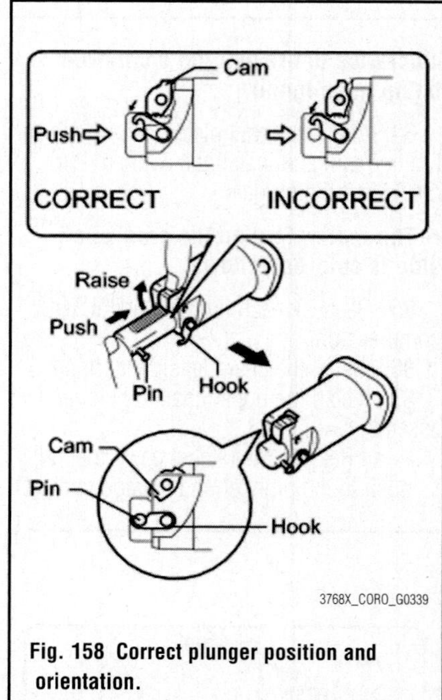

Fig. 158 Correct plunger position and orientation.

40. Install a new gasket and the chain tensioner with the 2 nuts. Tighten to 80 inch lbs. (9 Nm).

41. To complete installation, reverse remaining removal procedure.

ENGINE PERFORMANCE & EMISSION CONTROLS

ACCELERATOR PEDAL POSITION (APP) SENSOR

LOCATION

See Figures 159.

Refer to the accompanying illustration for sensor location

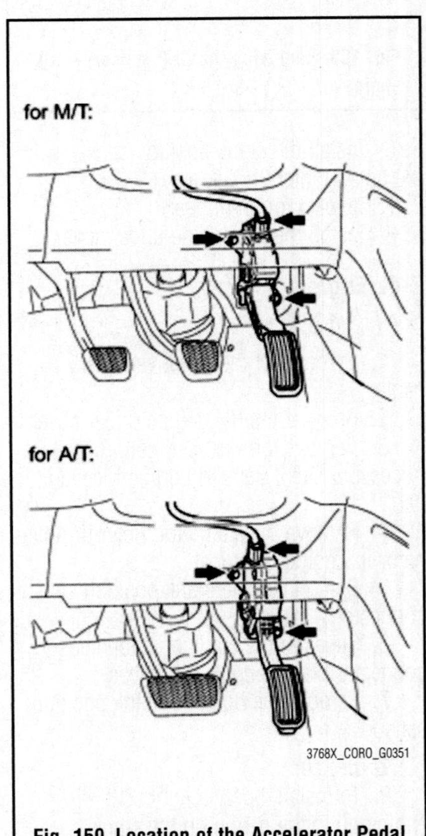

Fig. 159 Location of the Accelerator Pedal Position (APP) sensor

REMOVAL & INSTALLATION

1. Disconnect the negative battery cable.
2. Disconnect the Accelerator Pedal Position (APP) sensor connector.
3. Remove the 2 bolts and accelerator pedal..

To install:

4. To install, reverse the removal procedure.

CAMSHAFT POSITION (CMP) SENSOR

LOCATION

See Figures 160 and 161.

Refer to the accompanying illustrations for sensor location.

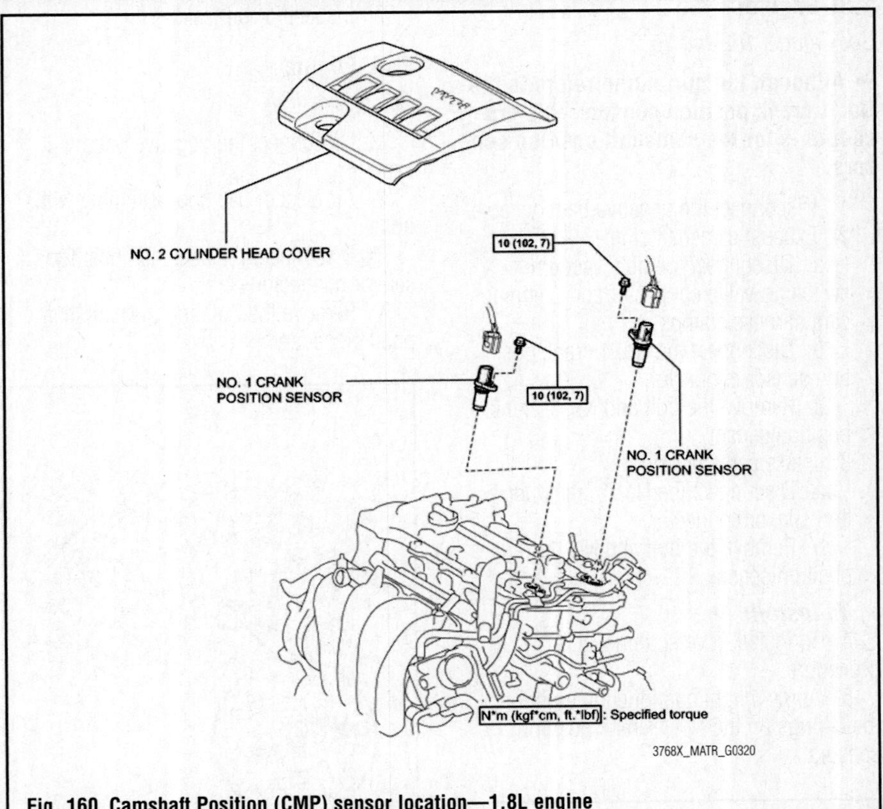

Fig. 160 Camshaft Position (CMP) sensor location—1.8L engine

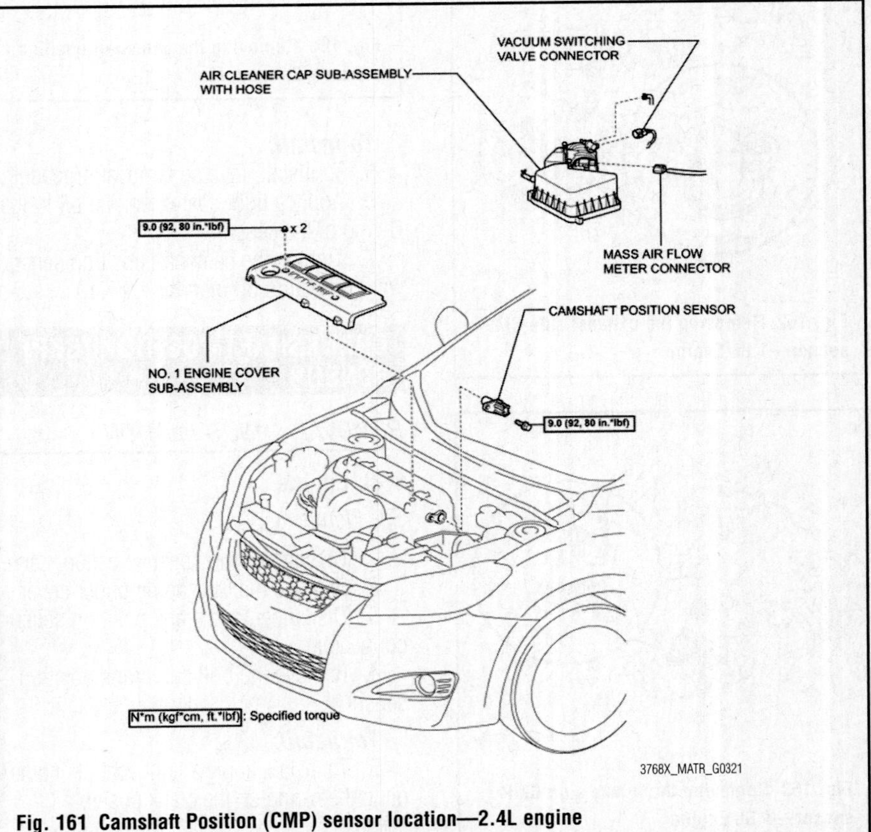

Fig. 161 Camshaft Position (CMP) sensor location—2.4L engine

REMOVAL & INSTALLATION

1.8L Engine

See Figures 162 and 163.

➡ **Although the part name refers to the No. 1 crank position sensors, this procedure is for the camshaft position sensors.**

1. Disconnect the negative battery cable.
2. Exhaust camshaft side:
   a. Disconnect the duty vacuum switching valve connector and 3 engine wire harness clamps.
   b. Disconnect the No. 1 crank position sensor connector.
   c. Remove the bolt and No. 1 crank position sensor.
3. Intake camshaft side:
   a. Disconnect the No. 1 crank position sensor connector.
   b. Remove the bolt and No. 1 crank position sensor.

To install:

4. To install, reverse removal procedure.
5. Apply a light coat of engine oil to the o-rings on the No. 1 crank position sensors.

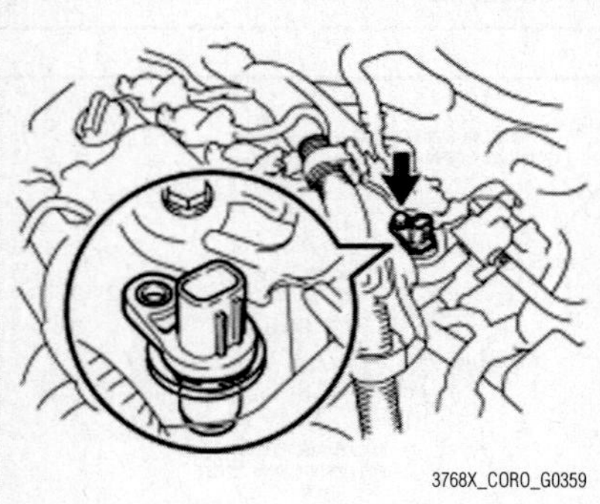

Fig. 162 Removing the exhaust side CMP sensor—1.8L Engine

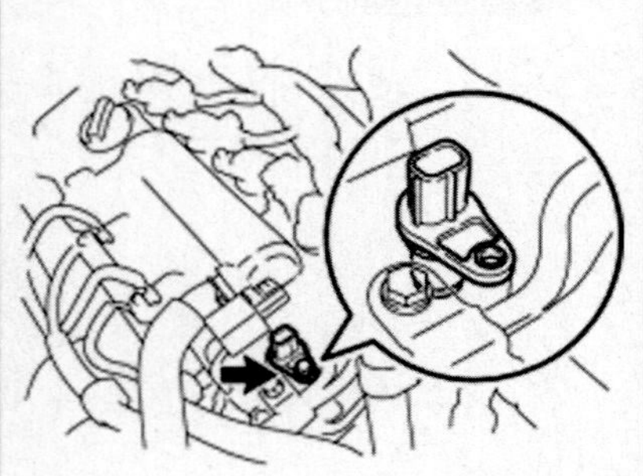

Fig. 163 Removing the intake side CMP sensor—1.8L Engine

6. Tighten the CMP sensor bolts to 7 ft. lbs. (10 Nm).
7. Inspect for oil leaks.

2.4L Engine

See Figure 164.

1. Disconnect the negative battery cable.
2. Air cleaner cap sub-assembly with hose.
3. Disconnect the camshaft position sensor connector.
4. Remove the bolt and the camshaft position sensor.

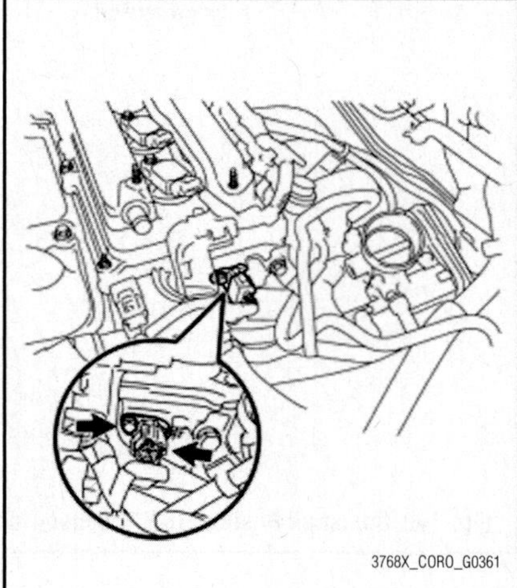

Fig. 164 Removing the camshaft position sensor—2.4L engine

To install:

5. To install, reverse removal procedure.
6. Apply a light coat of engine oil to the O-ring of the sensor.
7. Tighten the camshaft position sensor with the bolt to 80 inch lbs. (9 Nm).

CRANKSHAFT POSITION (CKP) SENSOR

REMOVAL & INSTALLATION

1.8L Engine

See Figure 165.

1. Disconnect the negative battery cable.
2. Remove the RH engine under cover.
3. Disconnect the crank position sensor connector.
4. Remove the bolt and crank position sensor.

To install:

5. To install, apply a light coat of engine oil to the o-ring on the crank position sensor.

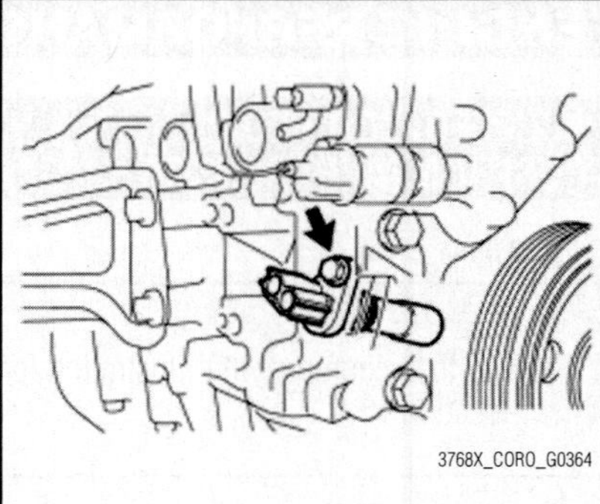

Fig. 165 Removing the CKP sensor—1.8L engine

6. Install the crank position sensor with the bolt. Tighten to 7 ft. lbs. (10 Nm).
7. Inspect for an oil leak.
8. Install the RH engine under cover.

2.4L Engine

See Figure 166.

1. Disconnect the negative battery cable.
2. Remove the RH engine under cover.
3. Remove the v-ribbed belt. Refer to Accessory Drive Belts in Engine Mechanical.
4. Remove the alternator. Refer to Alternator in Engine Electrical.
5. Disconnect the crank position sensor connector.
6. Separate the crank position sensor connector clamp and wire harness.
7. Remove the bolt and crank position sensor.

To install:

8. To install, apply a light coat of engine oil to the o-ring on the crank position sensor.

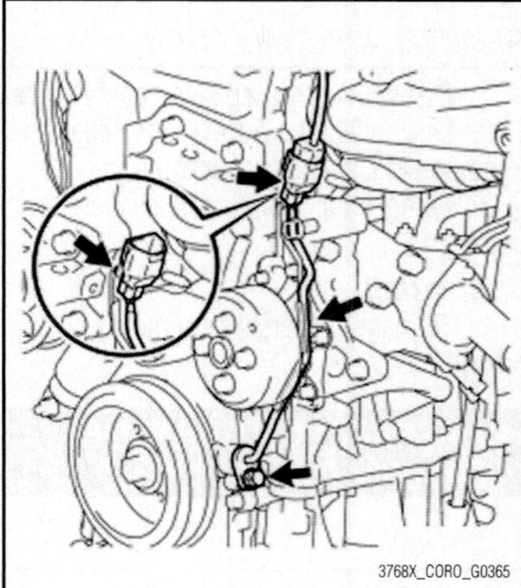

Fig. 166 Removing CKP sensor—2.4L engine

9. Install the crank position sensor with the bolt. Tighten to 80 inch lbs. (9 Nm).

10. To complete installation, reverse the remaining removal procedure.

11. Inspect for an oil leak.

12. Install the RH engine under cover.

ELECTRONIC CONTROL MODULE (ECM)

LOCATION

See Figures 167 and 168.

Refer to the accompanying illustrations for location.

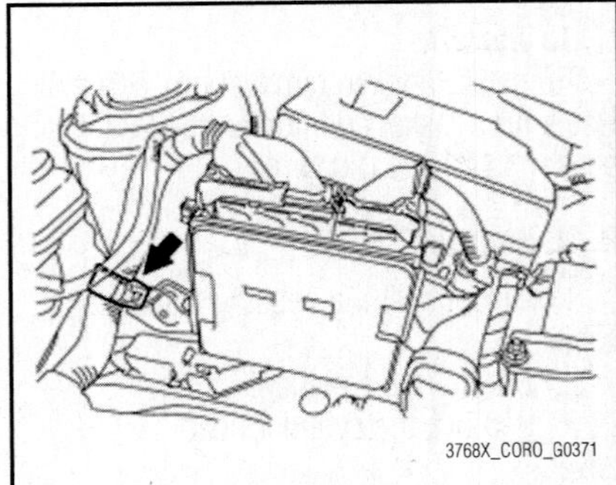

Fig. 167 Location of the ECM—1.8L engine

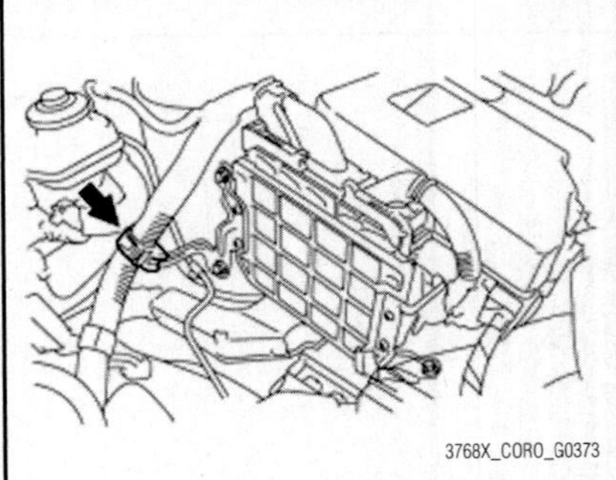

Fig. 168 Location of the ECM—2.4L engine

REMOVAL & INSTALLATION

1. Disconnect the negative battery cable.
2. Remove the air cleaner assembly.
3. Separate the wire harness clamp.
4. Disconnect the 2 ECM connectors. Push the locks on the 2 levers, then raise the levers, and disconnect the 2 ECM connectors.

➡After disconnecting the connectors, make sure that dirt, water or other foreign matter does not contact the connecting parts of the connectors.

5. Remove the 3 bolts and the ECM.
6. Remove the 4 screws and 2 ECM brackets.

To install:

7. To install, reverse the removal procedure.
8. Perform the VIN registration procedure and/or Automatic Transaxle Initialization.

Automatic Transaxle Initialization

➡The ECM memorizes the condition that the ECT controls the automatic transaxle assembly and engine assembly according to those characteristics. Therefore, when the automatic transaxle assembly, engine assembly, or ECM has been replaced, it is necessary to reset the memory so that the ECM can memorize the new information.

1. Turn the ignition switch off.
2. Connect the Techstream to the DLC3.
3. Turn the ignition switch to ON and the Techstream main switch on.
4. Enter the following menus: Powertrain / Engine and ECT / Utility / Reset Memory. Then, press "Next".

⁂ CAUTION

After performing the reset memory, be sure to perform the road test.

5. The ECM is learned by performing the road test.

VIN Registration

➡The Vehicle Identification Number (VIN) must be input into a replacement ECM.

➡The VIN is a 17-digit alphanumeric vehicle identification number. The Techstream is required to register the VIN.

➡This registration section consists of two parts: Read VIN and Write VIN.

1. Read VIN: This process allows the VIN stored in the ECM to be read in order to confirm that the two VINs, provided with the vehicle and stored in the vehicle's ECM, are the same.
2. Write VIN: This process allows the VIN to be input into the ECM. If the ECM is changed, or the ECM VIN and vehicle VIN do not match, the VIN can be registered, or overwritten in the ECM by following this procedure.
3. Read the VIN:
    a. Confirm the vehicle VIN.
    b. Connect the Techstream to the DLC3.
    c. Turn the ignition switch to ON.
    d. Turn the Techstream on.
    e. Enter the following menus: Powertrain / Engine and ECT / Utility / VIN / VIN Read.
4. Write the VIN:
    a. Confirm the vehicle VIN.
    b. Connect the Techstream to the DLC3.
    c. Turn the ignition switch to ON.
    d. Turn the Techstream on.
    e. Enter the following menus: Powertrain / Engine and ECT / Utility / VIN / VIN Write.

ROAD TEST

1. Perform the test at the ATF (Automatic Transmission Fluid) temperature 122 to 176 °F (50 to 80 °C) in the normal operation.
    a. D position test: Shift into the D position and fully depress the accelerator pedal and check the following points.
        - Check up-shift operation. Check that 1 → 2, 2 → 3, 3 → 4 and 4 → 5th up-shifts take place, and that the shift points conform to the automatic shift schedule.

➡5th Gear Up-shift Prohibition Control: Engine coolant temperature is 131°F (55°C) or less and vehicle speed is at 43 mph (70km/h) or less. ATF temperature is 28°F (-2°C) or less.

➡5th and 4th Gear Lock-up Prohibition Control: Brake pedal is depressed. Accelerator pedal is released. Engine coolant temperature is 140°F (60°C) or less.

- Check for shift shock and slip. Check for shock and slip at the 1 → 2, 2 → 3, 3 → 4, and 4→ 5th up-shifts.
- Check for abnormal noise and vibration. Check for abnormal noise and vibration when up-shifting from 1 → 2, 2 → 3, 3 → 4, and 4→ 5 while driving with the shift lever in the D position, and also check while driving in the lock-up condition. The check for the cause of abnormal noise and vibration must be done thoroughly as it could also be due to loss of balance in the differential, torque converter clutch, etc.
- Check kick-down operation.
- Check vehicle speeds when the 2nd to 1st, 3rd to 2nd, 4th to 3rd, and 5th to 4th kick-downs take place

while driving with the shift lever in the D position. Confirm that each speed is within the applicable vehicle speed range indicated in the automatic shift schedule.

- Check for abnormal shock and slip at kick-down.
- Check the lock-up mechanism: Drive in the D position (5th gear), at a steady speed (lock-up ON). Lightly depress the accelerator pedal and check that the engine speed does not change abruptly.

➡There is no lock-up function in the 1st, 2nd and 3rd gears. If there is a big jump in engine speed, there is no lock-up.

b. S position test: Shift to the S position, depress the accelerator pedal and check the following points:

- Check shift operation.
- While driving in the D position and 5th gear, shift into the S position and back to the D position. Check that the gear change 5 → 4down-shift and 4 → 5up-shift can be performed.
- With the shift lever in the S position (while the vehicle is stopped), shift into the "+" position to check that the shift position on the combination meter changes as follows: 1 → 2, 2 → 3, 3 → 4, and 4 → 5.
- While driving in the 4(S) position and 4th gear (at a vehicle speed of approximately 25 to 31mph (40 to 50 km/h)), shift into the "-" position and check if the 3rd gear down-shift occurs and the engine brake performs properly.
- While driving in the 3(S) position and 3rd gear (at a vehicle speed of approximately 19 to 25 mph (30 to 40 km/h)), shift into the "-" position and check if the 2nd gear down-shift occurs and the engine brake performs properly.
- While driving in the 2(S) position and 2nd gear (at a vehicle speed of approximately 12 to 19 mph (20 to 30 km/h)), shift into the "-" position and check if the 1st gear down-shift occurs and the engine brake performs properly.

➡Manual shift (S position) is prohibited under either of the following conditions: Down-shifting may cause engine overrun. The driver continuously down-shifts. (Down-shifting to 1st gear may not be performed.)

c. R position test: Shift into the R position, lightly depress the accelerator pedal, and check that the vehicle moves backward without any abnormal noise or vibration.

✲✲ CAUTION

Before conducting this test ensure that the test area is free from people and obstruction.

d. P position test: Stop the vehicle on a grade (more than 5°) and after shifting into the P position, release the parking brake. Then, check that the parking lock pawl holds the vehicle in place.

e. Uphill/downhill control function:

- Check that the gear does not up-shift to the 4th or 5th gear while the vehicle is driving uphill.
- Check that the gear automatically down-shifts from the 5th to 4th or from the 4th to 3rd gear when brake is applied while the vehicle is driving downhill.

ENGINE COOLANT TEMPERATURE (ECT) SENSOR

LOCATION

See Figures 169 and 170.

Refer to the accompanying illustrations for location.

REMOVAL & INSTALLATION

1.8L Engine

1. Disconnect the negative battery cable.
2. Drain engine coolant.
3. Remove the cylinder head cover.
4. Remove air cleaner cap sub-assembly.
5. Disconnect the Engine Coolant Temperature sensor (ECT) connector.
6. Remove the ECT sensor.

To install:

To install, reverse removal procedure. Use a new gasket and tighten the ECT sensor 14 ft. lbs. (20 Nm).

2.4L Engine

1. Disconnect the negative battery cable.
2. Drain engine coolant.
3. Remove the cylinder head cover.
4. Remove air cleaner cap sub-assembly with hose.
5. Remove air cleaner case sub-assembly.

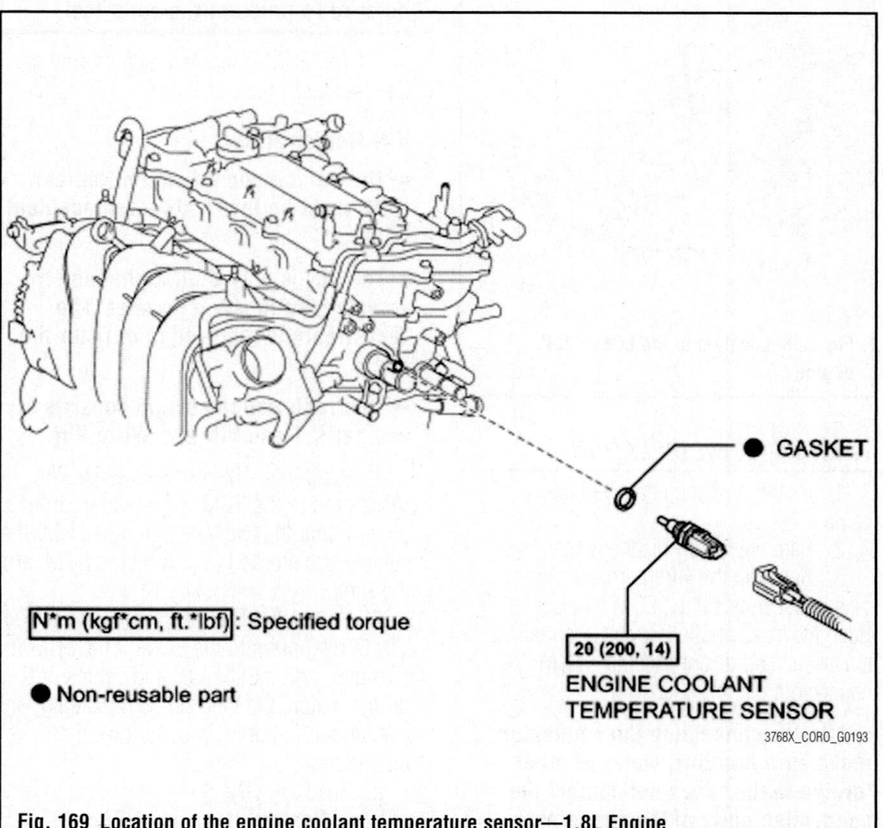

Fig. 169 Location of the engine coolant temperature sensor—1.8L Engine

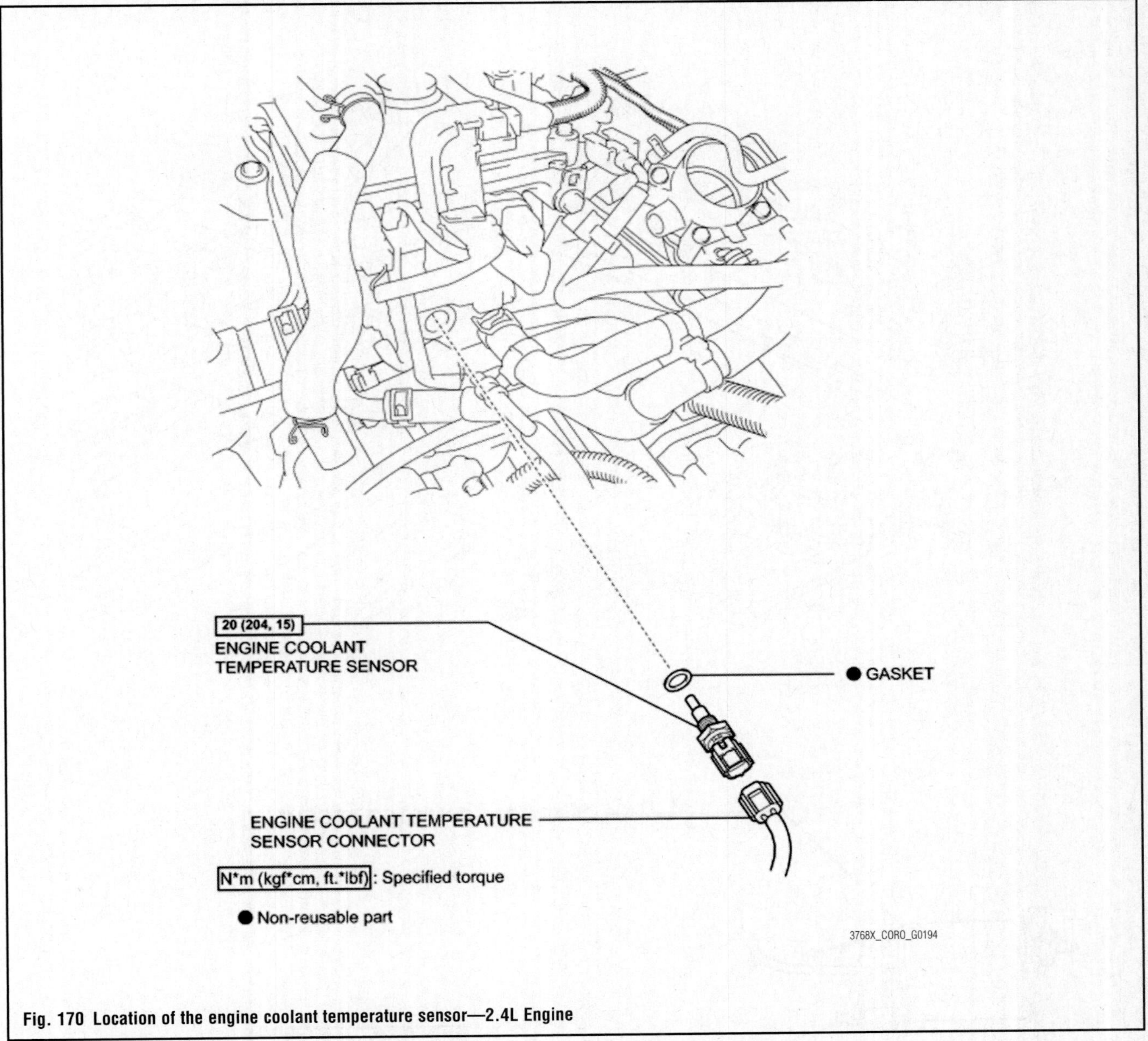

Fig. 170 Location of the engine coolant temperature sensor—2.4L Engine

6. Disconnect the Engine Coolant Temperature sensor (ECT) connector.
7. Remove the engine coolant temperature sensor and gasket.

To install:

To install, reverse removal procedure. Use a new gasket and tighten the ECT sensor 14 ft. lbs. (20 Nm).

EVAPORATIVE EMISSIONS (EVAP) CANISTER

LOCATION

See Figure 171.

Refer to the accompanying illustration for EVAP canister location.

REMOVAL & INSTALLATION

CAUTION

Observe all applicable safety precautions when working around fuel. Whenever servicing the fuel system, always work in a well ventilated area. Do not allow fuel spray or vapors to come in contact with a spark or open flame. Keep a dry chemical fire extinguisher near the work area. Always keep fuel in a container specifically designed for fuel storage; also, always properly seal fuel containers to avoid the possibility of fire or explosion.

1. Disconnect the fuel tank vent hose from the charcoal canister assembly.
2. Disconnect the vapor pressure sensor connector.
3. Disconnect the wire harness clamp.
4. Disconnect the purge line hose from the charcoal canister assembly.
5. Disconnect the charcoal canister filter sub-assembly from the charcoal canister assembly.
6. Remove the 3 bolts and charcoal canister.

To install:

7. To install, reverse the removal procedure.
8. Tighten the 3 bolts to 14 ft. lbs. (19 Nm).

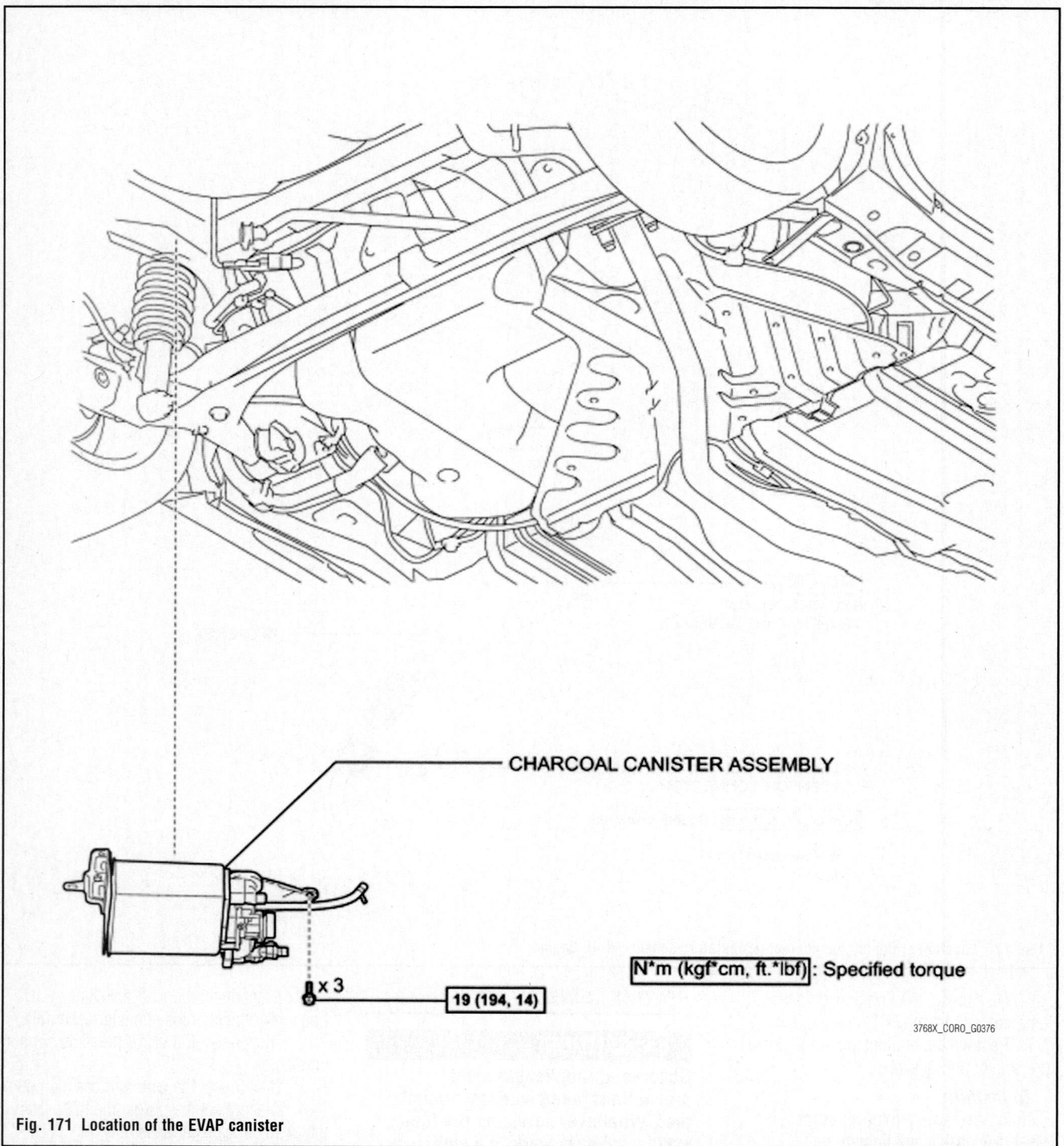

Fig. 171 Location of the EVAP canister

HEATED OXYGEN SENSOR (HO2S)

LOCATION

See Figures 172 and 173.

Refer to the accompanying illustrations for sensor location.

REMOVAL & INSTALLATION

See Figures 174 and 175.

1. Disconnect the negative battery cable.
2. Disconnect the heated oxygen sensor connector.
3. Remove the grommet and pull the sensor connector out of the cabin through the floor panel.
4. Remove the wire harness clamp bracket and disconnect the wire harness clamp.
5. Using a wrench or SST: 09224-00010, remove the heated oxygen sensor.

To install:

6. To install, reverse the removal procedure.
7. Tighten to 32 ft. lbs. (44 Nm).

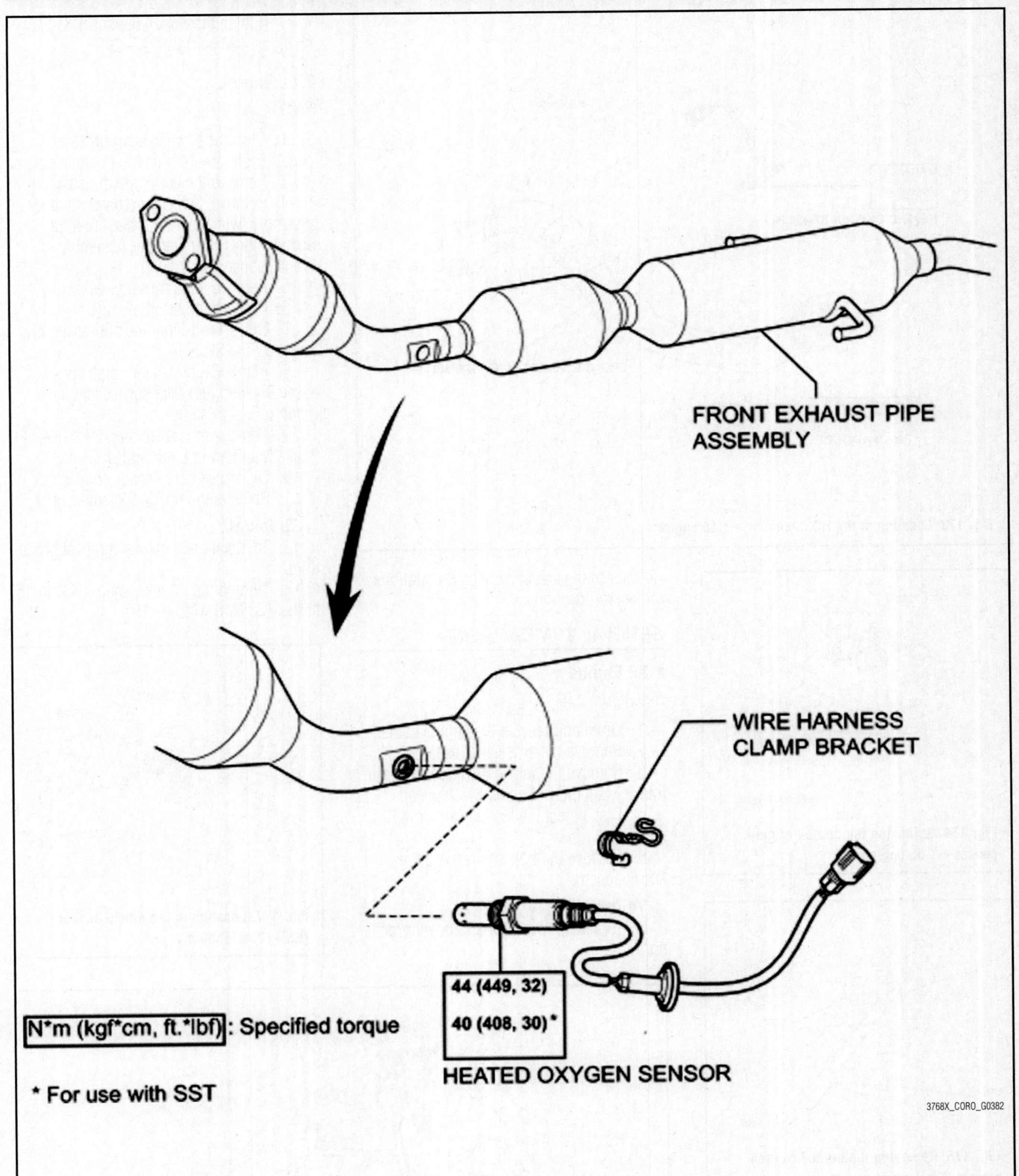

Fig. 172 Location of the HO2S sensor—1.8L engine

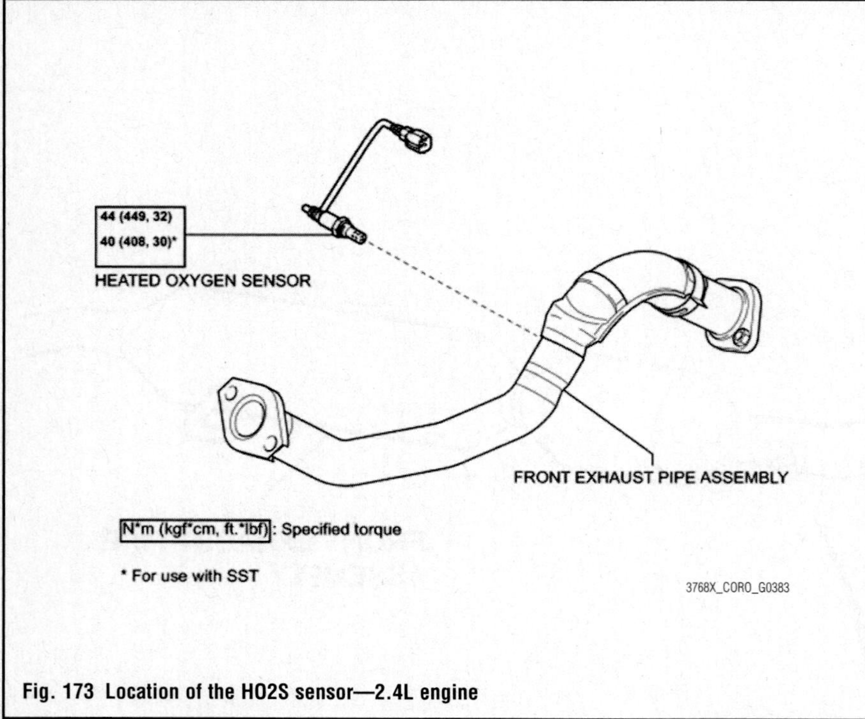

Fig. 173 Location of the HO2S sensor—2.4L engine

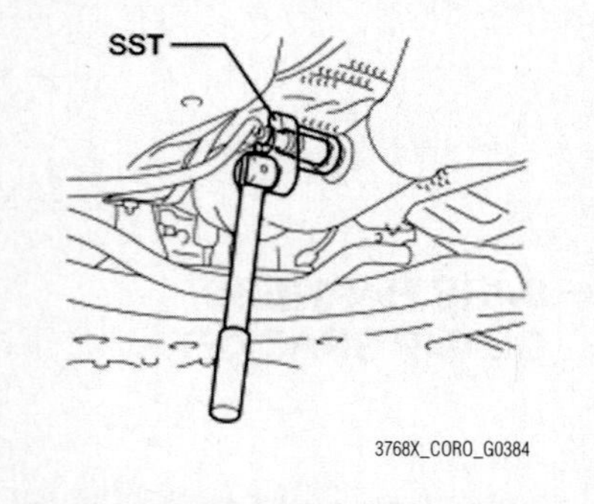

Fig. 174 Removing the heated oxygen sensor—1.8L engine

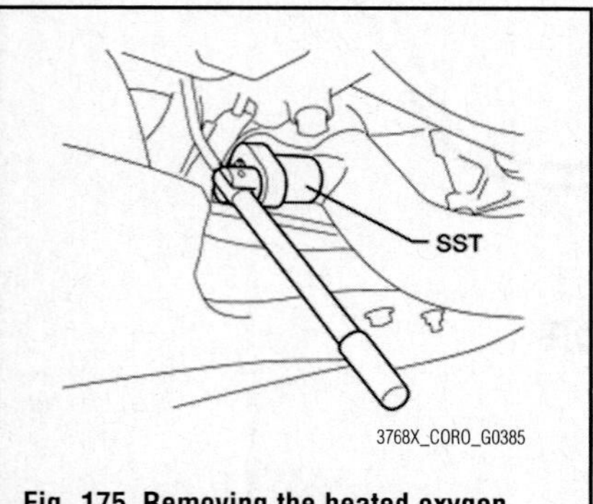

Fig. 175 Removing the heated oxygen sensor—2.4L engine

KNOCK SENSOR (KS)

LOCATION

See Figures 176 and 177.

Refer to the accompanying illustrations for sensor location.

REMOVAL & INSTALLATION

1.8L Engine

See Figure 176.

1. Drain and recycle the engine coolant.
2. Remove the air cleaner assembly.
3. Remove the intake manifold. Refer to Intake Manifold in Engine Mechanical.
4. Disconnect the Knock Sensor (KS) connector.
5. Remove the bolt and remove the knock sensor.

To install:

6. To install, reverse the removal procedure.
7. Tighten the Knock Sensor (KS) to 15 ft. lbs. (20 Nm) with +/- 20°.

2.4L Engine

See Figure 177.

1. Discharge fuel system pressure.
2. Drain and recycle the engine coolant.
3. Remove the wiper motor assembly.
4. Remove suspension tower damper assembly (with front strut bar). Refer to Suspension Tower Damper Assembly in Front Suspension.
5. Remove the air cleaner assembly.
6. Remove the throttle body.
7. Disconnect fuel tube sub-assembly. Refer to Fuel System.
8. Remove fuel delivery pipe sub-assembly with fuel tube sub-assembly. Refer to Fuel System.
9. Remove the intake manifold. Refer to Intake Manifold in Engine Mechanical.
10. Disconnect the sensor connector.
11. Remove the nut and knock sensor.

To install:

12. To install, reverse the removal procedure.
13. Tighten the Knock Sensor (KS) to 15 ft. lbs. (20 Nm) with +/- 10°.

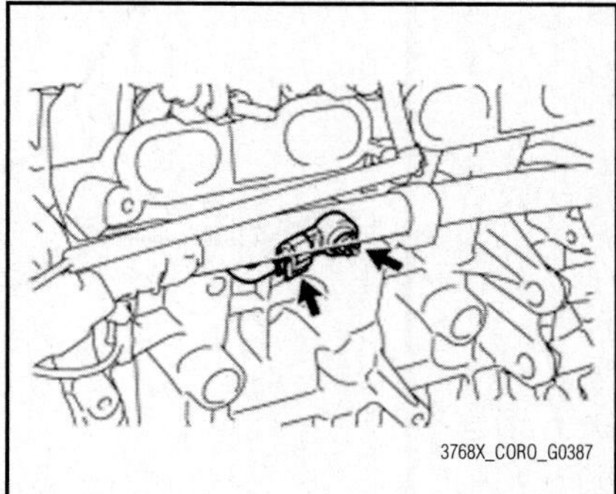

Fig. 177 Location of the Knock Sensor (KS)—2.4L engine

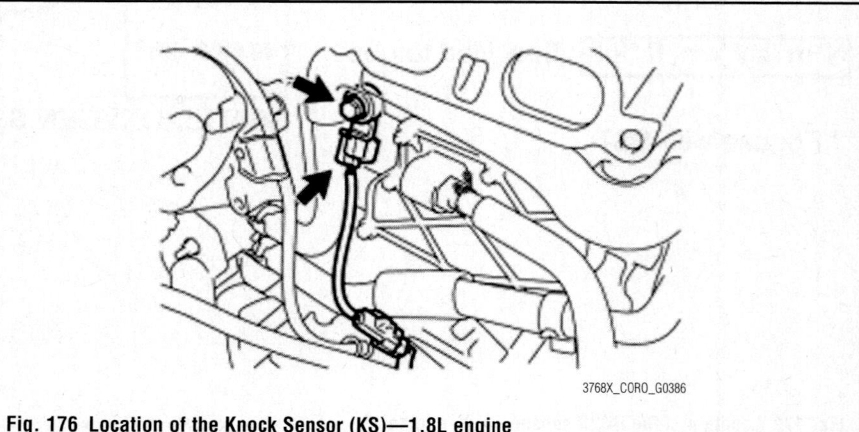

Fig. 176 Location of the Knock Sensor (KS)—1.8L engine

MALFUNCTION INDICATOR LIGHT (MIL)

RESET PROCEDURE

Using an ODB II scan tool, clear all DTC codes to reset malfunction indicator light.

MASS AIR FLOW (MAF) SENSOR

LOCATION

See Figures 178 and 179.

Refer to the accompanying illustrations for sensor location.

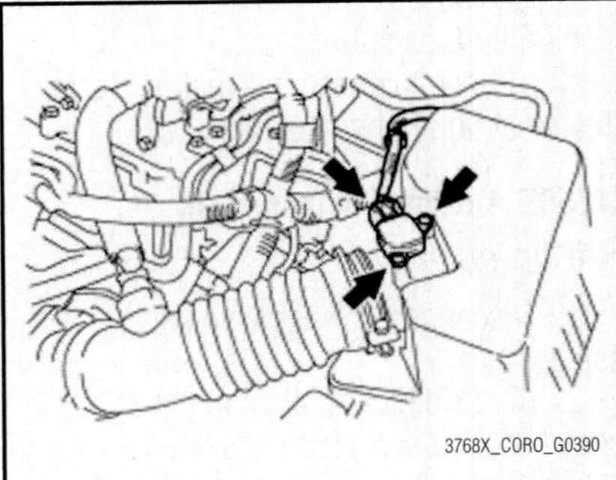

Fig. 178 Location of the MAF sensor—1.8L engine

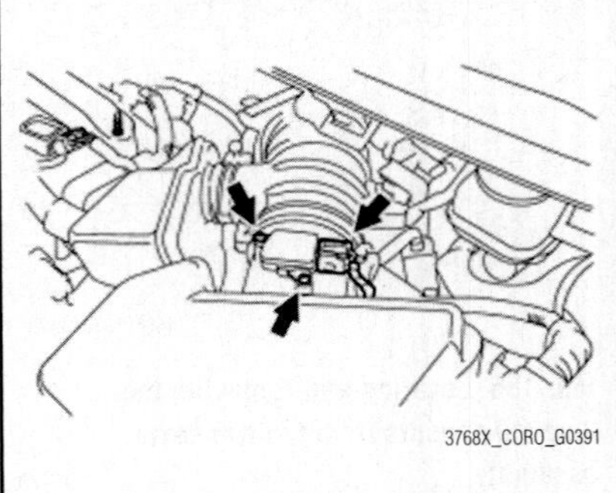

Fig. 179 Location of the MAF sensor—2.4L engine

REMOVAL & INSTALLATION

See Figures 178 and 179.

1. Disconnect the negative battery cable.
2. Disconnect the mass air flow meter connector.
3. Remove the 2 screws and the mass air flow meter.

➡Make sure that the O-ring is not cracked or does not jump out of position during installation.

To install:

4. To install, reverse the removal procedure.

POSITIVE CRANKCASE VENTILATION (PCV) VALVE

LOCATION

See Figure 180.

Refer to the accompanying illustration for location.

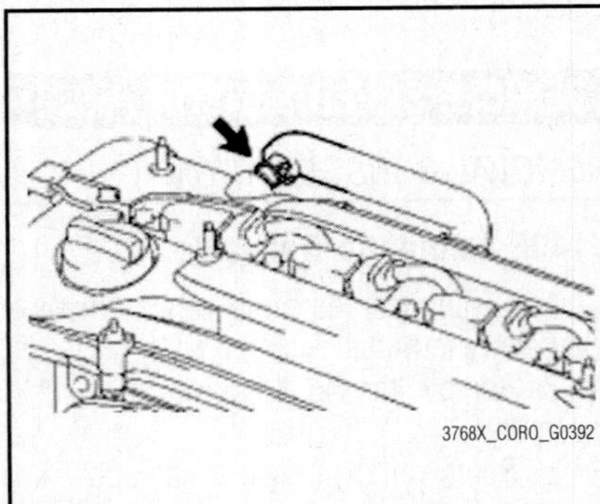

Fig. 180 Location of the PCV valve—2.4L engine

REMOVAL & INSTALLATION

1.8L Engine

1. Remove the intake manifold. Refer to Intake Manifold in Engine Mechanical.
2. Using a ball joint lock nut wrench (22 mm), remove the ventilation valve sub-assembly.

To install:

3. To install, reverse the removal procedure.
4. Apply adhesive to 2 or 3 threads of the ventilation valve sub-assembly.
5. Using a ball joint lock nut wrench (22 mm), install the ventilation valve sub-assembly. Tighten to:
   a. If tightening without a ball joint lock nut wrench: 15 ft. lbs. (20 Nm).
   b. If tightening with a ball joint lock nut wrench: 7 ft. lbs. (10 Nm).

2.4L Engine

1. Disconnect the ventilation hose from the ventilation valve sub-assembly.
2. Using a 22 mm deep socket wrench, remove the ventilation valve sub-assembly.

To install:

3. To install, reverse the removal procedure.
4. Apply adhesive to 2 or 3 threads of the ventilation valve. Use Toyota genuine adhesive 1324, three bond 1324 or equivalent.
5. Using a 22 mm deep socket wrench, install the ventilation valve and tighten to 14 ft. lbs. (19 Nm).
6. Inspect for an oil leak.

VARIABLE CAMSHAFT TIMING OIL CONTROL SOLENOID

LOCATION

See Figures 181 through 183.

Refer to the accompanying illustrations for location.

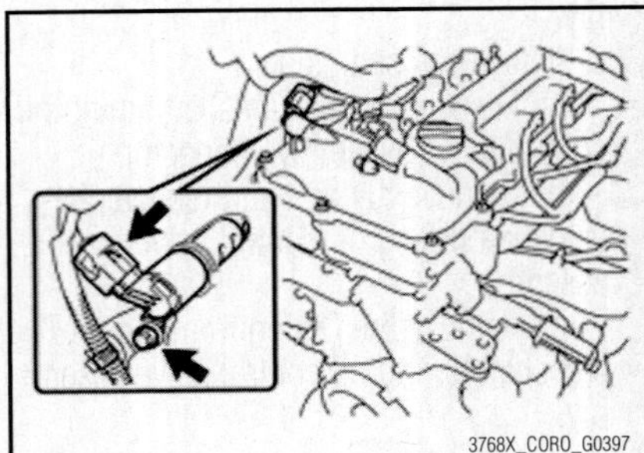

Fig. 181 Location of the camshaft oil control valve—Exhaust camshaft side—1.8L Engine

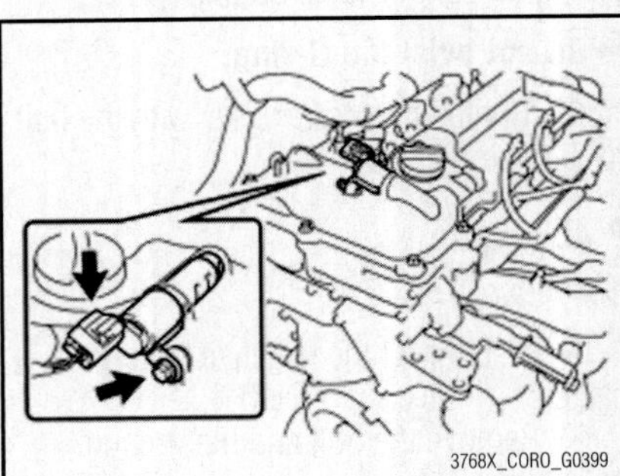

Fig. 182 Location of the camshaft oil control valve—Intake camshaft side—1.8L Engine

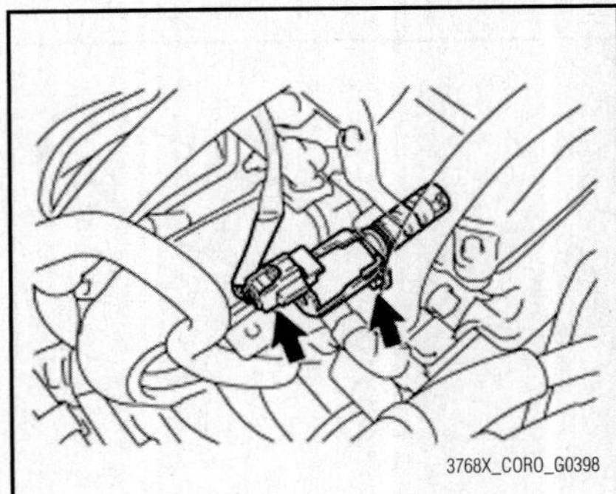

Fig. 183 Location of the camshaft oil control valve—2.4L Engine

REMOVAL & INSTALLATION

1.8L Engine

See Figure 184.

1. Disconnect the negative battery cable.
2. Exhaust camshaft side:
   a. Remove the bolt and the wire harness bracket.
   b. Disconnect the camshaft timing oil control valve assembly connector.
   c. Remove the bolt and the camshaft timing oil control valve assembly and separate the wire harness bracket.
   d. Remove the O-ring from the camshaft timing oil control valve assembly.
3. Intake camshaft side:
   a. Disconnect the camshaft timing oil control valve assembly connector.
   b. Remove the bolt and remove the camshaft timing oil control valve assembly.
   c. Remove the O-ring from the camshaft timing oil control valve assembly.

To install:

4. To install, reverse removal procedure.
5. Apply a light coat of engine oil to a new O-ring and install it onto the camshaft timing oil control valve assembly

➡**Do not twist the O-ring.**

6. Tighten harness bracket with the bolt to 7 ft. lbs. (10 Nm).

2.4L Engine

See Figure 184.

1. Disconnect the negative battery cable.
2. Remove the bolt and the vacuum hose clamp.
3. Disconnect the camshaft timing oil control valve connector.
4. Remove the bolt and the camshaft timing oil control valve assembly.

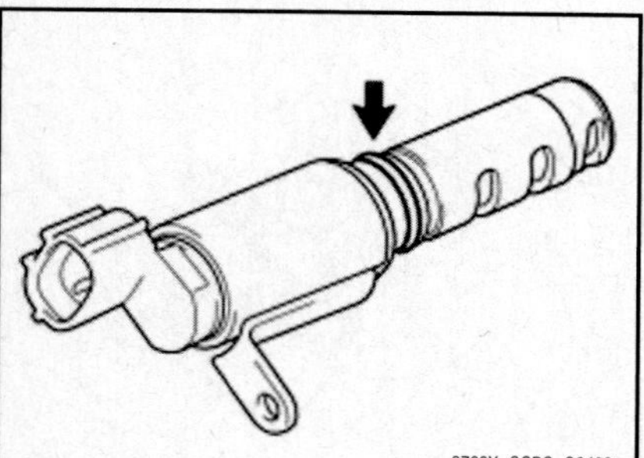

3768X_CORO_G0400

Fig. 184 Removing the O-ring from the camshaft timing oil control valve assembly

5. Remove the O-ring from the camshaft timing oil control valve assembly.

To install:

6. Apply a light coat of engine oil to a new O-ring and install it onto the camshaft timing oil control valve assembly. Tighten to 80 inch lbs. (9 Nm)
7. Install the vacuum hose clamp with the bolt and tighten to 75 inch lbs. (8.5 Nm).
8. Inspect for oil leaks.
9. Install the No. 1 engine cover sub assembly.

VEHICLE SPEED SENSOR (VSS)

REMOVAL & INSTALLATION

U140F Automatic Transaxle

1. Disconnect the cable from the negative battery terminal.
2. Remove the No. 1 engine cover sub assembly.
3. Remove the air cleaner cap sub assembly.
4. Remove the air cleaner case.
5. Remove the battery.
6. Remove the battery carrier.
7. Remove the transmission oil filler tube sub assembly.
   a. Remove the transmission oil level gauge sub assembly.
   b. Disconnect the transmission wire connector, park/ neutral position switch connector and 2 wire harness clamps.
   c. Remove the bolt and transmission oil filler tube sub-assembly.
   d. Remove the O-ring from the transmission oil filler tube sub-assembly.
8. Remove the speed sensor NT.
   a. Disconnect the sensor connector.
   b. Remove the bolt and sensor.
   c. Remove the o-ring from the sensor.
9. Remove the speed sensor NC.
   a. Disconnect the sensor connector.
   b. Remove the bolt and sensor.
   c. Remove the o-ring from the sensor.

To install:

10. Install the speed sensor NC.
    a. Coat a new O-ring with ATF WS and install it to the speed sensor.
    b. Install the speed sensor with the bolt. Tighten to 8 ft. lbs. (11 Nm).
    c. Connect the sensor connector.
11. Install the speed sensor NT.
    a. Coat a new O-ring with ATF WS and install it to the speed sensor.
    b. Install the speed sensor with the bolt. Tighten to 8 ft. lbs. (11 Nm).
    c. Connect the sensor connector.
12. Install the transmission oil filler tube sub assembly.
    a. Apply ATF to a new O-ring, and install it to the transmission oil filler tube sub-assembly.
    b. Install the transmission oil filler tube sub-assembly to the automatic transaxle with the bolt. Tighten to 49 inch lbs. (5.5 Nm).
    c. Connect the transmission wire connector, park/ neutral position switch connector and 2 wire harness clamps.
    d. Install the transmission oil level gauge sub-assembly to the transmission oil filler tube sub-assembly.
13. To complete the installation, reverse the remaining removal procedures.

U250E Automatic Transaxle

See Figures 185 and 186.

1. Disconnect the negative battery cable.
2. Remove the 2 bolts and the 2 speed sensors from the transaxle assembly.

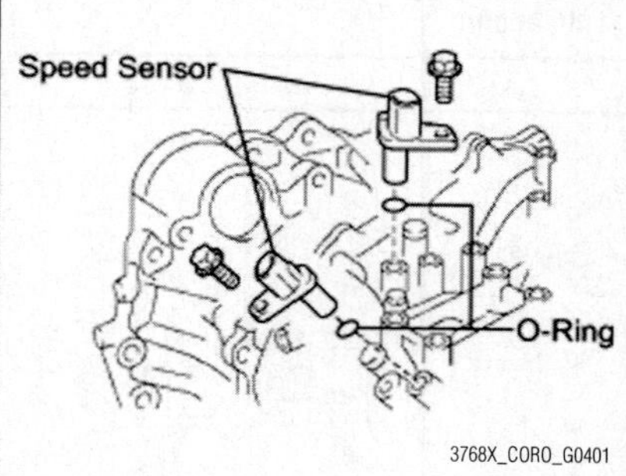

3768X_CORO_G0401

Fig. 185 Locating and removing the 2 speed sensors from the transaxle assembly

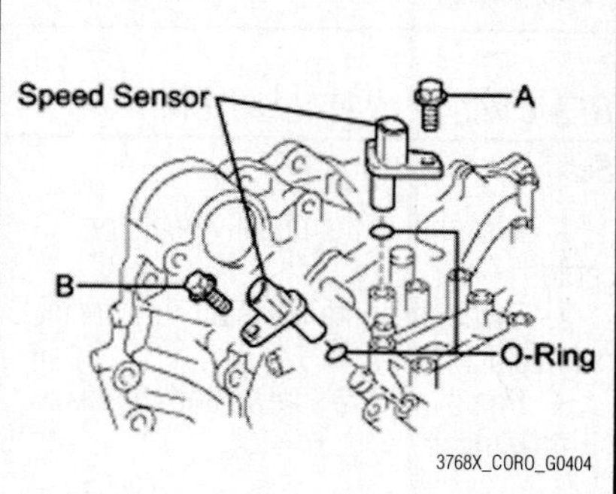

3768X_CORO_G0404

Fig. 186 Installing the VSS fasteners—U250E

3. Remove the 2 O-rings from the 2 speed sensors.

To install:

4. Coat 2 new O-rings with ATF, and install them to the 2 speed sensors.
5. Apply adhesive to the bolts threads. Toyota Genuine Adhesive 1344, Three Bond 1344 or Equivalent.
6. Install the 2 speed sensors to the transaxle case with the 2 bolts:
   a. Bolt A: 78 inch lbs. (9 Nm).
   b. Bolt B: 8 ft. lbs. (11 Nm).

U341E Automatic Transaxle

See Figure 187.

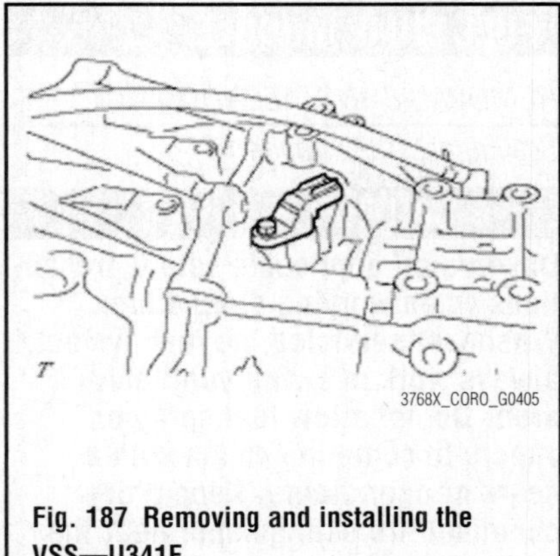

Fig. 187 Removing and installing the VSS—U341E

1. Disconnect the negative battery cable.
2. Disconnect the electrical connector
3. Remove the bolt and speed sensor from the transaxle case.

To install:

4. To install, coat a new O-ring with ATF, and install it to the speed sensor.
   a. Install the speed sensor to the transaxle case with the bolt. Tighten to 48 inch. lbs. (5.4 Nm).

FUEL — GASOLINE FUEL INJECTION SYSTEM

FUEL SYSTEM SERVICE PRECAUTIONS

Safety is the most important factor when performing not only fuel system maintenance, but any type of maintenance. Failure to conduct maintenance and repairs in a safe manner may result in serious personal injury or death. Work on a vehicle's fuel system components can be accomplished safely and effectively by adhering to the following rules and guidelines.

- To avoid the possibility of fire and personal injury, always disconnect the negative battery cable unless the repair or test procedure requires that battery voltage be applied.
- Always relieve the fuel system pressure prior to disconnecting any fuel system component (injector, fuel rail, pressure regulator, etc.) fitting or fuel line connection. Exercise extreme caution whenever relieving fuel system pressure to avoid exposing skin, face and eyes to fuel spray. Please be advised that fuel under pressure may penetrate the skin or any part of the body that it contacts.
- Always place a shop towel or cloth around the fitting or connection prior to loosening to absorb any excess fuel due to spillage. Ensure that all fuel spillage is quickly removed from engine surfaces. Ensure that all fuel-soaked cloths or towels are deposited into a flame-proof waste container with a lid.
- Always keep a dry chemical (Class B) fire extinguisher near the work area.
- Do not allow fuel spray or fuel vapors to come into contact with a spark or open flame.
- Always use a second wrench when loosening or tightening fuel line connection fittings. This will prevent unnecessary stress and torsion on fuel piping. Always follow the proper torque specifications.
- Always replace worn fuel fitting O-rings with new ones. Do not substitute fuel hose where rigid pipe is installed.

FUEL SYSTEM PRESSURE

RELIEVING

CAUTION

Observe all applicable safety precautions when working around fuel. Whenever servicing the fuel system, always work in a well ventilated area. Do not allow fuel spray or vapors to come in contact with a spark or open flame. Keep a dry chemical fire extinguisher near the work area. Always keep fuel in a container specifically designed for fuel storage; also, always properly seal fuel containers to avoid the possibility of fire or explosion.

CAUTION

Perform the following procedure to prevent fuel from spilling out before removing any fuel system parts. Pressure will still remain in the fuel lines even after performing the following procedure. When disconnecting a fuel line, cover it with a piece of cloth to prevent fuel from spraying or coming out.

1. Remove the rear seat cushion assembly.
2. Remove the rear floor service hole cover.
3. Disconnect the connector from the fuel pump assembly.
4. Start the engine. After the engine stops naturally, turn the ignition switch off.

➡Do not increase engine speed or drive the vehicle while waiting for the engine to stop naturally.

➡DTC P0171/25 (system too lean) may be set.

5. Crank the engine again and make sure that the engine does not start.
6. Remove the fuel tank cap and discharge the pressure from the fuel tank.
7. Disconnect the cable from the negative (-) battery terminal.
8. Connect the connector of the fuel pump assembly.
9. Install the rear floor service hole cover.
10. Install the rear seat cushion assembly.

FUEL FILTER

REMOVAL & INSTALLATION

The fuel filter is integrated in the fuel pump and must be replaced as a unit. Refer to Fuel Pump in this section.

FUEL LEVEL SENDING UNIT

REMOVAL & INSTALLATION

See Figure 188.

✲✲ CAUTION

Observe all applicable safety precautions when working around fuel. Whenever servicing the fuel system, always work in a well ventilated area. Do not allow fuel spray or vapors to come in contact with a spark or open flame. Keep a dry chemical fire extinguisher near the work area. Always keep fuel in a container specifically designed for fuel storage; also, always properly seal fuel containers to avoid the possibility of fire or explosion.

1. Discharge fuel system pressure.
2. Disconnect the negative battery cable.
3. Remove the fuel suction tube assembly with pump and gauge from the fuel tank. Refer to Fuel Pump Module in this section.
4. Disconnect the fuel sender gauge connector.

➡Do not damage the wire harness.

5. Release the lock and slide the fuel sender gauge assembly to remove it.

➡Make sure that the fuel sender gauge arm does not bend.

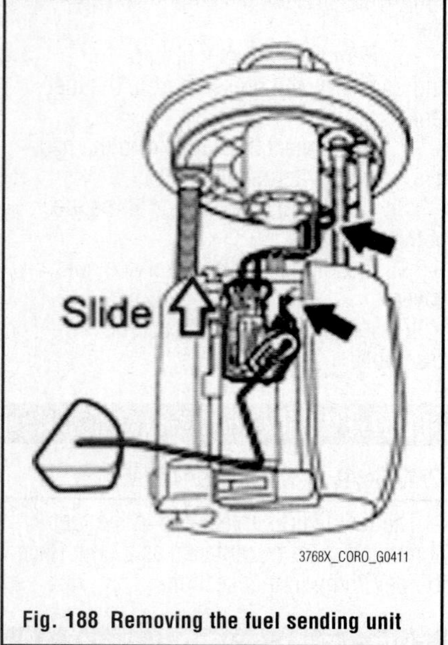

Fig. 188 Removing the fuel sending unit

To install:

6. To install, reverse removal procedure.
7. Install the fuel sender gauge assembly by sliding it downward.
8. Inspect for fuel leaks.

FUEL PUMP MODULE

REMOVAL & INSTALLATION

See Figures 189 through 194.

✲✲ CAUTION

Observe all applicable safety precautions when working around fuel. Whenever servicing the fuel system, always work in a well ventilated area. Do not allow fuel spray or vapors to come in contact with a spark or open flame. Keep a dry chemical fire extinguisher near the work area. Always keep fuel in a container specifically designed for fuel storage; also, always properly seal fuel containers to avoid the possibility of fire or explosion.

1. Remove the rear seat cushion assembly.
2. Remove the rear floor service hole cover.
3. Disconnect the connector from the fuel suction tube assembly.
4. Properly relieve the fuel system pressure.
5. Disconnect the negative battery cable.
6. Disconnect the fuel tank main tube:
    a. Remove the tube joint clip, and pull the fuel tube joint out of the plug of the fuel suction plate.

➡Check that there is no dirt or other foreign objects around the fuel tube joint before disconnecting it. Clean the joint if necessary.

- It is necessary to prevent mud or dirt from entering the joint. If mud or dirt gets in the joint, the O-rings may not seal properly.
- Only disconnect the joint by hand.
- Do not bend, kink or twist the nylon tubes.
- Protect the joint by covering it with a plastic bag.

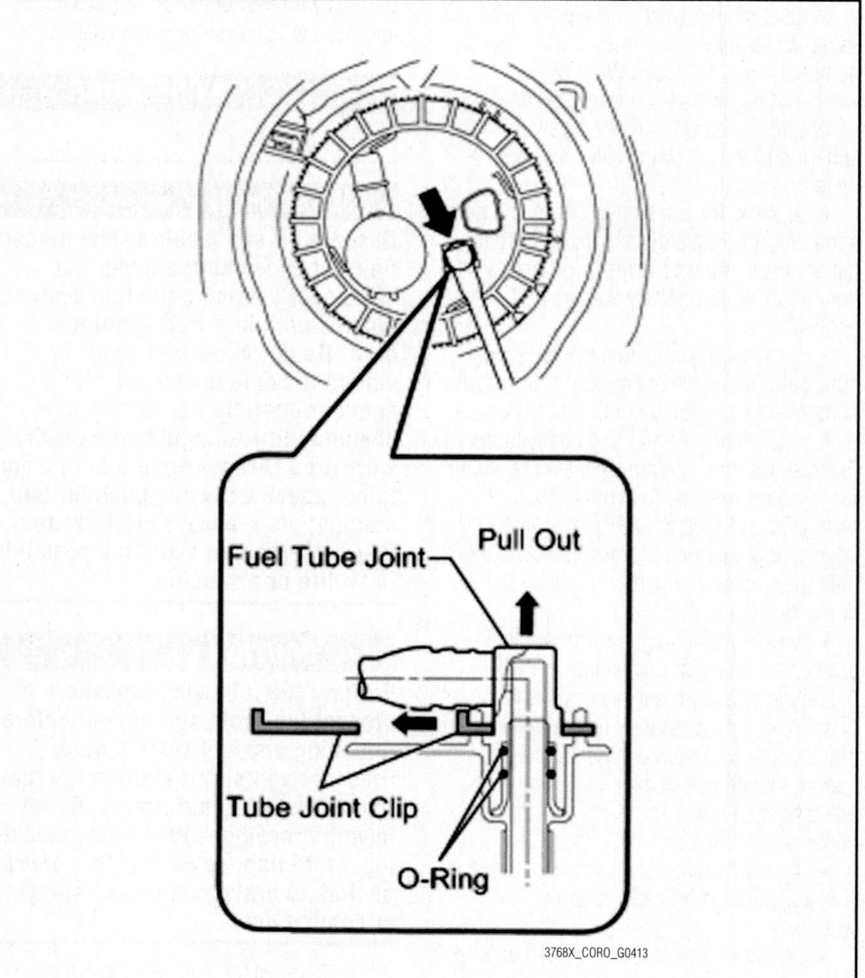

Fig. 189 Disconnecting the fuel tank main tube

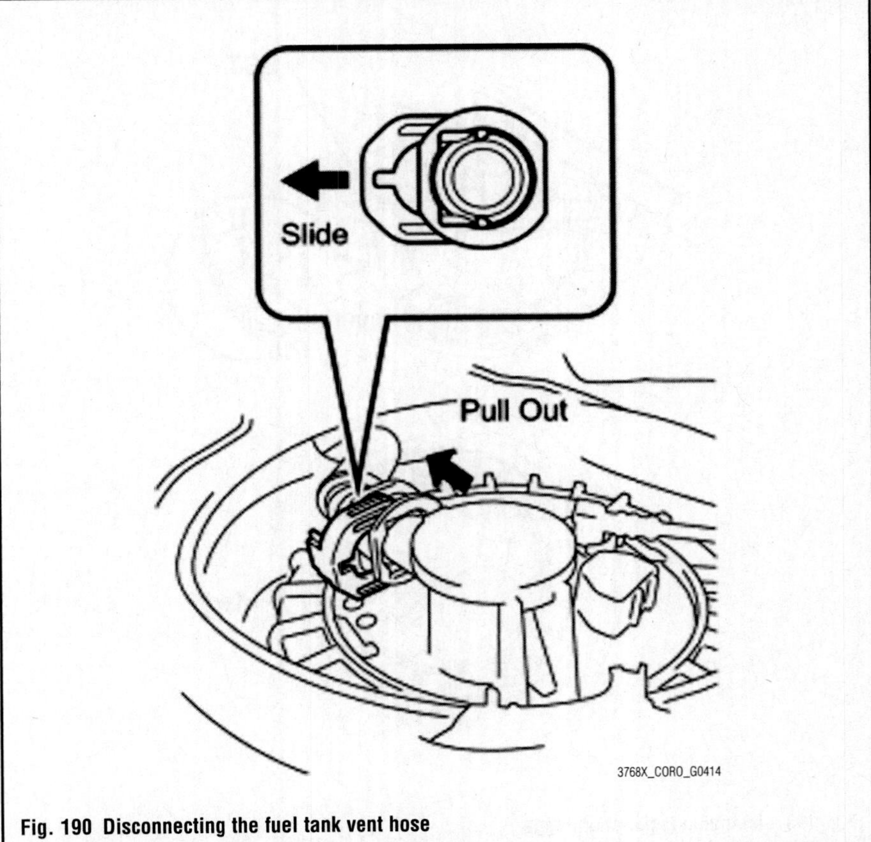

Fig. 190 Disconnecting the fuel tank vent hose

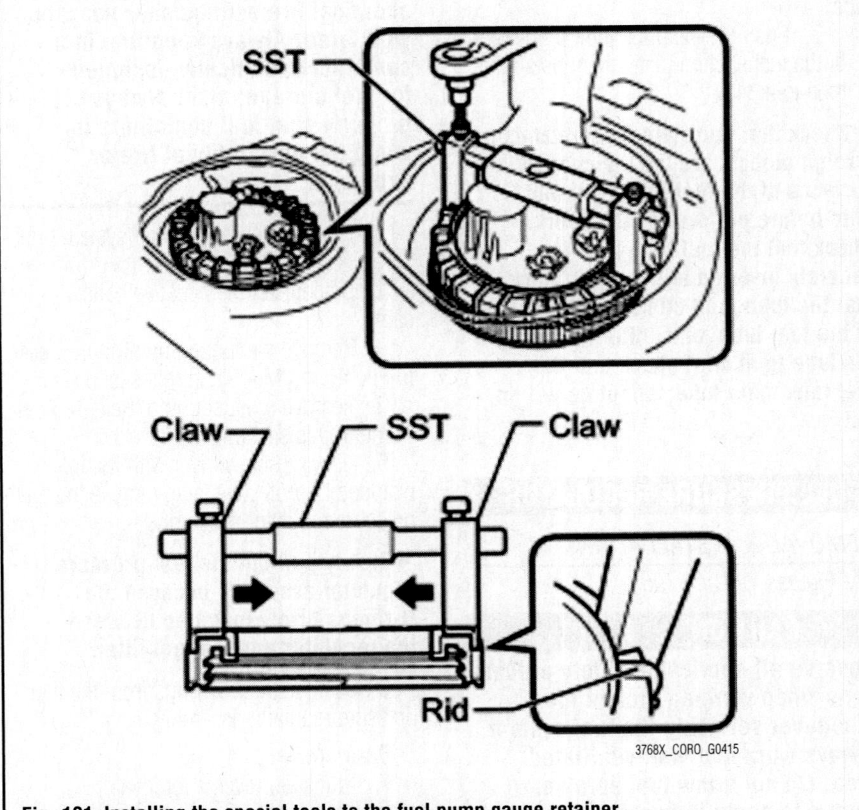

Fig. 191 Installing the special tools to the fuel pump gauge retainer

7. Disconnect the fuel tank vent hose:
   a. Pinch the retainer and pull the fuel tank vent connector out of the fuel tank to disconnect the fuel tank vent hose from the fuel suction plate.

➡Check that there is no dirt or other foreign objects around the fuel tube joint before disconnecting it. Clean the joint if necessary.

- It is necessary to prevent mud or dirt from entering the joint. If mud or dirt gets in the joint, the O-rings may not seal properly.
- Only disconnect the joint by hand.
- Do not bend, kink or twist the nylon tubes.
- Protect the joint by covering it with a plastic bag.
- If the pipe and connector are stuck, carefully try wiggling or pushing and pulling on the connector to release it. Pull the connector off the pipe carefully.

8. Remove the fuel pump gauge retainer:
   a. Using a 6 mm socket hexagon wrench, set SST 09808-14020 to the fuel pump gauge retainer.

➡Engage the SST claws securely with the fuel pump gauge retainer ribs to secure SST. Install SST while pressing the SST claws toward the fuel pump gauge retainer (towards the center of SST).

   b. Using SST 09808-14020, loosen the fuel pump gauge retainer.

➡Do not use any tools other than specified in this operation. Damage to the fuel pump gauge retainer or the fuel tank. Loosen the retainer by turning it counterclockwise while holding SST down. Do not allow the claw of the tank suction tube support to slip out of its groove on the fuel tank.

➡The ribs on the fuel pump gauge retainer can be fitted into the tips of SST.

9. Remove the fuel pump gauge retainer while holding the fuel suction tube assembly by hand.

➡Make sure the fuel level sending unit arm does not bend.

10. Remove the fuel pump module from the fuel tank.
11. Remove the gasket from the fuel tank.

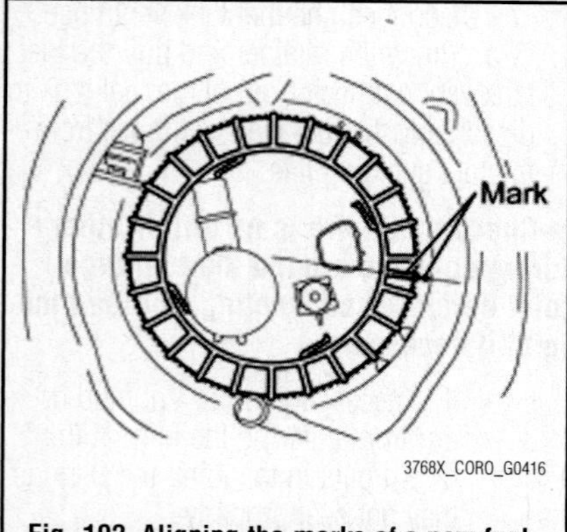

Fig. 192 Aligning the marks of a new fuel pump gauge retainer

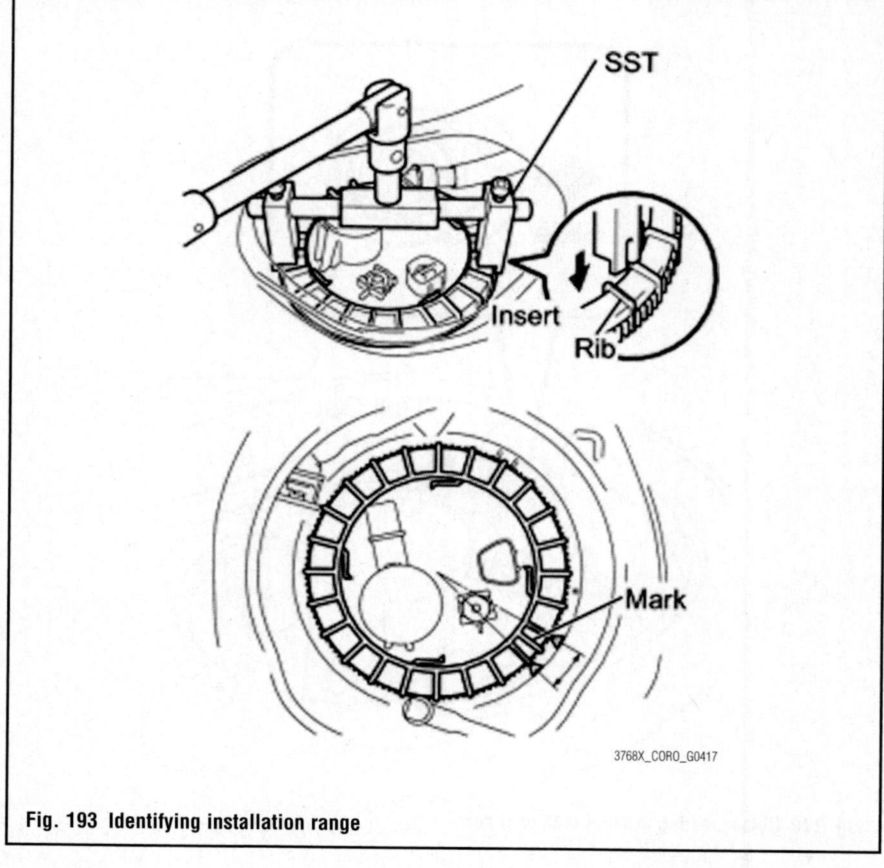

Fig. 193 Identifying installation range

To install:

12. Install a new gasket onto the fuel tank.
13. Set the fuel pump module in to the fuel tank.

➡Make sure the fuel level sending unit arm does not bend.

14. Install the fuel pump gauge retainer:
    a. Align the protrusion of the fuel suction tube assembly with the notch of the fuel tank.
    b. While holding the fuel suction tube assembly by hand, align the marks of a new fuel pump gauge retainer and fuel tank as shown in the illustration, then install the fuel pump gauge retainer.
    c. Using SST and your hand, tighten the fuel pump gauge retainer 2 revolutions so that the mark of the retainer comes within the range shown in the illustration.

➡Use the SST. Do not use any other tools such as a screwdriver.

➡Insert the notch of SST into the rib of the fuel pump gauge retainer.

15. Connect the fuel tank vent hose:
    a. Align the fuel tank vent connector with the pipe, then push in the fuel tank vent connector until the retainer makes a "click" sound to connect the fuel tank vent hose to the fuel suction plate.

➡Check that there are no scratches or foreign objects around the connecting surfaces of the fuel tank vent connector and pipe before performing this work. After connecting the fuel tank vent hose, check that the fuel tank vent hose is securely connected by pulling on the quick connector.

16. Connect the fuel tank main tube:
    a. Push the fuel tube joint in the plug of the fuel suction plate, then install the tube joint clip.

➡Check that there are no scratches or foreign objects around the connecting surfaces of the fuel tube joint and plug before performing this work. Check that the fuel tube joint is securely inserted to the end. Check that the tube joint clip is on the collar of the fuel tube joint. After installing the tube joint clip, check that the fuel tank main tube cannot be pulled out.

FUEL PRESSURE REGULATOR

REMOVAL & INSTALLATION

See Figures 195 and 196.

⁂ CAUTION

Observe all applicable safety precautions when working around fuel. Whenever servicing the fuel system, always work in a well ventilated area. Do not allow fuel spray or vapors to come in contact with a spark or open flame. Keep a dry chemical fire extinguisher near the work area. Always keep fuel in a container specifically designed for fuel storage; also, always properly seal fuel containers to avoid the possibility of fire or explosion.

1. Properly relieve the fuel system pressure.
2. Disconnect the negative battery cable.
3. Remove the fuel pump module. Refer to Fuel Pump Module in this section.
4. Remove the fuel pump. Refer to Fuel Pump in this section.
5. Using a screwdriver with its tip wrapped in protective tape, remove the fuel pressure regulator assembly.

➡Slowly pull out the fuel pressure regulator assembly because the O-ring is firmly installed between the regulator and the fuel filter.

6. Remove the 2 O-rings from the fuel pressure regulator assembly.

To install:

7. To install, reverse removal procedure.

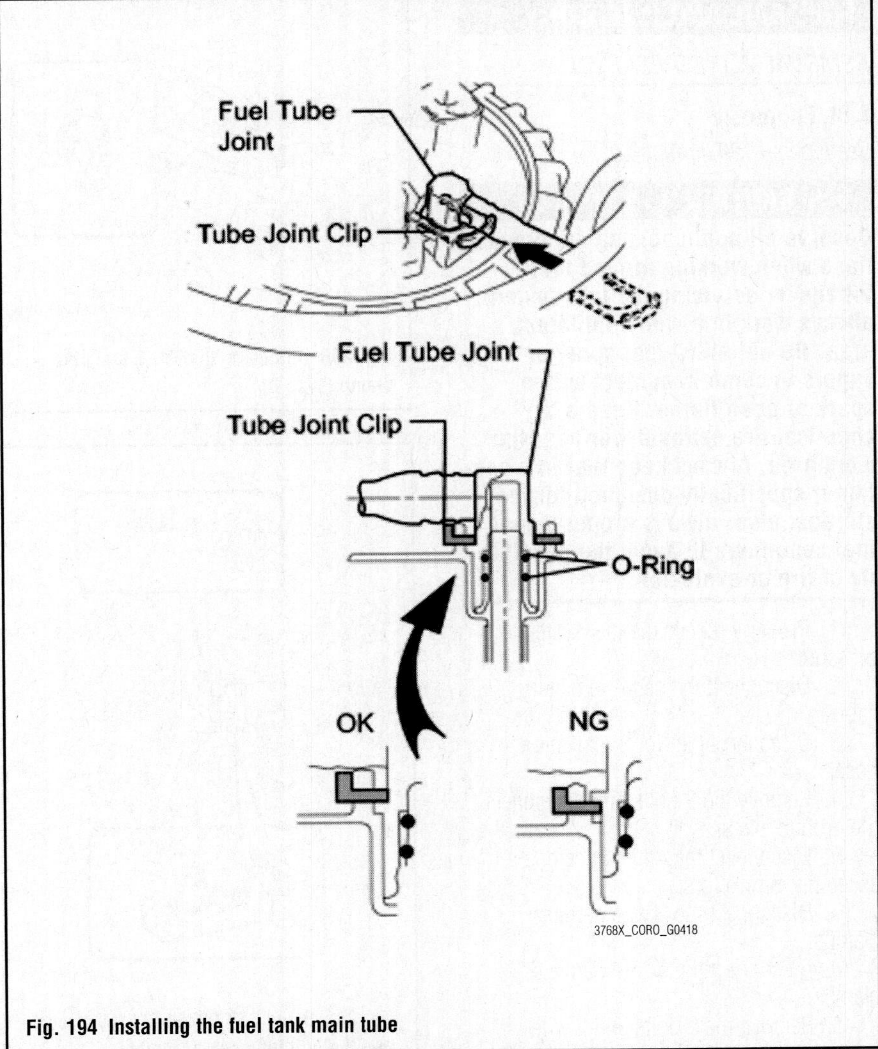

Fig. 194 Installing the fuel tank main tube

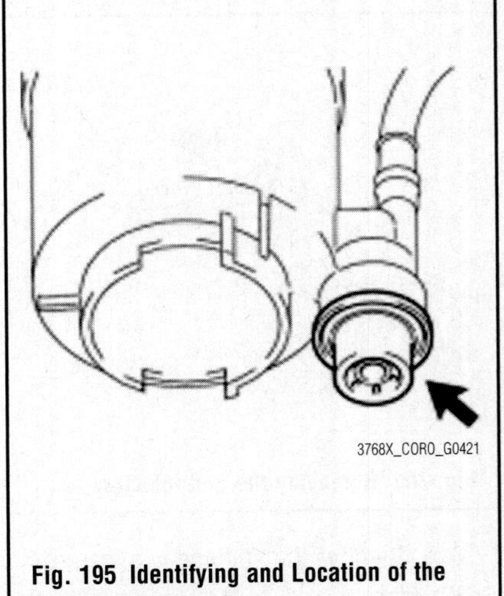

Fig. 195 Identifying and Location of the fuel pressure regulator

8. Apply gasoline to 2 new O-rings and then install them to the fuel pressure regulator assembly.

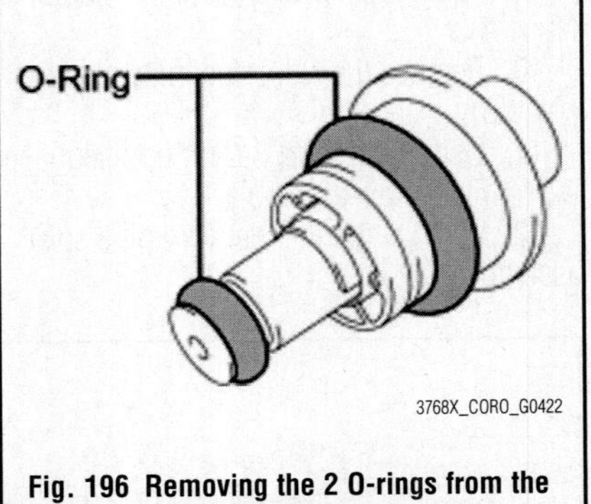

Fig. 196 Removing the 2 O-rings from the fuel pressure regulator assembly

FUEL PUMP

REMOVAL & INSTALLATION

See Figures 197 and 198.

✲✲ CAUTION

Observe all applicable safety precautions when working around fuel. Whenever servicing the fuel system, always work in a well ventilated area. Do not allow fuel spray or vapors to come in contact with a spark or open flame. Keep a dry chemical fire extinguisher near the work area. Always keep fuel in a container specifically designed for fuel storage; also, always properly seal fuel containers to avoid the possibility of fire or explosion.

1. Remove the fuel pump module. Refer to Fuel Pump Module in this section.
2. Release the claw and disconnect the fuel pump filter hose.
3. Remove the E-ring and separate the 2 claws, then remove the fuel pump sub-tank.

➡Do not separate the tube indicated in the illustration.

4. Disconnect the fuel pump harness connector. Do not damage the wire harness.
5. Using a screwdriver with its tip wrapped in protective tape, disengage the 2 claws and remove the No. 1 fuel suction support.
6. Using a screwdriver with its tip wrapped in protective tape, disengage the 5 claws, and remove the suction filter and fuel pump from the fuel filter.

➡Do not damage the fuel pump filter or fuel filter. Do not remove the suction filter. Do not use either the fuel pump or the suction filter if the suction filter is removed from the fuel pump.

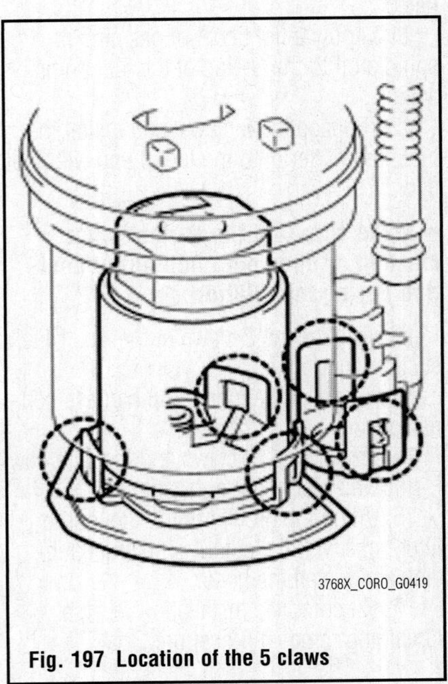

Fig. 197 Location of the 5 claws

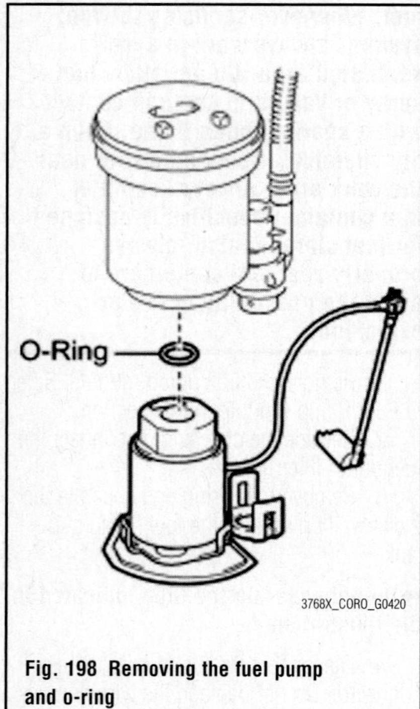

Fig. 198 Removing the fuel pump and o-ring

7. Disconnect the fuel pump connector.
8. Remove the O-ring from the fuel pump.

➡**Do not disassemble the fuel pump and the fuel filter because they are non-reusable parts.**

To install:

9. Apply a light coat of gasoline or spindle oil to a new O-ring, then install it into the fuel pump.
10. Connect the fuel pump harness connector.
11. Apply a light coat of gasoline or spindle oil to the O-ring of the fuel pump again.
12. Engage the 5 claws, and install the fuel pump filter onto the fuel pump with fuel filter.

➡**Make sure that the O-ring is not cracked or does not jump out of position during installation.**

13. Engage the 2 claws of the No. 1 fuel suction support.
14. Connect the fuel pump harness connector.
15. Engage the 2 claws and install a new E-ring and fuel pump sub-tank.
16. Align the groove of the fuel pump filter hose with the cutout of the fuel sub-tank and install the hose.
17. To complete installation reverse remaining removal procedure.

FUEL RAIL AND INJECTOR

REMOVAL & INSTALLATION

1.8L Engine

See Figures 199 through 202.

✲✲ CAUTION

Observe all applicable safety precautions when working around fuel. Whenever servicing the fuel system, always work in a well ventilated area. Do not allow fuel spray or vapors to come in contact with a spark or open flame. Keep a dry chemical fire extinguisher near the work area. Always keep fuel in a container specifically designed for fuel storage; also, always properly seal fuel containers to avoid the possibility of fire or explosion.

1. Properly relieve the fuel system pressure.
2. Disconnect the negative battery cable.
3. Disconnect the No. 2 ventilation hose.
4. Remove the 2 bolts and disconnect the ground wires.
5. Disconnect the 4 fuel injector assembly connectors.
6. Disconnect the 2 wire harness clamps.
7. Disconnect the 4 wire harness clamps.
8. Remove the 2 bolts and 2 wire harness brackets.
9. Remove the No. 2 fuel pipe clamp (Type A).
10. Remove the No. 2 fuel pipe clamp (Type B).
11. Using SST 09268-21010, disconnect the fuel tube sub-assembly.
12. Remove the fuel delivery pipe sub-assembly:

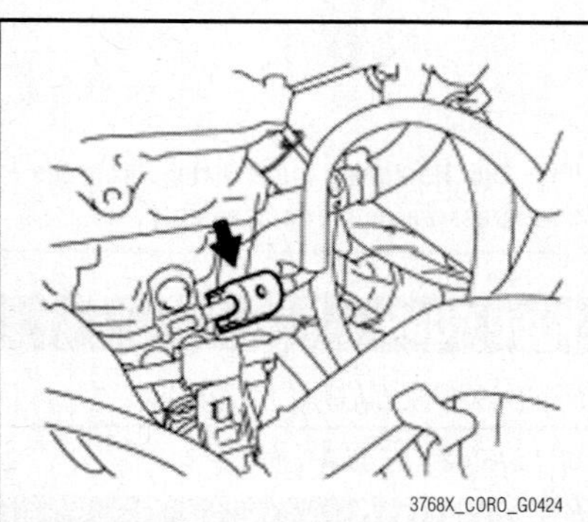

Fig. 199 Removing the No. 2 fuel pipe clamp (Type A)

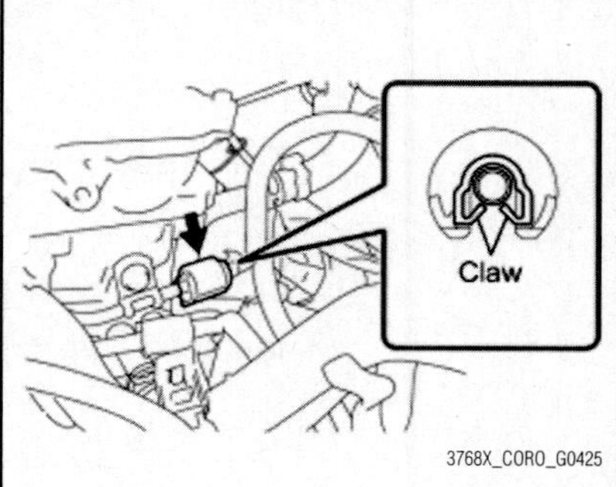

Fig. 200 Removing the No. 2 fuel pipe clamp (Type B)

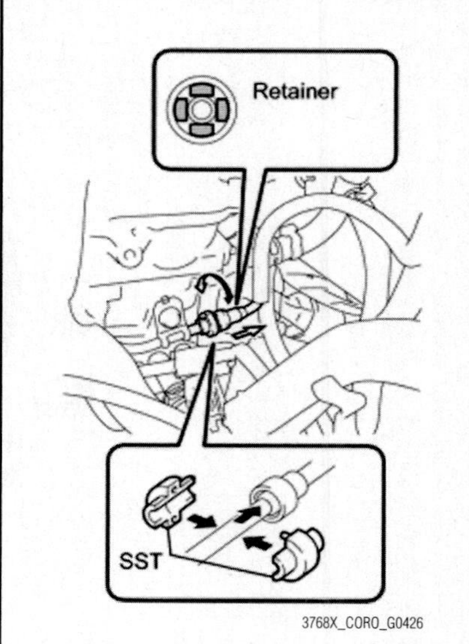

Fig. 201 Using SST 09268-21010, disconnect the fuel tube sub-assembly

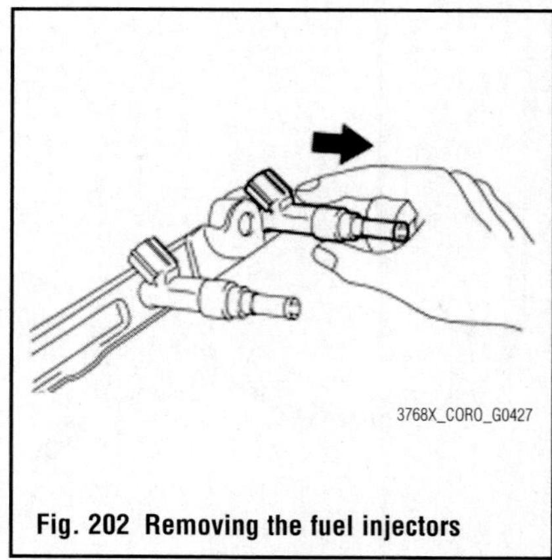

Fig. 202 Removing the fuel injectors

a. Remove the bolt and wire harness bracket.
b. Remove the 2 bolts.
c. Remove the bolt and the fuel delivery pipe sub-assembly.
d. Remove the 2 No. 1 delivery pipe spacers.

13. Pull the 4 fuel injector assemblies out of the fuel delivery pipe sub-assembly.

14. For reinstallation, attach a tag or label to the injector shaft.

➡**Prevent entry of foreign objects by covering the fuel injector with a plastic bag.**

15. Remove the O-rings from the fuel injector assemblies.

16. Remove the 4 injector vibration insulators.

To install:

17. Install 4 new injector vibration insulators to the 4 fuel injector assemblies.

18. Apply a light coat of gasoline or spindle oil to the contact surfaces of the O-rings of the fuel injector assemblies.

19. While turning the fuel injector assembly left and right, install it onto the fuel delivery pipe sub-assembly.

➡**Do not twist the O-ring. After installing the fuel injectors, check that they turn smoothly.**

20. Install the 2 No. 1 delivery pipe spacers onto the cylinder head.

➡**Install the No. 1 delivery pipe spacers in the correct direction.**

21. Install the fuel delivery pipe sub-assembly with the 4 fuel injector assemblies, then temporarily install the 2 bolts. Tighten the bolts to 15 ft. lbs. (21 Nm).

➡**Do not drop the fuel injectors when installing the fuel delivery pipe sub-assembly. Check that the fuel injector assemblies rotate smoothly after installing the fuel delivery pipe sub-assembly.**

22. Install the bolt to secure the fuel delivery pipe sub-assembly. Tighten the bolt to 15 ft. lbs. (21 Nm).

23. Install the wire harness bracket with the bolt. Tighten to 44 inch lbs. (5 Nm).

24. Insert the fuel tube sub-assembly connector into the fuel delivery pipe until a "click" sound can be heard.

➡**Check that there are no scratches or foreign matter around the contact surfaces of the fuel tube connector and pipe before performing this work. After connecting the fuel tube, check that the fuel tube connector and pipe are securely connected by pulling on them.**

25. Install a new No. 2 fuel pipe clamp (Type B).

26. Install a new No. 2 fuel pump clamp (Type A).

27. Install the 2 wire harness brackets with the 2 bolts. Tighten to 10 ft. lbs. (13 Nm).

28. Connect the 4 wire harness clamps.

29. Connect the 4 fuel injector assembly connectors.

30. Connect the 2 wire harness clamps.

31. Connect the ground wires with the 2 bolts.

32. Install the air cleaner assembly.

33. Connect the No. 2 ventilation hose.

34. Connect cable to negative battery terminal.

35. Inspect for fuel leaks.

2.4L Engine

See Figures 203 through 209.

CAUTION

Observe all applicable safety precautions when working around fuel. Whenever servicing the fuel system, always work in a well ventilated area. Do not allow fuel spray or vapors to come in contact with a spark or open flame. Keep a dry chemical fire extinguisher near the work area. Always keep fuel in a container specifically designed for fuel storage; also, always properly seal fuel containers to avoid the possibility of fire or explosion.

1. Properly relieve the fuel system pressure.

2. Disconnect the negative battery cable.

3. Remove the air cleaner assembly.

➡**Do not forcibly bend, kink or twist the fuel main tube.**

4. Remove the fuel tube from the fuel hose clamp.

5. Remove the fuel pipe clamp.

6. Wipe off any dirt on the fuel tube connector.

7. Hold the fuel tube connector, and then install SST 09268-21010.

8. Turn SST to align the retainer inside the fuel tube connector with the chamfered part of SST.

9. Insert SST into the fuel tube and hold it. Then push the fuel tube connector toward SST.

10. Mount the retainer of the fuel tube connector onto the chamfered part of SST.

11. Slide SST and fuel tube connector together towards the fuel tube until they make a "click" sound, and then disconnect the fuel tube.

12. Drain the fuel remaining inside the fuel tube.

13. Cover the fuel tube and fuel pipe with a plastic bag to protect the disconnected part.

14. Disconnect the No. 2 ventilation hose from the ventilation valve.

15. Remove the 2 wire harness clamps.

16. Disconnect the 4 fuel injector connectors.

17. Remove the 2 bolts, then remove the fuel delivery pipe together with the 4 fuel injectors.

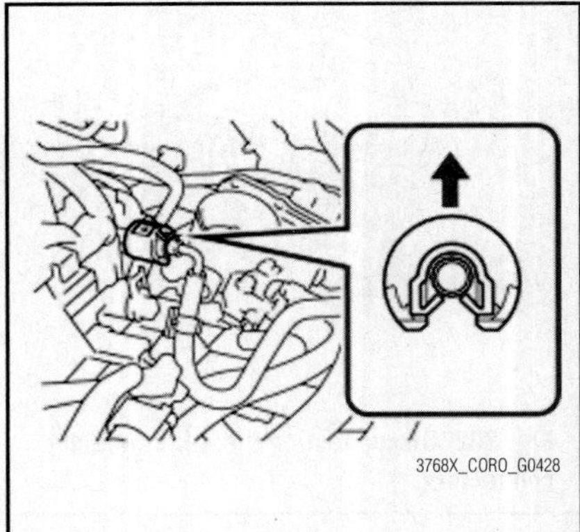

Fig. 203 Removing the fuel tube from the fuel hose clamp

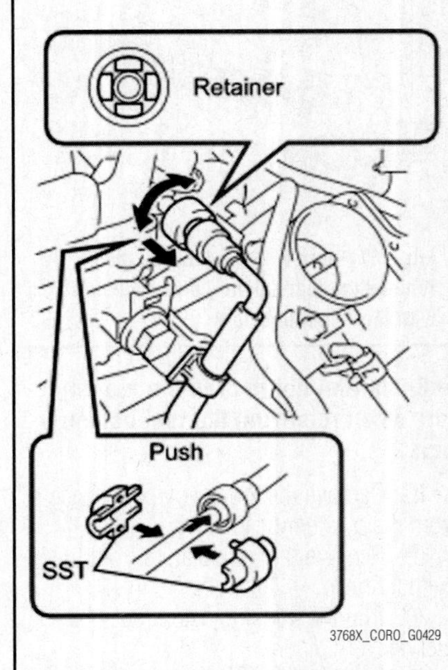

Fig. 204 Installing SST: 09268-21010 to the fuel tube connector

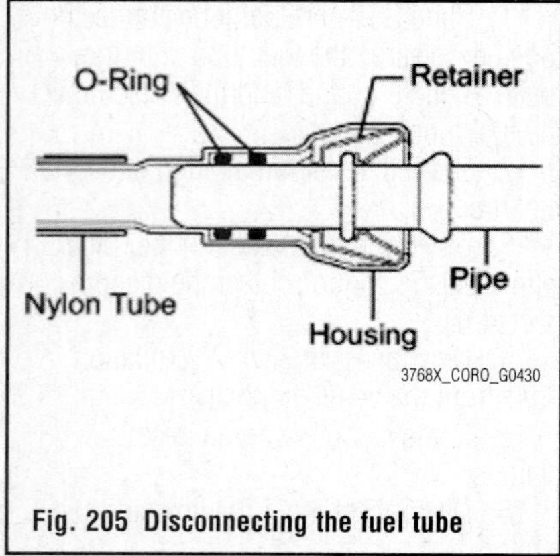

Fig. 205 Disconnecting the fuel tube

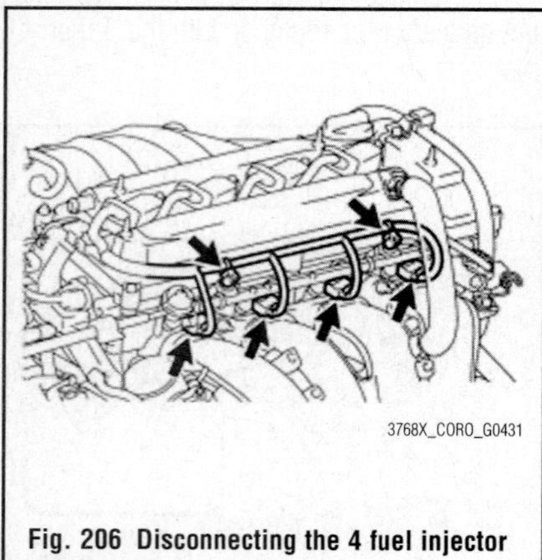

Fig. 206 Disconnecting the 4 fuel injector connectors

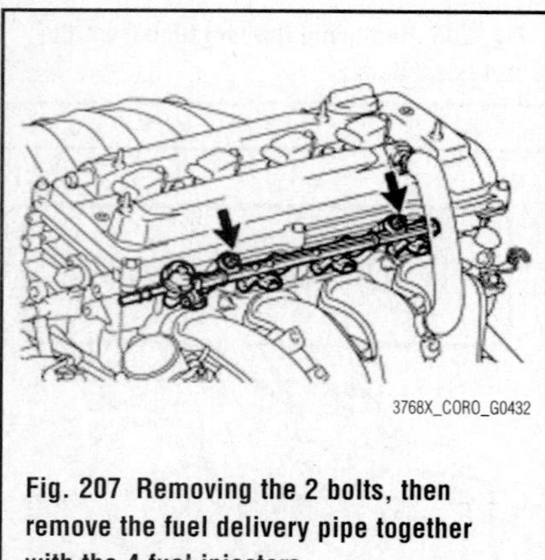

Fig. 207 Removing the 2 bolts, then remove the fuel delivery pipe together with the 4 fuel injectors

➡Be careful not to drop the fuel injectors when removing the fuel delivery pipe.

18. Remove the 2 delivery pipe spacers from the cylinder head.
19. Remove the 4 insulators from the cylinder head.
20. Pull the 4 fuel injectors out of the fuel delivery pipe.
21. Remove the 4 O-rings from the 4 fuel injectors.

To install:

22. Apply a light coat of gasoline or spindle oil to new O-rings, then install them to the fuel injectors.
23. Apply a light coat of gasoline or spindle oil to the part of the fuel delivery pipe which comes into contact with the O-ring of the fuel injector.
24. Apply a light coat of gasoline or spindle oil to the O-ring again, then install the right and left fuel injectors onto the fuel delivery pipe.

➡Make sure that the O-rings are not cracked or does not jump out of position during installation.

25. Check that the fuel injectors rotate smoothly. If the fuel injector does not rotate, replace the O-ring.
26. Install 4 new insulators to the cylinder head.
27. Install the 2 delivery pipe spacers onto the cylinder head.

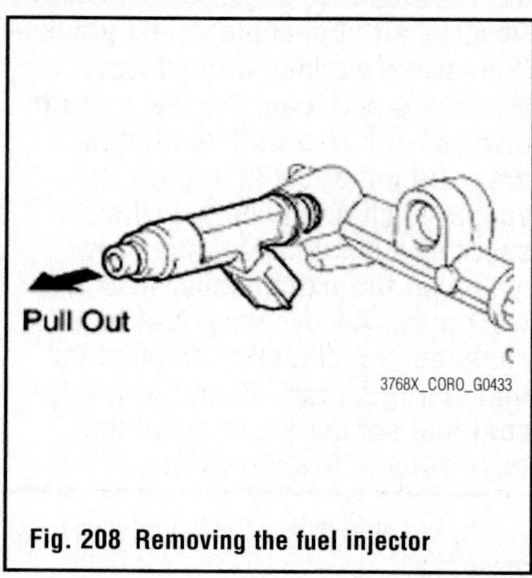

Fig. 208 Removing the fuel injector

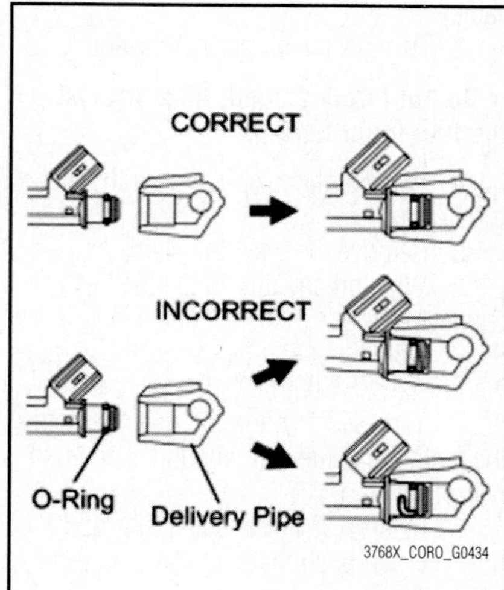

Fig. 209 Correct fuel injector installation position

28. Install the fuel delivery pipe together with the 4 fuel injectors, then temporarily tighten the 2 bolts.

➡Be careful not to drop the fuel injectors when installing the fuel delivery pipe.

29. Check that the fuel injector rotates smoothly.
30. If the fuel injector does not rotate smoothly, replace the O-ring.
31. Tighten the 2 fuel rail bolts to 15 ft. lbs. (20 Nm).
32. Connect the 4 fuel injector connectors.
33. Install the 2 wire harness clamps.
34. Connect the ventilation hose to the ventilation valve.

➡Make sure that the paint mark and hose clamp are at the correct angle when installing the hose.

35. Connect the fuel main tube.
36. Push the fuel tube connector until it makes a "click" sound.
37. Install the fuel pipe clamp.
38. Install the fuel tube to the fuel hose clamp.
39. Install the air cleaner assembly.
40. Connect cable to negative battery terminal.
41. Inspect for fuel leaks.

FUEL TANK

REMOVAL & INSTALLATION

See Figures 210 and 211.

✲✲ CAUTION

Observe all applicable safety precautions when working around fuel. Whenever servicing the fuel system, always work in a well ventilated area. Do not allow fuel spray or vapors to come in contact with a spark or open flame. Keep a dry chemical fire extinguisher near the work area. Always keep fuel in a container specifically designed for fuel storage; also, always properly seal fuel containers to avoid the possibility of fire or explosion.

1. Properly relieve the fuel system pressure.
2. Disconnect the negative battery cable.
3. Remove the fuel pump module. Refer to Fuel Pump Module in this section.

4. Drain fuel.

5. Remove front exhaust pipe assembly. Refer to Exhaust Manifold in Engine Mechanical.

6. Remove the 4 bolts and the No. 1 fuel tank protector.

7. Remove the 4 bolts, and separate the parking brake cables.

8. Disconnect the fuel tank vent hose.

9. Pull the fuel tank vent hose out of the pipe.

➡Check that there is no dirt or other foreign objects around the connector before disconnecting it. Clean the connector if necessary.

➡It is necessary to prevent mud or dirt from entering the connector. If mud or dirt gets in the connector, the O-rings may not seal properly.

➡Only disconnect the quick connector by hand.

➡Do not bend, kink or twist the nylon tubes.

➡Protect the connector by covering it with a plastic bag.

10. Disconnect breather tube fuel hose:

a. Pinch the retainer of the fuel tube connector, then pull the fuel tube connector out of the pipe.

➡Check that there is no dirt or other foreign objects around the connector before disconnecting it. Clean the connector if necessary.

➡It is necessary to prevent mud or dirt from entering the connector. If mud or dirt gets in the connector, the O-rings may not seal properly.

➡Only disconnect the quick connector by hand.

➡Do not bend, kink or twist the nylon tubes.

➡Protect the connector by covering it with a plastic bag.

➡If the pipe and connector are stuck, carefully try wiggling or pushing and pulling on the connector to release it. Pull the connector off the pipe carefully.

b. Separate the fuel breather tube fuel hose.

11. Disconnect the fuel tank main tube sub-assembly:

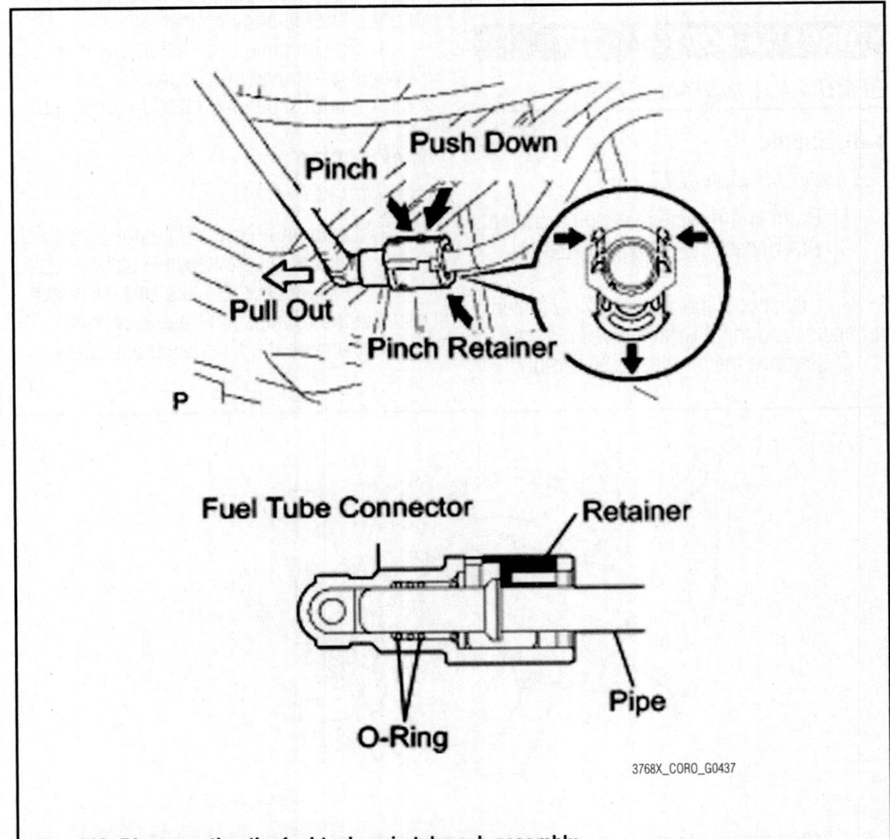

Fig. 210 Disconnecting the fuel tank main tube sub-assembly

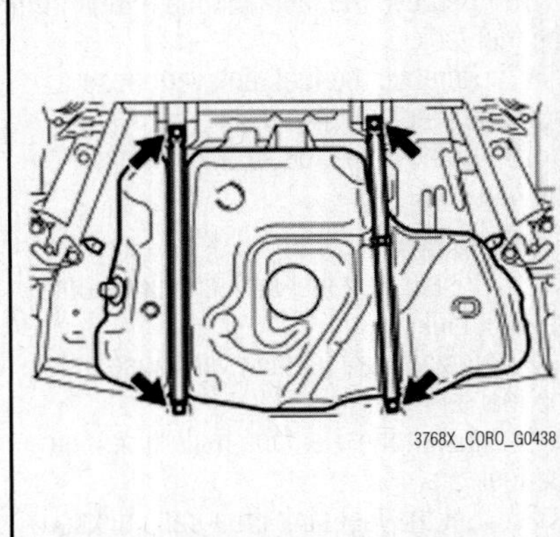

Fig. 211 Removing the 4 bolts, then remove the 2 No. 1 fuel tank bands

a. Pinch the tabs of the retainer of the fuel tube connector to disengage the lock claws and push it down as shown in the illustration.

b. Pull the fuel tank main tube out of the pipe.

➡Check that there is no dirt or other foreign objects around the connector before disconnecting it. Clean the connector if necessary.

➡It is necessary to prevent mud or dirt from entering the connector. If mud or dirt gets in the connector, the O-rings may not seal properly.

➡Only disconnect the quick connector by hand.

➡Do not bend, kink or twist the nylon tubes.

➡Protect the connector by covering it with a plastic bag.

➡If the pipe and connector are stuck, carefully try wiggling or pushing and pulling on the connector to release it. Pull the connector off the pipe carefully.

12. Disconnect fuel tank filler pipe sub-assembly:

a. Using a screwdriver, unfasten the claw. Then remove the fuel tank filler pipe cover from the fuel tank filler pipe.

b. Loosen the hose clamp bolt, then disconnect the fuel tank filler pipe hose from the fuel tank.

13. Hold the fuel tank using a transmission jack.

14. Remove the 4 bolts, then remove the 2 No. 1 fuel tank bands.

15. Operate the transmission jack, then remove the fuel tank.

16. Remove the fuel tank main tube from the fuel tank.
17. Remove the fuel tank vent hose from the fuel tank clamp.
18. Remove the fuel tank cushions from the fuel tank.

To install:

19. Install new fuel tank cushions onto the fuel tank.
20. Install the fuel tank vent hose onto the fuel tank clamp.
21. Install the fuel tank main tube onto the fuel tank.
22. Set the fuel tank on a transmission jack.
23. Operate the transmission jack, then install the fuel tank into the vehicle.
24. Install the 2 No. 1 fuel tank bands with the 4 bolts. Tighten to 29 ft. lbs. (39 Nm).
25. Connect the fuel tank filler pipe to the fuel tank.

➡Make sure that the hose clamp is facing in the correct direction when installing.

26. Engage the claw, then install the fuel tank filler pipe cover onto the fuel tank filler pipe.
27. Align the fuel tube connector with the pipe, then push the fuel tube connector in until it comes into contact with the seat to connect the fuel tank main tube to the pipe, then push the retainer up until the claws lock.

➡Check that there are no scratches or foreign objects around the connecting surfaces of the fuel tube connector and pipe before performing this work. After connecting the fuel tank main tube, check that the fuel tank main tube is securely connected by pulling on the fuel tube connector and pipe.

28. Connect the breather tube fuel hose to the clamp.
29. Align the fuel hose connector with the pipe, then push the fuel hose connector in until the retainer makes a "click" sound to connect the fuel tank breather tube fuel hose to the pipe.

➡Check that there are no scratches or foreign objects around the connecting surfaces of the fuel tube connector and pipe before performing this work. After connecting the fuel tank breather tube, check that the fuel pump tube is securely connected by pulling on the fuel tube connector and pipe.

30. Align the fuel tube connector with the pipe, then push the fuel tube connector in until it comes into contact with the seat to connect the fuel tank vent hose to the pipe.
31. Slide the retainer of the fuel tube connector to lock the claws

➡Check that there are no scratches or foreign objects around the connecting surfaces of the fuel tube connector and pipe before performing this work. After connecting the fuel tank breather tube, check that the fuel pump tube is securely connected by pulling on the fuel tube connector and pipe.

32. Install the fuel tank protector sub-assembly with the 4 bolts. Tighten to 49 inch lbs. (5.5 Nm).
33. Install the parking cables with the 4 bolts. Tighten to 49 inch lbs. (5.5 Nm).
34. To complete installation, reverse remaining removal procedure.
35. Inspect for fuel leaks.
36. Inspect for exhaust gas leak.

IDLE SPEED

ADJUSTMENT

Adjustment is not available or necessary.

THROTTLE BODY

REMOVAL & INSTALLATION

1.8L Engine

See Figures 212 and 213.

1. Drain and recycle the engine coolant.
2. Disconnect the negative battery cable.
3. Disconnect the mass air flow meter connector and the 2 wire harness clamps.
4. Remove the air cleaner assembly.

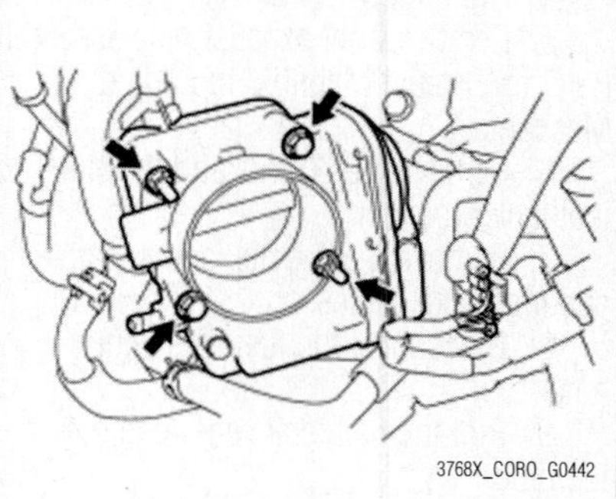

Fig. 213 Removing the 2 bolts and 2 nuts, and removing the throttle body assembly

5. Disconnect the throttle body connector and the 2 water by-pass hoses.
6. Remove the 2 bolts and 2 nuts, and remove the throttle body assembly.
7. Remove the gasket from the intake manifold.

To install:

8. Install a new gasket onto the intake manifold.
9. Install the throttle body assembly with the 2 bolts and 2 nuts. Tighten to 7 ft. lbs. (10 Nm).
10. Connect the 2 water by-pass hoses and the throttle body connector.
11. To complete installation, reverse remaining removal procedure.
12. Perform the initialization procedure.

2.4L Engine

See Figures 214 and 215.

1. Drain and recycle the engine coolant.
2. Disconnect the negative battery cable.
3. Disconnect the mass air flow meter connector and the 2 wire harness clamps.
4. Remove the air cleaner assembly.

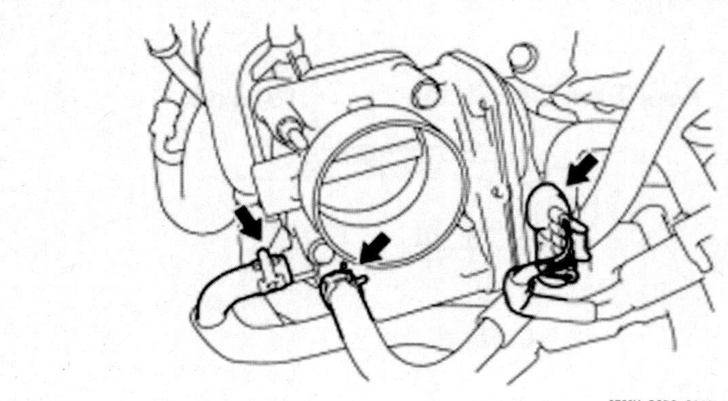

Fig. 212 Disconnecting the throttle body connector and the 2 water by-pass hoses

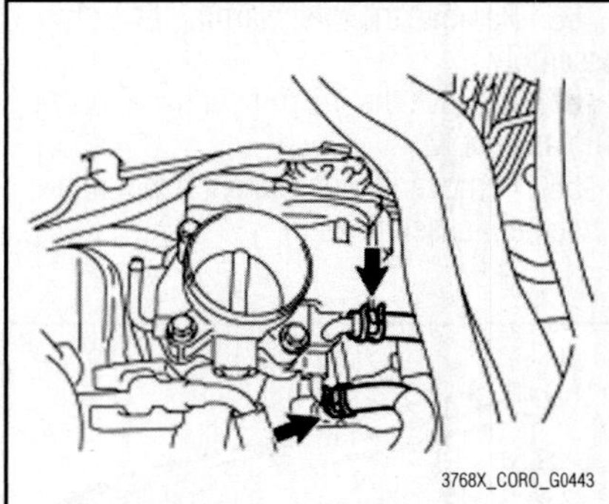
Fig. 214 Disconnecting the 2 water by-pass hoses

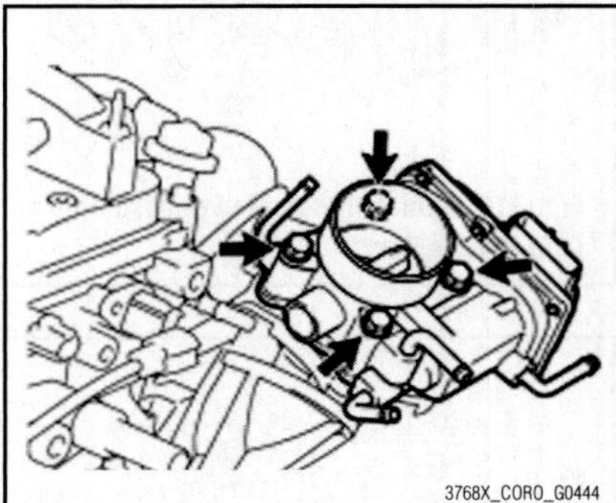
Fig. 215 Removing the 4 bolts and throttle body assembly

5. Disconnect the 2 water by-pass hoses.
6. Disconnect the throttle body assembly connector.
7. Disconnect the throttle body hose.
8. Remove the 4 bolts and throttle body assembly.
9. Remove the gasket from the intake manifold.

To install:

10. Install a new gasket onto the intake manifold.
11. Install the throttle body assembly with the 4 bolts. Tighten to 22 ft. lbs. (30 Nm).
12. Connect the 2 water by-pass hoses and the throttle body connector.
13. To complete installation, reverse remaining removal procedure.
14. Perform the initialization procedure.

INITIALIZATION

➡Be sure to perform this procedure after reassembling the throttle body assembly, removing and reinstalling any throttle body component or replacing the ECM.

1. Disconnect the cable from the negative (-) battery terminal. Wait at least 60 seconds and reconnect the cable.
2. Turn the ignition switch to ON without operating the accelerator pedal.

➡If the accelerator pedal is operated, perform the above steps again.

3. Connect the Techstream to the DLC3 and clear the DTC's.
4. Start the engine and check that the MIL is not illuminated and that the idle speed is within the specified range when the air conditioning is switched off after the engine is warmed up.
   a. Standard:
      - Condition: A/C switched off
      - Engine idle speed: 600 to 700 rpm

➡Be sure to perform this step with all accessories off. Make sure that the shift lever is in N or P.

5. Enter the following menus: Powertrain/ Engine and ECT/ Data List/ Throttle Pos. Sensor Output. Fully depress the accelerator pedal and check that the value is 60% or more.
6. Perform a road test and confirm that there are no abnormalities. Refer to Drive Train.

HEATING & AIR CONDITIONING SYSTEM

BLOWER MOTOR

REMOVAL & INSTALLATION

Front

See Figures 216 and 217.

1. Remove the No. 2 instrument panel under cover sub assembly (w/ instrument panel under cover).
2. Remove the blower motor sub assembly (w/o PTC heater).
   a. Disengage the clamp.
   b. Disconnect the connector.
   c. Remove the 3 screws and the blower motor sub assembly.

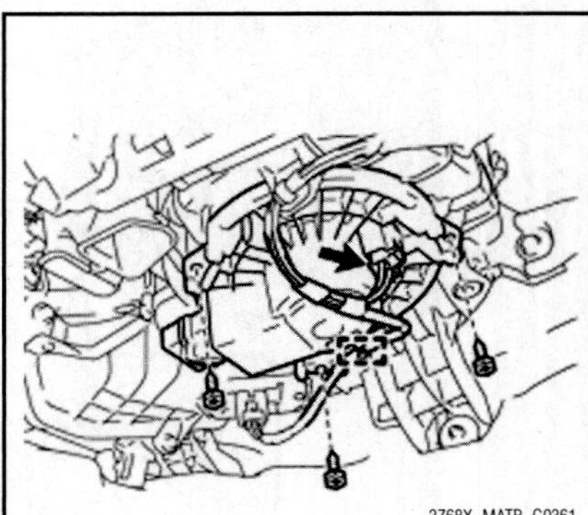
Fig. 216 Removing the blower motor sub assembly (w/o PTC heater)

Fig. 217 Removing the blower motor sub assembly (w/ PTC heater)

3. Remove the blower motor sub assembly (w/ PTC heater).
   a. Disengage the clamp.
   b. Remove the quick heater connector screw.
   c. Disconnect the connector.
   d. Remove the 3 screws and the blower motor sub-assembly.

To install:

4. To install, reverse the removal procedure.

Blower Unit

See Figures 218 through 220.

1. Turn the front wheels to face straight ahead.
2. Disconnect the cable from the negative battery terminal.

CAUTION

Wait at least 90 seconds after disconnecting the cable from the negative (-) battery terminal to disable the SRS system.

3. Recover the refrigerant from the refrigeration system.
4. Disconnect the suction pipe sub assembly.
5. Disconnect the liquid pipe sub assembly.
6. Disconnect the heater outlet water hose.
7. Disconnect the heater inlet water hose.
8. Remove the lower No. 3 steering wheel cover.
9. Remove the lower No. 2 steering wheel cover.
10. Remove the steering pad.
11. Remove the steering wheel assembly.
12. Remove the center instrument cluster finish panel sub assembly.
13. Remove the instrument cluster finish panel sub assembly.
14. Remove the meter hood sub assembly.
15. Remove the combination meter assembly.
16. Disconnect the front door opening trim weather-strip LH.
17. Remove the front pillar garnish LH.
18. Disconnect the front door opening trim weather strip RH.
19. Remove the front pillar garnish RH.
20. Remove the lower instrument panel finish panel assembly.
21. Remove the shift lever knob sub assembly.
22. Remove the center instrument cluster finish panel assembly.
23. Remove the glove compartment door assembly.
24. Remove the No. 1 instrument panel box door sub assembly.
25. Disconnect the instrument panel wire assembly.
26. Remove the upper instrument panel sub assembly.
27. Remove the radio receiver assembly with the bracket (w/o navigation system).
28. Remove the navigation receiver assembly with the bracket (w/ navigation system).
29. Remove the navigation antenna cord sub assembly (w/ navigation system).
30. Remove the air conditioning panel assembly.
31. Remove the front upper console box.
32. Remove the No. 2 switch hole base.
33. Remove the console box carpet.
34. Remove the front console box assembly.
35. Remove the front No. 1 console box insert.
36. Remove the front No. 2 console box insert.
37. Remove the lower center instrument panel finish panel.
38. Remove the No. 1 switch hole base.
39. Remove the lower instrument panel finish panel sub assembly.
40. Remove the front door scuff plate LH.
41. Remove the cowl side trim board LH.
42. Remove the No. 2 instrument panel under cover sub assembly (w/ instrument panel under cover).
43. Remove the front door scuff plate RH.
44. Remove the cowl side trim board RH.
45. Disconnect the shift lever assembly.
46. Remove the lower steering column cover.
47. Remove the upper steering column cover.
48. Remove the turn signal switch assembly with the spiral cable sub assembly.
49. Remove the radio wire sub assembly.
50. Remove the lower instrument panel sub assembly.
51. Remove the column hole cover silencer sheet.
52. Separate the No. 2 steering intermediate shaft assembly.
53. Remove the instrument panel sub reinforcement.
54. Remove the No. 2 air duct sub assembly.
55. Remove the transponder key amplifier (w/ engine immobilizer system).
56. Remove the stop light switch assembly.
57. Remove the steering post assembly.
58. Remove the power steering ECU assembly.
59. Remove the air conditioning amplifier assembly.
60. Remove the rear No. 3 air duct (w/ rear air duct).
61. Remove the rear No. 2 air duct (w/ rear air duct).
62. Remove the rear No. 1 air duct (w/ rear air duct).
63. Remove the center instrument panel register connector assembly.
64. Remove the No. 1 air duct sub assembly.
65. Remove the center instrument panel to cowl brace.
66. Remove the theft warning ECU assembly.
67. Remove the instrument panel brace sub assembly.
68. Remove the instrument panel reinforcement assembly.

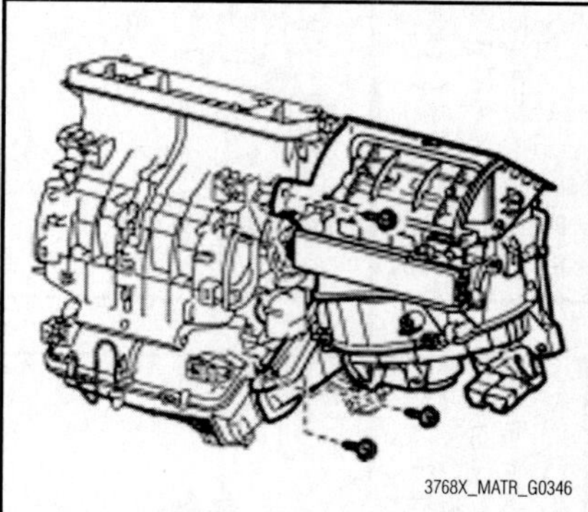

Fig. 218 Removing the blower assembly (w/o PTC heater)

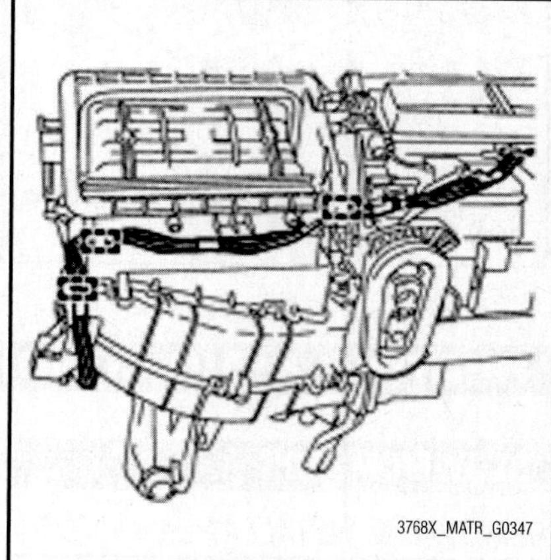

Fig. 219 Disengaging the clamps

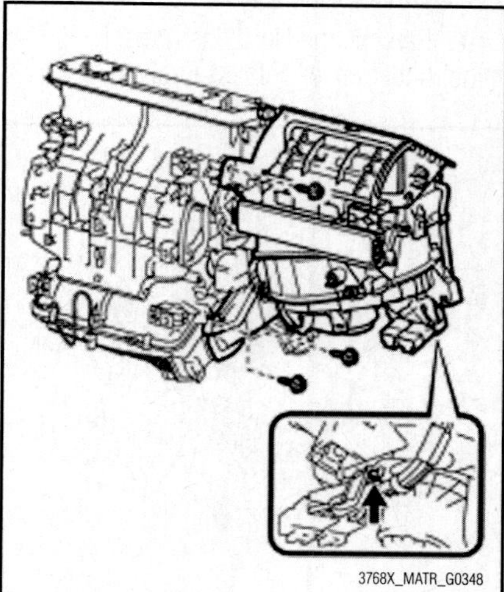

Fig. 220 Removing the blower assembly (w/ PTC heater)

69. Remove the air conditioning unit.

70. Remove the No. 3 air duct sub assembly.

71. Remove the blower assembly (w/o PTC heater).

a. Remove the 3 screws and the blower assembly.

72. Remove the blower assembly (w/ PTC heater).

a. Disengage each clamp.

b. Remove the screw and disengage the quick heater connector.

c. Remove the 3 screws and the blower assembly.

To install:

73. To install, reverse the removal procedure.

HEATER UNIT (PTC HEATER ASSEMBLY)

REMOVAL & INSTALLATION

See Figures 221 and 222.

1. Turn the front wheels to face straight ahead.

2. Disconnect the cable from the negative battery terminal.

**** CAUTION**

Wait at least 90 seconds after disconnecting the cable from the negative (-) battery terminal to disable the SRS system.

3. Recover the refrigerant from the refrigeration system.

4. Disconnect the suction pipe sub assembly.

5. Disconnect the liquid pipe sub assembly.

6. Disconnect the heater outlet water hose.

7. Disconnect the heater inlet water hose.

8. Remove the lower No. 3 steering wheel cover.

9. Remove the lower No. 2 steering wheel cover.

10. Remove the steering pad.

11. Remove the steering wheel assembly.

12. Remove the center instrument cluster finish panel sub assembly.

13. Remove the instrument cluster finish panel sub assembly.

14. Remove the meter hood sub assembly.

15. Remove the combination meter assembly.

16. Disconnect the front door opening trim weather-strip LH.

17. Remove the front pillar garnish LH.

18. Disconnect the front door opening trim weather-strip RH.

19. Remove the front pillar garnish RH.

20. Remove the lower instrument panel finish panel assembly.

21. Remove the shift lever knob sub assembly.

22. Remove the center instrument cluster finish panel assembly.

23. Remove the glove compartment door assembly.

24. Remove the No. 1 instrument panel box sub assembly.

25. Disconnect the instrument panel wire assembly.

26. Remove the radio receiver assembly with the bracket (w/o navigation system).

27. Remove the navigation receiver assembly with the bracket (w/ navigation system).

28. Remove the navigation antenna cord sub assembly (w/ navigation system).

29. Remove the air conditioning panel assembly.

30. Remove the front upper console box.

31. Remove the No. 2 switch hole base.

32. Remove the console box carpet.

33. Remove the front console box assembly.

34. Remove the front No. 1 console box insert.

35. Remove the front No. 2 console box insert.

36. Remove the lower center instrument panel finish panel.

37. Remove the No. 1 switch hole base.

38. Remove the lower instrument panel finish panel sub assembly.

39. Remove the front door scuff plate LH.

40. Remove the cowl side trim board LH.

41. Remove the No. 2 instrument panel under cover sub assembly (w/ instrument panel under cover).

42. Remove the front door scuff plate RH.

43. Remove the cowl side trim board RH.

44. Disconnect the shift lever assembly.

45. Remove the lower steering column cover.

46. Remove the upper steering column cover.

47. Remove the turn signal switch assembly with the spiral cable sub assembly.

48. Remove the radio wire sub assembly.

49. Remove the lower instrument panel sub assembly.

50. Remove the column hole cover silencer sheet.

51. Separate the No. 2 steering intermediate shaft assembly.

52. Remove the instrument panel sub reinforcement.

53. Remove the No. 2 air duct sub assembly.

54. Remove the transponder key amplifier (w/ engine immobilizer system).

55. Remove the stop light switch assembly.

56. Remove the steering post assembly.

57. Remove the power steering ECU assembly.

58. Remove the air conditioning amplifier assembly.

59. Remove the rear No. 3 air duct (w/ rear air duct).

60. Remove the rear No. 2 air duct (w/ rear air duct).

61. Remove the rear No. 1 air duct (w/ rear air duct).

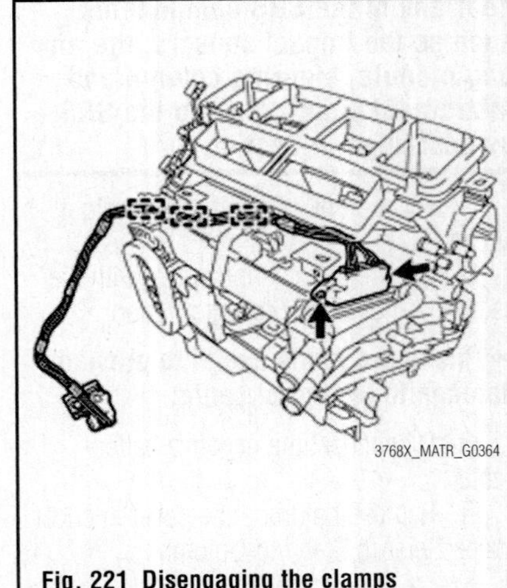

Fig. 221 Disengaging the clamps

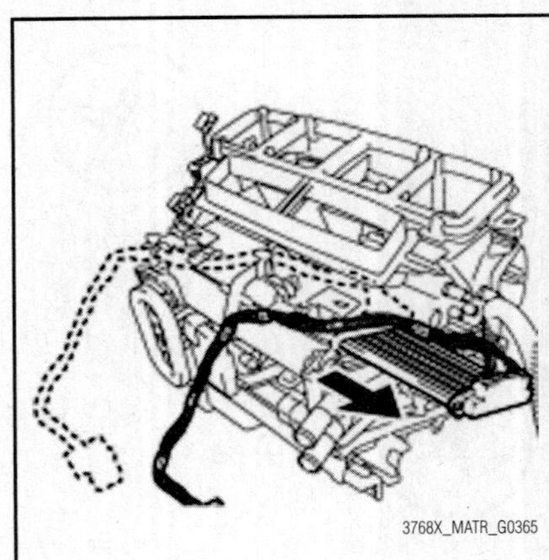

Fig. 222 Removing the quick heater assembly

62. Remove the center instrument panel register connector assembly.
63. Remove the No. 1 air duct sub assembly.
64. Remove the center instrument panel to cowl brace.
65. Remove the theft warning ECU assembly.
66. Remove the instrument panel brace sub assembly.
67. Remove the instrument panel reinforcement assembly.
68. Remove the air conditioning unit.
69. Remove the No. 3 air duct sub assembly.
70. Remove the blower assembly.
71. Remove the quick heater assembly.
   a. Disengage each clamp.
   b. Remove the 2 screws.
   c. Remove the quick heater assembly.

To install:

72. To install, reverse the removal procedure.

STEERING

POWER RACK & PINION STEERING GEAR

REMOVAL & INSTALLATION

2WD Vehicles

See Figures 223 through 227.

CAUTION

Some models covered by this manual may be equipped with a Supplemental Restraint System (SRS), which uses an air bag. Whenever working near any of the SRS components, such as the impact sensors, the air bag module, steering column and instrument panel, disable the SRS, as described in Section 6.

1. Place front wheels facing straight ahead.
2. Secure the steering wheel with the seat belt in order to prevent rotation.

➡This operation is useful to prevent damage to the spiral cable.

3. Disconnect the negative battery cable.
4. Remove column hole cover silencer sheet. Refer to Steering Column.

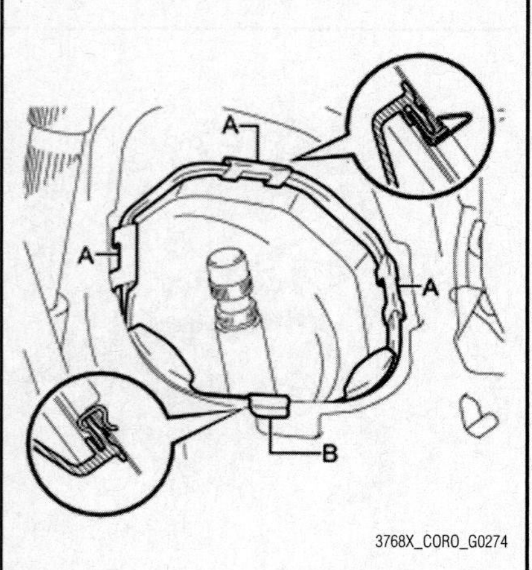

Fig. 223 Remove clips A and the No. 1 steering column hole cover sub-assembly

5. Separate no. 2 steering intermediate shaft assembly. Refer to Steering Column.
6. Remove clips A and the No. 1 steering column hole cover sub-assembly and disengage clip B from the body.
7. Remove the front wheels.
8. Remove the engine under covers.
9. Separate the tie rod end sub-assembly.
10. Separate front stabilizer link assembly.
11. Separate front lower suspension arms.
12. Remove front suspension crossmember sub-assembly. Refer to Front Suspension.
13. Remove the No. 1 steering column hole cover sub-assembly from the steering link assembly.
14. Put match marks on the steering intermediate shaft and the steering link assembly.
15. Remove the bolt and the steering intermediate shaft from the steering link assembly.
16. Remove the 4 bolts and steering link assembly from the front suspension crossmember sub-assembly.

➡Tape SST 09612-00012 before use.

17. Using SST 09612-00012, secure the steering link assembly in a vise.

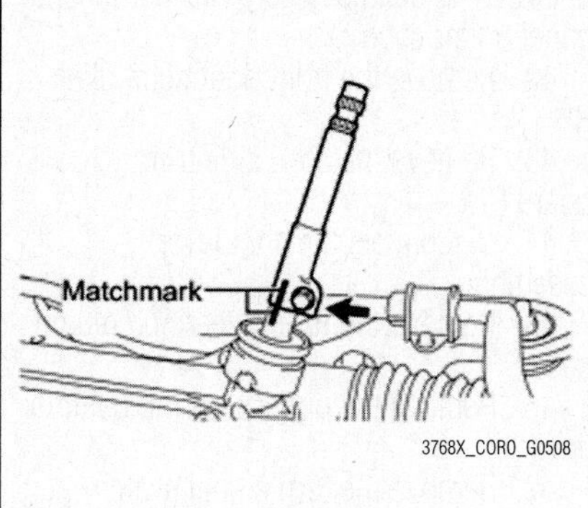

Fig. 224 Putting match marks on the steering intermediate shaft and the steering link assembly

18. Put matchmarks on the tie rod end sub-assembly LH and RH and steering gear assembly.
19. Remove the tie rod end sub-assembly and lock nuts.

To install:

20. Install the lock nut and the tie rod end sub-assembly to the steering gear assembly until the match marks are aligned.

➡After adjusting the toe-in, tighten the lock nut.

21. Install the steering link assembly to the front suspension crossmember sub-assembly with the 4 bolts:
   a. Temporarily tighten the bolts in order of (A), (B), (C), and then (D).

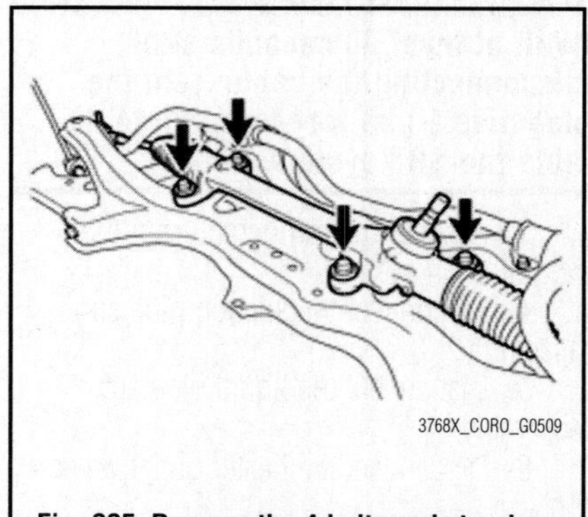

Fig. 225 Remove the 4 bolts and steering link assembly

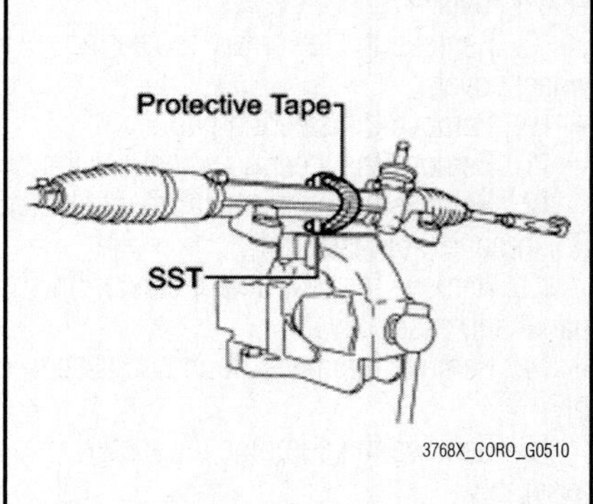

Fig. 226 Installing SST 09612-00012 to the steering link assembly

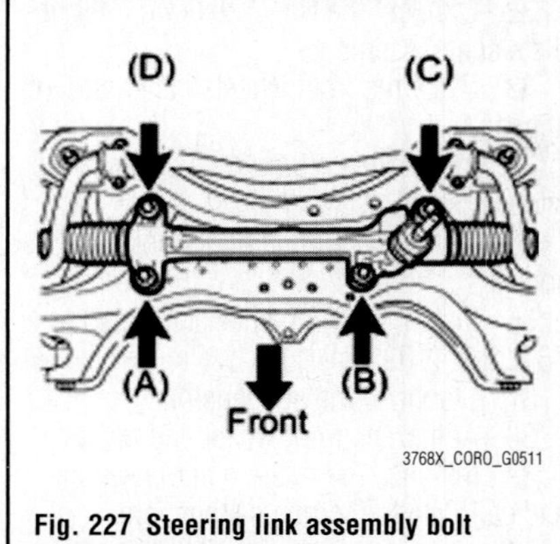

Fig. 227 Steering link assembly bolt tightening sequence

b. Tighten the bolts in order of (A), (B), (C), and then (D).

c. Tighten the bolts to 43 ft. lbs. (58 Nm).

22. Align the matchmarks and install the steering intermediate shaft to the steering link assembly.

23. Install the bolt. Tighten to 26 ft. lbs. (35 Nm).

24. To complete installation, reverse remaining removal procedure.

4WD Vehicles

See Figures 228 through 232.

1. Place the front wheels facing straight ahead.

2. Secure the steering wheel.

a. Secure the steering wheel with the seat belt in order to prevent rotation.

➡This operation is useful to prevent damage to the spiral cable.

3. Remove the column hole cover silencer shaft.

4. Separate the No. 2 steering intermediate shaft assembly.

5. Separate the No. 1 steering column hole cover sub assembly.

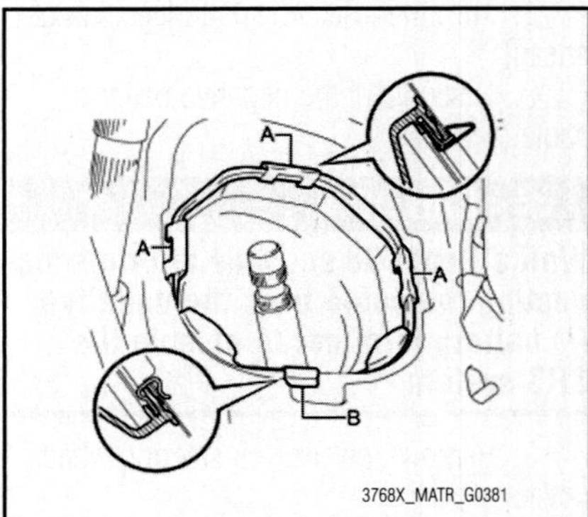

Fig. 228 Separating the No. 1 steering column hole cover sub assembly

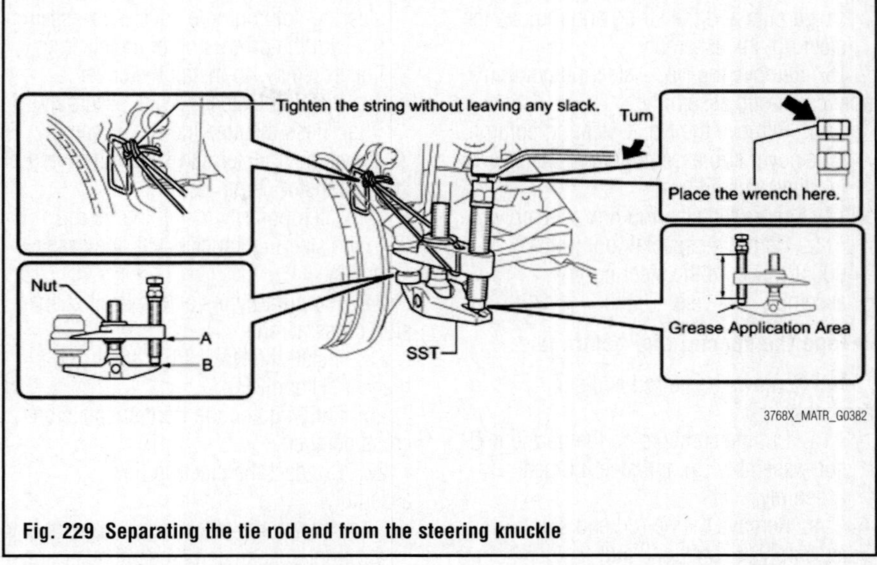

Fig. 229 Separating the tie rod end from the steering knuckle

a. Remove clips A and the No. 1 steering column hole cover sub-assembly and disengage clip B from the body.

➡Do not damage clips A and B.

6. Remove the front wheels.

7. Remove the engine under cover LH and RH.

8. Separate the tie rod end sub assembly LH.

a. Remove the cotter pin and nut.

b. Install the special tool (09960-20010).

➡Make sure that the upper ends of the tie rod end and SST are aligned.

c. Using the special tool, separate the tie rod end from the steering knuckle.

CAUTION

Apply grease to the bolt threads and the tip of the special tool.

➡Be sure to tighten the string firmly to secure SST to the steering knuckle to prevent SST from falling off.

➡Install SST with the center nut so that A and B are parallel. Otherwise, the dust cover may be damaged.

➡Be sure to place the wrench on the part indicated in the illustration.

➡Do not damage the front disc brake dust cover.

➡Do not damage the ball joint dust cover.

➡Do not damage the steering knuckle.

9. Separate the tie rod end sub assembly.

10. Separate the front lower suspension arm LH and RH.

11. Separate the steering link assembly.

a. Remove the 2 bolts and 2 nuts, and separate the steering link assembly from the front suspension crossmember sub-assembly.

➡Hold the nut when turning the bolt.

b. Install the stud bolt of each tie rod end to the steering knuckle and temporarily tighten the tie rod ends with the castle nuts.

c. Secure the steering link assembly to the transfer or stabilizer bar using a string or equivalent.

12. Loosen the front suspension crossmember.

13. Remove the No. 1 front stabilizer bracket LH and RH.

14. Remove the front suspension crossmember.

15. Remove the steering link assembly.

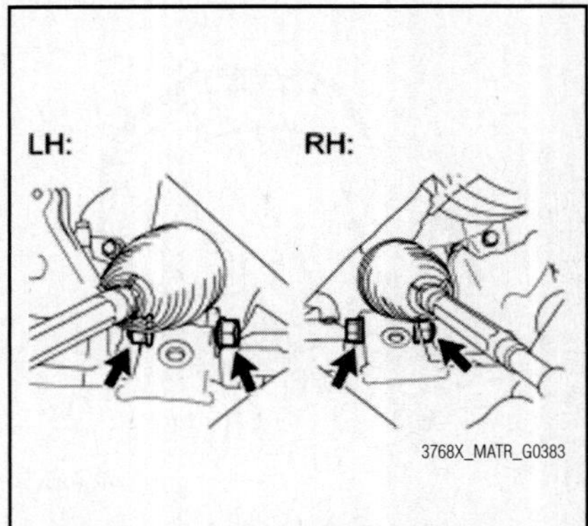

Fig. 230 Separating the steering link assembly

a. Remove the temporarily installed castle nuts and the string and remove the steering link assembly.

16. Remove the No. 1 steering column hole cover sub assembly.

a. Remove the No. 1 steering column hole cover sub assembly from the steering link assembly.

17. Secure the steering link assembly.

a. Using the special tool (09612-00012), secure the steering link assembly in a vise.

➡Tape the special tool before use.

18. Remove the tie rod end sub assembly LH.

a. Put matchmarks on the tie rod end sub-assembly LH and steering gear assembly.

b. Remove the tie rod end sub-assembly LH and lock nut.

19. Remove the tie rod end sub assembly RH.

a. Put matchmarks on the tie rod end sub-assembly RH and steering gear assembly.

b. Remove the tie rod end sub-assembly RH and lock nut.

To install:

20. Install the tie rod end sub assembly LH.

a. Install the lock nut and the tie rod end sub-assembly LH to the steering gear assembly until the matchmarks are aligned.

➡After adjusting the toe-in, tighten the lock nut.

21. Install the tie rod end sub assembly RH.

➡Use the same procedure for the LH side.

22. Install the No. 1 steering column hole cover sub assembly.

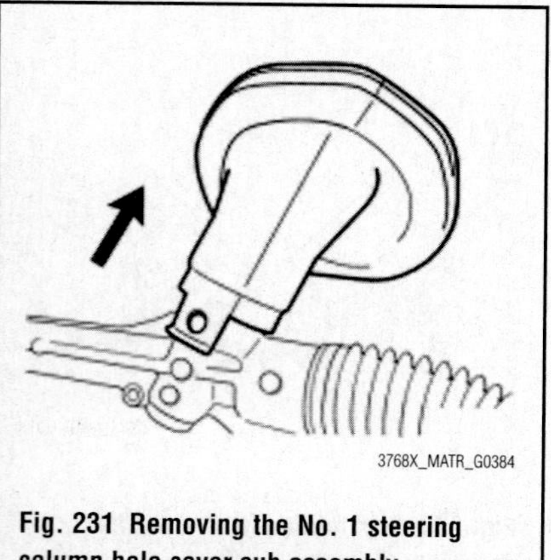

Fig. 231 Removing the No. 1 steering column hole cover sub assembly

a. Align the round hole in the No. 1 steering column hole cover sub-assembly with the protrusion of the steering link assembly to install the cover.

23. Install the steering link assembly.

a. Pass the steering link assembly above the transfer and secure it on top of the transfer using a string.

b. Temporarily install the tie rod to the steering knuckle with the castle nut.

24. Temporarily install the front suspension crossmember.

25. Install the No. 1 front stabilizer bracket LH and RH.

26. Fully tighten the front suspension crossmember.

27. Connect the steering link assembly.

a. Install the steering link assembly to the front suspension crossmember sub-assembly with the 2 bolts and 2 nuts. Tighten to 60 ft. lbs. (82 Nm).

28. Connect the front lower suspension arm LH and RH.

29. Connect the tie rod end sub assembly LH.

a. Connect the tie rod end sub-assembly LH to the steering knuckle with the nut. Tighten to 36 ft. lbs. (49 Nm).

➡Further tighten the nut up to 60° if the holes for the cotter pin are not aligned.

b. Install a new cotter pin.

30. Connect the tie rod end sub assembly RH.

31. Connect the No. 1 steering column hole cover sub assembly.

a. Engage clip B onto the body and install the No. 1 steering column hole cover sub-assembly onto the body with clips A.

➡Make sure that the lip of the No. 1 steering column hole cover sub-assembly is not damaged.

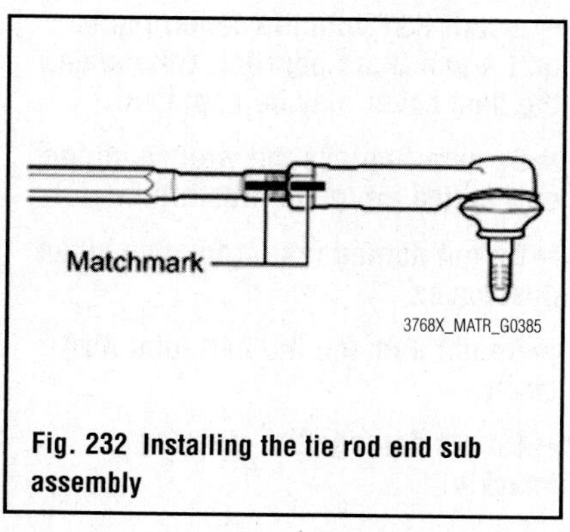

Fig. 232 Installing the tie rod end sub assembly

32. Connect the No. 2 steering intermediate shaft assembly.

33. Place the front wheels facing straight ahead.

34. Install the column hole cover silencer sheet.

35. Install the engine under cover LH and RH.

36. Install the front wheels. Tighten to 76 ft. lbs. (103 Nm).

37. Stabilizer the suspension.

38. Adjust the front wheel alignment.

39. Initialize the rotation angle sensor and calibrate the torque sensor zero point.

POWER STEERING PUMP

REMOVAL & INSTALLATION

➡The 2009–10 Matrix models do not use a power steering pump. They utilize a Power Steering Motor that works in conjunction with a power steering ECU. The power steering motor is integrated in to the steering column assembly and must be replaced as a unit. Refer to Power Steering Column Assembly in this section.

POWER STEERING COLUMN

REMOVAL & INSTALLATION

See Figures 233 through 239.

⁂ CAUTION

Some models covered by this manual may be equipped with a Supplemental Restraint System (SRS), which uses an air bag. Whenever working near any of the SRS components, such as the impact sensors, the air bag module, steering column and instrument panel, disable the SRS system.

1. Turn the front wheels to face straight ahead.

2. Disconnect the negative battery cable.

⁂ CAUTION

Wait at least 90 seconds after disconnecting the cable from the negative (-) battery terminal to disable the SRS system.

3. Remove lower no. 3 steering wheel cover.

4. Remove lower no. 2 steering wheel cover.

5. Remove the driver's side air bag.

6. Remove the steering wheel assembly.
7. Remove lower instrument panel finish panel sub-assembly.
8. Remove lower steering column cover:
   a. Push the right and left sides of the lower steering column cover, and disengage the 4 claws.
   b. Insert your fingers into the opening of the tilt lever of the lower steering column cover to disengage the claw. Spread the claw to disengage it.
   c. Using a screwdriver, insert the tip into each service hole to disengage the 2 claws and remove the lower steering column cover as shown in the illustration.
9. Disengage the claw and the 2 pins, and remove the upper steering column cover.
10. Disconnect the connectors from the turn signal switch assembly with spiral cable sub-assembly.
11. Use pliers to hold the clamp and raise the claw with a screwdriver. Remove the turn signal switch assembly with spiral cable sub-assembly from the steering column assembly.
12. Remove the upper instrument panel sub-assembly.
13. Turn back the floor carpet, and remove the 2 clips and column hole cover silencer sheet.
14. Put matchmarks on the No. 2 steering intermediate shaft assembly and the steering intermediate shaft.
15. Remove the bolt and separate the No. 2 steering intermediate shaft assembly from the steering intermediate shaft.
16. Remove the 2 bolts and instrument panel sub reinforcement.
17. Disengage the 2 claws and remove the No. 3 air duct sub-assembly.
18. Slide the transponder key amplifier to disengage the 2 claws.
19. Disconnect the connector and remove the transponder key amplifier.
20. Remove stop light switch assembly:
   a. Disconnect the connector.
   b. Turn the stop light switch assembly counterclockwise and remove the stop light switch assembly.
21. Remove steering post assembly (for 2ZR-FE):
   a. Separate the wire harness clamp from the power steering ECU assembly.
   b. Disconnect the 2 connectors from the power steering ECU assembly.
   c. Disconnect the connectors and disengage the wire harness clamps from the steering column assembly.

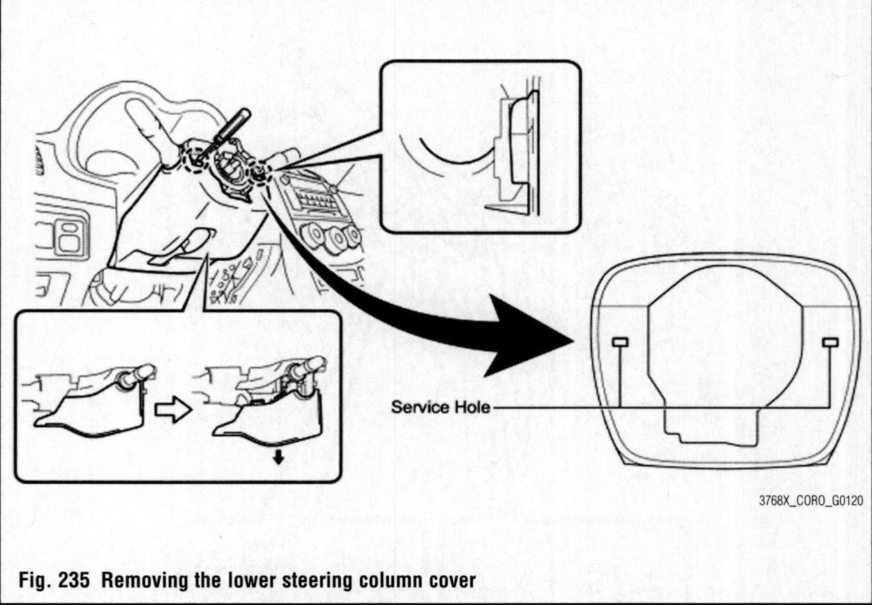

Fig. 235 Removing the lower steering column cover

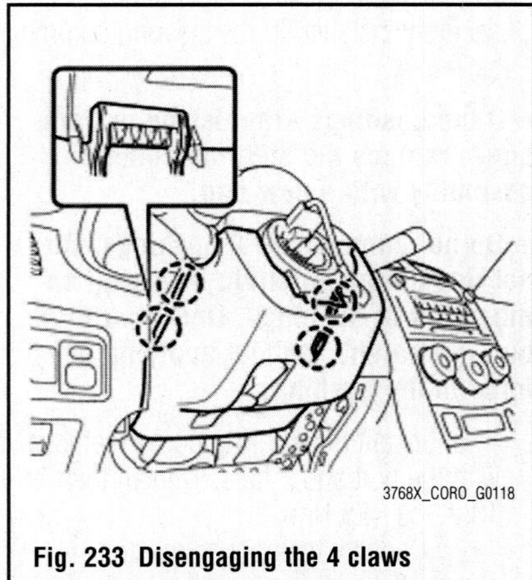

Fig. 233 Disengaging the 4 claws

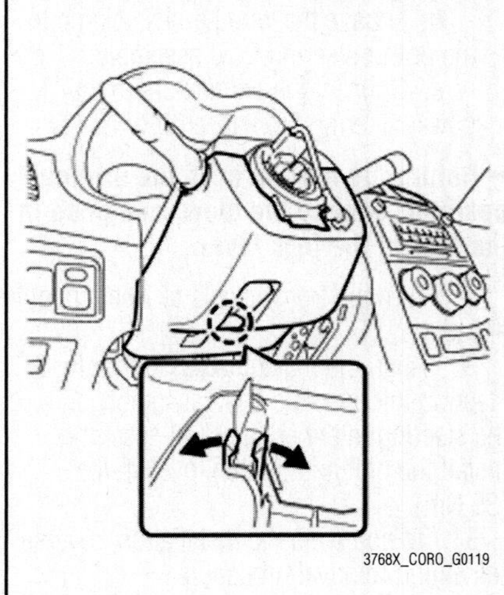

Fig. 234 Accessing the lower steering column cover to disengage the claw

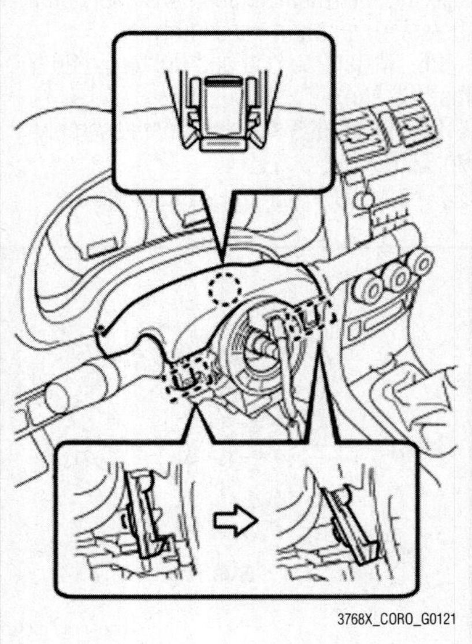

Fig. 236 Removing the upper steering column cover

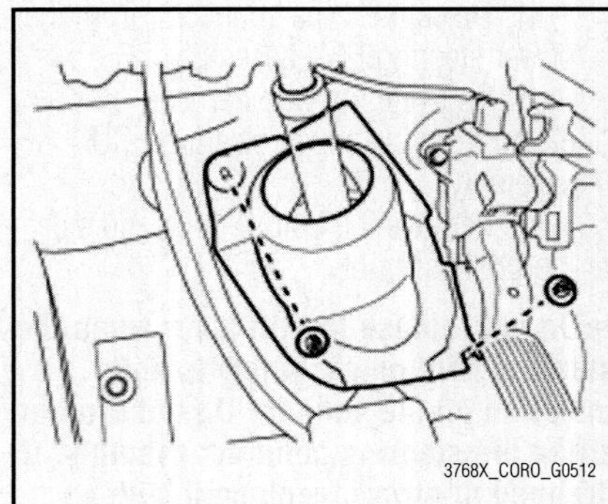

Fig. 237 Removing the 2 clips and column hole cover silencer sheet

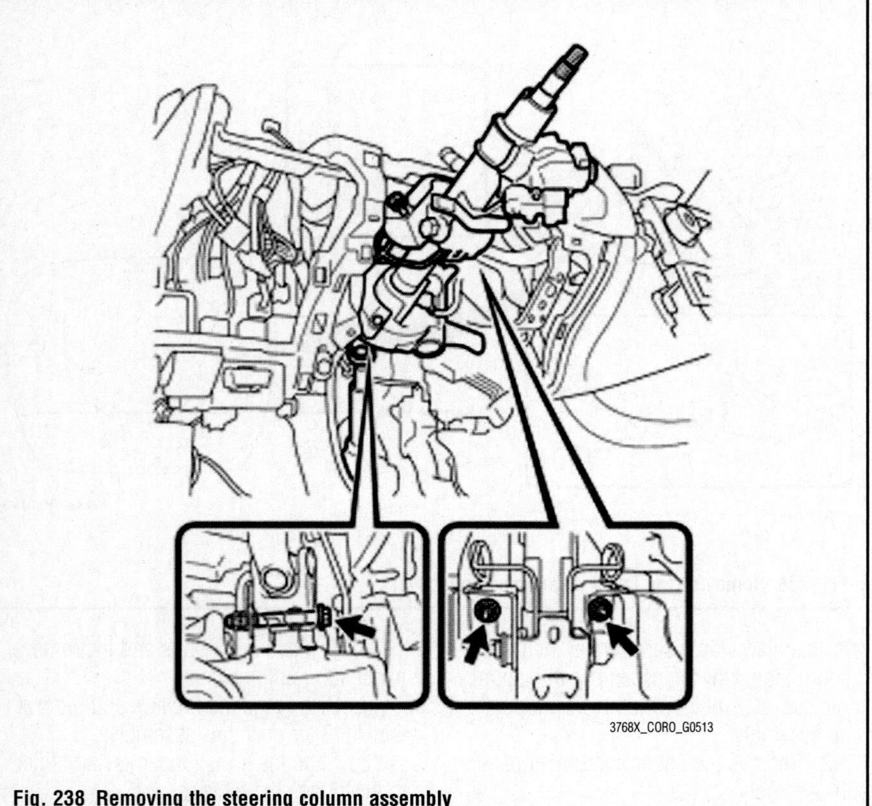

Fig. 238 Removing the steering column assembly

➡**Do not release the tilt lever when the steering column assembly is not installed on the vehicle. Do not drop or strike the steering column assembly. If dropped or struck, replace it with a new one.**

d. Remove the bolt, 2 nuts and steering post assembly.

22. Remove the steering post assembly (for 2AZ-FE):

a. Disconnect the connector from the power steering ECU assembly.

➡**Pull out the lock of the lock lever, disengage the claw, and turn the lock lever to disconnect the connector.**

b. Disconnect the connector from the power steering ECU assembly.

c. Separate the wire harness clamp from the power steering ECU assembly.

d. Remove the bolt, 2 nuts, and steering post assembly.

➡**Do not release the tilt lever when the steering column assembly is not installed on the vehicle. Do not drop or strike the steering column assembly. If dropped or struck, replace it with a new one.**

23. Put matchmarks on the No. 2 steering intermediate shaft assembly and the steering column assembly.

24. Remove the bolt and the No. 2 steering intermediate shaft assembly from the steering column assembly.

To install:

25. Align the matchmarks on the No. 2 steering intermediate shaft assembly and the steering column assembly.

26. Install the bolt and tighten to 26 ft. lbs. (35 Nm).

27. Install the steering post assembly (for 2ZR-FE):

a. Check that the 2 bushings are securely installed to the steering column assembly.

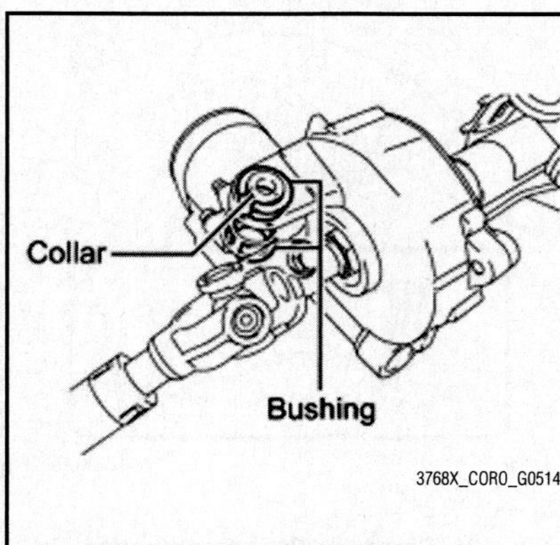

Fig. 239 Checking the 2 bushings

➡**If the bushings are missing or damaged, replace the steering column assembly with a new one.**

➡**Do not damage the 2 bushings. Do not line up the bolt hole by prying on the collar or bushings. Only install the bolt in straight, without applying any force to the bushings.**

b. Install the steering post assembly with the bolt and 2 nuts. Tighten them to 18 ft. lbs. (25 Nm).

c. Connect the connectors and engage the wire harness clamps to the steering column assembly.

d. Connect the 2 connectors to the power steering ECU assembly.

e. Install the wire harness clamp to the power steering ECU assembly.

28. Install the steering post assembly (for 2AZ-FE):

a. Check that the 2 bushings are securely installed to the steering column assembly.

➡**If the bushings are missing or damaged, replace the steering column assembly with a new one.**

➡**Do not damage the 2 bushings. Do not line up the bolt hole by prying on the collar or bushings. Only install the bolt in straight, without applying any force to the bushings.**

b. Install the steering post assembly with the bolt and 2 nuts. Tighten them to 18 ft. lbs. (25 Nm).

c. Connect the connectors and engage the wire harness clamps to the steering column assembly.

d. Engage the wire harness clamp to the power steering ECU assembly.

e. Connect the connectors to the power steering ECU assembly.

➡**Return the lock lever to the original position, engage the claw, and push in the lock of the lock lever.**

29. Turn the front wheels to face straight ahead.

30. Align the matchmarks on the No. 2 steering intermediate shaft assembly and the steering intermediate shaft assembly. Install the bolt and tighten to 26 ft. lbs. (35 Nm).

31. To complete the installation, reverse remaining removal procedure.

32. Calibrate the torque sensor zero point. Refer to Power Steering ECU.

LOWER BALL JOINT

REMOVAL & INSTALLATION

See Figures 240 through 242.

➡Use the same procedure for the LH side and RH side.

➡The following procedure listed is for the LH side.

1. Remove the front wheel.
2. Remove the front axle shaft nut.
3. Separate the front speed sensor.
4. Separate the front disc caliper assembly.
5. Remove the front disc.
6. Separate the tie rod end sub assembly.
7. Separate the front lower suspension arm.
   a. Remove the bolt and 2 nuts, and separate the front lower suspension arm from the lower ball joint.
8. Remove the front axle assembly.
9. Remove the front lower ball joint.
   a. Secure the front axle assembly in a vise.

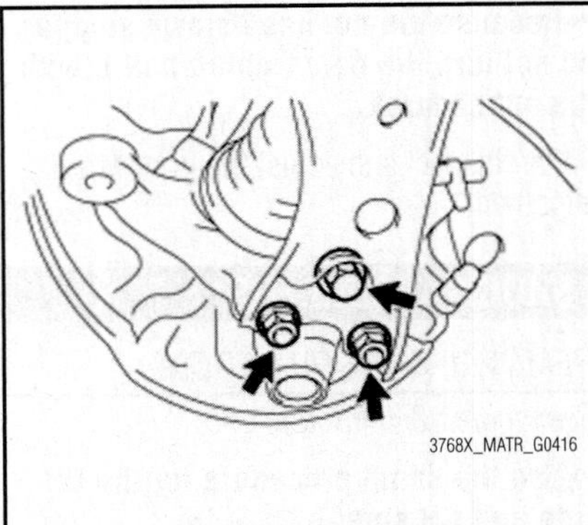

Fig. 240 Separating the front lower suspension arm

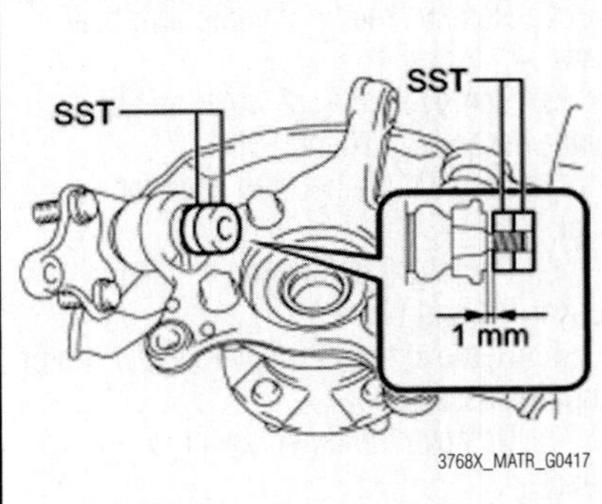

Fig. 241 Installing the special tools to the front lower ball joint

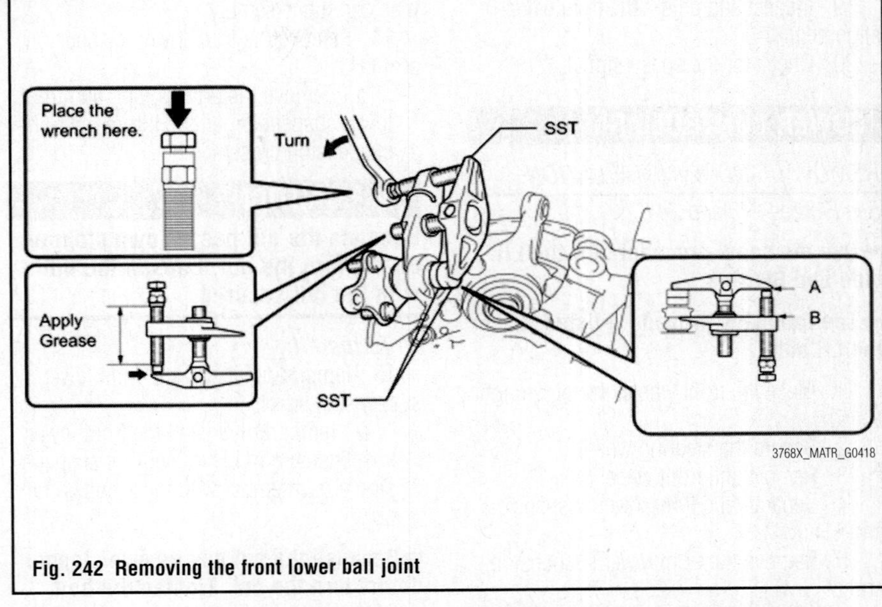

Fig. 242 Removing the front lower ball joint

➡When using a vise, do not over-tighten it.

   b. Remove the cotter pin and nut.
   c. Install the 2 special tools (09960-20010 or 09961-02050) to the front lower ball joint.

➡Check that the clearance measurement is between the special tool and the front axle assembly is 0.0394 inch (1 mm).

➡Use the 2 special tools of the same type.

   d. Using the special tool, remove the front lower ball joint from the front axle assembly.

CAUTION

Apply grease to the threads and end of the special tool bolt.

➡Install the special tool so that A and B are parallel.

➡Be sure to place a wrench on the part indicated in the illustration.

➡Do not damage the front lower ball joint dust cover.

➡Use 2 special tools of the same type.

To install:

10. Install the front lower ball joint.
    a. Secure the front axle assembly in a vise.

➡When using a vise, do not over-tighten it.

   b. Install the front lower ball join to the front axle assembly with the nut. For the 1.8L engine, tighten the nut to 76 ft. lbs. (103 Nm). For the 2.4L engine, tighten to 91 ft. lbs. (123 Nm).
   c. Install a new cotter pin.

➡Further tighten the nut up to 60° if the holes for the cotter pin are not aligned.

11. Install the front axle assembly.
12. Connect the front lower suspension arm.
    a. Install the front lower suspension arm to the lower ball joint with the bolt and 2 nuts. Tighten to 66 ft. lbs. (89 Nm).
13. Connect the tie rod end sub assembly.
14. Install the front disc.
15. Install the front disc brake caliper assembly.
16. Install the front speed senor.
    a. Install the front flexible hose and front speed sensor with the bolt. Tighten the bolt to 21 ft. lbs. (29 Nm).

➡Install the flexible hose and the speed sensor without twisting them.

   b. Install the front speed sensor onto the steering knuckle with the bolt. Tighten to 75 inch lbs. (8.5 Nm).
   c. Install the front speed sensor with the clamp to the front shock absorber.

17. Install the front axle shaft nut.

18. Install the front wheel. Tighten to 76 ft. lbs. (103 Nm).

19. Inspect and adjust the front wheel alignment.

20. Check for the speed signal.

LOWER CONTROL ARM

REMOVAL AND & INSTALLATION

See Figures 243 and 244.

➡Use the same procedure for the LH side and RH side.

➡The following procedure listed is for the LH side.

1. Place the front wheels facing straight ahead.
2. Secure the steering wheel.
3. Remove the front wheels.
4. Separate the front lower suspension arm LH (MT).
5. Remove the front lower suspension arm LH (MT).
    a. Remove the 2 bolts, nut, and front lower suspension arm LH from the front suspension crossmember.

➡Because the nut has its own stopper, do not turn the nut. Loosen bolt B with the nut secured.

6. Remove the column hole cover silencer sheet.
7. Separate the No. 2 steering intermediate shaft assembly.
8. Separating No. 1 steering column hole cover sub assembly.
9. Separate the tie rod end sub assembly LH.
10. Separate the tie rod end sub assembly RH.
11. Separate the front stabilizer link assembly LH and RH.

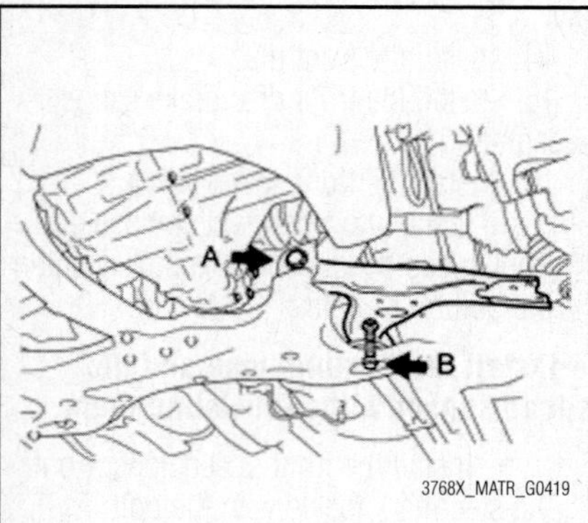

Fig. 243 Removing the front lower suspension arm LH (MT)

12. Separate the front lower suspension arm LH and RH.
13. Remove the front suspension crossmember sub assembly.
14. Remove the front lower suspension arm LH.
    a. Remove the 2 bolts, nut, and front lower suspension arm LH from the front suspension crossmember.

⁂ CAUTION

Because the nut has its own stopper, do not turn the nut. Loosen the bolt with the nut secured.

To install:

15. Temporarily install the front lower suspension arm LH.
    a. Temporarily install the front lower suspension arm LH to the front suspension crossmember with the 2 bolts and nut.

➡Because the nut has its own stopper, do not turn the nut. Tighten the bolt with the nut secured.

➡Fully tighten the 2 bolts after stabilizing the suspension.

16. Temporarily install the front lower suspension arm LH (MT).
    a. Temporarily install the front lower suspension arm LH to the front suspension crossmember with the 2 bolts and nut.

➡Because the nut has its own stopper, do not turn the nut. Tighten the bolt with the nut secured.

➡Fully tighten the 2 bolts after stabilizing the suspension.

17. Install the front suspension crossmember sub assembly.
18. Connect the front lower suspension arm LH and RH.
19. Connect the front stabilizer link assembly LH and RH.
20. Connect the tie rod end sub assembly LH and RH.
21. Connect the No. 1 steering column hole cover sub assembly.
22. Connect the No. 2 steering intermediate shaft assembly.
23. Place the front wheels facing straight ahead.
24. Install the column hole cover silencer sheet.
25. Install the front wheels. Tighten to 76 ft. lbs. (103 Nm).
26. Stabilize the suspension.
    a. Lower the vehicle and bounce it up and down several times to stabilize the front suspension. Raise the vehicle.

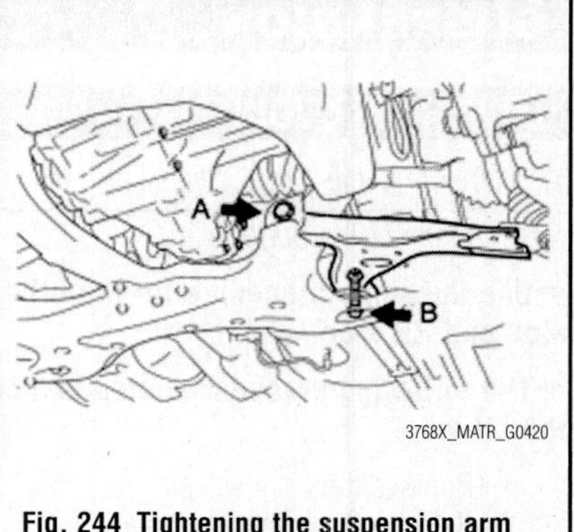

Fig. 244 Tightening the suspension arm bolts

27. Fully tighten the front lower suspension arm LH.
    a. Fully tighten bolt A to 101 ft. lbs. (137 Nm).

➡The final torque must be applied under standard vehicle height conditions.

    b. Fully tighten bolt B to 101 ft. lbs. (137 Nm).

➡The final torque must be applied under standard vehicle height conditions.

➡Because the nut has its own stopper, do not turn the nut. Tighten bolt B with the nut secured.

28. Inspect and adjust the front wheel alignment.

SHOCK ABSORBERS

REMOVAL & INSTALLATION

See Figures 245 through 250.

➡Use the same procedure for the LH side and RH side.

➡The following procedure listed is for the LH side.

1. Remove the front wheel.
2. Remove the front wiper arm head cap.
3. Remove the front wiper arm and blade assembly LH and RH.
4. Remove the hood to cowl top seal.
5. Remove the cowl top ventilator louver RH and LH.
6. Remove the windshield wiper motor and link assembly.
7. Remove the outer cowl top panel.
8. Separate the front stabilizer link assembly.
9. Separate the front speed sensor.

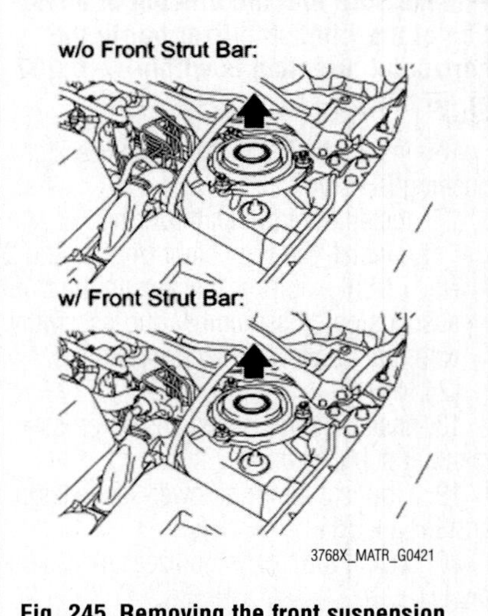

Fig. 245 Removing the front suspension dust cover

10. Remove the front suspension support dust cover.
    a. Remove the front suspension support dust cover.
11. Remove the shock absorber with the coil spring.
    a. Loosen the front support to front shock absorber nut of the front shock absorber.

➡Do not remove the front support to front shock absorber nut.

➡Loosen the nut only when the front shock absorber with coil spring needs to be disassembled.

b. Support the front axle using a jack and wooden block.

c. Remove the 2 bolts and 2 nuts, and separate the front shock absorber with coil spring (lower side) from the steering knuckle.

d. Remove the 3 nuts and front shock absorber with coil spring (upper side).

➡Make sure that the front speed sensor is completely separated from the front shock absorber with coil spring.

12. Secure the front shock absorber with the coil spring.
    a. Install the bolt and nut to the front shock absorber as shown in the illustration and secure the front shock absorber in a vise. "A" = 1.10 inch (28 mm)
13. Remove the front support to front shock absorber nut.
    a. Using the special tool (09727-30021), compress the front coil spring.

➡If the front coil spring is compressed at an angle, using 2 special tools will make the work easier.

b. Check that the front coil spring is fully compressed.

➡Do not use an impact wrench. It will damage the special tools.

c. Remove the front support to front shock absorber nut.

14. Remove the front suspension support sub assembly.
15. Remove the front suspension support dust seal.
16. Remove the front coil spring upper seat.
17. Remove the front coil spring upper insulator.
18. Remove the front coil spring.
19. Remove the front spring bumper.
20. Remove the front coil spring lower insulator.

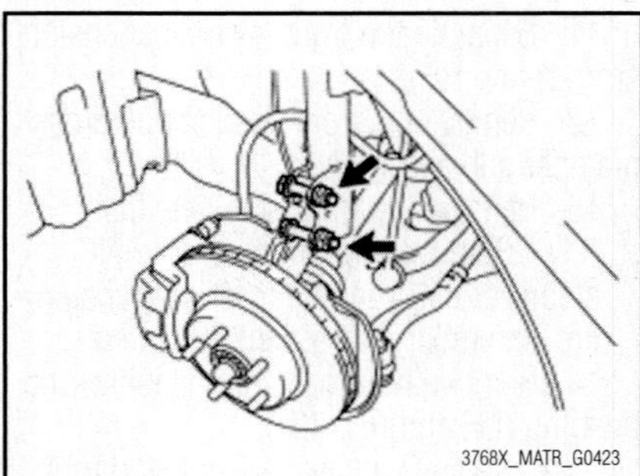

Fig. 247 Separating the front shock absorber with coil spring (lower side) from the steering knuckle

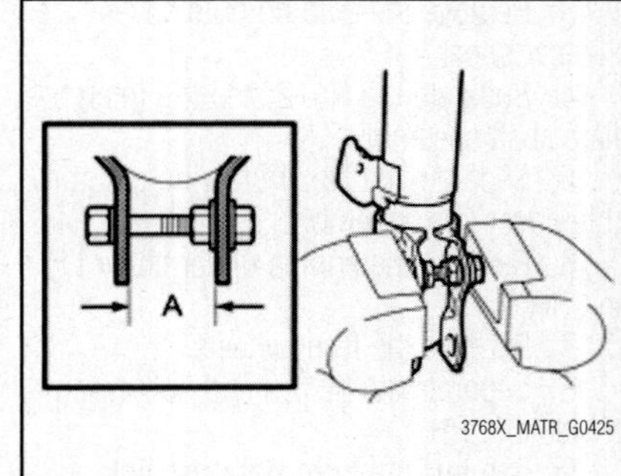

Fig. 249 Securing the front shock absorber with the coil spring

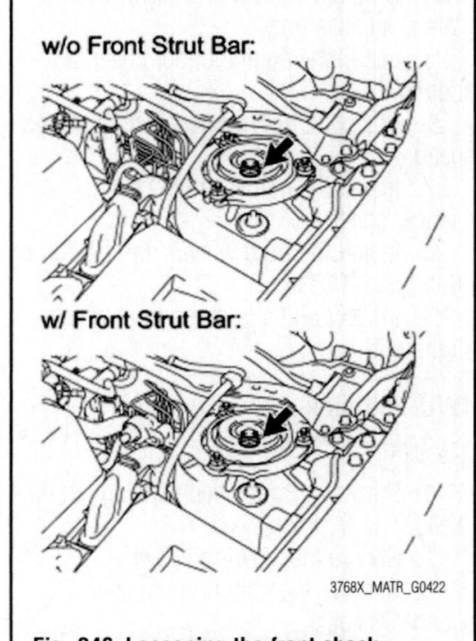

Fig. 246 Loosening the front shock absorber nut

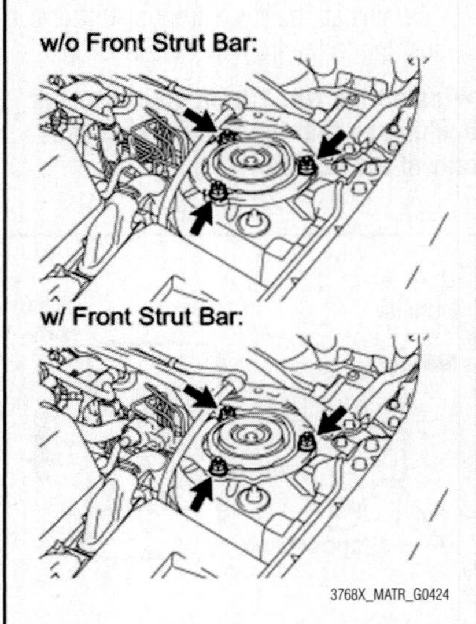

Fig. 248 Removing the shock absorber with the coil spring

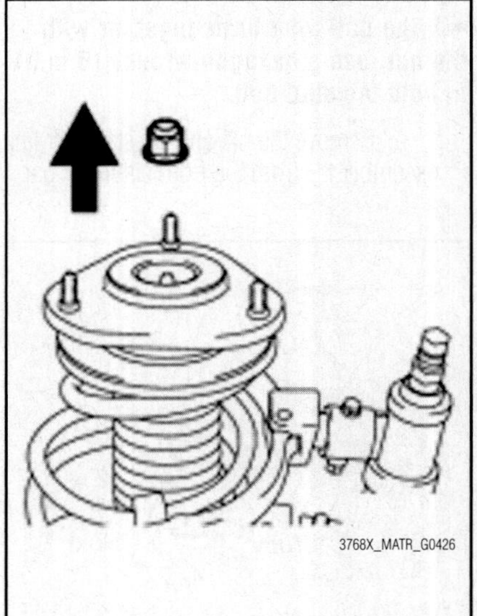

Fig. 250 Removing the front support to front shock absorber nut.

To install:

21. To install, reverse the removal procedure. Tighten the front shock absorber with the coil spring (upper side) nuts to 37 ft. lbs. (50 Nm) and (lower side) to 177 ft. lbs. (240 Nm). Tighten the front support to front shock absorber nut to 35 ft. lbs. (47 Nm).

STEERING KNUCKLE

REMOVAL & INSTALLATION

Refer to STEERING for removal and installation of the steering knuckle.

STABILIZER BAR

REMOVAL & INSTALLATION

2WD Vehicles

See Figures 251 through 253.

1. Place the front wheels facing straight ahead.
2. Secure the steering wheel.
3. Remove the column hole cover silencer sheet.
4. Separate the No. 2 steering intermediate shaft assembly.
5. Separate the No. 1 steering column hole cover sub assembly.
6. Remove the engine under cover LH and RH.
7. Remove the front wheels.
8. Separate the tie rod end sub assembly LH and RH.
9. Remove the front stabilizer link assembly LH.
   a. Remove the nut and separate the stabilizer link assembly LH from the front shock absorber with coil spring.

➡If the ball joint turns together with the nut, use a hexagon wrench (6 mm) to hold the stud bolt.

   b. Remove the nut and the stabilizer link assembly LH from the front stabilizer bar.

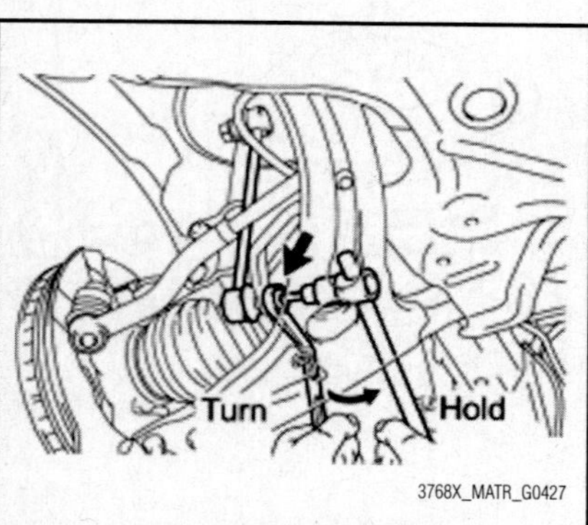

Fig. 251 Removing the stabilizer link assembly from the front stabilizer bar

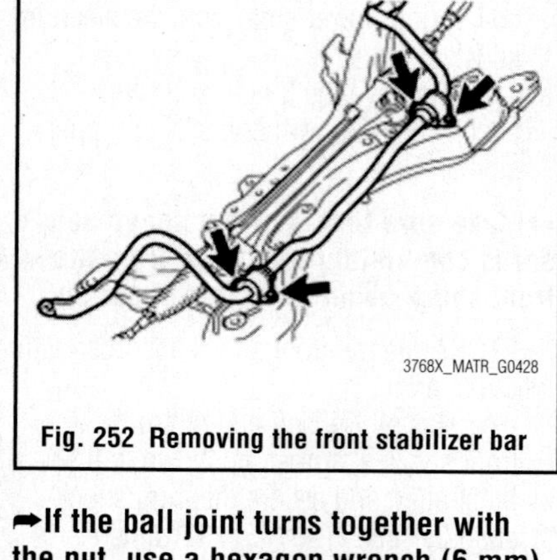

Fig. 252 Removing the front stabilizer bar

➡If the ball joint turns together with the nut, use a hexagon wrench (6 mm) to hold the stud bolt.

10. Remove the front stabilizer link assembly RH.

➡Perform the same procedure as the LH side.

11. Separate the front lower suspension arm LH and RH.
12. Remove the front suspension crossmember sub assembly.
13. Remove the front stabilizer bar.
   a. Remove the 4 bolts and front stabilizer bar from the front suspension crossmember sub-assembly.
14. Remove the No. 1 front stabilizer bar bushing (LH and RH side).
   a. Remove the No. 1 front stabilizer bar bushing from the front stabilizer bar.

To install:

15. Install the No. 1 front stabilizer bar bushing.
   a. Install the No. 1 front stabilizer bar bushing to the front stabilizer bar.

➡Install the No. 1 front stabilizer bar bushing so that the cutout faces the rear of the vehicle.

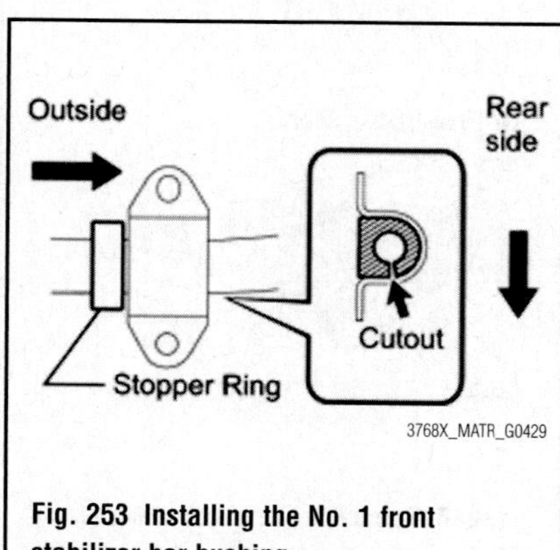

Fig. 253 Installing the No. 1 front stabilizer bar bushing

➡Make sure that the amount of deviation of the front stabilizer bar in the horizontal direction is within +/- 0.197 in. (5 mm).

16. Install the No. 1 front stabilizer bar bushing RH side.
17. Install the front stabilizer bar.
   a. Install the front stabilizer bar and 2 No. 1 front stabilizer brackets to the front suspension crossmember sub-assembly with the 4 bolts. Tighten to 18 ft. lbs. (24 Nm).
18. Install the front suspension crossmember sub assembly.
19. Connect the front lower suspension arm LH and RH.
20. Install the front stabilizer link assembly LH.
   a. Install the front stabilizer link assembly LH to the front shock absorber with coil spring with the nut. Tighten to 55 ft. lbs. (75 Nm).

➡If the ball joint turns together with the nut, use a hexagon wrench (6 mm) to hold the stud bolt.

   b. Install the front stabilizer link assembly LH to the front stabilizer bar with the nut. Tighten to 55 ft. lbs. (75 Nm).

➡If the ball joint turns together with the nut, use a hexagon wrench (6 mm) to hold the stud bolt.

21. Install the front stabilizer link assembly RH.
22. Connect the tie rod end sub assembly LH and RH.
23. Connect the No. 1 steering column hole cover sub assembly.
24. Connect the No. 2 steering intermediate shaft assembly.
25. Install the engine under cover LH and RH.
26. Place the front wheels facing straight ahead.
27. Install the column hole cover silencer sheet.
28. Install the front wheels and tighten to 76 ft. lbs. (103 Nm).
29. Inspect and adjust the front wheel alignment.

4WD Vehicles

See Figures 254 and 255.

1. Place the front wheels facing straight ahead.
2. Secure the steering wheel.
3. Remove the column hole cover silencer sheet.
4. Separate the No. 2 steering intermediate shaft assembly.

5. Separate the No. 1 steering column hole cover sub assembly.

6. Remove the engine under cover LH and RH.

7. Remove the front wheels.

8. Separate the tie rod end sub assembly LH and RH.

9. Loosen the front suspension crossmember sub assembly.

10. Remove the No. 1 front stabilizer bracket LH and RH.

11. Separate the steering link assembly.

12. Remove the front suspension crossmember sub assembly.

13. Remove the front stabilizer bar.

a. Remove the 2 nuts and the front stabilizer bar from the front stabilizer link.

➡If the ball joint turn together with the nut, use a hexagon wrench (6 mm) to hole the stud bolt.

14. Remove the front stabilizer link assembly LH.

a. Remove the nut and the stabilizer link assembly LH from the front shock absorber with coil spring.

➡If the ball joint turns together with the nut, use a hexagon wrench (6 mm) to hold the stud bolt.

15. Remove the front stabilizer link assembly RH.

➡Perform the same procedure as the LH side.

16. Remove the No. 1 front stabilizer bar bushing from the front stabilizer bar (LH and RH).

To install:

17. To install, reverse the removal procedure. Tighten the front stabilizer link and bar to 55 ft. lbs (74 Nm).

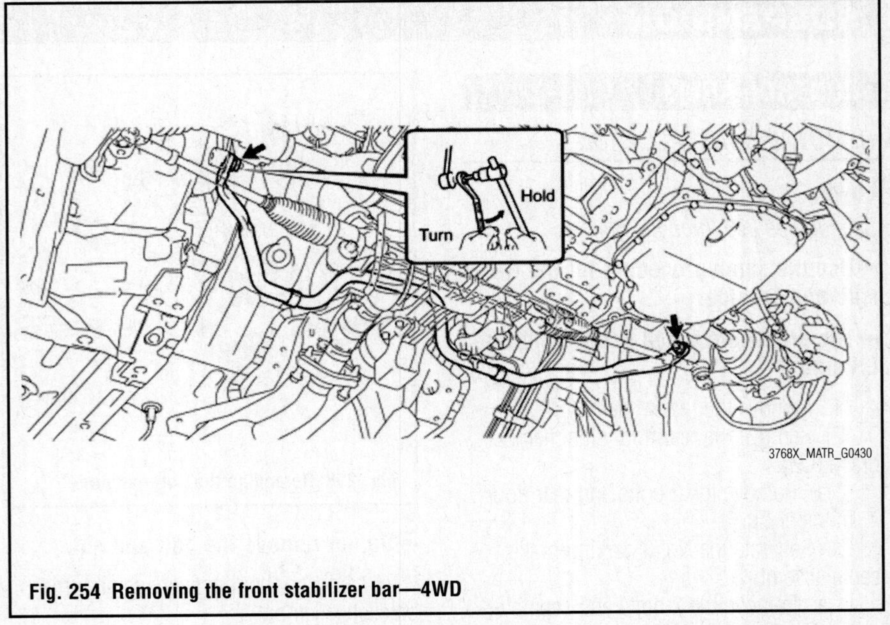

Fig. 254 Removing the front stabilizer bar—4WD

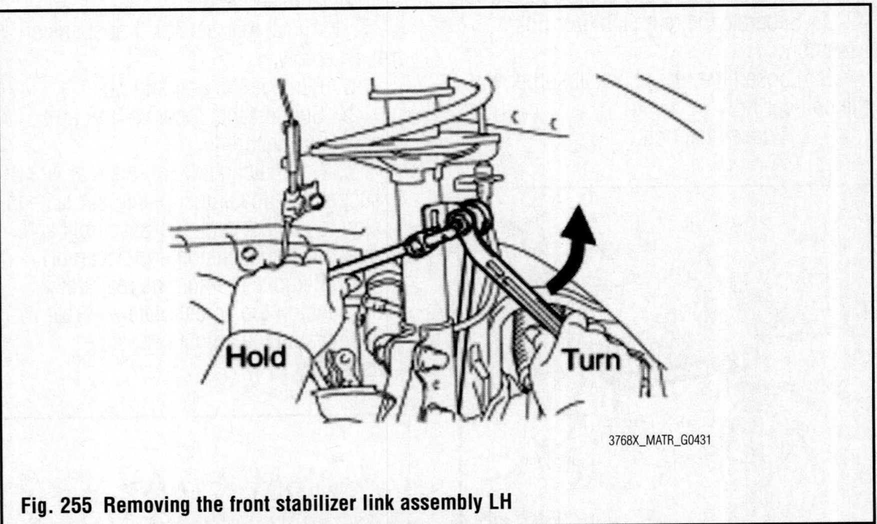

Fig. 255 Removing the front stabilizer link assembly LH

CONTROL ARMS

REMOVAL & INSTALLATION

Lower

See Figures 256 through 268.

➡Use the same procedure for the RH side and LH side.

➡The procedure listed below is for the LH side.

1. Remove the rear wheel.
2. Remove the rear floor side member brace (LH).
   a. Remove the 2 bolts and rear floor member brace.
3. Separate the No. 3 parking brake cable assembly.
   a. Remove the 2 bolts and separate the No. 3 parking brake cable assembly.
4. Separate the rear stabilizer link assembly.
5. Loosen the rear shock absorber with the coil spring.
   a. Loosen the bolt.

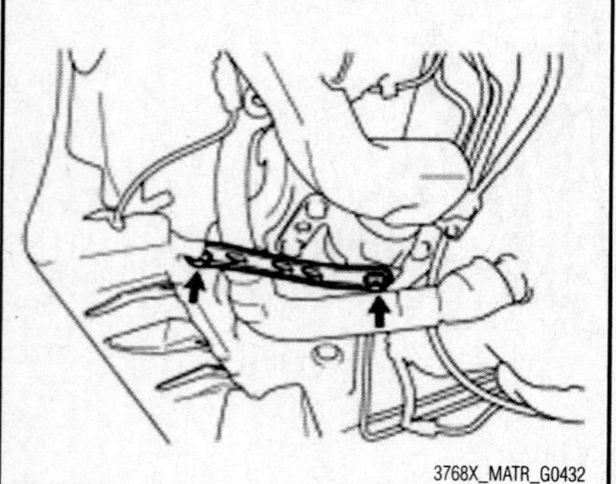

Fig. 256 Removing the rear floor side member brace

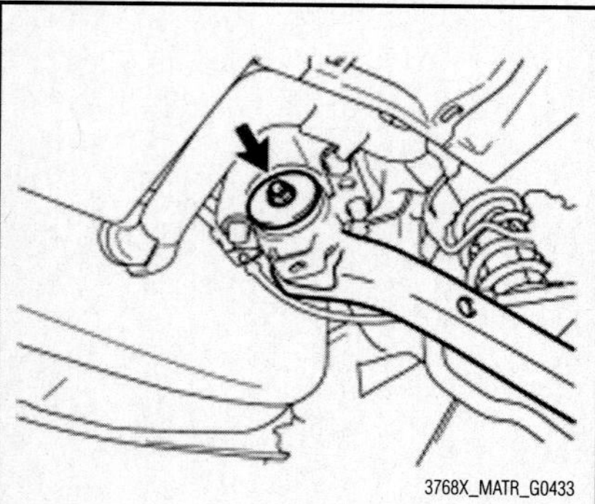

Fig. 257 Loosening the rear suspension arm bracket

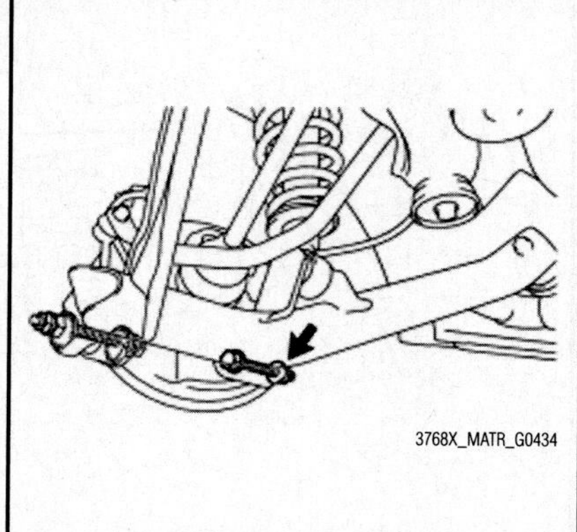

Fig. 258 Removing the bolt and nut

➡Do not remove the bolt and nut.

6. Loosen the rear suspension arm bracket assembly.
   a. Loosen the nut.
7. Remove the rear No. 1 suspension arm assembly.
   a. Remove the bolt and nut.
   b. Support the upper control arm assembly securely.
   c. Put matchmarks on the rear No. 2 suspension toe adjust plate, rear suspension toe adjust cam sub-assembly and rear No. 1 suspension arm assembly.
   d. Remove the nut, the rear No. 2 suspension toe adjust plate and rear suspension toe adjust cam sub-assembly.

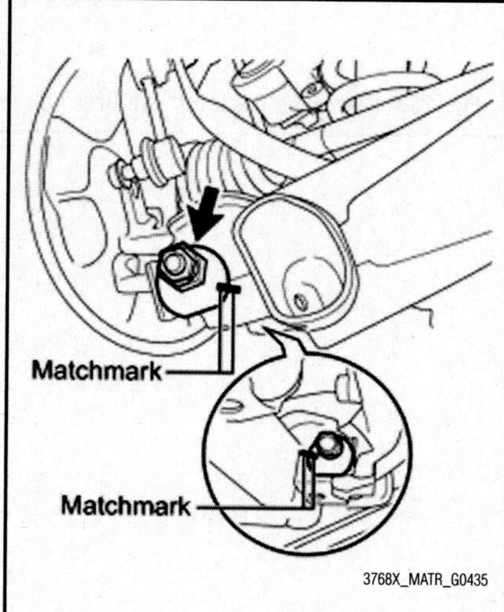

Fig. 259 Putting matchmarks on the rear No. 2 suspension toe adjustment plate, rear suspension toe adjust cam sub assembly and rear No. 1 suspension arm assembly

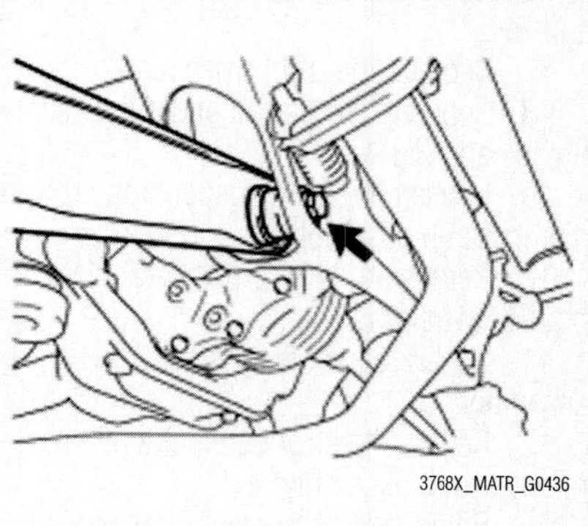

Fig. 260 Separating the rear part of the rear No. 1 suspension arm assembly

   e. Remove the bolt and nut, and disconnect the rear shock absorber with coil spring.
   f. Remove the bolt and nut, and separate the rear part of the rear No. 1 suspension arm assembly.

➡When removing the bolt, stop the nut from rotating and loosen the bolt.

   g. Remove the 3 bolts on the front part, and separate the front part of the rear No. 1 suspension arm assembly.
8. Remove the rear suspension arm bracket assembly.
9. Remove the rear suspension support stopper.
10. Remove the lower control arm bushing.
    a. Using the special tool (09632-36010), remove the lower control arm bushing.

To install:

11. Install the lower control arm bushing. "A": 0.591/0.0196 inch (15/0.5 mm).

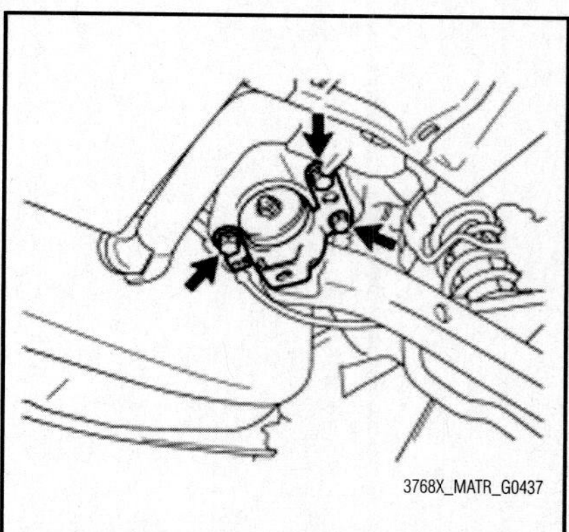

Fig. 261 Separating the front part of the rear No. 1 suspension arm assembly

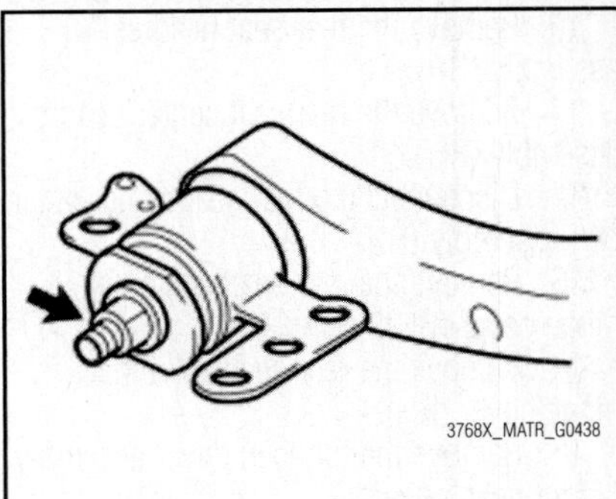

Fig. 262 Removing the rear suspension arm bracket assembly

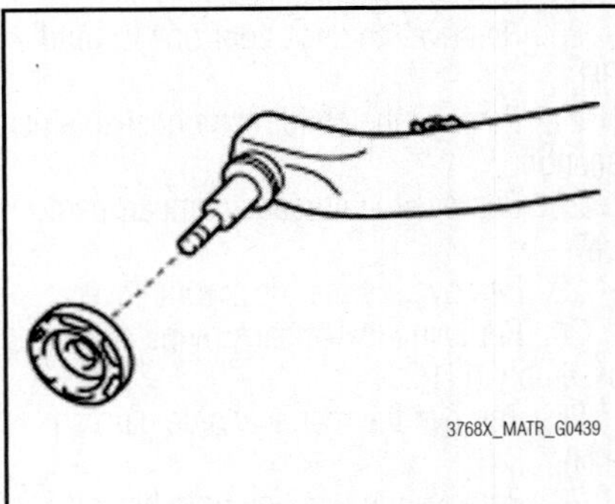

Fig. 263 Removing the rear suspension support stopper

➡Do not install the lower control arm bushing in the wrong direction.

➡Place the lower control arm bushing at the position in the illustration.

12. Install the rear suspension support stopper.

➡Do not install the rear suspension support stopper in the wrong direction.

13. Temporarily tighten the rear suspension arm bracket assembly with the nut.
14. Temporarily tighten the rear shock absorber with the coil spring with the nut and the bolt.
15. Temporarily tighten the rear No. 1 suspension arm assembly with the bolt and nut.

➡When installing the bolt, stop the nut from rotating and tighten the bolt.

a. Temporarily install the 3 bolts and fully tighten the 3 bolts in order from 1 to 3. Tighten to 48 ft. lbs. (65 Nm).
b. Temporarily tighten the rear axle assembly to the rear No. 1 suspension arm assembly with the bolt and the nut.
c. Insert the rear suspension toe adjust cam sub-assembly from the front side of the vehicle, and temporarily tighten the nut through the rear No. 2 suspension toe adjust plate.

16. Install the rear stabilizer link assembly.
17. Install the No. 3 parking brake cable assembly with the 2 bolts. Tighten to 53 inch lbs. (6 Nm).
18. Install the rear floor side member brace (LH) with the 2 bolts. Tighten to 22 ft. lbs. (30 Nm).
19. Stabilize the suspension.
20. Fully tighten the rear No. 1 suspension arm assembly.

a. Using the special tool, tighten the bolt. Using the special tool tighten to 64 ft. lbs. (87 Nm). Without the special tool tighten to 47 ft. lbs. (64 Nm).

➡Use a torque wrench with a fulcrum length of 425 mm (16.73 in.).

➡This torque value is effective when SST is parallel to the torque wrench.

➡When installing the bolt, stop the nut from rotating and tighten the bolt.

➡The final torque must be applied under standard vehicle height conditions.

b. Fully tighten the bolt to 103 ft. lbs. (140 Nm).

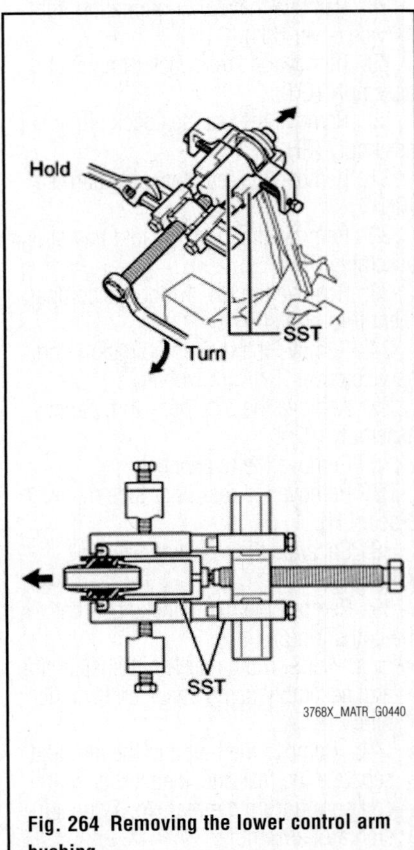

Fig. 264 Removing the lower control arm bushing

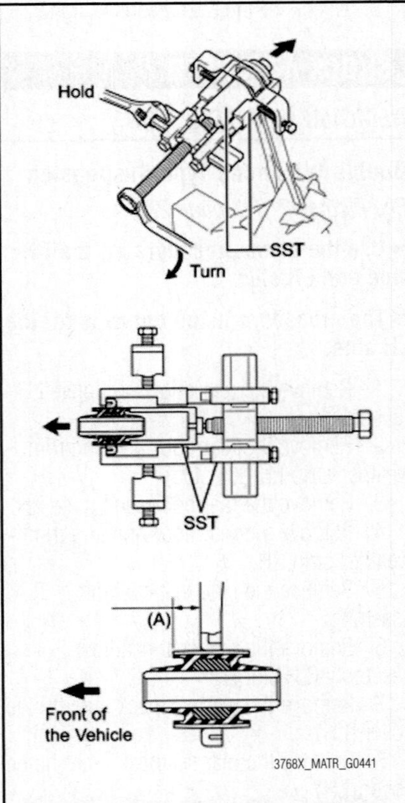

Fig. 265 Installing the lower control arm bushing

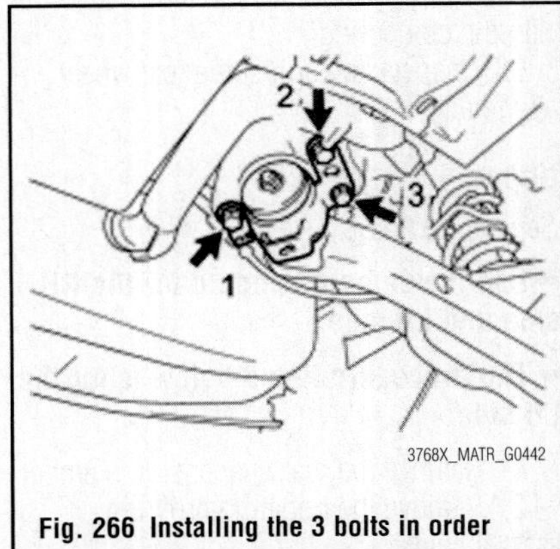

Fig. 266 Installing the 3 bolts in order

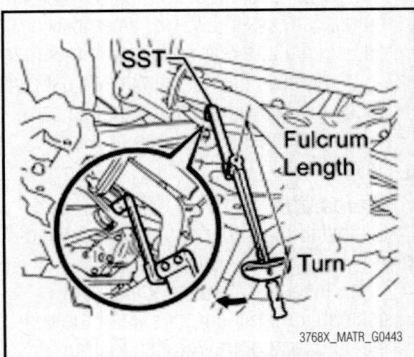

Fig. 267 Tightening the bolt with the special tool

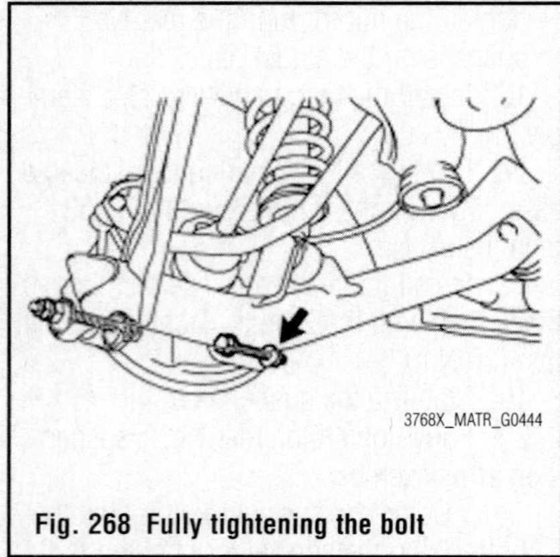

Fig. 268 Fully tightening the bolt

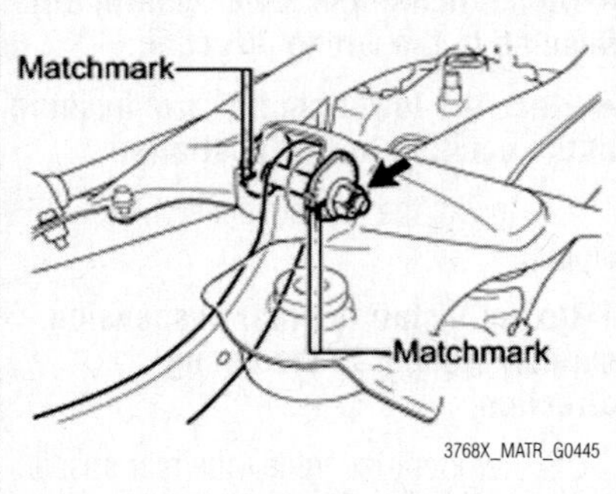

Fig. 269 Removing the upper control arm assembly LH

➡The final torque must be applied under standard vehicle height conditions.

c. Align the matchmarks on the rear No. 2 suspension toe adjust plate and fully tighten the nut. Tighten to 55 ft. lbs. (74 Nm).

➡The final torque must be applied under standard vehicle height conditions.

21. Fully tighten the rear suspension arm bracket assembly. Using the special tool tighten the nut to 81 ft. lbs. (110 Nm). Without the special tool tighten the nut to 60 ft. lbs. (81 Nm).
22. Fully tighten the rear shock absorber with the coil spring.
23. Inspect and adjust the rear wheel alignment.

Upper

See Figure 269.

➡Use the same procedure for the RH side and LH side.

➡The procedure listed below is for the LH side.

1. Remove the rear suspension member.
2. Remove the upper control arm assembly LH.
   a. Put matchmarks on the camber adjust cam assembly, No. 2 camber adjust cam and rear suspension member.
   b. Remove the nut, camber adjust cam assembly and No. 2 camber adjust cam.
   c. Remove the upper control arm assembly.

To install:

3. Install the upper control arm assembly LH.
   a. Insert the camber adjust cam assembly from the front side of the vehicle, and temporarily tighten the nut through the No. 2 camber adjust cam.
4. Install the rear suspension member.
5. Fully tighten the upper control arm assembly LH.
   a. Align the matchmarks on the No. 2 camber adjust cam and the camber adjust cam assembly.
   b. Fully tighten the nut to 55 ft. lbs. (74 Nm).
   c. Fully tighten the bolt and nut to 55 ft. lbs. (74 Nm).

➡When installing the bolt, stop the nut from rotating and tighten the bolt.

6. Fully tighten the upper control arm assembly RH and LH.
7. Inspect and adjust the rear wheel alignment.
8. Check for the speed sensor signal.

SHOCK ABSORBER

REMOVAL & INSTALLATION

Double Wishbone Type Suspension

See Figures 270 through 272.

➡Use the same procedure for the RH side and LH side.

➡The procedure listed below is for the LH side.

1. Remove the rear door scuff plate LH (LH side).
2. Remove the rear door opening trim weather-strip LH (LH side).
3. Remove the rear door scuff plate RH.
4. Remove the rear door opening trim weather-strip RH.
5. Remove the rear seat cushion assembly.
6. Remove the rear seat headrest assembly (LH side).
7. Remove the rear seatback outer hinge cover (LH).
8. Remove the rear seatback inner hinge cover (LH).
9. Remove the rear seatback assembly (LH).
10. Remove the rear seat headrest assembly (RH).
11. Remove the rear seat center headrest assembly (RH).
12. Disconnect the rear seat center outer belt assembly (RH).
13. Remove the rear seatback outer hinge cover (RH).
14. Remove the rear seatback inner hinge cover (RH).
15. Remove the tonneau cover assembly (w/ tonneau cover).
16. Remove the deck board assembly.
17. Remove the No. 2 deck board.
18. Remove the jack carrier assembly.
19. Remove the jack assembly.
20. Remove the deck floor box LH and RH.
21. Remove the child restraint seat tether anchor.
22. Remove the luggage compartment tray.
23. Remove the rear deck trim cover.
24. Remove the seatback hinge sub assembly (LH).
25. Remove the rear seat side garnish (LH).
26. Remove the luggage hold belt striker assembly (LH).
27. Remove the No. 1 luggage compartment trim hook (LH).
28. Remove the rear combination light service cover (LH).
29. Remove the rear deck trim panel assembly (LH).
30. Remove the rear seatback hinge sub assembly (RH).
31. Remove the rear seat side garnish (RH).
32. Remove the luggage hold belt striker assembly (RH).
33. Remove the No. 1 luggage compartment trim hook (RH).
34. Remove the rear combination light service cover (RH with woofer).
35. Remove the side deck trim panel assembly (RH).
36. Remove the rear wheel.
37. Remove the rear floor side member brace (LH).
38. Separate the rear stabilizer link assembly.
39. Remove the rear shock absorber with the coil spring.
   a. Support the rear No. 1 suspension arm assembly using a jack and wooden block.
   b. Remove the bolt and the nut, and separate the rear shock absorber with coil spring from the rear No. 1 suspension arm assembly.
   c. Remove the 3 nuts.

d. Remove the 3 bolts from the rear No. 1 suspension arm assembly.

e. Press the rear No. 1 suspension arm assembly down to the outside of the vehicle and remove the rear shock absorber with coil spring.

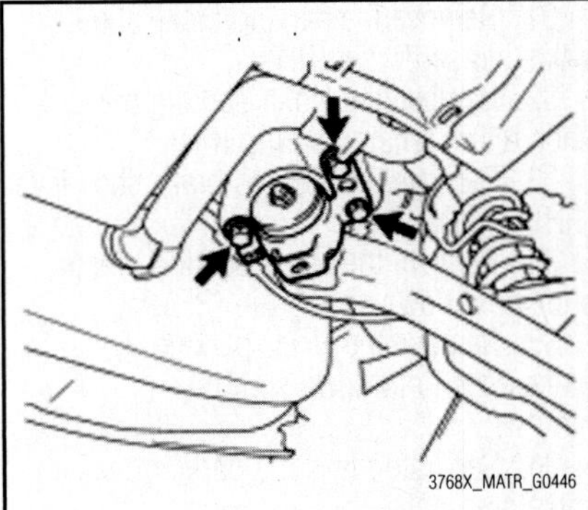

Fig. 270 Removing the 3 bolts from the rear No. 1 suspension arm assembly

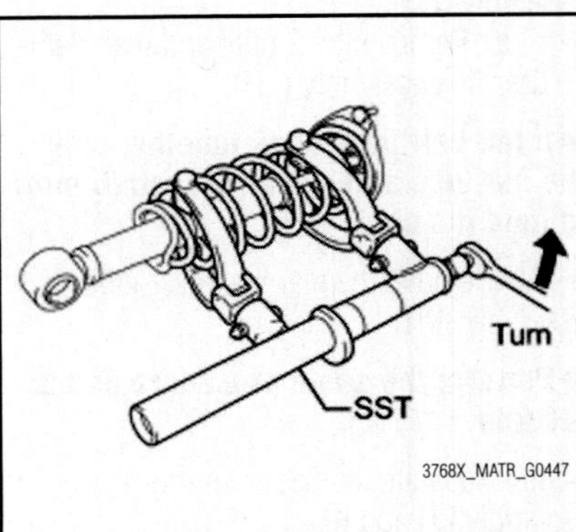

Fig. 271 Compressing the rear coil spring

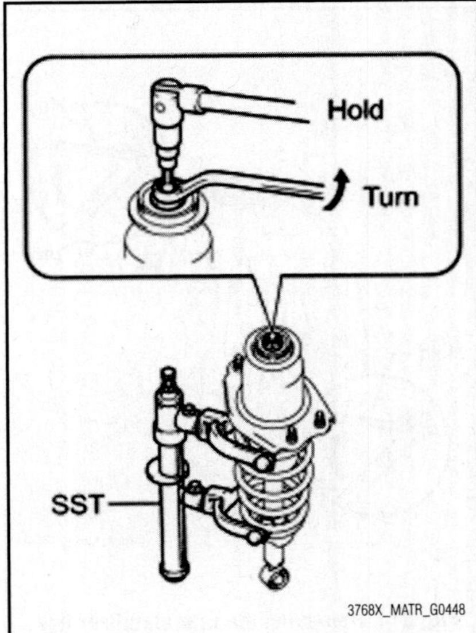

Fig. 272 Holding the rear shock absorber piston rod and removing the nut

40. Remove the rear No. 1 shock absorber cushion washer.

a. Using the special tool (09727-30021), compress the rear coil spring.

➡Do not use an impact wrench. It will damage the special tool.

b. Check that the rear coil spring is fully compressed.

c. Use a hexagon wrench (6 mm) to hold the rear shock absorber piston rod and remove the nut.

d. Remove the rear No. 1 shock absorber cushion washer.

41. Remove the rea4r suspension support.

42. Remove the rear spring front bracket sub assembly.

43. Remove the rear suspension support assembly.

44. Remove the rear coil spring upper insulator.

45. Remove the rear No. 1 spring bumper.

46. Remove the rear shock absorber assembly.

47. Remove the rear coil spring.

➡Do not use an impact wrench. It will damage the special tool.

To install:

48. To install, reverse the removal procedure. Tighten the shock absorber piston rod nut to 29 ft. lbs. (39 Nm). Tighten the shock absorber nuts to 59 ft. lbs. (80 Nm). Tighten the shock absorber bolts to 48 ft. lbs. (65 Nm).

Torsion Beam Type Suspension

See Figure 273.

➡Use the same procedure for the RH side and LH side.

➡The procedure listed below is for the LH side.

1. Remove the rear door scuff plate LH (LH side).
2. Remove the rear door opening trim weather-strip LH (LH side).
3. Remove the rear door scuff plate RH.
4. Remove the rear door opening trim weather-strip RH.
5. Remove the rear seat cushion assembly.
6. Remove the rear seat headrest assembly (LH side).
7. Remove the rear seatback outer hinge cover (LH).
8. Remove the rear seatback inner hinge cover (LH).
9. Remove the rear seatback assembly (LH).
10. Remove the rear seat headrest assembly (RH).
11. Remove the rear seat center headrest assembly (RH).
12. Disconnect the rear seat center outer belt assembly (RH).
13. Remove the rear seatback outer hinge cover (RH).
14. Remove the rear seatback inner hinge cover (RH).
15. Remove the tonneau cover assembly (w/ tonneau cover).
16. Remove the deck board assembly.
17. Remove the No. 2 deck board.
18. Remove the jack carrier assembly.
19. Remove the jack assembly.
20. Remove the deck floor box LH and RH.
21. Remove the child restraint seat tether anchor.
22. Remove the luggage compartment tray.
23. Remove the rear deck trim cover.
24. Remove the seatback hinge sub assembly (LH).
25. Remove the rear seat side garnish (LH).
26. Remove the luggage hold belt striker assembly (LH).
27. Remove the No. 1 luggage compartment trim hook (LH).
28. Remove the rear combination light service cover (LH).
29. Remove the rear deck trim panel assembly (LH).
30. Remove the rear seatback hinge sub assembly (RH).
31. Remove the rear seat side garnish (RH).
32. Remove the luggage hold belt striker assembly (RH).
33. Remove the No. 1 luggage compartment trim hook (RH).
34. Remove the rear combination light service cover (RH with woofer).

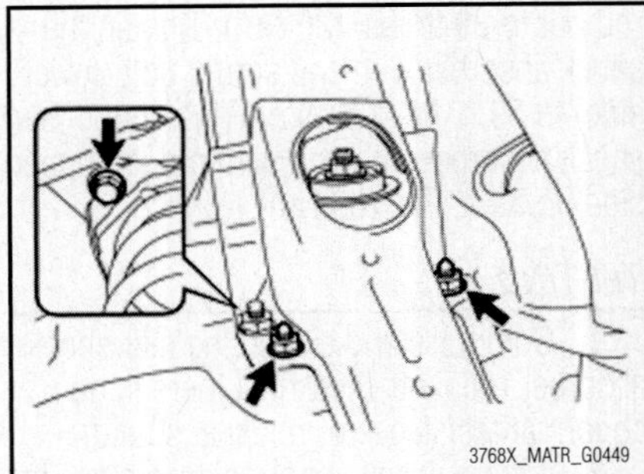

Fig. 273 Removing the 2 nuts from the rear shock absorber with coil spring (upper side)

35. Remove the side deck trim panel assembly (RH).
36. Remove the rear wheel.
37. Remove the rear shock absorber cushion retainer.
   a. Support the rear axle beam assembly using a jack and wooden block.
   b. Remove the nut and rear shock absorber cushion retainer.
38. Remove the rear shock absorber with the coil spring.
   a. Remove the 2 nuts from the rear shock absorber with coil spring (upper side).
   b. Remove the bolt (lower side) from the rear shock absorber with coil spring.
   c. Slowly lower the jack and remove the rear shock absorber with coil spring.
39. Remove the rear No. 1 shock absorber cushion washer.
   a. Using the special tool (09727-30021), compress the rear coil spring.
   b. Check that the rear coil spring is fully compressed.

➡Do not use an impact wrench. It will damage the special tool.

   c. Use a hexagon wrench (6 mm) to hold the rear shock absorber piston rod and remove the nut.
   d. Remove the rear No. 1 shock absorber cushion washer.
40. Remove the rear suspension support.
41. Remove the rear spring front bracket sub assembly.
42. Remove the rear suspension support assembly.
43. Remove the rear coil spring upper insulator.
44. Remove the rear No. 1 spring bumper.
45. Remove the rear shock absorber assembly.
46. Remove the rear coil spring.

To install:

47. To install, reverse the removal procedure. Tighten the rear shock absorber piston rod nut to 29 ft. lbs. (39 Nm). Tighten the shock absorber with coil spring bolt (lower side) to 59 ft. lbs. (80 Nm). Tighten the rear shock absorber with coil spring nuts (upper side) to 59 ft. lbs. (80 Nm).

TESTING

1. Compress and extend the rear shock absorber rod, and check that there is no abnormal resistance or unusual sound.
2. If there is any abnormality, replace the rear shock absorber assembly with a new one.

STABILIZER BAR

REMOVAL & INSTALLATION

Double Wishbone Type Suspension

See Figures 274 and 275.

1. Remove the rear wheels.
2. Remove the center exhaust pipe assembly.
3. Remove the propeller with the center bearing shaft assembly (4WD).
4. Drain the differential oil (4WD).
5. Remove the rear axle shaft nut LH (4WD).
6. Remove the rear axle shaft nut RH (4WD).
7. Remove the charcoal canister assembly (2WD).
8. Separate the rear disc brake caliper assembly LH.
9. Separate the rear disc brake caliper assembly RH.
10. Remove the rear disc LH and RH.
11. Disconnect the rear speed sensor wire LH and RH (2WD).
12. Separate the rear speed sensor wire LH and RH (2WD).
13. Separate the rear speed sensor LH and RH (4WD).
14. Remove the No. 3 parking brake show return tension spring (LH).
15. Remove the No. 2 parking brake shoe return tension spring (LH).
16. Remove the No. 1 parking brake shoe hold down compression spring (LH).
17. Remove the parking brake shoe strut (LH).
18. Remove the No. 2 parking brake shoe hold down compression spring (LH).
19. Separate the parking brake shoe lever (LH).
20. Remove the parking brake shoe adjusting screw set (LH).
21. Remove the No. 1 parking brake shoe return tension spring (LH).
22. Remove the parking brake shoe lever (LH).
23. Remove the No. 1 parking brake shoe hold down spring pin (LH).
24. Remove the No. 2 parking brake shoe hold down spring pin (LH).
25. Remove the No. 3 parking brake shoe return tension spring (RH).
26. Remove the No. 2 parking brake shoe return tension spring (RH).
27. Remove the No. 1 parking brake shoe hold down compression spring (RH).
28. Remove the parking brake shoe strut (RH).
29. Remove the No. 2 parking brake shoe hold down compression spring (RH).
30. Separate the parking brake shoe lever (RH).
31. Remove the parking brake shoe adjusting screw set (RH).
32. Remove the No. 1 parking brake shoe return tension spring (RH).
33. Remove the parking brake shoe lever (RH).
34. Remove the No. 1 parking brake shoe hold down spring pin (RH).
35. Remove the No. 2 parking brake shoe hold down spring pin (RH).
36. Separate the No. 3 parking brake cable assembly.
37. Separate the No. 2 parking brake cable assembly.
38. Remove the rear stabilizer link assembly (LH).
   a. Remove the 2 nuts and rear stabilizer link assembly (LH).

➡If the ball joint turns together with the nut, use a hexagon wrench (5 mm) to hold the stud.

39. Remove the rear stabilizer link assembly (RH).

➡Perform the same procedure as the LH side.

40. Separate the upper control arm assembly LH and RH.
41. Separate the rear No. 1 suspension arm assembly LH.

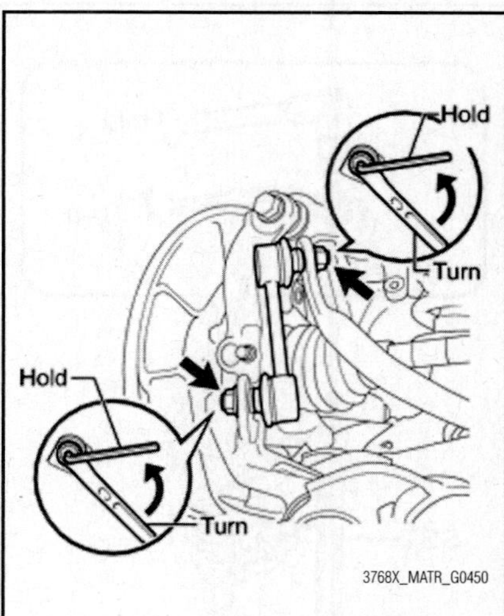

Fig. 274 Removing the rear stabilizer link assembly (LH)

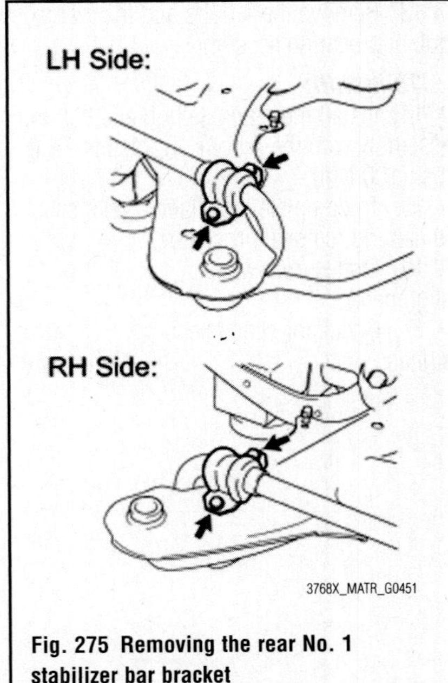

Fig. 275 Removing the rear No. 1 stabilizer bar bracket

42. Separate the rear No. 1 suspension arm assembly RH.
43. Remove the rear axle assembly LH and RH.
44. Remove the rear floor side member brace.
45. Loosen the rear shock absorber with the coil spring.
46. Remove the rear No. 1 suspension arm assembly LH and RH.
47. Remove the rear suspension member.
48. Remove the rear No. 1 stabilizer bar bracket.
 a. Remove the 4 bolts and 2 rear No. 1 stabilizer bar brackets from the rear suspension member.
49. Remove the rear stabilizer bushing.
 a. Remove the 2 rear stabilizer bushings from the rear stabilizer bar.
50. Remove the rear stabilizer bar.

To install:

51. To install, reverse the removal procedure. Tighten the rear No. 1 stabilizer bar bracket bolts to 13 ft. lbs. (18 Nm). Tighten the rear stabilizer link assembly nuts to 32 ft. lbs. (44 Nm). Check the speed sensor signal.

Torsion Bean Type Suspension

See Figures 276 through 279.

1. Remove the rear wheels.
2. Remove the rear axle beam damper.
3. Remove the rear stabilizer bar.

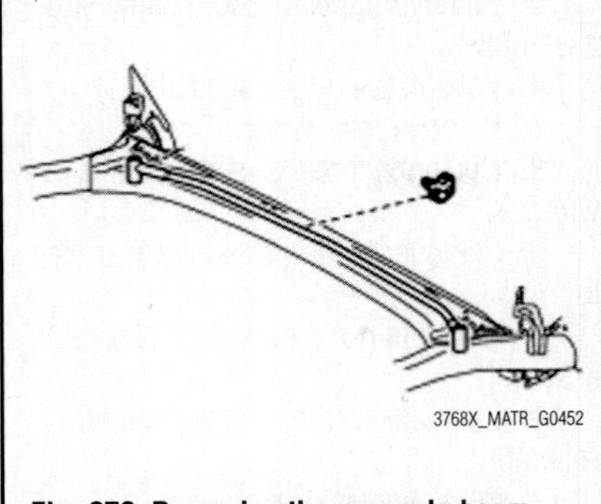

Fig. 276 Removing the rear axle beam damper

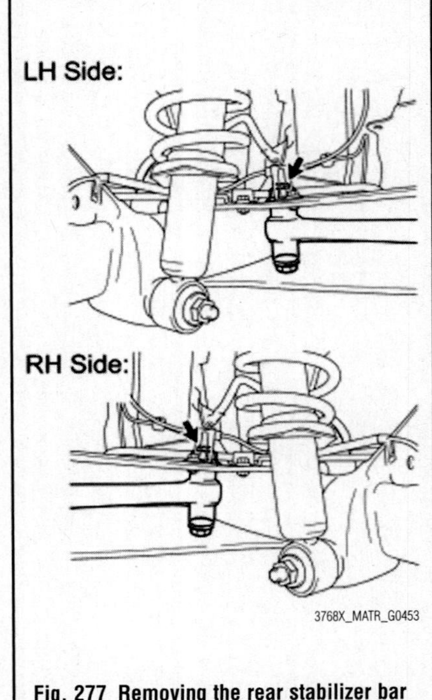

Fig. 277 Removing the rear stabilizer bar

 a. Remove the 2 bolts, 2 nuts and rear stabilizer bar.

➡Be sure to loosen the nuts.

To install:

4. Install the rear stabilizer bar.
 a. Check that the identification mark of the rear stabilizer bar is positioned on the left side of the vehicle.
 b. Install the rear stabilizer bar with the 2 bolts and 2 nuts. Tighten to 184 ft. lbs. (250 Nm).

➡Be sure to tighten the nuts.

➡When reinstalling the bolts, insert them from the upper side of the vehicle.

5. Install the rear axle beam damper.

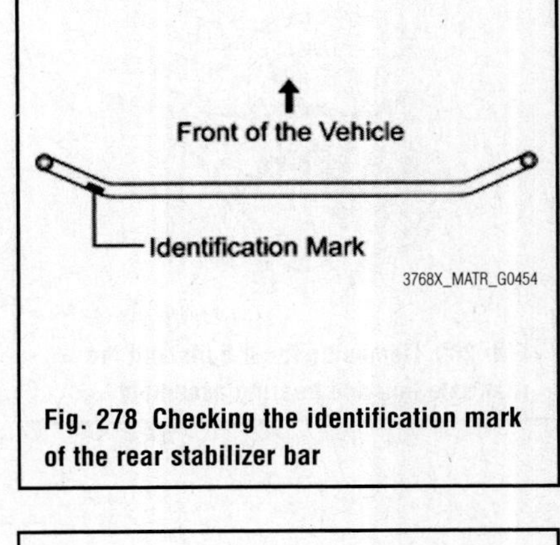

Fig. 278 Checking the identification mark of the rear stabilizer bar

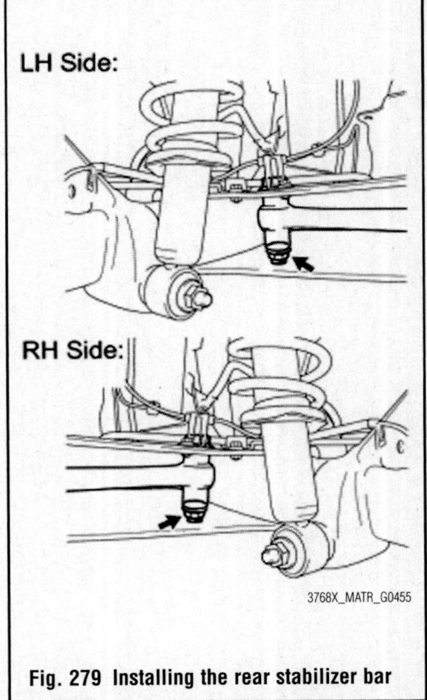

Fig. 279 Installing the rear stabilizer bar

 a. Install the rear axle beam damper to the center of the rear stabilizer bar.
6. Install the rear wheels. Tighten to 76 ft. lbs. (103 Nm).

WHEEL HUB & BEARING

REMOVAL & INSTALLATION

See Figure 280.

1. Disconnect cable from negative battery terminal.
2. Remove lower instrument panel finish panel.
3. Remove shift lever knob sub-assembly.
4. Remove center instrument cluster finish panel assembly.

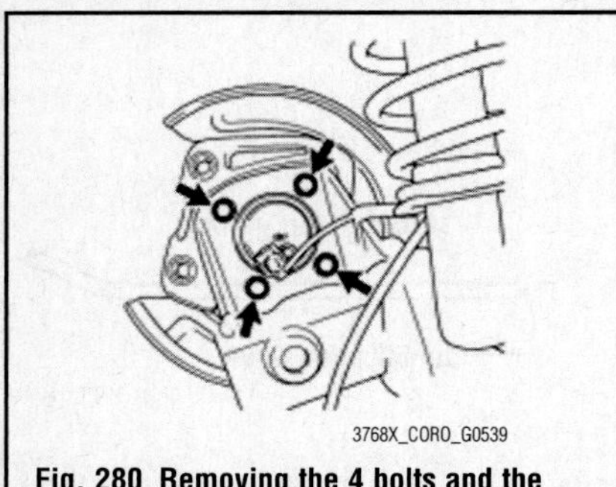

Fig. 280 Removing the 4 bolts and the rear axle hub and bearing assembly

5. Remove upper console panel sub-assembly.
6. Loosen parking brake cable.
7. Remove rear wheel.
8. Disconnect rear speed sensor wire.
9. Remove parking brake lever protector.
10. Separate no. 3 parking brake cable assembly.
11. Separate rear disc brake caliper assembly.
12. Remove rear disc.
13. Remove the 4 bolts and the rear axle hub and bearing assembly.

To install:

14. Install the rear axle hub and bearing assembly with the 4 bolts. Tighten to 74 ft. lbs. (100 Nm).
15. To complete installation, reverse remaining removal procedure.
16. Inspect rear wheel alignment.
17. Check for speed sensor signal.

13

TOYOTA

Prius

SPECIFICATIONS AND MAINTENANCE CHARTS

VEHICLE AND ENGINE IDENTIFICATION

Engine						
Code	Liters (cc)	Cu. in.	Cyl.	Fuel Sys.	Engine Type	Eng. Mfg.
1NZ-FXE	1.5 (1497)	91.4	4	SFI	DOHC	Toyota
2ZR-FXE	1.8 (1798)	109.7	4	SFI	DOHC	Toyota

Model Year	
Code	Year
9	2009
A	2010

SFI: Sequential Multiport Fuel Injection

DOHC: Dual Overhead Camshaft

3768X_PRIU_C0001

GENERAL ENGINE SPECIFICATIONS

Year	Engine ID/VIN	Engine Displacement Liters (cc)	Fuel System Type	Net Horsepower @ rpm	Net Torque @ rpm (ft. lbs.)	Bore x Stroke (in.)	Com-pression Ratio	Oil Pressure @ rpm
2009	1NZ-FXE	1.5 (1497)	SFI	110@5000	82@4200	2.95x3.33	13.0:1	22-80@2500
2010	2ZR-FXE	1.8 (1798)	SFI	98@5200	105@4000	3.17x3.48	13.0:1	21@2500

SFI: Sequential Multiport Fuel Injection

3768X_PRIU_C0002

GASOLINE ENGINE TUNE-UP SPECIFICATIONS

Year	Engine Displacement Liters	Engine ID/VIN	Spark Plugs Gap (in.)	Ignition Timing (deg.)		Fuel Pump (psi)	Idle Speed (rpm)		Valve Clearance	
				MT	AT		MT	AT	In.	Ex.
2009	1.5	1NZ-FXE	0.043	—	8-12B	44-50 ①	—	950-1050	0.007-0.009	0.011-0.013
2010	1.8	2ZR-FXE	0.043	—	8-12B	44-50 ①	—	950-1050	0.007-0.009	0.011-0.013

Note: The Vehicle Emission Control Information label often reflects specification changes made during production. The label figures must be used if they differ from those in this chart.

B: Before top dead center

① At idle

3768X_PRIU_C0003

CAPACITIES

Year	Model	Engine Displacement Liters	Engine ID/VIN	Engine Oil with Filter (qts.)	Transaxle (pts.) 5-Spd	Transaxle (pts.) Auto.	Fuel Tank (gal.)	Cooling System (qts.)
2009	Prius	1.5	1NZ-FXE	3.9	—	7.6 ①	11.9	②
2010	Prius	1.8	2ZR-FXE	4.4	—	7.6 ①	11.9	③

Note: All capacities are approximate. Add fluid gradualy and check to be sure a proper fluid level is obtained.

① Specification for Hybrid transaxle.

② Gasoline engine: 5.2 quarts

Electric motor, inverter and converter: 2.7 quarts

③ 7.7 quarts w/ exhaust heat recirculation system

6.8 quarts w/o exhaust heat recirculation system

3768X_PRIU_C0004

FLUID SPECIFICATIONS

Year	Model	Engine Displacement Liters	Engine ID/VIN	Engine Oil	Auto. Trans.	Manual Trans.	Drive Axle	Transfer Case	Power Steering Fluid	Brake Master Cylinder
2009	Prius	1.5	1NZ-FXE	5W-30	ATF WS	—	—	—	—	DOT 3
2010	Prius	1.8	2ZR-FXE	5W-30	ATF WS	—	—	—	—	DOT 3

DOT: Department Of Transpotation

3768X_PRIU_C0013

VALVE SPECIFICATIONS

Year	Engine ID/VIN	Engine Displacement Liters	Seat Angle (deg.)	Face Angle (deg.)	Spring Test Pressure (lbs. @ in.)	Spring Installed Height (in.)	Stem-to-Guide Clearance (in.) Intake	Stem-to-Guide Clearance (in.) Exhaust	Stem Diameter (in.) Intake	Stem Diameter (in.) Exhaust
2009	1NZ-FXE	1.5	45	45	40.5-44.5@ 0.9880	2.353	0.0001-0.0031	0.0012-0.0039	0.1957-0.1963	0.1955-0.1961
2010	2ZR-FXE	1.8	NA	NA	NA	NA	NA	NA	0.215-0.2160	0.215-0.2160

3768X_PRIU_C0005

CAMSHAFT AND BEARING SPECIFICATIONS CHART

All measurements are given in inches.

Year	Engine Displ. Liters	Engine VIN	Journal Dia.	Brg. Oil Clearance	Shaft End-play	Runout	Journal Bore	Lobe Lift Intake	Lobe Lift Exhaust
2009	1.5	1NZ-FXE	①	0.0001-0.0024	0.0016-0.0037	0.0012	NA	1.6657-1.6697	1.7341-1.7380
2010	1.8	2ZR-FXE	①	0.0001-0.0024	0.0016-0.0037	0.0012	NA	1.6657-1.6697	1.7341-1.7380

NA - Not available

① No 1: 1.3563-1.3569

All others: 0.9035-0.9041

3768X_PRIU_C0014

CRANKSHAFT AND CONNECTING ROD SPECIFICATIONS

All measurements are given in inches.

Year	Engine ID/VIN	Engine Displacement Liters	Crankshaft Main Brg. Journal Dia.	Crankshaft Main Brg. Oil Clearance	Crankshaft Shaft End-play	Crankshaft Thrust on No.	Connecting Rod Journal Diameter	Connecting Rod Oil Clearance	Connecting Rod Side Clearance
2009	1NZ-FXE	1.5	1.8106-1.8110	0.0004-0.0028	0.0035-0.0120	NA	1.5745-1.5748	0.0006-0.0024	0.0063-0.0142
2010	2ZR-FXE	1.8	1.7320-1.7323	0.0002-0.0004	0.0035-0.0120	NA	1.5745-1.5748	0.0005-0.0015	0.0063-0.0142

NA: Not Available

① Small end: 0.78787 - 0.78823 inch

Large end: Mark 1 = 1.85039 - 1.85070 inch

Mark 2 = 1.85074 - 1.85102 incl

Mark 3 = 1.85106 - 1.85133 incl

3768X_PRIU_C0006

PISTON AND RING SPECIFICATIONS

All measurements are given in inches.

Year	Engine ID/VIN	Engine Displacement Liters	Piston Clearance	Ring Gap Top Compression	Ring Gap Bottom Compression	Ring Gap Oil Control	Ring Side Clearance Top Compression	Ring Side Clearance Bottom Compression	Ring Side Clearance Oil Control
2009	1NZ-FXE	1.5	0.0018-0.0032	0.0079-0.0240	0.0118-0.0472	0.0039-0.0453	0.0008-0.0028	0.0008-0.0024	0.0008-0.0024
2010	2ZR-FXE	1.8	0.0017-0.0032	0.0079-0.0190	0.0118-0.0177	0.0039-0.0138	0.0008-0.0028	0.0008-0.0024	0.0008-0.0024

3768X_PRIU_C0007

TORQUE SPECIFICATIONS

All readings in ft. lbs.

Year	Engine ID/VIN	Engine Displ. Liters	Cylinder Head Bolts	Main Bearing Bolts	Rod Bearing Bolts	Crankshaft Damper Bolts	Flywheel Bolts	Manifold Intake	Manifold Exhaust	Spark Plugs	Oil Pan Drain Plug
2009	1NZ-FXE	1.5	①	②	③	95	36	15	20	13	13
2010	2ZR-FXE	1.8	⑤	⑥	③	95	36+90 deg.	21	15	15	27

① Step 1: 21 ft. lbs.
Step 2: turn head bolts 90 degrees
Step 3: turn head bolts 90 degrees

② Step 1: 16 ft. lbs.
Step 2: turn bearing cap bolts 90 degrees

③ Step 1: 11 ft. lbs.
Step 2: turn bearing cap bolts 90 degrees

④ Step 1: 62 ft. lbs.
Step 2: turn flywheel bolts 90 degrees

⑤ Step 1: 36 ft. lbs.
Step 2: turn head bolts 90 degrees
Step 3: turn head bolts 45 degrees

⑥ Step 1: 30 ft. lbs.
Step 2: turn bearing cap bolts 90 degrees

3768X_PRIU_C0008

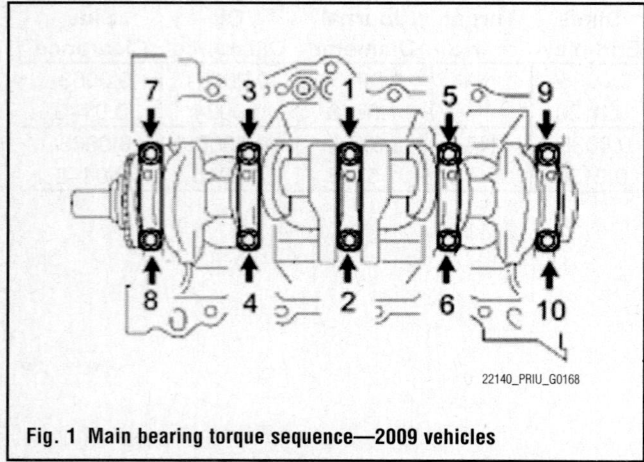

Fig. 1 Main bearing torque sequence—2009 vehicles

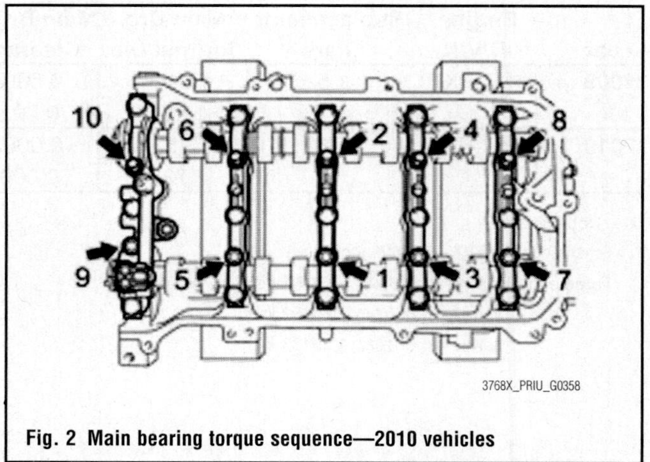

Fig. 2 Main bearing torque sequence—2010 vehicles

WHEEL ALIGNMENT

Year	Model		Caster Range (+/-Deg.)	Caster Preferred Setting (Deg.)	Camber Range (+/-Deg.)	Camber Preferred Setting (Deg.)	Toe-in (in.)	Steering Axis Inclination (Deg.)
2009	Prius	F	0.75	+3.17	0.75	-0.58	0+/-0.08	12.58+/-0.75
		R	—	—	0.50	-1.50	0.12+/-0.10	—
2010	Prius	F	0.75	+5.88	0.75	-0.22	0+/-0.08	12.27
		R	—	—	0.50	-1.48	0.12+/-0.10	—

3768X_PRIU_C0009

TIRE, WHEEL AND BALL JOINT SPECIFICATIONS

Year	Model	OEM Tires Standard	OEM Tires Optional	Tire Pressures (psi) Front	Tire Pressures (psi) Rear	Wheel Size	Ball Joint Inspection	Lug Nut (ft. lbs.)
2009	Prius	P185/65R15	None	35	33	6-JJ	①	76
2010	Prius	P195/65R15	None	35	32	6-JJ	①	76

OEM: Original Equipment Manufacturer

PSI: Pounds Per Square Inch

① Replace if any measurable movement is found.

3768X_PRIU_C0010

BRAKE SPECIFICATIONS

All measurements in inches unless noted

Year	Model	Brake Disc Original Thickness	Brake Disc Minimum Thickness	Brake Disc Maximum Runout	Brake Drum Diameter Original Inside Diameter	Brake Drum Diameter Max. Wear Limit	Brake Drum Diameter Max. Machine Diameter	Minimum Lining Thickness Front	Minimum Lining Thickness Rear	Brake Caliper Bracket bolts (ft. lbs.)	Brake Caliper Mounting bolts (ft. lbs.)
2009	Prius	0.866	0.787	0.002	7.874	7.913	7.913	0.039	0.039	81	25
2010	Prius	F: 0.984	F: 0.866	F: 0.002	NA	NA	NA	0.039	-	81	25
		R: 0.354	R: 0.295	R: 0.006	NA	NA	NA	-	0.039	81	25

3768X_PRIU_C0011

SCHEDULED MAINTENANCE INTERVALS
TOYOTA—PRIUS

TO BE SERVICED	TYPE OF SERVICE	VEHICLE MILEAGE INTERVAL (x1000)												
		5	10	15	20	25	30	35	40	45	50	55	60	65
Engine oil & filter	R		✓		✓		✓		✓		✓		✓	
Cabin air filter (solar power ventilation system)	S/I		✓		✓		✓		✓		✓		✓	
Cabin air filter (except solar power ventilation system)	S/I			✓			✓			✓			✓	
Driver's floor mat	S/I	✓	✓	✓	✓	✓	✓	✓	✓	✓	✓	✓	✓	✓
Hybrid transaxle fluid	S/I	✓	✓	✓	✓	✓	✓	✓	✓	✓	✓	✓	✓	✓
Drive axle boots	S/I	✓	✓	✓	✓	✓	✓	✓	✓	✓	✓	✓	✓	✓
Gear shift control operation	S/I	✓	✓	✓	✓	✓	✓	✓	✓	✓	✓	✓	✓	✓
Inspect & rotate tires	S/I	✓	✓	✓	✓	✓	✓	✓	✓	✓	✓	✓	✓	✓
Power steering system	S/I	✓	✓	✓	✓	✓	✓	✓	✓	✓	✓	✓	✓	✓
Suspension system	S/I	✓	✓	✓	✓	✓	✓	✓	✓	✓	✓	✓	✓	✓
Brake discs & pads	S/I	✓	✓	✓	✓	✓	✓	✓	✓	✓	✓	✓	✓	✓
Brake shoes & drums	S/I	✓	✓	✓	✓	✓	✓	✓	✓	✓	✓	✓	✓	✓
Brake hoses & pipes	S/I	✓		✓		✓		✓		✓		✓		✓
Brake fluid	S/I	✓	✓	✓	✓	✓	✓	✓	✓	✓	✓	✓	✓	✓
Brake pedal	S/I		✓		✓		✓		✓		✓		✓	
Cooling system, hoses & connections	S/I		✓		✓		✓		✓		✓		✓	
Fuel tank, cap & lines	S/I		✓		✓		✓		✓		✓		✓	
Air cleaner filter element	R				✓				✓				✓	
Engine coolant	R				✓				✓				✓	
Spark plugs	R				✓				✓				✓	
Drive belt	S/I				✓				✓				✓	
Engine inverter coolant	S/I			✓			✓			✓			✓	
Exhaust system	S/I			✓			✓			✓			✓	

R: Replace S/I: Service or Inspect

① Replace every 60,000 miles.

FREQUENT OPERATION MAINTENANCE (SEVERE SERVICE)

If a vehicle is operated under any of the following conditions it is considered severe service:

- Extremely dusty areas.

- 50% or more of the vehicle operation is in 32°C (90°F) or higher temperatures, or constant operation in temperatures below 0°C (32°F).

- Prolonged idling (vehicle operation in stop and go traffic).

- Frequent short running periods (engine does not warm to normal operating temperatures).

- Police, taxi, delivery usage or trailer towing usage.

Oil & oil filter: change every 3000 miles.

Brake discs & pads: service or inspect initially at 3000 miles, 6000 miles, & every 12,000 miles thereafter.

Brake hoses & pipes: service or inspect initially at 3000 miles, 6000 miles & every 12,000 miles thereafter.

Air cleaner filter element: service or inspect ever 3000 miles & replace every 30,000 miles (if not replaced previously).

Hybrid transaxle fluid: service or inspect every 6000 miles & replace every 15,000 miles (if not replaced previously).

Inspect & rotate tires: service or inspect every 6000 miles.

Power steering system: service or inspect every 6000 miles.

Steering system: service or inspect every 6000 miles.

Suspension system: service or inspect every 6000 miles.

Drive belts: service or inspect every 15,000 miles.

Exhaust system: service or inspect every 15,000 miles.

3768X_PRIU_C0012

BRAKES — INFORMATION AND PRECAUTIONS

ANTI-LOCK SYSTEMS

- Certain components within the ABS system are not intended to be serviced or repaired individually.
- Do not use rubber hoses or other parts not specifically specified for and ABS system. When using repair kits, replace all parts included in the kit. Partial or incorrect repair may lead to functional problems and require the replacement of components.
- Lubricate rubber parts with clean, fresh brake fluid to ease assembly. Do not use shop air to clean parts; damage to rubber components may result.
- Use only DOT 3 brake fluid from an unopened container.
- If any hydraulic component or line is removed or replaced, it may be necessary to bleed the entire system.
- A clean repair area is essential. Always clean the reservoir and cap thoroughly before removing the cap. The slightest amount of dirt in the fluid may plug an orifice and impair the system function. Perform repairs after components have been thoroughly cleaned; use only denatured alcohol to clean components. Do not allow ABS components to come into contact with any substance containing mineral oil; this includes used shop rags.
- The Anti-Lock control unit is a microprocessor similar to other computer units in the vehicle. Ensure that the ignition switch is **OFF** before removing or installing controller harnesses. Avoid static electricity discharge at or near the controller.
- If any arc welding is to be done on the vehicle, the control unit should be unplugged before welding operations begin.

DISC AND DRUM SYSTEMS

CAUTION

Dust and dirt accumulating on brake parts during normal use may contain asbestos fibers from production or aftermarket brake linings. Breathing excessive concentrations of asbestos fibers can cause serious bodily harm. Exercise care when servicing brake parts. Do not sand or grind brake lining unless equipment used is designed to contain the dust residue. Do not clean brake parts with compressed air or by dry brushing. Cleaning should be done by dampening the brake components with a fine mist of water, then wiping the brake components clean with a dampened cloth. Dispose of cloth and all residue containing asbestos fibers in an impermeable container with the appropriate label. Follow practices prescribed by the Occupational Safety and Health Administration (OSHA) and the Environmental Protection Agency (EPA) for the handling, processing, and disposing of dust or debris that may contain asbestos fibers.

BRAKES — BLEEDING THE BRAKE SYSTEM

BLEEDING PROCEDURE

Bleeding the brakes requires the use of a Toyota Techstream, or equivalent scan tool.

Follow tool manufacturer's procedures when bleeding the hydraulic brake system.

BRAKES — ANTI-LOCK BRAKE SYSTEM (ABS)

WHEEL SPEED SENSORS

REMOVAL & INSTALLATION

Front—2009 vehicles

See Figure 3.

1. Before servicing the vehicle, refer to the precautions.
2. Raise and safely support the vehicle securely on jackstands.
3. Remove front wheel.
4. Remove front fender liner.
5. Remove front speed sensor.
   a. Disconnect the sensor connector.
   b. Remove the 2 bolts and 2 sensor clamps from the body and front shock absorber.
   c. Remove the clip from the front shock absorber.
   d. Remove the bolt and sensor.

➡Do not allow foreign matter to attach to the tip or connecting portion of the speed sensor.

To install:

6. Install front speed sensor.

➡Do not twist the sensor wire harness during installation.

   a. Check that there is no foreign matter attached to the tip and connecting portion of the speed sensor.

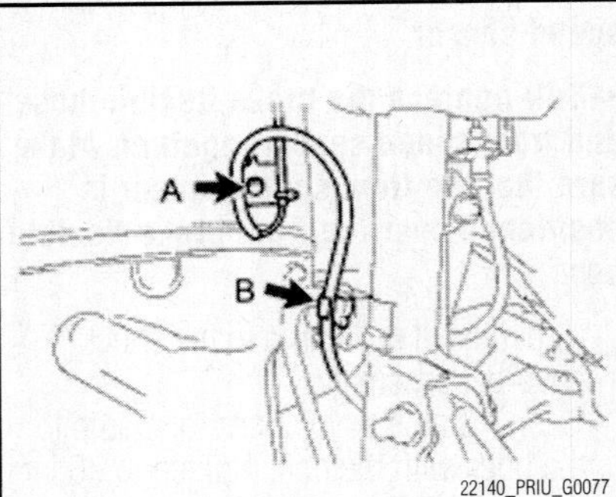

Fig. 3 Remove the 2 bolts (A and B) and 2 sensor clamps from the body and front shock absorber

   b. Clean the connecting portion of the sensor.
   c. Install the sensor with the bolt. Tighten to 71 inch lbs. (8 Nm).
   d. Install the clip to the front shock absorber.
   e. Install the sensor clamps to the body and front shock absorber and tighten bolt "A" to Tighten to 71 inch lbs. (8 Nm) and bolt "B" to 14 ft. lbs. (19 Nm).

➡Tighten the bolt "B" together with the brake flexible hose clamp and sensor clamp. The brake flexible hose clamp should be placed above.

   f. Connect the sensor connector.

7. Install front fender liner.
8. Install front wheel and lower the vehicle.
9. Check for proper speed sensor signal.

Front—2010 vehicles

See Figure 4.

➡While the battery is connected, even if the power switch is off, the brake

control system activates when the brake pedal is depressed or any door courtesy switch is turned on. Therefore, when servicing the brake system components, do not depress the brake pedal or open/ close the doors while the battery is connected.

➡Use the same procedure for the LH side and RH side.

➡The following procedure is for the LH side.

➡If the sensor rotor needs to be replaced, replace it together with the front axle hub and bearing assembly.

1. Remove the rear No. 2 floor board (separate type).
2. Remove the rear deck floor box.
3. Remove the rear No. 3 floor board.
4. Disconnect the cable from the negative battery terminal.

➡When disconnecting the cable, some systems need to be initialized after the cable is reconnected.

5. Remove the front wheel.
6. Remove the front fender liner.
   a. Remove the 12 clips, 4 screw, 3 grommets, and front fender liner.

➡Use the same procedure for the RH side and LH side.

7. Remove the front speed sensor.
   a. Remove the 2 clamps and disconnect the front speed sensor connector.
   b. Remove the bolt and No. 2 sensor clamp from the body.
   c. Remove the bolt, No. 1 sensor clamp and front brake flexible hose together from the shock absorber assembly.
   d. Remove the clamp from the shock absorber assembly.
   e. Remove the bolt and front speed sensor.

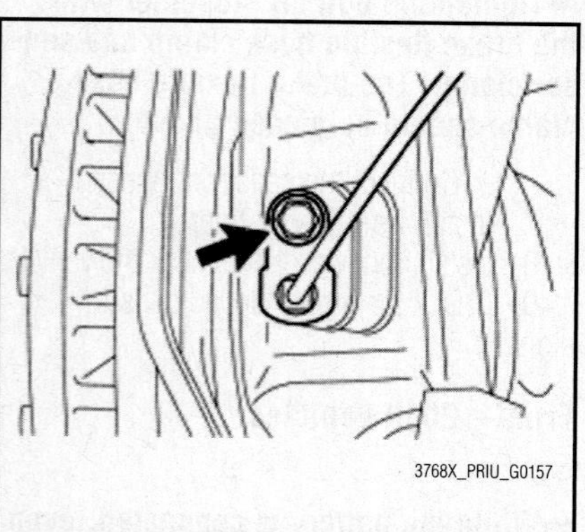

Fig. 4 Removing the front speed sensor

➡Prevent foreign matter from attaching to the front speed sensor tip.

➡Clean the speed sensor installation hole and the contact surfaces every time the front speed sensor is removed.

To install:

➡Use the same procedure for the LH side and RH side.

➡The following procedure is for the LH side.

➡If the sensor rotor needs to be replaced, replace it together with the front axle hub and bearing assembly.

8. Install the front speed sensor.
   a. Install the front speed sensor with the bolt. Tighten to 76 inch lbs. (8.5 Nm).

➡Prevent foreign matter from attaching to the front speed sensor tip.

➡Firmly insert the front speed sensor body into the knuckle before tightening the bolt.

➡After installing the front speed sensor to the knuckle, make sure that there is no clearance between the front speed sensor stay and knuckle. Also make sure that no foreign matter is stuck between the parts.

➡Before installing the clamp, firmly insert the points of the clamp into the installation holes.

   b. Temporarily install the front brake flexible hose.
   c. Install the front brake flexible hose and No. 1 sensor clamp together to the shock absorber with the bolt. Tighten to 14 ft. lbs. (19 Nm).

➡Do not twist the wire harness for the front speed sensor when installing the speed sensor.

➡Bolt tightens the brake flexible hose and front speed sensor together. Make sure that the front speed sensor is positioned over the front brake flexible hose.

   d. Install the clamp to the shock absorber assembly.
   e. Install the No. 2 sensor clamp to the body with the bolt. Tighten to 76 inch lbs. (8.5 Nm).
   f. Connect the speed sensor connector.
   g. Connect the 2 speed sensor wire harness clamps to the body.

9. Install the front fender liner with the 12 clips, 3 grommets and 4 screws.
10. Install the front wheel. Tighten to 76 ft. lbs. (103 Nm).
11. Connect the cable to the negative battery terminal.

➡When disconnecting the cable, some systems need to be initialized after the cable is reconnected.

12. Install the rear No. 3 floor board.
13. Install the rear deck floor box.
14. Install the rear No. 2 floor board (separate type).
15. Check for the speed sensor signal.

Rear—2009 vehicles

See Figures 5 through 7.

The rear wheel speed sensor is also known as the Skid Control Sensor.

1. Raise and safely support the vehicle securely on jackstands.
2. Remove rear wheel.
3. Remove rear brake drum.
4. Separate skid control sensor wire.
5. Remove rear axle hub and bearing assembly.
6. Remove skid control sensor.
   a. Install the 4 hub nuts to the 4 rear axle hub bolts. Placing an aluminum plate below the rear axle hub and bearing, place them into the vise.

➡Replace the rear axle hub and bearing if it is dropped or receives a strong shock.

   b. Using special tool, SST: 09520-00031 and 2 bolts, remove the sensor from the rear axle hub and bearing.
   c. Pull the sensor off straight, being

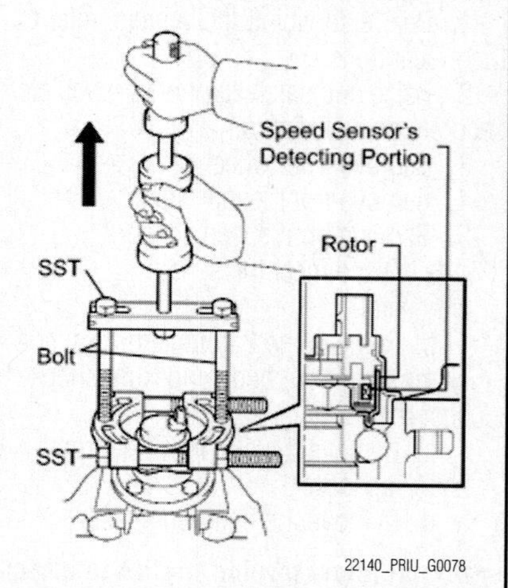

Fig. 5 Using the special tool, remove the sensor from the rear axle hub and bearing

careful not to make contact with the sensor rotor.

d. If the sensor rotor is damaged or deformed, replace the rear axle hub and bearing.

e. Do not scratch the area where the sensor contacts the rear axle hub and bearing.

f. Do not allow foreign matter to attach to the sensor rotor.

To install:

7. Install skid control sensor.

a. Wipe off sealant attached to the sensor's fitting surface with white gasoline.

➡Prevent foreign matter from attaching to the sensor rotor.

b. Install a new sensor to the rear axle hub and bearing. The sensor connector should be placed in the lowest position.

➡The narrow side of the rear axle hub and bearing should be placed below.

c. Using special tool SST: 09214-76011 and a press, press the speed sensor so that it becomes flush with the rear axle hub and bearing.

➡Do not use a hammer on the sensor.

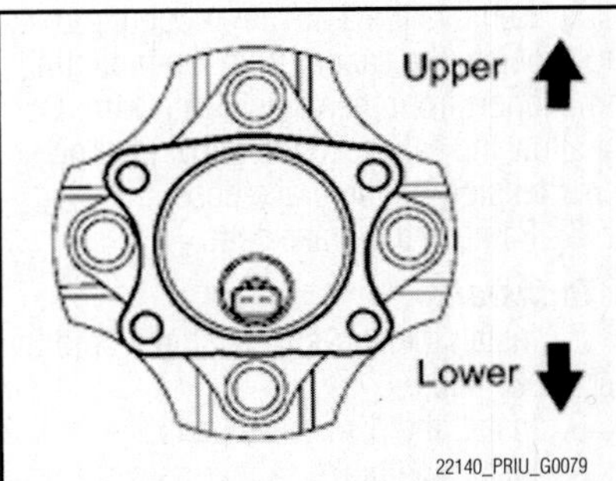

Fig. 6 Install the new sensor so that the sensor connector is placed in the lowest position

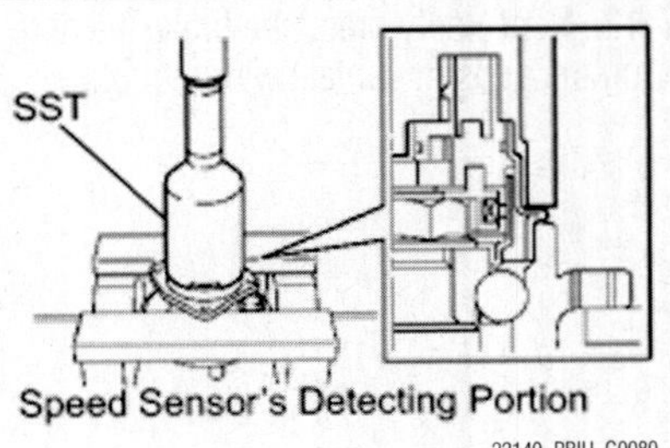

Fig. 7 Using the special tool and a press, install the speed sensor so that it becomes flush with the rear axle hub and bearing

d. Check that there is no foreign matter such as iron chips on the sensor's detecting portion.

e. Slowly press the sensor in straight.

f. Install rear axle hub and bearing assembly.

g. Install skid control sensor wire.

h. Install rear brake drum.

i. Install rear wheel and lower the vehicle.

j. Inspect rear wheel alignment.

k. Inspect the rear wheel alignment.

l. Check the speed sensor signal.

Rear—2010 vehicles

➡While the battery is connected, even if the power switch is off, the brake control system activates when the brake pedal is depressed or any door courtesy switch is turned on. Therefore, when servicing the brake system components, do not depress the brake pedal or open/ close the doors while the battery is connected.

➡When the brake pedal is first depressed after replacing the brake pads or pushing back the disc brake piston, DTC C1214 may be output. As there is no malfunction, clear the DTC.

➡Use the same procedure for the RH side and LH side.

➡The procedure listed below is for the LH side.

➡If the sensor rotor needs to be replaced, replace it together with the rear axle hub and bearing assembly.

➡The rear speed sensor is a component of the rear axle hub and bearing assembly. If the sensor malfunctions, replace the rear axle hub and bearing assembly.

1. Remove the rear No. 2 floor board (separate type).
2. Remove the rear deck floor box.
3. Remove the rear No. 3 floor board.
4. Disconnect the cable from the negative battery terminal.

➡When disconnecting the cable, some systems need to be initialized after the cable is reconnected.

5. Remove the rear wheel.
6. Remove the front door scuff plate LH.
7. Remove the cowl side trim sub assembly LH.
8. Remove the lower instrument panel finish panel assembly.
9. Loosen the parking brake cable.
10. Disconnect the rear speed sensor wire.

a. Using a screwdriver, disconnect the connector from the rear speed sensor.

➡Be careful not to damage the rear speed sensor.

11. Disconnect the No. 3 parking brake cable assembly.
12. Separate rear disc brake caliper assembly.
13. Remove the rear disc.
14. Remove the rear axle hub and bearing assembly.

➡The rear speed sensor is a component of the rear axle hub and bearing assembly. If the sensor malfunctions, replace the rear axle hub and bearing assembly.

To install:

15. Install the rear axle hub and bearing assembly.
16. Inspect the rear axle hub bearing looseness.
17. Inspect the rear axle hub runout.
18. Install the rear disc.
19. Install the rear disc brake caliper assembly.
20. Connect the No. 3 parking brake cable assembly.
21. Connect the rear speed sensor wire.

a. Connect the rear speed sensor wire connector to the rear speed sensor.

22. Adjust the parking brake lever travel.
23. Inspect the rear disc brake caliper operation lever and stopper clearance.
24. Install the lower instrument panel finish panel assembly.
25. Install the cowl side trim sub assembly LH.
26. Install the front door scuff plate LH.
27. Install the rear wheel. Tighten to 76 ft. lbs. (103 Nm).
28. Connect the cable to the negative battery terminal.

➡When disconnecting the cable, some systems need to be initialized after the cable is reconnected.

29. Install the rear No. 3 floor board.
30. Install the rear deck floor box.
31. Install the rear No. 2 floor board (separate type).
32. Check the speed sensor signal.

BRAKE CALIPER

REMOVAL & INSTALLATION

See Figures 8 and 9.

1. Before servicing the vehicle, refer to the precautions.
2. With the power switch **OFF** to prohibit brake control, remove the No. 1 and No. 2 motor relays.

➡If the pump motor operates while there is air remaining inside the brake actuator hose, the air will enter the actuator, resulting in difficulty in air bleeding.

3. Raise and safely support the front of the vehicle.
4. Remove the front wheel.
5. Remove the bolt and the gasket attaching the brake hose to the caliper.
6. Disconnect the brake hose, and plug the openings in the caliper and the brake hose to prevent fluid loss and contamination.
7. Hold the front disc brake caliper slide pin using a wrench.
8. Remove the 2 caliper mounting bolts and remove the caliper from the vehicle.
9. Remove the 2 disc brake pads from the front disc brake caliper mounting.
10. Remove the anti-squeal shims from the disc brake pads.
11. Remove the front disc brake pad support plates.
12. Remove the slide pins and bushing dust boots from the front disc brake caliper mounting, if necessary.
13. Remove the front disc brake caliper mounting from the steering knuckle, if necessary.

To install:

14. Install the front disc brake caliper mounting to the steering knuckle, if removed. Torque the 2 mounting bolts to 81 ft. lbs. (109 Nm).

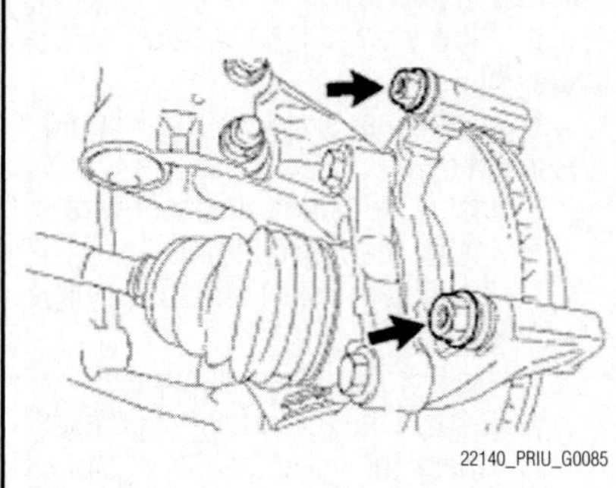

Fig. 8 Remove the front disc brake caliper slide pins from the mounting

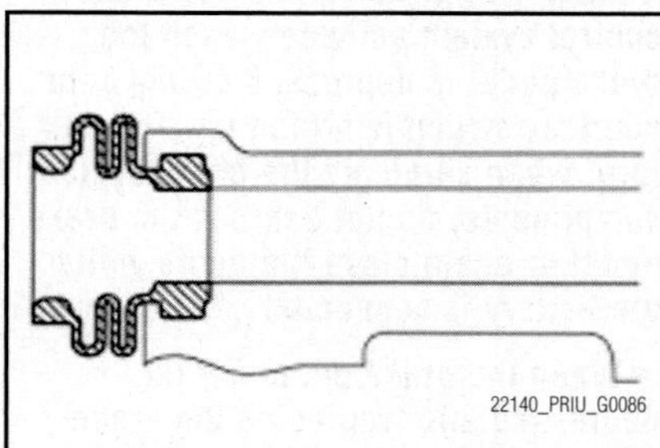

Fig. 9 Apply lithium soap base glycol grease to seal surface of 2 new bushing dust boots and install on to the front disc brake caliper mounting

15. Install new bushing dust boots and slide pins into the front disc brake caliper mounting, if necessary. Apply lithium soap base glycol grease to the sealing, sliding and fitting areas before installation.
16. Install the front disc brake pad support plates.
17. Install the anti-squeal shims onto the disc brake pads.
18. Install the disc brake pads onto the front disc brake caliper mounting.
19. Install the brake caliper onto the mounting along with the 2 caliper mounting bolts. Torque the 2 mounting bolts to 25 ft. lbs (34 Nm).
20. Connect the brake hose to the caliper with bolt and a new gasket. Tighten the caliper brake hose bolt to 24 ft. lbs. (33 Nm).
21. Fill the master cylinder to the proper level with clean brake fluid.
22. Bleed air from front and rear brake systems.
23. Recheck the fluid level.
24. Install the wheel and lower the vehicle.
25. Repeatedly press the brake pedal to bring the pads in contact with the rotor.
26. Check and clear the DTCs.

DISC BRAKE PADS

REMOVAL & INSTALLATION

1. Before servicing the vehicle, refer to the precautions.
2. Raise and safely support the front of the vehicle.
3. Remove the front wheel.
4. Hold the front disc brake caliper slide pin using a wrench.

➡Caliper removal is not necessary to service the brake pads.

5. Remove the 2 caliper mounting bolts and separate the caliper from the mounting and support from the vehicle with wire. Do not allow the caliper to hang low enough so to put tension on the brake hose.
6. Remove the brake pads.

To install:

7. Install the anti-squeal shims onto the disc brake pads.
8. Install the disc brake pads.
9. Install the brake caliper onto the mounting along with the 2 caliper mounting bolts. Torque the 2 mounting bolts to 25 ft. lbs (34 Nm).
10. Install the wheel.
11. Lower the vehicle.
12. Repeatedly press the brake pedal to bring the pads in contact with the rotor.

BRAKE CALIPER

REMOVAL & INSTALLATION

See Figures 10 and 11.

1. Before servicing the vehicle, refer to the precautions.

➡When the brake pedal is first depressed after replacing the brake pads or pushing back the disc brake piston, DTC C1214 may be output. As there is no malfunction, clear the DTC.

➡Use the same procedure for the LH side and RH side.

➡The following procedure is for the LH side.

2. Disable the brake control.
3. Remove the rear wheel.

➡If brake fluid leaks onto any painted surface, immediately wash it off.

4. Remove the front door scuff plate LH.
5. Remove the cowl side trim sub assembly LH.
6. Remove the lower instrument panel finish panel assembly.
7. Loosen the parking brake cable.
   a. Completely raise the parking brake pedal.
   b. Loosen the lock nut and adjusting nut to completely release the parking brake cable.
8. Disconnect the No. 3 parking brake cable assembly.
   a. Separate the No. 3 parking brake cable assembly from the rear disc brake caliper assembly.
   b. Separate the No. 3 parking brake cable assembly from the rear disc brake caliper assembly.

➡Insert an offset wrench (14 mm) at the base of the No. 3 parking brake cable assembly as shown in the illustration to disengage the clip. Pull out the No. 3 parking brake cable assembly from the rear disc brake caliper assembly.

9. Separate the rear flexible hose.
   a. Remove the union bolt and gasket, and separate the rear flexible hose from the rear disc brake caliper assembly.
10. Remove the rear disc brake caliper assembly.
   a. Hold the rear disc brake pad guide pin, and remove the 2 bolts and rear disc brake caliper assembly.

➡Remove the rear disc brake caliper assembly while holding both of the rear disc brake pads because the anti-squeal springs may fall off the rear disc brake pads.

To install:

11. Install the rear disc brake caliper assembly.
   a. To compensate for pad wear when reusing the pad, use SST to turn the piston to the position where the protrusion on the pad lines up properly with the piston groove.

➡Place the disc between the 2 brake pads and determine the piston return value.

   b. Hold the rear disc brake caliper pad guide pin, and install the rear disc brake caliper assembly to the rear disc brake caliper mounting with the 2 bolts. Tighten to 25 ft. lbs. (34 Nm).

➡Install the rear disc brake caliper assembly while holding both of the rear disc brake pads because the anti-squeal springs may fall off the rear disc brake pads.

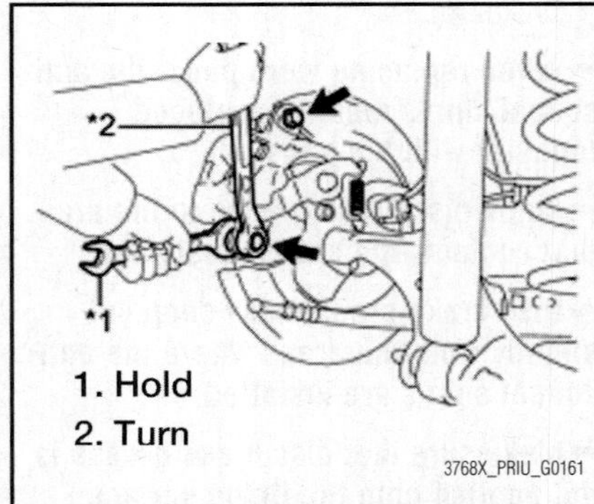

Fig. 10 Removing the rear disc brake caliper

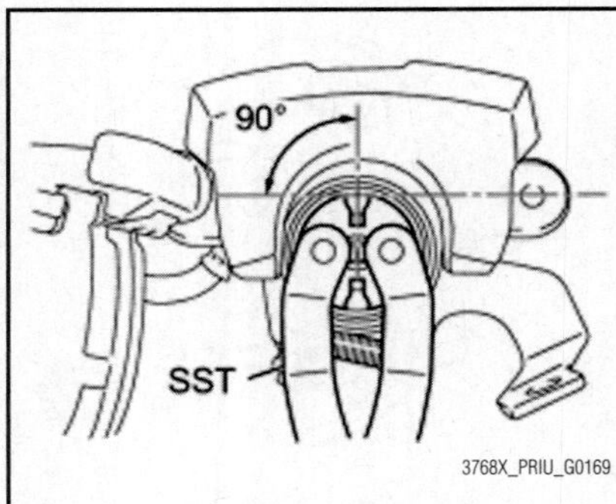

Fig. 11 Positioning the rear disc brake caliper

➡Be sure that the anti-squeal springs are installed to the rear disc brake pads.

12. Connect the rear flexible hose.
   a. Connect the rear flexible hose to the rear disc brake caliper assembly with a new union bolt and a new gasket. Tighten to 25 ft. lbs. (33 Nm).

➡Install the flexible hose lock securely into the lock hole in the disc brake caliper.

13. Connect the No. 3 parking brake cable assembly to the rear disc brake caliper assembly.
   a. Connect the No. 3 parking brake cable assembly to the rear disc brake caliper assembly.
14. Disconnect the cable from the negative battery terminal.

➡Perform this step only when the Techstream cannot prohibit brake control.

15. Connect the connector.
   a. Connect the connector to the brake booster with the master cylinder assembly.

➡Make sure that the connector can be connected smoothly. Do not allow water, oil or dirt to enter.

➡Make sure that the connector lock is locked securely.

16. Connect the cable to the negative battery terminal.
17. Bleed the brake line.
18. Perform the initialization and calibration of the linear solenoid valve.
19. Adjust the parking brake.
20. Install the lower instrument panel finish panel assembly.
21. Install the cowl side trim sub assembly LH.
22. Install the front door scuff plate LH.
23. Install the rear wheel. Tighten to 76 ft. lbs. (103 Nm).

DISC BRAKE PADS

REMOVAL & INSTALLATION

See Figures 12 through 16.

1. Before servicing the vehicle, refer to the precautions.

➡When the brake pedal is first depressed after replacing the brake pads or pushing back the disc brake piston, DTC C1214 may be output. As there is no malfunction, clear the DTC.

➡Use the same procedure for the LH side and RH side.

➡The following procedure is for the LH side.

2. Disable the brake control.
3. Remove the rear wheel.
4. Remove the rear brake caliper.
5. Remove the rear disc brake pad.
   a. Remove the 2 anti-squeal springs from the rear disc brake pads.
   b. Remove the 2 rear disc brake pads from the rear disc brake caliper mounting.
6. Remove the rear disc brake anti-squeal shim.
   a. Remove the rear No. 1 disc brake anti-squeal shim and rear No. 2 disc brake anti-squeal shim from each rear disc brake pad.
7. Remove the rear disc brake pad support plate.
   a. Remove the 2 rear disc brake pad support plates from the disc brake caliper mounting.

➡Each rear disc brake pad support plate has a different shape. Be sure to put an identification mark on each rear disc brake pad support plate so that it can be installed to its original position.

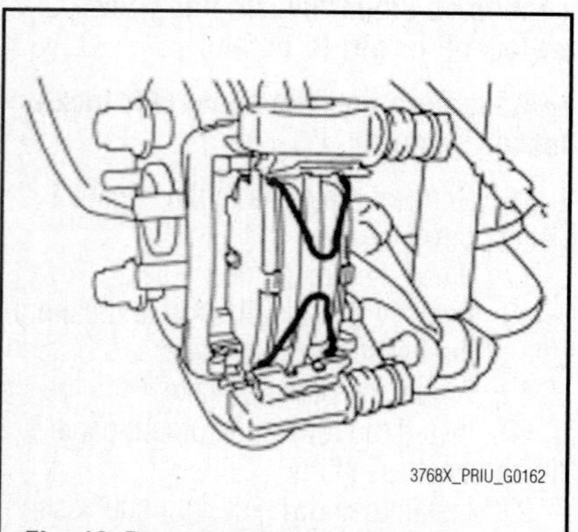

Fig. 12 Removing the rear disc pad

8. Remove the rear disc brake pad guide pin.
   a. Remove the 2 rear disc brake pad guide pins from the rear disc brake caliper mounting.

To install:

9. Install the rear disc brake pad guide pin.
   a. Apply a light layer of lithium soap base glycol grease to the sliding and sealing surfaces of the 2 rear disc brake pad guide pins.
   b. Install the 2 rear disc brake pad guide pins to the rear disc brake caliper mounting.
10. Install the rear disc brake pad support plate.
   a. Install the 2 rear disc brake pad support plates to the rear disc brake caliper mounting.

➡Be sure to install each rear disc brake pad support plate in the correct position and direction.

11. Install the rear disc brake anti-squeal shim.
   a. Apply disc brake grease to the back plate of the rear disc brake pads.
   b. Install the rear No. 1 disc brake anti-squeal shim and rear No. 2 disc brake anti-squeal shim to each rear disc brake pad.

➡When replacing worn pads, the anti-squeal shims must be replaced together with the pads.

➡Apply disc brake grease to the area that contacts the anti-squeal shim.

➡Disc brake grease may seep out slightly from the areas where the anti-squeal shims are installed.

➡Make sure that disc brake grease is not applied onto the lining surface.

12. Install the rear disc brake pad.

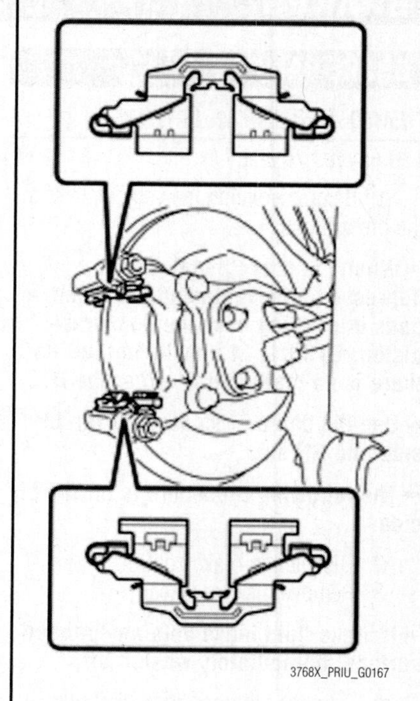

Fig. 15 Installing the rear disc brake pad support plate

   a. Install the 2 rear disc brake pads to the rear disc brake caliper mounting.

➡There should be no oil or grease on the friction surface of the disc brake pads or the rear disc.

   b. Install the 2 anti-squeal springs to the rear disc brake pads.

➡When replacing the rear disc brake pads with new ones make sure to replace the anti-squeal springs at the same time.

➡Be sure to install the anti-squeal springs into the rear disc brake pad installation holes as far as they will go.

13. Install the rear disc brake caliper.

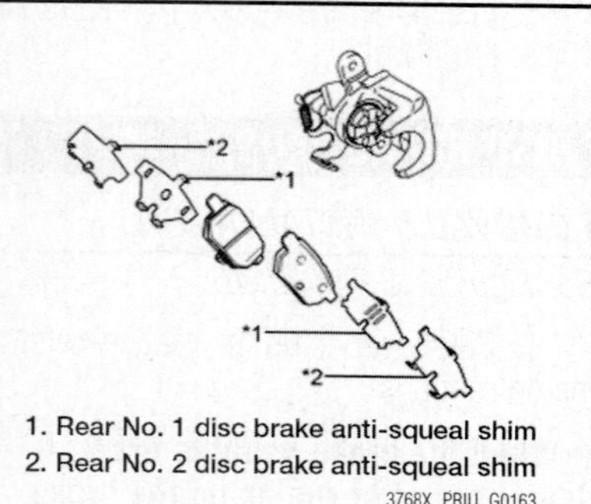

Fig. 13 Removing the rear disc brake anti-squeal shim

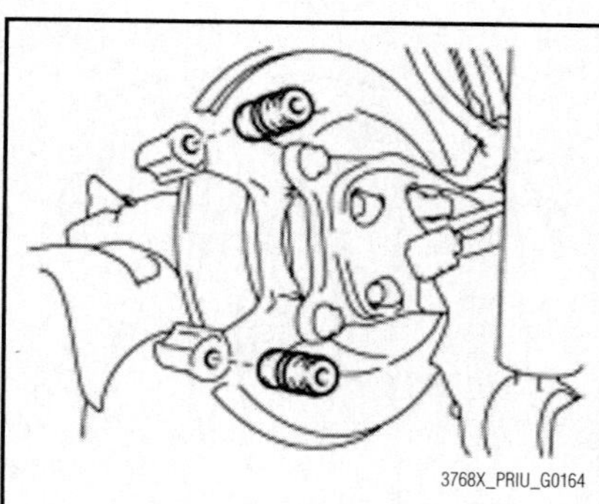

Fig. 14 Removing the rear disc brake pad guide pin

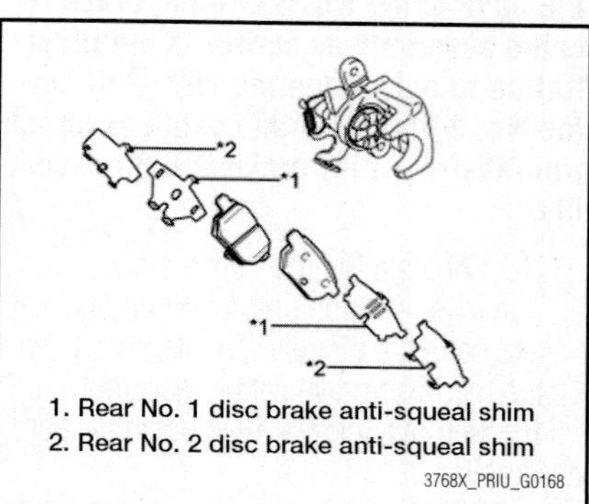

Fig. 16 Installing the rear disc brake anti-squeal shim

BRAKE DRUM

REMOVAL & INSTALLATION

See Figure 17.

1. Before servicing the vehicle, refer to the precautions.
2. Raise and safely support the rear of the vehicle.
3. Remove the rear wheels.
4. Put match marks on the rear brake drum and the axle hub.
5. Release the parking brake and remove the brake drum.
6. If the brake drum cannot be removed easily, do the following steps.
   a. Remove the plug and insert a screwdriver through the hole in the backing plate.
   b. Using another screwdriver, reduce the brake shoe adjuster by turning the adjusting wheel.

To install:

7. Aligning the match marks, install the brake drum.
8. Adjust the drum brake shoe clearance.
   a. Temporarily install 2 hub nuts.
   b. Remove the hole plug, and turn the adjuster to expand the shoe until the drum locks up.
   c. Rotate the adjuster back by 8 notches.
   d. Install the hole plug.
   e. Remove the 2 hub nuts.
9. Install the wheel. Torque the wheel lugs to 76 ft. lbs. (103 Nm).
10. Lower the vehicle.
11. Check brake fluid level in the reservoir and add to proper level with clean brake fluid, if necessary.

BRAKE SHOES

REMOVAL & INSTALLATION

See Figures 18 through 21.

1. Before servicing the vehicle, refer to the precautions.
2. Remove the rear wheels.
3. Remove the rear brake drums.
4. Disconnect the spring and parking brake shoe adjuster set from the brake shoes.
5. Remove the cup, shoe hold-down spring and pin.
6. Remove front brake shoe.
7. Remove the parking brake shoe adjuster set from the rear brake shoe.
8. Remove the automatic adjusting lever spring and automatic adjusting lever.
9. Remove anchor spring from the rear brake shoe.
10. Remove the cup, shoe hold-down spring and pin.
11. Using needle-nose pliers, disconnect the parking brake cable, and remove the rear brake shoe.
12. Using a screwdriver, remove the C-washer and parking brake lever.

To install:

13. Apply high-temperature grease to the shoe attached surface of the backing plate.

➡There should be no oil or grease adhering to the friction surfaces of the shoe lining and the drum.

14. Using needle-nose pliers, install the parking brake lever to the rear brake shoe with a new C-washer.
15. Using needle-nose pliers, connect the parking brake cable to the parking brake lever.
16. Install the rear brake shoe, pin, shoe hold-down spring and cup.
17. Install the automatic adjusting lever and automatic adjusting lever spring to the front brake shoe.
18. Apply high-temperature grease to the adjusting bolt and assemble the parking brake shoe adjuster set.
19. Install the rear brake shoe return spring to the adjuster set and install set.
20. Connect the anchor spring to the front and rear brake shoes.
21. Install the front brake shoe, pin, shoe hold-down spring and cup.
22. Connect the shoe return spring to the front and rear brake shoes.
23. Check to make sure that all the components of the brake shoe assembly are installed properly.
24. Install the brake drums.
25. Install the wheels.
26. Lower the vehicle.
27. Press the brake pedal 3–5 times to adjust the brake shoe clearance.

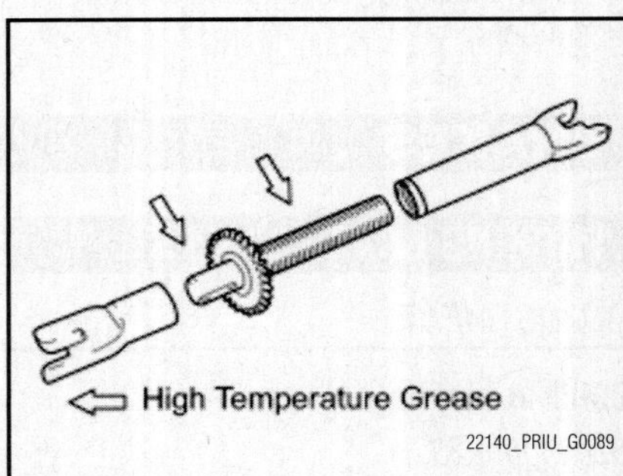

Fig. 19 Apply high temperature grease to the adjusting bolt and assemble the strut set

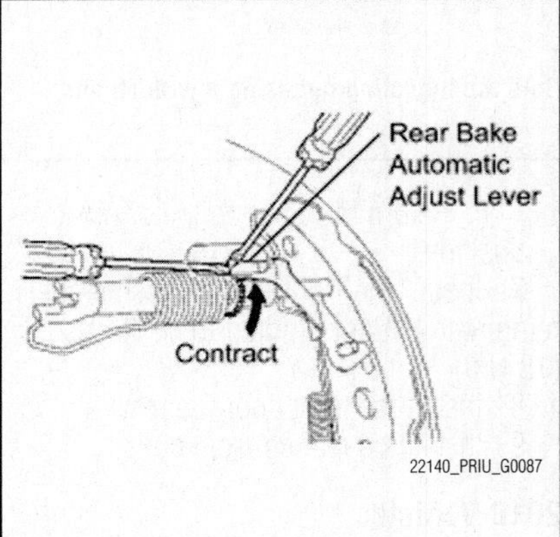

Fig. 17 Release the parking brake and remove the rear brake drum

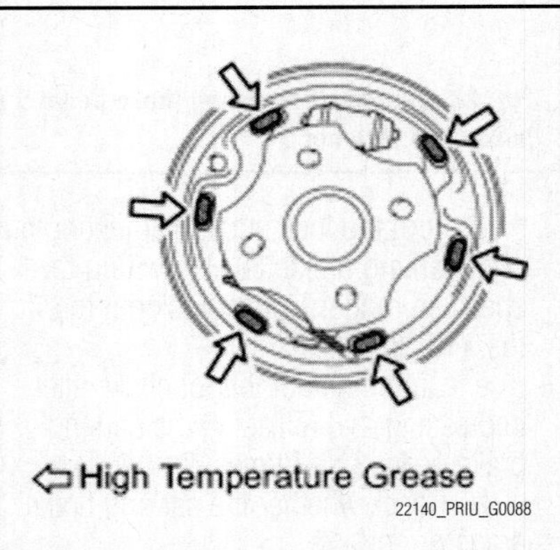

Fig. 18 Apply high temperature grease to the shoe attached surface of the backing plate

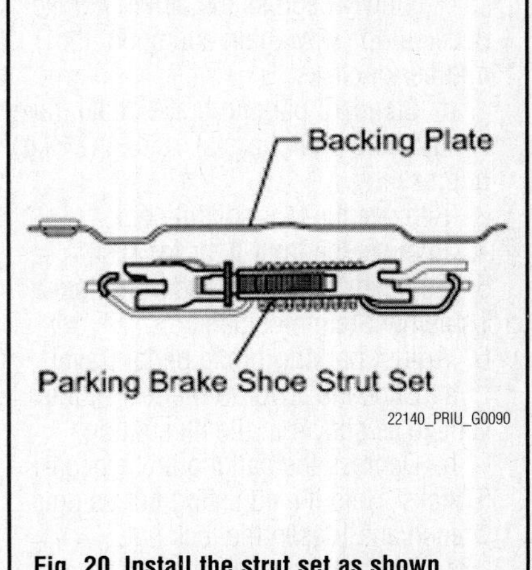

Fig. 20 Install the strut set as shown

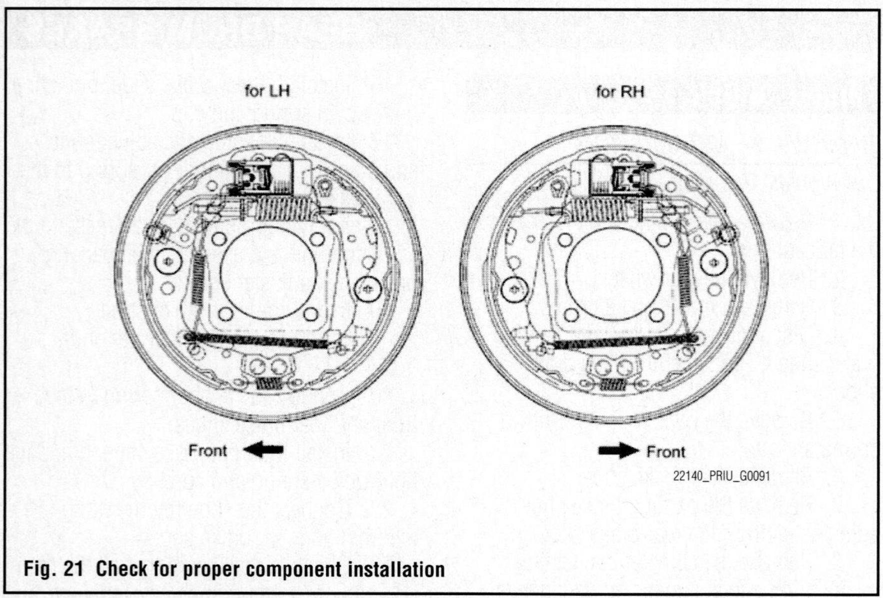

Fig. 21 Check for proper component installation

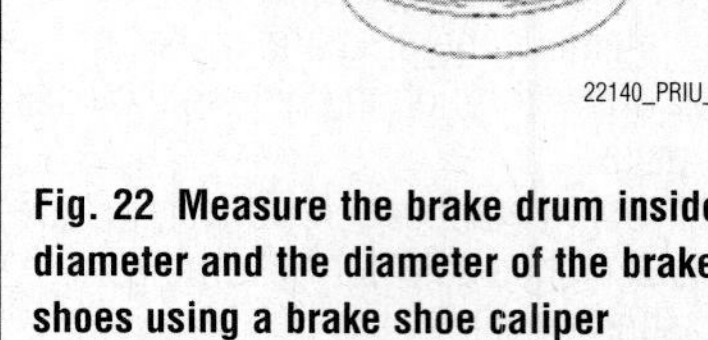

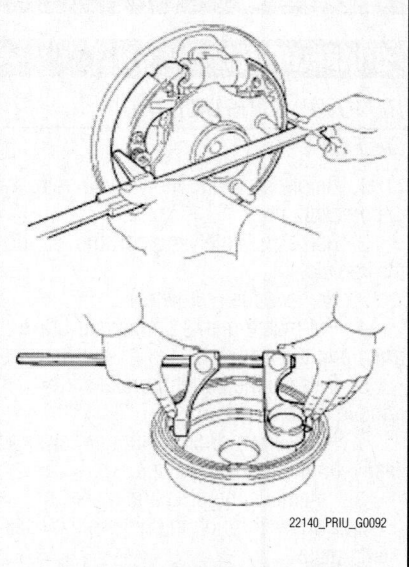

Fig. 22 Measure the brake drum inside diameter and the diameter of the brake shoes using a brake shoe caliper

ADJUSTMENT

See Figure 22.

1. With the brake drum off the vehicle:
   a. Measure the brake drum inside diameter and the diameter of the brake shoes using a brake shoe caliper.
   b. Check that the difference between the diameter is 0.024 in. (0.6 mm).
   c. Adjust the brake shoes as necessary.
2. With the brake drum installed on the vehicle:
   a. Remove the hole plug in the backing plate, and turn the adjuster to expand the shoe until the drum locks.
   b. Loosen the adjuster by 8 notches.
   c. Install the hole plug.

BRAKES

PARKING BRAKE

PARKING BRAKE CABLES

ADJUSTMENT

2009 Vehicles

See Figure 23.

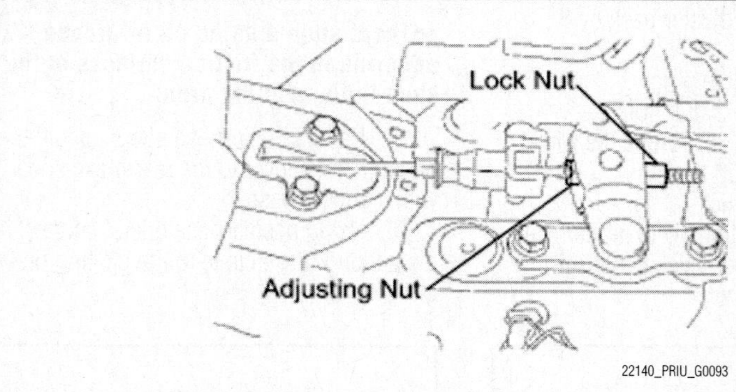

Fig. 23 Depress the parking brake pedal 5 clicks. Hold the adjusting nut using a wrench and loosen the lock nut

1. Adjust parking brake shoe clearance.
2. Check parking brake pedal travel.
   a. Firmly depress the parking brake pedal.
   b. Release the parking brake.
   c. Slowly depress the parking brake pedal all the way again and count the number of clicks.
   d. Standard parking brake pedal travel should be 6 to 9 clicks at 70 lbs. (31 kg) of pressure.
3. Remove the rear console box.
4. Remove the front floor footrest.
5. Remove the 8 bolts and front floor crossmember reinforcement.
6. Adjust parking brake pedal travel.
   a. Slide the cover to the rear of the vehicle as shown in the illustration.
   b. Depress the parking brake pedal 5 clicks. Hold the adjusting nut using a wrench and loosen the lock nut.
   c. Release the parking brake.
   d. Turn the lock nut and adjusting nut until parking brake travel is within the specified range. Tighten lock nut to 9 ft. lbs. (13 Nm).
   e. Count the number of clicks after depressing and releasing the parking brake pedal 3 to 4 times.
   f. Check whether the parking brake drags or not.
   g. When operating the parking brake pedal, check that the parking brake pedal indicator light comes on.
   h. Return the cover to the original position.
7. Install the front floor crossmember reinforcement and tighten to 13 ft. lbs. (18 Nm).
8. Install the front floor footrest.
9. Install the rear console box.

2010 Vehicles

See Figure 24.

1. Remove the lower instrument panel finish panel assembly.

Fig. 24 Releasing the parking brake cable

2. Completely release the parking brake pedal.

a. Loosen the lock nut and the adjusting nut to completely release the parking brake cable.

b. Turn the adjusting nut until the parking brake pedal travel is corrected to be within 8-11 notches at 67.5 ft. lbs. (300 N).

c. Using a wrench or an equivalent tool, hold the adjusting nut and tighten the lock nut to 48 inch lbs. (5.4 Nm).

d. Operate the parking brake pedal 3 to 4 times, and check the parking brake pedal travel.

e. Check whether the parking brake drags or not.

f. Install the lower instrument panel finish panel assembly

PARKING BRAKE SHOES

REMOVAL & INSTALLATION

The rear drum brake shoes serve as the parking brakes. Refer to Rear Drum Brakes, Brake Shoes, Removal & Installation.

ADJUSTMENT

The rear drum brake shoes serve as the parking brakes. Refer to Rear Drum Brakes, Brake Shoes, Adjustment.

CHASSIS ELECTRICAL

AIR BAG (SUPPLEMENTAL RESTRAINT SYSTEM)

GENERAL INFORMATION

These vehicles are equipped with an air bag system. The system must be disarmed before performing service on, or around, system components, the steering column, instrument panel components, wiring and sensors. Failure to follow the safety precautions and the disarming procedure could result in accidental air bag deployment, possible injury and unnecessary system repairs.

SERVICE PRECAUTIONS

CAUTION

Disconnect and isolate the battery negative cable before beginning any airbag system component diagnosis, testing, removal, or installation procedures. Allow system capacitor to discharge for two minutes before beginning any component service. This will disable the airbag system. Failure to disable the airbag system may result in accidental airbag deployment, personal injury, or death.

DISARMING THE SYSTEM

To avoid personal injury when working on vehicles equipped with an air bag, the negative battery cable must be disconnected and at least 90 seconds must elapse before working on the system. Failure to do so may result in deployment of the air bag.

ARMING THE SYSTEM

To rearm the air bag system, reconnect the battery cable (s).

CLOCKSPRING CENTERING

See Figure 25.

1. Before servicing the vehicle, refer to the precautions.
2. Install the spiral cable.

a. Check that the front wheels are facing straight ahead.

b. Set the turn signal switch to the neutral position.

➡Make sure that the turn signal switch is in the neutral position, as the pin of the turn signal switch may be snapped.

c. Install the spiral cable.

➡When replacing the spiral cable with a new one, remove the lock pin before installing the steering wheel.

d. Connect the connector to the spiral cable.

WARNING

When handling the airbag connector, do not damage the airbag wire harness.

3. Slowly rotate the spiral cable counterclockwise by hand until it feels firm.
4. Rotate the spiral cable clockwise approximately 2.5 turns to align the marks.
5. The spiral cable should rotate approximately 2.5 turns to the left and right from the center.

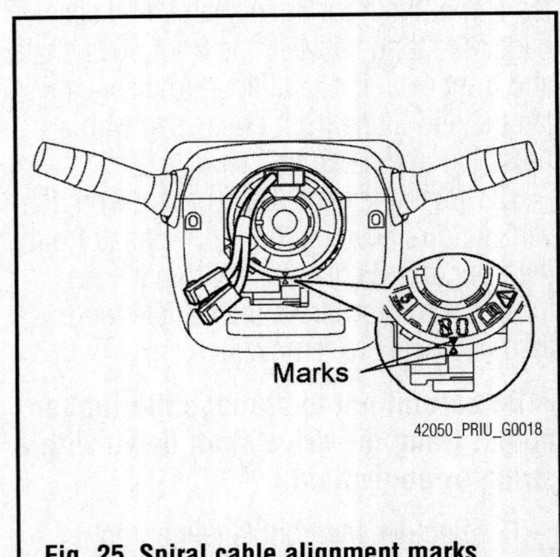

Fig. 25 Spiral cable alignment marks

DRIVE TRAIN

FRONT HALFSHAFT

REMOVAL & INSTALLATION

2009 Vehicles

See Figure 26.

1. Before servicing the vehicle, refer to the precautions.
2. Drain transaxle fluid.
3. After draining the fluid, tighten the drain plug, along with a new gasket to 29 ft. lbs. (39 Nm).
4. Raise and safely support the vehicle.
5. Remove the front wheel.
6. Unstake the axle hub nut and using a 30mm socket, remove the axle hub nut.
7. Remove the bolt and disconnect the speed sensor wire and flexible hose clamp from the strut.
8. Remove the bolt and front speed sensor from the steering knuckle.

➡Keep both the tip and installation part of the speed sensor free of foreign matter.

9. Remove the clip, castle nut and disconnect the tie rod end from the steering knuckle.
10. Remove the bolt, 2 nuts and disconnect the front lower ball joint from the front lower suspension arm.
11. Using a plastic-faced hammer, tap the end of the front drive shaft and disengage the fitting between the front drive shaft and front axle. If it is difficult to disengage, tap the end of the front drive shaft with a brass bar and hammer.
12. Push the front axle outward from the vehicle to remove the front drive shaft from the front axle. Be careful not to push the front axle outward from the vehicle more than necessary to remove it.

➡Be careful not to damage the rubber boots. Hang the drive shaft down with a string or equivalent.

13. Remove the front fender apron seal.
14. Hook the SST 09520-01010 and 09520-24010 claw in position to remove the front drive shaft.

CAUTION

Be careful not to damage the oil seal. Be careful not to damage the front drive shaft boot. Be careful not to drop the front drive shaft.

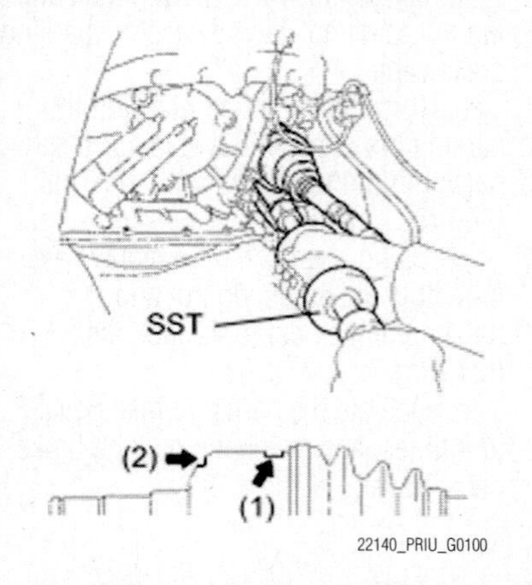

Fig. 26 Hook the special tool SST: 09520-01010 and 09520-24010 claws in position to remove the front drive shaft

15. Check for noticeable looseness when turning the joint up and down, left and right, and in the thrust direction.
16. Check for cracks, damage or grease leaks on the joint boot.

➡Carry the drive shaft levelly.

To install:

17. Apply ATF to the spline of the inboard joint.
18. Align the spline of the front drive shaft and insert the front drive shaft using a brass bar and hammer.

➡Face the snap ring cut area downward. Be careful not to damage the oil seal. Be careful not to damage the front drive shaft boot.

19. Install the front fender apron seal.
20. Install the front drive shaft dust cover (right drive shaft).
21. Push the front axle outward from the vehicle to align the spline of the front drive shaft with the front axle and insert.
22. Connect the front suspension lower arm to the front lower ball joint and tighten the bolt and 2 nuts to 66 ft. lbs. (89 Nm).
23. Connect the tie rod end to the steering knuckle and install it with the castle nut. Torque the castle nut to 36 ft. lbs. (49 Nm) and install a new cotter pin.

➡The cotter pin hole alignment should be done after tightening the castle nut up to 60 degrees beyond the torque specification.

24. Connect the front speed sensor wire and flexible hose clamp to the strut with the bolt.
25. Install the front speed sensor to the steering knuckle and tighten the bolt to 71 inch lbs. (8 Nm).
26. Using a 30mm socket wrench, install a new hub nut and tighten to 159 ft. lbs. (216 Nm).
27. Using a chisel and hammer, stake the hub nut.
28. Install the front wheel.
29. Refill the transaxle fluid to the proper level.
30. Connect the negative battery cable.
31. Perform the power window initialization procedure.
32. Check the front wheel alignment.
33. Check the ABS speed sensor signal.

2010 Vehicles

See Figures 27 through 29.

➡Use the same procedure for the RH side and LH side.

➡The procedure listed below is for the LH side.

1. Remove the front wheels.
2. Remove No. 1 engine under cover.
3. Remove the rear engine under cover LH.
4. Remove the rear engine under cover RH.
5. Drain the hybrid transaxle fluid.
6. Remove the front axle shaft nut.
   a. Using the special tool (09930-00010) and a hammer, release the staked part of the front axle shaft nut.

➡Loosen the staked part of the nut completely, otherwise the threads of the drive shaft may be damaged.

   b. While applying the brakes, remove the front axle shaft nut.
7. Separate the front speed sensor.
8. Separate the front flexible hose.
9. Separate the tie rod end sub assembly.
10. Separate the front stabilizer link assembly.
11. Separate the front No. 1 lower suspension arm sub assembly.
12. Separate the front drive shaft assembly.
13. Remove the front drive shaft assembly.
   a. Using the special tool (09520-00031, 09520-01010, 09521-00020), remove the front drive shaft assembly.

➡Do not damage the transaxle case oil seal.

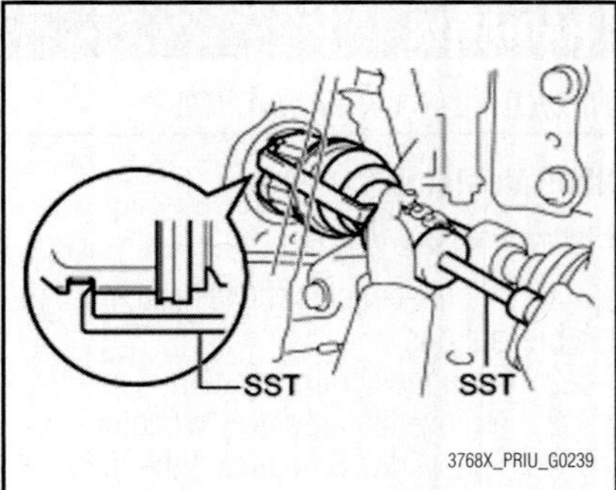

Fig. 27 Removing the front drive shaft assembly

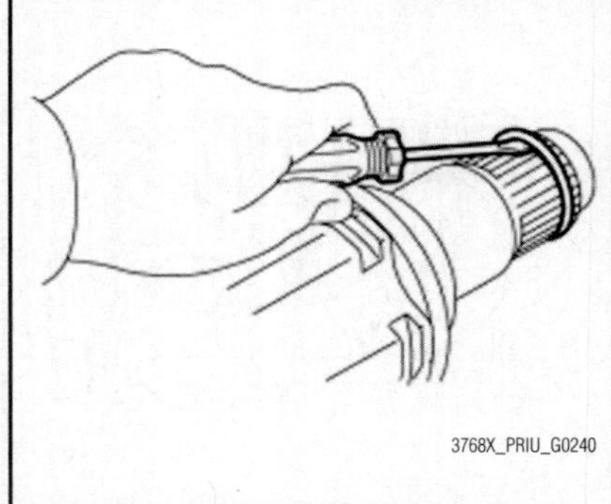

Fig. 28 Removing the front drive shaft hole snap ring

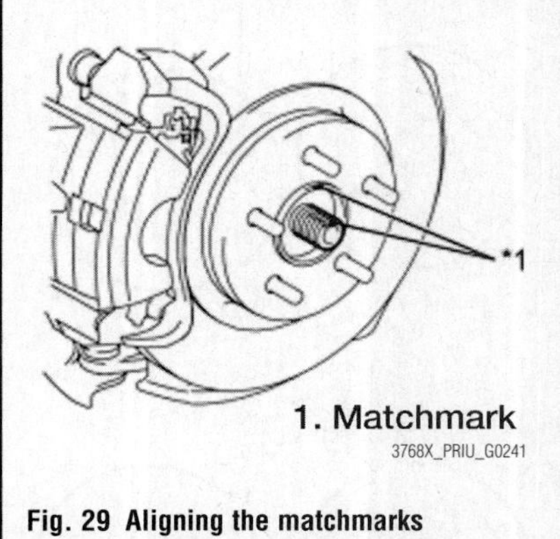

Fig. 29 Aligning the matchmarks

➡Do not damage the inboard joint boot.

➡Do not drop the front drive shaft assembly.

14. Remove the front drive shaft hole snap ring.
    a. Using a screwdriver, remove the front drive shaft hole snap ring.

To install:

15. Install the front drive shaft hole snap ring.
    a. Install a new front drive shaft hole snap ring to the front drive inboard joint assembly.

➡Face the end gap of the front drive inboard joint hole snap ring downward.

16. Install the front drive shaft assembly.
    a. Align the inboard joint splines, and using a brass bar and a hammer, install the front drive shaft assembly.

➡Face the end gap of the front drive shaft hole snap ring downward.

➡Do not damage the transaxle case oil seal.

➡Do not damage the inboard joint boot.

➡Make sure to center the front drive shaft assembly during installation to prevent damage to the front drive shaft hole snap ring.

➡Confirm whether the drive shaft is securely driven in by checking the reaction force and sound.

    b. Align the matchmarks and install the front drive shaft assembly to the front axle hub sub-assembly.
17. Connect the front No. 1 lower suspension arm sub assembly.
18. Connect the front stabilizer link assembly.
19. Connect the tie rod end sub assembly.
20. Install the front flexible hose.
21. Connect the front speed sensor.
22. Install the front axle shaft nut.
    a. Clean the threaded parts on the drive shaft and a new axle shaft nut using a non residue solvent.

➡Be sure to perform this work even when using a new drive shaft.

➡Keep the threaded parts free of oil and foreign matter.

    b. Using a socket wrench (30 mm), install the axle shaft nut. Tighten to 159 ft. lbs. (216 Nm).
    c. Using a chisel and hammer, stake the front axle shaft nut.
23. Add hybrid transaxle fluid.
24. Inspect hybrid transaxle fluid.
25. Install the front wheels. Tighten to 76 ft. lbs. (103 Nm).
26. Inspect and adjust front wheel alignment.
27. Install the rear engine under cover LH.
28. Install the rear engine under cover RH.
29. Install the No. 1 engine under cover.
30. Inspect the speed signal.

ENGINE COOLING

ENGINE FAN

REMOVAL & INSTALLATION

2009 Vehicles

See Figure 30.

1. Before servicing the vehicle, refer to the precautions.
2. Disconnect the negative battery cable.
3. Remove radiator support opening cover.
4. Remove the left engine under cover.
5. Remove the right engine under cover.
6. Drain the coolant in the radiator on the engine side.
7. After draining the coolant in the radiator on the engine side, remove the radiator drain cock plug.
8. Drain the coolant in the radiator on the hybrid side.
9. Remove front bumper cover.
10. Disconnect the radiator hoses
11. Disconnect the radiator reservoir tank hose.
12. Disconnect the inverter reservoir tank hose from the clamps.
13. Remove the 3 bolts and inverter bracket.
14. Remove the 2 bolts and cooler bracket.
15. Disconnect the horn connector.
16. Remove the 3 bolts and disconnect the hood lock control cable, then remove the hood lock assembly.
17. Remove the 5 bolts and radiator support.
18. Remove the hood lock control cable from the radiator support.
19. Disconnect the wire harness clamps shown in the illustration.
20. Remove the 4 bolts, then remove the fan with motor from the vehicle.

To install:

21. Set the fan with motor to the vehicle, then install it with the 4 bolts. Tighten to 66 inch lbs. (8 Nm).
22. Connect the 2 wire harness clamps.
23. Install the hood lock control cable to the radiator support.
24. Install the radiator support and tighten the bolts to 44 in. lbs (5.0 Nm).

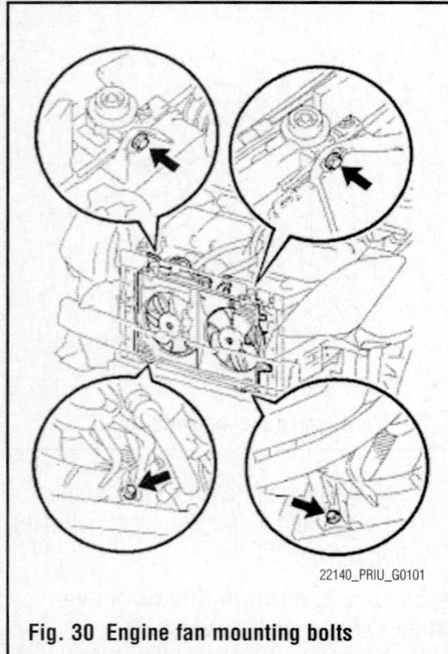
Fig. 30 Engine fan mounting bolts

25. Connect the hood lock control cable to the hood lock assembly.
26. Install the hood lock assembly with the 3 bolts.
27. Connect the horn connector.
28. Install the cooler bracket with the 2 bolts. Tighten bolt "A" to 15 ft. lbs. (20 Nm), and bolt "B" to 75 inch lbs. (9 Nm).
29. Install the inverter bracket with the 3 bolts.
30. Connect the inverter reservoir tank hose from the clamps.
31. Connect the radiator reservoir tank hose.
32. Connect the radiator hoses
33. Install front bumper cover.
34. Drain the coolant in the radiator on the hybrid side.
35. Install the radiator drain cock plug.
36. Drain the coolant in the radiator on the engine side.
37. Install the right engine under cover.
38. Install the left engine under cover.
39. Install radiator support opening cover.
40. Connect the negative battery cable.
41. Perform the power window initialization procedure.

2010 Vehicles

See Figures 31 and 32.

1. Remove the radiator.
2. Remove the fan.
   a. Remove the nut and fan.
3. Remove the No. 2 fan.

To install:

4. To install, reverse the removal procedure.

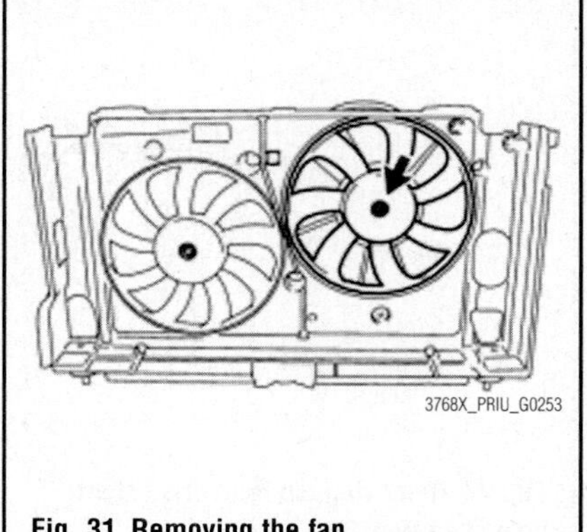
Fig. 31 Removing the fan

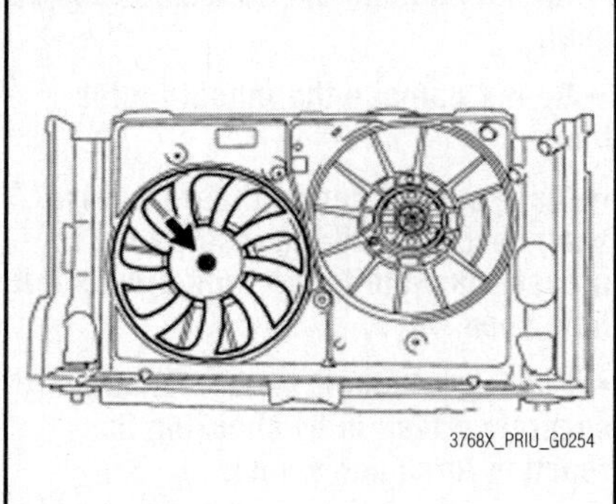
Fig. 32 Removing the No. 2 fan

RADIATOR

REMOVAL & INSTALLATION

2009 Vehicles

See Figures 33 and 34.

1. Before servicing the vehicle, refer to the precautions.
2. Drain the cooling system.
3. Remove fan assembly w/motor.
4. Remove the 3 bolts and the right upper radiator support.
5. Remove the 3 bolts and the left upper radiator support.
6. Remove the 3 bolts and the right lower radiator support.
7. Remove the 3 bolts and the left lower radiator support.
8. Remove the radiator assembly from the vehicle.

To install:

9. Install the radiator assembly to the vehicle.
10. Referring to the illustration, install the radiator supports.
11. Tighten bolts marked "A" to 44 in. lbs (5 Nm) and bolts marked "B" to 35 in. lbs (3.9 Nm).
12. Install the fan assembly.
13. Fill the cooling system.

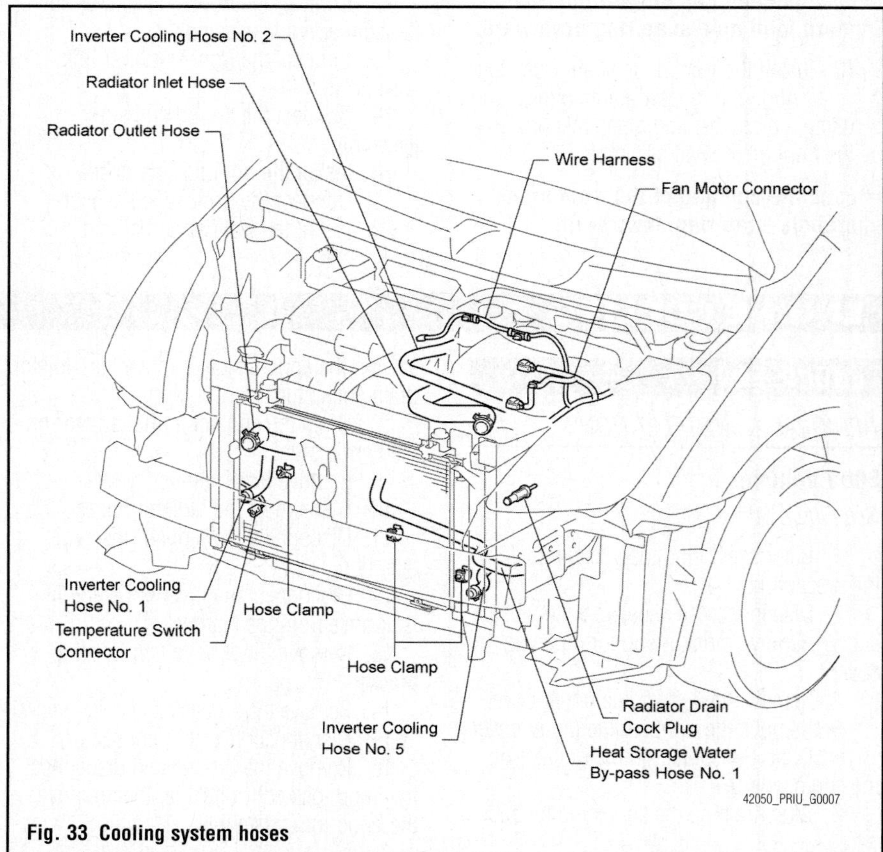

Fig. 33 Cooling system hoses

Fig. 34 Radiator support bolt identification

2010 Vehicles

See Figures 35 and 36.

1. remove the radiator support opening cover.
2. Remove the rear No. 2 floor board (separate type).
3. Remove the rear deck floor box.
4. Remove the rear No. 3 floor board.
5. Disconnect the cable from the negative battery terminal.

➡When disconnecting the cable, some systems need to be initialized after the cable is reconnected.

6. Remove the No. 1 engine under cover.
7. Drain the engine coolant.
8. Disconnect the No. 1 radiator hose from the radiator assembly.
9. Disconnect the No. 2 radiator hose from the radiator assembly.
10. Remove the front bumper assembly.
11. Remove the millimeter wave radar sensor assembly (w/dynamic radar cruise control system).
12. Remove the millimeter wave radar sensor bracket.
13. Remove the No. 1 inverter bracket.
14. Remove the hood lock support sub assembly.
    a. Disconnect the water by pass hose clamp from the radiator support RH.
    b. Disconnect the hood lock connector and hood lock control cable wire.
    c. Disconnect the 3 wire harness clamps.
    d. Disconnect the connector from the No. 2 cooling fan motor.
    e. Disconnect the 2 wire harness clamps and connector from the fan shroud and cooling fan motor.
15. Disconnect the 2 horn connectors.
    a. Remove the 2 bolts, radiator support RH and radiator support LH with the 2 cushions from the upper radiator support.
    b. Remove the 4 bolts and upper radiator support.
16. Remove the air cleaner cap sub assembly.
17. Remove the inlet air cleaner assembly.
18. Remove the air cleaner case.
19. Remove the No. 2 fan shroud.
    a. Disconnect the No. 1 water by pass hose from the radiator assembly.
    b. Disconnect the water by pass hose from the radiator assembly.
    c. Disconnect the 6 water by pass hose clamps from the No. 2 fan shroud.
    d. Remove the 2 bolts and No. 2 fan shroud from the radiator assembly.
20. Remove the radiator assembly with the fan shroud.

➡For vehicles with the air conditioning system, do not apply any excessive force to the cooler condenser assembly or pipe when removing the radiator assembly.

a. Remove the 2 lower radiator supports.
b. Remove the 2 bolts.
c. Remove the fan shroud from the radiator assembly.

To install:

21. Install the radiator assembly with the 2 bolts. Tighten to 62 inch lbs. (7 Nm).
    a. Install the fan shroud to the radiator assembly.

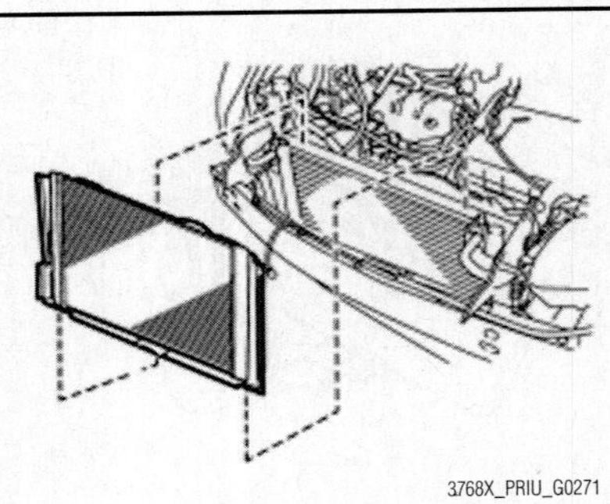

Fig. 35 Removing the radiator with the fan shroud

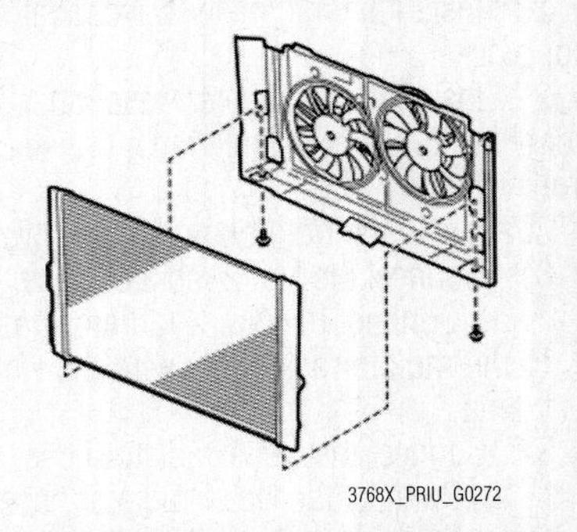

Fig. 36 Removing the fan shroud from the radiator assembly

b. Install the 2 lower radiator supports.
c. Install the radiator assembly with the fan shroud.

➡For vehicles with the air conditioning system, do not apply any excessive force to the cooler condenser assembly or pipe when removing the radiator assembly.

22. Install the No. 2 fan shroud with the 2 bolts. Tighten to 62 inch lbs. (7 Nm).
    a. Connect the 6 water by pass hose clamps to the No. 2 fan shroud.
    b. Connect the water by pass hose to the radiator assembly.
    c. Connect the No. 1 water by pass hose to the radiator assembly.
23. Install the air cleaner case.
24. Install the inlet air cleaner assembly.
25. Install the air cleaner cap sub assembly.
26. Install the hood lock support sub assembly.
    a. Install the upper radiator support with the 4 bolts. Tighten to 10 ft. lbs. (13 Nm).
    b. Install the 2 cushions to the radiator support RH and radiator support LH.
    c. Install the radiator support RH and radiator support LH with the 2 bolts. Tighten to 14 ft. lbs. (19 Nm).
    d. Connect the 2 horn connectors.
    e. Connect the 2 wire harness clamps and connector to the fan shroud and cooling fan motor.
    f. Connect the connector to the No. 2 cooling fan motor.
    g. Connect the hood lock connector and hood lock control cable wire.
    h. Connect the 3 wire harness clamps.
    i. Connect the water by pass hose clamp to the radiator support RH.
27. Install the No. 1 inverter bracket.

28. Install the millimeter wave radar sensor bracket.
29. Install the millimeter wave radar sensor assembly (w/ dynamic radar cruise control system).
30. Install the front bumper assembly.
31. Connect the No. 2 radiator hose.
    a. Connect the No. 2 radiator hose to the radiator assembly with the clamp.
32. Connect the No. 1 radiator hose.
    a. Connect the No. 1 radiator hose to the radiator assembly with the clamp.
33. Add engine coolant.
34. Inspect for coolant leaks.
35. Install the No. 1 engine under cover.
36. Connect the cable to the negative battery terminal.
37. Install the rear No. 3 floor board.
38. Install the rear deck floor box.
39. Install the rear No. 2 floor board (separate type).
40. Install the radiator support opening cover.
41. Add washer fluid (w/ headlight cleaner system).
42. Prepare the vehicle for fog light aim adjustment (w/ fog light).
43. Prepare for fog light aiming.
44. Inspect for fog light aiming.
45. Adjust the fog light aiming.

THERMOSTAT

REMOVAL & INSTALLATION

2009 Vehicles

See Figure 37.

1. Before servicing the vehicle, refer to the precautions.
2. Remove radiator support opening cover.
3. Remove the left engine under cover.
4. Remove the right engine under cover.
5. Drain engine coolant.
6. Remove the 2 nuts and water inlet with radiator outlet hose.
7. Remove the thermostat.
8. Remove the gasket from the thermostat.

To install:

9. Install a new gasket to the thermostat.
10. Install the thermostat so the jiggle valve faces upward.
11. Install the water inlet with radiator outlet hose with the 2 nuts and tighten to 9.0 Nm (80 in. lbs)
12. Add engine coolant.

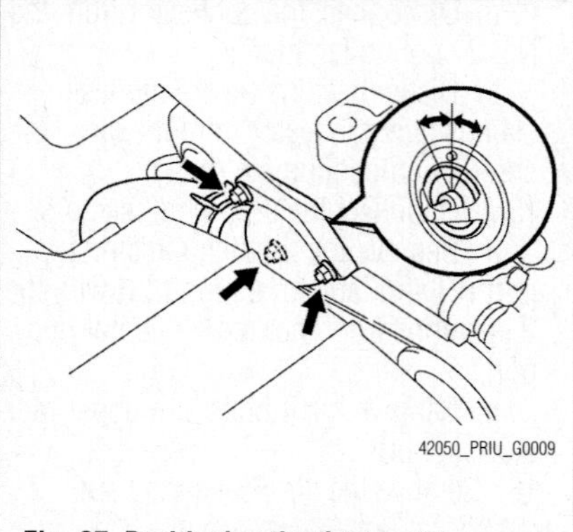

Fig. 37 Positioning the thermostat

13. Check for engine coolant leaks.
14. Install the right engine under cover.
15. Install the left engine under cover.
16. Install radiator support opening cover.

2010 Vehicles

See Figures 38 and 39.

1. Remove the No. 1 engine under cover.
2. Remove the inlet air cleaner assembly.
3. Remove the engine oil level dipstick guide sub assembly.
4. Drain the engine coolant.
5. Remove the water inlet with the thermostat sub assembly.
    a. Disconnect the No. 2 radiator hose and the No. 3 water by pass hose.

➡Do not apply force to the water inlet with thermostat sub-assembly when disconnecting the No. 3 water by-pass hose.

➡Do not damage the water inlet with thermostat sub-assembly.

➡When disconnecting the No. 3 water by-pass hose, pinch the hose clamp, rotate the hose and pull it straight off the pipe.

Fig. 38 Removing the water inlet with the thermostat sub assembly

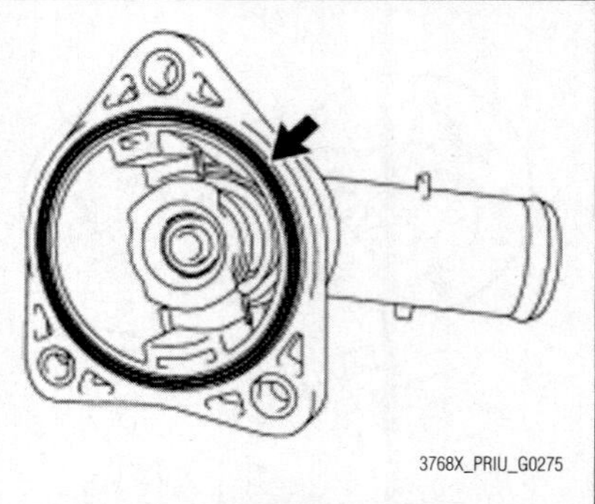

Fig. 39 Removing the gasket from the water inlet with the thermostat sub assembly

    b. Remove the 2 nuts, bolt and water inlet with thermostat sub assembly.
    c. Remove the gasket from the water inlet with thermostat sub assembly.

To install:

6. To install, reverse the removal procedure. Tighten the thermostat sub assembly to 7 ft. lbs. (10 Nm).

WATER PUMP

REMOVAL & INSTALLATION

2009 Vehicles

See Figure 40.

1. Before servicing the vehicle, refer to the precautions.
2. Disconnect the negative battery cable.
3. Drain the cooling system into a suitable container and tighten the drain plug.
4. Remove the radiator support opening cover, if necessary.
5. Remove the right engine under cover.
6. Remove the left engine under cover, if necessary.
7. Remove the accessory drive belt.
8. Remove the right engine mounting insulator, if necessary.
9. Remove the water pump pulley.
10. Remove the water pump mounting bolts and nuts.
11. Remove the water pump and gasket from engine block.

To install:

12. Install the water pump and new gasket to engine block.
13. Install the water pump mounting bolts and nuts. Tighten to 8 ft. lbs. (11 Nm).
14. Install the water pump pulley and mounting bolts. Tighten to 11 ft. lbs. (15 Nm).
15. Install the right engine mounting insulator, if necessary. Torque to 38 ft. lbs. (52 Nm).

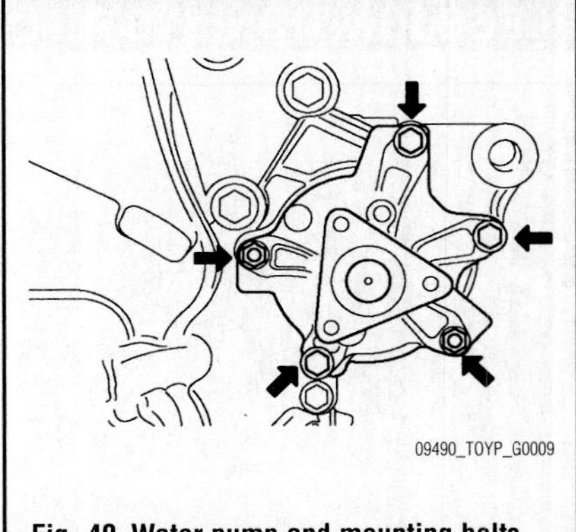

Fig. 40 Water pump and mounting bolts

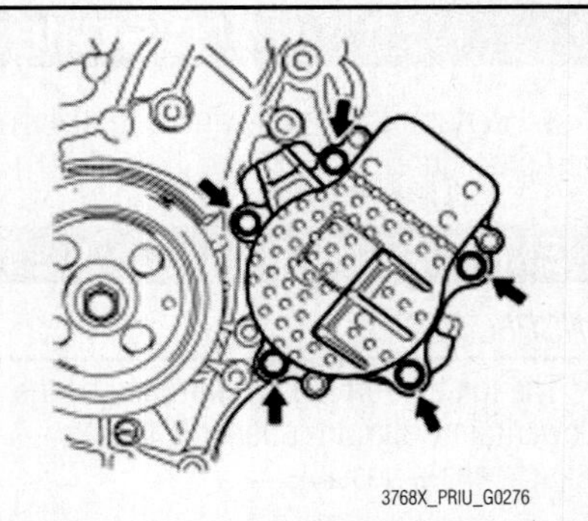

Fig. 41 Removing the water pump assembly front the timing chain cover

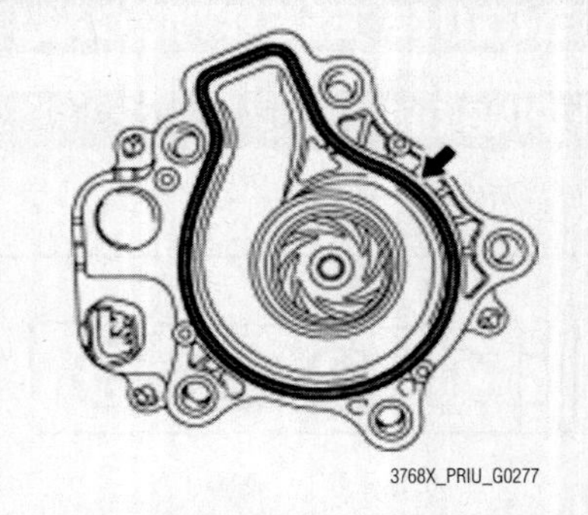

Fig. 42 Removing the water pump gasket from the water pump

16. Install the accessory drive belt.
17. Refill the engine cooling system.
18. Connect the negative battery cable.
19. Perform the power window initialization procedure.
20. Start the engine and top off the coolant as necessary.
21. Check the cooling system for leaks.
22. Install the right engine under cover
23. Install the left engine under cover, if necessary
24. Install the radiator support opening cover, if necessary

2010 Vehicles

See Figures 41 through 43.

1. Remove the No. 1 engine under cover.
2. Remove the inlet air cleaner assembly.
3. Drain the engine coolant.
4. Remove the water pump assembly.
    a. Disconnect the water pump connector from the water pump assembly.
    b. Remove the 5 bolts and water pump assembly from the timing chain cover.
    c. Remove the water pump gasket from the water pump.

To install:

5. To install, reverse the removal procedure. Tighten the water pump assembly A bolts to 19 ft. lbs. (26 Nm). Tighten the water pump B bolts to 15 ft. lbs. (21 Nm).

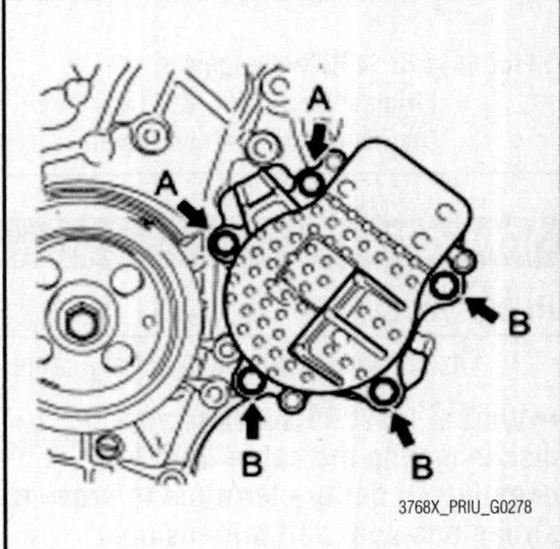

Fig. 43 Installing the water pump assembly to the timing chain cover

ENGINE ELECTRICAL

CHARGING SYSTEM

ALTERNATOR

The Toyota Prius, being a hybrid vehicle that utilizes both electric and gasoline (internal combustion) engine power for mobility, does not require (or come equipped with) an alternator as a part of its charging system. The Toyota Hybrid system replaces the alternator with a pair of electrical motor-generators, a computerized shunt system to control them, a mechanical power splitter that acts as a second differential, and a battery pack that serves as an energy reservoir.

FIRING ORDER

See Figure 44.

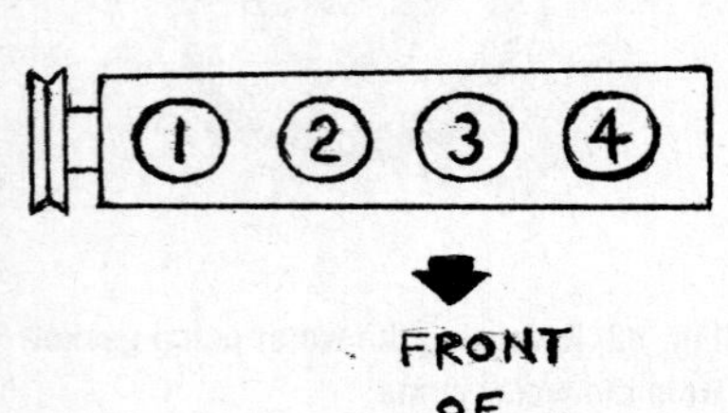

Fig. 44 1.5L I4 Hybrid engine Firing order: 1–3–4–2 Distributorless ignition system

IGNITION COIL

REMOVAL & INSTALLATION

1. Disconnect the negative battery cable.

➡Wait at least 90 seconds after disconnecting the cable from the negative (-) battery terminal to prevent airbag and seat belt pretensioner activation.

2. Disconnect the engine room relay block.
3. Disconnect the 4 ignition coil connectors.
4. Remove the 4 bolts and pull out the 4 ignition coils.

To install:

5. Install the 4 ignition coils with the 4 bolts. Tighten to 80 inch lbs. (9 Nm).
6. Connect the 4 ignition coil connectors.
7. Install the relay block with the 2 bolts. Tighten to 74 inch lbs. (8 Nm).
8. Connect the negative battery cable.
9. Perform the power window initialization procedure.

IGNITION TIMING

INSPECTION

The ignition timing is controlled by the Powertrain Control Module (PCM). No adjustment is possible.

SPARK PLUGS & IGNITION COIL

REMOVAL & INSTALLATION

2009 Vehicles

1. Remove the ignition coils.
2. Using a 16 mm plug wrench, remove the spark plugs.

To install:

3. Using a 16 mm plug wrench, install the spark plugs and tighten to 17.5 Nm (13 ft lbs)
4. Reinstall the ignition coils. Tighten to 80 inch lbs. (9 Nm).

2010 Vehicles

See Figures 45 and 46.

1. Remove the windshield wiper motor and link.
2. Remove the outer cowl top panel sub assembly.
3. Remove the No. 2 cylinder head cover.
4. Remove the ignition coil assembly.
   a. Disconnect the 4 ignition coil connectors.
   b. Remove the 4 bolts and 4 ignition coils.

➡When removing each ignition coil, do not damage the plug cap on the engine head cover opening or the upper edge of the spark plug tube.

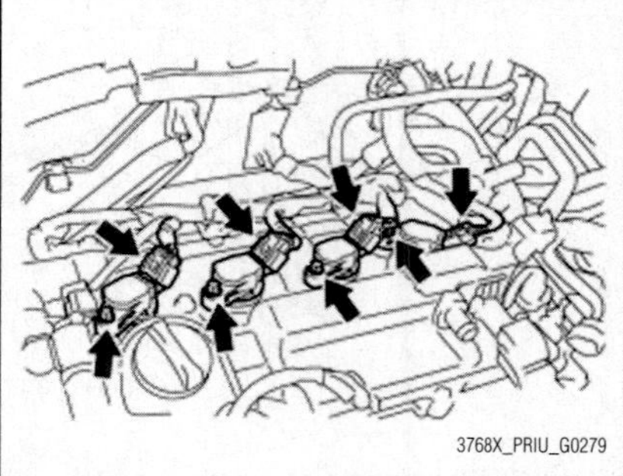

Fig. 45 Removing the ignition coil assembly

5. Remove the spark plugs.
   a. Using a 14 mm spark plug wrench, remove the 4 spark plugs.

To install:

6. To install, reverse the removal procedure. Tighten the spark plugs to 15 ft. lbs. (20 Nm). Tighten the ignition coils to 7 ft. lbs. (10 Nm).

➡When installing each ignition coil, do not damage the plug cap on the engine head cover opening or the upper edge of the spark plug tube.

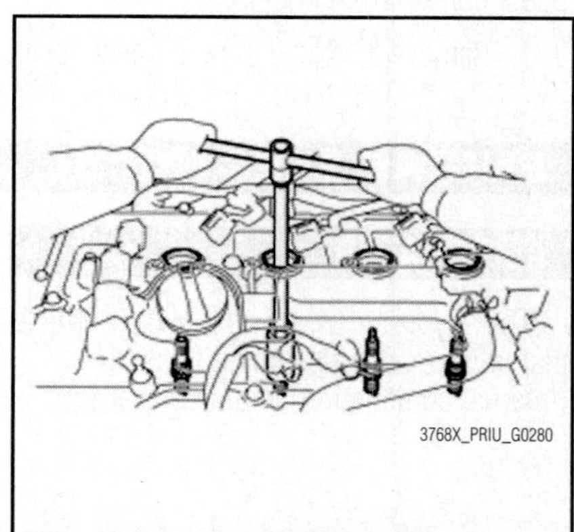

Fig. 46 Removing the spark plugs

ENGINE ELECTRICAL — STARTING SYSTEM

STARTER

The Toyota Prius, being a hybrid vehicle that utilizes both electric and gasoline (internal combustion) engine power for mobility, does not require (or come equipped with) a starter motor as a part of its starting system. The function of the starter motor is performed by a pair of electrical motor-generators, a computerized shunt system to control them, a mechanical power splitter that acts as a second differential, and a battery pack that serves as an energy reservoir.

ENGINE ELECTRICAL — HYBRID CONTROL SYSTEM

PRECAUTIONS

- Before inspecting the high-voltage system, take safety precautions to prevent electrical shocks, such as wearing insulated gloves and removing the service plug grip. After removing the service plug grip, put it in your pocket to prevent other technicians from reconnecting it while you are servicing the high-voltage system.
- Turning the power switch **ON (READY)** with the service plug grip removed could cause a malfunction. Therefore, do not turn the power switch **ON (READY)** unless instructed by the repair manual.
- After disconnecting the service plug grip, wait for at least 5 minutes before touching any of high-voltage connectors or terminals No. .
- At least 5 minutes are required to discharge the high-voltage condenser inside the inverter.
- Since liquid leakage may occur, wear protective goggles when checking inside the high voltage battery.
- Wear insulated gloves, turn the power switch **OFF**, and disconnect the negative terminal of the auxiliary battery before touching any of the orange-colored wires of the high-voltage system.
- Turn the power switch **OFF** before performing a resistance check.
- Turn the power switch **OFF** before disconnecting or reconnecting any connector.
- To install the service plug grip, the lever must be flipped and locked downward. Once it is locked in place, it turns the interlock switch **ON**. Make sure to lock it securely because if you leave it unlocked, the system will output a DTC pertaining to the interlock switch system.
- When the warning light is illuminated or the battery has been disconnected and reconnected, pressing the power switch may not start the system on the first attempt. If so, press the power switch again. With the power switches power mode changed to **ON (IG)**, disconnect the battery. If the key is not in the key slot during reconnection, DTC B2799 may be output.

BATTERY BLOWER

REMOVAL & INSTALLATION

2009 Vehicles

See Figures 47 and 48.

1. Disconnect the negative battery cable.

➡Wait at least 90 seconds after disconnecting the cable from the negative (-) battery terminal to prevent airbag and seat belt pretensioner activation.

2. Remove the rear deck trim cover.
3. Remove the tonneau cover.
4. Remove the rear seat cushion.
5. Remove the rear floor board.
6. Remove the rear side seat back frame.
7. Remove the deck trim side panel.
8. Remove the 3 clips and the inner quarter vent duct.
9. Remove the 3 nuts, then disconnect the battery blower assembly from the vehicle.

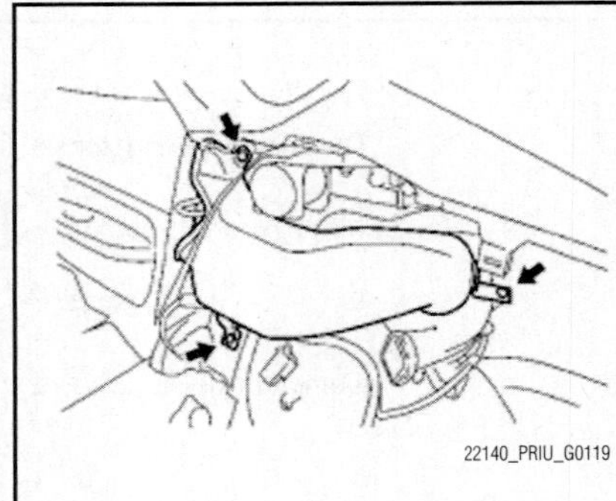

Fig. 47 Inner quarter vent duct attaching clips

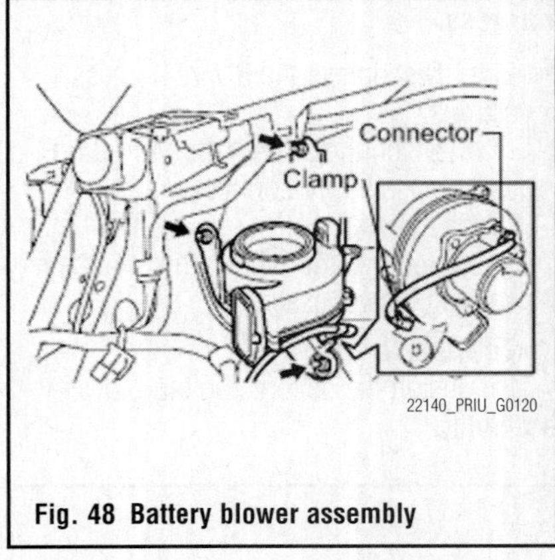

Fig. 48 Battery blower assembly

10. Remove the connector, clamp and battery blower.

To install:

11. Install the connector and clamp on the battery blower.
12. Install the battery blower assembly then the 3 nuts. Tighten to 44 inch lbs. (5 Nm).
13. Install the inner quarter vent duct with the 3 clips.
14. Install the deck trim side panel.
15. Install the rear side seat back frame.
16. Install the rear floor board.
17. Install the rear seat cushion.
18. Install the tonneau cover.
19. Install the rear deck trim cover.
20. Connect the negative battery cable.
21. Perform the power window initialization procedure.

2010 Vehicles

See Figures 49 and 50.

1. Remove the rear seat cushion assembly.
2. Remove the rear No. 1 floor board sub assembly.
3. Remove the rear No. 2 floor board sub assembly.
4. Remove the rear No. 1 floor board.
5. Remove the rear floor board spacer.
6. Remove the No. 1 hybrid battery exhaust duct.
7. Remove the rear door scuff plates (LH and RH).

8. Remove the rear side seat back assembly RH.
9. Remove the No. 1 hybrid battery intake duct.
10. Remove the battery cooling blower assembly.

➡Be sure not to touch the fan part of the battery cooling blower assemblies.

➡Do not lift the battery cooling blower assemblies using the wire harness.

a. Disconnect the 3 wire harness clamps.
b. Disconnect the battery cooling blower assembly connector and clamp.
c. Remove the 2 bolts, nut and battery cooling blower assembly.

To install:

11. Install the battery cooling blower assembly.

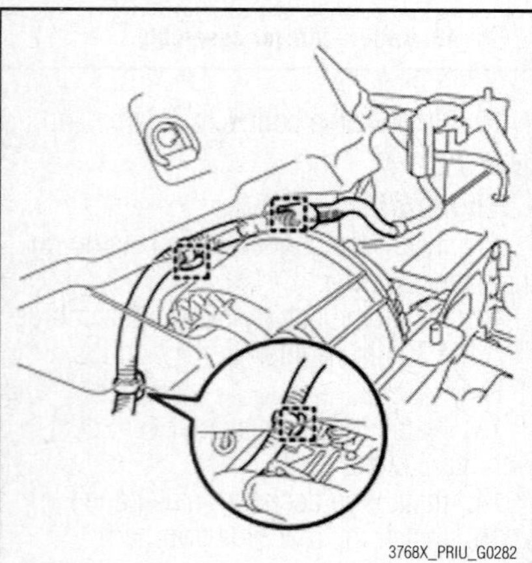

Fig. 49 Disconnecting the 3 wire harness clamps

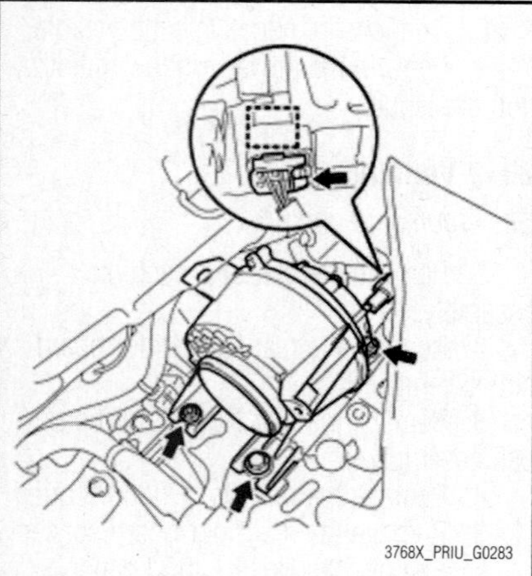

Fig. 50 Disconnecting the battery cooling blower assembly connector and clamp

➡Be sure not to touch the fan part of the battery cooling blower assemblies.

➡Do not lift the battery cooling blower assemblies using the wire harness.

a. install the battery cooling blower assembly with the 2 bolts and nut. Tighten to 66 inch lbs. (7.5 Nm).
b. Connect each battery cooling blower assembly connector and clamp.
c. Connect the 3 wire harness clamps.

12. Install the No. 1 hybrid battery intake duct.
13. Install the rear side seat back assembly RH.
14. Install the rear door scuff plates (LH and RH).
15. Install the No. 1 hybrid battery exhaust duct.
16. Install the rear floor board spacer.
17. Install the rear No. 1 floor board.
18. Install the rear No. 2 floor board sub assembly.
19. Install the rear No. 1 floor board sub assembly.
20. Install the rear seat cushion assembly.

BATTERY CONTROL UNIT

REMOVAL & INSTALLATION

2009 Vehicles

See Figure 51.

1. Disconnect the negative battery cable.

➡Wait at least 90 seconds after disconnecting the cable from the negative (-) battery terminal to prevent airbag and seat belt pretensioner activation.

2. Remove service plug grip.
3. Remove the rear deck trim cover.
4. Remove the tonneau cover.
5. Remove the rear seat cushion assembly.
6. Remove the rear No. 1 floor board.
7. Remove the rear side seat back frame
8. Remove the rear No. 4 floor board.
9. Remove the deck floor box.
10. Remove the deck trim side panel.
11. Remove the battery carrier bracket.
12. Remove the No. 6 battery carrier panel.
13. Remove the junction terminal.
14. Remove the battery ECU.

✱✱ CAUTION

Wear insulating gloves.

a. Disconnect the system main relay connector.

✱✱ CAUTION

Insulate the disconnected connectors with insulating tape.

b. Disconnect the battery ECU connector.
c. Disconnect the thermistor connector.

✱✱ CAUTION

Insulate the disconnected connectors with insulating tape.

d. Remove the clamp, then disconnect the No. 2 frame wire connector.
e. Remove the nut, then disconnect the aluminum shield wire of the main battery cable.

15. Remove the 2 screws and battery ECU.

To install:

16. Install battery ECU.

a. Install the battery ECU with the 2 screws. Tighten to 29 inch lbs. (3 Nm).
b. Install the aluminum shield wire of the main battery cable together with the battery ECU with a new nut. Tighten to Tighten to 29 inch lbs. (3 Nm).
c. Connect the clamp and No. 2 frame wire connector.

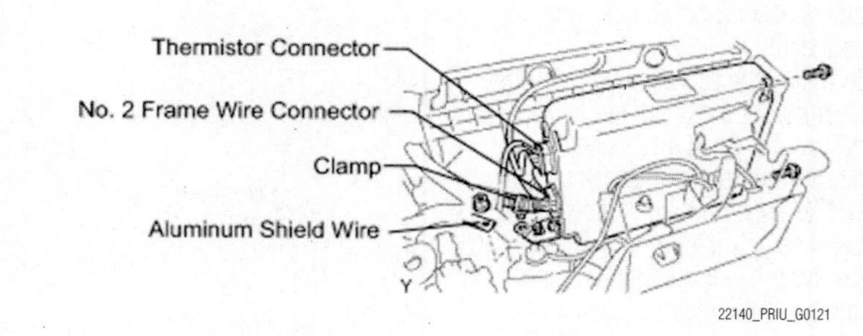

Fig. 51 Battery control unit connections

d. Connect the thermistor connector.

e. Connect the system main relay connector.

17. Install junction terminal.
18. Install No. 6 battery carrier panel.
19. Install battery carrier bracket.
20. Install deck trim side panel.
21. Install deck floor box.
22. Install rear No. 4 floor board.
23. Install rear side seat back frame.
24. Install rear No. 1 floor board.
25. Install rear seat cushion assembly.
26. Install tonneau cover.
27. Install rear deck trim cover.
28. Install service plug grip.
29. Connect the negative battery cable.
30. Perform the power window initialization procedure.

2010 Vehicles

See Figure 52.

1. Remove the hybrid battery junction block.
2. Remove the battery smart unit.

CAUTION

Be sure to wear insulated gloves and protective goggles.

a. Disconnect the 3 connectors.

b. Remove the 2 bolts and battery smart unit.

To install:

3. Install the battery smart unit.

CAUTION

Be sure to wear insulated gloves and protective goggles.

a. Install the batter smart unit with the 2 bolts. Tighten to 66 inch lbs. (7.5 Nm).

b. Connect the 3 connectors.

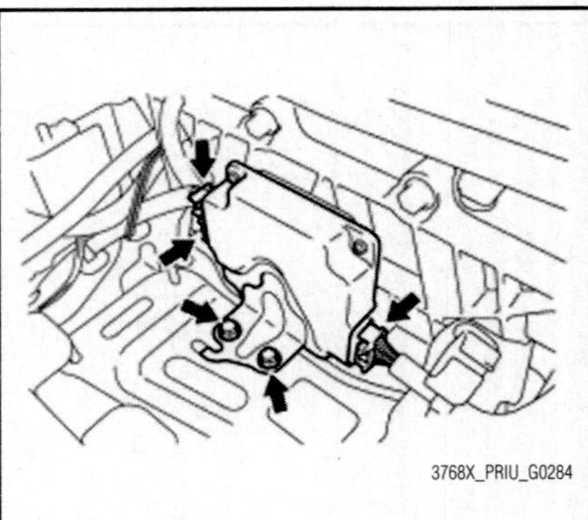

Fig. 52 Removing the battery smart unit

➡The connectors should be connected securely.

4. Install the hybrid battery junction block.

HIGH VOLTAGE BATTERY

REMOVAL & INSTALLATION

2009 Vehicles

See Figures 53 through 60.

1. Disconnect the negative battery cable.
2. Remove service plug grip.
3. Detach the 4 clips, then remove the deck trim cover.
4. Remove tonneau cover.
5. Remove rear seat cushion assembly.
6. Remove rear No. 1 floor board.
7. Remove rear side seat back frames.
8. Remove rear No. 4 floor board.
9. Remove deck floor boxes.
10. Remove deck trim side panels.
11. Remove battery carrier bracket.
12. Remove No. 2 inner quarter vent duct.
13. Remove battery bracket reinforcement.
14. Remove quarter vent duct.
15. Remove No. 6 battery carrier panel
16. Wearing insulating gloves, remove junction terminal.

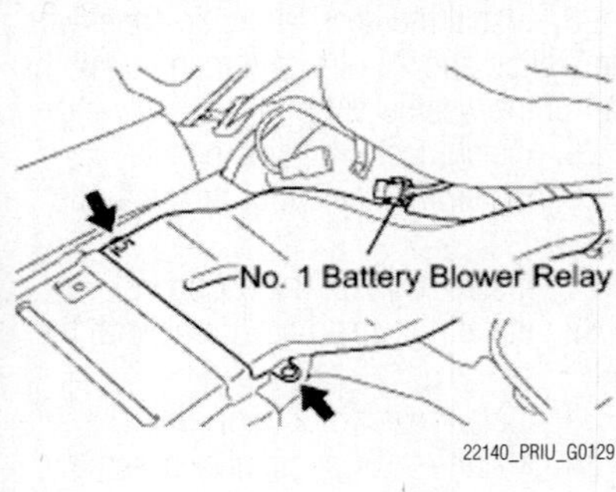

Fig. 53 Remove the 2 clips, then slide the vent duct to the battery side, then remove it

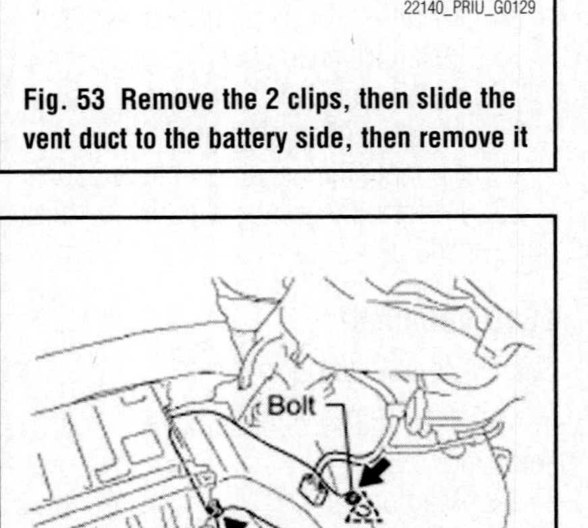

Fig. 54 Remove the bolt, clip and quarter vent duct

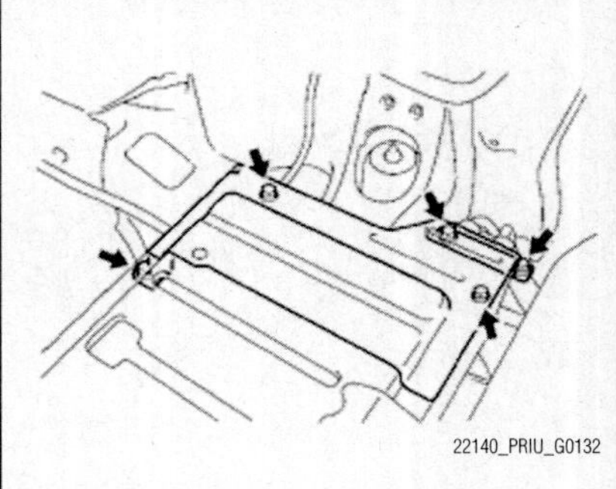

Fig. 55 Remove the 3 bolts, 2 nuts and battery carrier panel

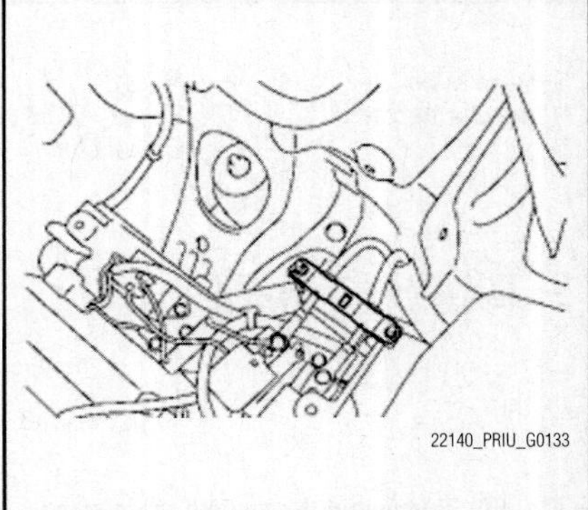

Fig. 56 Remove the junction terminal

17. Wearing insulating gloves, remove frame wire. Insulate the removed terminals with insulating tape.
18. Remove high voltage battery assembly.

a. Remove the earth bolt and 4 bolts shown in the illustration.

b. Disconnect the system main relay connector.

c. Wearing insulating gloves, disconnect the interlock connector.

d. Wearing insulating gloves, remove the clamp, then disconnect the battery ECU connector.

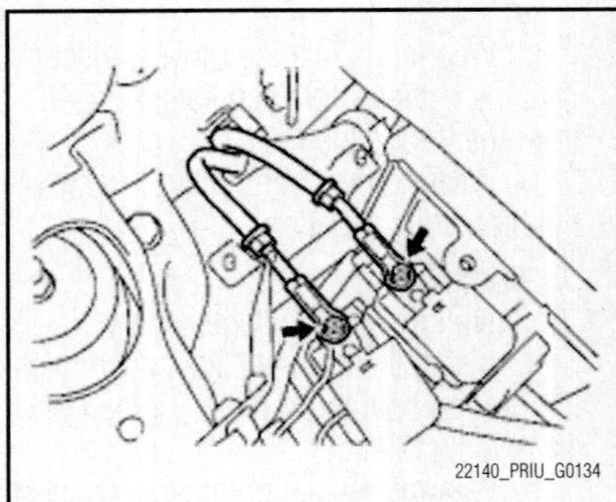

Fig. 57 Remove the wire frame and insulate the terminals No.

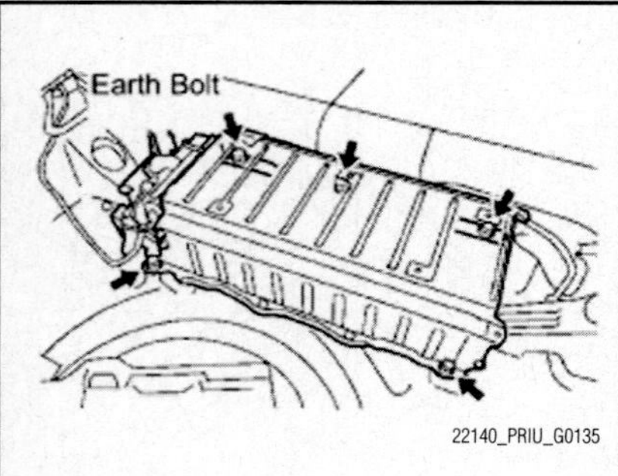

Fig. 58 Remove the ground bolt and 4 bolts shown from the high voltage battery

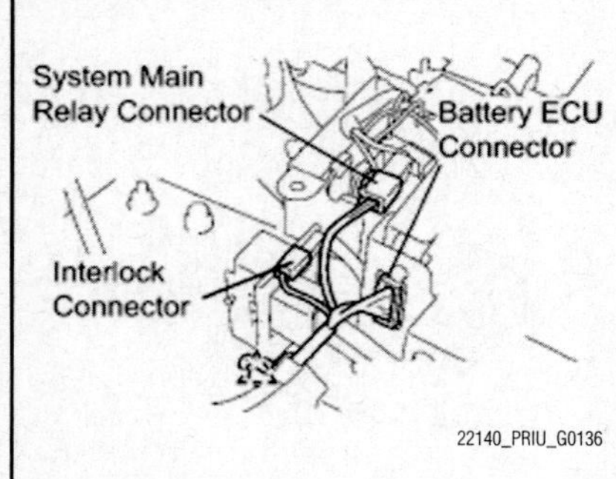

Fig. 59 Disconnect the system main relay connector

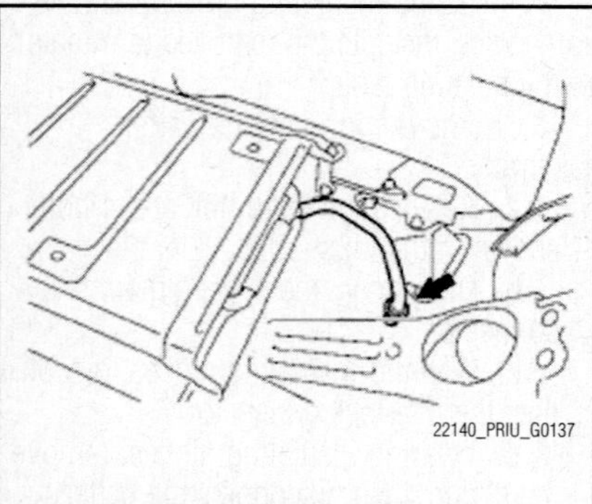

Fig. 60 Disconnect the battery room ventilation hose from the floor panel

e. Wearing insulating gloves, disconnect the battery room ventilation hose from the floor panel.

f. Wearing insulating gloves, remove the high voltage battery.

To install:

19. Install the high voltage battery.

a. Install the high voltage battery with the ground bolt and 4 bolts. Tighten to 14 ft. lbs. (19 Nm).

b. Wearing insulating gloves, connect the system main relay connector.

c. Wearing insulating gloves, connect the interlock connector.

d. Wearing insulating gloves, connect the battery ECU connector.

e. Wearing insulating gloves, connect the battery room ventilation hose to the floor panel.

20. Wearing insulating gloves, install frame wire on the No. 2 and No. 3 system main relay with 2 new nuts. Tighten to 50 inch lbs. (6 Nm). Connect the 2 clamps, then install the frame wire to the upper battery carrier.

21. Wearing insulating gloves, install junction terminal.

22. Install No. 6 battery carrier panel

23. Install the battery carrier panel with the bolt and 2 nuts. Tighten to 66 inch lbs. (8 Nm).

24. Install the quarter vent duct with the bolt and clip. Tighten to 35 inch lbs. (4 Nm).

25. Install the battery bracket reinforcement with the 7 bolts. Tighten to 21 ft. lbs. (28 Nm).

26. Insert the inner quarter vent duct until it touches the backside of the high voltage battery. Slide the fitting surface of the quarter vent duct inner downward, then attach it to the battery blower assembly.

27. Install the battery carrier bracket with the 7 bolts. Tighten to 21 ft. lbs. (28 Nm).

28. Install the deck trim side panel with the 7 clips and 2 bolts, and then install the part of the weather strip.

29. Install deck trim side panel.

30. Install the deck floor box with the clip.

31. Install rear No. 4 floor board.

32. Install the seat back frame with the bolt.

33. Install rear No. 1 floor board.

34. Install rear seat cushion assembly.

35. Install tonneau cover.

36. Install rear deck trim cover.

37. Install service plug grip.

38. Connect the negative battery cable.

39. Perform the power window initialization procedure.

2010 Vehicles

See Figures 61 through 84.

1. Refer to Precautions before servicing.

2. Read output DTC.

➡Check for DTCs and confirm that P0AA6 (Hybrid Battery Voltage System Isolation Fault) is not output before doing removal or installation inside the battery. If this DTC is output, perform troubleshooting first for this DTC.

3. Remove the rear No. 2 floor board (separate type).

4. Remove the rear deck floor box.

5. Remove the rear No. 3 floor board.

6. Disconnect the cable from the negative battery terminal.

➡When disconnecting the cable, some systems need to be initialized after the cable is reconnected.

7. Remove the service plug grip.

8. Remove the inverter terminal cover.

9. Check the terminal voltage.

10. Install the inverter terminal cover.

11. Remove the tonneau cover assembly (w/ tonneau cover).

12. Remove the rear seat cushion assembly.

13. Remove the rear No. 1 floor board sub assembly.

14. Remove the rear No. 2 floor board sub assembly.

15. Remove the rear No. 1 floor board.

16. Remove the rear door scuff plates (LH and RH.

17. Remove the rear side seat back assemblies (LH and RH).

18. Remove the rear No. 4 floor board.

19. Remove the deck floor box LH.

20. Remove the deck trim service hole cover.

21. Remove the rear deck trim cover.

22. Remove the luggage hold belt striker assembly (LH side).

23. Remove the tonneau cover holder cap (LH side).

24. Remove the deck trim side panel assembly LH.

25. Remove the tonneau cover holder cap (RH side).

26. Remove the luggage hold belt striker assembly (RH side).

27. Remove the deck trim side panel assembly RH.

28. Remove the rear floor board spacer.

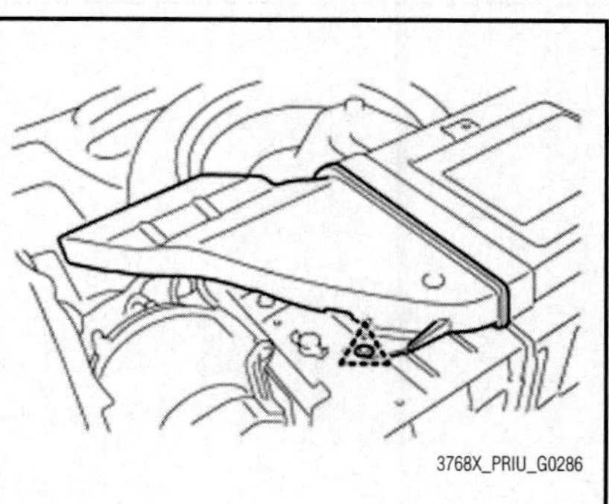

Fig. 61 Removing the No. 1 hybrid battery exhaust duct

a. Remove the 2 clips and rear floor board spacer.

29. Remove the No. 1 hybrid battery exhaust duct.

a. Remove the clip and No. 1 hybrid battery exhaust duct.

30. Remove the upper hybrid battery cover sub assembly.

CAUTION

Be sure to wear insulated gloves and protective goggles.

a. Using the service plug grip, remove the battery cover lock striker.

➡Insert the projection part of the service plug grip, and turn the button of the battery cover lock striker counterclockwise, and release the lock.

b. Remove the 4 nuts and upper hybrid battery cover sub assembly.

31. Remove the No. 1 hybrid battery intake duct.

a. Remove the 2 clips and No. 1 hybrid battery intake duct.

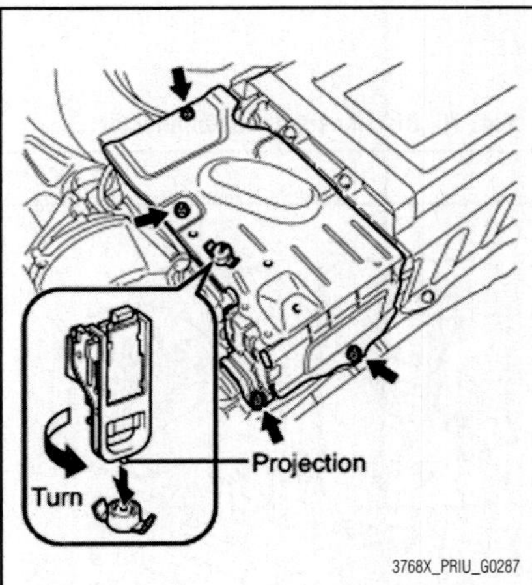

Fig. 62 Removing the upper hybrid battery cover sub assembly

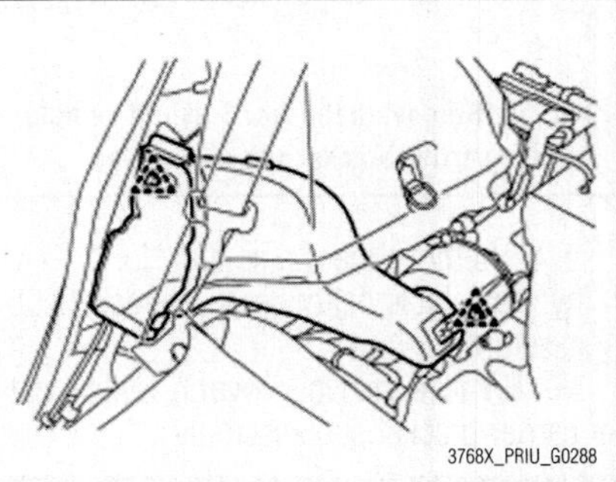

Fig. 63 Removing the No. 1 hybrid battery intake duct

32. Remove the battery cooling blower assembly.

➡Be sure not to touch the fan part of the battery cooling blower assemblies.

➡Do not lift the battery cooling blower assemblies using the wire harness.

33. Remove the No. 7 hybrid vehicle battery upper carrier bracket.

a. Disconnect the wire harness clamp.

b. Remove the bolt and No. 7 hybrid battery upper carrier bracket.

34. Remove the child restraint seat anchor bracket sub assembly LH.

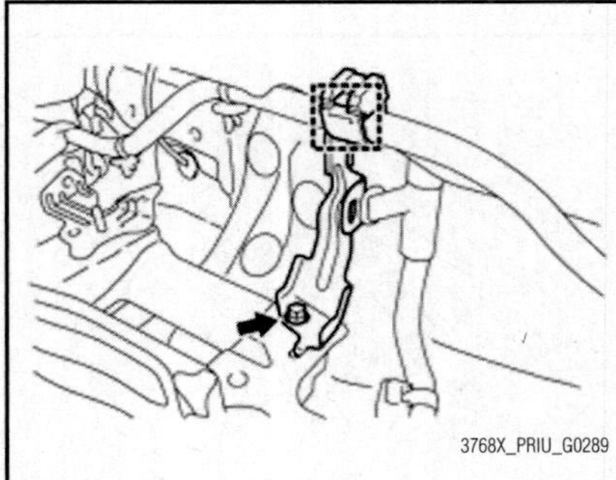

Fig. 64 Removing the No. 7 hybrid vehicle battery upper carrier bracket

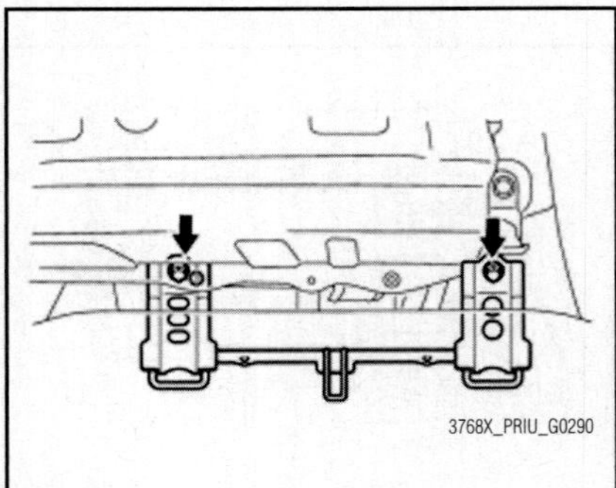

Fig. 65 Removing the child restraint seat anchor bracket sub assembly LH

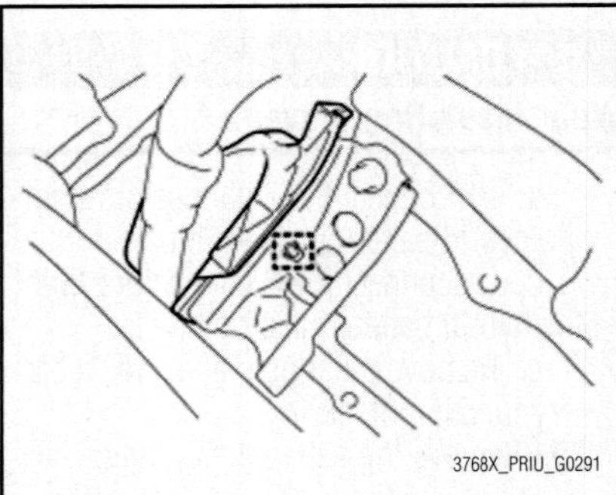

Fig. 66 Disconnecting the wire harness protector clamp

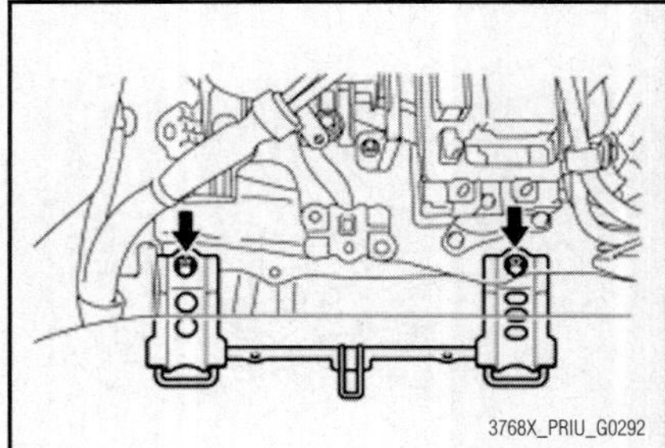

Fig. 67 Removing the 2 bolts and child restraint seat anchor bracket sub assembly RH

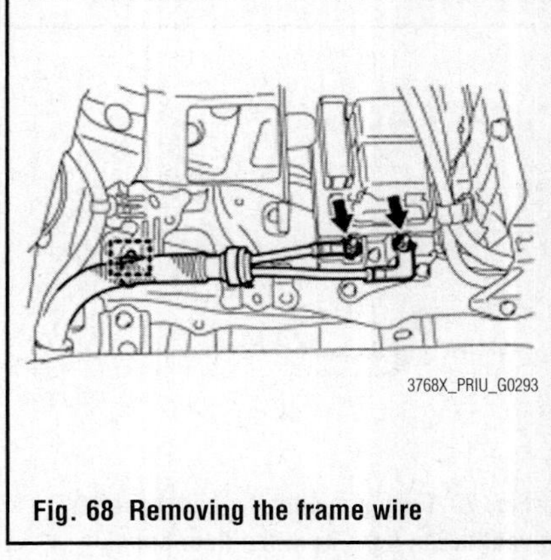

Fig. 68 Removing the frame wire

35. Remove the child restraint seat anchor bracket sub assembly RH.

a. Disconnect the wire harness protector clamp.

b. Remove the 2 bolts and child restraint seat anchor bracket sub-assembly RH.

36. Remove the frame wire.

CAUTION

Wear insulating gloves.

CAUTION

Insulate the removed terminals with insulating tape.

a. Remove the 2 nuts, then disconnect the frame wire from the hybrid battery junction block assembly.

b. Disconnect the clamp and frame wire.

37. Remove the HV battery assembly.

CAUTION

Wear insulating gloves.

➡Since the HV battery is very heavy, 2 people are needed to remove the HV battery. When removing the HV battery, do not damage the parts around it.

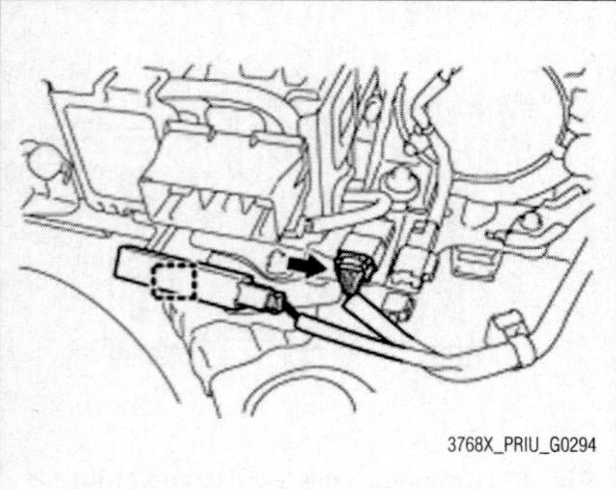

Fig. 69 Disconnecting the connector and electrical key oscillator clamp

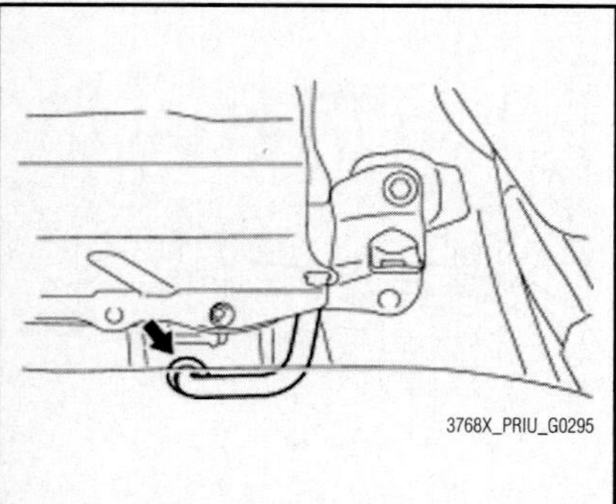

Fig. 70 Disconnecting the battery room ventilation hose from the floor panel

a. Disconnect the connector and electrical key oscillator clamp.

b. Disconnect the battery room ventilation hose from the floor panel.

c. Remove the 4 bolts shown in the illustration.

38. Remove the hybrid battery junction block.

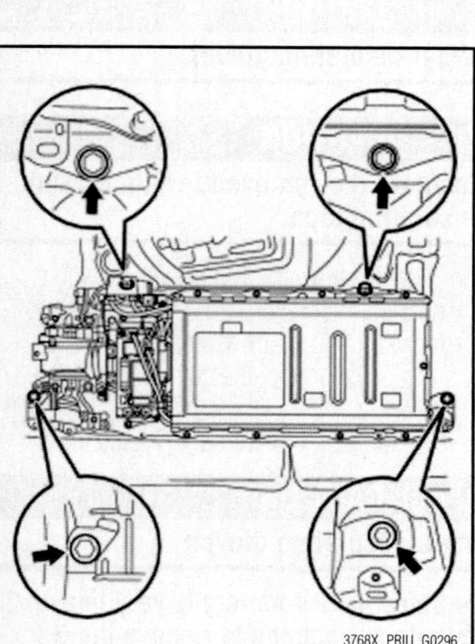

Fig. 71 Removing the 4 bolts

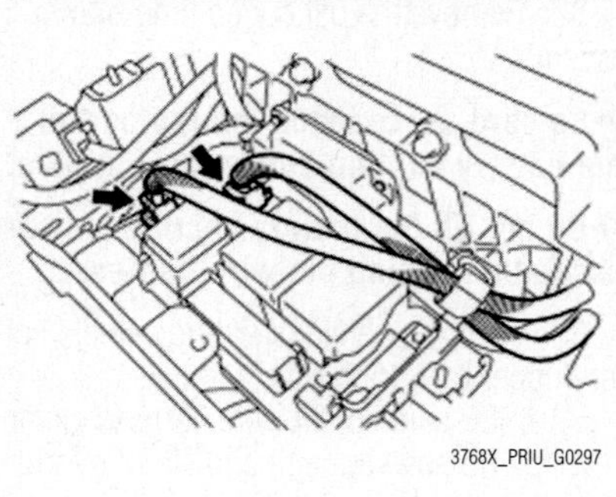

Fig. 72 Disconnecting the 2 connectors from the hybrid battery junction block

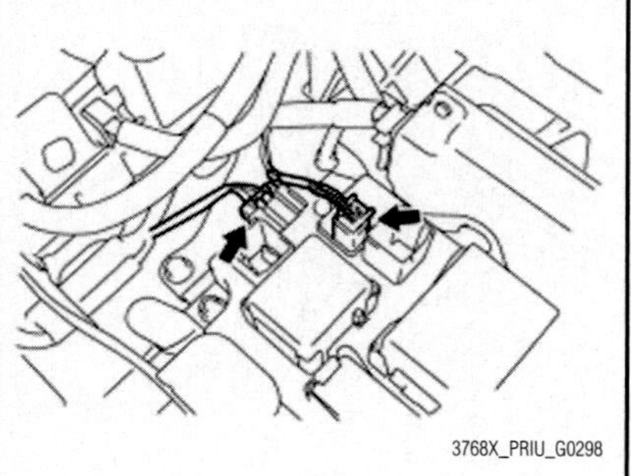

Fig. 73 Disconnecting 2 more connectors from the hybrid battery junction block

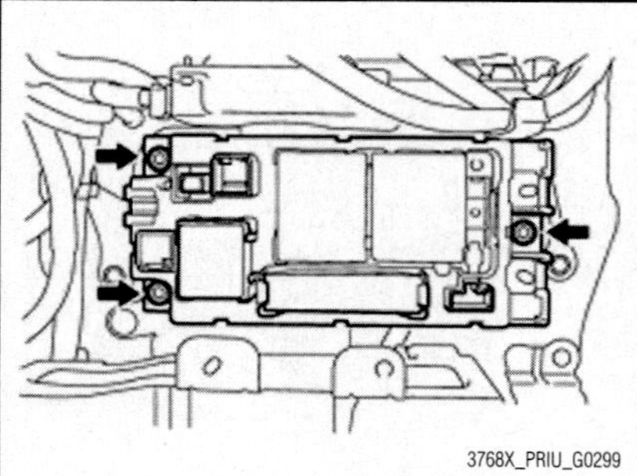

Fig. 74 Removing the 3 nuts and hybrid battery junction block

CAUTION

Wear insulating gloves.

a. Disconnect the 2 connectors from the hybrid battery junction block.

b. Disconnect the 2 connectors from the hybrid battery junction block.

c. Remove the 3 nuts and hybrid battery junction block.

39. Remove the battery smart unit.

CAUTION

Wear insulating gloves.

40. Remove the No. 1 hybrid vehicle battery carrier bracket assembly.

CAUTION

Wear insulating gloves.

a. Disconnect the connector.

b. Remove the bolt and EV battery plug.

c. Using the service plug grip, remove the battery cover lock striker.

➡Insert the projection part of the service plug grip, and turn the button of the battery cover lock striker counterclockwise, and release the lock.

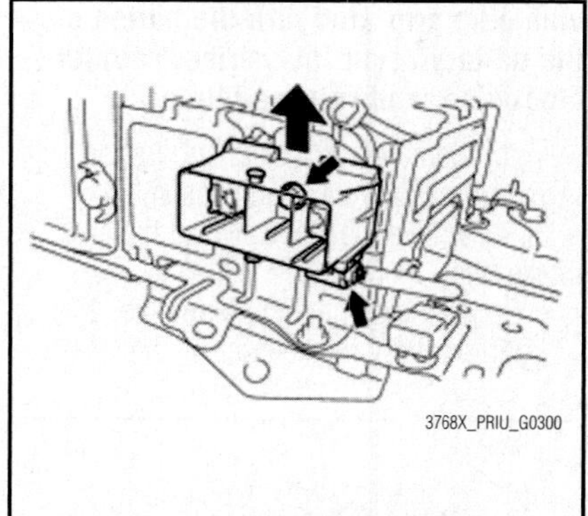

Fig. 75 Disconnecting the connector

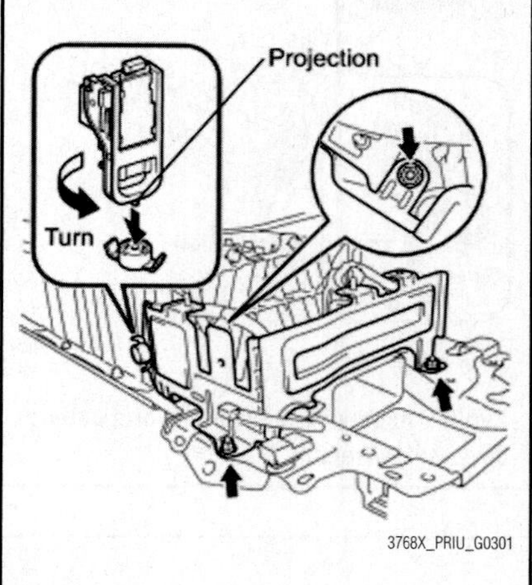

Fig. 76 Removing the No. 1 hybrid vehicle battery carrier bracket sub assembly

d. Remove the 3 nuts and No. 1 hybrid vehicle battery carrier bracket sub assembly.

41. Remove the No. 4 hybrid vehicle batter carrier bracket sub assembly.

CAUTION

Wear insulating gloves.

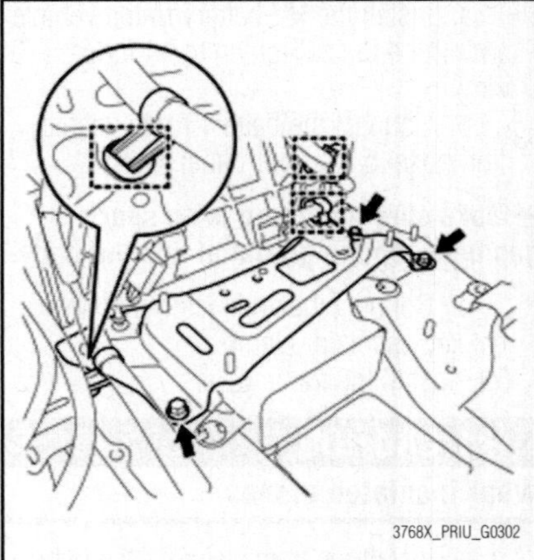
Fig. 77 Disconnecting the 3 wire harness clamps

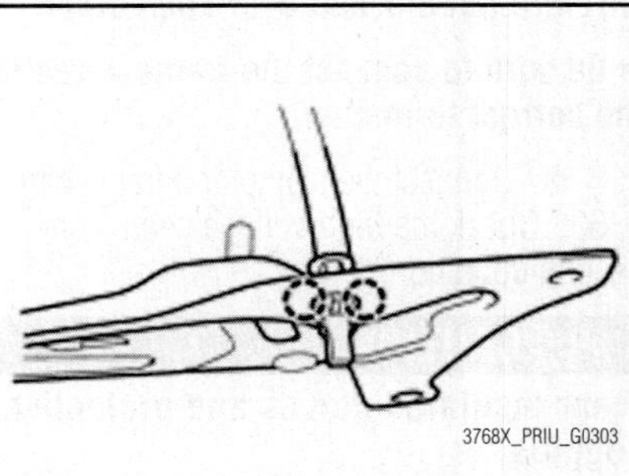
Fig. 78 Removing the HV battery thermistor

a. Disconnect the 3 wire harness clamps.

b. Remove the 3 bolts and No. 4 hybrid vehicle battery carrier bracket sub assembly.

c. Disengage the 2 claws and remove the HV battery thermistor.

42. Remove the No. 1 hybrid battery cover intake duct.

a. Remove the 2 clips and No. 1 hybrid battery cover intake duct.

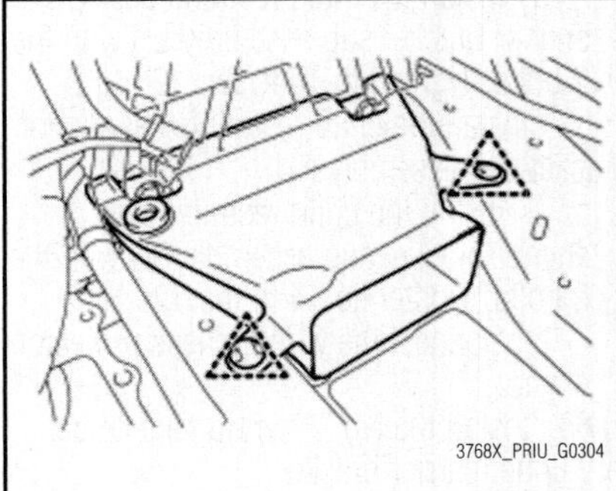
Fig. 79 Removing the No. 1 hybrid battery cover intake duct

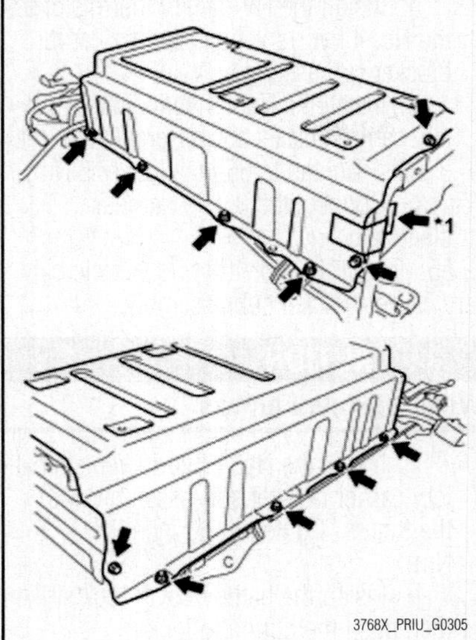

Fig. 80 Removing the upper hybrid battery cover sub assembly

43. Remove the upper hybrid battery cover sub assembly.

CAUTION

Be sure to wear insulated gloves and protective goggles.

a. Remove the tape (*1).

b. Remove the 5 bolts, 6 nuts and battery cover with the No. 1 hybrid battery shield sub assembly.

44. Remove the No. 1 hybrid battery packing.

CAUTION

Be sure to wear insulated gloves and protective goggles.

a. Remove the 2 clamps and No. 1 hybrid battery packing.

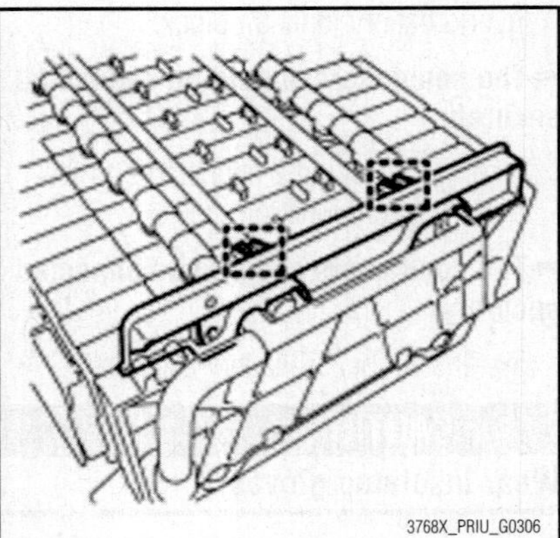
Fig. 81 Removing the No. 1 hybrid battery packing

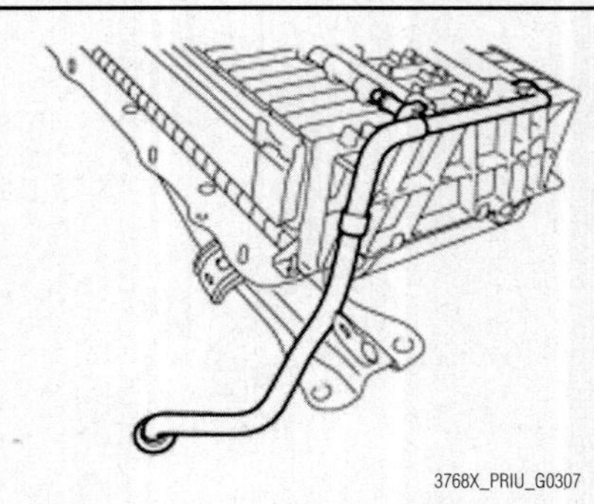
Fig. 82 Removing the hybrid battery hose assembly

45. Remove the hybrid battery hose assembly.

CAUTION

Be sure to wear insulated gloves and protective goggles.

a. Remove the hybrid battery hose assembly from the HV battery.

46. Remove the No. 1 hybrid vehicle battery hose.

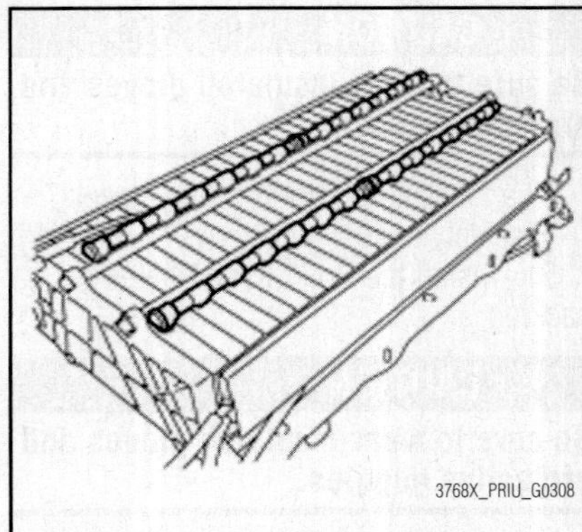
Fig. 83 Removing the 2 No. 1 hybrid vehicle battery hoses

CAUTION

Be sure to wear insulated gloves and protective goggles.

a. Remove the 2 No. 1 hybrid vehicle battery hoses from the HV battery.

47. Remove the No. 2 hybrid battery pack wire.

a. Disconnect the 2 clamps, then remove the No. 2 hybrid battery pack wire.

To install:

48. Install the No. 2 hybrid battery pack wire.

a. Connect the 2 clamps and No. 2 hybrid battery pack wire.

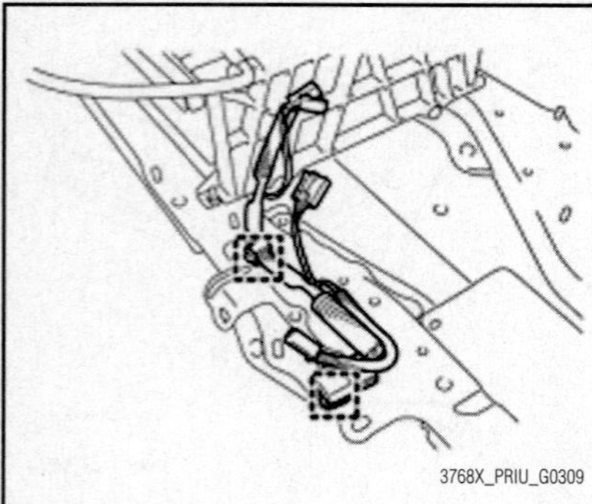

Fig. 84 Removing the No. 2 hybrid battery pack wire

49. Install the No. 1hybrid vehicle battery hose.

⁂ CAUTION

Be sure to wear insulated gloves and protective goggles.

a. Install the 2 No. 1 hybrid vehicle battery hoses to the HV battery.

50. Install the hybrid battery hose assembly.

⁂ CAUTION

Be sure to wear insulated gloves and protective goggles.

a. Install the hybrid battery hose assembly to the HV battery.

51. Install the No. 1 hybrid battery packing.

⁂ CAUTION

Be sure to wear insulated gloves and protective goggles.

a. Install the No. 1 hybrid battery packing with the 2 clamps.

52. Install the upper hybrid battery cover sub assembly.

⁂ CAUTION

Be sure to wear insulated gloves and protective goggles.

a. Install the battery cover and No. 1 hybrid battery shield sub assembly with the 5 bolts and 6 nuts. Tighten to 66 inch lbs. (7.5 Nm).

b. install the tape.

53. Install the No. 1hybrid battery cover intake duct with the 2 clips.

54. Install the No. 4 hybrid vehicle battery carrier bracket sub assembly.

⁂ CAUTION

Wear insulated gloves.

a. Install the HV battery thermistor to the No. 4 hybrid vehicle battery carrier bracket sub assembly.

b. Install the No. 4 hybrid vehicle battery carrier bracket sub assembly with the 3 bolts. Tighten to 66 inch lbs. (7.5 Nm).

c. Connect the 3 wire harness clamps.

55. Install the No. 1 hybrid vehicle battery carrier bracket sub assembly.

⁂ CAUTION

Wear insulated gloves.

a. Install the No. 1 hybrid vehicle battery carrier bracket sub-assembly with the 3 nuts. Tighten to 66 inch lbs. (7.5 Nm).

b. Install the battery cover lock striker, then push the button to lock it.

c. Install the electric vehicle battery plug assembly with the bolt. Tighten to 66 inch lbs. (7.5 Nm).

d. Connect the connector.

➡The connector should be connected securely.

56. Install the battery smart unit.

⁂ CAUTION

Wear insulated gloves.

a. Install the battery smart unit with the 2 bolts. Tighten to 66 inch lbs. (7.5 Nm).

b. Connect the 3 connectors.

➡The connectors should be connected securely.

57. Install the hybrid battery junction block.

a. Install the hybrid battery junction block with the 3 nuts. Tighten to 66 inch lbs. (7.5 Nm).

b. Connect the 2 connectors to the hybrid battery junction block.

➡The connectors should be connected securely.

c. Connect the 2 connectors to the hybrid battery junction block.

➡The connectors should be connected securely.

58. Install the HV battery assembly.

⁂ CAUTION

Wear insulated gloves.

➡Since the HV battery is very heavy, 2 people are needed to install the HV battery. When installing the HV battery, do not damage the parts around it.

a. Install the HV battery to the vehicle with the 4 bolts. Tighten to 14 ft. lbs. (19 Nm).

b. Connect the battery room ventilation hose to the floor panel.

➡Make sure that there is no space or gap between the grommet and the body.

c. Connect the connector and electrical key oscillator clamp.

59. Install the frame wire.

⁂ CAUTION

Wear insulated gloves.

a. Install the frame wire on the hybrid battery junction block assembly with the 2 nuts. Tighten to 80 inch lbs. (9 Nm).

➡Make sure that the ends of the frame wire are not crossed over each other.

➡Be sure to connect the frame wires to the correct terminals.

b. Connect the clamp and frame wire.

60. Check the high voltage cable connection condition.

⁂ CAUTION

Wear insulated gloves and protective goggles.

a. Check that each wire harness is being installed securely.

➡Make sure that the end of the frame wire are not crossover each other.

➡Be sure to connect the frame wire to the correct terminals.

➡The connectors should be connected securely.

➡The nuts should be fastened securely.

➡Make sure that the 4 plastic covers are engaged securely.

61. Install the child restraint seat anchor bracket sub assembly LH.

a. Install the child restraint seat anchor bracket sub assembly LH with the 2 bolts. Tighten to 14 ft. lbs. (20 Nm).

62. Install the child restraint seat anchor bracket sub assembly RH.

a. Install the child restraint seat anchor bracket sub assembly RH with the 2 bolts. Tighten to 14 ft. lbs. (20 Nm).

b. Connect the wire harness protector clamp.

63. Install the No. 7 hybrid vehicle battery upper carrier bracket.

a. Install the No. 7 hybrid battery upper carrier bracket with the bolt. Tighten to 66 inch lbs. (7.5 Nm).

64. Install the battery cooling blower assembly.

➡Be sure not to touch the fan part of the battery cooling blower assemblies.

➡Do not lift the battery cooling blower assemblies using the wire harness.

65. Install the No. 1 hybrid battery intake duct.

➡Ensure that the duct is installed securely.

66. Install the upper hybrid battery cover sub assembly with the 4 nuts. Tighten to 66 inch lbs. (7.5 Nm).
   a. Install the battery cover lock striker, then push the button to lock.
67. Install the No. 1 hybrid battery exhaust duct with the clip.

➡Ensure that the duct is installed securely.

68. Install the rear floor board spacer with the 2 clips.
69. Install the deck trim side panel assembly LH.
70. Install the tonneau cover holder cap (LH side).
71. Install the luggage hold belt striker assembly (LH side).
72. Install the deck trim side panel assembly RH.
73. Install the tonneau cover holder cap (RH side).
74. Install the luggage hold belt striker assembly (RH side).
75. Install the rear deck trim cover.
76. Install the deck trim service hole cover.
77. Install the deck floor box LH.
78. Install the rear No. 4 floor board.
79. Install the rear side seat back assembly LH.
80. Install the rear side seat back assembly RH.
81. Install the rear door scuff plate LH.
82. Install the rear door scuff plate RH.
83. Install the rear No. 1 floor board.
84. Install the rear No. 2 floor board sub assembly.
85. Install the rear No. 1 floor board sub assembly.
86. Install the rear seat cushion assembly.
87. Install the tonneau cover assembly (w/ tonneau cover).
88. Install the service plug grip.
89. Connect the cable to the negative battery terminal.

➡When disconnecting the cable, some systems need to be initialized after the cable is reconnected.

90. Install the rear No. 3 floor board.
91. Install the rear deck floor box.
92. Install the rear No. 2 floor board (separate type).

HYBRID VEHICLE CONTROL UNIT

REMOVAL & INSTALLATION

See Figure 85.

1. Disconnect the negative battery cable.

➡Wait at least 90 seconds after disconnecting the cable from the negative (-) battery terminal to prevent airbag and seat belt pretensioner activation.

2. Remove the No. 1 instrument panel register.
3. Remove the lower instrument panel finish panel.
4. Remove the upper instrument panel finish panel.
5. Remove the No. 3 instrument panel register.
6. Remove the No. 4 instrument panel register.
7. Remove the No. 2 instrument panel register.
8. Remove the multi-display assembly.
9. Remove the glove compartment door stopper.
10. Remove the glove compartment door assembly.
11. Remove the glove compartment door.
12. Remove the No. 1 instrument panel speaker panel.
13. Remove the front pillar garnish corner piece.
14. Remove the front pillar garnish.
15. Disconnect airbag connector.
16. Remove the instrument panel.

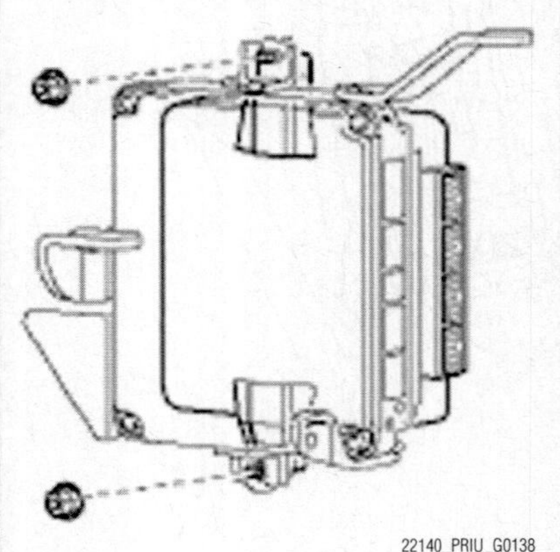

Fig. 85 Hybrid vehicle control unit

17. Remove the No. 3 heater to register duct.
18. Remove the ECM.
19. Remove the hybrid vehicle control ECU.

To install:

20. Install the hybrid vehicle control ECU with the 2 nuts. Tighten to 49 inch lbs. (6 Nm).
21. Install ECM.
22. Install No. 3 heater to register duct.
23. Install instrument panel.
24. Connect airbag connector.
25. Install front pillar garnish.
26. Install front pillar garnish corner piece.
27. Install No. 1 instrument panel speaker panel.
28. Install glove compartment door.
29. Install glove compartment door assembly.
30. Install glove compartment door stopper.
31. Install multi-display assembly.
32. Install No. 2 instrument panel register.
33. Install No. 4 instrument panel register.
34. Install No. 3 instrument panel register.
35. Install upper instrument panel finish panel.
36. Install lower instrument panel finish panel.
37. Install No. 1 instrument panel register.
38. Connect the negative battery cable.
39. Perform the power window initialization procedure.
40. Check SRS warning light.

INVERTER WITH CONVERTER

REMOVAL & INSTALLATION

2009 Vehicles

See Figures 86 and 87.

1. While wearing insulating gloves, remove the inverter with converter assembly.
   a. Remove the left and right engine under covers.
   b. Drain the high voltage coolant.
   c. Remove the radiator support opening cover.
   d. Remove the inverter cover.
   e. Verify that there is 0volt (s) at the inverter with converter, using a voltmeter utilizing a measuring range of DC 400volt (s) or more.
   f. Again, verify that there is 0volt (s) by measuring the voltage between the terminals of the three phase connector (U-V, V-W, U-W), using a voltmeter

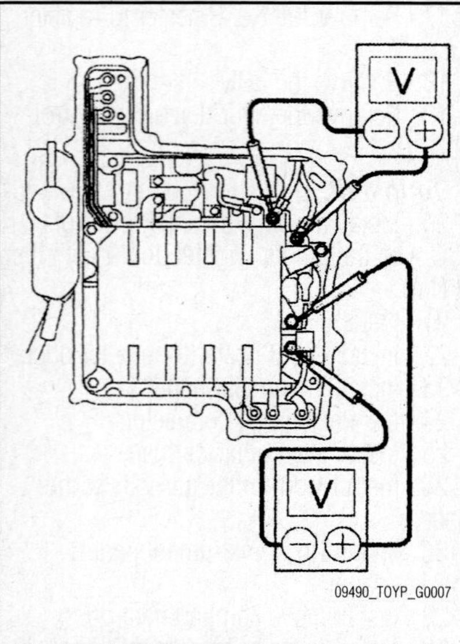

Fig. 86 Verifying that there is 0 volt (s) at the inverter with converter

utilizing a measuring range of DC 400volt (s) or more.

g. Disconnect the No. 1, No. 2 and No. 6 inverter cooling hoses.

h. Disconnect the No. 1 circuit breaker sensor by first moving the outer section away toward the wire side.

i. Disconnect the 2 frame wire connectors from the inverter with converter assembly and protect the electrode and connector parts with insulating tape.

j. Use a small screwdriver to lift up the green lock pin, then disconnect the connector for the air conditioning inverter.

k. Disconnect any remaining wiring connectors including the engine main wiring harness.

l. Remove the mounting bolts and disconnect the MG1 and MG2 power cables. Protect the connector parts with insulating tape.

m. Remove the 3 mounting bolts and the inverter with converter assembly.

To install:

2. While wearing insulating gloves, install the inverter with converter assembly.

a. Install the inverter with converter assembly and tighten the three mounting bolts to 16 ft. lbs. (21 Nm).

b. Connect the MG1 and MG2 power cables and tighten the mounting bolts to 71 inch lbs. (8 Nm).

c. Connect any remaining wiring connectors including the engine main wiring harness. Be sure to insert the grommet of the engine main wiring harness into the U-shaped groove of the inverter case.

d. Engage the connector for the air conditioning inverter and secure by pushing in the lock pin.

e. Connect the 2 frame wire connectors to the inverter with converter assembly.

f. Connect the No. 1 circuit breaker sensor.

g. Connect the No. 6, No. 2 and No. 1 inverter cooling hoses.

h. Install the inverter cover and tighten the mounting fasteners to 8 ft. lbs. (11 Nm).

i. Install the radiator support opening cover.

j. Install the left and right engine under covers.

2010 Vehicles

See Figures 88 through 108.

1. Refer to Precautions before servicing the vehicle.
2. Remove the No. 2 floor board (separate type).
3. Remove the rear deck floor box.
4. Remove the rear No. 3 floor board.
5. Disconnect the cable from the negative battery terminal.

➡When disconnecting the cable, some systems need to be initialized after the cable is reconnected.

6. Remove the service plug grip.
7. Remove the front spoiler cover (w/ front spoiler).
8. Remove the engine under cover (w/ cover).
9. Remove the No. 1 engine under cover.
10. Drain the coolant.
11. Remove the radiator support opening cover.
12. Remove the No. 1 inverter bracket.
13. Remove the 3 bolts and No. 1 inverter bracket.
14. Disconnect the engine room main wire.

a. Raise the lock lever and disconnect the inverter with the converter connector.

b. Disconnect the engine wire from the engine room main wire.

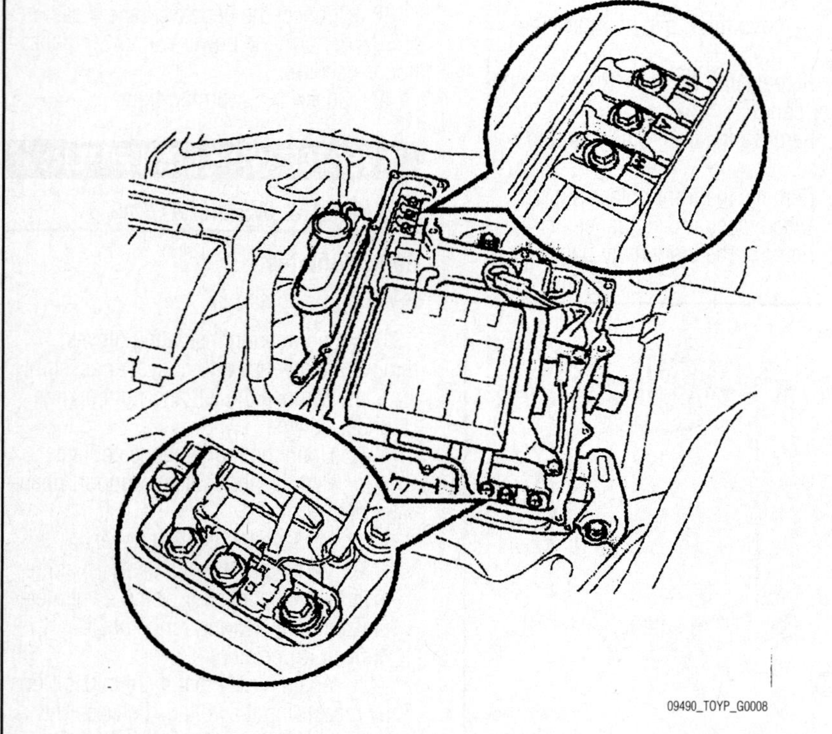

Fig. 87 Measuring voltage between the terminals of the three phase connector

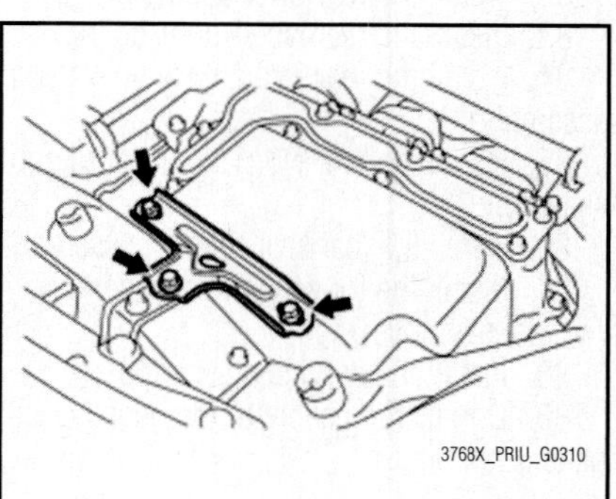

Fig. 88 Removing the 3 bolts and No. 1 inverter bracket

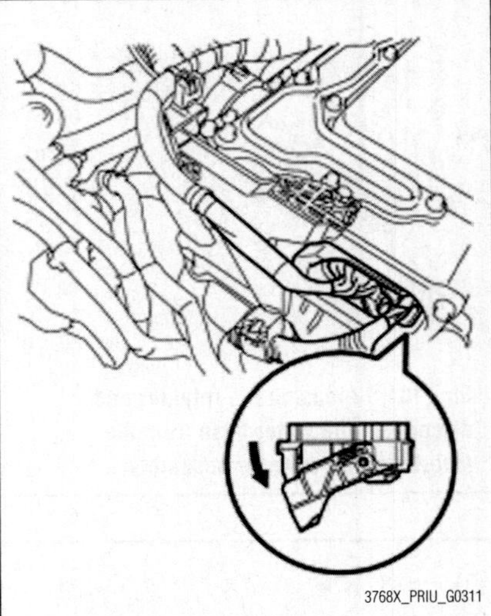
Fig. 89 Disconnecting the inverter with the converter connector

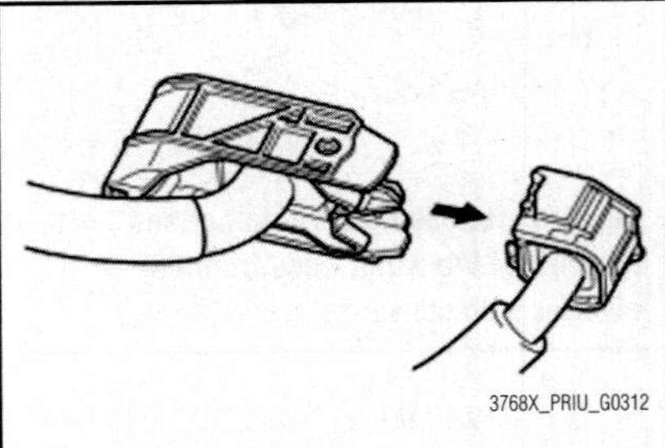
Fig. 90 Disconnecting the engine wire from the engine room main wire

c. Remove the bolt.

d. Remove the bolt, clamp and clip, and disconnect the engine room main wire.

15. Remove the inverter terminal cover.

CAUTION

Wear insulating gloves.

a. Remove the 9 bolts and inverter terminal cover.

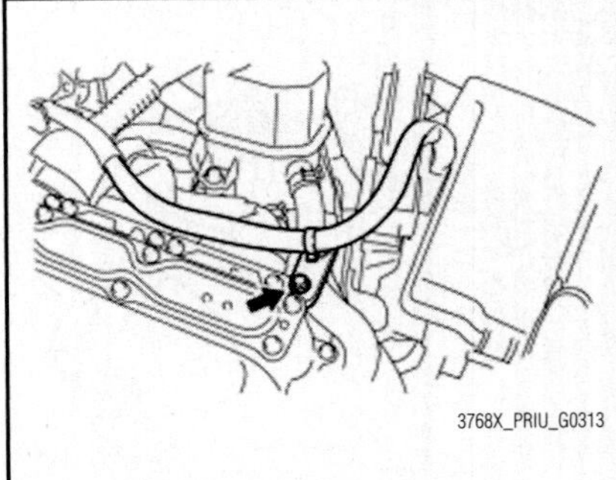
Fig. 91 Removing the bolt

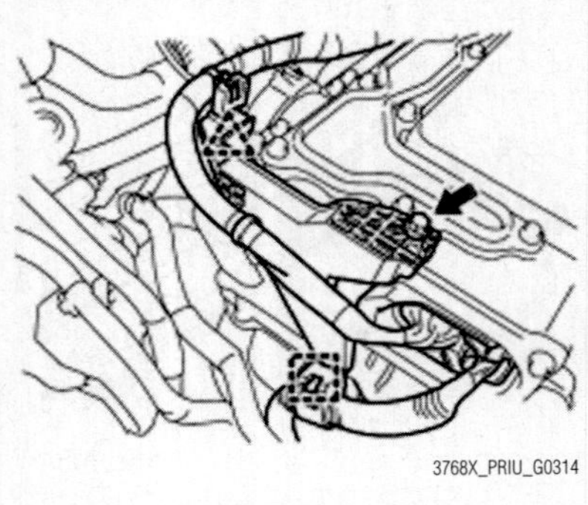
Fig. 92 Disconnecting the engine room main wire

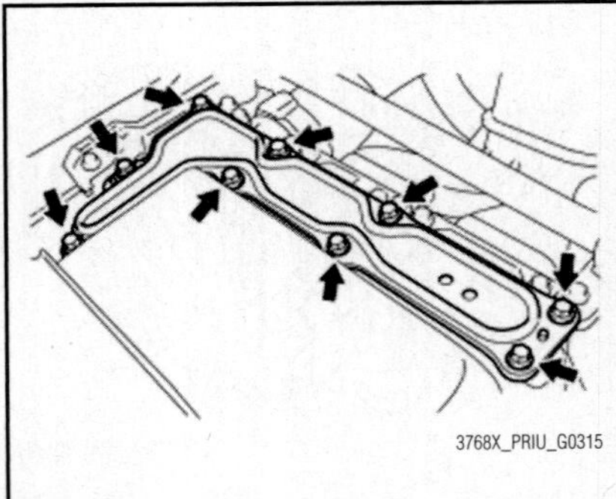
Fig. 93 Removing the inverter terminal cover

➡Make sure to pull the inverter terminal cover straight up, as a connector is connected to the bottom of the cover.

16. Check the terminal voltage.

CAUTION

Wear insulating gloves.

➡Do not allow any foreign objects or water to enter the inverter with the converter assembly.

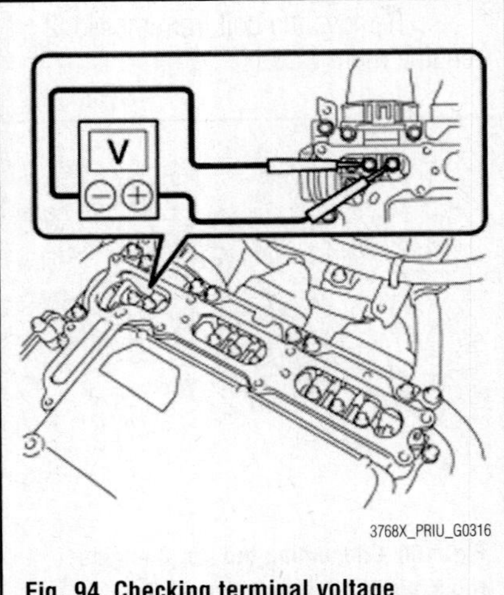

Fig. 94 Checking terminal voltage

a. Using a voltmeter, measure the voltage between the terminals of the 2 phase connectors. The standard voltage is 0 V.

➡Use measuring range of DC 750 V or more on the voltmeter.

17. Disconnect the frame wire.

CAUTION

Wear insulating gloves.

➡Insulate the removed terminals with insulating tape.

➡Cover the hole where the cable was connected with tape or equivalent (non-residue type) to prevent entry of foreign matter.

18. Disconnect the high voltage cable from the front transaxle.

CAUTION

Wear insulating gloves.

➡Insulate the removed terminals with insulating tape.

➡Cover the hole where the cable was connected with tape or equivalent (non-residue type) to prevent entry of foreign matter.

a. Remove the 5 bolts, and disconnect the high voltage cables of the generator (MG1) from the inverter with the converter assembly.

b. Turn back the wire harness cover and release the cable.

c. Remove the 5 bolts, and disconnect the high voltage cables of the motor

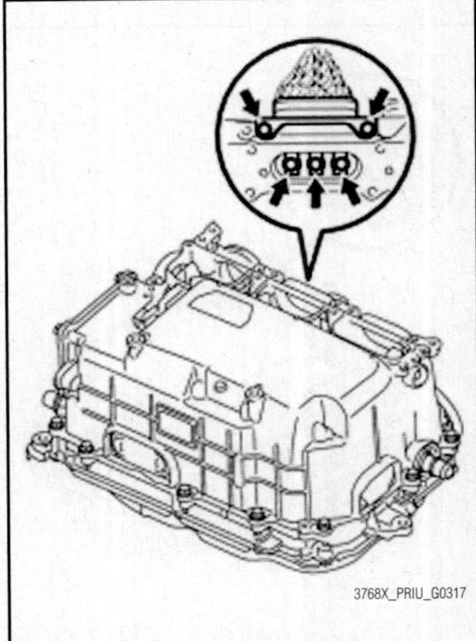
Fig. 95 Disconnecting the high voltage cables from the generator (MG1)

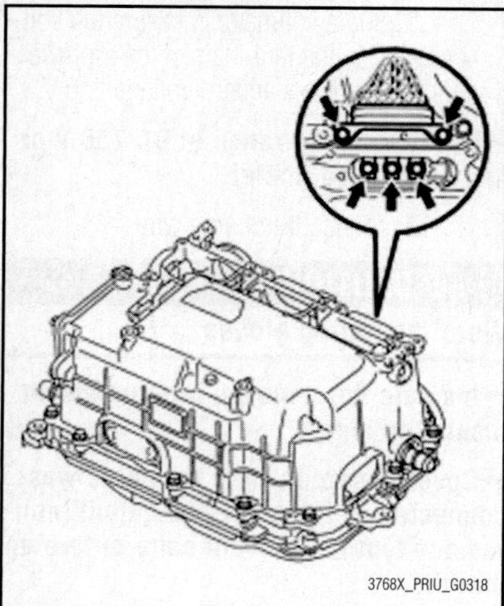

Fig. 96 Disconnecting the high voltage cables of the motor (MG2)

(MG2) from the inverter with converter assembly.

d. Disconnect the harness clamp.

19. Disconnect the No. 2 engine wire.

CAUTION

Wear insulating gloves.

➡Insulate the removed terminals with insulating tape.

➡Cover the hole where the cable was connected with tape or equivalent (non-residue type) to prevent entry of foreign matter.

a. Remove the 4 bolts, and disconnect the No. 2 engine wire (high voltage cables for the air conditioning compressor) from the inverter with converter assembly.

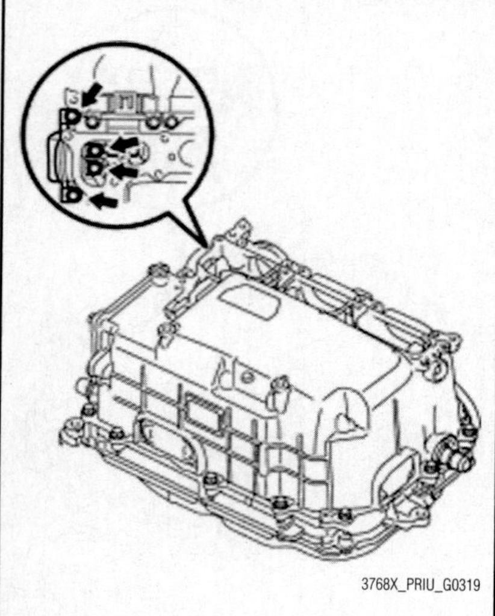

Fig. 97 Disconnecting the No. 2 engine wire

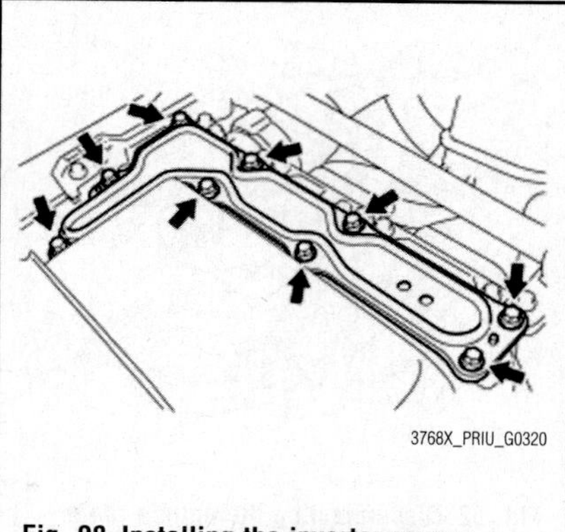

Fig. 98 Installing the inverter cover

Fig. 99 Removing the No. 1 relay block cover

b. Disconnect the harness clamp.

20. Install the inverter terminal cover.

a. Temporarily install the inverter terminal cover with the 9 bolts to prevent any foreign objects or water from entering the inverter with converter assembly.

21. Disconnect the No. 2 engine room wire.

a. Remove the relay block cover.

b. Release the 2 clamps, and remove the No. 1 relay block cover.

c. Remove the bolt from the No. 2 engine room wire.

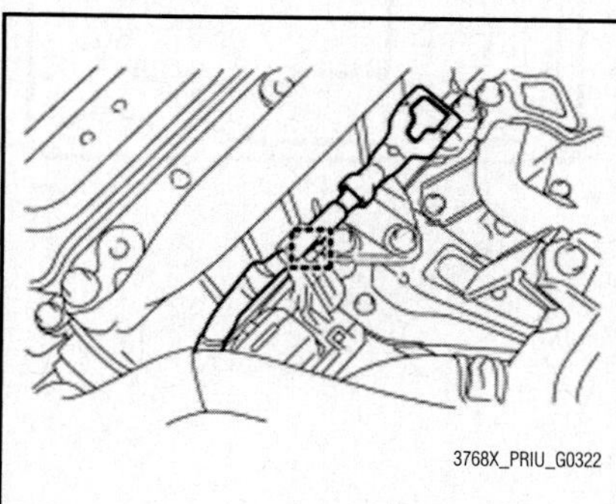

Fig. 100 Connecting the No. 2 engine room wire to the protector

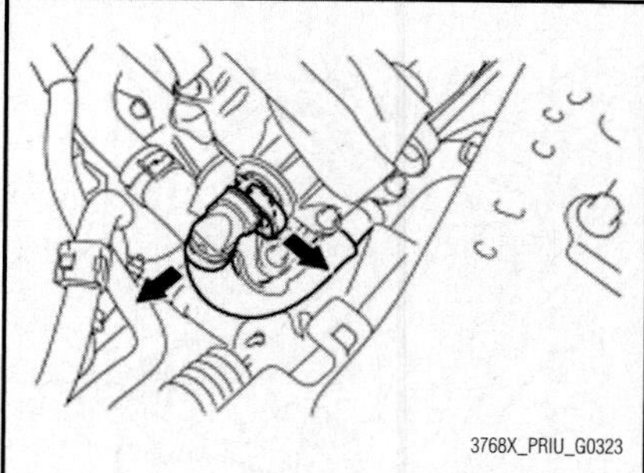

Fig. 101 Releasing the retainer and disconnect the water hose from the inverter with converter assembly

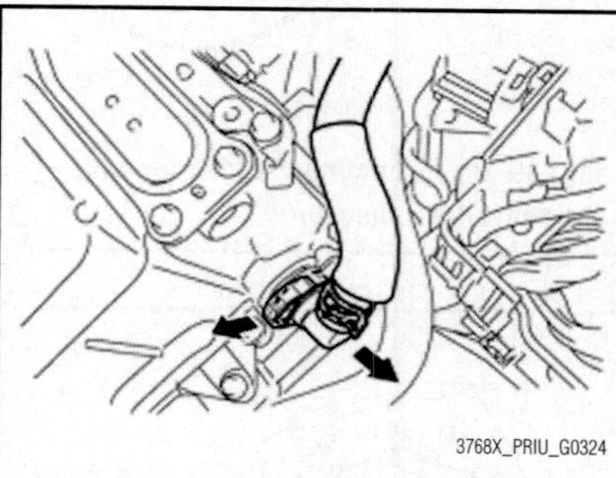

Fig. 102 Releasing the retainer and disconnect the water hose from the inverter with converter assembly

d. Release the 2 claws, and disconnect the No. 2 engine room wire.

e. Connect the No. 2 engine room wire to the protector.

22. Disconnect the water hose.

a. Release the retainer and disconnect the water hose from the inverter with converter assembly.

b. Release the retainer and disconnect the water hose from the inverter with converter assembly.

c. Disconnect the coolant hose from

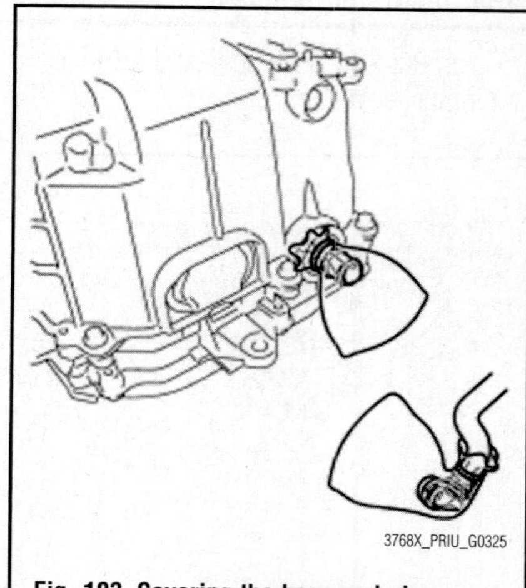

Fig. 103 Covering the hose and pipe

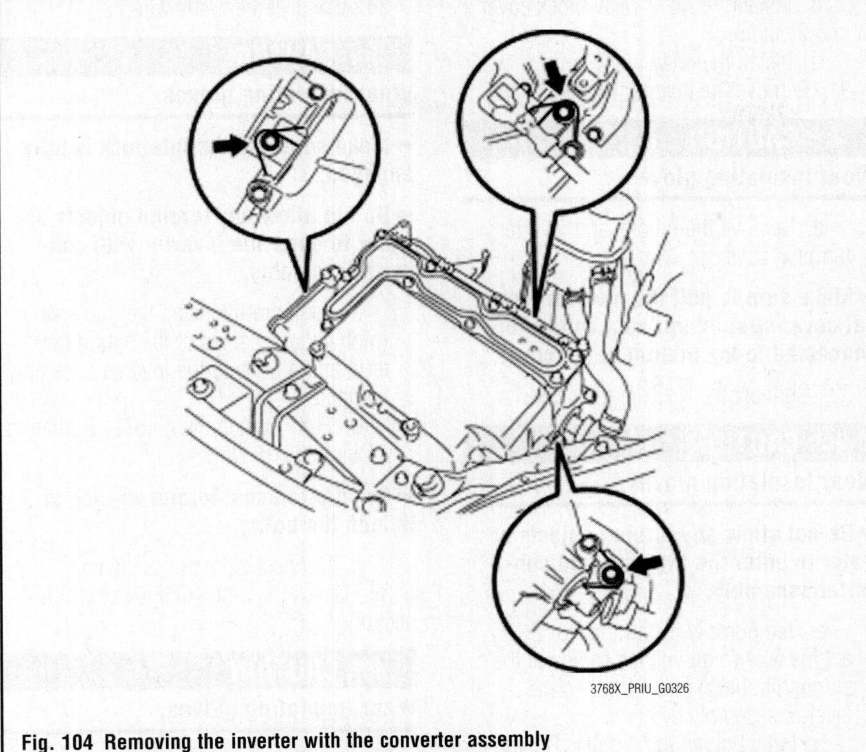
Fig. 104 Removing the inverter with the converter assembly

the inverter with converter assembly. Put a piece of cloth in the pipe and in the disconnected hose or cover the pipe and hose with plastic bags as shown in the illustration, so that foreign matter doesn't stick to the union or the inside of the connector and to prevent coolant from spilling near the inverter with converter assembly.

23. Remove the inverter with the converter assembly.

CAUTION

Wear insulating gloves.

a. Remove the 3 bolts and inverter with converter assembly.

➡Since the inverter with converter assembly is very heavy, 2 people are needed to remove the inverter with converter assembly. When removing the inverter with converter assembly, do not damage the parts around it.

➡To prevent damage, do not hold the inverter with converter assembly by the connectors.

➡To prevent damage due to static electricity, do not touch the terminals of the disconnected connectors.

24. Remove the motor cable bracket.

a. Remove the 2 bolts and motor cable bracket.

25. Remove the high voltage fuse.

CAUTION

Wear insulating gloves.

➡Perform this procedure only when replacement of the high voltage fuse is necessary.

a. Remove the 9 bolts and inverter terminal cover.

➡Make sure to pull the inverter terminal cover straight up, as a connector is connected to the bottom of the cover.

b. Remove the 2 bolts and high voltage fuse from the inverter with the converter assembly.

➡Do not allow any foreign objects or water to enter the inverter with converter assembly.

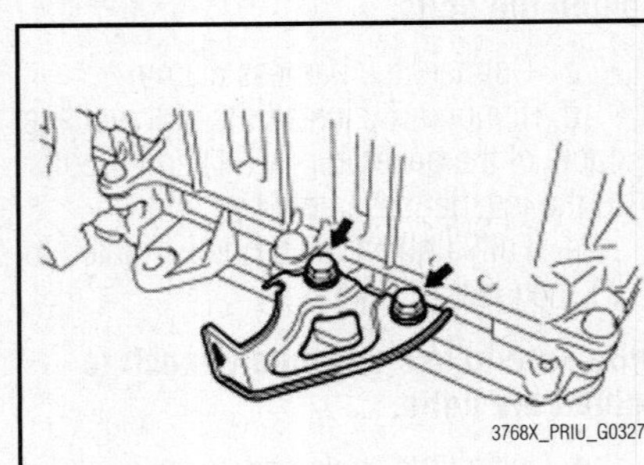
Fig. 105 Removing the motor cable bracket

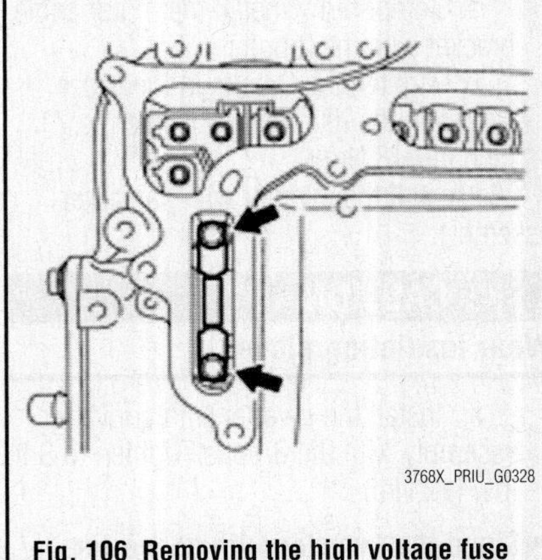
Fig. 106 Removing the high voltage fuse

c. Temporarily install the inverter terminal cover with the 9 bolts to prevent any foreign objects or water from entering the inverter with converter assembly.

To install:

26. Install the high voltage fuse.

CAUTION

Wear insulating gloves.

➡Perform this procedure only when replacement of the high voltage fuse is necessary.

a. Remove the 9 bolts and inverter terminal cover.

➡Make sure to pull the inverter terminal cover straight up, as a connector is connected to the bottom of the cover.

b. Install the high voltage fuse with the 2 bolts. Tighten to 35 inch lbs. (4 Nm).

➡Be sure to use a torque wrench to tighten the bolts.

c. Temporarily install the inverter terminal cover with the 9 bolts to prevent any foreign objects or water from entering the inverter with converter assembly.

27. Install the motor cable bracket.

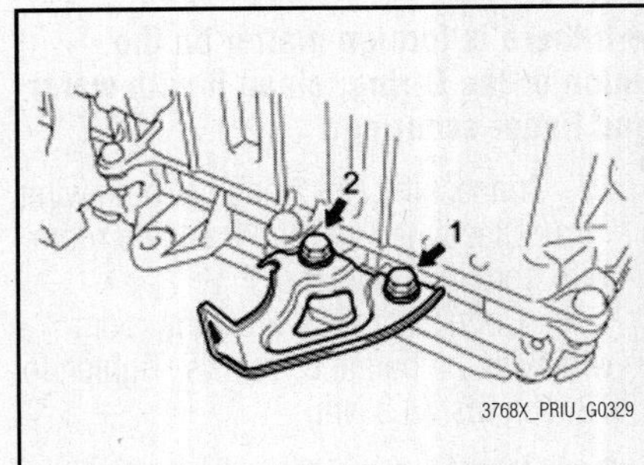

Fig. 107 Identifying the bolt tightening sequence

a. Temporarily install the motor cable bracket with the 2 bolts.

b. Tighten the 2 bolts in the order shown in the illustration. Tighten to 71 inch lbs. (8 Nm).

28. Install the inverter with converter assembly.

CAUTION

Wear insulating gloves.

a. Install the inverter with converter assembly with the 3 bolts. Tighten to 8 ft. lbs. (12 Nm).

➡Since the inverter with converter assembly is very heavy, 2 people are needed to install the inverter with converter assembly. When installing the inverter with converter assembly, do not damage the parts around it.

➡To prevent damage, do not hold the inverter with converter assembly by the connectors.

➡To prevent damage due to static electricity, do not touch the terminals of the disconnected connectors.

29. Connect the water hose.

a. Connect the water hose to the inverter with converter assembly and lock the hose with the retainer.

➡Insert the retainer until a click sound is heard.

➡Pull on the hose to confirm that the hose is securely connected.

➡If there is foreign matter on the union or the O-ring, clean it with water and finger scouring.

b. Connect the water hose to the inverter with converter assembly and lock the hose with the retainer.

➡Insert the retainer until a click sound is heard.

➡Pull on the hose to confirm that the hose is securely connected.

➡If there is foreign matter on the union or the O-ring, clean it with water and finger scouring.

30. Connect the No. 2 engine room wire.

a. Disconnect the No. 2 engine room wire from the protector.

b. Connect the No. 2 engine room wire with the bolt and 2 claws. Tighten to 73 inch lbs. (8.3 Nm).

➡Pass the No. 2 engine room wire under the two cooling hoses that pass beside the inverter.

c. Install the No. 1 relay block cover and 2 clamps.

d. Install the relay block cover.

31. Remove the inverter terminal cover.

CAUTION

Wear insulating gloves.

a. Remove the 9 bolts and inverter terminal cover.

➡Make sure to pull the inverter terminal cover straight up, as a connector is connected to the bottom of the cover.

32. Connect the No. 2 engine wire.

CAUTION

Wear insulating gloves.

➡Do not allow any foreign objects or water to enter the inverter with converter assembly.

a. Temporarily install the No. 2 engine wire (high voltage cables of the air conditioning) and 4 bolts to the inverter assembly by hand.

b. Fully tighten the 4 bolts. Tighten to 71 inch lbs. (8 Nm).

➡Be sure to use a torque wrench to tighten the bolts.

c. Connect the harness clamp.

33. Connect the high voltage cable of the front transaxle.

CAUTION

Wear insulating gloves.

➡Do not allow any foreign objects or water to enter the inverter with converter assembly.

a. Temporarily install the high voltage cable of the motor (MG2) and 5 bolts to the inverter assembly by hand.

b. Fully tighten the 5 bolts. Tighten to 71 inch lbs. (8 Nm).

➡Be sure to use a torque wrench to tighten the bolts.

c. Connect the harness clamp.

d. Temporarily install the high voltage cable of the generator (MG1) and 5 bolts to the inverter assembly by hand.

e. Fully tighten the 5 bolts. Tighten to 71 inch lbs. (8 Nm).

➡Be sure to use a torque wrench to tighten the bolts.

f. Install the cable and cover.

➡Close the cover so that the matchmarks are not visible.

34. Connect the frame wire.

CAUTION

Wear insulating gloves.

➡Make sure that the interlock is fully engaged.

➡Do not allow any foreign objects or water to enter the inverter with converter assembly.

a. Temporarily install the frame wire (high voltage cables of the hybrid battery) and 4 bolts to the inverter assembly by hand.

b. Fully tighten the 4 bolts. Tighten to 71 inch lbs. (8 Nm).

➡Be sure to use a torque wrench to tighten the bolts.

c. Connect the harness clamp.

35. Check the high voltage cable connection.

CAUTION

Wear insulating gloves.

➡Do not allow any foreign objects or water to enter the inverter with converter assembly.

a. Check that each connector and terminal is firmly installed.

➡Make sure that the bolts are fully tightened.

36. Install the inverter terminal cover.

CAUTION

Wear insulating gloves.

➡Make sure that the interlock is fully engaged.

➡Do not allow any foreign objects or water to enter the inverter with converter assembly.

a. Install the inverter terminal cover with the 9 bolts to the inverter with converter assembly. Tighten to 71 inch lbs. (8 Nm).

37. Install the engine room main wire.

➡Make sure that the interlock is fully engaged.

➡Do not allow any foreign objects or water to enter the inverter with converter assembly.

a. Install the bolt, clamp and clip, and connect the engine room main ware. Tighten to 9 ft. lbs. (12 Nm).

b. Install the bolt. Tighten to 9 ft. lbs. (12 Nm).

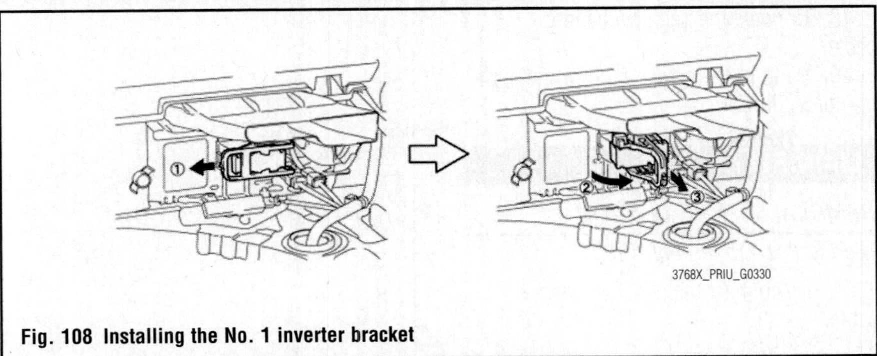

Fig. 108 Installing the No. 1 inverter bracket

c. Connect the engine wire to the engine room main wire.

d. Connect the connector to the inverter with converter assembly and lock the connector with the lock lever.

38. Install the No. 1 inverter bracket.

a. Temporarily install the No. 1 inverter bracket with the 3 bolts.

b. Tighten the 3 bolts in the order shown in the illustration. Tighten to 10 ft. lbs. (14 Nm).

39. Install the service plug grip.

40. Connect the cable to the negative battery terminal.

➡When disconnecting the cable, some systems need to be initialized after the cable is reconnected.

41. Install the rear No. 3 floor board.

42. Install the rear deck floor box.

43. Install the rear No. 2 floor board (separate type).

44. Add coolant.

45. Inspect for coolant leaks.

46. Install the No. 1 engine under cover.

47. Install the front spoiler cover (w/ front spoiler).

48. Install the engine under cover (w/ cover).

49. Install the radiator support opening cover.

MAIN BATTERY CABLE

REMOVAL & INSTALLATION

See Figures 109 through 114.

1. Disconnect the negative battery cable.

➡Wait at least 90 seconds after disconnecting the cable from the negative (-) battery terminal to prevent airbag and seat belt pretensioner activation.

2. Remove service plug grip.

3. Remove the high voltage battery from the vehicle.

4. Remove battery cover.

5. Wearing insulating gloves, remove No. 1 wire harness protector cover.

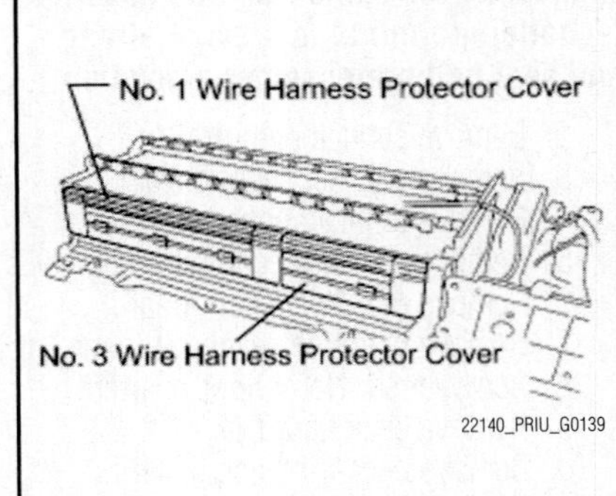

Fig. 109 Remove the wire harness protector cover

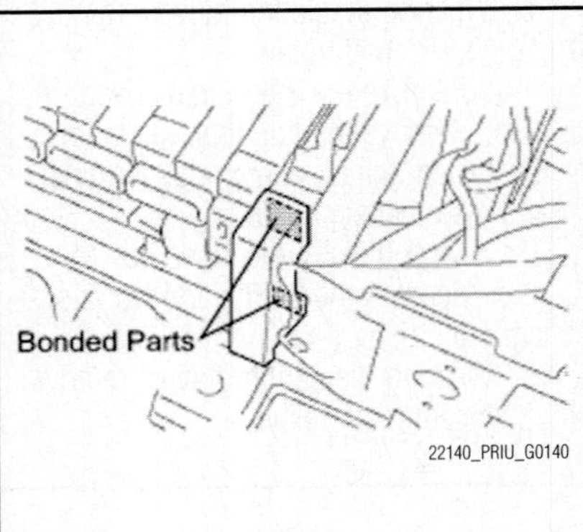

Fig. 110 Peel off the bonded parts, then remove the battery carrier cushion

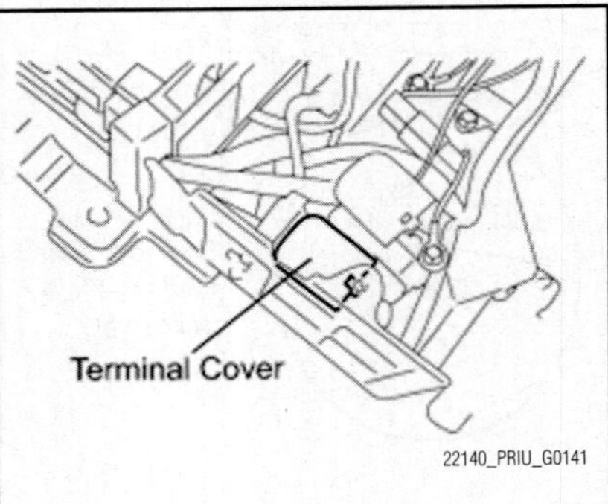

Fig. 111 Remove the main battery terminal cover shown

6. Wearing insulating gloves, remove No. 3 wire harness protector cover.

7. Wearing insulating gloves, peel off the bonded parts, then remove the battery carrier cushion.

8. Wearing insulating gloves, remove main battery cable.

a. Remove the terminal cover.

b. Remove the nut, then disconnect the aluminum shield wire.

c. Remove the nut, then disconnect the main battery cable from the No. 3 system main relay.

d. Remove the nut, then disconnect the main battery cable from the No. 2 frame wire.

e. Remove the main battery cable from the high voltage battery.

9. Wearing insulating gloves, remove No. 2 main battery cable.

a. Remove the terminal cover.

b. Remove the nut, then disconnect the system main relay terminal and No. 2 main battery cable from the No. 2 system main relay.

c. Remove the nut, then disconnect the No. 2 main battery cable from the frame wire.

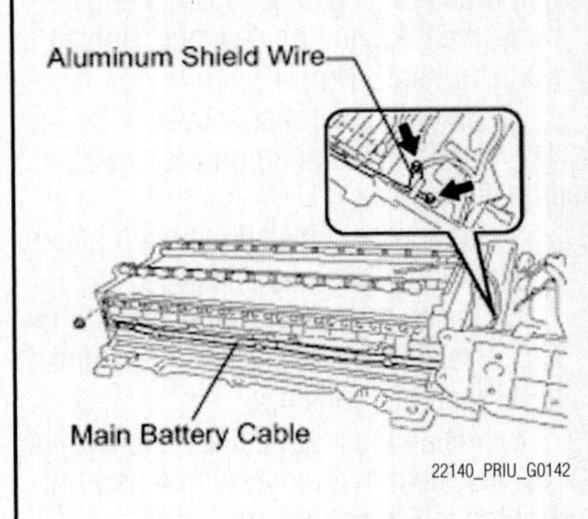

Fig. 112 Remove the nut, then disconnect the aluminum shield wire

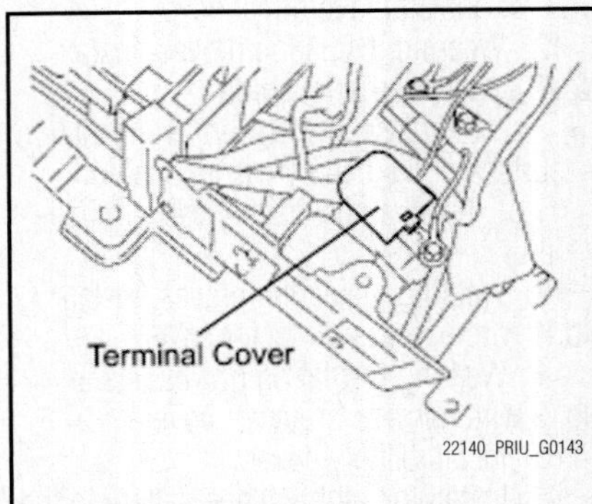

Fig. 113 Remove the No. 2 main battery cable terminal cover shown

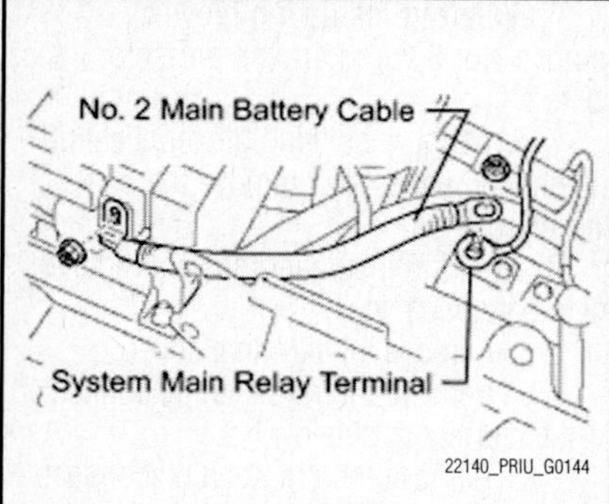

Fig. 114 Disconnect the system main relay terminal and No. 2 main battery cable from the No. 2 system main relay

d. Remove the No. 2 main battery cable from the high voltage battery.

To install:

10. Wearing insulating gloves, install No. 2 main battery cable.
   a. Temporarily install the No. 2 main battery cable to the high voltage battery.
   b. Install the No. 2 main battery cable to the No. 2 frame wire with a new nut. Tighten to 48 inch lbs. (5 Nm).
   c. Temporarily install the main battery cable and system main relay terminal, in that order, to the No. 2 system main relay, then tighten the new nut. Tighten to 50 inch lbs. (6 Nm).
   d. Install the terminal cover.
11. Wearing insulating gloves, install main battery cable.
   a. Temporarily install the main battery cable to the high voltage battery.
   b. Install the main battery cable to the No. 2 frame wire with a new nut. Tighten to 48 inch lbs. (5 Nm).
   c. Install the main battery cable to the No. 3 system main relay with a new nut. Tighten to 50 inch lbs. (6 Nm).
   d. Install the aluminum shield wire with a new nut. Tighten to 29 inch lbs. (3 Nm).
   e. Install the terminal cover.
12. Wearing insulating gloves, install No. 3 battery carrier cushion
   a. Degrease and clean the installation surface of the battery carrier cushion.
   b. Install a new battery carrier cushion.
13. Wearing insulating gloves, install No. 1 wire harness protector cover
14. Wearing insulating gloves, install No. 3 wire harness protector cover
15. Install battery cover
16. Install the high voltage battery to the vehicle.
17. Install service plug grip.
18. Connect the negative battery cable.
19. Perform the power window initialization procedure.

MAIN RELAY

REMOVAL & INSTALLATION

See Figures 115 through 119.

1. Disconnect the negative battery cable.

➡Wait at least 90 seconds after disconnecting the cable from the negative (-) battery terminal to prevent airbag and seat belt pretensioner activation.

2. Remove service plug grip.
3. Remove rear deck trim cover.
4. Remove tonneau cover.
5. Remove rear seat cushion assembly.
6. Remove rear No. 1 floor board.
7. Remove rear side seat back frame.
8. Remove rear No. 4 floor board.
9. Remove deck floor box.
10. Remove deck trim side panel.
11. Remove battery carrier bracket.
12. Remove No. 6 battery carrier panel.
13. Remove junction terminal.
14. Separate frame wire.
15. Wearing insulating gloves, remove No. 2 system main relay.
   a. Remove the 2 terminal covers.
   b. Remove the nut, then disconnect the system main battery relay terminal and main battery cable.
   c. Disconnect the connector.
   d. Remove the 2 nuts and No. 2 system main relay.
16. Wearing insulating gloves, remove No. 3 system main relay.

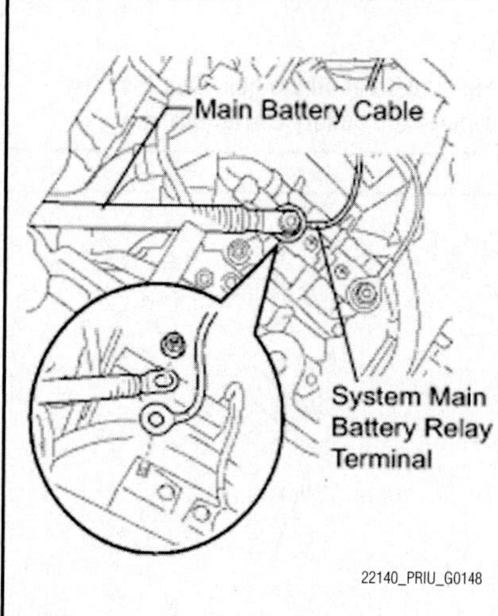

Fig. 115 System main battery relay terminal

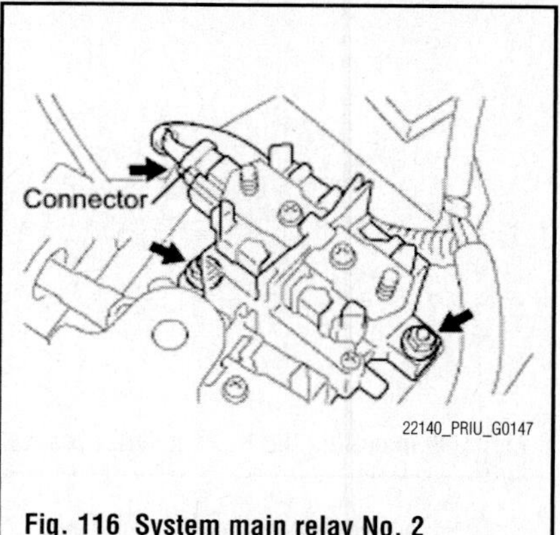

Fig. 116 System main relay No. 2

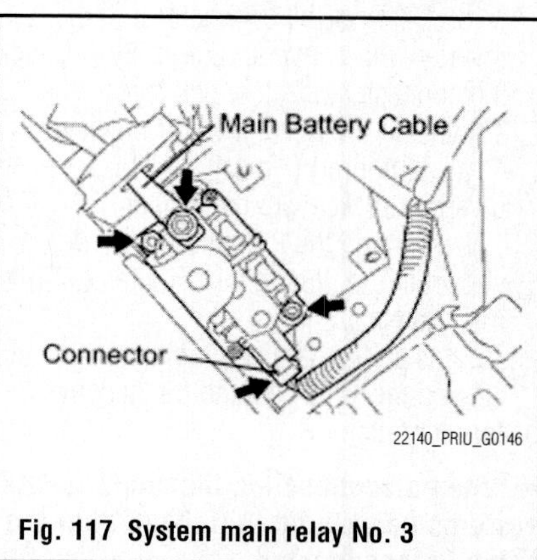

Fig. 117 System main relay No. 3

   a. Remove the nut, then disconnect the main battery cable.
   b. Remove the 2 nuts, then disconnect the No. 3 system main relay.
   c. Disconnect the connector and remove the No. 3 system main relay.
17. Wearing insulating gloves, remove No. 1 system main relay.
   a. Disconnect the connector.
   b. Remove the nut, then disconnect the ground terminal.
   c. Disconnect the 2 clamps.
   d. Remove the system main relay connector.
   e. Remove the 2 bolts and No. 1 system main relay.

To install:

18. Wearing insulating gloves, install No. 1 system main relay.
   a. Install the No. 1 system main relay with the 2 bolts. Tighten to 30 inch lbs. (3 Nm).
   b. Connect the system main relay connector.
   c. Connect the 2 clamps, then install the system main relay sub wiring harness to the upper battery carrier.
   d. Install the ground terminal with the nut. Tighten to 50 inch lbs. (6 Nm).
   e. Install the connector.

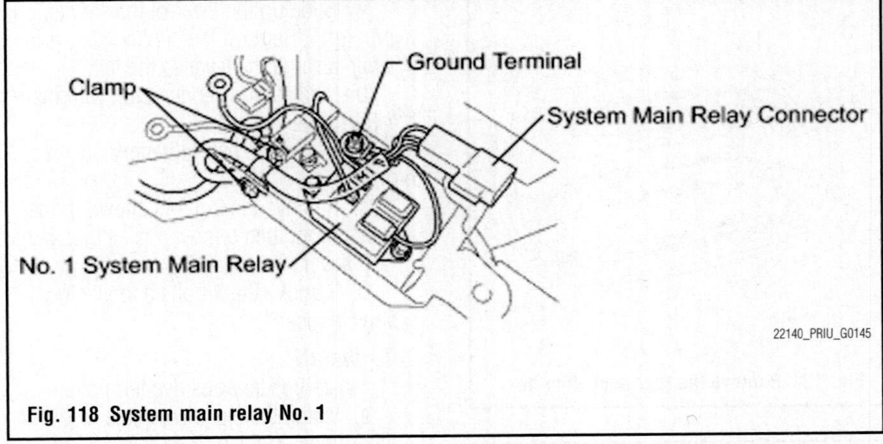

Fig. 118 System main relay No. 1

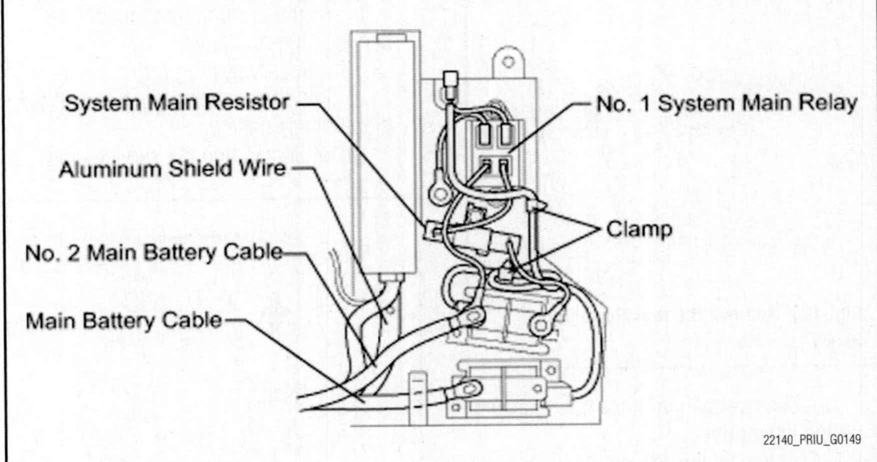

Fig. 119 Check that all the wire harnesses are correctly and securely connected on the upper battery carrier

19. Wearing insulating gloves, install No. 3 system main relay.
   a. Connect the connector, then temporarily install the No. 3 system main relay.
   b. Install the No. 3 system main relay with 2 new nuts. Tighten to 30 inch lbs. (3 Nm).
   c. Install the main battery cable with the nut. Tighten to 50 inch lbs. (6 Nm).
20. Wearing insulating gloves, install No. 2 system main relay.
   a. Install the No. 2 system main relay with the 2 nuts. Tighten to 30 inch lbs. (3 Nm).
   b. Connect the connector.
   c. Temporarily install the system main relay terminal and No. 2 main battery cable terminal, in that order, and tighten a new nut. Tighten to 50 inch lbs. (6 Nm).
   d. Install the 2 terminal covers.
21. Wearing insulating gloves, inspect contact condition.
   a. Check that all the wire harnesses are correctly and securely connected on the upper battery carrier.
   b. Make sure that the connectors and clamps are connected exactly the same as the illustration above. If any of them is different, correct it immediately.
22. Install frame wire.
23. Install junction terminal.
24. Install No. 6 battery carrier panel.
25. Install battery carrier bracket.
26. Install deck trim side panel.
27. Install deck floor box.
28. Install rear No. 4 floor board.
29. Install rear side seat back frame.
30. Install rear No. 1 floor board.
31. Install rear seat cushion assembly.
32. Install tonneau cover.
33. Install rear deck trim cover.
34. Install service plug grip
35. Connect the negative battery cable.
36. Perform the power window initialization procedure.

MAIN RESISTOR

REMOVAL & INSTALLATION

See Figure 120.

1. Disconnect the negative battery cable.

➡Wait at least 90 seconds after disconnecting the cable from the negative (-) battery terminal to prevent airbag and seat belt pretensioner activation.

2. Remove service plug grip.
3. Remove rear deck trim cover.
4. Remove tonneau cover.
5. Remove rear seat cushion assembly.
6. Remove rear No. 1 floor board.
7. Remove rear side seat back frame.
8. Remove rear No. 4 floor board.
9. Remove deck floor box.
10. Remove deck trim side panel.
11. Remove battery carrier bracket.
12. Remove No. 6 battery carrier panel.
13. Remove junction terminal.
14. Wearing insulating gloves, separate frame wire.
   a. Remove the nut, then disconnect the frame wire from the No. 2 system main relay.
   b. Insulate the removed terminals with insulating tape.
15. Wearing insulating gloves, remove system main resistor.
   a. Disconnect the connector from the No. 1 system main relay.
   b. Remove the bolt and system main resistor.

To install:

16. Wearing insulating gloves, install system main resistor.
   a. Install the system main resistor with the bolt. Tighten to 30 inch lbs. (3 Nm).
   b. Install the connector to the No. 1 system main relay.
17. Check that all the wire harnesses are correctly and securely connected on the upper battery carrier.

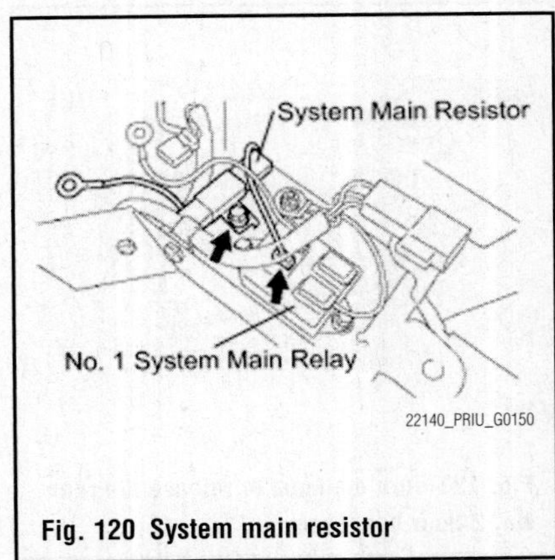

Fig. 120 System main resistor

18. Wearing insulating gloves, install frame wire.
   a. Temporarily install the system main resistor terminal and frame wire, in that order, and tighten a new nut. Tighten to 50 inch lbs. (6 Nm).
19. Install frame wire.
20. Install junction terminal.
21. Install No. 6 battery carrier panel.
22. Install battery carrier bracket.
23. Install deck trim side panel.
24. Install deck floor box.
25. Install rear No. 4 floor board.
26. Install rear side seat back frame.
27. Install rear No. 1 floor board.
28. Install rear seat cushion assembly.
29. Install tonneau cover.
30. Install rear deck trim cover.
31. Install service plug grip
32. Connect the negative battery cable.
33. Perform the power window initialization procedure.

SERVICE PLUG GRIP

REMOVAL & INSTALLATION

2009 Vehicles

See Figures 121 through 126.

**** WARNING**

After removing the service plug grip, do not operate the power switch as it may damage the hybrid vehicle control ECU.

1. Remove rear No. 2 floor board.
   a. As shown in the illustration, turn the knob to release the lock.
   b. Remove the rear floor board No. 2.
2. Remove rear deck floor box.
   a. remove the deck floor box rear shown in the illustration.
3. Remove rear No. 3 floor board.

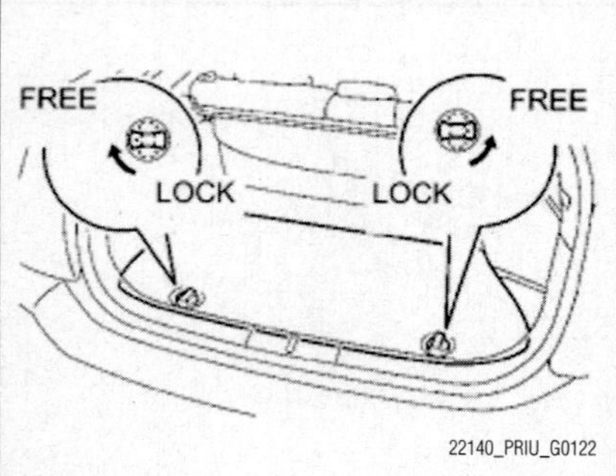

Fig. 121 Turn the knob to release the rear No. 2 floor board lock

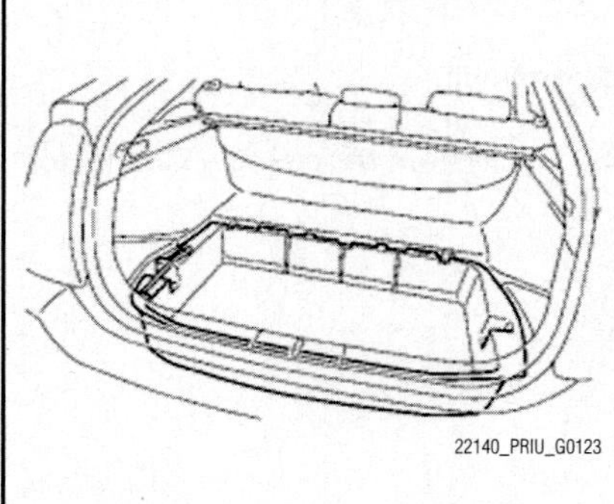

Fig. 122 Remove the rear deck floor box

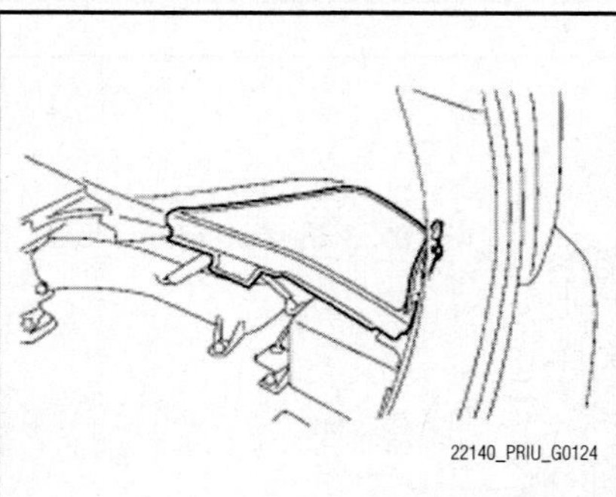

Fig. 123 Remove the rear No. 3 floor board

   a. Remove the rear floor board shown in the illustration.
4. Disconnect the 12 volt auxiliary (not hybrid high voltage) battery negative cable.
5. Remove service plug grip.

**** CAUTION**

Wear insulating gloves.

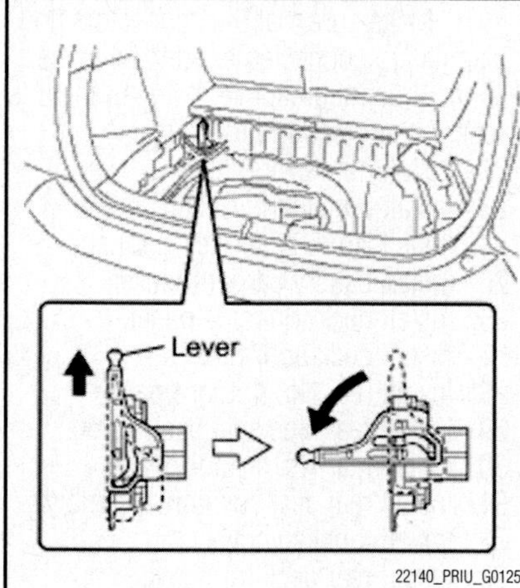

Fig. 124 Slide up the lever of the service plug grip. Remove the service plug grip while turning the lever to the left

   a. Slide up the lever of the service plug grip. Remove the service plug grip while turning the lever to the left.
   b. Insulate the service plug with insulating tape.
6. Remove the hybrid battery positive terminal.
   a. Remove the 2 joints shown in the illustration, then remove the service plug grip cover.
   b. Remove the 2 bolts and electric vehicle fuse.

To install:

7. Install battery positive terminal.
   a. Install the electric vehicle fuse with the 2 bolts. Tighten to 48 inch lbs. (5 Nm).
   b. Fit the 2 joints, then install the service plug grip cover.
8. Install service plug grip
   a. Join the service plug grip with the high voltage battery.
   b. While pushing the service plug

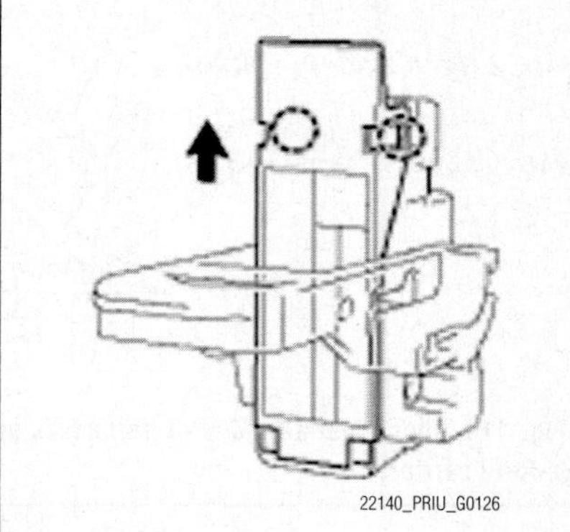

Fig. 125 Remove the 2 joints shown in the illustration, then remove the service plug grip cover

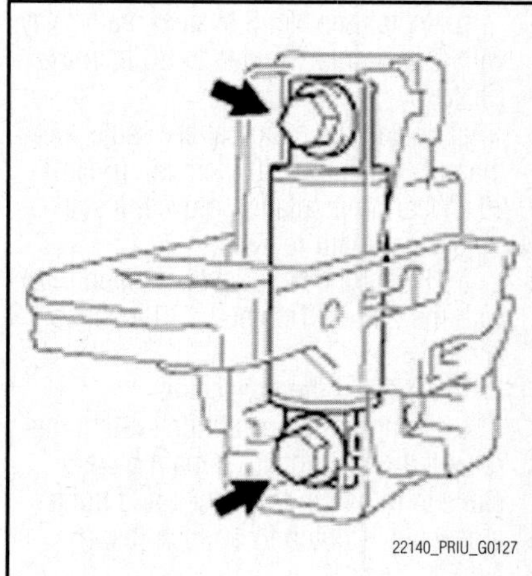

Fig. 126 Remove the 2 bolts and electric vehicle fuse

grip to the right, rotate the lever to the right.

c. Slide the lever down to lock the service plug grip in place.

➡The service plug grip must be locked. If not, it may cause DTCs to output.

⁂ WARNING

Do not operate the power switch when the service plug grip is not properly locked. It may damage the hybrid vehicle control ECU in rare cases.

9. Connect the negative battery cable.
10. Perform the power window initialization procedure.
11. Install rear No. 3 floor board.
12. Install rear deck floor box.
13. Install rear No. 2 floor board.

2010 Vehicles

See Figure 127.

1. Refer to Precautions before servicing the vehicle.
2. Remove the rear No. 2 floor board (separate type).
3. Remove the rear deck floor box.
4. Remove the rear No. 3 floor board.
5. Disconnect the cable from the negative battery terminal.

➡When disconnecting the cable, some systems need to be initialized after the cable is reconnected.

6. Remove the service plug grip.

⁂ CAUTION

Wear insulating gloves.

⁂ CAUTION

Remove the service plug grip to interrupt the high voltage circuit at the time of inspection or repair.

⁂ CAUTION

Keep the removed service plug grip in your pocket to prevent other technicians from accidentally reconnecting it while you are servicing the vehicle.

⁂ CAUTION

All the high voltage wiring connectors are colored in orange.

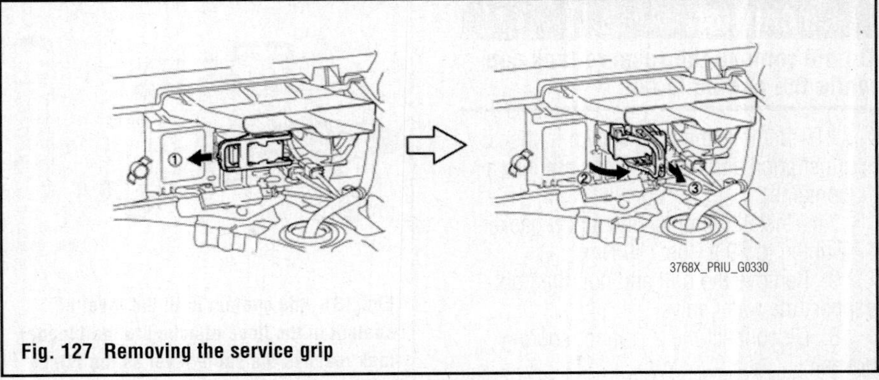

Fig. 127 Removing the service grip

a. Wear insulating gloves and remove the service plug grip after sliding up the lever of the service plug grip.

⁂ CAUTION

Keep the removed service plug grip in your pocket to prevent other technicians from accidentally reconnecting it while you are servicing the vehicle.

⁂ CAUTION

After removing the service plug grip, do not touch the high voltage connectors or terminals for 10 minutes.

➡Waiting for at least 10 minutes is required to discharge the high-voltage capacitor inside the inverter with converter assembly.

SUB RADIATOR

REMOVAL & INSTALLATION

2010 Vehicles

See Figure 128.

1. Remove the condenser with receiver assembly.

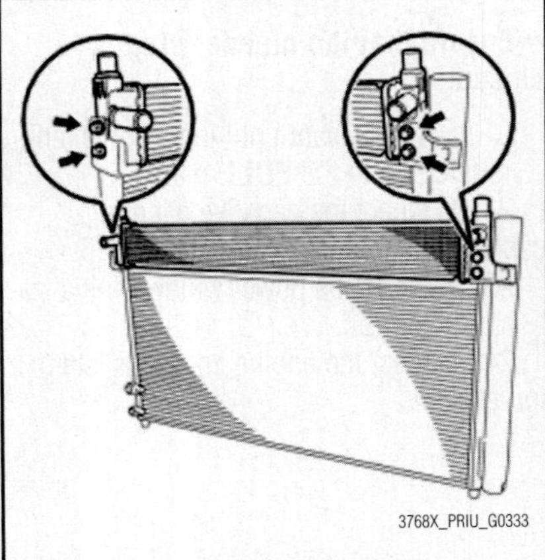

Fig. 128 Removing the sub radiator

2. Remove the radiator assembly.

a. Remove the 4 bolts and radiator assembly from the condenser assembly.

To install:

3. To install, reverse the removal procedure. Tighten the bolts to 80 inch lbs. (9 Nm).

WATER PUMP WITH MOTOR

REMOVAL & INSTALLATION

2009 Vehicles

See Figures 129 through 131.

1. Disconnect the negative battery cable.

➡Wait at least 90 seconds after disconnecting the cable from the negative (-) battery terminal to prevent airbag and seat belt pretensioner activation.

2. Disconnect front fender liner.
3. Remove center engine under cover.
4. Remove front spoiler cover.
5. Remove front bumper cover.
6. Remove front bumper energy absorber.
7. Remove left headlight assembly.
8. Drain coolant for the inverter.

a. Remove the transaxle-side reserve tank cap.

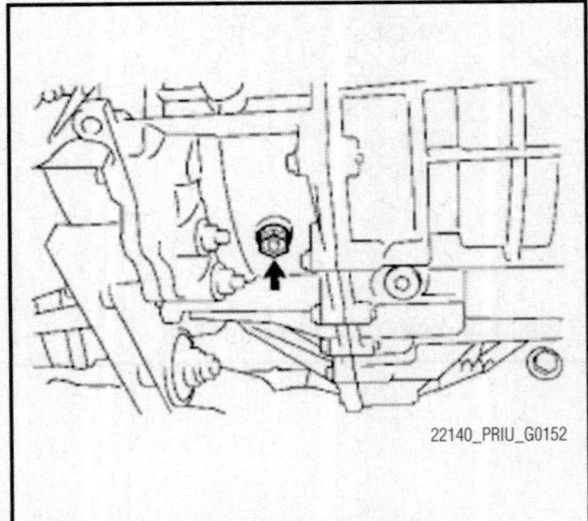

Fig. 129 Drain plug for inverter

⁂ CAUTION

Do not remove the reserve tank cap while the engine is hot.

b. Remove the plug shown in the illustration and drain the coolant into a container.

c. Install the plug with a new gasket. Tighten to 29 ft. lbs. (39 Nm).

9. Remove the bolt and nut, then disconnect the water valve.
10. Disconnect the 2 inverter cooling hoses.
11. Remove the bolt, then disconnect the water pump with motor assembly from the vehicle.
12. Disconnect the connector and remove the water pump with motor assembly.

To install:

13. Install the water pump with motor assembly. Tighten to 62 inch lbs. (7 Nm).
    a. Connect the connector.
    b. Connect the 2 inverter cooling hoses.
14. Install the water valve with the bolt and nut. Tighten to 62 inch lbs. (7 Nm).
15. Install left headlight assembly.
16. Install front bumper energy absorber.
17. Install front bumper cover.
18. Install front spoiler cover.
19. Install center engine under cover.
20. Install front fender liner.
21. Add coolant for the inverter.
    a. Loosen the bleeder plug and connect a hose.

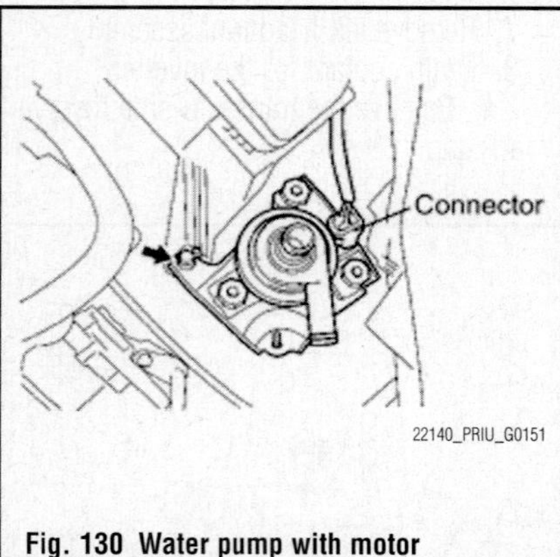

Fig. 130 Water pump with motor

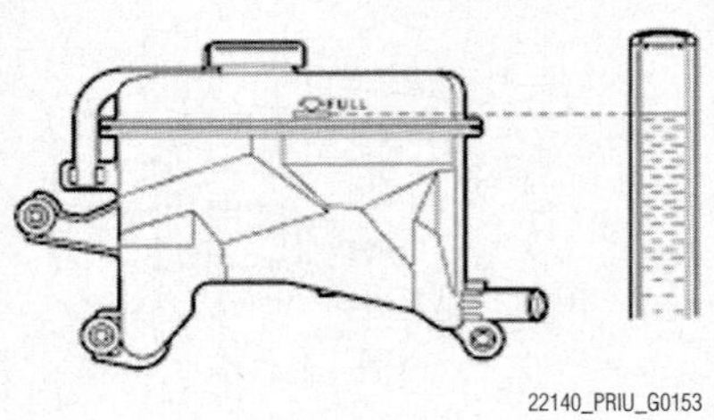

Fig. 131 Add coolant until the level of coolant in the hose attached to the bleeder tank reaches the same level as the FULL line of the reserve tank

b. Insert one end of the hose into the bleeder tank.

c. Add coolant until the level of coolant in the hose attached to the bleeder tank reaches the same level as the **FULL** line of the reserve tank.

d. Close the bleeder plug.

e. Turn the power switch **ON** and run the water pump for approximately 20 seconds.

f. Turn the power switch **OFF**

g. Loosen the bleeder plug and bleed the air from the transaxle.

h. Add coolant into the bleeder tank.

i. Repeat the previous three steps again.

j. Coolant system air bleeding is complete when water pump noise becomes softer and coolant circulation in reserve tank improves.

k. If air remains in the coolant system, the water pump noise becomes louder and the coolant circulation in the reserve tank becomes worse.

l. Turn the power switch **ON** and run the water pump for approximately 5 minutes after completing air bleeding of the coolant system.

➡Ensure that the bleeder plug is closed.

m. Add coolant until the reserve tank is filled up to the **FULL** mark.

22. Connect the negative battery cable.
23. Perform the power window initialization procedure.
24. Inspect the engine cooling system for leakage.

2010 Vehicles

See Figures 132 and 133.

1. Remove the front spoiler cover (w/ front spoiler).
2. Remove the engine under cover (w/ cover).
3. Remove the No. 1 engine under cover.
4. Drain the coolant (inverter).
5. Remove the inverter with converter assembly.
6. Remove the inverter reserve tank assembly.
7. Remove the inverter tray bracket.
8. Remove the water pump with the motor assembly.
    a. Disconnect the water hose and connector.
    b. Remove the 3 bolts and the water pump with motor assembly with the water pump bracket.

To install:

9. To install, reverse the removal procedure. Tighten the water pump bolts to 54 inch lbs. (6.1 Nm).

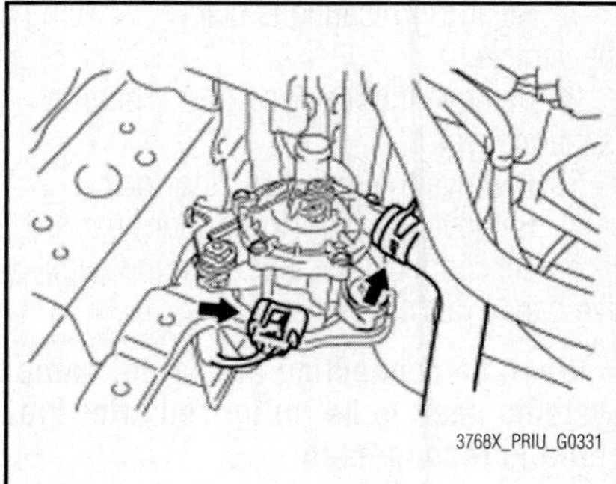

Fig. 132 Disconnecting the water hose and connector

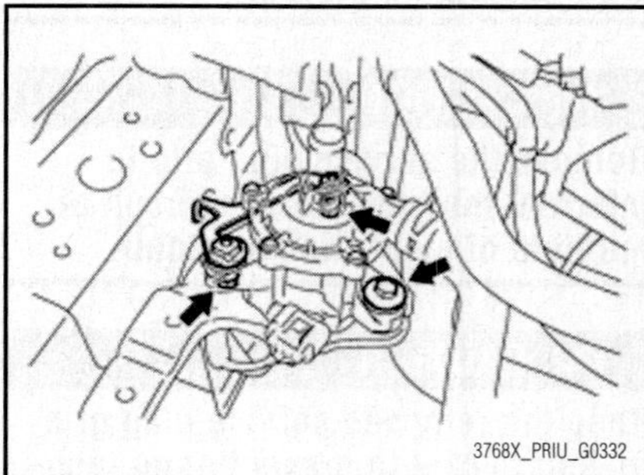

Fig. 133 Removing the 3 bolts and the water pump with motor assembly with the water pump bracket

ENGINE MECHANICAL

ACCESSORY DRIVE BELTS

ACCESSORY BELT ROUTING

See Figures 134 and 135.

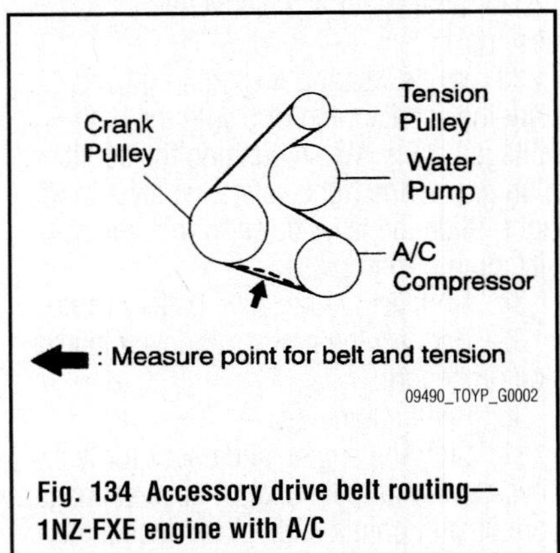

Fig. 134 Accessory drive belt routing—1NZ-FXE engine with A/C

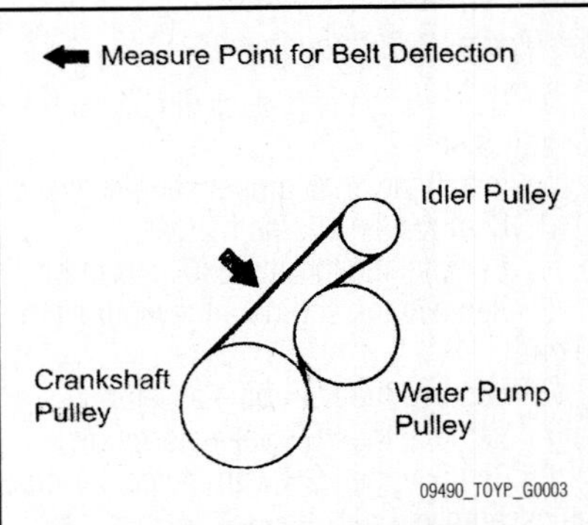

Fig. 135 Accessory drive belt routing—1NZ-FXE engine without A/C

INSPECTION

Inspect the drive belt for signs of glazing or cracking. A glazed belt will be perfectly smooth from slippage, while a good belt will have a slight texture of fabric visible. Cracks will usually start at the inner edge of the belt and run outward. All worn or damaged drive belts should be replaced immediately.

ADJUSTMENT

See Figure 136.

1. Using a belt tension gauge, measure the belt tension.
   a. New belt tension should be 99 to 121 ft. lbs. used belt tension should be 55 to 77 ft. lbs.

➡A "new belt" is a belt which has been used less than 5 minutes on a running engine. A "used belt" is a belt which has been used on a running engine for 5 minutes or more.

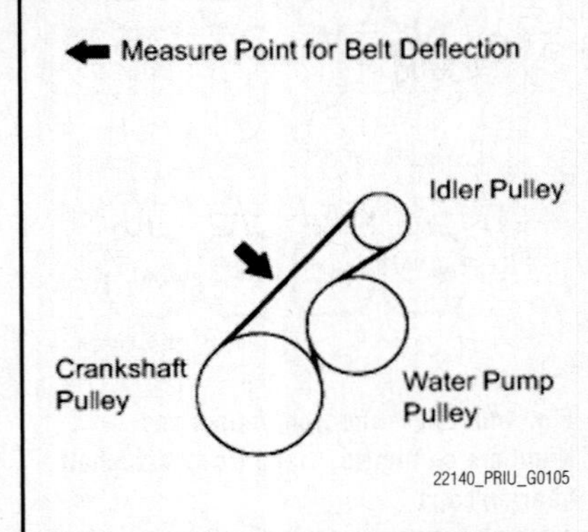

Fig. 136 Measuring point for belt deflection

   b. If tension is not as specified, adjust the belt by loosening nut "A", then turning adjust bolt "B" to provide the proper belt tension.
2. When not using a belt tension gauge, measure the belt deflection.
   a. Press the belt at the center of the longest span using 22 ft. lbs. (10 kg) of force.
   b. New belt tension should be 0.35 to 0.47 in. (9.0 to 12.0 mm), used belt tension should be 0.43 to 0.59 in. (11 to 15 mm)

REMOVAL & INSTALLATION

2009 Vehicles

See Figure 137.

1. Before servicing the vehicle, refer to the precautions.
2. Remove the engine under cover.
3. Loosen the clamp, then disconnect the air cleaner inlet from the air cleaner case.
4. Referring to the illustration, loosen nut "A", then turn adjust bolt "B" to relieve the V-ribbed belt tension.

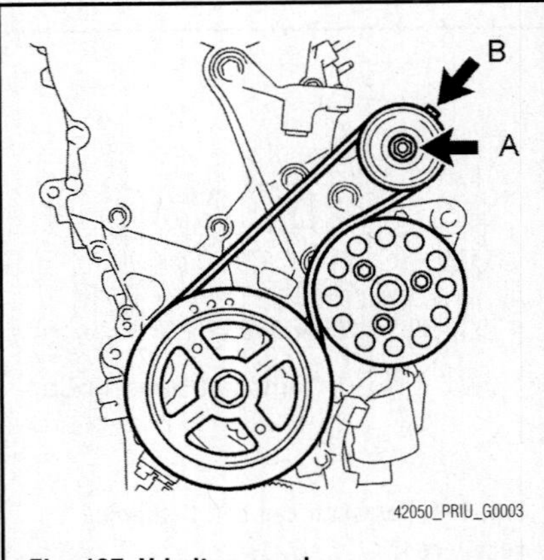

Fig. 137 V-belt removal

5. Remove the belt.

To install:

6. Install the V-ribbed belt to each pulley.

➡After installing the drive belt, check that it fits properly in the ribbed grooves. Check with your hands to confirm that the belt has not slipped out of the groove on the bottom of the crankshaft pulley. After installing a new belt, run the engine for approximately 5 minutes and then recheck the tension.

7. Turn adjust bolt "B" to adjust the V-ribbed belt tension.
8. Tighten nut "A" to 40 Nm (30 ft. lbs).

CAMSHAFT AND VALVE LIFTERS

REMOVAL & INSTALLATION

2009 Vehicles

See Figures 138 through 141.

1. Before servicing the vehicle, refer to the precautions.
2. Properly relieve the fuel system pressure.
3. Drain engine oil.
4. Disconnect the negative battery cable.

➡Wait at least 90 seconds after disconnecting the cable from the negative (-) battery terminal to prevent airbag and seat belt pretensioner activation.

5. While wearing insulating gloves, slide up the lever of the service plug grip and remove while turning the lever to the left. Be sure to insulate the service plug with insulating tape.

⁂ CAUTION

Do not touch the high voltage connectors and terminals for 5 minutes after removing the service plug grip.

6. While wearing insulating gloves, remove the inverter with converter assembly.
7. Remove timing chain.
8. Disconnect associated connectors and wire harnesses.
9. Remove the fuel pipe clamp.
10. Disconnect the fuel tube from the fuel delivery pipe. Even if the fuel tube is stuck and cannot be disconnected, do not use any tools. Push and pull the parts with

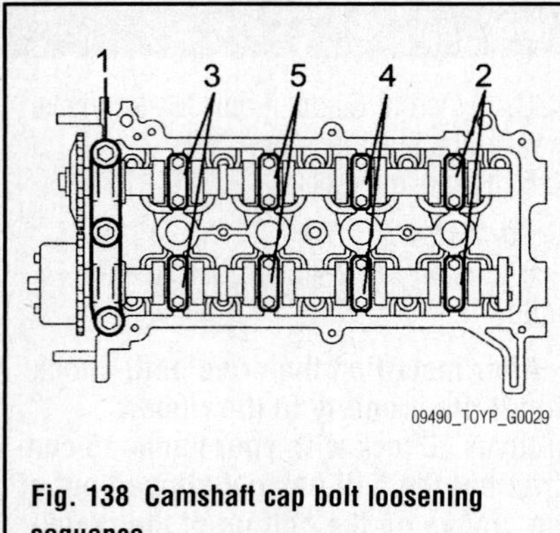

Fig. 138 Camshaft cap bolt loosening sequence

the quick connector pinched to disconnect the tube.

11. Cover the disconnected fuel tube and fuel delivery pipe with a plastic bag in order to prevent foreign objects from entering them.

12. Disconnect the radiator inlet hose from the cylinder head.

13. Disconnect the hoses from the cylinder head.

14. Remove the bolt and disconnect the No. 1 water by-pass pipe.

15. Disconnect the hoses, remove the bolt and disconnect the oil dipstick guide.

16. Remove the No. 1 and No. 2 camshaft bearing caps in the proper sequence, then remove the camshaft and No. 2 camshaft.

➡Uniformly loosen the bolts, keeping the camshaft level.

17. Remove the valve lifters. Keep the valve lifters in the correct order so that they can be returned to their original locations when reassembling.

To install:

18. Apply a light coat of engine oil to the valve lifter and install the valve lifter. Check that the valve lifter rotates smoothly by hand.

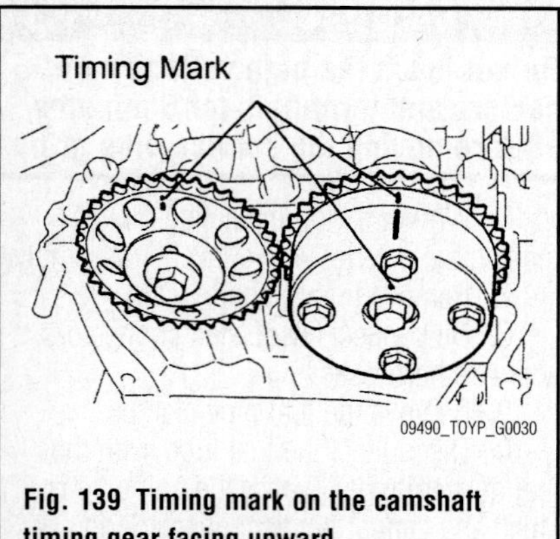

Fig. 139 Timing mark on the camshaft timing gear facing upward

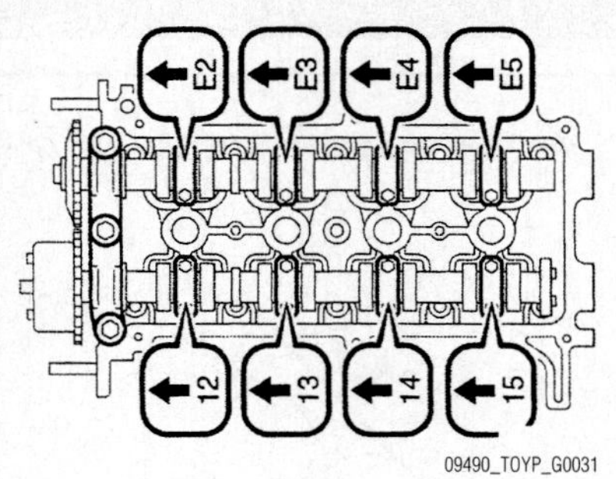

Fig. 140 Check the front marks and Numbers on the No. 1 and No. 2 camshaft bearing caps

➡If turning the camshaft with the chain removed, turn the crankshaft counterclockwise by 40 degrees from TDC/compression.

19. Apply engine oil to the cam and cylinder head journals.

20. Place the camshaft and No. 2 camshaft on the cylinder head with the timing mark on the camshaft timing gear facing upward.

21. Check the front marks and Numbers on the No. 1 and No. 2 camshaft bearing caps, then temporarily install them.

22. Uniformly tighten the No. 2 camshaft bearing caps in several steps in the proper sequence to 9.6 ft. lbs. (13 Nm).

23. Uniformly loosen the bolts, keeping the camshaft level.

24. Install the No. 1 camshaft bearing cap. Torque to 17 ft. lbs. (23 Nm).

25. Connect the water by-pass pipe with the bolt and tighten to 80 inch lbs. (9 Nm).

26. Connect the hoses.

27. Connect the radiator inlet hose.

28. Push the fuel main tube into the fuel delivery pipe until it makes a "click" sound. If the fuel tube is connected too tightly, apply a light coat of engine oil to the tip of the fuel delivery pipe. After connecting,

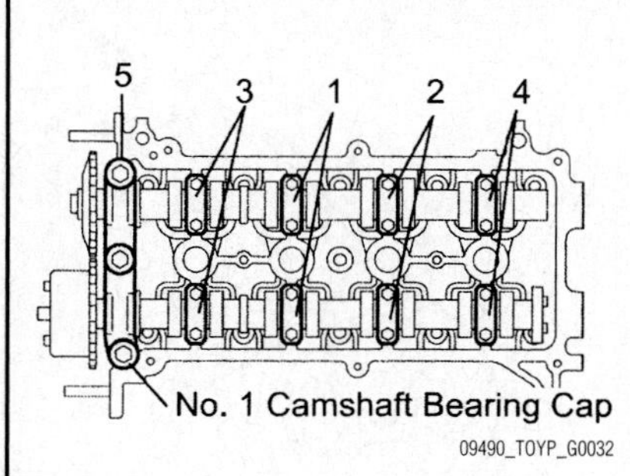

Fig. 141 Camshaft cap bolt tightening sequence

check that the fuel tube is securely connected by pulling it.

29. Install the fuel pipe clamp.

30. Connect all associated connectors and wire harnesses.

31. Install the timing chain.

32. While wearing insulating gloves, install the inverter with converter assembly.

33. While wearing insulating gloves, joint the service plug grip with the high voltage battery. While pushing the service plug grip to the right, rotate the lever to the right. Slide the lever down to lock the service plug grip in place.

34. Connect the negative battery cable.

35. Perform the power window initialization procedure.

36. Refill engine oil.

37. Start the engine and check for leaks, check for abnormal noises, shock slippage, correct shift points and smooth operation.

2010 Vehicles

See Figures 142 through 175.

1. Remove the engine assembly with the transaxle.
2. Install the engine on an engine stand.
3. Remove the engine hangers.
4. Remove the throttle body assembly.
5. Remove the engine oil level dipstick guide.
6. Remove the EGR pipe assembly.
7. Remove the EGR valve assembly.
8. Remove the EGR with the cooler pipe sub assembly.
9. Remove the intake manifold.
10. Remove the fuel vapor feed pipe.
11. Remove the fuel delivery pipe sub assembly.
12. Remove the No. 1 delivery pipe spacer.
13. Remove the fuel injector assembly.
14. Remove the ignition coil assembly.
15. Remove the cylinder head cover sub assembly.
16. Remove the cylinder head cover gasket.
17. Remove the spark plug tube gasket.
18. Set the No. 1 cylinder to TDC/Compression.
    a. Turn the crankshaft pulley until its notch and timing mark "0" of the timing chain cover are aligned.
    b. Check that timing marks on both the camshaft timing sprocket and camshaft timing gear are facing upward as shown in the illustration.
    c. If not, turn the crankshaft 1 complete revolution (360°) and align the marks.

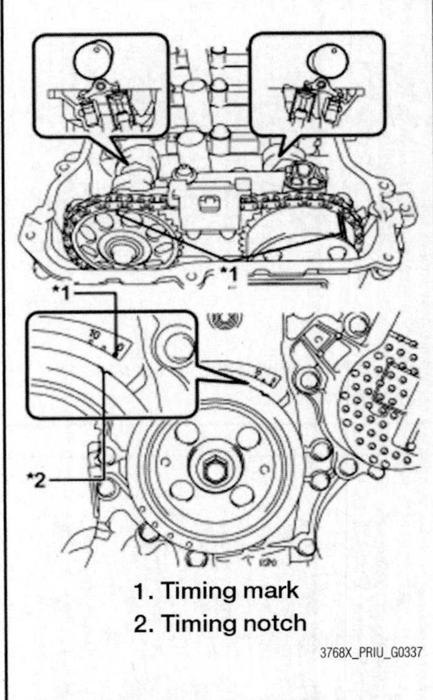

Fig. 142 Setting the No. 1 cylinder to TDC/Compression

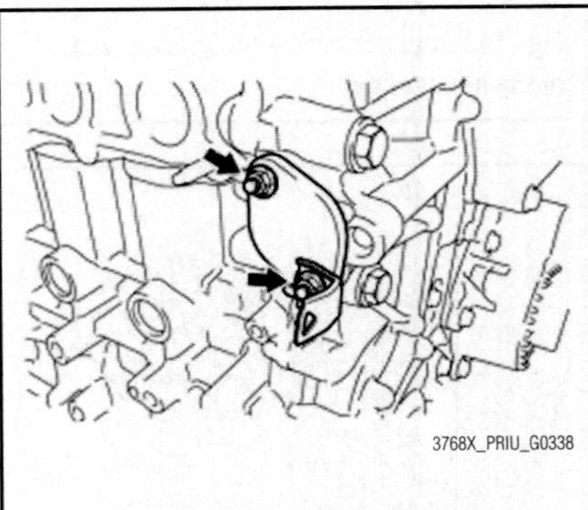

Fig. 143 Removing the No. 1 chain tensioner assembly

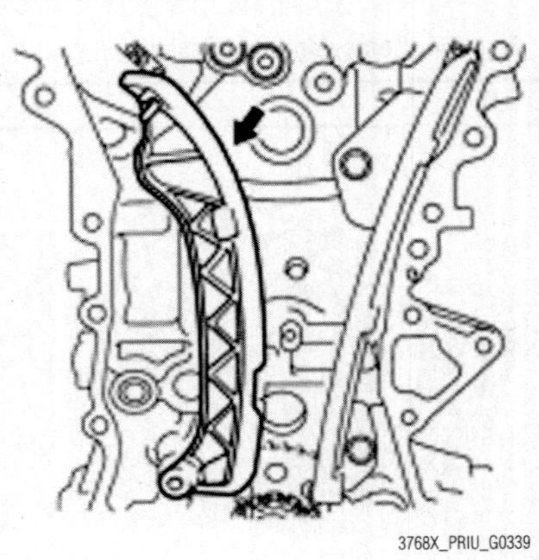

Fig. 144 Removing the chain tensioner slipper

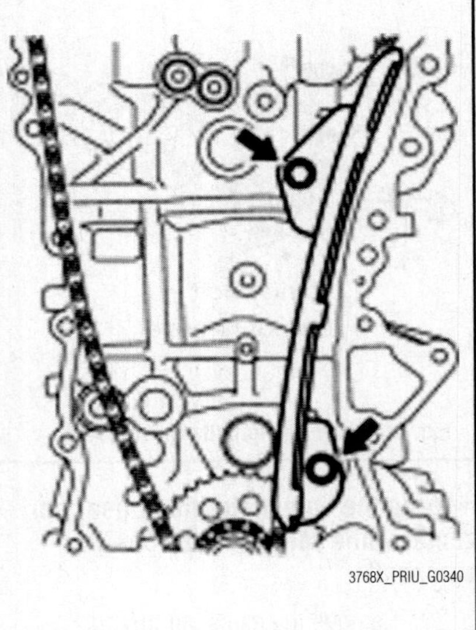

Fig. 145 Removing the 2 bolts and chain vibration damper

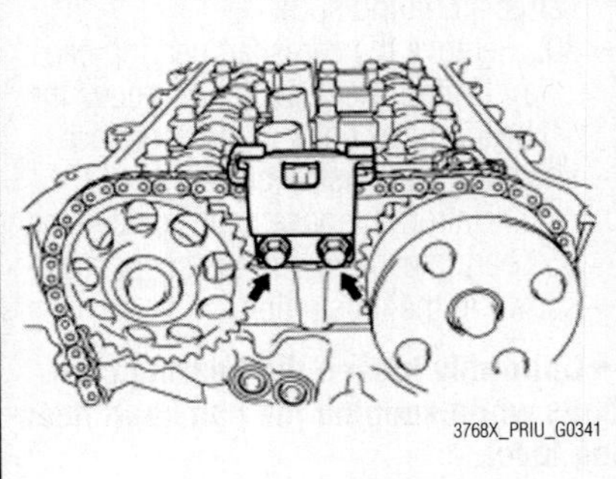

Fig. 146 Removing the No. 2 chain vibration damper

19. Remove the crankshaft pulley.
20. Remove the No. 1 chain tensioner assembly.
    a. Remove the 2 nuts, bracket, chain tensioner and gasket.

➡Do not turn the crankshaft without the No. 1 chain tensioner installed.

21. Remove the timing chain cover sub assembly.
22. Remove the timing chain cover oil seal.
23. Remove the chain tensioner slipper.
    a. Remove the chain tensioner slipper from the cylinder block.
24. Remove the 2 bolts and chain vibration damper.
25. Remove the 2 bolts and No. 2 chain vibration damper.
26. Remove the chain sub assembly.
    a. Hold the hexagonal portion of the camshaft with a wrench and turn the camshaft timing gear counterclockwise to loosen the chain between the camshaft timing gears.
    b. With the chain loosened, release the chain from the camshaft timing gear and place it on the camshaft timing gear.

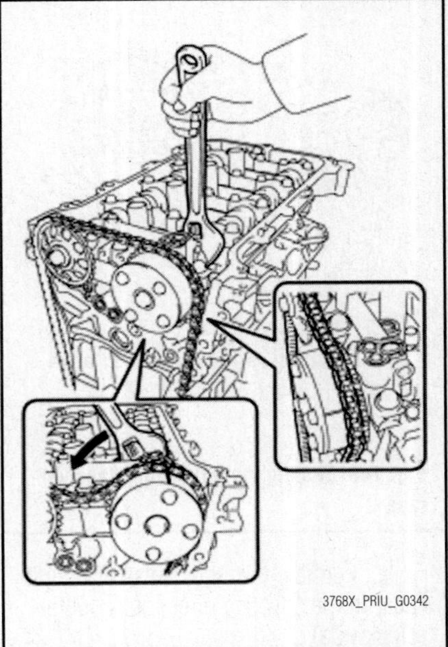

Fig. 147 Removing the chain sub assembly

➡Be sure to release the chain from the sprocket completely.

    c. Turn the camshaft clockwise to return it to the original position and remove the chain.
27. Inspect the camshaft timing gear assembly.
    a. Inspect the lock of the camshaft timing gear.
    b. After cleaning and degreasing the VVT oil hole on the intake side of the No. 1 camshaft bearing cap, completely seal the oil hole with adhesive tape or equivalent as shown in the illustration to prevent air from leaking.

➡Be sure to cover the oil hole completely because air leaks due to insufficient sealing will prevent the lock pin from being released.

    c. Prick a hole in the tape covering the oil hole as shown in the illustration.
    d. Apply approximately 22 psi (150 kPa) of air pressure to the hole pricked in procedure A to release the lock pin.

➡If air leaks out, reattach the adhesive tape.

➡Cover the oil hole with a piece of cloth when applying air pressure to prevent oil from spraying.

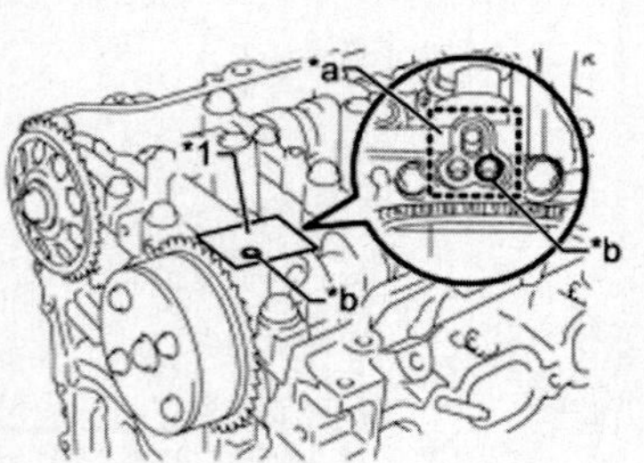

Fig. 148 Sealing the hole with adhesive tape

e. Forcibly turn the camshaft timing gear in the advance direction (counterclockwise).

➡Depending on the air pressure applied, the camshaft timing gear may turn in the advance direction without assistance.

f. Turn the camshaft timing gear within its movable range (26.5 to 28.5°) 2 or 3 times without turning it to the most retarded position. Make sure that the camshaft timing gear turns smoothly.

g. Remove the adhesive tape from the No. 1 camshaft bearing cap.

28. Remove the camshaft timing gear assembly.

a. Remove the flange bolt while holding the hexagonal portion of the camshaft with a wrench, and then remove the camshaft timing gear.

➡Before removing the camshaft timing gear, make sure that the lock pin has been released.

➡Be sure not to remove the other 4 bolts.

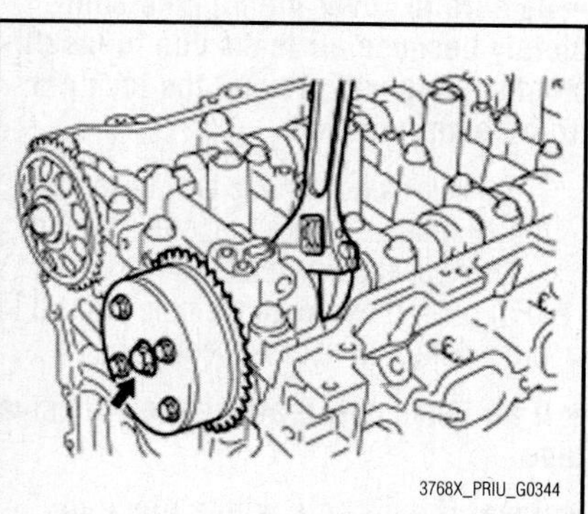

Fig. 149 Removing the flange bolt

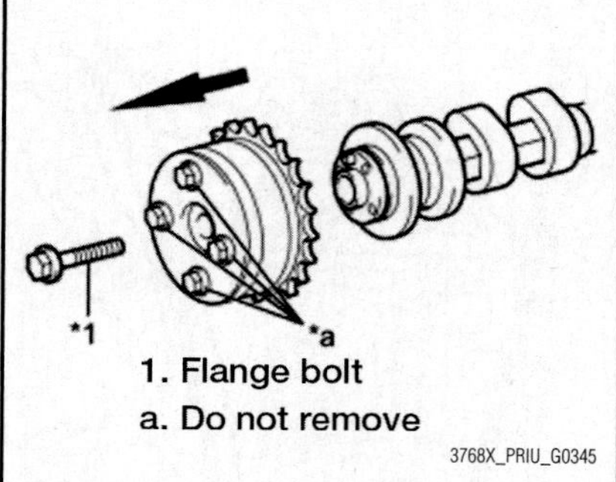

Fig. 150 Identifying bolts for removal

➡Keep the camshaft timing gear horizontal while removing it from the camshaft.

29. Remove the camshaft timing sprocket.

a. Remove the flange bolt while holding the hexagonal portion of the camshaft with a wrench, and then remove the camshaft timing sprocket.

30. Remove the camshaft bearing caps.

a. Uniformly loosen and remove the 10 bearing cap bolts in the sequence shown in the illustration.

b. Uniformly loosen and remove the 15 bearing cap bolts in the sequence shown in the illustration.

➡Uniformly loosen the bearing cap bolts while keeping the camshaft housing level.

c. Remove the 5 bearing caps.

➡Arrange the removed parts in the correct order.

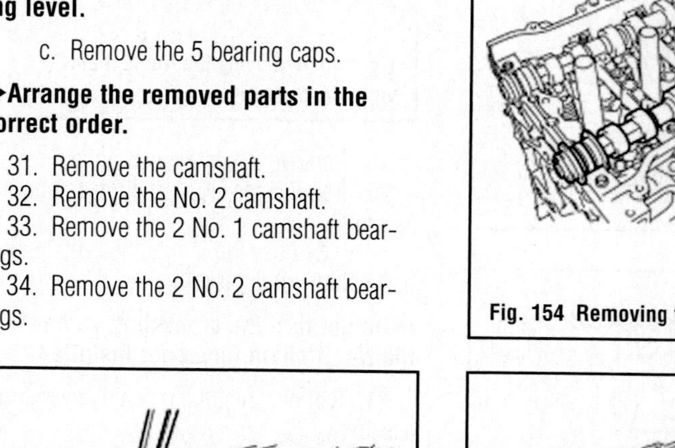

31. Remove the camshaft.

32. Remove the No. 2 camshaft.

33. Remove the 2 No. 1 camshaft bearings.

34. Remove the 2 No. 2 camshaft bearings.

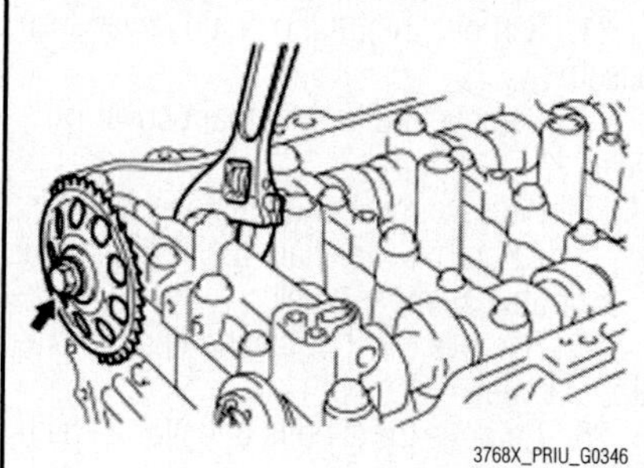

Fig. 151 Removing the camshaft timing sprocket

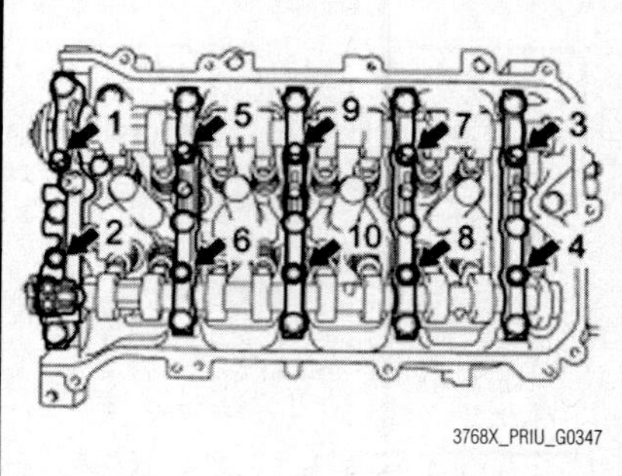

Fig. 152 Identifying the 10 bearing cap removal sequence

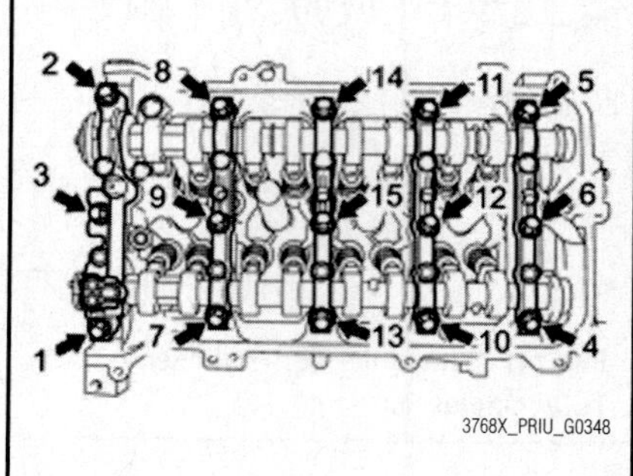

Fig. 153 Identifying the 15 bearing cap removal sequence

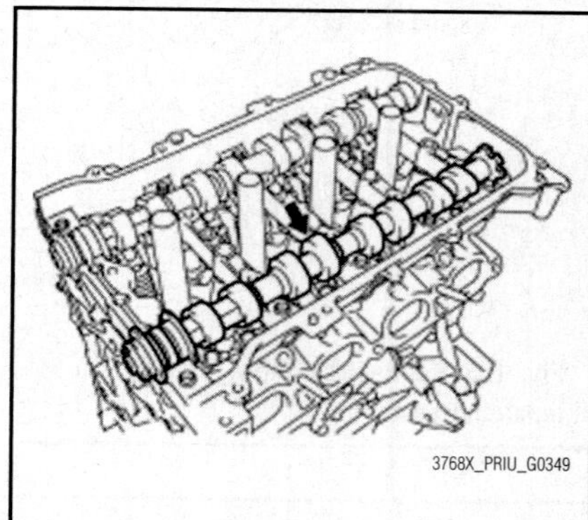

Fig. 154 Removing the camshaft

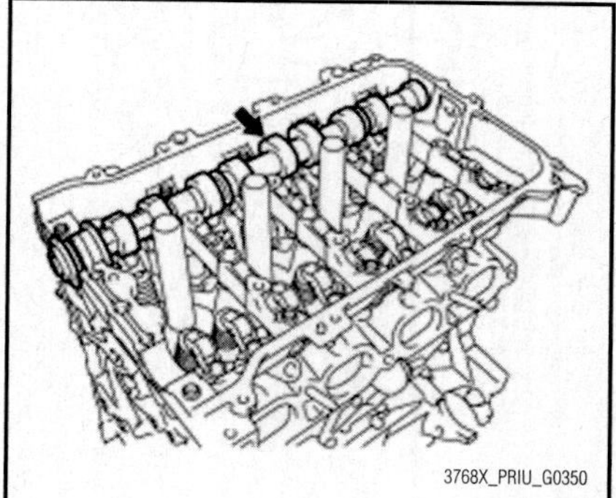

Fig. 155 Removing the No. 2 camshaft

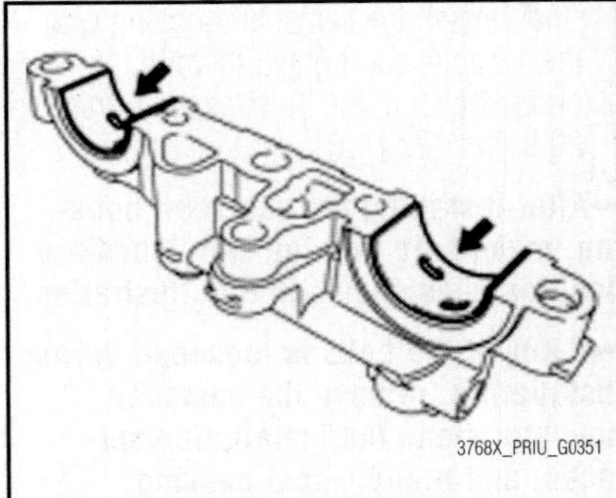

Fig. 156 Removing the 2 No. 1 camshaft bearings

Fig. 157 Removing the 2 No. 2 camshaft bearings

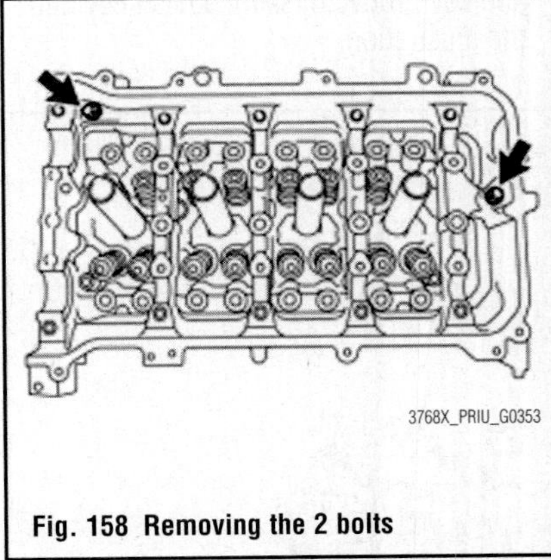

Fig. 158 Removing the 2 bolts

35. Remove the camshaft housing sub assembly.
   a. Remove the 2 bolts.
   b. Remove the camshaft housing by prying between the cylinder head and camshaft housing with a screwdriver.

➡Be careful not to damage the contact surfaces of the cylinder head and camshaft housing.

➡Tape (*1) the screwdriver tip before use.

To install:

36. Install the No. 1 camshaft bearing.

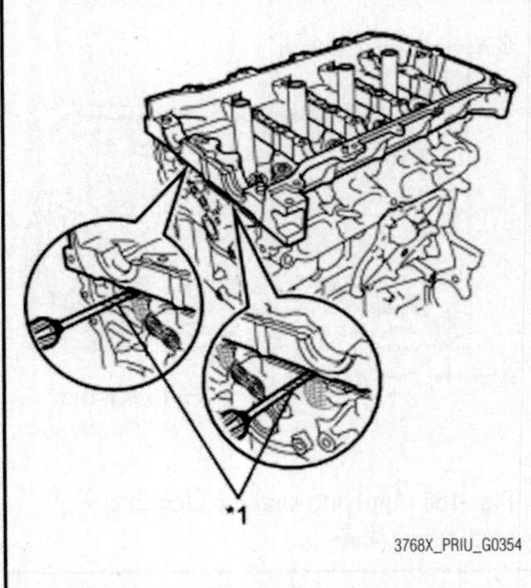

Fig. 159 Removing the camshaft housing

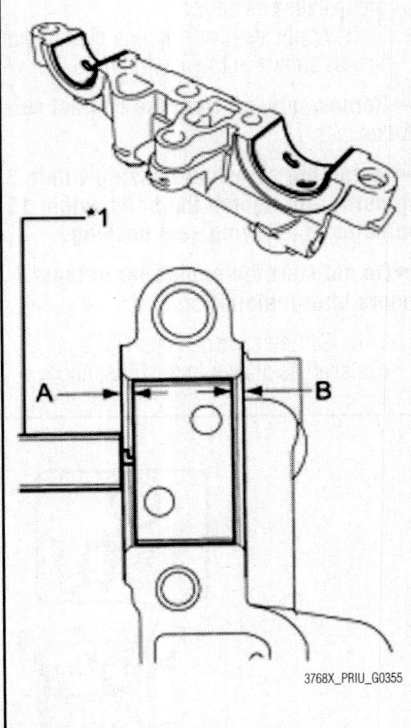

Fig. 160 Measuring the distance between the bearing cap edge and the camshaft bearing edge

   a. Clean both surfaces of the bearings.
   b. Install the 2 No. 1 camshaft bearings.
   c. Using a vernier caliper, measure the distance between the bearing cap edge and the camshaft bearing edge. The standard dimension is (A-B): 0.0276 inch (0.7 mm) or less.

➡Position the bearings to the center of the bearing cap by measuring the dimensions A and B.

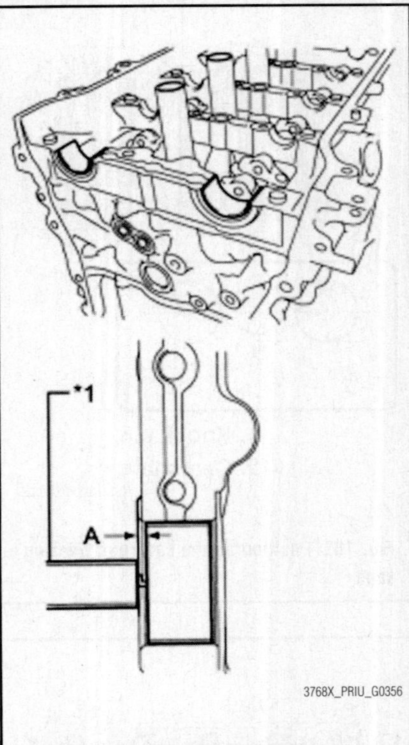

Fig. 161 Installing the No. 2 camshaft bearing

37. Install the No. 2 camshaft bearing.
   a. Clean both surfaces of the bearings.
   b. Install the 2 No. 2 camshaft bearings.
   c. Using a vernier caliper, measure the distance between the bearing cap edge and the camshaft bearing edge. The standard dimension for A is 0.0413-0.0689 inch (1.05-1.75 mm).

➡Positing the bearings to the center of the bearing cap by measuring dimension A.

38. Install the No. 2 camshaft.
   a. Clean the camshaft journals.
   b. Apply a light coat of engine oil to the camshaft journals, camshaft housings and bearing caps.
   c. Install the No. 2 camshaft to the camshaft housing.
39. Install the camshaft.
   a. Clean the camshaft journals.
   b. Apply a light coat of engine oil to the camshaft journals, camshaft housings and bearing caps.
   c. Install the camshaft to the camshaft housing.
40. Install the bearing cap.
   a. Apply engine oil to the camshaft journals, camshaft housings and bearing caps.

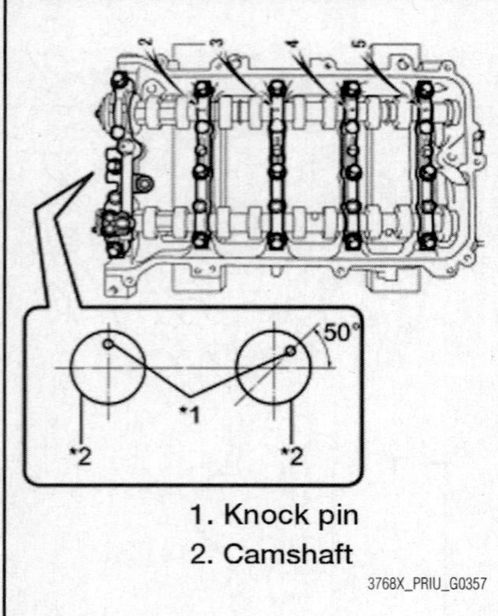

Fig. 162 Positioning the camshaft bearing caps

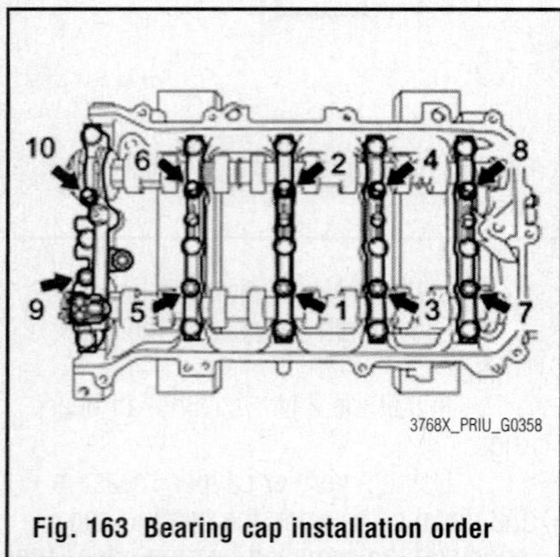

Fig. 163 Bearing cap installation order

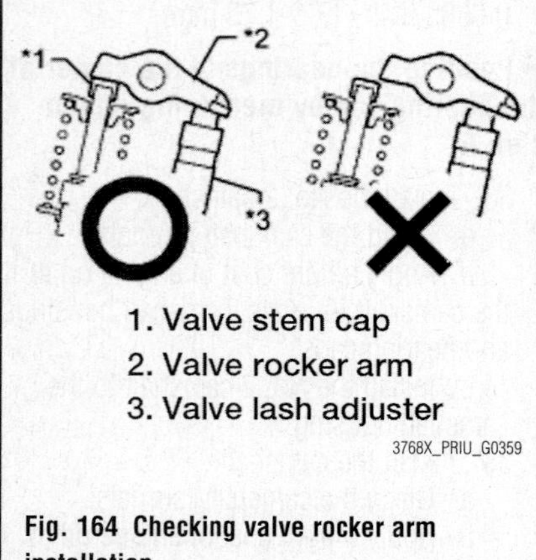

Fig. 164 Checking valve rocker arm installation

b. Make sure of the marks and numbers on the camshaft bearing caps and place them in each proper position and direction.

c. Tighten the 10 bolts in the order shown. Tighten to 12 ft. lbs. (16 Nm).

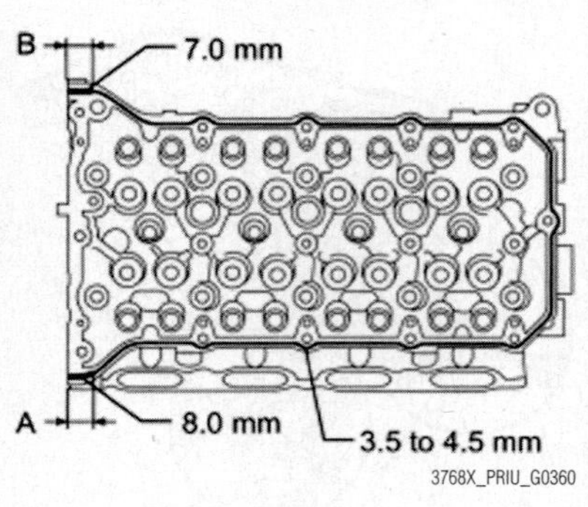

Fig. 165 Applying seal packing in a continuous line

41. Install the camshaft housing sub assembly.

a. Check that the valve rocker arms are installed as shown.

b. Apply seal packing in a continuous line as shown in the illustration.

➡Remove any oil from the contact surfaces.

➡Install the camshaft housing within 3 minutes and tighten the bolts within 10 minutes of applying seal packing.

➡Do not start the engine for at least 2 hours after installation.

c. Set the camshaft and No. 2 camshaft as shown in the illustration.

d. Install the camshaft housing with the 17 bolts and tighten them in the order shown in the illustration. Tighten to 20 ft. lbs. (27 Nm).

➡After installing the camshaft housing, make sure that the cam lobes are positioned as shown in the illustration.

➡If any of the bolts is loosened during installation, remove the camshaft housing, clean the installation surfaces, and reapply seal packing.

➡If the camshaft housing is removed because any of the bolts is loosened during installation, make sure that the previously applied seal packing does not enter any oil passages.

➡After installing the camshaft housing, wipe off any seal packing that seeped out from between the housing and cylinder head.

42. Install the camshaft timing sprocket.

a. Tighten the flange bolt with the camshaft timing sprocket secured in place. Tighten to 40 ft. lbs. (54 Nm).

43. Install the camshaft timing gear assembly.

a. Put the camshaft timing gear and camshaft together with the straight pin and key groove misaligned as shown in the illustration.

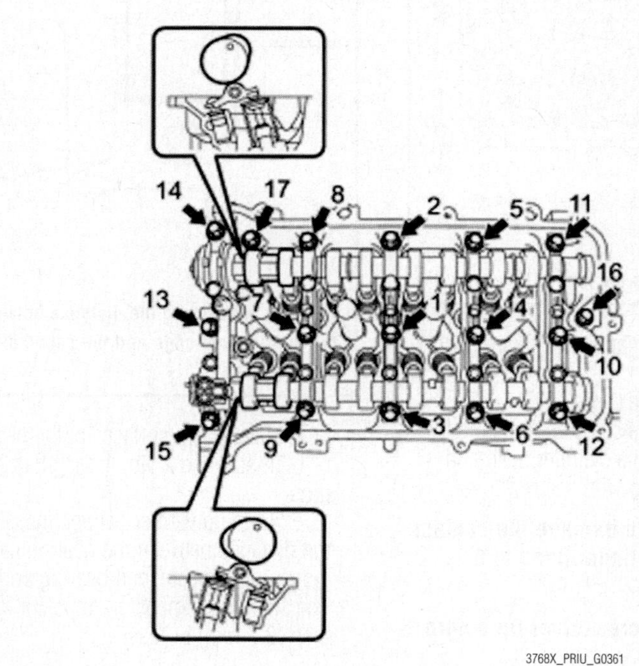

Fig. 166 Setting the camshaft and No. 2 camshaft and identifying the bolt tightening sequence

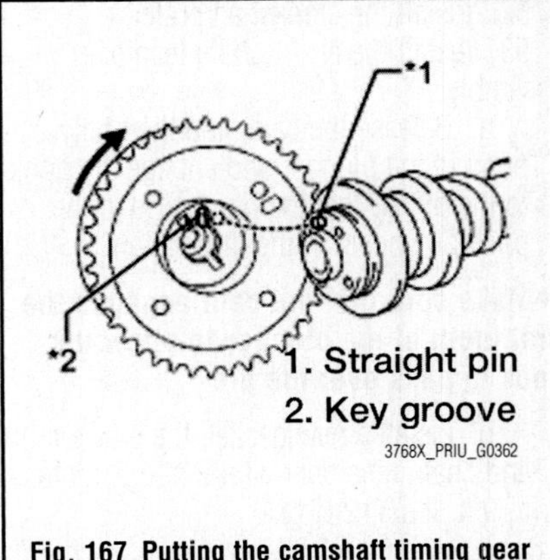

Fig. 167 Putting the camshaft timing gear and camshaft together

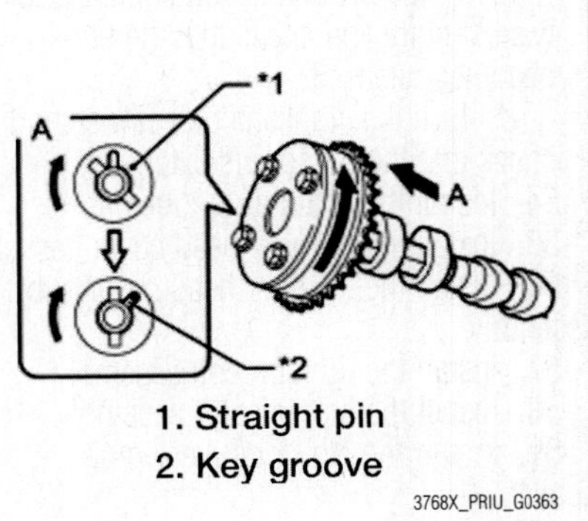

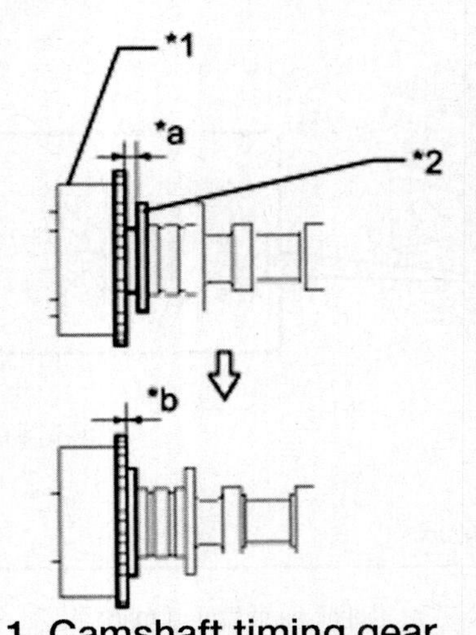

Fig. 169 Checking for clearance

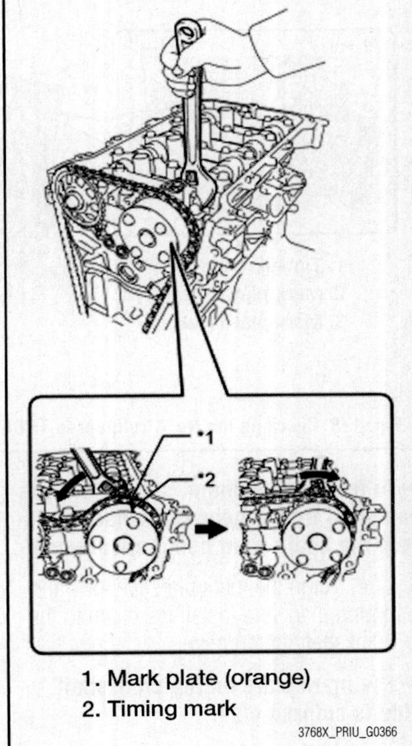

Fig. 171 Aligning the mark plate (orange) with the timing mark

Fig. 168 Turning the camshaft timing gear

➡Do not forcefully push in the camshaft timing gear. This may cause the camshaft straight pin tip to damage the installation surface of the camshaft timing gear.

b. Turn the camshaft timing gear as shown in the illustration while pushing it gently against the camshaft. Push further at the position where the pin fits into the groove.

➡Do not turn the camshaft timing gear in the retard direction (clockwise).

c. Check that there is no clearance between the camshaft timing gear and camshaft flange.

d. Tighten the flange bolt with the camshaft timing gear secured in place. Tighten to 40 ft. lbs. (54 Nm).

e. Check that the camshaft timing gear can move in the retard direction (clockwise) and is locked in the most retarded position.

44. Install the No. 1 chain vibration damper with the 2 bolts. Tighten to 15 ft. lbs. (21 Nm).

45. Set the No. 1 cylinder to TDC/Compression.

a. Temporarily install the crankshaft pulley bolt.

b. Turn the crankshaft to position the timing gear key to the top.

c. Check that the timing marks (*1) on the camshaft timing gear and camshaft timing sprocket are aligned.

d. Remove the crankshaft pulley bolt.

46. Install the chain sub assembly.

a. Align the mark plate (orange) with the timing mark as shown in the illustration and install the chain.

➡Be sure to position the mark plate at the front of the engine.

➡The mark plate on the camshaft side is colored orange.

➡Do not pass the chain around the sprocket of the camshaft timing gear. Only place it on the sprocket.

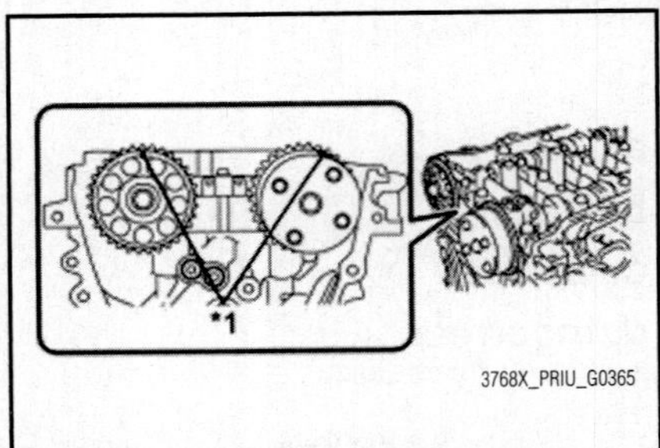

Fig. 170 Checking the timing marks

➡Pass the chain through the No. 1 vibration damper.

b. Hold the hexagonal portion of the camshaft with a wrench and turn the camshaft timing gear counterclockwise to align the mark plate (orange) and timing mark, and then install the chain.

c. Hold the hexagonal portion of the camshaft with a wrench and turn the camshaft timing gear clockwise.

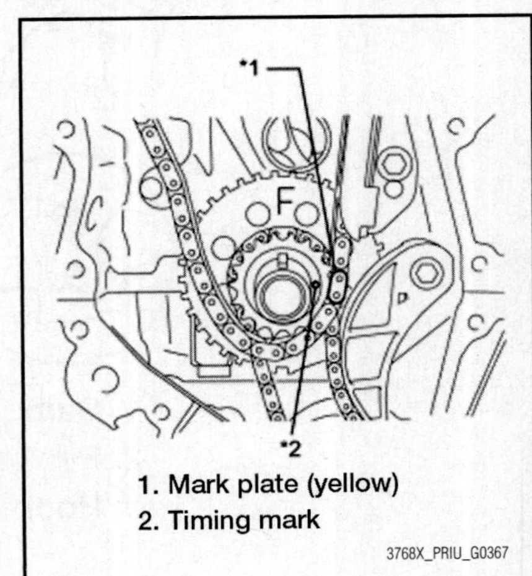

Fig. 172 Aligning the mark plate (yellow) and the timing mark

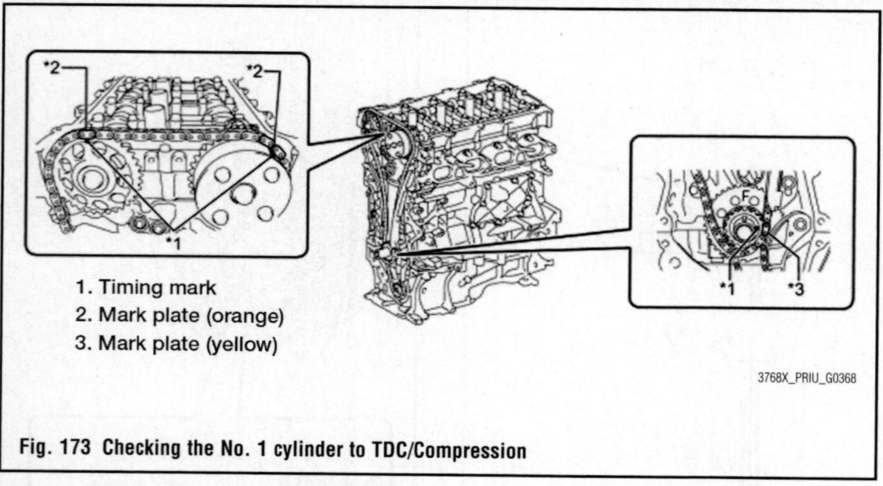

Fig. 173 Checking the No. 1 cylinder to TDC/Compression

➡To tension the chain, slowly turn the camshaft timing gear clockwise to prevent the chain from being misaligned.

d. Align the mark plate (yellow) and timing mark and install the chain to the crankshaft timing gear.

➡The mark plate on the crankshaft side is colored pink.

47. Check the No. 1 cylinder to TDC/Compression.

a. Check each timing mark at TDC/Compression.

48. Install the No. 2 chain vibration damper with the 2 bolts. Tighten to 7 ft. lbs. (10 Nm).

49. Install the chain tensioner slipper to the cylinder block.

50. Install the timing chain cover oil seal.

51. Install the timing chain cover sub assembly.

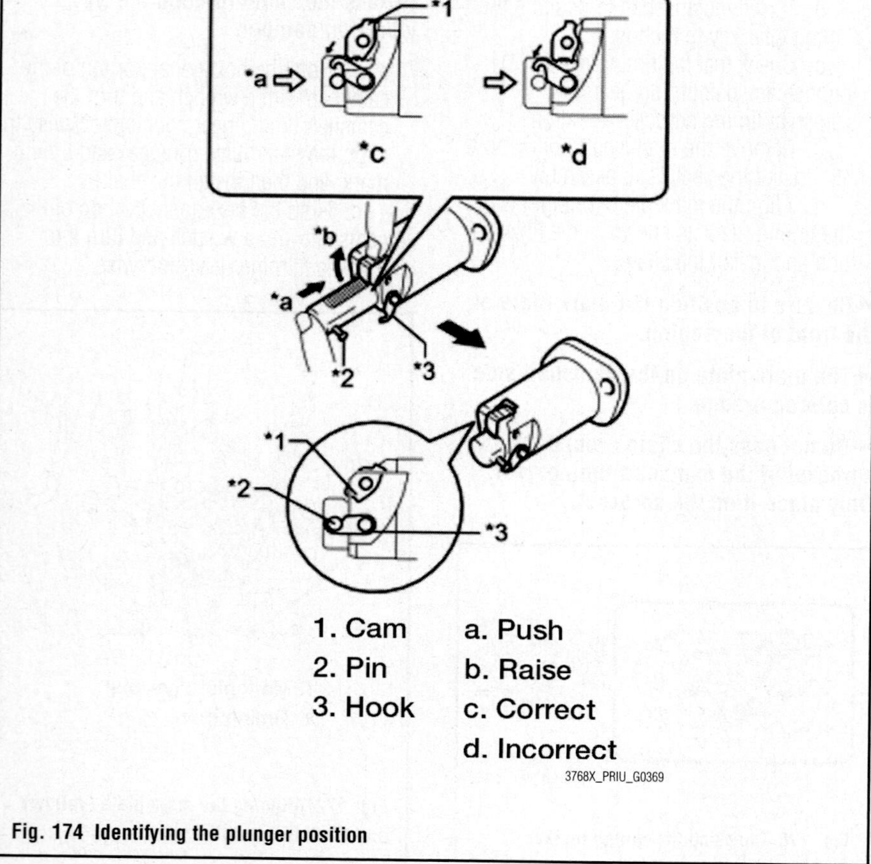

Fig. 174 Identifying the plunger position

52. Install the crankshaft pulley.

53. Install the No. 1 chain tensioner assembly.

a. Release the cam, and then fully push in the plunger and engage the hook to the pin so that the plunger is in the position shown in the illustration.

➡Make sure that the cam engages the first tooth of the plunger to allow the hook to pass over the pin.

b. Install a new gasket, the bracket and chain tensioner with 2 nuts. Tighten to 9 ft. lbs. (12 Nm).

➡If the hook releases the plunger while the chain tensioner is being installed, set the hook again.

c. Rotate the crankshaft counterclockwise slightly and check that the hook becomes released.

d. Turn the crankshaft clockwise and check that the plunger is extended.

54. Install the spark plug tube gasket.

55. Install the cylinder head cover gasket.

56. Install the cylinder head cover sub assembly.

57. Install the ignition coil assembly.

58. Install the fuel injector assembly.

59. Install the No. 1 delivery pipe spacer.

60. Install the fuel delivery pipe sub assembly.

61. Install the fuel vapor feed pipe.

62. Install the intake manifold.

63. Install the EGR with the cooler pipe sub assembly.

64. Install the EGR valve assembly.

65. Install the EGR pipe assembly.

66. Install the engine oil level dipstick guide.

67. Install the throttle body assembly.

68. Install the engine hangers.

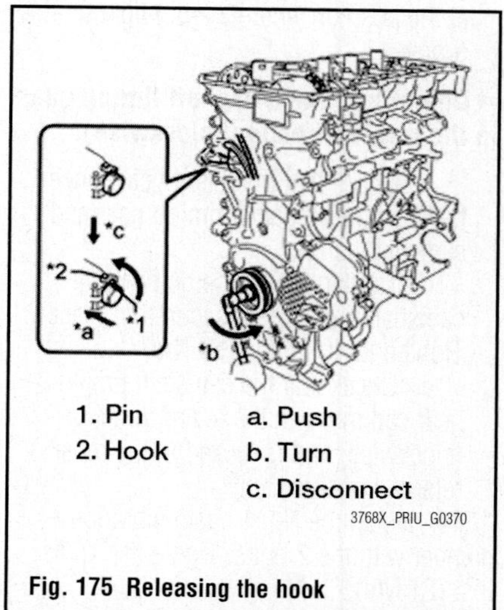

Fig. 175 Releasing the hook

69. Remove the engine from the engine stand.

70. Install the engine assembly with the transaxle.

CRANKSHAFT DAMPER

REMOVAL & INSTALLATION

See Figures 176 and 177.

1. Before servicing the vehicle, refer to the precautions.
2. Remove right engine under cover.
3. Loosen the clamp, and disconnect the air cleaner inlet from the air cleaner case.
4. Remove drive belt.
   a. Loosen the locknut.
   b. Turn adjust bolt, and then release the belt tension.
   c. Remove the belt.
5. Remove crankshaft pulley.
6. Using special tool SST: 09213-58013, hold the crankshaft pulley and loosen the crankshaft bolt.

➡When installing the special tool, be careful that the bolt which holds it does not interfere with the chain cover.

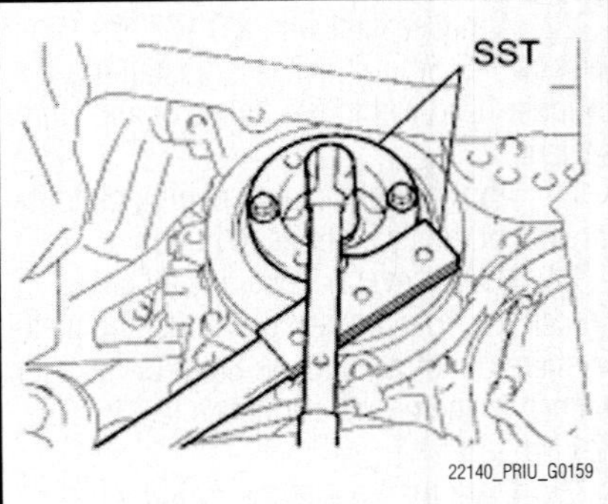

Fig. 176 Using the special tool, hold the crankshaft pulley and loosen the crankshaft bolt

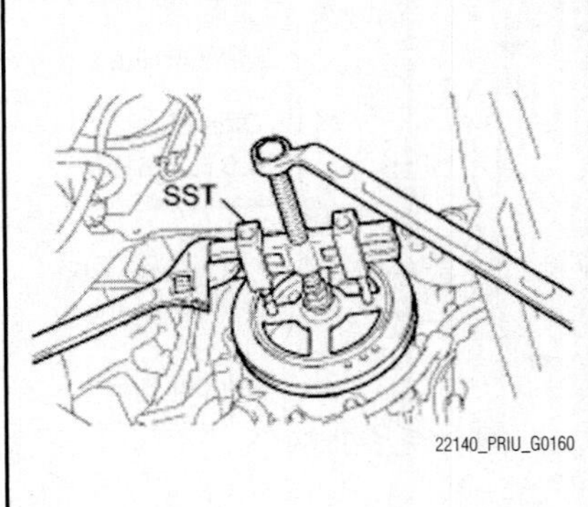

Fig. 177 Using the special tool, remove the crankshaft damper

   a. Loosen the crankshaft bolt until 2 to 3 threads of the bolt are tightened to the crankshaft.
   b. Using special tool SST:09950-50013, remove the crankshaft damper.
   c. Remove the crankshaft bolt.

To install:

7. Install crankshaft pulley.
   a. Align the hole of the crankshaft damper with the straight pin, then install the crankshaft damper.
   b. Using special tool SST: 09213-58013, hold the crankshaft damper and tighten the crankshaft bolt to 95 ft. lbs. (128 Nm).
8. Install drive belt.
   a. Temporarily install the belt on each pulley.
   b. Adjust drive belt tension.
   c. Turn adjust bolt to adjust the belt tension.
   d. Tighten the locknut to 30 ft. lbs. 40 Nm).
   e. Check drive belt tension.
9. Connect the air cleaner inlet to the air cleaner case and tighten the clamp to 27 inch lbs. (3 Nm).
10. Install right engine under cover.

CRANKSHAFT FRONT SEAL

REMOVAL & INSTALLATION

2009 Vehicles

See Figure 178.

1. Before servicing the vehicle, refer to the precautions.
2. Remove the crankshaft damper.
3. Using a knife, cut off the lip of the oil seal.
4. Using a screwdriver with the tip wrapped in tape, pry out the oil seal.

To install:

➡After removal, check if the crankshaft is damaged. If it is damaged, smooth the surface with 400-grit sandpaper.

5. Apply multi-purpose grease to the lip of a new oil seal. Keep the lip free of foreign objects.
6. Using seal installer and a hammer, tap in the oil seal until its surface is flush with the timing chain cover edge.

➡Be careful not to tap the oil seal at an angle.

7. Wipe any extra grease off the crankshaft.
8. Install the crankshaft damper.

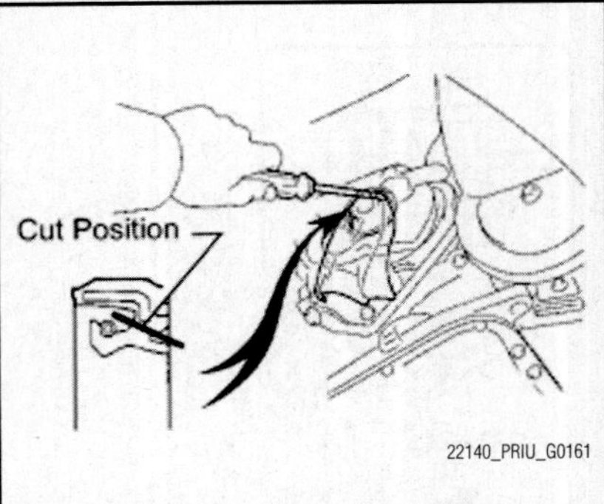

Fig. 178 Using a knife, cut off the lip of the oil seal

2010 Vehicles

See Figures 179 through 181.

1. Remove the front RH wheel.
2. Remove the rear engine under cover RH.
3. Remove the rear side rail reinforcement sub assembly RH.
4. Remove the crankshaft pulley.
   a. Using the special tool (09213-58014, 91551-80840 or 09330-00021), hold the pulley in place and loosen the pulley bolt.
   b. Using the special tool (09951-05010, 09952-05010, 09953-05020 or 09954-05021), remove the crankshaft pulley and pulley bolt.

➡If necessary, remove the pulley and pulley bolt using the special tool.

5. Remove the timing chain cover oil seal.
   a. Using a knife, cut off the lip of the oil seal.
   b. Using a screwdriver with the tip wrapped with tape, pry out the oil seal.

➡After removing, check the crankshaft for damage. If damage, smooth the surface with 400 grit sandpaper.

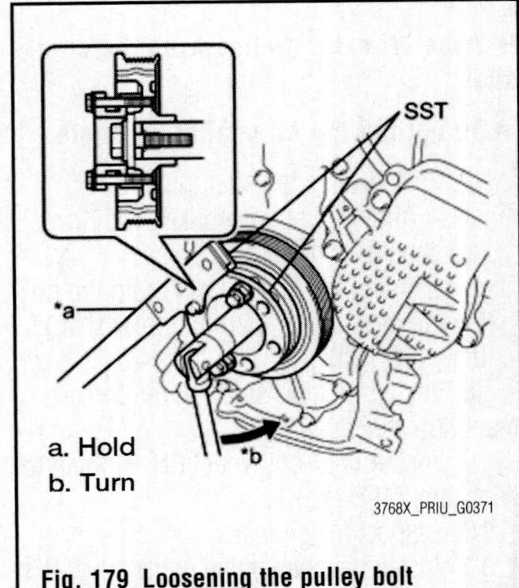

Fig. 179 Loosening the pulley bolt

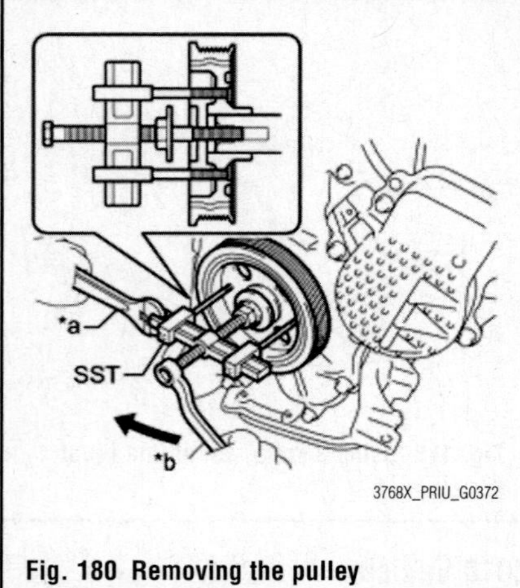

Fig. 180 Removing the pulley

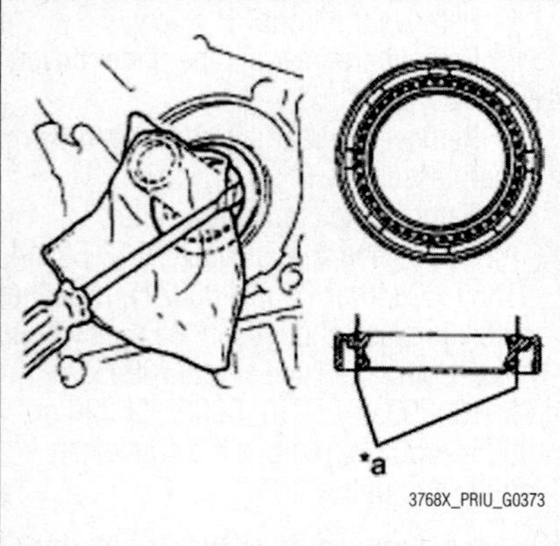

Fig. 181 Removing the timing chain cover oil seal

To install:

6. Install the timing chain cover oil seal.
   a. Apply MP grease to the lip of a new oil seal.

➡Keep the lip free from foreign matter.

   b. Using the special tool and a hammer, tap in the oil seal retainer edge.

➡Wipe off extra grease from the crankshaft.

➡Do not tap the oil seal at an angle.

7. Install the crankshaft pulley.
   a. Align the pulley set key with the key groove of the pulley.
   b. Using the special too, hold the pulley in place and tighten the bolt to 140 ft. lbs. (190 Nm).
8. Install the rear side rail reinforcement sub assembly RH.
9. Install the front wheel RH. Tighten to 76 ft. lbs. (103 Nm).
10. Inspect for oil leaks.
11. Install the rear engine under cover RH.

CYLINDER HEAD

REMOVAL & INSTALLATION

2009 Vehicles

See Figures 182 through 185.

1. Before servicing the vehicle, refer to the precautions.
2. Properly relieve the fuel system pressure.
3. Drain engine oil.
4. Drain transaxle and engine coolant.
5. Disconnect the negative battery cable.

➡Wait at least 90 seconds after disconnecting the cable from the negative (-) battery terminal to prevent airbag and seat belt pretensioner activation.

6. While wearing insulating gloves, slide up the lever of the service plug grip and remove while turning the lever to the left. Be sure to insulate the service plug with insulating tape.

⁂ CAUTION

Do not touch the high voltage connectors and terminals for 5 minutes after removing the service plug grip.

7. Remove the 2 bolts and 2 compression rings, then disconnect the front exhaust pipe assembly from the exhaust manifold.
8. While wearing insulating gloves, remove the inverter with converter assembly.
9. Remove timing chain.
10. Disconnect associated connectors and wire harnesses.
11. Remove the fuel pipe clamp.
12. Disconnect the fuel tube from the fuel delivery pipe. Even if the fuel tube is stuck and cannot be disconnected, do not use any tools. Push and pull the parts with the quick connector pinched to disconnect the tube.
13. Cover the disconnected fuel tube and fuel delivery pipe with a plastic bag in order to prevent foreign objects from entering them.
14. Disconnect the radiator inlet hose from the cylinder head.
15. Disconnect the hoses from the cylinder head as shown in the illustration.
16. Remove the bolt and disconnect the No. 1 water by-pass pipe.
17. Disconnect the hoses, remove the bolt and disconnect the oil dipstick guide.
18. Remove the camshafts.
19. Using an 8mm bi-hexagon wrench, loosen the cylinder head bolts in several steps in the proper sequence. Then remove the cylinder head bolts and washer.

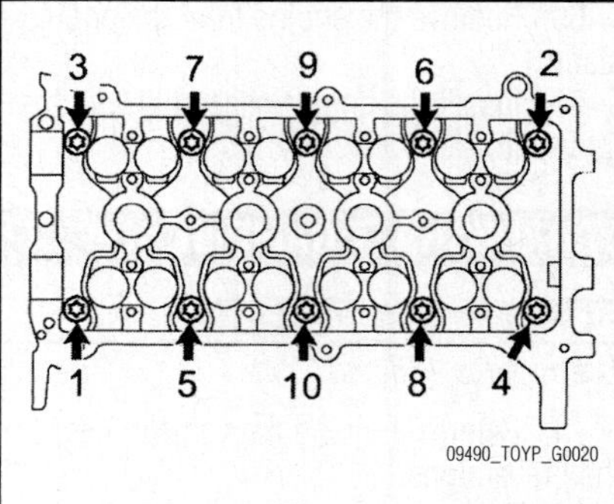

Fig. 182 Cylinder head bolt loosening sequence

⁂ CAUTION

When removing the bolt, do not drop the washer into the engine. Removing the cylinder head bolts in the wrong order may cause damage to the cylinder head.

20. Remove the cylinder head and gasket.

To install:

21. Install the cylinder head along with a new gasket.
22. Apply sealant with a 0.177–0.217 inch (4.5–5.5mm) thickness. Install the cylinder head within 3 minutes of applying sealant.
23. Apply a light coat of engine oil to the threads of the cylinder head bolts.
24. Using several steps, install and tighten the 10 cylinder head bolts and plate washers uniformly with an 8mm bi-hexagon wrench in the proper sequence to 21 ft. lbs. (29 Nm).
25. Mark the front of the cylinder head bolt with paint.
26. Retighten the cylinder head bolts by an additional 90 degrees and then another

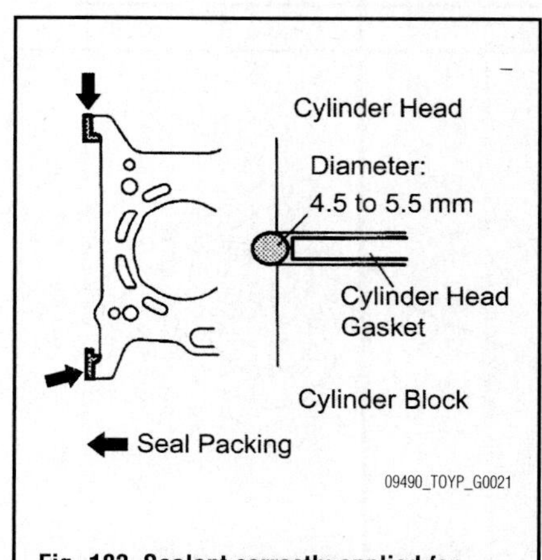

Fig. 183 Sealant correctly applied for cylinder head installation

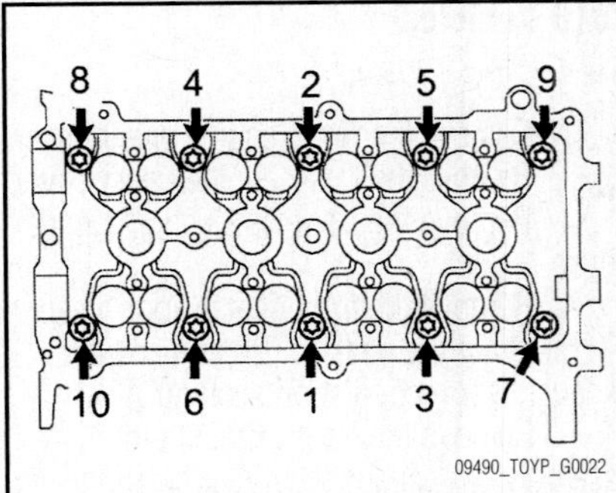

Fig. 184 Cylinder head bolt tightening sequence

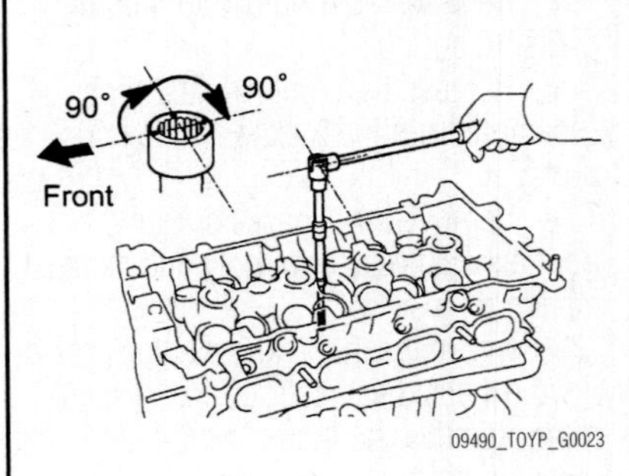

Fig. 185 Retightening the cylinder head bolts by an additional 90 degrees

90 degrees. Check that the paint mark is now 180 degrees opposite to the front.

27. Install the camshafts.
28. Install the oil dipstick guide with the bolt and tighten to 80 inch lbs. (9 Nm). Connect the hose.
29. Connect the water by-pass pipe with the bolt and tighten to 80 inch lbs. (9 Nm).
30. Connect the hoses.
31. Connect the radiator inlet hose.
32. Push the fuel main tube into the fuel delivery pipe until it makes a "click" sound. If the fuel tube is connected too tightly, apply a light coat of engine oil to the tip of the fuel delivery pipe. After connecting, check that the fuel tube is securely connected by pulling it.
33. Install the fuel pipe clamp.
34. Connect all associated connectors and wire harnesses.
35. Install the timing chain.
36. While wearing insulating gloves, install the inverter with converter assembly.
37. Install the front exhaust assembly.
38. While wearing insulating gloves, joint the service plug grip with the high voltage battery. While pushing the service plug grip to the right, rotate the lever to the right. Slide the lever down to lock the service plug grip in place.
39. Connect the negative battery cable.
40. Perform the power window initialization procedure.
41. Refill engine oil.
42. Refill transaxle and engine coolant.
43. Start the engine and check for leaks, check for abnormal noises, shock slippage, correct shift points and smooth operation.
44. Recheck transaxle and engine coolant.

2010 Vehicles

See Figures 186 and 187.

1. Remove the camshaft housing sub assembly.
2. Remove the No. 1 exhaust manifold heat insulator.
3. Remove the manifold stay.
4. Remove the exhaust manifold.
5. Remove the No. 1 valve rocker arm sub assembly.
6. Remove the valve lash adjuster assembly.
7. Remove the valve stem cap.
8. Remove the cylinder head sub assembly.
   a. Using a 10 mm bi-hexagon wrench, uniformly loosen and remove the 10 cylinder head bolts and 10 plate washers in several steps in the sequence shown in the illustration.

➡Be careful not to drop washers into the cylinder head.

➡Head warpage or cracking could result from removing the bolts in the wrong order.

   b. Using a screwdriver with its tip wrapped with tape, pry between the cylinder head and cylinder block, and remove the cylinder head.

➡Be careful not to damage the contact surfaces of the cylinder head and cylinder block.

To install:

➡The cylinder head bolts are tightened in 3 progressive steps.

   c. Place the cylinder head on the cylinder block.

➡Make sure that no oil is on the mounting surface of the cylinder head.

➡Place the cylinder head on the cylinder block gently in order not to damage the gasket with the bottom part of the head.

   d. Install the plate washers to the cylinder head bolts.
   e. Apply a light coat of engine oil to the threads and under the heads of the cylinder head bolts.
   f. Step 1: Using a 10 mm bi-hexagon wrench, install and uniformly tighten the 10 cylinder head bolts in several steps, in the sequence shown in the illustration. (Reverse the sequence shown). Tighten to 36 ft. lbs. (49 Nm).

➡Do not drop the plate washers into the cylinder head.

   g. Step 2: Mark each cylinder head bolt head with paint. Tighten the cylinder head bolts 90° in the sequence shown.
   h. Step 3: Tighten the cylinder head bolts another 45° in the same sequence as above. Check that the paint mark is now at a 135° angle to the front.

9. Install the valve stem cap.
10. Install the valve lash adjuster assembly.
11. Install the No. 1 valve rocker arm sub assembly.
12. Install the exhaust manifold.
13. Install the manifold stay.
14. Install the No. 2 exhaust manifold heat insulator.
15. Install the camshaft housing sub assembly.

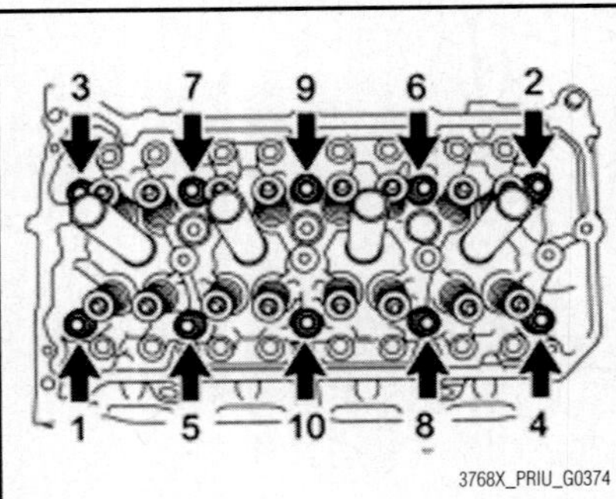

Fig. 186 Removing and installing the cylinder head sub assembly

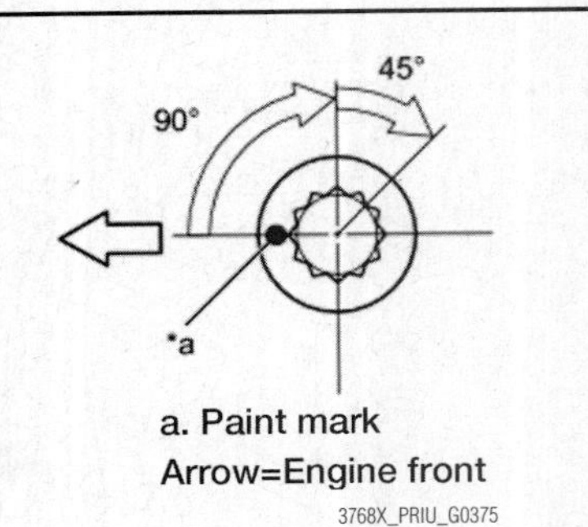

Fig. 187 Marking and tightening the cylinder head bolts an additional 90°

EXHAUST MANIFOLD

REMOVAL & INSTALLATION

2009 Vehicles

See Figures 188 and 189.

Toyota recommends this procedure be performed with the cylinder head removed. It may be possible to remove the exhaust manifold without removing the cylinder head.

***** CAUTION**

To avoid the danger of being burned, do not service the exhaust system while it is hot. Service should be performed only after the system cools down.

1. Before servicing the vehicle, refer to the precautions.
2. Properly relieve the fuel system pressure.
3. Disconnect the negative battery cable.

➡Wait at least 90 seconds after disconnecting the cable from the negative (-) battery terminal to prevent airbag and seat belt pretensioner activation.

4. While wearing insulating gloves, slide up the lever of the service plug grip and remove while turning the lever to the left. Be sure to insulate the service plug with insulating tape.

***** CAUTION**

Do not touch the high voltage connectors and terminals for 5 minutes after removing the service plug grip.

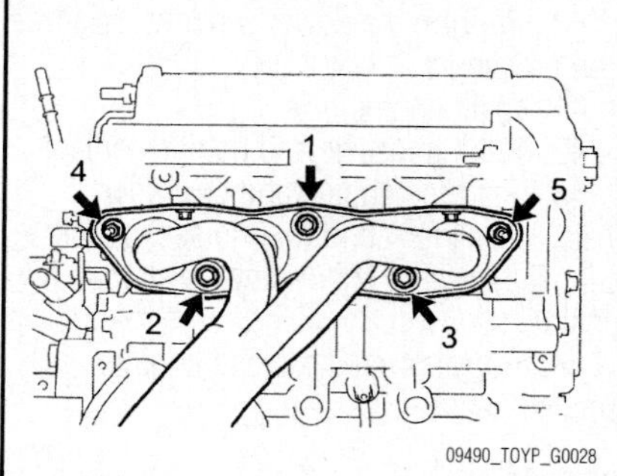

Fig. 189 Exhaust manifold nut and bolt tightening sequence

5. Remove the cylinder head.
6. Remove the 4 bolts and exhaust manifold insulator.
7. Remove the 3 bolts and 2 nuts, then remove the exhaust manifold.
8. Clean the sealing surfaces of the exhaust manifold and the cylinder head.

To install:

9. Install a new gasket, then install the exhaust manifold.
10. Tighten the 3 bolts and 2 nuts in the proper sequence to 20 ft. lbs. (27 Nm).
11. Install the exhaust manifold insulator with the 4 bolts and tighten to 71 inch lbs. (8 Nm).
12. Install the cylinder head.
13. Install the service plug.
14. Connect the negative battery cable.
15. Perform the power window initialization procedure.
16. Run the engine and check for exhaust leaks.

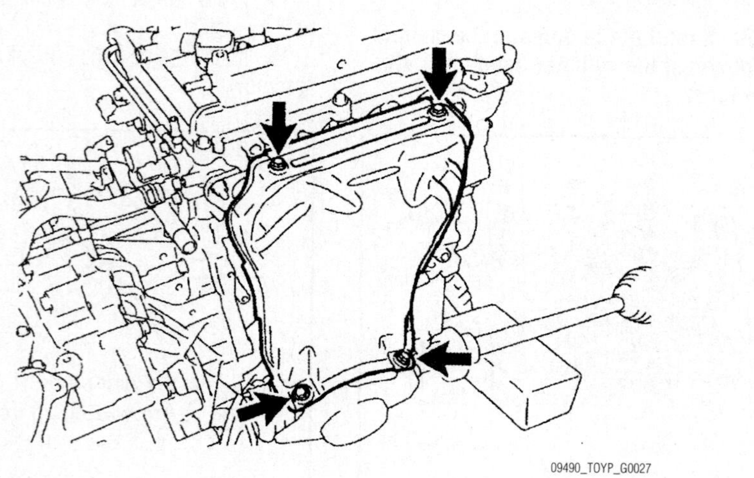

Fig. 188 Exhaust manifold insulator bolt locations

2010 Vehicles

See Figures 190 and 191.

1. Remove the No. 1 engine under cover.
2. Remove the No. 2 engine under cover.
3. Remove the front No. 3 engine under cover.
4. Remove the front center floor brace.
5. Drain the coolant (for engine with exhaust heat recirculation system).
6. Remove the front exhaust pipe assembly (w/ exhaust heat recirculation system).
   a. Disconnect the 3 clamps and oxygen sensor connector.
   b. Disconnect the 2 heater water hoses.
   c. Remove the 4 bolts and 4 compression springs.
   d. Remove the front exhaust pipe assembly from the 3 exhaust pipe supports.
   e. Remove the 2 gaskets from the front exhaust pipe assembly and exhaust manifold.
7. Remove the front exhaust pipe assembly (w/o exhaust heat recirculation system).
   a. Remove the 4 bolts and 4 compression springs.
   b. Remove the front exhaust pipe assembly from the 3 exhaust pipe supports.
   c. Remove the 2 gaskets from the front exhaust pipe assembly and exhaust manifold.
8. Remove the front No. 1 floor heat insulator.
   a. Remove the 3 nuts and No. 1 floor heat insulator.
9. Remove the No. 1 exhaust manifold heat insulator.
   a. Remove the 3 bolts and No. 1 exhaust manifold heat insulator.
10. Remove the air fuel ratio sensor.
11. Remove the manifold stay.
12. Remove the exhaust manifold.
    a. Remove the 7 nuts, exhaust manifold and 2 gaskets.

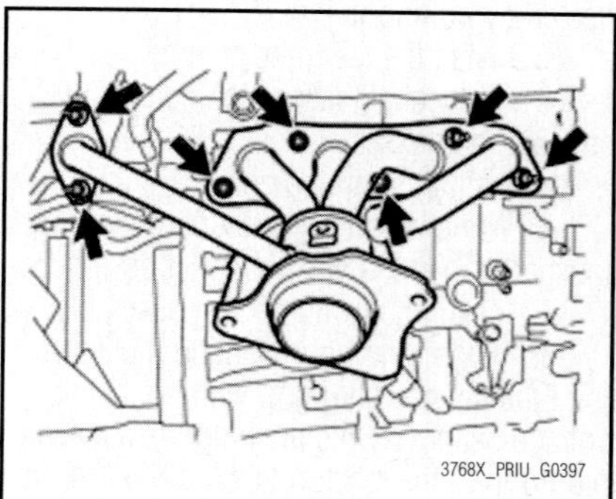

Fig. 190 Removing the exhaust manifold

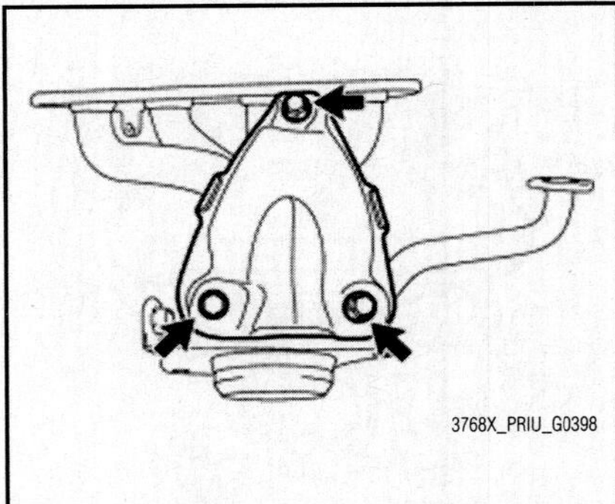

Fig. 191 Removing the No. 2 exhaust manifold heat insulator

13. Remove the No. 2 exhaust manifold heat insulator.
    a. Remove the 3 bolts and No. 2 exhaust manifold heat insulator.

To install:

14. Install the No. 2 exhaust manifold heat insulator with the 3 bolts. Tighten to 9 ft. lbs. (12 Nm).
15. Install the exhaust manifold.
    a. Install the 2 new gaskets and the exhaust manifold with the 7 nuts. Tighten to 15 ft. lbs. (21 Nm).
16. Install the manifold stay with the 3 bolts. Tighten to 32 ft. lbs. (43 Nm).
17. Install the air fuel ratio sensor.
18. Install the No. 1 exhaust manifold heat insulator with the 3 bolts. Tighten to 9 ft. lbs. (12 Nm).
19. Install the front No. 1 floor heat insulator with the 3 nuts. Tighten to 49 inch lbs. (5.5 Nm).
20. Install the front exhaust pipe assembly (w/ exhaust heat recirculation system).

➡When installing the water hose, ensure that the exhaust heat recirculation system is filled with coolant. Otherwise, the electric water pump may be damaged.

   a. Using a vernier caliper, measure the free length of the compression rings. Minimum (front) is 1.64 inch (41.5 mm). Minimum (rear) is 1.52 inch (38.5 mm).

➡If the free length is less than the minimum, replace the compression spring.

   b. Fully insert 2 new gaskets to the exhaust manifold and from exhaust pipe assembly.
   c. Using a plastic hammer and wooden block, tap in the new gaskets until its surface is flush with the exhaust manifold and from exhaust pipe assembly.
   d. Connect the front exhaust pipe assembly to the 3 exhaust pipe supports.
   e. Install the front exhaust pipe assembly with the 4 bolts and 4 compression springs. Tighten to 32 ft. lbs. (43 Nm).
   f. Connect the 2 heater water hoses.
   g. Connect the 3 clamps and oxygen sensor connector.

21. Install the front exhaust pipe assembly (w/o exhaust heat recirculation system).
    a. Using a vernier caliper, measure the free length of the compression springs. Minimum (front) is 1.64 inch (41.5 mm). Minimum (rear) is 1.52 inch (38.5 mm).

➡If the free length is less than the minimum, replace the compression spring.

   b. Fully insert 2 new gaskets to the exhaust manifold and front exhaust pipe assembly.
   c. Using a plastic hammer and wooden block, tap in the new gaskets until its surface is flush with the exhaust manifold and front exhaust pipe assembly.

➡Be careful with the installation direction of the gaskets.

➡Do not reuse the gaskets.

➡Do not damage the gaskets.

➡Do not push in the gasket by using the exhaust pipe when connecting it.

   d. Connect the front exhaust pipe assembly to the 3 exhaust pipe supports.
   e. Install the front exhaust pipe assembly with the 4 bolts and compression springs. Tighten to 32 ft. lbs. (43 Nm).

22. Add coolant (for engine with exhaust heat recirculation system).
23. Inspect for coolant leak (for engine with exhaust heat recirculation system).
24. Install the front center floor brace.
25. Install the front No. 3 engine under cover.
26. Install the No. 2 engine under cover.
27. Install the No. 1 engine under cover.
28. Inspect for exhaust gas leaks.

FLYWHEEL

REMOVAL & INSTALLATION

2009 Vehicles

See Figures 192 and 193.

1. Before servicing the vehicle, refer to the precautions.
2. Remove hybrid vehicle transaxle assembly.
3. Remove transmission input damper assembly.

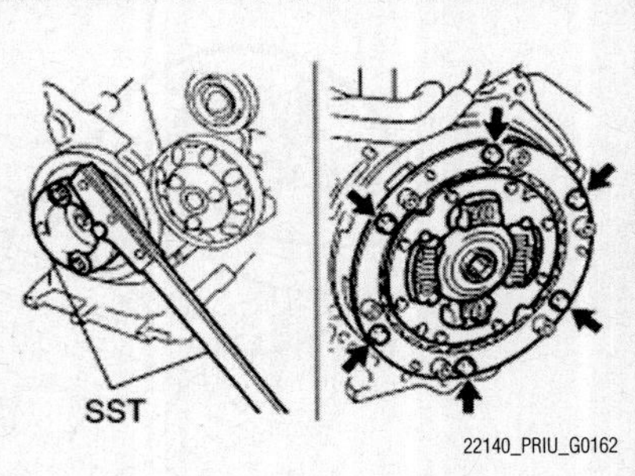

Fig. 192 Using the special tool, hold the crankshaft, remove the 6 bolts, then remove the input damper and input damper cover

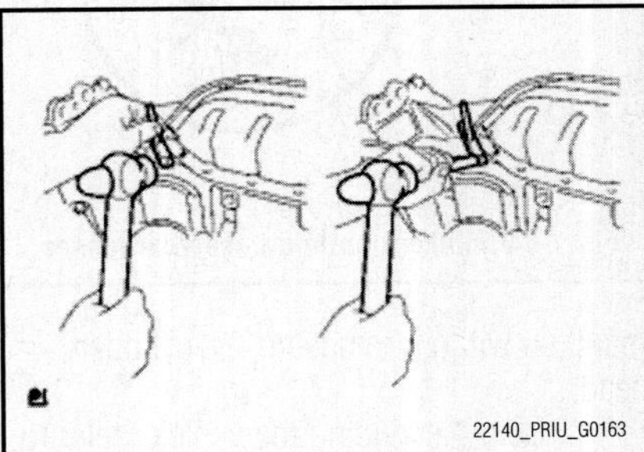

Fig. 193 Using the special tool, hold the crankshaft, remove the 6 bolts and flywheel

4. Remove the 6 bolts, then remove the input damper and input damper cover.
5. Remove the 6 bolts and flywheel.

To install:

6. Apply adhesive Part No. 08833-00070, THREE BOND 1324, or equivalent to the 2 or 3 threads of the bolt end.
7. Install the flywheel with the 6 bolts.
   a. 49 Nm (36 ft. lbs)
   b. Plus 90 degrees
8. Install the input damper and tighten the bolts to 20 Nm (15 ft. lbs)
9. Install hybrid vehicle transaxle assembly.

2010 Vehicles

Refer to REAR MAIN SEAL for removal and installation procedures.

INTAKE MANIFOLD

REMOVAL & INSTALLATION

2009 Vehicles

See Figures 194 and 195.

Toyota recommends this procedure be performed with the cylinder head removed. It may be possible to remove the intake

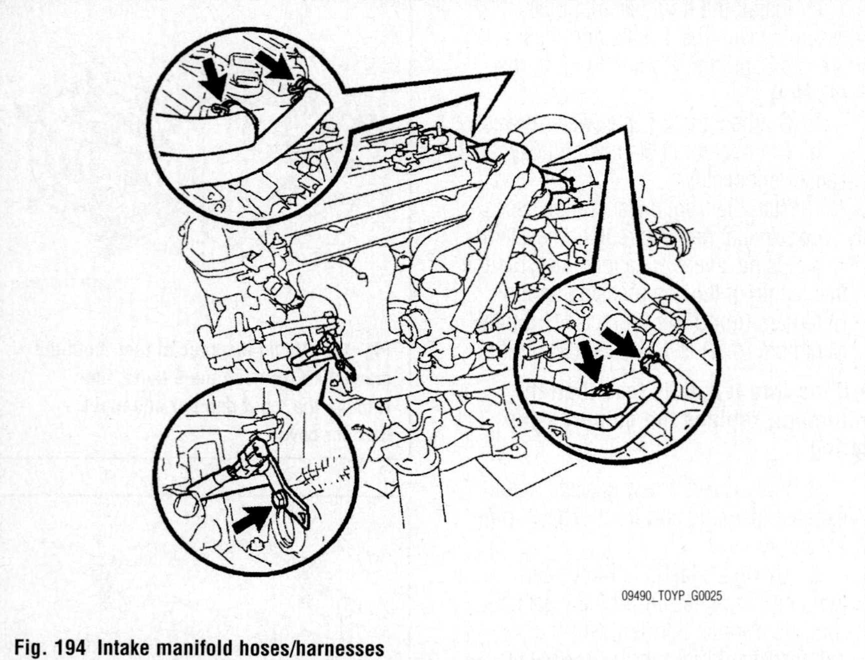

Fig. 194 Intake manifold hoses/harnesses

manifold without removing the cylinder head.

1. Before servicing the vehicle, refer to the precautions.
2. Disconnect the negative battery cable.

➡**Wait at least 90 seconds after disconnecting the cable from the negative (-) battery terminal to prevent airbag and seat belt pretensioner activation.**

3. Drain the engine coolant.
4. Remove the air cleaner assembly.
5. Remove the cylinder head.
6. Remove the oil dipstick guide.
7. Remove the bolt and knock control sensor with bracket.
8. Disconnect the wiring harness from the bracket.
9. Disconnect the ventilation hose.
10. Disconnect the water by-pass hose.

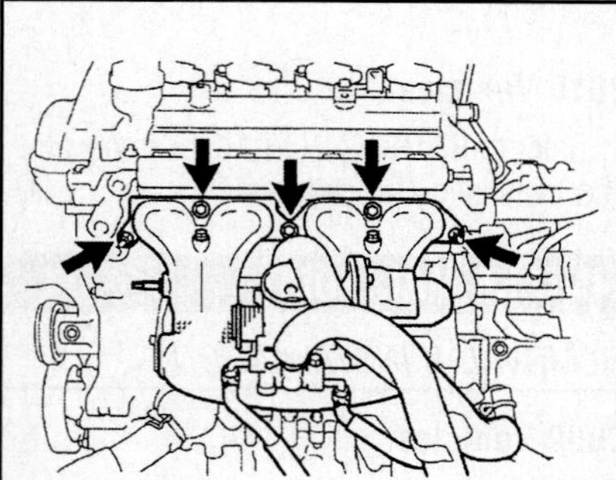

Fig. 195 Intake manifold mounting bolts and nuts

11. Remove the 3 bolts and 2 nuts, then remove the intake manifold and gasket.

To install:

12. Install a new gasket, then install the intake manifold with the bolts, nuts and brackets. Uniformly tighten the bolts and nuts in several passes to 15 ft. lbs. (20 Nm).
13. Connect the water by-pass hose.
14. Connect the ventilation hose.
15. Install the knock control sensor with bracket with the bolt and tighten to 80 inch lbs. (9 Nm).
16. Apply engine oil to a new O-ring, then install it to the dipstick guide.
17. Install the dipstick guide with the bolt and tighten to 80 inch lbs. (9 Nm).
18. Install the cylinder head.
19. Install the dipstick.
20. Install the air cleaner assembly.
21. Refill the engine coolant.
22. Connect the negative battery cable.
23. Perform the power window initialization procedure.
24. Start the engine and check for coolant leaks.

2010 Vehicles

See Figures 196 and 197.

1. Remove the throttle body assembly.
2. Remove the EGR pipe assembly.
3. Remove the manifold absolute pressure sensor.
4. Remove the vacuum switching valve assembly.

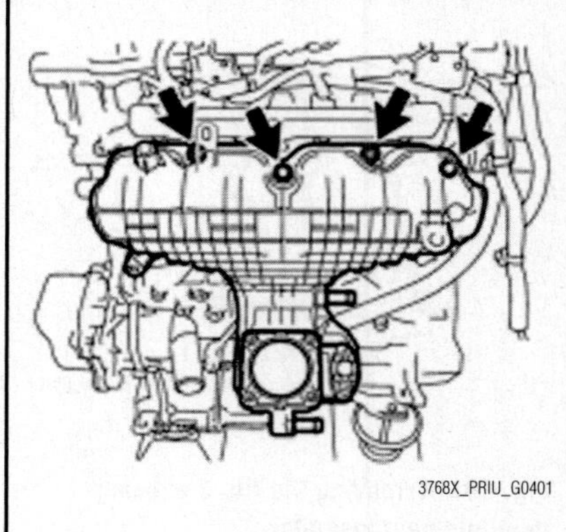

Fig. 196 Removing the 2 bolts, 2 nuts and intake manifold

5. Remove the intake manifold.
   a. Disconnect the 3 connectors and 2 wire harness clamps.
   b. Remove the engine oil level dipstick guide sub assembly.
   c. Disconnect the wire harness clamp, and then remove the 2 bolts and engine oil level dipstick guide sub-assembly.
   d. Remove the O-ring from the engine oil level dipstick guide sub-assembly.
   e. Disconnect the fuel vapor feed hose and ventilation hose.
   f. Remove the 2 bolts, 2 nuts and intake manifold.
   g. Remove the No. 1 intake manifold to head gasket.
6. Replace the stud bolt.

➡**If the stud bolt is deformed or the threads are damaged, replace it.**

   a. Using an E6 TORX® wrench, replace the 2 stud bolts. Tighten to 44 inch lbs. (5 Nm).

To install:

7. Install the intake manifold.
   a. Install a new No. 1 intake manifold to head gasket to the intake manifold.
   b. Install the intake manifold with the 2 bolts and 2 nuts. Tighten to 21 ft. lbs. (28 Nm).
   c. Connect the fuel vapor feed hose and ventilation hose.
   d. Install a new O-ring to the engine oil level dipstick guide sub-assembly.
   e. Apply a light coat of engine oil to the O-ring.
   f. Install the engine oil level dipstick guide sub-assembly with the 2 bolts and connect the wire harness clamp. Tighten bolt A to 21 ft. lbs. (28 Nm). Tighten bolt B to 15 ft. lbs. (21 Nm).
   g. Install the engine oil level dipstick.
   h. Connect the 2 wire harness clamps and 3 connectors.

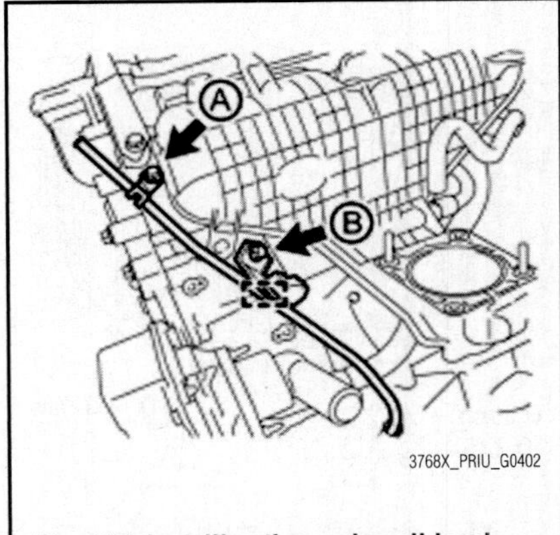

Fig. 197 Installing the engine oil level dipstick guide sub assembly

8. Install the vacuum switching valve assembly.
9. Install the manifold absolute pressure sensor.
10. Install the EGR pipe assembly.
11. Install the throttle body assembly.

OIL PAN

REMOVAL & INSTALLATION

2009 Vehicles

See Figures 198 through 203.

Toyota recommends this procedure be performed with the engine assembly removed. It may be possible to remove the oil pan without removing the engine assembly.

1. Before servicing the vehicle, refer to the precautions.
2. Disconnect the negative battery cable.

➡Wait at least 90 seconds after disconnecting the cable from the negative (-) battery terminal to prevent airbag and seat belt pretensioner activation.

3. Raise and safely support the vehicle securely on jackstands.
4. Drain engine oil.
5. Remove engine assembly.
6. Install engine to engine stand.
7. Remove the timing chain.
8. Remove the cylinder head.
9. Remove engine wire.
10. Remove the 2 nuts, bolt and water bypass pipe.
11. Remove the thermostat.
12. Remove the Knock Sensor (KS).
13. Remove the oil pressure switch.
14. Remove the engine coolant drain union.
15. Remove the oil filter.

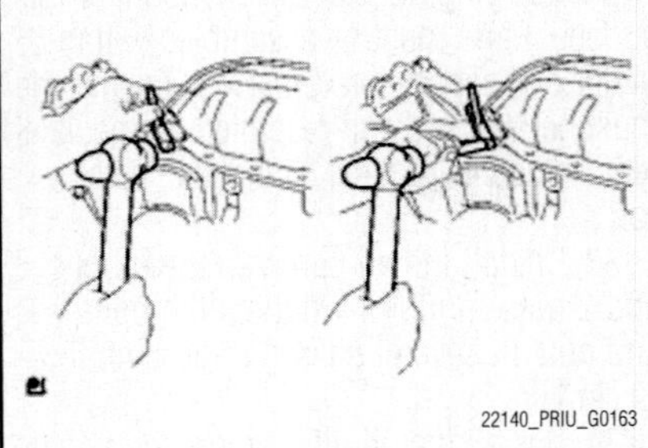

Fig. 198 Insert the blade of oil pan seal cutter between the oil pan and No. 2 oil pan, cut off the applied sealer and remove the No. 2 oil pan

16. Using a 12mm hexagon wrench, remove the oil filter union.
17. Remove the 9 bolts and 2 nuts of the No. 2 oil pan.
18. Insert the blade of SST 09032-00100 between the oil pan No. 1 and oil pan No. 2, and cut off applied sealer and remove the oil pan. Be careful not to the damage the oil pan contact surface of the oil pan No. 1. or the oil pan No. 2 flange.
19. Remove the bolt and 2 nuts, oil strainer and gasket.
20. Uniformly loosen and remove the 13 bolts, in several passes.
21. Using screwdriver remove the oil pan No. 1 by prying the portions between the cylinder block and oil pan No. 1.
22. Remove the 2 O-rings from the cylinder block.

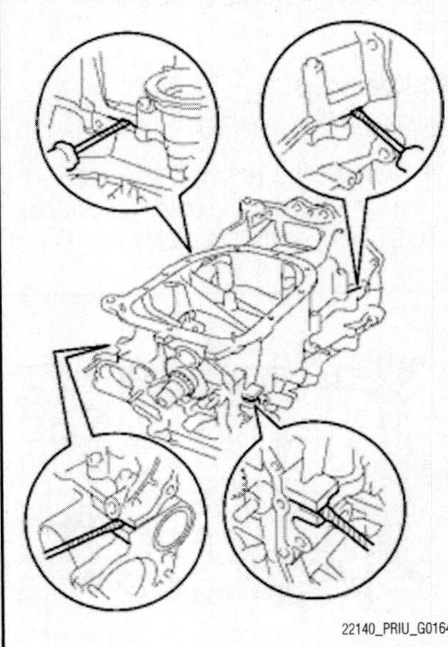

Fig. 199 Using a screwdriver, remove the oil pan by prying between the cylinder block and oil pan

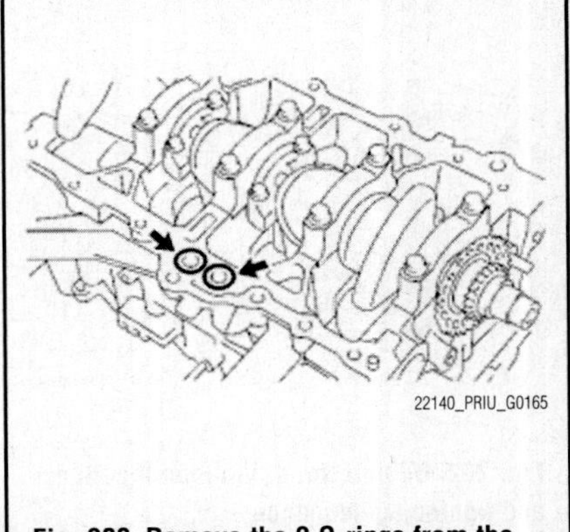

Fig. 200 Remove the 2 O-rings from the cylinder block

To install:

23. Remove any old sealant material and be careful not to drop any oil on the contact surface of the oil pan No. 1 and cylinder block. Using a razor blade and gasket scraper, remove all the old sealant material from the gasket surfaces and sealing grooves. Thoroughly clean all components to remove all the loose material. Using a non-residue solvent, clean both sealing surfaces.
24. Apply sealant to the oil pan No. 1 with a seal width of 0.08–0.12 inch (2–3mm). Avoid applying an excessive amount to the surface. Parts must be assembled within 3 minutes of application. Otherwise the material must be removed and reapplied. Immediately remove nozzle from the tube and reinstall cap.

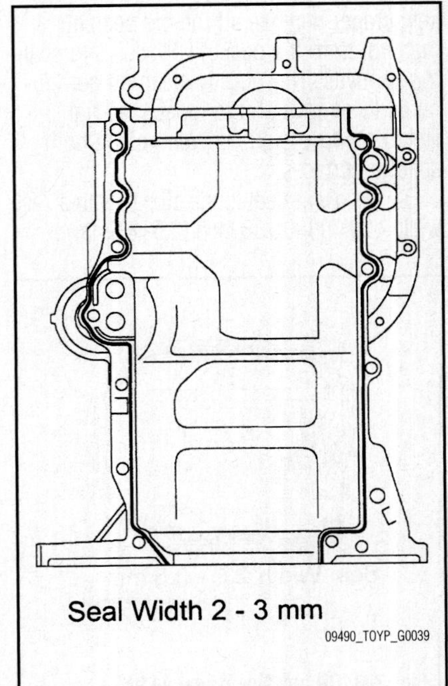

Fig. 201 Oil pan No. 1 seal width

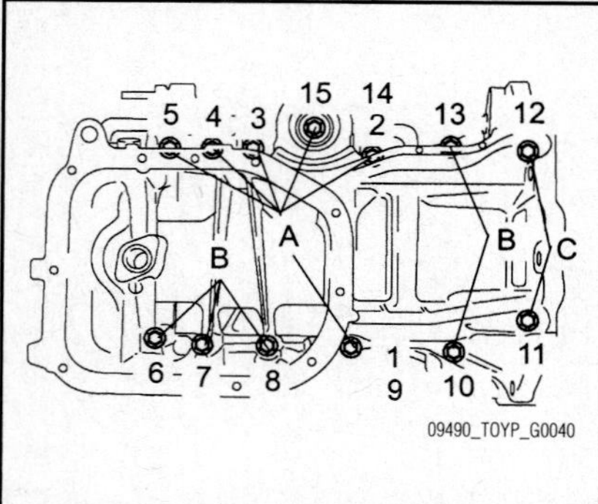

Fig. 202 Oil pan No. 1 bolt identification and tightening sequence

25. Install new O-rings to the cylinder block.

26. Using a plastic-faced hammer, lightly tap the oil pan No. 1 to ensure a proper fit.

27. Install and uniformly tighten the 13 bolts, in several passes, in sequence to 18 ft. lbs. (24 Nm).

28. Each bolt indicated in the illustration shown is the following length:

- Bolt A—1.929 inches (49mm)
- Bolt B—3.465 inches (88mm)
- Bolt C—5.669 inches (144mm)

29. Install rear crankshaft oil seal. Wipe sealant away from the contact surface of the cylinder block assembly and oil seal.

30. Install a new gasket, and oil strainer with the bolt and 2 nuts. Tighten the bolt and 2 nuts to 8 ft. lbs. (11 Nm).

31. Remove any old sealant material and be careful not to drop any oil on the contact surface of the main bearing cap and oil pan. Using a razor blade and gasket scraper, remove all the old sealant material from the gasket surfaces and sealing grooves. Thoroughly clean all components to remove all the loose material. Using a non-residue solvent, clean both sealing surfaces.

32. Apply sealant to the oil pan with a seal width of 0.098–0.138 inch (2.5–3.5mm).

Fig. 203 Oil pan No. 2 seal width

Avoid applying an excessive amount to the surface. Parts must be assembled within 3 minutes of application. Otherwise the material must be removed and reapplied. Immediately remove nozzle from the tube and reinstall cap.

33. Install the oil pan with the 9 bolts and 2 nuts. Uniformly tighten the bolts and nuts in several passes to 80 inch lbs. (9 Nm).

34. Install the oil filter union and tighten to 21 ft. lbs. (30 Nm).

35. Install the oil filter.

36. Apply adhesive to 2 or 3 threads and install the engine coolant drain union. Torque the union to 25 ft. lbs. (35 Nm) and after applying the specified torque, rotate the drain union clockwise until its drain port is facing downward.

37. Install the Knock Sensor (KS) and tighten to 29 ft. lbs. (39 Nm).

38. Install the oil pressure switch.

39. Install the thermostat.

40. Install the water bypass pipe and tighten bolts to 80 inch lbs. (9 Nm)

41. Install the engine wire.

42. Install the cylinder head.

43. Install the timing chain.

44. Remove the engine assembly from the engine stand.

45. Install the engine assembly into the vehicle.

46. Refill the engine with oil.

47. Connect the negative battery cable.

48. Perform the power window initialization procedure.

49. Start the engine and check for leaks.

2010 Vehicles

See Figures 204 and 205.

1. Remove the 10 bolts and 2 nuts.
   a. Insert the blade of the oil pan seal cutter between the crankcase and oil pan.

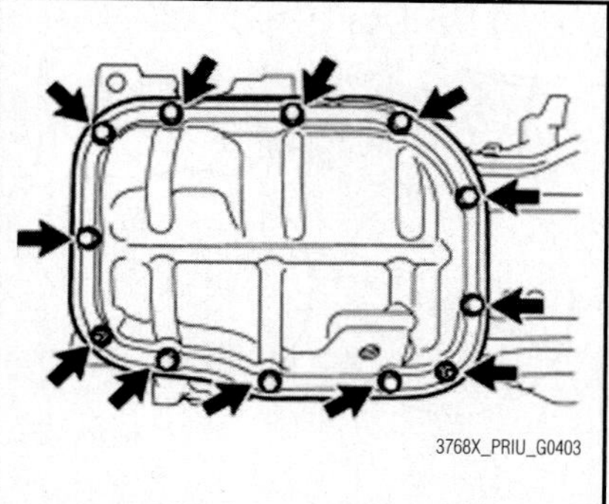

Fig. 204 Removing the oil pan bolts and nuts

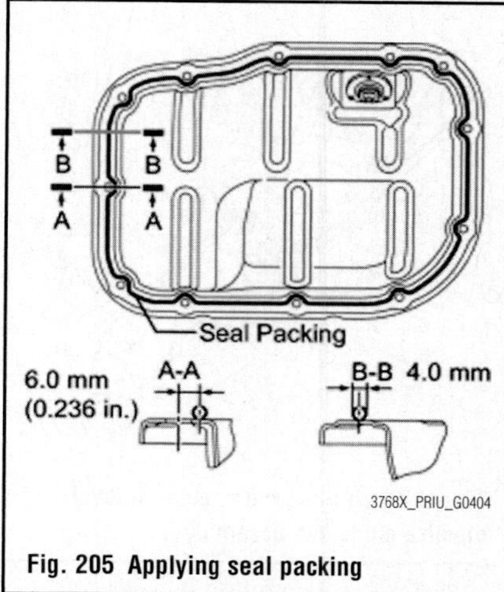

Fig. 205 Applying seal packing

Cut through the sealer and remove the oil pan.

➡Be careful not to damage the contact surfaces of the crankcase, chain, cover, and oil pan.

To install:

2. Install the No. 2 oil pan sub assembly.
   a. Remove any old packing material and be careful not to drop any oil on the contact surfaces of the cylinder block and oil pan.
   b. Apply a continuous bead of seal packing (0.157 inch (4 mm) diameter) as shown in the illustration.

➡Remove any oil from the contact surfaces.

➡Install the oil pan within 3 minutes after applying seal packing.

➡Do not start the engine for at least 2 hours after installing the oil pan.

   c. Install the No. 2 oil pan with the 10 bolts and 2 nuts. Tighten to 7 ft. lbs. (10 Nm).

OIL PUMP

REMOVAL & INSTALLATION

2009 Vehicles

See Figures 206 and 207.

1. Before servicing the vehicle, refer to the precautions.
2. Disconnect the negative battery cable.

➡Wait at least 90 seconds after disconnecting the cable from the negative (-) battery terminal to prevent airbag and seat belt pretensioner activation.

3. Drain the engine oil.
4. Remove the timing chain cover.

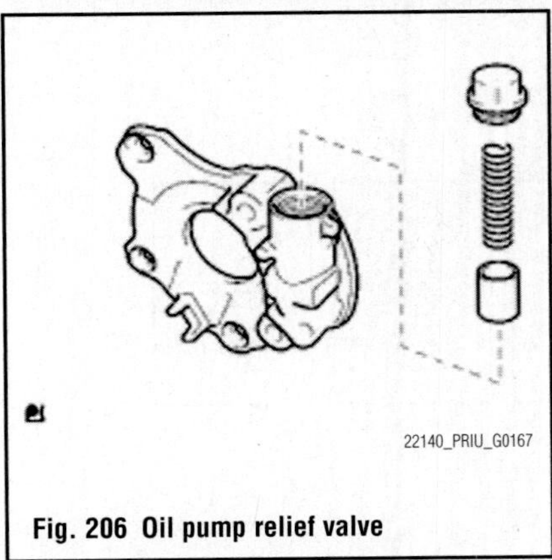

Fig. 206 Oil pump relief valve

5. Remove the 2 bolts, 3 screws and oil pump cover from the timing chain cover.

6. Remove the drive and driven rotors.

7. Remove the plug, spring and relief valve.

To install:

8. Insert the relief valve and spring into the pump body hole, and install the plug. Torque the plug to 18 ft. lbs. (25 Nm).

9. Place the drive and driven rotors into timing chain cover with the marks facing the oil pump cover side.

10. Install the oil pump cover to the timing chain cover with the 2 bolts and 3 screws. Torque the bolts to 78 inch lbs. (8.8 Nm), and the screws to 8 ft. lbs. (10.3 Nm).

11. Install the timing chain cover (refer to the timing chain procedure).

12. Refill engine with engine oil.

13. Connect the negative battery cable.

14. Perform the power window initialization procedure.

15. Start the engine and check the engine oil pressure.

16. Check that no leaks are present.

2010 Vehicles

See Figures 208 through 213.

1. Remove the engine assembly with the transaxle.
2. Install the engine on an engine stand.
3. Remove the engine hangers.
4. Remove the throttle body assembly.
5. Remove the engine oil level dipstick guide.
6. Remove the EGR pipe assembly.
7. Remove the EGR valve assembly.
8. Remove the EGR cooler sub assembly.
9. Remove the intake manifold.
10. Remove the fuel vapor feed pipe.
11. Remove the fuel delivery pipe sub assembly.
12. Remove the No. 1 delivery pipe spacer.
13. Remove the fuel injector assembly.
14. Remove the water inlet with the thermostat sub assembly.
15. Remove the ignition coil assembly.
16. Remove the cylinder head cover sub assembly.
17. Remove the cylinder head cover gasket.
18. Remove the spark plug tube gasket.
19. Set the No. 1 cylinder to TDC/Compression.
20. Remove the crankshaft pulley.
21. Remove the No. 1 chain tensioner assembly.
22. Remove the timing chain cover sub assembly.
    a. Remove the 3 bolts and engine mounting bracket RH.
    b. Remove the 4 bolts and oil filter bracket.
    c. Remove the 2 o-rings.
    d. Remove the 18 bolts and seal washer.
    e. Remove the timing chain cover by prying between the timing chain cover and cylinder head or cylinder block with a screwdriver.

➡Be careful not to damage the contact surfaces of the timing chain cover, cylinder block, and cylinder head.

➡Pry the timing chain cover out evenly in order to prevent damaging the knock pins.

➡Tape the screwdriver tip before use.

    f. Remove the 3 o-rings.

23. Remove the timing chain cover oil seal.
    a. Place the timing chain cover on wooden blocks.
    b. Using a screwdriver, pry out the oil seal.

➡Tape the screwdriver tip before use.

➡Do not damage the surface of the oil seal press fit hole.

24. Remove the chain tensioner slipper.
25. Remove the No. 1 chain vibration damper.
26. Remove the No. 2 chain vibration damper.
27. Remove the chain sub assembly.
28. Remove the crankshaft timing sprocket.
29. Remove the No. 2 chain sub assembly.
    a. Temporarily tighten the crankshaft pulley and crankshaft pulley bolt.
    b. Using the special tool (09213-58014, 91551-80840, or 09330-

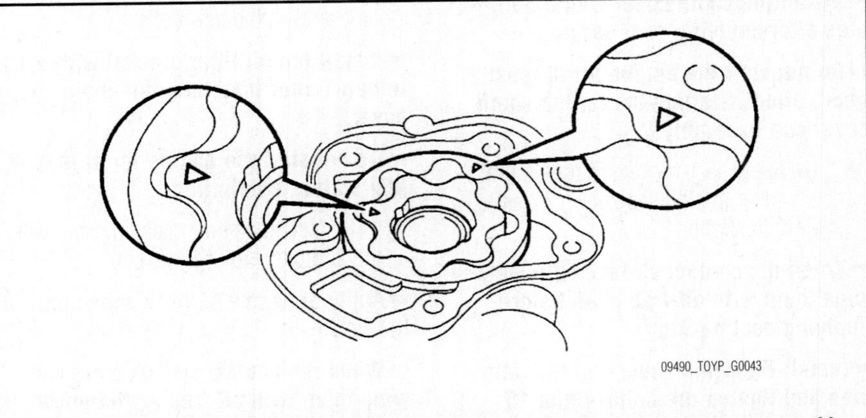

Fig. 207 Install the rotors into timing chain cover with the marks facing the oil pump cover side

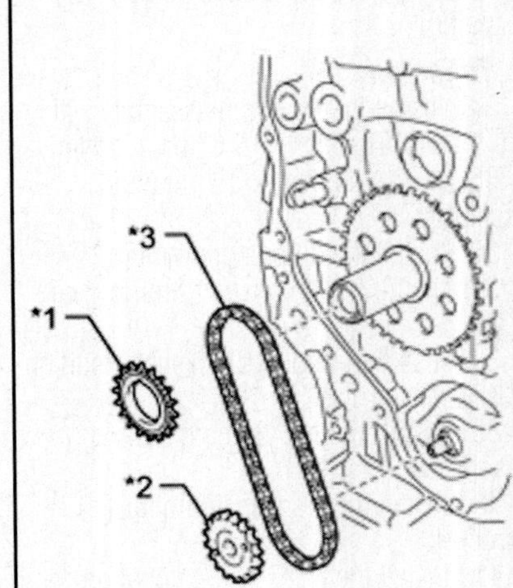

Fig. 208 Removing the oil pump drive gear, oil pump drive shaft gear and No. 2 chain sub assembly

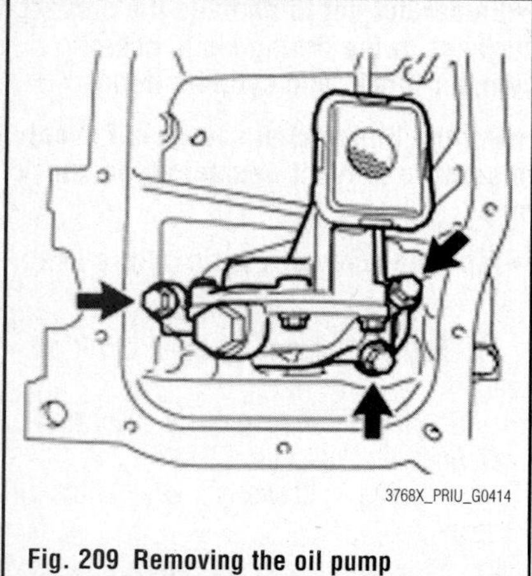

Fig. 209 Removing the oil pump

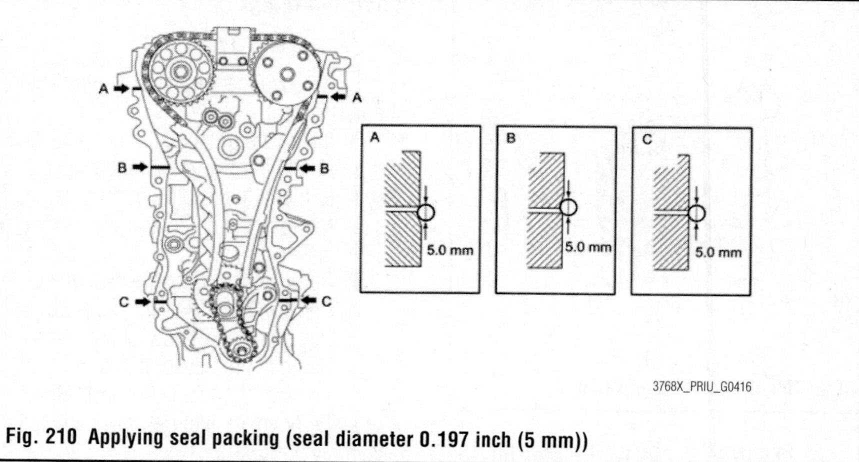

Fig. 210 Applying seal packing (seal diameter 0.197 inch (5 mm))

00021), remove the oil pump drive shaft sprocket nut while holding the crankshaft pulley.

c. Remove the special tool, crankshaft pulley, and crankshaft pulley bolt.

d. Remove the bolt, chain tensioner plate, and spring.

e. Remove the oil pump drive gear, oil pump drive shaft gear, and No. 2 chain sub assembly.

30. Remove the No. 1 Crankshaft Position (CKP) sensor plate.
31. Remove the oil pan drain plug.
32. Remove the No. 2 oil pan sub assembly.
33. Remove the oil pump assembly.

a. Remove the 3 bolts and oil pump.

To install:

34. Install the oil pump assembly with the 3 bolts. Tighten to 16 ft. lbs. (21 Nm).
35. Install the No. 2 oil pan sub assembly.
36. Install the oil pan drain plug.
37. Install the crankshaft timing gear key.
38. Install the No. 1 Crankshaft Position (CKP) sensor plate.
39. Install the No. 2 chain sub assembly.
40. Install the crankshaft timing sprocket.
41. Install the No. 1 chain vibration damper.
42. Set the No. 1 cylinder to TDC/Compression.
43. Install the chain sub assembly.
44. Check the No. 1 cylinder to TDC/Compression.
45. Install the chain tensioner slipper.
46. Install the No. 2 chain vibration damper.
47. Install the timing chain cover oil seal.

a. Using the special tool (09223-22010), tap in a new oil seal until its surface is flush with the timing chain cover edge.

➡Keep the lip free from foreign matter.

➡Do not tap on the oil seal at an angle.

➡Make sure that the oil seal edge does not stick out of the timing chain case.

➡Tap in the oil seal so that it is positioned within 1.0 mm from the edge of the timing chain case.

b. Apply MP grease to the lip of the oil seal.

48. Install the timing chain cover sub assembly.

a. Remove any old packing (FIPG) material and be careful not to drop any oil on the contact surfaces of the timing chain cover, cylinder head, and cylinder block.

b. Install 3 new o-rings.

c. Apply seal packing.

➡Remove any oil from the contact surfaces.

➡Install the chain cover within 3 minutes after applying seal packing.

➡Do not start the engine for at least 2 hours after installing the timing chain cover sub-assembly.

d. Apply seal packing to the timing chain cover in a line as shown in the following illustration.

➡When the contact surfaces are wet, wipe them with oil-free cloth before applying seal packing.

➡Install the chain cover within 3 minutes and tighten the bolts within 15 minutes after applying seal packing.

➡Do not start the engine for at least 2 hours after installation.

➡When the contact surfaces are wet, wipe them with oil-free cloth before applying seal packing.

➡Install the timing chain cover within 3 minutes and tighten the bolts within 10 minutes after applying seal packing.

➡After applying seal packing to the timing chain cover, install the engine mounting bracket and oil filter bracket within 10 minutes.

➡Do not add engine oil for at least 2 hours after installation.

e. Clean the bolt and fitting hole.

f. Install the timing chain cover.

g. Temporarily tighten the engine mounting bracket RH with the 3 bolts.

➡Install the mounting bracket within 10 minutes after installing the chain cover.

➡Do not start the engine for at least 2 hours after installation.

h. Install 2 new o-rings.

i. Temporarily tighten the oil filter bracket with the new bolts.

➡Install the oil filter bracket within 10 minutes after installing the chain cover.

➡Do not start the engine for at least 2 hours after installation.

j. Install the timing chain cover with the 25 bolts and seal washer.

➡Apply adhesive 1324 to screw part of the bolt F.

➡When the contact surfaces are wet, wipe them with oil-free cloth before applying seal packing.

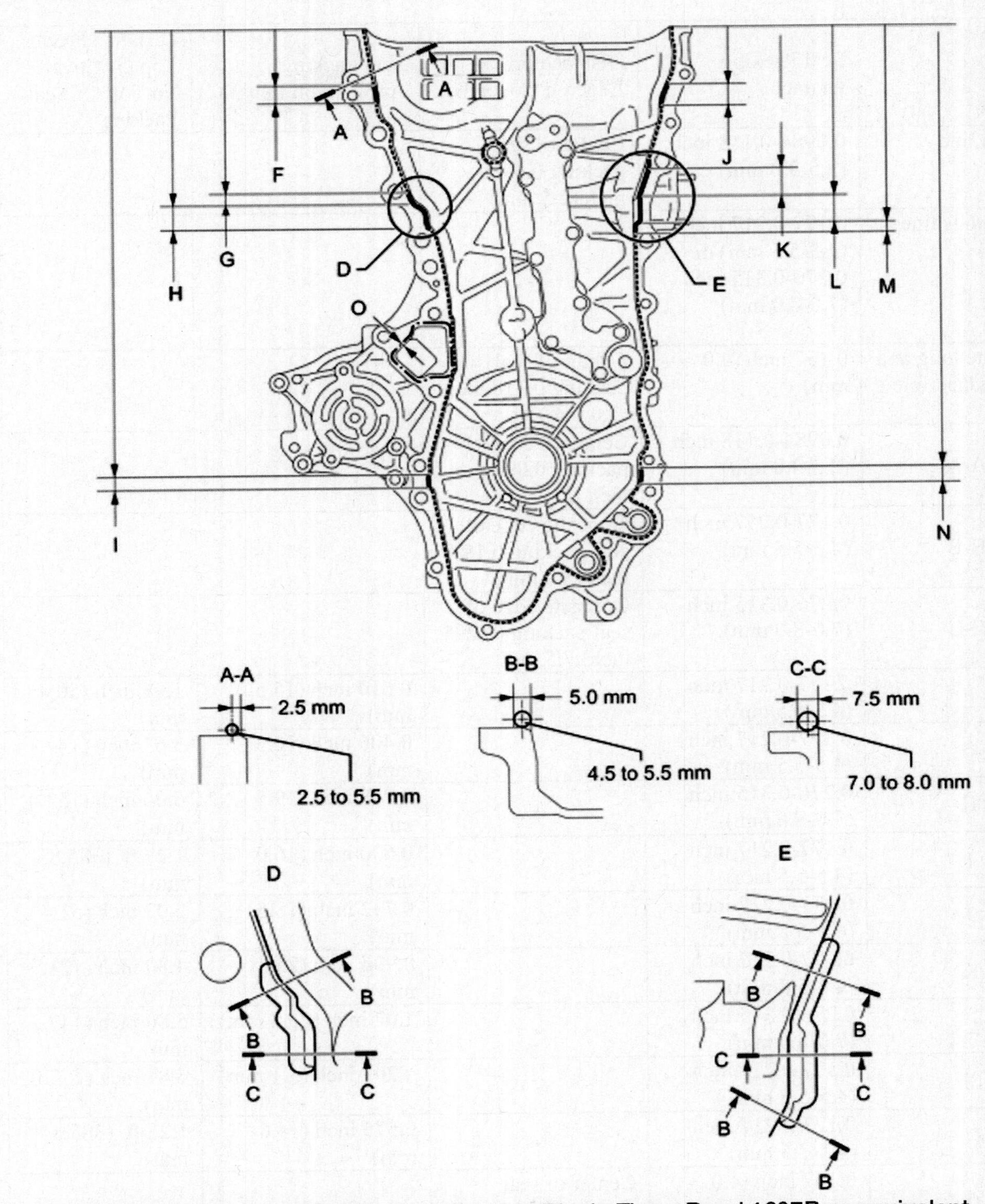

Fig. 211 Applying seal packing to the timing chain cover

Area	Seal Packing Diameter	Distance From Edge Of Cover To:	Seal Packing Application Length	Distance From Top Of Cover To Top Of Seal Packing
Dashed line	0.0984-0.118 inch (2.5-3.0 mm)	Center of seal packing 0.0984 inch (2.5 mm)	-	-
Continuous line	0.177-0.217 inch (4.5-5.5 mm) or 0.276-0.315 inch (7.0-8.0 mm)	-	-	-
Alternate long and short dashed line	0.157 inch (4.0 mm)	Center of seal packing 0.118 inch (3.0 mm)	-	-
A-A	0.0984-0.118 inch (2.5-3.0 mm)	Center of seal packing 0.0984 inch (2.5 mm)	-	-
B-B	0.177-0.217 inch (4.5-5.5 mm)	Opposite edge of seal packing 0.197 inch (5.0 mm)	-	-
C-C	0.276-0.315 inch (7.0-8.0 mm)	Opposite edge of seal packing (0.295 inch) 7.5 mm	-	-
F	0.177-0.217 inch (4.5-5.5 mm)	-	0.610 inch (15.5 mm)	1.98 inch (50.4 mm)
G	0.177-0.217 inch (4.5-5.5 mm)	-	0.406 inch (10.3 mm)	5.63 inch (143.1 mm)
H	0.276-0.315 inch (7.0-8.0 mm)	-	0.768 inch (19.5 mm)	6.04 inch (153.4 mm)
I	0.177-0.217 inch (4.5-5.5 mm)	-	0.630 inch (16.0 mm)	1.27 ft. (385.8 mm)
J	0.177-0.217 inch (4.5-5.5 mm)	-	0.732 inch (18.6 mm)	2.02 inch (51.4 mm)
K	0.177-0.217 inch (4.5-5.5 mm)	-	0.996 inch (25.3 mm)	4.80 inch (121.9 mm)
L	0.276-0.315 inch (7.0-8.0 mm)	-	1.02 inch (25.8 mm)	5.80 inch (147.2 mm)
M	0.177-0.217 inch (4.5-5.5 mm)	-	0.201 inch (5.1 mm)	6.81 inch (173.0 mm)
N	0.177-0.217 inch (4.5-5.5 mm)	-	0.575 inch (14.6 mm)	1.27 ft. (385.8 mm)
O	0.157 inch (4.0 mm)	Center of seal packing 0.118 inch (3.0 mm	-	-

3768X_PRIU_G0418

Fig. 212 Seal packing application specifications

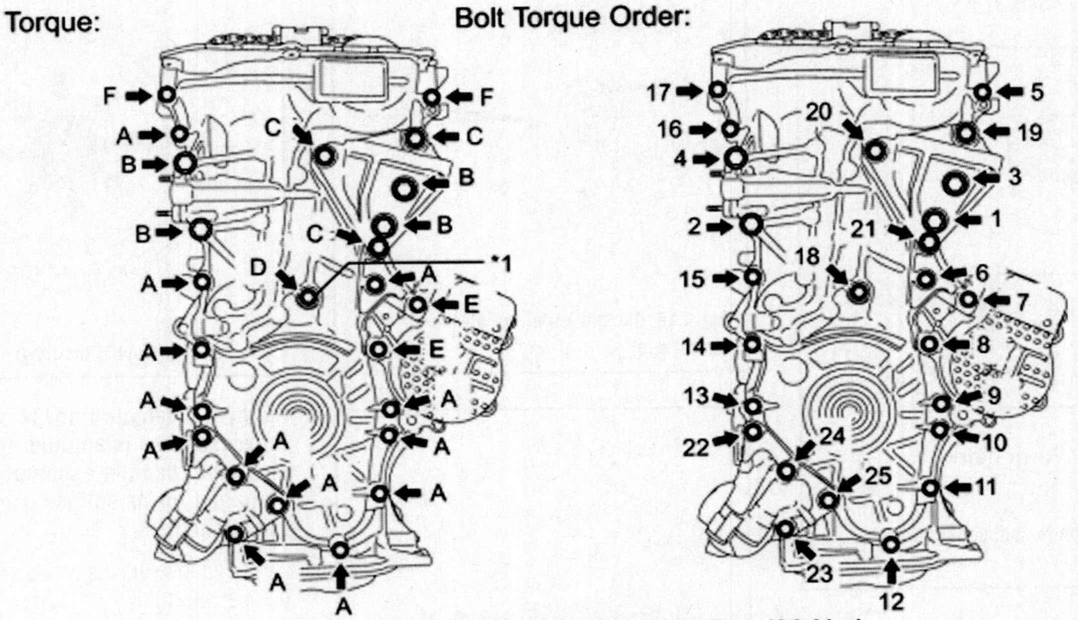

Torque:

Bolt Torque Order:

Bolt A, E, F - Torque: 19 ft. lbs. (26 Nm)

Bolt B, C - Torque: 28 ft. lbs. (51 Nm)

Bolt D - Torque: 7 ft. lbs. (10 Nm)

3768X_PRIU_G0419

Fig. 213 Installing the timing chain cover

➡Install the chain cover within 3 minutes and tighten the bolts within 15 minutes after applying the seal packing.

➡Do not add engine oil for at least 2 hours after installing the chain cover.

➡Do not start the engine for at least 2 hours after installing the chain cover.

49. Install the crankshaft pulley.
50. Install the No. 1 chain tensioner assembly.
51. Install the spark plug tube gasket.
52. Install the cylinder head cover gasket.
53. Install the cylinder head cover sub assembly.
54. Install the water inlet with the thermostat sub assembly.
55. Install the fuel injector assembly.
56. Install the No. 1 deliver pipe spacer.
57. Install the fuel delivery pipe sub assembly.
58. Install the No. 1 delivery pipe spacer.
59. Install the fuel delivery pipe sub assembly.
60. Install the fuel vapor feed pipe.
61. Install the intake manifold.
62. Install the EGR cooler sub assembly.
63. Install the EGR valve assembly.
64. Install the EGR pipe assembly.
65. Install the engine oil level dipstick guide.
66. Install the throttle body assembly.
67. Install the engine hangers.
68. Remove the engine on engine stand.
69. Install the engine assembly with the transaxle.

PISTON AND RING

POSITIONING

2009 Vehicles

See Figures 214 through 216.

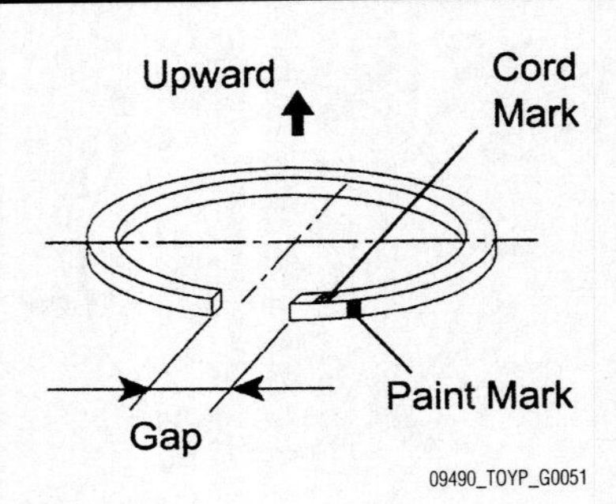

Fig. 214 Piston ring positioning and mark locations—1.5L Hybrid engine

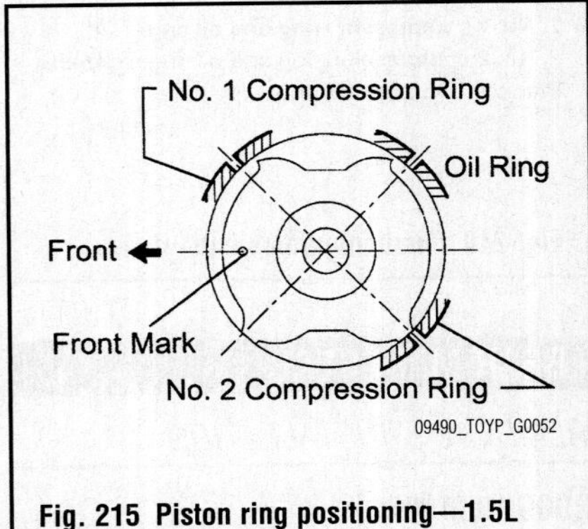

Fig. 215 Piston ring positioning—1.5L Hybrid engine

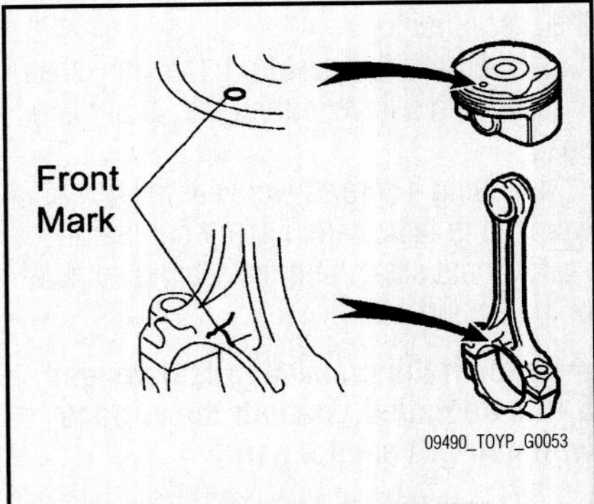

Fig. 216 Piston-to-connecting rod orientation—1.5L Hybrid engine

2010 Vehicles

See Figures 217 and 218.

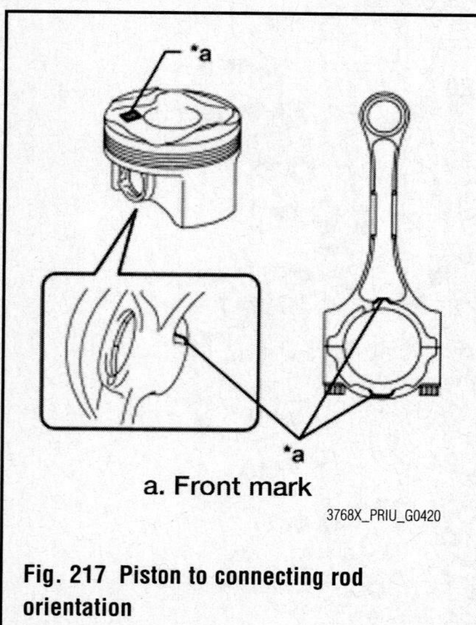

Fig. 217 Piston to connecting rod orientation

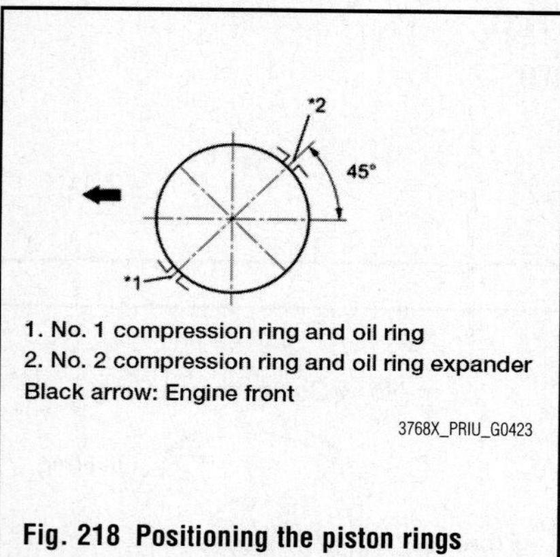

Fig. 218 Positioning the piston rings

REAR MAIN SEAL

REMOVAL & INSTALLATION

2009 Vehicles

See Figures 219 and 220.

1. Before servicing the vehicle, refer to the precautions.
2. Remove flywheel from the crankshaft.
3. Using a knife, cut off the lip of the oil seal.
4. Using a screwdriver with the tip wrapped in tape, carefully pry out the oil seal without scratching the sealing surface of the crankshaft.

➡Check if the crankshaft is damaged. If it is damaged, smooth the surface with 400-grit sandpaper.

To install:

5. Apply multi-purpose grease to the lip of the new seal.

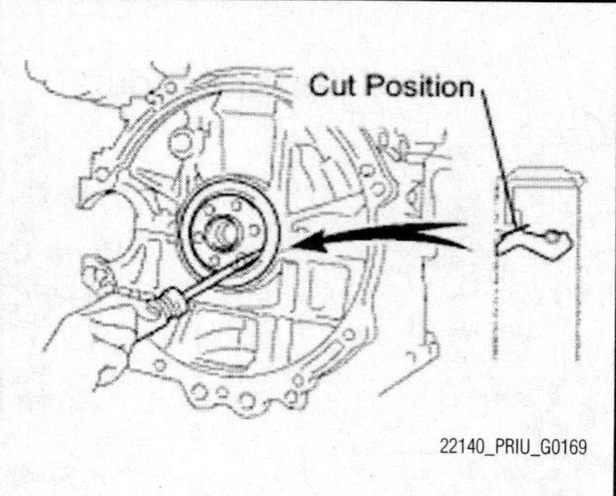

Fig. 219 Using a knife, cut off the lip of the oil seal

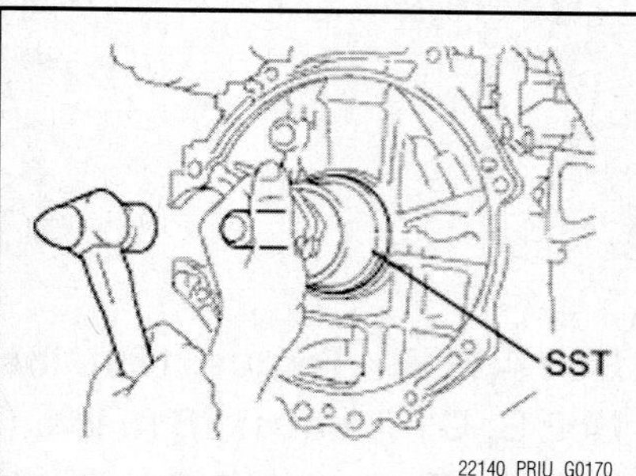

Fig. 220 Using a seal installer and a hammer, tap in the oil seal until its surface is flush with the rear oil seal retainer edge

➡Keep the lip free of foreign materials.

6. Install the seal in the retainer using a suitable seal driver. Wipe any extra grease off the crankshaft.
7. Install the flywheel.

2010 Vehicles

See Figures 221 through 224.

1. Remove the hybrid vehicle transaxle assembly.

➡Be careful not to apply excess force to the transmission input damper assembly when removing or installing the hybrid vehicle transaxle assembly. If excess force is applied, the transmission input damper assembly may be damaged, or its splines may become misaligned.

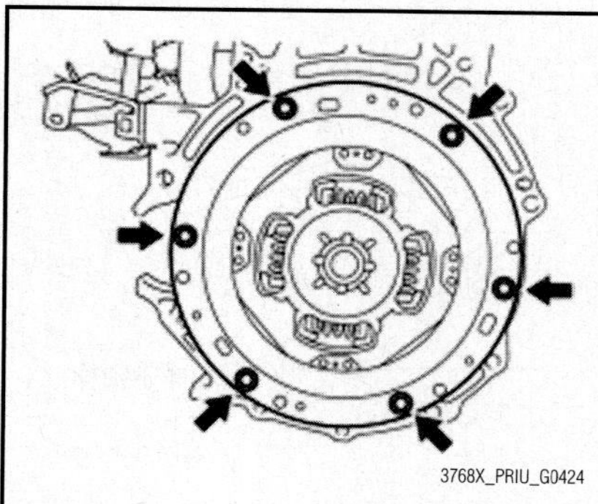

Fig. 221 Removing the transmission input damper assembly

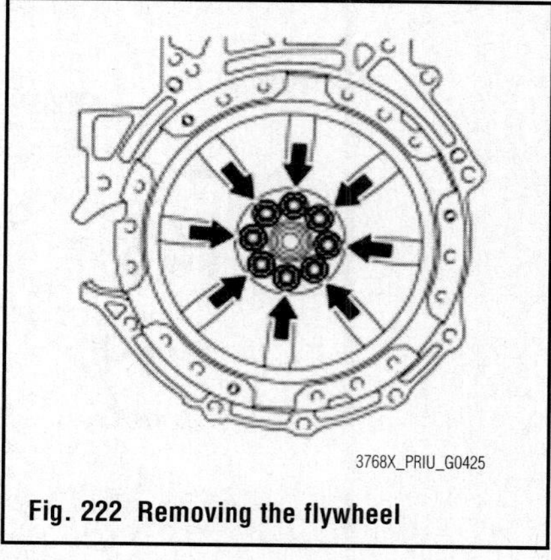

Fig. 222 Removing the flywheel

2. Remove the transmission input damper assembly.
   a. Using the special tool (09213-58014, 91551-80840 or 09330-00021), hold the crankshaft damper.
   b. Remove the 6 bolts and transmission input damper assembly from the flywheel sub assembly.
3. Remove the flywheel sub assembly.
   a. Using the special tool (09213-58014, 91551-80840 or 09330-00021), hold the crankshaft damper.
   b. Remove the 8 bolts and the flywheel.
4. Remove the rear engine oil seal.
   a. Using a knife, cut off the lip of the oil seal.
   b. Using a screwdriver with its tip wrapped with tape, pry out the oil seal.

➡After removing, check the crankshaft for damage. If damaged, smooth the surface with 400 grit sandpaper.

To install:

5. Install the rear engine oil seal.
   a. Apply MP grease to the lip of a new oil seal.

➡Keep the lip free from foreign matter.

   b. Using special tool (09223-15030, 09950-70010 or 09951-70010)

➡Wipe any extra grease off the crankshaft.

➡Do not tap the oil seal at an angle.

6. Install the flywheel sub assembly.
   a. Using the special tool (09213-58014, 91551-80840 or 09330-00021), hold the crankshaft.
   b. Apply adhesive to the 2 or 3 end threads of the bolts.

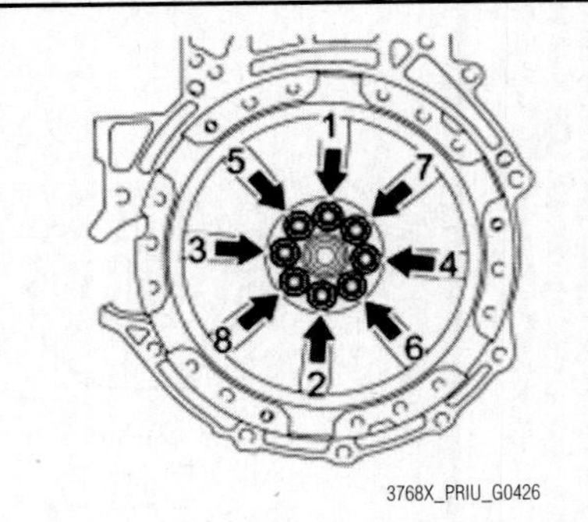

Fig. 223 Installing the flywheel sub assembly

c. Using several steps, uniformly install and tighten the 8 bolts in the sequence shown. Tighten to 36 ft. lbs. (49 Nm).

d. Mark the front of the bolts with paint.

e. Retighten the 8 bolts 90° in the same sequence.

f. Check that the paint marks are now at a 90° angle to the front.

g. Check that the crankshaft turns smoothly.

7. Install the transmission input damper assembly.

a. Using the special tool (09213-58014, 91551-80840, 09330-00021), hold the crankshaft.

b. Install the transmission input damper.

c. In several steps, uniformly install and tighten the 6 bolts in the sequence shown. Tighten to 22 ft. lbs. (30 Nm).

➡Take care not to insert the transmission input damper in a wrong direction.

8. Install the hybrid vehicle transaxle assembly.

➡Be careful not to apply excess force to the transmission input damper assembly when removing or installing the hybrid vehicle transaxle assembly. If excess force is applied, the transmission input damper assembly may be damaged, or its splines may become misaligned.

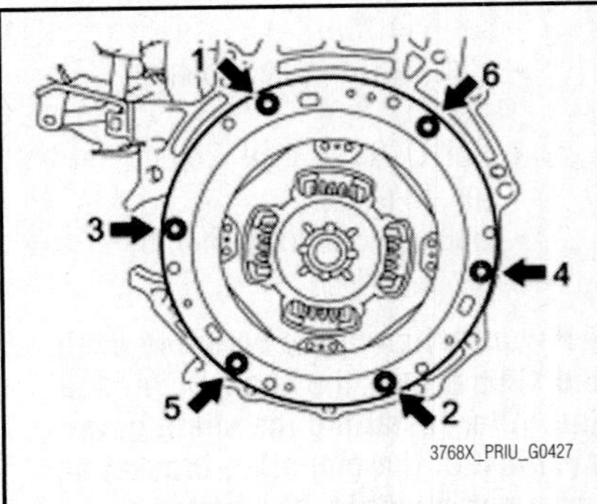

Fig. 224 Installing the transmission input damper

ROCKER ARMS

REMOVAL & INSTALLATION

2009 Vehicles

The 1.5L hybrid engine does not utilize rocker arms/shafts, the camshaft directly actuates the valves.

2010 Vehicles

See Figures 225 and 226.

1. Remove the camshafts.
2. Remove the camshaft housing straight pin.
3. Remove the No. 1 valve rocker arm sub assembly.

a. Remove the 16 valve rocker arms.

➡Arrange the removed parts in the correct order.

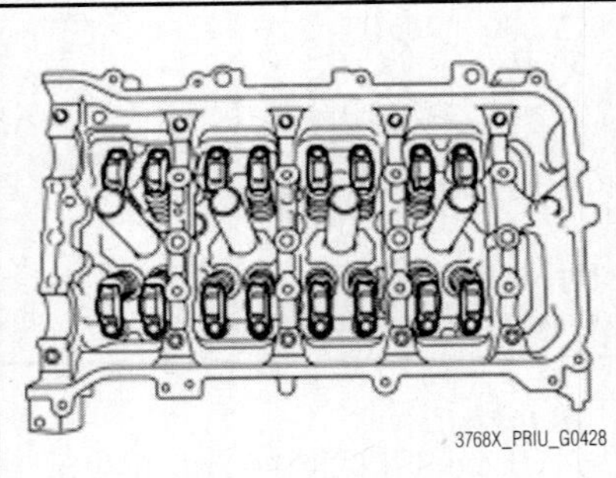

Fig. 225 Removing the No. 1 valve rocker arm sub assembly

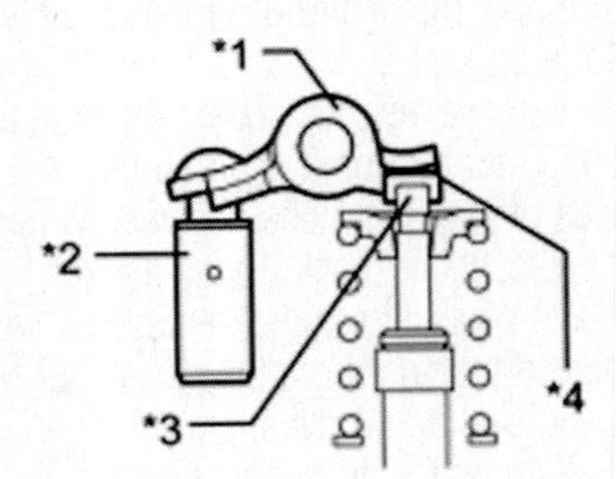

Fig. 226 Installing the valve rocker arms

To install:

4. Install the No. 1 valve rocker arm sub assembly.

a. Apply engine oil to the valve lash adjuster tips and valve stem cap ends.

b. Make sure that the No. 1 valve rocker arms are installed as shown.

5. Install the camshaft housing straight pin.
6. Install the camshafts.

TIMING CHAIN COVER AND SEAL

REMOVAL & INSTALLATION

2009 Vehicles

See Figures 227 through 233.

On this engine, the timing chain cover also functions as a housing for the oil pump and may be referred to as the oil pump or oil pump cover.

1. Before servicing the vehicle, refer to the precautions.
2. Properly relieve the fuel system pressure.
3. Disconnect the negative battery cable.

➡Wait at least 90 seconds after disconnecting the cable from the negative (-) battery terminal to prevent airbag and seat belt pretensioner activation.

4. While wearing insulating gloves, slide up the lever of the service plug grip and remove while turning the lever to the left. Be sure to insulate the service plug with insulating tape.

⁂ CAUTION

Do not touch the high voltage connectors and terminals for 5 minutes after removing the service plug grip.

5. Remove the outer front cowl top panel.
6. Remove right engine under cover.
7. Drain the engine coolant.
8. Remove the air cleaner assembly.
9. Remove the brake fluid level sensor connector.
10. Remove the brake fluid reservoir tank mounting bolts and suspend it with rope.
11. Remove the brake fluid reservoir tank bracket.
12. Remove the ignition connectors.
13. Remove the fuel injector connectors.
14. Remove the vacuum switching valve connectors.
15. Remove the Camshaft Position (CMP) sensor connector.

16. Remove the water temperature connector.

17. Remove the camshaft timing oil control valve connector.

18. Remove the air cleaner inlet hose.

19. Remove the engine coolant reservoir tank.

20. Remove the vacuum suction valve from engine mounting insulator.

21. Remove the accessory drive belt.

22. Place a floor jack under the engine to support it, with a block of wood between the engine and the jack.

23. Remove the right engine mounting insulator.

24. Remove the engine wiring from cylinder head cover.

25. Remove the ignition coils.

26. Remove the PCV hoses.

27. Remove the 7 bolts, 2 seal washers, 2 nuts, cylinder head cover and gasket.

28. Set the No. 1 cylinder to Top Dead Center/compression by turning the crankshaft pulley and aligning its groove with timing mark **"0"** of the timing chain cover.

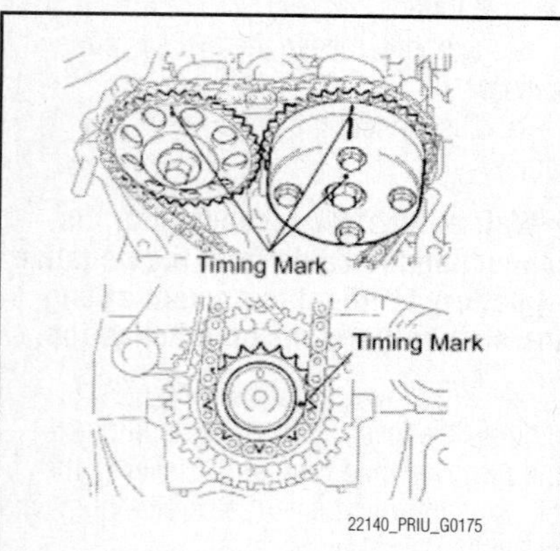

Fig. 227 Aligning both timing marks on the camshaft timing sprocket and valve timing controller assembly

29. Check that both timing marks on the camshaft timing sprocket and valve timing controller assembly are facing right up. If not, turn the crankshaft 1 revolution (360 degrees) and align the marks.

30. Remove the crankshaft pulley bolt and the crankshaft pulley.

31. Remove the Crankshaft Position (CKP) sensor.

32. Remove the right engine mounting bracket.

33. Remove the water pump.

34. Remove the oil control valve.

35. Remove the timing chain cover.

36. Using a screwdriver, remove the oil seal. Tape the screwdriver tip before use.

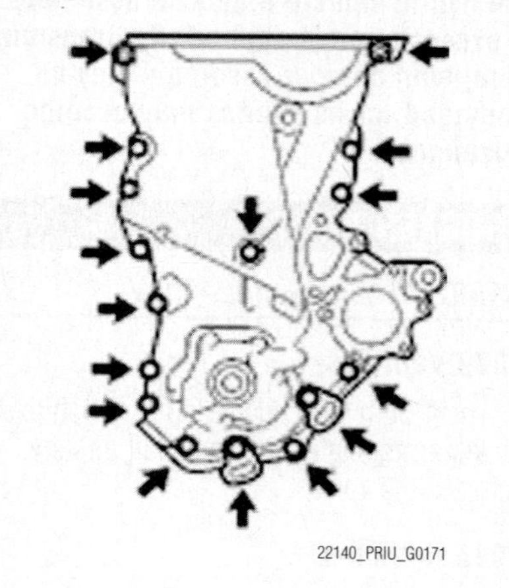

Fig. 228 Timing chain attaching bolt locations

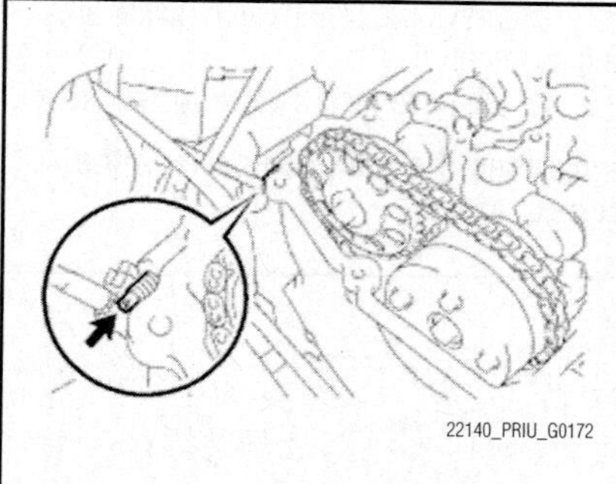

Fig. 229 Using an E8 Torx®socket wrench, remove the stud bolt

To install:

37. Using SST 09950-60010, 09950-70010 and a hammer, tap in a new oil seal until its surface is flush with the timing chain cover edge. Be careful not to tap the oil seal at an angle. Keep the lip free of foreign objects.

38. Apply multi-purpose grease to the lip of the oil seal.

39. Remove any old sealant material and be careful not to drop any oil on the contact surfaces of the timing chain cover, cylinder head and cylinder block. Using a razor blade and a gasket scraper, remove all the old sealant material from the gasket surfaces and sealing grooves. Thoroughly clean all components to remove all the loose material. Using a non-residue solvent, clean both sealing surfaces.

40. Apply sealant to the timing chain cover. Install a nozzle that has been cut to a 0.16–0.20 inch (4–5mm) opening. Sealant shall be accumulated in the groove to a depth of 0.10 inch (2.5mm) or more. Avoid applying an excessive amount to the surface. Parts must be assembled within 3 minutes of application. Otherwise the material must be removed and reapplied. Immediately remove nozzle from the tube and reinstall cap.

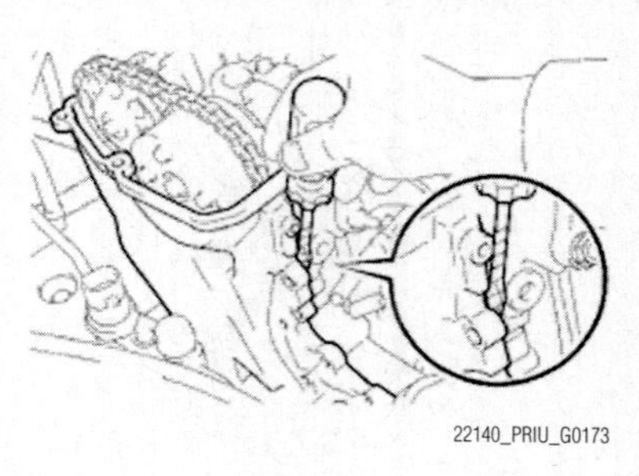

Fig. 230 Using a screwdriver with its tip wrapped in tape, remove the cover by prying out between the cylinder head and cylinder block

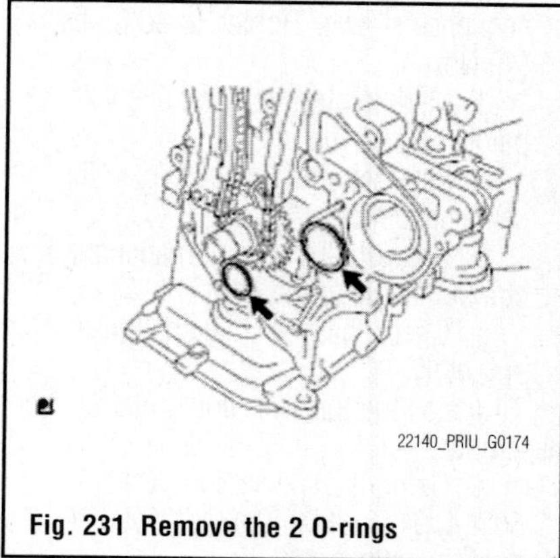

Fig. 231 Remove the 2 O-rings

41. Install 2 new O-rings to the cylinder block and oil pan No. 1.

42. Install the timing chain cover, new O-ring and water pump with the bolts and nuts. Uniformly tighten the bolts and nut in several passes to the following torque values:

- Bolt A, 0.787 inch (20mm)—18 ft. lbs. (24 Nm)
- Bolt B, 1.181 inch (30mm)—8 ft. lbs. (11 Nm)
- Bolt C, 1.378 inch (35mm)—8 ft. lbs. (11 Nm)
- Bolt D, 0.787 inch (20mm)—18 ft. lbs. (24 Nm)
- Bolt E, 1.378 inch (35mm)—18 ft. lbs. (24 Nm)

➡Pay attention not to wrap the chain and slipper over the chain cover seal line. After installing the chain cover, must install the mounting bracket and water pump within 15 minutes.

43. Apply sealant to threads of the engine mounting bracket mounting bolt, but

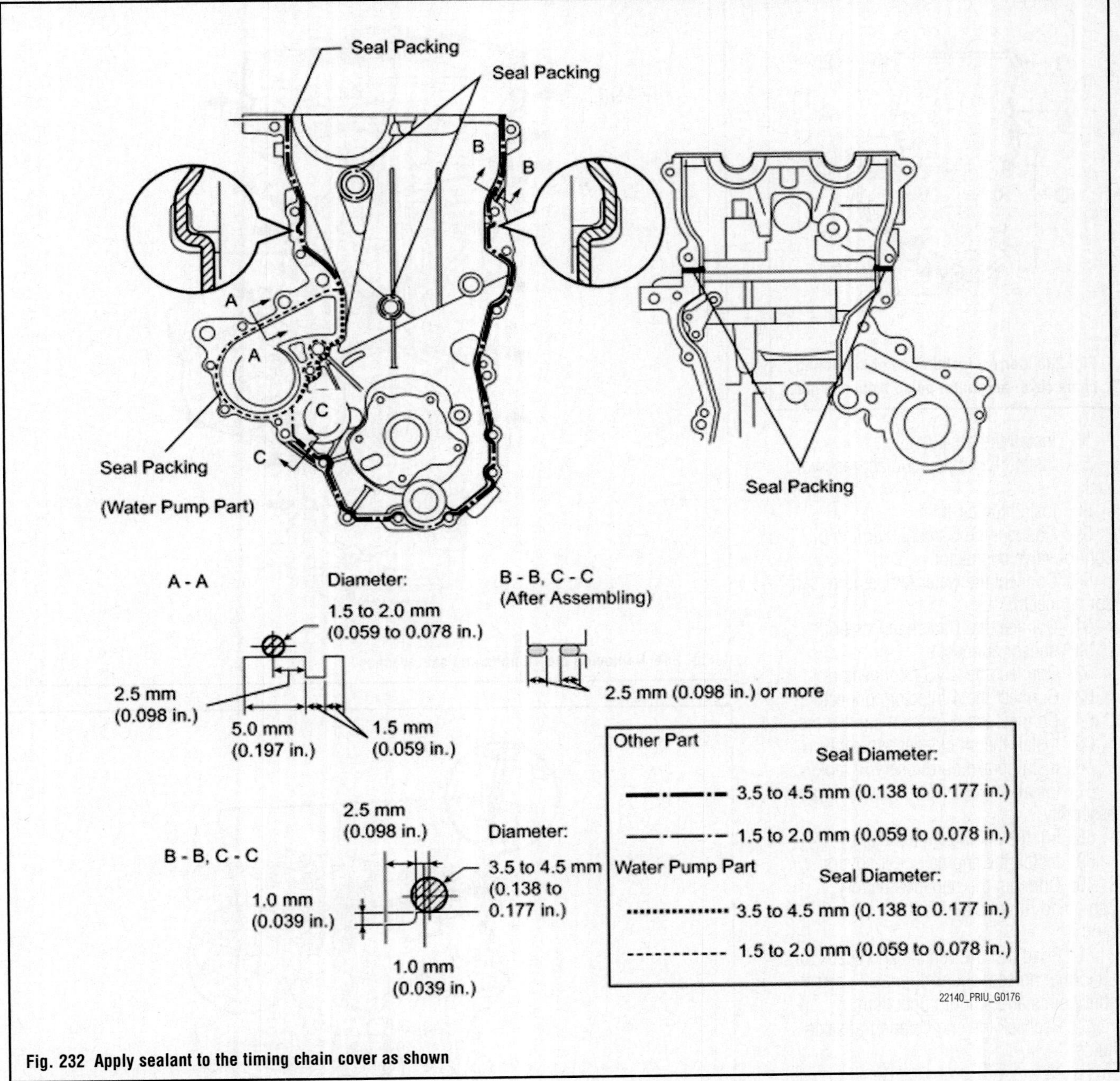

Fig. 232 Apply sealant to the timing chain cover as shown

do not apply sealant to 2 or 3 threads of the bolt end.

44. Install the right mounting bracket with the 4 bolts and tighten to 41 ft. lbs. (55 Nm).

45. Install the Crankshaft Position (CKP) sensor and tighten the bolt at the sensor to 66 inch lbs. (7.5 Nm), and the other bolts to 8 ft. lbs. (11 Nm).

46. Install the oil control valve and tighten to 66 inch lbs. (7.5 Nm).

47. Install the crankshaft pulley.

a. Clean the crankshaft pulley inside.

b. Install the pin to the crankshaft.

c. Align the hole in the crank pulley with the pin position and install the crank pulley.

d. Using SST 09213-70011 and 09330-00021, install the pulley bolt and tighten to 94 ft. lbs. (128 Nm).

48. Remove any old sealant material and apply sealant to 2 locations as shown in the illustration.

49. Install the gasket to the cylinder head cover.

➡Part must be assembled within 3 minutes of sealant application. Otherwise the material must be remove and reapplied.

50. Install the cylinder head cover and cable bracket with the 7 bolts, 2 seal washers and 2 nuts. Uniformly tighten the bolts and nuts, in the several passes, in the sequence to 7 ft. lbs. (10 Nm).

51. Connect the 2 PCV hoses to the cylinder head cover.

52. Connect the engine wire to cylinder head cover.

53. Install the ignition coils.

54. Install the RH engine mounting insulator with the 5 bolts and 2 nuts.

55. Install VSV to right engine mounting insulator

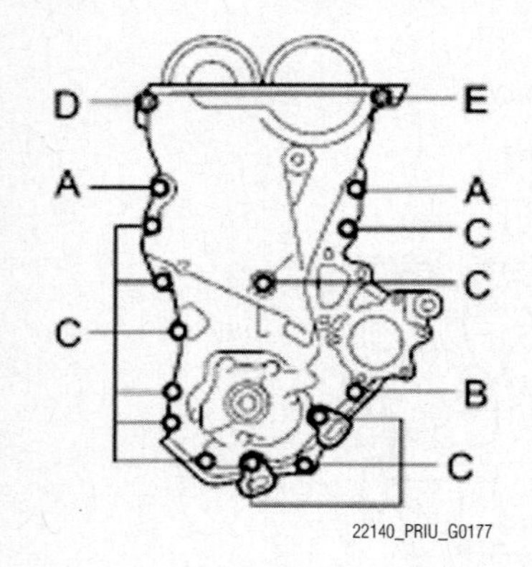

Fig. 233 Correct installation of the timing chain cover and water pump bolts and nuts

56. Install the drive belt.
57. Install the engine coolant reservoir tank.
58. Install the air inlet.
59. Connect the Camshaft timing oil control valve connector.
60. Connect the water temperature sensor connector.
61. Connect the Camshaft Position (CMP) sensor connector.
62. Connect the 2 VSV connectors.
63. Connect the 4 injector connectors.
64. Connect the 4 ignition connectors.
65. Install the air cleaner assembly.
66. Install brake fluid reservoir tank.
67. Install the outer front cowl top panel assembly.
68. Fill the engine with coolant.
69. Install the engine under covers.
70. Connect the negative battery cable and high voltage battery service plug.
71. Road test the vehicle and check for abnormal noises, shock slippage, correct shift points and smooth operation.
72. Recheck the engine and transaxle fluids.

2010 Vehicles

See Figures 234 through 241 and 246 through 252.

1. Before servicing the vehicle, refer to the precautions.
2. Remove the crankshaft pulley.
3. Remove the No. 1 chain tensioner assembly.
4. Remove the timing chain cover sub assembly.
   a. Remove the 3 bolts and engine mounting bracket RH.
   b. Remove the 4 bolts and oil filter bracket.
   c. Remove the 2 o-rings.

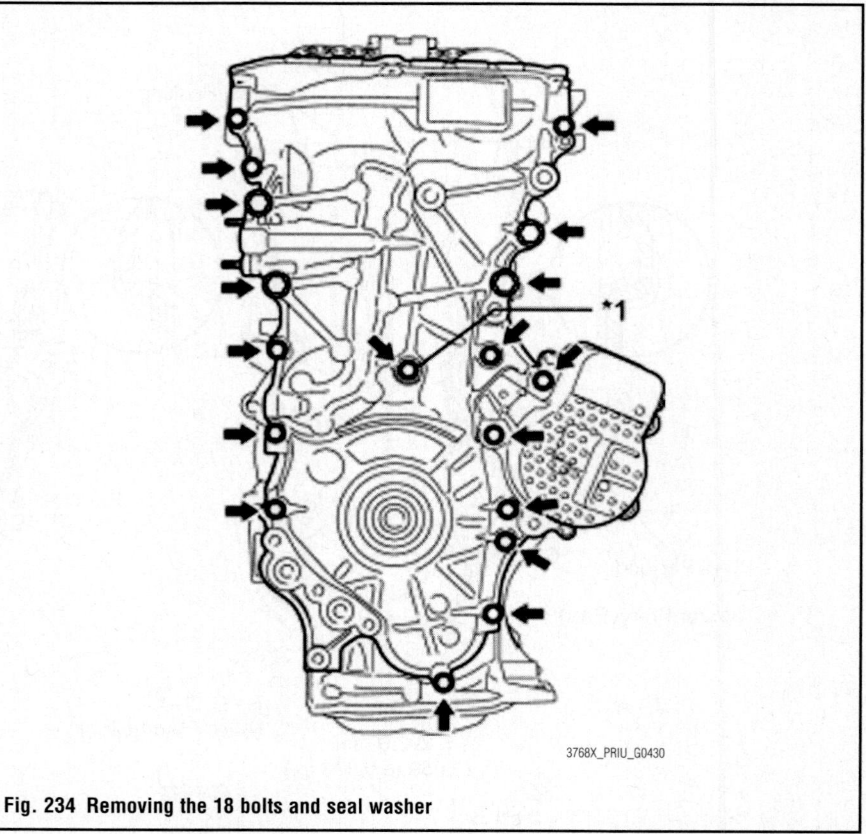

Fig. 234 Removing the 18 bolts and seal washer

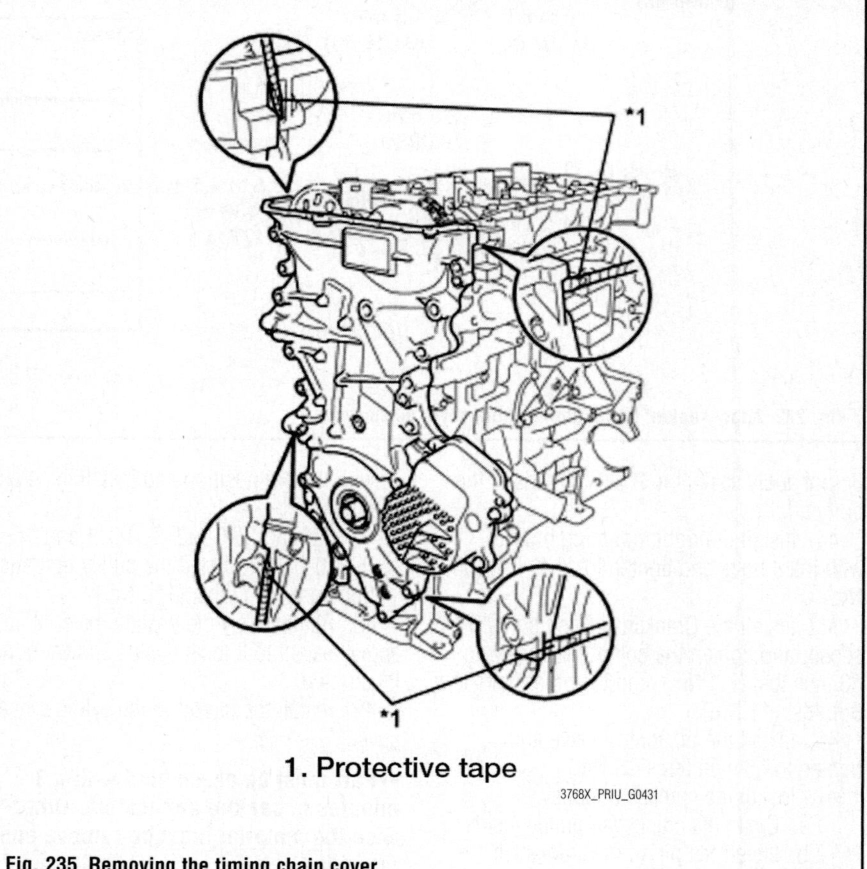

Fig. 235 Removing the timing chain cover

d. Remove the 18 bolts and seal washer (*1).

e. Remove the timing chain cover by prying between the timing chain cover and cylinder head or cylinder block with a screwdriver.

➡**Be careful not to damage the contact surfaces of the timing chain cover, cylinder block, and cylinder head.**

➡**Pry the timing chain cover out evenly in order to prevent damaging the knock pins.**

➡**Tape the screwdriver tip before use.**

f. Remove the 3 o-rings.

5. Remove the timing chain cover oil seal.

a. Place the timing chain cover on wooden blocks.

b. Using a screwdriver, pry out the oil seal.

➡**Tape the screwdriver tip before use.**

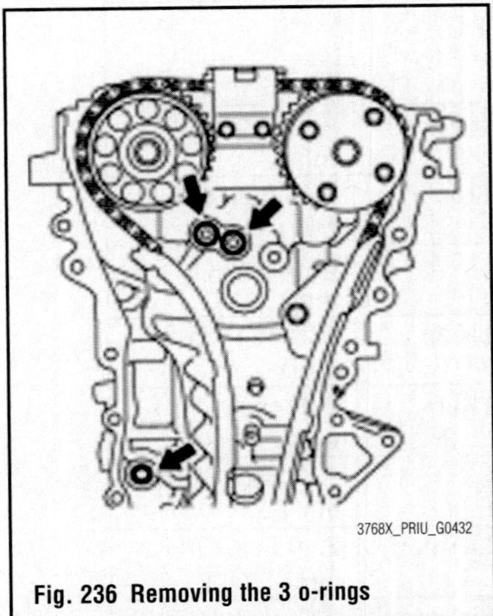

Fig. 236 Removing the 3 o-rings

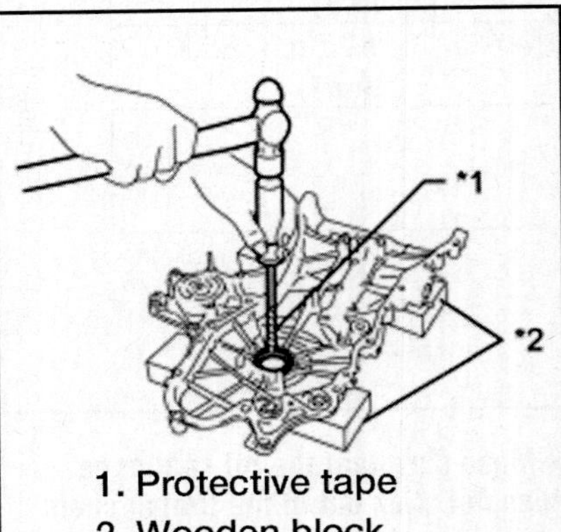

Fig. 237 Removing the timing chain cover oil seal

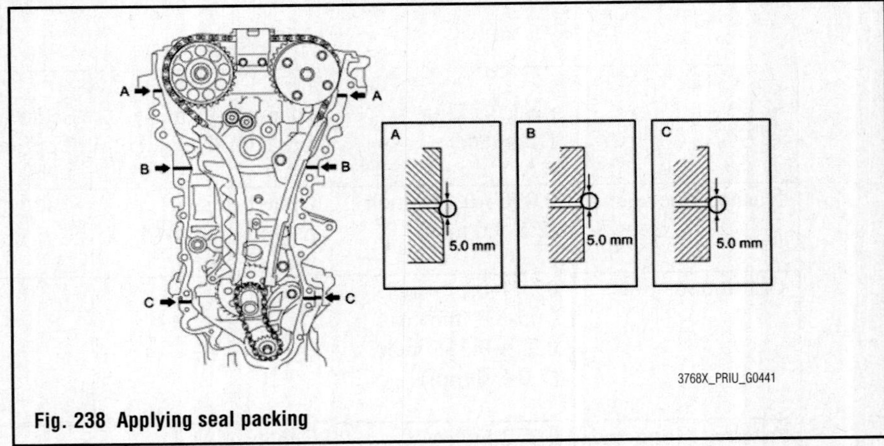

Fig. 238 Applying seal packing

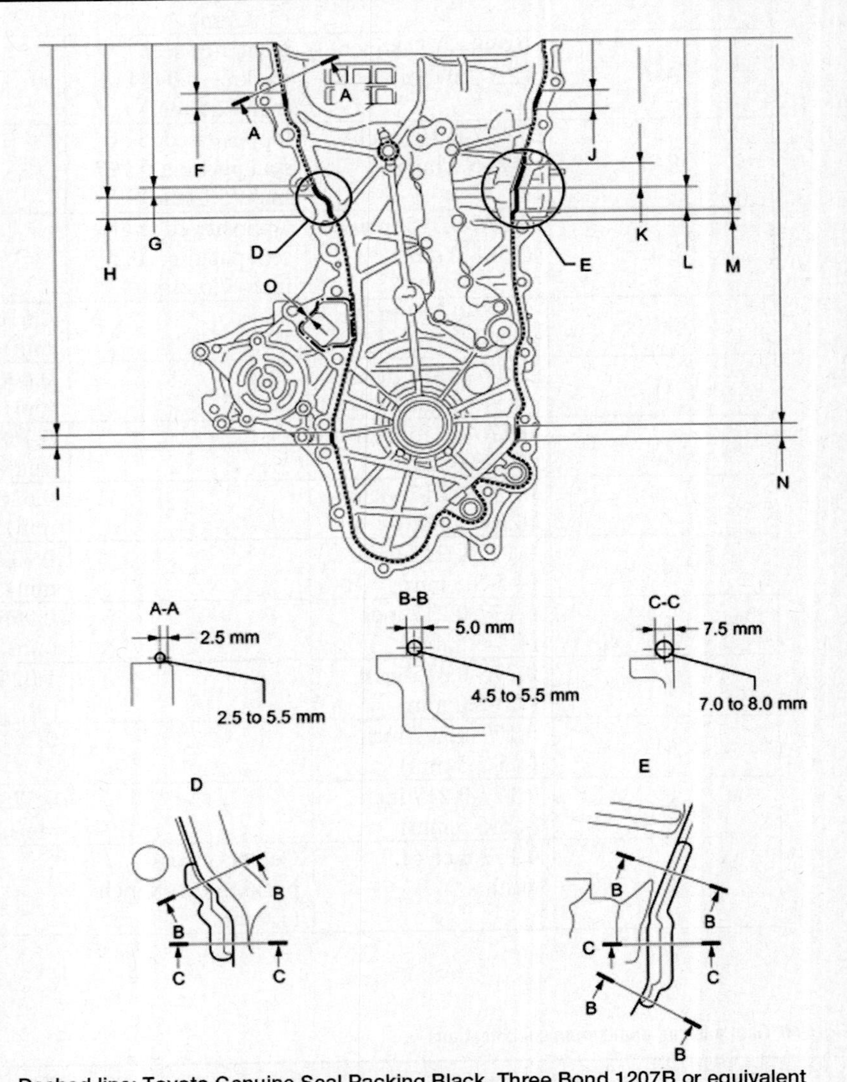

Fig. 239 Applying seal packing to the timing chain cover

Area	Seal Packing Diameter	Distance From Edge Of Cover To:	Seal Packing Application Length	Distance From Top Of Cover To Top Of Seal Packing
Dashed line	0.0984-0.118 inch (2.5-3.0 mm)	Center of seal packing 0.0984 inch (2.5 mm)	-	-
Continuous line	0.177-0.217 inch (4.5-5.5 mm) or 0.276-0.315 inch (7.0-8.0 mm)	-	-	-
Alternate long and short dashed line	0.157 inch (4.0 mm)	Center of seal packing 0.118 inch (3.0 mm)	-	-
A-A	0.0984-0.118 inch (2.5-3.0 mm)	Center of seal packing 0.0984 inch (2.5 mm)	-	-
B-B	0.177-0.217 inch (4.5-5.5 mm)	Opposite edge of seal packing 0.197 inch (5.0 mm)	-	-
C-C	0.276-0.315 inch (7.0-8.0 mm)	Opposite edge of seal packing (0.295 inch) 7.5 mm	-	-
F	0.177-0.217 inch (4.5-5.5 mm)	-	0.610 inch (15.5 mm)	1.98 inch (50.4 mm)
G	0.177-0.217 inch (4.5-5.5 mm)	-	0.406 inch (10.3 mm)	5.63 inch (143.1 mm)
H	0.276-0.315 inch (7.0-8.0 mm)	-	0.768 inch (19.5 mm)	6.04 inch (153.4 mm)
I	0.177-0.217 inch (4.5-5.5 mm)	-	0.630 inch (16.0 mm)	1.27 ft. (385.8 mm)
J	0.177-0.217 inch (4.5-5.5 mm)	-	0.732 inch (18.6 mm)	2.02 inch (51.4 mm)
K	0.177-0.217 inch (4.5-5.5 mm)	-	0.996 inch (25.3 mm)	4.80 inch (121.9 mm)
L	0.276-0.315 inch (7.0-8.0 mm)	-	1.02 inch (25.8 mm)	5.80 inch (147.2 mm)
M	0.177-0.217 inch (4.5-5.5 mm)	-	0.201 inch (5.1 mm)	6.81 inch (173.0 mm)
N	0.177-0.217 inch (4.5-5.5 mm)	-	0.575 inch (14.6 mm)	1.27 ft. (385.8 mm)
O	0.157 inch (4.0 mm)	Center of seal packing 0.118 inch (3.0 mm	-	-

3768X_PRIU_G0418

Fig. 240 Seal packing application specifications

➡Do not damage the surface of the oil seal press fit hole.

To install:

6. Install the timing chain cover oil seal.
   a. Using the special tool (0923-22010), tap in a new oil seal until its surface is flush with the timing chain cover edge.

➡Keep the lip free from foreign matter.

➡Do not tap on the oil seal at an angle.

➡Make sure that the oil seal edge does not stick out of the timing chain case.

➡Tap in the oil seal so that it is positioned within 1.0 mm from the edge of the timing chain case.

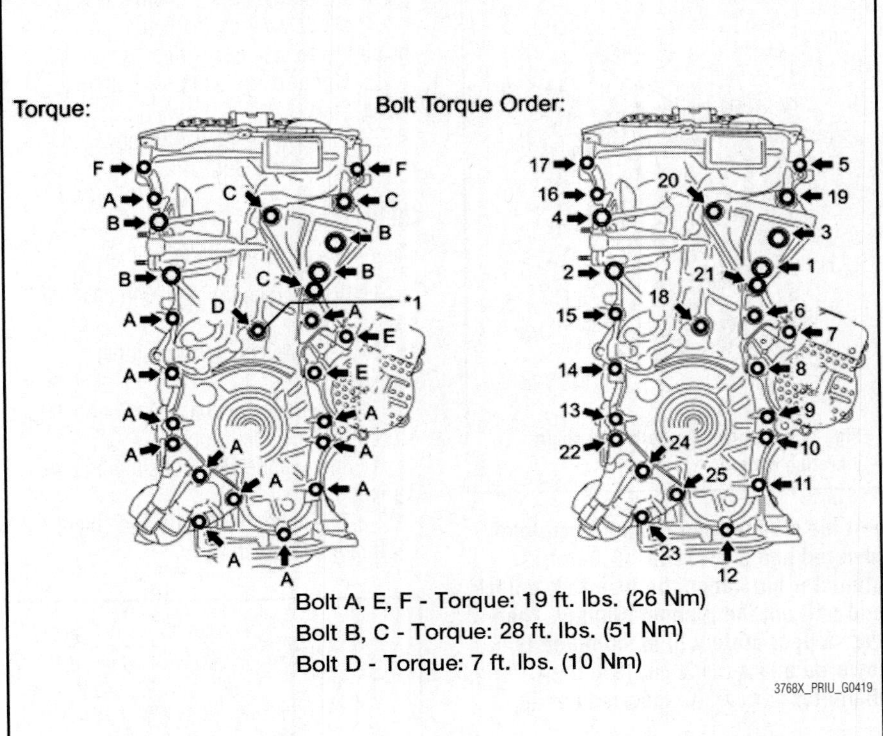

Fig. 241 Installing the timing chain cover

b. Apply MP grease to the lip of the oil seal.

7. Install the timing chain cover sub assembly.

a. Remove any old packing (FIPG) material and be careful not to drop any oil on the contact surfaces of the timing chain cover, cylinder head, and cylinder block.

b. Install 3 new o-rings.

c. Apply seal packing.

➡Remove any oil from the contact surfaces.

➡Install the chain cover within 3 minutes after applying seal packing.

➡Do not start the engine for at least 2 hours after installing the timing chain cover sub-assembly.

d. Apply seal packing to the timing chain cover in a line as shown in the following illustration.

➡When the contact surfaces are wet, wipe them with oil-free cloth before applying seal packing.

➡Install the chain cover within 3 minutes and tighten the bolts within 15 minutes after applying seal packing.

➡Do not start the engine for at least 2 hours after installation.

➡When the contact surfaces are wet, wipe them with oil-free cloth before applying seal packing.

➡Install the timing chain cover within 3 minutes and tighten the bolts within 10 minutes after applying seal packing.

➡After applying seal packing to the timing chain cover, install the engine mounting bracket and oil filter bracket within 10 minutes.

➡Do not add engine oil for at least 2 hours after installation.

e. Clean the bolt and fitting hole.

f. Install the timing chain cover.

g. Temporarily tighten the engine mounting bracket RH with the 3 bolts.

➡Install the mounting bracket within 10 minutes after installing the chain cover.

➡Do not start the engine for at least 2 hours after installation.

h. Install 2 new o-rings.

i. Temporarily tighten the oil filter bracket with the 4 bolts.

➡Install the oil filter bracket within 10 minutes after installing the chain cover.

➡Do not start the engine for at least 2 hours after installation.

j. Install the timing chain cover with the 25 bolts and seal washer as shown in the illustration.

➡Apply adhesive 1324 to screw part of the bolt F.

➡When the contact surfaces are wet, wipe them with oil-free cloth before applying seal packing.

➡Install the chain cover within 3 minutes and tighten the bolts within 15 minutes after applying the seal packing.

➡Do not add engine oil for at least 2 hours after installing the chain cover.

➡Do not start the engine for at least 2 hours after installing the chain cover.

TIMING CHAIN & SPROCKETS

REMOVAL & INSTALLATION

2009 Vehicles

See Figures 242 through 245.

1. Before servicing the vehicle, refer to the precautions.
2. Properly relieve the fuel system pressure.
3. Disconnect the negative battery cable.

➡Wait at least 90 seconds after disconnecting the cable from the negative (-) battery terminal to prevent airbag and seat belt pretensioner activation.

4. While wearing insulating gloves, slide up the lever of the service plug grip and remove while turning the lever to the left. Be sure to insulate the service plug with insulating tape.

⁂ CAUTION

Do not touch the high voltage connectors and terminals for 5 minutes after removing the service plug grip.

5. Remove the timing chain cover.
6. Remove the chain tensioner.

a. Lift up the stopper plate, then unlock the plunger.

b. Push in the plunger to the end with the plunger unlocked.

c. Lower the stopper plate with the plunger pushed to the end, then lock the plunger.

d. Insert a 3.0 mm (0.118 in.) diameter bar into the hole of the stopper plate with the plunger locked.

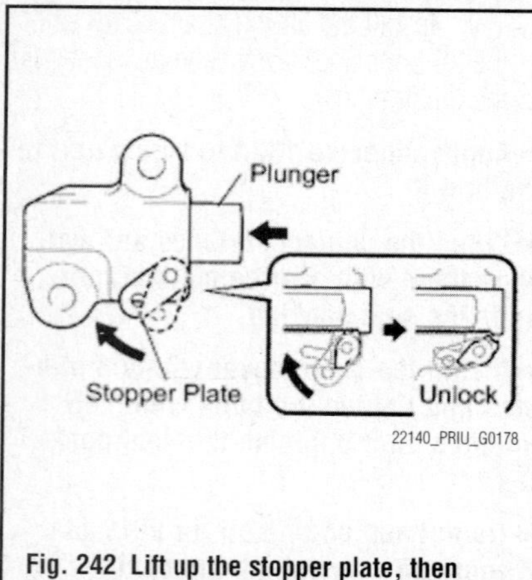

Fig. 242 Lift up the stopper plate, then unlock the plunger then push in the plunger to the end with the plunger unlocked

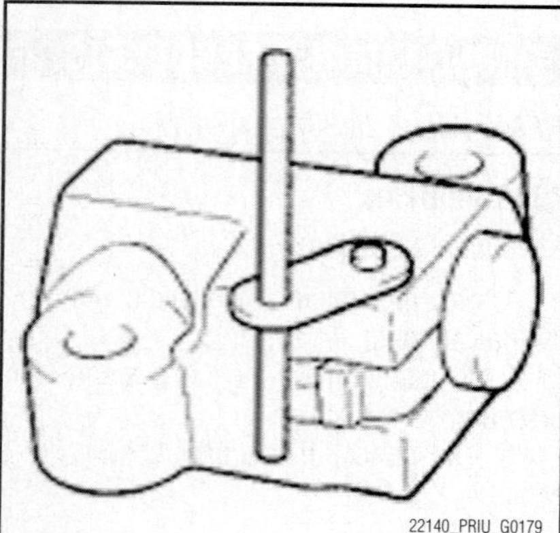

Fig. 243 Insert a 0.118 in. (3.0 mm) diameter bar into the hole of the stopper plate with the plunger locked

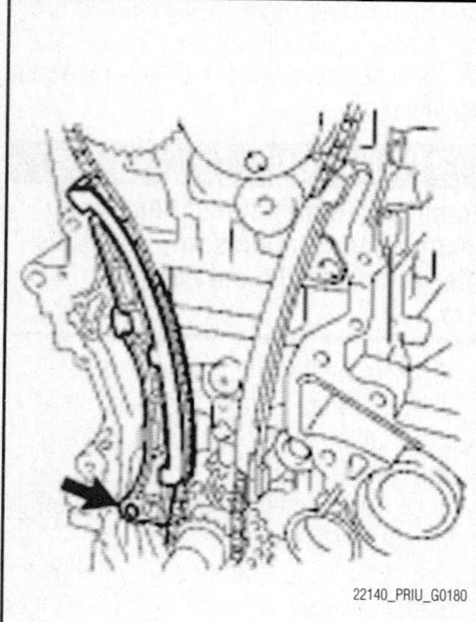

Fig. 244 Remove the bolt and tensioner slipper

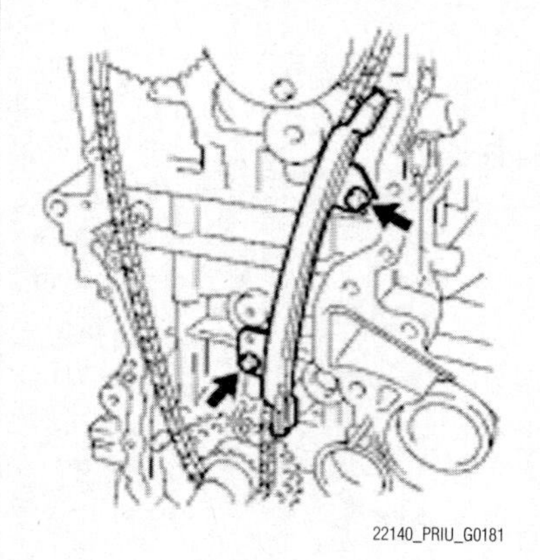

Fig. 245 Remove the bolts and chain vibration damper

➡If the stopper plate is not completely lowered and a 0.118 in. (3.0 mm) diameter bar cannot be inserted, unlock and pull out the plunger slightly. Then the stopper plate will be completely lowered and a 0.118 in. (3.0 mm) diameter bar can be inserted easily.

e. Remove the 2 bolts and chain tensioner.

7. Remove the chain tensioner slipper.
8. Remove the chain vibration damper.
9. Remove the timing chain.

To install:

10. Install the timing chain.
11. After setting the crankshaft at ATDC 40–140 degrees, set cams of intake and exhaust timing sprockets at ATDC 20 degrees and then the reset the crankshaft at ATDC 20 degrees.
12. Install the chain vibration damper with the 2 bolts and tighten to 80 inch lbs. (9 Nm).
13. Align the match marks of timing chain mark plate (Yellow), camshaft timing sprocket, camshaft timing gear and crankshaft timing sprocket to install the timing chain. To prevent the exhaust camshaft from spring back, turn it using a wrench and set it at the mark on a chain.
14. Install the chain tensioner slipper.
15. While rotating the lock plate of the tensioner up-ward, push in the plunger of the tensioner.
16. While rotating the lock plate of the tensioner down-ward, insert a bar of 0.098 inch (2.5mm) into the holes in the lock plate.
17. Install the chain tensioner with the 2 bolts and tighten to 80 inch lbs. (9 Nm).
18. Remove the bar from the chain tensioner.
19. Check the tension between the intake and exhaust camshaft timing sprocket.
20. Install the timing chain cover.
21. While wearing insulating gloves, install the service plug grip.
22. Connect the negative battery cable.
23. Perform the power window initialization procedure.

2010 Vehicles

See Figures 246 through 252.

1. Remove the timing chain cover and seal.
2. Remove the chain tensioner slipper.
   a. Remove the chain tensioner slipper from the cylinder block.
3. Remove the No. 1 chain vibration damper.
   a. Remove the 2 bolts and chain vibration damper.

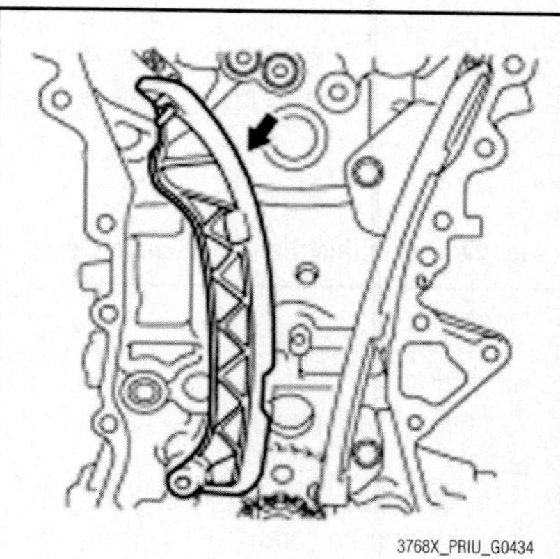

Fig. 246 Removing the chain tensioner slipper

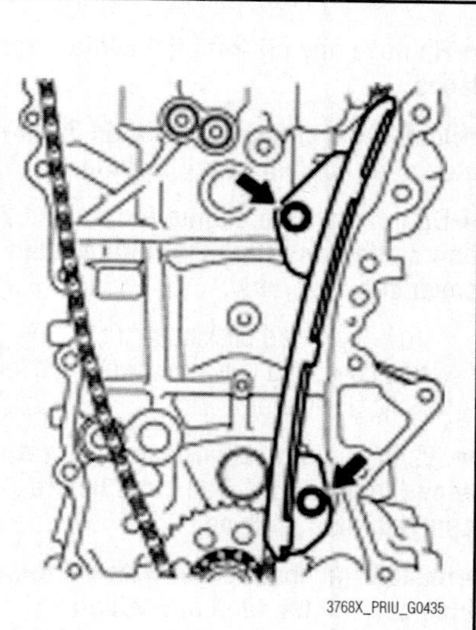

Fig. 247 Removing the No. 1 chain vibration damper

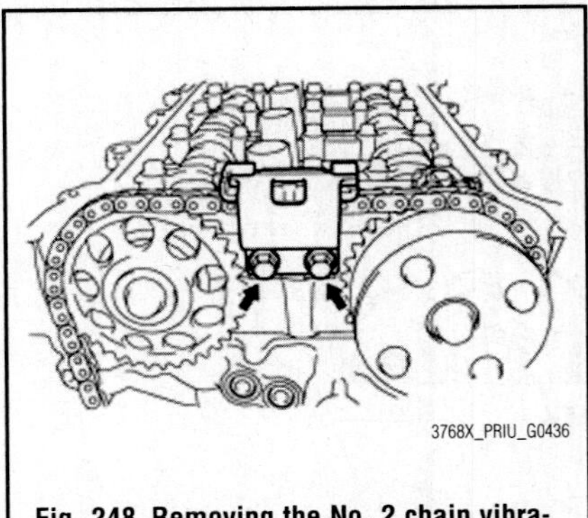

Fig. 248 Removing the No. 2 chain vibration damper

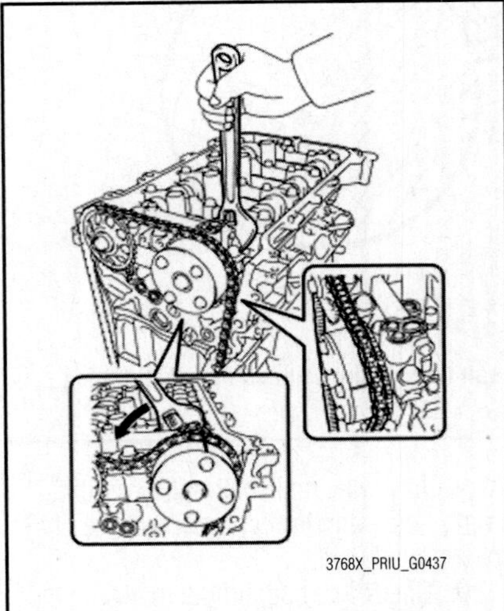

Fig. 249 Removing the chain sub assembly

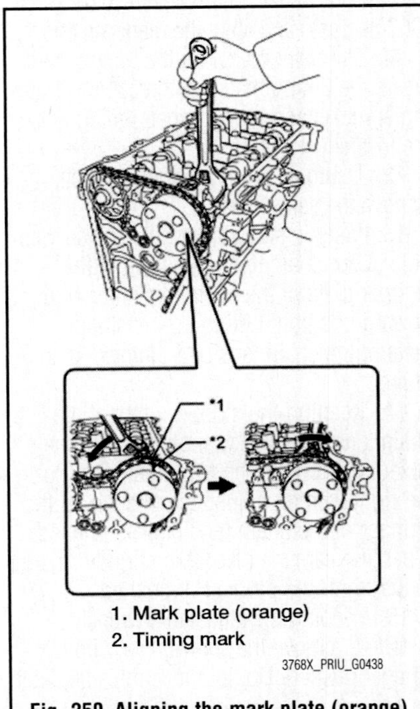

Fig. 250 Aligning the mark plate (orange) with the timing mark

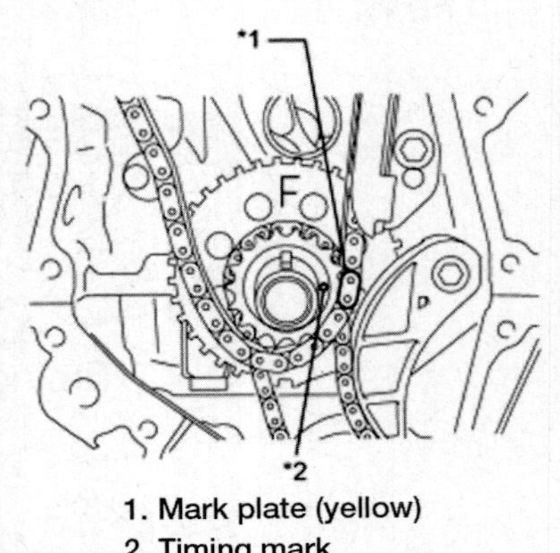

Fig. 251 Aligning the mark plate (yellow) with the timing mark

4. Remove the No. 2 chain vibration damper.

a. Remove the 2 bolts and No. 2 chain vibration damper.

5. Remove the chain sub assembly.

a. Hold the hexagonal portion of the camshaft with a wrench and turn the camshaft timing gear counterclockwise to loosen the chain between the camshaft timing gears.

b. With the chain loosened, release the chain from the camshaft timing gear and place it on the camshaft timing gear.

➡**Be sure to release the chain from the sprocket completely.**

c. Turn the camshaft clockwise to return it to the original position and remove the chain.

To install:

6. Install the chain sub assembly.

a. Align the mark plate (orange) with the timing mark to install the chain.

➡**Be sure to position the mark plate at the front of the engine.**

➡**The mark plate on the camshaft side is colored orange.**

➡**Do not pass the chain around the sprocket of the camshaft timing gear. Only place it on the sprocket.**

➡**Pass the chain through the No. 1 vibration damper.**

b. Hold the hexagonal portion of the camshaft with a wrench and turn the camshaft timing gear counterclockwise to align the mark plate (orange) and timing mark, and then install the chain.

c. Hold the hexagonal portion of the camshaft with a wrench and turn the camshaft timing gear clockwise.

➡**To tension the chain, slowly turn the camshaft timing gear clockwise to prevent the chain from being misaligned.**

d. Align the mark plate (yellow) and timing mark and install the chain to the crankshaft timing gear.

➡**The mark plate on the crankshaft side is colored pink.**

7. Check the No. 1 cylinder TDC/Compression.

a. Check each timing mark at TDC/Compression.

8. Install the No. 2 chain vibration damper.

a. Install the No. 2 chain vibration damper with the 2 bolts. Tighten to 7 ft. lbs. (10 Nm).

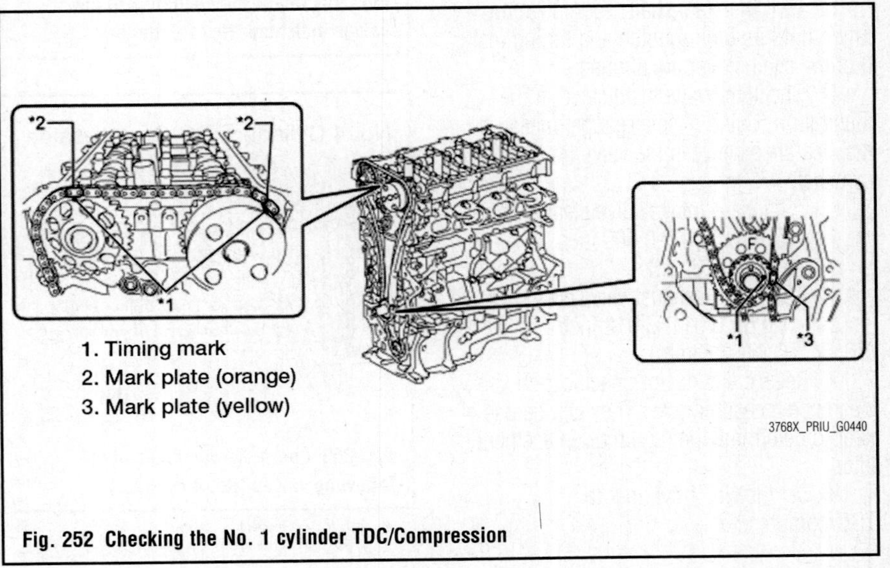

Fig. 252 Checking the No. 1 cylinder TDC/Compression

9. Install the chain tensioner slipper.
   a. Install the chain tensioner slipper to the cylinder block.
10. Install the timing chain cover oil seal.
11. Install the timing chain cover sub assembly.

VALVE LASH

ADJUSTMENT

See Figures 253 through 258.

➡Inspect and adjust the valve clearance when the engine is cold.

1. Before servicing the vehicle, refer to the precautions.
2. Set the No. 1 cylinder to TDC/compression.
   - Turn the crankshaft pulley until its timing notch and timing mark 0 of the chain cover are aligned.
   - Check that both timing marks on the camshaft timing sprocket and camshaft timing gear are facing upward as shown in the illustration. If not, turn the crankshaft 1 complete revolution (360 degrees) and align the marks as above.
3. Check the valves indicated in the illustration. Using a feeler gauge, measure the clearance between the valve lifter and camshaft.
   - Standard intake valve clearance (Cold): 0.007–0.009 inch (0.17–0.23mm)
   - Standard exhaust valve clearance (Cold): 0.011–0.013 inch (0.27–0.33mm)
4. Record any out-of-specification valve clearance measurements. They will be used later to determine the required replacement lifter.
5. Turn the crankshaft 1 complete revolution until its timing notch and timing mark 0 of the chain cover are aligned.
6. Check the valves indicated in the illustration. Using a feeler gauge, measure the clearance between the valve lifter and camshaft.
   - Standard intake valve clearance (Cold): 0.007–0.009 inch (0.17–0.23mm)
   - Standard exhaust valve clearance (Cold): 0.011–0.013 inch (0.27–0.33mm)
7. Record any out-of-specification valve clearance measurements. They will be used later to determine the required replacement lifter.
8. Set the No. 1 cylinder to TDC/compression.

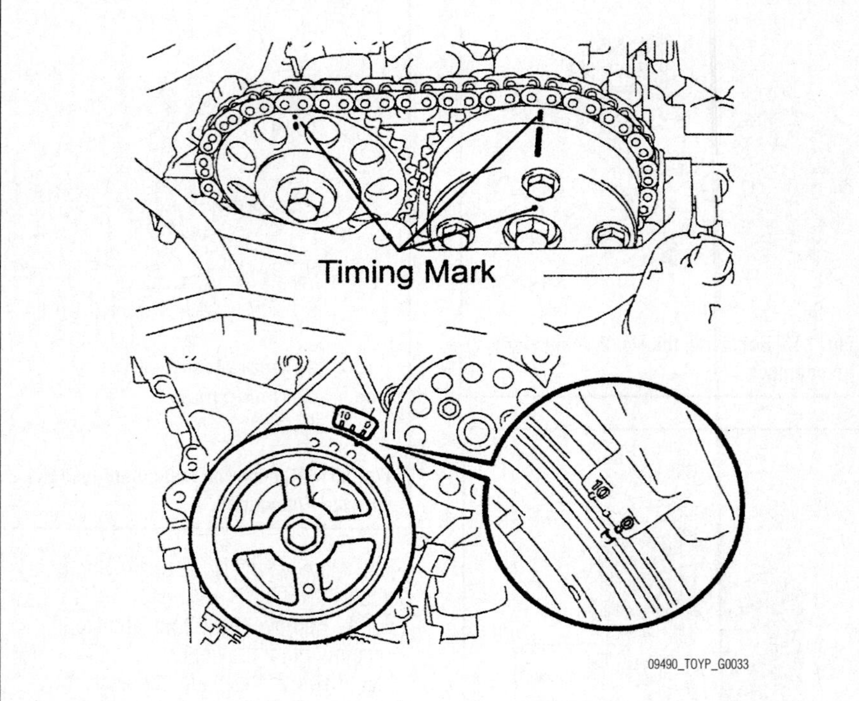

Fig. 253 Check that both timing marks on the camshaft timing sprocket and camshaft timing gear are facing upward

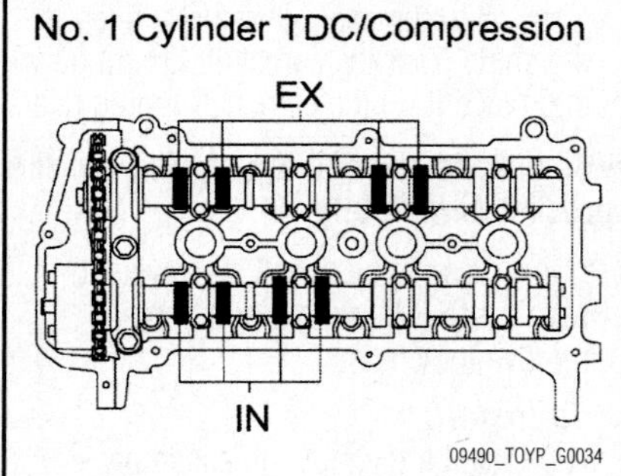

Fig. 254 Check the clearance of the valves indicated (No. 1 Cylinder)

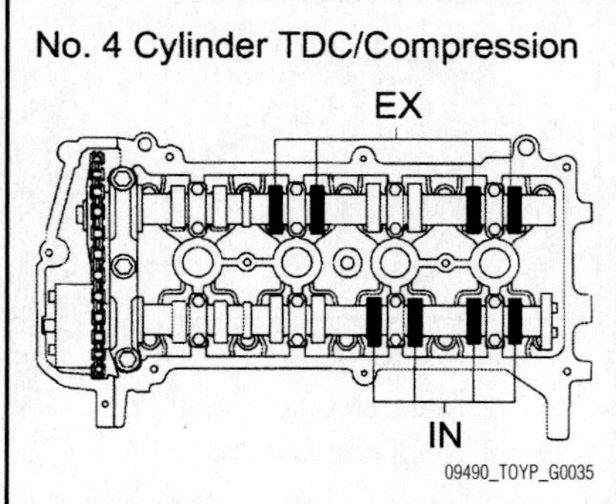

Fig. 255 Check the clearance of the following valves (No. 4 cylinder)

9. Turn the crankshaft pulley until its timing notch and timing mark 0 of the chain cover are aligned.
10. Check that both timing marks on the camshaft timing sprocket and valve timing controller assembly are facing upward. If not, turn the crankshaft 1 complete revolution (360 degrees) and align the marks as above.
11. Put paint marks on the timing chain where the timing marks of the camshaft timing sprocket and the camshaft timing gear are located.
12. Using an 8mm hexagon wrench, remove the screw plug.
13. Insert a screwdriver into the service hole of the chain tensioner to hold the stopper plate of the chain tensioner at an upward position. Lifting up the stopper plate of the chain tensioner unlocks the plunger.
14. Keeping the stopper plate of the chain tensioner lifted, slightly rotate the hexagonal lobe of the No. 2 camshaft to the right with an adjustable wrench so the plunger of the chain tensioner is pushed. When the camshaft No. 2 is slightly rotated to the right, the plunger is pushed.
15. Keeping the adjustable wrench installed, remove the screwdriver with the plunger pushed. Do not move the adjustable wrench.

➡**Removing the screwdriver lowers the stopper plate and locks the plunger.**

16. Insert a 0.118 inch (3.0mm) diameter bar into the hole of the stopper plate with the stopper plate of the chain tensioner lowered and locked. If the bar cannot be inserted into the hole of the stopper plate, rotate the No. 2 camshaft slightly to the left and right. Then that bar can be inserted easily.
17. Secure the bar with tape.
18. Hold the hexagonal lobe of the camshaft No. 2 with the adjustable wrench.
19. Using SST 09023-38400, loosen the bolt.
20. Using several steps, uniformly loosen and remove the 11 bearing cap bolts in the sequence shown in the illustration. Then remove the 5 bearing caps. Loosen each bolt uniformly, keeping the camshaft level.
21. Remove the flange bolt with the No. 2 camshaft lifted up. Then detach the No. 2 camshaft and the camshaft timing sprocket.
22. Using several steps, uniformly loosen and remove the 8 bearing cap bolts in the sequence shown in the illustration. Then remove the 4 bearing caps. Loosen each bolt uniformly, keeping the camshaft level.

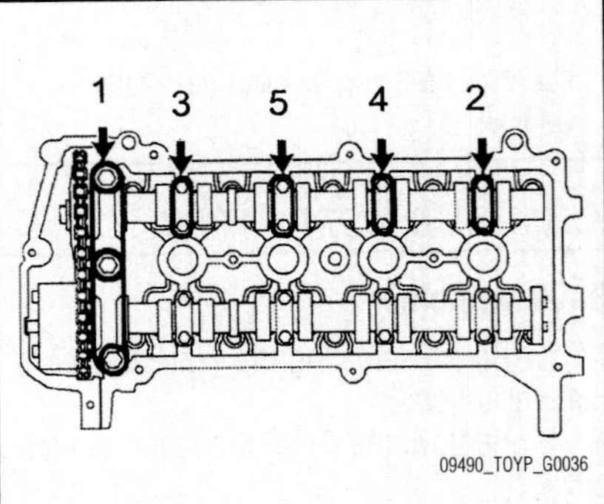

Fig. 256 Remove the 11 bearing cap bolts in the sequence

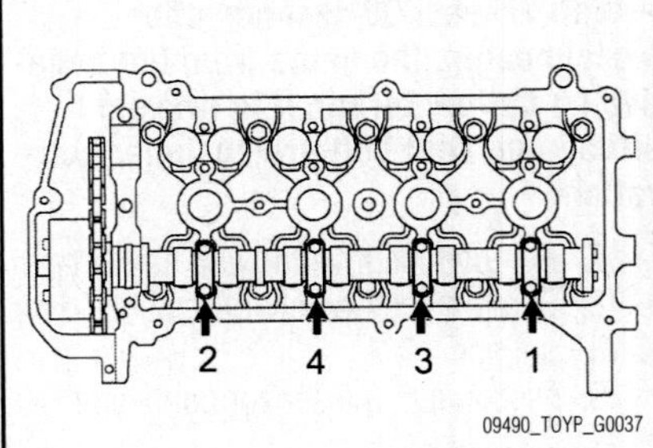

Fig. 257 Remove the 8 bearing cap bolts in the sequence

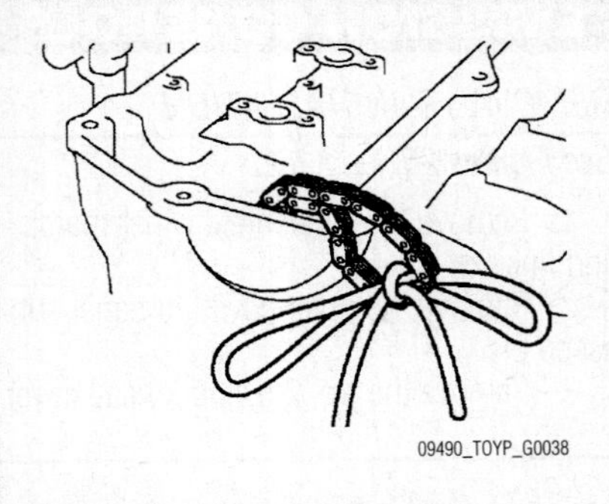

Fig. 258 Tie the timing chain with a string

23. Hold the timing chain with one hand, and remove the camshaft and the camshaft timing gear assembly.
24. Tie the timing chain with a string as shown in the illustration.

CAUTION

Be careful not to drop anything inside the timing chain cover.

25. Remove the valve lifters.
26. Using a micrometer, measure the thickness of the removed lifter.
27. Calculate the thickness of a new lifter so that the valve clearance comes within the specified value.
28. Select a new lifter with the thickness as close to the calculated values as possible.
- EXAMPLE: (Intake) Measured valve clearance equals 0.0158 inch (0.40mm)
- 0.0158 inch (0.40mm) minus 0.0079 inch (0.20mm) equals 0.0079 inch (0.20mm) (Measured minus Specification equals Excess clearance)
- Used lifter measurement equals 0.2067 inch (5.25mm)
- 0.0079 inch (0.20mm) plus 0.2067 inch (5.25mm) equals 0.2146 inch (5.45mm) (Excess clearance plus Used lifter equals Ideal new lifter)
- Closest new lifter equals 5.45 mm (0.2146 in.); select lifter (0.2150 inch (5.46mm))

➡**Lifters are available in 35 sizes in increments of 0.0008 inch (0.020mm), from 0.1992 inch (5.060mm) to 0.2260 inch (5.740mm).**

INSPECTION

See Figure 259.

➡**Keep the valve lash adjuster free from dirt and foreign matter.**

➡**Only use clean engine oil.**

1. Place the lash adjuster into a container full of new engine oil.
2. Insert the tip of the special tool (09276-75010) into the lash adjuster plunger and use the tip to press down on the check ball inside the plunger.
   a. Squeeze SST and the valve lash adjuster together to move the plunger up and down 5 to 6 times.
   b. Check the movement of the plunger and bleed the air. OK: Plunger moves up and down.

➡**When bleeding high-pressure air from the compression chamber, make sure that the tip of SST is actually pressing the check ball as shown in the illustration. If the check ball is not pressed, air bleeding is not possible.**

   c. After bleeding the air, remove SST. Then try to quickly and firmly press the plunger with your fingers. OK: Plunger can be pressed 3 times. If the plunger can still be compressed after pressing it 3 times, replace the valve lash adjuster with a new one.

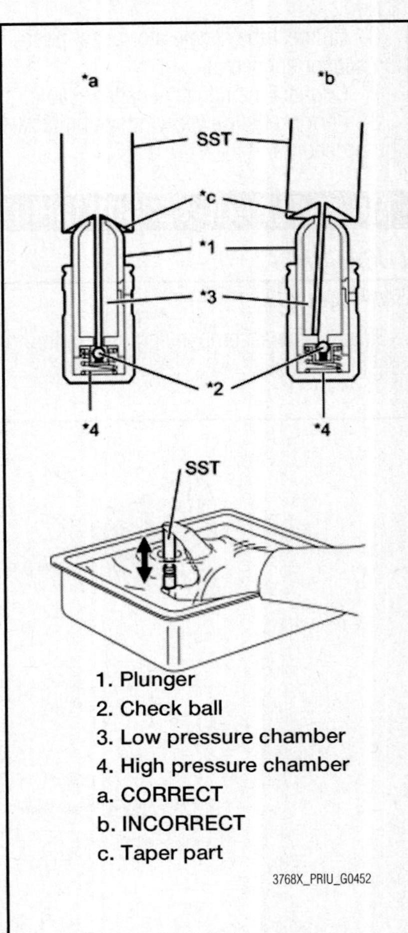

Fig. 259 Inspecting the lash adjuster

ACCELERATOR PEDAL POSITION (APP) SENSOR

LOCATION

The APP sensor is a integral part of the accelerator pedal assembly.

REMOVAL & INSTALLATION

1. Before servicing the vehicle, refer to the precautions.
2. Disconnect the negative battery cable.

➡Wait at least 90 seconds after disconnecting the cable from the negative (-) battery terminal to prevent airbag and seat belt pretensioner activation.

3. Disconnect the accelerator pedal position sensor connector.
4. Remove the 2 bolts, then remove the accelerator pedal rod.

To install:

5. Install the accelerator pedal rod with the 2 bolts. Tighten the bolts to 66 inch lbs. (8 Nm).
6. Connect the accelerator pedal position sensor connector.
7. Connect the negative battery cable.
8. Perform the power window initialization procedure.

AIR-FUEL RATIO SENSOR (AFR)

LOCATION

See Figure 260.

Refer to the accompanying illustration for sensor location.

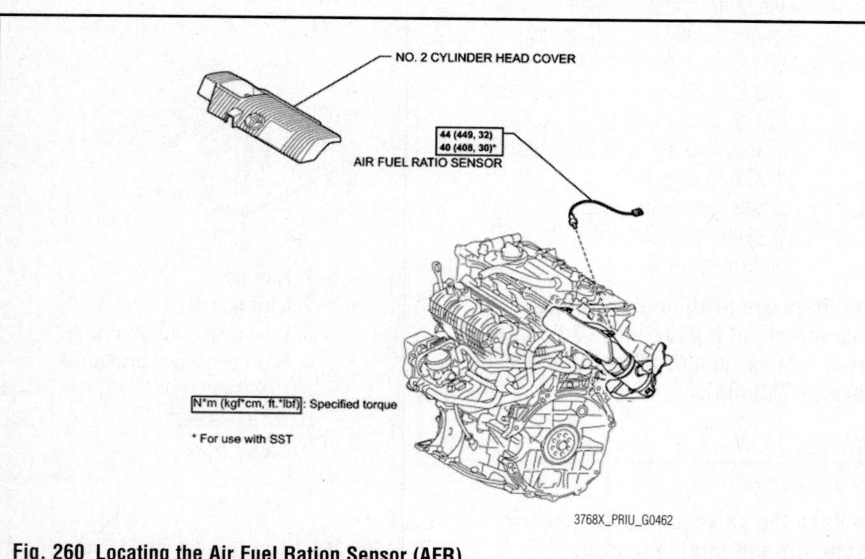

Fig. 260 Locating the Air Fuel Ration Sensor (AFR)

REMOVAL & INSTALLATION

See Figures 261 and 262.

1. Remove the windshield wiper motor and link.
2. Remove the outer cowl top panel sub assembly.
3. Remove the No. 2 cylinder head cover.

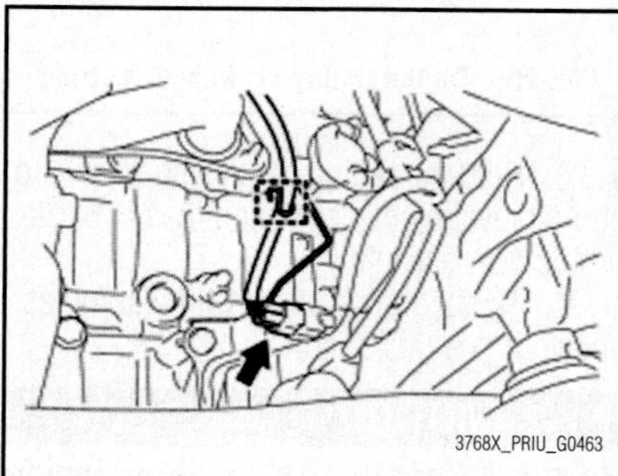

Fig. 261 Disconnecting the air fuel ratio sensor connector and clamp

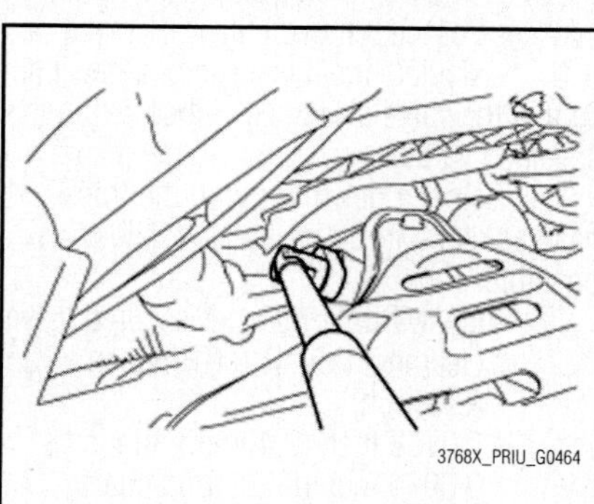

Fig. 262 Removing the air fuel ratio sensor

4. Remove the air fuel ratio sensor.
   a. Disconnect the air fuel ratio sensor connector and clamp.
   b. Using the special tool (09224-00010), remove the air fuel ratio sensor.

To install:

5. To install, reverse the removal procedure. Tighten the air fuel ratio sensor to 32 ft. lbs. (44 Nm). If using the special tool to torque, tighten to 30 ft. lbs. (40 Nm).

CAMSHAFT POSITION (CMP) SENSOR

LOCATION

See Figures 263 and 264.

Refer to the accompanying illustrations for sensor location.

Fig. 263 CMP sensor location—2009 vehicles

REMOVAL & INSTALLATION

2009 Vehicles

1. Before servicing the vehicle, refer to the precautions.
2. Remove the radiator support opening cover.
3. Remove the engine under cover.
4. Drain engine coolant.
5. Drain high voltage battery coolant.
6. Disconnect the negative battery cable.

➡Wait at least 90 seconds after disconnecting the cable from the negative (-) battery terminal to prevent airbag and seat belt pretensioner activation.

7. Remove the inverter with converter.
8. Remove the Camshaft Position (CMP) sensor.
9. Disconnect the sensor connector.

To install:

10. Install the Camshaft Position (CMP) sensor.

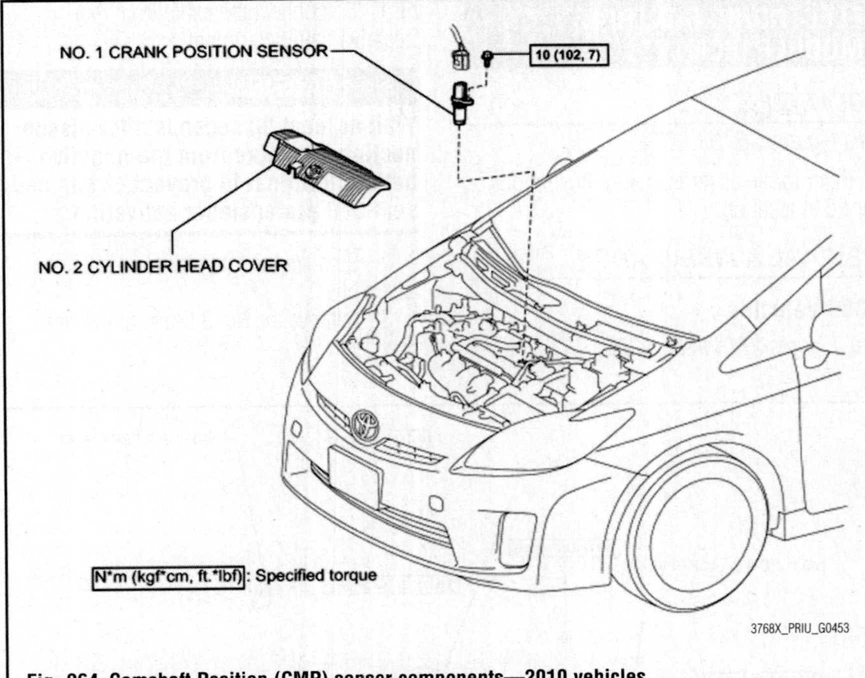

Fig. 264 Camshaft Position (CMP) sensor components—2010 vehicles

11. Install the sensor with the bolt. Tighten to 66 inch lbs. (8 Nm).
12. Connect the sensor connector.
13. Install the inverter with converter.
14. Install the inverter with converter.
15. Add high voltage battery coolant.
16. Add engine coolant.
17. Inspect for coolant leaks.
18. Install the radiator support opening cover.
19. Install the engine under cover.
20. Connect the negative battery cable.
21. Perform the power window initialization procedure.

2010 Vehicles

See Figure 265.

➡Although the part name refers to the No. 1 crank position sensors, this procedure is for the Camshaft Position (CMP) sensors.

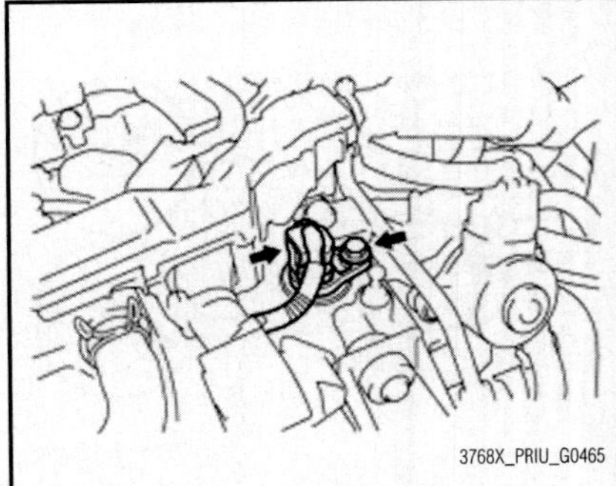

Fig. 265 Removing the No. 1 crank position sensor

1. Remove the No. 2 cylinder head cover.
2. Remove the No. 1 crank position sensor.
   a. Disconnect the No.1 crank position sensor connector.
   b. Remove the bolt and No. 1 crank position sensor.

To install:

➡Although the part name refers to the No. 1 crank position sensors, this procedure is for the Camshaft Position (CMP) sensors.

3. Install the No. 1 crank position sensor.
   a. Apply a light coat of engine oil to the o-rings on the No. 1 crank position sensor.
   b. Install the No. 1 crank position sensor with the bolt. Tighten to 7 ft. lbs. (10 Nm).
   c. Connect the No. 1 crank position sensor connector.
4. Inspect for oil leaks.
5. Install the No. 2 cylinder head cover.

CRANKSHAFT POSITION (CKP) SENSOR

LOCATION

See Figures 266 and 267.

Refer to the accompanying illustrations for sensor location.

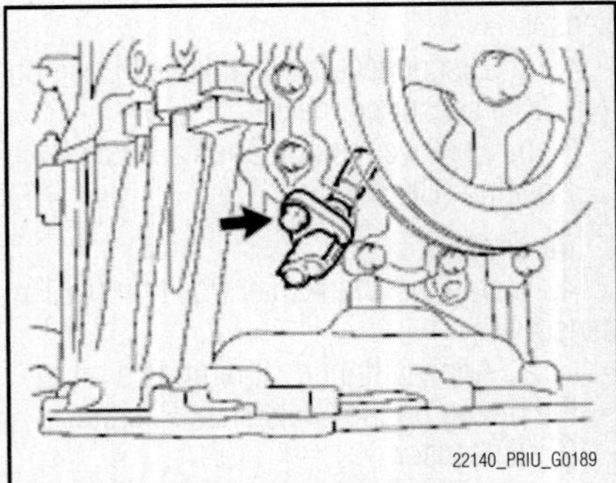

Fig. 266 CKP sensor location

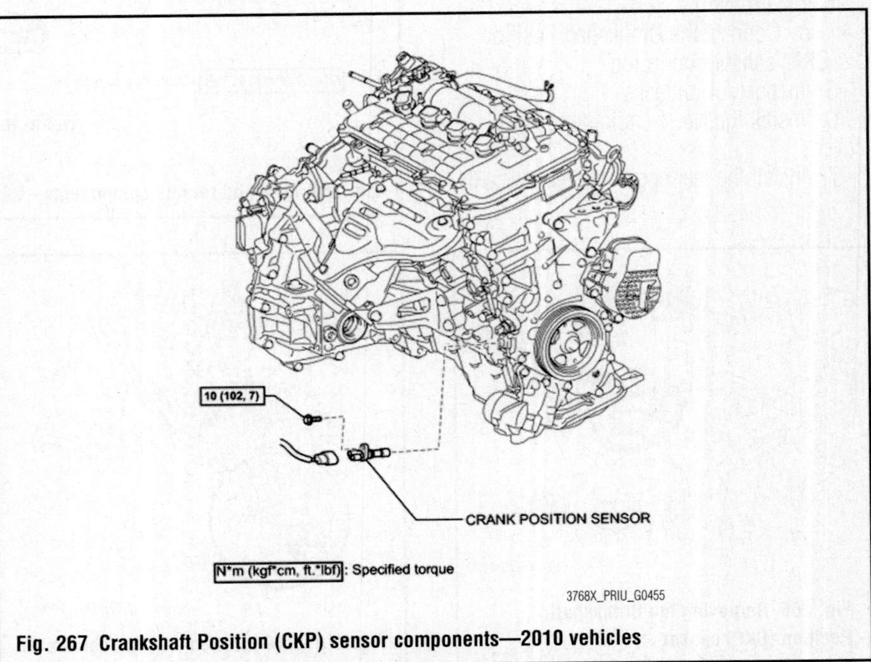

Fig. 267 Crankshaft Position (CKP) sensor components—2010 vehicles

REMOVAL & INSTALLATION

2009 Vehicles

1. Before servicing the vehicle, refer to the precautions.
2. Raise and safely support the vehicle securely on jackstands.
3. Remove engine under cover.
4. Disconnect the sensor connector.
5. Remove the bolt and sensor.

To install:

6. Install the sensor with the bolt. Tighten to 66 inch lbs. (8 Nm).
7. Connect the sensor connector.
8. Install engine under cover.
9. Lower the vehicle.

2010 Vehicles

See Figure 268.

1. Remove the rear engine under cover RH.
2. Remove the No. 1 engine under cover.
3. Remove the crank position sensor.
   a. Disconnect the Crankshaft Position (CKP) sensor connector.
   b. Remove the bolt and Crankshaft Position (CKP) sensor.

To install:

4. Install the Crankshaft Position (CKP) sensor.
   a. Apply a light coat of engine oil to the o-ring on the Crankshaft Position (CKP) sensor.
   b. Install the Crankshaft Position (CKP) sensor with the bolt. Tighten to 7 ft. lbs. (10 Nm).
   c. Connect the Crankshaft Position (CKP) sensor connector.
5. Inspect for oil leaks.
6. Install the No. 1 engine under cover.
7. Install the rear engine under cover RH.

ELECTRONIC CONTROL MODULE (ECM)

LOCATION

See Figures 269 and 270.

Refer to the accompanying illustrations for ECM location.

REMOVAL & INSTALLATION

2009 Vehicles

See Figures 271 through 275.

1. Disconnect the cable from the negative battery terminal.

CAUTION

Wait at least 90 seconds after disconnecting the cable from the negative (-) battery terminal to prevent airbag and seat belt pretensioner activation.

2. Remove the instrument panel sub assembly.
3. Remove the No. 3 heater to register duct.

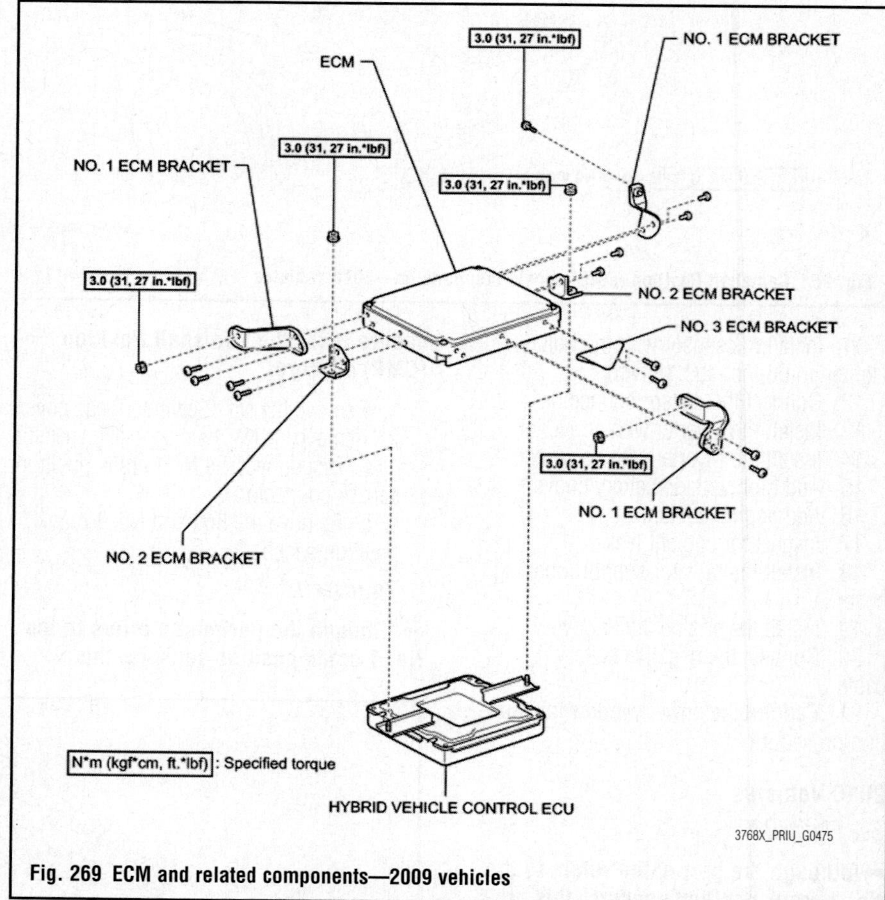

Fig. 269 ECM and related components—2009 vehicles

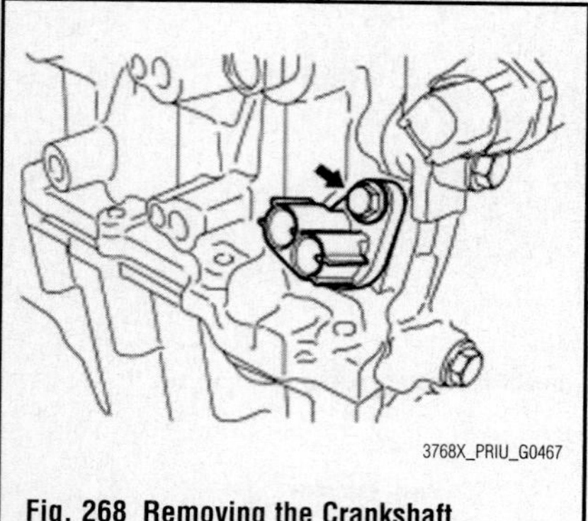

Fig. 268 Removing the Crankshaft Position (CKP) sensor

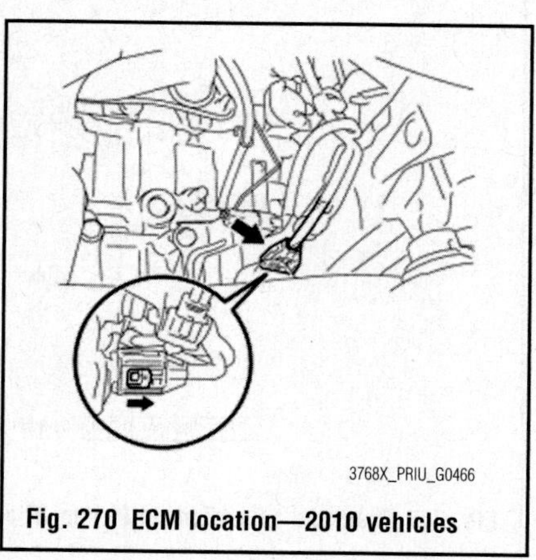

Fig. 270 ECM location—2010 vehicles

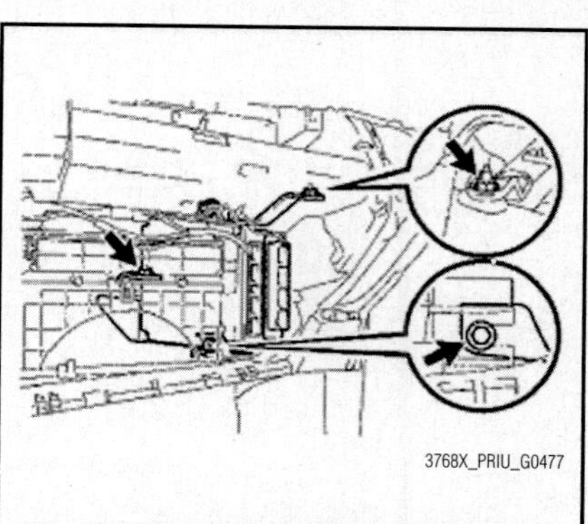

Fig. 271 Removing the ECM with the bracket

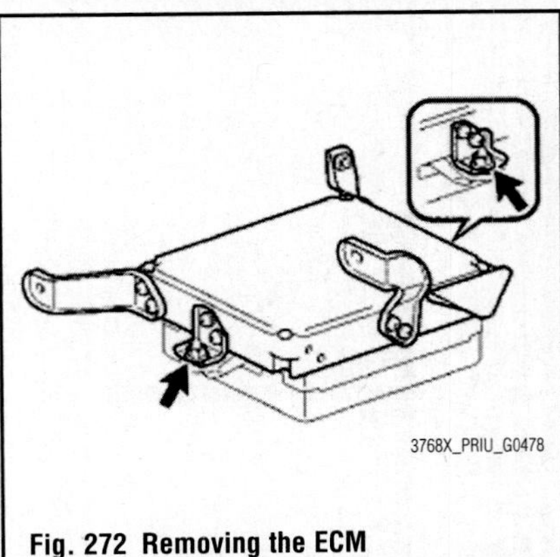

Fig. 272 Removing the ECM

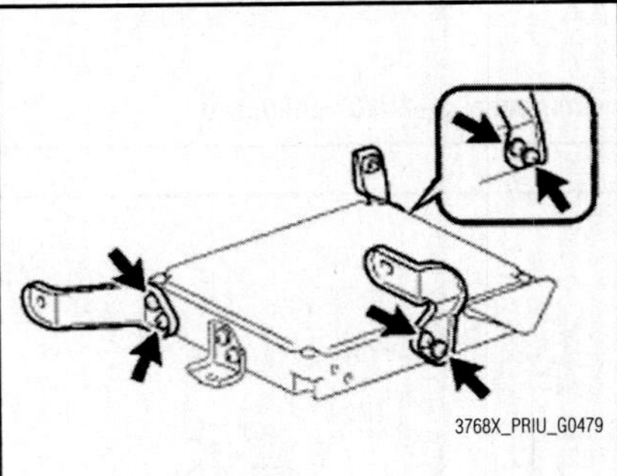

Fig. 273 Removing the No. 1 ECM brackets

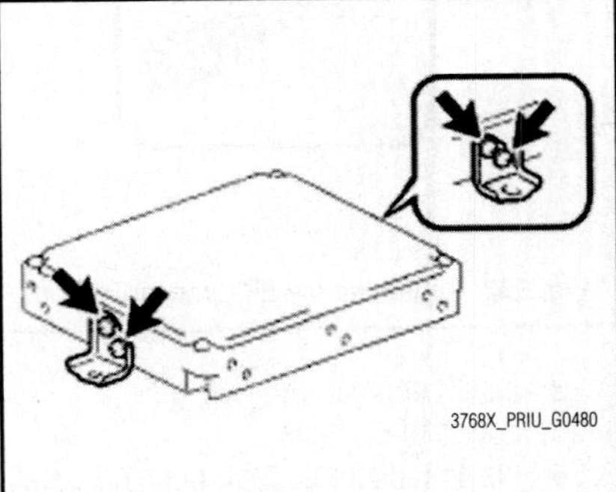

Fig. 274 Removing the No. 2 ECM brackets

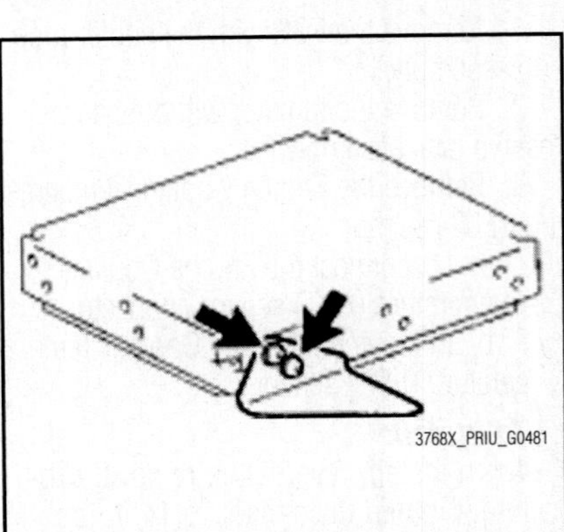

Fig. 275 Removing the No. 3 ECM bracket

4. Remove the ECM.
   a. Disconnect the 4 ECM connectors.
   b. Disconnect the 4 hybrid vehicle control ECU connectors.
   c. Remove the 2 nuts and bolt, and ECM with the bracket.
   d. Remove the 2 nuts and ECM.
   e. Remove the 6 screws and 3 No. 1 ECM brackets.
   f. Remove the 4 screws and 2 No. 2 ECM brackets.
   g. Remove the 2 screws and No. 3 ECM bracket.

To install:

5. To install, reverse the removal procedure. Tighten the ECM nuts and bolt to 27 inch lbs. (3 Nm).

2010 Vehicles

See Figures 276 through 280.

1. Remove the rear No. 2 floor board (separate type).
2. Remove the rear deck floor box.

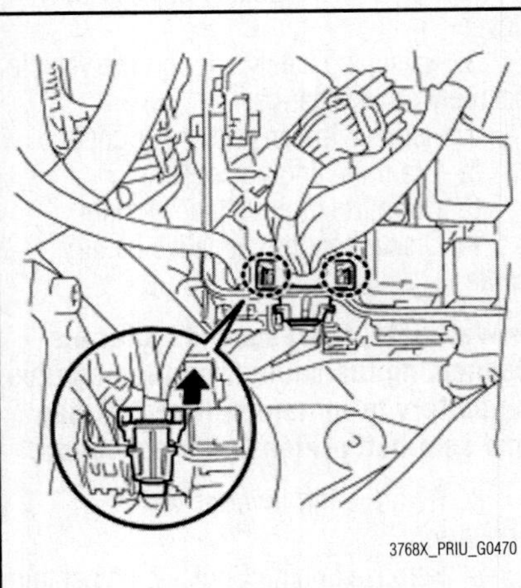

Fig. 276 Separating the engine room wire harness from the engine room relay block

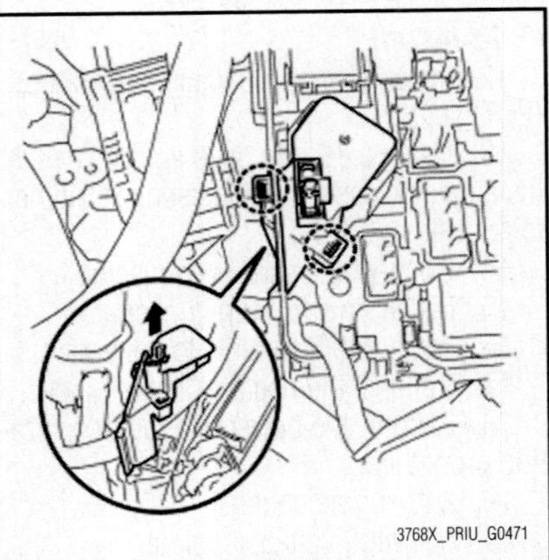

Fig. 277 Removing the side cover

3. Remove the rear No. 3 floor board.
4. Disconnect the cable from the negative battery terminal.

➡When disconnecting the cable, some systems need to be initialized after the cable is reconnected.

5. Separate the radiator without the cap reserve tank assembly.
   a. Remove the 2 bolts and separate the radiator without cap reserve tank assembly.
6. Remove the ECM.
   a. Remove the No. 1 relay block cover.
   b. Disconnect the 3 wire harness connectors.
   c. Disengage the 2 claws, push up the engine room wire harness and separate it from the engine room relay block.
   d. Disengage the 2 claws and push up the side cover to remove it.
   e. Loosen the bolt and separate the No. 2 engine room wire.

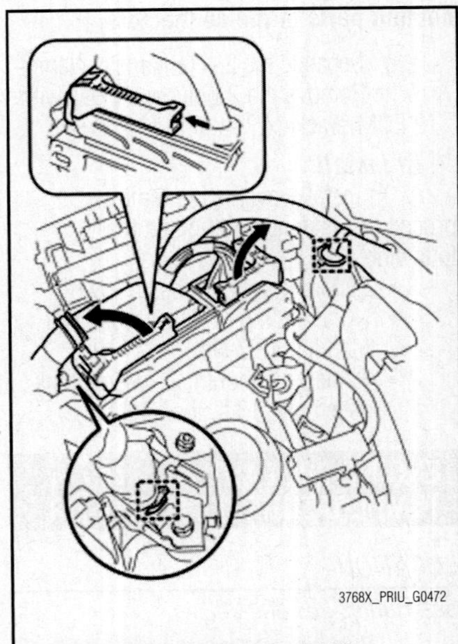

Fig. 278 Disconnecting the 2 ECM connectors

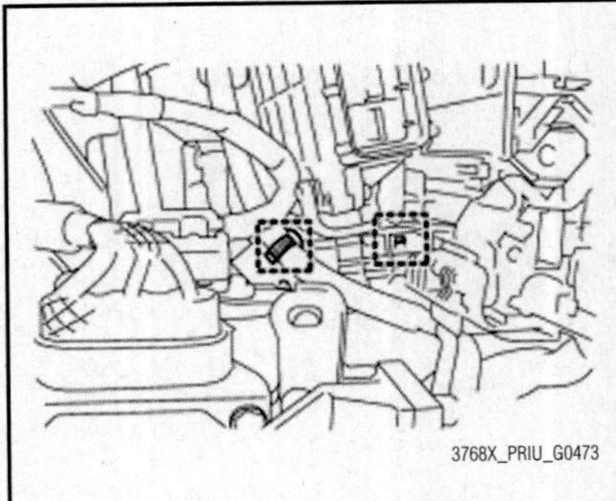

Fig. 279 Separating the wire harness clamps

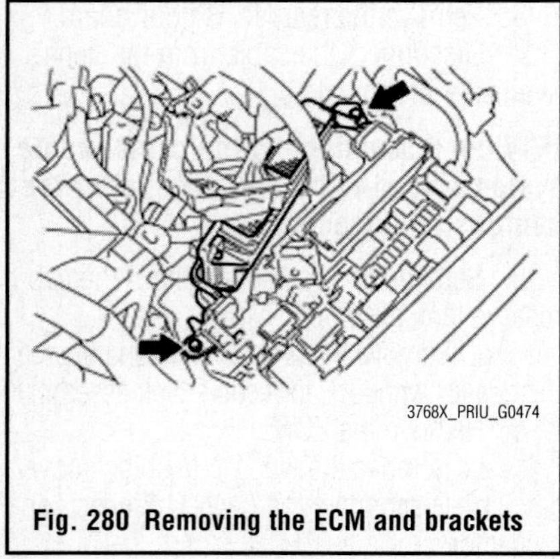

Fig. 280 Removing the ECM and brackets

f. Disconnect the 2 ECM connectors and wire harness clamps. Push the locks on the 2 levers, then raise the levers, and disconnect the 2 ECM connectors.

➡After disconnecting the connectors, make sure that dirt, water or other foreign matter does not contact the connecting parts of the connectors.

g. Separate the 2 wire harness clamps.
h. Remove the 2 bolts and ECM with 2 ECM brackets.

To install:

7. To install, reverse the removal procedure. Take note to tighten the following:
- ECM bolts: 10 ft. lbs. (13 Nm)
- No. 2 engine room wire bolt: 73 inch lbs. (8.3 Nm)
- Radiator without cap reserve tank assembly: 13 ft. lbs. (18 Nm)

ENGINE COOLANT TEMPERATURE (ECT) SENSOR

LOCATION

See Figures 281 and 282.

Refer to the accompanying illustrations for sensor location.

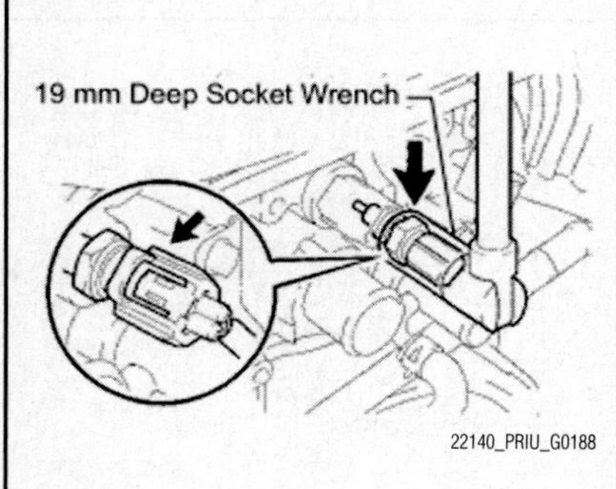

Fig. 281 ECT sensor location—2009 vehicles

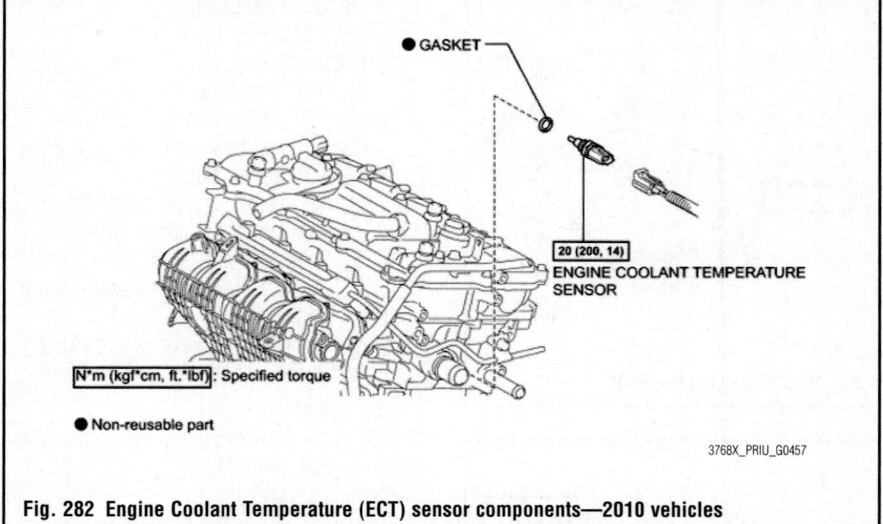

Fig. 282 Engine Coolant Temperature (ECT) sensor components—2010 vehicles

REMOVAL & INSTALLATION

2009 Vehicles

1. Before servicing the vehicle, refer to the precautions.
2. Remove the radiator support opening cover.
3. Raise and safely support the vehicle securely on jackstands.
4. Remove the engine under cover.
5. Drain the engine coolant.
6. Drain the high voltage coolant.
7. Disconnect the negative battery cable.

➡Wait at least 90 seconds after disconnecting the cable from the negative (-) battery terminal to prevent airbag and seat belt pretensioner activation.

8. Remove the inverter with converter.
9. Remove Engine Coolant Temperature (ECT) sensor.
10. Disconnect the sensor connector.
11. Using a 19 mm deep socket wrench, remove the sensor and gasket.

To install:

12. Install Engine Coolant Temperature (ECT) sensor.
13. Using a 19 mm deep socket wrench, install a new gasket and the sensor. Tighten to 15 ft. lbs. (20 Nm).
14. Connect the sensor connector.
15. Install inverter with converter.
16. Install the inverter with converter.
17. Connect the negative battery cable.
18. Perform the power window initialization procedure.
19. Add engine coolant.
20. Add high voltage coolant.
21. Inspect for coolant leaks.

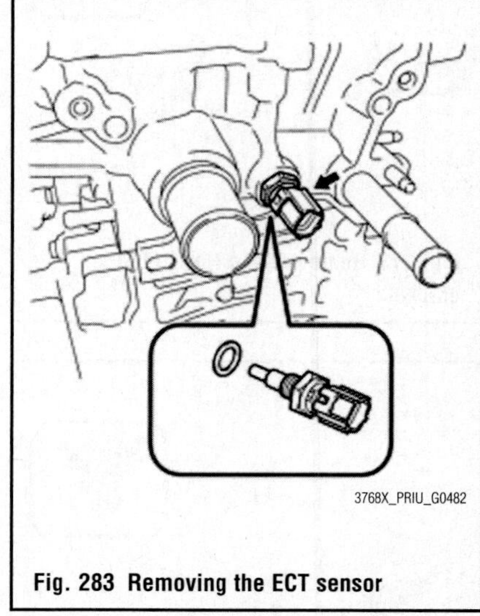

Fig. 283 Removing the ECT sensor

22. Install engine under cover.
23. Lower the vehicle.
24. Install radiator support opening cover.

2010 Vehicles

See Figure 283.

1. Remove the EGR with the cooler pipe sub assembly.
2. Remove the radiator without cap reserve tank assembly.
3. Remove the Engine Coolant Temperature (ECT) sensor.
   a. Disconnect the Engine Coolant Temperature (ECT) sensor connector.
   b. Remove the Engine Coolant Temperature (ECT) sensor.

To install:

4. To install, reverse the removal procedure. Tighten the sensor to 14 ft. lbs. (20 Nm).

EXHAUST GAS RECIRCULATION (EGR) VALVE

LOCATION

See Figures 284 and 285.

Refer to the accompanying illustrations for location.

REMOVAL & INSTALLATION

See Figures 286 through 289.

1. Remove the windshield wiper motor and link assembly.
2. Remove the cowl body mounting reinforcement LH.
3. Remove the outer cowl top panel sub assembly.
4. Drain the coolant (engine).

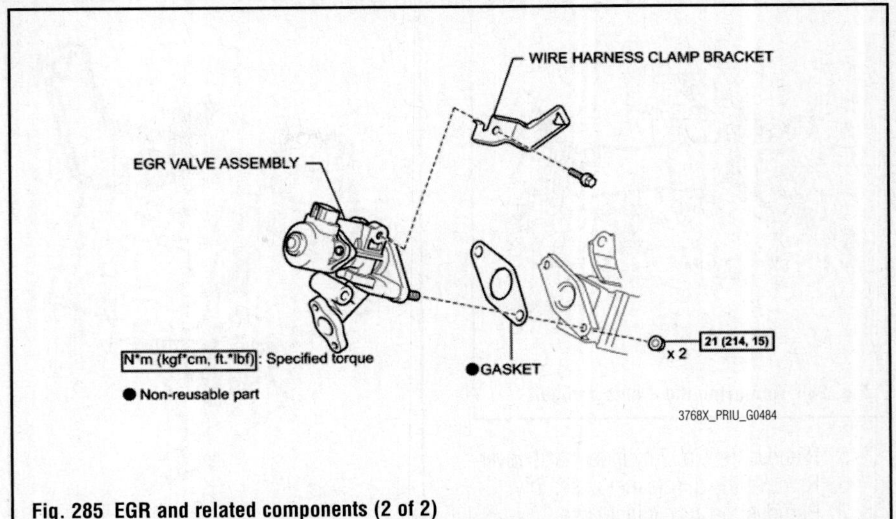

Fig. 285 EGR and related components (2 of 2)

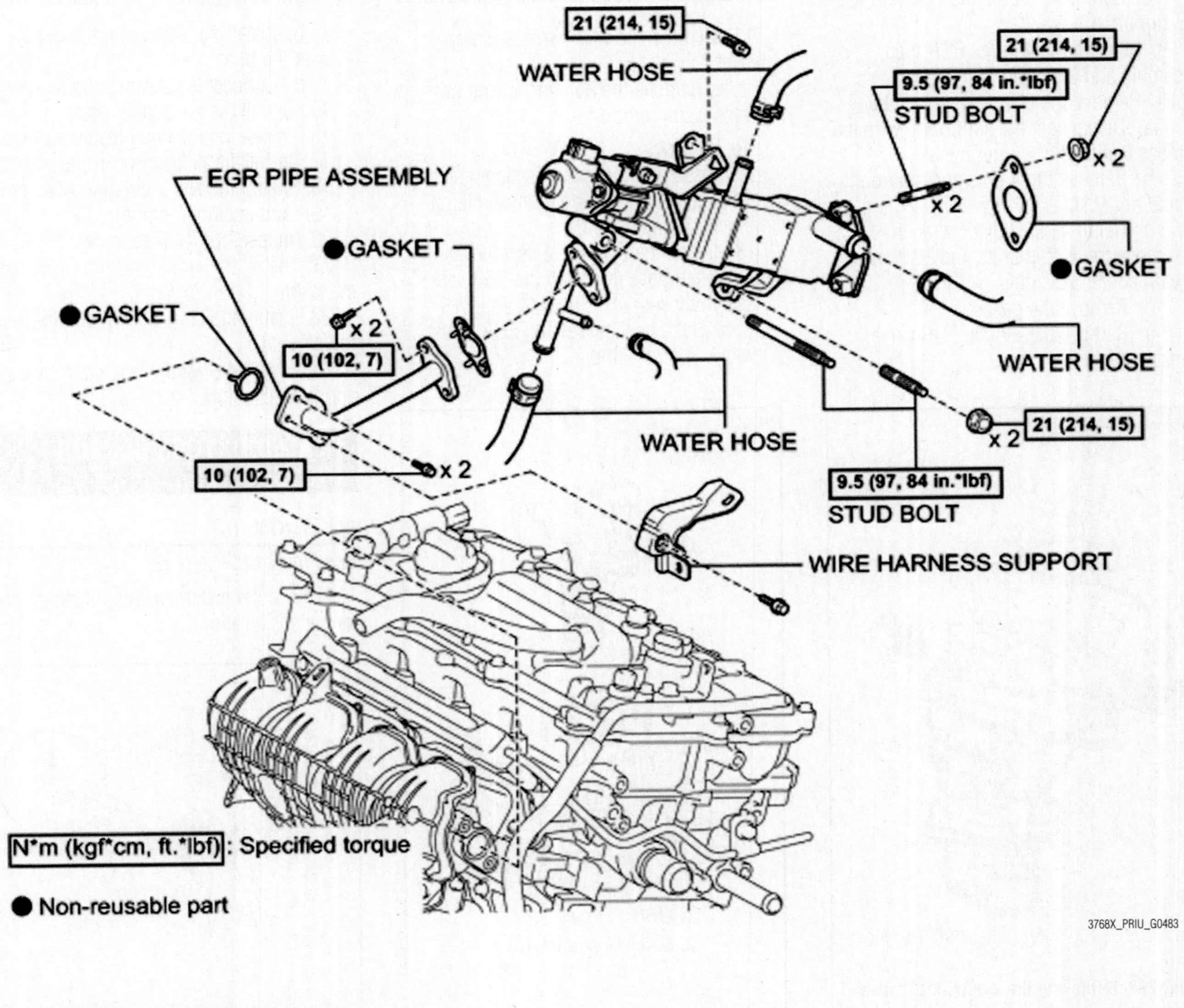

Fig. 284 EGR and related components (1 of 2)

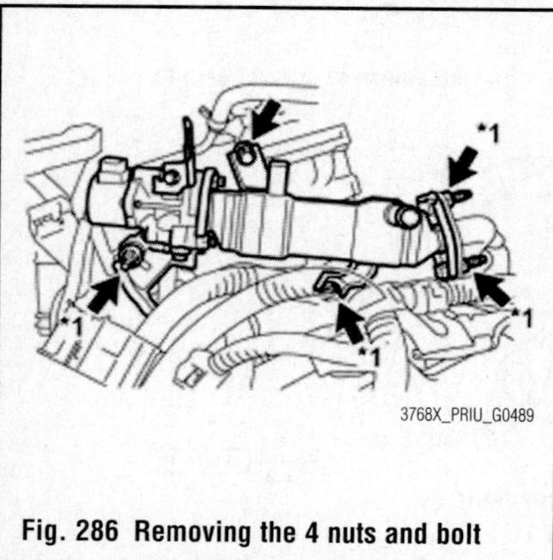

Fig. 286 Removing the 4 nuts and bolt

5. Remove the No. 2 cylinder head cover.
6. Remove the air cleaner assembly.
7. Remove the air cleaner case.
8. Remove the EGR pipe assembly.

a. Disconnect the 2 wire harness clamps.

b. Remove the bolt and wire harness support.

9. Remove the 4 bolts, EGR pipe assembly and 2 gaskets.
10. Remove the EGR valve assembly.

a. Disconnect the connector, wire harness clamp and 4 water hoses.

b. Remove the 4 nuts and stud bolt and nut (*1).

c. Using an E8 TORX® wrench, remove the 4 stud bolts and EGR valve with cooler assembly.

d. Remove the gasket.

e. Remove the 2 nuts, EGR valve assembly and gasket.

Fig. 288 Removing the wiring harness clamp bracket

11. Remove the wire harness clamp bracket.

a. Remove the bolt and wiring harness clamp bracket.

To install:

12. Install the wire harness clamp bracket.

a. Install the wiring harness clamp bracket with the bolt.

13. Install the EGR valve assembly.

a. Temporarily install the EGR valve assembly with the 2 nuts (A and B).

b. Set a new gasket and EGR valve with cooler assembly.

c. Using an E8 TORX® wrench, install the 4 stud bolts. Tighten to 84 inch lbs. (9.5 Nm).

d. Temporarily install the 2 nuts (C) and bolt.

e. Tighten nut (A) to 15 ft. lbs. (21 Nm).

f. Tighten the 3 nuts (B and C) and bolt to 15 ft. lbs. (21 Nm).

➡Make sure that all installation surfaces of the EGR valve with cooler assembly are in even contact when tightening the bolts and nuts.

g. Connect the connector, wire harness clamp and 4 water hoses.

14. Install the EGR pipe assembly.

a. Install 2 new gaskets and EGR pipe assembly with the 4 bolts. Tighten to 7 ft. lbs. (10 Nm).

➡Make sure to tighten the 4 bolts evenly until the flanges of the EGR pipe assembly contact with the installation surfaces.

b. Install the wire harness support with the bolt.

c. Connect the 2 wire harness clamps.

15. Install the air cleaner case.
16. Install the inlet air cleaner assembly.
17. Install the air cleaner cap sub assembly.
18. Install the No. 2 cylinder head cover.
19. Add coolant (engine).
20. Inspect for leaks (engine).
21. Install the outer cowl top panel sub assembly.
22. Install the cowl body mounting reinforcement LH.
23. Install the windshield wiper motor and link assembly.

HEATED OXYGEN SENSOR (HO2S)

LOCATION

See Figures 290 and 291.

Refer to the accompanying illustrations for sensor location.

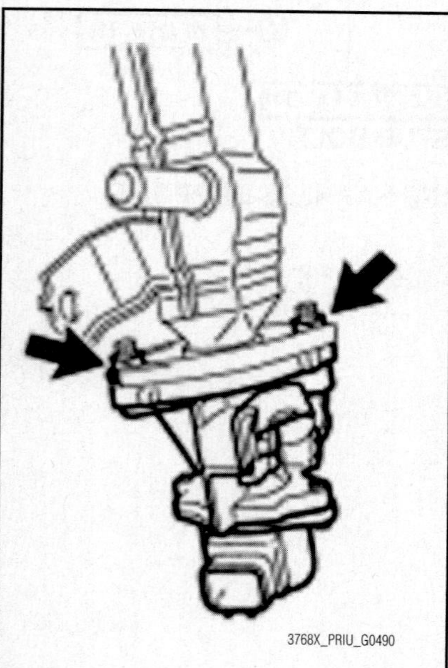

Fig. 287 Removing the 2 nuts, EGR valve assembly and gasket

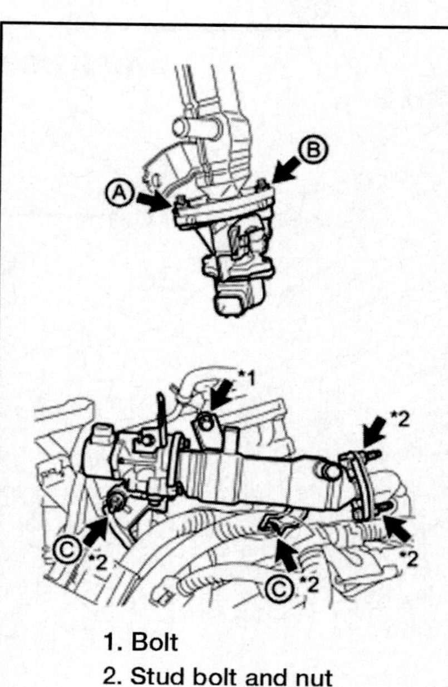

Fig. 289 Installing the EGR valve assembly

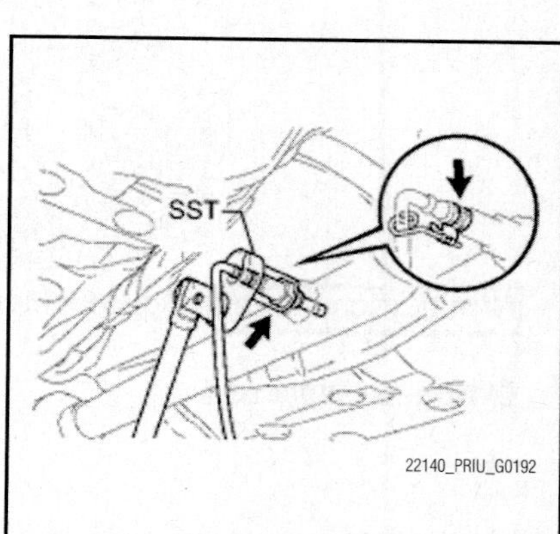

Fig. 290 HO2S location—2009 vehicles

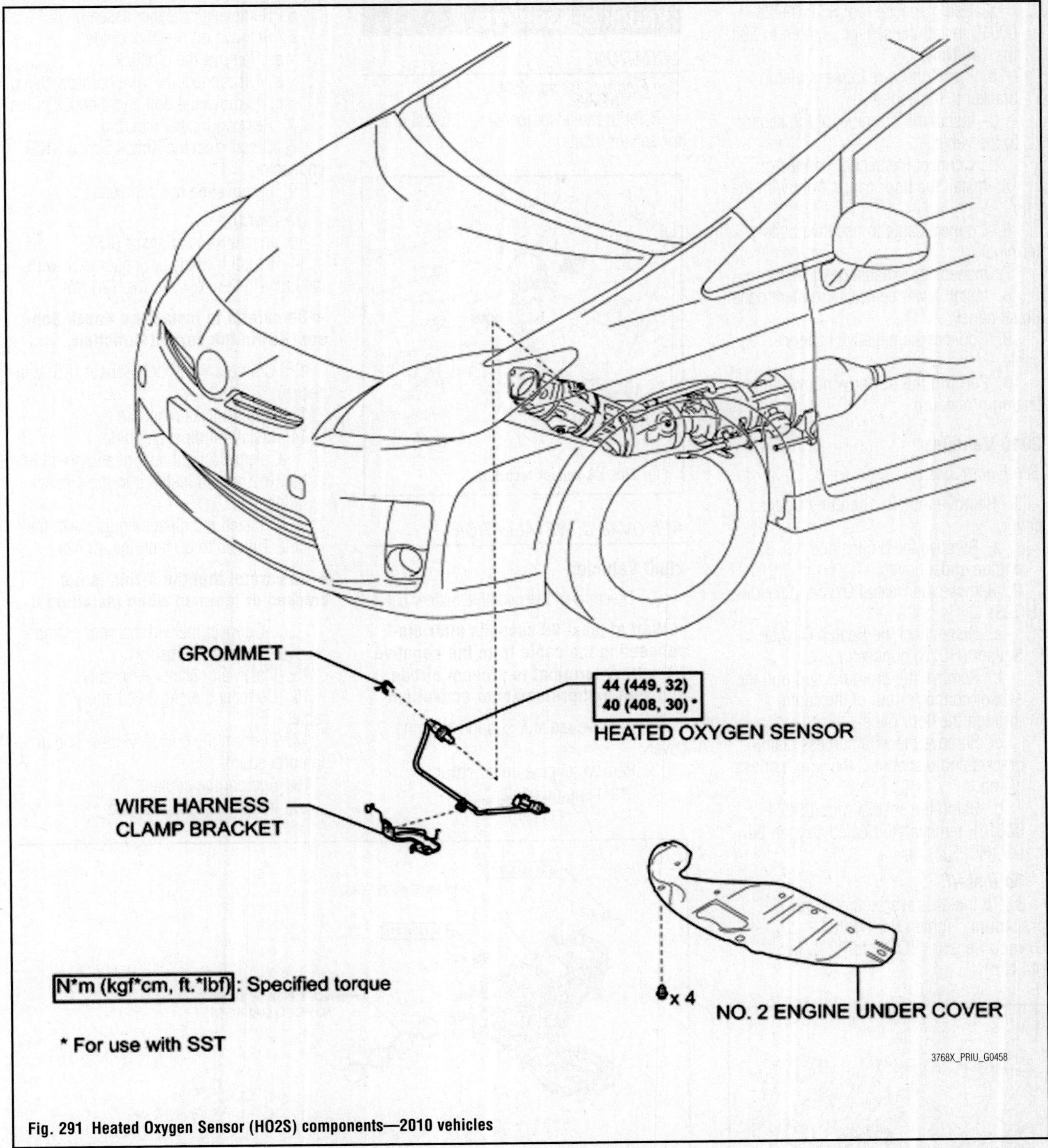

Fig. 291 Heated Oxygen Sensor (HO2S) components—2010 vehicles

REMOVAL & INSTALLATION

2009 Vehicles

1. Disconnect the negative battery cable.

➡**Wait at least 90 seconds after disconnecting the cable from the negative (-) battery terminal to prevent airbag and seat belt pretensioner activation.**

2. Remove lower center instrument panel finish panel.
3. Remove Heated Oxygen Sensor (HO2S).
   a. Using a clip remover, remove the clip.
   b. Fold back the floor carpet front.
   c. Disconnect the sensor connector.
   d. Remove the grommet of the sensor from the vehicle.
   e. Remove the wire harness clamp bracket from the sensor.
   f. Using special tool SST: 09224-00010 remove the sensor.

To install:

4. Install Heated Oxygen Sensor (HO2S).

a. Using special tool SST: 09224-00010 install the sensor. Tighten to 30ft. lbs. (40 Nm).

b. Install the wire harness clamp bracket to the sensor.

c. Install the grommet of the sensor to the vehicle.

d. Connect the sensor connector.

5. Install the floor carpet front with the clip.
6. Connect cable to negative battery terminal.
7. Inspect for exhaust gas leak.
8. Install lower center instrument panel finish panel.
9. Connect the negative battery cable.
10. Perform the power window initialization procedure.

2010 Vehicles

See Figure 292.

1. Remove the No. 2 engine under cover.

a. Remove the 3 clips and No. 3 engine under cover.

2. Remove the Heated Oxygen Sensor (HO2S).

a. Disconnect the Heated Oxygen Sensor (HO2S) connector.

b. Remove the grommet and pull the sensor connector out of the cabin through the floor panel.

c. Remove the wire harness clamp bracket and disconnect the wire harness clamp.

d. Using the special tool (09224-00010), remove the Heated Oxygen Sensor (HO2S).

To install:

3. To install, reverse the removal procedure. Tighten the Heated Oxygen Sensor (HO2S) to 32 ft. lbs. (44 Nm).

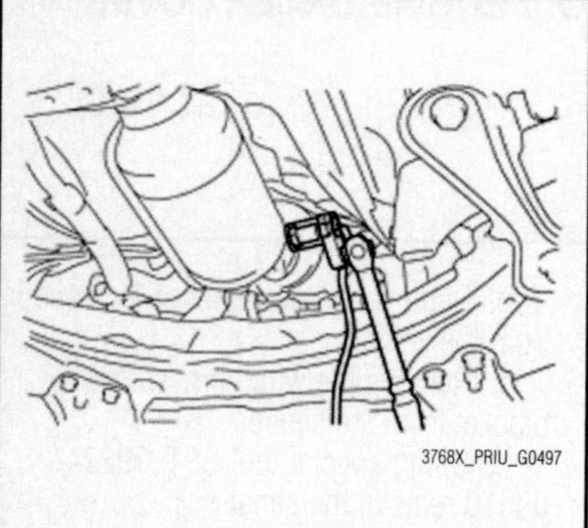

Fig. 292 Removing the Heated Oxygen Sensor (HO2S)

KNOCK SENSOR (KS)

LOCATION

See Figures 293 and 294.

Refer to the accompanying illustrations for sensor location.

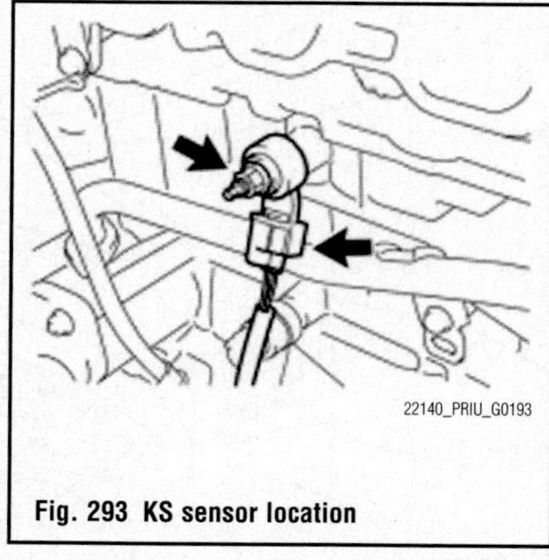

Fig. 293 KS sensor location

REMOVAL & INSTALLATION

2009 Vehicles

1. Disconnect the negative battery cable.

➡Wait at least 90 seconds after disconnecting the cable from the negative (-) battery terminal to prevent airbag and seat belt pretensioner activation.

2. Remove radiator support opening cover.
3. Remove engine under cover.
4. Drain engine coolant.
5. Remove air cleaner assembly.
6. Remove oil dipstick guide.

a. Remove the dipstick.

b. Disconnect the wire harness clamp.

c. Remove the bolt and dipstick guide.

7. Remove intake manifold.
8. Disconnect the Knock Sensor (KS) connector.
9. Remove the nut and sensor.

To install:

10. Install Knock Sensor (KS).
11. Install the Knock Sensor (KS) with the nut. Tighten to 15 ft. lbs. (20 Nm).

➡Be careful to install the Knock Sensor (KS) in the correct direction.

12. Connect the Knock Sensor (KS) connector.
13. Install intake manifold.
14. Install oil dipstick guide.

a. Apply a light coat of engine oil to a new o-ring and install it to the dipstick guide.

b. Install the dipstick guide with the bolt. Tighten to 80 inch lbs. (9 Nm).

➡Be careful that the o-ring is not cracked or jammed when installing it.

c. Connect the wire harness clamp.

d. Install the dipstick.

15. Install air cleaner assembly.
16. Connect the negative battery cable.
17. Perform the power window initialization procedure.
18. Add engine coolant.
19. Inspect for coolant leak.

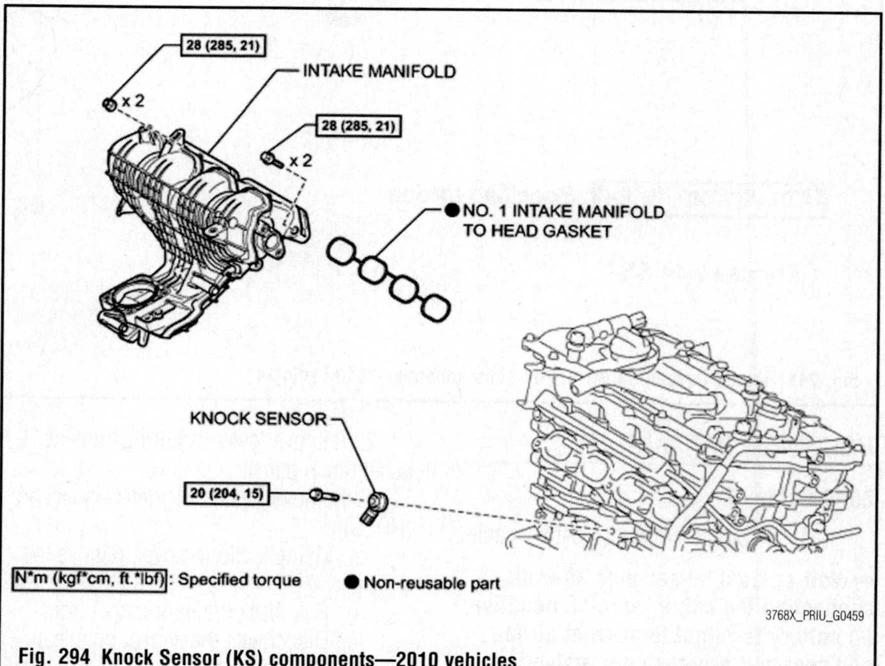

Fig. 294 Knock Sensor (KS) components—2010 vehicles

20. Install engine under cover.
21. Install radiator support opening cover.

2010 Vehicles

See Figures 295 and 296.

1. Remove the intake manifold.
2. Remove the Knock Sensor (KS).
   a. Disconnect the Knock Sensor (KS) connector.
   b. Remove the bolt and remove the Knock Sensor (KS).

To install:

3. Install the Knock Sensor (KS).

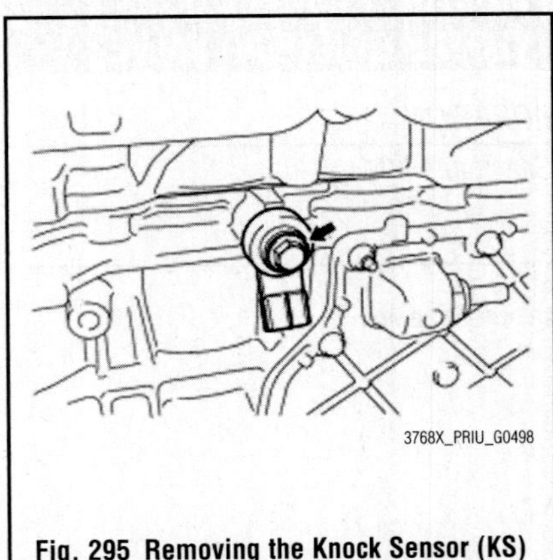

Fig. 295 Removing the Knock Sensor (KS)

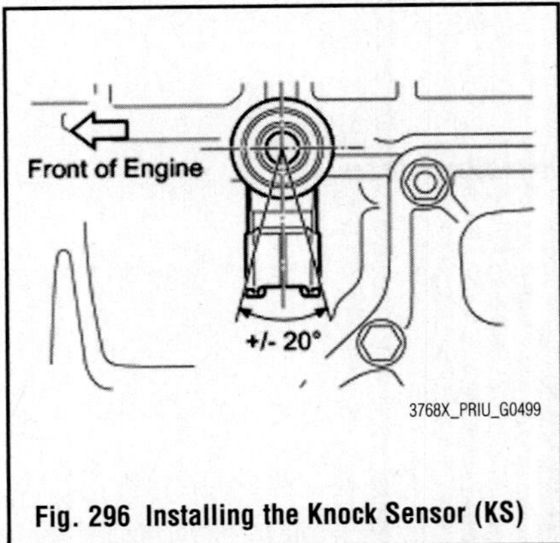

Fig. 296 Installing the Knock Sensor (KS)

   a. Install the Knock Sensor (KS) with the bolt. Tighten to 15 ft. lbs. (20 Nm).

➡Make sure that the knock control sensor is in the correct position.

   b. Connect the Knock Sensor (KS) connector.
4. Install the intake manifold.

MALFUNCTION INDICATOR LIGHT (MIL)

RESET PROCEDURE

See Figure 297.

Remove the EFI and ETCS fuses from the engine room relay block for more than 60 seconds, or disconnect the battery cable for more than 60 seconds.

➡After reconnecting the battery cable, perform the power window initialization procedure.

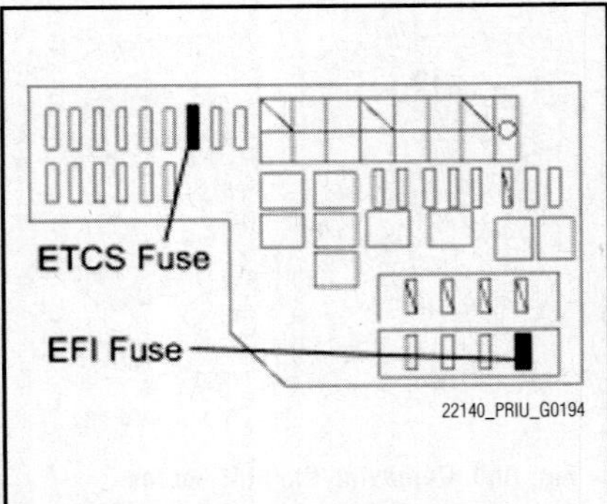

Fig. 297 EFI and ETCS fuse locations in the engine room relay block

MASS AIR FLOW (MAF) SENSOR

LOCATION

See Figures 298 and 299.

Refer to the accompanying illustrations for sensor location.

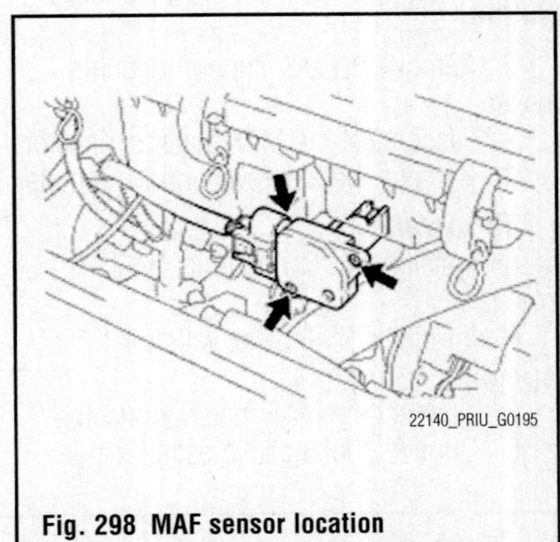

Fig. 298 MAF sensor location

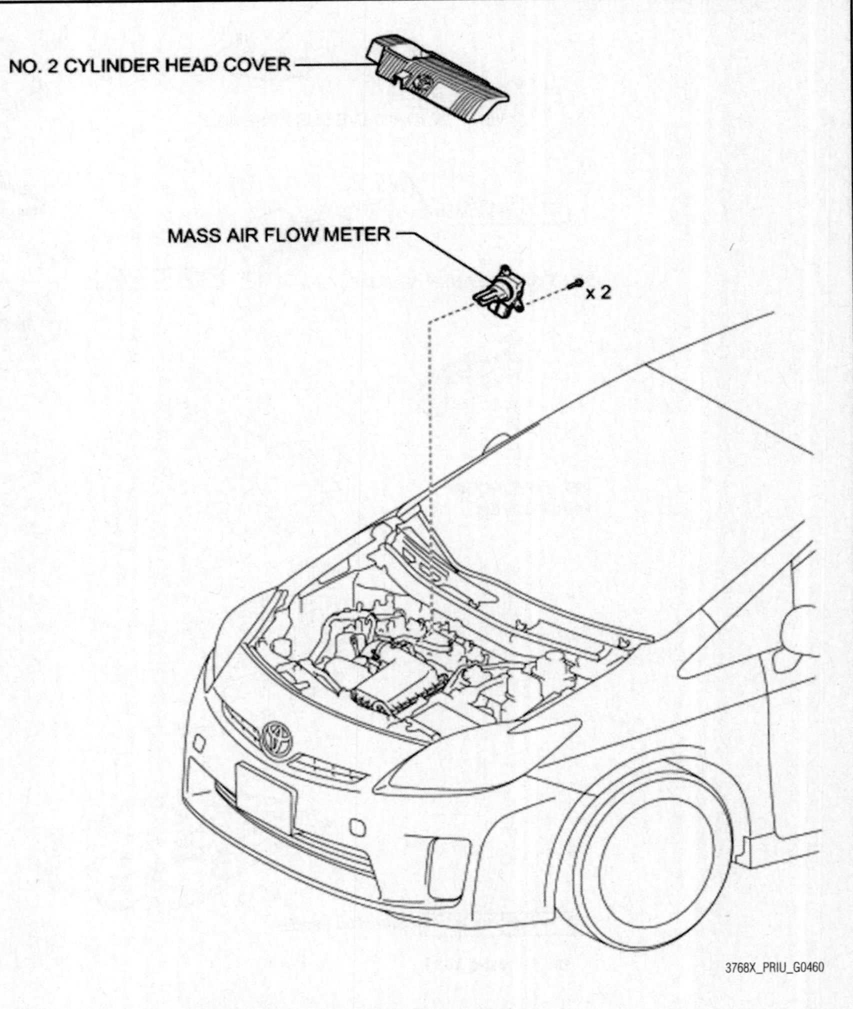

Fig. 299 Mass Air Flow (MAF) sensor components—2010 vehicles

REMOVAL & INSTALLATION

2009 Vehicles

1. Disconnect the negative battery cable.

➡Wait at least 90 seconds after disconnecting the cable from the negative (-) battery terminal to prevent airbag and seat belt pretensioner activation.

2. Remove radiator support opening cover.
3. Disconnect the MAF meter connector.
4. Remove the 2 screws and MAF meter.

To install:

5. Install a new O-ring to the MAF meter.
6. Install the MAF meter with the 2 screws.
7. Connect the MAF meter connector.
8. Connect the negative battery cable.
9. Perform the power window initialization procedure.

2010 Vehicles

See Figure 300.

1. Remove the No. 2 cylinder head cover.
2. Remove the MAF meter.

Fig. 300 Removing the MAF meter

a. Disconnect the MAF meter connector.

b. Remove the 2 screws and MAF meter.

To install:

3. Install the MAF meter.

a. Install the MAF meter with the 2 screws.

➡Make sure that the o-ring is not cracked or does not jump out of position during installation.

b. Connect the MAF meter connector.

4. Install the No. 2 cylinder head cover.

POSITIVE CRANKCASE VENTILATION (PCV) VALVE

LOCATION

See Figure 301.

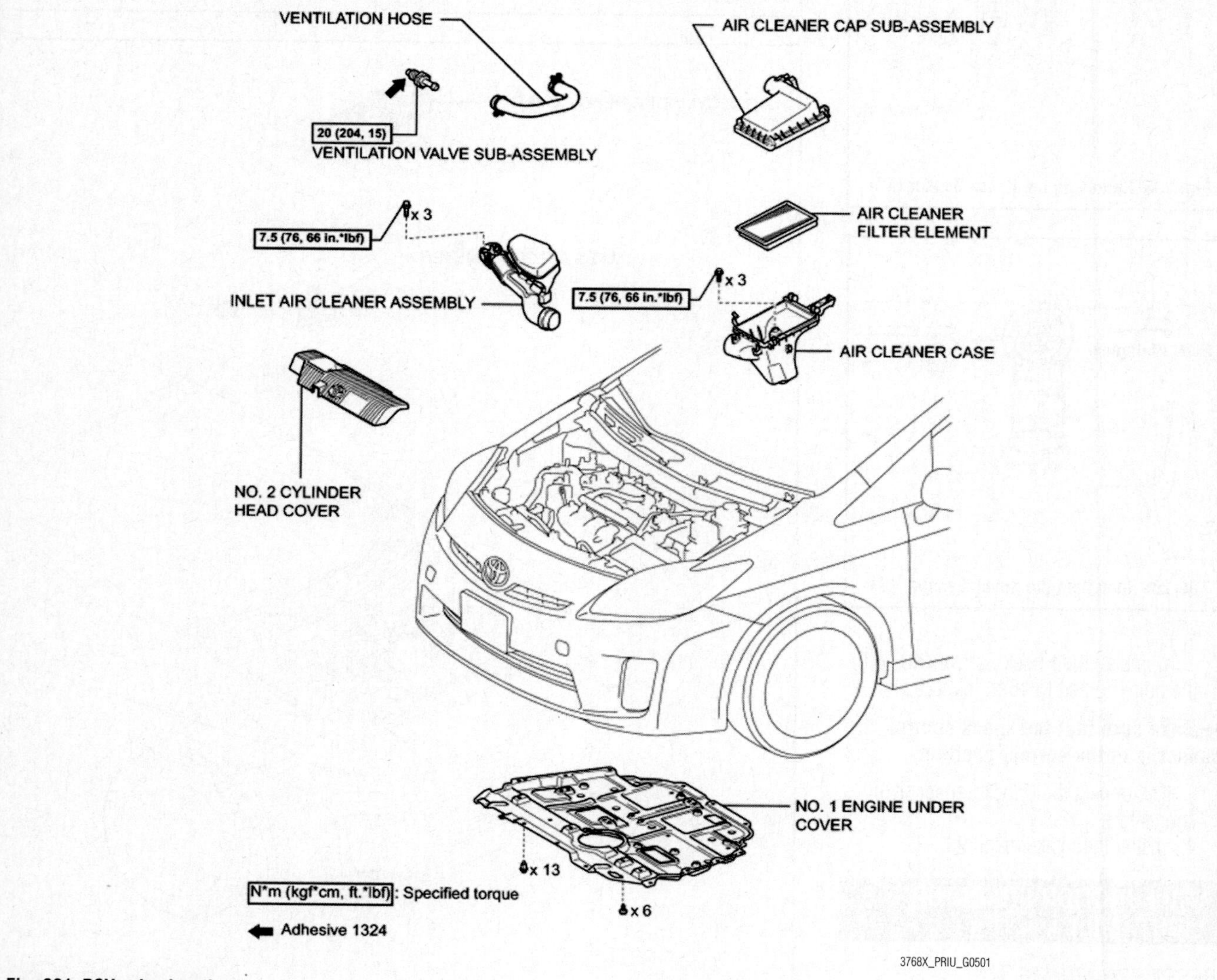

VENTILATION HOSE
AIR CLEANER CAP SUB-ASSEMBLY
20 (204, 15)
VENTILATION VALVE SUB-ASSEMBLY
x 3
7.5 (76, 66 in.*lbf)
AIR CLEANER FILTER ELEMENT
INLET AIR CLEANER ASSEMBLY
7.5 (76, 66 in.*lbf)
x 3
AIR CLEANER CASE
NO. 2 CYLINDER HEAD COVER
NO. 1 ENGINE UNDER COVER
x 13
N*m (kgf*cm, ft.*lbf): Specified torque
x 6
Adhesive 1324
3768X_PRIU_G0501

Fig. 301 PCV valve location

Refer to the accompanying illustration for location.

REMOVAL & INSTALLATION

See Figures 302 and 303.

1. Remove the No. 2 cylinder head cover.
2. Remove the air cleaner sap sub assembly.
3. Remove the inlet air cleaner assembly.
4. Remove the air cleaner case.
5. Remove the No. 1 engine under cover.
6. Remove the ventilation valve sub assembly.
   a. Remove the ventilation hose from the ventilation valve sub assembly.
   b. Using a ball joint lock nut wrench (22 mm), remove the ventilation valve sub assembly.

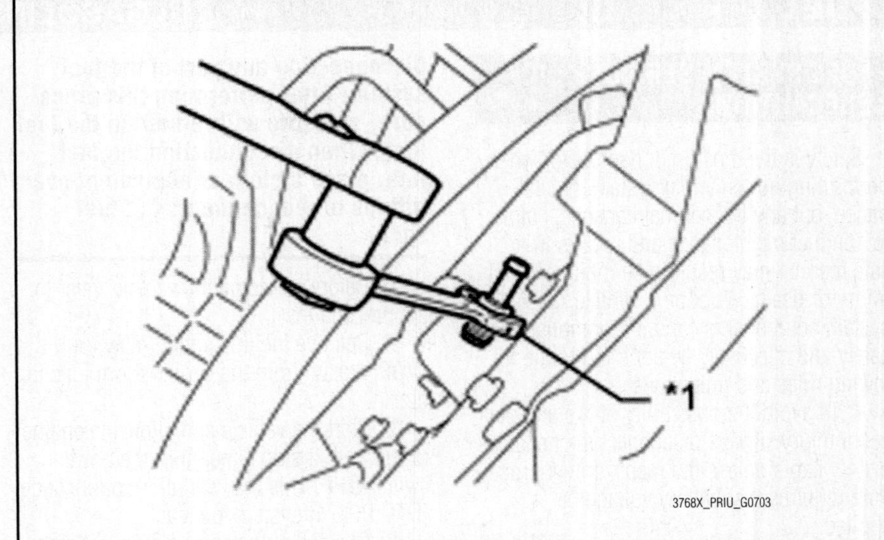

Fig. 303 Using a ball joint lock nut wrench (22 mm) to remove the ventilation valve subassembly

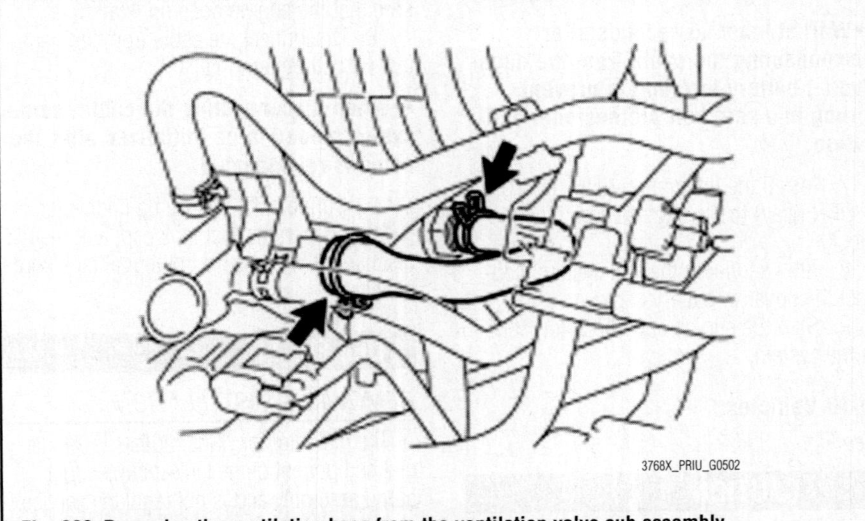

Fig. 302 Removing the ventilation hose from the ventilation valve sub assembly

To install:

7. Install the ventilation valve sub assembly.
   a. Apply adhesive to 2 or 3 threads of the ventilation valve sub assembly.
   b. Using a ball joint lock nut wrench (22 mm), install the ventilation valve sub assembly. Tighten to 15 ft. lbs. (20 Nm).
   c. Install the ventilation hose to the ventilation valve sub assembly.
8. Install the No. 1 engine under cover.
9. Install the air cleaner case.
10. Install the inlet air cleaner assembly.
11. Install the No. 2 cylinder head cover.

FUEL SYSTEM SERVICE PRECAUTIONS

Safety is the most important factor when performing not only fuel system maintenance, but any type of maintenance. Failure to conduct maintenance and repairs in a safe manner may result in serious personal injury or death. Work on a vehicle's fuel system components can be accomplished safely and effectively by adhering to the following rules and guidelines.

- To avoid the possibility of fire and personal injury, always disconnect the negative battery cable unless the repair or test procedure requires that battery voltage be applied.
- Always relieve the fuel system pressure prior to disconnecting any fuel system component (injector, fuel rail, pressure regulator, etc.) fitting or fuel line connection. Exercise extreme caution whenever relieving fuel system pressure to avoid exposing skin, face and eyes to fuel spray. Please be advised that fuel under pressure may penetrate the skin or any part of the body that it contacts.
- Always place a shop towel or cloth around the fitting or connection prior to loosening to absorb any excess fuel due to spillage. Ensure that all fuel spillage is quickly removed from engine surfaces. Ensure that all fuel-soaked cloths or towels are deposited into a flame-proof waste container with a lid.
- Always keep a dry chemical (Class B) fire extinguisher near the work area.
- Do not allow fuel spray or fuel vapors to come into contact with a spark or open flame.
- Always use a second wrench when loosening or tightening fuel line connection fittings. This will prevent unnecessary stress and torsion on fuel piping. Always follow the proper torque specifications.
- Always replace worn fuel fitting O-rings with new ones. Do not substitute fuel hose where rigid pipe is installed.

Before servicing the vehicle, make sure to also refer to the precautions in the beginning of this section as well.

FUEL SYSTEM PRESSURE

RELIEVING

2009 Vehicles

**** CAUTION**

The fuel system pressure relief procedure must be performed before disconnecting any part of the fuel system. After performing this procedure, pressure will remain in the fuel line. When disconnecting the fuel line, place a cloth or equivalent over fittings to reduce the risk of fuel spray.

1. Before servicing the vehicle, refer to the precautions.
2. Remove the integration relay (unit C: C/OPN relay) from the engine room junction block.
3. Start the vehicle and allow the engine to run until it stops, then turn the power switch **OFF**. This may set off a trouble code (DTC P0171: system too lean).
4. Check that the engine does not start.
5. Remove the fuel filler cap from the filler neck to release the fuel vapor pressure in the fuel tank.
6. Disconnect the negative battery cable.

➡Wait at least 90 seconds after disconnecting the cable from the negative (-) battery terminal to prevent airbag and seat belt pretensioner activation.

7. Install the integration relay (unit C: C/OPN relay) to the engine room junction block.
8. After servicing the fuel system, connect the negative battery cable.
9. Start the engine and check for leaks in the system.

2010 Vehicles

See Figure 304.

**** CAUTION**

Perform the following procedure to prevent fuel from spilling out before removing any fuel system parts.

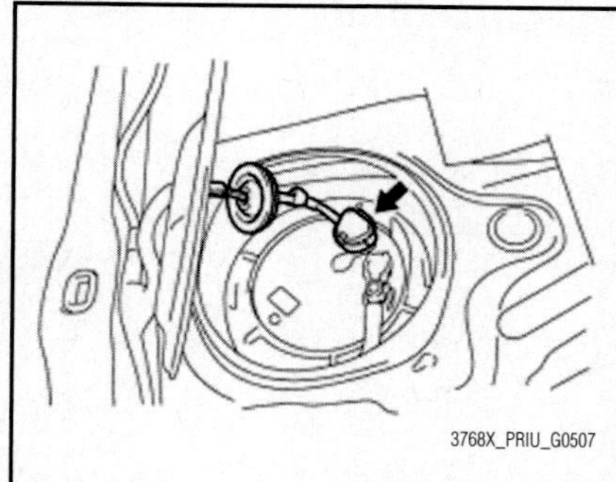

Fig. 304 Disconnecting the fuel pump connector

**** CAUTION**

Pressure will still remain in the fuel lines even after performing the following procedure. When disconnecting a fuel line, cover it with a piece of cloth to prevent fuel from spraying or coming out.

1. Remove the rear seat cushion assembly.
2. Remove the rear floor service hole cover.
3. Disconnect the fuel pump connector.
4. Put the vehicle in the "Inspection Mode".
5. Start the engine.
6. After the engine has stopped on its own, turn the power switch off.

➡DTCs P0171/25 may be detected.

7. Crank the engine again and make sure that the engine does not start.
8. Disconnect the cable from the negative (-) battery terminal.

➡When disconnecting the cable, some systems need to be initialized after the cable is reconnected.

9. Connect the fuel pump connector.
10. Loosen the fuel tank cap, and then discharge the pressure in the fuel tank completely.

FUEL FILTER

REMOVAL & INSTALLATION

The fuel filter for Prius models is an integral component of the tank-mounted fuel pump assembly and is not regularly serviced.

FUEL PUMP

REMOVAL & INSTALLATION

2009 Vehicles

1. Before servicing the vehicle, refer to the precautions.
2. Relieve the pressure from the fuel system.
3. Disconnect the negative battery cable.

➡Wait at least 90 seconds after disconnecting the cable from the negative (-) battery terminal to prevent airbag and seat belt pretensioner activation.

4. Remove the fuel tank.
5. Remove the trap canister with pump module.

a. Disconnect the VSV connector.

b. Remove the clamp from the fuel tank vent hose and canister hose.

c. Remove the fuel tank vent hose from the 2 fuel tube clamps.

d. Remove the 2 bolts and trap canister with pump module and disconnect the ground terminal of the fuel tank wire.

e. Remove the gasket from the fuel tank.

f. Remove the 2 clamps from the trap canister with pump module.

To install:

6. Install the trap canister with pump module.

a. Install a new gasket to the fuel tank.

b. Insert the trap canister with pump module to the fuel tank. Be careful that the gasket does not drop in the fuel tank.

c. Install the 2 clamps to the trap canister with pump module.

d. Install the trap canister with pump module and connect the ground terminal of the fuel tank wire with the 2 bolts. Tighten the 2 bolts to 53 inch lbs. (6 Nm).

e. Install the fuel tank vent hose to the 2 fuel tube clamps.

f. Install the clamp to the fuel tank vent hose and canister hose.

g. Connect the VSV connector.

7. Install the fuel tank.

8. Connect the negative battery cable.

9. Perform the power window initialization procedure.

10. Check for fuel leaks and proper fuel pressure.

2010 Vehicles

See Figures 305 through 309.

1. Remove the rear seat cushion assembly.

2. Remove the rear floor service hole cover.

a. Disconnect the fuel pump connector.

3. Discharge the fuel system pressure.

4. Remove the No. 2 floor board (separate type).

5. Remove the rear deck floor box.

6. Remove the rear No. 3 floor board.

7. Disconnect the cable from the negative battery terminal.

➡When disconnecting the cable, some systems need to be initialized after the cable is reconnected.

8. Remove the fuel tank main tube sub assembly.

a. Remove the tube joint clip and disengage the fuel tank main tube clamp, then pull the fuel tube joint out of the plug of the fuel suction tube assembly.

➡Check that there is no dirt or other foreign objects around the fuel tube joint before disconnecting it. Clean the joint if necessary.

➡It is necessary to prevent mud or dirt from entering the joint. If mud or dirt gets in the joint, the O-rings may not seal properly.

➡Only disconnect the joint by hand.

➡Do not bend, kink or twist the nylon tubes.

➡Protect the contact surfaces by covering it with a plastic bag.

9. Remove the fuel pump gauge retainer.

a. Using a 6 mm socket hexagon wrench, set the special tool (09808-14020) to the fuel pump gauge retainer.

➡Engage the special tool claws securely with the fuel pump gauge retainer holes to secure special tool.

➡Install special tool while pressing the special tool claws toward the fuel pump gauge retainer (towards the center of the special tool).

b. Using the special tool, loosen the fuel pump gauge retainer.

➡Do not use any tools other than specified in this operation. Damage to the fuel pump gauge retainer or the fuel tank may result.

➡Loosen the retainer by turning it counterclockwise while holding SST down. Do not allow the claw of the tank suction tube support to slip out of its groove on the fuel tank.

➡The holes on the fuel pump gauge retainer can be fitted into the tips of the special tool.

c. Remove the fuel pump gauge retainer while holding the fuel suction tube assembly by hand.

10. Remove the fuel suction tube assembly with the pump and gauge.

a. Remove the fuel suction tube assembly with pump and gauge from the fuel tank.

➡Make sure that the fuel sender gauge arm does not bend.

b. Remove the gasket from the fuel tank.

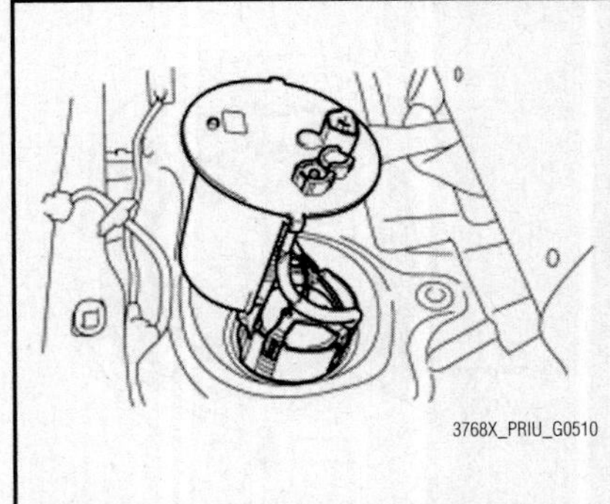

Fig. 305 Remove the fuel suction tube assembly with the pump gauge

To install:

11. Install the fuel suction tube assembly with the pump and gauge.

a. Install a new gasket onto the fuel tank.

b. Install the fuel suction tube assembly with pump and gauge to the fuel tank.

➡Make sure that the fuel sender gauge arm does not bend.

c. Align the protrusions of the fuel suction tube assembly with pump and gauge with the notches of the fuel tank.

➡Ensure the fuel suction tube gasket is in the correct position.

12. Install the fuel pump gauge retainer.

a. While holding the fuel suction tube assembly with pump and gauge by hand, position the fuel pump gauge retainer and tighten it lightly by hand.

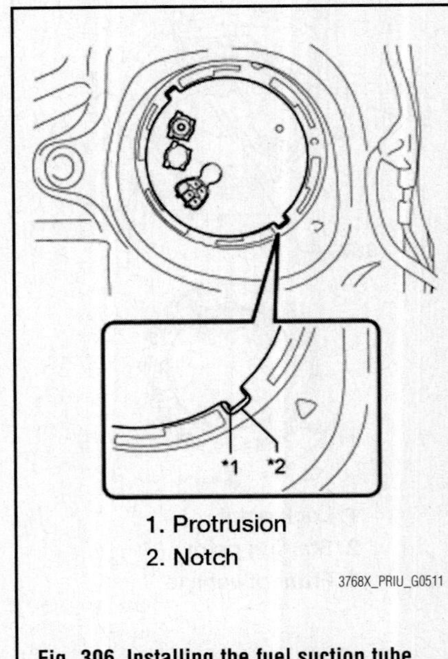

Fig. 306 Installing the fuel suction tube assembly with pump and gauge

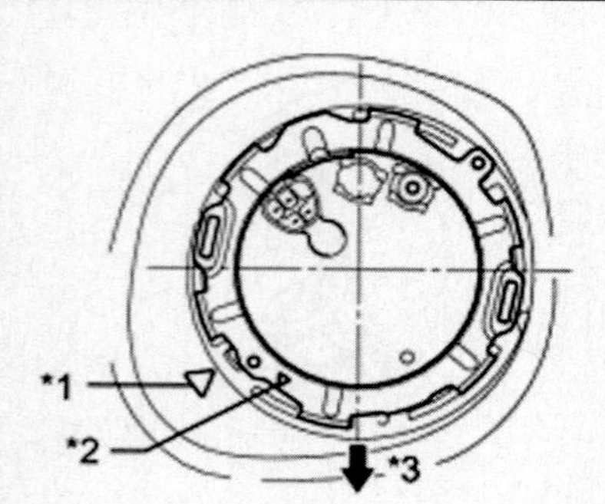

Fig. 307 Positioning the fuel pump gauge retainer

➡Check that the contact surface of the fuel tank retainer is not scratched or damaged, and prevent the entry of foreign objects.

b. Using a 6 mm socket hexagon wrench, set special tool (09808-14020) to the fuel pump gauge retainer.

➡Hold the fuel suction tube assembly upright by hand to ensure that the fuel suction tube gasket is not moved out of position.

➡Engage the special tool (09808-14020) claws securely with the fuel pump gauge retainer holes to secure special tool (09808-14020).

➡Install special tool (09808-14020) while pressing the special tool (09808-14020) claws toward the fuel pump gauge retainer (toward the center of special tool (09808-14020)).

c. Using the special tool (09808-14020), align the marks on the fuel tank and fuel pump gauge retainer.

➡Do not use any tools other than specified in this operation. Damage to the fuel pump gauge retainer or the fuel tank may result.

➡Tighten the retainer by turning it clockwise while holding SST down.

13. Connect the fuel tank main tube sub assembly.

a. Push the fuel tube joint in the plug of the fuel suction plate, then install the tube joint clip.

b. Connect the fuel pump connector.

14. Connect the cable to the negative battery terminal.

➡When disconnecting the cable, some systems need to be initialized after the cable is reconnected.

15. Inspect for fuel leaks.
16. Install the rear floor service hole cover with new butyl tape.
17. Install the rear seat cushion assembly.
18. Install the rear No. 3 floor board.
19. Install the rear deck floor board.
20. Install the rear No. 2 floor board (separate type).

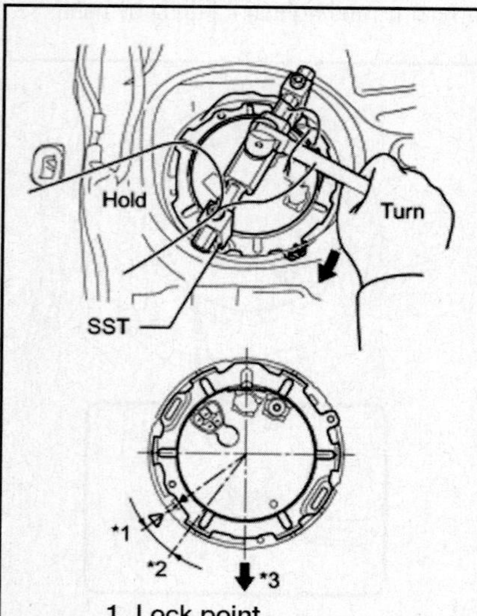

Fig. 308 Aligning the marks on the fuel tank and fuel tank retainer

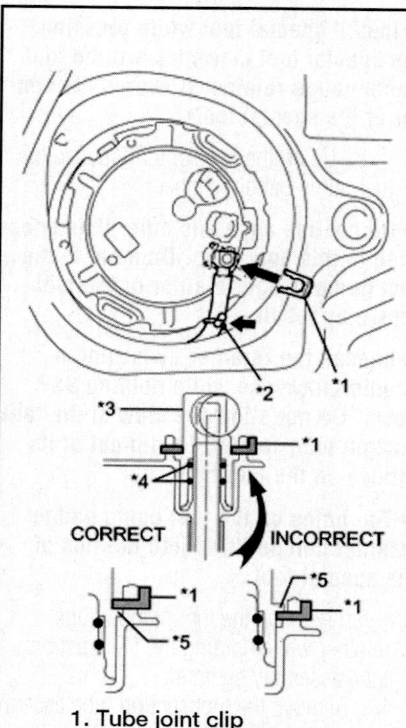

Fig. 309 Connecting the fuel tank main tube sub assembly

FUEL PRESSURE REGULATOR

REMOVAL & INSTALLATION

See Figure 310.

1. Remove the fuel suction tube assembly with the pump and gauge.
2. Remove the fuel sender gauge assembly.
3. Remove the fuel pump.
4. Remove the fuel pressure regulator assembly.

a. Using a screwdriver with its tip wrapped in protective tape, remove the fuel pressure regulator assembly.

➡Slowly pull out the fuel pressure regulator assembly because the o-ring is firmly installed between the regulator and the fuel filter.

b. Remove the 2 o-rings from the pressure regulator assembly.

To install:

5. Install the fuel pressure regulator assembly.

a. Apply gasoline to the 2 new o-rings and then install them to the fuel pressure regulator assembly.

b. Install the fuel pressure regulator assembly.

6. Install the fuel pump.
7. Install the fuel sender gauge assembly.
8. Install the fuel suction tube assembly with the pump and gauge.

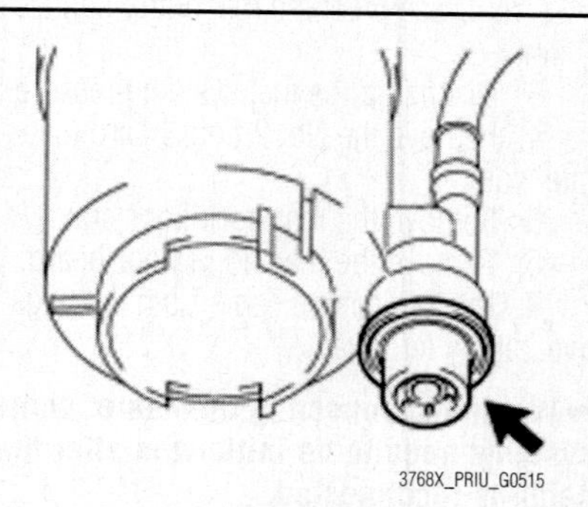

Fig. 310 Removing the fuel pressure regulator assembly

FUEL RAIL AND INJECTOR

REMOVAL & INSTALLATION

2009 Vehicles

See Figures 311 through 313.

1. Before servicing the vehicle, refer to the precautions.
2. Relieve the pressure from the fuel system.
3. Disconnect the negative battery cable.
4. Remove the windshield wipers and windshield wiper motor and link assembly, if necessary.
5. Remove the front outer cowl top panel, if necessary.
6. Remove the air cleaner assembly.
7. Remove the brake master cylinder reservoir.
   a. Disconnect the brake fluid level switch connector.
   b. Remove the 2 bolts.
   c. Disconnect the claw fitting, and then remove the brake master cylinder reservoir.
8. Remove the reservoir bracket.
   a. Remove the No. 2 fuel vapor feed hose from the hose clamp.
   b. Remove the connector clamp.
   c. Remove the wire harness clamp, the 3 bolts and reservoir bracket.
9. Remove the cylinder head cover.
10. Remove fuel delivery pipe.
   a. Disconnect the fuel tube from the fuel delivery pipe.
   b. Remove the No. 1 fuel pipe clamp.
   c. Pinch the retainer of the fuel tube connector, and then pull out the fuel tube connector to disconnect the fuel tube from the fuel delivery pipe.

CAUTION

Be careful not to drop the injectors when removing the delivery pipe.

   d. Disconnect the 4 injector connectors from injector.
   e. Remove the 3 bolts and delivery pipe together with the 4 injectors and fuel pipe.
   f. Remove the 2 spacers from the cylinder head.
   g. Pull out the 4 injectors from the delivery pipe.
   h. Remove the O-ring and grommet from each injector.

To install:

11. Install the grommet to each injector.
12. Apply a light coat of spindle oil or gasoline to new O-ring and install them to each injector.

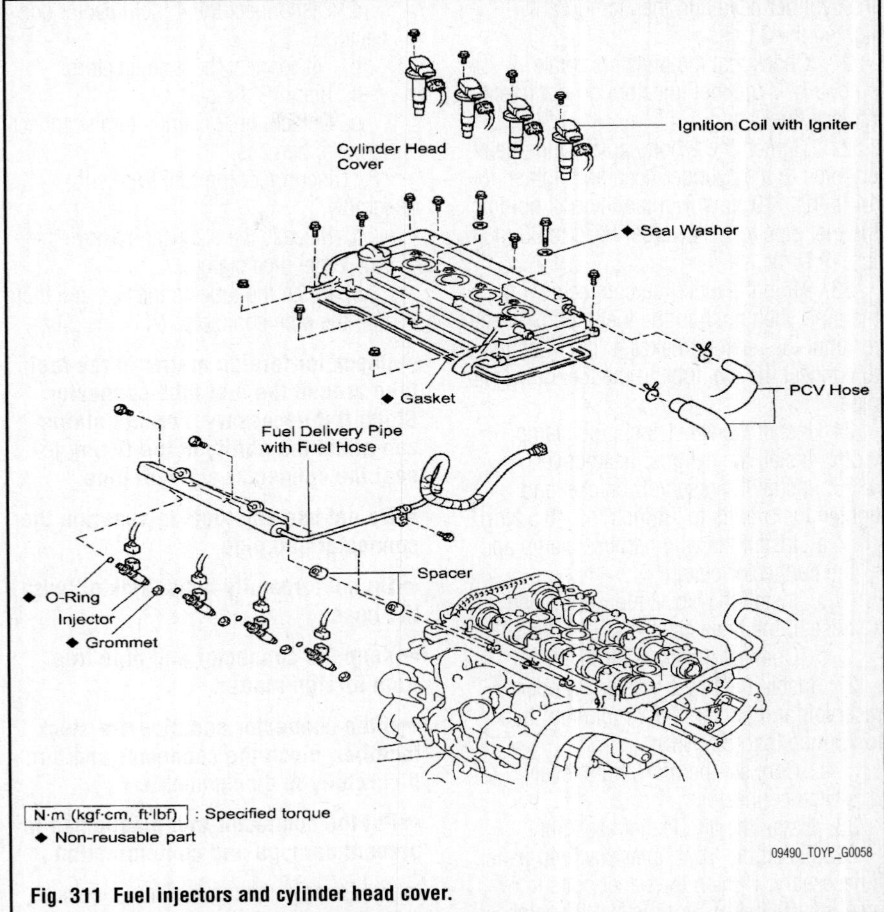

Fig. 311 Fuel injectors and cylinder head cover.

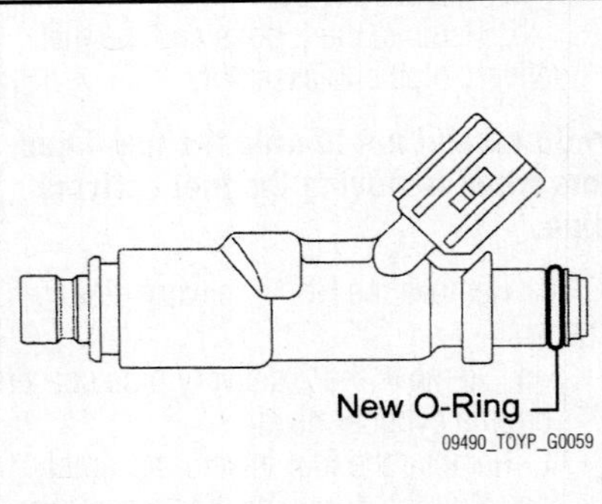

Fig. 312 Installation of new O-ring to the injectors.

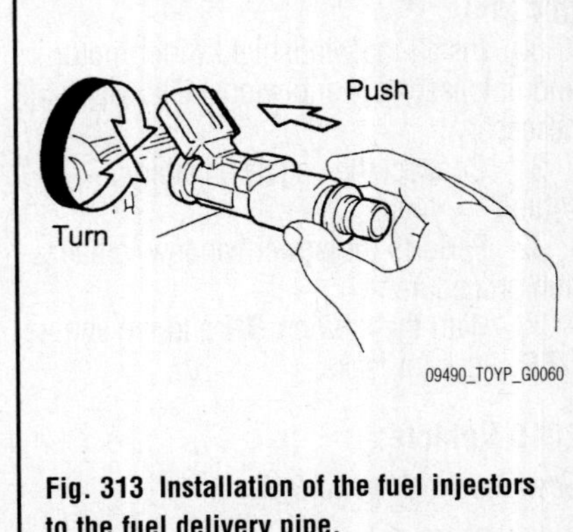

Fig. 313 Installation of the fuel injectors to the fuel delivery pipe.

13. Apply a light coat of spindle oil or gasoline to the surface of the fuel delivery pipe which contacts the O-ring of the fuel injector.
14. Apply a light coat of spindle oil or gasoline to the O-ring again, and install the fuel injector by turning it right and left while pushing it onto the fuel delivery pipe.

➡Be careful that the O-ring is not cracked or jammed when installing it.

15. Check that the fuel injector rotates smoothly. If the fuel injector does not rotate, replace the O-ring.
16. Install the 4 injectors.
17. Position the injector connector outward.
18. Install 4 new insulators to the cylinder head.
19. Install the 2 delivery pipe spacers to the cylinder head.
20. Place the delivery pipe and fuel pipe together with the 4 injectors in position on

the cylinder head and then temporarily tighten the 3 bolts.

21. Check that the injectors rotate smoothly. If the fuel injectors do not rotate, replace the O-ring.

22. Tighten the 2 bolts holding the delivery pipe to the cylinder head and tighten to 14 ft. lbs. (19 Nm). Tighten the bolt holding the fuel pipe to the cylinder head to 80 inch lbs. (9 Nm).

23. Align the fuel tube connector with the pipe, then push in the fuel tube connector until the retainer makes a "click" sound to connect the fuel tube to the fuel delivery pipe.

24. Install the No. 1 fuel pipe clamp.

25. Install the cylinder head cover.

26. Install the reservoir bracket and tighten the 3 bolts to 75 inch lbs. (8.5 Nm).

a. Install the wire harness clamp and the connector clamp.

b. Install the No. 2 fuel vapor feed hose to the hose clamp.

c. Connect the claw fitting.

27. Install the brake master cylinder reservoir and tighten the 2 mounting bolts to 75 inch lbs. (8.5 Nm).

a. Connect the brake fluid level switch connector.

28. Install the air cleaner assembly.

29. Install the front outer cowl top panel, if necessary. Tighten the panel bolts to 57 inch lbs. (6.4 Nm), and the No. 2 engine room relay block bolts to 74 inch lbs. (8.4 Nm).

30. Install the windshield wiper motor and link assembly and windshield wipers, if necessary.

31. Connect the negative battery cable.

32. Perform the power window initialization procedure.

33. With the ignition **ON** and the engine **OFF** check for leaks.

2010 Vehicles

See Figures 314 and 315.

1. Discharge the fuel system pressure.
2. Remove the rear No. 2 floor board (separate type).
3. Remove the rear deck floor box.
4. Remove the rear No. 3 floor board.
5. Disconnect the cable from the negative battery terminal.

➡When disconnecting the cable, some systems need to be initialized after the cable is reconnected.

6. Remove the EGR with the cooler pipe sub assembly.
7. Disconnect the engine wire.

a. Disconnect the 4 fuel injector connectors.

b. Disconnect the 4 connectors.

c. Remove the bolt.

d. Detach the 2 clamps to disconnect the wire harness.

8. Disconnect the fuel tube sub assembly.

a. Release the claw and remove the No. 1 fuel pipe clamp.

b. Pinch the tube connector, and then pull the tube connector off of the pipe.

➡Check for foreign matter in the fuel tube around the fuel tube connector. Clean it if necessary. Foreign matter can affect the ability of the O-ring to seal the connector and fuel pipe.

➡Do not use any tools to separate the connector and pipe.

➡Do not forcefully bend, kink or twist the hose.

➡Keep the connector and pipe free from foreign matter.

➡If the connector and pipe are stuck together, pinch the connector and turn it carefully to disconnect it.

➡Put the connector in a plastic bag to prevent damage and contamination.

9. Remove the fuel delivery pipe sub assembly.

a. Remove the bolt.

b. Remove the 2 bolts and the fuel delivery pipe sub assembly.

➡Be careful not to drop the fuel injectors when removing the fuel delivery pipe.

10. Remove the No. 1 delivery pipe spacer.

a. Remove the 2 delivery pipe spacers from the cylinder head.

11. Remove the fuel injector assembly.

a. Pull the 4 fuel injector assemblies out of the fuel delivery pipe sub assembly.

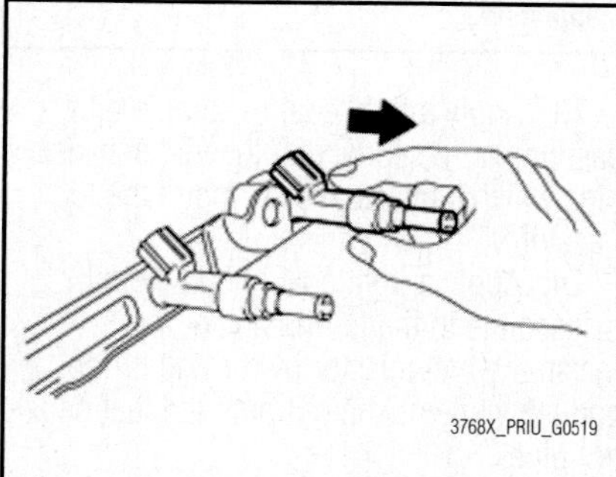

Fig. 314 Removing the fuel injector assemblies

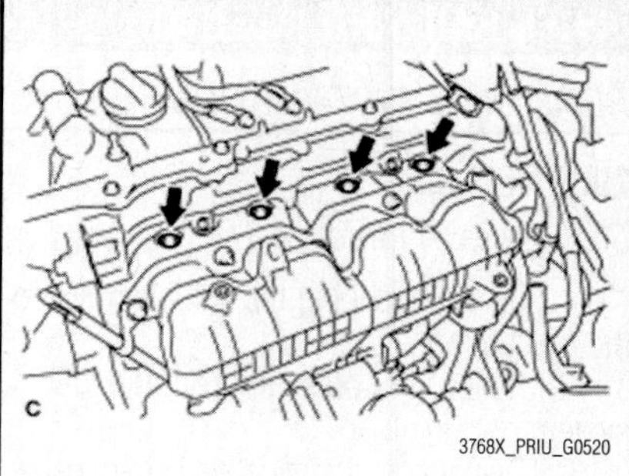

Fig. 315 Removing the injector vibration insulators

b. Remove the o-ring from each fuel injector.

c. For reinstallation, attach a tag or label to each injector shaft.

➡Prevent entry of foreign objects by covering the fuel injectors with plastic bags.

d. Remove the 4 injector vibration insulators.

To install:

12. Install the fuel injector assembly.

a. Install a new insulator and o-ring to each fuel injector assembly.

b. Apply a light coat of gasoline or spindle oil to the contact surfaces of the new o-ring on each fuel injector assembly.

c. While turning the fuel injector assembly left and right, install it onto the fuel delivery pipe sub-assembly.

➡Do not damage the fuel injector assembly or O-ring.

➡Do not twist the O-ring.

➡After installing each fuel injector, check that it turns smoothly. If not, replace the O-ring with a new one.

13. Install the No. 1 delivery pipe spacer.

a. Install the 2 No. 1 delivery pipe spacers onto the cylinder head.

➡Install the No. 1 delivery pipe spacers in the correct direction.

14. Install the fuel delivery pipe sub assembly.

a. Install the fuel deliver pipe sub assembly with the 4 fuel injector assemblies and install the 2 bolts. Tighten to 15 ft. lbs. (21 Nm).

➡Do not drop the fuel injectors when installing the fuel delivery pipe sub-assembly.

➡**Check that the fuel injector assemblies rotate smoothly after installing the fuel delivery pipe sub-assembly.**

b. Install the bolt to secure the fuel delivery pipe sub assembly. Tighten to 15 ft. lbs. (21 Nm).

15. Connect the fuel tube sub assembly.

a. Push the tube connector to the pipe until the tube connector makes a "click" sound.

➡**Before connecting the connector and fuel pipe, check that there is no damage or foreign matter on the connecting part of the fuel pipe.**

➡**After connecting the fuel tube connector and pipe, check that they are securely connected by trying to pull them apart.**

b. Engage the lock claw to install the No. 1 fuel pipe clamp.

16. Connect the engine wire.

a. Install the bolt and tighten to 7 ft. lbs. (10 Nm).

b. Connect the 4 fuel injector connectors.

c. Connect the 4 connectors.

d. Attach the 2 clamps to connect the wire harness.

17. Install the EGR with the cooler pipe sub assembly.

18. Install the rear No. 3 floor board.

19. Install the rear deck floor box.

20. Install the rear No. 2 floor board (separate type).Install the rear No. 2 floor board (separate type).

21. Connect the cable to the negative battery terminal.

22. Inspect for fuel leaks.

FUEL TANK

REMOVAL & INSTALLATION

2009 Vehicles

See Figures 316 through 328.

1. Before servicing the vehicle, refer to the precautions.

2. Relieve the pressure from the fuel system.

3. Disconnect the negative battery cable.

➡**Wait at least 90 seconds after disconnecting the cable from the negative (-) battery terminal to prevent airbag and seat belt pretensioner activation.**

4. Remove instrument panel finish panel lower center.

5. Remove the front floor panel brace and front exhaust pipe assembly.

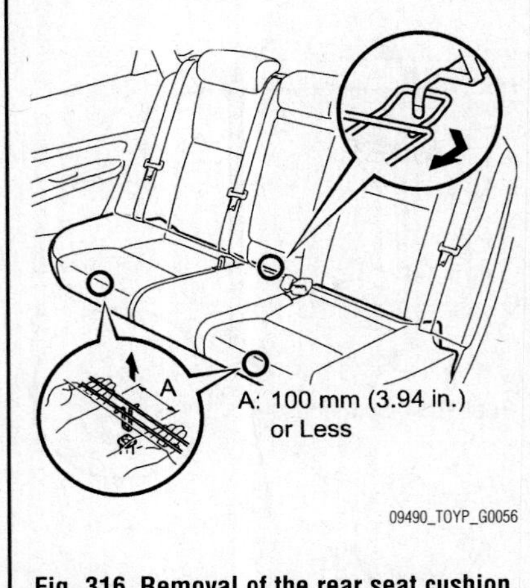

Fig. 316 Removal of the rear seat cushion assembly

6. Remove the rear seat cushion.

a. Detach the seat cushions 2 front hooks from the vehicle body by choosing a hook to detach first. Place your hands near one of the hooks, and then lift the seat cushion to detach the hook.

b. Repeat for the other hook.

c. Detach the seat cushions rear hook.

d. Remove the seat cushion.

7. Remove the butyl tape and rear floor service hole cover.

CAUTION

Remove dirt or foreign objects on the fuel line connectors before any disconnecting procedures. Do not allow any scratches or foreign objects on the parts when disconnecting them as the fuel connectors have O-rings that seal the pipes. Perform such work by hand. Do not use any tools. Do not forcibly bend, twist or turn the nylon tube. Protect the connecting part by covering it with a plastic bag after disconnecting the tube. If the connector and pipe are stuck, push and pull them to release them.

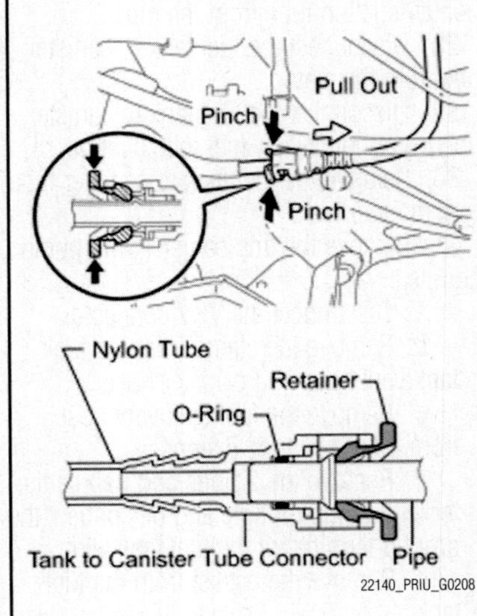

Fig. 317 Disconnect the fuel tank to canister tube from the pipe

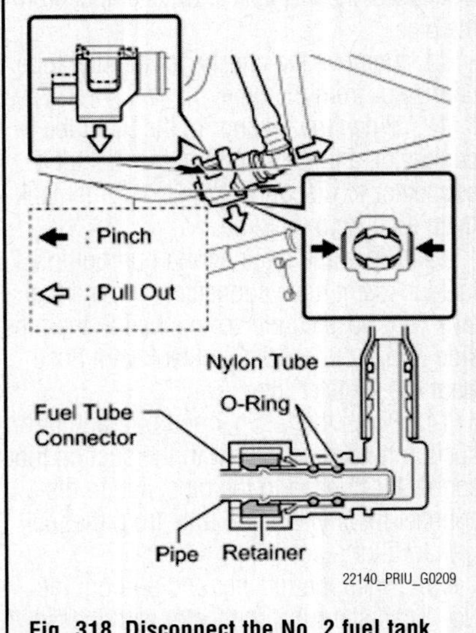

Fig. 318 Disconnect the No. 2 fuel tank main tube from the pipe

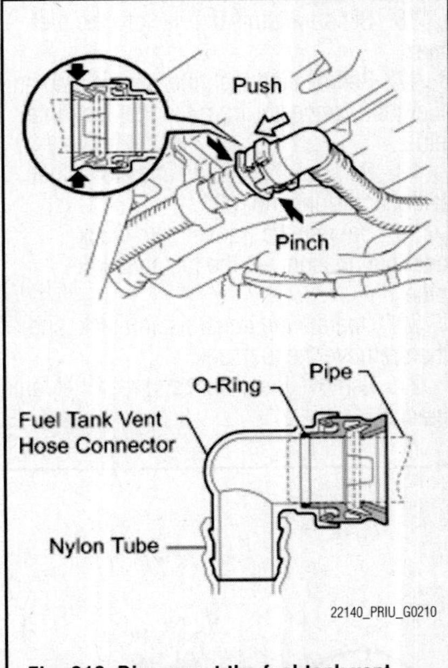

Fig. 319 Disconnect the fuel tank vent hose from the canister filter

8. Disconnect the fuel pump connector.

9. Disconnect the wire-to-wire connector.

10. Pinch the retainer of the fuel tube connector, then pull out the fuel tube connector to

disconnect the fuel tank to canister tube from the pipe.

11. Remove the checker of the fuel tube connector from the pipe.

12. Pinch the retainer of the fuel tube connector, and then pull out the fuel tube connector to disconnect the No. 2 fuel tank main tube from the pipe.

13. Pinch the retainer and pull out the fuel tank vent hose connector with the fuel tank vent hose connector pushed to the pipe side to disconnect the fuel tank vent hose from the canister filter.

14. Pinch the retainer and pull out the suction tube connector with the suction tube connector pushed to the pipe side to disconnect the fuel suction tube from the fuel tank to filler pipe.

15. Pinch the retainer and pull out the No. 1 canister tube connector with the No. 1 canister tube connector pushed to the pipe side to disconnect the No. 1 canister tube from the fuel tank to filler pipe.

16. Set a transmission jack to the fuel tank.

17. Remove the fuel filler pipe clamp and fuel tube connector from the fuel tank inlet pipe.

18. Remove the 4 bolts and No. 1 fuel tank band right and left.

19. Operate the transmission jack, and then disconnect the fuel tank inlet pipe.

20. Operate the transmission jack, and then remove the fuel tank.

21. Remove the 3 nuts and rear fuel tank bracket.

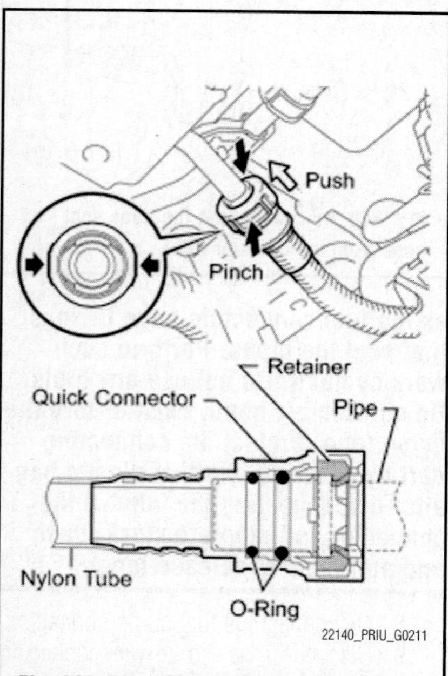

Fig. 320 Disconnect the fuel suction tube from the fuel tank to filler pipe

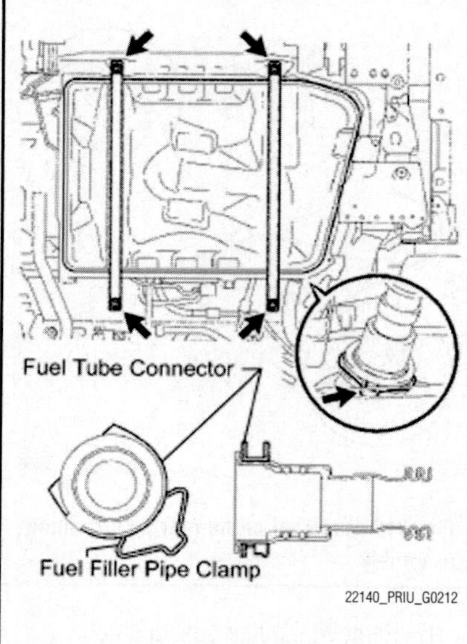

Fig. 321 Remove the fuel filler pipe clamp and fuel tube connector from the fuel tank inlet pipe

22. Disconnect the wire to wire connector from the rear fuel tank bracket.

23. Disconnect the No. 2 fuel tank main tube from the clamp.

24. Remove the checker of the main tube connector from the pipe.

25. Pinch the retainer of the main tube connector, then pull out the fuel tube connector to disconnect the No. 2 fuel tank main tube from the pipe.

26. Disconnect the fuel suction tube from the 2 No. 1 fuel tube clamps.

27. Pinch the retainer and pull out the suction tube connector with the quick connector pushed to the pipe side to disconnect the fuel suction tube from the pipe.

28. Disconnect the fuel tank to canister tube from the clamp.

29. Disconnect the fuel tank to canister tube from the 2 No. 1 fuel tube clamps.

30. Remove the fuel tank to canister tube from the fuel tank.

31. Remove the trap canister with pump module.

a. Disconnect the VSV connector.

b. Remove the clamp from the fuel tank vent hose and canister hose.

c. Remove the fuel tank vent hose from the 2 fuel tube clamps.

d. Remove the 2 bolts and trap canister with pump module and disconnect the ground terminal of the fuel tank wire.

e. Remove the gasket from the fuel tank.

f. Remove the 2 clamps from the trap canister with pump module.

32. Pinch the retainer and pull out the fuel tank vent hose connector with the fuel tank vent hose connector pushed to the fuel tank vent hose side to disconnect the fuel tank vent hose from the trap with outlet valve canister.

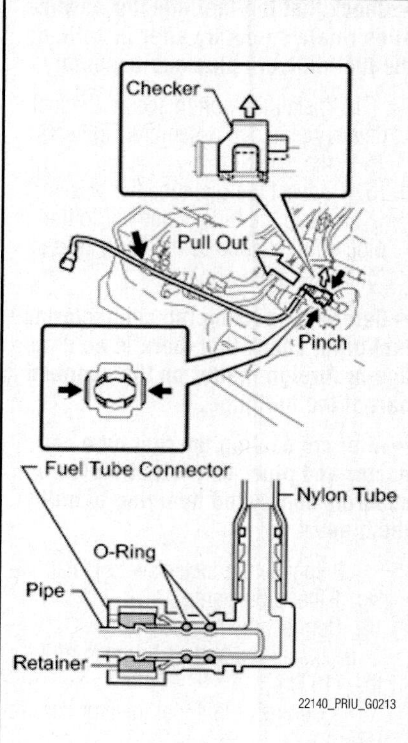

Fig. 322 Remove the checker of the main tube connector from the pipe

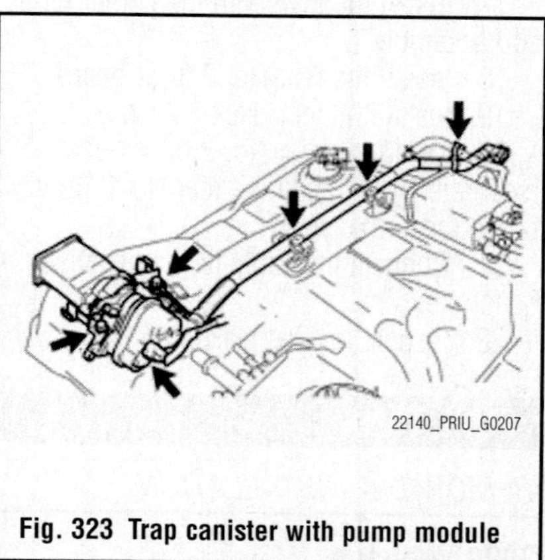

Fig. 323 Trap canister with pump module

33. Remove fuel tank wire.

a. Remove the clamp as shown in the illustration "A".

b. Disconnect the VSV connector as shown in the illustration "B".

c. Disconnect the vapor pressure sensor connector as shown in the illustration "C".

d. Remove the 3 wire harness clamps as shown in the illustration "D".

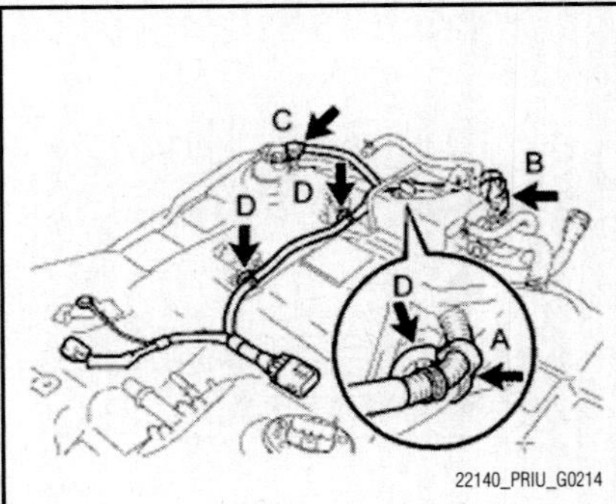

Fig. 324 Remove the clamp (A), disconnect the VSV connector (B), disconnect the vapor pressure sensor connector (C) and remove the 3 wire harness clamps (D).

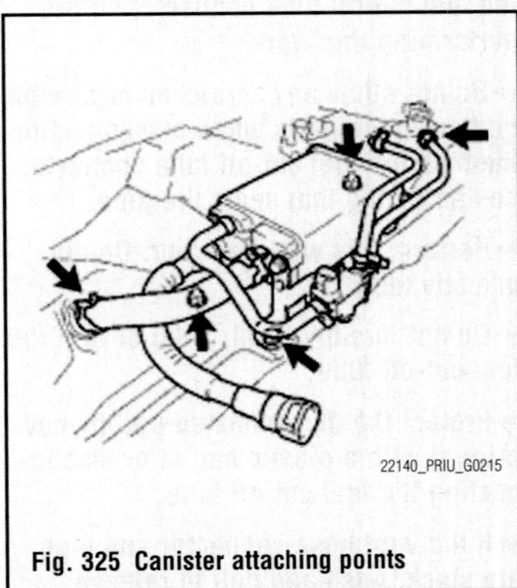

Fig. 325 Canister attaching points

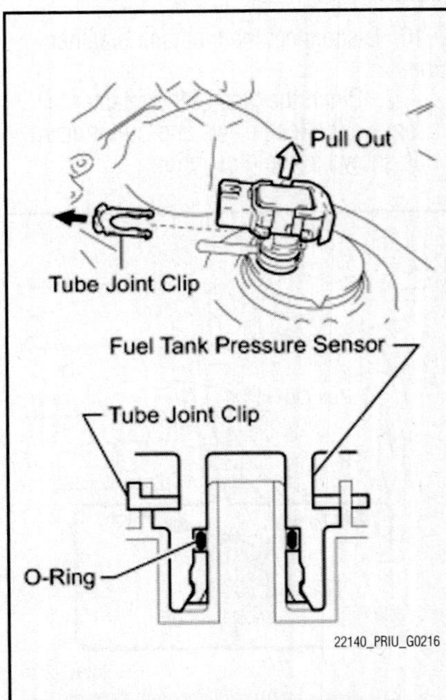

Fig. 326 Remove the tube joint clip, then pull out the fuel tank pressure sensor from the fuel tank retainer

34. Remove canister.
   a. Disconnect the canister hose from the fuel tank retainer.
   b. Disconnect the No. 1 canister outlet hose from the fuel tank.
   c. Remove the bolt, 2 nuts and canister.
   d. Remove the nut from the fuel tank.
35. Remove fuel tank pressure sensor.
   a. Remove the tube joint clip, then pull out the fuel tank pressure sensor from the fuel tank retainer.
36. Remove fuel tank retainer.
   a. Insert a clip remover between the fuel tank retainer and gasket, then remove the fuel tank retainer by lifting it little by little.

WARNING

The fuel tank retainer is made of resin and easily damaged if removed or installed forcibly. Handle the part correctly to ensure proper sealing.

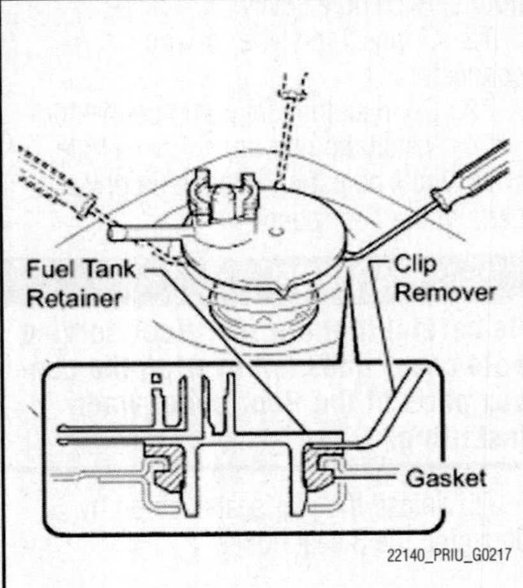

Fig. 327 Insert a clip remover between the fuel tank retainer and gasket, then remove the fuel tank retainer by lifting it little by little

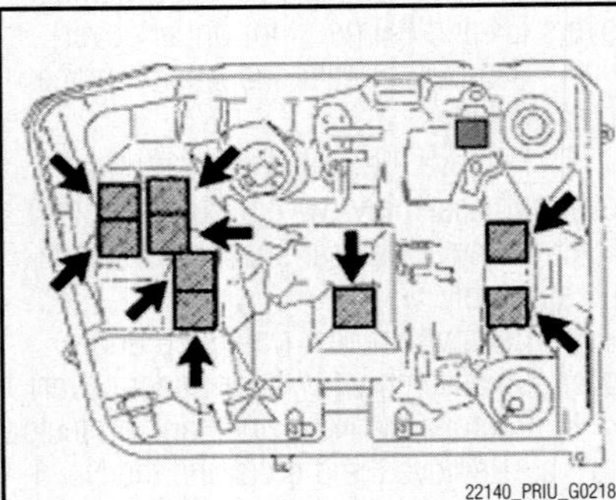

Fig. 328 Remove the 9 cushions from the fuel tank

After removing the fuel tank retainer, check that the contact surface of the fuel tank retainer on the fuel tank is not damaged.

   b. Remove the gasket from the fuel tank.
37. Remove the 2 clamps from the fuel tank.
38. Remove the 9 cushions from the fuel tank.

To install:

39. Install 9 new cushions to the fuel tank.
40. Install the 2 clamps to the fuel tank.
41. Install fuel tank retainer.
   a. Install a new gasket to the fuel tank.
   b. While being careful that the gasket does not drop in the fuel tank, insert the fuel tank retainer into the fuel tank so the protrusion of the fuel tank retainer is in the middle of the 2 convex pats of the fuel tank.
42. Push the fuel tank pressure sensor to the plug of the fuel tank retainer, and then install the tube joint clip.
43. Install canister.
   a. Install the nut to the fuel tank.
   b. Install the canister with the bolt and 2 nuts. Tighten to 53 inch lbs. (6 Nm).
   c. Connect the No. 1 canister outlet hose to the fuel tank.
   d. Connect the canister hose to the fuel tank retainer.
44. Install fuel tank wire.
   a. Install the 3 wire harness clamps.
   b. Connect the vapor pressure sensor connector.
   c. Connect the VSV connector.
   d. Install the clamp.
45. Install fuel tank vent hose.
   a. Align the fuel tank vent hose connector with the pipe, then push in the fuel tank vent hose connector until the retainer makes a "click" sound to install the fuel tank vent hose to the trap canister with pump module.
46. Install the trap canister with pump module.
   a. Install a new gasket to the fuel tank.
   b. Insert the trap canister with pump module to the fuel tank. Be careful that the gasket does not drop in the fuel tank.
   c. Install the 2 clamps to the trap canister with pump module.
   d. Install the trap canister with pump module and connect the ground terminal of the fuel tank wire with the 2 bolts. Tighten the 2 bolts to 53 inch lbs. (6 Nm).

e. Install the fuel tank vent hose to the 2 fuel tube clamps.

f. Install the clamp to the fuel tank vent hose and canister hose.

g. Connect the VSV connector.

47. Install the fuel tank to canister tube to the canister's hose.

48. Connect the fuel tank to canister tube to the 2 No. 1 fuel tube clamps.

CAUTION

Check that there are no scratches or foreign objects around any connected part of the fuel line connectors and pipes before these procedures. After connecting any fuel line tubes, check that the tube is securely connected by pulling on the connector.

49. Align the suction tube connector with the pipe, and then push in the suction tube connector until the retainer makes a "click" sound to install the fuel suction tube to the pipe.

50. Connect the fuel suction tube to the 2 No. 1 fuel tube clamps.

51. Align the main tube connector with the pipe, and then push in the main tube connector until the retainer makes a "click" sound to install the No. 2 fuel tank main tube to the pipe. Install the checker to the pipe.

52. Connect the No. 2 fuel tank main tube to the clamp.

53. Connect the connector clamp to the rear fuel tank bracket.

54. Install the rear fuel tank bracket with the 3 nuts and tighten to 53 inch lbs. (6 Nm).

55. Set the fuel tank to a transmission jack.

56. Operate the transmission jack, and then install the fuel tank to the vehicle.

57. Operate the transmission jack, and then connect the fuel tank inlet pipe.

58. Install the No. 1 fuel tank band right and left with the 4 bolts. Torque the 4 bolts to 29 ft. lbs. (39 Nm).

59. Install the fuel tube connector and fuel filler pipe clamp to the fuel tank inlet pipe.

60. Align the No. 1 canister tube connector with the pipe, and then push in the No. 1 canister tube connector until the retainer makes a "click" sound to connect the No. 1 canister tube to the fuel tank to filler pipe.

61. Align the suction tube connector with the pipe, and then push in the suction tube connector until the retainer makes a "click" sound to connect the fuel suction tube to the fuel tank to filler pipe.

62. Align the fuel tank vent hose connector with the pipe, and then push in the fuel tank vent hose connector until the retainer makes a "click" sound to connect the fuel tank vent hose to the canister filter.

63. Align the fuel tube connector with the pipe, and then push in the fuel tube connector until the retainer makes a "click" sound to connect the No. 2 fuel tank main tube to the pipe. Install the checker to the pipe.

64. Align the fuel tank to canister tube connector with the pipe, and then push in the fuel tank to canister tube connector until the retainer makes a "click" sound to connect the fuel tank to canister tube to the pipe.

65. Install the front exhaust pipe

66. Connect the negative battery cable.

67. Perform the power window initialization procedure.

68. Check for fuel and exhaust leaks.

69. Install the front floor panel brace.

70. Install the instrument panel finish panel lower center.

71. Attach new butyl tape to the rear floor service hole cover.

72. Connect the wire-to-wire connector.

73. Connect the fuel pump connector.

74. Install the rear floor service hole cover while adjusting it to the 3 convex parts of the floor panel.

WARNING

Be careful that the rear floor service hole cover does not overlap the convex parts of the floor panel when installing.

75. Install the rear seat cushion by engaging the 3 seat hooks.

2010 Vehicles

See Figures 329 through 335.

1. Remove the fuel suction tube assembly with the pump and gauge.

2. Drain the fuel.

3. Remove the rear floor side member covers LH and RH (w/ floor under cover).

4. Remove the rear suspension brace sub assembly.

5. Remove the rear floor step under cover sub assembly (w/ floor under cover).

6. Remove the rear suspension brace sub assembly.

7. Remove the rear floor step under cover sub assembly (w/ floor under cover).

8. Remove the No. 1 fuel tank protector.

a. Remove the 3 bolts and the No. 1 fuel tank protector.

9. Disconnect the fuel cut-off tube from the charcoal canister assembly.

Fig. 329 Removing the fuel tank protector

➡Do not remove the retainer.

➡Remove any dirt or foreign matter on the fuel cut-off tube connector before performing this work.

➡Do not allow any scratches or foreign matter on the parts when disconnecting them as the fuel cut-off tube connector has an O-ring that seals the pipe.

➡Perform this work by hand. Do not use any tools.

➡Do not forcibly bend, twist or turn the fuel cut-off tube.

➡Protect the disconnected part by covering it with a plastic bag after disconnecting the fuel cut-off tube.

➡If the vent hose connector and pipe are stuck, push and pull to release them.

10. Disconnect the fuel tank breather tube.

a. Pinch the tabs of the retainer to remove the lock claws and pull it down as shown in the illustration.

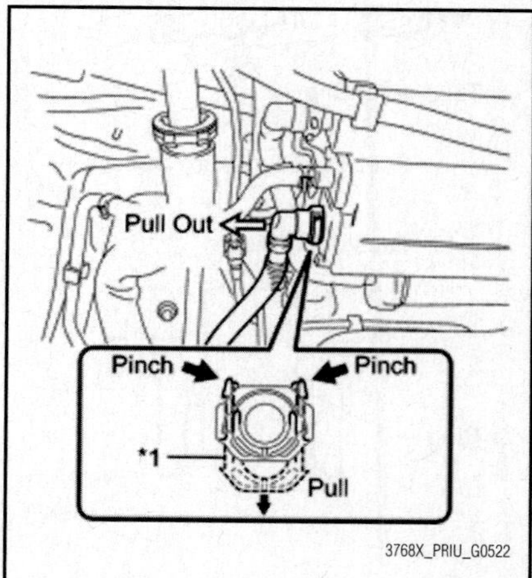

Fig. 330 Disconnecting the fuel cut off tube from the charcoal canister assembly

b. Pull out the fuel tank breather tube.

➡Check that there is no dirt or other foreign objects around the connector before this operation and clean the connector as necessary.

➡It is necessary to prevent mud or dirt from entering the connector. If mud or dirt gets in the connector, the O-rings may not seal properly.

➡Do not use any tools in this operation.

➡Do not bend, kink or twist the nylon tube. Protect the connector by covering it with a plastic bag.

➡When the pipe and connector are stuck, push and pull the connector to release and pull the connector out carefully.

11. Disconnect the fuel tank to filler pipe hose.
 a. Pull the tabs of the retainer to disengage the lock claws and pull it down.
 b. Pull out the fuel tank to filler pipe hose.

➡Check that there is no dirt or other foreign objects around the connector before this operation and clean the connector as necessary.

➡It is necessary to prevent mud or dirt from entering the connector. If mud or dirt gets in the connector, the O-rings may not seal properly.

1. Retainer
2. Quick connector
3. Pipe
4. O-ring

3768X_PRIU_G0523

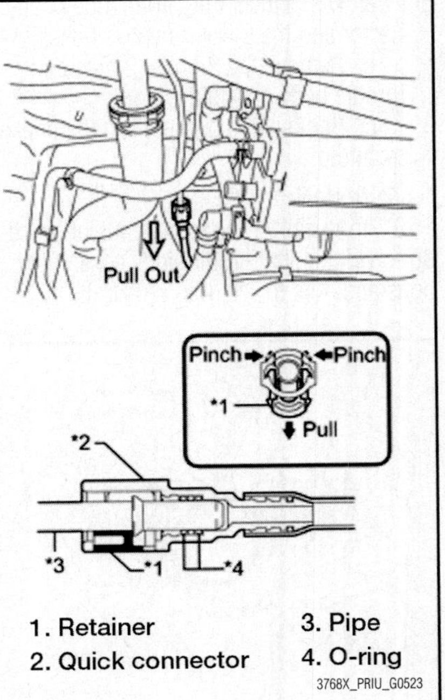

Fig. 331 Disconnecting the fuel tank breather tube

➡Do not use any tools in this operation.

➡Do not bend, kink or twist the nylon tube. Protect the connector by covering it with a plastic bag.

➡When the pipe and connector are stuck, push and pull the connector to release and pull the connector out carefully.

12. Remove the fuel tank main tube sub assembly.
 a. Pinch the tabs of the retainer to remove the lock claws and pull it down as shown in the illustration.
 b. Pull out and remove the fuel tank main tube sub-assembly.

➡Check that there is no dirt or other foreign objects around the connector before this operation and clean the connector as necessary.

➡It is necessary to prevent mud or dirt from entering the connector. If mud or dirt gets in the connector, the O-rings may not seal properly.

➡Do not use any tools in this operation.

➡Do not bend, kink or twist the nylon tube. Protect the connector by covering it with a plastic bag.

➡When the pipe and connector are stuck, push and pull the connector to release and pull the connector out carefully.

13. Remove the fuel tank assembly.

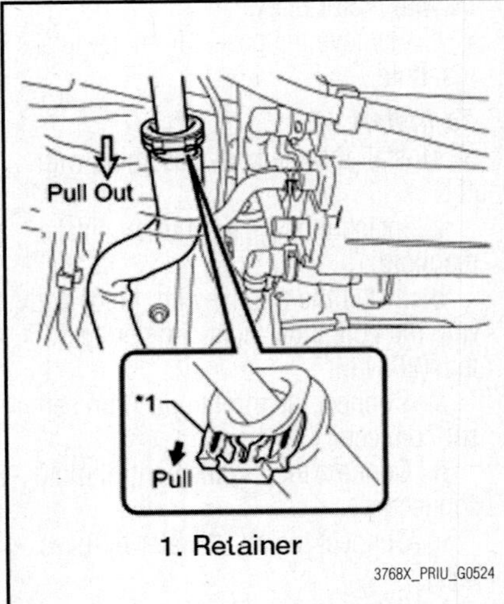

Fig. 332 Disconnecting the fuel tank to filler pipe hose

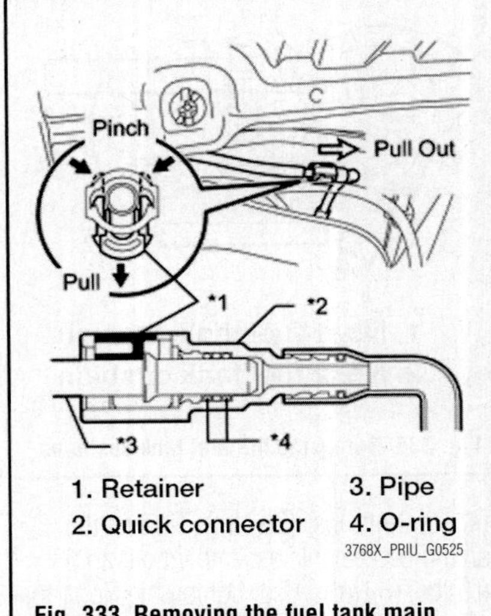

Fig. 333 Removing the fuel tank main tube sub assembly

 a. Remove the 2 bolts and disconnect the parking brake cable assembly.
 b. Support the fuel tank using an engine lifter.
 c. Remove the 4 set bolts of the 2 fuel tank bands.
 d. Lower the engine lifter to remove the fuel tank.

➡Slowly operate the engine lifter to lower the fuel tank.

➡Do not drop the fuel tank.

➡When removing the fuel tank, tilt it slightly to prevent it from interfering with the suspension arm or other surrounding parts.

14. Remove the fuel tank cushion.
 a. Remove the 3 No. 1 fuel tank cushions and 2 No. 2 fuel tank cushions.

To install:

15. To install, reverse the removal procedure. Take note to tighten the fuel tank band

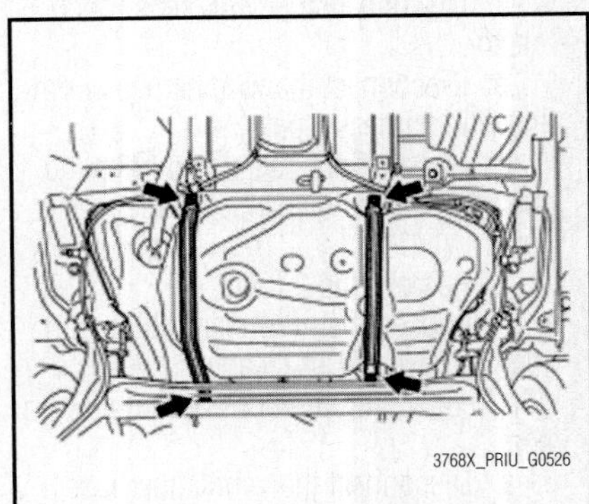

Fig. 334 Removing the bolts from the fuel tank bands

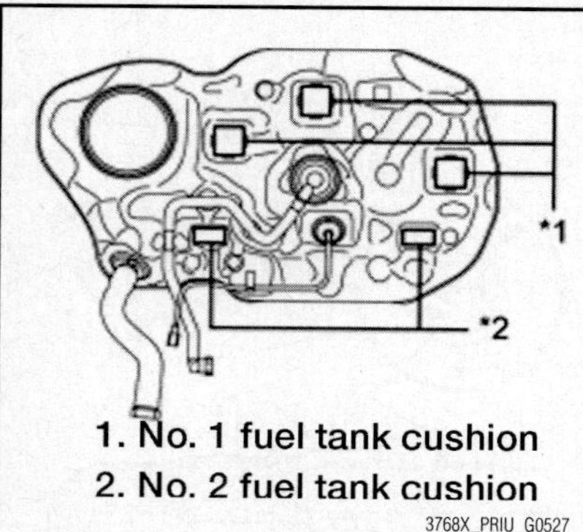

Fig. 335 Removing the fuel tank cushions

bolts to 29 ft. lbs. (39 Nm). Tighten the parking brake cable assembly bolts to 53 inch lbs. (6 Nm). Also, tighten the No. 1 fuel tank protector bolts to 49 inch lbs. (5.5 Nm).

IDLE SPEED

ADJUSTMENT

Idle speed is maintained by the Electronic Control Module (ECM). No adjustment is possible.

THROTTLE BODY

REMOVAL & INSTALLATION

2009 Vehicles

See Figures 336 and 337.

1. Disconnect the negative battery cable.

➡Wait at least 90 seconds after disconnecting the cable from the negative (-) battery terminal to prevent airbag and seat belt pretensioner activation.

2. Remove engine under cover.
3. Drain engine coolant.
4. Remove the 6 clips and radiator support opening cover.
5. Remove air cleaner assembly
   a. Disconnect the MAF sensor connector.
   b. Disconnect the wire harness from the wire harness clamp.
   c. Loosen the hose clamp bolt, and then disconnect the No. 1 air cleaner inlet.
   d. Remove the 2 bolts.
   e. Loosen the hose clamp bolt, and then remove the air cleaner.
6. Remove throttle body assembly with motor.
   a. Disconnect the ventilation hose.
   b. Disconnect the No. 2 ventilation hose.

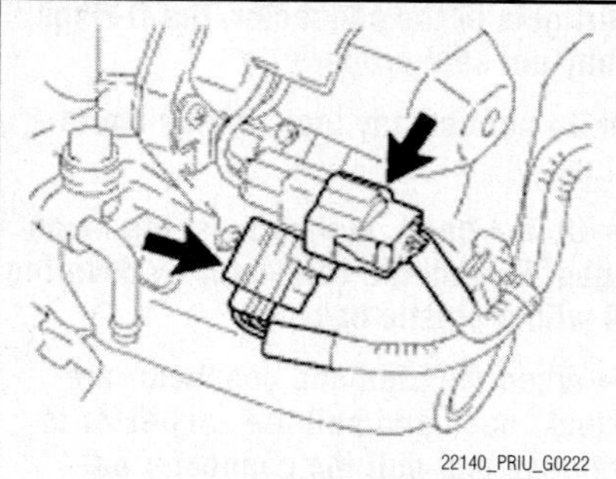

Fig. 336 Disconnect the throttle control motor and throttle position sensor connector

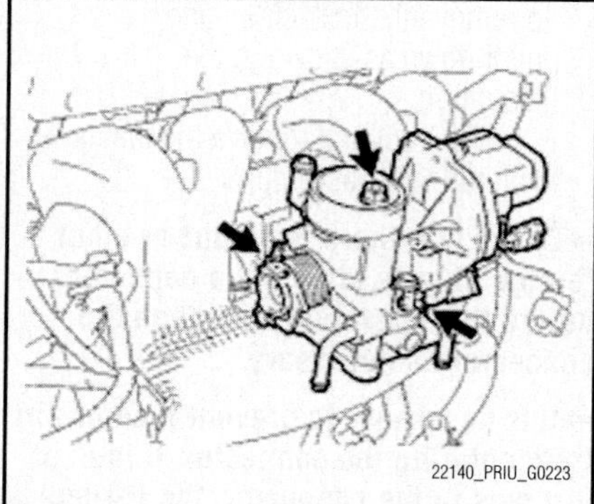

Fig. 337 Throttle body fastener locations

   c. Disconnect the No. 1 fuel vapor feed hose.
   d. Disconnect the water by-pass hose.
   e. Disconnect the No. 2 water by-pass hose.
   f. Disconnect the throttle control motor connector.
   g. Disconnect the throttle position sensor connector.
   h. Remove the bolt, 2 nuts and throttle with motor body.
   i. Remove the gasket from the intake manifold.

To install:

7. Install throttle body assembly with motor
   a. Install a new gasket to the intake manifold.
   b. Install the throttle with motor body with the bolt and 2 nuts. Tighten to 15 ft. lbs. (20 Nm).
   c. Connect the throttle position sensor connector.
   d. Connect the throttle control motor connector.
   e. Connect the No. 2 water by-pass hose.
   f. Connect the water by-pass hose.
   g. Connect the No. 1 fuel vapor feed hose.
   h. Connect the No. 2 ventilation hose.
   i. Connect the ventilation hose.
8. Install air cleaner assembly
   a. Install the air cleaner with the 2 bolts. Tighten to 62 inch lbs. (7 Nm).
   b. Tighten the hose clamp bolt to 27 inch lbs. (3 Nm).
   c. Connect the No. 1 air cleaner inlet, and tighten the hose clamp bolt to 27 inch lbs. (3 Nm).
   d. Connect the MAF sensor meter connector.
9. Connect the negative battery cable.
10. Perform the power window initialization procedure.
11. Add engine coolant
12. Inspect for coolant leaks.
13. Install engine under cover.
14. Install radiator support opening cover.

2010 Vehicles

See Figure 338.

1. Remove the No. 1 engine under cover.
2. Drain the engine coolant (engine).
3. Remove the No. 2 cylinder head cover.
4. Remove the air cleaner cap sub assembly.
5. Remove the inlet air cleaner assembly.
6. Remove the air cleaner case.
7. Remove the air cleaner hose assembly.
8. Remove the throttle body assembly.
   a. Disconnect the throttle body connector and the 2 water bypass hoses.
   b. Remove the 2 bolts, 2 nuts and throttle body assembly.
   c. Remove the gasket from the intake manifold.

To install:

9. To install, reverse the removal procedure. Tighten the throttle body assembly bolts and nuts to 7 ft. lbs. (10 Nm).

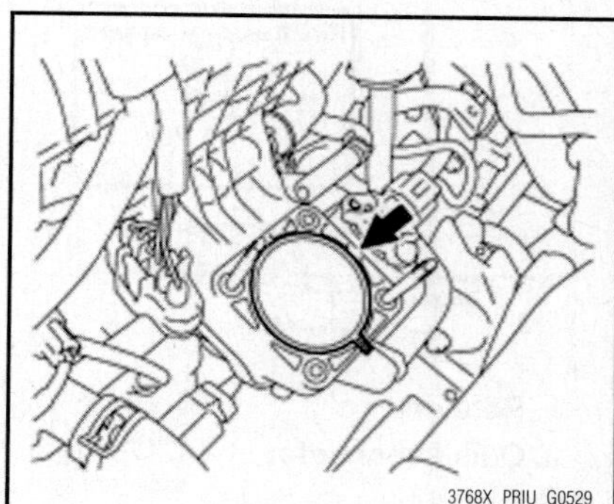

Fig. 338 Removing the gasket from the intake manifold

HEATING & AIR CONDITIONING SYSTEM

AIR CONDITIONING UNIT

REMOVAL & INSTALLATION

2009 Vehicles

See Figures 339 through 342.

1. Before servicing the vehicle, refer to the precautions.
2. Disconnect the negative battery cable.

➡Wait at least 90 seconds after disconnecting the cable from the negative (-) battery terminal to prevent airbag and seat belt pretensioner activation.

3. Place front wheels in the straight ahead position.
4. Drain the cooling system.
5. Recover refrigerant from A/C system.
6. Disconnect the suction hose.
   a. Check SST 09870-00015 installation direction. Set SST so that the stopper side is on the piping clamp lock side.
   b. Install SST on the piping clamp.
   c. Push down on special tool with your thumb while holding the pipe with both hands. Be careful not to bend the pipe.
   d. Pull special tool until the stopper touches the pipe.
   e. Raise SST stopper and remove the piping clamp with SST from the pipe.
   f. Remove the piping clamp from SST.
   g. Disconnect the suction hose by hand or using a screwdriver.
   h. Remove the 2 O-rings from the suction hose. Do not apply excessive force to the suction hose. Seal the opening of the disconnected part using vinyl tape to prevent moisture and foreign matter from entering it.

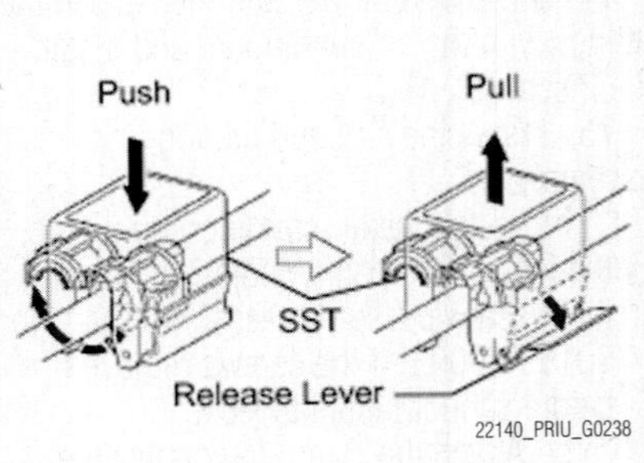

Fig. 339 Push down on special tool with your thumb while holding the pipe with both hands. Pull special tool until the stopper touches the pipe

7. Disconnect the cooler refrigerant liquid pipe to cooler unit by using the same procedures described for the suction hose utilizing SST 09870-00025.
8. Slide the clip and disconnect the heater water hoses. Do not apply excessive force to the water hoses. Prepare a drain pan or cloth for when the cooling water leaks.
9. Detach the 2 claws and 4 clips, and remove the instrument panel register.
10. Remove the 2 screws and disconnect the hood lock control cable.
11. Detach the 4 claws, 5 clips and disconnect all connectors and remove the instrument panel finish panel.
12. Detach the 3 claws, 4 clips and disconnect the connector and remove the instrument panel finish panel.
13. Detach the claw, 5 clips and disconnect the connector and remove the No. 3 instrument panel register.
14. Detach the 6 clips and remove the No. 4 instrument panel register.
15. Detach the 2 claws and 4 clips, and remove the No. 2 instrument panel register.
16. Remove the multi-display assembly.
17. Remove the glove compartment door stopper from the glove compartment door.
18. While pushing in the sides of the glove compartment door, open the door to release it from the 2 stoppers.
19. Open the door until it is horizontal.
20. Pull the glove compartment door toward the rear of the vehicle to detach the 2 hinges and remove the glove compartment door.
21. Remove the instrument panel cushion.
22. Remove the instrument cluster finish panel end.
23. Using a screwdriver, remove the 2 instrument panel hole covers.
24. Remove the 2 screws, detach the claw and 2 clips, and remove the glove compartment door.
25. Remove the No. 1 instrument panel speaker.
26. Remove the left and right front pillar garnish corner pieces.
27. Remove the left and right front pillar garnishes.
28. Disconnect the passenger airbag connector.
29. Remove the 3 screws, 2 bolts, pull up the instrument panel to detach the 6 claws and 3 clips.
30. Remove the instrument panel safety pad.
31. Remove the NO. 2 and No. 3 steering wheel lower covers.
32. Remove the air bag module.
33. Remove the steering wheel.
34. Remove the tilt lever bracket.
35. Remove the steering column cover.
36. Remove the spiral cable.
37. Remove left and right front door scuff plates.
38. Remove left and right cowl side trim boards.
39. Remove the 2 screws, detach the 4 clips and remove the instrument cluster finish panel.
40. Remove the No. 1 center cluster cushion.
41. Remove the radio receiver.
42. Using a clip remover, remove the clip from the instrument panel finish panel lower center.
43. Detach the 4 claws and 2 clips, and remove the instrument panel finish panel.
44. Detach the 6 claws and remove the glove compartment door lock cover.
45. Remove the 4 screws and glove compartment door lock.
46. Detach the 4 claws and remove the instrument panel finish panel lower.
47. Detach the 2 claws and remove the No. 1 instrument panel under cover.
48. Remove power steering ECU.
49. Remove the shift lever assembly lower.
50. Using a clip remover, remove the 8 clips from the duct and lower instrument panel.
51. Remove the 6 bolts, 2 screws, disconnect all connectors, and detach all clamps.
52. Remove the instrument panel lower.
53. Fold back the floor carpet so that the No. 3 air duct rear can be removed.
54. Detach the 10 claws and then remove the No. 3 air duct rear.
55. Remove the clip and No. 3 heater to register duct with the No. 2 side defroster nozzle duct.
56. Remove the No. 1 duct with the No. 1 side defroster nozzle duct.
57. Remove the No. 2 heater-to-register duct.
58. Remove the 2 clips, detach the 3 claws and then remove the defroster nozzle.
59. Remove the transaxle control ECU assembly.
60. Remove the ECM.
61. Remove the network gateway ECU.

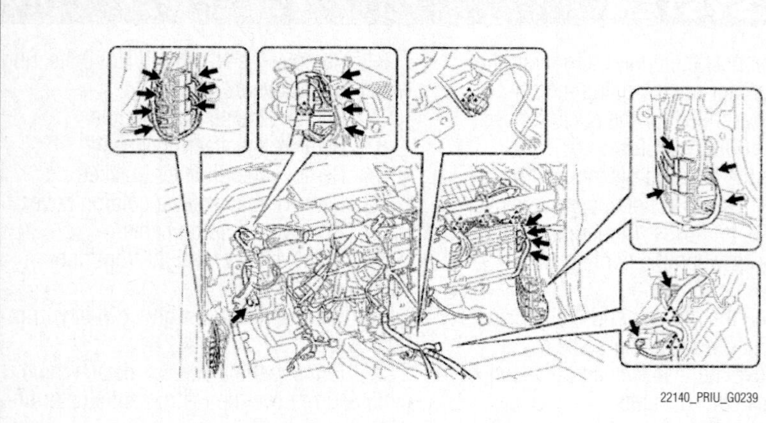

Fig. 340 Disconnect each connector and remove each clamp. Disconnect the wire harness

62. Detach the clamp and disconnect the harness.
63. Remove the 2 bolts, nut and instrument panel brace.
64. Remove the air conditioning amplifier assembly by disconnecting the connector and removing the 2 screws.
65. Separate steering column from under the lower instrument panel.
66. Disconnect each connector and remove each clamp. Disconnect the wire harness.
67. Remove the 7 bolts, 2 nuts and then remove the instrument panel reinforcement with the air conditioner unit.
68. Remove the air conditioning unit assembly.

a. Disconnect the 5 connectors.
b. Detach the 9 clamps and disconnect the wire harness.
c. Remove the bolt.
d. Detach the clamp and disconnect the junction connector.
e. Remove the 2 screws and air conditioning unit from the instrument panel reinforcement.

To install:

69. Install air conditioning tube and accessory assembly.
a. Sufficiently apply compressor oil ND-OIL 11 to 2 new O-rings and the fitting surface. Install the 2 O-rings to the air conditioning tube assembly.

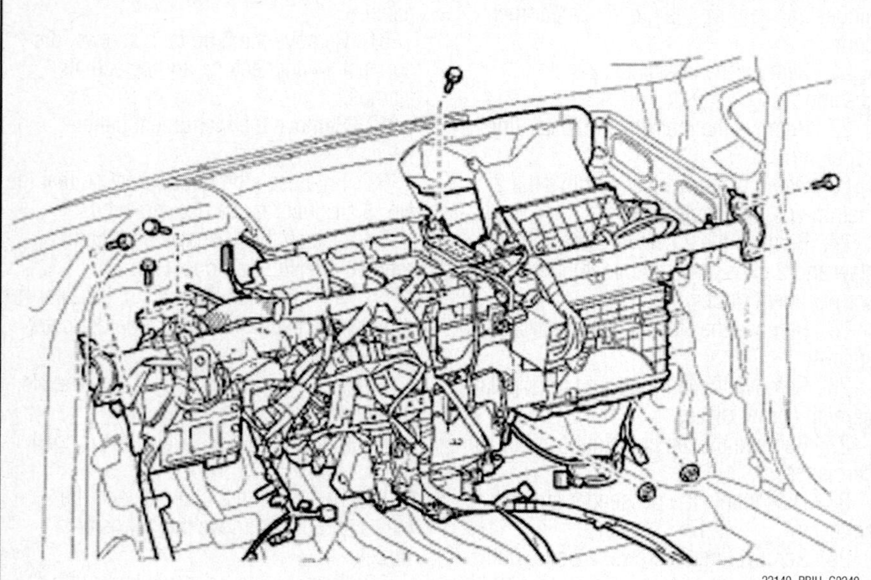

Fig. 341 Remove the 7 bolts, 2 nuts and then remove the instrument panel reinforcement with the air conditioner unit

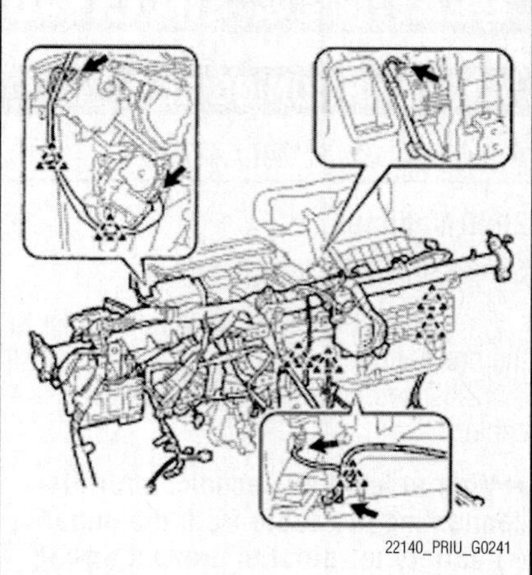

Fig. 342 Disconnect the 5 connectors. Detach the 9 clamps and disconnect the wire harness to remove the air conditioning unit.

✲✲ CAUTION

Do not use any compressor oil other than ND-OIL 11. If any compressor oil other than ND-OIL 11 is used, compressor motor insulation performance may decrease, resulting in a leakage of electric power.

b. Install the air conditioning tube to the No. 1 cooler evaporator, placing the cooler expansion valve between them. Using a 0.16 inch (4mm) hexagon wrench, install the 2 hexagon bolts and tighten to 30 inch lbs. (3.5 Nm).
c. Install the expansion cover with the 2 screws.
70. Install the air outlet control servo motor and attach the claw.
71. Attach the claw and install the 2 screws.
72. Install the air mix control servo motor with the 2 screws.
73. Install the defroster nozzle lower and attach the 2 claws.
74. Install the blower and then attach the fittings with the air conditioner and install the 2 screws.
75. Install the air conditioning unit assembly.
a. Install the air conditioning unit to the instrument panel reinforcement with the 2 screws. Use repair screws (part No. 90159-70003) if the screws removed before cannot be tightened.
b. Attach the clamp to connect the junction connector.
c. Install the bolt.
d. Attach the 9 clamps to connect the wire harness.
e. Connect the 5 connectors.

76. Install the instrument panel reinforcement with the 7 bolts.

77. Install the 2 nuts to the air conditioning unit and temporarily tighten them.

78. Connect each connector and install each clamp. Then connect the wire harness.

79. Install the air conditioning amplifier assembly with the 2 screws and engage the connector.

80. Place steering column into position underneath the lower instrument panel and tighten the mounting bolts.

81. Install the instrument panel brace with the 2 bolts and nut.

82. Attach the clamp to connect the harness.

83. Fully tighten the air conditioning unit with the 2 nuts. Tighten the left nut first, then the right nut second.

84. Install the network gateway ECU.

85. Install the ECM.

86. Install the transaxle control ECU assembly.

87. Install the defroster nozzle and attach the 3 claws.

88. Install the 2 clips.

89. Install the No. 2 heater-to-register duct.

90. Install the duct with the No. 1 side defroster nozzle duct.

91. Install the clip and then the duct with the No. 2 side defroster nozzle duct.

92. Attach the 10 claws to install the No. 3 air duct rear, then return the carpet to its original position.

93. Install the lower instrument panel.

94. Connect all the connectors and attach all the clamps.

95. Install the 2 screws, 6 bolts, 8 clips to the duct and lower instrument panel.

96. Install the lower shift lever assembly.

97. Install the power steering ECU.

98. Attach the 2 claws to install the No. 1 instrument panel under cover.

99. Attach the 4 claws to install the instrument panel finish panel.

100. Install the glove compartment door lock with the 4 screws.

101. Attach the 6 claws to install the glove compartment door lock cover.

102. Attach the 4 claws and 2 clips to install the instrument panel finish panel lower center. Install the clip.

103. Install the radio receiver.

104. Install the No. 1 center cluster cushion.

105. Attach the 4 claws and 2 clips to install the instrument cluster finish panel assembly center. Install the 2 screws.

106. Install left and right cowl side trim boards.

107. Install left and right front door scuff plates.

108. Install the spiral cable.

109. Install the steering column cover.

110. Install the tilt lever bracket.

111. Install the steering wheel and tighten the center nut to 37 ft. lbs. (50 Nm).

112. Install the air bag module.

113. Install the N0. 2 and No. 3 steering wheel lower covers.

114. Attach the 6 claws and 3 clips to install the instrument panel safety pad. Install the 2 bolts and 3 screws.

115. Connect the passenger airbag connector.

116. Install left and right front pillar garnishes.

117. Install left and right front pillar garnish corner pieces.

118. Install the No. 1 instrument panel speaker (w/ JBL Sound System).

119. Attach the claw and 2 clips to install the glove compartment door. Install the 2 screws.

120. Install the 2 instrument panel hole covers.

121. Install the instrument cluster finish panel end.

122. Install the instrument panel cushion.

123. Attach the 2 hinges to install the glove compartment door.

124. While pushing in the sides of the glove compartment door, engage it to the 2 stoppers.

125. Install glove compartment door stopper.

126. Install the multi-display assembly.

127. Attach the 2 claws and 4 clips to install the No. 2 instrument panel register.

128. Attach the 6 clips to install the No. 4 instrument panel register.

129. Connect the connector, attach the claw and 5 clips to install the No. 3 instrument panel register.

130. Connect the connector, attach the 3 claws and 4 clips to install the upper instrument panel finish panel.

131. Connect the connectors, attach the 4 claws and 5 clips to install the lower instrument panel finish panel.

132. Connect the hood lock control cable and install the 2 screws.

133. Attach the 2 claws and 4 clips to install the instrument panel register.

134. Connect the heater water hose A and slide the clip.

135. Connect the heater water hose B and slide the clip.

136. Connect the cooler refrigerant liquid pipe to cooler unit.

a. Remove the attached vinyl tape from the pipe's disconnected part.

b. Sufficiently apply compressor oil ND-OIL 11 to 2 new O-rings and the pipe's connecting part.

CAUTION

Do not use any compressor oil other than ND-OIL 11. If any compressor oil other than ND-OIL 11 is used, compressor motor insulation performance may decrease, resulting in a leakage of electric power.

c. Install the O-rings to the pipe.

d. Insert the pipe joint into the cooler unit fitting hole securely.

e. Using the piping clamp, connect the cooler unit refrigerant liquid pipe. Ensure that the piping clamp is securely engaged.

137. Connect the suction hose by using the same procedures described for the cooler unit refrigerant liquid pipe.

138. Refill the cooling system.

139. Evacuate and recharge the refrigerant.

140. Connect the negative battery cable.

141. Perform the power window initialization procedure.

142. Operate the engine to normal operating temperatures; then, check the climate control operation and check for coolant and refrigerant leaks.

143. Connect the negative battery cable.

144. Perform the power window initialization procedure.

145. Check the SRS warning light.

2010 Vehicles

See Figures 343 through 358.

1. Refer to the Precautions before servicing.

➡Before disconnecting the cable, set the air conditioning control switch to DEF-MODE. (for Automatic Air Conditioning System).

2. Recover the refrigerant from the refrigeration system.

3. Align the front wheels straight ahead.

4. Remove the rear No. 2 floor board (separate type).

5. Remove the rear deck floor box.

6. Remove the rear No. 3 floor board.

7. Disconnect the cable from the negative battery terminal.

CAUTION

Wait at least 90 seconds after disconnecting the cable from the negative (-) battery terminal to disable the SRS system.

➡When disconnecting the cable, some systems need to be initialized after the cable is reconnected

8. Remove the front wiper arm head cap.
9. Remove the front wiper arm and blade assemblies LH and RH.
10. Remove the cowl side ventilator sub assemblies LH and RH.
11. Remove the cowl top ventilator louver sub assembly.
12. Remove the windshield wiper motor and link assembly.
13. Remove the cowl body mounting reinforcement LH.
14. Remove the outer cowl top panel sub assembly.
15. Disconnect the suction pipe sub assembly.
    a. Remove the bolt and slide the hook connector.
    b. Disconnect the suction pipe assembly.
    c. Remove the O-ring from the suction pipe sub-assembly.

➡Seal the openings of the disconnected parts using vinyl tape to prevent entry of moisture and foreign matter.

16. Disconnect the air conditioning tube and accessory assembly.
    a. Disconnect the air conditioning tube and accessory assembly.
    b. Remove the O-ring from the air conditioning tube and accessory assembly.

➡Seal the openings of the disconnected parts using vinyl tape to prevent entry of moisture and foreign matter.

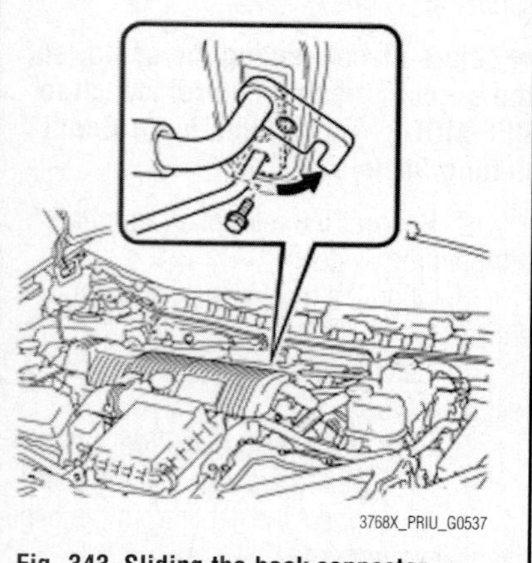

Fig. 343 Sliding the hook connector

17. Disconnect the inlet heater water hose.
    a. Using pliers, grip the claws of the clip and slide the clip to disconnect the outlet heater water hose.

➡Do not apply excessive force to the outlet heater water hose.

➡Prepare a drain pan or cloth in case the coolant leaks.

18. Disconnect the outlet heater water hose.
    a. Using pliers, grip the claws of the clip and slide the clip to disconnect the outlet heater water hose.

➡Do not apply excessive force to the outlet heater water hose.

➡Prepare a drain pan or cloth in case the coolant leaks.

19. Remove the integration control and panel assembly.
20. Remove the lower center instrument cluster finish panel sub assembly.

Fig. 344 Disconnecting the inlet heater water hose

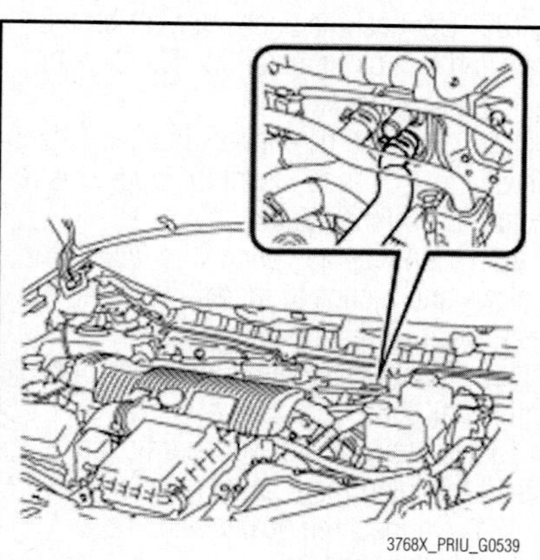

Fig. 345 Disconnecting the outlet heater water hose

21. Remove the instrument cluster finish panel garnish.
22. Remove the upper instrument panel finish panel sub assembly.
23. Remove the radio receiver with the bracket (w/o navigation system).
24. Remove the navigation receiver with the bracket (w/ navigation system).
25. Remove the front door scuff plate LH.
26. Remove the cowl side trim sub assembly LH.
27. Remove the lower instrument panel finish panel assembly.
28. Remove the No. 1 instrument panel register.
29. remove the No. 1 center instrument cluster finish panel.
30. Disconnect the front door opening trim weather-strip LH.
31. Remove the instrument panel finish panel end LH.
32. Remove the No. 1 side defroster nozzle.
33. Remove the No. 2 instrument panel register.
34. Remove the glove compartment door.
35. Remove the front door opening trim weather-strip RH.

➡Use the same procedure for the RH side and the LH side.

36. Remove the instrument panel finish panel end RH.
37. Remove the front pillar garnish LH.
38. Remove the front pillar garnish corner piece LH.
39. Remove the No. 1 instrument panel speaker panel sub assembly.
40. Remove the front No. 2 speaker assembly.
41. Remove the front pillar garnish RH.

➡Use the same procedure for the RH side and LH side.

42. Remove the front pillar garnish corner piece RH.
43. Remove the No. 2instrument panel speaker panel sub assembly.
44. Remove the front No. 2 speaker assembly.

➡Use the same procedure for the RH side and LH side.

45. Remove the center instrument cluster finish panel garnish.
46. Disconnect the instrument panel wire.
47. Remove the upper instrument panel assembly.

48. Remove the rear console box pocket (w/ power outlet socket).
49. Remove the upper console panel (w/ power outlet socket).
50. Remove the console box carpet.
51. Remove the rear console box assembly (w/ power outlet socket).
52. Remove the rear console box assembly (w/o power outlet socket).
53. Remove the electrical key oscillator.
54. Remove the No. 2 console box mounting bracket.
55. Remove the front No. 1 console box insert.
56. Remove the front No. 2 console box insert.
57. Remove the box bottom mat.
58. Separate the console box assembly.
59. Remove the air conditioning control assembly.
60. Remove the shift lock control unit assembly.
61. Remove the upper instrument panel finish panel assembly.
62. Remove the console box assembly.
63. Remove the No. 1 switch hole base.
64. Remove the lower No. 3 steering wheel cover.
65. remove the lower No. 2 steering wheel cover.
66. Remove the steering pad.
67. Remove the steering wheel assembly.
68. Remove the lower steering column cover.
69. Remove the upper steering column cover.
70. Remove the turn signal switch assembly with the spiral cable sub assembly.
71. Remove the No. 1 instrument panel under cover sub assembly.
72. Remove the drive side knee airbag assembly.
73. Remove the No. 2 instrument panel under cover sub assembly.
74. Remove the glove compartment door assembly.
75. Remove the front door scuff plate RH.

➡Use the same procedure for the RH side and the LH side.

76. Remove the cowl side trim board RH.

➡Use the same procedure for the RH side and the LH side.

77. Remove the No. 1 heater to register duct.
78. Remove the rear combination meter assembly.
79. Remove the power steering ECU assembly.
80. Remove the instrument panel safety pad plate sub assembly.
81. Remove the No. 2 antenna cord sub assembly.
82. Remove the lower instrument panel sub assembly.
83. Remove the driver side junction block assembly.
84. Remove the stop light switch assembly.
85. remove the stop light switch mounting adjuster.
86. Remove the brake pedal returning spring.
87. Remove the push rod pin.
88. Remove the brake pedal support assembly.
89. Remove the No. 2 air duct sub assembly.
90. Remove the column hole cover silencer sheet.
91. Separate the No. 2 steering intermediate shaft assembly.
92. Remove the steering post assembly.
93. Remove the rear No. 2 air duct (w/ rear air duct).
   a. Disengage each claw to open the 2 door scuff plate clamps as shown in the illustration.
   b. Disengage the clip and fastener.
   c. Disengage the 2 claws and turn back the floor carpet.
   d. Disengage the 2 claws and remove the rear No. 2 air duct.
94. Remove the rear No. 3 air duct (w/ rear air duct).
   a. Disengage each claw to open the 2 door scuff plate clamps as shown in the illustration.
   b. Disengage the clip and fastener.
   c. Disengage the 2 claws and turn back the floor carpet.
   d. Disengage the 2 claws and remove the rear No. 3 air duct.
95. Remove the rear No. 1 air duct (w/ rear air duct).

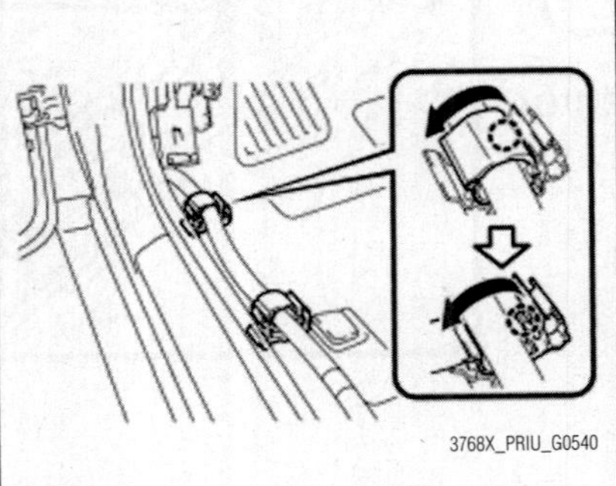

Fig. 346 Disengaging each claw to open the 2 door scuff plate clamps

Fig. 347 Disengaging each claw to open the 2 door scuff plate clamps

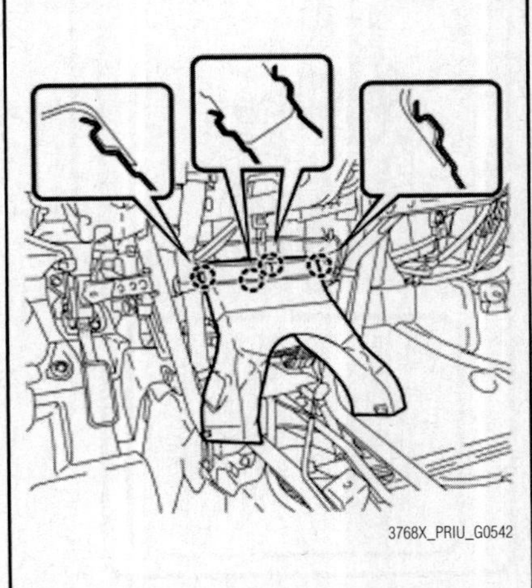

Fig. 348 Disengaging the 4 claws and remove the rear No. 1 air duct

   a. Disengage the 4 claws and remove the rear No. 1 air duct.
96. Remove the No. 3 side defroster nozzle duct.
   a. Remove the clip and No. 3 side defroster nozzle duct.
97. Remove the No. 1 instrument panel brace sub assembly.
   a. Check that the power switch is off.

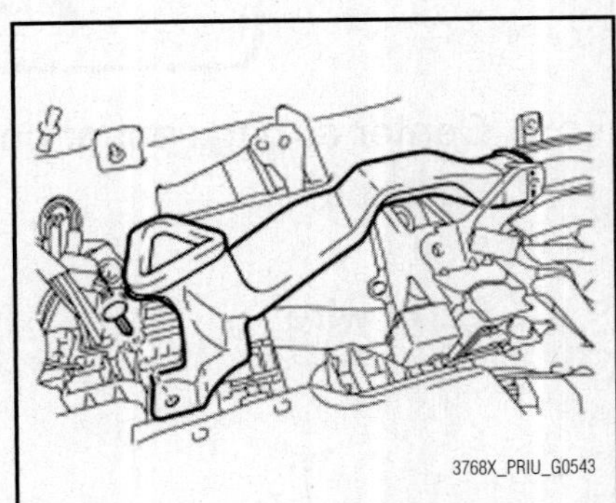

Fig. 349 Removing the clip and No. 3 side defroster nozzle duct

b. Check that the cable is disconnected from the negative (-) battery terminal.

CAUTION

Wait at least 90 seconds after disconnecting the cable from the negative (-) battery terminal to disable the SRS system.

c. Disconnect the center airbag sensor connectors from the center airbag sensor assembly as shown in the illustration.

When disconnecting any airbag connector, take care not to damage the airbag wire harness.

d. Remove the screw.

e. Remove the 2 bolts and disconnect the 2 earth wires.

f. Disconnect each connector.

g. Disengage each clamp and claw.

h. Remove the screw.

i. Remove the bolt, nut and No. 1 instrument panel brace sub-assembly.

98. Remove the No. 2 instrument panel brace sub assembly.

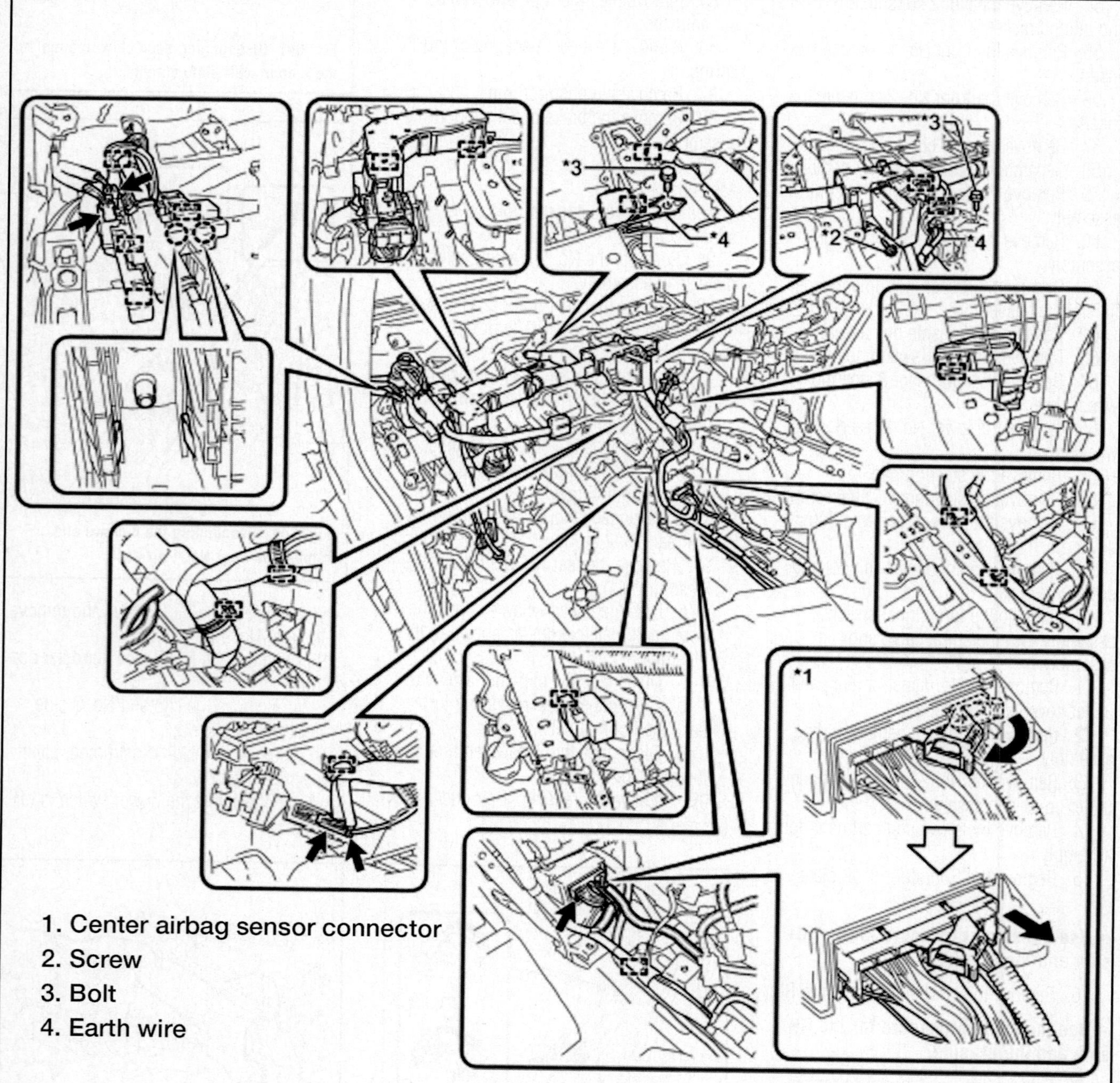

Fig. 350 Removing the No. 1 instrument panel brace sub assembly

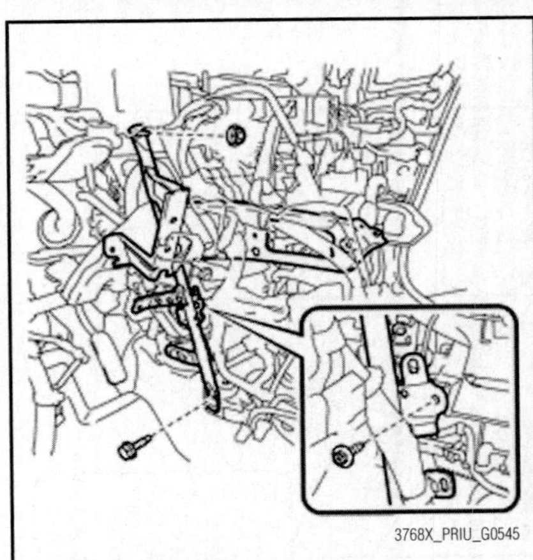

3768X_PRIU_G0545
Fig. 351 Removing the screw

a. Remove the 3 screws
b. Remove the bolt and disconnect the earth wire.
c. Disconnect each connector.
d. Disengage each clamp.
e. Remove the screw and 2 nuts from the computer integration box RH.
f. Remove the screw.
g. Remove the bolt, nut and No. 2 instrument panel brace sub-assembly.

99. Remove the defroster nozzle assembly
a. Disengage the 3 claws.
b. Remove the defroster nozzle assembly as shown in the illustration.

100. Remove the lower defroster nozzle assembly.

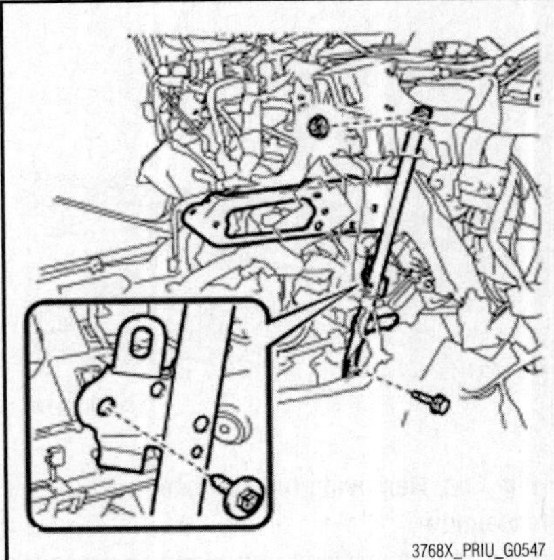

3768X_PRIU_G0547
Fig. 353 Removing the screw

*1
*2
*3
*4
*5
*6
1. Screw (A)
2. Bolt
3. Earth wire
4. Screw (B)
5. Nut
6. Computer integration box RH
3768X_PRIU_G0546
Fig. 352 Removing the No. 2 instrument panel brace sub assembly

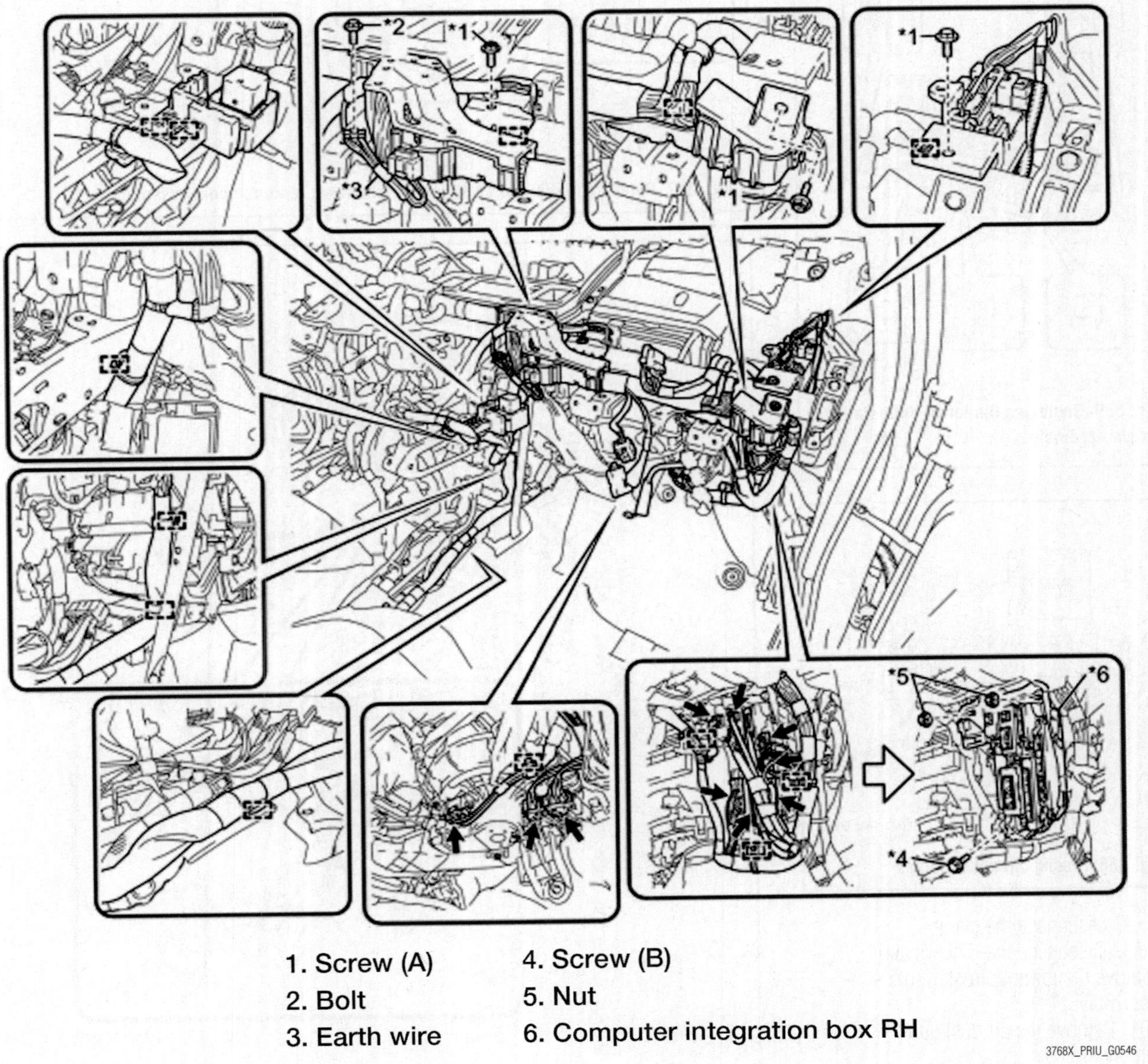

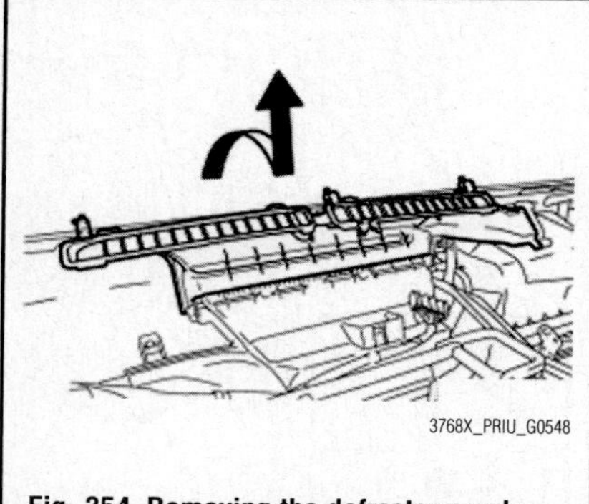

Fig. 354 Removing the defroster nozzle assembly

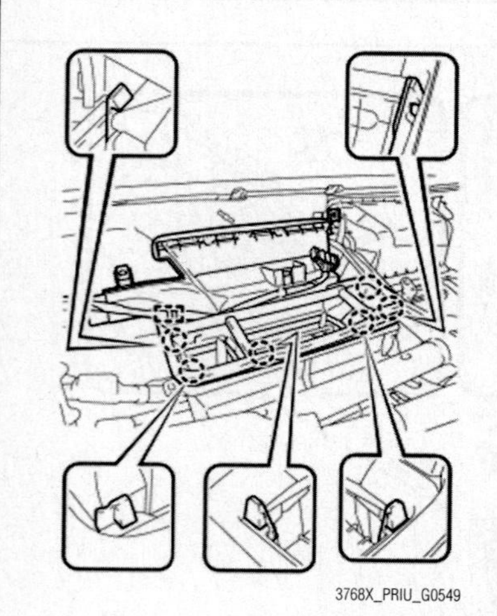

Fig. 355 Removing the lower defroster nozzle assembly

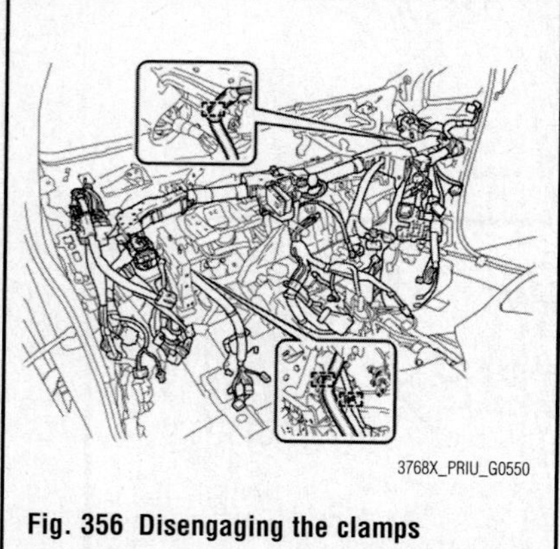

Fig. 356 Disengaging the clamps

a. Disengage the clamp.

b. Disengage the 6 claws and remove the lower defroster nozzle assembly.

101. Remove the air duct sub assembly.

a. Remove the 2 nuts and No. 3 air duct sub assembly.

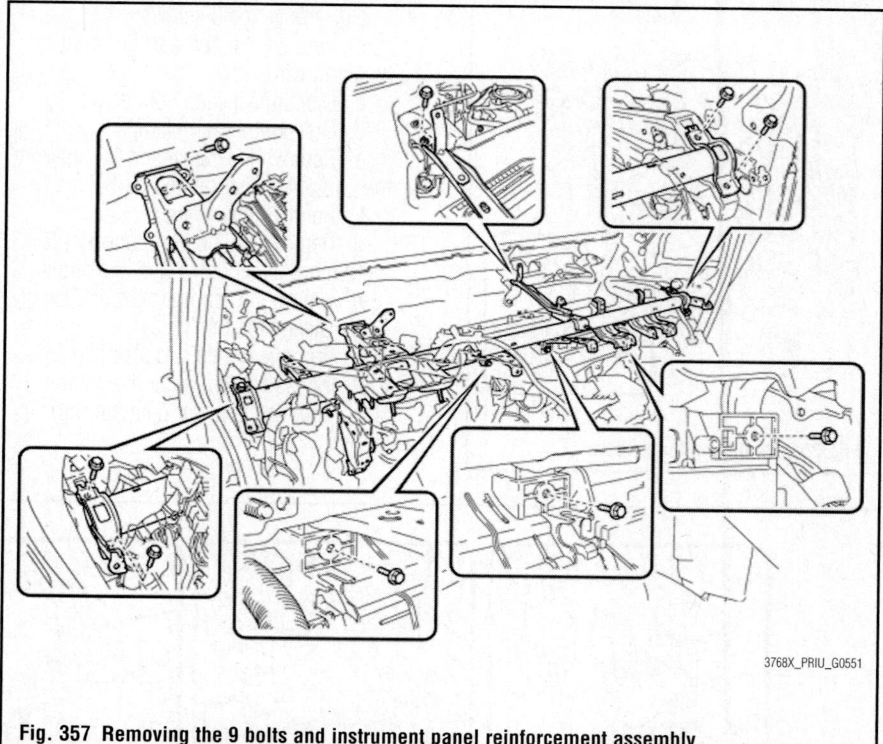

Fig. 357 Removing the 9 bolts and instrument panel reinforcement assembly

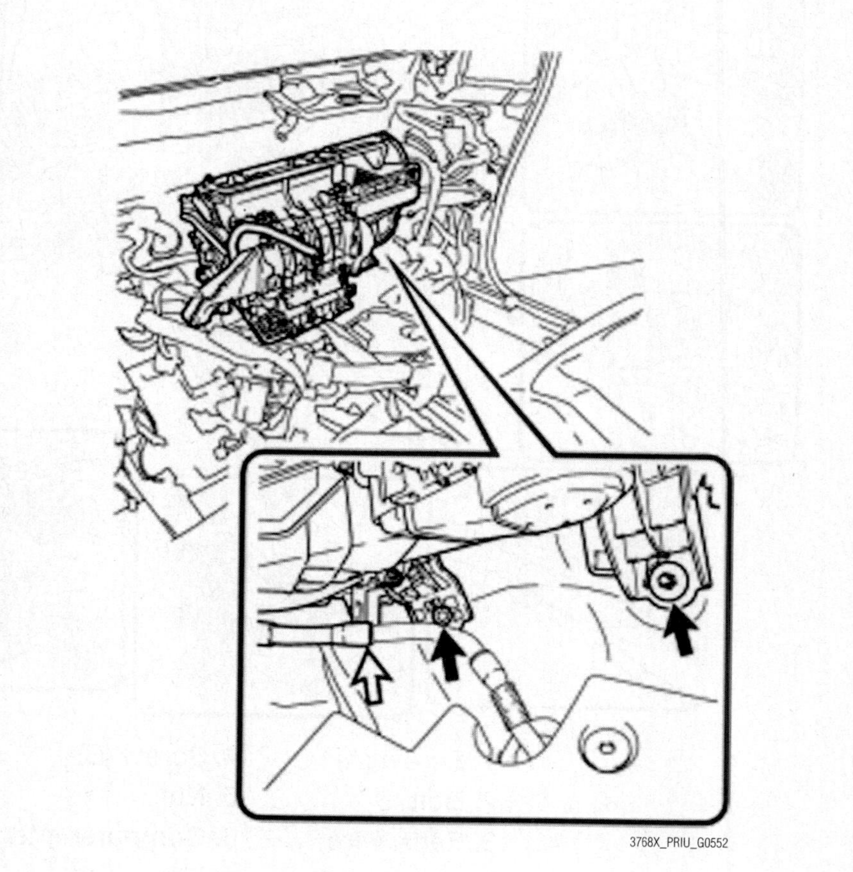

Fig. 358 Disengage the cooler drain hose

102. Remove the instrument panel reinforcement assembly.

a. Disengage each clamp.

b. Remove the 9 bolts and instrument panel reinforcement assembly.

103. Remove the air conditioning unit assembly.

➡Be sure to support the air conditioning unit assembly when removing it because failure to do so may cause the bracket of the air conditioning unit assembly to break.

➡When disassembling the air conditioning unit, eliminate static electricity by touching the vehicle body to prevent the components from being damaged.

a. Disengage the cooler drain hose.

b. Remove the bolt, nut and air conditioning unit assembly.

To install:

104. To install, reverse the removal procedure. Tighten the instrument panel reinforcement assembly bolts to 87 inch lbs. (9.8 Nm). Tighten the air conditioning unit assembly bolt and nut to 87 inch lbs. (9.8 Nm). Tighten the No. 3 air duct sub assembly nuts to 87 inch lbs. (9.8 Nm). Tighten the No. 2 instrument panel brace sub assembly to 35 inch lbs. (4 Nm). Tighten the ground wire to 74 inch lbs. (8.4 Nm). Tighten the No. 1 instrument panel brace sub assembly to 35 inch lbs. (4 Nm).

BLOWER MOTOR

REMOVAL & INSTALLATION

2009 Vehicles

See Figures 359 through 361.

1. Before servicing the vehicle, refer to the precautions.
2. Remove instrument panel with passenger airbag assembly.
3. Remove lower instrument panel.
4. Remove transmission control ECU assembly.
5. Remove ECM.
6. Remove network gateway ECU.
7. Remove No. 1 heater to register duct.
8. Remove air duct.
9. Remove blower assembly.
   a. Disconnect the 10 connectors.
   b. Detach the 4 clamps and disconnect the wire harness.
   c. Remove the 3 screws, nut and blower.
10. Remove No. 2 cooler wiring harness.

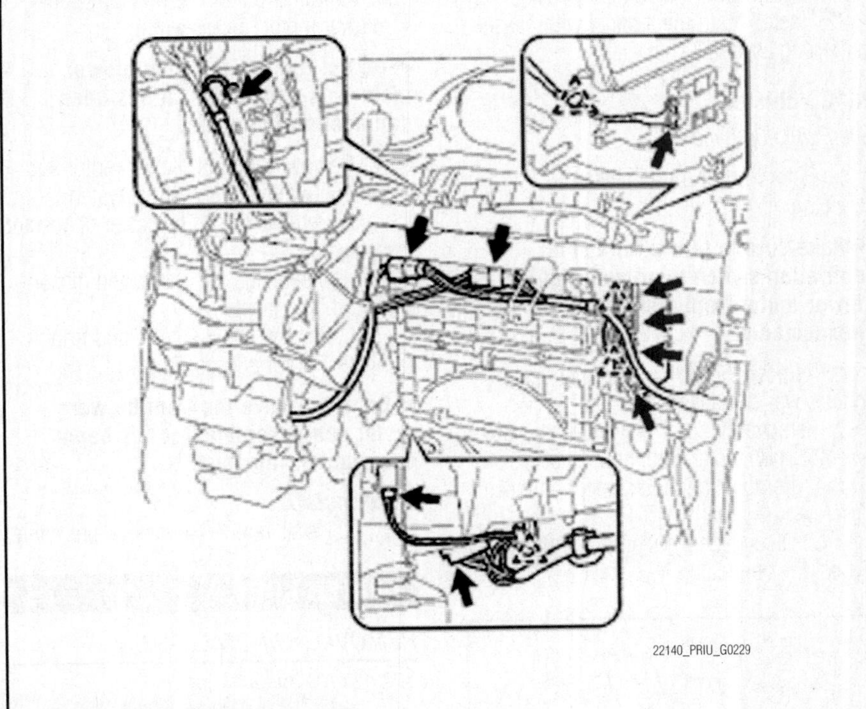

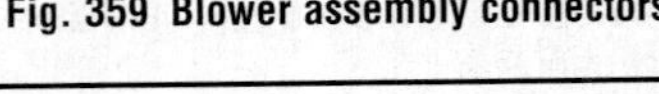
Fig. 359 Blower assembly connectors

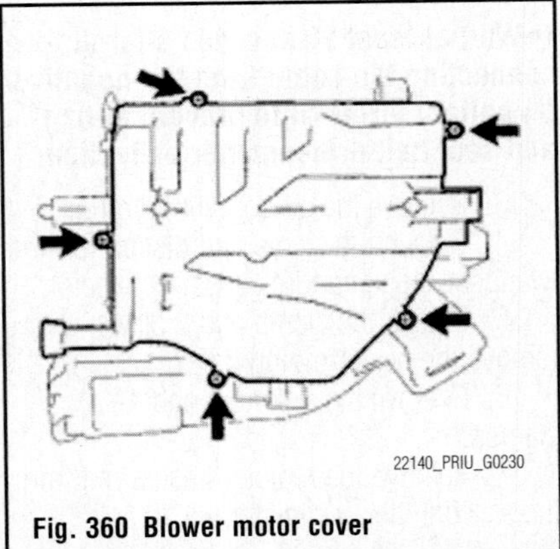

Fig. 360 Blower motor cover

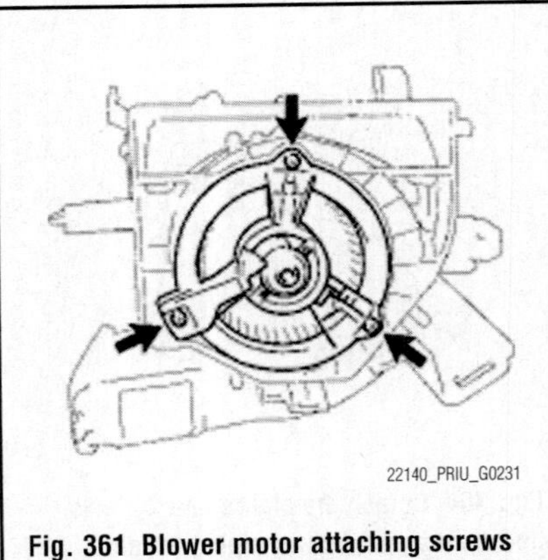

Fig. 361 Blower motor attaching screws

a. Remove the 5 screws and blower motor cover.

b. Remove the No. 2 cooler wiring harness from the blower with fan motor.

11. Remove blower with fan motor.
   a. Remove the 3 screws and blower with fan motor.

To install:

12. Install blower with fan motor.
   a. Install the blower with fan motor with the 3 screws.
13. Install No. 2 cooler wiring harness.
   a. Install the No. 2 cooler wiring harness to the blower with fan motor.
   b. Install the blower motor cover with the 5 screws.
14. Install blower assembly.
   a. Install the blower with the 3 screws.

➡15. Use a repair screw (part No. 90159-70003) if the screw removed before cannot be tightened.

a. Attach the 4 clamps to connect the wire harness.

b. Connect the connectors.

c. Install the 10 connectors of the transponder key ECU.

16. Install air duct.
   a. Install the air duct with the 2 screws.
17. Install No. 1 heater to register duct.
18. Install network gateway ECU.
19. Install ECU.
20. Install transmission control ECU assembly.

21. Install lower instrument panel.
22. Install instrument panel with passenger airbag assembly.

2010 Vehicles

See Figure 362 and 363.

1. Refer to Precautions before servicing.

➡Make sure to turn off the solar ventilation switch to prevent the blower motor from operating unexpectedly.

2. Remove the No. 2 instrument panel under cover sub assembly.
3. Remove the front blower motor sub assembly (w/o solar ventilation system).
   a. Remove the quick heater connector screw.
   b. Disengage the clamp and disconnect the connector.

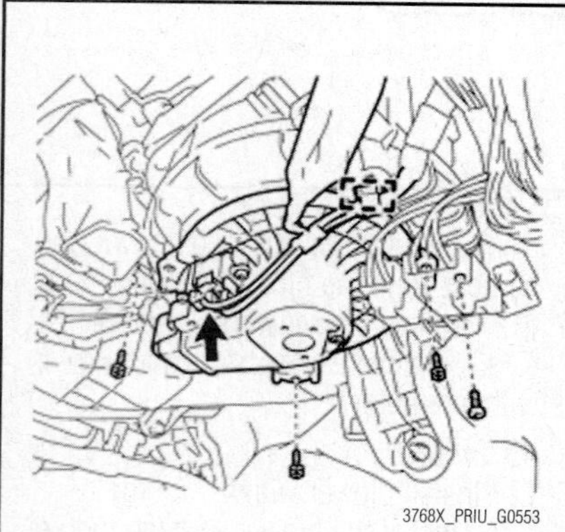

Fig. 362 Removing the front blower motor sub assembly (w/ solar ventilation system)

Fig. 363 Removing the front blower motor sub assembly (w/o solar ventilation system)

   c. Remove the 3 screws and front blower motor sub-assembly.

➡Do not remove the front blower motor sub-assembly if it has been damaged or impacted.

4. Remove the front blower motor sub assembly (w/ solar ventilation system).
   a. Remove the quick heater connector screw.
   b. Disengage the clamp and disconnect the connector.
   c. Remove the 3 screws and front blower motor sub-assembly.

➡Do not remove the front blower motor sub-assembly if it has been damaged or impacted.

To install:

5. To install, reverse the removal procedure.

HEATER CORE

REMOVAL & INSTALLATION

See Figures 364 and 365.

1. Before servicing the vehicle, refer to the precautions.
2. Disconnect the negative battery cable.

➡Wait at least 90 seconds after disconnecting the cable from the negative (-) battery terminal to prevent airbag and seat belt pretensioner activation.

3. Remove the air conditioning unit.
4. Detach the clamp and disconnect the evaporator temperature sensor connector.
5. Detach the clamp and 2 claws, and remove the heater piping cover.
6. Remove the 4 screws and 4 clamps.
7. Remove the radiator heater unit from the air conditioner radiator.

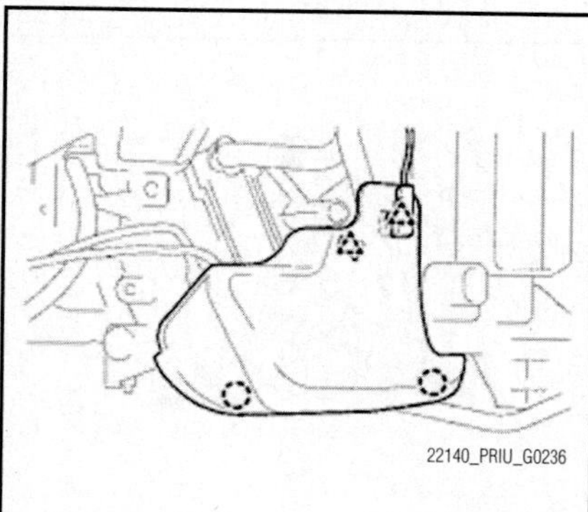

Fig. 364 Detach the clamp and 2 claws, and remove the heater piping cover

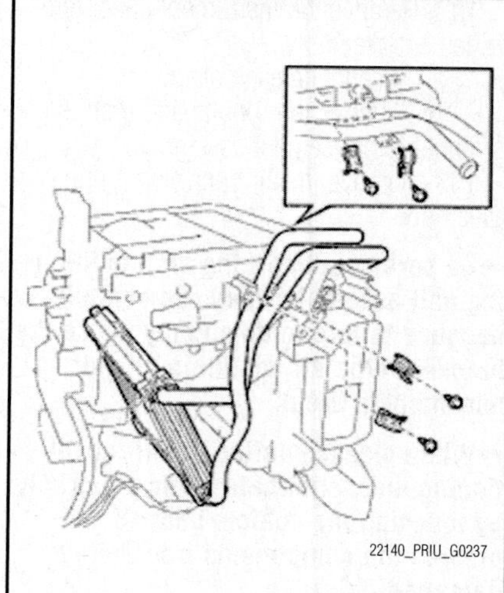

Fig. 365 Remove the radiator heater unit from the air conditioner radiator

➡Prepare a drain pan or cloth for when the cooling water leaks.

8. Remove the 2 screws and No. 1 air duct.
9. Remove the 4 screws and quick heater assembly.
10. Remove the 7 screws and No. 1 cooler evaporator from the heater case.
11. Remove the 2 O-rings from the No. 1 cooler evaporator.
12. Detach the 2 claws and remove the evaporator temperature sensor.

To install:

13. Install the air conditioning radiator to the radiator heater unit and install the 4 clamps and 4 screws.
14. Attach the 2 claws and clamp to install the heater piping cover.
15. Attach the clamp and connect the evaporator temperature sensor connector.
16. Install the cooler expansion valve to the No. 1 cooler evaporator.
17. Install the air conditioning unit.
18. Connect the negative battery cable.
19. Perform the power window initialization procedure.
20. Operate the engine to normal operating temperatures; then, check the climate control operation and check for coolant and refrigerant leaks.
21. Connect the negative battery cable.
22. Perform the power window initialization procedure.
23. Check the SRS warning light.

STEERING

POWER RACK & PINION STEERING GEAR

REMOVAL & INSTALLATION

2009 Vehicles

See Figures 366 and 367.

1. Before servicing the vehicle, refer to the precautions.
2. Place the front wheels in the straight ahead position.
3. Disconnect the negative battery cable.

➡Wait at least 90 seconds after disconnecting the cable from the negative (-) battery terminal to prevent airbag and seat belt pretensioner activation.

4. Fix the steering wheel with the seat belt in order to prevent rotation and damage to the spiral cable.
5. Remove engine under covers.
6. If equipped, remove the steering column hole cover sheet.
7. Put match marks in the sliding yoke and intermediate shaft.
8. Loosen the top bolt and remove the lower bolt to separate the sliding yoke.

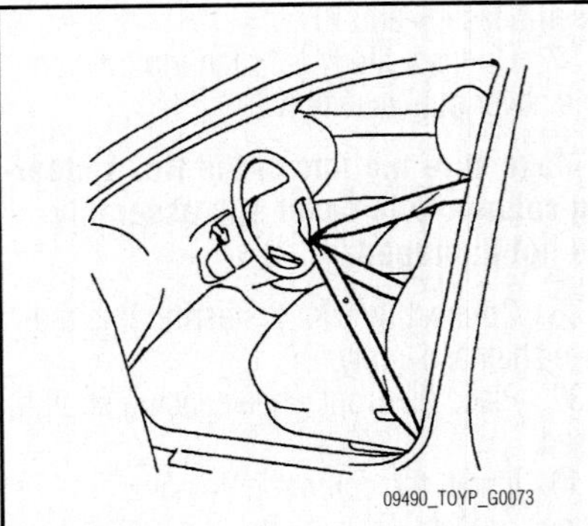

Fig. 366 Steering wheel in fixed position using seat belt

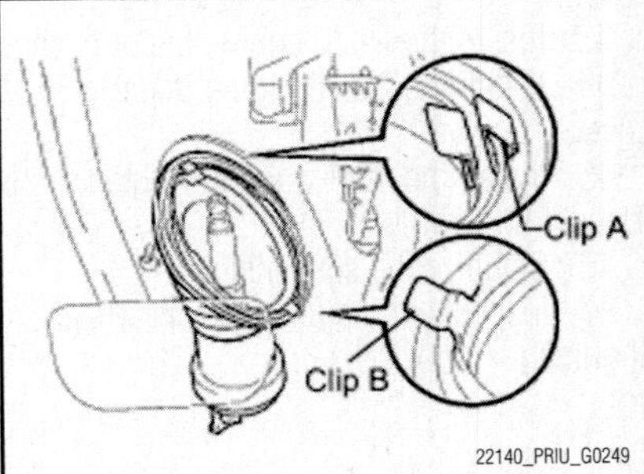

Fig. 367 Remove the clip labeled "A" and disconnect the hole cover from the body. Do not damage the clip labeled "B"

9. If necessary, remove the steering column hole cover from the body. Be careful not to damage the clips.
10. Remove the front wheels.
11. Disconnect the tie rod ends from the steering knuckles.
12. Disconnect the stabilizer links from the struts.
13. Disconnect the lower suspension arm from the lower ball joint.
14. Place match marks on the intermediate shaft assembly and power steering gear.
15. Remove the bolt and disconnect the intermediate shaft from the control valve shaft.
16. Disconnect the torque rod.
17. Remove the front suspension crossmember assembly.
18. Remove the stabilizer bar mounting brackets and the stabilizer bar.
19. Remove the 4 bolts and power steering gear assembly from the front suspension crossmember.

To install:

20. Install the steering column hole cover to the steering gear.
21. Install the power steering gear assembly with the 4 new bolts to the front suspension crossmember. Torque the bolts to 43 ft. lbs. (58 Nm).
22. Install the stabilizer bar and stabilizer bar mounting brackets. Torque the mounting bracket bolts to 14 ft. lbs. (19 Nm).
23. Install the front suspension crossmember. Torque the left and right front corner bolts to 83 ft. lbs. (113 Nm). Torque the left and right rear corner bolts to 116 ft. lbs. (157 Nm).
24. Connect the torque rod. Tighten the through bolt to 74 ft. lbs. (100 Nm).
25. Align the match marks and connect the intermediate shaft to the control valve shaft. Torque the bolt to 26 ft. lbs. (35 Nm).
26. Connect the left and right ball joints to the front lower control arms. Torque to 66 ft. lbs. (89 Nm).
27. Connect the stabilizer bar link to the strut. If the ball joint turns together with the nut, use a hexagon wrench to hold the stud. Torque to 55 ft. lbs. (74 Nm).
28. Connect the tie rod end to the steering knuckle. Torque the tie rod end nut to 36 ft. lbs. (49 Nm) and install a new cotter pin.
29. Install the front wheels.
30. Put the dust seal back to the engine compartment side.
31. Align the match marks on the intermediate shaft assembly and control valve shaft. Install and tighten the bolts to 26 ft. lbs. (35 Nm)
32. Install the steering column hole cover.
33. Align the match marks on the intermediate shaft assembly and sliding yoke. Install and tighten the bolts to 26 ft. lbs. (35 Nm)
34. If equipped, install the steering column hole cover sheet.
35. Install the engine under covers.
36. Remove the seat belt from the steering wheel.
37. Connect the negative battery cable.
38. Perform the power window initialization procedure.
39. Place the front wheels in the straight ahead position.
40. Check the front wheel alignment.

2010 Vehicles

See Figures 368 through 370.

1. Place the front wheels facing straight ahead.
2. Secure the steering wheel with the seat belt in order to prevent the rotation.

➡This operation is useful to prevent damage to the spiral cable.

3. Remove the column hole cover silencer sheet.
4. Separate the No. 2 steering intermediate shaft assembly.
5. Separate the No. 1 steering column hole cover sub assembly.

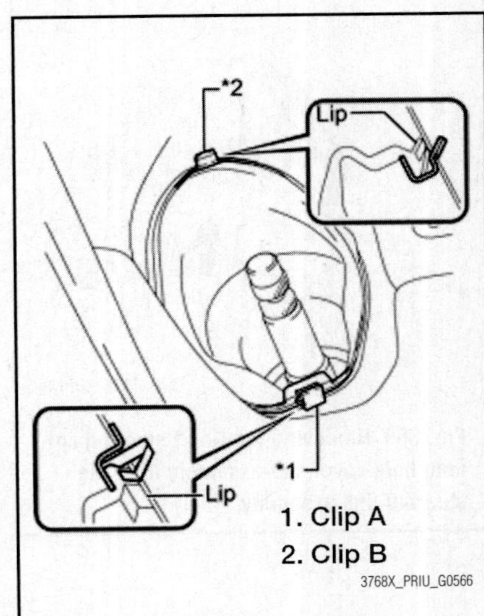

Fig. 368 Separating the No. 1 steering column hole cover sub assembly

a. Remove clip A, detach clip B from the body and disconnect the No. 1 steering column hole cover sub-assembly.

➡Do not damage the clips.

6. Remove the front wheels.
7. Remove the No. 1 engine under cover.
8. Remove the No. 2 engine under cover.
9. Remove the front No. 3 engine under cover.
10. Remove the rear engine under covers LH and RH.
11. Separate the front stabilizer link assemblies LH and RH.
12. Separate the tie rod end sub assembly LH.
13. Separate the tie rod end sub assembly RH.
14. Separate the front No. 1 lower suspension arm sub assemblies LH and RH.
15. Remove the front engine mounting bracket lower reinforcement.
16. Remove the rear side rail reinforcement sub assemblies LH and RH.
17. Remove the front suspension member rear braces LH and RH.
18. Remove the front suspension crossmember sub assembly.
19. Remove the No. 1 steering column hole cover sub assembly.
 a. Remove the No. 1 steering column hole cover sub-assembly from the steering link assembly.
20. Remove the steering intermediate shaft.

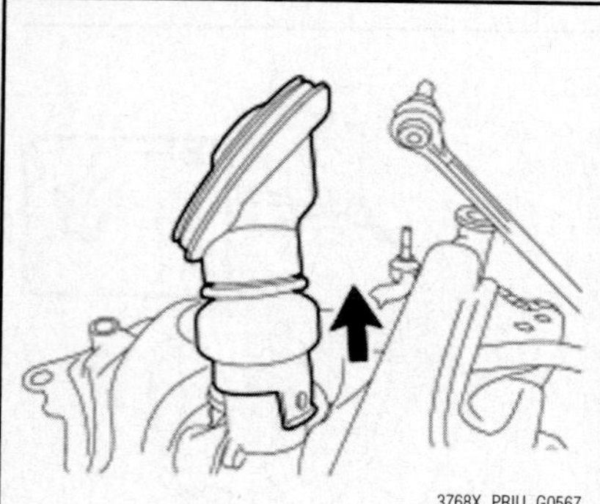

Fig. 369 Removing the No. 1 steering column hole cover sub-assembly from the steering link assembly

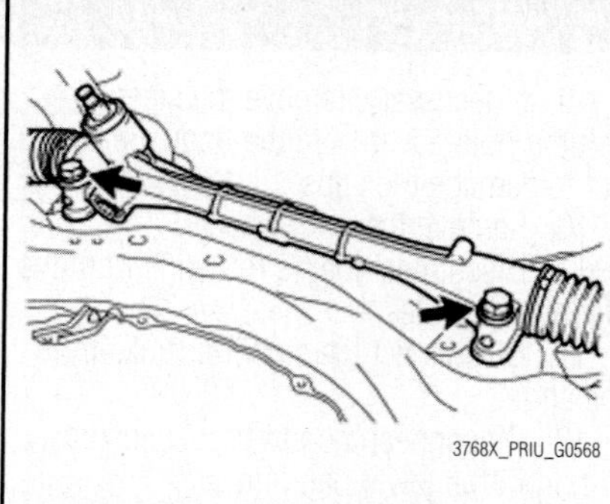

Fig. 370 Removing the steering link assembly

 a. Put matchmarks on the steering intermediate shaft and steering link assembly.
 b. Remove the bolt and steering intermediate shaft from the steering link assembly.
21. Remove the steering link assembly.
 a. Remove the 2 bolts, 2 nuts and steering link assembly from the front suspension crossmember sub-assembly.

➡Keep the nut from rotating while turning the bolt because the nut has its own stopper.

22. Secure the steering link assembly.
 a. Using the special tool (09612-00012), secure the steering link assembly in a vise.

➡Tape the special tool before use.

23. Remove the tie rod end sub assembly LH.
24. Remove the tie rod end sub assembly RH.

To install:

25. Install the tie rod end sub assemblies LH and RH.

➡After adjusting the toe in, tighten the lock nut.

26. Install the steering link assembly.
 a. Install the steering link assembly to the front suspension crossmember sub-assembly with the 2 bolts and 2 nuts. Tighten to 102 ft. lbs. (138 Nm).

➡Keep the nut from rotating while turning the bolt because the nut has its own stopper.

➡Make sure to tighten the bolts starting from the left side of the vehicle.

27. Install the steering intermediate shaft.
 a. Align the matchmarks and install the steering intermediate shaft to the steering link assembly.
 b. Install the bolt and tighten to 26 ft. lbs. (35 Nm).
28. Install the No. 1 steering column hole cover sub assembly.
 a. Align the round hole in the No. 1 steering column hole cover sub-assembly with the protrusion of the steering link assembly to install the cover.
29. Install the front suspension crossmember sub assembly.
30. Install the front suspension member rear brace LH.
31. Install the front suspension member rear brace RH.
32. Install the rear side rail reinforcement sub assemblies LH and RH.
33. Install the front engine mounting bracket lower reinforcement.
34. Connect the front No. 1 lower suspension arm sub assemblies LH and RH.
35. Connect the tie rod end sub assemblies LH and RH. Tighten to 36 ft. lbs. (49 Nm).

➡Further tighten the nut up to 60° if the holes for the clip are not aligned.

36. Connect the front stabilizer link assemblies LH and RH.
37. Connect the No. 1 steering column hole cover sub assembly.

➡Make sure the lips of the No. 1 steering column hole cover sub assembly are not damaged.

38. Connect the No. 2 steering intermediate shaft assembly.
39. Place the front wheels facing straight ahead.
40. Install the column hole cover silencer sheet.
41. Install the rear engine under covers LH and RH.
42. Install the front No. 3 engine under cover.
43. Install the No. 2 engine under cover.
44. Install the No. 1 engine under cover.
45. Install the front wheels. Tighten to 76 ft. lbs. (103 Nm).
46. Stabilize the suspension.
47. Inspect and adjust the front wheel alignment.

LOWER BALL JOINT

REMOVAL & INSTALLATION

2009 Vehicles

See Figures 371 and 372.

1. Before servicing the vehicle, refer to the precautions.
2. Remove the steering knuckle with the hub/bearing assembly. Refer to Front Suspension, Steering Knuckle, Removal & Installation.
3. Mount the steering knuckle in a vise.
4. Remove the cotter pin and nut.
5. Using SST 09628-62011, remove the lower ball joint.

To install:

6. Install the lower ball joint and tighten the nut to 52 ft. lbs. (71 Nm).
7. Install a new cotter pin. If the holes for the cotter pin are not aligned, tighten the nut further up to 60 degrees.
8. Install the steering knuckle with the hub/bearing assembly.
9. Check the ABS speed sensor signal.
10. Check and adjust the front wheel alignment.
11. Check for proper speed sensor signal.

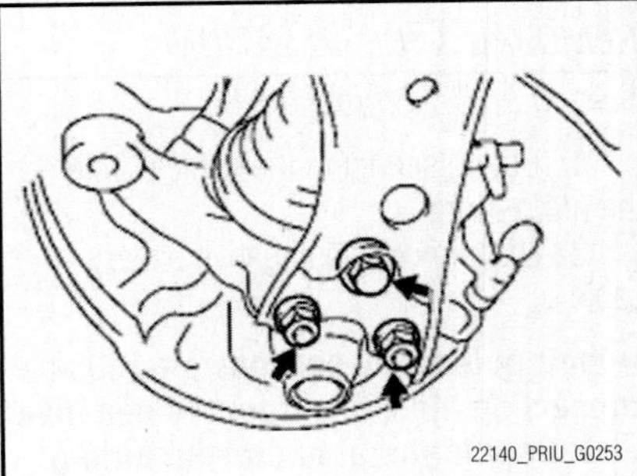

Fig. 371 Lower ball joint attaching bolt/nuts

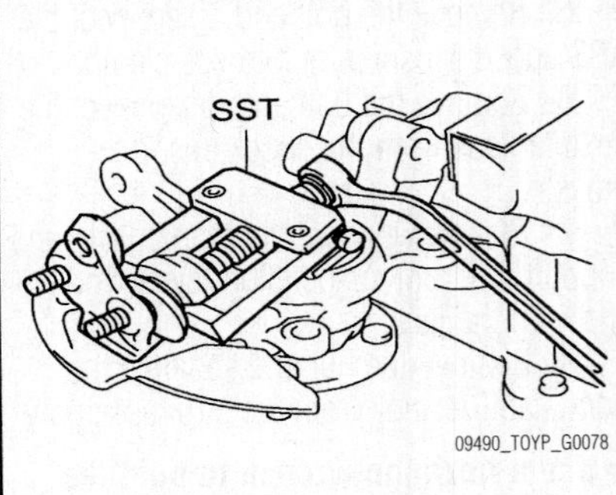

Fig. 372 Separate the ball joint from the steering knuckle.

2010 Vehicles

See Figures 373 and 374.

➡Use the same procedure for the LH side and RH side.

➡The procedure listed below is for the LH side.

1. Remove the front axle.
2. Remove the front lower ball joint assembly.
   a. Secure the front axle assembly in a vise.

➡When using a vise, do not over-tighten it.

   b. Remove the clip and nut.
   c. Install the special tool (09960-20010) to the front lower ball joint.

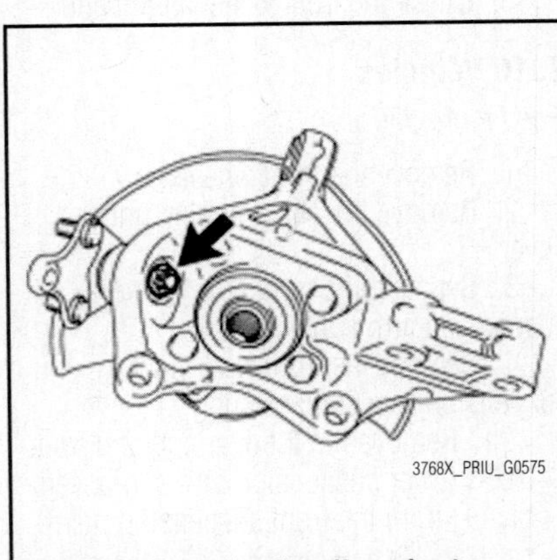

Fig. 373 Removing the clip and nut

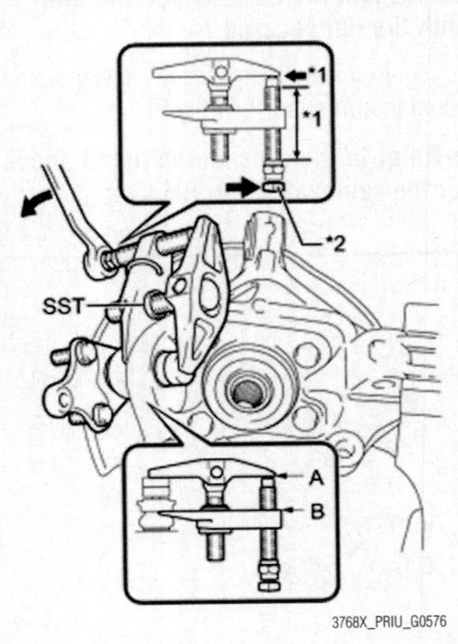

Fig. 374 Removing the front lower ball joint assembly

➡Check that the clearance measurement between SST and the front axle assembly is 1 mm (0.0394 in.).

   d. Using the special tool (09960-20010), remove the front lower ball joint assembly from the front axle assembly.

⁂ CAUTION

Apply grease to the threads and end of the SST bolt.

➡Install the special tool so that A and B are parallel.

➡Be sure to place the wrench on the part indicated in the illustration.

➡Do not damage the front lower ball joint dust cover.

To install:

3. Install the front lower ball joint assembly.
   a. Secure the front axle assembly in a vise.

➡When using a vise, do not over-tighten it.

   b. Install the front lower ball joint assembly to the front axle assembly with the nut. Tighten to 52 ft. lbs. (71 Nm).
   c. Install a new clip.

➡Further tighten the nut up to 60° if the holes for the cotter pin are not aligned.

4. Install the front axle assembly.

LOWER CONTROL ARM

REMOVAL & INSTALLATION

2009 Vehicles

See Figure 375.

1. Before servicing the vehicle, refer to the precautions.
2. Place front wheels facing straight ahead.
3. Disconnect the negative battery cable.
4. Raise and safely support the front of the vehicle. Let the control arms hang free.
5. Remove the front wheels
6. Remove the steering column hole cover silencer, if necessary.
7. Remove the engine under cover.
8. Remove the front exhaust pipe assembly, if necessary.
9. Remove the tie rod end cotter pin and nut.

10. Remove the tie rod end from the steering knuckle.

11. Remove the stabilizer bar link from the strut. if the ball joint turns together with the nut, use a hexagon wrench to hold the stud.

12. Remove the bolt and 2 nuts, and disconnect the lower suspension arm from the lower ball joint.

13. Remove the loosen the 2 lower suspension arm set bolts.

14. Remove the steering sliding yoke.

15. Remove the drive shafts, if necessary.

16. Support the suspension member with a transmission jack.

17. Remove the bolt and nut; disconnect the torque rod from the suspension member.

18. Remove the 4 bolts and disconnect the suspension member from the body.

19. Remove the 2 lower suspension arm set bolts and disconnect the lower suspension arm from suspension crossmember.

To install:

20. Install the front lower control arm to the suspension crossmember and temporarily tighten the front suspension lower No. 1 arm with the 2 bolts and nut.

21. Install the front suspension crossmember. Torque the left and right front corner bolts to 83 ft. lbs. (113 Nm). Torque the left and right rear corner bolts to 116 ft. lbs. (157 Nm).

22. Install the torque rod. Tighten the 2 nuts and 2 bolts, at the engine, to 74 ft. lbs. (100 Nm). Torque the 2 bolts, at the body side, to 41 ft. lbs. (56 Nm).

23. Install the left and right drive shafts.

24. Install the steering sliding yoke. Torque the pinch bolt to 26 ft. lbs. (35 Nm).

25. Install the left and right ball joints to the front lower control arms. Torque to 66 ft. lbs. (89 Nm).

26. Install the stabilizer bar link to the strut. If the ball joint turns together with the nut, use a hexagon wrench to hold the stud. Torque to 55 ft. lbs. (74 Nm).

27. Install the tie rod end to the steering knuckle. Torque the tie rod end nut to 36 ft. lbs. (49 Nm) and install a new clip.

28. Install the front wheels

29. Install the front exhaust pipe assembly, if necessary

30. Install the steering column hole cover silencer, if necessary

31. Lower the vehicle and bounce it up and down several times to stabilize the front suspension.

32. Fully tighten the 2 front lower control arm bolts. Torque to 101 ft. lbs. (137 Nm). Keep the nut from rotating while tightening the rear-side bolt. Lower the tires to the ground using a 4-post lift.

33. Connect the negative battery cable.

34. Perform the power window initialization procedure.

35. Check the front wheel alignment.

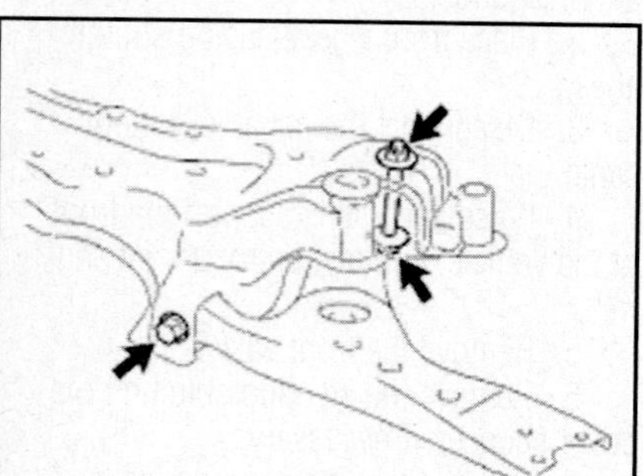

Fig. 375 Lower control arm attaching points

2010 Vehicles

See Figure 376.

1. Remove the front wheels.
2. Remove the No. 1 engine under cover.
3. Separate the front No. 1 lower suspension arm sub assembly LH.
4. Remove the front No. 1 lower suspension arm sub assembly LH.
   a. Remove the 2 bolts, nut, and front No. 1 lower suspension arm sub-assembly LH from the front suspension member.

➡Because the nut has its own stopper, do not turn the nut. Loosen the bolt with the nut secured

5. Remove the front No. 1 lower suspension arm sub assembly RH.

➡Refer to the procedures listed above for the removal of the RH side.

Fig. 376 Removing the front suspension arm sub assembly

To install:

6. Temporarily tighten the front No. 1 lower suspension arm sub assembly RH and LH.
   a. Temporarily install the front No. 1 lower suspension arm LH and RH to the front suspension crossmember with the 2 bolts and nut.

➡Because the nut has its own stopper, do not turn the nut. Tighten the bolt with the nut secured.

7. Connect the front No. 1 lower suspension arm sub assembly LH.
8. Install the front wheels. Tighten to 76 ft. lbs. (103 Nm).
9. Stabilize the suspension.
   a. Lower the vehicle.
   b. Press down on the vehicle several times to stabilize the suspension.
10. Fully tighten the front No. 1 lower suspension arm sub assemblies LH and RH. Tighten to 172 ft. lbs. (233 Nm).

➡Because the nut has its own stopper, do not turn the nut. Tighten the bolt with the nut secured.

11. Install the No. 1 engine under cover.
12. Inspect and adjust the front wheel alignment.

MACPHERSON STRUT

REMOVAL & INSTALLATION

See Figures 377 through 381.

1. Before servicing the vehicle, refer to the precautions.
2. Disconnect the negative battery cable.

➡Wait at least 90 seconds after disconnecting the cable from the negative (-) battery terminal to prevent airbag and seat belt pretensioner activation.

3. Remove front wheel.
4. Remove front wipers and front wiper motor link assembly.
5. Remove the bolt and disconnect the ABS speed sensor wire harness clamp.
6. Remove the bolt and disconnect the flexible hose from the shock absorber bracket.
7. Place a wooden block on a jack, and support the front suspension lower No. 1 arm with the jack.
8. Remove the nut and separate the front stabilizer link from the strut assembly.

➡Use a hexagon wrench to hold the stud if the ball joint turns together with the nut.

9. Remove the 2 nuts on the lower side of the front strut assembly, but keep the bolts inserted.

10. Remove the 3 nuts from the top of the strut assembly.

11. Lower the jack slowly. Remove the 2 bolts on the lower side and the front strut assembly.

➡Ensure that the speed sensor front is completely disconnected from the front shock absorber with coil spring.

12. Separate the coil spring from the strut assembly.

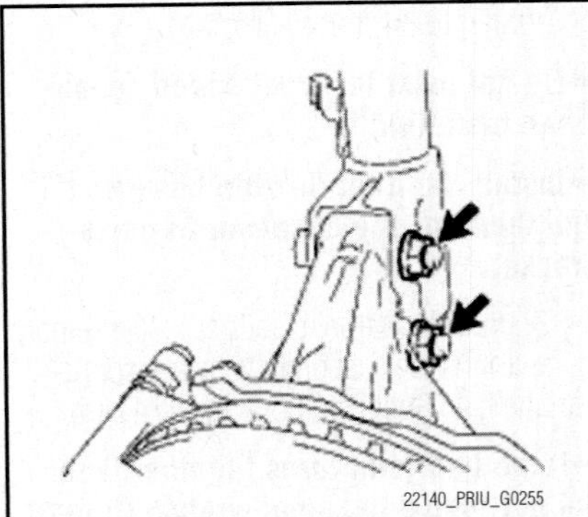

Fig. 377 Remove the 2 nuts on the lower side of the strut

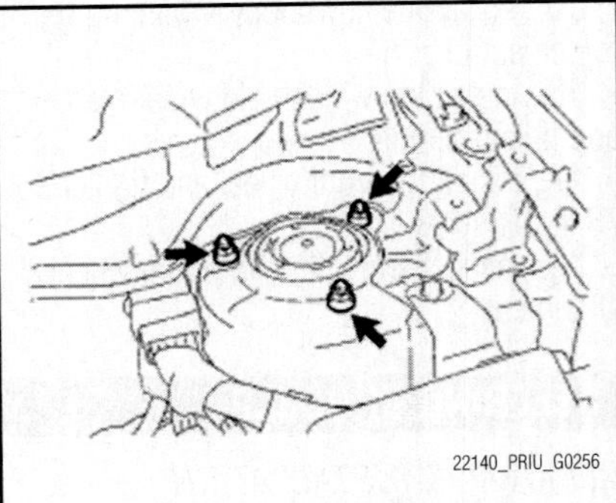

Fig. 378 Remove the 3 nuts on the strut tower

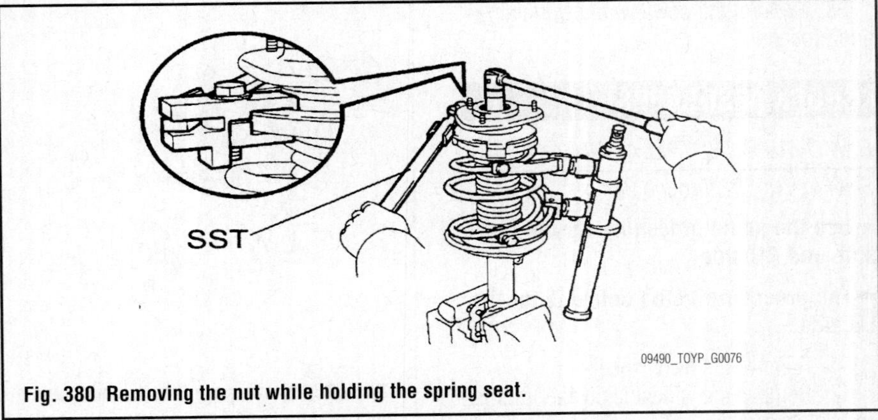

Fig. 380 Removing the nut while holding the spring seat.

a. Install 2 nuts and a bolt to the bracket at the lower side of the shock absorber and secure it in a vise.

b. Using SST 09727-30021, compress the coil spring.

c. Remove the cap from the suspension support.

d. Using SST 09729-22031 to hold the spring seat, remove the nut.

e. Remove the suspension support, dust seal, spring seat, upper insulator, coil spring, spring bumper and lower insulator.

To install:

13. Install the coil spring by assembling the strut assembly.

a. Install the lower insulator onto the strut.

b. Install the spring bumper to piston rod.

c. Using SST 09727-30021, compress the coil spring. Do not use an impact wrench. It will damage the SST.

d. Install the coil spring to the shock absorber, fitting the lower end of the coil spring into the gap of the spring lower seat.

e. Install the upper insulator.

f. Install the spring seat to the shock absorber with the arrow mark facing to the outside of the vehicle.

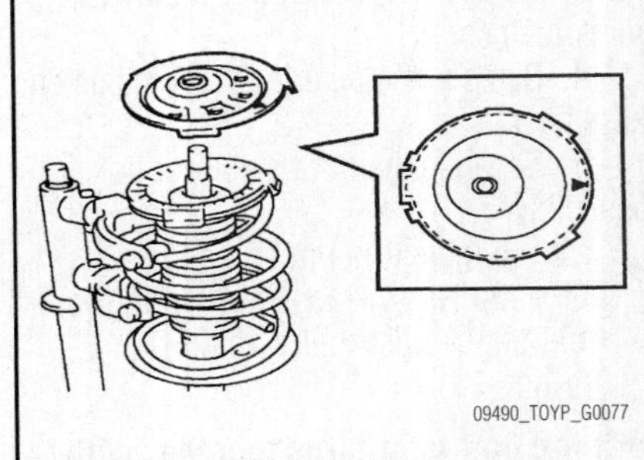

Fig. 381 Install the spring seat to the shock absorber with the arrow mark facing to the outside of the vehicle.

g. Install the dust seal and suspension support.

h. Using SST 09729-22031 to hold the suspension support, install a new nut and torque to 34 ft. lbs. (47 Nm).

i. Remove the tool.

j. Apply MP grease No.2 into the suspension support. Do not touch grease on rubber surface of upper support.

k. Install the cap.

14. Install the strut assembly to the steering knuckle and tighten the steering knuckle-to-strut assembly nuts and the bolts to 113 ft. lbs. (153 Nm)

15. Connect the stabilizer shaft link to strut assembly and tighten the stabilizer link-to-strut assembly nut to 55 ft. lbs. (74 Nm)

16. Install the flexible brake hose to the shock absorber bracket.

17. Connect the ABS speed sensor wire harness clamp.

18. Install the front wiper motor link assembly and the front wipers.

19. Install the front wheel.

20. Lower the vehicle.

21. Install the nuts securing the strut assembly to the body of the vehicle and tighten to 29 ft. lbs. (39 Nm).

22. Connect the negative battery cable.

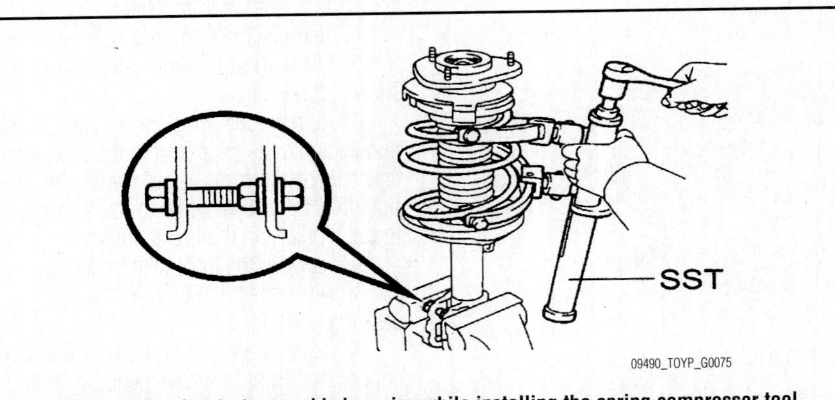

Fig. 379 Securing the strut assembly in a vise while installing the spring compressor tool.

23. Perform the power window initialization procedure.

SHOCK ABSORBERS

REMOVAL & INSTALLATION

See Figures 382 through 384.

➡Use the same procedure for the LH side and RH side.

➡The procedure listed below is for the LH side.

1. Remove the front wheel.
2. Remove the windshield wiper motor and link assembly.
3. Remove the cowl body mounting reinforcement LH.
4. Remove the outer cowl top panel sub assembly.
5. Separate the front stabilizer link assembly.
   a. Remove the nut and separate the stabilizer link assembly from the front shock absorber with coil spring.

➡If the ball joint turns together with the nut, use a hexagon wrench (6 mm) to hold the stud bolt.

6. Separate the front speed sensor.
   a. Remove the bolt and clamp, and separate the front speed sensor and front flexible hose from the front shock absorber with coil spring.

➡Be sure to separate the front speed sensor from the front shock absorber with coil spring completely.

7. Remove the front suspension support dust cover.
8. Remove the front shock absorber with the coil spring.
   a. Loosen the front support to front shock absorber nut of the front shock absorber.

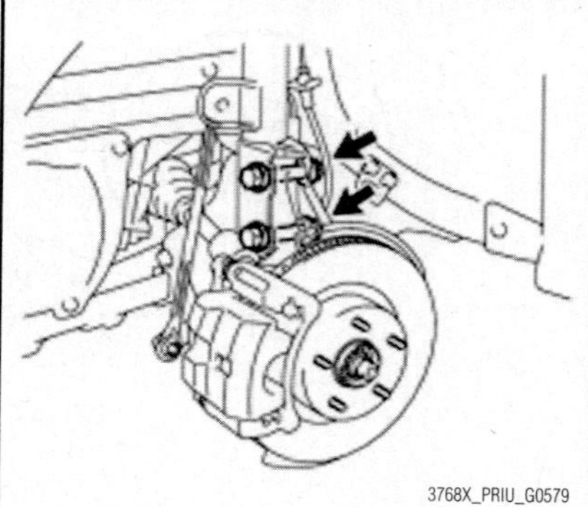

Fig. 383 Removing the 2 bolts and 2 nuts, and separate the front shock absorber with coil spring (lower side) from the steering knuckle

➡Do not remove the front support to front shock absorber nut.

➡Loosen the nut only when the front shock absorber with coil spring needs to be disassembled.

   b. Support the front axle using a jack and wooden block.
   c. Remove the 2 bolts and 2 nuts, and separate the front shock absorber with coil spring (lower side) from the steering knuckle.
   d. Remove the 3 nuts and front shock absorber with coil spring.

➡Make sure that the front speed sensor is completely separated from the front shock absorber with coil spring.

To install:

9. Install the front shock absorber with the coil spring.
   a. Install the front shock absorber with coil spring (upper side) with the 3 nuts. Tighten to 37 ft. lbs. (50 Nm).
   b. Install the front shock absorber with coil spring (lower side) to the steering knuckle with the 2 bolts and 2 nuts.

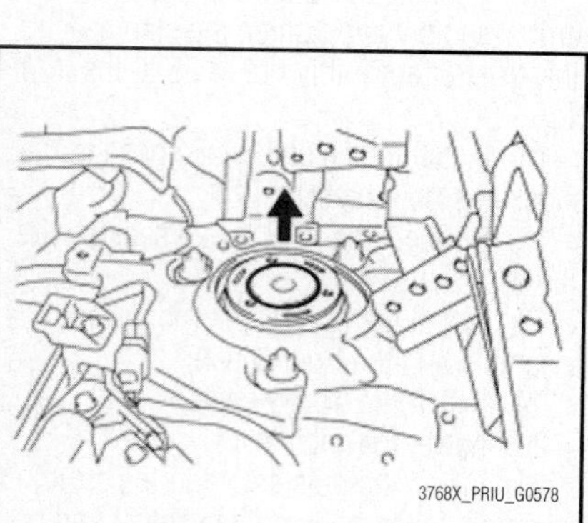

Fig. 382 Removing the front suspension support dust cover

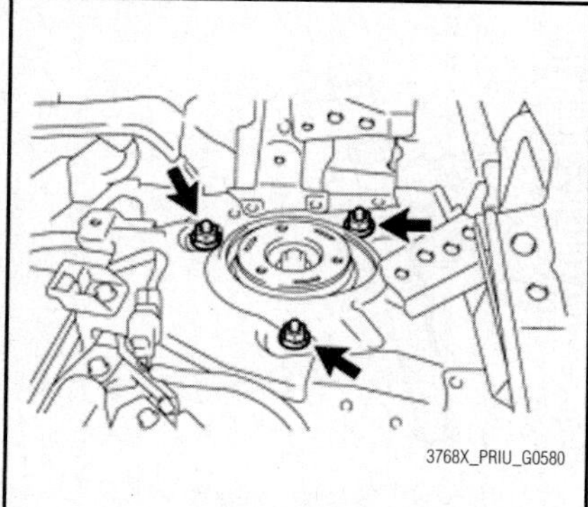

Fig. 384 Removing the 3 nuts

   c. Install the front shock absorber with coil spring (lower side) to the steering knuckle with the 2 bolts and 2 nuts. Tighten to 177 ft. lbs. (240 Nm).

➡While keeping the bolts from rotating, tighten the nuts.

10. Fully tighten the front support to front shock absorber nut to 35 ft. lbs. (47 Nm).
11. Install the front suspension support dust cover.
12. Install the front speed sensor.
   a. Install the front speed sensor and front flexible hose to the front shock absorber with the bolt and clamp. Tighten to 14 ft. lbs. (19 Nm).

➡Do not twist the front speed sensor when installing it.

➡Install the front flexible hose first and then the speed sensor harness bracket.

13. Install the front stabilizer link assembly to the front shock absorber with coil spring with the nut. Tighten to 55 ft. lbs. (74 Nm).

➡If the ball joint turns together with the nut, use a hexagon wrench (6 mm) to hole the stud bolt.

14. Install the outer cowl top panel sub assembly.
15. Install the cowl body mounting reinforcement LH.
16. Install the windshield wiper motor and link assembly.
17. Install the front wheel and tighten to 76 ft. lbs. (103 Nm).
18. Inspect and adjust the front wheel alignment.

STEERING KNUCKLE

REMOVAL & INSTALLATION

1. Before servicing the vehicle, refer to the precautions.
2. Raise and safely support the vehicle.
3. Remove the front wheel.
4. Remove the front axle hub nut.
5. Disconnect the front ABS speed sensor connector.
6. Remove the 2 caliper mounting bolts and separate the caliper from the mounting and support from the vehicle with wire. Do not allow the caliper to hang low enough as to put tension on the brake hose.
7. Remove the front disc brake rotor.
8. Disconnect the tie rod from the steering knuckle.
9. Remove the bolt and 2 nuts and disconnect the front lower suspension control arm from the lower ball joint.

10. Separate the front drive shaft from the hub/bearing and steering knuckle assembly.
11. Separate the steering knuckle from the strut assembly.

To install:

12. Install the steering knuckle to the strut assembly. Torque the nuts and bolts to 113 ft. lbs. (153 Nm)
13. Install the front drive shaft.
14. Install the lower ball joint to the front lower suspension control arm. Torque the bolt and 2 nuts to 66 ft. lbs. (89 Nm).
15. Connect the tie rod end to the steering knuckle. Torque the tie rod end nut to 36 ft. lbs. (49 Nm) and install a new clip.
16. Install the front disc brake rotor.
17. Install the front disc brake caliper.
18. Connect the front ABS speed sensor connector.
19. Install the front axle hub nut. Torque the nut to 159 ft. lbs. (216 Nm), then stake the hub nut, using a chisel and hammer.
20. Install the front wheel.
21. Inspect and adjust the front wheel alignment.
22. Check ABS speed sensor signal.

STABILIZER BAR

REMOVAL & INSTALLATION

2009 Vehicles

See Figures 385 and 386.

1. Place front wheels facing straight ahead.
2. Remove column hole cover silencer sheet.
3. Disconnect steering sliding yoke.
4. Disconnect No. 1 steering column hole cover.
5. Raise and safely support the vehicle securely on jackstands.
6. Remove front wheel.
7. Remove the 2 nuts and front stabilizer link.

➡Use a 6 mm hexagon wrench to hold the stud if the ball joint turns together with the nut.

8. Disconnect tie rod ends.
9. Remove floor panel brace front.
10. Remove front exhaust pipe assembly.
11. Remove the 2 bolts and the No. 1 stabilizer bracket from the front suspension crossmember.
12. Disconnect front stabilizer bracket.
13. Remove steering intermediate shaft.
14. Remove steering column hole cover.
15. Remove steering gear assembly.

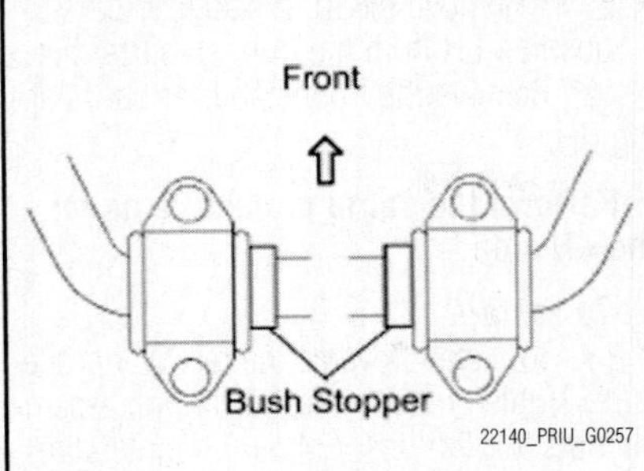

Fig. 385 Install the bushing to the outer side of the bushing stopper on the stabilizer bar

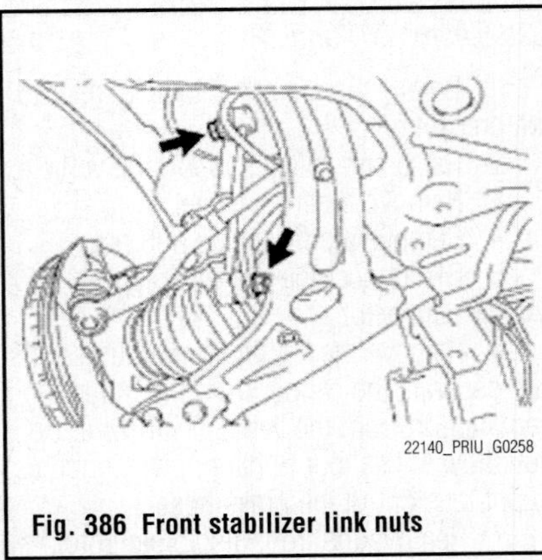

Fig. 386 Front stabilizer link nuts

16. Remove the 2 front stabilizer bar bushes from the stabilizer bar.
17. Remove the front stabilizer bar from the right side of the vehicle.

To install:

18. Insert the front stabilizer bar from the right side of the vehicle.
19. Install the bushing to the outer side of the bushing stopper on the stabilizer bar.

➡Place the cutout of the stabilizer bushing facing the rear side. Ensure the right and left deviation of the stabilizer bar is 0.20 in. (5 mm) or less.

20. Install steering gear assembly.
21. Install steering column hole cover.
22. Install steering intermediate shaft.
23. Install the stabilizer bracket the front suspension crossmember with the 2 bolts. Tighten to 14 ft. lbs. (19 Nm).
24. Connect front stabilizer bracket.
25. Install front exhaust pipe assembly.
26. Install floor panel brace front.
27. Install the front stabilizer link with the 2 nuts. Tighten to 55 ft. lbs. (74 Nm).
28. Install the tie rod ends.
29. Install front wheels and lower the vehicle.
30. Inspect and adjust the front wheel alignment

2010 Vehicles

See Figures 387 through 390.

1. Remove the front wheels.
2. Remove the No. 1 engine under cover.
3. Remove the No. 2 engine under cover.
4. Remove the front stabilizer link assembly LH.
   a. Remove the 2 nuts and front stabilizer link assemblies LH.

➡If the ball joint turns together with the nut, use a hexagon wrench (6 mm) to hold the stud bolt.

5. Remove the front stabilizer link assembly RH.

➡Perform the same procedure as for the LH side.

6. Separate the front No. 1 lower suspension arm sub assembly LH.
7. Remove the front No. 1 lower suspension arm sub assembly LH.
8. Remove the front suspension member front brace LH.
   a. Remove the 4 bolts and front suspension member front brace LH.
9. Remove the front suspension member front brace RH.

➡Perform the same procedure as for the LH side.

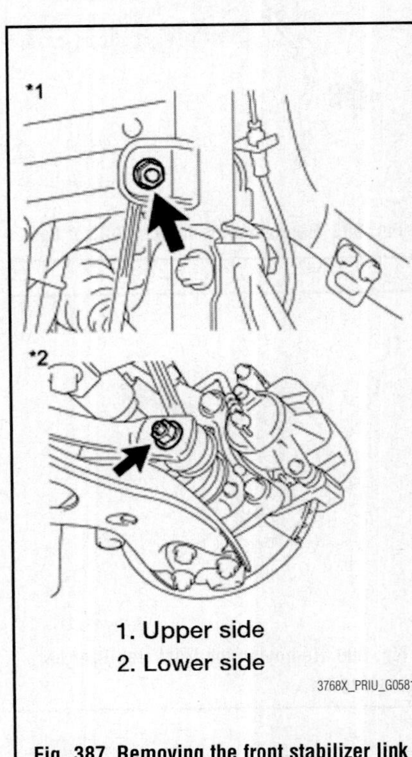

Fig. 387 Removing the front stabilizer link assembly LH

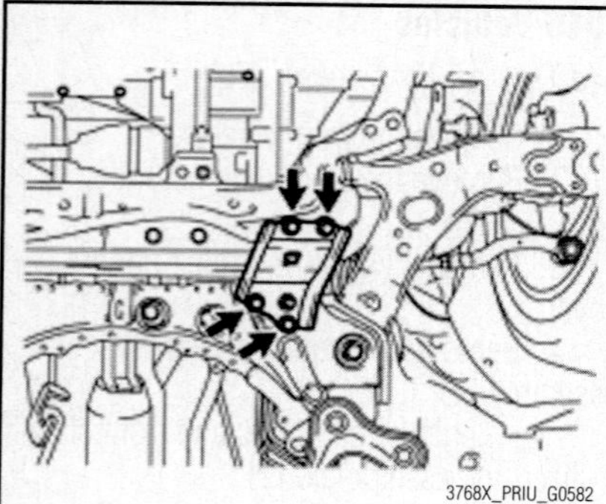

Fig. 388 Removing the front suspension member front brace LH

10. Remove the front stabilizer bar.
   a. Remove the front stabilizer bar with front stabilizer bar bushing from the front suspension crossmember sub-assembly.
11. Remove the front stabilizer bar bushing LH.

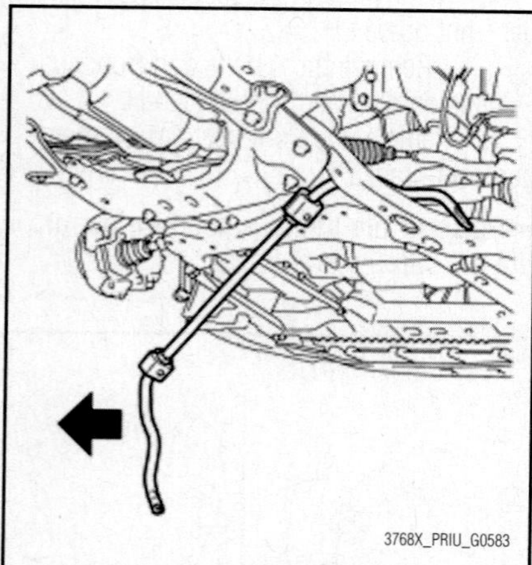

Fig. 389 Removing the front stabilizer bar

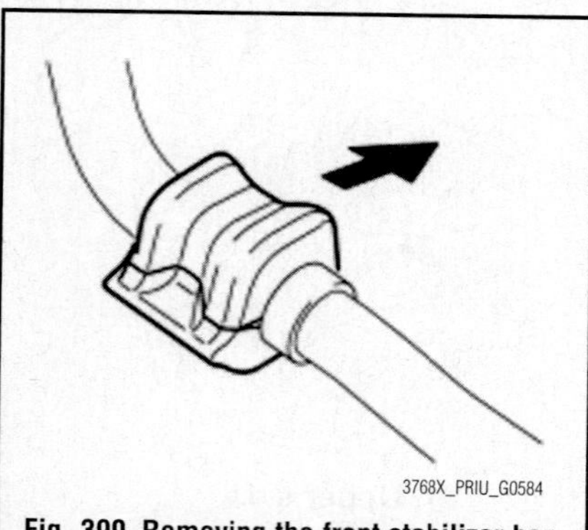

Fig. 390 Removing the front stabilizer bar bushing LH

   a. Remove the front stabilizer bar bushing LH from the front stabilizer bar.
12. Remove the front stabilizer bar bushing RH.

➡Perform the same procedure as for the LH side.

To install:

13. To install, reverse the removal procedure. Tighten the front stabilizer link assembly nuts to 55 ft. lbs. (74 Nm). Tighten the front wheels to 76 ft. lbs. (103 Nm).

WHEEL HUB & BEARING

REMOVAL & INSTALLATION

See Figures 391 and 392.

1. Before servicing the vehicle, refer to the precautions.
2. Raise and safely support the vehicle.
3. Remove the front wheel.
4. Remove the front axle hub nut.
5. Disconnect the front ABS speed sensor connector.
6. Remove the 2 caliper mounting bolts and separate the caliper from the mounting and support from the vehicle with wire. Do not allow the caliper to hang low enough as to put tension on the brake hose.
7. Remove the front disc brake rotor.
8. Disconnect the tie rod from the steering knuckle.
9. Remove the bolt and 2 nuts and disconnect the front lower suspension control arm from the lower ball joint.
10. Separate the front drive shaft from the hub/bearing and steering knuckle assembly.
11. Separate the steering knuckle from the strut assembly.
12. Mount the steering knuckle and hub/bearing assembly securely in a soft vice. Using a screwdriver, remove the dust deflector from the steering knuckle.
13. Remove the 4 mounting bolts and hub/bearing assembly along with the dust cover.

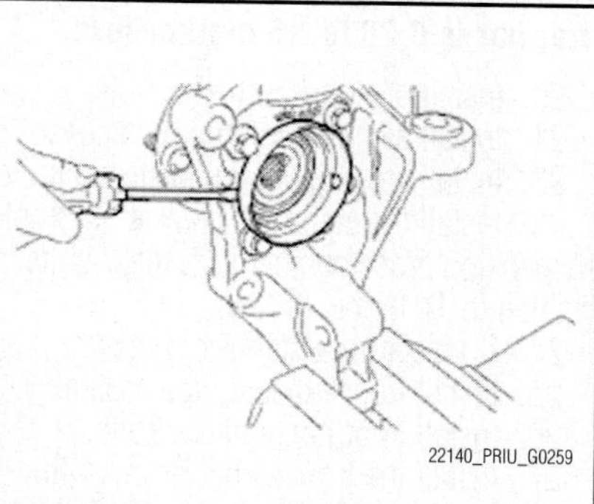

Fig. 391 Using a screwdriver, remove the deflector from the steering knuckle

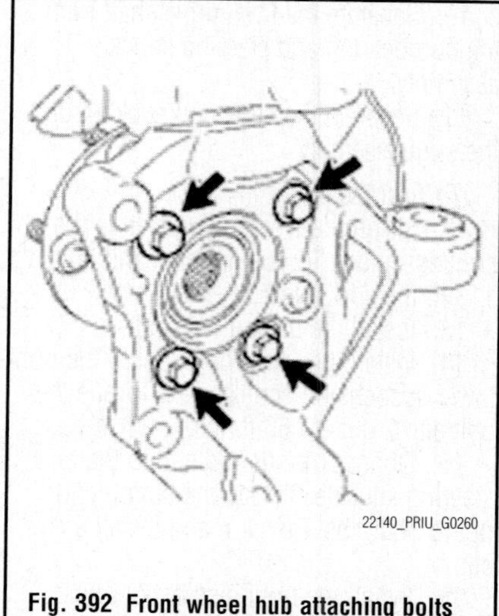

Fig. 392 Front wheel hub attaching bolts

To install:

14. Install the hub/bearing assembly and dust cover with the 4 mounting bolts.
15. Tighten the hub/bearing assembly mounting bolts to 41 ft. lbs. (56 Nm).
16. Using SST 09950-70010, 09608-320-10 and 09950-60020, press in a new dust deflector.
17. Install the steering knuckle to the strut assembly. Torque the nuts and bolts to 113 ft. lbs. (153 Nm)
18. Install the front drive shaft.
19. Install the lower ball joint to the front lower suspension control arm. Torque the bolt and 2 nuts to 66 ft. lbs. (89 Nm).
20. Connect the tie rod end to the steering knuckle. Torque the tie rod end nut to 36 ft. lbs. (49 Nm) and install a new clip.
21. Install the front disc brake rotor.
22. Install the front disc brake caliper.
23. Connect the front ABS speed sensor connector.
24. Install the front axle hub nut. Torque the nut to 159 ft. lbs. (216 Nm), then stake the hub nut, using a chisel and hammer.
25. Install the front wheel.
26. Inspect and adjust the front wheel alignment.
27. Check ABS speed sensor signal.

ADJUSTMENT

The front wheel bearings are a cartridge type design and cannot be adjusted.

COIL SPRING

REMOVAL & INSTALLATION

See Figures 393 through 397.

1. Remove the rear wheel.
2. Disconnect the rear speed sensor wire LH and RH.
3. Separate the rear speed sensor wire LH and RH.
   a. Remove the nut and separate the 2 clamps and rear speed sensor wire.
4. Separate the rear height control sensor sub assembly RH (w/ height control).
   a. Remove the bolt and separate the rear height control sensor sub-assembly RH from the rear axle beam assembly.
   b. Using a vinyl tape (*1), secure the rear height control sensor sub-assembly RH as shown in the illustration.
5. Separate the rear wheel house liner LH and RH.
   a. Remove the clip and turn back the rear wheel house liner LH to separate the rear wheel house liner.
6. Remove the rear coil spring LH.
   a. Loosen the 2 bolts.

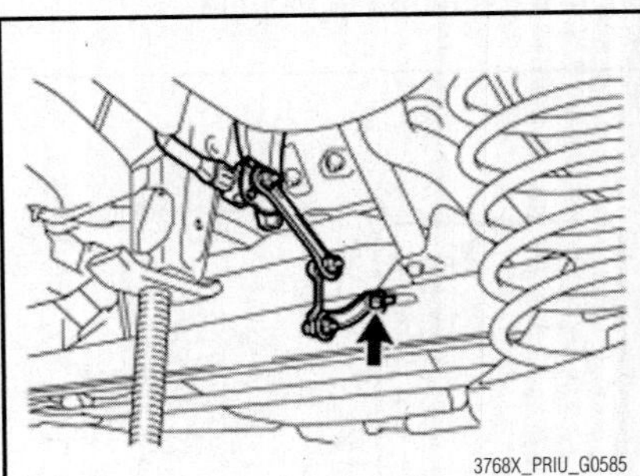

Fig. 393 Removing the bolt and separating the rear height control sensor sub-assembly RH from the rear axle beam assembly

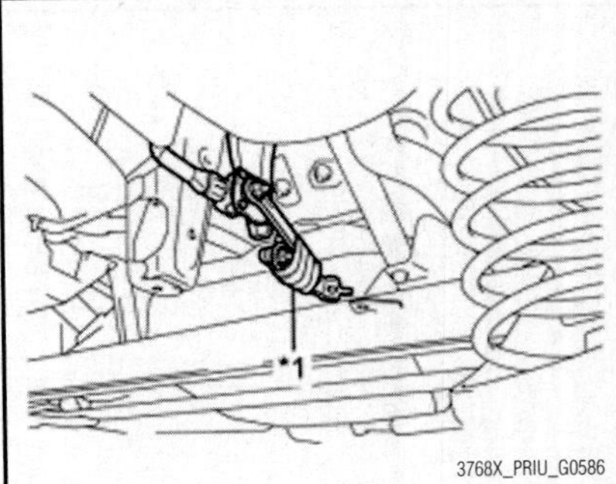

Fig. 394 Securing the height control sensor sub assembly

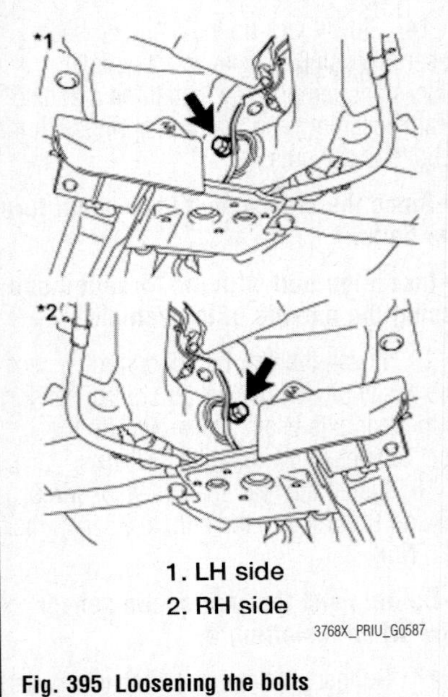

Fig. 395 Loosening the bolts

➡Do not remove the bolts.

   b. Support the spring seat of the rear axle beam assembly using 2 jacks and 2 wooden blocks.

CAUTION

Do not jack up the rear axle beam assembly too high as the vehicle may fall.

➡Support the rear shock absorber at a position where it compresses by approximately 0.787 to 1.18 inch (20 to 30 mm).

   c. Remove the 2 bolts while holding the 2 nuts and separate the rear axle beam assembly from the rear shock absorber assemblies LH and RH.

➡Since the stopper nuts are used, turn the bolts.

   d. Slowly lower the rear axle beam assembly using 2 jacks and 2 wooden blocks, and remove the rear coil spring LH.

➡When moving the rear axle beam assembly beyond full rebound, make sure that the rear axle beam assembly is not out of position for more than 10 minutes.

   e. Slowly jack up the rear axle beam assembly using 2 jacks and 2 wooden blocks, and temporarily tighten the rear axle beam assembly to the rear shock absorber assemblies LH and RH with the 2 bolts and 2 nuts.

➡Since the stopper nuts are used, turn the bolts.

7. Remove the rear coil spring RH.

➡Perform the same procedure as the LH side.

8. Remove the rear upper coil spring insulators LH and RH.
9. Remove the rear lower coil spring insulators LH and RH.

To install:

10. Install the rear upper coil spring insulators LH and RH.

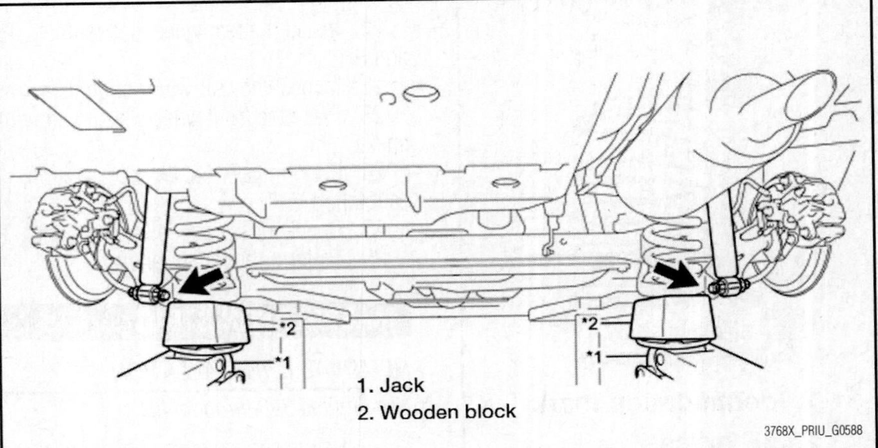

Fig. 396 Supporting the spring seat of the rear axle beam assembly using 2 jacks and 2 wooden blocks

a. Install the rear upper coil spring insulator to the rear coil spring.

➡**Install the rear upper coil spring insulator so that the dimension between the stopper and the upper end of the rear coil spring is 0.394 inch (10 mm) or less.**

11. Install the rear upper coil spring insulators LH and RH.
12. Install the rear lower coil spring insulators LH and RH.
13. Install the rear coil springs.
 a. Support the spring seat of the rear axle beam assembly using 2 jacks and 2 wooden blocks.
 b. Remove the 2 bolts while holding the 2 nuts and separate the rear axle beam assembly from the rear shock absorber assemblies LH and RH.

➡**Since the stopper nuts are used, turn the bolts.**

 c. Slowly lower the rear axle beam assembly using 2 jacks and 2 wooden blocks.
 d. Set the rear coil spring to the rear axle beam assembly.

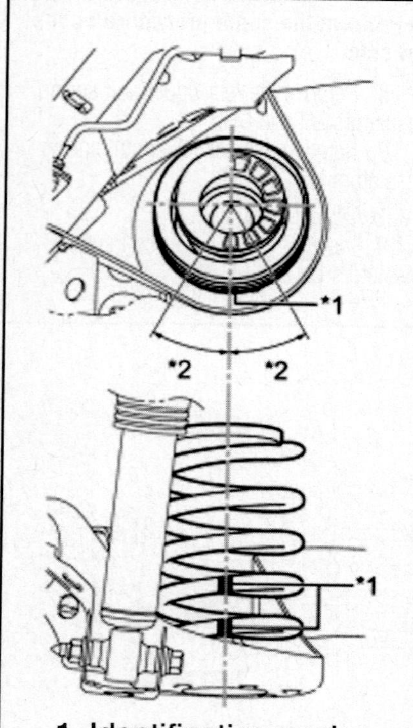

Fig. 397 Setting the rear coil spring

➡**Set the rear coil spring so that the identification marks are positioned as shown in the illustration.**

14. Slowly jack up the rear axle beam assembly using 2 jacks and 2 wooden blocks and temporarily install the rear axle beam assembly and rear coil spring with the 2 bolts and 2 nuts.

➡**Since the stopper nuts are used, turn the bolts.**

➡**Insert the bolt with the threaded end facing the outside of the vehicle.**

15. Install the rear height control sensor sub assembly RH (w/ height control sensor) to the rear axle beam assembly with the bolt. Tighten to 71 inch lbs. (8 Nm).
16. Install the rear speed sensor wires LH and RH. Tighten the nuts to 75 inch lbs. (8.5 Nm).

➡**Do not twist the rear speed sensor wire when installing it.**

17. Connect the rear speed sensor wires LH and RH.
18. Install the rear wheels. Tighten to 76 ft. lbs. (103 Nm).
19. Stabilize the suspension.
20. Fully tighten the rear axle beam assembly bolts to 100 ft. lbs. (135 Nm).

➡**The final torque must be applied under the standard vehicle height conditions.**

 a. Fully tighten the 2 bolts to 66 ft. lbs. (90 Nm).

➡**Since the stopper nut are used, turn the bolts.**

➡**The final torque must be applied under the standard vehicle height conditions.**

21. Install the rear wheel house liners LH and RH.
22. Inspect the rear wheel alignment.
23. Place the front wheels facing straight ahead.
24. Perform yaw rate and acceleration calibration.
25. Check for the speed sensor signal.
26. Perform the initialization.

MACPHERSON STRUT

REMOVAL & INSTALLATION

See Figures 398 through 401.

1. Before servicing the vehicle, refer to the precautions.
2. Remove the rear No. 2 floor board.
3. Remove the rear deck floor box.
4. Remove the rear deck trim cover.
5. Remove the tonneau cover assembly.
6. Remove the left rear seatback assembly.
7. Remove the rear No. 1 floor board.
8. Remove the left rear side seatback frame.
9. Remove the rear No. 4 floor board.
10. Remove the left deck floor box.
11. Remove the left deck trim side panel assembly
12. Remove the battery carrier bracket.
13. Remove the rear wheel (s).
14. Support the rear axle beam with a jack. Insert a wooden block between the jack and the rear axle beam to prevent damage.
15. Remove the 2 nuts from the rear strut assembly (upper side).
16. Remove the rear strut assembly (upper side) bolt from the under-side of the vehicle.
17. Remove the nut and spacer from the rear strut assembly (lower side).
18. Remove the rear strut assembly while slowly lowering the jack.

➡**Seat the jack so that no extra load is placed on the strut assembly on the opposite side of the vehicle.**

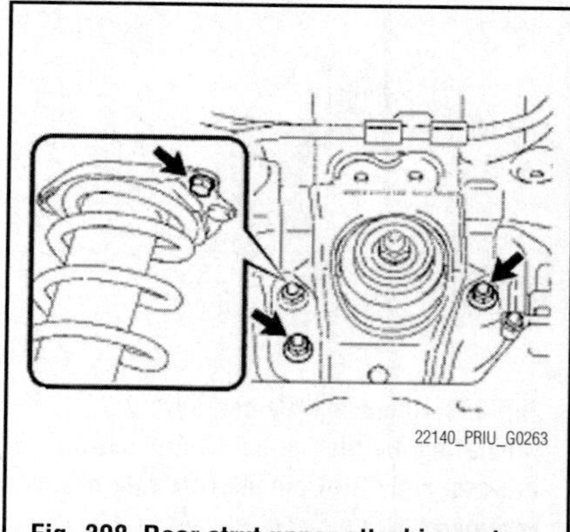

Fig. 398 Rear strut upper attaching nuts

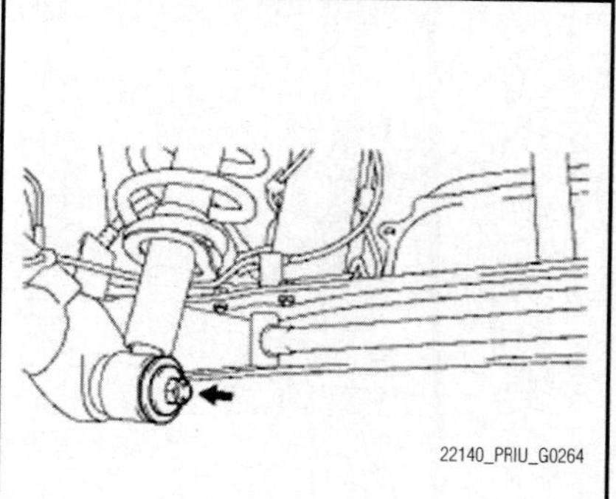

Fig. 399 Rear strut lower attaching nut

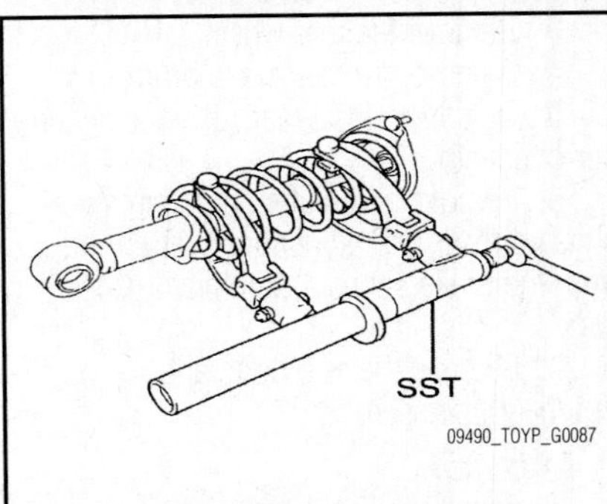

Fig. 400 Compressing the coil spring using a coil spring compressor.

19. Separate the coil spring from the strut assembly.

a. Use a 6mm socket hexagon wrench to secure the piston rod of the shock absorber and loosen the nut.

➡Do not remove the nut. Sufficiently insert the hexagon wrench.

b. Attach SST 09727-30021 to the coil spring so that the upper and lower hooks of the installed area are as wide as possible.

c. Compress the coil spring until it moves freely.

➡Do not use an impact wrench. It will damage SST.

d. Remove the nut.

e. Remove the No. 1 cushion washer, No. 1 cushion, rear spring front bracket, rear suspension support, rear coil spring insulator upper and rear No. 1 spring bumper. The shock absorber can be replaced without removing the No. 1 cushion and rear suspension support from the rear spring front bracket.

f. Release SST and remove it from the coil spring after removing the coil spring from the strut.

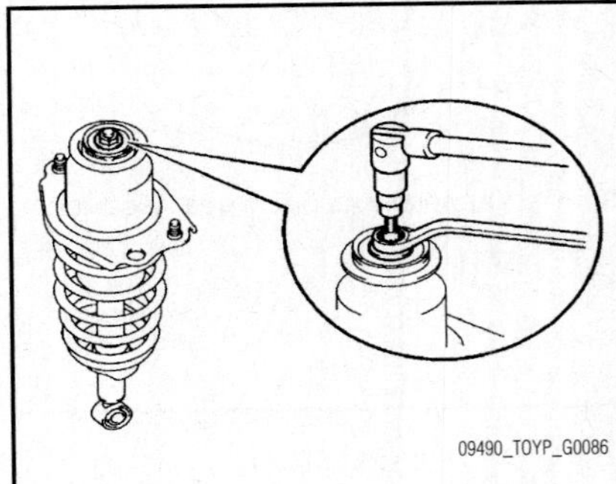

Fig. 401 Securing the piston rod of the shock absorber and loosening the nut.

To install:

20. Install the coil spring by assembling the strut assembly.

a. Using SST 09727-30021, compress the coil spring.

➡Do not use an impact wrench. It will damage SST.

b. Fit the coil spring end into the recessed part of the strut assembly lower seat.

c. Fit the rear coil spring insulator upper to the rear spring front bracket.

d. Install the rear No. 1 spring bumper, rear suspension support, rear spring front bracket, No. 1 cushion and No. 1 cushion washer.

➡Install the rear spring front bracket so that it is aligned with the strut lower bush, as shown in the illustration. Install the No. 1 cushion washer with the protruding portion facing down.

e. Use a 6mm socket hexagon wrench to fix the strut piston rod and tighten the nut to 41 ft. lbs. (56 Nm).

➡Sufficiently insert the hexagon wrench.

f. Release SST and remove it from the coil spring.

➡Do not use an impact wrench. It will damage SST. Remove SST while confirming the direction of the rear spring front bracket.

21. Install the rear strut assembly to the rear axle beam. Place the spacer and temporarily tighten the nut.

22. Install the rear strut assembly (upper side) to the vehicle by slowly raising the rear axle beam on a jack. Insert a wooden block between the jack and the rear axle beam to prevent damage.

23. Do not raise the rear axle beam more than necessary. Securely insert the rear spring front bracket stud bolt into the vehicle.

24. Tighten the bolt and 2 nuts of the rear strut assembly (upper side) to 59 ft. lbs. (80 Nm).

25. Install the rear wheel.

26. After lowering the vehicle, bounce the vehicle up and down to stabilize the rear suspension.

27. Fully tighten the rear strut assembly (lower side) installation nut to 59 ft. lbs. (80 Nm). Ensure the vehicle is lowered to the ground.

28. Install the battery carrier bracket.

29. Install the left deck trim side panel assembly.

30. Install the left deck floor box.

31. Install the rear No. 4 floor board.

32. Install the left rear side seatback frame.

33. Install the rear No. 1 floor board.

34. Install the left rear seatback assembly.

35. Install the tonneau cover assembly.

36. Install the rear deck trim cover.

37. Install the rear deck floor box

38. Install the rear No. 2 floor board.

39. Inspect the rear wheel alignment.

SHOCK ABSORBER

REMOVAL & INSTALLATION

See Figure 402.

➡Use the same procedure for the RH side and LH side.

➡The procedure listed below is for the LH side.

1. Remove the tonneau cover assembly (w/ tonneau cover).
2. Remove the rear No. 2 floor board (separate type).
3. Remove the rear deck floor box.
4. Remove the rear No. 4 floor board.
5. Remove the floor box LH.
6. Remove the rear No. 3 floor board (RH).
7. Remove the No. 1 floor board sub assembly.
8. Remove the No. 2 floor board sub assembly.
9. Remove the rear No. 1 floor board.
10. Remove the deck trim service hole cover.
11. Remove the rear deck trim cover.
12. Remove the rear door scuff plates LH and RH.
13. Remove the rear seat cushion assembly.
14. Remove the rear side seatback assembly LH.
15. Remove the luggage hold belt striker assembly LH.
16. Remove the tonneau cover holder cap LH.
17. Remove the deck trim side panel assembly LH.
18. Remove the rear side seatback assembly RH.
19. Remove the luggage hold belt striker assembly RH.
20. Remove the tonneau cover holder cap RH.
21. Remove the deck trim side panel assembly RH.
22. Remove the rear wheel.

23. Separate the rear height control sensor sub assembly RH (w/ height control sensor).

24. Remove the rear No. 1 shock absorber cushion washer.

a. Support the spring seat of the rear axle beam assembly using a jack and wooden block.

CAUTION

Do not jack up the rear axle beam assembly too high as the vehicle may fall.

➡Keep supporting the rear axle beam assembly with a jack until the installation of the rear shock absorber assembly has been completed.

➡Support the rear shock absorber at a position where it compresses by approximately 0.787 to 1.18 inch (20 to 30 mm).

b. Using a hexagon socket wrench, secure the rear shock absorber rod and remove the lock nut.

➡Securely insert the hexagon socket wrench to the rear shock absorber rod to prevent damage to the rear shock absorber assembly when removing the nut.

25. Remove the rear No. 1 shock absorber cushion washer.

26. Remove the rear suspension support.

27. Remove the rear shock absorber assembly.

a. Remove the bolt while holding the nut and remove the rear shock absorber assembly.

➡Since the stopper nut is used, turn the bolt.

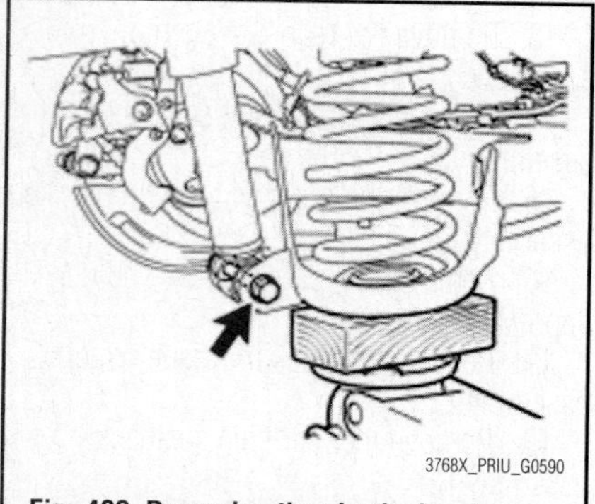

Fig. 402 Removing the shock absorber assembly

28. Remove the rear No. 1 spring bumper from the rear shock absorber assembly.

To install:

29. To install, reverse the removal procedure. Tighten the shock absorber lock nut to 18 ft. lbs. (25 Nm). Tighten the shock absorber assembly bolt to 66 ft. lbs. (90 Nm).

TESTING

Compress and extend the rear shock absorber rod, and check that there is no abnormal resistance or unusual sound during operation. If there is any abnormality, replace the shock absorber with a new one.

WHEEL HUB & BEARING

REMOVAL & INSTALLATION

See Figure 403.

1. Before servicing the vehicle, refer to the precautions.
2. Raise and safely support the vehicle.
3. Remove the rear wheel.
4. Remove the rear brake drum.
5. Disconnect the rear ABS speed sensor connector.
6. Remove automatic adjusting lever, shoe adjuster and adjusting lever spring from the brake shoes assembly, if necessary.
7. Remove the hub bolts and hub/bearing assembly.

To install:

8. Install the hub/bearing assembly.
9. Tighten the hub assembly bolts to 45 ft. lbs. (61 Nm).
10. Install adjusting lever spring, shoe adjuster and automatic adjusting lever from the brake shoes assembly, if removed earlier.
11. Connect the rear ABS speed sensor connector.
12. Install the rear brake drum.
13. Install the rear wheel.
14. Lower the vehicle.
15. Inspect the rear wheel alignment.
16. Check to ensure that the brakes are free from dragging and that proper braking is obtained.

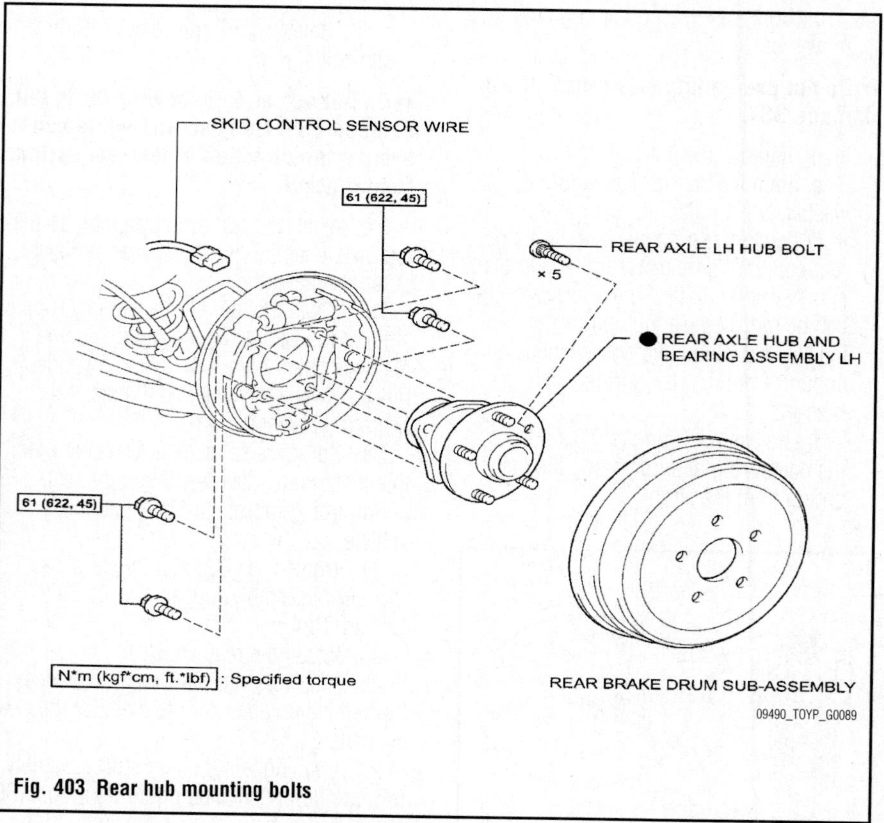

Fig. 403 Rear hub mounting bolts

TOYOTA

RAV4

14

SPECIFICATIONS AND MAINTENANCE CHARTS

ENGINE AND VEHICLE IDENTIFICATION

Engine							Model Year	
Code/VIN ①	Liters (cc)	Cu. In.	Cyl.	Fuel Sys.	Engine Type	Eng. Mfg.	Code ②	Year
2AR-FE	2.4 (2362)	144	4	SFI	DOHC	Toyota	9	2009
2GR-FE	3.5 (3498)	213	6	SFI	DOHC	Toyota	A	2010

SFI: Sequential Fuel Injection

DOHC: Double Overhead Camshaft

① 5th digit of the vehicle identification number (VIN)

② 10th digit of the Vehicle Identification Number (VIN)

3768X_RAV4_C0001

GENERAL ENGINE SPECIFICATIONS

Year	Model	Engine Displacement Liters	Engine Series Code/VIN	Net Horsepower @ rpm	Net Torque @ rpm (ft. lbs.)	Bore x Stroke (in.)	Com-pression Ratio	Oil Pressure @ rpm
2009	RAV4	2.4	2AR-FE	166@6000	165@4000	3.48x3.78	NA	①
	RAV4	3.5	2GR-FE	269@6200	246@4700	NA	NA	②
2010	RAV4	2.4	2AR-FE	166@6000	165@4000	3.48x3.78	NA	①
	RAV4	3.5	2GR-FE	269@6200	246@4700	NA	NA	②

NA: Not Available

① 4.3 psi or more at idle

② 11.6 psi or more at idle

3768X_RAV4_C0002

ENGINE TUNE-UP SPECIFICATIONS

Year	Engine Displacement Liters	Engine Code/VIN	Spark Plug Gap (in.)	Ignition Timing (deg.)	Fuel Pump (psi)	Idle Speed (rpm) MT	Idle Speed (rpm) AT	Valve Clearance Intake	Valve Clearance Exhaust
2009	2.4	2AR-FE	0.043	①	44-50	—	600-700	0.0075-0.0114	0.0150-0.0189
	3.5	2GR-FE	0.043	②	44-50	—	600-700	NA	NA
2010	2.4	2AR-FE	0.043	①	44-50	—	600-700	0.0075-0.0114	0.0150-0.0189
	3.5	2GR-FE	0.043	②	44-50	—	600-700	NA	NA

NOTE: The Vehicle Emission Control Information label often reflects specification changes made during production. The label figures must be used if they differ from those in this chart.

NA: Not Available

① 5-15 degrees BTDC when using intelligent tester tool.

8-12 degrees BTDC at idle. Connect SST when not using intelligent tester tool.

5-15 degrees BTDC at idle. Disconnect SST when not using intelligent tester tool.

② 8-12 degrees BTDC when using intelligent tester tool.

8-12 degrees BTDC at idle. Connect SST when not using intelligent tester tool.

5-15 degrees BTDC at idle. Disconnect SST when not using intelligent tester tool.

3768X_RAV4_C0003

CAPACITIES

Year	Model	Engine Displacement Liters	Engine Code/VIN	Engine Oil with Filter (qts.)	Transmission (pts.) 5-Spd	Transmission (pts.) Auto.*	Transfer Case (pts.)	Drive Axle Front (pts.)	Drive Axle Rear (pts.)	Fuel Tank (gal.)	Cooling System (qts.)
2009	RAV4	2.4	2AR-FE	4.5	—	7.4	1.0	—	1.0	15.9	7.2
	RAV4	3.5	2GR-FE	6.4	—	NA	1.0	—	1.0	15.9	①
2010	RAV4	2.4	2AR-FE	4.5	—	7.4	1.0	—	1.0	15.9	7.2
	RAV4	3.5	2GR-FE	6.4	—	NA	1.0	—	1.0	15.9	①

*After draining, add the following amounts, then fill to the cold full line.

NA: Not Available

① STD: 9.4
TWG: 9.8

3768X_RAV4_C0004

FLUID SPECIFICATIONS

Year	Model	Engine Displacement Liters	Engine Oil	Auto. Trans.	Drive Axle	Power Steering Fluid	Brake Master Cylinder
2009	RAV4	2.4	①	ATF World Standard	②	NA	DOT 3
	RAV4	3.5	5W-30	ATF World Standard	③	NA	DOT 3
2010	RAV4	2.4	①	ATF World Standard	②	NA	DOT 3
	RAV4	3.5	5W-30	ATF World Standard	③	NA	DOT 3

DOT: Department Of Transportation

NA: Not Available

Note: If specification disagrees with specification in owners manual, use specification in owners manaual

① 0W-20 above 40 degrees F. 5W-20 below 40 degrees F

② API GL-5 SAE 90 above 0 degrees F. API GL-5 SAE 80W-90 below 0 degrees

③ API GL-5 SAE 85W-90 above 0 degrees F. API GL-5 SAE 80W-90 below 0 degrees

3768X_RAV4_C0014

VALVE SPECIFICATIONS

Year	Engine Displacement Liters	Engine Code/VIN	Seat Angle (deg.)	Face Angle (deg.)	Spring Test Pressure (lbs. @ in.)	Spring Installed Height (in.)	Stem-to-Guide Clearance (in.) Intake	Stem-to-Guide Clearance (in.) Exhaust	Stem Diameter (in.) Intake	Stem Diameter (in.) Exhaust
2009	2.4	2AR-FE	45	44.5	NA	1.8670	0.0010-0.0024	0.0012-0.0026	0.2154-0.2159	0.2152-0.2158
	3.5	2GR-FE	45	44.5	NA	1.7898	0.0010-0.0024	0.0012-0.0026	0.2154-0.2159	0.2151-0.2157
2010	2.4	2AR-FE	45	44.5	NA	1.8670	0.0010-0.0024	0.0012-0.0026	0.2154-0.2159	0.2152-0.2158
	3.5	2GR-FE	45	44.5	NA	1.7898	0.0010-0.0024	0.0012-0.0026	0.2154-0.2159	0.2151-0.2157

NA: Not Available

3768X_RAV4_C0005

CAMSHAFT SPECIFICATIONS

All measurements in inches unless noted

Year	Engine Displacement Liters	Engine Code/VIN	Journal Dia.	Brg. Oil Clearance	Shaft End-play ①	Circle Runout	Lobe Height	
							Intake	Exhaust
2009	2.4	2AR-FE	②	NA	NA	NA	1.8624-1.8664	1.8104-1.8143
	3.5	2GR-FE	③	④	0.0031-0.0051	0.0016	1.7447-1.7487	1.7426-1.7465
2010	2.4	2AR-FE	②	NA	NA	NA	1.8624-1.8664	1.8104-1.8143
	3.5	2GR-FE	③	④	0.0031-0.0051	0.0016	1.7447-1.7487	1.7426-1.7465

① Thrust clearance

② Mark 1, 2 and 3: 1.4162-1.4167

③ No. 1: 1.4152-1.4157

All others: 1.0220-1.0226

④ No. 1: 0.0016-0.0031

All others: 0.0010-0.0024

3768X_RAV4_C0006

CRANKSHAFT AND CONNECTING ROD SPECIFICATIONS

All measurements are given in inches.

Year	Engine Displ. Liters	Engine Code/VIN	Crankshaft				Connecting Rod		
			Main Brg. Journal Dia.	Main Brg. Oil Clearance	Shaft End-play	Thrust on No.	Journal Diameter	Oil Clearance	Side Clearance
2009	2.4	2AR-FE	2.1649-2.1654	0.0006-0.0015	0.0016-0.0095	3	1.8894-1.8898	0.0012-0.0025	0.0063-0.0202
	3.5	2GR-FE	2.4011-2.4016	0.0010-0.0019	0.0016-0.0095	2	2.0863-2.0866	0.0018-0.0026	0.0059-0.0157
2010	2.4	2AR-FE	2.1649-2.1654	0.0006-0.0015	0.0016-0.0095	3	1.8894-1.8898	0.0012-0.0025	0.0063-0.02
	3.5	2GR-FE	2.4011-2.4016	0.0010-0.0019	0.0016-0.0095	2	2.0863-2.0866	0.0018-0.0026	0.0059-0.0157

3768X_RAV4_C0007

PISTON AND RING SPECIFICATIONS

All measurements are given in inches.

Year	Engine Displ. Liters	Engine Code/VIN	Piston Clearance	Ring Gap Top Comp.	Ring Gap Bottom Comp.	Ring Gap Oil Control	Ring Side Clearance Top Comp.	Ring Side Clearance Bottom Comp.	Ring Side Clearance Oil Control
2009	2.4	2AR-FE	0.0020-0.0029	0.0094-0.0122	0.0130-0.0169	0.0040-0.0119	0.0008-0.0028	0.0008-0.0024	0.0008-0.0028
	3.5	2GR-FE	0.0018-0.0020	0.0098-0.0138	0.0197-0.0413	0.0039-0.0157	0.0008-0.0028	0.0008-0.0024	0.0028-0.0059
2010	2.4	2AR-FE	0.0020-0.0029	0.0094-0.0122	0.0130-0.0169	0.0040-0.0119	0.0008-0.0028	0.0008-0.0024	0.0008-0.0028
	3.5	2GR-FE	0.0018-0.0020	0.0098-0.0138	0.0197-0.0413	0.0039-0.0157	0.0008-0.0028	0.0008-0.0024	0.0028-0.0059

3768X_RAV4_C0008

TORQUE SPECIFICATIONS

All readings in ft. lbs.

Year	Engine Displacement Liters	Engine Code/VIN	Cylinder Head Bolts	Main Bearing Bolts	Rod Bearing Bolts	Crankshaft Damper Bolts	Flywheel Bolts	Manifold Intake	Manifold Exhaust	Spark Plugs	Oil Pan Drain Plug
2009	2.4	2AR-FE	①	②	③	133	72	22	27	18	30
	3.5	2GR-FE	④	⑤	⑥	184	132	15	15	13	30
2010	2.4	2AR-FE	①	②	③	133	72	22	27	18	30
	3.5	2GR-FE	④	⑤	⑥	184	132	15	15	13	30

① Step 1: 52 ft. lbs.
Step 2: plus 90 degrees

② Step 1: 30 ft. lbs.
Step 2: plus 90 degrees

③ Step 1: 30 ft. lbs.
Step 2: plus 90 degrees

④ Step 1: 27 ft. lbs.
Step 2: plus 90 degrees
Step 3: plus 90 degrees
Bolt should be 22 ft. lbs. on Bank 2

⑤ Step 1: 45 ft. lbs.
Step 2: plus 90 degrees
Main bearing cap bolt: 38 ft. lbs.

⑥ Step 1: 25 ft. lbs.
Step 2: plus 90 degrees

3768X_RAV4_C0009

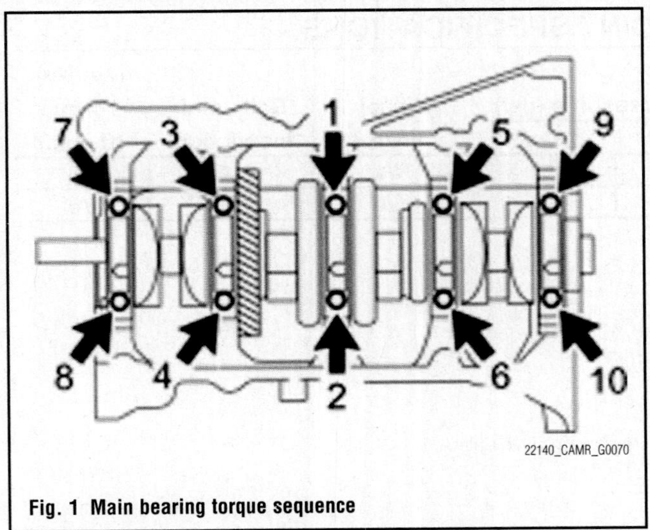

Fig. 1 Main bearing torque sequence

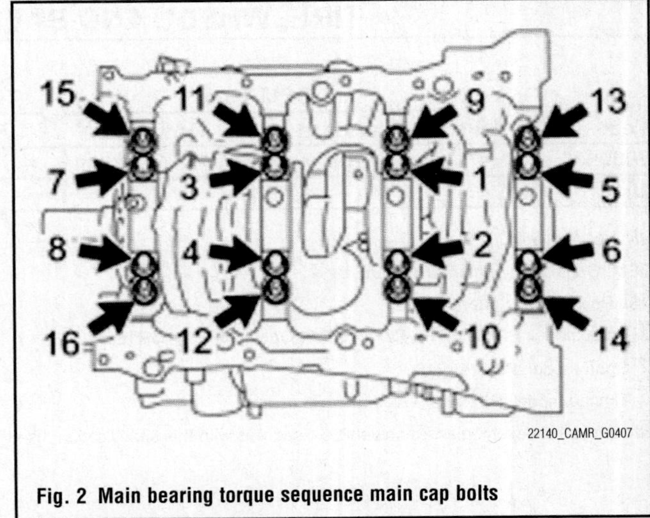

Fig. 2 Main bearing torque sequence main cap bolts

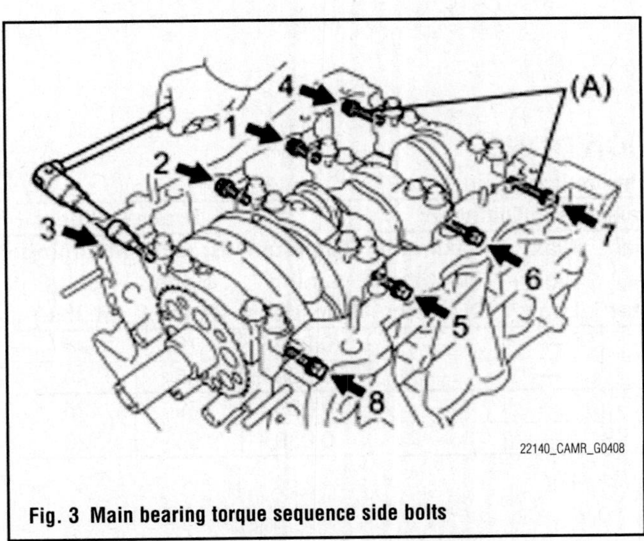

Fig. 3 Main bearing torque sequence side bolts

WHEEL ALIGNMENT

Year	Model		Caster Range (+/-Deg.)	Caster Preferred Setting (Deg.)	Camber Range (+/-Deg.)	Camber Preferred Setting (Deg.)	Toe-in (in.)	Steering Axis Inclination (Deg.)
2009	RAV4	Front	0.75	+5.72	0.75	-0.13	0+/-0.08	11.27+/-0.50
		Rear	NA	NA	0.75	-0.92	0.04+/-0.09	11.27+/-0.50
2010	RAV4	Front	0.75	+5.72	0.75	-0.13	0+/-0.08	11.27+/-0.50
		Rear	NA	NA	0.75	-0.92	0.04+/-0.09	11.27+/-0.50

3768X_RAV4_C0010

TIRE, WHEEL AND BALL JOINT SPECIFICATIONS

		OEM Tires		Tire Pressures (psi)		Wheel	Ball Joint	Lug Nut Torque
Year	Model	Standard	Optional	Front	Rear	Size	Inspection	(ft. lbs.)
2009	RAV4	①	①	②	②	NA	NA	76
2010	RAV4	①	①	②	②	NA	NA	76

NA: Not Available

OEM: Original Equipment Manufacturer

PSI: Pounds Per Square Inch

① Base model: P215/70R16, P225/65R17. Optional P235/55R18

Sport model: P235/55R18

Limited model: P225/65R17

② 32 PSI. However if placard on vehicle disagrees with this specification, use the specification on vehicle placard.

3768X_RAV4_C0011

BRAKE SPECIFICATIONS

All measurements in inches unless noted

			Brake Disc			Brake Drum Diameter				Brake Caliper	
			Original	Minimum	Maximum	Original Inside	Max. Wear	Maximum Machine	Minimum Lining	Bracket Bolts	Mounting Bolts
Year	Model		Thickness	Thickness	Runout	Diameter	Limit	Diameter	Thickness	(ft. lbs.)	(ft. lbs.)
2009	RAV4	F	①	②	0.0020	—	—	—	0.039	79	25
		R	0.472	0.413	0.0059	—	—	—	0.039	65	20
2010	RAV4	F	①	②	0.0020	—	—	—	0.039	79	25
		R	0.472	0.413	0.0059	—	—	—	0.039	65	20

F: Front

R: Rear

① 275 disc (15" disc): 0.984

296 disc (16" disc) : 1.102

② 275 disc (15" disc): 0.886

296 disc (16" disc): 0.984

3768X_RAV4_C0012

SCHEDULED MAINTENANCE INTERVALS

TOYOTA—RAV4

TO BE SERVICED	TYPE OF SERVICE	VEHICLE MILEAGE INTERVAL (x1000)																		
		5	10	15	20	25	30	35	40	45	50	55	60	65	70	75	80	85	90	95
Automatic transmission and differential fluid	S/I			✓			✓			✓			✓			✓			✓	
Ball joints and boots	S/I			✓			✓			✓			✓			✓			✓	
Brake system	S/I	✓	✓	✓	✓	✓	✓	✓	✓	✓	✓	✓	✓	✓	✓	✓	✓	✓	✓	✓
Charcoal canister	S/I												✓							
Drive belts	S/I						✓						✓						✓	
Driveshaft bushing	L						✓						✓						✓	
Engine coolant	R			✓			✓			✓			✓			✓			✓	
Engine oil & filter	R	✓	✓	✓	✓	✓	✓	✓	✓	✓	✓	✓	✓	✓	✓	✓	✓	✓	✓	✓
Exhaust pipes & mounts	S/I			✓			✓			✓			✓			✓			✓	
Fuel tank cap gasket	S/I						✓						✓						✓	
Halfshaft boots & flange bolts	S/I			✓			✓			✓			✓			✓			✓	
Limited slip differential fluid	R						✓						✓						✓	
Manual transmission and differential fluid	S/I						✓						✓						✓	
Platinum spark plugs	R												✓							
Propeller shaft bolts	S/I			✓			✓			✓			✓			✓			✓	
Rack and pinion assembly	S/I			✓			✓			✓			✓			✓			✓	
Tires (rotate)	S/I	✓	✓	✓	✓	✓	✓	✓	✓	✓	✓	✓	✓	✓	✓	✓	✓	✓	✓	✓
Transfer case and differential fluid	S/I			✓			✓			✓			✓			✓			✓	
Valves	S/I												✓							

R: Replace S/I: Service or Inspect L: Lubricate

FREQUENT OPERATION MAINTENANCE (SEVERE SERVICE)

If a vehicle is operated under any of the following conditions it is considered severe service:

- Towing a trailer or using a camper or car-top carrier.

- Repeated short trips of less than 5 miles in temperatures below freezing.

- Excessive idling or low-speed driving for long distances as in heavy commercial use, such as delivery, taxi or police cars.

- Operating on rough, muddy or salt-covered roads.

- Operating on unpaved or dusty roads.

Oil filter: service or inspect every 5000 miles or 4 months, whichever occurs first.

Brake linings and discs or drums: service or inspect every 5000 miles or 4 months, whichever occurs first.

Steering linkage: service or inspect every 5000 miles or 4 months, whichever occurs first.

Ball joints and boots: service or inspect every 5000 miles or 4 months, whichever occurs first.

Brake discs & pads (front): service or inspect every 6000 miles.

Halfshaft boots: service or inspect every 5000 miles or 4 months. Retighten the flange bolts, whichever occurs first.

Body chassis bolts and nuts: service or inspect every 5000 miles or 4 months, whichever occurs first.

Transmission and differential fluid: replace every 15,000 miles or 12 months, whichever occurs first.

Transfer case and differential fluid: replace every 15,000 miles or 12 months, whichever occurs first.

Timing belt: replace every 60,000 miles or 48 months, whichever occurs first.

3768X_RAV4_C0013

BRAKES — INFORMATION AND PRECAUTIONS

ANTI-LOCK SYSTEMS

- Certain components within the ABS system are not intended to be serviced or repaired individually.
- Do not use rubber hoses or other parts not specifically specified for and ABS system. When using repair kits, replace all parts included in the kit. Partial or incorrect repair may lead to functional problems and require the replacement of components.
- Lubricate rubber parts with clean, fresh brake fluid to ease assembly. Do not use shop air to clean parts; damage to rubber components may result.
- Use only DOT 3 brake fluid from an unopened container.
- If any hydraulic component or line is removed or replaced, it may be necessary to bleed the entire system.
- A clean repair area is essential. Always clean the reservoir and cap thoroughly before removing the cap. The slightest amount of dirt in the fluid may plug an orifice and impair the system function. Perform repairs after components have been thoroughly cleaned; use only denatured alcohol to clean components. Do not allow ABS components to come into contact with any substance containing mineral oil; this includes used shop rags.
- The Anti-Lock control unit is a microprocessor similar to other computer units in the vehicle. Ensure that the ignition switch is **OFF** before removing or installing controller harnesses. Avoid static electricity discharge at or near the controller.
- If any arc welding is to be done on the vehicle, the control unit should be unplugged before welding operations begin.

DISC AND DRUM SYSTEMS

CAUTION

Dust and dirt accumulating on brake parts during normal use may contain asbestos fibers from production or aftermarket brake linings. Breathing excessive concentrations of asbestos fibers can cause serious bodily harm. Exercise care when servicing brake parts. Do not sand or grind brake lining unless equipment used is designed to contain the dust residue. Do not clean brake parts with compressed air or by dry brushing. Cleaning should be done by dampening the brake components with a fine mist of water, then wiping the brake components clean with a dampened cloth. Dispose of cloth and all residue containing asbestos fibers in an impermeable container with the appropriate label. Follow practices prescribed by the Occupational Safety and Health Administration (OSHA) and the Environmental Protection Agency (EPA) for the handling, processing, and disposing of dust or debris that may contain asbestos fibers.

BRAKES — BLEEDING THE BRAKE SYSTEM

BLEEDING PROCEDURE

See Figure 4.

➡If any work is performed on the brake system or if air in the brake lines is suspected, bleed air from the brake system.

➡Wash off brake fluid immediately if it comes in contact with any painted surface.

1. Remove the cowl top ventilator louver bracket LH.
   a. Detach the 6 claws and remove the bracket.
2. Fill the reservoir with brake fluid.
   a. Set a brake fluid can upside down on the reservoir.

➡Make sure there is sufficient brake fluid in the can.

➡After adding brake fluid, make sure the reservoir is sufficient full.

3. Bleed the air from the brake master cylinder.

➡If the master cylinder has been disassembled or if the reservoir becomes empty, bleed air from the master cylinder.

   a. Using a union nut wrench (10 mm), disconnect the 2 brake lines from the master cylinder.
   b. Slowly depress and hole the brake pedal.
   c. Block the outer holes with your fingers, and release the pedal.
   d. Repeat the 2 previous steps 3 or 4 times.
   e. Using a union nut wrench (10 mm), connect the 2 brake lines to the master cylinder. Tighten to 11 ft. lbs. (15 Nm).

➡Use a torque wrench with a fulcrum length of 11.8 inch.

➡This torque value is effective when the union nut wrench is parallel to the torque wrench.

4. Bleed the air from the brake line.
   a. Remove the bleeder plug cap.
   b. Connect a vinyl tube to the bleeder plug.
   c. Depress the brake pedal several times, and then loosen the bleeder plug with the pedal depressed.
   d. When fluid stops coming out, immediately tighten the bleeder plug. Then release the pedal.
   e. Repeat the 2 previous steps until all the air in the brake fluid is gone.
   f. Tighten the bleeder plug. Tighten to 73 inch lbs. (8.3 Nm).
   g. Remove the vinyl tube and install the cap.

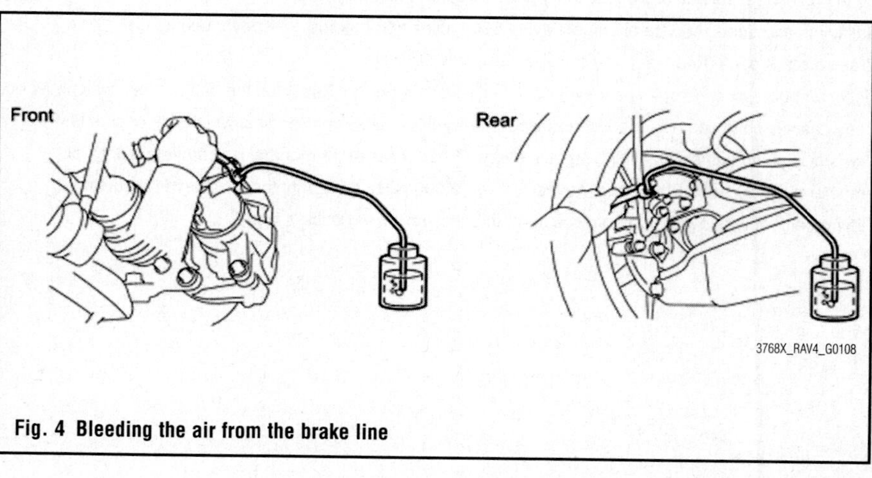

Fig. 4 Bleeding the air from the brake line

h. Bleed air from the brake line for each wheel by repeating the above procedures.

5. Bleed the air from the ABS and traction actuator assembly.

➡**After bleeding the air from the brake system, if the height or feel of the brake pedal is not correct, perform air bleeding of the brake actuator with the Techstream by performing the following procedure.**

a. Depress the brake pedal more than 20 times with the engine off.

b. Connect the Techstream to the DLC3, and turn the ignition switch to ON.

➡**DO NOT start the engine.**

c. Select AIR BLEEDING on the Techstream.

➡**Refer to the Techstream operator's manual for further details.**

Bleed the air from the suction line.

➡**Perform the bleeding for the right front wheel and right rear wheel.**

➡**Bleed the air by following the steps displayed on the Techstream.**

a. Connect a vinyl tube to the bleeder plug for the right front wheel.

b. Loosen the bleeder plug.

c. Operate the brake actuator to bleed the air using the Techstream.

➡**Do not depress the brake pedal at this time.**

➡**This operation stops automatically after 4 seconds.**

d. Check if the operation has stopped by referring to the Techstream display.

e. Tighten the bleeder plug.

f. Repeat the 4 previous steps until all air in the fluid is completely bled out.

g. Tighten the bleeder plug. Tighten to 73 inch lbs. (8.3 Nm).

h. Repeat all of the above procedures for the right rear wheel to bleed the air out of the suction line.

Bleed the air out of the pressure reduction line.

➡**Perform the bleeding for all 4 wheels.**

➡**Bleed the air by following the steps displayed on the Techstream.**

a. Connect a vinyl tube to one of the bleeder plugs.

b. Loosen the bleeder plug.

c. Using the Techstream, operate the brake actuator assembly, completely depress the brake pedal and hold it there.

➡**During this procedure, the pedal will feel heavy, but completely depress it so that the brake fluid comes out of the bleeder plug.**

➡**Depress and hold the brake pedal. Do not depress and release the pedal repeatedly.**

➡**The operation stops automatically after 4 seconds. When performing this procedure consecutively, wait at least 20 seconds between each operation.**

d. Tighten the bleeder plug, then release the brake pedal.

e. Repeat the 3 previous steps until all air in the fluid is completely bled out.

f. Tighten the bleeder plug. Tighten to 73 inch lbs. (8.3 Nm).

g. Repeat all of the above procedures for the other wheels to bleed the air out of the pressure reduction line.

6. Check the brake fluid level in the reservoir.

7. Install the cowl top ventilator louver bracket LH.

a. Attach the 6 claws to install the bracket.

WHEEL SPEED SENSORS

REMOVAL & INSTALLATION

➡**Use the same procedure for the RH and LH sides.**

➡**The procedures listed below are for the LH side.**

Front

See Figures 5 through 9.

1. Disconnect the cable from the negative battery terminal.

CAUTION

Wait at least 90 seconds after disconnecting the cable from the negative (-) battery terminal to disable the SRS system.

2. Remove the front wheel.
3. Remove the front fender liner LH.
4. Remove the front speed sensor LH.

a. Disconnect the sensor connector.

b. Remove the sensor clip (labeled A), bolt (labeled B) and sensor clamp (labeled C).

c. Remove the sensor clip (labeled D), bolt (labeled E) and sensor clamp (labeled F).

d. Remove the bolt and sensor body from the knuckle.

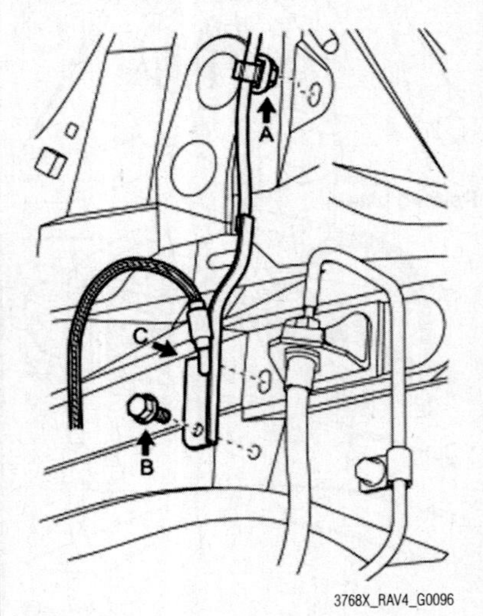

Fig. 5 Removing the sensor clip, bolt and sensor clamp (1 of 2)

➡**Keep the sensor tip and sensor installation hole free from foreign matter.**

To install:

5. Install the speed sensor front LH.

➡**To prevent interference with other parts, do not twist the painted line areas of the sensor wire when installing it.**

a. Set the sensor body into the knuckle, and then install the sensor with the bolt. Tighten to 75 inch lbs. (8.5 Nm).

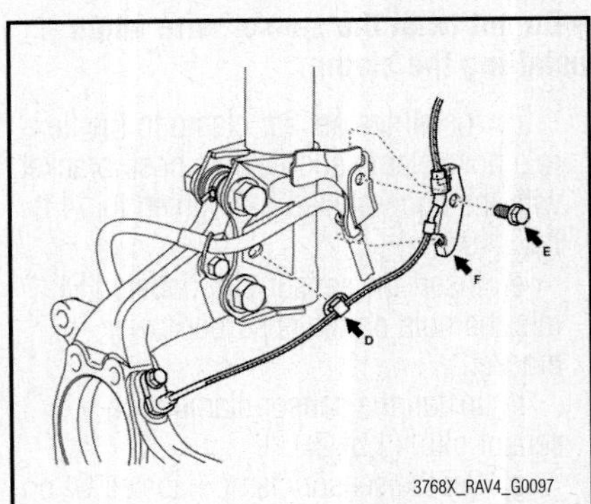

Fig. 6 Removing the sensor clip, bolt and sensor clamp (1 of 2)

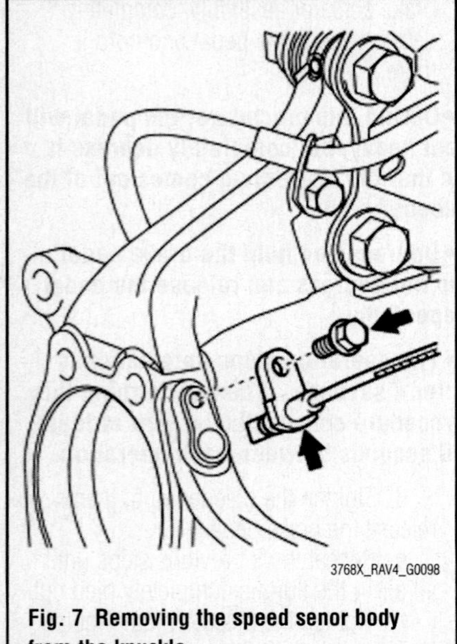
Fig. 7 Removing the speed senor body from the knuckle

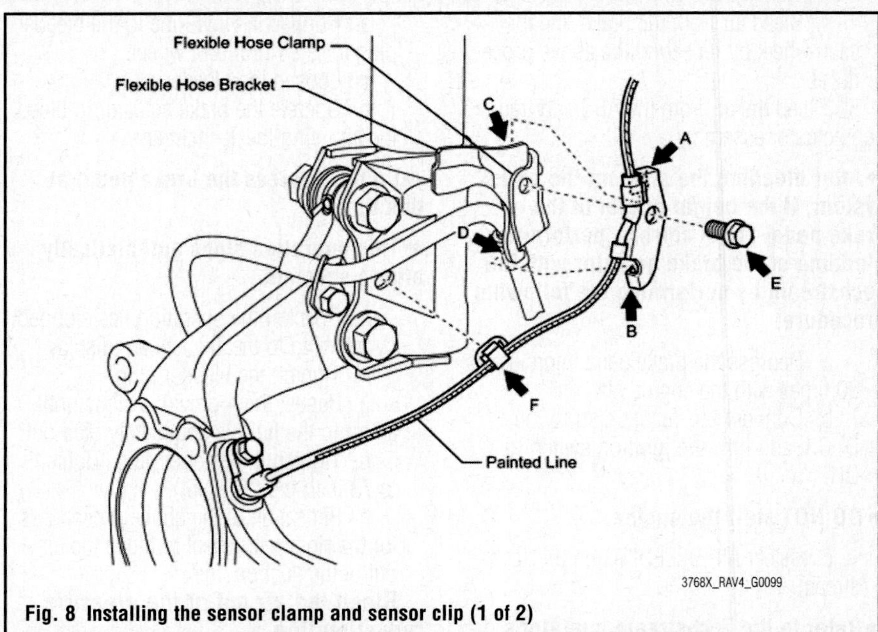

Fig. 8 Installing the sensor clamp and sensor clip (1 of 2)

➡**Keep the sensor tip and sensor installation hole free from foreign matter.**

➡**Firmly insert the sensor body into the knuckle before tightening the bolt.**

➡**After installing the sensor to the knuckle, make sure that there is no clearance between the sensor stay and knuckle. Also make sure that no foreign matter is stuck between the parts.**

➡**To prevent interference between the sensor and magnetic rotor, do not rotate the sensor body during or after the insertion of the sensor body to the knuckle.**

b. Install the sensor clamp and sensor clip as follows.

c. Simultaneously perform the following: 1) hang the hook part of the sensor clamp (labeled A) on the flexible hose bracket (labeled C); and 2) insert the hook part of the sensor clamp (labeled B) into the flexible hose bracket (labeled D).

➡**Do not twist the sensor wire when installing the clamp.**

d. Install the sensor clamp to the flexible hose clamp and flexible hose bracket with the bolt (labeled E). Tighten to 14 ft. lbs. (19 Nm).

e. Insert the sensor clip (labeled F) into the hole on the absorber lower bracket.

f. Install the sensor clamp and sensor clip. (1 of 2).

g. Set the sensor clamp (labeled G) on the side member, and then install the bolt (labeled H). Tighten to 75 inch lbs. (8.5 Nm).

➡**Do not twist the sensor wire when installing the clamp.**

h. Insert the sensor clip (labeled I) into the hole on the apron.

i. Connect the sensor connector.

6. Install the front fender liner LH.

➡**Install the fender liner so that the sensor wire harness passes beyond the fender liner installation clip towards the rear side of the vehicle.**

7. Install the front wheel. Tighten to 76 ft. lbs. (103 Nm).

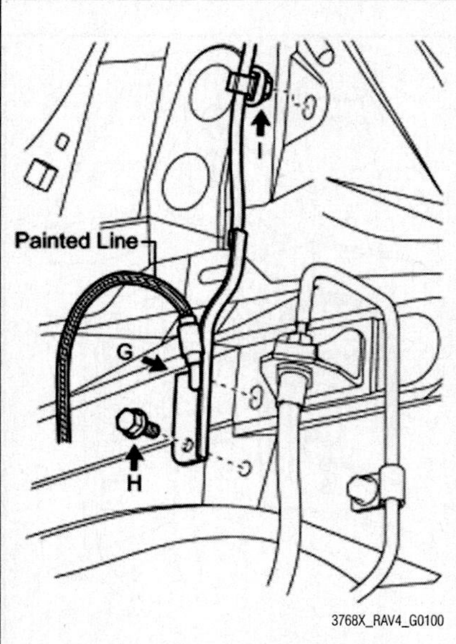

Fig. 9 Installing the sensor clamp and sensor clip (2 of 2)

8. Connect the cable to the negative battery terminal.

9. Check the ABS speed sensor signal.

Rear

See Figures 10 through 14.

1. Disconnect the cable from the negative battery terminal.

⁂ CAUTION

Wait at least 90 seconds after disconnecting the cable from the negative (-) battery terminal to disable the SRS system.

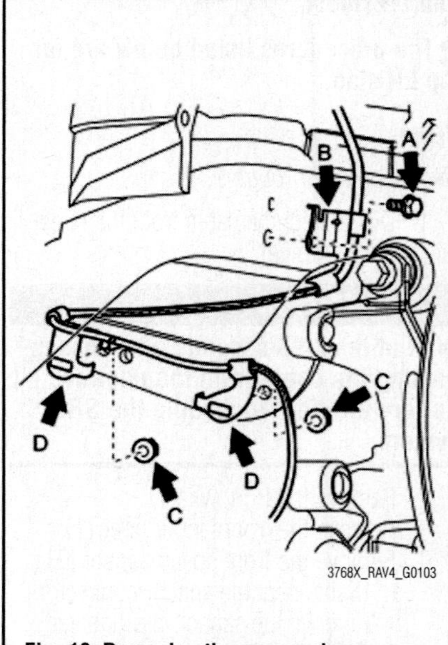

Fig. 10 Removing the sensor clamp (1 of 2)

2. Remove the rear wheel.

3. Remove the deck trim side panel assembly LH.

➡Refer to the procedures from the removal of the rear door scuff plate LH up until the removal of the deck trim side panel assembly LH.

4. Remove the rear speed sensor LH.

a. Disconnect the speed sensor connector.

b. Disconnect the grommet of the speed sensor wire from the hole of the wheel house.

c. Remove the bolt (labeled A) and sensor clamp (labeled B) from the side member.

d. Remove the 2 nuts (labeled C) and 2 sensor clamps (labeled D) from the upper arm.

e. Remove the nut (labeled E) and sensor clamp (labeled F) from the trailing arm.

f. Remove the bolt (labeled G) and sensor body (labeled H) from the carrier.

➡Keep the sensor tip and sensor installation hole free from foreign matter.

To install:

5. Install the rear speed sensor LH.

➡To prevent interference with other parts, do not twist the painted line areas of the sensor wire when installing it.

a. Install the sensor (labeled A) with the bolt (labeled B). Tighten to 75 inch lbs. (8.5 Nm).

3768X_RAV4_G0104

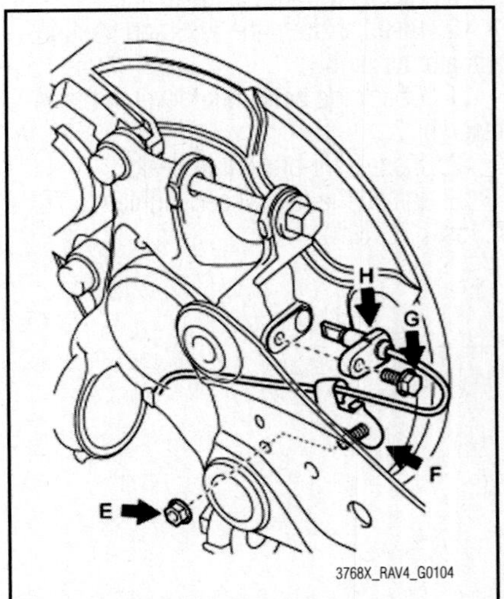

Fig. 11 Removing the sensor clamp (2 of 2)

➡Keep the sensor tip and sensor installation hole free from foreign matter.

➡To prevent interference with the bearing rotor, do not rotate the sensor body when inserting the sensor body or after inserting the sensor body.

b. Install the sensor clamp (labeled C) with the nut (labeled D). Tighten to 44 inch lbs. (5.0 Nm).

c. Install the 2 sensor clamps (labeled E) with the 2 nuts (labeled F).

➡Do not twist the sensor wire when installing the clamps.

d. Install the sensor clamp (labeled G) with the bolt (labeled H). Tighten to 75 inch lbs. (8.5 Nm).

➡Do not twist the sensor wire when installing the clamps.

e. Insert the connector and grommet into the inside of the vehicle through the passage hole in the wheel house.

➡Make sure the grommet band clamp remains on the outside of the vehicle.

f. Hold the grommet and pull it toward the outside of the vehicle. Then fix the grommet in place so that it is not tilted.

➡When pulling out the grommet, do not grip the sensor wire.

➡Fix the grommet in place within the range shown in the illustration.

g. Connect the speed sensor connector.

6. Install the deck trim side panel assembly LH.

➡Refer to the procedures from the installation of the deck trim side panel LH up until the installation of the rear door scuff plate LH.

7. Install the rear wheel. Tighten to 76 ft. lbs. (103 Nm).

8. Connect the cable to the negative battery terminal.

9. Check the speed sensor signal.

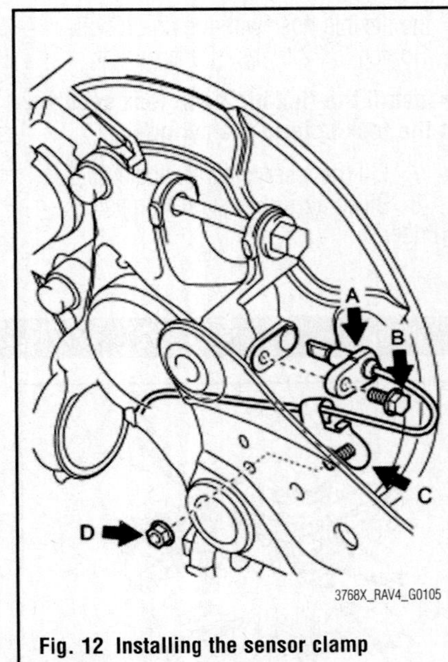

Fig. 12 Installing the sensor clamp (1 of 2)

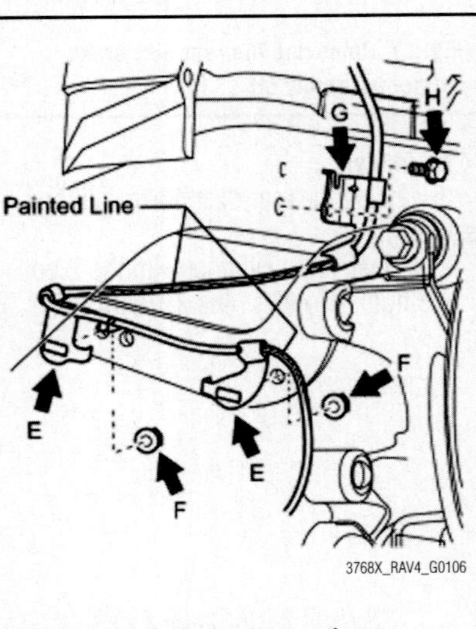

Fig. 13 Installing the sensor clamp (2 of 2)

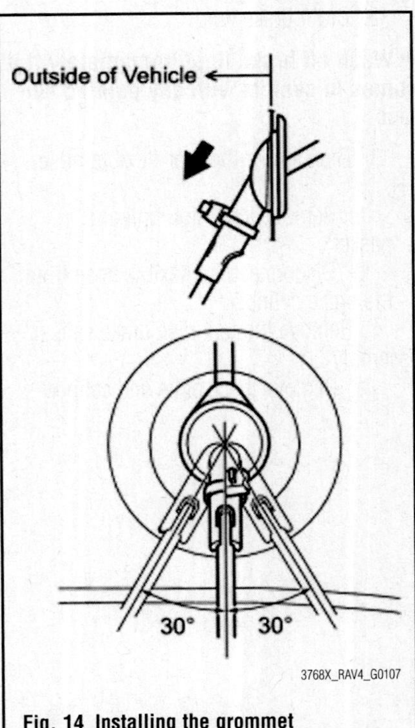

Fig. 14 Installing the grommet

BRAKES

FRONT DISC BRAKES

BRAKE CALIPER

REMOVAL & INSTALLATION

See Figures 15 and 16.

1. Remove the front wheel.
2. Drain the brake fluid.

➡Wash off brake fluid immediately if it comes in contact with any painted surface.

3. Disconnect the front flexible hose.
   a. Remove the union bolt and gasket.
   b. Disconnect the flexible hose from the brake caliper.
4. Remove the disc brake caliper assembly.
   a. Remove the 2 bolts and caliper.

To install:

5. Install the disc brake caliper assembly. Tighten the 2 bolts to 25 ft. lbs. (34 Nm).
6. Connect the front flexible hose.
   a. Install a new gasket and connect the flexible hose with the union bolt. Tighten to 22 ft. lbs. (29 Nm).

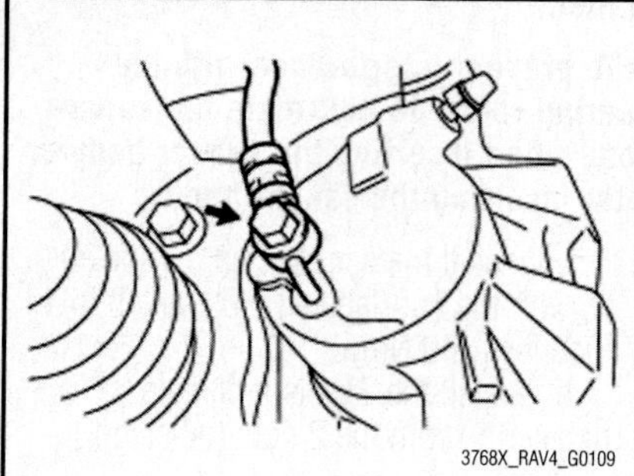
3768X_RAV4_G0109

Fig. 15 Disconnecting the flexible hose from the brake cylinder

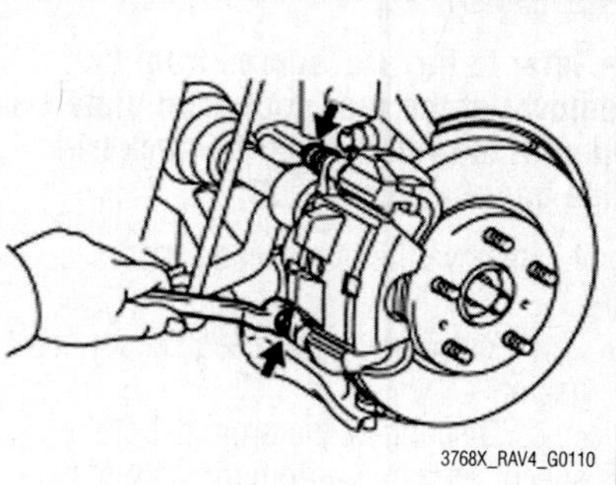
3768X_RAV4_G0110

Fig. 16 Removing the disc brake caliper assembly

➡Install the flexible hose lock securely in the lock hole in the cylinder.

7. Fill the reservoir with brake fluid.
8. Bleed air from the brake master cylinder.
9. Bleed air from the brake line.
10. Bleed air from the ABS and traction actuator assembly.
11. Check the brake fluid level in the reservoir.
12. Inspect for fluid leaks.
13. Install the front wheel. Tighten to 76 ft. lbs. (103 Nm).

BRAKES

REAR DISC BRAKES

BRAKE CALIPER

REMOVAL & INSTALLATION

See Figure 17.

1. Remove the rear wheel.
2. Drain brake fluid.

➡Wash off brake fluid immediately if it comes in contact with any painted surface.

3. Disconnect the rear flexible hose LH.
   a. Remove the union bolt and gasket.
   b. Disconnect the flexible hose from the brake cylinder.
4. Remove the rear disc brake caliper assembly LH.
   a. Remove the 2 bolts and caliper.

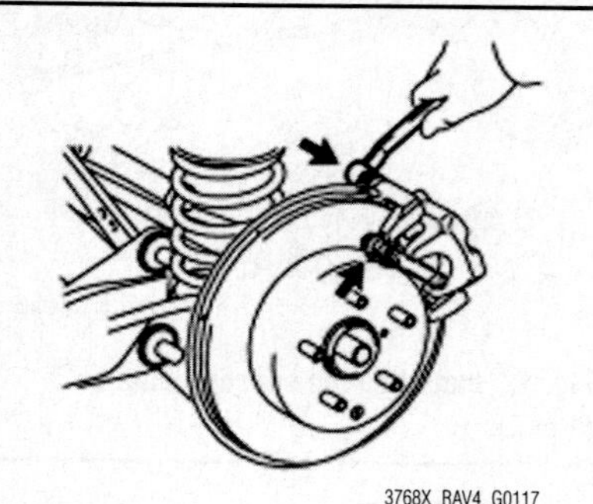
3768X_RAV4_G0117

Fig. 17 Removing the rear disc brake caliper assembly LH

To install:

5. Install the rear disc brake caliper assembly.
   a. Install the cylinder with the 2 bolts and tighten to 20 ft. lbs. (27 Nm).
6. Connect the rear flexible hose LH.
   a. Install a new gasket and connect the flexible hose with the union bolt. Tighten to 22 ft. lbs. (30 Nm).

➡Install the flexible hose lock securely in the lock hole in the cylinder.

7. Fill the reservoir with brake fluid.
8. Bleed air from the brake master cylinder.
9. Bleed air from the brake line.
10. Bleed air from the ABS and traction actuator assembly.
11. Check the brake fluid level in the reservoir.
12. Inspect for brake fluid leaks.
13. Install the rear wheel. Tighten to 76 ft. lbs. (103 Nm).

PARKING BRAKE SHOES

REMOVAL & INSTALLATION

See Figures 18 through 22.

➡Use the same procedures for the LH side and RH side.

➡The procedures listed below are for the LH side.

1. Remove the rear brake discs.
2. Remove the parking brake shoe return tension spring (upper side).

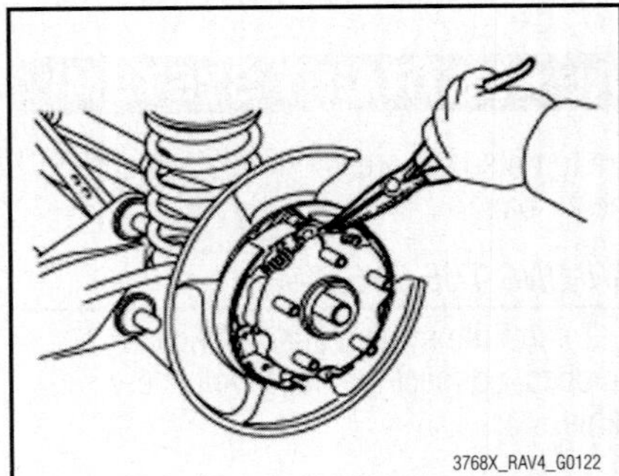

Fig. 18 Removing the parking brake shoe return tension spring (upper side)

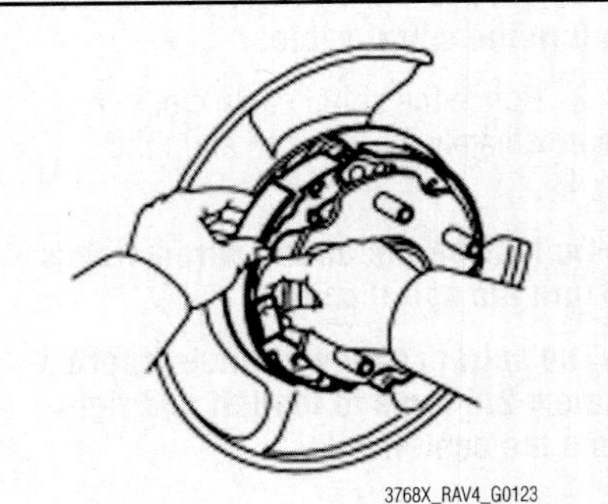

Fig. 19 Removing the No. 1 parking brake shoe assembly LH

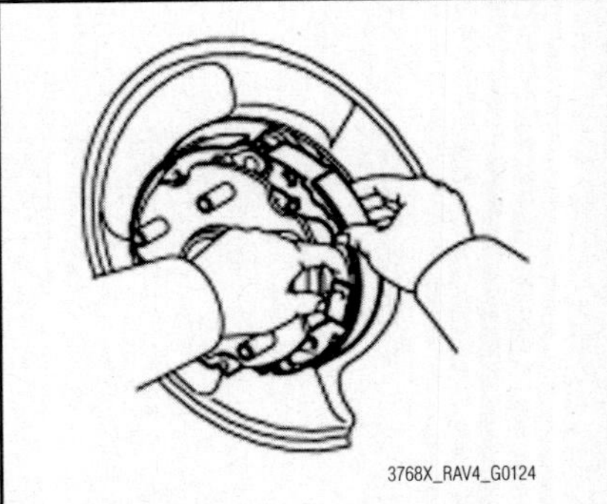

Fig. 20 Removing the No. 2 parking brake shoe assembly LH

High temperature grease

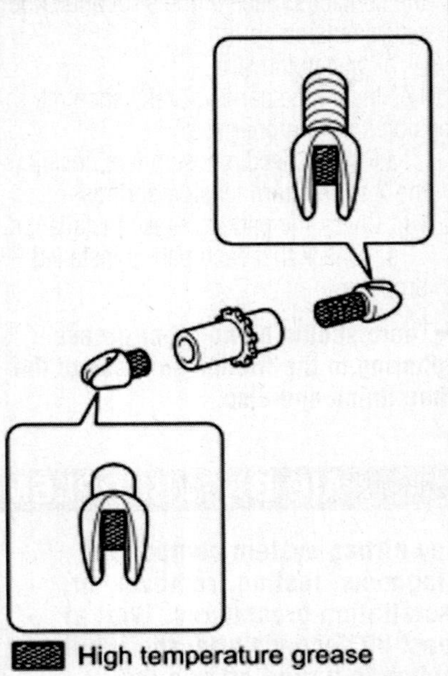

3768X_RAV4_G0125

Fig. 21 Applying high temperature grease to the show adjusting screw set

a. Using needle-nose pliers, remove the 2 shoe return tension springs.

3. Remove the parking brake shoe strut LH.
4. Remove the parking brake shoe return tension spring (lower side).
5. Using needle nose pliers, remove the tension spring.
6. Remove the parking brake shoe adjusting screw set.
7. Remove the No. 1 parking brake shoe assembly LH.
   a. Press the shoe hold down spring to remove the pin, shoe hold down spring and shoe.
8. Remove the No. 2 parking brake shoe assembly LH.
   a. Slide the shoe to remove the shoe hold down spring, pin and shoe.
   b. Disconnect the parking brake cable from the parking brake shoe lever.
9. Remove the parking brake shoe lever LH.

To install:

10. Apply high temperature grease to the areas of the backing plate that contact the shoes.
11. Install the parking brake shoe lever LH.
12. Install the No. 2 parking brake shoe assembly LH.
    a. Connect the parking brake cable to the parking brake shoe lever.
    b. Apply high temperature grease to the following contact surfaces of the shoe:
       - Shoe to anchor pin
       - Shoe to parking brake lever pin
       - Shoe to spring
    c. Install the shoe with the shoe hold down spring and pin.
13. Install the No. 1 parking brake shoe assembly.

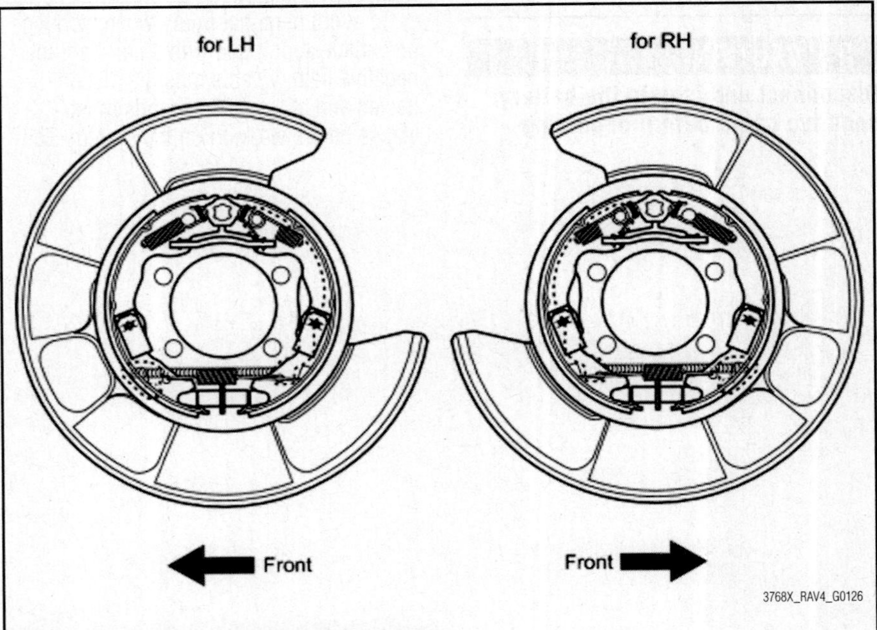

Fig. 22 Checking the parking brake installation

a. Apply high temperature grease to the following contact surfaces of the shoe:

- Shoe to anchor pin
- Shoe to spring

b. Install the shoe with the shoe hold-down spring and pin.

14. Install the parking brake shoe adjusting screw set.

a. Apply high temperature grease to the areas of the shoe adjusting screw set shown.

b. Install the screw set.

15. Install the parking brake shoe return tension spring (lower side).

a. Using needle nose pliers, install the tension spring.

16. Install the parking brake shoe strut.

a. Apply high temperature grease to the contact surface of the strut and shoe return tension spring.

b. Install the strut.

17. Install the parking brake shoe return tension spring (upper side).

a. Using needle nose pliers, install the 2 shoe return tension springs.

18. Check the parking brake installation.

a. Check that each part is installed properly.

➡There should be no oil or grease adhering to the friction surfaces of the shoe lining and disc.

19. Install the rear brake disc.

20. Adjust the parking brake shoe clearance.

ADJUSTMENT

1. Remove the shoe adjusting hole plug, and then turn the adjuster to expand the shoe adjuster until the disc locks.
2. Turn the shoe adjuster so that it contracts to a point where the disc can rotate smoothly. Standard is 8 notches.
3. Check that the shoe has no brake drag.

CHASSIS ELECTRICAL

AIR BAG (SUPPLEMENTAL RESTRAINT SYSTEM)

GENERAL INFORMATION

This vehicle is equipped with a Supplemental Restraint System (SRS). It consists of a driver airbag, front passenger airbag, driver side knee airbag, front seat side airbag and curtain shield airbag. Failure to carry out service operations in the correct sequence could cause the SRS to unexpectedly deploy during servicing, possibly leading to a serious accident. Further, if a mistake is made in servicing the SRS, it is possible that the SRS may fail to operate when required. Before performing servicing (including removal or installation of parts, inspection or replacement), be sure to read the following carefully, then follow the correct procedures indicated in the repair manual.

SERVICE PRECAUTIONS

⁂ CAUTION

Disconnect and isolate the battery negative cable before beginning any airbag system component diagnosis, testing, removal, or installation procedures. Wait at least 90 seconds after the ignition switch is turned off and the negative (-) terminal cable is disconnected from the battery before starting the operation. The SRS is equipped with a backup power source, so if work is started within 90 seconds after disconnecting the negative (-) terminal cable from the battery, the SRS may be deployed. Failure to disable the airbag system may result in accidental airbag deployment, personal injury, or death.

DISARMING THE SYSTEM

To avoid personal injury when working on vehicles equipped with an air bag, the negative battery cable must be disconnected and at least 90 seconds must elapse before working on the system. Failure to do so may result in deployment of the air bag.

ARMING THE SYSTEM

To arm the system after service is finished, connect the negative battery cable.

CLOCKSPRING CENTERING

1. Rotate the spiral cable counterclockwise by hand until it feels firm

➡Do not use the airbag wire harness to turn the spiral cable.

2. Rotate the spiral cable clockwise approximately 2.5 turns to align the marks.

➡Do not use the airbag wire harness to turn the spiral cable.

➡The spiral cable will rotate approximately 2.5 turns to the left and right from the center.

DRIVE TRAIN

TRANSFER CASE ASSEMBLY

REMOVAL & INSTALLATION

See Figures 23 and 24.

The transfer case is part of the engine/transaxle assembly and is removed with those units.

FRONT HALFSHAFT

REMOVAL & INSTALLATION

See Figures 25 and 26.

1. Before servicing the vehicle, refer to the Precautions Section.
2. Disconnect the negative battery cable.

➡Wait at least 90 seconds after disconnecting the negative battery cable before starting any repair work to prevent air bag and seat belt pretensioner activation.

3. Raise and support the vehicle safely.

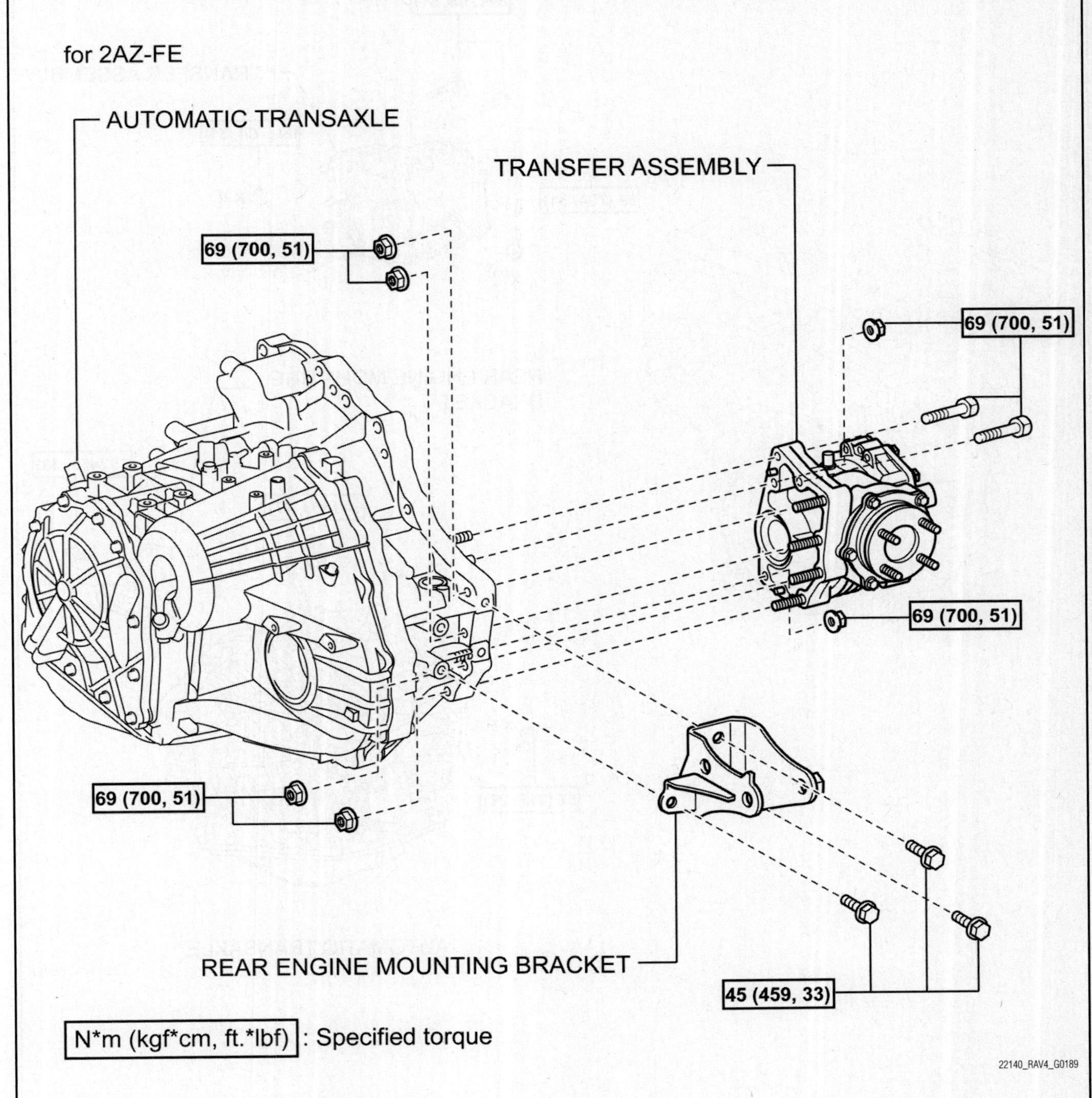

Fig. 23 Transfer case and related components—2AR-FE engine

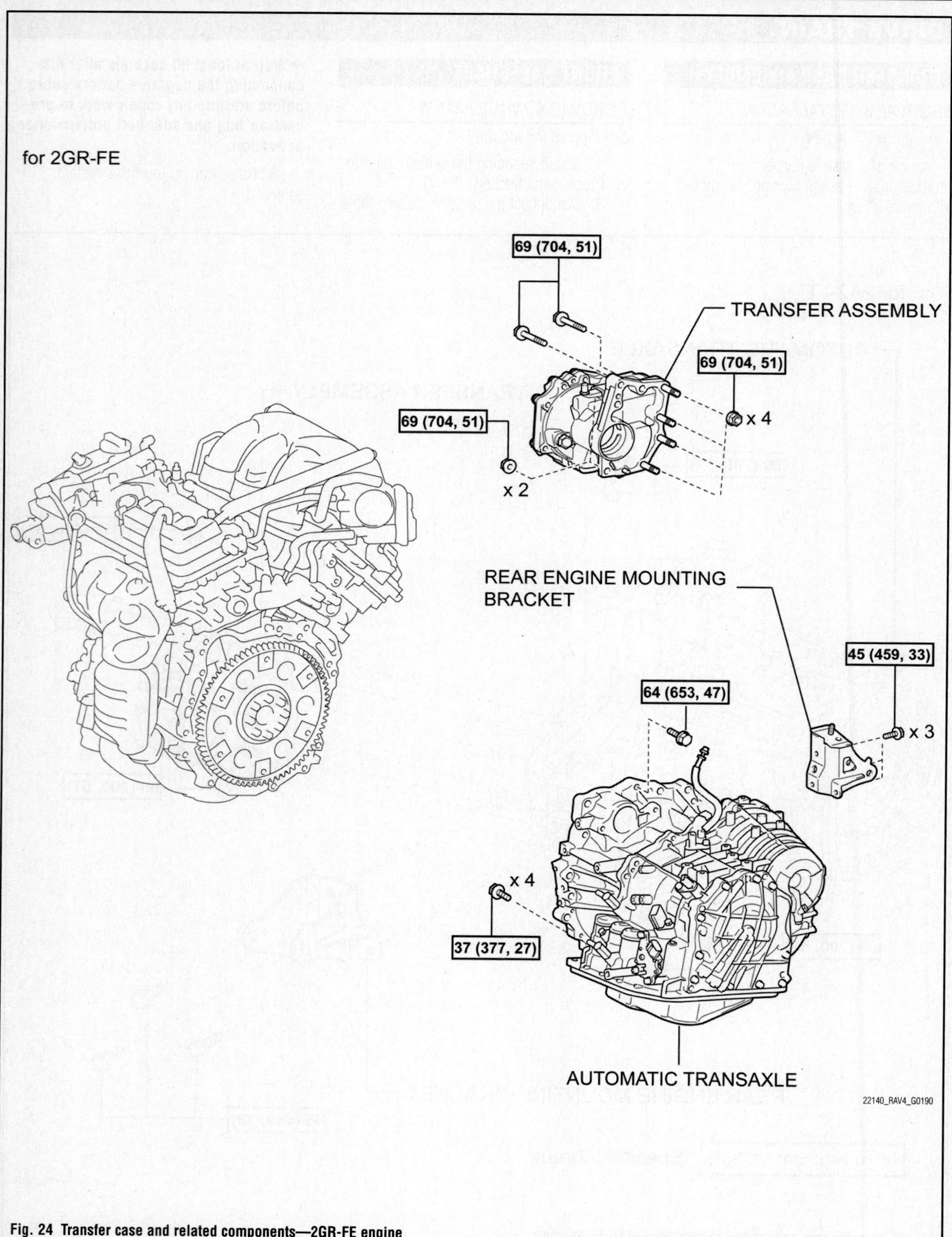

Fig. 24 Transfer case and related components—2GR-FE engine

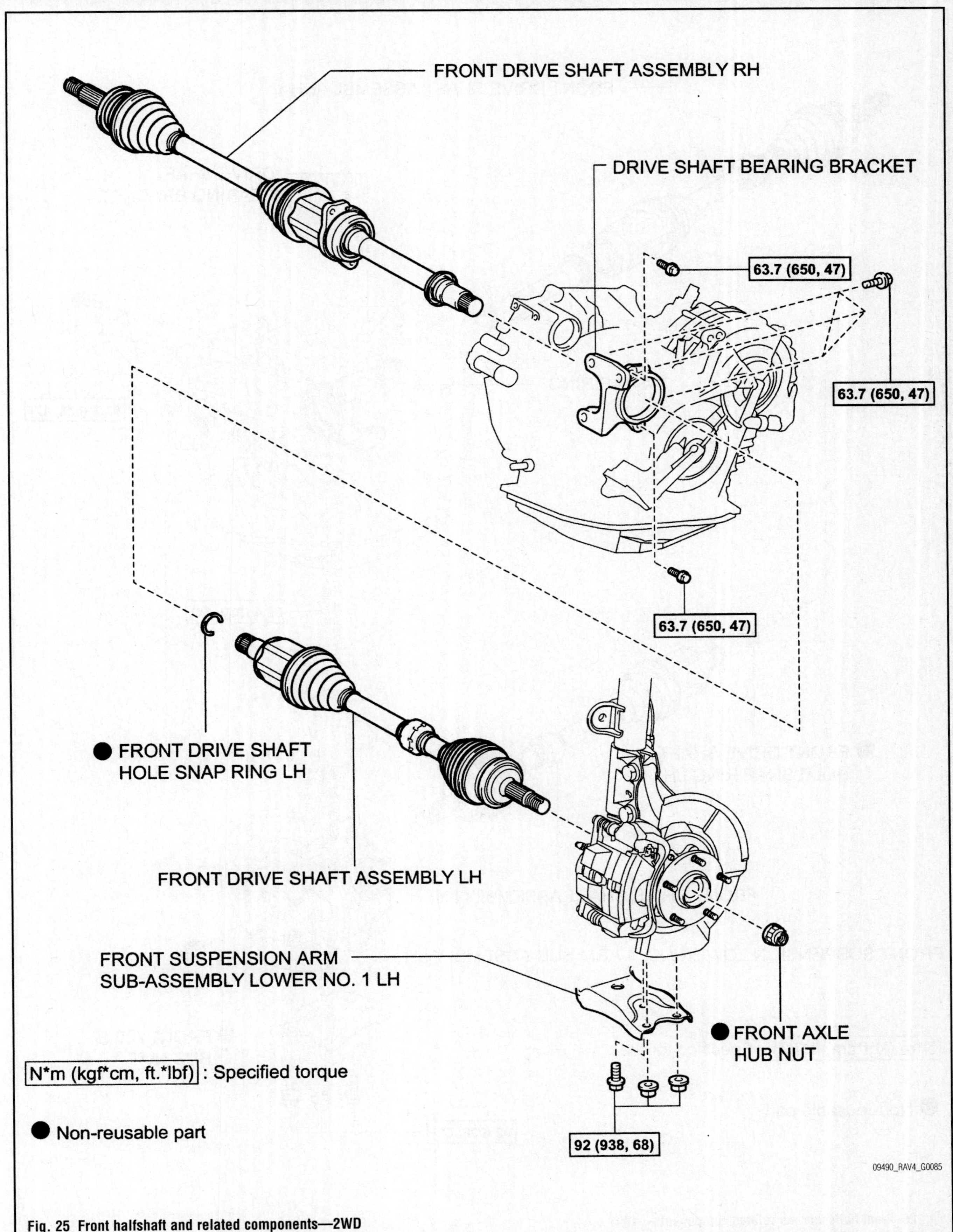

Fig. 25 Front halfshaft and related components—2WD

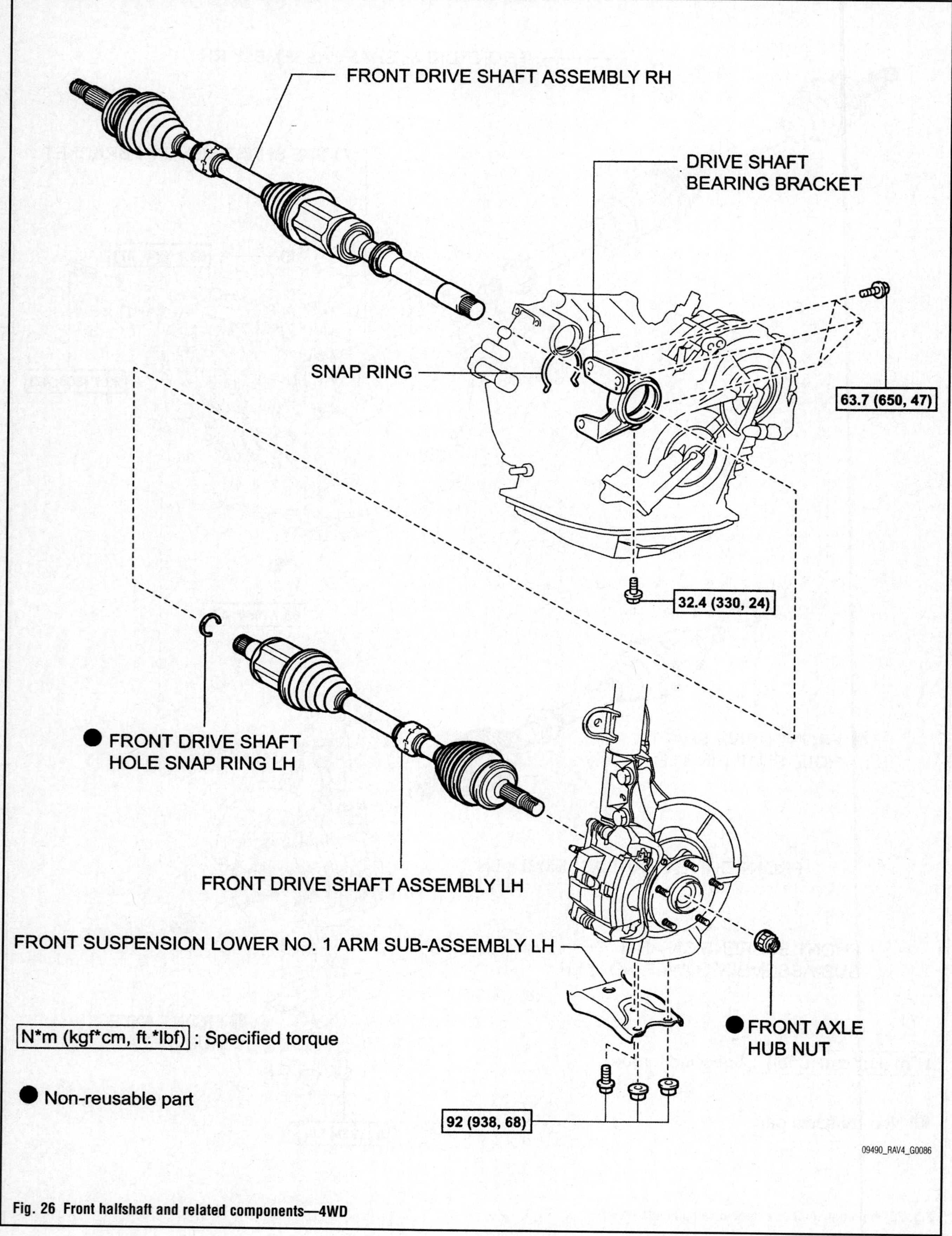

Fig. 26 Front halfshaft and related components—4WD

4. Remove the tire and wheel assembly. Remove the front axle hub nut.
5. Drain the transaxle fluid.
6. Disconnect the left and right speed sensors.
7. Remove the left and right brake calipers.
8. Disconnect the left and right front stabilizer link assemblies.
9. Disconnect the left and right front lower number one arm subassemblies.
10. Matchmark the halfshaft and the axle hub, left and right side.

➡Do not punch the marks.

11. Using a plastic hammer, disconnect the steering knuckle with the axle hub, left and right side.

➡Be careful not to damage the boot and speed sensor rotor. Do not push out excessively the halfshaft from the axle assembly.

12. Disconnect the left and right tie rod subassemblies.
13. On the left side, using tool SST09520-01010, or equivalent remove the front halfshaft.
14. On 2WD, to remove the right halfshaft remove the two bolts and pull out the halfshaft together with the halfshaft bearing case. Remove the halfshaft from the transaxle.

➡Be careful not to damage the boot and speed sensor rotor. Do not push out excessively the halfshaft from the axle assembly.

15. On 4WD, to remove the right halfshaft use a brass bar and hammer to remove the halfshaft.

➡Do not damage the oil seal, boot or allow the halfshaft to fall out.

16. Support the front axle assembly.

➡The hub bearing could be damaged if it is subjected to the vehicle weight. If it is necessary to place weight on the hub bearing, such as moving it when the halfshaft is removed, support it using too SST09608-16042, or equivalent.

To install:

17. Installation is the reverse of the removal procedure.
18. On vehicles manufactured thru 01/2006 tighten the front axle hub nut to 159 ft. lbs. On vehicles manufactured after 01/2006 and equipped with the 2.4L engine, tighten the axle hub nut to 159 ft. lbs. On vehicles manufactured after 01/2006 and equipped with the 3.5L engine, tighten the axle hub nut to 215 ft. lbs.
19. On 2WD, tighten the right side bearing bracket bolts to 47 ft. lbs.
20. On 4WD, tighten the right side bearing bracket bolts to 24 ft. lbs.
21. Check and adjust the wheel alignment, as required.
22. Be sure to fill the transaxle with the proper grade and type transaxle fluid.
23. Start the engine and check for leaks.

REAR AXLE HOUSING

REMOVAL & INSTALLATION

See Figures 27 through 29.

1. Disconnect the negative battery cable.

➡Wait at least 90 seconds after disconnecting the negative battery cable to prevent air bag and seat belt pretensioner activation.

2. Raise and safely support the vehicle.
3. Drain the rear differential fluid. Be sure to properly dispose of used fluid.
4. Remove the tire and wheel assembly.
5. Remove the fuel tank.
6. Remove the tailpipe assembly. Remove the center exhaust pipe assembly.
7. Remove the driveshaft with the center bearing.
8. Remove the two retaining bolts and remove the left side rear suspension member brace.
9. Remove the two retaining bolts and remove the right side rear suspension member brace.
10. Disconnect the harness clamp. Remove the breather tube. Disconnect the connector.
11. Properly support the differential carrier assembly with a transmission jack, or equivalent.
12. Remove the two bolts. See the illustration for location. Do not loosen the nuts, loosen the bolts.
13. Slowly lower the transmission jack and tilt the differential carrier assembly.
14. Position the tip of a suitable tool to the position on the rear halfshaft, as shown in the illustration. Using the ribbed part of the differential carrier as a fulcrum, disconnect the left and right halfshafts.

➡Do not scratch the rear halfshaft dust cover.

15. Remove the unit from the vehicle.
16. Remove the four rear differential carrier number one and number two supports from the rear differential carrier.
17. Remove the four bolts and remove the rear differential carrier support from the carrier.

To install:

18. Installation is the reverse of the removal procedure.
19. Tighten the four rear differential carrier support bolts to 72 ft. lbs.
20. Tighten the four rear differential number one and number two support bolts 41 ft. lbs.
21. Position the rear differential carrier in position. Install the rear halfshafts.
22. Align the splines of the halfshaft inboard joints and using a brass drift and hammer tap the left and right halfshafts.

➡Face the cutout section of the snapring downward.. Do not damage the oil seal during insertion. Do not strike the tip of the outboard joint with the hammer.

➡Determine whether or not the halfshaft is completely tapped in by checking for changes in sound or reaction force of the brass bar.

23. Slowly raise the assembly into position. Tighten the assembly retaining bolts to 63 ft. lbs and then to 103 ft. lbs. Tighten the bolts, not the nuts.
24. Continue the installation in the reverse order of the removal procedure.
25. Tighten the rear suspension crossmember bolts to 44 ft. lbs.
26. Be sure fill the rear differential carrier assembly with the proper grade and type fluid.
27. Be sure to check for fuel leaks, correct as required.
28. Be sure to check for exhaust leaks, correct as required.

REAR HALFSHAFT

REMOVAL & INSTALLATION

See Figures 30 and 31.

1. Before servicing the vehicle, refer to the Precautions Section.
2. Disconnect the negative battery cable.

➡Wait at least 90 seconds after disconnecting the negative battery cable before starting any repair work to prevent air bag and seat belt pretensioner activation.

3. Raise and support the vehicle safely.
4. Remove the tire and wheel assembly.
5. Drain the differential oil.
6. Remove the tailpipe assembly.
7. Remove the driveshaft with the center bearing.
8. Remove the rear axle shaft nut.
9. Support the rear differential using a suitable jack.
10. Fix the nuts in place and remove bolt

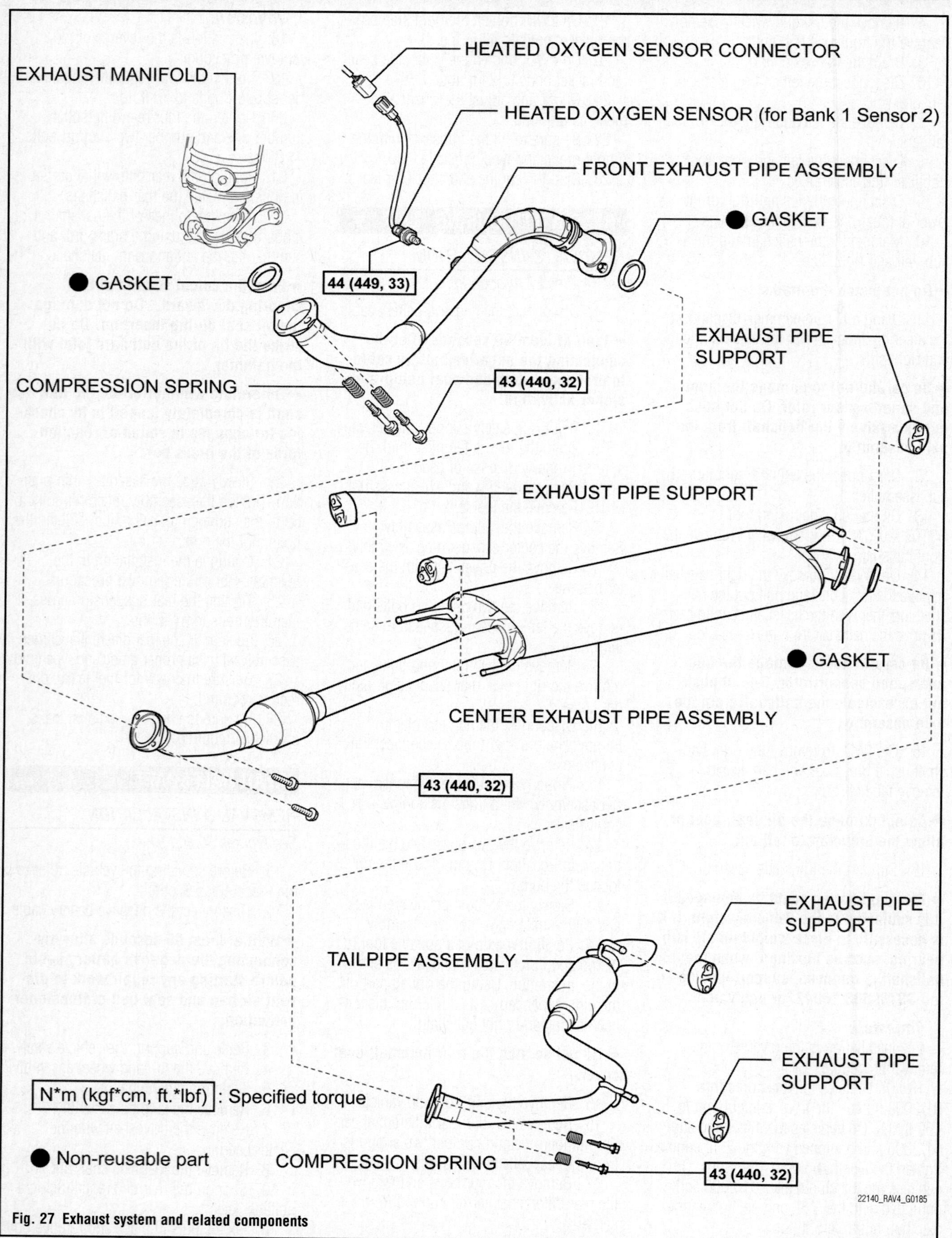

Fig. 27 Exhaust system and related components

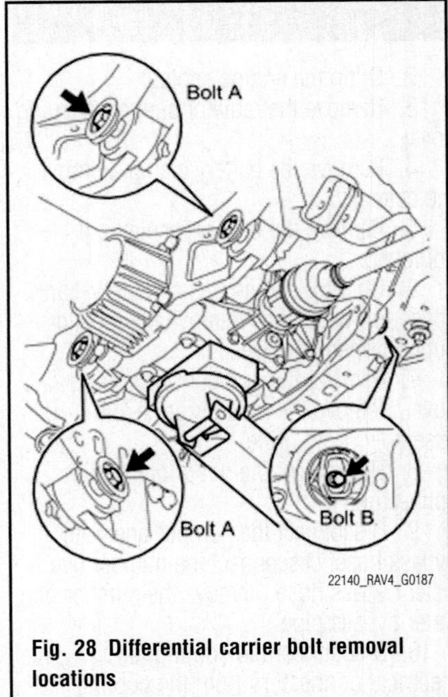

Fig. 28 Differential carrier bolt removal locations

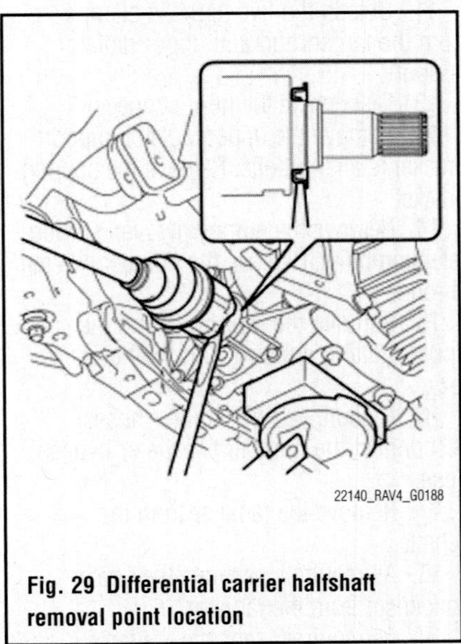

Fig. 29 Differential carrier halfshaft removal point location

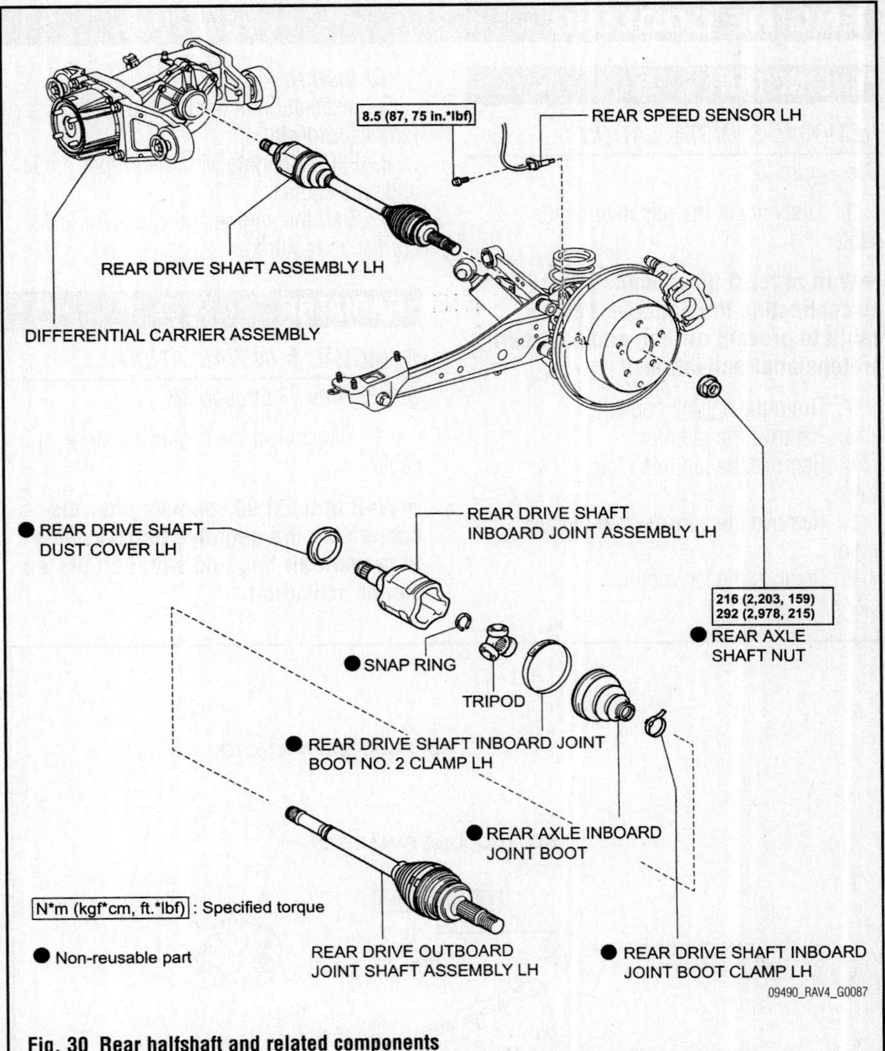

Fig. 30 Rear halfshaft and related components

"A" "B" and "C". Do not loosen the nuts, loosen the bolts. Slowly lower the jack and tilt the rear differential carrier, as shown in the illustration.

11. Using a suitable tool disconnect the left and right rear halfshafts from the differential carrier.

12. On 2WD, disconnect the skid control sensor wire.

13. On 4WD, disconnect the rear speed sensor, left side.

14. Put matchmarks on the halfshaft and the axle hub. Do not punch the marks.

15. Using a plastic faced hammer, separate the halfshaft from the axle hub.

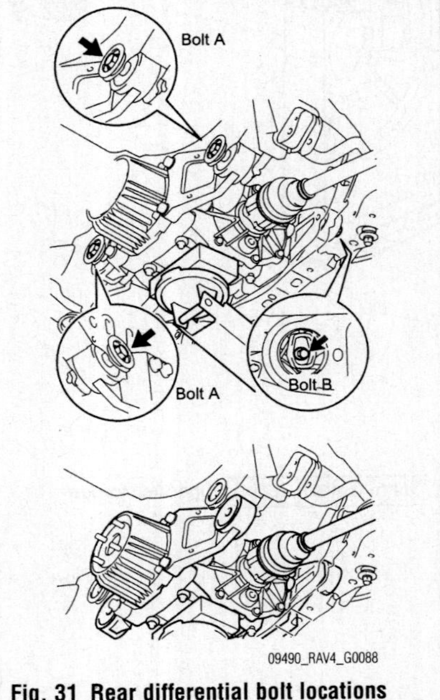

Fig. 31 Rear differential bolt locations

➡Be careful not to damage the boot and speed sensor rotor. Do not excessively push out the halfshaft from the axle.

16. Support the rear halfshaft assembly.

➡The hub bearing could be damaged if it is subjected to the vehicle weight. If it is necessary to place weight on the hub bearing, such as moving it when the halfshaft is removed, support it using too SST09608-16042, or equivalent.

To install:

17. Installation is the reverse of the removal procedure.

18. Tighten the differential carrier bolts to 63 ft. lbs. for bolt "A", 103 ft. lbs. for bolt "B".

19. Check and adjust the wheel alignment, as required.

20. Be sure to fill the differential with the proper grade and type fluid.

21. Start the engine and check for leaks.

ENGINE COOLING

ENGINE FAN

REMOVAL & INSTALLATION

See Figure 32.

1. Disconnect the negative battery cable.

➡Wait at least 90 seconds after disconnecting the negative battery cable to prevent air bag and seat belt pretensioner activation.

2. Drain the engine coolant.
3. Remove the radiator.
4. Remove the fan retaining nuts.
5. Remove the screws from the fan motor.
6. Remove the fan motor.

To install:

7. Installation is the reverse of the removal procedure.
8. Fill the engine with the proper grade and type coolant.
9. Start the engine and check for leaks, correct as required.

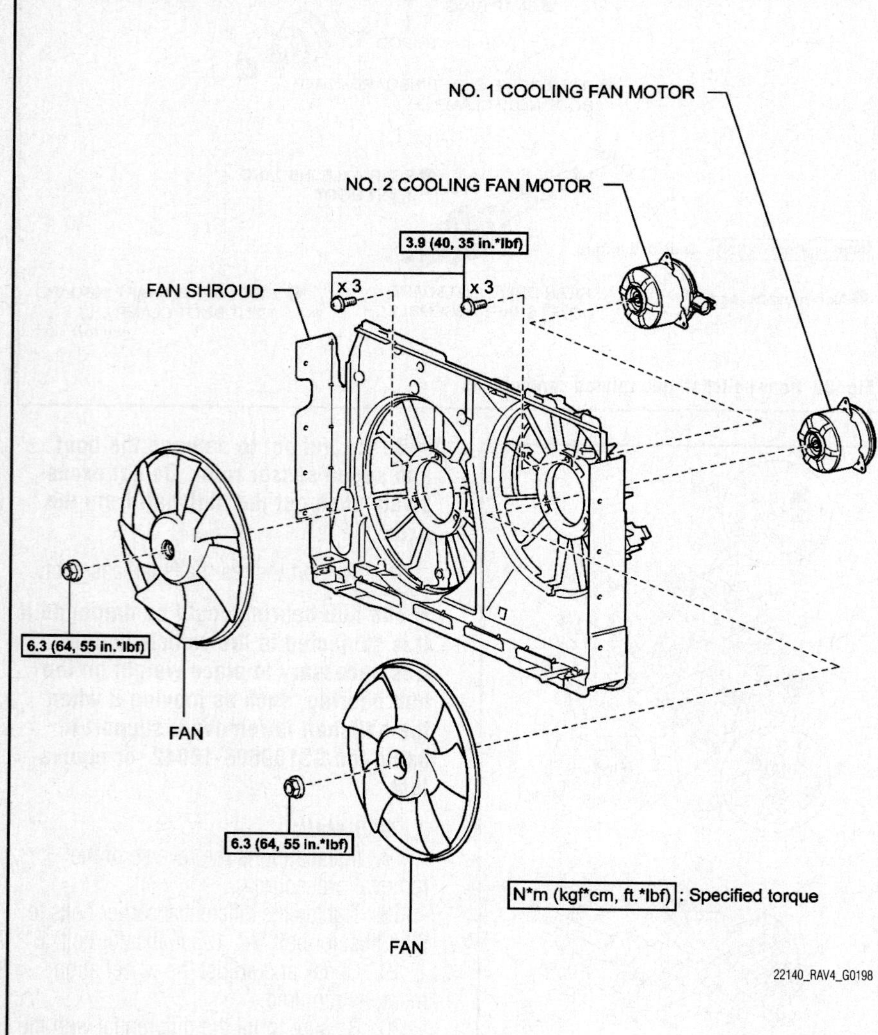

Fig. 32 Cooling fan motors and related components

RADIATOR

REMOVAL & INSTALLATION

See Figures 33 through 36.

1. Disconnect the negative battery cable.

➡Wait at least 90 seconds after disconnecting the negative battery cable to prevent air bag and seat belt pretensioner activation.

2. Drain the engine coolant.
3. Remove the radiator support opening cover.
4. Remove the battery clamp. Remove the battery.
5. Remove the battery insulator, if equipped.
6. On 3.5L engine, remove the V-bank cover sub-assembly. Remove the radiator support opening cover.
7. Remove the number one engine cover. Remove the radiator grille sub-assembly (front grille).
8. Disconnect the hood lock switch connector.
9. Disconnect the number one water bypass hose. Disconnect the number five water bypass hose. Remove the number one water bypass pipe.
10. Disconnect the cooling fan electrical connectors from the cooling fans.
11. Detach the five harness clamps from the fan shroud and upper radiator support.
12. Disconnect the horn connector.
13. Remove the upper radiator support bracket retaining bolts. Remove the support bracket.
14. Remove the number two fan shroud retaining bolts. Remove the number two fan shroud.
15. Remove the four bolts and the upper radiator support with the hood lock.
16. Disconnect the radiator hoses. Disconnect the number two water bypass hose.
17. Remove the radiator from the vehicle.
18. As required, separate the cooler condenser from the radiator.
19. As required, separate the fan shroud and cooling fans from the radiator assembly.

To install:

20. Installation is the reverse of the removal procedure.
21. Be sure to install hoses clamps as indicated in the illustration.
22. Be sure to install bypass hose clamps as indicated in the illustration.
23. Fill the engine with the proper grade and type coolant.
24. Start the engine and check for leaks, correct as required.

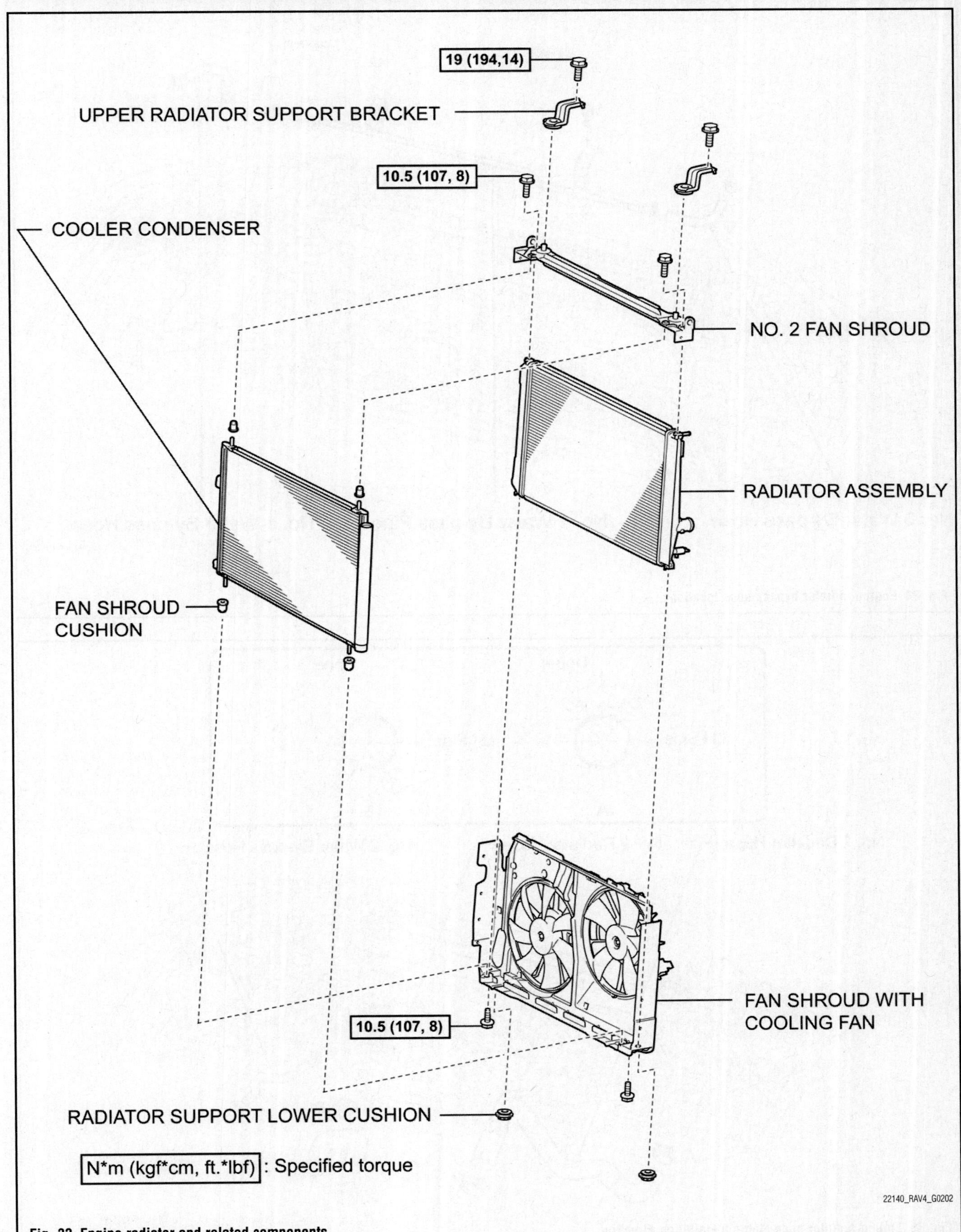

Fig. 33 Engine radiator and related components

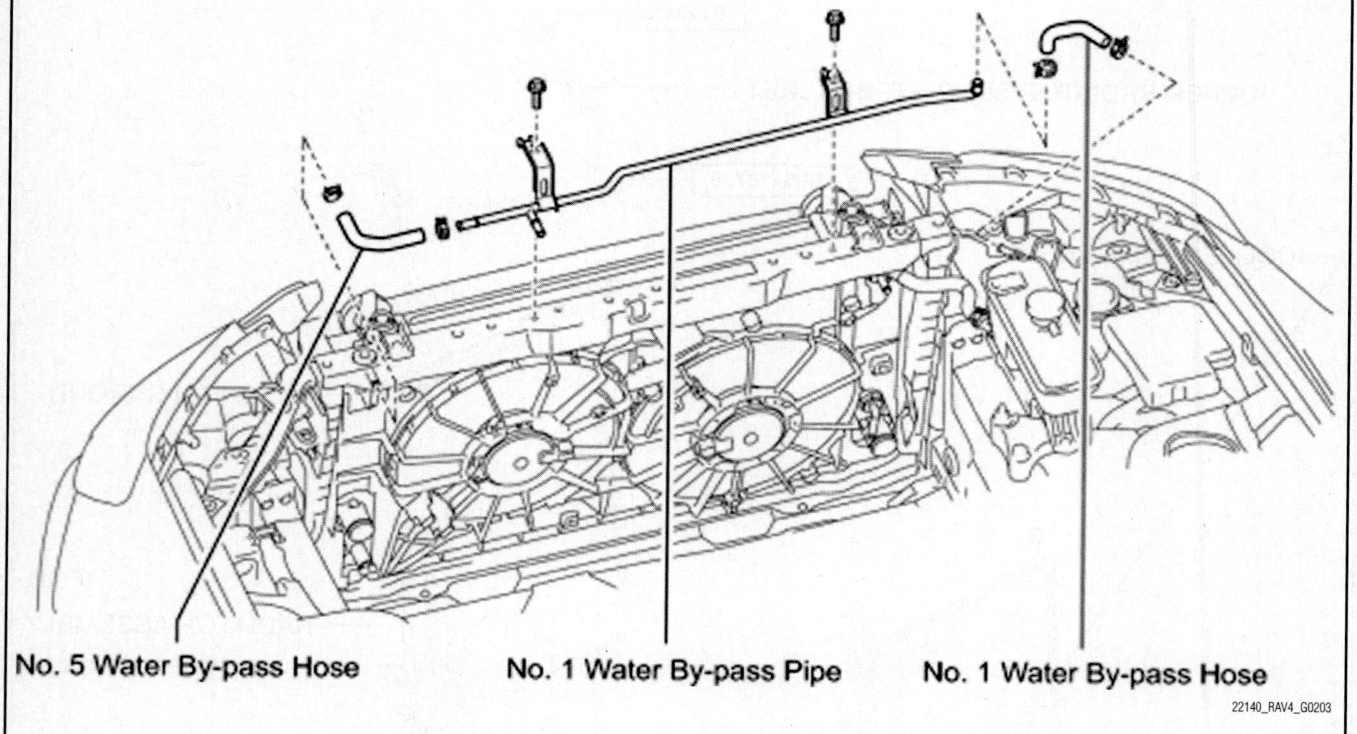

Fig. 34 Engine radiator bypass hose location

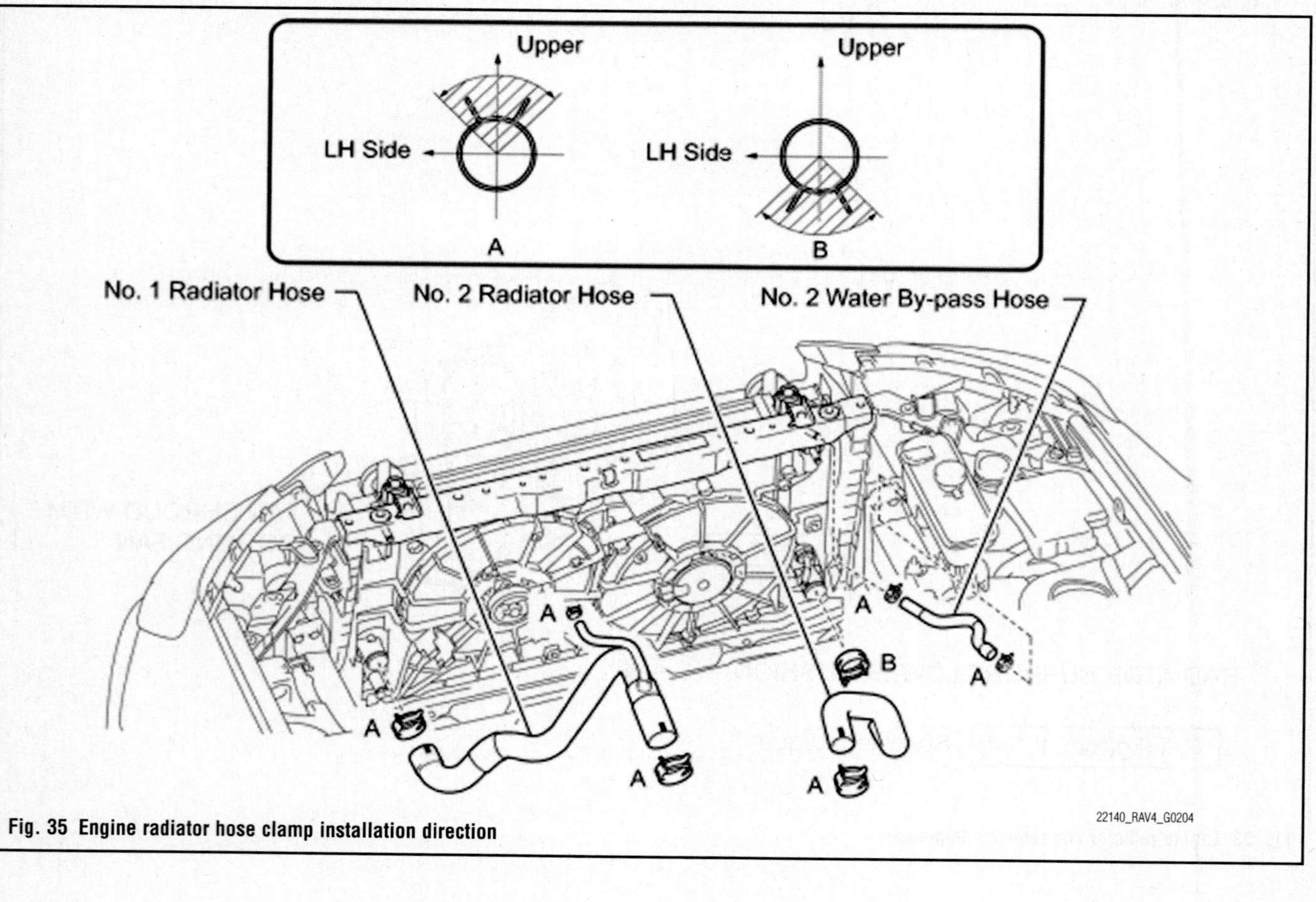

Fig. 35 Engine radiator hose clamp installation direction

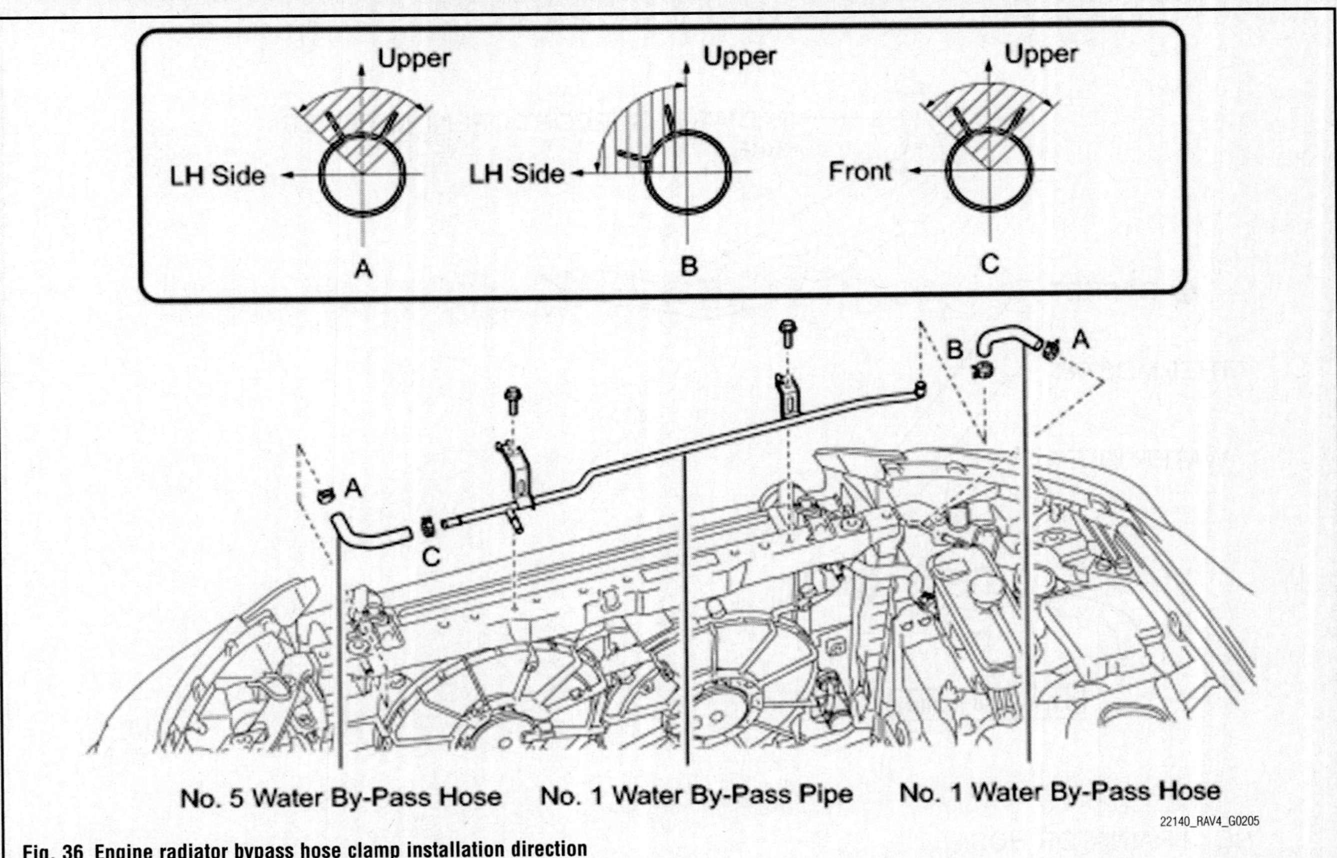

Fig. 36 Engine radiator bypass hose clamp installation direction

THERMOSTAT

REMOVAL & INSTALLATION

2.4L Engine

See Figures 37 and 38.

1. Disconnect the negative battery cable.

➡Wait at least 90 seconds after disconnecting the negative battery cable to prevent air bag and seat belt pretensioner activation.

2. Drain the engine coolant.
3. Remove the radiator support opening cover.
4. Disconnect the number two radiator hose.
5. Remove the water inlet retaining nuts. Remove the water inlet from the cylinder block.
6. Remove the thermostat from its mounting. Discard the gasket.

To install:

7. Installation is the reverse of the removal procedure.
8. Be sure to install a new gasket onto the thermostat.
9. Install the thermostat with the jiggle valve upward. The jiggle valve may be set within 10 degrees on either side of the assembly. See illustration.
10. Tighten the retaining nuts to 80 inch lbs.
11. Fill the engine with the proper grade and type coolant.
12. Start the engine and check for leaks, correct as required.

3.5L Engine

See Figures 38 through 42.

1. Disconnect the negative battery cable.

➡Wait at least 90 seconds after disconnecting the negative battery cable to prevent air bag and seat belt pretensioner activation.

2. Drain the engine coolant.
3. Remove the right tire and wheel assembly.
4. Remove the number one engine cover.
5. Remove the engine under cover rear, right side.
6. Remove the front suspension member reinforcement, right side.
7. Remove the V-bank cover assembly.
8. Remove the radiator support opening cover.
9. Remove the radiator reservoir tank assembly.
10. Remove the fan and alternator belt.
11. Remove the engine mounting insulator, right side five retaining bolts. Remove the component.
12. Remove the number one engine mounting bracket, left side.
13. Remove the bolt, number two idler pulley cover plate and number two idler pulley.
14. Disconnect the number two radiator hose.
15. Remove the water inlet housing retaining nuts. Remove the water inlet housing.
16. Remove the thermostat from its mounting. Discard the gasket.

To install:

17. Installation is the reverse of the removal procedure.
18. Be sure to install a new gasket onto the thermostat.
19. Install the thermostat with the jiggle valve upward. The jiggle valve may be set within 10 degrees on either side of the assembly. See illustration.
20. Tighten the retaining nuts to 7 ft. lbs.
21. Tighten the number two idler pulley bolt to 32 ft. lbs.

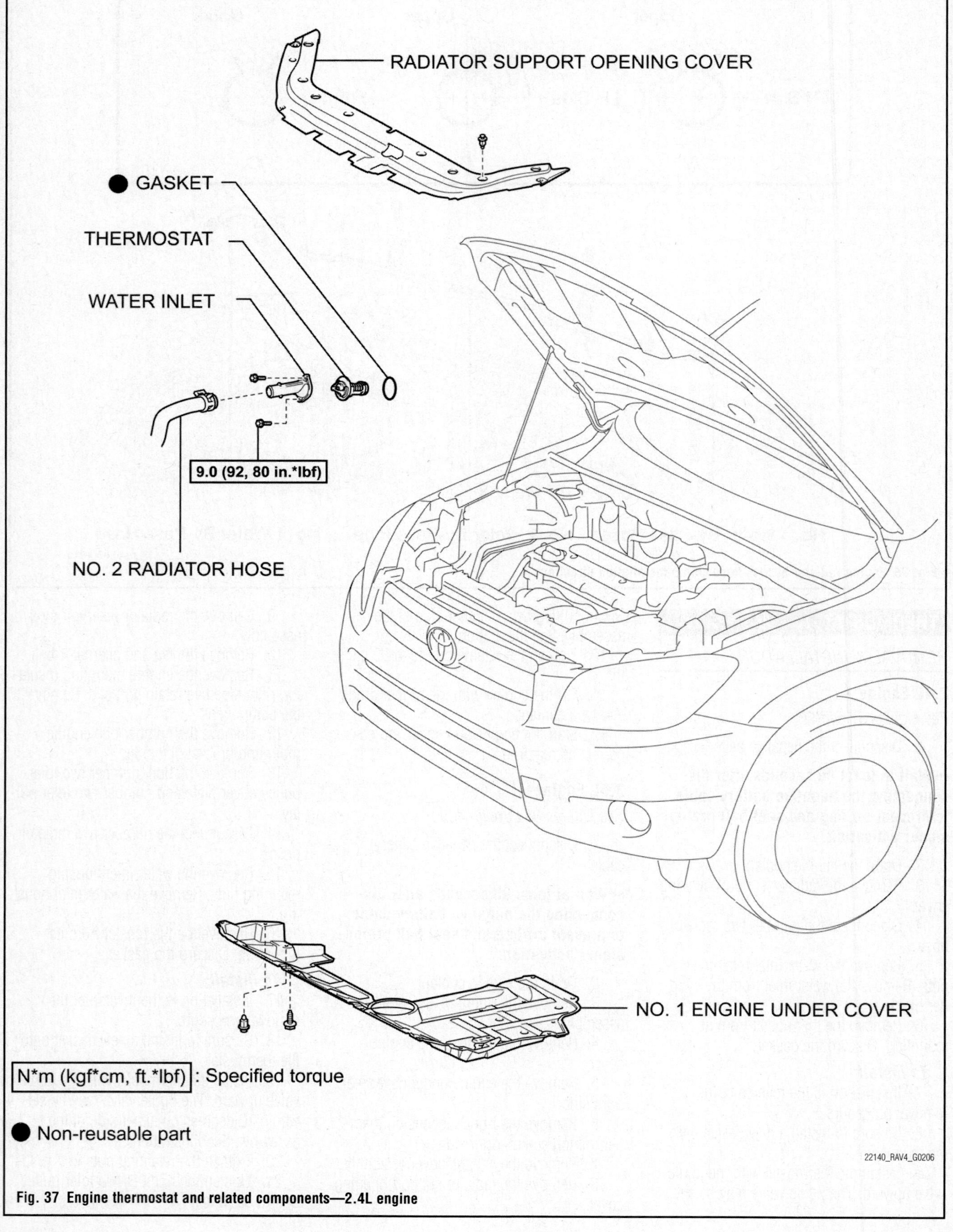

Fig. 37 Engine thermostat and related components—2.4L engine

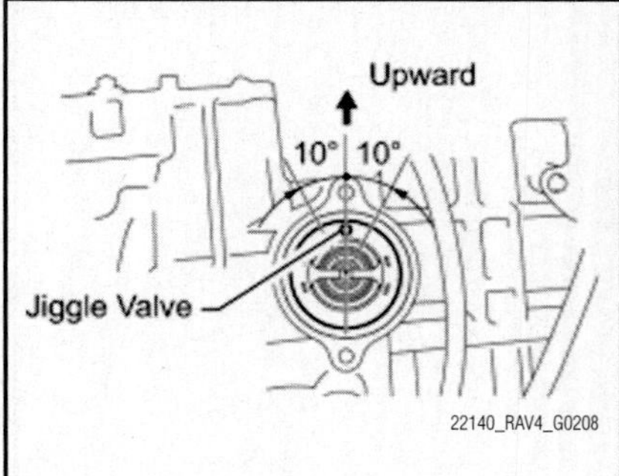

Fig. 38 Engine thermostat jiggle valve positioning—2.4L engine

22. Fill the engine with the proper grade and type coolant.

23. Start the engine and check for leaks, correct as required.

WATER PUMP

REMOVAL & INSTALLATION

2.4L Engine

See Figures 43 and 44.

1. Before servicing the vehicle, refer to the Precautions Section.
2. Disconnect the negative battery cable.

➡Wait at least 90 seconds after disconnecting the negative battery cable before starting any repair work to prevent air bag and seat belt pretensioner activation.

3. Remove the number one engine undercover.
4. Remove the front fender apron, right side.
5. Drain the cooling system. Remove the radiator support opening cover.
6. Remove the front suspension member reinforcement, right side.
7. Remove the fan and alternator drive belt. Remove the alternator.

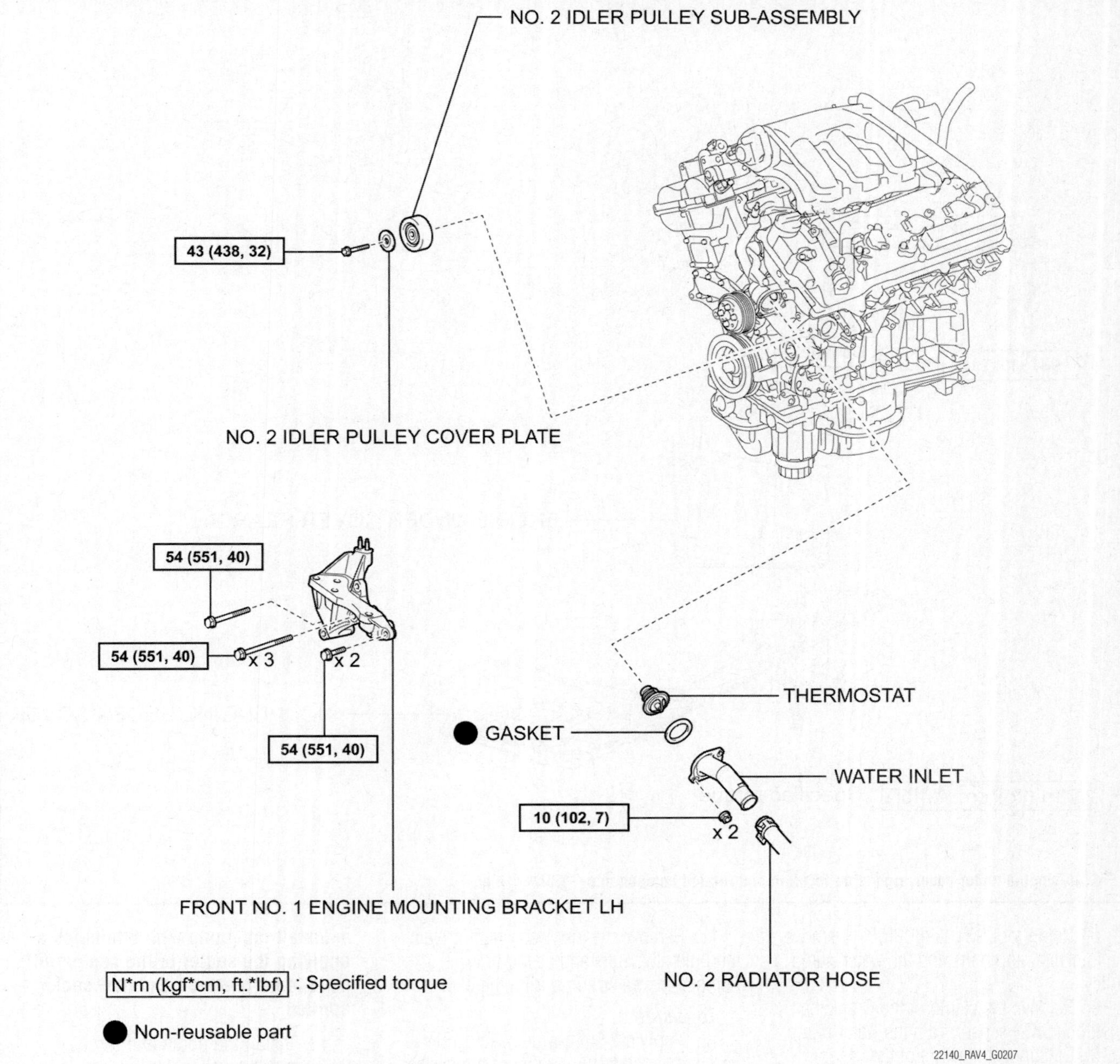

Fig. 39 Engine thermostat and related components—3.5L engine

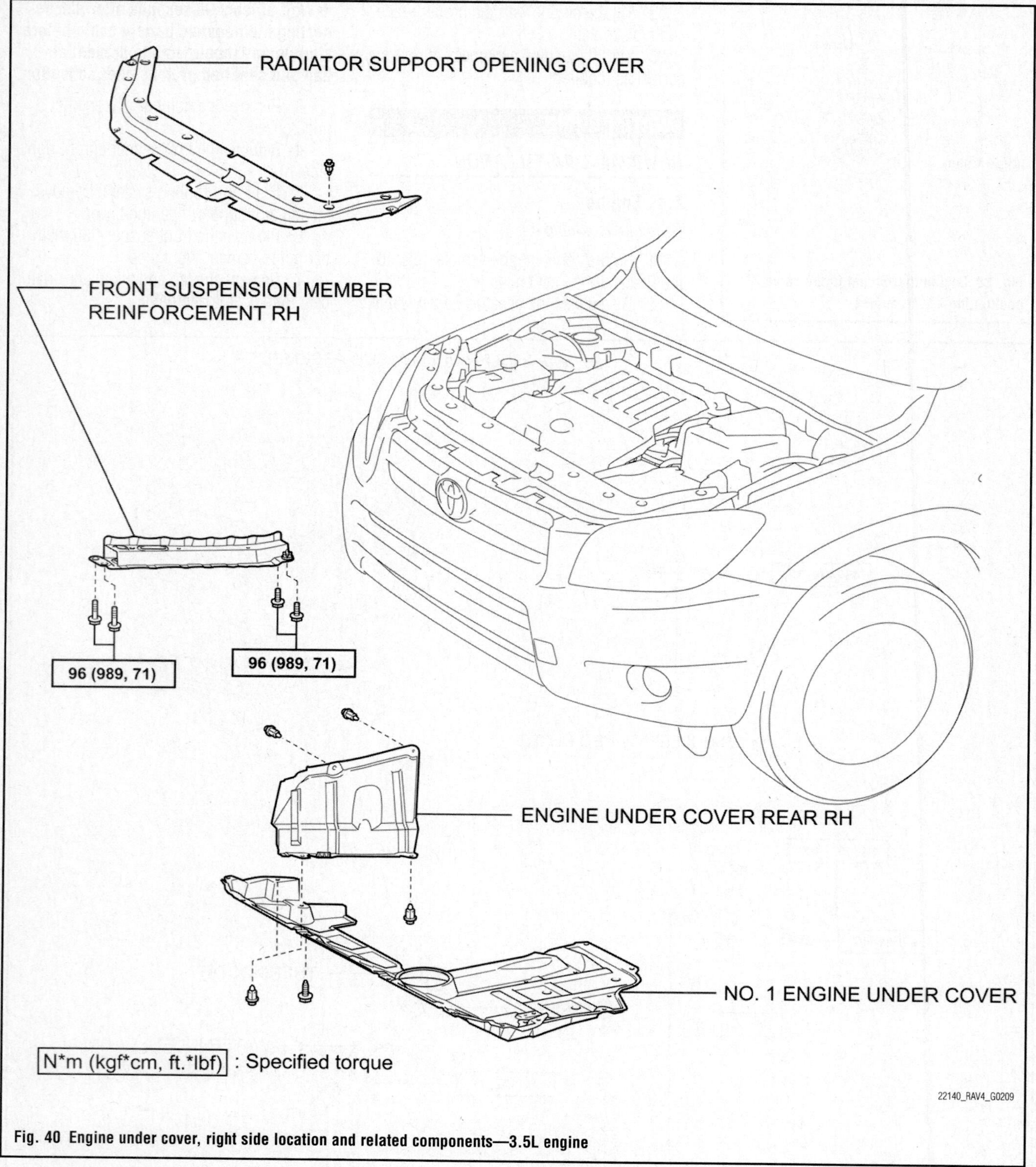

Fig. 40 Engine under cover, right side location and related components—3.5L engine

8. Using tool SST09960-10010 remove the four retaining bolts and the water pump pulley.
9. Remove the clamp of the Crankshaft Position (CKP) sensor from the water pump.
10. Disconnect the wire of the sensor from the clamp bracket.
11. Remove the four water pump retaining bolts, two nuts and clamp bracket. Remove the water pump from the engine.

To install:

12. Apply a 2.5mm wide bead of RTV gasket material to the pump sealing surface as shown.

➡Install the pump with 5 minutes of applying the sealer or the sealer will have to be removed and new sealer applied.

13. Install the pump and torque the nuts and bolts to 80 inch lbs. (9 Nm).

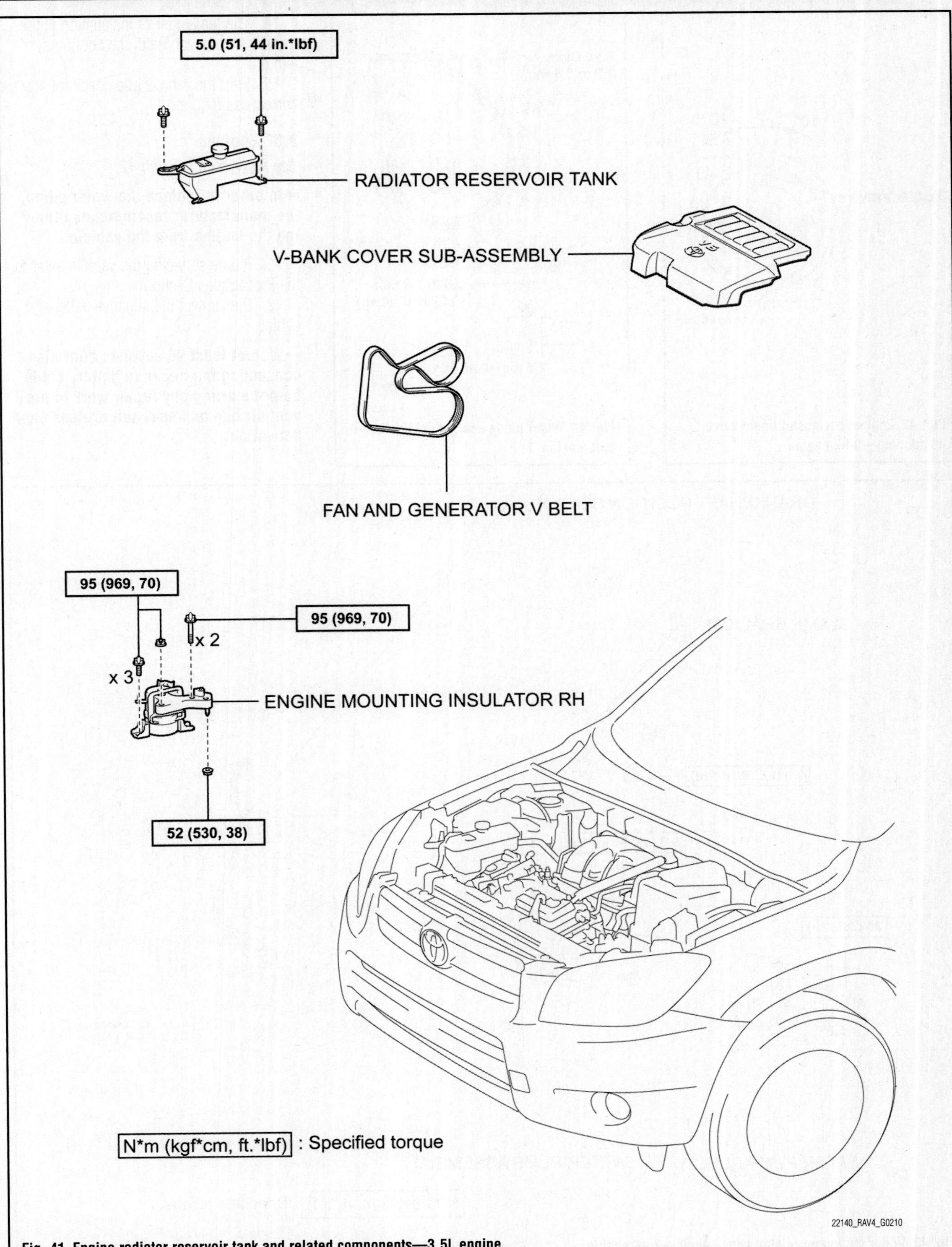

Fig. 41 Engine radiator reservoir tank and related components—3.5L engine

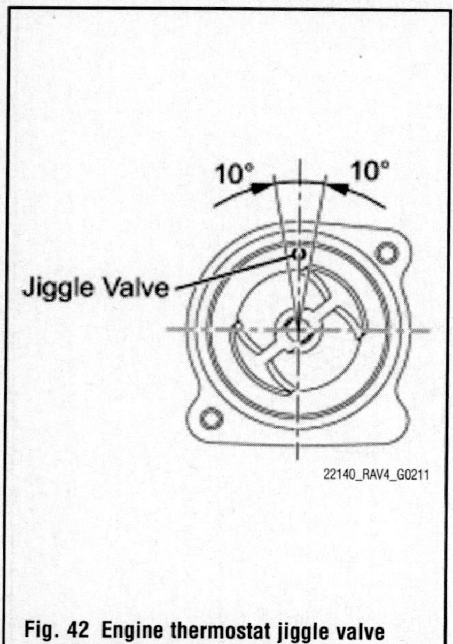

Fig. 42 Engine thermostat jiggle valve positioning—3.5L engine

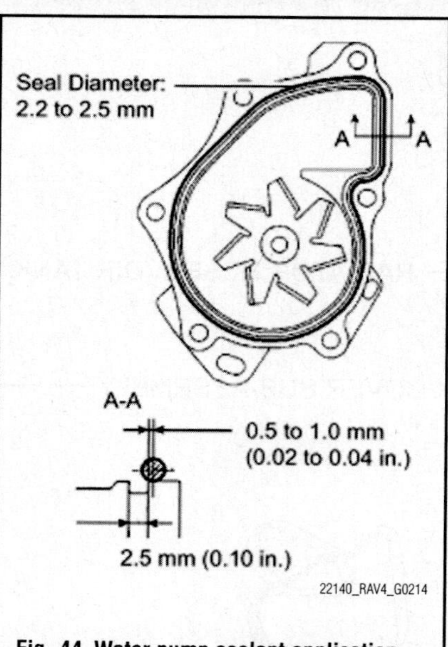

Fig. 44 Water pump sealant application—2.4L engine

14. The remainder of installation is the reverse of removal. Refill the cooling system.

15. Start the vehicle and check for leaks, correct ass required.

3.5L Engine

See Figures 45 through 47.

➡In order to replace the water pump, the manufacturer recommends removing the engine from the vehicle.

1. Before servicing the vehicle, refer to the Precautions Section.
2. Disconnect the negative battery cable.

➡Wait at least 90 seconds after disconnecting the negative battery cable before starting any repair work to prevent air bag and seat belt pretensioner activation.

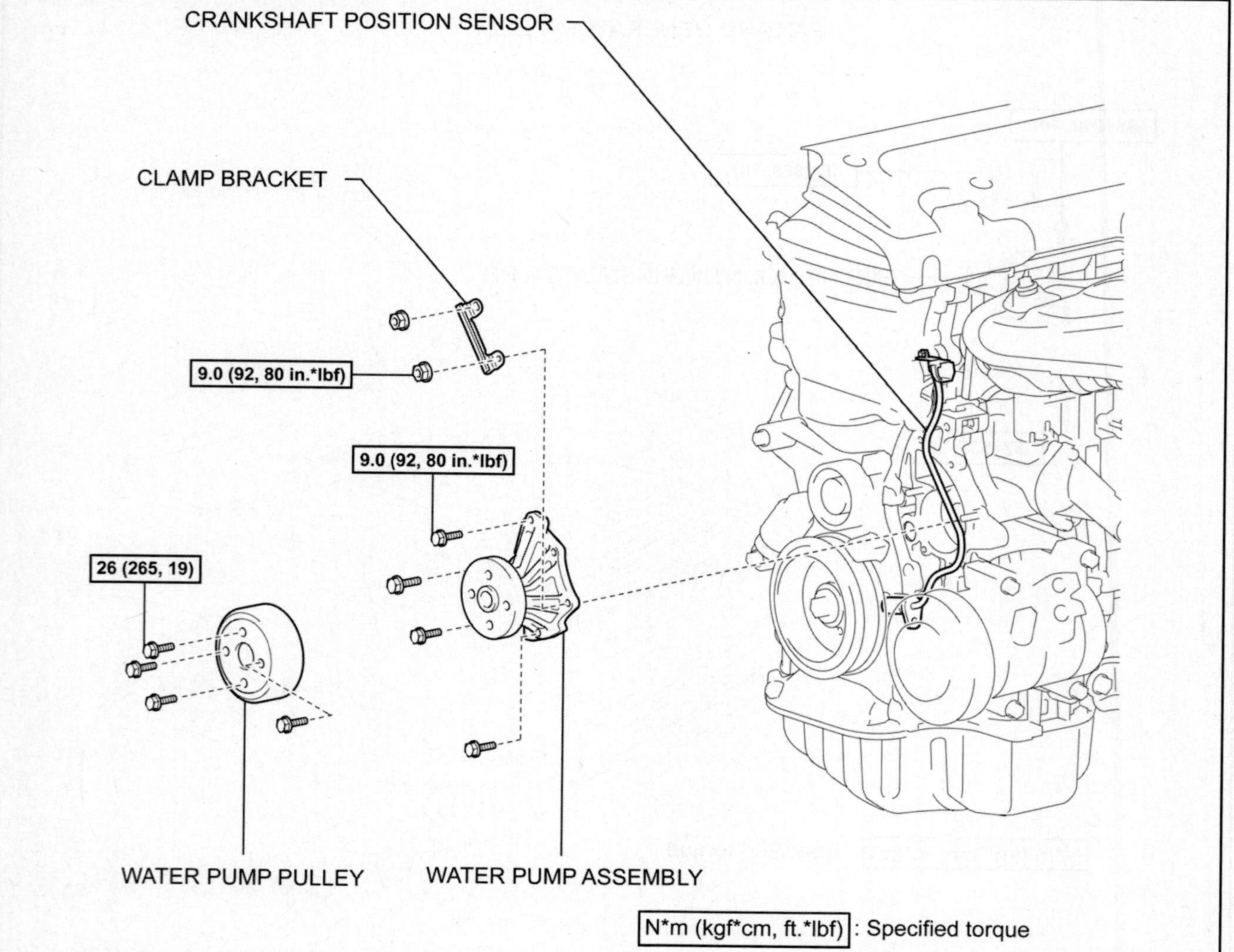

CRANKSHAFT POSITION SENSOR
CLAMP BRACKET
9.0 (92, 80 in.*lbf)
9.0 (92, 80 in.*lbf)
26 (265, 19)
WATER PUMP PULLEY
WATER PUMP ASSEMBLY
N*m (kgf*cm, ft.*lbf) : Specified torque
22140_RAV4_G0213

Fig. 43 Water pump and related components—2.4L engine

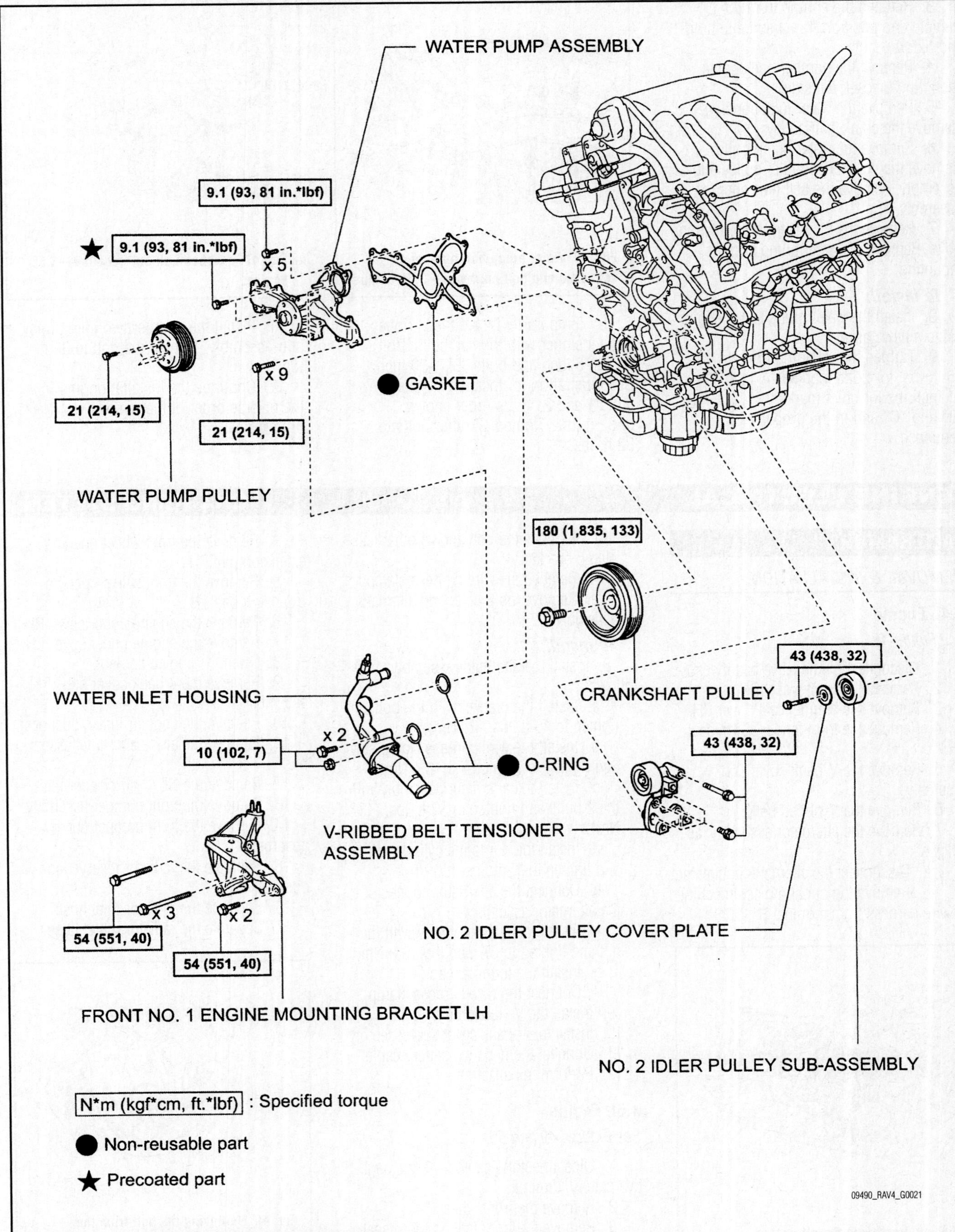

Fig. 45 Water pump and related components—3.5L engine

3. Remove the engine from the vehicle and position it in a suitable holding fixture.
4. Remove the number one engine mounting bracket, left side.
5. Remove the water inlet housing. Remove the crankshaft pulley.
6. Remove the water pump pulley. Remove the number two idler pulley sub-assembly. Remove the belt tensioner assembly.
7. Remove the 16 water pump retaining bolts. Remove the water pump from its mounting.

To install:

8. Install the water pump to the engine using a new gasket.
9. Tighten the retaining bolts to 15 ft. lbs. for bolts marked "A" and 81 inch lbs for bolts marked "B" and "C" and in the proper sequence.

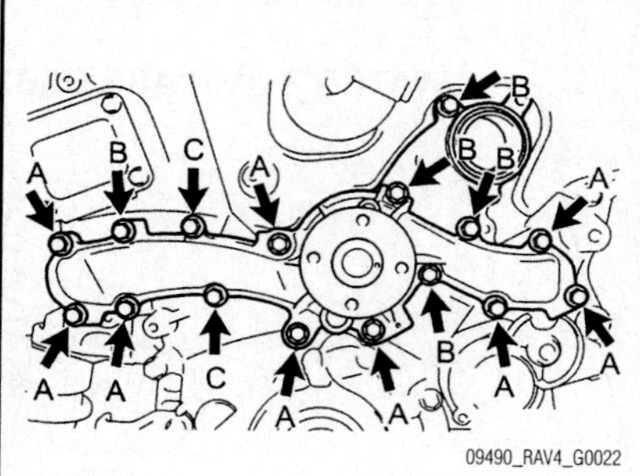

Fig. 46 Water pump bolt tightening sequence and bolt markings—3.5L engine

10. Temporarily install the V-ribbed belt tensioner with the five bolts. Bolt "A" 2.76 inch and bolts "B" 1.30 inch. See illustration for location. Tighten bolts 1 and 2, in the order shown, to 32 ft. lbs. Tighten all other bolts to 32 ft. lbs.

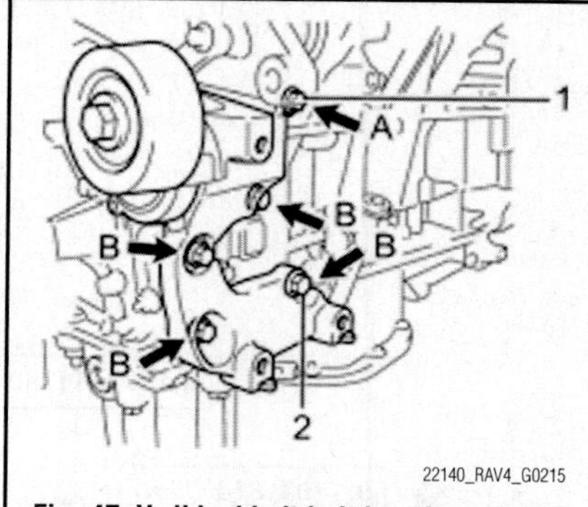

Fig. 47 V-ribbed belt bolt location—3.5L engine

11. Install the number two idler pulley sub-assembly. Tighten the bolt to 32 ft. lbs.
12. Continue the installation in the reverse order of the removal procedure.

ENGINE ELECTRICAL

CHARGING SYSTEM

ALTERNATOR

REMOVAL & INSTALLATION

2.4L Engine

See Figure 48.

1. Disconnect the negative battery cable.
2. Remove the front wheel RH.
3. Remove the engine front cover RH.
4. Remove the front fender apron seal RH.
5. Remove the V-bank cover sub-assembly.
6. Remove the V-ribbed belt.
7. Remove the alternator assembly, as follows:
   a. Disconnect the alternator connector.
   b. Remove the nut and disconnect the wire harness from terminal B.

3768X_CAMR_G0116

Fig. 48 Removing the alternator

   c. Remove the bolt and wire harness clamp bracket.
   d. Remove the wire harness clamps.
   e. Remove the 2 bolts and alternator assembly.

To install:

8. Install the alternator assembly, as follows:
   a. Install the bracket with the bolt and tighten to 15 ft. lbs. (20 Nm).
   b. Install the wire harness clamp stay and tighten to 74 inch lbs. (8.4 Nm).
   c. Install the alternator assembly with the 2 bolts and tighten to 32 ft. lbs. (43 Nm).
   d. Install the nut to the cylinder block and tighten to 15 ft. lbs. (20 Nm).
   e. Connect the alternator connector to the alternator assembly.
   f. Install the alternator wire with the nut and tighten to 87 inch lbs. (9.8 Nm).
   g. Install the terminal cap.
   h. Connect the wire harness clamp.
9. Install the V-ribbed belt.
10. Install the V-bank cover sub-assembly.
11. Connect the negative battery cable.
12. Perform initialization.

3.5L Engine

See Figures 49 and 50.

1. Disconnect the cable from the negative battery terminal.
2. Remove the front wheel.
3. Remove the front fender apron seal RH.
4. Remove the front wheel opening extension pad RH.
5. Remove the front wheel opening extension pad LH.
6. Remove the engine under cover RH.
7. Remove the engine under cover LH.
8. Drain the engine coolant.
9. Remove the v-bank cover sub assembly.
10. Remove the cool air intake duct seal.
11. Remove the air cleaner inlet assembly.
12. Remove the No. 1 air cleaner inlet.
13. Remove the front bumper assembly.
14. Remove the front bumper energy absorber.
15. Separate the radiator reserve tank hose.
16. Separate the radiator inlet hose.
17. Separate the radiator outlet hose.

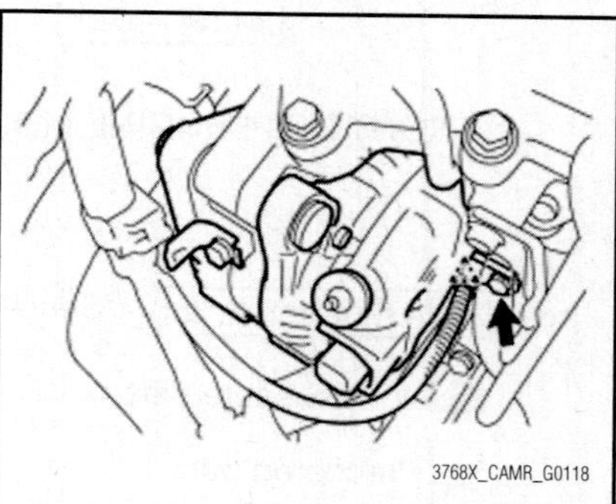

Fig. 49 Removing the bolt from the cylinder block

18. Separate the No. 1 oil cooler inlet hose.
19. Separate the No. 1 oil cooler outlet hose.
20. Remove the radiator support upper.
21. Remove the fan shroud.
22. Remove the radiator assembly.
23. Remove the v-ribbed belt.
24. Remove the alternator assembly.
    a. Remove the terminal cap.
    b. Remove the nut and disconnect the wire harness from terminal B.
    c. Disconnect the alternator connector from the alternator assembly.

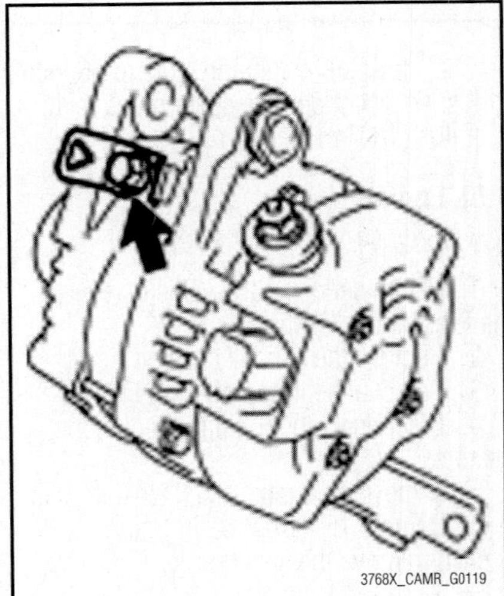

Fig. 50 Removing the wire harness clamp stay

    d. Disconnect the connector from the compressor and magnetic clutch.
    e. Disconnect the 2 wire harness clamps.
    f. Remove the 2 bolts.
    g. Remove the bolt from the cylinder block.
    h. Disconnect the wire harness clamp and remove the alternator assembly.
    i. Remove the bolt and wire harness clamp stay.
    j. Remove the bolt and bracket.

To install:

25. Install the alternator assembly.
    a. Install the bracket with the bolt. Tighten to 15 ft. lbs. (20 Nm).
    b. Install the wire harness clamp stay and tighten to 74 inch lbs. (8.4 Nm).
    c. Connect the wire harness clamp.
    d. Install the alternator assembly to the cylinder block with the bolt. Tighten to 15 ft. lbs. (20 Nm).
    e. Install the 2 bolts. Tighten to 32 ft. lbs. (43 Nm).
    f. Connect the alternator connector to the alternator assembly.
    g. Install the alternator wire with the nut and tighten to 87 inch lbs. (9.8 Nm).
    h. Install the terminal cap.
    i. Connect the 2 wire harness clamps.
    j. Connect the magnetic clutch connector to the compressor and magnetic clutch.
26. Install the v-ribbed belt.
27. Install the radiator assembly.
28. Install the fan shroud.
29. Install the radiator support upper.
30. Connect the No. 1 oil cooler outlet tube.
31. Connect the No. 1 oil cooler inlet tube.
32. Connect the radiator reserve tank hose.
33. Install the front bumper energy absorber.
34. Install the front bumper assembly.
35. Install the No. 1 air cleaner inlet.
36. Install the air cleaner cap sub assembly.
37. Install the air cleaner inlet assembly.
38. Connect the cable to the negative battery terminal.
39. Add engine coolant.
40. Check for engine coolant leaks.
41. Install the v-bank cover sub assembly.
42. Install the cool air intake duct seal.
43. Install the front fender apron seal RH.
44. Install the engine under cover RH.
45. Install the engine under cover LH.
46. Install the front wheel opening extension pad RH.
47. Install the front wheel opening extension pad LH.
48. Install the front wheel and tighten to 76 ft. lbs. (103 Nm).

ENGINE ELECTRICAL

IGNITION SYSTEM

FIRING ORDER

Firing order for the 2.4L engine: 1–3–4–2.

Firing order for 3.5L engine: 1–2–3–4–5–6

IGNITION COIL

REMOVAL & INSTALLATION

2.4L Engine

See Figure 51.

1. Before servicing the vehicle, refer to the Precautions Section.
2. Disconnect the negative battery cable.
3. Remove engine cover(s).
4. Disconnect the 4 ignition coil connectors. Remove the 4 bolts and 4 ignition coils.

To install:

5. To install, reverse removal procedure. Tighten ignition coil bolts to 80 inch lbs. (9 Nm).

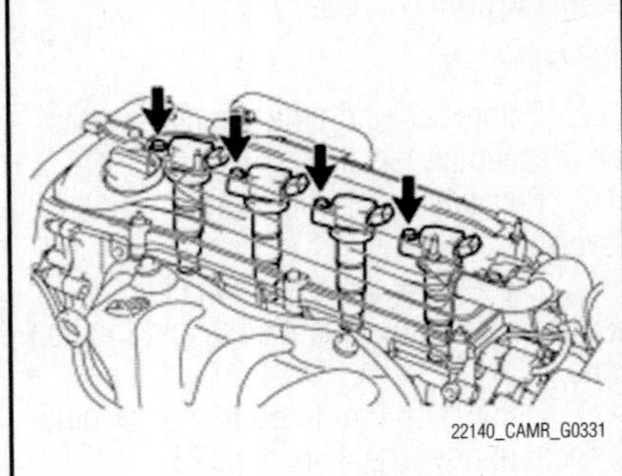

Fig. 51 Removing ignition coils—2.4L engines

3.5L Engine

See Figure 52.

1. Before servicing the vehicle, refer to the Precautions Section.
2. Disconnect the negative battery cable.
3. Drain and recycle the engine coolant.

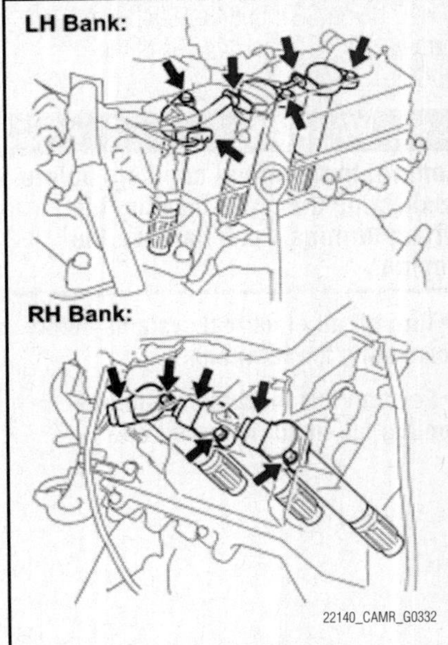

Fig. 52 Removing ignition coils—3.5L engines

4. Remove windshield wiper link assembly.
5. Remove cowl top panel outer sub-assembly.
6. Remove v-bank cover sub-assembly.
7. Remove air cleaner cap sub-assembly.
8. Remove intake air surge tank assembly.
9. Remove No. 1 surge tank stay by performing the following:
   a. Remove the bolt and disconnect the harness clamp.
   b. Remove the bolt and No. 1 surge tank stay.
10. Disconnect the 6 ignition coil connectors.
11. Remove the 6 bolts and 6 ignition coils.

To install:

12. To install, reverse removal procedure.
13. Tighten the following to specification:
   a. 6 ignition coil bolt: 10 ft. lbs. (10 Nm).
   b. No. 1 surge tank stay bolt: 15 ft. lbs. (21 Nm).
   c. No. 1 surge tank stay bolt and clamp: 62 inch lbs. (7 Nm).

IGNITION TIMING

INSPECTION

3.5L Engine

1. Warm up the engine.
2. Using SST: 09843-18040, connect terminals 13 (TC) and 4 (CG) of the DLC3.

⁂ WARNING

Confirm the terminal numbers before connecting them. Connecting the wrong terminals can damage the engine.

➡Turn off all electrical systems before connecting the terminals.

➡Perform this inspection after the cooling fan motor is turned off.

3. Remove the v-bank cover.
4. Pull out the red lead wire harness.
5. Connect the tester terminal of the timing light to the red lead wire.

➡Use a timing light which can detect the first signal.

6. Check the ignition timing at idle. Standard ignition timing: 8 to 12°BTDC at idle.

➡When checking the ignition timing, the transmission should be in the neutral position.

➡Run the engine at 1000 to 1300 rpm for 5 seconds, and then check that the engine rpm returns to idle speed.

7. Disconnect terminals 13 (TC) and 4 (CG) of the DLC3.
8. Check the ignition timing at idle. Standard ignition timing: 7 to 24°BTDC at idle.
9. Confirm that the ignition timing advances immediately when the engine rpm is increased.
10. Remove the timing light from the engine.

ADJUSTMENT

All engines are equipped with a Distributorless Ignition System (DIS). No timing adjustment is possible.

SPARK PLUGS

REMOVAL & INSTALLATION

2.4L Engine

See Figure 53.

1. Before servicing the vehicle, refer to the Precautions Section.
2. Remove the plastic engine cover.
3. Disconnect the 4 ignition coil connectors and remove the 4 bolts and ignition coils.
4. Using a 16 mm (0.63 in.) spark plug wrench, remove the 4 spark plugs.

To install:

5. Installation is the reverse of removal.

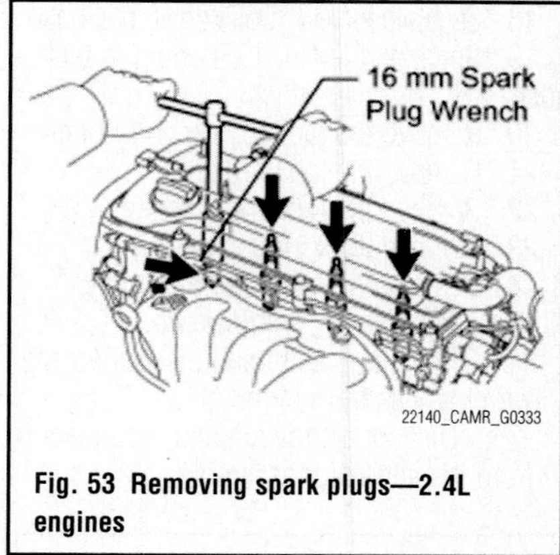

Fig. 53 Removing spark plugs—2.4L engines

   a. Torque the ignition coils to 66 inch lbs. (7.5 Nm) and the spark plugs to 13 ft. lbs (18 Nm).

3.5L Engine

See Figure 54.

1. Before servicing the vehicle, refer to the Precautions Section.
2. Remove the V-bank cover.
3. Remove the intake air surge tank.
4. Disconnect the 6 ignition coil connectors.
5. Remove the 6 bolts and 6 ignition coils.
6. Using a 16 mm (0.63 in.) plug wrench, remove the spark plugs.

To install:

7. Installation is the reverse of removal, noting the following:
   a. Torque the ignition coils to 66 inch lbs. (7.5 Nm) and the spark plugs to 13 ft. lbs (18 Nm).

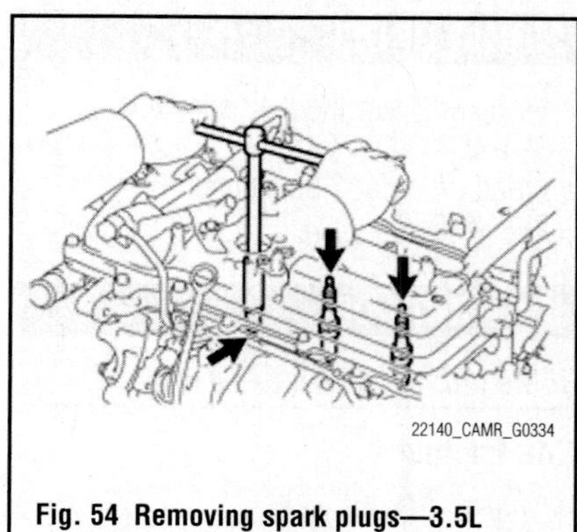

Fig. 54 Removing spark plugs—3.5L engines

STARTER

REMOVAL & INSTALLATION

2.4L Engine

See Figures 55 through 59.

1. Disconnect the cable from the negative battery terminal.
2. Remove the air cleaner inlet assembly.
3. Remove the air cleaner cap sub assembly.
4. Remove the air cleaner case sub assembly.
5. Remove the starter assembly (manual transaxle).
    a. Disconnect the terminal 50 connector from the starter assembly.
    b. Remove the nut and disconnect the wire harness from terminal 30.
    c. Remove the 3 bolts, clutch flexible hose bracket and starter assembly.
6. Remove the starter assembly (automatic transaxle).
    a. Disconnect the terminal 50 connector from the starter assembly.
    b. Remove the nut and disconnect the wire harness from terminal 30.
    c. Remove the 2 bolts and starter assembly.

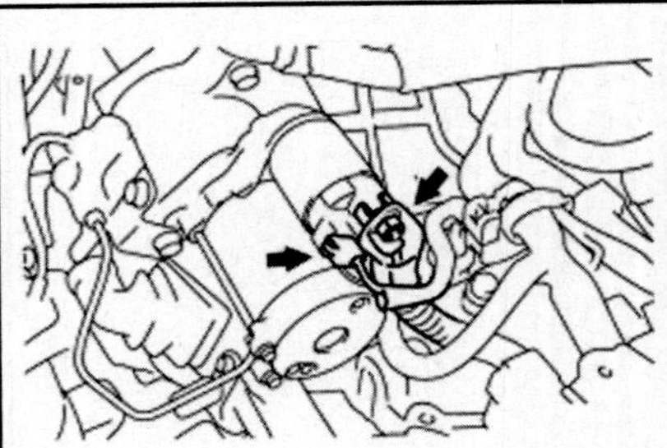

Fig. 55 Disconnecting the terminal 50 connector from the starter assembly (2.4L engine, manual transaxle)

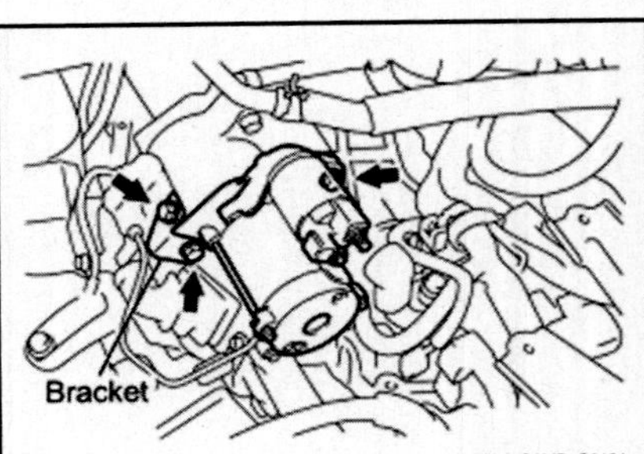

Fig. 56 Removing the clutch flexible hose bracket and starter assembly (manual transaxle)

To install:

7. Install the starter assembly (manual transaxle).
    a. Install the starter assembly and clutch flexible hose bracket with the 3 bolts. Tighten bolt A to 28 ft. lbs. (37 Nm). Tighten bolt B to 9 ft. lbs. (12 Nm).
    b. Connect the wire harness to terminal 30 and install the nut. Then, attach the terminal cap. Tighten to 87 inch lbs. (9.8 Nm).
    c. Connect the terminal 50 connector to the starter assembly.

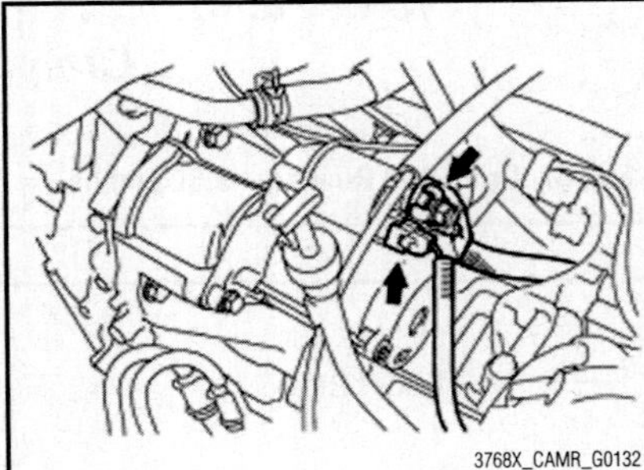

Fig. 57 Disconnecting the terminal 50 connector from the starter assembly (2.4L engine, automatic transaxle)

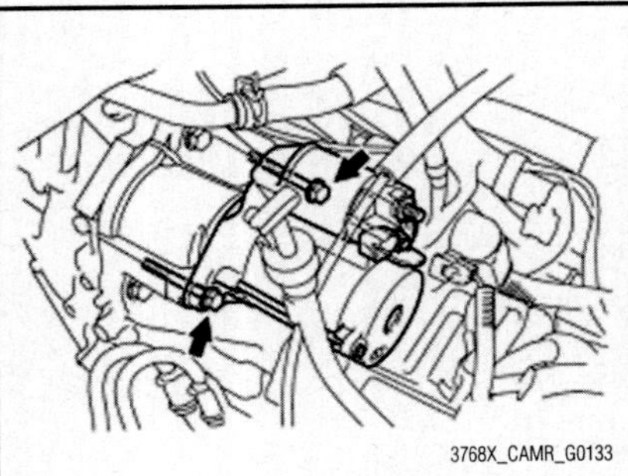

Fig. 58 Removing the starter assembly (2.4L engine, automatic transaxle)

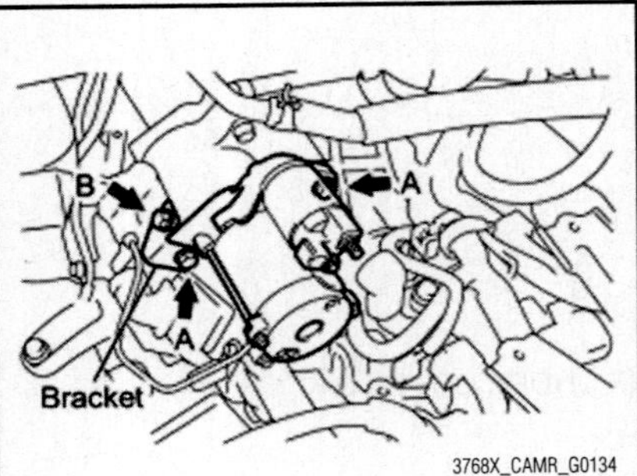

Fig. 59 Installing the starter assembly (manual transaxle)

8. Install the starter assembly (automatic transaxle).
    a. Install the starter assembly with the 2 bolts. Tighten to 28 ft. lbs. (37 Nm).
    b. Connect the wire harness to terminal 30 and install the nut. Then, attach the terminal cap. Tighten to 87 inch lbs. (9.8 Nm).
    c. Connect the terminal 50 connector to the starter assembly.
9. Install the air cleaner case sub assembly.
10. Install the air cleaner cap sub assembly.
11. Install the air cleaner inlet assembly.
12. Connect the cable to the negative battery terminal.

3.5L Engine

See Figure 60.

1. Before servicing the vehicle, refer to the Precautions Section.
2. Disconnect the negative battery cable.
3. Remove cool air intake duct seal.
4. Remove v-bank cover sub-assembly.
5. Remove air cleaner inlet assembly.
6. Remove air cleaner cap sub-assembly.
7. Remove air cleaner case sub-assembly.
8. Remove No. 1 air cleaner inlet.
9. Disconnect the terminal 50 connector from the starter assembly.
10. Remove the nut and disconnect the wire harness from terminal 30.
11. Remove the 2 bolts and starter assembly.

To install:

12. Install the starter assembly with the 2 bolts and tighten to 26 ft. lbs. (37 Nm).
13. Connect the wire harness to terminal 30 and install the nut and tighten to 87 inch lbs. (9.8 Nm).
14. Cover the nut with the cap.
15. Connect terminal 50 to the starter assembly.
16. To complete installation, reverse removal procedure.

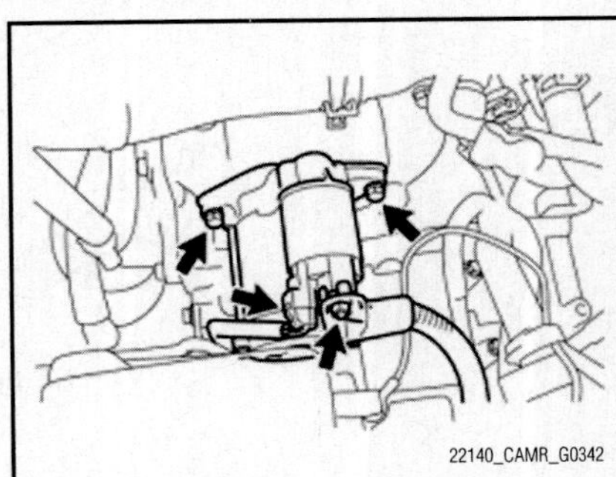

Fig. 60 Removing starter assembly—3.5L engine

ENGINE MECHANICAL

ACCESSORY DRIVE BELTS

ACCESSORY BELT ROUTING

See Figures 61 and 62.

INSPECTION

See Figure 63.

Visually check the V-ribbed belt for excessive wear, frayed cords, etc. If any defect has been found, replace the V-ribbed belt.

- Cracks on the rib side of a belt are considered acceptable. If the belt has chunks missing from the ribs, it should be replaced
- A "new belt" is a belt which has been used for less than 5 minutes with the engine running

Generator
Idler
Water Pump
Tensioner
Idler
A/C Compressor
Vane Pump
Crankshaft
22140_CAMR_G0007

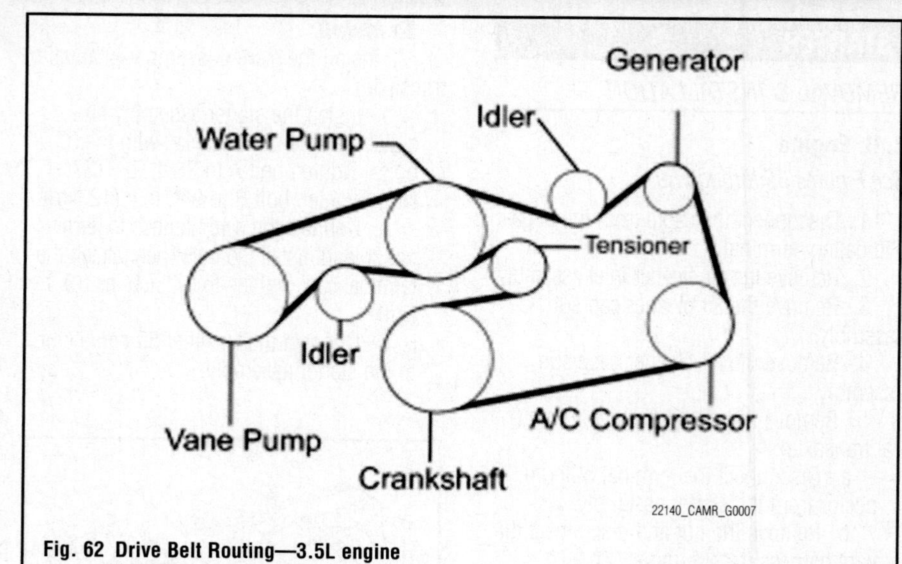

Fig. 62 Drive Belt Routing—3.5L engine

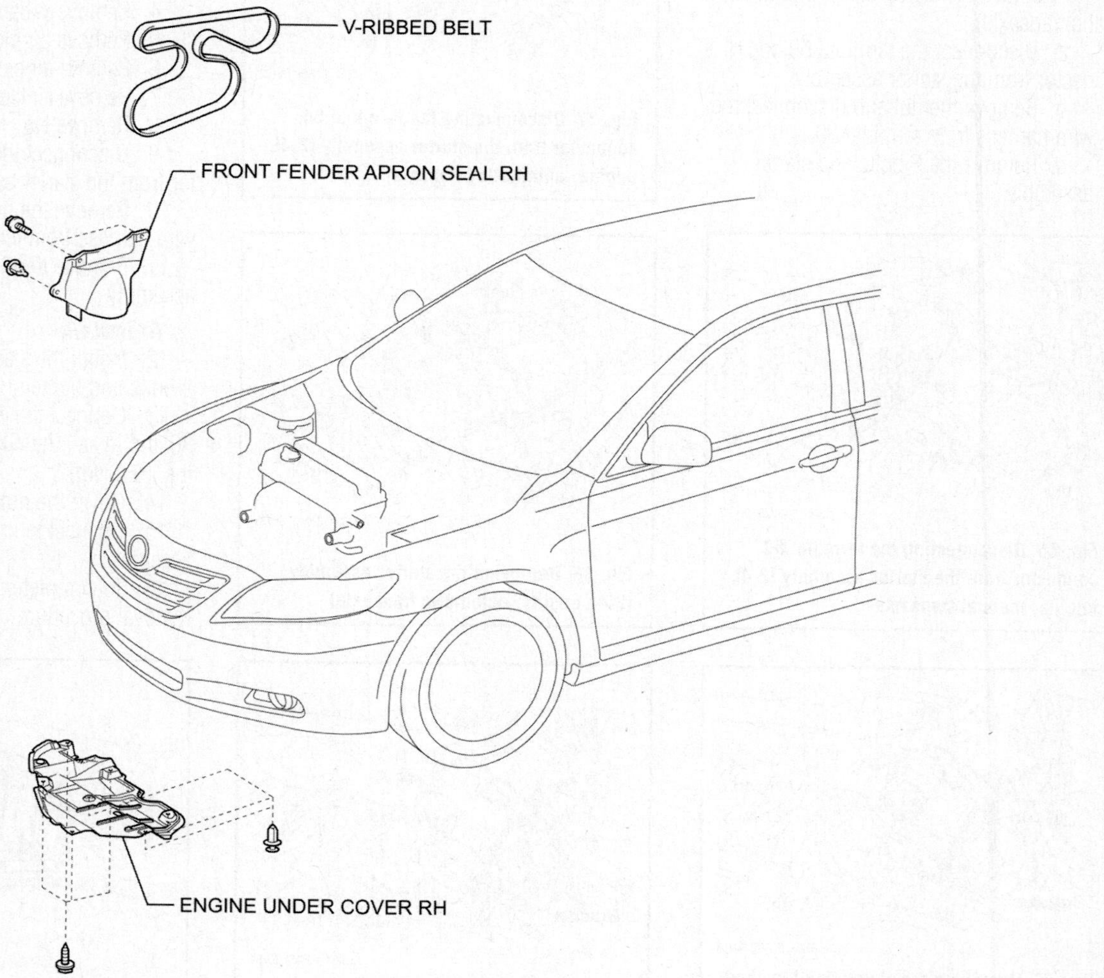

V-RIBBED BELT
FRONT FENDER APRON SEAL RH
ENGINE UNDER COVER RH
22140_CAMR_G0004

Fig. 61 Locating drive belt components—2.4L engine

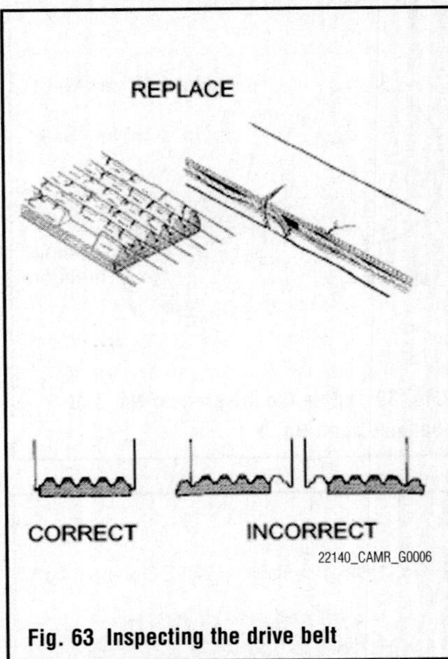

Fig. 63 Inspecting the drive belt

- A "used belt" is a belt which has been used for 5 minutes or more with the engine running

ADJUSTMENT

This vehicle is equipped with an auto-tensioner and cannot be adjusted.

REMOVAL & INSTALLATION

2.4L Engine

See Figure 61.

1. Before servicing the vehicle, refer to the Precautions Section.
2. Remove the right hand front wheel.
3. Remove the right hand engine under cover.
4. Remove the right hand front fender apron seal.

➡Before removing, take note of the following:

- Be sure to connect Special Tool: 09216—42010 and the tools so that they are in line during use
- When retracting the tensioner, turn it clockwise slowly for 3 seconds or more. Do not apply force rapidly
- After the tensioner is fully retracted, do not apply force any more than necessary

5. Using the Special Tool and a 19 mm socket wrench, loosen the v-ribbed belt tensioner arm clockwise, then remove the v-ribbed belt.
6. Remove the v-ribbed belt.

To install:

7. To install, reverse removal procedure.
8. After installing the V-ribbed belt, check that it fits properly in the ribbed grooves. Check to confirm that the belt has not slipped out of the grooves on the bottom of the crank pulley by hand.
9. Tighten the right hand front wheel to: 76 ft. lbs. (103 Nm).

3.5L Engine

See Figures 64 and 65.

1. Before servicing the vehicle, refer to the Precautions Section.
2. Remove the right hand front wheel.
3. Remove the right hand front fender apron seal.
4. Remove the V-bank cover sub-assembly.
5. Using Special Tool: 09249—63010, release the belt tension by turning the belt tensioner counterclockwise, and remove the V-ribbed belt from the belt tensioner.
6. While turning the belt tensioner counterclockwise, align with its holes and then insert the 5 mm bi-hexagon wrench into the holes to fix the V-ribbed belt tensioner.
7. Remove the v-ribbed belt.

To install:

8. To install, reverse removal procedure.
9. If it is difficult to install the V-ribbed belt, perform the following procedure:
   a. Put the V-ribbed belt on every pulley except the tensioner pulley.
   b. While releasing the belt tension by turning the belt tensioner counter-

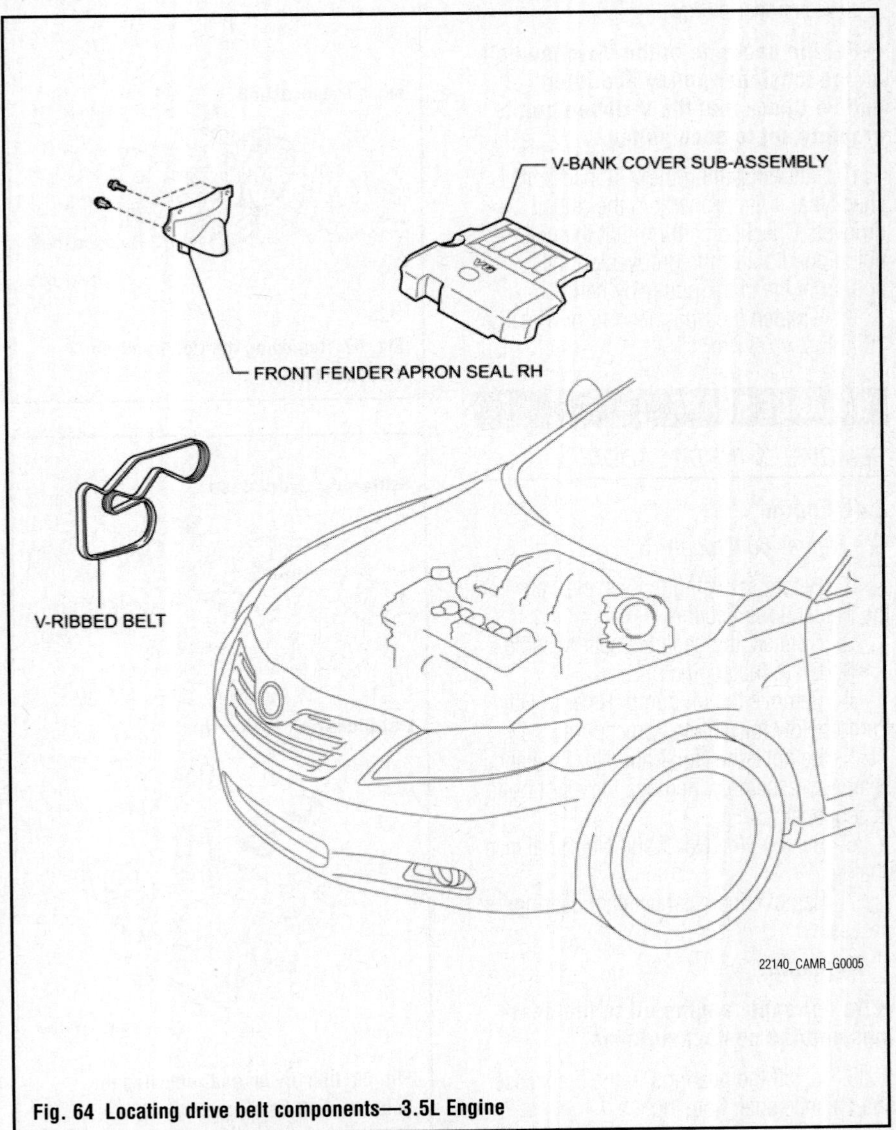

Fig. 64 Locating drive belt components—3.5L Engine

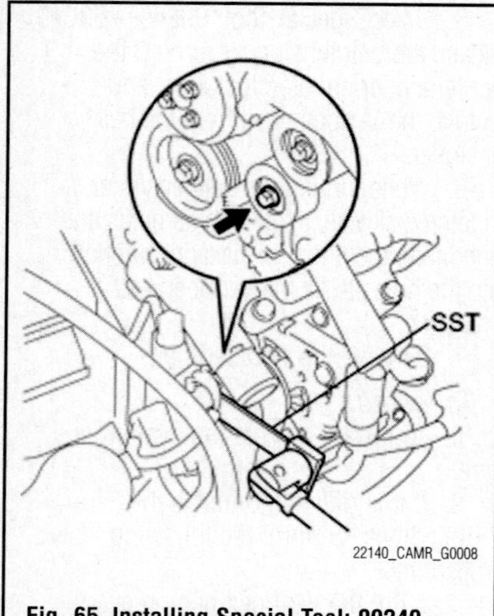

Fig. 65 Installing Special Tool: 09249—63010

clockwise, put the V-ribbed belt on the tensioner pulley.

➡**Put the backside of the V-ribbed belt on the tensioner pulley and idler pulley. Check that the V-ribbed belt is properly set to each pulley.**

10. After installing the V-ribbed belt, check that it fits properly in the ribbed grooves. Check to confirm that the belt has not slipped out of the grooves on the bottom of the crank pulley by hand.

11. Tighten the right hand front wheel to: 76 ft. lbs. (103 Nm).

BALANCE SHAFT

REMOVAL & INSTALLATION

2.4L Engine

See Figures 66 through 73.

1. Before servicing the vehicle, refer to the Precautions Section.
2. Connect the negative battery cable.
3. Drain the engine oil.
4. Remove the oil pump. Refer to Oil Pump below for removal procedure.
5. Remove the No. 1 and No. 2 balance shaft sub-assembly. Remove the eight bolts in sequence.
6. Remove the No. 1 and No. 2 balance shafts.
7. Remove the balance shaft bearings if necessary.

To install:

➡**Do not apply engine oil to the bearings and the contact surfaces.**

8. Install the bearings in the crankcase and balance shaft housing.

9. Apply a light coat of engine oil to the bearings.

➡**Confirm that the match marks on driven gears No. 1 and No. 2 are matched.**

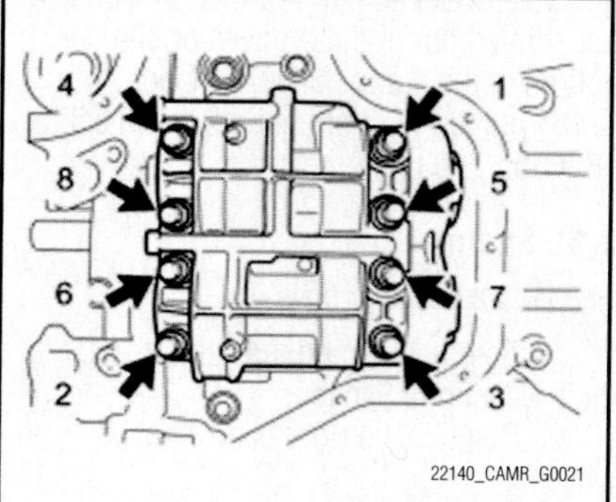

Fig. 66 Sub-assembly bolt removal sequence

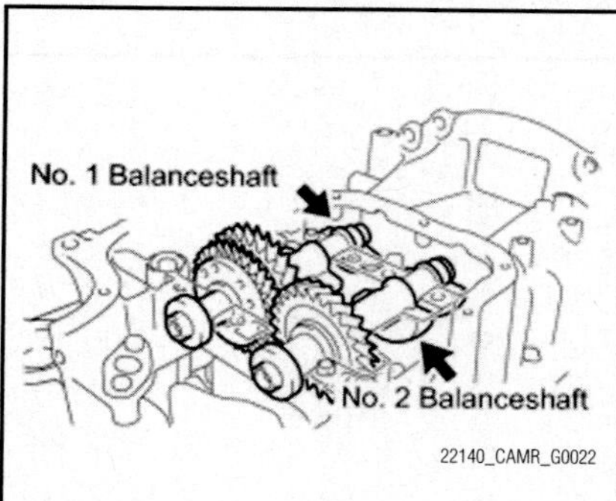

Fig. 67 Removing the No. 1 and No. 2 balance shafts

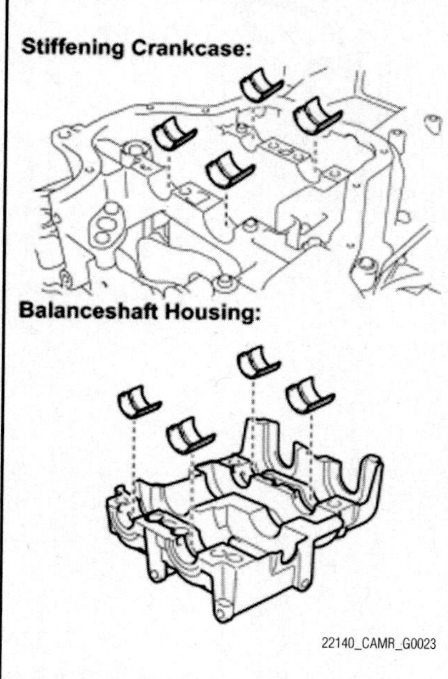

Fig. 68 Identifying and removing the balance shaft bearings

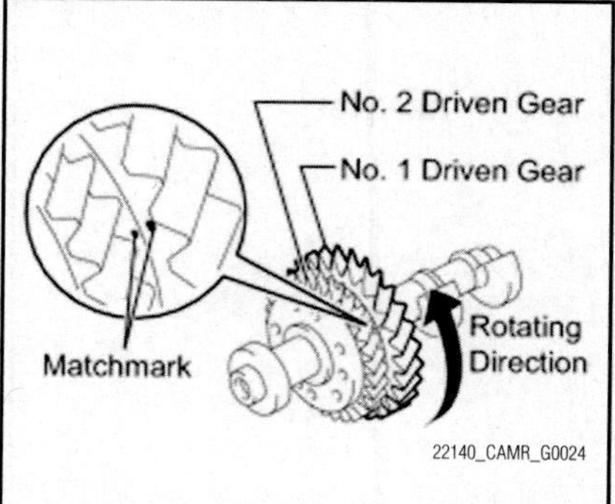

Fig. 69 Rotate the driven gear No. 1 of balance shaft No. 1

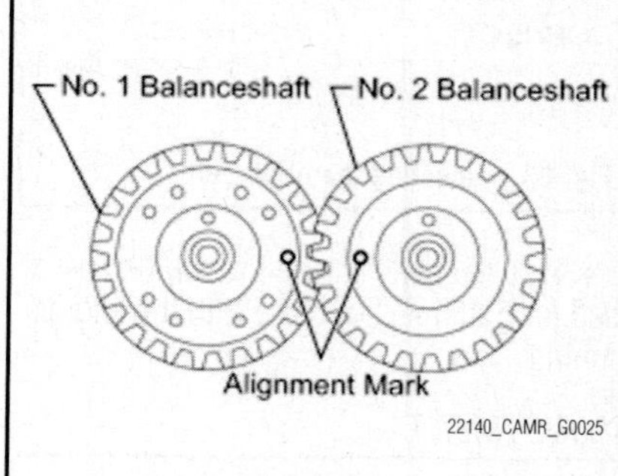

Fig. 70 Aligning marks of the No. 1 and No. 2 balance shafts

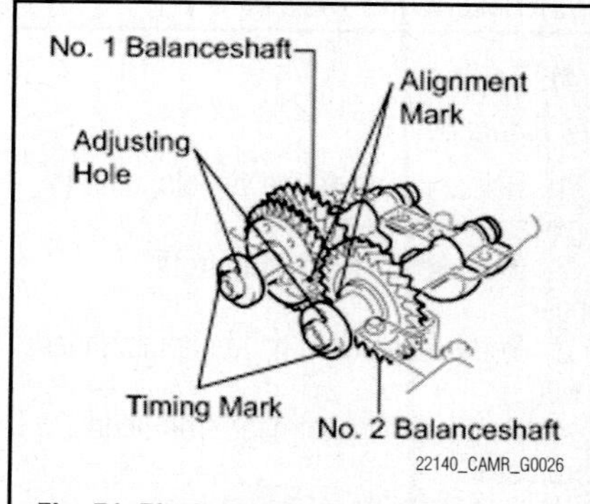

Fig. 71 Placing the No. 1 and No. 2 balance shafts on the crankcase

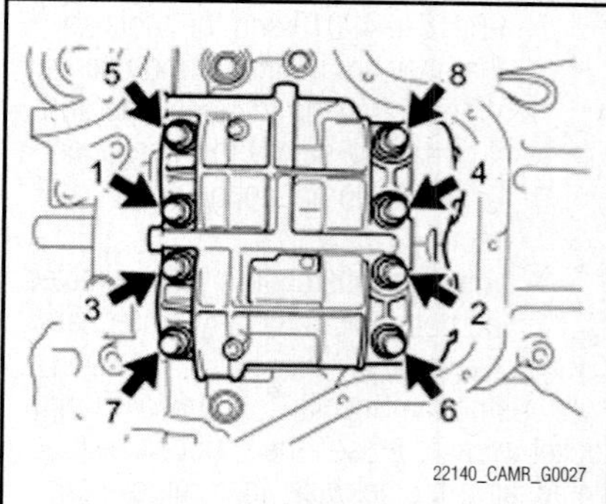

Fig. 72 Balance shaft housing bolt tightening sequence

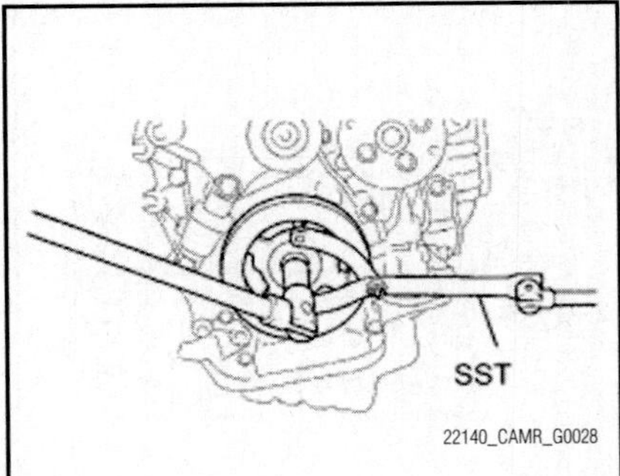

Fig. 73 Marking the bolt head and tightening procedure

10. Install No. 1 and No. 2 balance shaft sub-assembly. Rotate the driven gear No. 1 of balance shaft No. 1 in the rotating direction until it hits the stopper.
11. Align the alignment marks of the No. 1 and No. 2 balance shafts as shown.
12. Place the No. 1 and No. 2 balance shafts on the crankcase.
13. Apply a light coat of engine oil under the heads of the balance shaft housing bolts.
14. Install the balance shaft housing bolts. The balance shaft housing bolts should be tightened in 2 progressive steps as follows:
    a. Tighten the eight balance shaft housing bolts in sequence to: 16 ft. lbs. (22 Nm).
    b. Mark the front side of each balance shaft housing bolt head with paint. Retighten the bolts by 90°. Check that the paint marks are now at a 90°angle to the front.
15. To complete installation, reverse remaining removal procedure.
16. Check the engine for leaks.

CAMSHAFT AND TIMING GEAR

INSPECTION

2.4L Engine

➡Be careful not to damage the camshafts.

1. To inspect the No.1 camshaft, perform the following:
    a. Clamp the camshaft in a vise, and confirm that the camshaft timing gear is locked.
    b. Release the lock pin.

➡The 2 advance side paths are provided in the groove of the camshaft. Plug one of the paths with a rubber piece.

c. Cover the 4 oil paths of the cam journal with vinyl tape.

d. Break through the tape of the advance side path and the retard side path on the opposite side to the hole of the advance side path.

➡Cover the paths with a piece of cloth when applying pressure to keep oil from splashing.

e. Apply approximately 28 psi of air pressure to the two broken paths.

f. Check that the camshaft timing gear revolves in the advance direction when reducing the air pressure of the retard side path.

➡This operation releases the lock pin for the most retarded position.

➡Do not remove the air gun from the advance side path first. The gear may abruptly shift in the retard direction and break the lock pin.

g. When the camshaft timing gear reaches the most advanced position, remove the air gun from the retard side path and advance side path, in that order.

➡Do not use an air gun to check for smooth operation.

h. Rotate the camshaft timing gear within its movable range several times, but do not turn it to the most retarded position. Check that the gear rotates smoothly.

i. Check the lock in the most retarded position. Confirm that the camshaft timing gear is locked at the most retarded position.

j. To inspect the camshaft for runout, place the camshaft on V-blocks.

k. Using a dial indicator, measure the circle runout at the center journal. Maximum circle runout: 0.0012 inches (0.03 mm). If the circle runout is greater than the maximum, replace the camshaft.

l. Inspect the cam lobes by using a micrometer and measure the cam lobe height. Standard cam lobe height: 1.8624 to 1.8664 inches. (47.306 to 47.406 mm). Minimum cam lobe height: 1.8581 inches (47.196 mm).

m. If the cam lobe height is less than the minimum, replace the No.1 camshaft.

n. Inspect the camshaft journals by using a micrometer and measure the journal diameter. If the journal diameter is not as specified, check the oil clearance.

2. To inspect the No.2 Camshaft, perform the following:
    a. Place the camshaft on V-blocks.
    b. Using a dial indicator, measure the circle runout at the center journal. Maximum circle runout: 0.0012 inches. (0.03 mm). If the circle runout is greater than the maximum, replace the No. 2 camshaft.
    c. Inspect the cam lobes by using a micrometer and measure the cam lobe height. Standard cam lobe height: 1.8104 to 1.8143 inches. (45.983 to 46.083 mm). Minimum cam lobe height: 1.8060 inches (45.873 mm).
    d. If the cam lobe height is less than the minimum, replace the No.2 camshaft.
    e. Inspect the camshaft journals by using a micrometer and measure the journal diameter. If the journal diameter is not as specified, check the oil clearance.

3.5L Engine

1. To inspect the camshaft, place the camshaft on V-blocks.
2. Using a dial indicator, measure the circle runout at the center journal.
    a. Maximum circle runout: 0.0016 inches (0.04 mm). If the circle runout is greater than the maximum, replace the camshaft.

➡Check the oil clearance after replacing the camshaft.

3. Inspect the cam lobes by using a micrometer and measure the cam lobe height.
    a. Standard cam lobe height:
        - Intake: 1.7447 to 1.7487 inches (44.316 to 44.416 mm)
        - Exhaust: 1.7426 to 1.7465 inches. (44.262 to 44.362 mm)
    b. Maximum cam lobe height:
        - 1.7388 inches (44.166 mm)
        - 1.7367 inches (44.112 mm)
4. Inspect the camshaft journals by using a micrometer and measure the journal diameter. If the journal diameter is not as specified, check the oil clearance.
    a. No. journal: 1.4152 to 1.4157 inches (35.946 to 35.960 mm).
    b. Other journals: 1.0220 to 1.0226 inches (25.959 to 25. 975 mm).

REMOVAL & INSTALLATION

CAUTION

All models are equipped with a Supplemental Restraint System (SRS), which uses an air bag. Whenever working near any of the SRS components, such as the impact sensors, the air bag module, steering column and instrument panel, disable the SRS.

2.4L Engine

See Figures 74 through 85.

1. Before servicing the vehicle, refer to the Precautions Section.
2. Disconnect the negative battery cable.
3. Loosen the lug nuts on the front right hand wheel.
4. Apply the parking brake, block the rear wheels, then raise and safely support the front of the vehicle securely on jackstands.
5. Remove the front right hand wheel.
6. Remove the left and right hand under cover.
7. Remove the No.1 engine cover sub-assembly and 2 nuts.
8. Remove the ignition coil assembly.
9. Remove the cylinder head cover sub-assembly
10. Set No.1 Cylinder to TDC/Compression by performing the following:
    - Turn the crankshaft pulley until its groove and the timing mark "0" of the timing chain cover are aligned
    - Check that each timing mark of the camshaft timing gear and sprocket is aligned with each timing mark located on the No. 1 and No. 2 bearing caps as shown in the illustration. If not, turn the crankshaft by 1 revolution (360°) to align the timing marks

➡Do not turn the crankshaft without the chain tensioner.

11. Remove No.1 Chain tensioner assembly. Remove the 2 nuts, tensioner and gasket.
12. Remove the No. 2 camshaft by performing the following:
    a. While holding the camshaft with a wrench, loosen the camshaft timing set bolt.
    b. Using several steps, uniformly loosen and remove the 10 bearing cap bolts in the sequence shown in the illustration.
    c. While holding the No. 2 camshaft by hand, remove the camshaft timing sprocket set bolt.
    d. Remove the camshaft timing sprocket from the No. 2 camshaft with the timing chain wrapped on the sprocket.
    e. Remove the camshaft timing sprocket from the timing chain.
13. Remove the No. 1 camshaft by performing the following:
    a. Using several steps, uniformly loosen and remove the 10 bearing cap bolts in the sequence shown in the illustration.
    b. Remove the 5 bearing caps.
    c. Remove the camshaft and camshaft timing gear while holding the timing chain by hand.

➡Be careful not to drop anything inside the timing chain cover.

    d. Tie the timing chain with a string.
14. Remove the camshaft timing gear assembly by performing the following:
    a. Clamp the camshaft in a vise, and make sure that the camshaft timing gear does not rotate.
    b. Cover all the oil ports except the advance side port shown in the illustration with vinyl tape

➡Cover the paths with a shop rag or piece of cloth to avoid oil splashes.

➡Depending on the air pressure, the camshaft timing gear will turn to the advance angle side without applying force by hand. Also, if the pressure is difficult to apply because of air leakage from the port, the lock may be difficult to release.

    c. Apply air pressure of 14 psi to the oil path, then turn the camshaft timing gear in the advance direction (counterclockwise) by hand.
    d. Remove the flange bolt of the camshaft timing gear. Be sure not to remove the other four bolts. If planning to reuse the gear, be sure to release the straight pin lock before installing the gear.

To install:

15. Put the camshaft timing gear and camshaft together with the straight pin and key groove misaligned.

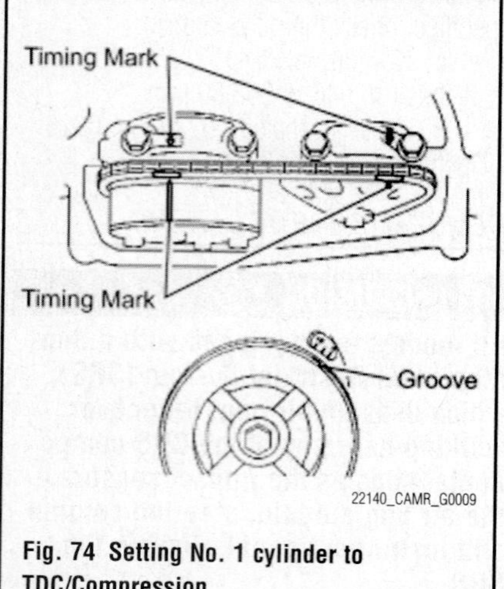

Fig. 74 Setting No. 1 cylinder to TDC/Compression

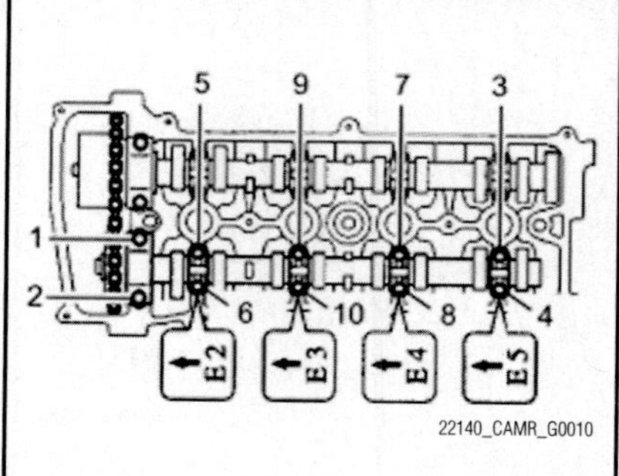

Fig. 75 No. 2 camshaft bearing cap bolt removal sequence

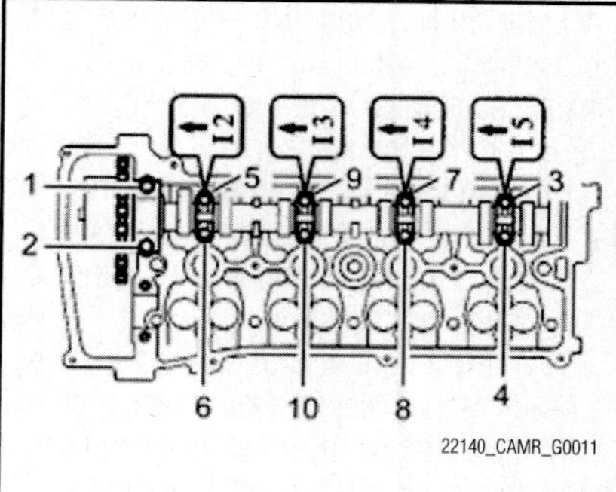

Fig. 76 No.1 camshaft bearing cap bolt removal sequence

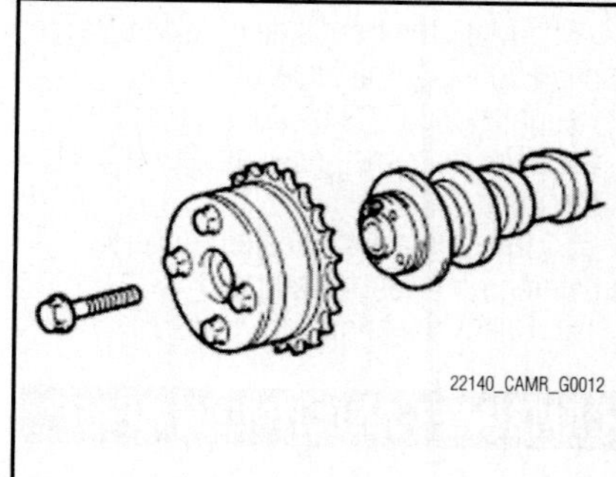

Fig. 77 Removing the flange bolt from the camshaft timing gear

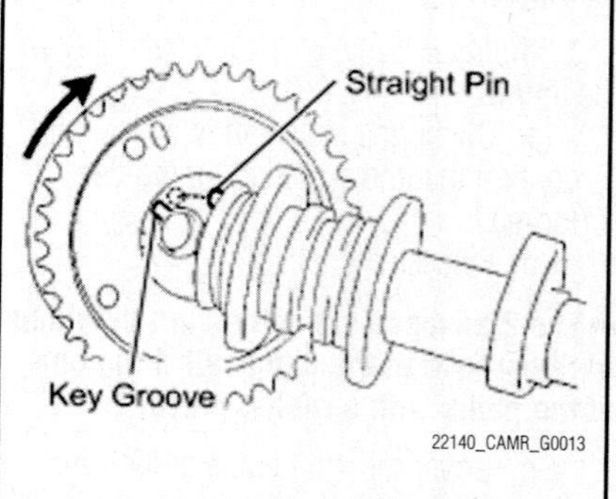

Fig. 78 No. 1 camshaft straight pin and key groove misaligned

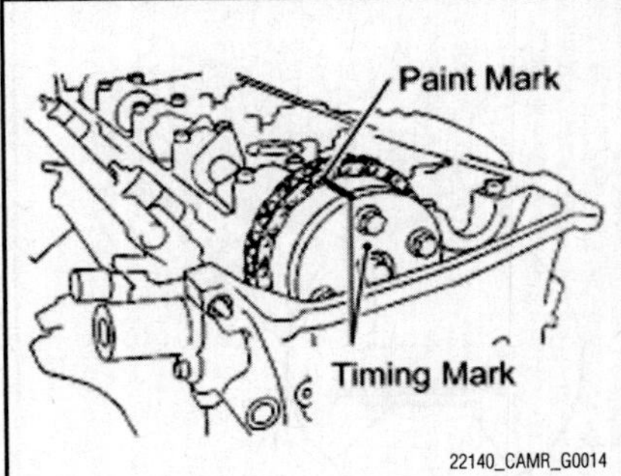

Fig. 79 Aligning No. 1 camshaft paint mark with the timing mark

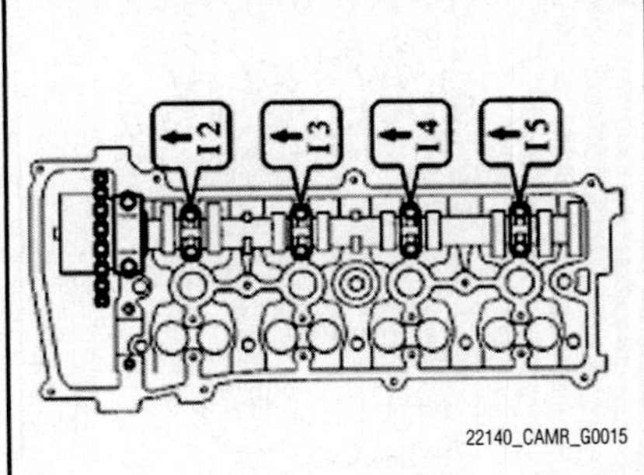

Fig. 80 Checking the No. 1 camshaft bearing cap front marks, numbers and order

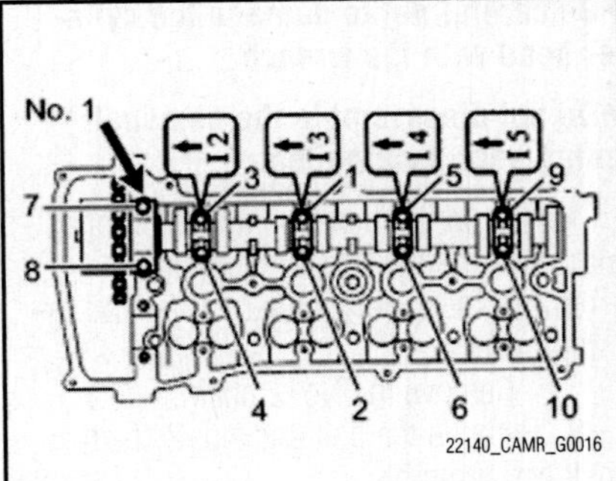

Fig. 81 No. 1 camshaft bearing cap tightening sequence

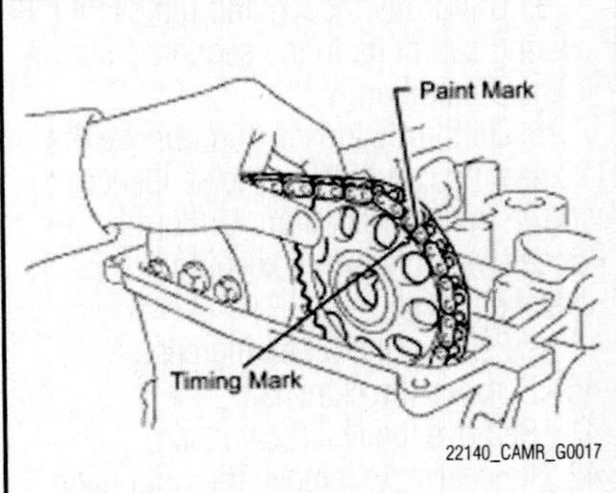

Fig. 82 Aligning No. 2 camshaft paint mark with the timing mark

➡**Be sure not to turn the camshaft timing gear to the retard angle side (the right angle).**

16. Turn the camshaft timing gear as shown in the illustration while pushing it gently against the camshaft. Push further at the position where the pin fits into the groove.

17. Check that there is no clearance between the gear and camshaft.

18. Tighten the flange bolt with the camshaft timing gear fixed in place and tighten to 40 ft. lbs. (54 Nm).

19. Check that the camshaft timing gear can move to the retard angle side (the right direction) and is locked in the most retarded position.

20. To install the No.1 camshaft, perform the following:

a. Apply a light coat of engine oil to the journal portion of the camshaft.

b. Install the timing chain onto the camshaft timing gear with the paint mark aligned with the timing mark in the camshaft timing gear.

c. Examine the front marks and numbers, and check that the order is as shown in the illustration below. Then install the bearing caps into the cylinder head.

d. Apply a light coat of engine oil to the threads and under the heads of the bearing cap bolts.

e. Using several steps, uniformly tighten the 10 bearing cap bolts in the sequence shown in the illustration. Tighten the following to specification:

- No. 1 Bearing cap: 22 ft. lbs. (30 Nm)
- No. 3 Bearing cap: 80 inch lbs. (9 Nm)

21. To install camshaft No. 2, perform the following:

a. Apply a light coat of engine oil to the journal portion of the No. 2 camshaft.

b. Put the No. 2 camshaft on the cylinder head with the paint mark of the chain aligned with the timing mark on the camshaft timing sprocket.

c. While holding the No. 2 camshaft by hand, temporarily tighten the camshaft timing sprocket set bolt.

d. Examine the front marks and numbers, and check that the order is as shown in the illustration. Then install the bearing caps onto the cylinder head.

e. Apply a light coat of engine oil to the threads and under the heads of the bearing cap bolts.

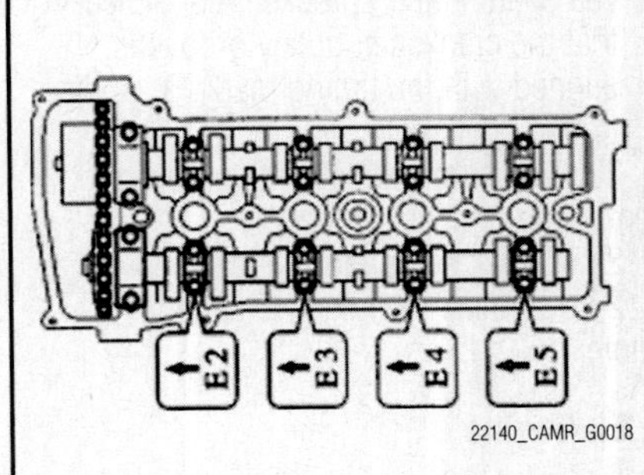

Fig. 83 Checking the No. 2 camshaft bearing cap front marks, numbers and order

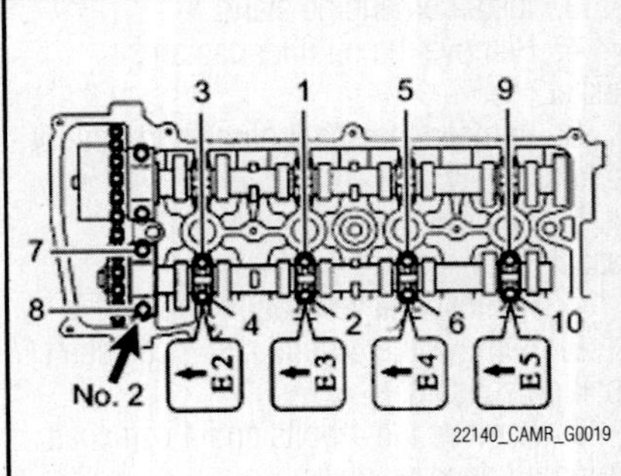

Fig. 84 No. 2 camshaft bearing cap tightening sequence

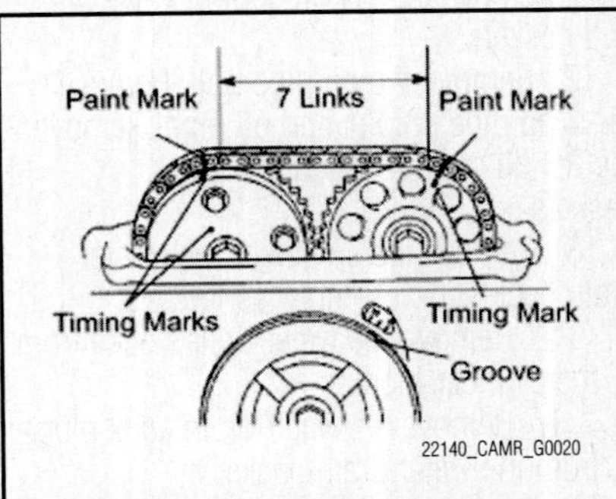

Fig. 85 Aligning paint marks on the chain with the timing marks on the camshaft timing gear and camshaft timing sprocket

f. Using several steps, uniformly tighten the 10 bearing cap bolts in the sequence shown in the illustration. Tighten the following to specification:

- No. 1 Bearing cap: 22 ft. lbs. (30 Nm)
- No. 3 Bearing cap: 80 inch lbs. (9 Nm)

g. While holding the camshaft with a wrench, tighten the camshaft timing sprocket set bolt and tighten to 40 ft. lbs. (54 Nm).

h. Check that the paint marks on the chain are aligned with the timing marks on the camshaft timing gear and camshaft timing sprocket. Also, check that the crankshaft pulley groove is aligned with the timing mark "0" of the timing chain cover.

22. To complete installation, reverse remaining removal procedure.

23. Check engine for oil leaks.

24. Connect the negative battery cable.

3.5L Engine

See Figures 86 through 105.

1. Before servicing the vehicle, refer to the Precautions Section.
2. Remove the engine assembly.
3. Install on engine stand.
4. Remove the oil filler cap and gasket.
5. Remove the spark plugs and ignition coil assembly.
6. Remove the drain plug and gasket.
7. Remove the ventilation valve.
8. Remove the 4 bolts and 4 camshaft position sensors.
9. Remove the 4 bolts and 4 camshaft timing oil control valves.
10. Remove the bolt and crankshaft position sensor.
11. Remove the 2 oil pipe unions and oil pipe. Remove the LH oil control valve filter and gaskets.
12. Remove the oil pipe bolt. Remove the 2 oil pipe unions and oil pipe. Remove the RH oil control valve filter and gaskets.
13. Remove the cylinder block water drain cock sub-assembly, as follows:
    a. Remove the water drain cocks from the cylinder block.
    b. Remove the water drain cock plugs from the water drain cocks.
14. Remove the oil filter.
15. Remove the crankshaft pulley, as follows:
    a. Using SST (SST: 09213-70011, SST: 09330-00021) or equivalent, loosen the crankshaft pulley bolt.
    b. Using SST (SST: 09950-50013) or equivalent, remove the crankshaft pulley bolt and crankshaft pulley.
16. Remove the 6 bolts and the left hand No. 1 front engine mounting bracket. Using Torx® socket wrench E8, remove the 2 stud bolts.
17. Remove the water inlet housing, as follows:
    a. Remove the 2 nuts, water inlet and thermostat.
    b. Remove the gasket.
    c. Remove the drain cock plug.
    d. Remove the drain cock.
    e. Remove the 2 stud bolts.
    f. Remove the 2 bolts, nut, and water inlet housing.
    g. Remove the 2 O-rings.
18. Remove the water outlet, as follows:
    a. Remove the 2 bolts, 4 nuts and water outlet.
    b. Remove the 2 gaskets and O-ring.
19. Remove the 12 bolts, valve cover (for Bank 1) and gasket, and then remove the 3 gaskets.
20. Remove the 12 bolts, valve cover (for Bank 2) and gasket, then remove the 3 gaskets.
21. Remove the No. 2 oil pan sub-assembly.
22. Remove the oil strainer sub-assembly.
23. Remove the oil pan sub-assembly.
24. Remove the No. 1 oil pan baffle plate.
25. Remove the engine rear oil seal, as follows:
    a. Remove the 6 bolts.
    b. Using a screwdriver with the tip taped, pry out the oil seal retainer. Be careful not to damage the oil seal retainer.

➡Be careful not to damage the engine rear oil seal retainer.

➡Tape the screwdriver tip before use.

26. Place the oil seal retainer on wooden blocks. Using a screwdriver and a hammer, tap out the oil seal.
27. Remove the water pump assembly.
28. Remove the timing chain cover sub-assembly.
29. Remove timing chain case oil seal.
30. Set the No. 1 cylinder to TDC/compression.
31. Remove the No. 1 chain tensioner assembly.
32. Remove the chain tensioner slipper.
33. Remove the chain sub-assembly.
34. Remove the idle sprocket assembly.
35. Remove the No. 1 chain vibration damper.
36. Remove crankshaft timing sprocket. Remove the 2 pulley set keys from the crankshaft.
37. Remove camshaft timing gears and No. 2 chain (for Bank 1), as follows:
    a. While raising the No. 2 chain tensioner, insert a pin of 0.039 in (1.0 mm) into the hole to fix the No. 2 chain tensioner.

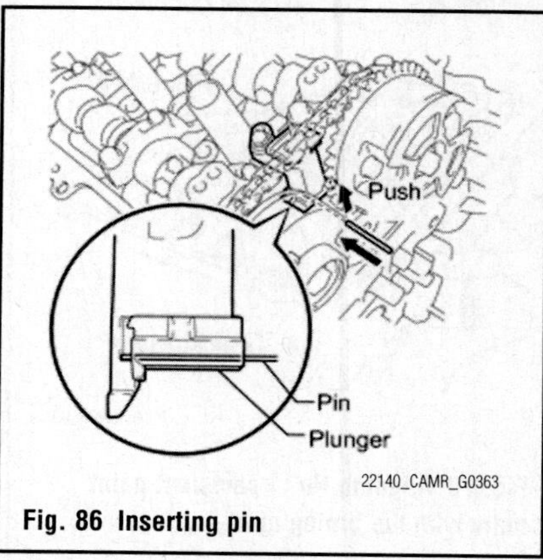

Fig. 86 Inserting pin

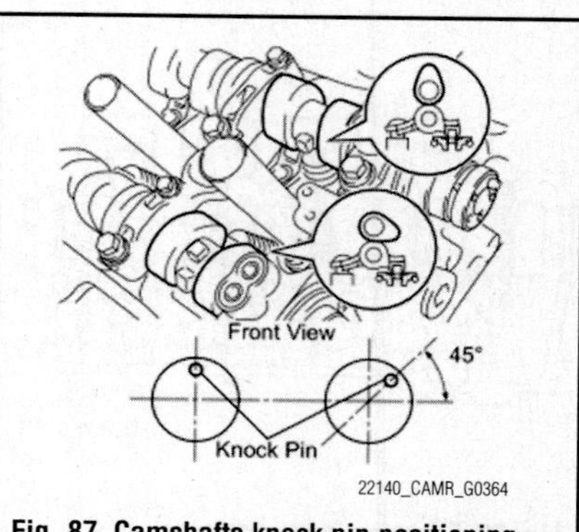

Fig. 87 Camshafts knock pin positioning—bank 1

➡Be careful not to damage the cylinder head with the wrench.

➡Do not disassemble the camshaft timing gear assemblies.

    b. Hold the hexagonal portion of the camshaft with a wrench, and remove the 2 bolts and 2 camshaft timing gears.
    c. Remove the No. 2 chain.
38. Remove the bolt and No. 2 chain tensioner assembly.
39. Remove camshaft bearing cap (for Bank 1), as follows:
    a. Check that the camshafts are positioned as shown in the illustration.
    b. Uniformly loosen and remove the 8 bearing cap bolts in the sequence shown in the illustration.
    c. Uniformly loosen and remove the 12 bearing cap bolts in the sequence shown in the illustration. Uniformly loosen the bolts while keeping the camshaft level.
    d. Remove the 5 bearing caps.
40. Remove the camshaft.
41. Remove the No. 2 camshaft.
42. If necessary, remove the right hand camshaft housing sub-assembly by prying between the cylinder head and the camshaft

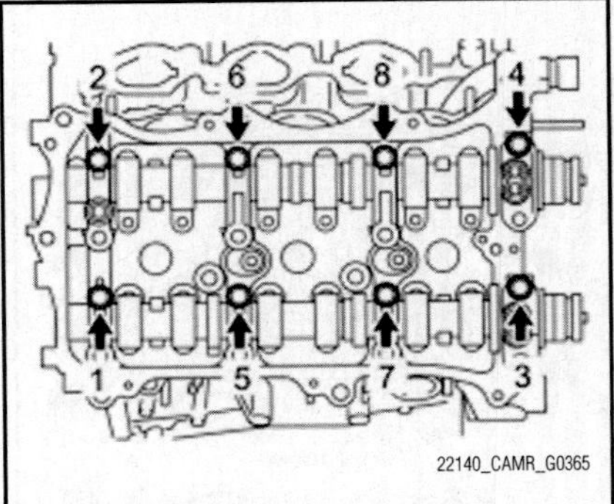

Fig. 88 8 bearing cap bolts removal sequence—bank 1

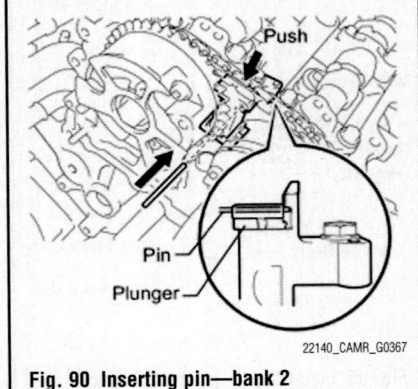

Push
Pin
Plunger
22140_CAMR_G0367

Fig. 90 Inserting pin—bank 2

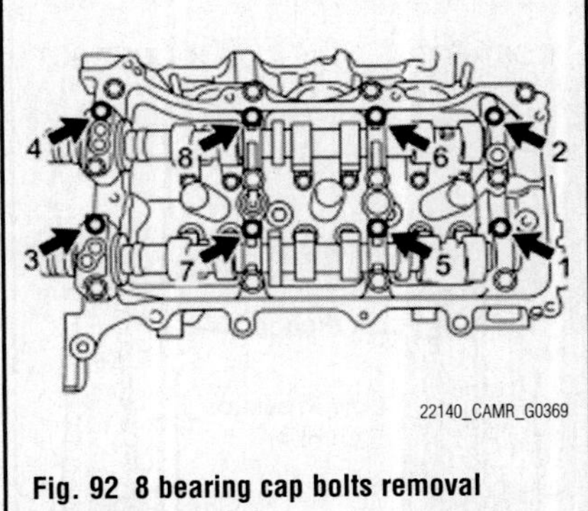

Fig. 92 8 bearing cap bolts removal sequence—bank 1

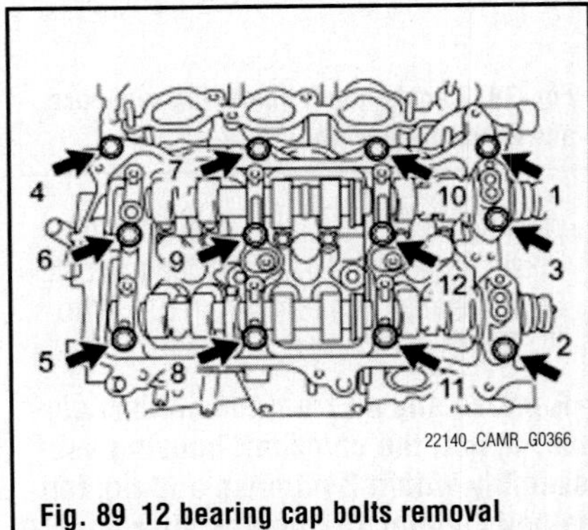

Fig. 89 12 bearing cap bolts removal sequence—bank 1

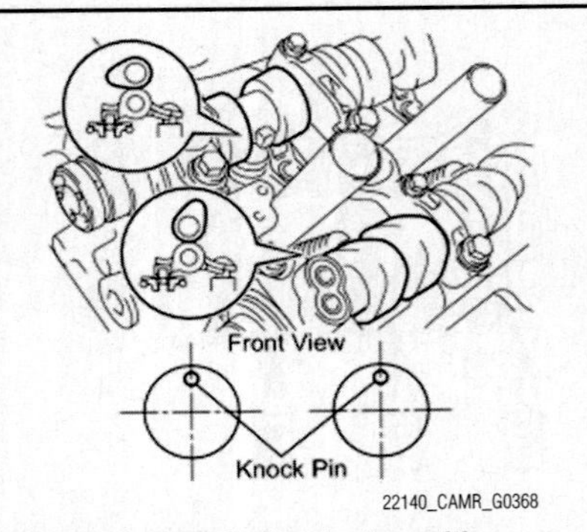

Fig. 91 Camshafts knock pin positioning—bank 2

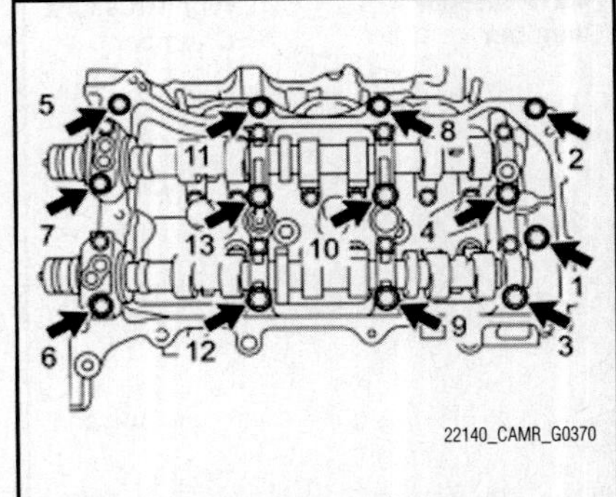

Fig. 93 13 bearing cap bolts removal sequence—bank 1

housing with a screwdriver with the tip taped.

➡Be careful not to damage the contact surfaces of the cylinder head and the camshaft housing.

43. Remove the camshaft timing gears and No. 2 chain (for Left-hand Bank), as follows:
 a. While pushing down the No. 3 chain tensioner, insert a pin of 1.0 mm (0.039 in.) into the hole to fix the No. 3 chain tensioner.
 b. Hold the hexagonal portion of the camshaft with a wrench, and remove the 2 bolts and 2 camshaft timing gears.

➡Be careful not to damage the cylinder head with the wrench.

➡Do not disassemble the camshaft timing gear assemblies.

 c. Remove the No. 2 chain.
44. Remove the bolt and No. 3 chain tensioner.
45. Remove the camshaft bearing cap (bank 2), as follows:
 a. Make sure that the knock pin of the camshaft is positioned as shown in the illustration.
 b. Uniformly loosen and remove the 8 bearing cap bolts in the sequence shown in the illustration.
 c. Uniformly loosen and remove the 13 bearing cap bolts in the sequence shown in the illustration. Loosen the bolts while keeping the camshaft level.
 d. Remove the 5 camshaft bearing caps.
46. Remove the No. 3 camshaft.
47. Remove the No. 4 camshaft.
48. If necessary, remove the left hand camshaft housing sub-assembly by prying between the cylinder head and the camshaft housing with a screwdriver with the tip taped.

➡Be careful not to damage the contact surfaces of the cylinder head and the camshaft housing.

49. Remove the No. 1 valve rocker arm sub-assembly, as follows:
 a. Remove the 24 valve rocker arms.

➡Arrange the removed parts in the correct order.

50. Remove the valve lash adjuster assembly, as follows:
 a. Remove the 24 valve lash adjusters from the cylinder head.

➡Arrange the removed parts in the correct order.

To install:

➡Keep the lash adjuster free of dirt and foreign objects. Only use clean engine oil.

51. Install the valve lash adjuster assembly, as follows:
 a. Place the lash adjuster into a container filled with engine oil. Insert the SST 09276-75010 tip into the lash adjuster's plunger and use the tip to press down on the check ball inside the plunger.
 b. Squeeze the SST and lash adjuster together to move the plunger up and down 5 to 6 times.
 c. Check the movement of the plunger and bleed the air. Make sure the plunger moves up and down.

➡When bleeding air from the high-pressure chamber, make sure that the tip of the SST is actually pressing the check ball as shown in the illustration. If the check ball is not pressed, air will not bleed.

 d. After bleeding the air, remove the SST. Then, try to press the plunger quickly and firmly with a finger. Make sure the plunger is very difficult to move.

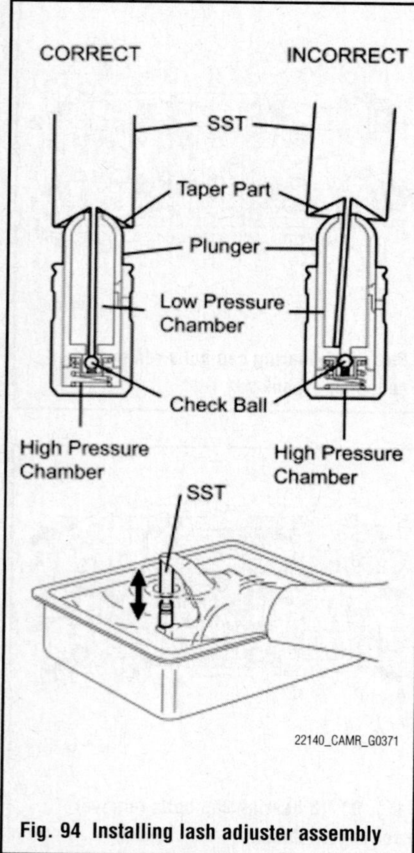

Fig. 94 Installing lash adjuster assembly

If the result is not as specified, replace the lash adjuster.

➡Install the lash adjuster to the same place it was removed from.

e. Install the lash adjusters.

52. Install the No. 1 valve rocker arm sub-assembly, as follows:

a. Apply engine oil to the lash adjuster tips and valve stem cap ends.

b. Make sure that the valve rocker arms are installed as shown in the illustration.

53. Install the right hand camshaft bearing cap, as follows:

a. Apply engine oil to the camshaft journals, camshaft housing and bearing caps.

b. Install the camshaft and No. 2 camshaft to the right camshaft housing.

c. Make sure of the marks and numbers on the camshaft bearing caps and place them in each proper position and direction.

d. Temporarily tighten the 8 bearing cap bolts to 7 ft. lbs. (10 Nm) in the order shown in the illustration.

54. Install the right hand camshaft housing sub-assembly, as follows:

a. Apply seal packing in a continuous line as shown in the illustration. Seal

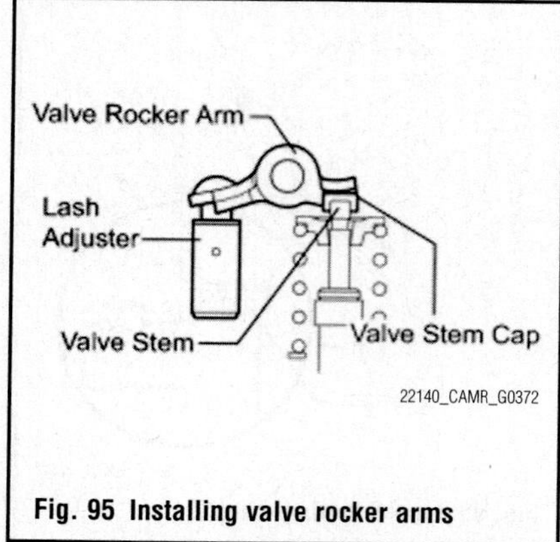

Fig. 95 Installing valve rocker arms

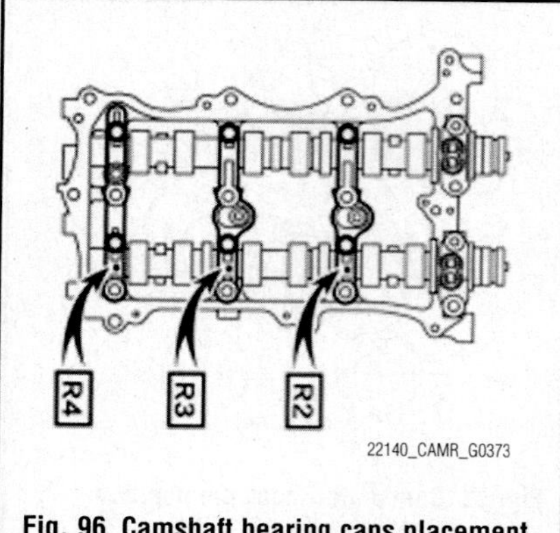

Fig. 96 Camshaft bearing caps placement

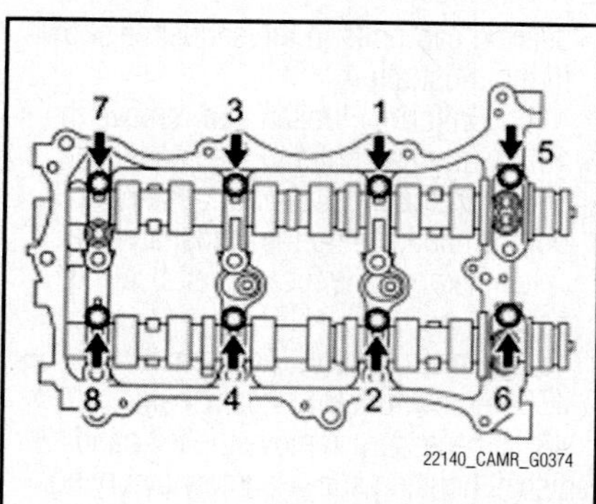

Fig. 97 Camshafts bearing cap tightening sequence

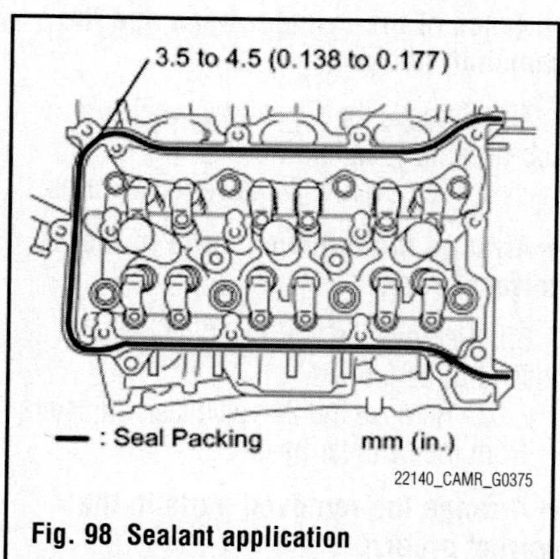

Fig. 98 Sealant application

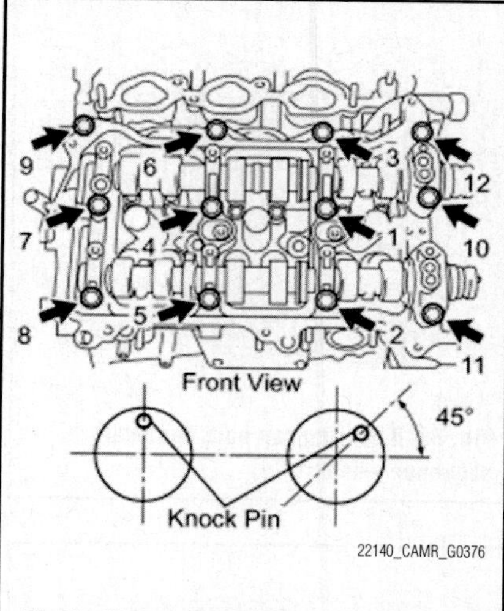

Fig. 99 Camshaft bolt tightening sequence and knock pin position

packing: Toyota Genuine Seal Packing Black, Three Bond 1207B or equivalent. Seal diameter: 3.5 to 4.5 mm (0.138 to 0.177 in.).

➡Remove any oil from the contact surface. Install the camshaft housing subassembly within 3 minutes and tighten the bolts within 15 minutes after applying sealant. Do not start the engine for at least 2 hours after installing.

b. Make sure that the knock pins of the camshafts are positioned as shown. Install the right hand camshaft housing and tighten the 12 bolts in the order shown in the illustration to 21 ft. lbs. (28 Nm).

➡When installing the camshaft housing RH, it is necessary to correctly position the camshafts as shown in the illustration. Failure to correctly position these parts may result in damage due to contact between the pistons and valves. If a camshaft is rotated with a piston at TDC, valve contact will occur.

➡If any of the bolts are loosened during installation, remove the camshaft housing, clean the installation surfaces, and reapply seal packing.

➡If the camshaft housing is removed because any of the bolts are loosened during installation, make sure that the previously applied seal packing does not enter any oil passages.

c. Tighten the 8 bolts to 12 ft. lbs. (16 Nm) in the order shown in the illustration.

55. Install the camshaft bearing cap (bank 2), as follows:

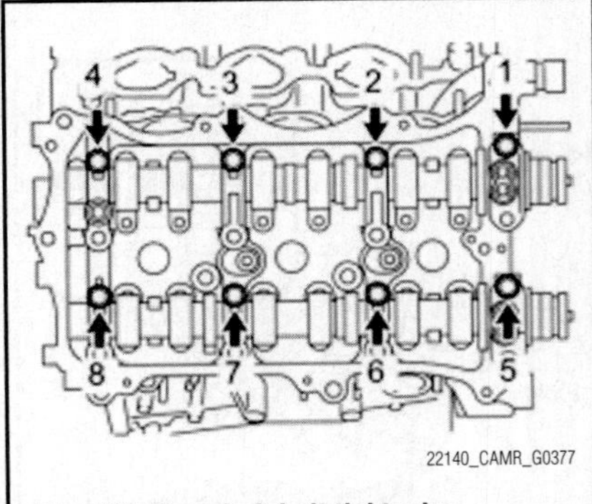

Fig. 100 Camshaft bolt tightening sequence

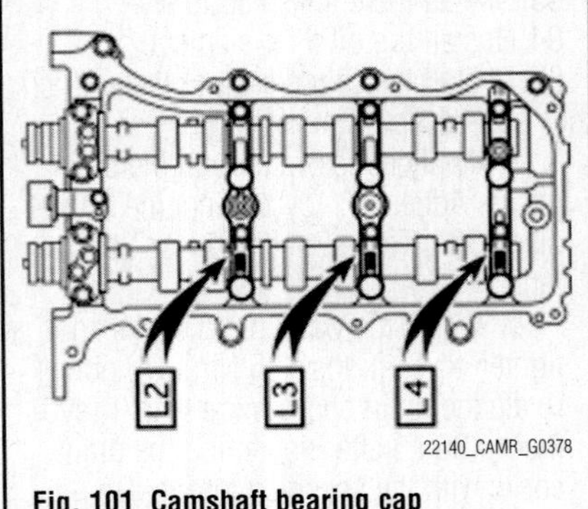

Fig. 101 Camshaft bearing cap positioning

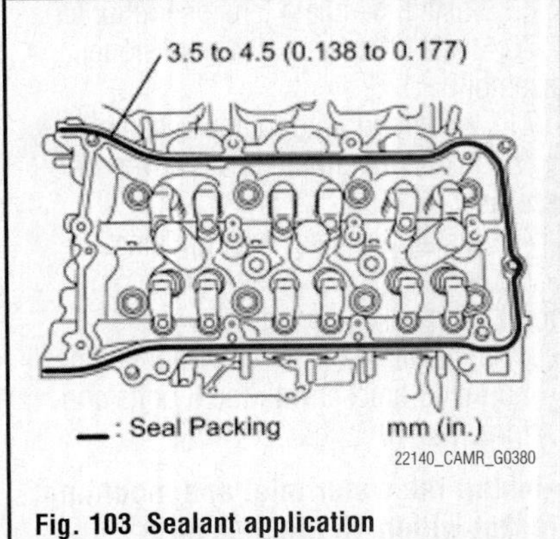

Fig. 103 Sealant application

a. Apply engine oil to the camshaft journals, camshaft housing and bearing caps.

b. Install the No. 3 camshaft and No. 4 camshaft to the left hand camshaft housing.

c. Make sure of the marks and numbers on the camshaft bearing caps and place them in each proper position and direction.

d. Temporarily tighten the 8 bolts in the order shown in the illustration to 7 ft. lbs. (10 Nm).

56. Make sure that the valve rocker arm is installed.

57. Install the left camshaft housing sub-assembly, as follows:

a. Apply seal packing in a continuous line as shown in the illustration. Seal packing: Toyota Genuine Seal Packing Black, Three Bond 1207B or equivalent. Seal diameter: 3.5 to 4.5 mm (0.138 to 0.177 in.).

➡Remove any oil from the contact surface. Install the camshaft housing sub-assembly within 3 minutes and tighten the bolts within 15 minutes after applying sealant. Do not start the engine for at least 2 hours after installing.

b. Make sure that the knock pins of the camshafts are positioned as shown. Install the left hand camshaft housing and tighten the 13 bolts in the order shown in the illustration to 21 ft. lbs. (28 Nm).

➡When installing the camshaft housing LH, it is necessary to correctly position the camshafts as shown in the illustration. Failure to correctly position these parts may result in damage due to contact between the pistons and valves. If a camshaft is rotated with a piston at TDC, valve contact will occur.

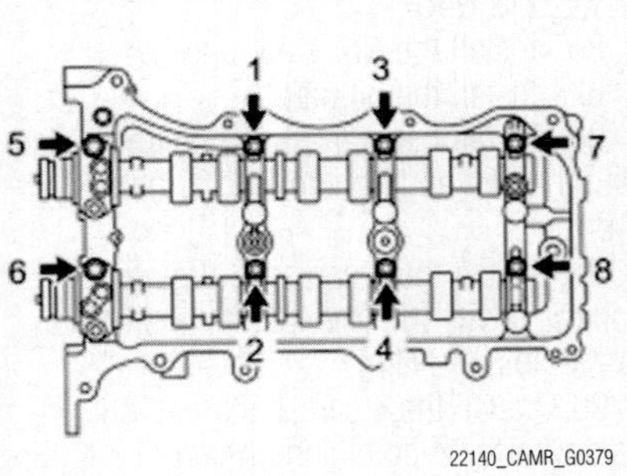

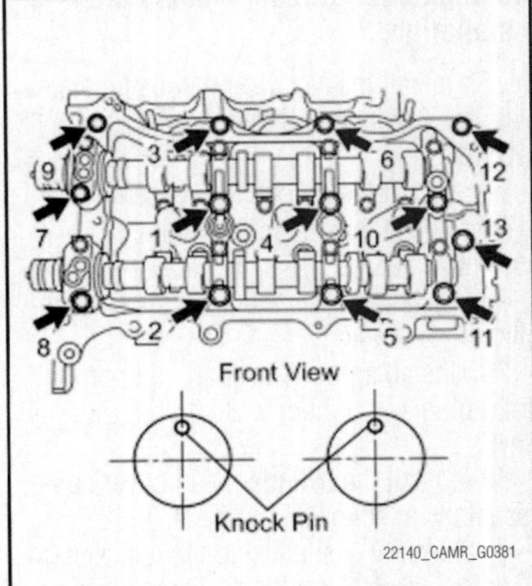

Fig. 102 Camshaft bolt tightening sequence

➡If any of the bolts are loosened during installation, remove the camshaft housing, clean the installation surfaces, and reapply seal packing.

➡If the camshaft housing is removed because any of the bolts are loosened during installation, make sure that the previously applied seal packing does not enter any oil passages.

c. Tighten the 8 bolts to 12 ft. lbs. (16 Nm) in the order shown in the illustration.

58. Install the No. 2 chain tensioner assembly with the bolt and tighten to 15 ft. lbs. (21 Nm).

59. While pushing in the tensioner, insert a pin of 1.0 mm (0.039 in.) diameter into the hole to hold it.

60. Install the camshaft timing gears and No. 2 chain (for Right-hand Bank).

61. Install the No. 3 chain tensioner assembly with the bolt and tighten to 15 ft. lbs. (21 Nm).

62. While pushing in the tensioner, insert a pin of 1.0 mm (0.039 in.) diameter into the hole to hold it.

63. Install the camshaft timing gears and No. 2 chain (for Left-hand Bank).

Fig. 104 Camshaft housing bolt tightening sequence and knock pin position

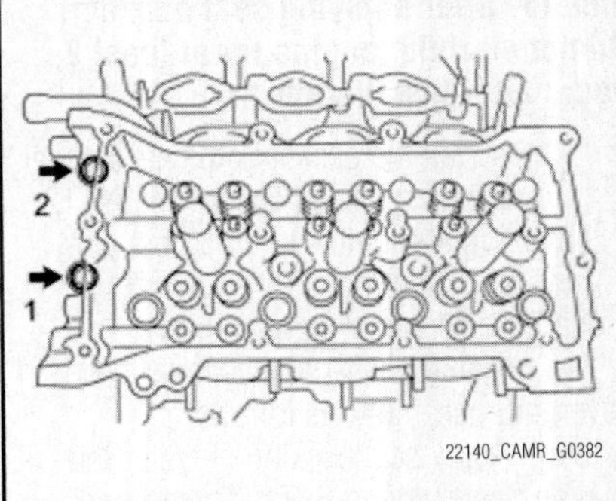

Fig. 105 Camshaft housing bolt tightening sequence

64. Install the No 1 chain vibration damper with the 2 bolts and tighten to 17 ft. lbs. (23 Nm).

65. Install the No 2 chain vibration damper.

66. Install the timing gear set keys and timing gear as shown in the illustration.

67. Install the idle sprocket assembly.

68. Install the chain sub-assembly.

69. Install the chain tensioner slipper.
70. Install the No. 1 chain tensioner assembly.
71. Install the water pump assembly.
72. Install the timing chain cover sub-assembly.
73. Install the water inlet housing.
74. Install the No. 1 left front engine mounting bracket, as follows:
  a. Install the No. 1 left front engine mounting bracket with the 6 bolts and tighten to 40 ft. lbs. (54 Nm).

➡Install the water inlet and mounting bracket within 15 minutes after installing the chain cover. Do not start the engine for at least 2 hours after installation.

75. Install the No. 1 oil pan baffle plate with the 7 bolts and tighten to 7 ft. lbs. (10 Nm).
76. Install the oil pan sub-assembly.
77. Install the oil strainer sub-assembly.
78. Install the No. 2 oil pan sub-assembly.
79. Install a new gasket and oil pan drain plug and tighten to 30 ft. lbs. (40 Nm).
80. Install the cylinder head cover sub-assembly, as follows:
  a. Apply seal packing (Toyota Genuine Seal Packing Black, Three Bond 1207B or equivalent).

➡Remove any oil from the contact surface. Install the crankcase within 3 minutes after applying seal packing. Do not start the engine for at least 2 hours after installation.

  b. Install the gasket to the head cover.
  c. Install the head cover with the 12 bolts. Tighten bolt A to 15 ft. lbs. (21 Nm), and other bolts to 7 ft. lbs. (10 Nm). Be certain to tighten bolt 1.
81. Install the left-hand cylinder head cover sub-assembly, as follows:
  a. Apply seal packing (Toyota Genuine Seal Packing Black, Three Bond 1207B or equivalent).

➡Remove any oil from the contact surface. Install the crankcase within 3 minutes after applying seal packing. Do not start the engine for at least 2 hours after installation.

  b. Install the gasket to the head cover.
  c. Install the head cover with the 14 bolts. Tighten bolt A to 15 ft. lbs. (21 Nm), and other bolts to 7 ft. lbs. (10 Nm). Be certain to tighten bolts 1 and 10.
82. Install water outlet.
83. Install the crankshaft pulley.
84. Install the oil filter element.
85. Install the cylinder block water drain cock sub-assembly, as follows:
  a. Apply adhesive around the drain cocks. Adhesive: Toyota Genuine Adhesive 1324, Three Bond 1324 or Equivalent.
  b. Install the water drain cocks and tighten to 18 ft. lbs. (25 Nm). Do not rotate the drain cocks more than 1 revolution (360°) after tightening the drain cocks with the specified torque. Do not loosen after setting correctly.
  c. Install the water drain cock plug to the water drain cocks and tighten to 9 ft. lbs. (13 Nm).
86. Install the No. 1 oil pipe.
87. Install the oil pipe.
88. Install the crankshaft position sensor with the bolt and tighten to 7 ft. lbs. (10 Nm).
89. Install the 4 camshaft timing oil control valves with the 4 bolts and tighten to 7 ft. lbs. (10 Nm).
90. Install the 4 camshaft position sensors with the 4 bolts and tighten to 7 ft. lbs. (10 Nm).
91. Install the ventilation valve sub-assembly, as follows:
  a. Apply adhesive (Toyota Genuine Adhesive 1324, Three Bond 1324 or equivalent) around the ventilation valve.
  b. Install the ventilation valve and tighten to 20 ft. lbs. (27 Nm).
92. Install the 6 spark plugs and the ignition coil assembly.
93. Install the oil filler cap sub-assembly.
94. Remove the engine stand.
95. Install the engine assembly.

Camshaft Bearing Replacement

To remove and replace the camshaft bearings, refer to the camshaft removal procedure.

CRANKSHAFT DAMPER

REMOVAL & INSTALLATION

See Figure 106.

1. Before servicing the vehicle, refer to the Precautions Section.
2. Remove the engine assembly.
3. Install on engine stand.
4. Remove the oil filler cap and gasket.
5. Remove the spark plugs and ignition coil assembly.
6. Remove the drain plug and gasket.
7. Remove the ventilation valve.

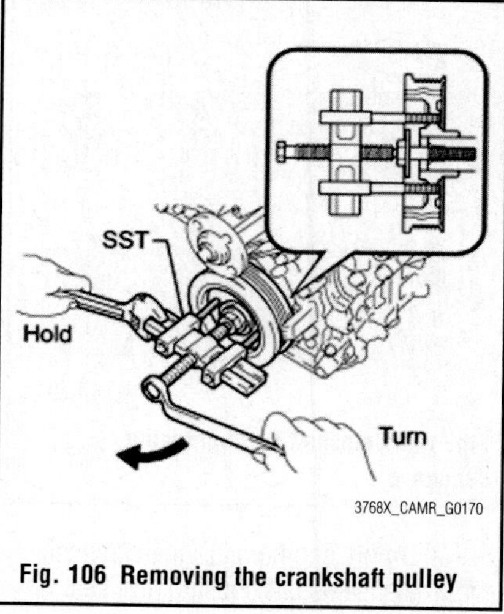

Fig. 106 Removing the crankshaft pulley

8. Remove the 4 bolts and 4 camshaft position sensors.
9. Remove the 4 bolts and 4 camshaft timing oil control valves.
10. Remove the bolt and crankshaft position sensor.
11. Remove the No. 1 oil pipe.
12. Remove the oil pipe.
13. Remove the cylinder block water drain cock sub-assembly.
14. Remove the oil filter element.
15. Remove the crankshaft pulley, as follows:
  a. Using SST (SST: 09213-70011, SST: 09330-00021) or equivalent, loosen the crankshaft pulley bolt.
  b. Using SST (SST: 09950-50013) or equivalent, remove the crankshaft pulley bolt and crankshaft pulley.

To install:

16. Install the crankshaft pulley, as follows:
  a. Align the pulley set key with the key groove of the pulley, and slide on the pulley.
  b. Using SST (SST: 09213-70011, SST: 09330-00021) or equivalent, install the pulley bolt and tighten to 184 ft. lbs. (250 Nm).
17. Install the oil filter element.
18. Install the cylinder block water drain cock sub-assembly.
19. Install the No. 1 oil pipe.
20. Install the oil pipe.
21. Install the crankshaft position sensor.
22. Install the camshaft timing oil control valve assembly.
23. Install the camshaft position sensor.
24. Install the ventilation valve sub-assembly.
25. Install the spark plugs and the ignition coil assembly.
26. Install the oil filler cap sub-assembly.

27. Remove the engine stand.
28. Install the engine assembly.

CRANKSHAFT FRONT SEAL

REMOVAL & INSTALLATION

2.4L Engine

See Figures 107 through 111.

1. Before servicing the vehicle, refer to the Precautions Section.
2. Disconnect the negative battery cable.
3. Drain the engine oil
4. Loosen the lug nuts on the left front wheel.
5. Apply the parking brake, block the rear wheels, then raise and safely support the front of the vehicle securely on jack stands.
6. Remove the left front wheel.
7. Remove the right hand front fender apron seal.
8. Remove the left and right hand engine under cover.
9. Remove the drive belt.
10. Remove the crankshaft pulley and take note of the following:

- For TMMK made crankshaft pulley, use Special Tool 09960-10010 to fix the pulley in place and to loosen the bolt. Use Special Tool 09950-40011 to remove the pulley and bolt.
- For TMC made crankshaft pulley, use Special Tools 09213-54015 and 09330-00021 to fix the pulley in place and to loosen the bolt. Use Special Tools 09950-50013 and 09950-40011 to remove the pulley and bolt.

11. Using a knife, cut off the oil seal lip. Using a screwdriver with the tip taped, pry out the oil seal.
12. After the removal, check the crankshaft for damage. If it is damaged, smooth the surface with 400-grit sandpaper.

To install:

13. Apply MP grease to a new oil seal lip. Keep the lip free from foreign matter.
14. Using Special Tool 09223-22010 and a hammer, tap in the oil seal until its surface is flush with the rear oil seal retainer edge.
15. For TMMK made pulleys, align the pulley set key with the key groove of the pulley. Using Special Tool 09960-10010, keep the pulley in place and tighten the bolt to: 125 ft. lbs. (170 Nm).
16. For TMC made pulleys, align the pulley set key with the key groove of the pulley. Using Special Tools 09213-54015 and 09330-00021, keep the pulley in place and tighten the bolt to 133 ft. lbs. (180 Nm).
17. To complete installation, reverse remaining removal procedure.
18. Check engine for oil leaks.

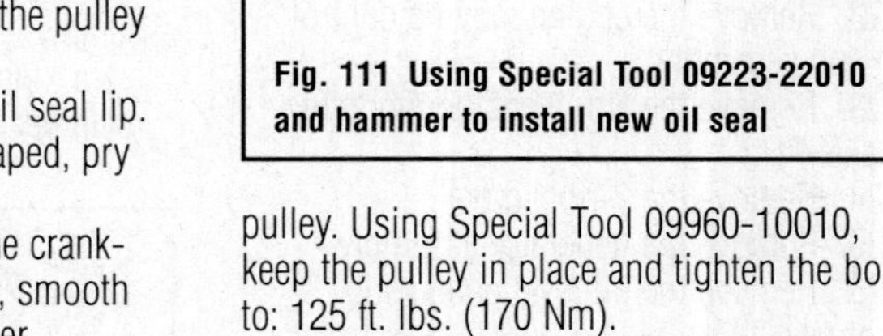

Fig. 111 Using Special Tool 09223-22010 and hammer to install new oil seal

3.5L Engine

See Timing Chain Cover and Seal.

CYLINDER HEAD

REMOVAL & INSTALLATION

2.4L Engine

See Figures 112 through 118.

1. Discharge the fuel system pressure.
2. Disconnect the cable from the negative battery terminal.
3. Remove the engine under covers.
4. Remove the front fender apron seal RH.
5. Remove the No. 1 engine cover sub assembly.
6. Drain the engine coolant.
7. Drain the engine oil.
8. Remove the windshield wiper link assembly.
9. Remove the cowl top panel outer sub assembly.
10. Remove the air cleaner inlet assembly.
11. Remove the air cleaner cap sub assembly.
12. Remove the air cleaner case sub assembly.
13. Remove the battery.
14. Remove the throttle body assembly.
15. Disconnect the fuel tube sub assembly.

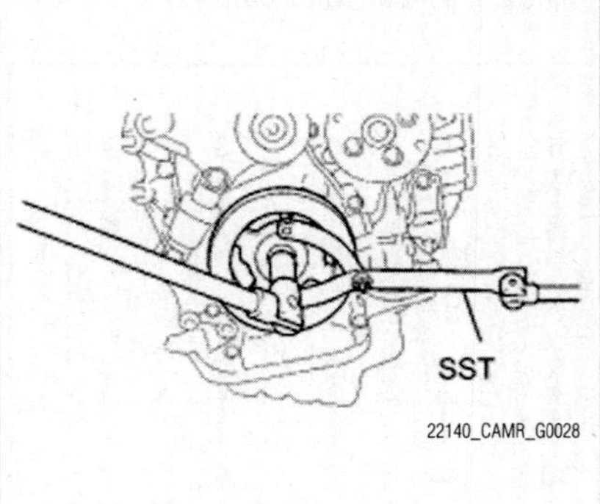

Fig. 107 Using Special Tool 09960-10010, fix the pulley in place and loosen the bolt—TMMK Pulley

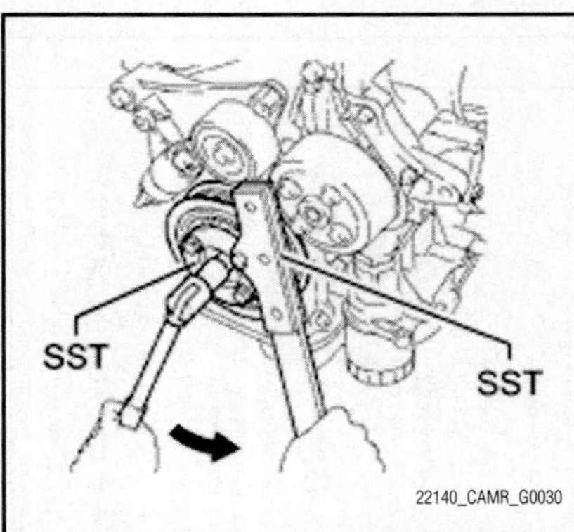
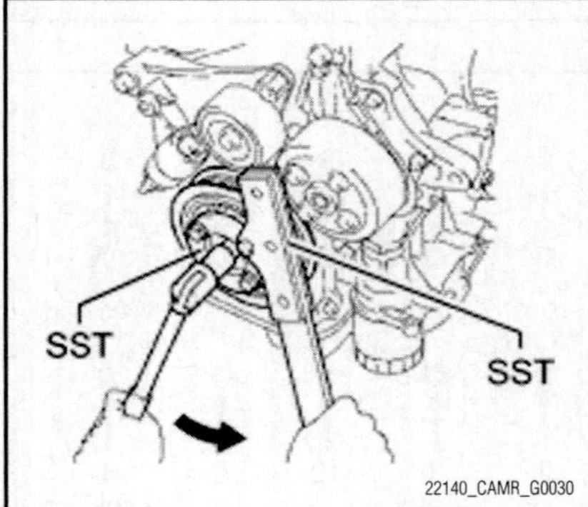

Fig. 109 Using Special Tools 09213-54015 and 09330-00021 to fix the pulley in place and to loosen the bolt—TMC Pulley

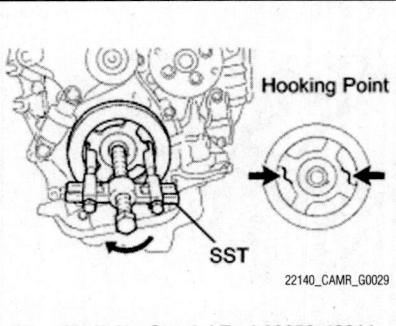

Fig. 108 Using Special Tool 09950-40011 to remove the pulley and bolt—TMMK Pulley

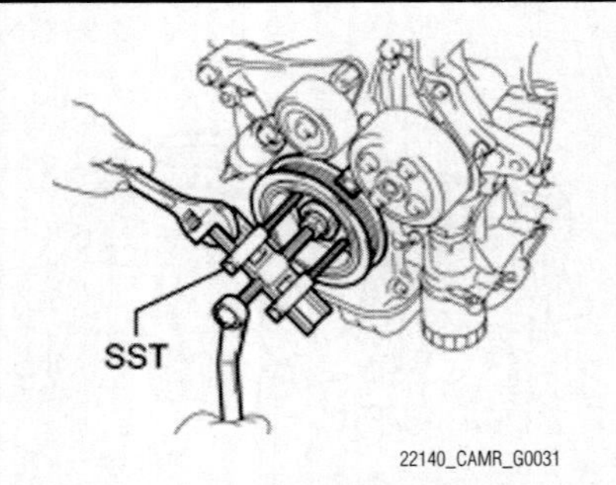

Fig. 110 Using Special Tools 09950-50013 and 09950-40011 to remove the pulley and bolt—TMC Pulley

16. Remove the fuel delivery pipe with the injectors.
17. Remove the intake manifold.
18. Remove the intake air control valve (PZEV).
19. Remove the No. 1 intake manifold insulator.
20. Remove the front exhaust pipe assembly.
21. Remove the No. 2 engine mounting stay RH.
22. Remove the engine moving control rod sub assembly.
23. Remove the No. 2 engine mounting bracket RH.
24. Remove the v-ribbed belt.
25. Remove the generator assembly.
26. Remove the oil level gauge sub assembly.
27. Remove the oil level gauge guide.
28. Remove the manifold stay.
29. Remove the No. 2 manifold stay.
30. Remove the exhaust manifold converter sub assembly.
31. Remove the No. 2 camshaft.
32. Remove the camshaft.
33. Remove the camshaft timing oil control valve assembly.
34. Disconnect the radiator hose inlet.
35. Disconnect the engine wire.
a. Disconnect the radio setting condenser connector.
b. Disconnect the engine oil pressure switch connector.
c. Disconnect the engine coolant temperature sensor connector.
d. Disconnect the camshaft position sensor connector.
e. Remove the bolt and ground cable.
36. Remove the No. 2 camshaft bearing.
37. Remove the cylinder head sub assembly.
a. Using several steps, uniformly loosen and remove the 10 cylinder head bolts and 10 plate washers with a 10 mm bi-hexagon wrench in the sequence shown in the illustration.

➡Head warpage or cracking could result from removing the bolts in the wrong order.

b. Using a screwdriver with its tip wrapped with tape, pry between the cylinder head and cylinder block, and remove the cylinder head.

➡Be careful not to damage the contact surfaces of the cylinder head and cylinder block.

38. Remove the cylinder head gasket.

To install:

39. Install the cylinder head gasket.
a. Place a new gasket on the cylinder block surface with the Lot No. stamp facing upward.

➡Remove any oil from the contact surfaces.

➡Make sure that the gasket is installed in the correct direction.

40. Install the cylinder head sub assembly.
a. Install the cylinder head on the cylinder block.

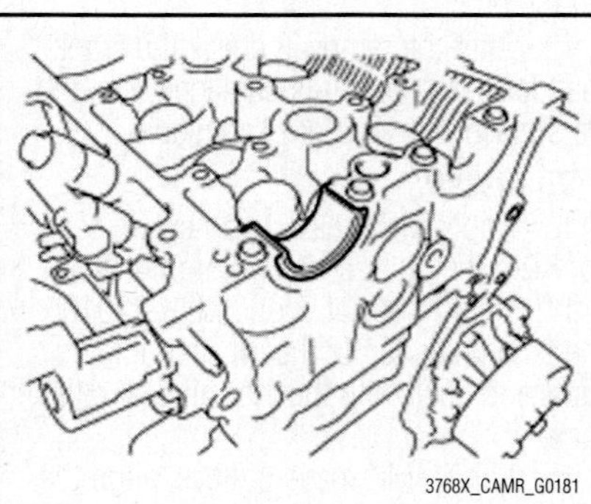

Fig. 112 Removing the No. 2 camshaft bearing

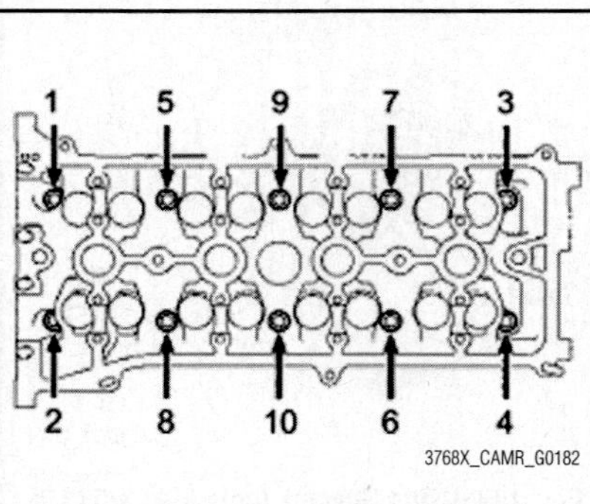

Fig. 113 Identifying the cylinder head bolt and plate washer removal sequence

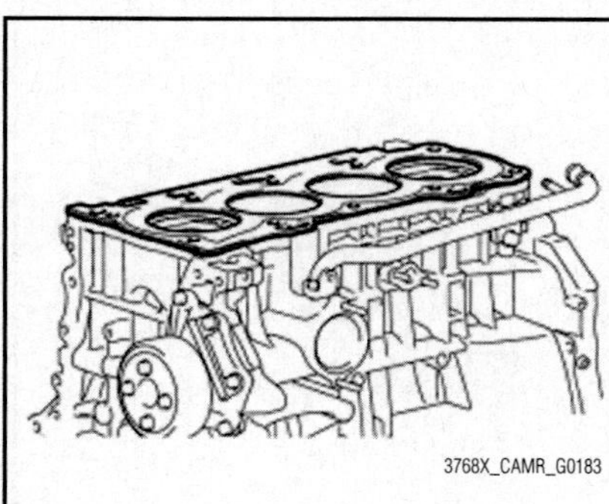

Fig. 114 Removing the cylinder head gasket

➡The cylinder head bolts are tightened in 2 progressive steps.

b. Apply a light coat of engine oil to the bolt threads and the area beneath the bolt heads that come in contact with the washers.
c. Install the bolts and plate washers to the cylinder head.

➡Do not drop the washers into the cylinder head.

d. Using several steps, uniformly install and tighten the 10 cylinder head set bolts and plate washers with a 10 mm bi-hexagon wrench in the order shown in the illustration. Tighten to 52 ft. lbs. (70 Nm).
e. Mark the front side of the cylinder head bolt with paint.
f. Retighten the cylinder head bolts 90° in the same sequence provided before.
g. Check that the paint mark is now at a 90° angle to the front.
41. Connect the engine wire.
a. Connect the ground cable with the bolt and tighten to 74 inch lbs. (8.4 Nm).
b. Connect the camshaft position sensor connector.
c. Connect the engine coolant temperature sensor connector.

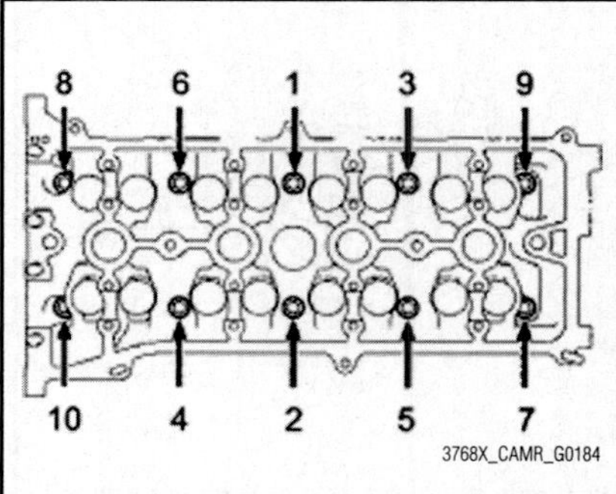

Fig. 115 Identifying the cylinder head set bolts and plate washer installation order

Fig. 116 Placing the camshafts on the cylinder head

Fig. 117 Examining front marks and numbers

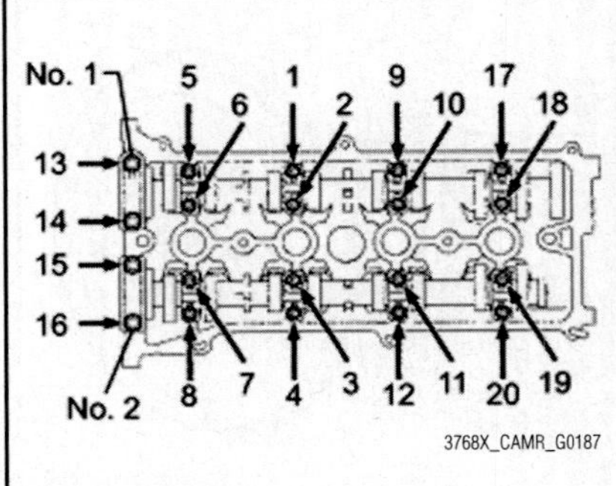

Fig. 118 Bearing cap installation sequence

d. Connect the engine oil pressure switch connector.

e. Connect the radio setting condenser connector.

42. Connect the radiator hose inlet.
43. Install the No. 2 camshaft bearing.
44. Install the camshaft timing oil control valve assembly.
45. Install the camshafts.

a. Apply a light coat of engine oil to the camshaft journals.

b. Place the 2 camshafts on the cylinder head with the No. 1 cam lobes facing the directions shown in the illustration.

c. Examine the front marks and numbers, and check that the order is as shown in the illustration. Then install the bearing caps onto the cylinder head.

d. Apply a light coat of engine oil to the threads and under the heads of the bearing cap bolts.

e. Using several steps, uniformly tighten the 20 bearing cap bolts in the sequence shown in the illustration. Tighten the No. 1 and No. 2 bearing caps to 22 ft. lbs. (30 Nm). Tighten the No. 3 bearing cap to 80 inch lbs. (9 Nm).

46. To complete installation, reverse the remaining removal procedure.

EXHAUST MANIFOLD

REMOVAL & INSTALLATION

2.4L Engine

See Figures 119 through 123.

1. Before servicing the vehicle, refer to the Precautions Section.
2. Disconnect the negative battery cable.
3. Remove engine cover.
4. Remove air cleaner assembly.

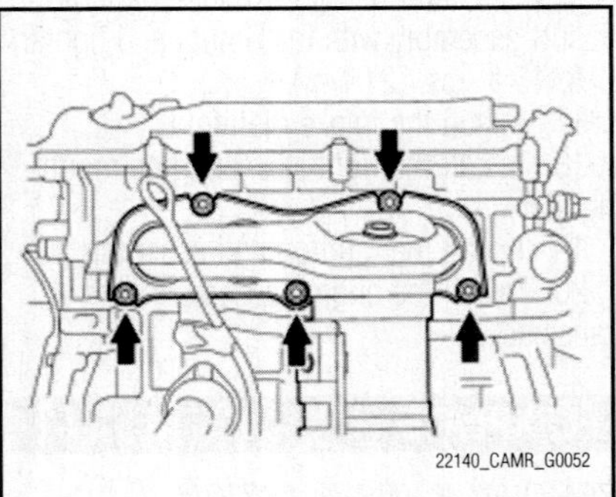

Fig. 119 Removing five nuts, exhaust manifold and gasket—PZEV vehicles

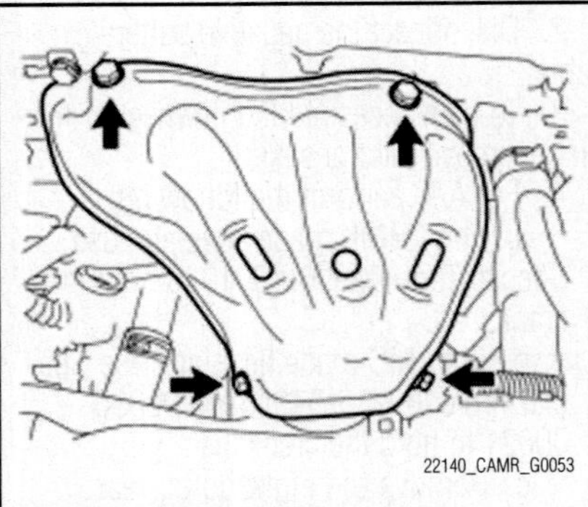

Fig. 120 Removing four bolts and insulator—except PZEV vehicles

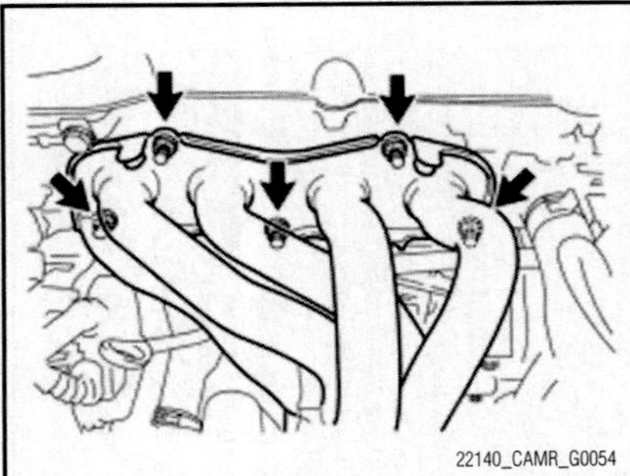

Fig. 121 Removing the five nuts, exhaust manifold and gasket—except PZEV vehicles

5. Remove manifold stays.
6. Disconnect the air-fuel ratio sensor connector.
7. Remove or disconnect remaining components from the exhaust manifold.
8. For PZEV vehicles, perform the following:

   a. Remove the five nuts, manifold converter and gasket.

9. For non-PZEV vehicles, perform the following:

   a. Remove the four bolts and insulator.

   b. Remove the five nuts, manifold converter and gasket.

10. Remove exhaust manifold from catalytic converter.

To install:

11. To install, reverse removal procedure.
12. Install new gaskets for exhaust manifolds.
13. For non-PZEV vehicles, tighten exhaust manifold bolts in sequence to: 27 ft. lbs. (37 Nm). Tighten the exhaust manifold heat insulator bolts to: 9 ft. lbs. (12 Nm).
14. For PZEV vehicles, tighten exhaust manifold bolts in sequence to: 27 ft. lbs. (37 Nm).
15. Tighten the exhaust manifold stays to 32 ft. lbs. (44 Nm).

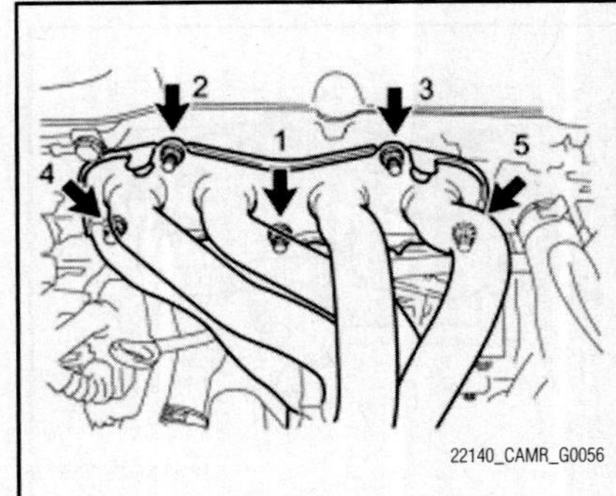

Fig. 122 Exhaust manifold tightening sequence—except PZEV vehicles

Fig. 123 Exhaust manifold tightening sequence—PZEV vehicles

3.5L Engine

See Figures 124 and 125.

1. Before servicing the vehicle, refer to the Precautions Section.
2. Remove the engine assembly with transaxle.
3. Secure the engine.
4. Remove the ignition coil assembly.
5. Remove the right No. 2 engine mounting stay.
6. Remove the intake manifold.
7. Remove the right exhaust manifold sub-assembly, as follows:
   a. Uniformly loosen and remove the 6 nuts.
   b. Remove the manifold and gasket.
8. Remove the oil level gauge guide sub-assembly.
9. Remove the bolt, nut and No. 2 manifold stay.
10. Remove the 3 bolts and No. 2 exhaust manifold heat insulator.
11. Remove the left exhaust manifold sub-assembly, as follows:
    a. Uniformly loosen and remove the 6 nuts.
    b. Remove the manifold and gasket.

To install:

12. Install the left exhaust manifold sub-assembly, as follows:
    a. Install a new gasket.
    b. Install the left exhaust manifold sub-assembly with the 6 nuts and tighten to 15 ft. lbs. (21 Nm).
13. Install the No. 2 exhaust manifold heat insulator with the 3 bolts and tighten to 75 inch lbs. (8.5 Nm).
14. Install the No. 2 manifold stay with the bolt and nut and tighten to 25 ft. lbs. (34 Nm).
15. Install the oil level gauge guide sub-assembly.
16. Install the right exhaust manifold sub-assembly, as follows:
    a. Install a new gasket.
    b. Install the right exhaust manifold sub-assembly with the 6 nuts and tighten to 15 ft. lbs. (21 Nm).
17. Install the intake manifold.
18. Install the right No. 2 engine mounting stay.
19. Install the ignition coil assembly.
20. Install the engine assembly with transaxle.

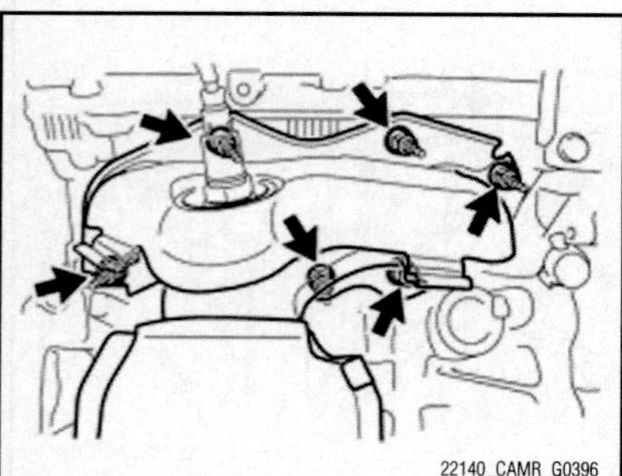

Fig. 124 Removing the exhaust manifold 6 nuts—Right

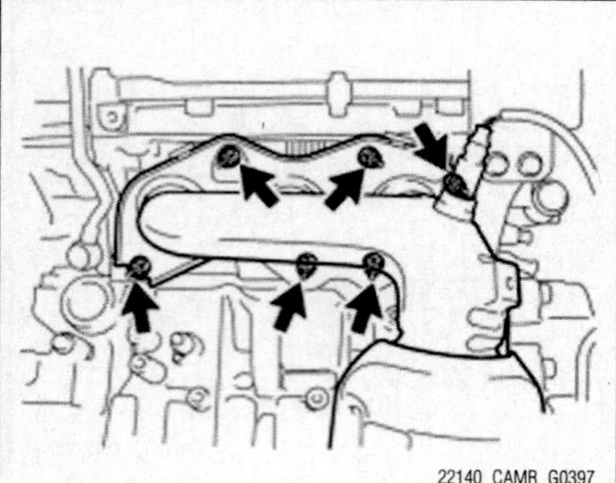

Fig. 125 Removing the exhaust manifold 6 nuts—Left

FLEXPLATE

REMOVAL & INSTALLATION

See Figures 126 through 129.

1. Before servicing the vehicle, refer to the Precautions Section.
2. Disconnect the negative battery cable.
3. To gain access to the flywheel or flexplate, remove the transaxle.
4. For A/T, perform the following:
   a. For TMMK made flexplate: use Special Tool: 09960-10010 to hold the crankshaft.
   b. For TMC made flexplate: use Special Tools: 09213-54015and 09330-00021 to hold the crankshaft.
   c. Remove the eight bolts, rear spacer, drive plate and front spacer.

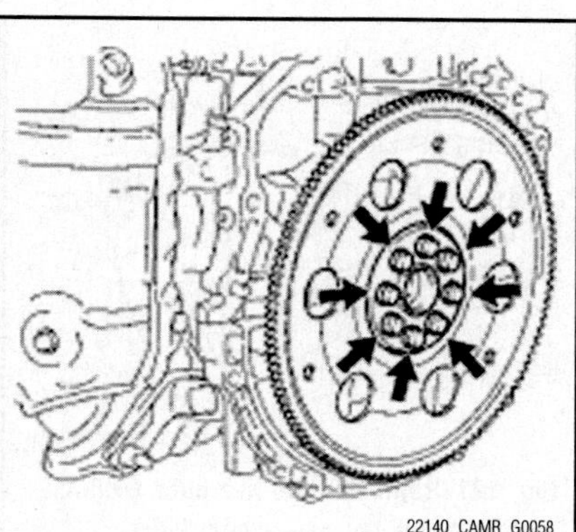

Fig. 126 Removing A/T flexplate bolts

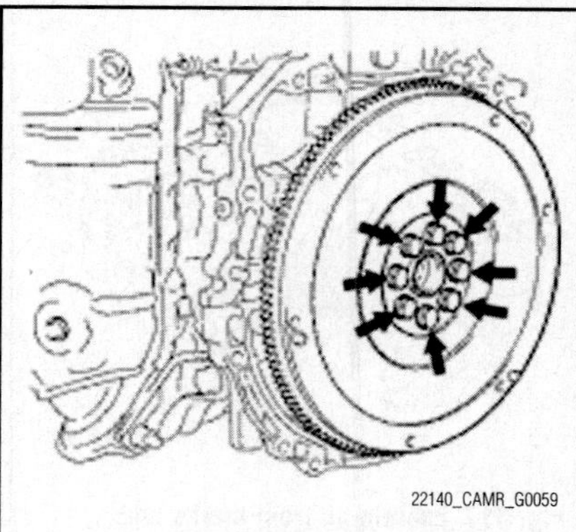

Fig. 127 Removing M/T flywheel bolts

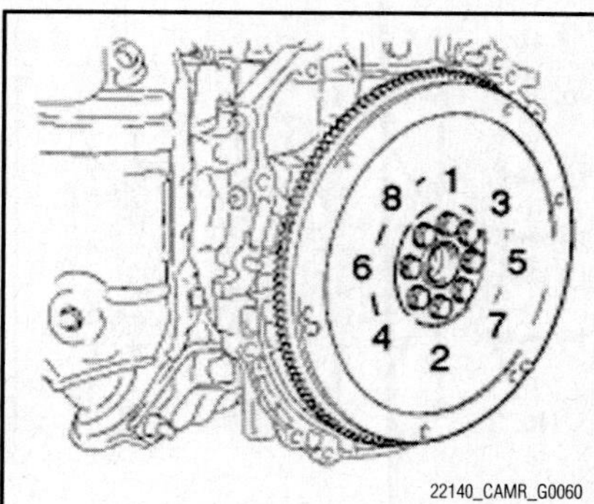

Fig. 128 M/T flywheel bolt tightening sequence

5. For M/T, perform the following:
   a. Remove the clutch cover assembly and clutch disc assembly.
   b. For TMMK made flywheels: use Special Tool: 09960-10010 to hold the crankshaft.
   c. For TMC made flywheels: use Special Tools: 09213-54015and 09330-00021 to hold the crankshaft.
   d. Remove the eight bolts and flywheel.

To install:

6. For M/T, perform the following:
   a. For TMMK made flywheels: use Special Tool: 09960-10010 to hold the crankshaft.
   b. For TMC made flywheels: use Special Tools: 09213-54015and 09330-00021 to hold the crankshaft.
   c. Clean the bolt and the bolt hole.
   d. Apply adhesive to 2 or 3 threads of the bolt end. Adhesive: Part No. 08833-00070, Three Bond or equivalent.
   e. Install the flywheel with the eight bolts. Uniformly tighten the 8 bolts in the sequence. Tighten to 96 ft. lbs. (130 Nm).
   f. Install clutch cover assembly and clutch disc assembly.
7. For M/T, perform the following:

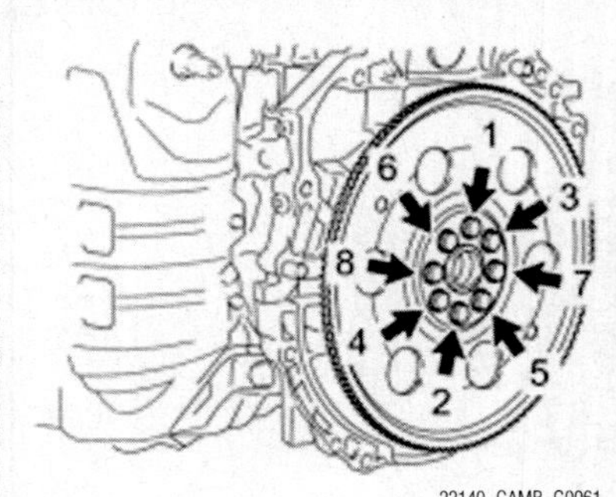

Fig. 129 A/T flywheel bolt tightening sequence

a. For TMMK made flywheel: use Special Tool: 09960-10010 to hold the crankshaft.

b. For TMC made flywheels: use Special Tools: 09213-54015and 09330-00021 to hold the crankshaft.

c. Clean the bolt and the bolt hole.

d. Apply adhesive to 2 or 3 threads of the bolt end. Adhesive: Part No. 08833-00070, Three Bond or equivalent.

e. Install the front spacer, drive plate and rear spacer with the 8 bolts. Uniformly tighten the eight bolts in the sequence. Tighten 72 ft. lbs. (98 Nm).

INTAKE MANIFOLD

REMOVAL & INSTALLATION

3.5L Engine

See Figure 130.

1. Before servicing the vehicle, refer to the Precautions Section.
2. Remove the engine assembly with transaxle.
3. Secure the engine.
4. Remove the ignition coil assembly.
5. Remove the right No. 2 engine mounting stay.

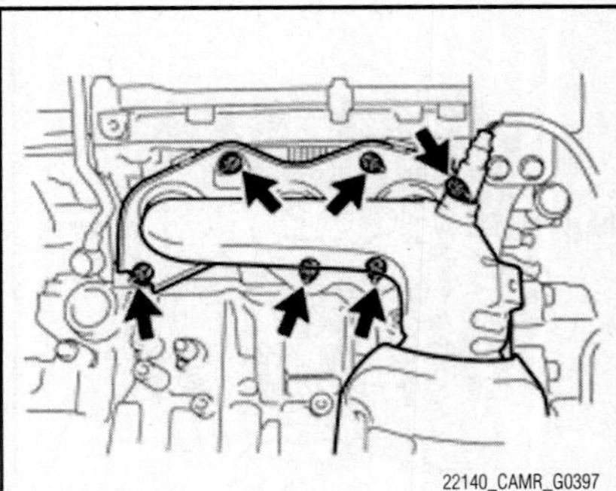

Fig. 130 Locating intake manifold bolts and nuts

6. Remove the intake manifold, as follows:

a. Uniformly loosen and remove the 6 bolts and 4 nuts.

b. Remove the intake manifold and 2 gaskets.

To install:

7. Install the intake manifold, as follows:

WARNING

DO NOTapplies oil to the intake manifold and cylinder head sub-assembly bolts.

a. Set a new gasket on each cylinder head.

➡Align the port holes of the gasket and cylinder head.

➡Make sure that the gasket is installed in the correct direction.

b. Set the intake manifold on the cylinder heads.

c. Install and tighten the 6 bolts and 4 nuts uniformly in several steps to 15 ft. lbs. (21 Nm).

8. Install the right No. 2 engine mounting stay.
9. Install the ignition coil assembly.
10. Install the engine assembly with transaxle.

OIL PAN

REMOVAL & INSTALLATION

2.4L Engine

See Figures 131 and 132

1. Before servicing the vehicle, refer to the Precautions Section.
2. Disconnect the negative battery cable.
3. Drain the engine oil.
4. Remove the oil pan drain plug and gasket.
5. Remove the oil pan 12 bolts and 2 nuts.

➡Be careful not to damage the contact surfaces of the crankcase, chain cover and oil pan.

6. Insert the blade of Special Tool: 09032-00100 between the crankcase and oil pan. Cut through the sealer and remove the oil pan.

To install:

7. Remove any old packing material and be careful not to drop any oil on the contact surfaces of the cylinder block and oil pan.

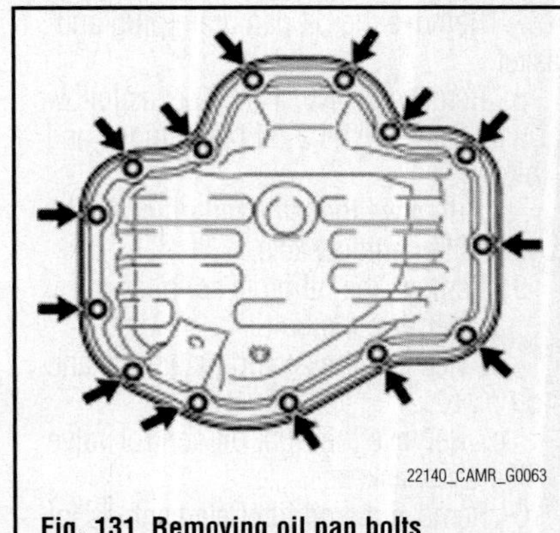

Fig. 131 Removing oil pan bolts

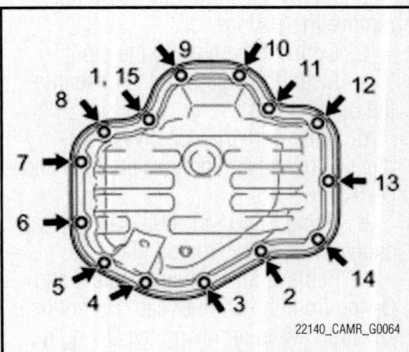

Fig. 132 Oil pan bolt and nut tightening sequence

8. Apply a continuous bead of seal packing (Diameter 0.118 to 0.157 inches (3.0 to 4.0 mm)). Use Toyota Genuine Seal Packing Block, Three Bond 1207B or Equivalent.

➡Remove any oil from the contact surfaces. Install the oil pan within 3 minutes after applying seal packing. Do not start the engine for at least 2 hours after installing.

9. Uniformly tighten the 12 bolts and 2 nuts in sequence. Tighten the bolts and nuts to 80 inch lbs. (9 Nm).
10. To complete installation, reverse remaining removal procedure.
11. Add oil and check for leaks.

3.5L Engine

See Figures 133 through 137.

1. Before servicing the vehicle, refer to the Precautions Section.
2. Drain the engine oil.
3. Remove the engine assembly with transaxle.
4. Secure the engine.
5. Remove the oil filler cap and gasket.
6. Remove the oil pan drain plug and gasket.

7. Remove the oil pan drain plug and gasket.

8. Remove the No. 1 oil pipe, as follows:

a. Remove the 2 oil pipe unions and oil pipe.

b. Remove the left hand oil control valve filter and gaskets.

9. Remove the oil pipe, as follows:

a. Remove the bolt.

b. Remove the 2 oil pipe unions and oil pipe.

c. Remove the right oil control valve filter and gaskets.

10. Remove the oil filter element, as follows:

a. Remove the drain plug. Do not remove the O-ring.

b. Connect the hose to the pipe.

c. Insert the pipe with the hose into the oil filter cap.

d. Make sure that the oil is completely drained and remove the pipe and O-ring.

e. Using SST (SST: 09228-06501) or equivalent, remove the oil filter cap.

f. Remove the oil filter element and O-ring from the oil filter cap. Do not use any tools when removing the O-ring to prevent the O-ring groove from being damaged.

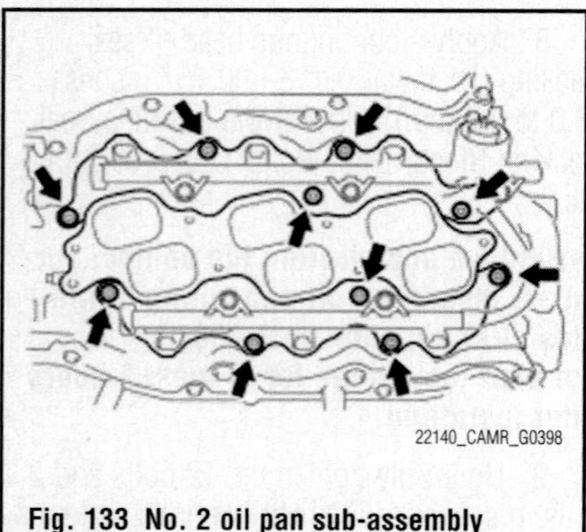

Fig. 133 No. 2 oil pan sub-assembly removal

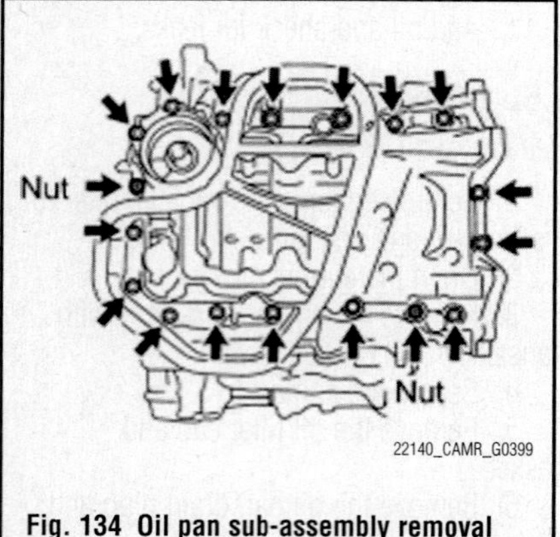

Fig. 134 Oil pan sub-assembly removal

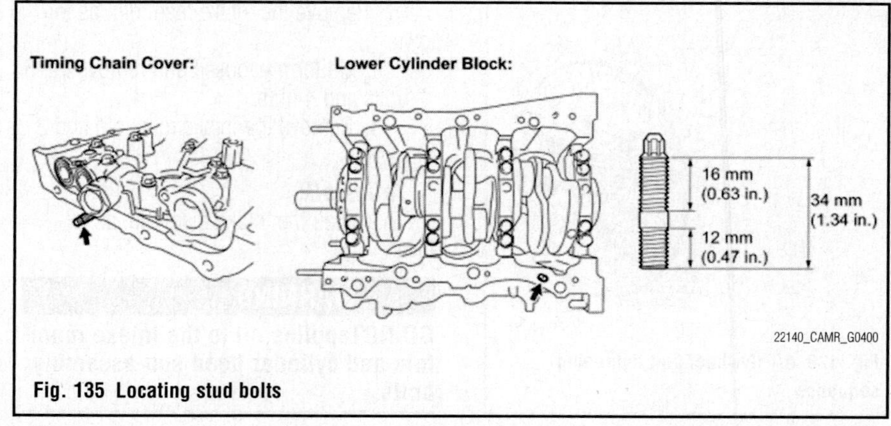

Fig. 135 Locating stud bolts

11. Remove the No. 2 oil pan sub-assembly, as follows:

a. Remove the 16 bolts and 2 nuts.

b. Insert the blade of SST (SST: 09032-00100) or equivalent tool between the oil pans. Cut through the applied sealer and remove the No. 2 oil pan sub-assembly.

➡Be careful not to damage the contact surfaces of the oil pans.

12. Remove the oil pan sub-assembly, as follows:

a. Remove the 16 bolts and 2 nuts.

➡Be sure to clean the bolts and stud bolts and check the threads for cracks or other damage.

b. Remove the oil pan by prying between the oil pan and cylinder block with a taped screwdriver.

➡Be careful not to damage the contact surfaces of the cylinder block and oil pan.

c. Remove the 2 O-rings.

To install:

13. Install the oil pan sub-assembly, as follows:

a. Using an E8 Torx® socket wrench, install the stud bolts as shown in the illustration. Tighten to 7 ft. lbs (10 Nm).

b. Install 2 new O-rings.

➡Remove any oil from the contact surface.

➡Install the oil pan within 3 minutes after applying seal packing.

➡Do not start the engine for at least 2 hours after installing.

c. Apply seal packing (Toyota Genuine Seal Packing Black, Three Bond 1207B or equivalent) in a continuous line as shown in the illustration. Seal diameter: 3.0 to 4.0 mm (0.118 to 0.156 in.).

d. Install the oil pan with the 16 bolts and 2 nuts and tighten to 7 ft. lbs (10 Nm), and 15 ft. lbs (21 Nm).

14. Install the No. 2 oil pan sub-assembly, as follows:

➡Remove any oil from the contact surface.

➡Install the No. 2 oil pan within 3 minutes after applying seal packing.

➡Do not start the engine for at least 2 hours after installing.

a. Using an E6 Torx® socket wrench, install the stud bolts as shown in the

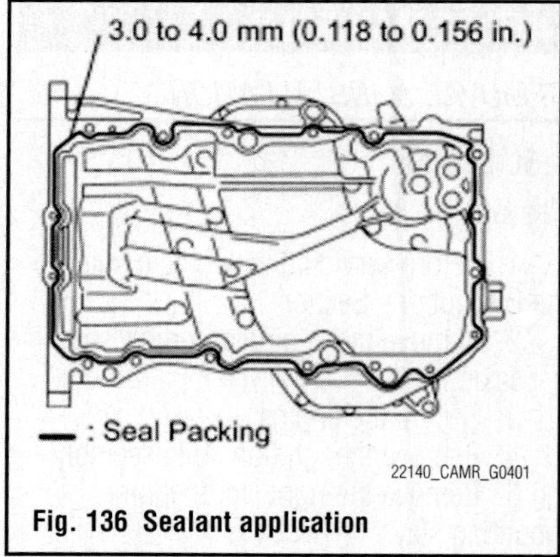

Fig. 136 Sealant application

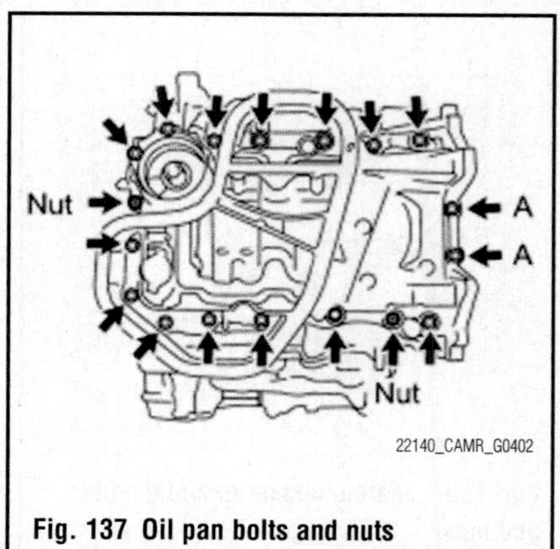

Fig. 137 Oil pan bolts and nuts

illustration and tighten to 35 inch lbs (4 Nm).

b. Apply seal packing (Toyota Genuine Seal Packing Black, Three Bond 1207B or equivalent) in a continuous line as shown in the illustration. Seal diameter: 3.0 to 4.0 mm (0.118 to 0.156 in.).

c. Install the No. 2 oil pan with the 16 bolts and 2 nuts and tighten to 7 ft. lbs (10 Nm).

15. Install the oil pan drain plug and a new gasket. Tighten to 30 ft. lbs (40 Nm).

16. Install the oil filter element, as follows:

a. Clean the inside of the oil filter cap, the threads and O-ring groove.

b. Apply a small amount of engine oil to a new O-ring and install it to the oil filter cap.

c. Set a new oil filter element to the oil filter cap.

d. Remove dirt or foreign matter from the installation surface and inside of the engine.

e. Apply a small amount of engine oil to the O-ring again and install the oil filter cap.

➡Be careful that the O-ring does not get caught between the parts. The O-ring must not be twisted on the groove.

f. Using SST (SST: 09228-06501) or equivalent, install the oil filter cap and tighten to 18 ft. lbs (25 Nm). Make sure that the oil filter is installed securely as shown in the illustration.

17. Install the oil filler cap sub-assembly.

18. Install the engine assembly with transaxle.

19. Check for oil leaks.

OIL PUMP

REMOVAL & INSTALLATION

2.4L Engine

See Figure 138.

1. Before servicing the vehicle, refer to the Precautions Section.

2. Disconnect the negative battery cable.

3. Drain the engine oil.

4. Remove or disconnect the following:

- Plastic engine cover and two nuts
- Front right wheel
- LH and RH engine under cover
- RH front fender apron seal
- Front exhaust pipe assembly
- RH engine mount stay
- Engine mount
- RH engine mounting bracket
- Drive belt
- Generator assembly
- Vane pump
- Ignition coil assembly
- Ventilation hoses
- Valve cover

5. Turn the crankshaft pulley until its groove and the timing mark "0" of the timing chain cover are aligned.

6. Check that each timing mark of the camshaft timing gear and sprocket is aligned with each timing mark located on the intake and exhaust bearing caps. If not, turn the crankshaft by 1 revolution (360°) to align the timing marks.

7. Remove or disconnect the following:

- Crankshaft pulley
- Crankshaft position sensor
- Oil pan
- Chain upper tensioner assembly

8. Install the No. 1 engine hanger (12281-28010) and No. 2 engine hanger (12282-28010) with the bolts (91512-61020) and tighten to: 28 ft. lbs. (38 Nm).

9. Remove or disconnect the following:

- Drive belt tensioner
- Engine mount insulator
- RH engine mount bracket
- Using an E10 Torx® socket, remove the stud bolt for the drive belt tensioner from the cylinder block
- Timing chain cover twelve bolts and two nuts

➡Be careful not to damage the contact surfaces of the timing chain cover, cylinder block and cylinder head. Tape the screwdriver tip before use.

- Timing chain cover by prying between the timing chain cover and cylinder head or cylinder block with a screwdriver

➡Tape the screwdriver tip before use.

- Using a screwdriver and a hammer, tap out the timing chain case oil seal
- Crankshaft position sensor plate

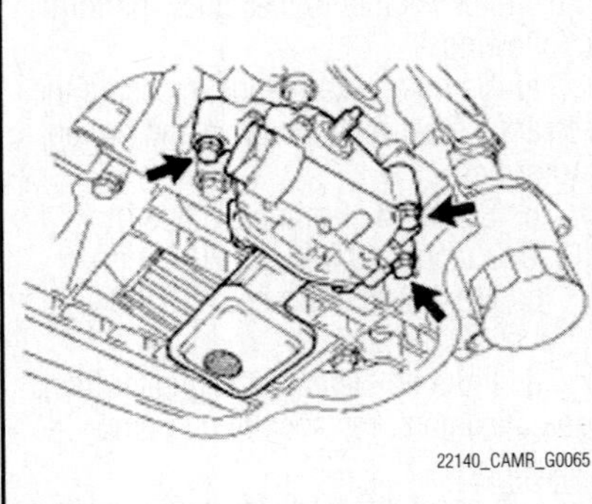

Fig. 139 Removing oil pump

- Chain tensioner slipper
- Chain vibration damper
- Timing chain guide
- Upper chain assembly
- Crankshaft timing sprocket
- Lower chain assembly
- Three bolts, oil pump and gasket

To install:

10. Install a new gasket and the oil pump with the 3 bolts. Tighten the bolts 14 ft. lbs. (19 Nm).

11. To complete installation, reverse removal procedure. Refer to the appropriate sections to install components correctly.

3.5L Engine

1. Before servicing the vehicle, refer to the Precautions Section.

2. Remove the engine assembly with transaxle.

3. Secure engine.

4. Remove the engine wire.

5. Remove the front frame assembly.

6. Remove the starter assembly.

7. Remove the automatic transaxle assembly.

8. Remove the oil level gauge guide sub-assembly.

9. Remove the right and left exhaust manifold sub-assemblies.

10. Remove the drive plate and ring gear sub-assembly.

11. Remove the No. 2 idler pulley sub-assembly.

12. Remove the V-ribbed belt tensioner assembly.

13. Remove the water pump pulley.

14. Remove the water inlet housing.

15. Remove the crankshaft pulley.

16. Remove the No. 2 oil pan sub-assembly.

17. Remove the oil strainer sub-assembly.

18. Remove the oil pan sub-assembly.

19. Remove the intake air surge tank assembly.

20. Remove the ignition coil assembly.

21. Remove the No. 1 and 2 oil pipes.

22. Remove the right and left cylinder head cover sub-assemblies.

23. Remove the timing chain or belt cover sub-assembly.

24. Remove the timing gear case or timing chain case oil seal, as follows:

a. Using a screwdriver with the tip taped, pry out the oil seal.

To install:

25. Install timing gear case or timing chain case oil seal, as follows:

a. Using SST (SST: 09316-60011) or equivalent tool, tap in a new oil seal until its surface is flush with the timing chain case edge.

➡**Keep the lip free from foreign matter.**

➡**Do not tap on the oil seal at an angle.**

➡**Make sure that the oil seal edge does not stick out of the timing chain case.**

b. Apply MP grease to the oil seal lip.

26. Install timing chain or belt cover sub-assembly.
27. Install the right and left cylinder head cover sub-assemblies.
28. Install the No. 1 and 2 oil pipes.
29. Install the ignition coil assembly.
30. Install the intake air surge tank assembly.
31. Install the oil pan sub-assembly.
32. Install the oil strainer sub-assembly.
33. Install the No. 2 oil pan sub-assembly.
34. Install the crankshaft pulley.
35. Install the water inlet housing.
36. Install the water pump pulley.
37. Install the V-ribbed belt tensioner assembly.
38. Install the No. 2 idler pulley sub-assembly.
39. Install the drive plate and ring gear sub-assembly.
40. Install the right and left exhaust manifold sub-assemblies.
41. Install the oil level gauge guide sub-assembly.
42. Install the automatic transaxle assembly.
43. Install the starter assembly.
44. Install the front frame assembly.
45. Install the engine wire.
46. Install the engine assembly with transaxle.

INSPECTION

2.4L Engine

See Figures 139 through 142.

1. Check the oil jet located above the timing chain sprocket for damage or clogging. If necessary, repair the cylinder block.

➡**If the valve does not fall smoothly, replace the relief valve. If necessary, replace the oil pump assembly.**

2. Coat the relief valve with engine oil, and then check that the valve falls smoothly into the valve hole under its own weight.
3. Install the oil pump rotor. Coat the drive rotor and driven rotor with engine oil.

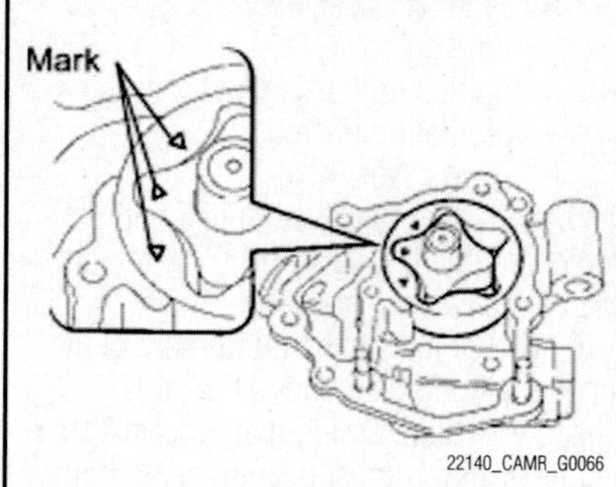

Fig. 139 Aligning the oil pump driven rotors and oil pump cover side marks

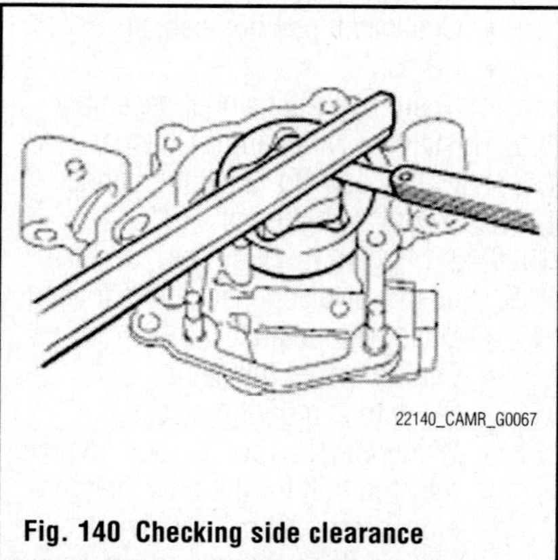

Fig. 140 Checking side clearance

4. Place the drive and driven rotors into the oil pump with the marks facing the pump covers side.
5. To check the side clearance, perform the following:
   a. Using a feeler gauge and precision straightedge, measure the clearance between the rotors and precision straightedge.
   b. Standard clearance: 0.0012 to 0.0033 inches (0.030 to 0.085 mm).
   c. Maximum clearance: 0.0063 inches (0.16 mm).
   d. If the side clearance is greater than the maximum, replace the oil pump assembly.
6. To check the tip clearance, perform the following:
   a. Using a feeler gauge, measure the clearance between the drive and driven rotor tips.
   b. Standard clearance: 0.0031 to 0.0063 inches (0.080 to 0.160 mm).
   c. Maximum clearance: 0.0138 inches (0.35 mm).
   d. If the tip clearance is greater than the maximum, replace the oil pump assembly.
7. To check the body clearance, perform the following:

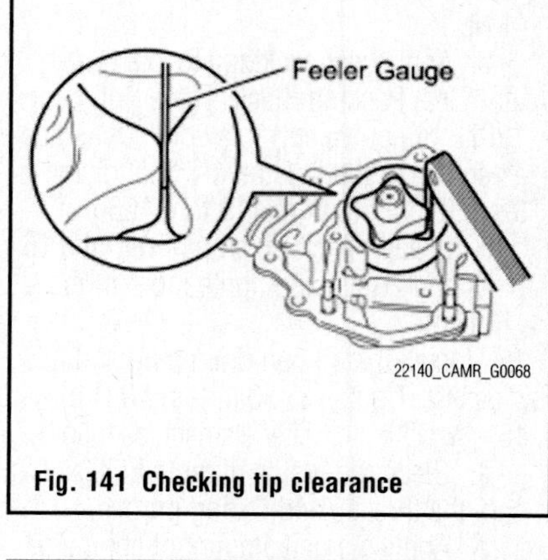

Fig. 141 Checking tip clearance

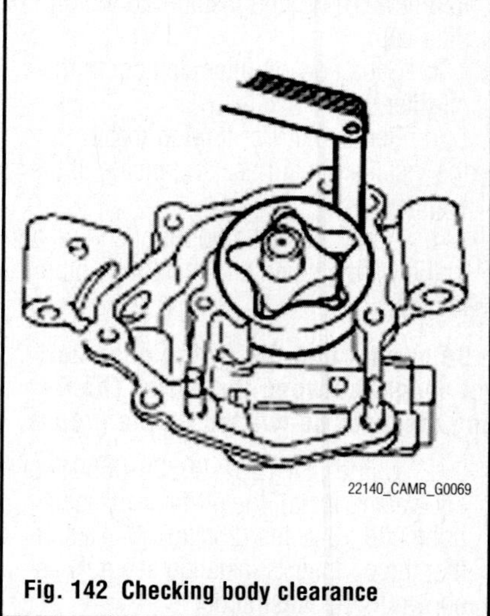

Fig. 142 Checking body clearance

   a. Using a feeler gauge, measure the clearance between the driven rotor and pump body.
   b. Standard clearance: 0.0039 to 0.0067 inches (0.100 to 0.170 mm).
   c. Maximum clearance: .0128 inches. (0.325 mm).
   d. If the body clearance is greater than the maximum, replace the oil pump assembly.
8. Inspect the lower chain assembly and replace as necessary.
9. Inspect the oil pump drive sprocket and replace as necessary.
10. Inspect the chain tensioner plate and replace as necessary.

3.5L Engine

See Figures 143 through 146.

1. Inspect the oil pump relief valve, as follows:
   a. Coat the relief valve with engine oil and check that it falls smoothly into the valve hole by its own weight. If the valve does not fall smoothly, replace the relief

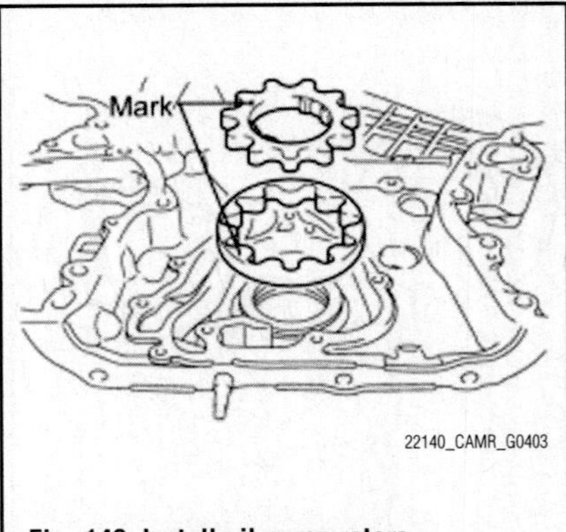

Fig. 143 Install oil pump rotors

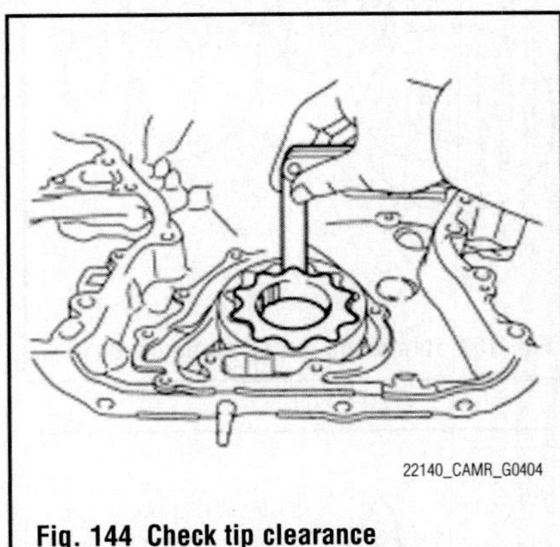

Fig. 144 Check tip clearance

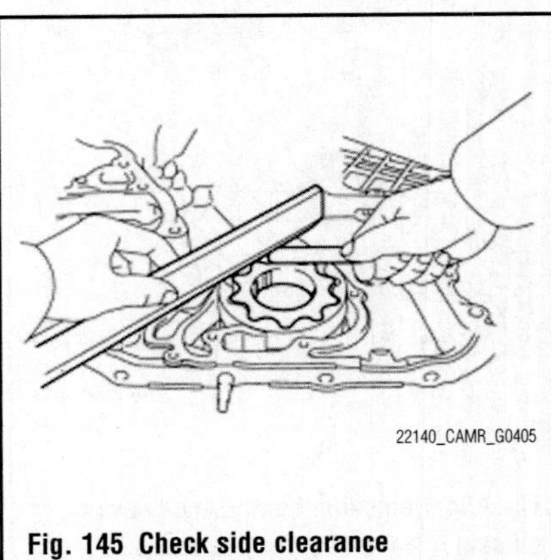

Fig. 145 Check side clearance

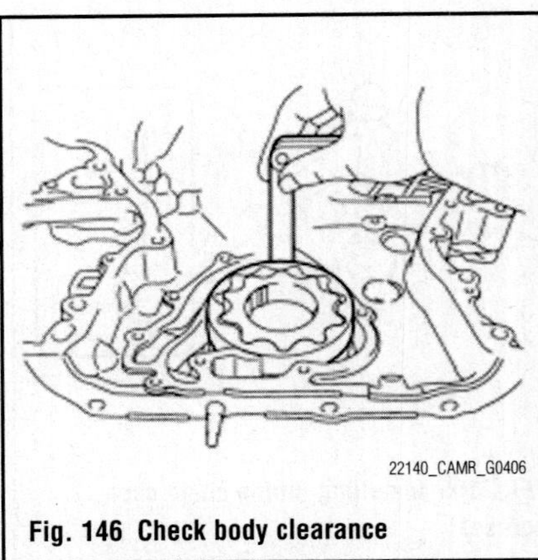

Fig. 146 Check body clearance

valve. If necessary, replace the oil pump assembly.

2. Inspect the oil pump rotor set, as follows:

a. Install the rotors to the timing chain cover with the rotors' marks outward. Check that the rotors rotate smoothly.

b. Check the tip clearance: using a feeler gauge, measure the clearance between the drive and driven rotor tips, as shown in the illustration. If the clearance is greater than the maximum, replace the drive and driven rotors. Standard: 0.0024 to 0.0063 inches (0.060 to 0.160 mm). Maximum: 0.0063 inches (0.16 mm).

c. Check the side clearance: using a feeler gauge and precision straightedge, measure the clearance between the rotors and precision straightedge, as shown in the illustration. If the side clearance is greater than the maximum, replace the timing chain cover sub-assembly. Standard: 0.0012 to 0.0035 inches (0.030 to 0.090 mm). Maximum: 0.0035 inches (0.090 mm).

d. Check the body clearance: using a feeler gauge, measure the clearance between the timing chain cover and driven rotor, as shown in the illustration. If the body clearance is greater than the maximum, replace the timing chain cover sub-assembly. Standard: 0.0098 to 0.0128 inches (0.250 to 0.325 mm). Maximum: 0.0128 inches (0.325 mm).

PISTON AND RING

POSITIONING

See Figure 147.

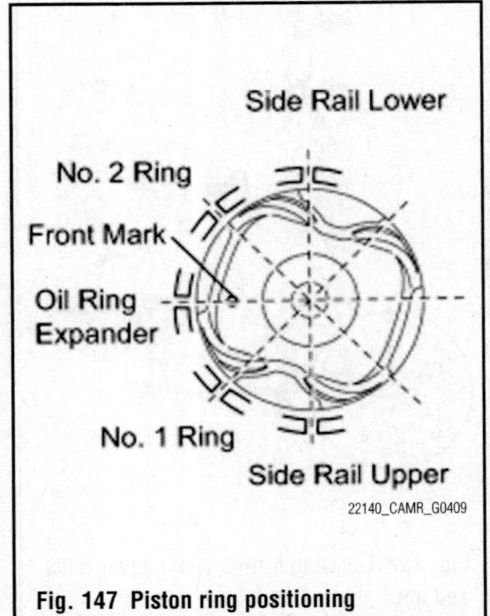

Fig. 147 Piston ring positioning

REAR MAIN SEAL

REMOVAL & INSTALLATION

2.4L Engine

See Figures 148 and 149.

1. Before servicing the vehicle, refer to the Precautions Section.
2. Disconnect the negative battery cable.
3. For A/T vehicles, separate the transaxle.
4. For M/T vehicles, separate the transaxle.
5. For A/T vehicles, remove the flexplate. Refer to Flexplate and Flywheel above.
6. For M/T vehicles, remove the flywheel. Refer to Flexplate and Flywheel above.
7. Using a knife, cut through the oil seal lip.
8. Using a screwdriver with its tip taped, pry out the oil seal.

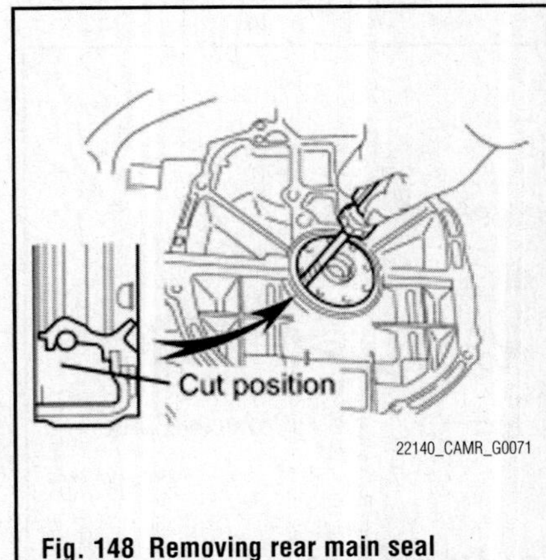

Fig. 148 Removing rear main seal

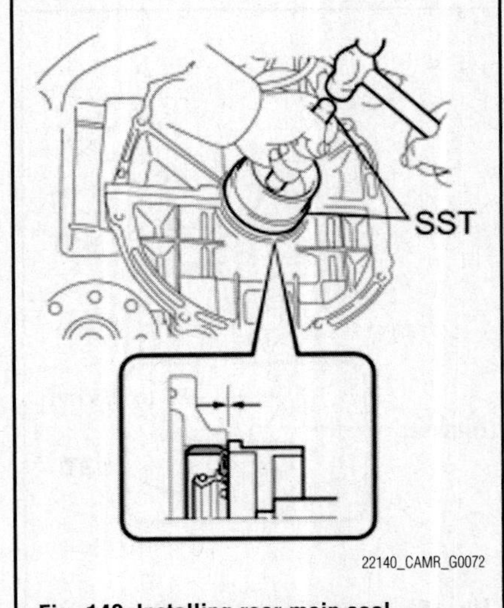

Fig. 149 Installing rear main seal

9. After the removal, check the crankshaft for damage. If it is damaged, smooth the surface with 400-grit sandpaper.

To install:

➡Keep the lip free from foreign matter.

10. Apply MP grease to a new oil seal lip.
11. Using Special Tools: 09223-15030, 09950-70010 and a hammer, tap in the oil seal until its surface is flush with the rear oil seal retainer edge. Wipe off extra grease from the crankshaft.
12. To complete installation, reverse remaining removal procedure.

3.5L Engine

See Figures 150 and 151.

1. Before servicing the vehicle, refer to the Precautions Section.
2. Remove the automatic transaxle assembly.
3. Remove the drive plate and ring gear sub-assembly.
4. Remove the rear main seal, as follows:

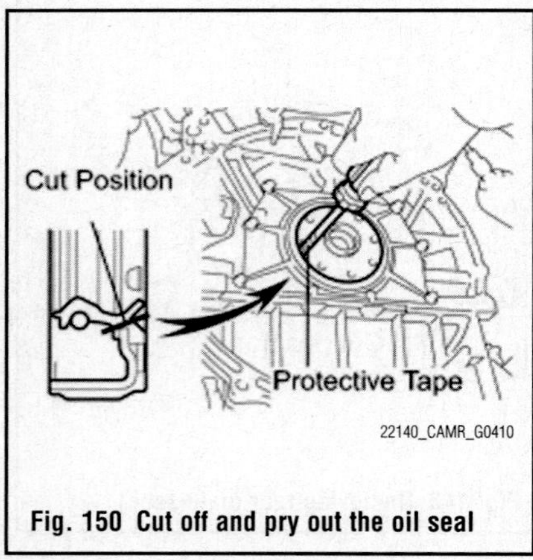

Fig. 150 Cut off and pry out the oil seal

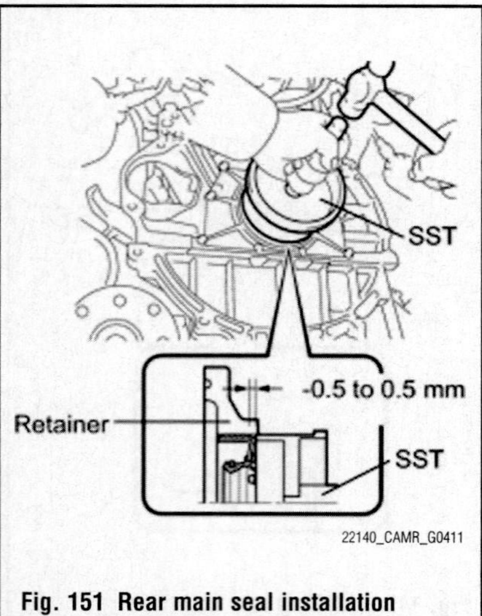

Fig. 151 Rear main seal installation

a. Using a knife, cut off the oil seal lip.

To install:

5. Apply MP grease to a new oil seal lip.
6. Using SST (SST: 09223-15030, SST: 09950-70010) or equivalent and a hammer, tap in the oil seal. Oil seal tap in depth: -0.020 to 0.020 in. (-0.5 to 0.5 mm)
7. Install the drive plate and ring gear sub-assembly.
8. Install automatic transaxle assembly.

ROCKER ARMS

REMOVAL & INSTALLATION

For 3.5L engines, see Camshafts.

TIMING CHAIN FRONT COVER

REMOVAL & INSTALLATION

2.4L Engine

See Figures 152 through 157

1. Before servicing the vehicle, refer to the Precautions Section.
2. Using an E10 Torx® socket, remove the stud bolt for the drive belt tensioner from the cylinder block.
3. Remove the 12 bolts and 2 nuts.

➡Be careful not to damage the contact surfaces of the timing chain cover, cylinder block and cylinder head.

4. Remove the timing chain cover by prying between the timing chain cover and cylinder head or cylinder block with a screwdriver. Tape the screwdriver tip before use.

Fig. 152 Locating timing chain cover bolts and nuts

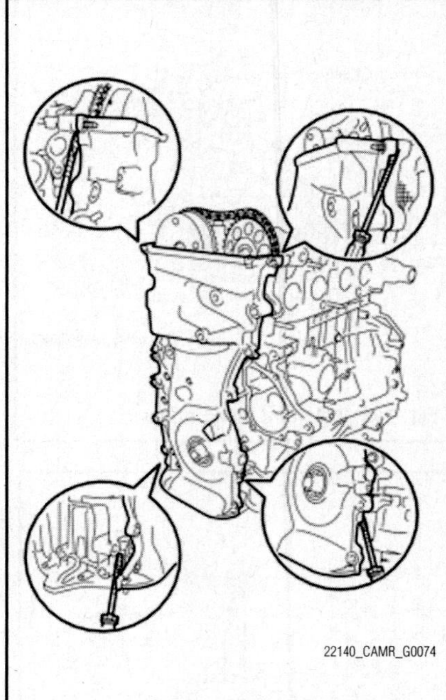

Fig. 153 Removing timing chain cover

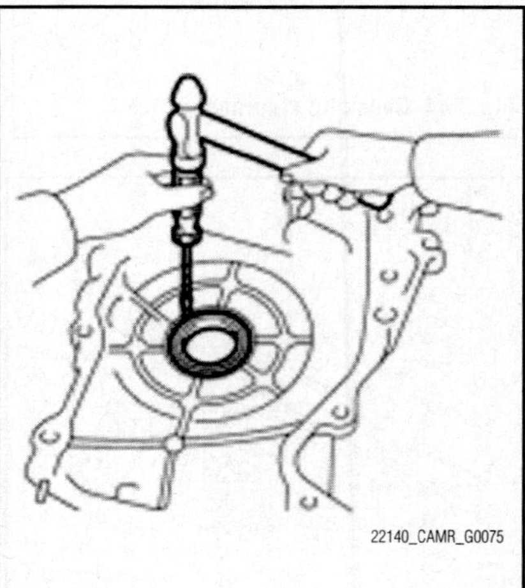

Fig. 154 Removing timing chain case oil seal

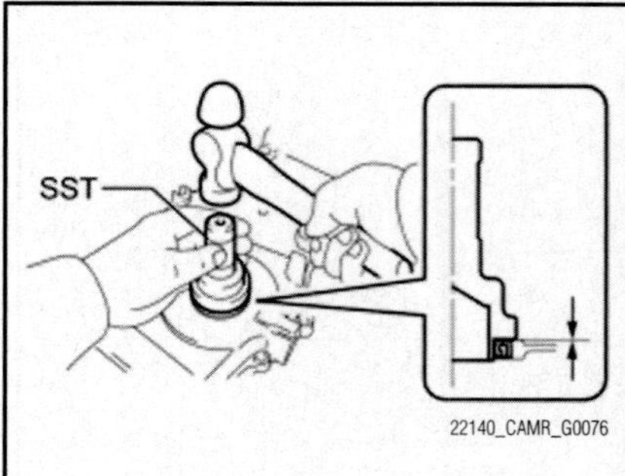

Fig. 155 Installing timing chain case oil seal

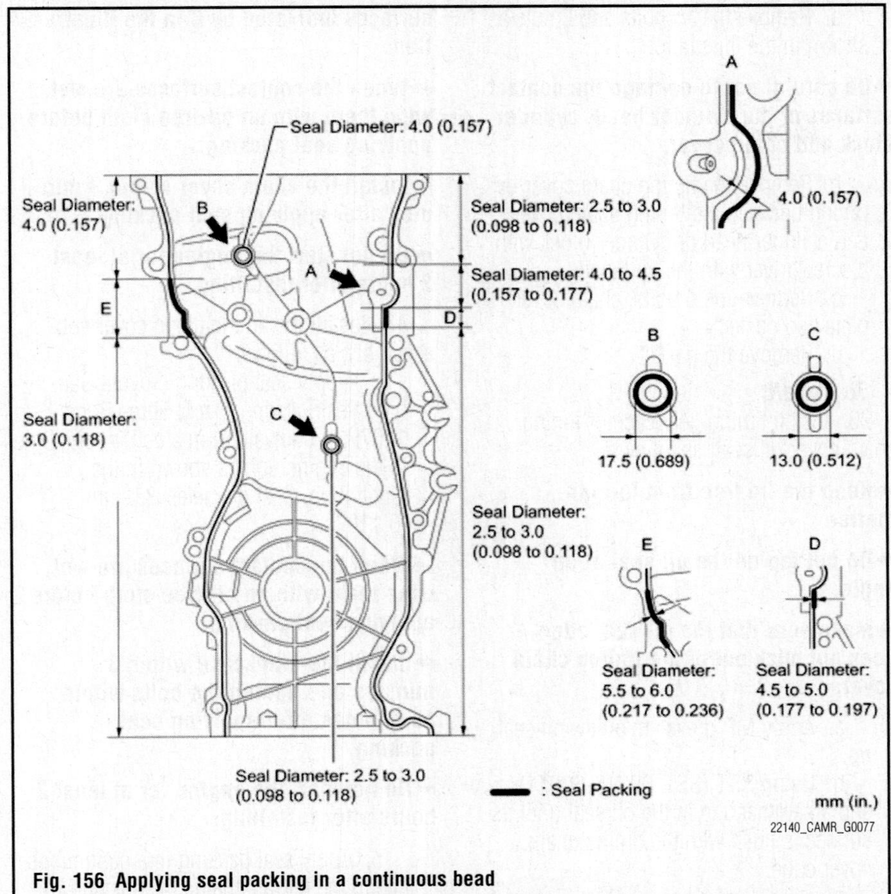

Fig. 156 Applying seal packing in a continuous bead

5. Using a screwdriver and a hammer, tap out the oil seal.

To install:

➡Keep the gap between the timing chain cover edge and the oil seal free of foreign matter.

6. Using Special Tool: 09223-22010, tap in a new oil seal until its surface is flush with the timing chain cover edge. Apply a light coat of MP grease to the lip of the oil seal.
7. Remove any old packing (FIPG) material and be careful not to drop any oil on the contact surfaces of the timing chain cover, cylinder head and cylinder block.
8. Apply Toyota Genuine Seal Packing Black, Three Bond 1207B or Equivalent seal packing in a diameter of 0.157 to 0.177 inches (4.0 to 4.5 mm).
9. Apply seal packing in a continuous bead as shown in the illustration below.

➡Remove any oil from the contact surface. Install the chain cover within 3 minutes after applying seal packing. Do not start the engine for at least 2 hours after installing.

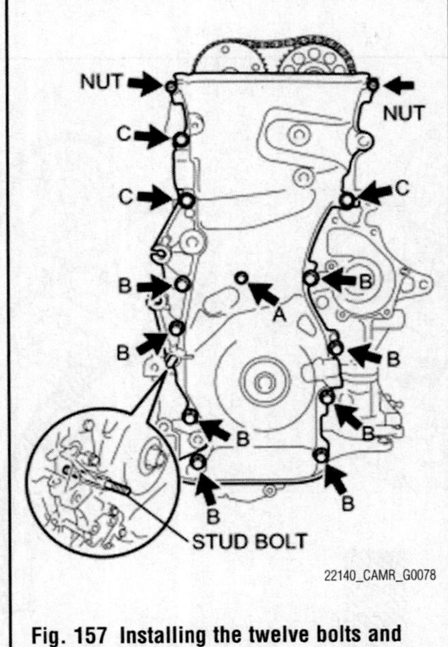

Fig. 157 Installing the twelve bolts and two nuts

10. Install the timing chain cover with the twelve bolts and two nuts in sequence to the following torque specification:
    - Bolt A length: 1.18 inches (30 mm) for 10 mm head: 80 inch lbs. (9 Nm)
    - Bolt B length: 1.18 inches (30 mm) for 12 mm head: 18 ft. lbs. (25 Nm)
    - Bolt C length: 1.57 inches (40 mm) for 14 mm head: 41 ft. lbs. (55 Nm)
    - Nut: 8 ft. lbs. (11 Nm)
11. Using a E10 Torx® socket, install the stud bolt to the drive belt tensioner and tighten to 16 ft. lbs. (22 Nm).

3.5L Engine

See Figures 158 through 163.

1. Before servicing the vehicle, refer to the Precautions Section.
2. Remove the engine assembly with transaxle.
3. Secure engine.
4. Remove the oil filler cap sub-assembly.
5. Remove the spark plugs and ignition coil assembly.
6. Remove the oil pan drain plug and gasket.
7. Remove the ventilation valve sub-assembly.
8. Remove the camshaft position sensor.
9. Remove the camshaft timing oil control valve assembly.
10. Remove crankshaft position sensor.
11. Remove the No. 1 oil pipe.
12. Remove the oil pipe.
13. Remove the cylinder block water drain cock sub-assembly.
14. Remove the oil filter.
15. Remove the crankshaft pulley.
16. Remove the left hand No. 1 front engine mounting bracket.
17. Remove the water inlet housing.
18. Remove the water outlet.
19. Remove the left-hand cylinder head cover sub-assembly and gasket.
20. Remove the cylinder head cover sub-assembly and gasket.

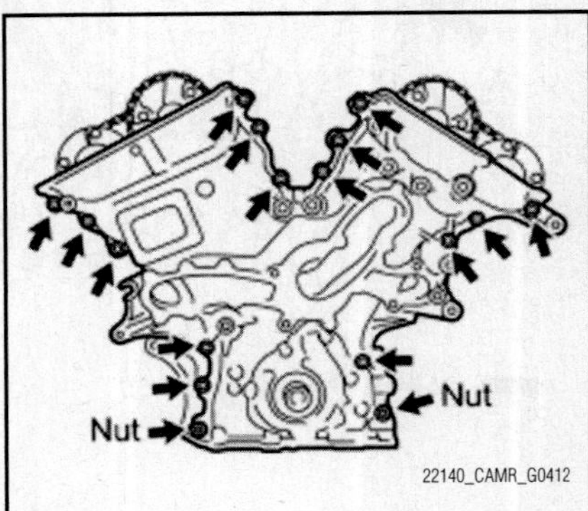

Fig. 158 Locating timing chain cover sub-assembly bolts and nuts

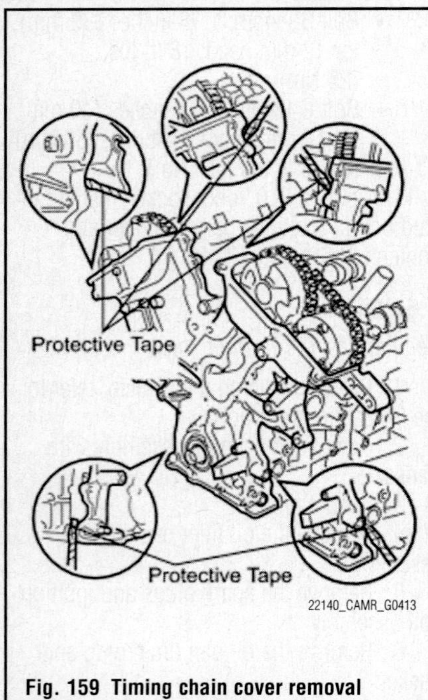

Fig. 159 Timing chain cover removal

21. Remove the No. 2 oil pan sub-assembly.
22. Remove the oil strainer sub-assembly.
23. Remove the oil pan sub-assembly.
24. Remove the water pump assembly.
25. Remove the timing chain cover sub-assembly, as follows:

a. Remove the 15 bolts and 2 nuts as shown in the illustration.

➡Be careful not to damage the contact surfaces of the cylinder head, cylinder block and chain cover.

b. Remove the timing chain cover by prying between the timing chain cover and cylinder head or cylinder block with a screwdriver with the tip taped.
c. Remove the 4 bolts, chain cover plate and gasket.
d. Remove the gasket.

To install:

26. Install timing gear case or timing chain cover oil seal, as follows:

➡Keep the lip free from foreign matter.

➡Do not tap on the oil seal at an angle.

➡Make sure that the oil seal edge does not stick out of the timing chain cover.

a. Apply MP grease to a new oil seal lip.
b. Using SST (SST: 09316-60011) and a hammer, tap in the oil seal until its surface is flush with the timing chain cover edge.

➡Be sure to clean and degrease the contact surfaces, especially the surfaces indicated by C in the illustration.

➡When the contact surfaces are wet, wipe them with an oil-free cloth before applying seal packing.

➡Install the chain cover within 3 minutes after applying seal packing.

➡Do not start the engine for at least 2 hours after installing.

27. Install the timing chain cover sub-assembly, as follows:

a. Apply seal packing (Toyota Genuine Seal Packing Black, Three Bond 1207B or equivalent) in a continuous line to the engine unit as shown in the illustration. Seal diameter: 3.0 mm (0.118 in.).

➡When the contact surfaces are wet, wipe them with an oil-free cloth before applying seal packing.

➡Install the crankcase within 3 minutes and tighten the bolts within 15 minutes after applying seal packing.

➡Do not start the engine for at least 2 hours after installing.

b. Apply seal packing in a continuous line to the timing chain cover as shown in the following illustration. Seal packing: Toyota Genuine Seal Packing Black,

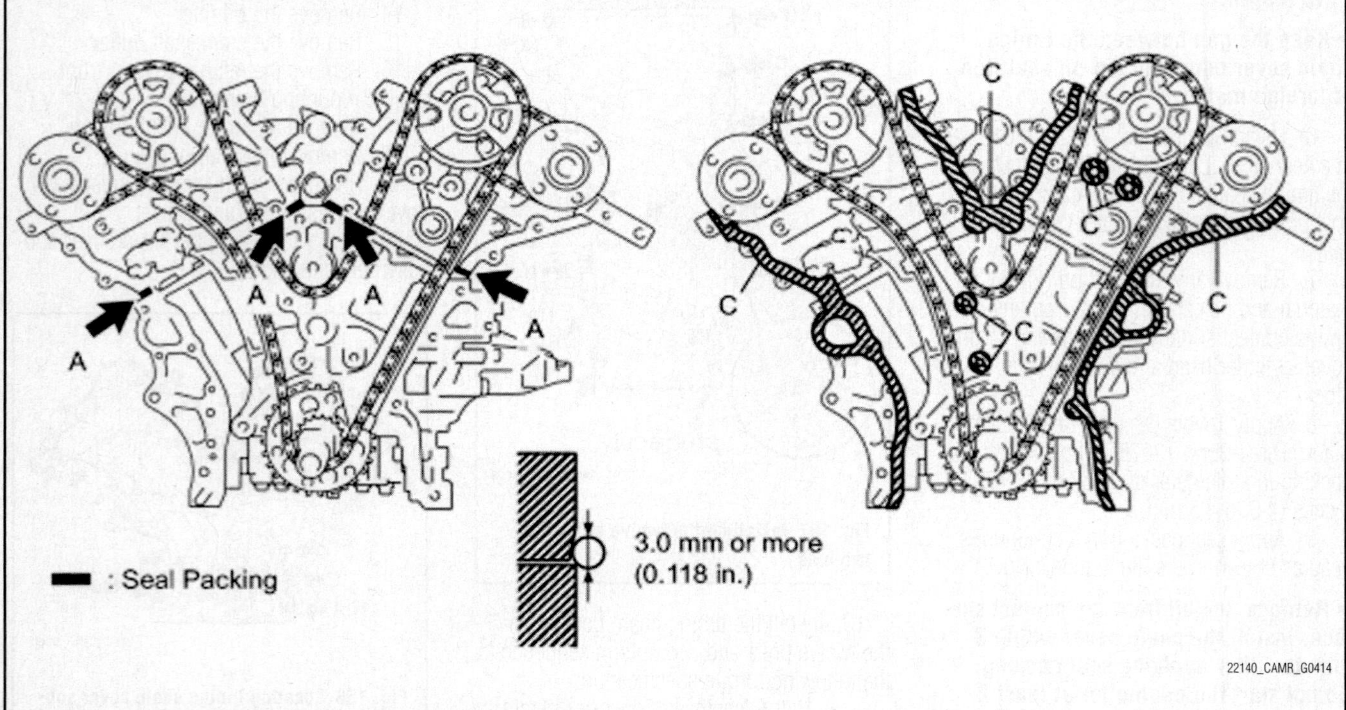

Fig. 160 Applying seal packing to timing chain cover sub-assembly

Three Bond 1207B or equivalent, Toyota Genuine Seal Packing Black, Three Bond 1282B, Three Bond 1282B or equivalent.

c. Install a new gasket.

d. Align the oil pump's drive rotor spline and the crankshaft as shown in the illustration. Install the spline and chain cover to the crankshaft.

e. Loosely install the timing chain cover with the 23 bolts and 2 nuts, but do not tighten the bolts and 2 nuts yet.

CAUTION

Make sure that there is no oil on the bolt and nut threads.

f. Fully tighten the bolts in this order: Area 1 and Area 2, tighten to 15 ft. lbs. (21 Nm).

g. Fully tighten the bolts in Area 3 to 15 ft. lbs. (21 Nm). Tighten the bolts and nuts in the order of upper to lower as shown in the illustration.

h. Fully tighten the bolts in Area 4 to 32 ft. lbs. (43 Nm), and to 15 ft. lbs. (21 Nm). Tighten the bolts and nuts in the order of lower to upper as shown in the illustration.

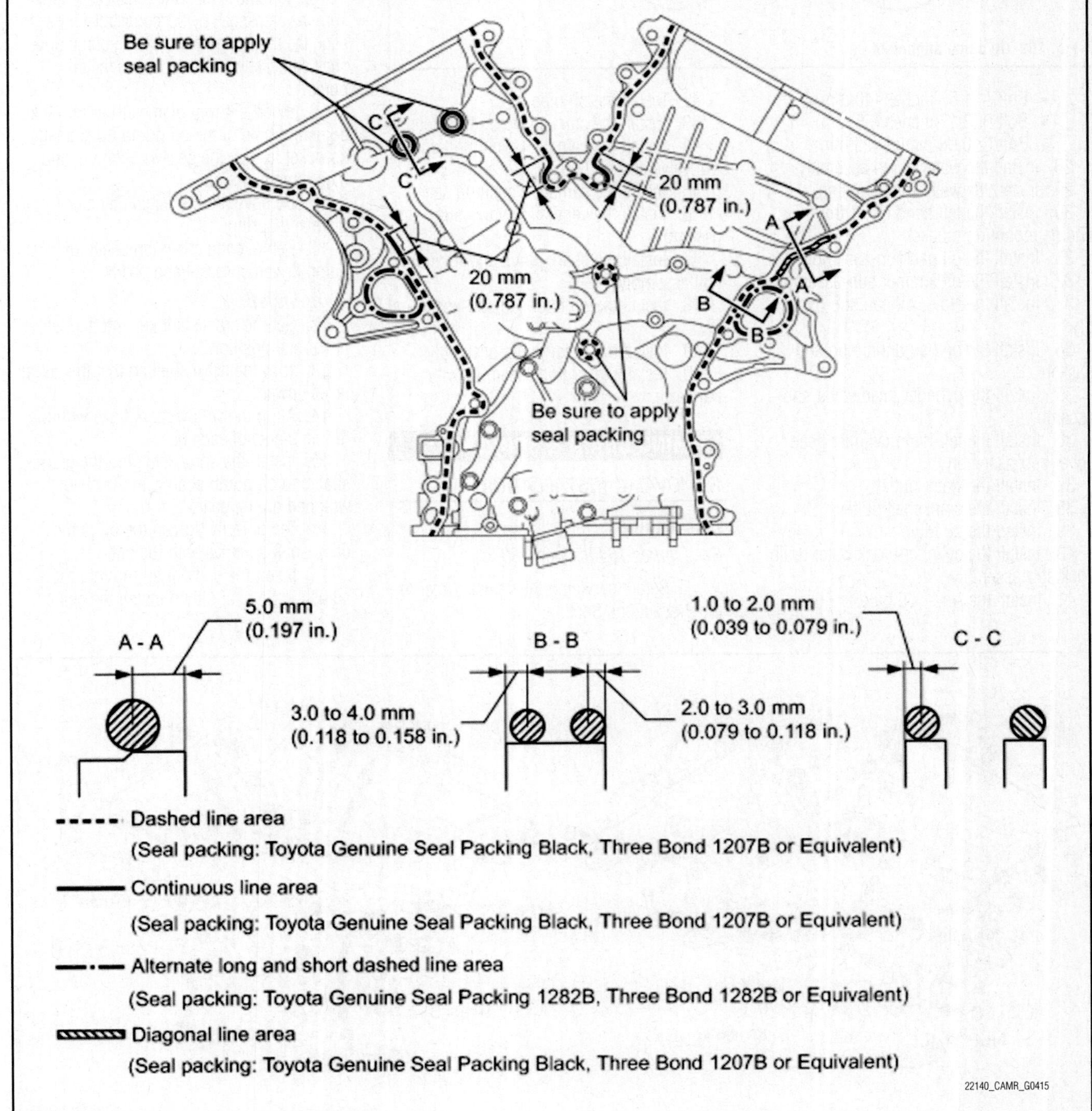

Fig. 161 Applying seal packing to timing chain cover

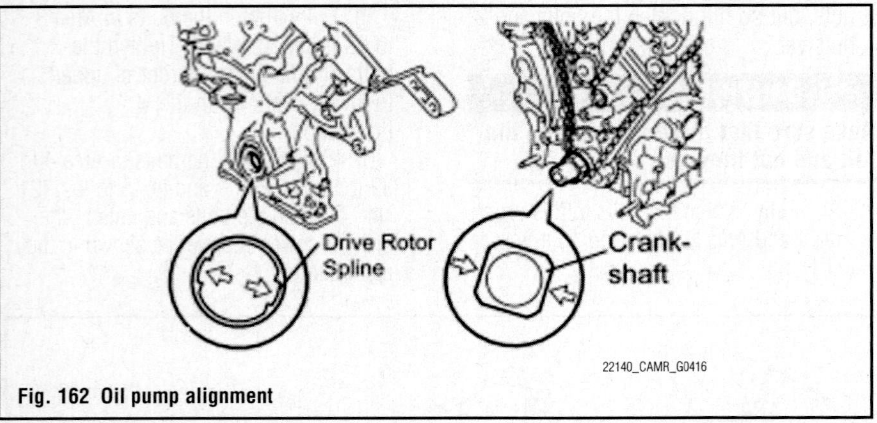

Fig. 162 Oil pump alignment

- Bolt A: 1.57 inches (40 mm)
- Bolt B: 2.17 inches (55 mm)
- Bolt C: 0.98 inches (25 mm)

28. Install the water pump assembly.
29. Install the water inlet housing.
30. Install the left hand No. 1 front engine mounting bracket.
31. Install the oil pan sub-assembly.
32. Install the oil strainer sub-assembly.
33. Install the No. 2 oil pan sub-assembly.
34. Install the oil pan drain plug and gasket.
35. Install the cylinder head cover sub-assembly.
36. Install the left-hand cylinder head cover sub-assembly.
37. Install the water outlet.
38. Install the crankshaft pulley.
39. Install the oil filter.
40. Install the cylinder block water drain cock sub-assembly.
41. Install the No. 1 oil pipe.
42. Install the oil pipe.
43. Install the crankshaft position sensor.
44. Install the camshaft timing oil control valve assembly.
45. Install the camshaft position sensor.
46. Install the ventilation valve sub-assembly.
47. Install the spark plugs and ignition coil assembly.
48. Install the oil filler cap sub-assembly.
49. Install the water pump assembly.
50. Install the engine assembly with transaxle.

TIMING CHAIN & SPROCKETS

REMOVAL & INSTALLATION

2.4L Engine

See Figures 164 through 183.

1. Before servicing the vehicle, refer to the Precautions Section.
2. Remove the crankshaft position sensor plate.
3. Remove the bolt and chain tensioner slipper.
4. Remove the two bolts and chain vibration damper.
5. Remove the bolt and timing chain guide.
6. Remove the upper chain assembly.
7. Remove the crankshaft timing sprocket.
8. To remove the lower chain assembly. Turn the crankshaft by 90°counterclockwise to align the adjusting hole of the oil pump drive shaft sprocket with the groove of the oil pump.
9. Insert a 4 mm diameter bar into the adjusting hole of the oil pump drive shaft sprocket to lock the gear in position, and then remove the nut.
10. Remove the bolt, chain tensioner plate and spring.
11. Remove the chain tensioner, oil pump driven sprocket and chain.

To install:

12. Set the crankshaft key into the left horizontal position.
13. Turn the drive shaft so that the cutout faces upward.
14. Align the yellow mark links with the timing marks of each gear.
15. Install the sprockets onto the crankshaft and oil pump shaft with the chain wrapped on the gears
16. Temporarily tighten the oil pump drive shaft sprocket with the nut.
17. Insert the damper spring into the adjusting hole, and then install the chain

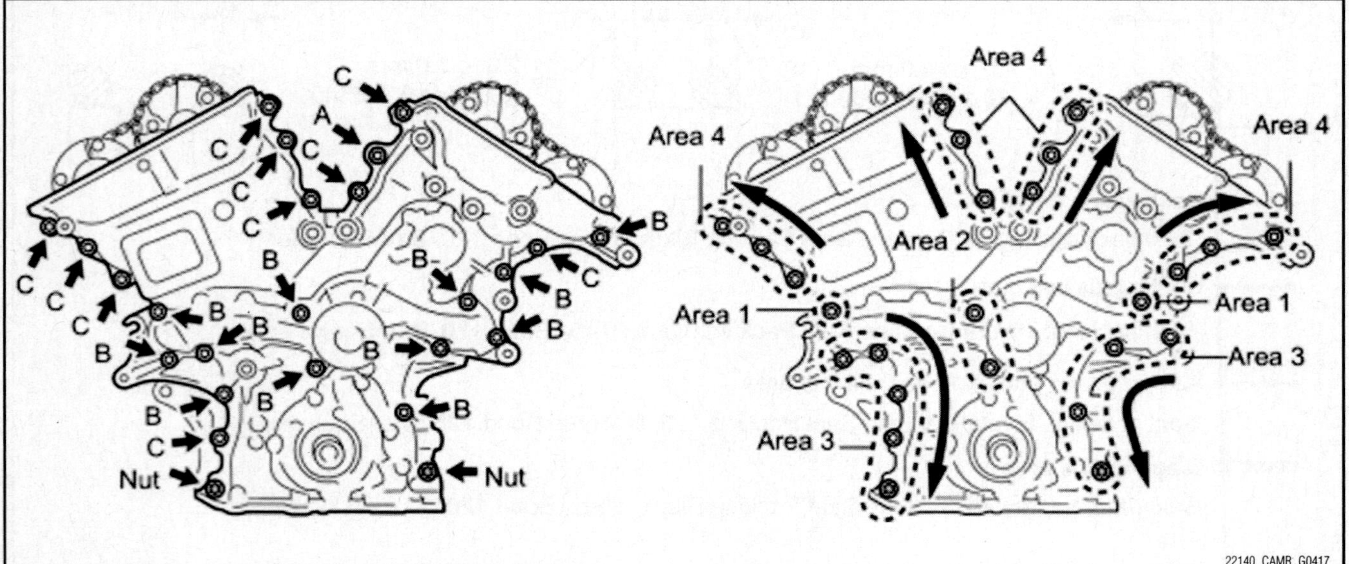

Fig. 163 Timing chain cover bolts and nuts tightening sequence

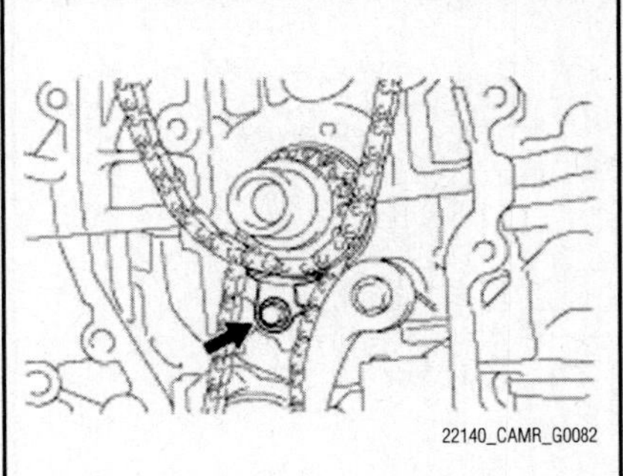

Fig. 164 Removing the bolt and timing chain guide

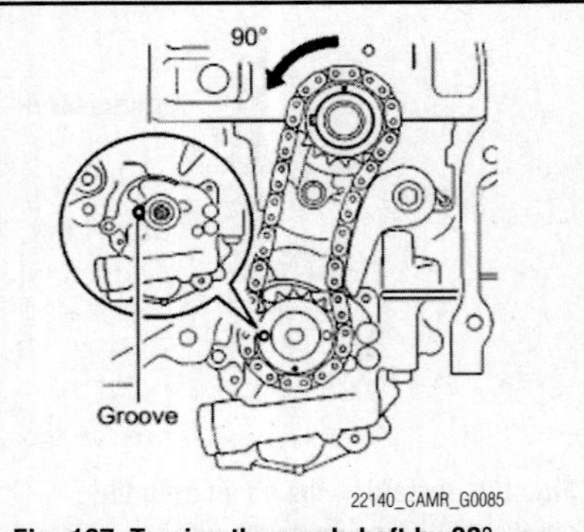

Fig. 167 Turning the crankshaft by 90° counterclockwise to align the adjusting hole

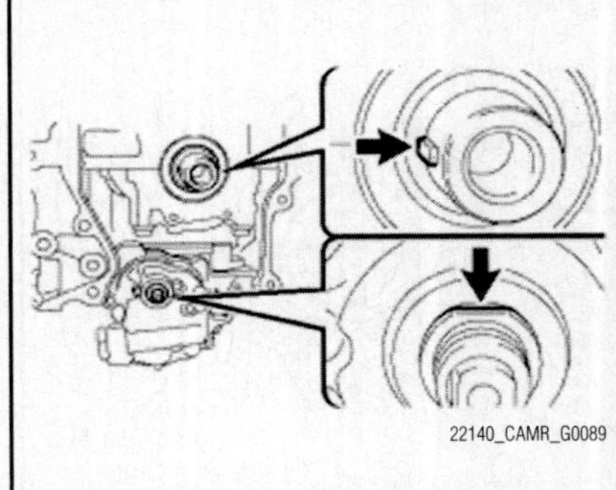

Fig. 171 Setting the crankshaft key and cutout face

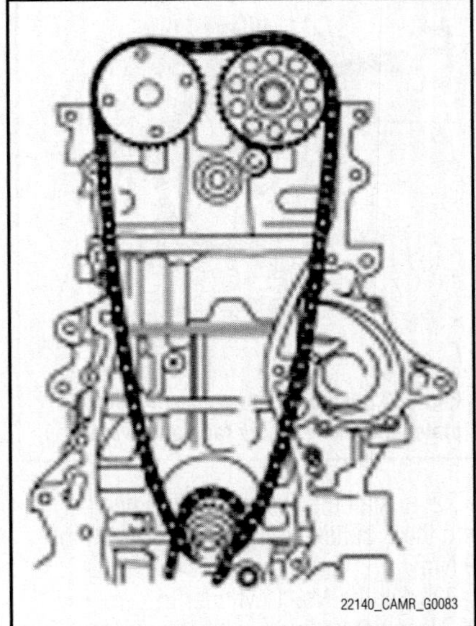

Fig. 165 Removing the upper chain assembly

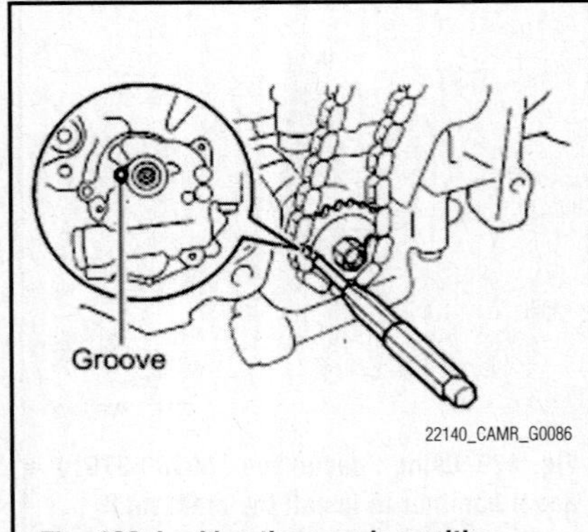

Fig. 168 Locking the gear in position

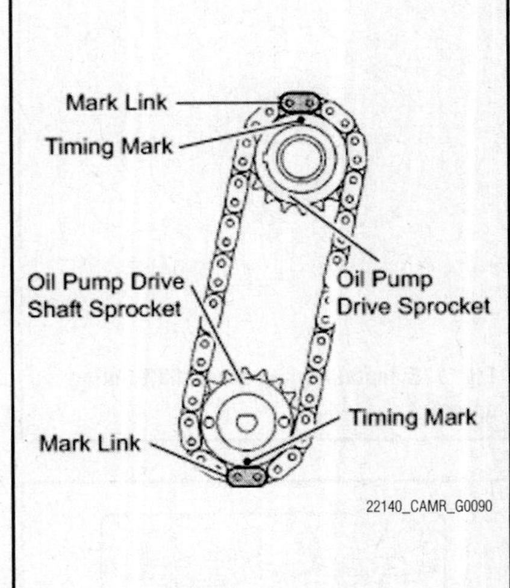

Fig. 172 Aligning the yellow mark links with the timing marks of each gear

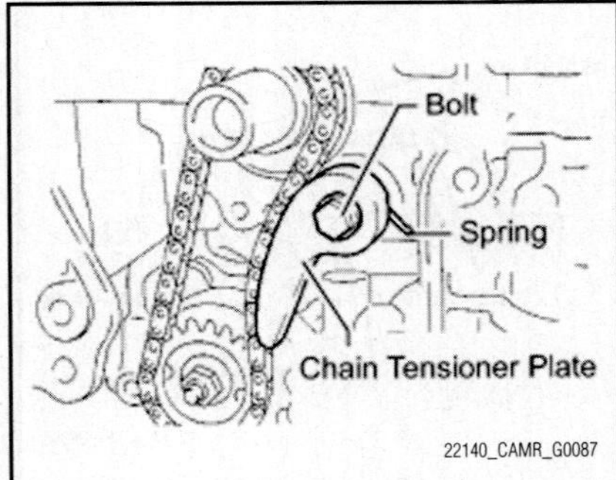

Fig. 169 Removing the bolt, chain tensioner plate and spring

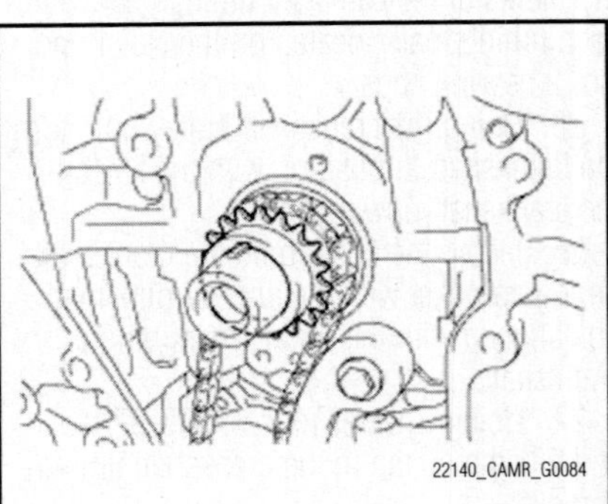

Fig. 166 Removing the crankshaft timing sprocket

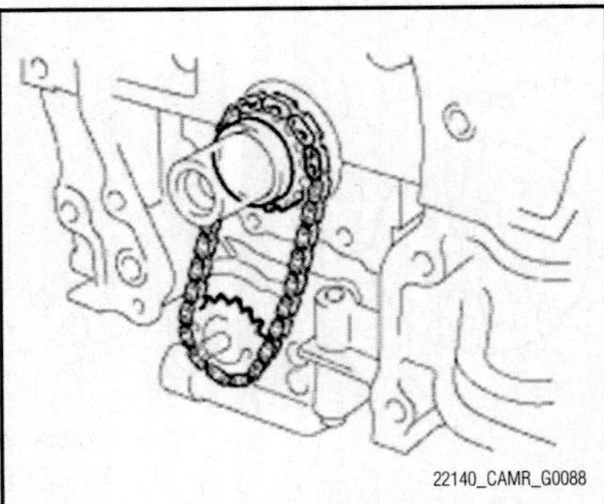

Fig. 170 Removing the chain tensioner, oil pump driven sprocket and chain

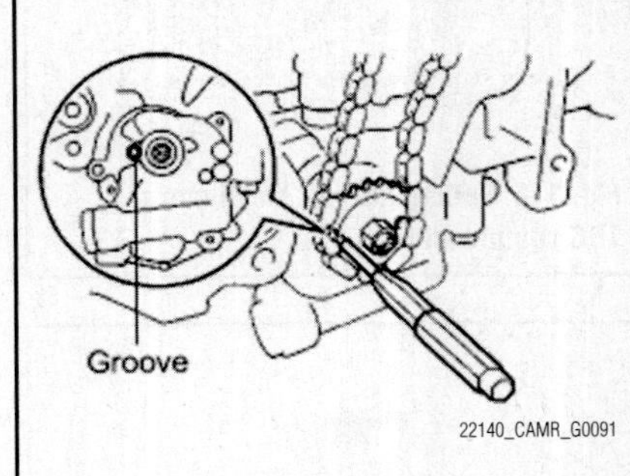

Fig. 173 Aligning the adjusting hole of the oil pump drive shaft sprocket

tensioner plate with the bolt and tighten to 9 ft. lbs. (12 Nm).

18. Align the adjusting hole of the oil pump drive shaft sprocket with the groove of the oil pump.

19. Insert a 4 mm diameter bar into the adjusting hole of the oil pump drive shaft gear to lock the gear in position, and then tighten the nut to 22 ft. lbs. (30 Nm).

20. Rotate the crankshaft clockwise by 90°, and align the crankshaft key to the top.

21. Install the crankshaft timing sprocket.

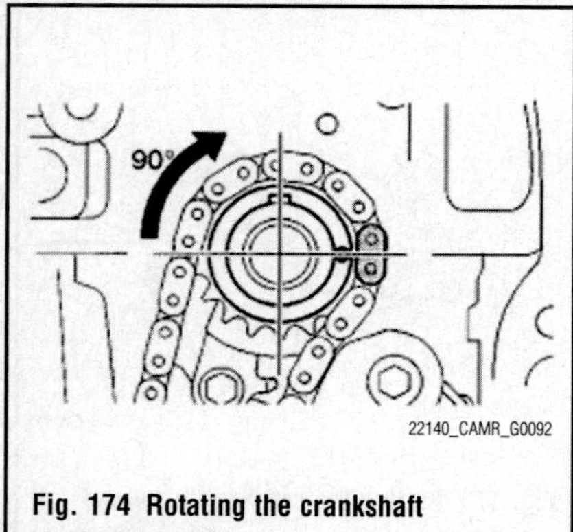

Fig. 174 Rotating the crankshaft clockwise by 90°

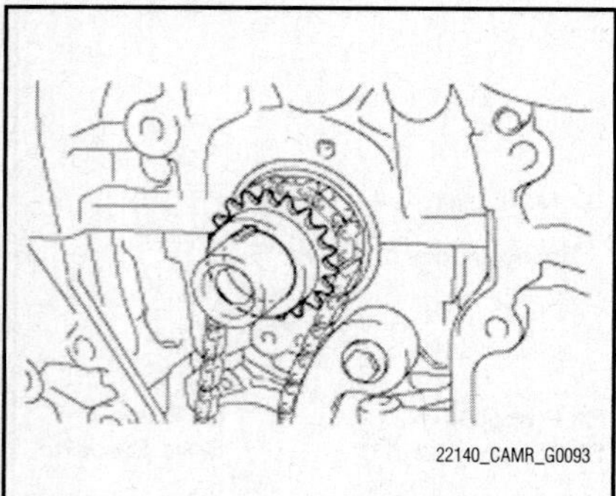

Fig. 175 Installing the crankshaft timing sprocket

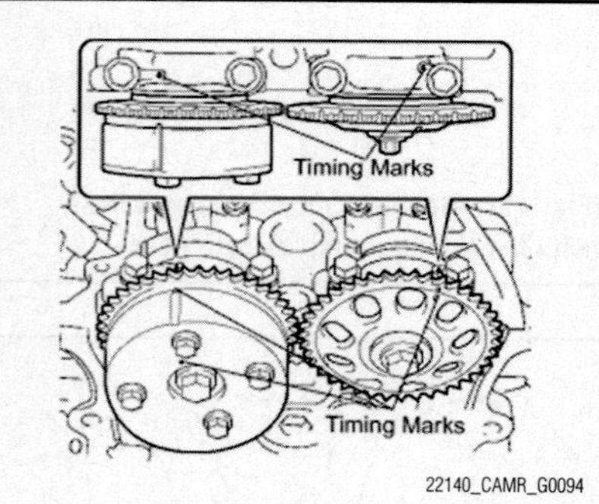

Fig. 176 Setting the No. 1 cylinder to TDC/compression

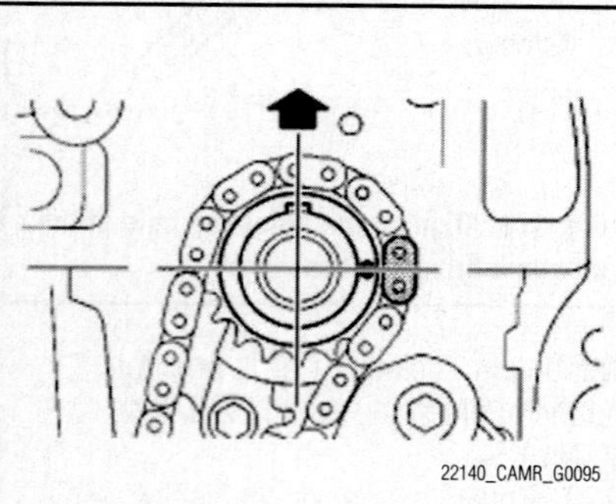

Fig. 177 Turning the crankshaft to position with the key on the crankshaft upward

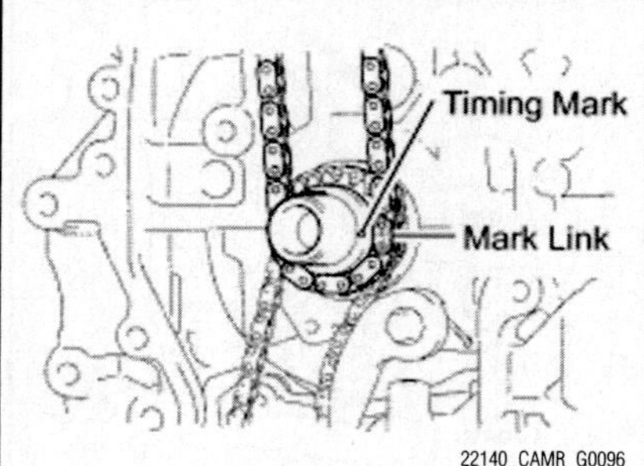

Fig. 178 Installing the chain onto the crankshaft timing sprocket with the gold or pink mark link

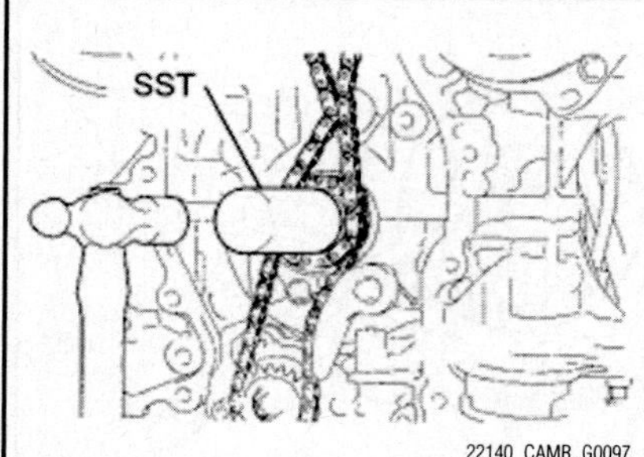

Fig. 179 Using Special Tool: 09309-37010 and a hammer to install the crankshaft timing sprocket

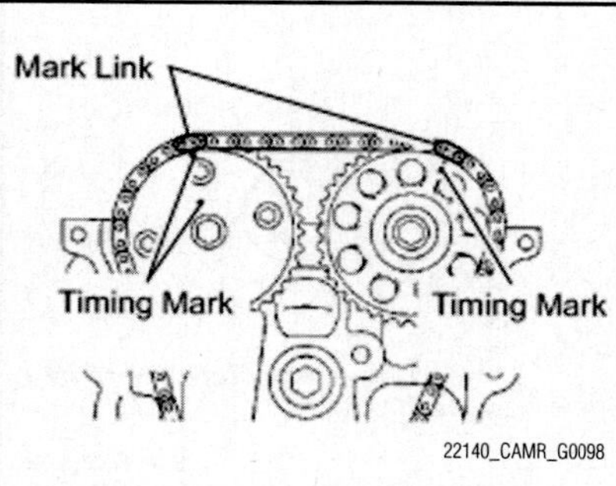

Fig. 180 Aligning the gold or yellow link with each timing mark located on the camshaft timing gear and sprocket

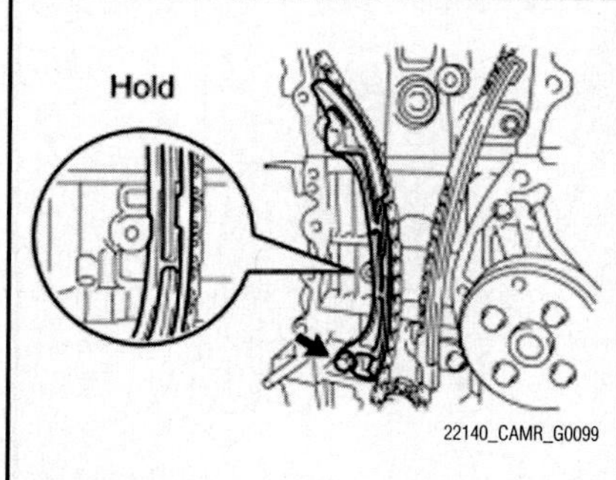

Fig. 181 Installing the chain tensioner slipper and bolt

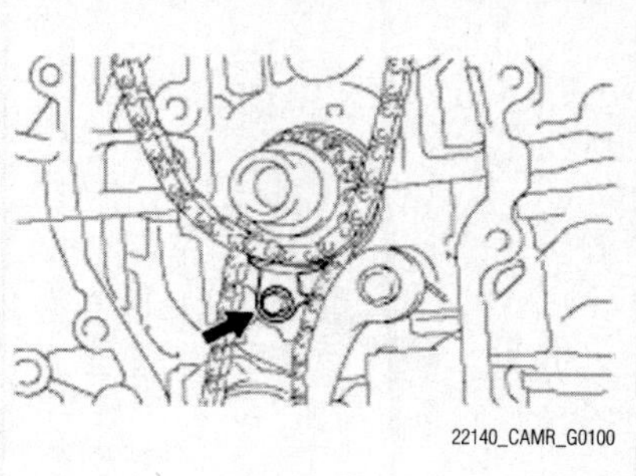

Fig. 182 Installing the timing chain guide and bolt

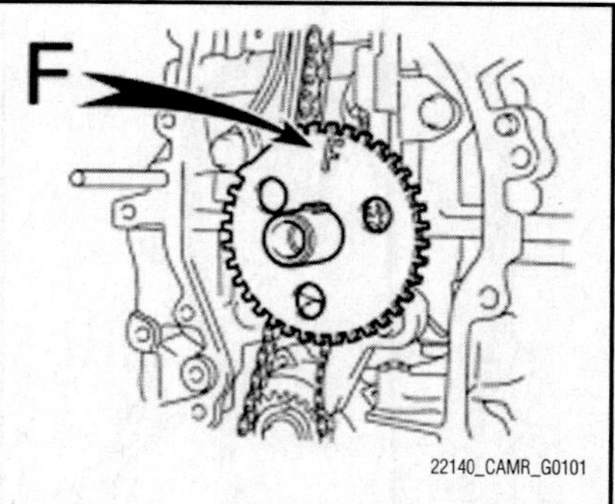

Fig. 183 Installing the crankshaft sensor plate with the "F" mark facing forward

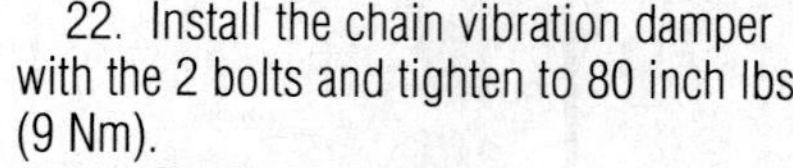

22. Install the chain vibration damper with the 2 bolts and tighten to 80 inch lbs. (9 Nm).
23. Set the No. 1 cylinder to TDC/compression.
24. Turn the camshafts with a wrench (using the hexagonal lobe) to align the timing marks of the camshaft timing gear with each timing mark located on the No. 1 and No. 2 bearing caps.
25. Using the crankshaft pulley bolt, turn the crankshaft to position with the key on the crankshaft upward.
26. Install the chain onto the crankshaft timing sprocket with the gold or pink mark link aligned with the timing mark on the crankshaft.
27. Using Special Tool: 09309-37010 and a hammer, tap in the crankshaft timing sprocket.
28. Align the gold or yellow link with each timing mark located on the camshaft timing gear and sprocket, then install the chain.
29. Install the chain tensioner slipper with the bolt and tighten to: 14 ft. lbs. (19 Nm).
30. Install the timing chain guide with the bolt and tighten to: 80 inch lbs. (9 Nm).

31. Install the crankshaft sensor plate with the "F" mark facing forward.

3.5L Engine

See Figures 184 through 195.

1. Before servicing the vehicle, refer to the Precautions Section.
2. Remove the engine assembly with transaxle.
3. Secure engine.
4. Remove the oil filler cap sub-assembly.
5. Remove the spark plugs and ignition coil assembly.
6. Remove the oil pan drain plug and gasket.
7. Remove the ventilation valve sub-assembly.
8. Remove the camshaft position sensor.
9. Remove the camshaft timing oil control valve assembly.
10. Remove crankshaft position sensor.
11. Remove the No. 1 oil pipe.
12. Remove the oil pipe.
13. Remove the cylinder block water drain cock sub-assembly.
14. Remove the oil filter.
15. Remove the crankshaft pulley.
16. Remove the left hand No. 1 front engine mounting bracket.
17. Remove the water inlet housing.
18. Remove the water outlet.
19. Remove the left-hand cylinder head cover sub-assembly and gasket.
20. Remove the cylinder head cover sub-assembly and gasket.
21. Remove the No. 2 oil pan sub-assembly.
22. Remove the oil strainer sub-assembly.
23. Remove the oil pan sub-assembly.
24. Remove the water pump assembly.
25. Remove the timing chain cover and seal.
26. Set no. 1 cylinder to TDC/compression, as follows:
    a. Temporarily tighten the pulley set bolt. Set the timing mark on the crank angle sensor plate to the RH block bore center line (TDC/compression).
    b. Check that the timing marks of the camshaft timing gears are aligned with the timing marks of the bearing cap as shown in the illustration. If not, turn the crankshaft 1 revolution (360°) and align the timing marks as above.
27. Remove the No. 1 chain tensioner assembly, as follows:
    a. Move the stopper plate upward to release the lock, and push the plunger deep into the tensioner.
    b. Move the stopper plate downward to set the lock, and insert a pin of _1.27 mm (0.05 in.) into the stopper plate's hole.
    c. Remove the 2 bolts and chain tensioner.
28. Remove the chain tensioner slipper.
29. Remove the chain sub-assembly, as follows:
    a. Turn the crankshaft counterclockwise 10° to loosen the chain of the crankshaft timing sprocket.
    b. Remove the pulley set bolt.
    c. Remove the chain from the crankshaft timing sprocket and place it on the crankshaft.
    d. Turn the camshaft timing gear assembly on the right hand bank clockwise (approximately 60°) and set it as shown in the illustration. Be sure to loosen the chain between the banks.
    e. Remove the chain.
30. Remove the idle sprocket assembly, as follows:
    a. Using a 10 mm hexagon wrench, remove the No. 2 idle gear shaft, sprocket and No. 1 idle gear shaft.
31. Remove the 2 bolts and the No. 1 chain vibration damper.
32. Remove the No. 2 chain vibration damper.
33. Remove the crankshaft timing sprocket, as follows:
    a. Remove the pulley set bolt.
    b. Remove the crankshaft timing gear from the crankshaft.

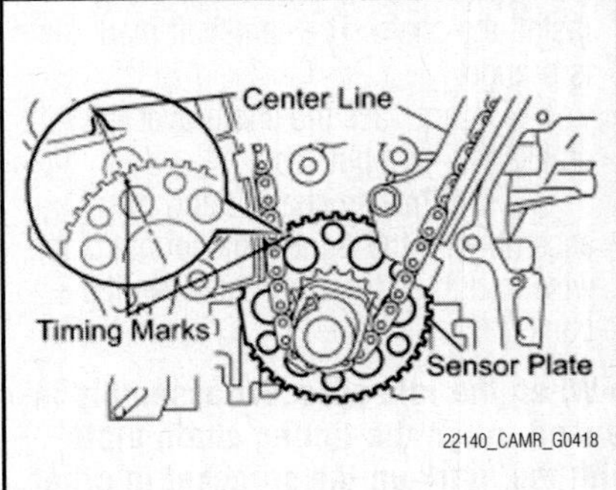

Fig. 184 Set the timing mark on the crank angle sensor plate

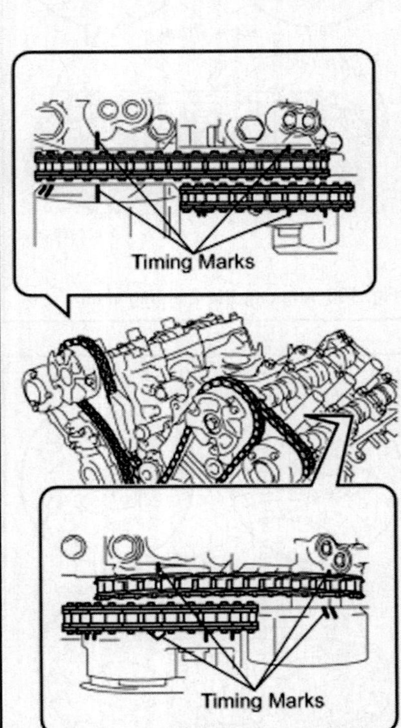

Fig. 185 Check that the timing marks of the camshaft timing gears

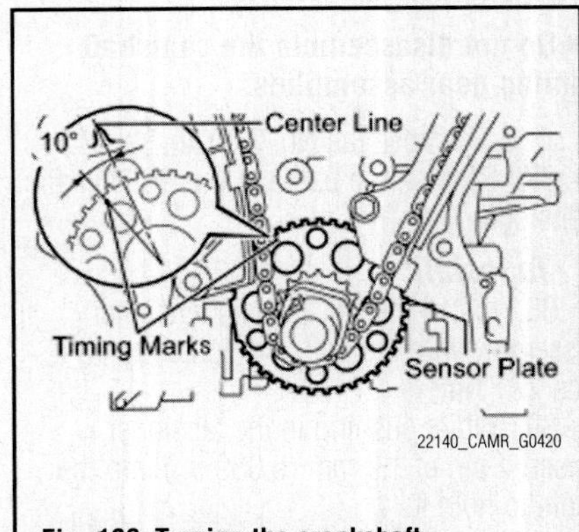

Fig. 186 Turning the crankshaft counterclockwise 10°

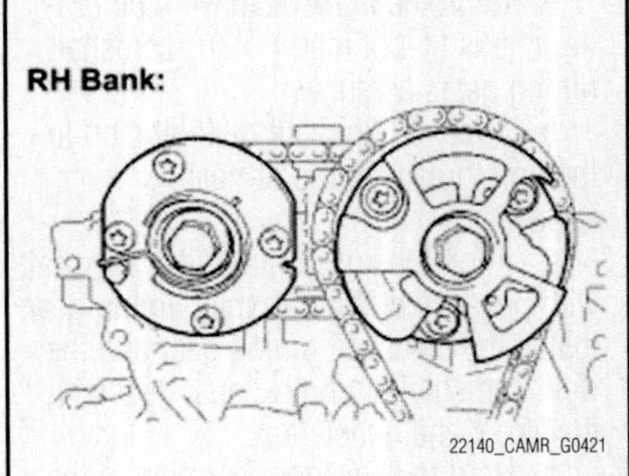

Fig. 187 Camshaft timing gear assembly positioning

c. Remove the 2 pulley set keys from the crankshaft.

34. Remove the camshaft timing gears and No. 2 chain (for Right-hand Bank), as follows:

a. While raising up the No. 2 chain tensioner, insert a pin of 1.0 mm (0.039 in.) into the hole to hold it.

b. Hold the hexagonal portion of the camshaft with a wrench, and remove the 2 bolts and 2 camshaft timing gears.

➡Be careful not to damage the cylinder head with the wrench.

➡Do not disassemble the camshaft timing gear assemblies.

c. Remove the No. 2 chain.

35. Remove the bolt and No. 2 chain tensioner.

36. Remove the camshaft timing gears and No. 2 chain (for Left-hand Bank), as follows:

a. While pushing down on the No. 3 chain tensioner, insert a pin of 1.0 mm (0.039 in.) into the hole to hold it.

b. Hold the hexagonal portion of the camshaft with a wrench, and remove the 2 bolts and 2 camshaft timing gears.

➡Be careful not to damage the cylinder head with the wrench.

➡Do not disassemble the camshaft timing gear assemblies.

c. Remove the No. 2 chain.

37. Remove the bolt and the No. 3 chain tensioner.

To install:

38. Install the No. 2 chain tensioner assembly with the bolt and tighten to 15 ft. lbs. (21 Nm).

39. While pushing in the tensioner, insert a pin of 1.0 mm (0.039 in.) into the hole to hold it.

40. Install the camshaft timing gears and No. 2 chain (for Right-hand Bank), as follows:

a. Align the mark plate with the timing marks (1-dot mark) of the camshaft timing gears as shown.

b. Apply a light coat of engine oil to the bolt threads and bolt-seating surface.

c. Align the knock pin of the camshaft with pin hole of the camshaft timing gear. Install the camshaft timing gear and the right camshaft timing exhaust gear with the No. 2 chain installed.

d. Hold the hexagonal portion of the camshaft with the wrench and tighten the two bolts to 74 ft. lbs. (100 Nm).

e. Remove the pin from the No. 2 chain tensioner.

41. Install the No. 3 chain tensioner assembly with the bolt and tighten to 15 ft. lbs. (21 Nm).

42. While pushing in the tensioner, insert a pin of 1.0 mm (0.039 in.) into the hole to hold it.

43. Install the camshaft timing gears and No. 2 chain (for Left-hand Bank), as follows:

a. Align the mark plate (yellow) with the timing marks (2-dot mark) of the camshaft timing gears as shown.

b. Apply a light coat of engine oil to the bolt threads and bolts seating surface.

c. Align the knock pin of the camshaft with pin hole of the camshaft timing gear. Install the camshaft timing gear and the left camshaft timing exhaust gear with the No. 2 chain installed.

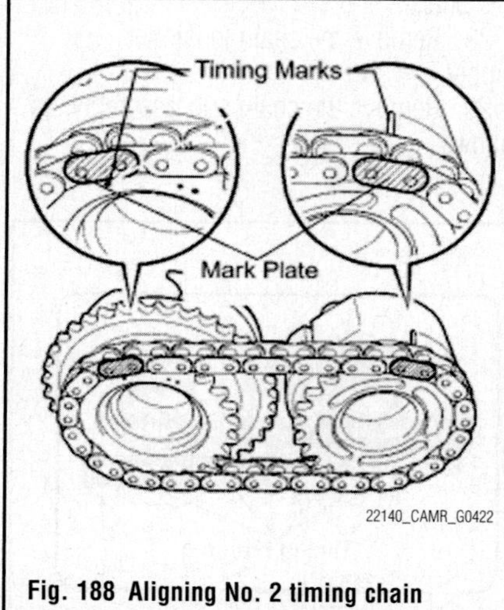

Fig. 188 Aligning No. 2 timing chain

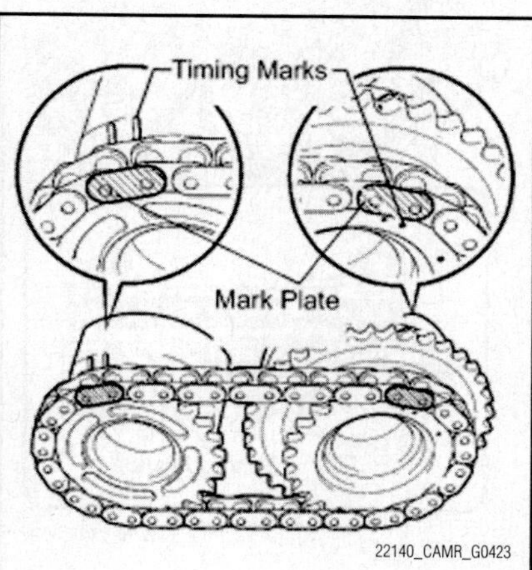

Fig. 189 Aligning No. 2 timing chain (bank 2)

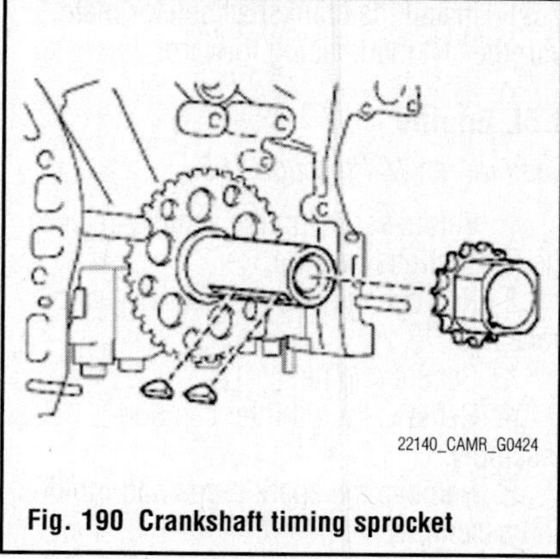

Fig. 190 Crankshaft timing sprocket

d. Hold the hexagonal portion of the camshaft with the wrench and tighten the two bolts to 74 ft. lbs. (100 Nm).

e. Remove the pin from the No 2 chain tensioner.

44. Install the No. 1 and 2 chain vibration dampers.

45. Install the timing gear set keys and crankshaft timing sprocket as shown in the illustration.

46. Install the idle sprocket assembly, as follows:

a. Apply a light coat of engine oil to the rotating surface of the No. 1 idle gear shaft.

b. Temporarily install the No. 1 idle gear shaft and idle sprocket with the No. 2 idle gear shaft while aligning the knock pin of the No. 1 idle gear with the knock pin groove of the cylinder block. Be careful of the idle gear direction.

c. Using a 10 mm hexagon wrench, tighten the No. 2 idle gear shaft to 44 ft. lbs. (60 Nm).

d. After installing the idle sprocket assembly, check that the idle sprocket turns smoothly.

47. Install the chain sub-assembly, as follows:

a. Align the mark plate and timing marks as shown in the illustration and install the chain. The camshaft mark plate is orange.

b. Do not pass the chain over the crankshaft, just put it on.

c. Turn the camshaft timing gear assembly on the right bank counterclockwise to tighten the chain between the banks.

➡When the idle sprocket assembly is reused, align the timing chain plate with the mark on the sprocket in order to tighten the chain between the banks.

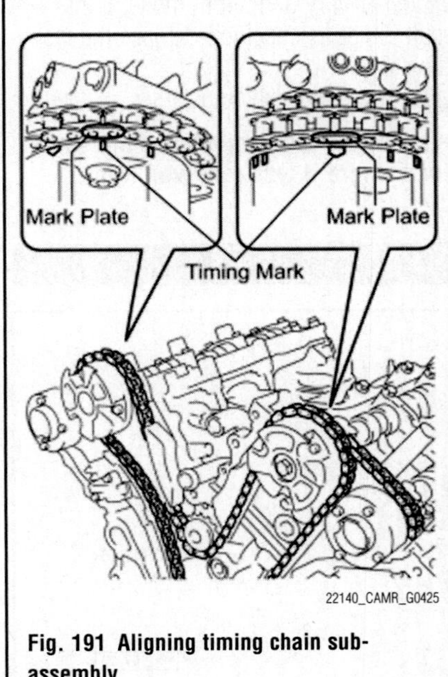

Fig. 191 Aligning timing chain sub-assembly

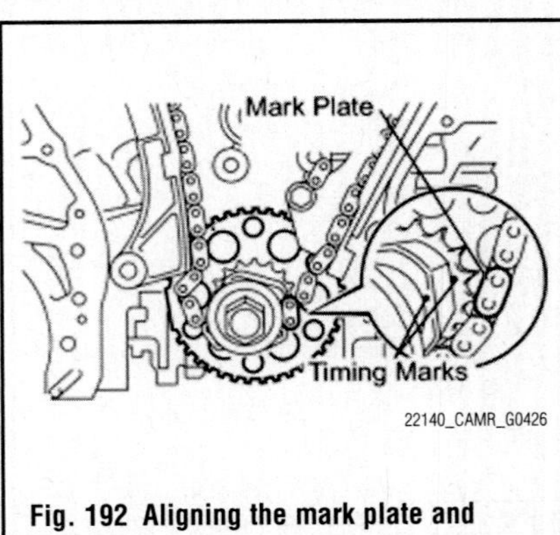

Fig. 192 Aligning the mark plate and timing mark

d. Align the mark plate and timing marks as shown in the illustration and install the chain onto the crankshaft timing sprocket. The crankshaft to mark plate is yellow.

e. Temporarily tighten the pulley set bolt.

f. Turn the crankshaft clockwise to set it to the right-hand block bore more centerline.

(TDC/compression).

48. Install the chain tensioner slipper.

49. Install the No. 1 chain tensioner assembly, as follows:

a. Move the stopper plate upward to release the lock, and push the plunger deep into the tensioner.

b. Move the stopper plate downward to set the lock, and insert a hexagon wrench into the hole of the stopper plate.

c. Install the No. 1 chain tensioner with the 2 bolts and tighten to 7 ft. lbs. (10 Nm).

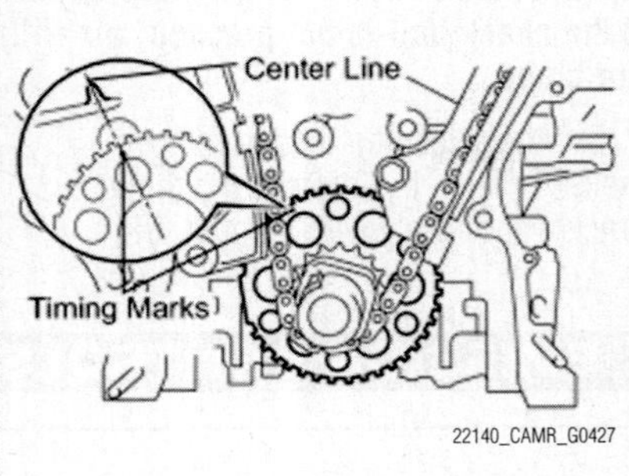

Fig. 193 Turning the crankshaft clockwise (TDC/compression)

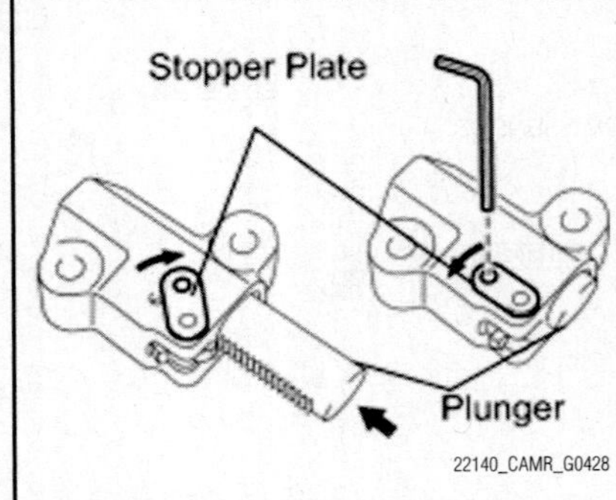

Fig. 194 Set chain tensioner plunger position

d. Remove the lock pin of the No. 1 chain tensioner. Check that each timing mark is aligned with the crankshaft at TDC/compression.

e. Remove the pulley set bolt.

50. Install timing chain cover and seal.

51. Install the water pump assembly.

52. Install the water inlet housing.

53. Install the left hand No. 1 front engine mounting bracket.

54. Install the oil pan sub-assembly.

55. Install the oil strainer sub-assembly.

56. Install the No. 2 oil pan sub-assembly.

57. Install the oil pan drain plug and gasket.

58. Install the cylinder head cover sub-assembly.

59. Install the left-hand cylinder head cover sub-assembly.

60. Install the water outlet.

61. Install the crankshaft pulley.

62. Install the oil filter.

63. Install the cylinder block water drain cock sub-assembly.

64. Install the No. 1 oil pipe.

65. Install the oil pipe.

66. Install the crankshaft position sensor.

67. Install the camshaft timing oil control valve assembly.

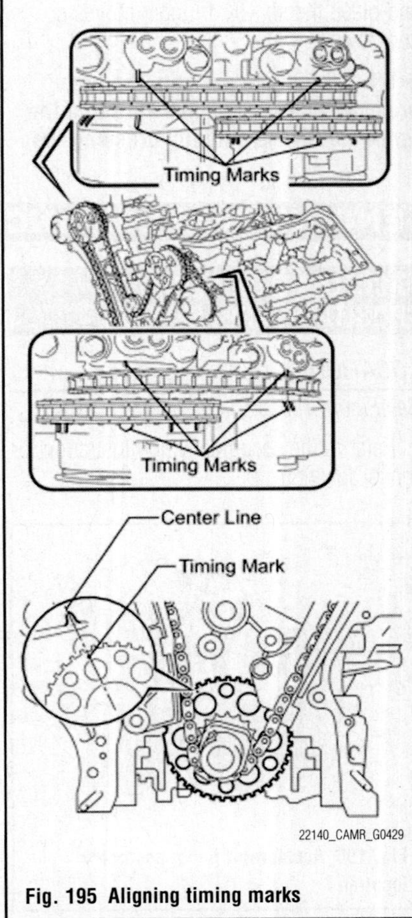

Fig. 195 Aligning timing marks

68. Install the camshaft position sensor.

69. Install the ventilation valve sub-assembly.

70. Install the spark plugs and ignition coil assembly.

71. Install the oil filler cap sub-assembly.

72. Install the engine assembly with transaxle.

VALVE LASH

ADJUSTMENT

➡**Keep the lash adjuster free of dirt and foreign objects.**

➡**Only use clean engine oil.**

1. Place the lash adjuster into a container filled with engine oil.

2. Insert the SST's (SST: 09276-75010) tip into the lash adjuster's plunger and use the tip to press down on the check ball inside the plunger.

3. Squeeze the SST and lash adjuster together to move the plunger up and down 5 to 6 times.

4. Check the movement of the plunger and bleed the air. OK: Plunger moves up and down.

➡**When bleeding air from the high-pressure chamber, make sure that the tip of the SST is actually pressing the check ball as shown in the illustration. If the check ball is not pressed, air will not bleed.**

5. After bleeding the air, remove the SST. Then, try to press the plunger quickly and firmly with a finger. OK: Plunger is very difficult to move. If the result is not as specified, replace the lash adjuster.

6. Install the lash adjusters.

➡**Install the lash adjuster to the same place where it was removed from.**

ENGINE PERFORMANCE & EMISSION CONTROLS

ACCELERATOR PEDAL POSITION (APP) SENSOR

LOCATION

See Figure 196.

Refer to the accompanying illustration for sensor location.

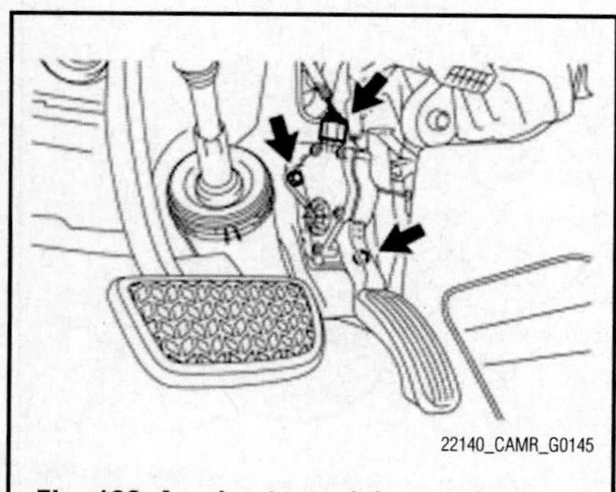

Fig. 196 Accelerator pedal assembly location

REMOVAL & INSTALLATION

See Figure 196.

1. Before servicing the vehicle, refer to the Precautions Section.
2. Remove left center floor carpet cover.
3. Disconnect the accelerator pedal connector.
4. Remove the 2 nuts and accelerator pedal assembly.

➡**Avoid physical shock to the accelerator pedal assembly.**

➡**Do not disassemble the accelerator pedal assembly.**

To install:

5. Installation is the reverse of the removal procedure. Tighten the accelerator pedal rod nuts to 43 inch lbs. (5.4 Nm).

CAMSHAFT POSITION (CMP) SENSOR

LOCATION

See Figure 197.

Refer to the accompanying illustration for sensor location.

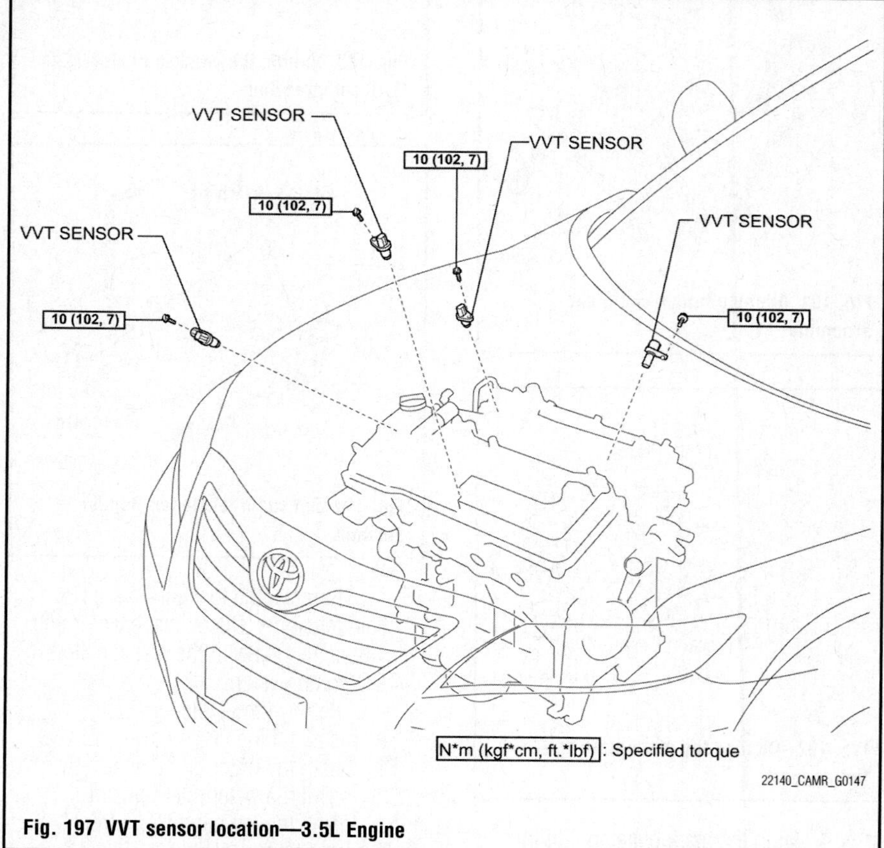

Fig. 197 VVT sensor location—3.5L Engine

REMOVAL & INSTALLATION

2.4L Engine

See Figure 198.

1. Before servicing the vehicle, refer to the Precautions Section.
2. Remove the plastic engine cover.
3. Remove air cleaner cap sub-assembly.
4. Disconnect the Camshaft Position (CMP) sensor connector.
5. Remove the bolt and Camshaft Position (CMP) sensor.

To install:

➡**Make sure that the O-ring is not cracked or jammed when installing it.**

6. Apply a light coat of engine oil to the O-ring of the sensor.
7. Install the Camshaft Position (CMP) sensor with the bolt and tighten to: 80 inch lbs. (9 Nm).

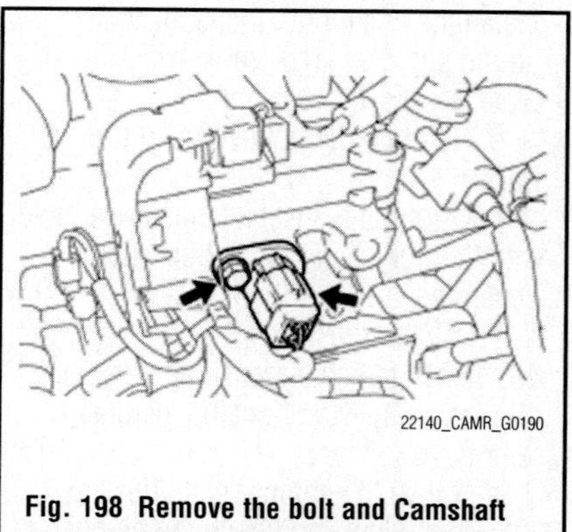

Fig. 198 Remove the bolt and Camshaft Position (CMP) sensor—2.4L Engine

8. To complete installation, reverse removal procedure.

3.5L Engine

See Figures 199 through 201.

1. Before servicing the vehicle, refer to the Precautions Section.
2. Drain and recycle the engine coolant.
3. Disconnect the negative battery cable.
4. Remove the V-bank cover sub-assembly.
5. Remove the windshield wiper link assembly.
6. Remove the front cowl top outside panel.
7. Remove the Intake camshaft VVT sensor (Bank 1), as follows:
    a. Disconnect the VVT sensor connector.
    b. Remove the bolt and VVT sensor.
8. Remove the Exhaust camshaft VVT sensor (Bank 1), as follows:
    a. Disconnect the VVT sensor connector.
    b. Remove the bolt and VVT sensor.
9. Remove the Exhaust camshaft VVT sensor (Bank 2), as follows:
    a. Disconnect the VVT sensor connector.
    b. Remove the bolt and VVT sensor.
10. Remove the Intake camshaft VVT sensor (Bank 2), as follows:

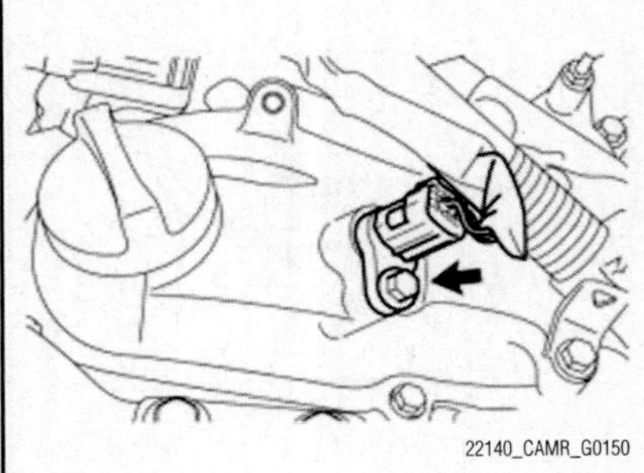

Fig. 201 Bank 2 intake camshaft VVT sensor

    a. Disconnect the VVT sensor connector.
    b. Remove the bolt and VVT sensor.

To install:

11. Install the VVT sensors and tighten to 7 ft. lbs. (10 Nm).
    a. Connect the VVT sensor connectors.
12. To complete installation, reverse the remaining removal procedure.

CRANKSHAFT POSITION (CKP) SENSOR

LOCATION

See Figures 202 and 203.

Refer to the accompanying illustrations for sensor location.

REMOVAL & INSTALLATION

2.4L Engine

See Figures 204 and 205.

1. Before servicing the vehicle, refer to the Precautions Section.
2. Disconnect the negative battery cable.
3. Remove the front right wheel.
4. Remove the right hand front fender apron.
5. Remove the drive belt.
6. Remove the alternator assembly.
7. Disconnect the Crankshaft Position (CKP) sensor connector.
8. Remove the connector clamp and wire harness clamp.
9. Remove the wire harness clamp bracket from the wire harness.
10. Remove the bolt, then remove the Crankshaft Position (CKP) sensor.

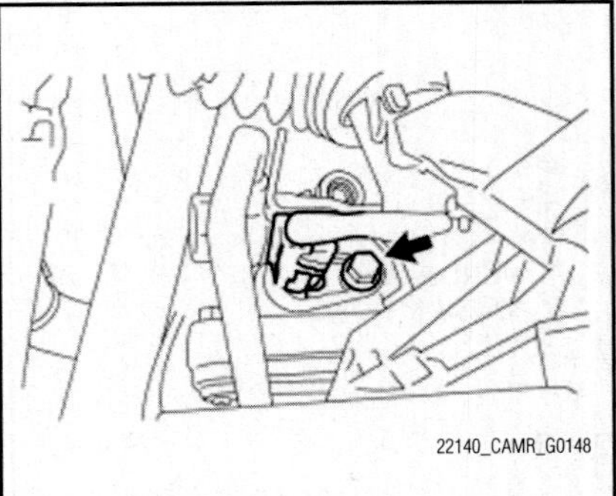

Fig. 201 Bank 1 intake camshaft VVT sensor

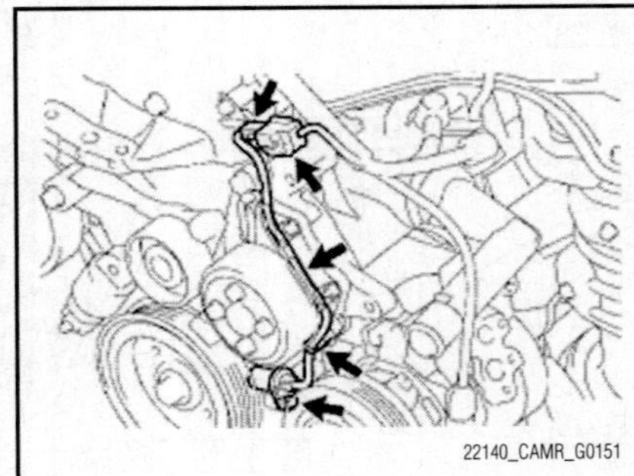

Fig. 202 Crankshaft Position (CKP) sensor location—2.4L engine

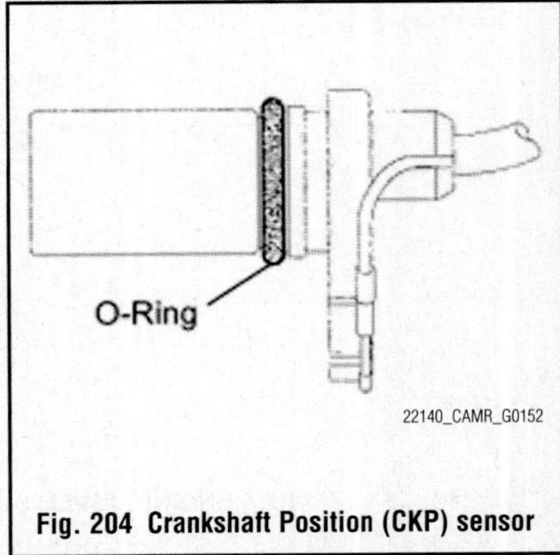

Fig. 204 Crankshaft Position (CKP) sensor

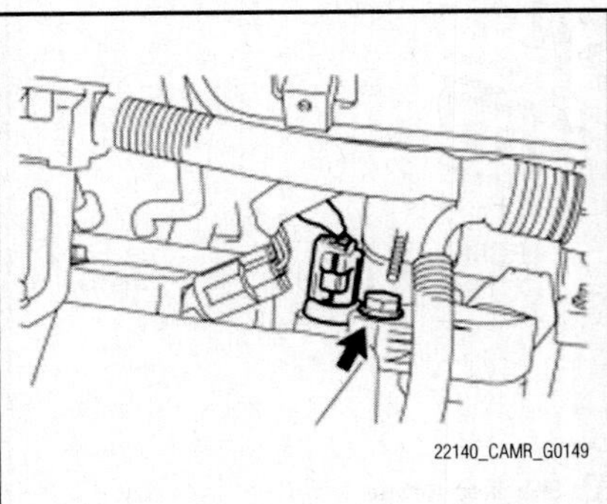

Fig. 200 Bank 1 exhaust camshaft VVT sensor

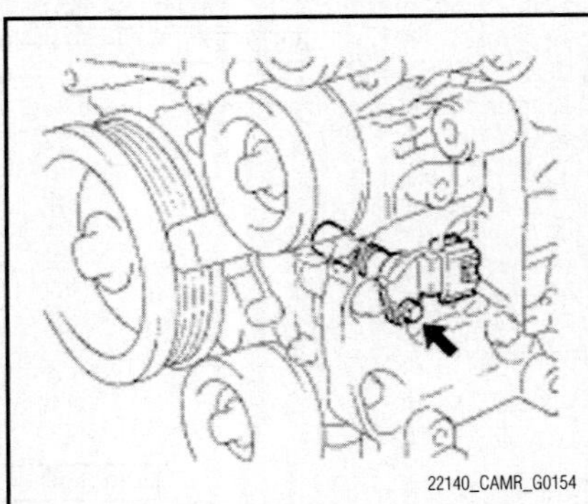

Fig. 203 Crankshaft Position (CKP) sensor location—3.5L engine

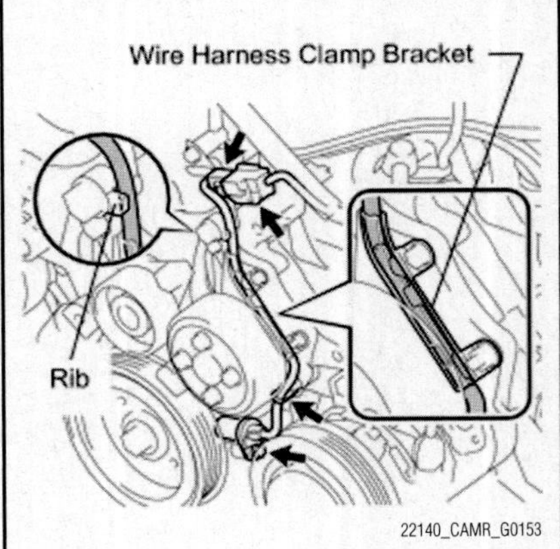

Fig. 205 Installing Crankshaft Position (CKP) sensor—2.4L Engine

To install:

11. Apply a light coat of engine oil to the O-ring on the Crankshaft Position (CKP) sensor.
12. Install the Crankshaft Position (CKP) sensor with the bolt and tighten to 7ft. lbs. (10 Nm).
13. Connect the Crankshaft Position (CKP) sensor connector.
14. The remainder of installation is the reverse of the removal procedure.

3.5L Engine

See Figure 206.

1. Before servicing the vehicle, refer to the Precautions Section.
2. Disconnect the negative battery cable.
3. Remove alternator assembly.

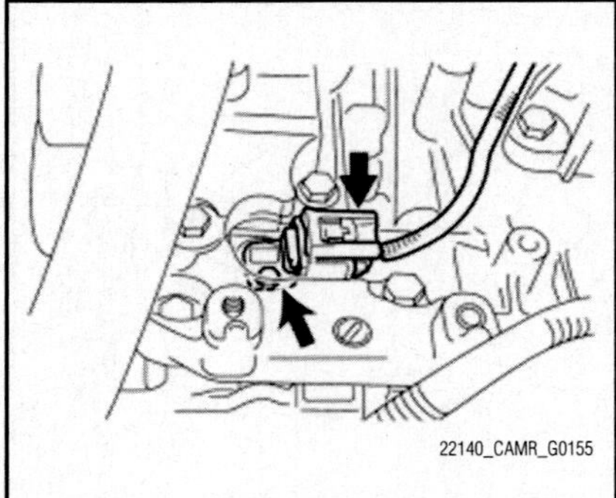

Fig. 206 Installing Crankshaft Position (CKP) sensor—3.5L Engine

4. Disconnect the cooler compressor assembly.
5. Remove the Crankshaft Position (CKP) sensor connector.
6. Remove the bolt, and then remove the Crankshaft Position (CKP) sensor.

To install:

7. Apply a light coat of engine oil to the O-ring on the Crankshaft Position (CKP) sensor.
8. Install the Crankshaft Position (CKP) sensor with the bolt and tighten to 7ft. lbs. (10 Nm).
9. Connect the Crankshaft Position (CKP) sensor connector.
10. The remainder of installation is the reverse of the removal procedure.

ELECTRONIC CONTROL MODULE (ECM)

LOCATION

See Figures 207 and 208.

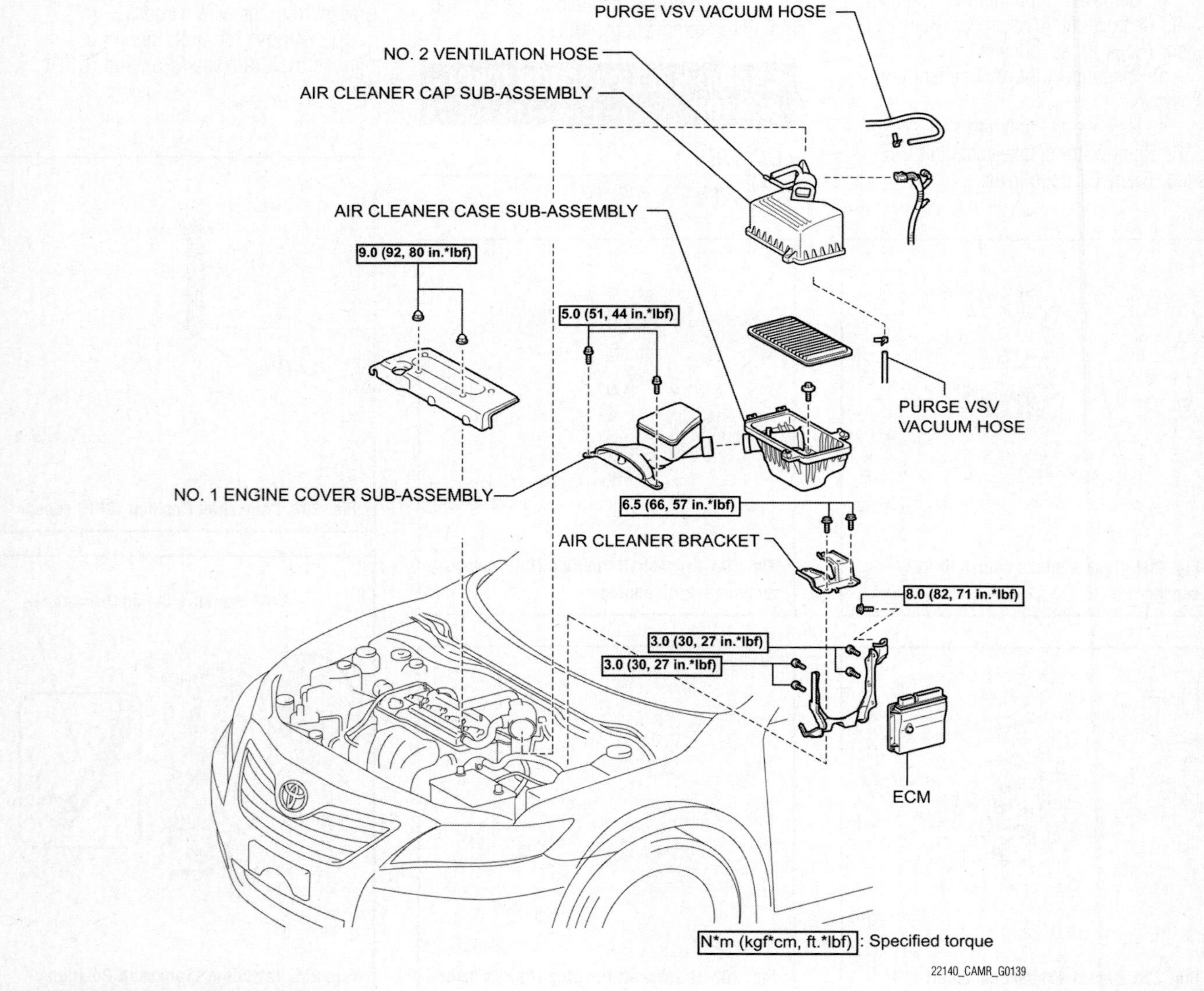

Fig. 207 ECM location—2.4L Engine

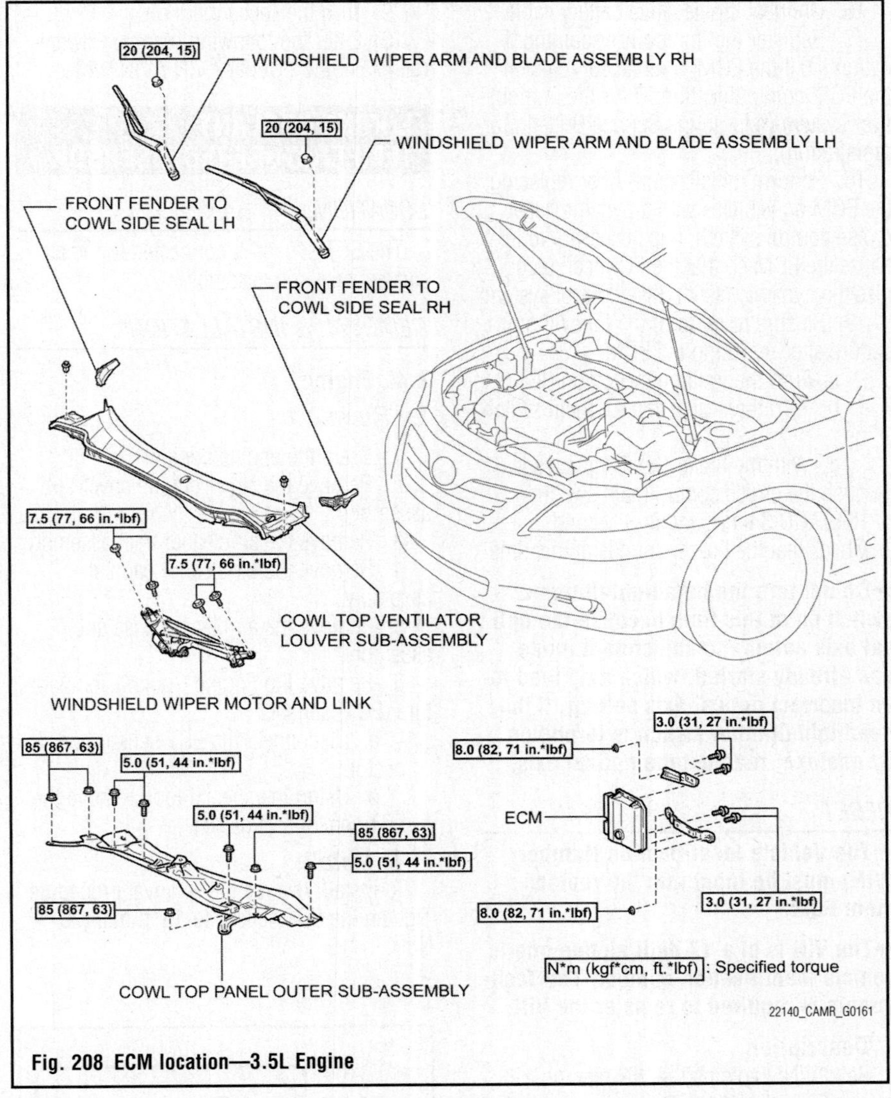

Fig. 208 ECM location—3.5L Engine

Refer to the accompanying illustrations for location.

REMOVAL & INSTALLATION

2.4L Engine

See Figure 209.

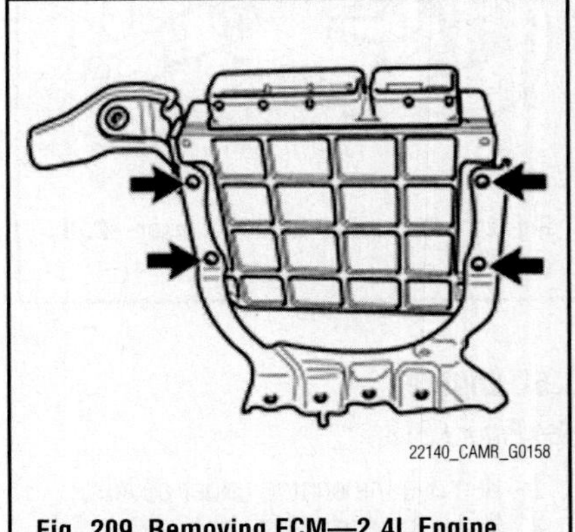

Fig. 209 Removing ECM—2.4L Engine

1. Before servicing the vehicle, refer to the Precautions Section.
2. Disconnect the negative battery cable.
3. Remove the plastic engine cover.
4. Remove the air cleaner inlet assembly.
5. Remove the air cleaner cap sub-assembly.
6. Remove the air cleaner case sub-assembly.
7. Remove the two bolts and air cleaner bracket.
8. Disconnect the two ECM connectors by raising the two levers. While pushing the locks on the two levers, disconnect the two ECM connectors.

➡After disconnecting the connectors, make sure that dirt, water or other foreign matter does not contact the connections of the connectors.

a. Remove the ECM with the bracket and three bolts.

b. Remove the four screws and ECM brackets.

To install:

9. Install the bracket to the ECM with the 4 screws, and tighten to 27 inch lbs. (3 Nm).
10. Attach the ECM with the three nuts and tighten to 71 inch lbs. (8 Nm).

➡Make sure that dirt, water or other foreign matter does not contact the connections of the connectors.

11. Connect the two ECM connectors and lower the two levers.
12. Install the ECM to the body.
13. Install the two bolts and air cleaner bracket.
14. Install the air cleaner case sub-assembly.
15. Install the air cleaner cap sub-assembly.
16. Install the air cleaner inlet assembly.
17. Remove the plastic engine cover.
18. Connect the negative battery cable.
19. Register the immobilizer communication ID. If the ECM is replaced, register the ECM communication ID for the immobilizer system (refer to the Service Bulletin for registration).
20. Perform initialization. After replacing the ECM on vehicles with a dynamic laser cruise control system, it is necessary to initialize the ECM so that the ECM can recognize the dynamic laser cruise control system.
21. Be sure to perform the following procedure after replacing the ECM:

a. Turn the ignition switch on (IG).

b. Turn the cruise control main switch on.

c. With the brake pedal depressed, push the cruise control main switch to RES/ACC 3 times within 3 seconds. Check that the buzzer sounds at this time.

➡Do not turn the headlight dimmer switch on at this time because the optical axis automatic adjustment mode has already started, which may lead to an incorrect optical axis setting. If the headlight dimmer switch is turned on by mistake, readjust the optical axis.

3.5L Engine

See Figure 210.

1. Before servicing the vehicle, refer to the Precautions Section.
2. Disconnect the negative battery cable.

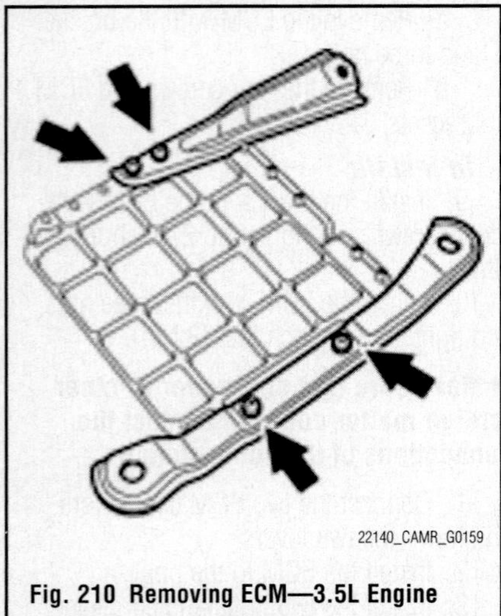

Fig. 210 Removing ECM—3.5L Engine

3. Remove both windshield wiper arm and blade assemblies.
4. Remove the cowl top ventilator louver sub-assembly.
5. Remove the windshield wiper motor and link assembly.
6. Remove the outer cowl top panel.
7. Remove the ECM, as follows:
   a. Remove the 3 nuts.
   b. Separate the ECM from the body. When separating the ECM, do not apply excessive force to the wire harness.
   c. Raise the 2 levers while pushing the locks on the 2 levers, and disconnect the 2 ECM connectors.

➡After disconnecting the connectors, make sure that dirt, water or other foreign matter does not contact the connections of the connectors.

   d. Remove the ECM.
   e. Remove the 4 screws and 2 ECM brackets.

To install:

8. Install the 2 ECM brackets with the 4 screws, and tighten to 27 inch lbs. (3 Nm).
9. Connect the 2 ECM connectors and lower the 2 levers.

➡Make sure that dirt, water or other foreign matter does not contact the connections of the connectors.

10. Install the ECM to the body.
11. Attach the ECM with the 3 nuts and tighten to 71 inch lbs. (8 Nm).
12. Install the outer cowl top panel.
13. Install the windshield wiper motor and link assembly.
14. Install the cowl top ventilator louver sub-assembly.
15. Install both windshield wiper arm and blade assemblies.
16. Connect the negative battery cable.
17. Register the immobilizer communication ID. If the ECM is replaced, register the ECM communication ID for the immobilizer system (refer to the Service Bulletin for registration).
18. Perform initialization. After replacing the ECM on vehicles with a dynamic laser cruise control system, it is necessary to initialize the ECM so that the ECM can recognize the dynamic laser cruise control system.
19. Be sure to perform the following procedure after replacing the ECM:
   a. Turn the ignition switch on (IG).
   b. Turn the cruise control main switch on.
   c. With the brake pedal depressed, push the cruise control main switch to RES/ACC 3 times within 3 seconds. Check that the buzzer sounds at this time.

➡Do not turn the headlight dimmer switch on at this time because the optical axis automatic adjustment mode has already started, which may lead to an incorrect optical axis setting. If the headlight dimmer switch is turned on by mistake, readjust the optical axis.

RESET

➡The Vehicle Identification Number (VIN) must be input into the replacement ECM.

➡The VIN is of a 17-digit alphanumeric vehicle identification number. The Techstream is required to register the VIN.

Description

Read VIN: Explains the VIN reading process in a flowchart. This process allows the VIN stored in the ECM to be read in order to confirm that the 2 VINs, provided with the vehicle and stored in the vehicle ECM, are the same.

Write VIN: Explains the VIN writing process in a flowchart. This process allows the VIN to be input into the ECM. If the ECM is replaced, or the ECM VIN and vehicle VIN do not match, the VIN can be registered, or overwritten in the ECM by following this procedure.

Read VIN

1. Confirm the vehicle VIN.
2. Connect the Techstream to the DLC3.
3. Turn the ignition switch to ON.
4. Turn the Techstream on.
5. Enter the following menus: Powertrain / Engine / Utility / VIN / VIN Read.

Write VIN

6. Confirm the vehicle VIN.
7. Connect the Techstream to the DLC3.
8. Turn the ignition switch to ON.
9. Turn the Techs ream on.
10. Enter the following menus: Powertrain / Engine / Utility / VIN / VIN Write.

ENGINE COOLANT TEMPERATURE (ECT) SENSOR

LOCATION

The ECT sensor is connected to the air cleaner case sub assembly.

REMOVAL & INSTALLATION

2.4L Engine

See Figure 211.

1. Drain the engine coolant.
2. Remove the No. 1 engine cover sub assembly.
3. Remove the air cleaner inlet assembly.
4. Remove the air cleaner cap sub assembly.
5. Remove the air cleaner case sub assembly.
6. Remove the Engine Coolant Temperature (ECT) sensor.
   a. Disconnect the ECT sensor connector.
   b. Using the special tool, remove the ECT sensor and gasket.

To install:

To install, reverse the removal procedure. Tighten the ECT sensor to 15 ft. lbs. (20 Nm).

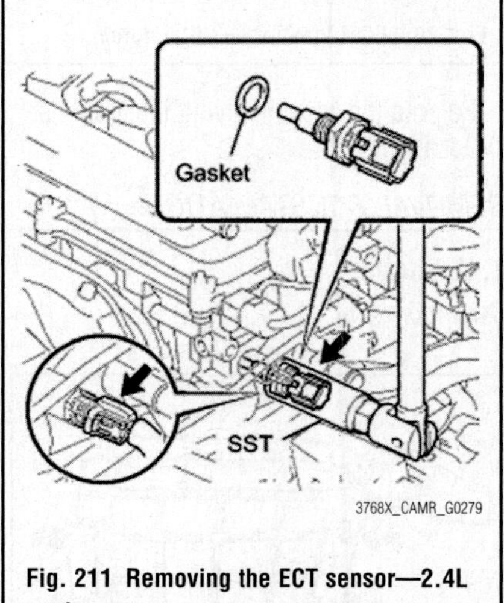

Fig. 211 Removing the ECT sensor—2.4L engine

3.5L Engine

See Figure 212.

1. Remove the engine under covers.
2. Drain the engine coolant.

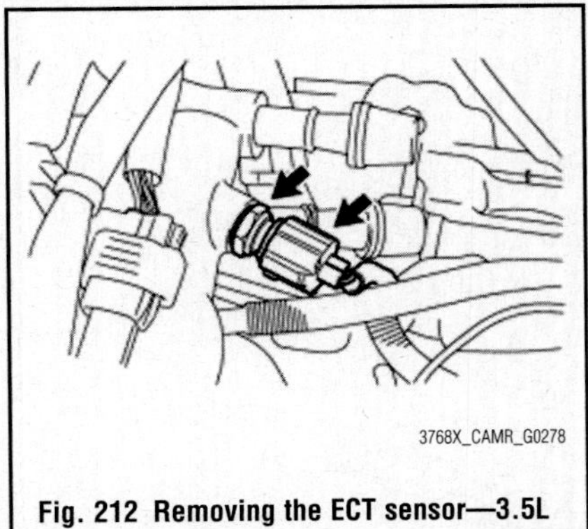

Fig. 212 Removing the ECT sensor—3.5L engine

3. Remove the v-bank cover sub assembly.
4. Remove the air cleaner inlet assembly.
5. Remove the air cleaner cap sub assembly.
6. Remove the air cleaner case sub assembly.
7. Remove the No. 1 air cleaner inlet.
8. Remove the Engine Coolant Temperature (ECT) sensor.
    a. Remove the ECT sensor connector.
    b. Remove the engine coolant temperature sensor.

To install:

To install, reverse the removal procedure. Tighten the ECT sensor to 15 ft. lbs. (20 Nm).

EVAPORATIVE EMISSIONS (EVAP) CANISTER

REMOVAL & INSTALLATION

2.4L Engine

See Figures 213 through 215.

1. Remove the fuel tank.
2. Remove the charcoal canister assembly.
    a. Disconnect the fuel tank vent hose from the charcoal canister. Push the connector deep inside. Pinch portion A. Pull out the connector.
    b. Disconnect the charcoal canister filter sub assembly from the charcoal canister. Push the connector deep inside. Pinch portion A. Pull out the connector.
    c. Disconnect the vapor pressure sensor connector.
    d. Disconnect the wire harness clamp.
    e. Disconnect the purge line hose from the charcoal canister.
    f. Remove the 2 bolts, clip and charcoal canister.

To install:

To install, reverse the removal procedure. Tighten the charcoal canister to 29 ft. lbs. (39 Nm).

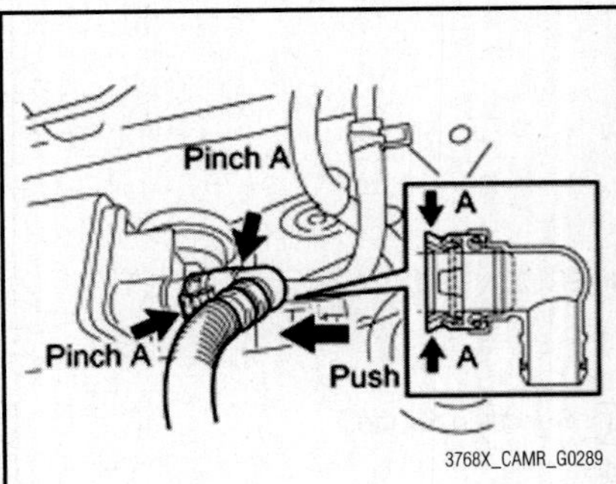

Fig. 213 Disconnecting the fuel tank vent hose from the charcoal canister

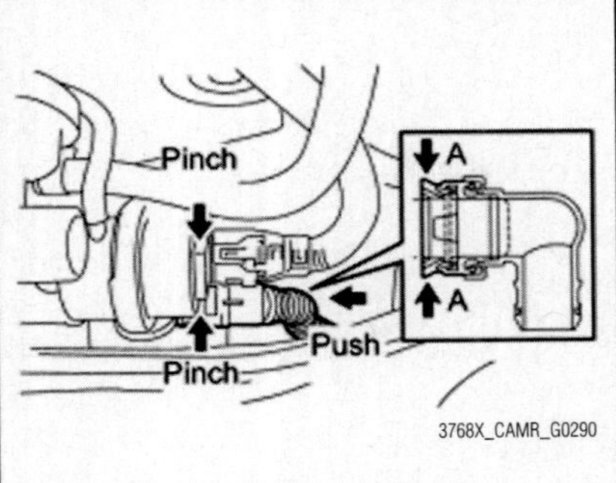

Fig. 214 Disconnecting the charcoal canister filter sub assembly

Fig. 215 Removing the charcoal canister

3.5L Engine

See Figures 216 through 218.

1. Remove the fuel tank.
2. Remove the charcoal canister assembly.
    a. Disconnect the fuel tank vent hose from the charcoal canister. Push the connector deep inside. Pinch portion A. Pull out the connector.
    b. Disconnect the charcoal canister filter sub assembly from the charcoal canister. Push the connector deep inside. Pinch portion A. Pull out the connector.

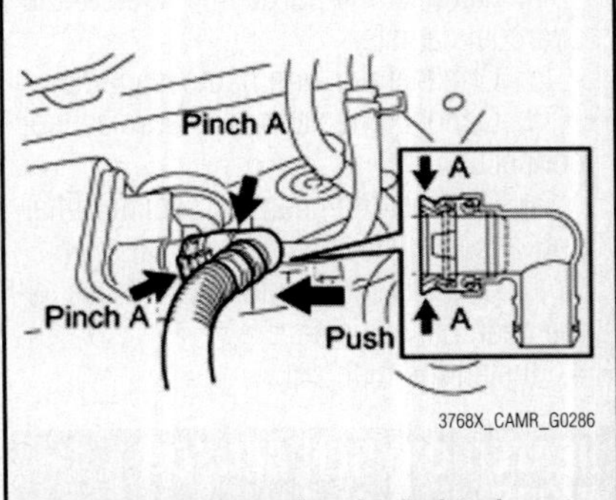

Fig. 216 Disconnecting the fuel tank vent hose from the charcoal canister

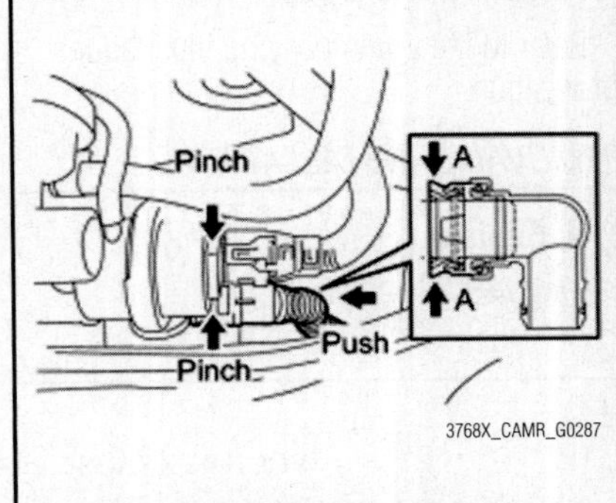

Fig. 217 Disconnecting the charcoal canister filter sub assembly

Fig. 218 Removing the charcoal canister

    c. Disconnect the vapor pressure sensor connector.
    d. Disconnect the wire harness clamp.
    e. Disconnect the purge line hose from the charcoal canister.
    f. Remove the 2 bolts, clip and the charcoal canister.

To install:

3. Install the charcoal canister assembly.
    a. Install the 2 bolts, clip and charcoal canister. Tighten to 29 ft. lbs. (39 Nm).

b. Connect the purge line hose to the charcoal canister.

c. Connect the wire harness clamp.

d. Connect the vapor pressure sensor connector.

e. Connect the charcoal canister filter sub-assembly to the charcoal canister.

f. Connect the fuel tank vent hose to the charcoal canister.

4. Install the fuel tank.

HEATED OXYGEN SENSOR (HO2S)

LOCATION

See Figures 219 and 220.

Refer to the accompanying illustrations for location.

REMOVAL & INSTALLATION

2.4L Engine

See Figures 221 and 222.

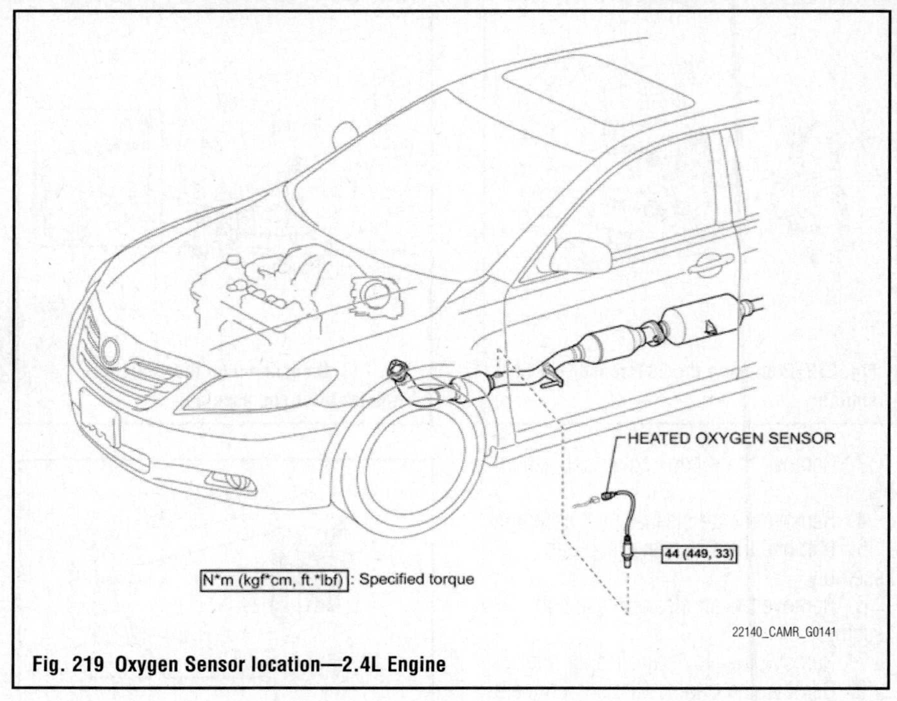

Fig. 219 Oxygen Sensor location—2.4L Engine

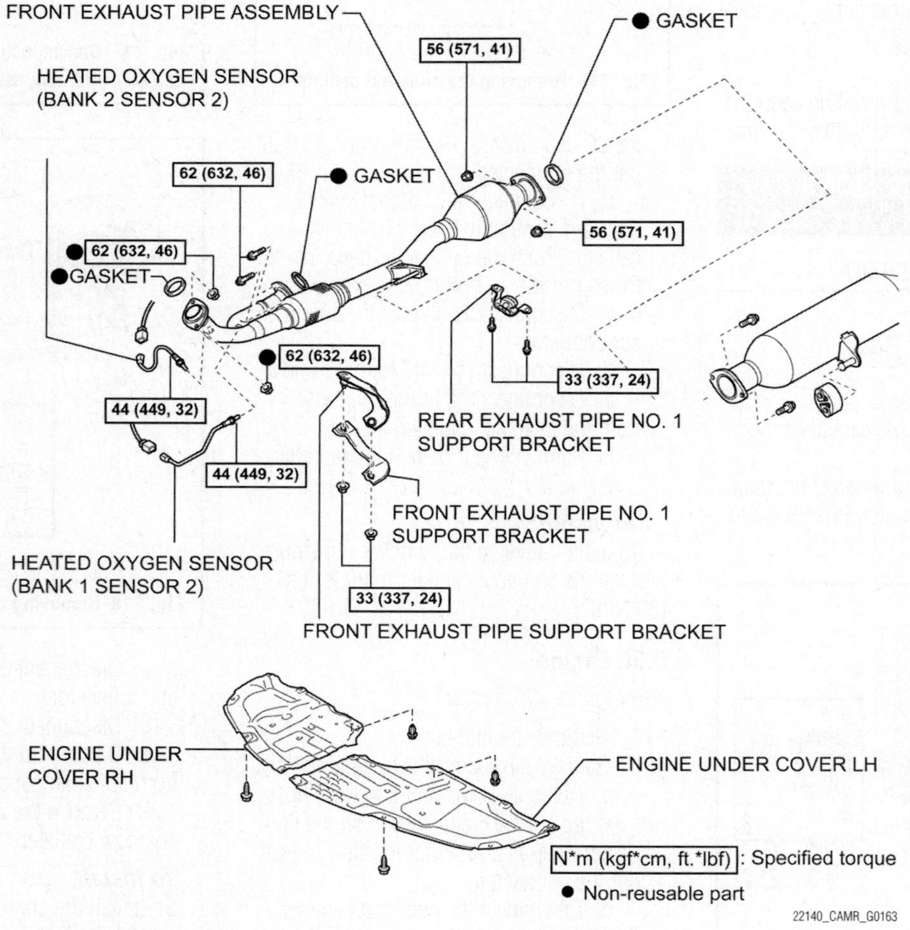

Fig. 220 Oxygen Sensor location—3.5L Engine

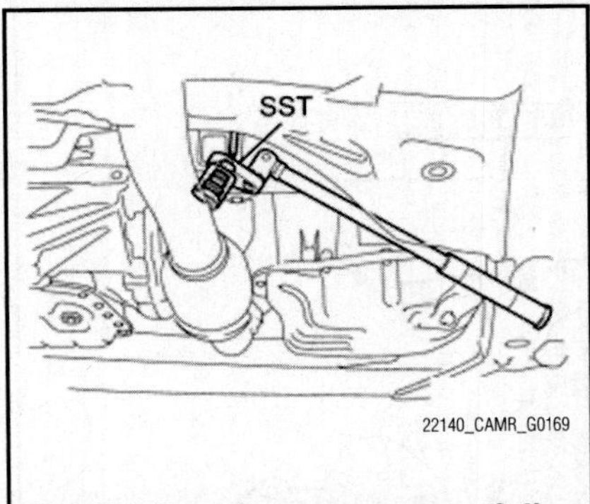

Fig. 221 Removing oxygen sensor—2.4L Engine

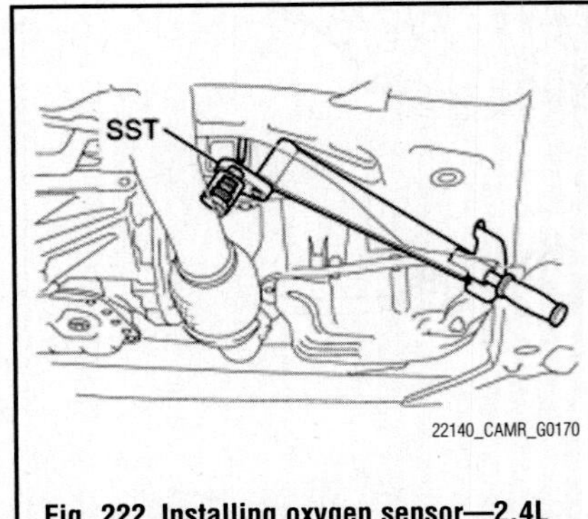

Fig. 222 Installing oxygen sensor—2.4L Engine

1. Before servicing the vehicle, refer to the Precautions Section.
2. Disconnect the oxygen sensor connectors.

➡Do not damage the heated oxygen sensor.

3. Using Special Tool: 09224-00010 or equivalent, remove the two oxygen sensors from the front pipe assembly.

To install:

4. Install the oxygen sensor to the front pipe assembly. Tighten to 32 ft. lbs. (44 Nm). Use a torque wrench with a fulcrum length of 300 mm (11.81 in.).
5. Connect the oxygen sensor connector.

3.5L Engine

See Figures 223 and 224.

1. Before servicing the vehicle, refer to the Precautions Section.
2. Remove the front exhaust pipe assembly.
3. Disconnect the 2 oxygen sensor connectors.
4. Using Special Tool: 09224-00010 or equivalent, remove the 2 oxygen sensors from the front pipe assembly.

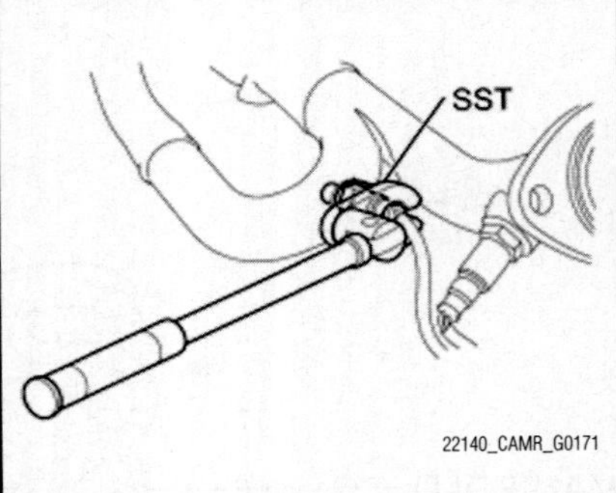

Fig. 223 Removing oxygen sensor—3.5L Engine

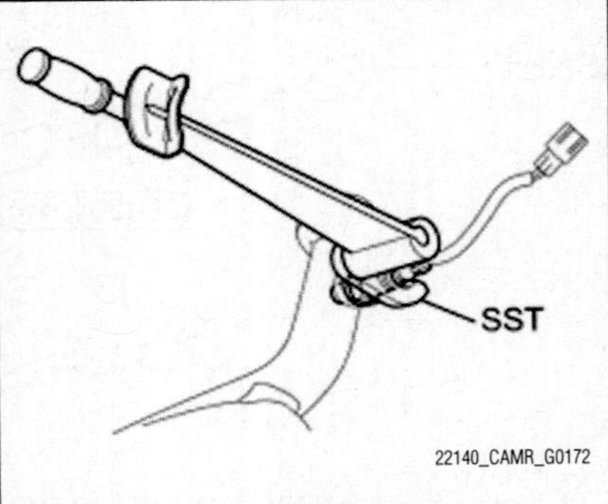

Fig. 224 Installing oxygen sensor—3.5L Engine

To install:

5. Install the 2 oxygen sensors to the front pipe assembly. Tighten to 32 ft. lbs. (44 Nm) and 30 ft. lbs. (40 Nm). Use a torque wrench with a fulcrum length of 300 mm (11.81 in.).
6. Connect the 2 oxygen sensor connectors.
7. Remove the front exhaust pipe assembly.

INTAKE AIR TEMPERATURE (IAT) SENSOR

LOCATION

The Intake Air Temperature (IAT) sensor is mounted on the Mass Air Flow (MAF) meter.

REMOVAL & INSTALLATION

See Mass Air Flow (MAF) meter.

KNOCK SENSOR (KS)

LOCATION

See Figures 225 and 226.

Refer to the accompanying illustrations for sensor location.

REMOVAL & INSTALLATION

2.4L Engine

See Figures 227 and 228.

⁂ CAUTION

Observe all applicable safety precautions when working around fuel. Whenever servicing the fuel system, always work in a well ventilated area. Do not allow fuel spray or vapors to come in contact with a spark or open flame. Keep a dry chemical fire extinguisher near the work area. Always keep fuel in a container specifically designed for fuel storage; also, always properly seal fuel containers to avoid the possibility of fire or explosion.

1. Before servicing the vehicle, refer to the Precautions Section.
2. Properly discharge the fuel system pressure.
3. Disconnect battery negative cable.
4. Remove plastic engine cover.
5. Drain and recycle the engine coolant.
6. Remove windshield wiper arms and blade assemblies.
7. Remove both front fender to cowl side seals.
8. Remove cowl top ventilator louver sub-assembly.
9. Remove windshield wiper motor and link.
10. Remove the four bolts, four nuts and cowl top panel outer sub-assembly.
11. Remove air cleaner cap sub-assembly.
12. Remove air cleaner case sub-assembly.
13. Remove throttle body.
14. Disconnect fuel tube.
15. Remove fuel delivery pipe with injector.
16. Disconnect the union to check valve hose from the brake booster.
17. Disconnect the camshaft timing oil control valve connector.
18. Remove the wire harness clamp.
19. Remove the union to check valve hose from the vacuum hose clamp.
20. Remove the 5 bolts, 2 nuts and intake manifold. Remove the gasket from the intake manifold.
21. Disconnect the knock sensor connector. Remove the nut and knock sensor.

To install:

➡Make sure that the knock sensor is in the correct position.

22. Install the knock control sensor with the nut and tighten to 15 ft. lbs. (20 Nm).

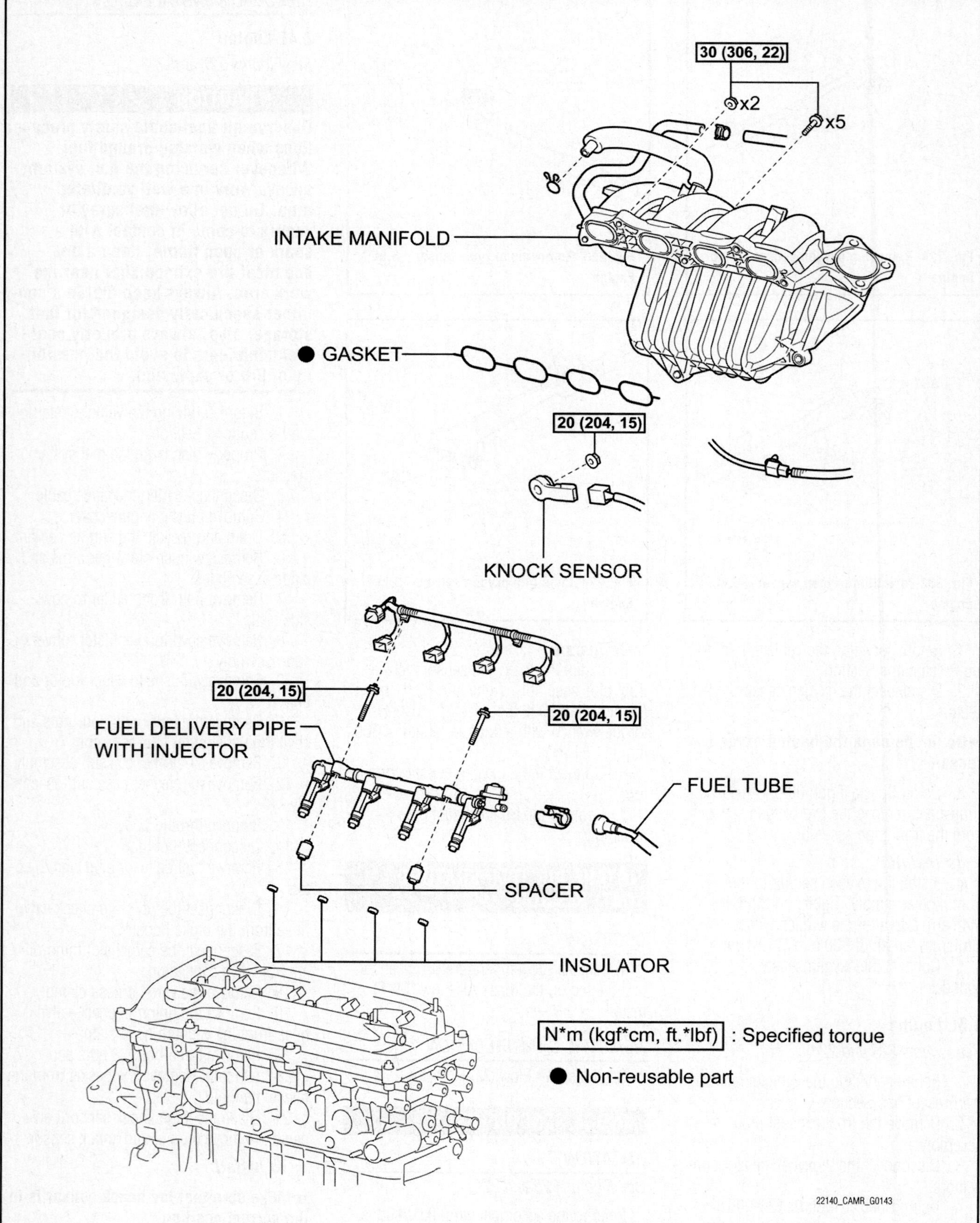

Fig. 225 Knock sensor location—2.4L Engine

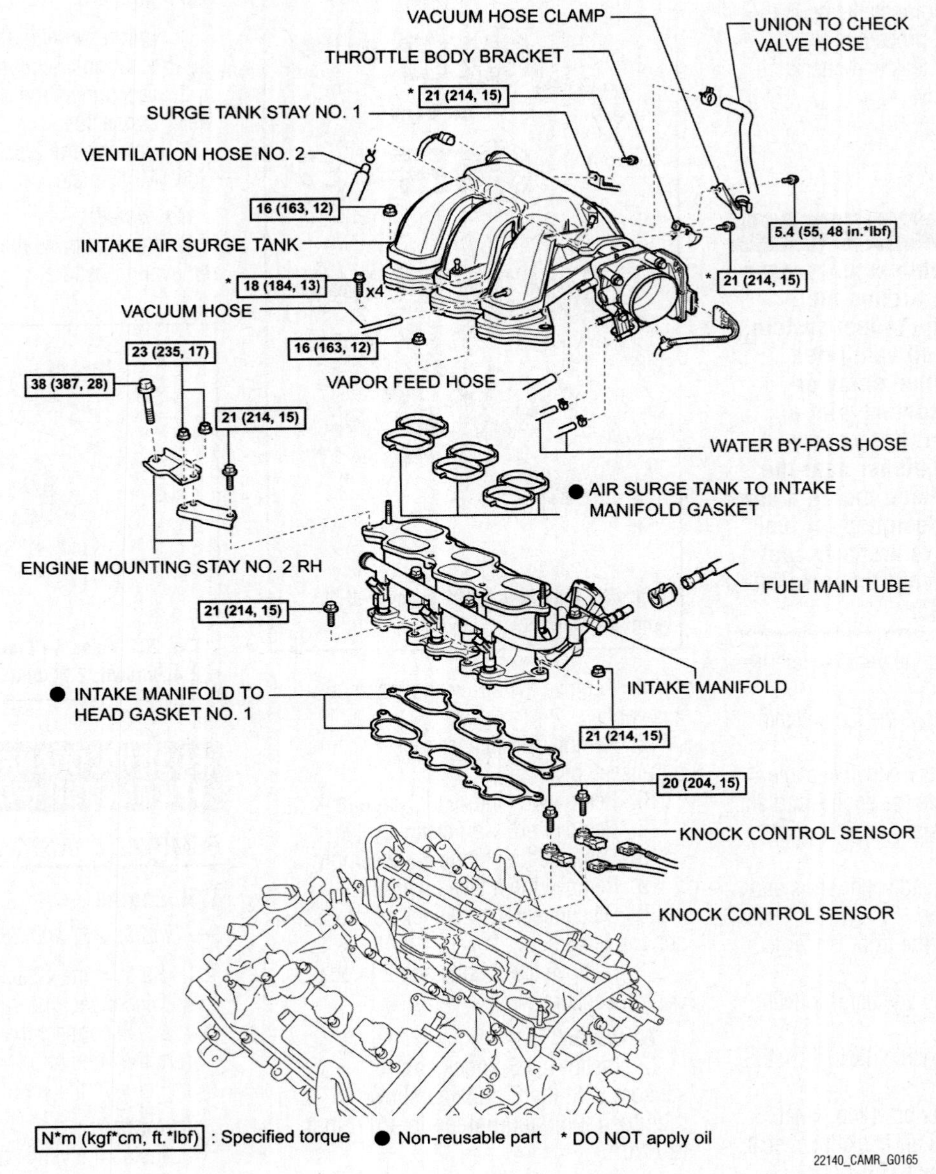

Fig. 226 Knock sensor location—3.5L Engine

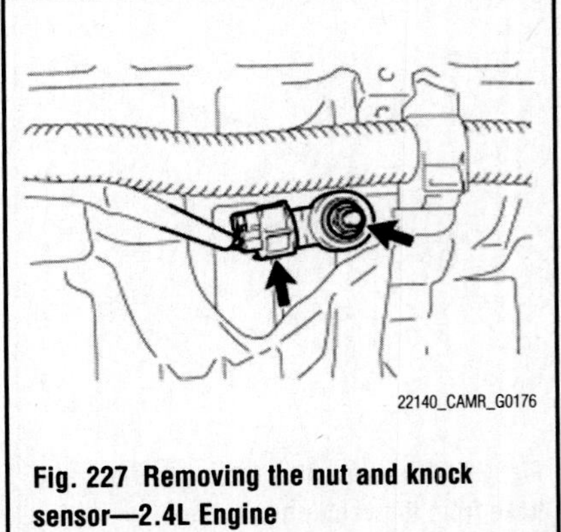

Fig. 227 Removing the nut and knock sensor—2.4L Engine

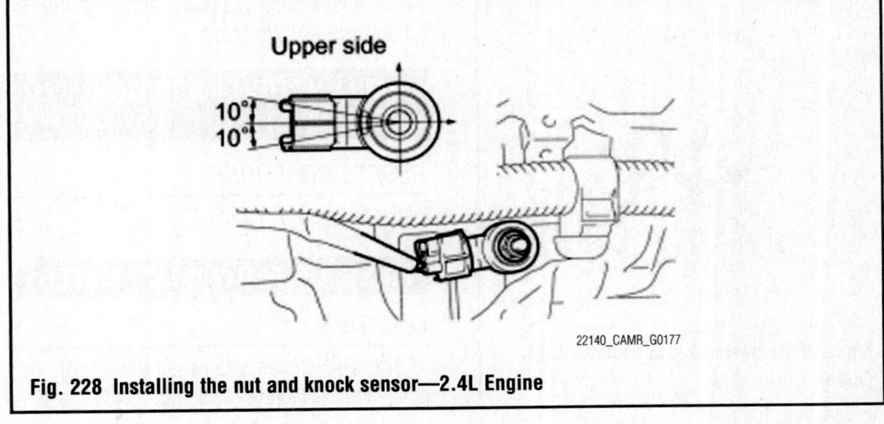

Fig. 228 Installing the nut and knock sensor—2.4L Engine

23. Connect the knock control sensor connector.
24. The remainder of installation is the reverse of the removal procedure.
25. Inspect for fuel leak and check the function of throttle body.

3.5L Engine

See Figures 229 and 230.

CAUTION

Observe all applicable safety precautions when working around fuel. Whenever servicing the fuel system, always work in a well ventilated area. Do not allow fuel spray or vapors to come in contact with a spark or open flame. Keep a dry chemical fire extinguisher near the work area. Always keep fuel in a container specifically designed for fuel storage; also, always properly seal fuel containers to avoid the possibility of fire or explosion.

1. Before servicing the vehicle, refer to the Precautions Section.
2. Properly discharge the fuel system pressure.
3. Disconnect battery negative cable.
4. Drain and recycle the engine coolant.
5. Remove both plastic engine under covers.
6. Remove windshield wiper arms and blade assemblies.
7. Remove both of the front fender to cowl side seals.
8. Remove cowl top ventilator louver sub-assembly.
9. Remove the windshield wiper motor and link assembly.
10. Remove the outer cowl top panel.
11. Remove the cool air intake duct seal.
12. Remove the V-bank cover sub-assembly.
13. Remove air cleaner inlet assembly.

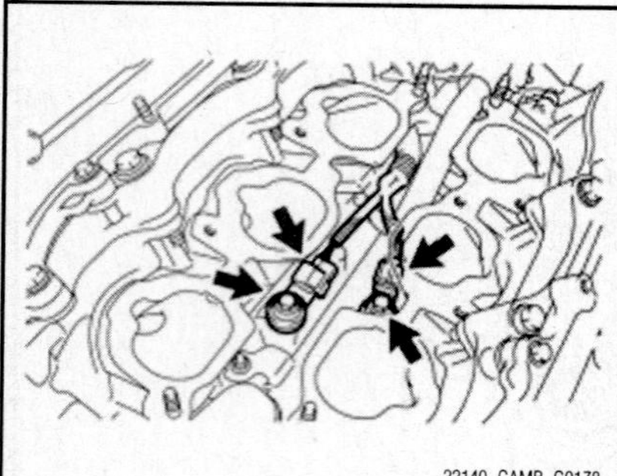

Fig. 229 Removing knock sensor—3.5L Engine

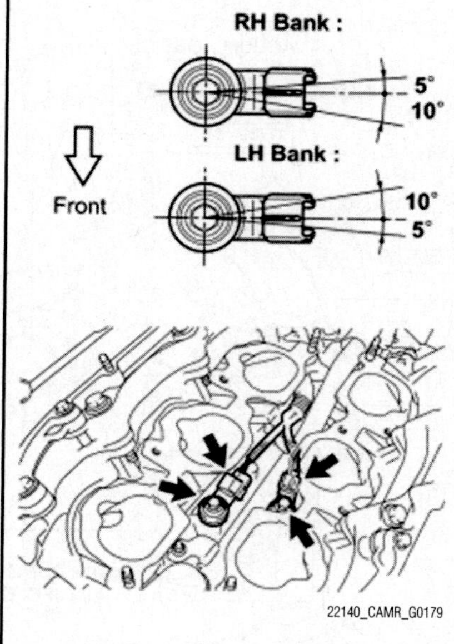

Fig. 230 Installing knock sensor—3.5L Engine

14. Remove the air cleaner cap sub-assembly.
15. Remove air cleaner case sub-assembly.
16. Remove the intake air surge tank.
17. Remove no. 1 air cleaner inlet.
18. Separate fuel tube sub-assembly.
19. Remove the intake manifold.
20. Disconnect the 2 knock control sensor connectors.
21. Remove the 2 bolts and 2 knock control sensors.

To install:

22. Install the 2 knock control sensors with the 2 bolts as shown in the illustration and tighten to 15 ft. lbs. (20 Nm).
23. Connect the 2 knock control sensor connectors.
24. The remainder of installation is the reverse of the removal procedure.
25. Inspect for fuel leak and check the function of throttle body.

MALFUNCTION INDICATOR LIGHT (MIL)

RESET PROCEDURE

Clear the DTC codes.

MASS AIR FLOW (MAF) METER

LOCATION

The MAF meter is between the throttle body and air cleaner housing.

REMOVAL & INSTALLATION

See Figure 231.

1. Before servicing the vehicle, refer to the Precautions Section.
2. Disconnect the Mass Air Flow (MAF) meter connector.
3. Remove the 2 screws and Mass Air Flow (MAF) meter.

To install:

4. Installation is the reverse of the removal procedure.

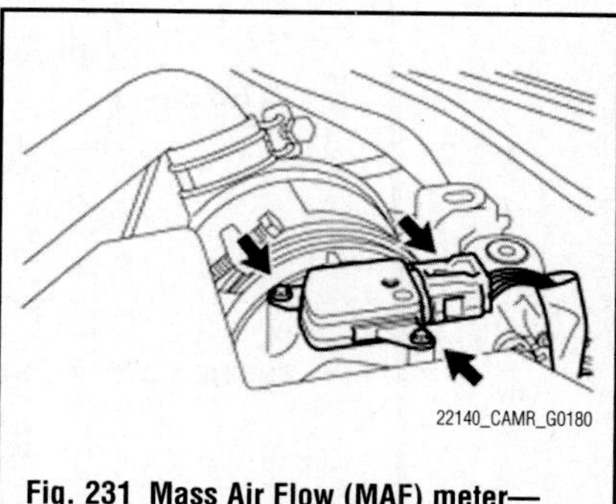

Fig. 231 Mass Air Flow (MAF) meter—2.4L shown, 2.5L and 3.5L are similar

POSITIVE CRANKCASE VENTILATION (PCV) VALVE

REMOVAL & INSTALLATION

3.5L Engine

See Figures 232 and 233.

1. Remove the v-bank cover sub assembly.
2. Disconnect the ventilation hose.
   a. Disconnect the ventilation hose from the ventilation valve.
3. Remove the ventilation valve.

To install:

4. Install the ventilation valve.
   a. Apply adhesive to 2 or 3 threads.
   b. Install the ventilation valve. Tighten to 20 ft. lbs. (27 Nm).

Fig. 232 Disconnecting the ventilation hose from the ventilation valve

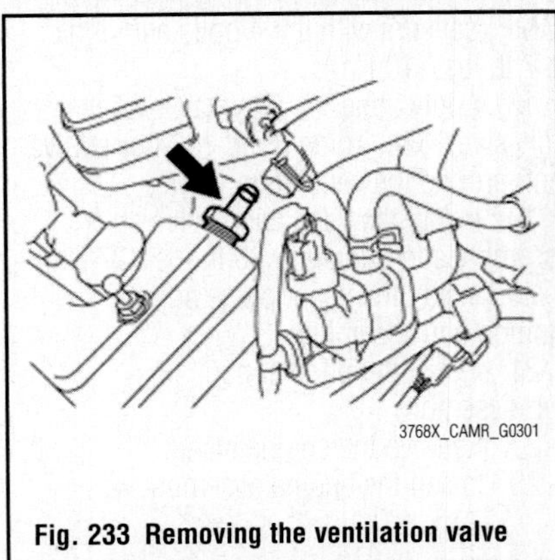
Fig. 233 Removing the ventilation valve

5. Connect the ventilation hose to the ventilation hose.
6. Install the v-bank cover sub assembly.

THROTTLE POSITION SENSOR (TPS)

LOCATION

See Figures 234 and 235.

Refer to the accompanying illustrations for sensor location.

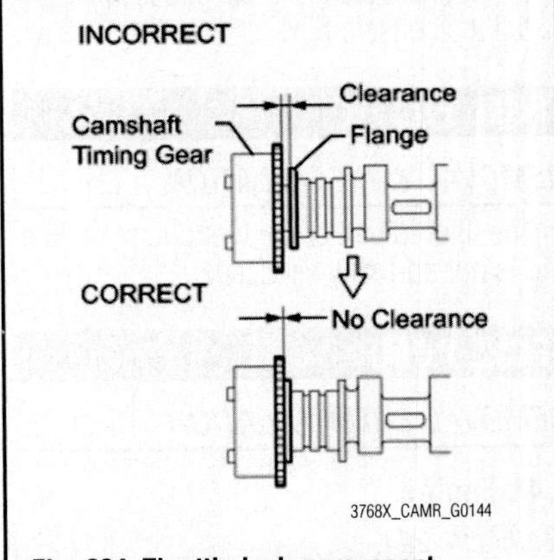

Fig. 234 Throttle body component locations including throttle position sensor—2.4L Engine

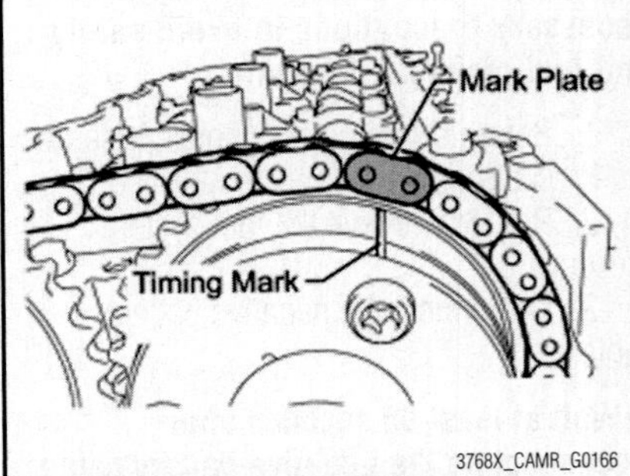

Fig. 235 Throttle body component locations including throttle position sensor—3.5L Engine

REMOVAL & INSTALLATION

Refer to the Throttle Body removal and installation procedures in the Fuel System Section.

VARIABLE CAMSHAFT TIMING OIL CONTROL SOLENOID

REMOVAL & INSTALLATION

2.4L Engine

See Figure 236.

1. Before servicing the vehicle, refer to the Precautions Section.
2. Remove the plastic engine cover.
3. Remove the bolt and disconnect the vacuum hose clamp.
4. Remove the clip nut.
5. Disconnect the camshaft timing oil control valve assembly connector.
6. Remove the bolt and camshaft timing oil control valve assembly.

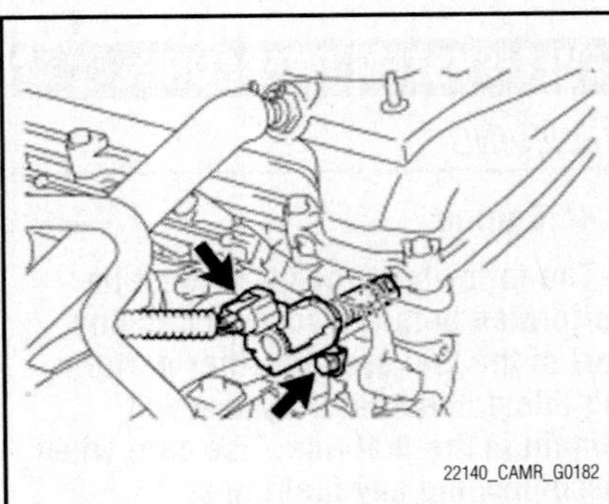
Fig. 236 Removing the camshaft timing oil control valve assembly—2.4L Engine

To install:

7. Apply a light coat of engine oil to an O-ring of the camshaft timing oil control valve assembly sensor.
8. Install the camshaft timing oil control valve assembly with the bolt and tighten to 80 inch lbs. (9 Nm).
9. To complete installation, reverse removal procedure.

3.5L Engine

See Figures 237 through 240.

1. Remove engine under cover.
2. Drain and recycle the engine coolant.
3. Remove windshield wiper arm and blade assembly.
4. Remove front fender to cowl side seal.
5. Remove cowl top ventilator louver sub-assembly.

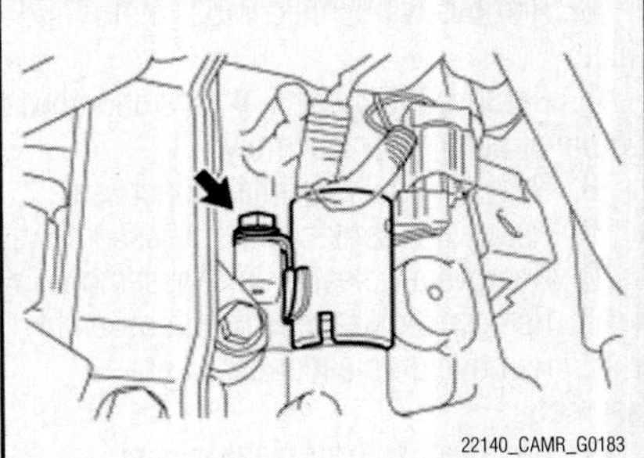
Fig. 237 Removing camshaft timing oil control valve assembly—bank 1, exhaust side—3.5L Engine

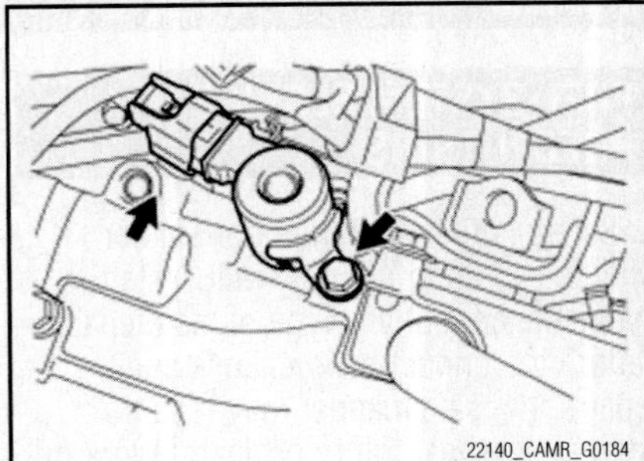
Fig. 238 Removing camshaft timing oil control valve assembly—bank 1, intake side—3.5L Engine

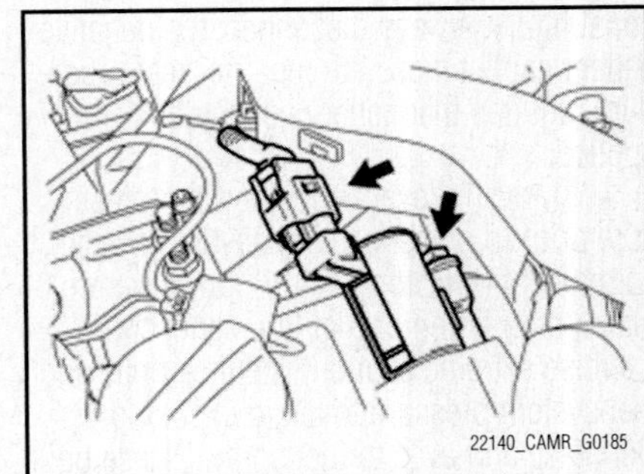
Fig. 239 Removing camshaft timing oil control valve assembly—bank 2, exhaust side—3.5L Engine

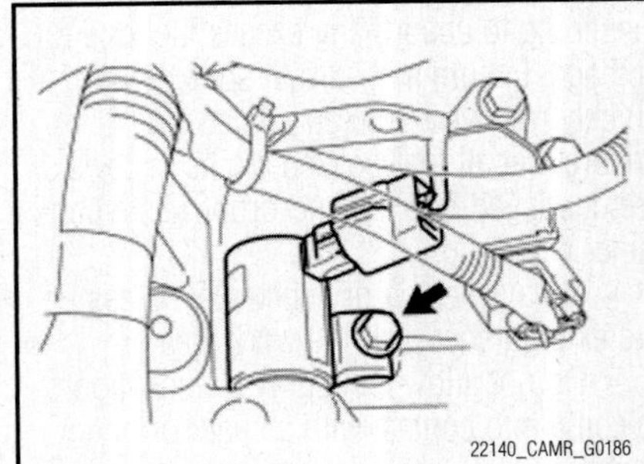
Fig. 240 Removing camshaft timing oil control valve assembly—bank 2, intake side—3.5L Engine

6. Remove windshield wiper motor and link.
7. Remove the 4 bolts, 4 nuts and cowl top panel outer sub-assembly.
8. Remove cools air intake duct seal.
9. Remove v-bank cover sub-assembly.
10. Remove air cleaner inlet assembly.
11. Remove air cleaner cap sub-assembly.
12. Remove air cleaner case sub-assembly.
13. Remove no. 1 air cleaner inlet.
14. Remove intake air surge tank.
15. Remove the four camshaft timing oil control valves by performing the following:
   a. Disconnect the camshaft timing oil control valve connector.
   b. Remove the bolt and camshaft timing oil control valve.
   c. Remove the O-ring from the camshaft timing oil control valve.

To install:

16. Install the camshaft timing oil control valve assembly, as follows:
   a. Install the 4 oil control valves with the 4 bolts and tighten to 7 ft. lbs. (10 Nm).
17. Install the 4 Camshaft Position (CMP) sensors with the 4 bolts and tighten to 7 ft. lbs. (10 Nm).
18. Apply adhesive (Toyota Genuine Adhesive 1324, Three Bond 1324 or equivalent) around the ventilation valve.
19. Install the ventilation valve sub-assembly and tighten to 20 ft. lbs. (27 Nm).
20. Install the spark plugs and the ignition coil assembly.
21. Install the oil filler cap sub-assembly.
22. Remove the engine stand.
23. Install the engine assembly.

FUEL

GASOLINE FUEL INJECTION SYSTEM

FUEL SYSTEM SERVICE PRECAUTIONS

Safety is the most important factor when performing not only fuel system maintenance, but any type of maintenance. Failure to conduct maintenance and repairs in a safe manner may result in serious personal injury or death. Work on a vehicle's fuel system components can be accomplished safely and effectively by adhering to the following rules and guidelines.

- To avoid the possibility of fire and personal injury, always disconnect the negative battery cable unless the repair or test procedure requires that battery voltage be applied.
- Always relieve the fuel system pressure prior to disconnecting any fuel system component (injector, fuel rail, pressure regulator, etc.) fitting or fuel line connection. Exercise extreme caution whenever relieving fuel system pressure to avoid exposing skin, face and eyes to fuel spray. Please be advised that fuel under pressure may penetrate the skin or any part of the body that it contacts.
- Always place a shop towel or cloth around the fitting or connection prior to loosening to absorb any excess fuel due to spillage. Ensure that all fuel spillage is quickly removed from engine surfaces. Ensure that all fuel-soaked cloths or towels are deposited into a flame-proof waste container with a lid.
- Always keep a dry chemical (Class B) fire extinguisher near the work area.
- Do not allow fuel spray or fuel vapors to come into contact with a spark or open flame.
- Always use a second wrench when loosening or tightening fuel line connection fittings. This will prevent unnecessary stress and torsion on fuel piping. Always follow the proper torque specifications.
- Always replace worn fuel fitting O-rings with new ones. Do not substitute fuel hose where rigid pipe is installed.

FUEL SYSTEM PRESSURE

RELIEVING

2.4L Engine

➡The following procedure must be performed before disconnecting any part of the fuel system. After performing this procedure, pressure will remain in the fuel line. Use care when disconnecting any fuel lines.

1. Before servicing the vehicle, refer to the Precautions Section.
2. Remove the console box.
3. Disconnect the connector.
4. Start the engine. After the engine has stopped, turn the ignition switch to the OFF position.

➡DTC P0171 (system lean) may set.

5. Check that the engine does not start.
6. Disconnect the negative battery cable.
7. Remove the fuel tank cap.
8. Connect the electrical connector. Install the console box.

3.5L Engine

➡The following procedure must be performed before disconnecting any part of the fuel system. After performing this procedure, pressure will remain in the fuel line. Use care when disconnecting any fuel lines.

1. Before servicing the vehicle, refer to the Precautions Section.
2. Remove the console box.
3. Disconnect the connector.
4. Start the engine. After the engine has stopped, turn the ignition switch to the OFF position.

➡DTC P0171/P0172 (system lean) may set.

5. Check that the engine does not start.
6. Disconnect the negative battery cable.
7. Remove the fuel tank cap.
8. Connect the electrical connector. Install the console box.

FUEL FILTER

REMOVAL & INSTALLATION

The filter is part of the fuel pump module and is not normally serviced.

FUEL PUMP

REMOVAL & INSTALLATION

2.4L Engine

See Figure 241.

➡The fuel tank must first be removed from the vehicle. Be sure to check and adjust the fuel level as required, before removing the fuel tank. Take all the necessary precautions to avoid safety and fuel disposal problems.

1. Before servicing the vehicle, refer to the Precautions Section.
2. Properly relieve the fuel system pressure.
3. Disconnect the negative battery cable.

➡Wait at least 90 seconds after disconnecting the negative battery cable before starting any repair work to prevent air bag and seat belt pretensioner activation.

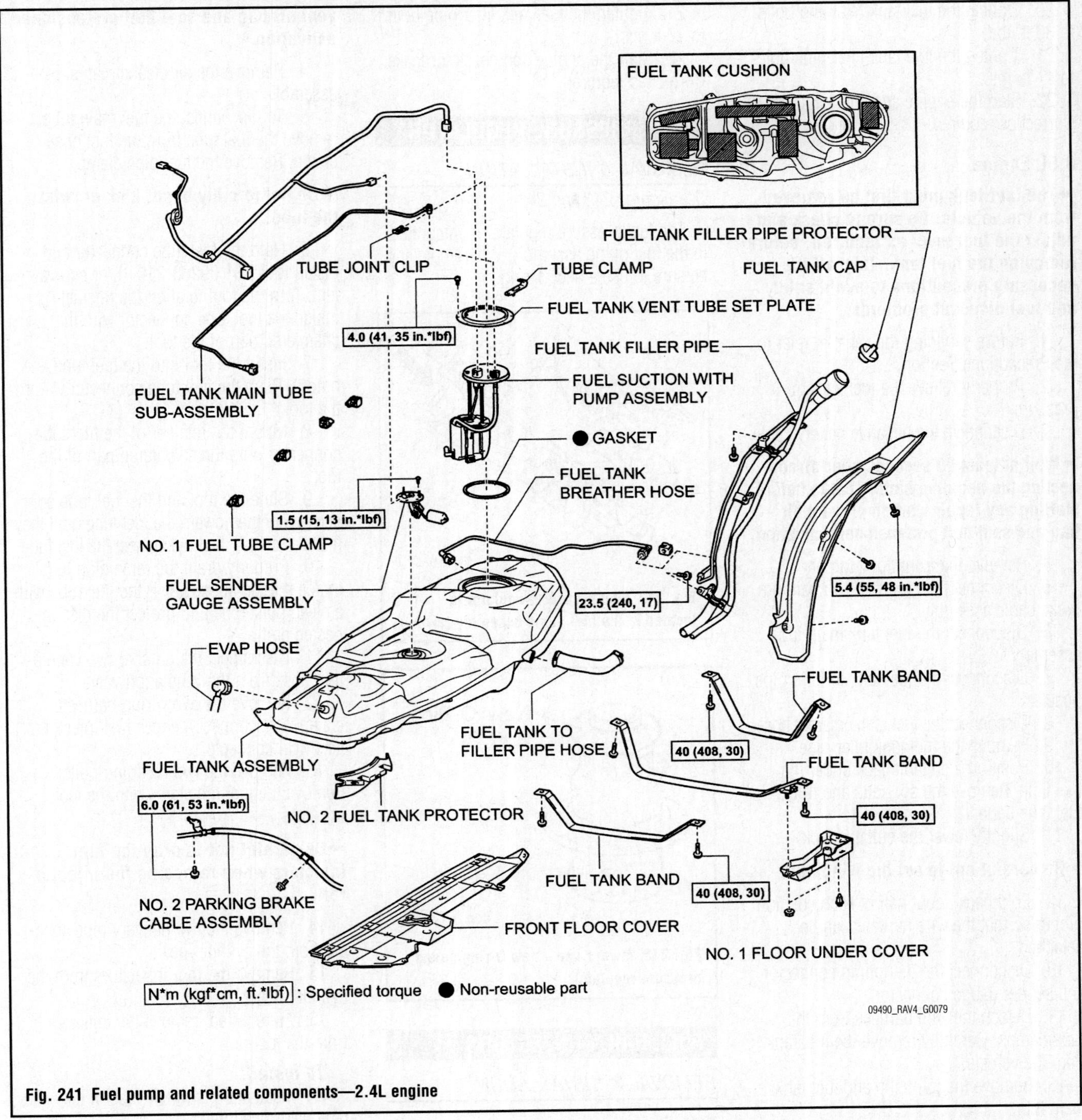

Fig. 241 Fuel pump and related components—2.4L engine

4. Remove the front floor carpet.
5. Disconnect the number two parking brake cable assembly.
6. Disconnect the fuel tank main tube assembly.
7. Disconnect the fuel tank to filler pipe hose.
8. Disconnect the fuel tank breather hose.
9. Remove the fuel tank filler pipe.
10. Position a suitable jack under the fuel tank. Remove the six bolts and three fuel tank bands.
11. Slightly lower the suitable jack.

➡Be careful not to cut the wing nuts.

12. Fold back about half of each cushion rubber so that the wire harness can be detached.
13. Disconnect the fuel pump connector and sender gauge connector.
14. Detach the wire harness from the four clamps. Carefully remove the fuel tank from the vehicle.
15. Remove the joint clip and fuel tank main tube.
16. Remove the eight bolts and the fuel tank vent tube set plate.
17. Disconnect the fuel hose and remove the fuel pump assembly from the fuel tank. Discard the gasket.

To install:

18. Installation is the reverse of the removal procedure.
19. Tighten the eight retaining bolts to 35 inch lbs, and in an alternating sequence.

20. Tighten the fuel tank retaining bolts to 30 ft. lbs.
21. Tighten the fuel tank filler pipe bolts to 17 ft. lbs.
22. Start the engine and check for leaks. Correct as required.

3.5L Engine

➡The fuel tank must first be removed from the vehicle. Be sure to check and adjust the fuel level as required, before removing the fuel tank. Take all the necessary precautions to avoid safety and fuel disposal problems.

1. Before servicing the vehicle, refer to the Precautions Section.
2. Properly relieve the fuel system pressure.
3. Disconnect the negative battery cable.

➡Wait at least 90 seconds after disconnecting the negative battery cable before starting any repair work to prevent air bag and seat belt pretensioner activation.

4. Remove the front floor carpet.
5. Disconnect the number two parking brake cable assembly.
6. Disconnect the fuel tank main tube assembly.
7. Disconnect the fuel tank to filler pipe hose.
8. Disconnect the fuel tank breather hose.
9. Remove the fuel tank filler pipe.
10. Position a suitable jack under the fuel tank. Remove the six bolts and three fuel tank bands.
11. Slightly lower the suitable jack.

➡Be careful not to cut the wing nuts.

12. Fold back about half of each cushion rubber so that the wire harness can be detached.
13. Disconnect the fuel pump connector and sender gauge connector.
14. Detach the wire harness from the four clamps. Carefully remove the fuel tank from the vehicle.
15. Remove the joint clip and fuel tank main tube.
16. Remove the eight bolts and the fuel tank vent tube set plate.
17. Disconnect the fuel hose and remove the fuel pump assembly from the fuel tank. Discard the gasket.

To install:

18. Installation is the reverse of the removal procedure.
19. Tighten the eight retaining bolts to 35 inch lbs, and in an alternating sequence.
20. Tighten the fuel tank retaining bolts to 30 ft. lbs.
21. Tighten the fuel tank filler pipe bolts to 17 ft. lbs.
22. Start the engine and check for leaks. Correct as required.

FUEL PRESSURE REGULATOR

REMOVAL & INSTALLATION

See Figures 242 and 243.

The fuel pressure regulator is integrated in the fuel pump assembly. When replacing, be sure to use a new O-ring.

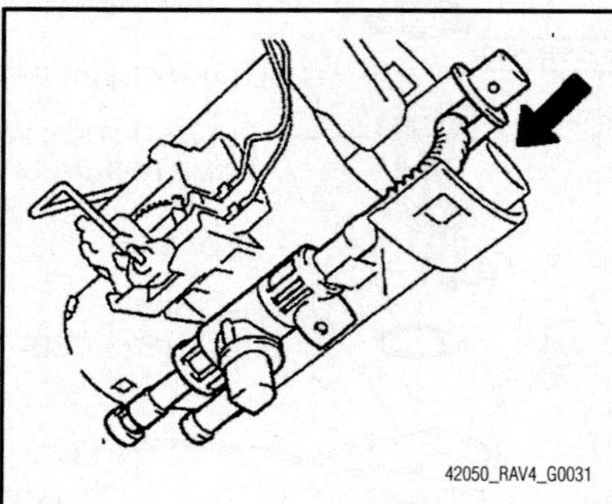

Fig. 242 Fuel pressure regulator is located in the fuel pump assembly (arrow)

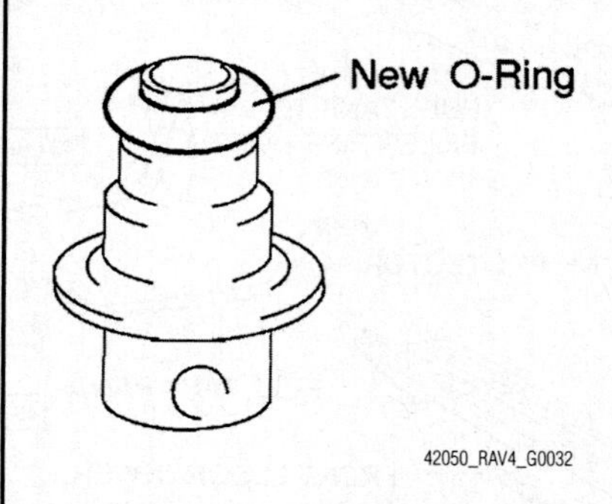

Fig. 243 Always use a new O-ring on the pressure regulator

FUEL RAIL & INJECTORS

REMOVAL & INSTALLATION

2.4L Engine

See Figures 244 through 247.

1. Before servicing the vehicle, refer to the Precautions Section.
2. Properly relieve the fuel system pressure.
3. Disconnect the negative battery cable.

➡Wait at least 90 seconds after disconnecting the negative battery cable before starting any repair work to prevent air bag and seat belt pretensioner activation.

4. Remove the air cleaner cap sub-assembly.
5. To disconnect the fuel main tube; remove the fuel tube from the fuel hose clamp. Remove the fuel pipe clamp.

➡Do not forcibly bend, kink or twist the tube.

6. Hold the fuel tube connector and install tool SST 09268-21010, or equivalent. Turn the tool to align the retainer inside the fuel tube connector with the chamfered part of the tool.
7. Insert the tool into the fuel tube and hold it. Push the fuel tube connector toward the tool,
8. Mount the retainer of the fuel tube connector onto the chamfered part of the tool.
9. Slide the tool and the fuel tube connector together toward the fuel tube until they make a "click", then disconnect the fuel tube.
10. Properly drain the remaining fuel inside the fuel tube. Cover the fuel tube with a clean plastic bag, to protect the disconnected part.
11. Disconnect the number two ventilation hose from the ventilation valve.
12. Remove the two wiring harness clamps. Disconnect the four fuel injector electrical connectors.
13. Remove the two retaining bolts. Remove the fuel rail along with the fuel injectors.

➡Be careful not to drop the fuel injectors when removing the injector rail.

14. Remove the two delivery pipe spacers from the cylinder head.
15. Remove the four insulators from the cylinder head.
16. As required, remove the injectors from the fuel rail.

To install:

17. Installation is the reverse of the removal procedure.
18. Installation is the reverse of removal.
19. Apply a light coat gasoline to the new fuel injector O-rings and install them on to the injectors.
20. Apply a light coat of gasoline to the part of the injector rail that contacts the O-ring of the injector.
21. Install the injectors on the fuel rail.

➡Be sure that the O-rings are not damaged or jammed when you are installing them.

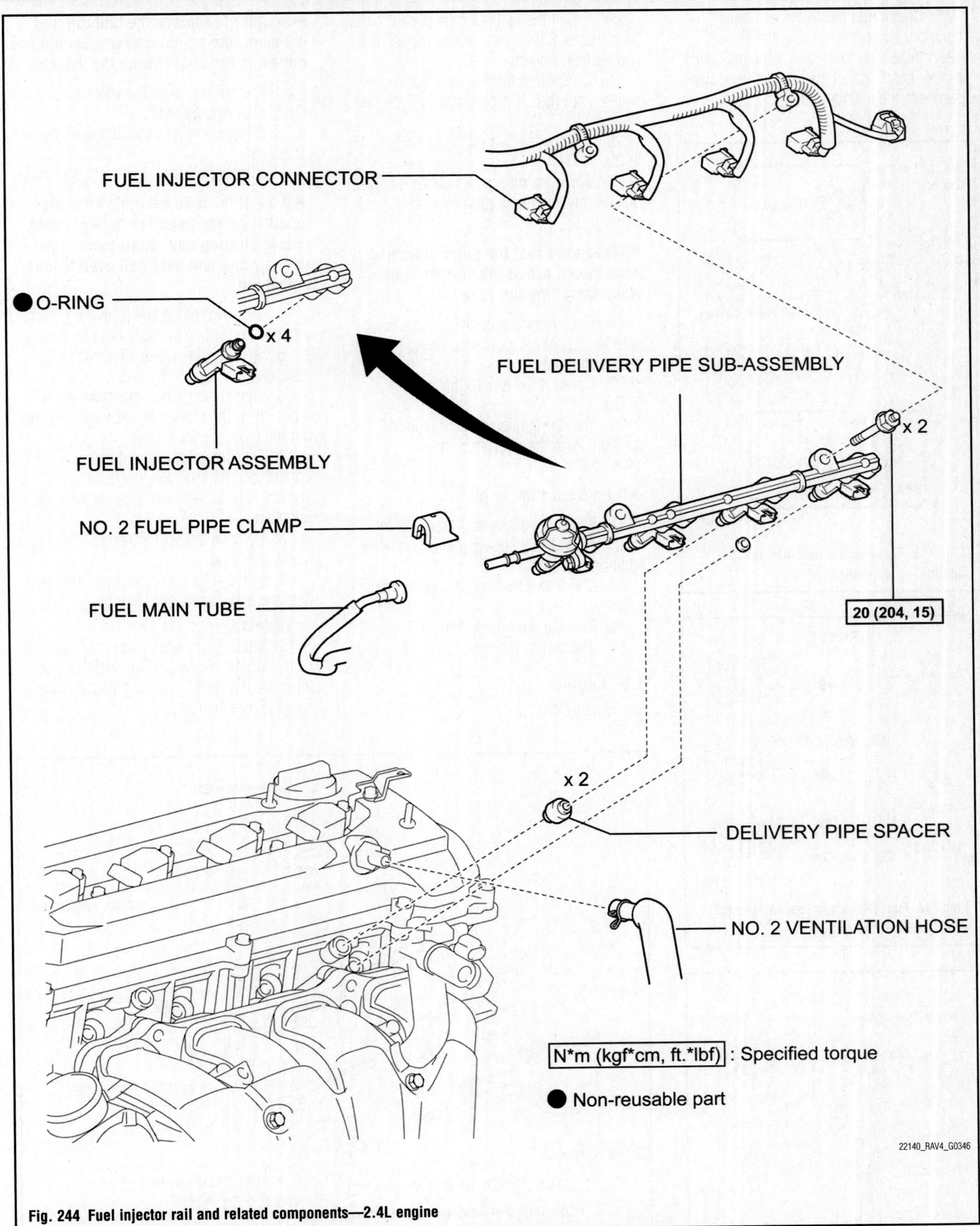

Fig. 244 Fuel injector rail and related components—2.4L engine

22. Check that the injector rotates freely. If it does not, replace the O-ring.

23. Install the four new insulators to the cylinder head. Install the two delivery pipe spacers to the cylinder head.

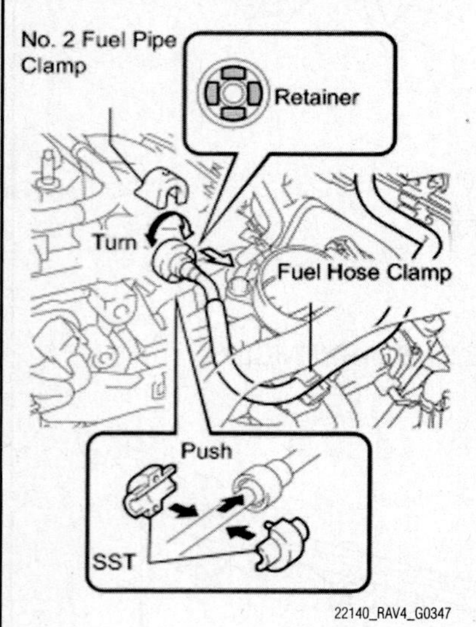

Fig. 245 Fuel injector fuel tube disconnection—2.4L engine

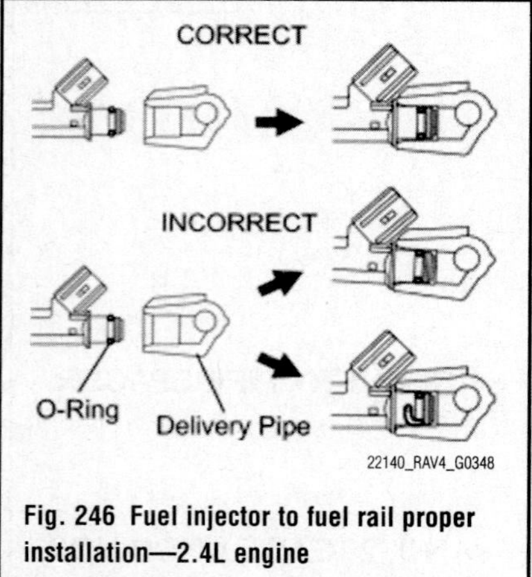

Fig. 246 Fuel injector to fuel rail proper installation—2.4L engine

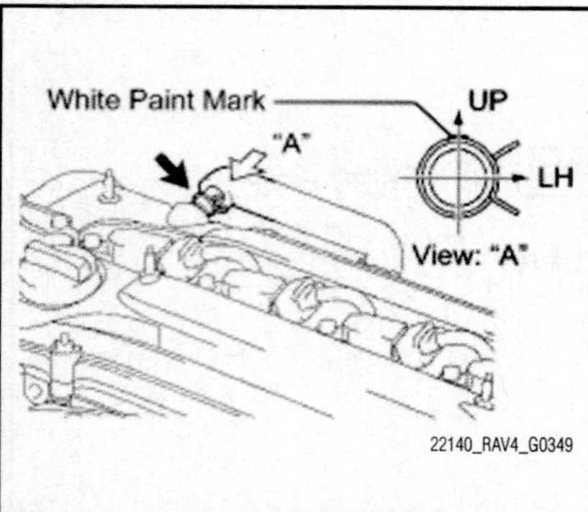

Fig. 247 Proper ventilation valve hose installation—2.4L engine

24. Install the fuel rail together with the fuel injectors, then temporarily tighten the two retaining bolts.

25. Check that the fuel injectors rotate smoothly. If it does not replace the O-ring.

26. Tighten the two retaining bolts to 15 ft. lbs.

27. When installing the number two ventilation, see the illustration for proper positioning.

➡Make sure that the paint mark and hose clamp are at the correct angle when installing the hose.

28. Once the installation is complete be sure to inspect for fuel leaks, using the Techstream tool.

29. Connect the tool to the DLC3.

30. Turn the ignition switch to the ON position. Push the Techstream main switch on.

➡Do not start the engine.

31. Select the active test mode on the tool. Refer to the tools instruction book for additional details.

32. Check that there are no fuel system leaks.

33. Turn the ignition switch OFF.

34. Disconnect the tool.

3.5L Engine

See Figure 248.

➡In order to replace the fuel rail and injectors, the manufacturer recommends removing the engine from the vehicle.

1. Before servicing the vehicle, refer to the Precautions Section.

2. Properly relieve the fuel system pressure.

3. Disconnect the negative battery cable.

➡Wait at least 90 seconds after disconnecting the negative battery cable before starting any repair work to prevent air bag and seat belt pretensioner activation.

4. Remove the engine from the vehicle and position it in a suitable holding fixture.

5. Remove the intake air surge tank assembly.

6. Disconnect the number two main tube. Pinch the tube connector and then pull out the pipe.

7. Disconnect the six injector connectors. Remove the five bolts and the fuel delivery rail together with the injectors.

8. Remove the six insulators from the intake manifold.

9. Pull out the injectors from the rail. Remove and discard the O-rings.

To install:

10. Installation is the reverse of removal.

11. Coat the new O-rings with gasoline. Push the injectors onto the pipes and make sure they rotate freely.

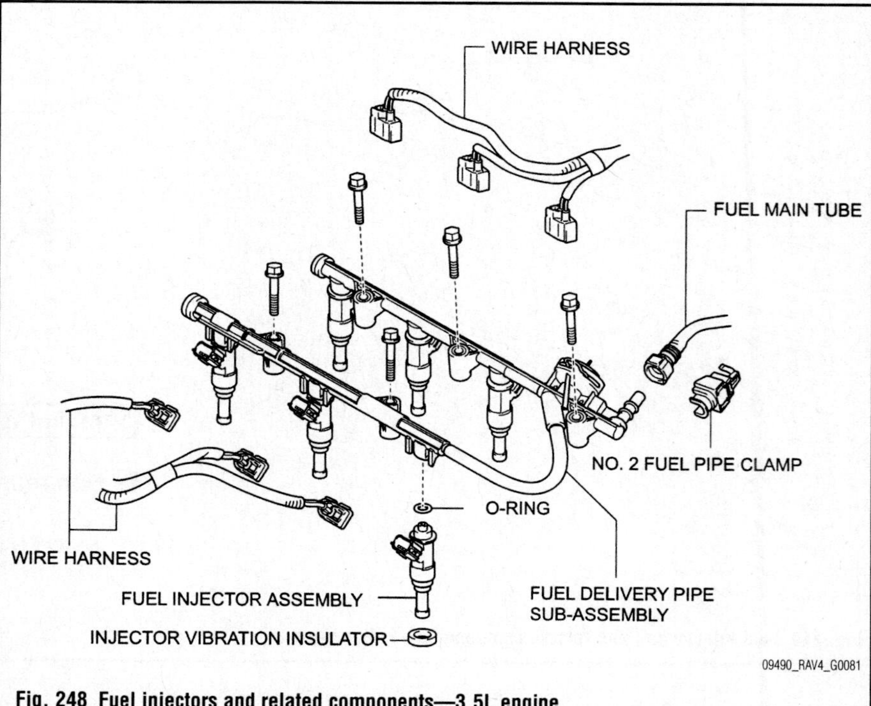

Fig. 248 Fuel injectors and related components—3.5L engine

12. Position the assembly onto the head and install the bolts finger tight. Make sure that the injectors still rotate freely. If not, replace the O-rings. Torque the bolts to 15 ft. lbs. (20 Nm).

13. Start the engine and check for leaks, correct as required.

FUEL TANK

REMOVAL & INSTALLATION

See Figures 249 through 253.

1. Before servicing the vehicle, refer to the Precautions Section.

2. Properly relieve the fuel system pressure.

3. Disconnect the negative battery cable.

➡Wait at least 90 seconds after disconnecting the negative battery cable before starting any repair work to

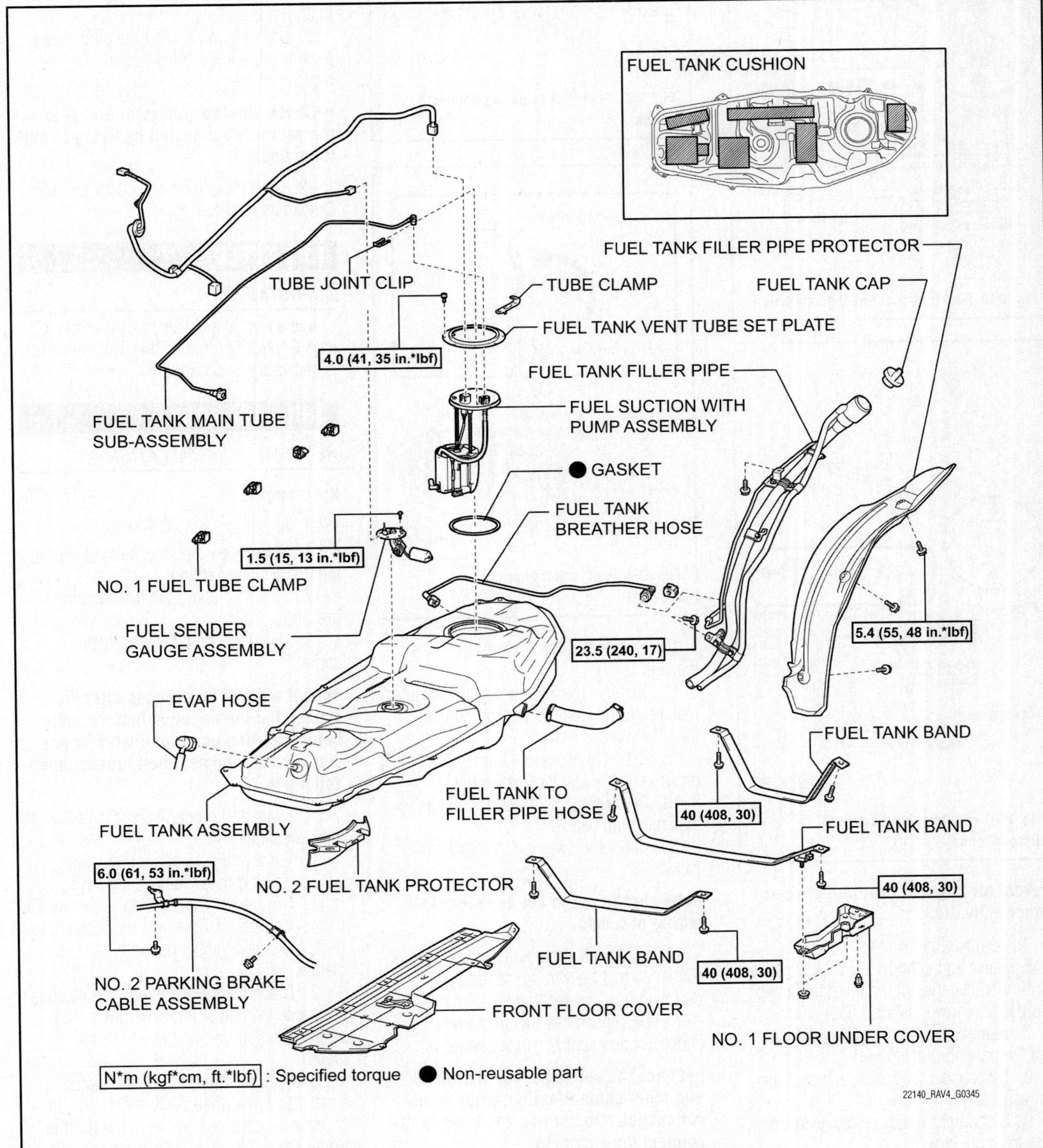

Fig. 249 Fuel tank and related components

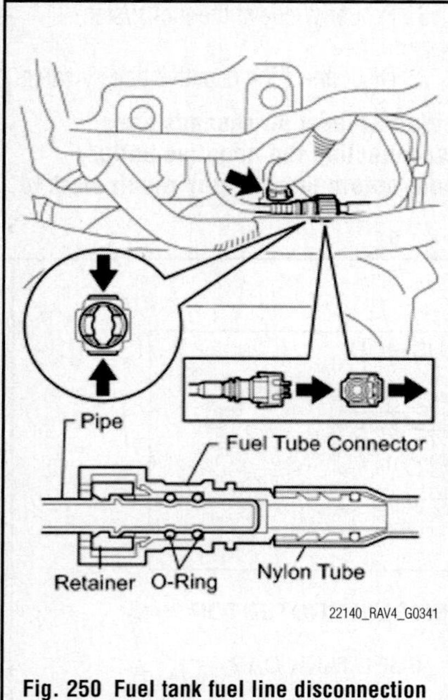

Fig. 250 Fuel tank fuel line disconnection

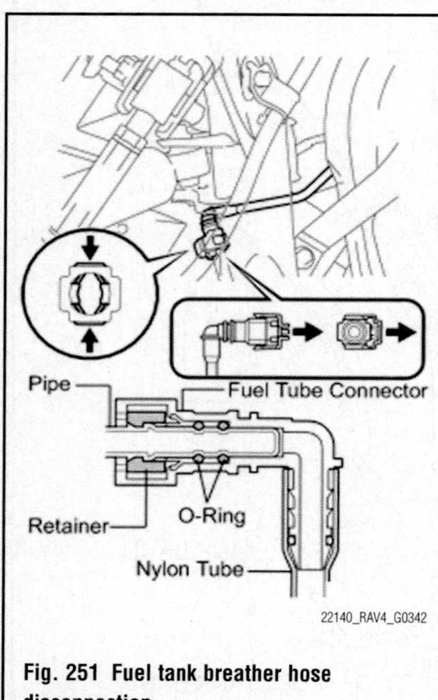

Fig. 251 Fuel tank breather hose disconnection

prevent air bag and seat belt pretensioner activation.

4. Remove the fuel cap.
5. Raise and support the vehicle safely.
6. Remove the nut, bolt and three clips in order to remove the front floor cover.
7. Remove the two bolts and disconnect the parking brake cover.
8. Disconnect the fuel tank main tube sub-assembly.
9. Disconnect the evaporative emission hose from the tank.

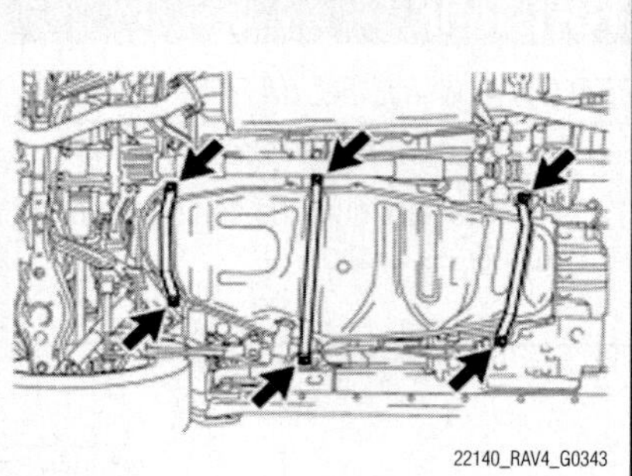

Fig. 252 Fuel tank retaining bolts and bands

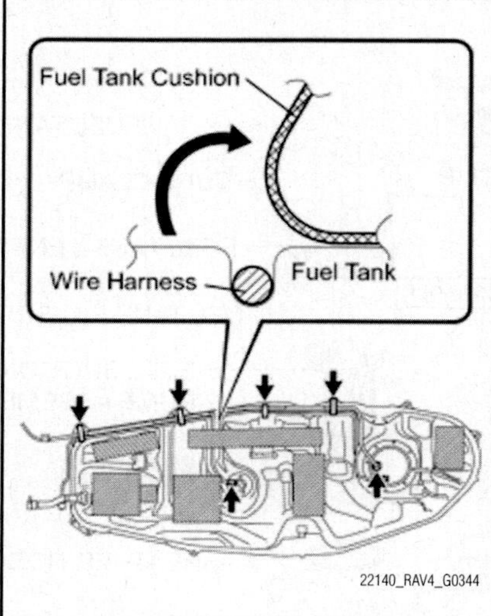

Fig. 253 Fuel tank rubber cushion removal

10. Disconnect the breather hose from the tank.
11. Remove the three bolts and the filler pipe protector. Remove the two bolts and the filler pipe.
12. Properly position a transmission jack underneath the fuel tank assembly.
13. Remove the six bolts and three fuel tank retaining bands.
14. Carefully lower the transmission jacks.

➡Be careful not to cut any electrical wiring or hoses.

15. Fold back about half of each cushion rubber so that the wire harness can be detached from the fuel tank.
16. Disconnect the fuel pump electrical connector and sender gauge connector.

➡Check the connector for dirt, mud and other contamination, prior to disconnection. Do not use any tools to disconnect the connector.

17. Detach the wire harness from the four clamps.
18. Remove the fuel tank from the vehicle.

To install:

19. Installation is the reverse of the removal procedure.
20. Tighten the six band retaining bolts to 30 ft. lbs.
21. When connecting tube connectors push firmly until a "click" sound is heard.

➡Check that the connector and pipe are securely connected by trying to pull them apart.

22. Start the engine and check for leaks. Correct as required.

IDLE SPEED

ADJUSTMENT

Idle speed is maintained by the Powertrain Control Module (PCM). No adjustment is necessary or possible.

THROTTLE BODY

REMOVAL & INSTALLATION

2.4L Engine

See Figures 254 through 258.

1. Before servicing the vehicle, refer to the Precautions Section.
2. Properly relieve the fuel system pressure.
3. Disconnect the negative battery cable.

➡Wait at least 90 seconds after disconnecting the negative battery cable before starting any repair work to prevent air bag and seat belt pretensioner activation.

4. Drain and properly dispose of the engine coolant.
5. Remove the number one engine cover retaining nuts. Remove the cover.
6. Disconnect the Mass Air Flow (MAF) meter connector. Disconnect the purge connector. Disconnect the four wire harness clamps.
7. Disconnect the number two ventilation hose from the air cleaner hose. Disconnect the purge line hose from the clamp.
8. Lock the number one air cleaner hose clamp and then disconnect the number one air cleaner hose from the throttle body.

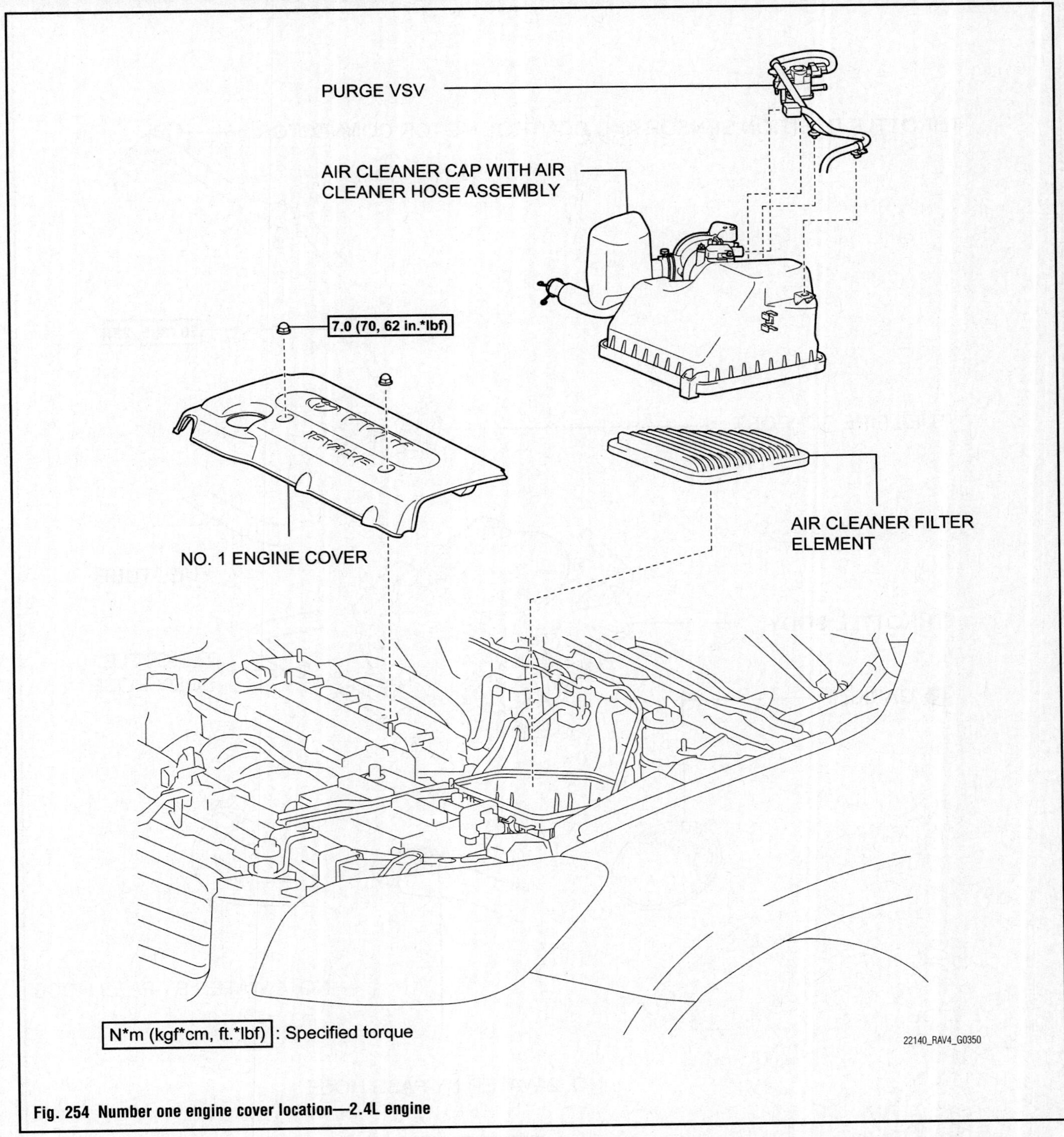

Fig. 254 Number one engine cover location—2.4L engine

9. Unfasten the two hook clamps and remove the air cleaner cap.
10. Remove the filter element from the case.
11. Disconnect the purge line hose from the throttle body.
12. Disconnect the water bypass hose from the throttle body. Disconnect the number two water bypass hose from the throttle body.
13. Disconnect the number one water bypass hose from the throttle body.
14. Disconnect the throttle position sensor and the control motor connector. Disconnect the wire harness clamp.
15. Disconnect the fuel tube from the clamp.
16. Remove the throttle body retaining bolts.
17. Remove the component from its mounting. Remove and discard the gasket from the intake manifold.

To install:

18. Installation is the reverse of removal.
19. Be sure to use a new gasket.
20. Tighten the retaining bolts to 22 ft. lbs.
21. Be sure that the air clean hose is properly installed.

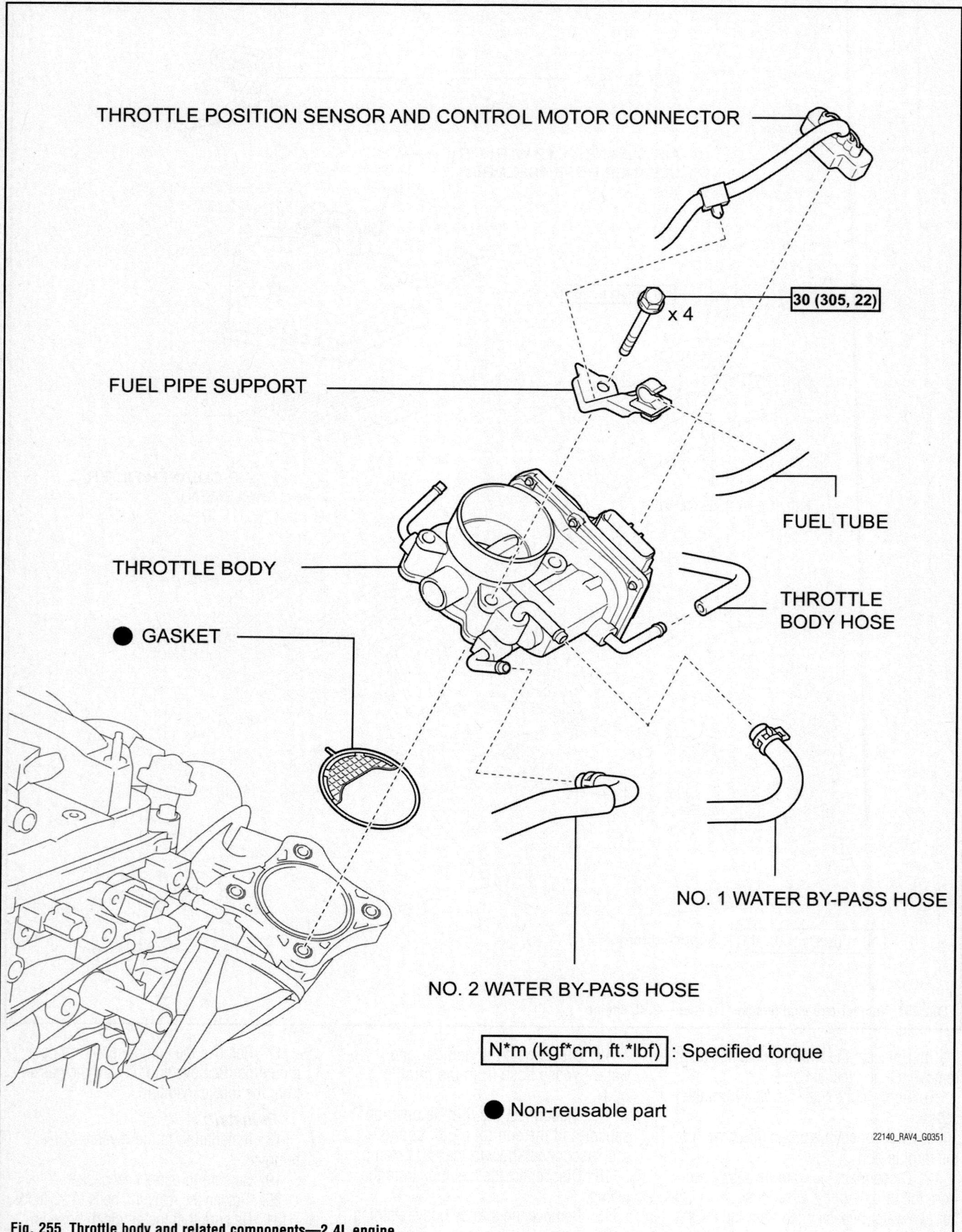

Fig. 255 Throttle body and related components—2.4L engine

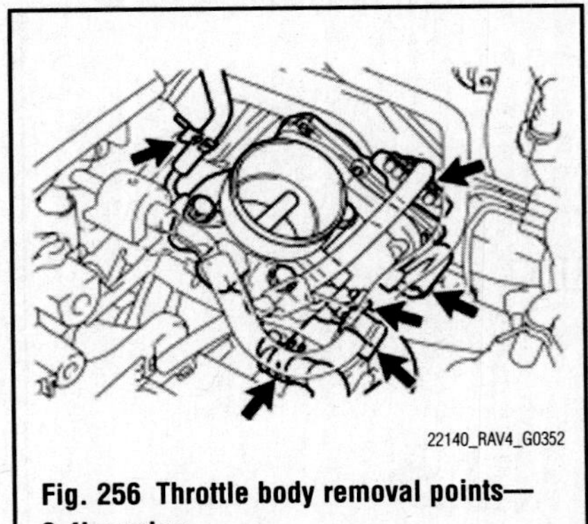

Fig. 256 Throttle body removal points—2.4L engine

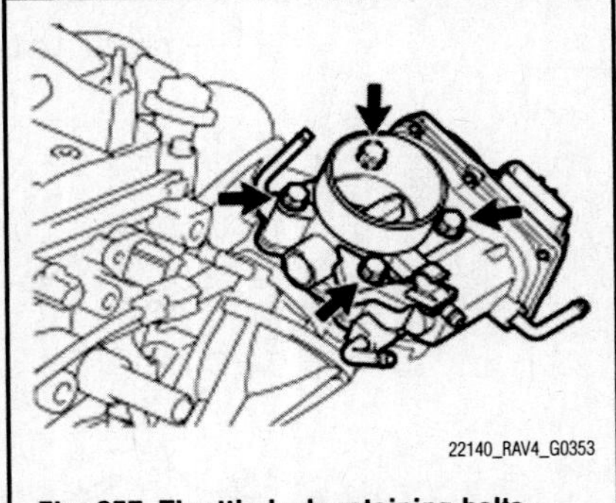

Fig. 257 Throttle body retaining bolts—2.4L engine

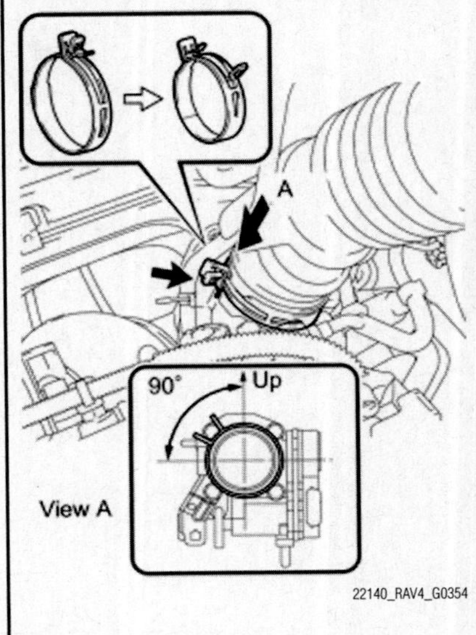

Fig. 258 Proper air cleaner hose installation—2.4L engine

22. Start the engine and check for leaks. Correct as required.

3.5L Engine

See Figures 259 through 261.

1. Before servicing the vehicle, refer to the Precautions Section.
2. Properly relieve the fuel system pressure.
3. Disconnect the negative battery cable.

➡Wait at least 90 seconds after disconnecting the negative battery cable before starting any repair work to

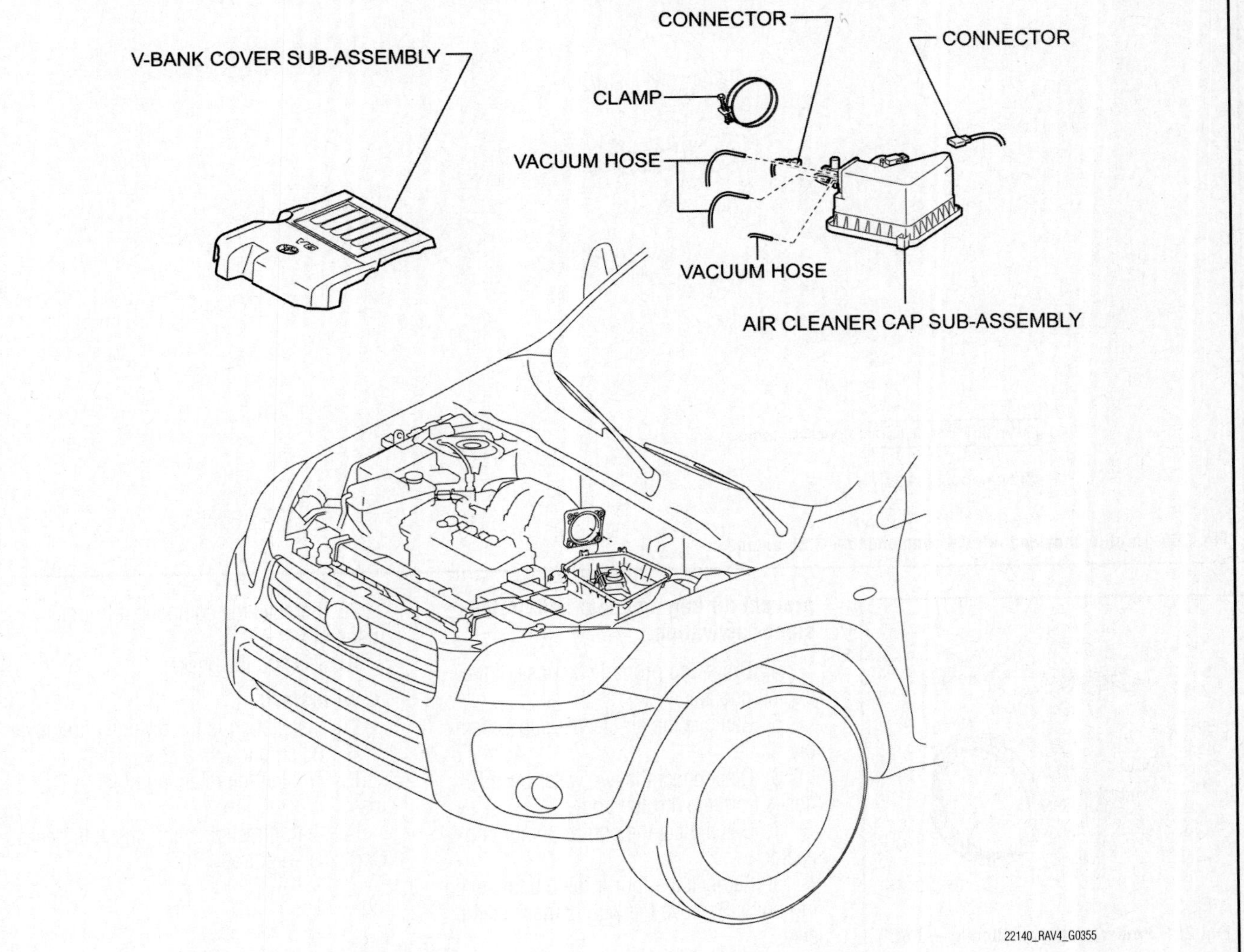

Fig. 259 V-bank cover location—3.5L engine

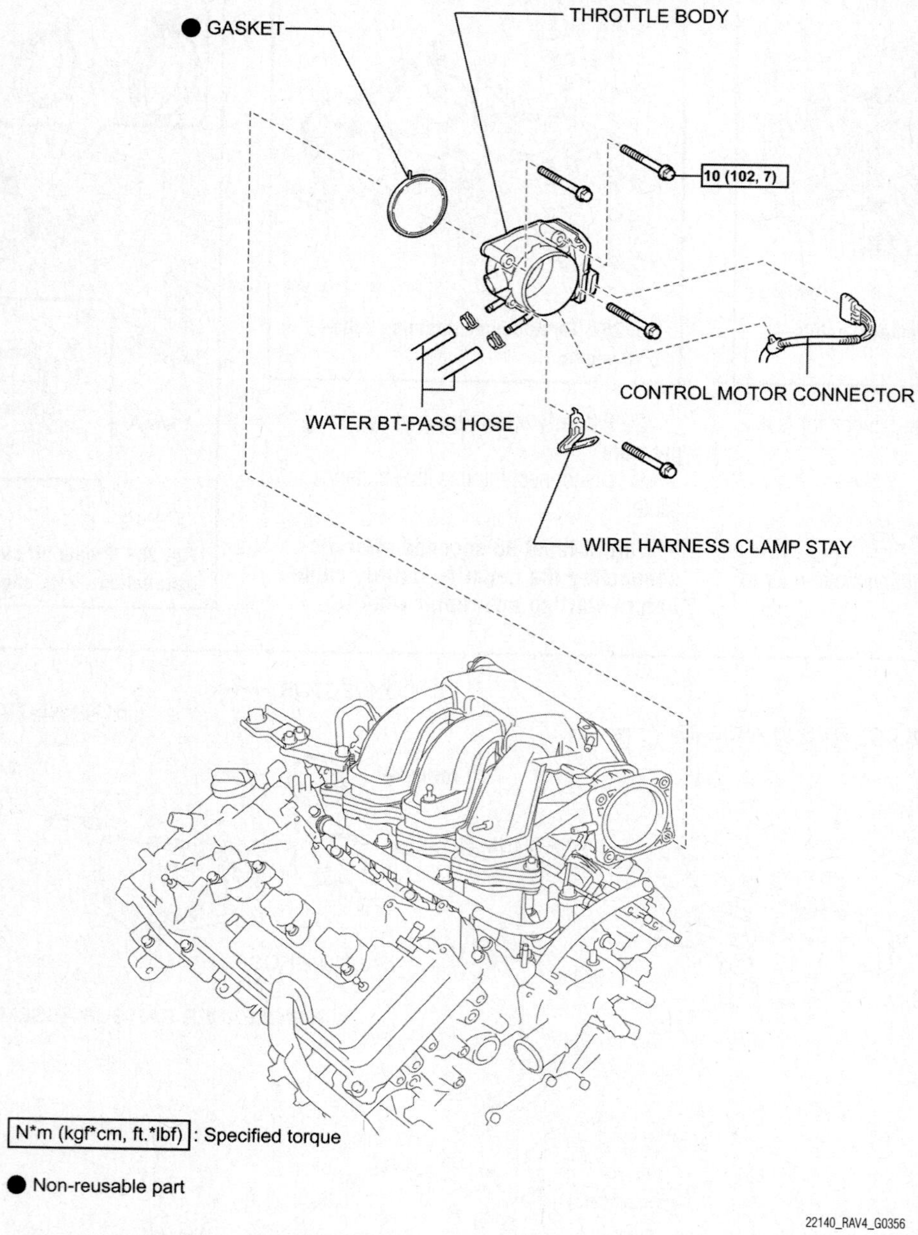

GASKET
THROTTLE BODY
10 (102, 7)
CONTROL MOTOR CONNECTOR
WATER BT-PASS HOSE
WIRE HARNESS CLAMP STAY
N*m (kgf*cm, ft.*lbf) : Specified torque
Non-reusable part
22140_RAV4_G0356

Fig. 260 Throttle body and related components—3.5L engine

Recess
22140_RAV4_G0357

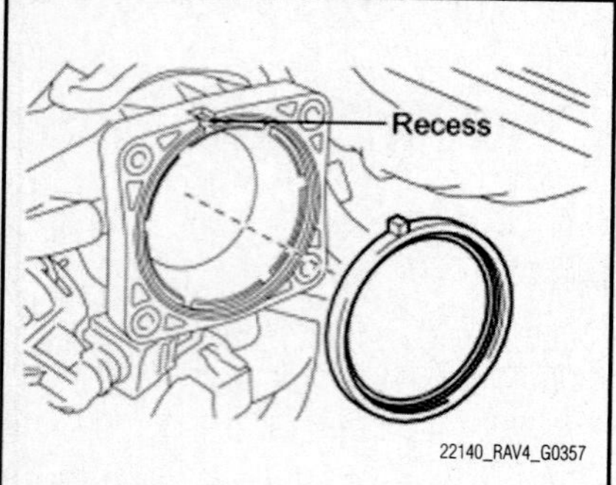

Fig. 261 Proper gasket positioning—3.5L engine

prevent air bag and seat belt pretensioner activation.

4. Drain and properly dispose of the engine coolant.
5. Remove the air cleaner sub-assembly.
6. Disconnect the two water bypass hoses from the throttle body.
7. Disconnect the control motor connector.
8. Remove the four throttle body retaining bolts. Remove the wire harness clamp stay.
9. Remove the component from its mounting.
10. Discard the gasket.

To install:

11. Installation is the reverse of removal.
12. Be sure to use a new gasket.
13. Tighten the retaining bolts to 7 ft. lbs.
14. Start the engine and check for leaks. Correct as required.

HEATING & AIR CONDITIONING SYSTEM

BLOWER MOTOR

REMOVAL & INSTALLATION

See Figures 262 and 263.

1. Before servicing the vehicle, refer to the Precautions Section.
2. Disconnect the negative battery cable.

➡Wait at least 90 seconds after disconnecting the negative battery cable before starting any repair work to prevent air bag and seat belt pretensioner activation.

3. Properly discharge the air conditioning system. Be sure to plug the refrigerant lines to avoid dirt and moisture from entering the system. Properly recycle the ac refrigerant.
4. Remove the upper instrument panel.
5. Remove the lower instrument panel.
6. Remove the air duct.
7. Remove the heater/air conditioning unit.
8. Remove the blower assembly.
9. Disconnect the electrical connector.
10. Remove the retaining screws.
11. Remove the component from its mounting.
12. Remove the blower motor retaining screws.
13. Remove the blower motor from its mounting.

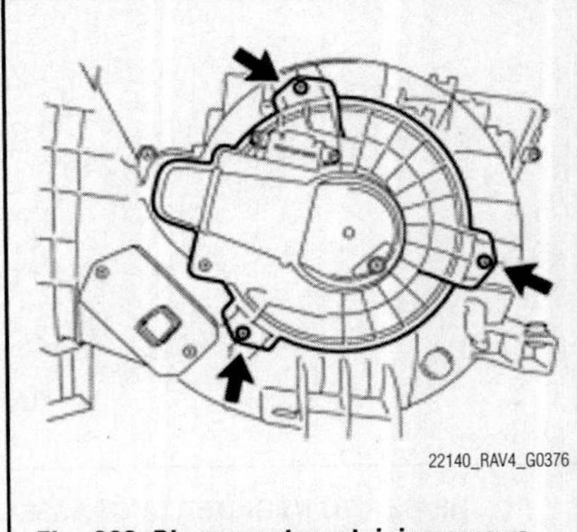

Fig. 263 Blower motor retaining screws

To install:

14. Installation is the reverse of the removal procedure.
15. Properly recharge the air conditioning system.
16. Check the SRS warning light for proper operation.

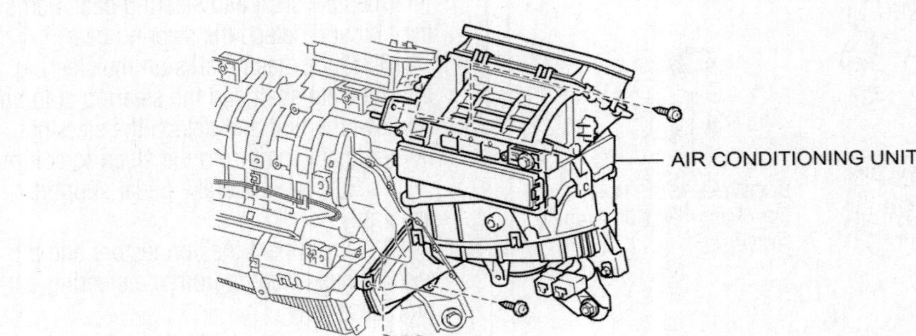

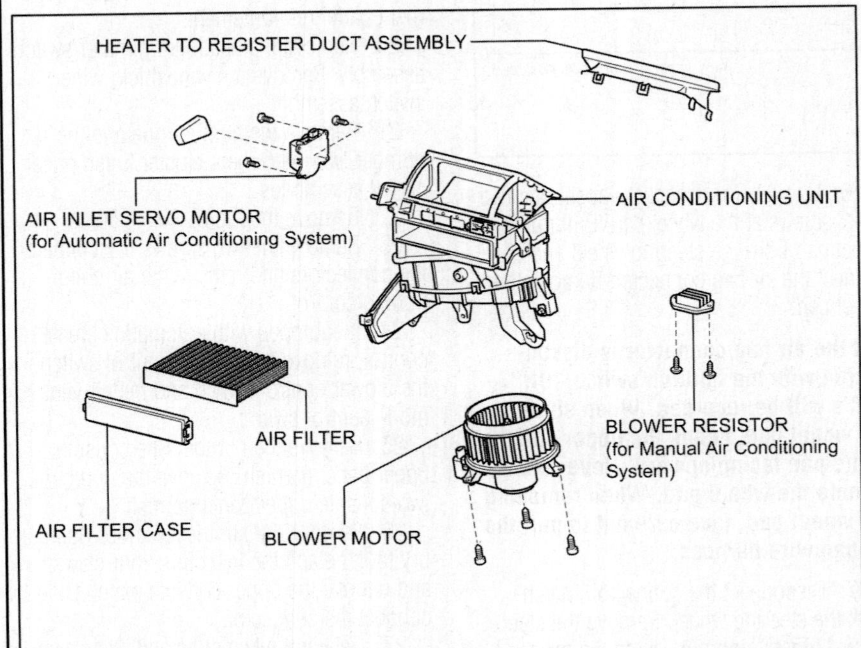

Fig. 262 Blower assembly and related components

BLOWER UNIT

REMOVAL & INSTALLATION

See Figure 264.

1. Before servicing the vehicle, refer to the Precautions Section.
2. Disconnect the negative battery cable.

➡Wait at least 90 seconds after disconnecting the negative battery cable before starting any repair work to prevent air bag and seat belt pretensioner activation.

3. Properly discharge the air conditioning system. Be sure to plug the refrigerant lines to avoid dirt and moisture from entering the system. Properly recycle the ac refrigerant.
4. Remove the upper instrument panel.
5. Remove the lower instrument panel.
6. Remove the air duct.
7. Remove the heater/air conditioning unit.
8. Remove the blower assembly.
9. Disconnect the electrical connector.
10. Remove the retaining screws.
11. Remove the component from its mounting.

To install:

12. Installation is the reverse of the removal procedure.
13. Properly recharge the air conditioning system.
14. Check the SRS warning light for proper operation.

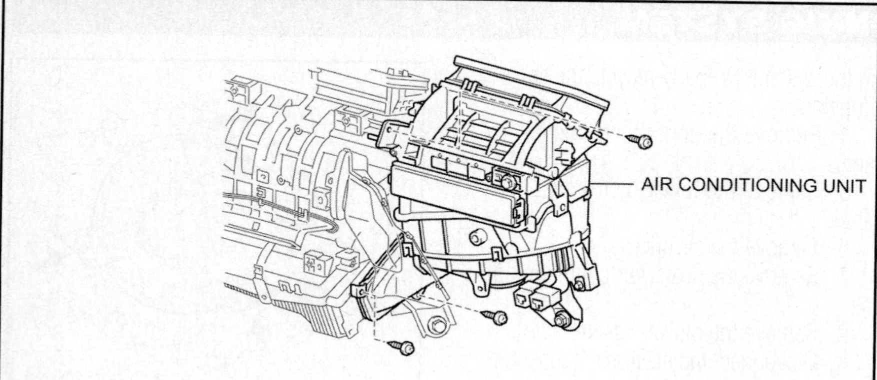

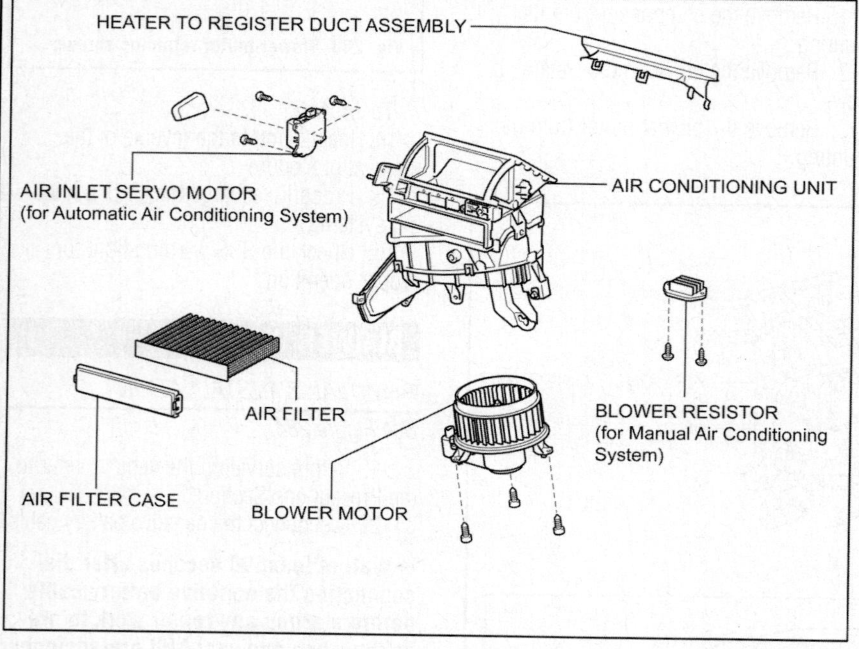

Fig. 264 Blower unit and related components

HEATER CORE

REMOVAL & INSTALLATION

See Figures 265 through 273.

1. Before servicing the vehicle, refer to the Precautions Section.
2. Position the front wheels in the straight ahead position.
3. Discharge the air conditioning system. Drain the engine coolant.
4. Disconnect the negative battery cable.

➡Wait at least 90 seconds after disconnecting the negative battery cable before starting any repair work to prevent air bag and seat belt pretensioner activation.

5. Disconnect and plug the refrigerant lines at the evaporator. Disconnect the heater hoses at the heater core.
6. Using a Torx® socket, loosen the two Torx® screws at the wheel pad. Pull out the wheel pad from the steering wheel and disconnect the air bag connector. Remove the wheel pad.

➡If the air bag connector is disconnected with the ignition switch "ON", DTC's will be recorded. When storing the wheel pad, keeps the upper surface of the pad facing upward. Never disassemble the wheel pad. When removing the wheel pad, take care not to pull the air bag wire harness.

7. Disconnect the connector. Matchmark the steering wheel. Remove the steering wheel retaining nut. Using the proper removal tool, remove the steering wheel.
8. Detach the four claws, release the tilt lever and remove the lower steering column cover. Detach the claw and remove the upper steering column cover.
9. Disconnect all connectors from the turn signal switch and the spiral cable. Remove the spiral cable.
10. Detach the clamp holding the combination switch in place. Remove the combination switch from the steering column.
11. Remove the instrument panel subassembly.
12. Disconnect the power steering motor wire harness and torque sensor wire harness clamps from the power steering ECU side. Disconnect the two steering column connectors from the power steering ECU.
13. Remove the lower instrument panel finish panel.
14. Turn back the driver's side carpet. Remove the two clips and remove the steering column hole cover silencer sheet.
15. Place matchmarks on the steering intermediate shaft and steering gear. Remove the bolt and detach the steering gear.
16. Place matchmarks on the steering intermediate shaft and the steering column. Remove the bolt and detach the steering intermediate shaft from the steering column.
17. Remove the brake pedal support assembly.
18. Disconnect the connectors and the wire harness clamps from the steering column assembly.
19. Remove the bolts and nuts and remove the steering column from the instrument panel reinforcement.
20. Remove the headlight dimmer switch assembly. Remove the windshield wiper switch assembly.
21. Remove the number one and the number two instrument cluster finish panel center assemblies.
22. Remove the radio.
23. Remove the two screws. Disconnect the connectors and remove the air conditioning control.
24. If equipped with automatic climate control, remove the air mix control switch, the blower control switch and the air vent mode control switch.
25. Remove the number one console upper panel garnish. Remove the number two console upper panel garnish.
26. Remove the shift lever knob. Using a pry tool, detach the two clips, four claws and remove the upper console panel. Disconnect the connector.
27. Detach the two clips and two claws. Disconnect the connectors and remove the switch base.
28. Remove the two screws. Detach the two clips and remove the cup holder box.

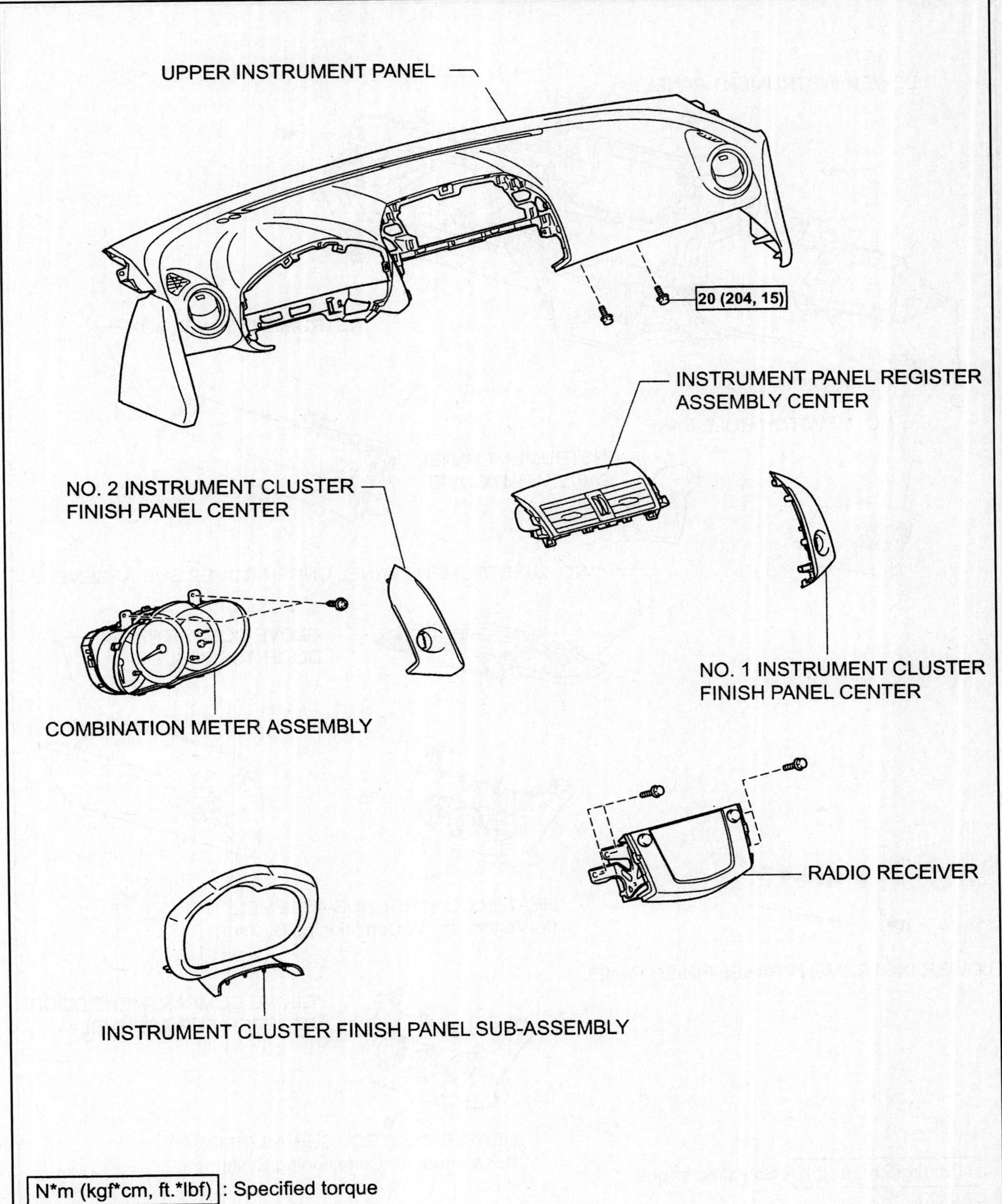

Fig. 265 Upper instrument panel and related components

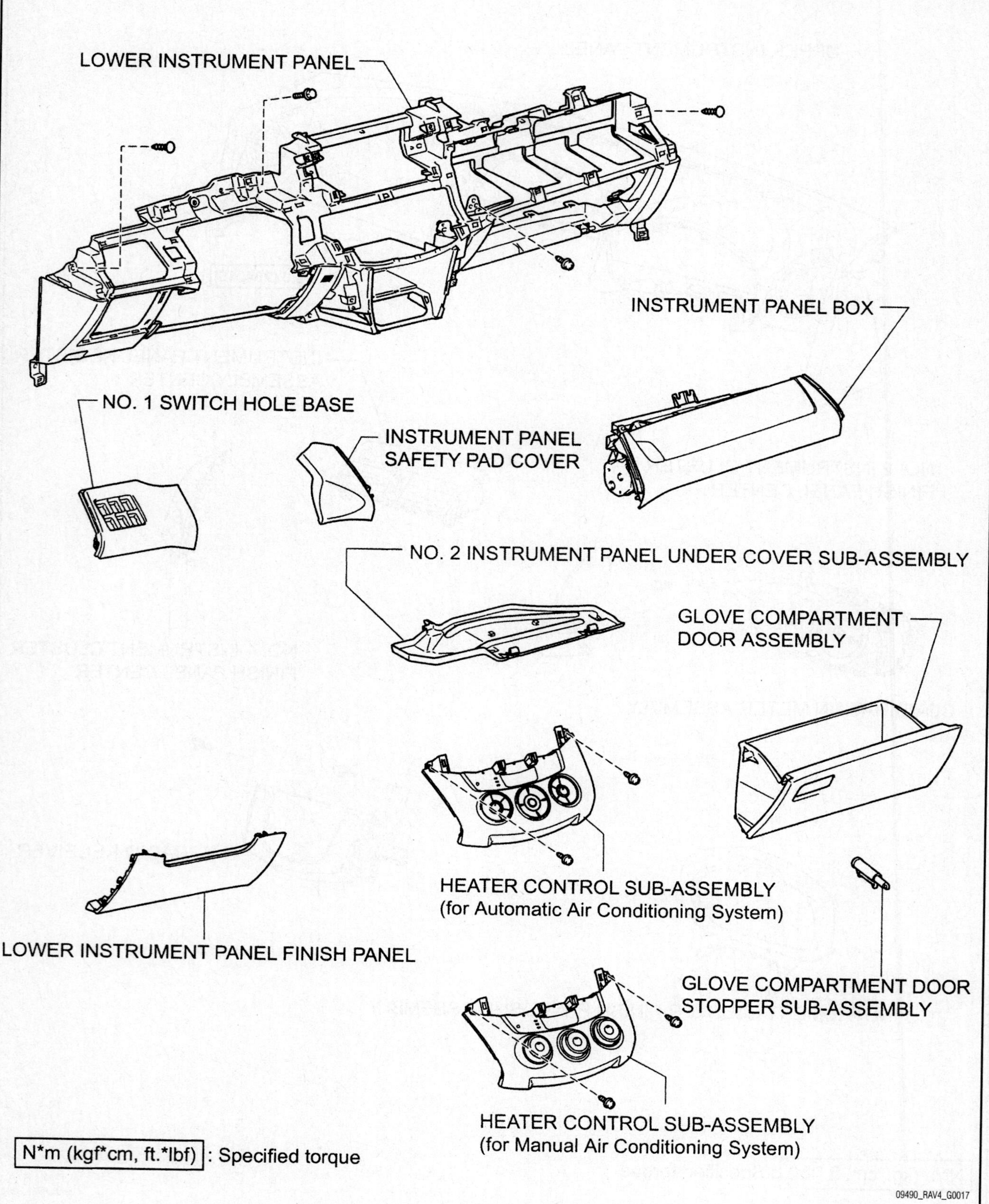

Fig. 266 Lower instrument panel and related components

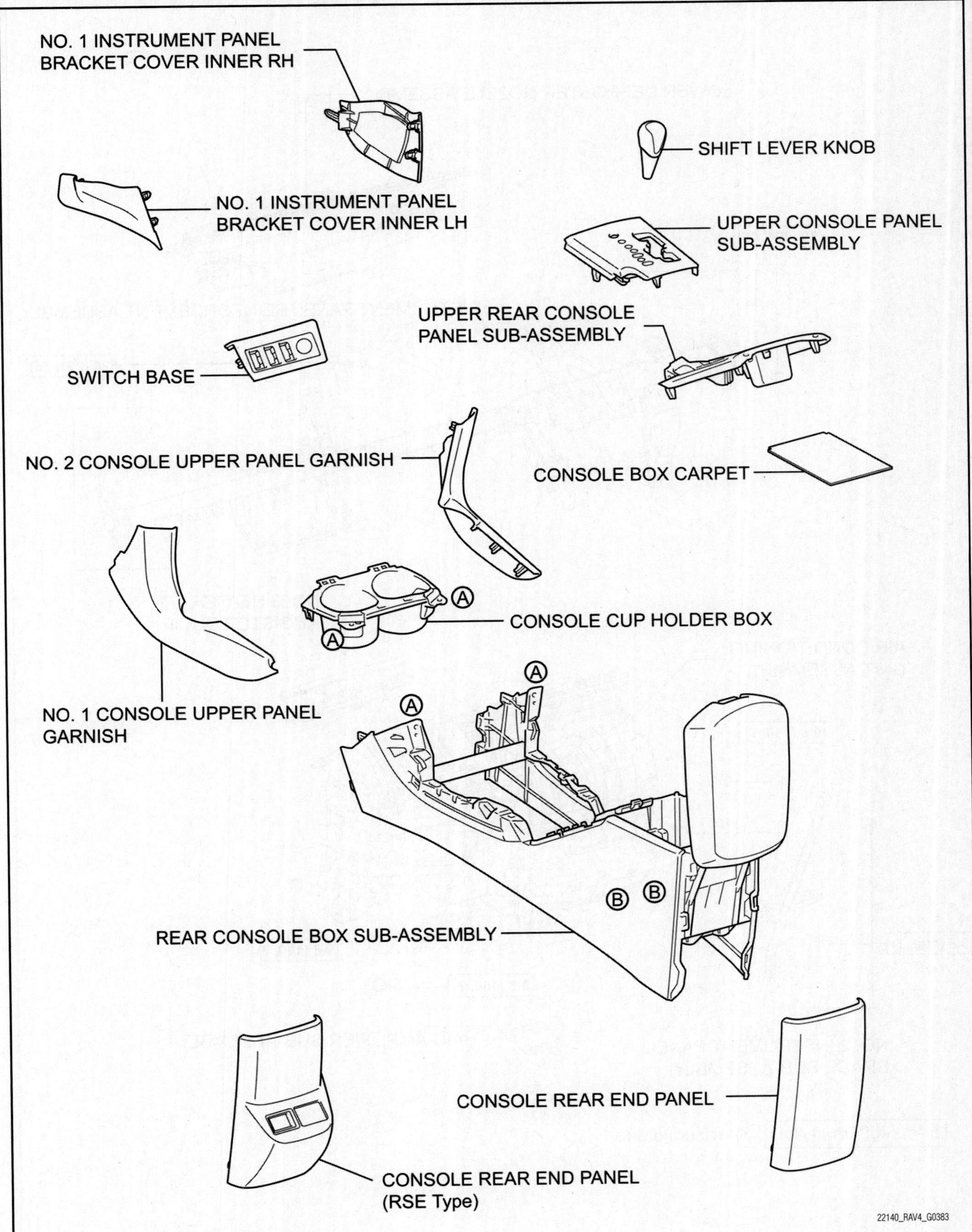

Fig. 267 Console assembly and related components

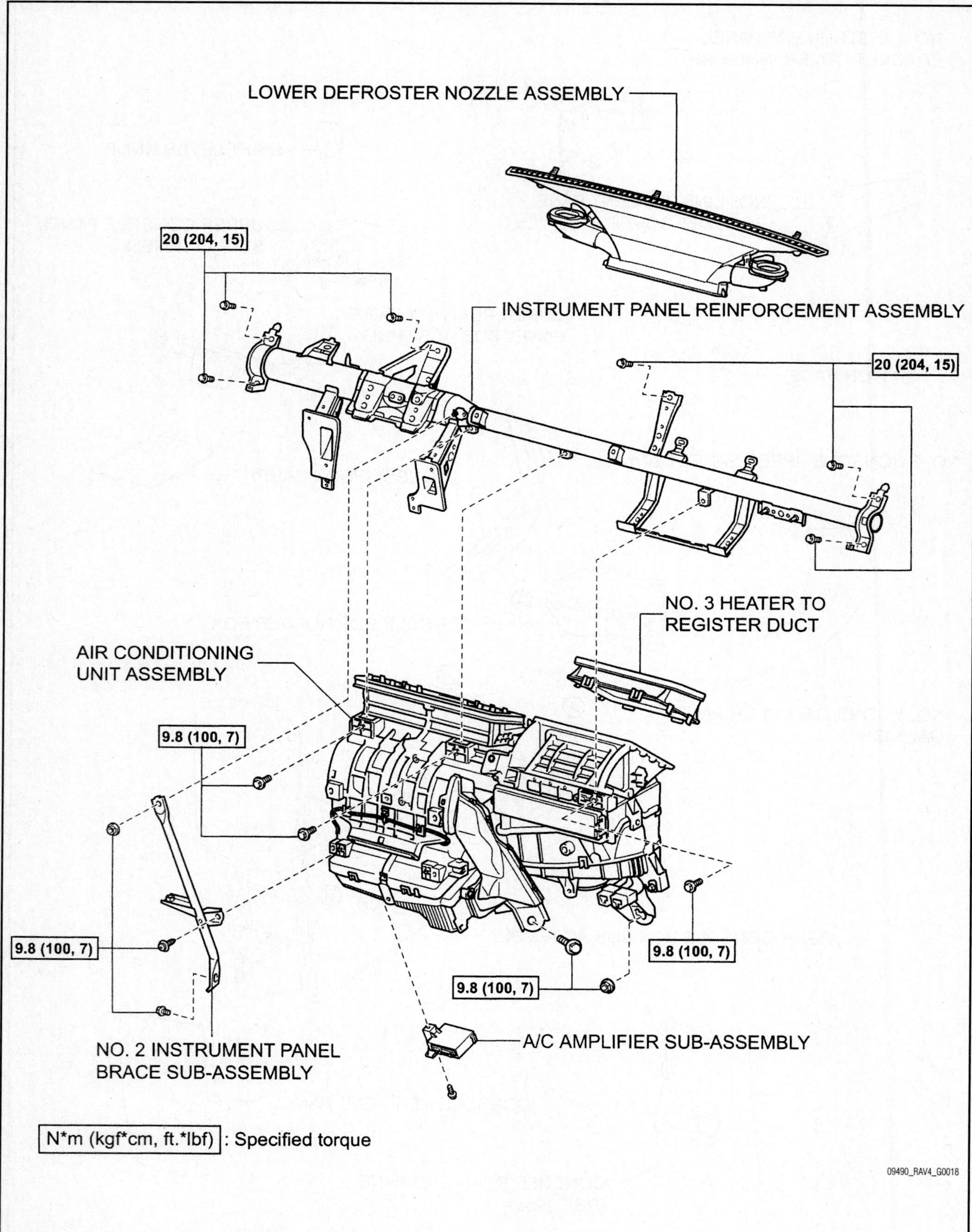

Fig. 268 Reinforcement panel and related components

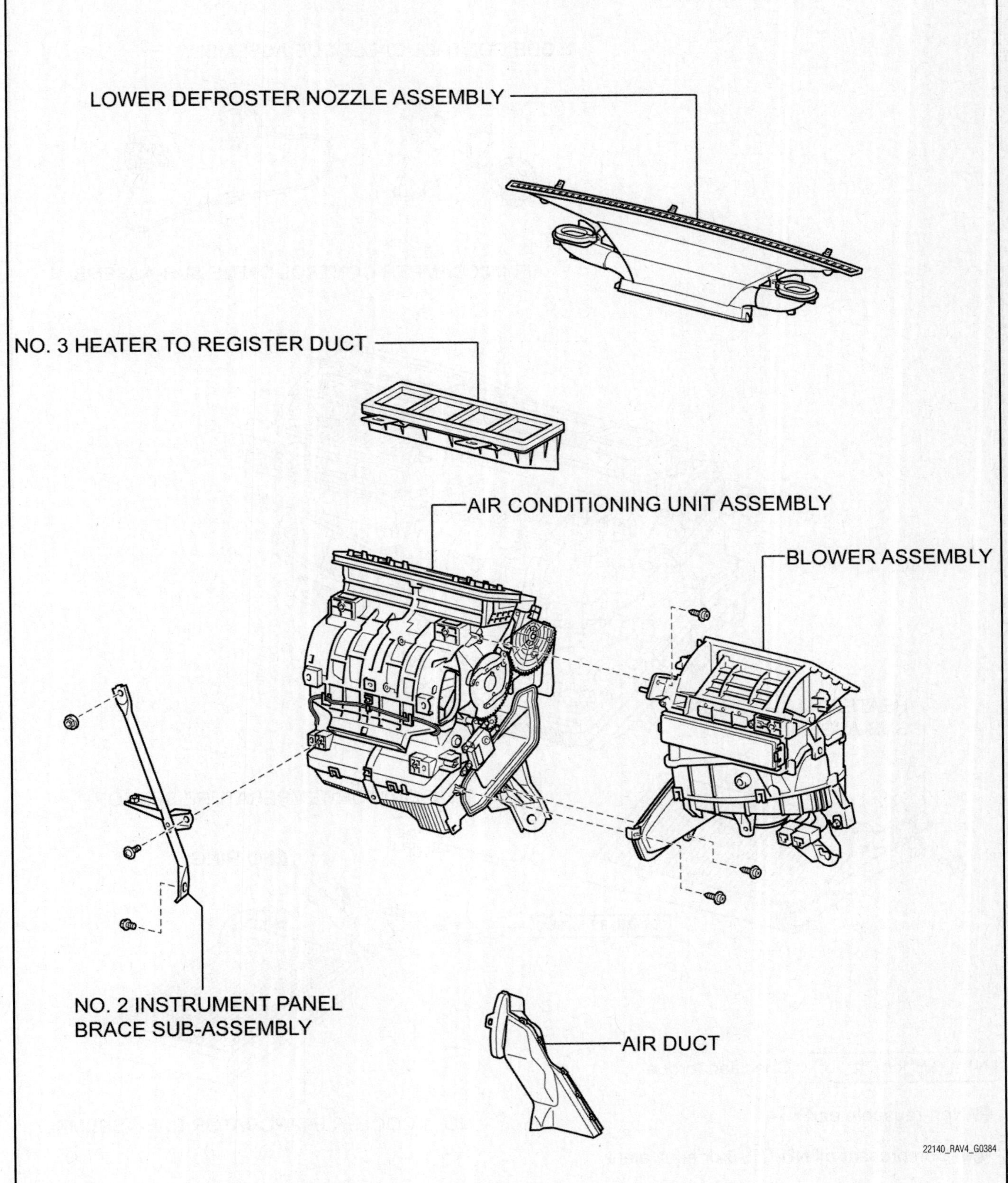

Fig. 269 Air conditioning/heater unit and related components

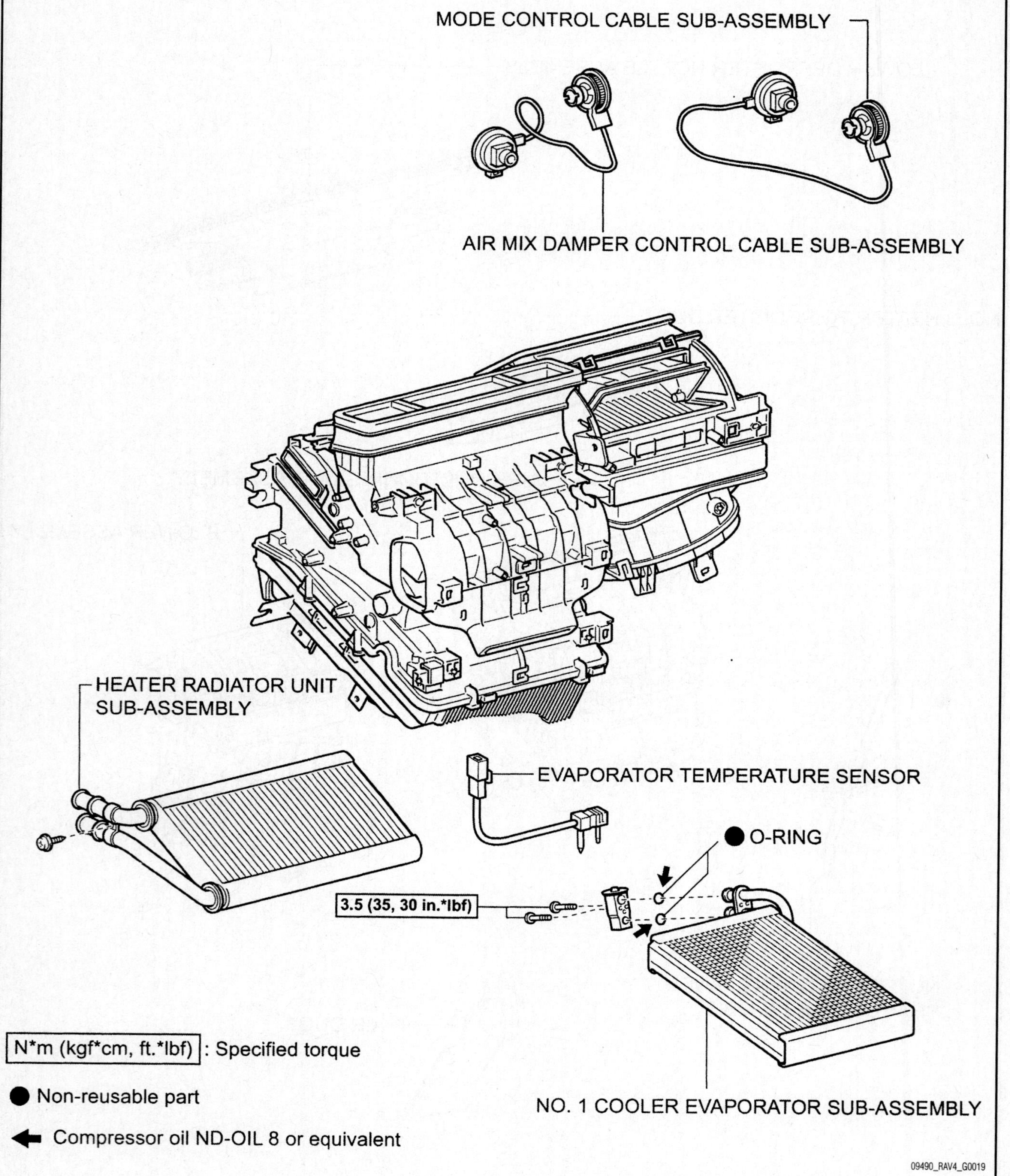

MODE CONTROL CABLE SUB-ASSEMBLY

AIR MIX DAMPER CONTROL CABLE SUB-ASSEMBLY

HEATER RADIATOR UNIT SUB-ASSEMBLY

EVAPORATOR TEMPERATURE SENSOR

● O-RING

3.5 (35, 30 in.*lbf)

N*m (kgf*cm, ft.*lbf) : Specified torque

● Non-reusable part

← Compressor oil ND-OIL 8 or equivalent

NO. 1 COOLER EVAPORATOR SUB-ASSEMBLY

09490_RAV4_G0019

Fig. 270 Manual air conditioning heater core and related components

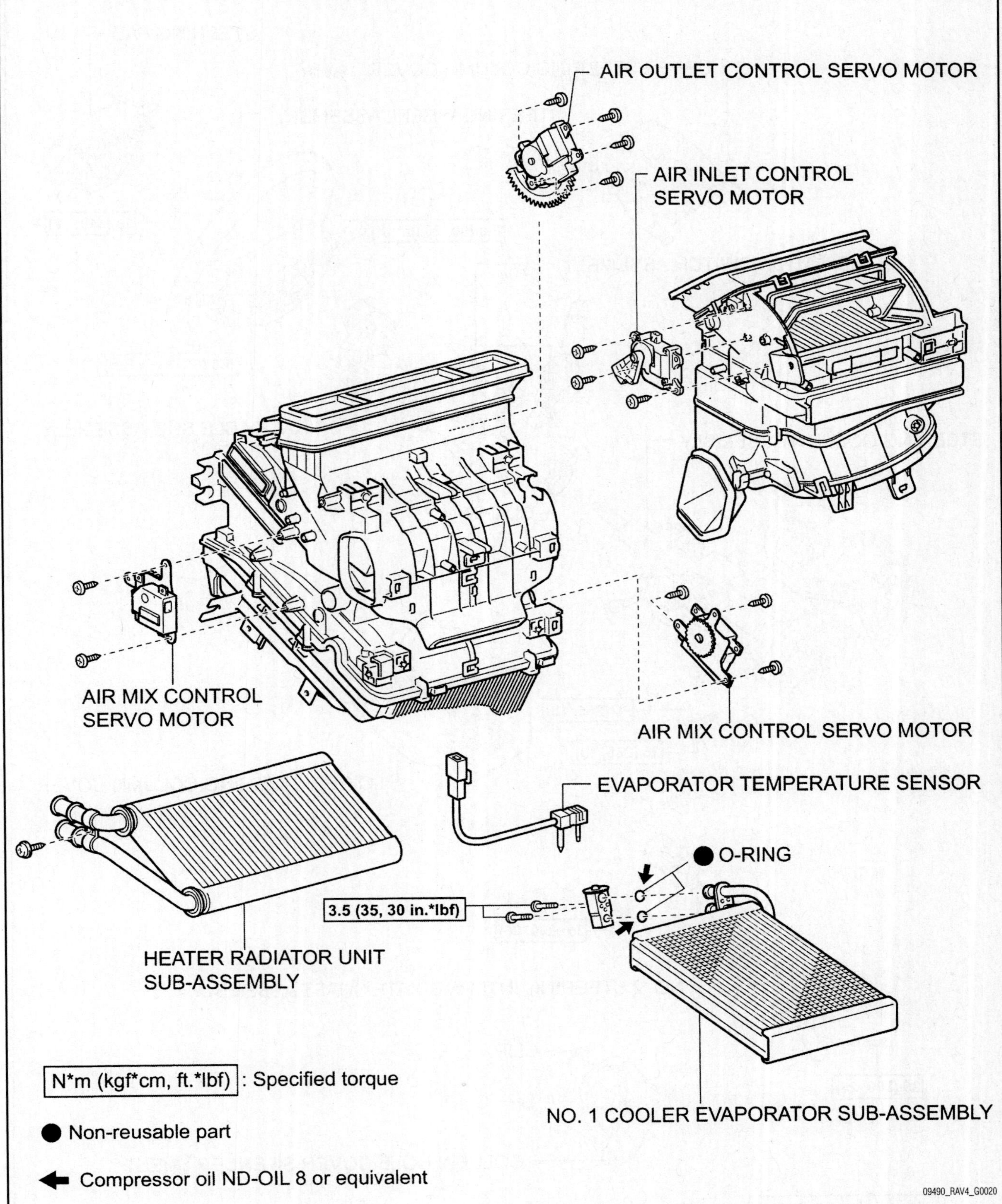

Fig. 271 Automatic air conditioning heater core and related components

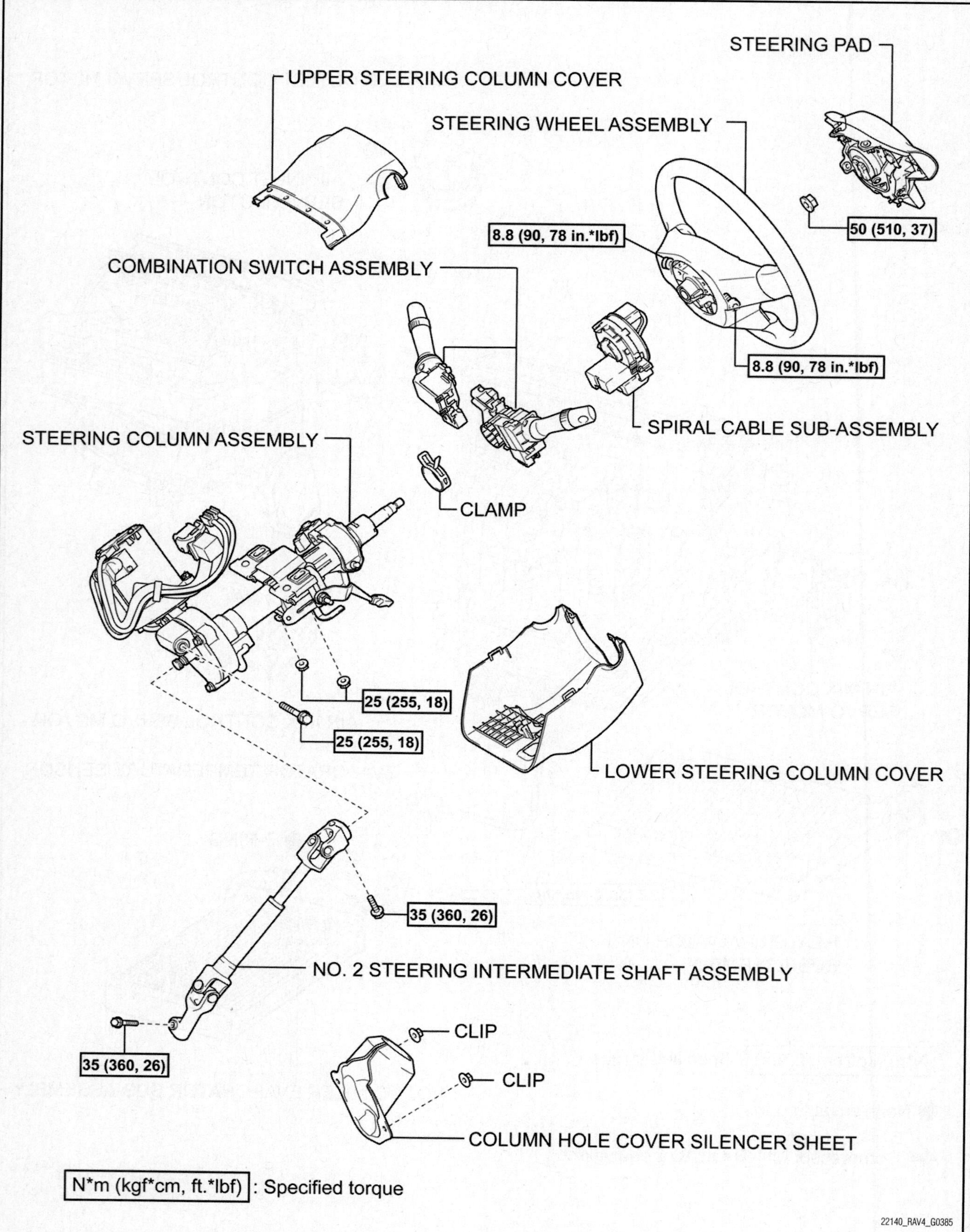

Fig. 272 Steering column retaining bolt locations

29. Using a pry tool, detach the four claws and two clips. Disconnect the connector and remove the upper rear console.

30. Using a pry tool, detach the six claws and remove the console rear end panel.

31. Detach the two claws and remove the right instrument panel bracket cover.

32. Detach the two claws and remove the left instrument panel bracket cover.

33. Remove the console box carpet. Remove the two bolts and two screws. Disconnect the connector and remove the rear console box.

34. Detach the two claws and remove the instrument panel under cover.

35. Remove the glove box door assembly.

36. Using a pry tool, detach the four clips. Disconnect the connectors and remove the number one switch hole base.

37. Using a pry tool, detach the four claws and remove the instrument panel safety pad cover.

38. Remove the screw. Using a pry tool, detach the six clips and remove the instrument panel box.

39. Disconnect the connectors. Remove the three nuts and remove the power steering ECU assembly.

40. Remove the two bolts, three screws and two clips. Disconnect the connectors and clamps. Remove the lower instrument panel from the vehicle.

41. Detach the three claws and remove the defroster nozzle assembly.

42. Fold back the carpet. Disconnect the clamp and disconnect the wire harness. Remove the bolt, nut, screw and instrument panel brace.

43. Disconnect the two clamps and disconnect the wire harness. Detach the three claws and remove the rear air duct.

44. Detach the two claws and remove the air duct.

45. Detach the two clamps. Disconnect the connector. Remove the screw and the air conditioning amplifier assembly. Remove the drain hose.

46. Disconnect the twelve retaining clamps. Remove the four bolts and disconnect the ground wire. Remove the six bolts. Remove the instrument panel reinforcement.

47. Remove the bolt, nut and the air conditioner unit assembly.

48. Remove the number three heater register duct. Remove the air duct. Remove the air outlet control servo motor. Remove the air mix control servo motor.

49. If equipped with manual air conditioning remove the mode control cable and the air mix damper control cable.

50. Remove the heater core from its mounting.

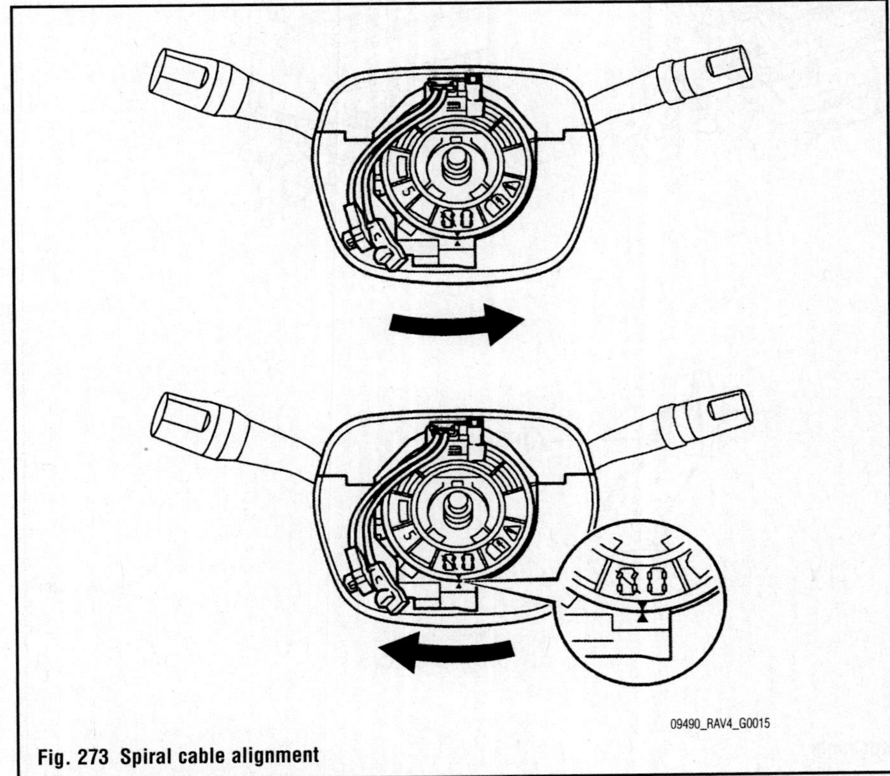

Fig. 273 Spiral cable alignment

To install:

51. Installation is the reverse of the removal procedure.

52. Tighten the steering column retaining bolts to 18 ft. lbs.

53. Tighten the sliding yoke to number intermediate shaft retaining bolt to 26 ft. lbs.

54. Tighten the sliding yoke to steering gear retaining bolt to 26 ft. lbs.

55. To center the spiral cable, check that the front wheels are in the straight ahead position. Turn the cable counterclockwise by hand until it becomes hard to turn. Then rotate the cable clockwise about 2.5 turns to align the marks.

➡The cable will rotate about 2.5 turns to either left or right of the center.

56. Tighten the steering wheel locknut to 37 ft. lbs.

57. Tighten the wheel pad retaining screws to 78 inch lbs.

58. Check the steering wheel center point.

59. Recharge the air conditioning system.

60. Fill the cooling system with the proper grade and type engine coolant.

61. Check the SRS warning light for proper operation.

62. Start the engine and check for leaks, correct as required.

HEATER/AIR CONDITIONING UNIT

REMOVAL & INSTALLATION

See Figures 274 through 280.

1. Before servicing the vehicle, refer to the Precautions Section.

2. Position the front wheels in the straight ahead position.

3. Discharge the air conditioning system. Drain the engine coolant.

4. Disconnect the negative battery cable.

➡Wait at least 90 seconds after disconnecting the negative battery cable before starting any repair work to prevent air bag and seat belt pretensioner activation.

5. Disconnect and plug the refrigerant lines at the evaporator. Disconnect the heater hoses at the heater core.

6. Using a Torx® socket, loosen the two Torx® screws at the wheel pad. Pull out the wheel pad from the steering wheel and disconnect the air bag connector. Remove the wheel pad.

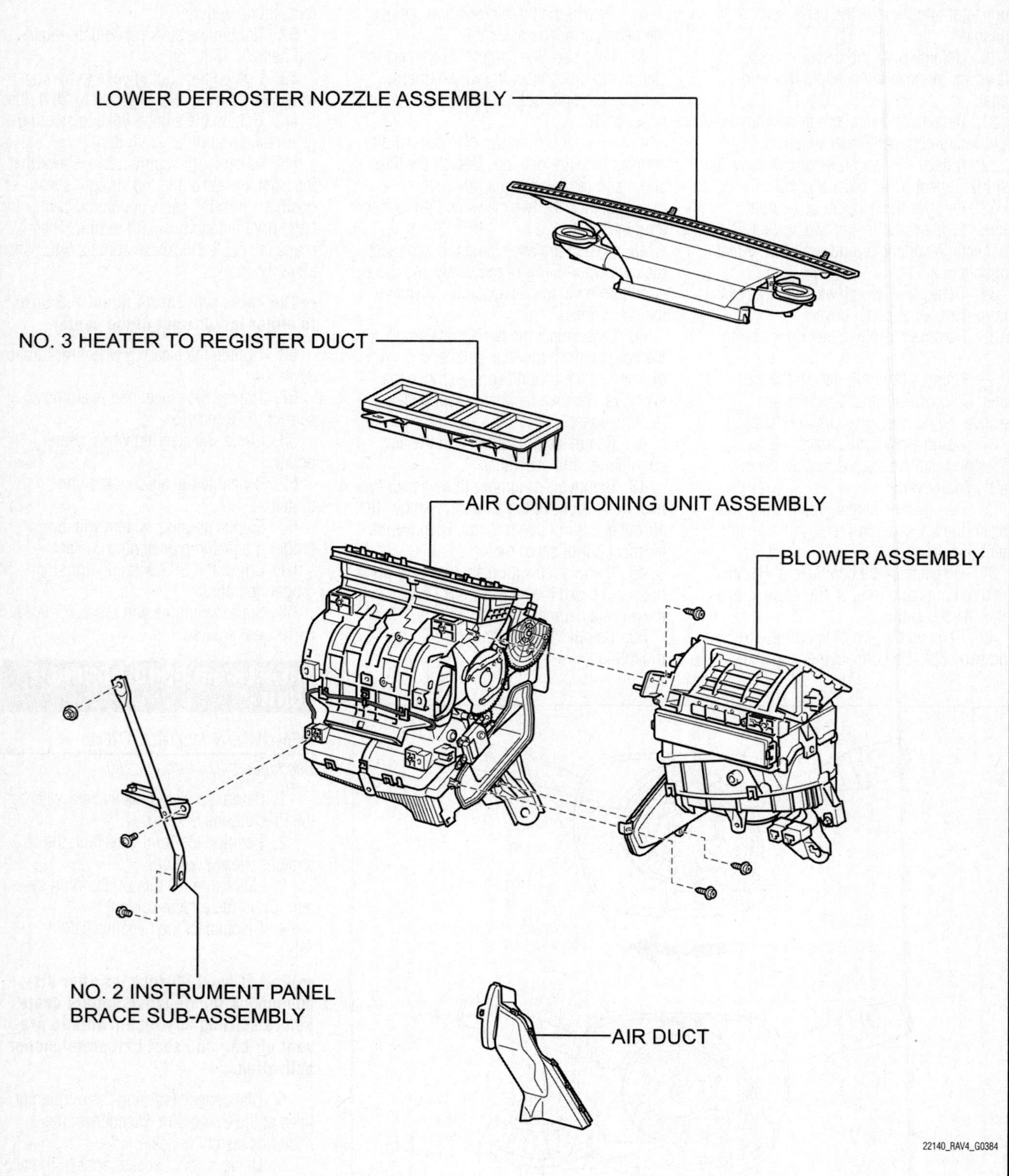

Fig. 274 Air conditioning/heater unit and related components

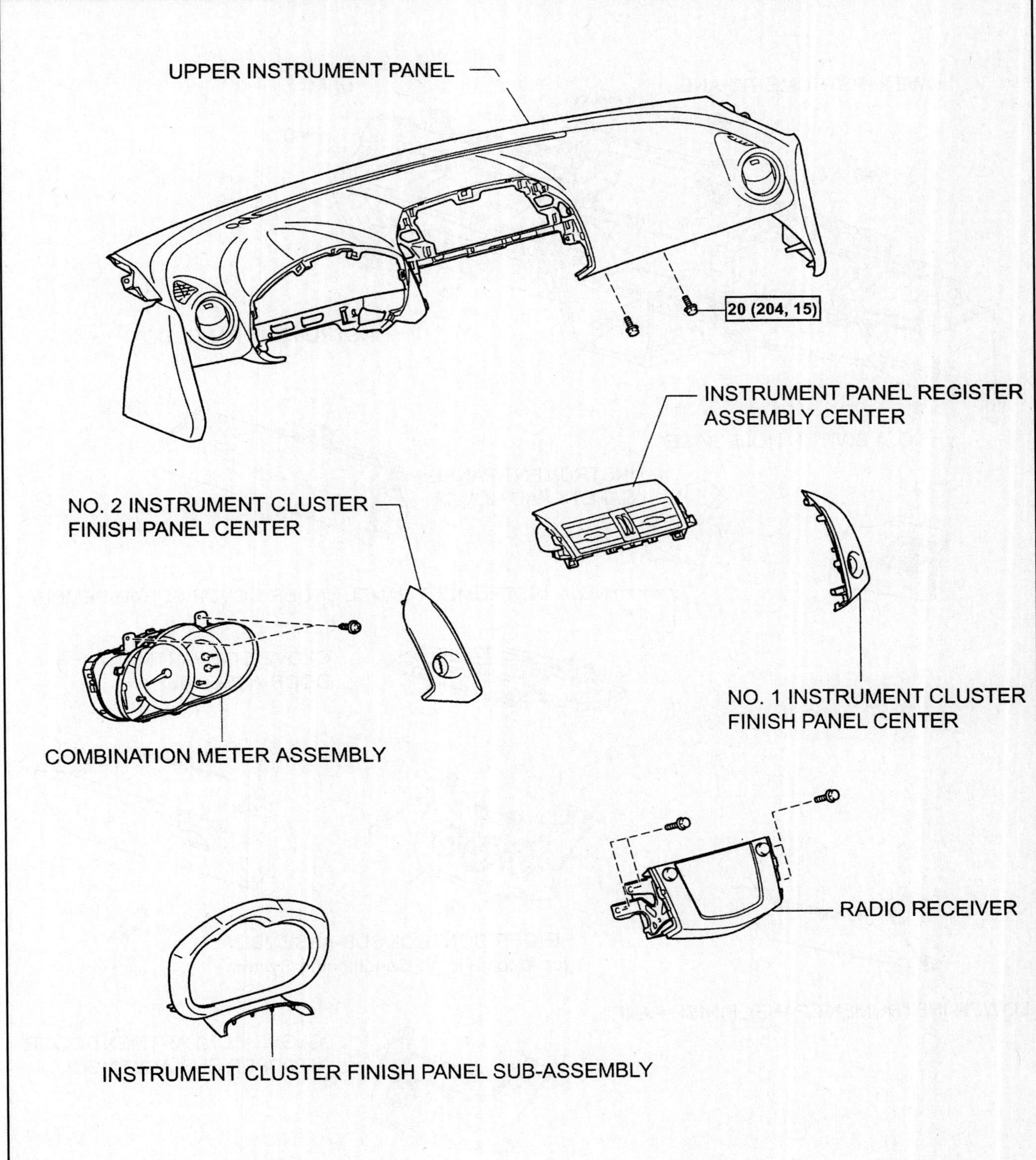

Fig. 275 Upper instrument panel and related components

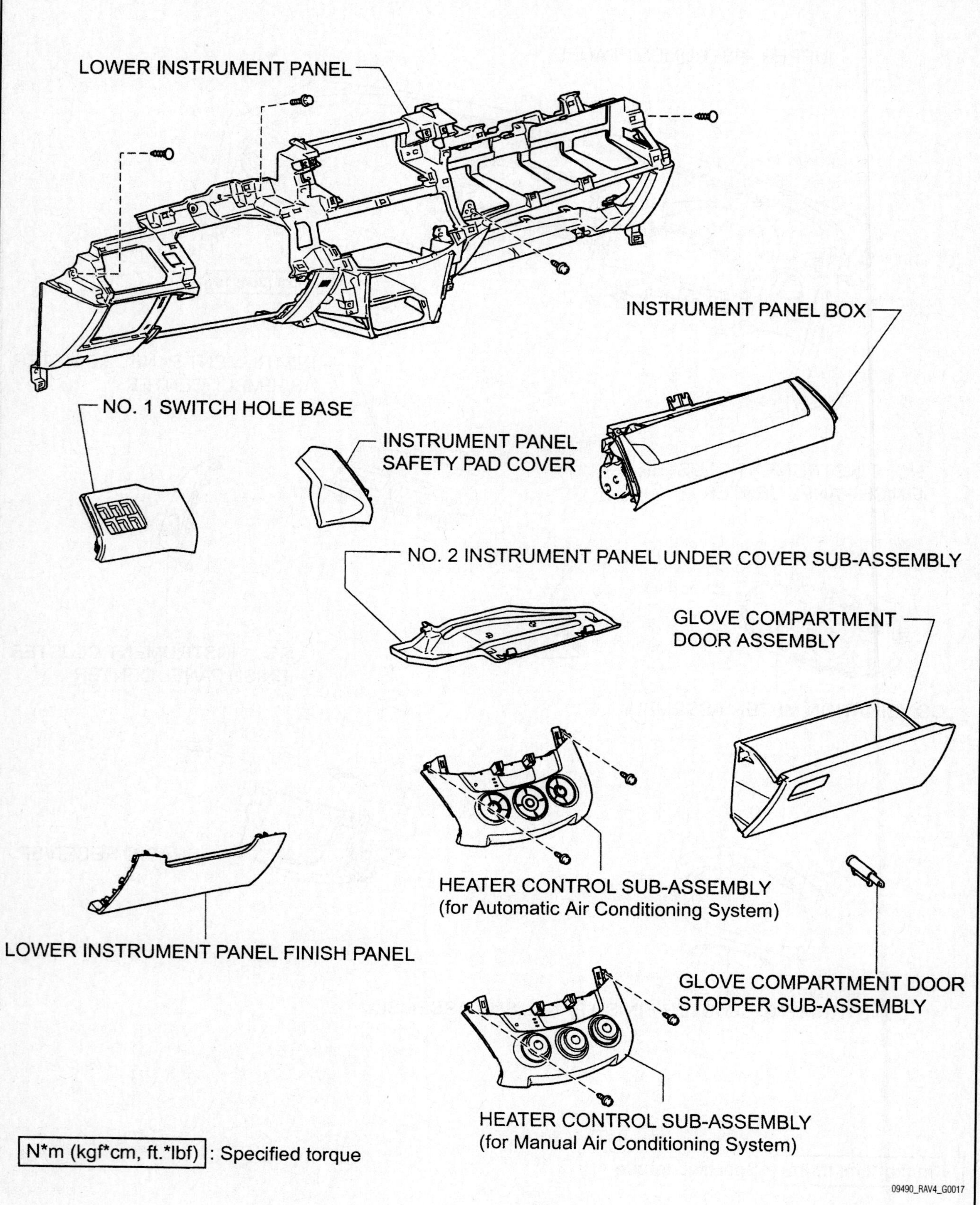

Fig. 276 Lower instrument panel and related components

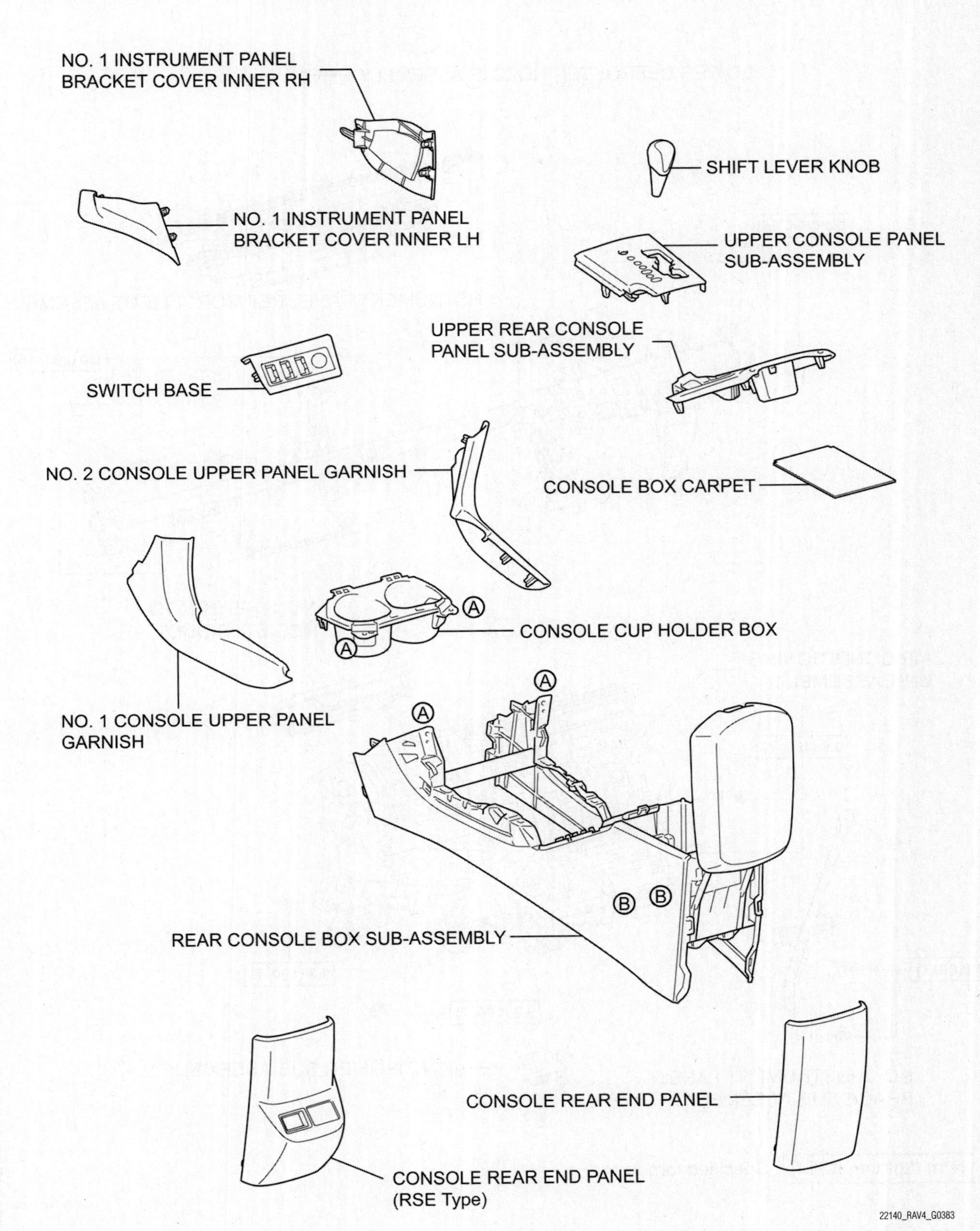

Fig. 277 Console assembly and related components

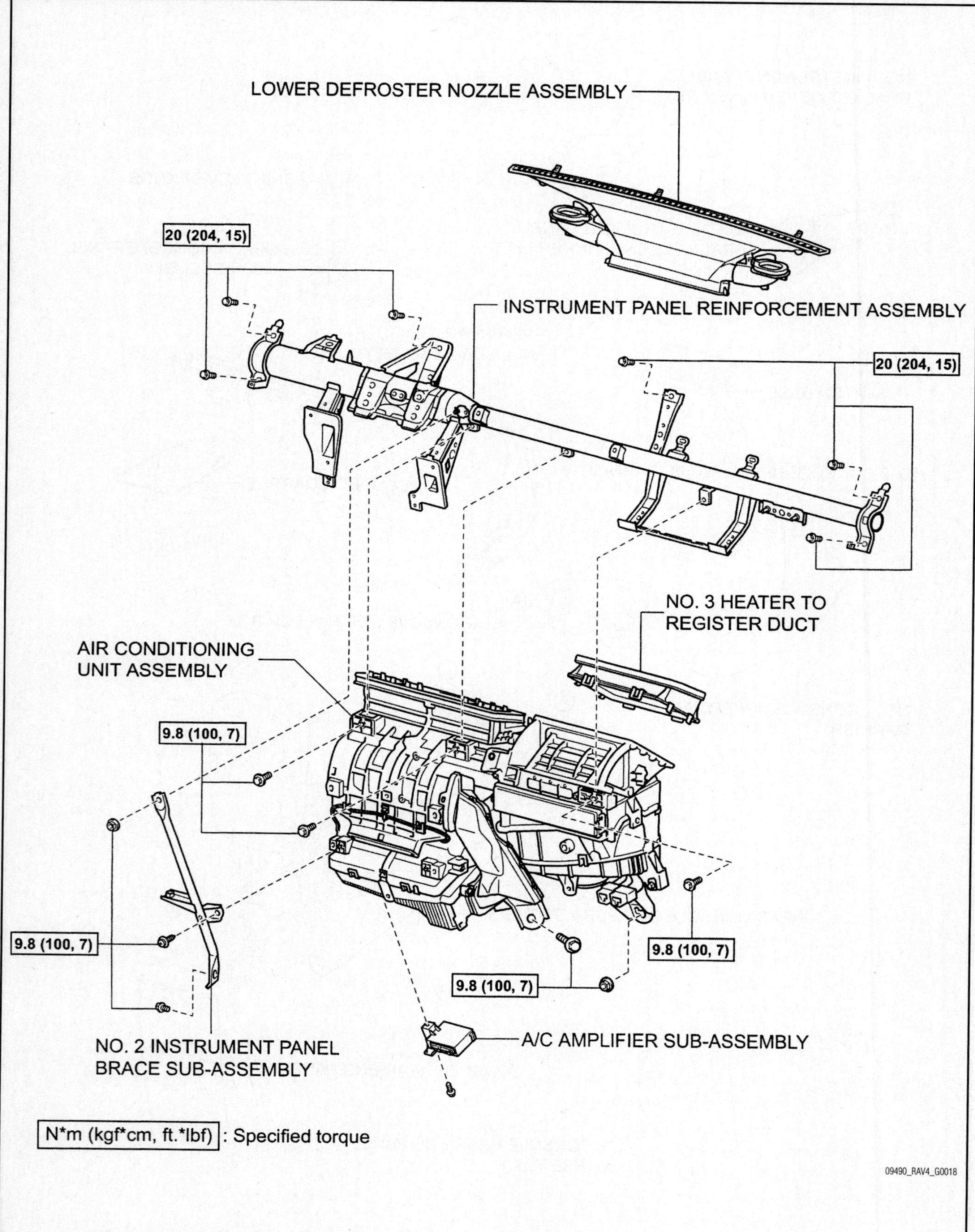

Fig. 278 Reinforcement panel and related components

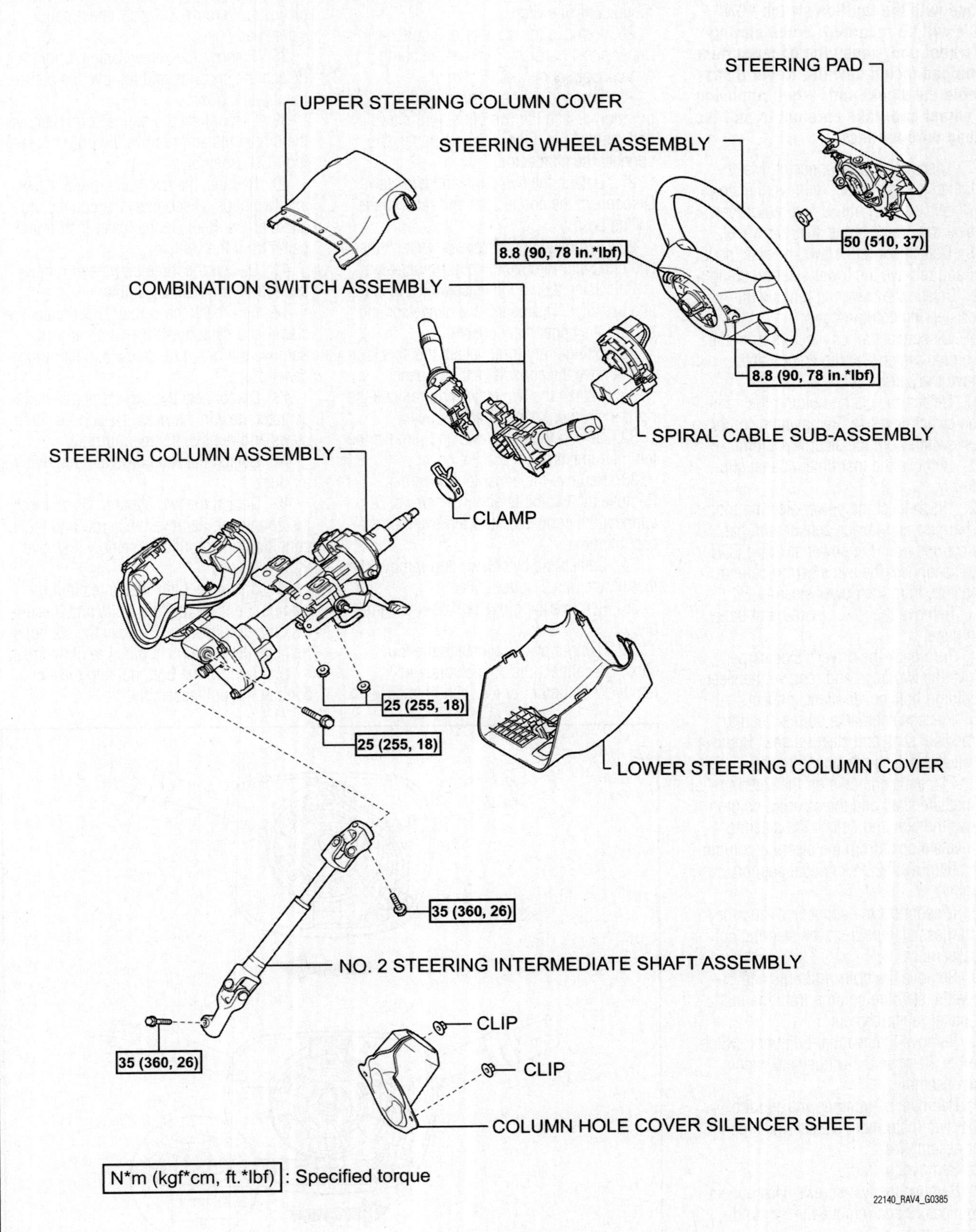

Fig. 279 Steering column retaining bolt locations

➡If the air bag connector is disconnected with the ignition switch "ON", DTC's will be recorded. When storing the wheel pad, keeps the upper surface of the pad facing upward. Never disassemble the wheel pad. When removing the wheel pad, take care not to pull the air bag wire harness.

7. Disconnect the connector. Matchmark the steering wheel. Remove the steering wheel retaining nut. Using the proper removal tool, remove the steering wheel.
8. Detach the four claws, release the tilt lever and remove the lower steering column cover. Detach the claw and remove the upper steering column cover.
9. Disconnect all connectors from the turn signal switch and the spiral cable. Remove the spiral cable.
10. Detach the clamp holding the combination switch in place. Remove the combination switch from the steering column.
11. Remove the instrument panel sub-assembly.
12. Disconnect the power steering motor wire harness and torque sensor wire harness clamps from the power steering ECU side. Disconnect the two steering column connectors from the power steering ECU.
13. Remove the lower instrument panel finish panel.
14. Turn back the driver's side carpet. Remove the two clips and remove the steering column hole cover silencer sheet.
15. Place matchmarks on the steering intermediate shaft and steering gear. Remove the bolt and detach the steering gear.
16. Place matchmarks on the steering intermediate shaft and the steering column. Remove the bolt and detach the steering intermediate shaft from the steering column.
17. Remove the brake pedal support assembly.
18. Disconnect the connectors and the wire harness clamps from the steering column assembly.
19. Remove the bolts and nuts and remove the steering column from the instrument panel reinforcement.
20. Remove the headlight dimmer switch assembly. Remove the windshield wiper switch assembly.
21. Remove the number one and the number two instrument cluster finish panel center assemblies.
22. Remove the radio.
23. Remove the two screws. Disconnect the connectors and remove the air conditioning control.
24. If equipped with automatic climate control, remove the air mix control switch, the blower control switch and the air vent mode control switch.
25. Remove the number one console upper panel garnish. Remove the number two console upper panel garnish.
26. Remove the shift lever knob. Using a pry tool, detach the two clips, four claws and remove the upper console panel. Disconnect the connector.
27. Detach the two clips and two claws. Disconnect the connectors and remove the switch base.
28. Remove the two screws. Detach the two clips and remove the cup holder box.
29. Using a pry tool, detach the four claws and two clips. Disconnect the connector and remove the upper rear console.
30. Using a pry tool, detach the six claws and remove the console rear end panel.
31. Detach the two claws and remove the right instrument panel bracket cover.
32. Detach the two claws and remove the left instrument panel bracket cover.
33. Remove the console box carpet. Remove the two bolts and two screws. Disconnect the connector and remove the rear console box.
34. Detach the two claws and remove the instrument panel under cover.
35. Remove the glove box door assembly.
36. Using a pry tool, detach the four clips. Disconnect the connectors and remove the number one switch hole base.
37. Using a pry tool, detach the four claws and remove the instrument panel safety pad cover.
38. Remove the screw. Using a pry tool, detach the six clips and remove the instrument panel box.
39. Disconnect the connectors. Remove the three nuts and remove the power steering ECU assembly.
40. Remove the two bolts, three screws and two clips. Disconnect the connectors and clamps. Remove the lower instrument panel from the vehicle.
41. Detach the three claws and remove the defroster nozzle assembly.
42. Fold back the carpet. Disconnect the clamp and disconnect the wire harness. Remove the bolt, nut, screw and instrument panel brace.
43. Disconnect the two clamps and disconnect the wire harness. Detach the three claws and remove the rear air duct.
44. Detach the two claws and remove the air duct.
45. Detach the two clamps. Disconnect the connector. Remove the screw and the air conditioning amplifier assembly. Remove the drain hose.
46. Disconnect the twelve retaining clamps. Remove the four bolts and disconnect the ground wire. Remove the six bolts. Remove the instrument panel reinforcement.
47. Remove the bolt, nut and the air conditioner unit assembly.

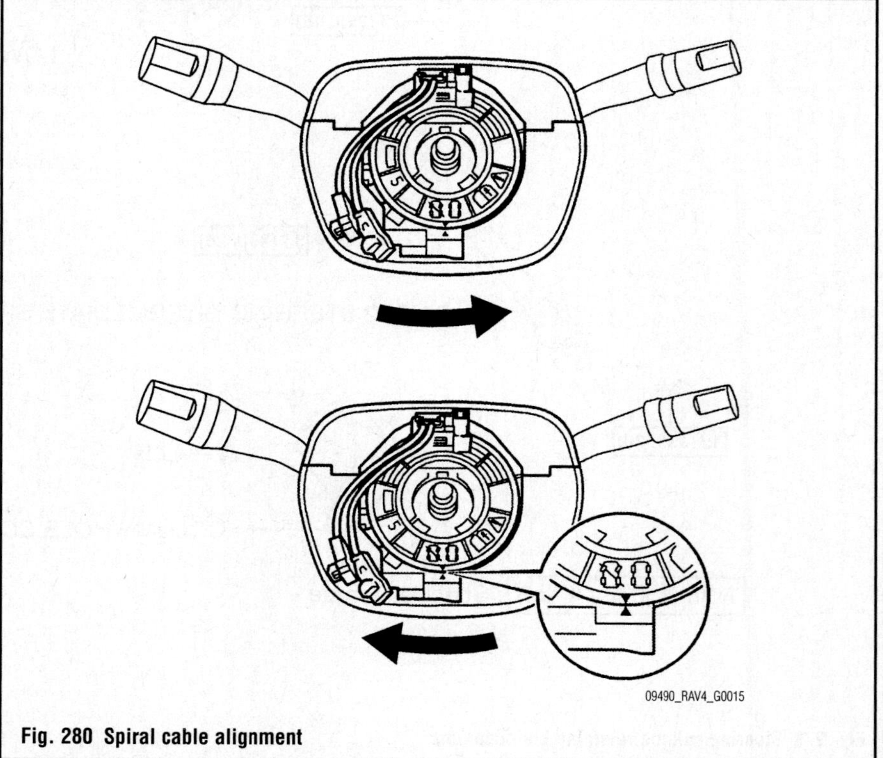

Fig. 280 Spiral cable alignment

To install:

48. Installation is the reverse of the removal procedure.

49. Tighten the steering column retaining bolts to 18 ft. lbs.

50. Tighten the sliding yoke to number intermediate shaft retaining bolt to 26 ft. lbs.

51. Tighten the sliding yoke to steering gear retaining bolt to 26 ft. lbs.

52. To center the spiral cable, check that the front wheels are in the straight ahead position. Turn the cable counterclockwise by hand until it becomes hard to turn. Then rotate the cable clockwise about 2.5 turns to align the marks.

➡The cable will rotate about 2.5 turns to either left or right of the center.

53. Tighten the steering wheel locknut to 37 ft. lbs.

54. Tighten the wheel pad retaining screws to 78 inch lbs.

55. Check the steering wheel center point.

56. Recharge the air conditioning system.

57. Fill the cooling system with the proper grade and type engine coolant.

58. Check the SRS warning light for proper operation.

59. Start the engine and check for leaks, correct as required.

STEERING

POWER RACK & PINION STEERING GEAR

REMOVAL & INSTALLATION

See Figures 281 through 283.

1. Before servicing the vehicle, refer to the Precautions Section.

2. Disconnect the negative battery cable.

➡Wait at least 90 seconds after disconnecting the negative battery cable before starting any repair work to prevent air bag and seat belt pretensioner activation.

3. Place the wheels in a straight-ahead position.

4. Raise and support the vehicle safely.

5. Remove the tire and wheel assemblies.

6. Disconnect the right and left tie rod ends, using tool SST09628-62011 or equivalent.

7. Remove the floor carpet. Remove the two clips and the column hole silencer cover.

8. Use the seat belt to position the steering wheel in order to avoid breakage of the spiral cable.

9. Matchmark the sliding yoke of the steering intermediate shaft. Remove the bolt and disconnect the sliding yoke.

10. Remove the bottom clip and detach the upper clip from the body and disconnect the number one steering column hole cover. Be careful not to damage the clips.

11. Remove the engine and transaxle assembly from the vehicle and position it in a suitable holding fixture.

12. Remove the clamp and disconnect the number one column hole cover from the steering gear.

13. Matchmark the intermediate shaft of the steering gear. Remove the bolt and disconnect the steering intermediate shaft from the steering gear.

14. Remove the two bolts, two nuts and the steering gear from the crossmember. Be sure to keep the nut from rotating while turning the bolt.

To install:

15. Installation is the reverse of the removal procedure.

16. Tighten the steering gear retaining bolts to 102 ft. lbs. Be sure to keep

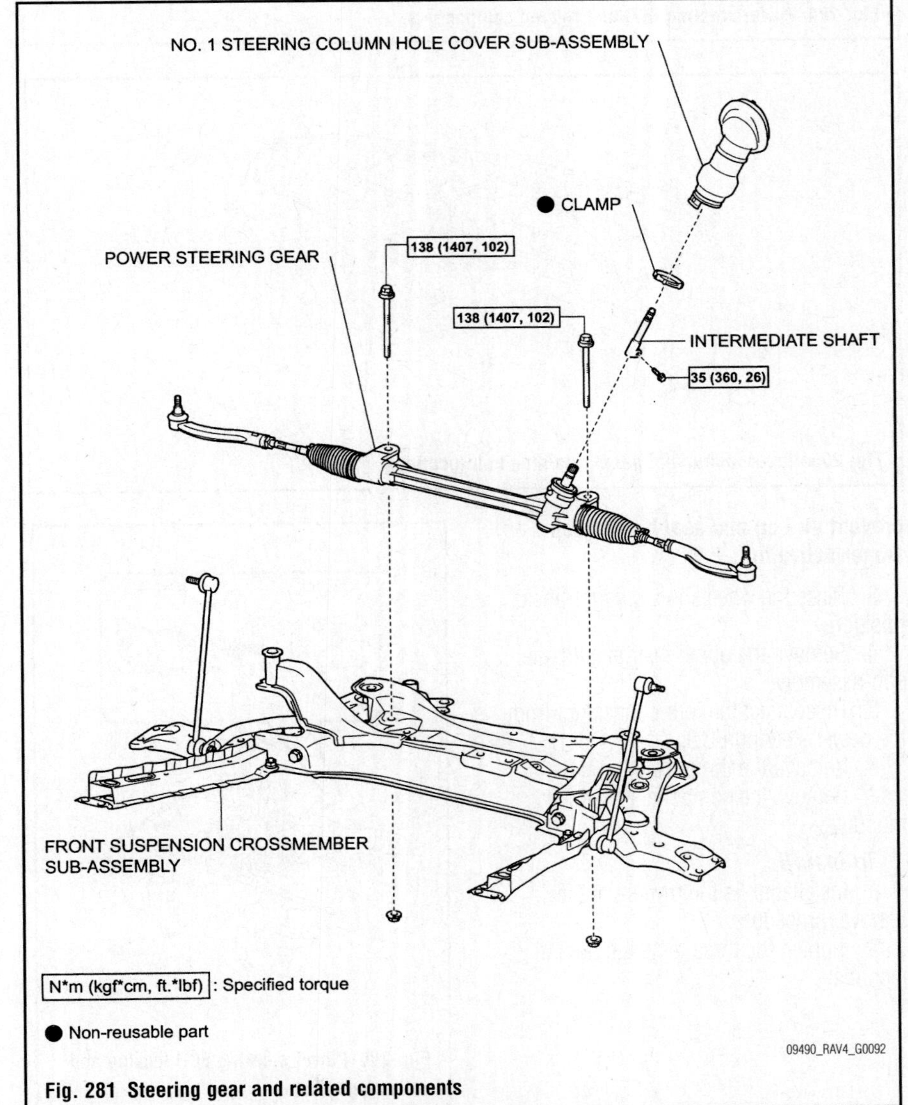

Fig. 281 Steering gear and related components

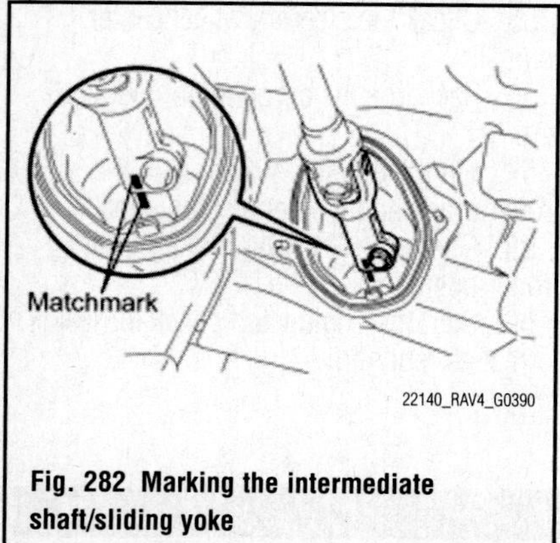

Fig. 282 Marking the intermediate shaft/sliding yoke

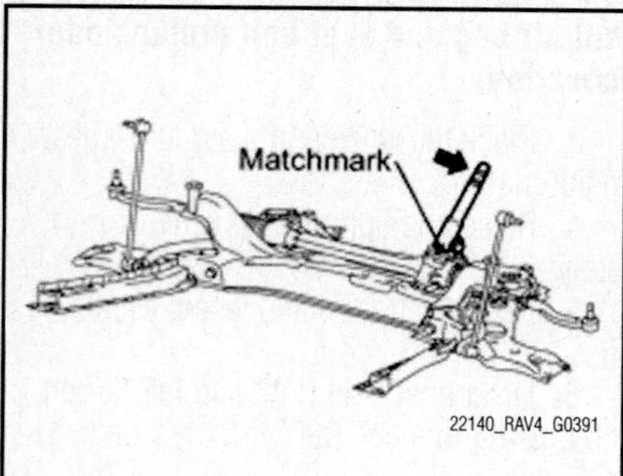

Fig. 283 Marking the intermediate shaft/steering gear

the nut from rotating while turning the bolt.

17. Tighten the intermediate shaft retaining bolt to 26 ft. lbs.
18. Tighten the sliding yoke retaining bolt to 26 ft. lbs.
19. Tighten the tie rod end castle nut to 36 ft. lbs. If the holes for the clip are not aligned, tighten the nut an additional 60 degrees. Be sure to use a new cotter pin.
20. Check and adjust the alignment, as required.

POWER STEERING ECU

REMOVAL & INSTALLATION

See Figures 284 through 286.

1. Before servicing the vehicle, refer to the Precautions Section.
2. Disconnect the negative battery cable.

➡Wait at least 90 seconds after disconnecting the negative battery cable before starting any repair work to prevent air bag and seat belt pretensioner activation.

3. Place the wheels in a straight-ahead position.
4. Remove the upper instrument panel sub-assembly.
5. Disconnect the four connectors from the power steering ECU.
6. Remove the three retaining nuts.
7. Remove the component from its mounting.

To install:

8. Installation is the reverse of the removal procedure.
9. Tighten the three retaining nuts to 75 inch lbs.

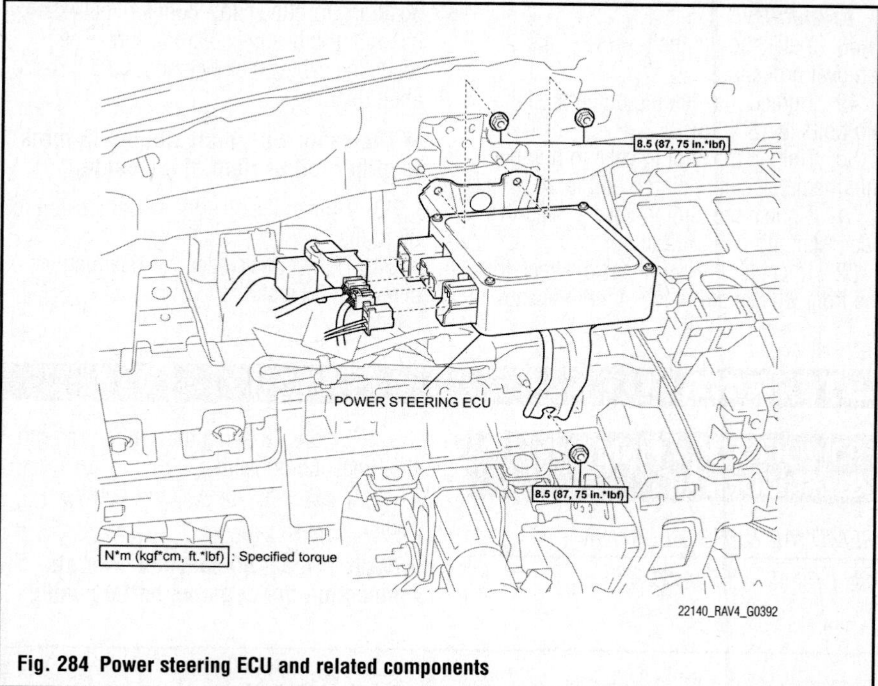

Fig. 284 Power steering ECU and related components

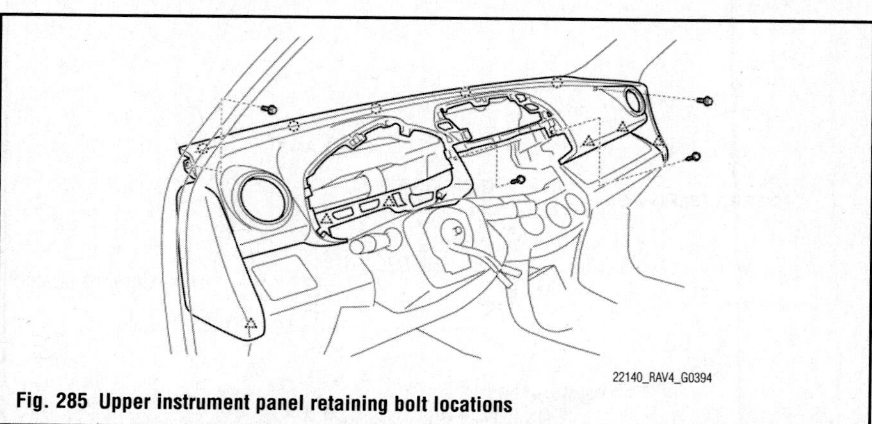

Fig. 285 Upper instrument panel retaining bolt locations

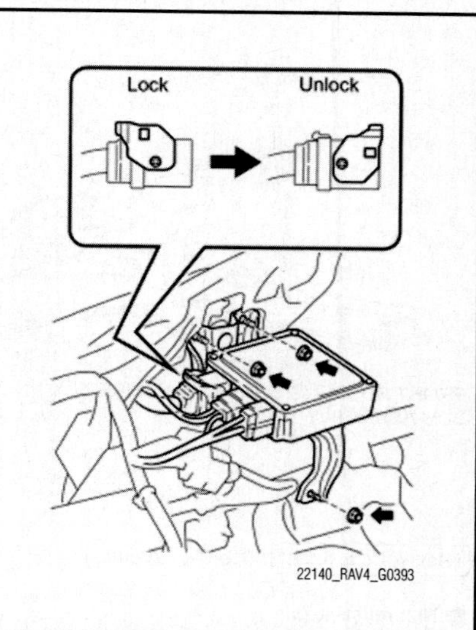

Fig. 286 Power steering ECU locking and unlocking tabs

LOWER BALL JOINT

REMOVAL & INSTALLATION

See Figures 287 and 288.

1. Before servicing the vehicle, refer to the Precautions Section.
2. Raise and support the vehicle safely.
3. Remove the tire and wheel assembly.
4. On 2WD, remove the front speed sensor.
5. Remove the front caliper. Remove the rotor.
6. Remove the front axle hub nut.
7. To disconnect the front suspension lower number one arm, remove the bolt and two nuts. Disconnect the lower arm from the ball joint.
8. Using the proper tools, disconnect the tie rod end.
9. Remove the steering knuckle with the axle hub, using the proper removal tools.
10. Remove the cotter pin and nut. Using tool SST09628-62011, remove the lower ball joint.

To install:

11. Installation is the reverse of the removal procedure.
12. Tighten the two ball joint to the lower arm nuts to 68 ft. lbs. (92 Nm).

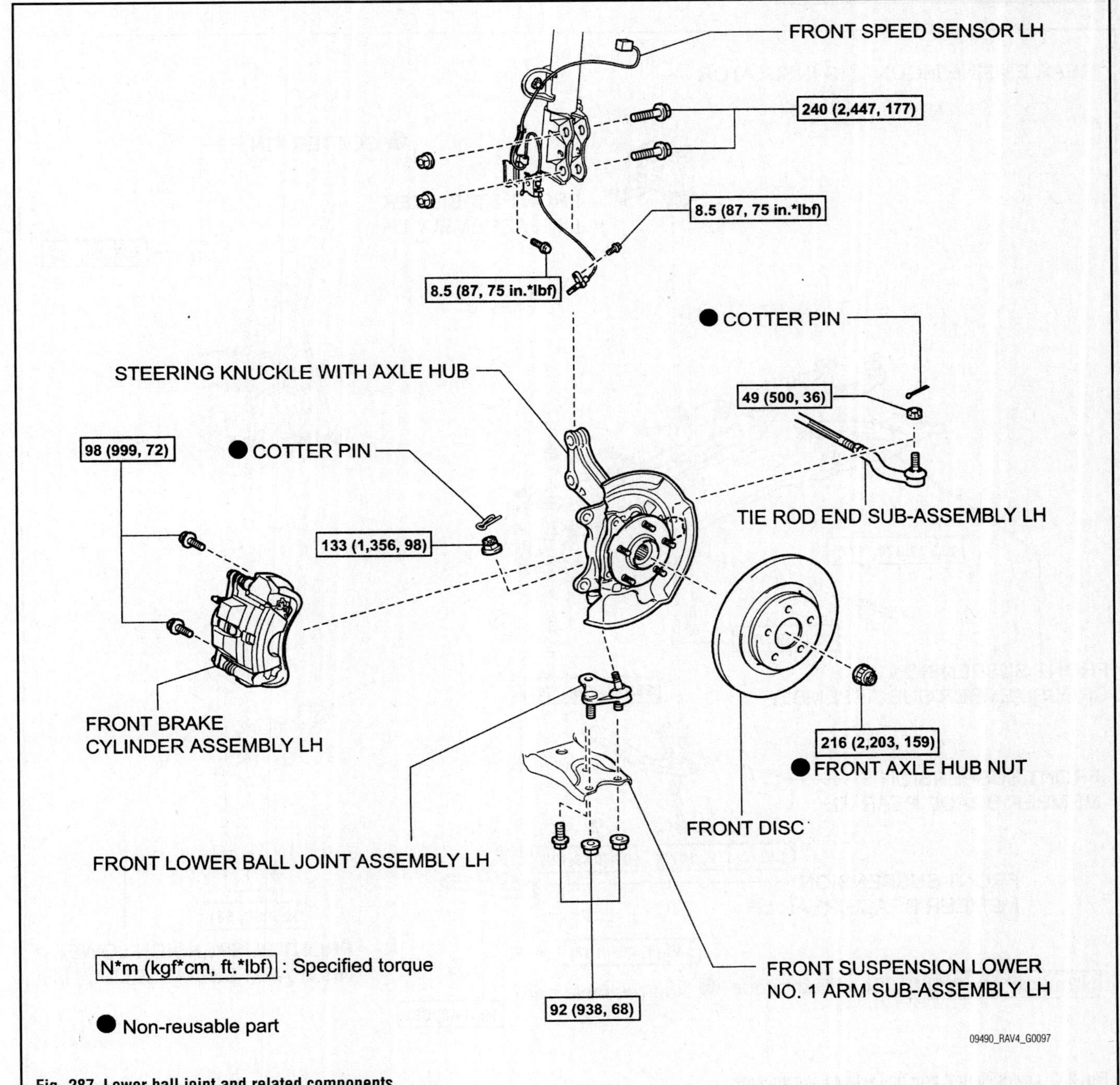

Fig. 287 Lower ball joint and related components

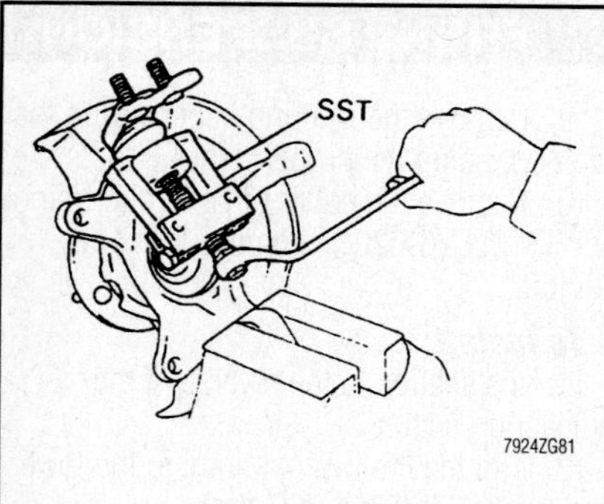

Fig. 288 Use a 2-jaw puller to remove the lower ball joint

13. Tighten the lower ball joint to the steering knuckle to 98 ft. lbs. (133 Nm).
14. Tighten the wheel lug nuts to 76 ft. lbs. (103 Nm).
15. Be sure to check and adjust the alignment, as required.

LOWER CONTROL ARM

REMOVAL & INSTALLATION

See Figures 289 and 290.

1. Before servicing the vehicle, refer to the Precautions Section.
2. Disconnect the negative battery cable.

➡Wait at least 90 seconds after disconnecting the negative battery cable before starting any repair work to prevent air bag and seat belt pretensioner activation.

3. Matchmark and remove the hood.
4. Raise and support the vehicle safely.
5. Remove the tire and wheel assembly.
6. Install engine hanger tools 12281-28010 and 12282-28010. Suspend the engine assembly using the proper engine removal tools.

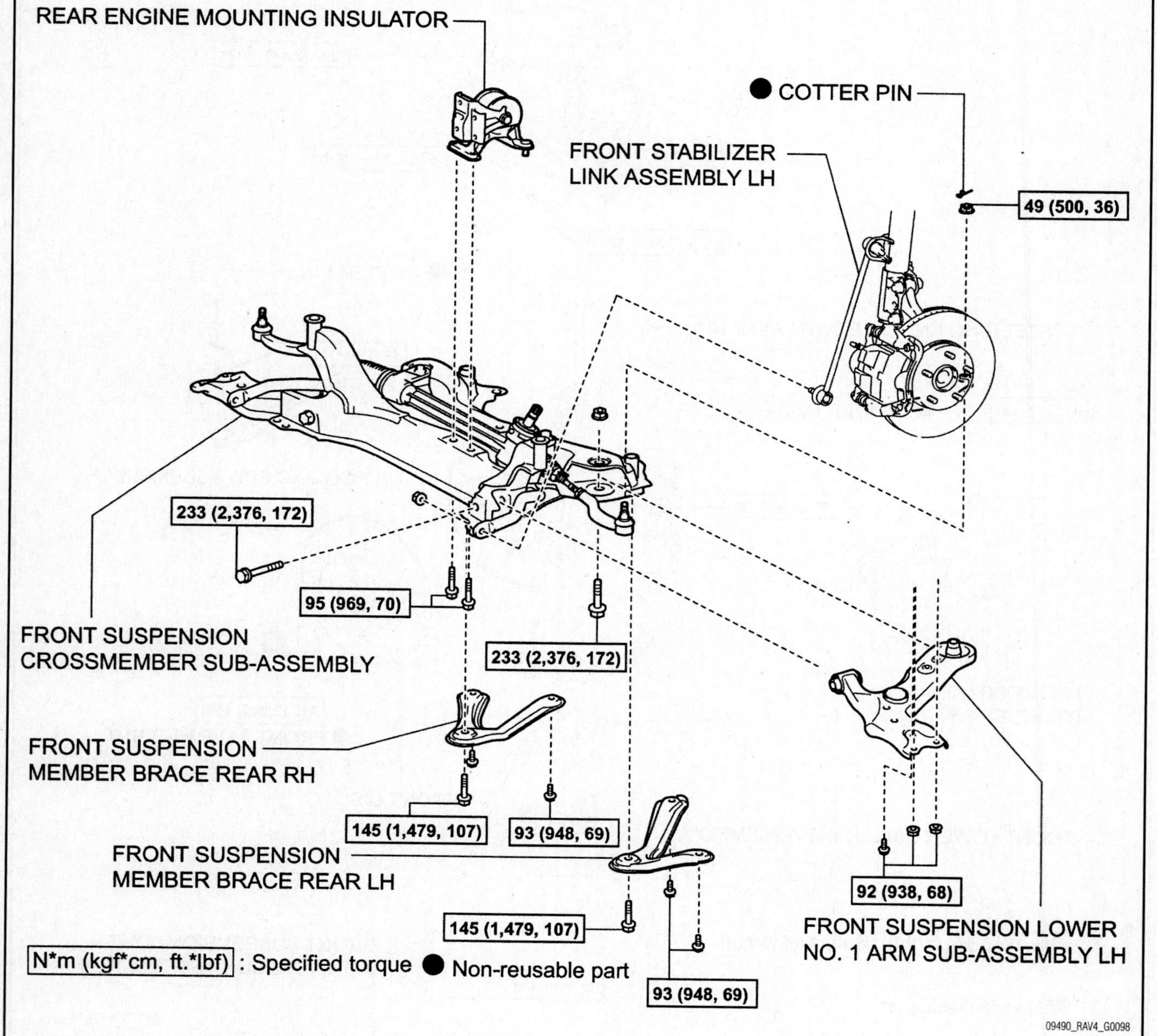

REAR ENGINE MOUNTING INSULATOR
● COTTER PIN
FRONT STABILIZER LINK ASSEMBLY LH
49 (500, 36)
233 (2,376, 172)
95 (969, 70)
FRONT SUSPENSION CROSSMEMBER SUB-ASSEMBLY
233 (2,376, 172)
FRONT SUSPENSION MEMBER BRACE REAR RH
145 (1,479, 107)
93 (948, 69)
FRONT SUSPENSION MEMBER BRACE REAR LH
92 (938, 68)
145 (1,479, 107)
FRONT SUSPENSION LOWER NO. 1 ARM SUB-ASSEMBLY LH
N*m (kgf*cm, ft.*lbf) : Specified torque ● Non-reusable part
93 (948, 69)
09490_RAV4_G0098

Fig. 289 Lower control arm and related components

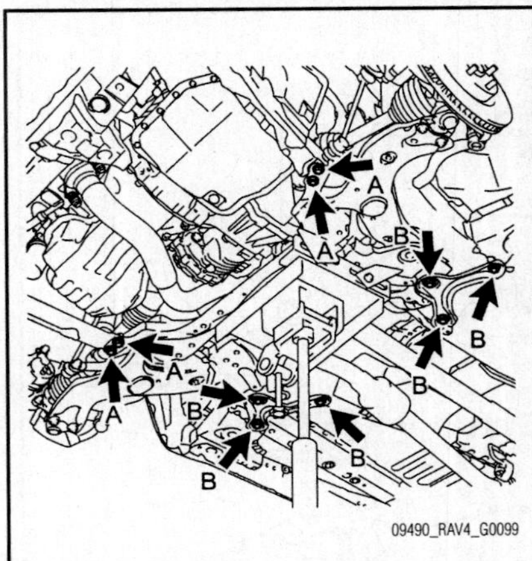

Fig. 290 Front crossmember bolt location and identification

7. Disconnect the front stabilizer links. Disconnect the front suspension lower number one arm assembly.

8. Remove the two nuts, two bolts and engine mounting rear insulator. Remove the bolt from the suspension member.

9. Support the crossmember, using a suitable jack.

10. Remove the four bolts "A" from the member reinforcement. Remove the six bolts "B" from the member reinforcement.

11. Carefully lower the jack and disconnect the crossmember from the vehicle.

12. Remove the bolt and nut from the suspension member (front). Remove the bolt and nut from the suspension member (rear). Remove the lower control arm.

To install:

13. Temporarily install the front suspension lower arm in its mounting.

14. Connect the front crossmember sub-assembly.

15. Install, but do not fully tighten, the four retaining bolts "A", the six retaining bolts "B", the suspension member with the bolt to the body and the rear mounting insulator bolts.

16. Install the front stabilizer link.

17. Connect the front suspension lower number one arm sub-assembly. Tighten the two bolts and nut to 68 ft. lbs.

18. Install the tire and wheel assembly.

19. To stabilize the suspension, lower the vehicle to ground height. Press down on the vehicle several times to stabilize the suspension.

20. Tighten the front crossmember sub-assembly retaining bolts to 64 ft. lbs. for bolt "A", 69 ft. lbs. (93 Nm) for bolt "B", 107 ft. lbs (145 Nm). for the bolt to the body and 70 ft. lbs. (95 Nm) for the rear engine insulator bolts.

21. Tighten the front suspension lower number one arm sub-assembly.

22. Continue the installation in the reverse order of the removal procedure.

23. Tighten the wheel lug nuts to 76 ft. lbs. (103 Nm).

24. Be sure to check and adjust the alignment, as required.

MACPHERSON STRUT

REMOVAL & INSTALLATION

See Figures 291 through 293.

1. Before servicing the vehicle, refer to the Precautions Section.

2. Raise and support the vehicle safely.

3. Remove the tire and wheel assembly.

4. Remove the front speed sensor.

5. Remove the stabilizer link assembly.

6. Remove the two bolts and disconnect the strut from the steering knuckle.

7. Remove the three strut upper retaining bolts.

8. Remove the strut from the vehicle.

To install:

9. Installation is the reverse of the removal procedure.

10. Tighten the three upper strut retaining nuts to 37 ft. lbs. (50 Nm).

11. Tighten the lower strut retaining bolts to 177 ft. lbs. (240 Nm).

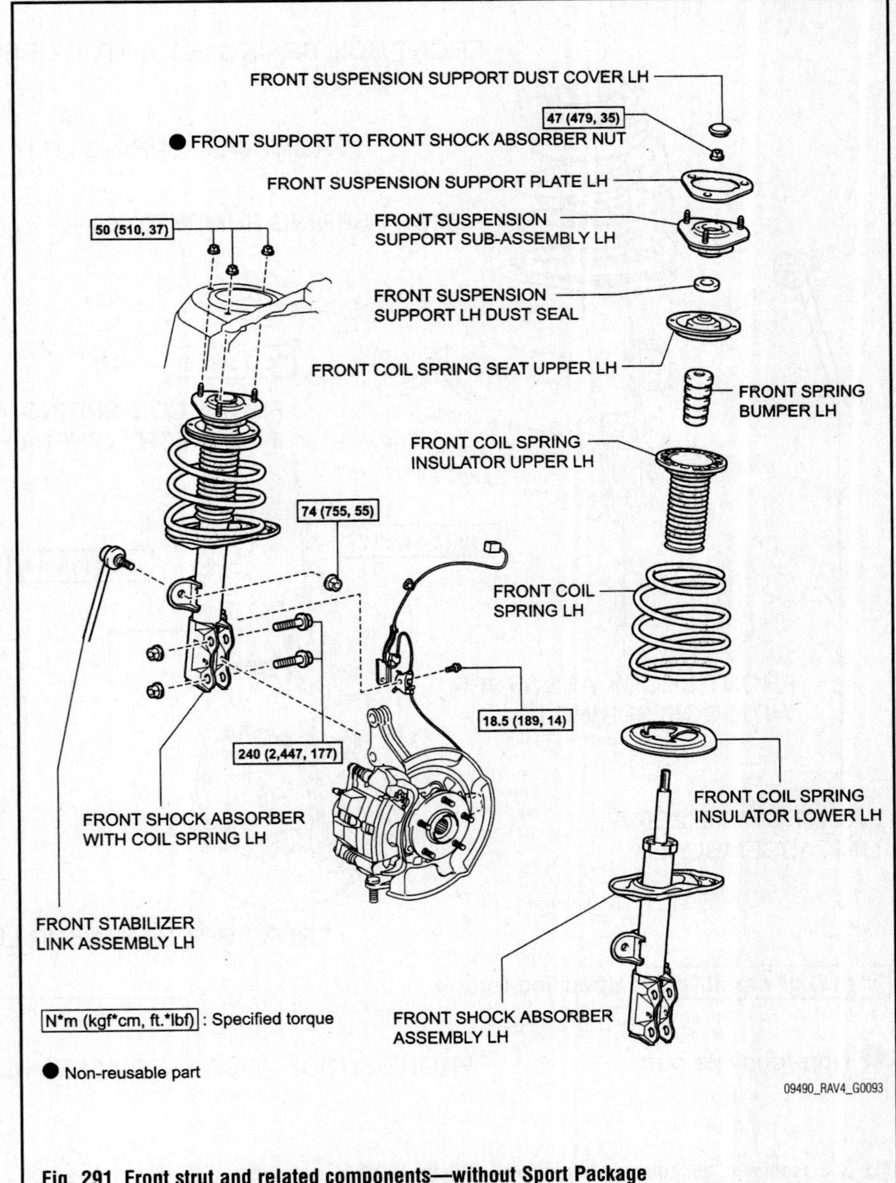

Fig. 291 Front strut and related components—without Sport Package

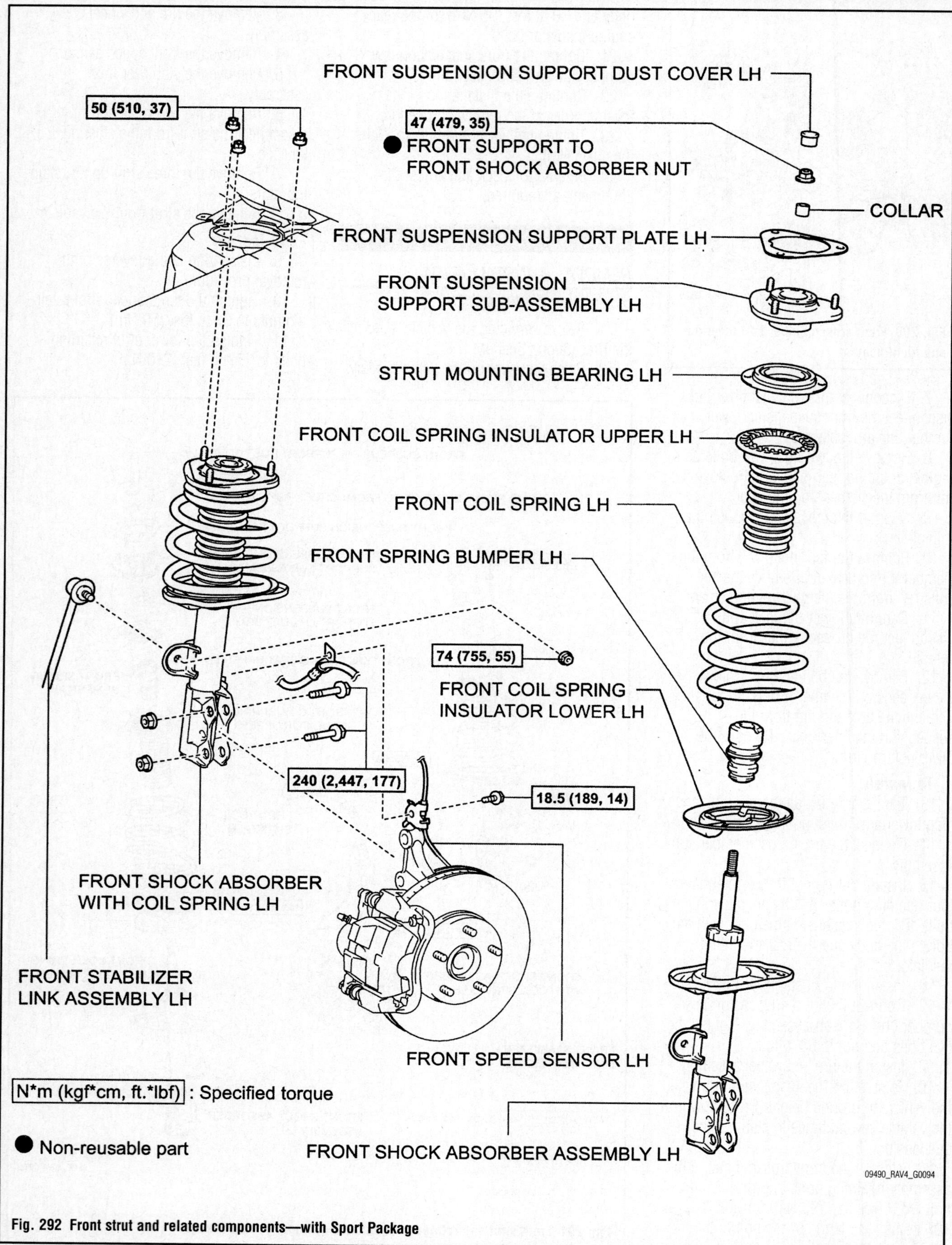

Fig. 292 Front strut and related components—with Sport Package

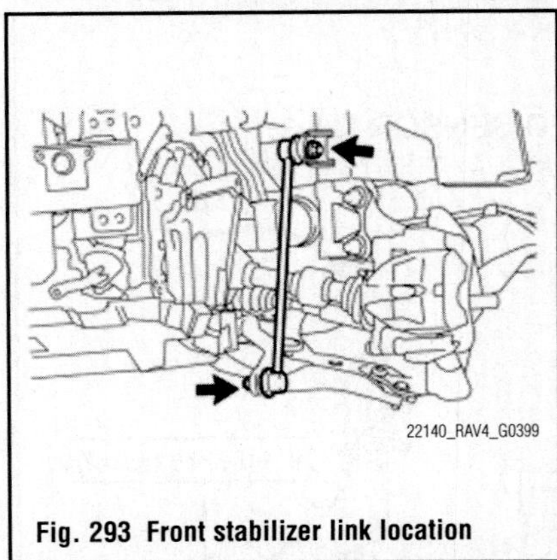

Fig. 293 Front stabilizer link location

12. Tighten the wheel lug nuts to 76 ft. lbs. (103 Nm).

13. To stabilize the suspension, lower the vehicle to ground height. Press down on the vehicle several times to stabilize the suspension.

14. Check and adjust the alignment, as required.

OVERHAUL

See Figures 294 through 297.

1. Before servicing the vehicle, refer to the Precautions Section.

2. Remove the strut from the vehicle.

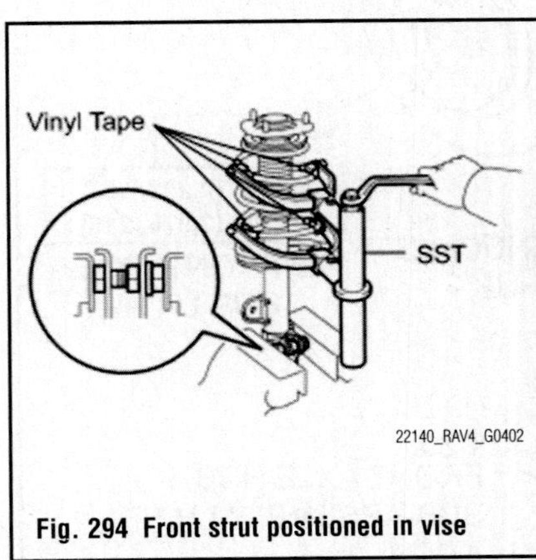

Fig. 294 Front strut positioned in vise

Fig. 295 Front strut disassembly tool installation

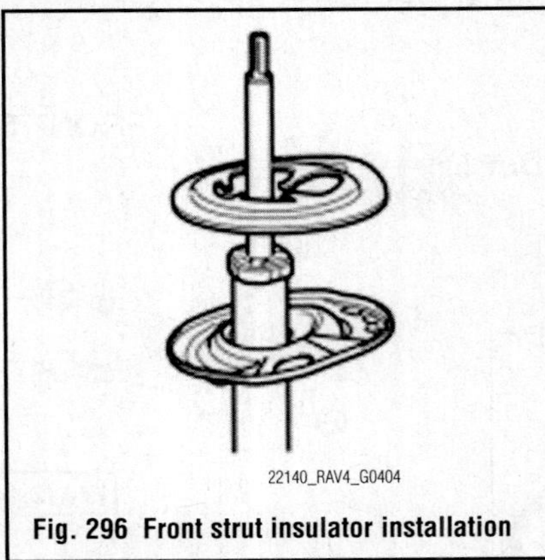

Fig. 296 Front strut insulator installation

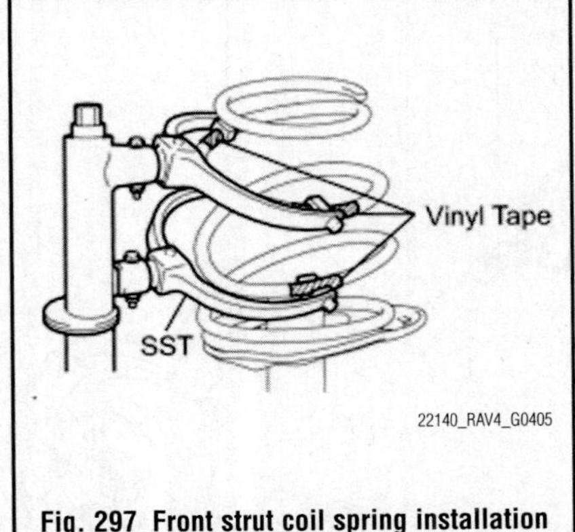

Fig. 297 Front strut coil spring installation

3. Install a nut/bolt to the bracket at the lower portion of the strut assembly and secure it in a vise.

4. Compress the coil spring with a spring compressor.

CAUTION

The proper tools must be used for this procedure. The spring on the strut is under high pressure and can cause serious injury if not properly removed and installed.

5. Remove or disconnect the following:
   - Center retaining nut, by holding the spring seat
   - Support, dust seal, spring seat, insulator and spring from the strut assembly

To install:

6. Install the spring bumper and lower insulator to the strut assembly.

7. Compress the coil spring and fit the lower end of the spring into the spring seat gap.

8. Install or connect the following:
   - Upper insulator, spring seat, dust seal, support and spring seat. Tighten the new retaining nut to 34 ft. lbs. (47 Nm).
   - Strut

9. If required, bleed the brake system and check for leaks.

10. Check and/or adjust the front wheel alignment.

STEERING KNUCKLE

REMOVAL & INSTALLATION

See Figures 298 and 299.

1. Before servicing the vehicle, refer to the Precautions Section.

2. Disconnect the negative battery cable.

➡Wait at least 90 seconds after disconnecting the negative battery cable before starting any repair work to prevent air bag and seat belt pretensioner activation.

3. Raise and support the vehicle safely.

4. Drain the transaxle fluid.

5. Remove the tire and wheel assembly. Remove the front axle hub nut.

6. Remove the front speed sensor. Remove the brake caliper. Remove the rotor.

7. Disconnect the tie rod end, using the proper tools.

8. Disconnect the front suspension number one lower arm sub-assembly.

9. Remove the two bolts and two nuts. Disconnect the strut from the steering knuckle.

10. Matchmark the halfshaft and the axle hub.

11. Remove the steering knuckle with the axle hub.

➡Be careful not to damage the boot and the speed sensor rotor. Do not excessively push out the halfshaft from the axle assembly.

12. Remove the four bolts and the axle hub from the steering knuckle. Remove the dust cover from the steering knuckle.

➡Do not place the hub and bearing's magnet rotor side so that it is facing downward. Do not allow the magnet rotor side to become damaged or contact foreign matter.

To install:

13. Installation is the reverse of the removal procedure.

14. Tighten the four axle hub bolts to 71 ft. lbs. (96 Nm).

15. Tighten the two steering knuckle to axle hub bolts to 177 ft. lbs. (240 Nm).

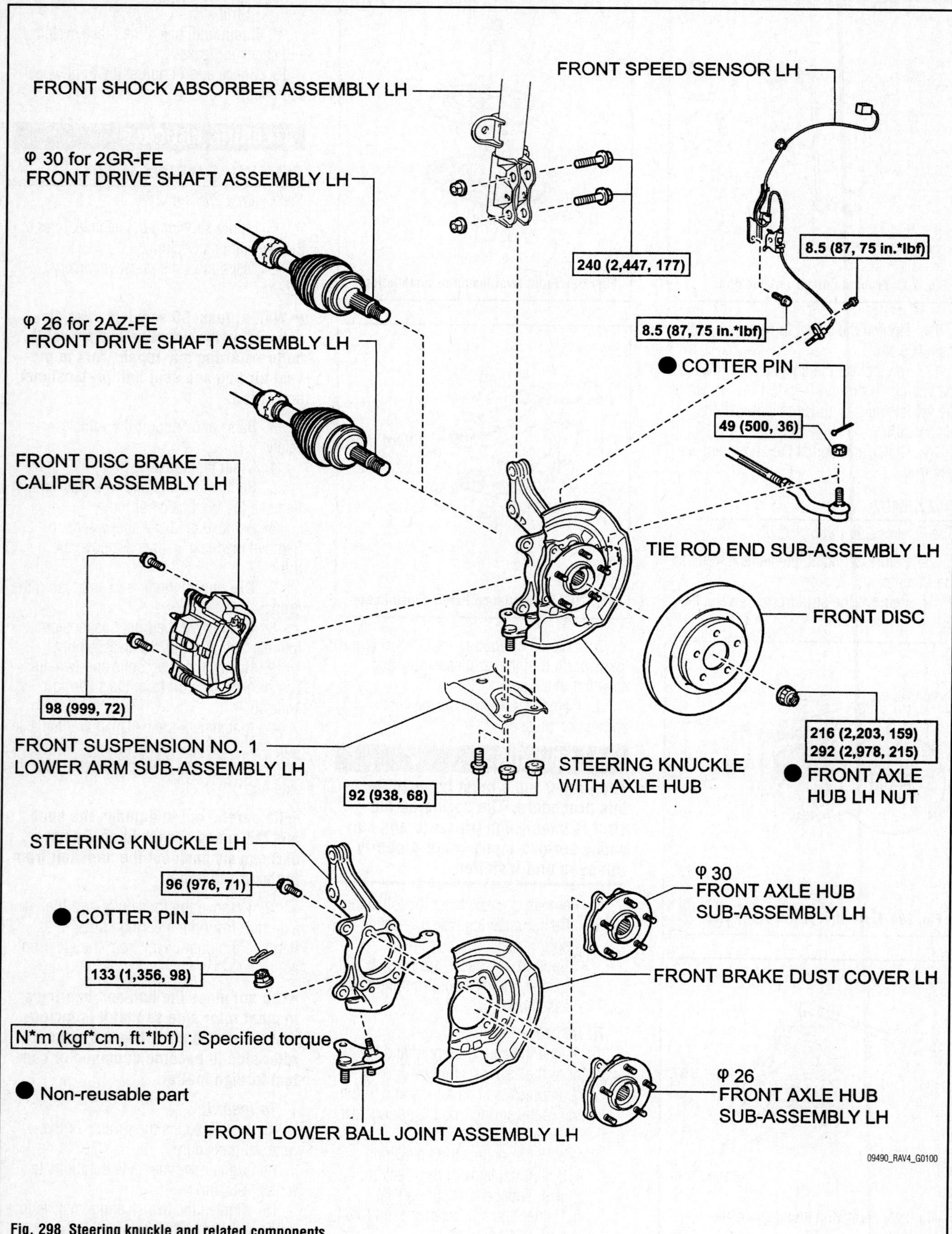

Fig. 298 Steering knuckle and related components

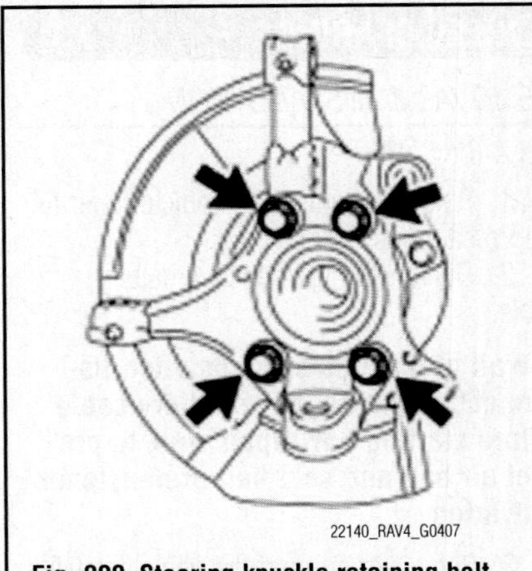

Fig. 299 Steering knuckle retaining bolt locations

16. Tighten the front axle hub nut to 159 ft. lbs. (215 Nm) for vehicles with the 2.4L engine, for vehicles equipped with the 3.5L engine tighten the nut to 215 ft. lbs. (291 Nm).
17. Be sure to refill the transaxle with the proper grade and type transaxle fluid.
18. Start the engine and check for leaks, correct as required.
19. Using the Techstream tool, or equivalent check the speed sensor signal.
20. Check and adjust the alignment, as required.

STABILIZER BAR

REMOVAL & INSTALLATION

See Figures 300 and 301.

1. Before servicing the vehicle, refer to the Precautions Section.
2. Raise and support the vehicle safely.
3. Remove the tire and wheel assembly.
4. Remove the stabilizer bar link retaining nuts.
5. Remove the four bolts and remove the left front suspension member brace.
6. Remove the four bolts and remove the right front suspension member brace.
7. Remove the stabilizer bar from the crossmember.

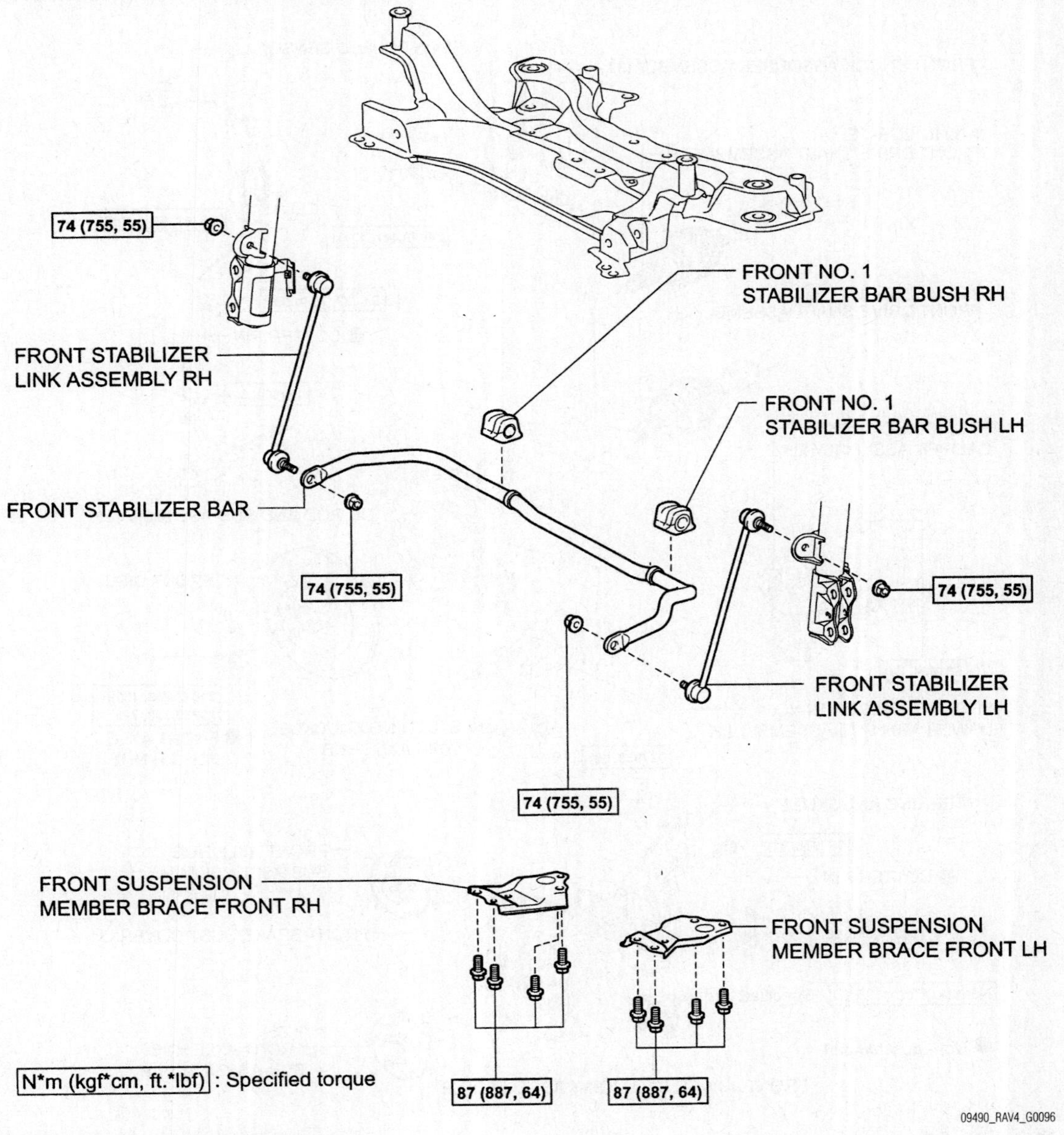

Fig. 300 Front stabilizer bar and related components

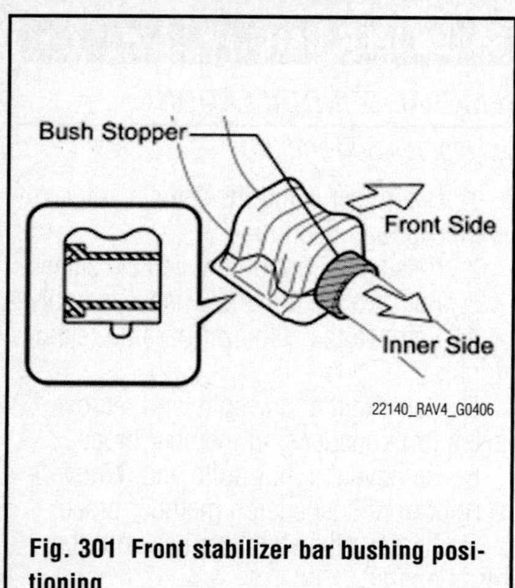

Fig. 301 Front stabilizer bar bushing positioning

8. Remove the bushings from the stabilizer bar.

To install:

9. Installation is the reverse order of the removal procedure.
10. Install the bushings to the inner side of each bushing stopper on the stabilizer bar.
11. Install the bushing with its slit facing the vehicle rear side.
12. Tighten the stabilizer bar link nuts to 55 ft. lbs. (74 Nm).
13. Tighten the front suspension member brace bolts to 64 ft. lbs. (87 Nm).
14. To stabilize the suspension, lower the vehicle to ground height. Press down on the vehicle several times to stabilize the suspension.

WHEEL BEARINGS

REMOVAL & INSTALLATION

See Figure 302.

1. Before servicing the vehicle, refer to the Precautions Section.
2. Disconnect the negative battery cable.

➡Wait at least 90 seconds after disconnecting the negative battery cable before starting any repair work to prevent air bag and seat belt pretensioner activation.

3. Raise and support the vehicle safely.
4. Drain the transaxle fluid.

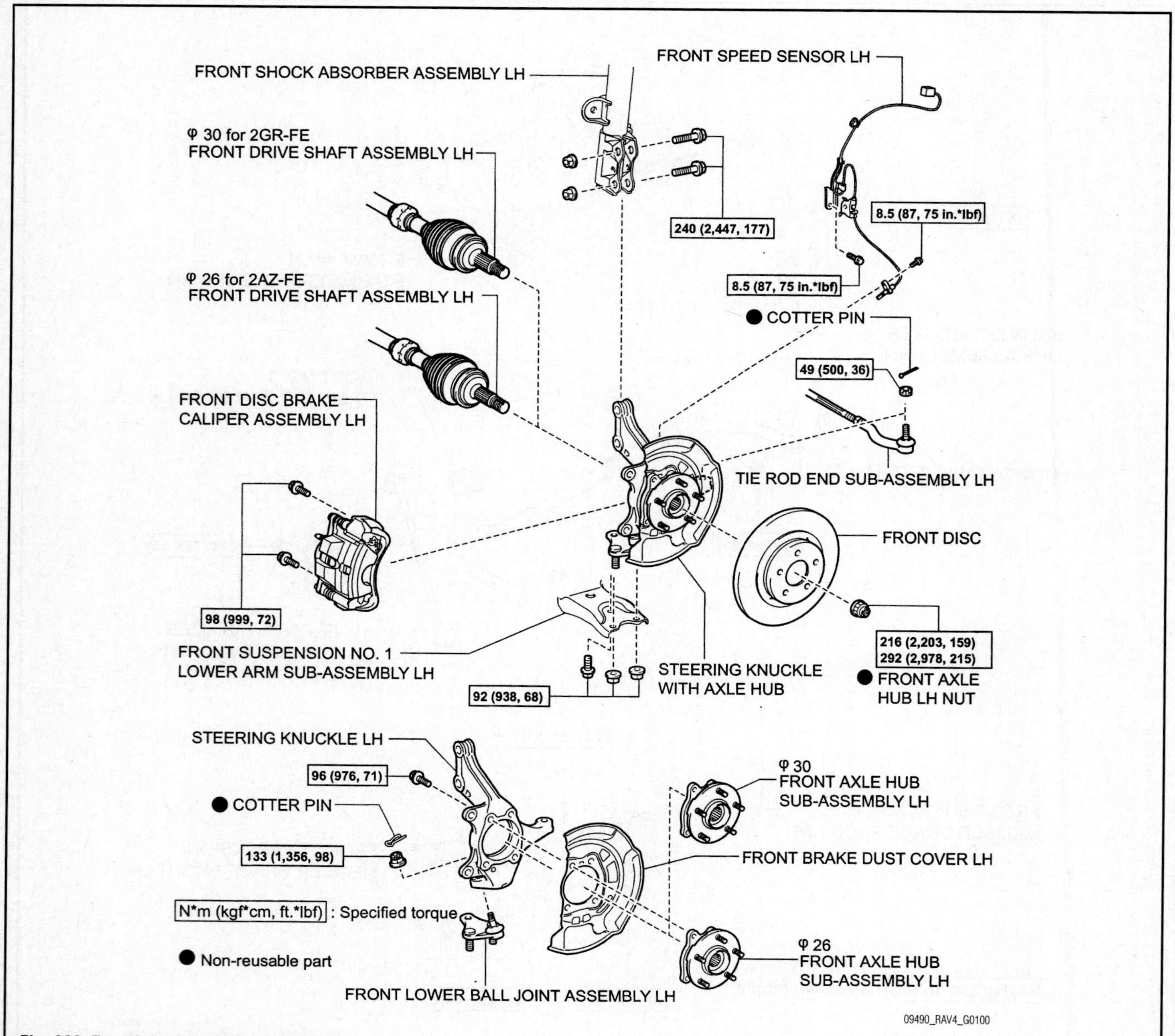

Fig. 302 Front hub and related components

5. Remove the tire and wheel assembly. Remove the front axle hub nut.

6. Remove the front speed sensor. Remove the brake caliper. Remove the rotor.

7. Disconnect the tie rod end, using the proper tools.

8. Disconnect the front suspension number one lower arm sub-assembly.

9. Remove the two bolts and two nuts. Disconnect the strut from the steering knuckle.

10. Matchmark the halfshaft and the axle hub.

11. Remove the steering knuckle with the axle hub.

➡Be careful not to damage the boot and the speed sensor rotor. Do not excessively push out the halfshaft from the axle assembly.

12. Remove the four bolts and the axle hub from the steering knuckle. Remove the dust cover from the steering knuckle.

➡Do not place the hub and bearing's magnet rotor side so that it is facing downward. Do not allow the magnet rotor side to become damaged or contact foreign matter.

To install:

13. Installation is the reverse of the removal procedure.

14. Tighten the four axle hub bolts to 71 ft. lbs. (96 Nm).

15. Tighten the two steering knuckle to axle hub bolts to 177 ft. lbs. (240 Nm).

16. Tighten the front axle hub nut to 159 ft. lbs. (215 Nm) for vehicles equipped with the 2.4L engine; for vehicles equipped with the 3.5L engine tighten the nut to 215 ft. lbs. (291 Nm).

17. Be sure to refill the transaxle with the proper grade and type transaxle fluid.

18. Start the engine and check for leaks, correct as required.

19. Using the Techstream tool, or equivalent check the speed sensor signal.

20. Check and adjust the alignment, as required.

ADJUSTMENT

See Figure 303.

1. Before servicing the vehicle, refer to the Precautions Section.

2. Disconnect the negative battery cable.

➡Wait at least 90 seconds after disconnecting the negative battery cable before starting any repair work to prevent air bag and seat belt pretensioner activation.

3. Raise and support the vehicle safely.

4. Remove the tire and wheel assembly.

5. Remove the brake caliper. Remove the rotor.

6. Using a dial indicator tool, check the backlash near the center of the axle hub. Maximum backlash should be 0.0020 inch.

7. If backlash is greater than specification, replace the bearing.

8. Using a dial indicator tool, check the deviation on the surface of the axle hub. Maximum backlash should be 0.0020 inch.

9. If backlash is greater than specification, replace the bearing.

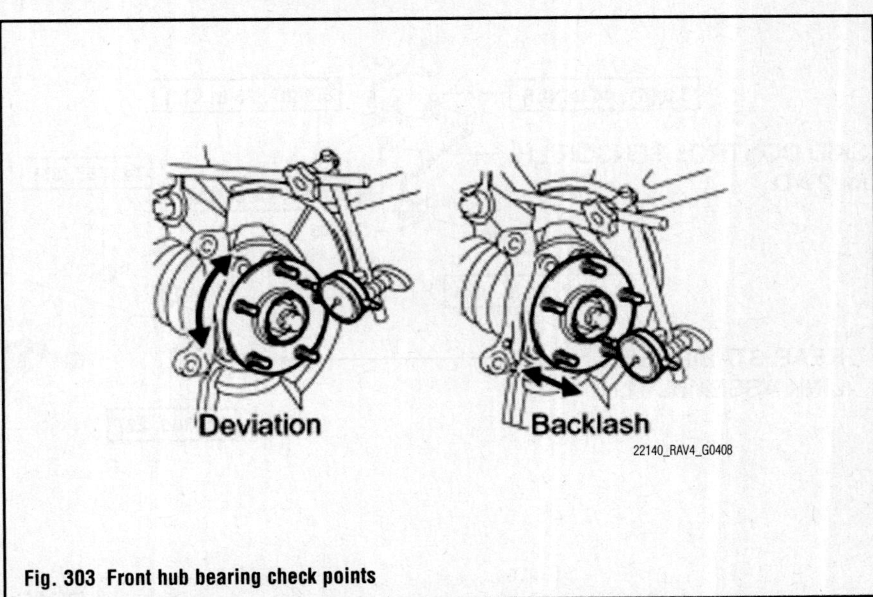

Fig. 303 Front hub bearing check points

COIL SPRING

REMOVAL & INSTALLATION

See Figure 304.

1. Before servicing the vehicle, refer to the Precautions Section.
2. Raise and support the vehicle safely.
3. Remove the tire and wheel assembly.
4. On 2WD, remove the skid control sensor wire.
5. On 4WD, remove the rear speed sensor wire.
6. Disconnect the number two parking brake cable assembly.
7. Disconnect the rear stabilizer link assembly.
8. To disconnect the rear suspension number two arm assembly, loosen the bolt from the suspension member side. Support

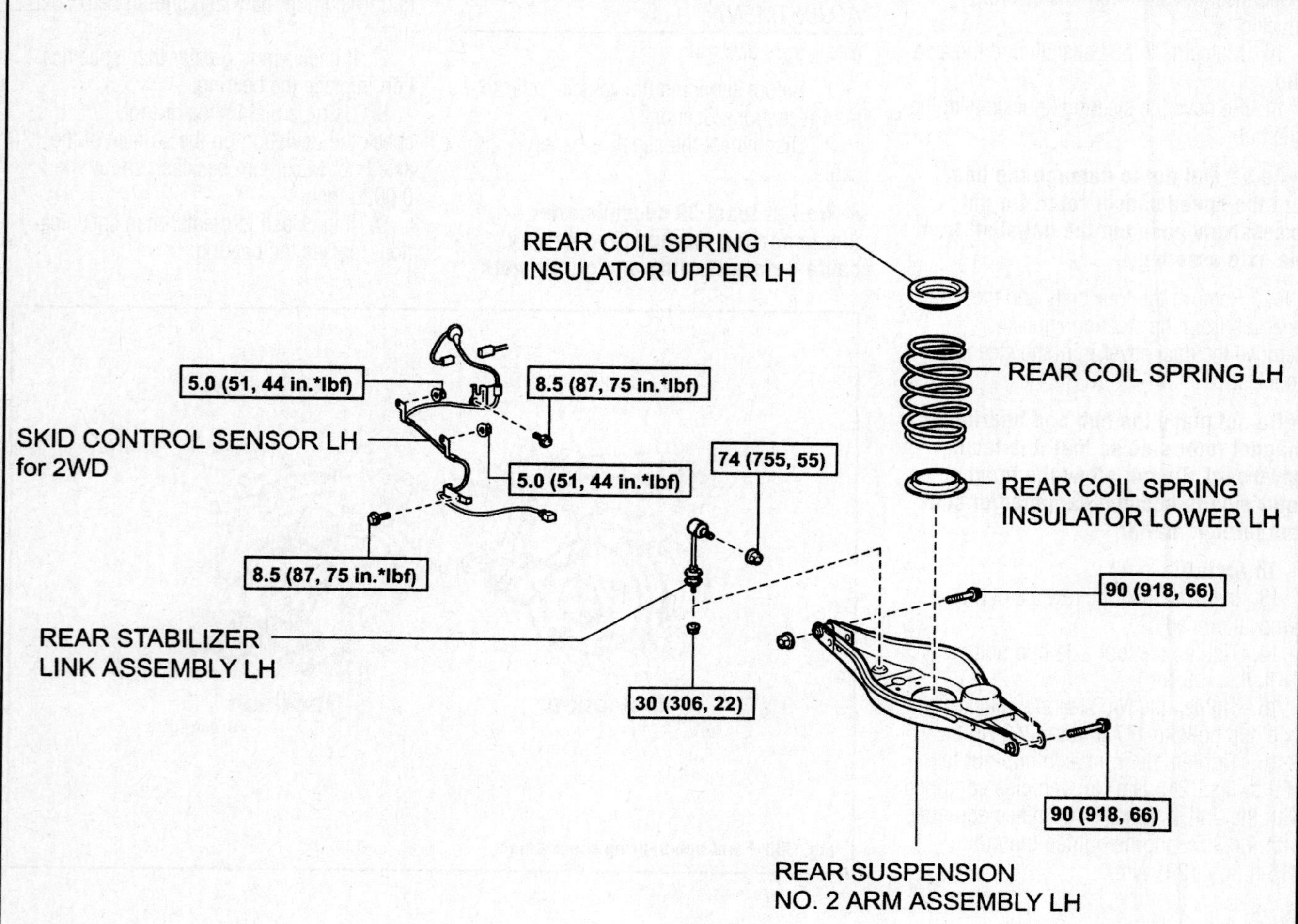

Fig. 304 Rear coil spring and related components

the number two suspension arm with a suitable jack.

➡**Do not remove the bolt, only loosen it.**

9. Remove the bolt and nut from the axle carrier side. Slowly lower the jack and disconnect the number two suspension arm from the axle carrier.

10. Remove the upper spring insulator. Remove the spring. Remove the lower insulator.

To install:

11. Installation is the reverse of the removal procedure.

12. Do not apply final tightening torque to the rear suspension number two arm assembly until the suspension is stabilized.

13. To stabilize the suspension, lower the vehicle to ground height. Press down on the vehicle several times to stabilize the suspension.

14. Check and adjust the alignment, as required.

CONTROL ARMS/LINKS

REMOVAL & INSTALLATION

See Figures 305 and 306.

1. Before servicing the vehicle, refer to the Precautions Section.
2. Raise and support the vehicle safely.
3. Remove the tire and wheel assembly.

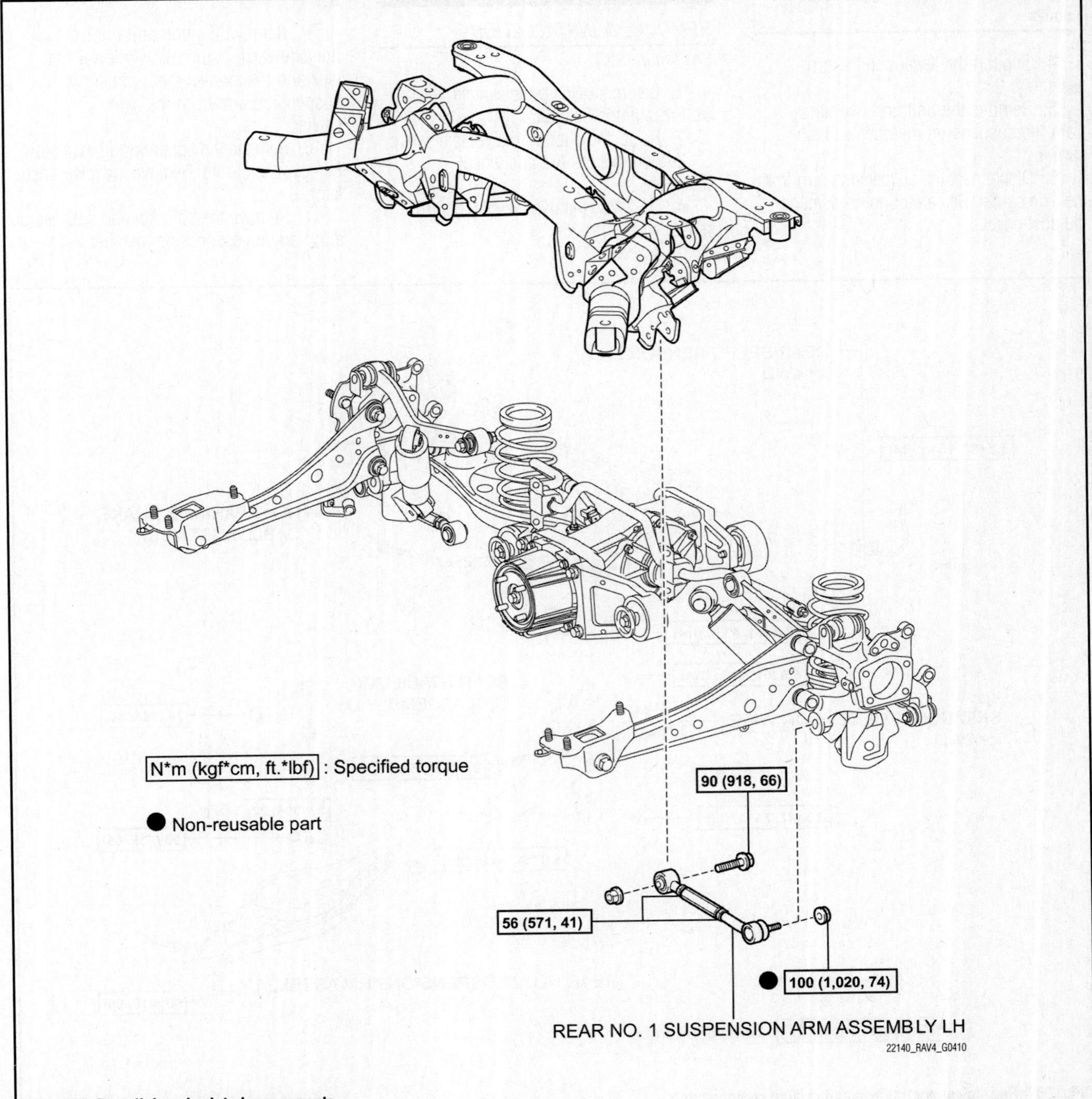

Fig. 305 Rear link and related components

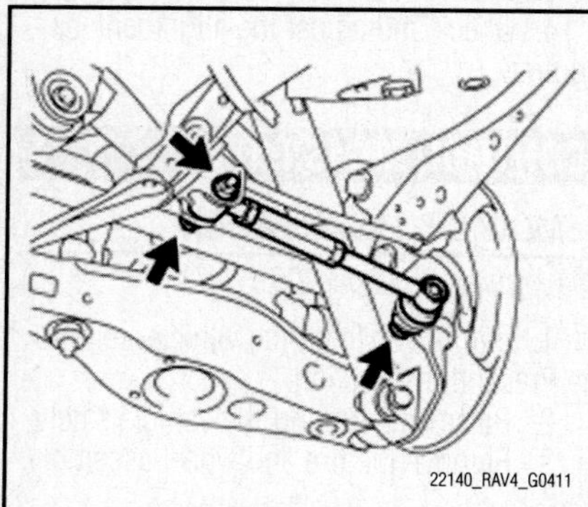

Fig. 306 Rear link retaining bolt location points

4. Support the lower control arm assembly.

5. Remove the bolt and two nuts from the suspension member and axle carrier.

6. Disconnect the suspension arm from the axle carrier. Be careful not to damage the dust cover.

To install:

7. Installation is the reverse of the removal procedure.

8. Do not apply final tightening torque to the rear suspension number two arm assembly until the suspension is stabilized.

9. Tighten the bolts to 66 ft. lbs. (89 Nm). Tighten the nut to 74 ft. lbs. (100 Nm).

10. Tighten the wheel lug nuts to 76 ft. lbs. (103 Nm).

LOWER CONTROL ARM

REMOVAL & INSTALLATION

See Figure 307.

1. Before servicing the vehicle, refer to the Precautions Section.

2. Raise and support the vehicle safely.

3. Remove the tire and wheel assembly.

4. On 2WD, remove the skid control sensor wire.

5. On 4WD, remove the rear speed sensor wire.

6. Disconnect the number two parking brake cable assembly.

7. Disconnect the rear stabilizer link assembly.

8. To disconnect the rear suspension number two arm assembly, loosen the bolt from the suspension member side. Support the number two suspension arm with a suitable jack.

➡Do not remove the bolt, only loosen it.

9. Remove the bolt and nut from the axle carrier side. Slowly lower the jack and disconnect the number two suspension arm from the axle carrier.

10. Remove the upper spring insulator. Remove the spring. Remove the lower insulator.

11. Remove the bolt, nut and suspension arm from the suspension member.

REAR SPEED SENSOR LH for 4WD
8.5 (87, 75 in.*lbf)
NO. 2 PARKING BRAKE CABLE ASSEMBLY
6.0 (61, 53 in.*lbf)
5.0 (51, 44 in.*lbf)
REAR STABILIZER LINK ASSEMBLY LH
SKID CONTROL SENSOR WIRE for 2WD
74 (755, 55)
8.5 (87, 75 in.*lbf)
30 (306, 22)
8.5 (87, 75 in.*lbf)
90 (918, 66)
5.0 (51, 44 in.*lbf)
REAR NO. 2 SUSPENSION ARM ASSEMBLY LH
90 (918, 66)
N*m (kgf*cm, ft.*lbf) : Specified torque
22140_RAV4_G0409

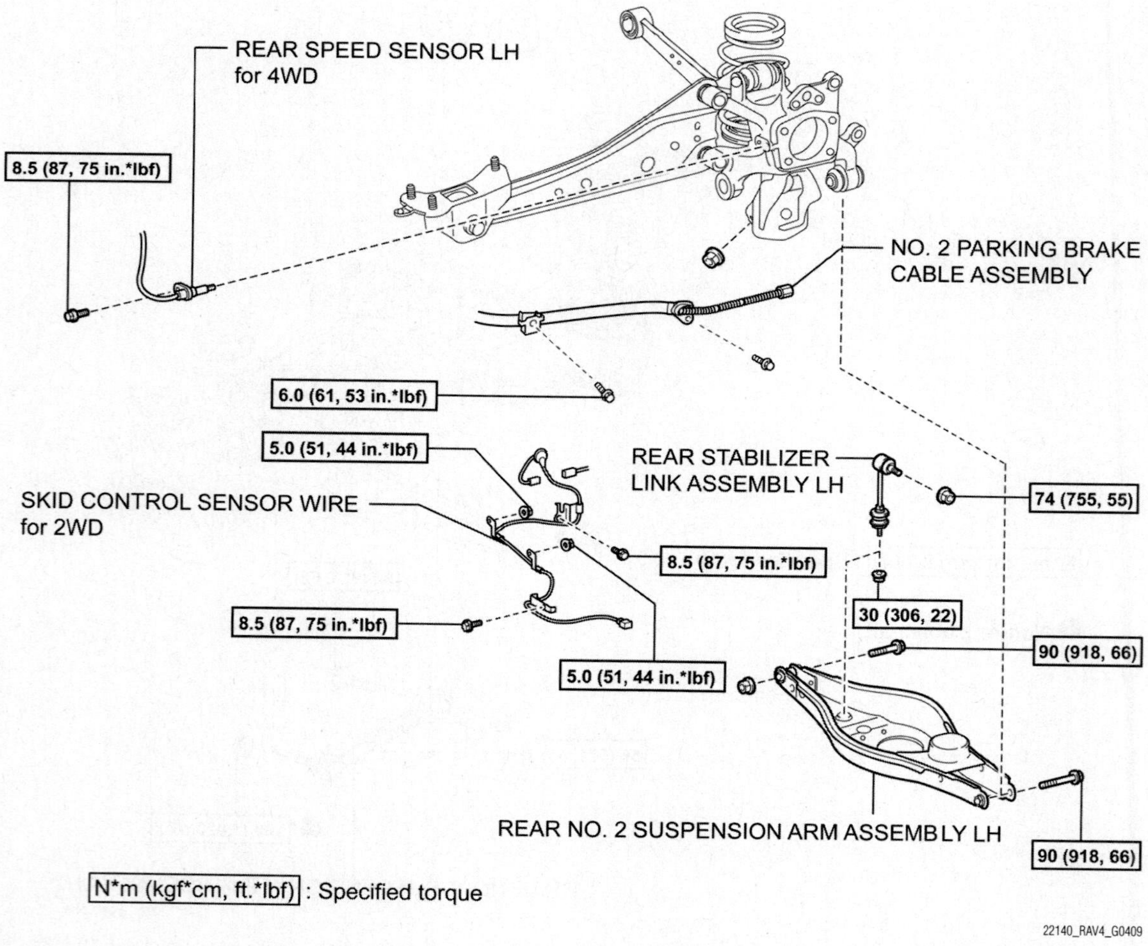

Fig. 307 Rear lower control arm and related components

To install:

12. Installation is the reverse of the removal procedure.
13. Do not apply final tightening torque to the rear suspension number two arm assembly until the suspension is stabilized.
14. To stabilize the suspension, lower the vehicle to ground height. Press down on the vehicle several times to stabilize the suspension.
15. Check and adjust the alignment, as required.

SHOCK ABSORBER

REMOVAL & INSTALLATION

See Figure 308.

1. Before servicing the vehicle, refer to the Precautions Section.
2. Disconnect the negative battery cable.

➡Wait at least 90 seconds after disconnecting the negative battery cable before starting any repair work to prevent air bag and seat belt pretensioner activation.

3. Raise and support the vehicle safely.
4. Remove the tire and wheel assembly.
5. Support the number two suspension arm, using a suitable jack.
6. Remove the bolt and two nuts from the suspension member and axle carrier.
7. Remove the two bolts and disconnect the shock absorber with the bracket.
8. Remove the nut and bolt from the shock absorber upper side.
9. Remove the shock absorber from the vehicle.

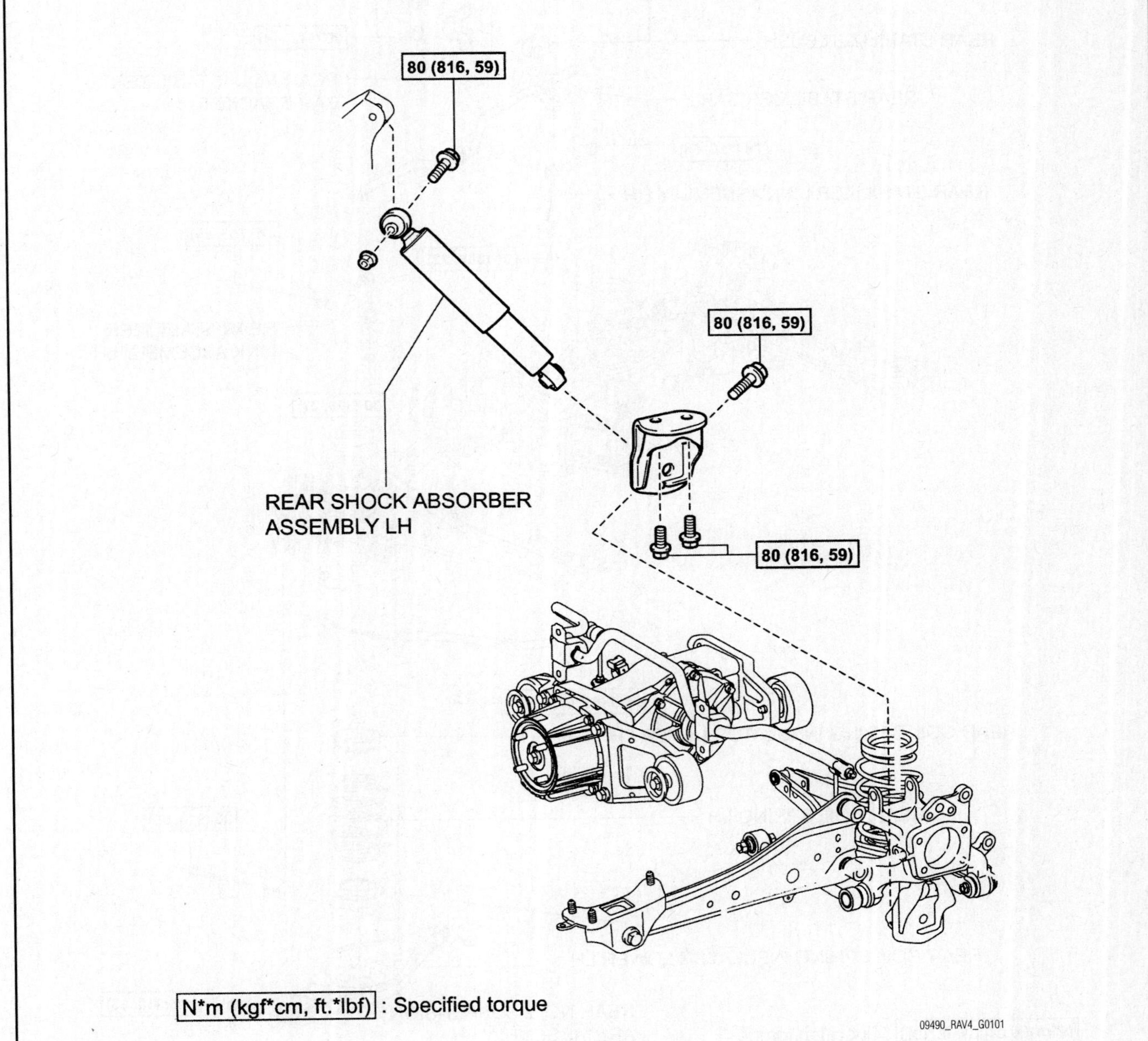

Fig. 308 Rear shock absorber and related components

To install:

10. Installation is the reverse of the removal procedure.
11. Do not apply final tightening torque until the suspension is stabilized.
12. To stabilize the suspension, lower the vehicle to ground height. Press down on the vehicle several times to stabilize the suspension.

STABILIZER BAR

REMOVAL & INSTALLATION

See Figure 309.

1. Before servicing the vehicle, refer to the Precautions Section.
2. Raise and support the vehicle safely.
3. Remove the tire and wheel assembly.
4. Remove the nut and disconnect the link from the suspension number two arm. Remove the nut and the link from the stabilizer bar.
5. Remove the rear number two suspension arms (lower control arm).
6. Remove the coil springs.
7. Remove the stabilizer bracket retaining bolts. Remove the stabilizer bar

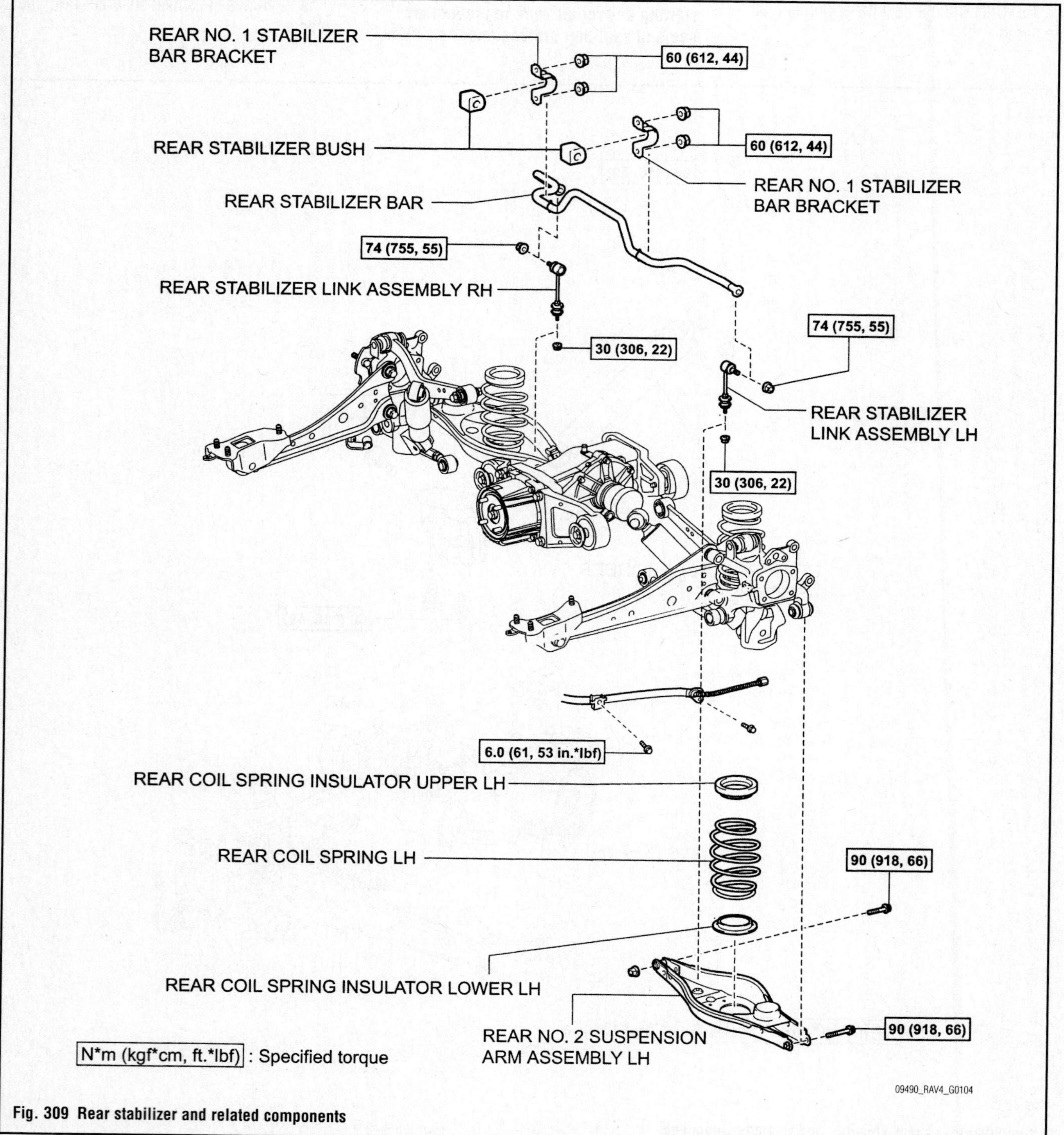

Fig. 309 Rear stabilizer and related components

from the vehicle. Remove the bushings from the bar.

➡**When removing the bar be sure not to damage the sensor wire, brake hose etc.**

To install:

8. Installation is the reverse of the removal procedure.
9. Tighten the stabilizer bar links to 55 ft. lbs. (74 Nm).
10. Tighten the stabilizer bracket and bushing bolts to 44 ft. lbs. (60 Nm).
11. Install each bushing to the outer side of the bushing stopper on each stabilizer bar. Install each bushing with its slit facing the vehicle front side.
12. Do not apply final tightening torque until the suspension is stabilized.
13. To stabilize the suspension, lower the vehicle to ground height. Press down on the vehicle several times to stabilize the suspension.
14. Check and adjust the alignment, as required.

UPPER CONTROL ARM

REMOVAL & INSTALLATION

See Figure 310.

1. Before servicing the vehicle, refer to the Precautions Section.
2. Raise and support the vehicle safely.
3. Remove the tire and wheel assembly.

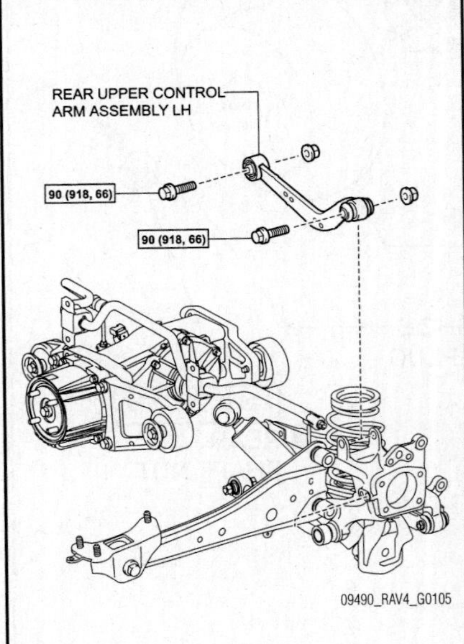

Fig. 310 Rear upper control arm and related components

4. On 2WD, remove the skid control sensor wire.
5. On 4WD, remove the rear speed sensor wire.
6. Remove the upper control arm retaining bolts. Remove the upper control arm from the vehicle.

To install:

7. Installation is the reverse of the removal procedure.
8. Do not apply final tightening torque to the component until the suspension is stabilized.
9. To stabilize the suspension, lower the vehicle to ground height. Press down on the vehicle several times to stabilize the suspension.
10. Check and adjust the alignment, as required.

WHEEL HUB & BEARING

REMOVAL & INSTALLATION

See Figures 311 through 313.

1. Before servicing the vehicle, refer to the Precautions Section.
2. Properly relieve the fuel system pressure.
3. Disconnect the negative battery cable.

➡**Wait at least 90 seconds after disconnecting the negative battery cable before starting any repair work to**

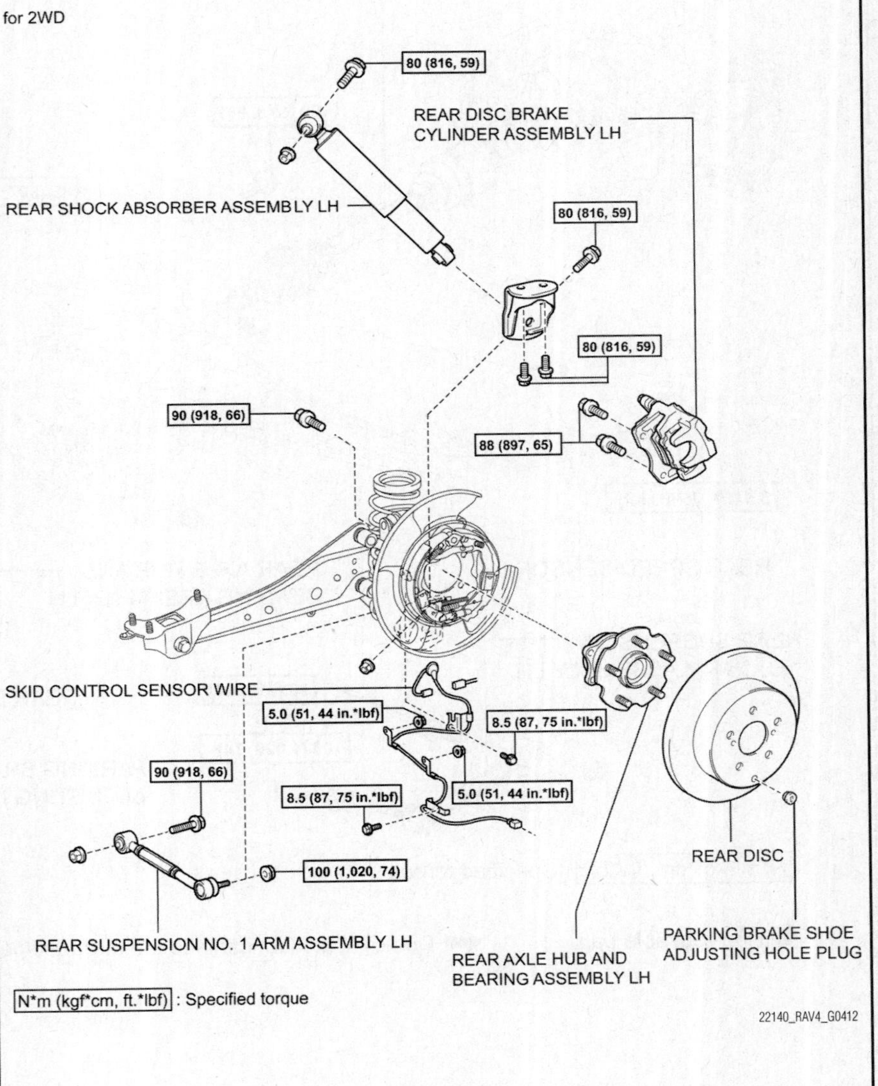

Fig. 311 Rear hub and related components—2WD vehicles

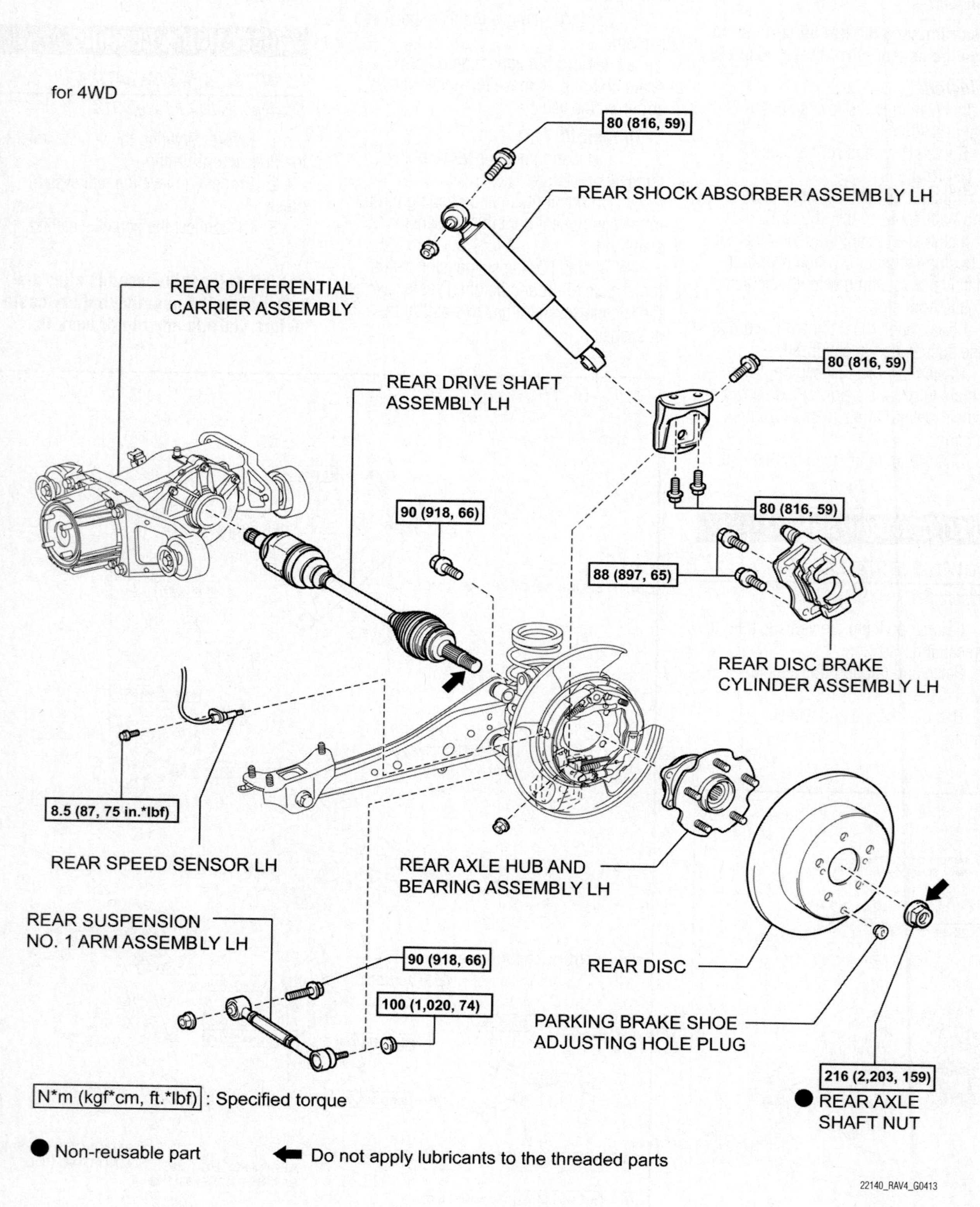

Fig. 312 Rear hub and related components—4WD vehicles

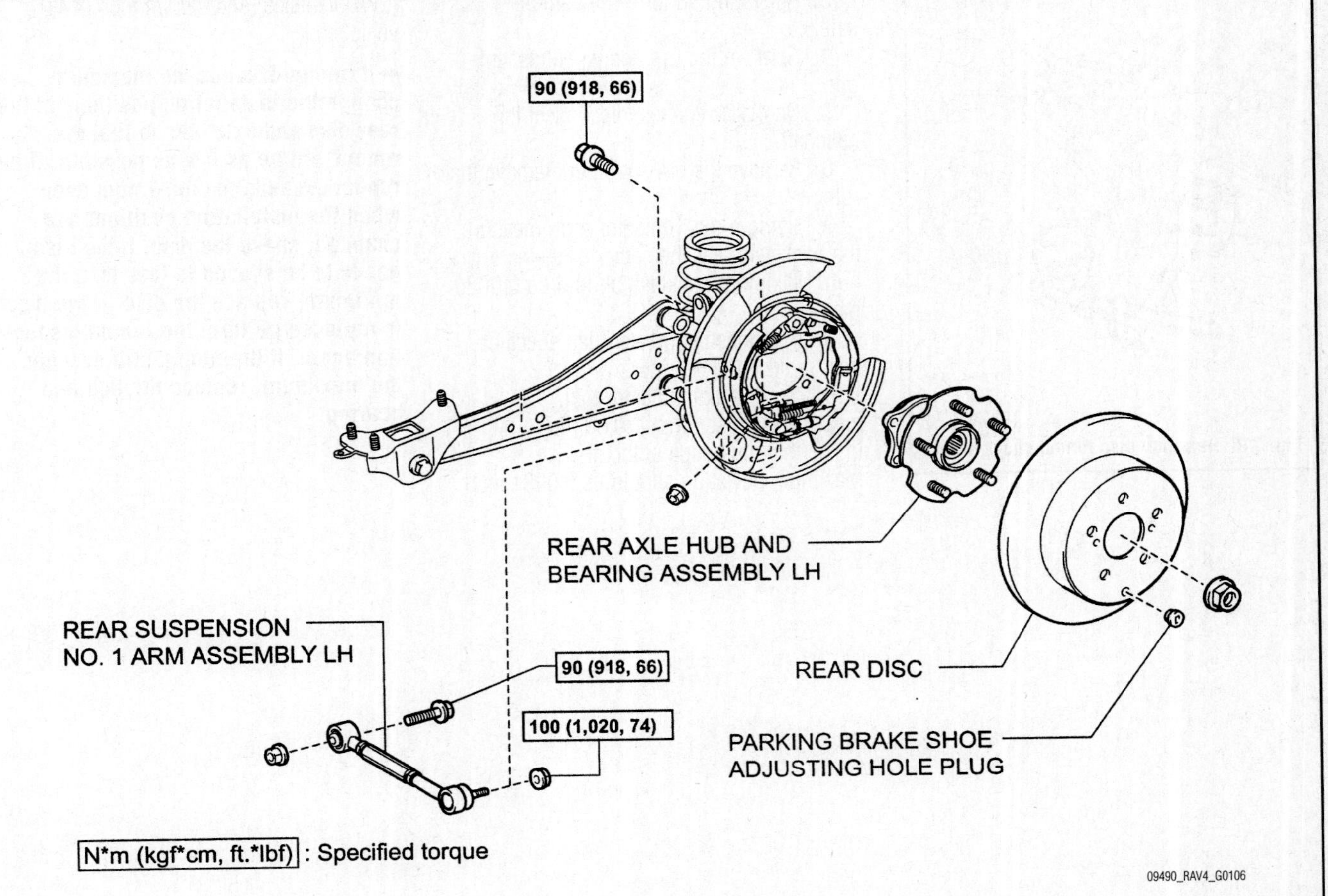

90 (918, 66)

REAR AXLE HUB AND BEARING ASSEMBLY LH

REAR SUSPENSION NO. 1 ARM ASSEMBLY LH

90 (918, 66)

100 (1,020, 74)

REAR DISC

PARKING BRAKE SHOE ADJUSTING HOLE PLUG

N*m (kgf*cm, ft.*lbf) : Specified torque

09490_RAV4_G0106

Fig. 313 Rear hub and bearing assembly

prevent air bag and seat belt pretensioner activation.

4. Raise and support the vehicle safely.
5. Remove the tire and wheel assembly.
6. On 4WD, remove the rear axle shaft nut.
7. Remove the caliper. Remove the rotor.
8. On 2WD remove the skid control sensor wire.
9. On 4WD, remove the rear speed sensor.
10. Remove the rear suspension number one arm.
11. Disconnect the shock absorber from the axle carrier.
12. On 2WD, remove the four bolts, and the axle hub and bearing from the axle carrier.
13. On 4WD, matchmark the halfshaft and the axle hub and bearing. Remove the four bolts, and the axle hub and bearing from the axle carrier.

➡Do not place the hub and bearing's magnet rotor side so that it is facing downward. Do not allow the magnet rotor side to become damaged or contact foreign matter.

To install:

14. Installation is the reverse of the removal procedure.
15. Tighten the four hub and bearing bolts to 68 ft. lbs. (92 Nm).
16. Do not apply final tightening torque to the component until the suspension is stabilized.
17. To stabilize the suspension, lower the vehicle to ground height. Press down on the vehicle several times to stabilize the suspension.
18. Check and adjust the alignment, as required.

ADJUSTMENT

See Figures 314 and 315.

1. Before servicing the vehicle, refer to the Precautions Section.
2. Disconnect the negative battery cable.

➡Wait at least 90 seconds after disconnecting the negative battery cable before starting any repair work to prevent air bag and seat belt pretensioner activation.

Fig. 314 Rear hub backlash check point

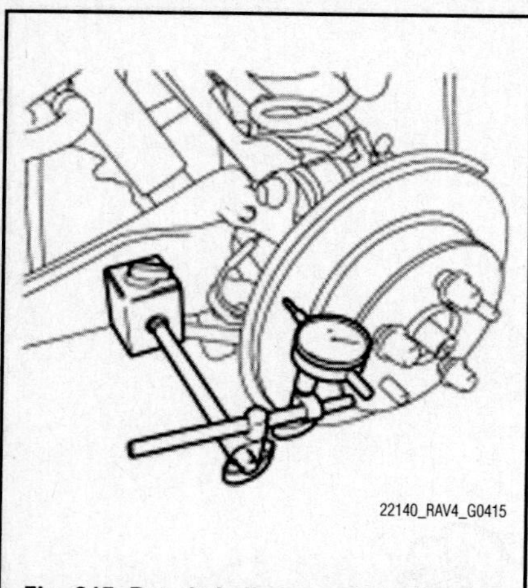

Fig. 315 Rear hub disc runout check point

3. Raise and support the vehicle safely.
4. Remove the tire and wheel assembly.
5. On 4WD vehicles, disconnect the halfshaft.
6. Remove the brake caliper. Remove the rotor.
7. Using a dial indicator tool, measure the backlash near the center of the axle hub. Maximum backlash should be 0.0020 inch.
8. If backlash is greater than specification, replace the bearing.
9. Using a dial indicator tool, measure the disc runout (0.39 inch) inside the outer edge of the disc. Maximum deviation should be 0.0031 inch (2WD vehicles) and 0.0024 inch (4WD vehicles).

➡If runout exceeds the maximum, change the installation positions of the rear disc and axle hub so that the runout will be as low as possible. If the runout exceeds the maximum even when the installation positions are changed, shave the disc. If the disc needs to be shaved to less than the minimum, replace the disc. If the disc is replaced perform the runout inspection again. If the runout still exceeds the maximum, replace the hub and bearing.

TOYOTA

Sequoia • Tundra

15

SPECIFICATIONS AND MAINTENANCE CHARTS

ENGINE AND VEHICLE IDENTIFICATION

Engine						
Code	Liters (cc)	Cu. In.	Cyl.	Fuel Sys.	Engine Type	Eng. Mfg.
1GR-FE/U	4.0 (3956)	241	6	MFI	DOHC	Toyota
1UR-FE/M	4.6 (4608)	282	8	SFI	DOHC	Toyota
2UZ-FE/T	4.7 (4664)	285	8	SFI	DOHC	Toyota
3UR-FE/V,Y	5.7 (5663)	346	8	SFI	DOHC	Toyota
3UR-FBE/W	5.7 (5663)	346	8	SFI	DOHC	Toyota

Model Year	
Code ①	Year
8	2008
9	2009
A	2010

SFI: Sequential Fuel Injection

MFI: Multi-port Fuel Injection

DOHC: Double Overhead Camshaft

① 10th digit of the VIN number

3768X_SEQU_C0001

GENERAL ENGINE SPECIFICATIONS

Year	Model	Engine Displacement Liters	Engine Series Code/ID	Net Horsepower @ rpm	Net Torque @ rpm (ft. lbs.)	Bore x Stroke (in.)	Com-pression Ratio	Oil Pressure @ rpm
2008	Tundra	4.0	1GR-FE/U	236@5200	266@4000	3.70x3.74	N/A	43-85@3000
		4.7	2UZ-FE/T	271@5400	313@3400	3.70x3.70	9.6:1	43-85@3000
		5.7	3UR-FE/V, Y	383@5600	403@3600	3.70x4.02	10.2:1	45-65@3000
	Sequoia	4.7	2UZ-FE/T	271@5400	313@3400	3.70x3.70	9.6:1	43-85@3000
		5.7	3UR-FE/V, Y	383@5600	403@3600	3.70x4.02	10.2:1	45-65@3000
2009	Tundra	4.0	1GR-FE/U	236@5200	266@4000	3.70x3.74	N/A	43-85@3000
		4.7	2UZ-FE/T	271@5400	313@3400	3.70x3.70	9.6:1	43-85@3000
		5.7	3UR-FE/V, Y	383@5600	403@3600	3.70x4.02	10.2:1	45-65@3000
		5.7	3UR-FE/W	383@5600	403@3600	3.70x4.02	10.2:1	45-65@3000
	Sequoia	4.7	2UZ-FE/T	271@5400	313@3400	3.70x3.70	9.6:1	43-85@3000
		5.7	3UR-FE/V, Y	383@5600	403@3600	3.70x4.02	10.2:1	45-65@3000
		5.7	3UR-FE/W	383@5600	403@3600	3.70x4.02	10.2:1	45-65@3000
2010	Tundra	4.0	1GR-FE/U	236@5200	266@4000	3.70x3.74	N/A	43-85@3000
		4.6	1UR-FE/M	310@5600	327@3400	3.70x3.27	10.2:1	32@2500
		5.7	3UR-FE/W	383@5600	403@3600	3.70x4.02	10.2:1	45-65@3000
		5.7	3UR-FE/V, Y	383@5600	403@3600	3.70x4.02	10.2:1	45-65@3000
	Sequoia	4.6	1UR-FE/M	310@5600	327@3400	3.70x3.27	10.2:1	32@2500
		5.7	3UR-FE/V, Y	383@5600	403@3600	3.70x4.02	10.2:1	45-65@3000
		5.7	3UR-FE/W	383@5600	403@3600	3.70x4.02	10.2:1	45-65@3000

N/A: Not Available

3768X_SEQU_C0002

ENGINE TUNE-UP SPECIFICATIONS

Year	Engine Displacement Liters	Engine Code/ID	Spark Plug Gap (in.)	Ignition Timing (deg.)	Fuel Pump (psi)	Idle Speed (rpm)		Valve Clearance	
						MT	AT	Intake	Exhaust
2008	4.0	1GR-FE/U	0.039-0.043	7-24B ①	40.8-41.7	N/A	650-750	0.0591-0.0098	0.0114-0.0154
	4.7	2UZ-FE/T	0.039-0.043	5-15	38-44	N/A	650-750	0.0591-0.0098	0.0114-0.0154
	5.7	3UR-FE/V, Y	0.039-0.043	8-12 ②	38-44	N/A	650-750	N/A	N/A
2009	4.0	1GR-FE/U	0.039-0.043	7-24B ①	40.8-41.7	N/A	650-750	0.0591-0.0098	0.0114-0.0154
	4.7	2UZ-FE/T	0.039-0.043	5-15	38-44	N/A	650-750	0.0591-0.0098	0.0114-0.0154
	5.7	3UR-FE/W	0.039-0.043	8-12 ②	38-44	N/A	650-750	N/A	N/A
	5.7	3UR-FE/V, Y	0.039-0.043	8-12 ②	38-44	N/A	650-750	N/A	N/A
2010	4.0	1GR-FE/U	0.039-0.043	8-12 ②	40.8-41.7	N/A	650-750	0.0591-0.0098	0.0114-0.0154
	4.6	1UR-FE/M	0.039-0.043	8-12 ②	41-42	N/A	650-750	N/A	N/A
	5.7	3UR-FE/W	0.039-0.043	8-12 ②	38-44	N/A	650-750	N/A	N/A
	5.7	3UR-FE/V, Y	0.039-0.043	8-12 ②	38-44	N/A	650-750	N/A	N/A

NOTE: The Vehicle Emission Control Information label often reflects specification changes made during production. The label figures must be used if they differ from those in this chart.

B: Before top dead center

N/A: Not Available

① With terminals TE1 and E1 connected to DLC1. Adjustments are not possible or necessary.

② With terminals TC and Cgof the DLC3 connected. Transmission in N and AC off.

3768X_SEQU_C0003

CAPACITIES

Year	Model	Engine Displacement Liters	Engine Code/ID	Engine Oil with Filter (qts.)	Transmission (qts.)		Transfer Case (qts.)	Drive Axle		Fuel Tank (gal.)	Cooling System (qts.)
					5-Spd	Auto.		Front (qts.)	Rear (qts.)		
2008	Tundra	4.0	1GR-FE/U	5.5	NA	3.2	1.20	2.20	①	26.4	10.1
		4.7	2UZ-FE/T	6.6	NA	3.2	1.20	2.20	①	26.4	10.3
		5.7	3UR-FE/V, Y	7.4	NA	3.2	1.20	2.20	②	26.4	13.7
	Sequoia	4.7	2UZ-FE/T	6.6	NA	3.2	1.43	2.17	1.64	26.4	13.3
		5.7	3UR-FE/V, Y	7.4	NA	3.2	1.43	2.17	1.64	26.4	③
2009	Tundra	4.0	1GR-FE/U	5.5	NA	3.2	1.20	2.20	①	26.4	10.1
		4.7	2UZ-FE/T	6.6	NA	3.2	1.20	2.20	①	26.4	10.3
		5.7	3UR-FE/W	7.4	NA	3.2	1.20	2.20	②	26.4	13.7
		5.7	3UR-FE/V, Y	7.4	NA	3.2	1.20	2.20	②	26.4	13.7
	Sequoia	4.7	2UZ-FE/T	6.6	NA	3.2	1.43	2.17	1.64	26.4	10.3
		5.7	3UR-FE/W	7.4	NA	3.2	1.43	2.17	1.64	26.4	③
		5.7	3UR-FE/V, Y	7.4	NA	3.2	1.43	2.17	1.64	26.4	③
2010	Tundra	4.0	1GR-FE/U	5.5	NA	3.6	NA	NA	①	26.4	10.1
		4.6	1UR-FE/M	7.4	NA	3.6	1.20	①	①	26.4	④
		5.7	3UR-FE/W	7.4	NA	4.4	1.20	②	②	26.4	④
		5.7	3UR-FE/V, Y	7.4	NA	4.4	1.20	②	②	26.4	④
	Sequoia	4.6	1UR-FE/M	7.4	NA	3.6	1.43	2.17	1.64	26.4	③
		5.7	3UR-FE/W	7.4	NA	4.4	1.43	2.17	1.64	26.4	③
		5.7	3UR-FE/V, Y	7.4	NA	4.4	1.43	2.17	1.64	26.4	③

NA: Not Available

① 4.0L, 4.6L & 4.7L standard bed: 4.3
4.0L, 4.6L & 4.7L long bed: 4.9

② 5.7L regular cab standard bed: 3.7
5.7L regular cab long bed: 3.8
5.7 double cab standard bed: 3.8
5.7L double cab long bed: 4.0
5.7L crew cab 2wd: 4.0
5.7L crew cab 4wd: 3.8

③ 14.5 w/o tow package
15.4 with tow package

④ 12.8 w/o tow packag
13.9 with tow package

3768X_SEQU_C0004

FLUID SPECIFICATIONS

Year	Model	Engine Displacement Liters	Engine ID/VIN	Engine Oil	Auto. Trans.	Power Steering Fluid	Brake Master Cylinder
2008	Tundra	4.0	1GR-FE/U	5W-30	①	②	DOT 3
		4.7	2UZ-FE/T	5W-30	①	②	DOT 3
		5.7	3UR-FE/V, Y	5W-20	①	②	DOT 3
	Sequoia	4.7	2UZ-FE/T	5W-30	①	②	DOT 3
		5.7	3UR-FE/V, Y	5W-20	①	②	DOT 3
2009	Tundra	4.0	1GR-FE/U	5W-30	①	②	DOT 3
		4.7	2UZ-FE/T	5W-30	①	②	DOT 3
		5.7	3UR-FE/W	5W-20	①	②	DOT 3
		5.7	3UR-FE/V, Y	5W-20	①	②	DOT 3
	Sequoia	4.7	2UZ-FE/T	5W-30	①	②	DOT 3
		5.7	3UR-FE/W	5W-20	①	②	DOT 3
		5.7	3UR-FE/V, Y	5W-20	①	②	DOT 3
2010	Tundra	4.0	1GR-FE/U	5W-30	③	②	DOT 3
		4.6	1UR-FE/M	5W-20	③	②	DOT 3
		5.7	3UR-FE/W	5W-20	③	②	DOT 3
		5.7	3UR-FE/V, Y	5W-20	③	②	DOT 3
	Sequoia	4.6	1UR-FE/M	5W-20	③	②	DOT 3
		5.7	3UR-FE/W	5W-20	③	②	DOT 3
		5.7	3UR-FE/V, Y	5W-20	③	②	DOT 3

DOT: Department Of Transportation

① A750E, A750F: ATF WS

② ATF Dexron II or III

③ A750E, A750F, AB60E, AB60F: ATF WS

3768X_SEQU_C0015

VALVE SPECIFICATIONS

Year	Engine Displ. Liters	Engine Code/ID	Seat Angle (deg.)	Face Angle (deg.)	Spring Test Pressure (lbs. @ in.)	Spring Installed Height (in.)	Stem-to-Guide Clearance (in.)		Stem Diameter (in.)	
							Intake	Exhaust	Intake	Exhaust
2008	4.0	1GR-FE/U	N/A	44.5	42-46@ 1.31	1.882	0.00094-0.00236	0.00118-0.00256	0.2154 0.2159	0.2152 0.2156
	4.7	2UZ-FE/T	45	44.5	47.2-50.7@ 1.31	N/A	0.00094-0.00236	0.00118-0.00256	0.2150-0.2160	0.2150-0.2160
	5.7	3UR-FE/V, Y	N/A	N/A	N/A	N/A	0.000984-0.00236	0.00118-0.00256	0.2150-0.2160	0.2150-0.2160
2009	4.0	1GR-FE/U	N/A	44.5	42-46@ 1.31	1.882	0.00094-0.00236	0.00118-0.00256	0.2154 0.2159	0.2152 0.2156
	4.7	2UZ-FE/T	45	44.5	47.2-50.7@ 1.31	N/A	0.00094-0.00236	0.00118-0.00256	0.2150-0.2160	0.2150-0.2160
	5.7	3UR-FE/W	N/A	N/A	N/A	N/A	0.000984-0.00236	0.00118-0.00256	0.2150-0.2160	0.2150-0.2160
	5.7	3UR-FE/V, Y	N/A	N/A	N/A	N/A	0.000984-0.00236	0.00118-0.00256	0.2150-0.2160	0.2150-0.2160
2010	4.0	1GR-FE/U	N/A	44.5	42-46@ 1.31	1.882	0.00094-0.00236	0.00118-0.00256	0.2154 0.2159	0.2152 0.2156
	4.6	1UR-FE/M	N/A	N/A	N/A	N/A	0.000984-0.00236	0.00118-0.00256	0.2150-0.2160	0.2150-0.2160
	5.7	3UR-FE/W	N/A	N/A	N/A	N/A	0.000984-0.00236	0.00118-0.00256	0.2150-0.2160	0.2150-0.2160
	5.7	3UR-FE/V, Y	N/A	N/A	N/A	N/A	0.000984-0.00236	0.00118-0.00256	0.2150-0.2160	0.2150-0.2160

N/A: Not Available

3768X_SEQU_C0005

CAMSHAFT SPECIFICATIONS

All measurements are given in inches.

Year	Engine Displacement Liters	Engine Code/ID	Journal Dia.	Brg. Oil Clearance	Shaft End-play ①	Circle Runout	Lobe Height	
							Intake	Exhaust
2008	4.0	1GR-FE/U	1.4160-1.4170	②	0.01570-0.03540	0.00236	1.7390-1.7420	1.75511-1.75905
	4.7	2UZ-FE/T	1.0610-1.0620	0.00118-0.00280	③	0.00118	1.6776-1.6815	1.67830-1.68230
	5.7	3UR-FE/V, Y	④	N/A	0.00315-0.00531	0.00157	1.7440-1.7500	1.74000-1.74600
2009	4.0	1GR-FE/U	1.4160-1.4170	②	0.01570-0.03540	0.00236	1.7390-1.7420	1.75511-1.75905
	4.7	2UZ-FE/T	1.0610-1.0620	0.00118-0.00280	③	0.00118	1.6776-1.6815	1.67830-1.68230
	5.7	3UR-FE/W	④	N/A	0.00315-0.00531	0.00157	1.7440-1.7500	1.74000-1.74600
	5.7	3UR-FE/V, Y	④	N/A	0.00315-0.00531	0.00157	1.7440-1.7500	1.74000-1.74600
2010	4.0	1GR-FE/U	1.4160-1.4170	②	0.01570-0.03540	0.00236	1.7390-1.7420	1.75511-1.75905
	4.6	1UR-FE/M	④	N/A	0.00315-0.00531	0.00157	1.7440-1.7500	1.74000-1.74600
	5.7	3UR-FE/W	④	N/A	0.00315-0.00531	0.00157	1.7440-1.7500	1.74000-1.74600
	5.7	3UR-FE/V, Y	④	N/A	0.00315-0.00531	0.00157	1.7440-1.7500	1.74000-1.74600

N/A: Not Available

① Thrust clearance

② No. 1 intake right side:0.000315-0.00150
No. 1 exhaust right side: 0.00157-0.00311
No. 1 intake left side: 0.00157-0.00311
Other journal: 0.000984-0.00244
No. 1 intake right side: 0.00276
Other journal: 0.00394
LH side journal: 0.00394

③ Intake side:0-0.00157. Exhaust side: 0.00118-0.00276

④ No. 1 1.1793-1.1799
Other journal: 1.0220-1.0226

3768X_SEQU_C0014

CRANKSHAFT AND CONNECTING ROD SPECIFICATIONS

All measurements are given in inches.

	Engine		Crankshaft				Connecting Rod		
Year	Engine Displ. Liters	Engine Code/ID	Main Brg. Journal Dia.	Main Brg. Oil Clearance	Shaft End-play	Thrust on No.	Journal Diameter	Oil Clearance	Side Clearance
2008	4.0	1GR-FE/U	2.8342-2.8346	0.00079-0.00118	NA	NA	①	②	②
	4.7	2UZ-FE/T	2.6373-2.6378	0.0016-0.0023	0.0008-0.0087	3	2.0465-2.0472	0.0011-0.0021	0.0063-0.0138
	5.7	3UR-FE/V, Y	2.6373-2.6378	③	0.0008-0.0087	3	2.3230-2.3235	0.000984-0.00197	0.00591-0.02170
2009	4.0	1GR-FE/U	2.8342-2.8346	0.00079-0.00118	NA	NA	①	②	②
	4.7	2UZ-FE/T	2.6373-2.6378	0.0016-0.0023	0.0008-0.0087	3	2.0465-2.0472	0.0011-0.0021	0.0063-0.0138
	5.7	3UR-FE/W	2.6373-2.6378	③	0.0008-0.0087	3	2.3230-2.3235	0.000984-0.00197	0.00591-0.02170
	5.7	3UR-FE/V, Y	2.6373-2.6378	③	0.0008-0.0087	3	2.3230-2.3235	0.000984-0.00197	0.00591-0.02170
2010	4.0	1GR-FE/U	2.8342-2.8346	0.00079-0.00118	NA	NA	①	②	②
	4.6	1UR-FE/M	2.6373-2.6378	③	0.0008-0.0087	3	NA	0.000984-0.00197	0.00591-0.02170
	5.7	3UR-FE/W	2.6373-2.6378	③	0.0008-0.0087	3	2.3230-2.3235	0.000984-0.00197	0.00591-0.02170
	5.7	3UR-FE/V, Y	2.6373-2.6378	③	0.0008-0.0087	3	2.3230-2.3235	0.000984-0.00197	0.00591-0.02170

NA: Not Available

① Rod bolt tension portion diameter: 0.2830-0.2870

② Rod oil clearance: 0.00102-0.00181. Bushing oil clearance: 0.000197-0.000433

③ Nos. 1 and 5: 0.000669-0.00118
All others: 0.000945-0.00146

3768X_SEQU_C0007

PISTON AND RING SPECIFICATIONS

All measurements are given in inches.

Year	Engine Displ. Liters	Engine Code/ID	Piston Clearance	Ring Gap Top Compression	Ring Gap Bottom Compression	Ring Gap Oil Control	Ring Side Clearance Top Compression	Ring Side Clearance Bottom Compression	Ring Side Clearance Oil Control
2008	4.0	1GR-FE/U	0.0031-0.0040	0.0118-0.0157	0.0157-0.0197	0.0039-0.0157	0.0008-0.0028	0.0008-0.0024	0.0028-0.0060
	4.7	2UZ-FE/T	0.0035-0.0044	0.0118-0.0197	0.0157-0.0256	0.0051-0.0189	0.0012-0.0031	0.0012-0.0028	SNUG
	5.7	3UR-FE/V, Y	0.0016-0.0024	0.0098-0.0138	0.0157-0.0197	0.0039-0.0157	0.0007-0.0028	0.0007-0.0027	0.0028-0.0057
2009	4.0	1GR-FE/U	0.0031-0.0040	0.0118-0.0157	0.0157-0.0197	0.0039-0.0157	0.0008-0.0028	0.0008-0.0024	0.0028-0.0060
	4.7	2UZ-FE/T	0.0035-0.0044	0.0118-0.0197	0.0157-0.0256	0.0051-0.0189	0.0012-0.0031	0.0012-0.0028	SNUG
	5.7	3UR-FE/W	0.0016-0.0024	0.0098-0.0138	0.0157-0.0197	0.0039-0.0157	0.0007-0.0028	0.0007-0.0027	0.0028-0.0057
	5.7	3UR-FE/V, Y	0.0016-0.0024	0.0098-0.0138	0.0157-0.0197	0.0039-0.0157	0.0007-0.0028	0.0007-0.0027	0.0028-0.0057
2010	4.0	1GR-FE/U	0.0031-0.0040	0.0118-0.0157	0.0157-0.0197	0.0039-0.0157	0.0008-0.0028	0.0008-0.0024	0.0028-0.0060
	4.6	1UR-FE/M	NA	0.00906-0.0130	0.0157-0.0197	0.00394-0.0157	0.000787-0.000276	0.000787-0.000236	0.00276-0.00571
	5.7	3UR-FE/W	0.0016-0.0024	0.0098-0.0138	0.0157-0.0197	0.0039-0.0157	0.0007-0.0028	0.0007-0.0027	0.0028-0.0057
	5.7	3UR-FE/V, Y	0.0016-0.0024	0.0098-0.0138	0.0157-0.0197	0.0039-0.0157	0.0007-0.0028	0.0007-0.0027	0.0028-0.0057

NA: Not Available

3768X_SEQU_C0008

TORQUE SPECIFICATIONS

All readings in ft. lbs.

Year	Engine Displacement Liters	Engine Code/ID	Cylinder Head Bolts	Main Bearing Bolts	Rod Bearing Bolts	Crankshaft Damper Bolts	Flywheel Bolts	Manifold Intake	Manifold Exhaust	Spark Plugs	Oil Pan Drain Plug
2008	4.0	1GR-FE/U	①	②	③	185	61	19	15	15	30
	4.7	2UZ-FE/T	④	⑤	③	181	⑥	13	32	15	29
	5.7	3UR-FE/V, Y	⑦	⑧	⑨	221	⑩	15	15	15	30
2009	4.0	1GR-FE/U	①	②	③	185	61	19	15	15	30
	4.7	2UZ-FE/T	④	⑤	③	181	⑥	13	32	15	29
	5.7	3UR-FE/W	⑦	⑧	⑨	221	⑩	15	15	15	30
	5.7	3UR-FE/V, Y	⑦	⑧	⑨	221	⑩	15	15	15	30
2010	4.0	1GR-FE/U	①	②	③	185	61	19	15	15	30
	4.6	1UR-FE/M	⑦	⑧	⑨	221	⑩	15	15	15	30
	5.7	3UR-FE/W	⑦	⑧	⑨	221	⑩	15	15	15	30
	5.7	3UR-FE/V, Y	⑦	⑧	⑨	221	⑩	15	15	15	30

① Right: 27 ft. lbs., then + 90 degrees, then +90 degrees
Left : (8 bolts) 27 ft. lbs., then + 90 degrees, then + 90 degrees. (2 bolts) 27 ft. lbs.

② (16 bolts): 45 ft. lbs., then +90 degrees
(8 bolts): 19 ft. lbs.

③ Step 1: 18 ft. lbs.
Step 2: Plus 90 degrees

④ Step 1: 24 ft. lbs.
Step 2: Plus 90 degrees
Step 3: Plus 90 degrees

⑤ Step 1: 20 ft. lbs.
Step 2: Plus 90 degrees

⑥ Step 1: 35 ft. lbs.
Step 2: Plus 90 degrees

⑦ Step 1: 27 ft. lbs.
Step 2: Plus 90 degrees
Step 3: Plus 90 degrees
12 mm head 15 ft. lbs.

⑧ Inside position 45 ft. lbs.
Step 1: Outside position 20 ft. lbs.
Step 2: Plus 90 degrees
Step 3: Cylinder block side position 33 ft. lbs.

⑨ Step 1: 30 ft. lbs.
Step 2: Plus 90 degrees

⑩ Step 1: 22 ft. lbs.
Step 2: Plus 90 degrees

3768X_SEQU_C0009

WHEEL ALIGNMENT

Year	Model	Caster Range (+/-Deg.)	Caster Preferred Setting (Deg.)	Camber Range (+/-Deg.)	Camber Preferred Setting (Deg.)	Toe-in (in.)	Steering Axis Inclination (Deg.)
2008	Tundra	NA	NA	NA	NA	NA	NA
	Sequoia	NA	NA	NA	NA	NA	NA
2009	Tundra	NA	NA	NA	NA	NA	NA
	Sequoia	NA	NA	NA	NA	NA	NA
2010	Tundra	NA	NA	NA	NA	NA	NA
	Sequoia	NA	NA	NA	NA	NA	NA

NA: Not available

3768X_SEQU_C0011

TIRE, WHEEL AND BALL JOINT SPECIFICATIONS

Year	Model	OEM Tires Standard	OEM Tires Optional	Tire Pressures (psi) Front	Tire Pressures (psi) Rear	Wheel Size	Ball Joint Inspection	Lug Nut Torque (ft. lbs.)
2008	Tundra	①	①	②	②	8J	③	④
	Sequoia	⑤	⑤	⑥	⑥	8J	②	④
2009	Tundra	①	①	②	②	8J	③	④
	Sequoia	⑤	⑤	⑥	⑥	8J	②	④
2010	Tundra	①	①	②	②	8J	③	④
	Sequoia	⑤	⑤	⑥	⑥	8J	②	④

OEM: Original Equipment Manufacturer

PSI: Pounds Per Square Inch

STD: Standard

OPT: Optional

NS: Not specified by manufacturer

NA: Not available

① P275/70R18, P275/65R18 or P275/55R20

② Upper: turning torque within 6-39 inch lbs.

② Front: 30, Rear: 33. Use specification on vehicle placard if different from one given.

③ Turning torque 89 inch lbs.

④ Steel wheel 154 ft. lbs. aluminum wheels 97 ft. lbs. If specification differs from one given, see owners manual.

⑤ P275/65R18 or P275/55R20

⑥ P275/65R18 tires: 33. P275/55R20 tires: front: 30, rear: 33. Use specification on vehicle placard if different from one given.

3768X_SEQU_C0012

BRAKE SPECIFICATIONS

All measurements in inches unless noted

			Brake Disc				Brake Caliper	
Year	Model		Original Thickness	Minimum Thickness	Maximum Runout	Minimum Lining Thickness	Bracket Bolts (ft. lbs.)	Mounting Bolts (ft. lbs.)
2008	Tundra	F	1.260	1.140	0.00197	0.469	—	73
		R	0.709	0.630	0.00787	0.472	70	65
	Sequoia	F	1.260	1.140	0.00197	0.469	—	73
		R	0.709	0.630	0.00590	0.472	70	65
2009	Tundra	F	1.260	1.140	0.00197	0.469	—	73
		R	0.709	0.630	0.00787	0.472	70	65
	Sequoia	F	1.260	1.140	0.00197	0.469	—	73
		R	0.709	0.630	0.00590	0.472	70	65
2010	Tundra	F	1.260	1.140	0.00197	0.469	—	73
		R	0.709	0.630	0.00787	0.472	70	65
	Sequoia	F	1.260	1.140	0.00197	0.469	—	73
		R	0.709	0.630	0.00590	0.472	70	65

F: Front

R: Rear

3768X_SEQU_C0010

SCHEDULED MAINTENANCE INTERVALS

TOYOTA—SEQUOIA, TUNDRA

TO BE SERVICED	TYPE OF SERVICE	VEHICLE MILEAGE INTERVAL (x1000)																		
		5	10	15	20	25	30	35	40	45	50	55	60	65	70	75	80	85	90	95
Automatic transmission and differential fluid	S/I			✓			✓			✓			✓			✓			✓	
Ball joints and boots	S/I			✓			✓			✓			✓			✓			✓	
Brake system	S/I			✓			✓			✓			✓			✓			✓	
Cabin filter	S/I			✓			✓			✓			✓			✓			✓	
Charcoal canister	S/I												✓							
Drive belts	S/I						✓						✓						✓	
Driveshaft bushing	L						✓						✓						✓	
Engine coolant	R						✓						✓						✓	
Engine oil & filter	R	✓	✓	✓	✓	✓	✓	✓	✓	✓	✓	✓	✓	✓	✓	✓	✓	✓	✓	✓
Exhaust system	S/I			✓			✓			✓			✓			✓			✓	
Fuel lines	S/I						✓						✓						✓	
Fuel tank cap gasket	S/I						✓						✓						✓	
Halfshaft boots & flange bolts	S/I			✓			✓			✓			✓			✓			✓	
Limited slip differential fluid	R						✓						✓						✓	
Differential fluid	S/I						✓						✓						✓	
Non-platinum spark plugs	R						✓						✓						✓	
Platinum spark plugs	R												✓							
Propeller shaft (4WD)	L			✓			✓			✓			✓			✓			✓	
Propeller shaft bolts	S/I			✓			✓			✓			✓			✓			✓	
Steering gear	S/I			✓			✓			✓			✓			✓			✓	
Steering linkage	S/I			✓			✓			✓			✓			✓			✓	
Tires (rotate)	S/I	✓	✓	✓	✓	✓	✓	✓	✓	✓	✓	✓	✓	✓	✓	✓	✓	✓	✓	✓
Valves	S/I												✓							

R: Replace S/I: Service or Inspect L: Lubricate

FREQUENT OPERATION MAINTENANCE (SEVERE SERVICE)

If a vehicle is operated under any of the following conditions it is considered severe service:

- Towing a trailer or using a camper or car-top carrier.

- Repeated short trips of less than 5 miles in temperatures below freezing.

- Excessive idling or low-speed driving for long distances as in heavy commercial use, such as delivery, taxi or police cars.

- Operating on rough, muddy or salt-covered roads.

- Operating on unpaved or dusty roads.

Oil and filter: service every 2500 miles or 3 months, whichever occurs first.

Brake linings and discs or drums: service or inspect every 5000 miles or 4 months, whichever occurs first.

Steering linkage: service or inspect every 5000 miles or 4 months, whichever occurs first.

Ball joints and boots: service or inspect every 5000 miles or 4 months, whichever occurs first.

Brake discs & pads (front): service or inspect every 6000 miles.

Halfshaft boots: service or inspect every 5000 miles or 4 months. Retighten the flange bolts, whichever occurs first.

Body chassis bolts and nuts: service or inspect every 5000 miles or 4 months, whichever occurs first.

Transmission and differential fluid: replace every 15,000 miles or 12 months, whichever occurs first.

Transfer case and differential fluid: replace every 15,000 miles or 12 months, whichever occurs first.

3768X_SEQU_C0013

BRAKES — INFORMATION AND PRECAUTIONS

ANTI-LOCK SYSTEMS

- Certain components within the ABS system are not intended to be serviced or repaired individually.
- Do not use rubber hoses or other parts not specifically specified for and ABS system. When using repair kits, replace all parts included in the kit. Partial or incorrect repair may lead to functional problems and require the replacement of components.
- Lubricate rubber parts with clean, fresh brake fluid to ease assembly. Do not use shop air to clean parts; damage to rubber components may result.
- Use only DOT 3 brake fluid from an unopened container.
- If any hydraulic component or line is removed or replaced, it may be necessary to bleed the entire system.
- A clean repair area is essential. Always clean the reservoir and cap thoroughly before removing the cap. The slightest amount of dirt in the fluid may plug an orifice and impair the system function. Perform repairs after components have been thoroughly cleaned; use only denatured alcohol to clean components. Do not allow ABS components to come into contact with any substance containing mineral oil; this includes used shop rags.
- The Anti-Lock control unit is a microprocessor similar to other computer units in the vehicle. Ensure that the ignition switch is **OFF** before removing or installing controller harnesses. Avoid static electricity discharge at or near the controller.
- If any arc welding is to be done on the vehicle, the control unit should be unplugged before welding operations begin.

DISC AND DRUM SYSTEMS

CAUTION

Dust and dirt accumulating on brake parts during normal use may contain asbestos fibers from production or aftermarket brake linings. Breathing excessive concentrations of asbestos fibers can cause serious bodily harm. Exercise care when servicing brake parts. Do not sand or grind brake lining unless equipment used is designed to contain the dust residue. Do not clean brake parts with compressed air or by dry brushing. Cleaning should be done by dampening the brake components with a fine mist of water, then wiping the brake components clean with a dampened cloth. Dispose of cloth and all residue containing asbestos fibers in an impermeable container with the appropriate label. Follow practices prescribed by the Occupational Safety and Health Administration (OSHA) and the Environmental Protection Agency (EPA) for the handling, processing, and disposing of dust or debris that may contain asbestos fibers.

BRAKES — BLEEDING THE BRAKE SYSTEM

BLEEDING PROCEDURE

Brake System

➡Immediately wash off any brake fluid that comes into contact with any painted surfaces. Depressing the brake pedal with the reservoir cap removed will cause the fluid to spray. When bleeding, maintain the amount of fluid in the reservoir between the MIN. and MAX. lines indicated on the master cylinder.

1. Remove the master cylinder cap.
2. Check and add fluid as required. Install the cap.
3. Disconnect the two brake lines from the master cylinder.
4. Slowly depress and hold the brake pedal.
5. Block the outer holes with your fingers and release the pedal.
6. Repeat the above three or four times.
7. Connect the brake lines to the master cylinder.
8. Tighten to 14 ft. lbs. (20 Nm), without a union nut wrench and 13 ft. lbs. (18 Nm), with a union wrench.
9. Be sure that the brake cylinder is full of fluid.
10. Depress the pedal several times and loosen the bleeder plug. With the brake pedal depressed, bleed fluid from the front calipers, (RH and LH).
11. After bleeding tighten the bleeder plug and release the pedal.
12. Depress the pedal several times and loosen the bleeder plug. With the brake pedal depressed, bleed fluid from the rear calipers, (RH and LH).
13. After bleeding tighten the bleeder plug and release the pedal.

➡After bleeding air from the system, if the height or feel of the pedal cannot be obtained, perform air bleeding of the actuator using the Techstream diagnostic tool, or equivalent. Follow the directions on the tool.

Brake Lines

➡Immediately wash off any brake fluid that comes into contact with any painted surfaces. Depressing the brake pedal with the reservoir cap removed will cause the fluid to spray. When bleeding, maintain the amount of fluid in the reservoir between the MIN. and MAX. lines indicated on the master cylinder.

1. Be sure that the brake cylinder is full of fluid.
2. Depress the pedal several times and loosen the bleeder plug. With the brake pedal depressed, bleed fluid from the front calipers, (RH and LH).
3. After bleeding tighten the bleeder plug and release the pedal.
4. Depress the pedal several times and loosen the bleeder plug. With the brake pedal depressed, bleed fluid from the rear calipers, (RH and LH).
5. After bleeding tighten the bleeder plug and release the pedal.
6. If the following symptoms occur, low or spongy pedal and pedal is depressed but braking is insufficient, bleed the brake system.
7. Bleed the VSC actuator assembly.

➡Depress the pedal more than twenty times with the engine off. Connect the diagnostic tool to the DLC3. Turn the ignition switch ON. Do not start the engine. Select AIR BLEEDING on the tool and follow the directions. Bleed the air out of the suction line.

WHEEL SPEED SENSORS

REMOVAL & INSTALLATION

Front

See Figures 1 through 3.

1. Before servicing the vehicle, refer to the Precautions Section.

➡If working near and/or around the SRS system and components, be sure to disable the SRS system. Tape the negative battery cable with insulating tape. Always disconnect the negative battery cable first.

⁂ CAUTION

To avoid personal injury when working on vehicles equipped with an air bag, the negative battery cable must be disconnected and at least 90 seconds must elapse before working on the system. Failure to do so may result in deployment of the air bag.

2. Disconnect the negative battery cable. Tape the cable with insulating tape.

➡When disconnecting the cable, some systems need to be initialized after the cable is reconnected. You will need the Techstream diagnostic scan tool or equivalent. Follow the directions on the tool.

3. Raise and support the vehicle safely.
4. Remove front wheel.
5. On Tundra, remove the skid control wire.
6. Disconnect speed sensor connector.
7. Remove the clips and 3 clamp bolts holding the sensor harness from the frame, upper arm and steering knuckle.
8. Remove the bolt and speed sensor.

To install:

➡Be sure to use new fasteners, as required.

9. Installation is the reverse of the removal procedure.
10. Tighten bolt to 8 ft. lbs (11 Nm) for Sequoia. For Tundra, refer to illustration.
11. On Tundra, be sure to install the bracket so that the rotation stopper touches the knuckle.
12. Inspect the sensor signal. Correct as required.

Rear

See Figures 4 and 5.

1. Before servicing the vehicle, refer to the Precautions Section.

➡If working near and/or around the SRS system and components, be sure to disable the SRS system. Tape the negative battery cable with insulating tape. Always disconnect the negative battery cable first.

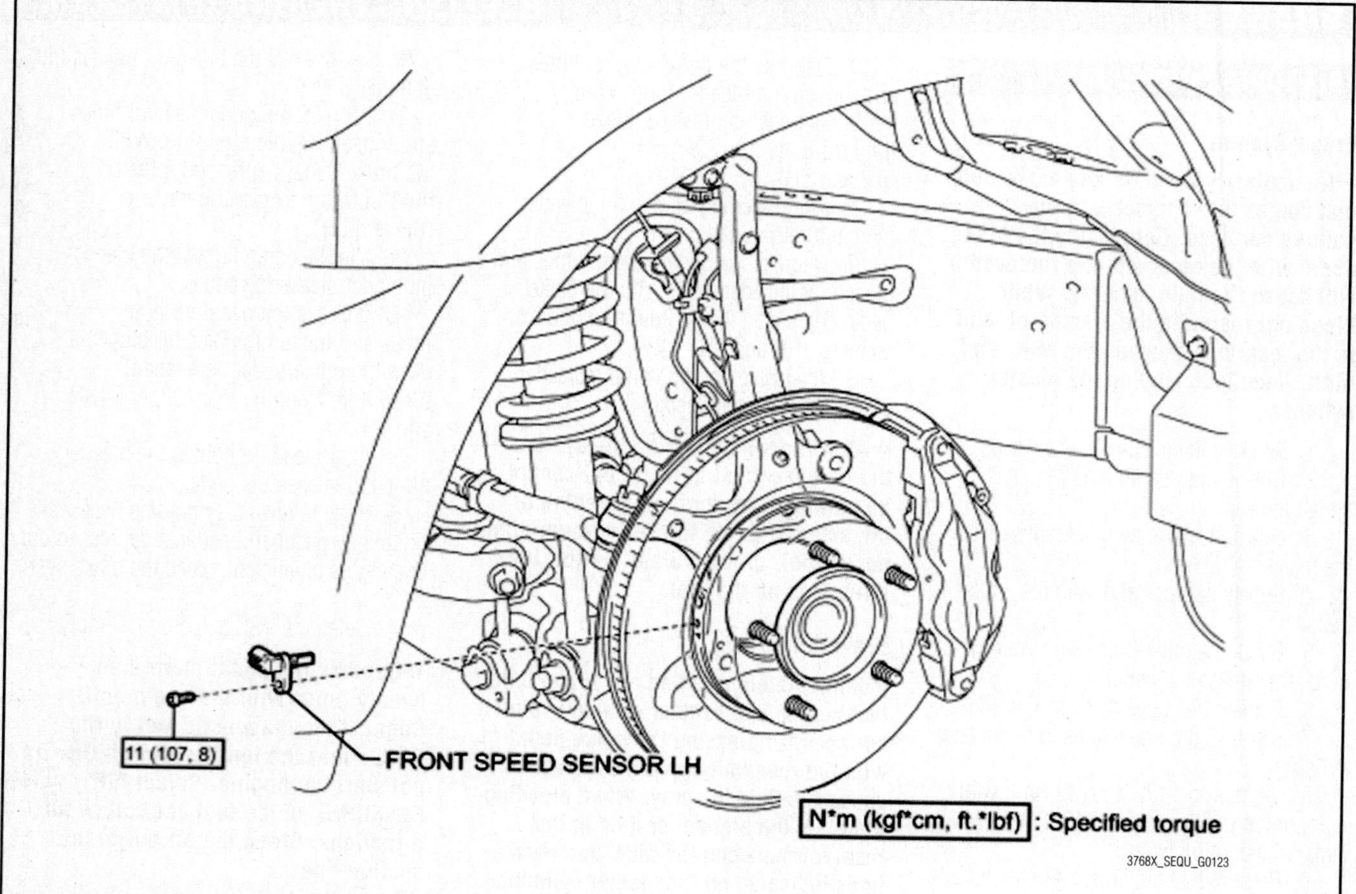

Fig. 1 Front wheel speed sensor and related components—Sequoia

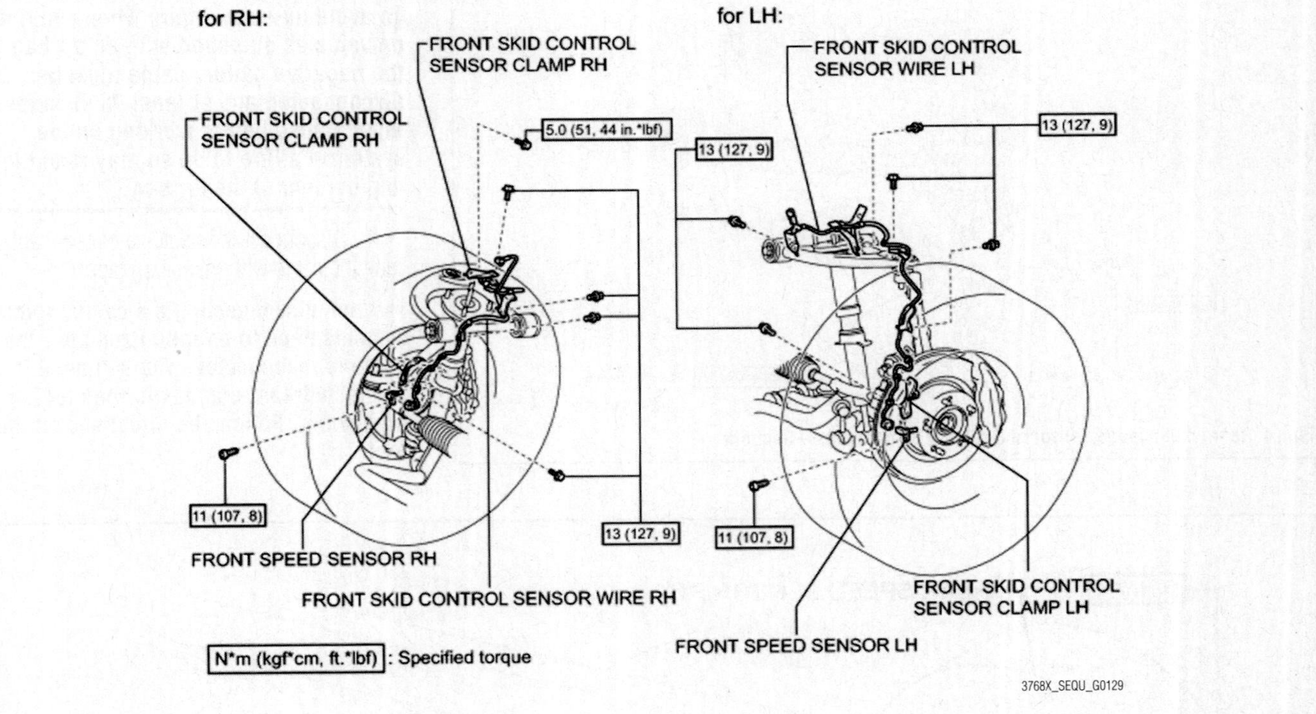

Fig. 2 Front wheel speed sensor and related components—Tundra

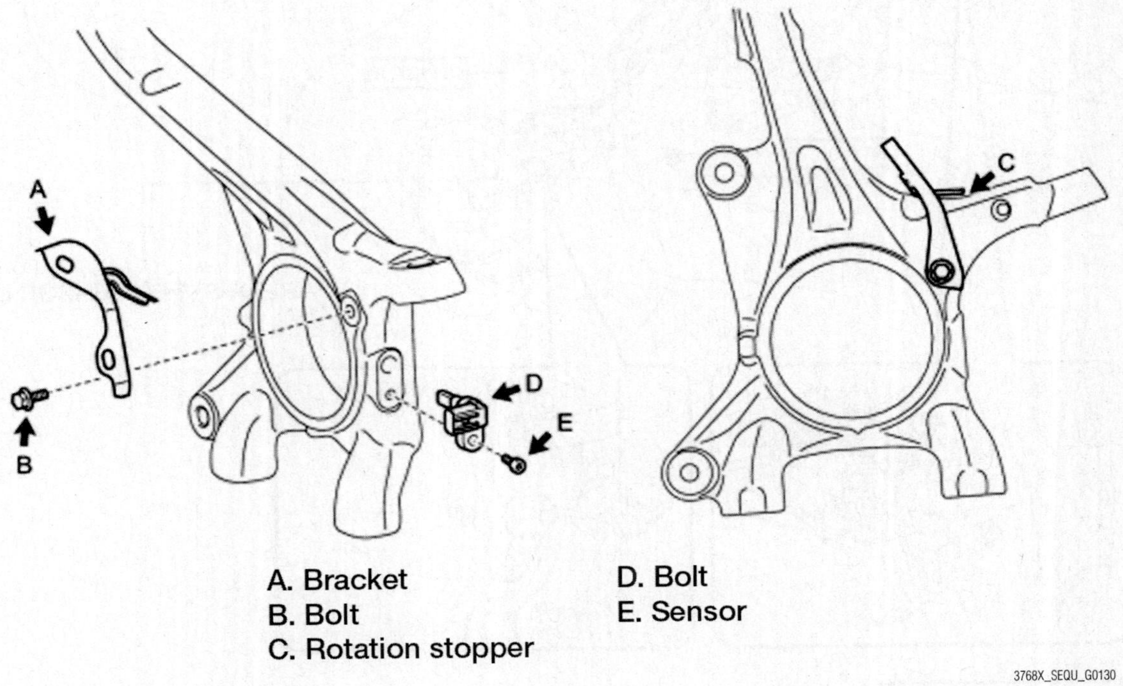

Fig. 3 Front wheel speed sensor bracket/stopper alignment—Tundra

8.3 (85, 73 in.*lbf)
REAR SPEED SENSOR LH
N*m (kgf*cm, ft.*lbf) : Specified torque
3768X_SEQU_G0124

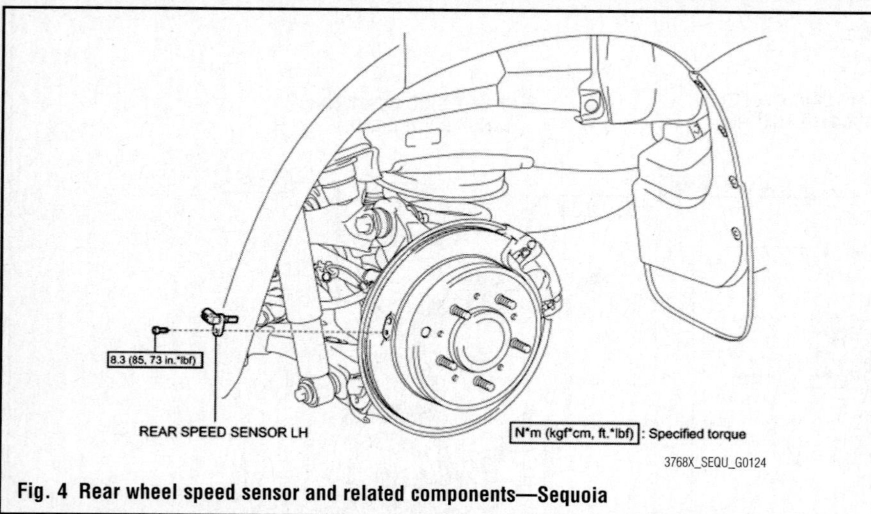

Fig. 4 Rear wheel speed sensor and related components—Sequoia

⁂ CAUTION

To avoid personal injury when working on vehicles equipped with an air bag, the negative battery cable must be disconnected and at least 90 seconds must elapse before working on the system. Failure to do so may result in deployment of the air bag.

2. Disconnect the negative battery cable. Tape the cable with insulating tape.

➡When disconnecting the cable, some systems need to be initialized after the cable is reconnected. You will need the Techstream diagnostic scan tool or equivalent. Follow the directions on the tool.

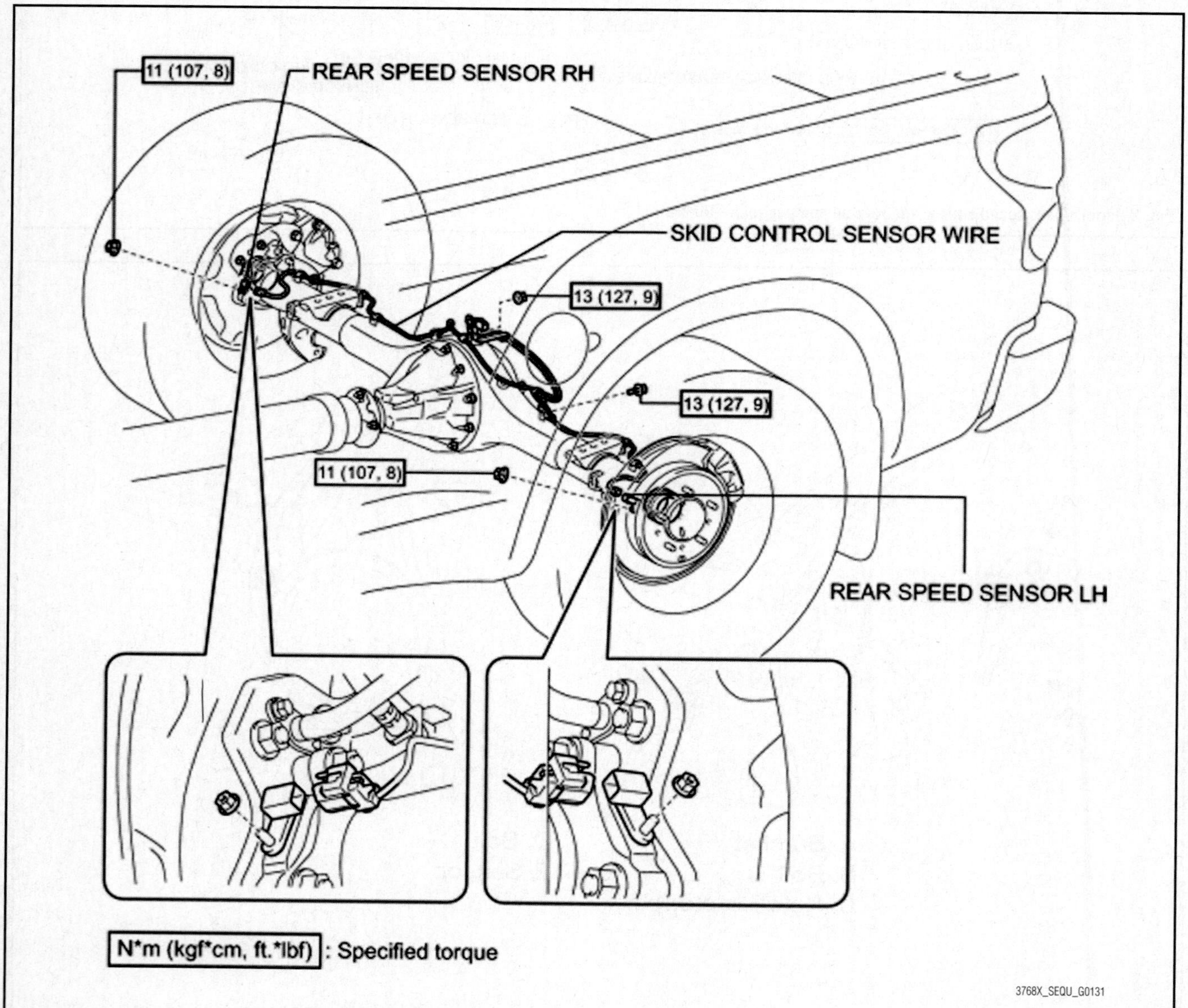

11 (107, 8)
REAR SPEED SENSOR RH
SKID CONTROL SENSOR WIRE
13 (127, 9)
13 (127, 9)
11 (107, 8)
REAR SPEED SENSOR LH
N*m (kgf*cm, ft.*lbf) : Specified torque
3768X_SEQU_G0131

Fig. 5 Rear wheel speed sensor and related components—Tundra

3. Raise and support the vehicle safely.
4. Remove the tire and wheel assembly.
5. Disconnect rear speed sensor connector.
6. Remove the bolt and speed sensor from its mounting.

To install:

➡Be sure to use new fasteners, as required.

7. Installation is the reverse of the removal procedure.
8. Tighten bolt to 73 inch lbs. (8.3 Nm), for Sequoia. On Tundra tighten bolt to 8 ft. lbs (11 Nm).
9. Inspect the sensor signal. Correct as required.

BRAKES

FRONT DISC BRAKES

BRAKE CALIPER

REMOVAL & INSTALLATION

See Figure 6.

1. Before servicing the vehicle, refer to the Precautions Section.

➡If working near and/or around the SRS system and components, be sure to disable the SRS system. Tape the negative battery cable with insulating tape. Always disconnect the negative battery cable first.

⁂ CAUTION

To avoid personal injury when working on vehicles equipped with an air bag, the negative battery cable must be disconnected and at least 90 seconds must elapse before working on the system. Failure to do so may result in deployment of the air bag.

2. Disconnect the negative battery cable. Tape the cable with insulating tape.

➡When disconnecting the cable, some systems need to be initialized after the cable is reconnected. You will need the Techstream diagnostic scan tool or equivalent. Follow the directions on the tool.

3. Raise and support the vehicle safely.
4. Remove the tire and wheel assembly.
5. To remove the front disc brake pad kit, remove the pin holddown clip. Remove the anti rattle spring. Remove the pads from the caliper. Remove the anti squeal shims.

➡The pin holddown clip can be used again if it has sufficient rebound, no deformation or wear and has all rust, dirt and foreign matter removed.

➡The anti rattle spring can be used again if it has sufficient rebound, no deformation or wear and has all rust, dirt and foreign matter removed.

6. Drain the brake fluid to an acceptable level.
7. Remove and plug the brake line hose.
8. Remove the caliper retaining bolts.
9. Remove the caliper from its mounting.

To install:

➡Be sure to use new fasteners, as required.

10. Installation is the reverse of the removal procedure.
11. Bleed the system.

DISC BRAKE PADS

REMOVAL & INSTALLATION

See Figure 6.

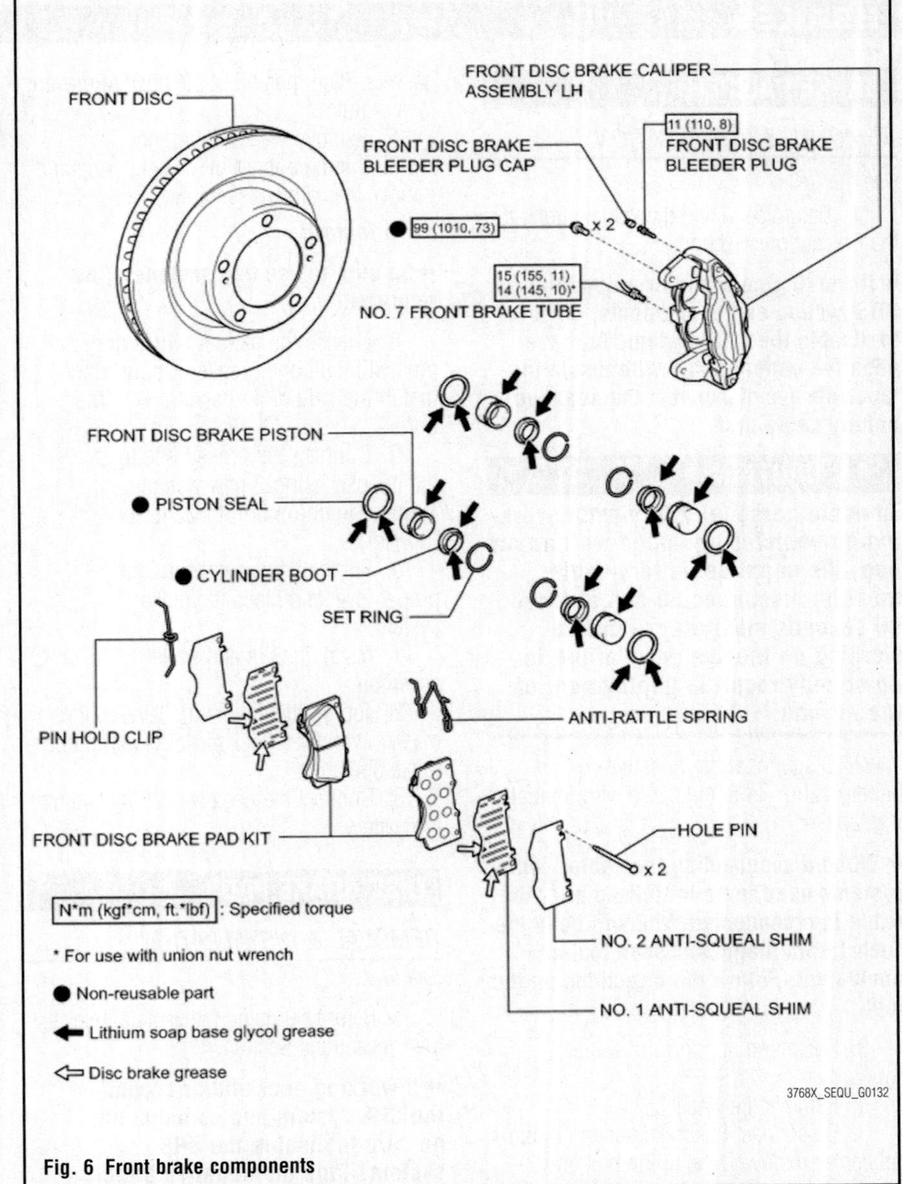

Fig. 6 Front brake components

1. Before servicing the vehicle, refer to the Precautions Section.

➡If working near and/or around the SRS system and components, be sure to disable the SRS system. Tape the negative battery cable with insulating tape. Always disconnect the negative battery cable first.

⁂ CAUTION

To avoid personal injury when working on vehicles equipped with an air bag, the negative battery cable must be disconnected and at least 90 seconds must elapse before working on the system. Failure to do so may result in deployment of the air bag.

2. Disconnect the negative battery cable. Tape the cable with insulating tape.

➡When disconnecting the cable, some systems need to be initialized after the cable is reconnected. You will need the Techstream diagnostic scan tool or equivalent. Follow the directions on the tool.

3. Raise and support the vehicle safely.
4. Remove the tire and wheel assembly.
5. To remove the front disc brake pad kit, remove the pin holddown clip. Remove the anti rattle spring. Remove the pads from the caliper. Remove the anti squeal shims.

➡The pin holddown clip can be used again if it has sufficient rebound, no deformation or wear and has all rust, dirt and foreign matter removed.

➡The anti rattle spring can be used again if it has sufficient rebound, no deformation or wear and has all rust, dirt and foreign matter removed.

To install:

➡Be sure to use new fasteners, as required.

6. Installation is the reverse of the removal procedure.

BRAKES

REAR DISC BRAKES

BRAKE CALIPER

REMOVAL & INSTALLATION

See Figure 7.

1. Before servicing the vehicle, refer to the Precautions Section.

➡If working near and/or around the SRS system and components, be sure to disable the SRS system. Tape the negative battery cable with insulating tape. Always disconnect the negative battery cable first.

⁂ CAUTION

To avoid personal injury when working on vehicles equipped with an air bag, the negative battery cable must be disconnected and at least 90 seconds must elapse before working on the system. Failure to do so may result in deployment of the air bag.

2. Disconnect the negative battery cable. Tape the cable with insulating tape.

➡When disconnecting the cable, some systems need to be initialized after the cable is reconnected. You will need the Techstream diagnostic scan tool or equivalent. Follow the directions on the tool.

3. Raise and support the vehicle safely.
4. Remove the wheels.
5. Disconnect the brake hose from the caliper by removing the union bolt and 2 gaskets. Plug the end of the hose to prevent loss of fluid.
6. Remove the 2 sliding pins.
7. Lift the bottom of the caliper up and remove the caliper assembly.

To install:

➡Be sure to use new fasteners, as required.

8. Grease the caliper slides and pins with silicone grease or equivalent. Install the caliper and secure with the bolts.
9. Connect the brake hose to the caliper, using 2 new washers. Torque the union bolt to 22 ft. lbs. (30 Nm).
10. Fill the brake system to the proper level and bleed the brake system.
11. Install the tire and wheel assembly.
12. Top off the brake fluid level in the master cylinder. Check for leaks and proper brake operation.
13. Connect the negative battery cable to the battery.

DISC BRAKE PADS

REMOVAL & INSTALLATION

See Figure 7.

1. Before servicing the vehicle, refer to the Precautions Section.

➡If working near and/or around the SRS system and components, be sure to disable the SRS system. Tape the negative battery cable with insulating tape. Always disconnect the negative battery cable first.

⁂ CAUTION

To avoid personal injury when working on vehicles equipped with an air bag, the negative battery cable must be disconnected and at least 90 seconds must elapse before working on the system. Failure to do so may result in deployment of the air bag.

2. Disconnect the negative battery cable. Tape the cable with insulating tape.

➡When disconnecting the cable, some systems need to be initialized after the cable is reconnected. You will need the Techstream diagnostic scan tool or equivalent. Follow the directions on the tool.

3. Raise the vehicle and support it safely.
4. Remove the wheels.
5. Remove the brake caliper and suspend it so the hose is not stretched.
6. Remove the brake pads, anti-squeal shim, pad support plates and wear indicators.

To install:

➡Be sure to use new fasteners, as required.

7. Before installing the new pads, check the disc thickness and disc runout.

8. Install the pad support plates.

9. Install the pad wear indicator plates on each pad.

10. Install the anti-squeal shim to the outer pad. Install the pads.

11. Install the brake caliper.

12. Install the wheels.

13. Apply the brake pedal several times.

14. Road-test the vehicle for proper operation.

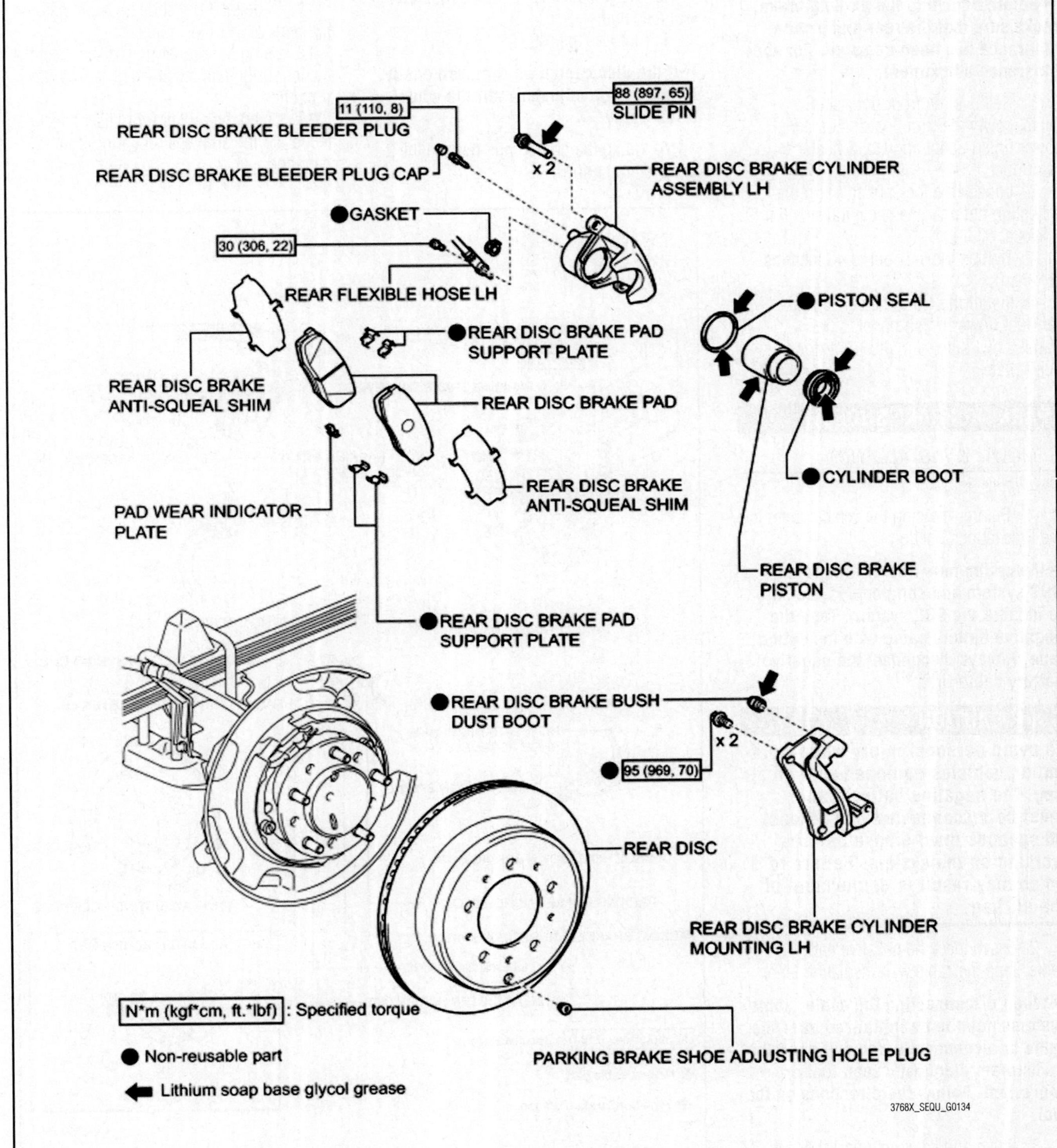

Fig. 7 Rear brake pads and related components—Tundra shown, Sequoia similar

PARKING BRAKE CABLES

ADJUSTMENT

➡Before adjusting the parking brake, make sure that the rear brake shoe clearance has been adjusted. For shoe clearance adjustment.

1. Remove the front door scuff plate, cowl side trim board, side panel, lower finish panel and No. 2 heater to register duct.
2. Loosen the lock nut and turn the adjusting nut until the pedal travel is 6 to 9 clicks at 67 ft lbs.
3. Tighten the lock nut to 48 inch lbs. (5.4 Nm).
4. Install the No. 2 heater to register duct, lower finish panel, side panel, cowl side trim and front door scuff plate.

PARKING BRAKE SHOES

REMOVAL & INSTALLATION

See Figure 8.

1. Before servicing the vehicle, refer to the Precautions Section.

➡If working near and/or around the SRS system and components, be sure to disable the SRS system. Tape the negative battery cable with insulating tape. Always disconnect the negative battery cable first.

CAUTION

To avoid personal injury when working on vehicles equipped with an air bag, the negative battery cable must be disconnected and at least 90 seconds must elapse before working on the system. Failure to do so may result in deployment of the air bag.

2. Disconnect the negative battery cable. Tape the cable with insulating tape.

➡When disconnecting the cable, some systems need to be initialized after the cable is reconnected. You will need the Techstream diagnostic scan tool or equivalent. Follow the directions on the tool.

3. Raise and safely support the vehicle.
4. Remove the rear wheel.
5. Remove the 2 mounting bolts and remove the disc brake assembly.
6. Suspend the disc brake securely and so the hose is not stretched.
7. Release the parking brake lever.
8. Place matchmarks on the rotor and rear axle hub.
9. Remove the rotor.

➡If the disc cannot be removed easily, turn the shoe adjuster until the wheel turns freely.

10. Using the proper tool, remove the 2 shoe return springs.

➡At the time of reassembly, install the strut with the spring facing forward.

11. Slide the front shoe toward outside and remove the shoe adjuster.
12. Using the proper tool, disconnect the anchor spring and tension spring from the front shoe.
13. Using the proper tool, disconnect the anchor spring and tension spring from the rear shoe.

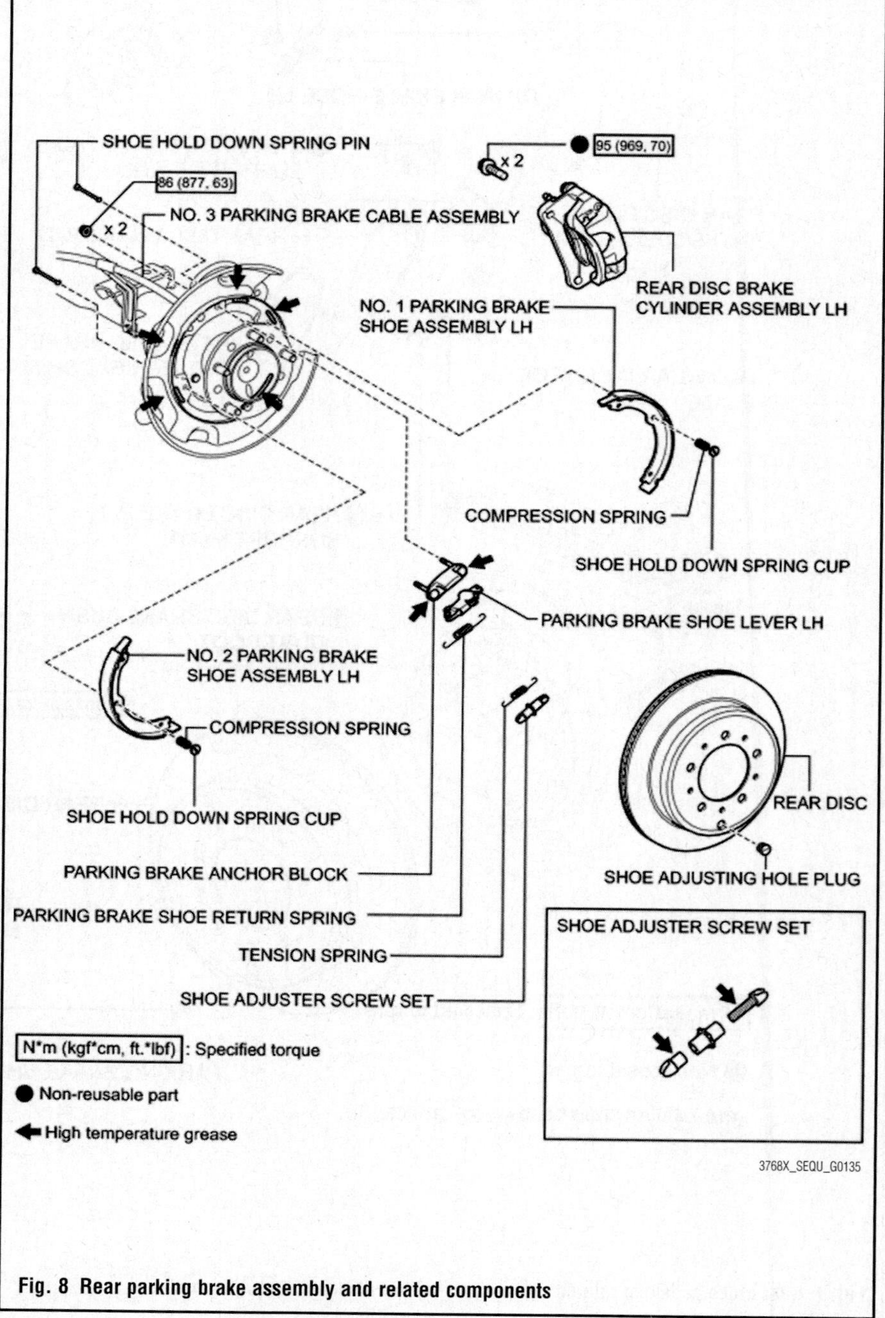

Fig. 8 Rear parking brake assembly and related components

To install:

➡Be sure to use new fasteners, as required.

14. Installation is the reverse of the removal procedure.
15. Be sure to apply high temperature grease to the shoe adjuster.
16. Adjust the parking brake.

ADJUSTMENT

1. Before servicing the vehicle, refer to the precautions in the beginning of this section.
2. Turn the adjuster and expand the shoes until the disc locks.
3. Return the adjuster 8 notches.
4. Depress the parking brake pedal with 147 N (33 lbs).
5. Drive the vehicle at about 50 km/h (31 mph) on a safe, level and dry road for about 400 meters (0.25 mile) in this condition.
6. Repeat this procedure 2 or 3 times.

CHASSIS ELECTRICAL

AIR BAG (SUPPLEMENTAL RESTRAINT SYSTEM)

GENERAL INFORMATION

All vehicles are equipped with an air bag system. The system must be disarmed before performing service on, or around, system components, the steering column, instrument panel components, wiring and sensors. Failure to follow the safety precautions and the disarming procedure could result in accidental air bag deployment, possible injury and unnecessary system repairs.

SERVICE PRECAUTIONS

⁂ CAUTION

Disconnect and isolate the battery negative cable before beginning any airbag system component diagnosis, testing, removal, or installation procedures. Allow system capacitor to discharge for two minutes before beginning any component service. This will disable the airbag system. Failure to disable the airbag system may result in accidental airbag deployment, personal injury, or death.

DISARMING THE SYSTEM

➡If working near and/or around the SRS system and components, be sure to disable the SRS system. Tape the negative battery cable with insulating tape. Always disconnect the negative battery cable first.

⁂ CAUTION

To avoid personal injury when working on vehicles equipped with an airbag, the negative battery cable must be disconnected and at least 90 seconds must elapse before working on the system. Failure to do so may result in deployment of the airbag.

ARMING THE SYSTEM

Connect the negative battery and wait 2 minutes before performing and work on the vehicle.

CLOCKSPRING CENTERING

See Figures 9 and 10.

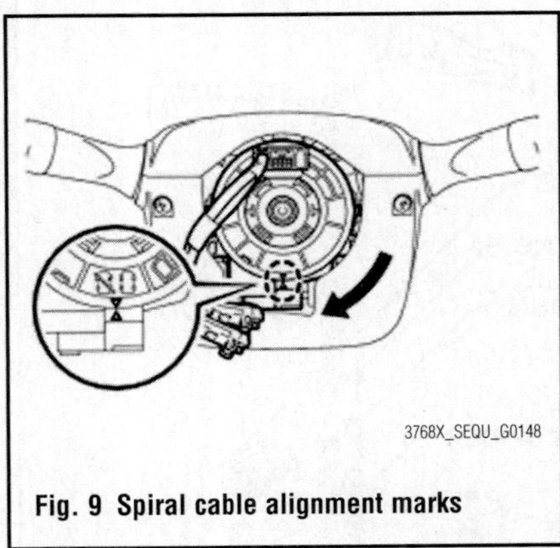

Fig. 9 Spiral cable alignment marks

When installing the spiral cable, check that the ignition switch is in the "OFF" position. Turn the cable counterclockwise by hand until it becomes hard to turn. Turn the cable clockwise about 21/2 turns to align the marks.

➡The cable will rotate about 21/2 turns both left and right from the center.

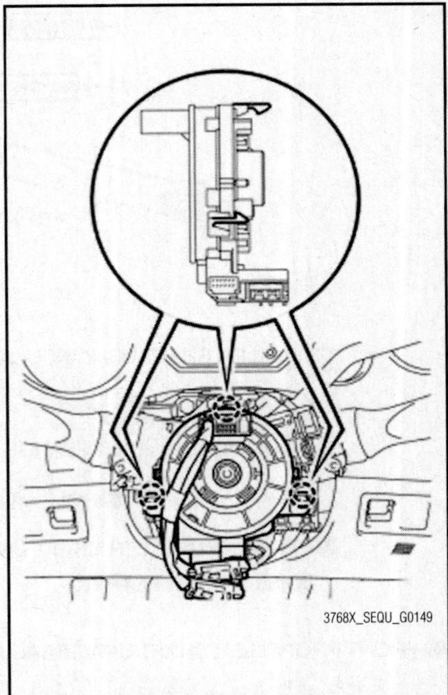

Fig. 10 Spiral cable installation points

DRIVE TRAIN

FRONT DRIVESHAFT

REMOVAL & INSTALLATION

See Figure 11.

1. Before servicing the vehicle, refer to the Precautions Section.

➡If working near and/or around the SRS system and components, be sure to disable the SRS system. Tape the negative battery cable with insulating tape. Always disconnect the negative battery cable first.

⁂ CAUTION

To avoid personal injury when working on vehicles equipped with an air bag, the negative battery cable must be disconnected and at least 90 seconds must elapse before working on the system. Failure to do so may result in deployment of the air bag.

2. Disconnect the negative battery cable. Tape the cable with insulating tape.

➡When disconnecting the cable, some systems need to be initialized after the cable is reconnected. You will need the Techstream diagnostic scan tool or equivalent. Follow the directions on the tool.

3. Raise and safely support the vehicle.
4. Remove the driveshaft heat insulator bracket sub assembly, if equipped.
5. Matchmark and remove the driveshaft retaining bolts.
6. Plug the transmission end to prevent fluid leakage.
7. Remove the component from the vehicle.

To install:

➡Be sure to use new fasteners, as required.

8. Installation is the reverse of the removal procedure.

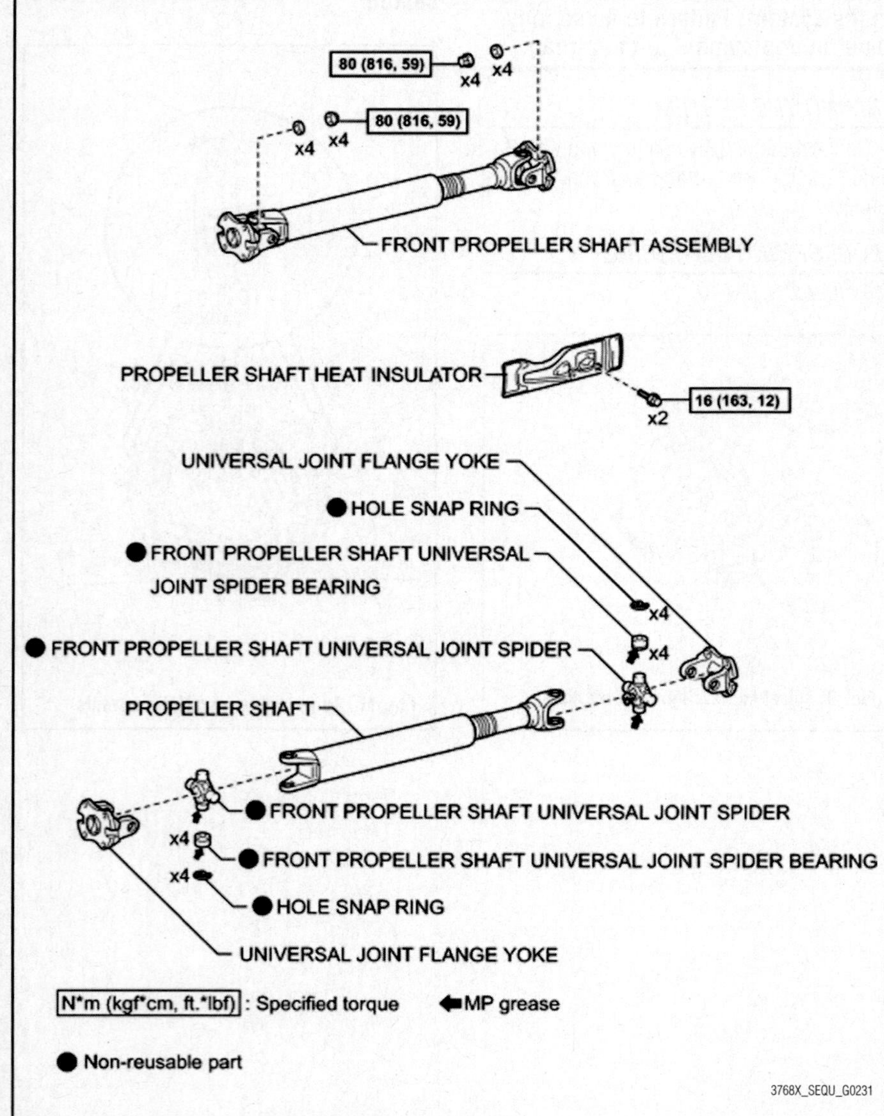

Fig. 11 Front driveshaft and related components

FRONT HALFSHAFTS

REMOVAL & INSTALLATION

Sequoia

See Figure 12.

1. Before servicing the vehicle, refer to the Precautions Section.

➡If working near and/or around the SRS system and components, be sure to disable the SRS system. Tape the negative battery cable with insulating tape. Always disconnect the negative battery cable first.

⁂ CAUTION

To avoid personal injury when working on vehicles equipped with an air bag, the negative battery cable must be disconnected and at least 90 seconds must elapse before working on the system. Failure to do so may result in deployment of the air bag.

2. Disconnect the negative battery cable. Tape the cable with insulating tape.

➡When disconnecting the cable, some systems need to be initialized after the cable is reconnected. You will need the Techstream diagnostic scan tool or equivalent. Follow the directions on the tool.

3. Drain the differential.
4. Remove the steering knuckle assembly.
5. Using tool SST09520-01010, and SST09520-24010 remove the halfshaft. Be careful not to damage the oil seal.

To install:

➡Be sure to use new fasteners, as required.

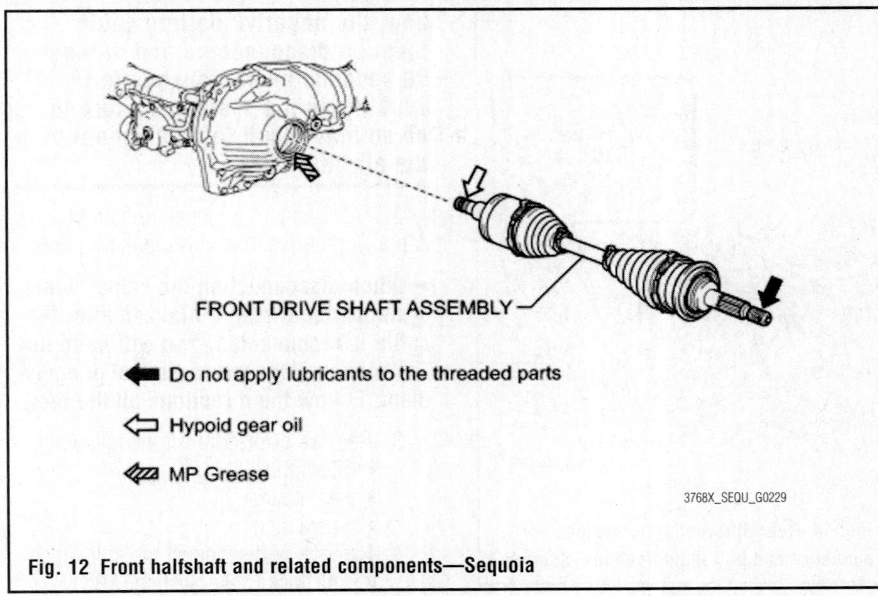

Fig. 12 Front halfshaft and related components—Sequoia

6. Installation is the reverse of the removal procedure.

Tundra

See Figure 13.

1. Before servicing the vehicle, refer to the Precautions Section.

➡If working near and/or around the SRS system and components, be sure to disable the SRS system. Tape the negative battery cable with insulating tape. Always disconnect the negative battery cable first.

✲✲ CAUTION

To avoid personal injury when working on vehicles equipped with an air bag, the negative battery cable must be disconnected and at least 90 seconds must elapse before working on the system. Failure to do so may result in deployment of the air bag.

2. Disconnect the negative battery cable. Tape the cable with insulating tape.

➡When disconnecting the cable, some systems need to be initialized after the cable is reconnected. You will need the Techstream diagnostic scan tool or equivalent. Follow the directions on the tool.

3. Remove or disconnect the following:
   - Front wheel
   - Under cover
4. Drain the differential oil.
5. Remove or disconnect the following:
   - Remove the 2 bolts and disconnect the disc brake caliper from the steering knuckle
   - Remove front disc
   - Grease cap ,4WD
   - Cotter pin and lock cap
   - Halfshaft locknut by applying the brakes
   - Remove the 4 bolts
   - For 4WD, using a plastic-faced hammer, tap out the front drive shaft from the front axle hub.
   - Remove the axle hub and dust cover from the steering knuckle
   - Remove the O-ring from the axle hub.

To install:

➡Be sure to use new fasteners, as required.

6. Install or connect the following:
   - After applying grease to the O-ring, install the O-ring to the axle hub

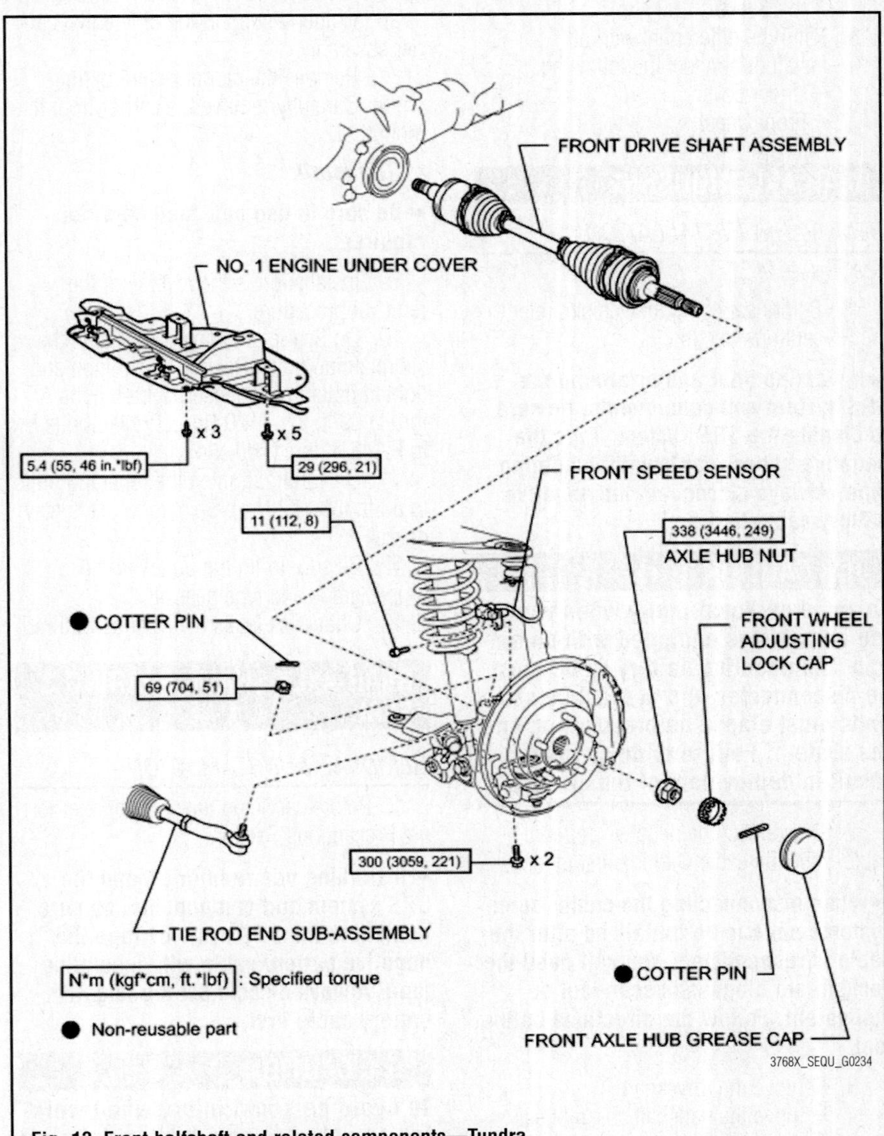

Fig. 13 Front halfshaft and related components—Tundra

- For 4WD, connect the front drive shaft to the front axle hub
- Install the dust cover and axle hub to the steering knuckle with the 4 bolts and tighten to 73 ft. lbs. (99 Nm)
- Clean the threaded parts on the drive shaft and axle hub nut using a non-residue solvent.

➡Be sure to perform this work for a new drive shaft. Keep the threaded parts free of oil and foreign objects.

7. Using a 39 mm socket wrench, install the hub nut and tighten to 249 ft. lbs. (338 Nm).
   - Lock cap and a new cotter pin
   - Grease cap
   - Front disc
   - Connect the front disc brake caliper and install 2 new bolts and tighten to 73 ft. lbs. (99 Nm)
8. Refill the differential with oil.
9. Install or connect the following:
   - Under cover
   - Front wheel

REAR AXLE HOUSING

REMOVAL & INSTALLATION

See Figure 14.

1. Before servicing the vehicle, refer to the Precautions Section.

➡If working near and/or around the SRS system and components, be sure to disable the SRS system. Tape the negative battery cable with insulating tape. Always disconnect the negative battery cable first.

✲✲ CAUTION

To avoid personal injury when working on vehicles equipped with an air bag, the negative battery cable must be disconnected and at least 90 seconds must elapse before working on the system. Failure to do so may result in deployment of the air bag.

2. Disconnect the negative battery cable. Tape the cable with insulating tape.

➡When disconnecting the cable, some systems need to be initialized after the cable is reconnected. You will need the Techstream diagnostic scan tool or equivalent. Follow the directions on the tool.

3. Remove the driveshaft.
4. Remove the halfshaft assemblies.
5. Drain the differential assembly.

Fig. 14 Rear differential tightening sequence and bolt identification—Sequoia

6. Properly support the differential carrier assembly.
7. Remove the carrier retaining bolts.
8. Carefully remove the carrier from its mounting.

To install:

➡Be sure to use new fasteners, as required.

9. Installation is the reverse of the removal procedure.
10. On Sequoia, temporarily install the six retaining bolts. Refer to illustration and tighten bolts in sequence. Tighten bolts A thru D 89 ft. lbs. (120 Nm). Tighten bolts E to F 148 ft. lbs. (200 Nm).
11. On Tundra, tighten the retaining bolts to 53 ft. lbs. (72 Nm). Be sure to use a new gasket.
12. Be sure to fill the unit with the proper grade and type gear oil.
13. Check for leaks. Correct as equired.

REAR AXLE SHAFT, BEARING & SEAL

REMOVAL & INSTALLATION

1. Before servicing the vehicle, refer to the Precautions Section.

➡If working near and/or around the SRS system and components, be sure to disable the SRS system. Tape the negative battery cable with insulating tape. Always disconnect the negative battery cable first.

✲✲ CAUTION

To avoid personal injury when working on vehicles equipped with an air bag, the negative battery cable must be disconnected and at least 90 seconds must elapse before working on the system. Failure to do so may result in deployment of the air bag.

2. Disconnect the negative battery cable. Tape the cable with insulating tape.

➡When disconnecting the cable, some systems need to be initialized after the cable is reconnected. You will need the Techstream diagnostic scan tool or equivalent. Follow the directions on the tool.

3. Remove or disconnect the following:
   - Rear wheel
   - Rear caliper
   - Brake disc
4. Remove or disconnect the following:
   - Anti-lock Brake System (ABS) speed sensor from the rear axle housing, if equipped
   - Parking brake cable
   - Parking brake shoe assemblies
   - Remove the 4 nuts and rear axle shaft together with the parking brake plate

✲✲ WARNING

Be careful not to damage the oil seal.

- O-ring from the rear axle housing
- Inner side oil seal using tool 09308-00010

To install:

➡Be sure to use new fasteners, as required.

5. Install or connect the following:
   - New O-ring to the rear axle
   - Install the rear axle shaft and parking brake plate with the 4 nuts and tighten to 44 ft. lbs. (60 Nm)
   - Anti-lock Brake System (ABS) speed sensor to the rear axle housing, if equipped
   - Parking brake cable assembly
   - Parking brake shoe assembly
   - Install rear disc
   - Adjust parking brake shoe clearance
   - Connect the rear disc brake cylinder and install 2 new bolts and tighten to 70 ft. lbs. (95 Nm)

➡Do not twist the flexible hose. Make sure that the bolts are free from damage and foreign matter. Do not over tighten the bolts.

- Rear wheel

REAR DRIVESHAFT

REMOVAL & INSTALLATION

See Figures 15 through 18.

1. Before servicing the vehicle, refer to the Precautions Section.

➡If working near and/or around the SRS system and components, be sure to disable the SRS system. Tape the negative battery cable with insulating tape. Always disconnect the negative battery cable first.

⁂ CAUTION

To avoid personal injury when working on vehicles equipped with an air bag, the negative battery cable must be disconnected and at least 90 seconds must elapse before working on the system. Failure to do so may result in deployment of the air bag.

2. Disconnect the negative battery cable. Tape the cable with insulating tape.

➡When disconnecting the cable, some systems need to be initialized after the cable is reconnected. You will need the Techstream diagnostic scan tool or equivalent. Follow the directions on the tool.

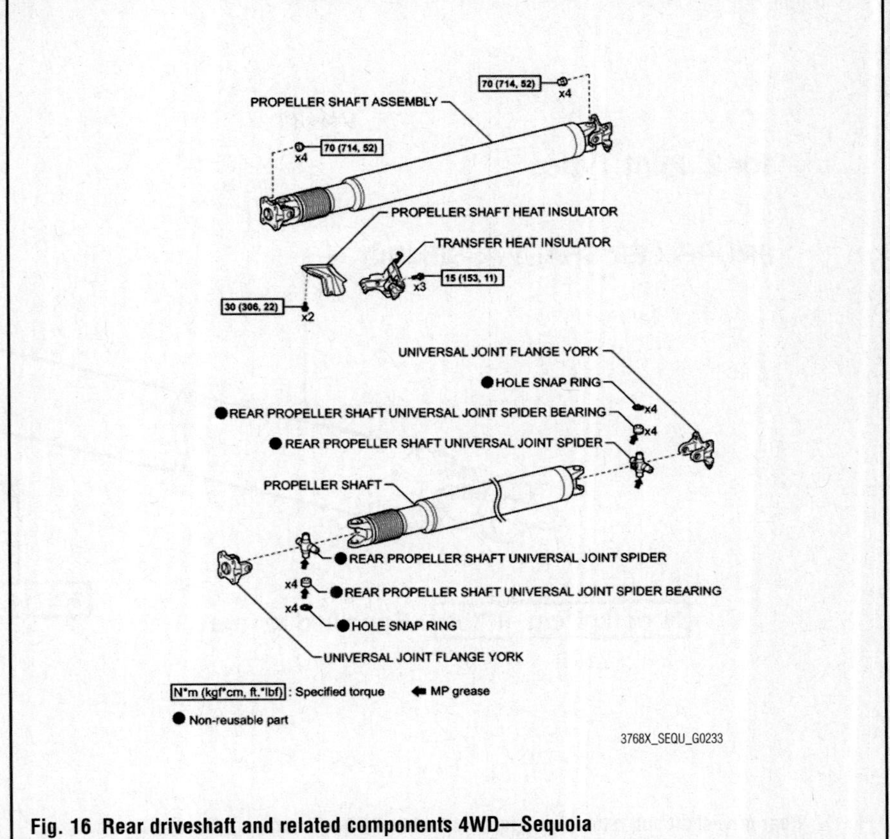

Fig. 16 Rear driveshaft and related components 4WD—Sequoia

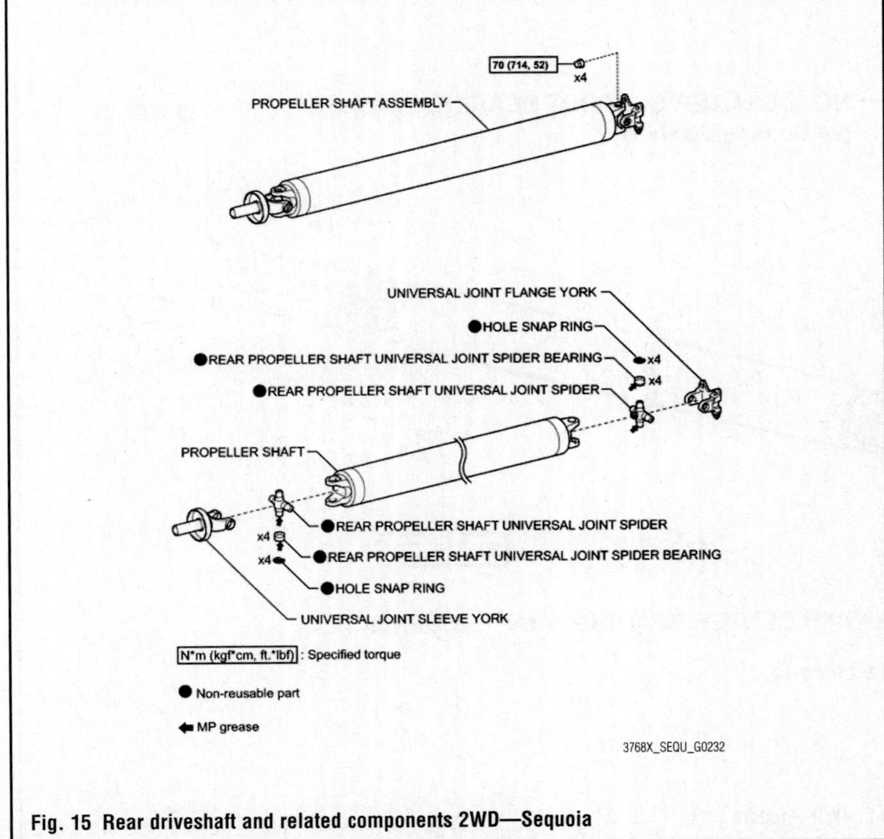

Fig. 15 Rear driveshaft and related components 2WD—Sequoia

3. Raise and safely support the vehicle.
4. Remove the driveshaft heat insulator bracket sub assembly, if equipped.
5. Matchmark and remove the driveshaft retaining bolts.
6. Plug the transmission end to prevent fluid leakage.
7. Remove the component from the vehicle.

To install:

➡Be sure to use new fasteners, as required.

8. Installation is the reverse of the removal procedure.

REAR PINION SEAL

REMOVAL & INSTALLATION

1. Before servicing the vehicle, refer to the precautions in the beginning of this section.
2. Drain the differential housing oil.
3. Remove the rear driveshaft.
4. Remove the companion flange, as follows:
   - Loosen the staked part of the nut, using a chisel and a hammer
   - Companion flange nut, using tool 09330-00021

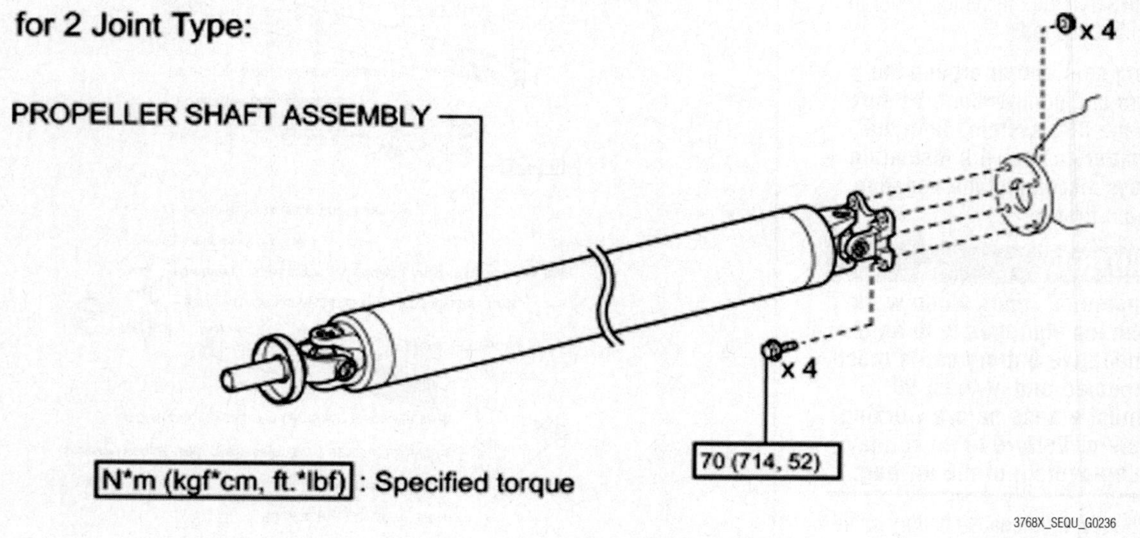

Fig. 17 Rear driveshaft and related components 1 of 2; 4WD—Tundra

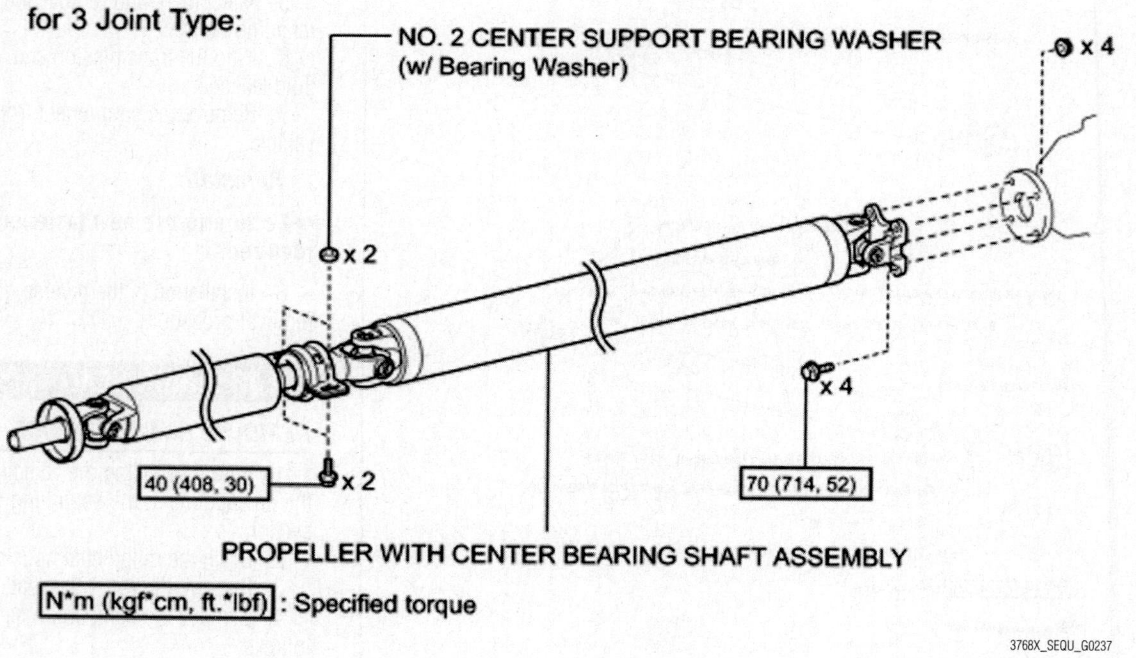

Fig. 18 Rear driveshaft and related components 2 of 2; 4WD—Tundra

- Companion flange, using tools 09950-30012 and 09955-03050
- Oil seal, using tool 09308-10010

To install:

5. Install the new oil seal until it is flush with the housing, using a plastic hammer and tools 09316-12010 and 09649-17010

➡Use vinyl tape to connect both oil seal installation tools.

6. Install the companion flange, as follows:
    - Companion flange
    - New nut, lubricated with hypoid gear oil
    - Torque the nut to 325 ft. lbs. (441 Nm), using tool 09330-00021.
7. Adjust the drive pinion preload
8. Rotate the drive pinion, using a torque wrench while tightening the flange nut to make sure the bearing preload is 10–14 inch lbs. (1.04–1.69 Nm) for a new bearing or 8–12 inch lbs. (0.85–01.37 Nm) for a used bearing. Tighten the flange nut to achieve the preload torque readings originally recorded.

CAUTION

Never loosen the pinion nut to reduce bearing preload.

9. Install or connect the following:
    - Drive pinion nut, stake it
    - Rear driveshaft.

TRANSFER CASE ASSEMBLY

REMOVAL & INSTALLATION

See Figure 19.

1. Before servicing the vehicle, refer to the Precautions Section.

➡If working near and/or around the SRS system and components, be sure to disable the SRS system. Tape the negative battery cable with insulating tape. Always disconnect the negative battery cable first.

CAUTION

To avoid personal injury when working on vehicles equipped with an air bag, the negative battery cable must be disconnected and at least 90 seconds must elapse before working on the system. Failure to do so may result in deployment of the air bag.

2. Disconnect the negative battery cable. Tape the cable with insulating tape.

➡When disconnecting the cable, some systems need to be initialized after the cable is reconnected. You will need the Techstream diagnostic scan tool or equivalent. Follow the directions on the tool.

3. Drain transfer case oil.
4. Remove the automatic transmission with transfer assembly.
5. Remove the 8 transfer adaptor rear mounting bolts.
6. Pull the transfer straight up and remove it from the transmission.

➡Take care not to damage the rear adaptor oil seal with the transfer input gear spline.

To install:

➡Be sure to use new fasteners, as required.

7. Install the transfer with the 8 bolts and tighten to 30 ft. lbs. (40 Nm).
8. Install the automatic transmission with transfer assembly.
9. Add transfer oil.
10. Check transfer oil.
11. Checks for leaks.

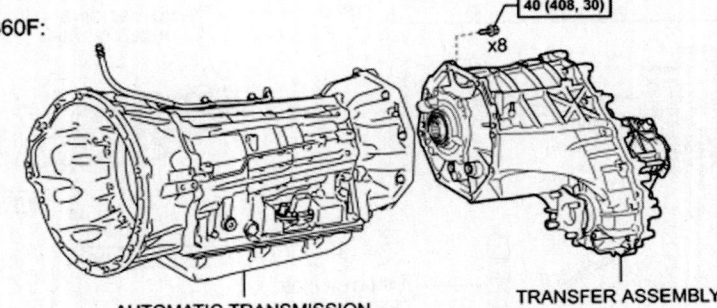

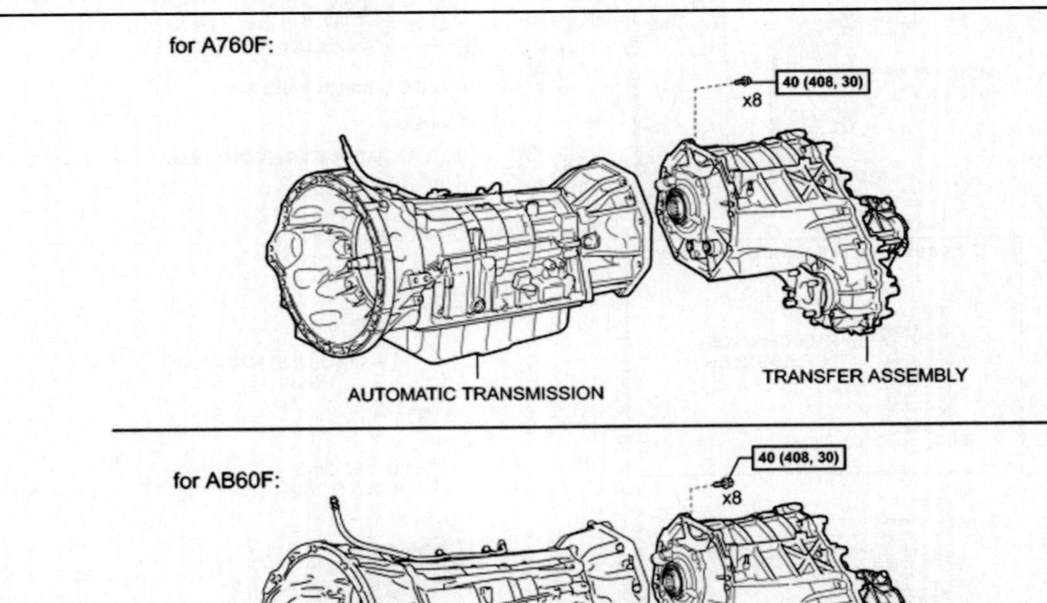

Fig. 19 Transfer case and related components

ENGINE COOLING

ENGINE FAN

REMOVAL & INSTALLATION

1. Before servicing the vehicle, refer to the Precautions Section.

➡If working near and/or around the SRS system and components, be sure to disable the SRS system. Tape the negative battery cable with insulating tape. Always disconnect the negative battery cable first.

⁂ CAUTION

To avoid personal injury when working on vehicles equipped with an air bag, the negative battery cable must be disconnected and at least 90 seconds must elapse before working on the system. Failure to do so may result in deployment of the air bag.

2. Disconnect the negative battery cable. Tape the cable with insulating tape.

➡When disconnecting the cable, some systems need to be initialized after the cable is reconnected. You will need the Techstream diagnostic scan tool or equivalent. Follow the directions on the tool.

3. Drain coolant.
4. Remover upper radiator hose.
5. Remove intake air connector assembly.
6. Remove fan shroud.
7. Loosen the 4 nuts holding the fluid coupling to the fan bracket.
8. Remove the drive belt.
9. Remove fan.

To install:

➡Be sure to use new fasteners, as required.

10. Installation is the reverse of the removal procedure.

RADIATOR

REMOVAL & INSTALLATION

See Figures 20 through 23.

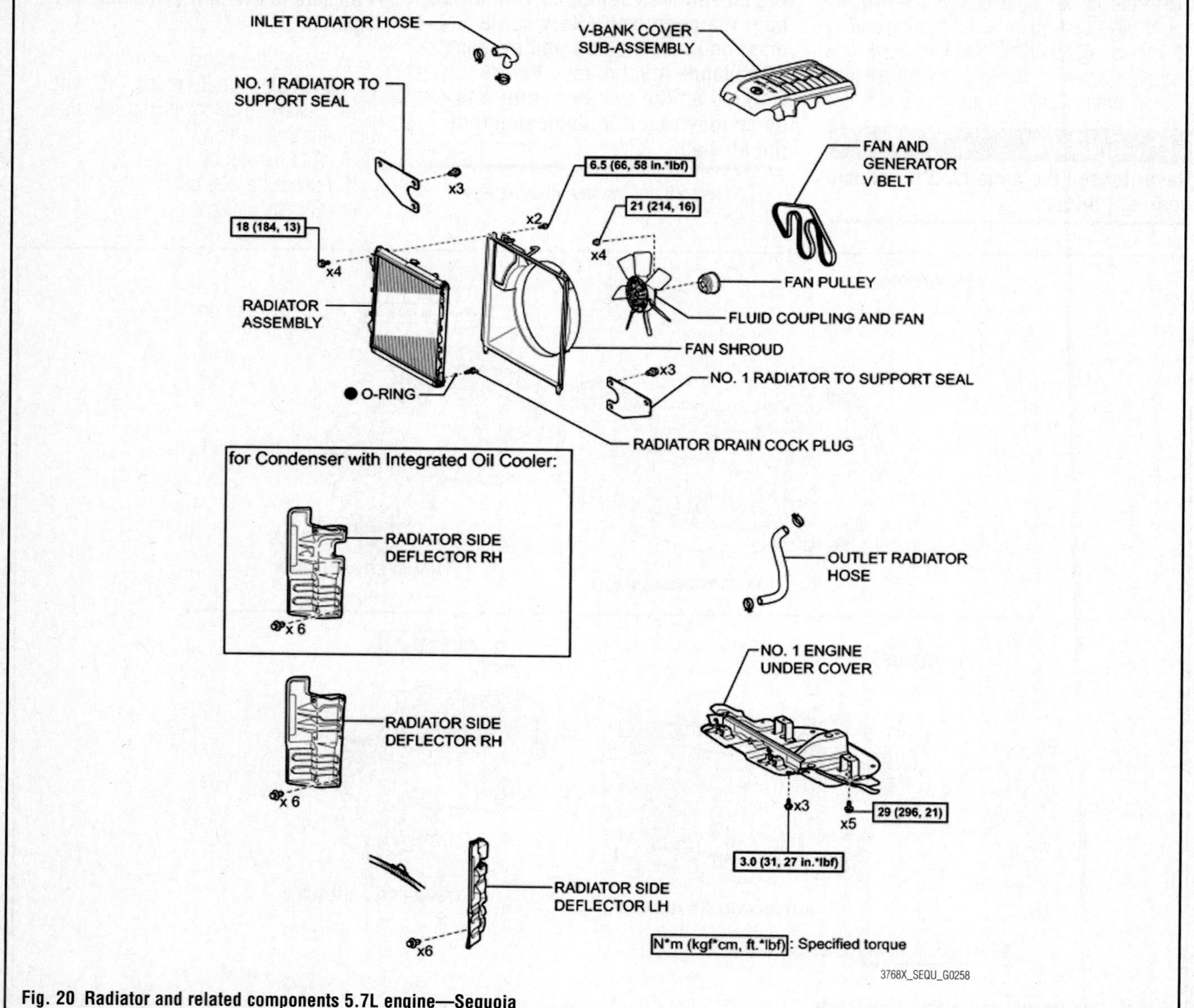

Fig. 20 Radiator and related components 5.7L engine—Sequoia

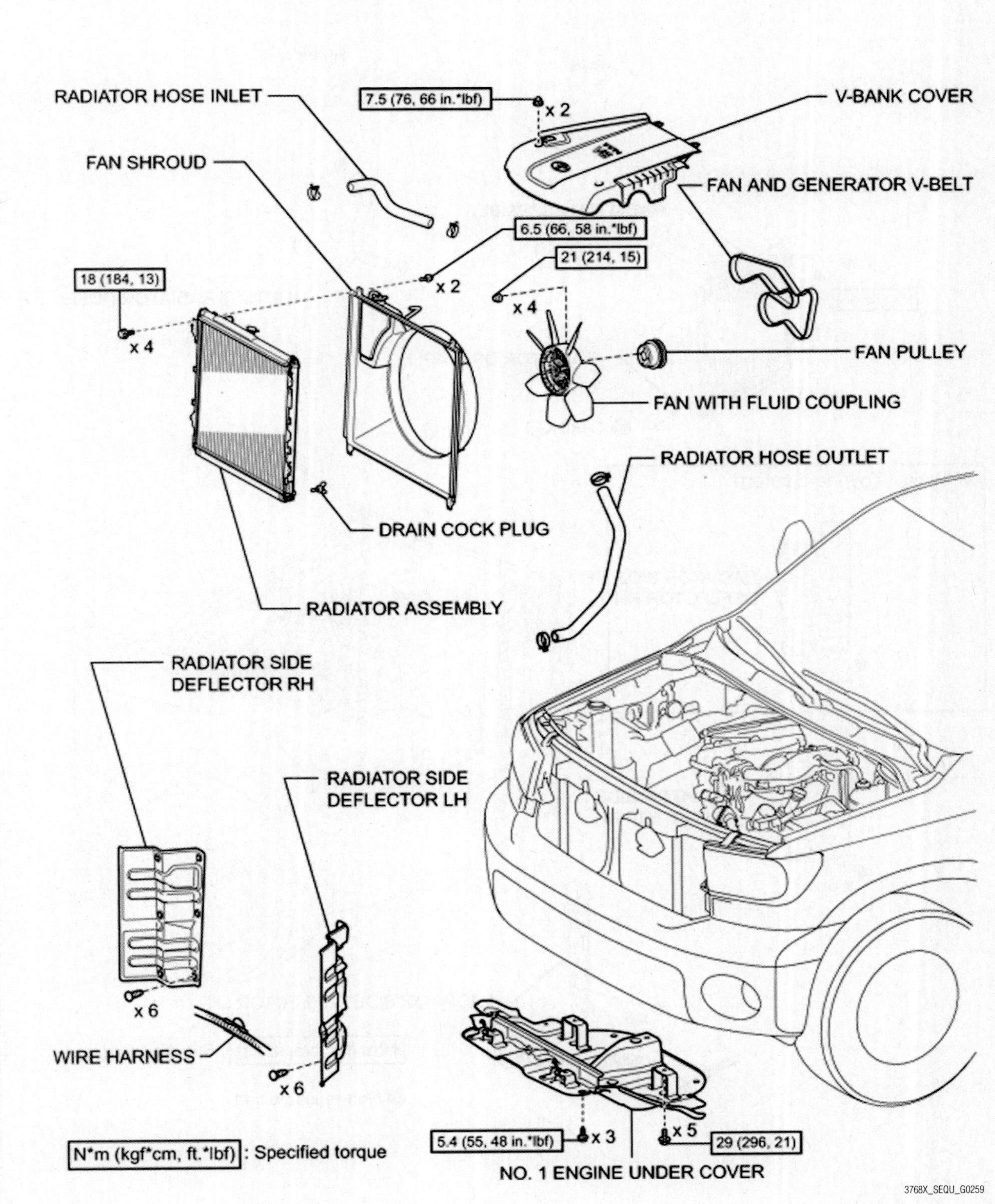

Fig. 21 Radiator and related components 4.0L engine—Tundra

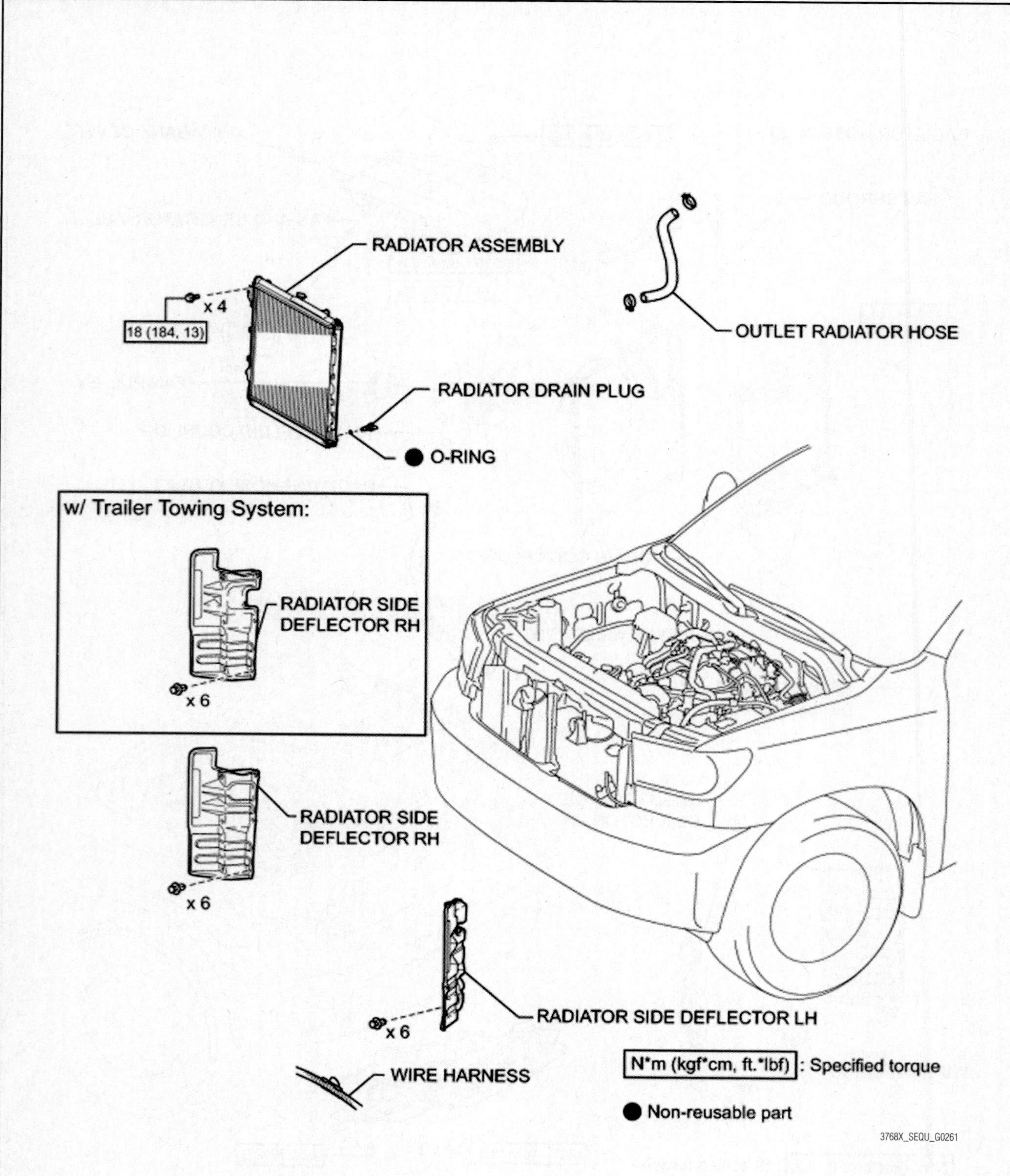

Fig. 22 Radiator and related components 4.6L engine—Tundra

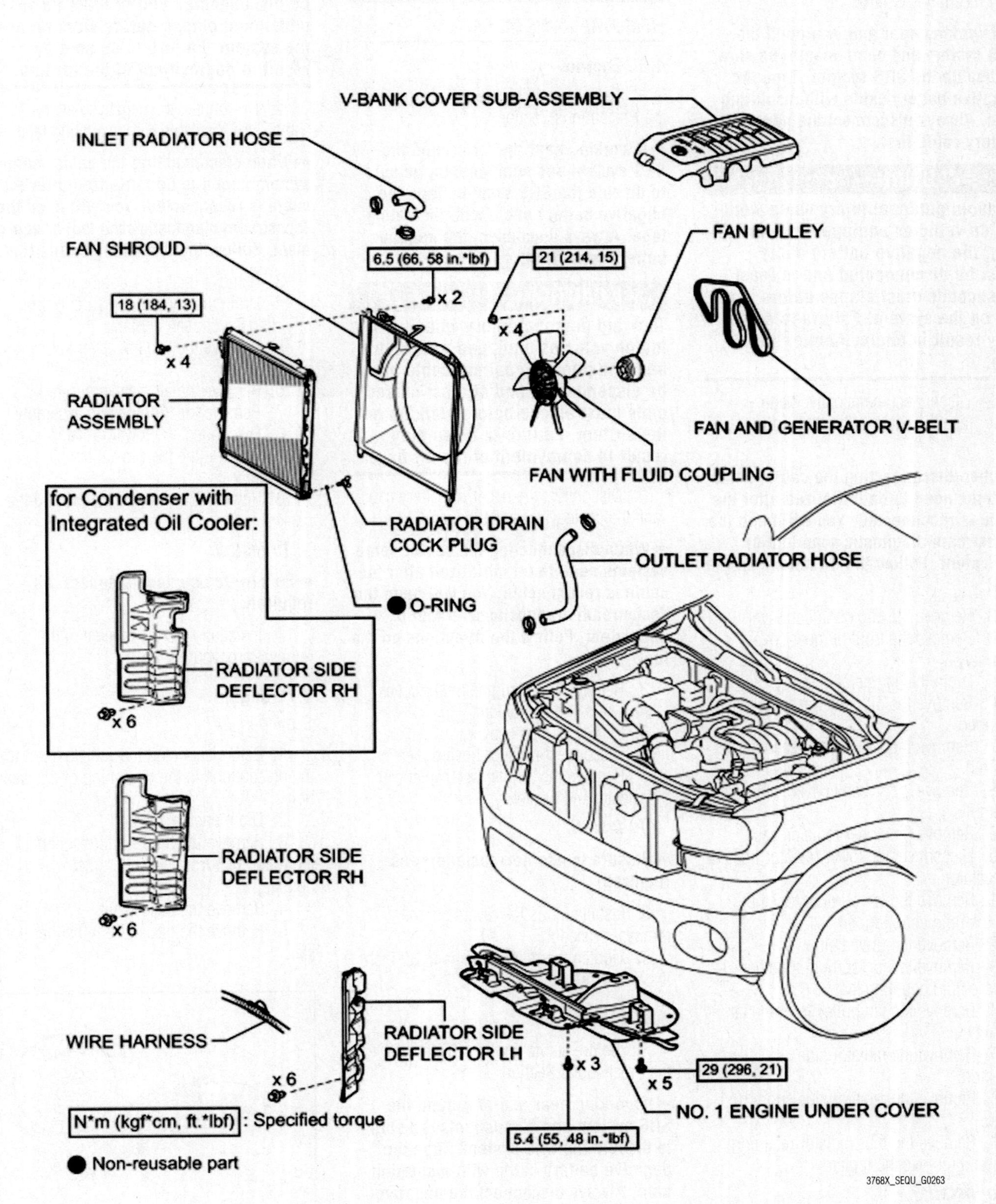

Fig. 23 Radiator and related components 5.7L engine 2 of 2—Tundra

1. Before servicing the vehicle, refer to the Precautions Section.

➡If working near and/or around the SRS system and components, be sure to disable the SRS system. Tape the negative battery cable with insulating tape. Always disconnect the negative battery cable first.

⁂ CAUTION

To avoid personal injury when working on vehicles equipped with an air bag, the negative battery cable must be disconnected and at least 90 seconds must elapse before working on the system. Failure to do so may result in deployment of the air bag.

2. Disconnect the negative battery cable. Tape the cable with insulating tape.

➡When disconnecting the cable, some systems need to be initialized after the cable is reconnected. You will need the Techstream diagnostic scan tool or equivalent. Follow the directions on the tool.

3. Remove the engine under cover.
4. Remove the front bumper cover assembly.
5. Drain the engine coolant.
6. Remove the oil cooler assembly, if equipped.
7. Remove the air cleaner assembly, as required.
8. Remove the V-bank cover assembly.
9. Remove the upper radiator hose.
10. Disconnect the lower radiator hose at the radiator.
11. Remove the fan shroud, together with the fluid coupling fan.
12. Remove the drive belt.
13. Remove the fan shroud, together with the fluid coupling fan.
14. Remove the fan pulley from the fan bracket.
15. Remove the radiator side deflectors.
16. Remove the radiator side support seals.
17. Remove the four radiator retaining bolts and remove the radiator.

To install:

➡Be sure to use new fasteners, as required.

18. Installation is the reverse of the removal procedure.

THERMOSTAT

REMOVAL & INSTALLATION

4.0L Engine

1. Before servicing the vehicle, refer to the Precautions Section.

➡If working near and/or around the SRS system and components, be sure to disable the SRS system. Tape the negative battery cable with insulating tape. Always disconnect the negative battery cable first.

⁂ CAUTION

To avoid personal injury when working on vehicles equipped with an air bag, the negative battery cable must be disconnected and at least 90 seconds must elapse before working on the system. Failure to do so may result in deployment of the air bag.

2. Disconnect the negative battery cable. Tape the cable with insulating tape.

➡When disconnecting the cable, some systems need to be initialized after the cable is reconnected. You will need the Techstream diagnostic scan tool or equivalent. Follow the directions on the tool.

3. Remove No. 1 engine under cover.
4. Drain engine coolant.
5. Remove V-bank cover.
6. Disconnect radiator hose outlet.
7. Remove the 3 nuts, water inlet with thermostat and gasket.

To install:

➡Be sure to use new fasteners, as required.

8. Installation is the reverse of the removal procedure.
9. Tighten the retaining nuts to 80 inch lbs. (9.0 Nm).

4.6L Engine

1. Before servicing the vehicle, refer to the Precautions Section.

➡If working near and/or around the SRS system and components, be sure to disable the SRS system. Tape the negative battery cable with insulating tape. Always disconnect the negative battery cable first.

⁂ CAUTION

To avoid personal injury when working on vehicles equipped with an air bag, the negative battery cable must be disconnected and at least 90 seconds must elapse before working on the system. Failure to do so may result in deployment of the air bag.

2. Disconnect the negative battery cable. Tape the cable with insulating tape.

➡When disconnecting the cable, some systems need to be initialized after the cable is reconnected. You will need the Techstream diagnostic scan tool or equivalent. Follow the directions on the tool.

3. Remove the No. 1 engine cover.
4. Drain the coolant. Be sure to properly dispose of used coolant.
5. Remove the V-bank cover sub assembly.
6. Remove the air cleaner assembly.
7. Remove the air tube sub assembly.
8. Disconnect the lower radiator hose.
9. Remove the thermostat housing retaining bolts.
10. Remove the thermostat. Discard the gasket.

To install:

➡Be sure to use new fasteners, as required.

11. Installation is the reverse of the removal procedure.

4.7L Engine

See Figure 24.

1. Before servicing the vehicle, refer to the precautions in the beginning of this section.
2. Drain engine coolant.
3. Remove the 3 nuts and disconnect the water inlet from the water inlet housing.
4. Remove the thermostat.
5. Remove the gasket from the thermostat.

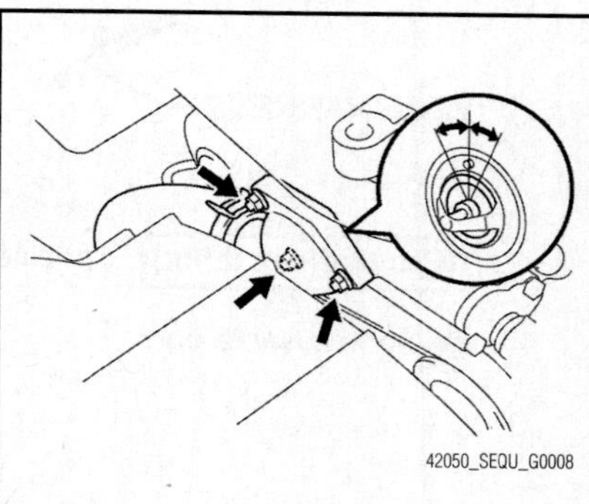

Fig. 24 Thermostat positioning and installation—4.7L engine

To install:

6. Install a new gasket to the thermostat.
7. Insert the thermostat into the water inlet housing with the jiggle valve facing straight upward.

➡The jiggle valve may be set within 30° of either side of the prescribed position.

8. Install the water inlet with the 3 nuts and tighten to 19 Nm (14 ft. lbs.).
9. Fill with engine coolant.
10. Start engine and check for coolant leaks.
11. Recheck engine coolant level.

5.7L Engine

See Figures 25 through 27.

1. Before servicing the vehicle, refer to the Precautions Section.

➡If working near and/or around the SRS system and components, be sure to disable the SRS system. Tape the negative battery cable with insulating tape. Always disconnect the negative battery cable first.

CAUTION

To avoid personal injury when working on vehicles equipped with an air bag, the negative battery cable must be disconnected and at least 90 seconds must elapse before working on the system. Failure to do so may result in deployment of the air bag.

2. Disconnect the negative battery cable. Tape the cable with insulating tape.

➡When disconnecting the cable, some systems need to be initialized after the cable is reconnected. You will need the Techstream diagnostic scan tool or equivalent. Follow the directions on the tool.

3. Remove No. 1 engine under cover.
4. Drain engine coolant.
5. Remove V-bank cover sub-assembly.
6. Disconnect outlet radiator hose.
7. Disconnect the No. 2 and No. 3 air hoses.
8. Disconnect the air pump's connector.
9. Disconnect the air pump connector clamp's holder.
10. Using a clip remover, detach the wire harness clamp.
11. Disconnect the No. 5 water by-pass hose.
12. Remove the air tube bracket's bolt.
13. Remove the 3 nuts, water inlet with thermostat and gasket.

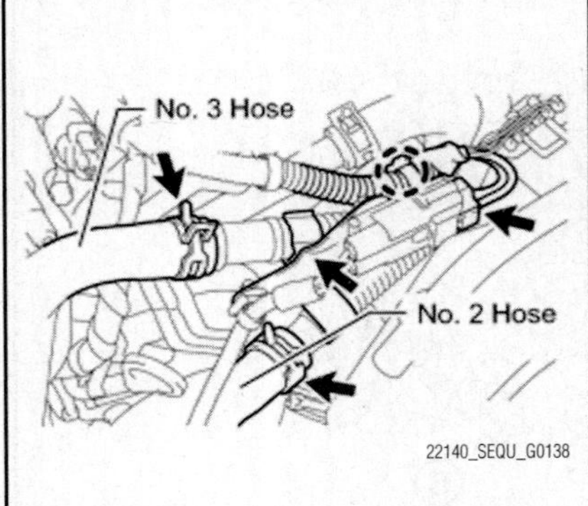

Fig. 25 Hose identification—5.7L engine

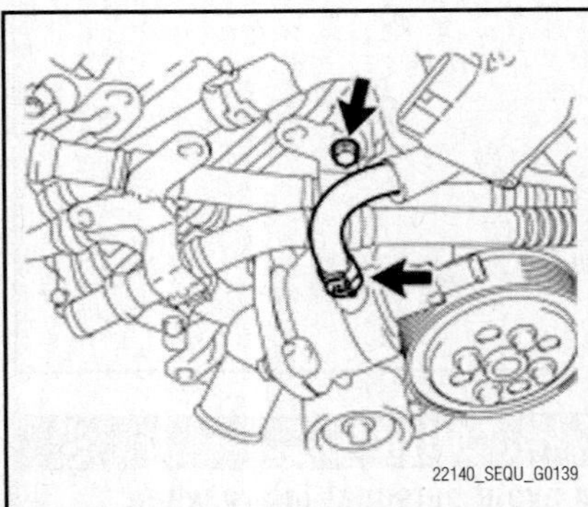

Fig. 26 By-pass Hose identification—5.7L engine

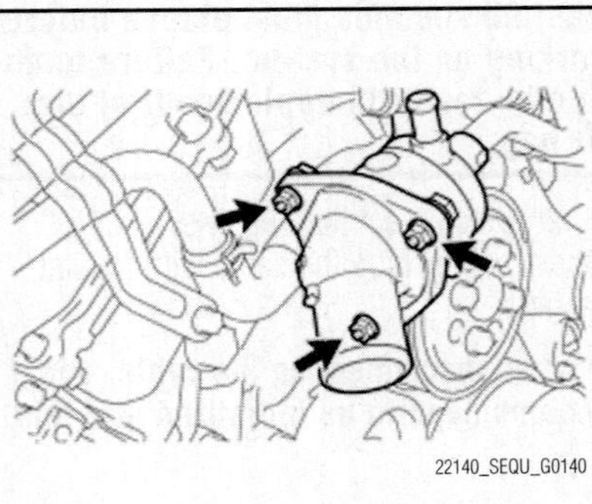

Fig. 27 Thermostat nut location—5.7L engine

To install:

➡Be sure to use new fasteners, as required.

14. Installation is the reverse of the removal procedure.

WATER PUMP

REMOVAL & INSTALLATION

4.0L Engine

See Figures 28 and 29.

1. Remove No. 1 engine under cover.
2. Drain engine coolant.
3. Remove V-bank cover.
4. Disconnect radiator hose inlet and outlet.
5. Remove fan shroud.
6. Remove No. 2 air cleaner hose.
7. Remove air cleaner assembly with element.
8. Disconnect the 2 oil cooler hoses.
9. Disconnect the 5 water by-pass hoses.
10. Remove the 5 bolts and water inlet.
11. Remove the gasket from the water outlet pipe.
12. Remove the 2 bolts, 2 cover plates and 2 idler pulleys.
13. Remove V-ribbed belt tensioner assembly.
14. Remove alternator assembly.
15. Remove the 17 bolts, water pump and gasket.

To install:

16. Install a new gasket and the water pump with the 17 bolts. Tighten 10 mm head bolts to 80 inch lbs. (9.0 Nm) and 12 mm head bolts to 17 ft. lbs. (23 Nm).
17. Install alternator assembly.
18. Install v-ribbed belt tensioner assembly.
19. Install the 2 idler pulleys and 2 cover plates with the 2 bolts an tighten to 29 ft. lbs. (39 Nm).
20. Install a new gasket to the water outlet pipe.
21. Install a new gasket to the water pump.
22. Apply soapy water to the gasket.
23. Install the water inlet with the 5 bolts and tighten to 80 inch lbs. (9.0 Nm).
24. Connect the 5 water by-pass hoses.
25. Connect the 2 oil cooler hoses.
26. Install fan shroud.
27. Connect radiator hose inlet and outlet.
28. Install air cleaner assembly with element.
29. Install No. 2 air cleaner hose.

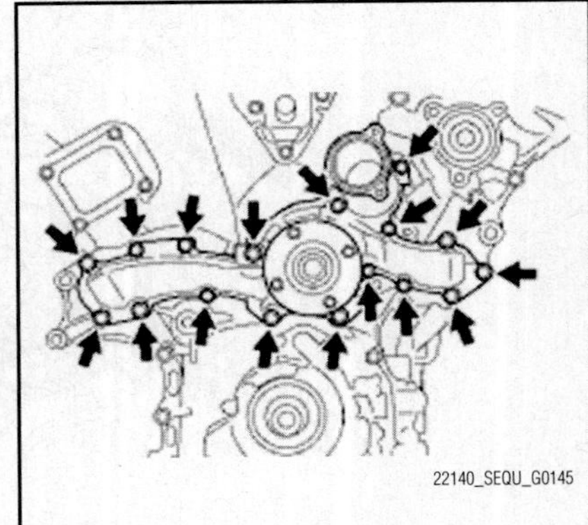

Fig. 28 Water pump bolt location

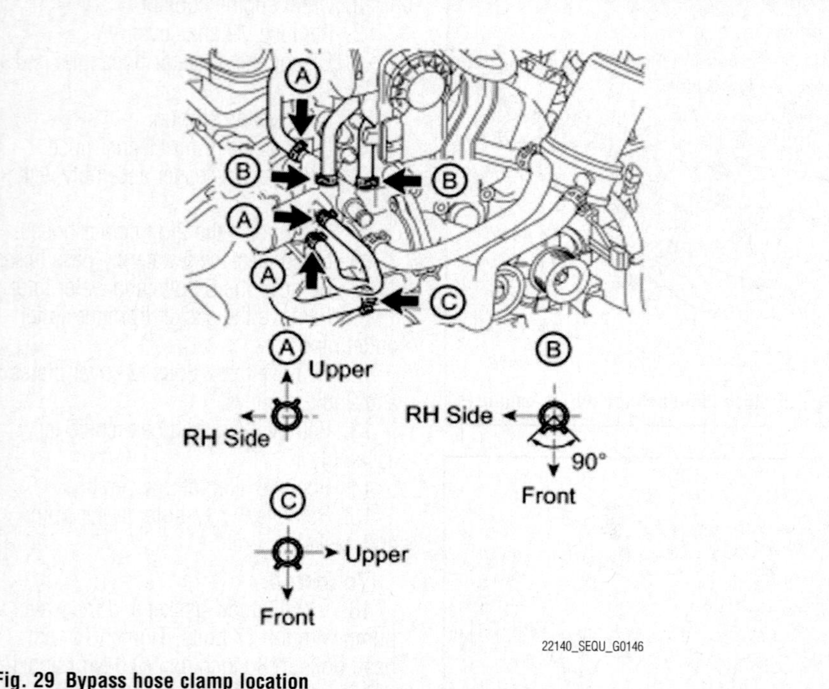

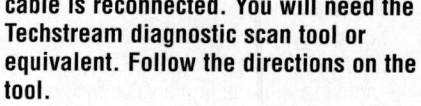

Fig. 29 Bypass hose clamp location

30. Add engine coolant.
31. Inspect for coolant leak.
32. Install V-bank cover.
33. Install No. 1 engine under cover.

4.6L Engine

See Figure 30.

1. Before servicing the vehicle, refer to the Precautions Section.

➡**If working near and/or around the SRS system and components, be sure to disable the SRS system. Tape the negative battery cable with insulating tape. Always disconnect the negative battery cable first.**

CAUTION

To avoid personal injury when working on vehicles equipped with an air bag, the negative battery cable must be disconnected and at least 90 seconds must elapse before working on the system. Failure to do so may result in deployment of the air bag.

2. Disconnect the negative battery cable. Tape the cable with insulating tape.

➡**When disconnecting the cable, some systems need to be initialized after the cable is reconnected. You will need the Techstream diagnostic scan tool or equivalent. Follow the directions on the tool.**

3. Drain the cooling system.
4. Remove the water inlet sub assembly.
5. Remove the top radiator hose.
6. Remove the fan shroud.
7. Remove the front fender apron seal, left side.
8. Disconnect the air conditioning compressor assembly. If the compressor has to be removed for clearance, be sure to properly discharge the system.
9. If equipped with oil cooler, remove the water bypass pipe.
10. Remove the No.1 water bypass hose.
11. Disconnect the No.8 water bypass hose.
12. Disconnect the No.5 water bypass hose.
13. Remove the water inlet housing.
14. Remove the water pump pulley, using the proper removal tools.
15. Remove the water pump retaining bolts.
16. Remove the water pump from its mounting. Discard the gasket.

To install:

➡**Be sure to use new fasteners, as required.**

17. Installation is the reverse of the removal procedure.
18. Be sure to use a new gasket.
19. Tighten the retaining bolts A to 35 ft. lbs. (47 Nm). Tighten the retaining bolts B to 17 ft. lbs. (23 Nm). Tighten the retaining bolts C to 15 ft. lbs. (20 Nm).
20. Be sure to fill the cooling system using the proper grade and type coolant.

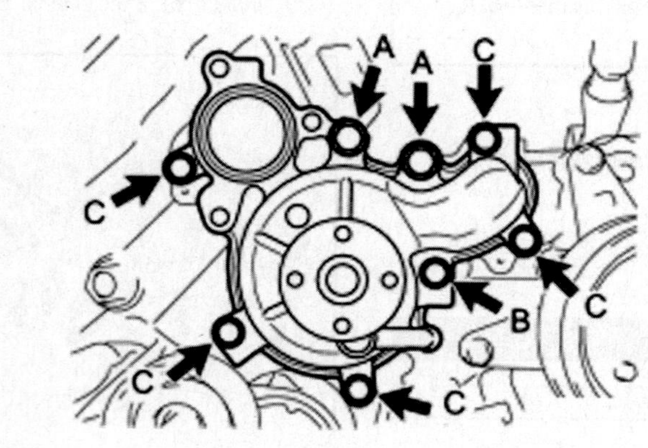

Fig. 30 Water pump bolt tightening sequence2—4.6L engine

4.7L Engine

See Figure 31.

1. Before servicing the vehicle, refer to the precautions in the beginning of this section.
2. Drain the cooling system.
3. Remove or disconnect the following:
   - Negative battery cable
   - Timing belt
   - No. 2 idler pulley
   - Radiator hose
   - Bypass hose
   - Water inlet housing assembly
   - Water pump

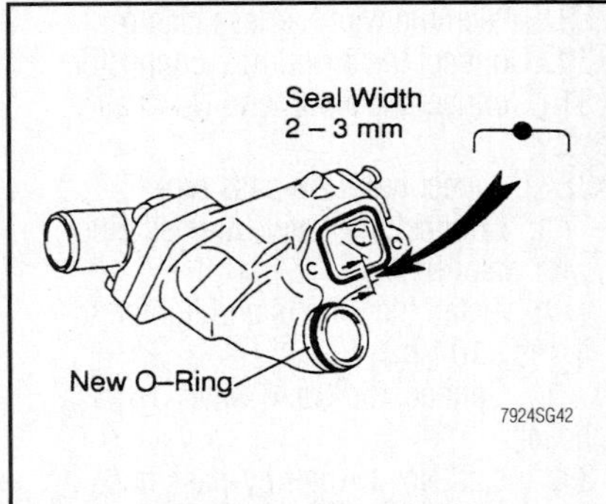

Fig. 31 Water inlet housing sealant application

To install:

4. Install or connect the following:
   - Water pump. Use a new gasket and tighten the bolts to 15 ft. lbs. (21 Nm). Tighten the stud bolt and nut to 13 ft. lbs. (18 Nm).
   - Water inlet housing assembly. Use a new O-ring and apply sealant as shown. Tighten the bolts to 13 ft. lbs. (18 Nm).
   - Bypass hose
   - Radiator hose
   - No. 2 idler pulley
   - Timing belt
   - Negative battery cable
5. Fill the cooling system.
6. Start the engine and check for leaks.

5.7L Engine

See Figures 32 and 33.

1. Before servicing the vehicle, refer to the Precautions Section.

➡If working near and/or around the SRS system and components, be sure to disable the SRS system. Tape the negative battery cable with insulating tape. Always disconnect the negative battery cable first.

⁂ CAUTION

To avoid personal injury when working on vehicles equipped with an air bag, the negative battery cable must be disconnected and at least 90 seconds must elapse before working on the system. Failure to do so may result in deployment of the air bag.

2. Disconnect the negative battery cable. Tape the cable with insulating tape.

➡When disconnecting the cable, some systems need to be initialized after the cable is reconnected. You will need the Techstream diagnostic scan tool or equivalent. Follow the directions on the tool.

3. Remove No. 1 engine under cover.
4. Drain engine coolant.
5. Remove V-bank cover sub-assembly.
6. Remove air cleaner assembly with element.
7. Remove air cleaner hose assembly.
8. Disconnect inlet radiator hose.
9. Remove fan shroud.
10. Remove No. 1 water by-pass hose.
11. Disconnect water by-pass pipe.
    a. Disconnect the No. 6 water by-pass hose.
    b. Remove the 3 bolts.
    c. Disconnect the water by-pass pipe with water hose.
12. Disconnect the No. 2 and No. 3 air hoses.
13. Disconnect the air pump's connector.
14. Disconnect the air pump connector clamp's holder.
15. Using a clip remover, detach the wire harness clamp.
16. Disconnect the No. 5 water by-pass hose.

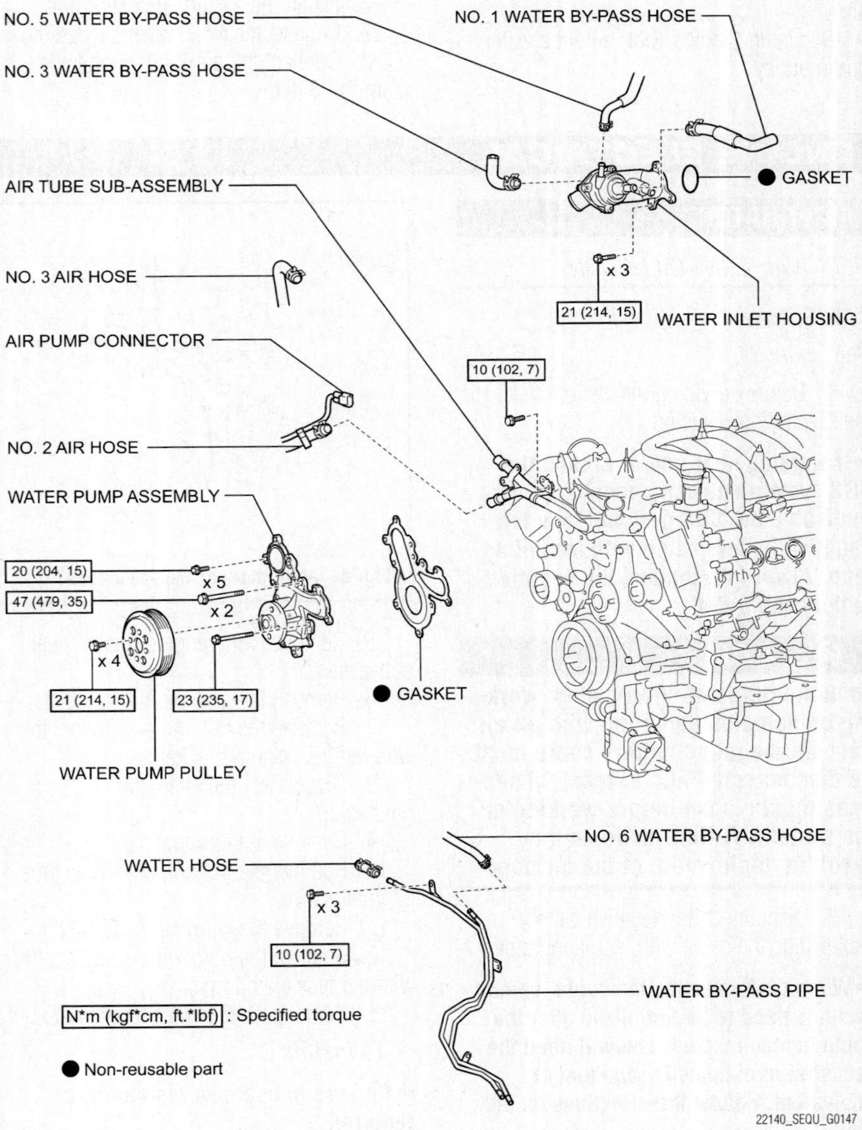

Fig. 32 Water pump components—5.7L engine

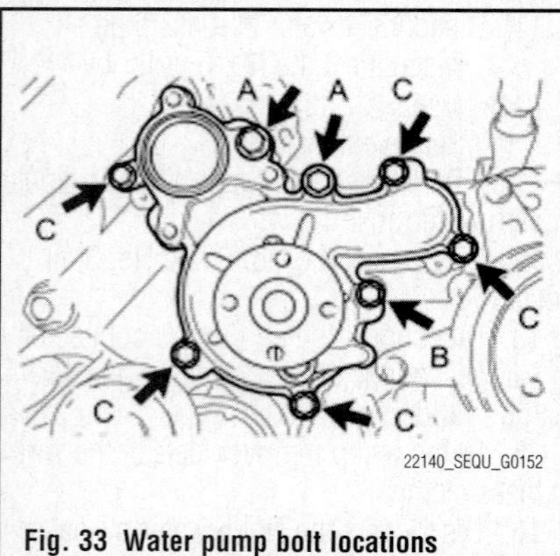

Fig. 33 Water pump bolt locations

17. Remove the air tube bracket's bolt.
18. Remove the 3 bolts, water inlet housing and No. 3 water by-pass hose.
19. Using Service Tool, hold the water pump pulley.
20. Remove the 4 bolts and water pump pulley.
21. Remove the 8 bolts, water pump and gasket.

To install:

➡Be sure to use new fasteners, as required.

22. Install a new gasket and the water pump with the 8 bolts and tighten: Bolt A to 35 ft. lbs. (47 Nm), bolt B to 17 ft. lbs. (23 Nm), and bolt C to 15 ft. lbs. (20 Nm).
23. Install the pulley with the 4 bolts and tighten to 15 ft. lbs. (21 Nm).
24. Install a new gasket to the water pump.
25. Install the No. 3 water by-pass hose and water inlet housing with the 3 bolts.
26. Install the air tube bracket's bolt.
27. Connect the No. 5 water by-pass hose.
28. Connect the air pump connector clamp's holder.
29. Attach the wire harness clamp.
30. Connect the air pump's connector.
31. Connect the No. 2 and No. 3 air hoses.
32. Connect water by-pass pipe.
    a. Connect the water by-pass pipe with water hose.
    b. Install the 3 bolts and tighten to 7 ft. lbs. (10 Nm).
    c. Connect the No. 6 water by-pass hose.
33. Install No. 1 water by-pass hose.
34. Install fan shroud.
35. Connect inlet and outlet radiator hose.
36. Install air cleaner hose assembly.
37. Install air cleaner assembly with element.
38. Add engine coolant.
39. Inspect for coolant leak.
40. Install V-bank cover sub-assembly.
41. Install No. 1 engine under cover.

ENGINE ELECTRICAL

CHARGING SYSTEM

ALTERNATOR

REMOVAL & INSTALLATION

4.0L Engine

See Figure 34.

1. Before servicing the vehicle, refer to the Precautions Section.

➡If working near and/or around the SRS system and components, be sure to disable the SRS system. Tape the negative battery cable with insulating tape. Always disconnect the negative battery cable first.

⁂ CAUTION

To avoid personal injury when working on vehicles equipped with an air bag, the negative battery cable must be disconnected and at least 90 seconds must elapse before working on the system. Failure to do so may result in deployment of the air bag.

2. Disconnect the negative battery cable. Tape the cable with insulating tape.

➡When disconnecting the cable, some systems need to be initialized after the cable is reconnected. You will need the Techstream diagnostic scan tool or equivalent. Follow the directions on the tool.

3. Remove battery.
4. Remove V-bank cover.

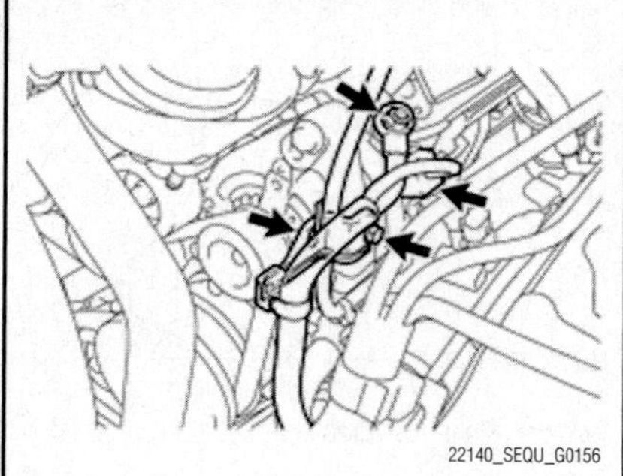

Fig. 34 Alternator connector location

5. Remove No. 1 engine under cover sub-assembly.
6. Remove fan and alternator V belt.
7. Remove the nut and disconnect the wire harness clamp bracket.
8. Disconnect the alternator connector.
9. Remove the terminal cap.
10. Remove the nut and disconnect the alternator wire.
11. Remove the wire harness clamp.
12. Remove the bolt and disconnect the wire harness clamp bracket.
13. Remove the 2 bolts and alternator.

To install:

➡Be sure to use new fasteners, as required.

14. Install the alternator with the 2 bolts and tighten to 32 ft. lbs. (43 Nm).
15. Install the wire harness clamp bracket with the bolt.
16. Install the wire harness clamp to the bracket.
17. Connect the alternator wire with the nut and tighten to 87 inch lbs. (9.8 Nm).
18. Install the terminal cap.
19. Connect the alternator connector.
20. Install the wire harness clamp bracket with the nut.
21. Install fan and alternator V belt.
22. Install No. 1 engine under cover sub-assembly.
23. Install V-bank cover.
24. Install battery.
25. Connect cable to negative battery terminal.

➡Perform initialization, if necessary.

4.6L Engine

➡If working near and/or around the SRS system and components, be sure to disable the SRS system. Tape the negative battery cable with insulating tape. Always disconnect the negative battery cable first.

⁂ CAUTION

To avoid personal injury when working on vehicles equipped with an air bag, the negative battery cable must be disconnected and at least 90 seconds must elapse before working on the system. Failure to do so may result in deployment of the air bag.

1. Disconnect the negative battery cable. Tape the cable with insulating tape.

➡**When disconnecting the cable, some systems need to be initialized after the cable is reconnected. You will need the Techstream diagnostic scan tool or equivalent. Follow the directions on the tool.**

2. Remove the fan and alternator V belt
3. Remove the air cleaner assembly.
4. Remove the front fender apron seal.
5. Disconnect the power steering pressure switch connector.
6. Disconnect the solenoid valve connector.
7. Disconnect the oil cooler tube sub assembly, if equipped.
8. Remove the alternator retaining bolts.
9. Disconnect the electrical connectors.
10. Remove the component from its mounting.

To install:

➡**Be sure to use new fasteners, as required.**

11. Installation is the reverse of the removal procedure.

4.7L Engine

See Figure 55.

1. Before servicing the vehicle, refer to the Precautions Section.

➡**If working near and/or around the SRS system and components, be sure to disable the SRS system. Tape the negative battery cable with insulating tape. Always disconnect the negative battery cable first.**

⁂ CAUTION

To avoid personal injury when working on vehicles equipped with an air bag, the negative battery cable must be disconnected and at least 90 seconds must elapse before working on the system. Failure to do so may result in deployment of the air bag.

2. Disconnect the negative battery cable. Tape the cable with insulating tape.

➡**When disconnecting the cable, some systems need to be initialized after the cable is reconnected. You will need the Techstream diagnostic scan tool or equivalent. Follow the directions on the tool.**

3. Drain the cooling system.
4. Remove or disconnect the following:
   - Negative battery cable
   - Accessory drive belt
   - Engine under cover
   - Power steering pump
   - Alternator harness connectors
   - Alternator

To install:

➡**Be sure to use new fasteners, as required.**

5. Install or connect the following:
   - Alternator. Tighten the fasteners to 29 ft. lbs. (39 Nm).
   - Alternator harness connectors
   - Power steering pump
   - Engine under cover
   - Accessory drive belt
   - Negative battery cable
6. Fill the cooling system.
7. Start the engine and check for leaks.

5.7L Engine

See Figures 35 and 36.

1. Before servicing the vehicle, refer to the Precautions Section.

➡**If working near and/or around the SRS system and components, be sure to disable the SRS system. Tape the negative battery cable with insulating tape. Always disconnect the negative battery cable first.**

⁂ CAUTION

To avoid personal injury when working on vehicles equipped with an air bag, the negative battery cable must be disconnected and at least 90 seconds must elapse before working on the system. Failure to do so may result in deployment of the air bag.

2. Disconnect the negative battery cable. Tape the cable with insulating tape.

➡**When disconnecting the cable, some systems need to be initialized after the cable is reconnected. You will need the Techstream diagnostic scan tool or equivalent. Follow the directions on the tool.**

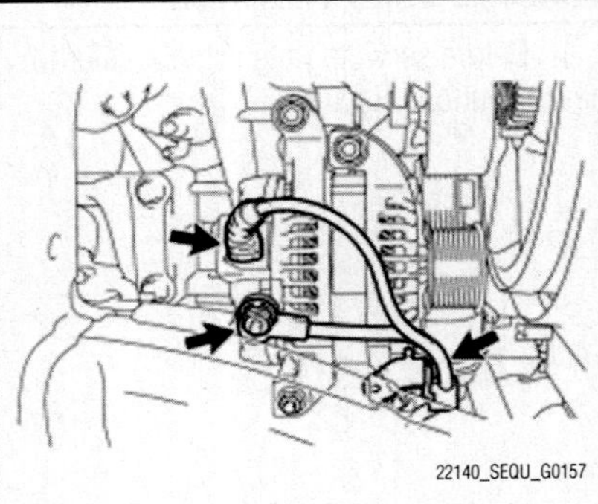

Fig. 35 Alternator connector location

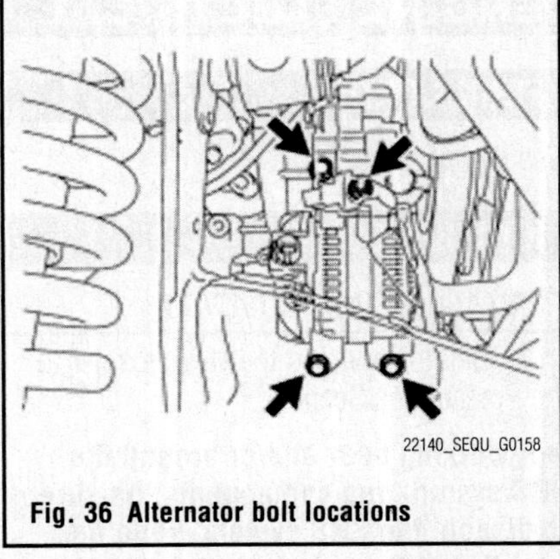

Fig. 36 Alternator bolt locations

3. Remove air cleaner hose assembly.
4. Remove air cleaner assembly.
5. Remove V-bank cover.
6. Remove No. 1 engine under cover sub-assembly.
7. Remove fan and alternator V belt.
8. Remove the 5 clips and apron seal.
9. Disconnect vane pump assembly.
10. Disconnect the alternator connector.
11. Remove the terminal cap and nut, and disconnect the alternator wire.
12. Remove the bolt and disconnect the wire harness bracket from the alternator.
13. Remove the 3 bolts and alternator.

To install:

➡**Be sure to use new fasteners, as required.**

14. Install the alternator with the 3 bolts and tighten to 32 ft. lbs. (43 Nm).
15. Connect the alternator wire with the nut and tighten to 87 inch lbs. (9.8 Nm).
16. Install the terminal cap.
17. Connect the alternator connector.
18. Install the wire harness clamp bracket with the nut.
19. Connect vane pump assembly.
20. Install the apron seal with the 5 clips.
21. Install fan and alternator V belt.
22. Install No. 1 engine under cover sub-assembly.
23. Install air cleaner assembly.
24. Install air cleaner hose assembly.
25. Install V-bank cover.
26. Install battery.
27. Connect cable to negative battery terminal.

➡**Perform initialization, if necessary.**

FIRING ORDER

See Figure 37.

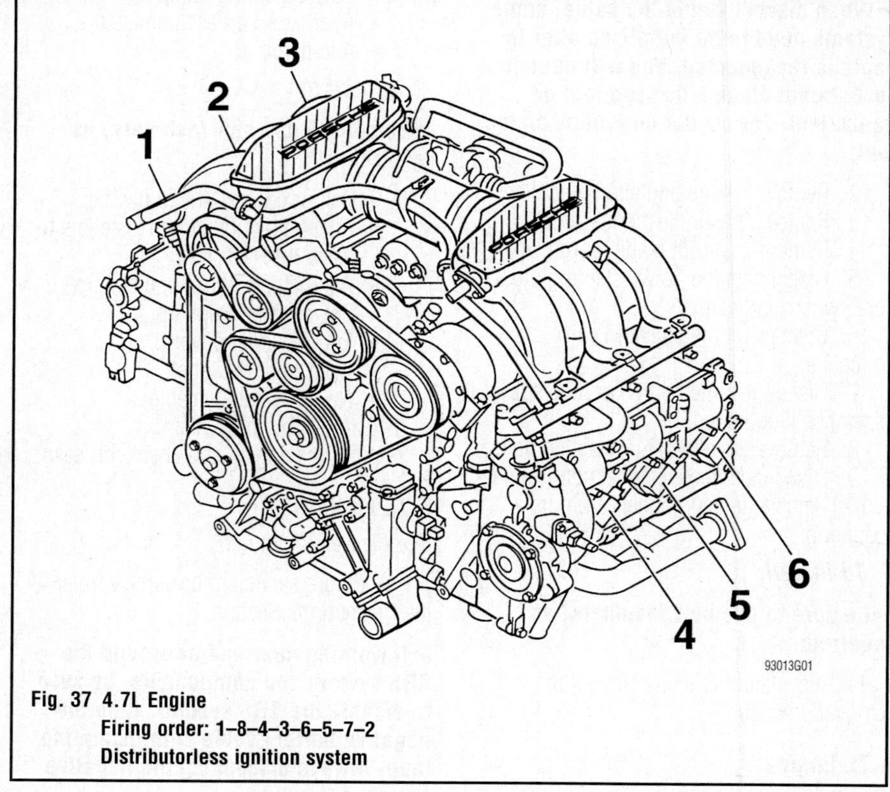

Fig. 37 4.7L Engine
Firing order: 1-8-4-3-6-5-7-2
Distributorless ignition system

IGNITION COIL

REMOVAL & INSTALLATION

1. Before servicing the vehicle, refer to the Precautions Section.

➡If working near and/or around the SRS system and components, be sure to disable the SRS system. Tape the negative battery cable with insulating tape. Always disconnect the negative battery cable first.

⁂ CAUTION

To avoid personal injury when working on vehicles equipped with an air bag, the negative battery cable must be disconnected and at least 90 seconds must elapse before working on the system. Failure to do so may result in deployment of the air bag.

2. Disconnect the negative battery cable. Tape the cable with insulating tape.

➡When disconnecting the cable, some systems need to be initialized after the cable is reconnected. You will need the Techstream diagnostic scan tool or equivalent. Follow the directions on the tool.

3. Remove the V bank cover assembly, as required.
4. Remove the air cleaner assembly, as required.
5. Remove the necessary components in order to gain access to the component.
6. Disconnect ignition coil (with igniter) connectors.
7. Remove the bolt, and pull out the ignition coil (with igniter). Remove the ignition coils (with igniter).

To install:

➡Be sure to use new fasteners, as required.

8. Installation is the reverse of the removal procedure.

IGNITION TIMING

ADJUSTMENT

The ignition timing is controlled by the Powertrain Control Module (PCM). No adjustment is necessary or possible.

SPARK PLUGS

REMOVAL & INSTALLATION

1. Before servicing the vehicle, refer to the Precautions Section.
2. Remove the ignition coils.
3. Using a 16 mm plug wrench, remove the spark plugs.
4. Clean the spark plugs.
5. If the electrode has traces of wet carbon, allow it to dry and then clean with a spark plug cleaner.
6. Check the spark plug for thread damage and insulator damage. If abnormal, replace the spark plug.
7. Adjust the spark plug electrode gap. Electrode gap for new spark plug is 1.0 to 1.1 mm (0.039 to 0.043 in.).
8. Using a 16 mm plug wrench, install the spark plugs and tighten to specification.
9. Reinstall the ignition coils.

STARTER

REMOVAL & INSTALLATION

4.0L Engine

See Figure 38.

1. Before servicing the vehicle, refer to the Precautions Section.

➡If working near and/or around the SRS system and components, be sure to disable the SRS system. Tape the negative battery cable with insulating tape. Always disconnect the negative battery cable first.

⁂ CAUTION

To avoid personal injury when working on vehicles equipped with an air bag, the negative battery cable must be disconnected and at least 90 seconds must elapse before working on the system. Failure to do so may result in deployment of the air bag.

2. Disconnect the negative battery cable. Tape the cable with insulating tape.

➡When disconnecting the cable, some systems need to be initialized after the cable is reconnected. You will need the Techstream diagnostic scan tool or equivalent. Follow the directions on the tool.

3. Disconnect cable from negative battery terminal.
4. Remove the terminal cap.
5. Remove the nut and disconnect the starter wire.
6. Disconnect the starter connector.
7. Remove the 2 bolts and starter assembly.

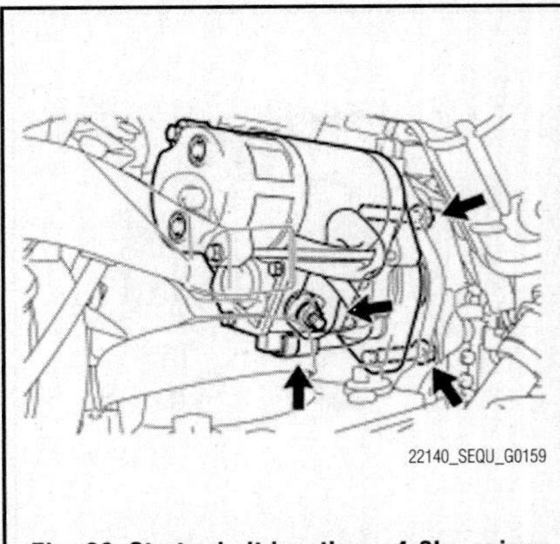

Fig. 38 Starter bolt location—4.0L engine

To install:

8. Install the starter with the 2 bolts and tighten to 43 ft. lbs. (58 Nm).
9. Connect the starter connector.
10. Connect the starter wire with the nut and tighten nut to 87 inch lbs. (9.8 Nm).
11. Install the terminal cap.
12. Connect cable to negative battery terminal.

➡Some systems need to be initialized after the cable is reconnected.

4.6L Engine

See Figure 39.

1. Before servicing the vehicle, refer to the Precautions Section.

➡If working near and/or around the SRS system and components, be sure to disable the SRS system. Tape the negative battery cable with insulating tape. Always disconnect the negative battery cable first.

⁂ CAUTION

To avoid personal injury when working on vehicles equipped with an air bag, the negative battery cable must be disconnected and at least 90 seconds must elapse before working on the system. Failure to do so may result in deployment of the air bag.

2. Disconnect the negative battery cable. Tape the cable with insulating tape.

➡When disconnecting the cable, some systems need to be initialized after the cable is reconnected. You will need the Techstream diagnostic scan tool or equivalent. Follow the directions on the tool.

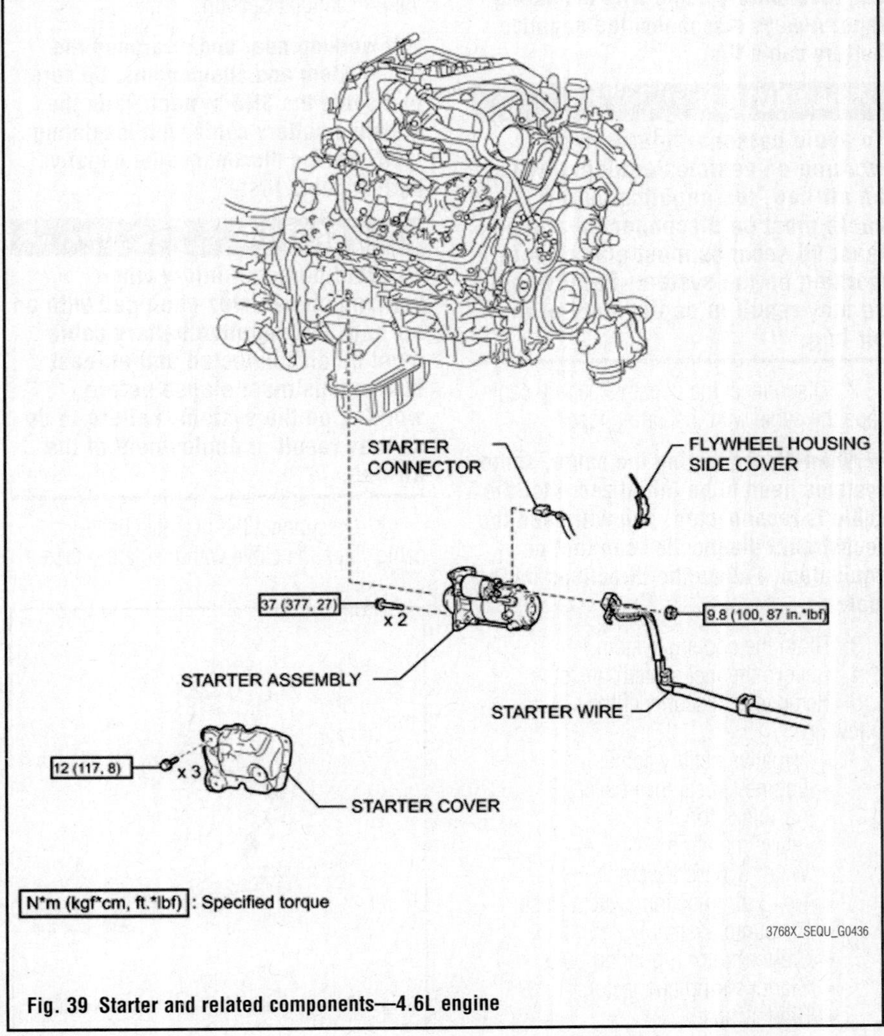

Fig. 39 Starter and related components—4.6L engine

3. Remove the exhaust manifold sub assembly.
4. Remove the starter cover.
5. Disconnect the electrical connector.
6. Remove the starter retaining bolts.
7. Remove the starter from its mounting.
8. Remove the flywheel side cover.

To install:

➡Be sure to use new fasteners, as required.

9. Installation is the reverse of the removal procedure.
10. Tighten the retaining bolts to 27 ft. lbs. (37 Nm).

4.7L Engine

See Figure 61.

1. Before servicing the vehicle, refer to the Precautions Section.

➡If working near and/or around the SRS system and components, be sure to disable the SRS system. Tape the negative battery cable with insulating tape. Always disconnect the negative battery cable first.

⁂ CAUTION

To avoid personal injury when working on vehicles equipped with an air bag, the negative battery cable must be disconnected and at least 90 seconds must elapse before working on the system. Failure to do so may result in deployment of the air bag.

2. Disconnect the negative battery cable. Tape the cable with insulating tape.

➡When disconnecting the cable, some systems need to be initialized after the cable is reconnected. You will need the Techstream diagnostic scan tool or equivalent. Follow the directions on the tool.

3. Drain the cooling system.
4. Relieve the fuel system pressure.
5. Remove or disconnect the following:
   - Negative battery cable
   - Engine appearance cover
   - Air intake tube
   - Intake manifold
   - Water bypass assembly
   - No. 4 air injection system hose
   - Air pump assembly
   - Starter motor mounting bolts
   - Starter wiring connectors
   - Starter motor

To install:

➡Be sure to use new fasteners, as required.

6. Install or connect the following:
   - Starter motor
   - Starter wiring connectors. Tighten the cable nut to 86 inch lbs. (10 Nm).
   - Starter motor mounting bolts. Tighten the bolts to 18 ft. lbs. (25 Nm).
   - Air pump assembly
   - No. 4 air injection system hose
   - Water bypass assembly
   - Intake manifold
   - Air intake tube
   - Engine appearance cover
   - Negative battery cable
7. Fill the cooling system.
8. Start the engine and check for leaks.

5.7L Engine

See Figures 40 through 46.

1. Before servicing the vehicle, refer to the Precautions Section.

➡If working near and/or around the SRS system and components, be sure to disable the SRS system. Tape the negative battery cable with insulating tape. Always disconnect the negative battery cable first.

⁂ CAUTION

To avoid personal injury when working on vehicles equipped with an air bag, the negative battery cable must be disconnected and at least 90 seconds must elapse before working on the system. Failure to do so may result in deployment of the air bag.

2. Disconnect the negative battery cable. Tape the cable with insulating tape.

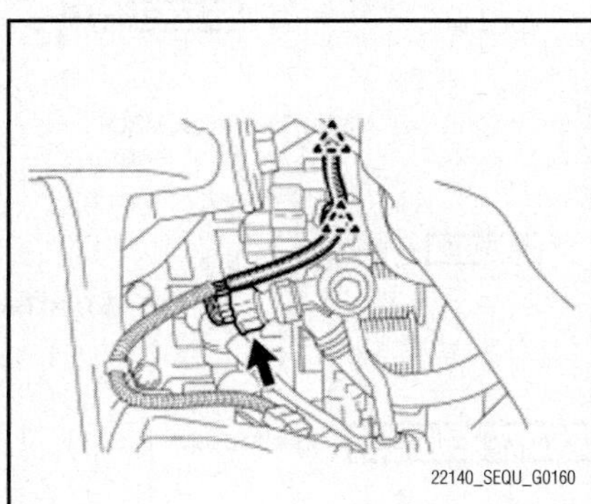

Fig. 40 Power steering connector location

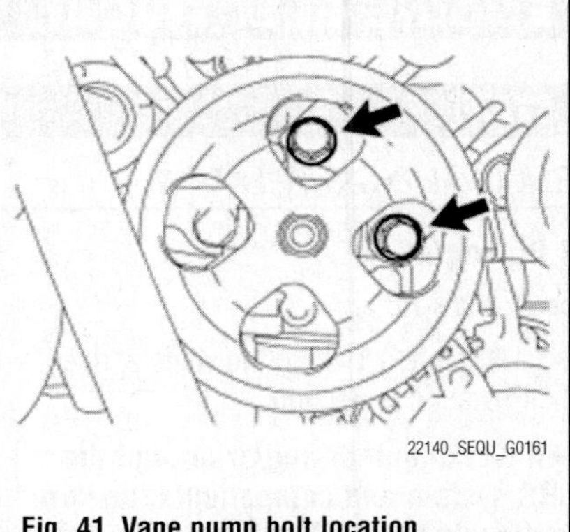

Fig. 41 Vane pump bolt location

➡When disconnecting the cable, some systems need to be initialized after the cable is reconnected. You will need the Techstream diagnostic scan tool or equivalent. Follow the directions on the tool.

3. Remove No. 1 engine under cover.
4. Remove V-bank cover.
5. Remove air cleaner hose assembly.
6. Remove air cleaner assembly.
7. Remove fan and alternator V belt.
8. Remove front fender apron seal RH.
9. Disconnect the 2 clamps and power steering oil pressure switch connector.
10. Remove the 2 bolts and disconnect the vane pump
11. Disconnect the alternator connector.
12. Remove the terminal cap and nut, and disconnect the alternator wire.
13. Remove the bolt and disconnect the wire harness bracket from the alternator.
14. Remove the 2 bolts, 2 nuts and alternator.
15. Remove the dipstick.
16. Remove the bolt and dipstick guide.
17. Remove the O-ring from the dipstick guide.

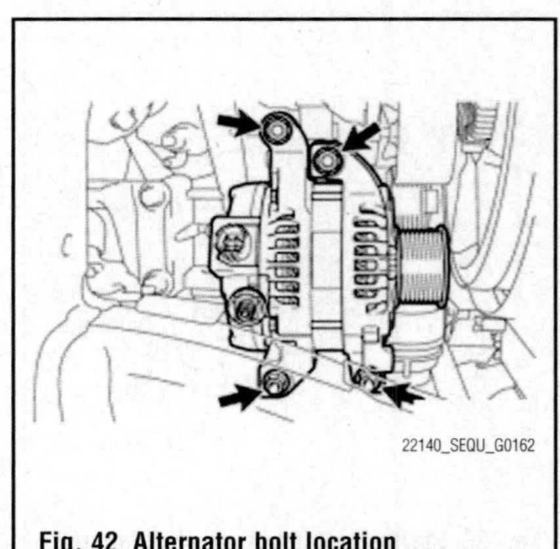

Fig. 42 Alternator bolt location

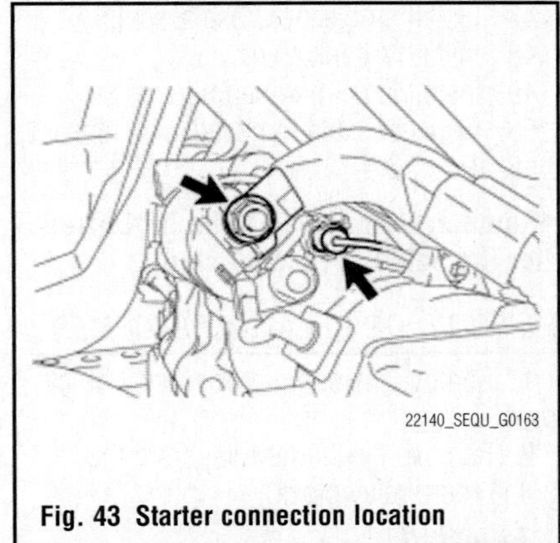

Fig. 43 Starter connection location

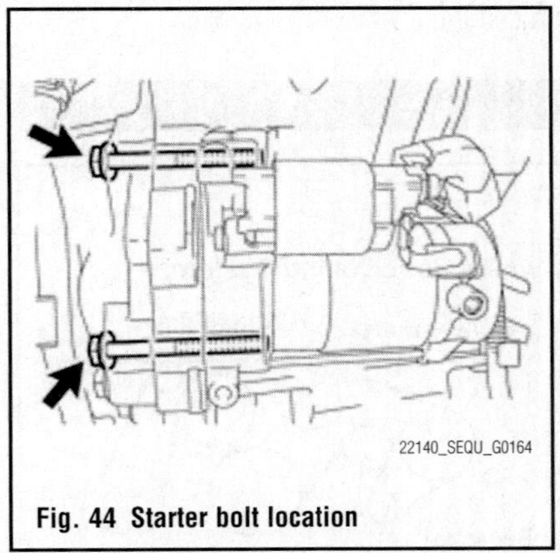

Fig. 44 Starter bolt location

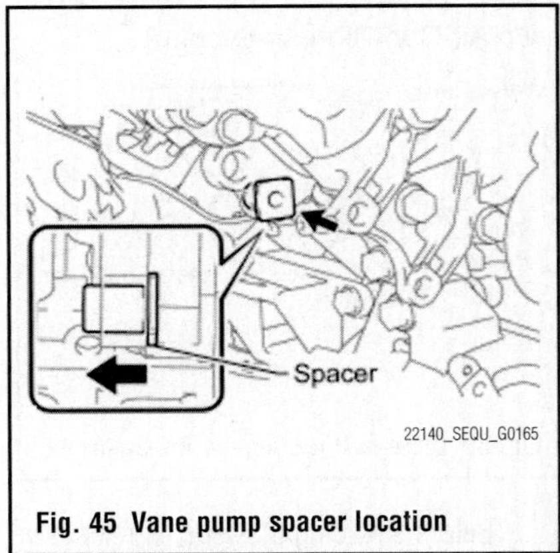

Fig. 45 Vane pump spacer location

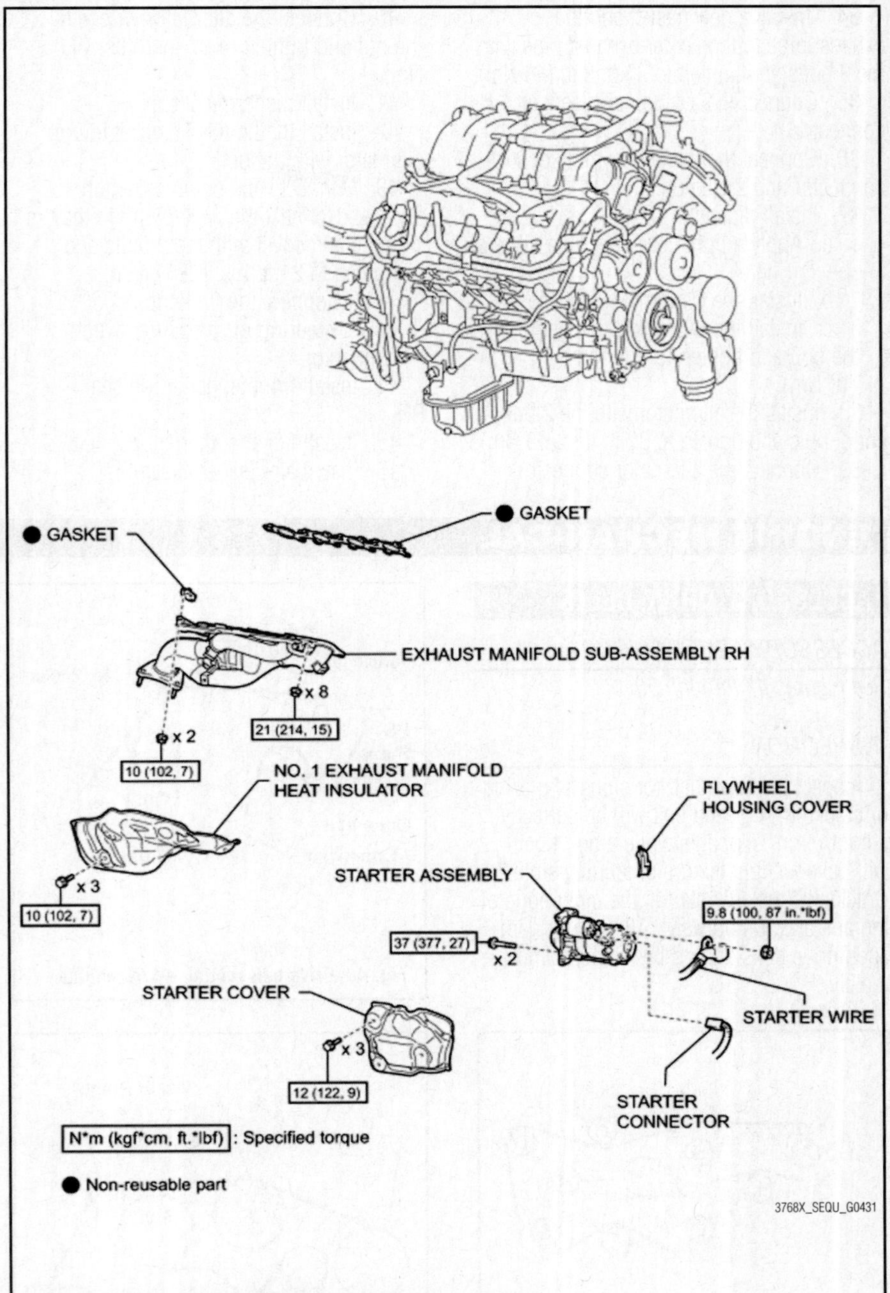

Fig. 46 Starter and related components—5.7L engine

18. Remove front exhaust pipe assembly.
 a. Disconnect the air fuel ratio sensor connector.
 b. Disconnect the heated oxygen sensor connector and 2 clamps.
 c. Remove the 2 bolts.
 d. Remove the 3 nuts and front exhaust pipe.
 e. Remove the 2 gaskets.
19. Remove No. 1 exhaust manifold heat insulator.
20. Remove exhaust manifold sub-assembly RH.
21. Remove the 3 bolts and starter cover.
22. Disconnect the starter connector.
23. Remove the nut and disconnect the starter wire.
24. Remove the 2 bolts and starter.
25. Remove the flywheel housing side cover.

To install:

➡Be sure to use new fasteners, as required.

26. Install the flywheel housing side cove.
27. Install the starter with the 2 bolts and tighten to 27 ft. lbs. (37 Nm).
28. Install the starter wire with the nut and tighten to 87 inch lbs. (9.8 Nm).
29. Connect the starter connector.
30. Install the starter cover with the 3 bolts and tighten to 9 ft. lbs. (12 Nm).
31. Install exhaust manifold sub-assembly RH.
32. Install No. 1 exhaust manifold heat insulator.
33. Install a new gasket and the exhaust pipe to the exhaust manifold RH with 3 new nuts and tighten to 40 ft. lbs. (54 Nm).

34. Install a new gasket and the exhaust pipe to the center exhaust pipe with the 2 bolts and tighten to 35 ft. lbs. (48 Nm).
35. Connect the air fuel ratio sensor connector.
36. Connect the heated oxygen sensor connector and 2 clamps.
37. Install dipstick.
  a. Apply a light coat of engine oil to a new O-ring.
  b. Install the O-ring to the guide.
  c. Install the dipstick guide with the bolt and tighten to 7 ft. lbs. (10 Nm).
38. Install the alternator with the 2 bolts and 2 nuts and tighten to 32 ft. lbs. (43 Nm).
39. Connect the alternator connector.
40. Connect the alternator wire with the nut and tighten to 87 inch lbs. (9.8 Nm).
41. Install the terminal cap.
42. Install the harness bracket to the alternator with the bolt.
43. Connect vane pump assembly.
  a. Connect the vane pump to the timing chain cover with the 2 bolts and tighten to 21 ft. lbs. (28 Nm).
  b. Connect the 2 clamps and power steering oil pressure switch connector.
44. Install front fender apron seal RH.
45. Install fan and alternator V belt.
46. Install air cleaner assembly.
47. Install air cleaner hose assembly.
48. Install V-bank cover.
49. Install No. 1 engine under cover.
50. Connect cable to negative battery terminal.

➡Some systems need to be initialized after the cable is reconnected.

SOLENOID OR RELAY REPLACEMENT

1. Remove the engine room relay block cover.
2. Remove the starter relay from the engine room relay block.

To install:

3. To install, reverse removal procedure.

ENGINE MECHANICAL

ACCESSORY DRIVE BELTS

ACCESSORY BELT ROUTING

See Figures 47 through 50.

INSPECTION

Inspect the drive belt for signs of glazing or cracking. A glazed belt will be perfectly smooth from slippage, while a good belt will have a slight texture of fabric visible. Cracks will usually start at the inner edge of the belt and run outward. All worn or damaged drive belts should be replaced immediately.

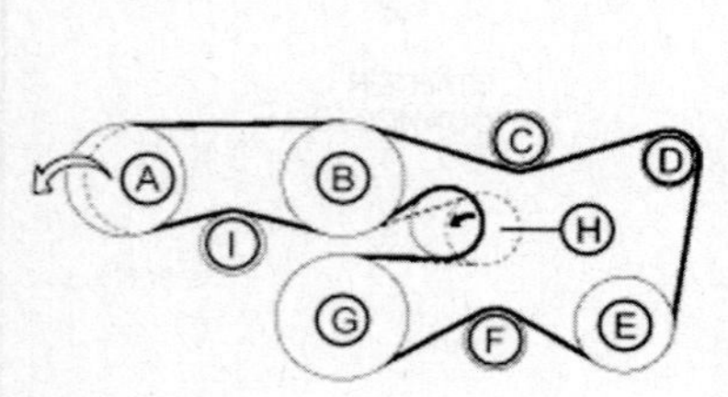

A: Vane Pump
B: Water Pump
C: No. 2 Idler
D: Generator
E: Cooler Compressor
F: No. 1 Idler
G: Crankshaft
H: V-Ribbed Belt Tensioner
I: No. 2 Idler

22140_SEQU_G0166

Fig. 47 Drive belt routing—4.0L engine

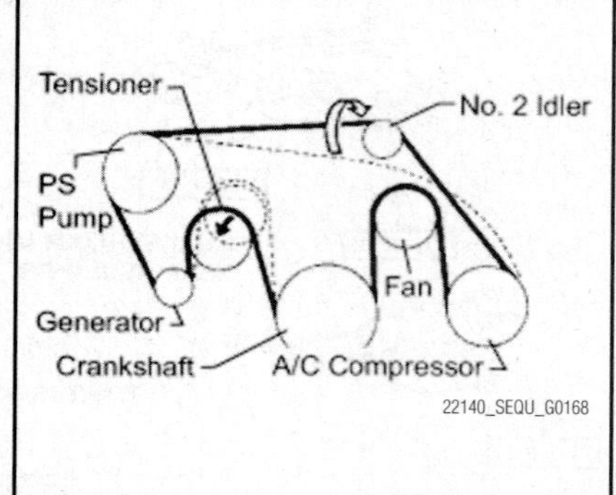

Fig. 48 Drive belt routing—4.7L engine

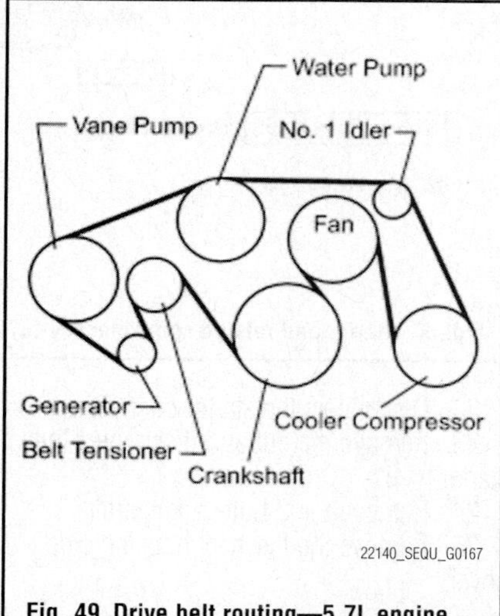

Fig. 49 Drive belt routing—5.7L engine

ADJUSTMENT

The belt does not require adjustment.

REMOVAL & INSTALLATION

See Figure 51.

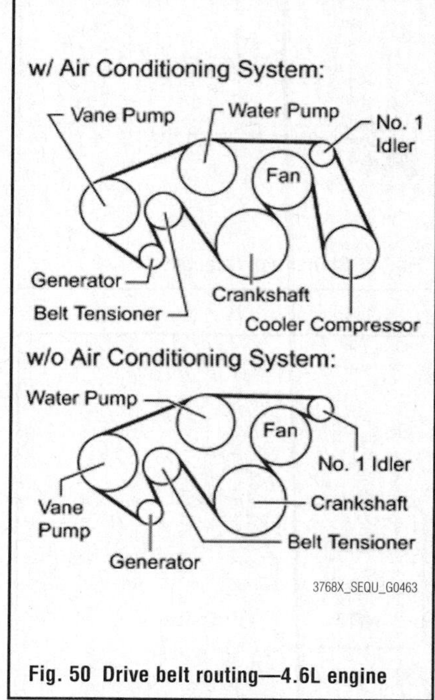

Fig. 50 Drive belt routing—4.6L engine

1. Before servicing the vehicle, refer to the Precautions Section.

➡If working near and/or around the SRS system and components, be sure to disable the SRS system. Tape the negative battery cable with insulating tape. Always disconnect the negative battery cable first.

CAUTION

To avoid personal injury when working on vehicles equipped with an air bag, the negative battery cable must be disconnected and at least 90 seconds must elapse before work-

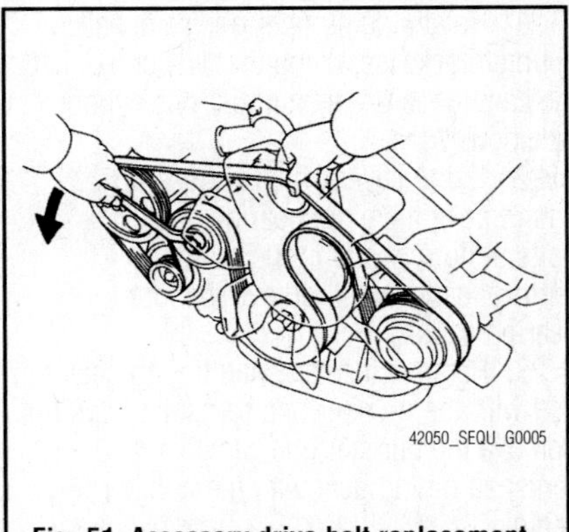

Fig. 51 Accessory drive belt replacement

ing on the system. Failure to do so may result in deployment of the air bag.

2. Disconnect the negative battery cable. Tape the cable with insulating tape.

➡When disconnecting the cable, some systems need to be initialized after the cable is reconnected. You will need the Techstream diagnostic scan tool or equivalent. Follow the directions on the tool.

3. Remove the V bank cover, as required.
4. Remove the air cleaner assembly, as required.
5. On 4.6L and 5.7L engines, while turning the belt tensioner counterclockwise, align the service hole for the belt tensioner and the belt tensioner fixing position. Insert a 0.197 inch bar into the service hole to hold the belt tensioner in place.
6. Loosen the drive belt tension by turning the drive belt tensioner counterclockwise, and remove the drive belt.

To install:

➡Be sure to use new fasteners, as required.

7. Installation is the reverse of the removal procedure.
8. On 4.6L and 5.7L engines, while turning the belt tensioner counterclockwise remove the pin.

CAMSHAFT AND VALVE LIFTERS

REMOVAL & INSTALLATION

4.0L Engine

Bank 1

See Figures 52 through 66.

1. Before servicing the vehicle, refer to the Precautions Section.

➡If working near and/or around the SRS system and components, be sure to disable the SRS system. Tape the negative battery cable with insulating tape. Always disconnect the negative battery cable first.

CAUTION

To avoid personal injury when working on vehicles equipped with an air bag, the negative battery cable must be disconnected and at least 90 seconds must elapse before working on the system. Failure to do so may result in deployment of the air bag.

2. Disconnect the negative battery cable. Tape the cable with insulating tape.

➡When disconnecting the cable, some systems need to be initialized after the cable is reconnected. You will need the Techstream diagnostic scan tool or equivalent. Follow the directions on the tool.

3. Drain the engine coolant. Remove the V bank cover.
4. Remove the air cleaner assembly.
5. Disconnect the two water bypass hoses. Disconnect the fuel vapor feed hose. Disconnect the ventilation hose. Disconnect the VSV connectors.
6. Disconnect the throttle body motor connector. Separate the three wire harness clamps and hose clamp.
7. Remove the two bolts and the throttle body bracket. Remove the bolt and the oil baffle plate. Remove the four bolts and the two serge tank stays.
8. Remove the two nuts. Remove the four bolts, intake air surge tank and gasket.
9. Remove the ignition coil assembly.
10. Remove the cylinder head cover retaining bolts. Remove the cylinder head cover.

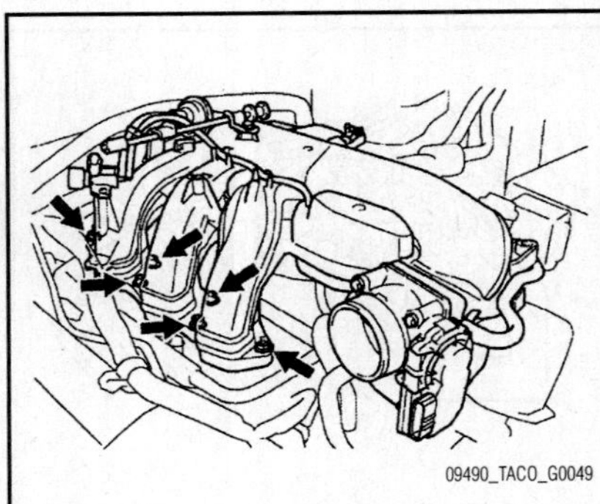

Fig. 52 Intake surge tank bolt locations—4.0L engine

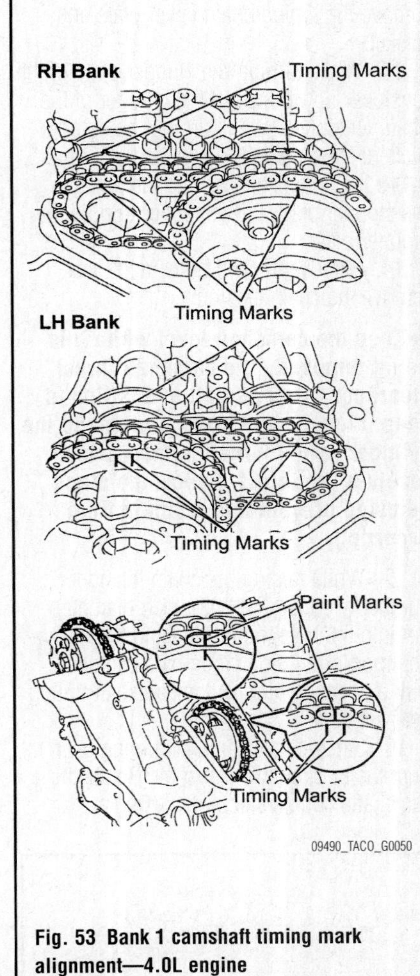

Fig. 53 Bank 1 camshaft timing mark alignment—4.0L engine

11. Turn the crankshaft pulley until its groove and the timing mark "0" of the timing chain cover are aligned. If not aligned at TDC of the compression stroke, turn the crankshaft one complete revolution, in the direction of rotation. Paint alignment marks on the number one chain links corresponding to the timing marks of the camshaft timing gears.

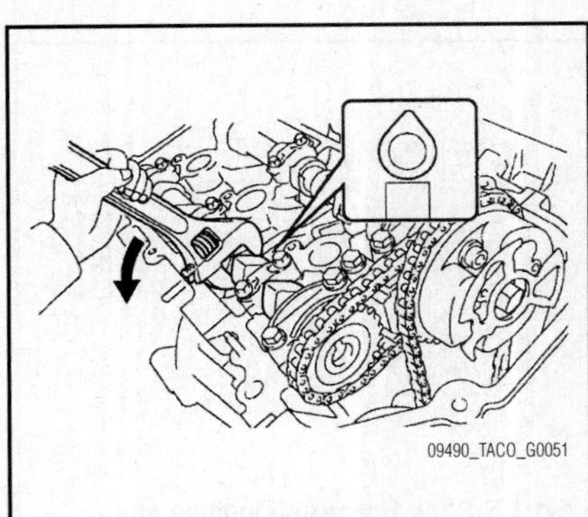

Fig. 54 Bank 1 camshaft lobe removal positioning—4.0L engine

12. Remove the four bolts, and then remove the timing chain cover plate and gasket.

13. While turning the stopper plate of the tensioner upward, push the plunger of the chain tensioner. While turning the stopper plate of the tensioner downward, insert a 0.118 inch diameter bar into the holes in the stopper plate and tensioner to hold the stopper plate.

14. Remove the two bolts, and then remove the chain tensioner.

➡Keep the camshaft level while it is being removed. The camshaft thrust clearance is very small and failing to keep it level could crack or damage the cylinder head journal surface, which receives the thrust. Follow the steps below to prevent this problem from occurring.

15. While raising the chain tensioner number two insert a 0.039 inch diameter pin into the hole to hold it. Hold the hexagonal portion of the number two camshaft with a wrench. Remove the camshaft timing gear set bolt.

16. Separate the camshaft timing gear from the number two camshaft. Rotate the camshafts counterclockwise, using a wrench, so that the cam lobes of the number one cylinder face in the direction shown.

17. Using several steps, loosen and remove the eight bearing cap bolts uniformly and in the proper removal sequence. Remove the four bearing caps and the number two camshaft.

18. Remove the number two chain tensioner bolt. Remove the number chain tensioner and camshaft timing gear.

➡Keep the camshaft level while it is being removed. The camshaft thrust clearance is very small and failing to keep it level could crack or damage the cylinder head journal surface, which receives the thrust. Follow the steps below to prevent this problem from occurring.

19. Hold the hexagonal portion of the number one camshaft, with a wrench. Loosen the camshaft timing gear set bolt.

➡Do not disassemble the camshaft timing gear assembly.

20. Slide the camshaft timing gear and separate the number one chain from the camshaft timing gear.

21. Rotate the number one camshaft counterclockwise, using the wrench so that the cam lobes on the number one cylinder face downward.

22. Using several steps, loosen and remove the eight bearing cap bolts uniformly and in the proper removal sequence. Remove the four bearing caps.

23. Remove the camshaft timing gear set bolt with the number one camshaft lifted up. Remove the number one camshaft and camshaft timing gear with the number two chain.

24. Tie the number one chain to the side. Be careful not to drop anything inside the timing chain cover.

To install:

➡Be sure to use new fasteners, as required.

➡Keep the camshaft level while it is being installed. The camshaft thrust clearance is very small and failing to keep it level could crack or damage the cylinder head journal surface, which receives the thrust.

25. Align the yellow mark link with the timing mark (1 dot mark) of the camshaft

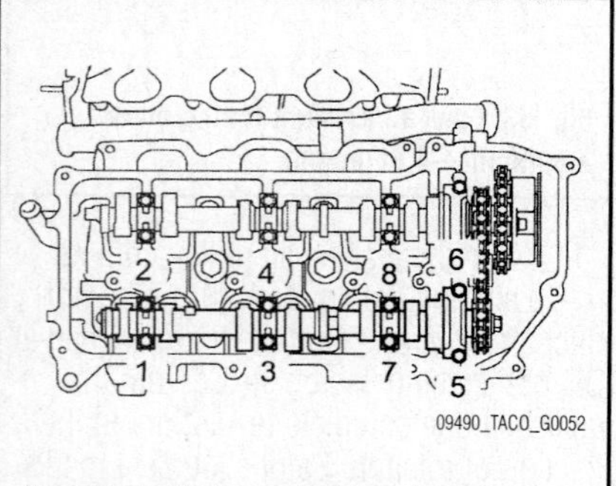

Fig. 55 Bank 1 camshaft number two bearing cap bolt removal sequence—4.0L engine

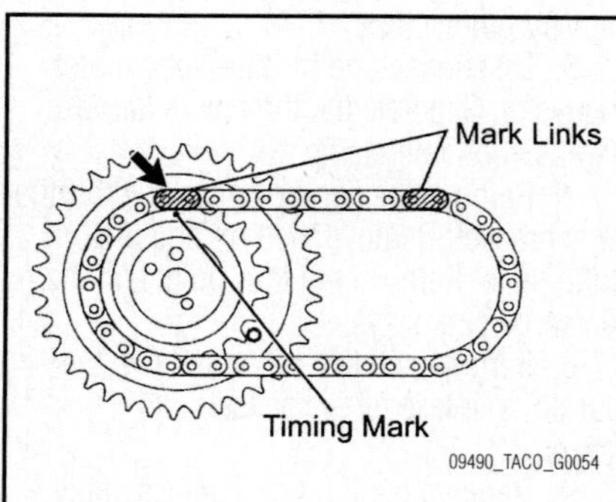

Fig. 57 Bank 1 camshaft number one alignment marks—4.0L engine

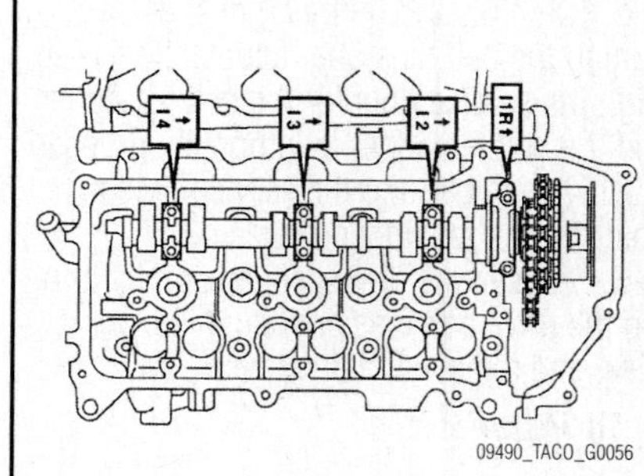

Fig. 59 Bank 1 camshaft number one bearing cap bolt installation sequence—4.0L engine

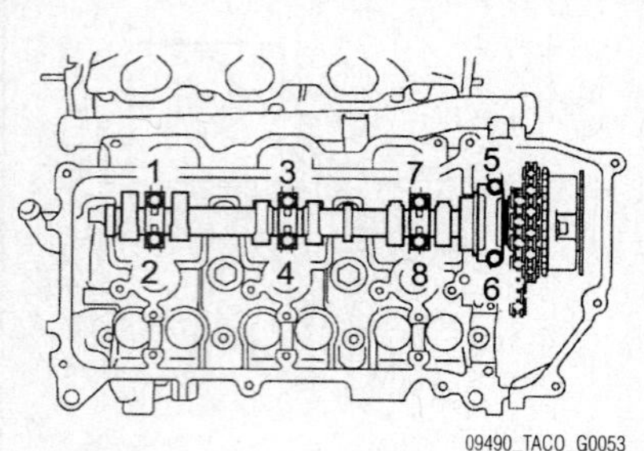

Fig. 56 Bank 1 camshaft number one bearing cap bolt removal sequence—4.0L engine

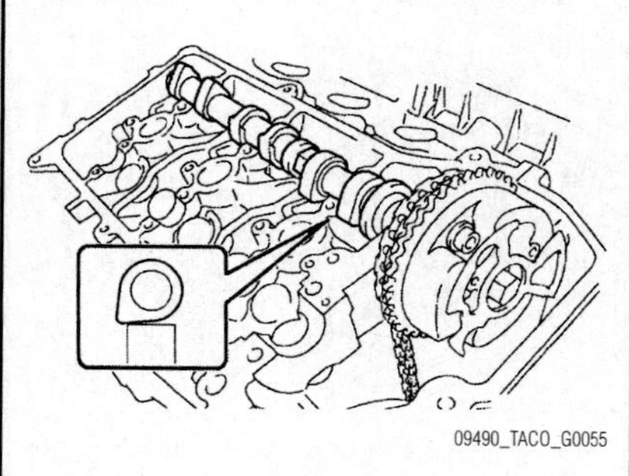

Fig. 58 Bank 1 camshaft number one lobe installation positioning—4.0L engine

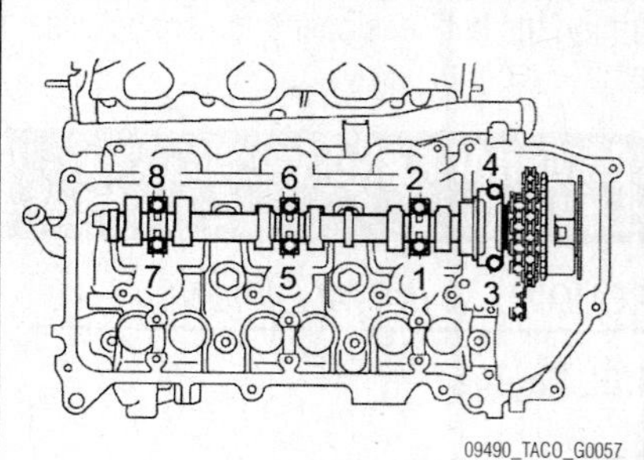

Fig. 60 Bank 1 camshaft number one bearing cap bolt torque sequence—4.0L engine

timing gear. Apply new engine oil to the thrust portion and journal of the camshafts.

26. Temporarily install the number one chain onto the number two chain of the camshaft timing gear.

27. Align the knock pin hole of the camshaft timing gear with the knock pin of the number one camshaft. Insert the number one camshaft into the camshaft timing gear.

28. Temporarily install the camshaft timing gear set bolt. Install the number one camshaft onto the right cylinder head with the cam lobes of the number one cylinder facing downward, as indicated in the illustration.

29. Install the four bearing caps, in the proper location. Apply a light coat of engine oil to the threads and under the heads of the cap bolts.

30. Using several steps, uniformly install and tighten the bearing cap bolts in the proper sequence to 80 inch lbs. for the 10mm bolts and 18 ft. lbs. for the 12mm bolts.

31. Rotate the number one camshaft clockwise, using a wrench so that the timing mark of the camshaft timing gear is aligned with the timing mark of the camshaft bearing cap.

32. Align the paint mark of the number one chain with the timing mark of the camshaft timing gear.

33. Hold the hexagonal portion of the number one camshaft with a wrench, and tighten the camshaft timing gear set bolt to 74 ft. lbs.

34. While pushing in on the number two chain tensioner, insert a 0.039 inch pin into the hole to hold it.

35. Temporarily install the camshaft timing gear and chain tensioner number two and align the yellow mark links with the timing marks (1 dot mark) on the camshaft timing gears. Tighten the bolt to 14 ft. lbs.

➡Keep the camshaft level while it is being installed. The camshaft thrust clearance is very small and failing to keep it level could crack or damage the cylinder head journal surface, which receives the thrust.

36. Install the number two camshaft onto the right cylinder head with the cam lobes of the number one cylinder facing upward, as indicated in the illustration.

37. Install the four bearing caps in the proper location. Apply a light coat of clean engine oil to the threads and under the heads of the bolts.

38. Using several steps, uniformly install and tighten the eight bearing cap bolts in sequence to 80 inch lbs, for the 10mm head bolts and 18 ft. lbs. for the 12mm head bolts.

39. Rotate the number two camshaft clockwise, using a wrench, so that the lock pin of the number two camshaft is aligned with the knock pin hole of the camshaft timing gear.

40. Hold the hexagonal portion of the number two camshaft, with a wrench, and install the camshaft timing gear set bolt and tighten it to 74 ft. lbs.

41. Remove the pin from the number two chain tensioner.

42. While turning the stopper plate of the tensioner clockwise, push in the plunger of the tensioner. While turning the stopper plate of the tensioner counterclockwise, insert a 0.138 inch bar into the holes in the stopper plate and tensioner to hold the stopper plate. Install the two chain tensioner bolts and tighten to 7.4 ft. lbs.

43. Remove the bar from the chain tensioner. Install a new gasket and the timing chain cover plate. Torque the bolts to 80 inch lbs.

44. Turn the crankshaft pulley two complete revolutions slowly until its groove and the timing mark "0" of the timing chain cover are aligned.

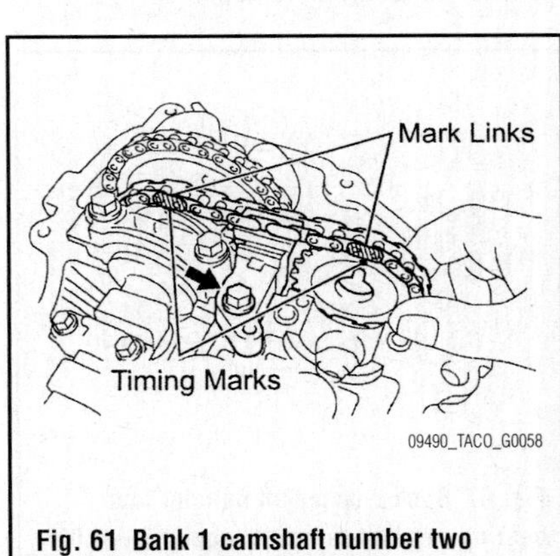

Fig. 61 Bank 1 camshaft number two alignment marks—4.0L engine

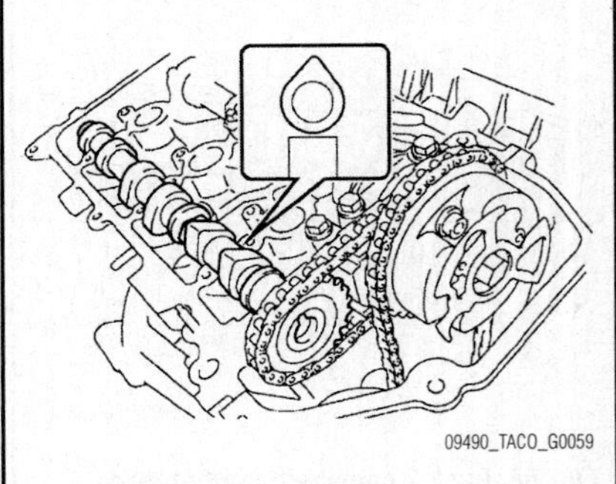

Fig. 62 Bank 1 camshaft number two lobe installation positioning—4.0L engine

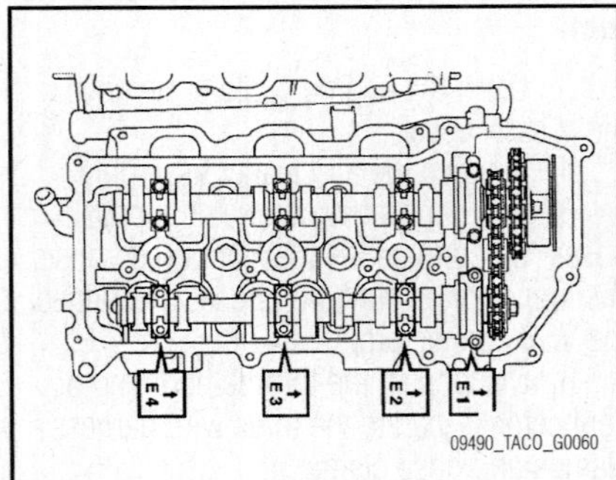

Fig. 63 Bank 1 camshaft number two bearing cap bolt installation sequence—4.0L engine

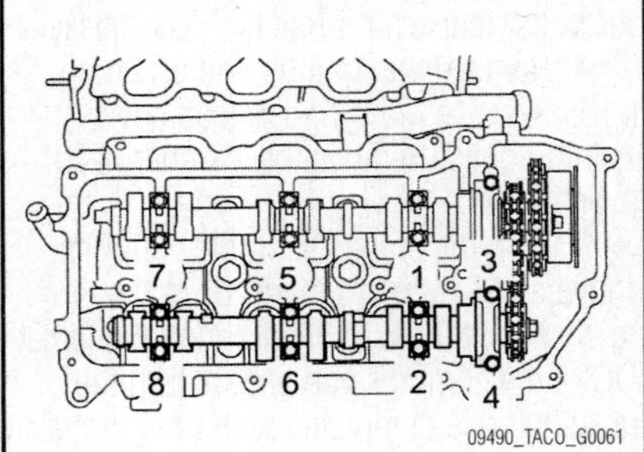

Fig. 64 Bank 1 camshaft number two bearing cap bolt torque sequence—4.0L engine

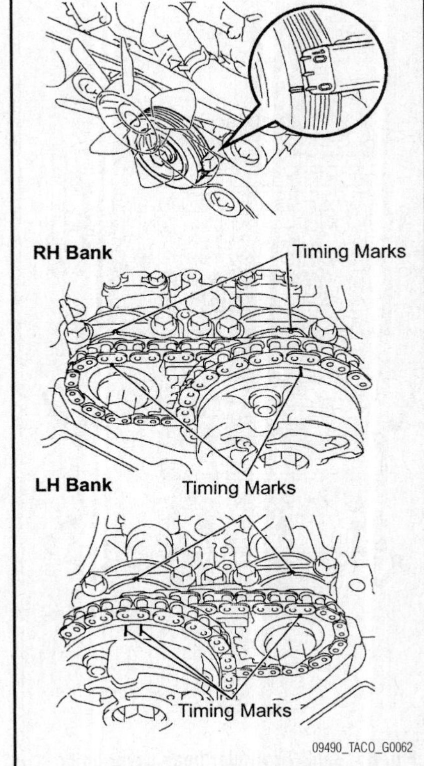

Fig. 65 Bank 1 and Bank 2 timing mark alignment—4.0L engine

45. Position the number one cylinder at TDC of the compression stroke. Inspect the valve clearance, adjust as required.

46. Apply a continuous bead (0.08–0.12 inch) of seal packing, part number 08826-00080 or equivalent, to the cylinder head as indicated in the illustration. Install the seal washers onto the bolts. Install the cylinder head cover bolts and nuts. Tighten bolts "A" to 7.4 ft. lbs. Tighten bolts "B" to 80 inch lbs. Tighten nuts to 80 inch lbs.

➡Be sure to remove any oil from the contact surfaces of the cylinder head cover and the cylinder head. Install the cover within three minutes after applying the seal packing. Tighten the bolts to specification within fifteen minutes after installing the cover. Do not add engine oil for at least two hours after installing the cover.

47. Continue the installation in the reverse order of the removal procedure.

48. Be sure to fill the engine with the proper grade and type engine coolant.

49. Be sure to fill the engine with the proper grade and type engine oil.

50. Start the engine and check for leaks. Correct as required.

51. Use the Techstream diagnostic tool, or equivalent and reprogram the required systems.

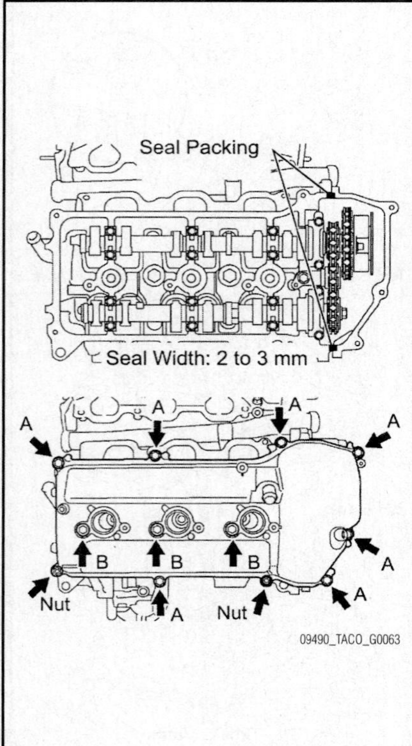

Fig. 66 Bank 1 cylinder head cover bolt sealant application and bolt identification—4.0L engine

Bank 2

See Figures 67 through 75.

1. Before servicing the vehicle, refer to the Precautions Section.

➡If working near and/or around the SRS system and components, be sure to disable the SRS system. Tape the negative battery cable with insulating tape. Always disconnect the negative battery cable first.

⁂ CAUTION

To avoid personal injury when working on vehicles equipped with an air bag, the negative battery cable must be disconnected and at least 90 seconds must elapse before working on the system. Failure to do so may result in deployment of the air bag.

2. Disconnect the negative battery cable. Tape the cable with insulating tape.

➡When disconnecting the cable, some systems need to be initialized after the cable is reconnected. You will need the Techstream diagnostic scan tool or equivalent. Follow the directions on the tool.

3. Drain the engine coolant. Remove the V bank cover.

4. Remove the air cleaner assembly.

5. Disconnect the two water bypass hoses. Disconnect the fuel vapor feed hose. Disconnect the ventilation hose. Disconnect the VSV connectors.

6. Disconnect the throttle body motor connector. Separate the three wire harness clamps and hose clamp.

7. Remove the two bolts and the throttle body bracket. Remove the bolt and the oil baffle plate. Remove the four bolts and the two serge tank stays.

8. Remove the two nuts. Remove the four bolts, intake air surge tank and gasket.

9. Remove the ignition coil assembly.

10. Remove the cylinder head cover retaining bolts. Remove the cylinder head cover.

11. Turn the crankshaft pulley until its groove and the timing mark "0" of the timing chain cover are aligned. If not aligned at TDC of the compression stroke, turn the crankshaft one complete revolution, in the direction of rotation. Paint alignment marks on the number one chain links corresponding to the timing marks of the camshaft timing gears.

12. While turning the stopper plate of the tensioner upward, push in the plunger of the chain tensioner. While turning the stopper plate of the tensioner downward, insert a 0.138 inch bar into the holes in the stopper plate and tensioner to hold the stopper plate. Remove the two bolts and then remove the number one chain tensioner assembly.

➡Never rotate the crankshaft with the chain tensioner removed. When rotating the camshaft with the tensioner removed, rotate the crankshaft counterclockwise forty degrees from TDC, first.

➡Keep the camshaft level while it is being removed. The camshaft thrust clearance is very small and failing to keep it level could crack or damage the cylinder head journal surface, which receives the thrust. Follow the steps below to prevent this problem from occurring.

13. While pushing down on chain tensioner number three insert a 0.039 inch diameter pin into the hole to hold it. Hold the hexagonal portion of the number four camshaft with a wrench. Remove the camshaft timing gear set bolt.

14. Separate the camshaft timing gear from the number four camshaft.

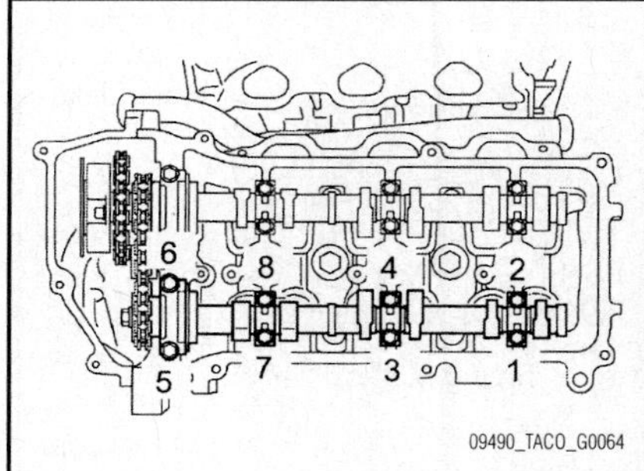

Fig. 67 Bank 2 camshaft number four bearing cap bolt removal sequence—4.0L engine

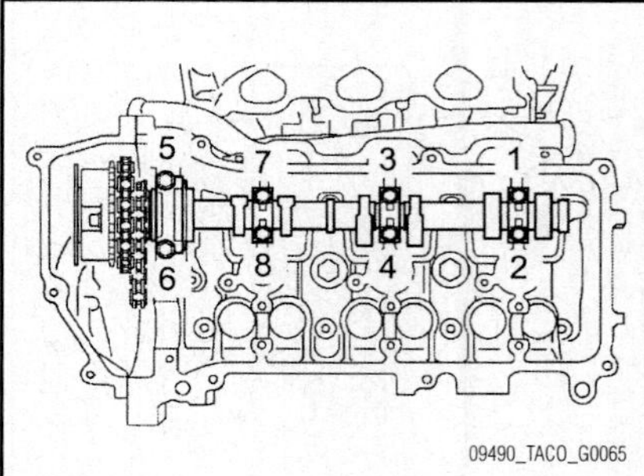

Fig. 68 Bank 2 camshaft number three bearing cap bolt removal sequence—4.0L engine

15. Using several steps, loosen and remove the eight bearing cap bolts uniformly and in the proper removal sequence. Remove the four bearing caps and the number four camshaft.

16. Remove the number three chain tensioner bolt. Remove the number chain tensioner and camshaft timing gear.

➡Keep the camshaft level while it is being removed. The camshaft thrust clearance is very small and failing to keep it level could crack or damage the cylinder head journal surface, which receives the thrust. Follow the steps below to prevent this problem from occurring.

17. Release the chain tension between the camshaft gear (LH bank) and the crankshaft timing gear by turning the crankshaft pulley counterclockwise slightly.

18. Hold the hexagonal portion of the number three camshaft, with a wrench. Loosen the camshaft timing gear set bolt.

➡Do not disassemble the camshaft timing gear assembly.

19. Slide the camshaft timing gear and separate the number one chain from the camshaft timing gear.

20. Using several steps, loosen and remove the eight bearing cap bolts uniformly and in the proper removal sequence. Remove the four bearing caps.

21. Remove the camshaft timing gear set bolt with the number three camshaft lifted up. Remove the number three camshaft and camshaft timing gear with the number two chain.

22. Tie the number one chain to the side. Be careful not to drop anything inside the timing chain cover.

To install:

➡Be sure to use new fasteners, as required.

➡Keep the camshaft level while it is being installed. The camshaft thrust clearance is very small and failing to keep it level could crack or damage the cylinder head journal surface, which receives the thrust.

23. Align the yellow mark link with the timing mark (2 dot mark) of the camshaft timing gear. Apply new engine oil to the thrust portion and journal of the camshafts.

24. Temporarily install the number one chain onto the number two chain of the camshaft timing gear.

25. Align the knock pin hole of the camshaft timing gear with the knock pin of the number three camshaft. Insert the number three camshaft into the camshaft timing gear.

26. Temporarily install the camshaft timing gear set bolt. Install the number three camshaft onto the left cylinder head with the cam lobes of the number two cylinder facing downward, as indicated in the illustration.

27. Install the four bearing caps, in the proper location. Apply a light coat of engine oil to the threads and under the heads of the cap bolts.

28. Using several steps, uniformly install and tighten the bearing cap bolts in the proper sequence to 80 inch lbs. for the 10mm bolts and 18 ft. lbs. for the 12mm bolts.

29. Rotate the number one camshaft clockwise, using a wrench so that the timing mark of the camshaft timing gear is aligned with the timing mark of the camshaft bearing cap.

30. Align the paint mark of the number one chain with the timing mark of the camshaft timing gear.

31. Hold the hexagonal portion of the number three camshaft with a wrench, and

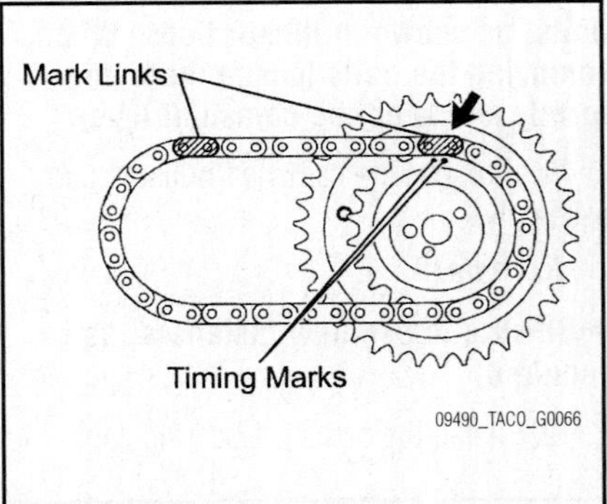

Fig. 69 Bank 2 camshaft number three alignment marks—4.0L engine

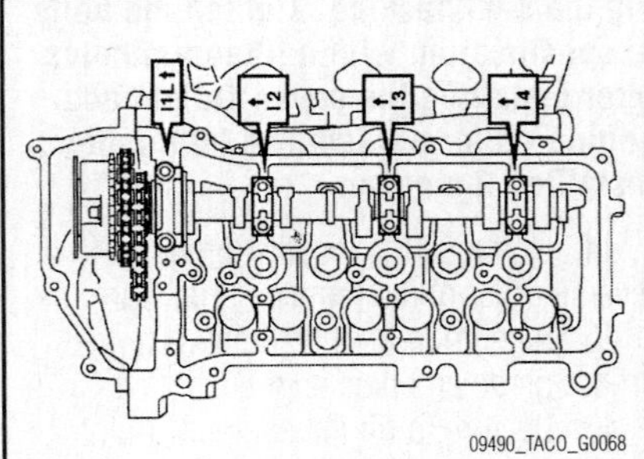

Fig. 71 Bank 2 camshaft number three bearing cap bolt installation sequence—4.0L engine

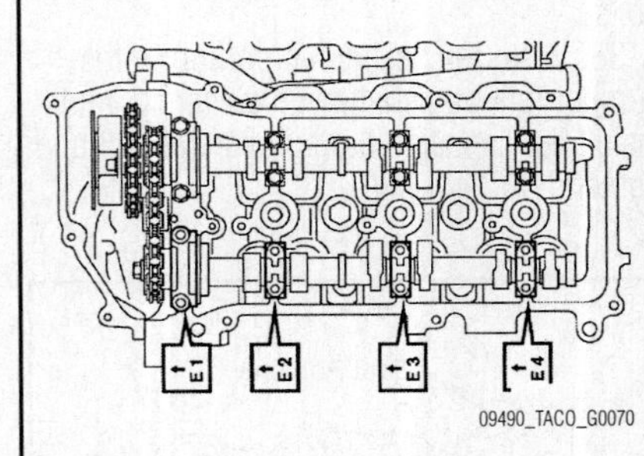

Fig. 73 Bank 2 camshaft number four bearing cap bolt installation sequence—4.0L engine

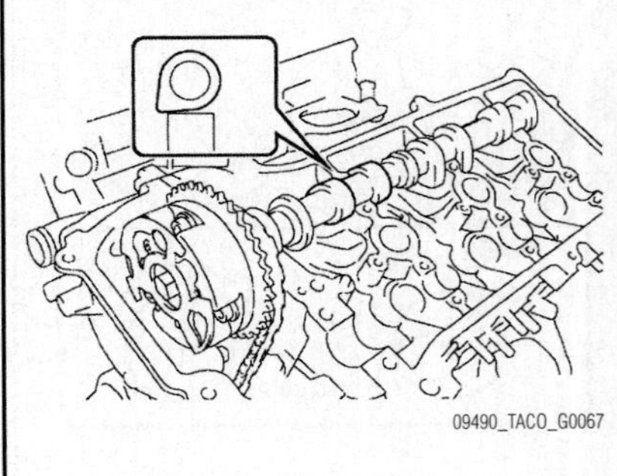

Fig. 70 Bank 2 camshaft number three lobe installation positioning—4.0L engine

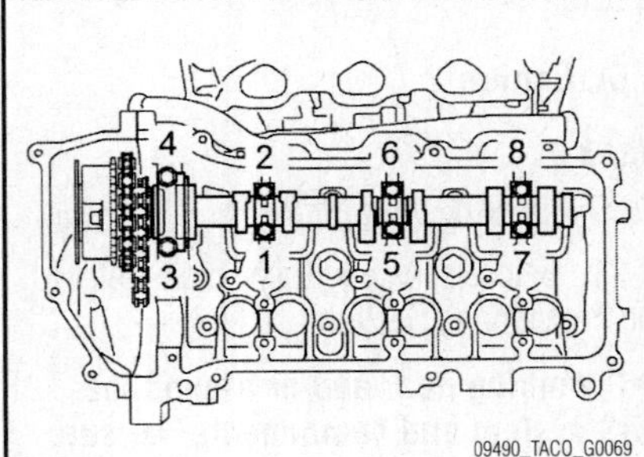

Fig. 72 Bank 2 camshaft number three bearing cap bolt torque sequence—4.0L engine

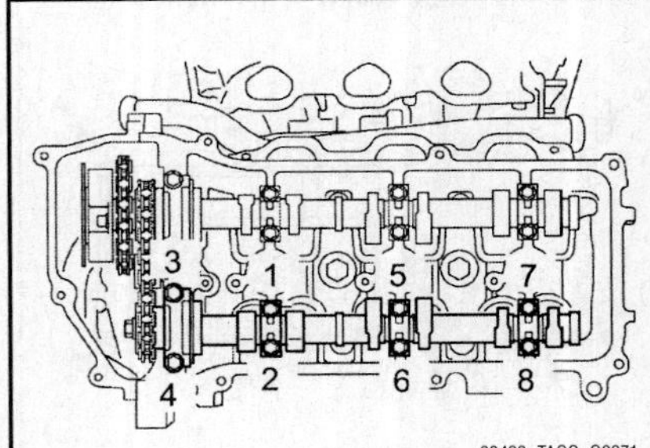

Fig. 74 Bank 2 camshaft number four bearing cap bolt torque sequence—4.0L engine

tighten the camshaft timing gear set bolt to 74 ft. lbs.

32. While pushing in on the number three chain tensioner, insert a 0.039 inch pin into the hole to hold it.

33. Temporarily install the camshaft timing gear and chain tensioner number three and align the yellow mark links with the timing marks (1 dot mark and 2 dot marks) on the camshaft timing gears. Tighten the bolt to 14 ft. lbs.

➡Keep the camshaft level while it is being installed. The camshaft thrust clearance is very small and failing to keep it level could crack or damage the cylinder head journal surface, which receives the thrust.

34. Align the knock pin hole in the camshaft timing gear with the knock pin of the number four camshaft, and insert the number four camshaft into the camshaft timing gear.

35. Temporarily install the camshaft timing gear set bolt.

36. Install the four bearing caps in the proper location. Apply a light coat of clean engine oil to the threads and under the heads of the bolts.

37. Using several steps, uniformly install and tighten the eight bearing cap bolts in sequence to 80 inch lbs, for the 10mm head bolts and 18 ft. lbs. for the 12mm head bolts.

38. Hold the hexagonal portion of the number four camshaft, with a wrench, and install the camshaft timing gear set bolt and tighten it to 74 ft. lbs.

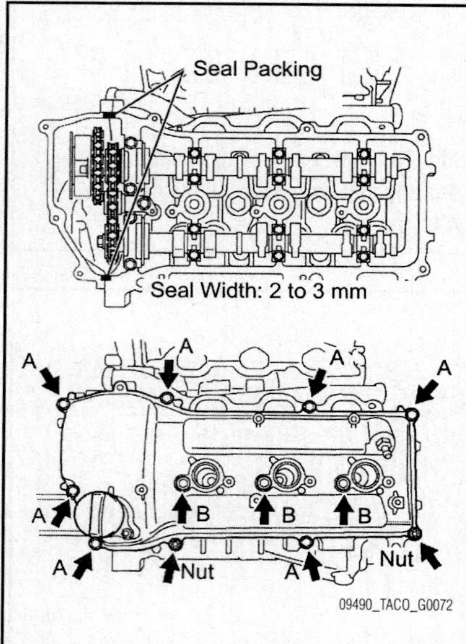

Fig. 75 Bank 2 cylinder head cover bolt sealant application and bolt identification—4.0L engine

39. Remove the pin from the number three chain tensioner.

40. Release the chain tension between the camshaft timing gear (RH bank) and the crankshaft timing gear by turning the crankshaft pulley clockwise slightly.

41. While turning the stopper plate of the tensioner clockwise, push in the plunger of the tensioner. While turning the stopper plate of the tensioner counterclockwise, insert a 0.138 inch bar into the holes in the stopper plate and tensioner to hold the stopper plate. Install the two chain tensioner bolts and tighten to 7.4 ft. lbs. Remove the bar from the chain tensioner.

42. Position the number one cylinder at TDC of the compression stroke. Inspect the valve clearance, adjust as required.

43. Apply a continuous bead (0.08–0.12 inch) of seal packing, part number 08826-00080 or equivalent, to the cylinder head as indicated in the illustration. Install the seal washers onto the bolts. Install the cylinder head cover bolts and nuts. Tighten bolts “A” to 7.4 ft. lbs. Tighten bolts “B” to 80 inch lbs. Tighten nuts to 80 inch lbs.

➡Be sure to remove any oil from the contact surfaces of the cylinder head cover and the cylinder head. Install the cover within three minutes after applying the seal packing. Tighten the bolts to specification within fifteen minutes after installing the cover. Do not add engine oil for at least two hours after installing the cover.

44. Continue the installation in the reverse order of the removal procedure.

45. Be sure to fill the engine with the proper grade and type engine coolant.

46. Be sure to fill the engine with the proper grade and type engine oil.

47. Start the engine and check for leaks. Correct as required.

48. Use the Techstream diagnostic tool, or equivalent and reprogram the required systems.

4.6L Engine

Bank 1

See Figures 76 through 89.

1. Before servicing the vehicle, refer to the Precautions Section.

➡If working near and/or around the SRS system and components, be sure to disable the SRS system. Tape the negative battery cable with insulating tape. Always disconnect the negative battery cable first.

⁂ CAUTION

To avoid personal injury when working on vehicles equipped with an air bag, the negative battery cable must be disconnected and at least 90 seconds must elapse before working on the system. Failure to do so may result in deployment of the air bag.

2. Disconnect the negative battery cable. Tape the cable with insulating tape.

➡When disconnecting the cable, some systems need to be initialized after the cable is reconnected. You will need the Techstream diagnostic scan tool or equivalent. Follow the directions on the tool.

3. Remove the timing chain cover.

4. Position the No. 1 cylinder at TDC on the compression stroke.

5. Remove the No. 1 chain tensioner assembly.

6. Remove the chain tensioner slipper.

7. Remove the No. 1 chain vibration damper.

8. Remove the No. 1 chain sub assembly.

9. Remove the No. 3 chain tensioner assembly.

10. Remove the camshaft bearing cap.

➡ Be sure that the knock pin is properly positioned. Remove retaining bolts, as shown in illustrations. When removing the bolts loosen them uniformly and keep the camshaft level.

11. Remove the camshaft housing sub assembly.

To install:

➡Be sure to use new fasteners, as required.

12. Install the camshaft bearing cap.

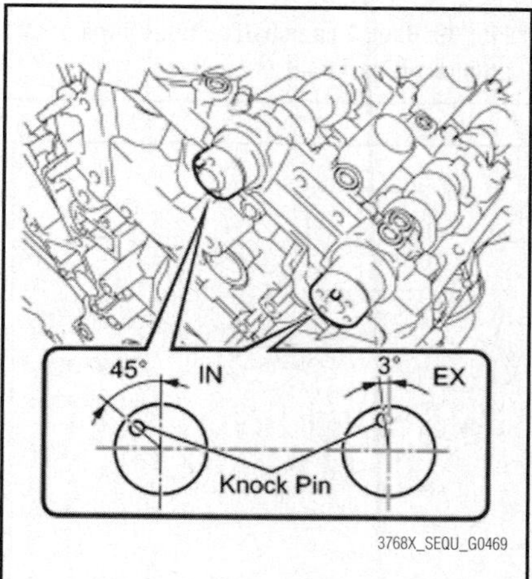

Fig. 76 Camshaft bearing cap removal Bank 1 (1 of 3)—4.6L engine

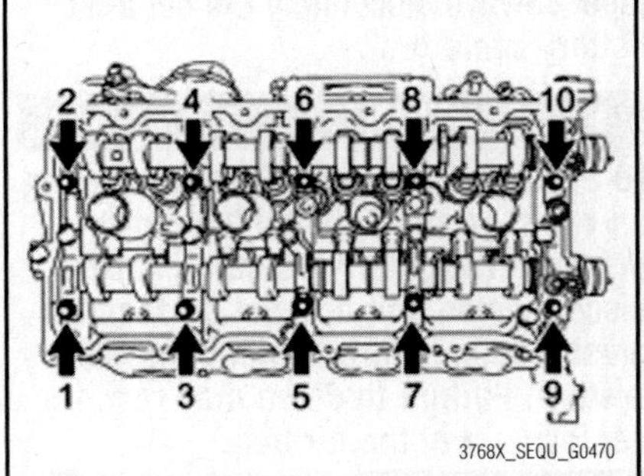

Fig. 77 Camshaft bearing cap removal Bank 1 (2 of 3)—4.6L engine

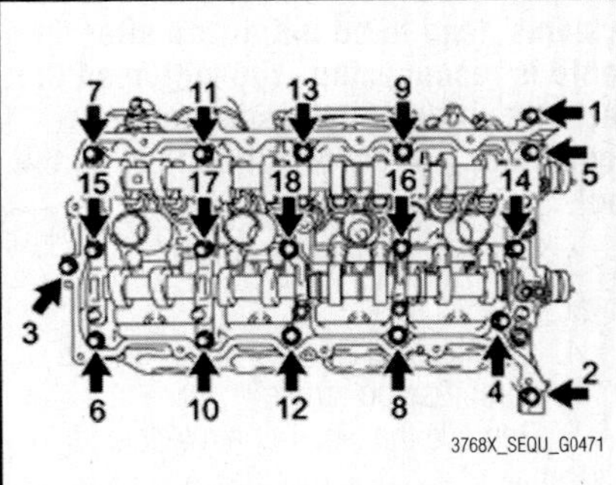

Fig. 78 Camshaft bearing cap removal Bank 1 (3 of 3)—4.6L engine

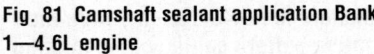

➡Apply a light coat of clean engine oil to the camshaft journals, housing and caps.

13. Install the No. 3 and No. 4 camshafts to the housing.
14. Temporarily install the ten bolts as shown in the illustration.
15. To install the camshaft housing sub assembly, install the rocker arms.
16. Apply seal packing in a continuous bead, see illustration.

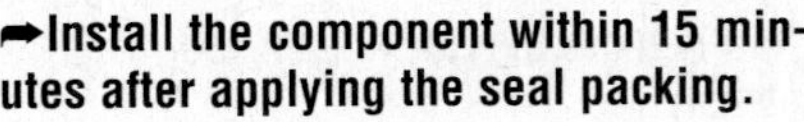

➡Install the component within 15 minutes after applying the seal packing.

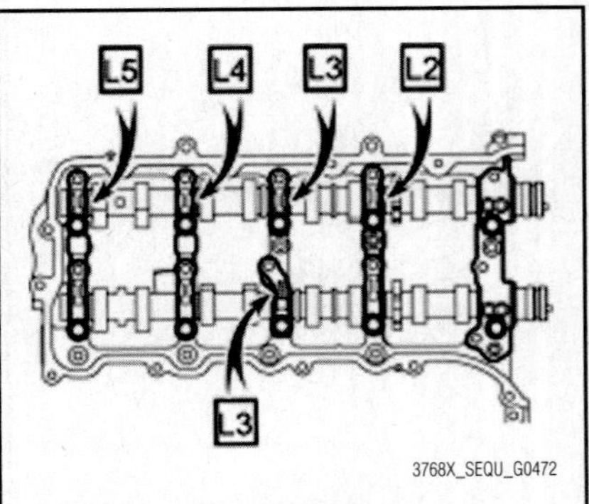

Fig. 79 Camshaft bearing cap installation Bank 1—4.6L engine

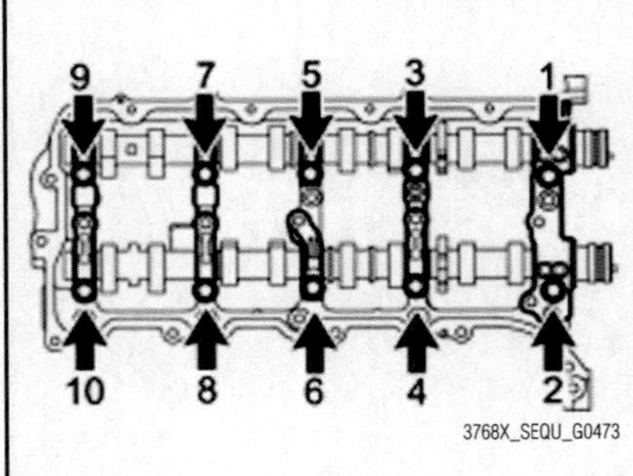

Fig. 80 Camshaft bearing bolt installation Bank 1—4.6L engine

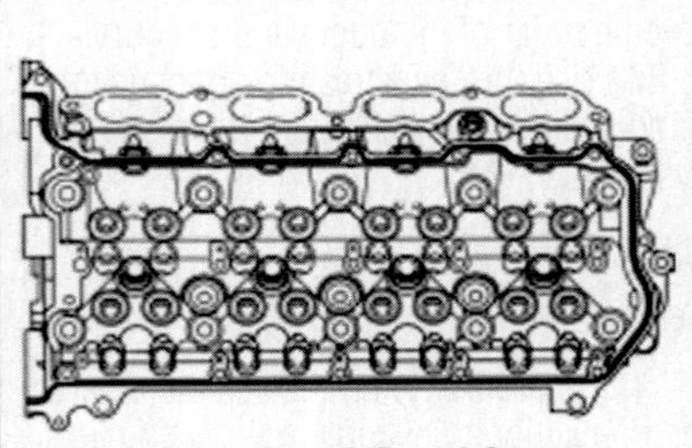

Fig. 81 Camshaft sealant application Bank 1—4.6L engine

17. Install the camshaft housing, and install the eighteen bolts as shown in the illustration. Tighten bolt A to 7 ft. lbs. (10 Nm). Tighten other bolts to 22 ft. lbs. (30 Nm). Be sure that the knock pin of the camshaft is properly positioned.

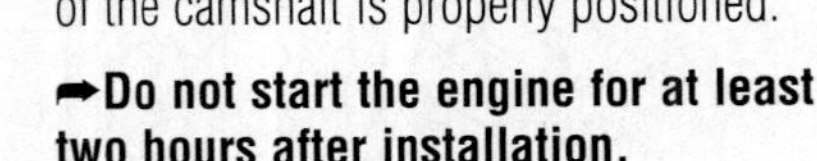

➡Do not start the engine for at least two hours after installation.

18. Tighten the ten bolts to 12 ft. lbs. (16 Nm), see illustration.

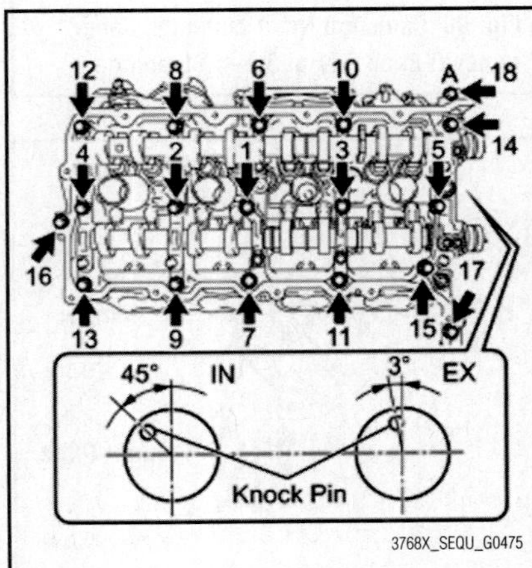

Fig. 82 Camshaft housing bolt (18) installation Bank 1—4.6L engine

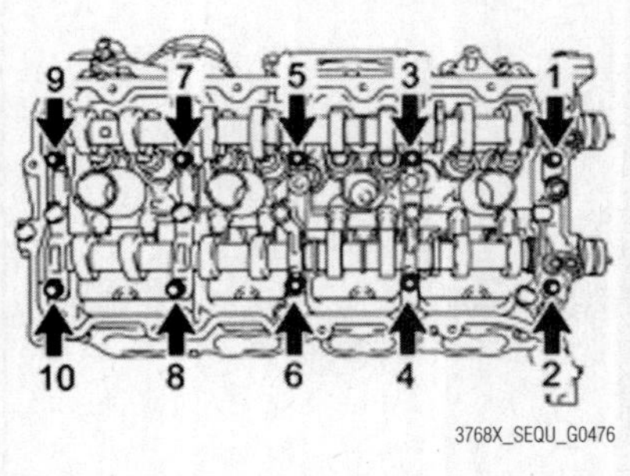

Fig. 83 Camshaft bolt (10) installation Bank 1—4.6L engine

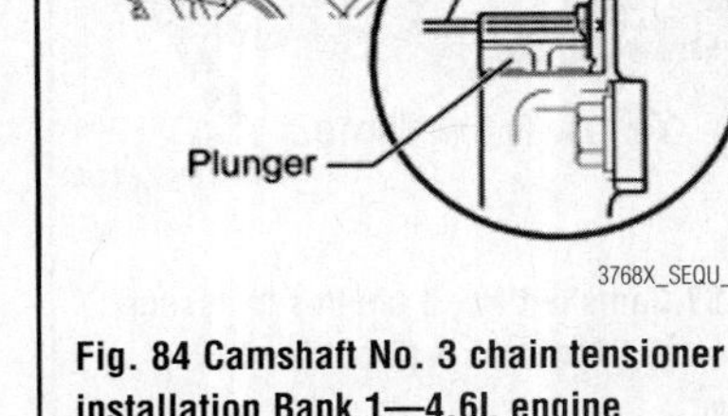

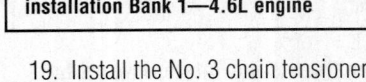

Fig. 84 Camshaft No. 3 chain tensioner installation Bank 1—4.6L engine

19. Install the No. 3 chain tensioner assembly.
20. Install the No. 1 chain sub assembly.
21. Install the No. 1 chain tensioner slipper.
22. Install the No. 1 chain tensioner assembly. Be sure to use a new gasket. Tighten the retaining bolts to 7 ft. lbs. (10 Nm).

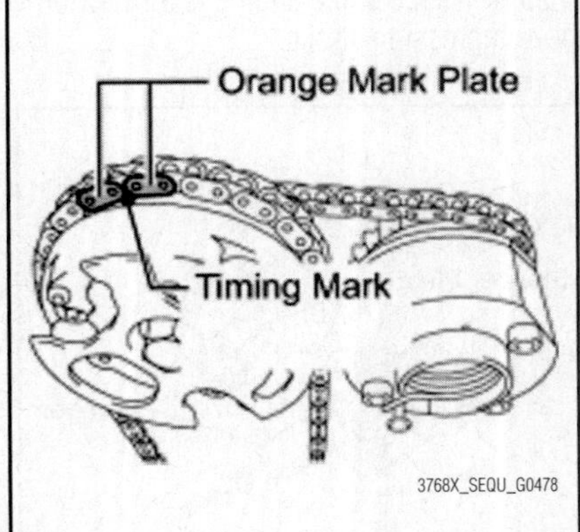

Fig. 85 Camshaft No. 1 chain sub assembly installation Bank 1 (1 of 3)—4.6L engine

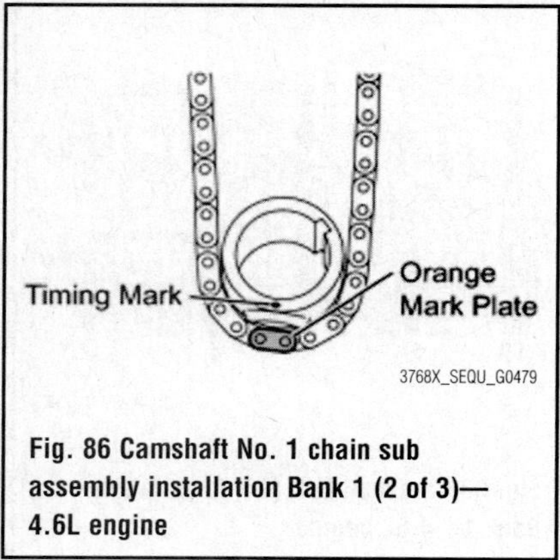

Fig. 86 Camshaft No. 1 chain sub assembly installation Bank 1 (2 of 3)—4.6L engine

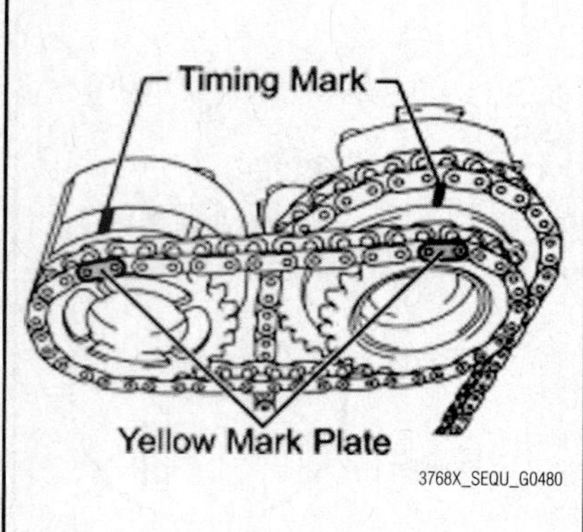

Fig. 87 Camshaft No. 1 chain sub assembly installation Bank 1 (3 of 3)—4.6L engine

➡Move the stopper plate upward to release the lock and push the plunger deep into the tensioner. Move the stopper plate downward to set the lock and insert a hex wrench into the hole of the stopper plate.

23. Install the No.1 chain vibration damper.
24. Tighten the camshaft timing gear. Tighten bolt to 74 ft. lbs. (100 Nm).
25. Check that the engine is at TDC, on the compression stroke.

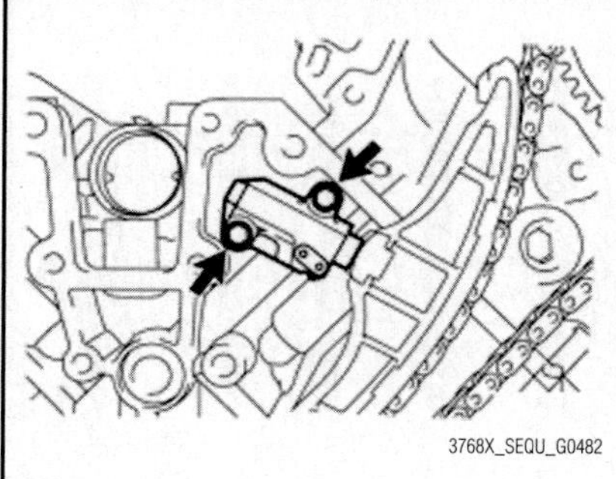

Fig. 89 Camshaft No. 1 chain tensioner installation Bank 1 (2 of 2)—4.6L engine

26. Continue the installation in the reverse order of the removal procedure.
27. Use the Techstream diagnostic tool, or equivalent and reprogram the required systems.

Bank 2

See Figures 90 through 109.

1. Before servicing the vehicle, refer to the Precautions Section.

➡If working near and/or around the SRS system and components, be sure to disable the SRS system. Tape the negative battery cable with insulating tape. Always disconnect the negative battery cable first.

CAUTION

To avoid personal injury when working on vehicles equipped with an air bag, the negative battery cable must be disconnected and at least 90 seconds must elapse before working on the system. Failure to do so may result in deployment of the air bag.

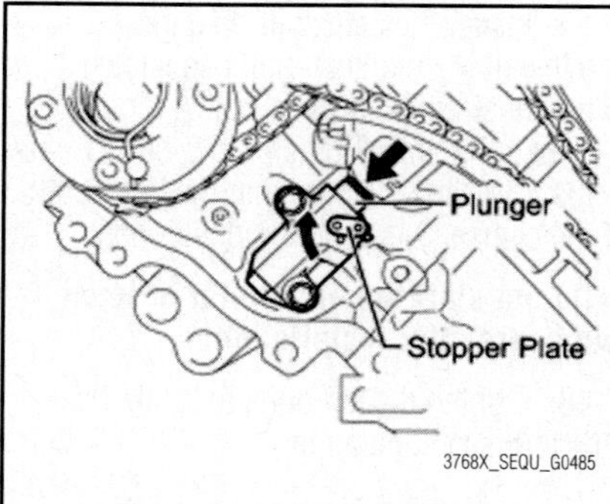

Fig. 90 Camshaft No. 1 chain tensioner removal Bank 2 (1 of 3)—4.6L engine

2. Disconnect the negative battery cable. Tape the cable with insulating tape.

➡When disconnecting the cable, some systems need to be initialized after the cable is reconnected. You will need the Techstream diagnostic scan tool or equivalent. Follow the directions on the tool.

3. Remove the timing chain cover.
4. Position the No. 1 cylinder at TDC on the compression stroke.
5. Remove the No. 1 chain tensioner assembly.
6. Remove the chain tensioner slipper.
7. Remove the No. 1 chain vibration damper.
8. Remove the No. 1 chain sub assembly.
9. Remove the No. 1 chain tensioner assembly.

➡Move the stopper plate upward to release the lock and push the plunger deep into the tensioner. Move the stopper plate downward to set the lock and insert a hex wrench into the hole of the stopper plate.

10. Remove the No. 1 chain tensioner slipper. Remove the No.1 vibration damper.
11. Remove the No. 1 chain sub assembly.

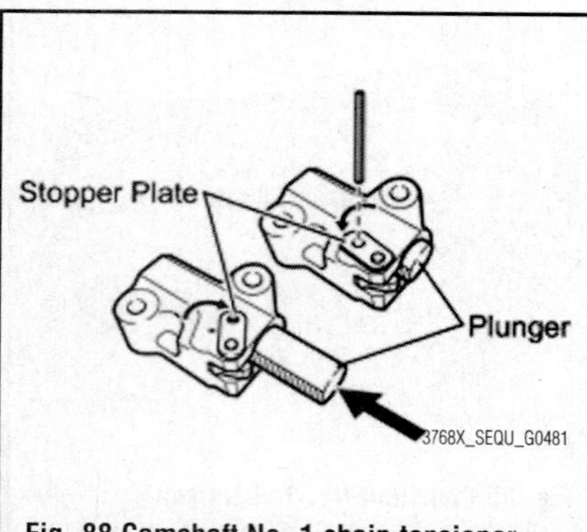

Fig. 88 Camshaft No. 1 chain tensioner installation Bank 1 (1 of 2)—4.6L engine

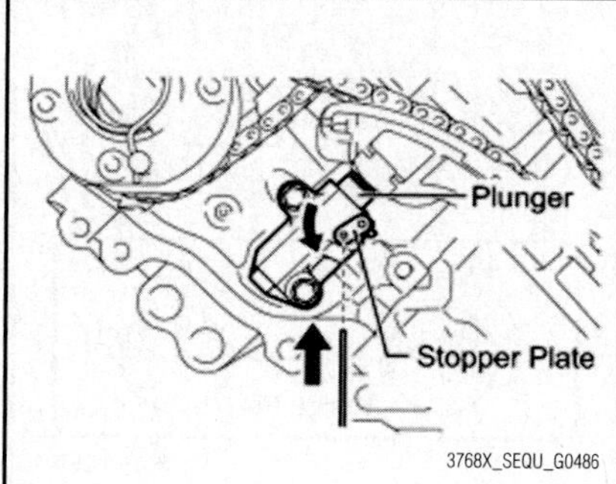

Fig. 91 Camshaft No. 1 chain tensioner removal Bank 2 (2 of 3)—4.6L engine

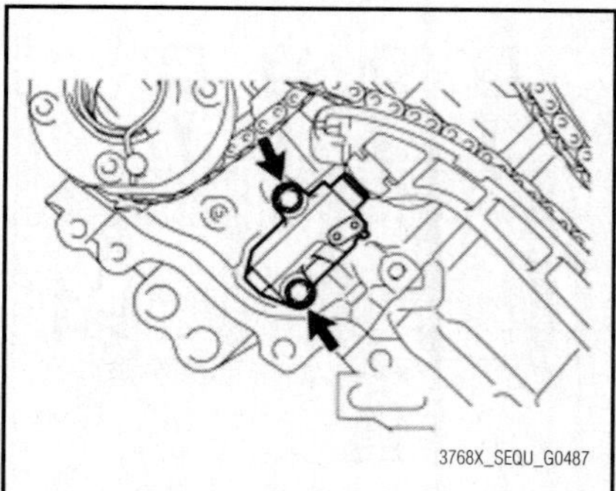

Fig. 92 Camshaft No. 1 chain tensioner removal Bank 2 (3 of 3)—4.6L engine

➡While raising up on the No. 2 chain tensioner, insert a 0.0394 inch pin into the hole to retain it in place.

12. Remove the No.1 and No.2 chains from the gear.
13. Remove the No. 2 chain tensioner assembly.
14. Remove the camshaft bearing cap.
15. Remove the camshaft housing sub assembly.

To install:

➡Be sure to use new fasteners, as required.

16. Install the camshaft bearing cap.

➡Apply a light coat of clean engine oil to the camshaft journals, housing and caps.

17. Install the No. 1 and No. 2 camshafts to the housing.
18. Temporarily install the ten bolts as shown in the illustration.
19. To install the camshaft housing sub assembly, install the rocker arms.
20. Apply seal packing in a continuous bead, see illustration.

➡Install the component within 15 minutes after applying the seal packing.

21. Install the camshaft housing, and install the eighteen bolts as shown in the illustration. Tighten bolt A to 7 ft. lbs. (10 Nm). Tighten other bolts to 22 ft. lbs. (30 Nm). Be sure that the knock pin of the camshaft is properly positioned.

➡Do not start the engine for at least two hours after installation.

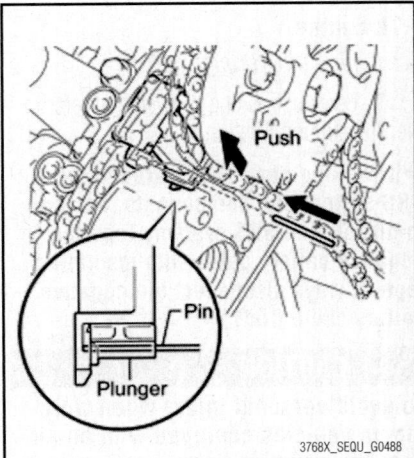

Fig. 93 Camshaft No. 1 chain sub assembly removal Bank 2—4.6L engine

22. Tighten the ten bolts to 12 ft. lbs. (16 Nm), see illustration.
23. Install the No. 2 chain tensioner assembly.
24. Install the No. 1 chain sub assembly.
25. Install the No. 1 chain vibration damper.
26. Install the No. 1 chain tensioner slipper.

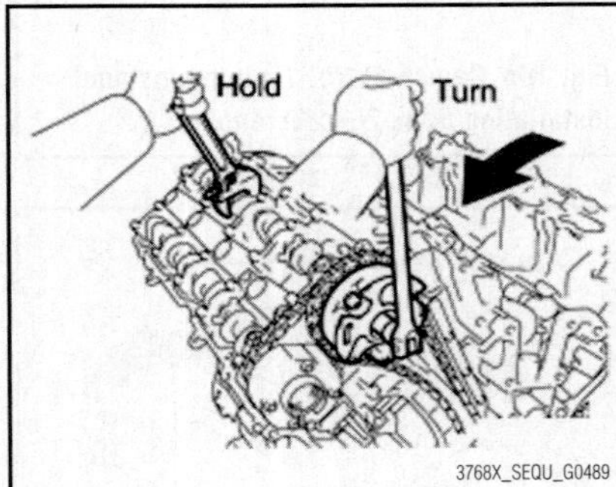

Fig. 94 Camshaft timing chain removal Bank 2 (1 of 2)—4.6L engine

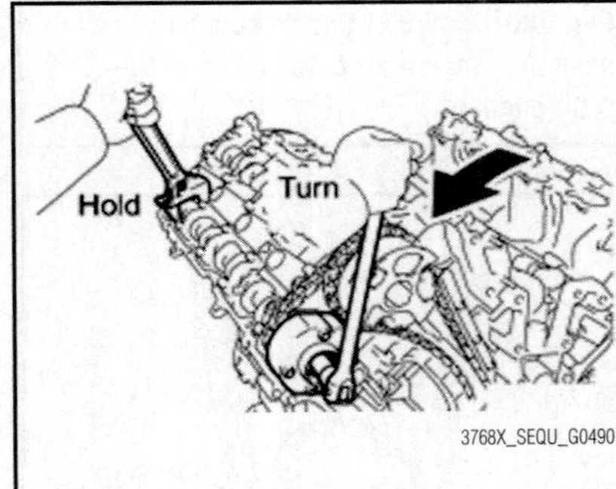

Fig. 95 Camshaft timing chain removal Bank 2 (2 of 2)—4.6L engine

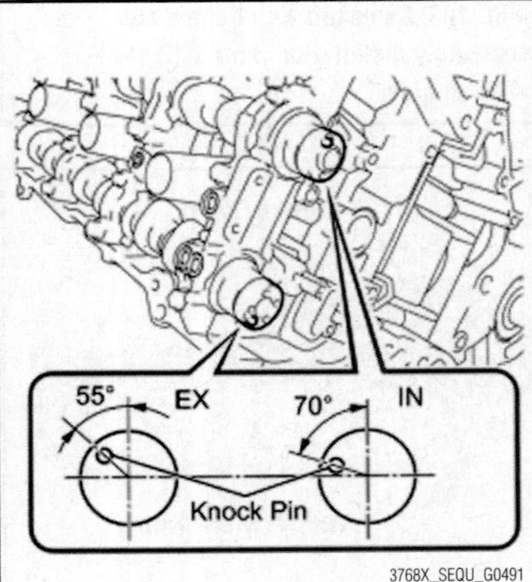

Fig. 96 Camshaft bearing cap removal Bank 2 (1 of 3)—4.6L engine

27. Install the No. 1 chain tensioner assembly. Be sure to use a new gasket. Tighten the retaining bolts to 7 ft. lbs. (10 Nm).

➡Move the stopper plate upward to release the lock and push the plunger deep into the tensioner. Move the stopper plate downward to set the lock and insert a hex wrench into the hole of the stopper plate.

28. Install the No. 1 chain sub assembly.
29. Install the No. 1 chain tensioner slipper.

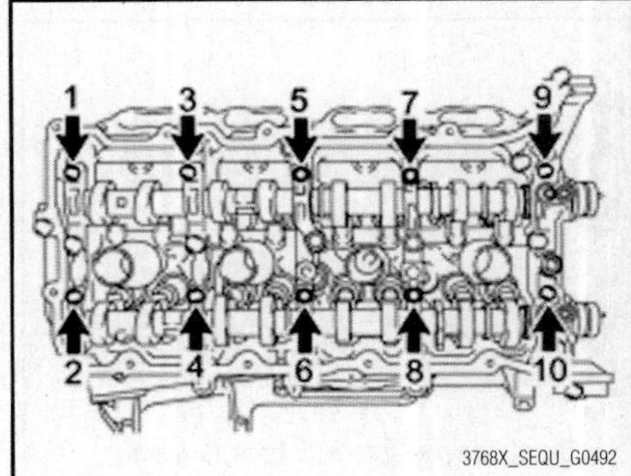

Fig. 97 Camshaft bearing cap removal Bank 2 (2 of 3)—4.6L engine

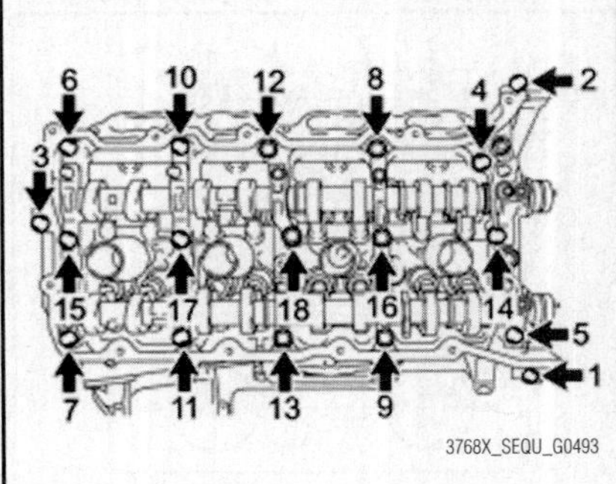

Fig. 98 Camshaft bearing cap removal Bank 2 (3 of 3)—4.6L engine

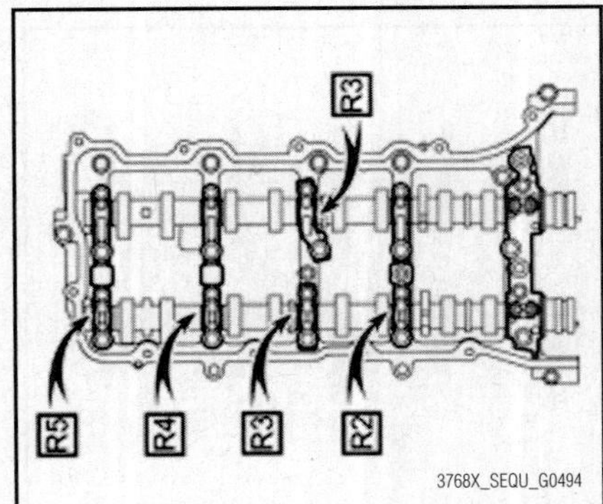

Fig. 99 Camshaft bearing cap installation Bank 2—4.6L engine

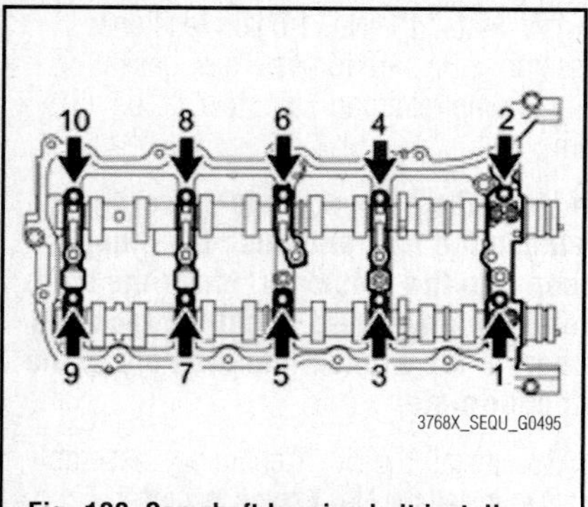

Fig. 100 Camshaft bearing bolt installation Bank 2—4.6L engine

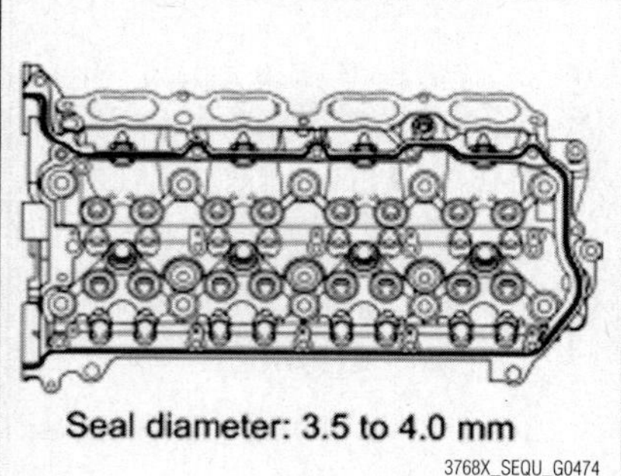

Fig. 101 Camshaft sealant application Bank 2—4.6L engine

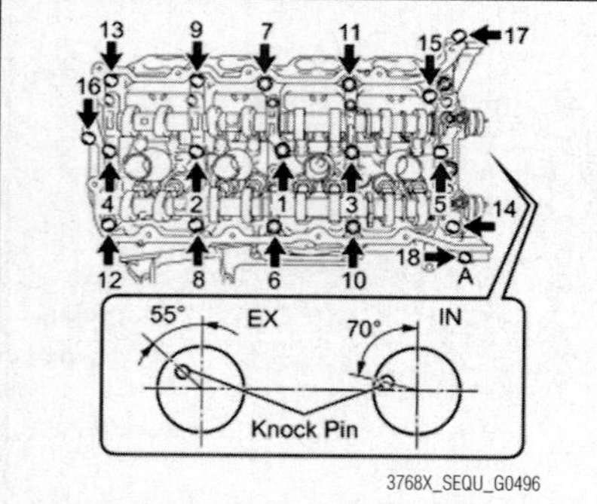

Fig. 102 Camshaft housing bolt (18) installation Bank 2—4.6L engine

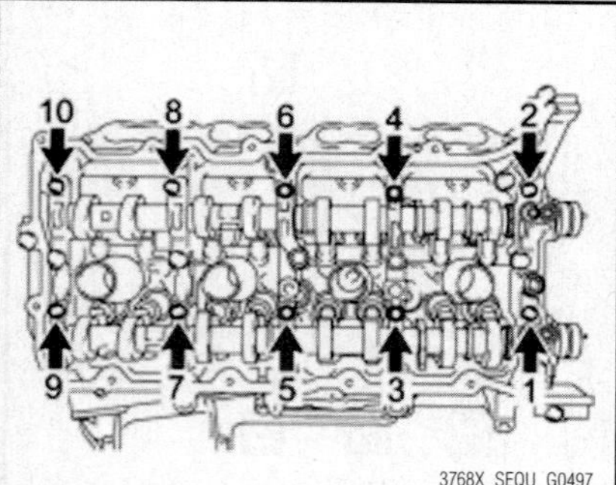

Fig. 103 Camshaft bolt (10) installation Bank 2—4.6L engine

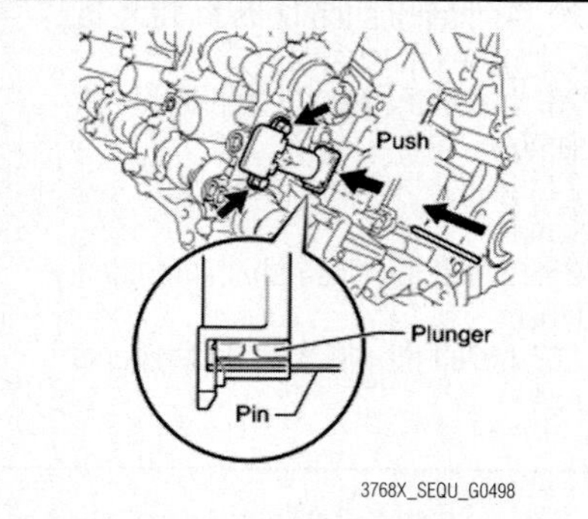

Fig. 104 Camshaft No. 1 chain tensioner installation Bank 2—4.6L engine

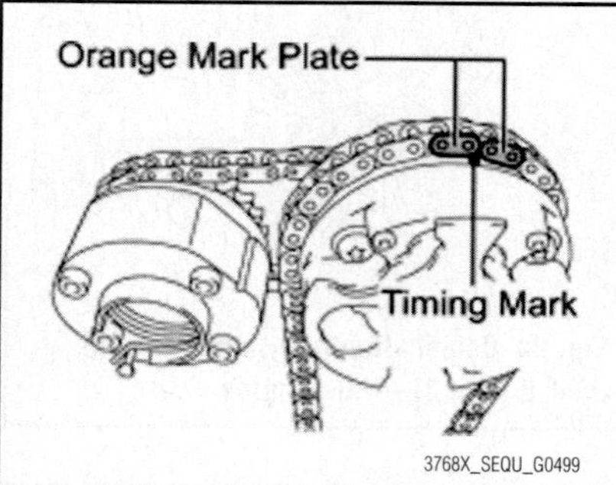

Fig. 105 Camshaft No. 1 chain sub assembly installation Bank 2 (1 of 3)—4.6L engine

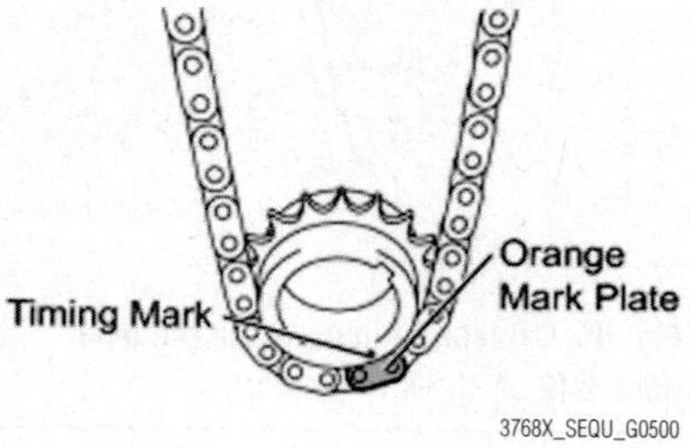

Timing Mark

Orange Mark Plate

3768X_SEQU_G0500

Fig. 106 Camshaft No. 1 chain sub assembly installation Bank 2 (2 of 3)—4.6L engine

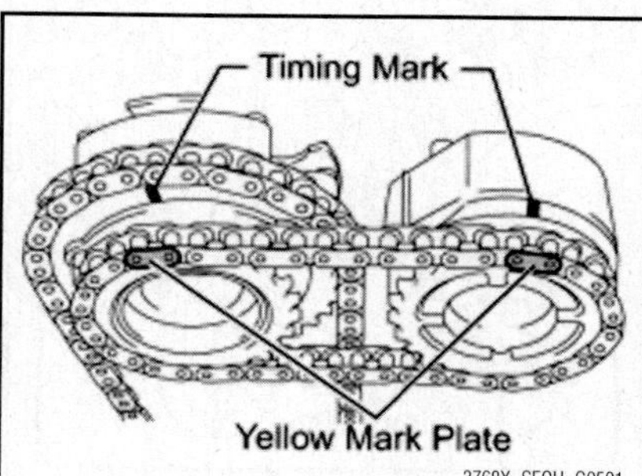

Fig. 107 Camshaft No. 1 chain sub assembly installation Bank 2 (3 of 3)—4.6L engine

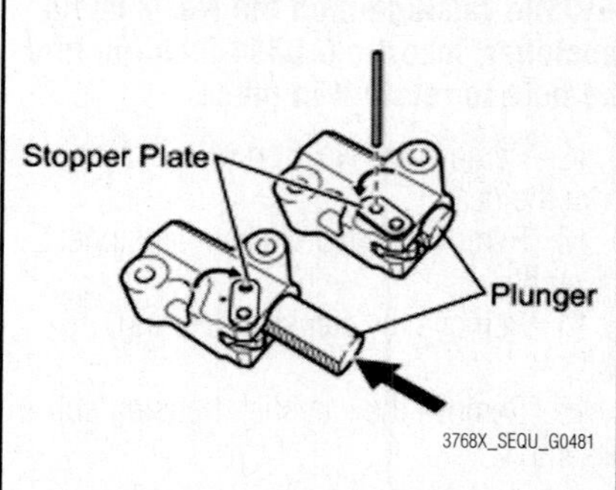

Fig. 108 Camshaft No. 1 chain tensioner installation Bank 2 (1 of 2)—4.6L engine

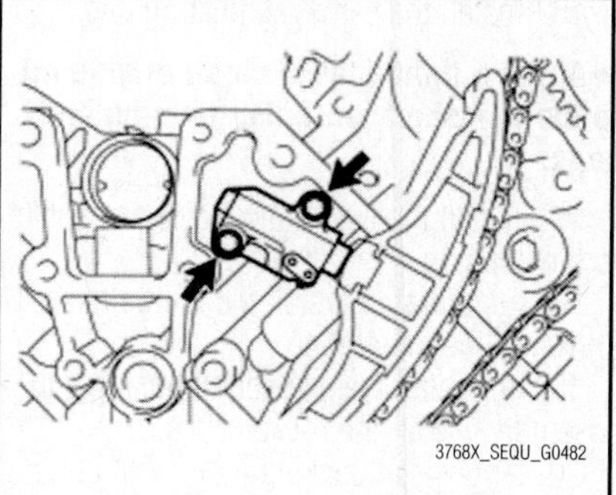

Fig. 109 Camshaft No. 1 chain tensioner installation Bank 2 (2 of 2)—4.6L engine

30. Install the No. 1 chain vibration damper.
31. Tighten the camshaft timing gear. Tighten bolt to 74 ft. lbs. (100 Nm).
32. Check that the engine is at TDC, on the compression stroke.
33. Continue the installation in the reverse order of the removal procedure.
34. Use the Techstream diagnostic tool, or equivalent and reprogram the required systems.

4.7L Engine

See Figures 110 through 120.

1. Before servicing the vehicle, refer to the Precautions Section.

➡If working near and/or around the SRS system and components, be sure to disable the SRS system. Tape the negative battery cable with insulating tape. Always disconnect the negative battery cable first.

⁂ CAUTION

To avoid personal injury when working on vehicles equipped with an air bag, the negative battery cable must be disconnected and at least 90 seconds must elapse before working on

the system. Failure to do so may result in deployment of the air bag.

2. Disconnect the negative battery cable. Tape the cable with insulating tape.

➡When disconnecting the cable, some systems need to be initialized after the cable is reconnected. You will need the Techstream diagnostic scan tool or equivalent. Follow the directions on the tool.

3. Drain the cooling system.
4. Relieve the fuel system pressure.
5. Remove or disconnect the following:
   - Engine under covers
   - Engine appearance cover
   - Air intake hose
   - Accessory drive belt
   - Cooling fan
   - Radiator
   - Idler pulley
   - Upper and middle timing belt covers
   - A/C compressor
   - Cooling fan bracket
   - Alternator
   - Accessory drive belt tensioner
6. Set the engine to Top Dead Center (TDC) with the camshaft sprocket timing marks aligned with the rear cover timing marks.
7. Rotate the crankshaft to 50 degrees After TDC as shown. The crankshaft pulley timing mark should align with the center of the No. 2 idler pulley bolt.
8. Remove or disconnect the following:
   - Crankshaft pulley
   - Lower timing cover
   - Timing belt.
   - Camshaft timing sprockets
   - Camshaft Position (CMP) sensor
   - Ignition coils
   - Valve cover
   - Timing belt rear covers
9. Rotate the right bank camshafts as necessary to access the exhaust camshaft sub-gear service bolt hole and install a 6mm x 1.0mm bolt.

➡Keep all valvetrain components in order for assembly.

10. Align the right bank camshaft 1 dot timing marks to a **10** degree angle as shown.
11. Loosen the bearing cap bolts in sequence and in several passes.
12. Remove the right bank camshafts.
13. Rotate the left bank camshafts as necessary to access the exhaust camshaft sub-gear service bolt hole and install a 6mm x 1.0mm bolt.
14. Align the left bank camshaft 2 dot timing marks as shown.
15. Loosen the bearing cap bolts in sequence and in several passes.
16. Remove the left bank camshafts.
17. Remove the valve lifters and shims.

To install:

➡Be sure to use new fasteners, as required.

18. Ensure that the crankshaft is at 50 degrees After TDC.

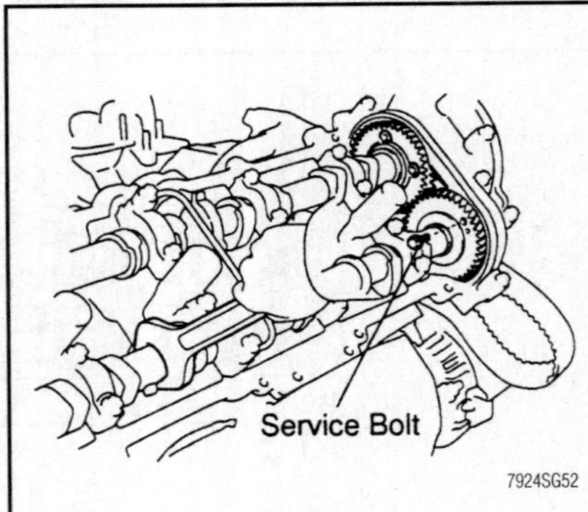

Fig. 111 Camshaft service bolt installation—4.7L engine

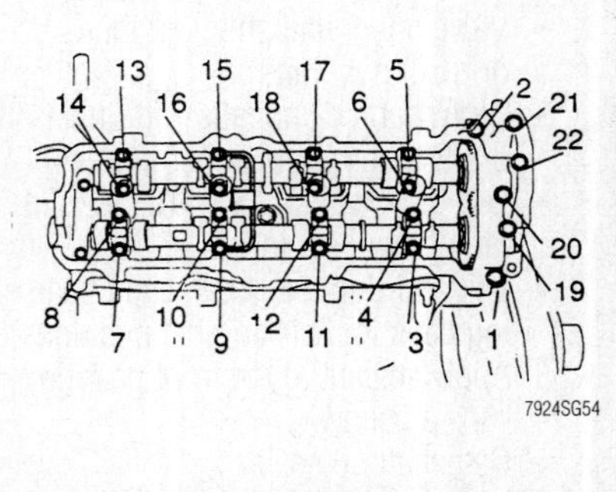

Fig. 113 Right bank camshaft bearing cap loosening sequence—4.7L engine

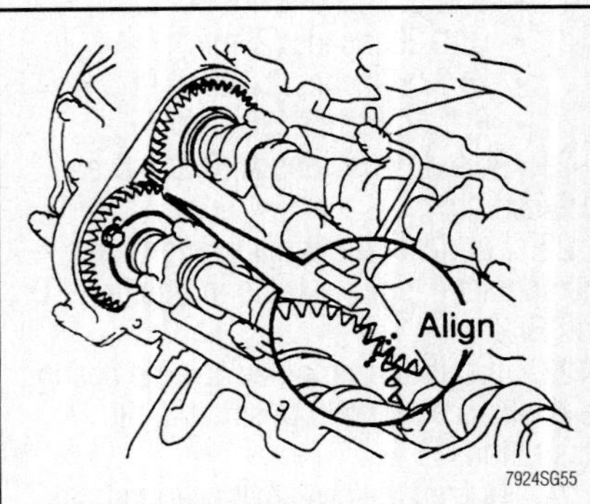

Fig. 114 Left bank camshaft timing mark (2 dot marks) alignment—4.7L engine

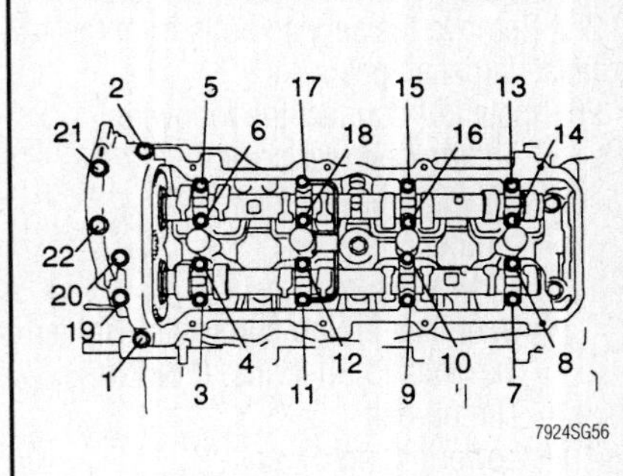

Fig. 115 Left bank camshaft bearing cap loosening sequence—4.7L engine

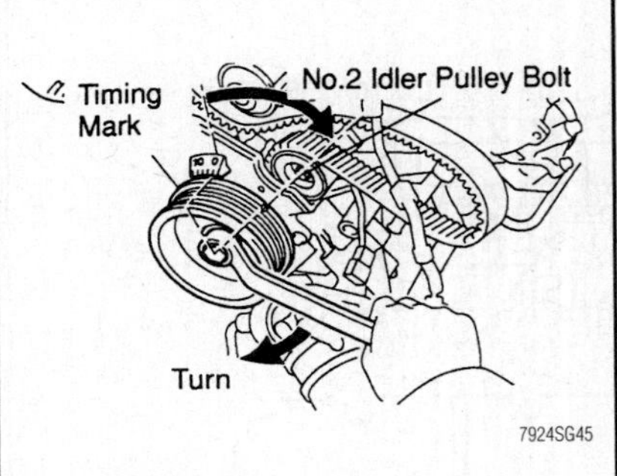

Fig. 110 Setting the crankshaft to 50 degrees ATDC—4.7L engine

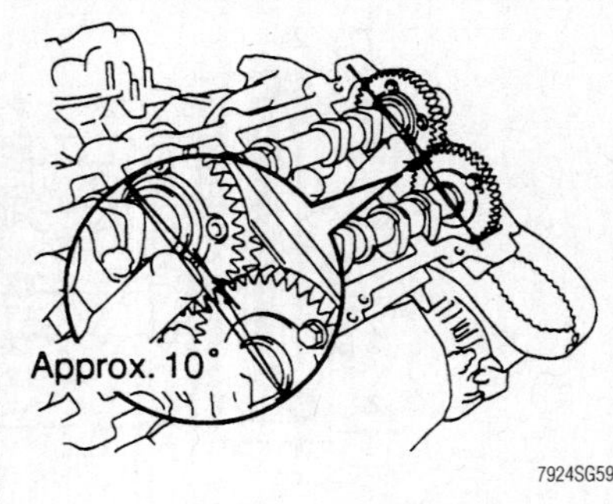

Fig. 112 Right bank camshaft timing mark (1 dot marks) alignment—4.7L engine

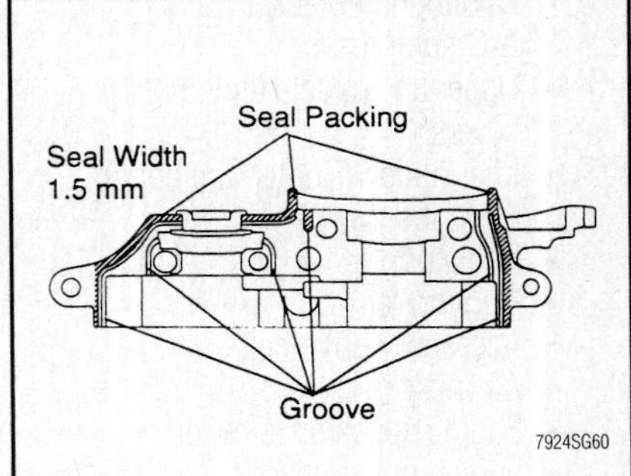

Fig. 116 Apply a 1.5mm bead of sealant to the front bearing caps—4.7L engine

19. Install or connect the following:
    - Valve lifters and shims in their original positions
    - Right bank camshafts with the 1 dot timing marks at 10 degrees
    - Left bank camshafts with the 2 dot timing marks aligned
    - Left and right bank camshaft bearing caps in their original positions. Apply sealant to the front bearing caps as shown.
    - Camshaft oil seals
20. The bearing cap bolts vary in length and are identified as follows:
    - A: 3.70 inches (94mm)
    - B: 2.83 inches (72mm)
    - C: 0.98 inches (25mm)
    - D: 2.05 inches (52mm)
    - E: 1.50 inches (38mm)
21. Bolts in positions **A**, **B** and **C** are installed dry.
22. Lubricate the threads and under the contact flange for bolts in positions **D** and **E**.
23. Install oil feed pipes and the bearing cap bolts according to position in the illustrations.
24. Tighten the camshaft bearing bolts in sequence and in several passes to the following specifications:
    - Bolt C: 66 inch lbs. (7.5 Nm)
    - All others: 12 ft. lbs. (16 Nm)
25. Remove the service bolts from the exhaust camshaft gears.
26. Install or connect the following:
    - Timing belt rear covers
    - Valve cover
    - Ignition coils
    - CMP sensor
    - Camshaft timing sprockets. Tighten the bolts to 80 ft. lbs. (108 Nm).
    - Timing belt
    - Lower timing cover
    - Crankshaft pulley. Tighten the bolt to 181 ft. lbs. (245 Nm).
    - Accessory drive belt tensioner
    - Alternator
    - Cooling fan bracket
    - A/C compressor
    - Upper and middle timing belt covers
    - Idler pulley. Tighten the bolt to 27 ft. lbs. (37 Nm).
    - Radiator
    - Cooling fan
    - Accessory drive belt
    - Air intake hose
    - Engine appearance cover
    - Engine under covers
    - Negative battery cable
27. Fill the cooling system.
28. Start the engine and check for leaks.

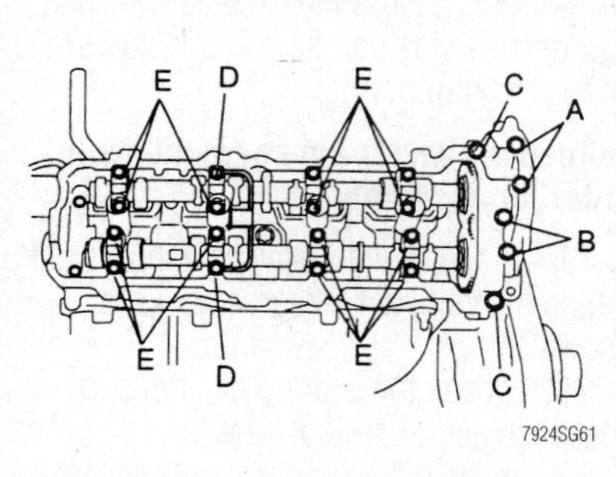

Fig. 117 Right bank bearing cap bolt location—4.7L engine

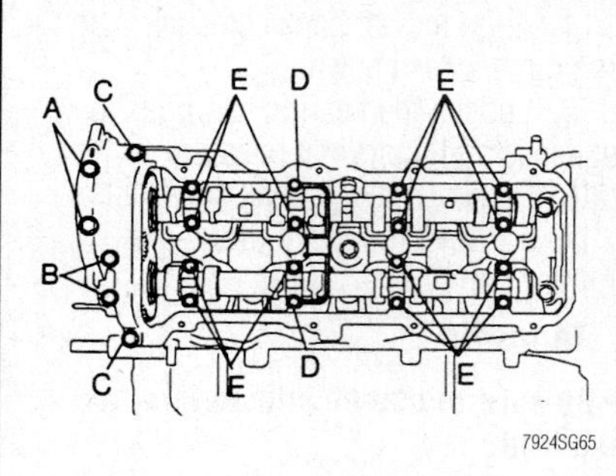

Fig. 118 Left camshaft bearing cap bolt locations—4.7L engine

29. Use the Techstream diagnostic tool, or equivalent and reprogram the required systems.

5.7L Engine

See Figures 121 through 139.

1. Before servicing the vehicle, refer to the Precautions Section.

➡If working near and/or around the SRS system and components, be sure to disable the SRS system. Tape the negative battery cable with insulating tape. Always disconnect the negative battery cable first.

✻✻ CAUTION

To avoid personal injury when working on vehicles equipped with an air bag, the negative battery cable must be disconnected and at least 90 seconds must elapse before working on the system. Failure to do so may result in deployment of the air bag.

2. Disconnect the negative battery cable. Tape the cable with insulating tape.

➡When disconnecting the cable, some systems need to be initialized after the cable is reconnected. You will need the

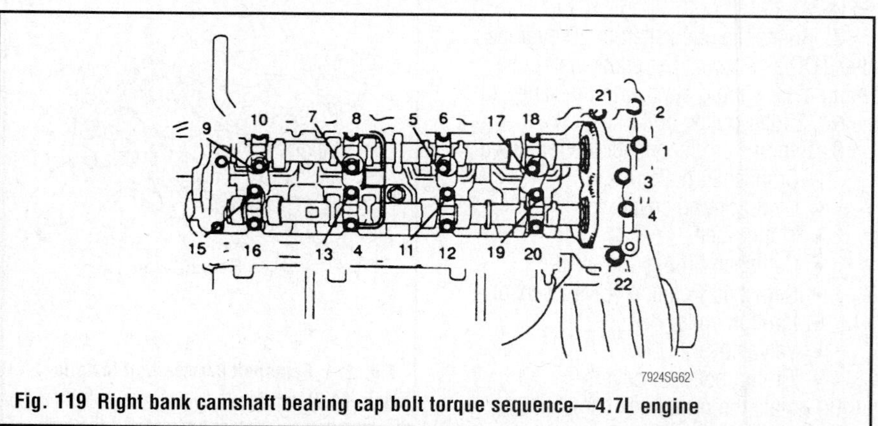

Fig. 119 Right bank camshaft bearing cap bolt torque sequence—4.7L engine

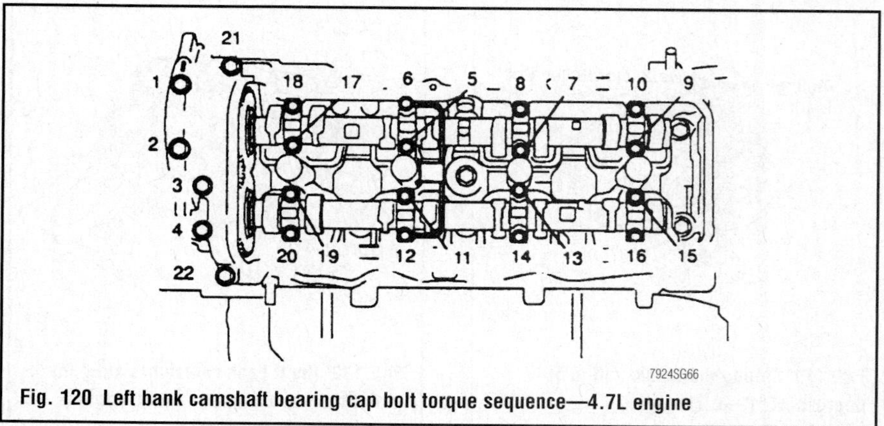

Fig. 120 Left bank camshaft bearing cap bolt torque sequence—4.7L engine

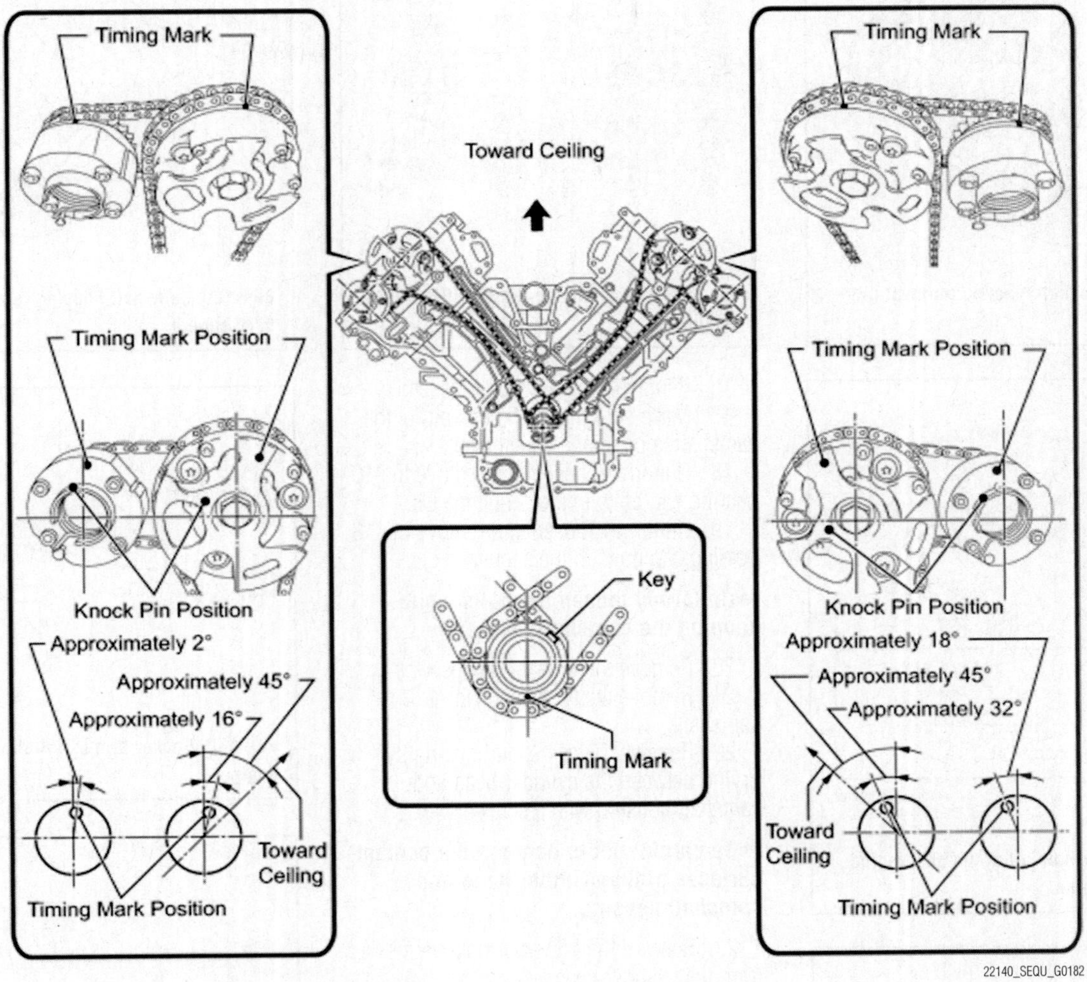

Fig. 121 Positioning engine to TDC—5.7L engine

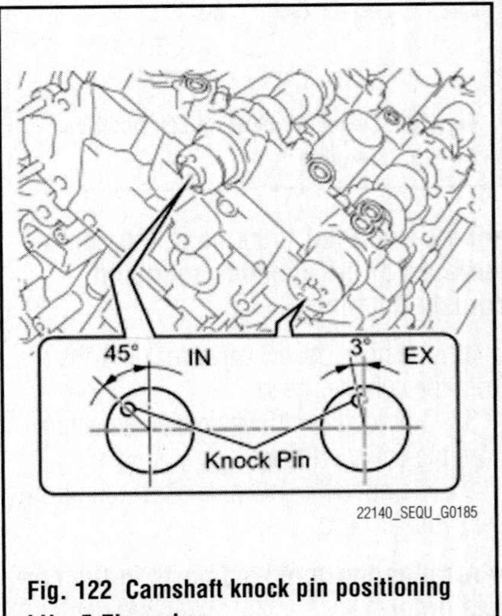

Fig. 122 Camshaft knock pin positioning LH—5.7L engine

Techstream diagnostic scan tool or equivalent. Follow the directions on the tool.

3. Discharge fuel system pressure.
4. Remove front exhaust pipe assembly.
5. Remove front fender apron seal.
6. Remove front fender apron seal rear.
7. Remove No. 2 steering intermediate shaft sub-assembly.

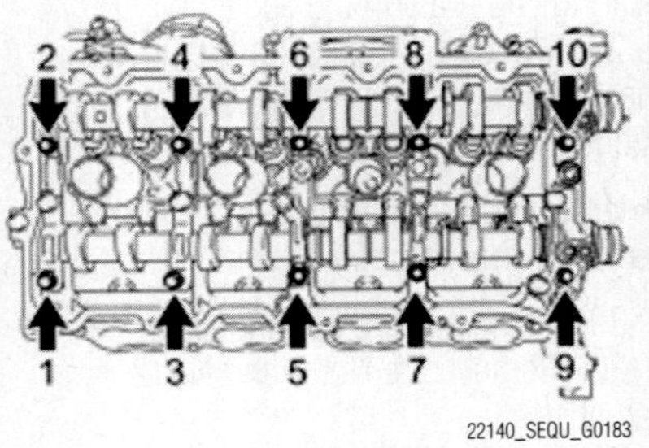

Fig. 123 Camshaft cap removal LH—5.7L engine

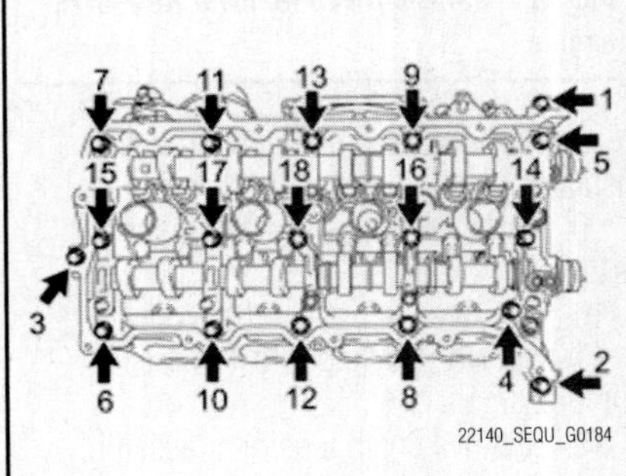

Fig. 124 Camshaft cap removal LH—5.7L engine

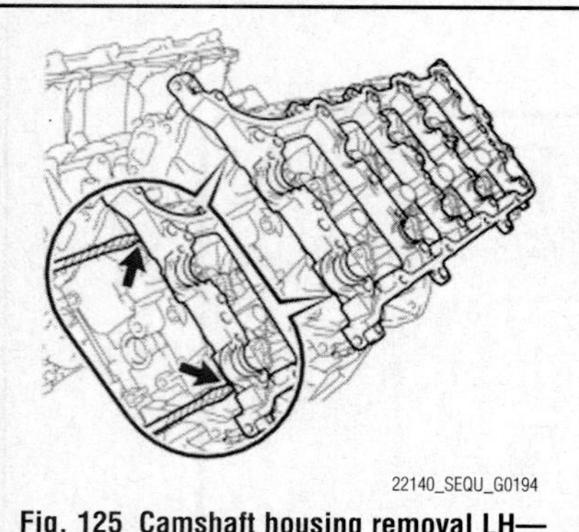

Fig. 125 Camshaft housing removal LH—5.7L engine

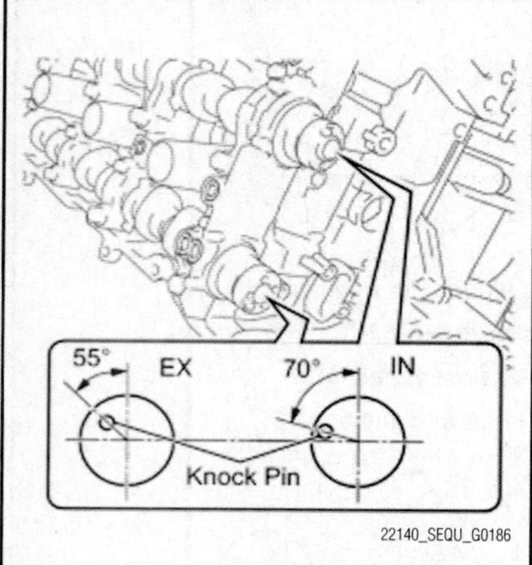

Fig. 126 Camshaft knock pin positioning RH—5.7L engine

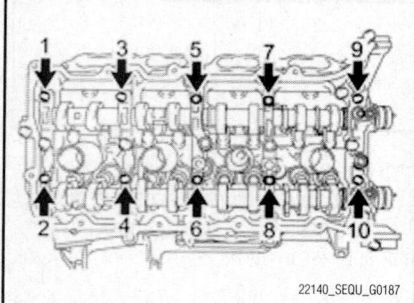

Fig. 127 Camshaft cap removal RH—5.7L engine

8. Remove exhaust manifold heat insulator.
9. Remove exhaust manifold sub-assembly.
10. Remove the timing chain cover.
11. Set No. 1 cylinder to TDC / compression.
12. Remove chain tensioner assembly.
13. Remove chain tensioner slipper.
14. Remove chain vibration damper.
15. Remove chain sub-assembly.

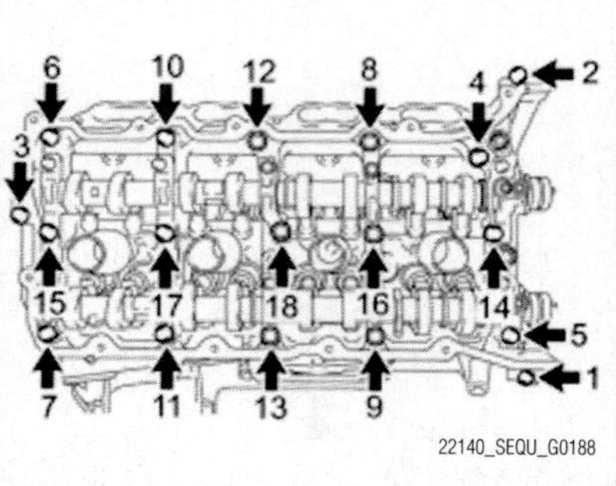

Fig. 128 Camshaft cap removal RH—5.7L engine

16. Remove chain tensioner assembly.
17. Make sure that the knock pin of the camshaft is positioned as shown.
18. Uniformly loosen and remove the 10 bearing cap bolts in the sequence LH.
19. Uniformly loosen and remove the 18 bearing cap bolts in the sequence LH.

➡Uniformly loosen the bolts while keeping the camshaft level.

20. Remove the 6 bearing caps.
21. Remove the No. 3 and No. 4 camshafts.
22. Remove the camshaft housing by prying between the cylinder head and camshaft housing with a screwdriver.

➡Be careful not to damage the contact surfaces of the cylinder head and camshaft housing.

23. Remove the 16 valve rocker arms from the cylinder head.
24. Remove the 16 valve lash adjusters from the cylinder head.
25. Remove the 16 valve stem caps from the cylinder head.

➡Arrange the removed parts in the correct order.

26. Make sure that the knock pin of the camshaft is positioned as shown in the illustration.
27. Uniformly loosen and remove the 10 bearing cap bolts in the sequence RH.
28. Uniformly loosen and remove the 18 bearing cap bolts in the sequence RH.

➡Uniformly loosen the bolts while keeping the camshaft level.

29. Remove the 6 bearing caps.
30. Remove the No. 1 and No. 2 camshafts.
31. Remove the camshaft housing by prying between the cylinder head and camshaft housing with a screwdriver.

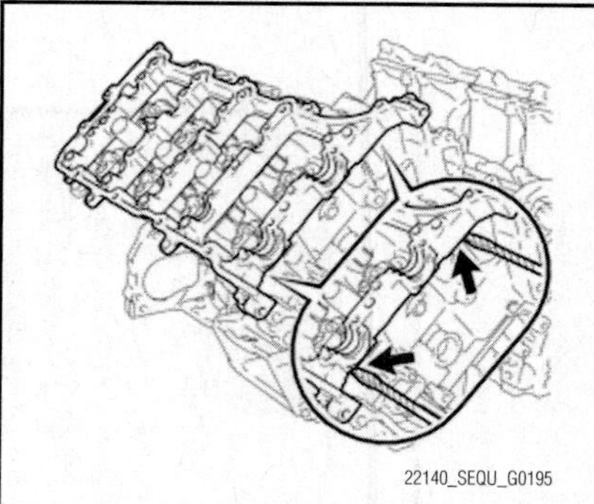

Fig. 129 Camshaft housing removal RH—5.7L engine

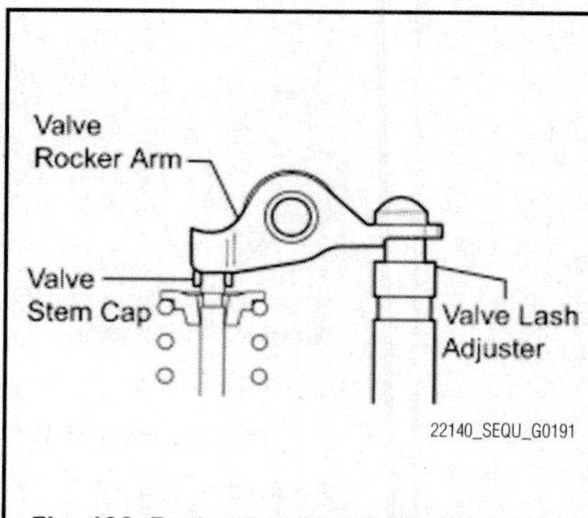

Fig. 130 Rocker arm installation—5.7L engine

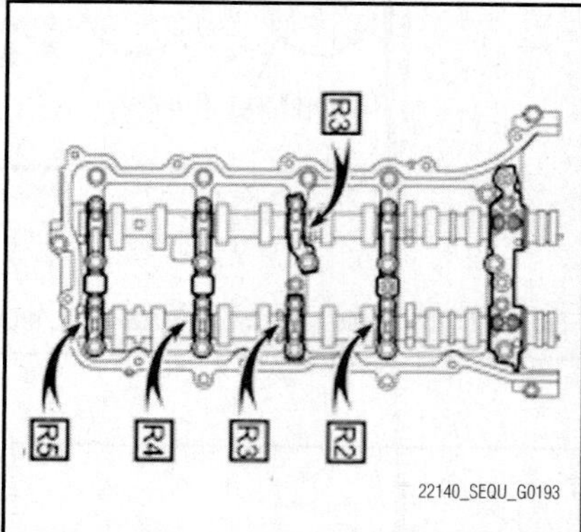

Fig. 131 Camshaft bearing cap position RH—5.7L engine

➡Be careful not to damage the contact surfaces of the cylinder head and camshaft housing.

32. Remove the 16 valve rocker arms from the cylinder head.
33. Remove the 16 valve lash adjusters from the cylinder head.
34. Remove the 16 valve stem caps from the cylinder head.

➡Arrange the removed parts in the correct order.

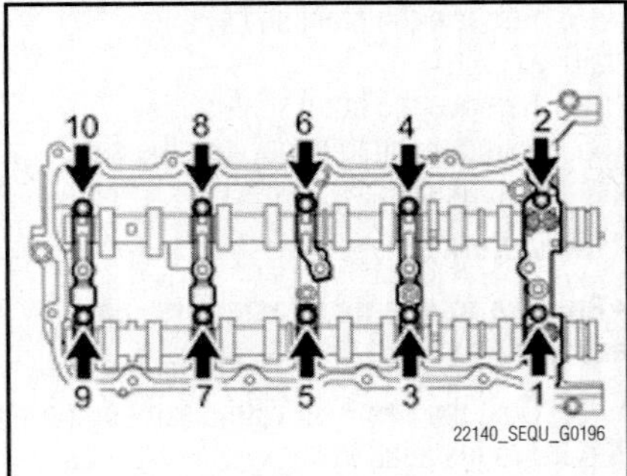

Fig. 132 Camshaft bearing cap tightening sequence RH—5.7L engine

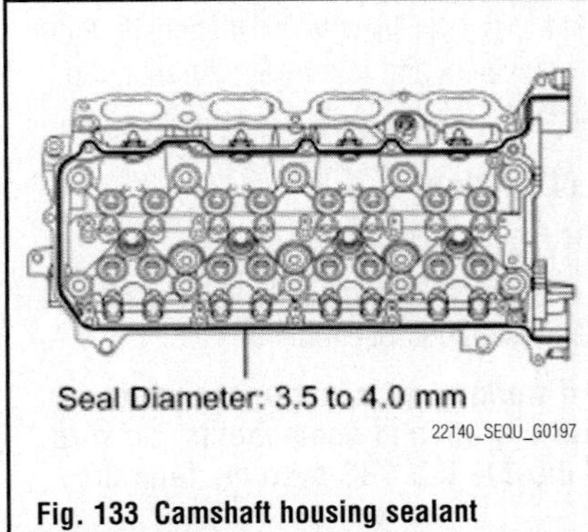

Fig. 133 Camshaft housing sealant position RH—5.7L engine

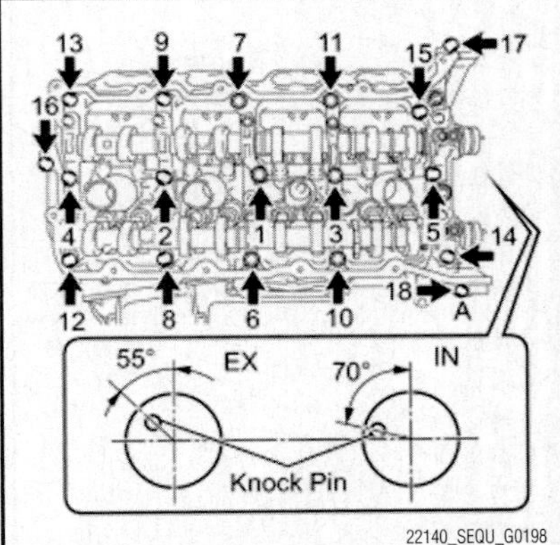

Fig. 134 Camshaft housing tightening sequence RH—5.7L engine

To install:

➡Be sure to use new fasteners, as required.

35. Install valve stem cap.
   a. Apply a light coat of engine oil to the valve stem caps.
   b. Install the 32 valve stem caps to the cylinder head.

➡Install the lash adjuster at the same place it was removed from.

36. Install the 32 valve rocker arms.

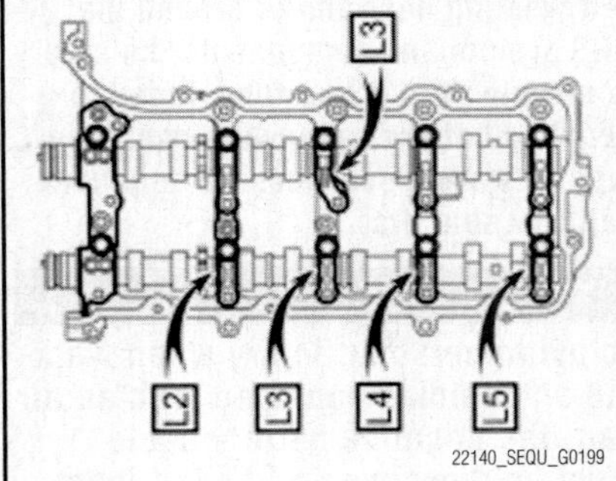

Fig. 135 Camshaft bearing cap position LH—5.7L engine

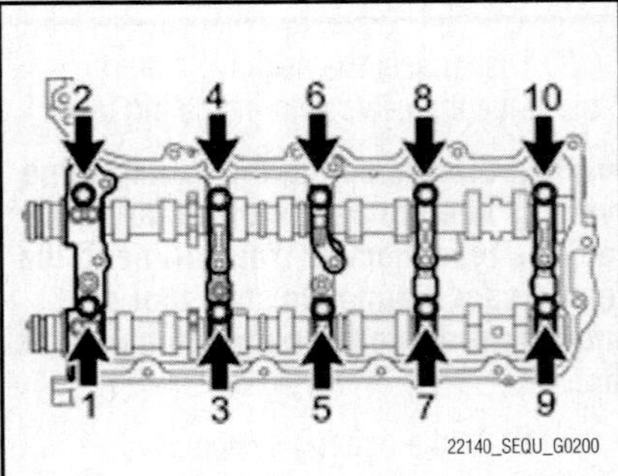

Fig. 136 Camshaft bearing cap tightening sequence LH—5.7L engine

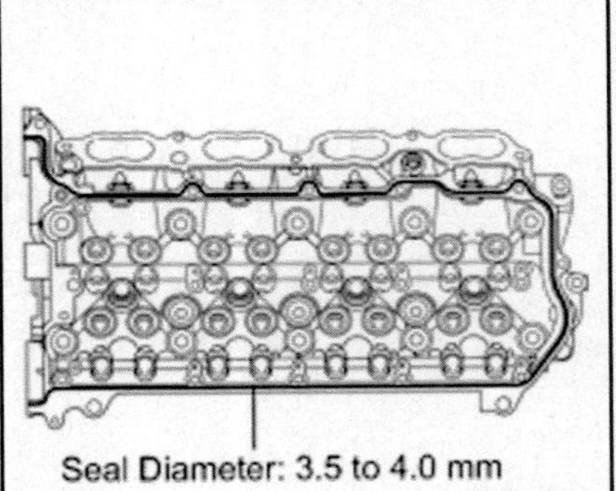

Fig. 137 Camshaft housing sealant position LH—5.7L engine

37. Apply a light coat of engine oil to the camshaft journals, camshaft housings and bearing caps.
38. Install the No. 1 and No. 2 camshafts to the camshaft housing.
39. Confirm the marks and numbers on the camshaft bearing caps and place them in their proper positions and directions.
40. Temporarily install the 10 bolts in the order shown in the illustration.
41. Make sure that the valve rocker arms are installed properly.
42. Apply seal packing in a continuous line.

➡Remove any oil from the contact surface. Install the camshaft housing within 3 minutes and tighten the bolts within 15 minutes after applying seal packing. Do not start the engine for at least 2 hours after the installation.

43. Install the camshaft housing, and install the 18 bolts in the order and tighten bolt "A" to 7 ft. lbs. (10 Nm), and all others to 22 ft. lbs. (30 Nm).

➡Make sure that the knock pin of the camshaft is positioned as shown in the illustration before installing the camshaft housing.

44. Tighten the 10 bolts in the order to 12 ft. lbs. (16 Nm).
45. Apply a light coat of engine oil to the camshaft journals, camshaft housings and bearing caps.
46. Install the No. 1 and No. 2 camshafts to the camshaft housing.
47. Confirm the marks and numbers on the camshaft bearing caps and place them in their proper positions and directions.
48. Temporarily install the 10 bolts in the order shown in the illustration.
49. Make sure that the valve rocker arms are installed properly.
50. Apply seal packing in a continuous line.

➡Remove any oil from the contact surface. Install the camshaft housing within 3 minutes and tighten the bolts within 15 minutes after applying seal packing. Do not start the engine for at least 2 hours after the installation.

51. Install the camshaft housing, and install the 18 bolts in the order and tighten bolt "A" to 7 ft. lbs. (10 Nm), and all others to 22 ft. lbs. (30 Nm).

➡Make sure that the knock pin of the camshaft is positioned as shown in the illustration before installing the camshaft housing.

52. Tighten the 10 bolts in the order to 12 ft. lbs. (16 Nm).
53. Install chain tensioner assembly.
54. Install chain sub-assembly.
55. Install chain tensioner slipper.
56. Install chain tensioner assembly.
57. Install chain vibration damper.
58. Tighten camshaft timing gear.
59. Check No. 1 cylinder to TDC.

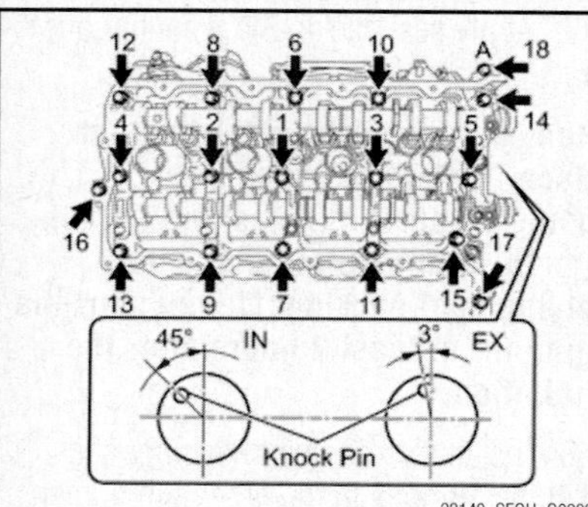

Fig. 138 Camshaft housing tightening sequence LH—5.7L engine

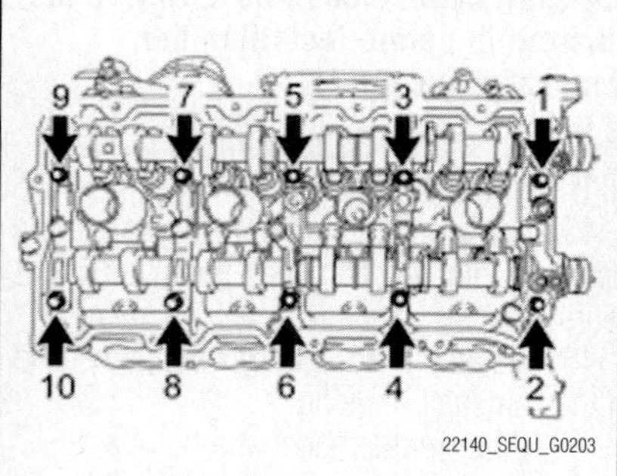

Fig. 139 Camshaft bearing cap tightening sequence LH—5.7L engine

60. Install the timing chain cover.
61. Install exhaust manifold sub-assembly.
62. Install exhaust manifold heat insulator.
63. Install No.2 steering intermediate shaft sub-assembly.
64. Install front fender apron seal rear.
65. Install front fender apron seal.
66. Install front exhaust pipe assembly.
67. Connect cable to negative battery terminal.
68. Use the Techstream diagnostic tool, or equivalent and reprogram the required systems.

CRANKSHAFT DAMPER

REMOVAL & INSTALLATION

1. See Crankshaft Front Seal procedure for damper removal.

CRANKSHAFT FRONT SEAL

REMOVAL & INSTALLATION

4.0L Engine

See Figure 140.

1. Before servicing the vehicle, refer to the Precautions Section.

➡If working near and/or around the SRS system and components, be sure to disable the SRS system. Tape the negative battery cable with insulating tape. Always disconnect the negative battery cable first.

⁂ CAUTION

To avoid personal injury when working on vehicles equipped with an air bag, the negative battery cable must be disconnected and at least 90 seconds must elapse before working on the system. Failure to do so may result in deployment of the air bag.

2. Disconnect the negative battery cable. Tape the cable with insulating tape.

➡When disconnecting the cable, some systems need to be initialized after the cable is reconnected. You will need the Techstream diagnostic scan tool or equivalent. Follow the directions on the tool.

3. Drain the cooling system.
4. Remove the radiator.
5. Remove the crankshaft pulley retaining bolt.
6. Remove the crankshaft pulley.
7. Using a suitable tool, pry the seal from its mounting.

To install:

➡Be sure to use new fasteners, as required.

8. Coat the new seal with clean engine oil prior to installation.
9. Using a seal installation tool, press the seal into position until its surface is flush with the timing chain edge cover.
10. Continue the installation in the reverse order of the removal procedure.
11. Use the Techstream diagnostic tool, or equivalent and reprogram the required systems.

4.6L Engine

See Figure 141.

1. Before servicing the vehicle, refer to the Precautions Section.

➡If working near and/or around the SRS system and components, be sure to disable the SRS system. Tape the

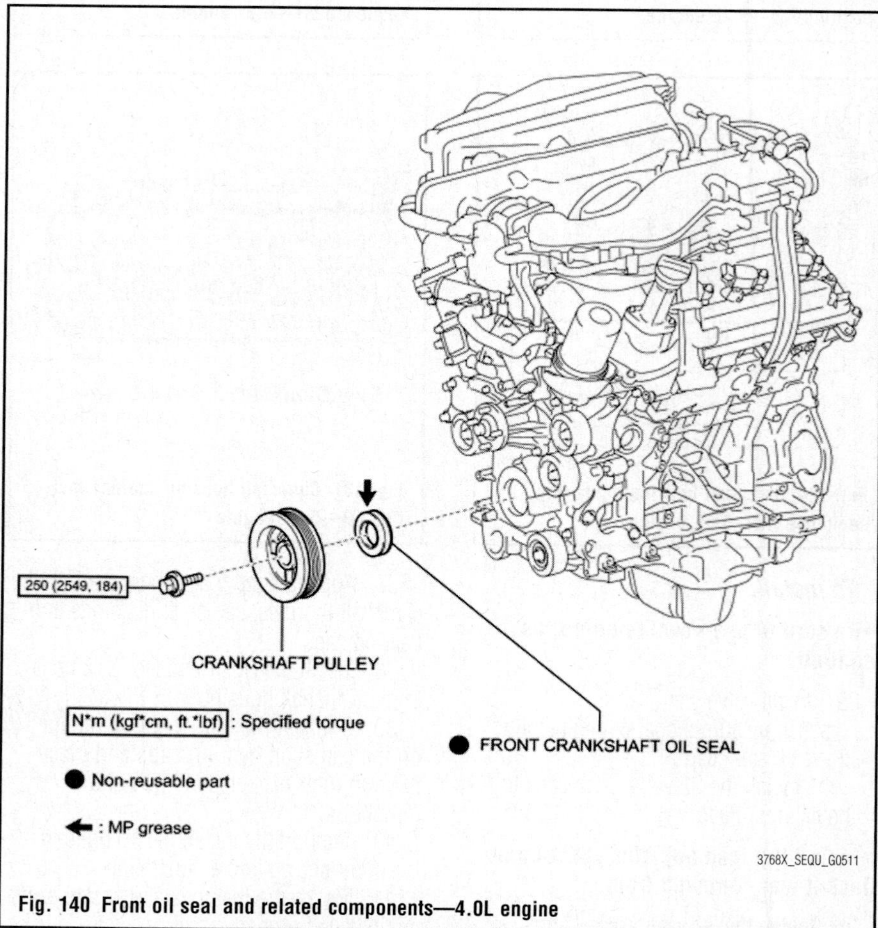

Fig. 140 Front oil seal and related components—4.0L engine

negative battery cable with insulating tape. Always disconnect the negative battery cable first.

CAUTION

To avoid personal injury when working on vehicles equipped with an air bag, the negative battery cable must be disconnected and at least 90 seconds must elapse before working on the system. Failure to do so may result in deployment of the air bag.

2. Disconnect the negative battery cable. Tape the cable with insulating tape.

➡When disconnecting the cable, some systems need to be initialized after the cable is reconnected. You will need the Techstream diagnostic scan tool or equivalent. Follow the directions on the tool.

3. Remove V-bank cover sub-assembly.
4. Remove No. 1 engine under cover.

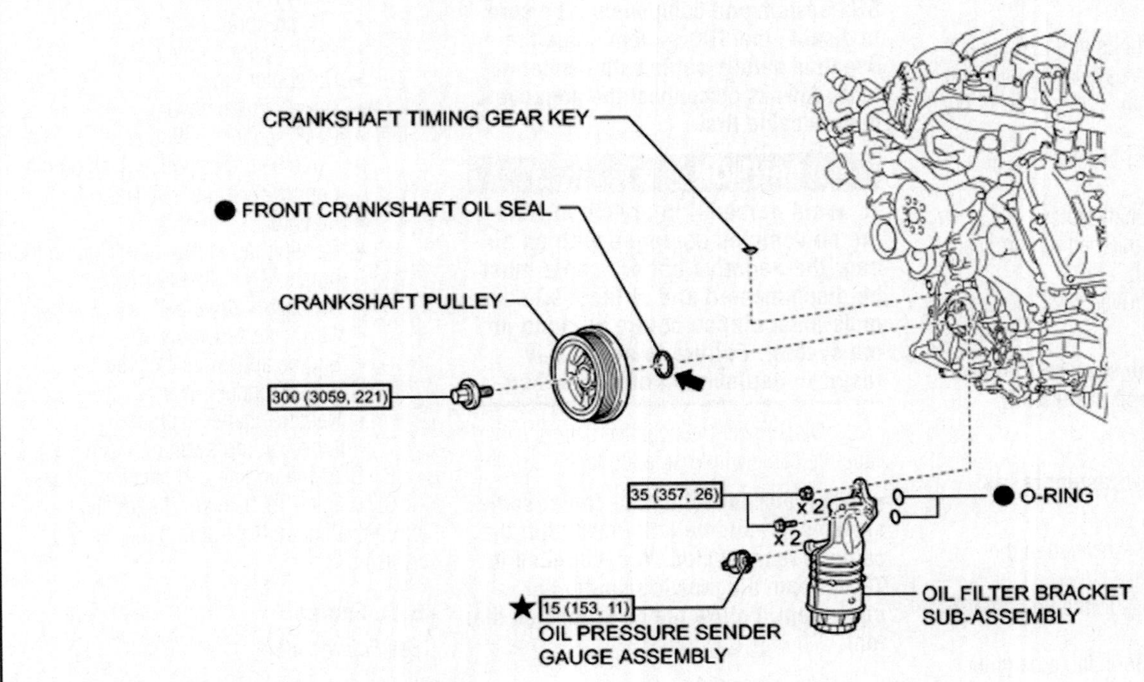

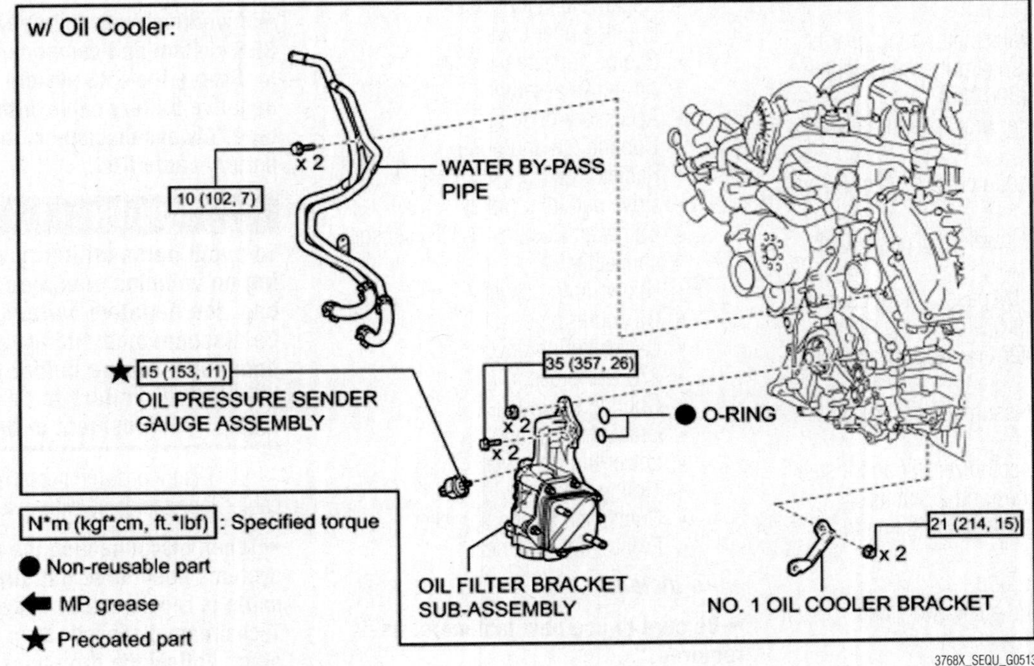

Fig. 141 Front oil seal and related components—4.6L engine

5. Drain engine coolant.
6. Remove inlet radiator hose.
7. Remove fan and alternator V belt.
8. Remove fan shroud.
9. Remove the 4 bolts, 2 stabilizer brackets and 2 stabilizer bushes. Then disconnect the stabilizer bar.
10. Remove oil pressure sender gauge assembly.
11. Remove oil filter bracket sub-assembly (w/o oil cooler).
12. Remove the 3 bolts and disconnect the 2 water by-pass hoses from the oil cooler.
13. Remove No. 1 oil cooler bracket (w/ oil cooler).
14. Remove oil filter bracket sub-assembly (w/ oil cooler).
15. Remove crankshaft pulley.
16. Remove the crankshaft timing gear key from the crankshaft.
17. Using a screwdriver, pry out the oil seal.

➡Do not damage the surface of the oil seal press fit hole and crankshaft.

To install:

➡Be sure to use new fasteners, as required.

18. Installation is the reverse of the removal procedure.
19. Apply MP grease to the lip of a new oil seal.
20. Using Service Tool and a hammer, tap in the oil seal to a depth between 0 to 1.0 mm (0 to 0.0394 in.) from the timing chain cover edge.
21. Install the crankshaft timing gear key.
22. Install crankshaft pulley and tighten bolt to 221 ft. lbs. (300 Nm).
23. Install oil filter bracket sub-assembly (w/oil cooler).
24. Install no. 1 oil cooler bracket (w/oil cooler).
25. Connect the 2 water by-pass hoses to the oil cooler.
26. Install the 3 by-pass pipe bolts and tighten to 7 ft. lbs. (10 Nm).
27. Install oil filter bracket sub-assembly (w/o oil cooler).
28. Install oil pressure sender gauge assembly.
29. Connect the stabilizer bar and install the 2 stabilizer bushes and 2 stabilizer brackets with the 4 bolts and tighten to 51 ft. lbs. (69 Nm).
30. Install fan shroud.
31. Install fan and alternator v belt.
32. Install inlet radiator hose.
33. Add engine coolant.
34. Inspect for leaks.
35. Install No. 1 engine under cover.
36. Install V-bank cover sub-assembly.
37. Use the Techstream diagnostic tool, or equivalent and reprogram the required systems.

4.7L Engine

1. Before servicing the vehicle, refer to the Precautions Section.

➡If working near and/or around the SRS system and components, be sure to disable the SRS system. Tape the negative battery cable with insulating tape. Always disconnect the negative battery cable first.

⁂ CAUTION

To avoid personal injury when working on vehicles equipped with an air bag, the negative battery cable must be disconnected and at least 90 seconds must elapse before working on the system. Failure to do so may result in deployment of the air bag.

2. Disconnect the negative battery cable. Tape the cable with insulating tape.

➡When disconnecting the cable, some systems need to be initialized after the cable is reconnected. You will need the Techstream diagnostic scan tool or equivalent. Follow the directions on the tool.

3. Drain the cooling system.
4. Remove or disconnect the following:
   - Negative battery cable
   - Engine under cover
   - Engine appearance cover
   - Air intake assembly
   - Accessory drive belt
   - Cooling fan and pulley
   - Radiator
   - Drive belt idler pulley
   - Camshaft Position (CMP) sensor connector
   - Upper timing covers
   - Oil cooler pipe
   - Center timing cover
   - A/C compressor
   - Cooling fan bracket
   - Crankshaft pulley
   - Lower timing cover
   - Timing belt.
   - Crankshaft timing sprocket
   - Front crankshaft seal

To install:

➡Be sure to use new fasteners, as required.

5. Install the oil seal so that it is flush with the oil pump housing.
6. Install or connect the following:
   - Crankshaft timing sprocket
   - Timing belt
   - Lower timing cover
   - Crankshaft pulley. Tighten the bolt to 181 ft. lbs. (245 Nm).
   - Cooling fan bracket. Tighten the 12mm bolts to 12 ft. lbs. (16 Nm) and the 14mm bolts to 24 ft. lbs. (32 Nm).
   - A/C compressor
   - Center timing cover
   - Oil cooler pipe
   - Upper timing covers
   - CMP sensor connector
   - Drive belt idler pulley. Tighten the bolt to 27 ft. lbs. (37 Nm).
   - Radiator
   - Cooling fan and pulley. Tighten the nuts to 16 ft. lbs. (21 Nm).
   - Accessory drive belt
   - Air intake assembly
   - Engine appearance cover
   - Engine under cover
   - Negative battery cable
7. Fill the cooling system.
8. Start the engine and check for leaks.
9. Use the Techstream diagnostic tool, or equivalent and reprogram the required systems.

5.7L Engine

See Figure 142.

1. Before servicing the vehicle, refer to the Precautions Section.

➡If working near and/or around the SRS system and components, be sure to disable the SRS system. Tape the negative battery cable with insulating tape. Always disconnect the negative battery cable first.

⁂ CAUTION

To avoid personal injury when working on vehicles equipped with an air bag, the negative battery cable must be disconnected and at least 90 seconds must elapse before working on the system. Failure to do so may result in deployment of the air bag.

2. Disconnect the negative battery cable. Tape the cable with insulating tape.

➡When disconnecting the cable, some systems need to be initialized after the cable is reconnected. You will need the Techstream diagnostic scan tool or equivalent. Follow the directions on the tool.

3. Remove V-bank cover sub-assembly.
4. Remove No. 1 engine under cover.

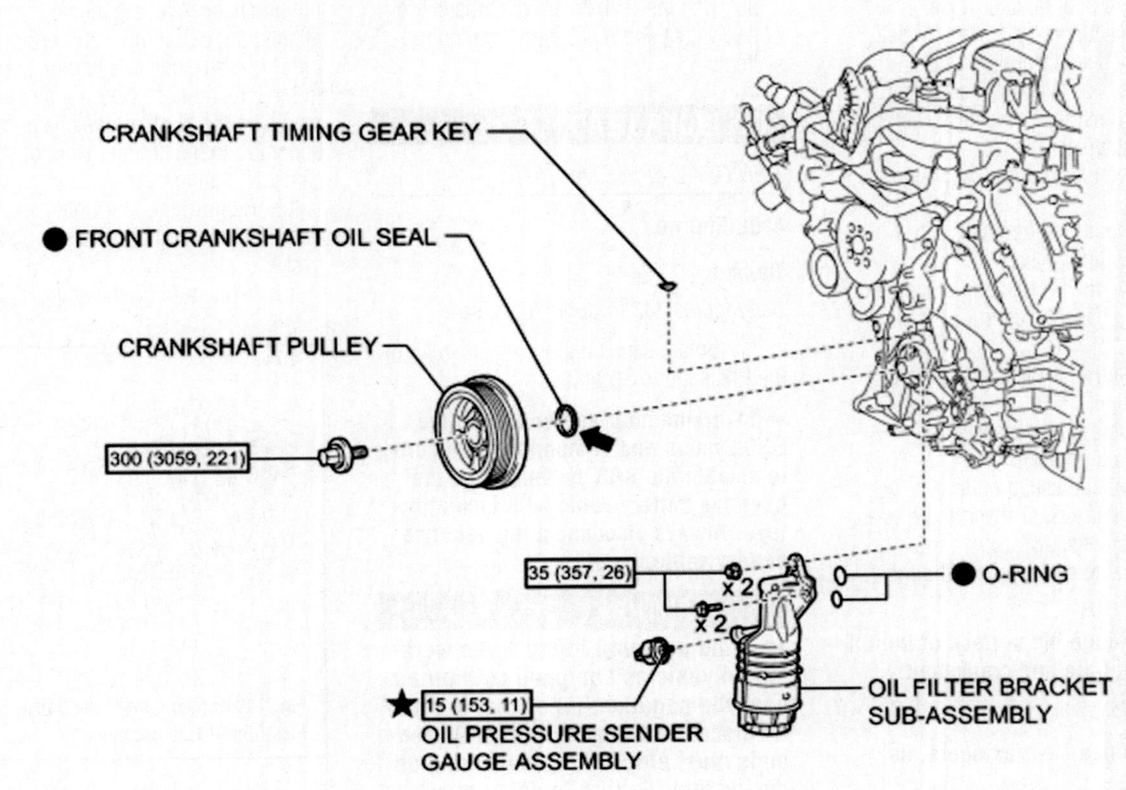

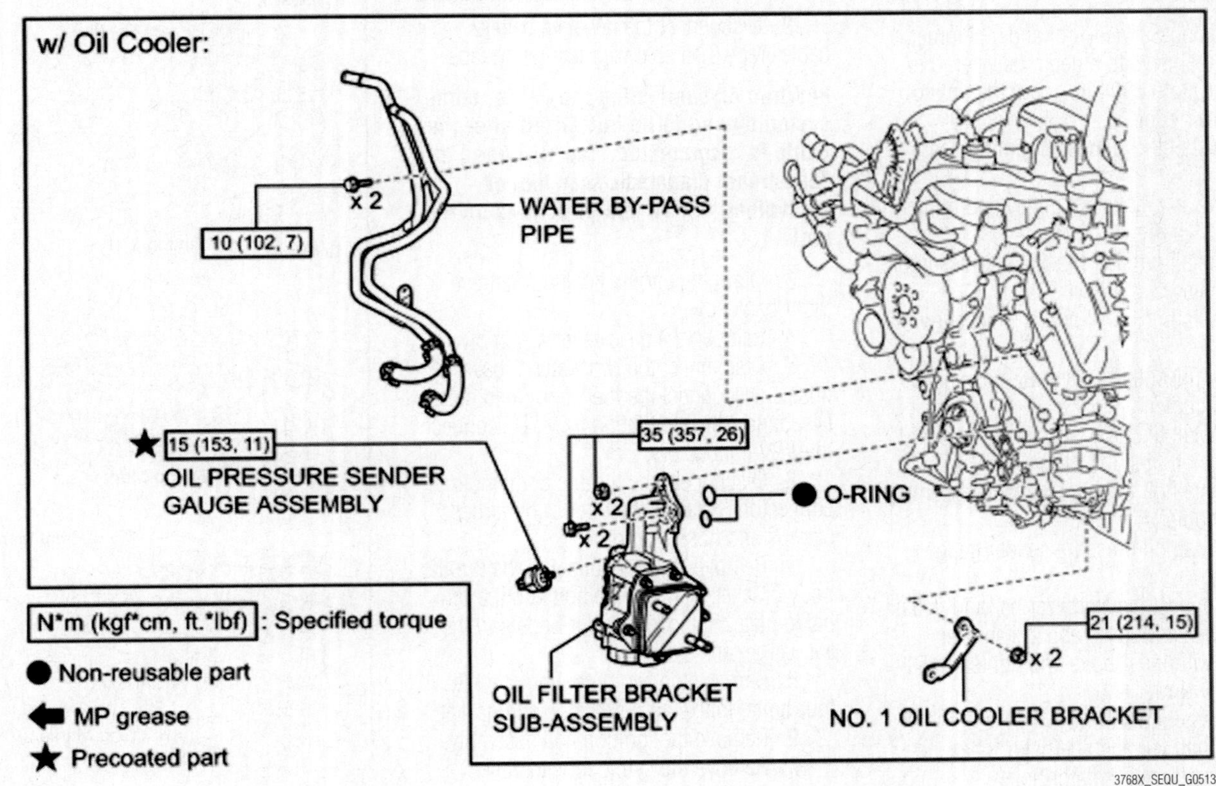

Fig. 142 Front oil seal and related components—5.7L engine

5. Drain engine coolant.
6. Remove inlet radiator hose.
7. Remove fan and alternator V belt.
8. Remove fan shroud.
9. Remove the 4 bolts, 2 stabilizer brackets and 2 stabilizer bushes. Then disconnect the stabilizer bar.
10. Remove oil pressure sender gauge assembly.
11. Remove oil filter bracket sub-assembly (w/o oil cooler).
12. Remove the 3 bolts and disconnect the 2 water by-pass hoses from the oil cooler.
13. Remove No. 1 oil cooler bracket (w/ oil cooler).
14. Remove oil filter bracket sub-assembly (w/ oil cooler).
15. Remove crankshaft pulley.
16. Remove the crankshaft timing gear key from the crankshaft.
17. Using a screwdriver, pry out the oil seal.

➡Do not damage the surface of the oil seal press fit hole and crankshaft.

To install:

➡Be sure to use new fasteners, as required.

18. Apply MP grease to the lip of a new oil seal.
19. Using Service Tool and a hammer, tap in the oil seal to a depth between 0 to 1.0 mm (0 to 0.0394 in.) from the timing chain cover edge.
20. Install the crankshaft timing gear key.
21. Install crankshaft pulley and tighten bolt to 221 ft. lbs. (300 Nm).
22. Install oil filter bracket sub-assembly (w/oil cooler).
23. Install no. 1 oil cooler bracket (w/oil cooler).
24. Connect the 2 water by-pass hoses to the oil cooler.
25. Install the 3 by-pass pipe bolts and tighten to 7 ft. lbs. (10 Nm).
26. Install oil filter bracket sub-assembly (w/o oil cooler).
27. Install oil pressure sender gauge assembly.
28. Connect the stabilizer bar and install the 2 stabilizer bushes and 2 stabilizer brackets with the 4 bolts and tighten to 51 ft. lbs. (69 Nm).
29. Install fan shroud.
30. Install fan and alternator v belt.
31. Install inlet radiator hose.
32. Add engine coolant.
33. Inspect for leaks.
34. Install No. 1 engine under cover.
35. Install V-bank cover sub-assembly.
36. Use the Techstream diagnostic tool, or equivalent and reprogram the required systems.

CYLINDER HEAD

REMOVAL & INSTALLATION

4.0L Engine

Bank 1

See Figures 143 through 160.

1. Before servicing the vehicle, refer to the Precautions Section.

➡If working near and/or around the SRS system and components, be sure to disable the SRS system. Tape the negative battery cable with insulating tape. Always disconnect the negative battery cable first.

CAUTION

To avoid personal injury when working on vehicles equipped with an air bag, the negative battery cable must be disconnected and at least 90 seconds must elapse before working on the system. Failure to do so may result in deployment of the air bag.

2. Disconnect the negative battery cable. Tape the cable with insulating tape.

➡When disconnecting the cable, some systems need to be initialized after the cable is reconnected. You will need the Techstream diagnostic scan tool or equivalent. Follow the directions on the tool.

3. Drain the engine coolant. Remove the V bank cover.
4. Remove the air cleaner assembly.
5. Disconnect the two water bypass hoses. Disconnect the fuel vapor feed hose. Disconnect the ventilation hose. Disconnect the VSV connectors.
6. Disconnect the throttle body motor connector. Separate the three wire harness clamps and hose clamp.
7. Remove the two bolts and the throttle body bracket. Remove the bolt and the oil baffle plate. Remove the four bolts and the two serge tank stays.
8. Remove the two nuts. Remove the four bolts, intake air surge tank and gasket.
9. Remove the ignition coil assembly.
10. Remove the cylinder head cover retaining bolts. Remove the cylinder head cover.
11. Turn the crankshaft pulley until its groove and the timing mark "0" of the timing chain cover are aligned. If not aligned at TDC of the compression stroke, turn the crankshaft one complete revolution, in the direction of rotation. Paint alignment marks on the number one chain links corresponding to the timing marks of the camshaft timing gears.
12. Remove the four bolts, and then remove the timing chain cover plate and gasket.
13. While turning the stopper plate of the

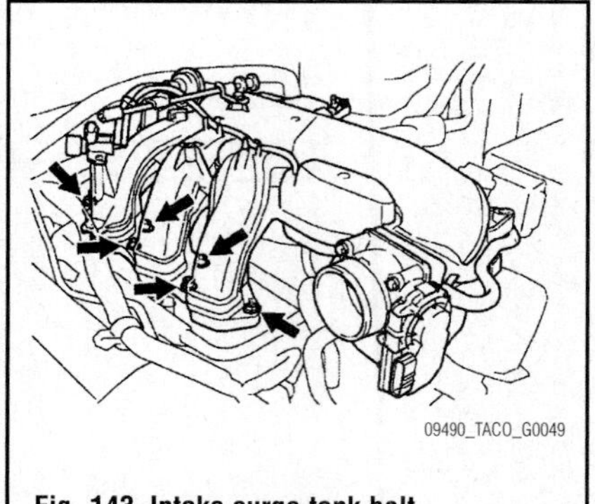

Fig. 143 Intake surge tank bolt locations—4.0L engine

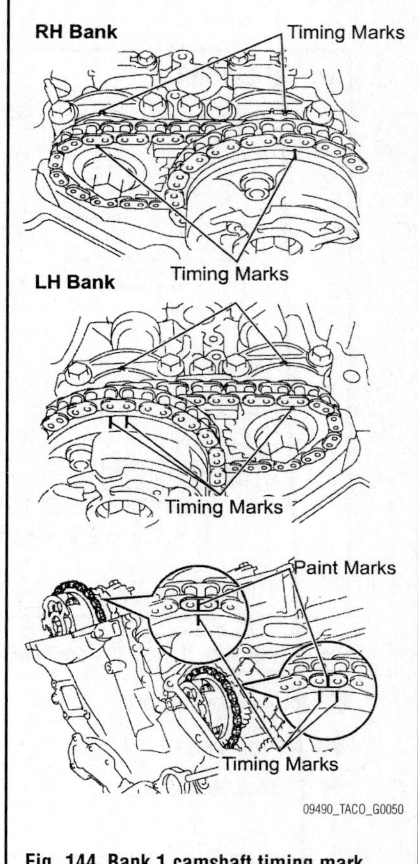

Fig. 144 Bank 1 camshaft timing mark alignment—4.0L engine

tensioner upward, push the plunger of the chain tensioner. While turning the stopper plate of the tensioner downward, insert a 0.118 inch diameter bar into the holes in the stopper plate and tensioner to hold the stopper plate.

14. Remove the two bolts, and then remove the chain tensioner.

➡**Keep the camshaft level while it is being removed. The camshaft thrust clearance is very small and failing to keep it level could crack or damage the cylinder head journal surface, which receives the thrust. Follow the steps below to prevent this problem from occurring.**

15. While raising the chain tensioner number two insert a 0.039 inch diameter pin into the hole to hold it. Hold the hexagonal portion of the number two camshaft with a wrench. Remove the camshaft timing gear set bolt.

16. Separate the camshaft timing gear from the number two camshaft. Rotate the camshafts counterclockwise, using a wrench, so that the cam lobes of the number one cylinder face in the direction shown.

17. Using several steps, loosen and remove the eight bearing cap bolts uniformly and in the proper removal sequence. Remove the four bearing caps and the number two camshaft.

18. Remove the number two chain tensioner bolt. Remove the number chain tensioner and camshaft timing gear.

➡**Keep the camshaft level while it is being removed. The camshaft thrust clearance is very small and failing to keep it level could crack or damage the cylinder head journal surface, which receives the thrust. Follow the steps below to prevent this problem from occurring.**

19. Hold the hexagonal portion of the number one camshaft, with a wrench. Loosen the camshaft timing gear set bolt.

➡**Do not disassemble the camshaft timing gear assembly.**

20. Slide the camshaft timing gear and separate the number one chain from the camshaft timing gear.

21. Rotate the number one camshaft counterclockwise, using the wrench so that the cam lobes on the number one cylinder face downward.

22. Using several steps, loosen and remove the eight bearing cap bolts uniformly and in the proper removal sequence. Remove the four bearing caps.

23. Remove the camshaft timing gear set bolt with the number one camshaft lifted up. Remove the number one camshaft and camshaft timing gear with the number two chain.

24. Tie the number one chain to the side. Be careful not to drop anything inside the timing chain cover.

25. Using a 10 mm bi-hexagon wrench, uniformly loosen the 8 cylinder head bolts in the sequence shown in the illustration. Remove the 8 cylinder head bolts and plate washers.

➡**Be careful not to drop the plate washers into the cylinder head. Cylinder head warpage or cracking could result from removing bolts in the incorrect order.**

26. Lift the cylinder head from the dowels on the cylinder block, and place the cylinder head on wooden blocks on a bench.

➡**Be careful not to damage the contact surfaces of the cylinder head and cylinder block.**

27. Remove cylinder head gasket.

To install:

➡**Be sure to use new fasteners, as required.**

28. Install cylinder head gasket.

a. Remove any old packing (FIPG) material and be careful not to drop any oil on the contact surfaces of the cylinder head and cylinder block.

b. Apply seal packing to a new cylinder head gasket.

➡**Remove any oil from the contact surface. Install the cylinder head gasket within 3 minutes after applying the seal packing. Do not add engine oil within 2 hours of installation.**

29. Place the cylinder head gasket on the cylinder block surface with the front face of the Lot No. stamp upward.

30. Place the cylinder head on the cylinder head gasket.

31. Install the 8 cylinder head bolts and plate washers.

a. Apply a light coat of engine oil to the threads and under the heads of the cylinder head bolts.

b. Install the plate washer to the cylinder head bolt.

32. Using a 10 mm bi-hexagon wrench, install and uniformly tighten the 8 cylinder head bolts to 27 ft. lbs. (36 Nm)

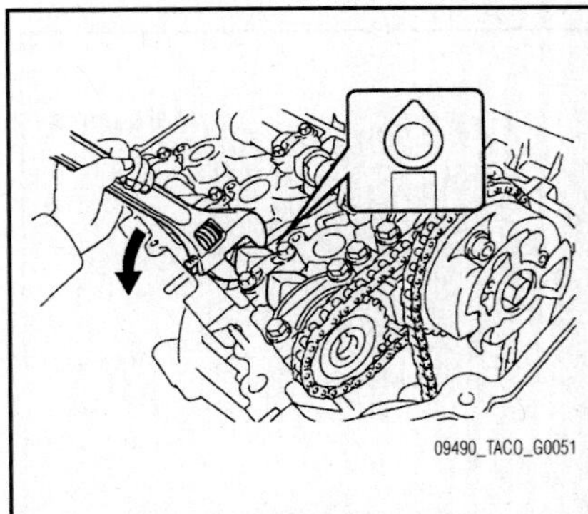

Fig. 145 Bank 1 camshaft lobe removal positioning—4.0L engine

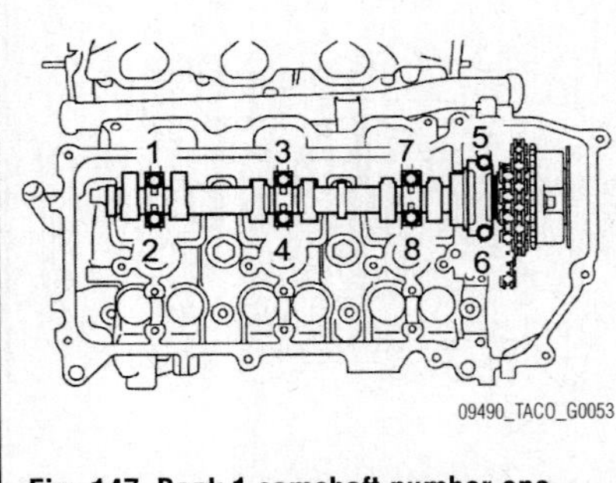

Fig. 147 Bank 1 camshaft number one bearing cap bolt removal sequence—4.0L engine

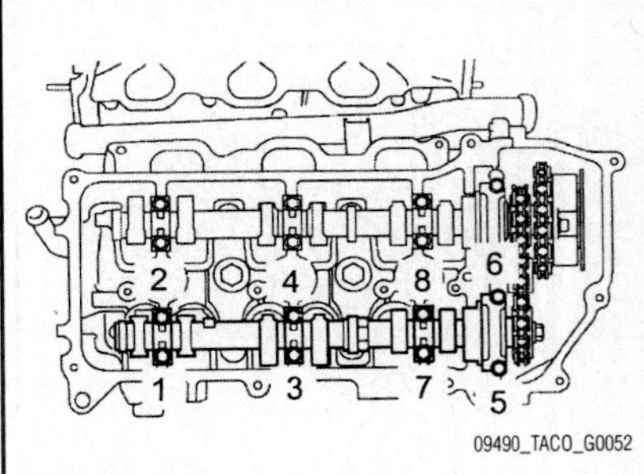

Fig. 146 Bank 1 camshaft number two bearing cap bolt removal sequence—4.0L engine

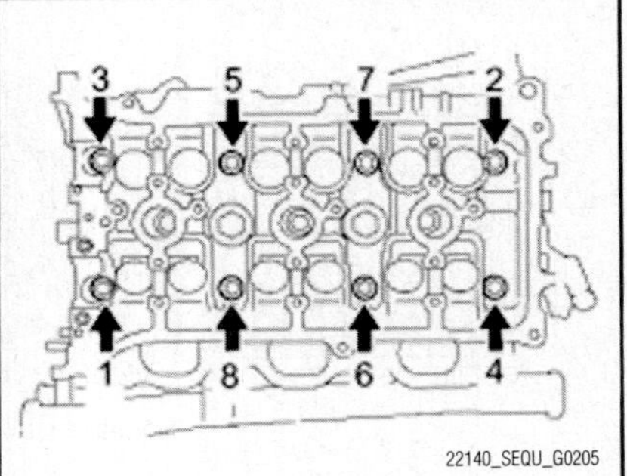

Fig. 148 Cylinder head removal sequence—4.0L engine

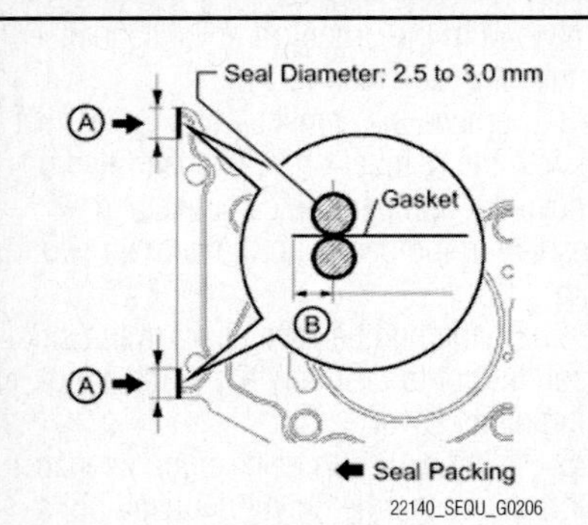

Fig. 149 Cylinder head gasket packing position—4.0L engine

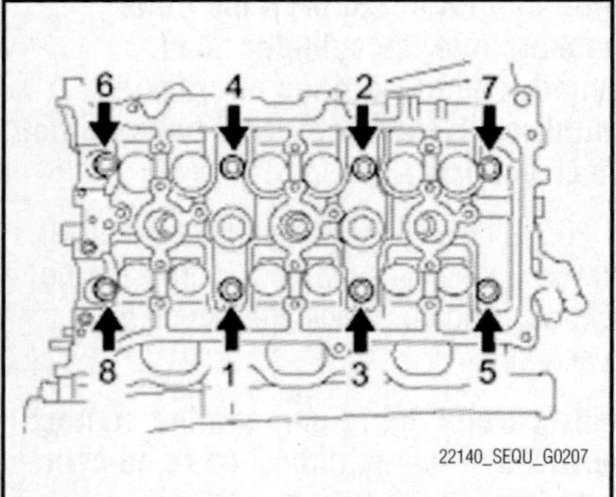

Fig. 150 Cylinder head bolt tightening sequence—4.0L engine

with the plate washers in several steps in sequence.

➡ **If any one of the cylinder head bolts does not meet the torque specification, replace the cylinder head bolt.**

33. Mark the cylinder head bolt heads with paint.

34. Tighten the cylinder head bolts another 90° in the sequence.

35. Tighten the cylinder head bolts another 90° in the sequence for a total of 180° from torque.

36. Check that the painted marks are now facing correctly.

37. Seal packing will seep out from the engine's front side. Thoroughly wipe off seeped out seal packing.

➡ **Keep the camshaft level while it is being installed. The camshaft thrust clearance is very small and failing to keep it level could crack or damage the cylinder head journal surface, which receives the thrust.**

38. Align the yellow mark link with the timing mark (1 dot mark) of the camshaft timing gear. Apply new engine oil to the thrust portion and journal of the camshafts.

39. Temporarily install the number one chain onto the number two chain of the camshaft timing gear.

40. Align the knock pin hole of the camshaft timing gear with the knock pin of the number one camshaft. Insert the number one camshaft into the camshaft timing gear.

41. Temporarily install the camshaft timing gear set bolt. Install the number one camshaft onto the right cylinder head with the cam lobes of the number one cylinder facing downward, as indicated in the illustration.

42. Install the four bearing caps, in the proper location. Apply a light coat of engine oil to the threads and under the heads of the cap bolts.

43. Using several steps, uniformly install and tighten the bearing cap bolts in the proper sequence to 80 inch lbs. for the 10mm bolts and 18 ft. lbs. for the 12mm bolts.

44. Rotate the number one camshaft clockwise, using a wrench so that the timing mark of the camshaft timing gear is aligned with the timing mark of the camshaft bearing cap.

45. Align the paint mark of the number one chain with the timing mark of the camshaft timing gear.

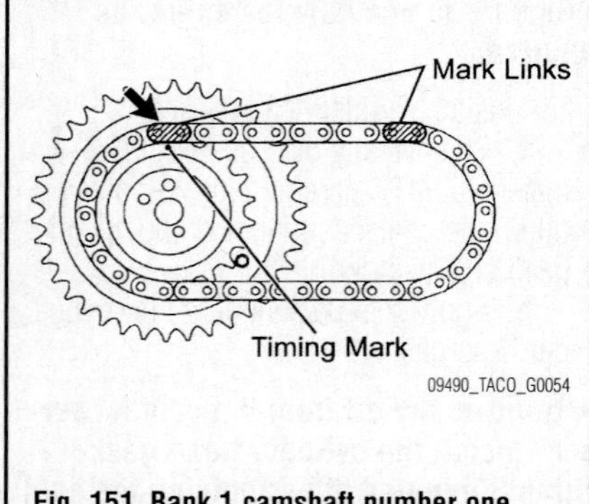

Fig. 151 Bank 1 camshaft number one alignment marks—4.0L engine

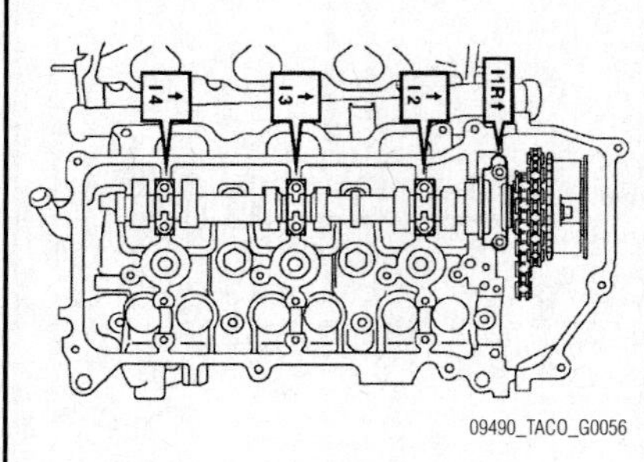

Fig. 153 Bank 1 camshaft number one bearing cap bolt installation sequence—4.0L engine

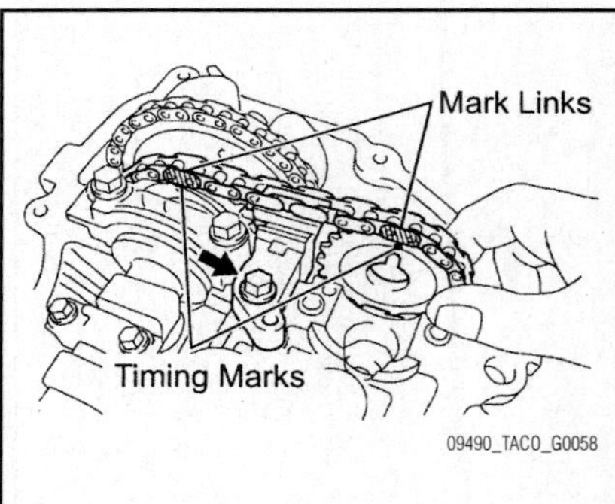

Fig. 155 Bank 1 camshaft number two alignment marks—4.0L engine

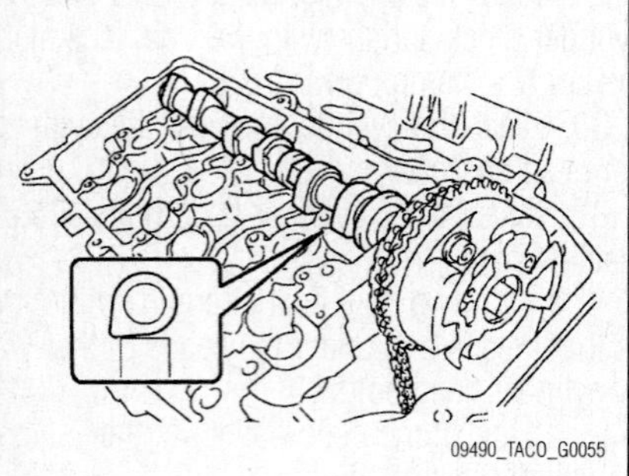

Fig. 152 Bank 1 camshaft number one lobe installation positioning—4.0L engine

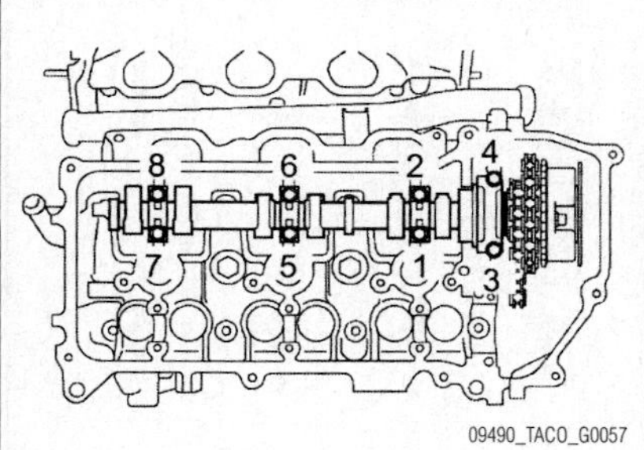

Fig. 154 Bank 1 camshaft number one bearing cap bolt torque sequence—4.0L engine

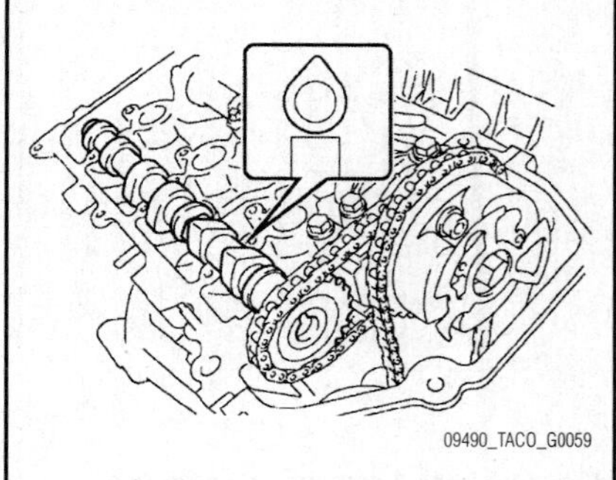

Fig. 156 Bank 1 camshaft number two lobe installation positioning—4.0L engine

46. Hold the hexagonal portion of the number one camshaft with a wrench, and tighten the camshaft timing gear set bolt to 74 ft. lbs.

47. While pushing in on the number two chain tensioner, insert a 0.039 inch pin into the hole to hold it.

48. Temporarily install the camshaft timing gear and chain tensioner number two and align the yellow mark links with the timing marks (1 dot mark) on the camshaft timing gears. Tighten the bolt to 14 ft. lbs.

➡Keep the camshaft level while it is being installed. The camshaft thrust clearance is very small and failing to keep it level could crack or damage the cylinder head journal surface, which receives the thrust.

49. Install the number two camshaft onto the right cylinder head with the cam lobes of the number one cylinder facing upward, as indicated in the illustration.

50. Install the four bearing caps in the proper location. Apply a light coat of clean engine oil to the threads and under the heads of the bolts.

51. Using several steps, uniformly install and tighten the eight bearing cap bolts in sequence to 80 inch lbs, for the 10mm head bolts and 18 ft. lbs. for the 12mm head bolts.

52. Rotate the number two camshaft clockwise, using a wrench, so that the lock pin of the number two camshaft is aligned with the knock pin hole of the camshaft timing gear.

53. Hold the hexagonal portion of the number two camshaft, with a wrench, and install the camshaft timing gear set bolt and tighten it to 74 ft. lbs.

54. Remove the pin from the number two chain tensioner.

55. While turning the stopper plate of the tensioner clockwise, push in the plunger of the tensioner. While turning the stopper plate of the tensioner counterclockwise, insert a 0.138 inch bar into the holes in the stopper plate and tensioner to hold the stopper plate. Install the two chain tensioner bolts and tighten to 7.4 ft. lbs.

56. Remove the bar from the chain tensioner. Install a new gasket and the timing chain cover plate. Torque the bolts to 80 inch lbs.

57. Turn the crankshaft pulley two complete revolutions slowly until its groove and the timing mark "0" of the timing chain cover are aligned.

58. Position the number one cylinder at TDC of the compression stroke. Inspect the valve clearance, adjust as required.

59. Apply a continuous bead (0.08–0.12 inch) of seal packing, part number 08826-00080 or equivalent, to the cylinder head as indicated in the illustration. Install the seal washers onto the bolts. Install the cylinder head cover bolts and nuts. Tighten bolts "A" to 7.4 ft. lbs. Tighten bolts "B" to 80 inch lbs. Tighten nuts to 80 inch lbs.

➡Be sure to remove any oil from the contact surfaces of the cylinder head cover and the cylinder head. Install the cover within three minutes after applying the seal packing. Tighten the bolts to specification within fifteen minutes after installing the cover. Do not add engine oil for at least two hours after installing the cover.

60. Continue the installation in the reverse order of the removal procedure.

61. Be sure to fill the engine with the proper grade and type engine coolant.

62. Be sure to fill the engine with the proper grade and type engine oil.

63. Start the engine and check for leaks. Correct as required.

64. Use the Techstream diagnostic tool, or equivalent and reprogram the required systems.

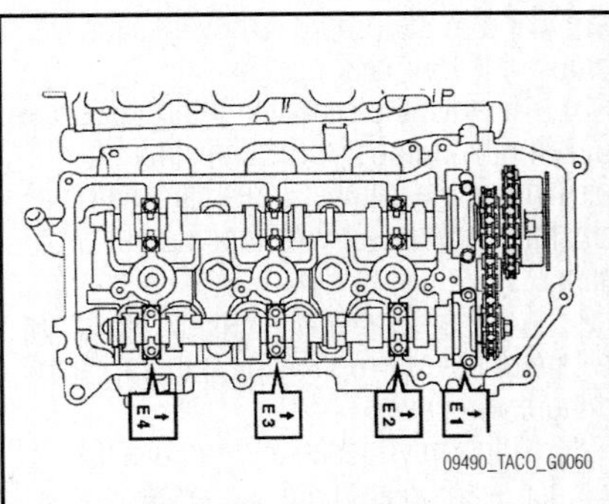

Fig. 157 Bank 1 camshaft number two bearing cap bolt installation sequence—4.0L engine

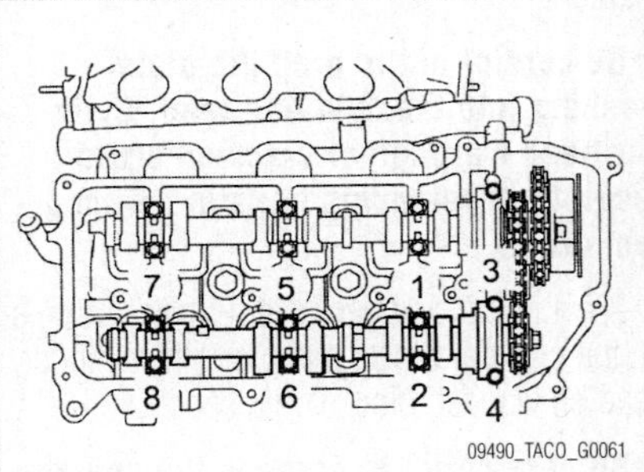

Fig. 158 Bank 1 camshaft number two bearing cap bolt torque sequence—4.0L engine

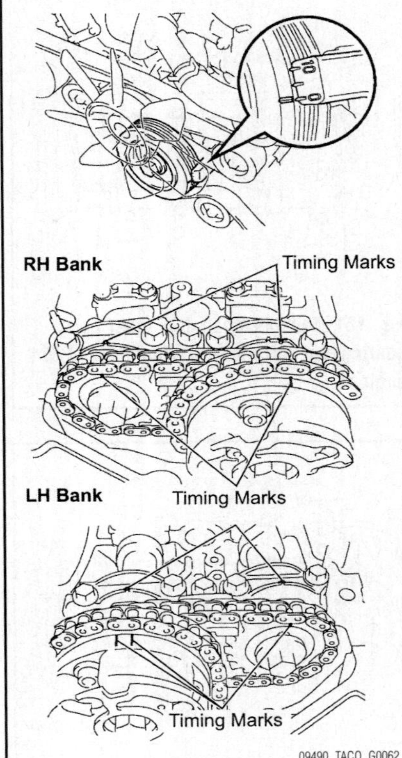

Fig. 159 Bank 1 and Bank 2 timing mark alignment—4.0L engine

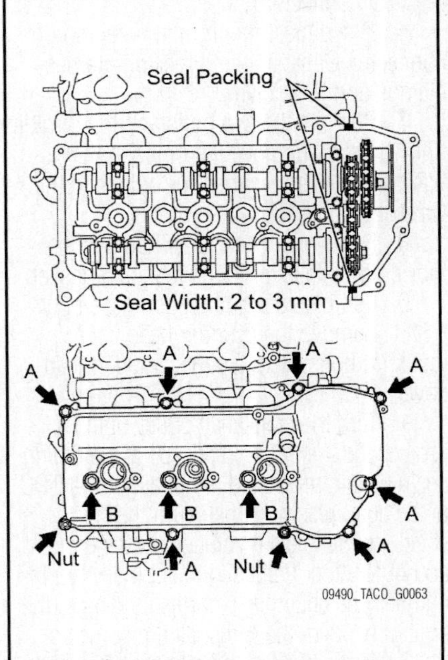

Fig. 160 Bank 1 cylinder head cover bolt sealant application and bolt identification—4.0L engine

Bank 2

See Figures 161 through 174.

1. Before servicing the vehicle, refer to the Precautions Section.

➡If working near and/or around the SRS system and components, be sure to disable the SRS system. Tape the negative battery cable with insulating tape. Always disconnect the negative battery cable first.

CAUTION

To avoid personal injury when working on vehicles equipped with an air bag, the negative battery cable must be disconnected and at least 90 seconds must elapse before working on the system. Failure to do so may result in deployment of the air bag.

2. Disconnect the negative battery cable. Tape the cable with insulating tape.

➡When disconnecting the cable, some systems need to be initialized after the cable is reconnected. You will need the Techstream diagnostic scan tool or equivalent. Follow the directions on the tool.

3. Drain the engine coolant. Remove the V bank cover.
4. Remove the air cleaner assembly.
5. Disconnect the two water bypass hoses. Disconnect the fuel vapor feed hose. Disconnect the ventilation hose. Disconnect the VSV connectors.
6. Disconnect the throttle body motor connector. Separate the three wire harness clamps and hose clamp.
7. Remove the two bolts and the throttle body bracket. Remove the bolt and the oil baffle plate. Remove the four bolts and the two serge tank stays.
8. Remove the two nuts. Remove the four bolts, intake air surge tank and gasket.
9. Remove the ignition coil assembly.
10. Remove the cylinder head cover retaining bolts. Remove the cylinder head cover.
11. Turn the crankshaft pulley until its groove and the timing mark "0" of the timing chain cover are aligned. If not aligned at TDC of the compression stroke, turn the crankshaft one complete revolution, in the direction of rotation. Paint alignment marks on the number one chain links corresponding to the timing marks of the camshaft timing gears.
12. While turning the stopper plate of the tensioner upward, push in the plunger of the chain tensioner. While turning the stopper plate of the tensioner downward, insert a 0.138 inch bar into the holes in the stopper plate and tensioner to hold the stopper plate. Remove the two bolts and then remove the number one chain tensioner assembly.

➡Never rotate the crankshaft with the chain tensioner removed. When rotating the camshaft with the tensioner removed, rotate the crankshaft counterclockwise forty degrees from TDC, first.

➡Keep the camshaft level while it is being removed. The camshaft thrust clearance is very small and failing to keep it level could crack or damage the cylinder head journal surface, which receives the thrust. Follow the steps below to prevent this problem from occurring.

13. While pushing down on chain tensioner number three insert a 0.039 inch diameter pin into the hole to hold it. Hold the hexagonal portion of the number four camshaft with a wrench. Remove the camshaft timing gear set bolt.
14. Separate the camshaft timing gear from the number four camshaft.
15. Using several steps, loosen and remove the eight bearing cap bolts uniformly and in the proper removal sequence. Remove the four bearing caps and the number four camshaft.

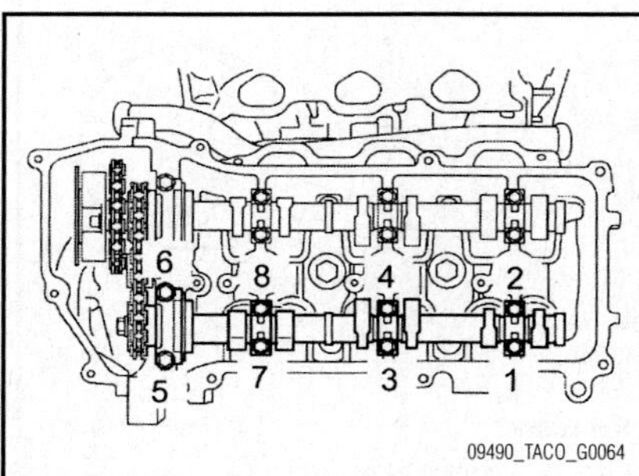

Fig. 161 Bank 2 camshaft number four bearing cap bolt removal sequence—4.0L engine

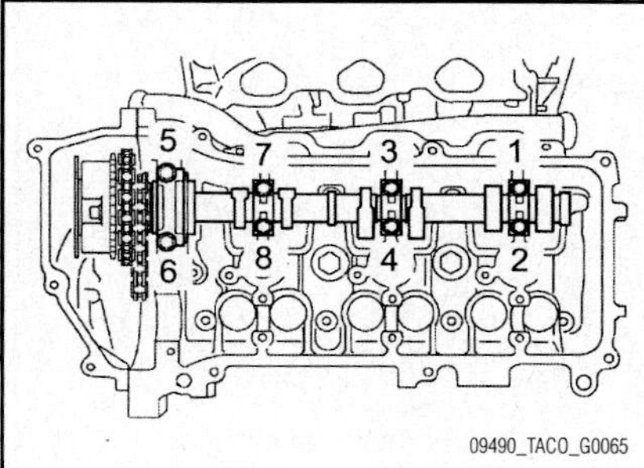

Fig. 162 Bank 2 camshaft number three bearing cap bolt removal sequence—4.0L engine

16. Remove the number three chain tensioner bolt. Remove the number chain tensioner and camshaft timing gear.

➡Keep the camshaft level while it is being removed. The camshaft thrust clearance is very small and failing to keep it level could crack or damage the cylinder head journal surface, which receives the thrust. Follow the steps below to prevent this problem from occurring.

17. Release the chain tension between the camshaft gear (LH bank) and the crankshaft timing gear by turning the crankshaft pulley counterclockwise slightly.
18. Hold the hexagonal portion of the number three camshaft, with a wrench. Loosen the camshaft timing gear set bolt.

➡Do not disassemble the camshaft timing gear assembly.

19. Slide the camshaft timing gear and separate the number one chain from the camshaft timing gear.
20. Using several steps, loosen and remove the eight bearing cap bolts uniformly and in the proper removal sequence. Remove the four bearing caps.
21. Remove the camshaft timing gear set bolt with the number three camshaft lifted up. Remove the number three camshaft and camshaft timing gear with the number two chain.
22. Tie the number one chain to the side. Be careful not to drop anything inside the timing chain cover.
23. Uniformly loosen and remove the 2 cylinder head bolts in the sequence.
24. Using a 10 mm bi-hexagon wrench, uniformly loosen the 8 cylinder head bolts in the sequence shown in the illustration. Remove the 8 cylinder head bolts and plate washers.

➡Be careful not to drop the plate washers into the cylinder head. Cylinder head warpage or cracking could result from removing bolts in the incorrect order.

25. Lift the cylinder head from the dowels on the cylinder block, and place the cylinder head on wooden blocks on a bench.

➡Be careful not to damage the contact surfaces of the cylinder head and cylinder block.

26. Remove cylinder head gasket.

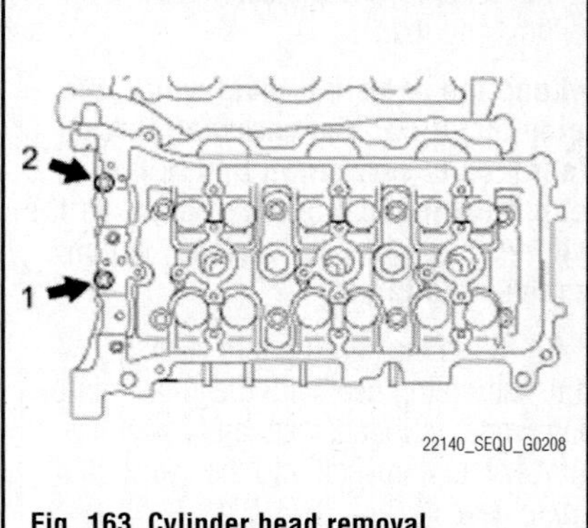

Fig. 163 Cylinder head removal sequence—4.0L engine

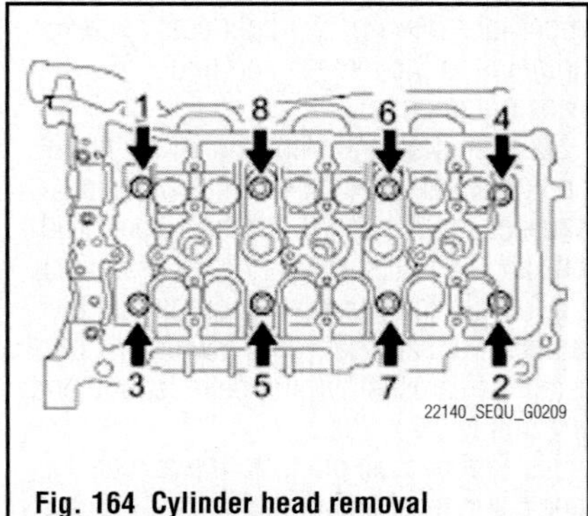

Fig. 164 Cylinder head removal sequence—4.0L engine

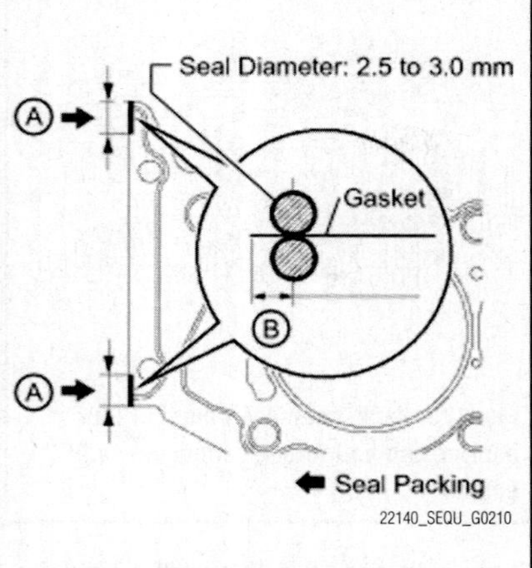

Fig. 165 Cylinder head gasket packing position—4.0L engine

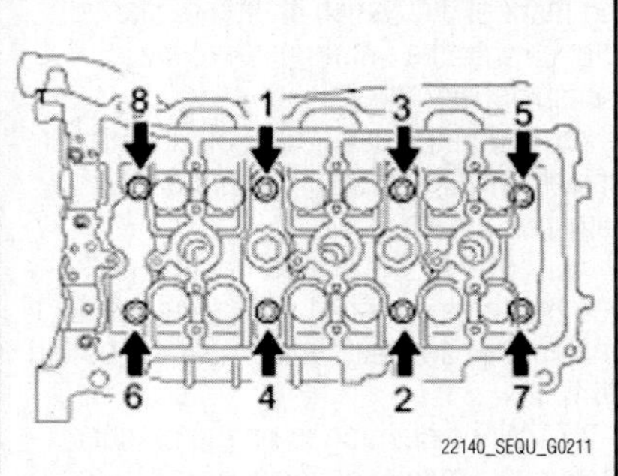

Fig. 166 Cylinder head bolt tightening sequence—4.0L engine

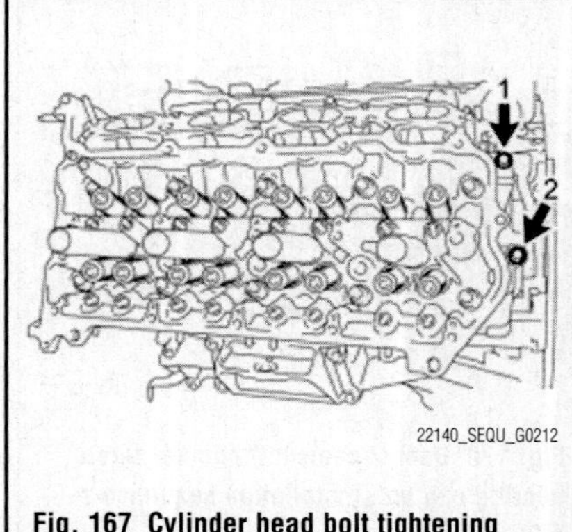

Fig. 167 Cylinder head bolt tightening sequence—4.0L engine

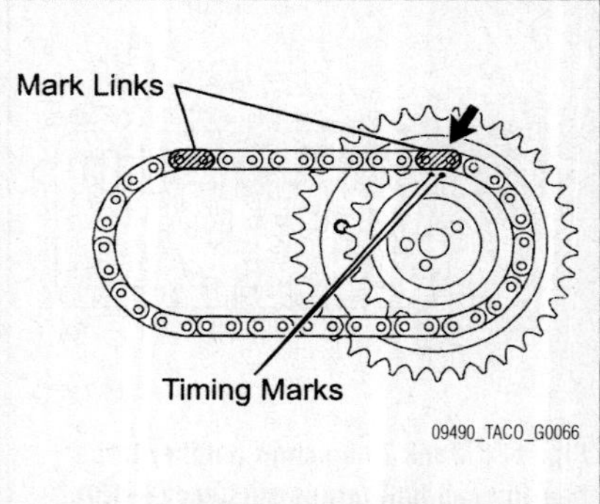

Fig. 168 Bank 2 camshaft number three alignment marks—4.0L engine

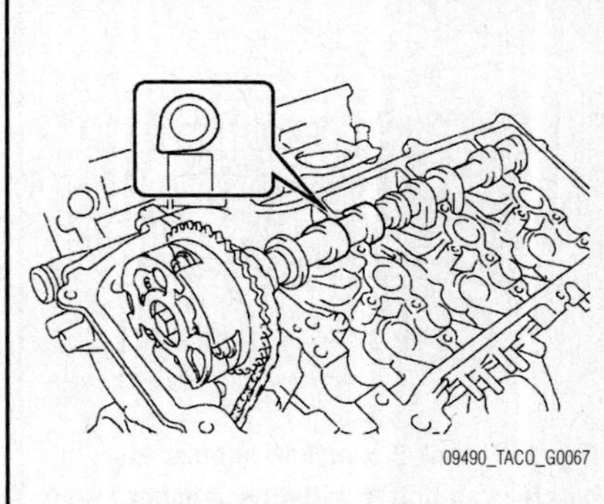

Fig. 169 Bank 2 camshaft number three lobe installation positioning—4.0L engine

To install:

➡Be sure to use new fasteners, as required.

27. Install cylinder head gasket.
   a. Remove any old packing (FIPG) material and be careful not to drop any oil on the contact surfaces of the cylinder head and cylinder block.
   b. Apply seal packing to a new cylinder head gasket.

➡Remove any oil from the contact surface. Install the cylinder head gasket within 3 minutes after applying the seal packing. Do not add engine oil within 2 hours of installation.

28. Place the cylinder head gasket on the cylinder block surface with the front face of the Lot No. stamp upward.
29. Place the cylinder head on the cylinder head gasket.
30. Install the 8 cylinder head bolts and plate washers.
   a. Apply a light coat of engine oil to the threads and under the heads of the cylinder head bolts
   b. Install the plate washer to the cylinder head bolt.
31. Using a 10 mm bi-hexagon wrench, install and uniformly tighten the 8 cylinder head bolts to 27 ft. lbs. (36 Nm) with the plate washers in several steps in sequence.

➡ If any one of the cylinder head bolts does not meet the torque specification, replace the cylinder head bolt.

32. Mark the cylinder head bolt heads with paint.
33. Tighten the cylinder head bolts another 90° in the sequence.
34. Tighten the cylinder head bolts another 90° in the sequence for a total of 180° from torque.
35. Check that the painted marks are now facing correctly.
36. Install the 2 bolts in the order and tighten to 22 ft. lbs. (30 Nm).
37. Seal packing will seep out from the engine's front side. Thoroughly wipe off seeped out seal packing.

➡Keep the camshaft level while it is being installed. The camshaft thrust clearance is very small and failing to keep it level could crack or damage the cylinder head journal surface, which receives the thrust.

38. Align the yellow mark link with the timing mark (2 dot mark) of the camshaft timing gear. Apply new engine oil to the thrust portion and journal of the camshafts.
39. Temporarily install the number one chain onto the number two chain of the camshaft timing gear.
40. Align the knock pin hole of the camshaft timing gear with the knock pin of the number three camshaft. Insert the num-

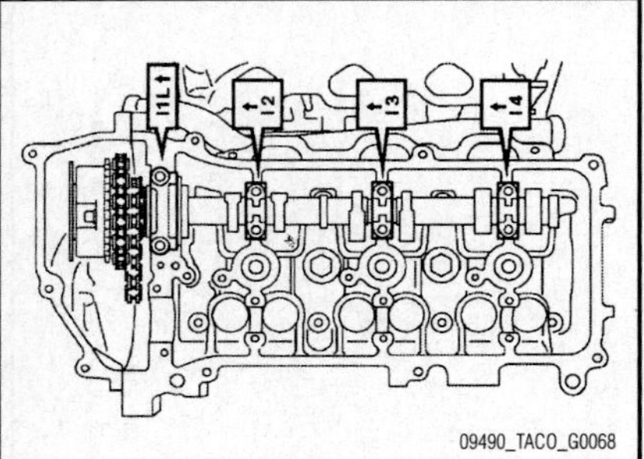

Fig. 170 Bank 2 camshaft number three bearing cap bolt installation sequence—4.0L engine

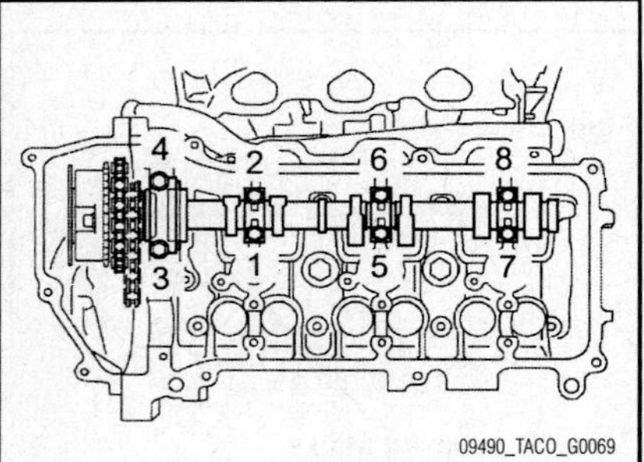

Fig. 171 Bank 2 camshaft number three bearing cap bolt torque sequence—4.0L engine

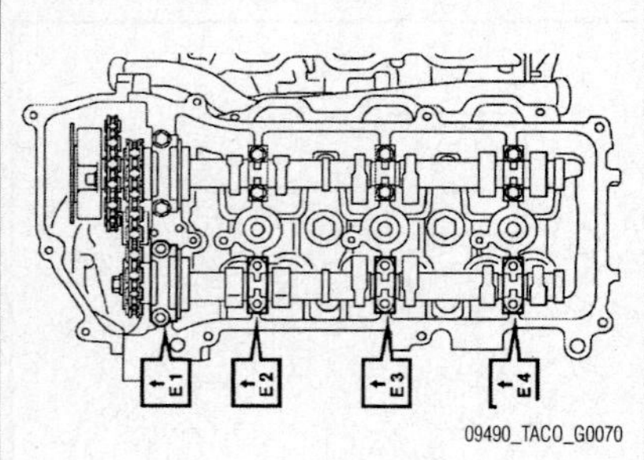

Fig. 172 Bank 2 camshaft number four bearing cap bolt installation sequence—4.0L engine

ber three camshaft into the camshaft timing gear.

41. Temporarily install the camshaft timing gear set bolt. Install the number three camshaft onto the left cylinder head with the cam lobes of the number two cylinder facing downward, as indicated in the illustration.

42. Install the four bearing caps, in the proper location. Apply a light coat of engine oil to the threads and under the heads of the cap bolts.

43. Using several steps, uniformly install and tighten the bearing cap bolts in the proper sequence to 80 inch lbs. for the

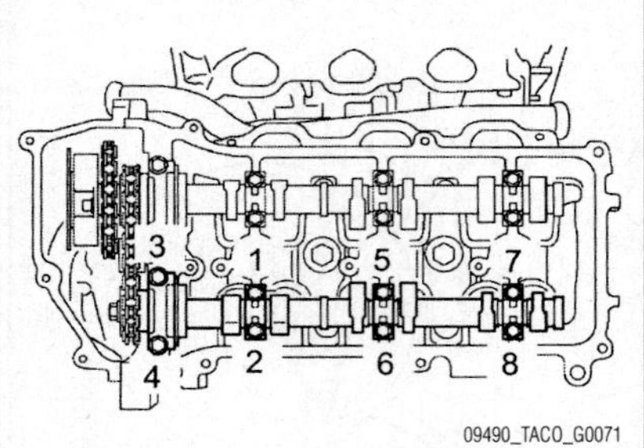

Fig. 173 Bank 2 camshaft number four bearing cap bolt torque sequence—4.0L engine

10mm bolts and 18 ft. lbs. for the 12mm bolts.

44. Rotate the number one camshaft clockwise, using a wrench so that the timing mark of the camshaft timing gear is aligned with the timing mark of the camshaft bearing cap.

45. Align the paint mark of the number one chain with the timing mark of the camshaft timing gear.

46. Hold the hexagonal portion of the number three camshaft with a wrench, and tighten the camshaft timing gear set bolt to 74 ft. lbs.

47. While pushing in on the number three chain tensioner, insert a 0.039 inch pin into the hole to hold it.

48. Temporarily install the camshaft timing gear and chain tensioner number three and align the yellow mark links with the timing marks (1 dot mark and 2 dot marks)

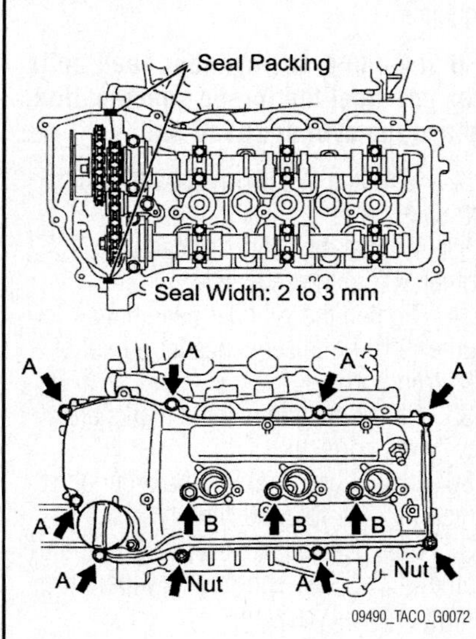

Fig. 174 Bank 2 cylinder head cover bolt sealant application and bolt identification—4.0L engine

on the camshaft timing gears. Tighten the bolt to 14 ft. lbs.

➡Keep the camshaft level while it is being installed. The camshaft thrust clearance is very small and failing to keep it level could crack or damage the cylinder head journal surface, which receives the thrust.

49. Align the knock pin hole in the camshaft timing gear with the knock pin of the number four camshaft, and insert the number four camshaft into the camshaft timing gear.

50. Temporarily install the camshaft timing gear set bolt.

51. Install the four bearing caps in the proper location. Apply a light coat of clean engine oil to the threads and under the heads of the bolts.

52. Using several steps, uniformly install and tighten the eight bearing cap bolts in sequence to 80 inch lbs, for the 10mm head bolts and 18 ft. lbs. for the 12mm head bolts.

53. Hold the hexagonal portion of the number four camshaft, with a wrench, and install the camshaft timing gear set bolt and tighten it to 74 ft. lbs.

54. Remove the pin from the number three chain tensioner.

55. Release the chain tension between the camshaft timing gear (RH bank) and the crankshaft timing gear by turning the crankshaft pulley clockwise slightly.

56. While turning the stopper plate of the tensioner clockwise, push in the plunger of the tensioner. While turning the stopper plate of the tensioner counterclockwise, insert a 0.138 inch bar into the holes in the stopper plate and tensioner to hold the stopper plate. Install the two chain tensioner bolts and tighten to 7.4 ft. lbs. Remove the bar from the chain tensioner.

57. Position the number one cylinder at TDC of the compression stroke. Inspect the valve clearance, adjust as required.

58. Apply a continuous bead (0.08–0.12 inch) of seal packing, part number 08826-00080 or equivalent, to the cylinder head as indicated in the illustration. Install the seal washers onto the bolts. Install the cylinder head cover bolts and nuts. Tighten bolts "A" to 7.4 ft. lbs. Tighten bolts "B" to 80 inch lbs. Tighten nuts to 80 inch lbs.

➡Be sure to remove any oil from the contact surfaces of the cylinder head cover and the cylinder head. Install the cover within three minutes after applying the seal packing. Tighten the bolts to specification within fifteen minutes after installing the cover. Do not add

engine oil for at least two hours after installing the cover.

59. Continue the installation in the reverse order of the removal procedure.
60. Be sure to fill the engine with the proper grade and type engine coolant.
61. Be sure to fill the engine with the proper grade and type engine oil.
62. Start the engine and check for leaks. Correct as required.
63. Use the Techstream diagnostic tool, or equivalent and reprogram the required systems.

4.6L Engine

Bank 1

See Figures 175 through 177.

1. Before servicing the vehicle, refer to the Precautions Section.

➡If working near and/or around the SRS system and components, be sure to disable the SRS system. Tape the negative battery cable with insulating tape. Always disconnect the negative battery cable first.

⁂ CAUTION

To avoid personal injury when working on vehicles equipped with an air bag, the negative battery cable must be disconnected and at least 90 seconds must elapse before working on the system. Failure to do so may result in deployment of the air bag.

2. Disconnect the negative battery cable. Tape the cable with insulating tape.

➡When disconnecting the cable, some systems need to be initialized after the cable is reconnected. You will need the Techstream diagnostic scan tool or equivalent. Follow the directions on the tool.

3. Remove the exhaust manifold sub assembly.
4. Remove the camshafts.
5. Remove the No. 1 valve rocker arm sub assembly.
6. Remove the 16 valve lash adjusters.
7. Remove the 16 valve stem caps from the cylinder head.
8. Remove bolts 1 and 2 from the head, see illustration. Remove the remaining bolts in the sequence shown in the illustration. Be sure to arrange the removed parts for installation separately.
9. Remove the cylinder head from its mounting.
10. Discard the gasket.

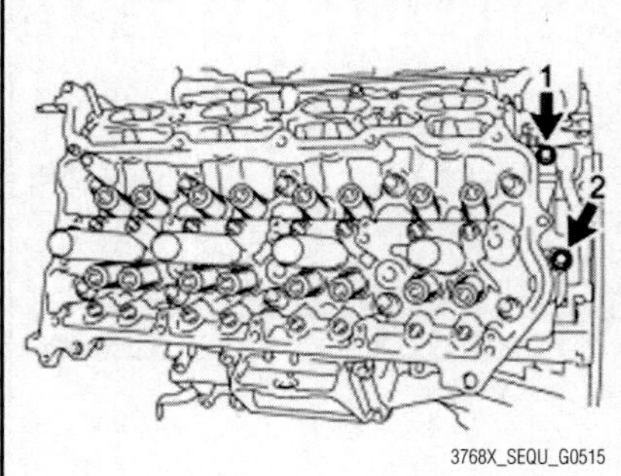

Fig. 175 Cylinder head bolt sequence removal Bank 1 (1 of 2)—4.6L engine

To install:

➡Be sure to use new fasteners, as required.

11. Installation is the reverse of the removal procedure.
12. Check for proper gasket positioning. Place the gasket on the head with the lot No. stamp upward. Be sure there is no oil on the gasket.
13. Tighten the bolts to specification and in the proper sequence. Lightly coat the bolt threads with clean engine oil.

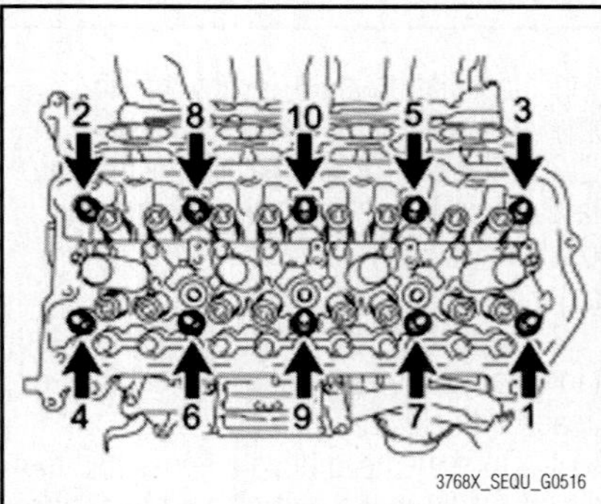

Fig. 176 Cylinder head bolt sequence removal Bank 1 (2 of 2)—4.6L engine

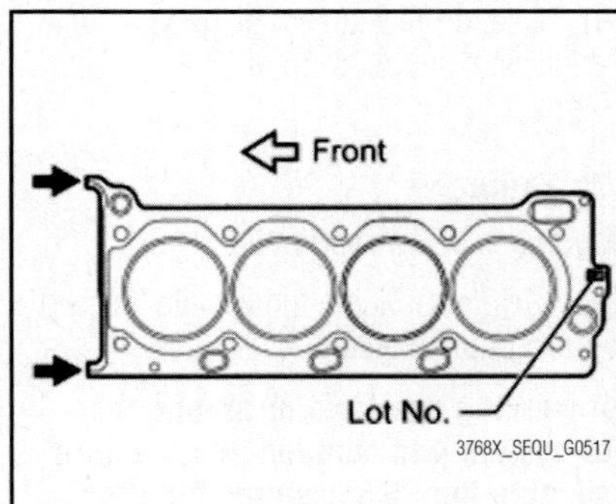

Fig. 177 Cylinder head gasket positioning Bank 1—4.6L engine

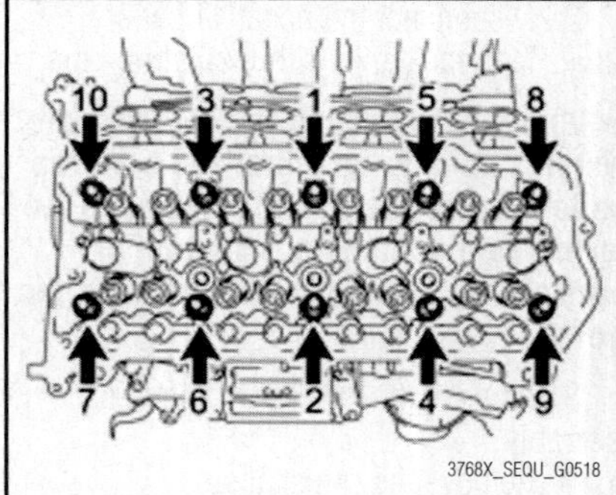

Fig. 178 Cylinder head gasket tightening sequence Bank 1—4.6L engine

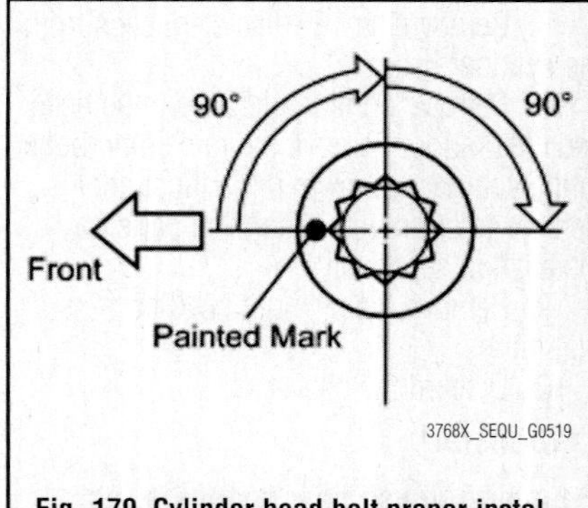

Fig. 179 Cylinder head bolt proper installation Bank 1 and Bank 2—4.6L engine

14. Once the head bolts are installed the painted mark (on the bolt) should be facing rearward.
15. Install and tighten the remaining two bolts.
16. Use the Techstream diagnostic tool, or equivalent and reprogram the required systems.

Bank 2

See Figures 180 through 183.

1. Before servicing the vehicle, refer to the Precautions Section.

➡If working near and/or around the SRS system and components, be sure to disable the SRS system. Tape the negative battery cable with insulating tape. Always disconnect the negative battery cable first.

⁂ CAUTION

To avoid personal injury when working on vehicles equipped with an air bag, the negative battery cable must be disconnected and at least 90 seconds must elapse before working on the system. Failure to do so may result in deployment of the air bag.

2. Disconnect the negative battery cable. Tape the cable with insulating tape.

➡When disconnecting the cable, some systems need to be initialized after the cable is reconnected. You will need the Techstream diagnostic scan tool or equivalent. Follow the directions on the tool.

3. Remove the exhaust manifold sub assembly.
4. Remove the camshafts.
5. Remove the No. 1 valve rocker arm sub assembly.
6. Remove the 16 valve lash adjusters.
7. Remove the 16 valve stem caps from the cylinder head.
8. Remove bolts 1 and 2 from the head, see illustration. Remove the remaining bolts in the sequence shown in the illustration. Be sure to arrange the removed parts for installation separately.
9. Remove the cylinder head from its mounting.
10. Discard the gasket.

To install:

➡Be sure to use new fasteners, as required.

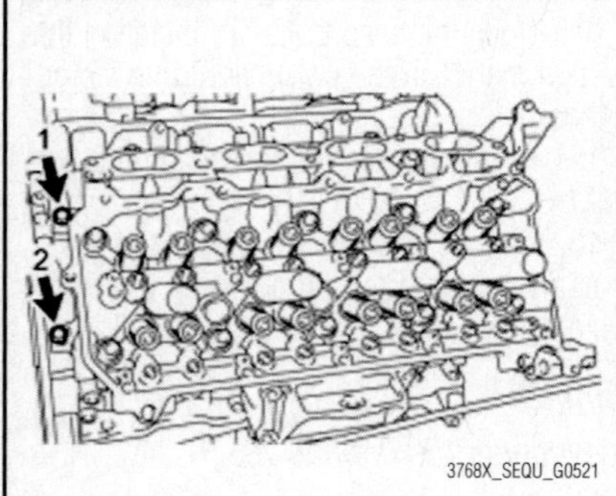

Fig. 180 Cylinder head bolt sequence removal Bank 2 (1 of 2)—4.6L engine

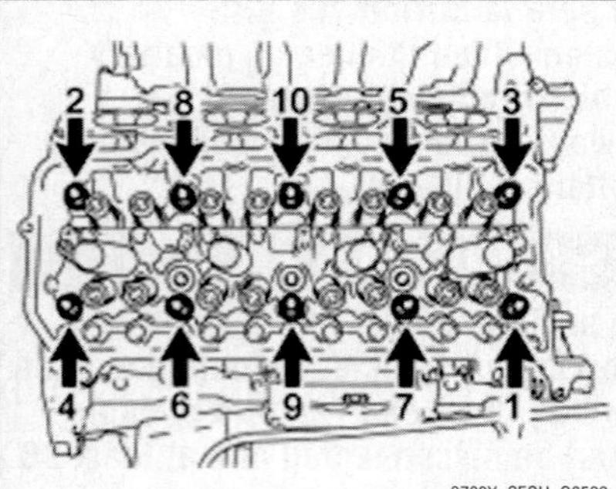

Fig. 181 Cylinder head bolt sequence removal Bank 2 (2 of 2)—4.6L engine

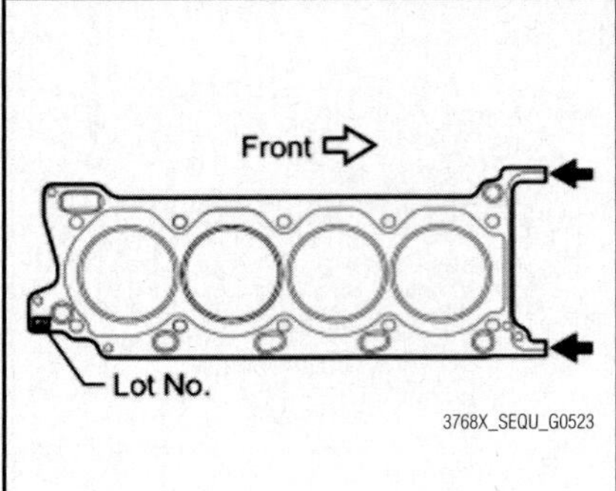

Fig. 182 Cylinder head gasket positioning Bank 2—4.6L engine

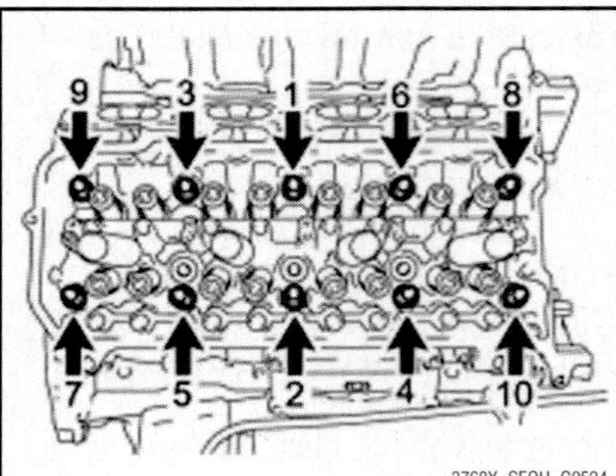

Fig. 183 Cylinder head gasket tightening sequence Bank 2—4.6L engine

11. Installation is the reverse of the removal procedure.
12. Check for proper gasket positioning. Place the gasket on the head with the lot No. stamp upward. Be sure there is no oil on the gasket.
13. Tighten the bolts to specification and in the proper sequence. Lightly coat the bolt threads with clean engine oil.
14. Once the head bolts are installed the painted mark (on the bolt) should be facing rearward.
15. Install and tighten the remaining two bolts.
16. Use the Techstream diagnostic tool, or equivalent and reprogram the required systems.

4.7L Engine

See Figures 184 through 186.

1. Before servicing the vehicle, refer to the Precautions Section.

➡If working near and/or around the SRS system and components, be sure to disable the SRS system. Tape the negative battery cable with insulating tape. Always disconnect the negative battery cable first.

CAUTION

To avoid personal injury when working on vehicles equipped with an air bag, the negative battery cable must be disconnected and at least 90 seconds must elapse before working on the system. Failure to do so may result in deployment of the air bag.

2. Disconnect the negative battery cable. Tape the cable with insulating tape.

➡When disconnecting the cable, some systems need to be initialized after the cable is reconnected. You will need the Techstream diagnostic scan tool or equivalent. Follow the directions on the tool.

3. Drain the cooling system.
4. Relieve the fuel system pressure.
5. Remove or disconnect the following:
   - Battery and tray
   - Engine appearance cover
   - Engine under covers
   - Air intake assembly
   - Accessory drive belt
   - A/C compressor and bracket
   - Cooling fan and bracket
   - Radiator
   - Idler pulley
   - Front covers
   - Timing belt.
   - Camshaft sprockets
   - Camshaft Position (CMP) sensor

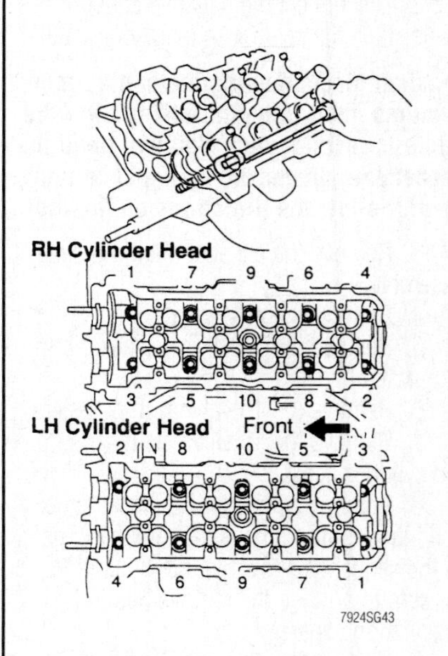

Fig. 184 Cylinder head loosening sequence—4.7L engine

- Power steering pump
- Exhaust front pipes
- Transmission dipstick tube
- Ignition coils
- Rear timing belt covers
- Fuel lines
- Intake manifold
- Water inlet housing assembly
- Front and rear water bypass joints
- Engine lifting eyes
- Oil dipstick tube
- Valve covers
- Camshafts
- Cylinder heads with the exhaust manifolds attached. Loosen the bolts in the sequence shown.

To install:

➡Be sure to use new fasteners, as required.

6. Install the cylinder heads with new gaskets. Tighten the bolts in sequence as follows:
    a. Step 1: 24 ft. lbs. (32 Nm).
    b. Step 2: Plus 180 degrees.
7. Install or connect the following:
    - Camshafts
    - Valve covers
    - Oil dipstick tube
    - Engine lifting eyes
    - Front and rear water bypass joints
    - Water inlet housing assembly
    - Intake manifold
    - Fuel lines
    - Rear timing belt covers
    - Ignition coils
    - Transmission dipstick tube

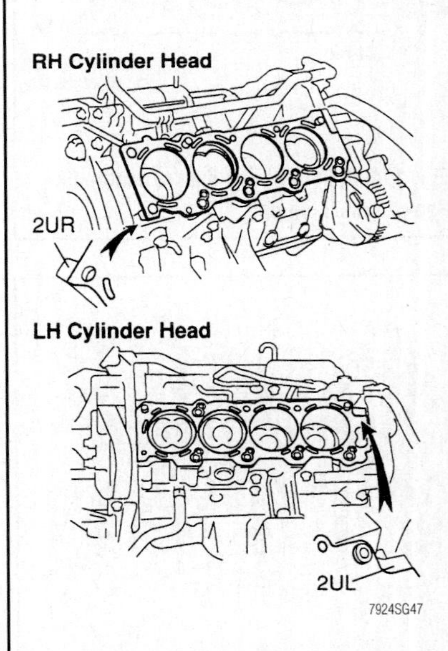

Fig. 185 Cylinder head gasket identification—4.7L engine

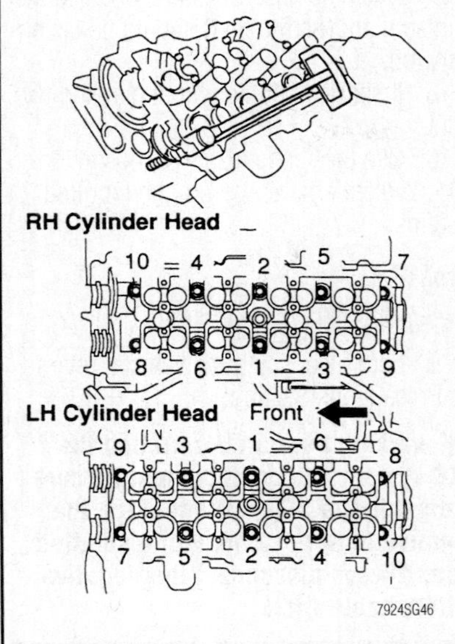

Fig. 186 Cylinder head torque sequence—4.7L engine

- Exhaust front pipes
- Power steering pump
- CMP sensor
- Camshaft sprockets
- Timing belt
- Front covers
- Idler pulley
- Radiator
- Cooling fan and bracket
- A/C compressor and bracket
- Accessory drive belt
- Air intake assembly
- Engine under covers
- Engine appearance cover
- Battery and tray

8. Fill the cooling system.
9. Start the engine and check for leaks.
10. Use the Techstream diagnostic tool, or equivalent and reprogram the required systems.

5.7L Engine

Bank 1

See Figures 187 through 191.

1. Before servicing the vehicle, refer to the Precautions Section.

➡If working near and/or around the SRS system and components, be sure to disable the SRS system. Tape the negative battery cable with insulating tape. Always disconnect the negative battery cable first.

⁂ CAUTION

To avoid personal injury when working on vehicles equipped with an air bag, the negative battery cable must be disconnected and at least 90 seconds must elapse before working on the system. Failure to do so may result in deployment of the air bag.

2. Disconnect the negative battery cable. Tape the cable with insulating tape.

➡When disconnecting the cable, some systems need to be initialized after the cable is reconnected. You will need the Techstream diagnostic scan tool or equivalent. Follow the directions on the tool.

3. Remove the exhaust manifold sub assembly.
4. Remove the camshafts.
5. Remove the No. 1 valve rocker arm sub assembly.
6. Remove the 16 valve lash adjusters.
7. Remove the 16 valve stem caps from the cylinder head.
8. Remove bolts 1 and 2 from the head, see illustration. Remove the remaining bolts in the sequence shown in the illustration. Be sure to arrange the removed parts for installation separately.
9. Remove the cylinder head from its mounting.
10. Discard the gasket.

To install:

➡Be sure to use new fasteners, as required.

11. Installation is the reverse of the removal procedure.
12. Check for proper gasket positioning. Place the gasket on the head with the lot No. stamp upward. Be sure there is no oil on the gasket.
13. Tighten the bolts to specification and in the proper sequence. Lightly coat the bolt threads with clean engine oil.

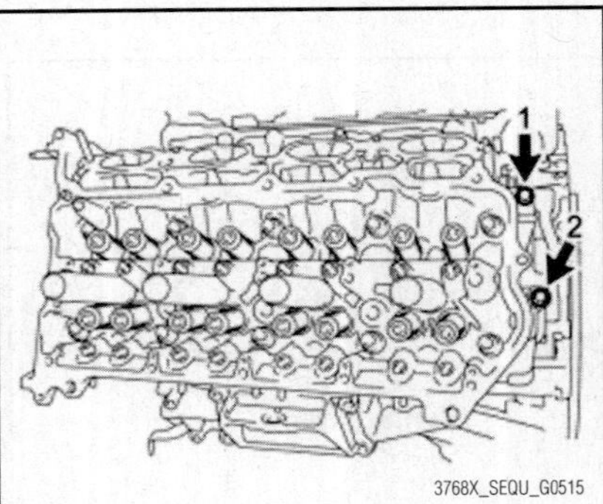

Fig. 187 Cylinder head bolt sequence removal Bank 1 (1 of 2)—5.7L engine

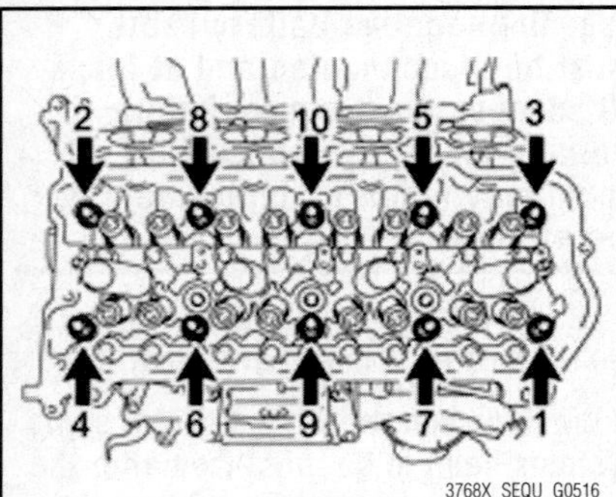

Fig. 188 Cylinder head bolt sequence removal Bank 1 (2 of 2)—5.7L engine

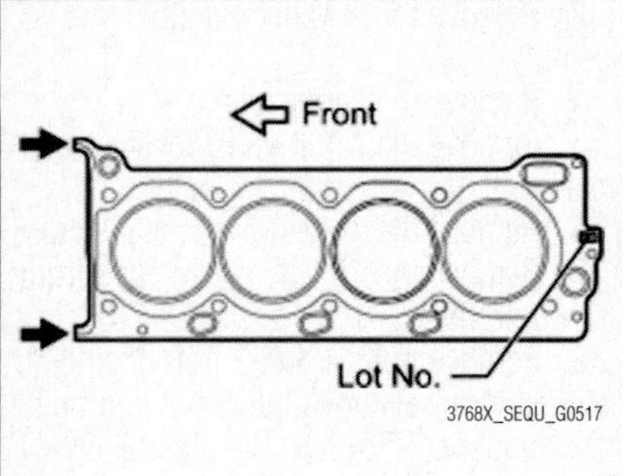

Fig. 189 Cylinder head gasket positioning Bank 1—5.7L engine

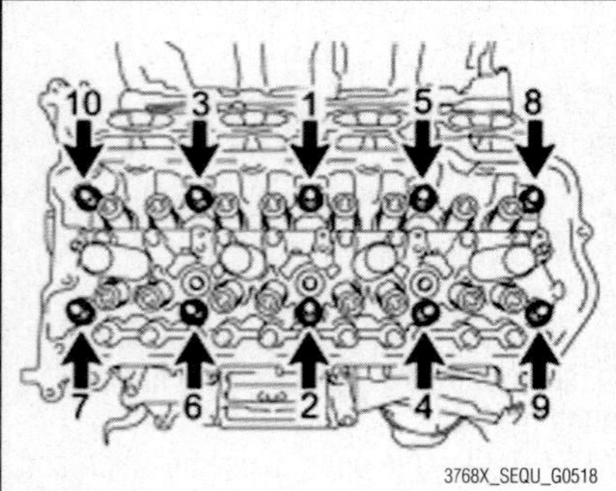

Fig. 190 Cylinder head gasket tightening sequence Bank 1—5.7L engine

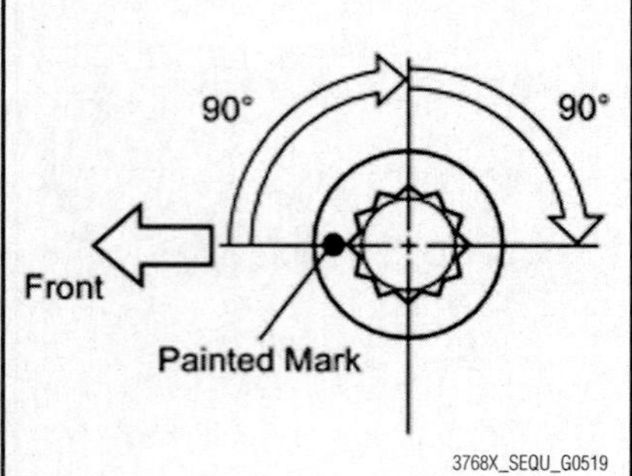

Fig. 191 Cylinder head bolt proper installation Bank 1 and Bank 2—5.7L engine

14. Once the head bolts are installed the painted mark (on the bolt) should be facing rearward.
15. Install and tighten the remaining two bolts.
16. Use the Techstream diagnostic tool, or equivalent and reprogram the required systems.

Bank 2

See Figures 192 through 195.

1. Before servicing the vehicle, refer to the Precautions Section.

➡If working near and/or around the SRS system and components, be sure to disable the SRS system. Tape the negative battery cable with insulating tape. Always disconnect the negative battery cable first.

⁂ CAUTION

To avoid personal injury when working on vehicles equipped with an air bag, the negative battery cable must be disconnected and at least 90 seconds must elapse before working on the system. Failure to do so may result in deployment of the air bag.

2. Disconnect the negative battery cable. Tape the cable with insulating tape.

➡When disconnecting the cable, some systems need to be initialized after the cable is reconnected. You will need the Techstream diagnostic scan tool or equivalent. Follow the directions on the tool.

3. Remove the exhaust manifold sub assembly.
4. Remove the camshafts.
5. Remove the No. 1 valve rocker arm sub assembly.
6. Remove the 16 valve lash adjusters.
7. Remove the 16 valve stem caps from the cylinder head.
8. Remove bolts 1 and 2 from the head, see illustration. Remove the remaining bolts in the sequence shown in the illustration. Be sure to arrange the removed parts for installation separately.
9. Remove the cylinder head from its mounting.
10. Discard the gasket.

To install:

➡Be sure to use new fasteners, as required.

11. Installation is the reverse of the removal procedure.
12. Check for proper gasket positioning. Place the gasket on the head with the lot

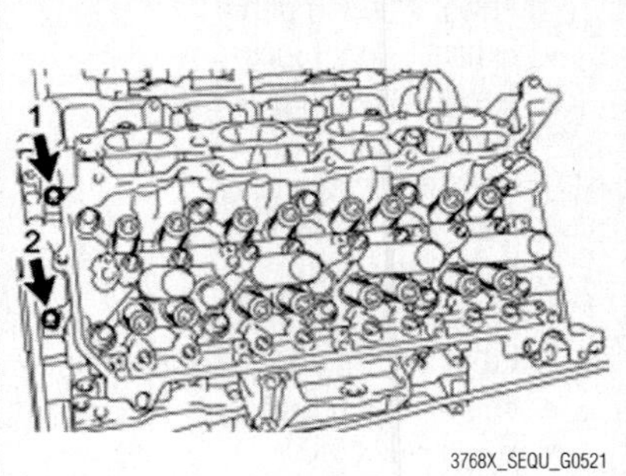

Fig. 192 Cylinder head bolt sequence removal Bank 2 (1 of 2)—5.7L engine

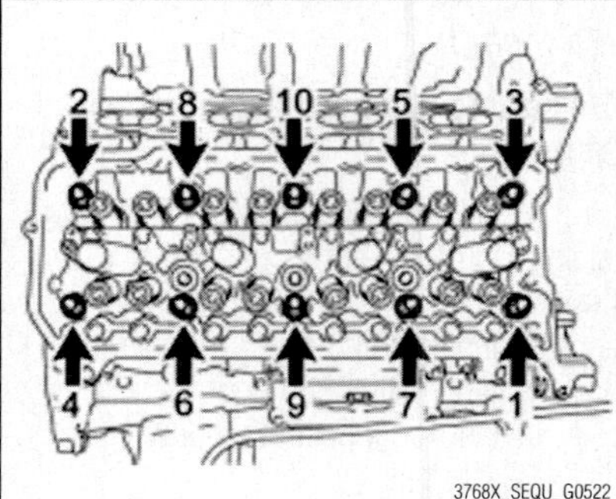

Fig. 193 Cylinder head bolt sequence removal Bank 2 (2 of 2)—5.7L engine

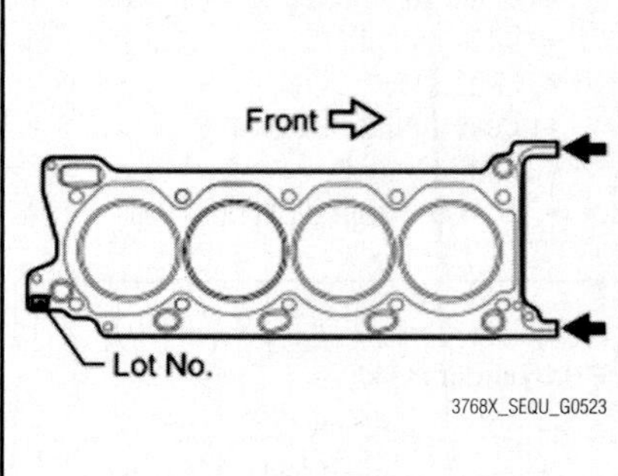

Fig. 194 Cylinder head gasket positioning Bank 2—5.7L engine

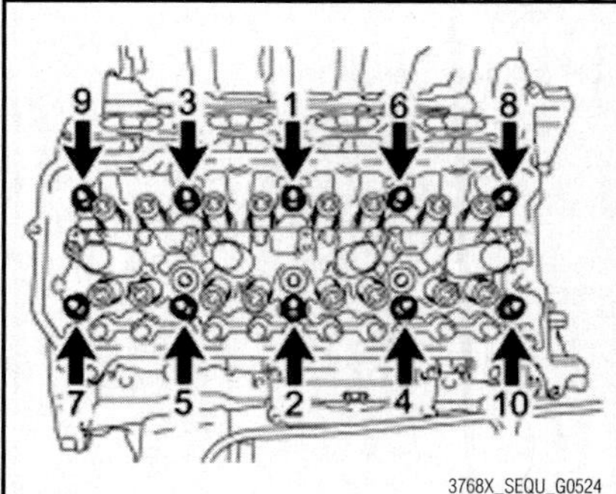

Fig. 195 Cylinder head gasket tightening sequence Bank 2—5.7L engine

No. stamp upward. Be sure there is no oil on the gasket.

13. Tighten the bolts to specification and in the proper sequence. Lightly coat the bolt threads with clean engine oil.

14. Once the head bolts are installed the painted mark (on the bolt) should be facing rearward.

15. Install and tighten the remaining two bolts.

16. Use the Techstream diagnostic tool, or equivalent and reprogram the required systems.

EXHAUST MANIFOLD

REMOVAL & INSTALLATION

4.0L Engine

See Figures 196 and 197.

1. Before servicing the vehicle, refer to the Precautions Section.

➡If working near and/or around the SRS system and components, be sure to disable the SRS system. Tape the negative battery cable with insulating tape. Always disconnect the negative battery cable first.

⁂ CAUTION

To avoid personal injury when working on vehicles equipped with an air bag, the negative battery cable must be disconnected and at least 90 seconds must elapse before working on the system. Failure to do so may result in deployment of the air bag.

2. Disconnect the negative battery cable. Tape the cable with insulating tape.

➡When disconnecting the cable, some systems need to be initialized after the cable is reconnected. You will need the Techstream diagnostic scan tool or equivalent. Follow the directions on the tool.

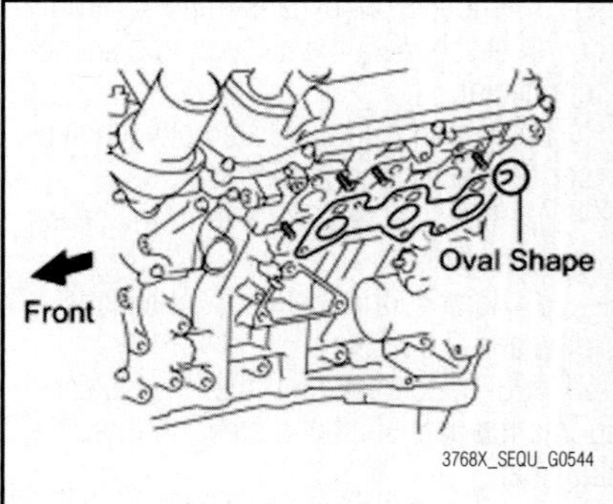

Fig. 196 Exhaust manifold gasket positioning left side—4.0L engine

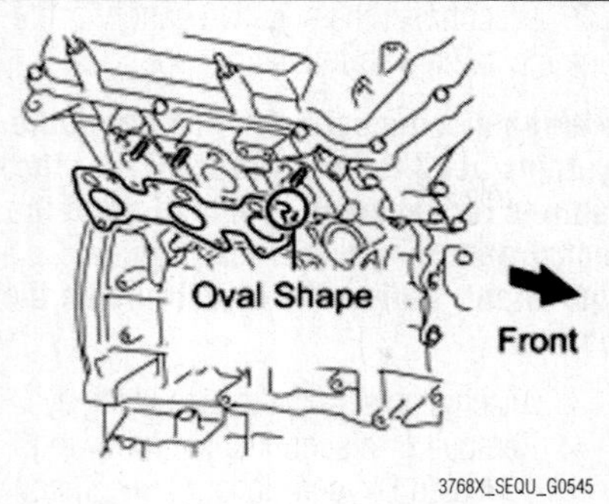

Fig. 197 Exhaust manifold gasket positioning right side—4.0L engine

3. Remove the air switching valve assembly.
4. Remove the No. 2 exhaust pipe assembly.
5. Remove the front exhaust pipe assembly.
6. Remove the manifold retaining nuts.
7. Remove the heat insulator, as required.
8. Remove the manifold from the vehicle.
9. Discard the gasket.

To install:

➡Be sure to use new fasteners, as required.

10. Installation is the reverse of the removal procedure.
11. Use the Techstream diagnostic tool, or equivalent and reprogram the required systems.

4.6L Engine

See Figures 198 through 201.

1. Before servicing the vehicle, refer to the Precautions Section.

➡If working near and/or around the SRS system and components, be sure to disable the SRS system. Tape the negative battery cable with insulating tape. Always disconnect the negative battery cable first.

⁂ CAUTION

To avoid personal injury when working on vehicles equipped with an air bag, the negative battery cable must be disconnected and at least 90 seconds must elapse before working on the system. Failure to do so may result in deployment of the air bag.

2. Disconnect the negative battery cable. Tape the cable with insulating tape.

➡When disconnecting the cable, some systems need to be initialized after the cable is reconnected. You will need the Techstream diagnostic scan tool or equivalent. Follow the directions on the tool.

3. Remove the front fender apron seals.
4. Remove the intake manifold assembly.
5. Remove the alternator.
6. Disconnect the compressor.

➡It may be necessary to properly discharge the system and remove the compressor, if unable to position it to the side without disconnect the refrigerant lines. If disconnecting the lines be sure to plug them after disconnection.

7. Remove the engine oil dipstick guide.
8. Remove the exhaust pipe assembly.
9. Remove the front driveshaft, if equipped with 4WD.
10. Remove the No. 3 and No. 4 floor heat insulators.
11. Remove the EGR pipe.
12. Remove the exhaust manifold heat insulators.
13. Remove the manifold retaining nuts. Discard the nuts.
14. Remove the manifold from the engine. Discard the gasket.

To install:

➡Be sure to use new fasteners, as required.

15. Installation is the reverse of the removal procedure.
16. Tighten nuts labeled A to 7 ft. lbs. (10 Nm). Tighten others to 15 ft. lbs. (21 Nm). See illustration.
17. Use the Techstream diagnostic tool, or equivalent and reprogram the required systems.

4.7L Engine

1. Before servicing the vehicle, refer to the Precautions Section.

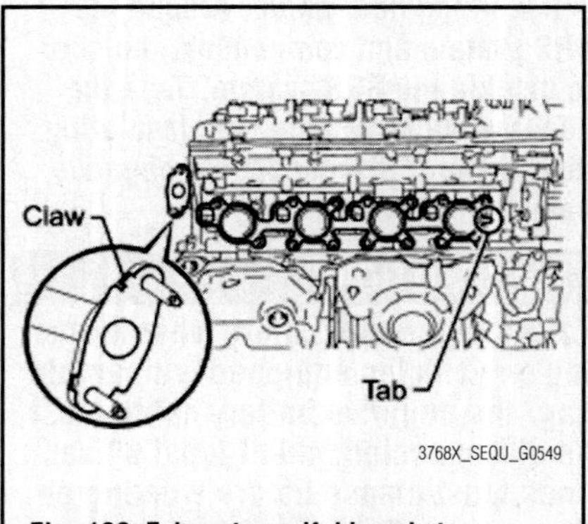

Fig. 198 Exhaust manifold gasket positioning left side—4.6L engine

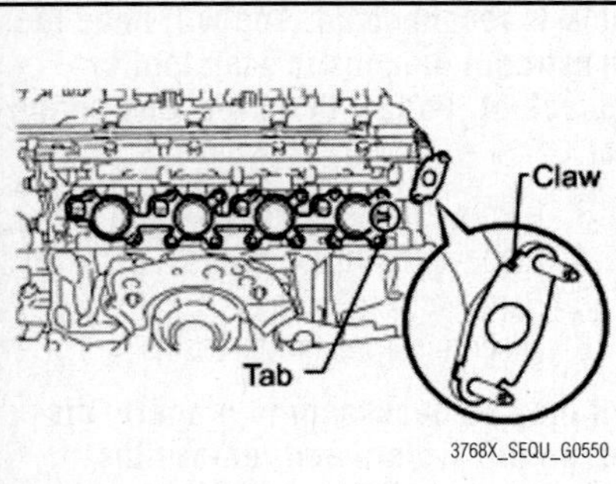

3768X_SEQU_G0550

Fig. 199 Exhaust manifold gasket positioning right side—4.6L engine

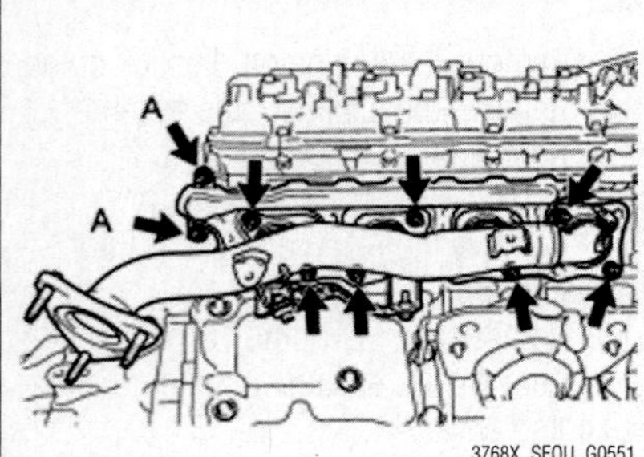

3768X_SEQU_G0551

Fig. 200 Exhaust manifold bolt location and tightening sequence left side—4.6L engine

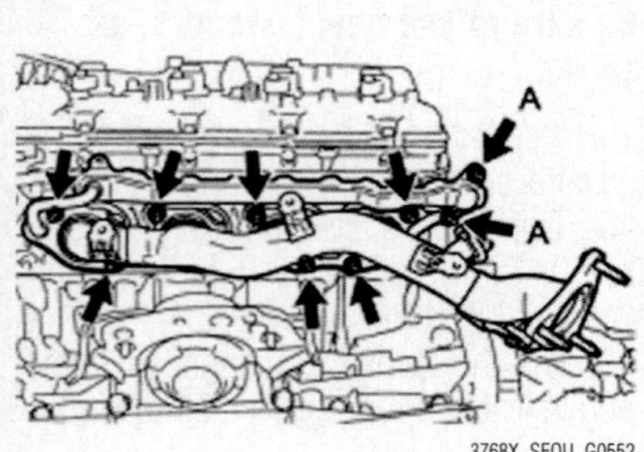

3768X_SEQU_G0552

Fig. 201 Exhaust manifold bolt location and tightening sequence right side—4.6L engine

➡If working near and/or around the SRS system and components, be sure to disable the SRS system. Tape the negative battery cable with insulating tape. Always disconnect the negative battery cable first.

⁂ CAUTION

To avoid personal injury when working on vehicles equipped with an air bag, the negative battery cable must be disconnected and at least 90 seconds must elapse before working on the system. Failure to do so may result in deployment of the air bag.

2. Disconnect the negative battery cable. Tape the cable with insulating tape.

➡When disconnecting the cable, some systems need to be initialized after the cable is reconnected. You will need the Techstream diagnostic scan tool or equivalent. Follow the directions on the tool.

3. Attach a hoist to the engine lifting eyes.
4. Remove or disconnect the following:
   - Heated Oxygen Sensor (HO_2S) connectors
   - Exhaust manifold heat shield
   - Exhaust front pipe
   - Motor mount
   - Motor mount bracket
   - Exhaust manifold

To install:

➡Be sure to use new fasteners, as required.

➡Use new exhaust manifold nuts for assembly.

5. Install or connect the following:
   - Exhaust manifold. Tighten the nuts to 32 ft. lbs. (44 Nm).
   - Motor mount bracket. Tighten the bolts to 27 ft. lbs. (36 Nm).
   - Motor mount. Tighten the fasteners to 22 ft. lbs. (30 Nm).
   - Exhaust front pipe. Tighten the nuts to 46 ft. lbs. (62 Nm).
   - Exhaust manifold heat shield
   - HO_2S connectors
   - Negative battery cable
6. Start the engine and check for leaks.
7. Use the Techstream diagnostic tool, or equivalent and reprogram the required systems.

5.7L Engine

See Figures 202 through 209.

1. Before servicing the vehicle, refer to the Precautions Section.

➡If working near and/or around the SRS system and components, be sure to disable the SRS system. Tape the negative battery cable with insulating tape. Always disconnect the negative battery cable first.

⁂ CAUTION

To avoid personal injury when working on vehicles equipped with an air bag, the negative battery cable must be disconnected and at least 90 seconds must elapse before working on the system. Failure to do so may result in deployment of the air bag.

2. Disconnect the negative battery cable. Tape the cable with insulating tape.

➡When disconnecting the cable, some systems need to be initialized after the cable is reconnected. You will need the Techstream diagnostic scan tool or equivalent. Follow the directions on the tool.

3. Remove No. 1 engine under cover.
4. Drain engine coolant.
5. Remove V-bank cover sub-assembly.
6. Remove air cleaner hose assembly.
7. Disconnect inlet and outlet radiator hose.
8. Remove fan and alternator v-belt.
9. Remove fan shroud.
10. Remove the 6 clips and fender apron seal.
11. Remove the 5 clips and fender apron seal.
12. Remove engine oil level dipstick guide.
13. Disconnect vane pump assembly.
14. Remove alternator assembly.
15. Disconnect cooler compressor assembly.
16. Remove the 2 bolts and heat insulator 4WD.
17. Remove front propeller shaft assembly (for 4WD).
18. Regular Cab Standard Deck:
    a. Disconnect the air fuel ratio sensor connector.
    b. Disconnect the heated oxygen sensor connector.
    c. Remove the 2 bolts, 5 nuts, No. 2 exhaust pipe and 2 gaskets.
19. Except Regular Cab Standard Deck:
    a. Disconnect the air fuel ratio sensor connector.
    b. Disconnect the heated oxygen sensor connector.
    c. Remove the 2 bolts, 3 nuts, No. 2 exhaust pipe and 2 gaskets.
    d. Disconnect the exhaust support.
20. Remove front exhaust pipe assembly.
    a. Disconnect the air fuel ratio sensor connector.
    b. Disconnect the heated oxygen sensor connector and 2 clamps.
    c. Remove the 2 bolts from the center exhaust pipe.
    d. Remove the 3 nuts, front exhaust pipe and 2 gaskets.
21. Put matchmarks on the No. 2 steering intermediate shaft and steering intermediate shaft.
    a. Remove the bolt and disconnect the No. 2 steering intermediate shaft from the steering intermediate shaft.

b. Put matchmarks on the No. 2 steering intermediate shaft and the power steering gear.

c. Remove the bolt and disconnect the No. 2 steering intermediate shaft from the power steering gear.

22. Remove the 3 bolts and heat insulator.

23. Remove the 10 nuts, exhaust manifold and 2 gaskets.

24. Remove the 3 bolts and heat insulator.

25. Remove the 10 nuts, exhaust manifold and 2 gaskets.

To install:

➡Be sure to use new fasteners, as required.

26. Install a new gasket to the cylinder head and a new gasket to the No. 2 air tube.

➡Install the exhaust manifold gasket with the gasket tab facing toward the front of the engine. Install the air tube gasket with the gasket's claws facing the tube side.

27. Temporarily install the exhaust manifold and then uniformly tighten 8 new nuts that are not labeled A.

28. Tighten the new nuts labeled A in the illustration to 7 ft. lbs. (10 Nm) and remaining nuts to 15 ft. lbs. (21 Nm).

29. Install a new gasket to the cylinder head and a new gasket to the No. 3 air tube.

➡Install the exhaust manifold gasket with the gasket tab facing toward the front of the engine. Install the air tube gasket with the gasket's claws facing the tube side.

30. Temporarily install the exhaust manifold and then uniformly tighten 8 new nuts that are not labeled A.

31. Tighten the new nuts labeled A in the illustration to 7 ft. lbs. (10 Nm) and remaining nuts to 15 ft. lbs. (21 Nm).

32. Install the heat insulator with the 3 bolts and tighten to 7 ft. lbs. (10 Nm).

33. Align the matchmarks and insert the No. 2 intermediate shaft into the intermediate shaft.

34. Align the matchmarks and insert the No. 2 intermediate shaft into the power steering gear.

35. Install the 2 bolt and tighten to 26 ft. lbs. (35 Nm).

36. Install front exhaust pipe assembly.

a. Install a new gasket and the front exhaust pipe to the exhaust manifold RH with 3 new nuts and tighten to 40 ft. lbs. (54 Nm).

b. Install a new gasket and the front exhaust pipe to the center exhaust pipe with the 2 bolts and tighten to 35 ft. lbs. (48 Nm).

c. Connect the air fuel ratio sensor connector.

d. Connect the heated oxygen sensor connector and 2 clamps.

37. Except Regular Cab Standard Deck:

a. Connect the front No. 2 exhaust pipe to the exhaust support.

b. Install a new gasket and the front No. 2 exhaust pipe to the exhaust manifold LH with 3 new nuts and tighten to 40 ft. lbs. (54 Nm).

c. Install a new gasket and the front No. 2 exhaust pipe to the center exhaust pipe with the 2 bolts and tighten to 35 ft. lbs. (48 Nm).

d. Connect the air fuel ratio sensor connector.

e. Connect the heated oxygen sensor connector.

38. Regular Cab Standard Deck :

a. Connect the front No. 2 exhaust pipe to the exhaust support.

b. Install a new gasket and the front No. 2 exhaust pipe to the exhaust

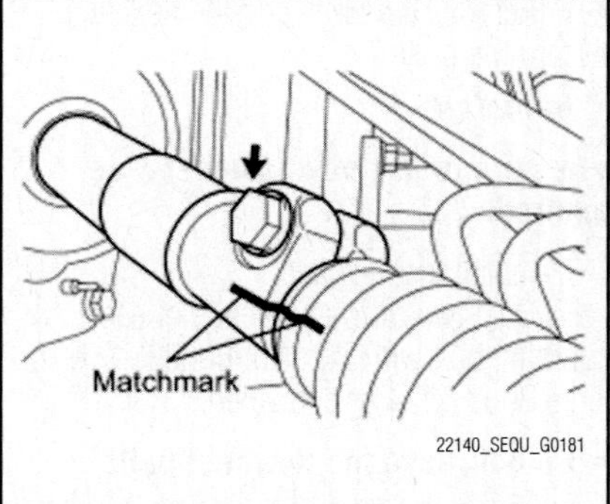

Fig. 202 Steering intermediate shaft to steering intermediate shaft—5.7L engine

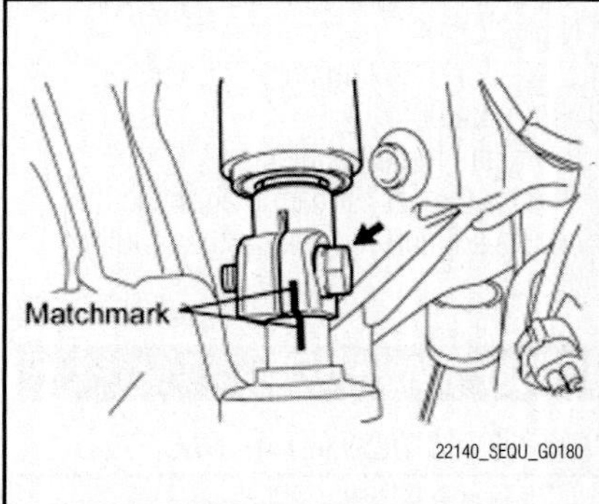

Fig. 203 Steering intermediate shaft to the power steering gear—5.7L engine

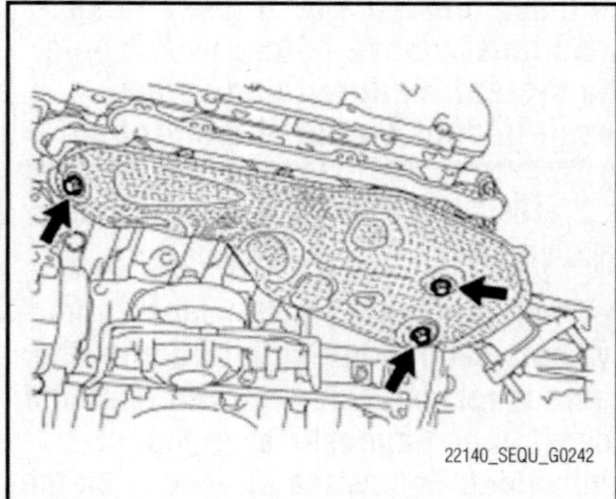

Fig. 204 Exhaust heat insulator No. 2—5.7L engine

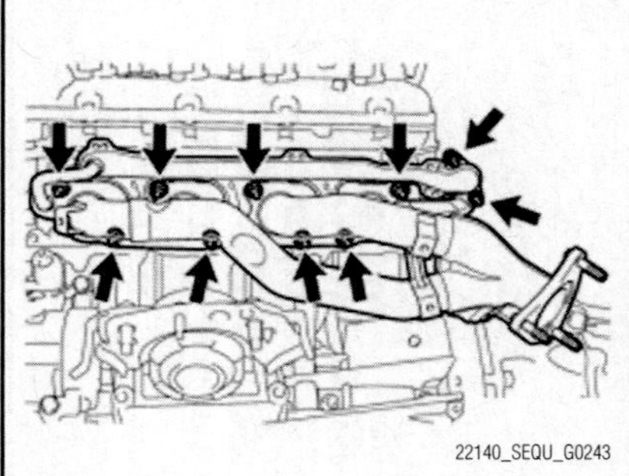

Fig. 205 Exhaust manifold LH—5.7L engine

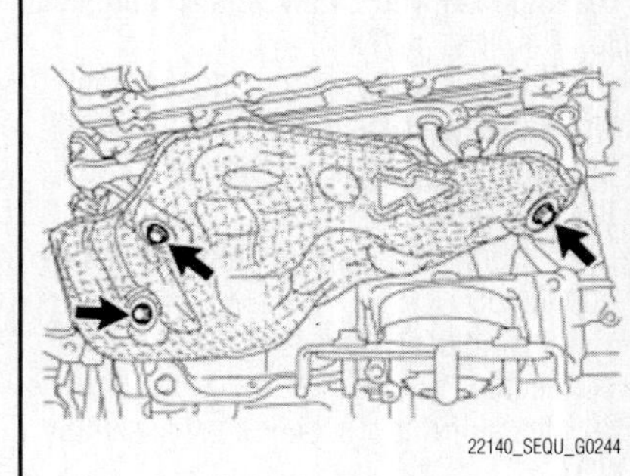

Fig. 206 Exhaust heat insulator No. 1—5.7L engine

Fig. 207 Exhaust manifold tightening sequence RH—5.7L engine

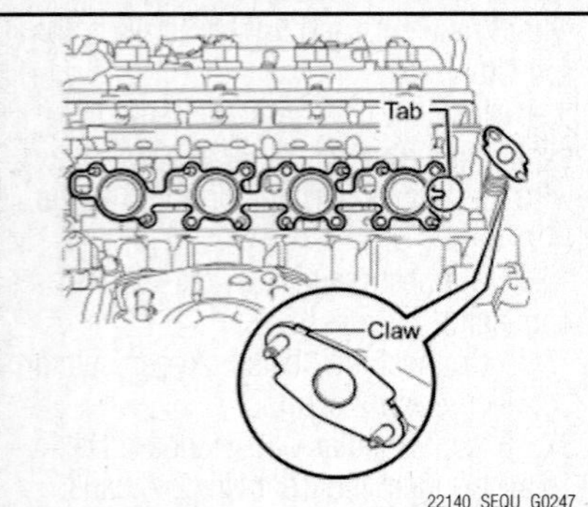

Fig. 208 Exhaust manifold gasket position LH—5.7L engine

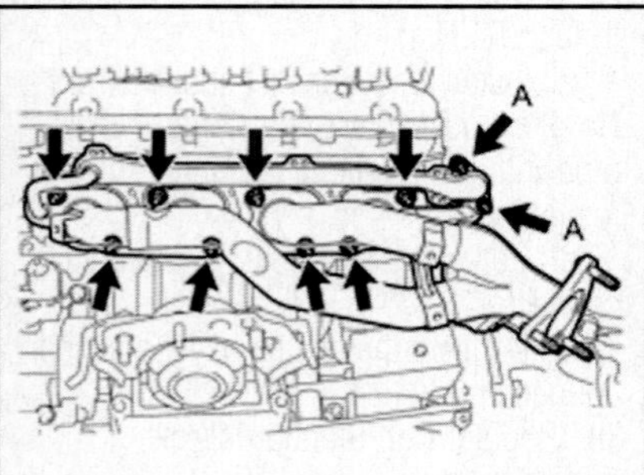

Fig. 209 Exhaust manifold tightening sequence LH—5.7L engine

manifold LH with 3 new nuts and tighten to 40 ft. lbs. (54 Nm).

c. Install a new gasket and the front No. 2 exhaust pipe to the center exhaust pipe with the 2 bolts and tighten to 35 ft. lbs. (48 Nm).

d. Connect the air fuel ratio sensor connector.

e. Connect the heated oxygen sensor connector.

39. Install front propeller shaft assembly (4WD).
40. Install the heat insulator with the 2 bolts (4WD).
41. Connect cooler compressor assembly.
42. Install alternator assembly.
43. Connect vane pump assembly.
44. Install engine oil level dipstick guide.
45. Install fan shroud.
46. Install fan and alternator v belt.
47. Connect outlet and inlet radiator hose.
48. Install air cleaner assembly.
49. Install air cleaner hose assembly.
50. Install front fender apron seal rear LH.
51. Install front fender apron seal LH.
52. Install front fender apron seal rear RH.
53. Install front fender apron seal RH.
54. Install No. 1 engine under cover.
55. Install V-bank cover sub-assembly.
56. Add engine coolant.
57. Inspect for exhaust gas leak.

➡If gas is leaking, tighten the areas necessary to stop the leak. Replace damaged parts as necessary.

58. Use the Techstream diagnostic tool, or equivalent and reprogram the required systems.

FLEXPLATE

REMOVAL & INSTALLATION

See Figures 210 through 212.

1. Before servicing the vehicle, refer to the Precautions Section.

➡If working near and/or around the SRS system and components, be sure to disable the SRS system. Tape the negative battery cable with insulating tape. Always disconnect the negative battery cable first.

CAUTION

To avoid personal injury when working on vehicles equipped with an air bag, the negative battery cable must be disconnected and at least 90 seconds must elapse before working on the system. Failure to do so may result in deployment of the air bag.

2. Disconnect the negative battery cable. Tape the cable with insulating tape.

➡When disconnecting the cable, some systems need to be initialized after the cable is reconnected. You will need the Techstream diagnostic scan tool or equivalent. Follow the directions on the tool.

3. Remove the transmission.

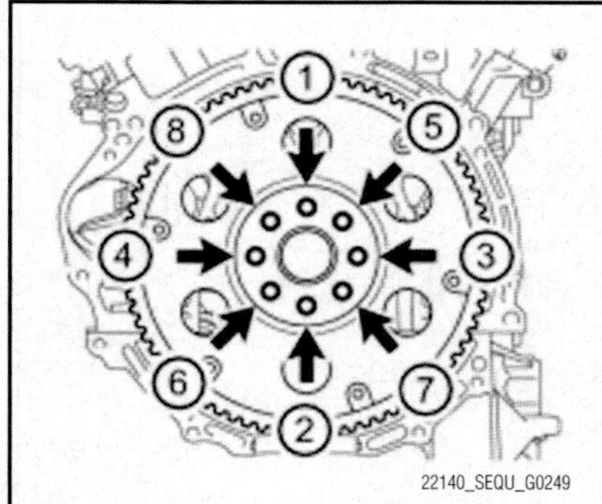

Fig. 210 Flexplate torque sequence—4.0L engine

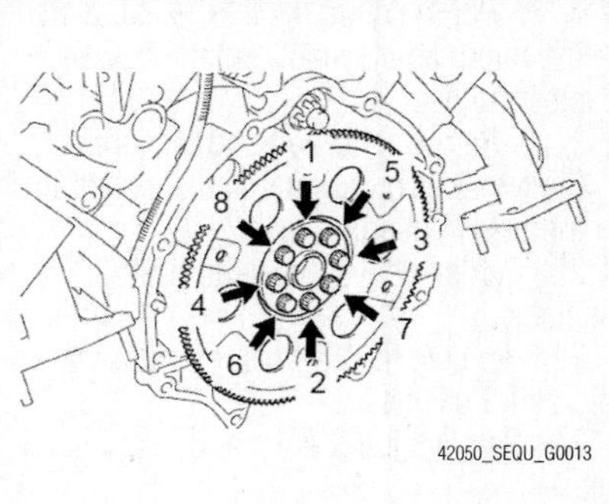

Fig. 211 Flexplate torque sequence—4.7L engine

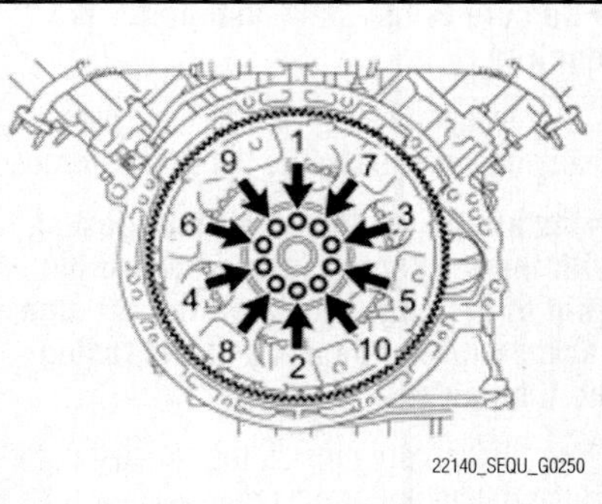

Fig. 212 Flexplate torque sequence—4.6L and 5.7L engines

4. Remove the bolts and the flexplate. Discard the bolts.

To install:

➡Be sure to use new fasteners, as required.

5. Install the flexplate.
6. Coat two or three threads of each mounting bolt with Part No. 008833-00070 Three Bond 1324 or equivalent.

➡ Do not reuse the flywheel bolts

7. Tighten the bolts in sequence as follows:

a. For 4.0L engine, 61 ft. lbs. (83 Nm)

b. For 4.7L engine, 36 ft. lbs. (49 Nm) Plus 90°.

c. For 5.7L engine, 22 ft. lbs. (30 Nm) plus 90°.

8. Install the transmission.
9. Use the Techstream diagnostic tool, or equivalent and reprogram the required systems.

INTAKE MANIFOLD

REMOVAL & INSTALLATION

4.0L Engine

See Figures 213 through 215.

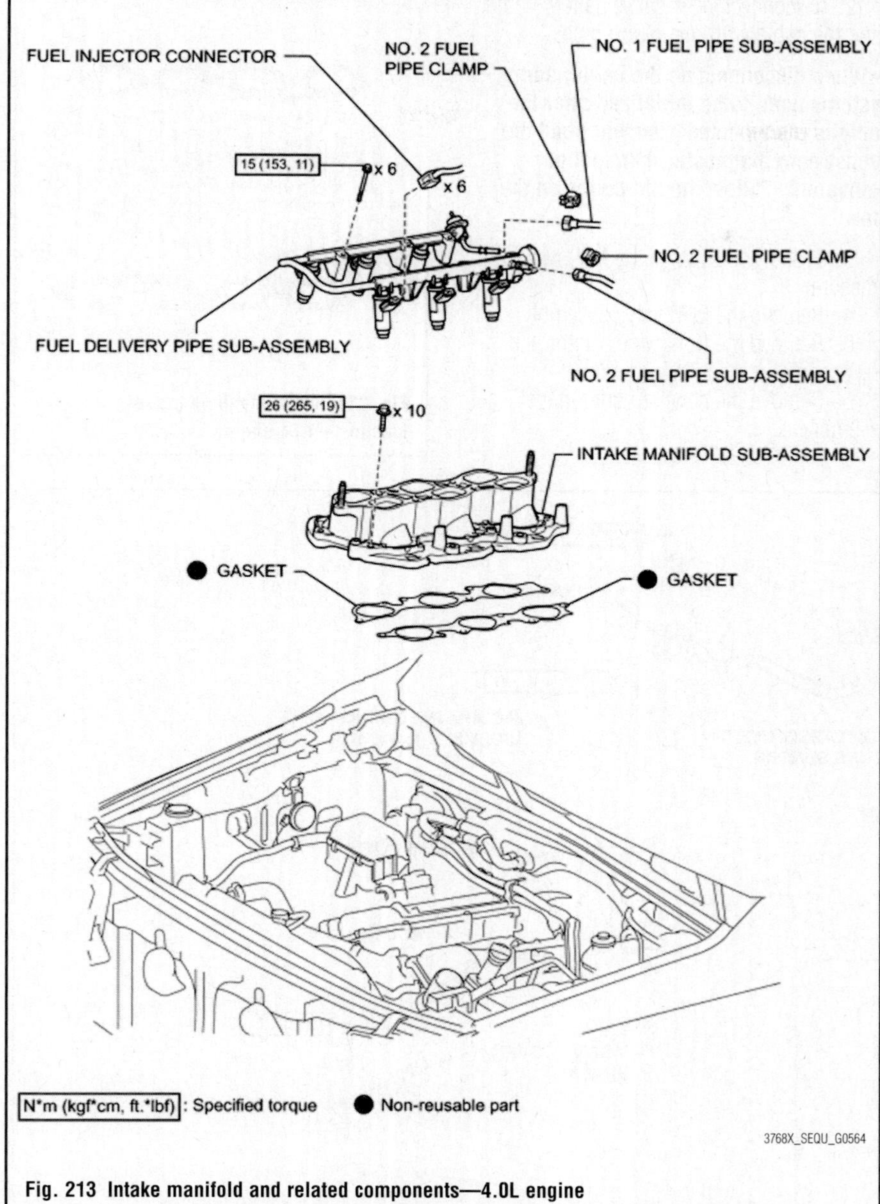

Fig. 213 Intake manifold and related components—4.0L engine

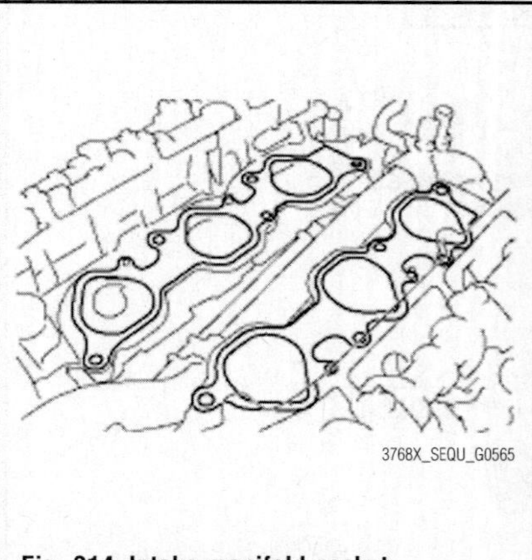

Fig. 214 Intake manifold gasket positioning—4.0L engine

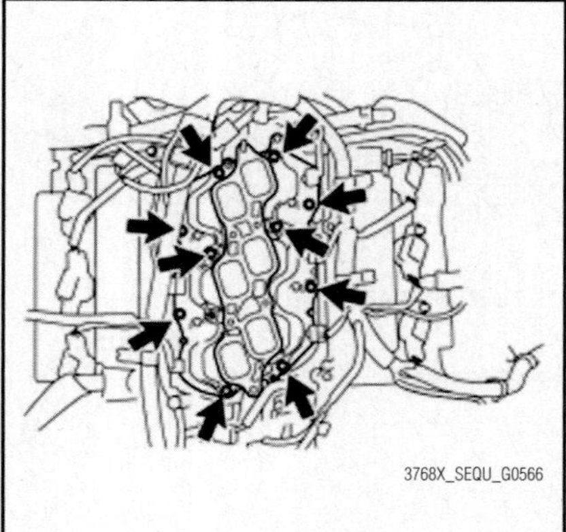

Fig. 215 Intake manifold bolt locations—4.0L engine

1. Before servicing the vehicle, refer to the Precautions Section.

➡If working near and/or around the SRS system and components, be sure to disable the SRS system. Tape the negative battery cable with insulating tape. Always disconnect the negative battery cable first.

CAUTION

To avoid personal injury when working on vehicles equipped with an air bag, the negative battery cable must be disconnected and at least 90 seconds must elapse before working on the system. Failure to do so may result in deployment of the air bag.

2. Disconnect the negative battery cable. Tape the cable with insulating tape.

➡When disconnecting the cable, some systems need to be initialized after the cable is reconnected. You will need the Techstream diagnostic scan tool or equivalent. Follow the directions on the tool.

3. Properly relieve the fuel system pressure.
4. Remove the engine under cover.
5. Drain the engine coolant. Be sure to properly dispose of used coolant.
6. Remove the front wiper motor and link assembly.
7. Remove the cowl top outer panel assembly.
8. Remove the V- bank cover. Remove the air cleaner assembly.
9. Remove the intake air surge tank. Discard the gasket.
10. Remove the fuel delivery pipe sub assembly.
11. Remove the intake manifold retaining bolts. Remove the intake manifold from the engine.
12. Discard the gasket.

To install:

➡Be sure to use new fasteners, as required.

13. Installation is the reverse of the removal procedure.
14. Be sure to use a new gasket.
15. Tighten the bolts to specification.
16. Use the Techstream diagnostic tool, or equivalent and reprogram the required systems.

4.6L Engine

See Figures 216 through 218.

1. Before servicing the vehicle, refer to the Precautions Section.

➡If working near and/or around the SRS system and components, be sure to disable the SRS system. Tape the negative battery cable with insulating tape. Always disconnect the negative battery cable first.

⁂ CAUTION

To avoid personal injury when working on vehicles equipped with an air bag, the negative battery cable must be disconnected and at least 90 seconds must elapse before working on the system. Failure to do so may result in deployment of the air bag.

2. Disconnect the negative battery cable. Tape the cable with insulating tape.

➡When disconnecting the cable, some systems need to be initialized after the cable is reconnected. You will need the Techstream diagnostic scan tool or equivalent. Follow the directions on the tool.

3. Properly discharge the fuel system pressure.
4. Remove the EGR valve assembly.
5. Remove the front wiper motor and link assembly.
6. Remove the cowl top outer panel assembly.

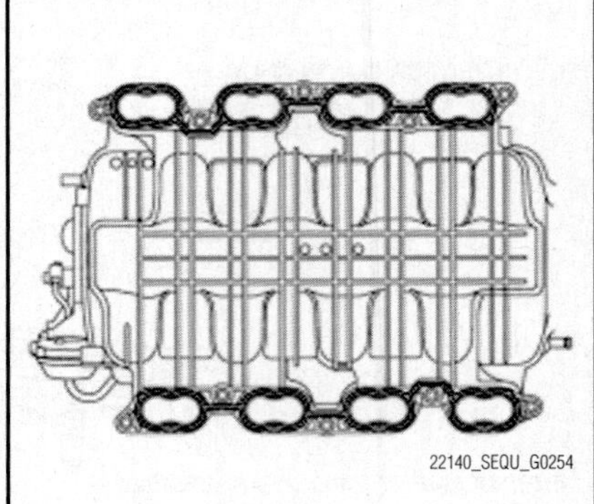

Fig. 217 Intake manifold gasket location—4.6L engine

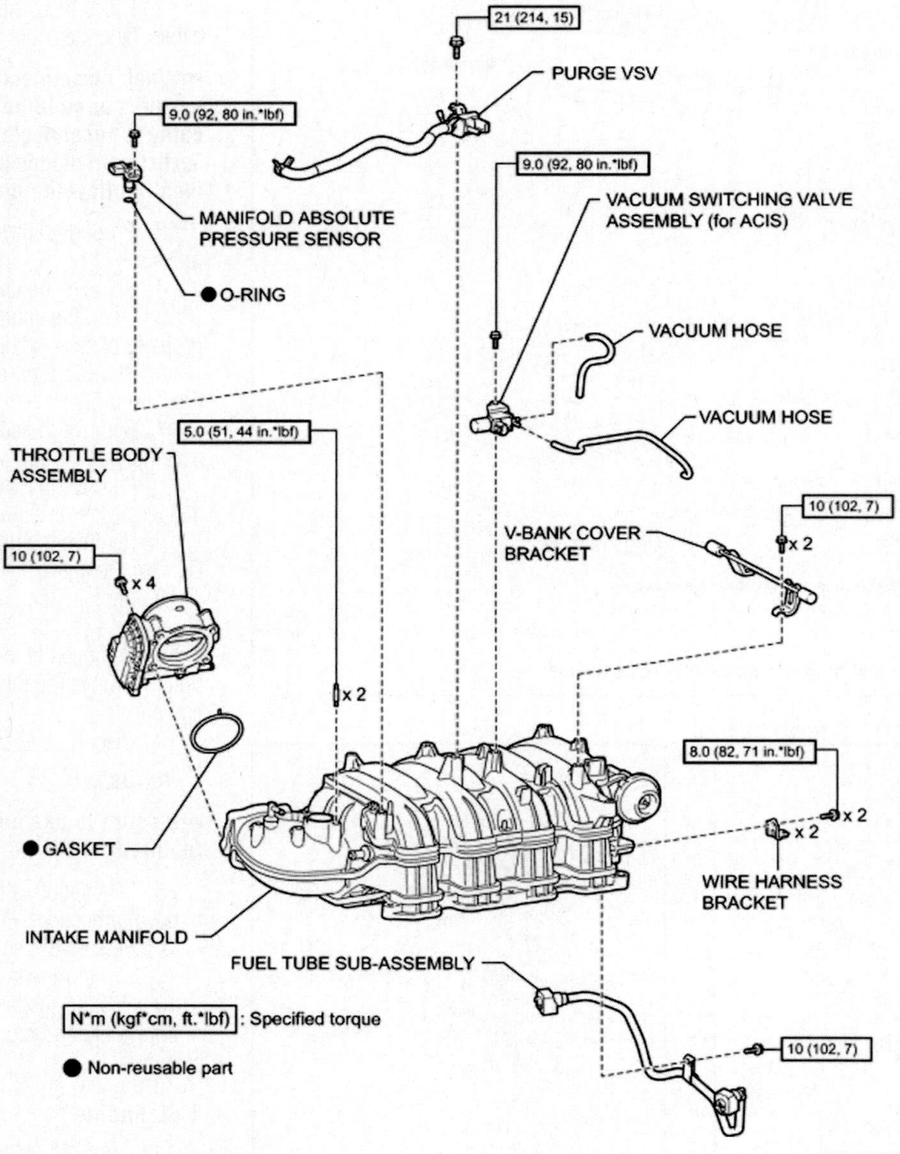

Fig. 216 Intake manifold and related components—4.6L engine

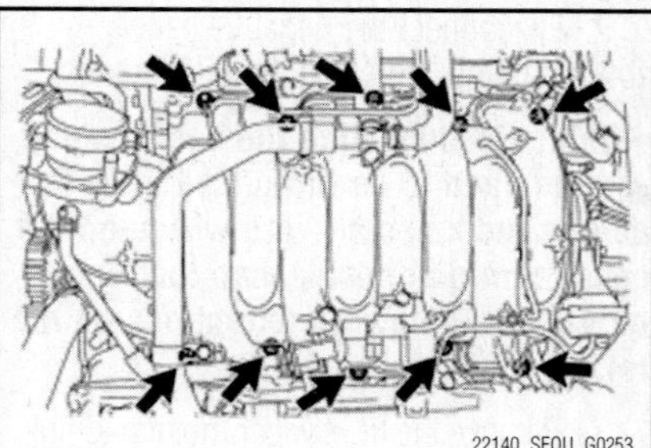

Fig. 218 Intake manifold bolt location—5.7L engine

7. Remove the ventilation hose.
8. Remove the air tube sub assembly, left side.
9. Disconnect the required electrical connectors.
10. Disconnect the required hoses.
11. Disconnect the fuel lines.
12. Remove the intake manifold retaining bolts.
13. Remove the intake manifold from its mounting.
14. Discard the gasket.

To install:

➡Be sure to use new fasteners, as required.

15. Installation is the reverse of the removal procedure.
16. Be sure to use a new gasket.
17. Use the Techstream diagnostic tool, or equivalent and reprogram the required systems.

4.7L Engine

See Figure 219.

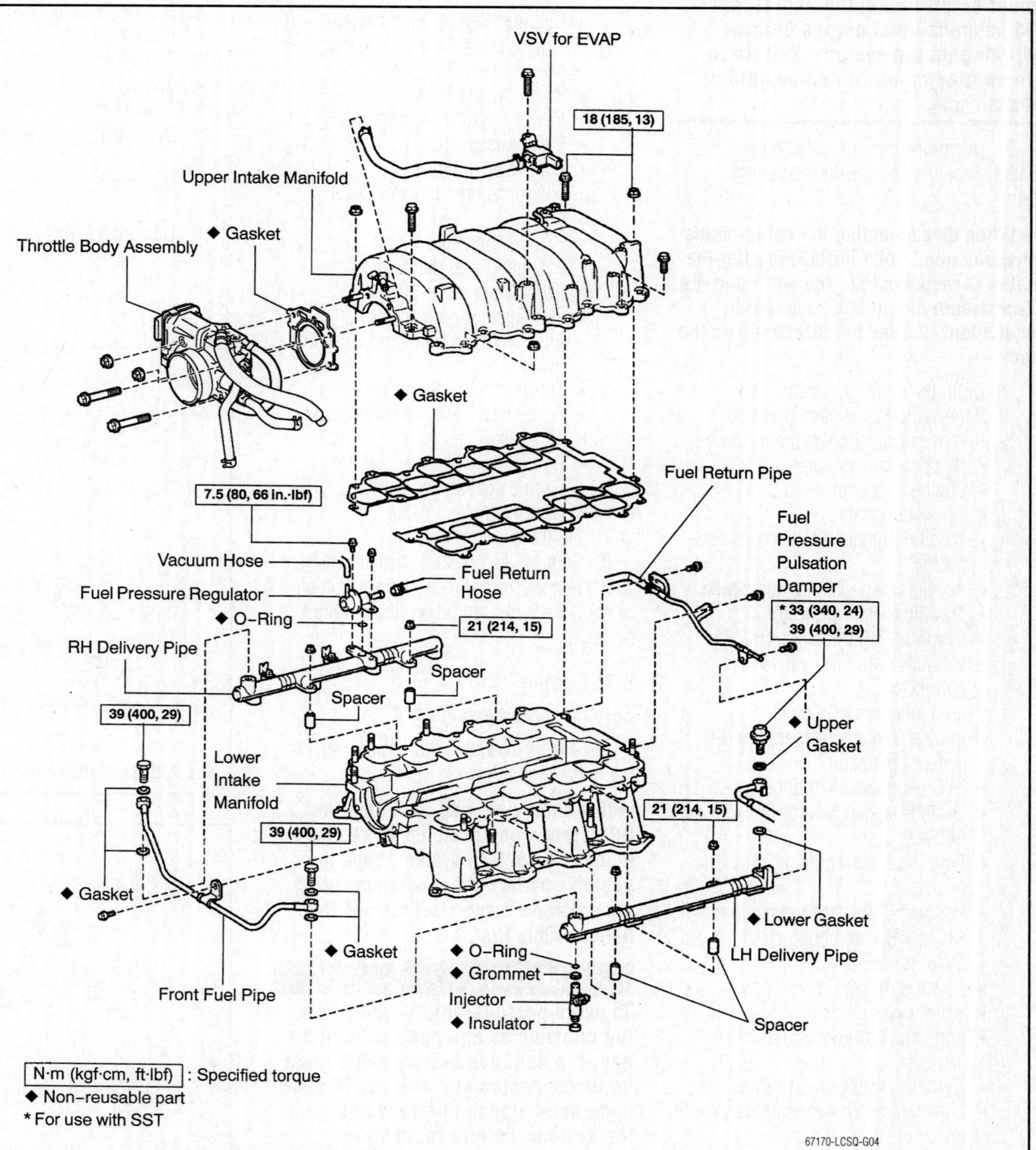

67170-LCSQ-G04

Fig. 219 Intake manifold and related components—4.7L engine

1. Before servicing the vehicle, refer to the Precautions Section.

➡**If working near and/or around the SRS system and components, be sure to disable the SRS system. Tape the negative battery cable with insulating tape. Always disconnect the negative battery cable first.**

CAUTION

To avoid personal injury when working on vehicles equipped with an air bag, the negative battery cable must be disconnected and at least 90 seconds must elapse before working on the system. Failure to do so may result in deployment of the air bag.

2. Disconnect the negative battery cable. Tape the cable with insulating tape.

➡**When disconnecting the cable, some systems need to be initialized after the cable is reconnected. You will need the Techstream diagnostic scan tool or equivalent. Follow the directions on the tool.**

3. Drain the cooling system.
4. Relieve the fuel system pressure.
5. Remove or disconnect the following:
   - Negative battery cable
   - Engine appearance cover
   - Accelerator cable
   - Throttle Position (TP) sensor connector
   - Accelerator pedal position sensor
   - Throttle motor connector
   - Evaporative Emissions (EVAP) vacuum switching valve connector
   - Fuel injector connectors
   - Engine Coolant Temperature (ECT) sensor connector
   - ETC gauge sender connector
   - Heated Oxygen Sensor (HO_2S) connectors
   - Fuel pressure regulator vacuum hose
   - Positive Crankcase Ventilation (PCV) valve and hose
   - EVAP hoses
   - Power steering vacuum hoses
   - Water bypass hose
   - Engine control wiring harness clamps
   - Cylinder head ground cables
   - Intake manifold wire harness protector
   - EVAP pipe
   - Engine appearance cover brackets
   - Intake manifold

To install:

➡**Be sure to use new fasteners, as required.**

6. Install or connect the following:
   - Intake manifold. Tighten the fasteners to 13 ft. lbs. (18 Nm).
   - Engine appearance cover brackets
   - EVAP pipe
   - Intake manifold wire harness protector
   - Cylinder head ground cables
   - Engine control wiring harness clamps
   - Water bypass hose
   - Power steering vacuum hoses
   - EVAP hoses
   - PCV valve and hose
   - Fuel pressure regulator vacuum hose
   - HO_2S connectors
   - ETC gauge sender connector
   - ECT sensor connector
   - Fuel injector connectors
   - EVAP vacuum switching valve connector
   - Throttle motor connector
   - Accelerator pedal position sensor
   - TP sensor connector
   - Accelerator cable
   - Engine appearance cover
   - Negative battery cable
7. Fill the cooling system.
8. Start the engine and check for leaks.
9. Use the Techstream diagnostic tool, or equivalent and reprogram the required systems.

5.7L Engine

See Figures 220 through 224.

1. Before servicing the vehicle, refer to the Precautions Section.

➡**If working near and/or around the SRS system and components, be sure to disable the SRS system. Tape the negative battery cable with insulating tape. Always disconnect the negative battery cable first.**

CAUTION

To avoid personal injury when working on vehicles equipped with an air bag, the negative battery cable must be disconnected and at least 90 seconds must elapse before working on the system. Failure to do so may result in deployment of the air bag.

2. Disconnect the negative battery cable. Tape the cable with insulating tape.

➡**When disconnecting the cable, some systems need to be initialized after the cable is reconnected. You will need the Techstream diagnostic scan tool or equivalent. Follow the directions on the tool.**

3. Remove the front wiper motor and link.
4. Disconnect the 2 washer hoses.
5. Remove the 7 bolts and outer panel.
6. Remove No. 1 engine under cover.
7. Drain engine coolant.
8. Remove V-bank cover sub-assembly.
9. Remove air cleaner hose assembly.
10. Disconnect the ventilation hose from the ventilation pipe of the cylinder head cover LH and RH.
11. Disconnect the 2 water by-pass hoses.
12. Disconnect the throttle body connector.
13. Disconnect the No. 1 ventilation hose.
14. Remove the No. 1 engine cover sub-assembly.

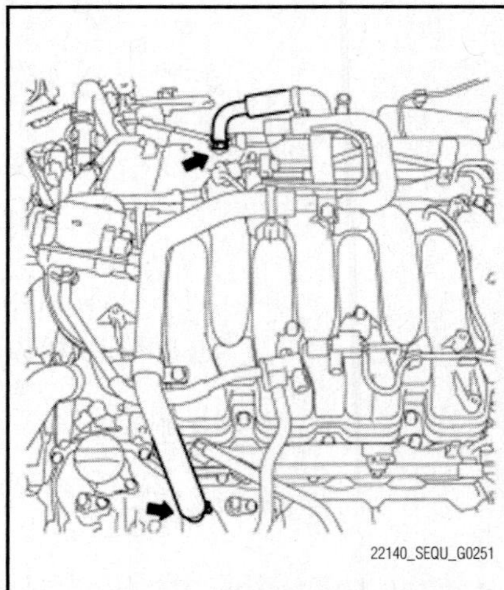

Fig. 220 Ventilation hose—5.7L engine

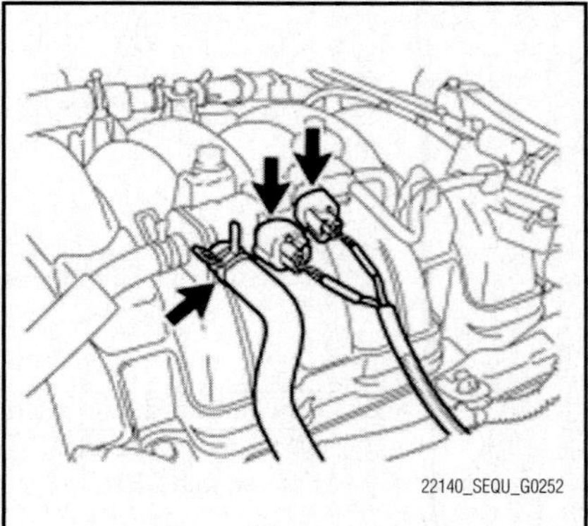

Fig. 221 VSV connector—5.7L engine

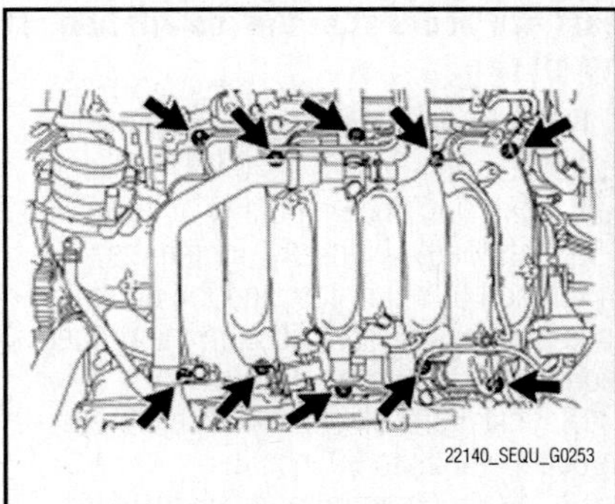

Fig. 222 Intake manifold bolt location—5.7L engine

15. Remove the No. 3 engine cover.
16. Disconnect the purge VSV connector.
17. Disconnect the purge line hose from the purge VSV.
18. Disconnect the vacuum switching valve connector (for ACIS).
19. Remove the 2 bolts.
20. Disconnect the No. 1 tube from the union to connector tube hose, and move the hose aside.
21. Disconnect the 3 wire clamps from the 3 wire brackets.
22. Remove the bolt and wire bracket from the intake manifold.
23. Remove the 2 nuts, 8 bolts, intake manifold and 2 gaskets.

To install:

➡Be sure to use new fasteners, as required.

24. Place 2 new gaskets on the intake manifold.
25. Place the intake manifold on the cylinder head.
26. Install and uniformly tighten the 8 bolts and 2 nuts to 15 ft. lbs. (21 Nm) in several steps.
27. Install the wire bracket to the intake manifold with the bolt.
28. Connect the 3 wire clamps to the 3 wire brackets.
29. Connect the No. 1 tube and install it to the intake manifold with the 2 bolts.
30. Connect the purge VSV connector.
31. Connect the purge line hose to the purge VSV.
32. Connect the vacuum switching valve connector (for ACIS).
33. Install the No. 1 engine cover sub-assembly.
34. Install the No. 3 engine cover.
35. Connect the No. 1 ventilation hose.
36. Connect the 2 water by-pass hoses.
37. Connect the throttle body connector.
38. Connect the ventilation hose to the ventilation pipe of the cylinder head cover LH and RH.
39. Install air cleaner hose assembly.
40. Install the outer panel with the 7 bolts.
41. Connect the 2 washer hoses.
42. Install the front wiper motor and link.
43. Add engine coolant.
44. Install V-bank cover sub-assembly.
45. Install No. 1 engine under cover.
46. Inspect for coolant leak.
47. Use the Techstream diagnostic tool, or equivalent and reprogram the required systems.

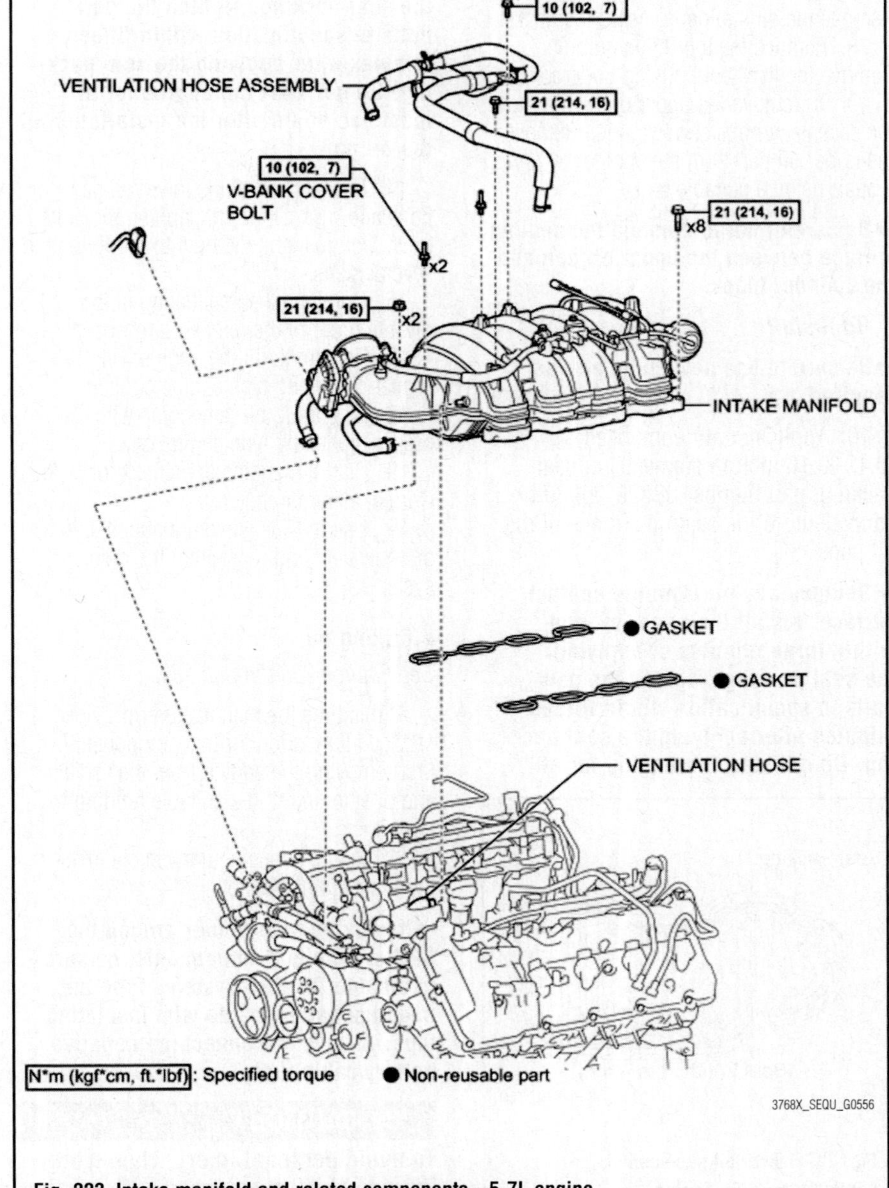

Fig. 223 Intake manifold and related components—5.7L engine

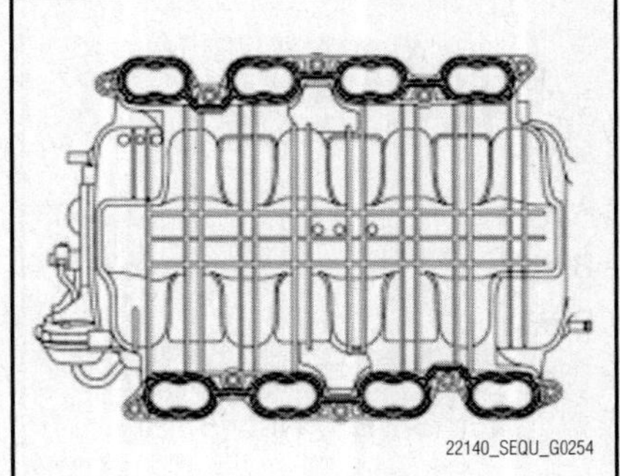

Fig. 224 Intake manifold gasket location—5.7L engine

OIL PAN

REMOVAL & INSTALLATION

4.0L Engine

See Figures 225 through 227.

1. Before servicing the vehicle, refer to the Precautions Section.

➡If working near and/or around the SRS system and components, be sure to disable the SRS system. Tape the negative battery cable with insulating tape. Always disconnect the negative battery cable first.

✻✻ CAUTION

To avoid personal injury when working on vehicles equipped with an air bag, the negative battery cable must be disconnected and at least 90 seconds must elapse before working on the system. Failure to do so may result in deployment of the air bag.

2. Disconnect the negative battery cable. Tape the cable with insulating tape.

➡When disconnecting the cable, some systems need to be initialized after the cable is reconnected. You will need the Techstream diagnostic scan tool or equivalent. Follow the directions on the tool.

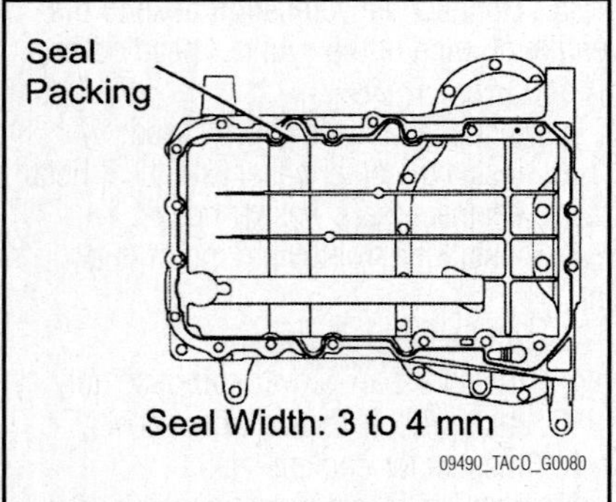

Fig. 225 Upper oil pan sealant application—4.0L engine

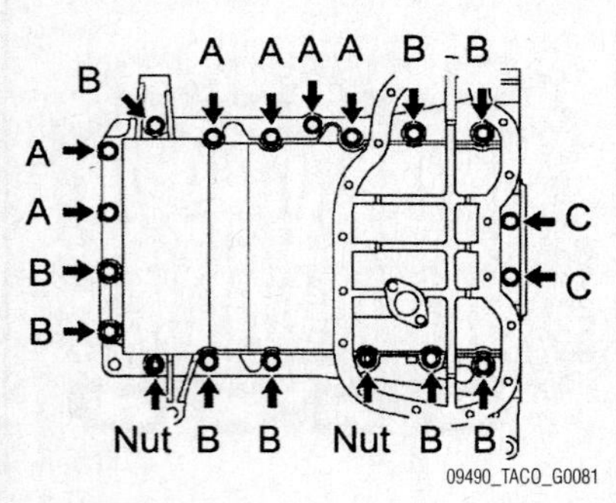

Fig. 226 Upper oil pan bolt torque sequence—4.0L engine

3. Raise and support the vehicle safely.
4. Remove the engine undercover. Drain the engine oil.
5. Remove the necessary components in order to gain access to the lower oil pan retaining bolts.
6. Remove the fifteen bolts and two nuts that retain the oil pan to the engine. Insert the blade of tool SST09032-00100 between the pans. Cut through the sealer and remove the lower oil pan from the engine.

➡Be careful not to damage the contact surface of the oil pans.

7. Remove the two bolts and nuts. Remove the oil strainer. Discard the gasket.
8. Remove the four housing bolts. Remove the flywheel housing undercover.
9. To remove the upper oil pan, remove the seventeen bolts and two nuts. Remove the upper oil pan from the engine, by prying it apart using a suitable tool.

➡Be careful not to damage the sealing surface between the upper oil pan and the cylinder block.

To install:

➡Be sure to use new fasteners, as required.

10. Apply a continuous bead (0.12–0.16 inch in diameter) of seal packing, part number 08826-00080 or equivalent, to the sealing surface of the oil pan.

➡Remove any oil from the contact surface. Install the upper oil pan within three minutes of applying the seal packing. Tighten the pan bolts to specification within fifteen minutes after applying the seal packing. Do not start the engine for at least two hours after the installation of the oil pan.

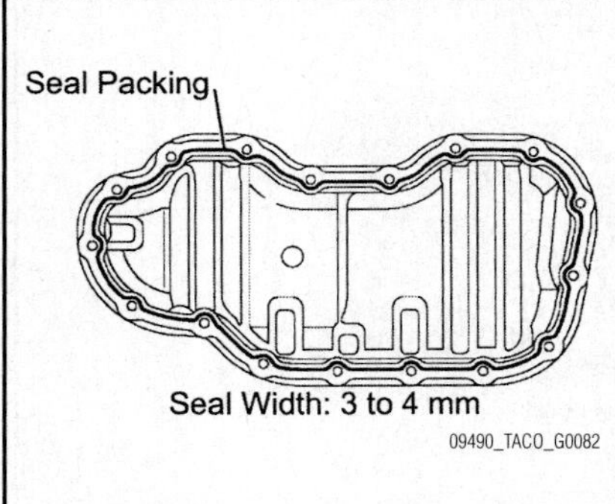

Fig. 227 Lower oil pan sealant application—4.0L engine

11. Loosely install the upper oil pan bolts and nuts. Bolt "A" is 0.98 inch long, bolt "B" is 1.77 inch long and bolt "C" is 0.55 inch long. Uniformly tighten the 14mm bolt to 7.0 ft. lbs, and the other bolts and nuts to 17 ft. lbs., in the proper sequence.
12. Install the oil strainer assembly. Torque the bolts to 80 inch lbs.
13. Apply a continuous bead (0.12–0.16 inch in diameter) of seal packing, part number 08826-00080 or equivalent, to the sealing surface of the oil pan.

➡Remove any oil from the contact surface. Install the lower oil pan within three minutes of applying the seal packing. Tighten the pan bolts to specification within fifteen minutes after applying the seal packing. Do not start the engine for at least two hours after the installation of the oil pan.

14. Loosely install the lower oil pan bolts and nuts. Uniformly tighten the bolts to 80 inch lbs, and the nuts to 7.0 ft. lbs., in several steps.
15. Continue the installation in the reverse order of the removal procedure.
16. Be sure to fill the engine with the proper grade and type engine coolant.
17. Be sure to fill the engine with the proper grade and type engine oil.
18. Start the engine and check for leaks. Correct as required.
19. Use the Techstream diagnostic tool, or equivalent and reprogram the required systems.

4.6L Engine

See Figures 228 through 231.

At this time the manufacturer provided service information for this component by first removing the engine from the vehicle and positioning it in a suitable holding fixture.

1. Before servicing the vehicle, refer to the Precautions Section.

➡If working near and/or around the SRS system and components, be sure to disable the SRS system. Tape the negative battery cable with insulating tape. Always disconnect the negative battery cable first.

✻✻ CAUTION

To avoid personal injury when working on vehicles equipped with an air

bag, the negative battery cable must be disconnected and at least 90 seconds must elapse before working on the system. Failure to do so may result in deployment of the air bag.

2. Disconnect the negative battery cable. Tape the cable with insulating tape.

➡When disconnecting the cable, some systems need to be initialized after the cable is reconnected. You will need the Techstream diagnostic scan tool or equivalent. Follow the directions on the tool.

3. Remove the engine from the vehicle and position it in a suitable holding fixture.
4. Remove the No. 2 pan attaching bolts.
5. Remove the No. 2 pan from the engine.
6. Discard the gasket.
7. Remove the No. 1 pan attaching bolts.
8. Remove the No. 1 pan from the engine.
9. Discard the gasket.

To install:

➡Be sure to use new fasteners, as required.

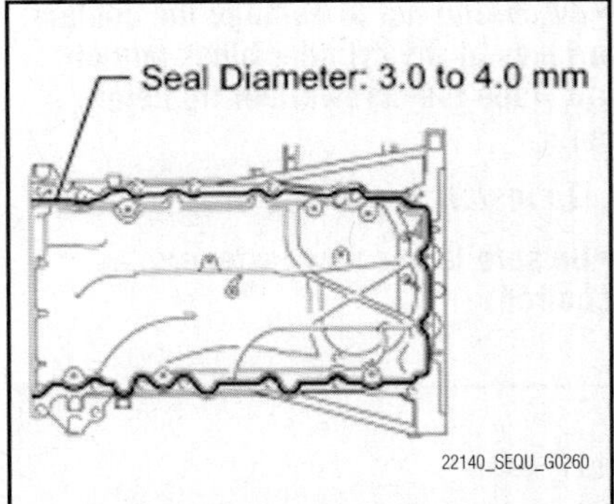

Fig. 228 Oil pan No. 1 sub-assembly packing —4.6L engine

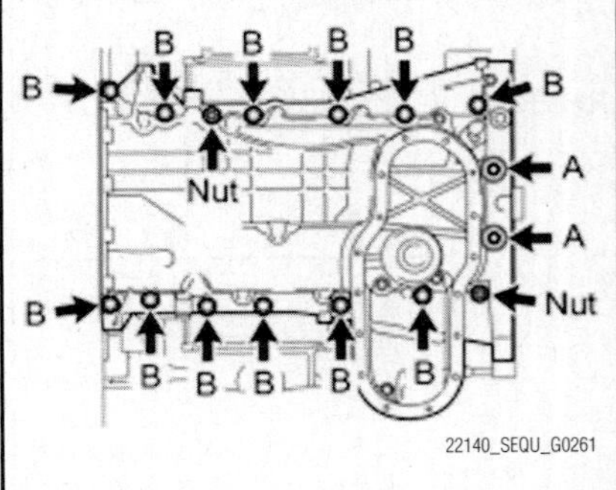

Fig. 229 Oil pan No. 1 sub-assembly tightening sequence—4.6L engine

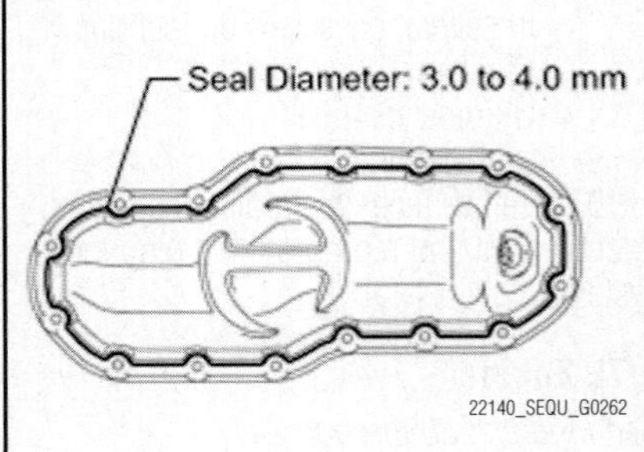

Fig. 230 Oil pan No. 2 sub-assembly packing—4.6L engine

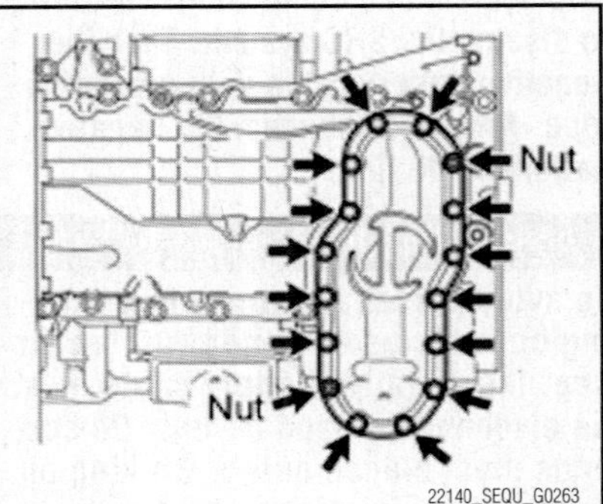

Fig. 231 Oil pan No. 1 sub-assembly tightening sequence—4.6L engine

10. Installation is the reverse of the removal procedure.

➡Remove any oil from the contact surface. Install the oil pan within 3 minutes and tighten the bolts and nuts within 15 minutes after applying seal packing. Do not start the engine for at least 2 hours after installing.

11. Install the oil pan with the 14 bolts and 2 nuts and tighten bolt A to 7 ft. lbs. (10 Nm), bolt B to 26 ft. lbs. (35 Nm) and nut to 26 ft. lbs. (35 Nm).
12. Apply seal packing in a continuous line as shown in the illustration.

➡Remove any oil from the contact surface. Install the oil pan within 3 minutes and tighten the bolts and nuts within 15 minutes after applying seal packing. Do not start the engine for at least 2 hours after installing.

13. Install the oil pan with the 14 bolts and 2 nuts and tighten bolts to 7 ft. lbs. (10 Nm).
14. Use the Techstream diagnostic tool, or equivalent and reprogram the required systems.

4.7L Engine

See Figures 232 through 234.

1. Before servicing the vehicle, refer to the Precautions Section.

➡If working near and/or around the SRS system and components, be sure to disable the SRS system. Tape the negative battery cable with insulating tape. Always disconnect the negative battery cable first.

⁂ CAUTION

To avoid personal injury when working on vehicles equipped with an air bag, the negative battery cable must be disconnected and at least 90 seconds must elapse before working on the system. Failure to do so may result in deployment of the air bag.

2. Disconnect the negative battery cable. Tape the cable with insulating tape.

➡When disconnecting the cable, some systems need to be initialized after the cable is reconnected. You will need the Techstream diagnostic scan tool or equivalent. Follow the directions on the tool.

3. Remove the engine from the vehicle and mount it on a stand.
4. Remove or disconnect the following:
   - Oil dipstick tube
   - Lower oil pan
   - Oil pan baffle
   - Upper oil pan

To install:

➡Be sure to use new fasteners, as required.

5. The upper oil pan bolts are different lengths and are identified as follows:
   - A: 0.79 inch (20mm) w/10mm head
   - B: 0.98 inch (25mm) w/12mm head
   - C: 2.36 inch (60mm) w/12mm head
   - D: 1.38 inch (35mm) w/10mm head
6. Apply silicone sealant to the upper oil pan as shown.
7. Install the upper oil pan and tighten the fasteners in several passes to the following specifications:
   - 10mm: 66 inch lbs. (7.5 Nm)
   - 12mm: 21 ft. lbs. (28 Nm)
8. Install or connect the following:
   - Oil pan baffle. Tighten the fasteners to 66 inch lbs. (7.5 Nm).

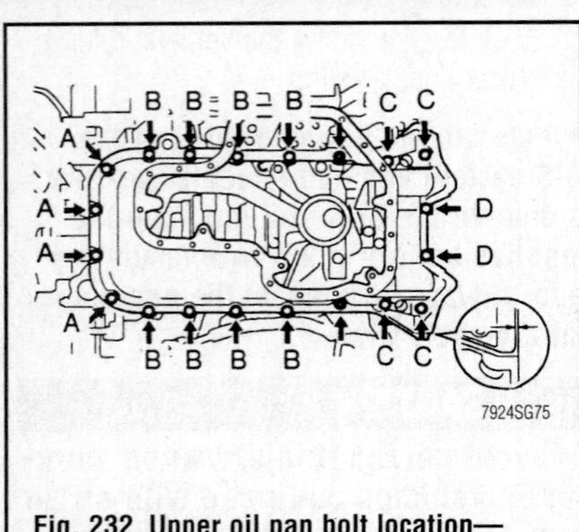

Fig. 232 Upper oil pan bolt location—4.7L engine

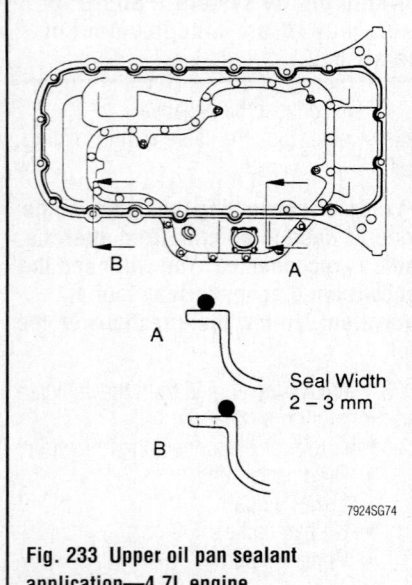

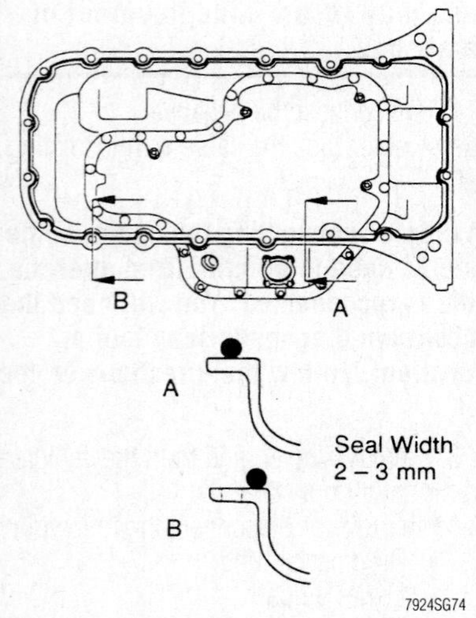

B
A
A
Seal Width
2 – 3 mm
B
7924SG74

Fig. 233 Upper oil pan sealant application—4.7L engine

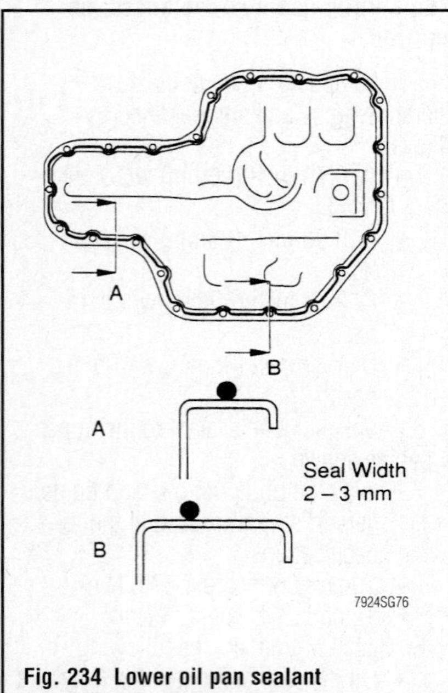

Fig. 234 Lower oil pan sealant application—4.7L engine

- Lower oil pan. Tighten the fasteners in several passes to 66 inch lbs. (7.5 Nm).
- Oil dipstick tube

9. Install the engine.
10. Use the Techstream diagnostic tool, or equivalent and reprogram the required systems.

5.7L Engine

See Figures 235 through 241.

1. Before servicing the vehicle, refer to the Precautions Section.

➡If working near and/or around the SRS system and components, be sure to disable the SRS system. Tape the negative battery cable with insulating tape. Always disconnect the negative battery cable first.

CAUTION

To avoid personal injury when working on vehicles equipped with an air bag, the negative battery cable must be disconnected and at least 90 seconds must elapse before working on the system. Failure to do so may result in deployment of the air bag.

2. Disconnect the negative battery cable. Tape the cable with insulating tape.

➡When disconnecting the cable, some systems need to be initialized after the cable is reconnected. You will need the Techstream diagnostic scan tool or equivalent. Follow the directions on the tool.

3. Disconnect the negative battery cable.
4. Raise and support the vehicle safely.
5. Remove the engine undercover. Drain the engine oil.
6. Remove the necessary components in order to gain access to the lower oil pan retaining bolts.
7. Remove the 14 bolts and 2 nuts.
8. Insert the blade of SST between the oil pans. Cut through the applied sealer and remove the No. 2 oil pan.

➡Be careful not to damage the contact surfaces of the oil pans.

9. Remove the 14 bolts and 2 nuts.

➡Be sure to clean the bolts and stud bolts, and check the threads for cracks or other damage.

10. Remove the oil pan by prying between the oil pan and cylinder block with a screwdriver.

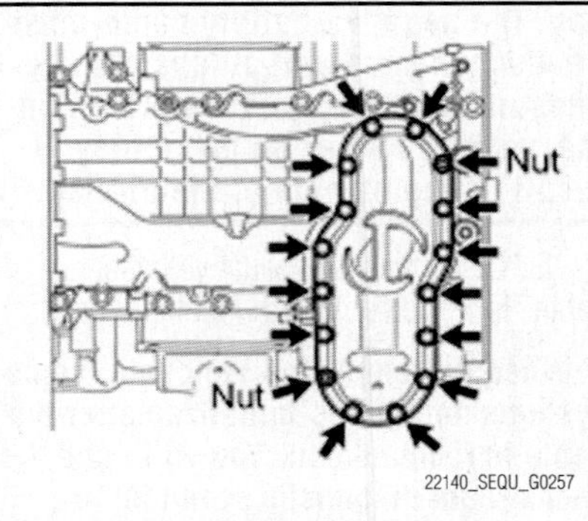

Fig. 235 Oil pan No. 2 sub-assembly—5.7L engine

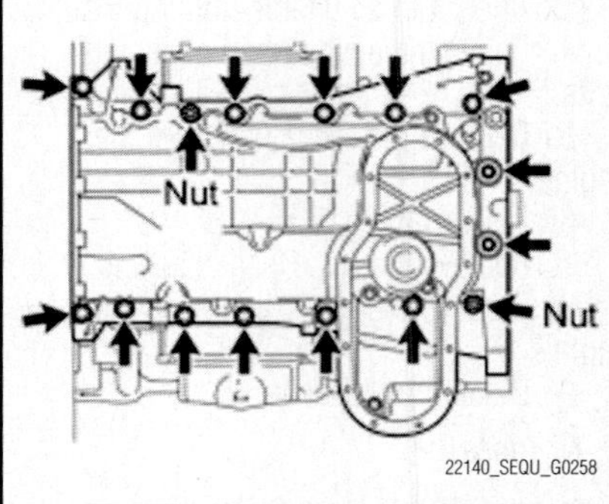

Fig. 236 Oil pan No. 1 sub-assembly—5.7L engine

➡Be careful not to damage the contact surfaces of the cylinder block and oil pan. Tape the screwdriver tip before use.

To install:

➡Be sure to use new fasteners, as required.

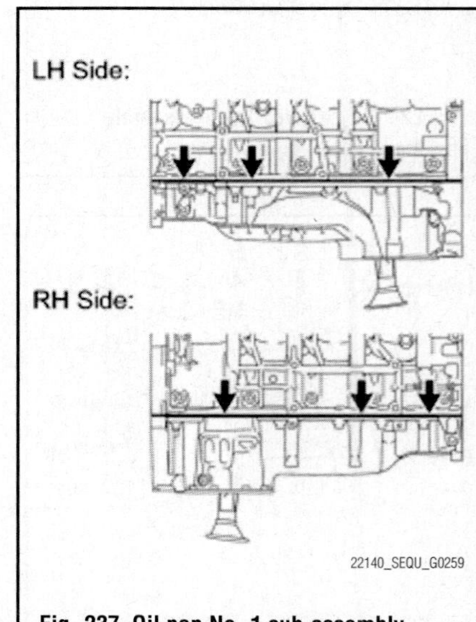

Fig. 237 Oil pan No. 1 sub-assembly removal—5.7L engine

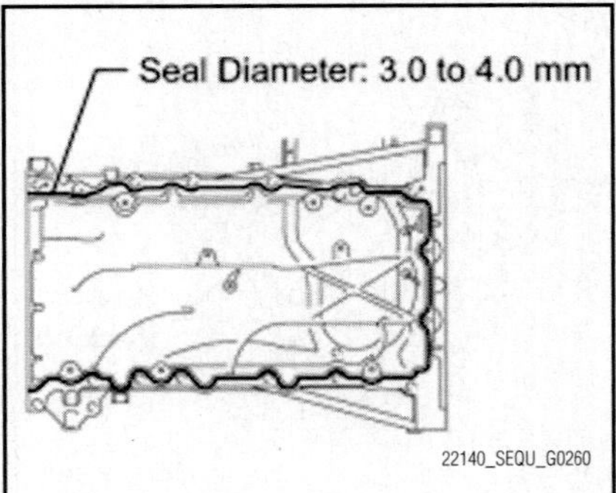

Fig. 238 Oil pan No. 1 sub-assembly packing—5.7L engine

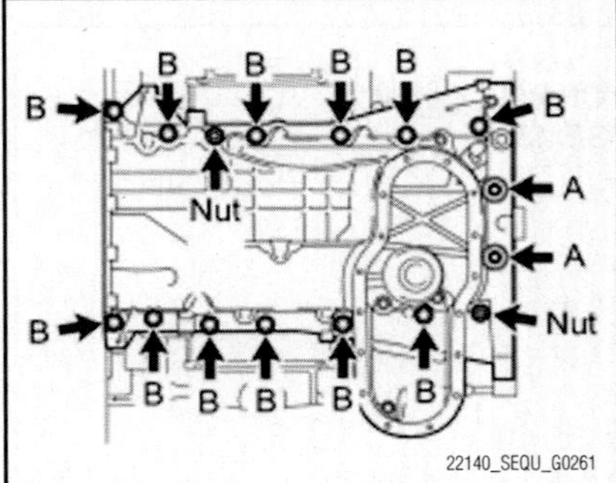

Fig. 239 Oil pan No. 1 sub-assembly tightening sequence—5.7L engine

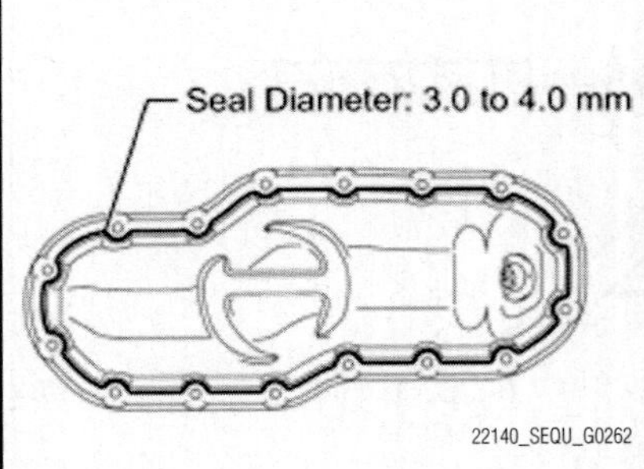

Fig. 240 Oil pan No. 2 sub-assembly packing—5.7L engine

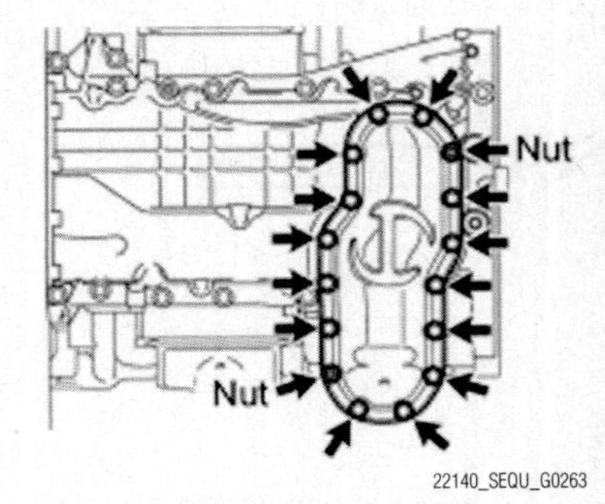

Fig. 241 Oil pan No. 1 sub-assembly tightening sequence—5.7L engine

11. Apply seal packing in a continuous line as shown in the illustration.

➡Remove any oil from the contact surface. Install the oil pan within 3 minutes and tighten the bolts and nuts within 15 minutes after applying seal packing. Do not start the engine for at least 2 hours after installing.

12. Install the oil pan with the 14 bolts and 2 nuts and tighten bolt A to 7 ft. lbs. (10 Nm), bolt B to 26 ft. lbs. (35 Nm) and nut to 26 ft. lbs. (35 Nm).
13. Apply seal packing in a continuous line as shown in the illustration.

➡Remove any oil from the contact surface. Install the oil pan within 3 minutes and tighten the bolts and nuts within 15 minutes after applying seal packing. Do not start the engine for at least 2 hours after installing.

14. Install the oil pan with the 14 bolts and 2 nuts and tighten bolts to 7 ft. lbs. (10 Nm).
15. Continue the installation in the reverse order of the removal procedure.
16. Be sure to fill the engine with the proper grade and type engine coolant.
17. Be sure to fill the engine with the proper grade and type engine oil.
18. Start the engine and check for leaks. Correct as required.
19. Use the Techstream diagnostic tool, or equivalent and reprogram the required systems.

OIL PUMP

REMOVAL & INSTALLATION

4.0L Engine

See Figures 242 through 247.

1. Before servicing the vehicle, refer to the Precautions Section.

➡If working near and/or around the SRS system and components, be sure to disable the SRS system. Tape the negative battery cable with insulating tape. Always disconnect the negative battery cable first.

CAUTION

To avoid personal injury when working on vehicles equipped with an air bag, the negative battery cable must be disconnected and at least 90 seconds must elapse before working on the system. Failure to do so may result in deployment of the air bag.

2. Disconnect the negative battery cable. Tape the cable with insulating tape.

➡When disconnecting the cable, some systems need to be initialized after the cable is reconnected. You will need the Techstream diagnostic scan tool or equivalent. Follow the directions on the tool.

3. Remove the engine and position it in a suitable holding fixture.
4. Remove the ignition coil assembly.
5. Remove the spark plugs.
6. Remove the oil dipstick guide.
7. Remove the No. 2 and No. 1 idler pulley sub assembly.
8. Disconnect the No. 2 oil cooler hose.
9. Remove the water inlet housing.
10. Remove the VVT sensor.
11. Remove the camshaft timing oil control valve assembly.
12. Remove the cylinder head covers.
13. Remove the crankshaft pulley.
14. Remove the No. 2 oil pan sub assembly retaining bolts. Remove the component. Discard the gasket.
15. Remove the oil strainer sub assembly.
16. Remove the No 1. oil pan sub assembly retaining bolts. Remove the component. Discard the gasket. Remove the O-ring from the cylinder block. Discard it.
17. Remove the oil filter bracket sub assembly.
18. Remove the timing cover retaining bolts.
19. Remove the timing cover. Discard the gasket.

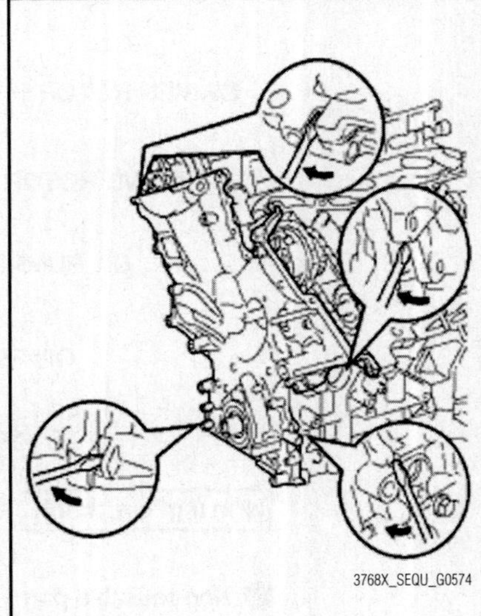

Fig. 242 Front cover removal—4.0L engine

➡When removing the cover carefully pry between the cover and the cylinder head or cylinder block with a suitable pry tool.

20. Remove the O-ring from the cylinder head, left side.
21. Remove and discard the oil seal.
22. Remove the oil pump outlet pipe. Discard the O-rings.
23. Remove the oil pump cover retaining bolts. Remove the cover. Discard the gasket.

To install:

➡Be sure to use new fasteners, as required.

24. Coat the oil pump relief valve with clean engine oil. Install the plug, using a new gasket and tighten to 36 ft. lbs.
25. Coat the oil pump gears with clean engine oil. Position the gears in the timing chain case cover with the identification marks facing oil pump cover side. Install the cover. Alternately tighten the bolts to 80 inch lbs. Install the oil pipe, tighten the bolts to 80 inch lbs.
26. Install a new front case oil seal. Install a new O-ring onto the left cylinder head.
27. To install the front cover, apply seal packing in a continuous bead in a diameter of 0.118–0.157 inch. See illustration.

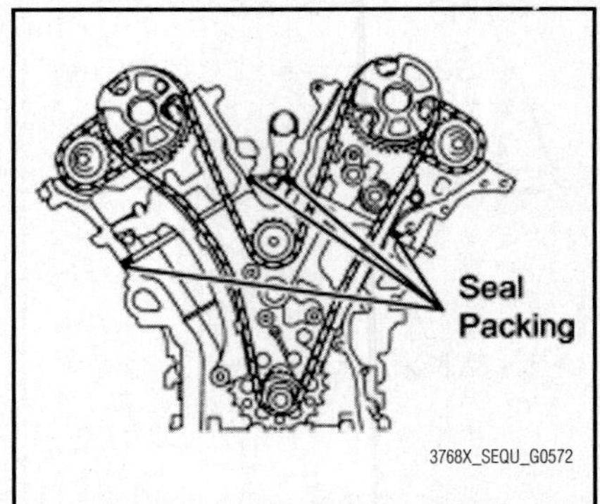

Fig. 244 Front cover sealant application point 1 of 2—4.0L engine

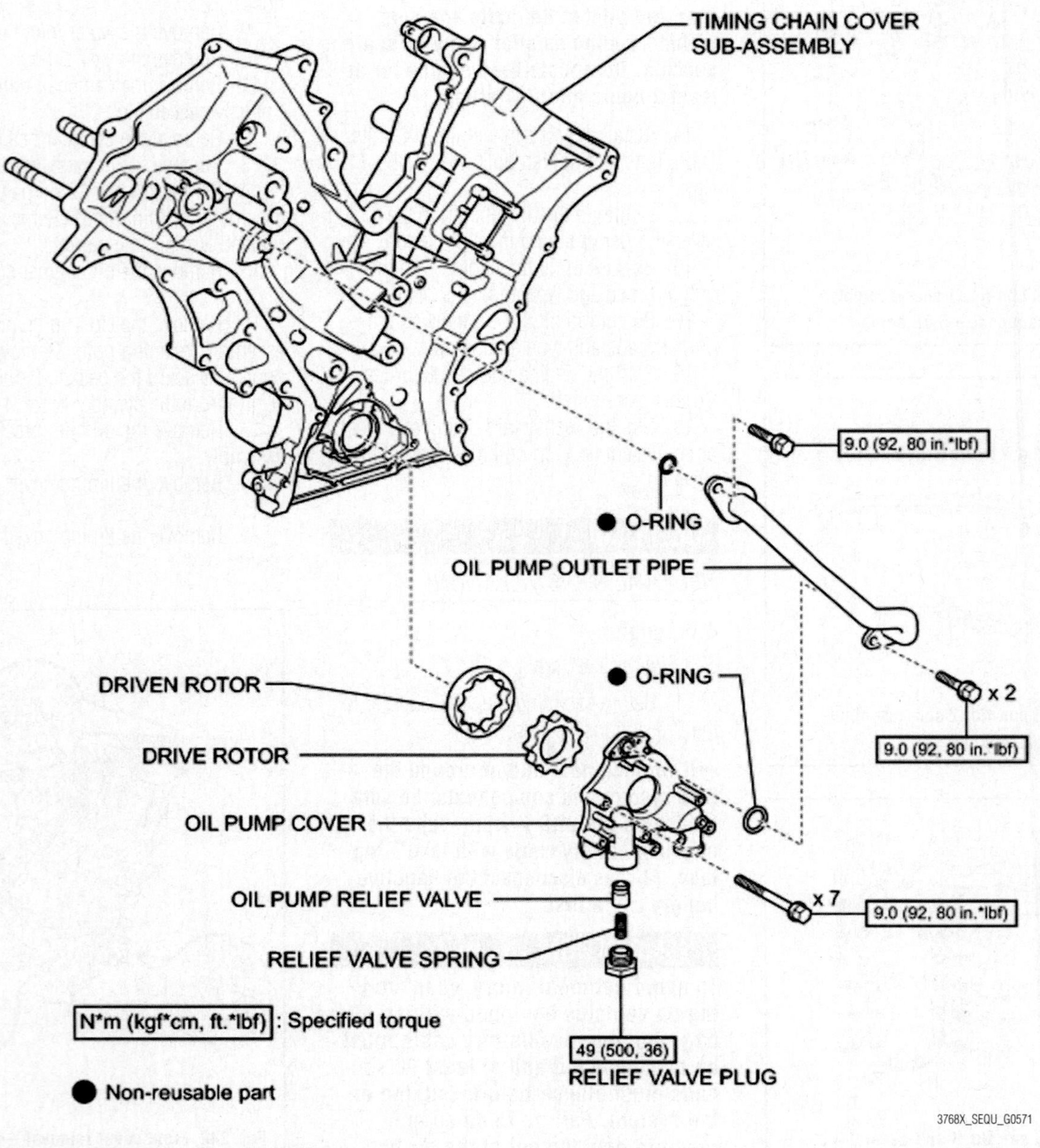

TIMING CHAIN COVER SUB-ASSEMBLY
9.0 (92, 80 in.*lbf)
● O-RING
OIL PUMP OUTLET PIPE
DRIVEN ROTOR
● O-RING
x 2
9.0 (92, 80 in.*lbf)
DRIVE ROTOR
OIL PUMP COVER
OIL PUMP RELIEF VALVE
x 7
9.0 (92, 80 in.*lbf)
RELIEF VALVE SPRING
N*m (kgf*cm, ft.*lbf) : Specified torque
49 (500, 36)
RELIEF VALVE PLUG
● Non-reusable part
3768X_SEQU_G0571

Fig. 243 Oil pump and related components—4.0L engine

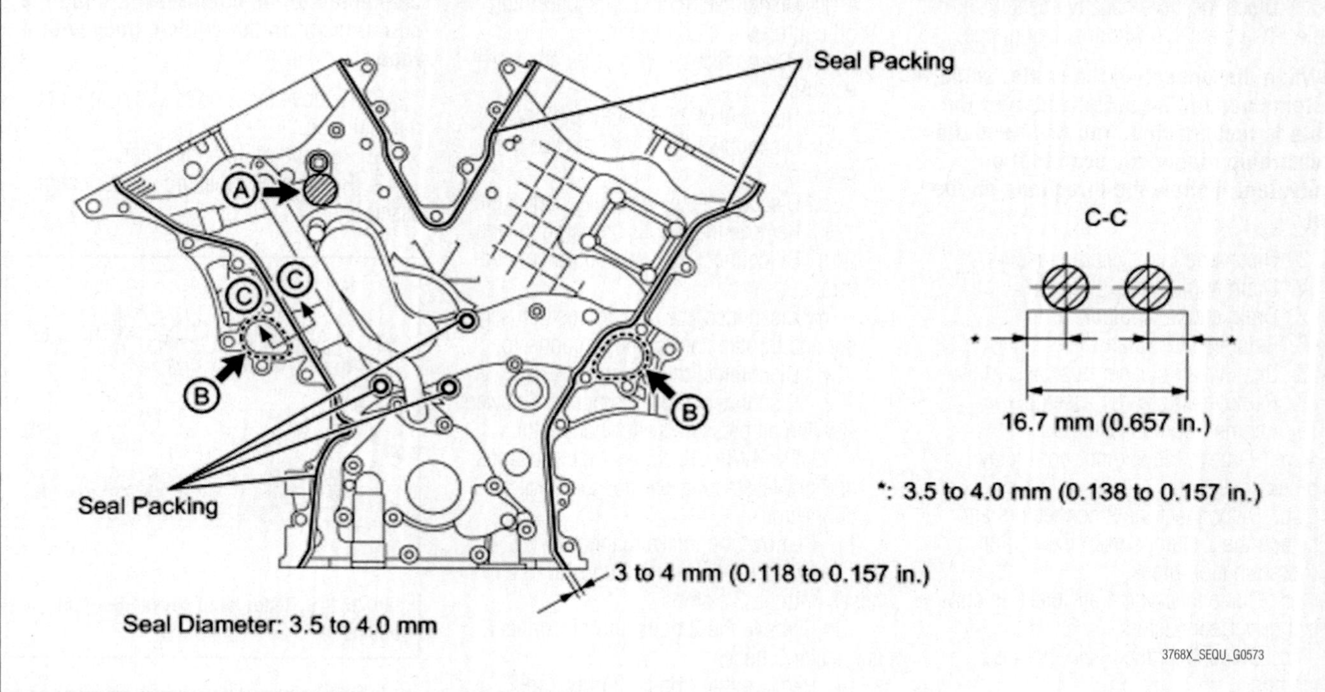

Fig. 245 Front cover sealant application point 2 of 2—4.0L engine

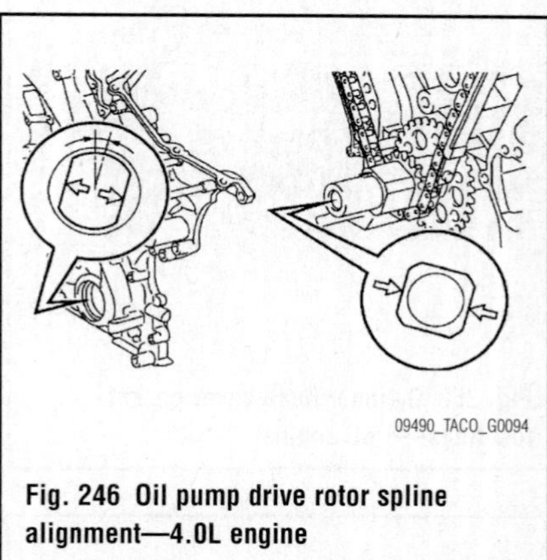

Fig. 246 Oil pump drive rotor spline alignment—4.0L engine

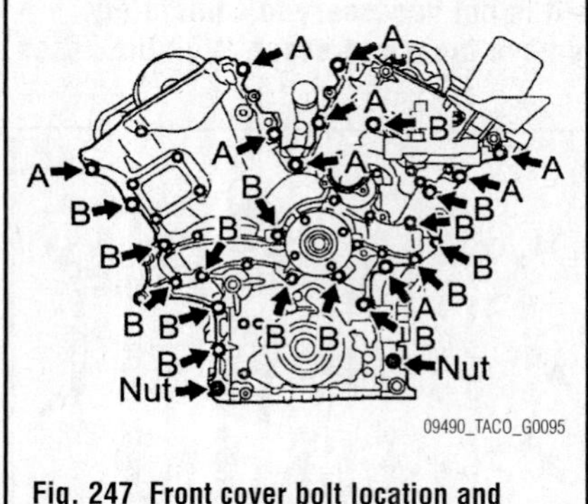

Fig. 247 Front cover bolt location and torque sequence—4.0L engine

28. Apply seal packing in a continuous bead in a diameter of 0.118–0.157 inch to the timing cover. Do not apply sealant to area A. See illustration.

➡Install the component within three minutes after applying the sealant. Tighten the bolts within fifteen minutes after applying the sealant. Do not start the engine for at least two hours after applying sealant.

29. Align the key way of the oil pump drive motor with the rectangular portion of the crankshaft timing gear and slide the timing chain case cover into place.
30. Install the timing chain case cover bolts. Tighten the bolts and nuts uniformly in several steps to 17 ft. lbs.
31. Continue the installation in the reverse order of the removal procedure.
32. When installing the cylinder head cover apply a continuous bead (0.08–0.12 inch) of seal packing, part number 08826-00080 or equivalent, to the cylinder head. Install the seal washers onto the bolts. Install the cylinder head cover bolts and nuts. Tighten bolts "A" to 7.4 ft. lbs. Tighten bolts "B" to 80 inch lbs. Tighten nuts to 80 inch lbs.

➡Be sure to remove any oil from the contact surfaces of the cylinder head cover and the cylinder head. Install the cover within three minutes after applying the seal packing. Tighten the bolts to specification within fifteen minutes after installing the cover. Do not add engine oil for at least two hours after installing the cover.

33. Be sure to fill the engine with the proper grade and type engine coolant.
34. Be sure to fill the engine with the proper grade and type engine oil.
35. Start the engine and check for leaks, correct as required.
36. Use the Techstream diagnostic tool, or equivalent and reprogram the required systems.

4.6L Engine

See Figures 248 through 264.

1. Before servicing the vehicle, refer to the Precautions Section.

➡If working near and/or around the SRS system and components, be sure to disable the SRS system. Tape the negative battery cable with insulating tape. Always disconnect the negative battery cable first.

✱✱ CAUTION

To avoid personal injury when working on vehicles equipped with an air bag, the negative battery cable must be disconnected and at least 90 seconds must elapse before working on the system. Failure to do so may result in deployment of the air bag.

2. Disconnect the negative battery cable. Tape the cable with insulating tape.

➡When disconnecting the cable, some systems need to be initialized after the cable is reconnected. You will need the Techstream diagnostic scan tool or equivalent. Follow the directions on the tool.

3. Discharge fuel system pressure.
4. Drain engine oil.
5. Drain engine coolant.
6. Remove the radiator.
7. Remove air cleaner hose assembly.
8. Remove air cleaner assembly.
9. Engine Room LH Side:
   a. Remove the engine room relay block cover.
   b. Disconnect the 2 connectors and detach the 2 clamps from the engine room junction block.
   c. Disconnect the 4 air injection control driver connectors.
   d. Disconnect the 2 wire harness clamps.
   e. Disconnect the injector connector.
   f. Disconnect the 4 ignition coil connectors.
   g. Disconnect the 2 VVT sensor connectors.
   h. Disconnect the 4 clamps.
   i. Remove the 2 bolts and ground wire.
   j. Disconnect the noise filter connector.
   k. Disconnect the engine coolant temperature sensor connector.
   l. Disconnect the 2 camshaft timing oil control valve connectors.
   m. Disconnect the camshaft position sensor connector.
   n. Disconnect the 3 clamps.
   o. Disconnect the cooler compressor connector.
10. Engine Room RH Side.
   a. Disconnect the 2 camshaft timing oil control valve connectors.
   b. Disconnect the 4 ignition coil connectors.
   c. Disconnect the injector connector.
   d. Disconnect the 2 VVT sensor connectors.
   e. Disconnect the noise filter connector.
   f. Remove the 2 bolts and ground wire.
   g. Disconnect the 2 air pump connectors.
   h. Disconnect the throttle position sensor and throttle control motor connector.
   i. Disconnect the 5 clamps.
   j. Disconnect the 2 clamps and power steering oil pressure switch connector.
   k. For 4WD: Disconnect the 2 clamps and power steering oil pressure switch connector.
11. Remove the intake manifold.
12. Disconnect the No. 2 and No. 3 air hoses and the 2 clamps.
13. Remove the 2 bolts and disconnect the heater 3 hoses.
14. Remove the 2 bolts, 2 nuts and 2 stud bolts, and disconnect the cooler compressor.

➡It is not necessary to completely remove the compressor. With the hoses connected to the compressor, hang the compressor on the vehicle body with a rope.

15. Remove the 2 bolts and disconnect the fuel tube.
16. Remove oil filter element.
17. Remove oil pressure sender gauge assembly.

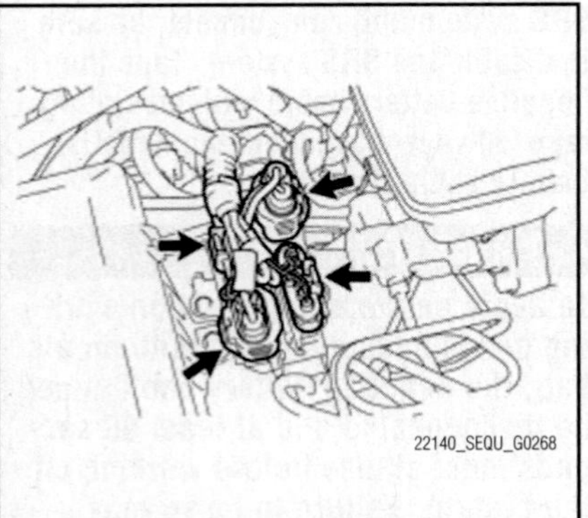
22140_SEQU_G0268

Fig. 248 Air injection control connectors—4.6L engine

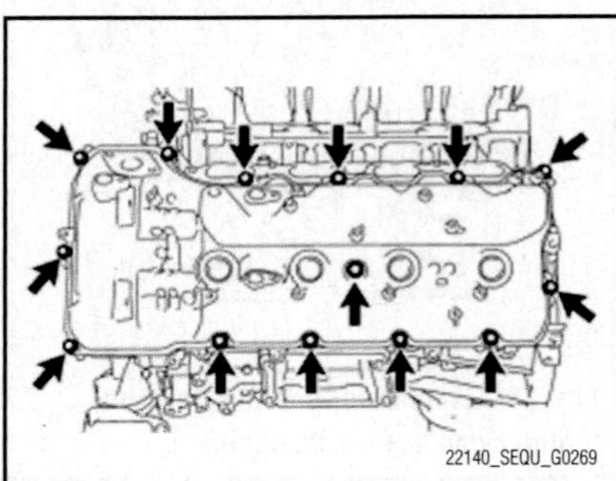
22140_SEQU_G0269

Fig. 249 Cylinder head cover LH—4.6L engine

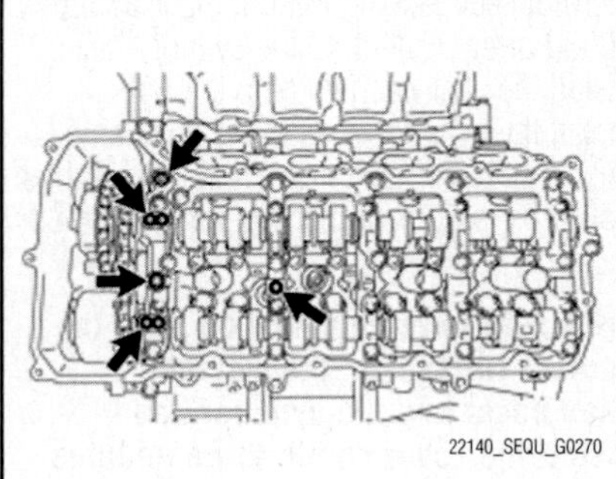
22140_SEQU_G0270

Fig. 250 Cylinder head cover gasket locations—4.6L engine

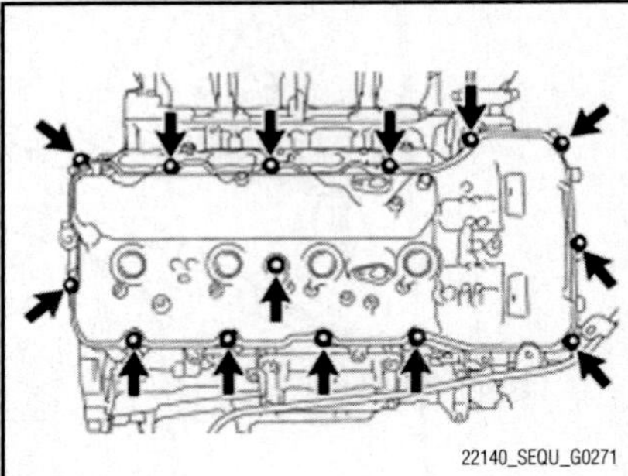
22140_SEQU_G0271

Fig. 251 Cylinder head cover RH—4.6L engine

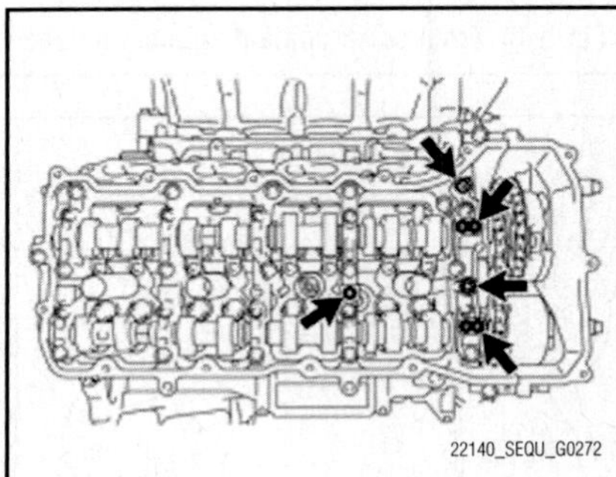
22140_SEQU_G0272

Fig. 252 Cylinder head cover gasket locations—4.6L engine

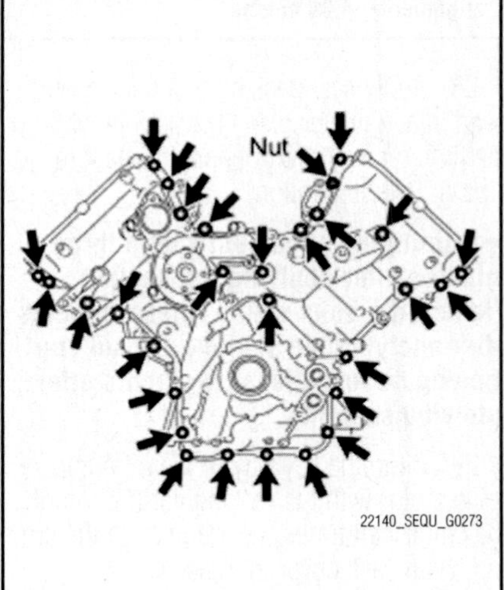

22140_SEQU_G0273

Fig. 253 Timing chain cover bolt locations—4.6L engine

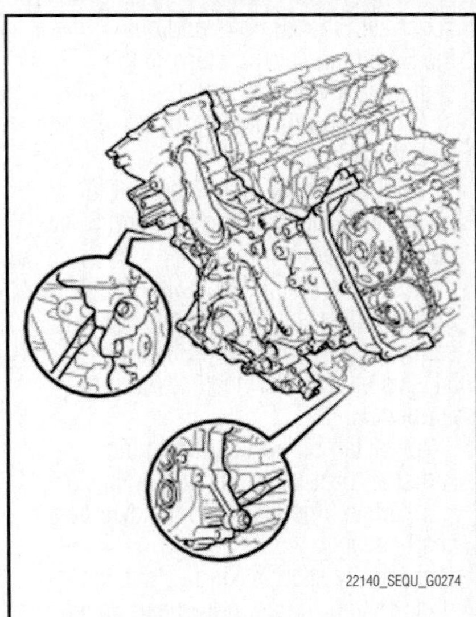

Fig. 254 Timing chain cover pry locations—4.6L engine

18. Disconnect the 4 hoses and remove the water by-pass pipe.
19. Remove No. 1 oil cooler bracket (w/ oil cooler).
20. Remove the 2 bolts, 2 nuts and oil filter bracket.
21. Remove the 2 O-rings.
22. Remove engine oil level dipstick guide.
23. Remove the 2 bolts and disconnect the vane pump.
24. Disconnect the alternator connector.
25. Remove the terminal cap and nut, and disconnect the alternator wire.
26. Remove the bolt and disconnect the wire harness bracket from the alternator.
27. Remove the 2 bolts, 2 nuts and alternator.
28. Remove the 2 stud bolts.
29. Remove the No. 1 water by-pass hose by disconnect the hose from the water inlet housing and front water by-pass joint.
30. Remove the bolt and disconnect the air tube.
31. Remove the 2 bolts and hoses and water by-pass pipe.
32. Disconnect the No. 2 water by-pass hose from the water by-pass joint.
33. Remove the 4 nuts, water by-pass joint and 2 gaskets.
34. Remove No.1 and No. 2 engine covers.
35. Remove the bolt, disconnect the 2 hoses, and remove the air tube.
36. Remove the 3 bolts, water inlet housing and gasket.
37. Remove the 4 bolts and water pump pulley.
38. Remove the bolt and idler pulley.
39. Remove the 4 bolts and fan bracket.
40. Remove the standard bolt, 6 mm hexagon wrench bolt and belt tensioner.
41. Remove the 8 bolts and 8 ignition coils.

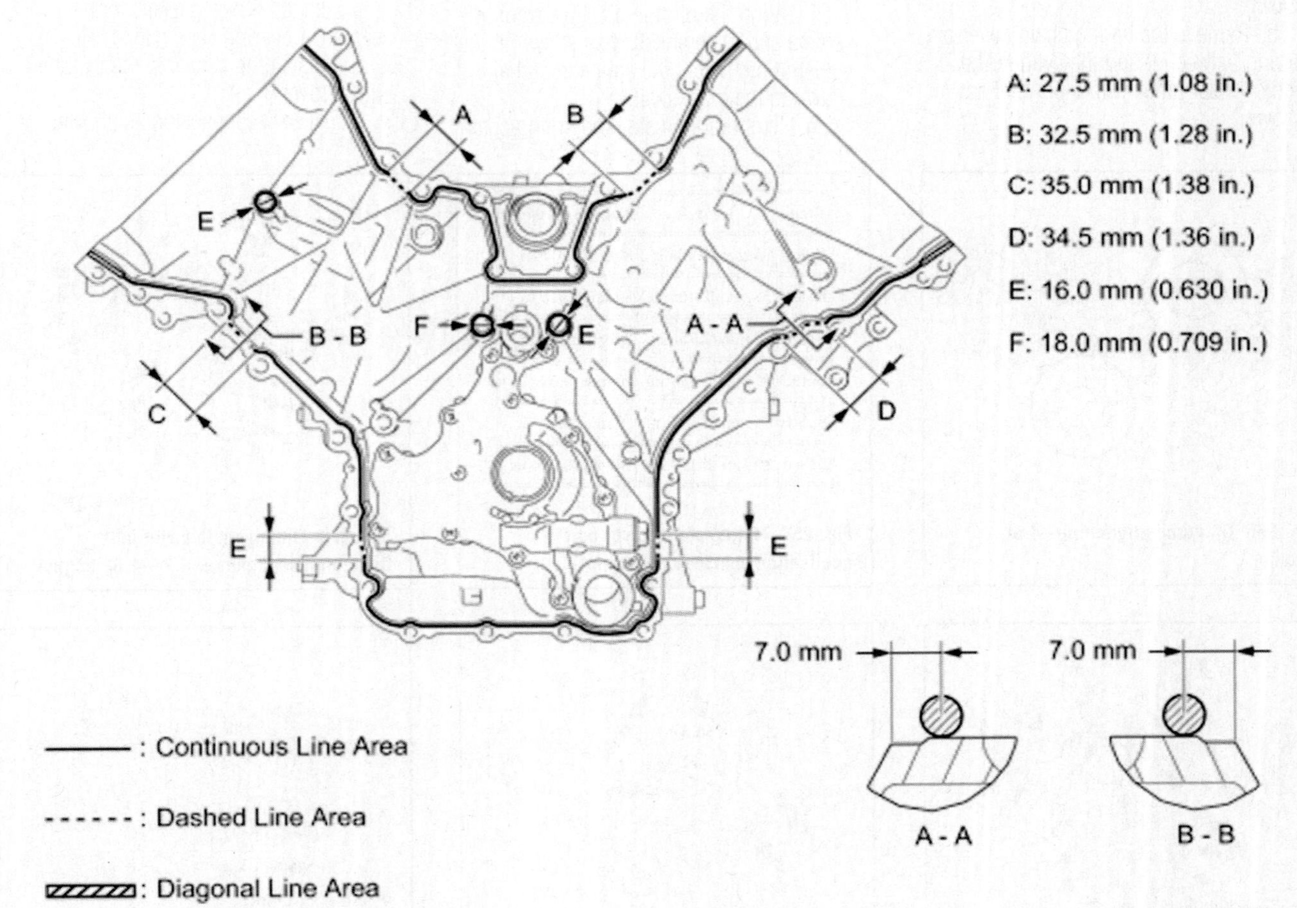

A: 27.5 mm (1.08 in.)
B: 32.5 mm (1.28 in.)
C: 35.0 mm (1.38 in.)
D: 34.5 mm (1.36 in.)
E: 16.0 mm (0.630 in.)
F: 18.0 mm (0.709 in.)

7.0 mm A - A
7.0 mm B - B

———— : Continuous Line Area
- - - - - - : Dashed Line Area
▨▨▨ : Diagonal Line Area

22140_SEQU_G0275

Fig. 255 Timing chain cover sealant locations—4.6L engine

42. Remove cylinder head cover sub-assembly LH.
   a. Remove the 14 bolts, seal washer, cylinder head cover and gasket.
   b. Remove the 5 gaskets from the camshaft bearing caps (No. 2, No. 3).
43. Remove cylinder head cover sub-assembly RH.
   a. Remove the bolt and noise filter.
   b. Remove the 14 bolts, seal washer, cylinder head cover and gasket.
   c. Remove the 5 gaskets from the camshaft bearing caps (No. 1, No. 3).
44. Remove crankshaft pulley.
   a. Loosen the crankshaft pulley set bolt.
   b. Partially install the pulley set bolt to the crankshaft until 2 or 3 threads are engaged.
   c. Remove the crankshaft pulley.
45. Remove the bolt and disconnect the wire harness bracket.
46. Remove timing chain cover sub-assembly.
   a. Remove the 28 bolts and nut shown.
   b. Remove the timing chain cover by prying between it and the cylinder head and cylinder block with a pry tool as shown.
47. Remove the oil pump gasket from the cylinder block.
48. Remove the O-ring from the oil pan.
49. Remove the water inlet pipe.
50. Remove the 2 O-rings from the water inlet pipe.

To install:

➡Be sure to use new fasteners, as required.

51. Apply soapy water to 2 new O-rings and install them to the inlet pipe.
52. Install the inlet pipe to the No. 1 heat exchanger cover.
53. Install timing chain cover sub-assembly.
   a. Apply a light coat of engine oil to a new oil pump gasket.
   b. Install the oil pump gasket.
   c. Apply a light coat of engine oil to a new O-ring.
   d. Install the O-ring.
   e. Apply seal packing in a continuous line to the timing chain cover as shown.
   f. Align the oil pump's drive rotor spline and the crankshaft as shown in the illustration. Install the spline and chain cover to the crankshaft.
   g. Temporarily install the timing chain cover with the 28 bolts and nut. Tighten the 5 bolts in several steps in the sequence shown to 35 ft. lbs. (47 Nm).
   h. Temporarily install the fan bracket with the 4 bolts.
   i. Temporarily install the belt tensioner with the standard bolt and 6 mm hexagon wrench bolt.
   j. Tighten the 23 bolts labeled 12 to 35 and nut in several steps in the sequence to 17 ft. lbs. (23 Nm).
54. Install cylinder head cover sub-assembly LH.
   a. Install 5 new gaskets to the camshaft bearing caps (No. 2, No. 3).
   b. Install the gasket to the cylinder head cover.
   c. Apply seal packing as shown.
   d. Install the cylinder head cover washer with a new seal and the 14 bolts and tighten bolt A to 15 ft. lbs. (21 Nm) and remaining bolts to 9 ft. lbs. (12 Nm).
55. Install cylinder head cover sub-assembly RH.
   a. Install 5 new gaskets to the camshaft bearing caps (No. 1, No. 3).
   b. Install the gasket to the cylinder head cover.
   c. Apply seal packing as shown.

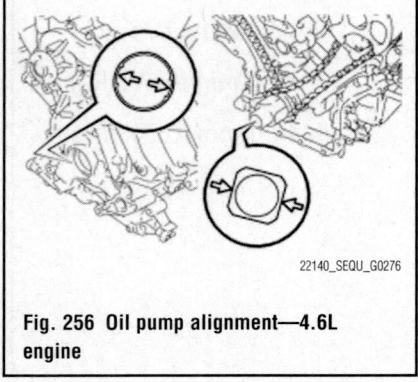

22140_SEQU_G0276

Fig. 256 Oil pump alignment—4.6L engine

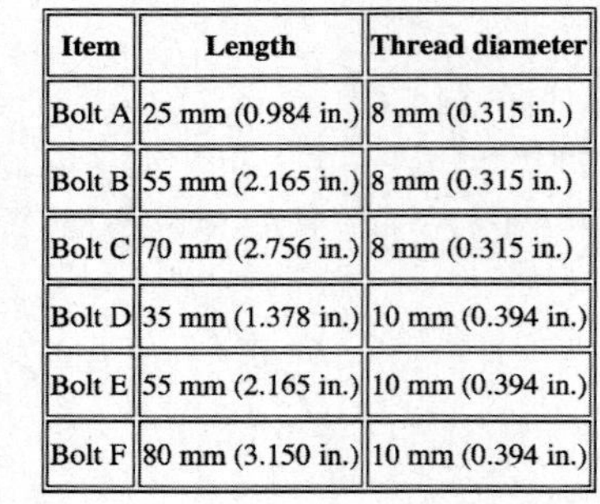

Item	Length	Thread diameter
Bolt A	25 mm (0.984 in.)	8 mm (0.315 in.)
Bolt B	55 mm (2.165 in.)	8 mm (0.315 in.)
Bolt C	70 mm (2.756 in.)	8 mm (0.315 in.)
Bolt D	35 mm (1.378 in.)	10 mm (0.394 in.)
Bolt E	55 mm (2.165 in.)	10 mm (0.394 in.)
Bolt F	80 mm (3.150 in.)	10 mm (0.394 in.)

22140_SEQU_G0278

Fig. 258 Timing chain cover bolt application chart—4.6L engine

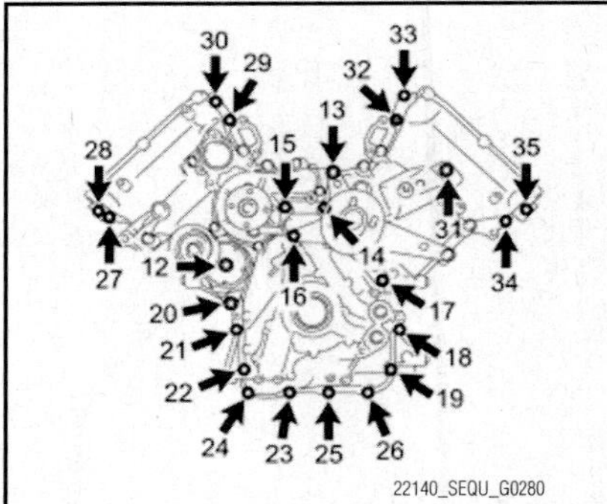

Fig. 260 Timing chain cover bolt tightening sequence No. 2—4.6L engine

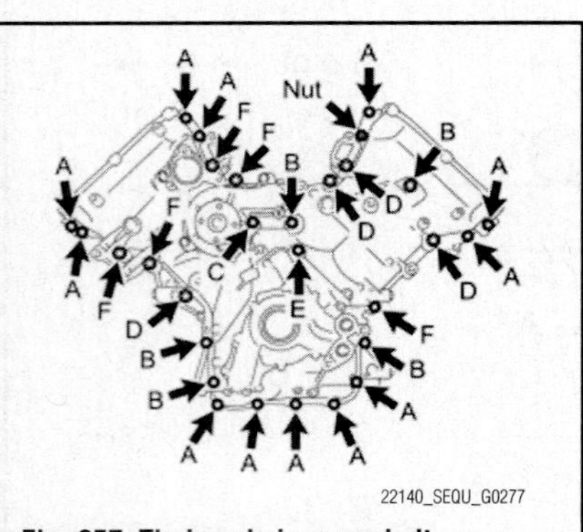

Fig. 257 Timing chain cover bolt locations—4.6L engine

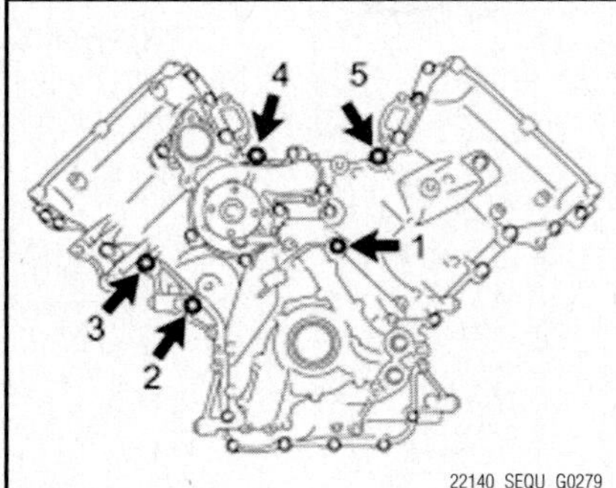

Fig. 259 Timing chain cover bolt tightening sequence No. 1—4.6L engine

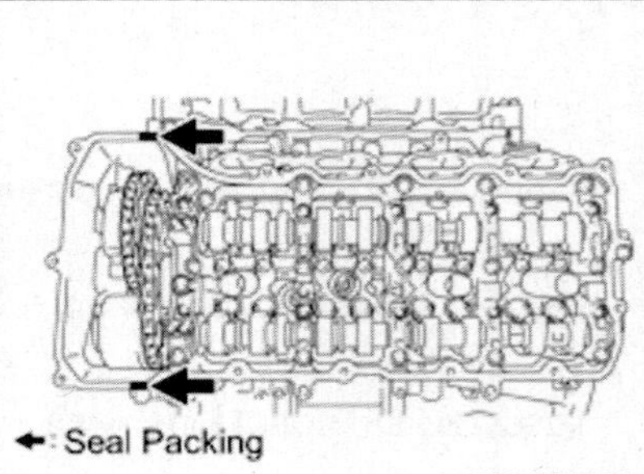

Fig. 261 Cylinder head cover packing location LH—4.6L engine

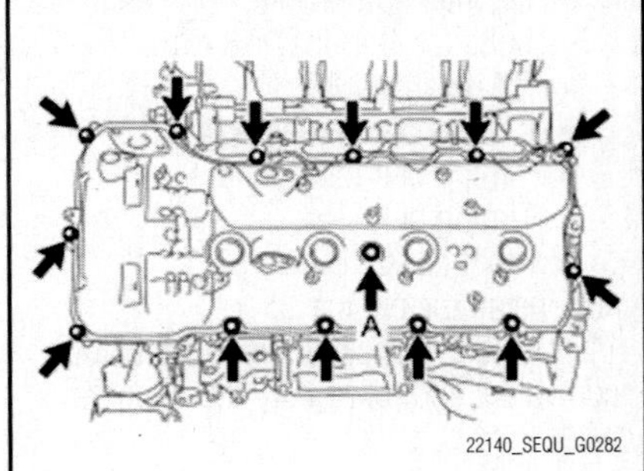

Fig. 262 Cylinder head cover bolt tightening sequence LH—4.6L engine

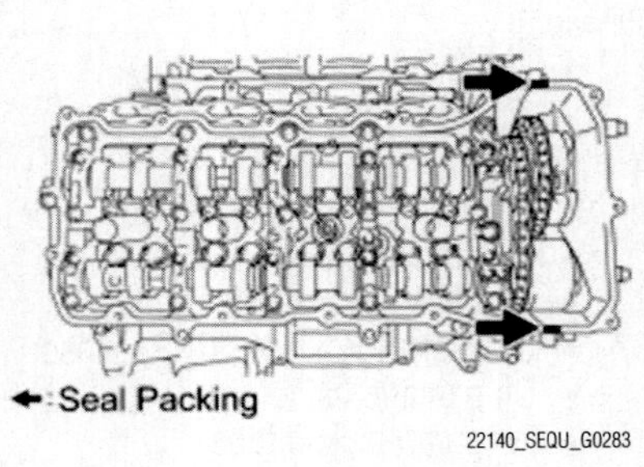

Fig. 263 Cylinder head cover packing location RH—4.6L engine

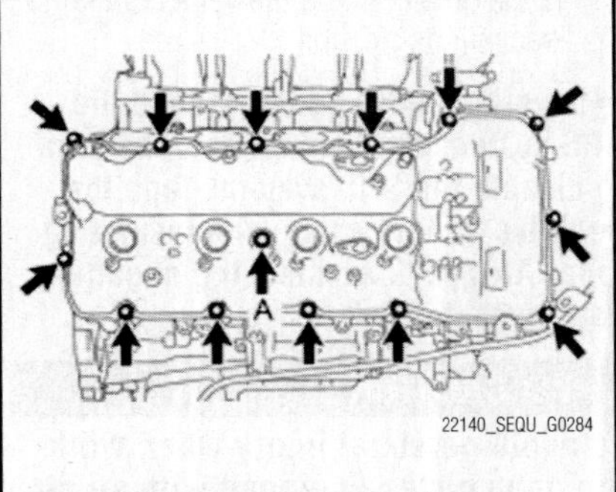

Fig. 264 Cylinder head cover bolt tightening sequence RH—4.6L engine

d. Install the cylinder head cover washer with a new seal and the 14 bolts and tighten bolt A to 15 ft. lbs. (21 Nm) and remaining bolts to 9 ft. lbs. (12 Nm).

e. Install the noise filter to the cylinder head cover with the bolt and tighten to 62 inch lbs. (7.0 Nm).

56. Install the 8 ignition coils with the 8 bolts and tighten to 7 ft. lbs. (10 Nm).

57. Align the pulley set key with the key groove of the pulley, and slide on the pulley and tighten to 221 ft. Lbs. (300 Nm).

58. Connect the bracket to the timing chain cover with the bolt.

59. Install the idler pulley with the bolt and tighten to 32 ft. lbs. (43 Nm).

60. Temporarily install the water pump pulley with the 4 bolts and tighten to 15 ft. lbs. (21 Nm).

61. Install a new water cover gasket to the timing chain cover.

62. Install the water inlet with the 3 bolts and tighten to 15 ft. lbs. (21 Nm).

63. Connect the 2 air tube hoses.

64. Connect the 2 hoses.

65. Install No. 1 and No 2 engine covers.

66. Install 2 new gaskets and the water by-pass joint with the 4 nuts and tighten to 15 ft. lbs. (21 Nm).

67. Connect the No. 2 water by-pass hose to the water by-pass joint.

68. Install the water by-pass pipe.

69. Connect the air tube.

70. Install the No. 1 water by-pass hose by connecting the hose to the water inlet housing and front water by-pass joint.

71. Install the 2 stud bolts.

72. Install the alternator with the 2 bolts and 2 nuts and tighten to 32 ft. lbs. (43 Nm).

73. Connect the alternator wire with the nut and tighten to 87 inch lbs. (9.8 Nm).

74. Connect the wire harness bracket to the alternator.

75. Connect the vane pump to the timing chain cover with the 2 bolts and tighten to 21 ft. lbs. (28 Nm).

76. Install engine oil level dipstick guide.

77. Install oil filter bracket (w/o Oil Cooler).

a. Apply a light coat of engine oil to 2 new O-rings.

b. Install the 2 O-rings to the timing chain cover.

c. Install the oil filter bracket with the 2 bolts and 2 nuts and tighten to 26 ft. lbs. (35 Nm).

78. Install No. 1 oil cooler bracket.

79. Install No. 2 water by-pass pipe.

80. Install oil pressure sender gauge assembly.

81. Install oil filter element.

82. Connect No. 2 fuel tube sub-assembly.

83. Install the cooler compressor with the 2 stud bolts and tighten to 7 ft. lbs. (10 Nm).

84. Install the 2 bolts and 2 nuts and tighten to 18 ft. lbs. (235 Nm).

85. Connect the 3 heater hoses.

86. Connect the No. 2 and No. 3 air hoses.

87. Connect the 2 clamps and air pump wire harness.

88. Install the intake manifold.

89. Engine Room RH Side:

a. Connect the 5 clamps.

b. Connect the throttle position sensor and throttle control motor connector.

c. Connect the 2 air pump connectors.

d. Install the ground wire with the 2 bolts.

e. Connect the noise filter connector.

f. Connect the 2 VVT sensor connectors.

g. Connect the injector connector.

h. Connect the 4 ignition coil connectors.

i. Connect the 2 camshaft timing oil control valve connectors.

j. Connect the 2 clamps and power steering oil pressure switch connector.

k. For 4WD: Connect the clamp and junction connector.

90. Engine Room LH Side:

a. Connect the cooler compressor connector.

b. Connect the 3 clamps.

c. Connect the camshaft position sensor connector.

d. Connect the 2 camshaft timing oil control valve connectors.

e. Connect the engine coolant temperature sensor connector.

f. Connect the 4 clamps.

g. Install the ground wire with the 2 bolts.

h. Connect the noise filter connector.

i. Connect the 2 VVT sensor connectors.

j. Connect the 4 ignition coil connectors.

k. Connect the injector connector.

l. Connect the 2 wire harness clamps.

m. Connect the 4 air injection control driver connectors.

n. Connect the 2 connectors and attach the 2 clamps to the engine room junction block.

o. Install the engine room relay block cover.

91. Install air cleaner assembly.

92. Install air cleaner hose assembly.

93. Install the radiator.

94. Add engine oil.

95. Add engine coolant.

96. Connect cable to negative battery terminal.

97. Inspect for leaks.

98. Check engine oil level.

99. Use the Techstream diagnostic tool, or equivalent and reprogram the required systems.

4.7L Engine

See Figures 265 through 267.

1. Before servicing the vehicle, refer to the Precautions Section.

➡If working near and/or around the SRS system and components, be sure to disable the SRS system. Tape the negative battery cable with insulating tape. Always disconnect the negative battery cable first.

CAUTION

To avoid personal injury when working on vehicles equipped with an air bag, the negative battery cable must be disconnected and at least 90 seconds must elapse before working on the system. Failure to do so may result in deployment of the air bag.

2. Disconnect the negative battery cable. Tape the cable with insulating tape.

➡When disconnecting the cable, some systems need to be initialized after the cable is reconnected. You will need the Techstream diagnostic scan tool or equivalent. Follow the directions on the tool.

3. Remove the engine from the vehicle and mount it on a stand.

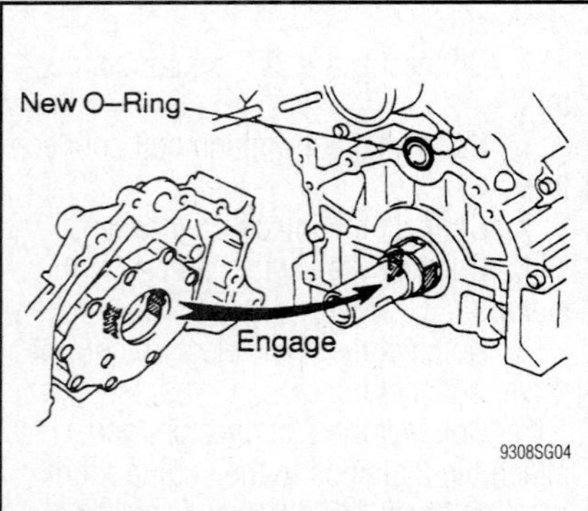

Fig. 265 Location of the O-ring seal—4.7L engine

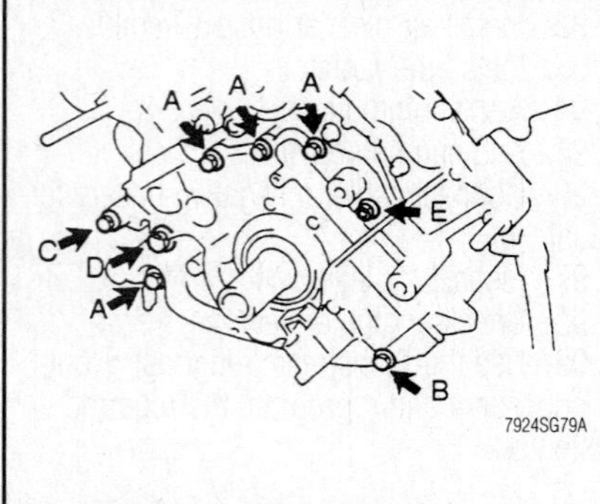

Fig. 266 Oil pump bolt location—4.7L engine

4. Remove or disconnect the following:
   - Front cover
   - Timing belt.
   - Timing belt idler pulleys
   - Crankshaft timing sprocket
   - Oil dipstick tube
   - Oil filter and bracket
   - Crankshaft Position (CKP) sensor
   - Oil pan and baffle
   - Oil pump pickup tube
   - Oil pump

To install:

5. The upper oil pan bolts are different lengths and are identified as follows:
   - A: 1.38 inch (35mm) w/12mm head
   - B: 1.97 inch (50mm) w/12mm head
   - C: 4.17 inch (106mm) w/12mm head
   - D: 1.57 inch (40mm) w/14mm head
   - E: 1.18 inch (30mm) w/6mm hex head
6. Install a new O-ring on the engine block.
7. Apply silicone sealant to the oil pump housing as shown.
8. Install the oil pump. Tighten the bolts in several passes to the following specifications:
   - 12mm: 11 ft. lbs. (15.5 Nm)
   - 14mm: 22 ft. lbs. (30.5 Nm)
   - 6mm Hex: 11 ft. lbs. (15.5 Nm)
9. Install or connect the following:
   - Oil pump pickup tube. Tighten the bolts to 66 inch lbs. (7.5 Nm).
   - Oil pan and baffle
   - CKP sensor
   - Oil filter and bracket. Tighten the bolts to 13 ft. lbs. (18 Nm).
   - Oil dipstick tube
   - Crankshaft timing sprocket
   - Timing belt idler pulleys
   - Timing belt
   - Front cover
10. Install the engine
11. Use the Techstream diagnostic tool, or equivalent and reprogram the required systems.

Fig. 267 Oil pump housing sealant application—4.7L engine

5.7L Engine

See Figures 268 through 284.

1. Before servicing the vehicle, refer to the Precautions Section.

➡If working near and/or around the SRS system and components, be sure to disable the SRS system. Tape the negative battery cable with insulating tape. Always disconnect the negative battery cable first.

CAUTION

To avoid personal injury when working on vehicles equipped with an air bag, the negative battery cable must be disconnected and at least 90 seconds must elapse before working on the system. Failure to do so may result in deployment of the air bag.

2. Disconnect the negative battery cable. Tape the cable with insulating tape.

➡When disconnecting the cable, some systems need to be initialized after the cable is reconnected. You will need the Techstream diagnostic scan tool or equivalent. Follow the directions on the tool.

3. Discharge fuel system pressure.
4. Drain engine oil.
5. Drain engine coolant.

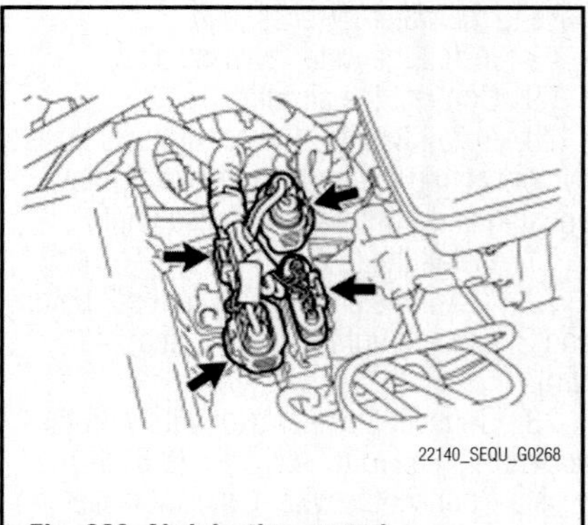

Fig. 268 Air injection control connectors—5.7L engine

6. Remove the radiator.
7. Remove air cleaner hose assembly.
8. Remove air cleaner assembly.
9. Engine Room LH Side:
   a. Remove the engine room relay block cover.
   b. Disconnect the 2 connectors and detach the 2 clamps from the engine room junction block.
   c. Disconnect the 4 air injection control driver connectors.
   d. Disconnect the 2 wire harness clamps.
   e. Disconnect the injector connector.
   f. Disconnect the 4 ignition coil connectors.
   g. Disconnect the 2 VVT sensor connectors.
   h. Disconnect the 4 clamps.
   i. Remove the 2 bolts and ground wire.
   j. Disconnect the noise filter connector.
   k. Disconnect the engine coolant temperature sensor connector.
   l. Disconnect the 2 camshaft timing oil control valve connectors.
   m. Disconnect the camshaft position sensor connector.
   n. Disconnect the 3 clamps.
   o. Disconnect the cooler compressor connector.
10. Engine Room RH Side.
   a. Disconnect the 2 camshaft timing oil control valve connectors.
   b. Disconnect the 4 ignition coil connectors.
   c. Disconnect the injector connector.
   d. Disconnect the 2 VVT sensor connectors.
   e. Disconnect the noise filter connector.
   f. Remove the 2 bolts and ground wire.
   g. Disconnect the 2 air pump connectors.
   h. Disconnect the throttle position sensor and throttle control motor connector.
   i. Disconnect the 5 clamps.
   j. Disconnect the 2 clamps and power steering oil pressure switch connector.
   k. For 4WD: Disconnect the 2 clamps and power steering oil pressure switch connector.
11. Remove the intake manifold.
12. Disconnect the No. 2 and No. 3 air hoses and the 2 clamps.
13. Remove the 2 bolts and disconnect the heater 3 hoses.
14. Remove the 2 bolts, 2 nuts and 2 stud bolts, and disconnect the cooler compressor.

➡It is not necessary to completely remove the compressor. With the hoses connected to the compressor, hang the compressor on the vehicle body with a rope.

15. Remove the 2 bolts and disconnect the fuel tube.
16. Remove oil filter element.
17. Remove oil pressure sender gauge assembly.
18. Disconnect the 4 hoses and remove the water by-pass pipe.
19. Remove No. 1 oil cooler bracket (w/ oil cooler).
20. Remove the 2 bolts, 2 nuts and oil filter bracket.
21. Remove the 2 O-rings.
22. Remove engine oil level dipstick guide.
23. Remove the 2 bolts and disconnect the vane pump.
24. Disconnect the alternator connector.
25. Remove the terminal cap and nut, and disconnect the alternator wire.
26. Remove the bolt and disconnect the wire harness bracket from the alternator.
27. Remove the 2 bolts, 2 nuts and alternator.
28. Remove the 2 stud bolts.
29. Remove the No. 1 water by-pass hose by disconnect the hose from the water inlet housing and front water by-pass joint.
30. Remove the bolt and disconnect the air tube.
31. Remove the 2 bolts and hoses and water by-pass pipe.
32. Disconnect the No. 2 water by-pass hose from the water by-pass joint.
33. Remove the 4 nuts, water by-pass joint and 2 gaskets.
34. Remove No.1 and No. 2 engine covers.
35. Remove the bolt, disconnect the 2 hoses, and remove the air tube.
36. Remove the 3 bolts, water inlet housing and gasket.
37. Remove the 4 bolts and water pump pulley.
38. Remove the bolt and idler pulley.
39. Remove the 4 bolts and fan bracket.
40. Remove the standard bolt, 6 mm hexagon wrench bolt and belt tensioner.
41. Remove the 8 bolts and 8 ignition coils.
42. Remove cylinder head cover subassembly LH.
   a. Remove the 14 bolts, seal washer, cylinder head cover and gasket.

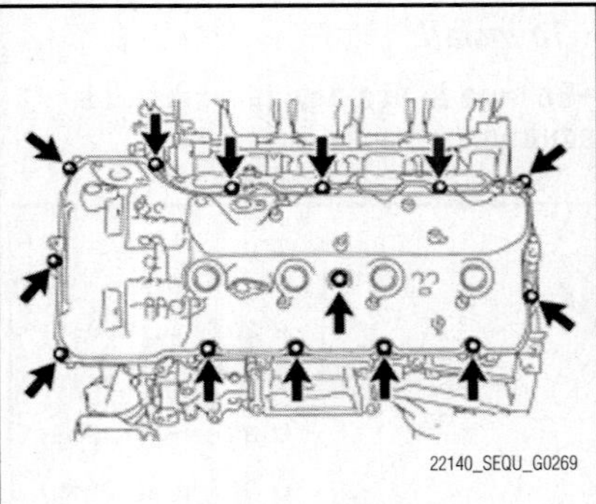

Fig. 269 Cylinder head cover LH—5.7L engine

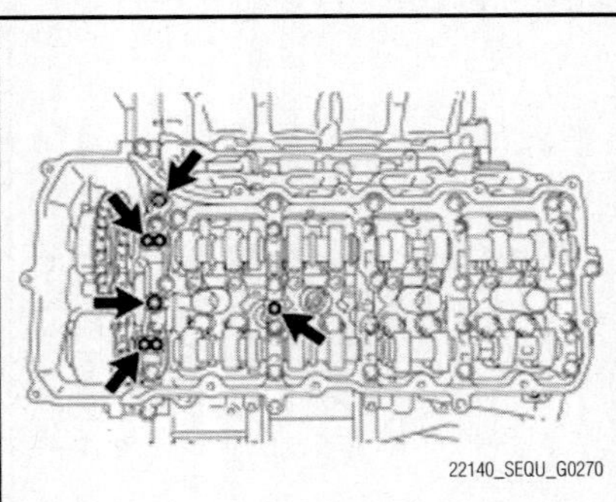

Fig. 270 Cylinder head cover gasket locations—5.7L engine

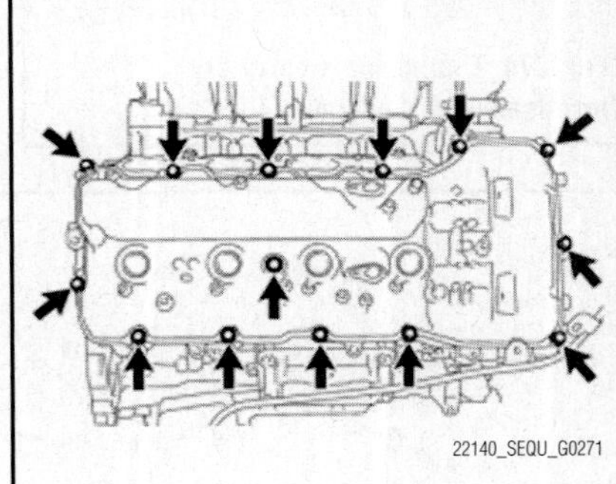

Fig. 271 Cylinder head cover RH—5.7L engine

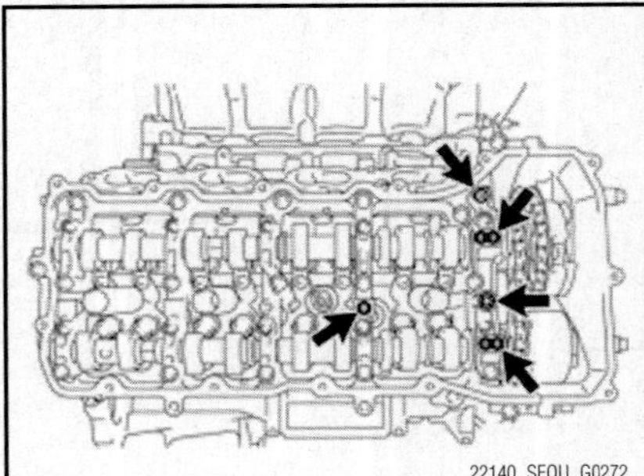

Fig. 272 Cylinder head cover gasket locations—5.7L engine

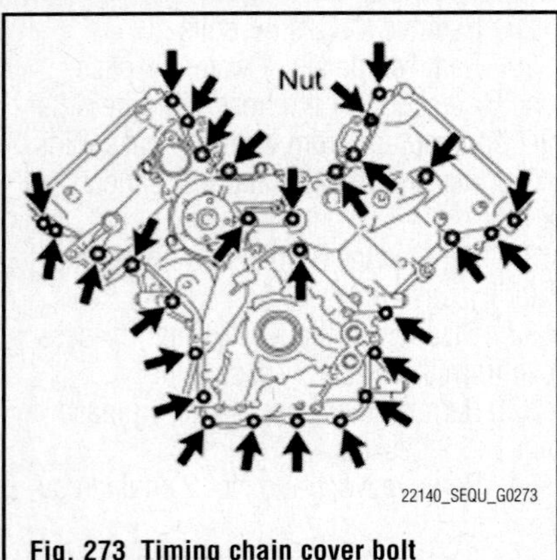

Fig. 273 Timing chain cover bolt locations—5.7L engine

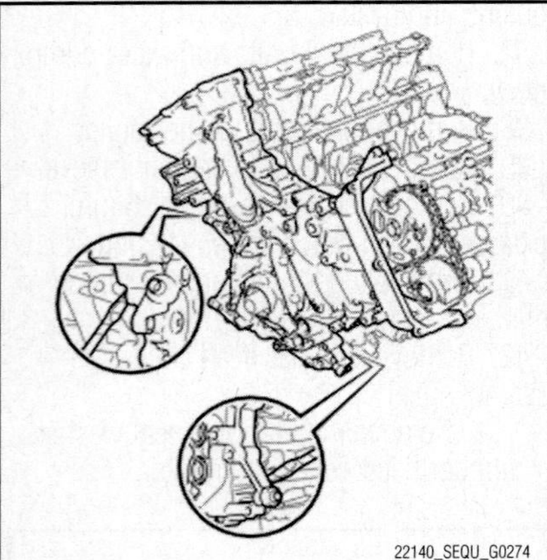

Fig. 274 Timing chain cover pry locations—5.7L engine

b. Remove the 5 gaskets from the camshaft bearing caps (No. 2, No. 3).

43. Remove cylinder head cover sub-assembly RH.

a. Remove the bolt and noise filter.

b. Remove the 14 bolts, seal washer, cylinder head cover and gasket.

c. Remove the 5 gaskets from the camshaft bearing caps (No. 1, No. 3).

44. Remove crankshaft pulley.

a. Loosen the crankshaft pulley set bolt.

b. Partially install the pulley set bolt to the crankshaft until 2 or 3 threads are engaged.

c. Remove the crankshaft pulley.

45. Remove the bolt and disconnect the wire harness bracket.

46. Remove timing chain cover sub-assembly.

a. Remove the 28 bolts and nut shown.

b. Remove the timing chain cover by prying between it and the cylinder head and cylinder block with a pry tool as shown.

47. Remove the oil pump gasket from the cylinder block.

48. Remove the O-ring from the oil pan.

49. Remove the water inlet pipe.

50. Remove the 2 O-rings from the water inlet pipe.

To install:

➡Be sure to use new fasteners, as required.

51. Apply soapy water to 2 new O-rings and install them to the inlet pipe.

52. Install the inlet pipe to the No. 1 heat exchanger cover.

53. Install timing chain cover sub-assembly.

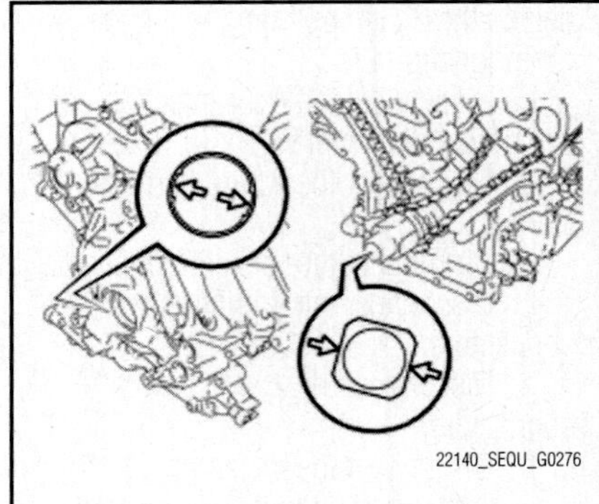

Fig. 276 Oil pump alignment—5.7L engine

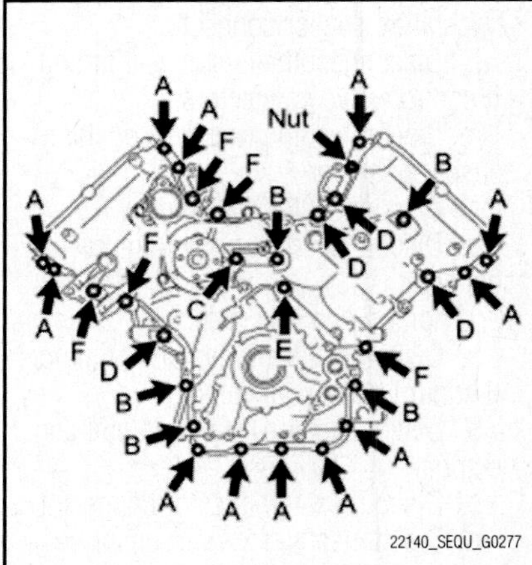

Fig. 277 Timing chain cover bolt locations—5.7L engine

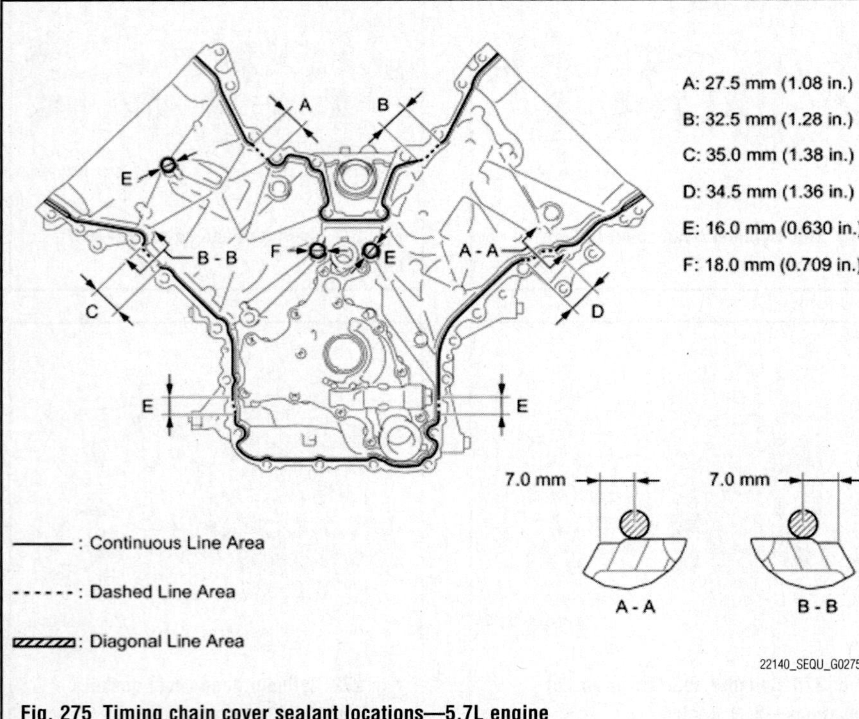

Fig. 275 Timing chain cover sealant locations—5.7L engine

Item	Length	Thread diameter
Bolt A	25 mm (0.984 in.)	8 mm (0.315 in.)
Bolt B	55 mm (2.165 in.)	8 mm (0.315 in.)
Bolt C	70 mm (2.756 in.)	8 mm (0.315 in.)
Bolt D	35 mm (1.378 in.)	10 mm (0.394 in.)
Bolt E	55 mm (2.165 in.)	10 mm (0.394 in.)
Bolt F	80 mm (3.150 in.)	10 mm (0.394 in.)

22140_SEQU_G0278

Fig. 278 Timing chain cover bolt application chart—5.7L engine

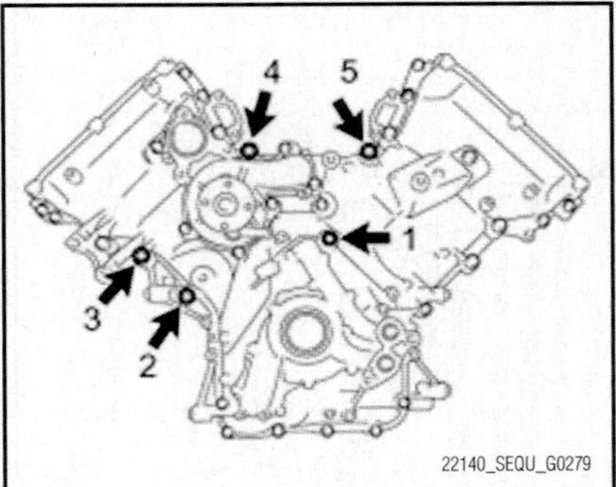

Fig. 279 Timing chain cover bolt tightening sequence No. 1—5.7L engine

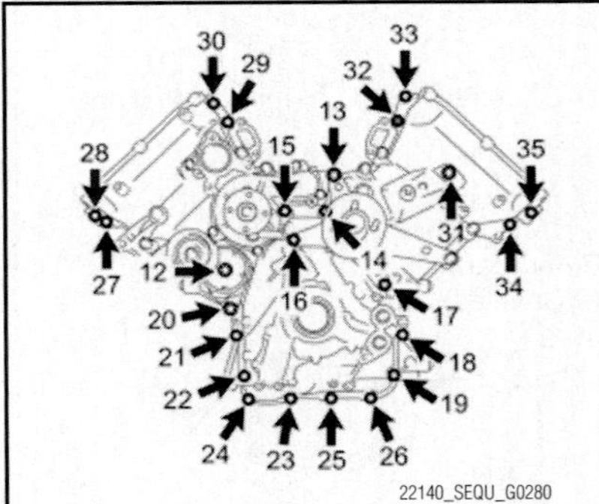

Fig. 280 Timing chain cover bolt tightening sequence No. 2—5.7L engine

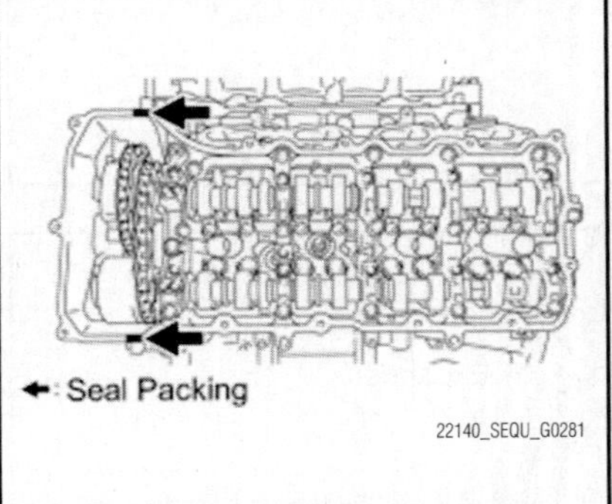

Fig. 281 Cylinder head cover packing location LH—5.7L engine

a. Apply a light coat of engine oil to a new oil pump gasket.

b. Install the oil pump gasket.

c. Apply a light coat of engine oil to a new O-ring.

d. Install the O-ring.

e. Apply seal packing in a continuous line to the timing chain cover as shown.

f. Align the oil pump's drive rotor spline and the crankshaft as shown in the illustration. Install the spline and chain cover to the crankshaft.

g. Temporarily install the timing chain cover with the 28 bolts and nut. Tighten the 5 bolts in several steps in the sequence shown to 35 ft. lbs. (47 Nm).

h. Temporarily install the fan bracket with the 4 bolts.

i. Temporarily install the belt tensioner with the standard bolt and 6 mm hexagon wrench bolt.

j. Tighten the 23 bolts labeled 12 to 35 and nut in several steps in the sequence to 17 ft. lbs. (23 Nm).

54. Install cylinder head cover sub-assembly LH.

a. Install 5 new gaskets to the camshaft bearing caps (No. 2, No. 3).

b. Install the gasket to the cylinder head cover.

c. Apply seal packing as shown.

d. Install the cylinder head cover washer with a new seal and the 14 bolts and tighten bolt A to 15 ft. lbs. (21 Nm) and remaining bolts to 9 ft. lbs. (12 Nm).

55. Install cylinder head cover sub-assembly RH.

a. Install 5 new gaskets to the camshaft bearing caps (No. 1, No. 3).

b. Install the gasket to the cylinder head cover.

c. Apply seal packing as shown.

d. Install the cylinder head cover washer with a new seal and the 14 bolts and tighten bolt A to 15 ft. lbs. (21 Nm) and remaining bolts to 9 ft. lbs. (12 Nm).

e. Install the noise filter to the cylinder head cover with the bolt and tighten to 62 inch lbs. (7.0 Nm).

56. Install the 8 ignition coils with the 8 bolts and tighten to 7 ft. lbs. (10 Nm).

57. Align the pulley set key with the key groove of the pulley, and slide on the pulley and tighten to 221 ft. Lbs. (300 Nm).

58. Connect the bracket to the timing chain cover with the bolt.

59. Install the idler pulley with the bolt and tighten to 32 ft. lbs. (43 Nm).

60. Temporarily install the water pump pulley with the 4 bolts and tighten to 15 ft. lbs. (21 Nm).

61. Install a new water cover gasket to the timing chain cover.

62. Install the water inlet with the 3 bolts and tighten to 15 ft. lbs. (21 Nm).

63. Connect the 2 air tube hoses.

64. Connect the 2 hoses.

65. Install No. 1 and No 2 engine covers.

66. Install 2 new gaskets and the water by-pass joint with the 4 nuts and tighten to 15 ft. lbs. (21 Nm).

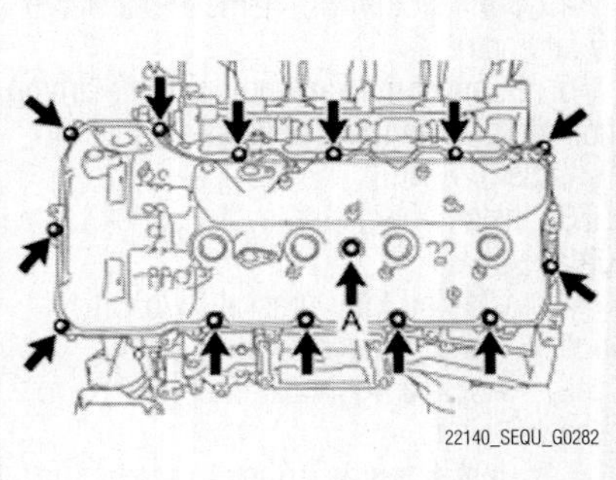

Fig. 282 Cylinder head cover bolt tightening sequence LH—5.7L engine

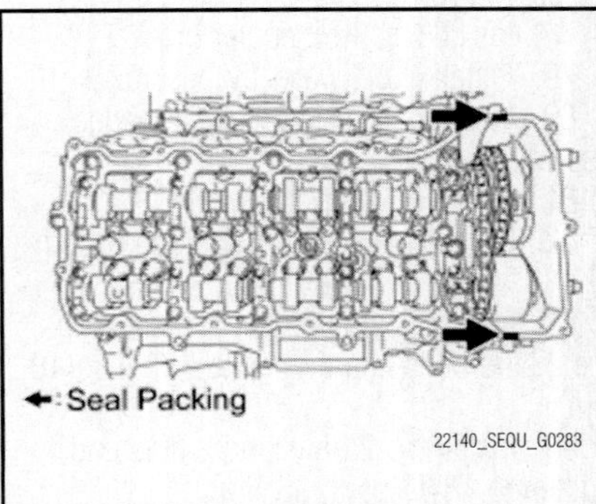

Fig. 283 Cylinder head cover packing location RH—5.7L engine

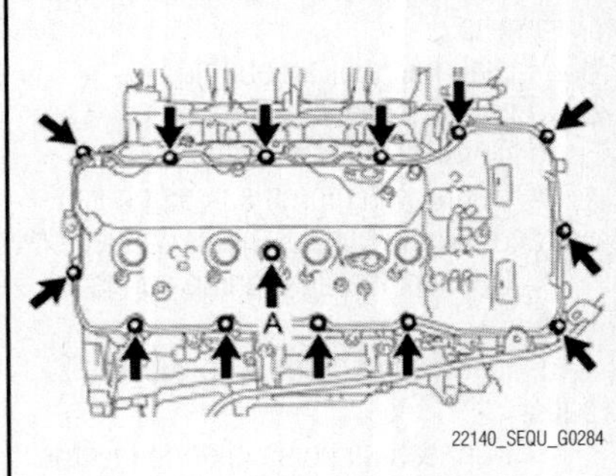

Fig. 284 Cylinder head cover bolt tightening sequence RH—5.7L engine

67. Connect the No. 2 water by-pass hose to the water by-pass joint.

68. Install the water by-pass pipe.

69. Connect the air tube.

70. Install the No. 1 water by-pass hose by connecting the hose to the water inlet housing and front water by-pass joint.

71. Install the 2 stud bolts.

72. Install the alternator with the 2 bolts and 2 nuts and tighten to 32 ft. lbs. (43 Nm).

73. Connect the alternator wire with the nut and tighten to 87 inch lbs. (9.8 Nm).

74. Connect the wire harness bracket to the alternator.
75. Connect the vane pump to the timing chain cover with the 2 bolts and tighten to 21 ft. lbs. (28 Nm).
76. Install engine oil level dipstick guide.
77. Install oil filter bracket (w/o Oil Cooler).
 a. Apply a light coat of engine oil to 2 new O-rings.
 b. Install the 2 O-rings to the timing chain cover.
 c. Install the oil filter bracket with the 2 bolts and 2 nuts and tighten to 26 ft. lbs. (35 Nm).
78. Install No. 1 oil cooler bracket.
79. Install No. 2 water by-pass pipe.
80. Install oil pressure sender gauge assembly.
81. Install oil filter element.
82. Connect No. 2 fuel tube sub-assembly.
83. Install the cooler compressor with the 2 stud bolts and tighten to 7 ft. lbs. (10 Nm).
84. Install the 2 bolts and 2 nuts and tighten to 18 ft. lbs. (235 Nm).
85. Connect the 3 heater hoses.
86. Connect the No. 2 and No. 3 air hoses.
87. Connect the 2 clamps and air pump wire harness.
88. Install the intake manifold.
89. Engine Room RH Side:
 a. Connect the 5 clamps.
 b. Connect the throttle position sensor and throttle control motor connector.
 c. Connect the 2 air pump connectors.
 d. Install the ground wire with the 2 bolts.
 e. Connect the noise filter connector.
 f. Connect the 2 VVT sensor connectors.
 g. Connect the injector connector.
 h. Connect the 4 ignition coil connectors.
 i. Connect the 2 camshaft timing oil control valve connectors.
 j. Connect the 2 clamps and power steering oil pressure switch connector.
 k. For 4WD: Connect the clamp and junction connector.
90. Engine Room LH Side:
 a. Connect the cooler compressor connector.
 b. Connect the 3 clamps.
 c. Connect the camshaft position sensor connector.
 d. Connect the 2 camshaft timing oil control valve connectors.
 e. Connect the engine coolant temperature sensor connector.
 f. Connect the 4 clamps.
 g. Install the ground wire with the 2 bolts.
 h. Connect the noise filter connector.
 i. Connect the 2 VVT sensor connectors.
 j. Connect the 4 ignition coil connectors.
 k. Connect the injector connector.
 l. Connect the 2 wire harness clamps.
 m. Connect the 4 air injection control driver connectors.
 n. Connect the 2 connectors and attach the 2 clamps to the engine room junction block.
 o. Install the engine room relay block cover.
91. Install air cleaner assembly.
92. Install air cleaner hose assembly.
93. Install the radiator.
94. Add engine oil.
95. Add engine coolant.
96. Connect cable to negative battery terminal.
97. Inspect for leaks.
98. Check engine oil level.
99. Use the Techstream diagnostic tool, or equivalent and reprogram the required systems.

PISTON AND RING

POSITIONING

See Figures 285 through 287.

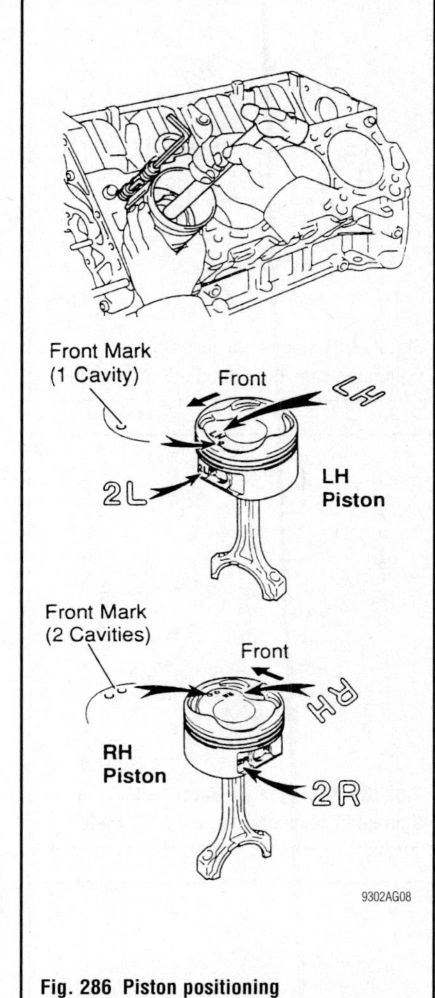

Fig. 286 Piston positioning

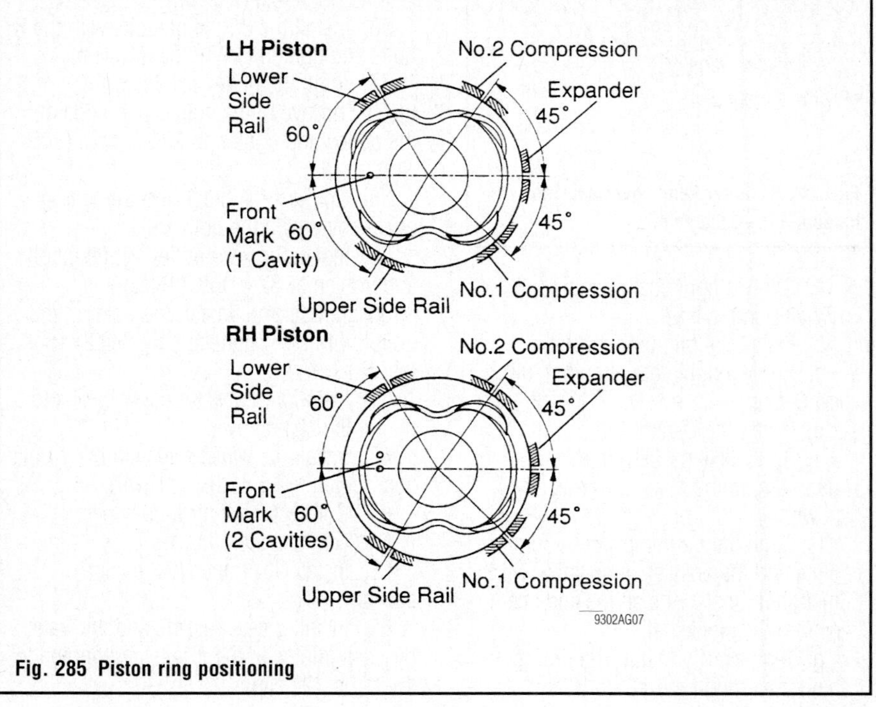

Fig. 285 Piston ring positioning

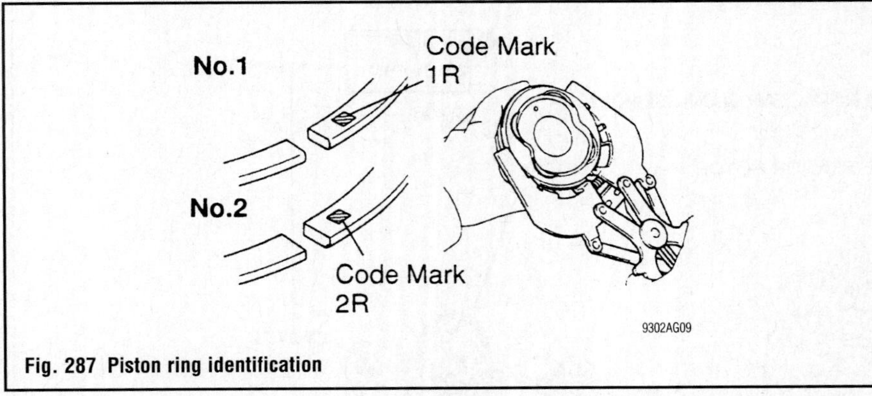

Fig. 287 Piston ring identification

REAR MAIN SEAL

REMOVAL & INSTALLATION

See Figures 288 through 290.

1. Before servicing the vehicle, refer to the Precautions Section.

➡If working near and/or around the SRS system and components, be sure to disable the SRS system. Tape the negative battery cable with insulating tape. Always disconnect the negative battery cable first.

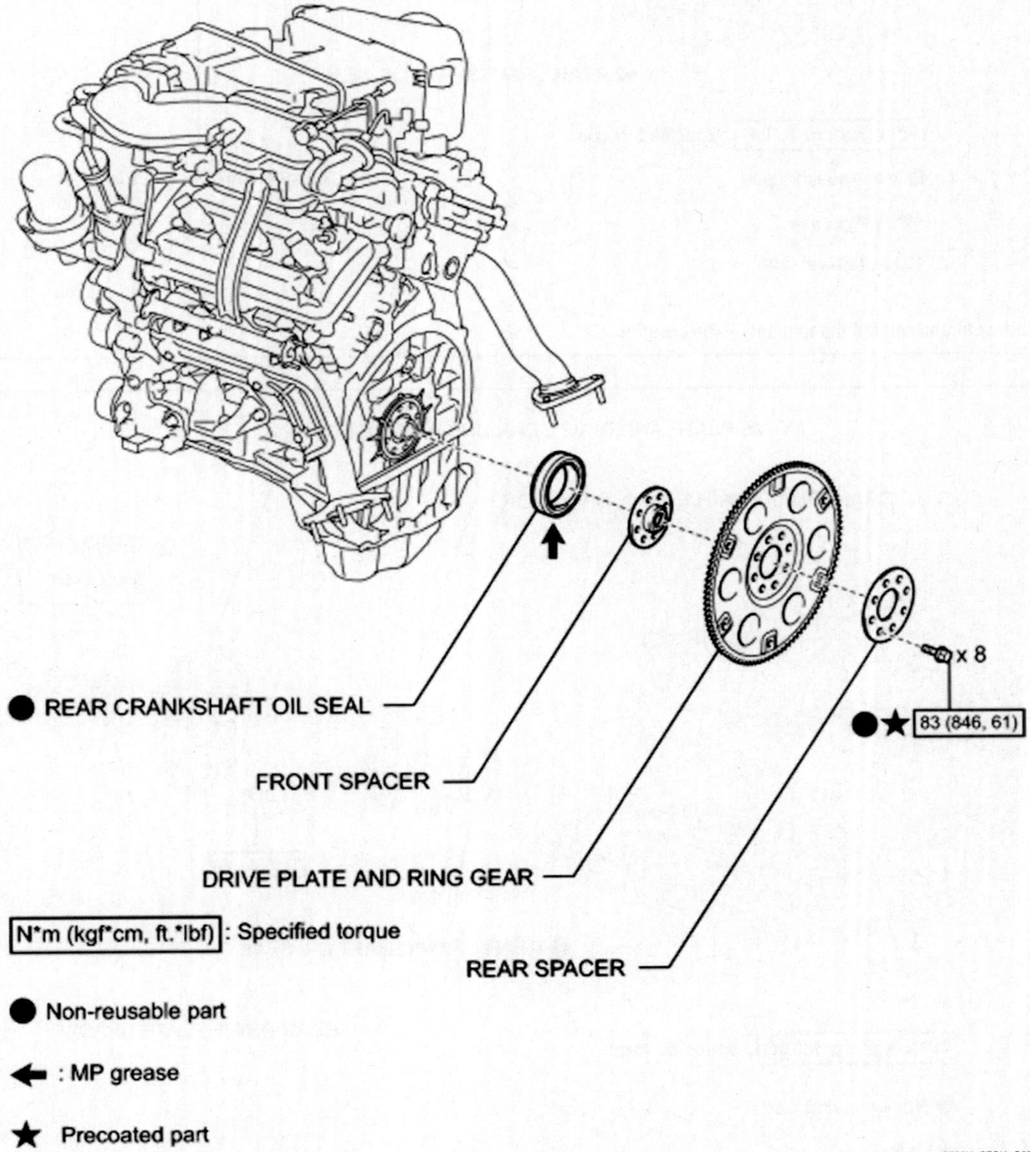

Fig. 288 Rear main seal and related components—4.0L engine

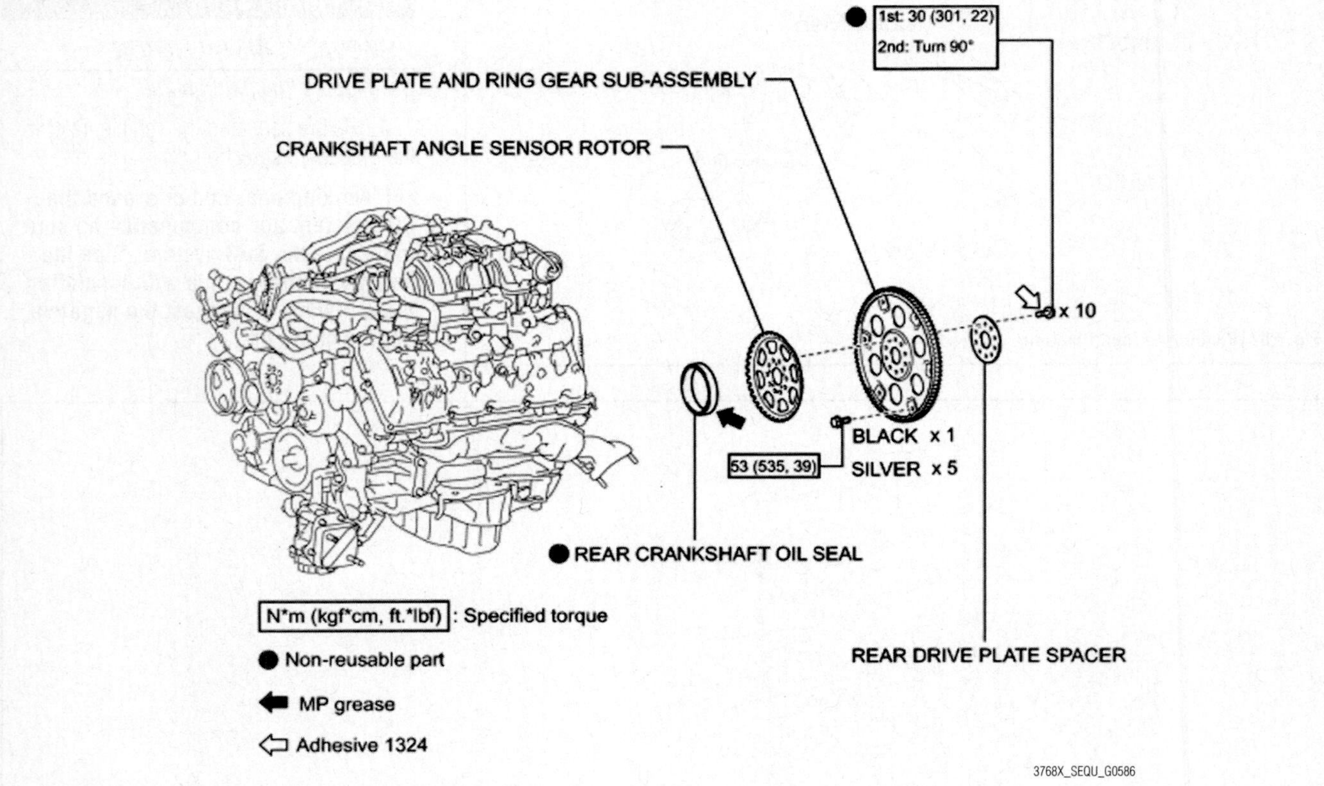

Fig. 289 Rear main seal and related components—4.6L engine

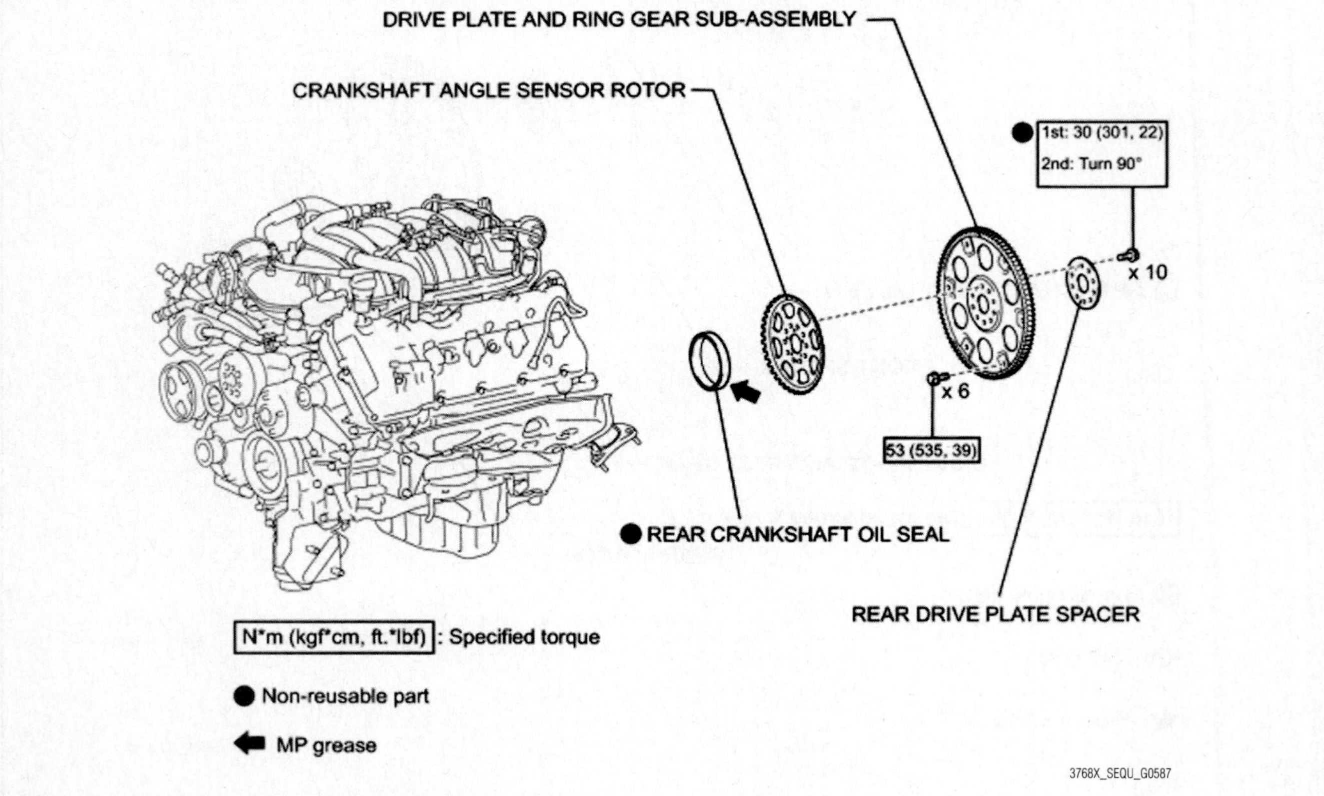

Fig. 290 Rear main seal and related components—5.7L engine

⁂ CAUTION

To avoid personal injury when working on vehicles equipped with an air bag, the negative battery cable must be disconnected and at least 90 seconds must elapse before working on the system. Failure to do so may result in deployment of the air bag.

2. Disconnect the negative battery cable. Tape the cable with insulating tape.

➡When disconnecting the cable, some systems need to be initialized after the cable is reconnected. You will need the Techstream diagnostic scan tool or equivalent. Follow the directions on the tool.

3. Remove the transmission and flywheel from the vehicle.
4. Cut off the rubber lip portion of the seal with a sharp knife.
5. Pry out the oil seal.

To install:

➡Be sure to use new fasteners, as required.

6. Installation is the reverse of the removal procedure.
7. Use the Techstream diagnostic tool, or equivalent and reprogram the required systems.

TIMING BELT FRONT COVER

REMOVAL & INSTALLATION

See Timing Belt and Sprockets.

TIMING BELT & SPROCKETS

REMOVAL & INSTALLATION

4.7L Engine

See Figures 291 through 298.

1. Before servicing the vehicle, refer to the Precautions Section.

➡If working near and/or around the SRS system and components, be sure to disable the SRS system. Tape the negative battery cable with insulating tape. Always disconnect the negative battery cable first.

⁂ CAUTION

To avoid personal injury when working on vehicles equipped with an air bag, the negative battery cable must be disconnected and at least 90 seconds must elapse before working on the system. Failure to do so may result in deployment of the air bag.

2. Disconnect the negative battery cable. Tape the cable with insulating tape.

➡When disconnecting the cable, some systems need to be initialized after the cable is reconnected. You will need the Techstream diagnostic scan tool or equivalent. Follow the directions on the tool.

3. Raise and safely support the vehicle.
4. Remove the oil pan protector and the engine under cover.
5. Drain the cooling system and store the coolant for refilling purposes.
6. Lower the vehicle and remove the battery clamp cover.
7. From the top of the engine, remove the fuel return hose, the engine cover nuts/bolts and the cover.
8. Remove the air cleaner and the intake air connector assembly.
9. Remove the cooling fan pulley by performing the following procedures:
    a. Loosen the 4 fan clutch-to-fan pulley nuts.
    b. Using a box-end wrench on the serpentine drive belt tensioner bolt, rotate the tensioner counterclockwise and remove the drive belt.

➡The serpentine drive belt tensioner bolt is a left-hand thread.

    c. Remove the fan clutch-to-fan pulley nuts, the fan, the clutch assembly and the fan pulley.
10. Remove the radiator by performing the following procedures:
    a. Disconnect the upper, lower and reservoir hoses from the radiator.
    b. Disconnect and plug the automatic transmission oil cooler at the radiator. Disconnect the automatic transmission oil cooler hoses from the fan shroud clamp.
    c. Remove the radiator reservoir tank.
    d. Remove the fan shroud-to-radiator bolts and the shroud.
    e. Remove the 2 upper radiator-to-chassis nuts.
    f. Remove the middle radiator-to-chassis nut/bolts and brackets.
    g. Carefully, lift the radiator from the vehicle.
11. Remove the serpentine drive belt idler pulley bolt, cover plate and pulley.
12. Remove the right side (No. 3) timing belt cover.
13. Remove the left side (No. 3) timing belt cover by performing the following procedures:
    a. Disconnect the engine wire from both wire clamps.
    b. Disconnect the camshaft position sensor wire from the wire clamp on the left-side (No.3) timing belt over.
    c. Disconnect the sensor connector from the connector bracket.
    d. Disconnect the sensor connector.
    e. Remove the wire grommet from the left-side (No. 3) timing belt cover.
    f. Remove the oil cooler tube bolts and tube.
14. Remove the middle (No. 2) timing belt cover bolts and cover.
15. Remove the cooling fan bracket nuts/bolts and bracket.

➡If reusing the timing belt, make sure that there are 3 installation marks on the belt; if there are none, install them.

16. Using the Crankshaft Pulley Holding tool 09213-70010, Bolt tool 90105-08076 and Companion Flange Holding tool 09330-00021, or equivalent, loosen the crankshaft pulley bolt.
17. Position the No. 1 cylinder to approximately 50 degrees After Top Dead Center (ATDC) of the compression stroke by performing the following procedures:
    a. Rotate the crankshaft pulley (CLOCKWISE) to align its groove with the timing mark "0" on the lower (No. 1) timing belt cover.
    b. Check that the camshaft sprocket timing marks are aligned with the rear timing belt plate marks; if not, rotate the crankshaft 1 revolution (360 degrees).
    c. Rotate the crankshaft pulley approximately 50 degrees (CLOCKWISE) and align the crankshaft pulley timing mark between the centers of the crankshaft pulley bolt and the idler pulley bolt.

⁂ WARNING

If the timing belt is disengaged, having the crankshaft pulley in the wrong angle can cause the valve to come into contact with the piston when removing the camshaft pulley.

18. Remove the crankshaft pulley bolt.

➡If reusing the timing belt and the installation marks have disappeared, place new installation marks on the timing belt to match the camshaft timing sprocket marks.

➡To avoid meshing the timing sprocket and the timing belt, secure one with a string; then, place matchmarks on the timing belt and the right-side camshaft timing sprocket.

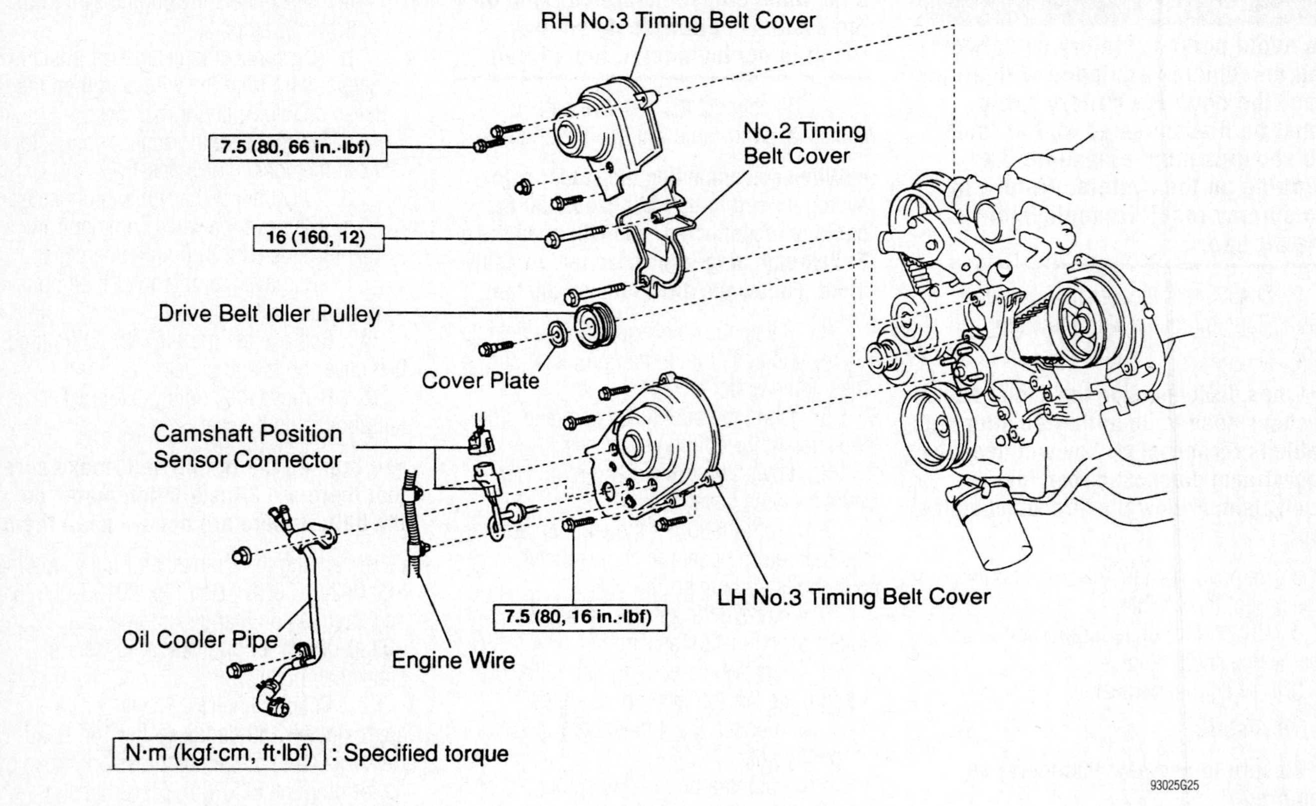

Fig. 291 Exploded view of upper timing belt covers

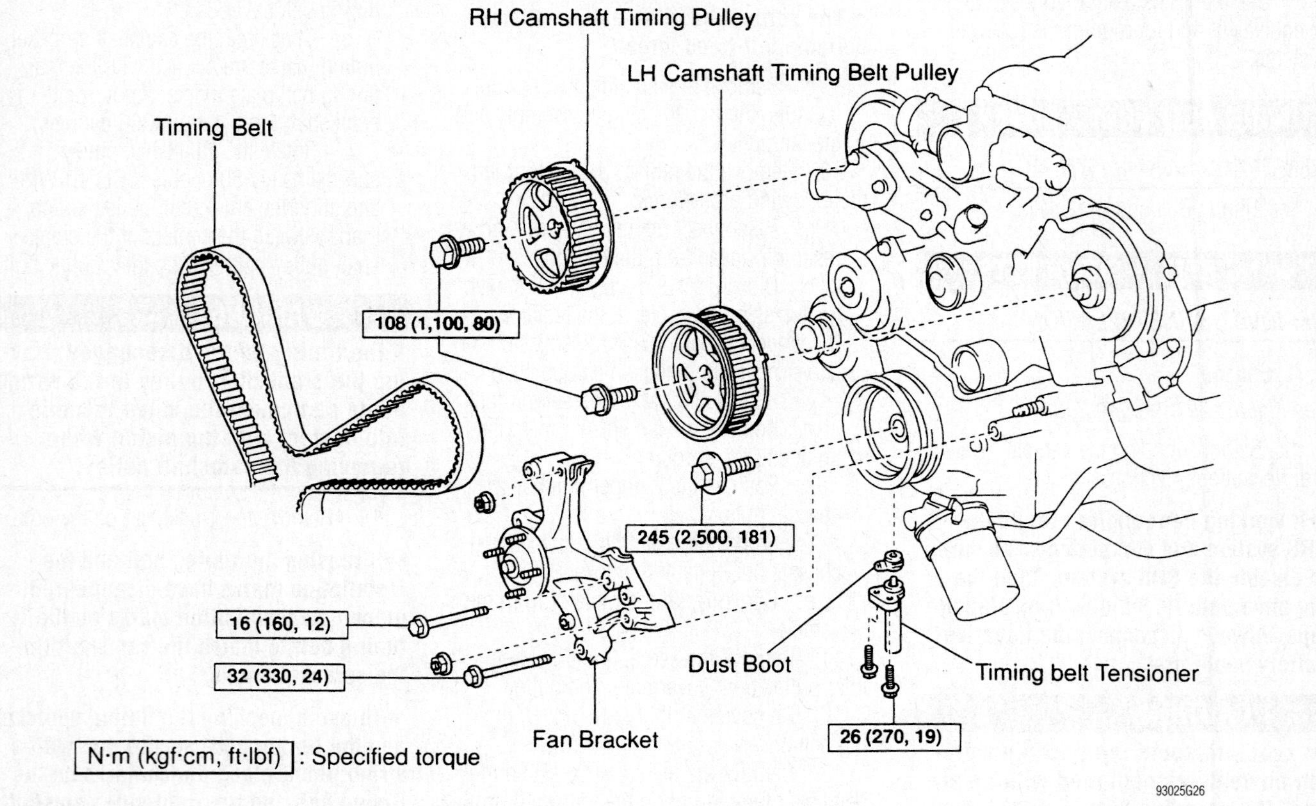

Fig. 292 Exploded view of upper timing sprockets and components

19. Remove the timing belt tensioner bolts and the tensioner.

20. Using the Camshaft Holding tool 09960-10010, or equivalent, slightly turn the left-side camshaft sprocket clockwise to loosen the tension spring. Then, disconnect the timing belt from the camshaft sprockets.

21. Remove the alternator by performing the following procedures:

a. Disconnect the electrical connector from the alternator.

b. Remove the rubber cap/nut and disconnect the battery wire from the alternator.

c. Disconnect the wire clamp from the alternator cord clip.

d. Remove the alternator-to-engine nuts/bolts and the alternator.

22. Remove the serpentine drive belt tensioner nuts/bolts and the tensioner.

23. Using the Crankshaft Puller Assembly tool 09950-50012, or equivalent, press the crankshaft pulley from the crankshaft.

⁂ WARNING

DO NOT rotate the crankshaft pulley.

24. Remove the lower (No. 1) timing belt cover bolts and the cover.

25. Remove the timing belt guide, spacer and the timing belt.

To install:

➡Be sure to use new fasteners, as required.

➡With the timing belt removed, this is a perfect opportunity to inspect and/or replace the water pump.

26. Inspect the timing belt tensioner by performing the following procedures:

a. Inspect the seal for leakage; if leakage is suspected, replace the tensioner.

b. Using both hands to hold the tensioner facing upward, strongly press the pushrod against a solid surface. If the pushrod moves, replace the tensioner.

⁂ WARNING

Never hold the tensioner with the pushrod facing downward.

c. Measure the pushrod protrusion from the housing end, it should be 0.413–0.453 in. (10.5–11.5mm). If the protrusion is not as specified, replace the tensioner.

27. Temporarily install the timing belt by performing the following procedures:

a. Align the timing belt's installation mark with the crankshaft timing sprocket.

b. Install the timing belt on the crankshaft timing sprocket, the No. 1 idler pulley and the No. 2 idler pulley.

28. Install the gasket to the timing belt cover spacer and install the cover spacer.

29. Install the timing belt guide with the cup side facing outward.

30. Install the lower (No. 1) timing belt cover.

31. Install the crankshaft pulley by performing the following procedures:

a. Align the crankshaft pulley with the crankshaft key.

b. Using the Crankshaft Installer tool 09223-46011, or equivalent, and a hammer, tap the crankshaft pulley into position.

32. Install the serpentine drive belt tensioner and torque the tensioner-to-engine bolts to 12 ft. lbs. (16 Nm).

➡To install the serpentine drive belt tensioner, use a bolt 4.18 in. (106mm) in length.

33. Check that the crankshaft pulley's timing mark is aligned with the centers of the idler pulley and crankshaft pulley bolts.

34. Install the alternator and torque the alternator-to-engine nuts/bolts to 29 ft. lbs.

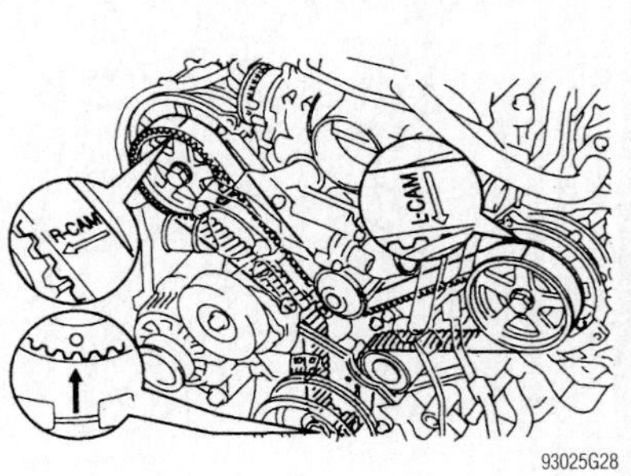

93025G28

Fig. 293 Alignment of timing belt with the timing sprockets

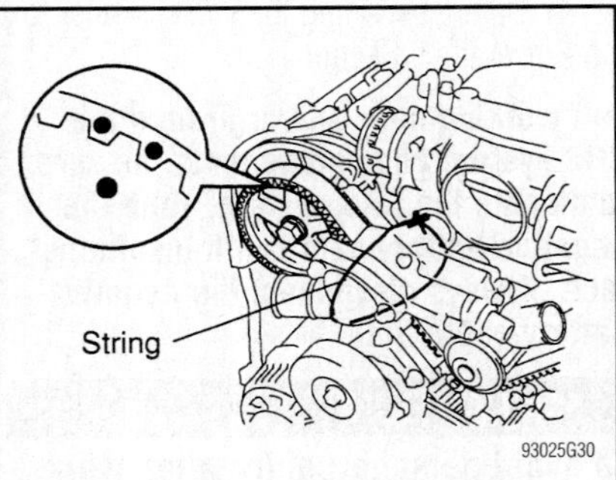

Fig. 295 Securing the timing belt with string and matchmarking the camshaft with the timing belt

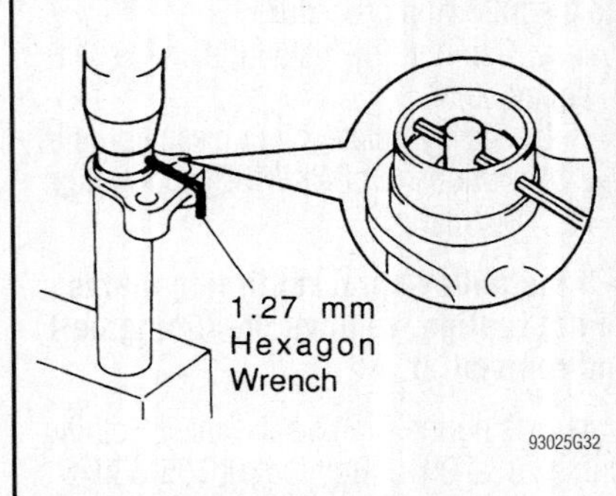

Fig. 297 Securing the timing belt tensioner pushrod

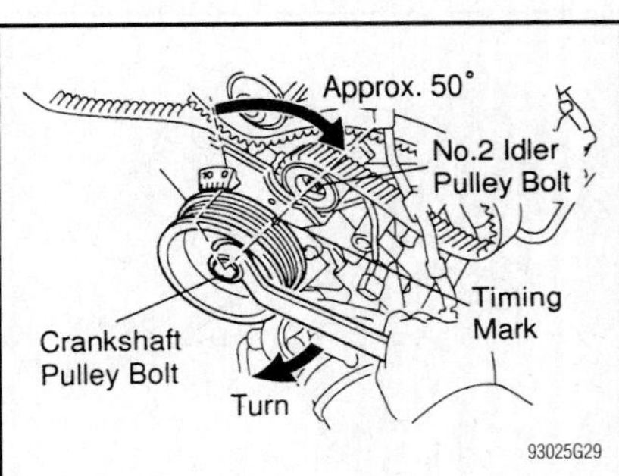

Fig. 294 Aligning of crankshaft pulley timing mark with the center line of the crankshaft pulley bolt and the idler pulley bolt

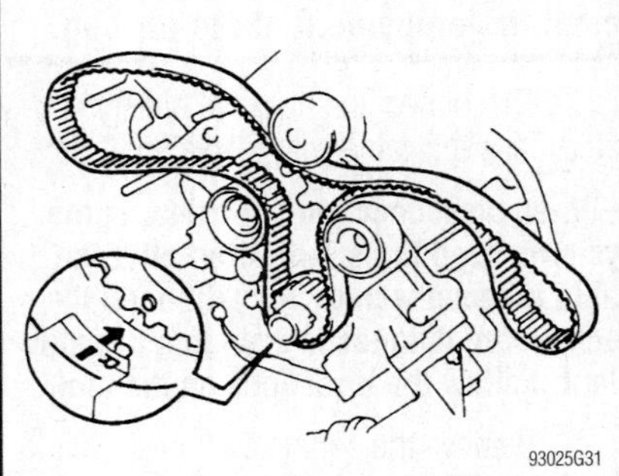

Fig. 296 Installing the timing belt on the crankshaft sprocket

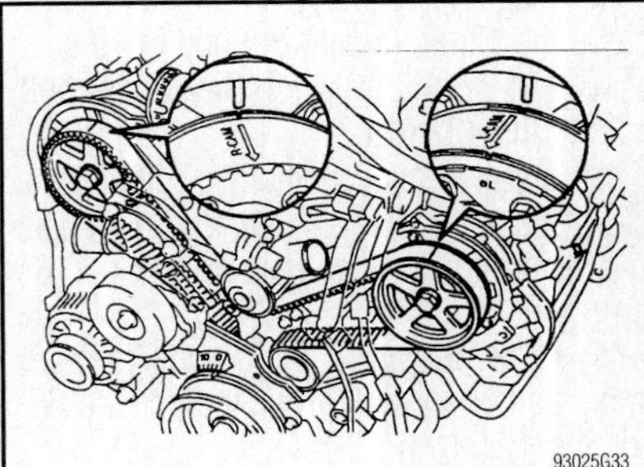

Fig. 298 Checking the TDC alignment marks after rotating the crankshaft 2 revolutions

(39 Nm). Connect the alternator's electrical connectors and clip.

35. Install the timing belt to the left-side camshaft by performing the following procedures:

a. Rotate the left-side camshaft pulley to align the timing belt installation mark with the camshaft sprocket's timing mark and slide the belt onto the camshaft timing sprocket.

b. Using the Camshaft Holding tool 09960-10010, or equivalent, slightly turn the left-side camshaft sprocket counterclockwise to place tension on the timing belt between the crankshaft sprocket and the camshaft sprocket.

36. Rotate the right-side camshaft pulley to align the timing belt installation mark with the camshaft sprocket's timing mark and slide the belt onto the camshaft timing sprocket.

37. Using a vertical press, slowly press the pushrod into the housing using 200–2205 lbs. (981–9807 N) until the holes align, then, install a 1.27mm Allen® wrench to secure the pushrod and release the press. Install the dust boot on the tensioner housing.

38. Install the timing belt tensioner and torque the bolts to 19 ft. lbs. (26 Nm).

39. Using a pair of pliers, remove the Allen® wrench from the tensioner housing.

40. Check the valve timing by performing the following procedure:

a. Temporarily install the crankshaft pulley bolt.

b. Slowly, rotate the crankshaft pulley 2 revolutions (CLOCKWISE) and realign the TDC marks.

➡If the pulley/sprocket timing marks do not realign, remove the timing belt and reinstall it.

41. Using the Crankshaft Pulley Holding tool 09213-70010, Bolt tool 90105-08076 and Companion Flange Holding tool 09330-00021, or equivalent, torque the crankshaft pulley bolt to 181 ft. lbs. (245 Nm).

42. Install the cooling fan bracket and torque the 12mm (head size) bolt to 12 ft. lbs. (16 Nm) and the 14mm (head size) bolt to 24 ft. lbs. (32 Nm).

43. Install the air conditioning compressor.

44. Install the middle (No. 2) timing belt cover and torque the bolts to 12 ft. lbs. (16 Nm).

45. Install the upper right-side (No. 3) timing belt cover and torque the bolts to 66 inch lbs. (7.5 Nm).

46. Install the upper left-side (No. 3) timing belt cover by performing the following procedures:

a. Install the oil cooler tube and bolt.

b. Feed the Camshaft Position Sensor (CPS) through the left-side (No. 3) timing belt cover hole.

c. Install the left-side (No. 3) timing belt cover and torque the bolts to 66 inch lbs. (7.5 Nm).

d. Install the wire grommet to the left-side (No. 3) timing belt cover.

e. Install the sensor connector to the connector bracket and connect the sensor connector.

f. Install the sensor wire and the engine wire to the clamps on the left-side (No. 3) timing belt cover.

47. Install the drive belt idler pulley and cover plate; then, torque the pulley bolt to 27 ft. lbs. (37 Nm).

48. To complete the installation, reverse the removal procedures.

49. Refill the cooling system and connect the negative battery cable.

50. Use the Techstream diagnostic tool, or equivalent and reprogram the required systems.

TIMING CHAIN FRONT COVER

REMOVAL & INSTALLATION

4.0L Engine

See Figures 299 through 303.

1. Before servicing the vehicle, refer to the Precautions Section.

➡If working near and/or around the SRS system and components, be sure to disable the SRS system. Tape the negative battery cable with insulating tape. Always disconnect the negative battery cable first.

⁂ CAUTION

To avoid personal injury when working on vehicles equipped with an air bag, the negative battery cable must be disconnected and at least 90 seconds must elapse before working on the system. Failure to do so may result in deployment of the air bag.

2. Disconnect the negative battery cable. Tape the cable with insulating tape.

➡When disconnecting the cable, some systems need to be initialized after the cable is reconnected. You will need the Techstream diagnostic scan tool or equivalent. Follow the directions on the tool.

3. Remove the engine and position it in a suitable holding fixture.

4. Remove the ignition coil assembly.

5. Remove the spark plugs.

6. Remove the oil dipstick guide.

7. Remove the No. 2 and No. 1 idler pulley sub assembly.

8. Disconnect the No. 2 oil cooler hose.

9. Remove the water inlet housing.

10. Remove the VVT sensor.

11. Remove the camshaft timing oil control valve assembly.

12. Remove the cylinder head covers.

13. Remove the crankshaft pulley.

14. Remove the No. 2 oil pan sub assembly retaining bolts. Remove the component. Discard the gasket.

15. Remove the oil strainer sub assembly.

16. Remove the No 1. oil pan sub assembly retaining bolts. Remove the component. Discard the gasket. Remove the O-ring from the cylinder block. Discard it.

17. Remove the oil filter bracket sub assembly.

18. Remove the timing cover retaining bolts.

19. Remove the timing cover. Discard the gasket.

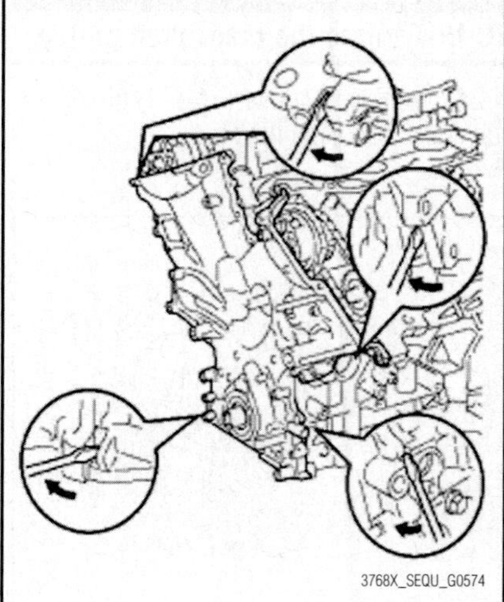

Fig. 299 Front cover removal—4.0L engine

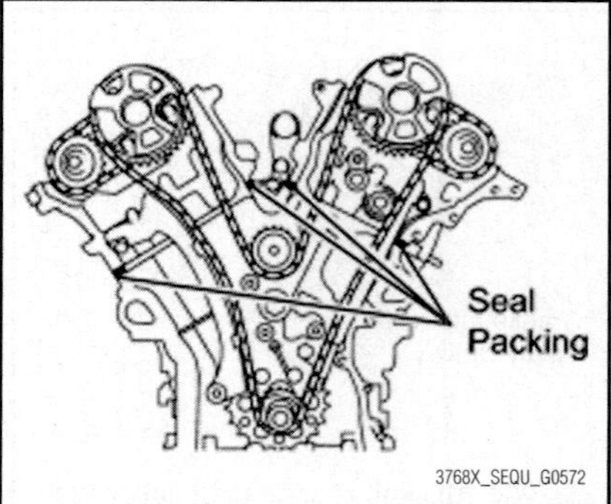

Fig. 300 Front cover sealant application point 1 of 2—4.0L engine

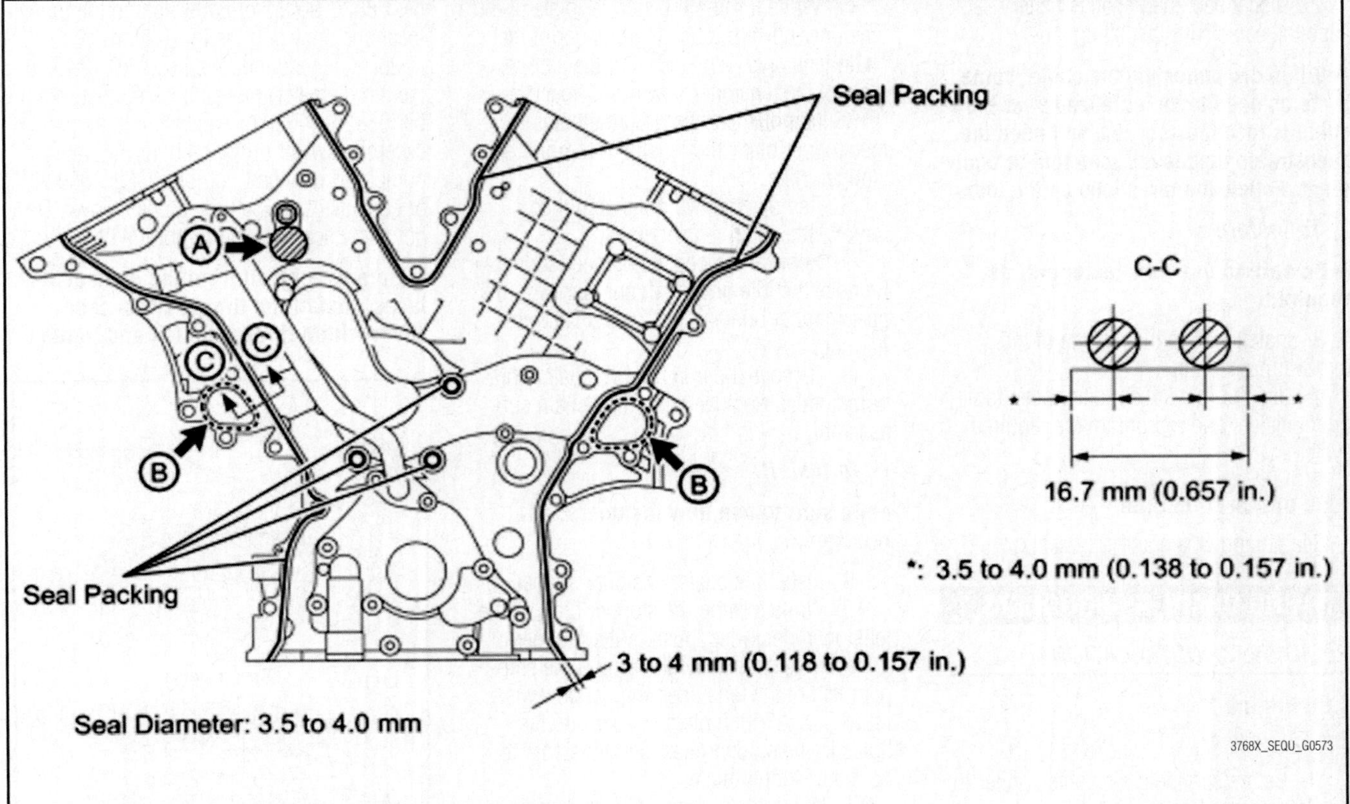

Fig. 301 Front cover sealant application point 2 of 2—4.0L engine

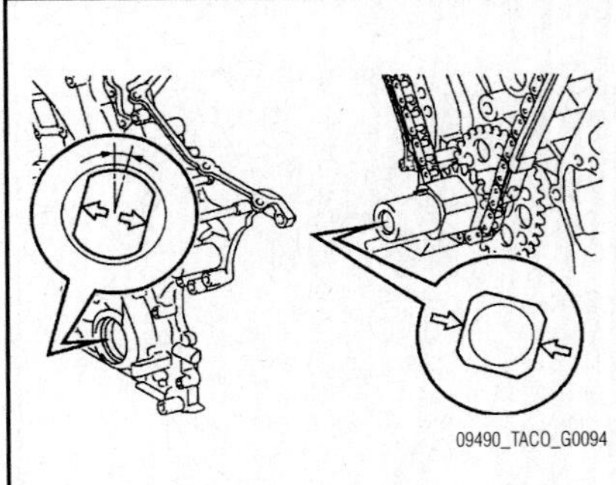

Fig. 302 Oil pump drive rotor spline alignment—4.0L engine

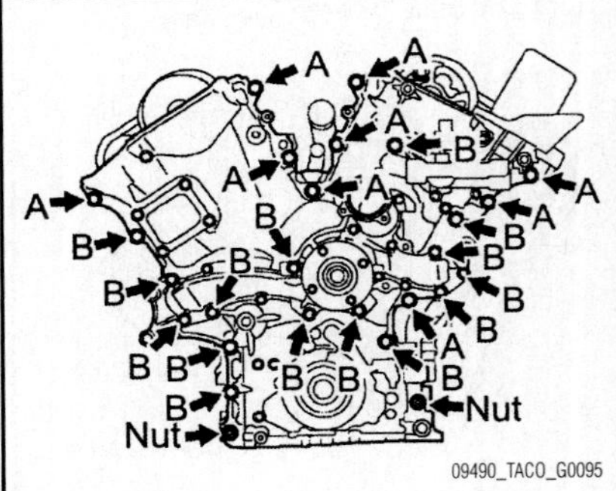

Fig. 303 Front cover bolt location and torque sequence—4.0L engine

➡When removing the cover carefully pry between the cover and the cylinder head or cylinder block with a suitable pry tool.

To install:

➡Be sure to use new fasteners, as required.

20. To install the front cover, remove any old packing material.
21. Install a new O-ring to the cylinder head, Bank 2.
22. Apply seal packing in a continuous bead in a diameter of 0.118–0.157 inch. See illustration.
23. Apply seal packing in a continuous bead in a diameter of 0.118–0.157 inch to the timing cover. Do not apply sealant to area A. See illustration.

➡Install the component within three minutes after applying the sealant. Tighten the bolts within fifteen minutes after applying the sealant. Do not start the engine for at least two hours after applying sealant.

24. Align the key way of the oil pump drive motor with the rectangular portion of the crankshaft timing gear and slide the timing chain case cover into place.
25. Install the timing chain case cover bolts. Tighten the bolts and nuts uniformly in several steps to 17 ft. lbs.
26. Continue the installation in the reverse order of the removal procedure.
27. Use the Techstream diagnostic tool, or equivalent and reprogram the required systems.

4.6L Engine

See Figure 133.

1. Before servicing the vehicle, refer to the Precautions Section.

➡If working near and/or around the SRS system and components, be sure to disable the SRS system. Tape the negative battery cable with insulating tape. Always disconnect the negative battery cable first.

CAUTION

To avoid personal injury when working on vehicles equipped with an air bag, the negative battery cable must be disconnected and at least 90 seconds must elapse before working on the system. Failure to do so may result in deployment of the air bag.

2. Disconnect the negative battery cable. Tape the cable with insulating tape.

➡When disconnecting the cable, some systems need to be initialized after the cable is reconnected. You will need the Techstream diagnostic scan tool or equivalent. Follow the directions on the tool.

To install:

➡Be sure to use new fasteners, as required.

3. Installation is the reverse of the removal procedure.
4. Use the Techstream diagnostic tool, or equivalent and reprogram the required systems.

4.6L and 5.7L Engine

See Timing Chain and Sprockets.

TIMING CHAIN & SPROCKETS

REMOVAL & INSTALLATION

4.0L Engine

See Figure 304.

1. Before servicing the vehicle, refer to the Precautions Section.

➡If working near and/or around the SRS system and components, be sure to disable the SRS system. Tape the negative battery cable with insulating tape. Always disconnect the negative battery cable first.

CAUTION

To avoid personal injury when working on vehicles equipped with an air bag, the negative battery cable must be disconnected and at least 90 seconds must elapse before working on the system. Failure to do so may result in deployment of the air bag.

2. Disconnect the negative battery cable. Tape the cable with insulating tape.

➡When disconnecting the cable, some systems need to be initialized after the cable is reconnected. You will need the Techstream diagnostic scan tool or equivalent. Follow the directions on the tool.

3. Remove the front cover.
4. Using the crankshaft pulley set bolt, turn the crankshaft to align the crankshaft set key with the timing line of the cylinder block. If not aligned at TDC of the compression stroke, turn the crankshaft one complete revolution, in the direction of rotation
5. While turning the stopper plate of the tensioner upward, push the plunger of the chain tensioner. While turning the stopper plate of the tensioner downward, insert a 0.138 inch diameter bar into the holes in the stopper plate and tensioner to hold the stopper plate.
6. Remove the two bolts, and then remove the chain tensioner.
7. Remove the chain tensioner slipper. Remove the idle gear shaft number two, idle gear number one and idle gear shaft number one.
8. Remove the number two chain vibration damper. Remove the timing chain subassembly.

To install:

➡Be sure to use new fasteners, as required.

9. Install the chain tensioner slipper.
10. While turning the stopper plate of the tensioner clockwise, push in the plunger of the chain tensioner. While turning the stopper plate of the tensioner counterclockwise, insert a 0.138 inch diameter bar into the holes in the stopper plate and tensioner to hold the stopper plate.
11. Install the chain tensioner. Tighten the bolts to 7.4 ft. lbs.
12. Position the engine at TDC on the compression stroke. Align the camshaft timing gears and bearing caps. Using the crankshaft pulley set bolt, align the crankshaft set key with the timing line of the cylinder.
13. Align the yellow mark line with the timing mark of the crankshaft timing link. Align the orange mark links with the timing marks of the camshaft timing gears, and install the chain.
14. Install the number two chain vibration damper.
15. Apply a light coat of clean engine oil to the rotating surface of the idle gear shaft number one.
16. Temporarily install the idle gear shaft number one together with idle gear shaft number two, while aligning the knock pin of idle gear shaft number one with the knock pin groove of the cylinder block.

➡Be careful of the idle gear direction.

17. Tighten the idle gear shaft number two to 44 ft. lbs. Remove the bar from the chain tensioner.
18. Install a new front case oil seal. Install a new O-ring onto the left cylinder head.
19. Apply continuous beads (0.12–0.16 inch in diameter) of seal packing, part number 08826-00080 or equivalent to the four locations shown in the illustration.
20. Apply continuous beads (0.12–0.16 inch in diameter) of seal packing, part number 08826-00080 or equivalent to all parts except the water pump part: for the water pump part use, part number 08826-00080 or equivalent, to the timing chain cover. Do not apply seal packing to portion "A".

➡Remove any oil from the contact surfaces. Install the timing chain case cover within three minutes and tighten

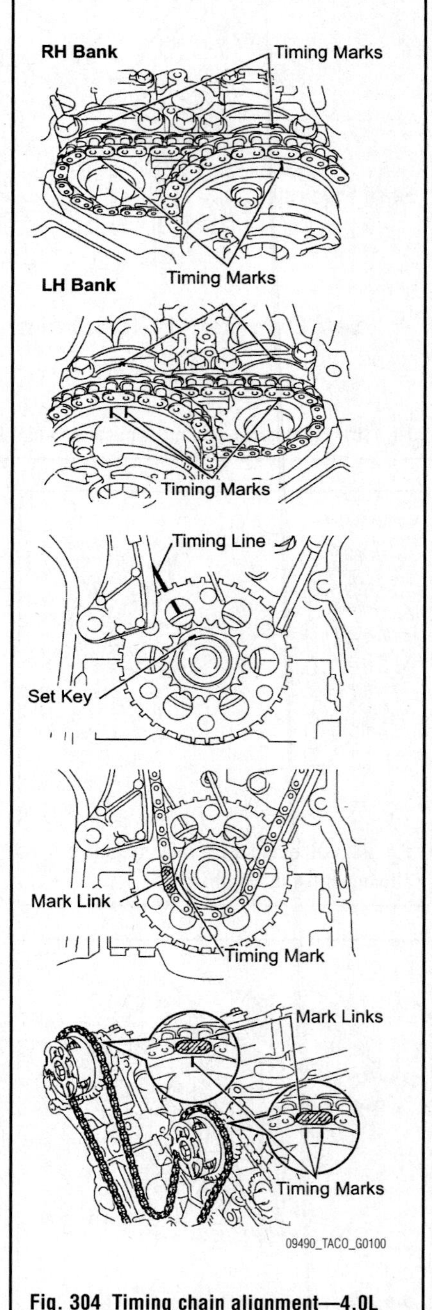

Fig. 304 Timing chain alignment—4.0L engine

the bolts within fifteen minutes of applying the seal packing.

21. Align the key way of the oil pump drive motor with the rectangular portion of the crankshaft timing gear and slide the timing chain case cover into place.
22. Install the timing chain case cover bolts. Tighten the bolts and nuts uniformly in several steps to 17 ft. lbs.

➡Do not wrap the chain and slipper over the timing chain case cover seal line.

23. Continue the installation in the reverse order of the removal procedure.
24. When installing the cylinder head cover apply a continuous bead (0.08–0.12 inch) of seal packing, part number 08826-00080 or equivalent, to the cylinder head. Install the seal washers onto the bolts. Install the cylinder head cover bolts and nuts. Tighten bolts "A" to 7.4 ft. lbs. Tighten bolts "B" to 80 inch lbs. Tighten nuts to 80 inch lbs.

➡Be sure to remove any oil from the contact surfaces of the cylinder head cover and the cylinder head. Install the cover within three minutes after applying the seal packing. Tighten the bolts to specification within fifteen minutes after installing the cover. Do not add engine oil for at least two hours after installing the cover.

25. Be sure to fill the engine with the proper grade and type engine coolant.
26. Be sure to fill the engine with the proper grade and type engine oil.
27. Start the engine and check for leaks, correct as required
28. Use the Techstream diagnostic tool, or equivalent and reprogram the required systems.

4.6L Engine

See Figures 305 through 338.

1. Before servicing the vehicle, refer to the Precautions Section.

➡If working near and/or around the SRS system and components, be sure to disable the SRS system. Tape the negative battery cable with insulating tape. Always disconnect the negative battery cable first.

CAUTION

To avoid personal injury when working on vehicles equipped with an air bag, the negative battery cable must be disconnected and at least 90 seconds must elapse before working on the system. Failure to do so may result in deployment of the air bag.

2. Disconnect the negative battery cable. Tape the cable with insulating tape.

➡When disconnecting the cable, some systems need to be initialized after the cable is reconnected. You will need the Techstream diagnostic scan tool or equivalent. Follow the directions on the tool.

3. Discharge fuel system pressure.
4. Drain engine oil.
5. Drain engine coolant.
6. Remove the radiator.
7. Remove air cleaner hose assembly.
8. Remove air cleaner assembly.
9. Engine Room LH Side:
   a. Remove the engine room relay block cover.
   b. Disconnect the 2 connectors and detach the 2 clamps from the engine room junction block.
   c. Disconnect the 4 air injection control driver connectors.
   d. Disconnect the 2 wire harness clamps.
   e. Disconnect the injector connector.
   f. Disconnect the 4 ignition coil connectors.
   g. Disconnect the 2 VVT sensor connectors.
   h. Disconnect the 4 clamps.
   i. Remove the 2 bolts and ground wire.
   j. Disconnect the noise filter connector.
   k. Disconnect the engine coolant temperature sensor connector.
   l. Disconnect the 2 camshaft timing oil control valve connectors.
   m. Disconnect the camshaft position sensor connector.

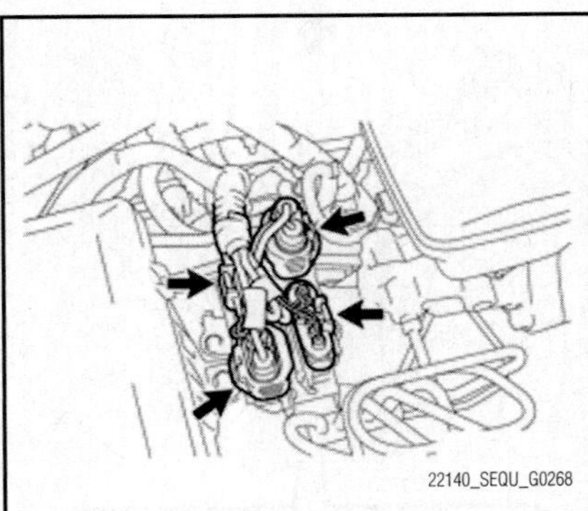

Fig. 305 Air injection control connectors—4.6L engine

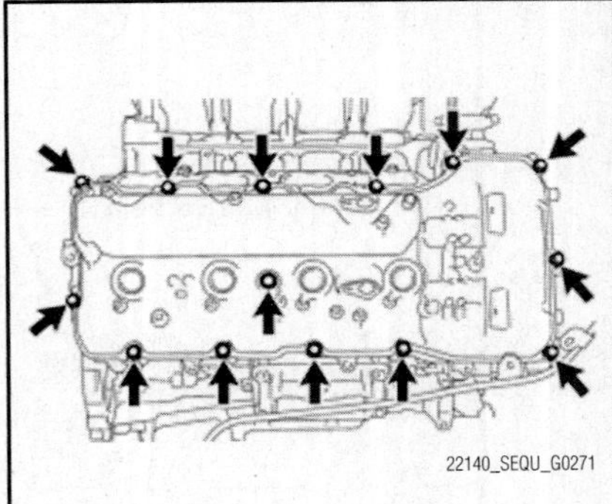

Fig. 308 Cylinder head cover RH—4.6L engine

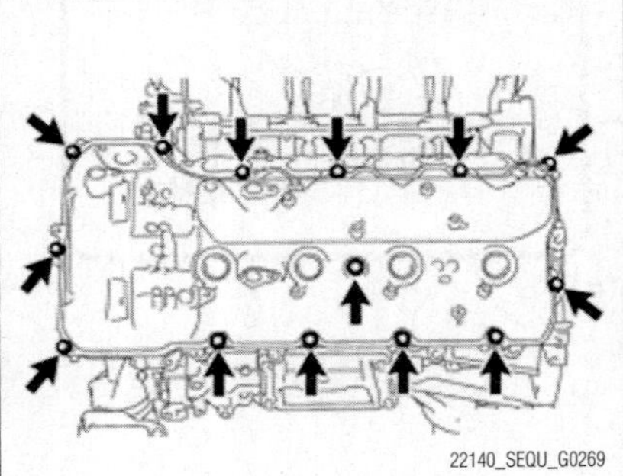

Fig. 306 Cylinder head cover LH—4.6L engine

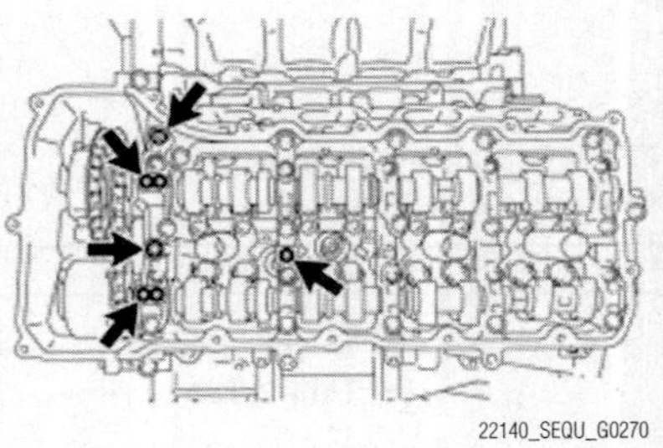

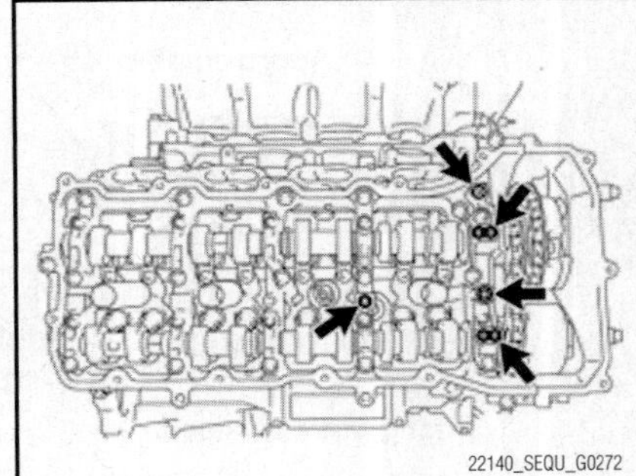

Fig. 309 Cylinder head cover gasket locations—4.6L engine

Fig. 307 Cylinder head cover gasket locations—4.6L engine

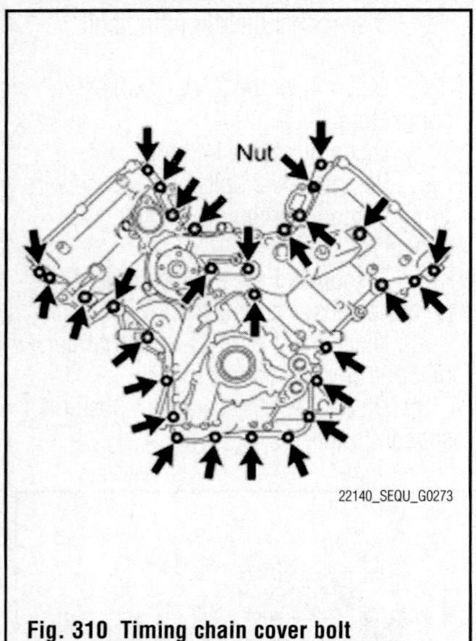

Fig. 310 Timing chain cover bolt locations—4.6L engine

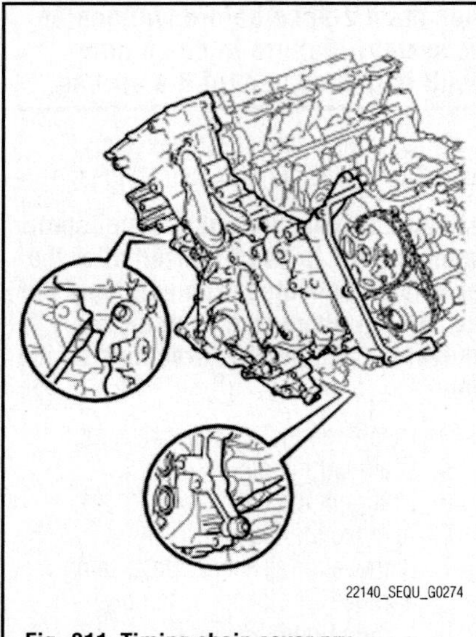

Fig. 311 Timing chain cover pry locations—4.6L engine

n. Disconnect the 3 clamps.

o. Disconnect the cooler compressor connector.

10. Engine Room RH Side.

a. Disconnect the 2 camshaft timing oil control valve connectors.

b. Disconnect the 4 ignition coil connectors.

c. Disconnect the injector connector.

d. Disconnect the 2 VVT sensor connectors.

e. Disconnect the noise filter connector.

f. Remove the 2 bolts and ground wire.

g. Disconnect the 2 air pump connectors.

h. Disconnect the throttle position sensor and throttle control motor connector.

i. Disconnect the 5 clamps.

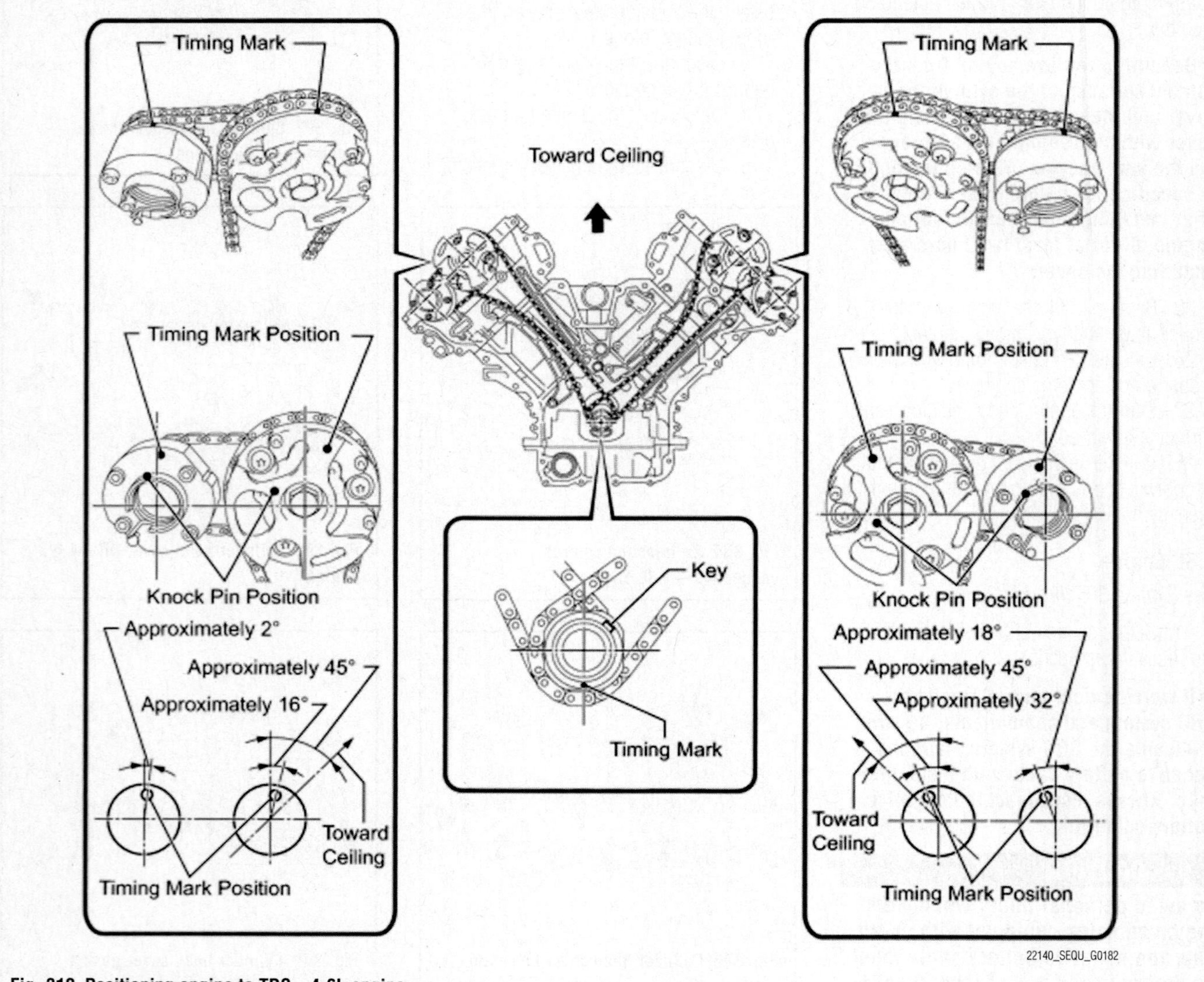

Fig. 312 Positioning engine to TDC—4.6L engine

j. Disconnect the 2 clamps and power steering oil pressure switch connector.

k. For 4WD: Disconnect the 2 clamps and power steering oil pressure switch connector.

11. Remove the intake manifold.
12. Disconnect the No. 2 and No. 3 air hoses and the 2 clamps.
13. Remove the 2 bolts and disconnect the heater 3 hoses.
14. Remove the 2 bolts, 2 nuts and 2 stud bolts, and disconnect the cooler compressor.

➡It is not necessary to completely remove the compressor. With the hoses connected to the compressor, hang the compressor on the vehicle body with a rope.

15. Remove the 2 bolts and disconnect the fuel tube.
16. Remove oil filter element.
17. Remove oil pressure sender gauge assembly.
18. Disconnect the 4 hoses and remove the water by-pass pipe.
19. Remove No. 1 oil cooler bracket (w/ oil cooler).
20. Remove the 2 bolts, 2 nuts and oil filter bracket.
21. Remove the 2 O-rings.
22. Remove engine oil level dipstick guide.
23. Remove the 2 bolts and disconnect the vane pump.
24. Disconnect the alternator connector.
25. Remove the terminal cap and nut, and disconnect the alternator wire.
26. Remove the bolt and disconnect the wire harness bracket from the alternator.
27. Remove the 2 bolts, 2 nuts and alternator.
28. Remove the 2 stud bolts.
29. Remove the No. 1 water by-pass hose by disconnect the hose from the water inlet housing and front water by-pass joint.
30. Remove the bolt and disconnect the air tube.
31. Remove the 2 bolts and hoses and water by-pass pipe.
32. Disconnect the No. 2 water by-pass hose from the water by-pass joint.
33. Remove the 4 nuts, water by-pass joint and 2 gaskets.
34. Remove No.1 and No. 2 engine covers .
35. Remove the bolt, disconnect the 2 hoses, and remove the air tube.
36. Remove the 3 bolts, water inlet housing and gasket.
37. Remove the 4 bolts and water pump pulley.
38. Remove the bolt and idler pulley.
39. Remove the 4 bolts and fan bracket.
40. Remove the standard bolt, 6 mm hexagon wrench bolt and belt tensioner.
41. Remove the 8 bolts and 8 ignition coils.
42. Remove cylinder head cover sub-assembly LH.

a. Remove the 14 bolts, seal washer, cylinder head cover and gasket.

b. Remove the 5 gaskets from the camshaft bearing caps (No. 2, No. 3).

43. Remove cylinder head cover sub-assembly RH.

a. Remove the bolt and noise filter.

b. Remove the 14 bolts, seal washer, cylinder head cover and gasket.

c. Remove the 5 gaskets from the camshaft bearing caps (No. 1, No. 3).

44. Remove crankshaft pulley.

a. Loosen the crankshaft pulley set bolt.

b. Partially install the pulley set bolt to the crankshaft until 2 or 3 threads are engaged.

c. Remove the crankshaft pulley.

45. Remove the bolt and disconnect the wire harness bracket.

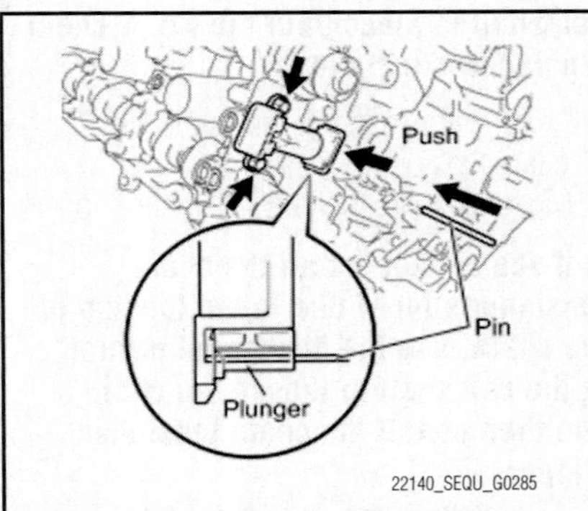

Fig. 313 Timing chain tensioner assembly No. 2—4.6L engine

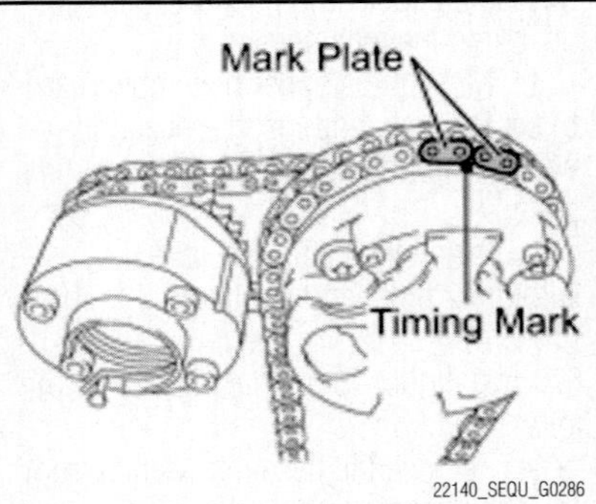

Fig. 314 Timing chain positioning RH—4.6L engine

46. Remove timing chain cover sub-assembly.

a. Remove the 28 bolts and nut shown.

b. Remove the timing chain cover by prying between it and the cylinder head and cylinder block with a screwdriver as shown

47. Remove the oil pump gasket from the cylinder block.
48. Remove the O-ring from the oil pan.
49. Remove the water inlet pipe.
50. Remove the 2 O-rings from the water inlet pipe.
51. Set No. 1 cylinder to TDC / compression.
52. Remove chain tensioner assembly.
53. Remove chain tensioner slipper.
54. Remove chain vibration damper.
55. Remove chain sub-assembly.
56. Remove chain tensioner assembly.

To install:

➡Be sure to use new fasteners, as required.

57. Check No. 1 cylinder to TDC.
58. Install No.2 chain tensioner assembly. While raising up the No. 2 chain tensioner, insert a pin of 1.0 mm (0.0394 in.) into the hole to fix it in place.

Fig. 315 Timing chain positioning RH—4.6L engine

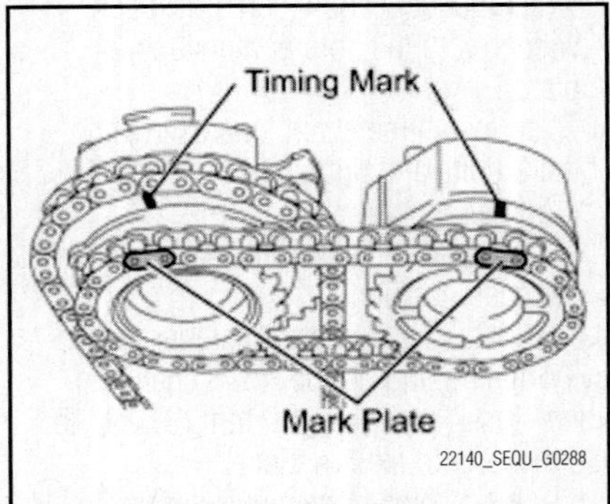

Fig. 316 Timing chain positioning RH—4.6L engine

59. Install No. 1 chain sub-assembly RH.

a. Align the No. 1 chain's orange mark plates with the camshaft timing gear's timing mark, and attach the chain to the gear as shown.

b. Align the No. 1 chain's orange mark plate with the crankshaft timing gear's timing mark, and attach the chain to the gear as shown.

c. Align the No. 2 chain's mark plates (yellow) with the timing marks of the camshaft timing gear and camshaft timing exhaust gear, and attach the No. 2 chain to the gears as shown.

➡The crankshaft timing gear and camshaft exhaust gear will be installed with the No. 1 and No. 2 chains connected to the gears.

d. Install the crankshaft timing sprocket to the crankshaft.

e. Align and attach the knock pin of the No. 1 camshaft with the pin hole of the camshaft timing gear.

f. Using the hexagonal portion of the No. 2 camshaft, align and attach the knock pin of the No. 2 camshaft with the pin hole of the camshaft timing exhaust gear.

g. Remove the pin from the No. 2 chain tensioner.

60. Install the vibration damper with the 2 bolts.

61. Install No. 1 chain tensioner slipper.

➡If you cannot install the chain tensioner slipper due to the tension of the chain, use the hexagonal portion of the camshaft to loosen the chain, and then install the chain tensioner slipper

62. Install No. 1 chain tensioner assembly.

a. Move the stopper plate upward to release the lock, and push the plunger deep into the tensioner.

b. Move the stopper plate downward to set the lock, and insert a hexagon wrench into the hole of the stopper plate.

c. Install the chain tensioner with the 2 bolts and tighten to 7 ft. lbs. (10 Nm).

d. Remove the hexagon wrench from the chain tensioner.

63. Install No.3 chain tensioner assembly. While raising up the No. 3 chain tensioner, insert a pin of _1.0 mm (0.0394 in.) into the hole to fix it in place.

64. Install No. 1 chain sub-assembly LH.

a. Align the No. 1 chain's orange mark plates with the camshaft timing gear's timing mark, and attach the chain to the gear as shown.

b. Align the No. 1 chain's orange mark plate with the crankshaft timing gear's timing mark, and attach the chain to the gear as shown

c. Align the No. 2 chain's mark plates (yellow) with the timing marks of the camshaft timing gear and camshaft timing exhaust gear, and attach the No. 2 chain to the gears as shown

➡The crankshaft timing gear and camshaft exhaust gear will be installed with the No. 1 and No. 2 chains connected to the gears.

d. Install the crankshaft timing sprocket to the crankshaft.

e. Align and attach the knock pin of the No. 3 camshaft with the pin hole of the camshaft timing gear.

f. Using the hexagonal portion of the No. 4 camshaft, align and attach the knock pin of the No. 4 camshaft with the pin hole of the camshaft timing exhaust gear.

➡Because the gears' timing mark positions may shift due to looseness of the No. 1 chain, use the hexagonal portion of the camshaft to hold the No. 3 camshaft in place until the No. 1 chain tensioner is installed.

g. Remove the pin from the No. 2 chain tensioner.

65. Install No. 1 chain tensioner slipper.

➡If you cannot install the chain tensioner slipper due to the tension of the chain, use the hexagonal portion of the camshaft to loosen the chain, and then install the chain tensioner slipper.

66. Install No. 1 chain tensioner assembly.

a. Move the stopper plate upward to release the lock, and push the plunger deep into the tensioner.

b. Move the stopper plate downward to set the lock, and insert a hexagon wrench into the hole of the stopper plate.

c. Install the chain tensioner with the 2 bolts and tighten to 7 ft. lbs. (10 Nm).

67. Install the vibration damper with the 2 bolts

a. Remove the hexagon wrench from the chain tensioner.

68. Tighten camshaft timing gears LH.

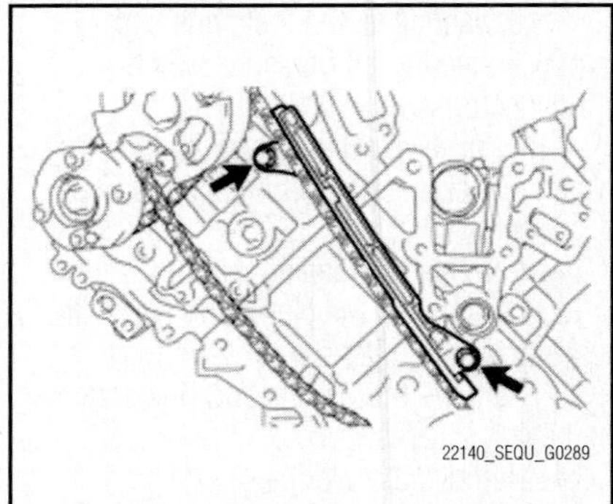

Fig. 317 Timing chain vibrations damper positioning RH—4.6L engine

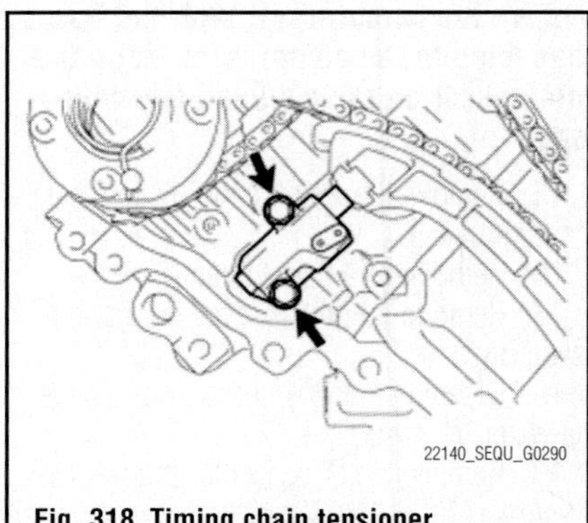

Fig. 318 Timing chain tensioner positioning RH—4.6L engine

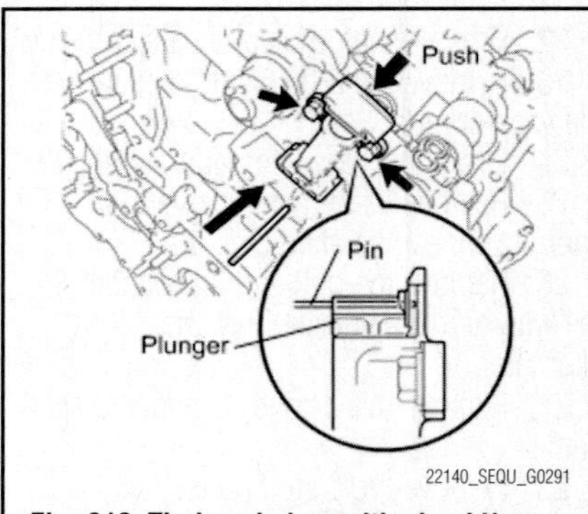

Fig. 319 Timing chain positioning LH—4.6L engine

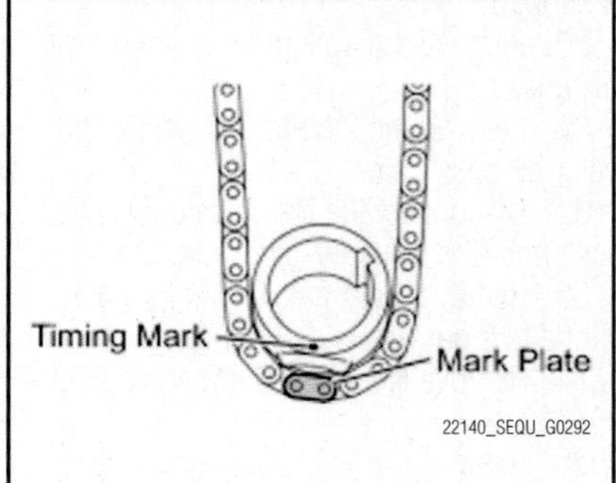

Fig. 320 Timing chain positioning LH—4.6L engine

a. Using a wrench to hold the hexagonal portion of the No. 3 camshaft, tighten the camshaft timing gear with the bolt and tighten to 74 ft. lbs. (100 Nm).

b. Using a wrench to hold the hexagonal portion of the No. 4 camshaft, tighten the camshaft timing gear with the bolt and tighten to 74 ft. lbs. (100 Nm).

69. Tighten camshaft timing gears RH.

a. Using a wrench to hold the hexagonal portion of the No. 1 camshaft, tighten the camshaft timing gear with the bolt and tighten to 74 ft. lbs. (100 Nm).

b. Using a wrench to hold the hexagonal portion of the No. 2 camshaft, tighten the camshaft timing gear with the bolt and tighten to 74 ft. lbs. (100 Nm).

70. Check No. 1 cylinder to TDC / compression.

a. Temporarily install the pulley set bolt.

b. Rotate the crankshaft clockwise, and check that the timing marks on the crankshaft timing gear and camshaft timing gears are as shown.

c. Remove the crankshaft pulley bolt.

71. Apply soapy water to 2 new O-rings and install them to the inlet pipe.

72. Install the inlet pipe to the No. 1 heat exchanger cover.

73. Install timing chain cover sub-assembly.

a. Apply a light coat of engine oil to a new oil pump gasket.

b. Install the oil pump gasket.

c. Apply a light coat of engine oil to a new O-ring.

d. Install the O-ring.

e. Apply seal packing in a continuous line to the timing chain cover as shown.

f. Align the oil pump's drive rotor spline and the crankshaft as shown in the illustration. Install the spline and chain cover to the crankshaft.

g. Temporarily install the timing chain cover with the 28 bolts and nut.

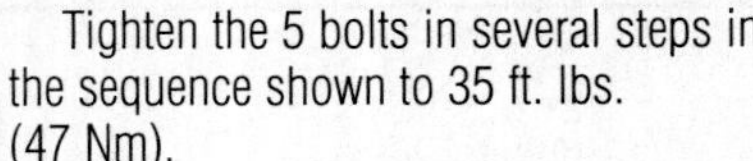

Tighten the 5 bolts in several steps in the sequence shown to 35 ft. lbs. (47 Nm).

h. Temporarily install the fan bracket with the 4 bolts.

i. Temporarily install the belt tensioner with the standard bolt and 6 mm hexagon wrench bolt.

j. Tighten the 23 bolts labeled 12 to 35 and nut in several steps in the sequence to 17 ft. lbs. (23 Nm).

74. Install cylinder head cover sub-assembly LH.

a. Install 5 new gaskets to the camshaft bearing caps (No. 2, No. 3).

b. Install the gasket to the cylinder head cover.

c. Apply seal packing as shown.

d. Install the cylinder head cover washer with a new seal and the 14 bolts and tighten bolt A to 15 ft. lbs. (21 Nm) and remaining bolts to 9 ft. lbs. (12 Nm).

75. Install cylinder head cover sub-assembly RH.

a. Install 5 new gaskets to the camshaft bearing caps (No. 1, No. 3).

b. Install the gasket to the cylinder head cover.

c. Apply seal packing as shown.

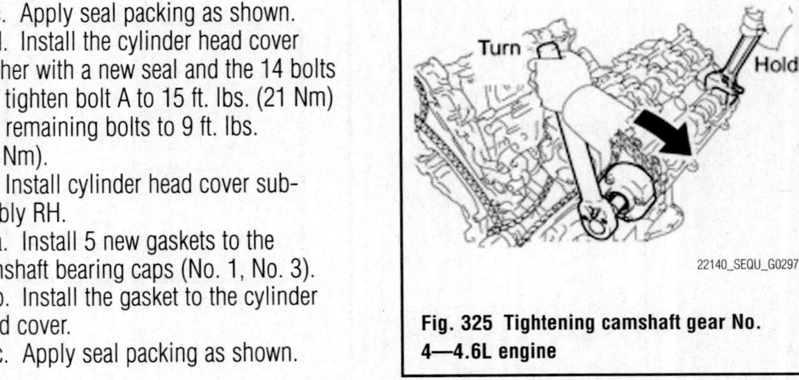

Fig. 324 Tightening camshaft gear No. 3—4.6L engine

Fig. 325 Tightening camshaft gear No. 4—4.6L engine

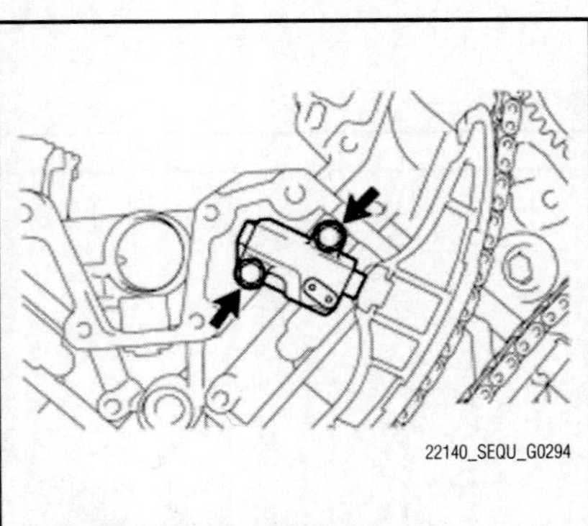

Fig. 322 Timing chain tensioner positioning LH—4.6L engine

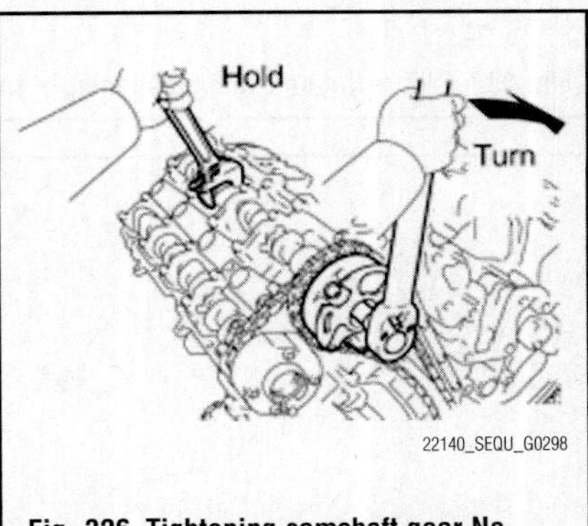

Fig. 326 Tightening camshaft gear No. 1—4.6L engine

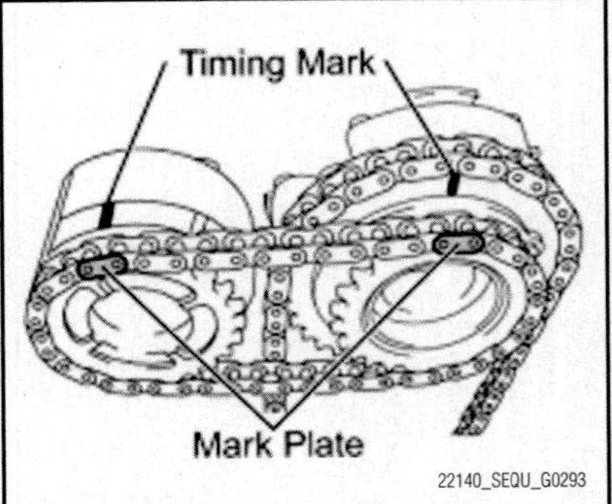

Fig. 321 Timing chain positioning LH—4.6L engine

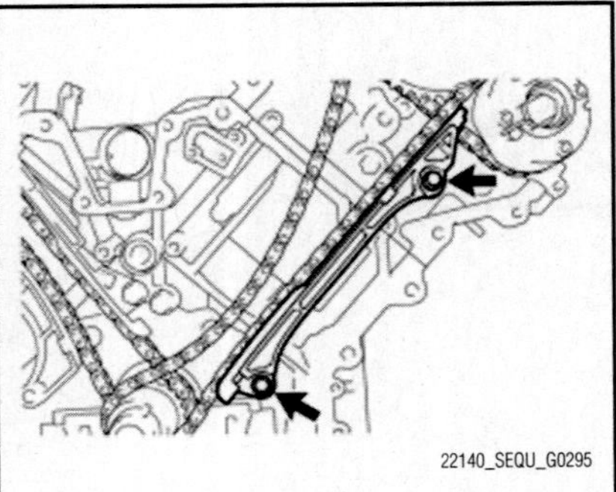

Fig. 323 Timing chain vibrations damper positioning RH—4.6L engine

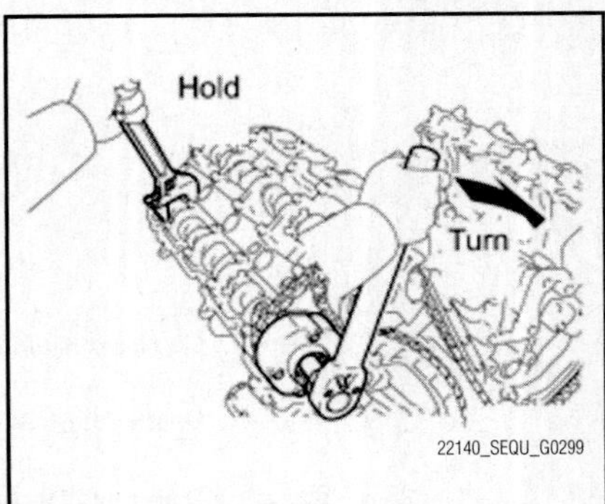

Fig. 327 Tightening camshaft gear No. 2—4.6L engine

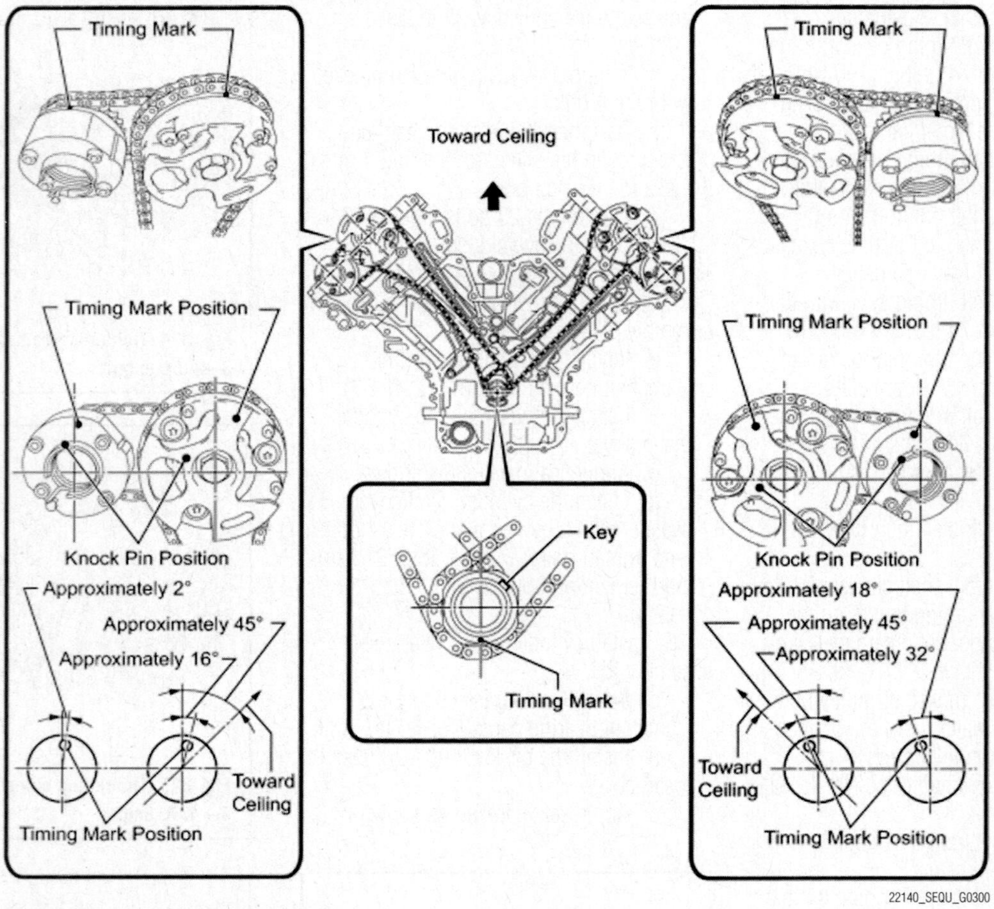

22140_SEQU_G0300

Fig. 328 Proper timing chain positioning—4.6L engine

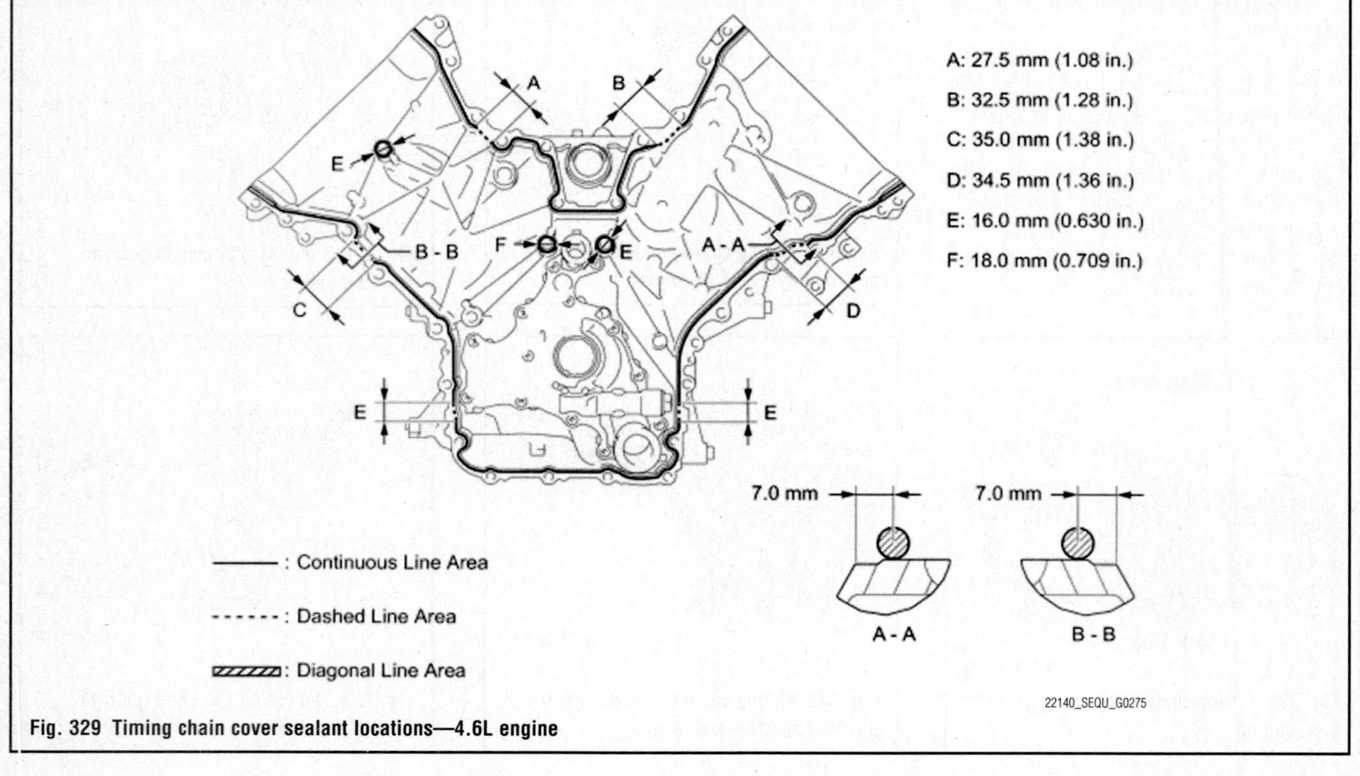

Fig. 329 Timing chain cover sealant locations—4.6L engine

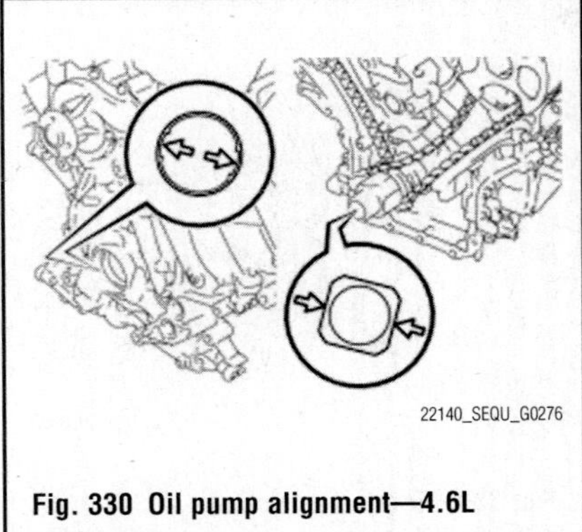

Fig. 330 Oil pump alignment—4.6L engine

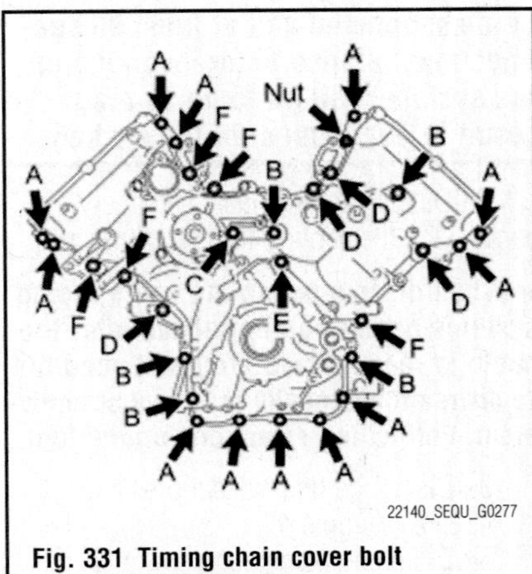

Fig. 331 Timing chain cover bolt locations—4.6L engine

d. Install the cylinder head cover washer with a new seal and the 14 bolts and tighten bolt A to 15 ft. lbs. (21 Nm) and remaining bolts to 9 ft. lbs. (12 Nm).

e. Install the noise filter to the cylinder head cover with the bolt and tighten to 62 inch lbs. (7.0 Nm).

76. Install the 8 ignition coils with the 8 bolts and tighten to 7 ft. lbs. (10 Nm).

Item	Length	Thread diameter
Bolt A	25 mm (0.984 in.)	8 mm (0.315 in.)
Bolt B	55 mm (2.165 in.)	8 mm (0.315 in.)
Bolt C	70 mm (2.756 in.)	8 mm (0.315 in.)
Bolt D	35 mm (1.378 in.)	10 mm (0.394 in.)
Bolt E	55 mm (2.165 in.)	10 mm (0.394 in.)
Bolt F	80 mm (3.150 in.)	10 mm (0.394 in.)

22140_SEQU_G0278

Fig. 332 Timing chain cover bolt application chart—4.6L engine

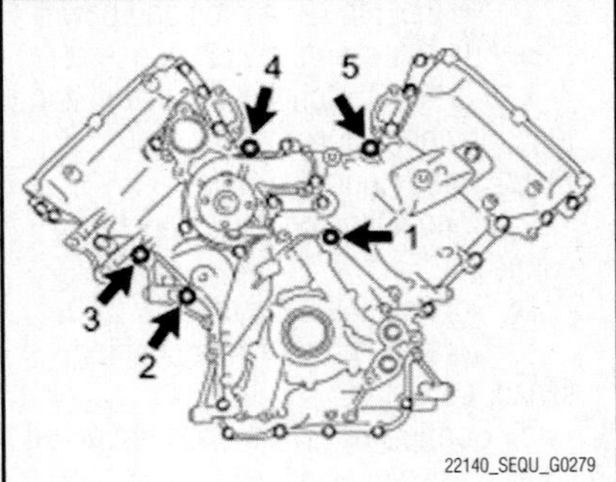

Fig. 333 Timing chain cover bolt tightening sequence No. 1—4.6L engine

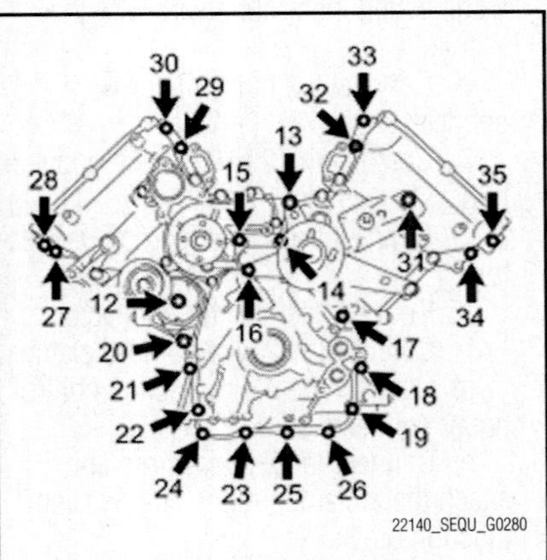

Fig. 334 Timing chain cover bolt tightening sequence No. 2—4.6L engine

77. Align the pulley set key with the key groove of the pulley, and slide on the pulley and tighten to 221 ft. Lbs. (300 Nm).

78. Connect the bracket to the timing chain cover with the bolt.

79. Install the idler pulley with the bolt and tighten to 32 ft. lbs. (43 Nm).

80. Temporarily install the water pump pulley with the 4 bolts and tighten to 15 ft. lbs. (21 Nm).

81. Install a new water cover gasket to the timing chain cover.

82. Install the water inlet with the 3 bolts and tighten to 15 ft. lbs. (21 Nm).

83. Connect the 2 air tube hoses.

84. Connect the 2 hoses.

85. Install No. 1 and No 2 engine covers.

86. Install 2 new gaskets and the water by-pass joint with the 4 nuts and tighten to 15 ft. lbs. (21 Nm).

87. Connect the No. 2 water by-pass hose to the water by-pass joint.

88. Install the water by-pass pipe.

89. Connect the air tube.

90. Install the No. 1 water by-pass hose by connecting the hose to the water inlet housing and front water by-pass joint.

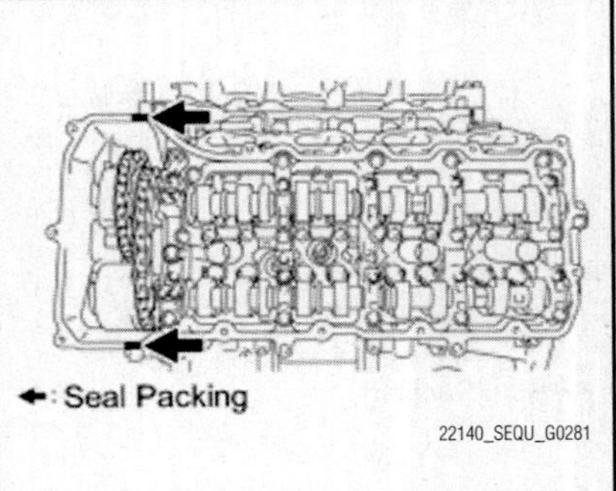

Fig. 335 Cylinder head cover packing location LH—4.6L engine

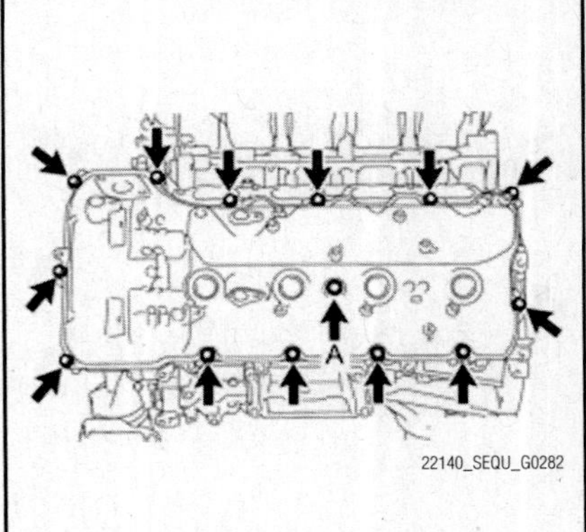

Fig. 336 Cylinder head cover bolt tightening sequence LH—4.6L engine

91. Install the 2 stud bolts.

92. Install the alternator with the 2 bolts and 2 nuts and tighten to 32 ft. lbs. (43 Nm).

93. Connect the alternator wire with the nut and tighten to 87 inch lbs. (9.8 Nm).

94. Connect the wire harness bracket to the alternator.

95. Connect the vane pump to the timing chain cover with the 2 bolts and tighten to 21 ft. lbs. (28 Nm).

96. Install engine oil level dipstick guide.

97. Install oil filter bracket (w/o Oil Cooler).

a. Apply a light coat of engine oil to 2 new O-rings.

b. Install the 2 O-rings to the timing chain cover.

c. Install the oil filter bracket with the 2 bolts and 2 nuts and tighten to 26 ft. lbs. (35 Nm).

98. Install No. 1 oil cooler bracket.

99. Install No. 2 water by-pass pipe.

100. Install oil pressure sender gauge assembly.

101. Install oil filter element.

102. Connect No. 2 fuel tube sub-assembly.

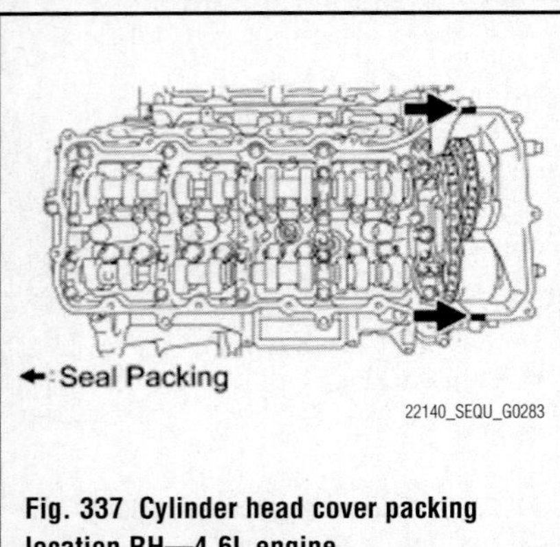

Fig. 337 Cylinder head cover packing location RH—4.6L engine

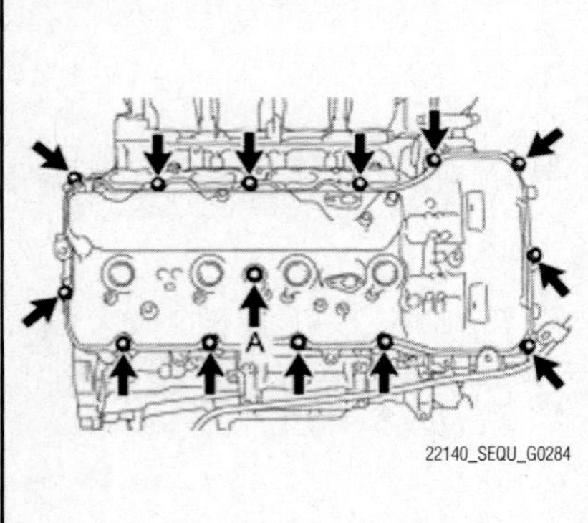

Fig. 338 Cylinder head cover bolt tightening sequence RH—4.6L engine

103. Install the cooler compressor with the 2 stud bolts and tighten to 7 ft. lbs. (10 Nm).

104. Install the 2 bolts and 2 nuts and tighten to 18 ft. lbs. (235 Nm).

105. Connect the 3 heater hoses.

106. Connect the No. 2 and No. 3 air hoses.

107. Connect the 2 clamps and air pump wire harness.

108. Install the intake manifold.

109. Engine Room RH Side:

a. Connect the 5 clamps.

b. Connect the throttle position sensor and throttle control motor connector.

c. Connect the 2 air pump connectors.

d. Install the ground wire with the 2 bolts.

e. Connect the noise filter connector.

f. Connect the 2 VVT sensor connectors.

g. Connect the injector connector.

h. Connect the 4 ignition coil connectors.

i. Connect the 2 camshaft timing oil control valve connectors.

j. Connect the 2 clamps and power steering oil pressure switch connector.

k. For 4WD: Connect the clamp and junction connector.

110. Engine Room LH Side:

a. Connect the cooler compressor connector.

b. Connect the 3 clamps.

c. Connect the camshaft position sensor connector.

d. Connect the 2 camshaft timing oil control valve connectors.

e. Connect the engine coolant temperature sensor connector.

f. Connect the 4 clamps.

g. Install the ground wire with the 2 bolts.

h. Connect the noise filter connector.

i. Connect the 2 VVT sensor connectors.

j. Connect the 4 ignition coil connectors.

k. Connect the injector connector.

l. Connect the 2 wire harness clamps.

m. Connect the 4 air injection control driver connectors.

n. Connect the 2 connectors and attach the 2 clamps to the engine room junction block.

o. Install the engine room relay block cover.

111. Install air cleaner assembly.

112. Install air cleaner hose assembly.

113. Install the radiator.

114. Add engine oil.

115. Add engine coolant.

116. Connect cable to negative battery terminal.

117. Inspect for leaks.

118. Check engine oil level.

119. Use the Techstream diagnostic tool, or equivalent and reprogram the required systems.

5.7L Engine

See Figures 339 through 372.

1. Before servicing the vehicle, refer to the Precautions Section.

➡If working near and/or around the SRS system and components, be sure to disable the SRS system. Tape the negative battery cable with insulating tape. Always disconnect the negative battery cable first.

CAUTION

To avoid personal injury when working on vehicles equipped with an air bag, the negative battery cable must be disconnected and at least 90 seconds must elapse before working on the system. Failure to do so may result in deployment of the air bag.

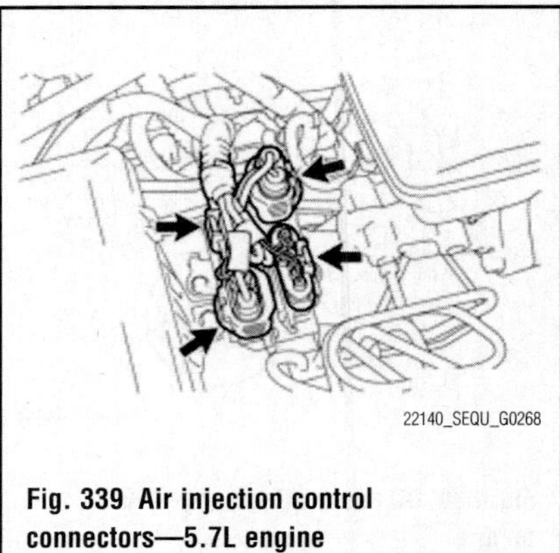

Fig. 339 Air injection control connectors—5.7L engine

2. Disconnect the negative battery cable. Tape the cable with insulating tape.

➡When disconnecting the cable, some systems need to be initialized after the cable is reconnected. You will need the Techstream diagnostic scan tool or equivalent. Follow the directions on the tool.

3. Discharge fuel system pressure.
4. Drain engine oil.
5. Drain engine coolant.
6. Remove the radiator.
7. Remove air cleaner hose assembly.
8. Remove air cleaner assembly.
9. Engine Room LH Side:

a. Remove the engine room relay block cover.

b. Disconnect the 2 connectors and detach the 2 clamps from the engine room junction block.

c. Disconnect the 4 air injection control driver connectors.

d. Disconnect the 2 wire harness clamps.

e. Disconnect the injector connector.

f. Disconnect the 4 ignition coil connectors.

g. Disconnect the 2 VVT sensor connectors.

h. Disconnect the 4 clamps.

i. Remove the 2 bolts and ground wire.

j. Disconnect the noise filter connector.

k. Disconnect the engine coolant temperature sensor connector.

l. Disconnect the 2 camshaft timing oil control valve connectors.

m. Disconnect the camshaft position sensor connector.

n. Disconnect the 3 clamps.

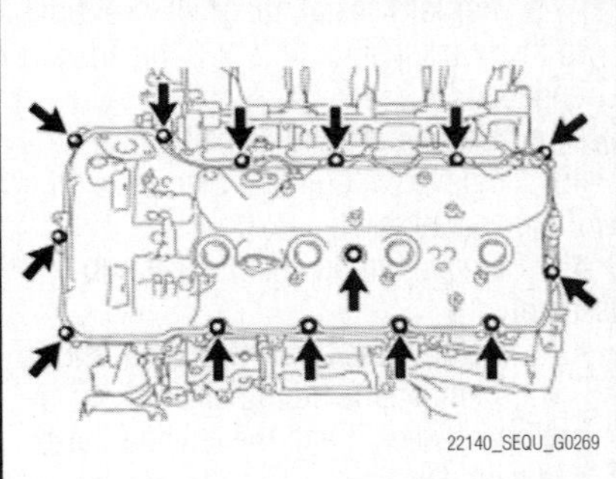

Fig. 340 Cylinder head cover LH—5.7L engine

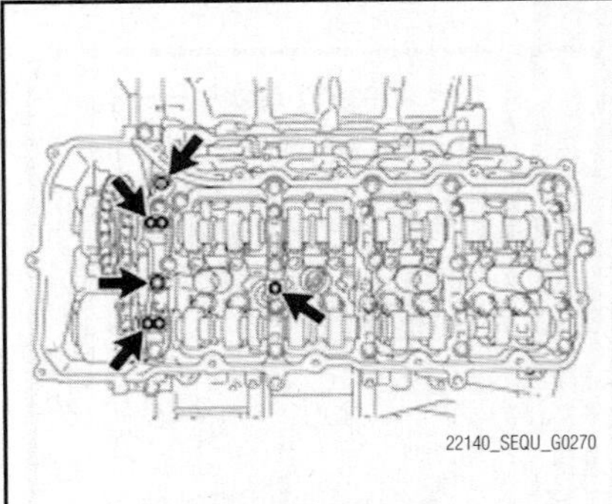

Fig. 341 Cylinder head cover gasket locations—5.7L engine

o. Disconnect the cooler compressor connector.

10. Engine Room RH Side.

a. Disconnect the 2 camshaft timing oil control valve connectors.

b. Disconnect the 4 ignition coil connectors.

c. Disconnect the injector connector.

d. Disconnect the 2 VVT sensor connectors.

e. Disconnect the noise filter connector.

f. Remove the 2 bolts and ground wire.

g. Disconnect the 2 air pump connectors.

h. Disconnect the throttle position sensor and throttle control motor connector.

i. Disconnect the 5 clamps.

j. Disconnect the 2 clamps and power steering oil pressure switch connector.

k. For 4WD: Disconnect the 2 clamps and power steering oil pressure switch connector.

11. Remove the intake manifold.

12. Disconnect the No. 2 and No. 3 air hoses and the 2 clamps.

13. Remove the 2 bolts and disconnect the heater 3 hoses.

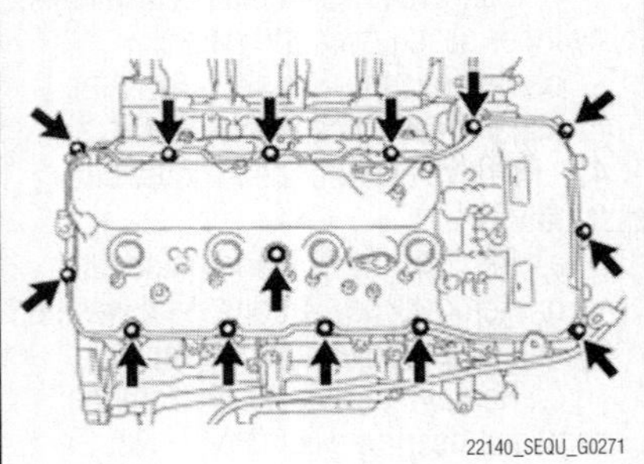

Fig. 342 Cylinder head cover RH—5.7L engine

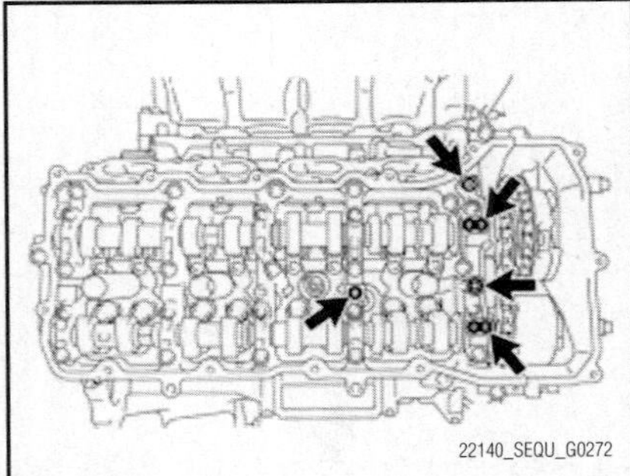
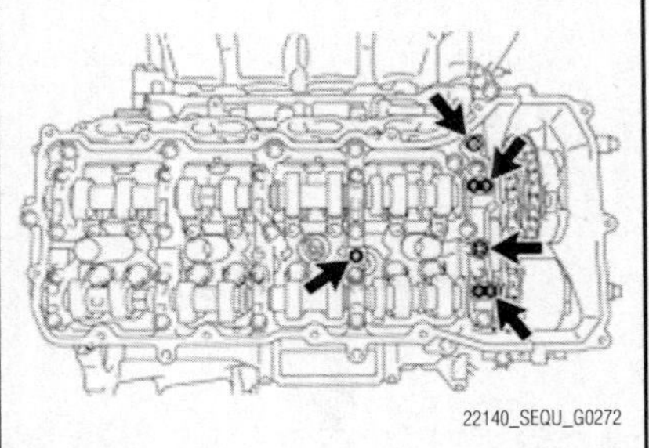

Fig. 343 Cylinder head cover gasket locations—5.7L engine

14. Remove the 2 bolts, 2 nuts and 2 stud bolts, and disconnect the cooler compressor.

➡It is not necessary to completely remove the compressor. With the hoses connected to the compressor, hang the compressor on the vehicle body with a rope.

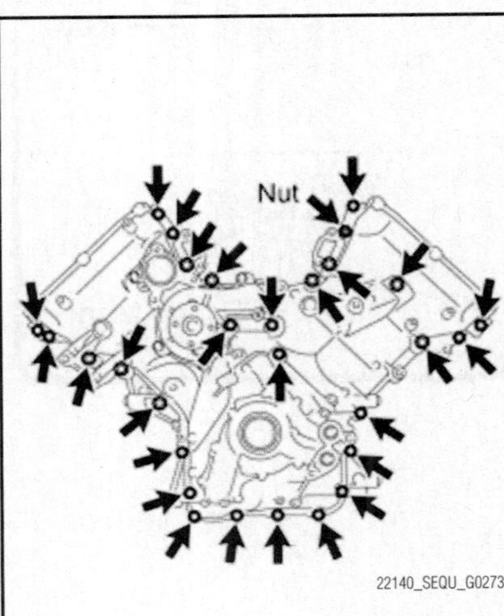

Fig. 344 Timing chain cover bolt locations—5.7L engine

15. Remove the 2 bolts and disconnect the fuel tube.

16. Remove oil filter element.

17. Remove oil pressure sender gauge assembly.

18. Disconnect the 4 hoses and remove the water by-pass pipe.

19. Remove No. 1 oil cooler bracket (w/ oil cooler).

20. Remove the 2 bolts, 2 nuts and oil filter bracket.

21. Remove the 2 O-rings.

22. Remove engine oil level dipstick guide.

23. Remove the 2 bolts and disconnect the vane pump.

24. Disconnect the alternator connector.

25. Remove the terminal cap and nut, and disconnect the alternator wire.

26. Remove the bolt and disconnect the wire harness bracket from the alternator.

27. Remove the 2 bolts, 2 nuts and alternator.

28. Remove the 2 stud bolts.

29. Remove the No. 1 water by-pass hose by disconnect the hose from the water inlet housing and front water by-pass joint.

30. Remove the bolt and disconnect the air tube.

31. Remove the 2 bolts and hoses and water by-pass pipe.

32. Disconnect the No. 2 water by-pass hose from the water by-pass joint.

33. Remove the 4 nuts, water by-pass joint and 2 gaskets.

34. Remove No.1 and No. 2 engine covers .

35. Remove the bolt, disconnect the 2 hoses, and remove the air tube.

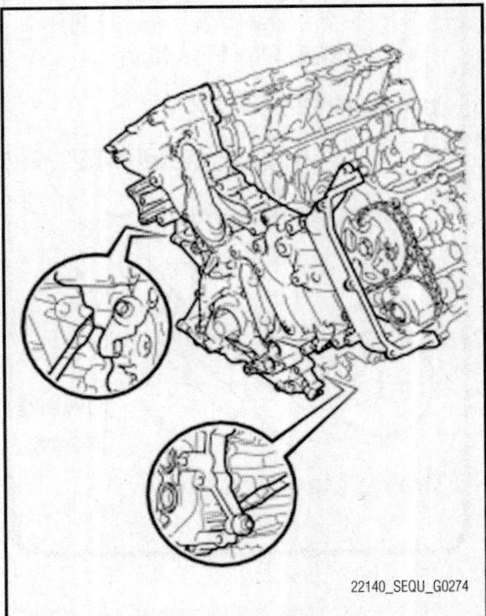

Fig. 345 Timing chain cover pry locations—5.7L engine

36. Remove the 3 bolts, water inlet housing and gasket.
37. Remove the 4 bolts and water pump pulley.
38. Remove the bolt and idler pulley.
39. Remove the 4 bolts and fan bracket.
40. Remove the standard bolt, 6 mm hexagon wrench bolt and belt tensioner.
41. Remove the 8 bolts and 8 ignition coils.
42. Remove cylinder head cover sub-assembly LH.
   a. Remove the 14 bolts, seal washer, cylinder head cover and gasket.
   b. Remove the 5 gaskets from the camshaft bearing caps (No. 2, No. 3).
43. Remove cylinder head cover sub-assembly RH.
   a. Remove the bolt and noise filter.
   b. Remove the 14 bolts, seal washer, cylinder head cover and gasket.
   c. Remove the 5 gaskets from the camshaft bearing caps (No. 1, No. 3).
44. Remove crankshaft pulley.
   a. Loosen the crankshaft pulley set bolt.
   b. Partially install the pulley set bolt to the crankshaft until 2 or 3 threads are engaged.
   c. Remove the crankshaft pulley.
45. Remove the bolt and disconnect the wire harness bracket.
46. Remove timing chain cover sub-assembly.
   a. Remove the 28 bolts and nut shown.
   b. Remove the timing chain cover by prying between it and the cylinder head and cylinder block with a screwdriver as shown

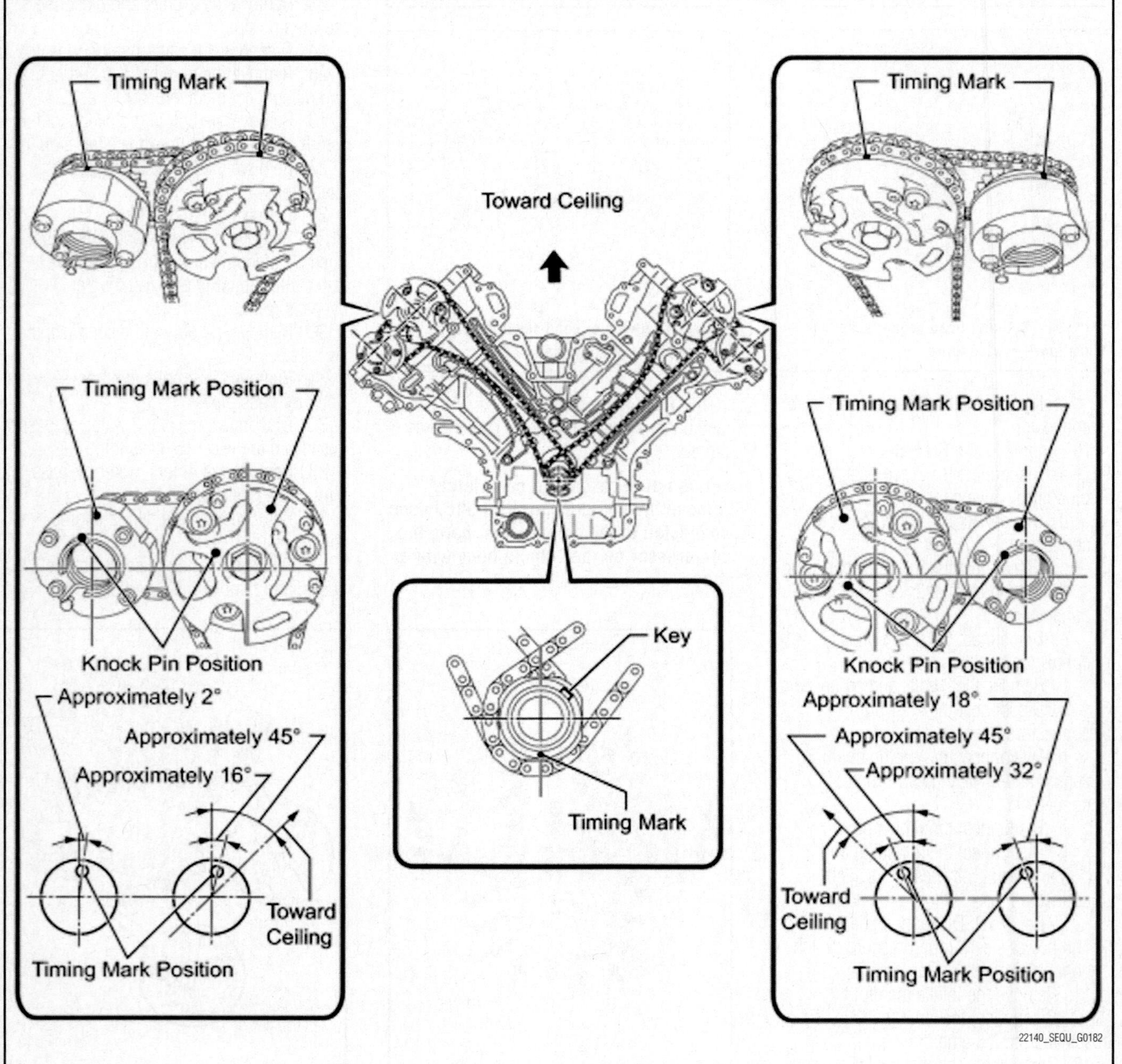

Fig. 346 Positioning engine to TDC—5.7L engine

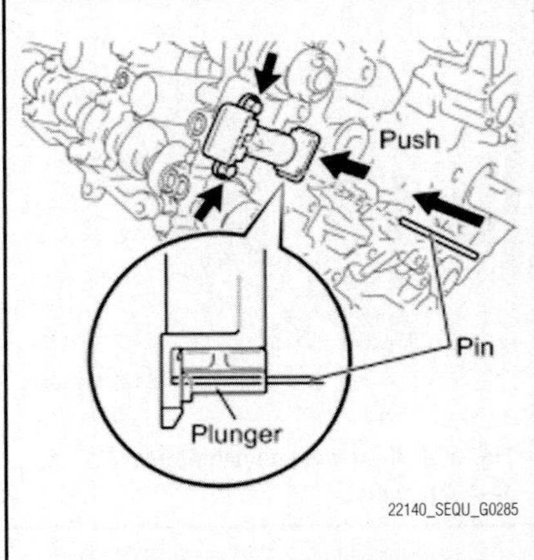

Fig. 347 Timing chain tensioner assembly No. 2—5.7L engine

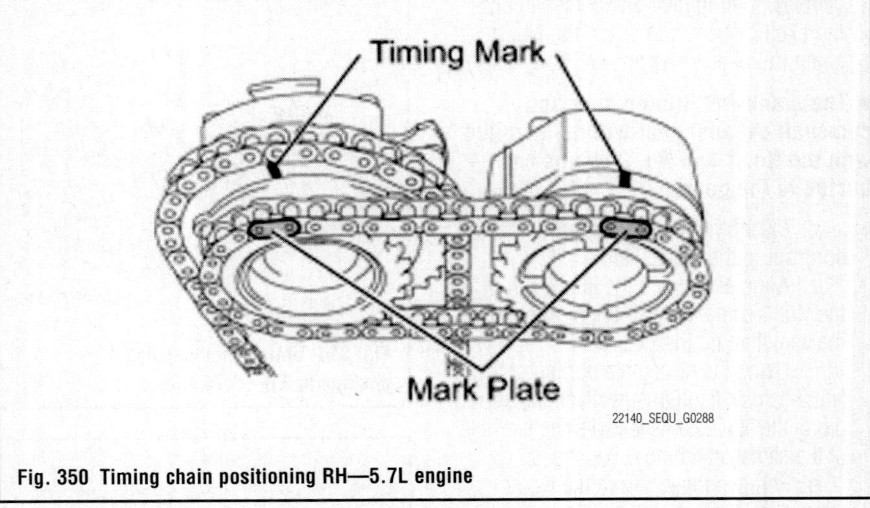

Fig. 350 Timing chain positioning RH—5.7L engine

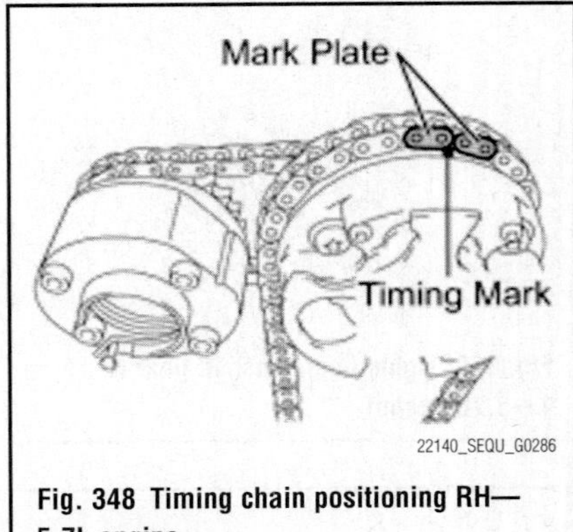

Fig. 348 Timing chain positioning RH—5.7L engine

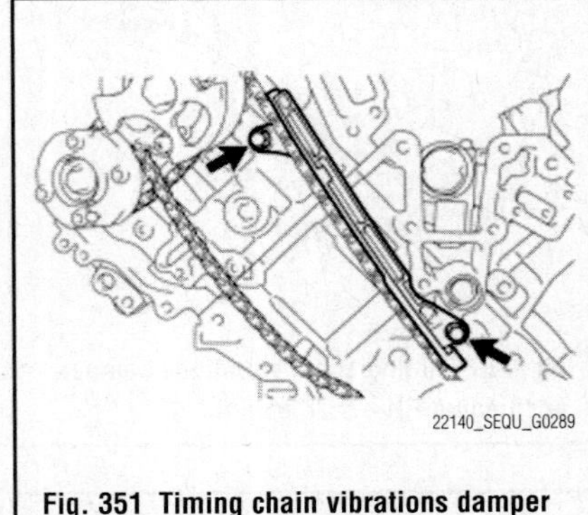

Fig. 351 Timing chain vibrations damper positioning RH—5.7L engine

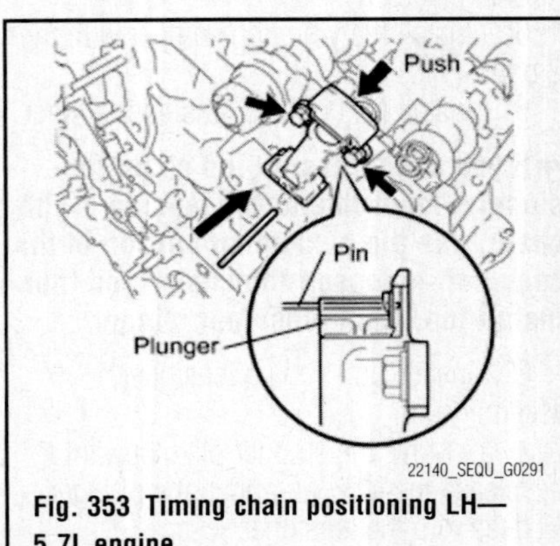

Fig. 353 Timing chain positioning LH—5.7L engine

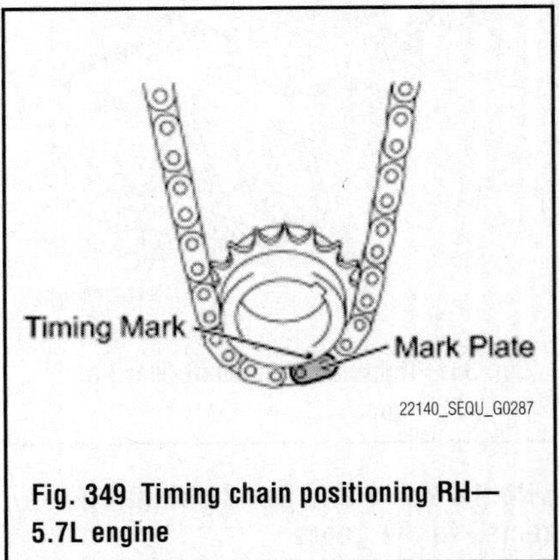

Fig. 349 Timing chain positioning RH—5.7L engine

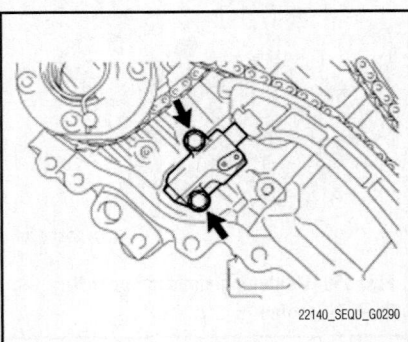

Fig. 352 Timing chain tensioner positioning RH—5.7L engine

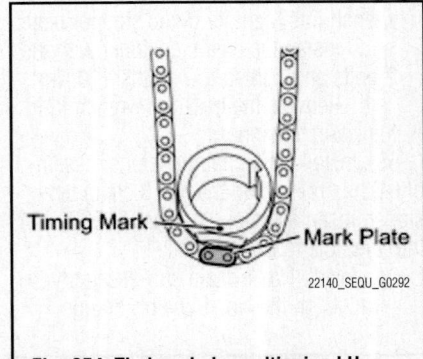

Fig. 354 Timing chain positioning LH—5.7L engine

47. Remove the oil pump gasket from the cylinder block.
48. Remove the O-ring from the oil pan.
49. Remove the water inlet pipe.
50. Remove the 2 O-rings from the water inlet pipe.
51. Set No. 1 cylinder to TDC / compression.
52. Remove chain tensioner assembly.
53. Remove chain tensioner slipper.
54. Remove chain vibration damper.
55. Remove chain sub-assembly.
56. Remove chain tensioner assembly.

To install:

➡Be sure to use new fasteners, as required.

57. Check No. 1 cylinder to TDC.
58. Install No.2 chain tensioner assembly. While raising up the No. 2 chain tensioner, insert a pin of ϕ1.0 mm (0.0394 in.) into the hole to fix it in place.
59. Install No. 1 chain sub-assembly RH.
   a. Align the No. 1 chain's orange mark plates with the camshaft timing gear's timing mark, and attach the chain to the gear as shown.
   b. Align the No. 1 chain's orange mark plate with the crankshaft timing gear's timing mark, and attach the chain to the gear as shown.
   c. Align the No. 2 chain's mark plates (yellow) with the timing marks of the

camshaft timing gear and camshaft timing exhaust gear, and attach the No. 2 chain to the gears as shown.

➡The crankshaft timing gear and camshaft exhaust gear will be installed with the No. 1 and No. 2 chains connected to the gears.

d. Install the crankshaft timing sprocket to the crankshaft.

e. Align and attach the knock pin of the No. 1 camshaft with the pin hole of the camshaft timing gear.

f. Using the hexagonal portion of the No. 2 camshaft, align and attach the knock pin of the No. 2 camshaft with the pin hole of the camshaft timing exhaust gear.

g. Remove the pin from the No. 2 chain tensioner.

60. Install the vibration damper with the 2 bolts.

61. Install No. 1 chain tensioner slipper.

➡If you cannot install the chain tensioner slipper due to the tension of the chain, use the hexagonal portion of the camshaft to loosen the chain, and then install the chain tensioner slipper

62. Install No. 1 chain tensioner assembly.

a. Move the stopper plate upward to release the lock, and push the plunger deep into the tensioner.

b. Move the stopper plate downward to set the lock, and insert a hexagon wrench into the hole of the stopper plate.

c. Install the chain tensioner with the 2 bolts and tighten to 7 ft. lbs. (10 Nm).

d. Remove the hexagon wrench from the chain tensioner.

63. Install No.3 chain tensioner assembly. While raising up the No. 3 chain tensioner, insert a pin of _1.0 mm (0.0394 in.) into the hole to fix it in place.

64. Install No. 1 chain sub-assembly LH.

a. Align the No. 1 chain's orange mark plates with the camshaft timing gear's timing mark, and attach the chain to the gear as shown.

b. Align the No. 1 chain's orange mark plate with the crankshaft timing gear's timing mark, and attach the chain to the gear as shown

c. Align the No. 2 chain's mark plates (yellow) with the timing marks of the camshaft timing gear and camshaft timing exhaust gear, and attach the No. 2 chain to the gears as shown

➡The crankshaft timing gear and camshaft exhaust gear will be installed with the No. 1 and No. 2 chains connected to the gears.

d. Install the crankshaft timing sprocket to the crankshaft.

e. Align and attach the knock pin of the No. 3 camshaft with the pin hole of the camshaft timing gear.

f. Using the hexagonal portion of the No. 4 camshaft, align and attach the knock pin of the No. 4 camshaft with the pin hole of the camshaft timing exhaust gear.

➡Because the gears' timing mark positions may shift due to looseness of the No. 1 chain, use the hexagonal portion

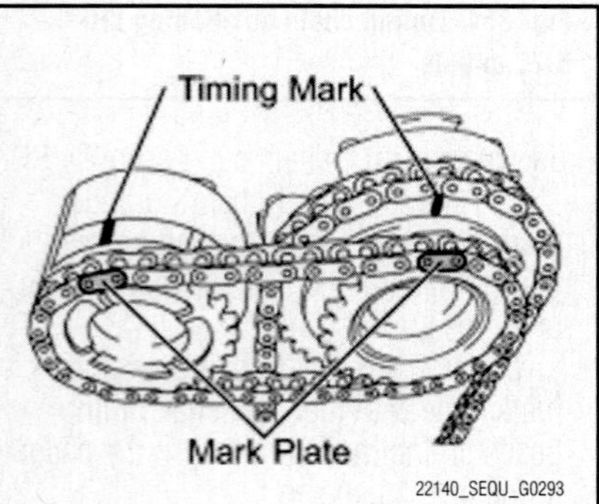

Fig. 355 Timing chain positioning LH—5.7L engine

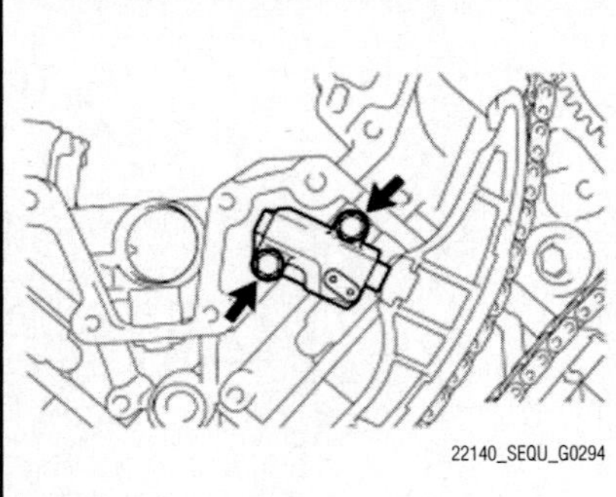

Fig. 356 Timing chain tensioner positioning LH—5.7L engine

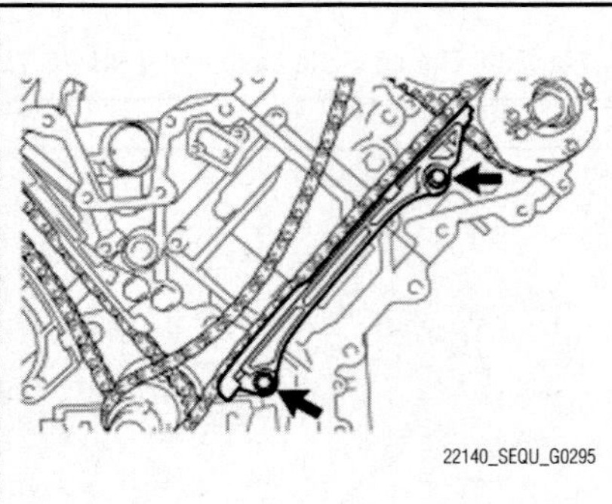

Fig. 357 Timing chain vibrations damper positioning RH—5.7L engine

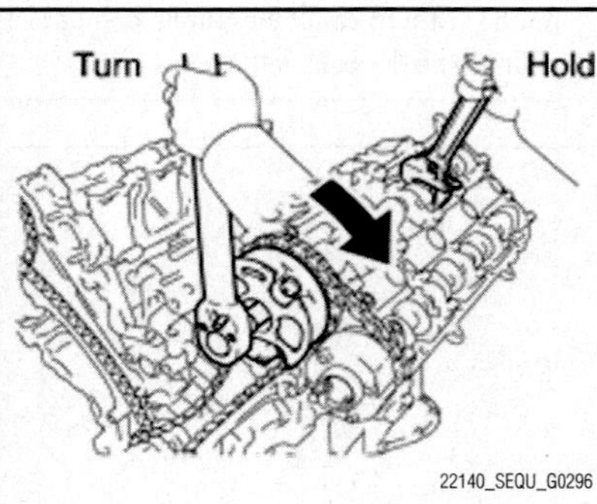

Fig. 358 Tightening camshaft gear No. 3—5.7L engine

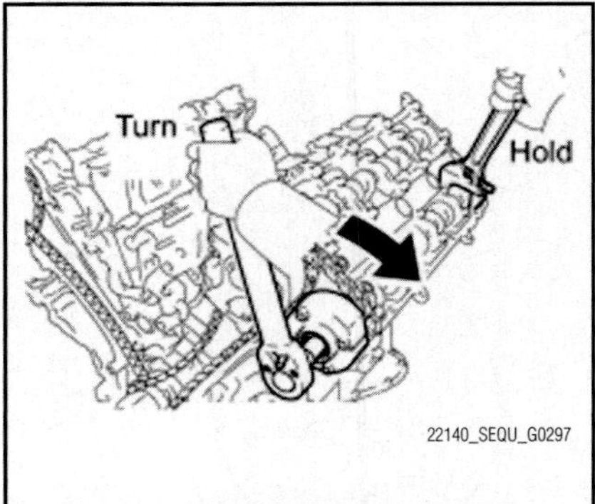

Fig. 359 Tightening camshaft gear No. 4—5.7L engine

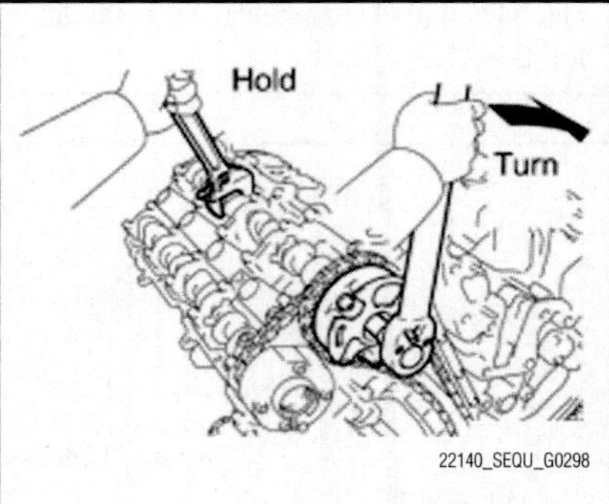

Fig. 360 Tightening camshaft gear No. 1—5.7L engine

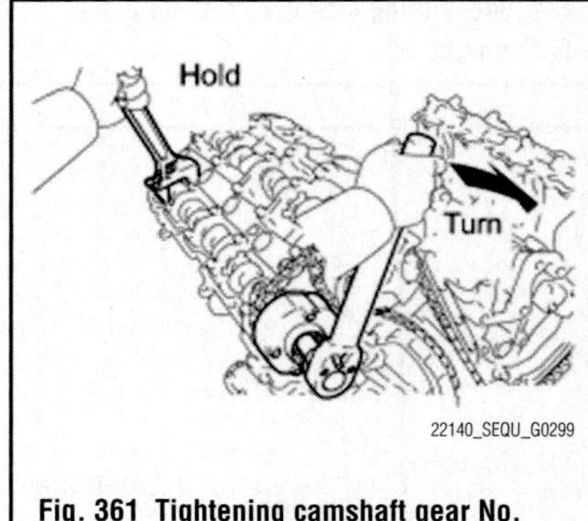

Fig. 361 Tightening camshaft gear No. 2—5.7L engine

of the camshaft to hold the No. 3 camshaft in place until the No. 1 chain tensioner is installed.

g. Remove the pin from the No. 2 chain tensioner.

65. Install No. 1 chain tensioner slipper.

➡If you cannot install the chain tensioner slipper due to the tension of the chain, use the hexagonal portion of the camshaft to loosen the chain, and then install the chain tensioner slipper.

66. Install No. 1 chain tensioner assembly.

a. Move the stopper plate upward to release the lock, and push the plunger deep into the tensioner.

b. Move the stopper plate downward to set the lock, and insert a hexagon wrench into the hole of the stopper plate.

c. Install the chain tensioner with the 2 bolts and tighten to 7 ft. lbs. (10 Nm).

67. Install the vibration damper with the 2 bolts

a. Remove the hexagon wrench from the chain tensioner.

68. Tighten camshaft timing gears LH.

a. Using a wrench to hold the hexagonal portion of the No. 3 camshaft, tighten the camshaft timing gear with the bolt and tighten to 74 ft. lbs. (100 Nm).

b. Using a wrench to hold the hexagonal portion of the No. 4 camshaft, tighten the camshaft timing gear with the bolt and tighten to 74 ft. lbs. (100 Nm).

69. Tighten camshaft timing gears RH.

a. Using a wrench to hold the hexagonal portion of the No. 1 camshaft, tighten the camshaft timing gear with the bolt and tighten to 74 ft. lbs. (100 Nm).

b. Using a wrench to hold the hexagonal portion of the No. 2 camshaft, tighten the camshaft timing gear with the bolt and tighten to 74 ft. lbs. (100 Nm).

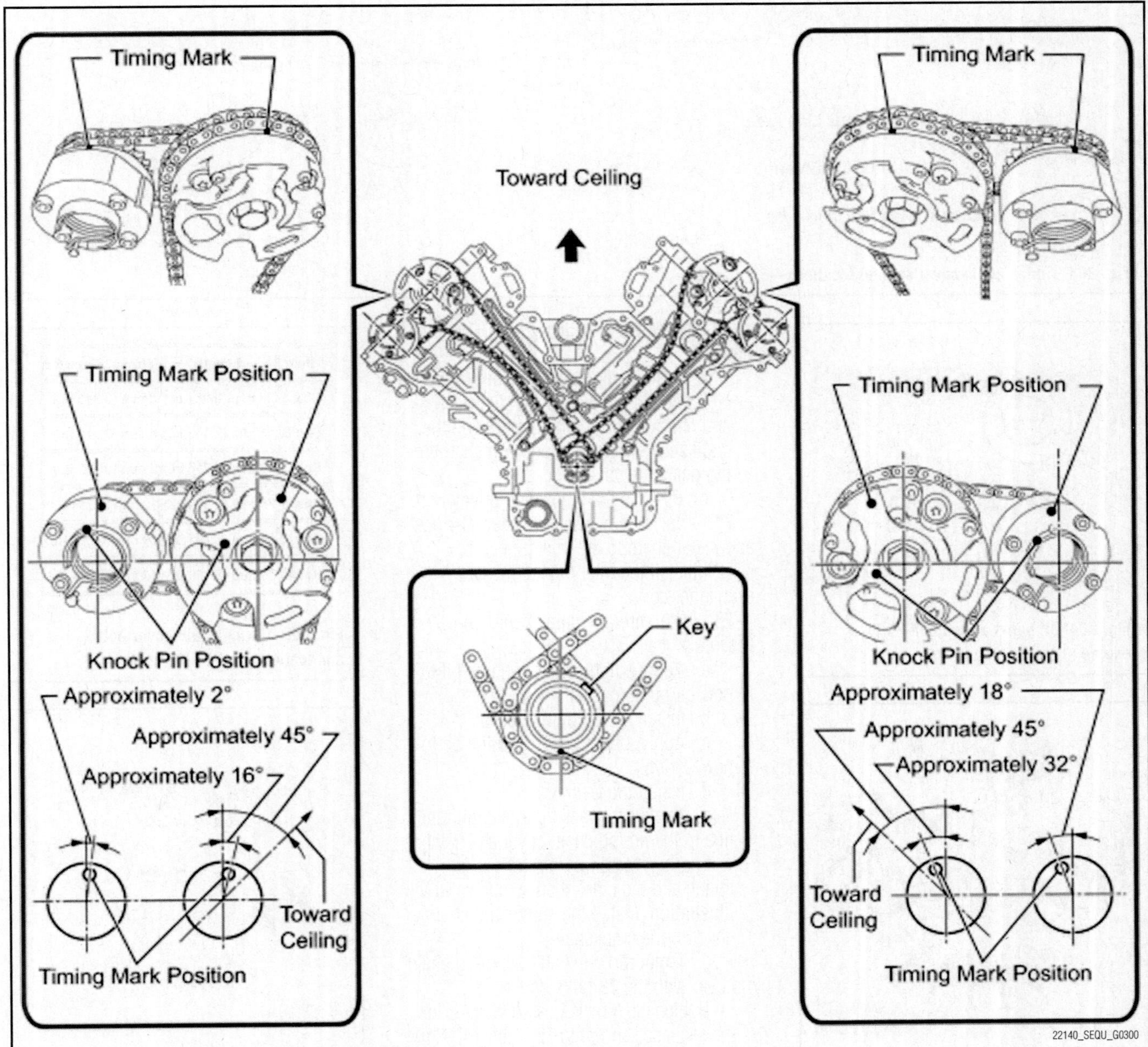

Fig. 362 Proper timing chain positioning—5.7L engine

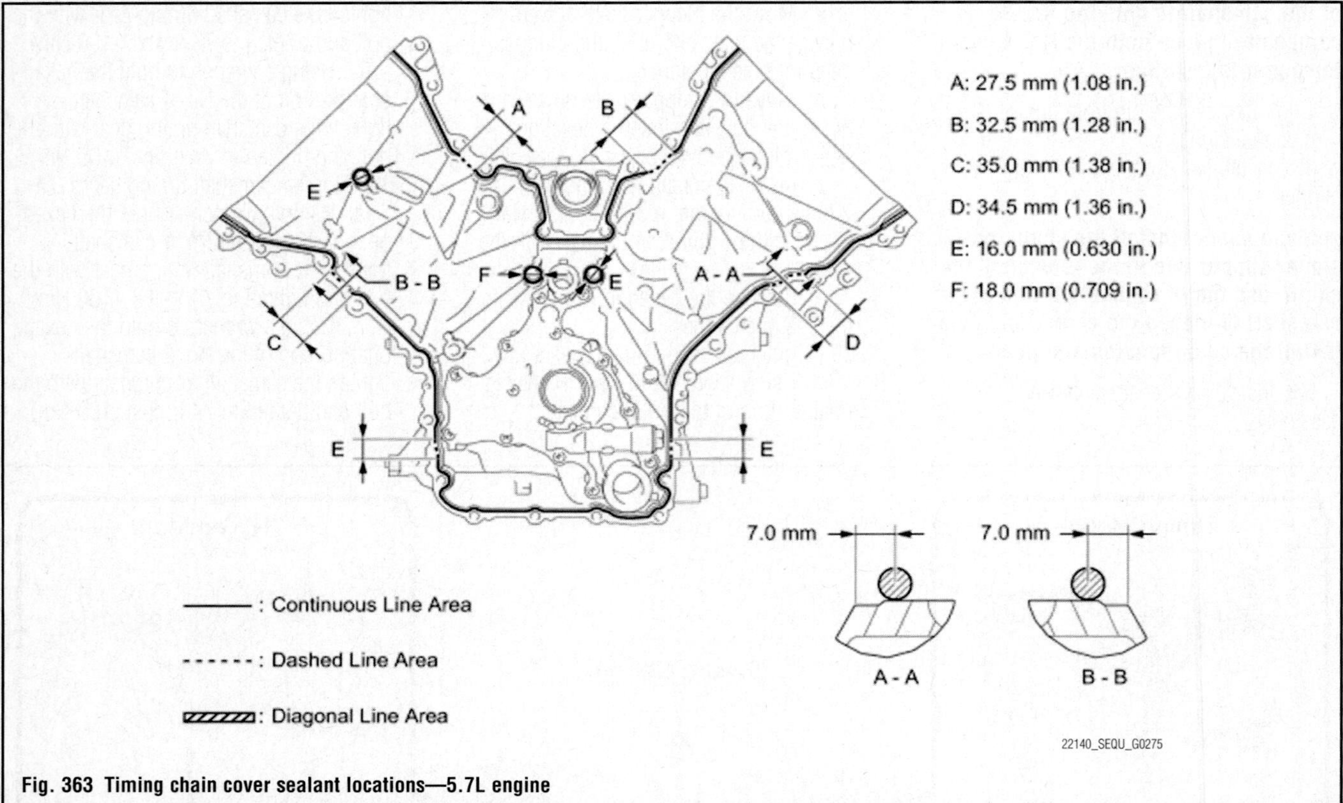

Fig. 363 Timing chain cover sealant locations—5.7L engine

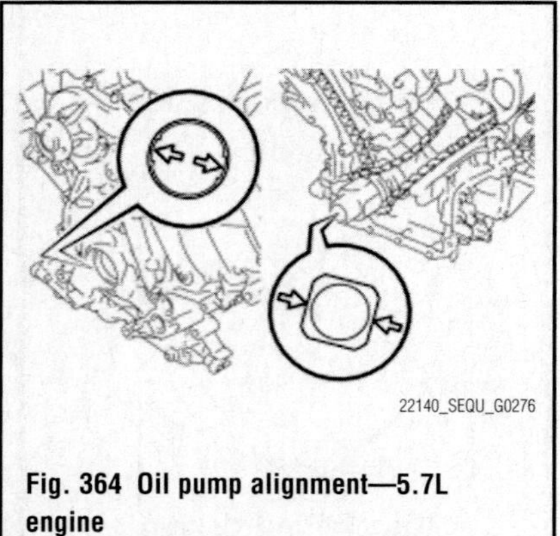

Fig. 364 Oil pump alignment—5.7L engine

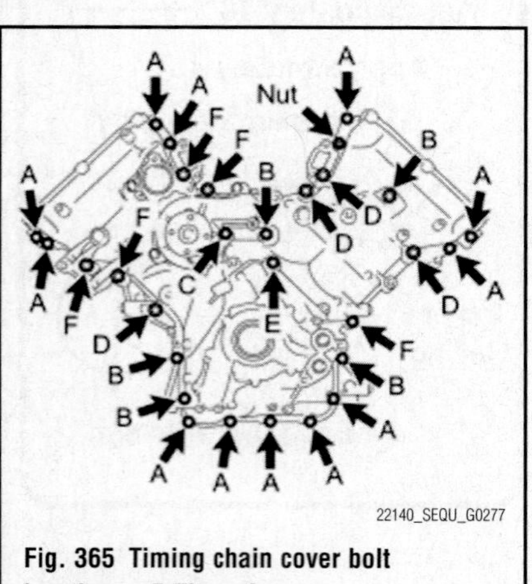

Fig. 365 Timing chain cover bolt locations—5.7L engine

70. Check No. 1 cylinder to TDC / compression.

a. Temporarily install the pulley set bolt.

b. Rotate the crankshaft clockwise, and check that the timing marks on the crankshaft timing gear and camshaft timing gears are as shown.

c. Remove the crankshaft pulley bolt.

71. Apply soapy water to 2 new O-rings and install them to the inlet pipe.

72. Install the inlet pipe to the No. 1 heat exchanger cover.

73. Install timing chain cover sub-assembly.

a. Apply a light coat of engine oil to a new oil pump gasket.

b. Install the oil pump gasket.

c. Apply a light coat of engine oil to a new O-ring.

d. Install the O-ring.

e. Apply seal packing in a continuous line to the timing chain cover as shown.

f. Align the oil pump's drive rotor spline and the crankshaft as shown in the illustration. Install the spline and chain cover to the crankshaft.

g. Temporarily install the timing chain cover with the 28 bolts and nut.

Tighten the 5 bolts in several steps in the sequence shown to 35 ft. lbs. (47 Nm).

h. Temporarily install the fan bracket with the 4 bolts.

Item	Length	Thread diameter
Bolt A	25 mm (0.984 in.)	8 mm (0.315 in.)
Bolt B	55 mm (2.165 in.)	8 mm (0.315 in.)
Bolt C	70 mm (2.756 in.)	8 mm (0.315 in.)
Bolt D	35 mm (1.378 in.)	10 mm (0.394 in.)
Bolt E	55 mm (2.165 in.)	10 mm (0.394 in.)
Bolt F	80 mm (3.150 in.)	10 mm (0.394 in.)

22140_SEQU_G0278

Fig. 366 Timing chain cover bolt application chart—5.7L engine

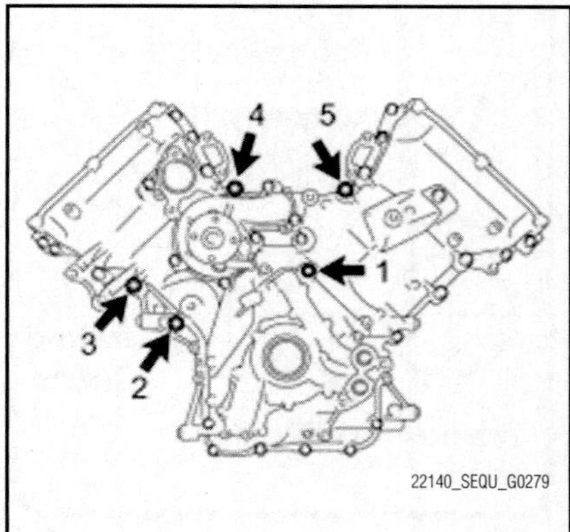

Fig. 367 Timing chain cover bolt tightening sequence No. 1—5.7L engine

i. Temporarily install the belt tensioner with the standard bolt and 6 mm hexagon wrench bolt.

j. Tighten the 23 bolts labeled 12 to 35 and nut in several steps in the sequence to 17 ft. lbs. (23 Nm).

74. Install cylinder head cover sub-assembly LH.

a. Install 5 new gaskets to the camshaft bearing caps (No. 2, No. 3).

b. Install the gasket to the cylinder head cover.

c. Apply seal packing as shown.

d. Install the cylinder head cover washer with a new seal and the 14 bolts and tighten bolt A to 15 ft. lbs. (21 Nm) and remaining bolts to 9 ft. lbs. (12 Nm).

75. Install cylinder head cover sub-assembly RH.

a. Install 5 new gaskets to the camshaft bearing caps (No. 1, No. 3).

b. Install the gasket to the cylinder head cover.

c. Apply seal packing as shown.

d. Install the cylinder head cover washer with a new seal and the 14 bolts and tighten bolt A to 15 ft. lbs. (21 Nm) and remaining bolts to 9 ft. lbs. (12 Nm).

e. Install the noise filter to the cylinder head cover with the bolt and tighten to 62 inch lbs. (7.0 Nm).

76. Install the 8 ignition coils with the 8 bolts and tighten to 7 ft. lbs. (10 Nm).

77. Align the pulley set key with the key groove of the pulley, and slide on the pulley and tighten to 221 ft. Lbs. (300 Nm).

78. Connect the bracket to the timing chain cover with the bolt.

79. Install the idler pulley with the bolt and tighten to 32 ft. lbs. (43 Nm).

80. Temporarily install the water pump pulley with the 4 bolts and tighten to 15 ft. lbs. (21 Nm).

81. Install a new water cover gasket to the timing chain cover.

82. Install the water inlet with the 3 bolts and tighten to 15 ft. lbs. (21 Nm).

83. Connect the 2 air tube hoses.

84. Connect the 2 hoses.

85. Install No. 1 and No 2 engine covers.

86. Install 2 new gaskets and the water by-pass joint with the 4 nuts and tighten to 15 ft. lbs. (21 Nm).

87. Connect the No. 2 water by-pass hose to the water by-pass joint.

88. Install the water by-pass pipe.

89. Connect the air tube.

90. Install the No. 1 water by-pass hose by connecting the hose to the water inlet housing and front water by-pass joint.

91. Install the 2 stud bolts.

92. Install the alternator with the 2 bolts and 2 nuts and tighten to 32 ft. lbs. (43 Nm).

93. Connect the alternator wire with the nut and tighten to 87 inch lbs. (9.8 Nm).

94. Connect the wire harness bracket to the alternator.

95. Connect the vane pump to the timing chain cover with the 2 bolts and tighten to 21 ft. lbs. (28 Nm).

96. Install engine oil level dipstick guide.

97. Install oil filter bracket (w/o Oil Cooler).

a. Apply a light coat of engine oil to 2 new O-rings.

b. Install the 2 O-rings to the timing chain cover.

c. Install the oil filter bracket with the 2 bolts and 2 nuts and tighten to 26 ft. lbs. (35 Nm).

98. Install No. 1 oil cooler bracket.

99. Install No. 2 water by-pass pipe.

100. Install oil pressure sender gauge assembly.

101. Install oil filter element.

102. Connect No. 2 fuel tube sub-assembly.

103. Install the cooler compressor with the 2 stud bolts and tighten to 7 ft. lbs. (10 Nm).

104. Install the 2 bolts and 2 nuts and tighten to 18 ft. lbs. (235 Nm).

105. Connect the 3 heater hoses.

106. Connect the No. 2 and No. 3 air hoses.

107. Connect the 2 clamps and air pump wire harness.

108. Install the intake manifold.

109. Engine Room RH Side:

a. Connect the 5 clamps.

b. Connect the throttle position sensor and throttle control motor connector.

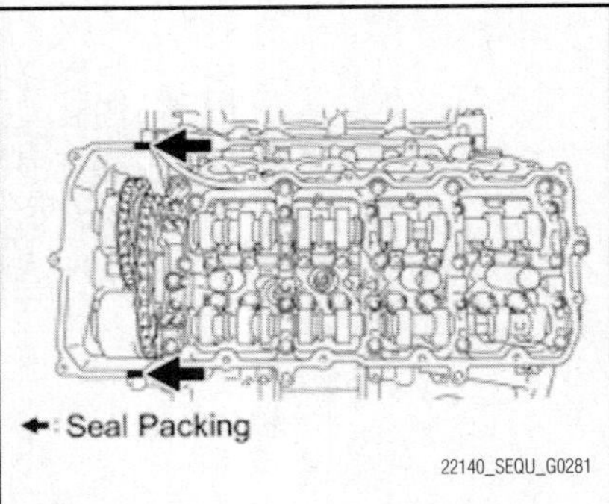

Fig. 369 Cylinder head cover packing location LH—5.7L engine

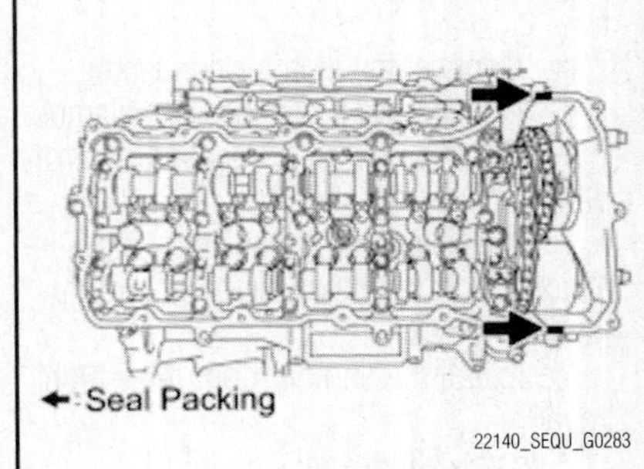

Fig. 371 Cylinder head cover packing location RH—5.7L engine

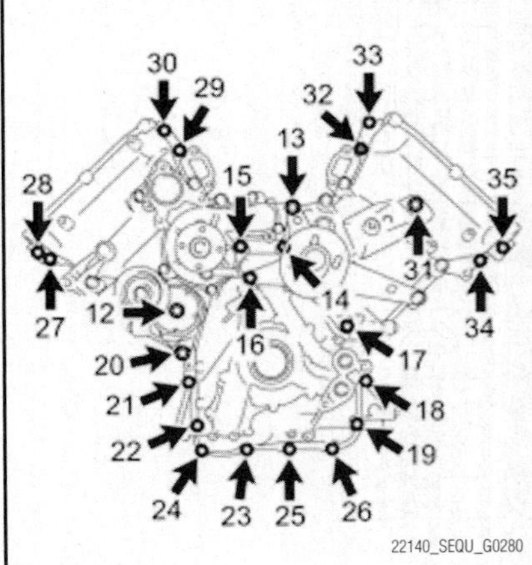

Fig. 368 Timing chain cover bolt tightening sequence No. 2—5.7L engine

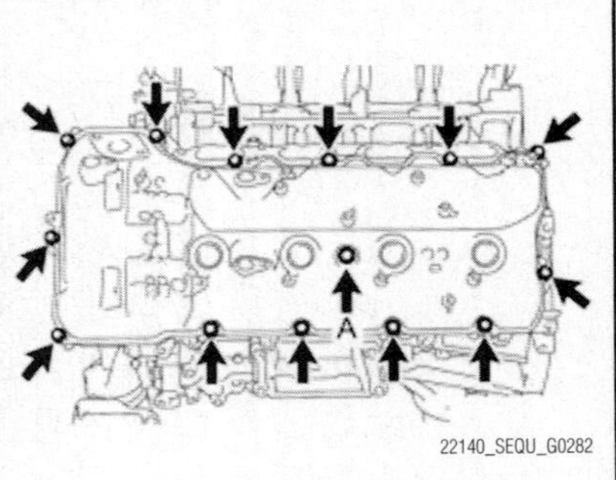

Fig. 370 Cylinder head cover bolt tightening sequence LH—5.7L engine

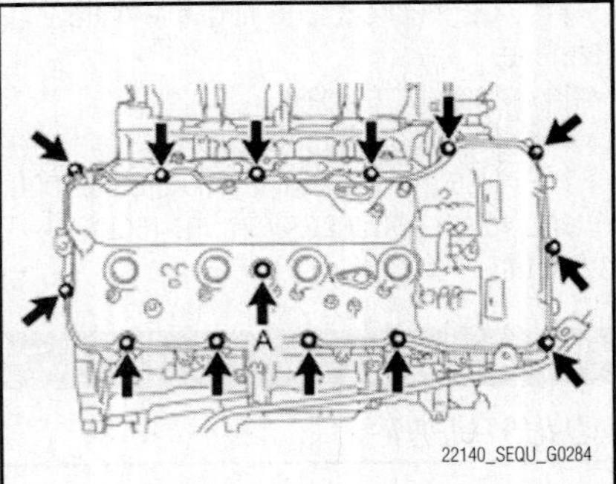

Fig. 372 Cylinder head cover bolt tightening sequence RH—5.7L engine

c. Connect the 2 air pump connectors.

d. Install the ground wire with the 2 bolts.

e. Connect the noise filter connector.

f. Connect the 2 VVT sensor connectors.

g. Connect the injector connector.

h. Connect the 4 ignition coil connectors.

i. Connect the 2 camshaft timing oil control valve connectors.

j. Connect the 2 clamps and power steering oil pressure switch connector.

k. For 4WD: Connect the clamp and junction connector.

110. Engine Room LH Side:

a. Connect the cooler compressor connector.

b. Connect the 3 clamps.

c. Connect the camshaft position sensor connector.

d. Connect the 2 camshaft timing oil control valve connectors.

e. Connect the engine coolant temperature sensor connector.

f. Connect the 4 clamps.

g. Install the ground wire with the 2 bolts.

h. Connect the noise filter connector.

i. Connect the 2 VVT sensor connectors.

j. Connect the 4 ignition coil connectors.

k. Connect the injector connector.

l. Connect the 2 wire harness clamps.

m. Connect the 4 air injection control driver connectors.

n. Connect the 2 connectors and attach the 2 clamps to the engine room junction block.

o. Install the engine room relay block cover.

111. Install air cleaner assembly.

112. Install air cleaner hose assembly.

113. Install the radiator.

114. Add engine oil.

115. Add engine coolant.

116. Connect cable to negative battery terminal.

117. Inspect for leaks.

118. Check engine oil level.

119. Use the Techstream diagnostic tool, or equivalent and reprogram the required systems.

VALVE LASH

ADJUSTMENT

4.0L Engine

See Figures 373 through 375.

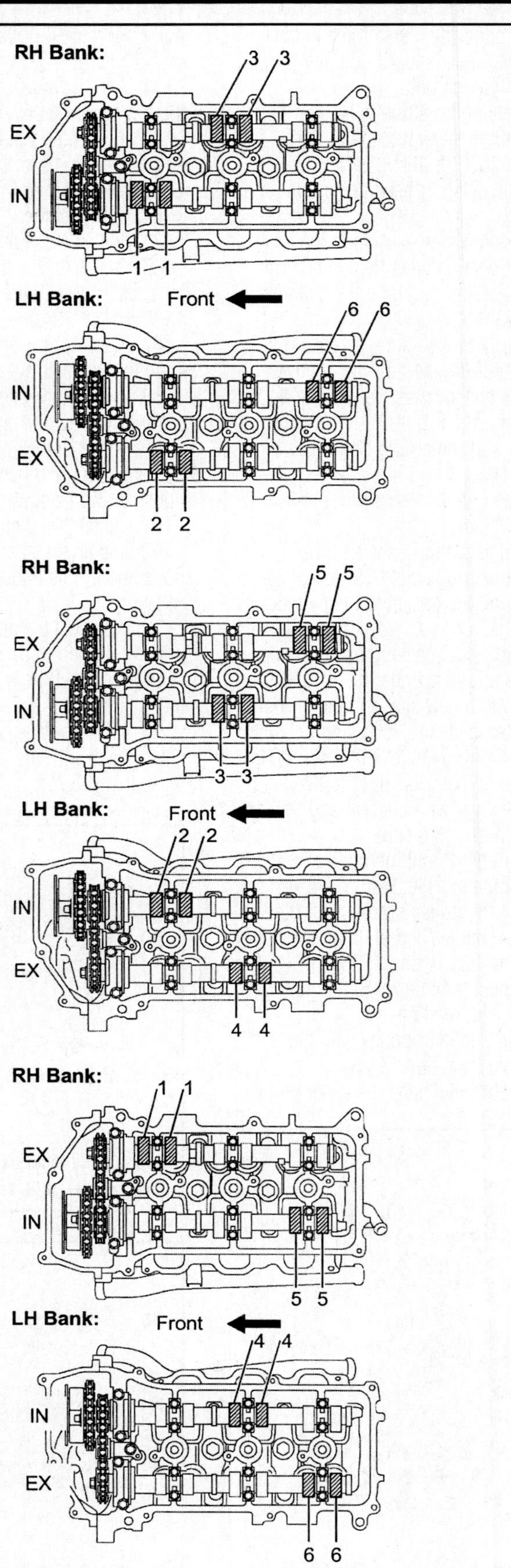

Fig. 373 Valve clearance location—4.0L engine

	5.060 (0.1992)	5.080 (0.2000)	5.100 (0.2008)	5.120 (0.2016)	5.140 (0.2024)	5.160 (0.2031)	5.180 (0.2039)	5.200 (0.2047)	5.210 (0.2051)	5.220 (0.2055)	5.230 (0.2059)	5.240 (0.2063)	5.250 (0.2067)	5.260 (0.2071)	5.270 (0.2075)	5.280 (0.2079)	5.290 (0.2083)	5.300 (0.2087)	5.310 (0.2091)	5.320 (0.2094)	5.330 (0.2098)	5.340 (0.2102)	5.350 (0.2106)	5.360 (0.2110)	5.370 (0.2114)	5.380 (0.2118)	5.390 (0.2122)	5.400 (0.2126)
0.000 - 0.020 (0.0000 - 0.0008)													06	08	08	10	10	12	12	14	14	16	16	18	18	20	20	22
0.021 - 0.040 (0.0008 - 0.0016)											06	08	08	10	10	12	12	14	14	16	16	18	18	20	20	22	22	24
0.041 - 0.060 (0.0016 - 0.0024)									06	08	08	10	10	12	12	14	14	16	16	18	18	20	20	22	22	24	24	26
0.061 - 0.080 (0.0024 - 0.0031)								08	08	10	10	12	12	14	14	16	16	18	18	20	20	22	22	24	24	26	26	28
0.081 - 0.100 (0.0032 - 0.0039)							08	10	10	12	12	14	14	16	16	18	18	20	20	22	22	24	24	26	26	28	28	30
0.101 - 0.120 (0.0040 - 0.0047)						08	10	12	12	14	14	16	16	18	18	20	20	22	22	24	24	26	26	28	28	30	30	32
0.121 - 0.140 (0.0048 - 0.0055)					08	10	12	14	14	16	16	18	18	20	20	22	22	24	24	26	26	28	28	30	30	32	32	34
0.141 - 0.149 (0.0056 - 0.0059)				06	08	10	12	14	16	16	18	18	20	20	22	22	24	24	26	26	28	28	30	30	32	32	34	34
0.251 - 0.270 (0.0099 - 0.0106)	12	14	16	18	20	22	24	26	28	28	30	30	32	32	34	34	36	36	38	38	40	40	42	42	44	44	46	46
0.271 - 0.290 (0.0107 - 0.0114)	14	16	18	20	22	24	26	28	30	30	32	32	34	34	36	36	38	38	40	40	42	42	44	44	46	46	48	48
0.291 - 0.310 (0.0115 - 0.0122)	16	18	20	22	24	26	28	30	32	32	34	34	36	36	38	38	40	40	42	42	44	44	46	46	48	48	50	50
0.311 - 0.330 (0.0122 - 0.0130)	18	20	22	24	26	28	30	32	34	34	36	36	38	38	40	40	42	42	44	44	46	46	48	48	50	50	52	52
0.331 - 0.350 (0.0130 - 0.0138)	20	22	24	26	28	30	32	34	36	36	38	38	40	40	42	42	44	44	46	46	48	48	50	50	52	52	54	54
0.351 - 0.370 (0.0138 - 0.0146)	22	24	26	28	30	32	34	36	38	38	40	40	42	42	44	44	46	46	48	48	50	50	52	52	54	54	56	56
0.371 - 0.390 (0.0146 - 0.0154)	24	26	28	30	32	34	36	38	40	40	42	42	44	44	46	46	48	48	50	50	52	52	54	54	56	56	58	58
0.391 - 0.410 (0.0154 - 0.0161)	26	28	30	32	34	36	38	40	42	42	44	44	46	46	48	48	50	50	52	52	54	54	56	56	58	58	60	60
0.411 - 0.430 (0.0162 - 0.0169)	28	30	32	34	36	38	40	42	44	44	46	46	48	48	50	50	52	52	54	54	56	56	58	58	60	60	62	62
0.431 - 0.450 (0.0170 - 0.0177)	30	32	34	36	38	40	42	44	46	46	48	48	50	50	52	52	54	54	56	56	58	58	60	60	62	62	64	64
0.451 - 0.470 (0.0178 - 0.0185)	32	34	36	38	40	42	44	46	48	48	50	50	52	52	54	54	56	56	58	58	60	60	62	62	64	64	66	66
0.471 - 0.490 (0.0185 - 0.0193)	34	36	38	40	42	44	46	48	50	50	52	52	54	54	56	56	58	58	60	60	62	62	64	64	66	66	68	68
0.491 - 0.510 (0.0193 - 0.0201)	36	38	40	42	44	46	48	50	52	52	54	54	56	56	58	58	60	60	62	62	64	64	66	66	68	68	70	70
0.511 - 0.530 (0.0201 - 0.0209)	38	40	42	44	46	48	50	52	54	54	56	56	58	58	60	60	62	62	64	64	66	66	68	68	70	70	72	72
0.531 - 0.550 (0.0209 - 0.0217)	40	42	44	46	48	50	52	54	56	56	58	58	60	60	62	62	64	64	66	66	68	68	70	70	72	72	74	
0.551 - 0.570 (0.0217 - 0.0224)	42	44	46	48	50	52	54	56	58	58	60	60	62	62	64	64	66	66	68	68	70	70	72	72	74			
0.571 - 0.590 (0.0225 - 0.0232)	44	46	48	50	52	54	56	58	60	60	62	62	64	64	66	66	68	68	70	70	72	72	74					
0.591 - 0.610 (0.0233 - 0.0240)	46	48	50	52	54	56	58	60	62	62	64	64	66	66	68	68	70	70	72	72	74							
0.611 - 0.630 (0.0241 - 0.0248)	48	50	52	54	56	58	60	62	64	64	66	66	68	68	70	70	72	72	74									
0.631 - 0.650 (0.0248 - 0.0256)	50	52	54	56	58	60	62	64	66	66	68	68	70	70	72	72	74											
0.651 - 0.670 (0.0256 - 0.0264)	52	54	56	58	60	62	64	66	68	68	70	70	72	72	74													
0.671 - 0.690 (0.0264 - 0.0272)	54	56	58	60	62	64	66	68	70	70	72	72	74															
0.691 - 0.710 (0.0272 - 0.0280)	56	58	60	62	64	66	68	70	72	72	74																	
0.711 - 0.730 (0.0280 - 0.0287)	58	60	62	64	66	68	70	72	74																			
0.731 - 0.750 (0.0288 - 0.0295)	60	62	64	66	68	70	72																					
0.751 - 0.770 (0.0296 - 0.0303)	62	64	66	68	70	72																						
0.771 - 0.790 (0.0304 - 0.0311)	64	66	68	70	72																							
0.791 - 0.810 (0.0311 - 0.0319)	66	68	70	72																								
0.811 - 0.830 (0.0319 - 0.0327)	68	70	72																									
0.831 - 0.850 (0.0327 - 0.0335)	70	72																										
0.851 - 0.870 (0.0335 - 0.0343)	72																											

	5.410 (0.2130)	5.420 (0.2134)	5.430 (0.2138)	5.440 (0.2142)	5.450 (0.2146)	5.460 (0.2150)	5.470 (0.2154)	5.480 (0.2157)	5.490 (0.2161)	5.500 (0.2165)	5.510 (0.2169)	5.520 (0.2173)	5.530 (0.2177)	5.540 (0.2181)	5.550 (0.2185)	5.560 (0.2189)	5.570 (0.2193)	5.580 (0.2197)	5.590 (0.2201)	5.600 (0.2205)	5.620 (0.2213)	5.640 (0.2220)	5.660 (0.2228)	5.680 (0.2236)	5.700 (0.2244)	5.720 (0.2252)	5.740 (0.2260)
0.000 - 0.020 (0.0000 - 0.0008)	22	24	24	26	26	28	28	30	30	32	32	34	34	36	36	38	38	40	40	42	44	46	48	50	52	54	56
0.021 - 0.040 (0.0008 - 0.0016)	24	26	26	28	28	30	30	32	32	34	34	36	36	38	38	40	40	42	42	44	46	48	50	52	54	56	58
0.041 - 0.060 (0.0016 - 0.0024)	26	28	28	30	30	32	32	34	34	36	36	38	38	40	40	42	42	44	44	46	48	50	52	54	56	58	60
0.061 - 0.080 (0.0024 - 0.0031)	28	30	30	32	32	34	34	36	36	38	38	40	40	42	42	44	44	46	46	48	50	52	54	56	58	60	62
0.081 - 0.100 (0.0032 - 0.0039)	30	32	32	34	34	36	36	38	38	40	40	42	42	44	44	46	46	48	48	50	52	54	56	58	60	62	64
0.101 - 0.120 (0.0040 - 0.0047)	32	34	34	36	36	38	38	40	40	42	42	44	44	46	46	48	48	50	50	52	54	56	58	60	62	64	66
0.121 - 0.140 (0.0048 - 0.0055)	34	36	36	38	38	40	40	42	42	44	44	46	46	48	48	50	50	52	52	54	56	58	60	62	64	66	68
0.141 - 0.149 (0.0056 - 0.0059)	36	36	38	38	40	40	42	42	44	44	46	46	48	48	50	50	52	52	54	54	56	58	60	62	64	66	68
0.251 - 0.270 (0.0099 - 0.0106)	48	48	50	50	52	52	54	54	56	56	58	58	60	60	62	62	64	64	66	66	68	70	72				
0.271 - 0.290 (0.0107 - 0.0114)	50	50	52	52	54	54	56	56	58	58	60	60	62	62	64	64	66	66	68	68	70	72					
0.291 - 0.310 (0.0115 - 0.0122)	52	52	54	54	56	56	58	58	60	60	62	62	64	64	66	66	68	68	70	70	72						
0.311 - 0.330 (0.0122 - 0.0130)	54	54	56	56	58	58	60	60	62	62	64	64	66	66	68	68	70	70	72	72							
0.331 - 0.350 (0.0130 - 0.0138)	56	56	58	58	60	60	62	62	64	64	66	66	68	68	70	70	72	72	74								
0.351 - 0.370 (0.0138 - 0.0146)	58	58	60	60	62	62	64	64	66	66	68	68	70	70	72	72	74										
0.371 - 0.390 (0.0146 - 0.0154)	60	60	62	62	64	64	66	66	68	68	70	70	72	72	74												
0.391 - 0.410 (0.0154 - 0.0161)	62	62	64	64	66	66	68	68	70	70	72	72	74														
0.411 - 0.430 (0.0162 - 0.0169)	64	64	66	66	68	68	70	70	72	72	74																
0.431 - 0.450 (0.0170 - 0.0177)	66	66	68	68	70	70	72	72	74																		
0.451 - 0.470 (0.0178 - 0.0185)	68	68	70	70	72	72	74																				
0.471 - 0.490 (0.0185 - 0.0193)	70	70	72	72	74																						
0.491 - 0.510 (0.0193 - 0.0201)	72	72	74																								
0.511 - 0.530 (0.0201 - 0.0209)	74																										

06	5.060 (0.1992)	30	5.300 (0.2087)	54	5.540 (0.2181)
08	5.080 (0.2000)	32	5.320 (0.2094)	56	5.560 (0.2189)
10	5.100 (0.2008)	34	5.340 (0.2102)	58	5.580 (0.2197)
12	5.120 (0.2016)	36	5.360 (0.2110)	60	5.600 (0.2205)
14	5.140 (0.2024)	38	5.380 (0.2118)	62	5.620 (0.2213)
16	5.160 (0.2031)	40	5.400 (0.2126)	64	5.640 (0.2220)
18	5.180 (0.2039)	42	5.420 (0.2134)	66	5.660 (0.2228)
20	5.200 (0.2047)	44	5.440 (0.2142)	68	5.680 (0.2236)
22	5.220 (0.2055)	46	5.460 (0.2150)	70	5.700 (0.2244)
24	5.240 (0.2063)	48	5.480 (0.2157)	72	5.720 (0.2252)
26	5.260 (0.2071)	50	5.500 (0.2165)	74	5.740 (0.2260)
28	5.280 (0.2079)	52	5.520 (0.2173)		

09490_TACO_G0074

Fig. 374 Shim selection chart, part one—4.0L engine

	5.060 (0.1992)	5.080 (0.2000)	5.100 (0.2008)	5.120 (0.2016)	5.140 (0.2024)	5.160 (0.2031)	5.180 (0.2039)	5.200 (0.2047)	5.210 (0.2051)	5.220 (0.2055)	5.230 (0.2059)	5.240 (0.2063)	5.250 (0.2067)	5.260 (0.2071)	5.270 (0.2075)	5.280 (0.2079)	5.290 (0.2083)	5.300 (0.2087)	5.310 (0.2091)	5.320 (0.2094)	5.330 (0.2098)	5.340 (0.2102)	5.350 (0.2106)	5.360 (0.2110)	5.370 (0.2114)	5.380 (0.2118)	5.390 (0.2122)	5.400 (0.2126)	5.410 (0.2130)	5.420 (0.2134)	5.430 (0.2138)	5.440 (0.2142)	5.450 (0.2146)	5.460 (0.2150)	5.470 (0.2154)	5.480 (0.2157)	5.490 (0.2161)	5.500 (0.2165)	5.510 (0.2169)	5.520 (0.2173)	5.530 (0.2177)	5.540 (0.2181)	5.550 (0.2185)	5.560 (0.2189)	5.570 (0.2193)	5.580 (0.2197)	5.590 (0.2201)	5.600 (0.2205)	5.620 (0.2213)	5.640 (0.2220)	5.660 (0.2228)	5.680 (0.2236)	5.700 (0.2244)	5.720 (0.2252)	5.740 (0.2260)
0.000 - 0.020 (0.0000 - 0.0008)																											06	08	08	10	10	12	12	14	14	16	16	18	18	20	20	22	22	24	24	26	26	28	30	32	34	36	38	40	42
0.021 - 0.040 (0.0008 - 0.0016)																									06	08	08	10	10	12	12	14	14	16	16	18	18	20	20	22	22	24	24	26	26	28	28	30	32	34	36	38	40	42	44
0.041 - 0.060 (0.0016 - 0.0024)																							06	08	08	10	10	12	12	14	14	16	16	18	18	20	20	22	22	24	24	26	26	28	28	30	30	32	34	36	38	40	42	44	46
0.061 - 0.080 (0.0024 - 0.0031)																					06	08	08	10	10	12	12	14	14	16	16	18	18	20	20	22	22	24	24	26	26	28	28	30	30	32	32	34	36	38	40	42	44	46	48
0.081 - 0.100 (0.0032 - 0.0039)																			06	08	08	10	10	12	12	14	14	16	16	18	18	20	20	22	22	24	24	26	26	28	28	30	30	32	32	34	34	36	38	40	42	44	46	48	50
0.101 - 0.120 (0.0040 - 0.0047)																	06	08	08	10	10	12	12	14	14	16	16	18	18	20	20	22	22	24	24	26	26	28	28	30	30	32	32	34	34	36	36	38	40	42	44	46	48	50	52
0.121 - 0.140 (0.0048 - 0.0055)															06	08	08	10	10	12	12	14	14	16	16	18	18	20	20	22	22	24	24	26	26	28	28	30	30	32	32	34	34	36	36	38	38	40	42	44	46	48	50	52	54
0.141 - 0.160 (0.0056 - 0.0063)													06	08	08	10	10	12	12	14	14	16	16	18	18	20	20	22	22	24	24	26	26	28	28	30	30	32	32	34	34	36	36	38	38	40	40	42	44	46	48	50	52	54	56
0.161 - 0.180 (0.0063 - 0.0071)											06	08	08	10	10	12	12	14	14	16	16	18	18	20	20	22	22	24	24	26	26	28	28	30	30	32	32	34	34	36	36	38	38	40	40	42	42	44	46	48	50	52	54	56	58
0.181 - 0.200 (0.0071 - 0.0079)									06	08	08	10	10	12	12	14	14	16	16	18	18	20	20	22	22	24	24	26	26	28	28	30	30	32	32	34	34	36	36	38	38	40	40	42	42	44	44	46	48	50	52	54	56	58	60
0.201 - 0.220 (0.0079 - 0.0087)								08	08	10	10	12	12	14	14	16	16	18	18	20	20	22	22	24	24	26	26	28	28	30	30	32	32	34	34	36	36	38	38	40	40	42	42	44	44	46	46	48	50	52	54	56	58	60	62
0.221 - 0.240 (0.0087 - 0.0094)							08	10	10	12	12	14	14	16	16	18	18	20	20	22	22	24	24	26	26	28	28	30	30	32	32	34	34	36	36	38	38	40	40	42	42	44	44	46	46	48	48	50	52	54	56	58	60	62	64
0.241 - 0.260 (0.0095 - 0.0102)						08	10	12	12	14	14	16	16	18	18	20	20	22	22	24	24	26	26	28	28	30	30	32	32	34	34	36	36	38	38	40	40	42	42	44	44	46	46	48	48	50	50	52	54	56	58	60	62	64	66
0.261 - 0.280 (0.0103 - 0.0110)					08	10	12	14	14	16	16	18	18	20	20	22	22	24	24	26	26	28	28	30	30	32	32	34	34	36	36	38	38	40	40	42	42	44	44	46	46	48	48	50	50	52	52	54	56	58	60	62	64	66	68
0.281 - 0.289 (0.0111 - 0.0114)				06	08	10	12	14	16	16	18	18	20	20	22	22	24	24	26	26	28	28	30	30	32	32	34	34	36	36	38	38	40	40	42	42	44	44	46	46	48	48	50	50	52	52	54	54	56	58	60	62	64	66	68
0.391 - 0.410 (0.0154 - 0.0161)	12	14	16	18	20	22	24	26	28	28	30	30	32	32	34	34	36	36	38	38	40	40	42	42	44	44	46	46	48	48	50	50	52	52	54	54	56	56	58	58	60	60	62	62	64	64	66	66	68	70	72				
0.411 - 0.430 (0.0162 - 0.0169)	14	16	18	20	22	24	26	28	30	30	32	32	34	34	36	36	38	38	40	40	42	42	44	44	46	46	48	48	50	50	52	52	54	54	56	56	58	58	60	60	62	62	64	64	66	66	68	68	70	72					
0.431 - 0.450 (0.0170 - 0.0177)	16	18	20	22	24	26	28	30	32	32	34	34	36	36	38	38	40	40	42	42	44	44	46	46	48	48	50	50	52	52	54	54	56	56	58	58	60	60	62	62	64	64	66	66	68	68	70	70	72						
0.451 - 0.470 (0.0178 - 0.0185)	18	20	22	24	26	28	30	32	34	34	36	36	38	38	40	40	42	42	44	44	46	46	48	48	50	50	52	52	54	54	56	56	58	58	60	60	62	62	64	64	66	66	68	68	70	70	72	72							
0.471 - 0.490 (0.0185 - 0.0193)	20	22	24	26	28	30	32	34	36	36	38	38	40	40	42	42	44	44	46	46	48	48	50	50	52	52	54	54	56	56	58	58	60	60	62	62	64	64	66	66	68	68	70	70	72	72	74								
0.491 - 0.510 (0.0193 - 0.0201)	22	24	26	28	30	32	34	36	38	38	40	40	42	42	44	44	46	46	48	48	50	50	52	52	54	54	56	56	58	58	60	60	62	62	64	64	66	66	68	68	70	70	72	72	74										
0.511 - 0.530 (0.0201 - 0.0209)	24	26	28	30	32	34	36	38	40	40	42	42	44	44	46	46	48	48	50	50	52	52	54	54	56	56	58	58	60	60	62	62	64	64	66	66	68	68	70	70	72	72	74												
0.531 - 0.550 (0.0209 - 0.0217)	26	28	30	32	34	36	38	40	42	42	44	44	46	46	48	48	50	50	52	52	54	54	56	56	58	58	60	60	62	62	64	64	66	66	68	68	70	70	72	72	74														
0.551 - 0.570 (0.0217 - 0.0224)	28	30	32	34	36	38	40	42	44	44	46	46	48	48	50	50	52	52	54	54	56	56	58	58	60	60	62	62	64	64	66	66	68	68	70	70	72	72	74																
0.571 - 0.590 (0.0225 - 0.0232)	30	32	34	36	38	40	42	44	46	46	48	48	50	50	52	52	54	54	56	56	58	58	60	60	62	62	64	64	66	66	68	68	70	70	72	72	74																		
0.591 - 0.610 (0.0233 - 0.0240)	32	34	36	38	40	42	44	46	48	48	50	50	52	52	54	54	56	56	58	58	60	60	62	62	64	64	66	66	68	68	70	70	72	72	74																				
0.611 - 0.630 (0.0241 - 0.0248)	34	36	38	40	42	44	46	48	50	50	52	52	54	54	56	56	58	58	60	60	62	62	64	64	66	66	68	68	70	70	72	72	74																						
0.631 - 0.650 (0.0248 - 0.0256)	36	38	40	42	44	46	48	50	52	52	54	54	56	56	58	58	60	60	62	62	64	64	66	66	68	68	70	70	72	72	74																								
0.651 - 0.670 (0.0256 - 0.0264)	38	40	42	44	46	48	50	52	54	54	56	56	58	58	60	60	62	62	64	64	66	66	68	68	70	70	72	72	74																										
0.671 - 0.690 (0.0264 - 0.0272)	40	42	44	46	48	50	52	54	56	56	58	58	60	60	62	62	64	64	66	66	68	68	70	70	72	72	74																												
0.691 - 0.710 (0.0272 - 0.0280)	42	44	46	48	50	52	54	56	58	58	60	60	62	62	64	64	66	66	68	68	70	70	72	72	74																														
0.711 - 0.730 (0.0280 - 0.0287)	44	46	48	50	52	54	56	58	60	60	62	62	64	64	66	66	68	68	70	70	72	72	74																																
0.731 - 0.750 (0.0288 - 0.0295)	46	48	50	52	54	56	58	60	62	62	64	64	66	66	68	68	70	70	72	72	74																																		
0.751 - 0.770 (0.0296 - 0.0303)	48	50	52	54	56	58	60	62	64	64	66	66	68	68	70	70	72	72	74																																				
0.771 - 0.790 (0.0304 - 0.0311)	50	52	54	56	58	60	62	64	66	66	68	68	70	70	72	72	74																																						
0.791 - 0.810 (0.0311 - 0.0319)	52	54	56	58	60	62	64	66	68	68	70	70	72	72	74																																								
0.811 - 0.830 (0.0319 - 0.0327)	54	56	58	60	62	64	66	68	70	70	72	72	74																																										
0.831 - 0.850 (0.0327 - 0.0335)	56	58	60	62	64	66	68	70	72	72	74																																												
0.851 - 0.870 (0.0335 - 0.0343)	58	60	62	64	66	68	70	72	74																																														
0.871 - 0.890 (0.0343 - 0.0350)	60	62	64	66	68	70	72																																																
0.891 - 0.910 (0.0351 - 0.0358)	62	64	66	68	70	72																																																	
0.911 - 0.930 (0.0359 - 0.0366)	64	66	68	70	72																																																		
0.931 - 0.950 (0.0367 - 0.0374)	66	68	70	72																																																			
0.951 - 0.970 (0.0374 - 0.0382)	68	70	72																																																				
0.971 - 0.990 (0.0382 - 0.0390)	70	72																																																					
0.991 - 1.010 (0.0390 - 0.0398)	72																																																						

06	5.060 (0.1992)	30	5.300 (0.2087)	54	5.540 (0.2181)
08	5.080 (0.2000)	32	5.320 (0.2094)	56	5.560 (0.2189)
10	5.100 (0.2008)	34	5.340 (0.2102)	58	5.580 (0.2197)
12	5.120 (0.2016)	36	5.360 (0.2110)	60	5.600 (0.2205)
14	5.140 (0.2024)	38	5.380 (0.2118)	62	5.620 (0.2213)
16	5.160 (0.2031)	40	5.400 (0.2126)	64	5.640 (0.2220)
18	5.180 (0.2039)	42	5.420 (0.2134)	66	5.660 (0.2228)
20	5.200 (0.2047)	44	5.440 (0.2142)	68	5.680 (0.2236)
22	5.220 (0.2055)	46	5.460 (0.2150)	70	5.700 (0.2244)
24	5.240 (0.2063)	48	5.480 (0.2157)	72	5.720 (0.2252)
26	5.260 (0.2071)	50	5.500 (0.2165)	74	5.740 (0.2260)
28	5.280 (0.2079)	52	5.520 (0.2173)		

09490_TACO_G0075

Fig. 375 Shim selection chart, part two—4.0L engine

1. Before servicing the vehicle, refer to the Precautions Section.

➡If working near and/or around the SRS system and components, be sure to disable the SRS system. Tape the negative battery cable with insulating tape. Always disconnect the negative battery cable first.

CAUTION

To avoid personal injury when working on vehicles equipped with an air bag, the negative battery cable must be disconnected and at least 90 seconds must elapse before working on the system. Failure to do so may result in deployment of the air bag.

2. Disconnect the negative battery cable. Tape the cable with insulating tape.

➡When disconnecting the cable, some systems need to be initialized after the cable is reconnected. You will need the Techstream diagnostic scan tool or equivalent. Follow the directions on the tool.

3. Drain the engine coolant. Remove the V bank cover.
4. Remove the air cleaner assembly.
5. Disconnect the two water bypass hoses. Disconnect the fuel vapor feed hose. Disconnect the ventilation hose. Disconnect the VSV connectors.
6. Disconnect the throttle body motor connector. Separate the three wire harness clamps and hose clamp.
7. Remove the two bolts and the throttle body bracket. Remove the bolt and the oil baffle plate. Remove the four bolts and the two surge tank stays.
8. Remove the two nuts. Remove the four bolts, intake air surge tank and gasket.
9. Remove the ignition coil assembly.
10. Remove the cylinder head cover retaining bolts. Remove the cylinder head cover.
11. Turn the crankshaft pulley until its groove and the timing mark "0" of the timing chain cover are aligned. If not aligned at TDC of the compression stroke, turn the crankshaft one complete revolution, in the direction of rotation.
12. Using a feeler gauge, check and record the valve clearance on the following valves: right bank, exhaust number three and intake number one, left bank exhaust number two and intake number six.
13. Rotate the crankshaft 240 degrees clockwise and using a feeler gauge, check and record the following valves: right bank, exhaust number five and intake number three, left bank exhaust number four and intake number two.
14. Rotate the crankshaft 240 degrees clockwise and using a feeler gauge, check and record the following valves: right bank, exhaust number one and intake number five, left bank exhaust number six and intake number four.
15. If adjustment is required, position the engine at TDC on the compression stroke.
16. Place paint marks on the number one chain links corresponding to the timing marks of the camshaft timing gears.
17. Remove the chain tensioner assembly number one. Remove the number two camshaft.
18. Remove the chain tensioner assembly number two. Remove the camshaft.
19. Remove the number four camshaft subassembly. Remove the chain tensioner assembly number three.
20. Remove the number three camshaft subassembly.
21. Remove the valve lifters.
22. Determine the replacement adjusting shim size according to the following formula or use the adjusting shim charts.
23. Using a micrometer, measure the thickness of the removed shim. Calculate the thickness of a new shim so that the valve clearance comes within the specified value.
   - T: Thickness of the removed shim
   - A: Measured valve clearance
   - N: Thickness of the new shim

   a. Intake: N = T + A
   b. Exhaust: N = T + A
24. Select a new lifter with a thickness as close as possible to the calculated value.
25. Install removed components in the reverse order of the removal procedure.
26. When installing the cylinder head cover, apply a continuous bead (0.08–0.12 inch) of seal packing, part number 08826-00080 or equivalent, to the cylinder head as indicated in the illustration. Install the seal washers onto the bolts. Install the cylinder head cover bolts and nuts. Tighten bolts "A" to 7.4 ft. lbs. Tighten bolts "B" to 80 inch lbs. Tighten nuts to 80 inch lbs.

➡Be sure to remove any oil from the contact surfaces of the cylinder head cover and the cylinder head. Install the cover within three minutes after applying the seal packing. Tighten the bolts to specification within fifteen minutes after installing the cover. Do not add engine oil for at least two hours after installing the cover.

27. Continue the installation in the reverse order of the removal procedure.
28. Be sure to fill the engine with the proper grade and type engine coolant.
29. Be sure to fill the engine with the proper grade and type engine oil.
30. Start the engine and check for leaks. Correct as required.
31. Use the Techstream diagnostic tool, or equivalent and reprogram the required systems.

4.6L Engine

Valve lash is not adjustable for this engine.

4.7L Engine

See Figures 376 and 377.

1. Before servicing the vehicle, refer to the Precautions Section.

➡If working near and/or around the SRS system and components, be sure to disable the SRS system. Tape the negative battery cable with insulating tape. Always disconnect the negative battery cable first.

CAUTION

To avoid personal injury when working on vehicles equipped with an air bag, the negative battery cable must be disconnected and at least 90 seconds must elapse before working on the system. Failure to do so may result in deployment of the air bag.

2. Disconnect the negative battery cable. Tape the cable with insulating tape.

➡When disconnecting the cable, some systems need to be initialized after the cable is reconnected. You will need the Techstream diagnostic scan tool or equivalent. Follow the directions on the tool.

➡Measure valve clearance with the engine cold.

3. Before servicing the vehicle, refer to the precautions in the beginning of this section.
4. Drain the cooling system.
5. Remove or disconnect the following:
   - Negative battery cable
   - Ignition coils
   - Valve covers
6. Set the engine to the top of the compression stroke with the valves closed for the cylinder to be measured.
7. Check the valve clearance. The valve clearance specifications are as follows:
   - Intake: 0.006–0.010 in. (0.15–0.25mm)
   - Exhaust: 0.010–0.014 in. (0.25–0.35mm)

Measured clearance mm (in.) \ Installed shim thickness mm (in.)	2.000 (0.0787)	2.020 (0.0795)	2.040 (0.0803)	2.060 (0.0811)	2.080 (0.0819)	2.100 (0.0827)	2.120 (0.0835)	2.140 (0.0843)	2.160 (0.0850)	2.180 (0.0858)	2.200 (0.0866)	2.210 (0.0870)	2.220 (0.0874)	2.230 (0.0878)	2.240 (0.0882)	2.250 (0.0886)	2.260 (0.0890)	2.270 (0.0894)	2.280 (0.0898)	2.290 (0.0902)	2.300 (0.0906)	2.310 (0.0909)	2.320 (0.0913)	2.330 (0.0917)	2.340 (0.0921)	2.350 (0.0925)	2.360 (0.0929)	2.370 (0.0933)	2.380 (0.0937)	2.390 (0.0941)	2.400 (0.0945)	2.410 (0.0949)	2.420 (0.0953)	2.430 (0.0957)	2.440 (0.0961)	2.450 (0.0965)	2.460 (0.0969)	2.470 (0.0972)	2.480 (0.0976)	2.490 (0.0980)	2.500 (0.0984)	2.510 (0.0988)	2.520 (0.0992)	2.530 (0.0996)	2.540 (0.1000)	2.550 (0.1004)	2.560 (0.1008)	2.570 (0.1012)	2.580 (0.1016)	2.590 (0.1020)	2.600 (0.1024)	2.620 (0.1031)	2.640 (0.1039)	2.660 (0.1047)	2.680 (0.1055)	2.700 (0.1063)	2.720 (0.1071)	2.740 (0.1079)	2.760 (0.1087)	2.780 (0.1094)	2.800 (0.1102)
0.000-0.030 (0.0000-0.0012)							00	00	00	00	02	02	04	04	06	06	08	08	10	10	12	12	14	14	16	16	18	18	20	20	22	22	24	24	26	26	28	28	30	30	32	32	34	34	36	36	38	38	40	40	42	44	46	48	50	52	54	56	58	60	62
0.031-0.050 (0.0012-0.0020)						00	00	00	00	02	04	06	06	08	08	10	10	12	12	14	14	16	16	18	18	20	20	22	22	24	24	26	26	28	28	30	30	32	32	34	34	36	36	38	38	40	40	42	42	44	44	46	48	50	52	54	56	58	60	62	64
0.051-0.070 (0.0020-0.0028)					00	00	00	00	02	04	06	08	08	10	10	12	12	14	14	16	16	18	18	20	20	22	22	24	24	26	26	28	28	30	30	32	32	34	34	36	36	38	38	40	40	42	42	44	44	46	46	48	50	52	54	56	58	60	62	64	66
0.071-0.090 (0.0028-0.0035)				00	00	00	00	02	04	06	08	10	10	12	12	14	14	16	16	18	18	20	20	22	22	24	24	26	26	28	28	30	30	32	32	34	34	36	36	38	38	40	40	42	42	44	44	46	46	48	48	50	52	54	56	58	60	62	64	66	68
0.091-0.110 (0.0036-0.0043)			00	00	00	00	02	04	06	08	10	12	12	14	14	16	16	18	18	20	20	22	22	24	24	26	26	28	28	30	30	32	32	34	34	36	36	38	38	40	40	42	42	44	44	46	46	48	48	50	50	52	54	56	58	60	62	64	66	68	70
0.111-0.130 (0.0044-0.0051)		00	00	00	00	02	04	06	08	10	12	14	14	16	16	18	18	20	20	22	22	24	24	26	26	28	28	30	30	32	32	34	34	36	36	38	38	40	40	42	42	44	44	46	46	48	48	50	50	52	52	54	56	58	60	62	64	66	68	70	72
0.131-0.149 (0.0052-0.0059)		00	00	00	02	04	06	08	10	12	14	16	16	18	18	20	20	22	22	24	24	26	26	28	28	30	30	32	32	34	34	36	36	38	38	40	40	42	42	44	44	46	46	48	48	50	50	52	52	54	54	56	58	60	62	64	66	68	70	72	74
0.150-0.250 (0.0059-0.0098)																																																													
0.251-0.270 (0.0099-0.0106)	06	08	10	12	14	16	18	20	22	24	26	28	28	30	30	32	32	34	34	36	36	38	38	40	40	42	42	44	44	46	46	48	48	50	50	52	52	54	54	56	56	58	58	60	60	62	62	64	64	66	66	68	70	72	74	76	78	80	80	80	
0.271-0.290 (0.0107-0.0114)	08	10	12	14	16	18	20	22	24	26	28	30	30	32	32	34	34	36	36	38	38	40	40	42	42	44	44	46	46	48	48	50	50	52	52	54	54	56	56	58	58	60	60	62	62	64	64	66	66	68	68	70	72	74	76	78	80	80	80		
0.291-0.310 (0.0115-0.0122)	10	12	14	16	18	20	22	24	26	28	30	32	32	34	34	36	36	38	38	40	40	42	42	44	44	46	46	48	48	50	50	52	52	54	54	56	56	58	58	60	60	62	62	64	64	66	66	68	68	70	70	72	74	76	78	80	80	80			
0.311-0.330 (0.0122-0.0130)	12	14	16	18	20	22	24	26	28	30	32	34	34	36	36	38	38	40	40	42	42	44	44	46	46	48	48	50	50	52	52	54	54	56	56	58	58	60	60	62	62	64	64	66	66	68	68	70	70	72	72	74	76	78	80	80	80				
0.331-0.350 (0.0130-0.0138)	14	16	18	20	22	24	26	28	30	32	34	36	36	38	38	40	40	42	42	44	44	46	46	48	48	50	50	52	52	54	54	56	56	58	58	60	60	62	62	64	64	66	66	68	68	70	70	72	72	74	74	76	78	80	80	80					
0.351-0.370 (0.0138-0.0146)	16	18	20	22	24	26	28	30	32	34	36	38	38	40	40	42	42	44	44	46	46	48	48	50	50	52	52	54	54	56	56	58	58	60	60	62	62	64	64	66	66	68	68	70	70	72	72	74	74	76	76	78	80	80	80						
0.371-0.390 (0.0146-0.0154)	18	20	22	24	26	28	30	32	34	36	38	40	40	42	42	44	44	46	46	48	48	50	50	52	52	54	54	56	56	58	58	60	60	62	62	64	64	66	66	68	68	70	70	72	72	74	74	76	76	78	78	80	80	80							
0.391-0.410 (0.0154-0.0161)	20	22	24	26	28	30	32	34	36	38	40	42	42	44	44	46	46	48	48	50	50	52	52	54	54	56	56	58	58	60	60	62	62	64	64	66	66	68	68	70	70	72	72	74	74	76	76	78	78	80	80	80	80								
0.411-0.430 (0.0162-0.0169)	22	24	26	28	30	32	34	36	38	40	42	44	44	46	46	48	48	50	50	52	52	54	54	56	56	58	58	60	60	62	62	64	64	66	66	68	68	70	70	72	72	74	74	76	76	78	78	80	80	80	80	80									
0.431-0.450 (0.0170-0.0177)	24	26	28	30	32	34	36	38	40	42	44	46	46	48	48	50	50	52	52	54	54	56	56	58	58	60	60	62	62	64	64	66	66	68	68	70	70	72	72	74	74	76	76	78	78	80	80	80	80	80	80										
0.451-0.470 (0.0178-0.0185)	26	28	30	32	34	36	38	40	42	44	46	48	48	50	50	52	52	54	54	56	56	58	58	60	60	62	62	64	64	66	66	68	68	70	70	72	72	74	74	76	76	78	78	80	80	80	80	80	80	80											
0.471-0.490 (0.0185-0.0193)	28	30	32	34	36	38	40	42	44	46	48	50	50	52	52	54	54	56	56	58	58	60	60	62	62	64	64	66	66	68	68	70	70	72	72	74	74	76	76	78	78	80	80	80	80	80	80	80													
0.491-0.510 (0.0193-0.0201)	30	32	34	36	38	40	42	44	46	48	50	52	52	54	54	56	56	58	58	60	60	62	62	64	64	66	66	68	68	70	70	72	72	74	74	76	76	78	78	80	80	80	80	80	80	80															
0.511-0.530 (0.0201-0.0209)	32	34	36	38	40	42	44	46	48	50	52	54	54	56	56	58	58	60	60	62	62	64	64	66	66	68	68	70	70	72	72	74	74	76	76	78	78	80	80	80	80	80	80	80																	
0.531-0.550 (0.0209-0.0217)	34	36	38	40	42	44	46	48	50	52	54	56	56	58	58	60	60	62	62	64	64	66	66	68	68	70	70	72	72	74	74	76	76	78	78	80	80	80	80	80	80	80																			
0.551-0.570 (0.0217-0.0224)	36	38	40	42	44	46	48	50	52	54	56	58	58	60	60	62	62	64	64	66	66	68	68	70	70	72	72	74	74	76	76	78	78	80	80	80	80	80	80	80																					
0.571-0.590 (0.0225-0.0232)	38	40	42	44	46	48	50	52	54	56	58	60	60	62	62	64	64	66	66	68	68	70	70	72	72	74	74	76	76	78	78	80	80	80	80	80	80	80																							
0.591-0.610 (0.0233-0.0240)	40	42	44	46	48	50	52	54	56	58	60	62	62	64	64	66	66	68	68	70	70	72	72	74	74	76	76	78	78	80	80	80	80	80	80	80																									
0.611-0.630 (0.0241-0.0248)	42	44	46	48	50	52	54	56	58	60	62	64	64	66	66	68	68	70	70	72	72	74	74	76	76	78	78	80	80	80	80	80	80	80																											
0.631-0.650 (0.0248-0.0256)	44	46	48	50	52	54	56	58	60	62	64	66	66	68	68	70	70	72	72	74	74	76	76	78	78	80	80	80	80	80	80	80																													
0.651-0.670 (0.0256-0.0264)	46	48	50	52	54	56	58	60	62	64	66	68	68	70	70	72	72	74	74	76	76	78	78	80	80	80	80	80	80	80																															
0.671-0.690 (0.0264-0.0272)	48	50	52	54	56	58	60	62	64	66	68	70	70	72	72	74	74	76	76	78	78	80	80	80	80	80	80	80																																	
0.691-0.710 (0.0272-0.0280)	50	52	54	56	58	60	62	64	66	68	70	72	72	74	74	76	76	78	78	80	80	80	80	80	80	80																																			
0.711-0.730 (0.0280-0.0287)	52	54	56	58	60	62	64	66	68	70	72	74	74	76	76	78	78	80	80	80	80	80	80	80																																					
0.731-0.750 (0.0288-0.0295)	54	56	58	60	62	64	66	68	70	72	74	76	76	78	78	80	80	80	80	80	80	80																																							
0.751-0.770 (0.0296-0.0303)	56	58	60	62	64	66	68	70	72	74	76	78	78	80	80	80	80	80	80	80																																									
0.771-0.790 (0.0304-0.0311)	58	60	62	64	66	68	70	72	74	76	78	80	80	80	80	80	80	80																																											
0.791-0.810 (0.0311-0.0319)	60	62	64	66	68	70	72	74	76	78	80	80	80	80	80	80																																													
0.811-0.830 (0.0319-0.0327)	62	64	66	68	70	72	74	76	78	80	80	80	80	80																																															
0.831-0.850 (0.0327-0.0335)	64	66	68	70	72	74	76	78	80	80	80	80																																																	
0.851-0.870 (0.0335-0.0343)	66	68	70	72	74	76	78	80	80	80																																																			
0.871-0.890 (0.0343-0.0350)	68	70	72	74	76	78	80	80	80																																																				
0.891-0.910 (0.0351-0.0358)	70	72	74	76	78	80	80	80																																																					
0.911-0.930 (0.0359-0.0366)	72	74	76	78	80	80	80																																																						
0.931-0.950 (0.0367-0.0374)	74	76	78	80	80	80																																																							
0.951-0.970 (0.0374-0.0382)	76	78	80	80	80																																																								
0.971-0.990 (0.0382-0.0390)	78	80	80	80																																																									
0.991-1.010 (0.0390-0.0398)	80	80	80																																																										
1.011-1.030 (0.0398-0.0406)	80	80																																																											
1.031-1.050 (0.0406-0.0413)	80																																																													

Intake valve clearance (Cold):
0.15 – 0.25 mm (0.006 – 0.010 in.)

EXAMPLE:
The 2.300 mm (0.0906 in.) shim is installed, and the measured clearance is 0.440 mm (0.0173 in.). Replace the 2.300 mm (0.0906 in.) shim with a No. 54 shim.

New shim thickness mm (in.)

Shim No.	Thickness	Shim No.	Thickness	Shim No.	Thickness
00	2.000 (0.0787)	28	2.280 (0.0898)	56	2.560 (0.1008)
02	2.020 (0.0795)	30	2.300 (0.0906)	58	2.580 (0.1016)
04	2.040 (0.0803)	32	2.320 (0.0913)	60	2.600 (0.1024)
06	2.060 (0.0811)	34	2.340 (0.0921)	62	2.620 (0.1031)
08	2.080 (0.0819)	36	2.360 (0.0929)	64	2.640 (0.1039)
10	2.100 (0.0827)	38	2.380 (0.0937)	66	2.660 (0.1047)
12	2.120 (0.0835)	40	2.400 (0.0945)	68	2.680 (0.1055)
14	2.140 (0.0843)	42	2.420 (0.0953)	70	2.700 (0.1063)
16	2.160 (0.0850)	44	2.440 (0.0961)	72	2.720 (0.1071)
18	2.180 (0.0858)	46	2.460 (0.0969)	74	2.740 (0.1079)
20	2.200 (0.0866)	48	2.480 (0.0976)	76	2.760 (0.1087)
22	2.220 (0.0874)	50	2.500 (0.0984)	78	2.780 (0.1094)
24	2.240 (0.0882)	52	2.520 (0.0992)	80	2.800 (0.1102)
26	2.260 (0.0890)	54	2.540 (0.1000)		

7924SG71

Fig. 376 Intake valve clearance shim selection chart—4.7L engine

Installed shim thickness mm (in.) (shim sizes 2.000–2.290 mm)

Measured clearance mm (in.)	2.000 (0.0787)	2.020 (0.0795)	2.040 (0.0803)	2.060 (0.0811)	2.080 (0.0819)	2.100 (0.0827)	2.120 (0.0835)	2.140 (0.0843)	2.160 (0.0850)	2.180 (0.0858)	2.200 (0.0866)	2.210 (0.0870)	2.220 (0.0874)	2.230 (0.0878)	2.240 (0.0882)	2.250 (0.0886)	2.260 (0.0890)	2.270 (0.0894)	2.280 (0.0898)	2.290 (0.0902)
0.000–0.030 (0.0000–0.0012)												00	00	00	00	00	00	00	00	00
0.031–0.050 (0.0012–0.0020)											00	00	00	00	00	00	00	02	02	04
0.051–0.070 (0.0020–0.0028)										00	00	00	00	00	00	02	02	04	04	06
0.071–0.090 (0.0028–0.0035)									00	00	00	00	00	02	02	04	04	06	06	08
0.091–0.110 (0.0036–0.0043)								00	00	00	00	02	02	04	04	06	06	08	08	10
0.111–0.130 (0.0044–0.0051)							00	00	00	00	02	04	04	06	06	08	08	10	10	12
0.131–0.150 (0.0052–0.0059)						00	00	00	00	02	04	06	06	08	08	10	10	12	12	14
0.151–0.170 (0.0059–0.0067)					00	00	00	00	02	04	06	08	08	10	10	12	12	14	14	16
0.171–0.190 (0.0067–0.0075)				00	00	00	00	02	04	06	08	10	10	12	12	14	14	16	16	18
0.191–0.210 (0.0075–0.0083)			00	00	00	00	02	04	06	08	10	12	12	14	14	16	16	18	18	20
0.211–0.230 (0.0083–0.0091)		00	00	00	00	02	04	06	08	10	12	14	14	16	16	18	18	20	20	22
0.231–0.249 (0.0091–0.0098)		00	00	00	02	04	06	08	10	12	14	16	16	18	18	20	20	22	22	24
0.250–0.350 (0.0098–0.0138)																				
0.351–0.370 (0.0138–0.0146)	06	08	10	12	14	16	18	20	22	24	26	28	28	30	30	32	32	34	34	36
0.371–0.390 (0.0146–0.0154)	08	10	12	14	16	18	20	22	24	26	28	30	30	32	32	34	34	36	36	38
0.391–0.410 (0.0154–0.0161)	10	12	14	16	18	20	22	24	26	28	30	32	32	34	34	36	36	38	38	40
0.411–0.430 (0.0162–0.0169)	12	14	16	18	20	22	24	26	28	30	32	34	34	36	36	38	38	40	40	42
0.431–0.450 (0.0170–0.0177)	14	16	18	20	22	24	26	28	30	32	34	36	36	38	38	40	40	42	42	44
0.451–0.470 (0.0178–0.0185)	16	18	20	22	24	26	28	30	32	34	36	38	38	40	40	42	42	44	44	46
0.471–0.490 (0.0185–0.0193)	18	20	22	24	26	28	30	32	34	36	38	40	40	42	42	44	44	46	46	48
0.491–0.510 (0.0193–0.0201)	20	22	24	26	28	30	32	34	36	38	40	42	42	44	44	46	46	48	48	50
0.511–0.530 (0.0201–0.0209)	22	24	26	28	30	32	34	36	38	40	42	44	44	46	46	48	48	50	50	52
0.531–0.550 (0.0209–0.0217)	24	26	28	30	32	34	36	38	40	42	44	46	46	48	48	50	50	52	52	54
0.551–0.570 (0.0217–0.0224)	26	28	30	32	34	36	38	40	42	44	46	48	48	50	50	52	52	54	54	56
0.571–0.590 (0.0225–0.0232)	28	30	32	34	36	38	40	42	44	46	48	50	50	52	52	54	54	56	56	58
0.591–0.610 (0.0233–0.0240)	30	32	34	36	38	40	42	44	46	48	50	52	52	54	54	56	56	58	58	60
0.611–0.630 (0.0241–0.0248)	32	34	36	38	40	42	44	46	48	50	52	54	54	56	56	58	58	60	60	62
0.631–0.650 (0.0248–0.0256)	34	36	38	40	42	44	46	48	50	52	54	56	56	58	58	60	60	62	62	64
0.651–0.670 (0.0256–0.0264)	36	38	40	42	44	46	48	50	52	54	56	58	58	60	60	62	62	64	64	66
0.671–0.690 (0.0264–0.0272)	38	40	42	44	46	48	50	52	54	56	58	60	60	62	62	64	64	66	66	68
0.691–0.710 (0.0272–0.0280)	40	42	44	46	48	50	52	54	56	58	60	62	62	64	64	66	66	68	68	70
0.711–0.730 (0.0280–0.0287)	42	44	46	48	50	52	54	56	58	60	62	64	64	66	66	68	68	70	70	72
0.731–0.750 (0.0288–0.0295)	44	46	48	50	52	54	56	58	60	62	64	66	66	68	68	70	70	72	72	74
0.751–0.770 (0.0296–0.0303)	46	48	50	52	54	56	58	60	62	64	66	68	68	70	70	72	72	74	74	76
0.771–0.790 (0.0304–0.0311)	48	50	52	54	56	58	60	62	64	66	68	70	70	72	72	74	74	76	76	78
0.791–0.810 (0.0311–0.0319)	50	52	54	56	58	60	62	64	66	68	70	72	72	74	74	76	76	78	78	80
0.811–0.830 (0.0319–0.0327)	52	54	56	58	60	62	64	66	68	70	72	74	74	76	76	78	78	80	80	80
0.831–0.850 (0.0327–0.0335)	54	56	58	60	62	64	66	68	70	72	74	76	76	78	78	80	80	80	80	80
0.851–0.870 (0.0335–0.0343)	56	58	60	62	64	66	68	70	72	74	76	78	78	80	80	80	80	80	80	80
0.871–0.890 (0.0343–0.0350)	58	60	62	64	66	68	70	72	74	76	78	80	80	80	80	80	80	80		
0.891–0.910 (0.0351–0.0358)	60	62	64	66	68	70	72	74	76	78	80	80	80	80	80	80				
0.911–0.930 (0.0359–0.0366)	62	64	66	68	70	72	74	76	78	80	80	80	80	80						
0.931–0.950 (0.0367–0.0374)	64	66	68	70	72	74	76	78	80	80	80	80								
0.951–0.970 (0.0374–0.0382)	66	68	70	72	74	76	78	80	80	80										
0.971–0.990 (0.0382–0.0390)	68	70	72	74	76	78	80	80	80											
0.991–1.010 (0.0390–0.0398)	70	72	74	76	78	80	80	80												
1.011–1.030 (0.0398–0.0406)	72	74	76	78	80	80	80													
1.031–1.050 (0.0406–0.0413)	74	76	78	80	80	80														
1.051–1.070 (0.0414–0.0421)	76	78	80	80	80															
1.071–1.090 (0.0422–0.0429)	78	80	80	80																
1.091–1.110 (0.0430–0.0437)	80	80	80																	
1.111–1.130 (0.0437–0.0445)	80	80																		
1.131–1.150 (0.0445–0.0453)	80																			

Installed shim thickness mm (in.) (shim sizes 2.300–2.490 mm)

Measured clearance mm (in.)	2.300 (0.0906)	2.310 (0.0909)	2.320 (0.0913)	2.330 (0.0917)	2.340 (0.0921)	2.350 (0.0925)	2.360 (0.0929)	2.370 (0.0933)	2.380 (0.0937)	2.390 (0.0941)	2.400 (0.0945)	2.410 (0.0949)	2.420 (0.0953)	2.430 (0.0957)	2.440 (0.0961)	2.450 (0.0965)	2.460 (0.0969)	2.470 (0.0972)	2.480 (0.0976)	2.490 (0.0980)
0.000–0.030 (0.0000–0.0012)	02	02	04	04	06	06	08	08	10	10	12	12	14	14	16	16	18	18	20	20
0.031–0.050 (0.0012–0.0020)	04	06	06	08	08	10	10	12	12	14	14	16	16	18	18	20	20	22	22	24
0.051–0.070 (0.0020–0.0028)	06	08	08	10	10	12	12	14	14	16	16	18	18	20	20	22	22	24	24	26
0.071–0.090 (0.0028–0.0035)	08	10	10	12	12	14	14	16	16	18	18	20	20	22	22	24	24	26	26	28
0.091–0.110 (0.0036–0.0043)	10	12	12	14	14	16	16	18	18	20	20	22	22	24	24	26	26	28	28	30
0.111–0.130 (0.0044–0.0051)	12	14	14	16	16	18	18	20	20	22	22	24	24	26	26	28	28	30	30	32
0.131–0.150 (0.0052–0.0059)	14	16	16	18	18	20	20	22	22	24	24	26	26	28	28	30	30	32	32	34
0.151–0.170 (0.0059–0.0067)	16	18	18	20	20	22	22	24	24	26	26	28	28	30	30	32	32	34	34	36
0.171–0.190 (0.0067–0.0075)	18	20	20	22	22	24	24	26	26	28	28	30	30	32	32	34	34	36	36	38
0.191–0.210 (0.0075–0.0083)	20	22	22	24	24	26	26	28	28	30	30	32	32	34	34	36	36	38	38	40
0.211–0.230 (0.0083–0.0091)	22	24	24	26	26	28	28	30	30	32	32	34	34	36	36	38	38	40	40	42
0.231–0.249 (0.0091–0.0098)	24	26	26	28	28	30	30	32	32	34	34	36	36	38	38	40	40	42	42	44
0.250–0.350 (0.0098–0.0138)																				
0.351–0.370 (0.0138–0.0146)	36	38	38	40	40	42	42	44	44	46	46	48	48	50	50	52	52	54	54	56
0.371–0.390 (0.0146–0.0154)	38	40	40	42	42	44	44	46	46	48	48	50	50	52	52	54	54	56	56	58
0.391–0.410 (0.0154–0.0161)	40	42	42	44	44	46	46	48	48	50	50	52	52	54	54	56	56	58	58	60
0.411–0.430 (0.0162–0.0169)	42	44	44	46	46	48	48	50	50	52	52	54	54	56	56	58	58	60	60	62
0.431–0.450 (0.0170–0.0177)	44	46	46	48	48	50	50	52	52	54	54	56	56	58	58	60	60	62	62	64
0.451–0.470 (0.0178–0.0185)	46	48	48	50	50	52	52	54	54	56	56	58	58	60	60	62	62	64	64	66
0.471–0.490 (0.0185–0.0193)	48	50	50	52	52	54	54	56	56	58	58	60	60	62	62	64	64	66	66	68
0.491–0.510 (0.0193–0.0201)	50	52	52	54	54	56	56	58	58	60	60	62	62	64	64	66	66	68	68	70
0.511–0.530 (0.0201–0.0209)	52	54	54	56	56	58	58	60	60	62	62	64	64	66	66	68	68	70	70	72
0.531–0.550 (0.0209–0.0217)	54	56	56	58	58	60	60	62	62	64	64	66	66	68	68	70	70	72	72	74
0.551–0.570 (0.0217–0.0224)	56	58	58	60	60	62	62	64	64	66	66	68	68	70	70	72	72	74	74	76
0.571–0.590 (0.0225–0.0232)	58	60	60	62	62	64	64	66	66	68	68	70	70	72	72	74	74	76	76	78
0.591–0.610 (0.0233–0.0240)	60	62	62	64	64	66	66	68	68	70	70	72	72	74	74	76	76	78	78	80
0.611–0.630 (0.0241–0.0248)	62	64	64	66	66	68	68	70	70	72	72	74	74	76	76	78	78	80	80	80
0.631–0.650 (0.0248–0.0256)	64	66	66	68	68	70	70	72	72	74	74	76	76	78	78	80	80	80	80	80
0.651–0.670 (0.0256–0.0264)	66	68	68	70	70	72	72	74	74	76	76	78	78	80	80	80	80	80	80	80
0.671–0.690 (0.0264–0.0272)	68	70	70	72	72	74	74	76	76	78	78	80	80	80	80	80	80	80		
0.691–0.710 (0.0272–0.0280)	70	72	72	74	74	76	76	78	78	80	80	80	80	80	80	80				
0.711–0.730 (0.0280–0.0287)	72	74	74	76	76	78	78	80	80	80	80	80	80	80						
0.731–0.750 (0.0288–0.0295)	74	76	76	78	78	80	80	80	80	80	80	80								
0.751–0.770 (0.0296–0.0303)	76	78	78	80	80	80	80	80	80	80										
0.771–0.790 (0.0304–0.0311)	78	80	80	80	80	80	80	80												
0.791–0.810 (0.0311–0.0319)	80	80	80	80	80	80														
0.811–0.830 (0.0319–0.0327)	80	80	80	80																
0.831–0.850 (0.0327–0.0335)	80	80																		

Installed shim thickness mm (in.) (shim sizes 2.500–2.800 mm)

Measured clearance mm (in.)	2.500 (0.0984)	2.510 (0.0988)	2.520 (0.0992)	2.530 (0.0996)	2.540 (0.1000)	2.550 (0.1004)	2.560 (0.1008)	2.570 (0.1012)	2.580 (0.1016)	2.590 (0.1020)	2.600 (0.1024)	2.620 (0.1031)	2.640 (0.1039)	2.660 (0.1047)	2.680 (0.1055)	2.700 (0.1063)	2.720 (0.1071)	2.740 (0.1079)	2.760 (0.1087)	2.780 (0.1094)	2.800 (0.1102)
0.000–0.030 (0.0000–0.0012)	22	22	24	24	26	26	28	28	30	30	32	34	36	38	40	42	44	46	48	50	52
0.031–0.050 (0.0012–0.0020)	24	26	26	28	28	30	30	32	32	34	34	36	38	40	42	44	46	48	50	52	54
0.051–0.070 (0.0020–0.0028)	26	28	28	30	30	32	32	34	34	36	36	38	40	42	44	46	48	50	52	54	56
0.071–0.090 (0.0028–0.0035)	28	30	30	32	32	34	34	36	36	38	38	40	42	44	46	48	50	52	54	56	58
0.091–0.110 (0.0036–0.0043)	30	32	32	34	34	36	36	38	38	40	40	42	44	46	48	50	52	54	56	58	60
0.111–0.130 (0.0044–0.0051)	32	34	34	36	36	38	38	40	40	42	42	44	46	48	50	52	54	56	58	60	62
0.131–0.150 (0.0052–0.0059)	34	36	36	38	38	40	40	42	42	44	44	46	48	50	52	54	56	58	60	62	64
0.151–0.170 (0.0059–0.0067)	36	38	38	40	40	42	42	44	44	46	46	48	50	52	54	56	58	60	62	64	66
0.171–0.190 (0.0067–0.0075)	38	40	40	42	42	44	44	46	46	48	48	50	52	54	56	58	60	62	64	66	68
0.191–0.210 (0.0075–0.0083)	40	42	42	44	44	46	46	48	48	50	50	52	54	56	58	60	62	64	66	68	70
0.211–0.230 (0.0083–0.0091)	42	44	44	46	46	48	48	50	50	52	52	54	56	58	60	62	64	66	68	70	72
0.231–0.249 (0.0091–0.0098)	44	46	46	48	48	50	50	52	52	54	54	56	58	60	62	64	66	68	70	72	74
0.250–0.350 (0.0098–0.0138)																					
0.351–0.370 (0.0138–0.0146)	56	58	58	60	60	62	62	64	64	66	66	68	70	72	74	76	78	80	80	80	
0.371–0.390 (0.0146–0.0154)	58	60	60	62	62	64	64	66	66	68	68	70	72	74	76	78	80	80	80		
0.391–0.410 (0.0154–0.0161)	60	62	62	64	64	66	66	68	68	70	70	72	74	76	78	80	80	80			
0.411–0.430 (0.0162–0.0169)	62	64	64	66	66	68	68	70	70	72	72	74	76	78	80	80	80				
0.431–0.450 (0.0170–0.0177)	64	66	66	68	68	70	70	72	72	74	74	76	78	80	80	80					
0.451–0.470 (0.0178–0.0185)	66	68	68	70	70	72	72	74	74	76	76	78	80	80	80						
0.471–0.490 (0.0185–0.0193)	68	70	70	72	72	74	74	76	76	78	78	80	80	80							
0.491–0.510 (0.0193–0.0201)	70	72	72	74	74	76	76	78	78	80	80	80	80								
0.511–0.530 (0.0201–0.0209)	72	74	74	76	76	78	78	80	80	80	80	80									
0.531–0.550 (0.0209–0.0217)	74	76	76	78	78	80	80	80	80	80	80										
0.551–0.570 (0.0217–0.0224)	76	78	78	80	80	80	80	80	80	80											
0.571–0.590 (0.0225–0.0232)	78	80	80	80	80	80	80	80													
0.591–0.610 (0.0233–0.0240)	80	80	80	80	80	80															
0.611–0.630 (0.0241–0.0248)	80	80	80	80																	
0.631–0.650 (0.0248–0.0256)	80	80																			

Exhaust valve clearance (Cold):
0.25 – 0.35 mm (0.010 – 0.014 in.)

EXAMPLE:
The 2.300 mm (0.0906 in.) shim is installed, and the measured clearance is 0.440 mm (0.0173 in.). Replace the 2.300 mm (0.0906 in.) shim with a No. 44 shim.

New shim thickness mm (in.)

Shim No.	Thickness	Shim No.	Thickness	Shim No.	Thickness
00	2.000 (0.0787)	28	2.280 (0.0898)	56	2.560 (0.1008)
02	2.020 (0.0795)	30	2.300 (0.0906)	58	2.580 (0.1016)
04	2.040 (0.0803)	32	2.320 (0.0913)	60	2.600 (0.1024)
06	2.060 (0.0811)	34	2.340 (0.0921)	62	2.620 (0.1031)
08	2.080 (0.0819)	36	2.360 (0.0929)	64	2.640 (0.1039)
10	2.100 (0.0827)	38	2.380 (0.0937)	66	2.660 (0.1047)
12	2.120 (0.0835)	40	2.400 (0.0945)	68	2.680 (0.1055)
14	2.140 (0.0843)	42	2.420 (0.0953)	70	2.700 (0.1063)
16	2.160 (0.0850)	44	2.440 (0.0961)	72	2.720 (0.1071)
18	2.180 (0.0858)	46	2.460 (0.0969)	74	2.740 (0.1079)
20	2.200 (0.0866)	48	2.480 (0.0976)	76	2.760 (0.1087)
22	2.220 (0.0874)	50	2.500 (0.0984)	78	2.780 (0.1094)
24	2.240 (0.0882)	52	2.520 (0.0992)	80	2.800 (0.1102)
26	2.260 (0.0890)	54	2.540 (0.1000)		

7924SG72

Fig. 377 Exhaust valve clearance shim selection chart—4.7L engine

8. Record the measurements for each valve.

9. When all valve clearances have been measured, remove the camshafts.

10. Remove the valve shims and measure them. Note this measurement along with the clearance measurement recorded earlier.

11. Using the valve clearance and shim thickness measurements, find replacement shims in the Adjusting Shim Selection charts.

12. Install or connect the following:

- Replacement valve shims
- Camshafts
- Valve covers
- Ignition coils
- Negative battery cable

13. Fill the cooling system.

14. Start the engine and check for leaks.

15. Use the Techstream diagnostic tool, or equivalent and reprogram the required systems.

5.7L Engine

Valve lash is not adjustable for this engine.

INSPECTION

Check for engine noise and vibration. Check valve noise or rough idle. Judge whether the valve clearance needs to be adjusted. If valve noise is to loud or engine idle is to rough, inspect the valve clearance.

ENGINE PERFORMANCE & EMISSION CONTROLS

CAMSHAFT POSITION (CMP) SENSOR

LOCATION

See Figure 378.

REMOVAL & INSTALLATION

4.0L Engine

1. Before servicing the vehicle, refer to the Precautions Section.

➡If working near and/or around the SRS system and components, be sure to disable the SRS system. Tape the negative battery cable with insulating tape. Always disconnect the negative battery cable first.

⁂ CAUTION

To avoid personal injury when working on vehicles equipped with an air bag, the negative battery cable must be disconnected and at least 90 seconds must elapse before working on the system. Failure to do so may result in deployment of the air bag.

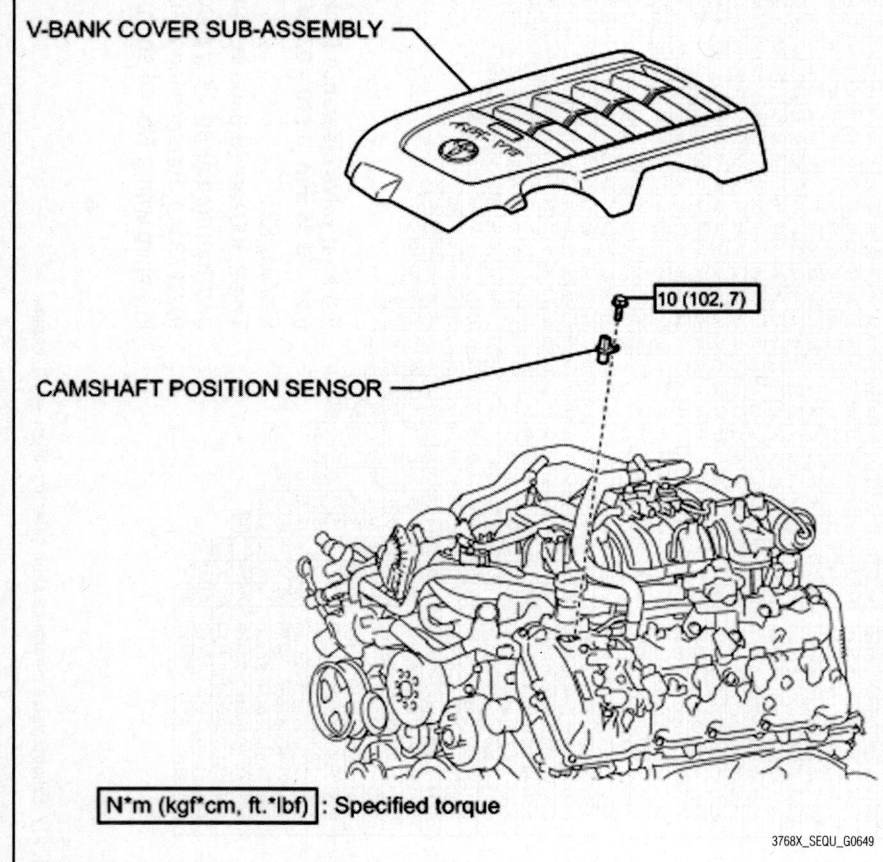

Fig. 378 Camshaft position sensor location

2. Disconnect the negative battery cable. Tape the cable with insulating tape.

➡When disconnecting the cable, some systems need to be initialized after the cable is reconnected. You will need the Techstream diagnostic scan tool or equivalent. Follow the directions on the tool.

3. Remove V-bank cover sub-assembly.

4. Disconnect the sensor connector.

5. Remove the bolt and sensor.

To install:

➡Be sure to use new fasteners, as required.

6. Install the sensor with the bolt and tighten to 7 ft. lbs. (10 Nm).

7. Connect the sensor connector.

8. Install V-bank cover sub-assembly.

9. Use the Techstream diagnostic tool, or equivalent and reprogram the required systems.

4.6L Engine

1. Before servicing the vehicle, refer to the Precautions Section.

➡If working near and/or around the SRS system and components, be sure to disable the SRS system. Tape the negative battery cable with insulating tape. Always disconnect the negative battery cable first.

⁂ CAUTION

To avoid personal injury when working on vehicles equipped with an air bag, the negative battery cable must be disconnected and at least 90 seconds must elapse before working on the system. Failure to do so may result in deployment of the air bag.

2. Disconnect the negative battery cable. Tape the cable with insulating tape.

➡When disconnecting the cable, some systems need to be initialized after the cable is reconnected. You will need the Techstream diagnostic scan tool or equivalent. Follow the directions on the tool.

3. Remove V-bank cover sub-assembly.
4. Disconnect the sensor connector.
5. Remove the bolt and sensor.

To install:

➡Be sure to use new fasteners, as required.

6. Install the sensor with the bolt and tighten to 7 ft. lbs. (10 Nm).
7. Connect the sensor connector.
8. Install V-bank cover sub-assembly.
9. Use the Techstream diagnostic tool, or equivalent and reprogram the required systems.

4.7L Engine

1. Before servicing the vehicle, refer to the Precautions Section.

➡If working near and/or around the SRS system and components, be sure to disable the SRS system. Tape the negative battery cable with insulating tape. Always disconnect the negative battery cable first.

⁂ CAUTION

To avoid personal injury when working on vehicles equipped with an air bag, the negative battery cable must be disconnected and at least 90 seconds must elapse before working on the system. Failure to do so may result in deployment of the air bag.

2. Disconnect the negative battery cable. Tape the cable with insulating tape.

➡When disconnecting the cable, some systems need to be initialized after the cable is reconnected. You will need the Techstream diagnostic scan tool or equivalent. Follow the directions on the tool.

3. Drain engine coolant.
4. Remove V-bank cover sub-assembly.
5. Remove fan and alternator v belt.
6. Remove oil cooler pipe.
7. Remove timing belt cover sub-assembly No. 3 LH.
8. Disconnect the camshaft position sensor connector.
9. Remove the bolt, stud bolt and camshaft position sensor.

To install:

➡Be sure to use new fasteners, as required.

10. Install the camshaft position sensor with the bolt and stud bolt. Tighten bolt to 66 inch lbs. (7.5 Nm).
11. Reconnect the camshaft position sensor connector.
12. Install timing belt cover sub-assembly No. 3 LH.
13. Install oil cooler pipe.
14. Install fan and alternator v belt.
15. Add engine coolant.
16. Check for engine coolant leaks.
17. Install V-bank cover sub-assembly.
18. Use the Techstream diagnostic tool, or equivalent and reprogram the required systems.

5.7L Engine

1. Before servicing the vehicle, refer to the Precautions Section.

➡If working near and/or around the SRS system and components, be sure to disable the SRS system. Tape the negative battery cable with insulating tape. Always disconnect the negative battery cable first.

⁂ CAUTION

To avoid personal injury when working on vehicles equipped with an air bag, the negative battery cable must be disconnected and at least 90 seconds must elapse before working on the system. Failure to do so may result in deployment of the air bag.

2. Disconnect the negative battery cable. Tape the cable with insulating tape.

➡When disconnecting the cable, some systems need to be initialized after the cable is reconnected. You will need the Techstream diagnostic scan tool or equivalent. Follow the directions on the tool.

3. Remove V-bank cover sub-assembly.
4. Disconnect the sensor connector.
5. Remove the bolt and sensor.

To install:

➡Be sure to use new fasteners, as required.

6. Install the sensor with the bolt and tighten to 7 ft. lbs. (10 Nm).
7. Connect the sensor connector.
8. Install V-bank cover sub-assembly.
9. Use the Techstream diagnostic tool, or equivalent and reprogram the required systems.

CRANKSHAFT POSITION (CKP) SENSOR

LOCATION

See Figures 379 through 382.

REMOVAL & INSTALLATION

4.0L Engine

1. Before servicing the vehicle, refer to the Precautions Section.

➡If working near and/or around the SRS system and components, be sure to disable the SRS system. Tape the negative battery cable with insulating tape. Always disconnect the negative battery cable first.

⁂ CAUTION

To avoid personal injury when working on vehicles equipped with an air bag, the negative battery cable must be disconnected and at least 90 seconds must elapse before working on the system. Failure to do so may result in deployment of the air bag.

2. Disconnect the negative battery cable. Tape the cable with insulating tape.

➡When disconnecting the cable, some systems need to be initialized after the cable is reconnected. You will need the Techstream diagnostic scan tool or equivalent. Follow the directions on the tool.

3. Remove no. 1 engine under cover.
4. Remove fan and alternator v belt.
5. Remove alternator assembly.
6. Disconnect cooler compressor assembly.
7. Disconnect the sensor connector.
8. Remove the bolt and sensor.

To install:

➡Be sure to use new fasteners, as required.

9. Install crankshaft position sensor.
10. Install the sensor with the bolt. Tighten bolt to 57 inch lbs. (6.5 nm).
11. Connect the sensor connector.
12. Reverse removal procedure.
13. Connect cable to negative battery terminal.
14. Use the Techstream diagnostic tool, or equivalent and reprogram the required systems.

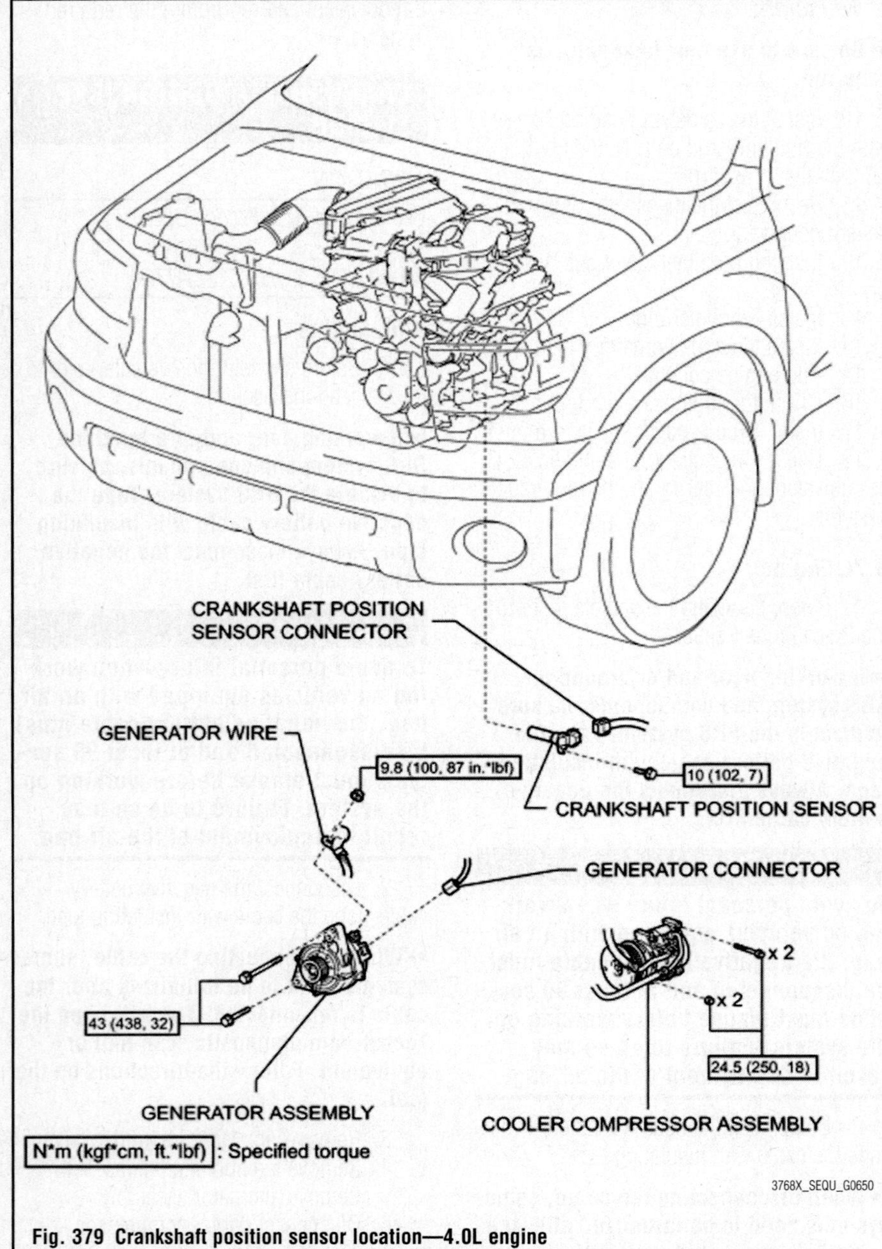

Fig. 379 Crankshaft position sensor location—4.0L engine

4.6L Engine

1. Before servicing the vehicle, refer to the Precautions Section.

➡If working near and/or around the SRS system and components, be sure to disable the SRS system. Tape the negative battery cable with insulating tape. Always disconnect the negative battery cable first.

⁂ CAUTION

To avoid personal injury when working on vehicles equipped with an air bag, the negative battery cable must be disconnected and at least 90 seconds must elapse before working on the system. Failure to do so may result in deployment of the air bag.

2. Disconnect the negative battery cable. Tape the cable with insulating tape.

➡When disconnecting the cable, some systems need to be initialized after the cable is reconnected. You will need the Techstream diagnostic scan tool or equivalent. Follow the directions on the tool.

3. Remove the sensor protector.
4. Disconnect the electrical connector.
5. Remove the retaining bolts.
6. Remove the component from its mounting.

To install:

➡Be sure to use new fasteners, as required.

7. Installation is the reverse of the removal procedure.
8. Install the sensor with the bolt and tighten to 7 ft. lbs. (10 Nm)
9. Use the Techstream diagnostic tool, or equivalent and reprogram the required systems.

4.7L Engine

1. Before servicing the vehicle, refer to the Precautions Section.

➡If working near and/or around the SRS system and components, be sure to disable the SRS system. Tape the negative battery cable with insulating tape. Always disconnect the negative battery cable first.

⁂ CAUTION

To avoid personal injury when working on vehicles equipped with an air bag, the negative battery cable must be disconnected and at least 90 seconds must elapse before working on the system. Failure to do so may result in deployment of the air bag.

2. Disconnect the negative battery cable. Tape the cable with insulating tape.

➡When disconnecting the cable, some systems need to be initialized after the cable is reconnected. You will need the Techstream diagnostic scan tool or equivalent. Follow the directions on the tool.

3. Remove no. 1 engine under cover.
4. Disconnect the sensor connector.
5. Remove the bolt and sensor.

To install:

➡Be sure to use new fasteners, as required.

6. Install crankshaft position sensor.
7. Install the sensor with the bolt. Tighten bolt to 57 inch lbs. (6.5 Nm).
8. Connect the sensor connector.
9. Install no. 1 engine under cover.
10. Connect cable to negative battery terminal.
11. Use the Techstream diagnostic tool, or equivalent and reprogram the required systems.

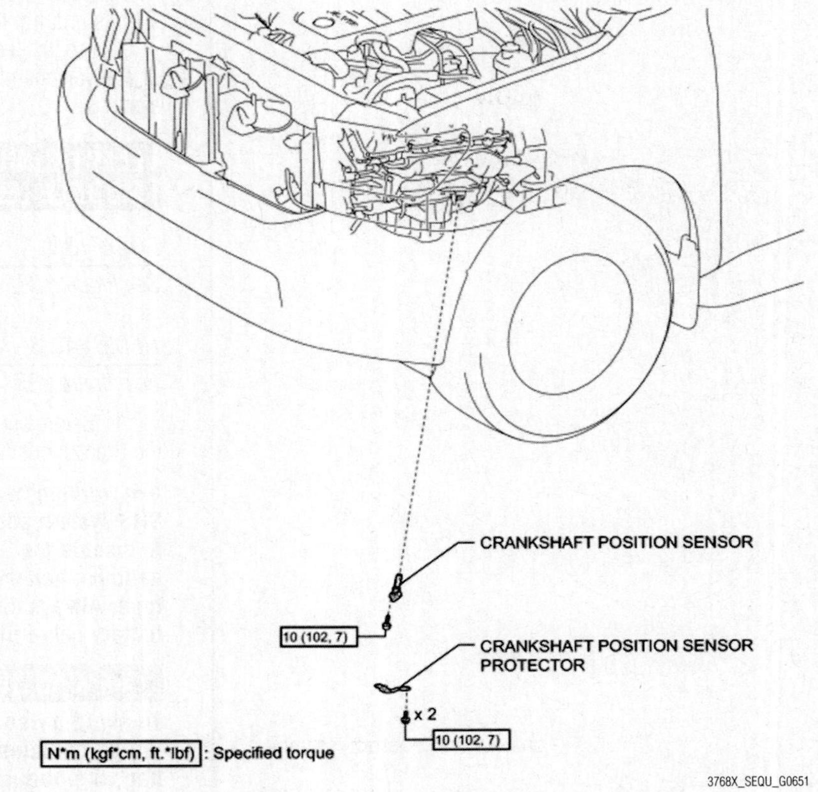

Fig. 380 Crankshaft position sensor location—4.6L engine

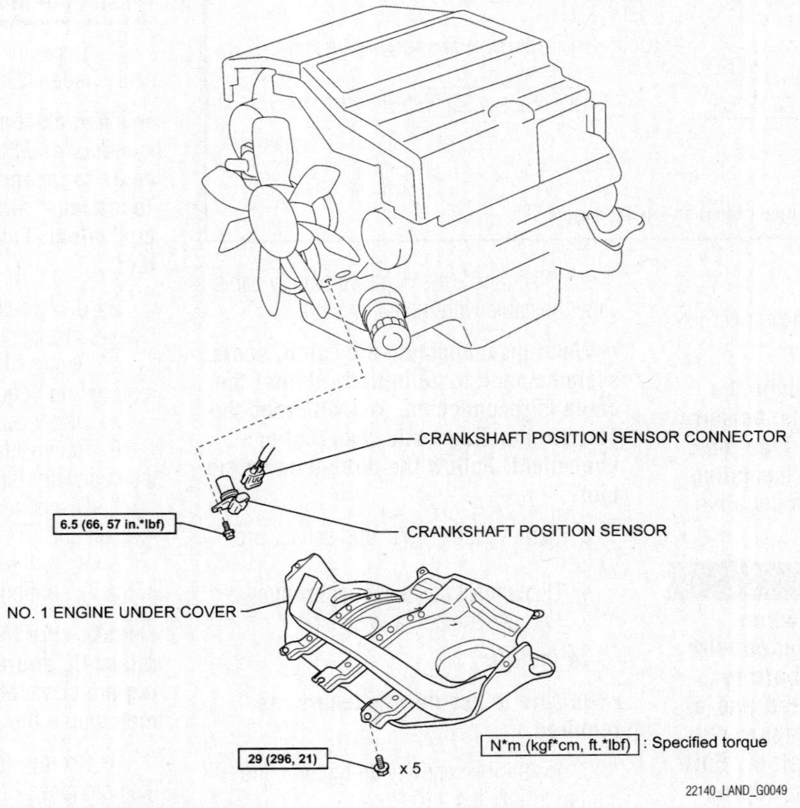

Fig. 381 Crankshaft position sensor location—4.7L engine

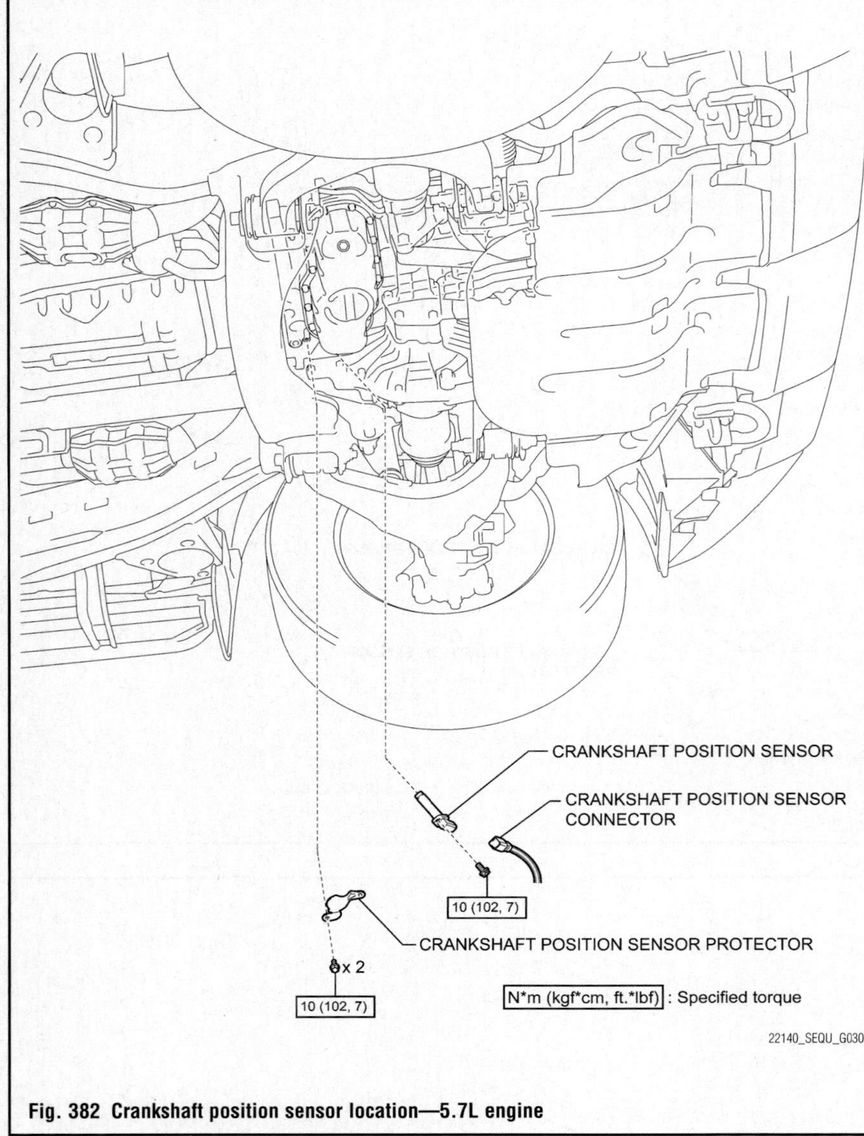

Fig. 382 Crankshaft position sensor location—5.7L engine

5.7L Engine

1. Before servicing the vehicle, refer to the Precautions Section.

➡If working near and/or around the SRS system and components, be sure to disable the SRS system. Tape the negative battery cable with insulating tape. Always disconnect the negative battery cable first.

⁂ CAUTION

To avoid personal injury when working on vehicles equipped with an air bag, the negative battery cable must be disconnected and at least 90 seconds must elapse before working on the system. Failure to do so may result in deployment of the air bag.

2. Disconnect the negative battery cable. Tape the cable with insulating tape.

➡When disconnecting the cable, some systems need to be initialized after the cable is reconnected. You will need the Techstream diagnostic scan tool or equivalent. Follow the directions on the tool.

3. Remove the 2 bolts and sensor protector
4. Disconnect the sensor connector.
5. Remove the bolt and sensor

To install:

➡Be sure to use new fasteners, as required.

6. Install the sensor with the bolt and tighten to 7 ft. lbs. (10 Nm)
7. Connect the sensor connector
8. Install the sensor protector with the 2 bolts and tighten to 7 ft. lbs. (10 Nm)
9. Use the Techstream diagnostic tool, or equivalent and reprogram the required systems.

ELECTRONIC CONTROL MODULE (ECM)

LOCATION

See Figures 383 and 384.

REMOVAL & INSTALLATION

See Figure 385.

1. Before servicing the vehicle, refer to the Precautions Section.

➡If working near and/or around the SRS system and components, be sure to disable the SRS system. Tape the negative battery cable with insulating tape. Always disconnect the negative battery cable first.

⁂ CAUTION

To avoid personal injury when working on vehicles equipped with an air bag, the negative battery cable must be disconnected and at least 90 seconds must elapse before working on the system. Failure to do so may result in deployment of the air bag.

2. Disconnect the negative battery cable. Tape the cable with insulating tape.

➡When disconnecting the cable, some systems need to be initialized after the cable is reconnected. You will need the Techstream diagnostic scan tool or equivalent. Follow the directions on the tool.

3. On 4.6L engine remove the V- back cover sub assembly.
4. Remove the air cleaner assembly, as required on Sequoia.
5. Disconnect the connector holder block.
6. Remove the retaining bolts and move the connector block.
7. Disconnect the wire harness clamp, 4.0L engine.
8. Raise the retaining levers, as shown in the illustration.

➡Make sure that the lock lever is raised 90 degrees, before disconnecting the connectors. Failure to do so may cause the connectors to break.

9. Remove the retaining bolts. Remove the component from its mounting. Discard the gasket.

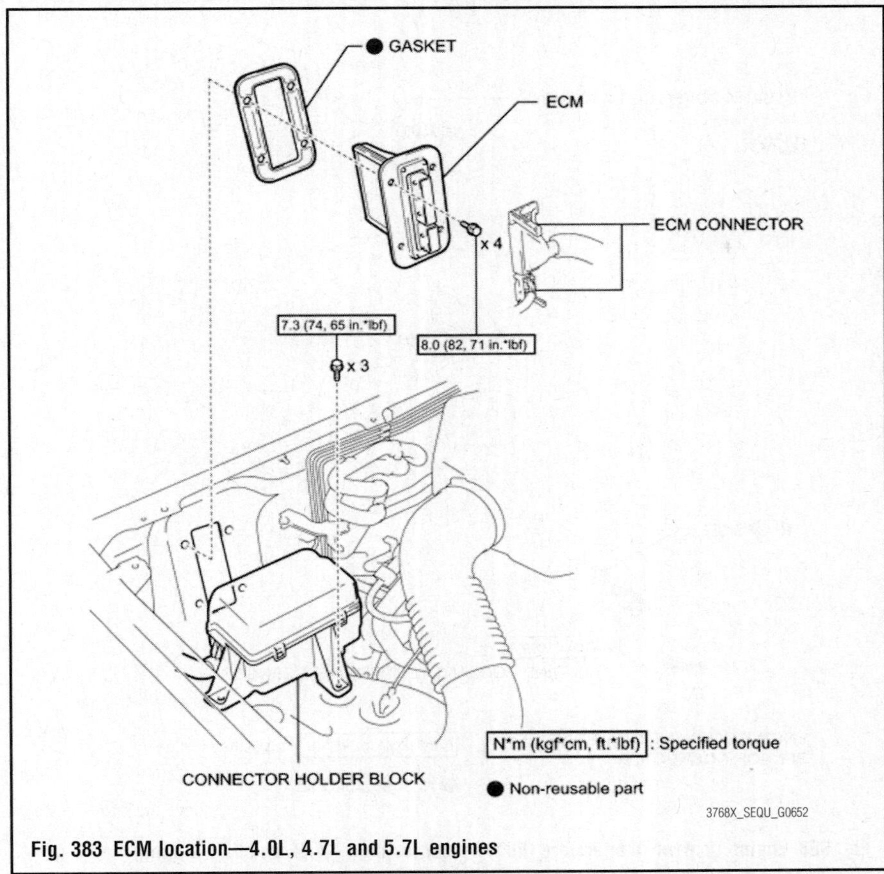

Fig. 383 ECM location—4.0L, 4.7L and 5.7L engines

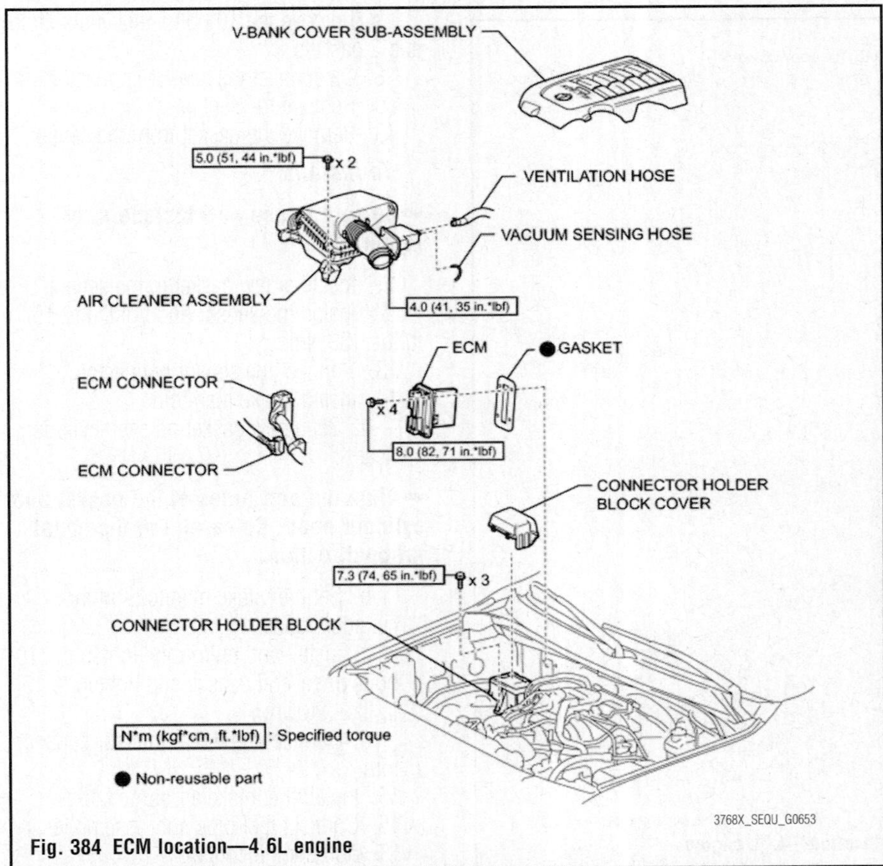

Fig. 384 ECM location—4.6L engine

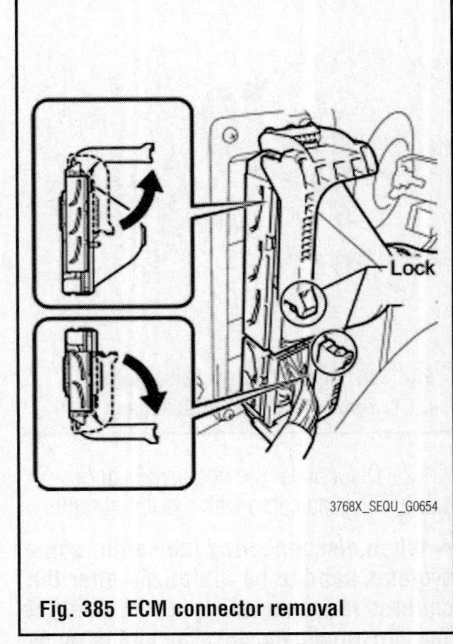

Fig. 385 ECM connector removal

To install:

➡Be sure to use new fasteners, as required.

10. Installation is the reverse of the removal procedure.
11. Use the Techstream diagnostic tool, or equivalent and reprogram the required systems.

ENGINE COOLANT TEMPERATURE (ECT) SENSOR

LOCATION

See Figures 386 through 388.

REMOVAL & INSTALLATION

4.0L Engine

1. Before servicing the vehicle, refer to the Precautions Section.

➡If working near and/or around the SRS system and components, be sure to disable the SRS system. Tape the negative battery cable with insulating tape. Always disconnect the negative battery cable first.

⁂ CAUTION

To avoid personal injury when working on vehicles equipped with an air bag, the negative battery cable must be disconnected and at least 90 seconds must elapse before working on the system. Failure to do so may result in deployment of the air bag.

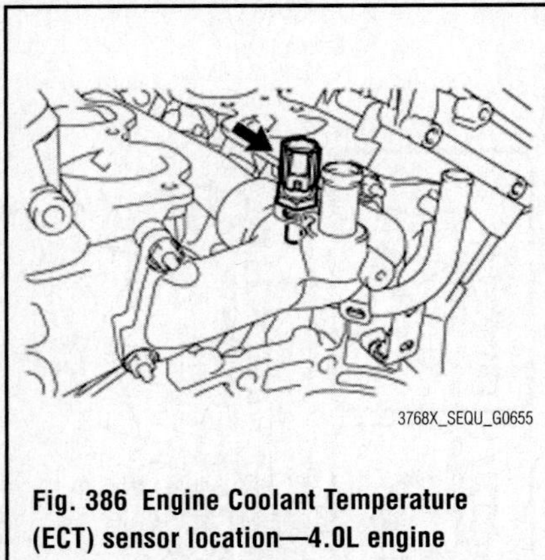

Fig. 386 Engine Coolant Temperature (ECT) sensor location—4.0L engine

2. Disconnect the negative battery cable. Tape the cable with insulating tape.

➡**When disconnecting the cable, some systems need to be initialized after the cable is reconnected. You will need the Techstream diagnostic scan tool or equivalent. Follow the directions on the tool.**

3. Disconnect cable from negative battery terminal.
4. Remove the intake air surge tank.
5. Disconnect No. 1 and No.1 2 fuel pipe sub-assemblies.
6. Disconnect the 6 fuel injector connectors.

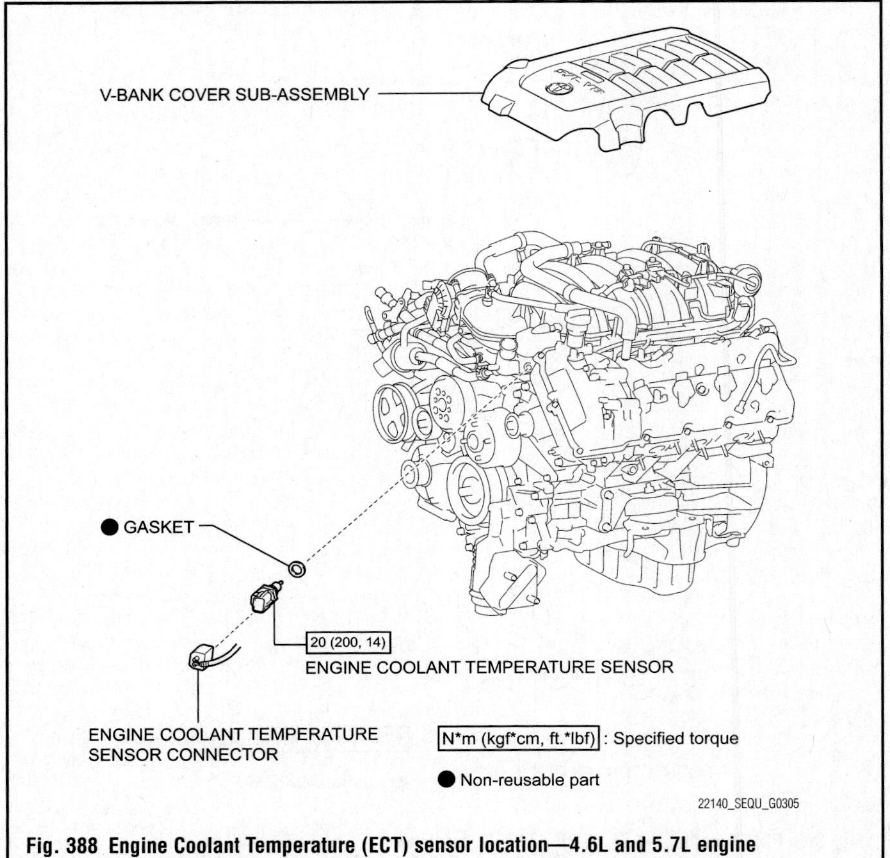

Fig. 388 Engine Coolant Temperature (ECT) sensor location—4.6L and 5.7L engine

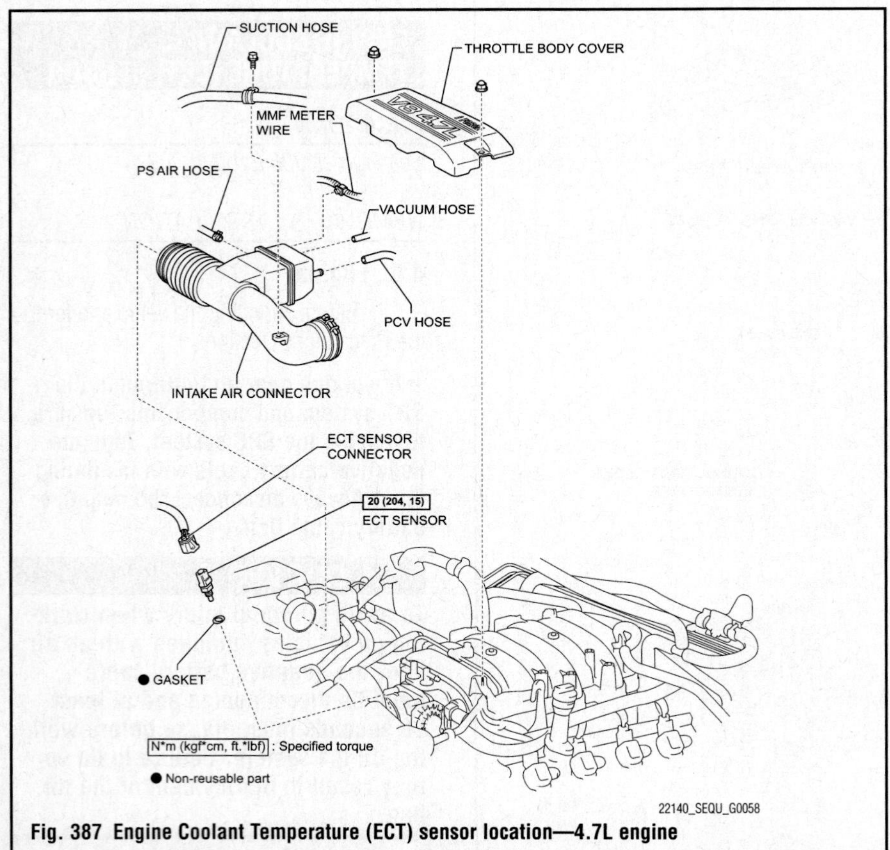

Fig. 387 Engine Coolant Temperature (ECT) sensor location—4.7L engine

7. Remove the 10 bolts, intake manifold and 2 gaskets.
8. Disconnect the sensor connector.
9. Remove the sensor.
10. Remove the gasket from the sensor.

To install:

➡**Be sure to use new fasteners, as required.**

11. Install a new gasket to the sensor.
12. Install the sensor and tighten to 15 ft. lbs. (20 Nm).
13. Connect the sensor connector.
14. Install intake manifold.
    a. Set a new gasket on each cylinder head.

➡**Align the port holes of the gasket and cylinder head. Be careful of the installation direction.**

    b. Set the intake manifold on the cylinder heads.
    c. Install and uniformly tighten the 10 bolts in several passes and tighten to 19 ft. lbs. (26 Nm).
    d. Connect the 6 fuel injector connectors.
15. Install the intake air surge tank.
16. Connect fuel pipe sub-assemblies.
17. Add engine coolant.

18. Connect cable to negative battery terminal.
19. Inspect for engine coolant leak.
20. Use the Techstream diagnostic tool, or equivalent and reprogram the required systems.

4.6L Engine

1. Before servicing the vehicle, refer to the Precautions Section.

➡If working near and/or around the SRS system and components, be sure to disable the SRS system. Tape the negative battery cable with insulating tape. Always disconnect the negative battery cable first.

CAUTION

To avoid personal injury when working on vehicles equipped with an air bag, the negative battery cable must be disconnected and at least 90 seconds must elapse before working on the system. Failure to do so may result in deployment of the air bag.

2. Disconnect the negative battery cable. Tape the cable with insulating tape.

➡When disconnecting the cable, some systems need to be initialized after the cable is reconnected. You will need the Techstream diagnostic scan tool or equivalent. Follow the directions on the tool.

3. Remove No. 1 engine under cover.
4. Remove V-bank cover sub-assembly.
5. Drain engine coolant.
6. Disconnect the sensor connector.
7. Remove the sensor.
8. Remove the gasket from the sensor.

To install:

➡Be sure to use new fasteners, as required.

9. Install a new gasket to the sensor.
10. Install the sensor and tighten to 14 ft. lbs. (20 Nm).
11. Connect the sensor connector.
12. Add engine coolant.
13. Inspect for coolant leak.
14. Install V-bank cover sub-assembly.
15. Install No. 1 engine under cover.
16. Use the Techstream diagnostic tool, or equivalent and reprogram the required systems.

4.7L Engine

See Figure 389.

1. Before servicing the vehicle, refer to the Precautions Section.

➡If working near and/or around the SRS system and components, be sure to disable the SRS system. Tape the negative battery cable with insulating tape. Always disconnect the negative battery cable first.

CAUTION

To avoid personal injury when working on vehicles equipped with an air bag, the negative battery cable must be disconnected and at least 90 seconds must elapse before working on the system. Failure to do so may result in deployment of the air bag.

2. Disconnect the negative battery cable. Tape the cable with insulating tape.

➡When disconnecting the cable, some systems need to be initialized after the cable is reconnected. You will need the Techstream diagnostic scan tool or equivalent. Follow the directions on the tool.

3. Drain engine coolant.
4. Remove v-bank cover sub-assembly.
5. Remove intake air connector pipe.
6. Remove throttle body.
7. Disconnect the sensor connector.
8. Remove the sensor.
9. Remove the gasket from the sensor.

To install:

➡Be sure to use new fasteners, as required.

10. Install a new gasket to the sensor.
11. Install the sensor. Tighten to 15 ft. lbs. (20 Nm).
12. Install throttle body.
13. Install intake air connector pipe.
14. Install v-bank cover sub-assembly.
15. Connect cable to negative battery terminal.

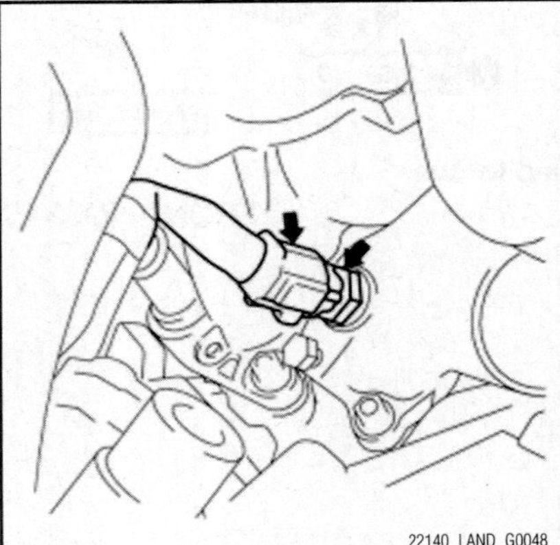

Fig. 389 Removing coolant temperature sensor—4.7L engine

16. Add engine coolant.
17. Check for engine coolant leaks.
18. Use the Techstream diagnostic tool, or equivalent and reprogram the required systems.

5.7L Engine

1. Before servicing the vehicle, refer to the Precautions Section.

➡If working near and/or around the SRS system and components, be sure to disable the SRS system. Tape the negative battery cable with insulating tape. Always disconnect the negative battery cable first.

CAUTION

To avoid personal injury when working on vehicles equipped with an air bag, the negative battery cable must be disconnected and at least 90 seconds must elapse before working on the system. Failure to do so may result in deployment of the air bag.

2. Disconnect the negative battery cable. Tape the cable with insulating tape.

➡When disconnecting the cable, some systems need to be initialized after the cable is reconnected. You will need the Techstream diagnostic scan tool or equivalent. Follow the directions on the tool.

3. Remove No. 1 engine under cover.
4. Remove V-bank cover sub-assembly.
5. Drain engine coolant.
6. Disconnect the sensor connector.
7. Remove the sensor.
8. Remove the gasket from the sensor.

To install:

➡Be sure to use new fasteners, as required.

9. Install a new gasket to the sensor.
10. Install the sensor and tighten to 14 ft. lbs. (20 Nm).
11. Connect the sensor connector.
12. Add engine coolant.
13. Inspect for coolant leak.
14. Install V-bank cover sub-assembly.
15. Install No. 1 engine under cover.
16. Use the Techstream diagnostic tool, or equivalent and reprogram the required systems.

HEATED OXYGEN SENSOR (HO2S)

LOCATION

See Figures 390 through 393.

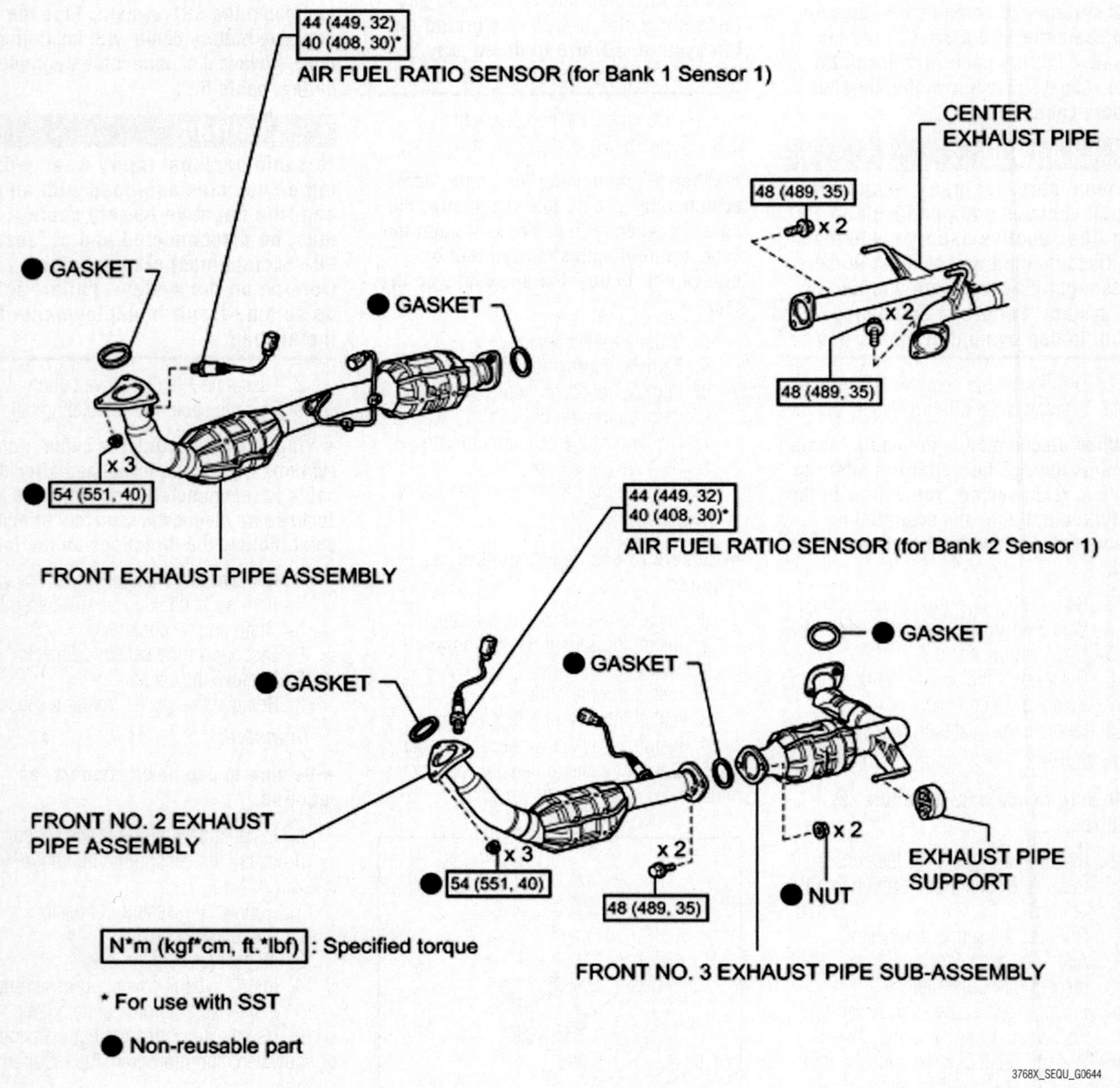

Fig. 390 Heated oxygen sensor location—4.0L engine

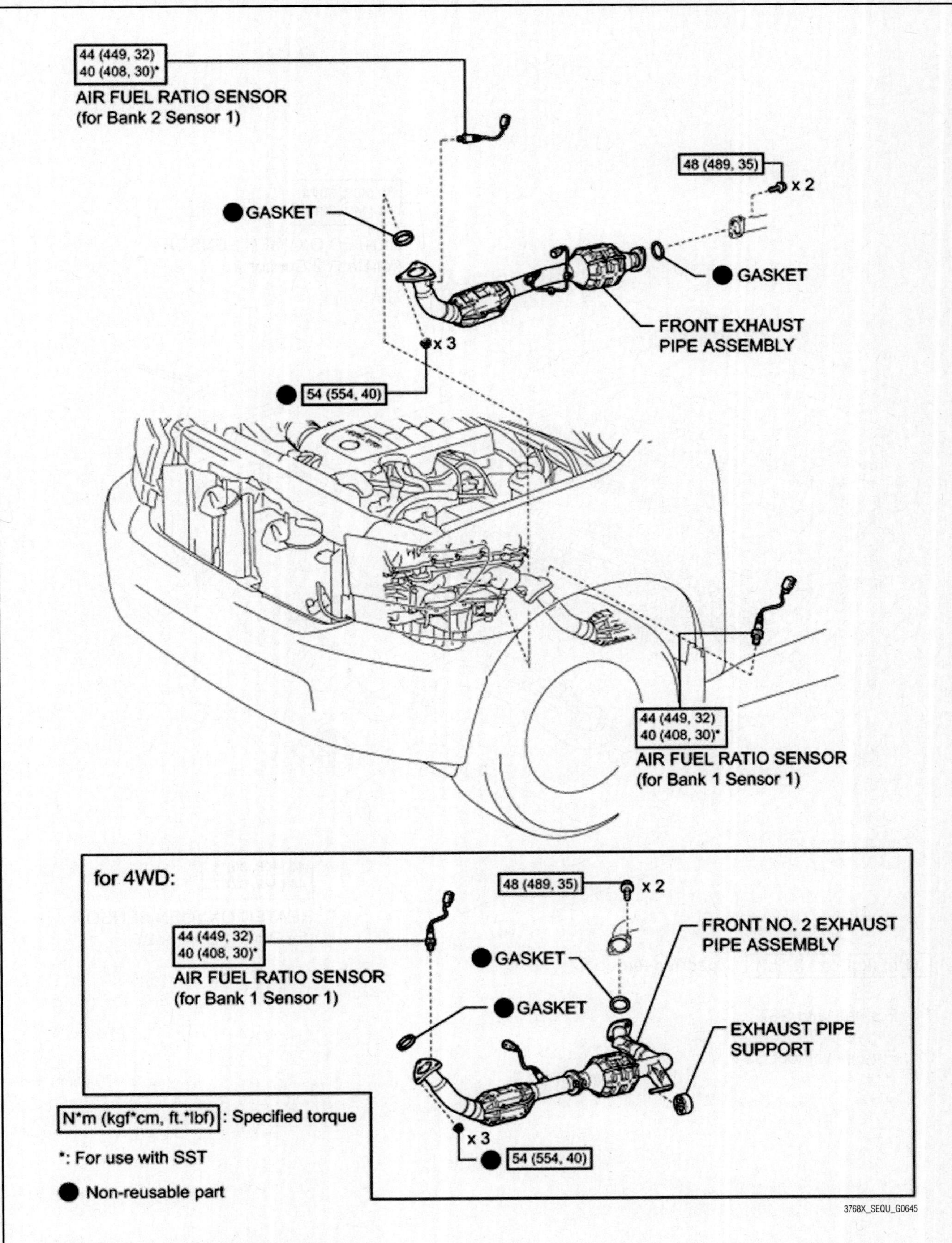

Fig. 391 Heated oxygen sensor location—4.6L engine

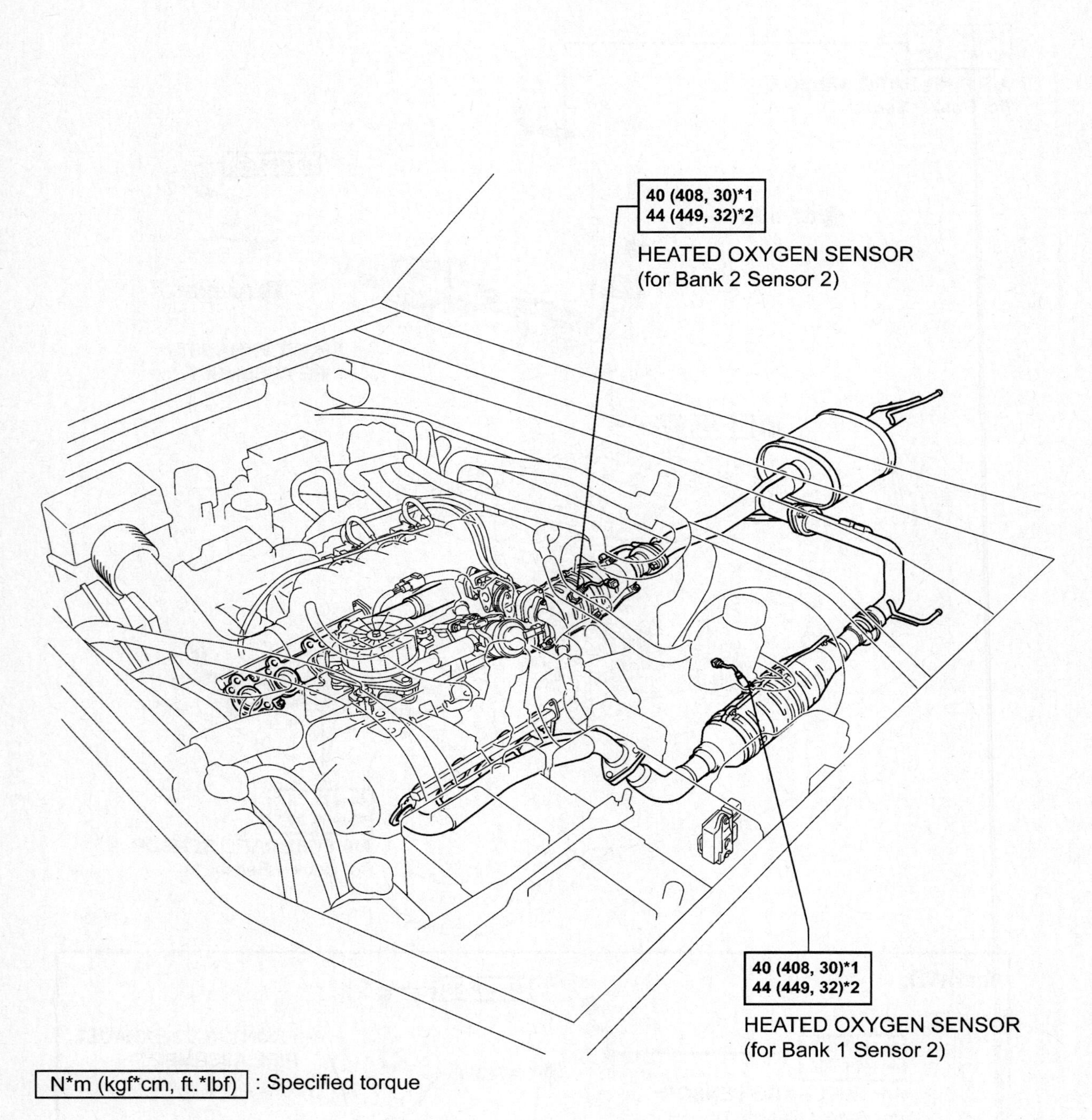

Fig. 392 Heated oxygen sensor location—4.7L engine

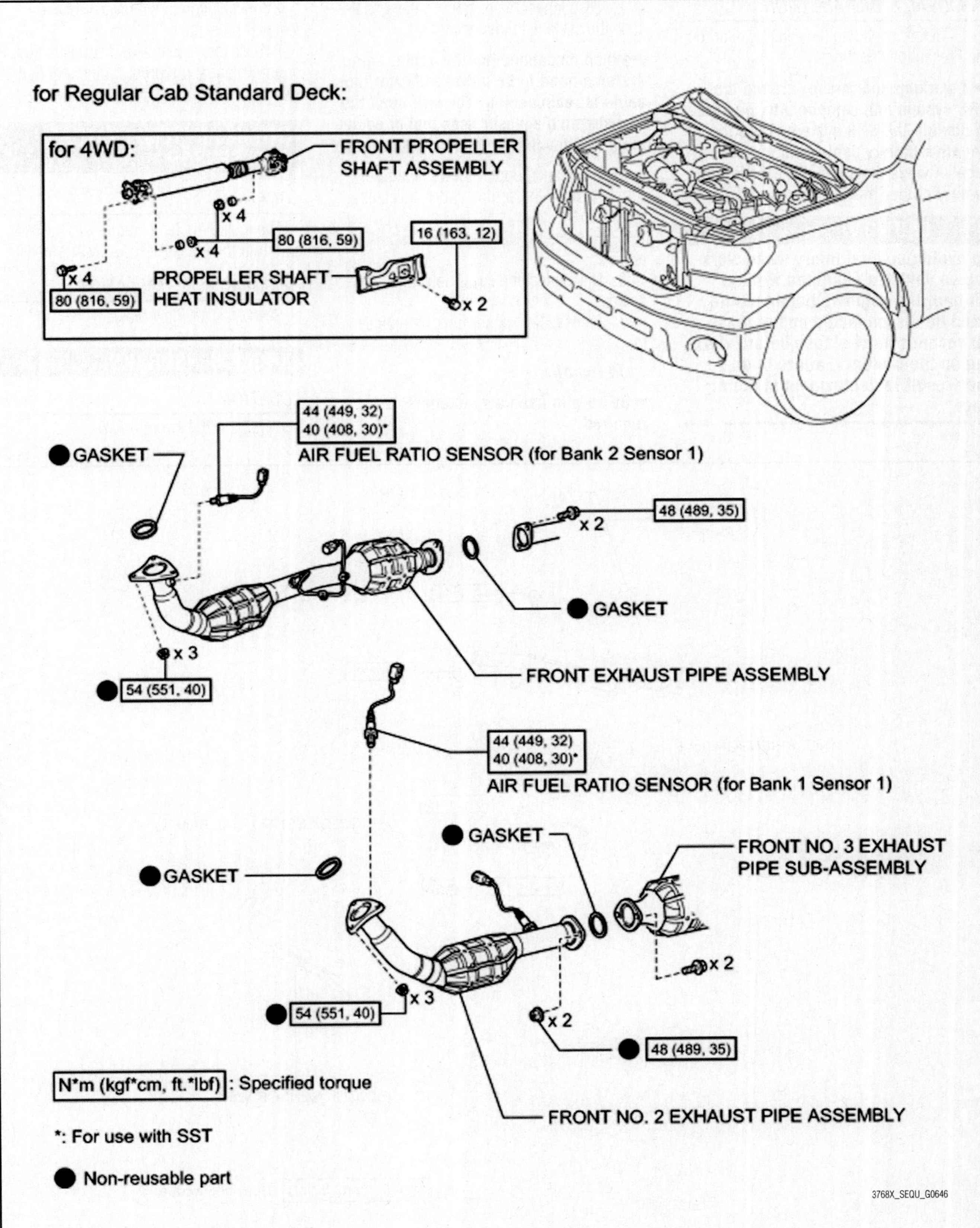

Fig. 393 Heated oxygen sensor location em dash>5.7L engine

REMOVAL & INSTALLATION

1. Before servicing the vehicle, refer to the Precautions Section.

➡If working near and/or around the SRS system and components, be sure to disable the SRS system. Tape the negative battery cable with insulating tape. Always disconnect the negative battery cable first.

⁂ CAUTION

To avoid personal injury when working on vehicles equipped with an air bag, the negative battery cable must be disconnected and at least 90 seconds must elapse before working on the system. Failure to do so may result in deployment of the air bag.

2. Disconnect the negative battery cable. Tape the cable with insulating tape.

➡When disconnecting the cable, some systems need to be initialized after the cable is reconnected. You will need the Techstream diagnostic scan tool or equivalent. Follow the directions on the tool.

3. Raise and support the vehicle safely.
4. Remove the engine under cover, as required.
5. Remove the heat insulator, as required.
6. Disconnect the sensor electrical connector.
7. Remove the sensor from its mounting.

To install:

➡Be sure to use new fasteners, as required.

8. Installation is the reverse of the removal procedure.
9. Use the Techstream diagnostic tool, or equivalent and reprogram the required systems.

INTAKE AIR TEMPERATURE (IAT) SENSOR

LOCATION

The Intake Air Temperature (IAT) sensor, built into the Mass Air Flow (MAF) meter.

REMOVAL & INSTALLATION

See Mass Air Flow Meter.

KNOCK SENSOR (KS)

LOCATION

See Figures 394 through 396.

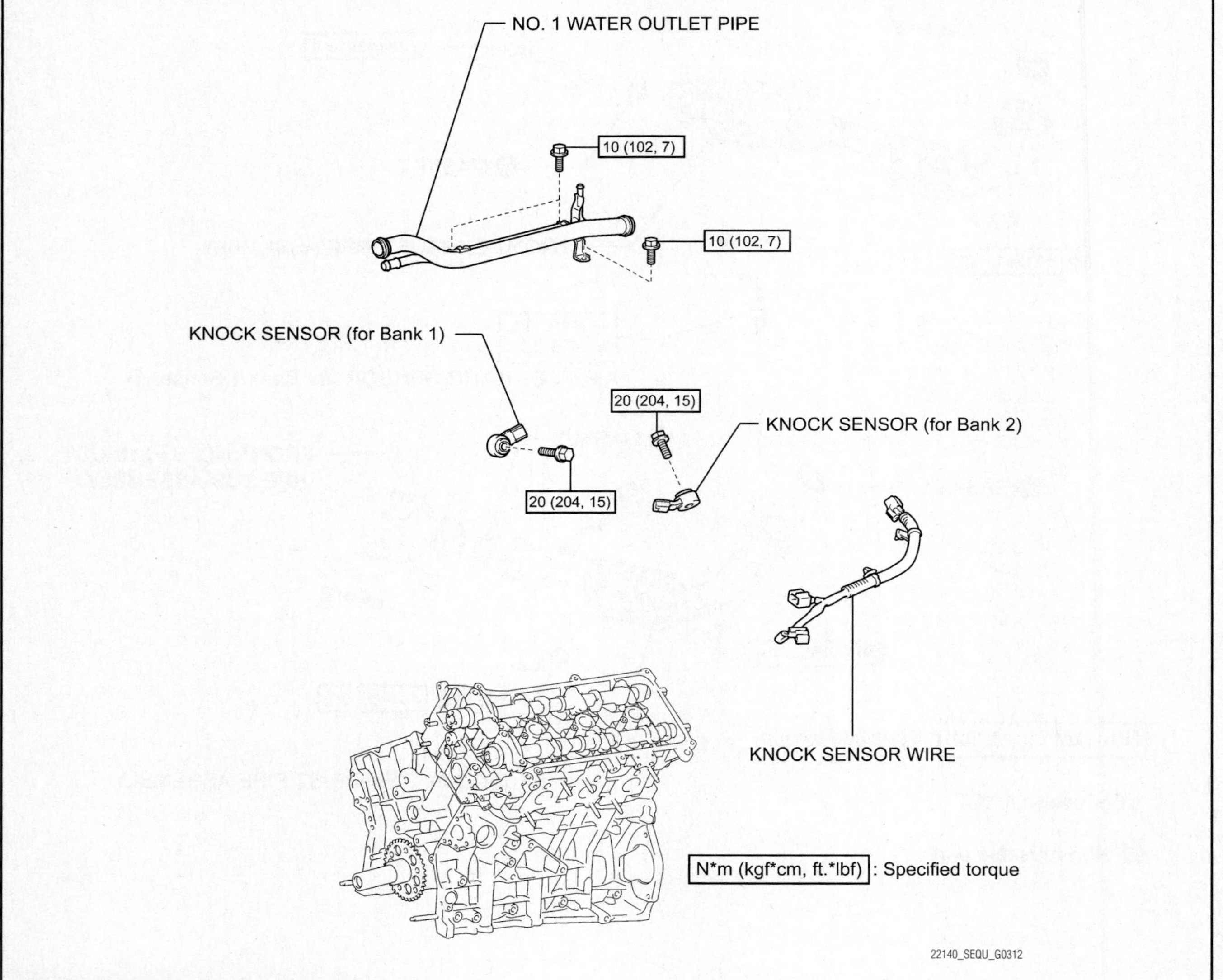

Fig. 394 Knock sensor location—4.0L engine

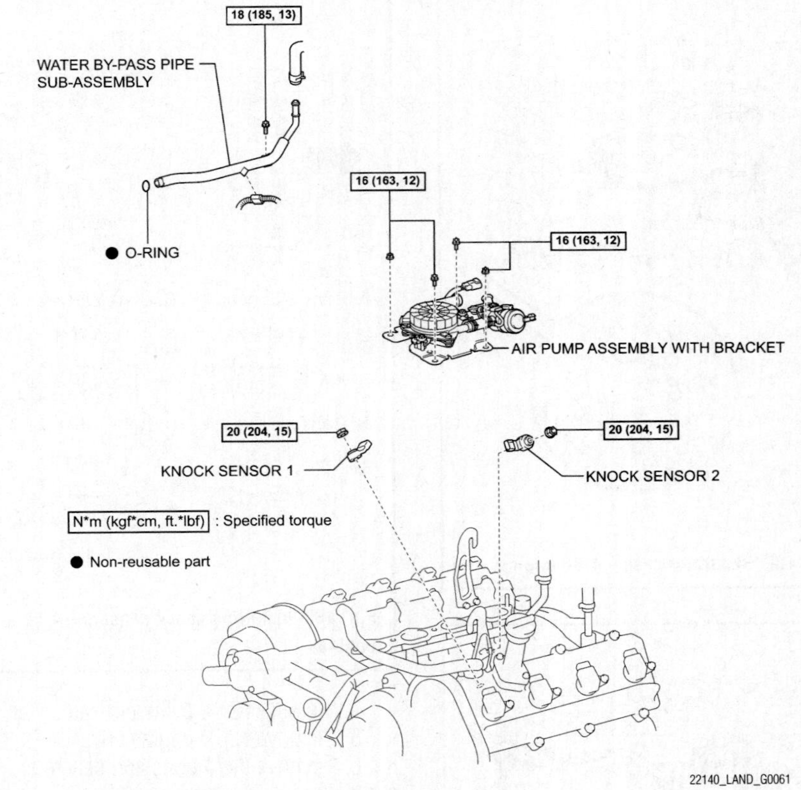

Fig. 395 Knock sensor location—4.7L engine

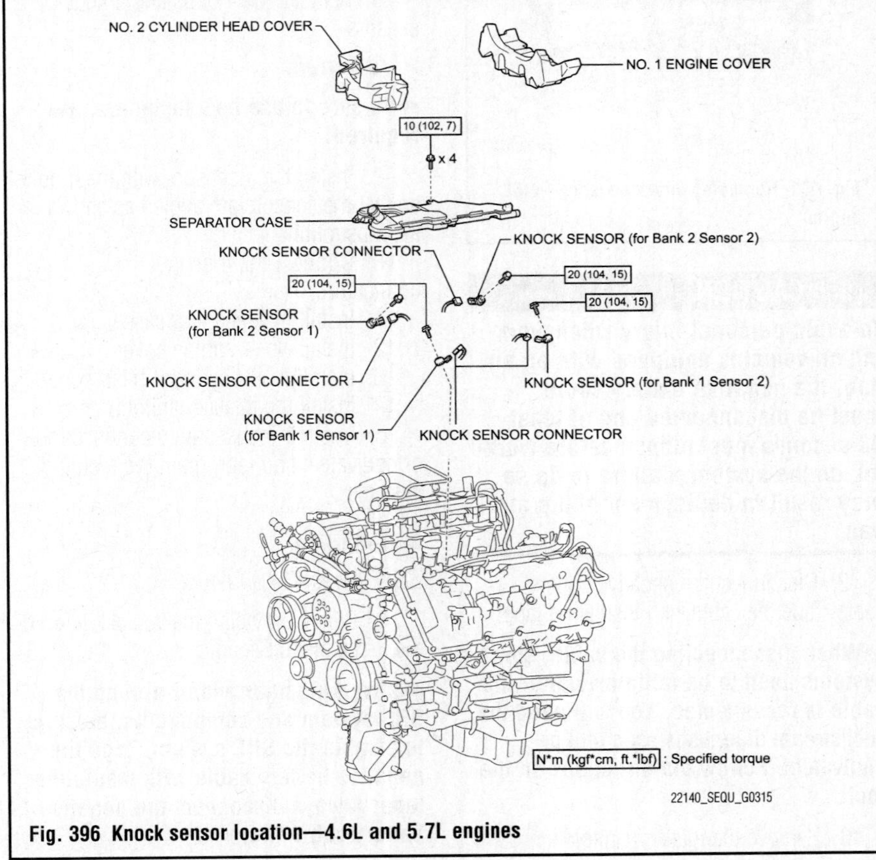

Fig. 396 Knock sensor location—4.6L and 5.7L engines

REMOVAL & INSTALLATION

4.0L Engine

See Figures 397 through 399.

1. Before servicing the vehicle, refer to the Precautions Section.

➡If working near and/or around the SRS system and components, be sure to disable the SRS system. Tape the negative battery cable with insulating tape. Always disconnect the negative battery cable first.

CAUTION

To avoid personal injury when working on vehicles equipped with an air bag, the negative battery cable must be disconnected and at least 90 seconds must elapse before working on the system. Failure to do so may result in deployment of the air bag.

2. Disconnect the negative battery cable. Tape the cable with insulating tape.

➡When disconnecting the cable, some systems need to be initialized after the cable is reconnected. You will need the Techstream diagnostic scan tool or equivalent. Follow the directions on the tool.

3. Remove the cylinder head RH.
4. Remove the 4 wire harness clamps [A].

Remove the 3 bolts [B] and water outlet pipe.

5. Disconnect the 2 sensor connectors.
6. Remove the 2 bolts and 2 sensors.

To install:

➡Be sure to use new fasteners, as required.

7. Install the 2 sensors with the 2 bolts and tighten to 15 ft. lbs. (20 Nm) as shown.
8. Connect the 2 sensor connectors.

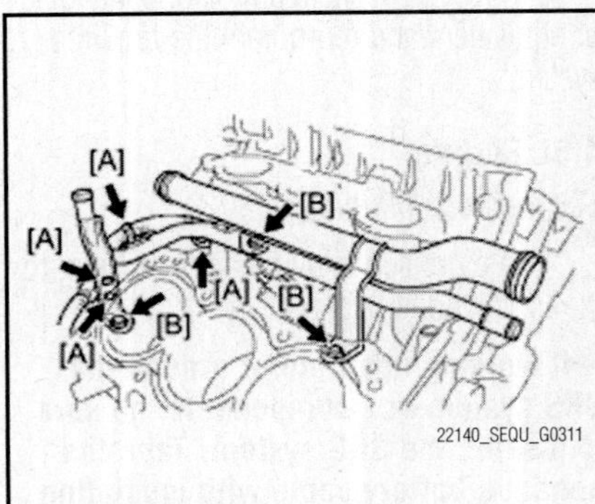

Fig. 397 Water pipe removal—4.0L engine

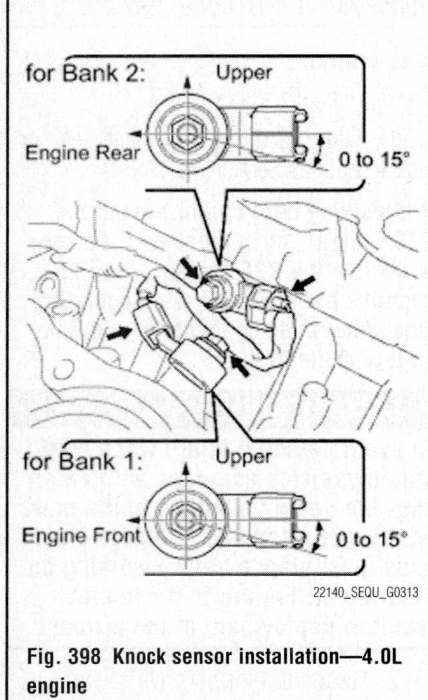

Fig. 398 Knock sensor installation—4.0L engine

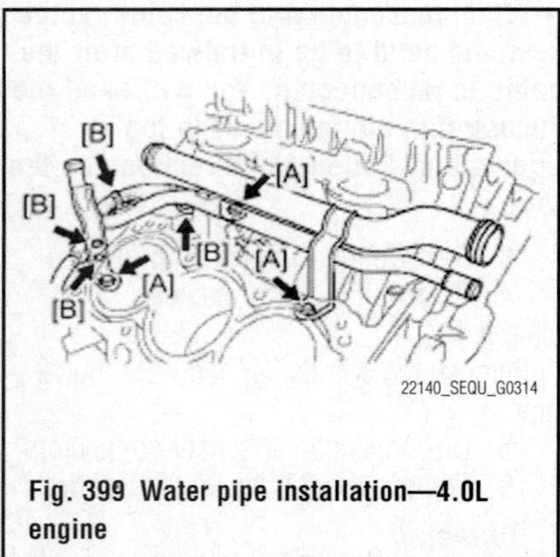

Fig. 399 Water pipe installation—4.0L engine

9. Install the 3 bolts [A] and water outlet pipe.
10. Install the 4 wire harness clamps [B].
11. Install the cylinder head RH.
12. Use the Techstream diagnostic tool, or equivalent and reprogram the required systems.

4.6L Engine

See Figures 400 through 402.

1. Before servicing the vehicle, refer to the Precautions Section.

➡If working near and/or around the SRS system and components, be sure to disable the SRS system. Tape the negative battery cable with insulating tape. Always disconnect the negative battery cable first.

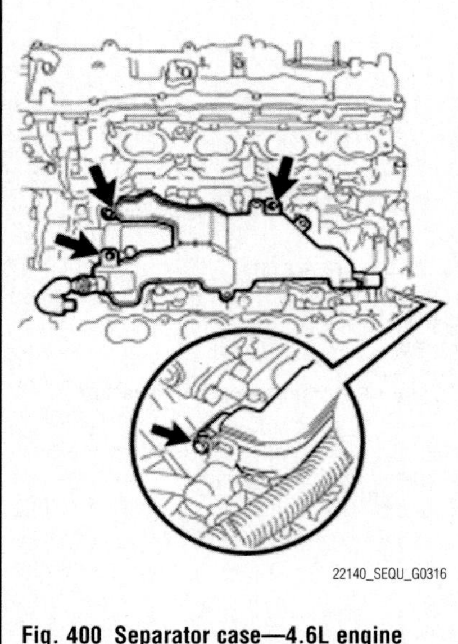

Fig. 400 Separator case—4.6L engine

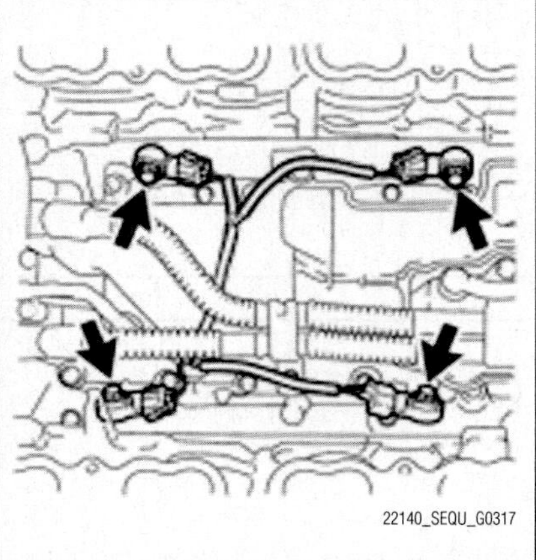

Fig. 401 Removing knock sensors—4.6L engine

⁂ CAUTION

To avoid personal injury when working on vehicles equipped with an air bag, the negative battery cable must be disconnected and at least 90 seconds must elapse before working on the system. Failure to do so may result in deployment of the air bag.

2. Disconnect the negative battery cable. Tape the cable with insulating tape.

➡When disconnecting the cable, some systems need to be initialized after the cable is reconnected. You will need the Techstream diagnostic scan tool or equivalent. Follow the directions on the tool.

3. Remove the intake manifold.

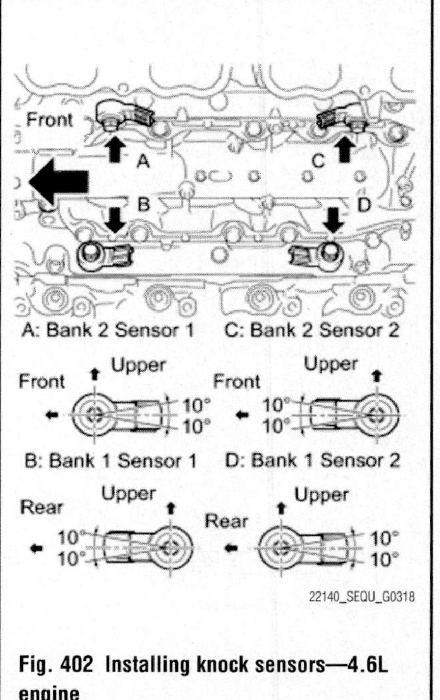

Fig. 402 Installing knock sensors—4.6L engine

4. Remove No. 2 cylinder head cover.
5. Remove No. 1 engine cover.
6. Remove the 4 bolts and separator case.
7. Disconnect the 4 knock sensor connectors.
8. Remove the 4 bolts and 4 knock sensors.

To install:

➡Be sure to use new fasteners, as required.

9. Install the 4 sensors with the 4 bolts so that the sensors are angled as shown in the illustration.
10. Connect the 4 sensor connectors.
11. Install the separator case.
12. Install No. 1 engine cover.
13. Install No. 2 cylinder head cover.
14. Install the intake manifold.
15. Use the Techstream diagnostic tool, or equivalent and reprogram the required systems.

4.7L Engine

See Figures 403 and 404.

1. Before servicing the vehicle, refer to the Precautions Section.

➡If working near and/or around the SRS system and components, be sure to disable the SRS system. Tape the negative battery cable with insulating tape. Always disconnect the negative battery cable first.

⁂ CAUTION

To avoid personal injury when working on vehicles equipped with an air bag, the negative battery cable must be disconnected and at least 90 seconds must elapse before working on the system. Failure to do so may result in deployment of the air bag.

2. Disconnect the negative battery cable. Tape the cable with insulating tape.

➡When disconnecting the cable, some systems need to be initialized after the cable is reconnected. You will need the Techstream diagnostic scan tool or equivalent. Follow the directions on the tool.

3. Discharge fuel system pressure.
4. Drain engine coolant.
5. Remove v-bank cover sub-assembly.
6. Disconnect the vacuum hoses (for the power steering idle-up and fuel pressure regulator) and ventilation hose.
7. Remove the air cleaner hose assembly.
8. Disconnect fuel hose.
9. Disconnect fuel hose no.2.
10. Disconnect the throttle control connector.
11. Disconnect the purge VSV connector.
12. Disconnect the 8 injector connectors.
13. Disconnect the ECT sensor connector.
14. Disconnect the 8 ignition coil connectors.
15. Disconnect the 2 VSV connectors for the air injection system.
16. Disconnect the 8 ignition coil connectors.
17. Disconnect the 2 air fuel ratio sensor connectors.
18. Disconnect the vacuum hose [A] from the fuel pressure regulator.
19. Disconnect the PCV hoses [B] from the PCV valve on the LH cylinder head.
20. Disconnect the EVAP hose (from the charcoal canister) [C] from the VSV for the EVAP.
21. Disconnect the 2 vacuum hoses [D] from the VSV for the air injection system.
22. Disconnect the 2 water by-pass hoses from the throttle body.
23. Disconnect the 2 wire clamps from the wire clamp bracket on the RH delivery pipe.
24. Remove the bolt and nut holding the engine wire protector from the intake manifold and cylinder head.
25. Remove the 2 bolts and ground cables from the RH and LH cylinder heads.
26. Remove the bolt and V-bank cover bracket from the intake manifold.

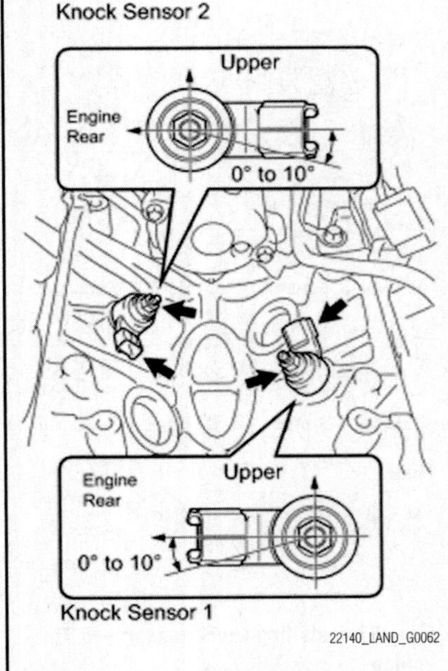

Fig. 403 Installing knock sensor—4.7L engine

27. Disconnect the engine wire from the engine hanger and wire bracket.
28. Remove the bolt and wire bracket from the intake manifold.
29. Remove the 6 bolts, 4 nuts, intake manifold assembly and 2 gaskets.
30. Remove air pump assembly with bracket.
31. Remove knock sensor.
32. Disconnect the 2 knock sensor connectors.

To install:

➡Be sure to use new fasteners, as required.

33. Install the 2 knock sensors with the 2 nuts as shown in the illustration and tighten nuts to 15 ft. lbs. (20 Nm).
34. Connect the 2 knock sensor connectors.

Fig. 404 Intake manifold bolts—4.7L engine

35. Place 2 new gaskets on the intake manifold.
36. Place the intake manifold on the cylinder heads.
37. Install and uniformly tighten the 6 bolts and 4 nuts in several steps to 13 ft. lbs. (18 Nm).
38. Install the V-bank cover bracket to the intake manifold.
39. Install the wire bracket to the intake manifold with the bolt.
40. Connect the engine wire to the engine hanger and wire bracket.
41. Connect the wire protector to the intake manifold and cylinder heads with the bolt and nut.
42. Install the 2 ground cables with the 2 bolts to the RH and LH cylinder heads.
43. Connect the 2 water by-pass hoses to the throttle body.
44. Connect the 2 wire clamps to the wire clamp bracket on the RH delivery pipe.
45. Connect the vacuum hose to the fuel pressure regulator.
46. Connect the PCV hose to the PCV valve on the LH cylinder head.
47. Connect the EVAP hose (from the charcoal canister) to the purge VSV.
48. Connect the 2 vacuum hoses to the VSV for the air injection system.
49. Connect the throttle control connector.
50. Connect the 2 VSV connectors for the air injection system.
51. Connect the purge VSV connector.
52. Connect the 8 injector connectors.
53. Connect the ECT sensor connector.
54. Connect the 8 ignition coil connectors.
55. Connect the 2 air fuel ratio sensor connectors.
56. Install fuel hose no.2.
57. Install fuel hose.
58. Install v-bank cover sub-assembly.
59. Add engine coolant.
60. Check for engine coolant leaks.
61. Check for fuel leaks.
62. Use the Techstream diagnostic tool, or equivalent and reprogram the required systems.

5.7 Engine

See Figures 405 through 407.

1. Before servicing the vehicle, refer to the Precautions Section.

➡If working near and/or around the SRS system and components, be sure to disable the SRS system. Tape the negative battery cable with insulating tape. Always disconnect the negative battery cable first.

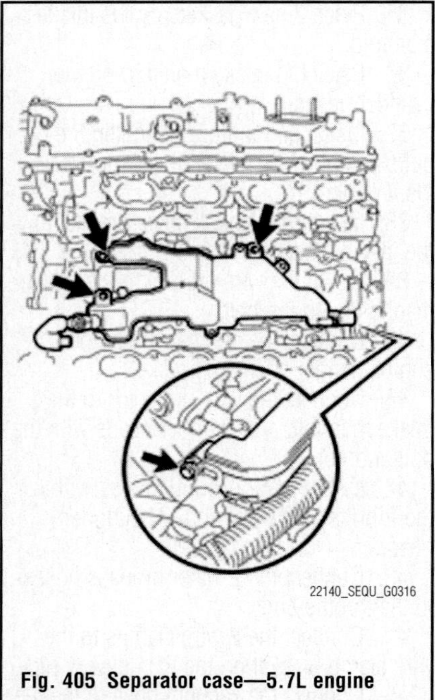

Fig. 405 Separator case—5.7L engine

CAUTION

To avoid personal injury when working on vehicles equipped with an air bag, the negative battery cable must be disconnected and at least 90 seconds must elapse before working on the system. Failure to do so may result in deployment of the air bag.

2. Disconnect the negative battery cable. Tape the cable with insulating tape.

➡When disconnecting the cable, some systems need to be initialized after the cable is reconnected. You will need the Techstream diagnostic scan tool or equivalent. Follow the directions on the tool.

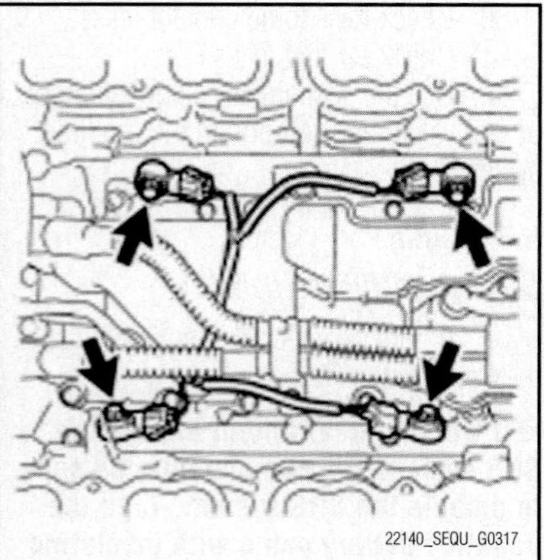

Fig. 406 Removing knock sensors—5.7L engine

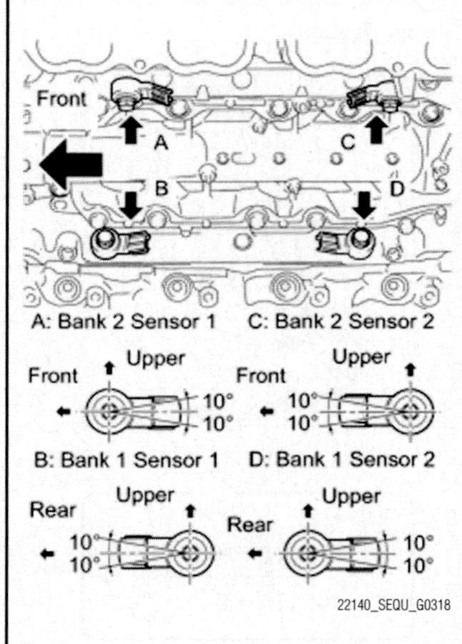

Fig. 407 Installing knock sensors—5.7L engine

3. Remove the intake manifold.
4. Remove No. 2 cylinder head cover.
5. Remove No. 1 engine cover.
6. Remove the 4 bolts and separator case.
7. Disconnect the 4 knock sensor connectors.
8. Remove the 4 bolts and 4 knock sensors.

To install:

➡Be sure to use new fasteners, as required.

9. Install the 4 sensors with the 4 bolts so that the sensors are angled as shown in the illustration.
10. Connect the 4 sensor connectors.
11. Install the separator case.
12. Install No. 1 engine cover.
13. Install No. 2 cylinder head cover.
14. Install the intake manifold.
15. Use the Techstream diagnostic tool, or equivalent and reprogram the required systems.

MASS AIR FLOW (MAF) SENSOR

LOCATION

The MAF sensor is located in the air intake snorkel.

REMOVAL & INSTALLATION

1. Before servicing the vehicle, refer to the Precautions Section.

➡If working near and/or around the SRS system and components, be sure to disable the SRS system. Tape the negative battery cable with insulating tape. Always disconnect the negative battery cable first.

CAUTION

To avoid personal injury when working on vehicles equipped with an air bag, the negative battery cable must be disconnected and at least 90 seconds must elapse before working on the system. Failure to do so may result in deployment of the air bag.

2. Disconnect the negative battery cable. Tape the cable with insulating tape.

➡When disconnecting the cable, some systems need to be initialized after the cable is reconnected. You will need the Techstream diagnostic scan tool or equivalent. Follow the directions on the tool.

3. Disconnect connector.
4. Remove attaching screws an remove MAF.

To install:

➡Be sure to use new fasteners, as required.

5. Installation is the reverse of the removal procedure.
6. Use the Techstream diagnostic tool, or equivalent and reprogram the required systems.

MANIFOLD ABSOLUTE PRESSURE (MAP) SENSOR

LOCATION

See Figure 408.

REMOVAL & INSTALLATION

See Figure 408.

1. Before servicing the vehicle, refer to the Precautions Section.

➡If working near and/or around the SRS system and components, be sure to disable the SRS system. Tape the negative battery cable with insulating tape. Always disconnect the negative battery cable first.

CAUTION

To avoid personal injury when working on vehicles equipped with an air bag, the negative battery cable must

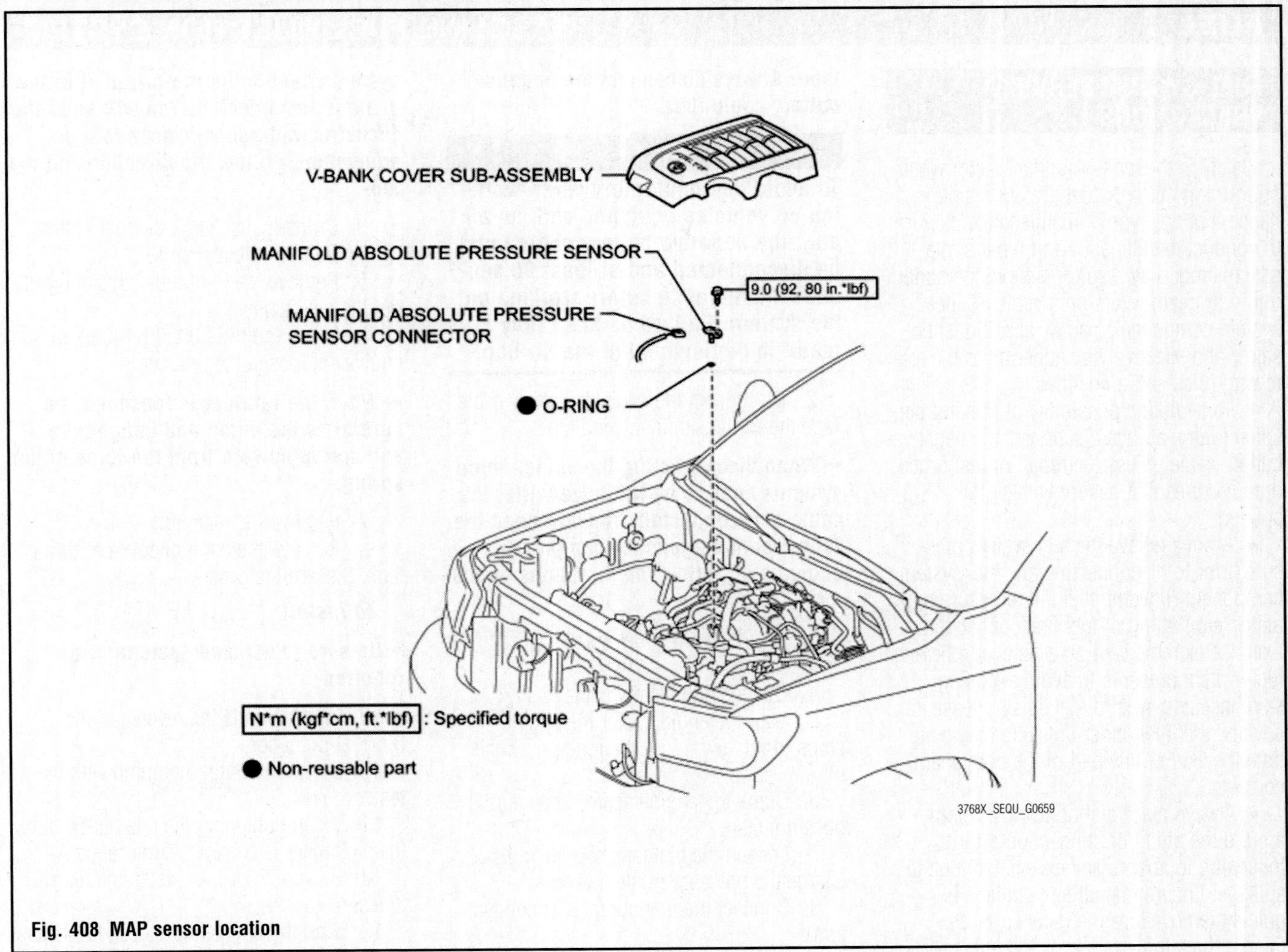

Fig. 408 MAP sensor location

be disconnected and at least 90 seconds must elapse before working on the system. Failure to do so may result in deployment of the air bag.

2. Disconnect the negative battery cable. Tape the cable with insulating tape.

➡When disconnecting the cable, some systems need to be initialized after the cable is reconnected. You will need the Techstream diagnostic scan tool or equivalent. Follow the directions on the tool.

3. Remove the V- bank cover assembly.
4. Disconnect the electrical connector.
5. Remove the sensor from its mounting.
6. Discard the O-ring.

To install:

➡Be sure to use new fasteners, as required.

7. Installation is the reverse of the removal procedure.
8. Use the Techstream diagnostic tool, or equivalent and reprogram the required systems.

FUEL SYSTEM SERVICE PRECAUTIONS

Safety is the most important factor when performing not only fuel system maintenance, but any type of maintenance. Failure to conduct maintenance and repairs in a safe manner may result in serious personal injury or death. Work on a vehicle's fuel system components can be accomplished safely and effectively by adhering to the following rules and guidelines.

- To avoid the possibility of fire and personal injury, always disconnect the negative battery cable unless the repair or test procedure requires that battery voltage be applied.
- Always relieve the fuel system pressure prior to disconnecting any fuel system component (injector, fuel rail, pressure regulator, etc.) fitting or fuel line connection. Exercise extreme caution whenever relieving fuel system pressure to avoid exposing skin, face and eyes to fuel spray. Please be advised that fuel under pressure may penetrate the skin or any part of the body that it contacts.
- Always place a shop towel or cloth around the fitting or connection prior to loosening to absorb any excess fuel due to spillage. Ensure that all fuel spillage is quickly removed from engine surfaces. Ensure that all fuel-soaked cloths or towels are deposited into a flame-proof waste container with a lid.
- Always keep a dry chemical (Class B) fire extinguisher near the work area.
- Do not allow fuel spray or fuel vapors to come into contact with a spark or open flame.
- Always use a second wrench when loosening or tightening fuel line connection fittings. This will prevent unnecessary stress and torsion on fuel piping. Always follow the proper torque specifications.
- Always replace worn fuel fitting O-rings with new ones. Do not substitute fuel hose where rigid pipe is installed.

FUEL SYSTEM PRESSURE

RELIEVING

1. Before servicing the vehicle, refer to the Precautions Section.

➡If working near and/or around the SRS system and components, be sure to disable the SRS system. Tape the negative battery cable with insulating tape. Always disconnect the negative battery cable first.

⁂ CAUTION

To avoid personal injury when working on vehicles equipped with an air bag, the negative battery cable must be disconnected and at least 90 seconds must elapse before working on the system. Failure to do so may result in deployment of the air bag.

2. Disconnect the negative battery cable. Tape the cable with insulating tape.

➡When disconnecting the cable, some systems need to be initialized after the cable is reconnected. You will need the Techstream diagnostic scan tool or equivalent. Follow the directions on the tool.

3. Disconnect the fuel pump ECU connectors.
4. Connect the battery cable.
5. Start the engine, after the engine stops (on its own) turn the ignition switch off.
6. Crank the engine again, to be sure it does not start.
7. Loosen the fuel tank cap, then discharge the pressure in the fuel tank.
8. Connect the fuel pump ECU connectors.

FUEL LEVEL SENDING UNIT

REMOVAL & INSTALLATION

1. Before servicing the vehicle, refer to the Precautions Section.

➡If working near and/or around the SRS system and components, be sure to disable the SRS system. Tape the negative battery cable with insulating tape. Always disconnect the negative battery cable first.

⁂ CAUTION

To avoid personal injury when working on vehicles equipped with an air bag, the negative battery cable must be disconnected and at least 90 seconds must elapse before working on the system. Failure to do so may result in deployment of the air bag.

2. Disconnect the negative battery cable. Tape the cable with insulating tape.

➡When disconnecting the cable, some systems need to be initialized after the cable is reconnected. You will need the Techstream diagnostic scan tool or equivalent. Follow the directions on the tool.

3. Properly discharge the fuel system.
4. Remove the fuel tank.
5. Remove the joint tube clips and pull out the tank tubes.
6. Using tool SST:09808-14020, or equivalent loosen the retainer.

➡When the retainer is loosened, be careful as the pump and gauge tube will spring upward from the force of the spring.

7. Remove the retainer.
8. Remove the component from the tank. Discard the gasket.

To install:

➡Be sure to use new fasteners, as required.

9. Installation is the reverse of the removal procedure.
10. Be sure to align the pump with the tank groove.
11. When tightening the retainer be sure the alignment is correct (within range A).
12. Be sure to proper install the return tubes.
13. Start the engine and check for leaks.
14. Correct as required.

FUEL PUMP MODULE

REMOVAL & INSTALLATION

1. Before servicing the vehicle, refer to the Precautions Section.

➡If working near and/or around the SRS system and components, be sure to disable the SRS system. Tape the negative battery cable with insulating tape. Always disconnect the negative battery cable first.

⁂ CAUTION

To avoid personal injury when working on vehicles equipped with an air bag, the negative battery cable must be disconnected and at least 90 seconds must elapse before working on the system. Failure to do so may result in deployment of the air bag.

2. Disconnect the negative battery cable. Tape the cable with insulating tape.

➡**When disconnecting the cable, some systems need to be initialized after the cable is reconnected. You will need the Techstream diagnostic scan tool or equivalent. Follow the directions on the tool.**

3. Properly discharge the fuel system.
4. Remove the fuel tank.
5. Remove the joint tube clips and pull out the tank tubes.
6. Using tool SST:09808-14020, or equivalent loosen the retainer.

➡**When the retainer is loosened, be careful as the pump and gauge tube will spring upward from the force of the spring.**

7. Remove the retainer.
8. Remove the component from the tank. Discard the gasket.

To install:

➡**Be sure to use new fasteners, as required.**

9. Installation is the reverse of the removal procedure.
10. Be sure to align the pump with the tank groove.
11. When tightening the retainer be sure the alignment is correct (within range A).
12. Be sure to proper install the return tubes.
13. Start the engine and check for leaks.
14. Correct as required.

FUEL RAIL AND INJECTOR

REMOVAL & INSTALLATION

4.0L Engine

See Figure 409.

1. Before servicing the vehicle, refer to the Precautions Section.

➡**If working near and/or around the SRS system and components, be sure to disable the SRS system. Tape the negative battery cable with insulating tape. Always disconnect the negative battery cable first.**

CAUTION

To avoid personal injury when working on vehicles equipped with an air bag, the negative battery cable must be disconnected and at least 90 seconds must elapse before working on the system. Failure to do so may result in deployment of the air bag.

2. Disconnect the negative battery cable. Tape the cable with insulating tape.

➡**When disconnecting the cable, some systems need to be initialized after the cable is reconnected. You will need the Techstream diagnostic scan tool or equivalent. Follow the directions on the tool.**

3. Relieve the fuel system pressure.
4. Drain the engine coolant. Remove the V bank cover.
5. Remove the air cleaner assembly.
6. Remove the front cowl top outer panel sub assembly.
7. Remove the intake manifold.
8. Disconnect and plug the fuel line hoses.
9. Disconnect the fuel injector connectors. Remove the six bolts and the fuel rail together with the injectors.
10. Pull the injectors out of the fuel rail.

To install:

➡**Be sure to use new fasteners, as required.**

11. Installation is the reverse of the removal procedure.
12. Be sure to use new gaskets and O-rings, as required.
13. Be sure to fill the cooling system with the proper grade and type engine coolant.
14. Start the engine and check for leaks, correct as required.

4.7L Engine

See Figure 410.

1. Before servicing the vehicle, refer to the Precautions Section.

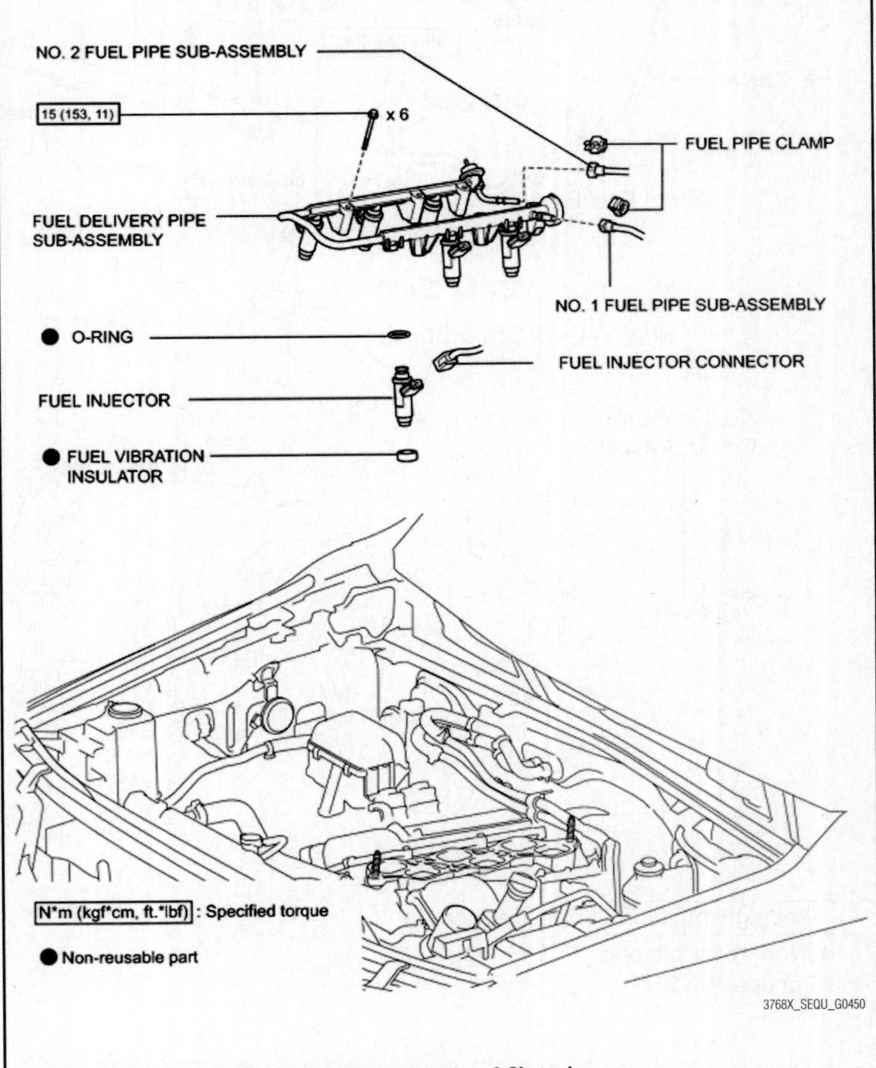

Fig. 409 Fuel injector rail and related components—4.0L engine

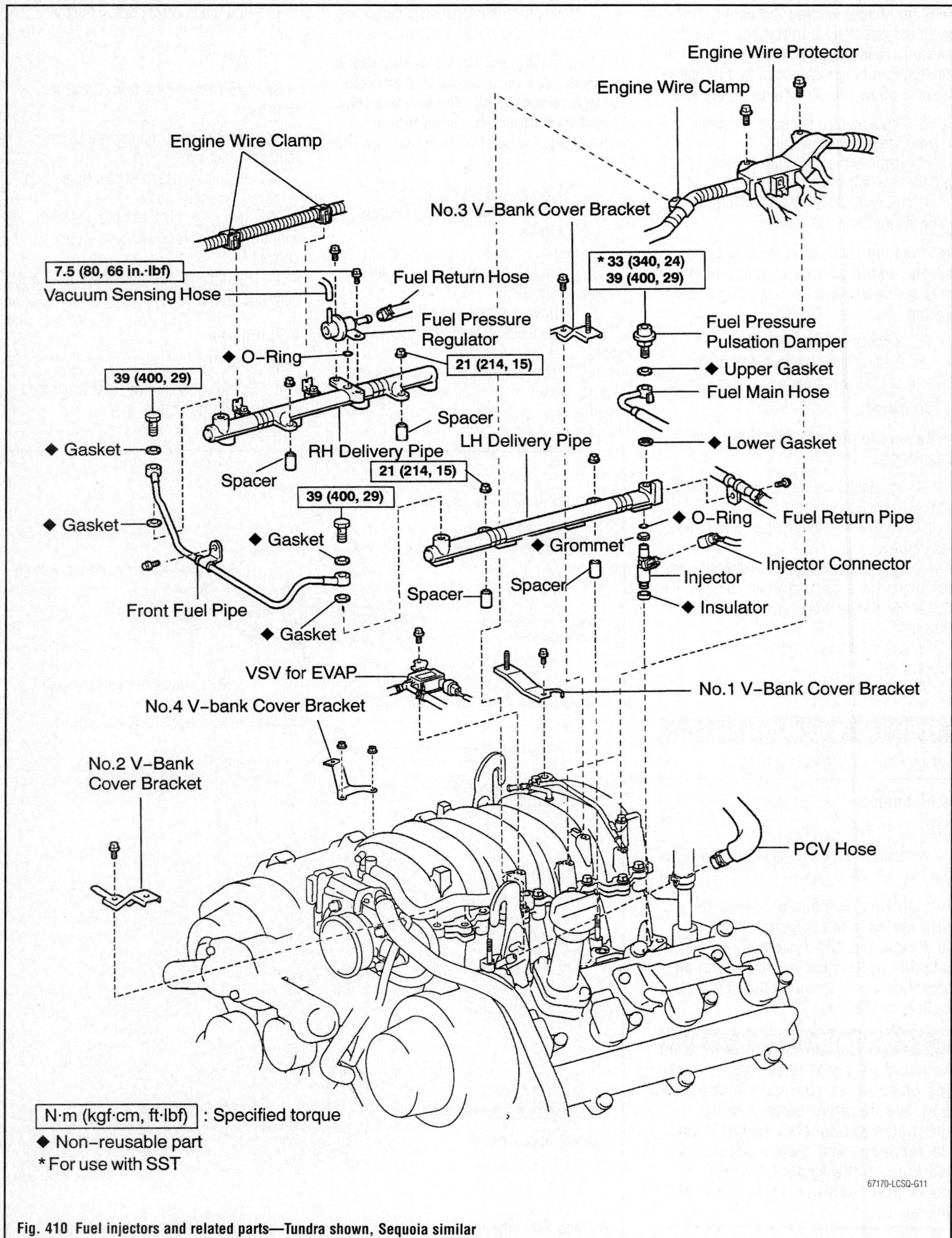
Engine Wire Protector
Engine Wire Clamp
Engine Wire Clamp
No.3 V-Bank Cover Bracket
7.5 (80, 66 in.·lbf)
Vacuum Sensing Hose
Fuel Return Hose
* 33 (340, 24)
39 (400, 29)
Fuel Pressure
Regulator
Fuel Pressure
Pulsation Damper
◆ O-Ring
21 (214, 15)
◆ Upper Gasket
39 (400, 29)
Fuel Main Hose
Spacer
LH Delivery Pipe
◆ Gasket
RH Delivery Pipe
◆ Lower Gasket
21 (214, 15)
Spacer
39 (400, 29)
◆ O-Ring
Fuel Return Pipe
◆ Gasket
◆ Gasket
◆ Grommet
Injector Connector
Injector
Spacer
Spacer
Front Fuel Pipe
◆ Insulator
◆ Gasket
VSV for EVAP
No.1 V-Bank Cover Bracket
No.4 V-bank Cover Bracket
No.2 V-Bank
Cover Bracket
PCV Hose
N·m (kgf·cm, ft·lbf) : Specified torque
◆ Non-reusable part
* For use with SST
67170-LCSQ-G11

Fig. 410 Fuel injectors and related parts—Tundra shown, Sequoia similar

➡If working near and/or around the SRS system and components, be sure to disable the SRS system. Tape the negative battery cable with insulating tape. Always disconnect the negative battery cable first.

⁂ CAUTION

To avoid personal injury when working on vehicles equipped with an air bag, the negative battery cable must be disconnected and at least 90 seconds must elapse before working on the system. Failure to do so may result in deployment of the air bag.

2. Disconnect the negative battery cable. Tape the cable with insulating tape.

➡When disconnecting the cable, some systems need to be initialized after the cable is reconnected. You will need the Techstream diagnostic scan tool or equivalent. Follow the directions on the tool.

3. Relieve the fuel system pressure.
4. Remove or disconnect the following:
   - Engine appearance cover
   - Air intake tube
   - Fuel lines
   - Fuel pulsation damper
   - Fuel pressure regulator vacuum line
   - Accelerator cable and bracket
   - Positive Crankcase Ventilation (PCV) valve and hose
   - Evaporative Emissions (EVAP) vacuum switching valve
   - Engine appearance cover brackets
   - Fuel injector harness connectors
   - Engine harness protector
   - Fuel supply manifold crossover pipe
   - Fuel supply manifolds with injectors attached
   - Fuel injectors

To install:

5. Install the fuel injectors to the supply manifold with new O-ring seals and new grommets.
6. Install new injector insulators to the intake manifold.
7. Install or connect the following:
   - Fuel supply manifolds with injectors attached. Tighten the bolts to 66 inch lbs. (7.5 Nm).
   - Fuel supply manifold crossover pipe. Tighten the bolts to 29 ft. lbs. (39 Nm).
   - Engine harness protector
   - Fuel injector harness connectors
   - Engine appearance cover brackets
   - EVAP vacuum switching valve
   - PCV valve and hose
   - Accelerator cable and bracket
   - Fuel pressure regulator vacuum line
   - Fuel pulsation damper
   - Fuel lines
   - Air intake tube
   - Engine appearance cover
   - Negative battery cable
8. Start the engine and check for leaks.

4.6L Engine

See Figure 411.

1. Properly discharge the fuel system.
2. Disconnect cable from negative battery terminal.
3. Remove the EGR valve bracket.
4. Disconnect the No. 2 fuel tube from the fuel pressure regulator.
5. Disconnect the No. 1 fuel tube from the fuel delivery pipes.
6. Disconnect the fuel tube from the fuel delivery pipe LH.
7. Disconnect the ventilation hose and No. 6 wire harness connector.
8. Remove the 2 bolts and fuel delivery pipe RH.
9. Remove the 2 delivery pipe spacers and 4 insulators from the intake manifold.
10. Disconnect the No. 7 wire harness connector.

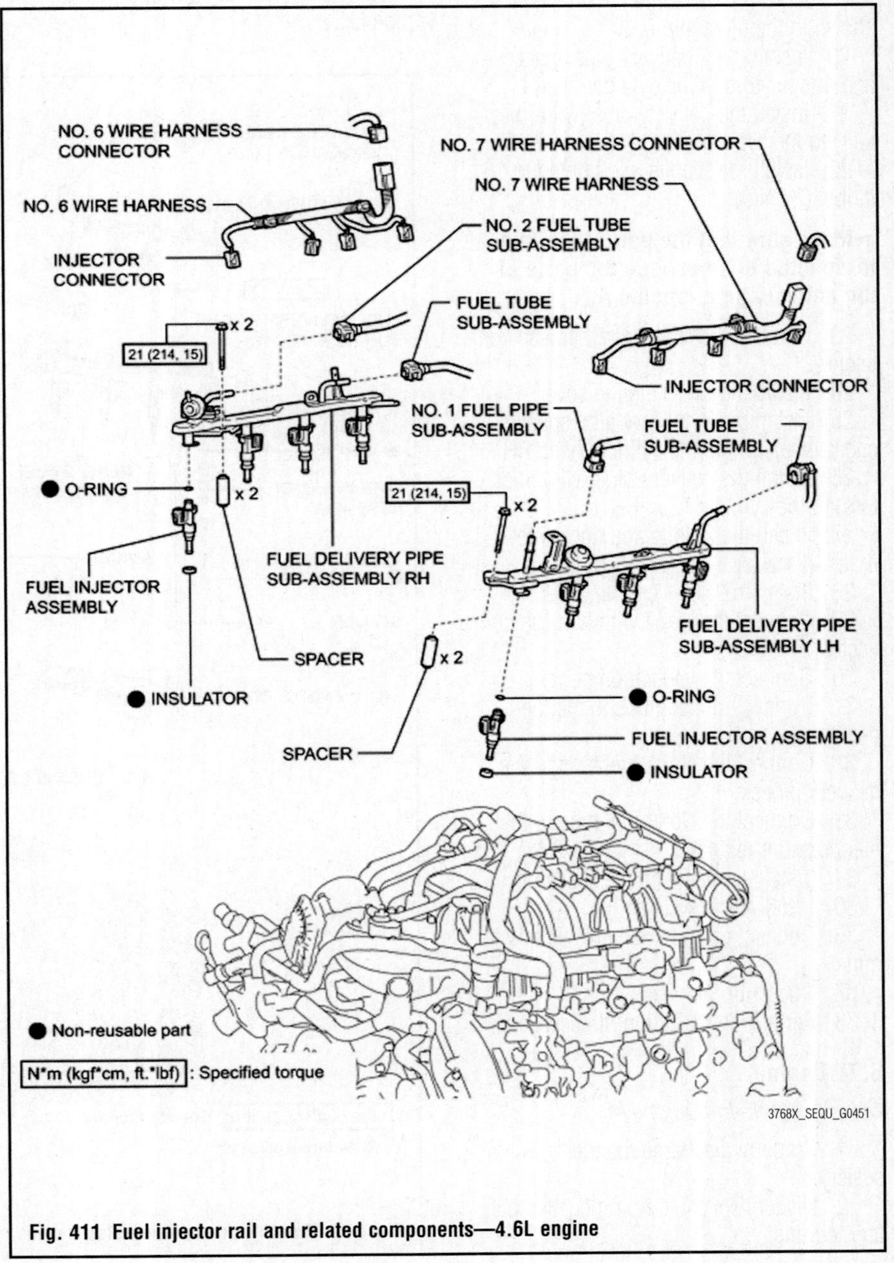

Fig. 411 Fuel injector rail and related components—4.6L engine

11. Remove the No. 3 engine cover.
12. Remove the 2 bolts and fuel delivery pipe LH.
13. Remove the 2 delivery pipe spacers and 4 insulators from the intake manifold.
14. Remove the fuel injector from the fuel delivery pipe, and then disconnect the injector connector.

To install:

15. Attach the 2 clamps to install the No. 6 wire harness to the delivery pipe.
16. Attach the 3 clamps to install the No. 7 wire harness to the delivery pipe LH.
17. Apply gasoline or spindle oil to a new O-ring and install it to the injector.
18. Connect the injector connector.
19. Check that each injector is installed to the delivery pipe facing the direction shown in the illustration.
20. Install the 2 delivery pipe spacers and 4 insulators to the cylinder head LH.
21. Install the delivery pipe (with injectors) to the cylinder head LH.
22. Install the 2 bolts and tighten to 15 ft. lbs. (21 Nm).

➡Make sure that the part of the injector labeled B is between the parts of the delivery pipe labeled A.

23. Connect the No. 6 wire harness connector.
24. Install the No. 3 engine cover.
25. Install the 2 delivery pipe spacers and 4 insulators to the cylinder head RH.
26. Install the delivery pipe (with injectors) to the cylinder head RH.
27. Install the 2 bolts and tighten to 15 ft. lbs. (21 Nm).
28. Install the No. 1 engine cover.
29. Connect the No. 7 wire harness connector.
30. Connect the ventilation hose.
31. Connect the fuel tube to the fuel delivery pipe LH.
32. Connect the No. 1 fuel tube to the delivery pipes.
33. Connect the No. 2 fuel tube to the fuel pressure regulator.
34. Install air cleaner hose assembly.
35. Install V-bank cover.
36. Connect cable to negative battery terminal.
37. Check for fuel leak.
38. Perform initialization, if necessary.

5.7L Engine

See Figures 464 through 414.

1. Properly discharge the fuel system.
2. Disconnect cable from negative battery terminal.
3. Remove V-bank cover. Remove air cleaner hose assembly.
4. Disconnect the No. 2 fuel tube from the fuel pressure regulator.
5. Disconnect the No. 1 fuel tube from the fuel delivery pipes.
6. Disconnect the fuel tube from the fuel delivery pipe LH.
7. Disconnect the ventilation hose and No. 6 wire harness connector.
8. Remove the 2 bolts and fuel delivery pipe RH.
9. Remove the 2 delivery pipe spacers and 4 insulators from the intake manifold.
10. Disconnect the No. 7 wire harness connector.
11. Remove the No. 3 engine cover.
12. Remove the 2 bolts and fuel delivery pipe LH.
13. Remove the 2 delivery pipe spacers and 4 insulators from the intake manifold.
14. Remove the fuel injector from the fuel delivery pipe, and then disconnect the injector connector.

To install:

15. Attach the 2 clamps to install the No. 6 wire harness to the delivery pipe.
16. Attach the 3 clamps to install the No. 7 wire harness to the delivery pipe LH.
17. Apply gasoline or spindle oil to a new O-ring and install it to the injector.
18. Connect the injector connector.
19. Check that each injector is installed to the delivery pipe facing the direction shown in the illustration.
20. Install the 2 delivery pipe spacers and 4 insulators to the cylinder head LH.

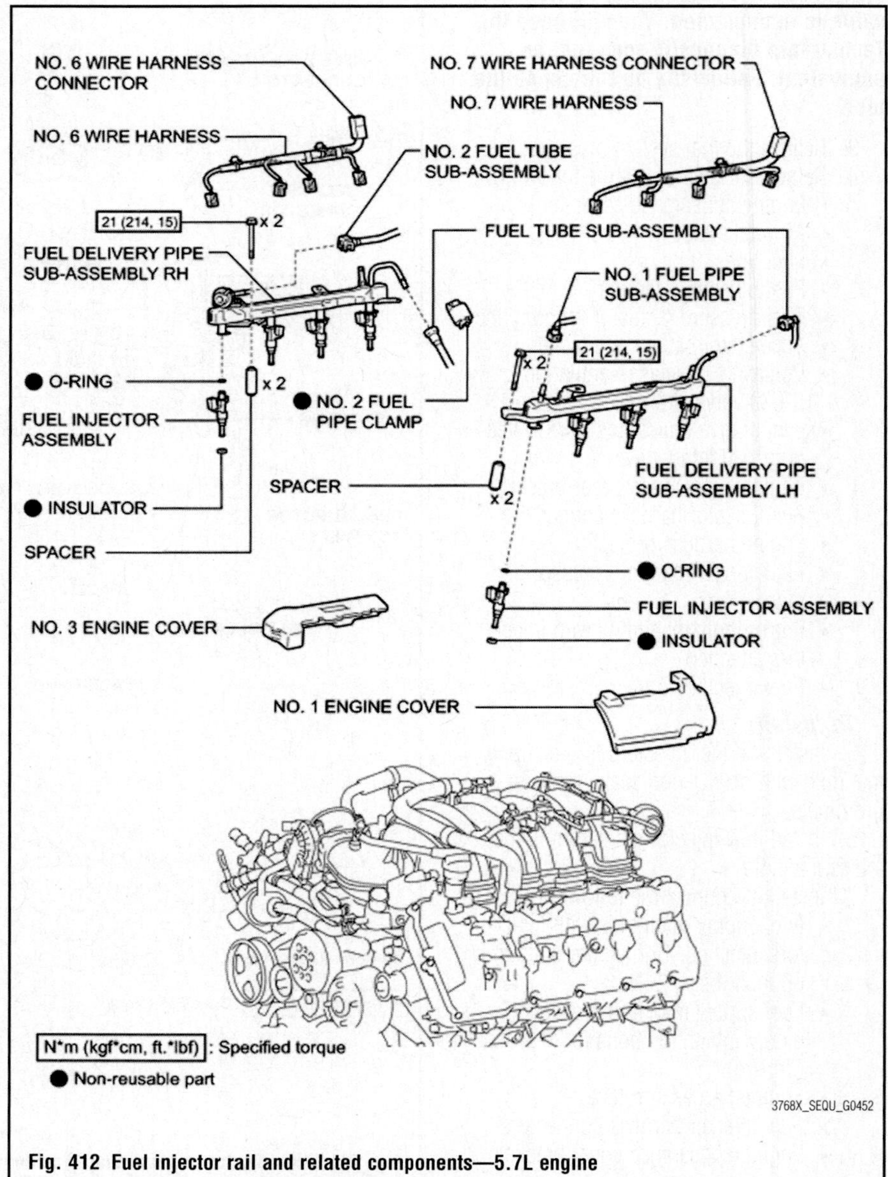

Fig. 412 Fuel injector rail and related components—5.7L engine

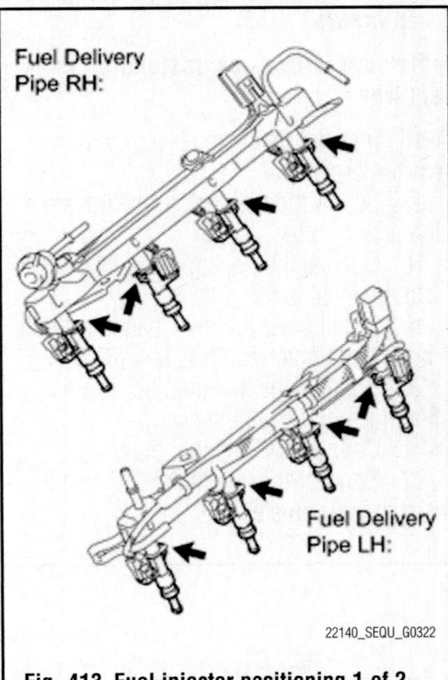

Fig. 413 Fuel injector positioning 1 of 2

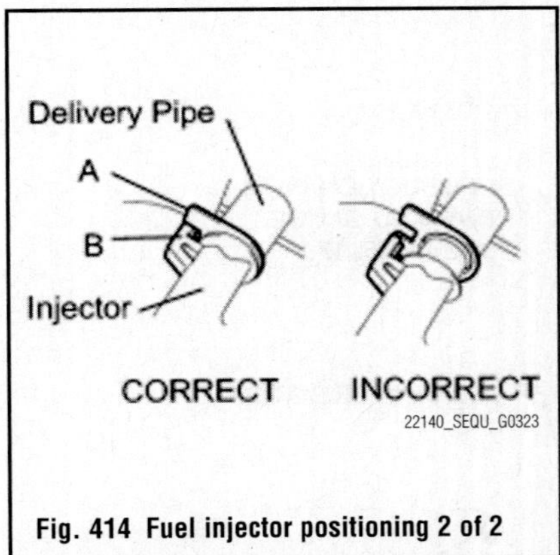

Fig. 414 Fuel injector positioning 2 of 2

21. Install the delivery pipe (with injectors) to the cylinder head LH.
22. Install the 2 bolts and tighten to 15 ft. lbs. (21 Nm).

➡Make sure that the part of the injector labeled B is between the parts of the delivery pipe labeled A.

23. Connect the No. 6 wire harness connector.
24. Install the No. 3 engine cover.
25. Install the 2 delivery pipe spacers and 4 insulators to the cylinder head RH.
26. Install the delivery pipe (with injectors) to the cylinder head RH.
27. Install the 2 bolts and tighten to 15 ft. lbs. (21 Nm).
28. Install the No. 1 engine cover.
29. Connect the No. 7 wire harness connector.
30. Connect the ventilation hose.
31. Connect the fuel tube to the fuel delivery pipe LH.
32. Connect the No. 1 fuel tube to the delivery pipes.
33. Connect the No. 2 fuel tube to the fuel pressure regulator.
34. Install air cleaner hose assembly.
35. Install V-bank cover.
36. Connect cable to negative battery terminal.
37. Check for fuel leak.
38. Perform initialization, if necessary.

FUEL TANK

REMOVAL & INSTALLATION

See Figures 415 and 416.

1. Before servicing the vehicle, refer to the Precautions Section.

➡If working near and/or around the SRS system and components, be sure to disable the SRS system. Tape the negative battery cable with insulating tape. Always disconnect the negative battery cable first.

⁂ CAUTION

To avoid personal injury when working on vehicles equipped with an air bag, the negative battery cable must be disconnected and at least 90 seconds must elapse before working on the system. Failure to do so may result in deployment of the air bag.

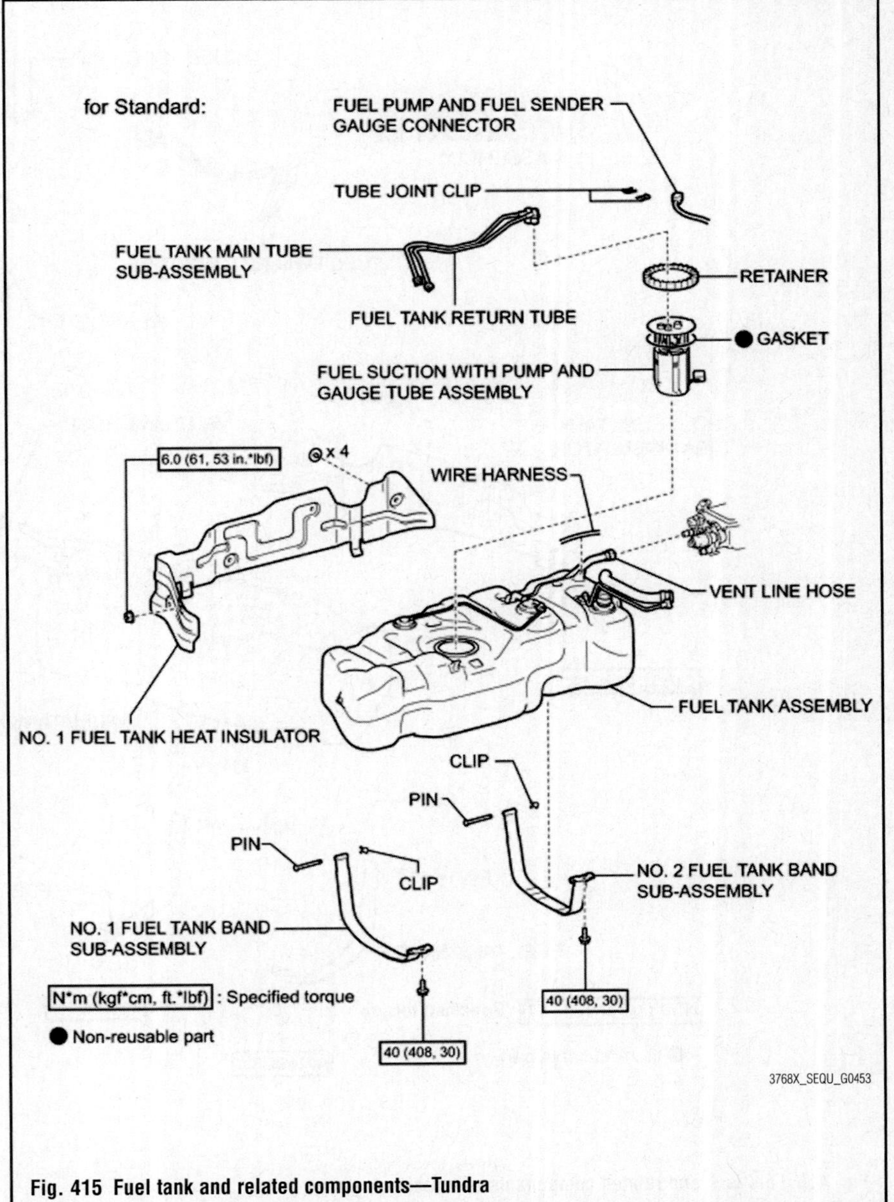

Fig. 415 Fuel tank and related components—Tundra

2. Disconnect the negative battery cable. Tape the cable with insulating tape.

➡When disconnecting the cable, some systems need to be initialized after the cable is reconnected. You will need the Techstream diagnostic scan tool or equivalent. Follow the directions on the tool.

3. Drain the fuel tank to an acceptable fuel level.
4. Discharge fuel system pressure.
5. Disconnect fuel pump connector.
6. Remove the 2 bolts, 2 nuts and fuel tank protector.
7. Disconnect fuel main tube and return tube.
8. Push the connector deep into the charcoal canister to release the locking tab.
9. Pinch and pull out the connector.
10. Loosen the hose clamp bolt and disconnect the fuel inlet hose from the fuel filler pipe.
11. Disconnect the breather tube.
12. Properly position a suitable jack under the fuel tank.
13. Remove the 2 bolts and disconnect the 2 fuel tank bands from the fuel tank.
14. Operate the jack and carefully lower the tank, check to see that nothing is interfering with the removal (wires, hoses etc).
15. Lower the jack and remove the fuel tank.

To install:

➡Be sure to use new fasteners, as required.

16. Installation is the reverse of the removal procedure.
17. Set up the fuel tank to the transmission jack.
18. Operate the transmission jack and install the fuel tank.
19. Install the 2 fuel tank bands with the 2 bolts and tighten to 45 ft. lbs. (62 Nm).
20. Connect the fuel inlet hose to the filler pipe and install the clamp.
21. Connect the breather tube.
22. Connect the fuel tank vent hose to the charcoal canister.

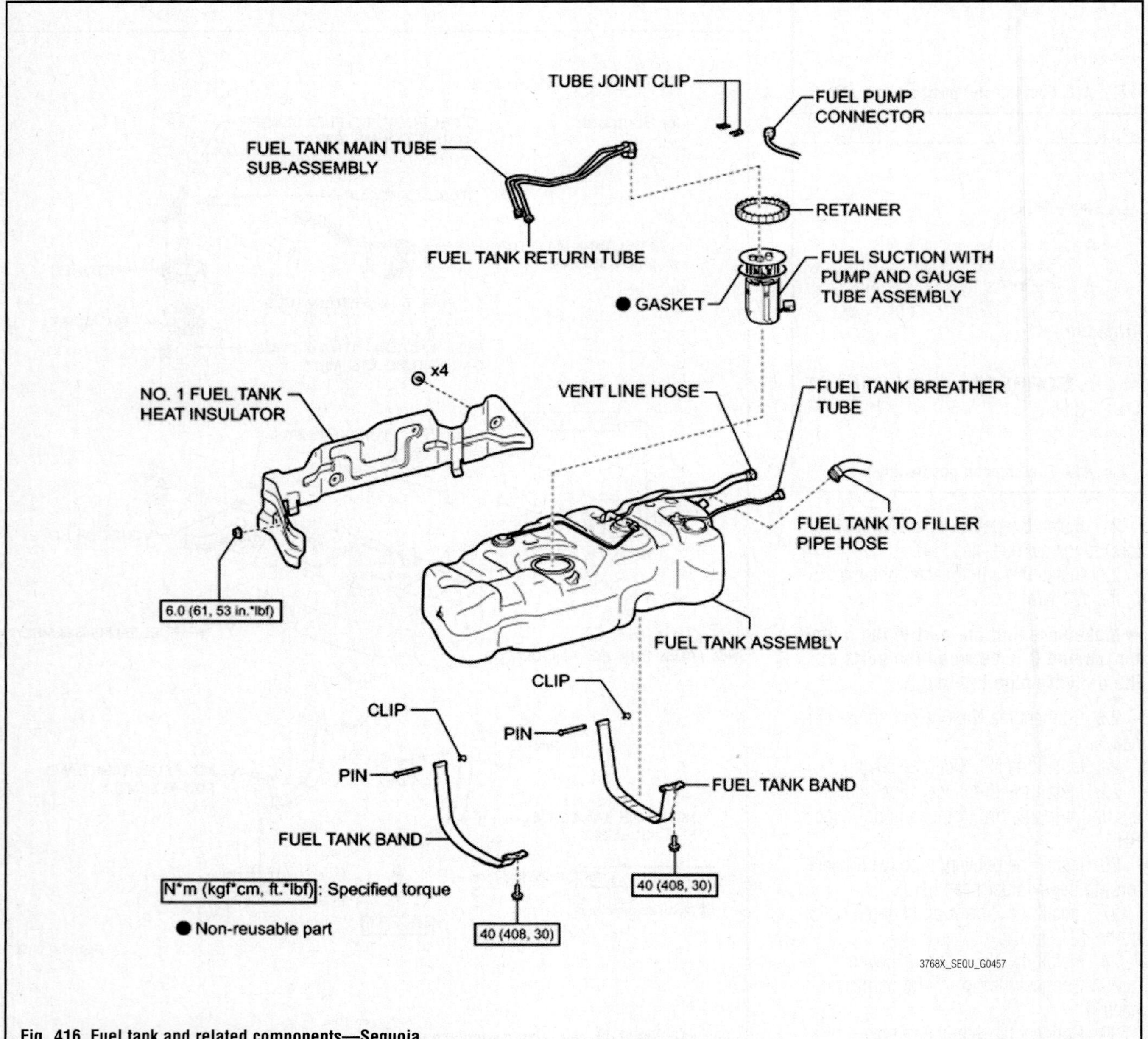

Fig. 416 Fuel tank and related components—Sequoia

23. Connect fuel tank main tube and return tube.
24. Install the fuel tank protector with the 2 bolts and 2 nuts.
25. Connect fuel pump connector.
26. Check for fuel leaks.
27. Install spare tire.

IDLE SPEED

ADJUSTMENT

Idle speed is maintained by the Powertrain Control Module (PCM). No adjustment is necessary or possible.

THROTTLE BODY

REMOVAL & INSTALLATION

See Figure 417.

1. Before servicing the vehicle, refer to the Precautions Section.

➡If working near and/or around the SRS system and components, be sure to disable the SRS system. Tape the negative battery cable with insulating tape. Always disconnect the negative battery cable first.

CAUTION

To avoid personal injury when working on vehicles equipped with an air bag, the negative battery cable must be disconnected and at least 90 seconds must elapse before working on the system. Failure to do so may result in deployment of the air bag.

2. Disconnect the negative battery cable. Tape the cable with insulating tape.

➡When disconnecting the cable, some systems need to be initialized after the cable is reconnected. You will need the Techstream diagnostic scan tool or equivalent. Follow the directions on the tool.

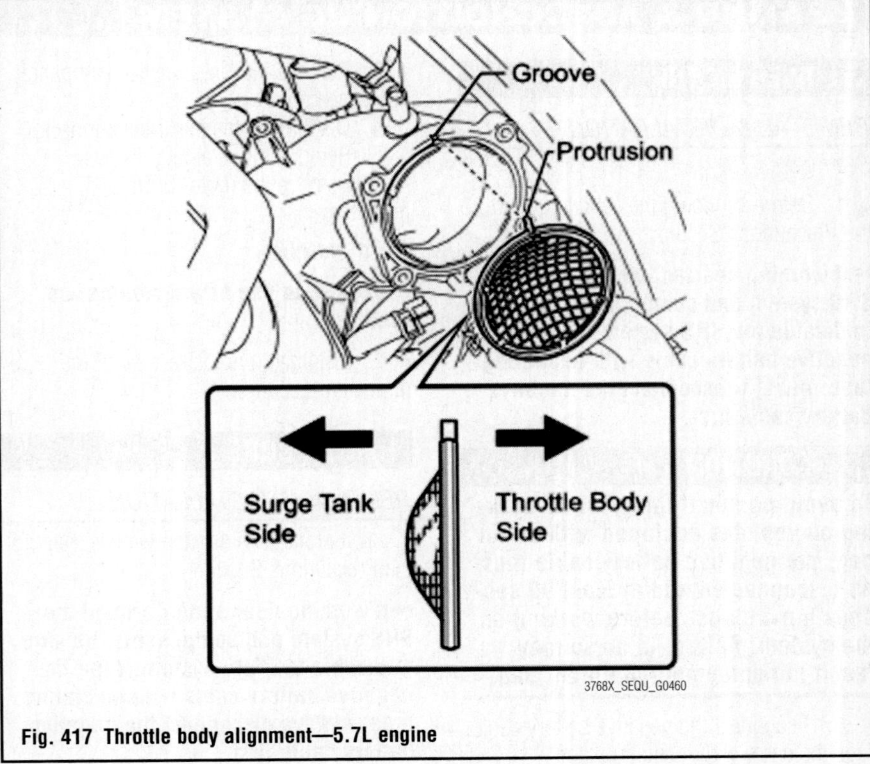

Fig. 417 Throttle body alignment—5.7L engine

➡On 4.0L and 5.7L engines the throttle body is removed along with the throttle body motor.

3. Remove the engine undercover, if equipped.
4. Drain the engine coolant. Be sure to properly dispose of used coolant.
5. Remove the throttle body sub cover assembly, 4.7L engine.
6. Remove the V- bank cover assembly.
7. Remove the air cleaner assembly.
8. On 4.0L engine remove the wiper motor assembly. Remove the front cowl top sub panel.
9. Disconnect the coolant by pass hoses.
10. Disconnect the electrical connector.
11. Remove the component retaining bolts.
12. Remove the component from its mounting. Discard the gasket.

To install:

➡Be sure to use new fasteners, as required.

13. Installation is the reverse of the removal procedure.
14. On 4.6L and 5.7L engines tighten the retaining bolts to 7 ft. lbs. (10 Nm). On 4.0L engine tighten to 8 ft. lbs. (11 Nm). On 4.7L engine tighten to 10 ft. lbs. (14 Nm).
15. Be sure to fill the cooling system with the proper grade and type coolant.
16. Start the engine and check for leaks. Correct as required.
17. Using the Techstream diagnostic tool, or equivalent, perform the initialization procedure. Follow the directions on the tool.

HEATING & AIR CONDITIONING SYSTEM

BLOWER MOTOR

REMOVAL & INSTALLATION

See Figure 418.

1. Before servicing the vehicle, refer to the Precautions Section.

➡If working near and/or around the SRS system and components, be sure to disable the SRS system. Tape the negative battery cable with insulating tape. Always disconnect the negative battery cable first.

✲✲ CAUTION

To avoid personal injury when working on vehicles equipped with an air bag, the negative battery cable must be disconnected and at least 90 seconds must elapse before working on the system. Failure to do so may result in deployment of the air bag.

2. Disconnect the negative battery cable. Tape the cable with insulating tape.

➡When disconnecting the cable, some systems need to be initialized after the cable is reconnected. You will need the Techstream diagnostic scan tool or equivalent. Follow the directions on the tool.

3. Remove the No. 2 instrument panel sub assembly.
4. Disconnect the electrical connector.
5. Remove the retaining screws.
6. Remove the motor from its mounting.

To install:

➡Be sure to use new fasteners, as required.

7. Installation is the reverse of the removal procedure.

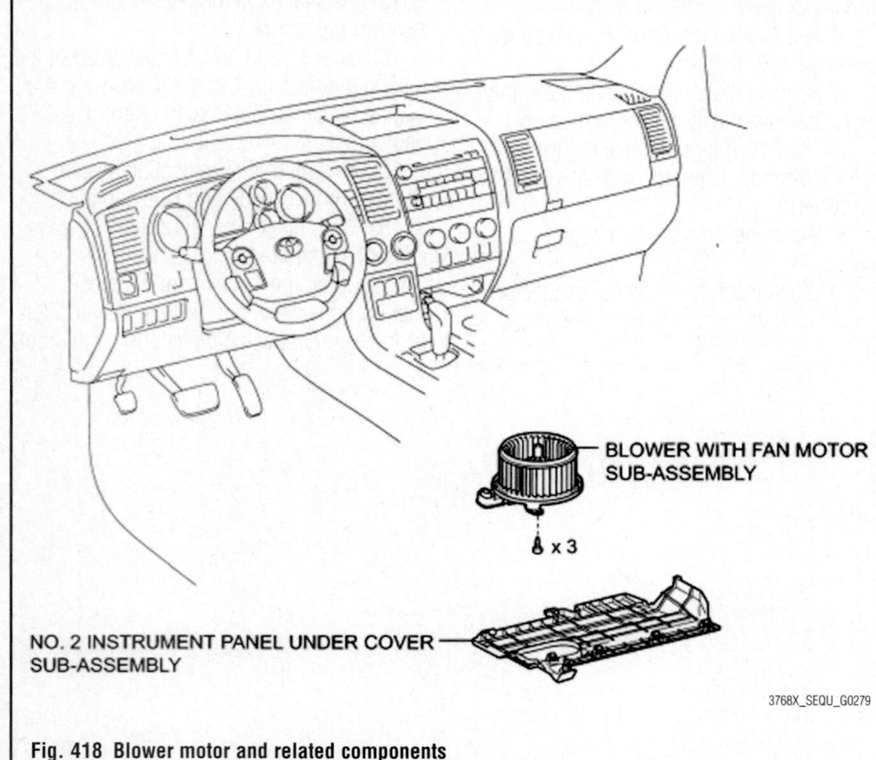

Fig. 418 Blower motor and related components

HEATER CORE

REMOVAL & INSTALLATION

1. Before servicing the vehicle, refer to the Precautions Section.

➡If working near and/or around the SRS system and components, be sure to disable the SRS system. Tape the negative battery cable with insulating tape. Always disconnect the negative battery cable first.

✲✲ CAUTION

To avoid personal injury when working on vehicles equipped with an air bag, the negative battery cable must be disconnected and at least 90 seconds must elapse before working on the system. Failure to do so may result in deployment of the air bag.

2. Disconnect the negative battery cable. Tape the cable with insulating tape.

➡When disconnecting the cable, some systems need to be initialized after the cable is reconnected. You will need the Techstream diagnostic scan tool or equivalent. Follow the directions on the tool.

3. Remove the heater/cooling unit from the vehicle.
4. Disassemble the unit as required to remove the heater core.

To install:

➡Be sure to use new fasteners, as required.

5. Installation is the reverse of the removal procedure.
6. Be sure to refill the cooling system, with the proper grade and type coolant.
7. Be sure to properly recharge the system.
8. Be sure to adjust the spiral cable, as required.
9. Check the SRS system, for proper operation.
10. Using the Techstream diagnostic tool or equivalent, reprogram the required systems.

HEATER/COOLING UNIT

REMOVAL & INSTALLATION

See Figures 419 through 426.

1. Before servicing the vehicle, refer to the Precautions Section.

➡If working near and/or around the SRS system and components, be sure to disable the SRS system. Tape the negative battery cable with insulating tape. Always disconnect the negative battery cable first.

✲✲ CAUTION

To avoid personal injury when working on vehicles equipped with an air bag, the negative battery cable must be disconnected and at least 90 seconds must elapse before working on the system. Failure to do so may result in deployment of the air bag.

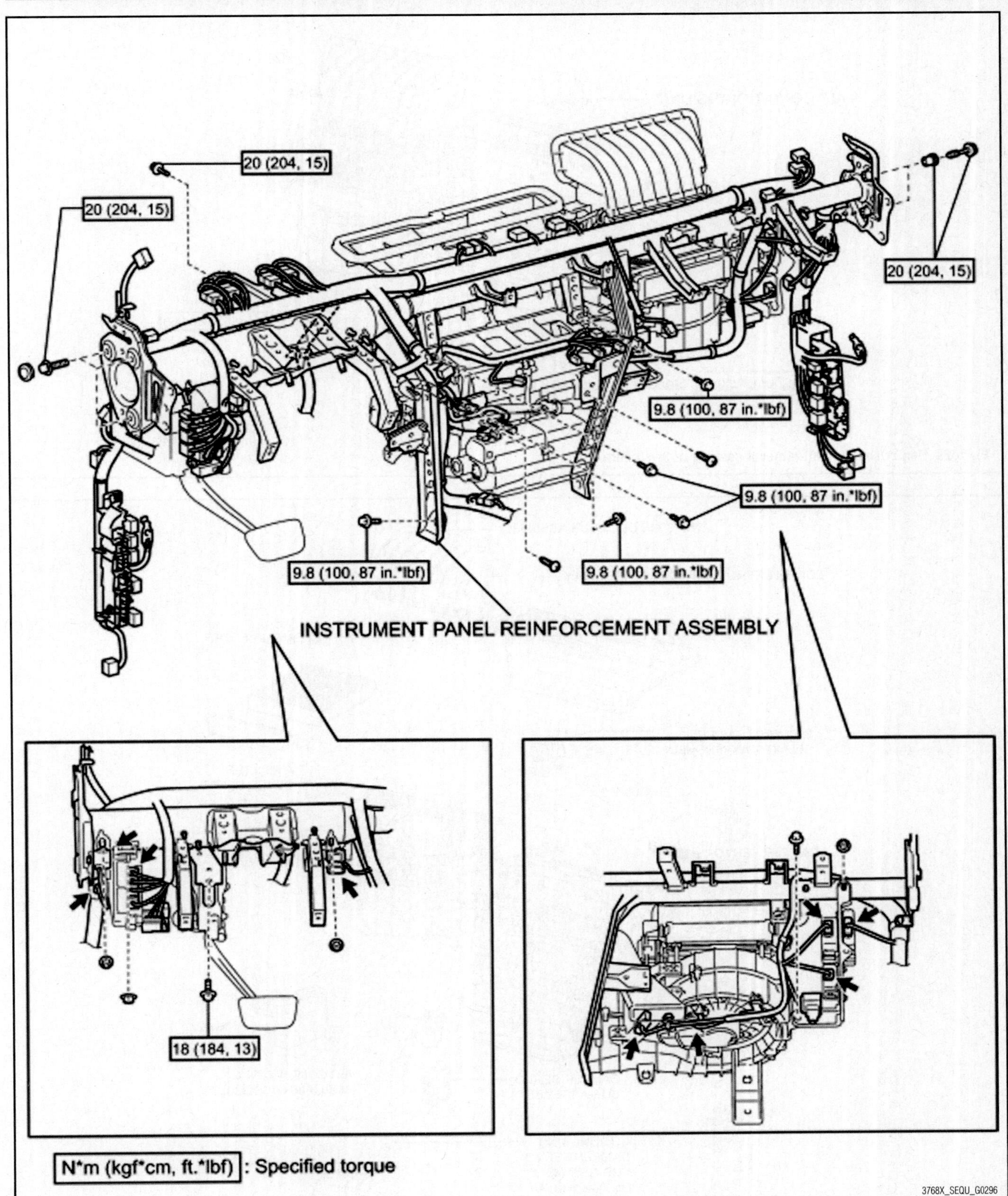

Fig. 419 Heater/cooling unit removal points 1 of 3—Tundra

AIR CONDITIONING UNIT
5.4 (55, 48 in.*lbf)
N*m (kgf*cm, ft.*lbf) : Specified torque
3768X_SEQU_G0297

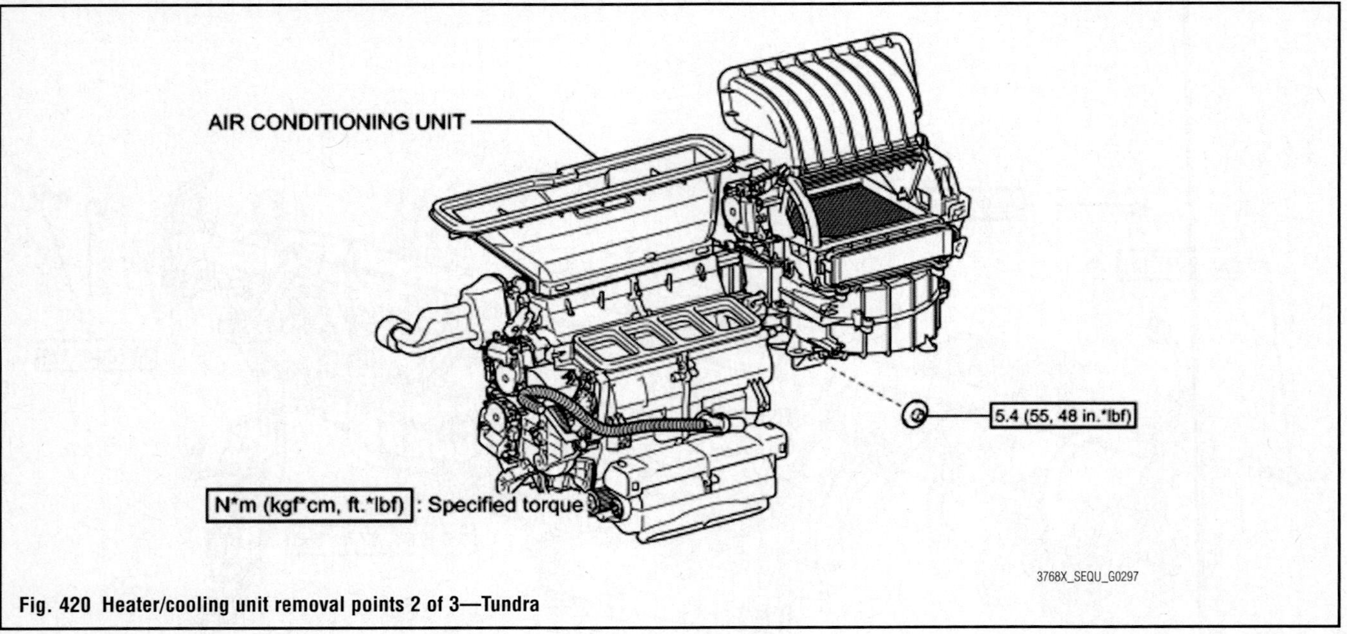

Fig. 420 Heater/cooling unit removal points 2 of 3—Tundra

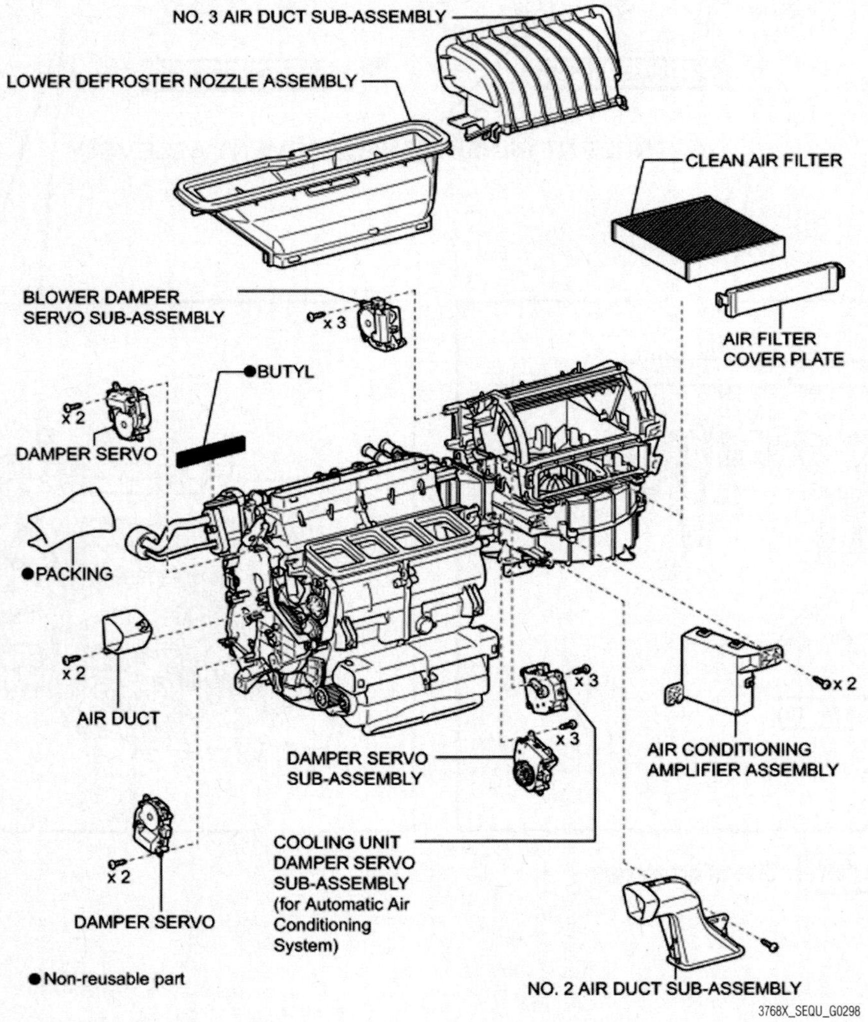

NO. 3 AIR DUCT SUB-ASSEMBLY
LOWER DEFROSTER NOZZLE ASSEMBLY
CLEAN AIR FILTER
BLOWER DAMPER
SERVO SUB-ASSEMBLY
x 3
AIR FILTER
COVER PLATE
●BUTYL
x 2
DAMPER SERVO
●PACKING
x 2
AIR DUCT
x 3
x 3
x 2
DAMPER SERVO
SUB-ASSEMBLY
AIR CONDITIONING
AMPLIFIER ASSEMBLY
COOLING UNIT
DAMPER SERVO
SUB-ASSEMBLY
(for Automatic Air
Conditioning
System)
x 2
DAMPER SERVO
●Non-reusable part
NO. 2 AIR DUCT SUB-ASSEMBLY
3768X_SEQU_G0298

Fig. 421 Heater/cooling unit removal points 3 of 3—Tundra

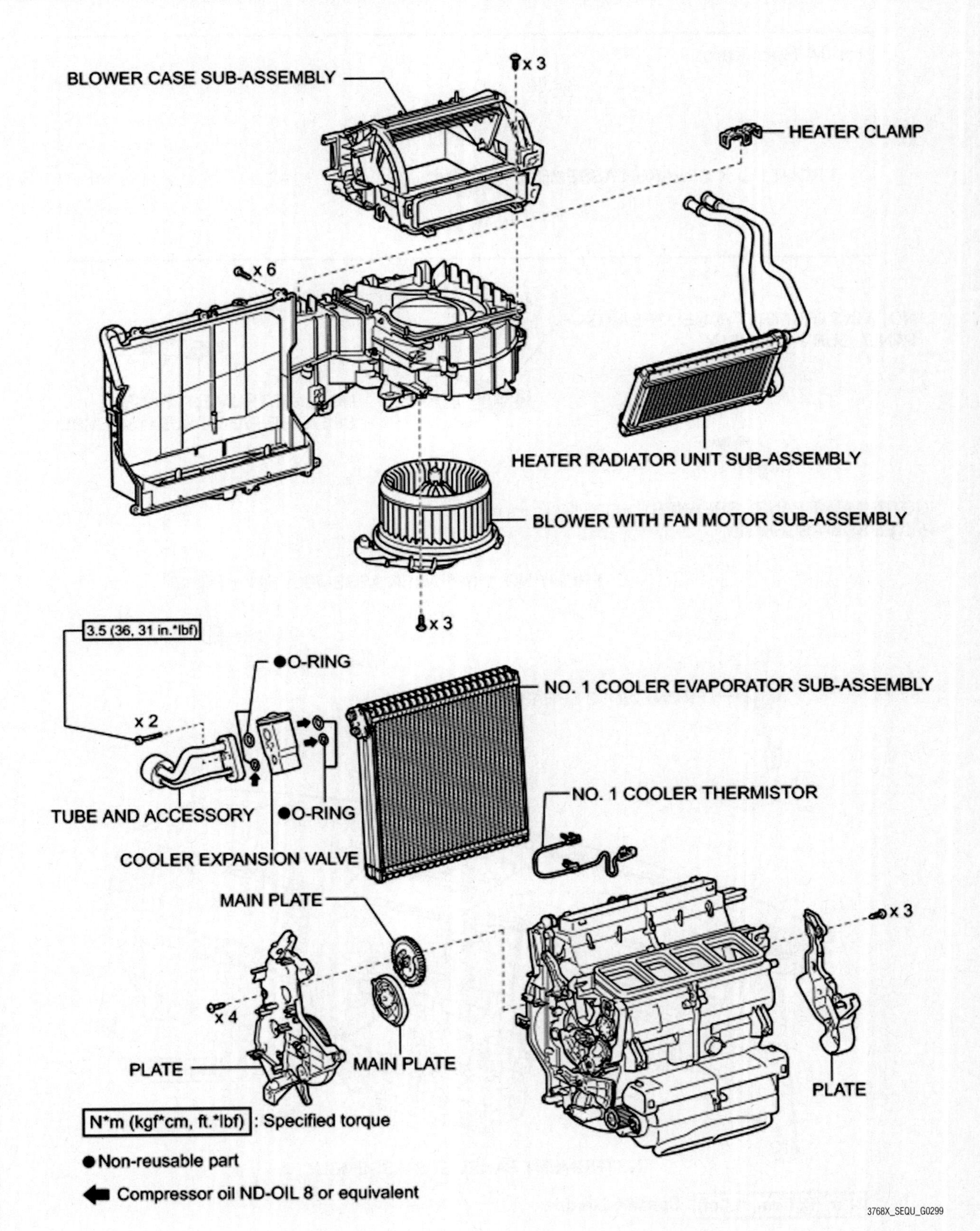

Fig. 422 Heater/cooling unit and related components—Tundra

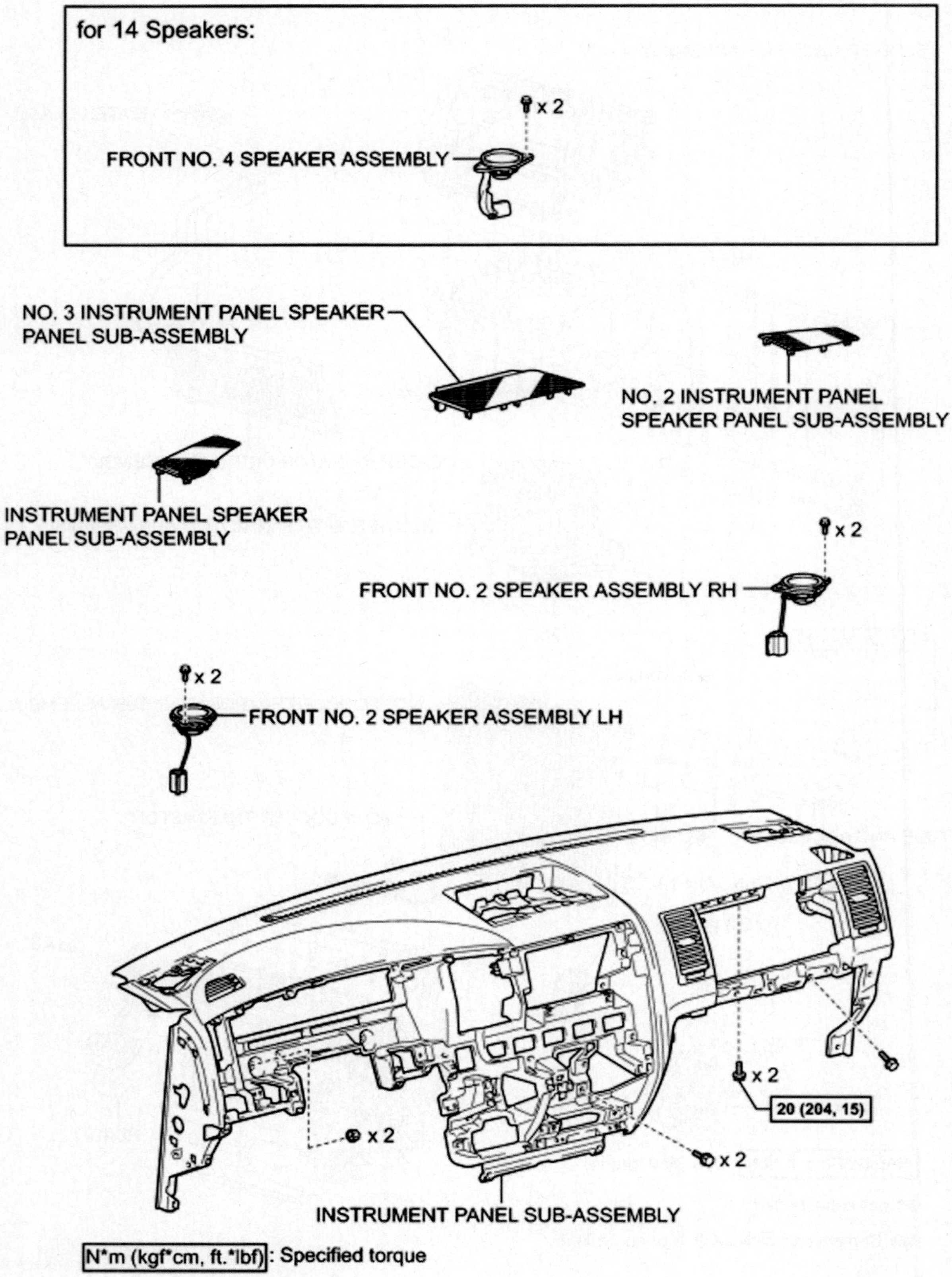

Fig. 423 Heater/cooling unit removal points 10 of 12—Sequoia

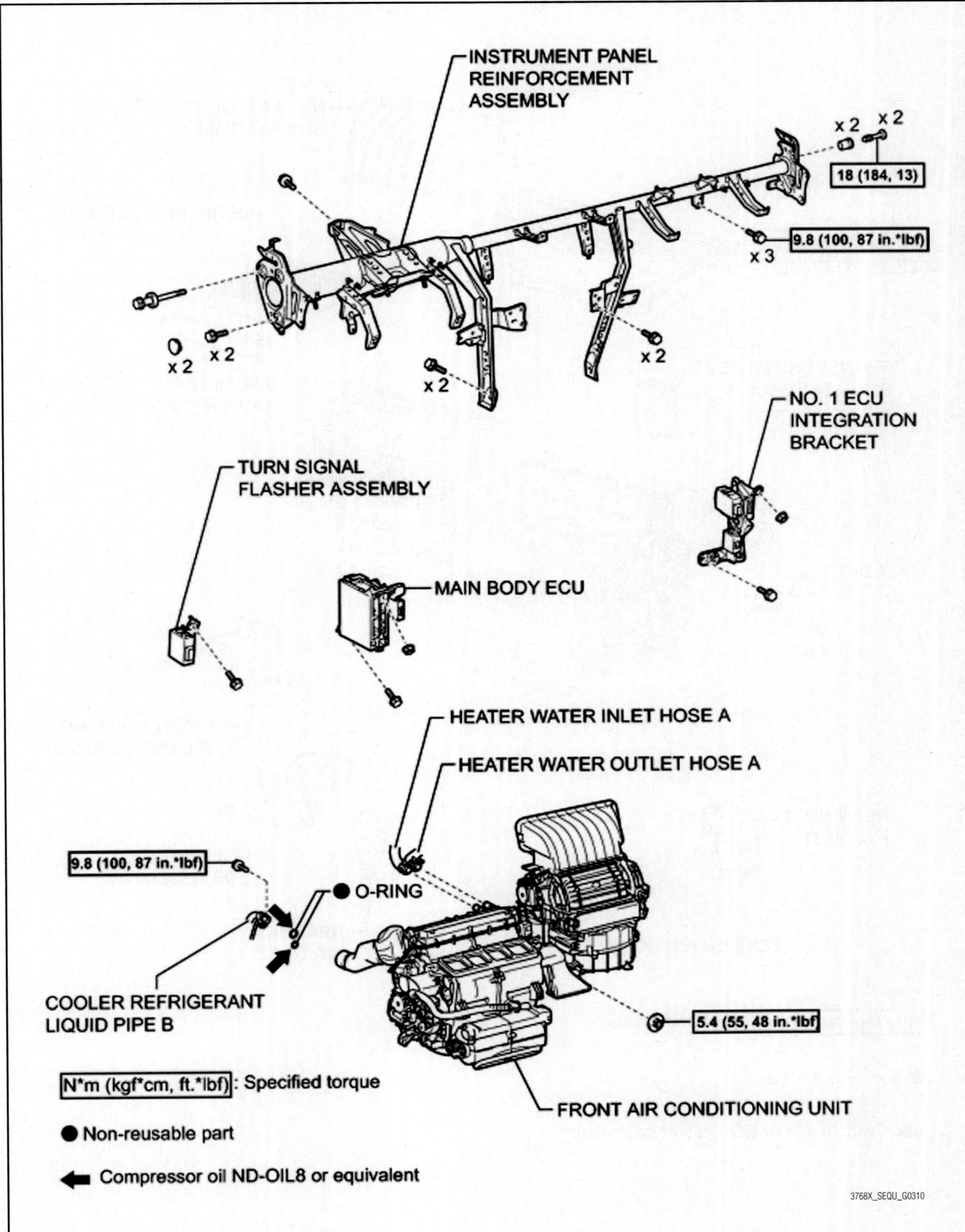

Fig. 424 Heater/cooling unit removal points 11 of 12—Sequoia

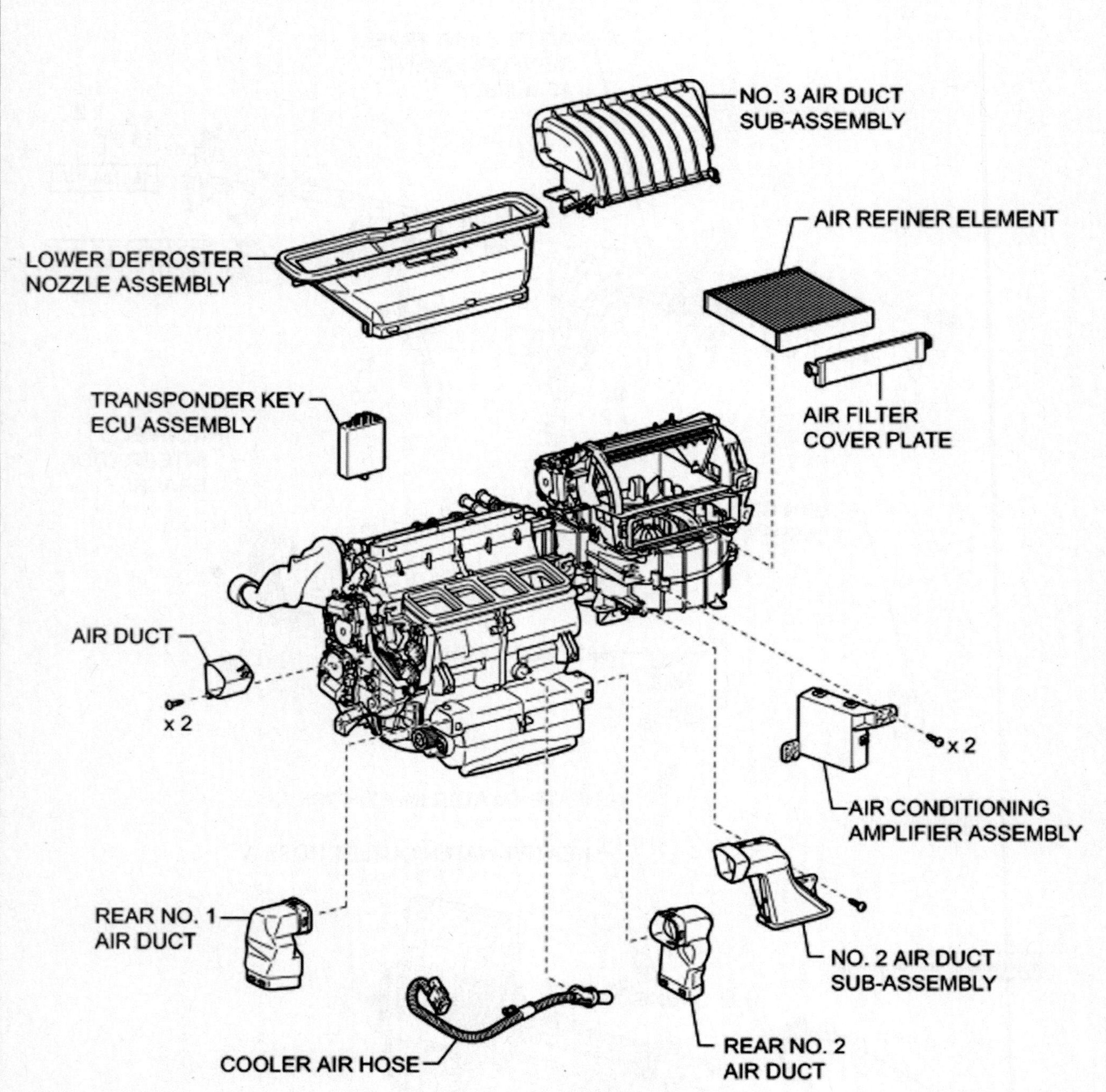

Fig. 425 Heater/cooling unit removal points 12 of 12—Sequoia

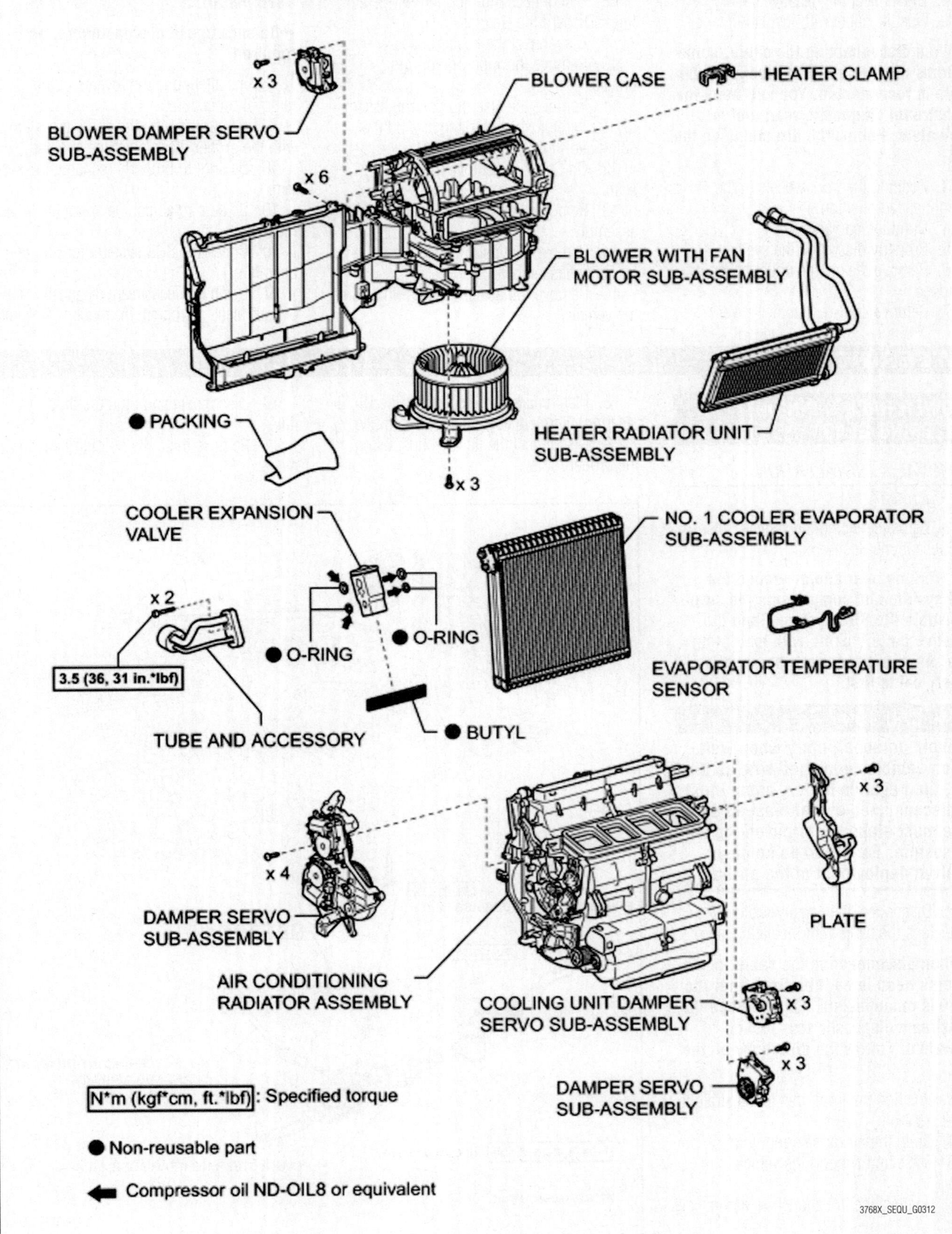

Fig. 426 Heater/cooling unit and related components—Sequoia

2. Disconnect the negative battery cable. Tape the cable with insulating tape.

➡**When disconnecting the cable, some systems need to be initialized after the cable is reconnected. You will need the Techstream diagnostic scan tool or equivalent. Follow the directions on the tool.**

3. Position the front wheels in the straight ahead position.
4. Drain the coolant.
5. Properly discharge the system.
6. Remove the engine under cover, if equipped.
7. Remove the No. 2 cooler cover.
8. Disconnect and plug the refrigerant lines. Discard the O-rings.
9. Disconnect the heater hoses.
10. Remove the wiper motor and bracket.
11. Remove the steering column assembly. Be careful when working around the spiral cable.
12. On Sequoia, remove the front seats.
13. Remove the instrument panel sub assembly.
14. Remove the instrument panel reinforcement assembly.
15. Remove the heater/cooling unit from the vehicle.

To install:

➡**Be sure to use new fasteners, as required.**

16. Installation is the reverse of the removal procedure.
17. Be sure to refill the cooling system, with the proper grade and type coolant.
18. Be sure to properly recharge the system.
19. Be sure to adjust the spiral cable, as required.
20. Check the SRS system, for proper operation.
21. Using the Techstream diagnostic tool or equivalent, reprogram the required systems.

STEERING

POWER RACK & PINION STEERING GEAR

REMOVAL & INSTALLATION

See Figure 427.

1. Before servicing the vehicle, refer to the Precautions Section.

➡**If working near and/or around the SRS system and components, be sure to disable the SRS system. Tape the negative battery cable with insulating tape. Always disconnect the negative battery cable first.**

CAUTION

To avoid personal injury when working on vehicles equipped with an air bag, the negative battery cable must be disconnected and at least 90 seconds must elapse before working on the system. Failure to do so may result in deployment of the air bag.

2. Disconnect the negative battery cable. Tape the cable with insulating tape.

➡**When disconnecting the cable, some systems need to be initialized after the cable is reconnected. You will need the Techstream diagnostic scan tool or equivalent. Follow the directions on the tool.**

3. Position the front tires in the straight ahead position.
4. Drain the power steering fluid.
5. Raise and support the vehicle safely.
6. Remove the tire and wheel assemblies.
7. Remove the engine from the vehicle.
8. Matchmark and disconnect the No. 2 steering intermediate shaft sub assembly.
9. Disconnect the tie rod ends, using the proper tools.
10. Disconnect and plug the fluid lines.
11. Remove the bolts holding lines in place.

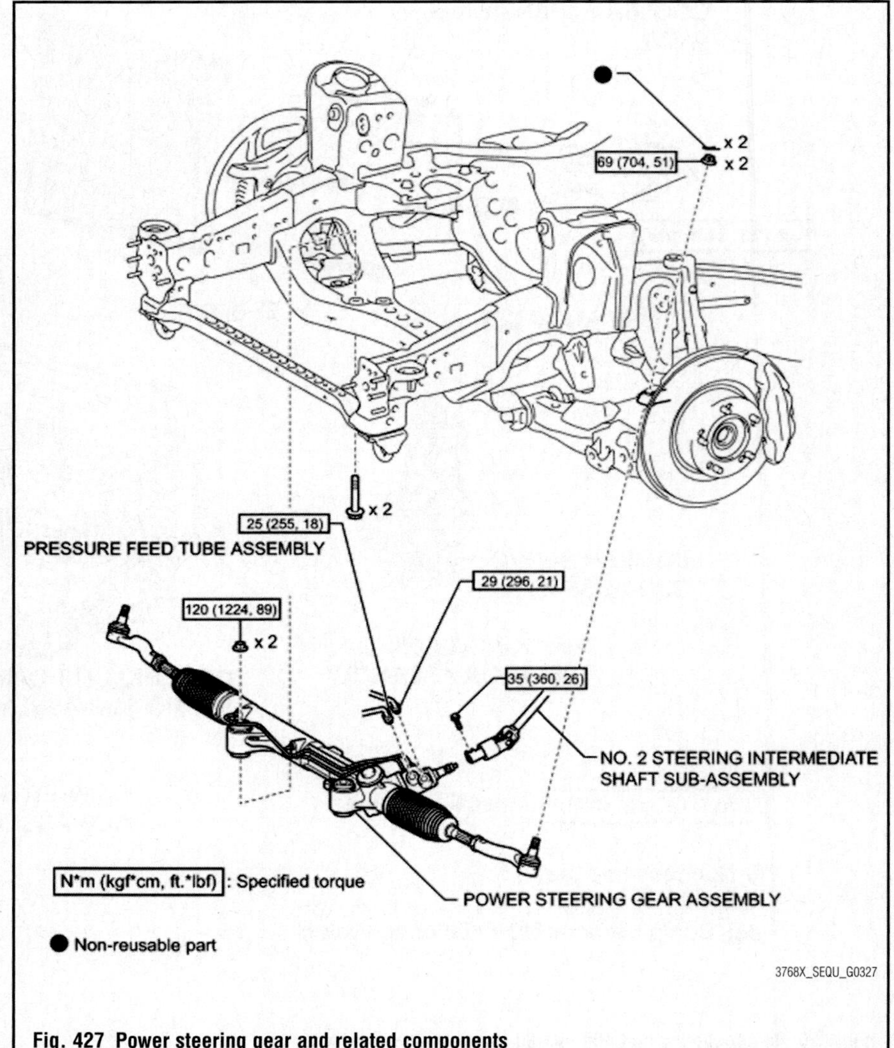

Fig. 427 Power steering gear and related components

12. Remove the gear retaining bolts.
13. Remove the gear assembly from its mounting.

To install:

➡Be sure to use new fasteners, as required.

14. Installation is the reverse of the removal procedure.
15. Tighten the gear retaining bolts to 89 ft. lbs. (120 Nm).
16. Use the matchmarks when installing the No.2 steering intermediate shaft to the steering gear. Tighten the retaining bolt to 26 ft. lbs. (35 Nm).
17. Adjust the front end alignment, as required.
18. Be sure to fill the power steering system, with the correct grade and type fluid.

POWER STEERING PUMP

REMOVAL & INSTALLATION

4.0L Engine

See Figure 428.

1. Before servicing the vehicle, refer to the Precautions Section.

➡If working near and/or around the SRS system and components, be sure to disable the SRS system. Tape the negative battery cable with insulating tape. Always disconnect the negative battery cable first.

✱✱ CAUTION

To avoid personal injury when working on vehicles equipped with an air bag, the negative battery cable must be disconnected and at least 90 seconds must elapse before working on the system. Failure to do so may result in deployment of the air bag.

2. Disconnect the negative battery cable. Tape the cable with insulating tape.

➡When disconnecting the cable, some systems need to be initialized after the cable is reconnected. You will need the Techstream diagnostic scan tool or equivalent. Follow the directions on the tool.

3. Remove No. 1 engine under cover.
4. Remove fan and alternator V belt.
5. Drain power steering fluid.
6. Disconnect suction hose.
7. Disconnect power steering oil pressure switch connector.
8. Remove the bolt and disconnect the pressure feed tube.

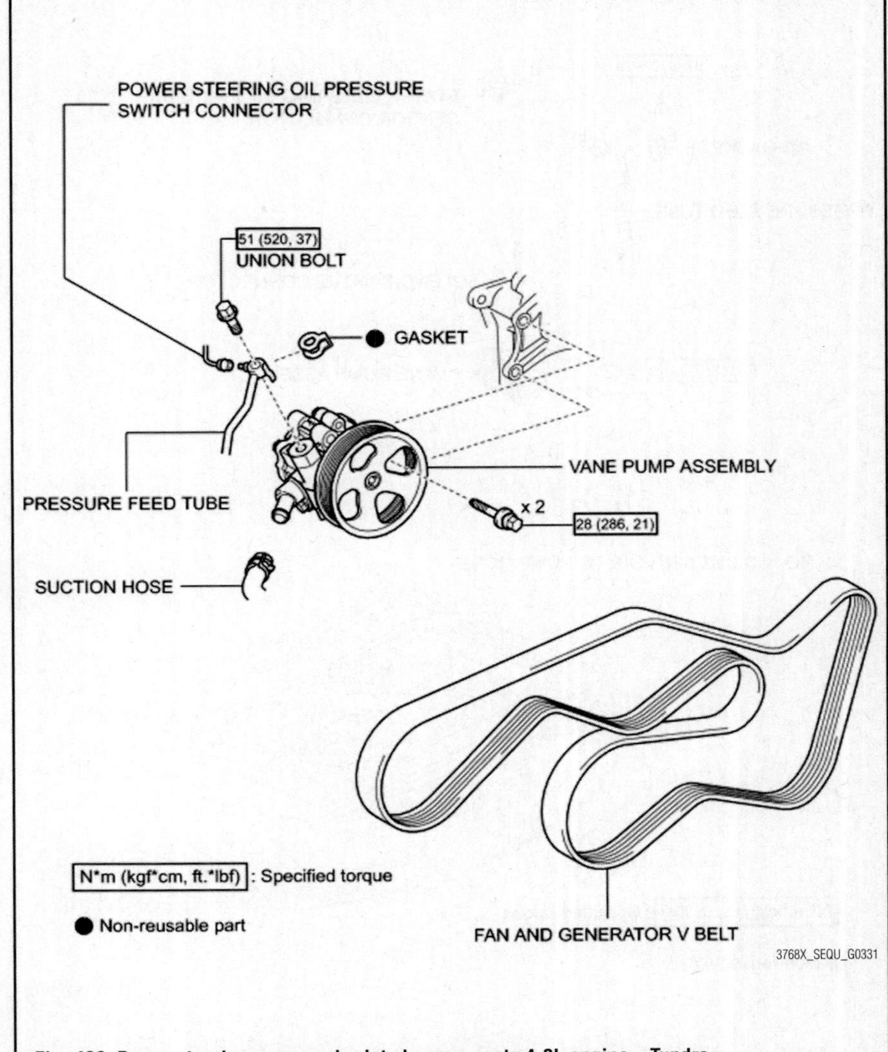

Fig. 428 Power steering pump and related components 4.0L engine—Tundra

9. Remove the gasket.
10. Remove the 2 bolts and vane pump.

To install:

➡Be sure to use new fasteners, as required.

11. Install the vane pump with the 2 bolts and tighten to 21 ft. lbs. (28 Nm).
12. Install a new gasket to the pressure feed tube.
13. Install the pressure feed tube with the union bolt and tighten to 38 ft lbs. (51 Nm).
14. Connect power steering oil pressure switch connector.
15. Connect the suction hose with the clip.
16. Install fan and alternator V belt.
17. Add power steering fluid.
18. Bleed power steering fluid.
19. Inspect for power steering fluid leak.
20. Install No. 1 engine under cover.
21. Connect cable to negative battery terminal.
22. Perform initialization, if necessary.

4.6L Engine

See Figures 429 and 430.

1. Before servicing the vehicle, refer to the Precautions Section.

➡If working near and/or around the SRS system and components, be sure to disable the SRS system. Tape the negative battery cable with insulating tape. Always disconnect the negative battery cable first.

✱✱ CAUTION

To avoid personal injury when working on vehicles equipped with an air bag, the negative battery cable must be disconnected and at least 90 seconds must elapse before working on the system. Failure to do so may result in deployment of the air bag.

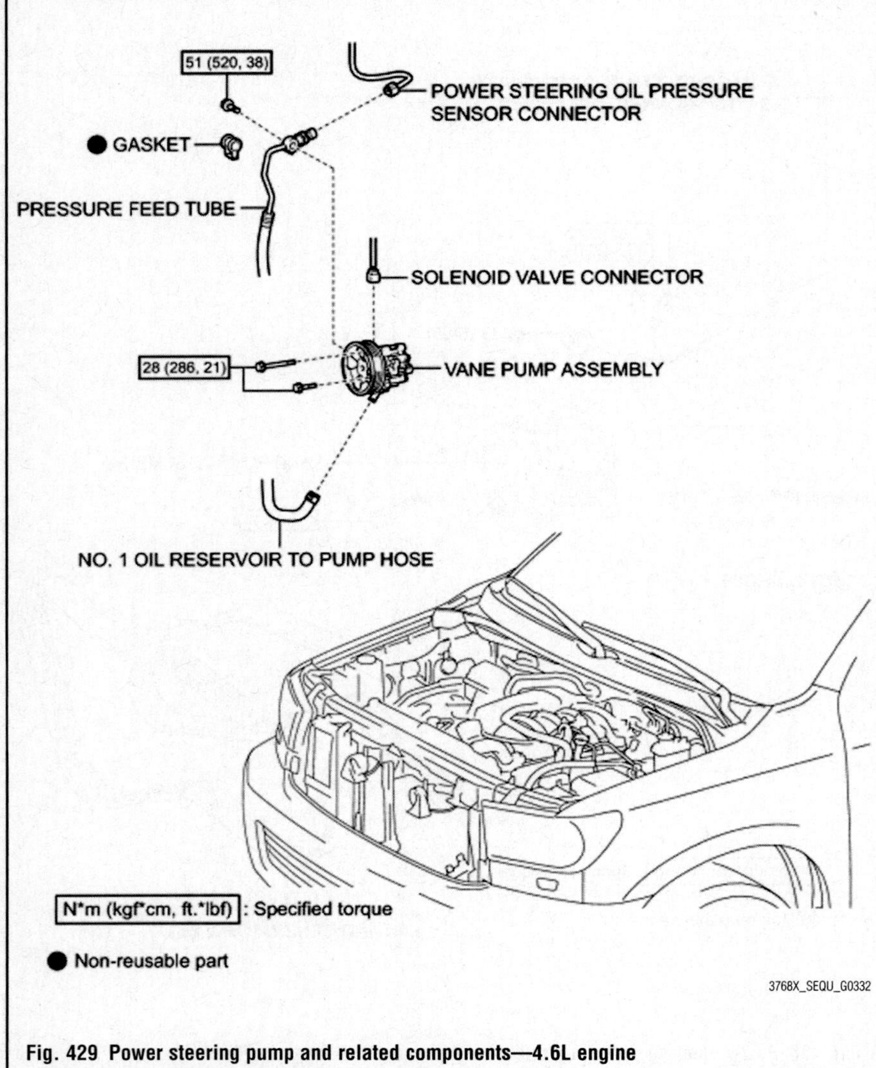

Fig. 429 Power steering pump and related components—4.6L engine

2. Disconnect the negative battery cable. Tape the cable with insulating tape.

➡**When disconnecting the cable, some systems need to be initialized after the cable is reconnected. You will need the Techstream diagnostic scan tool or equivalent. Follow the directions on the tool.**

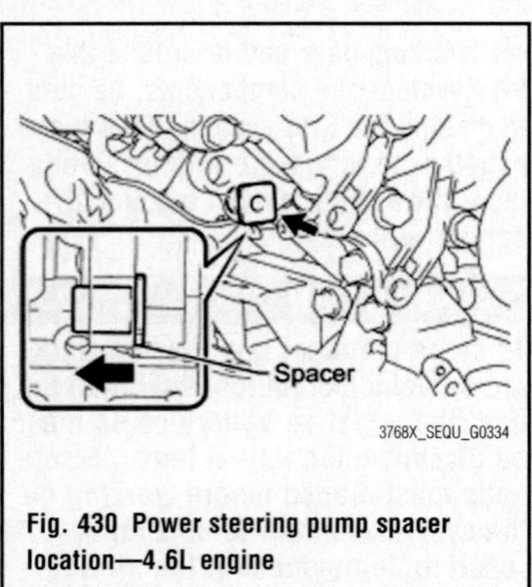

Fig. 430 Power steering pump spacer location—4.6L engine

3. Remove the drive belt.
4. Drain the cooling system. Remove the radiator.
5. Separate the harness bracket.
6. Drain the power steering system. Be sure to properly dispose of used fluid.
7. Disconnect and plug the fluid lines.
8. Disconnect the pressure sensor.
9. Disconnect the solenoid valve connector.
10. Remove the pump retaining bolts.
11. Remove the pump from its mounting.

To install:

➡**Be sure to use new fasteners, as required.**

12. Installation is the reverse of the removal procedure.
13. Move the space so that the pump can be installed.
14. Tighten the retaining bolts to 21 ft. lbs. (29 Nm).
15. Be sure to fill the cooling system, with the proper grade and type coolant.
16. Be sure to fill the power steering system, with the proper grade and type fluid.

4.7 Engine

See Figure 431.

1. Before servicing the vehicle, refer to the Precautions Section.

➡**If working near and/or around the SRS system and components, be sure to disable the SRS system. Tape the negative battery cable with insulating tape. Always disconnect the negative battery cable first.**

⁂ CAUTION

To avoid personal injury when working on vehicles equipped with an air bag, the negative battery cable must be disconnected and at least 90 seconds must elapse before working on the system. Failure to do so may result in deployment of the air bag.

2. Disconnect the negative battery cable. Tape the cable with insulating tape.

➡**When disconnecting the cable, some systems need to be initialized after the cable is reconnected. You will need the Techstream diagnostic scan tool or equivalent. Follow the directions on the tool.**

3. Disconnect the MAF meter connector.
4. Disconnect the hoses.

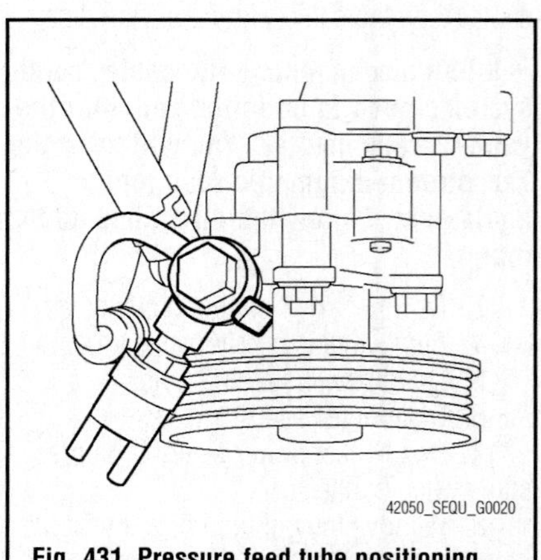

Fig. 431 Pressure feed tube positioning

5. Remove the clamp.
6. Remove the 3 bolts and air cleaner assembly with air cleaner hose connected.
7. Loosen the drive belt tension by turning the drive belt tensioner counter-clockwise, and remove the drive belt.
8. Remove the 2 clips and disconnect the 2 vacuum hoses.
9. Remove the clip and disconnect the return hose.
10. Remove the union bolt and gasket, disconnect the pressure feed tube.
11. Remove the 2 bolts, nut, stud bolt and power steering pump assembly.

To install:

➡Be sure to use new fasteners, as required.

12. Install the power steering pump assembly with the stud bolt.
13. Tighten the stud bolt to 22 Nm (16 ft. lbs.).
14. Install the 2 bolts and nut and tighten them to 44 Nm (33 ft. lbs.).
15. Install a new gasket and the union bolt on the pressure feed tube.

➡Make sure that the stopper of the pressure feed tube contacts the power steering pump body as shown in the illustration.

16. Tighten the union bolt to 46.5 Nm (34 ft. lbs.)
17. Connect the return hose with the clip.
18. Connect the 2 vacuum hoses and install the 2 clips.
19. Loosen the drive belt tension by turning the drive belt tensioner counter-clockwise, and install the belt.
20. Install the air cleaner assembly with air cleaner hose and the 3 bolts.
21. Install the clamp.
22. Connect the MAF meter connector.
23. Fill with power steering fluid and bleed the system.

5.7L Engine

See Figure 432.

1. Before servicing the vehicle, refer to the Precautions Section.

➡If working near and/or around the SRS system and components, be sure to disable the SRS system. Tape the negative battery cable with insulating tape. Always disconnect the negative battery cable first.

⁂ CAUTION

To avoid personal injury when working on vehicles equipped with an air bag, the negative battery cable must be disconnected and at least 90 seconds must elapse before working on the system. Failure to do so may result in deployment of the air bag.

2. Disconnect the negative battery cable. Tape the cable with insulating tape.

➡When disconnecting the cable, some systems need to be initialized after the cable is reconnected. You will need the Techstream diagnostic scan tool or equivalent. Follow the directions on the tool.

3. Remove the drive belt.
4. Remove the V- bank sub assembly.
5. Remove the air cleaner assembly.
6. Drain the power steering system. Be sure to properly dispose of used fluid.
7. Disconnect and plug the fluid lines.
8. Disconnect the pressure sensor.
9. Remove the pump retaining bolts.
10. Remove the pump from its mounting.

To install:

➡Be sure to use new fasteners, as required.

11. Installation is the reverse of the removal procedure.
12. Move the space so that the pump can be installed.

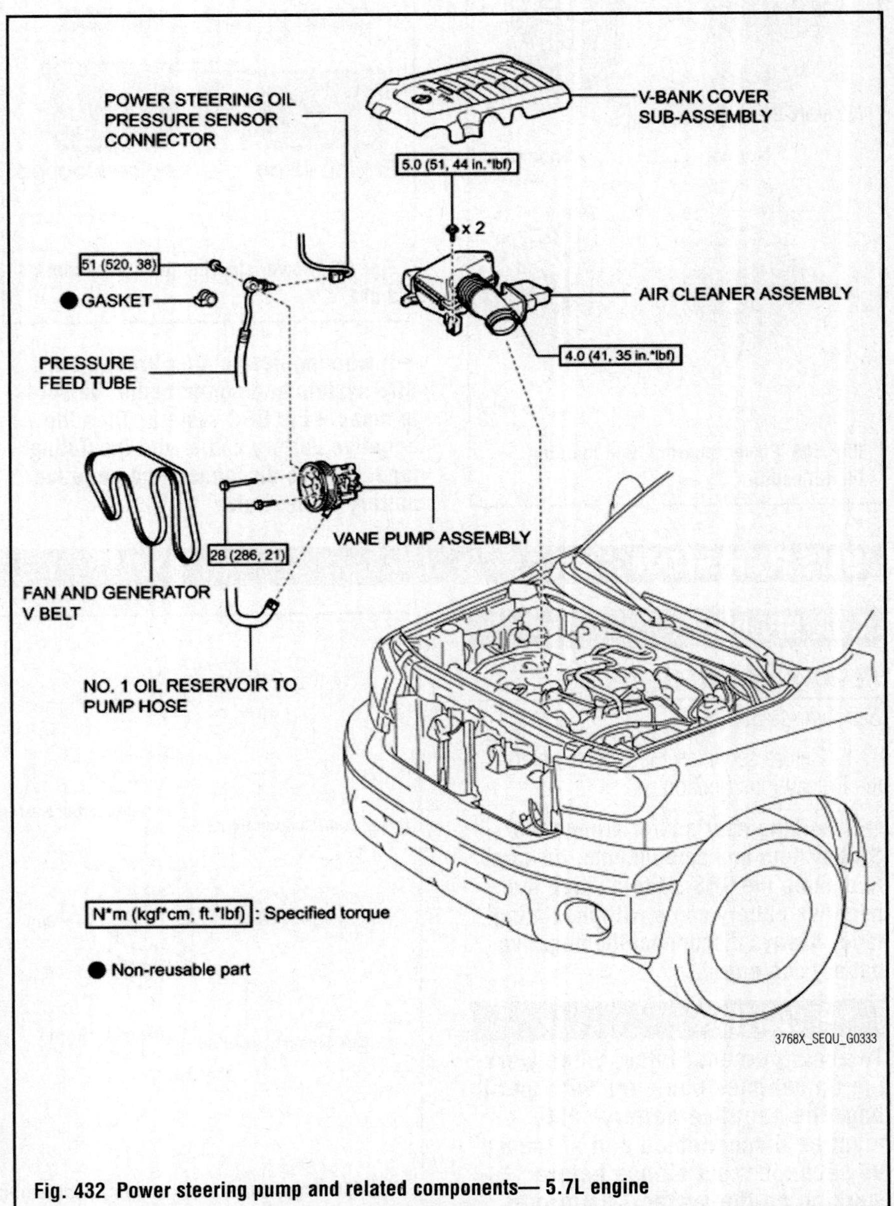

Fig. 432 Power steering pump and related components— 5.7L engine

13. Tighten the retaining bolts to 21 ft. lbs. (28 Nm).
14. Be sure to fill the cooling system, with the proper grade and type coolant.
15. Be sure to fill the power steering system, with the proper grade and type fluid.

BLEEDING

See Figures 433 through 435.

1. Before servicing the vehicle, refer to the Precautions Section.

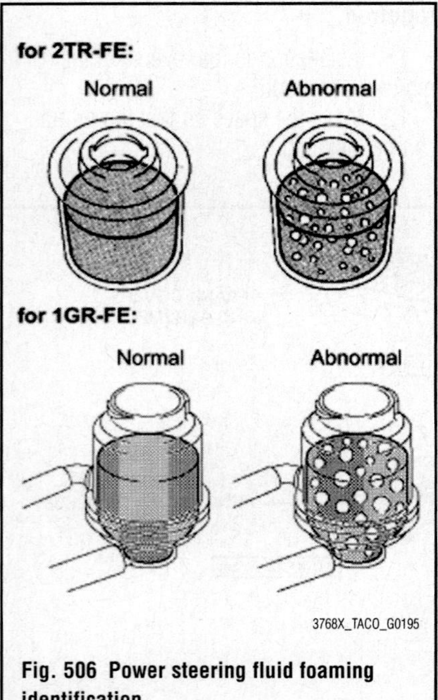

Fig. 506 Power steering fluid foaming identification

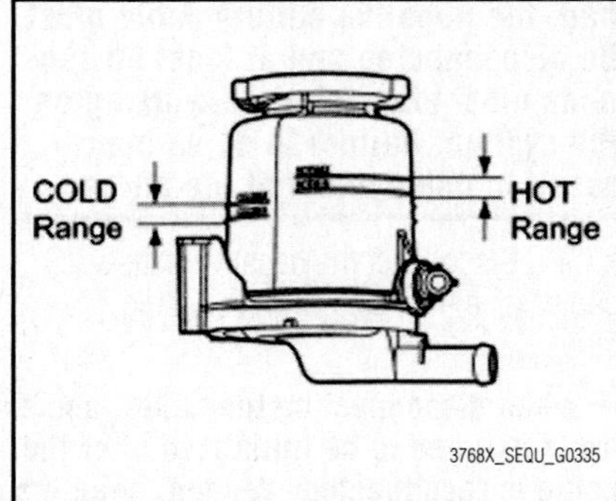

Fig. 434 Power steering fluid level check 1 of 2

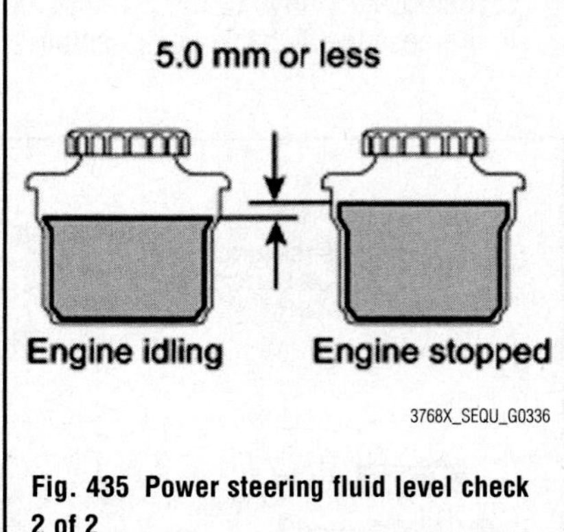

Fig. 435 Power steering fluid level check 2 of 2

➡If working near and/or around the SRS system and components, be sure to disable the SRS system. Tape the negative battery cable with insulating tape. Always disconnect the negative battery cable first.

CAUTION

To avoid personal injury when working on vehicles equipped with an air bag, the negative battery cable must be disconnected and at least 90 seconds must elapse before working on the system. Failure to do so may result in deployment of the air bag.

2. Disconnect the negative battery cable. Tape the cable with insulating tape.

➡When disconnecting the cable, some systems need to be initialized after the cable is reconnected. You will need the Techstream diagnostic scan tool or equivalent. Follow the directions on the tool.

3. Check the fluid level.
4. Raise and properly support the vehicle.
5. With the engine stopped, turn the wheels slowly from stop to stop several times.
6. Lower the vehicle.
7. Start the engine. Run at idle for a few minutes.
8. With the engine idling turn the steering wheel to the left or right full lock position and keep it there for two or three seconds. Repeat for the other side.
9. Repeat the above step several times.
10. Stop the engine. Check for foaming.
11. If foaming exists, repeat the procedure.
12. With the engine stopped check the fluid level.
13. Adjust as necessary. Be sure to use the proper grade and type fluid.

SUSPENSION — FRONT SUSPENSION

COIL SPRING

REMOVAL & INSTALLATION

See Figure 436.

1. Before servicing the vehicle, refer to the Precautions Section.

➡If working near and/or around the SRS system and components, be sure to disable the SRS system. Tape the negative battery cable with insulating tape. Always disconnect the negative battery cable first.

CAUTION

To avoid personal injury when working on vehicles equipped with an air bag, the negative battery cable must be disconnected and at least 90 seconds must elapse before working on the system. Failure to

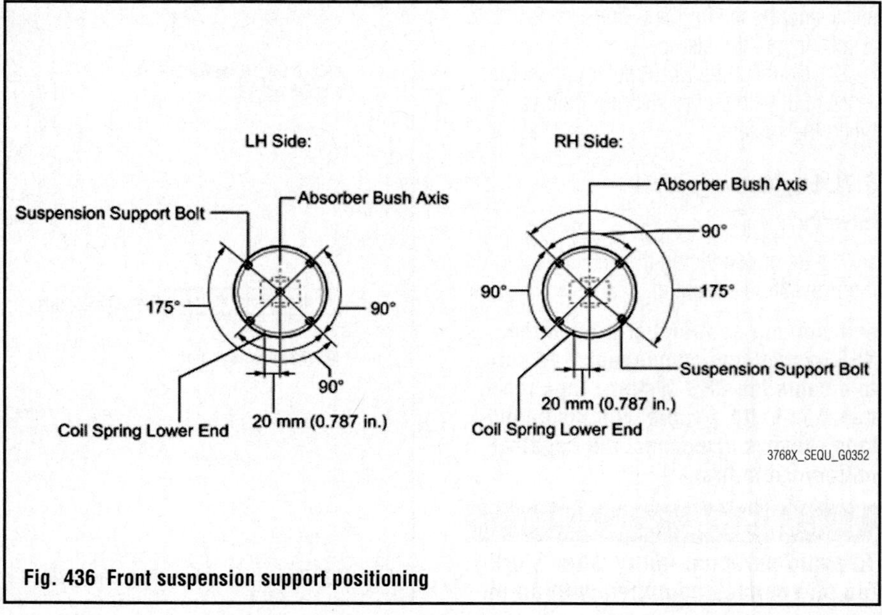

Fig. 436 Front suspension support positioning

do so may result in deployment of the air bag.

2. Disconnect the negative battery cable. Tape the cable with insulating tape.

➡When disconnecting the cable, some systems need to be initialized after the cable is reconnected. You will need the Techstream diagnostic scan tool or equivalent. Follow the directions on the tool.

3. Remove the strut.
4. Using a compressor, compress the coil spring.

➡A compressor with a force of 2,860 lbs. (12,740 N) or more must be used. Make sure that the suspension support is free from the spring. Do not compress the spring more than necessary. Do not position the spring with the upper end towards you.

5. Remove the center nut.
6. Remove the retainers, bushing support and spring.
7. Assembly is the reverse of disassembly. See the accompanying illustration for correct positioning of the suspension support and spring. Torque the nut to 18 ft. lbs. (25 Nm).

CONTROL LINKS

REMOVAL & INSTALLATION

See Figure 437.

1. Before servicing the vehicle, refer to the Precautions Section.

➡If working near and/or around the SRS system and components, be sure to disable the SRS system. Tape the negative battery cable with insulating tape. Always disconnect the negative battery cable first.

⁂ CAUTION

To avoid personal injury when working on vehicles equipped with an air bag, the negative battery cable must be disconnected and at least 90 seconds must elapse before working on the system. Failure to do so may result in deployment of the air bag.

2. Disconnect the negative battery cable. Tape the cable with insulating tape.

➡When disconnecting the cable, some systems need to be initialized after the cable is reconnected. You will need the Techstream diagnostic scan tool or

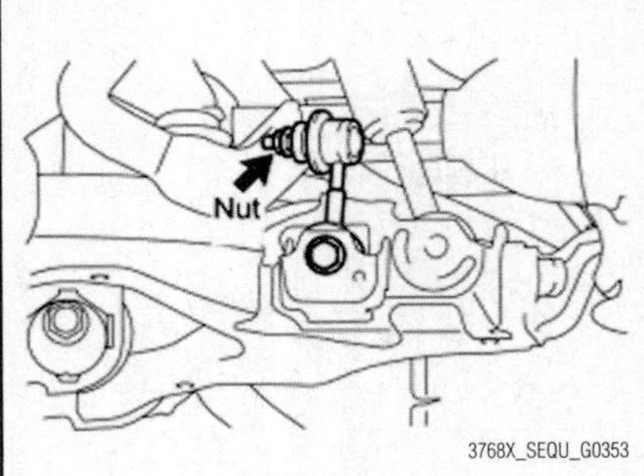

Fig. 437 Front suspension stabilizer link and related components

equivalent. Follow the directions on the tool.

3. Raise and support the vehicle safely.
4. Remove the tire and wheel assembly.
5. Remove the bolt, nut and stabilizer link from its mounting.

To install:

➡Be sure to use new fasteners, as required.

6. Installation is the reverse of the removal procedure.
7. Temporarily install the component.
8. Stabilize the suspension.
9. Tighten the retaining nut to 111 ft. lbs. (150 Nm).
10. Tighten the retaining bolt to 89 ft. lbs. (120 Nm).

LOWER BALL JOINT

REMOVAL & INSTALLATION

1. Before servicing the vehicle, refer to the Precautions Section.

➡If working near and/or around the SRS system and components, be sure to disable the SRS system. Tape the negative battery cable with insulating tape. Always disconnect the negative battery cable first.

⁂ CAUTION

To avoid personal injury when working on vehicles equipped with an air bag, the negative battery cable must be disconnected and at least 90 seconds must elapse before working on the system. Failure to do so may result in deployment of the air bag.

2. Disconnect the negative battery cable. Tape the cable with insulating tape.

➡When disconnecting the cable, some systems need to be initialized after the cable is reconnected. You will need the Techstream diagnostic scan tool or equivalent. Follow the directions on the tool.

3. Remove the lower control arm.
4. Remove the cotter pin and nut. Discard the cotter pin.
5. Using a ball joint removal tool, remove the ball joint.

To install:

➡Be sure to use new fasteners, as required.

6. Installation is the reverse of the removal procedure.
7. Tighten the retaining bolts to 221 ft. lbs. (300 Nm).
8. Be sure to use a new cotter pin.
9. Correct and adjust alignment, as required.

LOWER CONTROL ARM

REMOVAL & INSTALLATION

See Figures 438 and 439.

1. Before servicing the vehicle, refer to the Precautions Section.

➡If working near and/or around the SRS system and components, be sure to disable the SRS system. Tape the negative battery cable with insulating tape. Always disconnect the negative battery cable first.

⁂ CAUTION

To avoid personal injury when working on vehicles equipped with an air bag, the negative battery cable must be disconnected and at least 90 seconds must elapse before working on the system. Failure to do so may result in deployment of the air bag.

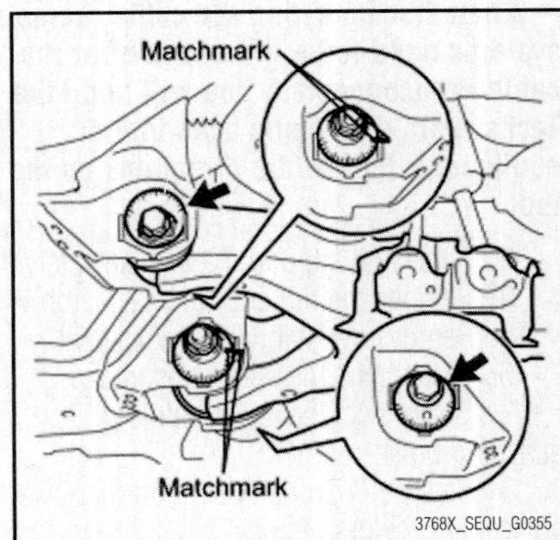

Fig. 438 Front suspension lower control arm matchmark points

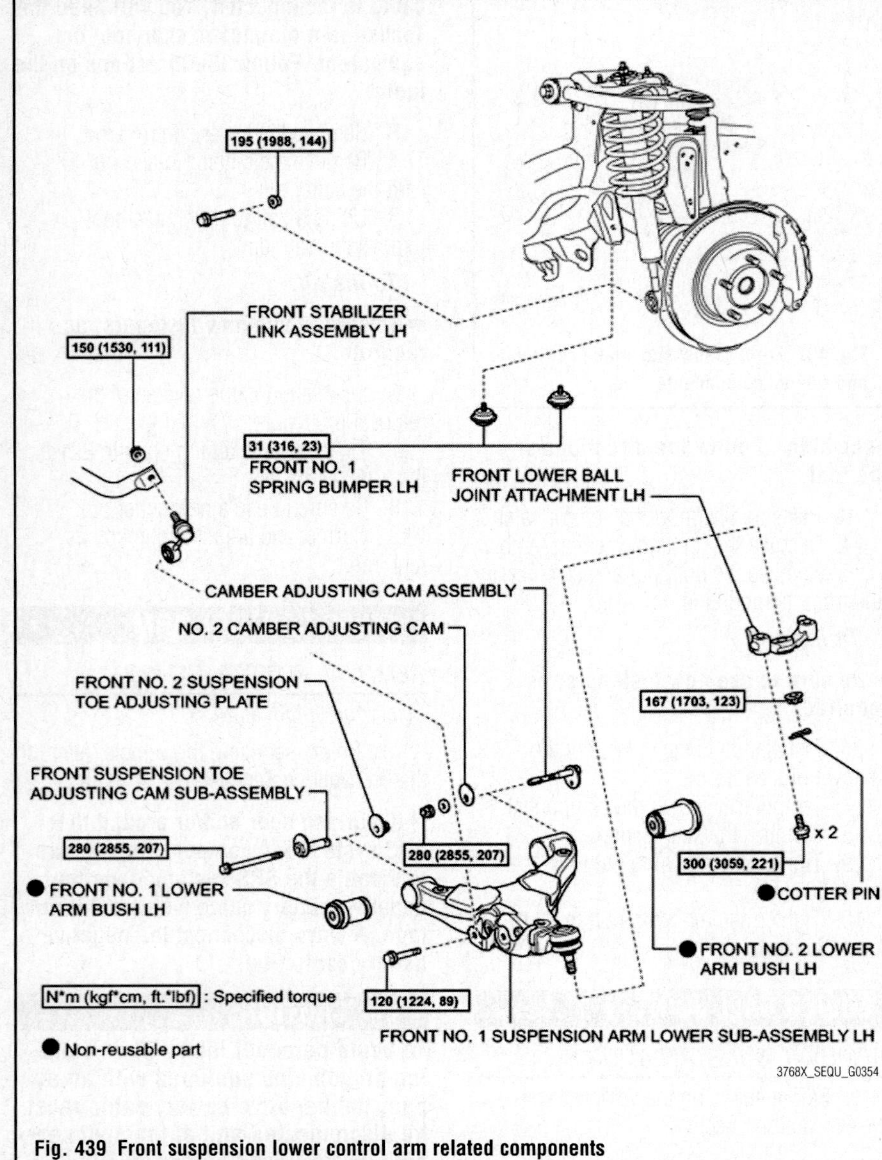

Fig. 439 Front suspension lower control arm related components

2. Disconnect the negative battery cable. Tape the cable with insulating tape.

➡**When disconnecting the cable, some systems need to be initialized after the cable is reconnected. You will need the Techstream diagnostic scan tool or equivalent. Follow the directions on the tool.**

3. Raise and support the vehicle safely.
4. Remove the tire and wheel assembly.
5. Remove the stabilizer link assembly.
6. Disconnect the shock absorber.
7. Disconnect the lower ball joint attaching bolts.
8. Be sure to properly support the lower control arm before removal.
9. Matchmark and remove the No. 1 suspension arm sub assembly.
10. As required, remove the cotter pin and nut from the ball joint. Install the proper ball joint removal tool and remove the ball joint.

To install:

➡**Be sure to use new fasteners, as required.**

11. Installation is the reverse of the removal procedure.
12. Temporarily install the component.
13. Stabilize the suspension.
14. Tighten the No. 1 suspension arm sub assembly bolt and nut to 207 ft. lbs. (280 Nm).
15. Tighten the shock bolt to 144 ft. lbs. (195 Nm).
16. Check and adjust the alignment, as required.

SHOCK ABSORBERS

REMOVAL & INSTALLATION

See Figures 440 through 442.

1. Before servicing the vehicle, refer to the Precautions Section.

➡**If working near and/or around the SRS system and components, be sure to disable the SRS system. Tape the negative battery cable with insulating tape. Always disconnect the negative battery cable first.**

⁂ CAUTION

To avoid personal injury when working on vehicles equipped with an air bag, the negative battery cable must be disconnected and at least 90 seconds must elapse before working on the system. Failure to do so may result in deployment of the air bag.

2. Disconnect the negative battery cable. Tape the cable with insulating tape.

➡**When disconnecting the cable, some systems need to be initialized after the cable is reconnected. You will need the Techstream diagnostic scan tool or equivalent. Follow the directions on the tool.**

3. Raise and support the vehicle safely.
4. Remove the tire and wheel assembly.
5. Remove the shock absorber control actuator, if equipped.
6. Remove the suspension control bracket, if equipped.
7. Remove the lower shock retaining nut.
8. Remove the upper shock retaining bolts.
9. Remove the component from the vehicle.

To install:

➡**Be sure to use new fasteners, as required.**

10. Installation is the reverse of the removal procedure.
11. Temporarily install the component.
12. Stabilize the suspension.
13. Tighten the lower bolt to 144 ft. lbs. (195 Nm).
14. Tighten the upper bolts to 33 ft. lbs. (45 Nm), in the sequence shown in the illustration.

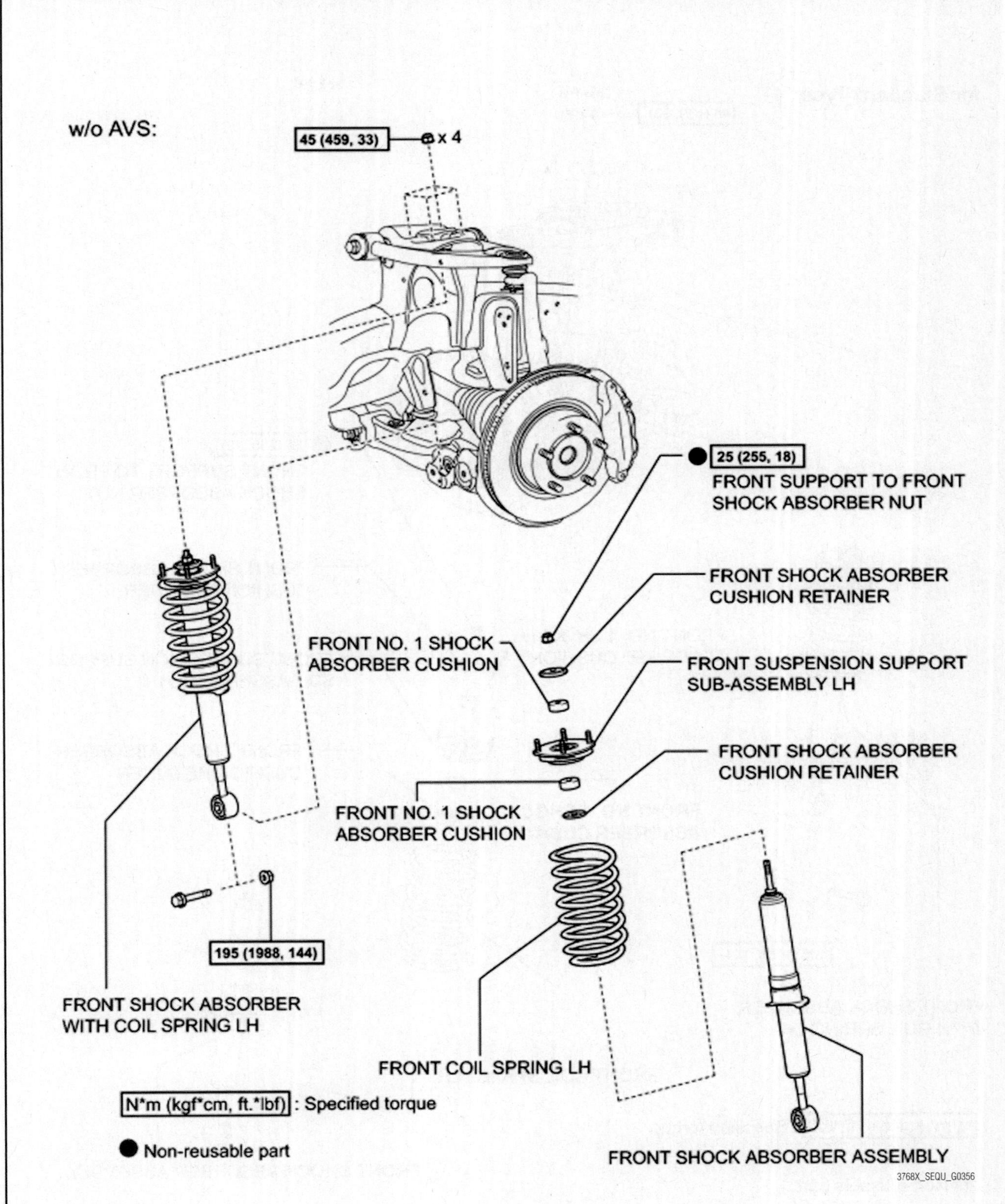

Fig. 440 Front shock absorber and related components—Sequoia without AVS shown; with AVS similar

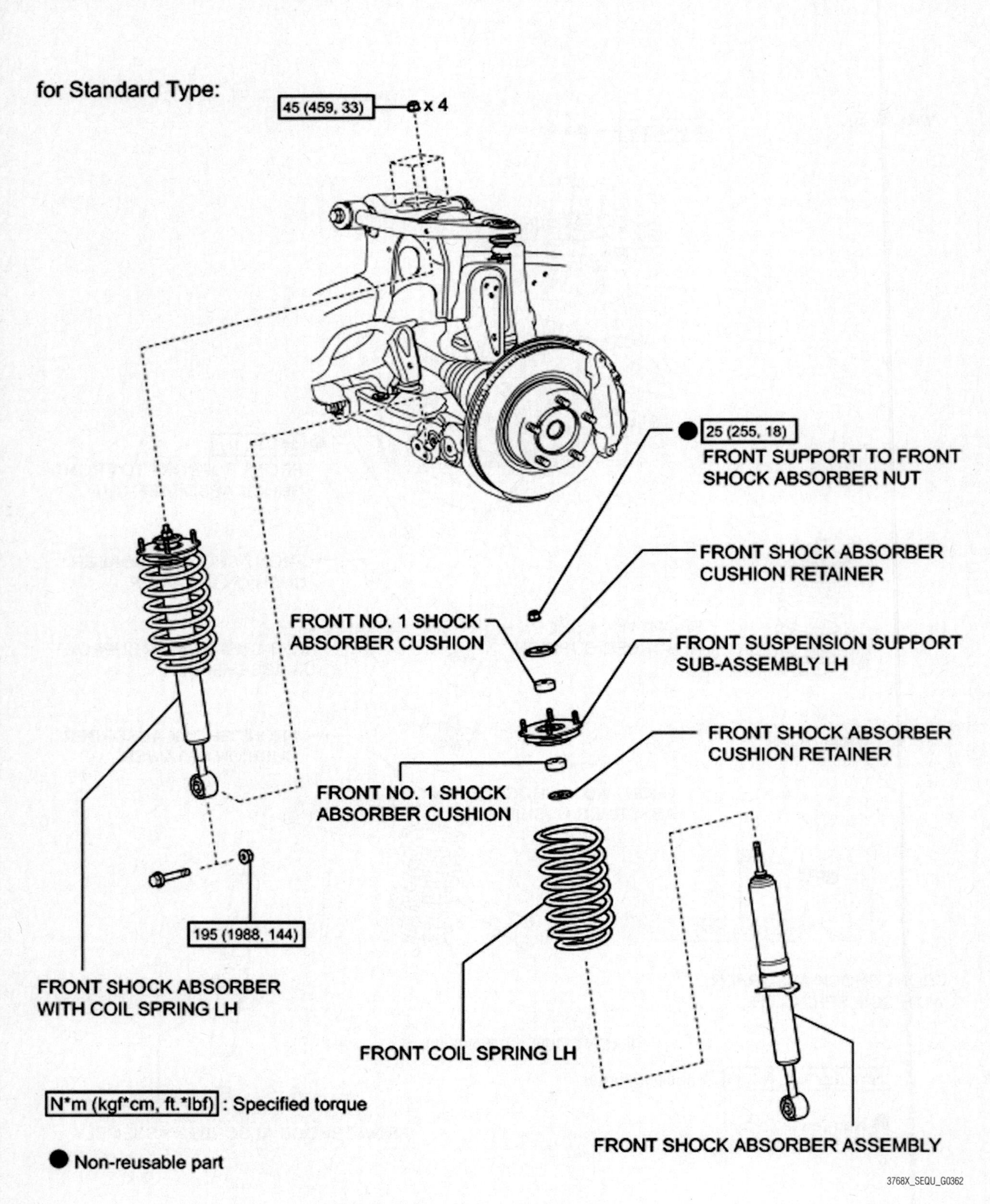

Fig. 441 Front shock absorber and related components—Tundra with standard package shown; off road package similar

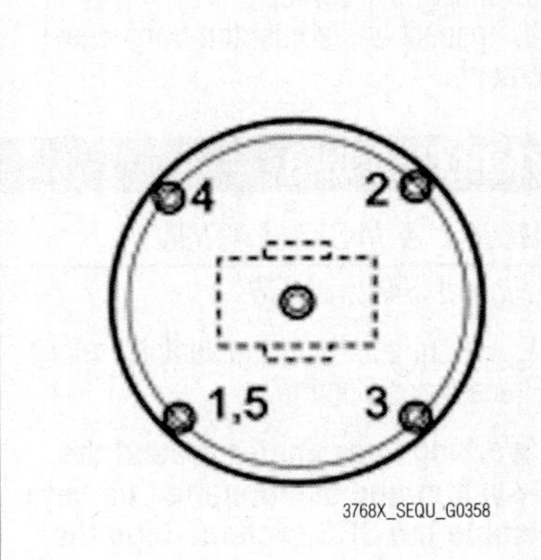

Fig. 442 Front shock absorber upper bolt tightening sequence

STEERING KNUCKLE

REMOVAL & INSTALLATION

See Figures 443 and 444.

1. Before servicing the vehicle, refer to the Precautions Section.

➡If working near and/or around the SRS system and components, be sure to disable the SRS system. Tape the negative battery cable with insulating tape. Always isconnect the negative battery cable first.

⁂ CAUTION

To avoid personal injury when working on vehicles equipped with an air bag, the negative battery cable must be disconnected and at least 90 seconds must elapse before working on the system. Failure to do so may result in deployment of the air bag.

2. Disconnect the negative battery cable. Tape the cable with insulating tape.

➡When disconnecting the cable, some systems need to be initialized after the cable is reconnected. You will need the Techstream diagnostic scan tool or equivalent. Follow the directions on the tool.

3. Remove the front axle hub.
4. Disconnect the speed sensor from the steering knuckle.
5. Disconnect tie rod end sub-assembly.

FRONT SUSPENSION UPPER ARM ASSEMBLY LH
FRONT SPEED SENSOR LH
29 (296, 21)
x2
99 (1010, 73)
13 (127, 9)
13 (127, 9)
FRONT DISC BRAKE CALIPER ASSEMBLY LH
110 (1122, 81)
CLIP
TIE ROD END SUB-ASSEMBLY LH
COTTER PIN
11 (107, 8)
x2
69 (704, 51)
FRONT LOWER BALL JOINT ATTACHMENT LH
300 (3059, 221)
FRONT AXLE HUB SUB-ASSEMBLY LH
FRONT DISC
for 4WD:
338 (3446, 249)
FRONT AXLE HUB NUT
FRONT WHEEL ADJUSTING LOCK CAP
COTTER PIN
FRONT AXLE HUB GREASE CAP
N*m (kgf*cm, ft.*lbf) : Specified torque
● Non-reusable part
Do not apply lubricants to the threaded parts
3768X_SEQU_G0364

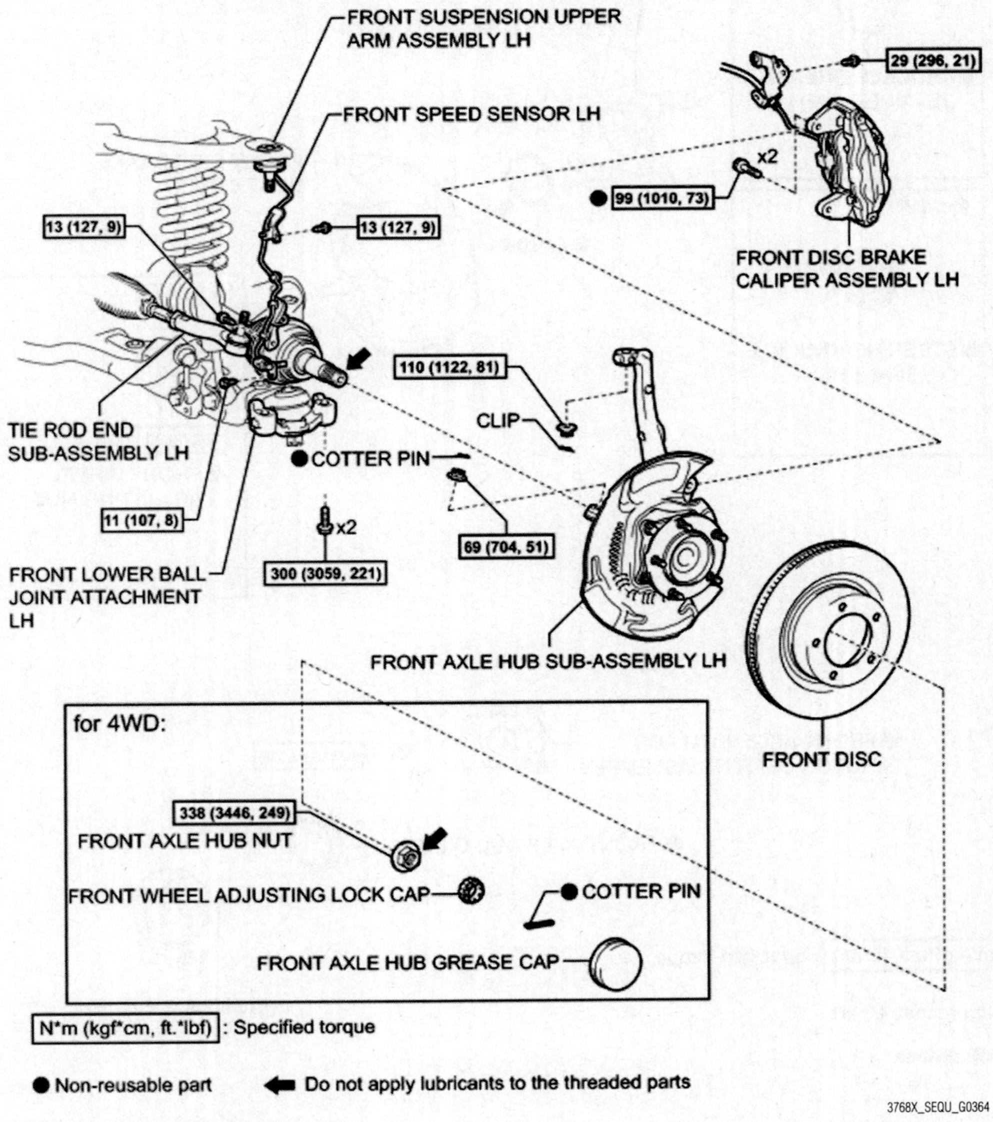

Fig. 443 Front steering knuckle and related components 1 of 2

a. Remove the cotter pin and nut.

b. Disconnect the tie rod end LH from the steering knuckle.

6. Disconnect front lower ball joint attachment.

7. Support the front suspension lower arm with a jack.

8. Remove the clip and the nut.

9. Disconnect the upper ball joint from the steering knuckle.

10. Remove the steering knuckle.

To install:

➡Be sure to use new fasteners, as required.

11. Install the front suspension upper arm to the steering knuckle with the nut and tighten to specification.

12. Install a new clip.

➡If the holes for the clip are not aligned, tighten the nut up to another 60°.

13. Connect front lower ball joint attachment.

14. Connect tie rod end sub-assembly.

15. Install the front axle hub and tighten nut to specification.

16. Connect the front speed sensor to the steering knuckle.

17. Attach the clip to the steering knuckle.

18. Install front wheel.

19. Inspect and adjust the front wheel alignment.

STABILIZER BAR

REMOVAL & INSTALLATION

See Figures 445 and 446.

1. Before servicing the vehicle, refer to the Precautions Section.

➡If working near and/or around the SRS system and components, be sure to disable the SRS system. Tape the

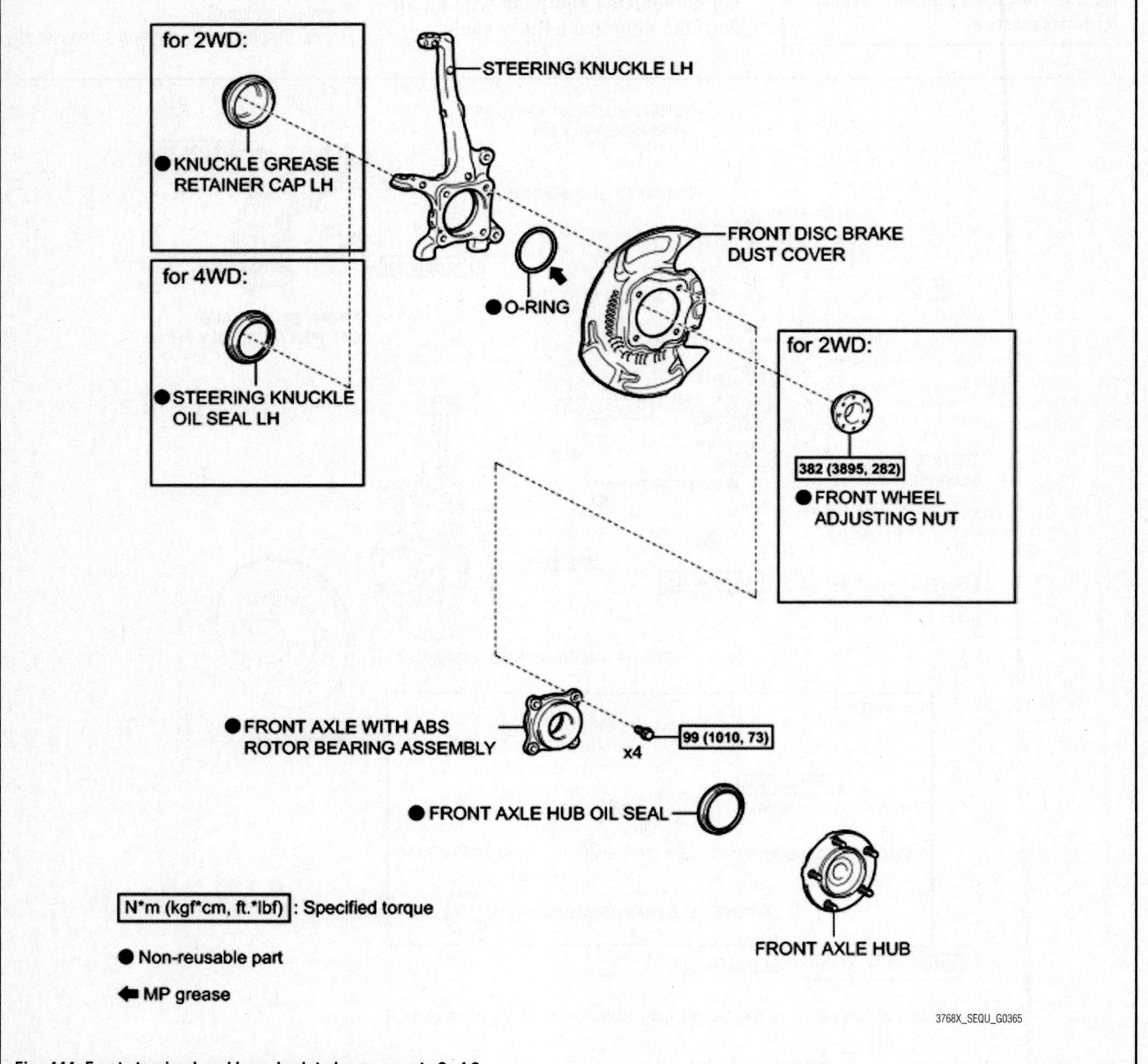

Fig. 444 Front steering knuckle and related components 2 of 2

negative battery cable with insulating tape. Always disconnect the negative battery cable first.

CAUTION

To avoid personal injury when working on vehicles equipped with an air bag, the negative battery cable must be disconnected and at least 90 seconds must elapse before working on the system. Failure to do so may result in deployment of the air bag.

2. Disconnect the negative battery cable. Tape the cable with insulating tape.

➡When disconnecting the cable, some systems need to be initialized after the cable is reconnected. You will need the Techstream diagnostic scan tool or equivalent. Follow the directions on the tool.

3. Remove the nut and disconnect the stabilizer bar links from the lower suspension arm.

➡If the ball joint turns together with the nut, use a hexagon (6 mm) wrench to hold the stud.

4. Remove the 2 bolts, nuts and stabilizer bar with the cushions and brackets.
5. Remove the 2 brackets and cushions from the stabilizer bar.
6. Hold the stabilizer bar link, and remove the nut.
7. Remove the stabilizer bar link, 2 retainers and bushings from each end of the stabilizer bar.

To install:

➡Be sure to use new fasteners, as required.

8. Temporarily install the component.
9. Stabilize the suspension.
10. Install the 2 bushings, retainers and stabilizer bar link to the stabilizer bar.
11. Hold the stabilizer bar link, and install a new nut and tighten to 14 ft. lbs. (19 Nm).
12. Install the 2 bushings with their cutout facing to the rearward of the stabilizer bar.

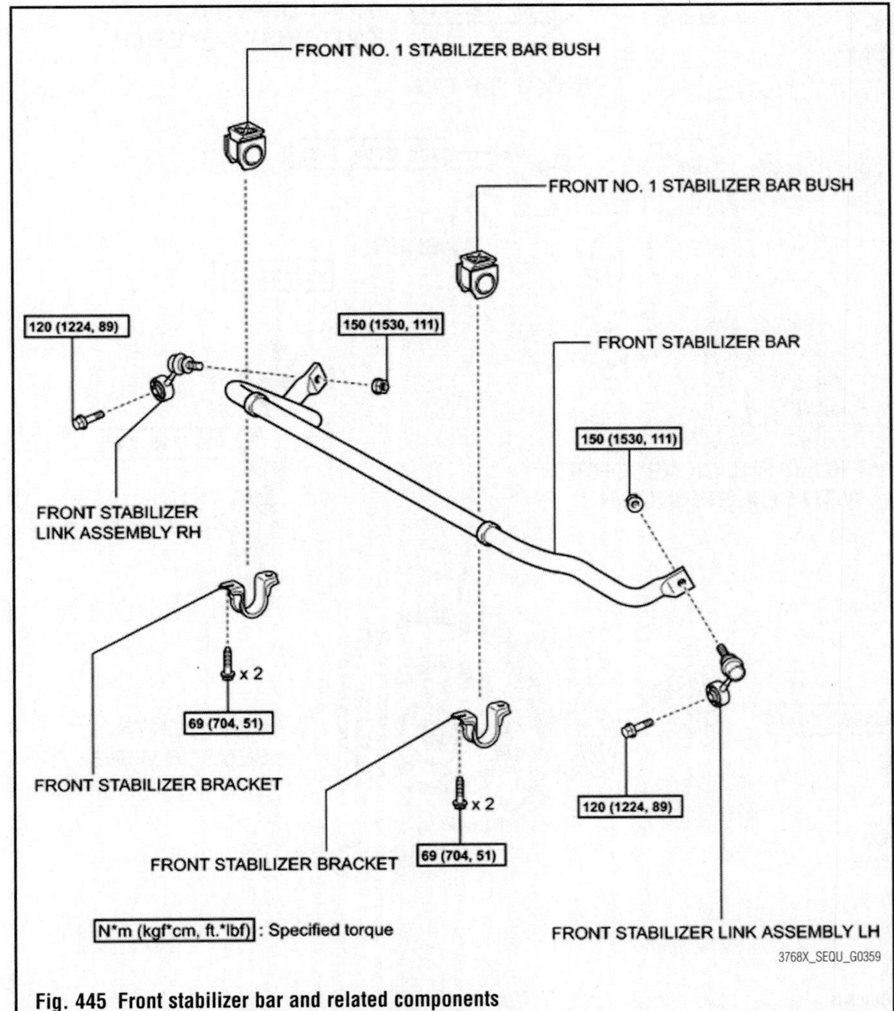

Fig. 445 Front stabilizer bar and related components

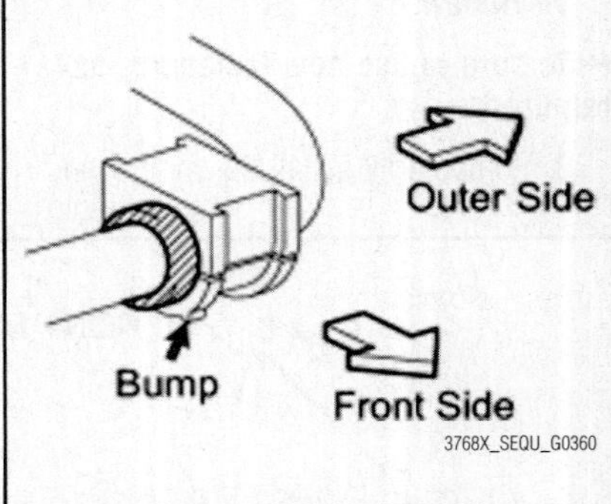

Fig. 446 Front stabilizer bar bushing positioning

13. Install the stabilizer bar and 2 brackets with the nuts and bolts and tighten to 27 ft. lbs. (37 Nm).
14. Connect front stabilizer links to suspension lower arm and tighten to 51 ft. lbs. (69 Nm).

➡If the ball joint turns together with the nut, use a hexagon (6 mm) wrench to hold the nut.

UPPER BALL JOINT

REMOVAL & INSTALLATION

1. Before servicing the vehicle, refer to the Precautions Section.

➡If working near and/or around the SRS system and components, be sure to disable the SRS system. Tape the negative battery cable with insulating tape. Always disconnect the negative battery cable first.

CAUTION

To avoid personal injury when working on vehicles equipped with an air bag, the negative battery cable must be disconnected and at least 90 seconds must elapse before working on the system. Failure to do so may result in deployment of the air bag.

2. Disconnect the negative battery cable. Tape the cable with insulating tape.

➡When disconnecting the cable, some systems need to be initialized after the cable is reconnected. You will need the Techstream diagnostic scan tool or equivalent. Follow the directions on the tool.

3. Remove the lower control arm.
4. Using the proper removal tools, separate the ball joint from the control arm.

To install:

➡Be sure to use new fasteners, as required.

5. Temporarily install the component.
6. Stabilize the suspension.
7. Tighten the retaining bolts to specification.
8. Check and adjust the alignment, as required.

UPPER CONTROL ARM

REMOVAL & INSTALLATION

See Figure 447.

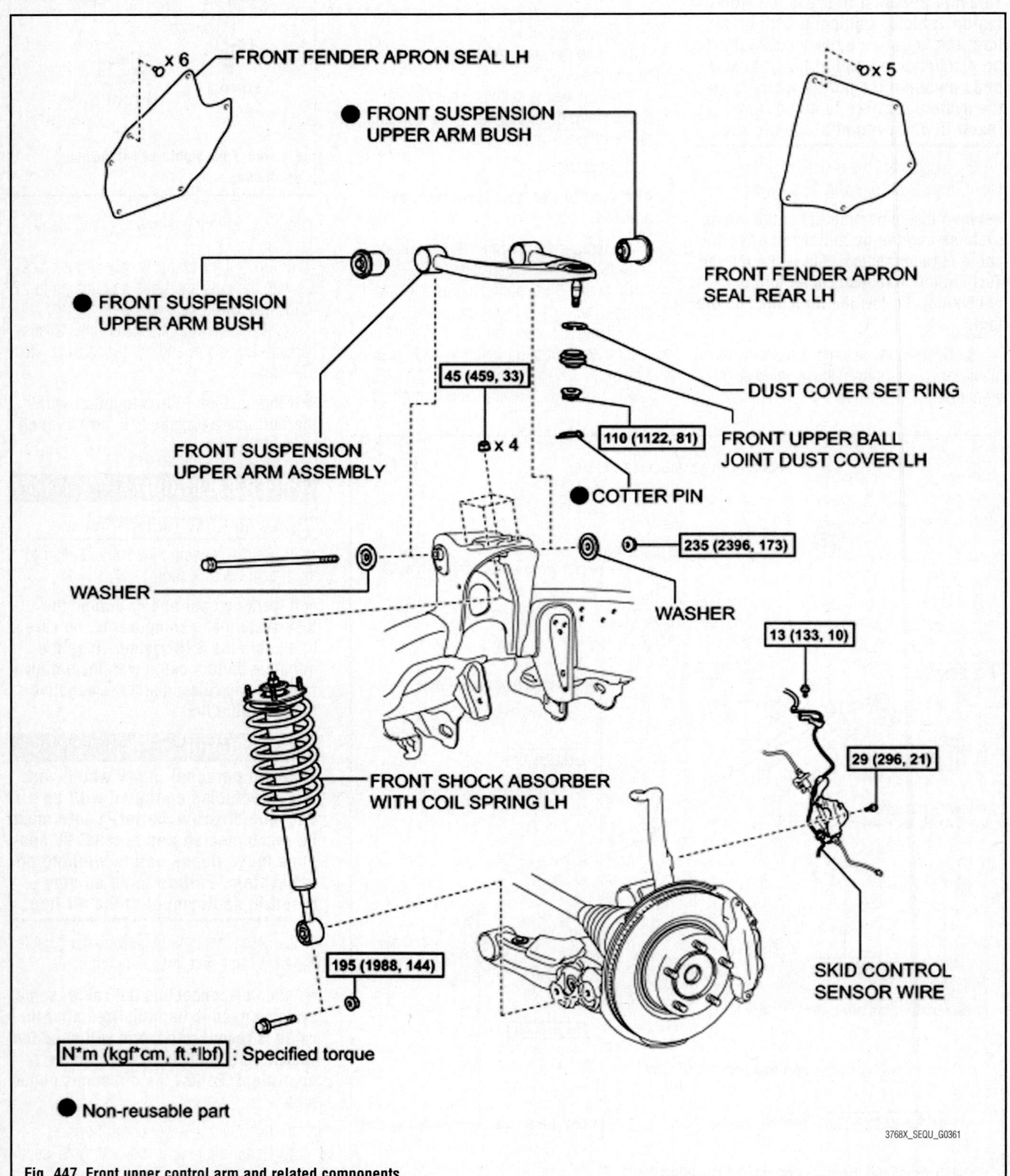

Fig. 447 Front upper control arm and related components

1. Before servicing the vehicle, refer to the Precautions Section.

➡**If working near and/or around the SRS system and components, be sure to disable the SRS system. Tape the negative battery cable with insulating tape. Always disconnect the negative battery cable first.**

⁂ CAUTION

To avoid personal injury when working on vehicles equipped with an air bag, the negative battery cable must be disconnected and at least 90 seconds must elapse before working on the system. Failure to do so may result in deployment of the air bag.

2. Disconnect the negative battery cable. Tape the cable with insulating tape.

➡**When disconnecting the cable, some systems need to be initialized after the cable is reconnected. You will need the Techstream diagnostic scan tool or equivalent. Follow the directions on the tool.**

3. Raise and support the vehicle safely.
4. Remove the tire and wheel assembly.
5. Remove the shock absorber.
6. Disconnect the skid control sensor wire.
7. Remove the upper ball joint cotter pin and nut. Discard the cotter pin.
8. Disconnect the upper ball joint from the steering knuckle. Be sure to properly support the lower control arm assembly.
9. Remove the fender apron seals.
10. Remove the component retaining bolts.
11. Remove the component from its mounting.

To install:

➡**Be sure to use new fasteners, as required.**

12. Temporarily install the component.
13. Stabilize the suspension.
14. Tighten all bolts to specification, refer to illustration.
15. Check and adjust the alignment, as required.

WHEEL HUB & BEARING

REMOVAL & INSTALLATION

1. Before servicing the vehicle, refer to the Precautions Section.

➡**If working near and/or around the SRS system and components, be sure to disable the SRS system. Tape the negative battery cable with insulating tape. Always disconnect the negative battery cable first.**

⁂ CAUTION

To avoid personal injury when working on vehicles equipped with an air bag, the negative battery cable must be disconnected and at least 90 seconds must elapse before working on the system. Failure to do so may result in deployment of the air bag.

2. Disconnect the negative battery cable. Tape the cable with insulating tape.

➡**When disconnecting the cable, some systems need to be initialized after the cable is reconnected. You will need the Techstream diagnostic scan tool or equivalent. Follow the directions on the tool.**

3. Raise and support the vehicle safely. Remove the tire and wheel assembly.
4. Remove the bolt and disconnect the brake tube bracket from the steering knuckle.
5. Remove the 2 bolts and disconnect the disc brake caliper from the steering knuckle.
6. Remove front disc.
7. Remove front axle hub grease cap (4WD).
   a. Remove the cotter pin and adjusting lock cap.
   b. Using a 39 mm socket wrench, remove the axle hub nut.
8. Remove front axle hub sub-assembly (4WD).
   a. Remove the 4 bolts.
   b. Using a plastic-faced hammer, tap out the front drive shaft from the front axle hub.
   c. Remove the O-ring from the axle hub.
9. Remove front axle hub sub-assembly (2WD).
   a. Remove the 4 bolts.
   b. Remove the axle hub and dust cover from the steering knuckle.
   c. Remove the O-ring from the axle hub.

To install:

➡**Be sure to use new fasteners, as required.**

10. Install front axle hub sub-assembly (2WD).
    a. Apply MP grease to a new O-ring.
    b. Install the O-ring to the axle hub.
    c. Install the dust cover and axle hub to the steering knuckle with the 4 bolts and tighten to 73 ft. lbs. (99 Nm).
11. Install front axle hub sub-assembly (4WD).
    a. Apply MP grease to a new O-ring.
    b. Install the O-ring to the axle hub.
    c. Connect the front drive shaft to the front axle hub.
    d. Install the dust cover and axle hub to the steering knuckle with the 4 bolts and tighten to 73 ft. lbs. (99 Nm).
12. Install front axle hub nut (4WD).
    a. Clean the threaded parts on the drive shaft and axle hub nut using a non-residue solvent.
    b. Using a 39 mm socket wrench, install the hub nut and tighten to 249 ft. lbs. (338 Nm).
    c. Install the front wheel adjusting lock cap and a new cotter pin.
13. Inspect the front axle hub.
14. Install the axle hub grease cap.
15. Install front disc.
16. Connect the front disc brake caliper and install 2 new bolts and tighten to 73 ft. lbs. (99 Nm).
17. Connect the brake tube bracket to the steering knuckle with the bolt and tighten to 21 ft. lbs. (29 Nm).
18. Install front wheel.
19. Check the speed sensor signal.

COIL SPRING

REMOVAL & INSTALLATION

Sequoia

See Figures 448 through 450.

1. Before servicing the vehicle, refer to the Precautions Section.

➡If working near and/or around the SRS system and components, be sure to disable the SRS system. Tape the negative battery cable with insulating tape. Always disconnect the negative battery cable first.

⁂ CAUTION

To avoid personal injury when working on vehicles equipped with an air bag, the negative battery cable must be disconnected and at least 90 seconds must elapse before working on the system. Failure to do so may result in deployment of the air bag.

2. Disconnect the negative battery cable. Tape the cable with insulating tape.

➡When disconnecting the cable, some systems need to be initialized after the cable is reconnected. You will need the Techstream diagnostic scan tool or equivalent. Follow the directions on the tool.

3. Raise and support the vehicle safely.
4. Remove the shock absorber.
5. Loosen the three upper control arm nuts. Do not remove them.
6. Insert a piece of wood between the jack and the No. 2 lower arm and raise the arm.
7. Using a spring compressor tool, compress the spring.

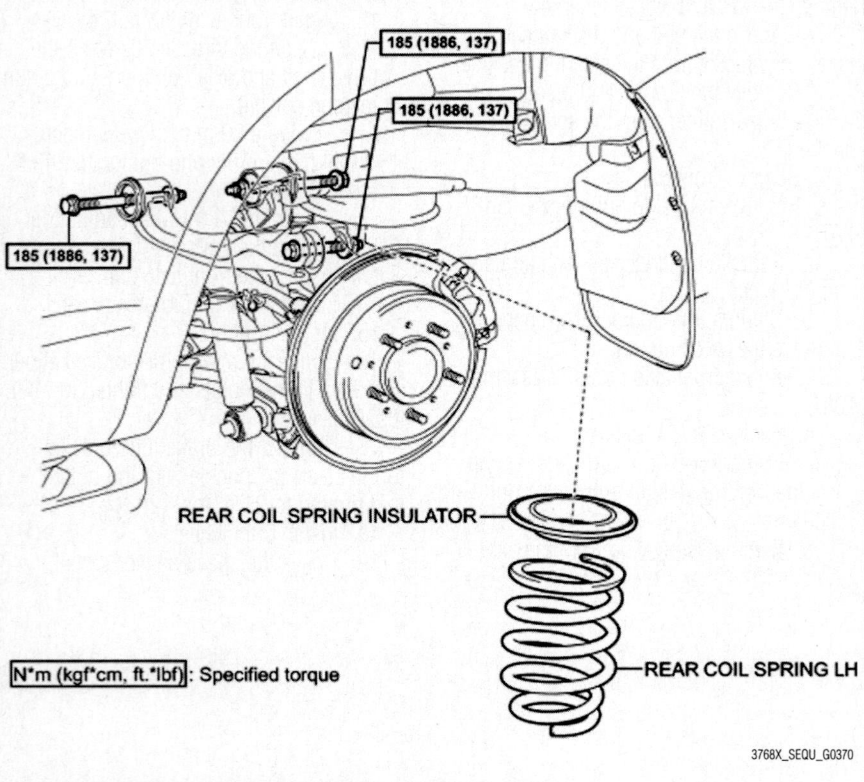

Fig. 448 Rear coil spring and related components—Sequoia

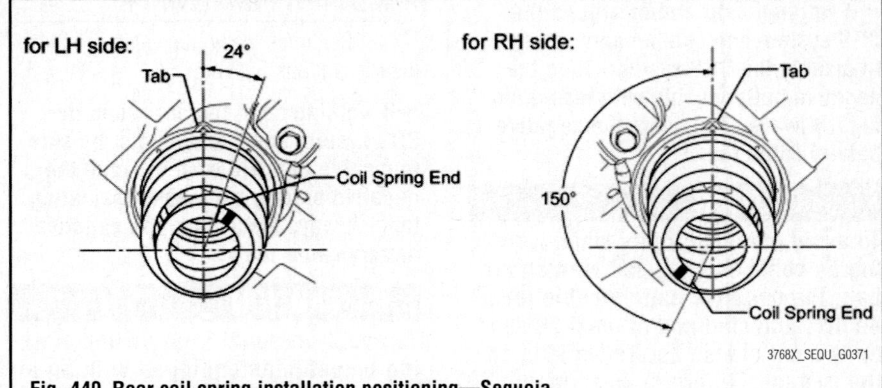

Fig. 449 Rear coil spring installation positioning—Sequoia

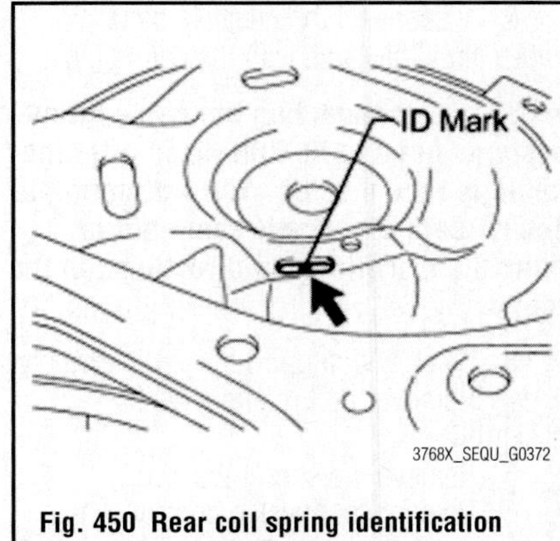

Fig. 450 Rear coil spring identification mark alignment—Sequoia

8. Gently lower the jack and remove the component from its mounting.

To install:

➡Be sure to use new fasteners, as required.

9. Installation is the reverse of the removal procedure.
10. Be sure that the spring is properly installed. Refer to illustration for proper alignment.
11. Be sure that the spring identification mark is visible thru the drain hole of the No. 2 lower arm.
12. Check and adjust alignment, as required.

CONTROL LINKS

REMOVAL & INSTALLATION

Sequoia

1. Before servicing the vehicle, refer to the Precautions Section.

➡If working near and/or around the SRS system and components, be sure

to disable the SRS system. Tape the negative battery cable with insulating tape. Always disconnect the negative battery cable first.

⁂ CAUTION

To avoid personal injury when working on vehicles equipped with an air bag, the negative battery cable must be disconnected and at least 90 seconds must elapse before working on the system. Failure to do so may result in deployment of the air bag.

2. Disconnect the negative battery cable. Tape the cable with insulating tape.

➡When disconnecting the cable, some systems need to be initialized after the cable is reconnected. You will need the Techstream diagnostic scan tool or equivalent. Follow the directions on the tool.

3. Raise and safely support the vehicle safely.
4. Remove the tire and wheel assemblies.
5. Remove the tail exhaust pipe assembly.
6. Remove the driveshaft.
7. Remove the stabilizer link retaining bolts. Remove the stabilizer link.

To install:

➡Be sure to use new fasteners, as required.

8. Installation is the reverse of the removal procedure.
9. Temporarily install the component.
10. Stabilize the suspension.

LEAF SPRING

REMOVAL & INSTALLATION

Tundra

See Figure 451.

1. Before servicing the vehicle, refer to the Precautions Section.

➡If working near and/or around the SRS system and components, be sure to disable the SRS system. Tape the negative battery cable with insulating tape. Always disconnect the negative battery cable first.

⁂ CAUTION

To avoid personal injury when working on vehicles equipped with an air bag, the negative battery cable must be disconnected and at least 90 seconds must elapse before working on the system. Failure to do so may result in deployment of the air bag.

2. Disconnect the negative battery cable. Tape the cable with insulating tape.

➡When disconnecting the cable, some systems need to be initialized after the cable is reconnected. You will need the Techstream diagnostic scan tool or equivalent. Follow the directions on the tool.

3. Raise and safely support the vehicle safely.
4. Remove the tire and wheel assemblies.
5. Lower the axle housing until the leaf spring tension is free. Keep it at this position.
6. Remove the spare tire.
7. Disconnect the No. 3 parking brake assembly.
8. Support the rear axle housing with a jack and wooden block. Remove the bolt and nut and disconnect the shock absorber assembly from the axle housing.
9. Remove the U bolt nuts and washers. Remove the U bolt.
10. Remove the spring bumper nut and washer and bolt from the front side.
11. Remove the nuts, washers and shackle outer plate from the rear side.
12. Remove the component from the vehicle.

To install:

➡Be sure to use new fasteners, as required.

13. Installation is the reverse of the removal procedure.
14. Temporarily install the component.
15. When installing the rear spring bumper sub assembly, align the leaf spring center bolt and nut with the installation hole of the rear axle housing and spring bumper so that they are not loose.
16. Stabilize the suspension.

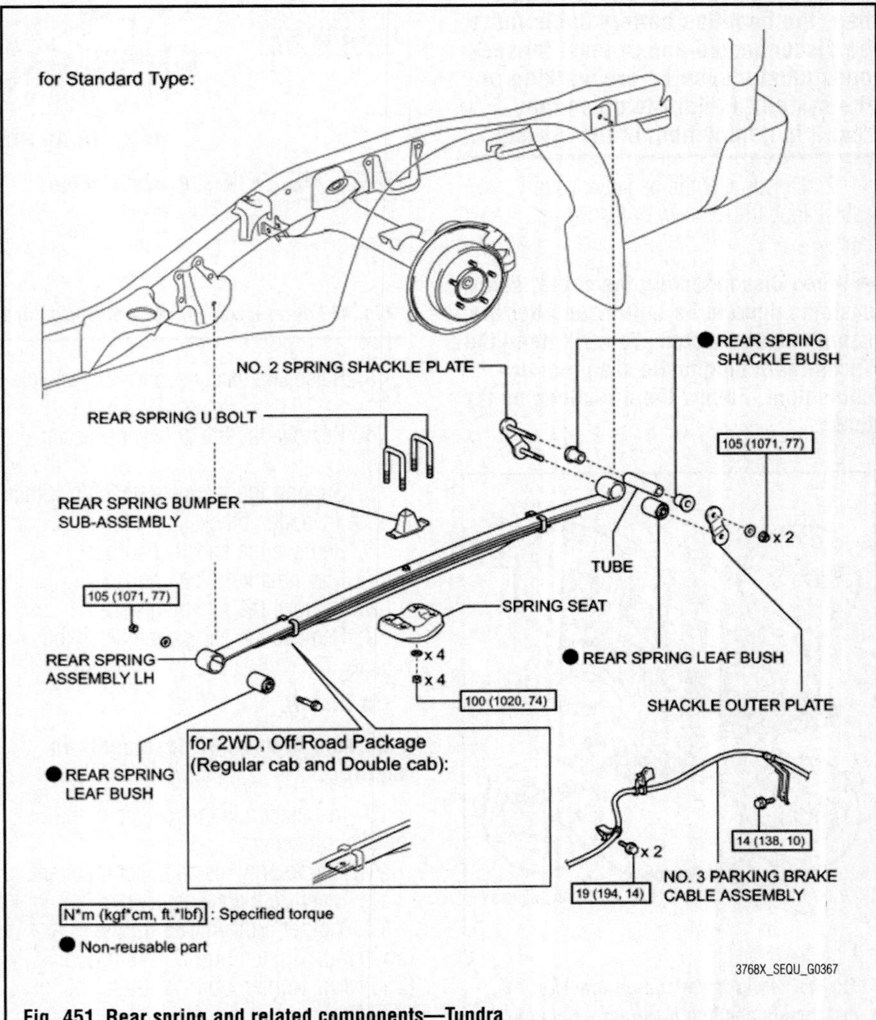

Fig. 451 Rear spring and related components—Tundra

17. When tightening the U bolts, tighten them in a criss cross pattern.
18. Tighten all bolts to specification, refer to illustration.

LOWER CONTROL ARMS

REMOVAL & INSTALLATION

Sequoia

No. 1 Arm

See Figures 452 and 453.

1. Before servicing the vehicle, refer to the Precautions Section.

➡If working near and/or around the SRS system and components, be sure to disable the SRS system. Tape the negative battery cable with insulating tape. Always disconnect the negative battery cable first.

⁂ CAUTION

To avoid personal injury when working on vehicles equipped with an air bag, the negative battery cable must be disconnected and at least 90 seconds must elapse before working on the system. Failure to do so may result in deployment of the air bag.

2. Disconnect the negative battery cable. Tape the cable with insulating tape.

➡When disconnecting the cable, some systems need to be initialized after the cable is reconnected. You will need the Techstream diagnostic scan tool or equivalent. Follow the directions on the tool.

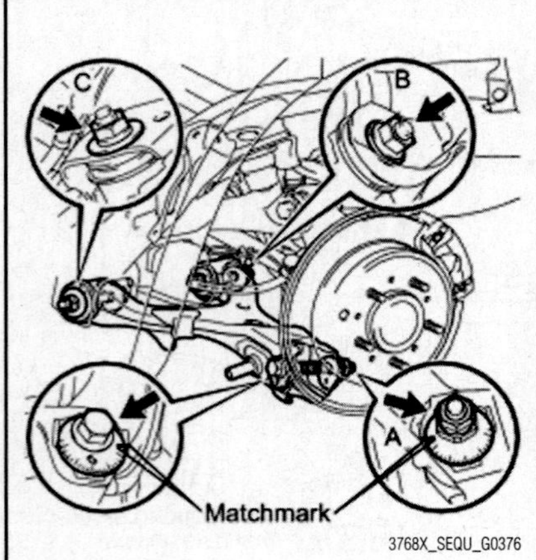

Fig. 452 Rear lower control arm (No. 1) matchmark and bolt locations—Sequoia

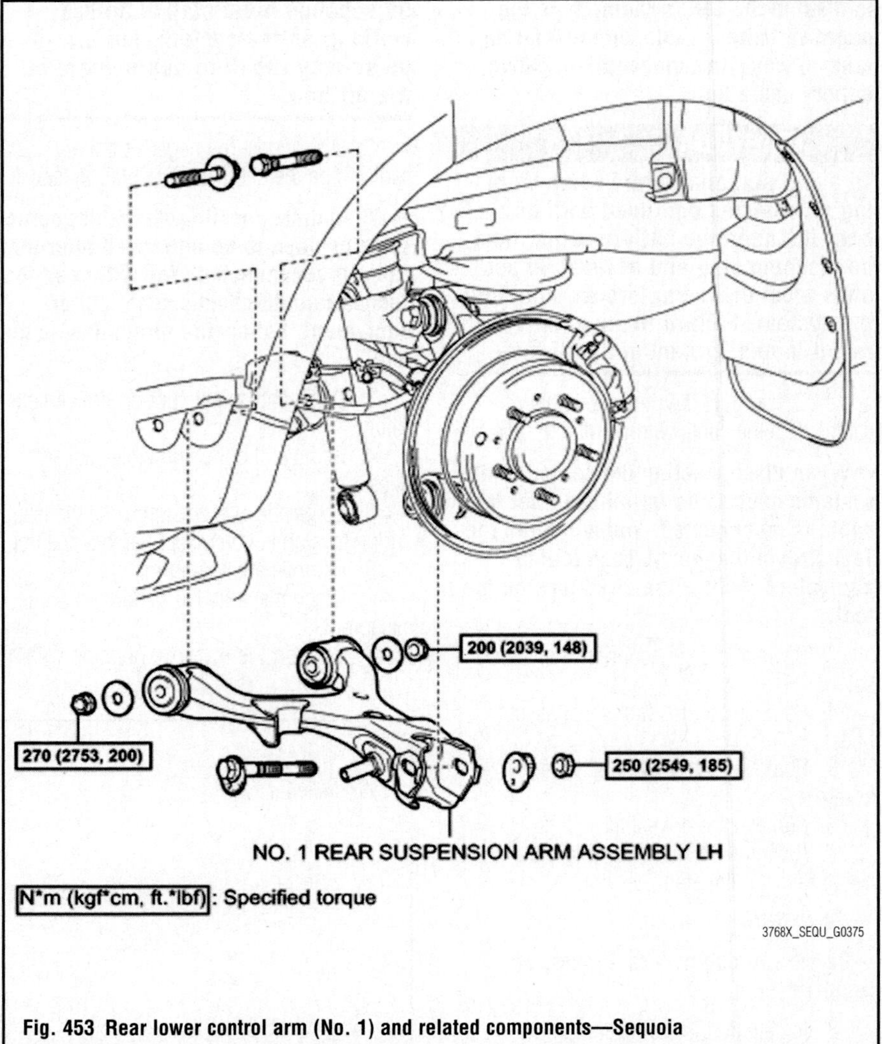

Fig. 453 Rear lower control arm (No. 1) and related components—Sequoia

3. Raise and safely support the vehicle safely.
4. Remove the tire and wheel assemblies.
5. Support the rear axle housing with a jack and wooden block.
6. Remove the shock absorber.
7. Matchmark the component.
8. Remove the retaining bolts.
9. Remove the component from the vehicle.

To install:

➡Be sure to use new fasteners, as required.

10. Installation is the reverse of the removal procedure.
11. Temporarily install the component.
12. Stabilize the suspension.
13. Tighten Bolt A: 185 ft. lbs. (250 Nm). Tighten bolt B: 148 ft. lbs. (200 Nm). Tighten bolt C: 200 ft. lbs. (270 Nm).
14. Check and adjust alignment, as required.

No. 2 Arm

See Figures 454 and 455.

1. Before servicing the vehicle, refer to the Precautions Section.

➡If working near and/or around the SRS system and components, be sure to disable the SRS system. Tape the negative battery cable with insulating tape. Always disconnect the negative battery cable first.

⁂ CAUTION

To avoid personal injury when working on vehicles equipped with an air bag, the negative battery cable must be disconnected and at least 90 seconds must elapse before working on the system. Failure to do so may result in deployment of the air bag.

2. Disconnect the negative battery cable. Tape the cable with insulating tape.

➡When disconnecting the cable, some systems need to be initialized after the cable is reconnected. You will need the Techstream diagnostic scan tool or equivalent. Follow the directions on the tool.

3. Raise and safely support the vehicle safely.
4. Remove the tire and wheel assemblies.
5. Support the rear axle housing with a jack and wooden block.
6. Remove the spring.
7. Remove the pneumatic cylinder, if equipped.
8. Remove the cotter pin and nut. Discard the cotter pin.
9. Using the proper removal tool, separate the rear suspension arm.
10. Matchmark the component.
11. Remove the retaining bolts.
12. Remove the component from the vehicle.

To install:

➡Be sure to use new fasteners, as required.

13. Installation is the reverse of the removal procedure.
14. Temporarily install the component.
15. Stabilize the suspension.
16. Tighten the joint retaining nut to specification.
17. Tighten the control arm retaining bolts to 89 ft. lbs. (120 Nm).
18. Check and adjust alignment, as required.

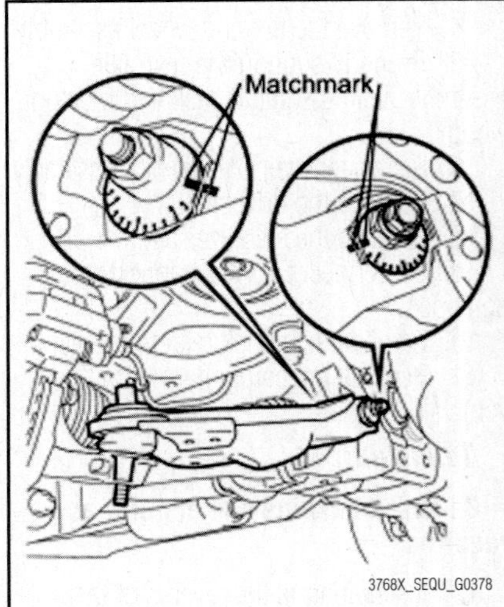

Fig. 454 Rear lower control arm (No. 2) matchmark locations—Sequoia

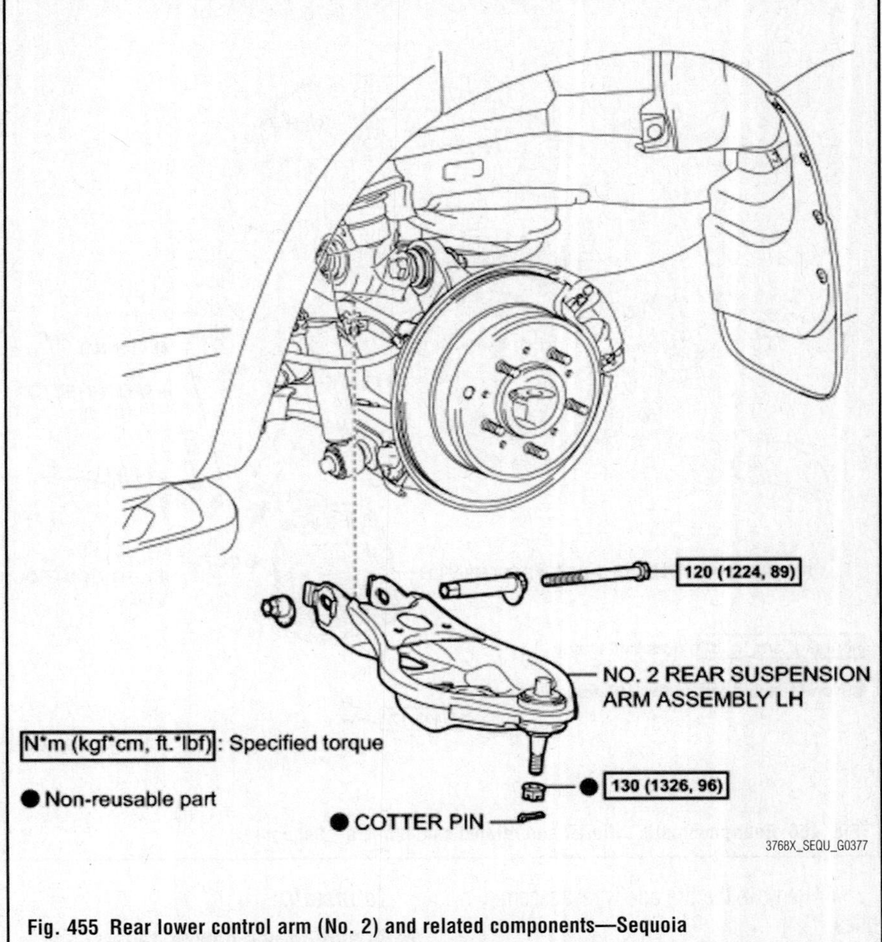

Fig. 455 Rear lower control arm (No. 2) and related components—Sequoia

PNEUMATIC CYLINDER

REMOVAL & INSTALLATION

Sequoia

See Figures 456 through 458.

1. Before servicing the vehicle, refer to the Precautions Section.

➡If working near and/or around the SRS system and components, be sure to disable the SRS system. Tape the negative battery cable with insulating tape. Always disconnect the negative battery cable first.

CAUTION

To avoid personal injury when working on vehicles equipped with an air bag, the negative battery cable must be disconnected and at least 90 seconds must elapse before working on the system. Failure to do so may result in deployment of the air bag.

2. Disconnect the negative battery cable. Tape the cable with insulating tape.

➡When disconnecting the cable, some systems need to be initialized after the cable is reconnected. You will need the Techstream diagnostic scan tool or equivalent. Follow the directions on the tool.

➡Do not extend the cylinder during or after removal. Do not put air in the cylinder after removal. Do not remove the cylinder without first releasing the air pressure. Do not remove the bolt from the bottom of the cylinder, if the cylinder is not filled with air.

➡When the cylinder is fitted on the vehicle, do not remove the absorber. Doing so will cause the No.2 control arm to move beyond its normal stroke in the rebound direction.

➡When performing this procedure be sure to support the vehicle frame and maintain the No. 2 lower arm height at a position lower than the normal vehicle height.

3. Raise and safely support the vehicle safely.

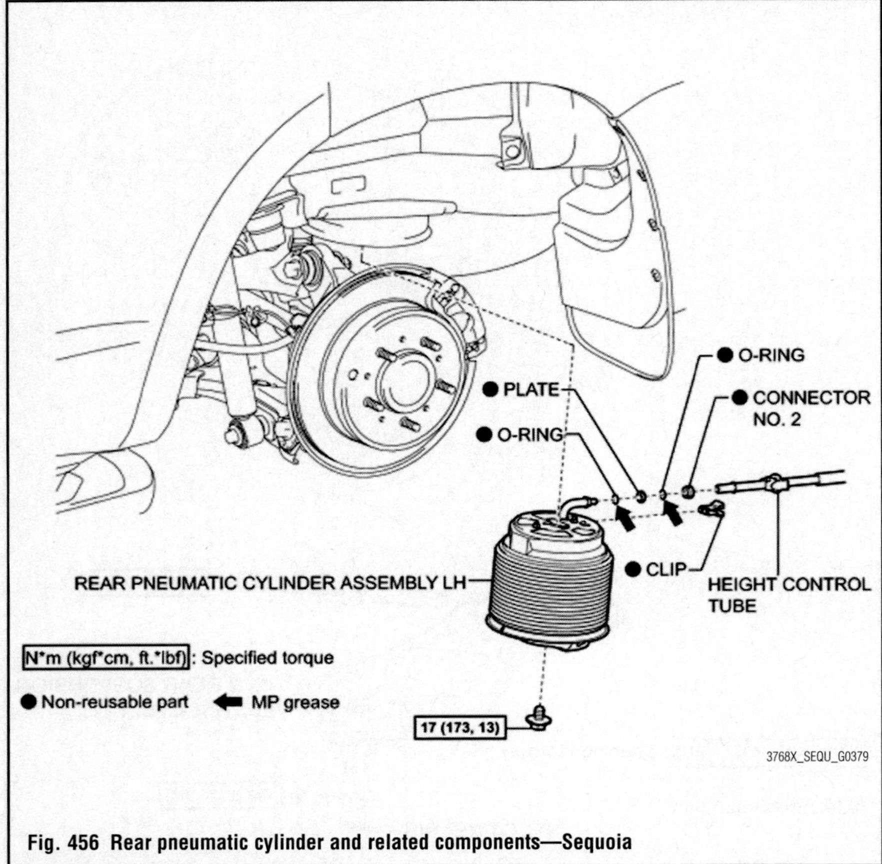

Fig. 456 Rear pneumatic cylinder and related components—Sequoia

4. Remove the tire and wheel assemblies.
5. Insert a piece of wood between the jack and the No.2 lower arm and raise the arm.
6. Remove the bolt. Remove the clip.
7. Disconnect the height control tube.
8. Gently lower the jack and remove the component.

To install:

➡Be sure to use new fasteners, as required.

9. Installation is the reverse of the removal procedure.
10. Align the cylinder, see illustration.
11. lower the lift that is supporting the frame to allow the tire to touch the floor.

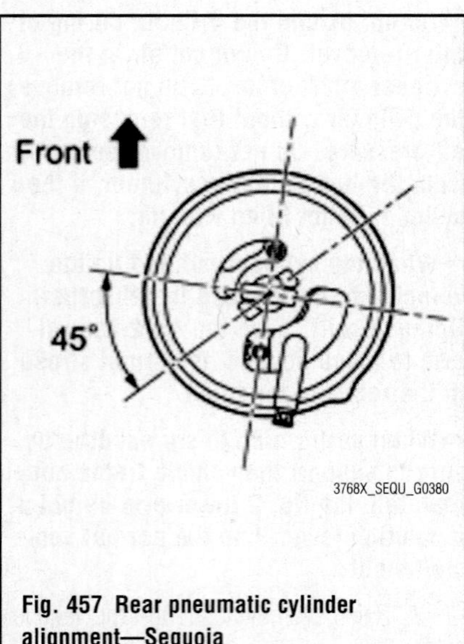

Fig. 457 Rear pneumatic cylinder alignment—Sequoia

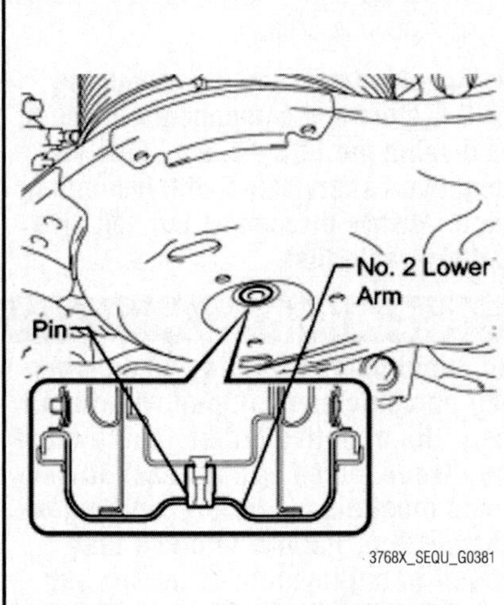

Fig. 458 Rear pneumatic cylinder proper piston pin alignment—Sequoia

Check that the piston side of the cylinder pin is in the center hole of the No. 2 control arm.

12. Start the engine and fill the cylinder with air. Install the bolt. Tighten to 13 ft. lbs.

SHOCK ABSORBER

REMOVAL & INSTALLATION

Sequoia

See Figure 459.

1. Before servicing the vehicle, refer to the Precautions Section.

➡If working near and/or around the SRS system and components, be sure to disable the SRS system. Tape the negative battery cable with insulating tape. Always disconnect the negative battery cable first.

⁂ CAUTION

To avoid personal injury when working on vehicles equipped with an air bag, the negative battery cable must be disconnected and at least 90 seconds must elapse before working on the system. Failure to do so may result in deployment of the air bag.

2. Disconnect the negative battery cable. Tape the cable with insulating tape.

➡When disconnecting the cable, some systems need to be initialized after the cable is reconnected. You will need the Techstream diagnostic scan tool or equivalent. Follow the directions on the tool.

3. Raise and safely support the vehicle.
4. Remove the tire and wheel assembly.
5. Properly support the rear axle assembly, with a suitable jack and block of wood.
6. Remove the tail exhaust pipe assembly.
7. Remove the driveshaft.
8. Remove the stabilizer bar.
9. Disconnect the AVS connector, if equipped.
10. Remove the shock retaining bolts.
11. Remove the component from the vehicle.

To install:

➡Be sure to use new fasteners, as required.

12. Installation is the reverse of the removal procedure.
13. Temporarily install the component.
14. Stabilize the suspension.

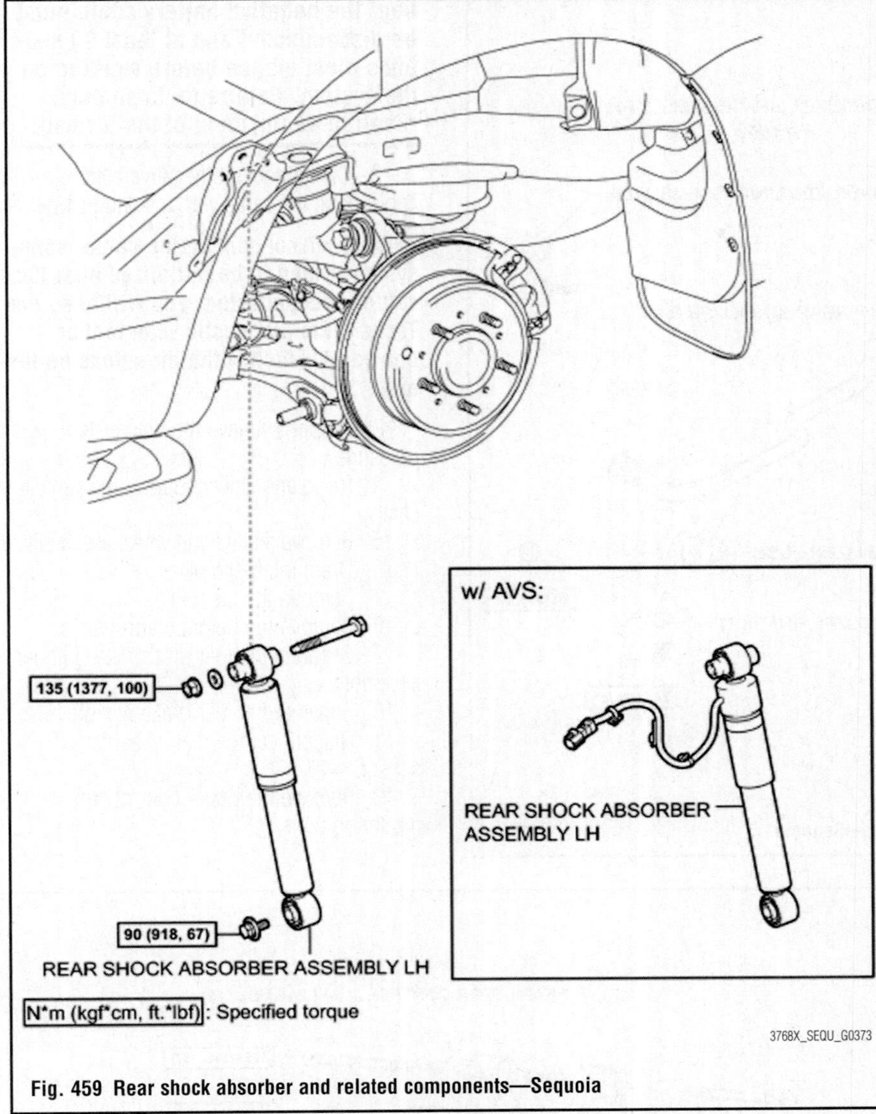

Fig. 459 Rear shock absorber and related components—Sequoia

15. Tighten the bolts to specification.
16. Check and adjust alignment, as required.

Tundra

1. Before servicing the vehicle, refer to the Precautions Section.

➡If working near and/or around the SRS system and components, be sure to disable the SRS system. Tape the negative battery cable with insulating tape. Always disconnect the negative battery cable first.

⁂ CAUTION

To avoid personal injury when working on vehicles equipped with an air bag, the negative battery cable must be disconnected and at least 90 seconds must elapse before working on the system. Failure to do so may result in deployment of the air bag.

2. Disconnect the negative battery cable. Tape the cable with insulating tape.

➡When disconnecting the cable, some systems need to be initialized after the cable is reconnected. You will need the Techstream diagnostic scan tool or equivalent. Follow the directions on the tool.

3. Raise and safely support the vehicle safely.
4. Remove the tire and wheel assemblies.
5. Lower the axle housing until the leaf spring tension is free. Keep it at this position.
6. Support the rear axle housing with a jack and wooden block. Remove the bolt and nut and disconnect the shock absorber assembly from the axle housing.
7. Remove the upper shock retaining bolts.
8. Remove the component from the vehicle.

To install:

➡Be sure to use new fasteners, as required.

9. Installation is the reverse of the removal procedure.
10. Temporarily install the component.
11. Stabilize the suspension.
12. Tighten the lower bolt and nut to 66 ft. lbs. (90 Nm).
13. Tighten the upper bolts and nuts to 21 ft. lbs. (28 Nm).

STABILIZER BAR

REMOVAL & INSTALLATION

Sequoia

See Figures 00 and 460.

1. Before servicing the vehicle, refer to the Precautions Section.

➡If working near and/or around the SRS system and components, be sure to disable the SRS system. Tape the negative battery cable with insulating tape. Always disconnect the negative battery cable first.

⁂ CAUTION

To avoid personal injury when working on vehicles equipped with an air bag, the negative battery cable must be disconnected and at least 90 seconds must elapse before working on the system. Failure to do so may result in deployment of the air bag.

2. Disconnect the negative battery cable. Tape the cable with insulating tape.

➡When disconnecting the cable, some systems need to be initialized after the cable is reconnected. You will need the Techstream diagnostic scan tool or equivalent. Follow the directions on the tool.

3. Raise and safely support the vehicle safely.
4. Remove the tire and wheel assemblies.
5. Remove the tail exhaust pipe assembly.
6. Remove the driveshaft.
7. Remove the stabilizer link retaining bolts. Remove the stabilizer link.
8. Remove the bushing bracket retaining bolts.
9. Remove the bushing brackets.

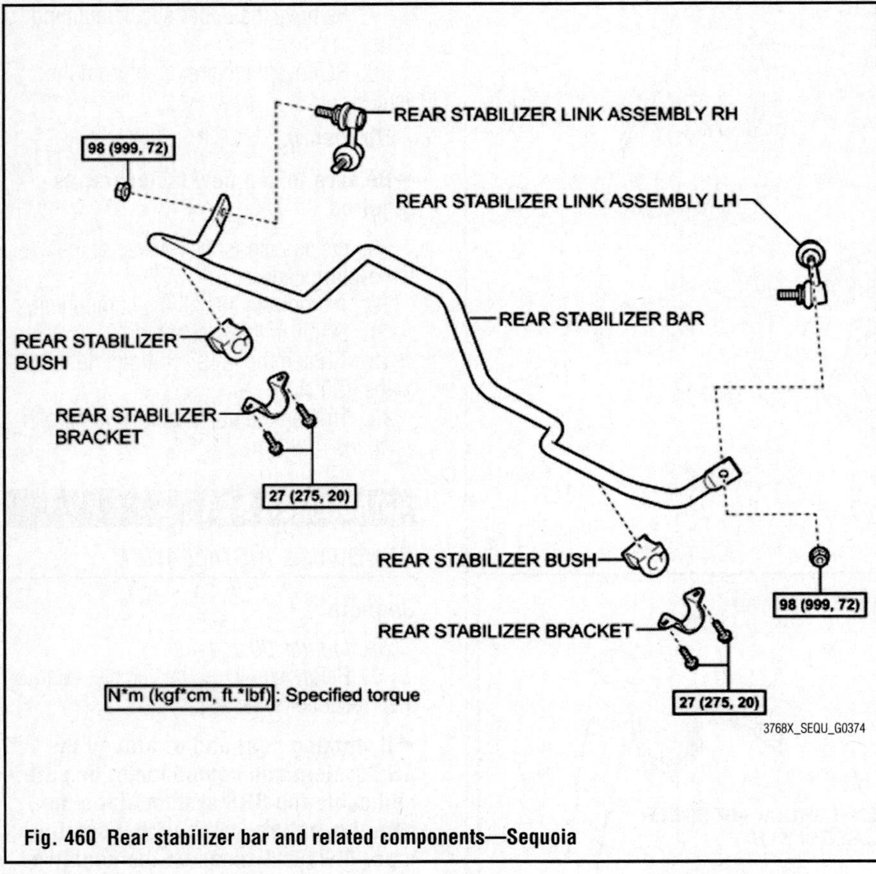

Fig. 460 Rear stabilizer bar and related components—Sequoia

10. Remove the stabilizer bar from its mounting.

To install:

➡**Be sure to use new fasteners, as required.**

11. Installation is the reverse of the removal procedure.
12. Temporarily install the component.
13. Stabilize the suspension.

UPPER CONTROL ARMS

REMOVAL & INSTALLATION

Sequoia

See Figure 461.

1. Before servicing the vehicle, refer to the Precautions Section.

➡**If working near and/or around the SRS system and components, be sure to disable the SRS system. Tape the negative battery cable with insulating tape. Always disconnect the negative battery cable first.**

⁂ CAUTION

To avoid personal injury when working on vehicles equipped with an air bag, the negative battery cable must be disconnected and at least 90 seconds must elapse before working on the system. Failure to do so may result in deployment of the air bag.

2. Disconnect the negative battery cable. Tape the cable with insulating tape.

➡**When disconnecting the cable, some systems need to be initialized after the cable is reconnected. You will need the Techstream diagnostic scan tool or equivalent. Follow the directions on the tool.**

3. Properly relieve the fuel system pressure.
4. Raise and safely support the vehicle safely.
5. Remove the tire and wheel assemblies.
6. Drain the fuel tank.
7. Remove the fuel tank.
8. Remove the height control sensor.
9. Separate the rear disc brake cylinder assembly.
10. Separate the rear brake flexible hose.
11. Disconnect the rear wheel speed sensor.
12. Remove the upper control arm mounting bolts.

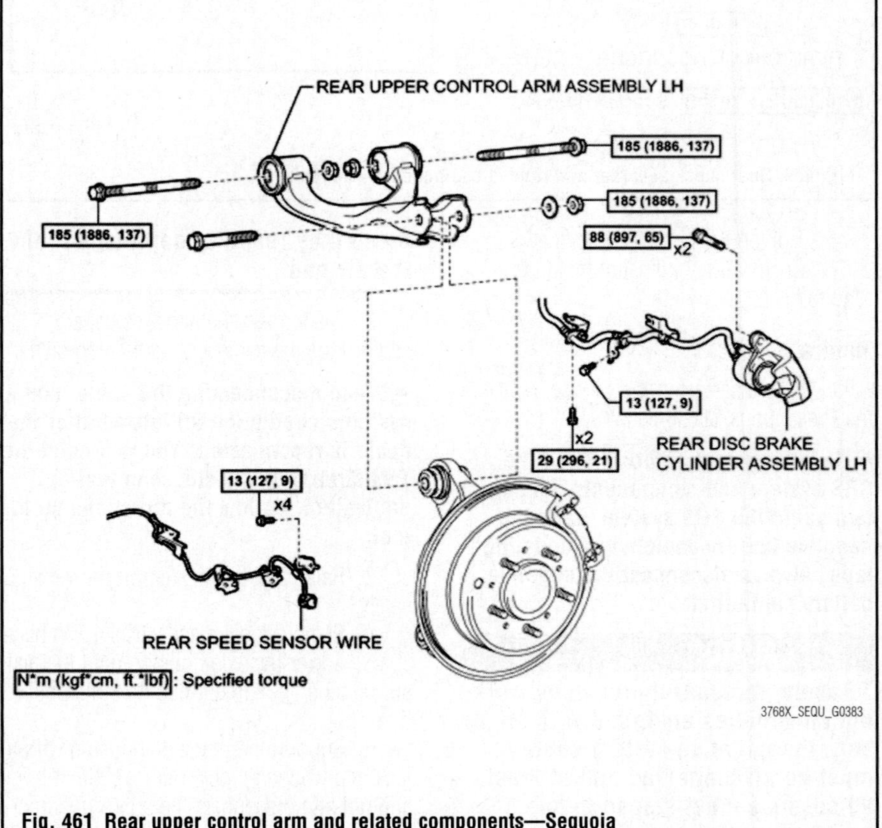

Fig. 461 Rear upper control arm and related components—Sequoia

13. Remove the component from the vehicle.

To install:

➡Be sure to use new fasteners, as required.

14. Installation is the reverse of the removal procedure.
15. Temporarily install the component.
16. Stabilize the suspension.

WHEEL BEARINGS

REMOVAL & INSTALLATION

Tundra

See Figures 462 through 467.

1. Disconnect cable from negative battery terminal.
2. Remove rear wheel.
3. Drain brake fluid.
4. Disconnect rear brake flexible hose.
5. Remove the clip.
6. Remove the 2 bolts and disconnect rear disc brake cylinder.
7. Remove rear disc.
8. Remove parking brake shoe return tension spring.
9. Remove parking brake shoe assembly.
10. Remove parking brake shoe lever sub-assembly.
11. Disconnect parking brake cable assembly.
12. Disconnect rear speed sensor.
13. Remove the 4 nuts and rear axle shaft together with the parking brake plate.
14. Remove the O-ring.
15. Using a snap ring expander, remove the snap ring.
16. Press out the rear axle shaft.
17. Remove the rear axle bearing inner retainer from the axle hub.
18. Remove the rear axle shaft washer from the axle hub.
19. Grind the rear axle bearing inner race surface using a grinder, then remove it with a chisel.
20. Temporarily install 4 nuts to the housing bolts.
21. Using a hammer, remove the 4 housing bolts and rear axle hub and bearing.
22. Remove the nuts.
23. Remove the 5 hub bolts.
24. Remove the deflector and deflector gasket from the rear axle shaft.

To install:

25. Install a new deflector gasket and deflector to the rear axle shaft.
26. Temporarily install a washer and nut to 5 new hub bolts.

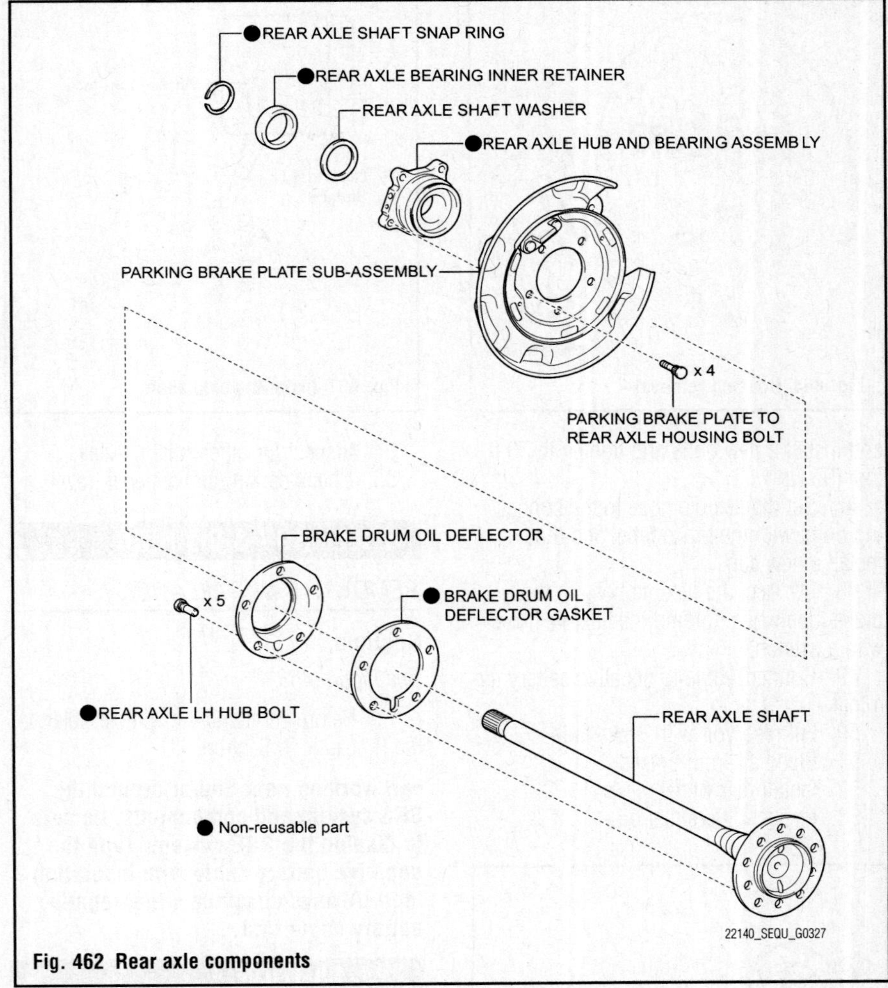

Fig. 462 Rear axle components

27. Install the hub bolts by tightening each nut.
28. Removed the washer and nut from each hub bolt.
29. Position the parking brake plate on a new rear axle hub and bearing.
30. Using 2 socket wrenches and a press, press in the 4 housing bolts.
31. Install the washer and a new retainer to the axle hub.

➡Install the washer with its tapered surface facing downward. Install the retainer with its chamfered surface facing downward.

32. Press in the rear axle shaft.
33. Using a snap ring expander, install a new snap ring.
34. Apply MP grease to the lip of the oil seal.
35. Install the O-ring to the axle shaft.
36. Install the rear axle shaft and parking brake plate with the 4 nuts and tighten to 44 ft. lbs. (60 Nm).
37. Inspect the rear axle shaft.
38. Connect rear speed sensor.
39. Connect parking brake cable assembly.
40. Install parking brake shoe lever sub-assembly.
41. Install parking brake shoe assemblies.
42. Install parking brake shoe return tension spring.
43. Check parking brake installation.
44. Install rear disc.
45. Adjust parking brake shoe clearance.
46. Connect the rear disc brake cylinder

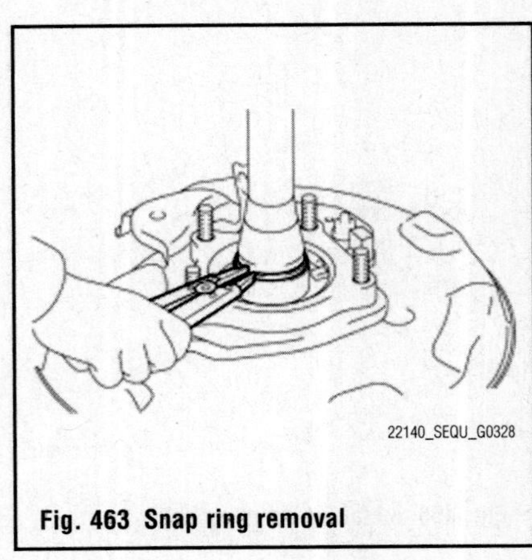

Fig. 463 Snap ring removal

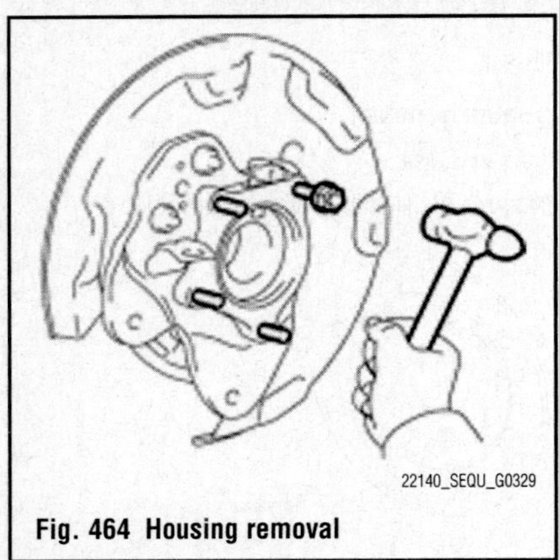

Fig. 464 Housing removal

and install 2 new bolts and tighten to 70 ft. lbs. (95 Nm).

47. Set the flexible hose to the connecting point with the brake tube, and then install a new clip.
48. Connect the flexible hose to the brake tube while holding the flexible hose with a wrench.
49. Connect cable to negative battery terminal.
50. Fill reservoir with brake fluid.
51. Bleed air from brakes.
52. Install rear wheel.
53. Inspect differential oil.

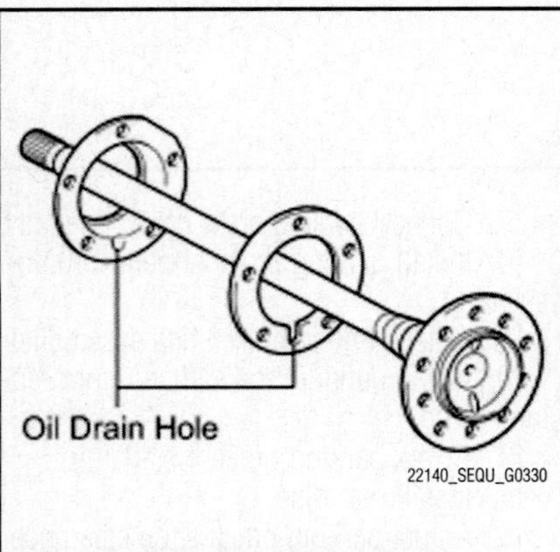

Fig. 465 Deflector gasket position

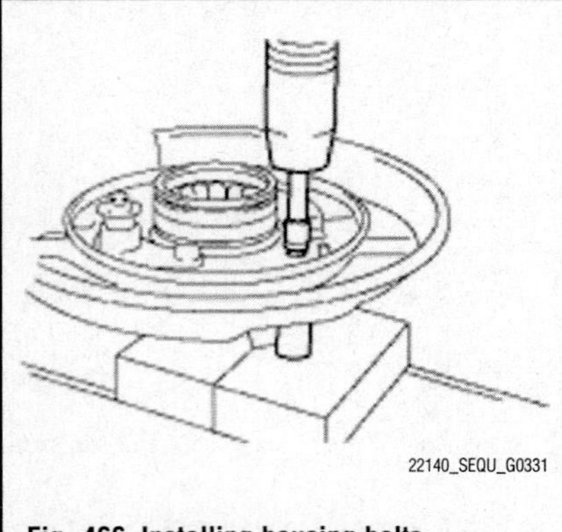

Fig. 466 Installing housing bolts

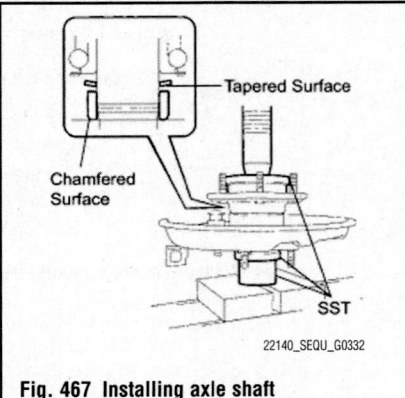

Fig. 467 Installing axle shaft

54. Inspect for differential oil leak.
55. Check parking brake pedal travel.

WHEEL HUB & BEARING

REMOVAL & INSTALLATION

Sequoia

See Figure 468.

1. Before servicing the vehicle, refer to the Precautions Section.

➡If working near and/or around the SRS system and components, be sure to disable the SRS system. Tape the negative battery cable with insulating tape. Always disconnect the negative battery cable first.

⁂ CAUTION

To avoid personal injury when working on vehicles equipped with an air bag, the negative battery cable must be disconnected and at least 90 seconds must elapse before working on the system. Failure to do so may result in deployment of the air bag.

2. Disconnect the negative battery cable. Tape the cable with insulating tape.

➡When disconnecting the cable, some systems need to be initialized after the cable is reconnected. You will need the Techstream diagnostic scan tool or equivalent. Follow the directions on the tool.

3. Raise and safely support the vehicle.
4. Remove the tire and wheel assembly.
5. Remove the coil spring, if equipped.
6. Remove the pneumatic cylinder, if equipped.
7. Remove the parking brake assembly.
8. Remove the rear speed sensor.
9. Remove the rear hub grease cap.
10. Remove the cotter pin. Remove the axle shaft nut. Discard the cotter pin.
11. Remove the cotter pin and nut. Using the proper removal tools, separate the No.2 rear suspension arm assembly.
12. Matchmark the No. 1 rear suspension arm and adjust cam.
13. Remove the bolts and nuts. Tap out the front halfshaft from the rear axle carrier sub assembly.
14. Remove the bolts and hub and bearing from the axle carrier.

To install:

➡Be sure to use new fasteners, as required.

15. Installation is the reverse of the removal procedure.
16. Be sure to use a new oil seal.
17. Tighten the retaining bolts to 64 ft. lbs. (86 Nm).

ADJUSTMENT

Adjustment is not possible.

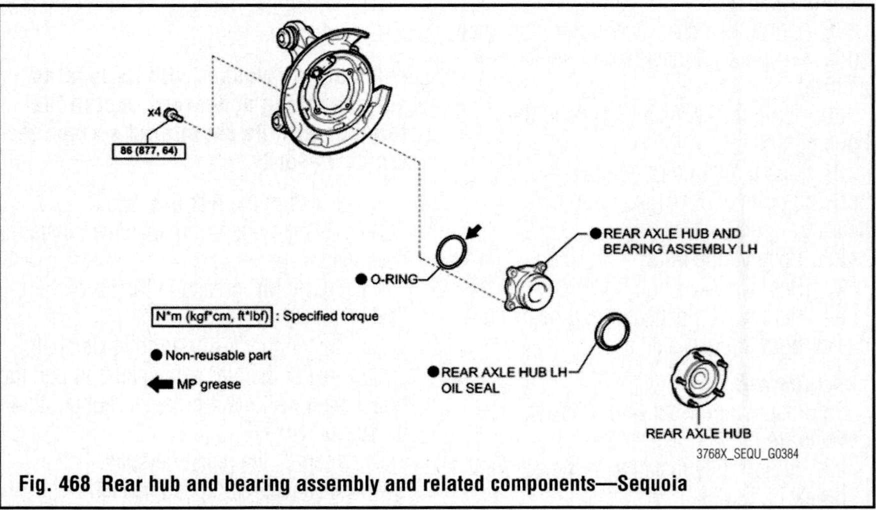

Fig. 468 Rear hub and bearing assembly and related components—Sequoia

TOYOTA

Sienna

16

SPECIFICATIONS AND MAINTENANCE CHARTS

ENGINE AND VEHICLE IDENTIFICATION

Engine							Model Year	
Code ①	Liters (cc)	Cu. In.	Cyl.	Fuel Sys.	Engine Type	Eng. Mfg.	Code ②	Year
2GR-FE	3.5 (3456)	NA	6	SFI	DOHC	Toyota	9	2009
							A	2010

MFI: Multi-port Fuel Injection

DOHC: Double Overhead Camshaft

① Stamped on the left side of the engine block

② 10th digit of the Vehicle Identification Number (VIN)

3768X_SIEN_C0001

GENERAL ENGINE SPECIFICATIONS

Year	Model	Engine Displacement Liters	Engine Series ID	Net Horsepower @ rpm	Net Torque @ rpm (ft. lbs.)	Bore x Stroke (in.)	Com-pression Ratio	Oil Pressure @ rpm
2009	Sienna	3.5	2GR-FE	268@6200	248@4700	3.70x3.27	NA	43-78@3000
2010	Sienna	3.5	2GR-FE	268@6200	248@4700	3.70x3.27	NA	43-78@3000

3768X_SIEN_C0002

ENGINE TUNE-UP SPECIFICATIONS

Year	Engine Displacement Liters	Engine ID	Spark Plug Gap (in.)	Ignition Timing (deg.)	Fuel Pump (psi)	Idle Speed (rpm)	Valve Clearance	
							Intake	Exhaust
2009	3.5	2GR-FE	0.039-0.043	①	44-50	550-650	0.006-0.010	0.010-0.014
2010	3.5	2GR-FE	0.039-0.043	①	44-50	550-650	0.006-0.010	0.010-0.014

NOTE: The Vehicle Emission Control Information label often reflects specification changes made during production.

The label figures must be used if they differ from those in this chart.

B: Before top dead center

① With terminal TC and CG of DLC3 connected: 8-12 degrees BTDC

With terminal TC and CG of DLC3 disconnected: 7-24 degrees BTDC

3768X_SIEN_C0003

CAPACITIES

Year	Model	Engine Displacement Liters	Engine ID	Engine Oil with Filter (qts.)	Transmission (pts.) Auto.	Transfer Case (pts.)	Drive Axle Rear (pts.)	Fuel Tank (gal.)	Cooling System (qts.)
2009	Sienna	3.5	2GR-FE	5.0	①	2.0	2.0	20.0	12.4
2010	Sienna	3.5	2GR-FE	5.0	①	2.0	2.0	20.0	12.4

① 2WD: 7.4 pts.
4WD: 7.6 pts.

3768X_SIEN_C0004

FLUID SPECIFICATIONS

Year	Model	Engine Displ. Liters	Engine Oil	Auto. Trans.	Drive Axle Rear ①	Transfer Case	Brake Master Cylinder	Cooling System
2009	Sienna	3.5	5W-30	ATF WS	80W-90	80W-90	DOT 3	SLLC ②
2010	Sienna	3.5	5W-30	ATF WS	80W-90	80W-90	DOT 3	SLLC ②

NE: Not Equipped

DOT: Department Of Transpotation

① Oil grade: Hypoid gear oil API GL-5

② Toyota Super Long Life Coolant

3768X_SIEN_C0014

VALVE SPECIFICATIONS

Year	Engine Displacement Liters	Engine ID	Seat Angle (deg.)	Face Angle (deg.)	Spring Test Pressure (lbs. @ in.)	Spring Installed Height (in.)	Stem-to-Guide Clearance (in.)		Stem Diameter (in.)	
							Intake	Exhaust	Intake	Exhaust
2009	3.5	2GR-FE	45	40.5	41.9-46.3@ 1.437	1.331	0.0010-0.0024	0.0012-0.0026	0.2154-0.2159	0.2152 0.2157
2010	3.5	2GR-FE	45	40.5	41.9-46.3@ 1.437	1.331	0.0010-0.0024	0.0012-0.0026	0.2154-0.2159	0.2152 0.2157

3768X_SIEN_C0005

CAMSHAFT AND BEARING SPECIFICATIONS

All measurements are given in inches.

Year	Engine Displacement Liters	Engine VIN	Journal Diameter	Brg. Oil Clearance	Shaft End-play	Runout	Journal Bore	Lobe Lift Intake	Lobe Lift Exhaust
2009	3.5	2GR-FE	①	②	NA	0.0016	NA	1.7447-1.7487	1.7426-1.7465
2010	3.5	2GR-FE	①	②	NA	0.0016	NA	1.7447-1.7487	1.7426-1.7465

NA: Not Available

① No. 1 journal: 1.4152 to 1.4157 inches

Other journals: 1.0220 to 1.0226 inches

② No. 1 journal: 0.0016 to 0.0031 inches

Other journals: 0.00098 to 0.0024 inches

3768X_SIEN_C0013

CRANKSHAFT AND CONNECTING ROD SPECIFICATIONS

All measurements are given in inches.

Year	Engine Displacement Liters	Engine ID	Crankshaft Main Brg. Journal Dia.	Crankshaft Main Brg. Oil Clearance	Crankshaft Shaft End-play	Crankshaft Thrust on No.	Connecting Rod Journal Diameter	Connecting Rod Oil Clearance	Connecting Rod Side Clearance
2009	3.5	2GR-FE	2.4011-2.4016	①	0.0016-0.0095	2	2.0863-2.0866	0.0015-0.0026	0.0059-0.0118
2010	3.5	2GR-FE	2.4011-2.4016	①	0.0016-0.0095	2	2.0863-2.0866	0.0015-0.0026	0.0059-0.0118

① Journals 1 and 4: 0.0006 - 0.0013 in.

Journals 2 and 3: 0.0010 - 0.0018 in.

3768X_SIEN_C0006

PISTON AND RING SPECIFICATIONS

All measurements are given in inches.

Year	Engine Displ. Liters	Engine ID	Piston Clearance	Ring Gap Top Comp.	Ring Gap Bottom Comp.	Ring Gap Oil Control	Ring Side Clearance Top Comp.	Ring Side Clearance Bottom Comp.	Ring Side Clearance Oil Control
2009	3.5	2GR-FE	0.0013-0.0023	0.0118-0.0157	0.0197-0.0236	0.0059-0.0157	0.0012-0.0031	0.0008-0.0024	0.0012-0.0043
2010	3.5	2GR-FE	0.0013-0.0023	0.0118-0.0157	0.0197-0.0236	0.0059-0.0157	0.0012-0.0031	0.0008-0.0024	0.0012-0.0043

3768X_SIEN_C0007

TORQUE SPECIFICATIONS

All readings in ft. lbs.

Year	Engine Displacement Liters	Engine ID	Cylinder Head Bolts	Main Bearing Bolts	Rod Bearing Bolts	Crankshaft Damper Bolts	Flywheel Bolts	Manifold Intake	Manifold Exhaust	Spark Plugs	Oil Pan Drain Plug
2009	3.5	2GR-FE	①	②	③	162	61	11	36	18	33
2010	3.5	2GR-FE	①	②	③	162	61	11	36	18	33

① Step 1: 40 ft. lbs.
Step 2: Plus 90 degrees
Recessed bolt: 13 ft. lbs.

② 6-point bolts: 20 ft. lbs.
12-point bolts:
Step 1: 16 ft. lbs.
Step 2: Plus 90 degrees

③ Step 1: 18 ft. lbs.
Step 2: Plus 90 degrees

3768X_SIEN_C0008

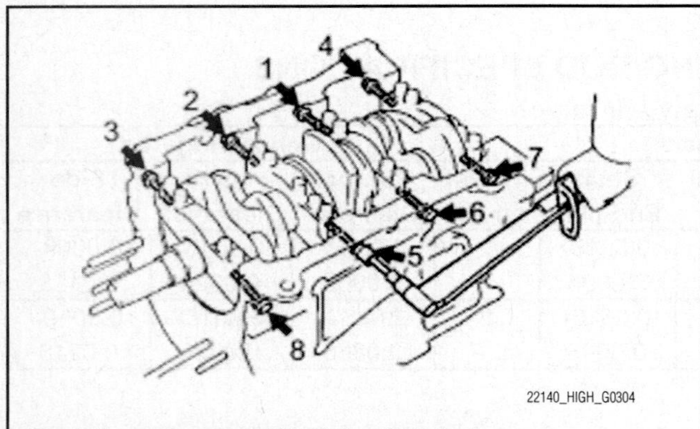

Fig. 1 Torque Sequence Side Bolts—3.5L Engines

WHEEL ALIGNMENT

Year	Model		Caster Range (+/-Deg.)	Caster Preferred Setting (Deg.)	Camber Range (+/-Deg.)	Camber Preferred Setting (Deg.)	Toe-in (in.)	Inside Wheel Angle (Deg.)
2009	Sienna FWD	F	0.75	2.80	0.75	-0.28	0+/-0.08	42.75
		R	NA	NA	0.50	-1.37	0.11+/-0.12	NA
	4WD	F	0.75	2.67	0.75	-0.28	0+/-0.08	42.7
		R	NA	NA	0.50	-1.42	0.05+/-0.12	NA
2010	Sienna FWD	F	0.75	2.80	0.75	-0.28	0+/-0.08	42.75
		R	NA	NA	0.50	-1.37	0.11+/-0.12	NA
	4WD	F	0.75	2.67	0.75	-0.28	0+/-0.08	42.7
		R	NA	NA	0.50	-1.42	0.05+/-0.12	NA

NA: Not Available

3768X_SIEN_C0009

TIRE, WHEEL AND BALL JOINT SPECIFICATIONS

Year	Model	OEM Tires Standard	OEM Tires Optional	Tire Pressures (psi) Front	Tire Pressures (psi) Rear	Wheel Size	Ball Joint Inspection	Lug Nut Torque (ft. lbs.)
2009	Sienna	P215/65R16 96T	P225/60R17 98T	35	35	6.5	①	76
2010	Sienna	P215/65R16 96T	P225/60R17 98T	35	35	6.5	①	76

OEM: Original Equipment Manufacturer

NA: Information not available

PSI: Pounds Per Square Inch

① Ball joint turning torque should be 8.7-30 inch lbs.

3768X_SIEN_C0010

BRAKE SPECIFICATIONS

All measurements in inches unless noted

Year	Model		Brake Disc Original Thickness	Brake Disc Minimum Thickness	Brake Disc Maximum Runout	Brake Drum Diameter Original Inside Diameter	Brake Drum Diameter Maximum Machine Diameter	Minimum Lining Thickness	Brake Caliper Bracket Bolts (ft. lbs.)	Brake Caliper Mounting Bolts (ft. lbs.)
2009	Sienna	F	1.102	1.024	0.0020	NE	NE	0.039	79	25
		R	0.472	0.413	0.0039	10.00	10.08	0.039	65	25
2010	Sienna	F	1.102	1.024	0.0020	NE	NE	0.039	79	25
		R	0.472	0.413	0.0039	10.00	10.08	0.039	65	25

F: Front

R: Rear

NE: Not Equipped

3768X_SIEN_C0011

SCHEDULED MAINTENANCE INTERVALS

TOYOTA—SIENNA

		VEHICLE MILEAGE INTERVAL (x1000)																	
TO BE SERVICED	**TYPE OF SERVICE**	**5**	**10**	**15**	**20**	**25**	**30**	**35**	**40**	**45**	**50**	**55**	**60**	**65**	**70**	**75**	**80**	**85**	**90**
Automatic transmission and differential fluid	S/I			✓			✓			✓			✓			✓			✓
Ball joints and boots	S/I			✓			✓			✓			✓			✓			✓
Brake linings, discs/drums, lines & hoses	S/I			✓			✓			✓			✓			✓			✓
Charcoal canister	S/I												✓						
Drive belts	S/I			✓			✓			✓			✓			✓			✓
Engine coolant	R						✓						✓						✓
Engine oil & filter	R	✓	✓	✓	✓	✓	✓	✓	✓	✓	✓	✓	✓	✓	✓	✓	✓	✓	✓
Exhaust pipes & mounts	S/I			✓			✓			✓			✓			✓			✓
Fuel lines & connections, fuel tank vapor vent system hoses, fuel tank band	S/I						✓			✓			✓			✓			✓
Fuel tank cap gasket	S/I						✓						✓						✓
Halfshaft boots & flange bolts	S/I			✓			✓			✓			✓			✓			✓
Non-platinum spark plugs	R						✓						✓						✓
Platinum spark plugs	R												✓						
Rotate tires		✓	✓	✓	✓	✓	✓	✓	✓	✓	✓	✓	✓	✓	✓	✓	✓	✓	✓
Rack and pinion assembly	S/I			✓			✓			✓			✓			✓			✓
Steering linkage	S/I			✓			✓			✓			✓			✓			✓
Valves	S/I												✓						

R: Replace S/I: Service or Inspect L: Lubricate

FREQUENT OPERATION MAINTENANCE (SEVERE SERVICE)

If a vehicle is operated under any of the following conditions it is considered severe service:

- Towing a trailer or using a camper or car-top carrier.

- Repeated short trips of less than 5 miles in temperatures below freezing.

- Excessive idling or low-speed driving for long distances as in heavy commercial use, such as delivery, taxi or police cars.

- Operating on rough, muddy or salt-covered roads.

- Operating on unpaved or dusty roads.

Oil filter: service or inspect every 5000 miles or 4 months, whichever occurs first.

Brake linings and discs or drums: service or inspect every 5000 miles or 4 months, whichever occurs first.

Steering linkage: service or inspect every 5000 miles or 4 months, whichever occurs first.

Ball joints and boots: service or inspect every 5000 miles or 4 months, whichever occurs first.

Brake discs & pads (front): service or inspect every 6000 miles.

Halfshaft boots: service or inspect every 5000 miles or 4 months. Retighten the flange bolts, whichever occurs first.

Body chassis bolts and nuts: service or inspect every 5000 miles or 4 months, whichever occurs first.

Transmission and differential fluid: replace every 15,000 miles or 12 months, whichever occurs first.

Timing belt: replace every 60,000 miles or 48 months, whichever occurs first.

3768X_SIEN_C0012

BRAKES — INFORMATION AND PRECAUTIONS

ANTI-LOCK SYSTEMS

- Certain components within the ABS system are not intended to be serviced or repaired individually.
- Do not use rubber hoses or other parts not specifically specified for and ABS system. When using repair kits, replace all parts included in the kit. Partial or incorrect repair may lead to functional problems and require the replacement of components.
- Lubricate rubber parts with clean, fresh brake fluid to ease assembly. Do not use shop air to clean parts; damage to rubber components may result.
- Use only DOT 3 brake fluid from an unopened container.
- If any hydraulic component or line is removed or replaced, it may be necessary to bleed the entire system.
- A clean repair area is essential. Always clean the reservoir and cap thoroughly before removing the cap. The slightest amount of dirt in the fluid may plug an orifice and impair the system function. Perform repairs after components have been thoroughly cleaned; use only denatured alcohol to clean components. Do not allow ABS components to come into contact with any substance containing mineral oil; this includes used shop rags.
- The Anti-Lock control unit is a microprocessor similar to other computer units in the vehicle. Ensure that the ignition switch is **OFF** before removing or installing controller harnesses. Avoid static electricity discharge at or near the controller.
- If any arc welding is to be done on the vehicle, the control unit should be unplugged before welding operations begin.

DISC AND DRUM SYSTEMS

CAUTION

Dust and dirt accumulating on brake parts during normal use may contain asbestos fibers from production or aftermarket brake linings. Breathing excessive concentrations of asbestos fibers can cause serious bodily harm. Exercise care when servicing brake parts. Do not sand or grind brake lining unless equipment used is designed to contain the dust residue. Do not clean brake parts with compressed air or by dry brushing. Cleaning should be done by dampening the brake components with a fine mist of water, then wiping the brake components clean with a dampened cloth. Dispose of cloth and all residue containing asbestos fibers in an impermeable container with the appropriate label. Follow practices prescribed by the Occupational Safety and Health Administration (OSHA) and the Environmental Protection Agency (EPA) for the handling, processing, and disposing of dust or debris that may contain asbestos fibers.

BRAKES — BLEEDING THE BRAKE SYSTEM

BLEEDING PROCEDURE

See Figure 2.

➡**Do not reuse the drained fluid. Use only clean DOT 3 Brake Fluid from an unopened container.**

WARNING

Make sure no dirt or other foreign matter is allowed to contaminate the brake fluid.

WARNING

Do not spill brake fluid on the vehicle, it may damage the paint; if brake fluid does contact the paint, wash it off immediately with water.

1. The reservoir on the master cylinder must be at the MAX level mark at the start of the bleeding procedure and checked after bleeding each brake caliper. Add fluid as required.
2. Make sure the brake fluid level in the reservoir is at the MAX level line.
3. Slide a piece of clear plastic hose over the first bleed screw, and submerge the other end in a container of new brake fluid.
4. Have someone slowly pump the brake pedal several times, then apply steady pressure.
5. Starting at the left-front, loosen the brake bleed screw to allow air to escape from the system. Then tighten the bleed screw securely. Bleed the right rear, right front and left rear using the same procedure.
6. Repeat the procedure for each wheel until air bubbles no longer appear in the fluid.
7. Refill the master cylinder reservoir to the MAX level line.

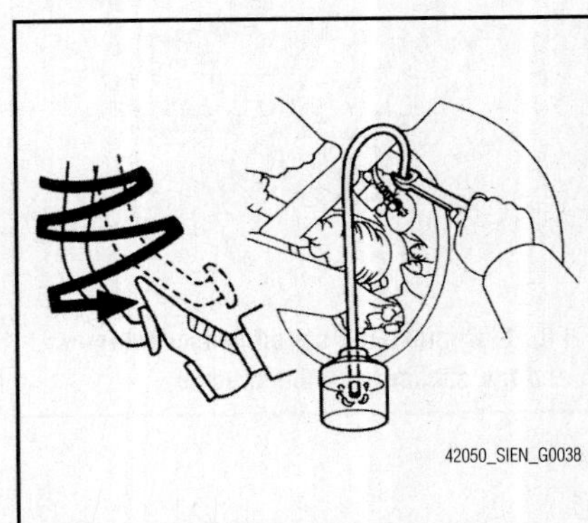

Fig. 2 Bleeding the brake system—Sienna

WHEEL SPEED SENSORS

REMOVAL & INSTALLATION

Front

See Figure 3.

1. Before servicing the vehicle, refer to the precautions section.
2. Remove the front wheel.
3. Remove the fender liner screws and the liner.
4. Unplug the connector from the speed sensor.
5. Remove the sensor harness and clamp from the body.
6. Remove the speed sensor retainers and the sensor from the knuckle.

To install:

7. Install the speed sensor and tighten the fasteners to 71 inch lbs. (8 Nm).
8. Route the hose, install the harness clamp and tighten the bolt to body to 71 inch lbs. (8 Nm) and the bolt to shock absorber to 14 ft. lbs. (19 Nm).
9. Attach the speed sensor connector and install fender liner and the wheel.

Fig. 3 Remove the speed sensor retainers and the sensor from the knuckle

Rear

2WD Vehicles

See Figure 4.

1. Before servicing the vehicle, refer to the precautions section.
2. Remove the rear wheel.
3. Unplug the connector from the rear speed sensor.
4. Remove the brake drum, if equipped.
5. Remove the brake rotor, if equipped.
6. Remove the rear wheel bearing/hub assembly.
7. Mount the rear wheel bearing/hub assembly in a vise.
8. Using a hammer and pin punch, drive out the 2 pins and remove the attachments from the tool illustrated.
9. Using the tool illustrated and two 12mm bolts with a pitch of 1.5mm, remove the rear speed sensor from the hub.

To install:

10. Clean the sensor and hub mating surfaces, place the sensor on the hub as illustrated.
11. Using the tool illustrated and a press, install the sensor into the hub making sure it goes in straight and slow.
12. Install the sensor with the rear axle hub assembly.
13. Install the brake rotor, if equipped.

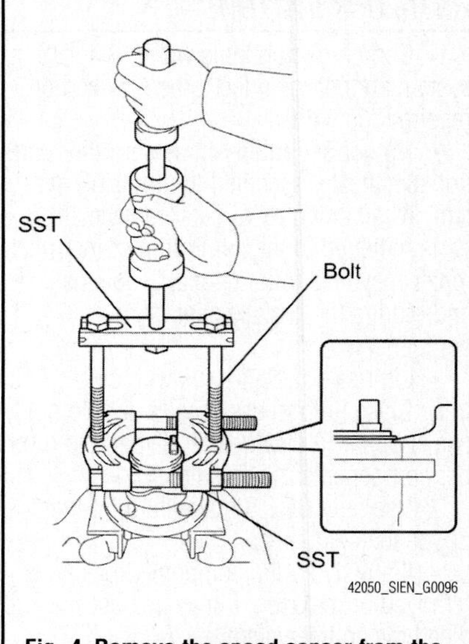

Fig. 4 Remove the speed sensor from the axle hub—2 Wheel Drive

14. Install the brake drum, if equipped.
15. Connect the sensor connector and install the wheel.

4WD Vehicles

See Figure 3.

1. Before servicing the vehicle, refer to the precautions section.
2. Remove the rear wheel.
3. Remove the quarter trim panel.
4. Unplug the connector from the speed sensor and pull out the sensor wire with the grommet.
5. Remove the clamp bolts.
6. Remove the bolts and speed sensor assembly.

To install:

7. Installation is the reverse of removal. Tighten the sensor bolts to 71 inch lbs. (8 Nm).

CAUTION

Dust and dirt accumulating on brake parts during normal use may contain asbestos fibers from production or aftermarket brake linings. Breathing excessive concentrations of asbestos fibers can cause serious bodily harm. Exercise care when servicing brake parts. Do not sand or grind brake lining unless equipment used is designed to contain the dust residue. Do not clean brake parts with compressed air or by dry brushing. Cleaning should be done by dampening the brake components with a fine mist of water, then wiping the brake components clean with a dampened cloth. Dispose of cloth and all residue containing asbestos fibers in an impermeable container with the appropriate label. Follow practices prescribed by the Occupational Safety and Health Administration (OSHA) and the Environmental Protection Agency (EPA) for the handling, processing, and disposing of dust or debris that may contain asbestos fibers.

BRAKE CALIPER

REMOVAL & INSTALLATION

See Figure 5.

1. Before servicing the vehicle, refer to the precautions section.

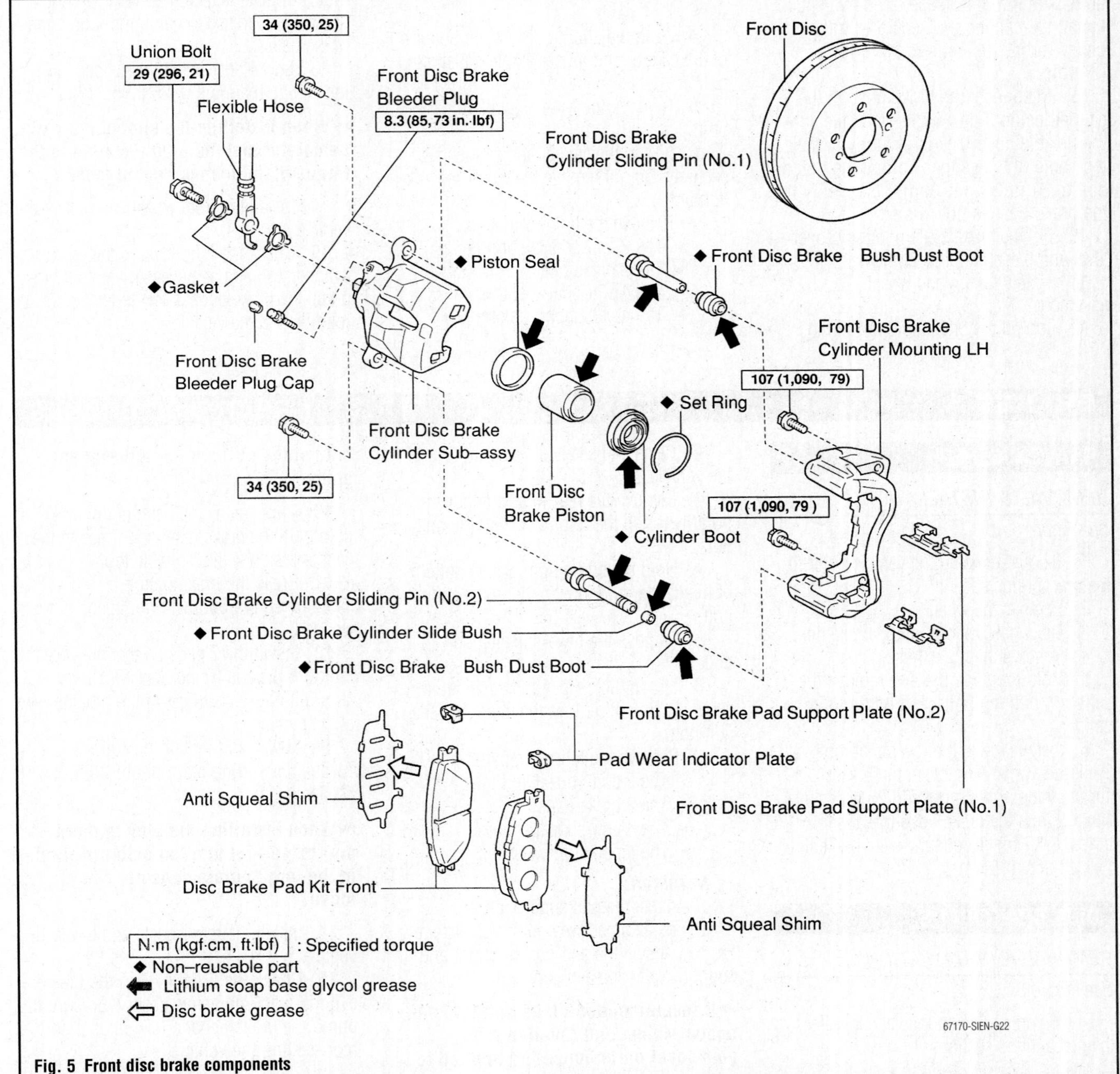

Fig. 5 Front disc brake components

2. Disconnect the negative battery cable from the battery.
3. Raise and support the vehicle safely.
4. Remove the wheels.
5. Disconnect the brake hose from the caliper by removing the union bolt and 2 gaskets. Plug the end of the hose to prevent loss of fluid.
6. Remove the bolts that attach the caliper to the torque plate.
7. Lift the bottom of the caliper up and remove the caliper assembly.

To install:

8. Grease the caliper slides and bolts with lithium grease or equivalent. Install the caliper and secure with the bolts. Torque the bolts to 25 ft. lbs. (34 Nm).
9. Reconnect the brake hose to the caliper, using 2 new washers. Make sure the flexible hose lock is securely in the lock hole of the caliper. Torque the union bolt to 21 ft. lbs. (29 Nm). Also, verify that the brake hose is not twisted.
10. Fill the brake system to the proper level and bleed the brake system.
11. Install the tire and wheel assembly.
12. Top off the brake fluid level in the master cylinder. Check for leaks and proper brake operation.
13. Connect the negative battery cable to the battery.

DISC BRAKE PADS

REMOVAL & INSTALLATION

See Figure 5.

1. Before servicing the vehicle, refer to the precautions section.
2. Raise and safely support the front of the vehicle.
3. Remove the front wheels and temporarily fasten the rotor disc with the hub nuts.
4. Hold the sliding pin on the bottom of the caliper and loosen the installation bolt.
5. Remove the lower installation bolt.
6. Lift up the caliper and suspend it securely. Do not remove the upper installation bolt.
7. Remove the following parts:
   - The 2 anti-squeal springs.
   - The 2 brake pads.
   - The 4 anti-squeal shims.
   - The 4 pad support plates.

To install:

8. Install the pad support plates.
9. Install a pad wear indicator plate to the pad. Install the anti-squeal shims and support plates to each pad.

➡It recommended that a suitable anti-squeal compound be applied to both sides of the inner anti-squeal shim.

10. Draw out a small amount of brake fluid from the brake reservoir. Press in the caliper piston with a suitable tool.
11. Press the brake piston in carefully so the boot will not become wedged.
12. Install the 2 pads so that the wear indicator plate is facing upward. Do not allow oil or grease to get in the rubbing face of the pads.
13. Lower and install the caliper. Torque the sliding main pin to 25 ft. lbs. (34 Nm).

➡When installing the sliding main pin, be careful that the plug installed in the torque plate does not come loose.

14. Install the front wheels and lower the vehicle.
15. Check the fluid level in the master cylinder and add as necessary. Be sure to pump the brake pedal a few times before road-testing the vehicle.

BRAKES — REAR DISC BRAKES

BRAKE CALIPER

REMOVAL & INSTALLATION

See Figure 6.

1. Before servicing the vehicle, refer to the precautions section.
2. Remove the wheel.
3. Disconnect and plug the brake line.
4. Remove the brake hose.
5. Hold the slide pin and remove the 2 caliper mounting bolts. Lift off the caliper.
6. Installation is the reverse of removal. Refill the system and bleed the brakes. Torque the mounting bolts to 25 ft. lbs. (34 Nm). Torque the brake hose-to-caliper to 17 ft. lbs. (23 Nm). Torque the steel brake line to 11 ft. lbs. (15 Nm).

DISC BRAKE PADS

REMOVAL & INSTALLATION

See Figure 8.

1. Before servicing the vehicle, refer to the precautions section.
2. Raise and safely support the rear of the vehicle.
3. Remove the rear wheels and temporarily fasten the rotor disc with the hub nuts.
4. Hold the sliding pin on the bottom of the caliper and loosen the installation bolt.
5. Remove the lower installation bolt.
6. Lift up the caliper and suspend it securely. Do not remove the upper installation bolt.
7. Remove the following parts:
   - The 2 anti-squeal springs.
   - The 2 brake pads.
   - The 4 anti-squeal shims.
   - The 4 pad support plates.

To install:

8. Install the pad support plates.
9. Install a pad wear indicator plate to the pad. Install the anti-squeal shims and support plates to each pad.

➡It recommended that a suitable anti-squeal compound (available at your local parts house) be applied to both sides of the inner anti-squeal shim.

10. Draw out a small amount of brake fluid from the brake reservoir. Press in the caliper piston with a suitable tool.
11. Press the brake piston in carefully so the boot will not become wedged.
12. Install the 2 pads so that the wear indicator plate is facing upward. Do not allow oil or grease to get in the rubbing face of the pads.
13. Lower and install the caliper. Torque the sliding main pin to 25 ft. lbs. (34 Nm).

➡When installing the sliding main pin, be careful that the plug installed in the torque plate does not come loose.

14. Install the rear wheels and lower the vehicle.
15. Check the fluid level in the master cylinder and add as necessary. Be sure to pump the brake pedal a few times before road-testing the vehicle.

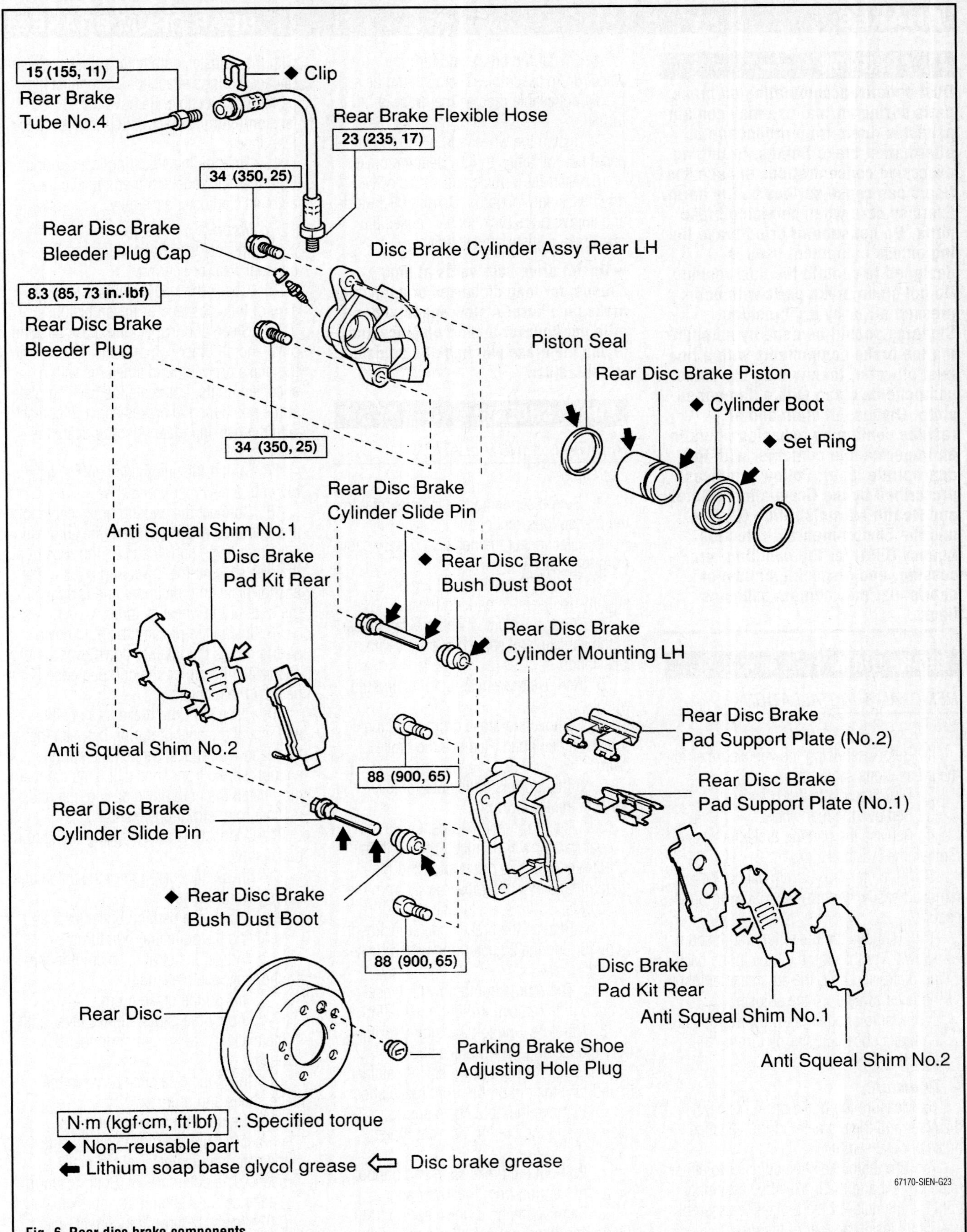

15 (155, 11)
Rear Brake Tube No.4
◆ Clip
Rear Brake Flexible Hose
23 (235, 17)
34 (350, 25)
Rear Disc Brake Bleeder Plug Cap
Disc Brake Cylinder Assy Rear LH
8.3 (85, 73 in.·lbf)
Rear Disc Brake Bleeder Plug
Piston Seal
Rear Disc Brake Piston
◆ Cylinder Boot
◆ Set Ring
34 (350, 25)
Rear Disc Brake Cylinder Slide Pin
Anti Squeal Shim No.1
Disc Brake Pad Kit Rear
◆ Rear Disc Brake Bush Dust Boot
Rear Disc Brake Cylinder Mounting LH
Rear Disc Brake Pad Support Plate (No.2)
Anti Squeal Shim No.2
88 (900, 65)
Rear Disc Brake Pad Support Plate (No.1)
Rear Disc Brake Cylinder Slide Pin
◆ Rear Disc Brake Bush Dust Boot
88 (900, 65)
Disc Brake Pad Kit Rear
Anti Squeal Shim No.1
Rear Disc
Parking Brake Shoe Adjusting Hole Plug
Anti Squeal Shim No.2
N·m (kgf·cm, ft·lbf) : Specified torque
◆ Non–reusable part
Lithium soap base glycol grease
Disc brake grease
67170-SIEN-G23

Fig. 6 Rear disc brake components

CAUTION

Dust and dirt accumulating on brake parts during normal use may contain asbestos fibers from production or aftermarket brake linings. Breathing excessive concentrations of asbestos fibers can cause serious bodily harm. Exercise care when servicing brake parts. Do not sand or grind brake lining unless equipment used is designed to contain the dust residue. Do not clean brake parts with compressed air or by dry brushing. Cleaning should be done by dampening the brake components with a fine mist of water, then wiping the brake components clean with a dampened cloth. Dispose of cloth and all residue containing asbestos fibers in an impermeable container with the appropriate label. Follow practices prescribed by the Occupational Safety and Health Administration (OSHA) and the Environmental Protection Agency (EPA) for the handling, processing, and disposing of dust or debris that may contain asbestos fibers.

BRAKE DRUM

REMOVAL & INSTALLATION

See Figure 7.

1. Before servicing the vehicle, refer to the precautions section.
2. Drain the brake fluid.
3. Remove the rear wheel.
4. Release the parking brake lever, and remove the rear brake drum.
5. If the rear brake drum cannot be removed easily, perform the following procedure:
   a. Remove the hole plug and insert a screwdriver through the hole in the backing plate, and hold the automatic adjusting lever away from the adjuster.
   b. Using another screwdriver, reduce the brake shoe adjuster by turning the adjusting wheel.

To install:

6. Measure the brake drum inside diameter and the diameter of the installed brake shoe assembly.
7. Turn the brake shoe adjuster to either retract or expand the brake shoe assembly until the diameter of the brake she assembly is slightly less than that of the drum.
8. Install the drum, and turn the wheel/drum assembly. There should be a hardly noticeable drag as the drum is turned.
9. Install the wheel and pump the brake pedal several times to fully seat the brakes.
10. Start the vehicle and very slowly drive back and forwards. Driving backwards and applying the brake pedal causes the rear brake adjuster to self adjust.

➡Do not drive backwards at high speeds, for long distances or apply the brakes too hard. A slow speed under 1 mile per hour for a short distance and plying the brake will help the adjuster to self adjust.

BRAKE SHOES

REMOVAL & INSTALLATION

See Figure 7.

1. Before servicing the vehicle, refer to the precautions section.
2. Disconnect the negative battery cable from the battery.
3. Loosen the rear wheel lug nuts slightly. Release the parking brake.
4. Block the front wheels, raise the rear of the vehicle, and safely support it with jackstands.
5. Remove the wheel lug nuts and the wheel.
6. Remove the brake drum retaining screws, if equipped. Remove the brake drum.
7. If the drum is difficult to remove, perform the following:
   a. Insert the end of a bent wire (a coat hanger will do nicely) through the hole in the brake drum and hold the automatic adjusting lever away from the adjuster.
   b. Reduce the brake shoe adjustment by turning the adjuster bolt with a brake tool.
   c. The drum should now be loose enough to remove without much effort.
8. Carefully unhook the return spring from the leading (front) brake shoe.
9. Press the hold down spring retainer in and turn the pin on the front brake shoe.
10. Remove the hold down spring, retainers and the pin for the front brake shoe.
11. Pull out the brake shoe and unhook the anchor spring from the lower edge.
12. Remove the hold down spring from the trailing (rear) shoe. Pull the shoe out with the adjuster, automatic adjuster assembly and springs attached. Disconnect the parking brake cable. Remove the tension/return and anchor springs from the rear shoe.
13. Unhook the adjusting lever spring from the rear shoe and then remove the automatic adjuster assembly.

To install:

14. Inspect the shoes for signs of unusual wear or scoring.
15. Check the wheel cylinder for any sign of fluid seepage or frozen pistons.
16. Clean and inspect the brake backing plate and all other components. Check that the brake drum inner diameter is within specified limits. Lubricate the backing plate at the positions the brakes come in contact with the backing plate. Also lubricate the anchor plate.
17. Mount the automatic adjuster assembly onto a new rear brake shoe.
18. Connect the parking brake cable to the rear shoe and then install the automatic adjusting lever, spring and E-ring. Position the rear shoe so the lower end rides in the anchor plate and the upper end is against the boot of the wheel cylinder.
19. Install the pin and the hold down spring. Press the retainer down over the pin and rotate the pin so the crimped edge is held by the retainer.
20. Place the front brake into position and install the anchor spring between the front and rear shoes. Stretch the spring enough so the front shoe will fit as the rear did. Install the hold down spring, pin and retainer to the front brake shoe.
21. Connect the return spring to the front brake shoe.
22. Check the operation of the automatic adjuster mechanism:
    a. Apply the parking brake lever and verifying the adjusting bolt turns.
    b. Adjust the strut to where it is the shortest possible length.
    c. Install the brake drum.
    d. Apply the parking brake lever until the clicking sound can no longer be heard.
23. Check the clearance between the brake shoes and drum:
    a. Remove the brake drum.
    b. Measure the brake drum inside diameter and diameter of the brake shoes. The difference is "Shoe-to-drum clearance" and should be approximately 0.024 inch (0.6mm). If incorrect, check the parking brake system.

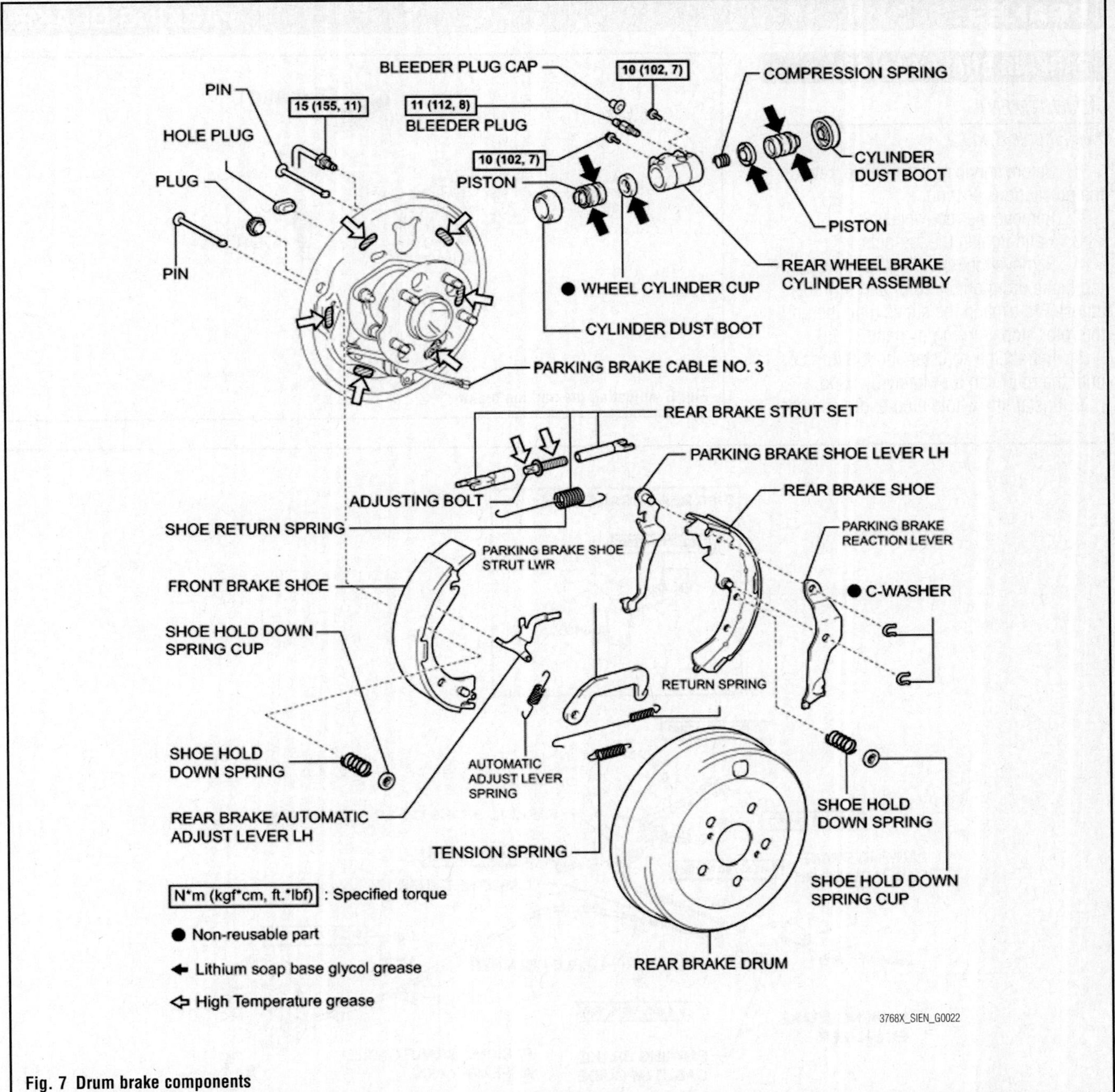

Fig. 7 Drum brake components

➡A special brake caliper tool is required to gauge the brake drum inside diameter and shoe-to-drum clearance. However it is not required to perform brake shoe adjustment.

24. Install the brake drum.
25. Adjust the brake pedal until a slight drag is felt when the drum is spun by hand.
26. Pull the parking lever all the way up until a clicking sound can no longer be heard. Check the clearance between brake shoes and brake drum.
27. Install the rear wheels, tighten the wheel lug nuts and lower the vehicle.
28. Retighten the wheel lug nuts and pump the brake pedal a few times before moving the vehicle. Adjust the rear brakes again if necessary.
29. Check the level of brake fluid in the master cylinder, then perform a test drive.
30. Connect the negative battery cable to the battery.

ADJUSTMENT

1. Before servicing the vehicle, refer to the precautions section.
2. Remove the rear wheels.
3. Measure the brake drum inside diameter and the diameter of the installed brake shoe assembly.
4. Turn the brake shoe adjuster to either retract or expand the brake shoe assembly until the diameter of the brake she assembly is slightly less than that of the drum.
5. Install the drum, and turn the adjuster to expand the shoes until the drum locks.
6. Back off the adjuster 15 notches and install the hole plug.

PARKING BRAKE CABLES

ADJUSTMENT

See Figures 8 and 9.

1. Before servicing the vehicle, refer to the precautions section.
2. Remove the rear wheel.
3. Hand tighten the lug nuts.
4. Remove the hole plug from the parking brake drum on the rotor and turn the adjuster to expand the shoes until the until the rotor stops turning by hand.
5. Retract the adjuster about 8 turns or until the rotor can turn freely by hand.
6. Install the whole plug and rear wheel.

Expand
Contract
42050_SIEN_G0094

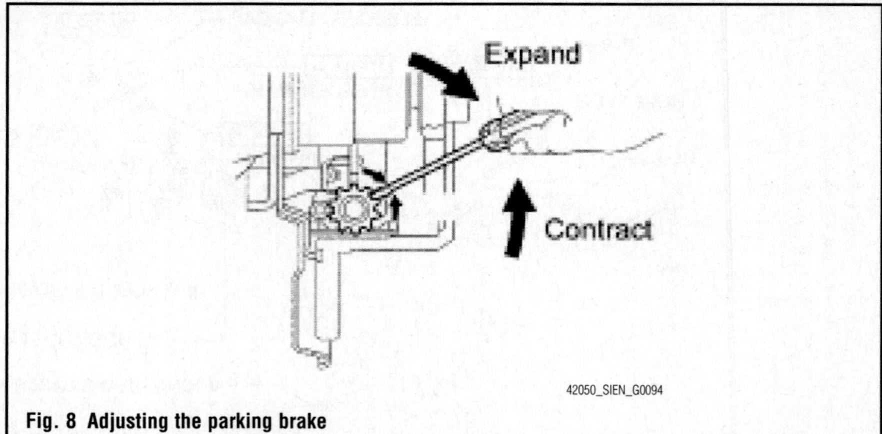

Fig. 8 Adjusting the parking brake

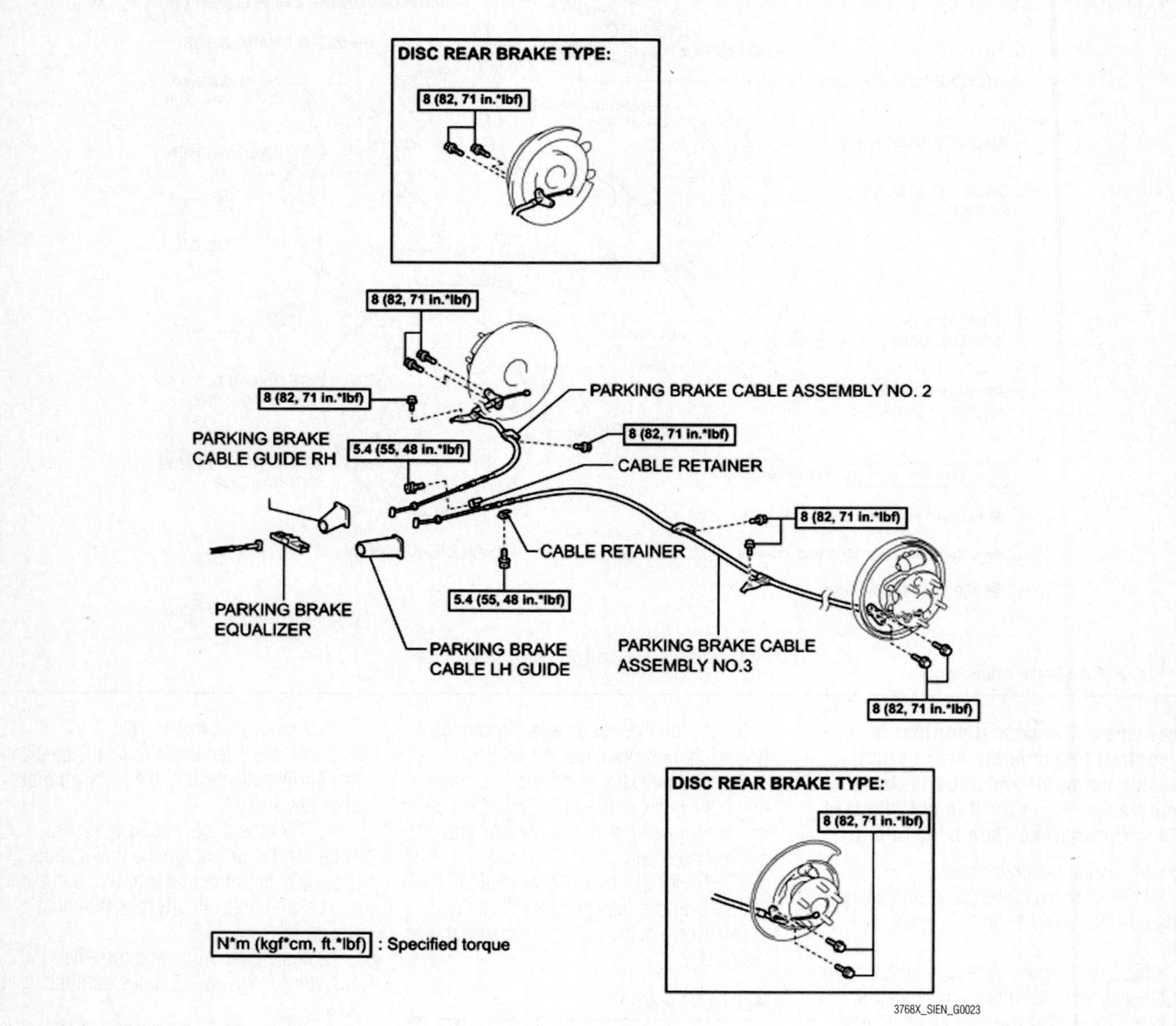

DISC REAR BRAKE TYPE:
8 (82, 71 in.*lbf)
8 (82, 71 in.*lbf)
8 (82, 71 in.*lbf)
PARKING BRAKE CABLE ASSEMBLY NO. 2
PARKING BRAKE CABLE GUIDE RH
5.4 (55, 48 in.*lbf)
8 (82, 71 in.*lbf)
CABLE RETAINER
8 (82, 71 in.*lbf)
CABLE RETAINER
5.4 (55, 48 in.*lbf)
PARKING BRAKE EQUALIZER
PARKING BRAKE CABLE LH GUIDE
PARKING BRAKE CABLE ASSEMBLY NO.3
8 (82, 71 in.*lbf)
DISC REAR BRAKE TYPE:
8 (82, 71 in.*lbf)
N*m (kgf*cm, ft.*lbf) : Specified torque
3768X_SIEN_G0023

Fig. 9 Parking brake cable and related components

PARKING BRAKE SHOES

REMOVAL & INSTALLATION

See Figure 10.

1. Before servicing the vehicle, refer to the precautions section.
2. Raise and safely support the vehicle.
3. Remove the wheels.
4. Remove the brake caliper.
5. Use a suitable piece of wire to suspend the caliper assembly from the upper control arm. This will prevent the weight of the caliper from being supported by the brake flex hose which will damage the hose.
6. Remove the caliper bracket bolts and remove the bracket.
7. Place match marks on the rotor and hub.
8. Remove the rotor. If the disc cannot be removed easily, turn the parking brake adjuster until the rotor turns freely and remove the rotor.
9. Use needle nosed pliers to remove the left hand No.1 parking brake shoe return springs.
10. Disconnect the tension spring from the left hand No.1 parking brake shoe.

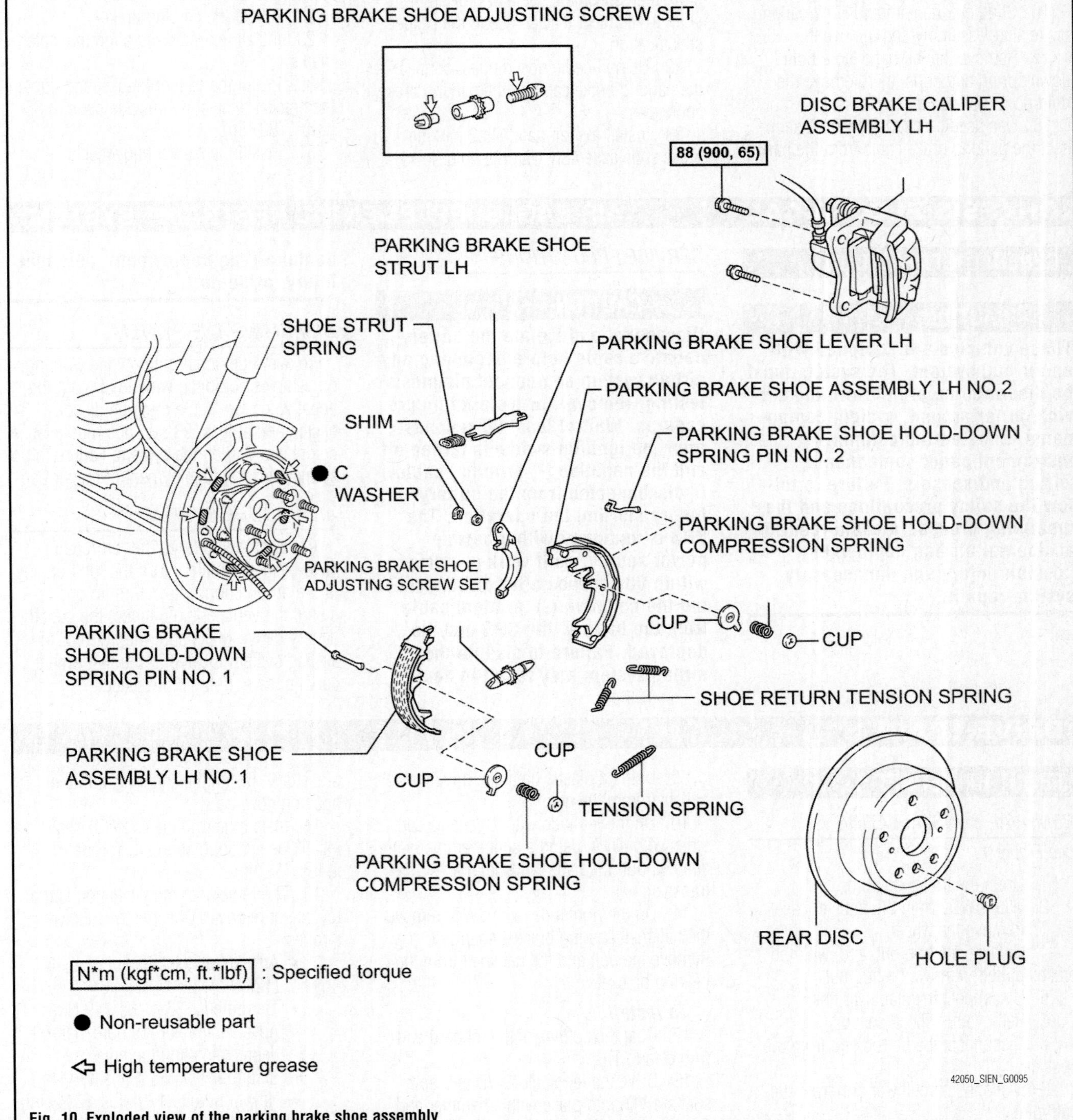

Fig. 10 Exploded view of the parking brake shoe assembly

11. Slide out the left hand No.1 parking brake shoe assembly and remove it.
12. Remove the left hand No.1 hold down spring and cups, then remove the hold down spring pin.
13. Remove the left hand No.1 parking brake shoe adjuster assembly.
14. Remove the left hand No.1 parking brake shoe strut.
15. Disconnect the return spring from the left hand No.2 parking brake shoe and remove the tension spring.
16. Slide out the left hand No.2 parking brake shoe assembly and remove it.
17. Remove the left hand No.2 hold down spring and cups, then remove the hold down spring pin.
18. Use needle nosed pliers to disconnect the parking brake cable from the parking brake shoe lever and the left hand No.2 parking brake shoe.

To install:

19. Apply high temperature grease to the surfaces where the parking brake shoe and backing plate contact.
20. Install the shoe lever and shim to the rear shoe using a new C-washer.
21. Measure the distance between the C-washer and shoe, the clearance should be 0.0138 inch (0.35mm). If the clearance is more than specified, use a 3, 6 0r 9mm shim to reduce the clearance to the proper specification.
22. Using needle nosed plies, connect the parking brake cable to the parking brake shoe lever.
23. Install the left hand No.2 parking brake shoe assembly with the hold down compression spring pin, compression spring and cups.
24. Apply high temperature grease to the adjuster assembly moving components and install the adjuster assembly.
25. Install the parking brake shoe strut.
26. Install the left hand No.1 parking brake shoe assembly with the hold down compression spring pin, compression spring and cups.
27. Install the tension spring and 2 return springs.
28. Adjust the parking brake.
29. Install the rotor and align the match marks.
30. Install the caliper bracket and tighten the bracket to steering knuckle bolts to 65 ft. lbs. (88 Nm).
31. Install the caliper and wheels.

CHASSIS ELECTRICAL

AIR BAG (SUPPLEMENTAL RESTRAINT SYSTEM)

GENERAL INFORMATION

CAUTION

These vehicles are equipped with an air bag system. The system must be disarmed before performing service on, or around, system components, the steering column, instrument panel components, wiring and sensors. Failure to follow the safety precautions and the disarming procedure could result in accidental air bag deployment, possible injury and unnecessary system repairs.

SERVICE PRECAUTIONS

CAUTION

Disconnect and isolate the battery negative cable before beginning any airbag system component diagnosis, testing, removal, or installation procedures. Wait at least 90 seconds after the ignition switch is turned off and the negative (-) terminal cable is disconnected from the battery before starting the operation. The SRS is equipped with a backup power source, so if work is started within 90 seconds after disconnecting the negative (-) terminal cable from the battery, the SRS may be deployed. Failure to disable the airbag system may result in accidental airbag deployment, personal injury, or death.

DISARMING THE SYSTEM

To avoid personal injury when working on vehicles equipped with an air bag, the negative battery cable must be disconnected and at least 90 seconds must elapse before working on the system. Failure to do so may result in deployment of the air bag.

ARMING THE SYSTEM

Connect the battery. After all repairs have been completed, turn the ignition key to the on position. Make sure no one is inside the vehicle and connect the negative battery cable. Make sure the light for the air bag system located in the instrument panel does not stay illuminated.

DRIVE TRAIN

FRONT HALFSHAFT

REMOVAL & INSTALLATION

See Figure 11.

1. Drain the transaxle fluid.
2. With 4WD, drain the transfer case.
3. Remove the wheel.
4. Unstake the hub nut, and, with the brake applied, remove the hub nut.
5. Disconnect the stabilizer link.
6. Remove the speed sensor.
7. Disconnect the tie rod end from the knuckle.
8. Disconnect the lower arm from the ball joint.
9. Using a plastic hammer, drive the halfshaft from the hub.
10. On the left side with 2WD and both sides with 4WD, using a slide hammer with adapter, pull the halfshaft from the transaxle.
11. On the right side, with 2WD, remove the halfshaft bearing bracket snapring. Remove the bolt and the halfshaft from the bearing bracket.

To install:

12. Coat the splines of the inboard end with clean ATF.
13. Drive the left shaft (2WD and both shafts 4WD) into place with a hammer and brass drift. Install the snapring with the opening downward.
14. Install the right side (2WD) shaft. Install the snapring and bolt. Torque to 24 ft. lbs. (32 Nm).
15. The remainder of installation is the reverse of removal. Observe the following torques:
    - Arm-to-ball joint: 94 ft. lbs. (127 Nm)
    - Tie rod end: 36 ft. lbs. (49 Nm). Advance the nut no more than 60 degrees to align the hole.
    - Stabilizer link: 55 ft. lbs. (74 Nm)
    - New hub nut: 217 ft. lbs. (294 Nm). Stake the nut.

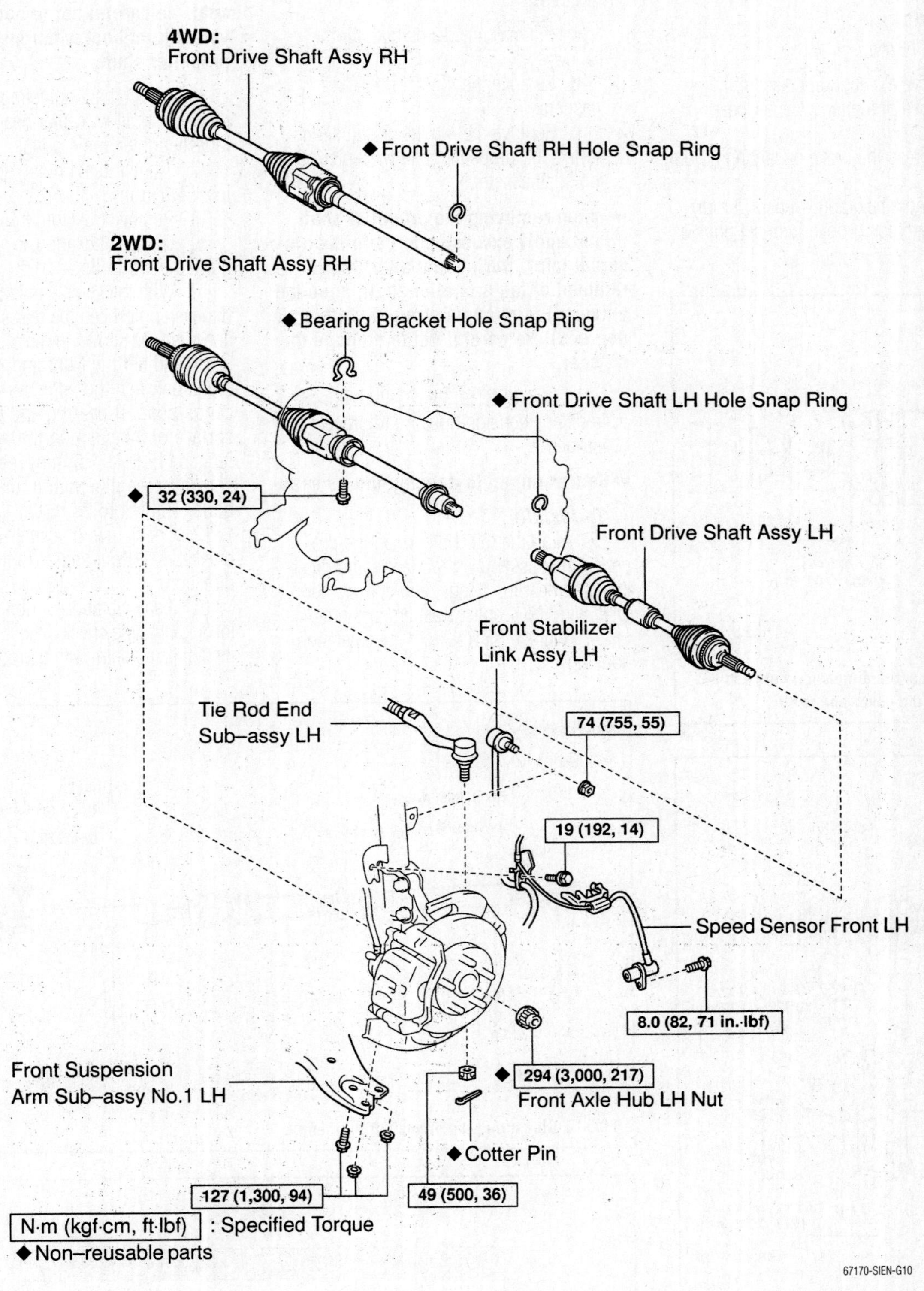

Fig. 11 Front halfshaft exploded view

REAR DRIVESHAFT

REMOVAL & INSTALLATION

4WD Models

See Figures 12 through 15.

1. Remove exhaust pipe assembly.
2. Remove propeller w/center bearing shaft assembly:
   a. Depress the brake pedal and hold it down.
   b. Using a hexagon wrench (6 mm), loosen the cross groove joint set bolts a half turn.

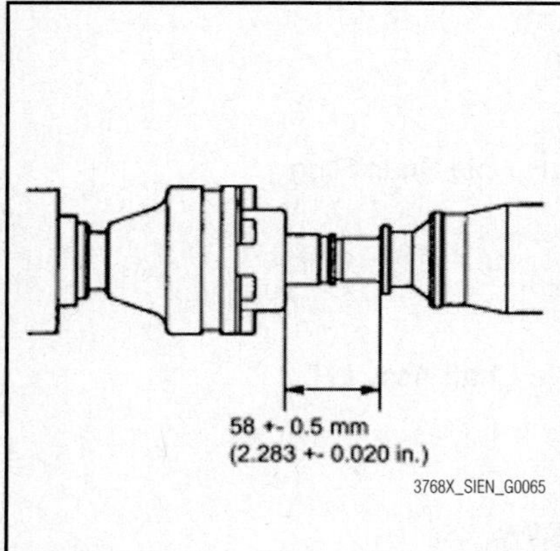

Fig. 12 Adjust the dimension between the rear side of the cover and shaft

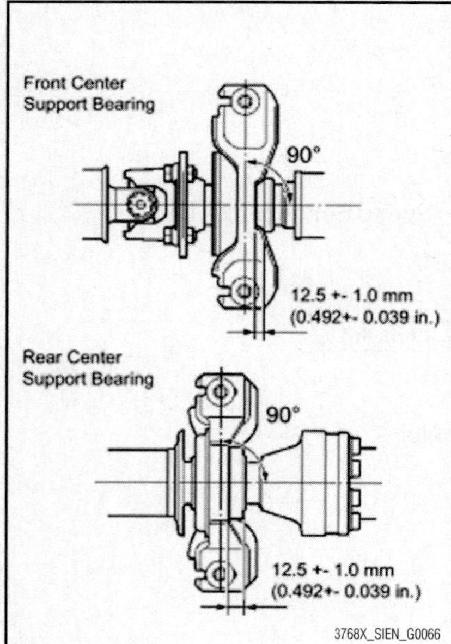

Fig. 13 Adjust the front and rear dimensions between the edge surface of the center support bearing and the edge surface of the cushion

➡Place a cloth in the inside of the universal joint cover so that the boot does not touch the inside of the universal joint cover.

   c. Put matchmarks on both the flanges.
   d. Remove the 4 nuts, bolts and washers.
   e. Remove the 4 bolts, 2 adjusting shims and propeller shaft w/ center bearing shaft assembly.

➡When removing the propeller shaft, do not apply excessive force to the universal joint. During and after the removal of the propeller shaft, keep the universal joint angle straight (within 15 degrees). Be careful not to damage the oil seal.

   f. Insert SST(s) 09325-20010 in the transfer extension housing to prevent oil leakage.

➡Be careful not to damage the oil seal.

To install:

3. Align the matchmarks on the propeller shaft assembly rear flange and differential companion flange, and connect the shaft with the 4 bolts, washers and nuts.
4. Remove SST(s) from the extension housing.
5. Insert the yoke into the extension housing.

➡Be careful not to damage the oil seal. Be careful not to damage the universal joint boot when installing the propeller shaft.

6. Install the 2 adjusting shims and propeller shaft w/ center bearing, and loosely tighten the 4 bolts.
7. Tighten the 4 nuts. Tighten to 54 ft. lbs. (74 Nm).
8. Remove the cloth from the joint.
9. Using a hexagon wrench (6 mm), tighten the 6 bolts to 21 ft. lbs. (29 Nm).
10. With the vehicle unloaded, adjust the dimension between the rear side of the cover and shaft, as shown in the illustration.
11. Adjust the front and rear dimensions between the edge surface of the center support bearing and the edge surface of the cushion to 0.492 +- 0.039 inches (12.5 +- 1.0 mm) respectively as shown in the illustration, then torque the bolts. Tighten to 27 ft. lbs. (37 Nm).
12. Check that the center line of the bracket is at right angles to the shaft axial direction.
13. If any vibration or noise occurs, perform joint angle check as follows and replace the adjusting shim with a proper one:

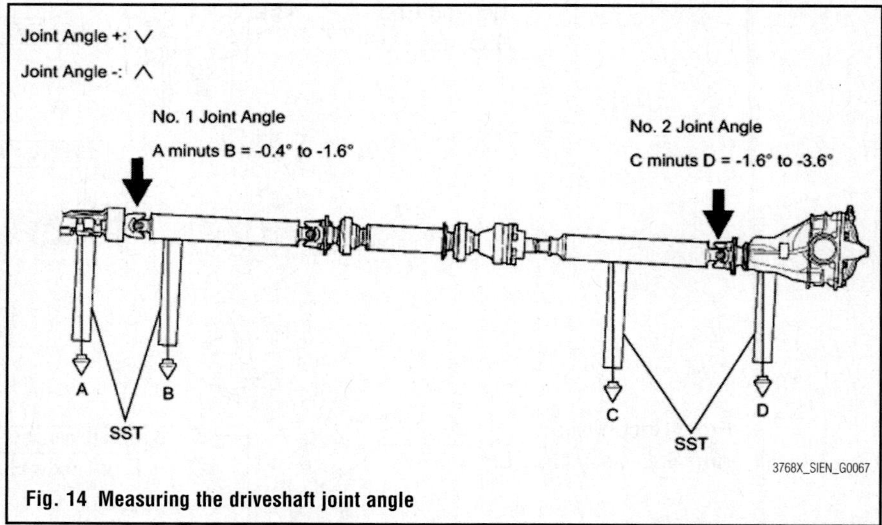

Fig. 14 Measuring the driveshaft joint angle

THICKNESS MM (IN.)	THICKNESS MM (IN.)
3.2 (0.126)	4.5 (0.177)
6.5 (0.256)	9.0 (0.354)

3768X_SIEN_G0068

Fig. 15 Center support bearing adjusting shim thickness

a. Turn the propeller shaft several times by hand to stabilize the center support bearings.

b. Using a jack, raise and lower the differential to stabilize the differential mounting cushion.

c. Remove the transfer dynamic damper.

d. Using SST(s), measure the installation angle of the transfer extension housing (A) and front propeller shaft (B).

- No. 1 joint angle: A minus B = -0.4° to -1.6°

e. Using SST(s), measure the installation angle of the rear propeller shaft (C) and rear differential (D).

- No. 2 joint angle: C minus D = -1.6° to -3.6°

14. If the measured angle is not within the specification, adjust with the center support bearing adjusting shim.

15. Install the transfer dynamic damper. Tighten to 19 ft. lbs. (26 Nm).

16. Install exhaust pipe assembly.

REAR HALFSHAFT

REMOVAL & INSTALLATION

See Figure 16.

1. Remove the wheel.
2. Remove the tail pipe.
3. Remove the speed sensor.
4. Unstake and remove the axle shaft nut.
5. Matchmark the halfshaft and differential side gear.
6. Remove the 4 nuts and washers and remove the shaft.
7. Installation is the reverse of removal. Torque the 4 nuts t 41 ft. lbs. (56 Nm). Torque the axle shaft nut to 159 ft. lbs. (216 Nm). Stake the nut.

REAR PINION SEAL

REMOVAL & INSTALLATION

See Figures 17 through 22.

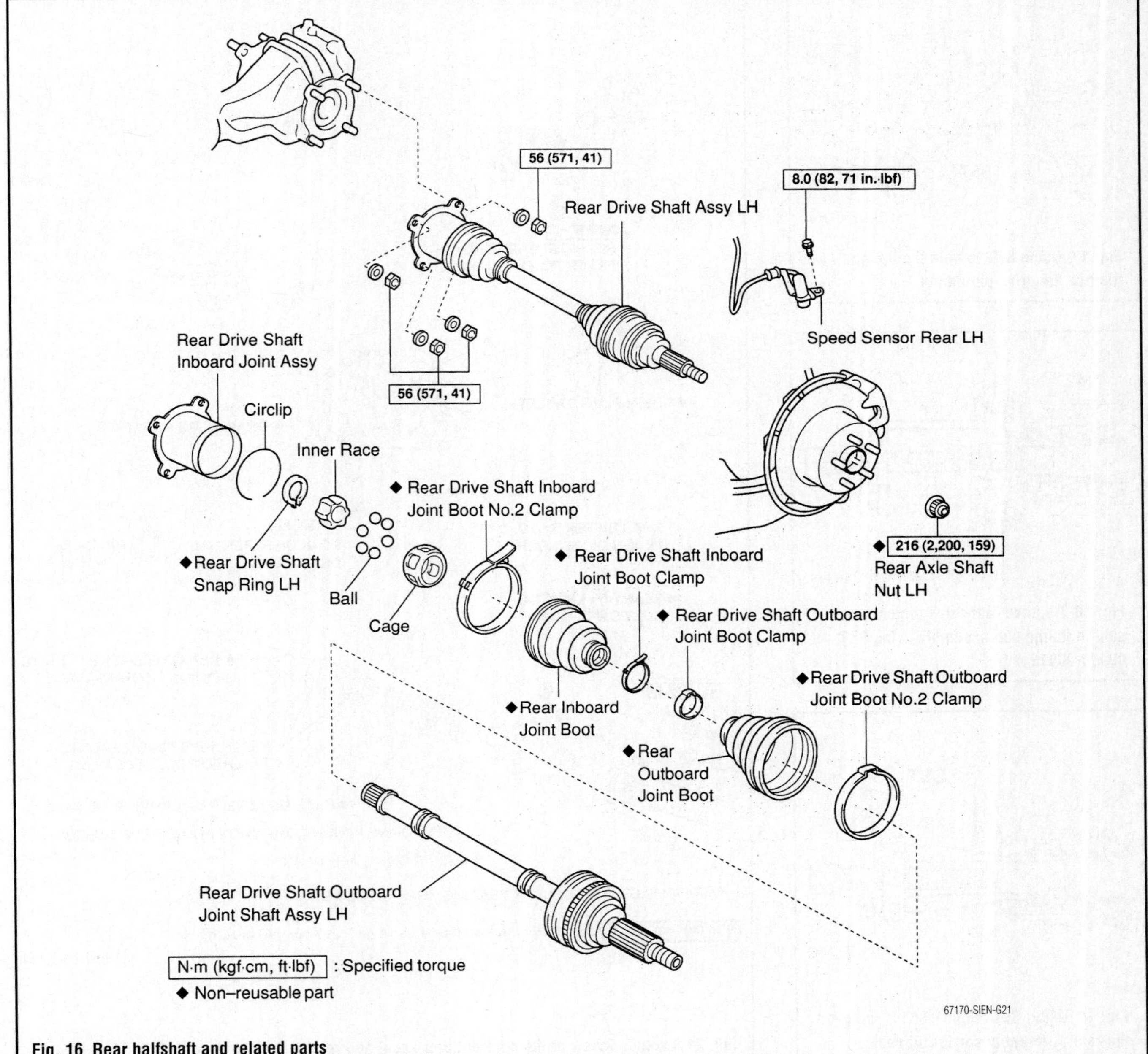

Fig. 16 Rear halfshaft and related parts

1. Remove front exhaust pipe assembly.
2. Remove driveshaft with center bearing shaft assembly.
3. Using a hexagon wrench (10 mm), remove the filler plug and gasket.
4. Using a hexagon wrench (10 mm), remove the drain plug and gasket, and drain the oil.
5. Remove rear drive pinion nut:
   a. Using a chisel and a hammer, unstake the staked part of the nut.
   b. Using SST 09330-00021 to hold the flange, remove the nut.
6. Remove rear drive pinion companion flange sub-assembly using SST: 09950-30012.
7. Using SST 09308-10010, remove the rear differential carrier oil seal.
8. Remove rear differential drive pinion oil slinger.
9. Using SST 09556-22010, remove the front tapered roller bearing.
10. Remove rear differential drive pinion bearing spacer.

To install:

11. Install a new bearing spacer.
12. Install the tapered roller bearing.
13. Install the oil slinger, as shown in the illustration.

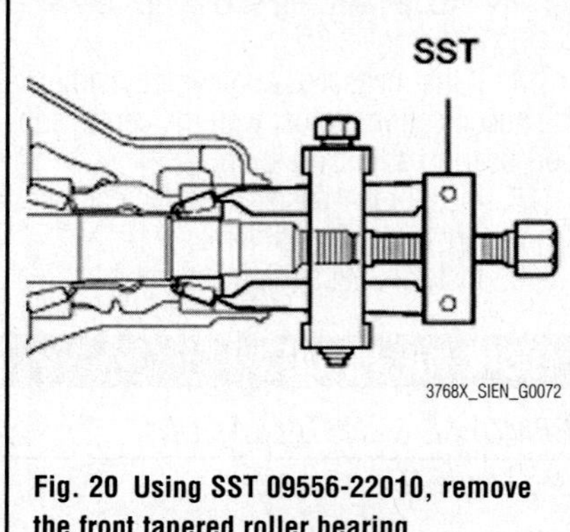

Fig. 20 Using SST 09556-22010, remove the front tapered roller bearing

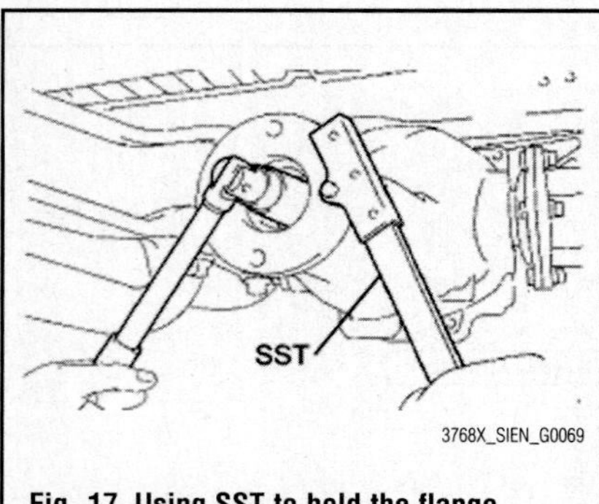

Fig. 17 Using SST to hold the flange, remove the drive pinion nut

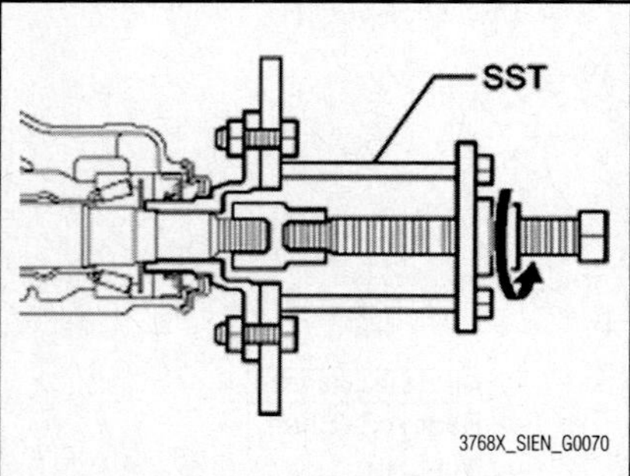

Fig. 18 Remove rear drive pinion companion flange sub-assembly using SST: 09950-30012

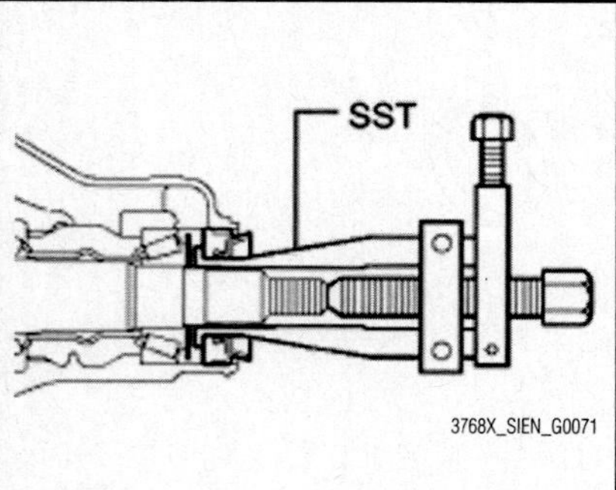

Fig. 19 Using SST 09308-10010, remove the rear differential carrier oil seal

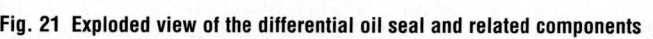

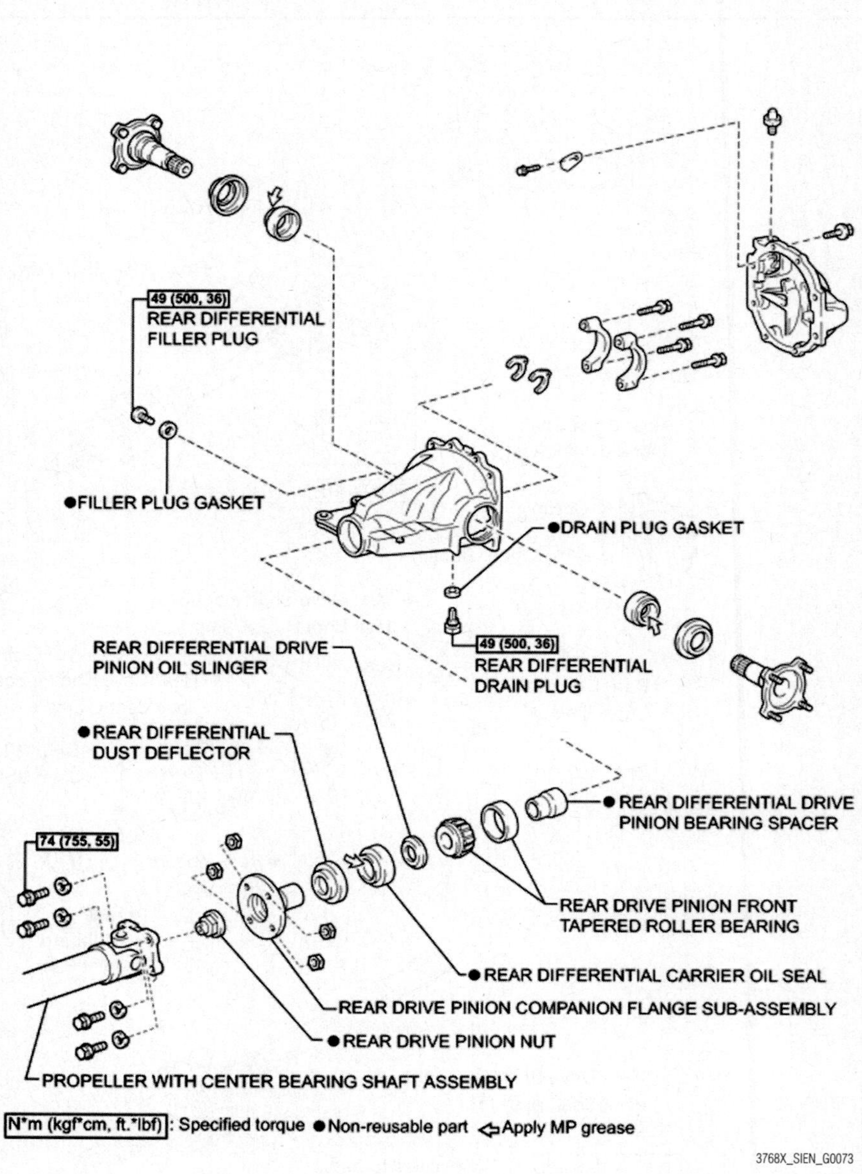

Fig. 21 Exploded view of the differential oil seal and related components

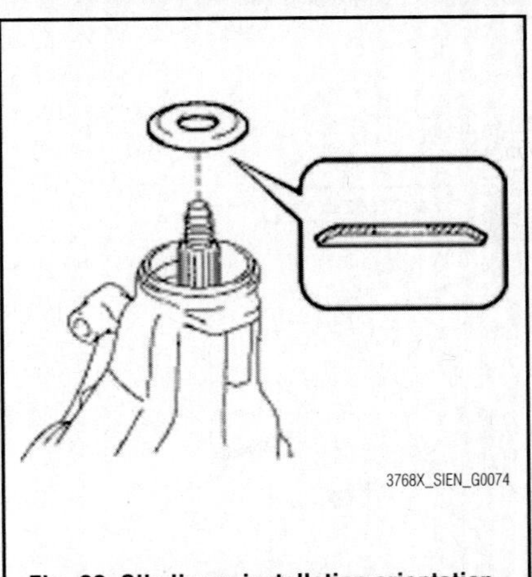

Fig. 22 Oil slinger installation orientation

14. Using SST 09554-22010 and a hammer, install a new oil seal.
15. Coat MP grease to the oil seal lip.
16. Using SST 09950-30012, install the companion flange on the shaft.

➡Apply hypoid gear oil to the SST center bolt tip and threads before use.

17. Install rear drive pinion nut:
a. Coat the threads of a new nut with hypoid gear oil LSD.
b. Using SST 09330-00021 to hold the flange, torque the nut. Tighten to 80 ft. lbs. (108 Nm).
18. Inspect differential drive pinion preload. Using a chisel and a hammer, stake the drive pinion nut.
19. Using a hexagon wrench (10 mm), install the filler plug with a new gasket. Tighten to 36 ft. lbs. (49 Nm).
20. Fill the rear differential carrier assembly with hypoid gear oil.
21. Inspect the differential oil level.
22. Using a hexagon wrench (10 mm), install the filler plug with a new gasket. Tighten to 36 ft. lbs. (49 Nm).
23. Install propeller with center bearing shaft assembly.
24. Fully tighten propeller with center bearing shaft assembly.
25. To complete the installation, reverse the remaining removal procedure.

ENGINE COOLING

ENGINE FAN

REMOVAL & INSTALLATION

See Figure 23.

1. Before servicing the vehicle, refer to the precautions section.
2. Disconnect the negative battery cable.
3. Drain and recycle the engine coolant.
4. Disconnect the radiator hoses.
5. Disconnect the oil cooler hoses from the oil cooler pipes.
6. Remove the front bumper cover.
7. Remove the air cleaner assembly.
8. Release the 4 claws from the hood lock protector and remove the protector.
9. Unfasten the 3 bolts and remove the hood lock assembly.
10. Unplug the horn connectors, unfasten the radiator upper support bolts and remove the support.
11. Disconnect the coolant reservoir hose and unplug the fan motor connector.

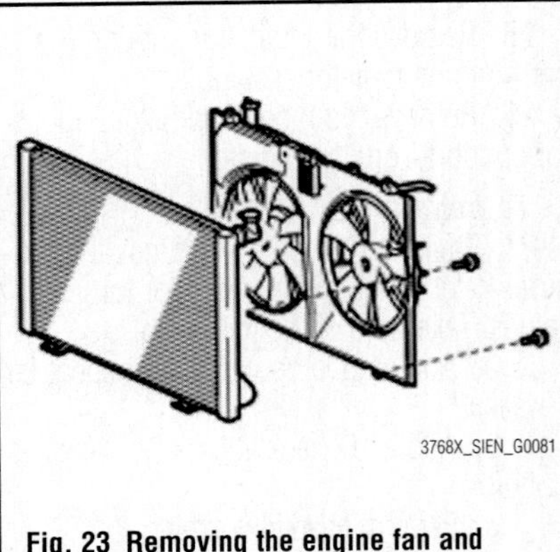

Fig. 23 Removing the engine fan and shroud

12. Remove the 4 bolts and the fan motor assembly.
13. Remove the radiator.

To install:

14. Install the fan motor assembly, attach the motor connector and the coolant reservoir hose.
15. Install the radiator upper support and bolts. Attach the horn connectors.
16. Install the hood lock assembly.
17. install the hood lock protector.
18. Install the air cleaner assembly.
19. Install the front bumper cover.
20. Connect the oil cooler hoses from the oil cooler pipes.
21. Connect the radiator hoses.
22. Fill the cooling system.
23. Connect the negative battery cable.

RADIATOR

REMOVAL & INSTALLATION

See Figures 24 through 30.

1. Before servicing the vehicle, refer to the precautions section.
2. Disconnect the negative battery cable.
3. Drain and recycle the engine coolant.
4. Remove no. 2 air cleaner inlet.
5. Remove battery.
6. Remove front bumper assembly:
a. Remove the 4 screws and separate the fender liner from the front bumper assembly.
b. Remove the 8 screws and separate the engine under cover from the front bumper assembly.
c. Remove the 5 clips.
d. Disengage the claws on the left and right sides of the front bumper and pull the front bumper toward the front of the vehicle.

➡Apply protective tape to the bottom part of the front fender to prevent it from being damaged. Disconnect the connectors if necessary.

7. Remove front bumper energy absorber.
8. Disconnect the radiator reserve tank hose or pipe from the radiator.
9. Disconnect both connections of the No. 1 radiator hose from the radiator.
10. Disconnect the No. 2 radiator hose from the radiator.
11. Disengage the 4 claws and remove the hood lock release lever protector.
12. Remove hood lock assembly:
a. With Hood Courtesy Switch: Disconnect the connector.
b. Remove the 3 bolts.
c. Disconnect the cable and remove the hood lock assembly.
13. Remove hood lock support sub-assembly:
a. Disconnect the ambient temperature sensor connector.
b. Detach the clamp.
c. Disconnect the low pitched horn and high pitched horn connectors.
d. Remove the 2 bolts and hood lock support sub-assembly.
14. Disconnect cooling fan ECU connector:
a. Disconnect the cooling fan ECU connector.
b. Detach the clamp.
15. Remove the 4 bolts and radiator upper support sub-assembly.
16. Disconnect no. 2 oil cooler outlet tube sub-assembly:

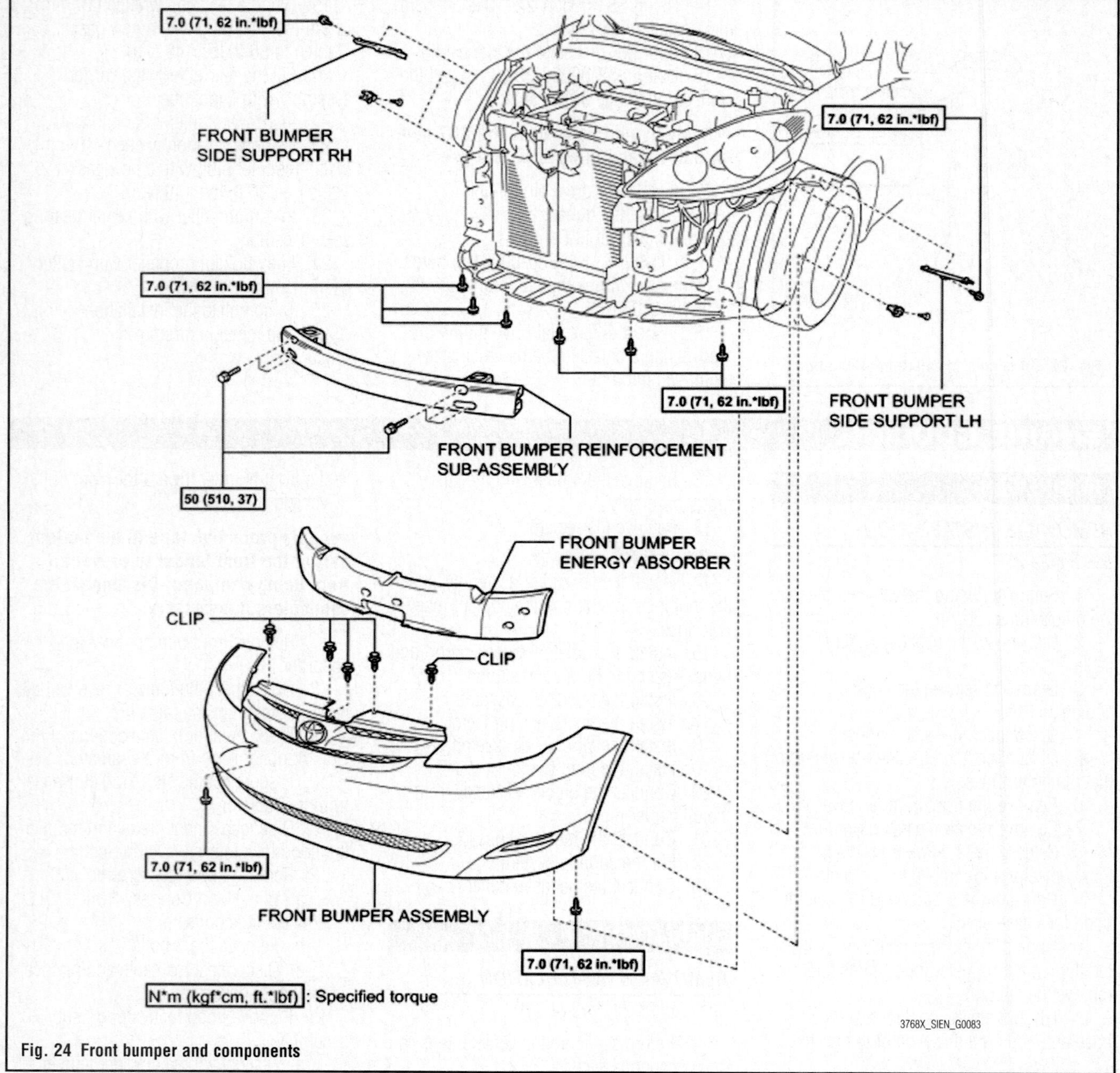

Fig. 24 Front bumper and components

a. Disconnect the No. 1 oil cooler inlet and No. 1 oil cooler outlet hoses from the No. 2 oil cooler outlet tube sub-assembly.

b. Disconnect the No. 1 transmission oil cooler hoses from the radiator.

c. Remove the 2 bolts and No. 2 oil cooler outlet tube sub-assembly.

17. Remove the headlight assembly.

18. Disengage the 3 claws and remove the radiator side deflector RH.

19. Remove the 3 nuts and headlight bracket RH.

20. Remove the two bolts and the pressure feed tube assembly.

21. Remove the 2 radiator support cushions from the No. 1 radiator support.

22. Remove no. 1 radiator support:

a. Detach the clamp.

b. Remove the 6 bolts and No. 1 radiator support.

23. Remove radiator assembly with fan shroud and fan motor:

a. Remove the 2 bolts and separate the condenser assembly from the radiator assembly with fan shroud and fan motor.

b. Remove the radiator assembly with fan shroud and fan motor from the body.

24. Remove the 2 bolts and fan shroud with fan motor from the radiator.

25. Remove the 2 radiator support lowers from the radiator assembly.

26. Remove the 2 bolts and No. 2 radiator support from the radiator.

To install:

27. To install, reverse the removal procedure. Refer to the illustrations for torque values. Take note of the following:

a. Add engine coolant and inspect for leaks.

b. Add and check automatic transaxle fluid.

c. Perform headlight aiming procedure.

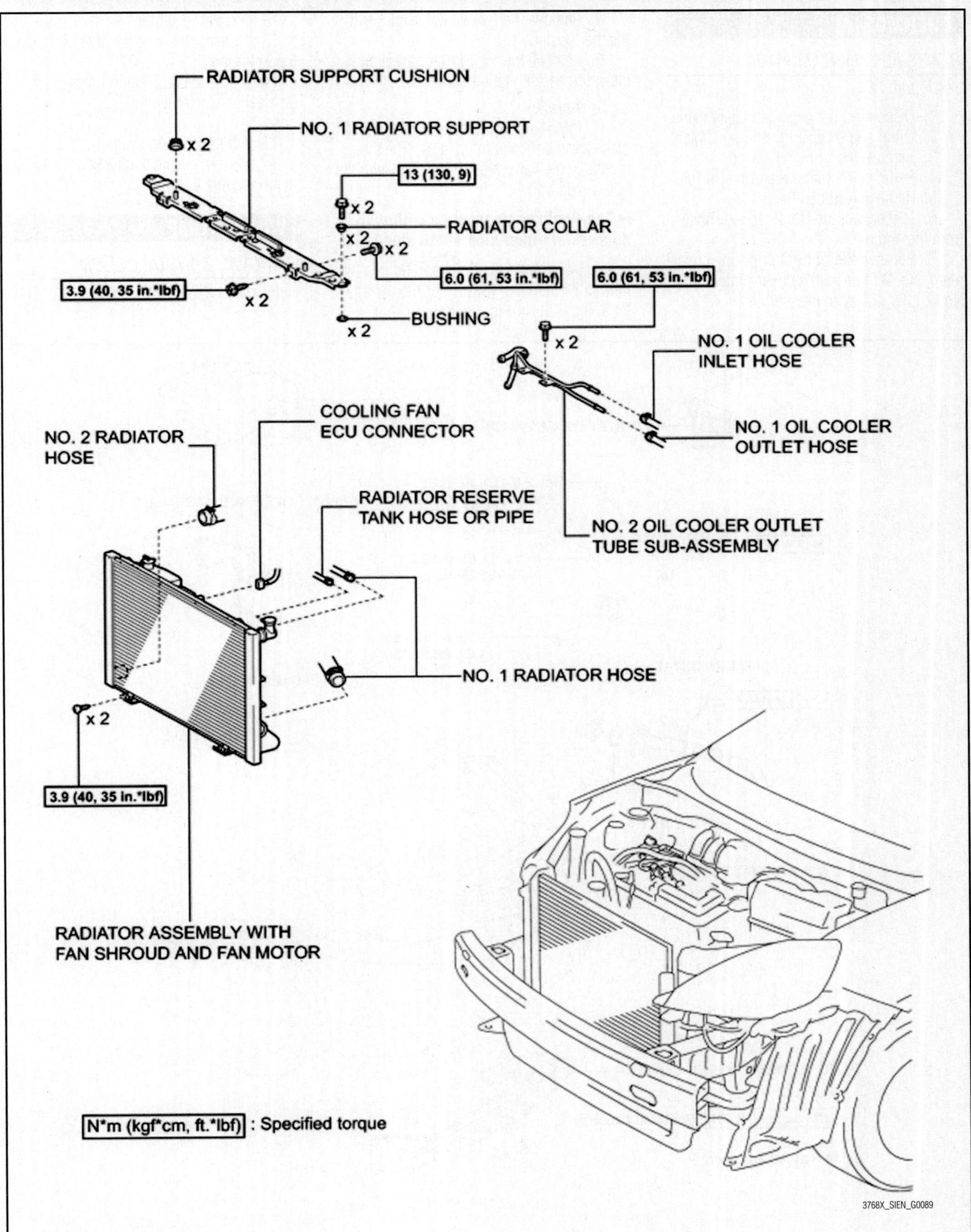

Fig. 25 Radiator and related components

THERMOSTAT

REMOVAL & INSTALLATION

See Figure 26.

1. Disconnect the negative battery cable.
2. Drain and recycle the engine coolant.
3. Remove the front RH wheel.
4. Remove front fender apron seal RH.
5. Remove v-ribbed belt.
6. Disconnect the No. 2 radiator hose from the engine.
7. Remove the bolt, idler pulley cover plate, No. 2 idler pulley cover plate and No. 2 idler pulley sub-assembly.
8. Remove the 2 nuts, water inlet and the thermostat.
9. Remove the thermostat gasket and clean the gasket mating surfaces.

To install:

10. Install a new thermostat gasket then place the thermostat in position. The thermostat jiggle valve must be aligned with the upper stud bolt.

➡The jiggle valve must be within 15 degrees of either side of the stud bolt.

11. Install the water inlet and tighten the nuts to 71 inch lbs. (8 Nm) for the nut and 14 ft. lbs. (20 Nm) for the bolt.
12. Install the starter.
13. Connect the wire harness clamp to the water inlet.
14. Install the air cleaner bracket, air cleaner assembly and hose.
15. Fill the cooling system
16. Start the vehicle, check for leaks and recheck coolant level.

WATER PUMP

REMOVAL & INSTALLATION

See Figures 27 and 28.

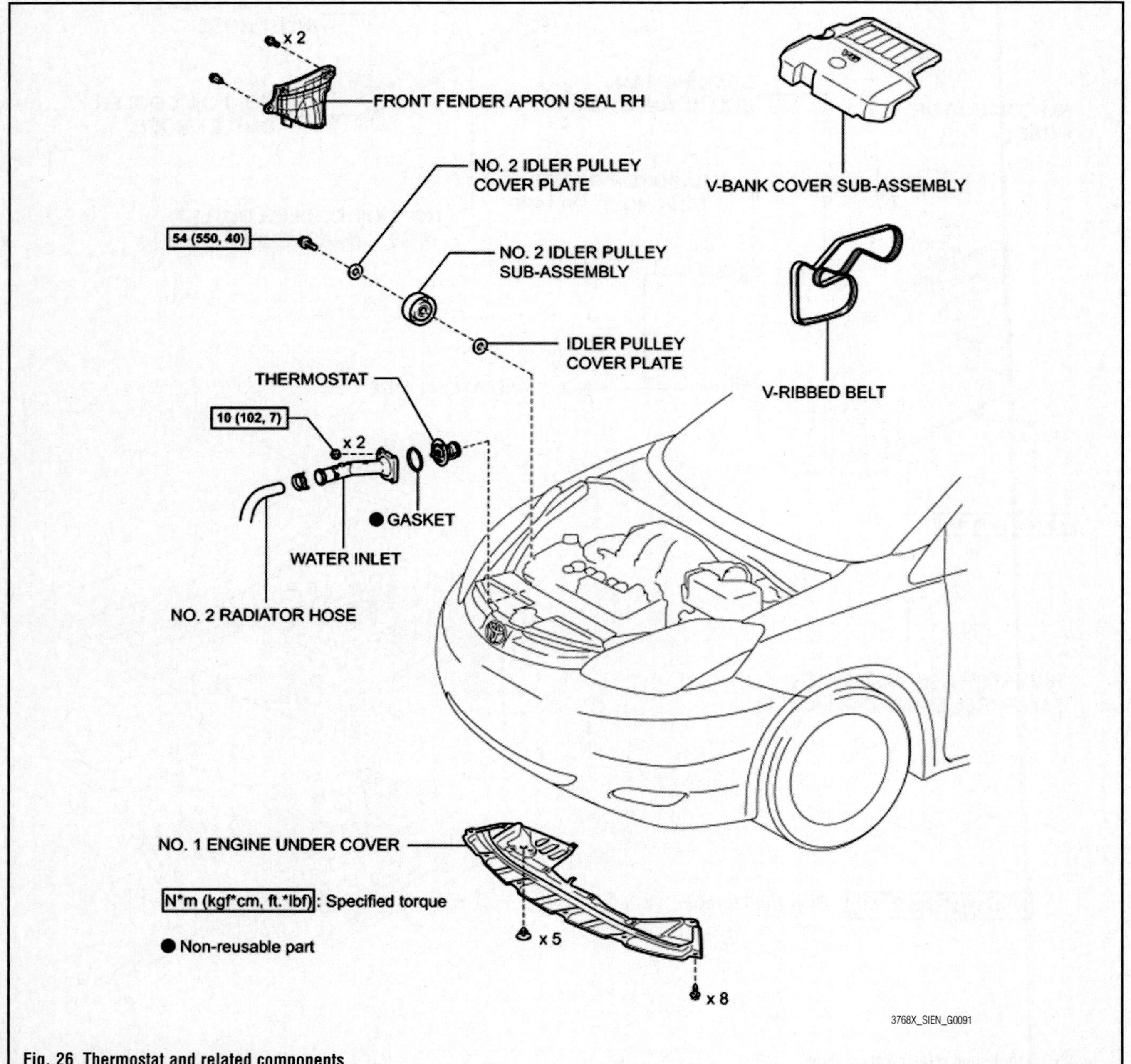

Fig. 26 Thermostat and related components

1. Before servicing the vehicle, refer to the precautions section.
2. Properly relieve the fuel system pressure.
3. Disconnect the negative battery cable.
4. Drain and recycle the engine coolant.
5. Remove engine assembly with transaxle.
6. Secure the engine.
7. Remove the alternator.
8. Remove the A/C compressor.
9. Remove no. 1 engine front mounting bracket.
10. Remove the 2 bolts, 2 idler pulley cover plates, 2 No. 2 idler pulley cover plates and 2 No. 2 idler pulley sub-assemblies.
11. Remove the 5 bolts and V-ribbed belt tensioner assembly.
12. Using SST, hold the water pump pulley. Remove the 4 bolts and water pump pulley.
13. Remove water inlet housing:
    a. Disconnect the water hose.
    b. Remove the 2 bolts, nut and water inlet housing.
    c. Remove the water inlet housing gasket and water outlet pipe O-ring.
14. Remove the 16 bolts, water pump assembly and water pump gasket.

To install:

15. Install a new water pump gasket and the water pump assembly with the 16 bolts.
    a. Bolt A: Tighten to 15 ft. lbs. (21 Nm).
    b. Bolt B: Tighten to 8 ft. lbs. (11 Nm).

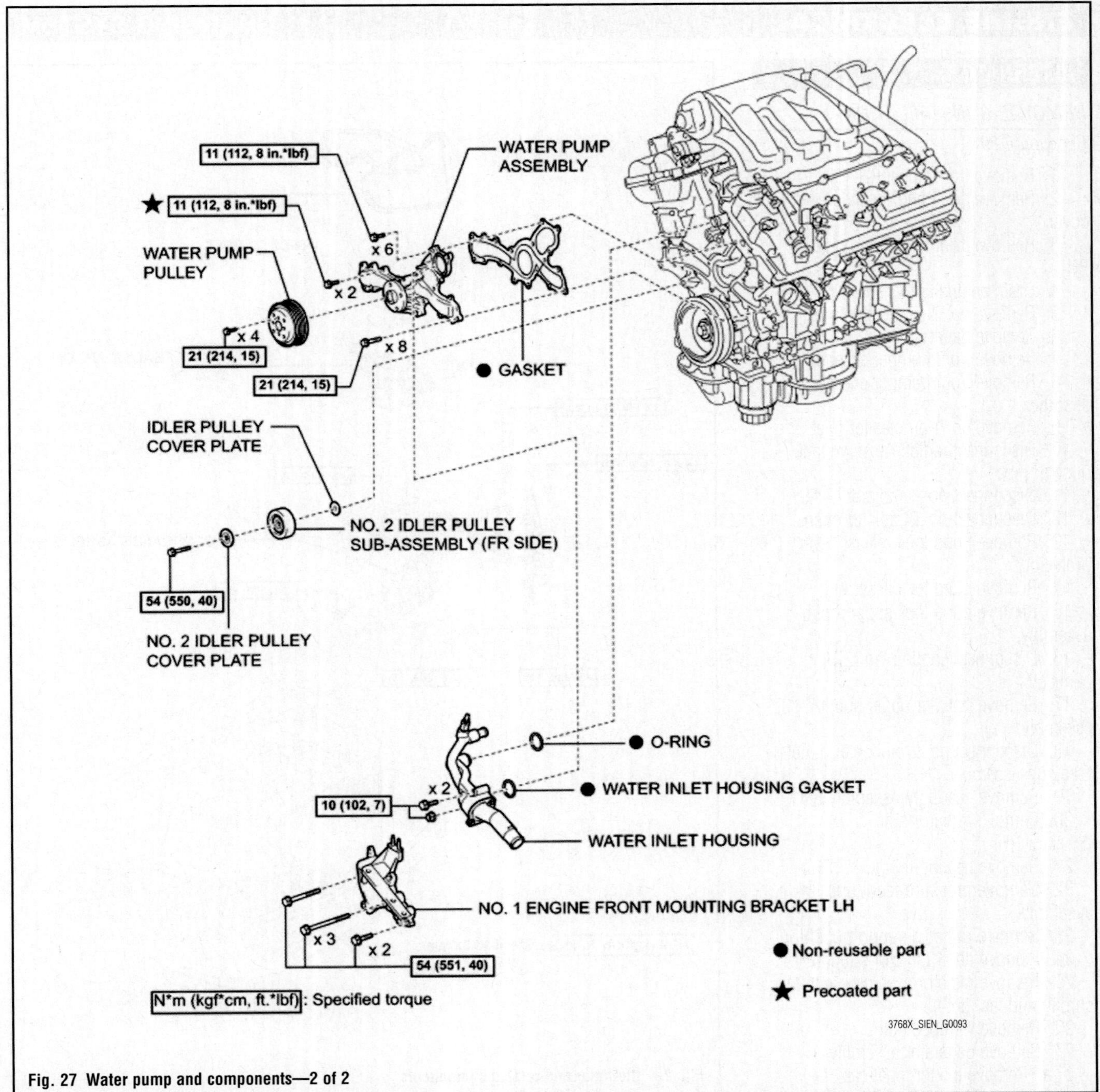

Fig. 27 Water pump and components—2 of 2

➡Make sure that there is no oil on the threads of bolts A. Be sure to replace 2 bolts C with new ones or reuse them after applying adhesive 1344. Adhesive: Toyota Genuine Adhesive 1344, Three Bond 1344 or equivalent.

16. Install the remaining components in the reverse order of removal.
17. Fill the cooling system and check for leaks.

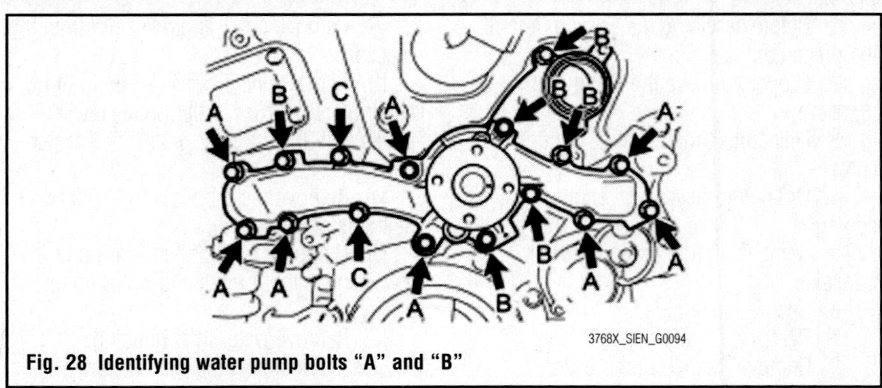

Fig. 28 Identifying water pump bolts "A" and "B"

ENGINE ELECTRICAL

CHARGING SYSTEM

ALTERNATOR

REMOVAL & INSTALLATION

See Figure 29.

1. Remove front wheel RH.
2. Remove no. 1 engine under cover.
3. Remove front fender apron seal RH.
4. Drain engine coolant.
5. Remove no. 2 air cleaner inlet.
6. Remove battery.
7. Remove front bumper assembly.
8. Remove front bumper energy absorber.
9. Remove no. 1 air cleaner inlet.
10. Disconnect radiator reserve tank hose or pipe.
11. Disconnect no. 1 radiator hose.
12. Disconnect no. 2 radiator hose.
13. Remove hood lock release lever protector.
14. Remove hood lock assembly.
15. Remove hood lock support sub-assembly.
16. Disconnect cooling fan ECU connector.
17. Remove radiator upper support sub-assembly.
18. Disconnect no. 2 oil cooler outlet tube sub-assembly.
19. Remove headlight assembly RH.
20. Remove radiator side deflector RH.
21. Remove headlight bracket RH.
22. Remove pressure feed tube assembly.
23. Remove radiator support cushion.
24. Remove no. 1 radiator support.
25. Remove radiator assembly with fan shroud and fan motor.
26. Remove v-ribbed belt
27. Remove generator assembly.
    a. Remove the terminal cap.

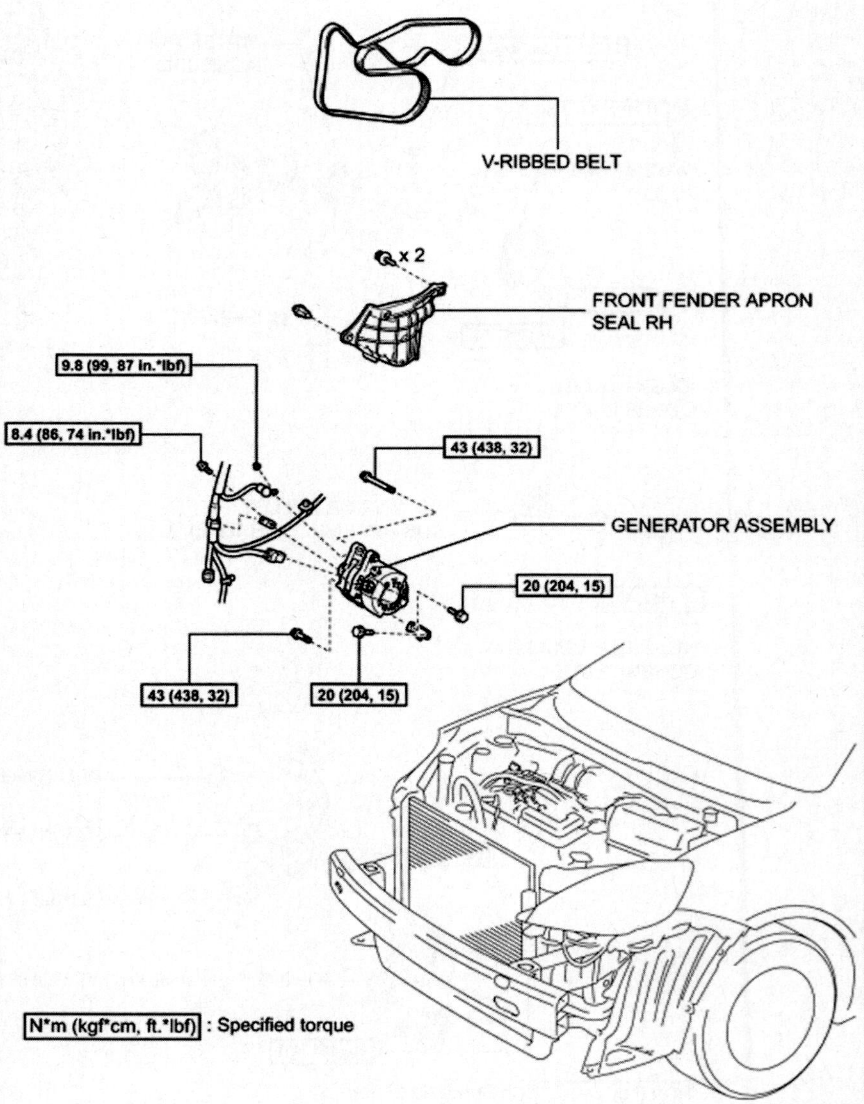

Fig. 29 Alternator and related components

b. Remove the nut and disconnect the wire harness from terminal B.

c. Disconnect the generator connector from the generator assembly.

d. Disconnect the connector from the compressor and magnetic clutch.

e. Disconnect the 2 wire harness clamps.

f. Remove the 2 bolts.

g. Remove the bolt from the cylinder block.

h. Disconnect the wire harness clamp and remove the generator assembly.

28. To install, reverse the removal procedure. Refer to the illustration for torque values.

ENGINE ELECTRICAL

IGNITION SYSTEM

FIRING ORDER

See Figure 30.

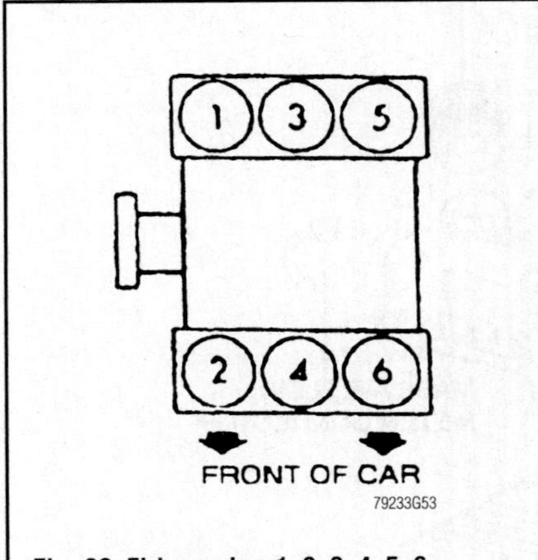

Fig. 30 Firing order: 1–2–3–4–5–6 Distributorless ignition system

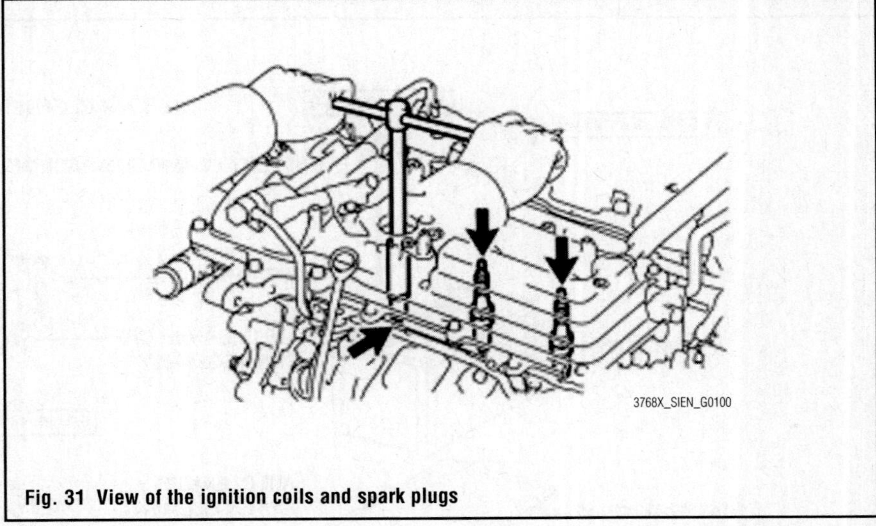

Fig. 31 View of the ignition coils and spark plugs

IGNITION COIL

REMOVAL & INSTALLATION

See Figure 40.

1. Disconnect cable from negative battery terminal.
2. Drain engine coolant.
3. Remove front wiper arm head cap.
4. Remove front wiper arm RH.
5. Remove front wiper arm LH.
6. Remove cowl top ventilator louver sub-assembly.
7. Remove windshield wiper motor and link assembly.
8. Remove no. 1 cowl top to cowl brace inner.
9. Remove cowl top panel sub-assembly outer front.
10. Remove no. 1 engine under cover.
11. Remove v-bank cover sub-assembly.
12. Remove air cleaner cap sub-assembly.
13. Remove intake air surge tank assembly.
14. Remove no. 1 surge tank stay.
    a. Remove the 2 bolts and nut and disconnect the 2 harness clamps.
    b. Remove the bolt and No. 1 surge tank stay.
15. Remove ignition coil assembly:
    a. Disconnect the 6 ignition coil connectors.
    b. Remove the 6 bolts and 6 ignition coils.
16. To install, reverse the removal procedure. Tighten the ignition coil bolts to 7 ft. lbs. (10 Nm).

IGNITION TIMING

The ignition timing is controlled by the Powertrain Control Module (PCM). No adjustment is necessary or possible.

SPARK PLUGS

REMOVAL & INSTALLATION

See Figure 31.

1. Remove the ignition coils.
2. Remove the 6 spark plugs.
3. To install, tighten to 13 ft. lbs. (18 Nm).

STARTER

REMOVAL & INSTALLATION

See Figure 32.

1. Before servicing the vehicle, refer to the precautions section.
2. Remove the battery and tray.
3. Remove the air cleaner assembly and inlet tubes.
4. Remove the air cleaner bracket.
5. Remove the wiring from the starter.
6. Remove the 2 bolts and lower the starter.
7. Installation is the reverse of removal. Torque the starter bolts to 26 ft. lbs. (37 Nm).

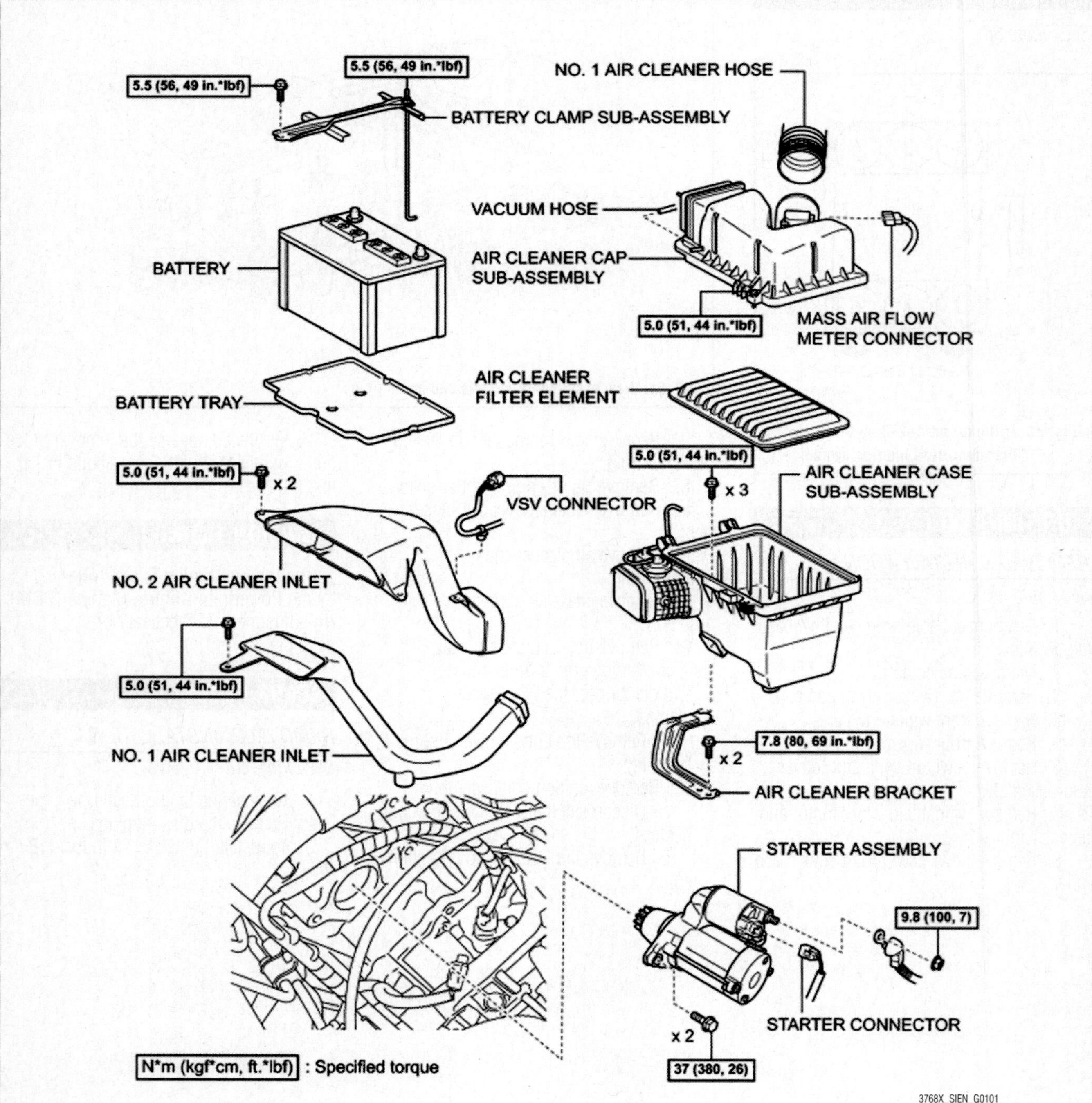

Fig. 32 Removing the starter and related components

ENGINE MECHANICAL

ACCESSORY DRIVE BELTS

ACCESSORY BELT ROUTING

See Figure 33.

Refer to accompanying illustration for belt routing.

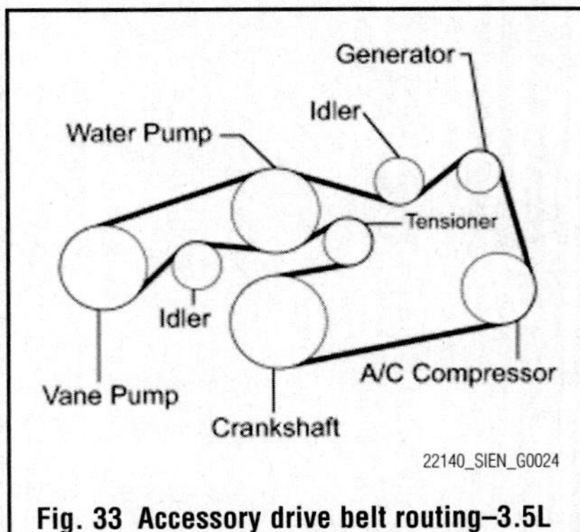

Fig. 33 Accessory drive belt routing–3.5L engine

INSPECTION

See Figure 34.

Inspect the drive belt for signs of glazing or cracking. A glazed belt will be perfectly smooth from slippage, while a good belt will have a slight texture of fabric visible. Cracks will usually start at the inner edge of the belt and run outward. All worn or damaged drive belts should be replaced immediately.

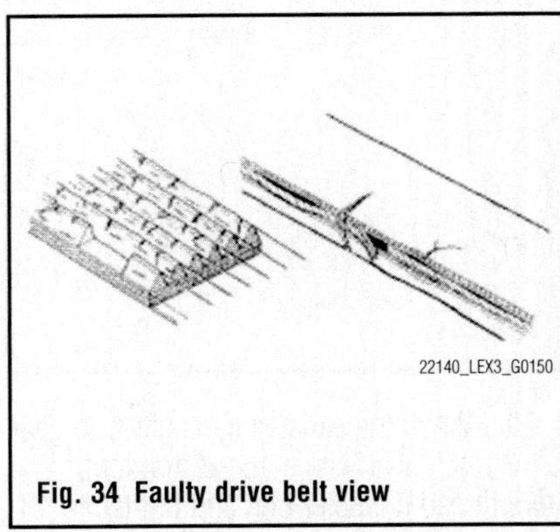

Fig. 34 Faulty drive belt view

ADJUSTMENT

These engines are equipped with automatic belt tensioners. Adjusting the belt tension is not possible or necessary.

REMOVAL & INSTALLATION

See Figures 35 and 36.

1. Before servicing the vehicle, refer to the precautions section.

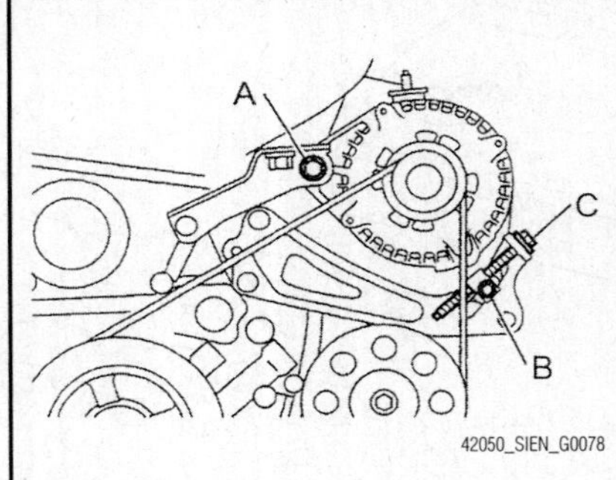

Fig. 35 A/C compressor to crankshaft pulley belt bolt locations

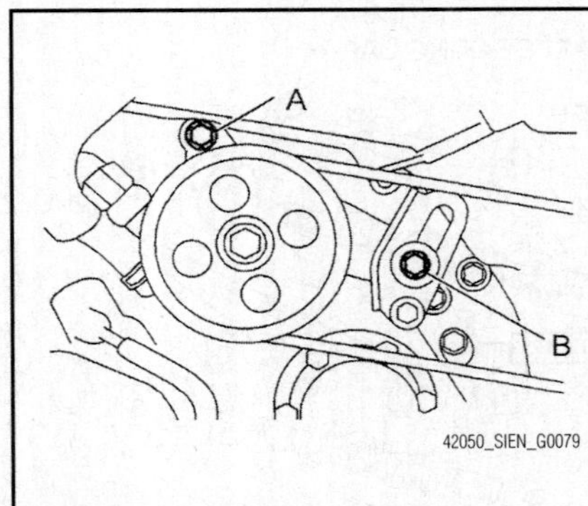

Fig. 36 Vane pump pulley belt bolt locations

2. Remove the right hand front wheel and apron seal.
3. Remove the A/C compressor to crankshaft pulley belt by loosening bolts A and B, then loosen adjuster bolt C and remove the belt.
4. Remove the vane pump belt by loosening bolts A and B, then remove the belt.

To install:

5. Install the vane pump belt and tighten the bolts to 32 ft. lbs. (44 Nm).
6. Install the A/C compressor to crankshaft pulley belt over the pulleys. Tighten adjuster bolt C to adjust the belt tension.
7. Tighten bolt A to 43 ft. lbs. (58 Nm) and bolt B to 13 ft. lbs. (18 Nm).
8. Install the apron seal and front wheel. Tighten the wheel nuts to 76 ft. lbs. (103 Nm)

CAMSHAFT AND VALVE LIFTERS

INSPECTION

1. Place the camshaft on V-blocks.
2. Using a dial indicator, measure the circle runout at the center journal.
3. Maximum circle runout: 0.04 mm (0.0016 in.)
4. If the circle runout is greater than the maximum, replace the camshaft.

REMOVAL & INSTALLATION

Left Side

See Figures 37 through 49.

1. Before servicing the vehicle, refer to the precautions section.
2. Properly relieve the fuel system pressure.
3. Drain and recycle the engine coolant.
4. Disconnect the negative battery cable.
5. Remove the right hand front wheel.
6. Remove the wiper arms.
7. Remove the cowl panel.
8. Remove the wiper link assembly.
9. Remove the cowl top to inner brace.
10. Remove the cowl top cover outer sub assembly.
11. Remove the radiator inlet hose.
12. Remove the ignition coil.
13. Remove the cylinder head cover.
14. Remove the apron seal.
15. Remove the A/C compressor to crankshaft pulley belt.
16. Remove the vane pump belt.
17. Remove the engine moving control rod bolts and rod.
18. Remove the number 2 engine mounting stay.
19. Remove the number 2 alternator bracket and the alternator belt adjusting bar.
20. Remove the crankshaft pulley.
21. Remove the timing belt covers 1 and 2. Refer to the timing belt removal and installation procedure.
22. Remove the right hand engine mount bracket.
23. Remove the timing belt guide number 2. Refer to the timing belt removal and installation procedure.
24. Remove the timing belt.
25. Remove the timing belt guide number 2 idler sub assembly. Refer to the timing belt removal and installation procedure.
26. Remove the camshaft pulley and lower timing cover. Refer to the timing belt removal and installation procedure.
27. Remove the timing belt number 3 cover. Refer to the timing belt removal and installation procedure.
28. Align the timing marks of the camshaft drive and driven gears by using a wrench to turn the camshaft.

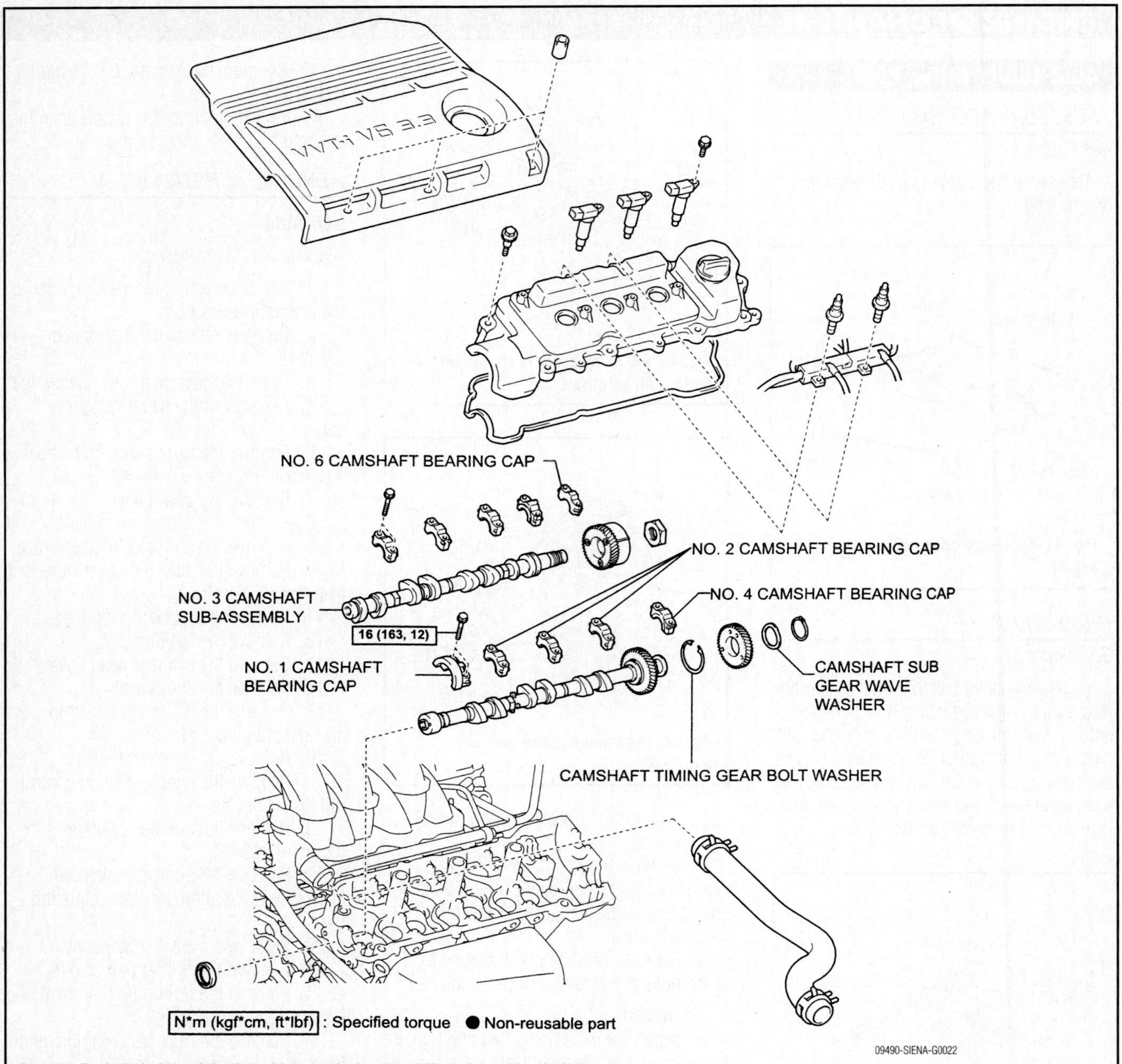

N*m (kgf*cm, ft*lbf) : Specified torque ● Non-reusable part

09490-SIENA-G0022

Fig. 37 Exploded view of the left side camshaft assemblies; right side similar

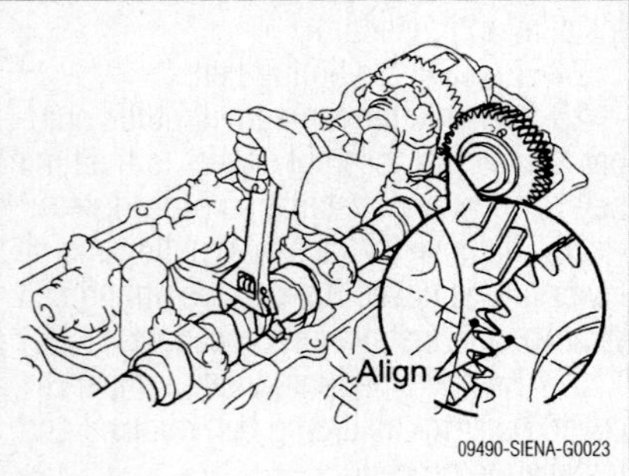

Fig. 38 Align the timing marks of the left side camshaft drive and driven gears by using a wrench to turn the camshaft

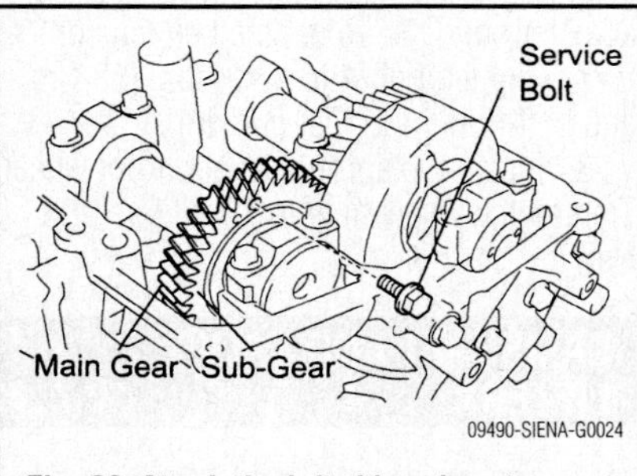

Fig. 39 Attach the left side exhaust camshaft sub gear to the main gear with a 16–20mm long x 6mm thread diameter bolt

29. Attach the exhaust camshaft sub gear to the main gear with a 16–20mm long x 6mm thread diameter bolt and tighten to 48 inch lbs. (5 Nm).

➡When removing the camshaft, be sure the torsional spring force of the sub gear has been taken up by the service bolt.

30. Mark the bearing caps prior to removal so they can be installed in their original positions.

31. Loosen the number 3 camshaft cap bolts in sequence using several steps.

32. Loosen the number 4 camshaft cap bolts in sequence using several steps.

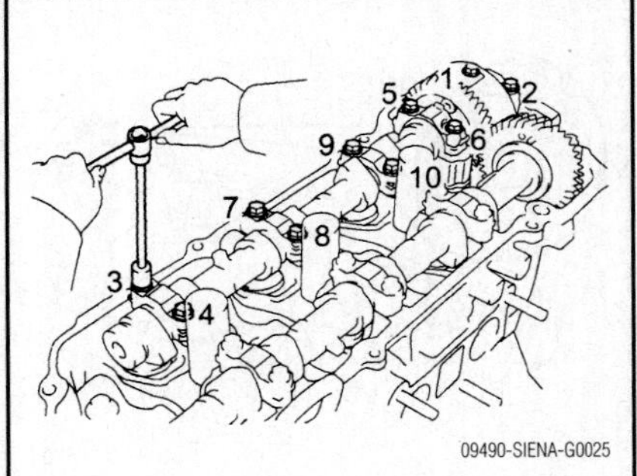

Fig. 40 Loosen the number 3 camshaft cap bolts in sequence

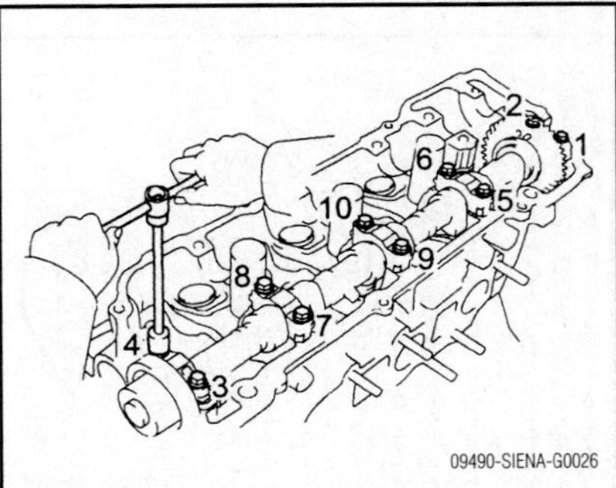

Fig. 41 Loosen the number 4 camshaft cap bolts in sequence

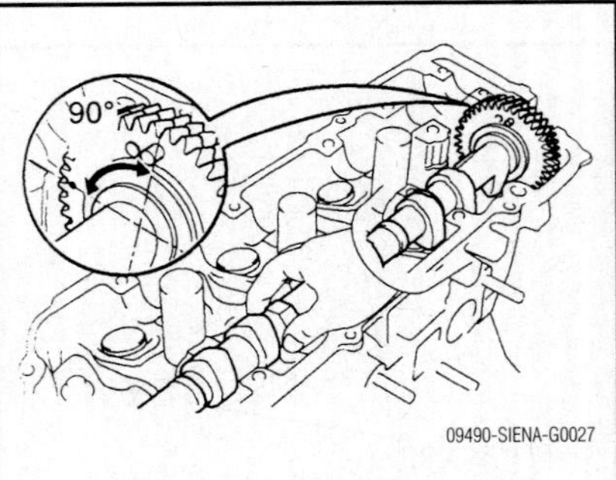

Fig. 42 Place the number 4 camshaft at a 90 degree angle of timing on the cylinder head

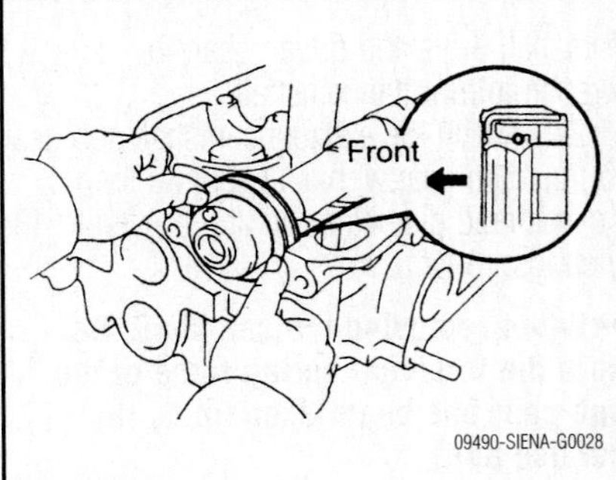

Fig. 43 Make sure to not turn the seal lip over and insert the seal until it stops

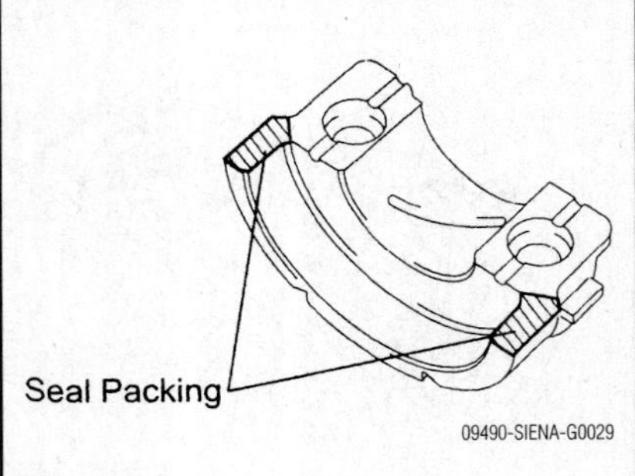

Fig. 44 Apply seal packing to the number 1 bearing cap at the locations shown

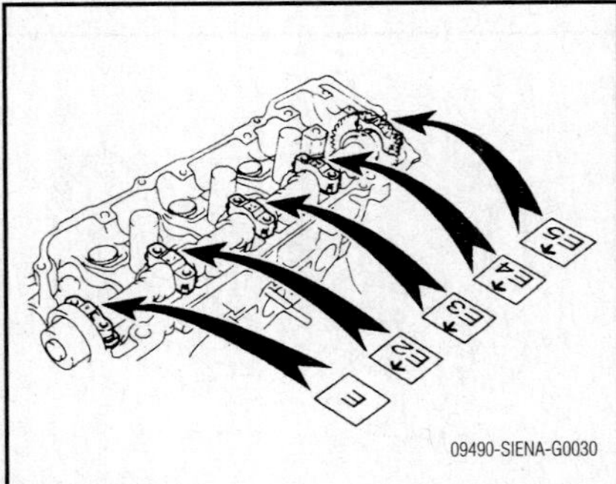

Fig. 45 Install the number 4 camshaft bearing caps in their original locations

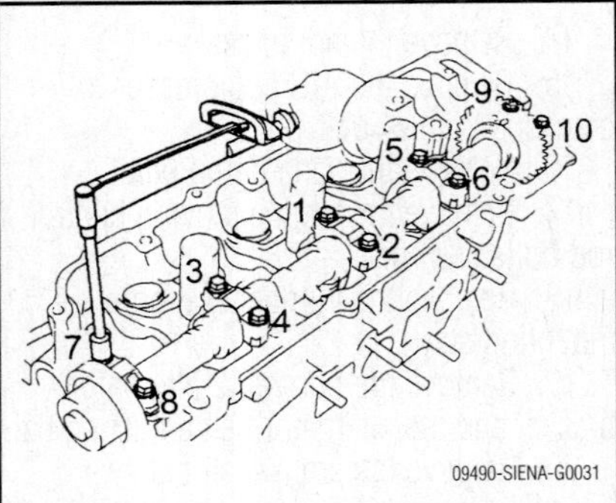

Fig. 46 Tighten the number 4 camshaft bearing caps in this sequence

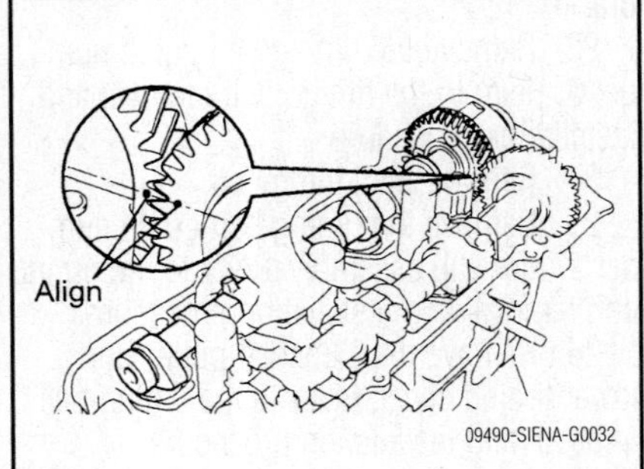

Fig. 47 Align the timing marks on the number 3 camshaft as shown

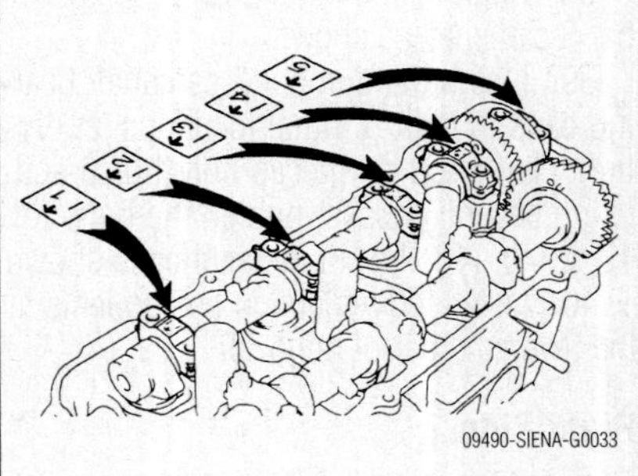

Fig. 48 Install the number 3 camshaft bearing caps in their original locations

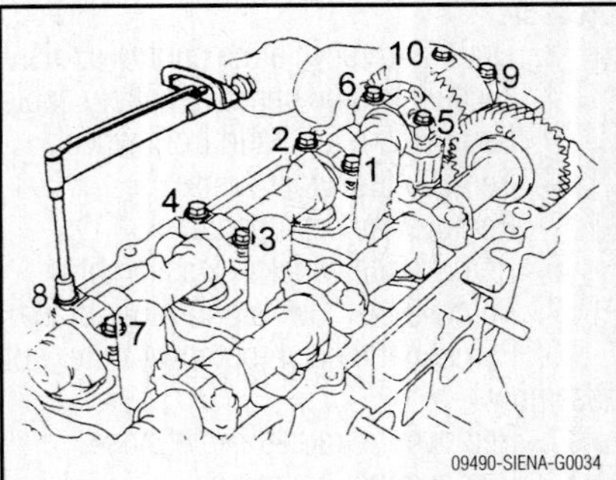

Fig. 49 Tighten the number 3 camshaft bearing caps in this sequence

To install:

The camshaft has a small thrust clearance, make sure to keep the camshaft level during installation to avoid damage to the camshaft and cylinder head.

33. Apply clean engine oil to the camshaft thrust and journal locations.
34. Place the number 4 camshaft at a 90 degree angle of timing on the cylinder head as illustrated.
35. Apply a multipurpose grease to a new oil seal lip and install the seal. Make sure to not turn the seal lip over and insert the seal until it stops. Remove any packing material from the seal surface.
36. Apply seal packing to the number 1 bearing cap as illustrated. Install the cap within 5 minutes of applying the packing. Do not let the seal come into contact with engine oil until at least two hours after it has been installed.
37. Install the number 4 camshaft bearing caps in their original locations, apply a light coat of oil to the cap bolt thread and using several passes; tighten the bolts to 12 ft. lbs. (16 Nm) in the sequence shown.

38. Align the timing marks on the number 3 camshaft as shown.

39. Install the number 3 camshaft bearing caps in their original locations, apply a light coat of oil to the cap bolt thread and using several passes; tighten the bolts to 12 ft. lbs. (16 Nm) in the sequence shown.

40. Install the remaining components in the reverse order of removal.

Right Side

See Figures 50 through 61.

1. Before servicing the vehicle, refer to the precautions section.
2. Properly relieve the fuel system pressure.
3. Drain and recycle the engine coolant.
4. Disconnect the negative battery cable.
5. Remove the left hand front wheel.
6. Remove the wiper arms.
7. Remove the cowl panel.
8. Remove the wiper link assembly.
9. Remove the cowl top to inner brace.
10. Remove the cowl top cover outer sub assembly.
11. Remove the radiator inlet hose.
12. Remove the ignition coil.
13. Remove the cylinder head cover.

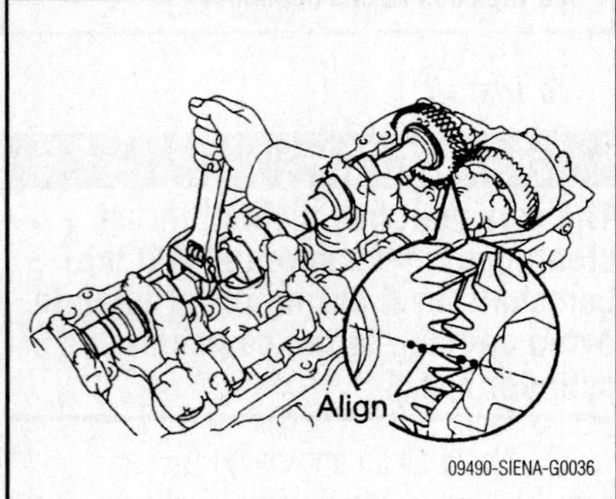

Fig. 50 Align the timing marks of the right side camshaft drive and driven gears by using a wrench to turn the camshaft

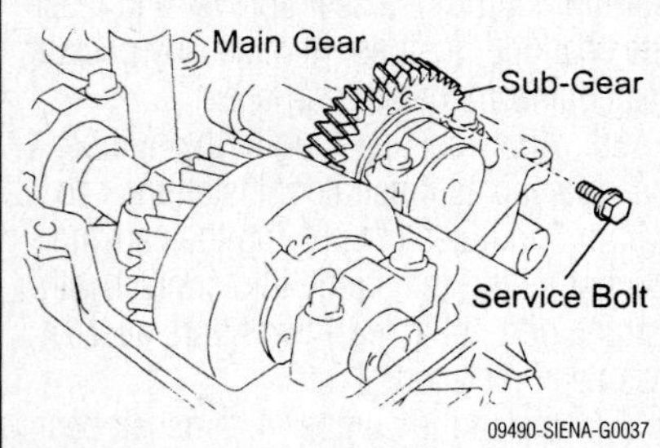

Fig. 51 Attach the right side exhaust camshaft sub gear to the main gear with a 16–20mm long x 6mm thread diameter bolt

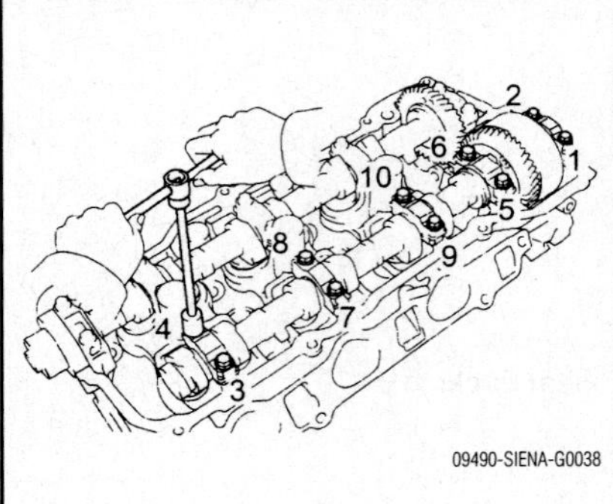

Fig. 52 Loosen the number 1 camshaft cap bolts in sequence

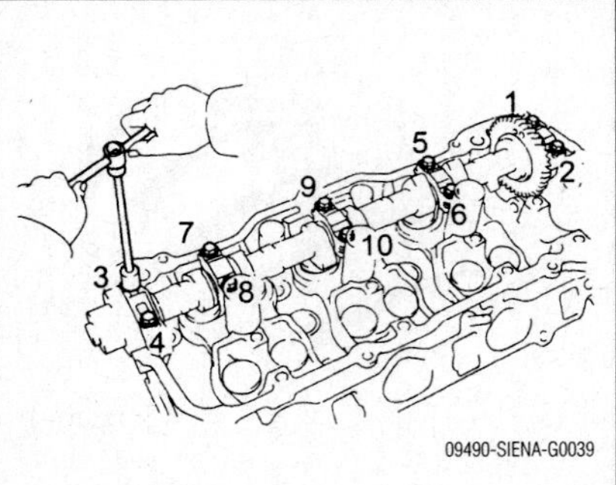

Fig. 53 Loosen the number 2 camshaft cap bolts in sequence

14. Remove the apron seal.
15. Remove the A/C compressor to crankshaft pulley belt.
16. Remove the vane pump belt.
17. Remove the engine moving control rod bolts and rod.
18. Remove the number 2 engine mounting stay.
19. Remove the number 2 alternator bracket and the alternator belt adjusting bar.
20. Remove the crankshaft pulley.
21. Remove the timing belt covers 1 and 2. Refer to the timing belt removal and installation procedure.
22. Remove the right hand engine mount bracket.
23. Remove the timing belt guide number 2. Refer to the timing belt removal and installation procedure.
24. Remove the timing belt.
25. Remove the timing belt guide number 2 idler sub assembly. Refer to the timing belt removal and installation procedure.
26. Remove the camshaft pulley and lower timing cover. Refer to the timing belt removal and installation procedure.
27. Remove the timing belt number 3 cover. Refer to the timing belt removal and installation procedure.
28. Align the timing marks of the camshaft drive and driven gears by using a wrench to turn the camshaft.

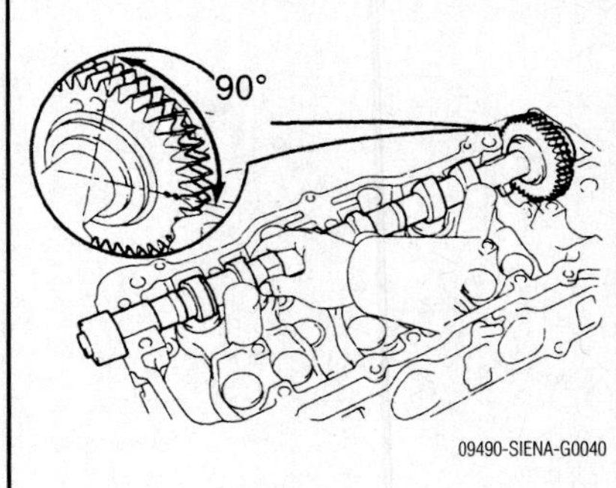

Fig. 54 Place the number 2 camshaft at a 90 degree angle of timing on the cylinder head

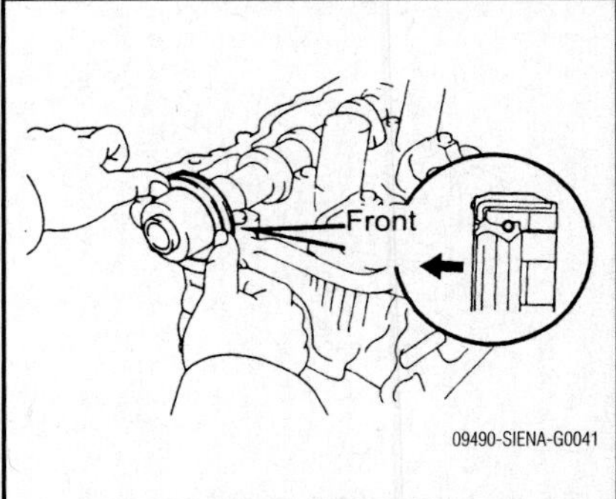

Fig. 55 Make sure to not turn the seal lip over and insert the seal until it stops

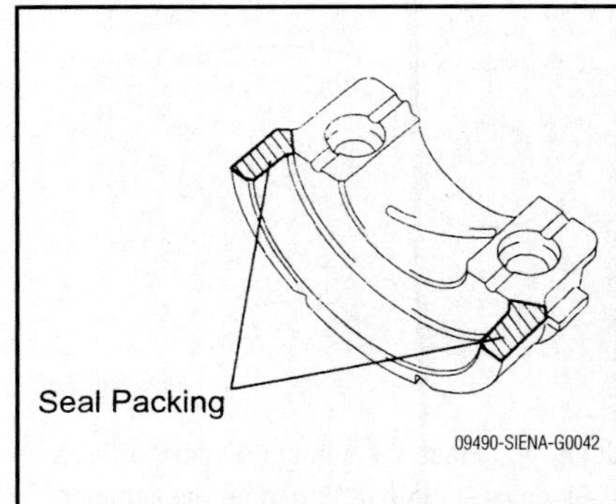

Fig. 56 Apply seal packing to the number 1 bearing cap as illustrated

29. Attach the exhaust camshaft sub gear to the main gear with a 16–20mm long x 6mm thread diameter bolt and tighten to 48 inch lbs. (5 Nm).

➡When removing the camshaft, be sure the torsional spring force of the sub gear has been taken up by the service bolt.

30. Mark the bearing caps prior to removal so they can be installed in their original positions.

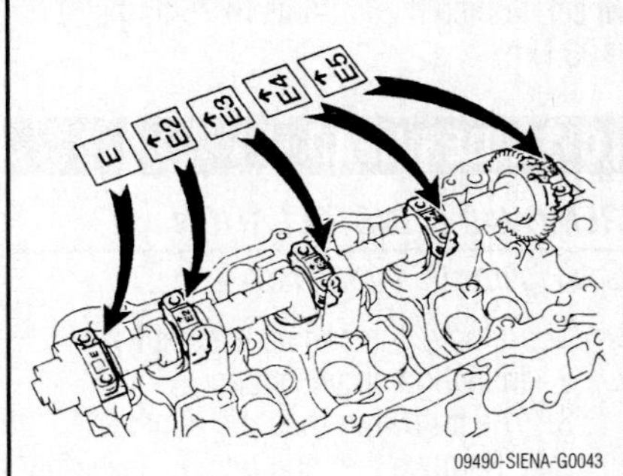

Fig. 57 Install the number 2 camshaft bearing caps in their original locations

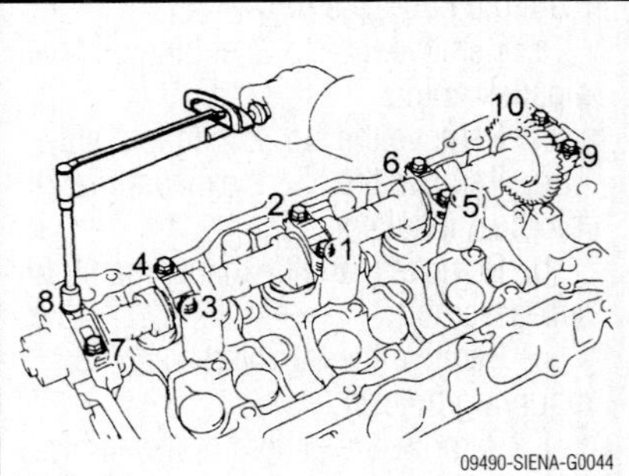

Fig. 58 Tighten the number 2 camshaft bearing caps in this sequence

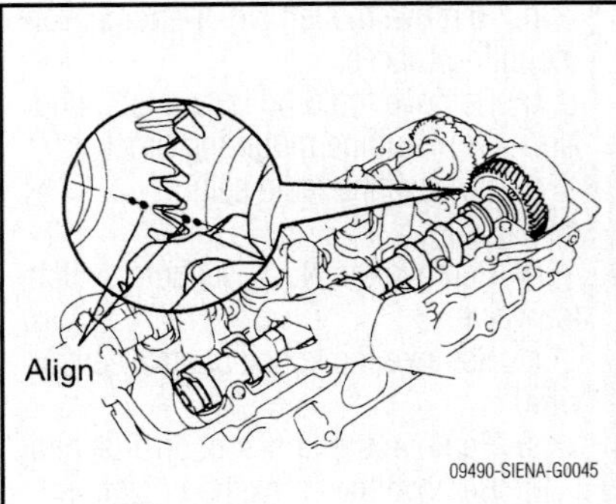

Fig. 59 Align the timing marks on the number 1 camshaft as shown

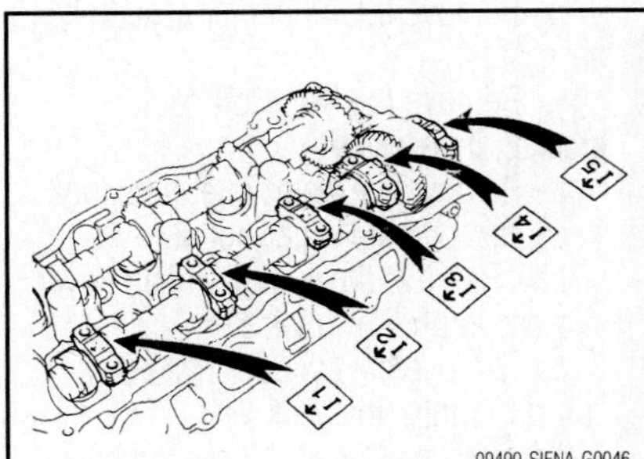

Fig. 60 Install the number 1 camshaft bearing caps in their original locations

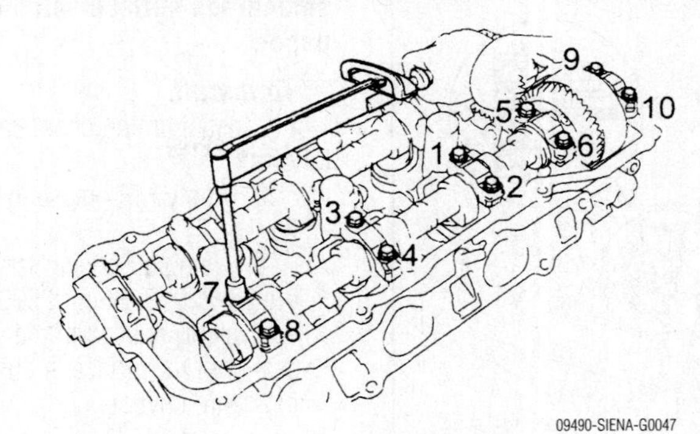

Fig. 61 Tighten the number 1 camshaft bearing caps in this sequence

31. Loosen the number 1 camshaft cap bolts in sequence using several steps.
32. Loosen the number 2 camshaft cap bolts in sequence using several steps.

To install:

WARNING

The camshaft has a small thrust clearance, make sure to keep the camshaft level during installation to avoid damage to the camshaft and cylinder head.

33. Apply clean engine oil to the camshaft thrust and journal locations.
34. Place the number 2 camshaft at a 90 degree angle of timing on the cylinder head as illustrated.
35. Apply a multipurpose grease to a new oil seal lip and install the seal. Make sure to not turn the seal lip over and insert the seal until it stops. Remove any packing material from the seal surface.
36. Apply seal packing to the number 1 bearing cap as illustrated. Install the cap within 5 minutes of applying the packing. Do not let the seal come into contact with engine oil until at least two hours after it has been installed.
37. Install the number 2 camshaft bearing caps in their original locations, apply a light coat of oil to the cap bolt thread and using several passes; tighten the bolts to 12 ft. lbs. (16 Nm) in the sequence shown.
38. Align the timing marks on the number 1 camshaft as shown.
39. Install the number 1 camshaft bearing caps in their original locations, apply a light coat of oil to the cap bolt thread and using several passes; tighten the bolts to 12 ft. lbs. (16 Nm) in the sequence shown.
40. Install the remaining components in the reverse order of removal.

CATALYTIC CONVERTER

REMOVAL & INSTALLATION

The catalytic converter is integrated with the exhaust manifold and the exhaust pipe. Refer to Exhaust Manifold in this section.

CRANKSHAFT DAMPER

REMOVAL & INSTALLATION

See Figures 62 and 63.

1. Before servicing the vehicle, refer to the precautions section.
2. Properly relieve the fuel system pressure.
3. Drain and recycle the engine coolant.
4. Disconnect the negative battery cable.
5. Remove the right hand front wheel.
6. Remove the wiper arms.
7. Remove the cowl panel.
8. Remove the wiper link assembly.
9. Remove the cowl top to inner brace.
10. Remove the cowl top cover outer sub assembly.
11. Remove the right side apron seal.
12. Remove the A/C compressor to crankshaft pulley belt.
13. Remove the vane pump belt.
14. Remove the number 2 engine mounting stay.

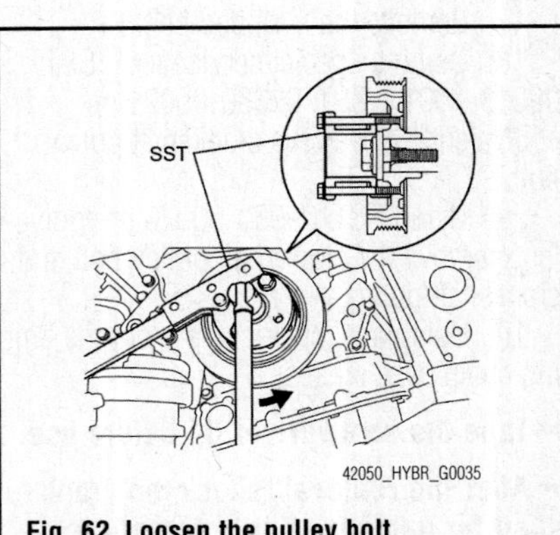

Fig. 62 Loosen the pulley bolt

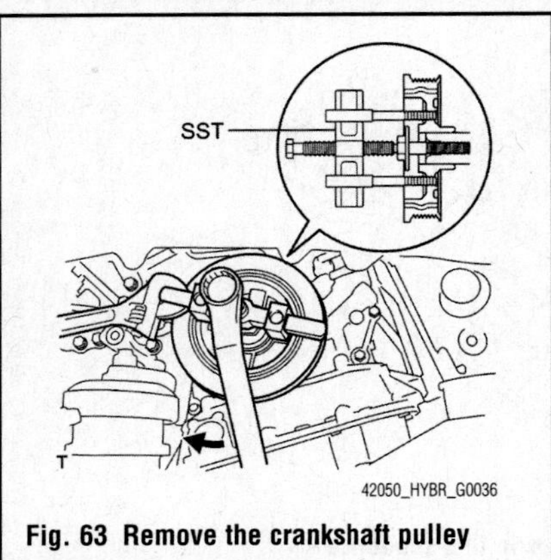

Fig. 63 Remove the crankshaft pulley

15. Remove the number 2 alternator bracket and the alternator belt adjusting bar.
16. Remove the crankshaft pulley.

To install:

17. Install the crankshaft pulley.
18. Using the Crankshaft Pulley Holding tool 09213-54015, Bolt tool 91651-60855 and Companion Flange Holding tool 09330-00021, or equivalent, install the crankshaft pulley bolt and torque the bolt to 162 ft. lbs. (220 Nm).
19. To complete the installation, reverse the removal procedures.
20. Connect the negative battery cable.
21. Start the engine and check for leaks.

CRANKSHAFT FRONT SEAL

REMOVAL & INSTALLATION

See Figures 64 and 65.

1. Before servicing the vehicle, refer to the precautions section.
2. Raise and support the vehicle.
3. Remove the right front wheel.
4. Remove the engine under cover assembly.
5. Remove the right front fender liner.
6. Remove the right front fender apron seal.
7. Remove the V-ribbed belt.
8. Using a special service tool (SST: 09213-70011, SST: 09330-00021 or equivalent), loosen the crankshaft pulley bolt.
9. Using SST: 09950-50013 or equivalent, remove the crankshaft pulley bolt and crankshaft pulley.
10. Using a screwdriver, pry out the timing chain case oil seal.

➡Tape the screwdriver tip before use.

➡After the removal, check the crankshaft for damage. If it is damaged, smooth the surface with 400-grit sandpaper.

To install:

11. Install timing chain case oil seal, as follows:
   a. Apply MP grease to a new oil seal lip.
   b. Using a Special Service Tool (SST: 09223-22010, SST: 09506-35010 or equivalent) and a hammer, tap in the oil seal until its surface is flush with the timing chain cover edge.

➡Keep the lip free of foreign matter.

➡Do not tap the oil seal at an angle.

12. Install the crankshaft pulley, as follows:
   a. Align the pulley set key with the key groove of the pulley, and slide on the pulley.
   b. Using SST: 09213-70011, SST: 09330-00021, or equivalent, install the pulley bolt. For 2.7L engines, tighten to 192 ft. lbs. (260 Nm). For 3.5L engines, tighten to 184 ft. lbs. (250 Nm).
13. The remainder of installation is the reverse of removal. When installing the wheel, tighten the lug nuts to 76 ft. lbs. (103 Nm).

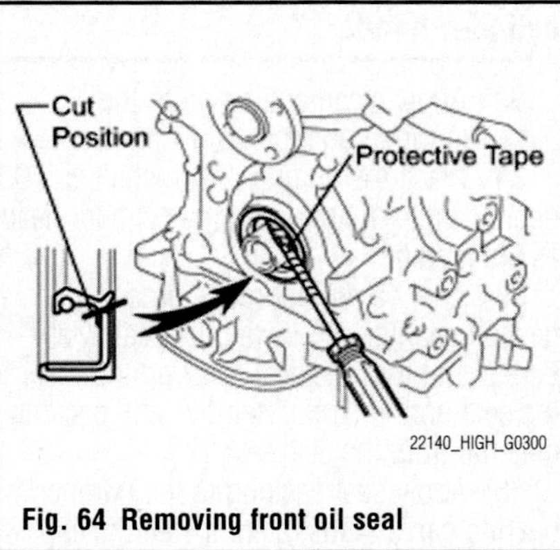

Fig. 64 Removing front oil seal

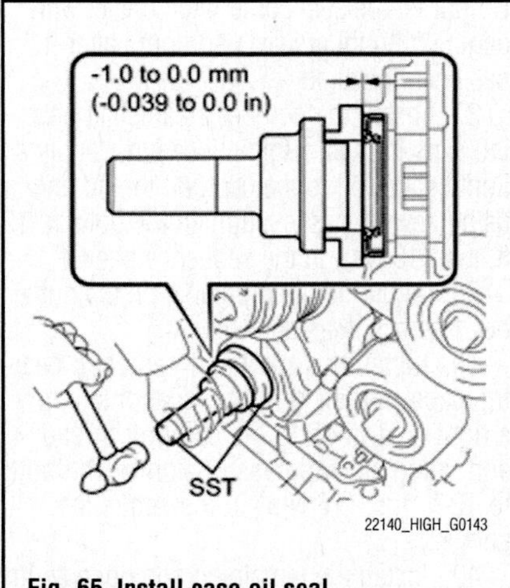

Fig. 65 Install case oil seal

CYLINDER HEAD

REMOVAL & INSTALLATION

See Figures 66 through 87.

1. Remove the engine assembly.
2. Install on engine stand.
3. The following must be removed:
   a. Remove the ignition coil assembly.
   b. Remove right hand No. 2 engine mounting stay.
   c. Remove the intake manifold.
   d. Remove the right hand exhaust manifold sub-assembly.
   e. Remove the No. 2 engine oil level dipstick guide.
   f. Remove the No. 2 manifold stay.
   g. Remove the No. 2 exhaust manifold heat insulator.
   h. Remove the left exhaust manifold sub-assembly.
   i. Remove the transverse engine mounting bracket.
   j. Remove the generator assembly.
   k. Remove the V-ribbed belt tensioner assembly.
   l. Remove the No. 2 timing gear cover.
   m. Remove the No. 2 idler pulley sub-assembly.
   n. Remove the left No. 1 engine front mounting bracket.
   o. Remove the 6 bolts and left hand No. 1 front engine mounting bracket.
   p. Remove the radio setting condenser.
   q. Remove the No. 1 vacuum switching valve.
   r. Remove the knock control sensor wire.
   s. Remove the knock control sensor.
   t. Remove the crankshaft position sensor.
   u. Remove the No. 1 oil pipe.
   v. Remove the oil pipe.
   w. Remove the crankshaft pulley.
   x. Remove the oil cooler assembly, if necessary.
   y. Remove the No. 1 oil cooler bracket, if necessary.
   z. Remove the water inlet housing.
   aa. Remove the water outlet.
   bb. Remove the cylinder head cover sub-assembly (for bank 1).
   cc. Remove the cylinder head cover sub-assembly (for bank 2).
   dd. Remove the No. 2 oil pan sub-assembly.
   ee. Remove the oil strainer sub-assembly.

ff. Remove the oil pan sub-assembly.

gg. Remove the timing chain cover sub-assembly.

hh. Remove the timing chain case oil seal.

ii. Set the No. 1 cylinder to TDC/compression.

jj. Remove the No. 1 chain tensioner assembly.

kk. Remove the chain tensioner slipper.

ll. Remove the chain sub-assembly.

mm. Remove the idle sprocket assembly.

nn. Remove the camshaft timing gears and No. 2 chain (for bank 1).

oo. While raising the No. 2 chain tensioner assembly, insert a pin of 1.0 mm (0.039 in.) diameter into the hole to fix the No. 2 chain tensioner assembly.

pp. Hold the hexagonal portion of the camshaft with a wrench, and remove the 2 bolts and 2 camshaft timing gear assemblies.

➡Be careful not to damage the cylinder head with the wrench. Do not disassemble the camshaft timing gear assemblies.

qq. Remove the No. 2 chain assembly.

rr. Remove the bolt and No. 2 chain tensioner assembly.

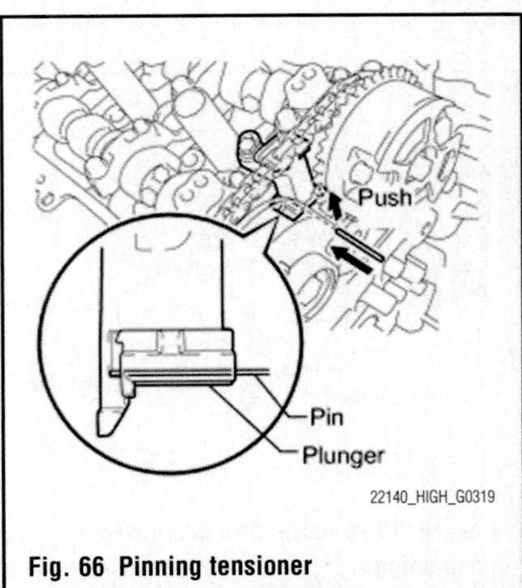

Fig. 66 Pinning tensioner

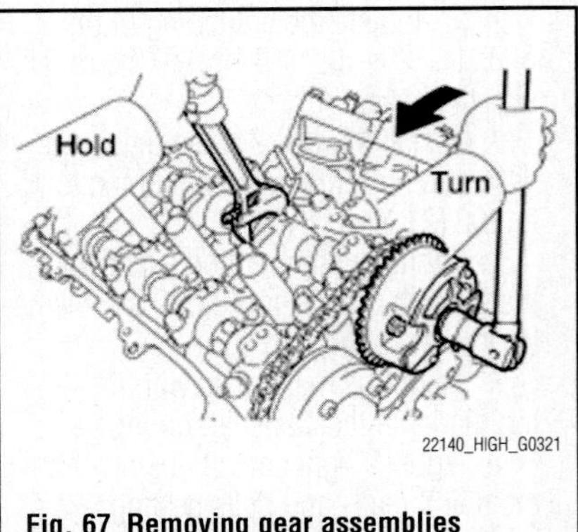

Fig. 67 Removing gear assemblies

ss. Check that the camshafts are positioned as shown in the illustration.

tt. Uniformly loosen and remove the 8 bearing cap bolts in several steps and in the sequence shown in the illustration.

uu. Uniformly loosen and remove the 12 bearing cap bolts in several steps and in the sequence shown in the illustration.

➡Uniformly loosen the bolts while keeping the camshaft level.

vv. Remove the 5 camshaft bearing caps.

ww. Remove the camshaft.

xx. Remove the No. 2 camshaft.

yy. Remove the right camshaft housing sub-assembly by prying between the cylinder head and right camshaft housing sub-assembly with a screwdriver.

➡Be careful not to damage the contact surfaces of the cylinder head and the right camshaft housing sub-assembly.

4. Remove the 24 valve lash adjuster assemblies from the cylinder head.

➡Arrange the removed parts in the correct order.

5. Uniformly loosen and remove the 2 cylinder head set bolts in several steps and in the sequence shown in the illustration.

➡Be careful not to drop washers into the cylinder head. Cylinder head warpage or cracking could result from removing bolts in an incorrect order. Be sure to keep separate the removed parts for each installation position.

6. Using a 10 mm bi-hexagon wrench, uniformly loosen the 8 bolts in the sequence shown in the illustration. Remove the 8 cylinder head bolts and plate washers.

7. Remove the cylinder head sub-assembly.

8. Remove the No. 2 cylinder head gasket.

To install:

9. Place the No. 2 cylinder head gasket on the cylinder block surface with the Lot No. stamp upward.

➡Gently lower the cylinder head in order not to damage the gasket with the bottom part of the head.

10. Place the cylinder head on the cylinder block.

➡Be careful not to allow oil to adhere to the bottom part of the cylinder head.

11. Apply a light coat of engine oil to the threads and under the heads of the cylinder head bolts.

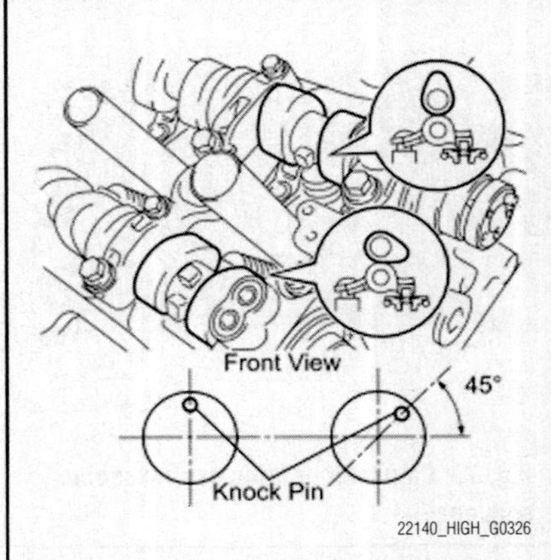

Fig. 68 Positioning camshafts for bearing cap removal

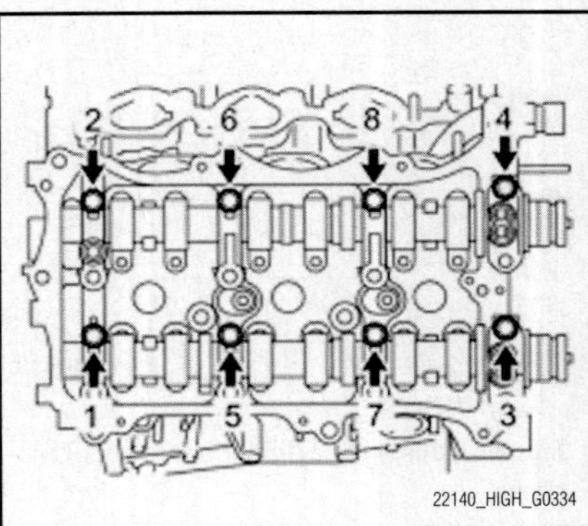

Fig. 69 Camshafts bearing cap removal sequence

➡The cylinder head bolts are tightened in 3 progressive steps.

a. Step 1: Using a 10 mm bi-hexagon wrench, install and uniformly tighten the 8 cylinder head bolts with the plate washers to 27 ft. lbs. (36 Nm) in several steps in the sequence shown in the illustration.

b. Step 2: Mark the cylinder head bolt head with paint. Tighten the cylinder head bolts another 90°.

c. Step 3: Tighten the cylinder head bolts an additional 90°.

12. Tighten the 2 bolts to 22 ft. lbs. (30 Nm) in the order shown in the illustration.

13. Install the 12 valve stem caps.

➡Keep the lash adjuster free of dirt and foreign objects. Only use clean engine oil.

14. Place the lash adjuster into a container filled with engine oil.

a. Insert Service Tool's tip into the lash adjuster's plunger and use the tip to press down on the check ball inside the plunger.

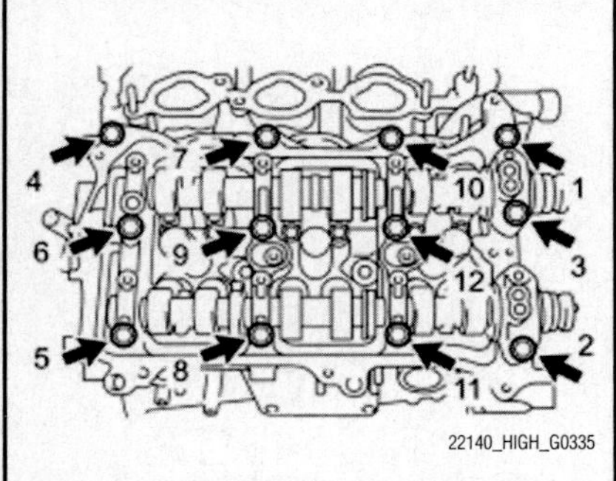

Fig. 70 Camshaft bearing cap loosening sequence

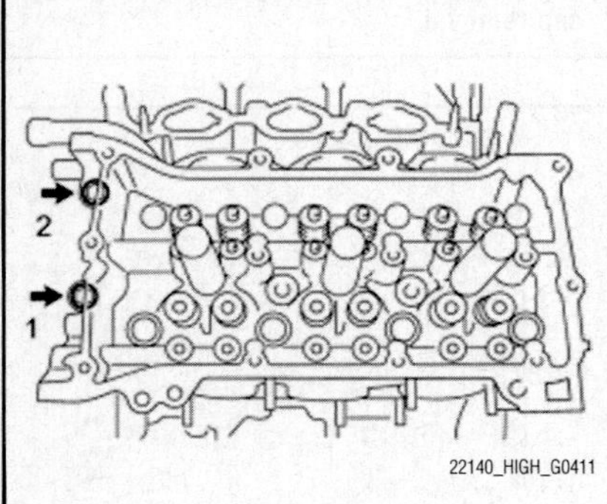

Fig. 71 Loosening cylinder head bolts LH shown

b. Squeeze Service Tool and lash adjuster together to move the plunger up and down 5 to 6 times.

c. Check the movement of the plunger and bleed the air. Confirm that the plunger moves up and down freely.

d. When bleeding air from the high-pressure chamber, make sure that the tip of SST is actually pressing the check ball as shown in the illustration. If the check ball is not pressed, air will not bleed.

e. After bleeding the air, remove Service Tool. Then, try to press the plunger quickly and firmly with by hand. Confirm that the plunger is very difficult to move.

f. If the results are not as specified, replace the defective lash adjuster.

g. Install the lash adjusters.

➡Install the lash adjuster to the same place where it was removed from.

15. Install the following:

a. Apply engine oil to the camshaft journals, right camshaft housing sub-assembly, and camshaft bearing caps.

b. Install the camshaft and No. 2 camshaft to the right camshaft housing sub-assembly.

c. Make sure of the marks and numbers

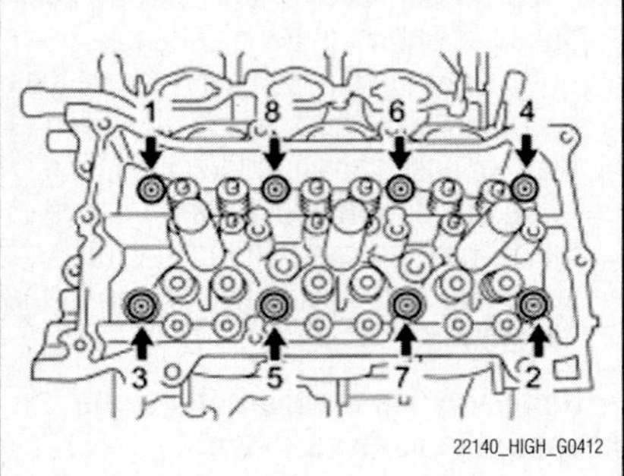

Fig. 72 Cylinder head bolt loosening sequence LH shown

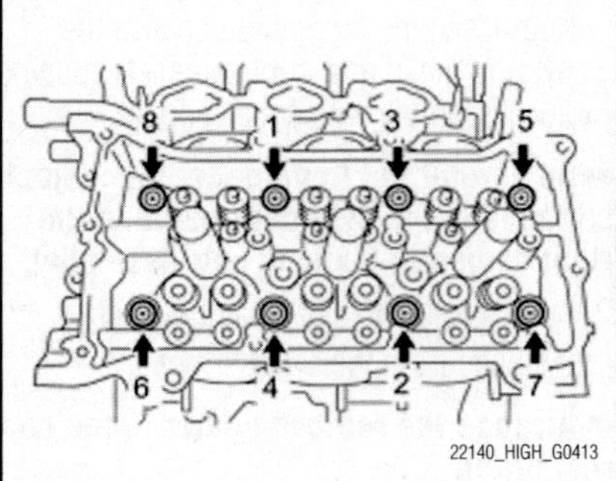

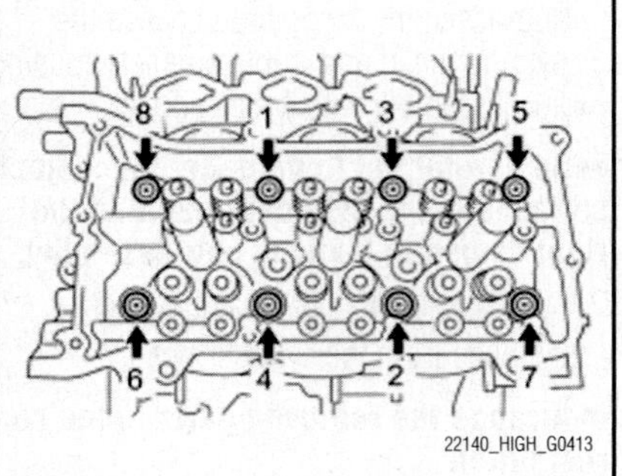

Fig. 73 Cylinder head bolt tightening sequence LH shown

on the camshaft bearing caps and place them in each proper position and direction.

d. Temporarily tighten the 8 bearing cap bolts to 7 ft. lbs. (10 Nm) in the order shown in the illustration.

e. Make sure that the No. 1 valve rocker arm sub-assembly is installed as shown in the illustration.

f. Apply seal packing in a continuous line as shown in the illustration.

➡Remove any oil from the contact surface. Install the camshaft housing sub-assembly RH within 3 minutes. Do not start the engine for at least 2 hours after installing.

g. Install the right camshaft housing sub-assembly and tighten the 12 bolts to 21 ft. lbs. (28 Nm) in the order shown in the illustration.

➡When installing the camshaft housing RH, it is necessary to correctly position the camshafts as shown in the removal illustration. If the camshaft housing sub-assembly is removed because any of the bolts are loosened during installation, make sure that the previously applied seal packing does not enter any oil passages.

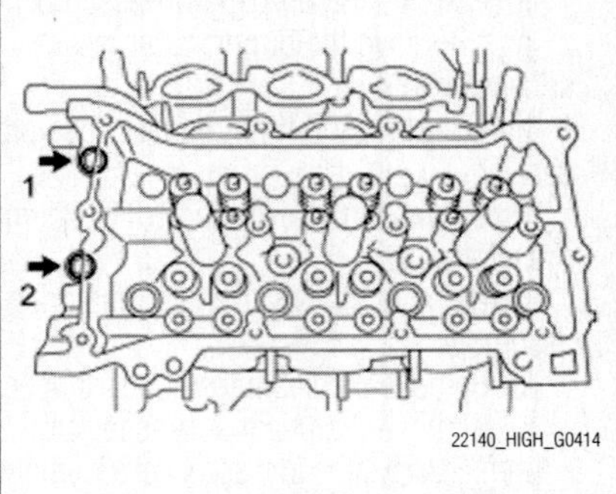

Fig. 74 Tightening cylinder head bolts LH shown

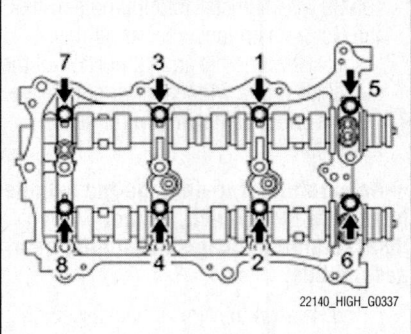

Fig. 75 Camshafts bearing cap tightening sequence

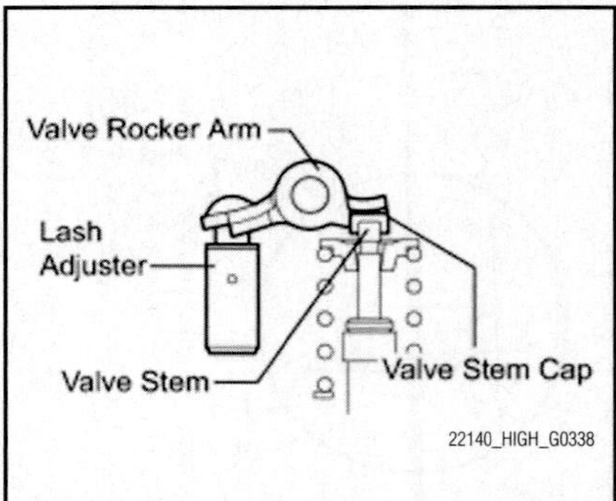

Fig. 76 Valve rocker arm sub-assembly positioning

h. Complete the tightening of the 8 bolts to 12 ft. lbs. (16 Nm) in the order shown above.

i. Install the No. 2 chain tensioner assembly with the bolt and tighten to 15 ft. lbs. (21 Nm).

j. While pushing in the tensioner, insert a pin of 1.0 mm (0.039 in.) diameter into the hole to fix it.

k. Align the mark plate with the timing marks of the camshaft timing gear.

l. Apply a light coat of engine oil to the bolt threads and bolts seating surface.

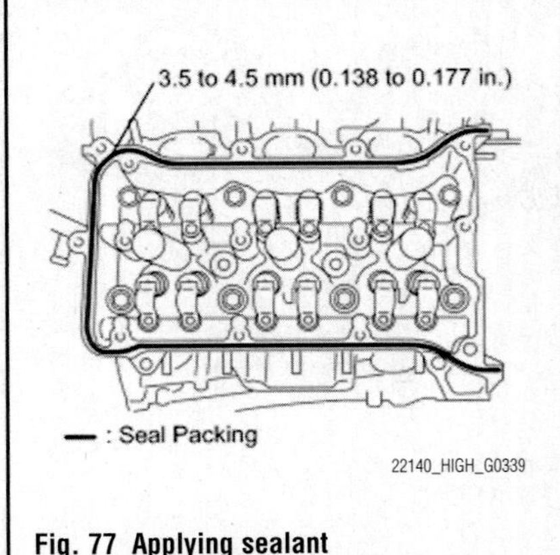

Fig. 77 Applying sealant

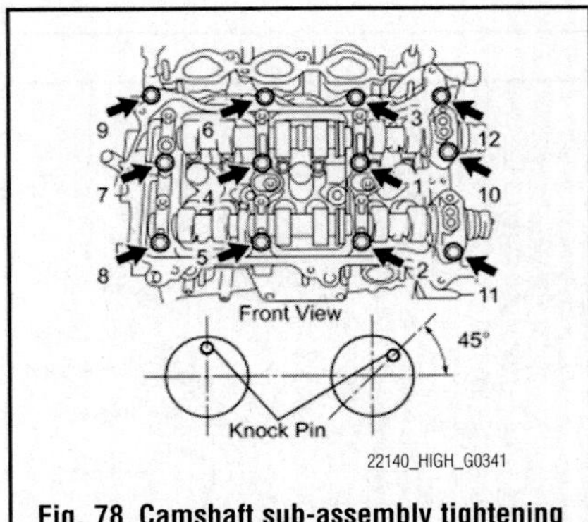

Fig. 78 Camshaft sub-assembly tightening sequence

m. Align the knockpin of the camshaft with pinhole of the camshaft timing gear assembly. Install the camshaft timing gear assembly and camshaft timing exhaust timing gear assembly and

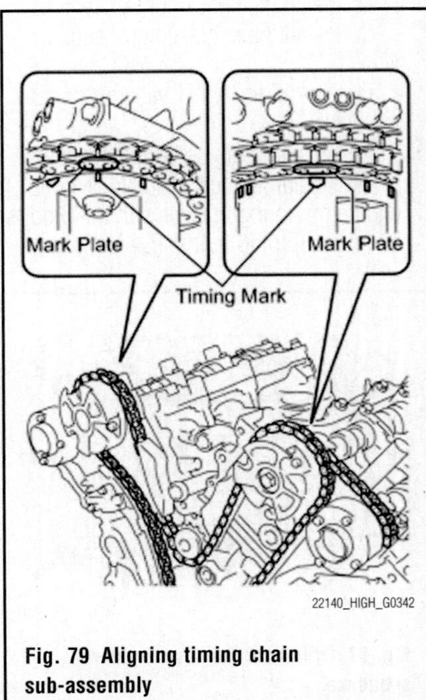

Fig. 79 Aligning timing chain sub-assembly

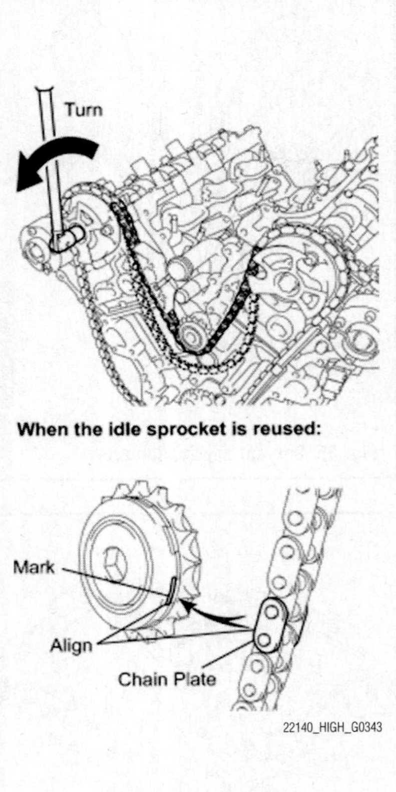

Fig. 80 Tightening timing chain

camshaft timing exhaust gear with the No. 2 chain sub-assembly installed.

n. Hold the hexagonal portion of the camshaft with the wrench and tighten the two bolts and camshaft timing gear assemblies to 74 ft. lbs. (100 Nm).

o. Remove the pan from the No 2 chain tensioner assembly.

p. Install idle sprocket assembly and tighten to 44 ft. lbs. (60 Nm).

q. Install chain sub-assembly.

r. Align the mark plate and timing marks as shown in the illustration and install the chain.

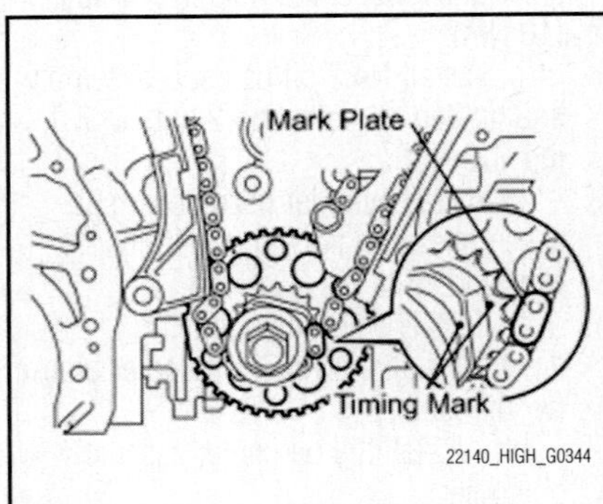

Fig. 81 Installing timing chain on crankshaft

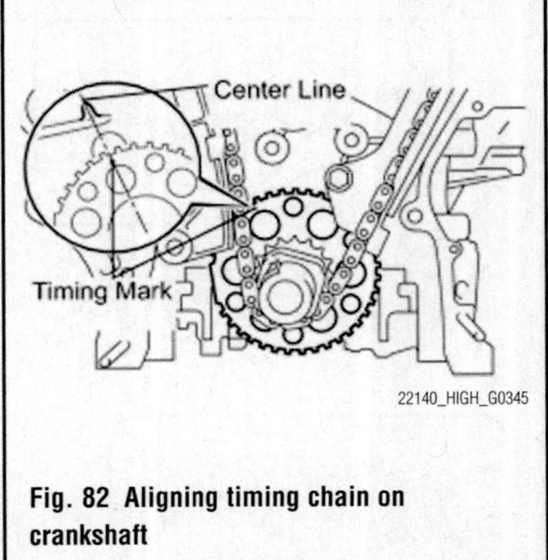

Fig. 82 Aligning timing chain on crankshaft

➡The camshaft mark plates are orange.

s. Do not pass the chain over the crankshaft, just temporarily place it on the crankshaft.

t. Turn the camshaft timing gear assembly on bank 1 counterclockwise to tighten the chain between the banks.

➡When the idle sprocket assembly is reused, align the timing chain plate with the mark where the plate has been in order to tighten the chain between the banks.

u. Align the mark plate and timing marks as shown in the illustration and install the chain onto the crankshaft timing sprocket. The crankshaft to mark plate is yellow.

v. Turn the crankshaft clockwise to set it to the right-hand block bore more centerline. (TDC Compression)

w. Install chain tensioner slipper.

x. Move the stopper plate upward to release the lock, and push the plunger deep into the tensioner.

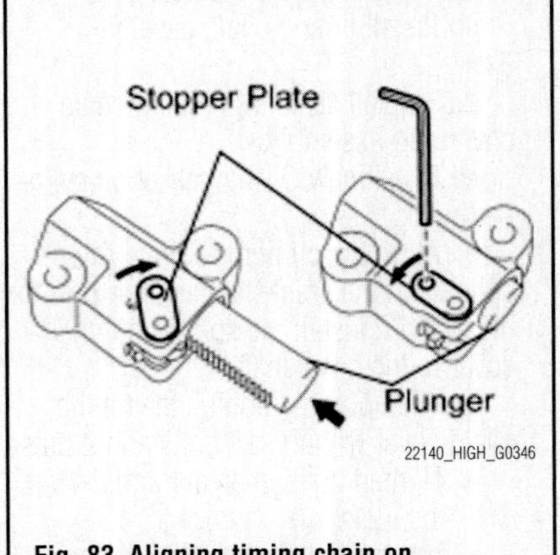

Fig. 83 Aligning timing chain on crankshaft

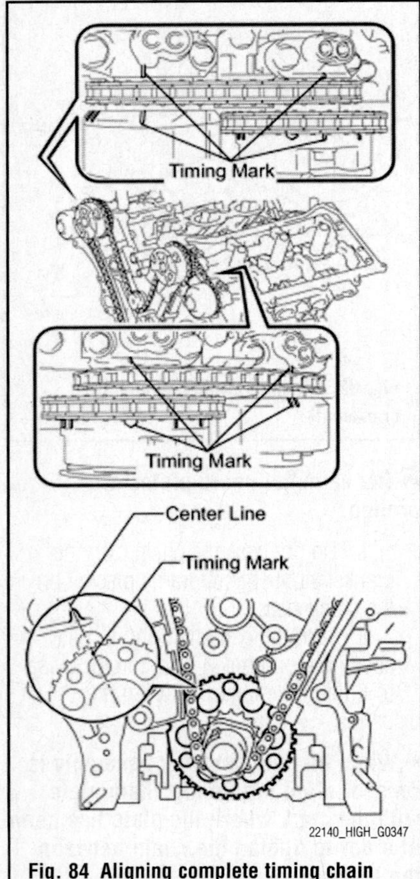

Fig. 84 Aligning complete timing chain

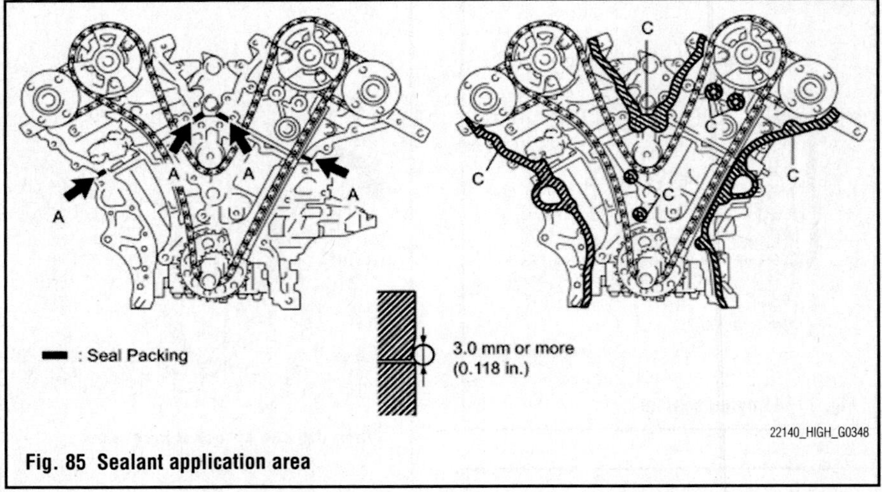

Fig. 85 Sealant application area

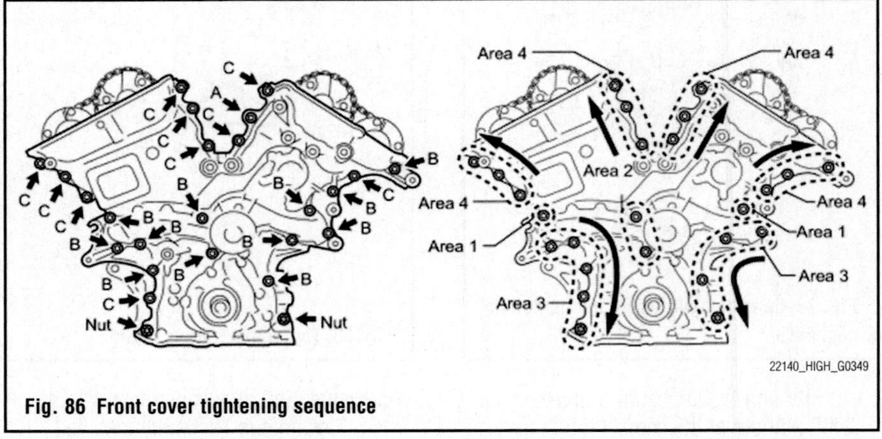

Fig. 86 Front cover tightening sequence

y. Move the stopper plate downward to set the loss, and insert a hexagon wrench into the hole of the stopper plate.

z. Install No. 1 chain tensioner assembly and tighten bolts to 7 ft. lbs. (10 Nm).

aa. Remove the hexagon wrench from the No. 1 chain tensioner assembly. Check that the each timing mark is aligned with the crankshaft at TDC compression.

bb. Remove the pulley set bolt.

cc. Install timing chain case oil seal.

dd. Install sealant to timing chain cover sub-assembly.

ee. Install new O ring gasket on cylinder block.

ff. Align the oil pump's drive rotor spline and the crankshaft as shown in the illustration. Install the spline and chain cover to the crankshaft.

gg. Temporarily tighten the timing chain cover with the 23 bolts and 2 nuts.

- Tighten bolts in area 1 and 2 15 ft. lbs. (21 Nm)
- Tighten bolt in area 3 15 ft. lbs. (21 Nm)

➡First tighten the upper bolts and nuts followed by the lower bolts and nuts as shown.

- Tighten bolt in area 4 32 ft. lbs. (43 Nm)
- Tighten bolt in area 4 15 ft. lbs. (21 Nm)

hh. Install oil pan subassembly and tighten 16 bolts and 2 nuts to 7 ft. lbs. (10 Nm) and 15 ft lbs. (21 Nm).

ii. Install oil strainers sub-assembly and tighten bolts and nuts to 7 ft. lbs. (10 Nm).

jj. Install No.2 oil pan sub-assembly and tighten 16 bolts and 2 nuts to 7 ft. lbs. (10 Nm).

kk. Install cylinder head cover sub-assemblies and tighten to 7 ft. lbs. (10 Nm) and 15 ft. lbs. (21 Nm).

ll. Install water inlet housing

mm. Install the No. 1 oil cooler bracket (w/ oil cooler).

nn. Install the oil cooler assembly (w/ oil cooler).

oo. Install the crankshaft pulley.

pp. Install the oil pipe.

qq. Install the No. 1 oil pipe.

rr. Install the crankshaft position sensor.

ss. Install the knock control sensor.

tt. Install the knock control sensor wire.

uu. Install the No. 1 vacuum switching valve.

vv. Install radio setting condenser.

ww. Install the left hand No. 1 front engine mounting bracket with the 6 bolts and tighten to 40 ft. lbs. (54 Nm).

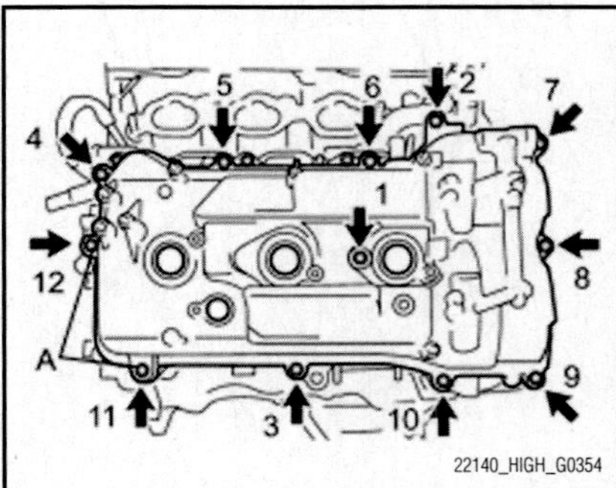

Fig. 87 Cylinder head cover tightening sequence

➡Install the water inlet and mounting bracket within 15 minutes after installing the chain cover. Do not

xx. start the engine for at least 2 hours after installation.

yy. Install the No. 2 idler pulley sub-assembly.

zz. Install the No. 2 timing gear cover.

53. Install the V-ribbed belt tensioner assembly.
54. Install the generator assembly.
55. Install the transverse engine mounting bracket.
56. Install left hand exhaust manifold sub-assembly and tighten to 15 ft. lbs. (21 Nm).
57. Install the No. 2 exhaust manifold heat insulator.
58. Install the No. 2 manifold stay.
59. Install the No. 2 engine oil level dipstick guide.
60. Install right hand exhaust manifold sub-assembly and tighten to 15 ft. lbs. (21 Nm).
61. Install the intake manifold and tighten the 6 bolts and 4 nuts uniformly in several steps to 15 ft. lbs. (21 Nm).
62. Install the right hand No. 2 engine mounting stay.
63. Install the ignition coil assembly.
64. Remove the engine stand.
65. Install the engine assembly.

EXHAUST MANIFOLD

REMOVAL & INSTALLATION

See Figures 88 through 91.

1. Before servicing the vehicle, refer to the precautions section.
2. Remove the right front wheel.
3. Remove the V-bank cover sub-assembly.

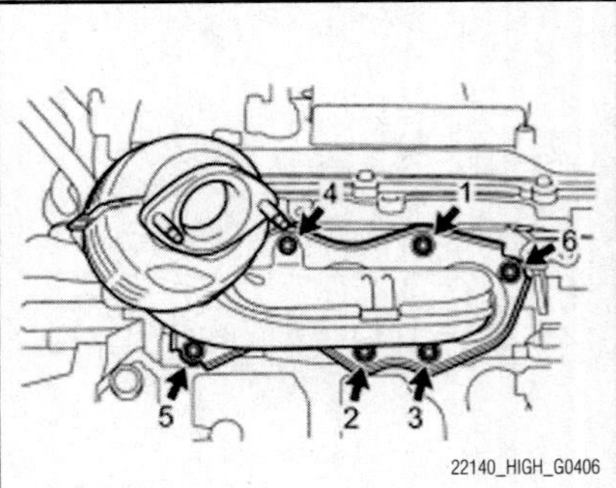

Fig. 88 Right exhaust manifold sub-assembly tightening sequence—3.5L engine

4. Remove the engine under cover assembly.
5. Remove the No. 1 and 2 engine under covers.
6. Drain engine coolant.
7. Disconnect the No. 1 radiator hose.
8. Remove the radiator reserve tank assembly, as follows:
   a. Disconnect the hose.
   b. Remove the 2 bolts and the radiator reserve tank assembly.
9. Remove the No. 2 oil level dipstick guide.

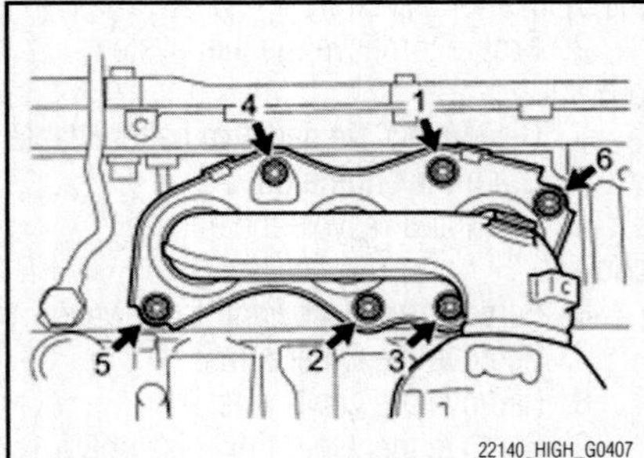

Fig. 89 Left exhaust manifold sub-assembly tightening sequence—3.5L engine

Fig. 90 No. 2 manifold stay tightening sequence—3.5L engine

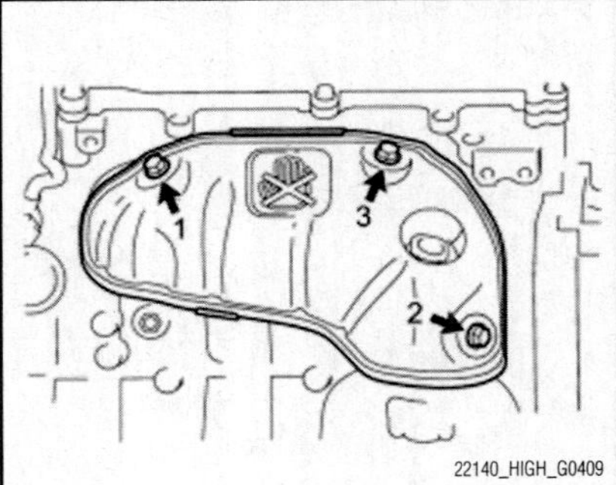

Fig. 91 No. 2 exhaust manifold heat insulator tightening sequence—3.5L engine

10. Remove the air fuel ratio sensor (for Bank 2 Sensor 1).
11. Remove the 3 bolts and No. 2 exhaust manifold heat insulator.
12. For 4WD vehicles, remove the propeller with center bearing shaft assembly.
13. Remove the tail exhaust pipe assembly.
14. Remove the center exhaust pipe assembly.
15. Remove the front No. 3 exhaust pipe sub-assembly.
16. Remove the front exhaust pipe assembly.
17. Remove the bolt, nut and No. 2 manifold stay.
18. Remove the 6 nuts and left hand exhaust manifold sub-assembly.
19. Remove the gasket.
20. Remove the bolt, nut and manifold stay.
21. Remove the right hand exhaust manifold sub-assembly, as follows:
    a. Disconnect the air fuel ratio sensor (for bank 1 sensor 1) connector and remove the clamp.
    b. Remove the 6 nuts and the right hand exhaust manifold sub-assembly.
    c. Remove the gasket.
22. Remove the air fuel ratio sensor (for Bank 1 Sensor 1).

To install:

23. Install the air fuel ratio sensor (for Bank 1 Sensor 1).
24. Install the right hand exhaust manifold sub-assembly, as follows:
    a. Install a new gasket.
    b. Install the right hand exhaust manifold sub-assembly by tightening the 6 nuts in the order shown to 15 ft. lbs (21 Nm).
    c. Connect the air fuel ratio sensor (for Bank 1 Sensor 1) connector and install the clamp.
25. Install the manifold stay with the bolt and nut and tighten to 25 ft. lbs (34 Nm), 26 ft. lbs (35 Nm).
26. Install the left hand exhaust manifold sub-assembly, as follows:
    a. Install a new gasket.
    b. Install the left hand exhaust manifold sub-assembly by tightening the 6 nuts in the order shown to 15 ft. lbs (21 Nm).
27. Install the No. 2 manifold stay by tightening the bolt and nut in the order shown to 25 ft. lbs (34 Nm).
28. Install the front exhaust pipe assembly.
29. Install the front No. 3 exhaust pipe sub-assembly.

30. Install the center exhaust pipe assembly.
31. Install the tail exhaust pipe assembly.
32. For 4WD vehicles, temporarily tighten the propeller with center bearing shaft assembly.
33. For 4WD vehicles, fully tighten the propeller with center bearing shaft assembly.
34. Install the No. 2 exhaust manifold heat insulator by tightening the 3 bolts in the order shown to 75 inch lbs (8.5 Nm).
35. Install the air fuel ratio sensor (for Bank 2 Sensor 1).
36. Install the No. 2 oil level dipstick guide.
37. Install the radiator reserve tank assembly with the 2 bolts and tighten to 48 inch lbs (5.4 Nm).
38. Connect the hose.
39. Connect the No. 1 radiator hose.
40. Add engine coolant.
41. Inspect for coolant leak.
42. Inspect for gas leak, and repair as necessary.
43. For 4WD vehicles, inspect and adjust transfer oil.
44. Install the No. 1 and 2 engine under covers.
45. Install the engine under cover assembly.
46. Install the V-bank cover sub-assembly.
47. Install the right front wheel.

FLEXPLATE

REMOVAL & INSTALLATION

See Figure 92.

1. Before servicing the vehicle, refer to the precautions section.
2. Remove the transmission assembly.
3. Hold the crankshaft and unfasten the flywheel bolts.
4. Remove the flywheel.

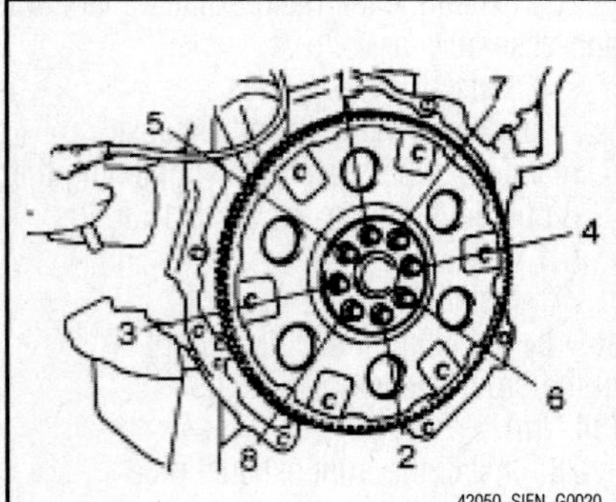

Fig. 92 Flywheel torque sequence—Sienna

To install:

5. Apply anti-lock compound to 2 or 3 threads of the flywheel bolts.
6. Install the front spacer, flywheel and rear plate.
7. Hold the crankshaft and tighten the bolts in a star pattern to 61 ft. lbs. (83 Nm).
8. Install the transaxle assembly.

INTAKE MANIFOLD

REMOVAL & INSTALLATION

See Figures 93 through 96.

1. Before servicing the vehicle, refer to the precautions section.
2. Properly relieve the fuel system pressure.
3. Disconnect the negative battery cable.
4. Drain the engine oil.
5. Drain and recycle the engine coolant.
6. Remove the right hand front wheel.
7. Remove the wiper arms.
8. Remove the cowl panel.
9. Remove the wiper link assembly.
10. Remove the cowl top to inner brace.
11. Remove the cowl top cover outer sub assembly.
12. Remove the engine cover.
13. Remove the air cleaner assembly.
14. Remove the emission control valve set.
15. Remove the air surge tank.
16. Disconnect the fuel pipe sub assembly.

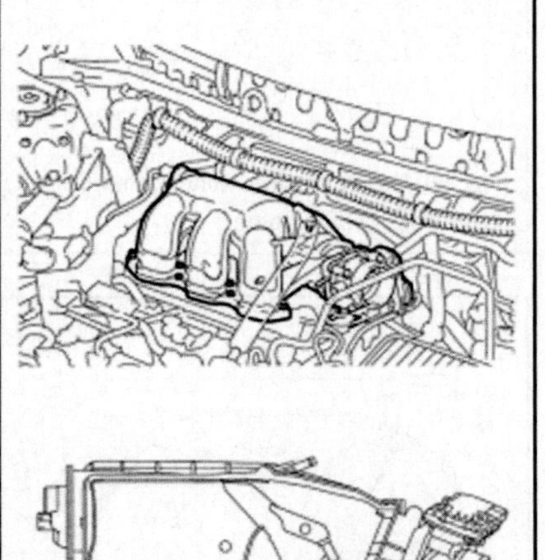

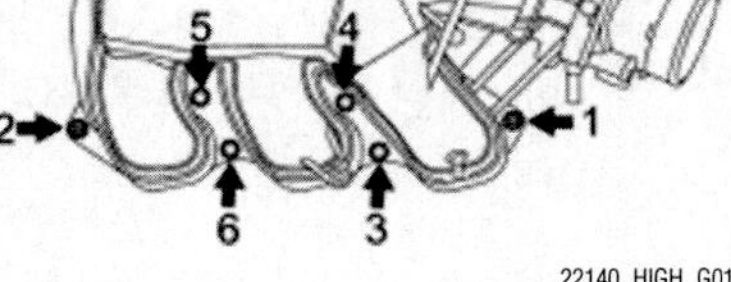

Fig. 93 Intake air surge tank assembly bolt removal sequence—3.5L engine

17. Disconnect the heater hose inlet pipe.
18. Remove the nut and ground cable.
19. Disconnect the fuel injector connectors.
20. Remove the intake manifold nuts and bolts in the sequence shown using several passes.
21. Remove the intake manifold.

To install:

22. Install the intake manifold. Tighten the retainers using several passes to 11 ft. lbs. (15 Nm).

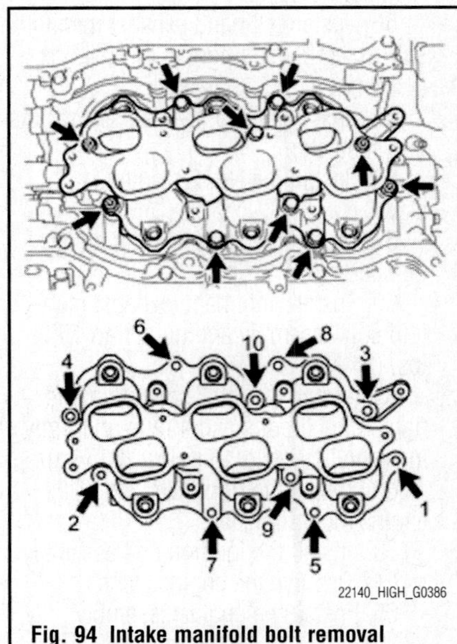

Fig. 94 Intake manifold bolt removal sequence—3.5L engine

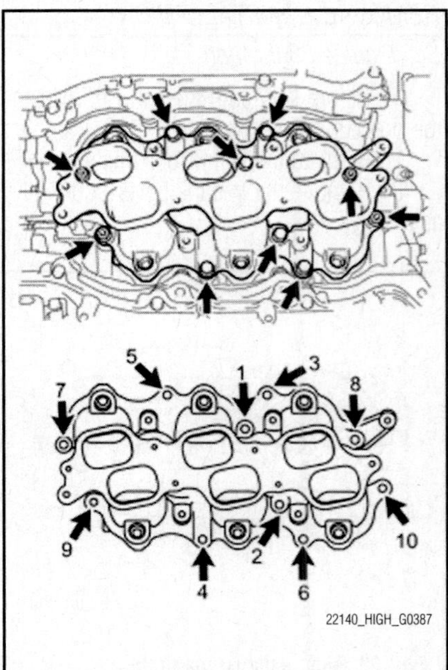

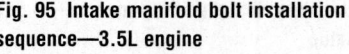

Fig. 95 Intake manifold bolt installation sequence—3.5L engine

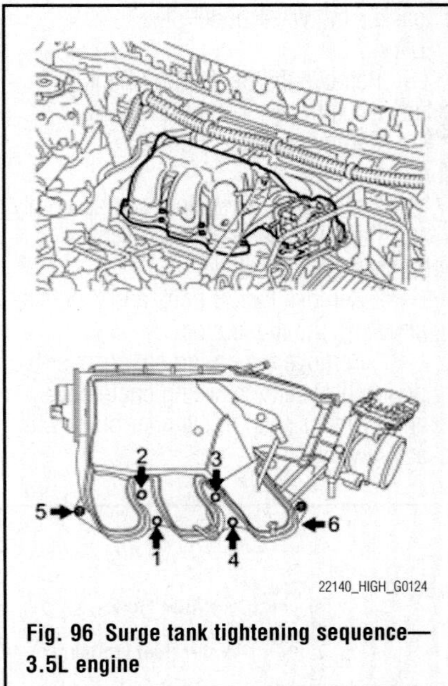

Fig. 96 Surge tank tightening sequence—3.5L engine

23. Install the remaining components in the reverse order of removal.

OIL PAN

REMOVAL & INSTALLATION

See Figures 97 through 101.

1. Before servicing the vehicle, refer to the precautions section.
2. Drain the engine oil.
3. Remove the No. 2 oil pan sub-assembly, as follows:
    a. Remove the 16 bolts and 2 nuts.
    b. Insert the blade of oil pan seal cutter between the oil pans. Cut through the applied sealer and remove the No. 2 oil pan sub-assembly.

➡Be careful not to damage the contact surfaces of the oil pans.

    c. Using a Torx® socket wrench E6, remove the 2 stud bolts.
4. Remove the oil strainer sub-assembly, as follows:
    a. Remove the bolt, 2 nuts, oil strainer sub-assembly and gasket.
    b. Using a Torx® socket wrench E6, remove the 2 stud bolts.
5. Remove the oil pan sub-assembly, as follows:
    a. Remove the 16 bolts and 2 nuts.

➡Be sure to clean the bolts and stud bolts and check the threads for cracks or other damage.

    b. Remove the oil pan sub-assembly by prying between the oil pan sub-assembly and cylinder block sub-assembly with a screwdriver.

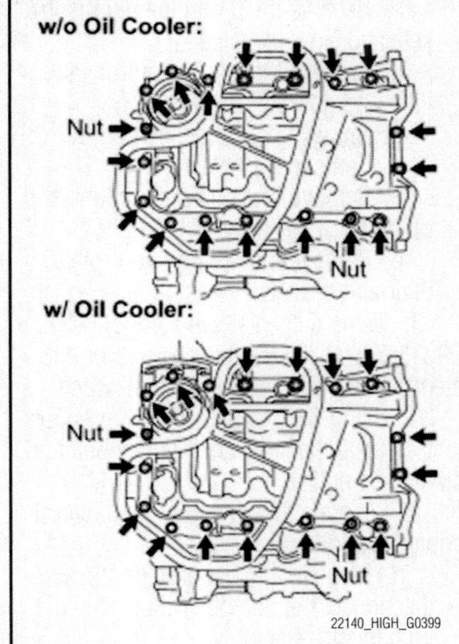

Fig. 97 Oil pan sub-assembly bolts and nuts—3.5L engine

➡Be careful not to damage the contact surfaces of the cylinder block and oil pans.

➡Tape the screwdriver tip before use.

    c. Remove the 2 O-rings.
    d. Using a Torx® socket wrench E8, remove the 2 stud bolts.

To install:

6. Install the oil pan sub-assembly, as follows:
    a. When replacing a stud bolt, install it by using an E8 Torx® socket wrench. Tighten to 7 ft. lbs. (10 Nm).

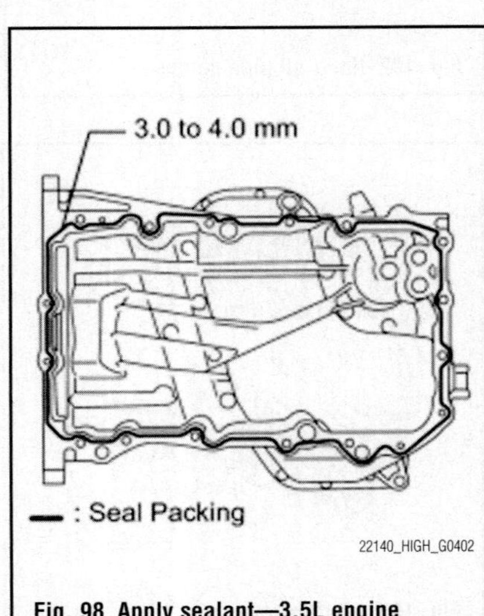

Fig. 98 Apply sealant—3.5L engine

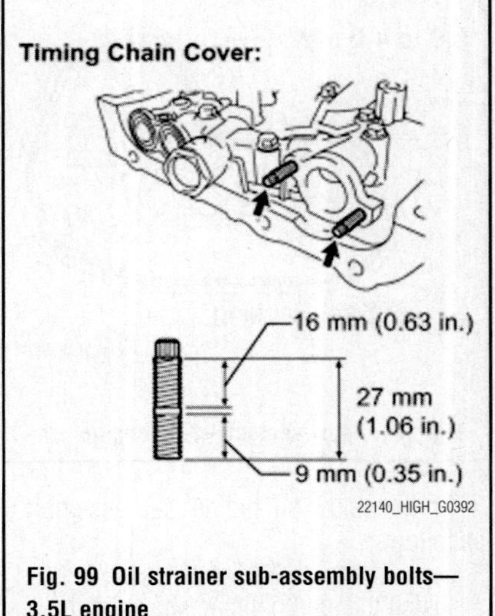

Fig. 99 Oil strainer sub-assembly bolts—3.5L engine

    b. Install 2 new O-rings.
    c. Apply seal packing in a continuous line as shown in the illustration. Seal packing: Toyota Genuine Seal Packing Black, Three Bond 1207B or equivalent. Seal diameter: 3.0 to 4.0 mm (0.118 to .0156 inches).

➡Remove any oil from the contact surface.

➡Install the oil pan within 3 minutes after applying seal packing.

➡Do not start the engine for at least 2 hours after installing.

    d. Install the oil pan with the 16 bolts and 2 nuts and tighten to 7 ft. lbs. (10 Nm), 15 ft. lbs. (21 Nm).

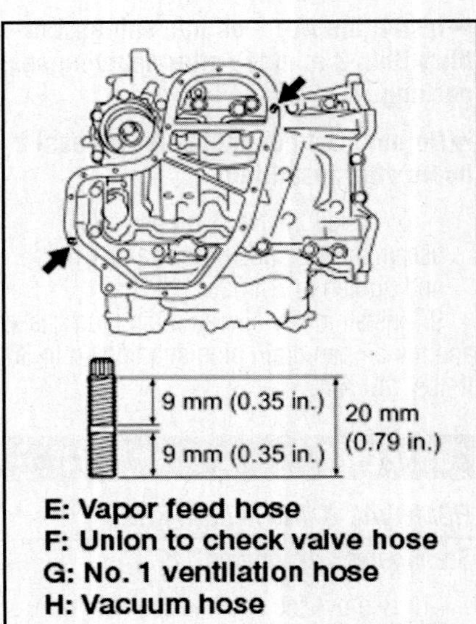

Fig. 100 No. 2 oil pan sub-assembly—3.5L engine

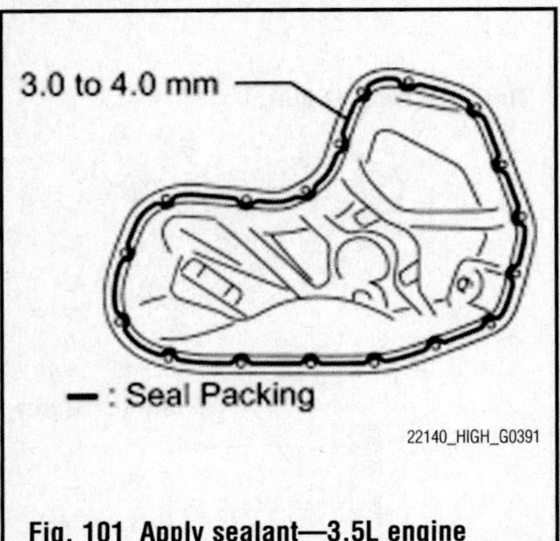

Fig. 101 Apply sealant—3.5L engine

7. Install the oil strainer sub-assembly, as follows:
   a. Using an E6 Torx® socket, install the stud bolts as shown in the illustration and tighten to 35 inch lbs. (4 Nm).
   b. Install a new gasket and the oil strainer sub-assembly with the bolt and 2 nuts and tighten to 7 ft. lbs. (10 Nm).
8. Install the No. 2 oil pan sub-assembly, as follows:
   a. Using an E6 Torx® socket, install the stud bolts as shown in the illustration and tighten to 35 inch lbs. (4 Nm).
   b. Apply seal packing in a continuous line as shown in the illustration. Seal packing: Toyota Genuine Seal Packing Black, Three Bond 1207B or equivalent. Seal diameter: 3.0 to 4.0 mm (0.118 to .0156 inches).

➡Remove any oil from the contact surface.

➡Install the No. 2 oil pan sub-assembly within 3 minutes after applying seal packing.

➡Do not start the engine for at least 2 hours after installing.

   c. Install the No. 2 oil pan sub-assembly with the 16 bolts and 2 nuts and tighten to 7 ft. lbs. (10 Nm).
9. Install a new oil pan drain plug gasket and the oil pan drain plug and tighten to 30 ft. lbs. (40 Nm).

OIL PUMP

REMOVAL & INSTALLATION

See Figures 102 through 112.

1. Remove the engine assembly with transaxle.
2. Secure the engine.
3. Remove the No. 1 oil pipe, as follows:
   a. Remove the 2 oil pipe unions, gaskets and No. 1 oil pipe.
   b. Remove the left oil control valve filter and gaskets.
4. Remove the oil pipe, as follows:
   a. Remove the bolt.
   b. Remove the 2 oil pipe unions and oil pipe.
   c. Remove the right oil control valve filter and gaskets.
5. Using a special service tool (SST: 09213-70011, SST: 09213-70011 or equivalent), loosen the crankshaft pulley bolt.
6. Using SST: 09950-50013 or equivalent, remove the crankshaft pulley bolt and crankshaft pulley.
7. Separate the oil cooler pipe (w/ oil cooler), as follows:
   a. Remove the bolt and 2 nuts, and disconnect the oil cooler pipe from the oil pan sub-assembly.
   b. Remove the gasket from the oil pan sub-assembly.
8. Remove the water inlet housing, as follows:
   a. Disconnect the water hose.
   b. Remove the 2 bolts, nut and water inlet housing.
   c. Remove the water inlet housing gasket and water outlet pipe O-ring.
9. Remove the Bank 1 and Bank 2 cylinder head cover sub-assemblies.
10. Remove the No. 2 oil pan sub-assembly.
11. Remove the oil strainer sub-assembly.
12. Remove the timing chain cover sub-assembly, as follows:
    a. Remove the 23 bolts and 2 nuts as shown in the illustration.
    b. Remove the timing chain cover by prying between the timing chain cover and cylinder head or cylinder block with a screwdriver.

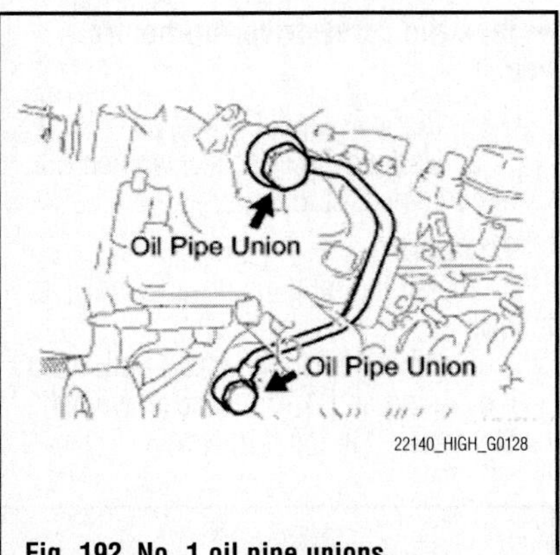

Fig. 192 No. 1 oil pipe unions

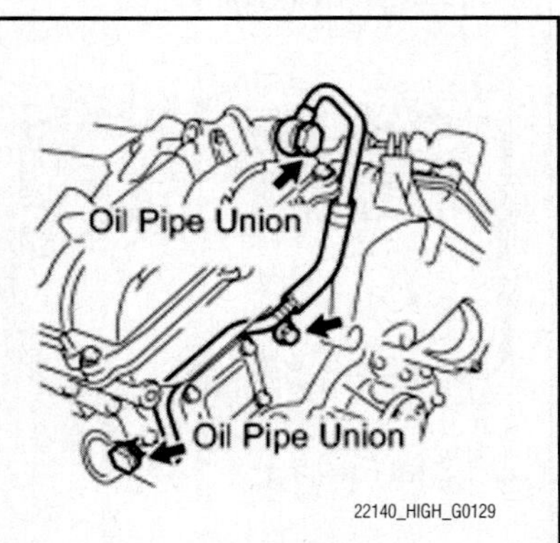

Fig. 103 Oil pipe unions

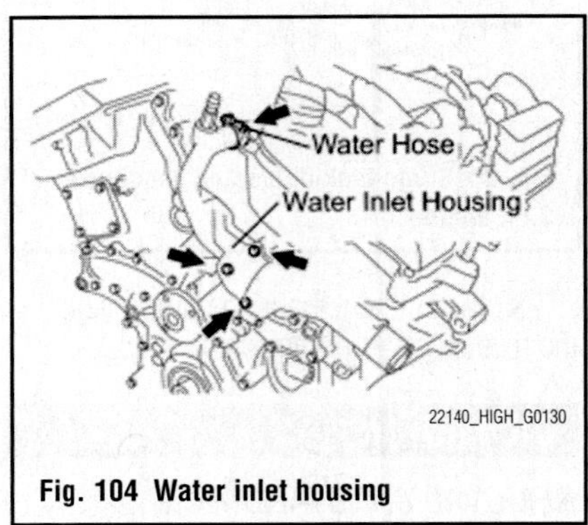

Fig. 104 Water inlet housing

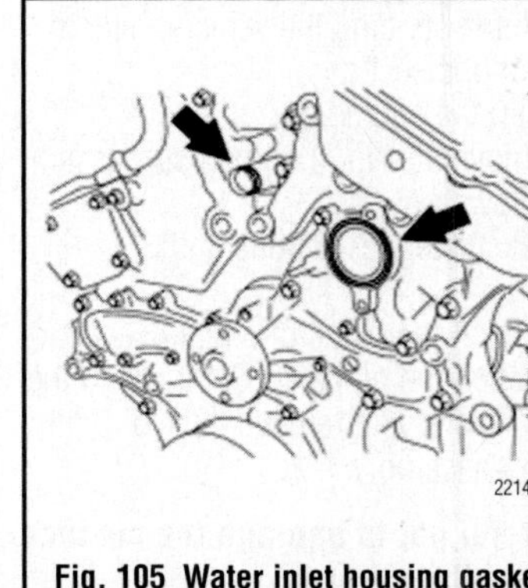

Fig. 105 Water inlet housing gasket and water outlet pipe O-ring

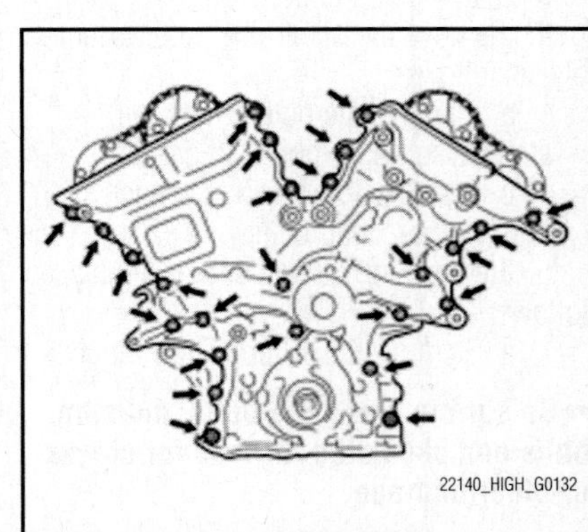

Fig. 106 Timing chain cover sub-assembly bolts and nuts

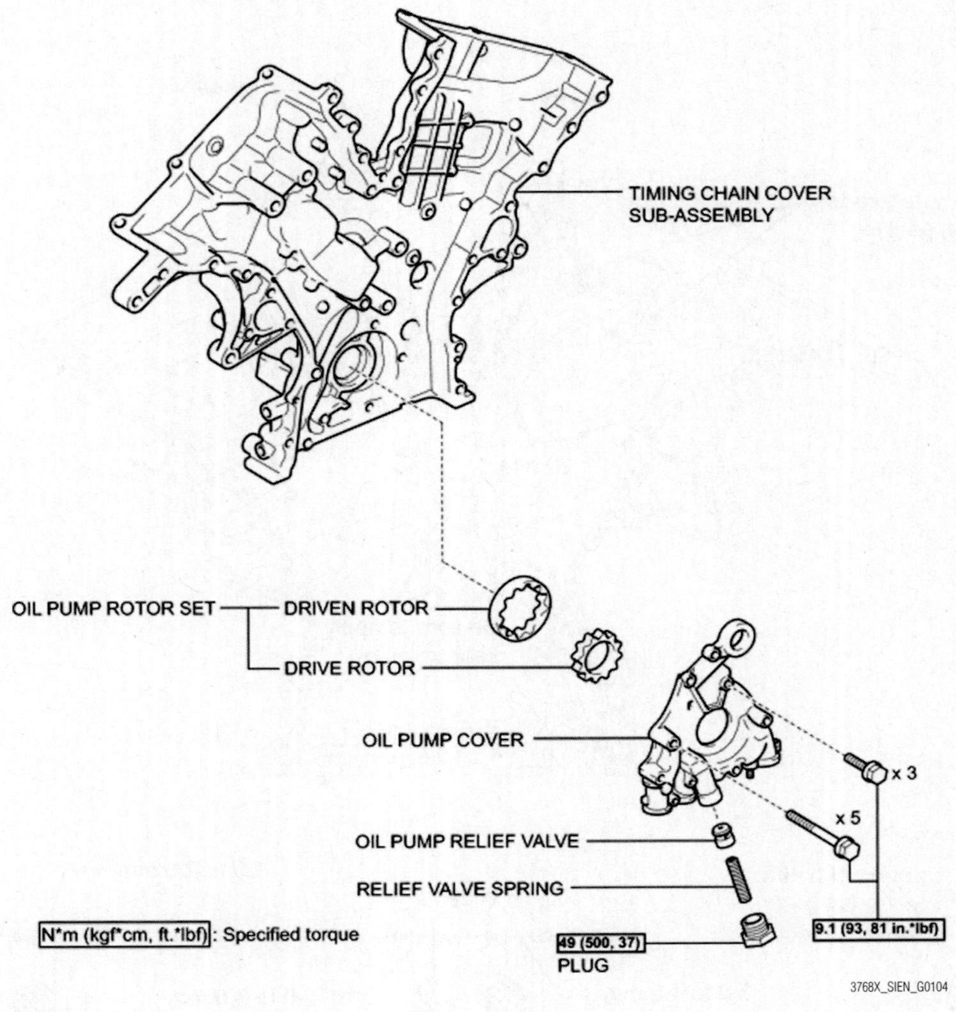

TIMING CHAIN COVER
SUB-ASSEMBLY
OIL PUMP ROTOR SET
DRIVEN ROTOR
DRIVE ROTOR
OIL PUMP COVER
x 3
x 5
OIL PUMP RELIEF VALVE
RELIEF VALVE SPRING
N*m (kgf*cm, ft.*lbf): Specified torque
49 (500, 37)
PLUG
9.1 (93, 81 in.*lbf)
3768X_SIEN_G0104

Fig. 107 Removing the oil pump

C
C
C
C
C
: Seal Packing
3.0 mm or more
22140_HIGH_G0389

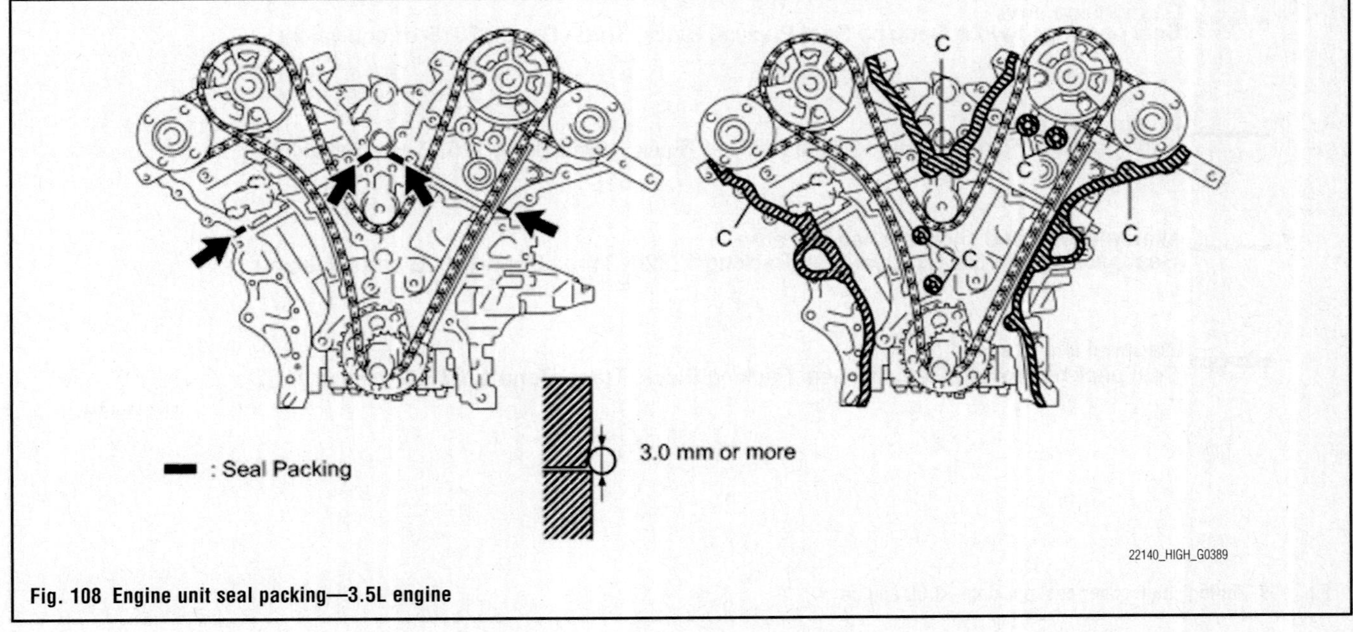

Fig. 108 Engine unit seal packing—3.5L engine

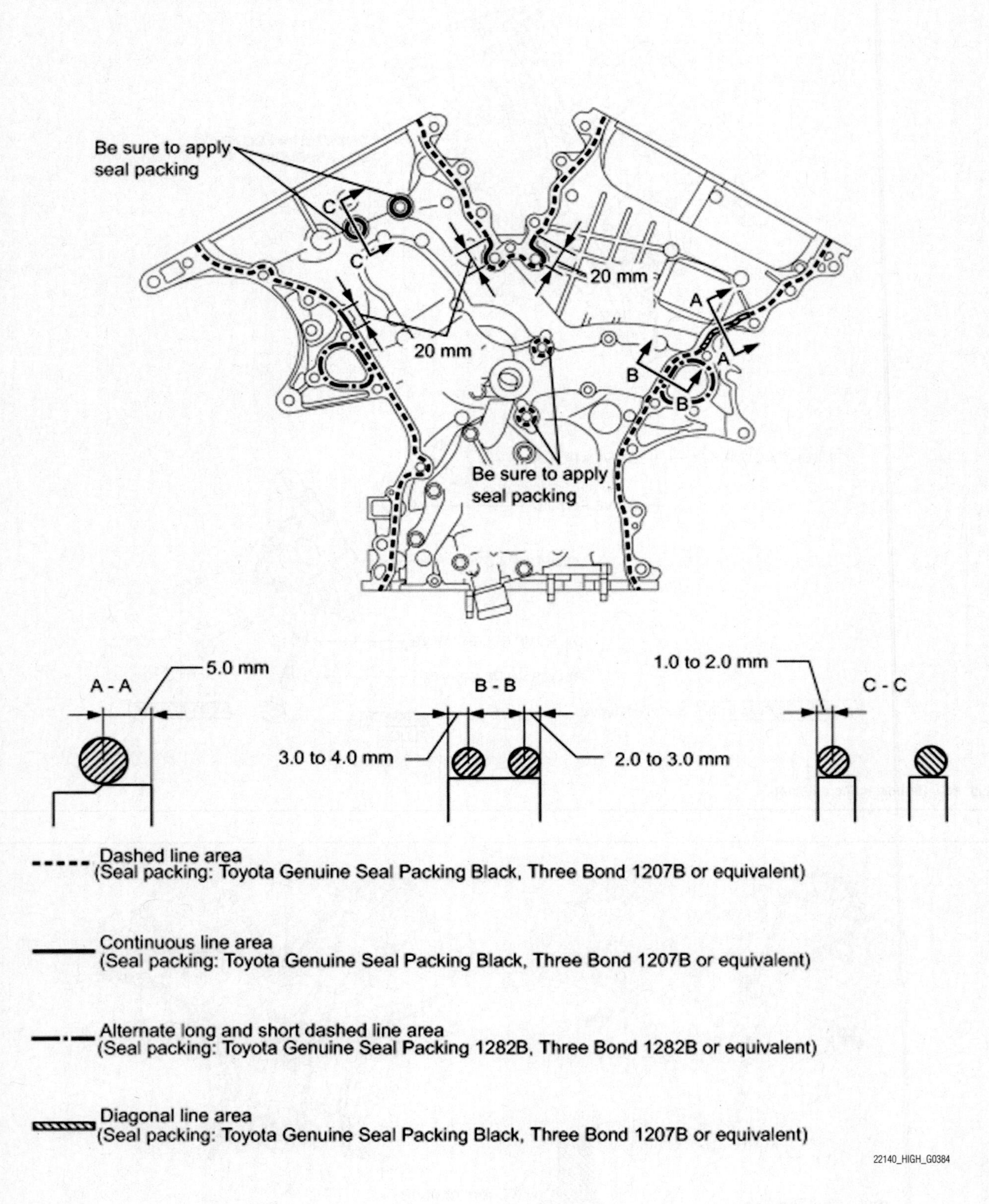

Fig. 109 Timing chain cover seal packing—3.5L engine

Area	Seal Packing Diameter	Application Position from Inside Seal Line
Continuous Line Area	4.5 mm (0.177 in.) or more	3.0 to 4.0 mm (0.118 to 0.158 in.)
Alternate Long and Short Dashed Line Area	3.5 mm (0.138 in.) or more	2.0 to 3.0 mm (0.079 to 0.118 in.)
Dashed Line Area	3.5 mm (0.138 in.) or more	3.0 to 4.0 mm (0.118 to 0.158 in.)
Diagonal Line Area	6.0 mm (0.236 in.) or more	5.0 mm (0.197 in.)

22140_HIGH_G0380

Fig. 110 Seal packing specifications—3.5L engine

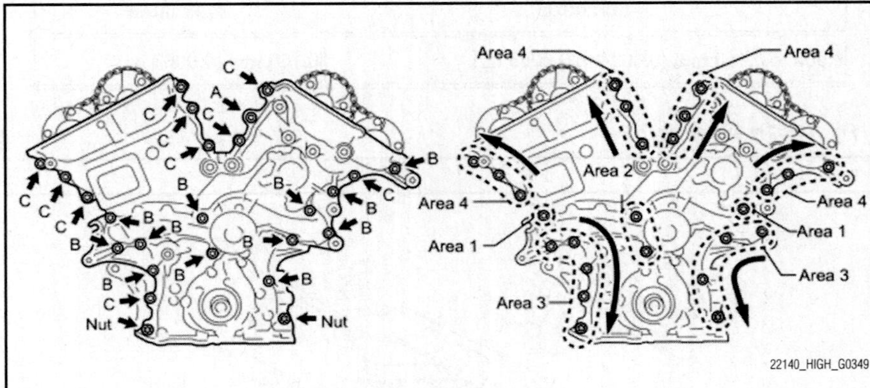

Fig. 111 Timing chain cover tightening sequence

Item	Length
Bolt A	40 mm (1.57 in.)
Bolt B	55 mm (2.17 in.)
Bolt C	25 mm (0.98 in.)

22140_HIGH_G0133

Fig. 112 Timing chain cover bolt length

➡Be careful not to damage the contact surfaces of the cylinder head, cylinder block and chain cover.

➡Tape the screwdriver tip before use.

c. Remove the gasket.

13. Using a screwdriver, pry out and remove the timing chain oil seal.

➡Tape the screwdriver tip before use.

To install:

14. Install the timing chain case oil seal, as follows:

a. Using SST (SST: 09223-22010, SST: 09506-35010) tap in a new oil seal until its surface is flush with the timing chain case edge.

➡Keep the lip free from foreign matter.

➡Make sure that the oil seal edge does not stick out of the timing chain case.

➡Do not tap on the oil seal at an angle.

15. Install the timing chain cover sub-assembly, as follows:

a. Apply seal packing in a continuous line to the engine unit as shown in the illustration. Seal packing: Toyota Seal Packing Black, Three Bond 1207B or equivalent. Seal diameter: 3.0 mm (0.118 in.).

➡Be sure to clean and degrease the contact surfaces, especially the surfaces indicated by C in the illustration.

➡If the contact surfaces are wet, wipe them with an oil-free cloth before applying seal packing.

➡Install the chain cover sub-assembly within 3 minutes after applying seal packing.

➡Do not start the engine for at least 2 hours after installing the chain cover sub-assembly.

b. Apply seal packing in a continuous line to the timing chain cover as shown in the illustration. Seal packing: Toyota Seal Packing Black, Three Bond 1207B, Three Bond 1282B or equivalent.

➡If the contact surfaces are wet, wipe them with an oil-free cloth before applying seal packing.

➡Install the chain cover sub-assembly within 3 minutes and tighten the bolts within 15 minutes after applying seal packing.

➡Do not start the engine for at least 2 hours after installing the chain cover sub-assembly.

c. Apply seal packing as follows:

d. Install a new gasket.

e. Align the oil pump's drive rotor spline and the crankshaft as shown in the illustration. Install the spline and chain cover to the crankshaft.

f. Temporarily tighten the timing chain cover with the 23 bolts and 2 nuts.

➡Make sure that there is no oil on the bolt threads.

g. Fully tighten the bolts in area 1 and area 2 (from top to bottom as shown) to 15 ft. lbs. (21 Nm).

h. Fully tighten the bolts and nuts in area 3 (from top to bottom as shown) to 15 ft. lbs. (21 Nm).

➡Tighten the bolts and nuts from top to bottom as shown in the illustration.

i. Fully tighten the bolts in area 4 (from bottom to top as shown) to 32 ft. lbs. (43 Nm), 15 ft. lbs. (21 Nm).

j. Install the oil pan sub-assembly.

k. Install the oil strainer sub-assembly.

l. Install the No. 2 oil pan sub-assembly.

m. Install the cylinder head cover sub-assembly.

16. Install the water inlet housing, as follows:

a. Install a new water inlet housing No. 1 gasket and water outlet pipe O-ring.

b. Install the water inlet housing with the 2 bolts and nut and tighten to 7 ft. lbs. (10 Nm).

➡**Be careful not to allow the O-ring to get caught between the parts.**

c. Connect the water hose.

17. Install the oil cooler pipe (w/oil cooler), as follows:

a. Install a new gasket to the oil pan sub-assembly.

b. Install the oil cooler pipe with the bolt and 2 nuts and tighten to 15 ft. lbs. (21 Nm).

18. Install the crankshaft pulley.
19. Install oil pipe.
20. Install the No. 1 oil pipe.
21. Install the engine hangers.
22. Remove engine stand.
23. Install engine assembly with transaxle.

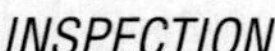

INSPECTION

See Figures 113 through 119.

1. Inspect the oil pump relief valve, as follows:

a. Coat the relief valve with engine oil and check that it falls smoothly into the valve hole by its own weight. If the valve does not fall smoothly, replace the relief valve. If necessary, replace the oil pump assembly.

2. Inspect the oil pump rotor set, as follows:

a. Install the rotors to the timing chain cover with the rotors' marks facing up. Check that the rotors rotate smoothly.

b. Check the tip clearance: using a feeler gauge, measure the clearance between the drive and driven rotor tips, as shown in the illustration. If the clearance is greater than the maximum, replace the drive and driven rotors.

c. Check the side clearance: using a feeler gauge and precision straightedge, measure the clearance between the rotors and precision straightedge, as shown in the illustration. If the side clearance is greater than the maximum, replace the timing chain cover sub-assembly.

d. Check the body clearance: using a feeler gauge, measure the clearance between the timing chain cover and driven rotor, as shown in the illustration. If the body clearance is greater than the maximum, replace the timing chain cover sub-assembly.

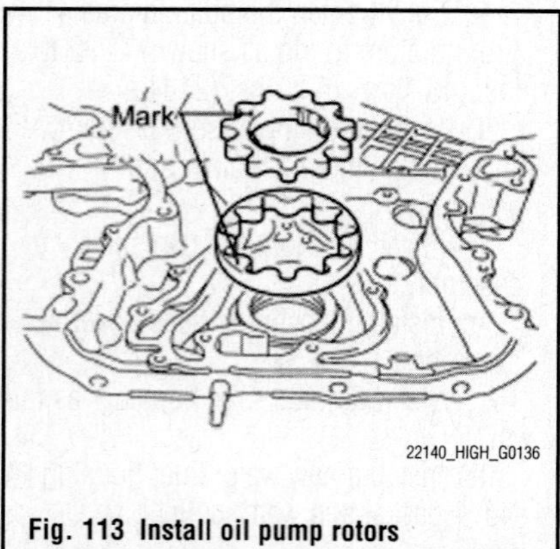

Fig. 113 Install oil pump rotors

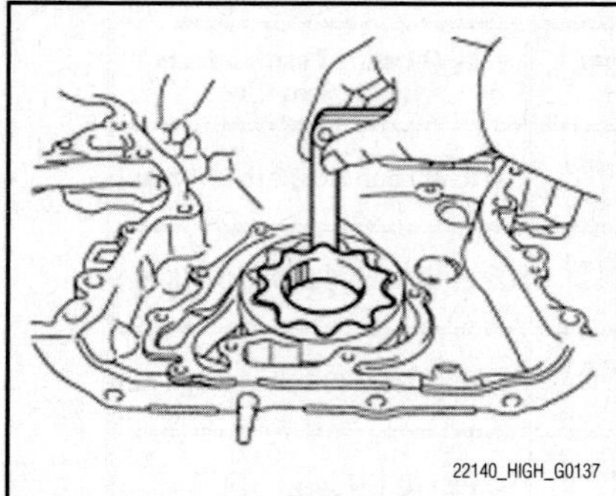

Fig. 114 Check tip clearance

Standard	Maximum
0.060 to 0.160 mm (0.0024 to 0.0063 in.)	0.160 mm (0.0063 in.)

22140_HIGH_G0138

Fig. 115 Tip clearance

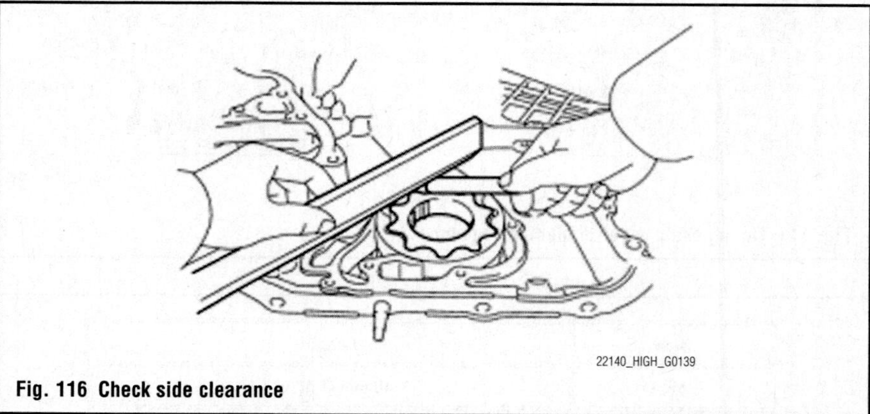

Fig. 116 Check side clearance

Standard	Maximum
0.030 to 0.090 mm (0.0012 to 0.0035 in.)	0.090 mm (0.0035 in.)

22140_HIGH_G0140

Fig. 117 Side clearance

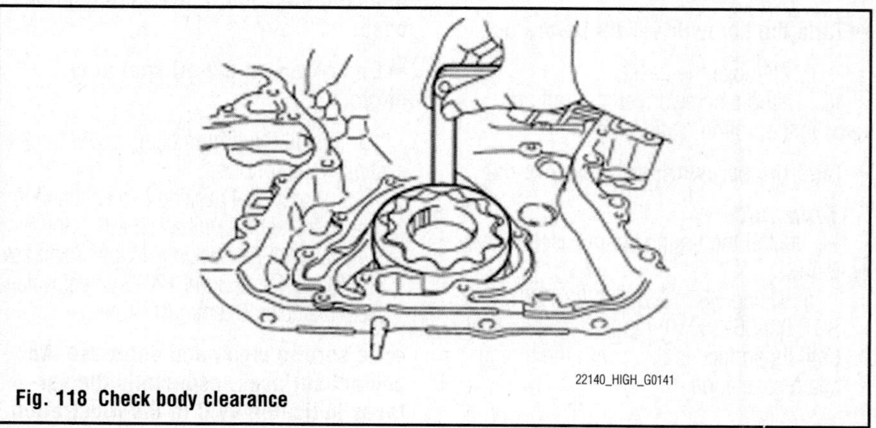

Fig. 118 Check body clearance

Standard	Maximum
0.250 to 0.325 mm (0.0098 to 0.0128 in.)	0.325 mm (0.0128 in.)

22140_HIGH_G0142

Fig. 119 Body clearance

PISTON AND RING

POSITIONING

See Figure 120.

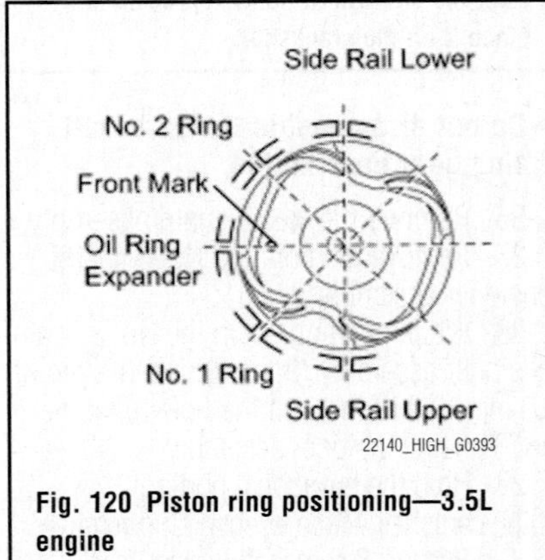

Fig. 120 Piston ring positioning—3.5L engine

REAR MAIN SEAL

REMOVAL & INSTALLATION

See Figures 121 through 124.

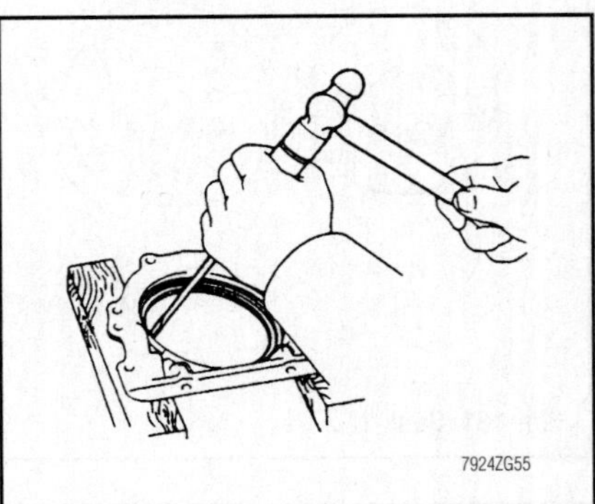

Fig. 121 Carefully tap the old seal from the retainer

If the rear oil seal retainer is not installed to the block, use a tapered ended screwdriver and hammer to remove the oil seal. Apply multi-purpose grease to the new oil seal lip. Using a seal driver, tap the seal into place. Be careful not to install it slantwise.

1. Before servicing the vehicle, refer to the precautions section.

If the rear oil seal retainer is installed on the cylinder block, using a knife, cut off the lip of the seal. Using a taped ended prytool, pry the old seal out of the retainer. Inspect the oil seal lip contacting surface of the crankshaft for cracks or damage. Apply multipurpose grease to the new oil seal, then tap the seal in place with a seal installer. Be careful not to install the seal slantwise.

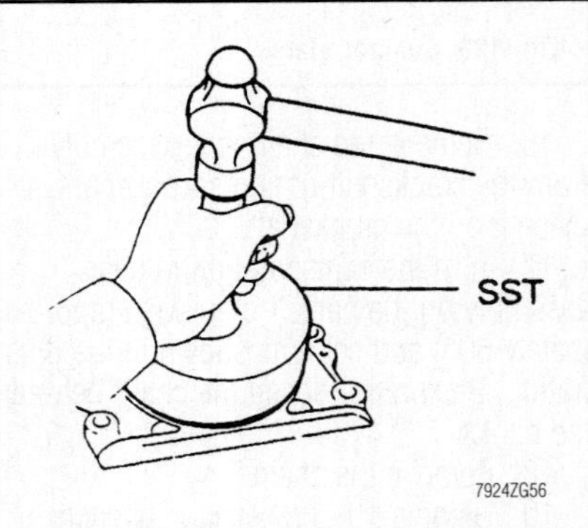

Fig. 122 Use the proper sized driver to seat the seal

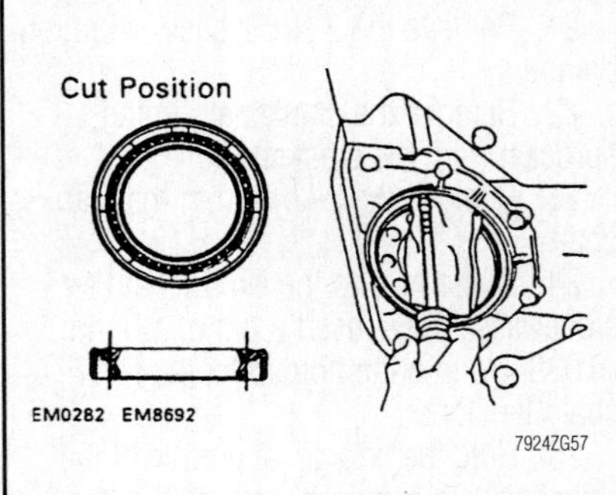

Fig. 123 Cut off the oil seal lip, then pry the seal out of the retaining plate

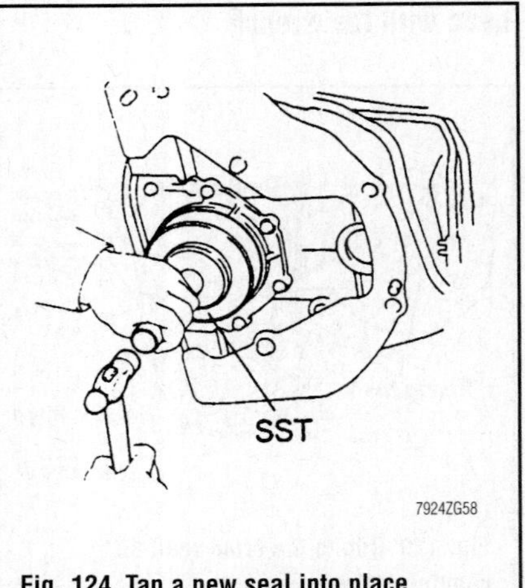

Fig. 124 Tap a new seal into place

TIMING CHAIN FRONT COVER

REMOVAL & INSTALLATION

Refer to the timing chain and sprocket procedure.

TIMING CHAIN & SPROCKETS

REMOVAL & INSTALLATION

See Figures 125 through 147.

1. Before servicing the vehicle, refer to the Precautions section.
2. Remove the engine from the vehicle and mount it on an engine stand.
3. Remove the 15 bolts and 2 nuts as shown in the illustration.
4. Remove the timing chain cover sub-assembly by prying between the timing chain cover and cylinder head sub-assembly or cylinder block sub-assembly with a prytool.

➡Be careful not to damage the contact surfaces of the cylinder head, cylinder block and chain cover.

5. Remove the 4 bolts, chain cover plate and chain cover plate gasket.
6. Temporarily tighten the pulley set bolt.
7. Set the timing mark on the crank angle sensor plate to the RH block bore center line (TDC / compression).
8. Check that the timing marks of the camshaft timing gears are aligned with those of the bearing cap as shown in the illustration.
9. If not, turn the crankshaft 1 revolution (360°) and align the timing marks as above.

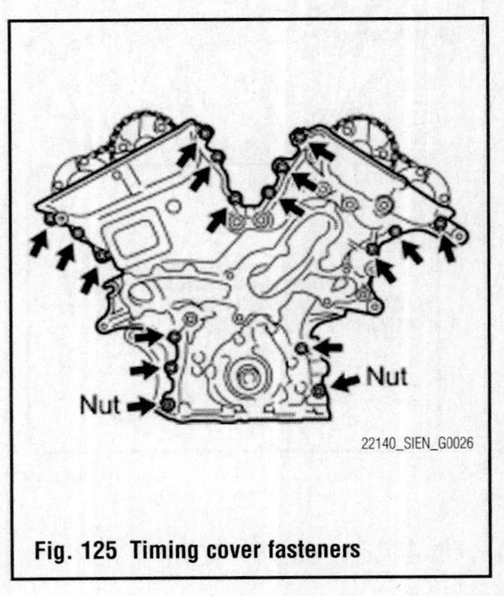

Fig. 125 Timing cover fasteners

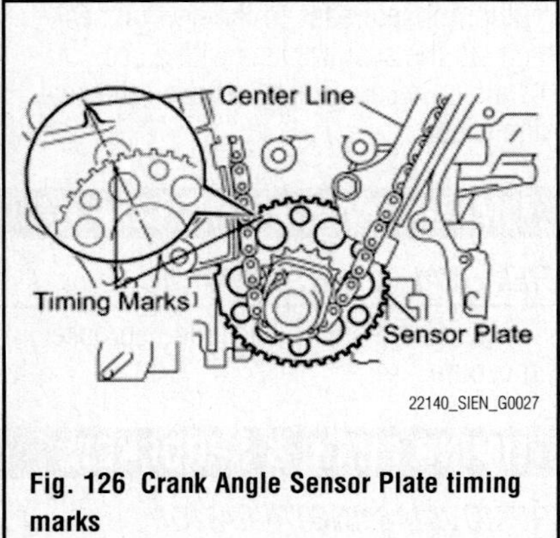

Fig. 126 Crank Angle Sensor Plate timing marks

10. Move the stopper plate upward to release the lock, and push the plunger deep into the tensioner.

11. Move the stopper plate downward to set the lock, and insert a pin of 1.27 mm (0.05 in.) into the stopper plate's hole.

12. Remove the 2 bolts and No. 1 chain tensioner assembly.

13. Remove the chain tensioner slipper.

14. Turn the crankshaft counterclockwise 10° to loosen the chain sub-assembly of the crankshaft timing sprocket.

15. Remove the pulley set bolt.

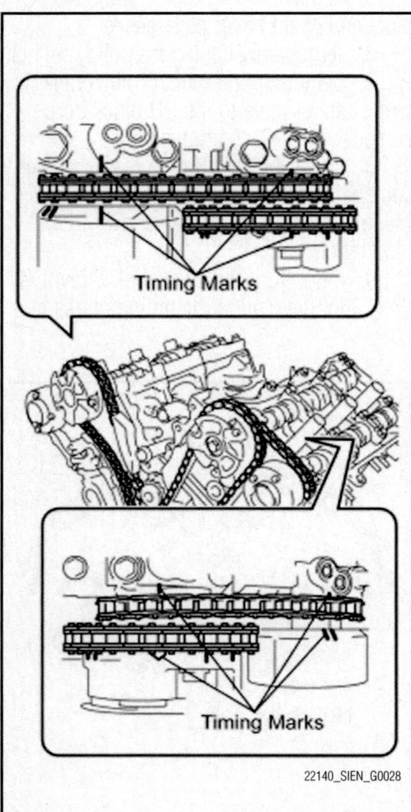

Fig. 127 Camshaft timing marks

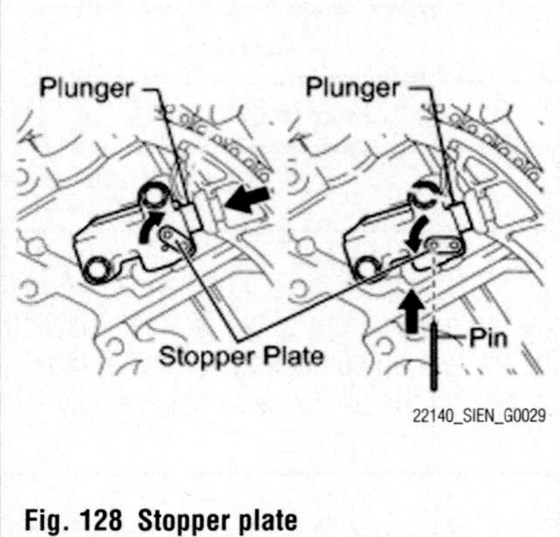

Fig. 128 Stopper plate

16. Remove the chain sub-assembly from the crankshaft timing sprocket and place it on the crankshaft.

17. Turn the camshaft timing gear assembly on the bank 1 clockwise (approximately 60°) and set it as shown in the illustration. Be sure to loosen the chain between the banks.

18. Remove the chain.

19. Using a 10 mm hexagon wrench, remove the No. 2 idle gear shaft, idle sprocket assembly and No. 1 idle gear shaft.

20. Remove the 2 bolts and No. 1 chain vibration damper.

21. Remove the 2 No. 2 chain vibration dampers.

22. Remove the crankshaft timing sprocket from the crankshaft.

23. Remove the 2 keys from the crankshaft.

24. While raising the No. 2 chain tensioner assembly, insert a pin of 1.0 mm (0.039 in.) into the hole to fix the No. 2 chain tensioner.

25. Hold the hexagonal portion of the camshaft with a wrench, and remove the 2 bolts and 2 camshaft timing gear assemblies.

➡Be careful not to damage the cylinder head with the wrench.

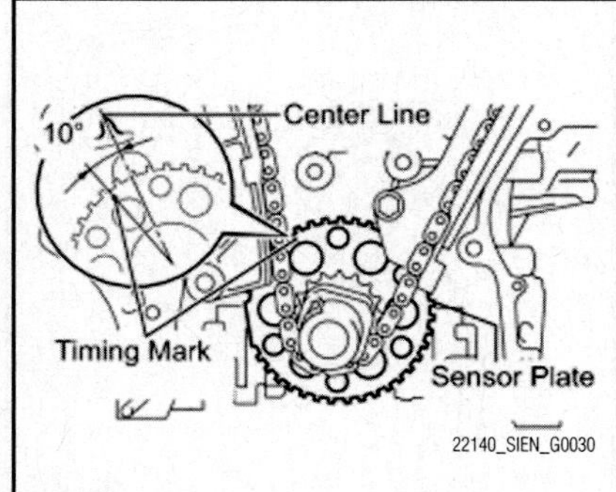

Fig. 129 Rotate the crankshaft 10° counterclockwise

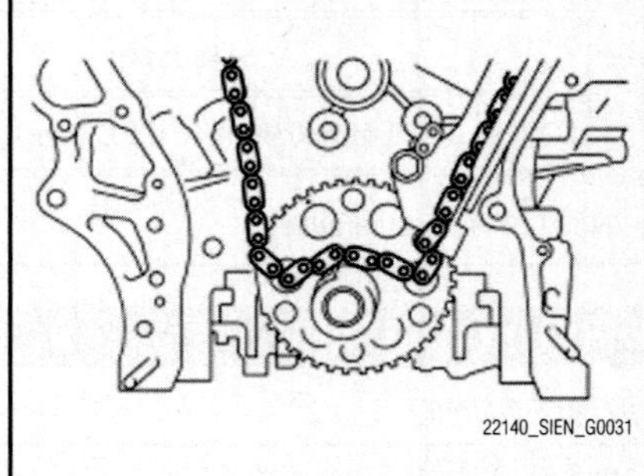

Fig. 130 Remove the chain sub-assembly from the crankshaft timing sprocket and place it on the crankshaft

➡Do not disassemble the camshaft timing gear assemblies.

26. Remove the No. 2 chain assembly.

27. Remove the bolt and No. 2 chain tensioner assembly.

28. While pushing down the No. 3 chain tensioner assembly (Bank 2), insert a pin of 1.0 mm (0.039 in.) into the hole to fix the No. 3 chain tensioner assembly.

29. Hold the hexagonal portion of the camshaft with a wrench, and remove the 2 bolts and 2 camshaft timing gear assemblies.

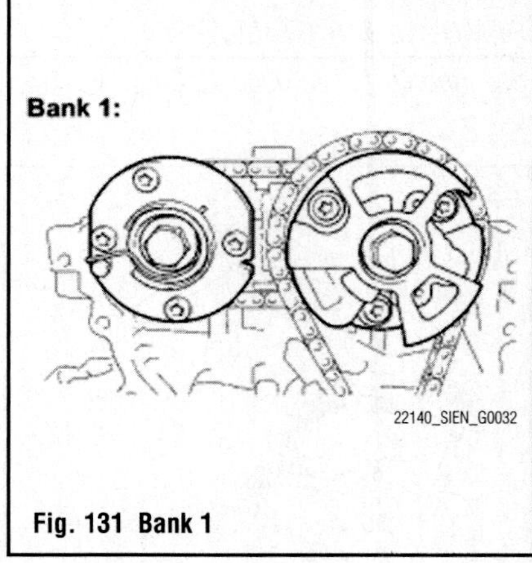

Fig. 131 Bank 1

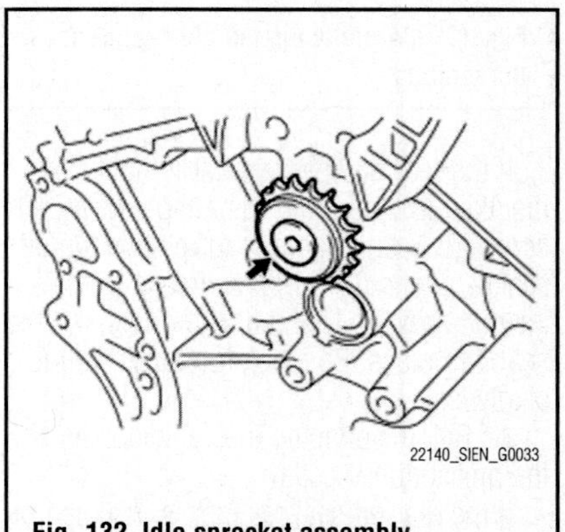

Fig. 132 Idle sprocket assembly

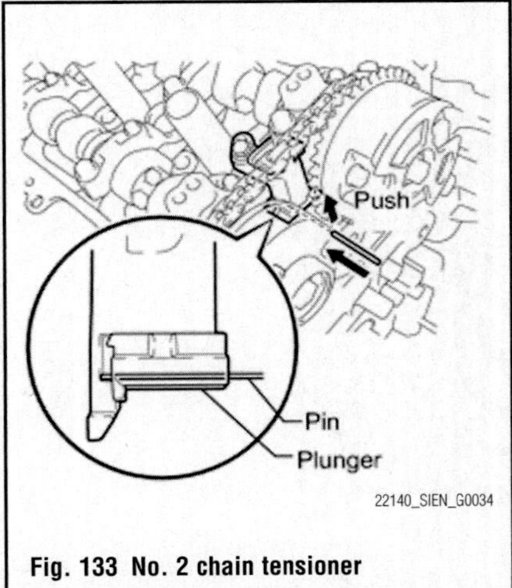

Fig. 133 No. 2 chain tensioner

➡**Be careful not to damage the cylinder head with the wrench.**

➡**Do not disassemble the camshaft timing gear assemblies.**

30. Remove the No. 2 chain sub-assembly.
31. Remove the bolt and No. 3 chain tensioner assembly.

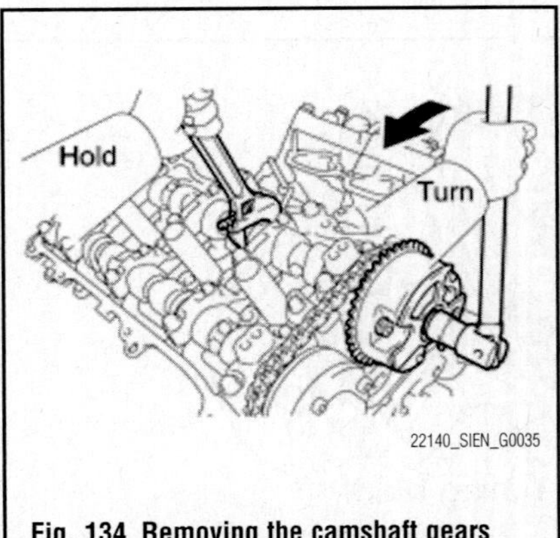

Fig. 134 Removing the camshaft gears

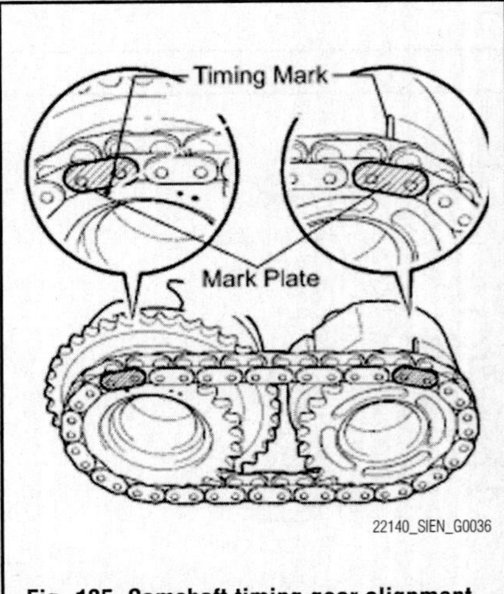

Fig. 135 Camshaft timing gear alignment

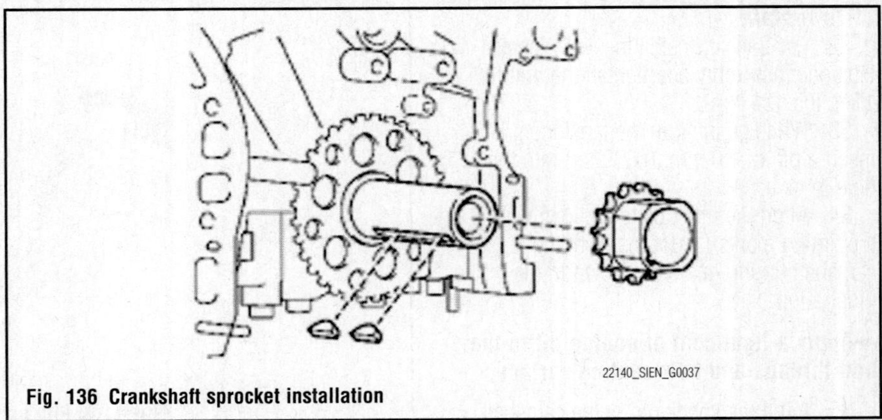

Fig. 136 Crankshaft sprocket installation

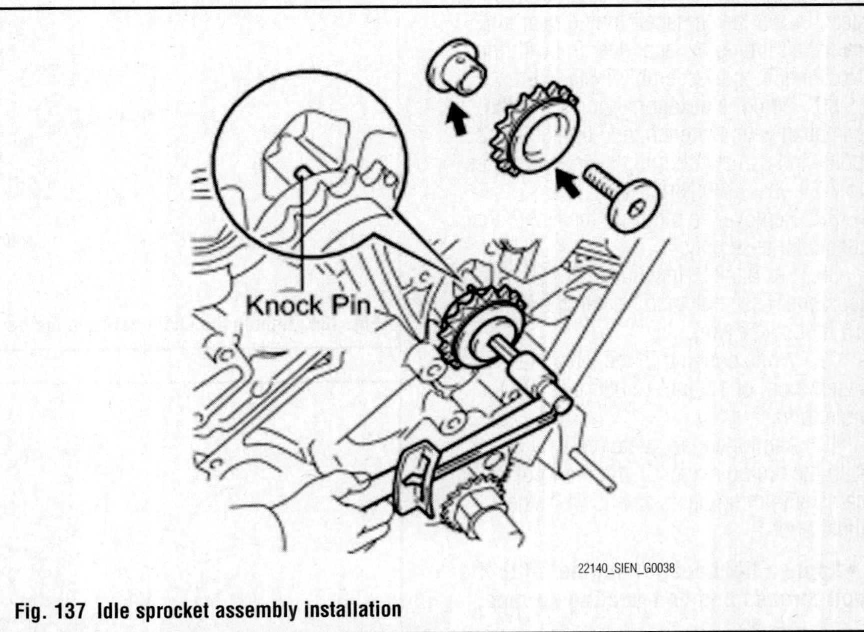

Fig. 137 Idle sprocket assembly installation

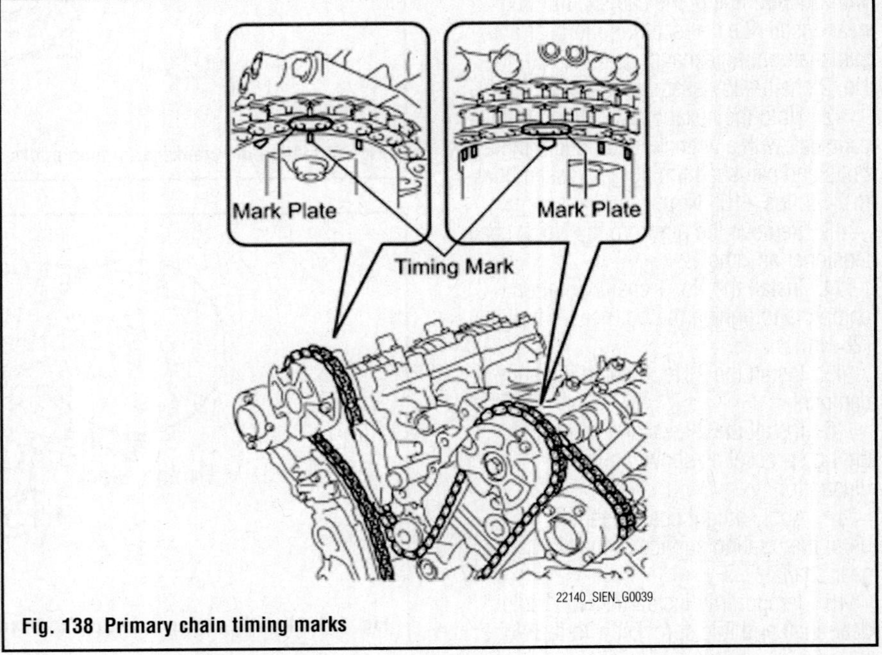

Fig. 138 Primary chain timing marks

To install:

32. For Bank 1, install the No. 2 chain tensioner assembly and tighten the bolt to 15 ft. lbs. (21 Nm).
33. While pushing in the tensioner, insert a pin of 1.0 mm (0.039 in.) into the hole to fix it.
34. Align the mark plate (yellow) with the timing marks (1-dot mark) of the camshaft timing gears as shown in the illustration.

➡**Apply a light coat of engine oil to the bolt threads and bolt-seating surface.**

35. Align the knock pin of the camshaft with the pin hole of the camshaft timing gear. Install the camshaft timing gear and camshaft timing exhaust gear RH with the No. 2 chain sub-assembly installed.
36. Hold the hexagonal portion of the camshaft with a wrench, and tighten the 2 bolts and camshaft timing gear assemblies to 74 ft. lbs. (100 Nm).
37. Remove the pin from the No. 2 chain tensioner assembly.
38. For Bank 2, install the No. 2 chain tensioner assembly and tighten the bolt to 15 ft. lbs. (21 Nm).
39. While pushing in the tensioner, insert a pin of 1.0 mm (0.039 in.) into the hole to fix it.
40. Align the mark plate (yellow) with the timing marks (1-dot mark) of the camshaft timing gears as shown in the illustration.

➡**Apply a light coat of engine oil to the bolt threads and bolt-seating surface.**

41. Align the knock pin of the camshaft with the pin hole of the camshaft timing gear. Install the camshaft timing gear and camshaft timing exhaust gear LH with the No. 2 chain sub-assembly installed.
42. Hold the hexagonal portion of the camshaft with a wrench, and tighten the 2 bolts and camshaft timing gear assemblies to 74 ft. lbs. (100 Nm).
43. Remove the pin from the No. 2 chain tensioner assembly.
44. Install the No. 1 chain vibration damper and tighten the bolts to 17 ft. lbs. (23 Nm).
45. Install the 2 No. 2 chain vibration dampers.
46. Install the 2 keys and crankshaft timing sprocket as shown in the illustration.
47. Apply a light coat of engine oil to the rotating surface of the No. 1 idle gear shaft.
48. Temporarily install the No. 1 idle gear shaft and idle sprocket with the No. 2

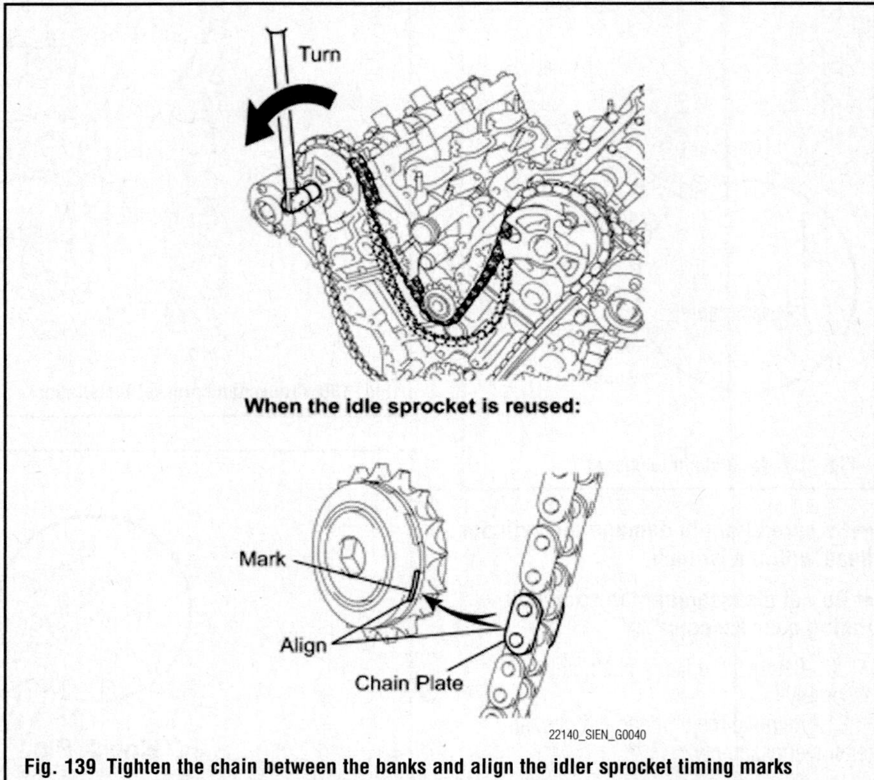

Fig. 139 Tighten the chain between the banks and align the idler sprocket timing marks

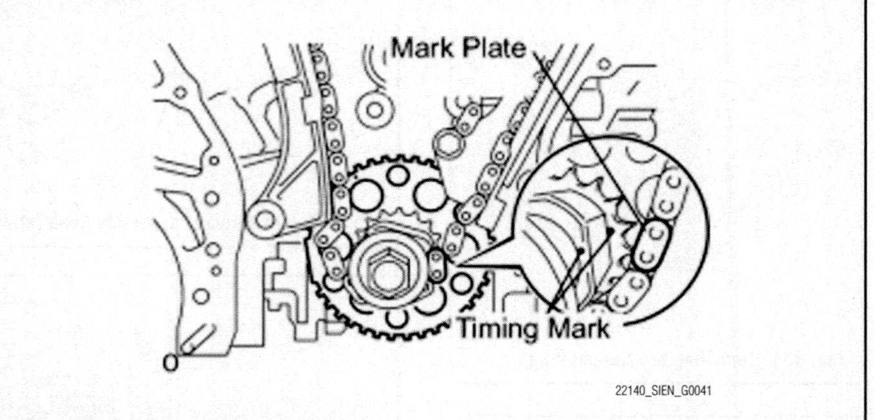

Fig. 140 Align the crankshaft timing marks

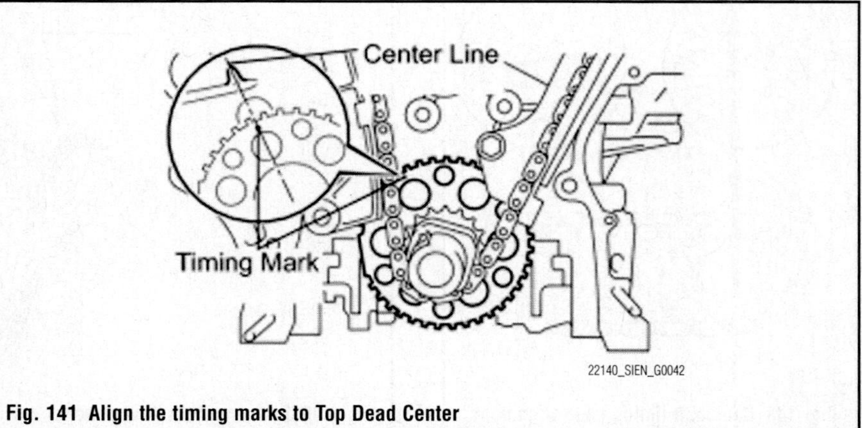

Fig. 141 Align the timing marks to Top Dead Center

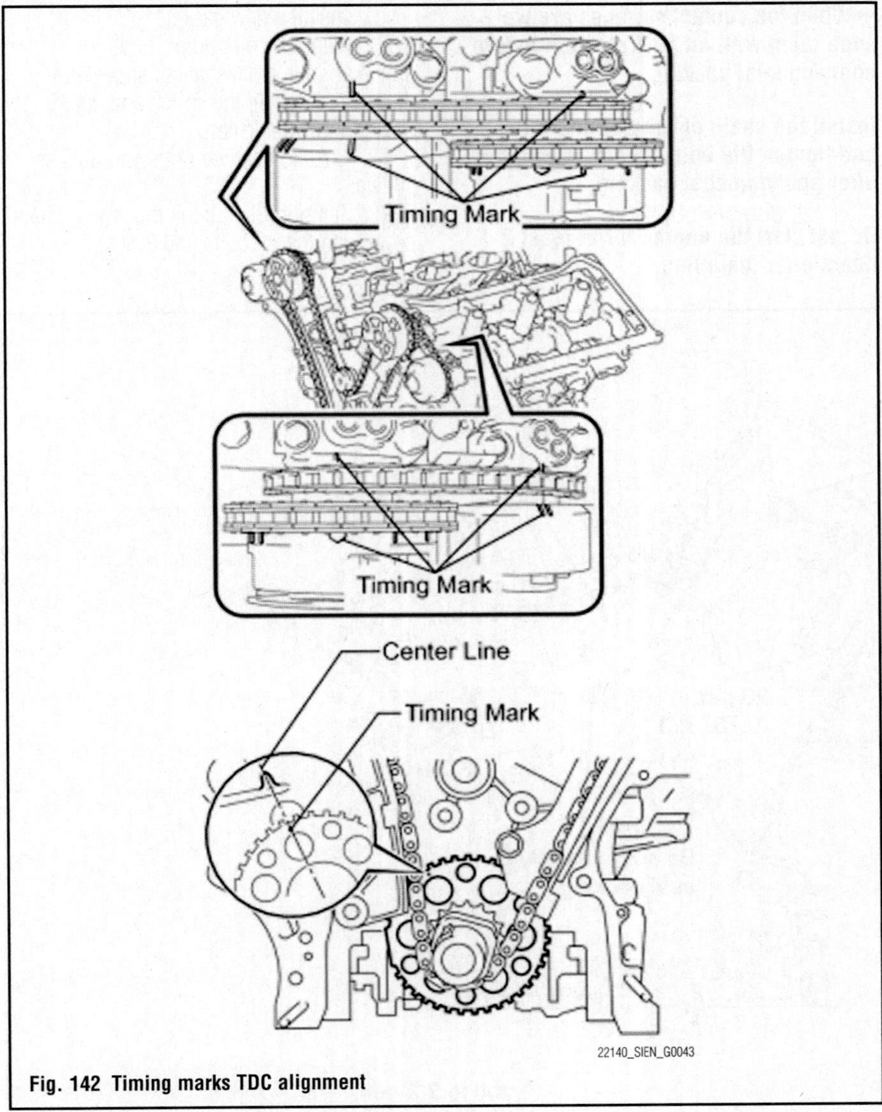

Fig. 142 Timing marks TDC alignment

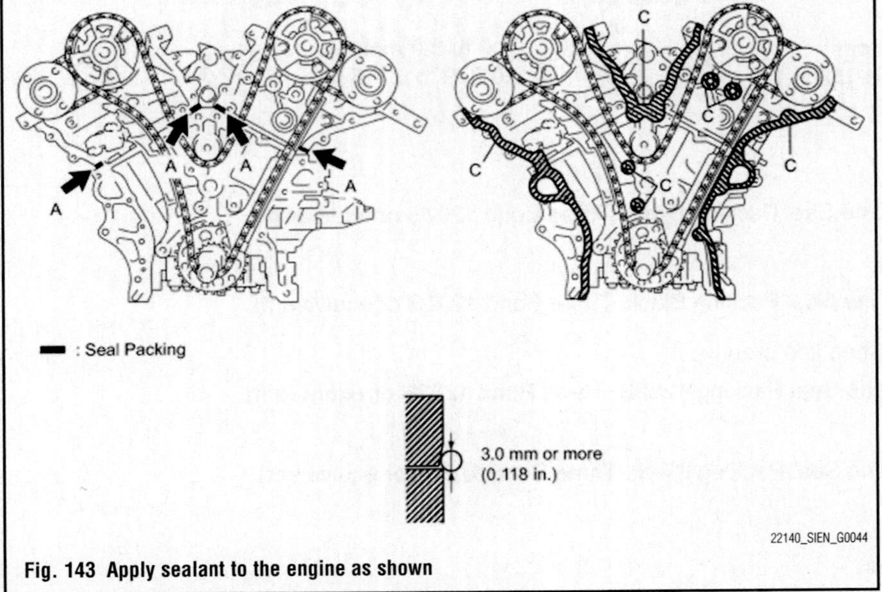

Fig. 143 Apply sealant to the engine as shown

idle gear shaft while aligning the knock pin of the No. 1 idle gear with the knock pin groove of the cylinder block.

➡Be careful of the idle gear direction.

49. Check that no foreign objects are on the No. 1 and No. 2 idle gear shafts.
50. Using a 10 mm hexagon wrench, tighten the No. 2 idle gear shaft to 44 ft. lbs. (60 Nm).
51. After installing the idle sprocket assembly, check that the idle sprocket turns smoothly.
52. Align the mark plate and timing marks as shown in the illustration and install the chain.

➡The camshaft mark plate is orange.

53. Do not pass the chain over the crankshaft, just put it on it.
54. Turn the camshaft timing gear assembly on the bank 1 counterclockwise to tighten the chain between the banks.
55. When the idle sprocket is reused, align the chain plate with the mark where the plate had been in order to tighten the chain between the banks.
56. Align the mark plate and timing marks as shown in the illustration and install the chain onto the crankshaft timing sprocket.

➡The crankshaft mark plate is yellow.

57. Temporarily tighten the pulley set bolt.
58. Turn the crankshaft clockwise to set it to the RH block bore center line (TDC / compression).
59. Install the chain tensioner slipper.
60. Move the stopper plate upward to release the lock, and push the plunger deep into the tensioner.
61. Move the stopper plate downward to set the lock, and insert a hexagon wrench into the hole of the stopper plate.
62. Install the chain tensioner and tighten the bolts to 84 inch lbs. (10 Nm).
63. Remove the hexagon wrench of the chain tensioner. Check that each timing mark is aligned with the crankshaft at the TDC / compression.
64. Remove the pulley set bolt.
65. Apply seal packing in a continuous line to the engine unit as shown in the following illustration. Use Toyota Genuine Seal Packing Black, Three Bond 1207B or equivalent
66. Be sure to clean and degrease the contact surfaces, especially the surfaces indicated by C in the illustration.
67. When the contact surfaces are wet,

wipe them with an oil-free cloth before applying seal packing.

68. Install the chain cover within 3 minutes.

➡**Do not start the engine for at least 2 hours after installing.**

69. Apply seal packing in a continuous line to the timing chain cover as shown in the following illustration.

➡**When the contact surfaces are wet, wipe them with an oil-free cloth before applying seal packing.**

Install the chain cover within 3 minutes and tighten the bolts within 15 minutes after applying seal packing.

Do not start the engine for at least 2 hours after installing.

70. Install a new gasket.

71. Align the oil pump's drive rotor spline and the crankshaft as shown in the illustration. Install the spline and chain cover to the crankshaft.

72. Install the cover bolts and nuts as shown.

73. Temporarily tighten the timing chain cover with the 23 bolts and 2 nuts.

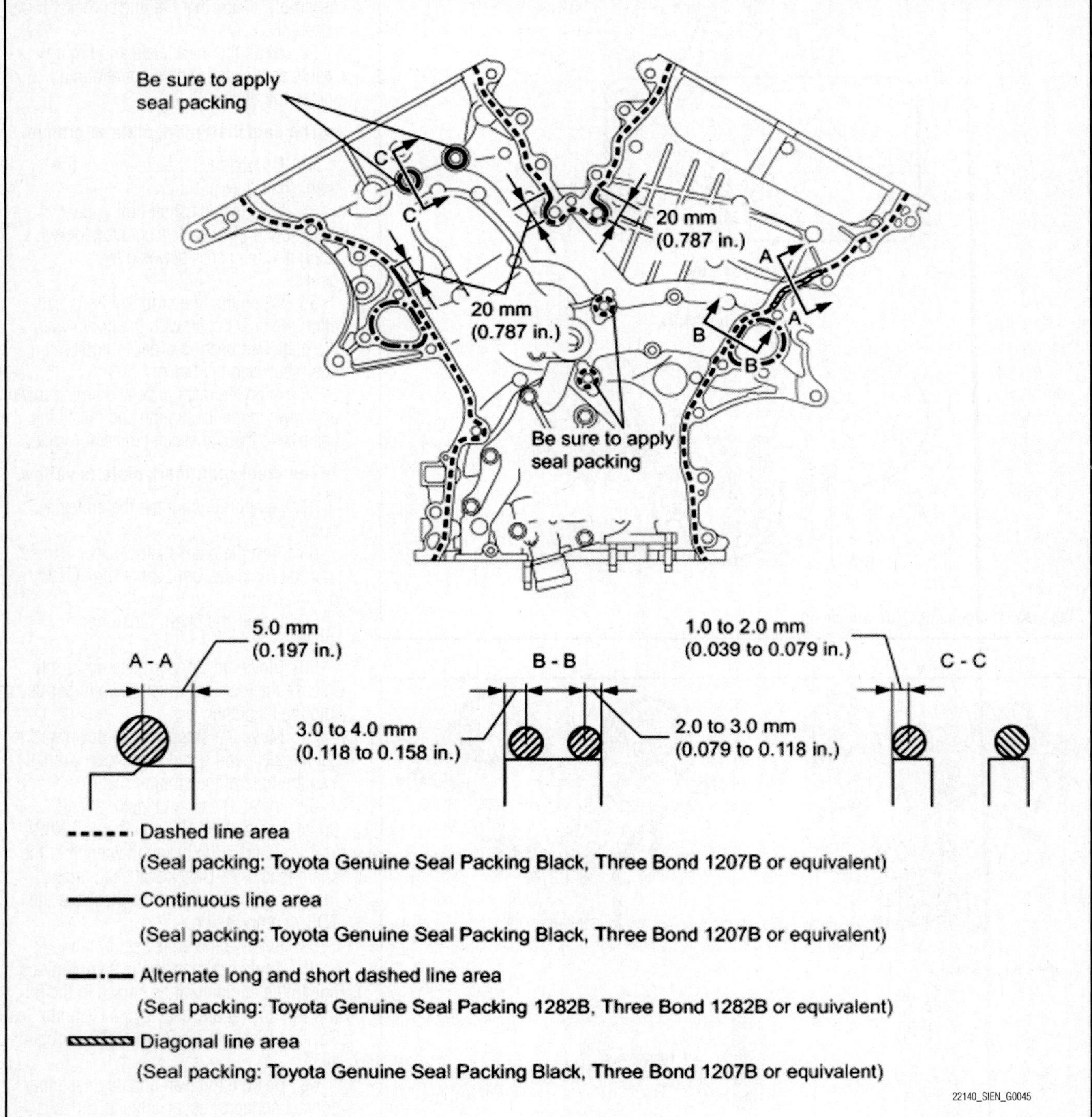

Fig. 144 Apply sealant to the timing cover as shown

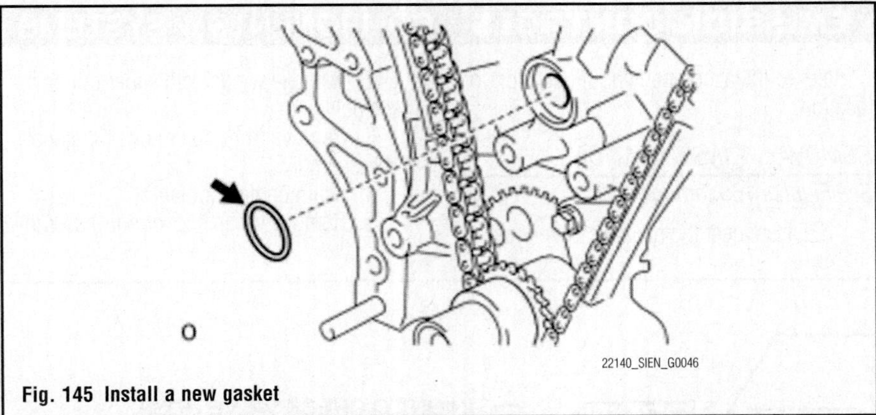

Fig. 145 Install a new gasket

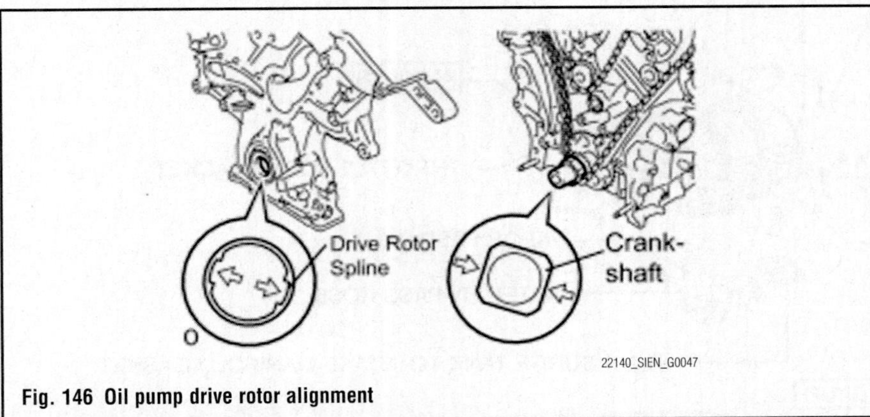

Fig. 146 Oil pump drive rotor alignment

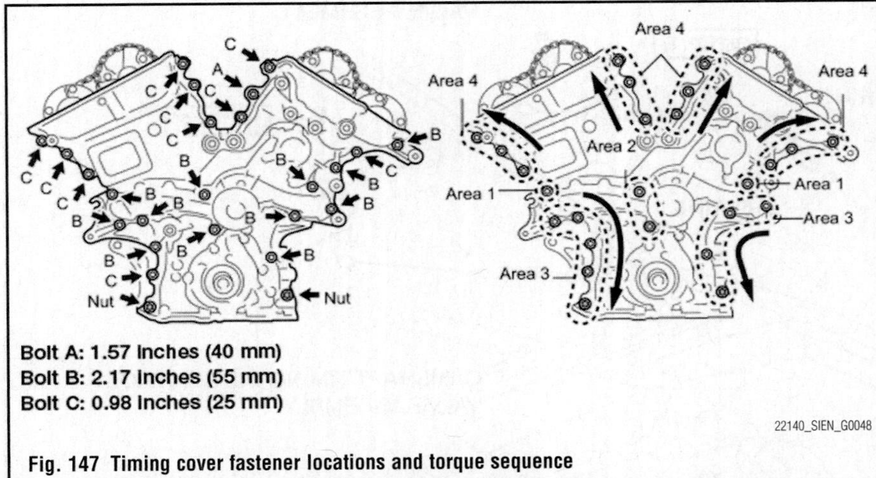

Fig. 147 Timing cover fastener locations and torque sequence

➡**Make sure that there is no oil on the bolt and nut threads.**

74. Fully tighten the bolts as follows:
   a. Step 1: Area 1 and Area 2 to 15 ft. lbs. (21 Nm)
   b. Step 2: Fully tighten the bolts in Area 3 to 15 ft. lbs. (21 Nm)
   c. Step 3: Tighten bolt A to 32 ft. lbs. (43 Nm)
   d. Step 4: Tighten bolts in Area 4 to 15 ft. lbs. (21 Nm)
75. Install the engine to the vehicle. Check for leaks and proper operation.

VALVE LASH

ADJUSTMENT

No adjustment is necessary on these engines.

INSPECTION

See Figure 148.

➡**Keep the adjuster free from dirt and foreign matter. Use only clean engine oil.**

1. Place the lash adjuster into a container full of new engine oil.
2. Insert the tip of SST: 09276-75010 into the lash adjuster plunger and use the tip to press down on the check ball inside the plunger.
3. Squeeze SST and the lash adjuster together to move the plunger up and down 5 to 6 times.
4. Check the movement of the plunger and bleed air. Plunger is OK if it moves up and down.

➡**When bleeding high-pressure air from the compression chamber, make sure that the tip of SST is actually pressing the check ball as shown in the illustration. If the check ball is not pressed, air will not bleed.**

5. After bleeding the air, remove SST. Then try to quickly and firmly press the plunger with your fingers. Plunger is OK if it can be pressed 3 times.
6. If the plunger can still be compressed after pressing it 3 times, replace the valve lash adjuster with a new one.

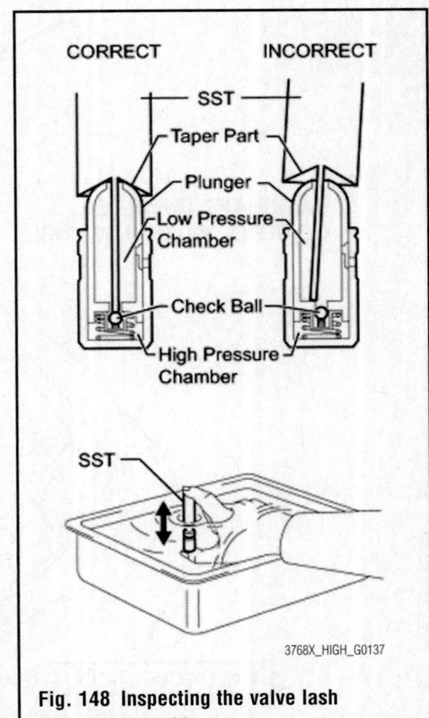

Fig. 148 Inspecting the valve lash adjuster assembly

CAMSHAFT OIL CONTROL VALVE

LOCATION

See Figure 149.

Refer to the accompanying illustration for location.

REMOVAL & INSTALLATION

See Figures 150 through 153.

1. Disconnect the negative battery cable.
2. Remove windshield wiper motor assembly.
3. Remove front outer cowl top panel sub-assembly.
4. Drain engine coolant.
5. Remove v-bank cover sub-assembly.

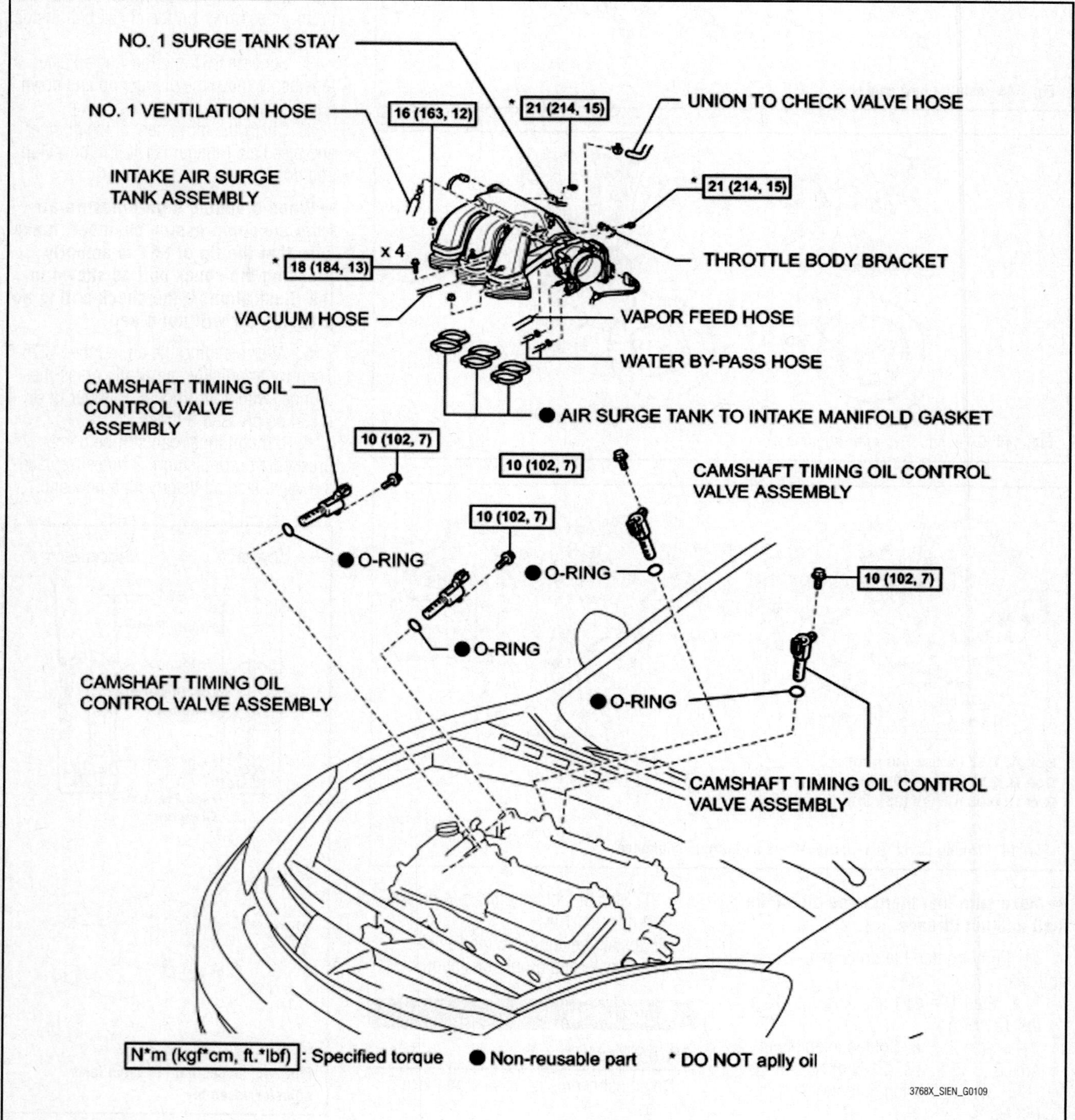

Fig. 149 Camshaft Oil Control Valve location

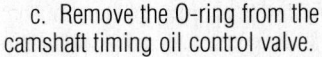

6. Remove no. 2 air cleaner inlet.
7. Remove no. 1 air cleaner inlet.
8. Remove air cleaner cap sub-assembly
9. Remove air cleaner case sub-assembly.
10. Remove intake air surge tank assembly.
11. Remove camshaft timing oil control valve assembly (for bank 1 exhaust side):
   a. Disconnect the camshaft timing oil control valve assembly connector.
   b. Remove the bolt and camshaft timing oil control valve assembly.
   c. Remove the O-ring from the camshaft timing oil control valve.
12. Remove camshaft timing oil control valve assembly (for bank 1 intake side):
   a. Disconnect the camshaft timing oil control valve assembly connector.
   b. Remove the bolt and camshaft timing oil control valve assembly.
   c. Remove the O-ring from the camshaft timing oil control valve.
13. Remove camshaft timing oil control valve assembly (for bank 2 exhaust side):
   a. Disconnect the camshaft timing oil control valve assembly connector.
   b. Remove the bolt and camshaft timing oil control valve assembly.
   c. Remove the O-ring from the camshaft timing oil control valve.
14. Remove camshaft timing oil control valve assembly (for bank 2 intake side):
   a. Disconnect the camshaft timing oil control valve assembly connector.
   b. Remove the bolt and camshaft timing oil control valve assembly.
   c. Remove the O-ring from the camshaft timing oil control valve.
15. To install, reverse the removal procedure.

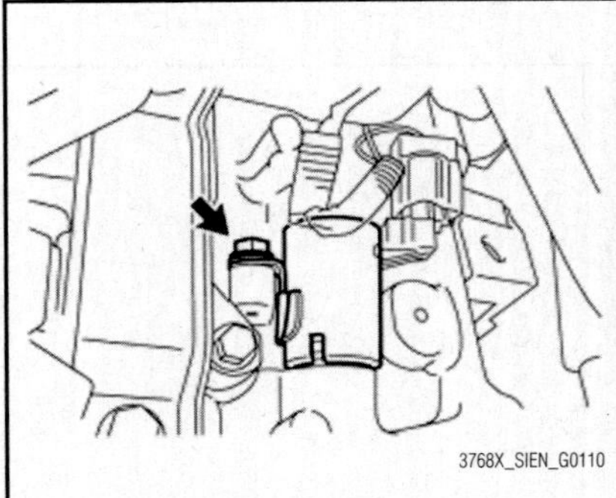

Fig. 150 Remove camshaft timing oil control valve assembly (for bank 1 exhaust side)

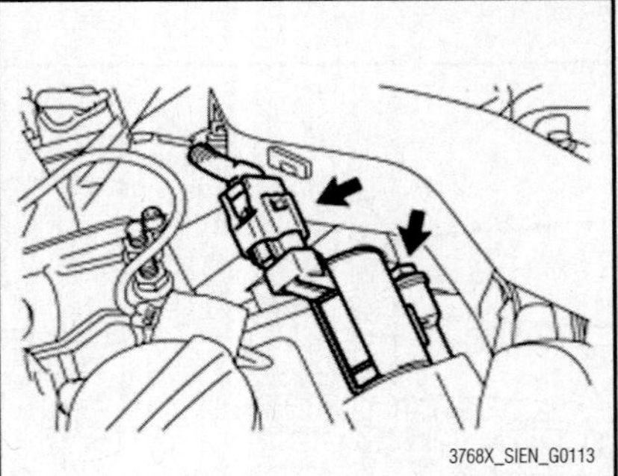

Fig. 153 Remove camshaft timing oil control valve assembly (for bank 2 intake side)

CRANKSHAFT POSITION (CKP) SENSOR

LOCATION

See Figure 154.

Refer to the accompanying illustration for sensor location.

REMOVAL & INSTALLATION

See Figure 155.

1. Remove the A/C compressor.
2. Disconnect the Crankshaft Position (CKP) sensor connector.
3. Remove the bolt, and then remove the crankshaft position sensor.

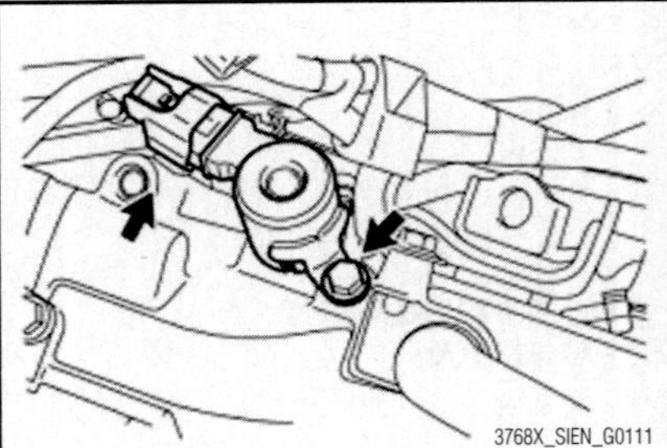

Fig. 151 Remove camshaft timing oil control valve assembly (for bank 1 intake side)

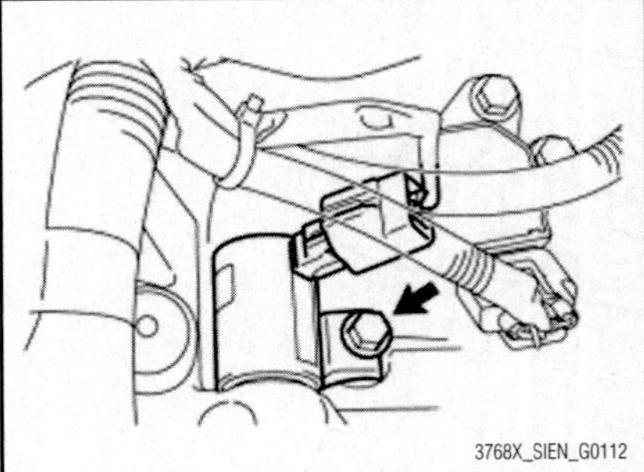

Fig. 152 Remove camshaft timing oil control valve assembly (for bank 2 exhaust side)

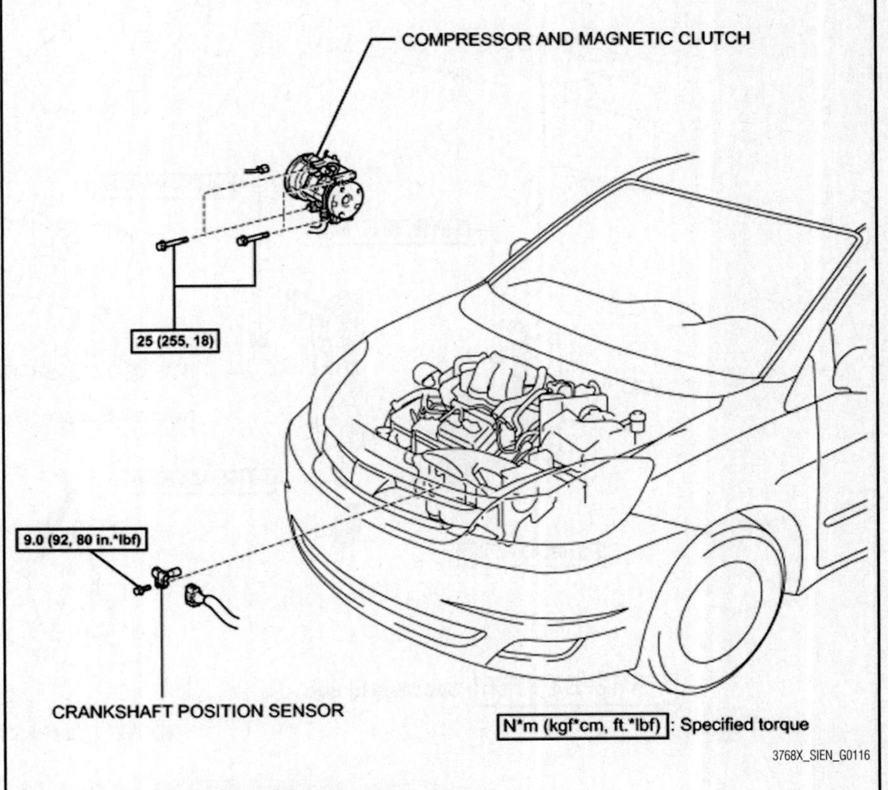

Fig. 154 Crankshaft Position (CKP) Sensor location

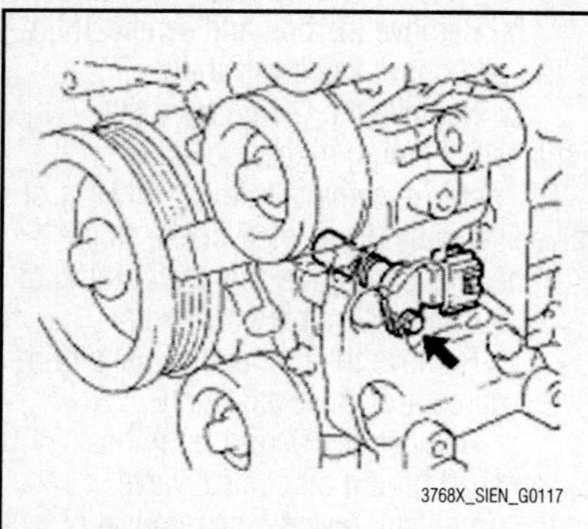

Fig. 155 Remove the bolt, and then remove the crankshaft position sensor

4. To install, apply a light coat of engine oil to the O-ring on the crankshaft position sensor.

ELECTRONIC CONTROL MODULE (ECM)

LOCATION

See Figure 156.

Refer to the accompanying illustration for location.

REMOVAL & INSTALLATION

See Figures 157 and 158.

1. Remove glove compartment door assembly.
2. Disengage the 2 claws and 4 clips, and remove the No. 2 instrument panel box.
3. Remove ECM:
   a. Disconnect the 5 ECM connectors.
   b. Remove the 2 nuts and ECM.
   c. If equipped with the 10 speaker system:
      - Disconnect the 5 ECM connectors and 2 stereo components amplifier connectors.
      - Remove the 4 nuts and the ECM with the stereo components amplifier.

5.5 (56, 49 in.*lbf)
5.5 (56, 49 in.*lbf)
NO. 2 INSTRUMENT PANEL BOX
ECM
STEREO COMPONENTS AMPLIFIER
(with 10 Speaker System)
5.5 (56, 49 in.*lbf)
5.5 (56, 49 in.*lbf)
N*m (kgf*cm, ft.*lbf) : Specified torque
GLOVE COMPARTMENT DOOR ASSEMBLY
3768X_SIEN_G0118

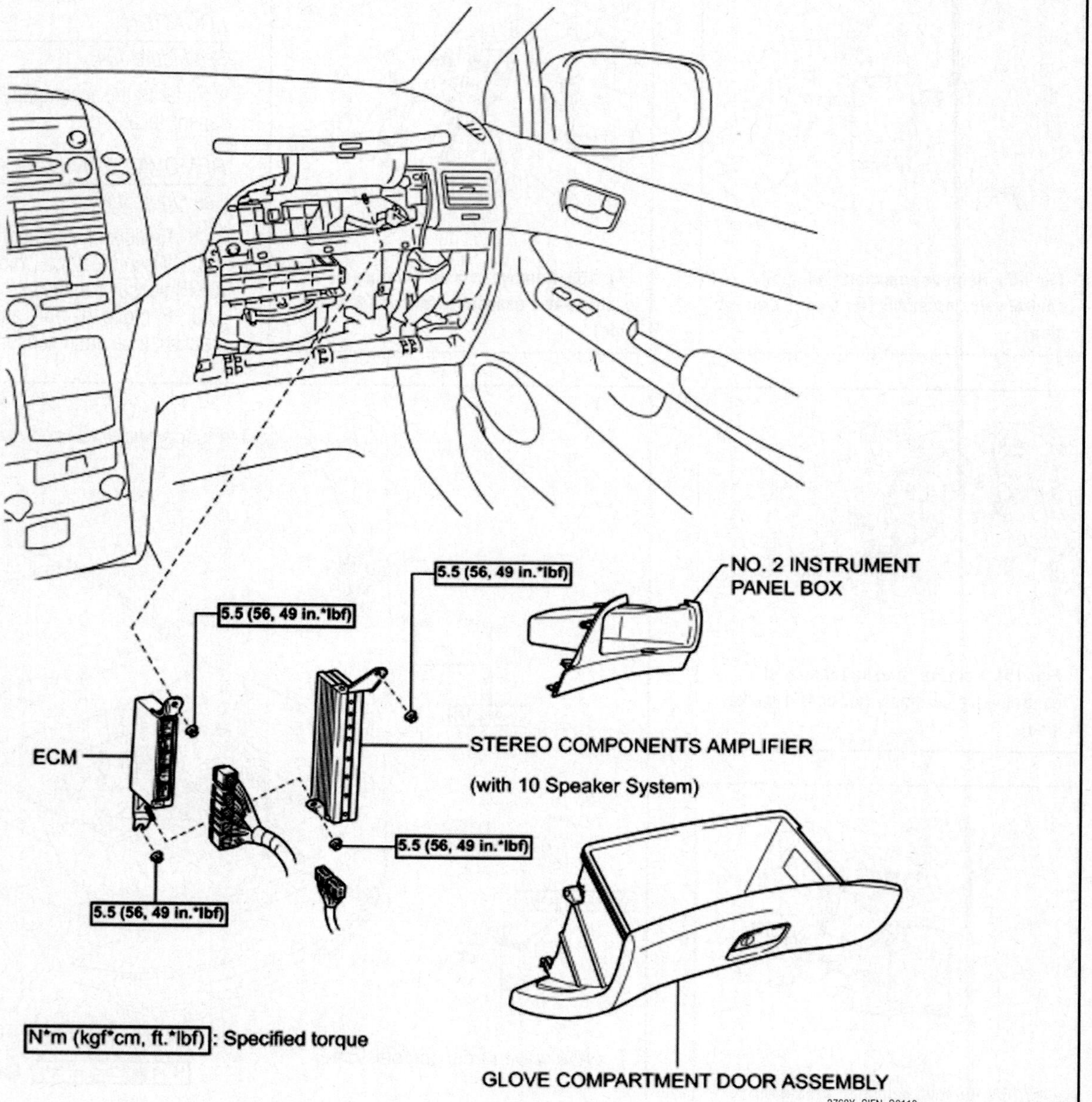

Fig. 156 ECM location

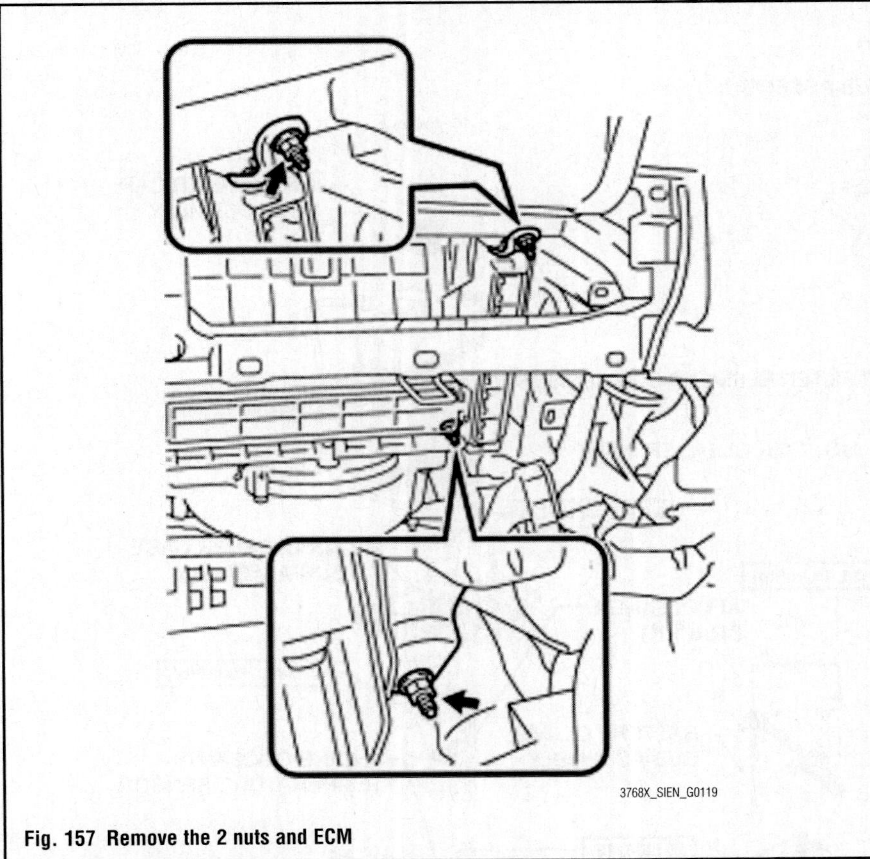

Fig. 157 Remove the 2 nuts and ECM

Fig. 158 Remove the 4 nuts and the ECM with the stereo components amplifier—10 speaker system

4. To install, reverse the installation procedure.

5. Perform ECU initialization/recognition procedure.

ECM INITIALIZATION REGISTRATION

➡After replacing the distance control ECU, it is necessary to initialize the distance control ECU so that the ECU can recognize the specification of the vehicle.

1. Be sure to perform the following procedures after replacing the distance control ECU.
   a. Turn the ignition switch to the on position.
   b. Turn the cruise main switch on.
   c. With the brake pedal depressed, push the cruise control main switch to RES/ACC 3 times within 3 seconds. Check that the buzzer sounds at this time.

➡Do not turn the headlight dimmer switch on at this time because the optical axis automatic adjustment mode has already started, which may lead to an incorrect optical axis setting. If the headlight dimmer switch is turned on by mistake, readjust the optical axis.

   d. Perform any of the following to cancel beam axis adjustment mode.
      - Turn the cruise control main switch off.
      - Turn the ignition switch off.
      - Accelerate the vehicle to 10 km/h or more.

ENGINE COOLANT TEMPERATURE (ECT) SENSOR

LOCATION

See Figure 159.

Refer to the accompanying illustration for sensor location.

REMOVAL & INSTALLATION

1. Drain and recycle the engine coolant.
2. Remove v-bank cover sub-assembly.
3. Remove no. 2 air cleaner inlet.
4. Remove no. 1 air cleaner inlet.
5. Remove air cleaner cap sub-assembly.

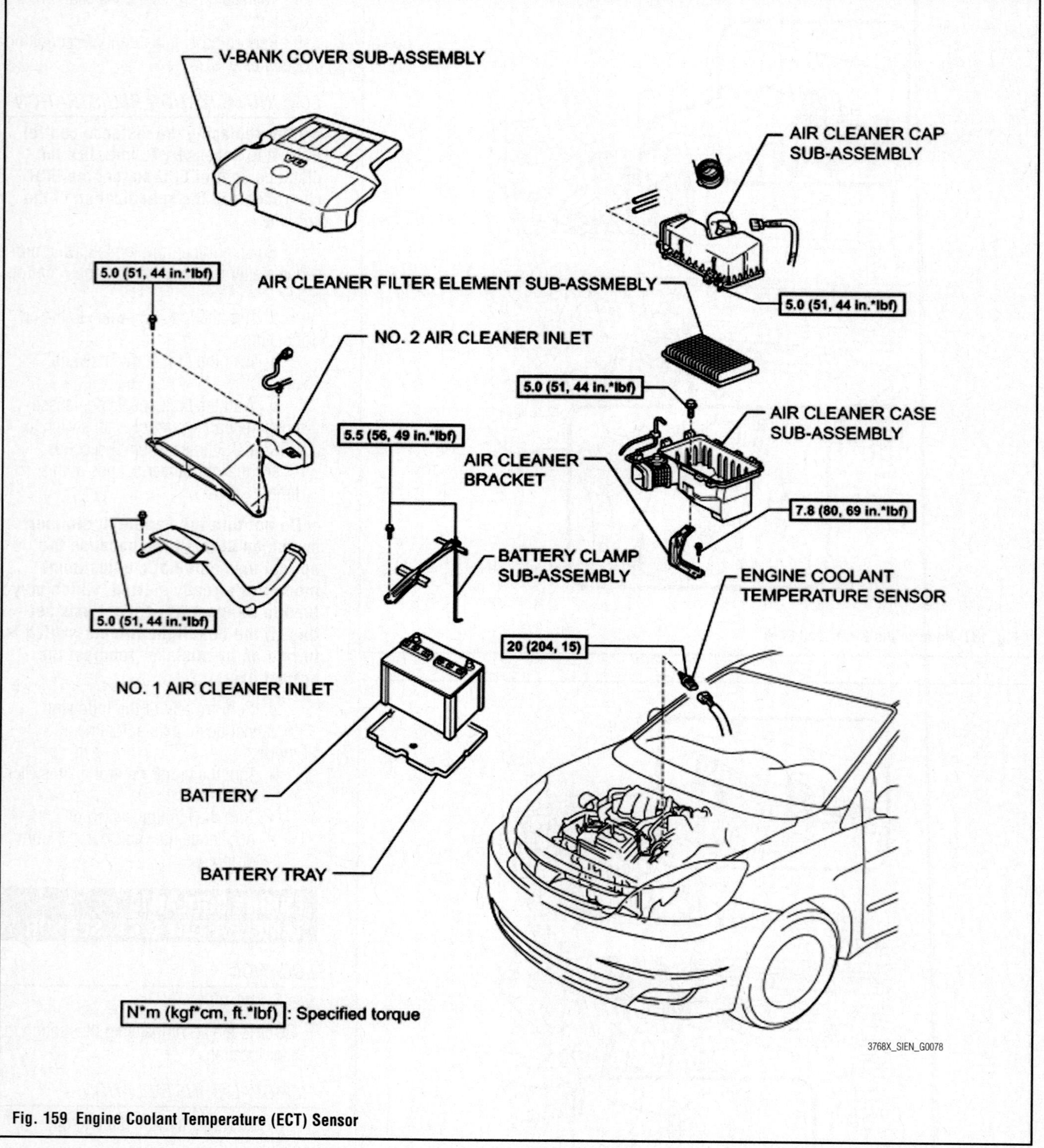

Fig. 159 Engine Coolant Temperature (ECT) Sensor

6. Disconnect the sensor connector.
7. Remove the sensor using a deep 19 mm socket.

To install:

8. Install the sensor with a new gasket.
9. Tighten the sensor to 14 ft. lbs. (20 Nm).
10. Attach the electrical connector and replenish the cooling system.
11. Connect the negative battery cable.
12. Start the vehicle and check for leaks

EVAPORATIVE EMISSIONS (EVAP) CANISTER

LOCATION

See Figure 160.

Refer to the accompanying illustration for location.

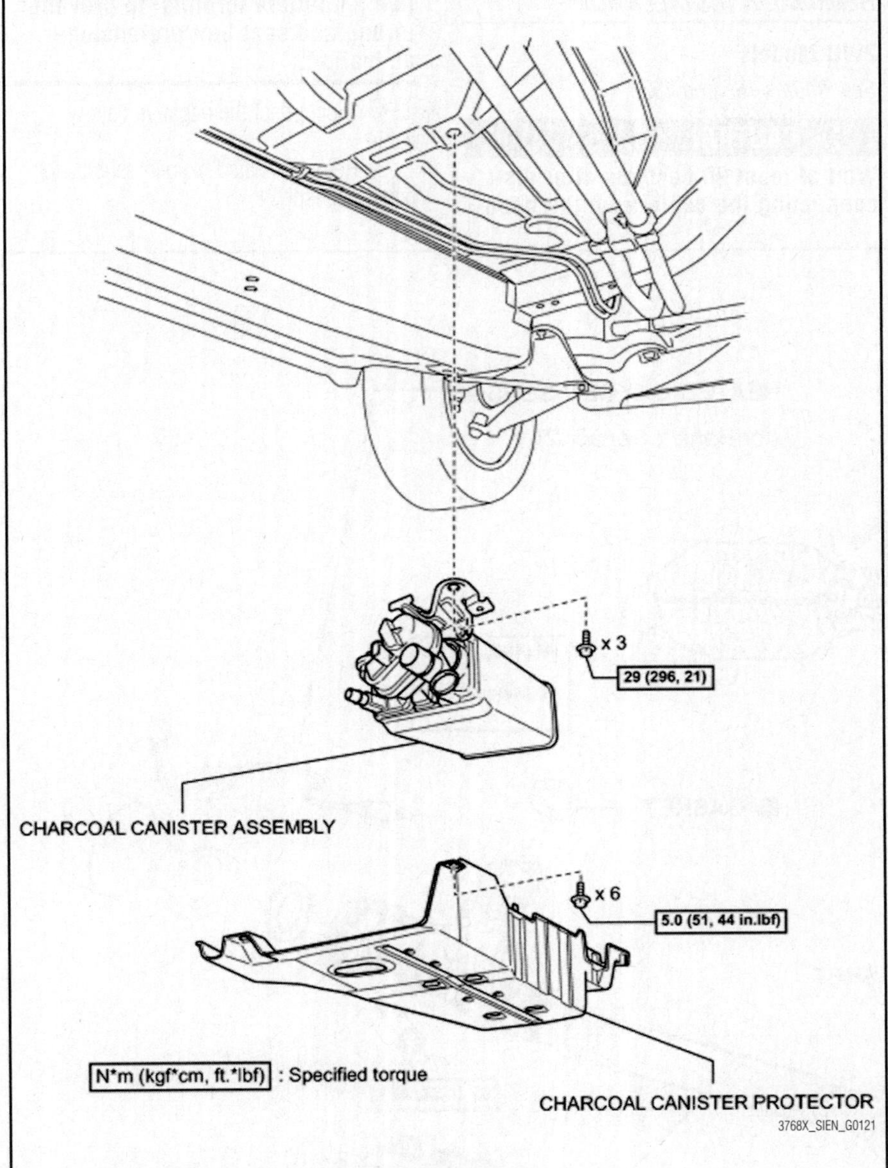

Fig. 160 Evaporative Emissions (EVAP) Canister location

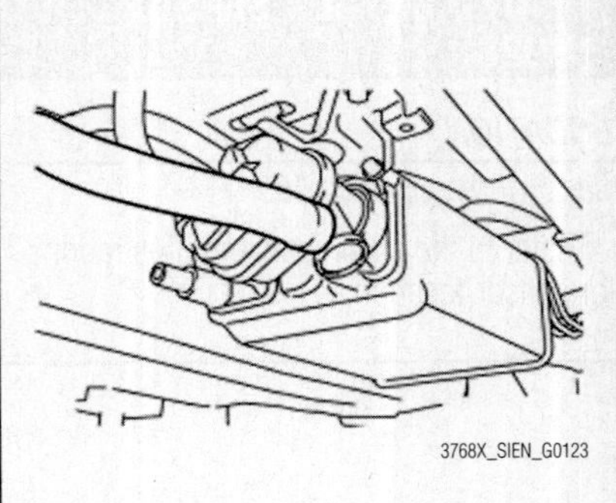

Fig. 162 Disconnect the charcoal canister filter sub-assembly from the charcoal canister

Fig. 163 Remove the 3 bolts and charcoal canister

REMOVAL & INSTALLATION

See Figures 161 through 163.

1. Remove the 6 bolts and charcoal canister protector.
2. Disconnect the fuel tank vent hose from the charcoal canister:
    a. Push the connector deep inside.
    b. Pinch portion A.
    c. Pull out the connector.
3. Disconnect the charcoal canister filter sub-assembly from the charcoal canister.
    a. Push the connector deep inside.
    b. Pinch portion A.
    c. Pull out the connector.
4. Disconnect the vapor pressure sensor connector.
5. Disconnect the wire harness clamp.
6. Disconnect the purge line hose from the charcoal canister.
7. Remove the 3 bolts and charcoal canister.
8. To install, reverse the removal procedure. Tighten the canister nuts to 21 ft. lbs. (29 Nm).

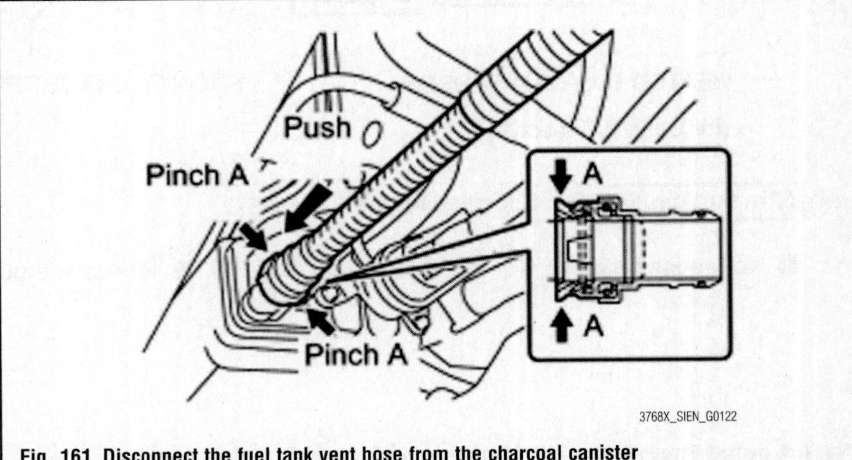

Fig. 161 Disconnect the fuel tank vent hose from the charcoal canister

HEATED OXYGEN SENSOR (HO2S)

LOCATION

See Figures 164 and 165.

Refer to the accompanying illustrations for sensor location.

REMOVAL & INSTALLATION

2WD Models

See Figures 166 and 167.

⁂ CAUTION

Wait at least 90 seconds after disconnecting the cable from the negative (-) battery terminal to prevent airbag and seat belt pretensioner activation.

1. Disconnect the negative battery cable.
2. Remove heated oxygen sensor (for bank 1 sensor 2):

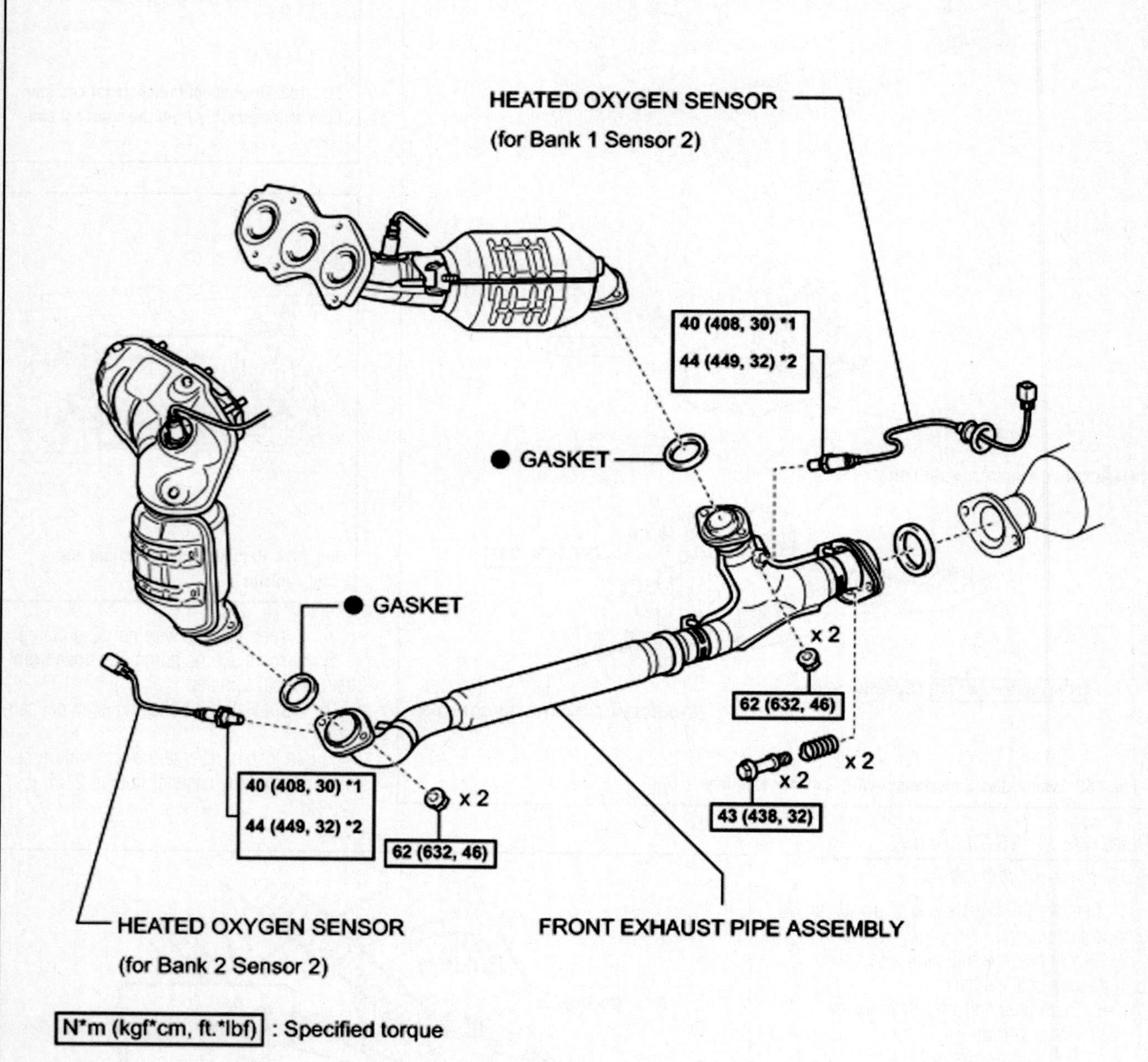

Fig. 164 Heated Oxygen Sensor (HO2S) location—2WD models

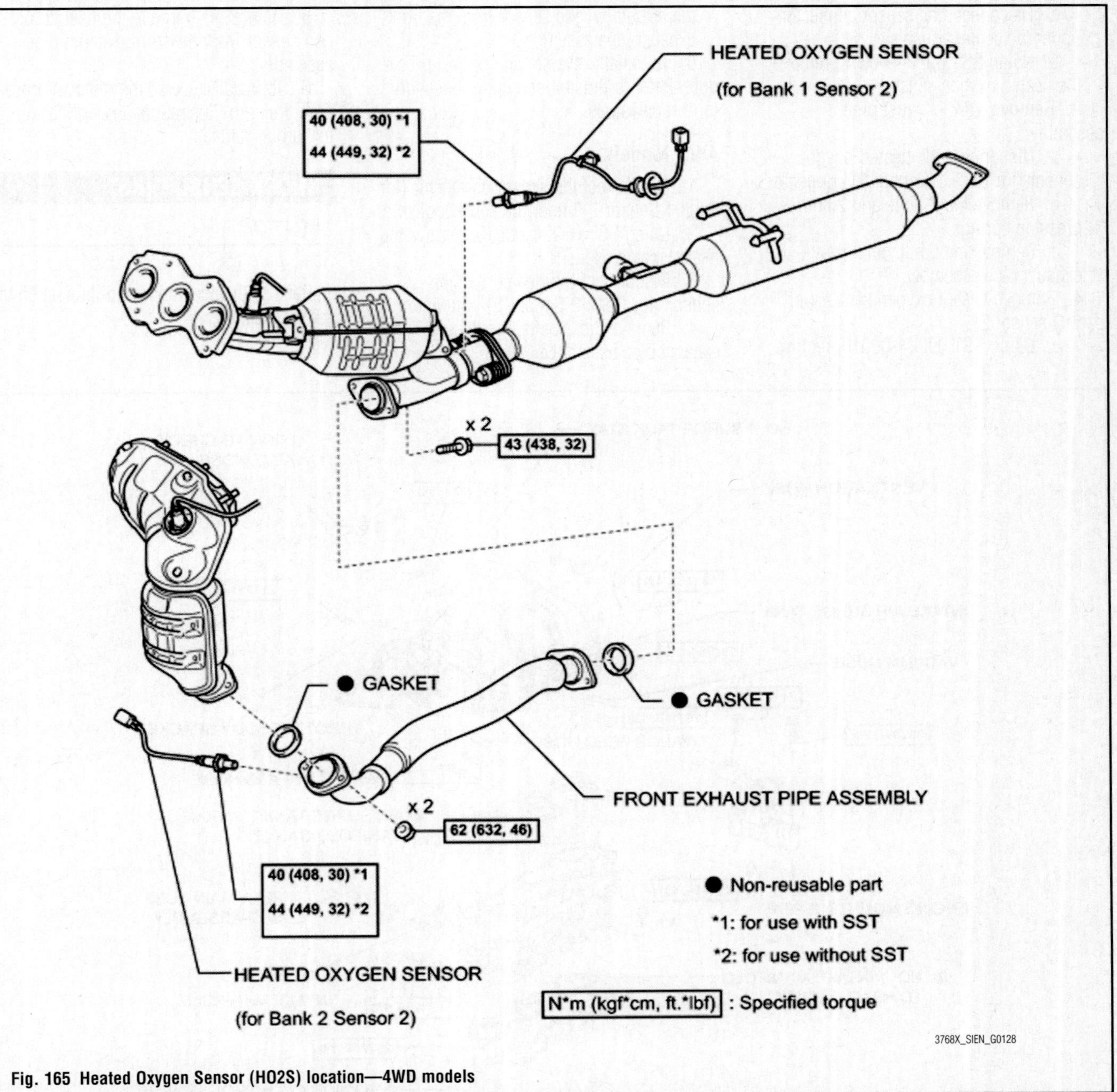

HEATED OXYGEN SENSOR
(for Bank 1 Sensor 2)
40 (408, 30) *1
44 (449, 32) *2
x 2
43 (438, 32)
● GASKET
● GASKET
FRONT EXHAUST PIPE ASSEMBLY
x 2
62 (632, 46)
40 (408, 30) *1
44 (449, 32) *2
HEATED OXYGEN SENSOR
(for Bank 2 Sensor 2)
● Non-reusable part
*1: for use with SST
*2: for use without SST
N*m (kgf*cm, ft.*lbf) : Specified torque
3768X_SIEN_G0128

Fig. 165 Heated Oxygen Sensor (HO2S) location—4WD models

SST
3768X_SIEN_G0129

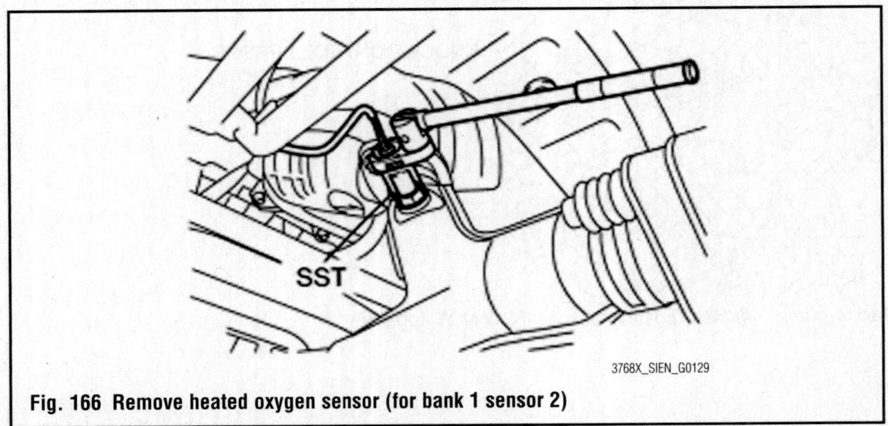

Fig. 166 Remove heated oxygen sensor (for bank 1 sensor 2)

SST
3768X_SIEN_G0130

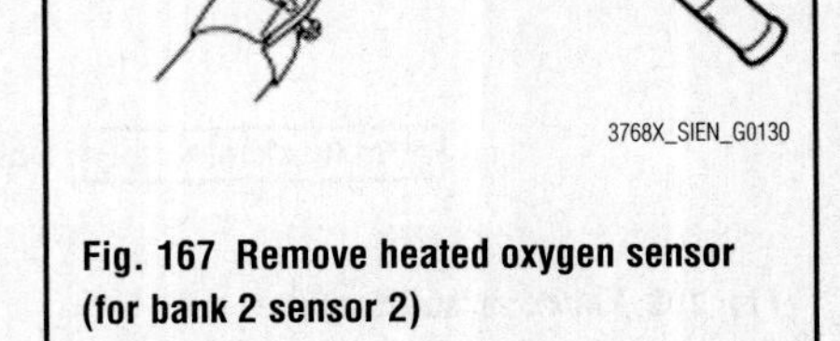

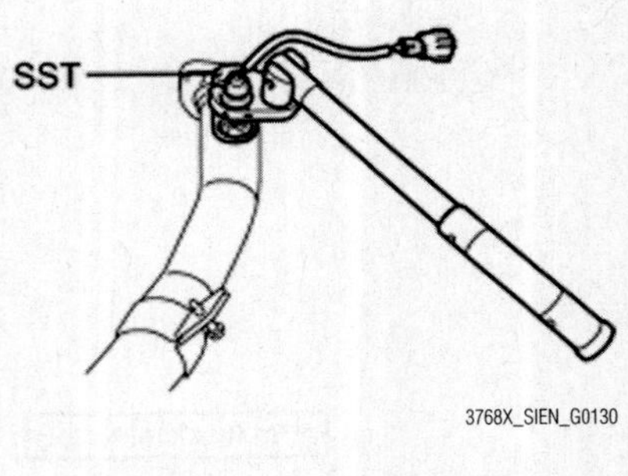

Fig. 167 Remove heated oxygen sensor (for bank 2 sensor 2)

a. Disconnect the sensor connector under the center console.

b. Using SST 09224-00010, remove the heated oxygen sensor.

3. Remove front exhaust pipe assembly:

a. Disconnect the heated oxygen sensor (for Bank 2 Sensor 2) connector.

b. Remove the 2 bolts and 2 compression springs.

c. Remove the 6 nuts and front exhaust pipe assembly.

4. Remove heated oxygen sensor (for bank 2 sensor 2):

a. Using SST 09224-00010, remove the heated oxygen sensor from the front exhaust pipe assembly.

5. To install, reverse the removal procedure. Refer to illustrations under Location for torque values.

4WD Models

1. Disconnect the sensor connector (for Bank 1 Sensor 2) under the center console.
2. Using SST 09224-00010, remove the heated oxygen sensor.
3. Disconnect the heated oxygen sensor (for Bank 2 Sensor 2) connector.
4. Remove the 2 bolts, 4 nuts and front exhaust pipe assembly.
5. Using SST, remove the heated oxygen sensor from the front exhaust pipe assembly.
6. To install, reverse the removal procedure. Refer to illustrations under Location for torque values.

KNOCK SENSOR (KS)

LOCATION

See Figure 168.

Refer to the accompanying illustration for sensor location.

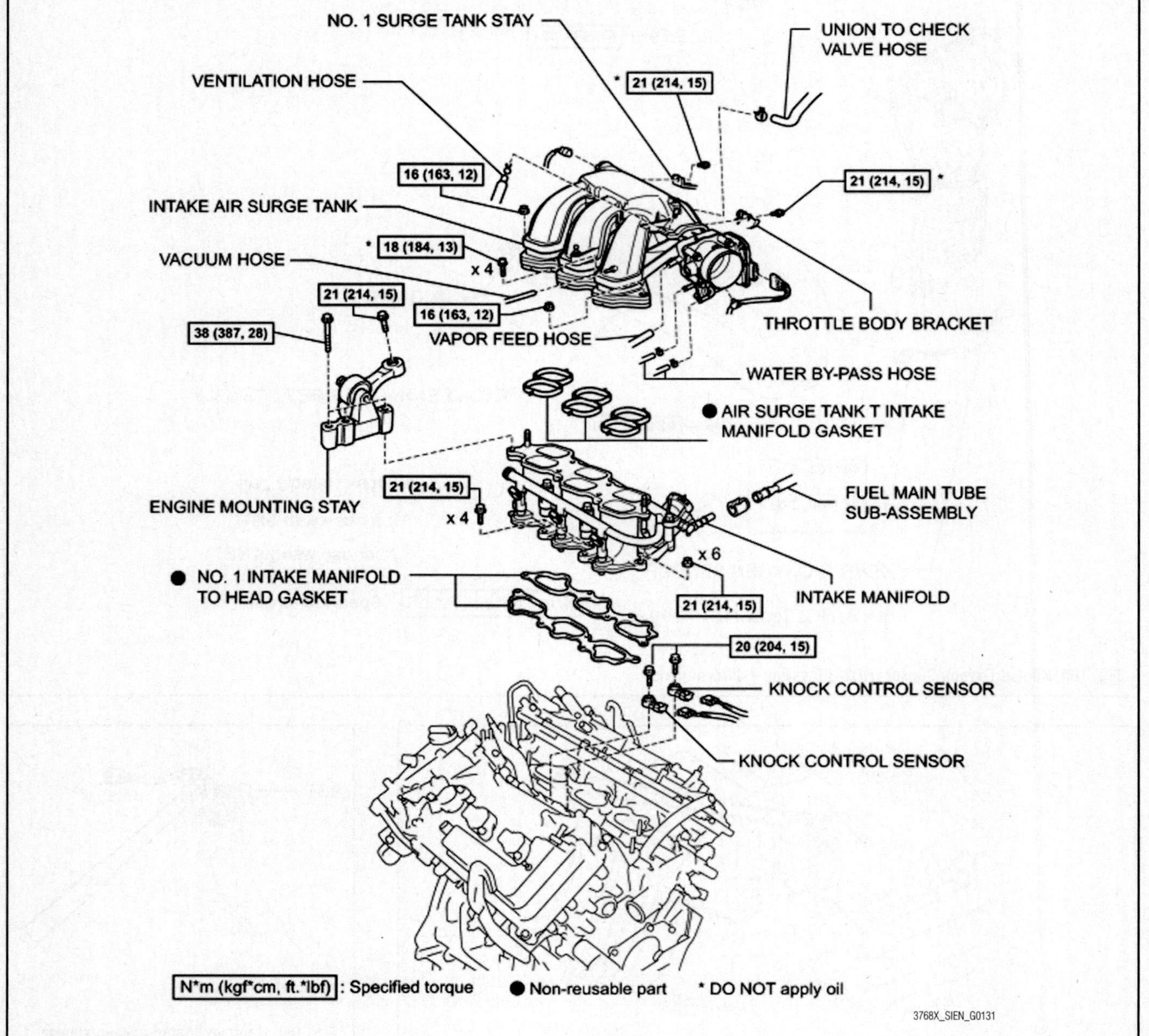

Fig. 168 Knock sensor location

REMOVAL & INSTALLATION

See Figure 169.

1. Remove the intake manifold.
2. Remove the throttle body.

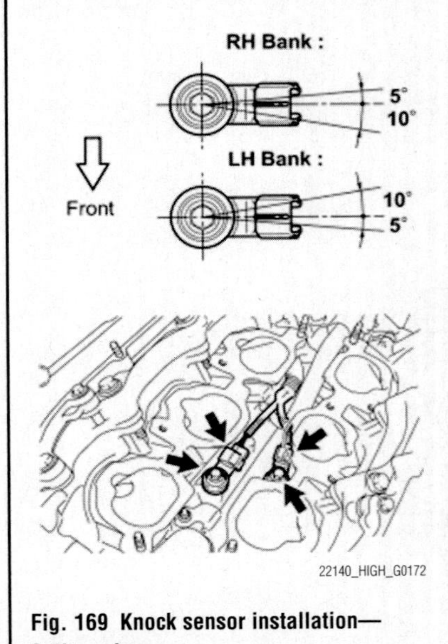

Fig. 169 Knock sensor installation—3.5L engine

3. Remove knock control sensor:
   a. Disconnect the 2 knock sensor connectors.
   b. Remove the 2 bolts and then remove the 2 knock control sensors.

To install:

4. Install the 2 knock control sensors with the 2 bolts as shown in the illustration, and tighten to 15 ft. lbs. (20 Nm).
5. Connect the 2 knock control sensor connectors.
6. Install the intake manifold.

MALFUNCTION INDICATOR LIGHT (MIL)

RESET PROCEDURE

Using a ODB II scan tool, clear the DTC codes.

MASS AIR FLOW (MAF) SENSOR

LOCATION

See Figure 170.

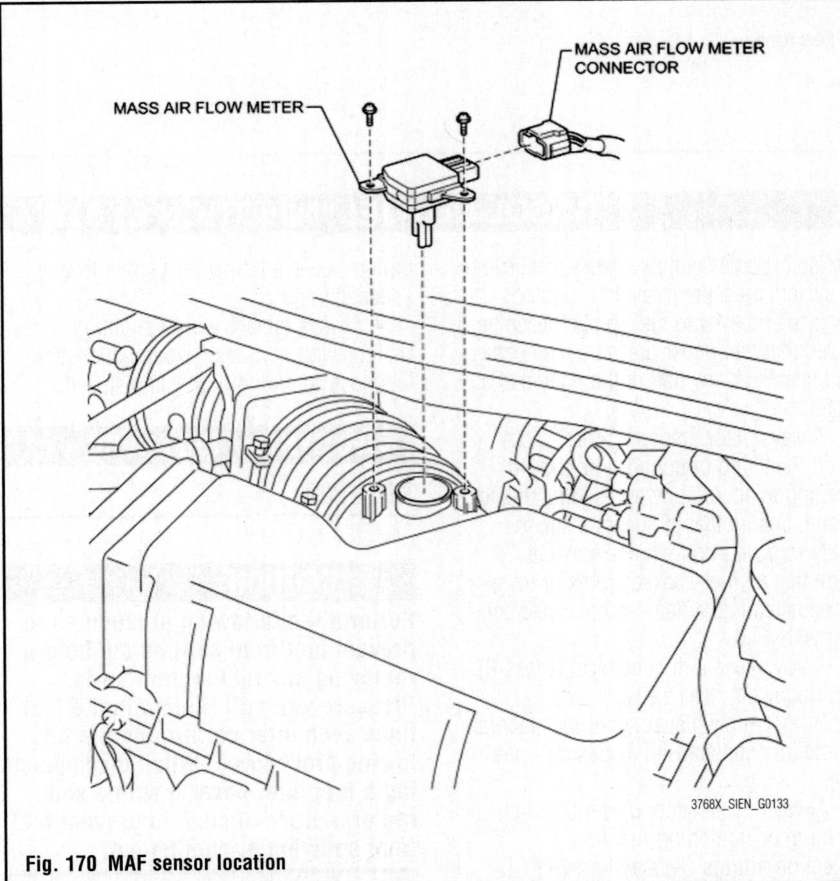

Fig. 170 MAF sensor location

Refer to the accompanying illustration for sensor location.

REMOVAL & INSTALLATION

See Figure 171.

1. Disconnect the negative battery cable.
2. Disconnect the mass air flow meter connector.
3. Remove the 2 screws and mass air flow meter.
4. To install, reverse the removal procedure.

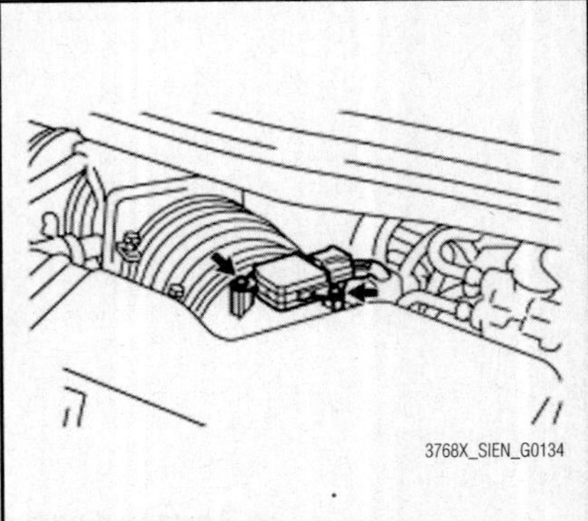

Fig. 171 Remove the 2 screws and mass air flow meter

VEHICLE SPEED SENSOR (VSS)

REMOVAL & INSTALLATION

See Figure 172.

1. Remove the battery.
2. Remove the air cleaner assembly.
3. Remove speed sensor (NT sensor):
   a. Disconnect the speed sensor connector.
   b. Remove the bolt and speed sensor.
4. Remove speed sensor (NC sensor):
   a. Disconnect the speed sensor connector.
   b. Remove the bolt and speed sensor.
5. To install, reverse the removal procedure. Take note of the following:
   a. Coat the VSS sensor O-ring with ATF.
   b. Tighten the VSS sensor bolts to: 8 ft. lbs. (11 Nm).

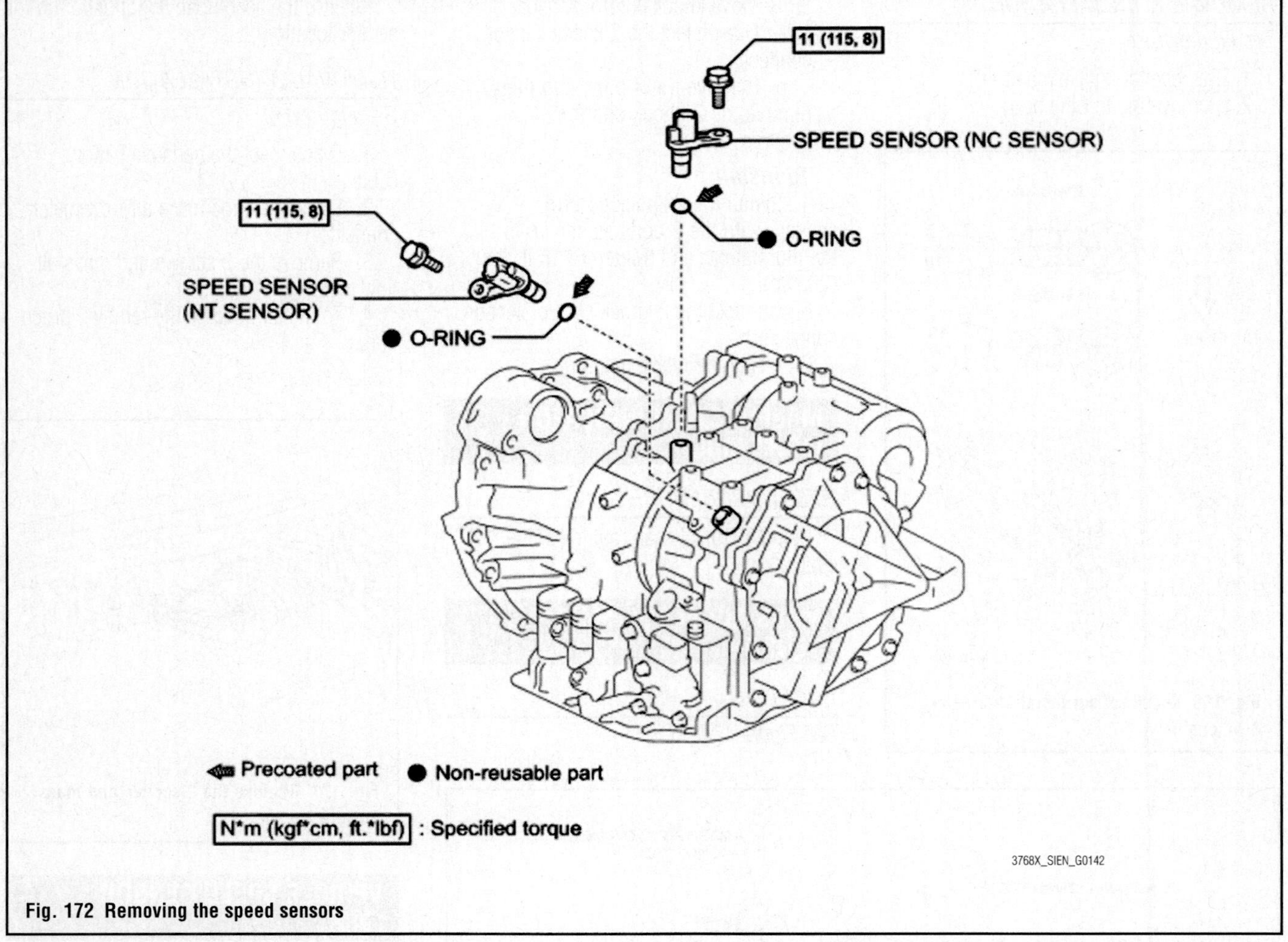

Fig. 172 Removing the speed sensors

FUEL — GASOLINE FUEL INJECTION SYSTEM

FUEL SYSTEM SERVICE PRECAUTIONS

Safety is the most important factor when performing not only fuel system maintenance, but any type of maintenance. Failure to conduct maintenance and repairs in a safe manner may result in serious personal injury or death. Work on a vehicle's fuel system components can be accomplished safely and effectively by adhering to the following rules and guidelines.

- To avoid the possibility of fire and personal injury, always disconnect the negative battery cable unless the repair or test procedure requires that battery voltage be applied.
- Always relieve the fuel system pressure prior to disconnecting any fuel system component (injector, fuel rail, pressure regulator, etc.) fitting or fuel line connection. Exercise extreme caution whenever relieving fuel system pressure to avoid exposing skin, face and eyes to fuel spray. Please be advised that fuel under pressure may penetrate the skin or any part of the body that it contacts.
- Always place a shop towel or cloth around the fitting or connection prior to loosening to absorb any excess fuel due to spillage. Ensure that all fuel spillage is quickly removed from engine surfaces. Ensure that all fuel-soaked cloths or towels are deposited into a flame-proof waste container with a lid.
- Always keep a dry chemical (Class B) fire extinguisher near the work area.
- Do not allow fuel spray or fuel vapors to come into contact with a spark or open flame.
- Always use a second wrench when loosening or tightening fuel line connection fittings. This will prevent unnecessary stress and torsion on fuel piping. Always follow the proper torque specifications.
- Always replace worn fuel fitting O-rings with new ones. Do not substitute fuel hose where rigid pipe is installed.

FUEL SYSTEM PRESSURE

RELIEVING

See Figure 173.

✲✲ CAUTION

Perform the following procedures to prevent fuel from spilling out before removing any fuel system parts. Pressure will still remain in the fuel lines even after performing the following procedures. When disconnecting a fuel line, cover it with a shop rag or a piece of cloth to prevent fuel from spraying or coming out.

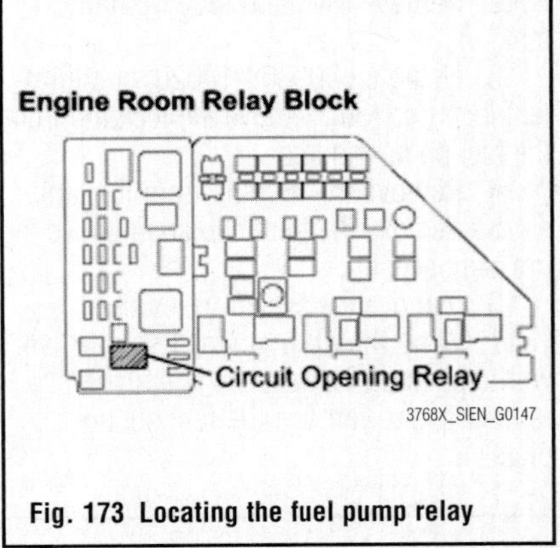

Fig. 173 Locating the fuel pump relay

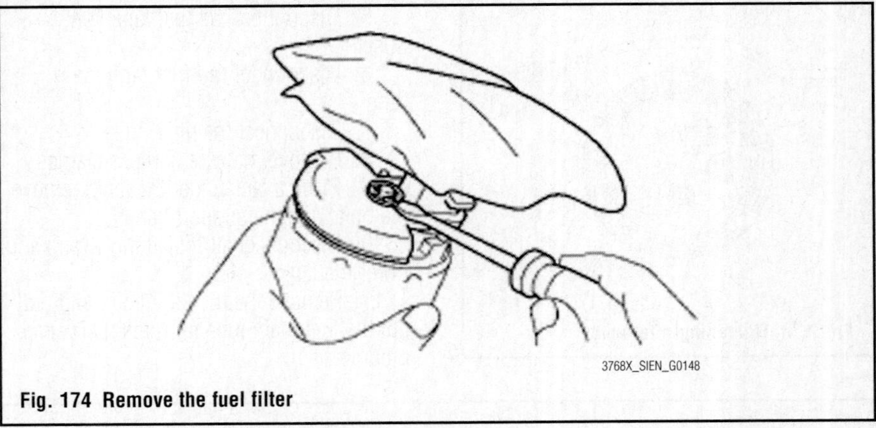

Fig. 174 Remove the fuel filter

CAUTION

Observe all applicable safety precautions when working around fuel. Whenever servicing the fuel system, always work in a well ventilated area. Do not allow fuel spray or vapors to come in contact with a spark or open flame. Keep a dry chemical fire extinguisher near the work area. Always keep fuel in a container specifically designed for fuel storage; also, always properly seal fuel containers to avoid the possibility of fire or explosion.

1. Remove the relay block cover.
2. Remove the FUEL PUMP relay.
3. Start the engine.
4. After the engine stops, turn the ignition switch off.

➡DTC P0171/25 (fuel problem) may be detected.

5. Crank the engine again. Check that the engine does not start.
6. Remove the fuel tank cap to discharge pressure from the fuel tank.
7. Disconnect the cable from the negative (-) battery terminal.
8. Install the FUEL PUMP relay.

FUEL FILTER

REMOVAL & INSTALLATION

See Figure 174.

1. Remove the fuel pump module from the fuel tank.
2. Using a small screwdriver, pry out the clip.
3. Remove the fuel filter assembly from the fuel pump.
4. To install, reverse the removal procedure.
5. Install a new clip.

FUEL LEVEL SENDING UNIT

REMOVAL & INSTALLATION

See Figure 175.

1. Before servicing the vehicle, refer to the precautions section.
2. Disconnect the negative battery cable.
3. Relieve the fuel system pressure.
4. Remove the fuel pump.
5. Unplug the sender electrical connector.
6. Unlock the sender gauge and slide it off the pump assembly.
7. Installation is the reverse of removal.

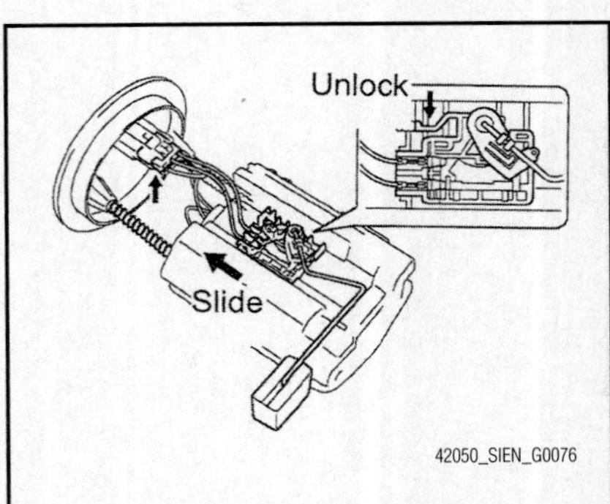

Fig. 175 Removing the fuel pump sender assembly

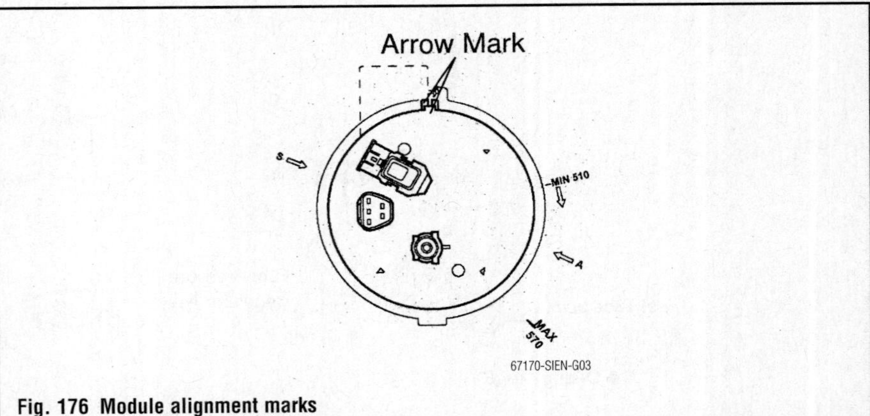

Fig. 176 Module alignment marks

FUEL PUMP MODULE

REMOVAL & INSTALLATION

See Figures 176 through 179.

1. Before servicing the vehicle, refer to the precautions section.
2. Discharge the fuel system pressure.
3. Remove the charcoal canister cover.
4. Remove the fuel tank filler hose cover.

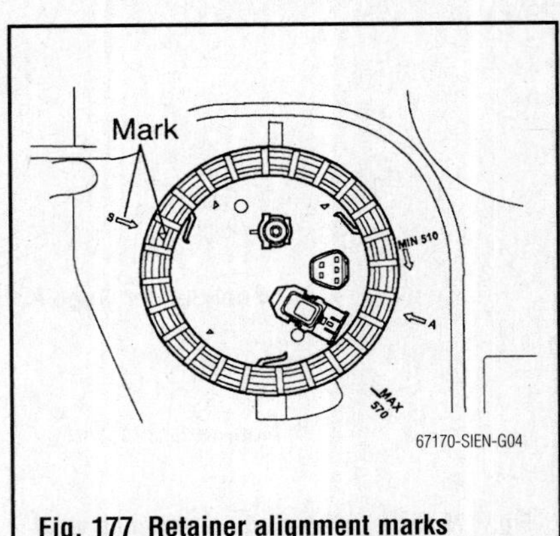

Fig. 177 Retainer alignment marks

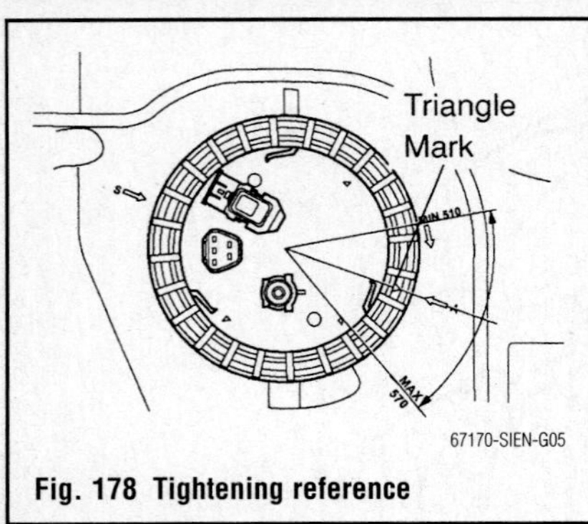

Fig. 178 Tightening reference

5. Disconnect the fuel tank vent hose.
6. Disconnect the fuel tank main tube.
7. Disconnect the filler hose.
8. Remove the wire harness clamps.
9. Place a jack under the tank, remove the bolts and the support bands.
10. Disconnect all remaining wiring and lower the tank.
11. Remove the tube joint clip and pull out the fuel main tube from the fuel pump module.
12. Remove the main tube from the tank.
13. Using tool 09808-14020, or equivalent lock ring tool, remove the lockring from the fuel pump module.
14. Remove the module from the tank.
15. Remove the joint clip and remove the pressure sensor.
16. Remove the sender assembly.
17. Wrap the tip of a small screwdriver with tape and disconnect the 4 snap retainers, and remove the fuel suction plate.

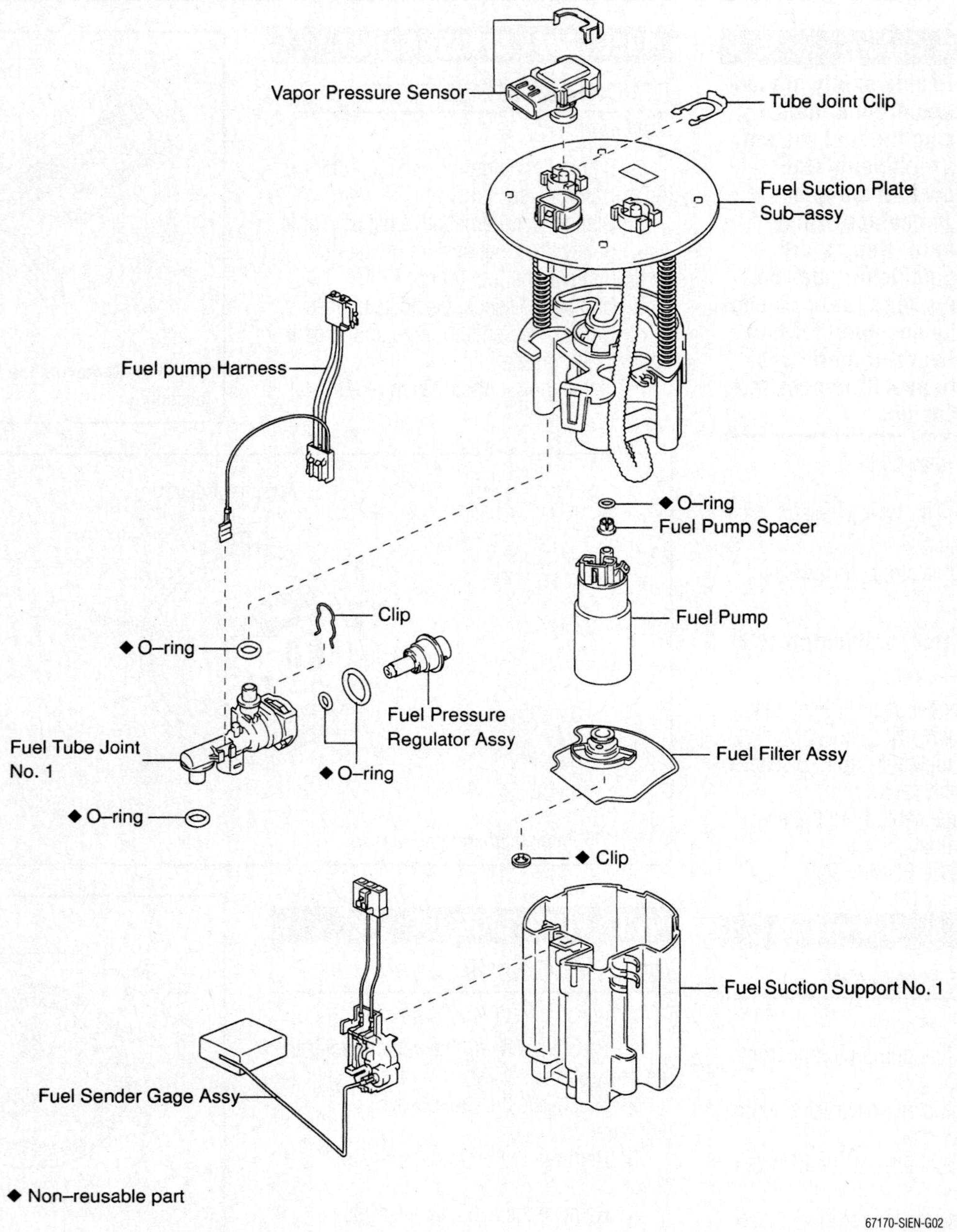

Vapor Pressure Sensor
Tube Joint Clip
Fuel Suction Plate Sub–assy
Fuel pump Harness
◆ O–ring
Fuel Pump Spacer
Fuel Pump
Clip
◆ O–ring
Fuel Pressure Regulator Assy
Fuel Tube Joint No. 1
◆ O–ring
Fuel Filter Assy
◆ O–ring
◆ Clip
Fuel Suction Support No. 1
Fuel Sender Gage Assy
◆ Non–reusable part
67170-SIEN-G02

Fig. 179 Fuel pump module exploded view

18. Disconnect the snap retainers, disconnect the connector, and remove the fuel pump.

19. Remove the O-ring and spacer from the pump.

20. Installation is the reverse of removal. Use a new O-ring coated with clean gasoline.

➡**Prior to assembly, all new parts must be stored at room temperature for a minimum of 12 hours.**

21. Make sure that the gasket groove is clean. Use a new gasket. Align the arrow on the fuel suction tube and the tank suction support.

22. Align the marks on the fuel pump module retainer and the fuel tank.

23. Position the retainer on the module and push down. Hold the module and turn the retainer by hand, one complete turn.

➡**Make sure that the anti-rotation tab is in the groove during tightening. The "S" arrow on the fuel tank indicates "0" degrees position. Make sure that the retainer isn't cross-threaded.**

24. Using the special tool, torque the retainer to 59–67 ft. lbs. (80–90 Nm). The triangle mark on the retainer should be about 1½ turn from the start.

FUEL RAIL AND INJECTOR

REMOVAL & INSTALLATION

See Figure 180.

1. Before servicing the vehicle, refer to the precautions section.
2. Relieve the fuel system pressure.
3. Disconnect the negative battery cable.
4. Drain the coolant.
5. Remove the wiper arms.
6. Remove the wiper motor.
7. Remove the cowl tops.
8. Remove the V-bank cover.
9. Remove the air cleaner assembly.
10. Remove the emission control valve set.
11. Remove the upper intake manifold (intake air surge tank). Discard the gasket.
12. Remove the fuel pipe sub-assembly.
    a. Remove the No. 2 fuel pipe clamp.
    b. Pinch the tube connector and pull out the fuel pipe.

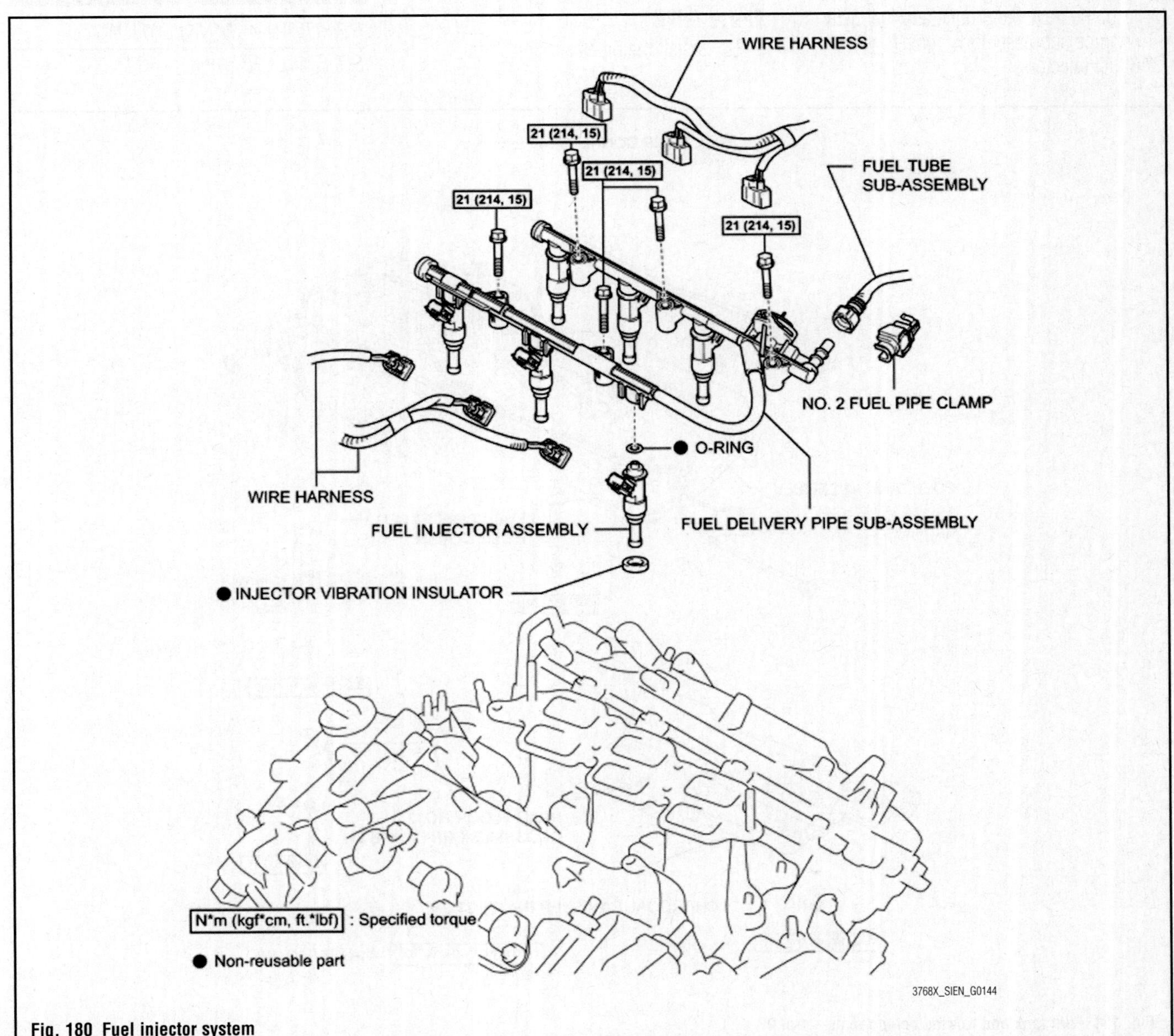

Fig. 180 Fuel injector system

➡Check that there is no dirt or other foreign objects around the connector before disconnecting it. Clean the connector as necessary. It is necessary to prevent mud or dirt from entering the quick connector. If mud or dirt gets in the connector, the O-rings may not seal properly. Only disconnect the quick connector by hand. Do not bend, kink or twist the nylon tubes. Protect the connector by covering it with a plastic bag. If the pipe and the connector are stuck, carefully try wiggling or pushing and pulling on the connector to release it. Pull the connector off the pipe carefully.

13. Remove fuel injector assembly:
    a. Disconnect the 6 fuel injector connectors.
    b. Remove the 5 bolts and fuel delivery pipe sub-assembly together with the 6 fuel injectors.

➡Be careful not to drop the fuel injectors when removing the fuel delivery pipe sub-assembly.

    c. Remove the 6 injector vibration insulators from the intake manifold.
    d. Pull out the fuel injectors from the fuel delivery pipe sub-assembly.
    e. Remove the 6 O-rings from the fuel injectors.

To install:

14. Install new O-rings on each injector. Apply a light coating of gasoline to the O-rings and mating points on the pipes.
15. Using a twisting motion, install the injectors on the pipes.

➡Be careful to avoid twisting the O-rings. After installation, check that the injectors turn smoothly. If not, use new O-rings.

16. Install the pipes and injectors.
17. Loosely install the bolts and make sure that the injectors still turn freely. If not, replace the O-rings.
18. Torque the bolts to 84 inch lbs. (10 Nm).
19. The remainder of installation is the reverse of removal. Observe the following torques:
    - Fuel line union bolt: 24 ft. lbs. (33 Nm)
    - Pulsation damper: 24 ft. lbs. (33 Nm)
    - Fuel feed pipe: 14 ft. lbs. (20 Nm)
    - Upper intake manifold (air surge tank): 21 ft. lbs. (28 Nm)
    - Upper intake manifold stays: 14 ft. lbs. (20 Nm)

FUEL TANK

REMOVAL & INSTALLATION

See Figures 181 through 184.

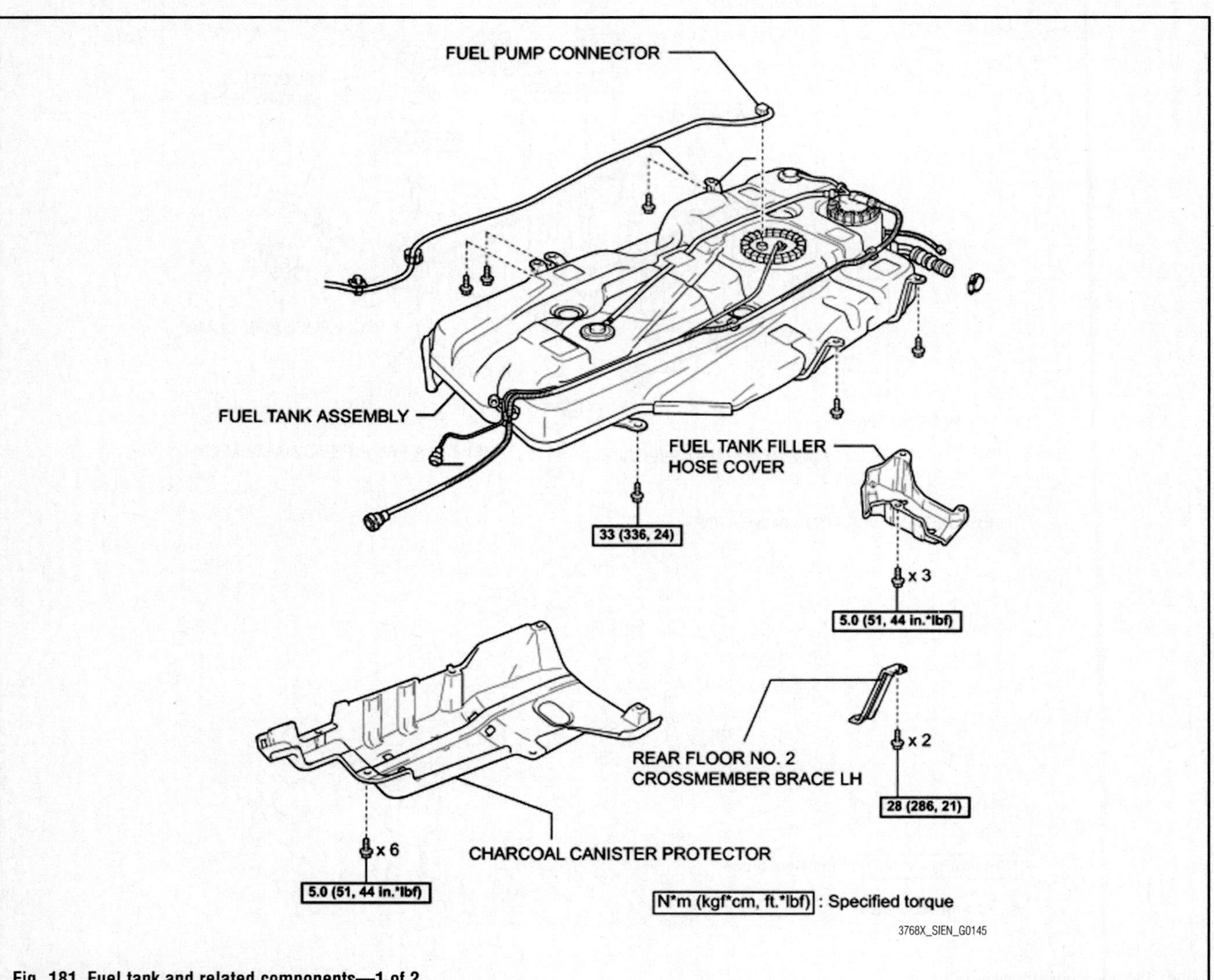

Fig. 181 Fuel tank and related components—1 of 2

1. Discharge fuel system pressure.
2. Remove charcoal canister protector.
3. Remove the 2 bolts and the rear floor No. 2 crossmember brace LH.
4. Remove the 3 bolts and the fuel tank filler hose cover.
5. Remove fuel tank assembly:
   a. Disconnect the fuel tank vent hose. Deeply push the connector to release the locking tab. Pinch and pull out the connector.
   b. Disconnect the fuel tank main tube sub-assembly. Pinch the tube connector and then pull out the fuel tank main tube sub-assembly.

➡Check if there is any dirt or mud around the connector before this operation and remove the dirt as necessary. Be careful of mud because the quick connector has an O-ring which seals the pipe and connector that can be contaminated. Do not use any tool in this operation. Do not bend or twist the nylon tube. Protect the connector by covering it with a vinyl or plastic bag. When the pipe and the connector are stuck, push and pull the connector to release and pull the connector out carefully.

   c. Disconnect the fuel tank to filler pipe hose. Loosen the hose clamp bolt and disconnect the fuel tank to filler pipe hose.
   d. Disconnect the fuel tube. Pinch the tube connector and then pull out the fuel tank main tube sub-assembly.

➡Check if there is any dirt or mud around the connector before this operation and remove the dirt as necessary. Be careful of mud because the quick connector has an O-ring which seals the pipe and connector that can be contaminated. Do not use any tool in this operation. Do not bend or twist the nylon tube. Protect the connector by covering it with a vinyl or plastic bag. When the pipe and the connector are stuck, push and pull the connector to release and pull the connector out carefully.

   e. Remove the 2 wire harness clamps.
   f. Set up a transmission jack under the fuel tank assembly.
   g. Remove the 6 set bolts of the 3 fuel tank bands.
   h. While operating the transmission jack, unfasten the 4 claws for the wire harness and disconnect the fuel pump connector, then remove the fuel tank and the 3 fuel tank bands from the vehicle.

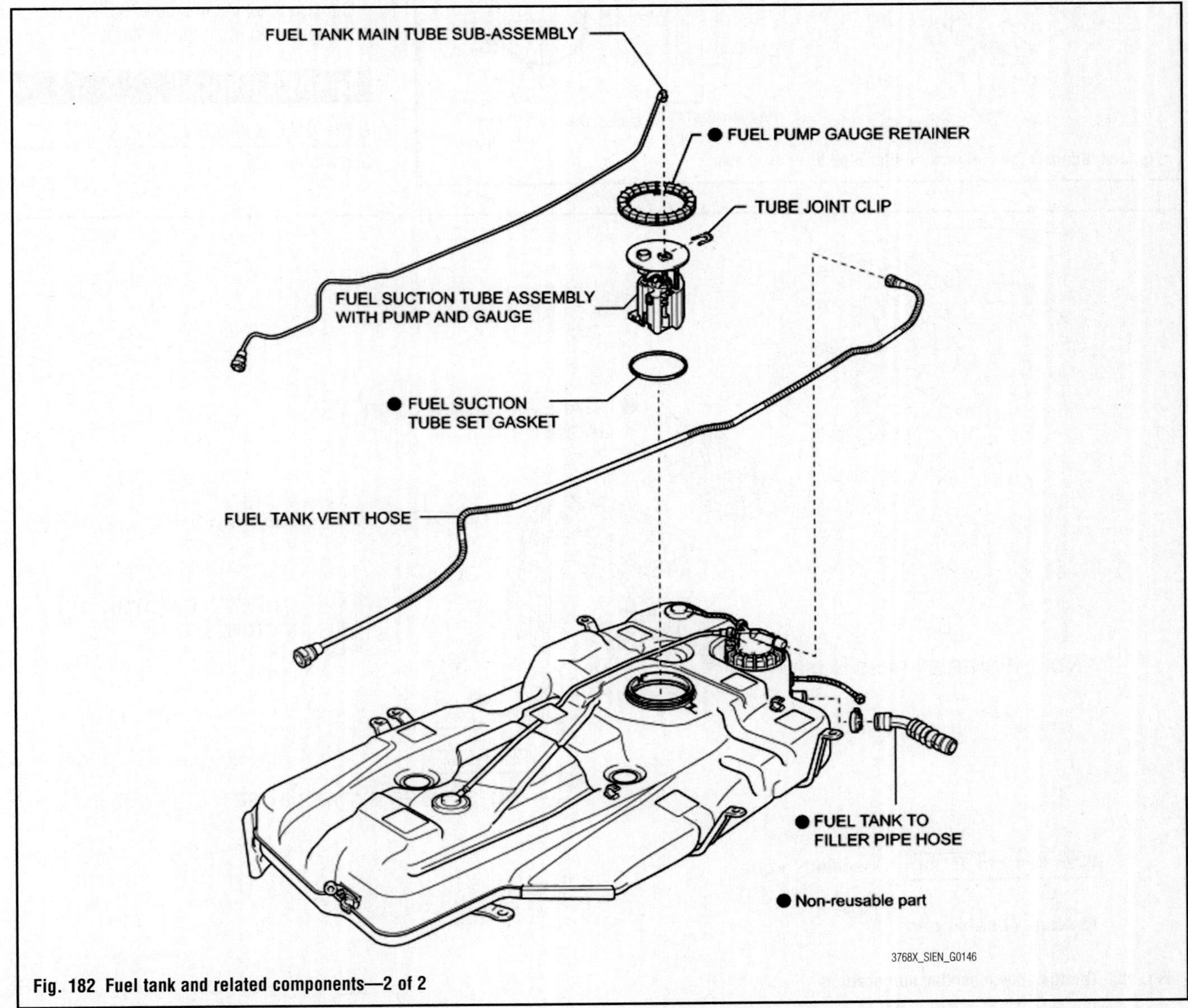

Fig. 182 Fuel tank and related components—2 of 2

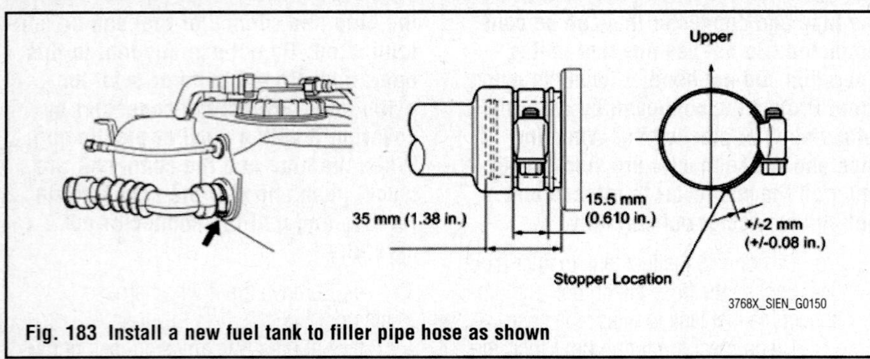

Fig. 183 Install a new fuel tank to filler pipe hose as shown

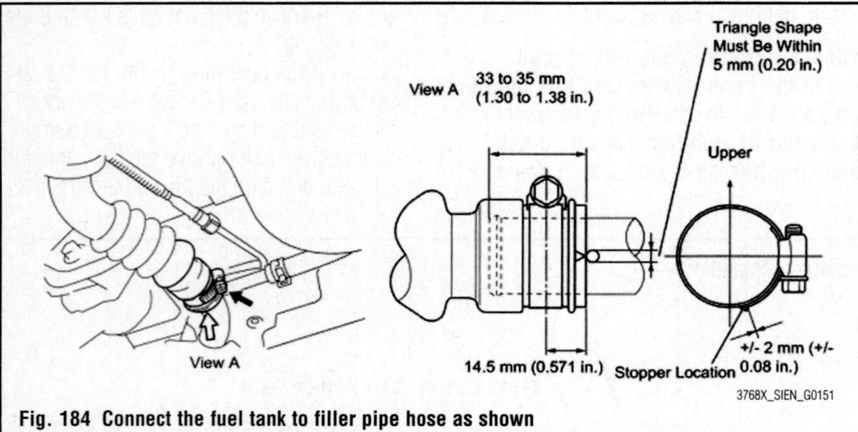

Fig. 184 Connect the fuel tank to filler pipe hose as shown

6. Remove fuel tank main tube sub-assembly.
7. Remove fuel suction tube assembly with pump and gauge.
8. Drain fuel.
9. Loosen the hose clamp bolt and remove the fuel tank to filler pipe hose from the fuel tank.
10. Remove fuel tank vent hose:
    a. Remove the hose of the fuel tank vent hose from the 3 hose clamps.
    b. Remove the fuel tank vent hose.
    c. Deeply push the connector to release the locking tab. Pinch and pull out the connector.

To install:

11. To install, reverse the removal procedure while paying attention to the following:
    a. Install a new fuel tank to filler pipe hose as shown in the illustration.
12. Connect the fuel tank to filler pipe hose as shown in the illustration.

THROTTLE BODY

REMOVAL & INSTALLATION

See Figure 185.

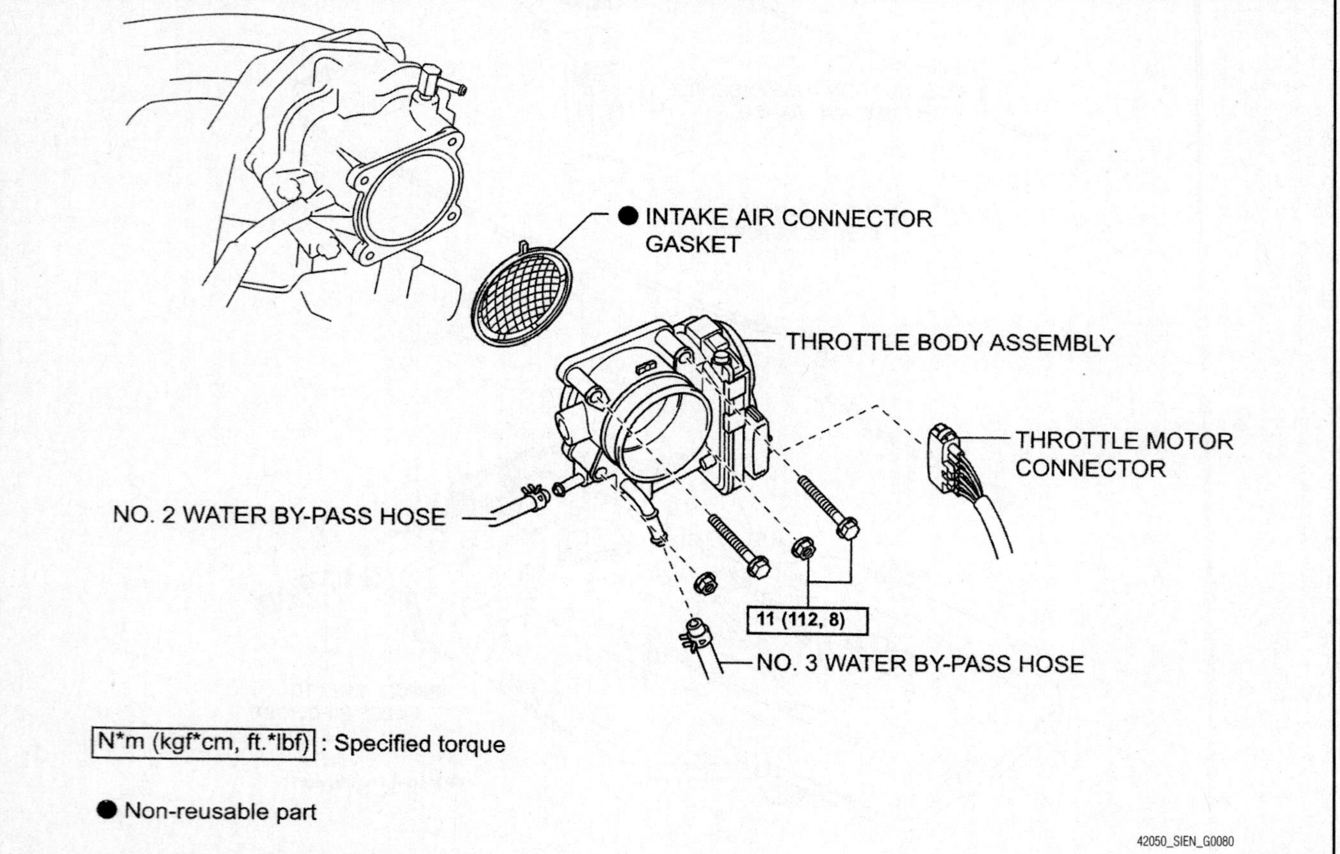

● INTAKE AIR CONNECTOR GASKET
THROTTLE BODY ASSEMBLY
THROTTLE MOTOR CONNECTOR
NO. 2 WATER BY-PASS HOSE
11 (112, 8)
NO. 3 WATER BY-PASS HOSE
N*m (kgf*cm, ft.*lbf) : Specified torque
● Non-reusable part
42050_SIEN_G0080

Fig. 185 Throttle body mounting nut locations

➡There are many hoses and electrical connectors that are either removed or disconnected to facilitate the throttle body removal. It is a good idea to mark component connections and locations prior to removal to avoid confusion during installation.

1. Before servicing the vehicle, refer to the precautions section.
2. Discharge the fuel system pressure.
3. Drain the cooling system.
4. Remove the engine cover.
5. Remove the air cleaner assembly as follows:
    a. Disconnect the mass air flow sensor, tag and unplug the 3 vacuum hoses.
    b. Tag and disconnect the fuel vapor feed hose, ventilation hose and remove the air cleaner cap bolts.
    c. Remove the hose from the cap and remove the cap with the hose.
    d. Remove the air filter.
6. Disconnect the throttle body motor connector.
7. Disconnect the water by pass hoses from the throttle body.
8. Loosen the throttle body nuts and nuts, remove the throttle body and the gasket.
9. Clean the gasket mating surfaces.

To install:

10. Install a new gasket and the throttle body. Tighten the throttle body nuts and bolts to 8 ft. lbs. (11 Nm).
11. Attach the hoses and electrical connectors.
12. Install the air cleaner assembly.
13. Install the remaining components.
14. Fill the cooling system.
15. Start the vehicle and check for leaks.

HEATING & AIR CONDITIONING SYSTEM

BLOWER MOTOR

REMOVAL & INSTALLATION

See Figure 186.

1. Before servicing the vehicle, refer to the precautions section.
2. Disconnect the negative battery cable.
3. Remove the instrument panel assembly.
4. Remove the clip and the heater to foot No. 1 duct.

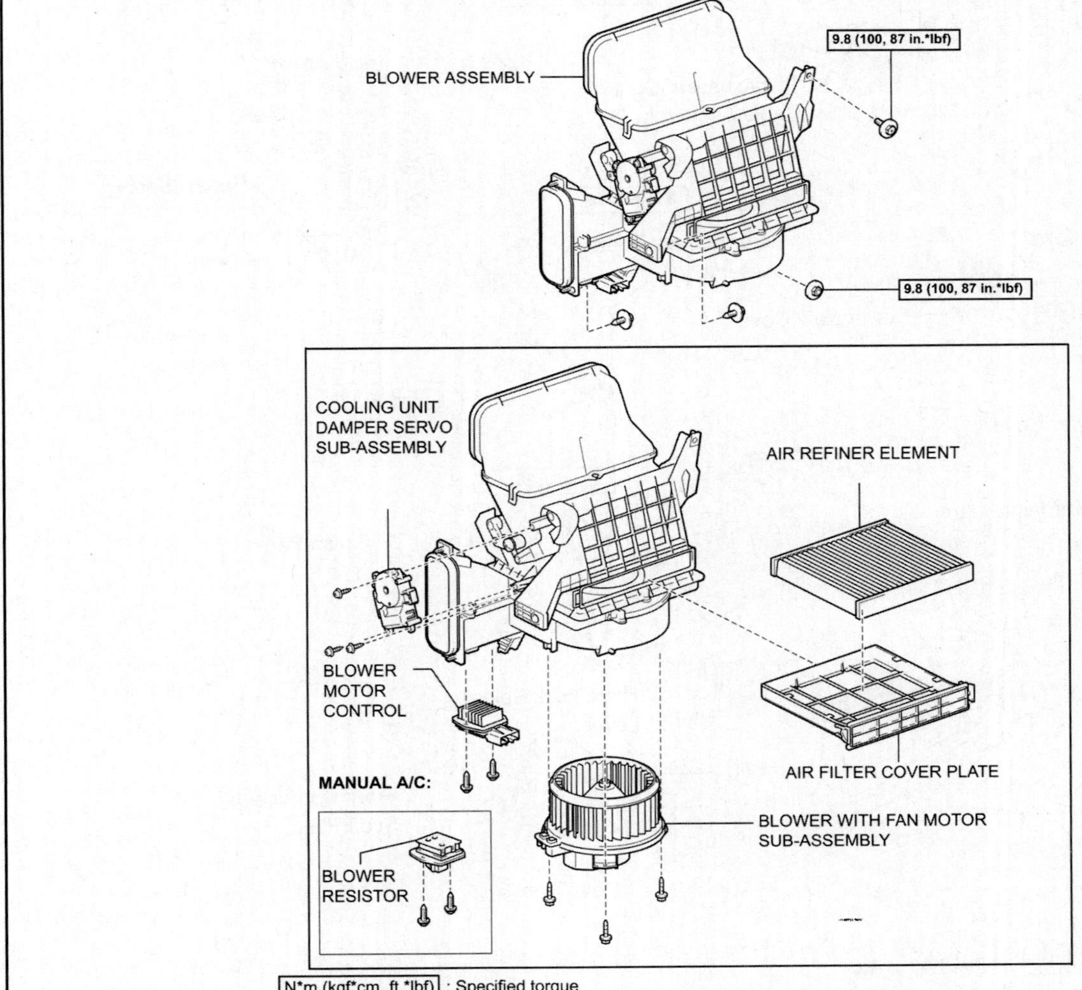

Fig. 186 Exploded view of the front blower motor assembly

5. Remove the ECM.
6. Remove the stereo amplifier assembly.
7. Release the clamps, unfasten the nuts and the wiring harness.
8. Remove the bolt, screws, nut and the blower assembly.
9. Remove the 3 screws and the blower motor.

To install:

10. Install the blower motor and tighten the screws.
11. Install the blower assembly, tighten the fasteners and attach the wiring connector.
12. Install the stereo amplifier assembly.
13. Install the ECM.
14. Install the heater to foot No. 1 duct and clip.
15. Install the instrument panel assembly.
16. Connect the negative battery cable.

HEATER CORE

REMOVAL & INSTALLATION

See Figures 187 through 189.

1. Before servicing the vehicle, refer to the precautions section.
2. Disconnect the negative battery cable.
3. Drain the cooling system into a clean container for reuse.
4. Disconnect the heater hoses from the heater core.

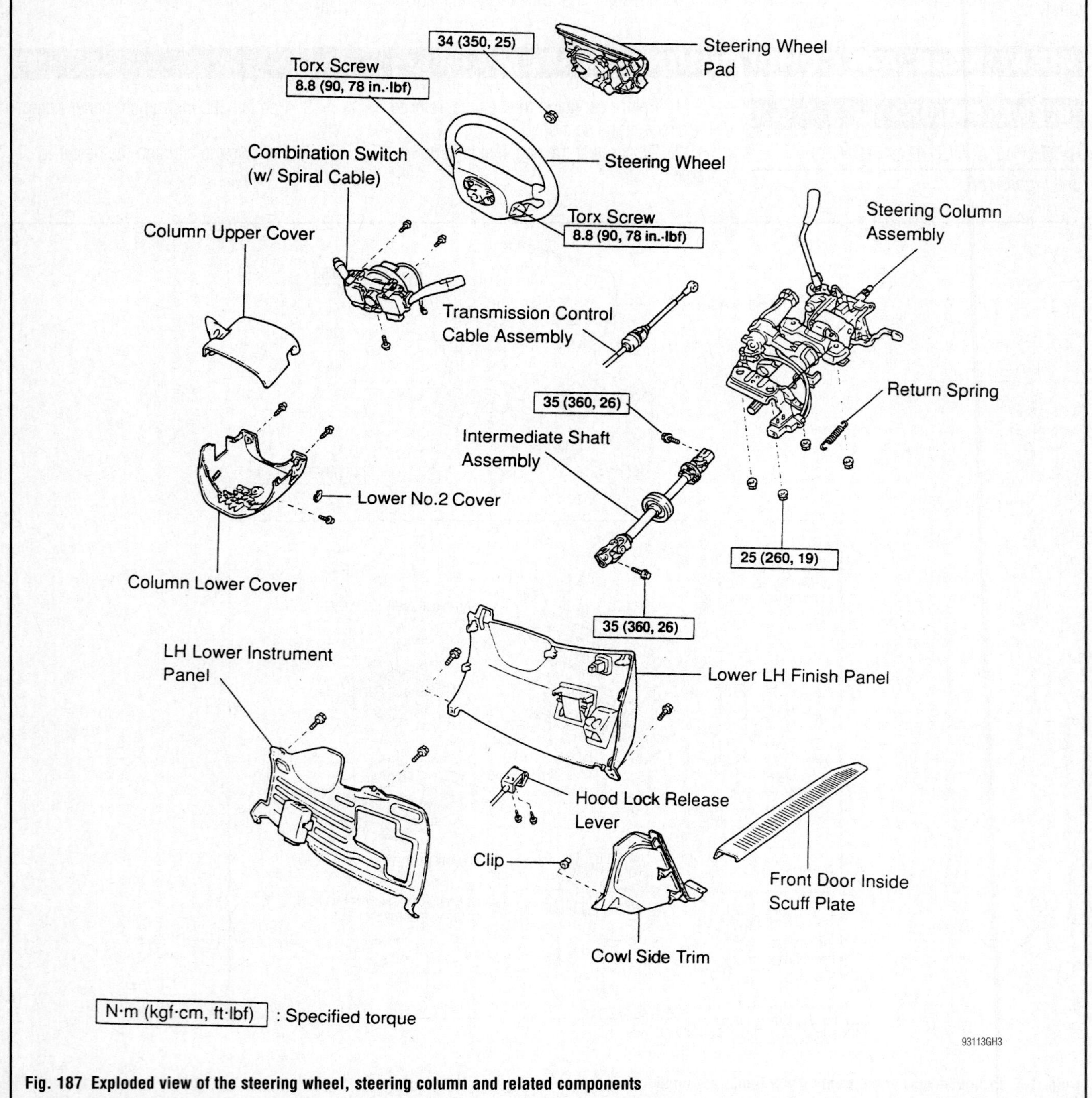

Fig. 187 Exploded view of the steering wheel, steering column and related components

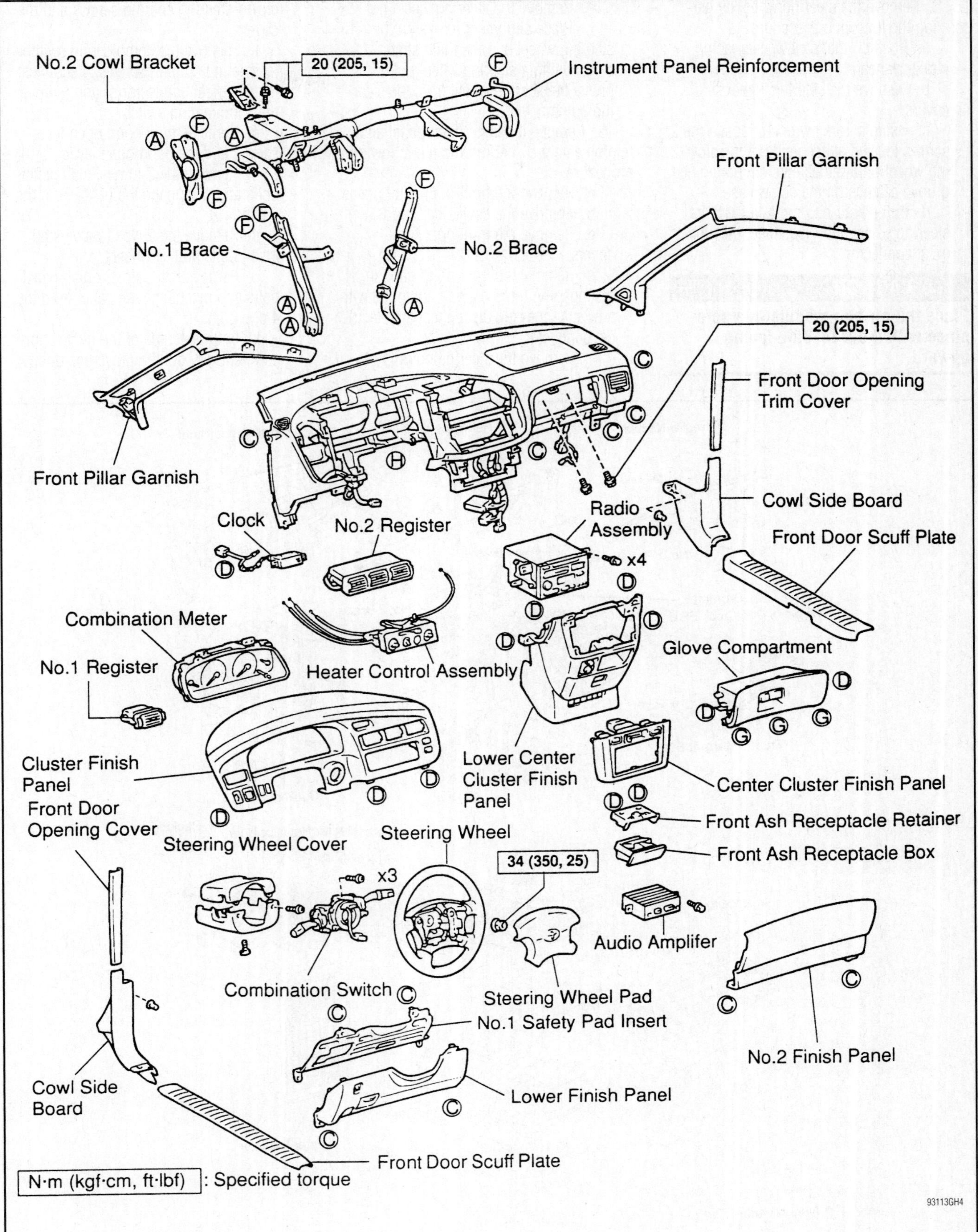

No.2 Cowl Bracket
20 (205, 15)
Instrument Panel Reinforcement
Front Pillar Garnish
No.1 Brace
No.2 Brace
20 (205, 15)
Front Door Opening Trim Cover
Front Pillar Garnish
Cowl Side Board
Radio Assembly
x4
Front Door Scuff Plate
Clock
No.2 Register
Combination Meter
No.1 Register
Heater Control Assembly
Glove Compartment
Cluster Finish Panel
Lower Center Cluster Finish Panel
Center Cluster Finish Panel
Front Door Opening Cover
Front Ash Receptacle Retainer
Steering Wheel Cover
Steering Wheel
Front Ash Receptacle Box
34 (350, 25)
x3
Audio Amplifer
Combination Switch
Steering Wheel Pad
No.1 Safety Pad Insert
No.2 Finish Panel
Cowl Side Board
Lower Finish Panel
Front Door Scuff Plate
N·m (kgf·cm, ft·lbf) : Specified torque
93113GH4

Fig. 188 Exploded view of the instrument panel and related components

5. Remove the steering wheel by performing the following procedure:

a. Position the front wheels facing straight-ahead.

b. Remove the steering wheel side covers.

c. Using a Torx® wrench, loosen the 2 screws located at each side of the steering wheel until the screw's circumference groove catches on the screw case.

d. Pull the air bag module from the steering wheel and disconnect the electrical connector.

CAUTION

Place the air bag module in a safe place with the front side facing upward.

e. Remove the steering wheel nut.

f. Place alignment marks on the steering wheel and the main shaft.

g. Using a steering wheel puller, press the steering wheel from the steering column.

6. Remove the instrument panel and reinforcement by performing the following procedure:

a. Remove the front door scuff plates.

b. Remove the cowl side boards.

c. Remove the front door trim covers.

d. Remove the front pillar garnish by disengaging the 5 clips. If equipped with a tweeter speaker, disconnect the electrical connector.

e. Remove the steering column covers-to-steering column screws and the covers.

f. Remove the combination switch-to-steering column screws, disconnect the electrical connector(s) and remove the combination switch.

g. Remove the 2 hood open lever screws and the hood open lever.

h. Remove the 2 lower finish panel bolts and disengage the panel from the 3 clips.

i. Remove the 2 No. 1 safety pad insert bolts and the insert.

j. Remove the 2 No. 2 finish panel bolts and disengage the panel from the 4 clips.

k. In the left side of the glove compartment, pry out the glove box door

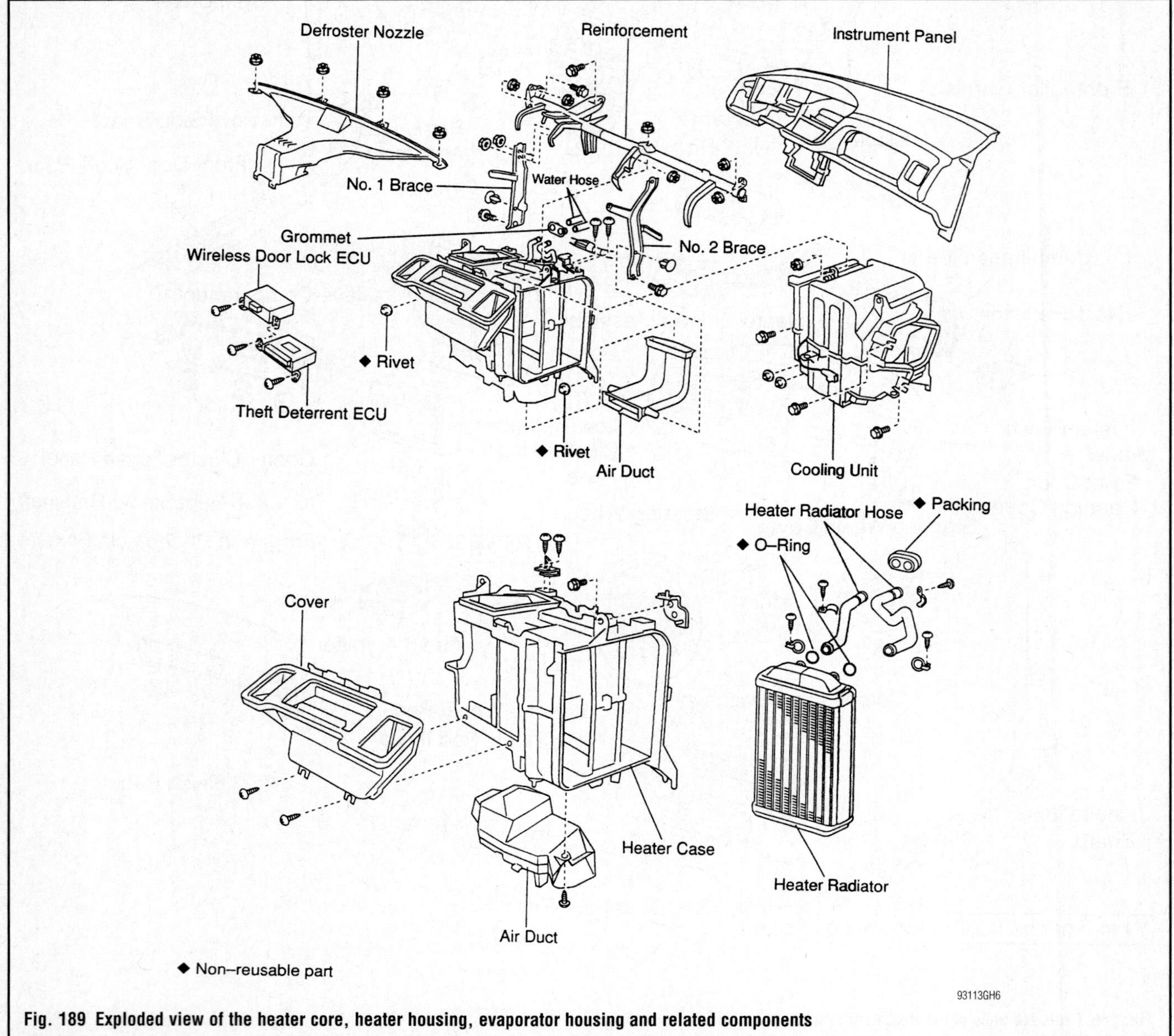

Fig. 189 Exploded view of the heater core, heater housing, evaporator housing and related components

finish plate and disconnect the air bag module connector.

l. Remove the glove box 3 nuts and 2 screws and the glove box.

m. Remove the center cluster finish panel by disengaging the claw (bottom center) and 4 clips (one at each corner).

n. Remove the ashtray, the 2 ashtray receptacle box screws.

o. Remove the 4 lower center cluster finish panel screws and disconnect the connector.

p. Remove the clock, the No. 1 and No. 2 registers from the panel.

q. Remove the 3 cluster finish panel screws, disengage the 8 clips and remove the panel.

r. Remove the combination meter.

s. Remove the radio assembly.

t. Remove the heater control assembly.

u. Remove 2 passenger's side air bag module bolts; then, disconnect and remove the air bag module.

CAUTION

Place the air bag module in a safe place with the front side facing upward.

v. Remove the instrument panel-to-chassis 5 bolts and nut.

w. Remove the audio amplifier.

x. Remove the No. 1 and No. 2 braces.

y. Remove the No. 2 cowl brace.

z. Remove the instrument panel reinforcement.

7. Remove the evaporator housing by performing the following procedure:

a. Discharge and recover the air conditioning system refrigerant.

b. In the engine compartment, remove the refrigerant lines-to-cowl connector bolts; then, disconnect the lines and discard the O-rings.

c. Disconnect the electrical connector at the evaporator housing.

d. Disconnect the wiring harness clamp.

e. Remove the evaporator housing-to-chassis 2 rivets, 3 bolts and nut.

f. Remove the evaporator housing.

8. Remove the 4 defroster nozzle nuts and the nozzle.

9. Disconnect and remove the theft deterrent and the wireless door lock ECUs.

10. Release the 2 air duct claws and the air duct.

11. Remove the 2 heater housing-to-chassis rivets and the heater housing.

➡When installing the heater housing, use new screws in place of the rivets.

12. Remove the heater core-to-heater housing cover.

13. Remove both heater core screws and clamps; then, remove the heater core.

To install:

14. Install the heater core and both heater core screws and clamps.

15. Install the heater core-to-heater housing cover.

➡When installing the heater housing, use new screws in place of the rivets.

16. Install the heater housing-to-chassis and the 2 heater housing screws.

17. Release the air duct and the air duct claws.

18. Connect and install the theft deterrent and the wireless door lock ECUs.

19. Install the defroster nozzle and the 4 nozzle nuts.

20. Install the evaporator housing by performing the following procedure:

a. Install the evaporator housing.

b. Install the evaporator housing-to-chassis 2 rivets, 3 bolts and nut.

c. Connect the wiring harness clamp.

d. Connect the electrical connector at the evaporator housing.

e. In the engine compartment, use new O-rings and install the refrigerant lines-to-cowl connector and install the bolts.

21. Install the instrument panel and reinforcement by performing the following procedure:

a. Install the instrument panel reinforcement.

b. Install the No. 2 cowl brace.

c. Install the No. 1 and No. 2 braces.

d. Install the audio amplifier.

e. Install the instrument panel-to-chassis 5 bolts and nut.

f. Connect and install the air bag module and the 2 passenger's side air bag module bolts.

g. Install the heater control assembly.

h. Install the radio assembly.

i. Install the combination meter.

j. Install the cluster finish panel, engage the 8 clips and install the panel screws.

k. Install the No. 1 and No. 2 registers and the clock to the panel.

l. Connect the lower center cluster finish panel connector and install the 4 lower center cluster finish panel screws.

m. Install the 2 ashtray receptacle box screws and the ashtray.

n. Install the center cluster finish panel by engaging the 4 clips (1 at each corner) and the claw (bottom center).

o. Install the glove box and the glove box 3 nuts and 2 screws.

p. In the left side of the glove compartment, connect the air bag module connector and install the glove box door finish plate.

q. Install the No. 2 finish panel, engage the 4 panel clips and install the 3 panel bolts.

r. Install the No. 1 safety pad insert and the 2 insert bolts.

s. Install the finish panel, engage the 3 finish panel clips and install 2 lower finish panel bolts.

t. Install the hood open lever and the 2 hood open lever screws.

u. Install the combination switch, connect the electrical connector(s) and install the combination switch-to-steering column screws.

v. Install the steering column covers and the covers-to-steering column screws.

w. Install the front pillar garnish by engaging the 5 clips. If equipped with a tweeter speaker, connect the electrical connector.

x. Install the front door trim covers.

y. Install the cowl side boards.

z. Install the front door scuff plates.

22. Install the steering wheel by performing the following procedure:

a. Install the steering wheel to the steering column.

b. Align the steering wheel-to-main shaft marks.

c. Install the steering wheel nut and torque the nut to 25 ft. lbs. (34 Nm).

d. Install the air bag module to the steering wheel and connect the electrical connector.

e. Using a Torx® wrench, tighten the steering wheel screws to 78 inch lbs. (8.8 Nm).

f. Install the steering wheel side covers.

23. Connect the heater hoses to the heater core.

24. Refill the cooling system.

25. Connect the negative battery cable.

26. Evacuate and charge the air conditioning system.

27. Run the engine to normal operating temperatures; then, check the climate control operation and check for leaks.

STEERING

POWER RACK & PINION STEERING GEAR

REMOVAL & INSTALLATION

See Figure 190.

1. Place the wheels in a straight-ahead position.
2. Remove the wheels.
3. Matchmark and remove the intermediate shaft from the gear.
4. Remove the steering column hole cover from the dash panel.
5. Disconnect the tie rod ends.
6. Disconnect the stabilizer bar.
7. Disconnect the pressure and return lines.
8. Remove the 2 gear mounting bolts.
9. Matchmark the intermediate shaft extension and control valve shaft.
10. Remove the intermediate shaft extension bolt and remove the gear.

To install:

11. Place the gear in position. Align the matchmarks and connect the intermediate extension shaft. Torque to 27 ft. lbs. (36 Nm).

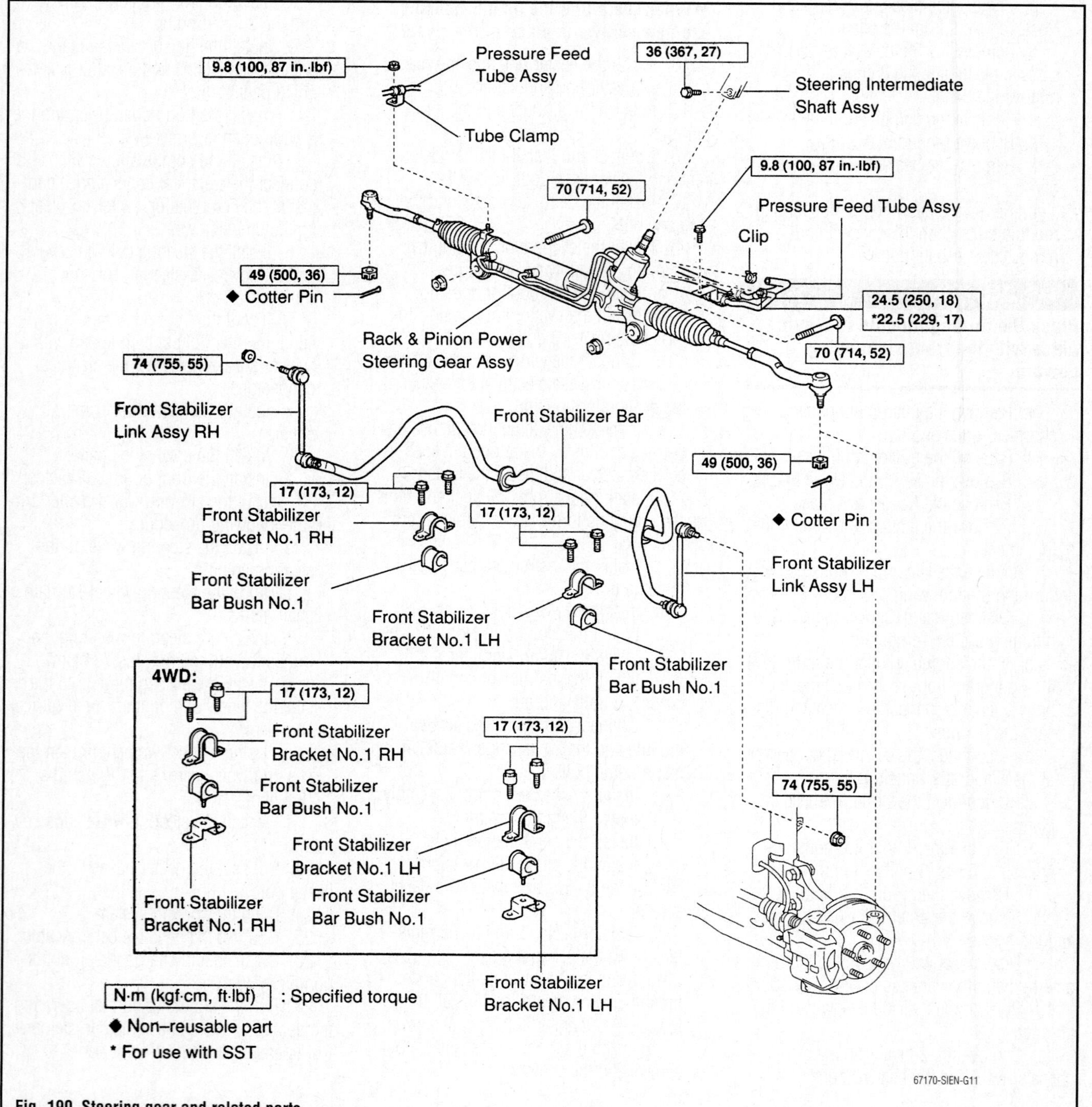

Fig. 190 Steering gear and related parts

12. Install the column cover plate and clamp.
13. Install the gear mounting bolts. Torque to 52 ft. lbs. (70 Nm).
14. Connect the stabilizer bar. Torque to 12 ft. lbs. (17 Nm).
15. Connect the tie rods ends. Torque to 36 ft. lbs. (49 Nm).
16. Connect the intermediate shaft. Torque to 26 ft. lbs. (36 Nm).
17. Check the alignment.

POWER STEERING PUMP

REMOVAL & INSTALLATION

See Figures 191 and 192.

1. Before servicing the vehicle, refer to the precautions section.
2. Remove the right hand front wheel.
3. Drain the power steering fluid.
4. Remove the right side fender apron.
5. Disconnect the clip and disconnect oil reservoir to No. 1 pump hose.
6. Disconnect the connector from the oil pressure switch and remove the switch.
7. Disconnect the pressure feed tube from the steering gear assembly using a 24mm wrench to hold the port union, then remove the union bolt and gasket. Disconnect the feed tube assembly.
8. Remove the drive belt.
9. Remove the pump bolts and the pump.

To install:

10. Install the pump assembly and hand tighten the 2 bolts.
11. Install the drive belt and adjust the belt tension.
12. Tighten the bolts to 32 ft. lbs. (44 Nm).
13. Install a new gasket and the pressure feed tube using the union bolt. Tighten the union bolt to 38 ft. lbs. (52 Nm)

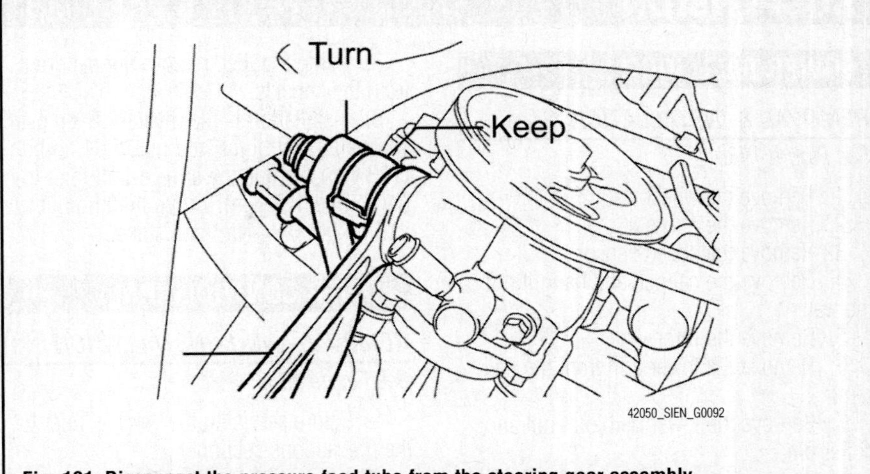

Fig. 191 Disconnect the pressure feed tube from the steering gear assembly

➡Make sure the stopper of the tube touches the front bracket as illustrated.

14. Install the oil pressure switch and tighten to 15 ft. lbs. (21 Nm).
15. Attach the oil pressure switch connector.
16. Connect the return hose and clip.
17. Install the right side fender apron.
18. Install the right hand front wheel.

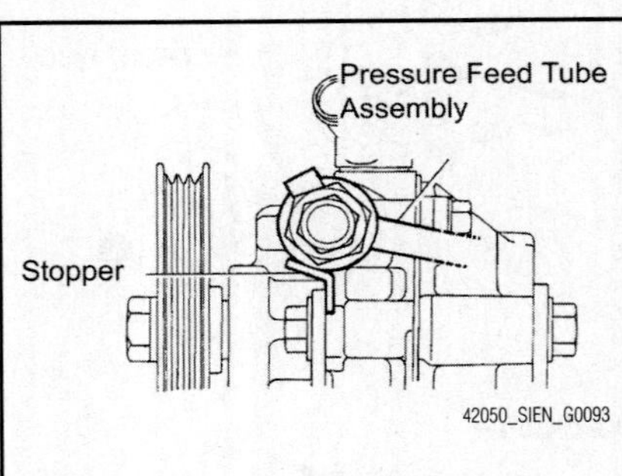

Fig. 192 Make sure the stopper of the pressure feed tube touches the front bracket

19. Fill the power steering fluid reservoir.
20. Bleed the power steering system.

BLEEDING

1. Before servicing the vehicle, refer to the precautions section.
2. Check the fluid level and top off as needed.
3. Jack up the front of the vehicle and support it with safety stands.
4. With the engine OFF, turn the wheel from lock to lock slowly several times.
5. Lower the vehicle and start the engine and allow to idle for a few minutes.
6. With the engine warm and at idle, turn the steering wheel to the left or right lock position and hold it there for 2 to 3 seconds , then turn the wheel to the opposite side lock and hold for 2 to 3 seconds.
7. Repeat these last two steps several times.
8. Turn the vehicle off and check the fluid level.
9. if there is a lot of foam in the reservoir, check the system for leaks and repair, then bleed the system again.

LOWER BALL JOINT

REMOVAL & INSTALLATION

See Figure 193.

1. Remove the wheel.
2. Remove the hub nut.
3. Remove the speed sensor.
4. Remove the caliper, and hang it out of the way.
5. Remove the rotor.
6. Remove the lower arm from the ball joint.
7. Remove the lower ball joint nut and cotter pin.
8. Using a puller, remove the ball joint from the knuckle.
9. Installation is the reverse of removal. Torque the ball joint stud nut to 91 ft. lbs. (123 Nm). Torque the arm-to-ball joint to 94 ft. lbs. (127 Nm).Torque the hub nut to 217 ft. lbs. (294 Nm) and stake it.

LOWER CONTROL ARM

REMOVAL AND & INSTALLATION

See Figure 193.

1. Before servicing the vehicle, refer to the precautions section.
2. Remove the engine and transaxle assembly.
3. Remove the nuts and the transverse engine mount.
4. Remove the two bolts on the front of the lower arm. Remove the nut and bolt, then remove the arm.

To install:

5. Install the arm assembly, install the nut and bolt and tighten to 152 ft. lbs. (206 Nm).
6. Install the two front side bolts and tighten to 148 ft. lbs. (200 Nm).
7. Install the engine mount and tighten the nuts to 64 ft. lbs. (87 Nm).
8. Install the engine and transaxle assembly.

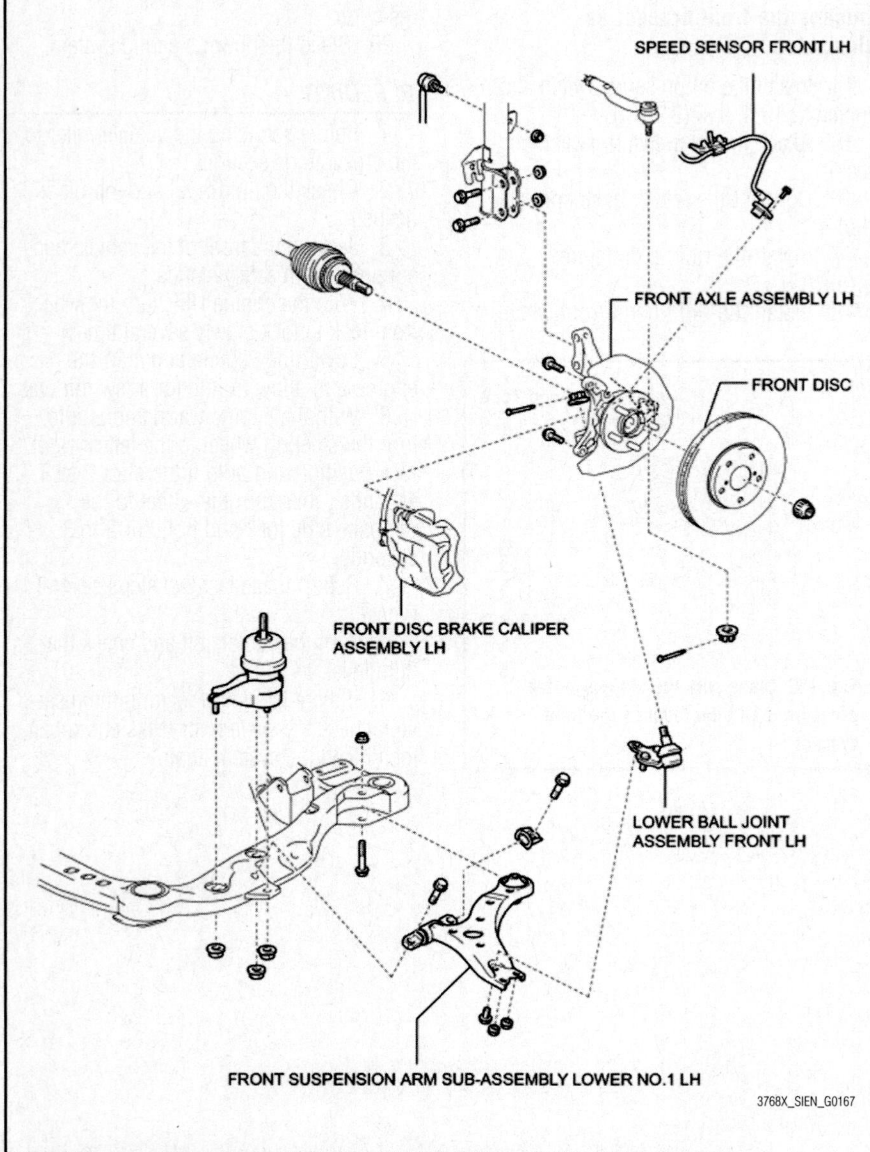

Fig. 193 Lower ball joint, control arm and related components—Front suspension

STEERING KNUCKLE

REMOVAL & INSTALLATION

See Figures 194 and 195.

1. Before servicing the vehicle, refer to the precautions section.
2. Remove or disconnect the following:
    - Front wheels
    - Fender apron seal
3. Check the bearing backlash and axle hub deviation, as follows:
    a. Remove the 2 brake caliper set bolts.
    b. Hang the caliper using stiff wire on the shock absorber assembly.
    c. Remove the rotor.
    d. Place a dial indicator near the center of the axle hub and check the backlash in the bearing shaft direction.
    e. Backlash maximum should read 0.0020 inch (0.05mm). If greater than specified, replace the bearing.
    f. Using the dial indicator, check the deviation at the surface of the axle hub outside and hub bolt. Maximum is 0.0020 inch (0.05mm). If greater than specified, replace the axle hub.
4. Install the rotor and caliper assembly.
5. Remove or disconnect the following:
    - Cotter pin (discard it) and lock cap off the center hub nut
    - Driveshaft locknut, by applying the front brakes
    - Tie rod end, from the steering knuckle
    - Left and right stabilizer end brackets, from the lower arms

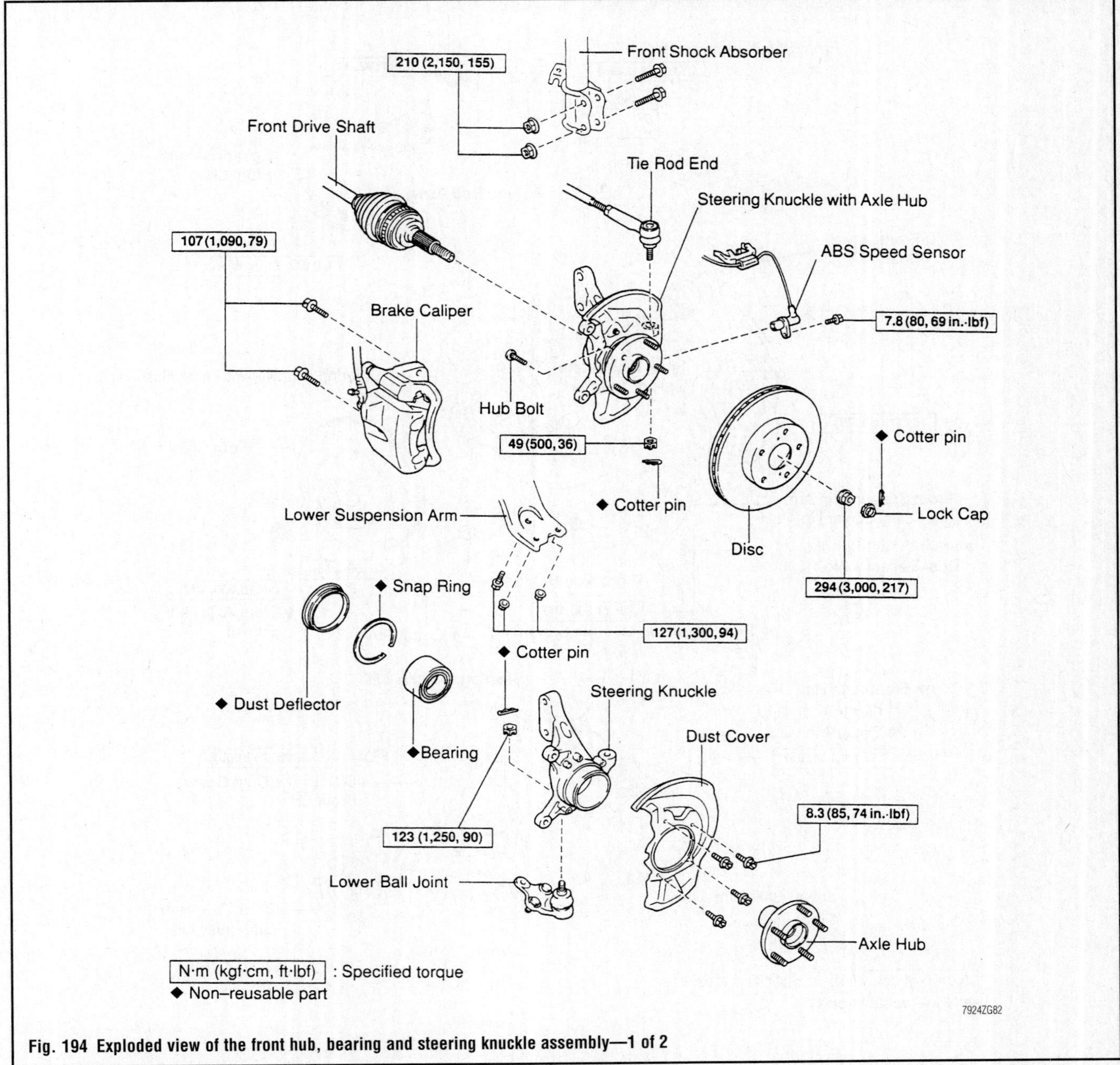

Fig. 194 Exploded view of the front hub, bearing and steering knuckle assembly—1 of 2

- Both nuts and the lower arm from the ball joint
- Driveshaft from the axle hub. Secure the shaft aside using wire.

WARNING

Be careful not to damage the shaft boot or Anti-lock Brake System (ABS) sensor rotor.

- Both brake caliper mounting bolts and the caliper.

➡Support caliper from the vehicle using wire.

- Brake rotor
- Sensor from the steering knuckle, if equipped with ABS
- Both nuts from the lower end of the shock
- Steering knuckle and hub assembly

To install:

6. Install or connect the following:
   - Steering knuckle/hub assembly and temporarily install the lower shock bolts
   - Lower ball joint to the lower arm. Tighten the bolt and nuts to 94 ft. lbs. (127 Nm).
   - Tie rod to the knuckle. Tighten the nut to 36 ft. lbs. (49 Nm).
   - New cotter pin
   - Tighten the lower shock nuts to 156 ft. lbs. (211 Nm).
   - Both side stabilizer end brackets to the lower arm.
   - Front ABS sensor. Tighten it to 69 inch lbs. (8 Nm).
   - Front brake rotor and caliper.
   - Driveshaft locknut, by applying the brakes. Tighten it to 217 ft. lbs. (294 Nm).
   - Lock cap and new cotter pin
   - Front fender apron seal

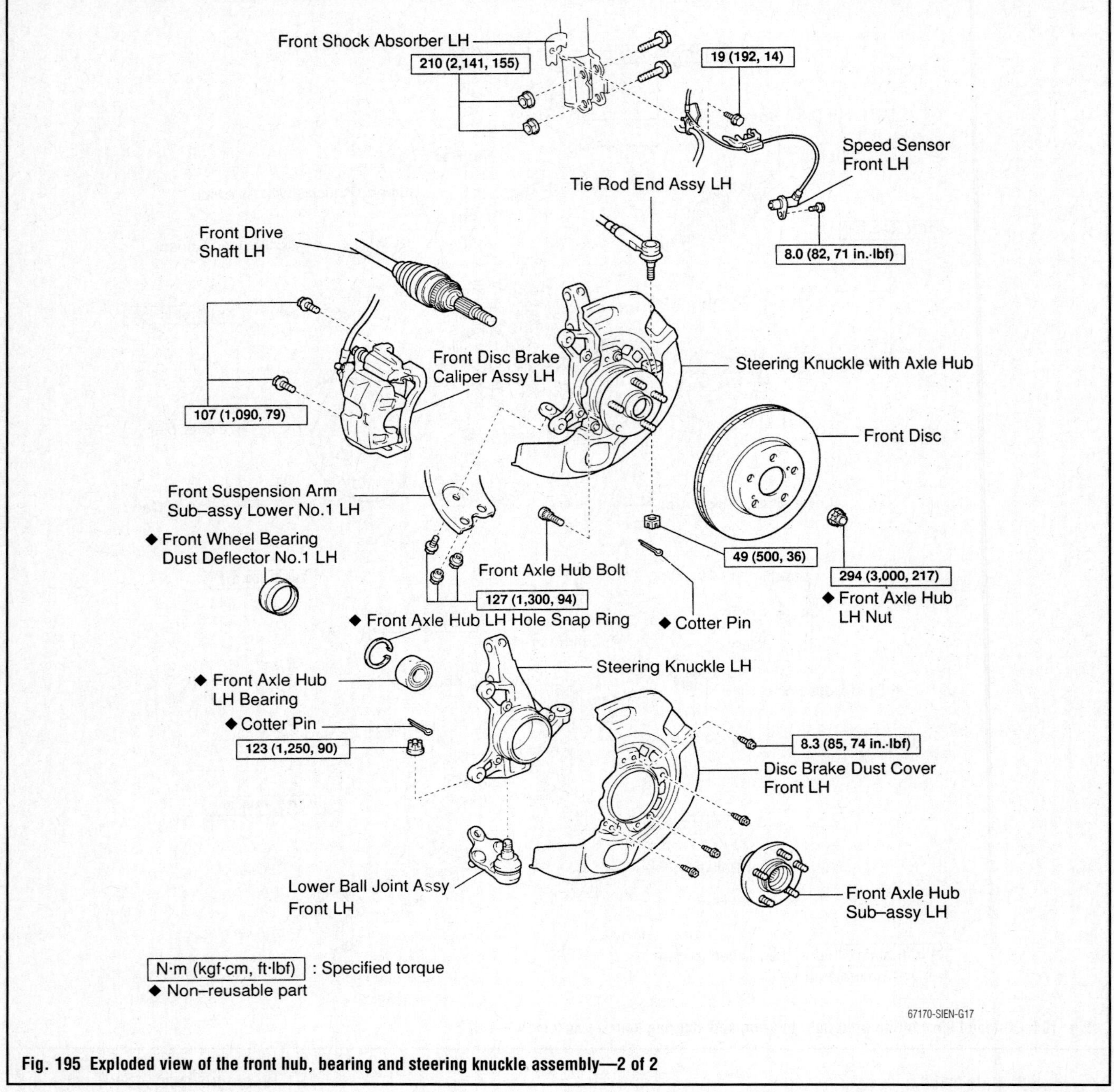

Fig. 195 Exploded view of the front hub, bearing and steering knuckle assembly—2 of 2

- Front wheel. Tighten the lug nuts to 76 ft. lbs. (103 Nm).

STRUT

REMOVAL & INSTALLATION

See Figure 196.

1. Remove the wheel.
2. Remove the wiper arms.
3. Remove the wiper motor.
4. Remove the top cowl.
5. Remove the stabilizer link from the strut.
6. Loosen the strut rod locknut. Don't remove it.
7. Remove the brake hose bracket from the strut.
8. Remove the lower strut bolts.
9. Remove the 3 upper strut nuts.
10. Remove the strut.

To install:

11. Installation is the reverse of removal. Observe the following torques:
    - Upper nuts: 59 ft. lbs. (80 Nm)
    - Lower nuts/bolts: 155 ft. lbs. (210 Nm)
    - Strut rod locknut: 36 ft. lbs. (49 Nm)
    - Stabilizer link: 55 ft. lbs. (74 Nm)

OVERHAUL

1. Before servicing the vehicle, refer to the precautions section.
2. Remove or disconnect the following:
    - Wheel

➡If equipped, be careful not to damage the oil seal, driveshaft boot and/or speed sensor rotor when removing the steering knuckle.

- Shock absorber (strut assembly)

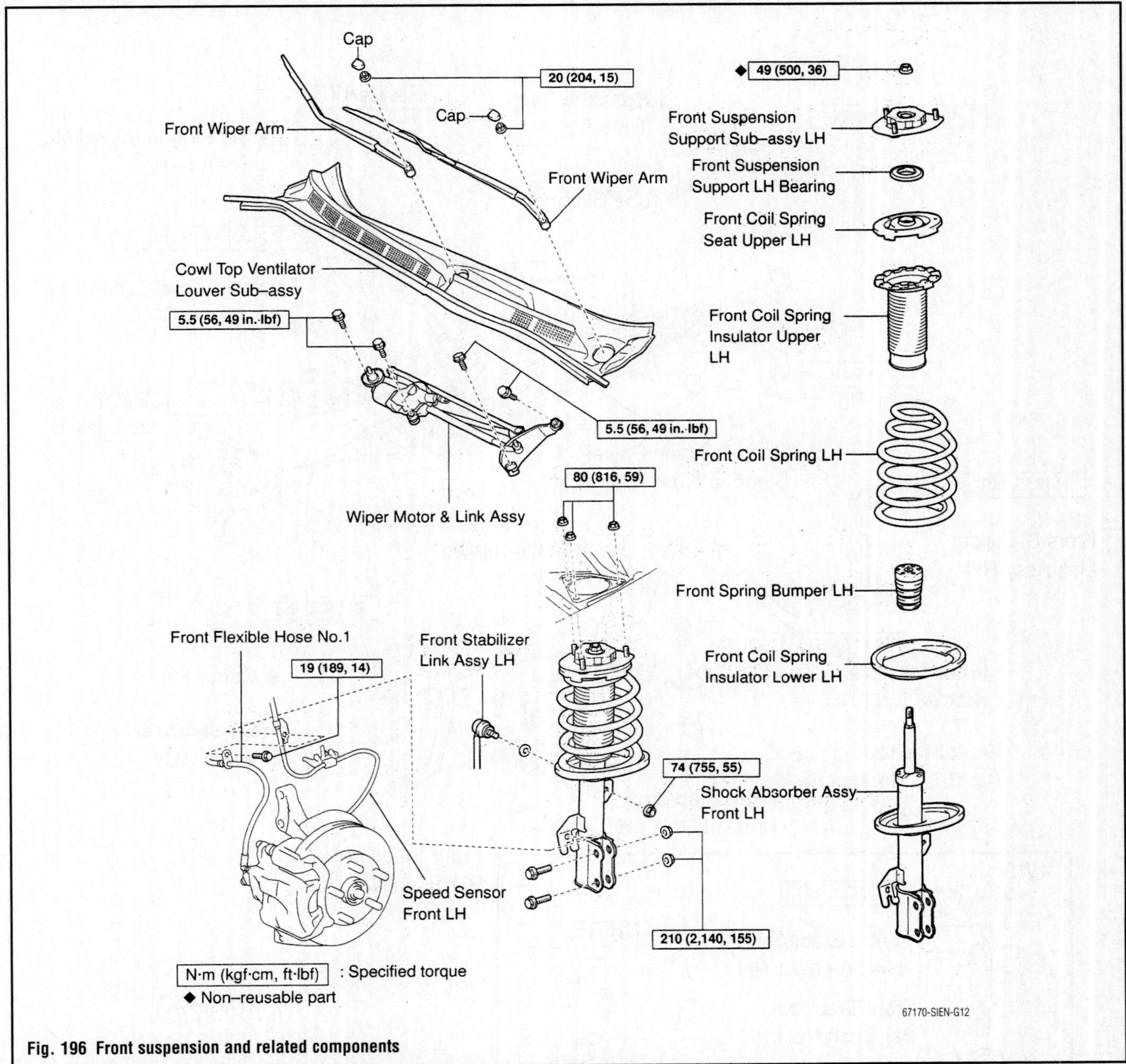

Fig. 196 Front suspension and related components

3. Install a nut/bolt to the bracket at the lower portion of the strut assembly and secure it in a vise.
4. Compress the coil spring with a spring compressor.

CAUTION

The proper tools must be used for this procedure. The spring on the strut is under high pressure and can cause serious injury if not properly removed and installed.

5. Remove or disconnect the following:
   - Center retaining nut, by holding the spring seat
   - Support, dust seal, spring seat, insulator and spring from the strut assembly

To install:

6. Install the spring bumper and lower insulator to the strut assembly.
7. Compress the coil spring and fit the lower end of the spring into the spring seat gap.
8. Install or connect the following:
   - Upper insulator, spring seat, dust seal, support and spring seat. Tighten the new retaining nut to 36 ft. lbs. (49 Nm).
9. Rotate the spring seat so that the OUT mark of the spring seat faces the outside of the vehicle.
   - Strut
   - Wheel
10. If required, bleed the brake system and check for leaks.
11. Check and/or adjust the front wheel alignment.

STABILIZER BAR & LINKS

REMOVAL & INSTALLATION

2-Wheel Drive

See Figures 197 and 198.

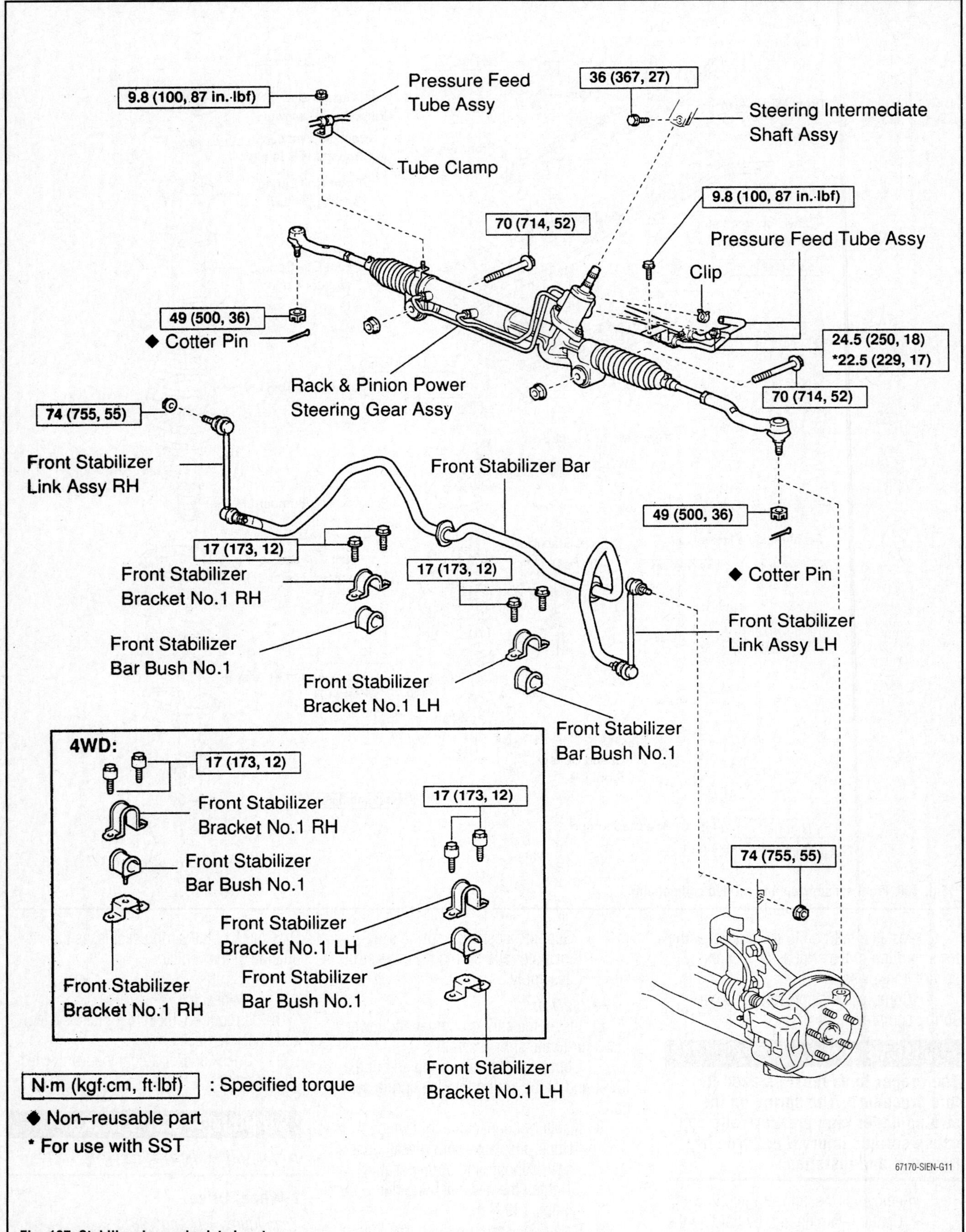

Fig. 197 Stabilizer bar and related parts

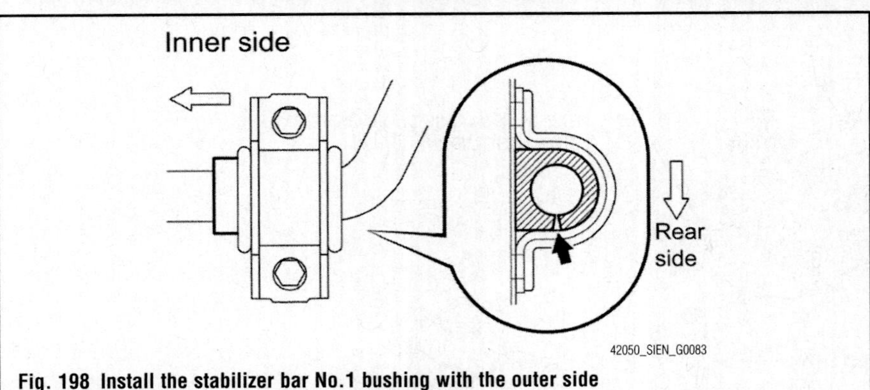

Fig. 198 Install the stabilizer bar No.1 bushing with the outer side of the bushing stopper on the stabilizer bar—2WD

1. Before servicing the vehicle, refer to the precautions section.
2. Remove the wheels.
3. Remove the stabilizer bar links on both sides.
4. Remove the bolts for the stabilizer bar brackets. Remove the brackets from both sides.
5. Disconnect the tie rod ends from the steering knuckle.
6. Disconnect the power steering pressure tube.
7. Separate the steering intermediate shaft.
8. Remove the power steering gear assembly.
9. Remove the stabilizer bar No.1 bushing and the stabilizer bar.
10. Remove the power steering gear set bolts and nuts, then remove the stabilizer bar from the left side.

To install:

11. Install the stabilizer bar.
12. Install the stabilizer bar No.1 bushing with the outer side of the bushing stopper on the stabilizer bar as illustrated.
13. Install the steering intermediate shaft and the power steering gear, then connect the pressure feed tube.
14. Attach the tie rod end, tighten the nut to 36 ft. lbs. (49 Nm) and install a new cotter pin.
15. Install the brackets and bushings. Tighten the bolts to 12 ft. lbs. (17 Nm).
16. Install the stabilizer bar links and tighten the nuts to 55 ft. lbs. (74 Nm).
17. Install the wheels and tighten to 76 ft. lbs. (103 Nm).
18. Perform a wheel alignment check.

4-Wheel Drive

See Figures 199 and 200.

1. Before servicing the vehicle, refer to the precautions section.
2. Remove the wheels.
3. Remove the stabilizer bar links on both sides.
4. Remove the exhaust pipe assembly.
5. Remove the bolts for the stabilizer bar brackets. Remove the brackets from both sides.
6. Disconnect the tie rod ends from the steering knuckle.
7. Disconnect the power steering pressure tube.
8. Separate the steering intermediate shaft.
9. Remove the power steering gear assembly.
10. Remove the exhaust manifold heat shield, manifold stay and manifold converter sub-assembly.
11. Remove the stabilizer bar No.1 bushing and the stabilizer bar.
12. Remove the power steering gear set bolts and nuts, then remove the stabilizer bar from the left side.

To install:

13. Install the stabilizer bar.

➡Install the bushings with the slit facing the rear of the vehicle.

14. Install the stabilizer bar No.1 bushing with the outer side of the bushing stopper on the stabilizer bar as illustrated.
15. Remove the exhaust manifold converter sub-assembly, manifold stay and heat shield.
16. Install the steering intermediate shaft and the power steering gear, then connect the pressure feed tube.
17. Attach the tie rod end, tighten the nut to 36 ft. lbs. (49 Nm) and install a new cotter pin.
18. Install the brackets and bushings. Tighten the bolts to 12 ft. lbs. (17 Nm).
19. Install the stabilizer bar links and tighten the nuts to 55 ft. lbs. (74 Nm).
20. Install the exhaust pipe assembly.
21. Install the wheels and tighten to 76 ft. lbs. (103 Nm).
22. Perform a wheel alignment check.

WHEEL HUB & BEARING

REMOVAL & INSTALLATION

See Figures 194 and 195.

1. Before servicing the vehicle, refer to the precautions section.
2. Remove or disconnect the following:
   - Front wheels
   - Fender apron seal
3. Check the bearing backlash and axle hub deviation, as follows:
   a. Remove the 2 brake caliper set bolts.
   b. Hang the caliper using stiff wire on the shock absorber assembly.
   c. Remove the rotor.
   d. Place a dial indicator near the center of the axle hub and check the backlash in the bearing shaft direction.
   e. Backlash maximum should read 0.0020 inch (0.05mm). If greater than specified, replace the bearing.
   f. Using the dial indicator, check the deviation at the surface of the axle hub outside and hub bolt. Maximum is 0.0020 inch (0.05mm). If greater than specified, replace the axle hub.
4. Install the rotor and caliper assembly.
5. Remove or disconnect the following:
   - Cotter pin (discard it) and lock cap off the center hub nut
   - Driveshaft locknut, by applying the front brakes
   - Tie rod end, from the steering knuckle
   - Left and right stabilizer end brackets, from the lower arms
   - Both nuts and the lower arm from the ball joint
   - Driveshaft from the axle hub. Secure the shaft aside using wire.

⁂ WARNING

Be careful not to damage the shaft boot or Anti-lock Brake System (ABS) sensor rotor.

   - Both brake caliper mounting bolts and the caliper.

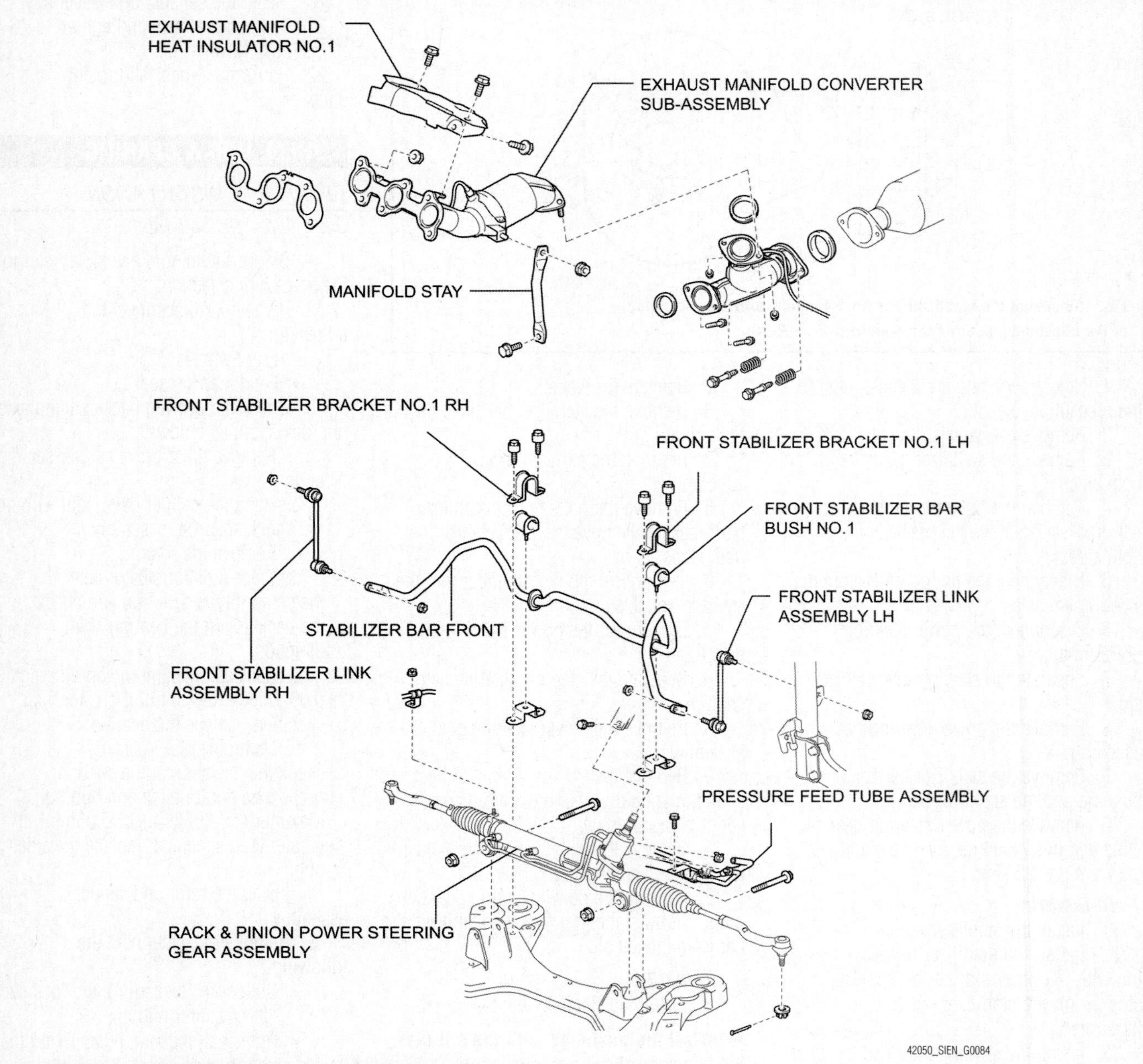

EXHAUST MANIFOLD HEAT INSULATOR NO.1
EXHAUST MANIFOLD CONVERTER SUB-ASSEMBLY
MANIFOLD STAY
FRONT STABILIZER BRACKET NO.1 RH
FRONT STABILIZER BRACKET NO.1 LH
FRONT STABILIZER BAR BUSH NO.1
FRONT STABILIZER LINK ASSEMBLY LH
STABILIZER BAR FRONT
FRONT STABILIZER LINK ASSEMBLY RH
PRESSURE FEED TUBE ASSEMBLY
RACK & PINION POWER STEERING GEAR ASSEMBLY
42050_SIEN_G0084

Fig. 199 Sway bar and related components—4 wheel drive Sienna

Inner side
Front side
Slit
Rear side
42050_SIEN_G0085

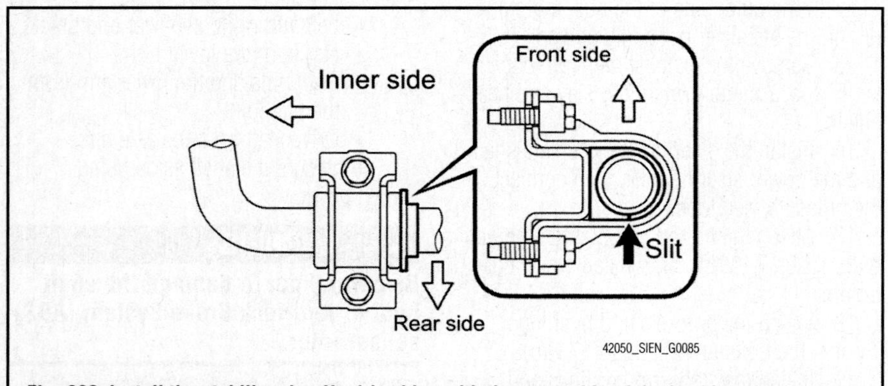

Fig. 200 Install the stabilizer bar No.1 bushing with the outer side of the bushing stopper on the stabilizer bar—4WD

➡Support caliper from the vehicle using wire.

- Brake rotor
- Sensor from the steering knuckle, if equipped with ABS
- Both nuts from the lower end of the shock
- Steering knuckle and hub assembly

6. Clamp the steering knuckle in a vise with soft jaws to protect the knuckle.
7. Remove or disconnect the following:

- Dust deflector from the hub, using a screwdriver

- Bearing inner oil seal, by prying it from the knuckle
- Snapring from the knuckle bore
- Dust deflector from the steering knuckle
- Axle hub, by pulling it from the dust deflector, using a 2-armed mechanical puller
- Inner (inside) bearing race from the bearing, using the puller
- Sensor control rotor from the axle hub, using Torx® wrench
- Outer bearing race, using the puller
- Outer bearing seal, using the puller

8. Position the inner (outside) race inside the bearing.

9. Using a brass rod, tap the bearing from the steering knuckle.

To install:

10. Clean all the oil seal and bearing seating surfaces with a clean, dry rag.

11. Install or connect the following:
    - Bearing into the bore, using a Bearing Driver tool 09608-32010 and a press
    - New outer oil seal, driving it into the steering knuckle, by inserting the seal side lip into the factory tool
    - Brake disc cover to the steering knuckle with the bolts

12. Apply multi-purpose grease between the oil seal lip, oil seal and bearing.

13. Install or connect the following:
    - Hub, by pressing it into the knuckle
    - New snapring into the knuckle
    - New oil seal, by pressing it into the knuckle once lubricated with multi-purpose grease
    - Dust deflector, by pressing it into the knuckle.

⁂ WARNING

Align the speed sensor holes in the dust deflector and steering knuckle, if equipped with ABS.

- Ball joint to the steering knuckle. Tighten the bolts to 94 ft. lbs. (127 Nm).
- Steering knuckle/hub assembly and temporarily install the lower shock bolts
- Lower ball joint to the lower arm. Tighten the bolt and nuts to 94 ft. lbs. (127 Nm).
- Tie rod to the knuckle. Tighten the nut to 36 ft. lbs. (49 Nm).
- New cotter pin
- Tighten the lower shock nuts to 156 ft. lbs. (211 Nm).
- Both side stabilizer end brackets to the lower arm. Tighten the fasteners to 55 ft. lbs. (74 Nm).
- Front ABS sensor. Tighten it to 69 inch lbs. (8 Nm).
- Front brake rotor and caliper.
- Driveshaft locknut, by applying the brakes. Tighten it to 217 ft. lbs. (294 Nm).
- Lock cap and new cotter pin
- Front fender apron seal
- Front wheel. Tighten the lug nuts to 76 ft. lbs. (103 Nm).

SUSPENSION

REAR SUSPENSION

COIL SPRING

REMOVAL & INSTALLATION

See Figure 201.

1. Before servicing the vehicle, refer to the precautions section.

2. Remove or disconnect the following:
    - Shock absorbers
    - Coil springs

To install:

3. Install or connect the following:
    - Coil springs
    - Raise the axle beam enough to apply tension on the springs
    - Shock absorbers

3768X_SIEN_G0168

Fig. 201 Removing the coil spring

SHOCK ABSORBER

REMOVAL & INSTALLATION

See Figures 202 through 204.

1. Remove the wheel.
2. Remove the shock absorber cap.
3. Support the axle with a jack.
4. Remove the upper locknut, retainer and bushing.
5. Remove the lower nut and remove the shock.
6. Remove the spring bumper from the shock.
7. Installation is the reverse of removal. Torque the upper end nut to 22 ft. lbs. (30 Nm). Install the lower end nut loosely. Raise the axle to load the shock. For 2WD models, the shock absorber length should be 9.22 in. (234mm); for 4WD, it should be 10.16 in. (258mm), then tighten the lower end nut to 85 ft. lbs. (115 Nm). If you can't reach the nut in this position, support the rear axle and place 198 lbs. (90 kg) in the trunk.

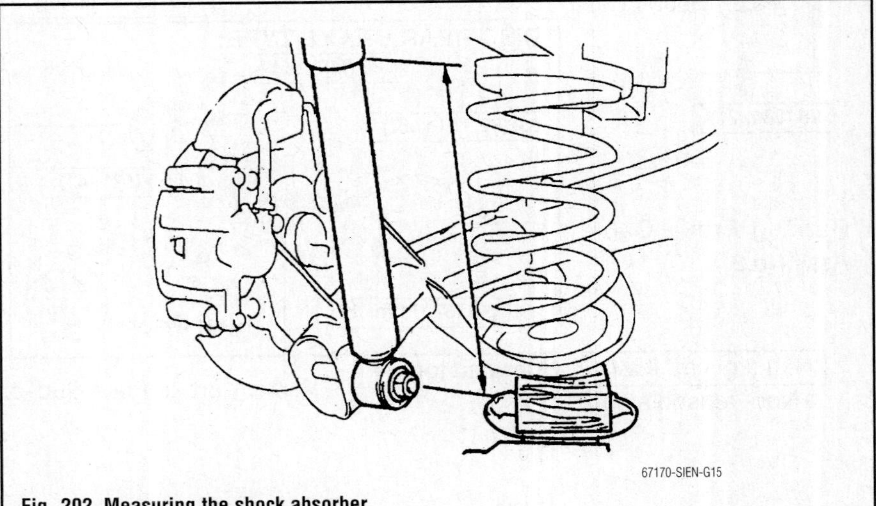

Fig. 202 Measuring the shock absorber

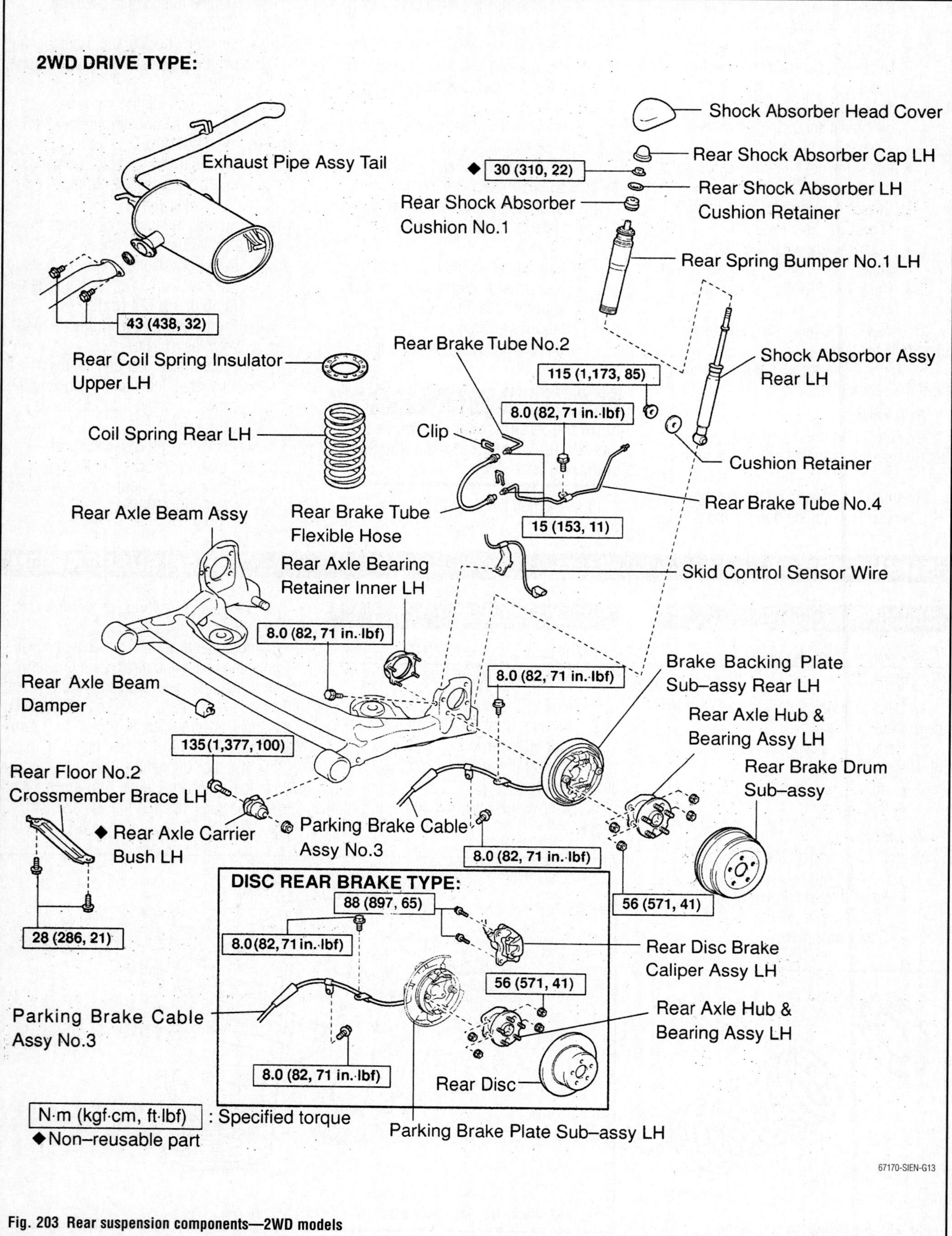
2WD DRIVE TYPE:
Shock Absorber Head Cover
Rear Shock Absorber Cap LH
Exhaust Pipe Assy Tail
◆ 30 (310, 22)
Rear Shock Absorber LH Cushion Retainer
Rear Shock Absorber Cushion No.1
Rear Spring Bumper No.1 LH
43 (438, 32)
Rear Brake Tube No.2
Rear Coil Spring Insulator Upper LH
115 (1,173, 85)
Shock Absorbor Assy Rear LH
8.0 (82, 71 in.·lbf)
Coil Spring Rear LH
Clip
Cushion Retainer
Rear Brake Tube No.4
Rear Axle Beam Assy
Rear Brake Tube Flexible Hose
15 (153, 11)
Rear Axle Bearing Retainer Inner LH
Skid Control Sensor Wire
8.0 (82, 71 in.·lbf)
Brake Backing Plate Sub–assy Rear LH
Rear Axle Beam Damper
8.0 (82, 71 in.·lbf)
Rear Axle Hub & Bearing Assy LH
135 (1,377, 100)
Rear Brake Drum Sub–assy
Rear Floor No.2 Crossmember Brace LH
◆ Rear Axle Carrier Bush LH
Parking Brake Cable Assy No.3
8.0 (82, 71 in.·lbf)
DISC REAR BRAKE TYPE:
88 (897, 65)
56 (571, 41)
28 (286, 21)
8.0 (82, 71 in.·lbf)
Rear Disc Brake Caliper Assy LH
56 (571, 41)
Parking Brake Cable Assy No.3
Rear Axle Hub & Bearing Assy LH
8.0 (82, 71 in.·lbf)
Rear Disc
N·m (kgf·cm, ft·lbf) : Specified torque
◆ Non–reusable part
Parking Brake Plate Sub–assy LH
67170-SIEN-G13

Fig. 203 Rear suspension components—2WD models

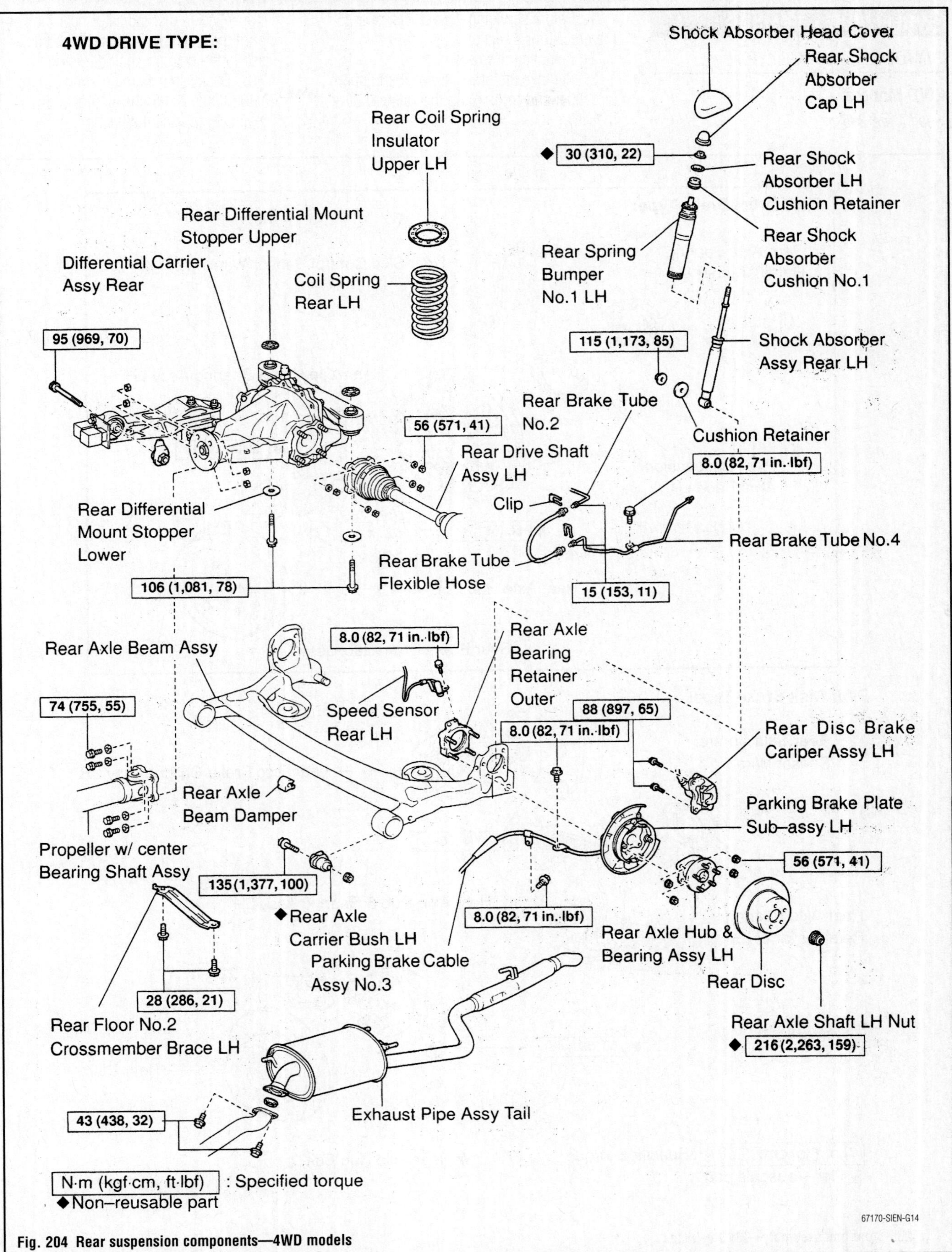
4WD DRIVE TYPE:
Shock Absorber Head Cover
Rear Shock Absorber Cap LH
Rear Coil Spring Insulator Upper LH
◆ 30 (310, 22)
Rear Shock Absorber LH Cushion Retainer
Rear Differential Mount Stopper Upper
Differential Carrier Assy Rear
Coil Spring Rear LH
Rear Spring Bumper No.1 LH
Rear Shock Absorber Cushion No.1
95 (969, 70)
115 (1,173, 85)
Shock Absorber Assy Rear LH
Rear Brake Tube No.2
56 (571, 41)
Cushion Retainer
Rear Drive Shaft Assy LH
8.0 (82, 71 in.·lbf)
Clip
Rear Differential Mount Stopper Lower
Rear Brake Tube No.4
Rear Brake Tube Flexible Hose
106 (1,081, 78)
15 (153, 11)
Rear Axle Bearing Retainer Outer
Rear Axle Beam Assy
8.0 (82, 71 in.·lbf)
Speed Sensor Rear LH
74 (755, 55)
88 (897, 65)
8.0 (82, 71 in.·lbf)
Rear Disc Brake Cariper Assy LH
Rear Axle Beam Damper
Parking Brake Plate Sub–assy LH
Propeller w/ center Bearing Shaft Assy
56 (571, 41)
135 (1,377, 100)
◆ Rear Axle Carrier Bush LH
8.0 (82, 71 in.·lbf)
Rear Axle Hub & Bearing Assy LH
Parking Brake Cable Assy No.3
Rear Disc
28 (286, 21)
Rear Floor No.2 Crossmember Brace LH
Rear Axle Shaft LH Nut
◆ 216 (2,263, 159)
43 (438, 32)
Exhaust Pipe Assy Tail
N·m (kgf·cm, ft·lbf) : Specified torque
◆ Non–reusable part
67170-SIEN-G14

Fig. 204 Rear suspension components—4WD models

WHEEL HUB & BEARING

REMOVAL & INSTALLATION

2WD Models

See Figure 205.

1. Before servicing the vehicle, refer to the precautions section.
2. Remove the wheel.
3. Remove the brake drum (if equipped) or caliper and rotor. Hang the caliper out of the way.

   a. Place a dial indicator near the center of the axle hub and check the backlash in the bearing shaft direction.

   b. Backlash maximum should read 0.0020 inch (0.05mm). If greater than specified, replace the bearing.

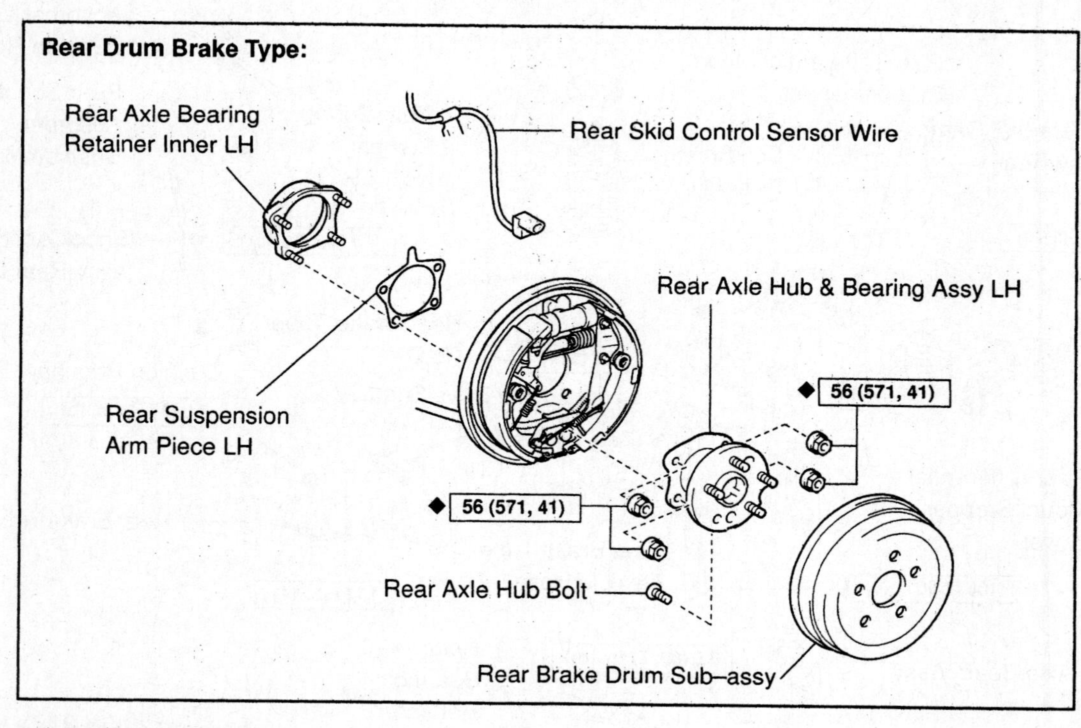

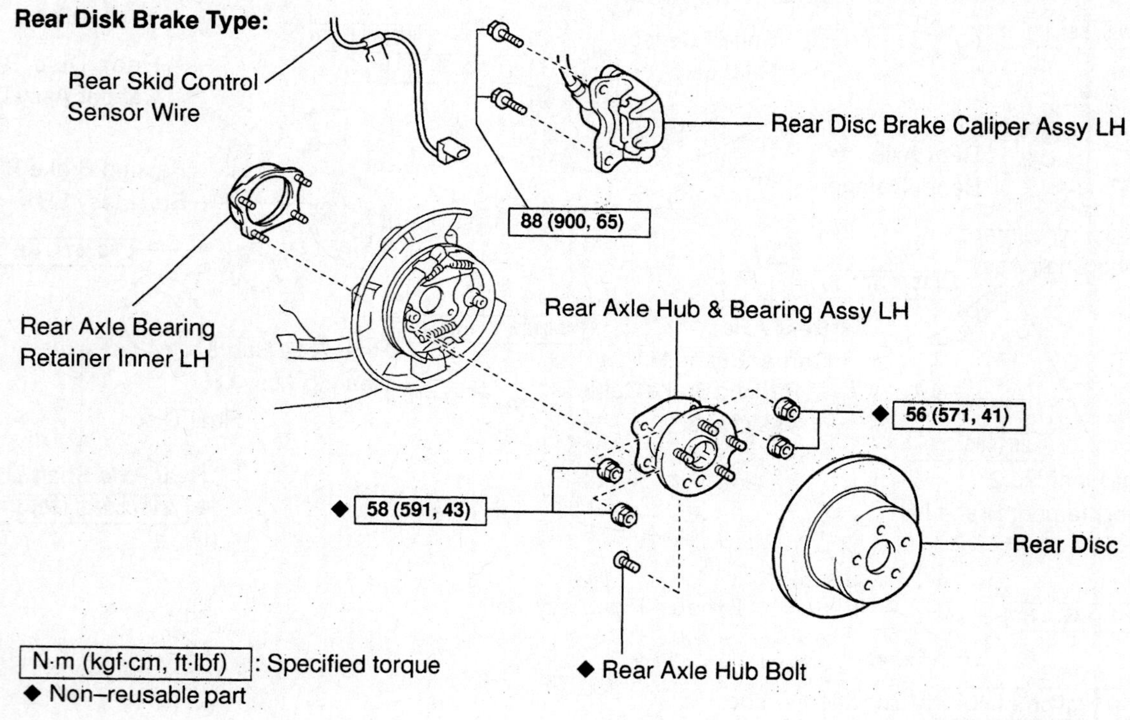

Fig. 205 Rear hub assembly—2WD vehicles

c. Using the dial indicator, check the deviation at the surface of the axle hub outside and hub bolt. Maximum is 0.0020 inch (0.05mm). If greater than specified, replace the axle hub.

4. Remove the ABS sensor wire.

5. Remove the 4 bolts and the hub/bearing assembly.

6. Installation is the reverse of removal. Torque the hub bolts to 41 ft. lbs. (56 Nm).

4WD Vehicles

See Figure 206.

1. Before servicing the vehicle, refer to the precautions section.

2. Remove the wheel.

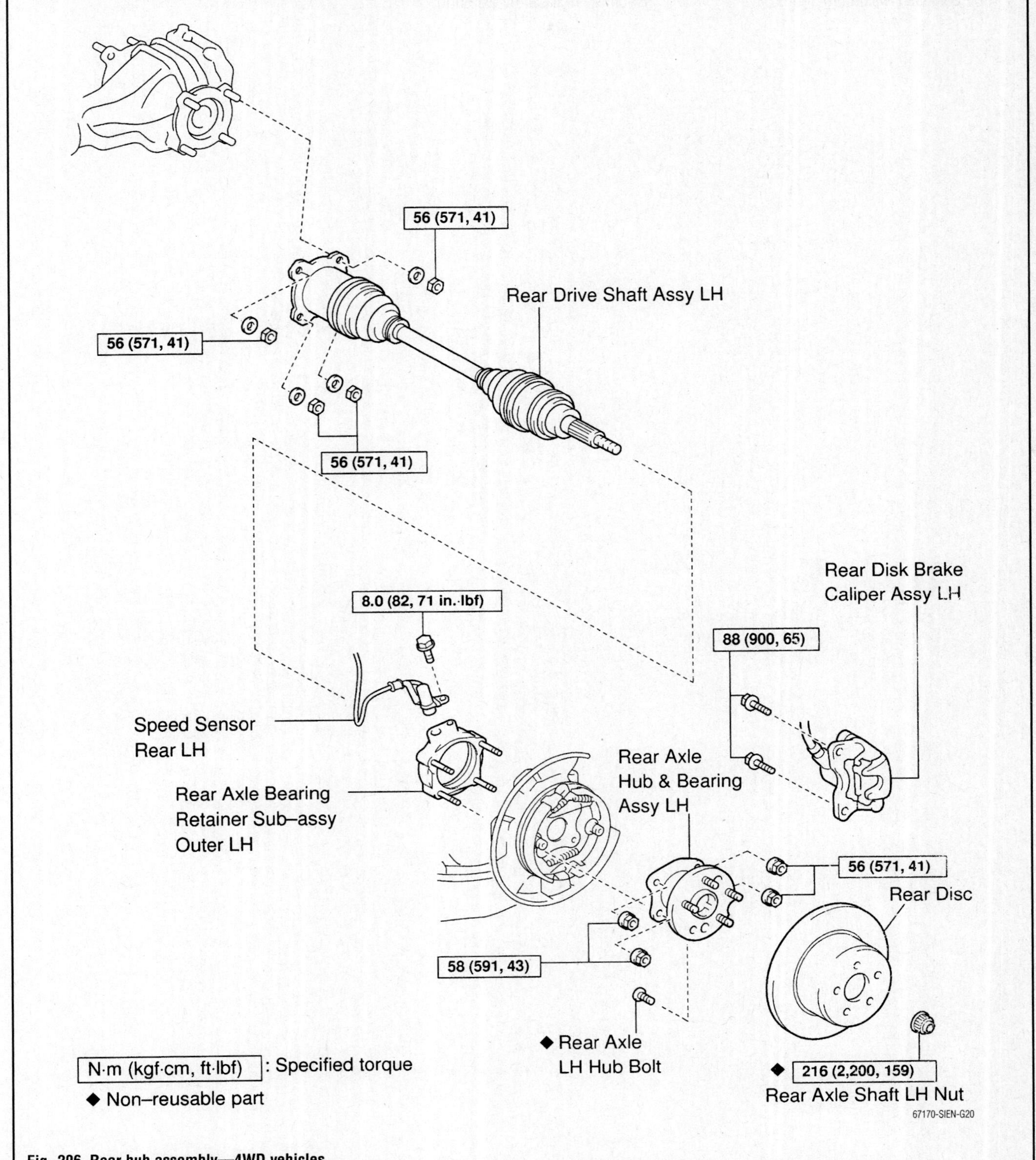

Fig. 206 Rear hub assembly—4WD vehicles

3. Unstake and remove the axle shaft nut.

4. Remove the caliper and rotor. Hang the caliper out of the way.

  a. Place a dial indicator near the center of the axle hub and check the backlash in the bearing shaft direction.

  b. Backlash maximum should read 0.0020 inch (0.05mm). If greater than specified, replace the bearing.

  c. Using the dial indicator, check the deviation at the surface of the axle hub outside and hub bolt. Maximum is 0.0020 inch (0.05mm). If greater than specified, replace the axle hub.

5. Remove the halfshaft.

6. Remove the ABS sensor wire.

7. Remove the 4 bolts and the hub/bearing assembly.

8. Installation is the reverse of removal. Torque the hub bolts to 41 ft. lbs. (56 Nm).

TOYOTA

Tacoma

17

SPECIFICATIONS AND MAINTENANCE CHARTS

ENGINE AND VEHICLE IDENTIFICATION

Engine							Model Year	
Code/ID	Liters (cc)	Cu. In.	Cyl.	Fuel Sys.	Engine Type	Eng. Mfg.	Code ①	Year
2TR-FE	2.7 (2693)	164	4	MFI	DOHC	Toyota	9	2009
1GR-FE	4.0 (3956)	241	6	MFI	DOHC	Toyota	A	2010

MFI: Multi-port Fuel Injection

DOHC: Double Overhead Camshaft

① 10th digit of the VIN number

3768X_TACO_C0001

GENERAL ENGINE SPECIFICATIONS

Year	Model	Engine Displacement Liters	Engine Series Code/ID	Net Horsepower @ rpm	Net Torque @ rpm (ft. lbs.)	Bore x Stroke (in.)	Com-pression Ratio	Oil Pressure @ rpm
2009	Tacoma	2.7	2TR-FE	159@5200	236@5200	3.74x3.74	NA	43-85@3000
		4.0	1GR-FE	236@3800	266@4000	3.70x3.74	NA	23-75@3000
2010	Tacoma	2.7	2TR-FE	159@5200	236@5200	3.74x3.74	NA	43-85@3000
		4.0	1GR-FE	236@3800	266@4000	3.70x3.74	NA	23-75@3000

NA: Not Available

3768X_TACO_C0002

ENGINE TUNE-UP SPECIFICATIONS

Year	Engine Displacement Liters	Engine Code/ID	Spark Plug Gap (in.)	Ignition Timing (deg.)	Fuel Pump (psi)	Idle Speed (rpm)		Valve Clearance	
						MT	AT	Intake	Exhaust
2009	2.7	2TR-FE	0.039-0.043	3-7B ①	40.8-41.7	600-700	600-700	②	②
	4.0	1GR-FE	0.039-0.043	7-24B ①	40.8-41.7	650-750	650-750	0.006-0.010	0.011-0.015
2010	2.7	2TR-FE	0.039-0.043	3-7B ①	40.8-41.7	600-700	600-700	②	②
	4.0	1GR-FE	0.039-0.043	7-24B ①	40.8-41.7	650-750	650-750	0.006-0.010	0.011-0.015

NOTE: The Vehicle Emission Control Information label often reflects specification changes made during production.

The label figures must be used if they differ from those in this chart.

B: Before top dead center

① With terminals TC and CG of the DLC3 connected, 2.7L engine. With Terminals TC and CG of DLC disconnected, 4.0L engine.

② Automatic adjustment

3768X_TACO_C0003

CAPACITIES

Year	Model	Engine Displacement Liters	Engine Code/ID	Engine Oil with Filter (qts.)	Transmission (pts.) 5-Spd	Transmission (pts.) Auto.	Transfer Case (pts.)	Drive Axle Front (pts.)	Drive Axle Rear (pts.)	Fuel Tank (gal.)	Cooling System (qts.)
2009	Tacoma	2.7	2TR-FE	6.1	①	②	2.2	3.2	③	21.1	④
		4.0	1GR-FE	⑤	3.8	②	2.2	3.2	③	21.1	⑥
2010	Tacoma	2.7	2TR-FE	6.1	①	②	2.2	3.2	⑦	21.1	④
		4.0	1GR-FE	⑤	3.8	②	2.2	3.2	⑦	21.1	⑥

NOTE: If specifications disagree with owners manual, use owners manual specifications.

① 2WD: 5.4
4WD: 4.6

② A340E: 4.2
A750E: 3.6

③ 2WD: 7
4WD: 6

④ MT: 9.2
AT: 9.1

⑤ 2WD, except PreRunner: 4.8
4WD and PreRunner: 5.5

⑥ MT: 10.3
AT: 10.4

⑦ 2WD: 7.3
4WD: 6.2

3768X_TACO_C0004

FLUID SPECIFICATIONS

Year	Model	Engine Displacement Liters	Engine ID/VIN	Engine Oil	Auto. Trans.	Manual Trans.	Power Steering Fluid	Brake Master Cylinder
2009	Tacoma	2.7	2TR-FE	5W-20	①	②	③	DOT 3
		4.0	1GR-FE	5W-30	①	②	③	DOT 3
2010	Tacoma	2.7	2TR-FE	5W-20	①	②	③	DOT 3
		4.0	1GR-FE	5W-30	①	②	③	DOT 3

DOT: Department Of Transportation

① 340E: ATF T-IV. A750E ATF WS

② API GL-4 or GL-5 75W-90

③ ATF Dexron II or III

3768X_TACO_C0014

VALVE SPECIFICATIONS

Year	Engine Displacement Liters	Engine Code/ID	Seat Angle (deg.)	Face Angle (deg.)	Spring Test Pressure (lbs. @ in.)	Spring Installed Height (in.)	Stem-to-Guide Clearance (in.)		Stem Diameter (in.)	
							Intake	Exhaust	Intake	Exhaust
2009	2.7	2TR-FE	45	NA	NA	1.9106	0.0010-0.0024	0.0012-0.0026	0.2154-0.2159	0.2151-0.2157
	4.0	1GR-FE	NA	44.5	41.9-46.3@ 1.311	1.882	0.0010-0.0024	0.0012-0.0026	0.2154-0.2159	0.2152-0.2158
2010	2.7	2TR-FE	45	NA	NA	1.9106	0.0010-0.0024	0.0012-0.0026	0.2154-0.2159	0.2151-0.2157
	4.0	1GR-FE	NA	44.5	41.9-46.3@ 1.311	1.882	0.0010-0.0024	0.0012-0.0026	0.2154-0.2159	0.2152-0.2158

NA: Not Available

3768X_TACO_C0005

CAMSHAFT SPECIFICATIONS

All measurements in inches unless noted

Year	Engine Displacement Liters	Engine Code/ID	Journal Dia.	Brg. Oil Clearance	Shaft End-play ①	Circle Runout	Lobe Height	
							Intake	Exhaust
2009	2.7	2TR-FE	②	③	0.0039-0.0090	0.0012	1.6872-1.6911	1.6872 1.6911
	4.0	1GR-FE	④	⑤	0.0160-0.0350	0.0024	⑥	⑥
2010	2.7	2TR-FE	②	③	0.0039-0.0090	0.0012	1.6872-1.6911	1.6872 1.6911
	4.0	1GR-FE	④	⑤	0.0160-0.0350	0.0024	⑥	⑥

NA: Not Available

① Thrust clearance

② No. 1: 1.4153-1.4159
All others: 1.0614-1.0620

③ No. 1: 0.0014-0.0029
All others: 0.0010-0.0024

④ No. 1: 1.4162-1.4167
All others: 0.9039-0.9045

⑤ No. 1: 0.0016-0.0031
All others: 0.0010-0.0024

⑥ No. 1 camshaft: 1.7389-1.7428
No. 2 camshaft: 1.7551-1.7591
No. 3 camshaft (sub assembly): 1.7389-1.7428
No. 2 camshaft (sub assembly): 1.7551-1.7591
No. 2 camshaft: 1.7551-1.7591

3768X_TACO_C0006

CRANKSHAFT AND CONNECTING ROD SPECIFICATIONS

All measurements are given in inches.

Year	Engine Displacement Liters	Engine Code/ID	Crankshaft Main Brg. Journal Dia.	Crankshaft Main Brg. Oil Clearance	Crankshaft Shaft End-play	Crankshaft Thrust on No.	Connecting Rod Journal Diameter	Connecting Rod Oil Clearance	Connecting Rod Side Clearance
2009	2.7	2TR-FE	①	②	0.0008-0.0087	NA	NA	0.0009-0.0019	0.0059-0.0138
	4.0	1GR-FE	2.8342-2.8346	0.0007-0.0012	NA	NA	NA	0.0010-0.0018	0.0059-0.0118
2010	2.7	2TR-FE	①	②	0.0008-0.0087	NA	NA	0.0009-0.0019	0.0059-0.0138
	4.0	1GR-FE	2.8342-2.8346	0.0007-0.0012	NA	NA	NA	0.0010-0.0018	0.0059-0.0118

NA: Not Available

① No. 3: 2.3615-2.3620
All others: 2.3619-2.3622

② No. 3: 0.0012-0.0022
All others: 0.0009-0.0019

3768X_TACO_C0007

PISTON AND RING SPECIFICATIONS

All measurements are given in inches.

Year	Engine Displacement Liters	Engine Code/ID	Piston Clearance	Ring Gap Top Compression	Ring Gap Bottom Compression	Ring Gap Oil Control	Ring Side Clearance Top Compression	Ring Side Clearance Bottom Compression	Ring Side Clearance Oil Control
2009	2.7	2TR-FE	0.0007-0.0020	0.0087-0.0134	0.0177-0.0224	0.0039-0.0157	0.0008-0.0030	0.0008-0.0026	0.0008-0.0028
	4.0	1GR-FE	0.0031-0.0040	0.0118-0.0157	0.0157-0.0197	0.0039-0.0157	0.0008-0.0028	0.0008-0.0024	0.0028-0.0060
2010	2.7	2TR-FE	0.0007-0.0020	0.0087-0.0134	0.0177-0.0224	0.0039-0.0157	0.0008-0.0030	0.0008-0.0026	0.0008-0.0028
	4.0	1GR-FE	0.0031-0.0040	0.0118-0.0157	0.0157-0.0197	0.0039-0.0157	0.0008-0.0028	0.0008-0.0024	0.0028-0.0060

3768X_TACO_C0008

TORQUE SPECIFICATIONS

All readings in ft. lbs.

Year	Engine Displacement Liters	Engine Code/ID	Cylinder Head Bolts	Main Bearing Bolts	Rod Bearing Bolts	Crankshaft Damper Bolts	Flywheel Bolts	Manifold Intake	Manifold Exhaust	Spark Plugs	Oil Pan Drain Plug
2009	2.7	2TR-FE	①	②	③	192	④	18	27	13	28
	4.0	1GR-FE	⑤	⑥	③	185	61	19	16	15	30
2010	2.7	2TR-FE	①	②	③	192	④	18	27	13	28
	4.0	1GR-FE	⑤	⑥	③	185	61	19	16	15	30

① Step 1: 29 ft. lbs.
Step 2: Plus 90 degrees
Step 3: Plus 90 degrees

② Step 1: 29 ft. lbs.
Step 2: Plus 90 degrees

③ Step 1: 18 ft. lbs.
Step 2: Plus 90 degrees

④ AT: 55 ft. lbs.
MT: 20 ft. lbs. plus 90 degrees

⑤ Step 1: 27 ft. lbs.
Step 2: Plus 180 degrees
Left side 14mm bolt: 22 ft. lbs.

⑥ 12 pointed head: 45 ft. lbs. Then plus 90 degrees
12mm head: 18 ft. lbs.

3768X_TACO_C0009

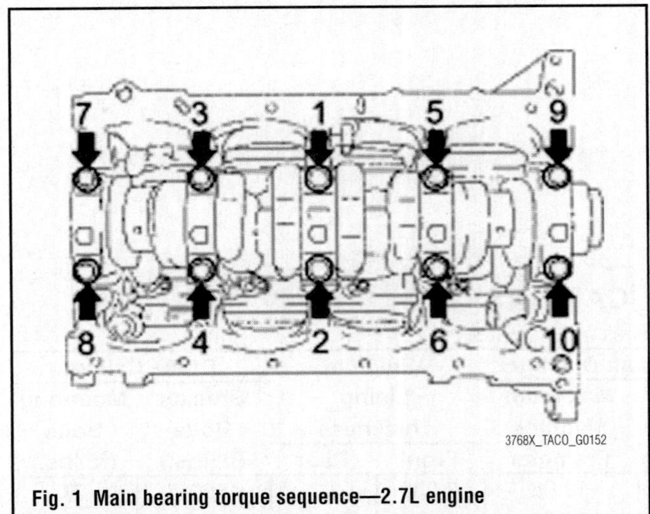

Fig. 1 Main bearing torque sequence—2.7L engine

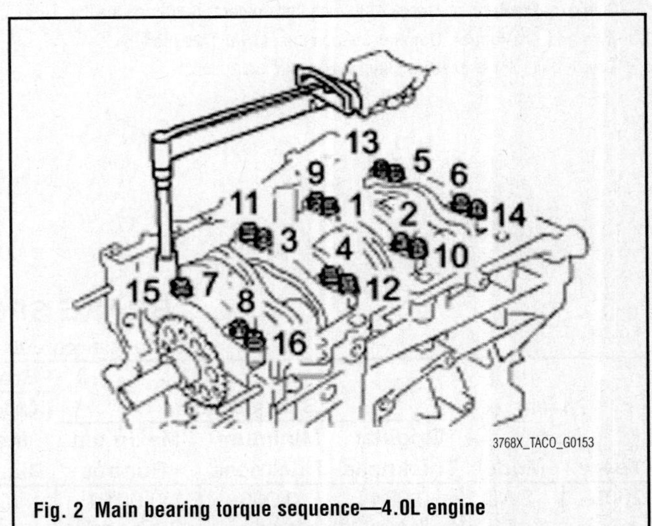

Fig. 2 Main bearing torque sequence—4.0L engine

WHEEL ALIGNMENT

Year	Model	Caster Range (+/-Deg.)	Caster Preferred Setting (Deg.)	Camber Range (+/-Deg.)	Camber Preferred Setting (Deg.)	Toe-in (in.)	Steering Axis Inclination (Deg.)
2009	2WD exc. PreRunner	0.75	+0.67	0.75	0	0.06+/-0.08	10.00
	4WD & PreRunner	0.75	0.30	0.75	+1.62	0.06+/-0.08	10.40
2010	2WD exc. PreRunner	0.75	+0.67	0.75	0	0.06+/-0.08	10.00
	4WD & PreRunner	0.75	0.30	0.75	+1.62	0.06+/-0.08	10.40

All alignment figures based on nominal ride height and standard tires

3768X_TACO_C0010

TIRE, WHEEL AND BALL JOINT SPECIFICATIONS

Year	Model	OEM Tires		Tire Pressures (psi)		Wheel	Ball Joint	Lug Nut Torque
		Standard	Optional	Front	Rear	Size	Inspection ①	(ft. lbs.)
2009	Tacoma	P215/70R15	None	30	33	NA	②	83
		P245/75R16	None	30	30	NA	②	83
		P265/70R16	None	29	32	NA	②	83
		P265/65R17	None	29	29	NA	②	83
		P255/45R18	None	35	35	NA	②	83
2010	Tacoma	P215/70R15	None	30	33	NA	②	83
		P245/75R16	None	30	30	NA	②	83
		P265/70R16	None	29	32	NA	②	83
		P265/65R17	None	29	29	NA	②	83
		P255/45R18	None	35	35	NA	②	83

OEM: Original Equipment Manufacturer

PSI: Pounds Per Square Inch

NOTE: If specification differs from driver's door placard, use data on placard.

NA: Information not available

① Torque required in inch lbs. to rotate ball joint when removed from the knuckle

② 2wd exc. PreRunner: Upper 4-30 inch lbs.; lower 0.8-30 inch lbs.
4wd and PreRunner: Upper 6-39 inch lbs.; lower 1-22 inch lbs.
Lower ball joint excessive play, all models: 0.020 inch

3768X_TACO_C0011

BRAKE SPECIFICATIONS

All measurements in inches unless noted

Year	Model	Brake Disc			Brake Drum Diameter		Minimum Lining Thickness		Brake Caliper	
		Original Thickness	Minimum Thickness	Maximum Runout	Original Inside Diameter	Maximum Machine Diameter	Front	Rear	Bracket Bolts (ft. lbs.)	Mounting Bolts (ft. lbs.)
2009	2WD	0.984	0.906	0.0020	10.00	10.08	0.039	—	80	27
	①	1.102	1.024	0.0020	10.00	10.08	—	0.039	91	—
2010	2WD	0.984	0.906	0.0020	10.00	10.08	0.039	—	80	27
	①	1.102	1.024	0.0020	10.00	10.08	—	0.039	91	—

① 4WD and PreRunner

3768X_TACO_C0012

SCHEDULED MAINTENANCE INTERVALS

TOYOTA—TACOMA

TO BE SERVICED	TYPE OF	VEHICLE MILEAGE INTERVAL (x1000)																		
		5	10	15	20	25	30	35	40	45	50	55	60	65	70	75	80	85	90	95
Air filter	R												✓							
Air filter (cabin)	R				✓				✓				✓				✓			
Automatic transmission and differential fluid	S/I			✓			✓			✓			✓			✓			✓	
Ball joints and boots	S/I			✓			✓			✓			✓			✓			✓	
Brake linings, discs/drums, lines & hoses	S/I	✓	✓	✓	✓	✓	✓	✓	✓	✓	✓	✓	✓	✓	✓	✓	✓	✓	✓	✓
Charcoal canister	S/I												✓							
Drive belts	S/I						✓						✓						✓	
Driveshaft bushing (4WD)	L						✓						✓						✓	
Engine coolant	R												✓						✓	
Engine oil & filter	R	✓	✓	✓	✓	✓	✓	✓	✓	✓	✓	✓	✓	✓	✓	✓	✓	✓	✓	✓
Exhaust pipes & mounts	S/I			✓			✓			✓			✓			✓			✓	
Fuel lines & connections, fuel tank vapor vent system hoses, fuel tank band	S/I						✓						✓						✓	
Fuel tank cap gasket	S/I						✓						✓						✓	
Halfshaft boots & flange bolts	S/I			✓			✓			✓			✓			✓			✓	
Limited slip differential fluid	R						✓						✓						✓	
Manual transmission and differential fluid	S/I						✓						✓						✓	
Non-platinum spark plugs	R						✓						✓						✓	
Platinum spark plugs	R																			
Propeller shaft (4WD)	L			✓			✓			✓			✓			✓			✓	
Propeller shaft bolts	S/I			✓			✓			✓			✓			✓			✓	
Rack and pinion assembly	S/I			✓			✓			✓			✓			✓			✓	
Rear wheel bearing	L						✓						✓						✓	
Rotate tires	S/I	✓	✓	✓	✓	✓	✓	✓	✓	✓	✓	✓	✓	✓	✓	✓	✓	✓	✓	✓
Steering linkage	S/I			✓			✓			✓			✓			✓			✓	
Valves	S/I												✓							

R: Replace S/I: Service or Inspect L: Lubricate

FREQUENT OPERATION MAINTENANCE (SEVERE SERVICE)

If a vehicle is operated under any of the following conditions it is considered severe service:

- Towing a trailer or using a camper or car-top carrier.

- Repeated short trips of less than 5 miles in temperatures below freezing.

- Excessive idling or low-speed driving for long distances as in heavy commercial use, such as delivery, taxi or police cars.

- Operating on rough, muddy or salt-covered roads.

- Operating on unpaved or dusty roads.

Oil filter: service or inspect every 5000 miles or 4 months, whichever occurs first.

Brake linings and discs or drums: service or inspect every 5000 miles or 4 months, whichever occurs first.

Steering linkage: service or inspect every 5000 miles or 4 months, whichever occurs first.

Ball joints and boots: service or inspect every 5000 miles or 4 months, whichever occurs first.

Brake discs & pads (front): service or inspect every 5000 miles.

Halfshaft boots: service or inspect every 5000 miles or 4 months. Retighten the flange bolts, whichever occurs first.

Body chassis bolts and nuts: service or inspect every 5000 miles or 4 months, whichever occurs first.

Transmission and differential fluid: replace every 60,000 miles or 72 months, whichever occurs first.

Transfer case and differential fluid: replace every 15,000 miles or 18 months, whichever occurs first.

Timing belt: replace every 60,000 miles or 48 months, whichever occurs first.

Lubricate driveshaft 4WD every 5,000 miles

Retorque driveshaft bolts 4WD every 5,000 miles

Inspect engine air filter at 15,000 miles or 18 months, than every 5,000 miles

3768X_TACO_C0013

BRAKES — INFORMATION AND PRECAUTIONS

ANTI-LOCK SYSTEMS

- Certain components within the ABS system are not intended to be serviced or repaired individually.
- Do not use rubber hoses or other parts not specifically specified for and ABS system. When using repair kits, replace all parts included in the kit. Partial or incorrect repair may lead to functional problems and require the replacement of components.
- Lubricate rubber parts with clean, fresh brake fluid to ease assembly. Do not use shop air to clean parts; damage to rubber components may result.
- Use only DOT 3 brake fluid from an unopened container.
- If any hydraulic component or line is removed or replaced, it may be necessary to bleed the entire system.
- A clean repair area is essential. Always clean the reservoir and cap thoroughly before removing the cap. The slightest amount of dirt in the fluid may plug an orifice and impair the system function. Perform repairs after components have been thoroughly cleaned; use only denatured alcohol to clean components. Do not allow ABS components to come into contact with any substance containing mineral oil; this includes used shop rags.
- The Anti-Lock control unit is a microprocessor similar to other computer units in the vehicle. Ensure that the ignition switch is **OFF** before removing or installing controller harnesses. Avoid static electricity discharge at or near the controller.
- If any arc welding is to be done on the vehicle, the control unit should be unplugged before welding operations begin.

DISC AND DRUM SYSTEMS

CAUTION

Dust and dirt accumulating on brake parts during normal use may contain asbestos fibers from production or aftermarket brake linings. Breathing excessive concentrations of asbestos fibers can cause serious bodily harm. Exercise care when servicing brake parts. Do not sand or grind brake lining unless equipment used is designed to contain the dust residue. Do not clean brake parts with compressed air or by dry brushing. Cleaning should be done by dampening the brake components with a fine mist of water, then wiping the brake components clean with a dampened cloth. Dispose of cloth and all residue containing asbestos fibers in an impermeable container with the appropriate label. Follow practices prescribed by the Occupational Safety and Health Administration (OSHA) and the Environmental Protection Agency (EPA) for the handling, processing, and disposing of dust or debris that may contain asbestos fibers.

BRAKES — BLEEDING THE BRAKE SYSTEM

BLEEDING PROCEDURE

HYDRAULIC BRAKE BOOSTER

Booster With Accumulator Assembly

➡Immediately wash off any brake fluid that comes into contact with any painted surfaces. Depressing the brake pedal with the reservoir cap removed will cause the fluid to spray. When bleeding, maintain the amount of fluid in the reservoir between the Min. and Max. lines.

1. Before servicing the vehicle, refer to the Precautions section.
2. Fill the reservoir with the proper grade and type brake fluid.
3. Turn the ignition switch to ON, and wait until the pump motor has stopped.
4. Turn the ignition switch OFF, depress the brake pedal more than twenty times.
5. Repeat the above steps about five times.
6. Turn the ignition switch ON, and check that the pump stops after approximately 8–14 seconds. If the pump does not stop, repeat the above step.

Master Cylinder Solenoid

➡Immediately wash off any brake fluid that comes into contact with any painted surfaces. Depressing the brake pedal with the reservoir cap removed will cause the fluid to spray. When bleeding, maintain the amount of fluid in the reservoir between the Min. and Max. lines.

1. Before servicing the vehicle, refer to the Precautions section.
2. Connect the Techstream diagnostic tool to the DLC3.
3. Turn the ignition switch to the ON position.
4. Select "Active Test" mode on the tool.
5. Connect the vinyl tube to the wheel cylinder.
6. Loosen the bleeder plug.
7. Select "TRAC Solenoid (SRMF&SRMR) to drive the solenoids and bleed air from the wheel cylinder.
8. Repeat the above step until all air is bled out.
9. When air is completely bled out of the brake fluid thru the bleeder plug, tighten the plug.
10. Repeat the above on the other line.
11. Turn the ignition switch OFF.
12. Turn the ignition switch ON.
13. Clear the DTC codes, as required.

Front Brake Line

➡Immediately wash off any brake fluid that comes into contact with any painted surfaces. Depressing the brake pedal with the reservoir cap removed will cause the fluid to spray. When bleeding, maintain the amount of fluid in the reservoir between the Min. and Max. lines.

1. Before servicing the vehicle, refer to the Precautions section.
2. Turn the ignition switch ON and wait until the pump motor stops.
3. Connect the vinyl tube to the brake caliper.
4. Depress the brake pedal several times, then loosen the bleeder plug with the pedal held down.
5. At the point when the fluid stops coming out, tighten the bleeder plug, then release the brake pedal.
6. Repeat until all the air in the fluid has been bled out.
7. Repeat the above procedures to bleed the other brake line.

Rear Brake Line

➡Immediately wash off any brake fluid that comes into contact with any painted surfaces. Depressing the brake pedal with the reservoir cap removed will cause the fluid to spray. When bleeding, maintain the amount of fluid in the reservoir between the Min. and Max. lines.

1. Before servicing the vehicle, refer to the Precautions section.
2. Connect the vinyl tube to the wheel cylinder.
3. Depress the brake pedal, hold it, and then loosen the bleeder plug.
4. Loosen the bleeder plug and release air.
5. When the air is completely bled out of the brake fluid through the bleeder plug, tighten the bleeder plug.
6. Repeat the above procedures to bleed the other brake line.

VACUUM BRAKE BOOSTER

Master Cylinder

➡Immediately wash off any brake fluid that comes into contact with any painted surfaces. Depressing the brake pedal with the reservoir cap removed will cause the fluid to spray. When bleeding, maintain the amount of fluid in the reservoir between the Min. and Max. lines.

1. Before servicing the vehicle, refer to the Precautions section.
2. Fill reservoir with dot3 brake fluid.
3. Using SST 09023-00101, disconnect the brake lines from the master cylinder.
4. Slowly depress the brake pedal and hold it there.
5. Block the outer holes with your fingers, and release the brake pedal.
6. Repeat 3 or 4 times.

Brake Lines

➡Immediately wash off any brake fluid that comes into contact with any painted surfaces. Depressing the brake pedal with the reservoir cap removed will cause the fluid to spray. When bleeding, maintain the amount of fluid in the reservoir between the Min. and Max. lines.

1. Before servicing the vehicle, refer to the Precautions section.
2. Connect the vinyl tube to the bleeder plug.
3. Depress the brake pedal several times, then loosen the bleeder plug with the pedal held down.
4. At the point where the fluid stops coming out, tighten the bleeder plug, then release the brake pedal.
5. Repeat until all the air in the fluid has been bled out.
6. Repeat the above procedure to bleed the air out of the brake line for each wheel.
7. Check the fluid level and add fluid if necessary.

BRAKES — ANTI-LOCK BRAKE SYSTEM (ABS)

WHEEL SPEED SENSORS

REMOVAL & INSTALLATION

Front

See Figure 3.

1. Before servicing the vehicle, refer to the Precautions Section.

➡If working near and/or around the SRS system and components, be sure to disable the SRS system. Tape the negative battery cable with insulating tape. Always disconnect the negative battery cable first.

⁂ CAUTION

To avoid personal injury when working on vehicles equipped with an airbag, the negative battery cable must be disconnected and at least 90 seconds must elapse before working on the system. Failure to do so may result in deployment of the airbag.

2. Disconnect the negative battery cable. Tape the cable with insulating tape.
3. Raise and safely support the vehicle.
4. Remove front wheel.
5. Disconnect the speed sensor connector.
6. Remove the bolt and front speed sensor.

➡Do not attach any foreign matter to the sensor tip. Ensure that no foreign matter enters the sensor installation part.

To install:

➡Be sure to use new fasteners, as required.

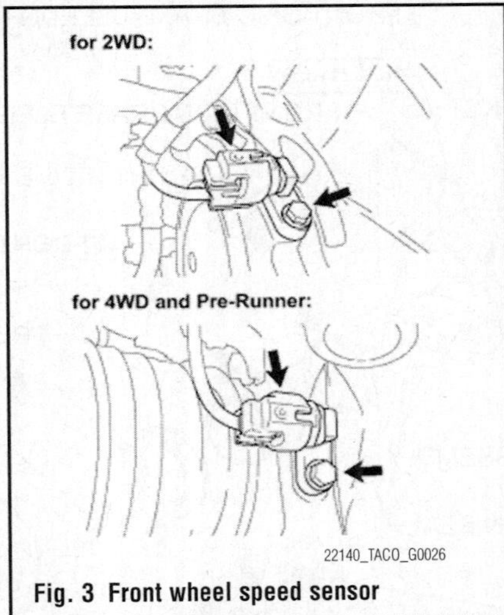

Fig. 3 Front wheel speed sensor

7. Installation is the reverse of the removal procedure.
8. Tighten the bolt to 73 inch lbs. (8.3 Nm).

Rear

See Figure 4.

1. Before servicing the vehicle, refer to the Precautions Section.

➡If working near and/or around the SRS system and components, be sure to disable the SRS system. Tape the negative battery cable with insulating tape. Always disconnect the negative battery cable first.

⁂ CAUTION

To avoid personal injury when working on vehicles equipped with an airbag, the negative battery cable must be disconnected and at least 90 seconds must elapse before working on the system. Failure to do so may result in deployment of the airbag.

2. Disconnect the negative battery cable. Tape the cable with insulating tape.
3. Raise and safely support the vehicle.
4. Remove rear wheel.
5. Disconnect the speed sensor connector.
6. Remove the nut and rear speed sensor.

➡Do not attach any foreign matter to the sensor tip. Ensure that no foreign matter enters the sensor installation part.

To install:

➡Be sure to use new fasteners, as required.

7. Installation is the reverse of the removal procedure.
8. Tighten the nut to 71 inch lbs. (8 Nm).

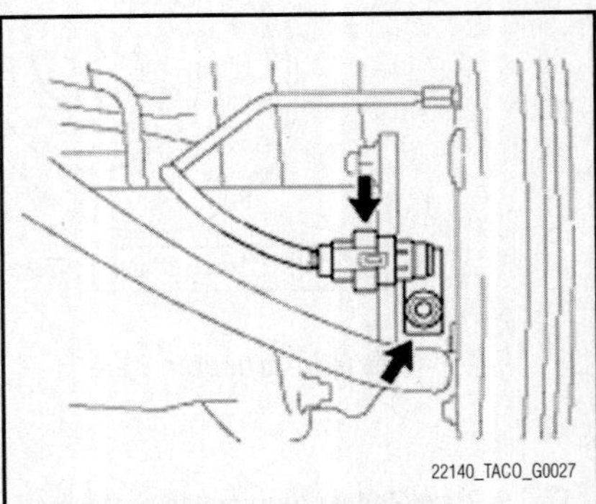

Fig. 4 Rear wheel speed sensor

BRAKE CALIPER

REMOVAL & INSTALLATION

See Figures 5 and 6.

1. Before servicing the vehicle, refer to the Precautions Section.
2. Disconnect the negative battery cable from the battery.
3. Raise and support the vehicle safely.
4. Remove the wheels.
5. Disconnect the brake hose from the caliper by removing the union bolt and 2 gaskets. Plug the end of the hose to prevent loss of fluid.
6. Remove the bolts that attach the caliper to its mounting.
7. Lift the bottom of the caliper up and remove the caliper assembly.

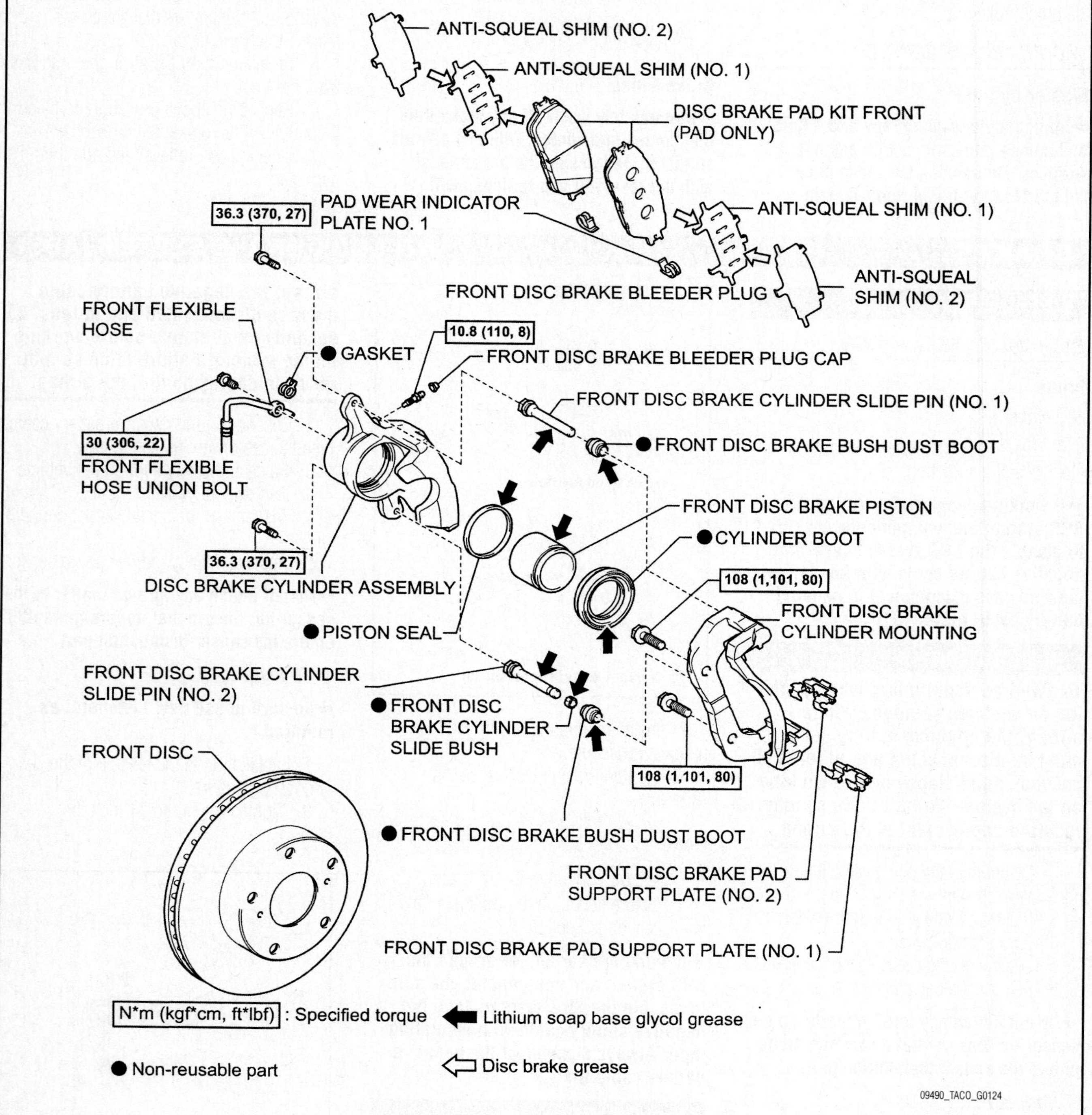

Fig. 5 Brake caliper and related components—2WD

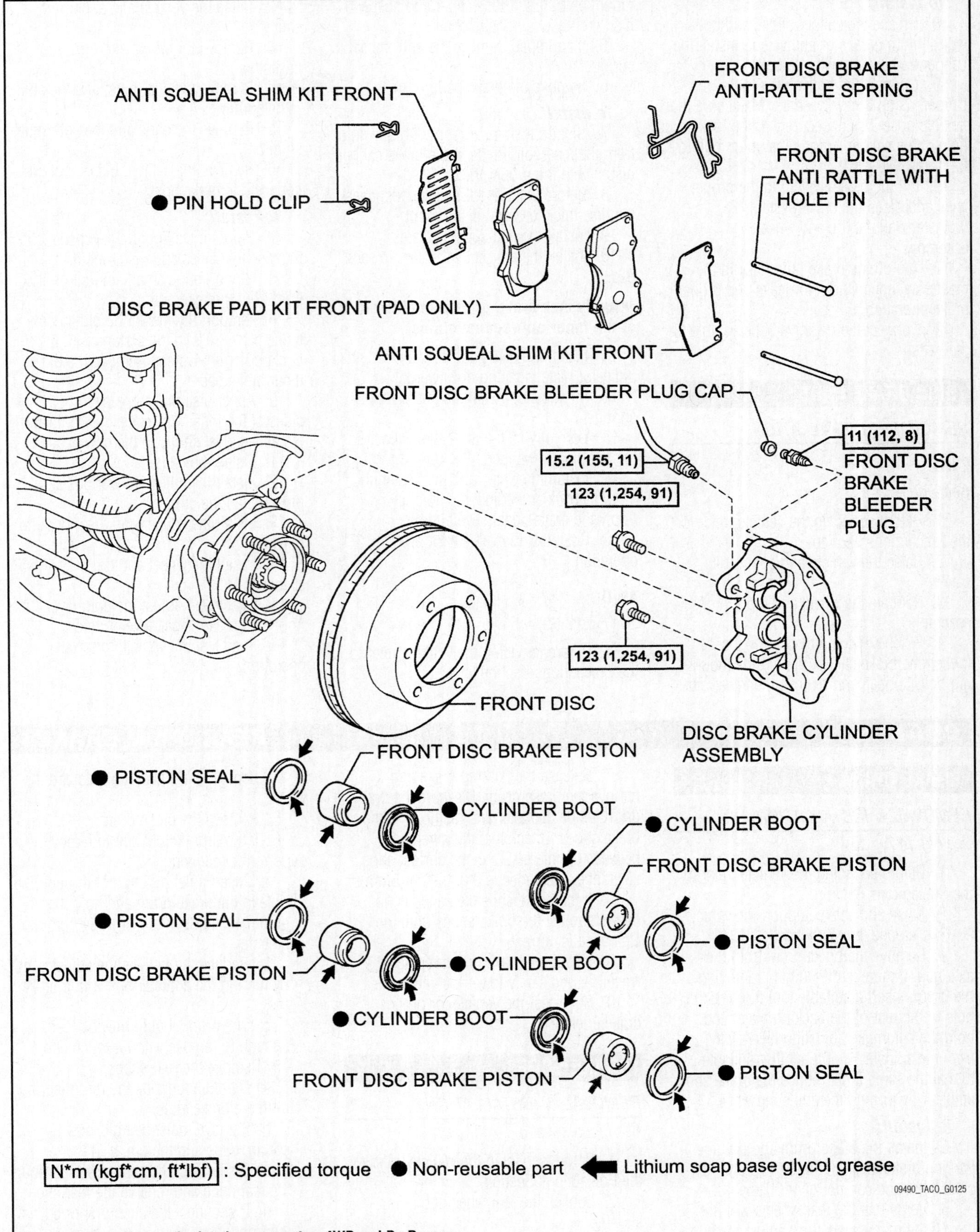

Fig. 6 Brake caliper and related components—4WD and PreRunner

To install:

8. Grease the caliper slides and bolts with lithium grease or equivalent. Install the caliper and secure with the bolts.
9. Connect the brake hose to the caliper, using 2 new washers. Make sure the flexible hose lock is securely in the lock hole of the caliper. Torque the union bolt to 22 ft. lbs. (30 Nm).
10. Fill the brake system to the proper level and bleed the brake system.
11. Install the tire and wheel assembly.
12. Top off the brake fluid level in the master cylinder. Check for leaks and proper brake operation.
13. Connect the negative battery cable to the battery.

DISC BRAKE PADS

REMOVAL & INSTALLATION

2WD

See Figure 5.

1. Before servicing the vehicle, refer to the Precautions Section.
2. Raise the vehicle and support it safely.
3. Remove the wheel and tire assembly.
4. When servicing the front pads, loosen the brake caliper upper side mounting bolt. Loosen and remove the lower side mounting bolt. Lift the caliper and suspend it so the hose is not stretched.
5. If equipped, remove the anti-squeal spring.
6. Remove the brake pads.

To install:

7. Siphon a small amount of brake fluid from the reservoir. Press in the brake caliper piston with the proper tool.
8. Before installing the new pads, check the disc thickness and disc runout.
9. Install the pad support plates.
10. Install the anti-squeal shims to each pad.

➡Apply disc brake grease to both sides of the inner anti-squeal shims.

11. Install the disc pads so the wear indicator plate is facing downward.
12. If removed, install the anti-squeal springs.
13. Carefully install the brake caliper so the boot is not wedged.
14. Install the wheel and tire assembly.
15. Check and adjust the fluid level. Apply the brake pedal several times.
16. Road test the vehicle for proper operation.

4WD

See Figure 6.

1. Before servicing the vehicle, refer to the Precautions Section.
2. Raise the vehicle and support it safely.
3. Remove the wheel and tire assembly.
4. Remove the clip, pins, and the anti-rattle spring.
5. Remove the pads and the anti-squeal shims.
6. Remove the caliper, but do not disconnect the brake hose.

To install:

7. Before installing the new pads, check the disc thickness and disc runout.
8. Siphon out a small amount of brake fluid from the reservoir.
9. Temporarily install the old inner brake pad. Press in the pistons with a C-clamp or equivalent. Remove the old inner brake pad.
10. Apply disc brake grease to both sides of the inner anti-squeal shim. Install the anti-squeal shims to the new pads.
11. Install the pads.
12. Install the anti-rattle springs and pins. Install the clip.
13. Install the caliper and the mounting bolts.
14. Install the wheel and tire assembly.
15. Check and adjust the fluid level. Apply the brake pedal several times.
16. Road test the vehicle for proper operation.

BRAKES — REAR DRUM BRAKES

BRAKE DRUM

REMOVAL & INSTALLATION

See Figure 7.

1. Before servicing the vehicle, refer to the Precautions Section.
2. Raise and safely support the vehicle.
3. Remove the rear wheel(s).
4. Remove the brake drum from the axle hub. If there is difficulty in removing the drum, insert a suitable tool through the hole in the rear of the backing plate, and hold the automatic adjusting lever away from the adjuster. Using another suitable tool at the same time, reduce the brake shoe adjuster by turning the adjusting wheel.

To install:

5. Install the brake drum and pull the parking brake lever all the way up until a clicking sound can no longer be heard.
6. Verify that the rear wheels will not turn. If the rear wheels turn, adjust the parking brake cable as necessary.
7. Release the parking brake and remove the brake drum. Measure the brake drum inside diameter and diameter of the brake shoes. Check that the difference between the diameters is the correct shoe clearance. Clearance is 0.020 inch (5mm).
8. If the brake shoe clearance is not correct, adjust the brake shoes until the clearance is correct.
9. Install the brake drum, replace the wheel(s), and safely lower the vehicle.
10. Road-test the vehicle for proper brake operation.

BRAKE SHOES

REMOVAL & INSTALLATION

See Figures 7 and 8.

1. Before servicing the vehicle, refer to the Precautions Section.
2. Loosen the rear wheel lug nuts slightly.
3. Raise and support the vehicle safely.
4. Remove the wheel lug nuts and the wheel.
5. Remove the brake drum.
6. If the drum is difficult to remove, perform the following:
   a. Insert a flat prying tool through the hole in the brake drum and hold the automatic adjusting lever away from the adjuster.
   b. Reduce the brake shoe adjustment by turning the adjuster bolt with a brake tool.
   c. The drum should now be loose enough to remove without much effort.
7. Remove the rear shoe.
   a. Carefully unhook the return spring from the brake shoe.
   b. Remove the shoe hold-down spring, cups and the pin.
   c. Disconnect the anchor spring from the rear shoe and remove the rear shoe.
   d. Disconnect the anchor spring from the front shoe.
8. Remove the front shoe.

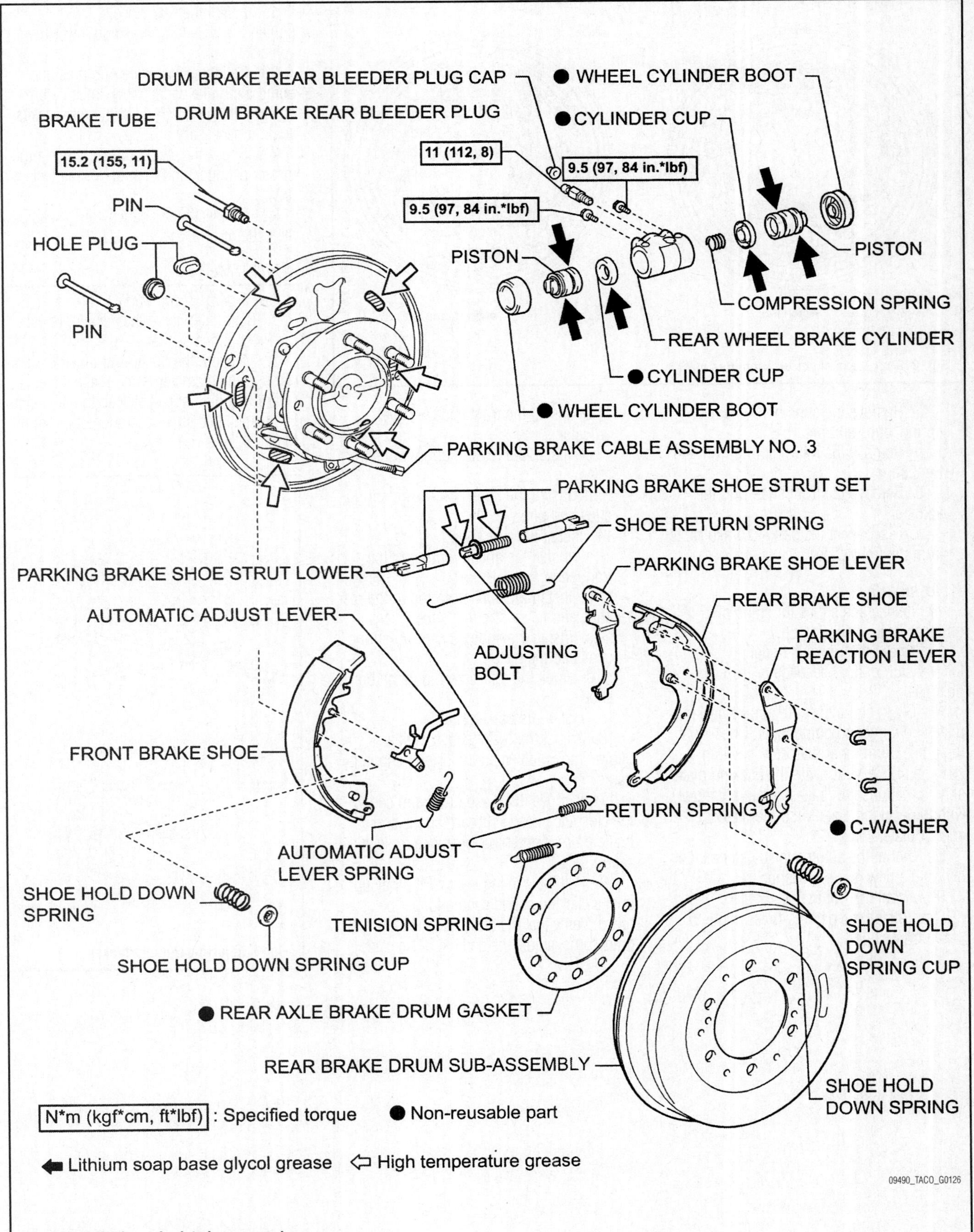

Fig. 7 Rear brake and related components

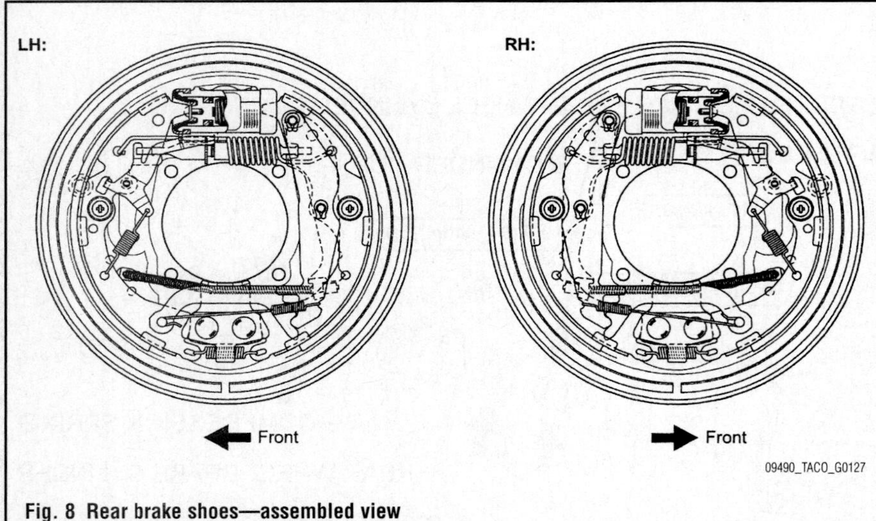

Fig. 8 Rear brake shoes—assembled view

a. Remove the shoe hold-down spring, cups and pin.

b. Remove the return spring from the front shoe.

c. Remove the front shoe with the adjuster.

d. Disconnect the parking brake cable from the front shoe.

To install:

9. Inspect the shoes for signs of unusual wear or scoring.

10. Check the wheel cylinder for any sign of fluid seepage or frozen pistons.

11. Clean and inspect the brake backing plate and all other components. Check that the brake drum inner diameter is within specified limits. Lubricate the backing plate at the positions the brakes come in contact with the backing plate. Also lubricate the anchor plate.

12. Mount the automatic adjuster assembly onto a new rear brake shoe.

13. Install the front shoe.

a. Install the parking brake cable to the front shoe.

b. Install the front shoe with the adjuster.

c. Install the return spring to the front shoe.

d. Install the shoe hold-down spring, cups and pin.

14. Install the rear shoe.

a. Install the anchor spring to the front shoe.

b. Install the anchor spring to the rear shoe and install the rear shoe.

c. Install the shoe hold-down spring, cups and the pin.

d. Hook the return spring to the brake shoe.

15. Install the brake drum.

16. Adjust the brake shoes until a slight drag is felt when the drum is spun by hand.

17. Remove the brake drum and check the clearance between brake shoes and brake drum. Adjust the clearance to specification.

18. Pull the parking lever all the way up until a clicking sound can no longer be heard. Verify that the drum doesn't turn. If the drum turns, adjust the parking brake cable.

19. Install the rear wheels, tighten the wheel lug nuts and lower the vehicle.

20. Retighten the wheel lug nuts and pump the brake pedal a few times before moving the vehicle. Adjust the rear brakes again if necessary.

21. Check the level of brake fluid in the master cylinder, and then perform a test drive.

22. Connect the negative battery cable to the battery.

ADJUSTMENT

See Figures 9 and 000.

1. Before servicing the vehicle, refer to the Precautions section.

2. Measure the brake drum inside diameter and the diameter of the brake shoes. Check that difference between the diameters is the specified shoe clearance of 0.5 mm (0.020 in.).

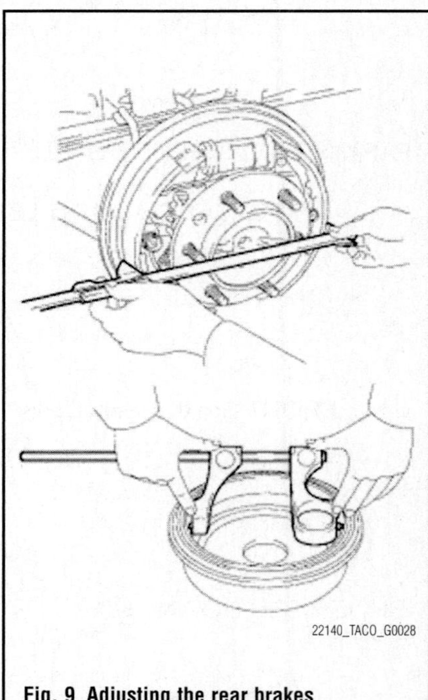

Fig. 9 Adjusting the rear brakes

PARKING BRAKE CABLES

ADJUSTMENT

Cable Adjustment

1. Before servicing the vehicle, refer to the Precautions section.
2. Turn the adjuster and expand the shoes until the disc locks.
3. Return the adjuster 8 notches.
4. Depress the parking brake pedal with 147 N (33 lbs).
5. Drive the vehicle at about 50 km/h (31 MPH) on a safe, level and dry road for about 400 meters (0.25 mile) in this condition.
6. Repeat this procedure 2 or 3 times.

PARKING BRAKE SHOES

REMOVAL & INSTALLATION

The rear drum brake shoes serve as the parking brakes.

GENERAL INFORMATION

All vehicles are equipped with an SRS system. The system must be disarmed before performing service on, or around, system components, the steering column, instrument panel components, wiring and sensors. Failure to follow the safety precautions and the disarming procedure could result in accidental airbag deployment, possible injury and unnecessary system repairs.

SERVICE PRECAUTIONS

**** CAUTION**

Disconnect and isolate the battery negative cable before beginning any airbag system component diagnosis, testing, removal, or installation procedures. Allow system capacitor to discharge for two minutes before beginning any component service. This will disable the airbag system. Failure to disable the airbag system may result in accidental airbag deployment, personal injury, or death.

DISARMING THE SYSTEM

➡If working near and/or around the SRS system and components, be sure to disable the SRS system. Tape the negative battery cable with insulating tape. Always disconnect the negative battery cable first.

**** CAUTION**

To avoid personal injury when working on vehicles equipped with an airbag, the negative battery cable must be disconnected and at least 90 seconds must elapse before working on the system. Failure to do so may result in deployment of the airbag.

ARMING THE SYSTEM

Reconnect the negative battery cable. Wait 2 minutes before performing any service.

CLOCKSPRING CENTERING

See Figure 10.

When installing the spiral cable, check that the ignition switch is in the "OFF" position. Turn the cable counterclockwise by hand until it becomes hard to turn. Turn the cable clockwise about 2 1/2 turns to align the marks.

➡The cable will rotate about 2½ turns both left and right from the center.

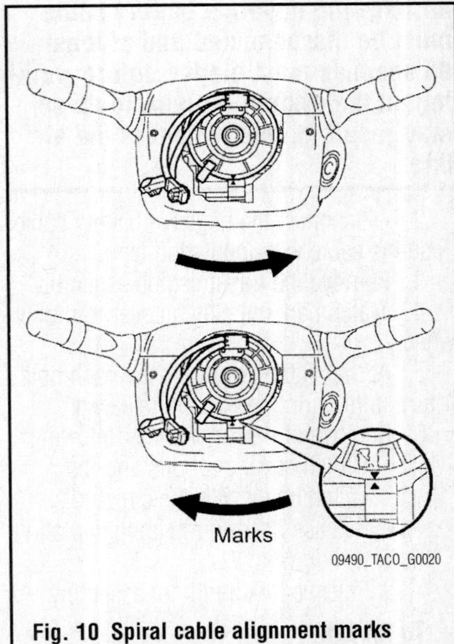

Fig. 10 Spiral cable alignment marks

DRIVE TRAIN

CLUTCH DRIVEN DISC & PRESSURE PLATE

REMOVAL & INSTALLATION

See Figure 11.

1. Before servicing the vehicle, refer to the Precautions Section.

➡If working near and/or around the SRS system and components, be sure to disable the SRS system. Tape the negative battery cable with insulating tape. Always disconnect the negative battery cable first.

**** CAUTION**

To avoid personal injury when working on vehicles equipped with an air bag, the negative battery cable must be disconnected and at least 90 seconds must elapse before working on the system. Failure to do so may result in deployment of the air bag.

2. Disconnect the negative battery cable. Tape the cable with insulating tape.
3. Remove the transmission assembly.
4. Matchmark the clutch cover to the flywheel.
5. At the clutch cover, loosen each bolt 1 turn until spring tension is released.
6. Remove or disconnect the following:
   - Clutch cover set bolts and the clutch cover with the clutch disc.
   - Release bearing retaining clip and withdraw it
   - Release fork and boot assembly

To install:

➡Be sure to use new fasteners, as required.

7. Install or connect the following:
   - Clutch disc onto the flywheel, using a clutch disc alignment tool
   - Clutch cover, position it onto the flywheel and if reusing the old pressure plate, align the matchmarks.
   - Clutch cover. Tighten the bolts in a crisscross pattern to specification.
8. Lubricate the release fork pivot and contact points, the release bearing, bearing hub and input shaft spline surfaces with a suitable molybdenum disulfide lithium based or multi-purpose grease.
9. Install or connect the following:
   - Boot, release fork, hub and the bearing assemblies
   - Transmission
   - Negative battery cable

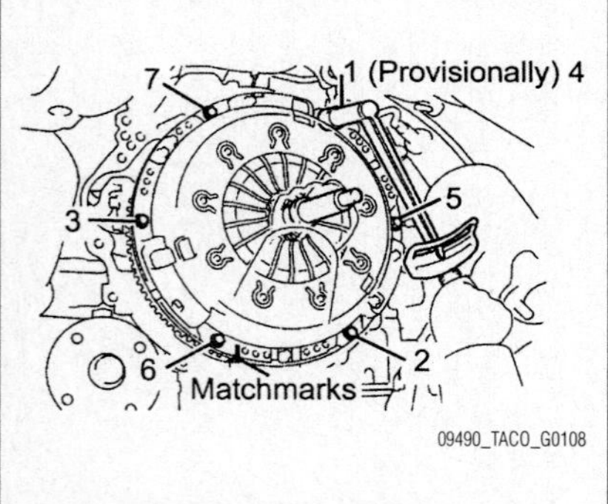

Fig. 11 Clutch cover bolt tightening sequence

ADJUSTMENTS

Clutch Pedal

See Figures 12 and 13.

1. Before servicing the vehicle, refer to the Precautions Section.

➡If working near and/or around the SRS system and components, be sure to disable the SRS system. Tape the negative battery cable with insulating tape. Always disconnect the negative battery cable first.

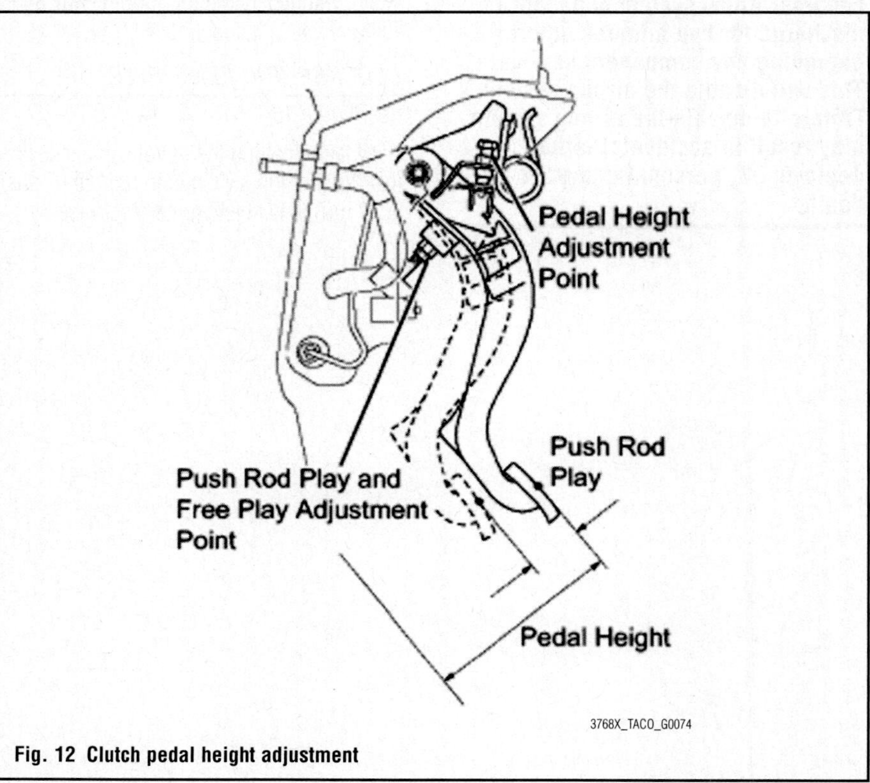

Fig. 12 Clutch pedal height adjustment

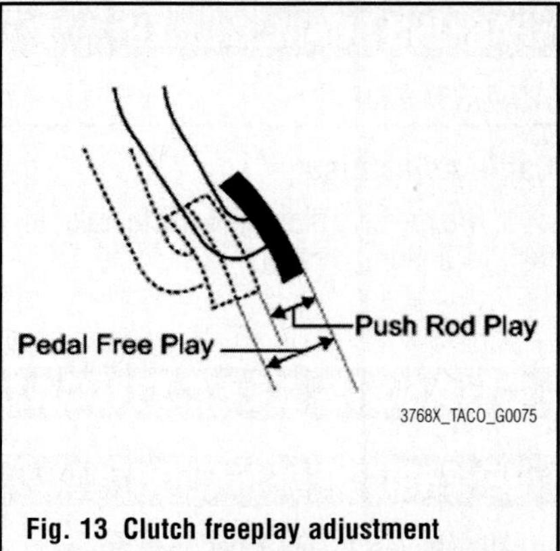

Fig. 13 Clutch freeplay adjustment

**** CAUTION**

To avoid personal injury when working on vehicles equipped with an air bag, the negative battery cable must be disconnected and at least 90 seconds must elapse before working on the system. Failure to do so may result in deployment of the air bag.

2. Disconnect the negative battery cable. Tape the cable with insulating tape.
3. Lift the floor carpet.
4. Check that the pedal height is correct. Specification should be 6.634–7.028 inch. See illustration.

5. To adjust, loosen the locknut and turn the stopper until adjustment is correct. Tighten locknut to 18 ft. lbs. (25 Nm).
6. To check freeplay, depress the pedal until resistance is felt. Specification should be 0.197—0.591 inch. See illustration.
7. Adjust as required.

HYDRAULIC SYSTEM BLEEDING

BLEEDING PROCEDURE

1. Before servicing the vehicle, refer to the Precautions Section.
2. Fill the clutch reservoir with brake fluid. Check the reservoir level frequently and add fluid as needed.
3. Connect one end of a vinyl tube to the bleeder plug on the slave cylinder and submerge the other end into a clear container half-filled with brake fluid.
4. Slowly pump the clutch pedal several times.
5. Have an assistant hold the clutch pedal down and loosen the bleeder plug until fluid and/or air starts to run out of the bleeder plug. Close the bleeder plug while the pedal is held to the floor.
6. Repeat Steps 2 and 3 until all the air bubbles are removed from the system.
7. Tighten the bleeder plug when all the air is gone.
8. Refill the master cylinder to the proper level as required.
9. Check the system for leaks.

TRANSFER CASE ASSEMBLY

REMOVAL & INSTALLATION

See Figure 14.

1. Before servicing the vehicle, refer to the Precautions Section.

➡If working near and/or around the SRS system and components, be sure to disable the SRS system. Tape the negative battery cable with insulating tape. Always disconnect the negative battery cable first.

⁂ CAUTION

To avoid personal injury when working on vehicles equipped with an air bag, the negative battery cable must be disconnected and at least 90 seconds must elapse before working on the system. Failure to do so may result in deployment of the air bag.

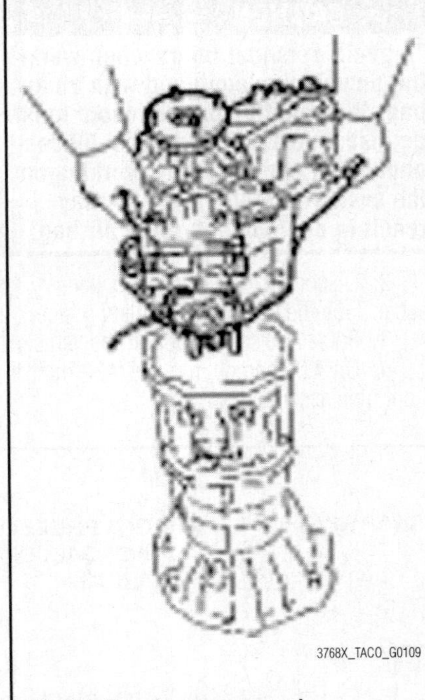

Fig. 14 Transfer case removal

2. Disconnect the negative battery cable. Tape the cable with insulating tape.
3. Drain the transfer case oil.
4. Remove the four bolts and remove the transfer case lower protector.
5. Remove the transmission assembly.
6. Remove the eight transfer adaptor rear mounting bolts. Pull the transfer case straight up and remove it from the transmission.

➡Take care not to damage the adaptor rear oil seal with the transfer case input gear spline.

To install:

➡Be sure to use new fasteners, as required.

7. Installation is the reverse of the removal procedure.
8. Tighten the eight transfer case mounting bolts to 17 ft. lbs. (23 Nm).
9. Be sure to fill the transfer case with the proper grade and type transmission fluid.
10. Start the engine and check for leaks, correct as required.

FRONT CV-JOINT

OVERHAUL

The outboard joint is replaced with halfshaft; no overhaul is possible or necessary.

Inboard Joint

1. Before servicing the vehicle, refer to the Precautions Section.
2. Remove the halfshaft from the vehicle.
3. Remove the large clamp from the inboard joint.
4. Remove the small clamp, using side cutters, from the inboard joint.
5. Slide the inboard joint boot toward the outboard joint.
6. Matchmark the inboard joint to the halfshaft.
7. Remove the inboard joint housing from the halfshaft.
8. Remove the snapring from the end of the halfshaft.
9. Matchmark the halfshaft to the tri-pot joint.
10. Remove the tri-pot from the halfshaft, using a brass bar and a hammer.

⁂ WARNING

Do not tap on the tri-pot joint.

11. Remove the inboard and outboard boots from the halfshaft.

⁂ WARNING

Do not disassemble the outboard joint.

To assemble:

12. Wrap vinyl tape around the halfshaft splines to prevent damaging the boots.
13. Install the outboard and inboard boots to the halfshaft with the small end clamps.
14. Assemble the tri-pot joint to the halfshaft with the beveled side facing the outboard joint and align the matchmarks.
15. Install the tri-pot joint, using a brass bar and a hammer.

⁂ WARNING

Do not tap on the roller.

16. Install the snapring.
17. Lubricate the outboard joint with ½ of the grease supplied with the kit.
18. Assemble the boot to the outboard joint
19. Assemble the inboard joint housing to the halfshaft by aligning the matchmarks.
20. Temporarily install the boot onto the tri-pot housing.
21. Make sure the boots are positioned in the shaft grooves.
22. Install a new inboard joint clamp.
23. Crimp the large clamp so that the crimp clearance is 0.039–0.197 in. (1.0–5.0mm).
24. Install the halfshaft.

FRONT DRIVESHAFT

REMOVAL & INSTALLATION

See Figures 15 and 16.

1. Before servicing the vehicle, refer to the Precautions Section.

➡**If working near and/or around the SRS system and components, be sure to disable the SRS system. Tape the negative battery cable with insulating tape. Always disconnect the negative battery cable first.**

⁂ CAUTION

To avoid personal injury when working on vehicles equipped with an air bag, the negative battery cable must be disconnected and at least 90 seconds must elapse before working on the system. Failure to do so may result in deployment of the air bag.

2. Disconnect the negative battery cable. Tape the cable with insulating tape.
3. Raise and safely support the vehicle.
4. On 4.0L engine, remove the exhaust front pipe assembly No. 2.
5. Remove the driveshaft heat insulator bracket sub assembly.
6. Matchmark and remove the driveshaft retaining bolts.
7. Remove the component from the vehicle.

To install:

➡**Be sure to use new fasteners, as required.**

8. Installation is the reverse of the removal procedure.
9. When replacing the spider bearing make sure that the grease fitting assembly

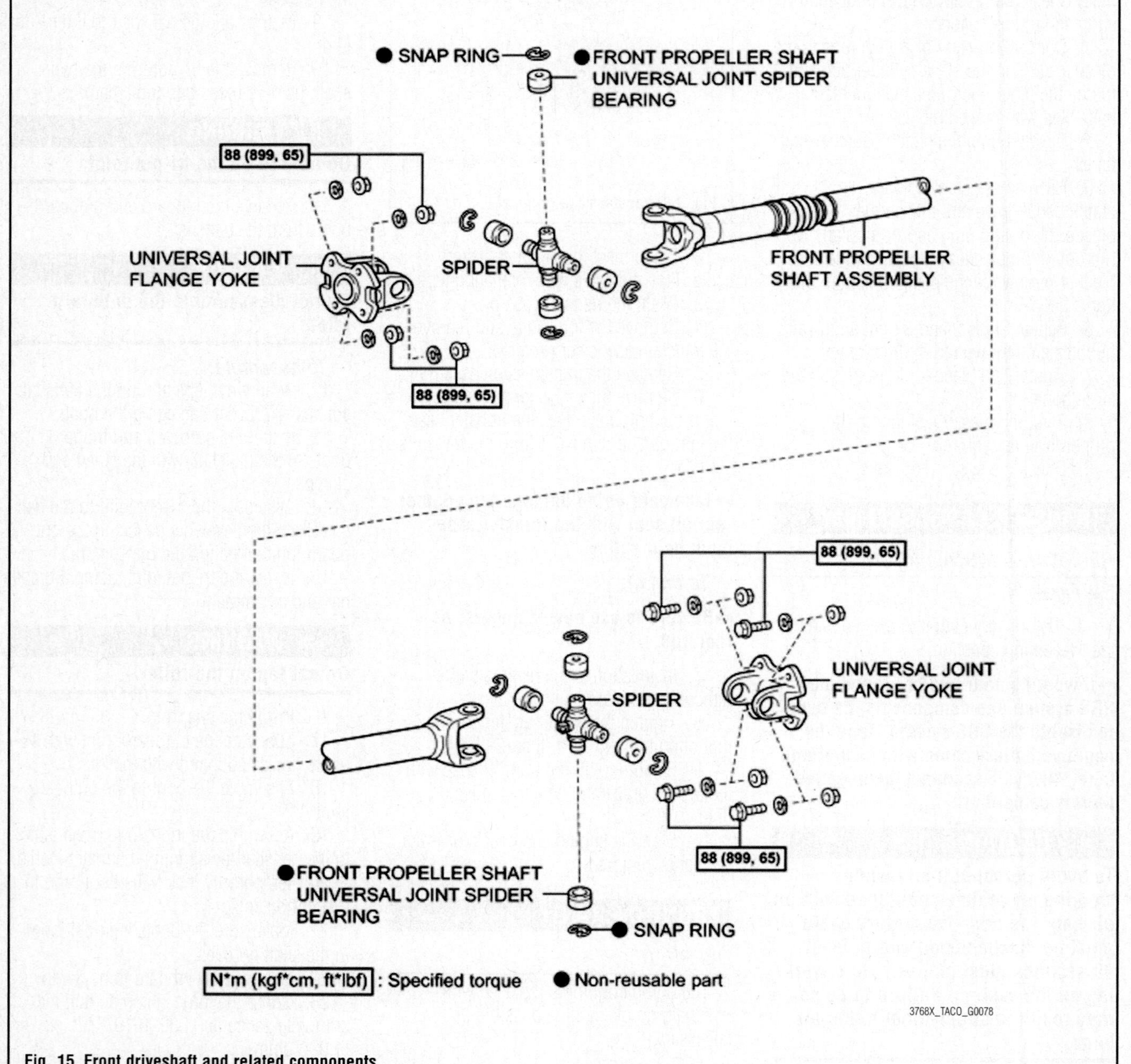

Fig. 15 Front driveshaft and related components

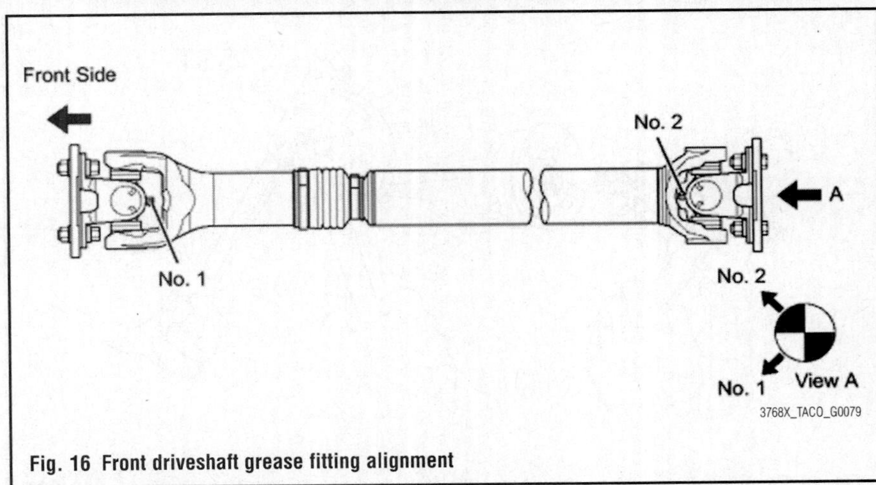

Fig. 16 Front driveshaft grease fitting alignment

hole is facing in the direction shown in the illustration.

10. Tighten the retaining bolts to 65 ft. lbs. (88 Nm).

FRONT HALFSHAFT

REMOVAL & INSTALLATION

See Figure 17.

1. Before servicing the vehicle, refer to the Precautions Section.

➡If working near and/or around the SRS system and components, be sure to disable the SRS system. Tape the negative battery cable with insulating tape. Always disconnect the negative battery cable first.

⁂ CAUTION

To avoid personal injury when working on vehicles equipped with an air bag, the negative battery cable must be disconnected and at least 90 seconds must elapse before working on the system. Failure to do so may result in deployment of the air bag.

2. Disconnect the negative battery cable. Tape the cable with insulating tape.
3. Raise and support the vehicle safely. Remove the tire and wheel assembly.
4. Drain the differential.
5. Remove the bolt and separate the front speed sensor. Disengage the two clamps. Remove the bolt and separate the speed sensor wire harness from the steering knuckle.
6. Remove the cotter pin and nut. Using tool SST09628-62011 or equivalent, separate the tie rod end from the steering knuckle.
7. Using a suitable tool and hammer, remove the front axle hub grease cap. Remove the cotter pin and adjusting cap. Remove the front axle hub nut.
8. Remove the two bolts and separate the front lower ball joint attachment front from the steering knuckle.
9. Using tool SST09520-01010, and SST09520-24010 remove the halfshaft. Be careful not to damage the oil seal.

To install:

➡Be sure to use new fasteners, as required.

10. Coat the spline of the inboard joint shaft with clean ATF.
11. Align the shaft splines and install the halfshaft.

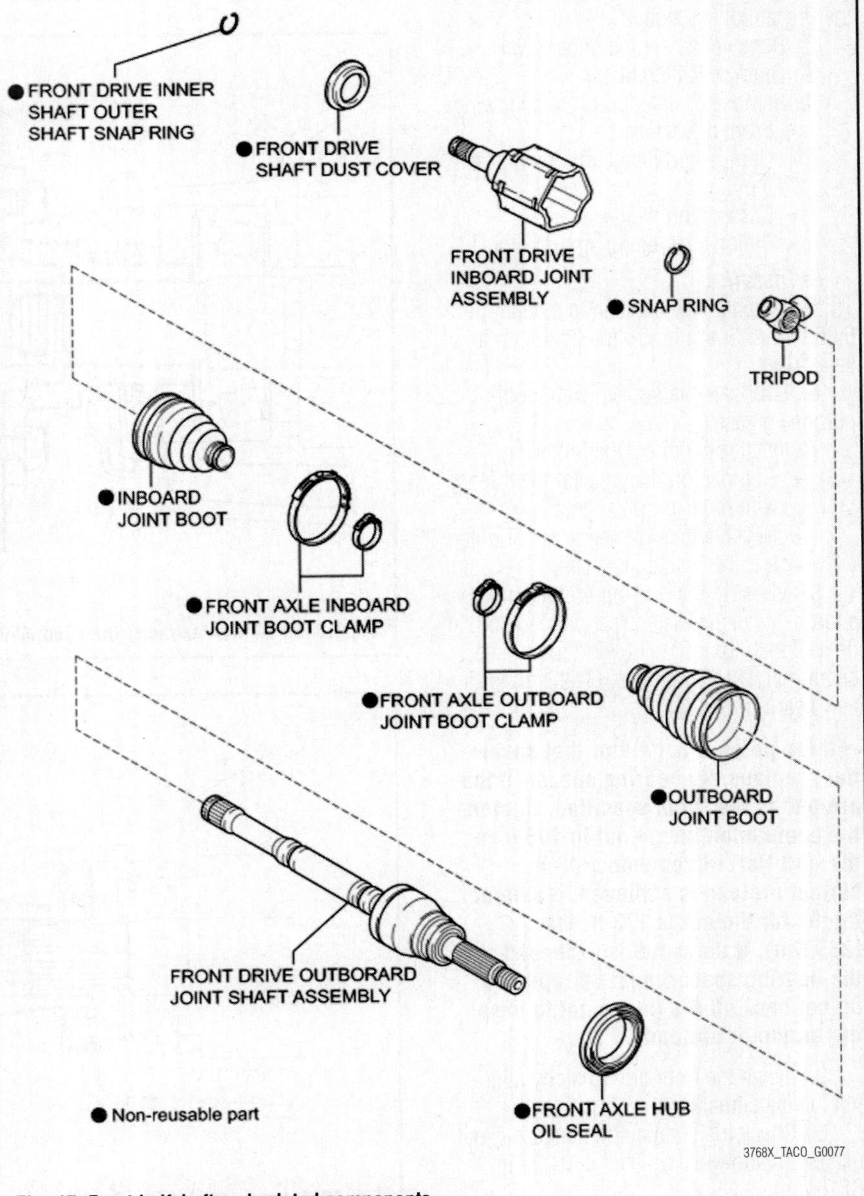

Fig. 17 Front halfshaft and related components

➡Set the snapring with the opening side facing downward. Be careful not to damage the oil seal.

12. Continue the installation in the reverse order of the removal procedure.
13. Tighten the hub nut to 173 ft. lbs. (234 Nm).
14. Be sure to fill the differential with the proper type and grade lubricant.
15. Check and correct leaks, as required.
16. Check and adjust the alignment, as required.

FRONT PINION SEAL

REMOVAL & INSTALLATION

See Figures 18 through 20.

1. Before servicing the vehicle, refer to the Precautions Section.
2. Remove the engine undercover.
3. Drain the differential oil.
4. Remove or disconnect the following:
   - Front driveshaft
   - Companion flange nut, by unstaking it
   - Companion flange
   - Pinion seal, using an extractor

To install:

5. Install a new oil seal, to a depth of 0.059 in. (1.5mm) below the lip, using a seal driver.
6. Lubricate the seal lip with multi-purpose grease.
7. Install or connect the following:
   - Companion flange, coat the threads with multi-purpose grease
   - New companion flange nut. Tighten it to 89 ft. lbs. (120 Nm).
8. Measure the bearing preload, using a torque wrench. The correct preload should be 5–9 inch lbs. (0.6–1.0 Nm) for a used bearing or 10–17 inch lbs. (1–2 Nm) for a new bearing.

➡If the preload is greater that specified, replace the bearing spacer. If the preload is less than specified, tighten the companion flange nut in 108 inch lbs. (13 Nm) increments until the correct preload is achieved. Maximum torque for the nut is 173 ft. lbs. (235 Nm). If the value is exceeded, the bearing spacer must be replaced; do not back off the flange nut to lower the torque or preload.

9. Install the front driveshaft by aligning the matchmarks.
10. Check the companion flange run-out; maximum allowable run-out is 0.003 in. (0.10mm).

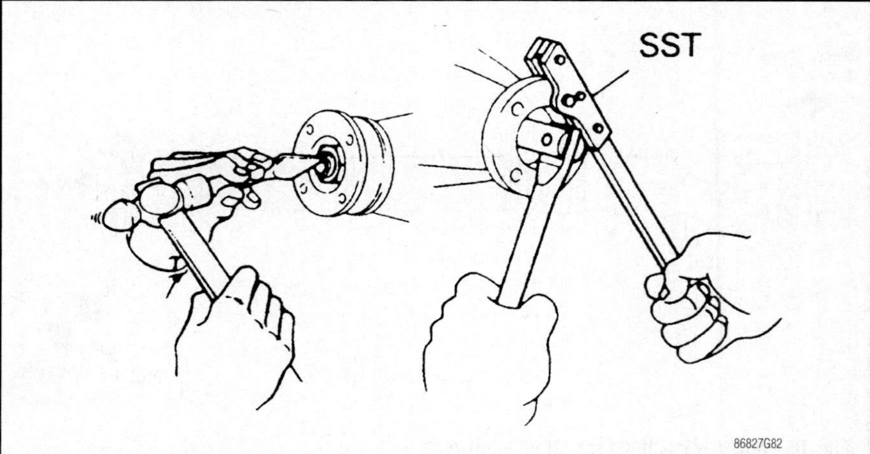

Fig. 18 Using a chisel and hammer, loosen the staked part of the nut. Hold the flange with SST 09950-30010 or equivalent and remove the nut

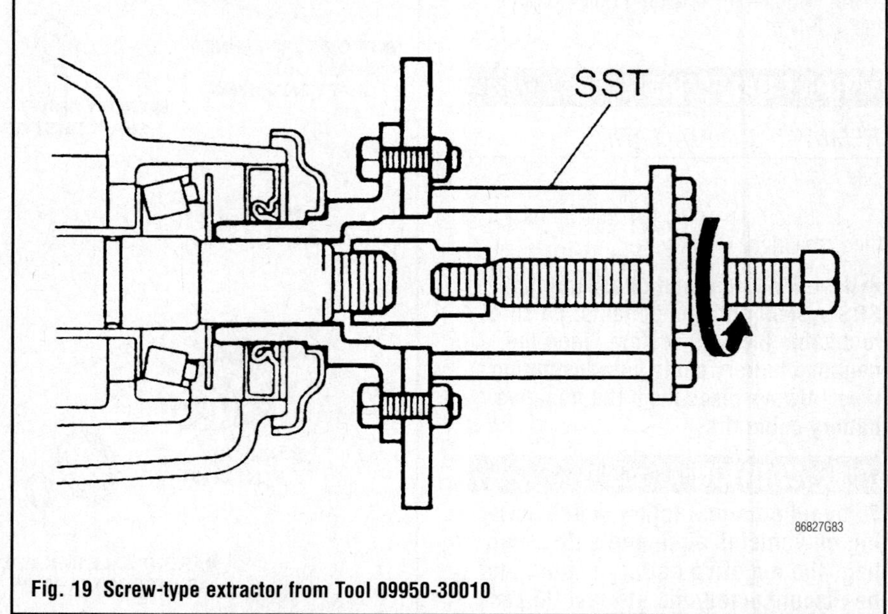

Fig. 19 Screw-type extractor from Tool 09950-30010

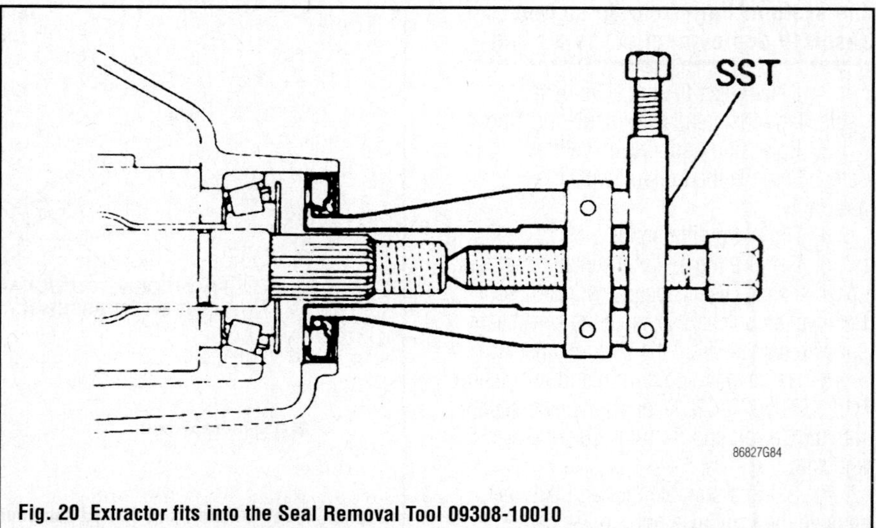

Fig. 20 Extractor fits into the Seal Removal Tool 09308-10010

11. Stake the pinion flange nut.
12. Refill the differential with oil.

REAR AXLE SHAFT, BEARING & SEAL

REMOVAL & INSTALLATION

See Figure 21.

1. Before servicing the vehicle, refer to the Precautions Section.

➡If working near and/or around the SRS system and components, be sure to disable the SRS system. Tape the negative battery cable with insulating tape. Always disconnect the negative battery cable first.

⁂ CAUTION

To avoid personal injury when working on vehicles equipped with an air bag, the negative battery cable must be disconnected and at least 90 seconds must elapse before working on the system. Failure to do so may result in deployment of the air bag.

2. Disconnect the negative battery cable. Tape the cable with insulating tape.
3. Drain the brake fluid.
4. Raise and support the vehicle safely. Remove the tire and wheel assembly.
5. Remove the brake drum. Remove the brake shoes.
6. Remove the rear speed sensor.
7. Remove the two bolts and disconnect the parking brake cable from the backing plate. Disconnect the brake line at the brake backing plate.

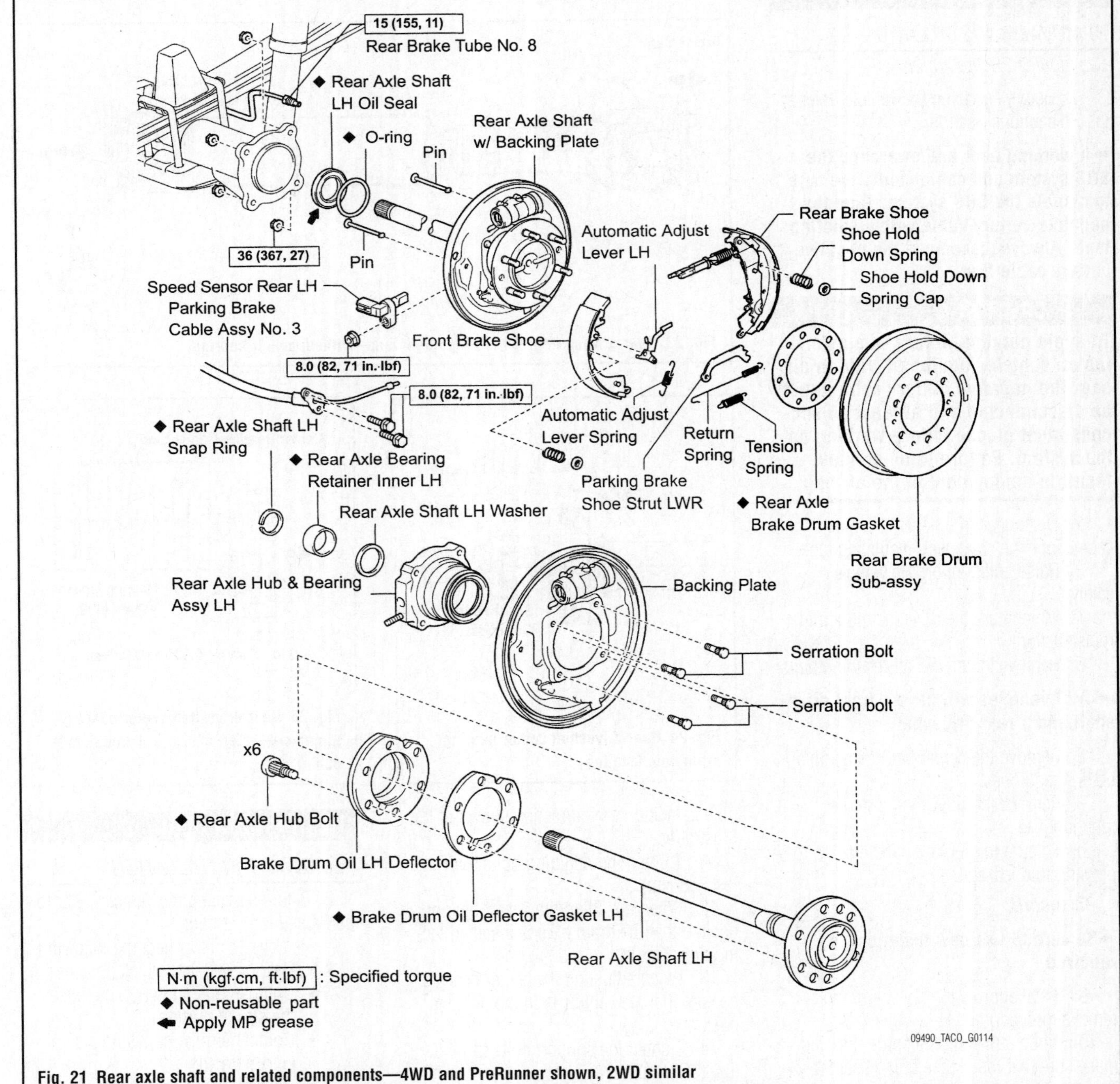

Fig. 21 Rear axle shaft and related components—4WD and PreRunner shown, 2WD similar

8. Remove the four nuts and the rear axle shaft and backing plate. Remove the O-ring.

9. Remove the rear axle shaft oil seal using tool SST09308-00010.

To install:

➡Be sure to use new fasteners, as required.

10. Installation is the reverse of the removal procedure.

11. Be sure to fill the master cylinder with the proper grade and type brake fluid.

12. Bleed the brakes, as required.

REAR DRIVESHAFT

REMOVAL & INSTALLATION

See Figures 22 through 25.

1. Before servicing the vehicle, refer to the Precautions Section.

➡If working near and/or around the SRS system and components, be sure to disable the SRS system. Tape the negative battery cable with insulating tape. Always disconnect the negative battery cable first.

✲✲ CAUTION

To avoid personal injury when working on vehicles equipped with an air bag, the negative battery cable must be disconnected and at least 90 seconds must elapse before working on the system. Failure to do so may result in deployment of the air bag.

2. Disconnect the negative battery cable. Tape the cable with insulating tape.

3. Raise and support the vehicle safely.

4. Matchmark the driveshaft to aid in reinstallation.

5. Remove the driveshaft retaining bolts.

➡4WD vehicles will have a front driveshaft and a rear driveshaft.

6. Remove the center bearing retaining bolts, if equipped.

7. Remove the driveshaft from its mounting.

8. Be sure to plug the transmission to prevent fluid leakage.

To install:

➡Be sure to use new fasteners, as required.

9. Installation is the reverse of the removal procedure.

10. When replacing the spider bearing make sure that the grease fitting assembly hole is facing in the direction shown in the illustration.

11. Tighten the retaining bolts to 65 ft. lbs. (88 Nm).

12. When installing the center bearing be sure that the drain hole is installed facing downwards.

13. When adjusting the center bearing be sure that the vehicle is in the unladen position.

14. Tighten the support bolts to 27 ft. lbs. (36 Nm).

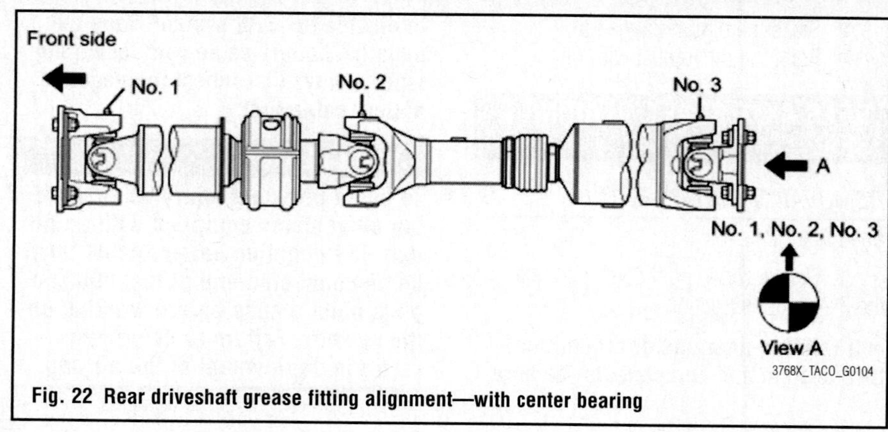

Fig. 22 Rear driveshaft grease fitting alignment—with center bearing

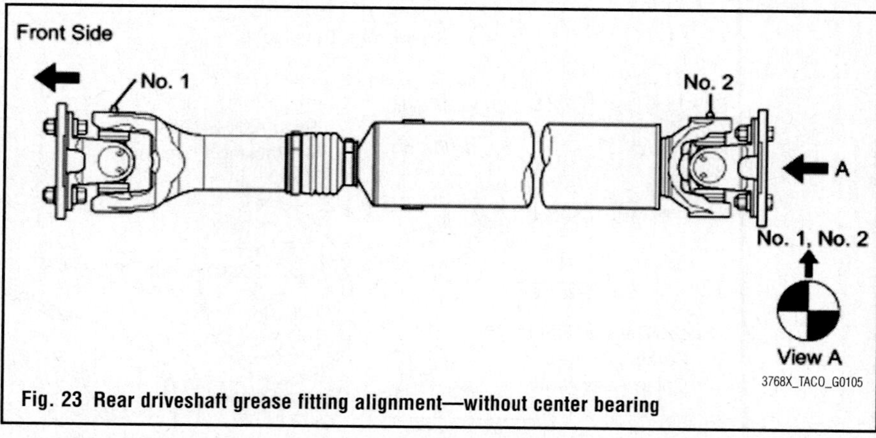

Fig. 23 Rear driveshaft grease fitting alignment—without center bearing

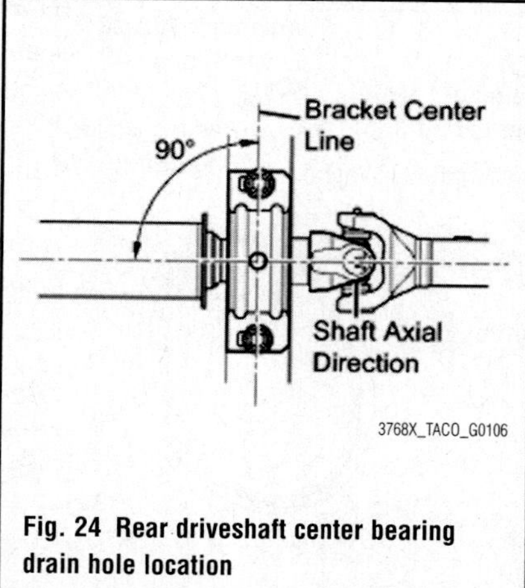

Fig. 24 Rear driveshaft center bearing drain hole location

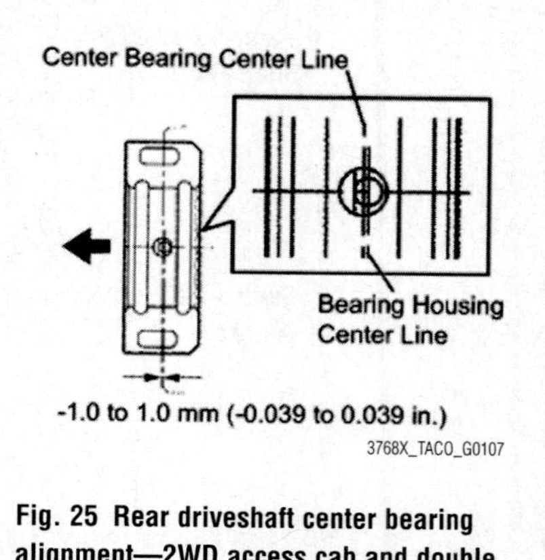

Fig. 25 Rear driveshaft center bearing alignment—2WD access cab and double cab

REAR PINION SEAL

REMOVAL & INSTALLATION

1. Before servicing the vehicle, refer to the Precautions Section.

2. Remove or disconnect the following:
   - Rear driveshaft by matchmarking it
   - Companion flange nut, by loosen the staked portion
   - Companion flange, using a screw-type extractor
   - Oil seal, using an extractor

To install:

3. Install a new oil seal, to a depth of 0.039 in. (1.0mm) below the lip, using a seal driver.
4. Lubricate the seal lip with multi-purpose grease.
5. Install or connect the following:
   - Companion flange, coat the threads with multi-purpose grease
   - New companion flange nut. Tighten it to 109 ft. lbs. (147 Nm).
6. Measure the bearing preload, using a torque wrench. The correct preload should be 8–11 inch lbs. (0.9–1.2 Nm) for a 2 spider gear differential or to 4–7 inch lbs. (0.4–0.8 Nm) for a 4 spider gear differential.

➡If the preload is greater that specified, replace the bearing spacer. If the preload is less than specified, tighten the companion flange nut in 9 ft. lbs. (13 Nm) increments until the correct preload is achieved. Maximum torque for the nut is 325 ft. lbs. (441 Nm). If the value is exceeded, the bearing spacer must be replaced; do not back off the flange nut to lower the torque or preload.

7. Stake the pinion flange nut.
8. Install the rear driveshaft by aligning the matchmarks.

ENGINE COOLING

RADIATOR

REMOVAL & INSTALLATION

2.7L Engines

See Figure 26.

1. Before servicing the vehicle, refer to the Precautions Section.

➡If working near and/or around the SRS system and components, be sure to disable the SRS system. Tape the negative battery cable with insulating tape. Always disconnect the negative battery cable first.

⁂ CAUTION

To avoid personal injury when working on vehicles equipped with an air bag, the negative battery cable must be disconnected and at least 90 seconds must elapse before working on the system. Failure to do so may result in deployment of the air bag.

2. Disconnect the negative battery cable. Tape the cable with insulating tape.

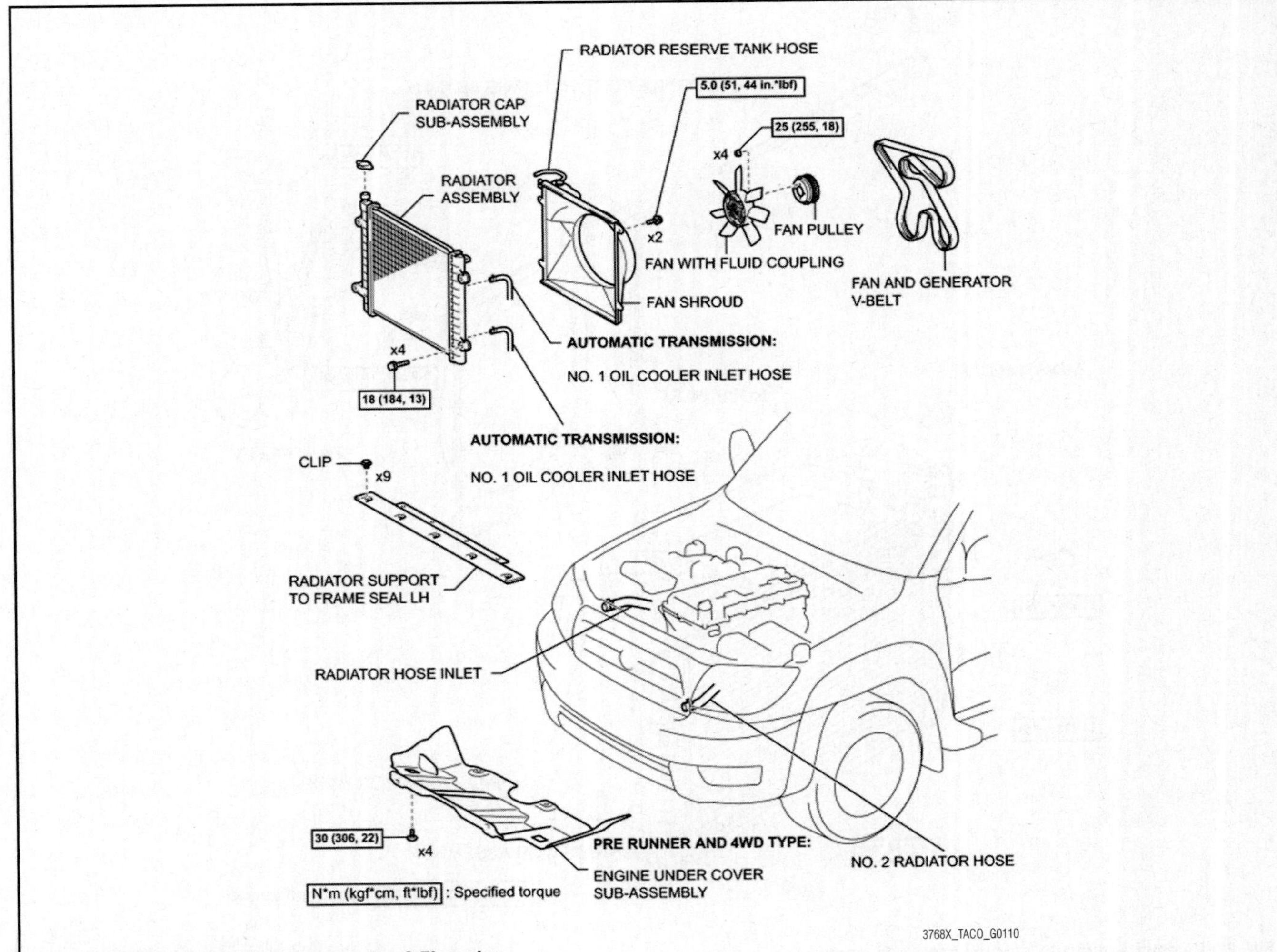

Fig. 26 Radiator and related components—2.7L engine

3. Remove the 4 bolts, then remove the engine under cover no. 1.
4. Drain engine coolant.
5. Disengage the 9 clips, then remove the radiator support to frame seal LH.
6. Remove the fan and generator V belt.
7. Disconnect the radiator reserve tank hose from the radiator.
8. Disconnect the oil cooler inlet and outlet hoses from the fan shroud.
9. Remove the 4 nuts, then remove the fan shroud and fan with fluid coupling together.

➡Make sure that the fan shroud and fan with fluid coupling do not make any contact with the radiator when they are removed.

10. Remove the fan pulley.
11. Disconnect the radiator hose inlet.
12. Disconnect the radiator hose No. 2.
13. Disconnect the oil cooler inlet and outlet hoses.
14. Remove the 4 bolts, then remove the radiator.

To install:

➡Be sure to use new fasteners, as required.

15. Install the radiator with the 4 bolts and tighten to 13 ft. lbs. (18 Nm)
16. Connect the oil cooler inlet and outlet hoses.
17. Connect the radiator hose NO. 2.
18. Connect the radiator hose inlet.
19. Install the fan pulley.
20. Install the fan shroud and fan with fluid coupling together.

➡Make sure that the fan shroud and fan with fluid coupling do not make any contact with the radiator when they are installed.

21. Install the fan with fluid coupling with the 4 nuts and tighten to 18 ft. lbs. (25 Nm).
22. Connect the oil cooler inlet and outlet hoses to the fan shroud.
23. Connect the radiator reserve tank hose to the radiator.
24. Install the fan generator V belt.
25. Install the radiator support to frame seal LH with the 9 clips.
26. Connect cable to negative battery terminal.
27. Add engine coolant.
28. Check for engine coolant leakage.
29. Install the engine under cover No. 1 with the 4 bolts and tighten to 30 Nm (22 ft. lbs.)

4.0L Engines

See Figure 27.

1. Before servicing the vehicle, refer to the Precautions Section.

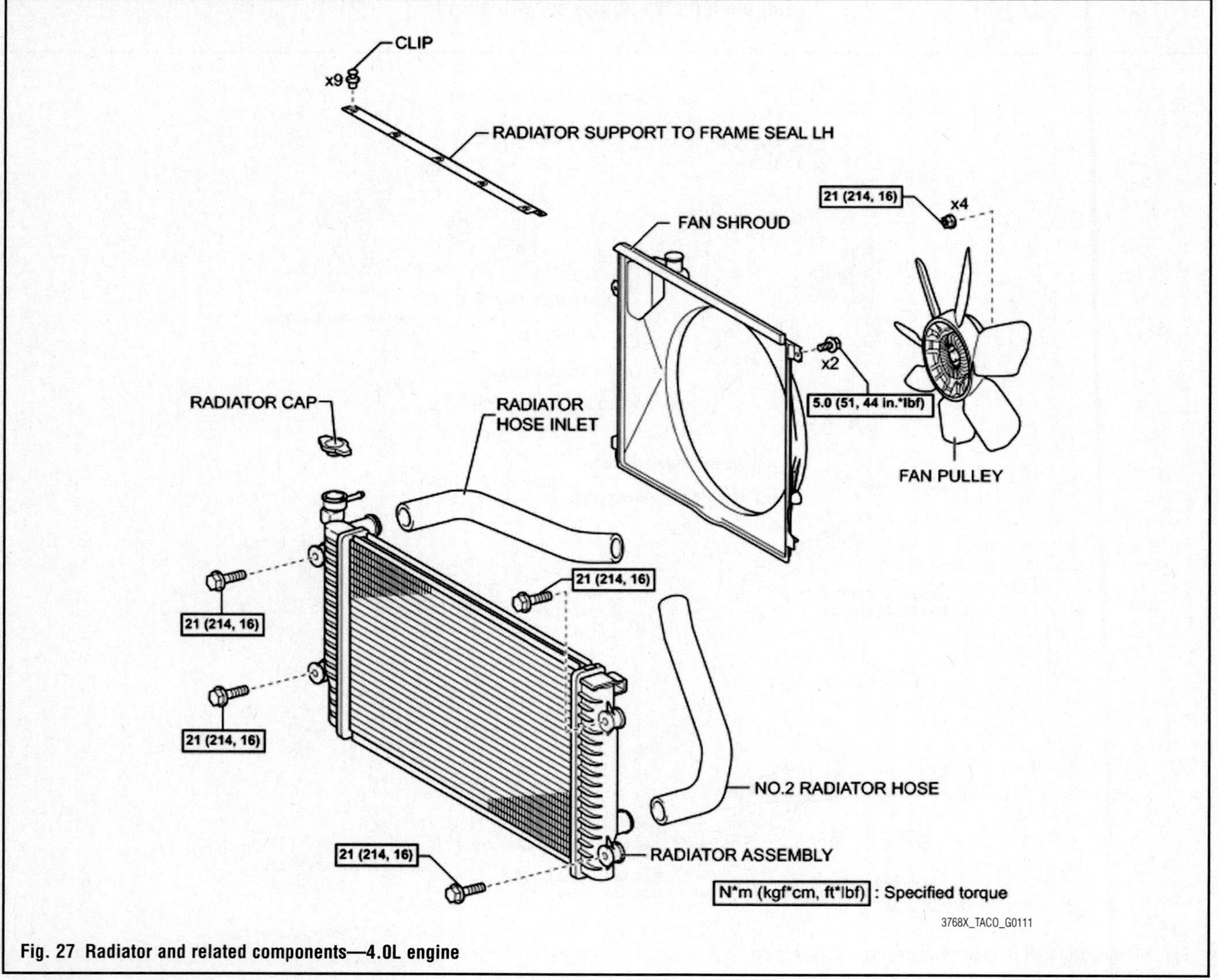

Fig. 27 Radiator and related components—4.0L engine

➡If working near and/or around the SRS system and components, be sure to disable the SRS system. Tape the negative battery cable with insulating tape. Always disconnect the negative battery cable first.

✻✻ CAUTION

To avoid personal injury when working on vehicles equipped with an air bag, the negative battery cable must be disconnected and at least 90 seconds must elapse before working on the system. Failure to do so may result in deployment of the air bag.

2. Disconnect the negative battery cable. Tape the cable with insulating tape.
3. Drain engine coolant.
4. Disengage the 9 clips, then remove the radiator support to frame seal.
5. Disconnect the hose from the radiator reserve tank.
6. Remove the 2 cooler hoses from the clamp.
7. Remove the 4 nuts from the fan pulley.
8. Remove the fan and generator V belt.
9. Remove the 2 bolts from the fan shroud.
10. Remove the fan pulley and fan shroud.
11. Disconnect the 2 radiator hoses.
12. Disconnect the 2 oil cooler hoses.
13. Remove radiator hose inlet.
14. Remove radiator hose no. 2.
15. Remove the 4 bolts, then remove the radiator.

To install:

➡Be sure to use new fasteners, as required.

16. Install the radiator with the 4 bolts and tighten to 13 ft. lbs. (18 Nm).
17. Install radiator hose no. 2.
18. Install radiator hose inlet.
19. Install radiator assembly.
20. Connect the 2 oil cooler hoses.
21. Provisionally install the fan shroud together with the fan pulley.
22. Install the fan and generator V belt.
23. Install the fan pulley with the 4 nuts and tighten to 16 ft. lbs. (22 Nm).
24. Install the fan shroud with the 2 bolts and tighten to 44 inch lbs. (5 Nm).
25. Install the 2 cooler hoses into the clamp.
26. Connect the hose to the radiator reserve tank.
27. Install radiator support to frame seal lh.
28. Connect cable to negative battery terminal.
29. Add engine coolant.
30. Check for engine coolant leakage.

THERMOSTAT

REMOVAL & INSTALLATION

2.7L Engines

See Figure 28.

1. Before servicing the vehicle, refer to the Precautions Section.

➡If working near and/or around the SRS system and components, be sure to disable the SRS system. Tape the negative battery cable with insulating tape. Always disconnect the negative battery cable first.

✻✻ CAUTION

To avoid personal injury when working on vehicles equipped with an air bag, the negative battery cable must be disconnected and at least 90 seconds must elapse before working on the system. Failure to do so may result in deployment of the air bag.

2. Disconnect the negative battery cable. Tape the cable with insulating tape.
3. Drain engine coolant.
4. Remove water inlet.
5. Remove the bolt and 2 nuts, then remove the water inlet.
6. Remove the gasket from the timing chain cover.
7. Remove the thermostat from the timing chain cover.

To install:

➡Be sure to use new fasteners, as required.

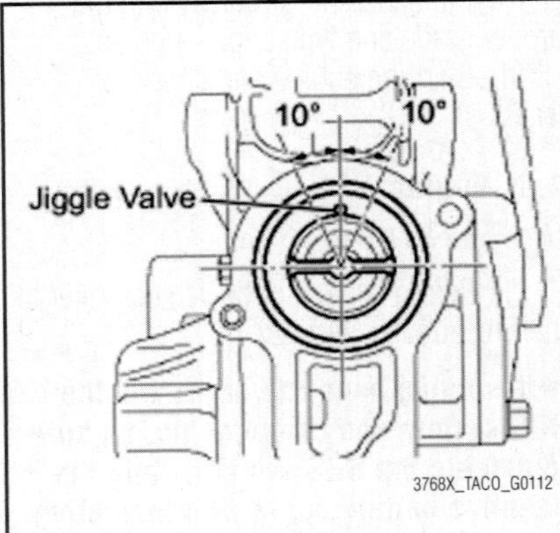

Fig. 28 Jiggle valve alignment—2.7L engine

8. Install a new gasket onto the thermostat.
9. Install the thermostat with the jiggle valve upward.

➡The jiggle valve may be set within 10° to either side of vertical position.

10. Install a new gasket and the water inlet with the bolt and 2 nuts. Tighten to 15 ft. lbs. (20 Nm).
11. Add engine coolant.
12. Check for engine coolant leakage.

4.0L Engines

1. Before servicing the vehicle, refer to the Precautions Section.

➡If working near and/or around the SRS system and components, be sure to disable the SRS system. Tape the negative battery cable with insulating tape. Always disconnect the negative battery cable first.

✻✻ CAUTION

To avoid personal injury when working on vehicles equipped with an air bag, the negative battery cable must be disconnected and at least 90 seconds must elapse before working on the system. Failure to do so may result in deployment of the air bag.

2. Disconnect the negative battery cable. Tape the cable with insulating tape.
3. Drain engine coolant.
4. Remove V-bank cover.
5. Disconnect radiator hose no. 2
6. Remove water inlet w/thermostat
7. Remove the 3 nuts, then remove the water inlet with thermostat and gasket.

To install:

➡Be sure to use new fasteners, as required.

8. Install a new gasket onto the water inlet with thermostat.
9. Install the water inlet with thermostat with the 3 nuts and tighten to 80 inch lbs. (9 Nm)
10. Connect radiator hose no. 2.
11. Connect cable to negative battery terminal.
12. Add engine coolant.
13. Check for engine coolant leakage.
14. Install the V-bank cover with the 2 nuts.

WATER PUMP

REMOVAL & INSTALLATION

2.7L Engines

See Figures 29 and 30.

1. Before servicing the vehicle, refer to the Precautions Section.

➡If working near and/or around the SRS system and components, be sure to disable the SRS system. Tape the negative battery cable with insulating tape. Always disconnect the negative battery cable first.

⁂ CAUTION

To avoid personal injury when working on vehicles equipped with an air bag, the negative battery cable must be disconnected and at least 90 seconds must elapse before working on the system. Failure to do so may result in deployment of the air bag.

2. Disconnect the negative battery cable. Tape the cable with insulating tape.
3. On 4WD and PreRunner, remove the four retaining bolts and remove the number one engine undercover subassembly.
4. Drain the engine coolant.
5. Remove the radiator support to frame seal, left side.
6. Remove the fan shroud. Remove the alternator.
7. Remove the three bolts, and remove the belt tensioner assembly.
8. Remove the water pump retaining bolts. Remove the water pump from the engine.

To install:

9. Clean all gasket mounting surfaces.
10. Using a new gasket install the water pump to the engine. Tighten bolts "A" to 15 ft. lbs. (20 Nm). Tighten bolts "B" to 80 inch lbs. (9 Nm).
11. Install the belt tensioner assembly. Tighten bolt "B" to 30 ft. lbs. (40 Nm). Tighten bolt "A" to 16 ft. lbs. (22 Nm). Tighten bolt "C" to 32 ft. lbs. (43 Nm).

➡Check that the bolt holes on the belt tensioner and timing chain cover are aligned, prior to installing bolt "C".

12. Continue the installation in the reverse order of the removal procedure.
13. Fill the cooling system with the proper grade and type engine coolant.
14. Start the engine and check for leaks.

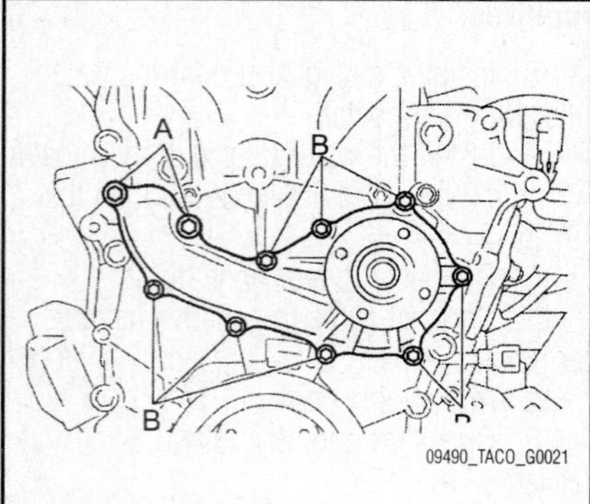

Fig. 29 Water pump bolt identification—2.7L engine

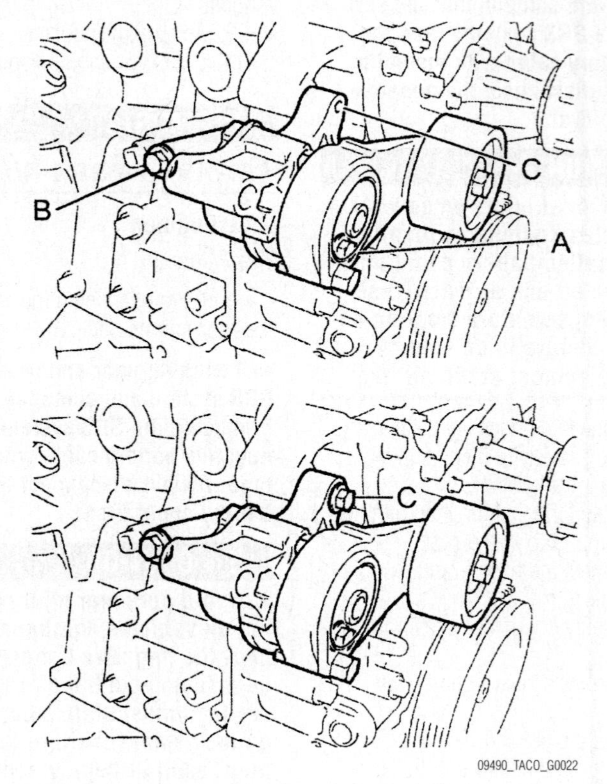

Fig. 30 Belt tensioner bolt identification and location—2.7L engine

4.0L Engines

See Figure 31.

1. Before servicing the vehicle, refer to the Precautions Section.

➡If working near and/or around the SRS system and components, be sure to disable the SRS system. Tape the negative battery cable with insulating tape. Always disconnect the negative battery cable first.

⁂ CAUTION

To avoid personal injury when working on vehicles equipped with an air bag, the negative battery cable must be disconnected and at least 90 seconds must elapse before working on the system. Failure to do so may result in deployment of the air bag.

2. Disconnect the negative battery cable. Tape the cable with insulating tape.

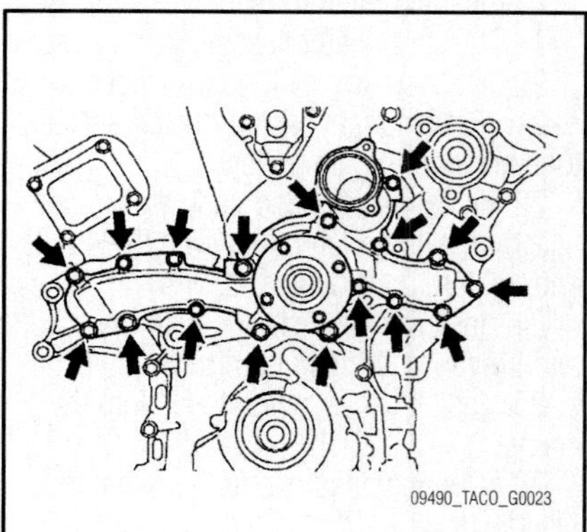

Fig. 31 Water pump bolt identification—4.0L engine

3. On 4WD and PreRunner, remove the four retaining bolts and remove the number one engine undercover subassembly.

4. Drain the engine coolant.

5. Remove the radiator support to frame seal, left side.

6. Remove the V bank cover.

7. Remove the fan shroud. Remove the air cleaner assembly.

8. Disconnect the two oil cooler hoses (with oil cooler) and remove the water inlet.

9. Disconnect the radiator hoses. Disconnect the five water bypass hoses.

10. Remove the five bolts and the water inlet. Remove the O-ring from the water outlet pipe. Remove the gasket from the water pump.

11. Remove the two bolts and remove the number two idler pulley subassembly.

12. Remove the alternator.

13. Remove the mounting bolt and separate the air conditioning compressor suction hose subassembly. Disconnect the air condition compressor connector. Remove the four bolts and separate the compressor from the belt tensioner assembly.

14. Remove belt tensioner assembly.

15. Remove the water pump retaining bolts. Remove the water pump from the engine.

➡Be sure to use new fasteners, as required.

To install:

16. Clean all gasket mounting surfaces.

17. Using a new gasket install the water pump to the engine. Tighten bolts "A" to 80 inch lbs. (9 Nm). Tighten bolts "B" to 17 ft. lbs. (23 Nm).

18. Install the belt tensioner assembly.

19. Continue the installation in the reverse order of the removal procedure.

20. Fill the cooling system with the proper grade and type engine coolant.

21. Start the engine and check for leaks.

ENGINE ELECTRICAL

CHARGING SYSTEM

ALTERNATOR

REMOVAL & INSTALLATION

See Figures 32 and 33.

On some vehicles, the alternator is mounted very low on the engine. It may be necessary to remove the gravel shield and work from beneath the vehicle in order to gain access to the alternator. Replacing the alternator while the engine is cold is recommended.

1. Before servicing the vehicle, refer to the Precautions Section.

➡If working near and/or around the SRS system and components, be sure to disable the SRS system. Tape the negative battery cable with insulating tape. Always disconnect the negative battery cable first.

⁂ CAUTION

To avoid personal injury when working on vehicles equipped with an air bag, the negative battery cable must be disconnected and at least 90 seconds must elapse before working on the system. Failure to do so may result in deployment of the air bag.

2. Disconnect the negative battery cable. Tape the cable with insulating tape.

3. Remove the V bank cover, if equipped.

4. Remove the radiator support to frame seal, left side.

5. Remove the radiator fan shroud.

6. Disconnect the alternator harness wiring.

7. Remove the drive belt.

8. Remove the alternator retaining bolts.

9. Remove the alternator from the vehicle.

To install:

➡Be sure to use new fasteners, as required.

10. Installation is the reverse of the removal procedure.

11. Tighten the alternator retaining bolts to 32 ft. lbs. (43 Nm.).

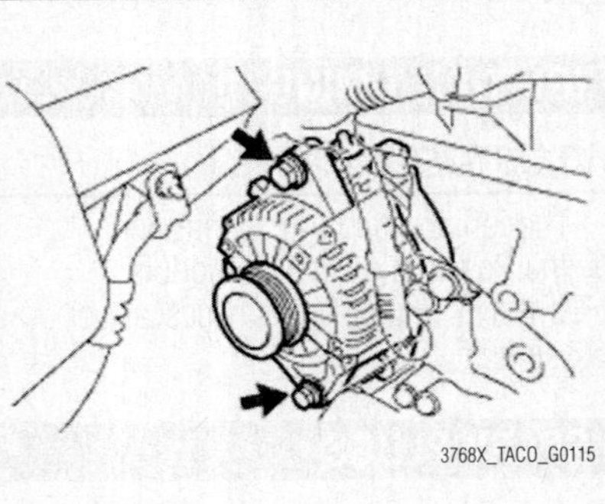

Fig. 32 Alternator mounting bolt locations—2.7L engine

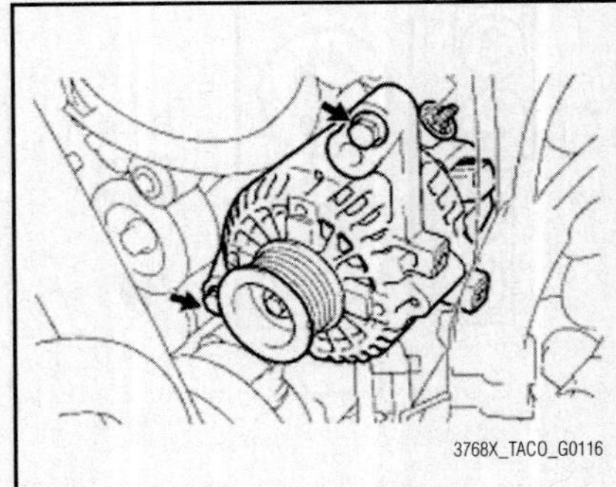

Fig. 33 Alternator mounting bolt locations—4.0L engine

FIRING ORDER

See Figures 34 and 35.

IGNITION COIL

REMOVAL & INSTALLATION

➡It is a good idea to remove and reinstall the coils one at a time to prevent the coils being installed out of order.

1. Disconnect the negative battery cable.
2. Disconnect the spark plug wire from the coil.
3. Disconnect the electrical connectors from the coil.
4. Remove the coil bolts and the coils.

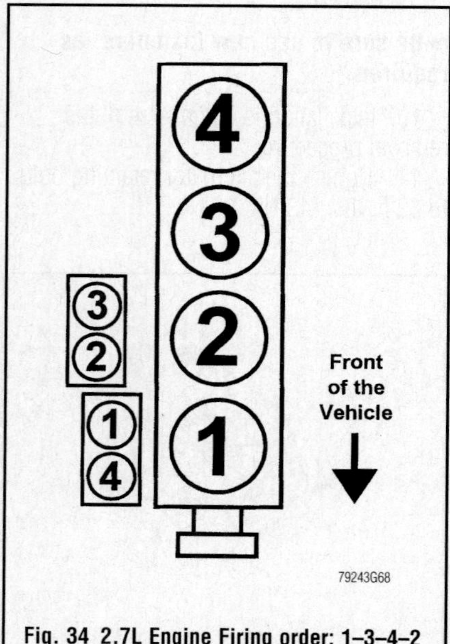

Fig. 34 2.7L Engine Firing order: 1–3–4–2 Distributorless ignition system

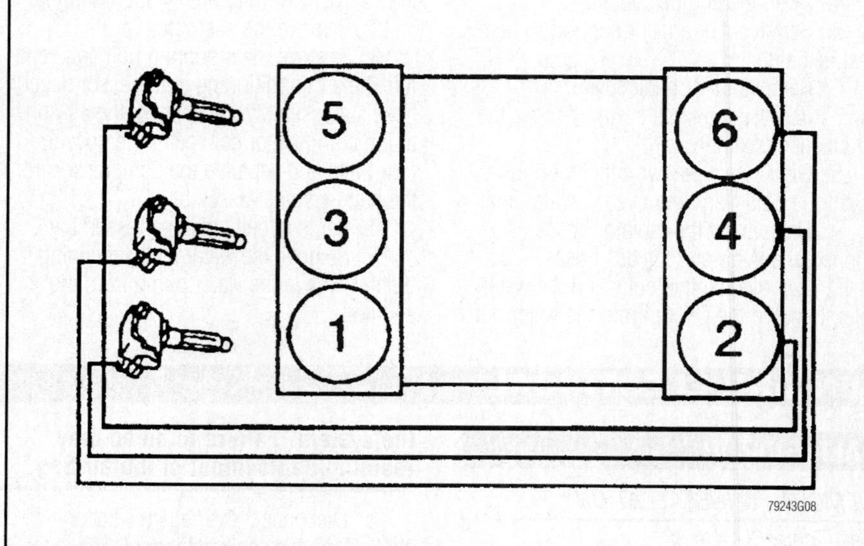

Fig. 35 4.0L EngineFiring order: 1–2–3–4–5–6 Distributorless ignition system

To install:

5. Install the coil and tighten the bolts to 69 inch lbs. (8 Nm).
6. Connect the wiring and spark plug wires
7. Connect the negative battery cable.

IGNITION TIMING

ADJUSTMENT

The ignition timing is controlled by the Powertrain Control Module (PCM). No adjustment is necessary or possible.

SPARK PLUGS

REMOVAL & INSTALLATION

1. Remove the ignition coils.
2. Using a 16 mm plug wrench, remove the spark plugs.
3. Clean the spark plugs.
4. If the electrode has traces of wet carbon, allow it to dry and then clean with a spark plug cleaner.
5. Check the spark plug for thread damage and insulator damage. If abnormal, replace the spark plug.
6. Adjust the spark plug electrode gap. Electrode gap for new spark plug is 1.0 to 1.1 mm (0.039 to 0.043 in.).
7. Using a 16 mm plug wrench, install the spark plugs and tighten to 13 ft. lbs. (17.5 Nm).
8. Reinstall the ignition coils.

STARTER

REMOVAL & INSTALLATION

2.7L Engines

See Figure 36.

1. Before servicing the vehicle, refer to the Precautions Section.

➡If working near and/or around the SRS system and components, be sure to disable the SRS system. Tape the negative battery cable with insulating tape. Always disconnect the negative battery cable first.

⁂ CAUTION

To avoid personal injury when working on vehicles equipped with an air bag, the negative battery cable must be disconnected and at least 90 seconds must elapse before working on the system. Failure to do so may result in deployment of the air bag.

2. Disconnect the negative battery cable. Tape the cable with insulating tape.

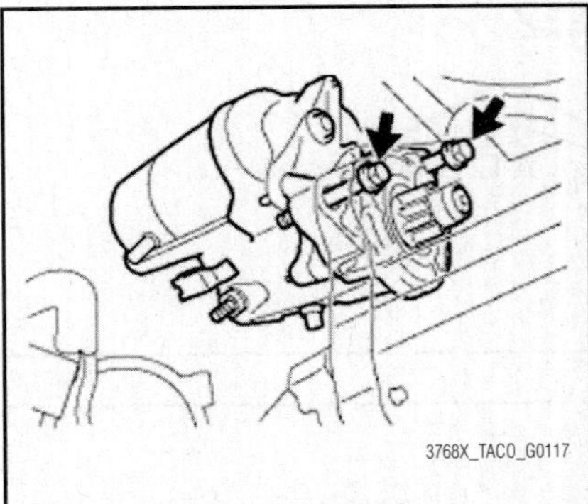

Fig. 36 Starter mounting bolt locations—2.7L engine

3. Raise and support the vehicle, as required.
4. Remove the terminal cap. Disconnect the electrical connections.
5. Disconnect the positive battery cable.
6. Remove the starter retaining bolts. Remove the starter from the vehicle.

To install:

➡Be sure to use new fasteners, as required.

7. Installation is the reverse of the removal procedure.
8. Tighten the retaining bolts to 27 ft. lbs. (36 Nm).

4.0L Engines

See Figure 37.

1. Before servicing the vehicle, refer to the Precautions Section.

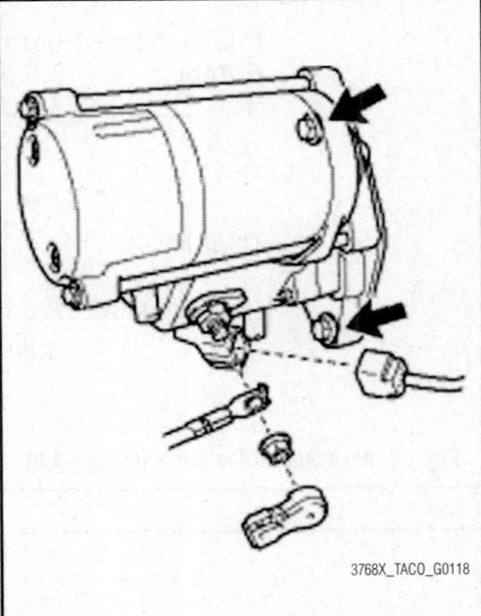

Fig. 37 Starter mounting bolt locations—4.0L engine

➡If working near and/or around the SRS system and components, be sure to disable the SRS system. Tape the negative battery cable with insulating tape. Always disconnect the negative battery cable first.

⁂ CAUTION

To avoid personal injury when working on vehicles equipped with an air bag, the negative battery cable must be disconnected and at least 90 seconds must elapse before working on the system. Failure to do so may result in deployment of the air bag.

2. Disconnect the negative battery cable. Tape the cable with insulating tape.
3. Remove the engine undercover assembly.
4. On 2WD and PreRunner, remove the number two manifold stay.
5. On 4WD vehicles, remove the number two exhaust front pipe assembly. Remove the five clips and then remove the front fender splash shield, left side. Remove the number two steering intermediate shaft.
6. Disconnect the starter electrical connectors.
7. Disconnect the positive battery cable.
8. Remove the starter retaining bolts. Remove the starter from the vehicle.

To install:

➡Be sure to use new fasteners, as required.

9. Installation is the reverse of the removal procedure.
10. Tighten the retaining bolts to 27 ft. lbs. (36 Nm).

ENGINE MECHANICAL

ACCESSORY DRIVE BELTS

ACCESSORY BELT ROUTING

See Figures 38 and 39.

INSPECTION

Inspect the drive belt for signs of glazing or cracking. A glazed belt will be perfectly smooth from slippage, while a good belt will have a slight texture of fabric visible. Cracks will usually start at the inner edge of the belt and run outward. All worn or damaged drive belts should be replaced immediately.

REMOVAL & INSTALLATION

See Figures 38 through 40.

1. Before servicing the vehicle, refer to the Precautions Section.

➡If working near and/or around the SRS system and components, be sure to disable the SRS system. Tape the negative battery cable with insulating tape. Always disconnect the negative battery cable first.

⁂ CAUTION

To avoid personal injury when working on vehicles equipped with an air bag, the negative battery cable must be disconnected and at least 90 seconds must elapse before working on the system. Failure to do so may result in deployment of the air bag.

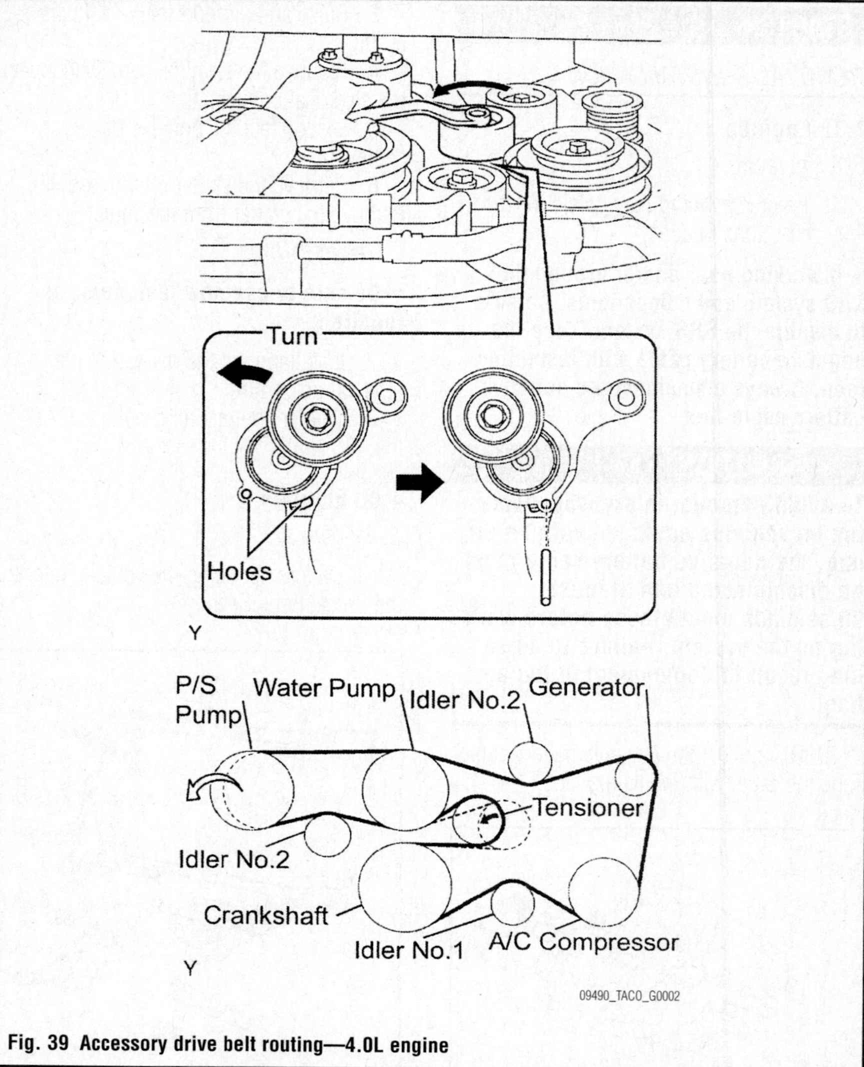

Fig. 39 Accessory drive belt routing—4.0L engine

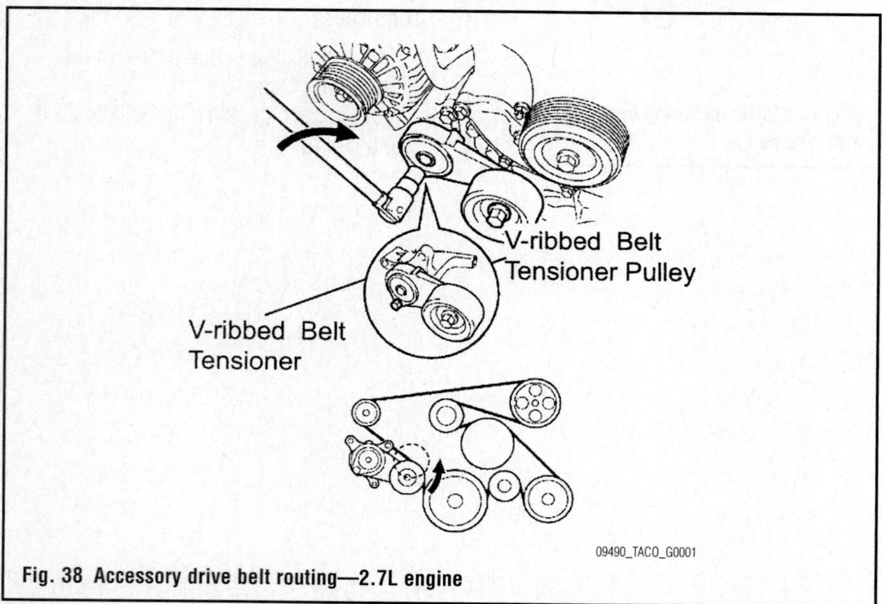

Fig. 38 Accessory drive belt routing—2.7L engine

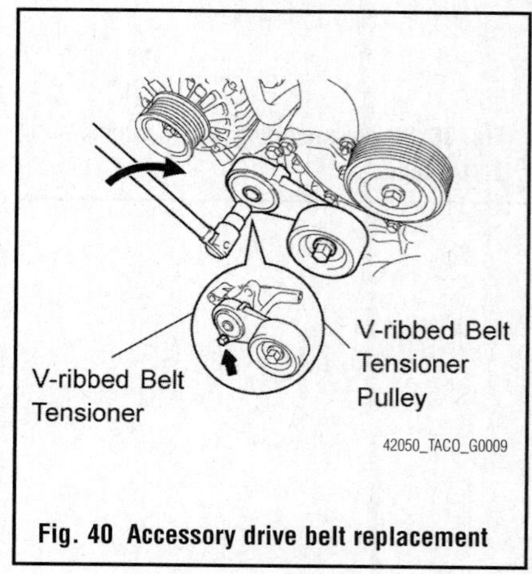

Fig. 40 Accessory drive belt replacement

2. Disconnect the negative battery cable. Tape the cable with insulating tape.
3. Loosen the drive belt tension by turning the drive belt tensioner

counterclockwise, and remove the drive belt.

To install:

➡**Be sure to use new fasteners, as required.**

4. Installation is the reverse of the removal procedure.

CAMSHAFT AND VALVE LIFTERS

INSPECTION

1. Before servicing the vehicle, refer to the Precautions Section.
2. Remove the camshaft from the engine.
3. Check the camshaft bearing journals for damage and binding.
4. If the journals are binding, check the cylinder head for damage.
5. Check the cylinder head for clogged oil holes.
6. Check the camshaft surface for abnormal wear and damage. Replace the camshaft, as required.
7. Measure the camshaft lobe surface and replace the camshaft if not within specification.
8. Measure the camshaft journal diameter and replace the camshaft if not within specification.
9. Measure the camshaft run out and replace the camshaft if not within specification.

REMOVAL & INSTALLATION

2.7L Engine

See Figures 41 through 48.

1. Before servicing the vehicle, refer to the Precautions Section.

➡**If working near and/or around the SRS system and components, be sure to disable the SRS system. Tape the negative battery cable with insulating tape. Always disconnect the negative battery cable first.**

⁂ CAUTION

To avoid personal injury when working on vehicles equipped with an air bag, the negative battery cable must be disconnected and at least 90 seconds must elapse before working on the system. Failure to do so may result in deployment of the air bag.

2. Disconnect the negative battery cable. Tape the cable with insulating tape.
3. On vehicles equipped with 4WD, remove the engine undercover subassembly.
4. Drain the engine coolant. Remove the radiator support to frame seal, left side. Remove the fan shroud.
5. Remove the air cleaner cap subassembly. Remove the intake air connector.
6. Disconnect the ignition coil connectors. Disconnect the throttle body motor connector. Disconnect the VSV connector.
7. Disconnect the Camshaft Position (CMP) sensor connector. Disconnect the engine wire harness clamps. Remove the ignition coils. Disconnect the PCV hose.
8. Remove the cylinder head cover retaining bolts. Remove the cylinder head cover.
9. Remove the two timing chain guide bolts. Remove the timing chain guide. Remove the O-ring.
10. Position the number one cylinder at TDC on the compression stroke.

➡**Turn the crankshaft pulley clockwise to align the timing mark notch with the timing mark "0". Paint marks on the timing chain plates that align with the timing marks on the camshaft timing gear.**

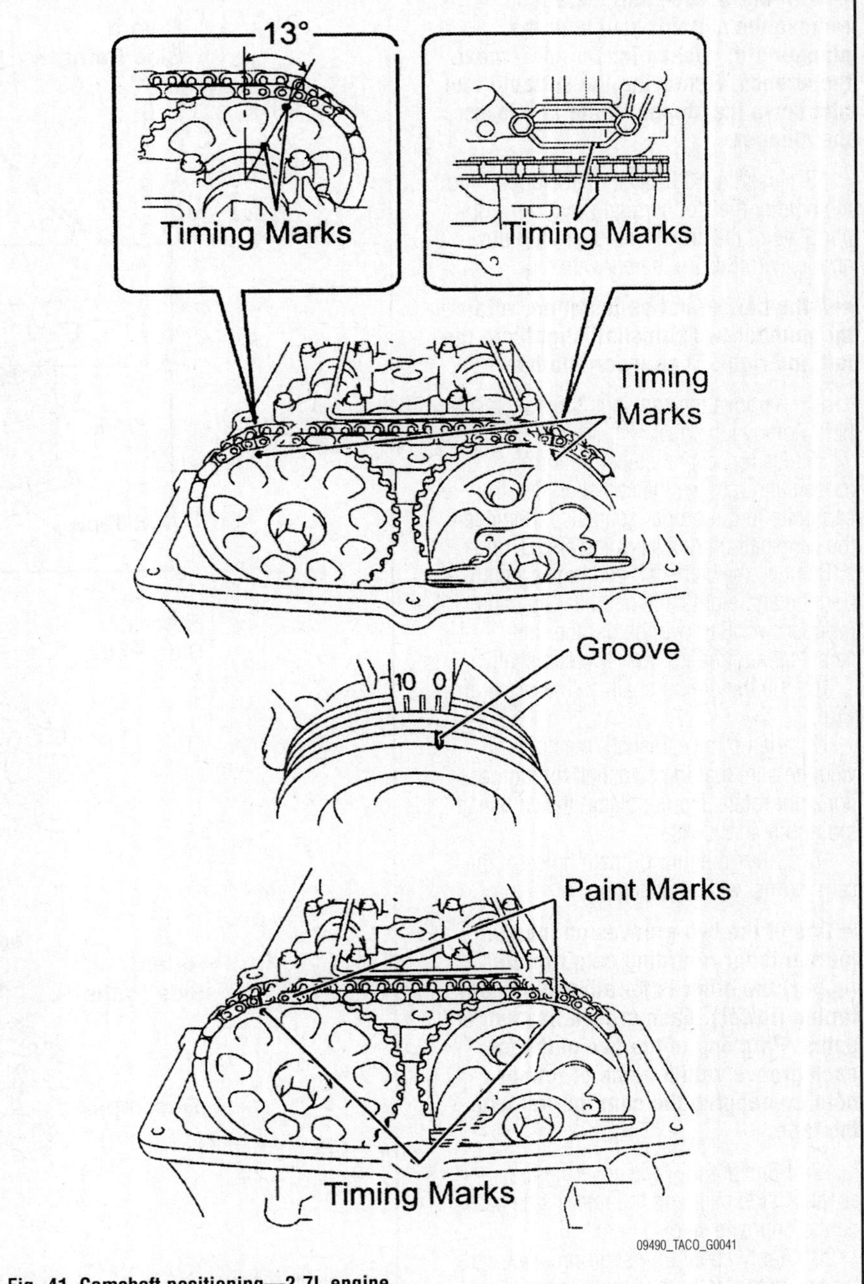

Fig. 41 Camshaft positioning—2.7L engine

11. Hold the hexagonal lobe of the number two camshaft, with a suitable tool. Loosen the bolt. Remove the head straight screw plug. Insert a suitable tool into the service hole of the chain tensioner to hold the stopper plate of the chain tensioner lifted up.

➡Lifting up the stopper plate of the chain tensioner unlocks the plunger.

12. While keeping the stopper plate of the chain tensioner lifted up, slightly rotate the hexagonal lobe of the number two camshaft clockwise so that the plunger of the chain tensioner is pushed. Be careful not to damage the camshaft oil delivery pipe.

➡With the wrench still installed, remove the suitable tool with the plunger still pushed in. Do not remove the wrench. Removing the suitable tool lifts down the stopper plate and locks the plunger.

13. Insert a 0.118 inch diameter bar into the hole of the stopper plate with the stopper plate of the chain tensioner lifter down and locked. Secure the bar with tape.

➡If the bar cannot be installed, rotate the number two camshaft slightly to the left and right. Then insert the bar.

14. Remove the camshaft timing gear bolt. Remove the gear.

15. Using several steps, uniformly loosen and remove the camshaft bearing cap bolts in the proper sequence. Remove the camshaft oil delivery pipe and O-ring.

Remove the camshaft bearing cap number one and eight camshaft bearing caps number two. Remove the number one camshaft and the number two camshaft.

16. Tie the timing chain with a piece of wire.

17. Clamp the camshaft in a soft jaw vise, be sure that the camshaft timing gear does not rotate. Do not clamp the camshaft too tightly in the vise.

18. Cover the four oil path holes of the cam journal with vinyl tape.

➡One of the two grooves on the cam journal is for retarding cam timing (upper) the other is for advancing cam timing (lower). Each groove has two oil paths. Plug one of the two paths for each groove with a piece of rubber before wrapping the cam journal with the tape.

19. Puncture the tape covering the advance side path and the retard side path on the opposite side.

20. Apply about 29 psi air pressure into the two paths, from the two punctures.

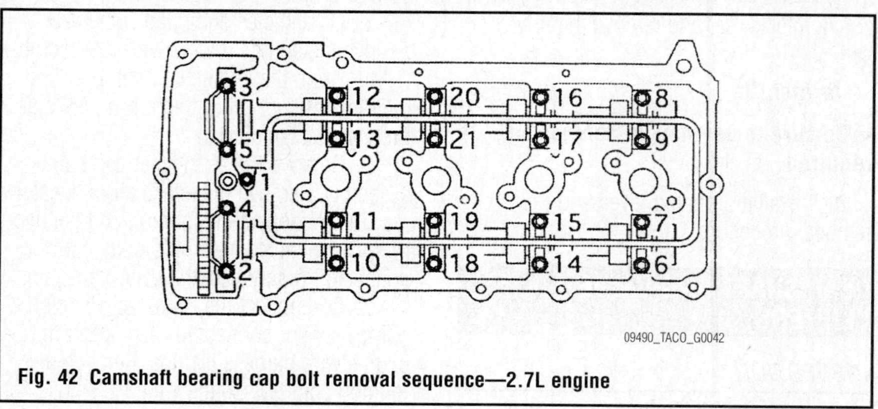

Fig. 42 Camshaft bearing cap bolt removal sequence—2.7L engine

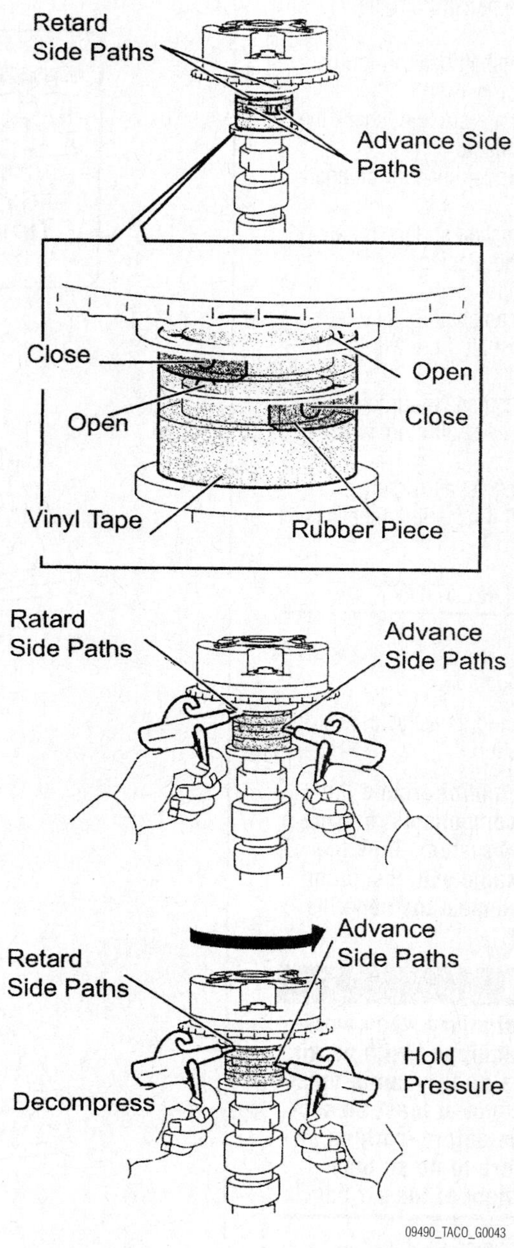

Fig. 43 Camshaft timing gear removal—2.7L engine

When applying air pressure, cover the paths with a shop rag to prevent oil splashes.

21. Confirm that the camshaft timing gear revolves in the advance direction, when reducing the air pressure on the retard side.

➡**The lock pin is released and the camshaft timing groove revolves in the advance direction.**

22. When the camshaft timing gear reaches the most advanced position, release the air pressure on the retard side path, then release the air pressure on the advance side path.

➡**If the air pressure on the advance path is released first, the camshaft timing gear assembly occasionally shifts in the retard direction abruptly. This may damage the lock pin. Be sure to release the air pressure on the retard side first.**

23. Remove the fringe bolt of the camshaft timing gear.

➡**Do not remove the other three bolts.**

To install:

➡**Be sure to use new fasteners, as required.**

24. Clean all surfaces.
25. Put the camshaft timing gear and the camshaft together by aligning the key groove and the straight pin.
26. Gently press the gear against the camshaft and turn the gear. Push further at the position where the pin fits the groove.

➡**Be sure not to turn the camshaft timing gear to the retard angle side (to the right angle).**

27. Check that there is no clearance between the gear's fringe and the camshaft. Tighten the fringe bolt to 58 ft. lbs. (78 Nm).
28. Check that the camshaft timing gear can move to the retard side (the right angle) and is locked in the extreme retard position.
29. Check that the valve rocker arm is correctly installed. Apply clean engine oil to the camshaft's cam portion and the cylinder head journals.
30. Install the chain onto the camshaft timing gear, with the painted mark of the link aligned with the timing mark of the camshaft timing gear.
31. Position the two camshafts in there mounting on the engine, see illustration.

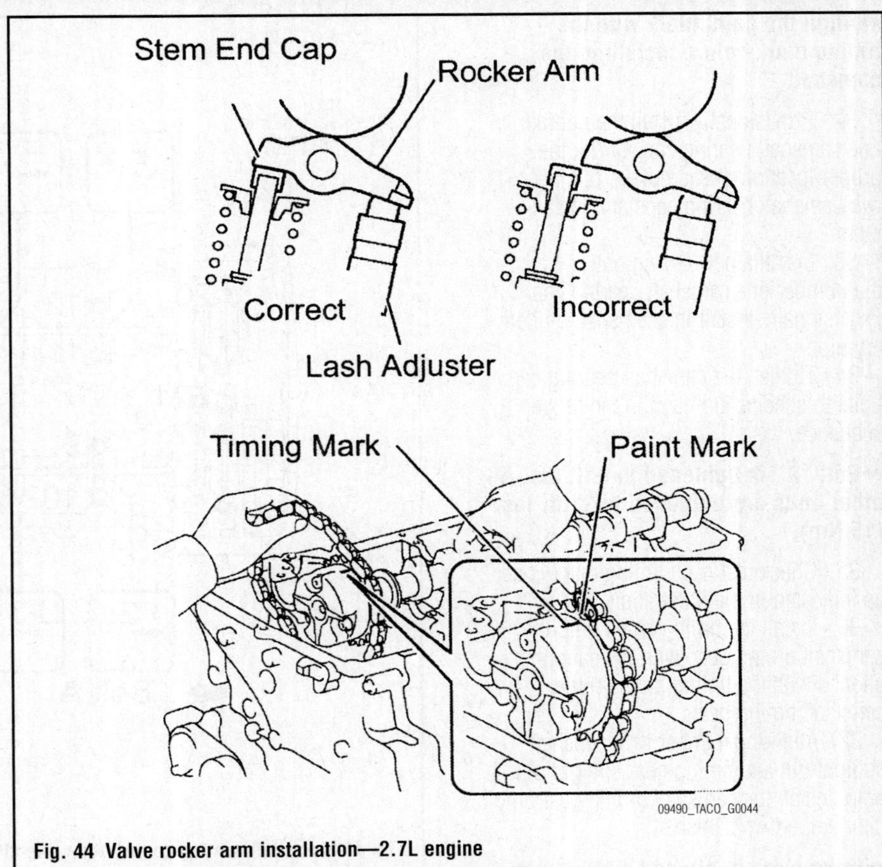

Fig. 44 Valve rocker arm installation—2.7L engine

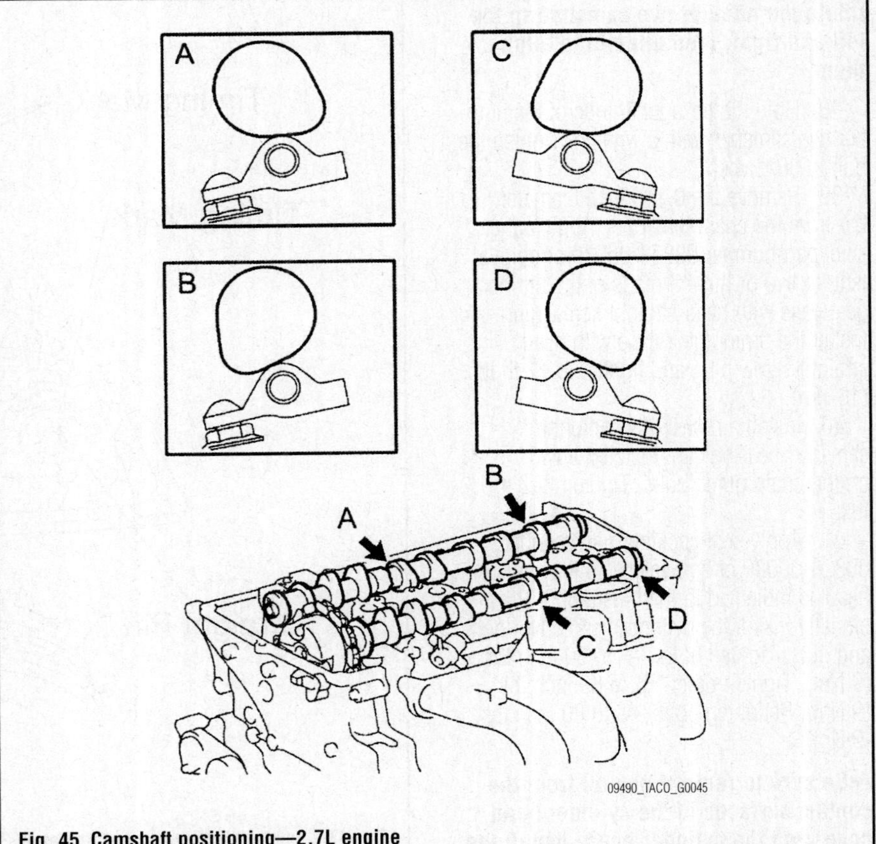

Fig. 45 Camshaft positioning—2.7L engine

➡Align the paint mark with the timing mark before installing the camshaft.

32. Provisionally install the number one camshaft bearing cap. Check the proper location of each of the number two camshaft bearing caps and install them.

33. Install a new O-ring onto the number one camshaft bearing cap. Provisionally install the camshaft oil delivery pipe.

34. Tighten the camshaft bearing cap bolts to specification and in the proper sequence.

➡Bolt "A" is tightened to 9 ft. lbs. All other bolts are tightened to 11 ft. lbs. (15 Nm).

35. Check that each timing mark is set as indicated in the illustration.

36. Install the timing chain onto the camshaft timing gear, with the paint mark aligned with the timing mark on the camshaft timing gear.

37. Align the number two camshaft straight pin and timing gear straight pin hole. Install the camshaft timing gear onto the number two camshaft.

➡If the straight pin and straight pin hole are difficult to align, slightly rotate the number two camshaft to the left and right, then attempt to align them.

38. Hold the hexagonal lobe of the number two camshaft with a wrench. Tighten the bolt to 58 ft. lbs.

39. Remove the 0.118 inch diameter bar from the chain tensioner. Apply adhesive, part number 08833-00070 or equivalent, to two or three threads of the timing gear case with head straight screw plug. Install the timing gear case with head straight screw plug and tighten to 12 ft. lbs. (16 Nm).

40. Install a new O-ring onto the camshaft bearing cap. Install the two timing chain guide bolts. Tighten to 7 ft. lbs.

41. Apply seal packing, part number 08826-00080 or equivalent, to the cylinder head as indicated in the illustration. Provisionally install the cylinder head cover bolts and nuts. Tighten bolts "A" to 80 inch lbs. (9 Nm). Tighten bolts "B" to 80 inch lbs. (9 Nm). Retighten bolts "A" to 80 inch lbs. (9 Nm).

➡Be sure to remove any oil from the contact surfaces of the cylinder head cover and the cylinder head. Install the

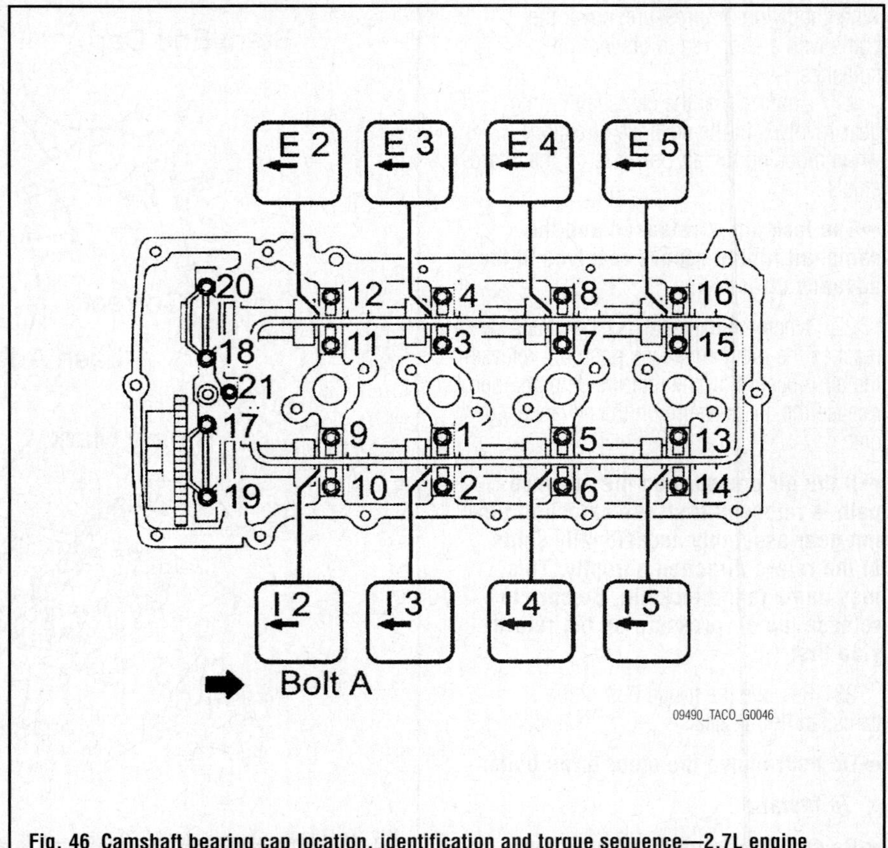

Fig. 46 Camshaft bearing cap location, identification and torque sequence—2.7L engine

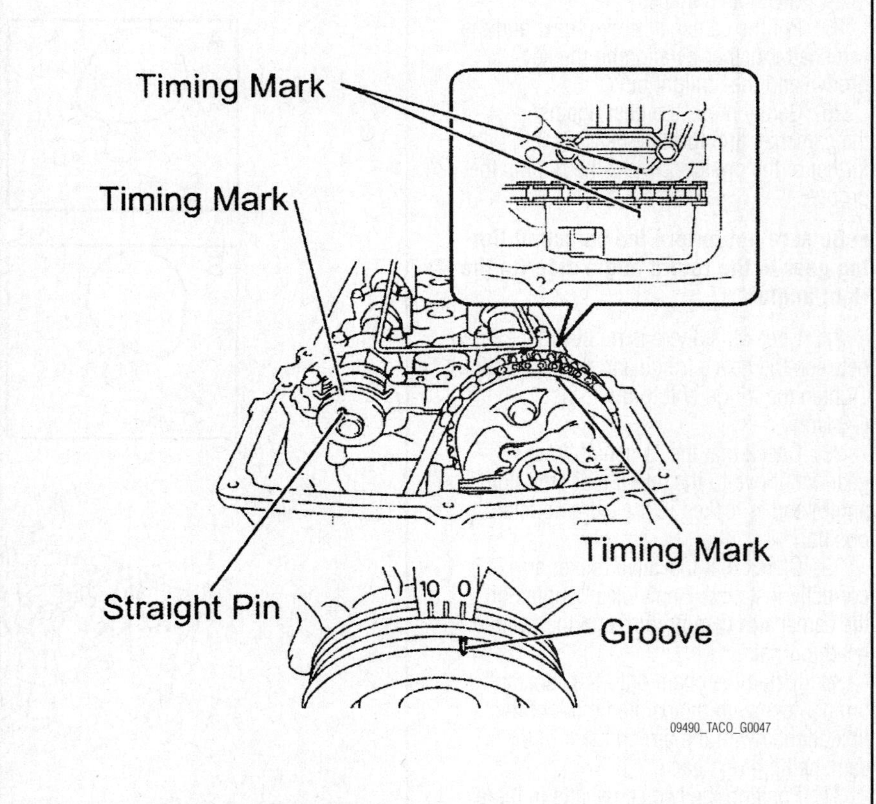

Fig. 47 Camshaft timing mark alignment—2.7L engine

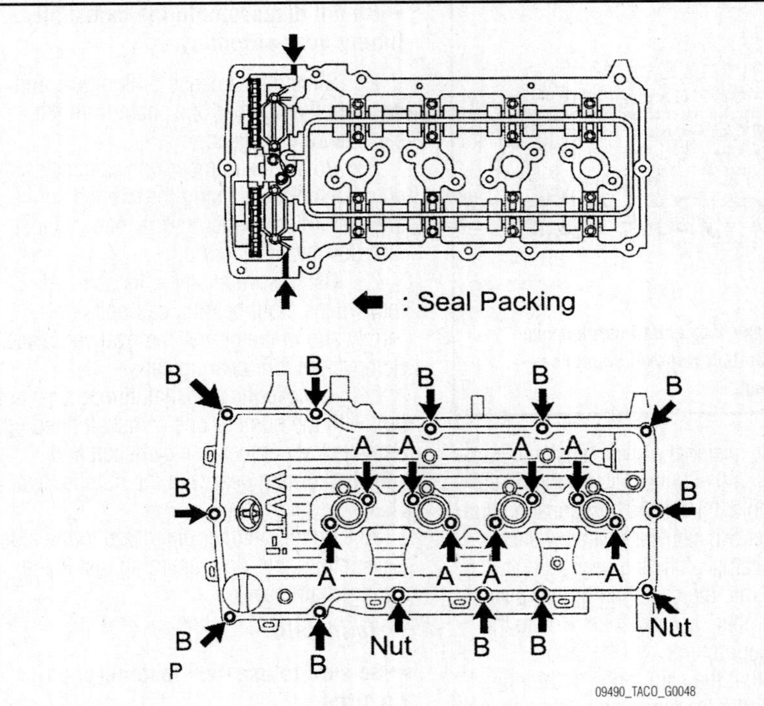

Fig. 48 Cylinder head cover bolt sealant application and bolt identification—2.7L engine

cover within three minutes after applying the seal packing. Do not add engine oil for at least two hours after installing the cover.

42. Continue the installation in the reverse order of the removal procedure.
43. Be sure to fill the engine with the proper grade and type engine coolant.
44. Be sure to fill the engine with the proper grade and type engine oil.
45. Start the engine and check for leaks. Correct as required.

4.0L Engine

Bank 1

See Figures 49 through 63.

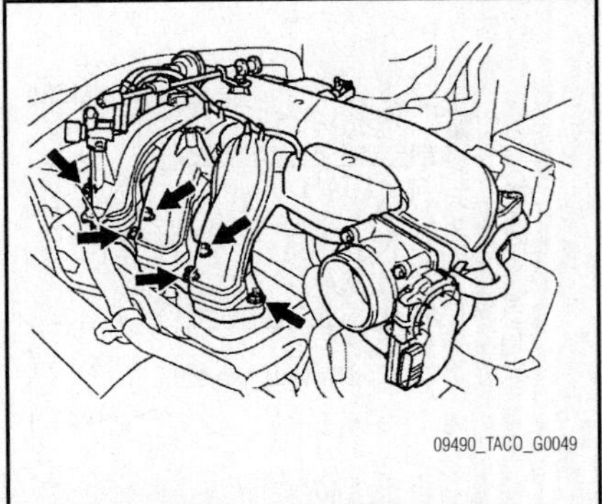

Fig. 49 Intake surge tank bolt locations—4.0L engine

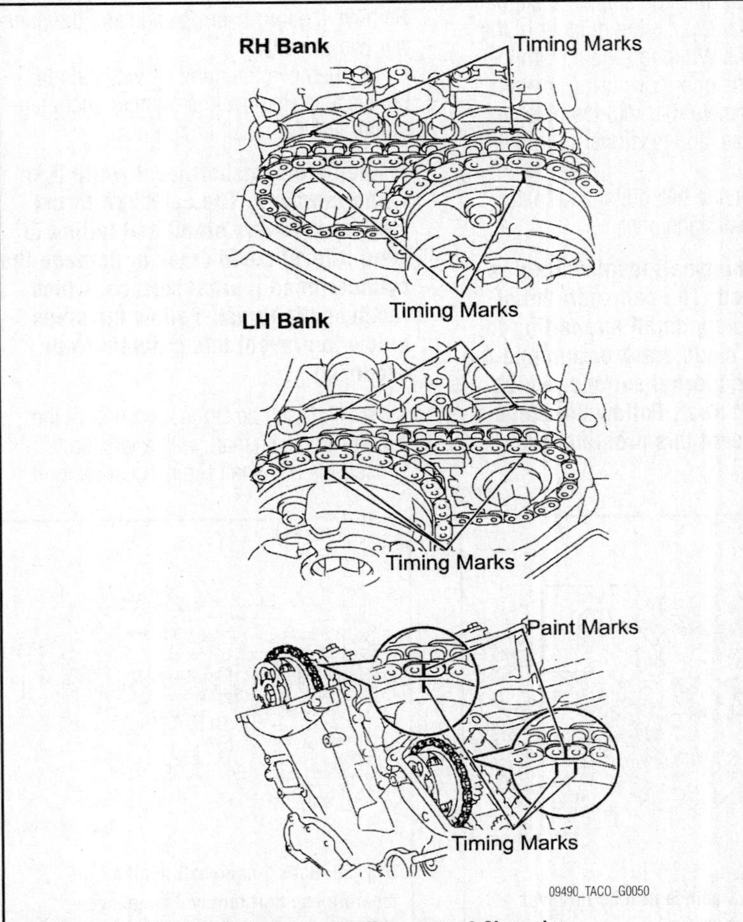

Fig. 50 Bank 1 camshaft timing mark alignment—4.0L engine

1. Before servicing the vehicle, refer to the Precautions Section.

➡If working near and/or around the SRS system and components, be sure to disable the SRS system. Tape the negative battery cable with insulating tape. Always disconnect the negative battery cable first.

⁂ CAUTION

To avoid personal injury when working on vehicles equipped with an air bag, the negative battery cable must be disconnected and at least 90 seconds must elapse before working on the system. Failure to do so may result in deployment of the air bag.

2. Disconnect the negative battery cable. Tape the cable with insulating tape.
3. Drain the engine coolant. Remove the V bank cover.
4. Remove the air cleaner assembly.
5. Disconnect the two water bypass hoses. Disconnect the fuel vapor feed hose. Disconnect the ventilation hose. Disconnect the VSV connectors.

6. Disconnect the throttle body motor connector. Separate the three wire harness clamps and hose clamp.

7. If equipped with manual transmission, remove the nut, then separate the clutch flexible hose bracket from the surge tank stay.

8. Remove the two bolts and the throttle body bracket. Remove the bolt and the oil baffle plate. Remove the four bolts and the two serge tank stays.

9. Remove the two nuts. Remove the four bolts, intake air surge tank and gasket.

10. Remove the ignition coil assembly.

11. Remove the cylinder head cover retaining bolts. Remove the cylinder head cover.

12. Turn the crankshaft pulley until its groove and the timing mark "0" of the timing chain cover are aligned. If not aligned at TDC of the compression stroke, turn the crankshaft one complete revolution, in the direction of rotation. Paint alignment marks on the number one chain links corresponding to the timing marks of the camshaft timing gears.

13. Remove the four bolts, and then remove the timing chain cover plate and gasket.

14. While turning the stopper plate of the tensioner upward, push the plunger of the chain tensioner. While turning the stopper plate of the tensioner downward, insert a 0.118 inch diameter bar into the holes in the stopper plate and tensioner to hold the stopper plate.

15. Remove the two bolts, and then remove the chain tensioner.

➡Keep the camshaft level while it is being removed. The camshaft thrust clearance is very small and failing to keep it level could crack or damage the cylinder head journal surface, which receives the thrust. Follow the steps below to prevent this problem from occurring.

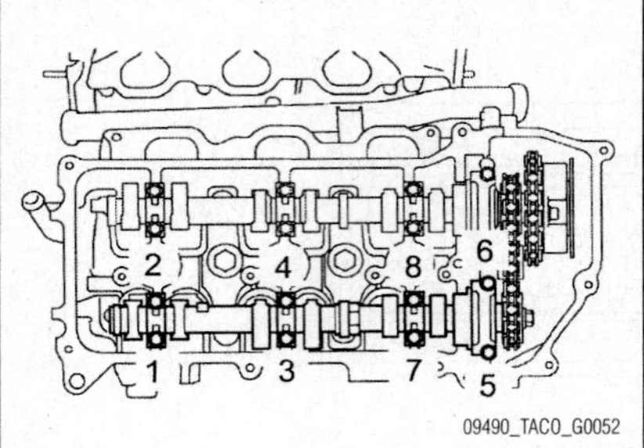

Fig. 52 Bank 1 camshaft number two bearing cap bolt removal sequence—4.0L engine

16. While raising the chain tensioner number two insert a 0.039 inch diameter pin into the hole to hold it. Hold the hexagonal portion of the number two camshaft with a wrench. Remove the camshaft timing gear set bolt.

17. Separate the camshaft timing gear from the number two camshaft. Rotate the camshafts counterclockwise, using a wrench, so that the cam lobes of the number one cylinder face in the direction shown.

18. Using several steps, loosen and remove the eight bearing cap bolts uniformly and in the proper removal sequence. Remove the four bearing caps and the number two camshaft.

19. Remove the number two chain tensioner bolt. Remove the number chain tensioner and camshaft timing gear.

➡Keep the camshaft level while it is being removed. The camshaft thrust clearance is very small and failing to keep it level could crack or damage the cylinder head journal surface, which receives the thrust. Follow the steps below to prevent this problem from occurring.

20. Hold the hexagonal portion of the number one camshaft, with a wrench. Loosen the camshaft timing gear set bolt.

➡Do not disassemble the camshaft timing gear assembly.

21. Slide the camshaft timing gear and separate the number one chain from the camshaft timing gear.

22. Rotate the number one camshaft counterclockwise, using the wrench so that the cam lobes on the number one cylinder face downward.

23. Using several steps, loosen and remove the eight bearing cap bolts uniformly and in the proper removal sequence. Remove the four bearing caps.

24. Remove the camshaft timing gear set bolt with the number one camshaft lifted up. Remove the number one camshaft and camshaft timing gear with the number two chain.

25. Tie the number one chain to the side. Be careful not to drop anything inside the timing chain cover.

To install:

➡Be sure to use new fasteners, as required.

➡Keep the camshaft level while it is being installed. The camshaft thrust clearance is very small and failing to keep it level could crack or damage the cylinder head journal surface, which receives the thrust.

26. Align the yellow mark link with the timing mark (1 dot mark) of the camshaft timing gear. Apply new engine oil to the thrust portion and journal of the camshafts.

27. Temporarily install the number one chain onto the number two chain of the camshaft timing gear.

28. Align the knock pin hole of the camshaft timing gear with the knock pin of the number one camshaft. Insert the number one camshaft into the camshaft timing gear.

29. Temporarily install the camshaft timing gear set bolt. Install the number one

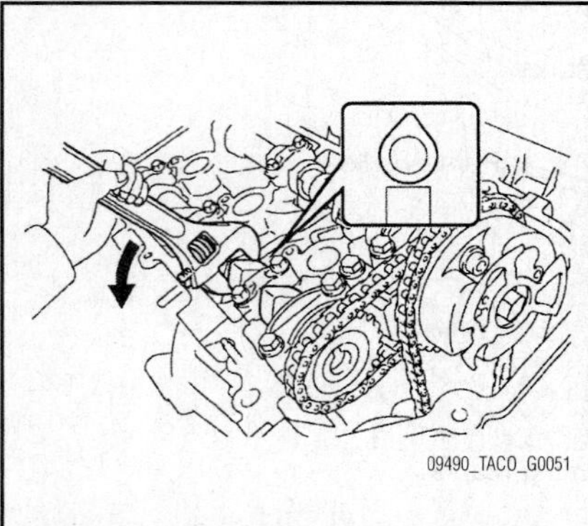

Fig. 51 Bank 1 camshaft lobe removal positioning—4.0L engine

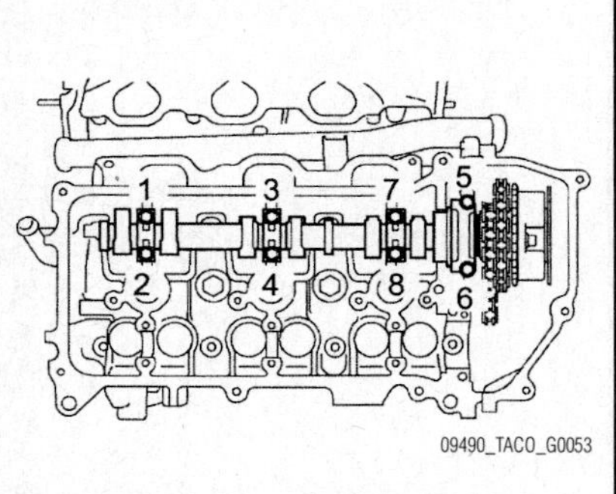

Fig. 53 Bank 1 camshaft number one bearing cap bolt removal sequence—4.0L engine

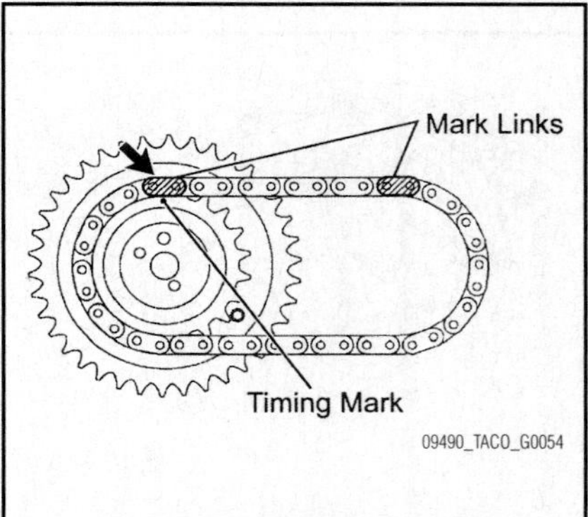

Fig. 54 Bank 1 camshaft number one alignment marks—4.0L engine

camshaft onto the right cylinder head with the cam lobes of the number one cylinder facing downward, as indicated in the illustration.

30. Install the four bearing caps, in the proper location. Apply a light coat of engine oil to the threads and under the heads of the cap bolts.

31. Using several steps, uniformly install and tighten the bearing cap bolts in the proper sequence to 80 inch lbs. (9 Nm) for the 10mm bolts and 18 ft. lbs. (24 Nm) for the 12mm bolts.

32. Rotate the number one camshaft clockwise, using a wrench so that the timing mark of the camshaft timing gear is aligned with the timing mark of the camshaft bearing cap.

33. Align the paint mark of the number one chain with the timing mark of the camshaft timing gear.

34. Hold the hexagonal portion of the number one camshaft with a wrench, and tighten the camshaft timing gear set bolt to 74 ft. lbs.

35. While pushing in on the number two chain tensioner, insert a 0.039 inch pin into the hole to hold it.

36. Temporarily install the camshaft timing gear and chain tensioner number two and align the yellow mark links with the timing marks (1 dot mark) on the camshaft timing gears. Tighten the bolt to 14 ft. lbs. (19 Nm).

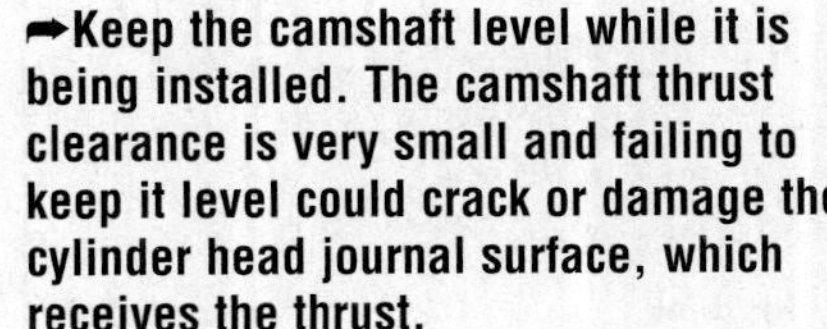

➡Keep the camshaft level while it is being installed. The camshaft thrust clearance is very small and failing to keep it level could crack or damage the cylinder head journal surface, which receives the thrust.

37. Install the number two camshaft onto the right cylinder head with the cam lobes of the number one cylinder facing upward, as indicated in the illustration.

38. Install the four bearing caps in the proper location. Apply a light coat of clean engine oil to the threads and under the heads of the bolts.

39. Using several steps, uniformly install and tighten the eight bearing cap bolts in sequence to 80 inch lbs. (9 Nm), for the 10mm head bolts and 18 ft. lbs. (25 Nm) for the 12mm head bolts.

40. Rotate the number two camshaft clockwise, using a wrench, so that the lock pin of the number two camshaft is aligned with the knock pin hole of the camshaft timing gear.

41. Hold the hexagonal portion of the number two camshaft, with a wrench, and install the camshaft timing gear set bolt and tighten it to 74 ft. lbs. (100 Nm).

42. Remove the pin from the number two chain tensioner.

43. While turning the stopper plate of the tensioner clockwise, push in the plunger of the tensioner. While turning the stopper plate of the tensioner counterclockwise, insert a 0.138 inch bar into the holes in the stopper plate and tensioner to hold the stopper plate. Install the two chain tensioner bolts and tighten to 7.4 ft. lbs. (10 Nm).

44. Remove the bar from the chain tensioner. Install a new gasket and the timing chain cover plate. Torque the bolts to 80 inch lbs.

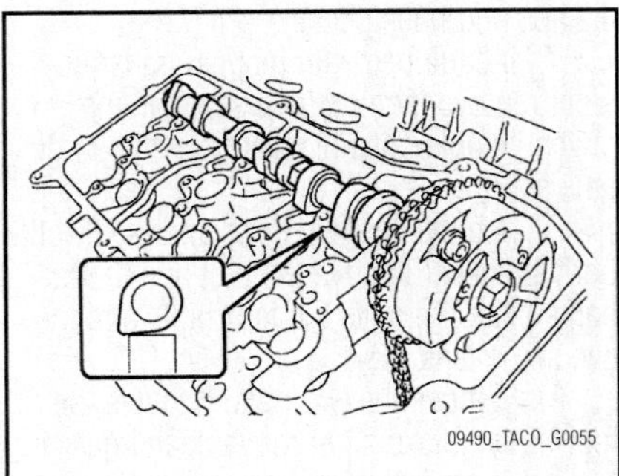

Fig. 55 Bank 1 camshaft number one lobe installation positioning—4.0L engine

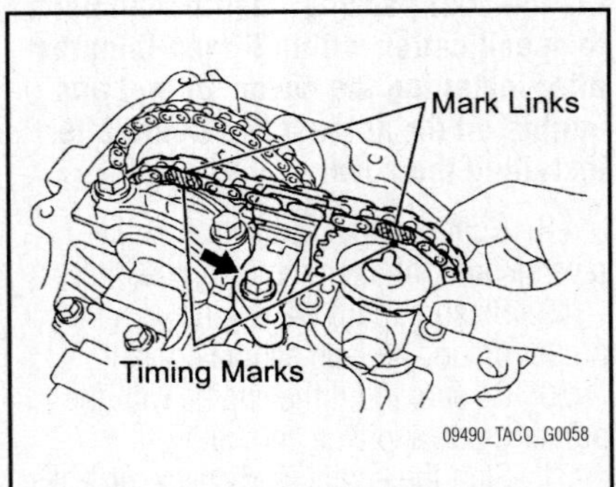

Fig. 58 Bank 1 camshaft number two alignment marks—4.0L engine

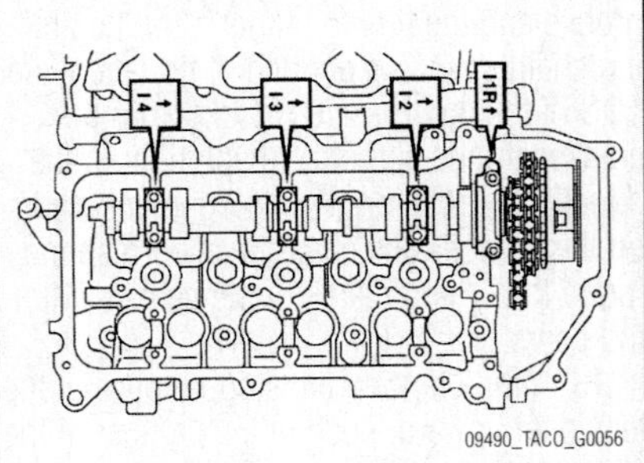

Fig. 56 Bank 1 camshaft number one bearing cap bolt installation sequence—4.0L engine

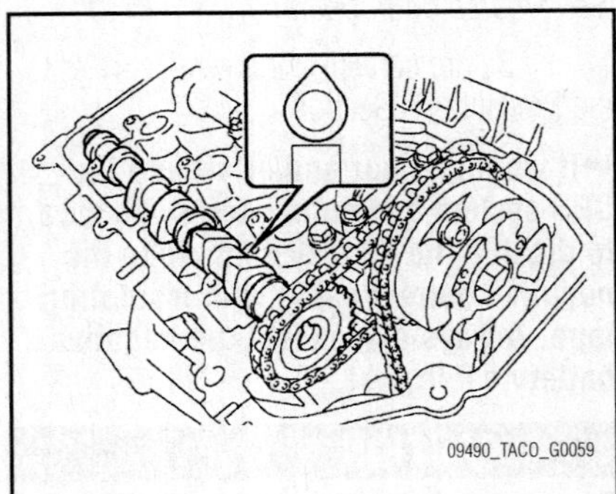

Fig. 59 Bank 1 camshaft number two lobe installation positioning—4.0L engine

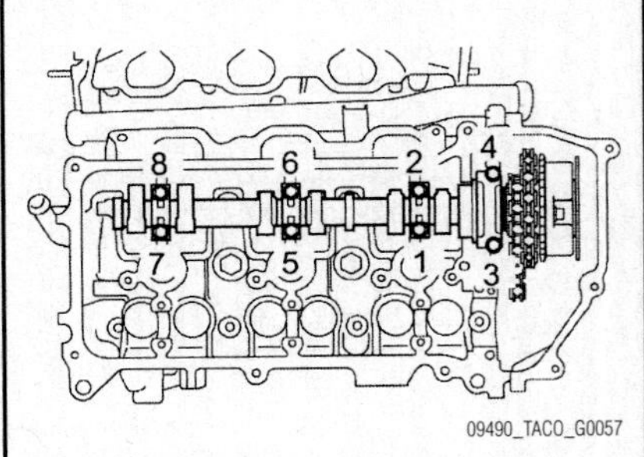

Fig. 57 Bank 1 camshaft number one bearing cap bolt torque sequence—4.0L engine

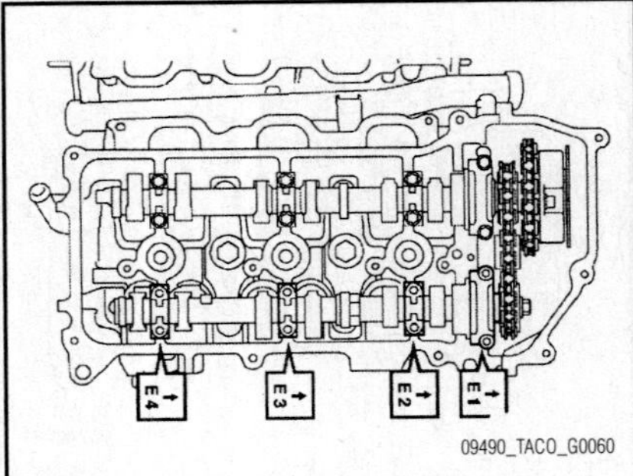

Fig. 60 Bank 1 camshaft number two bearing cap bolt installation sequence—4.0L engine

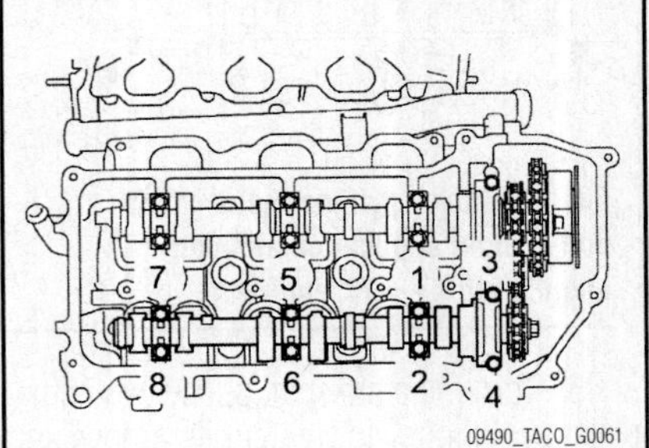

Fig. 61 Bank 1 camshaft number two bearing cap bolt torque sequence—4.0L engine

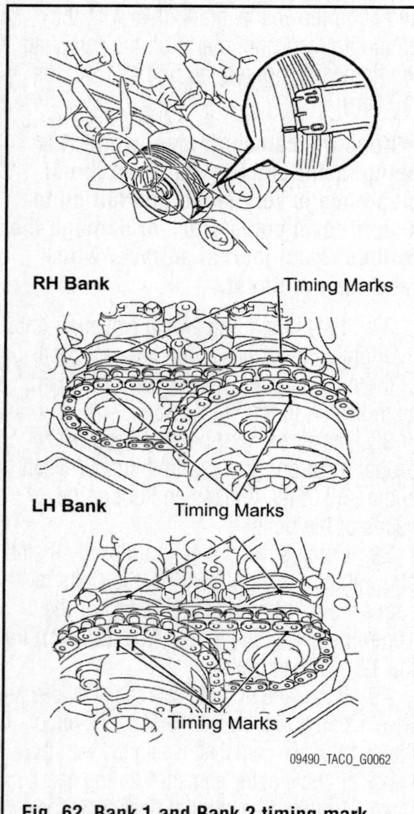

Fig. 62 Bank 1 and Bank 2 timing mark alignment—4.0L engine

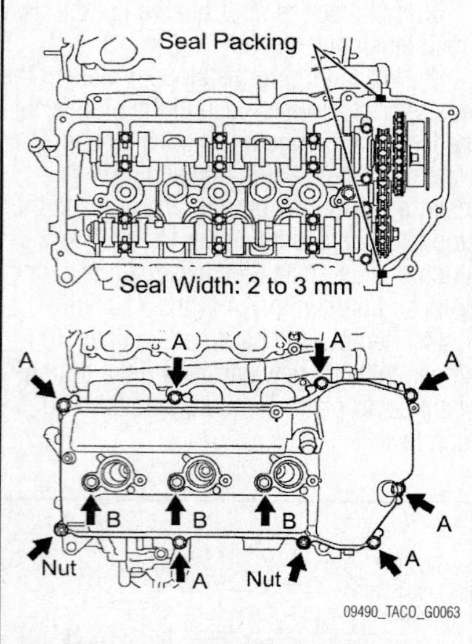

Fig. 63 Bank 1 cylinder head cover bolt sealant application and bolt identification—4.0L engine

45. Turn the crankshaft pulley two complete revolutions slowly until its groove and the timing mark "0" of the timing chain cover are aligned.

46. Position the number one cylinder at TDC of the compression stroke. Inspect the valve clearance, adjust as required.

47. Apply a continuous bead (0.08–0.12 inch) of seal packing, part number 08826-00080 or equivalent, to the cylinder head as indicated in the illustration. Install the seal washers onto the bolts. Install the cylinder head cover bolts and nuts. Tighten bolts "A" to 7.4 ft. lbs. (10 Nm). Tighten bolts "B" to 80 inch lbs. (9 Nm). Tighten nuts to 80 inch lbs. (9 Nm).

➡Be sure to remove any oil from the contact surfaces of the cylinder head cover and the cylinder head. Install the cover within three minutes after applying the seal packing. Tighten the bolts to specification within fifteen minutes after installing the cover. Do not add engine oil for at least two hours after installing the cover.

48. Continue the installation in the reverse order of the removal procedure.

49. Be sure to fill the engine with the proper grade and type engine coolant.

50. Be sure to fill the engine with the proper grade and type engine oil.

51. Start the engine and check for leaks. Correct as required.

Bank 2

See Figures 64 through 72.

1. Before servicing the vehicle, refer to the Precautions Section.

➡If working near and/or around the SRS system and components, be sure to disable the SRS system. Tape the negative battery cable with insulating tape. Always disconnect the negative battery cable first.

CAUTION

To avoid personal injury when working on vehicles equipped with an air bag, the negative battery cable must be disconnected and at least 90 seconds must elapse before working on the system. Failure to do so may result in deployment of the air bag.

2. Disconnect the negative battery cable. Tape the cable with insulating tape.

3. Drain the engine coolant. Remove the V bank cover.

4. Remove the air cleaner assembly.

5. Disconnect the two water bypass hoses. Disconnect the fuel vapor feed hose. Disconnect the ventilation hose. Disconnect the VSV connectors.

6. Disconnect the throttle body motor connector. Separate the three wire harness clamps and hose clamp.

7. If equipped with manual transmission, remove the nut, then separate the clutch flexible hose bracket from the surge tank stay.

8. Remove the two bolts and the throttle body bracket. Remove the bolt and the oil baffle plate. Remove the four bolts and the two serge tank stays.

9. Remove the two nuts. Remove the four bolts, intake air surge tank and gasket.

10. Remove the ignition coil assembly.

11. Remove the cylinder head cover retaining bolts. Remove the cylinder head cover.

12. Turn the crankshaft pulley until its groove and the timing mark "0" of the timing chain cover are aligned. If not aligned at TDC of the compression stroke, turn the crankshaft one complete revolution, in the direction of rotation. Paint alignment marks on the number one chain links corresponding to the timing marks of the camshaft timing gears.

13. While turning the stopper plate of the tensioner upward, push in the plunger of the chain tensioner. While turning the stopper plate of the tensioner downward, insert a 0.138 inch bar into the holes in the stopper plate and tensioner to hold the stopper plate.

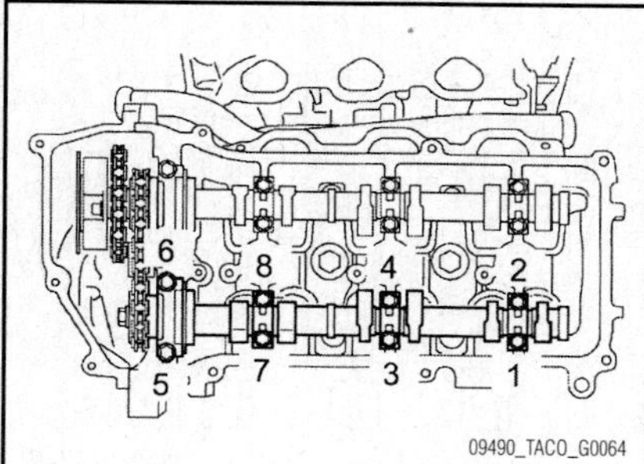

Fig. 64 Bank 2 camshaft number four bearing cap bolt removal sequence—4.0L engine

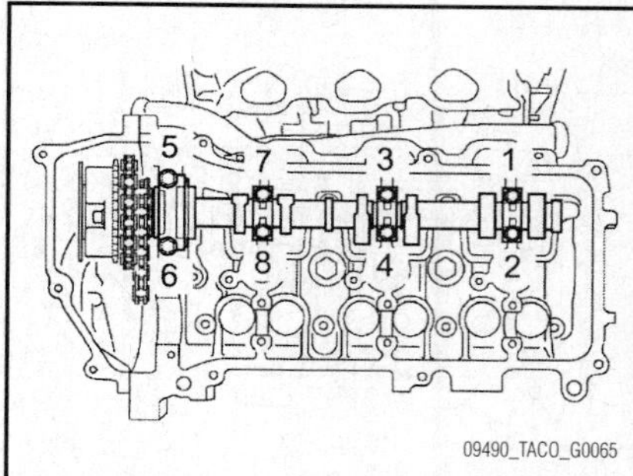

Fig. 65 Bank 2 camshaft number three bearing cap bolt removal sequence—4.0L engine

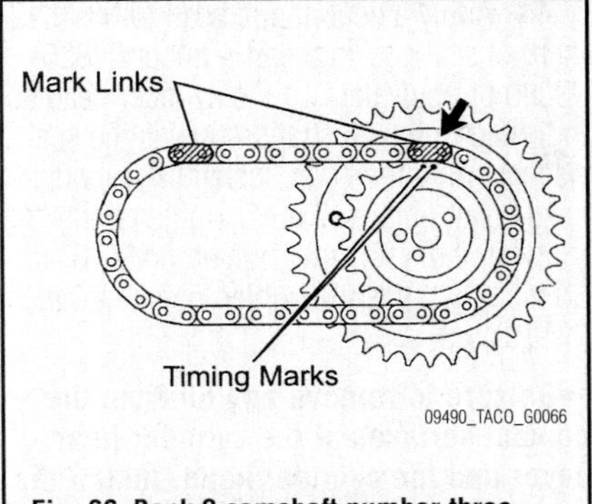

Fig. 66 Bank 2 camshaft number three alignment marks—4.0L engine

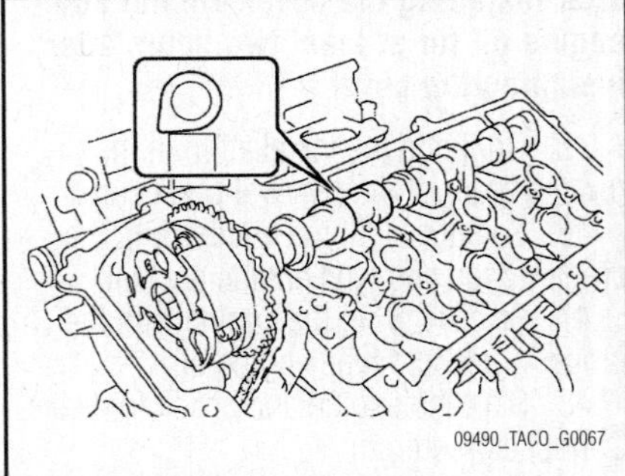

Fig. 67 Bank 2 camshaft number three lobe installation positioning—4.0L engine

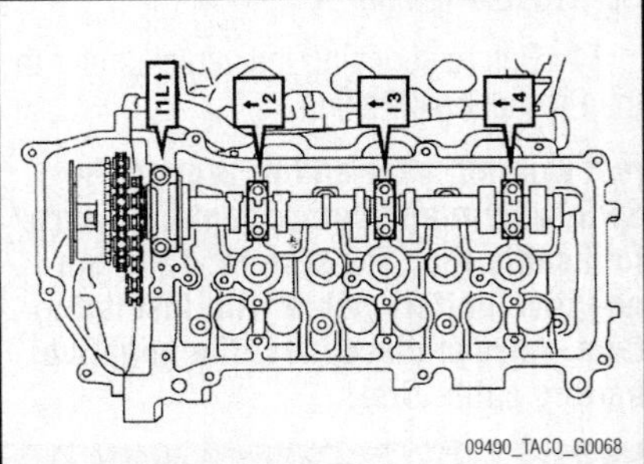

Fig. 68 Bank 2 camshaft number three bearing cap bolt installation sequence—4.0L engine

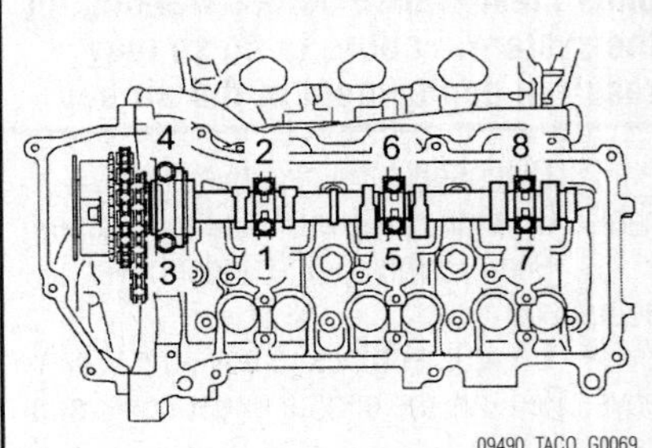

Fig. 69 Bank 2 camshaft number three bearing cap bolt torque sequence—4.0L engine

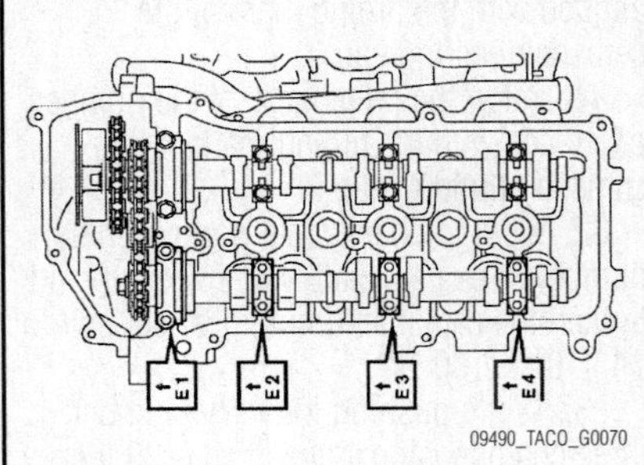

Fig. 70 Bank 2 camshaft number four bearing cap bolt installation sequence—4.0L engine

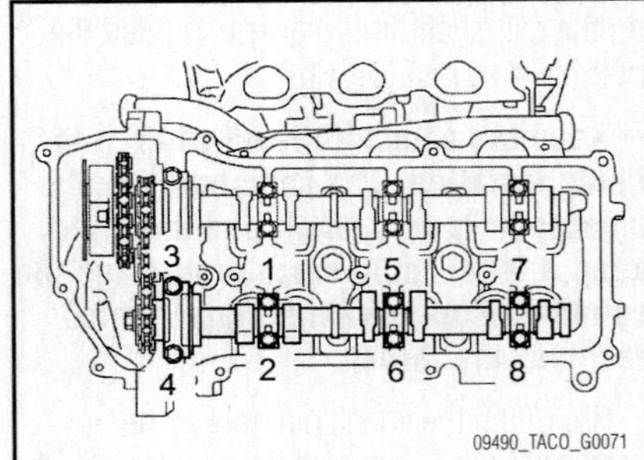

Fig. 71 Bank 2 camshaft number four bearing cap bolt torque sequence—4.0L engine

Remove the two bolts and then remove the number one chain tensioner assembly.

➡Never rotate the crankshaft with the chain tensioner removed. When rotating the camshaft with the tensioner removed, rotate the crankshaft counterclockwise forty degrees from TDC, first.

➡Keep the camshaft level while it is being removed. The camshaft thrust clearance is very small and failing to keep it level could crack or damage the cylinder head journal surface, which receives the thrust. Follow the steps below to prevent this problem from occurring.

14. While pushing down on chain tensioner number three insert a 0.039 inch diameter pin into the hole to hold it. Hold the hexagonal portion of the number four camshaft with a wrench. Remove the camshaft timing gear set bolt.

15. Separate the camshaft timing gear from the number four camshaft.

16. Using several steps, loosen and remove the eight bearing cap bolts uniformly and in the proper removal sequence. Remove the four bearing caps and the number four camshaft.

17. Remove the number three chain tensioner bolt. Remove the number chain tensioner and camshaft timing gear.

➡Keep the camshaft level while it is being removed. The camshaft thrust clearance is very small and failing to keep it level could crack or damage the cylinder head journal surface, which receives the thrust. Follow the steps below to prevent this problem from occurring.

18. Release the chain tension between the camshaft gear (LH bank) and the crank-

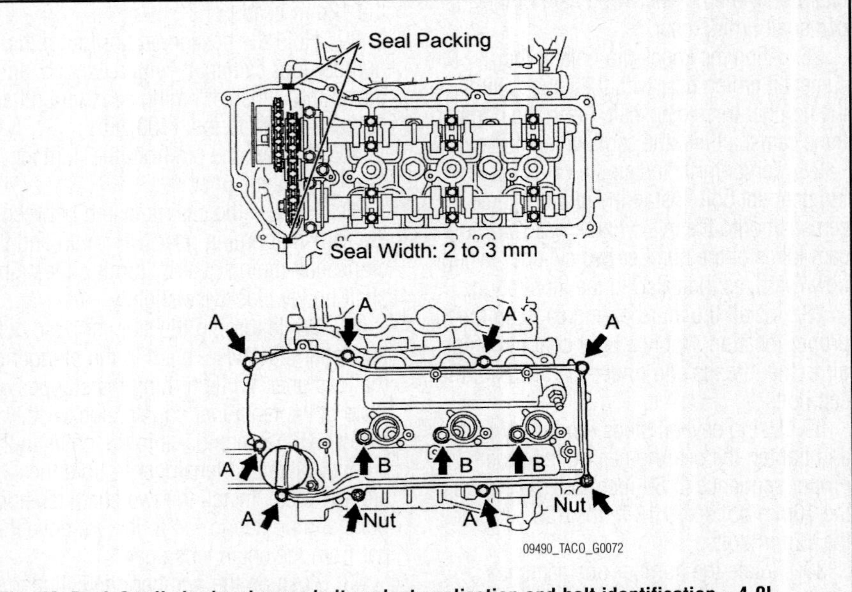

Fig. 72 Bank 2 cylinder head cover bolt sealant application and bolt identification—4.0L engine

shaft timing gear by turning the crankshaft pulley counterclockwise slightly.

19. Hold the hexagonal portion of the number three camshaft, with a wrench. Loosen the camshaft timing gear set bolt.

➡Do not disassemble the camshaft timing gear assembly.

20. Slide the camshaft timing gear and separate the number one chain from the camshaft timing gear.

21. Using several steps, loosen and remove the eight bearing cap bolts uniformly and in the proper removal sequence. Remove the four bearing caps.

22. Remove the camshaft timing gear set bolt with the number three camshaft lifted up. Remove the number three camshaft and camshaft timing gear with the number two chain.

23. Tie the number one chain to the side. Be careful not to drop anything inside the timing chain cover.

To install:

➡Be sure to use new fasteners, as required.

➡Keep the camshaft level while it is being installed. The camshaft thrust clearance is very small and failing to keep it level could crack or damage the cylinder head journal surface, which receives the thrust.

24. Align the yellow mark link with the timing mark (2 dot mark) of the camshaft timing gear. Apply new engine oil to the thrust portion and journal of the camshafts.

25. Temporarily install the number one chain onto the number two chain of the camshaft timing gear.

26. Align the knock pin hole of the camshaft timing gear with the knock pin of the number three camshaft. Insert the number three camshaft into the camshaft timing gear.

27. Temporarily install the camshaft timing gear set bolt. Install the number three camshaft onto the left cylinder head with the cam lobes of the number two cylinder facing downward, as indicated in the illustration.

28. Install the four bearing caps, in the proper location. Apply a light coat of engine oil to the threads and under the heads of the cap bolts.

29. Using several steps, uniformly install and tighten the bearing cap bolts in the proper sequence to 80 inch lbs. (9 Nm) for the 10mm bolts and 18 ft. lbs. (25 Nm) for the 12mm bolts.

30. Rotate the number one camshaft clockwise, using a wrench so that the timing mark of the camshaft timing gear is aligned with the timing mark of the camshaft bearing cap.

31. Align the paint mark of the number one chain with the timing mark of the camshaft timing gear.

32. Hold the hexagonal portion of the number three camshaft with a wrench, and tighten the camshaft timing gear set bolt to 74 ft. lbs. (100 Nm)

33. While pushing in on the number three chain tensioner, insert a 0.039 inch pin into the hole to hold it.

34. Temporarily install the camshaft timing gear and chain tensioner number three and align the yellow mark links with the timing marks (1 dot mark and 2 dot marks) on the camshaft timing gears. Tighten the bolt to 14 ft. lbs. (19 Nm).

➡Keep the camshaft level while it is being installed. The camshaft thrust clearance is very small and failing to keep it level could crack or damage the cylinder head journal surface, which receives the thrust.

35. Align the knock pin hole in the camshaft timing gear with the knock pin of the number four camshaft, and insert the number four camshaft into the camshaft timing gear.

36. Temporarily install the camshaft timing gear set bolt.

37. Install the four bearing caps in the proper location. Apply a light coat of clean engine oil to the threads and under the heads of the bolts.

38. Using several steps, uniformly install and tighten the eight bearing cap bolts in sequence to 80 inch lbs. (9 Nm), for the 10mm head bolts and 18 ft. lbs. (25 Nm) for the 12mm head bolts.

39. Hold the hexagonal portion of the number four camshaft, with a wrench, and install the camshaft timing gear set bolt and tighten it to 74 ft. lbs. (100 Nm).

40. Remove the pin from the number three chain tensioner.

41. Release the chain tension between the camshaft timing gear (RH bank) and the crankshaft timing gear by turning the crankshaft pulley clockwise slightly.

42. While turning the stopper plate of the tensioner clockwise, push in the plunger of the tensioner. While turning the stopper plate of the tensioner counterclockwise, insert a 0.138 inch bar into the holes in the stopper plate and tensioner to hold the stopper plate. Install the two chain tensioner bolts and tighten to 7.4 ft. lbs. Remove the bar from the chain tensioner.

43. Position the number one cylinder at TDC of the compression stroke. Inspect the valve clearance, adjust as required.

44. Apply a continuous bead (0.08–0.12 inch) of seal packing, part number 08826-00080 or equivalent, to the cylinder head as indicated in the illustration. Install the seal washers onto the bolts. Install the cylinder head cover bolts and nuts. Tighten bolts "A" to 7.4 ft. lbs. (10 Nm). Tighten bolts "B" to 80 inch lbs. (9 Nm). Tighten nuts to 80 inch lbs. (9 Nm).

➡Be sure to remove any oil from the contact surfaces of the cylinder head cover and the cylinder head. Install the cover within three minutes after applying the seal packing. Tighten the bolts to specification within fifteen minutes after installing the cover. Do not add engine oil for at least two hours after installing the cover.

45. Continue the installation in the reverse order of the removal procedure.

46. Be sure to fill the engine with the proper grade and type engine coolant.

47. Be sure to fill the engine with the proper grade and type engine oil.

48. Start the engine and check for leaks. Correct as required.

CRANKSHAFT DAMPER

REMOVAL & INSTALLATION

See Figures 73 and 74.

1. Before servicing the vehicle, refer to the Precautions Section.

➡If working near and/or around the SRS system and components, be sure to disable the SRS system. Tape the negative battery cable with insulating tape. Always disconnect the negative battery cable first.

***** CAUTION**

To avoid personal injury when working on vehicles equipped with an air bag, the negative battery cable must be disconnected and at least 90 seconds must elapse before working on the system. Failure to do so may result in deployment of the air bag.

2. Disconnect the negative battery cable. Tape the cable with insulating tape.

3. Remove the engine under cover, if equipped.

4. On 4.0L engines remove the V bank cover. Remove the engine under cover sub assembly No 1.

5. Drain the coolant. Be sure to properly dispose of used coolant.

6. Remove the radiator support to frame seal, left side.

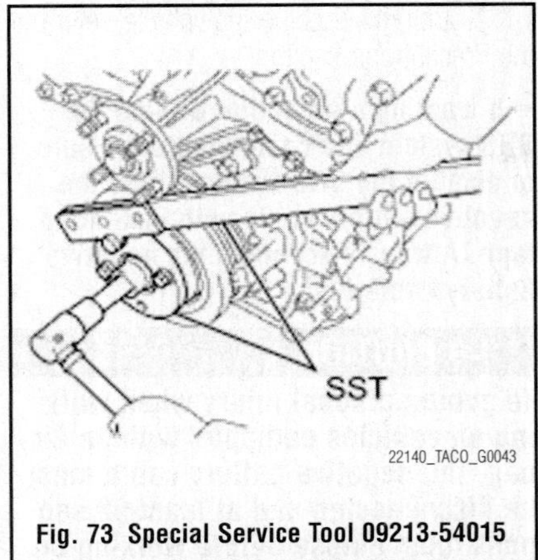

Fig. 73 Special Service Tool 09213-54015

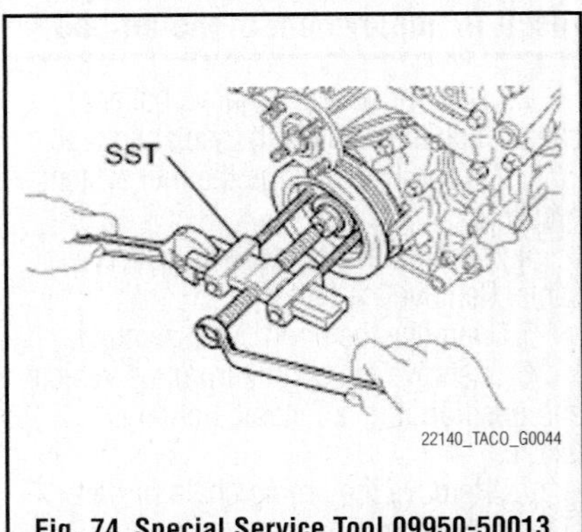

Fig. 74 Special Service Tool 09950-50013

7. Remove the fan shroud.
8. Remove the accessory drive belt.
9. Using SST 09213-54015, fix the pulley and loosen the pulley bolt.
10. Using the pulley set bolt and SST 09950-50013, remove the crankshaft pulley.
11. Discard the bolt.

To install:

➡Be sure to use new fasteners, as required.

12. Installation is the reverse of the removal procedure. Tighten the crankshaft bolt to 192 ft. lbs (260 Nm). Be sure to replace the bolt.
13. Be sure to fill the cooling system with the proper grade and type coolant.
14. Start the engine and check for leaks. Correct as required.

CRANKSHAFT FRONT SEAL

REMOVAL & INSTALLATION

See Figure 75.

1. Before servicing the vehicle, refer to the Precautions Section.

➡If working near and/or around the SRS system and components, be sure to disable the SRS system. Tape the negative battery cable with insulating tape. Always disconnect the negative battery cable first.

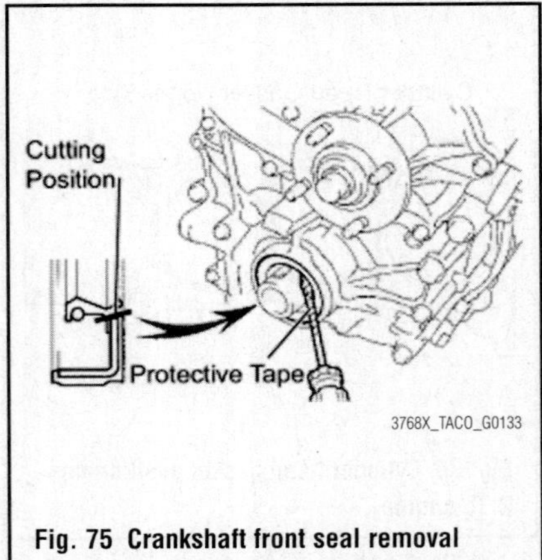

Fig. 75 Crankshaft front seal removal

⁂ CAUTION

To avoid personal injury when working on vehicles equipped with an air bag, the negative battery cable must be disconnected and at least 90 seconds must elapse before working on the system. Failure to do so may result in deployment of the air bag.

2. Disconnect the negative battery cable. Tape the cable with insulating tape.
3. Remove the crankshaft pulley. Discard the bolt.
4. Using a cutter knife cut off the lip of the seal.
5. Using a suitable tool, with its tip wrapped in tape, pry out the seal from its mounting.

➡Be careful not to damage the crankshaft, with the suitable tool.

To install:

➡Be sure to use new fasteners, as required.

6. Installation is the reverse of the removal procedure.
7. Be sure to fill the cooling system with the proper grade and type coolant.
8. Start the engine and check for leaks. Correct as required.

CYLINDER HEAD

REMOVAL & INSTALLATION

2.7L Engines

See Figures 76 through 79.

➡The engine must first be removed from the vehicle.

1. Before servicing the vehicle, refer to the Precautions Section.

➡If working near and/or around the SRS system and components, be sure to disable the SRS system. Tape the negative battery cable with insulating tape. Always disconnect the negative battery cable first.

⁂ CAUTION

To avoid personal injury when working on vehicles equipped with an air bag, the negative battery cable must be disconnected and at least 90 seconds must elapse before working on the system. Failure to do so may result in deployment of the air bag.

2. Disconnect the negative battery cable. Tape the cable with insulating tape.
3. Properly discharge the fuel system pressure.
4. Remove the hood subassembly.
5. Remove the engine from the vehicle and position it in a suitable holding fixture.
6. Remove the intake air connector. Remove the alternator.
7. Remove the number one exhaust manifold heat insulator. Remove the air switching valve assembly.
8. Remove the exhaust manifold.
9. Remove the belt tensioner assembly. Remove the number one idler pulley subassembly. Remove the idle pulley assembly and bracket.
10. Remove the Crankshaft Position (CKP) sensor. Remove the Camshaft Position (CMP) sensor.
11. Remove the number one intake manifold.
12. Remove the cylinder head cover. Remove the crankshaft pulley. Remove the oil gauge subassembly.
13. Remove the number two oil pan subassembly. Remove the oil strainer. Remove the oil pap subassembly.
14. Remove the timing chain cover. Remove the timing chain guide. Remove the number one chain tensioner assembly.
15. Remove the chain tensioner slipper. Remove the number one chain vibration damper. Remove the chain subassembly.
16. Remove the camshaft. Remove the valve rocker arms.
17. Remove the valve lash adjuster assembly.
18. Disconnect the water hoses. Disconnect the engine coolant temperature sensor wire. Remove the bolts and separate the wire from the harnesses. Remove the bolt, and then separate the wire harness bracket.

19. Loosen the cylinder head retaining bolts in several steps and in the proper removal sequence. Remove the bolts.

➡Be careful not to drop the washers into the cylinder head. Head warpage and cracking could result in removing the bolts in the wrong order.

To install:

➡Be sure to use new fasteners, as required.

20. Before installing, thoroughly clean the gasket mating surfaces and check for warpage.

21. Apply continuous beads of seal packing, part number 08826-00080 or equivalent, to the cylinder block upper side and cylinder head gasket upper side. Bead width should be 0.15–0.28 inch.

➡Remove any oil from the contact surface. Install the cylinder head gasket within three minutes after applying the seal packing. Install the cylinder head bolts within fifteen minutes after applying the seal packing. Do not fill the engine with oil for at least four hours.

➡Be sure to position the cylinder head gasket on the engine block with the lot number stamp facing upward.

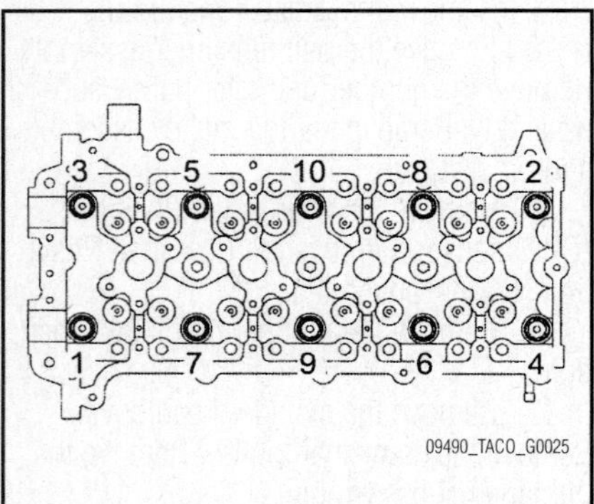

Fig. 76 Cylinder head bolt loosening sequence—2.7L engine

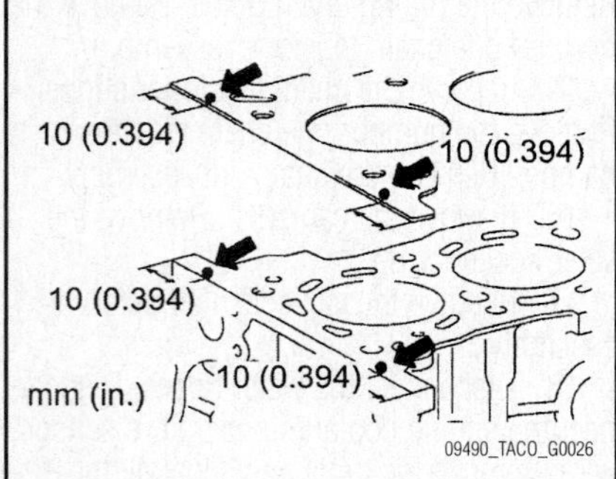

Fig. 77 Cylinder head gasket sealant application—2.7L engine

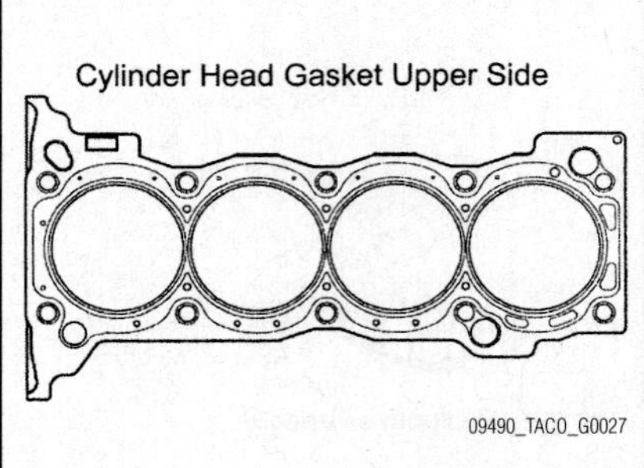

Fig. 78 Cylinder head gasket positioning—2.7L engine

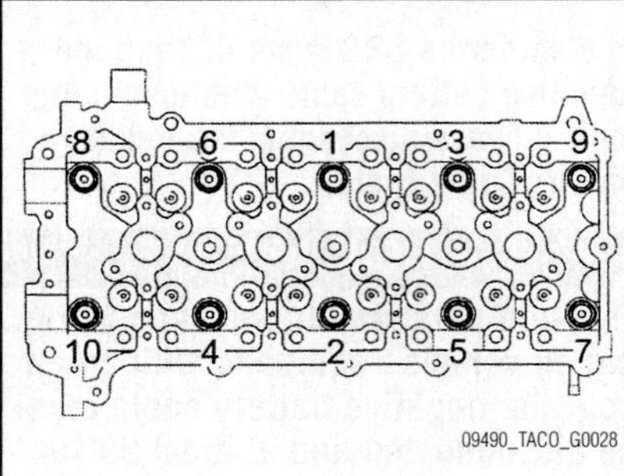

Fig. 79 Cylinder head bolt torque sequence—2.7L engine

➡Be sure that the head gasket is installed in the correct direction. Position the gasket gently in order to avoid damaging the gasket with the bottom part of the cylinder head.

22. Apply a light coat of clean engine oil to the cylinder head bolts. Tighten the cylinder head bolts, in two successive steps, to specification and in the proper sequence.

23. Continue the installation in the reverse of the removal procedure.

24. Be sure to fill the cooling system with the proper grade and type engine coolant.

25. Be sure to fill the engine oil with the proper grade and type oil.

➡Do not fill the engine with oil for at least four hours after the cylinder head has been installed.

26. Start the engine and check for leaks. Correct as required.

4.0L Engines

2WD Vehicles–Bank 1

See Figures 80 through 83.

➡The engine must first be removed from the vehicle.

1. Before servicing the vehicle, refer to the Precautions Section.

➡If working near and/or around the SRS system and components, be sure to disable the SRS system. Tape the negative battery cable with insulating tape. Always disconnect the negative battery cable first.

CAUTION

To avoid personal injury when working on vehicles equipped with an air bag, the negative battery cable must be disconnected and at least 90 seconds must elapse before working on the system. Failure to do so may result in deployment of the air bag.

2. Disconnect the negative battery cable. Tape the cable with insulating tape.

3. Properly discharge the fuel system pressure.

4. Disconnect the positive battery cable. Remove the battery.

5. Remove the hood subassembly.

6. Remove the engine from the vehicle and position it in a suitable holding fixture.

7. Remove the timing chain or belt cover subassembly.

8. Remove the chain subassembly.

9. Remove the number one cool inlet. Remove the front exhaust pipe assembly. Remove the three bolts and remove the exhaust manifold stay.

10. Disconnect the air fuel ratio sensor connector. Remove the six nuts; remove the exhaust manifold and gasket.

11. Disconnect the number one fuel pipe subassembly. Disconnect the number two fuel pipe subassembly.

12. Disconnect the fuel injector connectors. Remove the ten intake manifold retaining bolts. Remove the intake manifold from the engine.

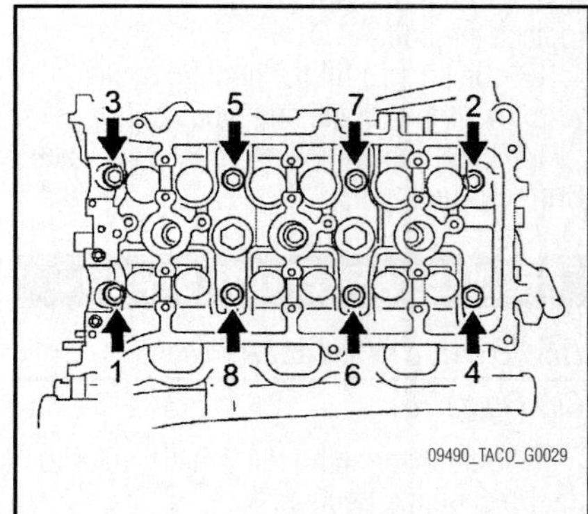

Fig. 80 Bank 1 cylinder head bolt loosening sequence—4.0L engine

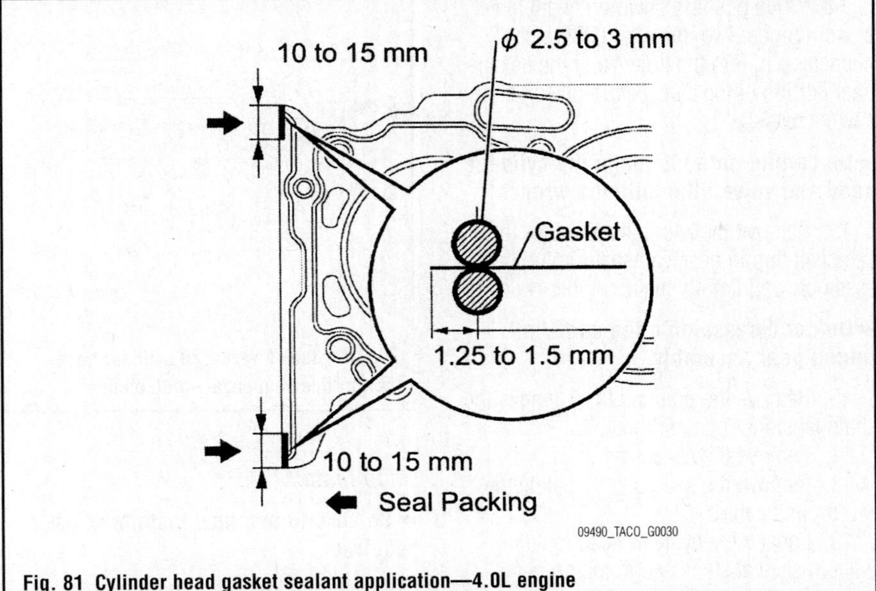

Fig. 81 Cylinder head gasket sealant application—4.0L engine

13. Disconnect the engine coolant temperature sensor connector. Disconnect the heater hose. Remove the two bolts and four nuts, and then remove the water bypass joint RR and two gaskets. Remove the O-ring from the water outlet hose.
14. While raising the chain tensioner number two, insert a 0.039 inch diameter pin, into the hole. Hold the hexagonal portion of the camshaft in place, using a wrench.

➡Be careful not to damage the cylinder head and valve lifter with the wrench.

15. Remove the two bolts. Remove the camshaft timing gear, camshaft timing gear assembly and timing chain number two.

➡Do not disassemble the camshaft timing gear assembly.

16. Remove the bolt and then remove the chain tensioner number two.
17. Remove the camshafts.
18. Remove the two bolts. Separate the two ground cables.
19. Loosen the cylinder head retaining bolts in several steps and in the proper removal sequence. Remove the bolts.

➡Be careful not to drop the washers into the cylinder head. Head warpage and cracking could result in removing the bolts in the wrong order.

20. Lift the cylinder head from the dowels on the cylinder block and remove it from the engine.

To install:

➡Be sure to use new fasteners, as required.

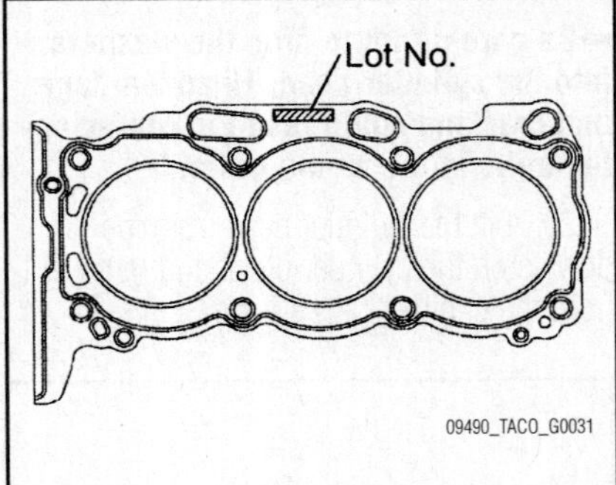

Fig. 82 Bank 1 cylinder head gasket positioning—4.0L engine

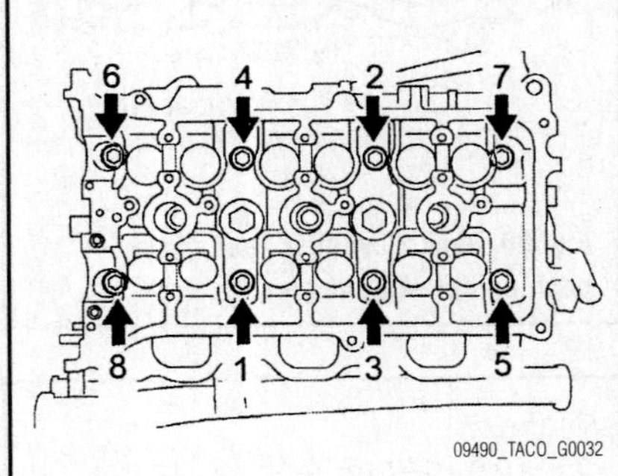

Fig. 83 Bank 1 cylinder head bolt torque sequence—4.0L engine

21. Before installing, thoroughly clean the gasket mating surfaces and check for warpage. Remove any old packing material. Be careful not to drop any oil on the contact surfaces of the cylinder head or engine block.
22. Apply continuous beads of seal packing, part number 08826-00080 or equivalent, to the cylinder head gasket. Bead width should be 0.098–0.118 inch.

➡Install the cylinder head gasket within three minutes after applying the seal packing. Install the cylinder head bolts within fifteen minutes after applying the seal packing.

➡Be sure to position the cylinder head gasket on the engine block with the lot number stamp facing upward.

➡Be sure that the head gasket is installed in the correct direction. Position the gasket gently in order to avoid damaging the gasket with the bottom part of the cylinder head.

23. Apply a light coat of clean engine oil to the cylinder head bolts. Tighten the cylinder head bolts, in two successive steps, to specification and in the proper sequence.

➡Replace defective bolts, as required.

24. Continue the installation in the reverse of the removal procedure.
25. Be sure to fill the cooling system with the proper grade and type engine coolant.
26. Be sure to fill the engine oil with the proper grade and type oil.

➡Do not fill the engine with oil for at least four hours after the cylinder head has been installed.

27. Start the engine and check for leaks. Correct as required.

2WD Vehicles–Bank 2

See Figures 84 through 88.

➡The engine must first be removed from the vehicle.

1. Before servicing the vehicle, refer to the Precautions Section.

➡If working near and/or around the SRS system and components, be sure to disable the SRS system. Tape the negative battery cable with insulating tape. Always disconnect the negative battery cable first.

CAUTION

To avoid personal injury when working on vehicles equipped with an air bag, the negative battery cable must be disconnected and at least 90 seconds must elapse before working on the system. Failure to do so may result in deployment of the air bag.

2. Disconnect the negative battery cable. Tape the cable with insulating tape.

3. Properly discharge the fuel system pressure.

4. Disconnect the positive battery cable. Remove the battery.

5. Remove the hood subassembly.

6. Remove the engine from the vehicle and position it in a suitable holding fixture.

7. Remove the timing chain or belt cover subassembly.

8. Remove the chain subassembly.

9. Remove the number one cool inlet. Remove the number two front exhaust pipe assembly. Remove the three bolts and remove the exhaust manifold stay.

10. Disconnect the air fuel ratio sensor connector. Remove the six nuts; remove the exhaust manifold and gasket.

11. Disconnect the fuel injector connectors. Remove the ten intake manifold retaining bolts. Remove the intake manifold from the engine.

12. Disconnect the engine coolant temperature sensor connector. Disconnect the heater hose. Remove the two bolts and four nuts, and then remove the water bypass joint RR and two gaskets. Remove the O-ring from the water outlet hose.

13. Remove the two bolts and then remove the chain vibration damper number one.

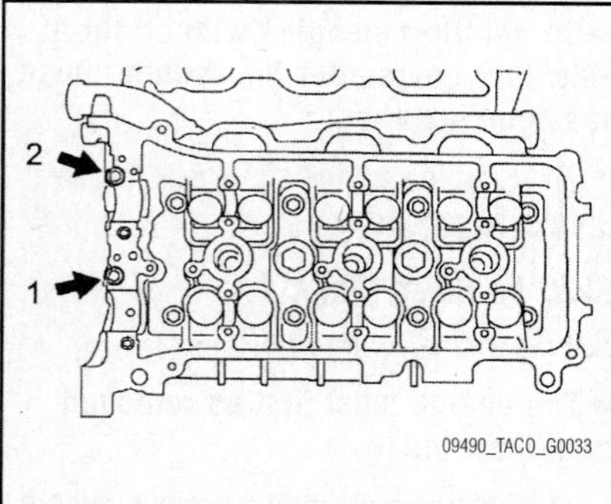

Fig. 84 Bank 2 recessed cylinder head bolt loosening sequence—4.0L engine

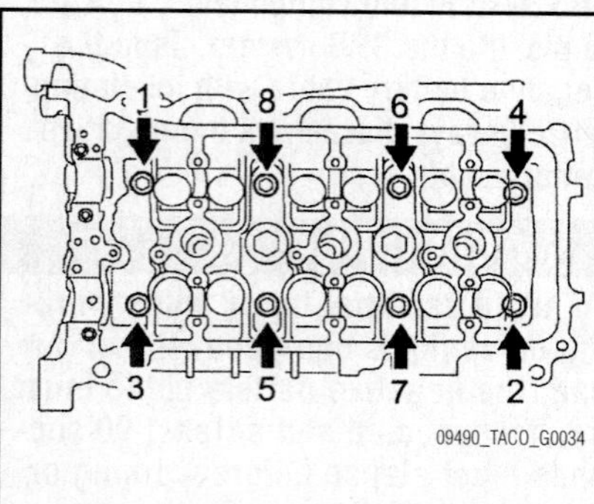

Fig. 85 Bank 2 cylinder head bolt loosening sequence—4.0L engine

14. While pushing down on chain tensioner number two, insert a 0.039 inch diameter pin, into the hole. Hold the hexagonal portion of the camshaft in place, using a wrench.

➡**Be careful not to damage the cylinder head and valve lifter with the wrench.**

15. Remove the two bolts. Remove the camshaft timing gear, camshaft timing gear assembly and timing chain number two.

➡**Do not disassemble the camshaft timing gear assembly.**

16. Remove the bolt and then remove the chain tensioner number three.

17. Remove the camshafts.

18. Remove the two bolts. Separate the two ground cables.

19. Loosen the cylinder head retaining bolts in several steps and in the proper removal sequence. Remove the bolts.

➡**Be careful not to drop the washers into the cylinder head. Head warpage and cracking could result in removing the bolts in the wrong order.**

20. Lift the cylinder head from the dowels on the cylinder block and remove it from the engine.

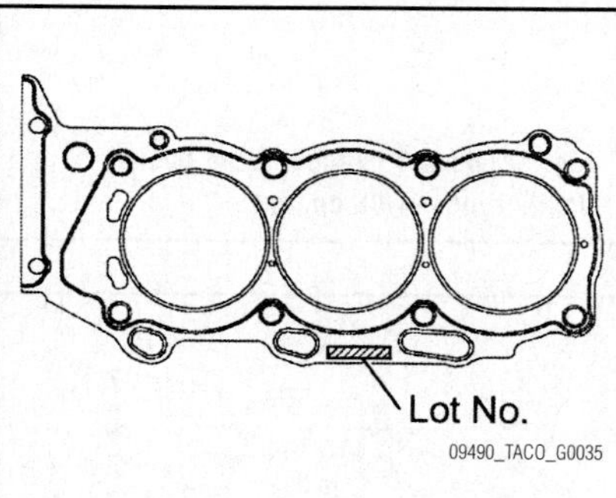

Fig. 86 Bank 2 cylinder head gasket positioning—4.0L engine

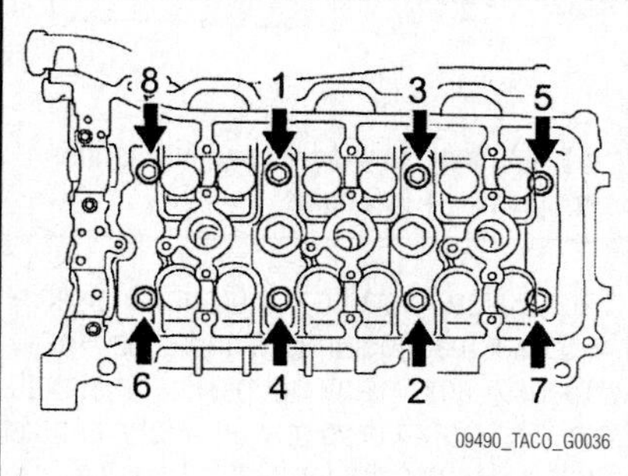

Fig. 87 Bank 2 Cylinder head bolt torque sequence—4.0L engine

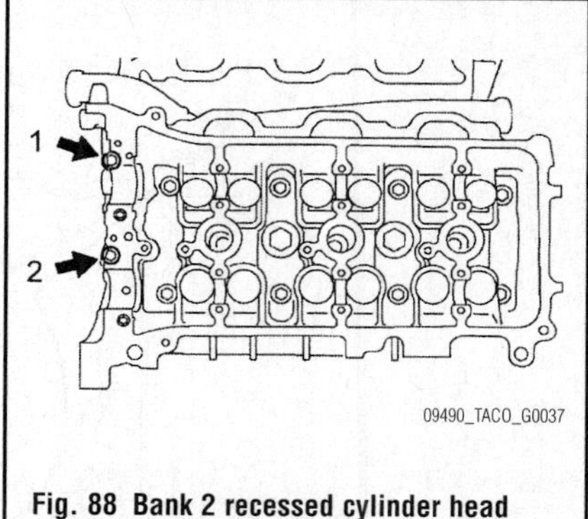

Fig. 88 Bank 2 recessed cylinder head bolt torque sequence—4.0L engine

To install:

➡**Be sure to use new fasteners, as required.**

21. Before installing, thoroughly clean the gasket mating surfaces and check for warpage. Remove any old packing material. Be careful not to drop any oil on the contact surfaces of the cylinder head or engine block.

22. Apply continuous beads of seal packing, part number 08826-00080 or equivalent, to the cylinder head gasket. Bead width should be 0.098–0.118 inch.

➡**Install the cylinder head gasket within three minutes after applying the seal packing. Install the cylinder head bolts within fifteen minutes after applying the seal packing.**

23. Be sure to position the cylinder head gasket on the engine block with the lot number stamp facing upward.

➡**Be sure that the head gasket is installed in the correct direction. Position the gasket gently in order to avoid damaging the gasket with the bottom part of the cylinder head.**

24. Apply a light coat of clean engine oil to the cylinder head bolts. Tighten the cylinder head bolts, in two successive steps, to specification and in the proper sequence.

➡**Replace defective bolts, as required.**

25. Continue the installation in the reverse of the removal procedure.

26. Be sure to fill the cooling system with the proper grade and type engine coolant.

27. Be sure to fill the engine oil with the proper grade and type oil.

➡**Do not fill the engine with oil for at least four hours after the cylinder head has been installed.**

28. Start the engine and check for leaks. Correct as required.

4WD Vehicles And PreRunner–Bank 1

➡The engine must first be removed from the vehicle.

1. Before servicing the vehicle, refer to the Precautions Section.

➡If working near and/or around the SRS system and components, be sure to disable the SRS system. Tape the negative battery cable with insulating tape. Always disconnect the negative battery cable first.

CAUTION

To avoid personal injury when working on vehicles equipped with an air bag, the negative battery cable must be disconnected and at least 90 seconds must elapse before working on the system. Failure to do so may result in deployment of the air bag.

2. Disconnect the negative battery cable. Tape the cable with insulating tape.
3. Properly discharge the fuel system pressure.
4. Disconnect the positive battery cable. Remove the battery.
5. Remove the power steering gear assembly.
6. Remove the front differential carrier assembly.
7. Remove the timing chain or belt cover subassembly.
8. Remove the chain subassembly.
9. Remove the number one cool inlet. Remove the front exhaust pipe assembly. Remove the three bolts and remove the exhaust manifold stay.
10. Disconnect the air fuel ratio sensor connector. Remove the six nuts; remove the exhaust manifold and gasket.
11. Disconnect the number one fuel pipe subassembly. Disconnect the number two fuel pipe subassembly.
12. Disconnect the fuel injector connectors. Remove the ten intake manifold retaining bolts. Remove the intake manifold from the engine.
13. Disconnect the engine coolant temperature sensor connector. Disconnect the heater hose. Remove the two bolts and four nuts, and then remove the water bypass joint RR and two gaskets. Remove the O-ring from the water outlet hose.
14. While raising the chain tensioner number two, insert a 0.039 inch diameter pin, into the hole. Hold the hexagonal portion of the camshaft in place, using a wrench.

➡Be careful not to damage the cylinder head and valve lifter with the wrench.

15. Remove the two bolts. Remove the camshaft timing gear, camshaft timing gear assembly and timing chain number two.

➡Do not disassemble the camshaft timing gear assembly.

16. Remove the bolt and then remove the chain tensioner number two.
17. Remove the camshafts.
18. Remove the two bolts. Separate the two ground cables.
19. Loosen the cylinder head retaining bolts in several steps and in the proper removal sequence. Remove the bolts.

➡Be careful not to drop the washers into the cylinder head. Head warpage and cracking could result in removing the bolts in the wrong order.

20. Lift the cylinder head from the dowels on the cylinder block and remove it from the engine.

To install:

➡Be sure to use new fasteners, as required.

21. Before installing, thoroughly clean the gasket mating surfaces and check for warpage. Remove any old packing material. Be careful not to drop any oil on the contact surfaces of the cylinder head or engine block.
22. Apply continuous beads of seal packing, part number 08826-00080 or equivalent, to the cylinder head gasket. Bead width should be 0.098–0.118 inch.

➡Install the cylinder head gasket within three minutes after applying the seal packing. Install the cylinder head bolts within fifteen minutes after applying the seal packing.

➡Be sure to position the cylinder head gasket on the engine block with the lot number stamp facing upward.

➡Be sure that the head gasket is installed in the correct direction. Position the gasket gently in order to avoid damaging the gasket with the bottom part of the cylinder head.

23. Apply a light coat of clean engine oil to the cylinder head bolts. Tighten the cylinder head bolts, in two successive steps, to specification and in the proper sequence.

➡Replace defective bolts, as required.

24. Continue the installation in the reverse of the removal procedure.
25. Be sure to fill the cooling system with the proper grade and type engine coolant.
26. Be sure to fill the engine oil with the proper grade and type oil.

➡Do not fill the engine with oil for at least four hours after the cylinder head has been installed.

27. Start the engine and check for leaks. Correct as required.

4WD Vehicles And PreRunner–Bank 2

➡The engine must first be removed from the vehicle.

1. Before servicing the vehicle, refer to the Precautions Section.

➡If working near and/or around the SRS system and components, be sure to disable the SRS system. Tape the negative battery cable with insulating tape. Always disconnect the negative battery cable first.

CAUTION

To avoid personal injury when working on vehicles equipped with an air bag, the negative battery cable must be disconnected and at least 90 seconds must elapse before working on the system. Failure to do so may result in deployment of the air bag.

2. Disconnect the negative battery cable. Tape the cable with insulating tape.
3. Properly discharge the fuel system pressure.
4. Disconnect the positive battery cable. Remove the battery.
5. Remove the power steering gear assembly.
6. Remove the front differential carrier assembly.
7. Remove the timing chain or belt cover subassembly.
8. Remove the chain subassembly.
9. Remove the number one cool inlet. Remove the number two front exhaust pipe assembly. Remove the three bolts and remove the exhaust manifold stay.

10. Disconnect the air fuel ratio sensor connector. Remove the six nuts; remove the exhaust manifold and gasket.

11. Disconnect the fuel injector connectors. Remove the ten intake manifold retaining bolts. Remove the intake manifold from the engine.

12. Disconnect the engine coolant temperature sensor connector. Disconnect the heater hose. Remove the two bolts and four nuts, and then remove the water bypass joint RR and two gaskets. Remove the O-ring from the water outlet hose.

13. Remove the two bolts and then remove the chain vibration damper number one.

14. While pushing down on chain tensioner number two, insert a 0.039 inch diameter pin, into the hole. Hold the hexagonal portion of the camshaft in place, using a wrench.

➡Be careful not to damage the cylinder head and valve lifter with the wrench.

15. Remove the two bolts. Remove the camshaft timing gear, camshaft timing gear assembly and timing chain number two.

➡Do not disassemble the camshaft timing gear assembly.

16. Remove the bolt and then remove the chain tensioner number three.

17. Remove the camshafts.

18. Remove the two bolts. Separate the two ground cables.

19. Loosen the cylinder head retaining bolts in several steps and in the proper removal sequence. Remove the bolts.

➡Be careful not to drop the washers into the cylinder head. Head warpage and cracking could result in removing the bolts in the wrong order.

20. Lift the cylinder head from the dowels on the cylinder block and remove it from the engine.

To install:

➡Be sure to use new fasteners, as required.

21. Before installing, thoroughly clean the gasket mating surfaces and check for warpage. Remove any old packing material. Be careful not to drop any oil on the contact surfaces of the cylinder head or engine block.

22. Apply continuous beads of seal packing, part number 08826-00080 or equivalent, to the cylinder head gasket. Bead width should be 0.098–0.118 inch.

➡Install the cylinder head gasket within three minutes after applying the seal packing. Install the cylinder head bolts within fifteen minutes after applying the seal packing.

➡Be sure to position the cylinder head gasket on the engine block with the lot number stamp facing upward.

➡Be sure that the head gasket is installed in the correct direction. Position the gasket gently in order to avoid damaging the gasket with the bottom part of the cylinder head.

23. Apply a light coat of clean engine oil to the cylinder head bolts. Tighten the cylinder head bolts, in two successive steps, to specification and in the proper sequence.

➡Replace defective bolts, as required.

24. Continue the installation in the reverse of the removal procedure.

25. Be sure to fill the cooling system with the proper grade and type engine coolant.

26. Be sure to fill the engine oil with the proper grade and type oil.

➡Do not fill the engine with oil for at least four hours after the cylinder head has been installed.

27. Start the engine and check for leaks. Correct as required.

EXHAUST MANIFOLD

REMOVAL & INSTALLATION

2.7L Engines

See Figure 89.

1. Before servicing the vehicle, refer to the Precautions Section.

➡If working near and/or around the SRS system and components, be sure to disable the SRS system. Tape the negative battery cable with insulating tape. Always disconnect the negative battery cable first.

CAUTION

To avoid personal injury when working on vehicles equipped with an air bag, the negative battery cable must be disconnected and at least 90 seconds must elapse before working on the system. Failure to do so may result in deployment of the air bag.

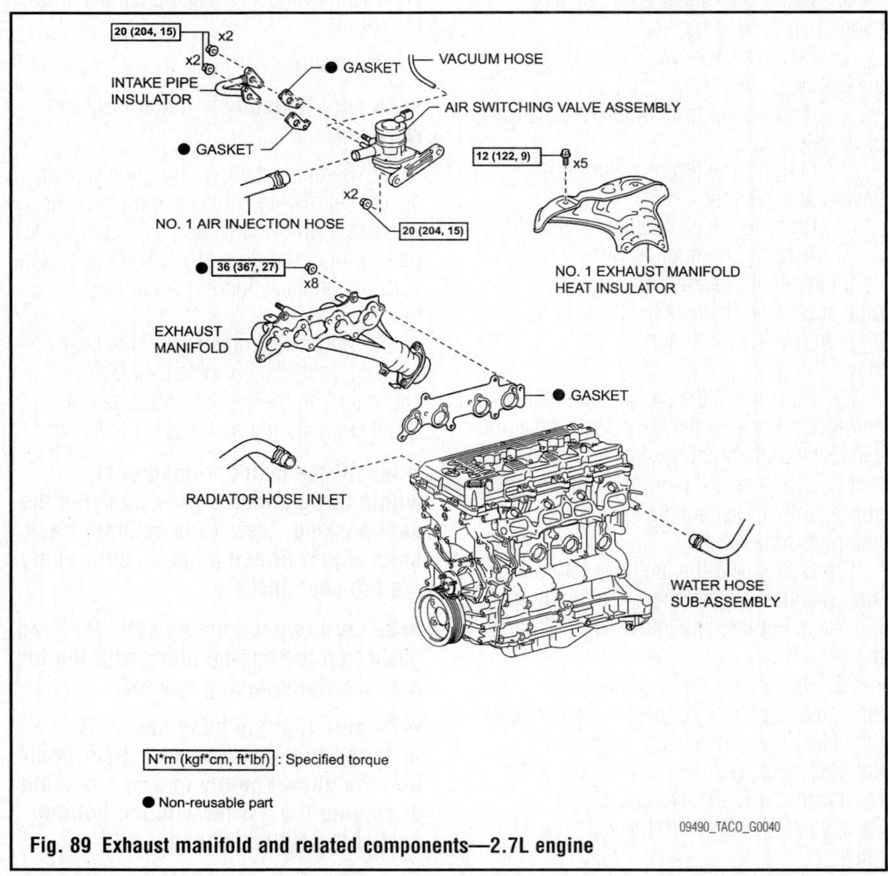

Fig. 89 Exhaust manifold and related components—2.7L engine

2. Disconnect the negative battery cable. Tape the cable with insulating tape.
3. Disconnect the exhaust manifold to exhaust flange nuts.
4. Remove the necessary components to gain access to the exhaust manifold retaining bolts.
5. Remove the exhaust manifold retaining nuts. Discard the nuts. Remove the exhaust manifold from the engine.

To install:

➡Be sure to use new fasteners, as required.

6. Installation is the reverse of the removal procedure.
7. Be sure to use new gaskets. Be sure to use new retaining nuts.

4.0L Engines

See Figure 90.

1. Before servicing the vehicle, refer to the Precautions Section.

➡If working near and/or around the SRS system and components, be sure to disable the SRS system. Tape the negative battery cable with insulating tape. Always disconnect the negative battery cable first.

⁂ CAUTION

To avoid personal injury when working on vehicles equipped with an air bag, the negative battery cable must be disconnected and at least 90 seconds must elapse before working on the system. Failure to do so may result in deployment of the air bag.

2. Disconnect the negative battery cable. Tape the cable with insulating tape.
3. Disconnect the exhaust manifold to exhaust flange nuts.
4. Remove the necessary components to gain access to the exhaust manifold retaining bolts.
5. Remove the exhaust manifold retaining nuts. Discard the nuts. Remove the exhaust manifold from the engine.

To install:

➡Be sure to use new fasteners, as required.

6. Installation is the reverse of the removal procedure.
7. Be sure to use new gaskets. Be sure to use new retaining nuts.

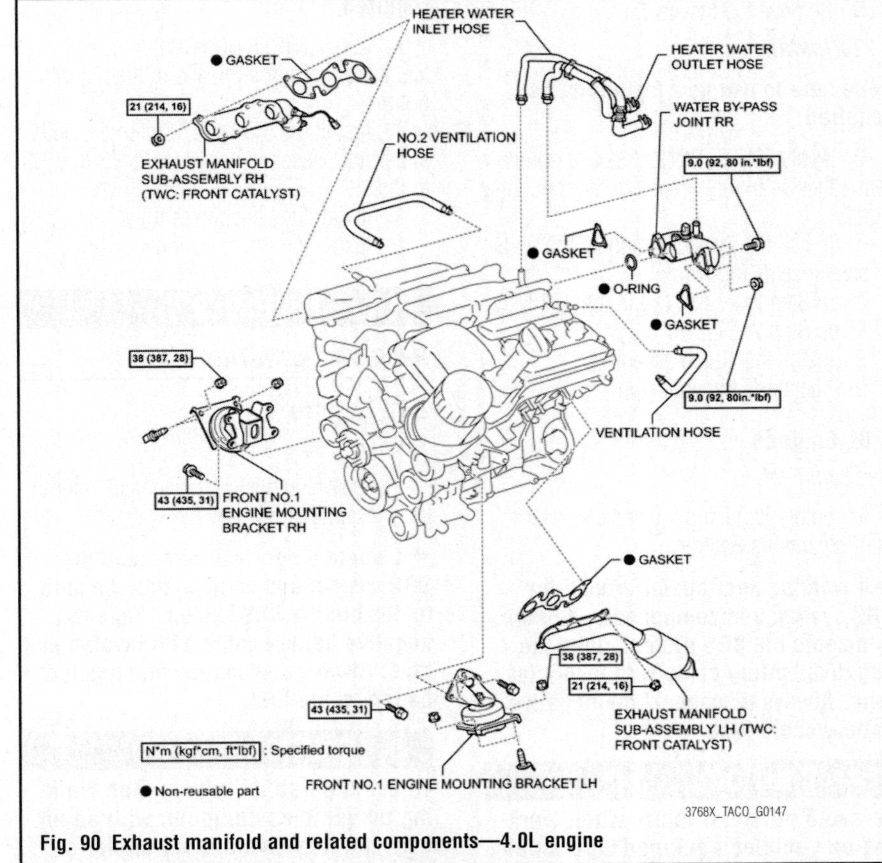

Fig. 90 Exhaust manifold and related components—4.0L engine

FLEXPLATE

REMOVAL & INSTALLATION

2.7L Engines

See Figure 91.

1. Before servicing the vehicle, refer to the Precautions Section.

➡If working near and/or around the SRS system and components, be sure to disable the SRS system. Tape the negative battery cable with insulating tape. Always disconnect the negative battery cable first.

⁂ CAUTION

To avoid personal injury when working on vehicles equipped with an air bag, the negative battery cable must be disconnected and at least 90 seconds must elapse before working on the system. Failure to do so may result in deployment of the air bag.

2. Disconnect the negative battery cable. Tape the cable with insulating tape.
3. Remove the transmission.
4. Remove the flexplate.

To install:

➡Be sure to use new fasteners, as required.

5. Apply THREE BOND 1324 or equivalent adhesive to the threads of the mounting bolts.
6. Install the flexplate. Tighten the bolts in several passes and in sequence to 74 Nm (55 ft. lbs.).
7. Install the transmission.

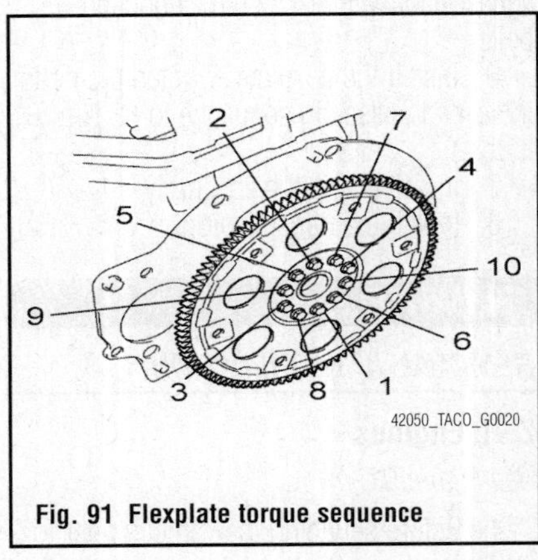

Fig. 91 Flexplate torque sequence

4.0L Engine

See Figure 92.

1. Before servicing the vehicle, refer to the Precautions Section.

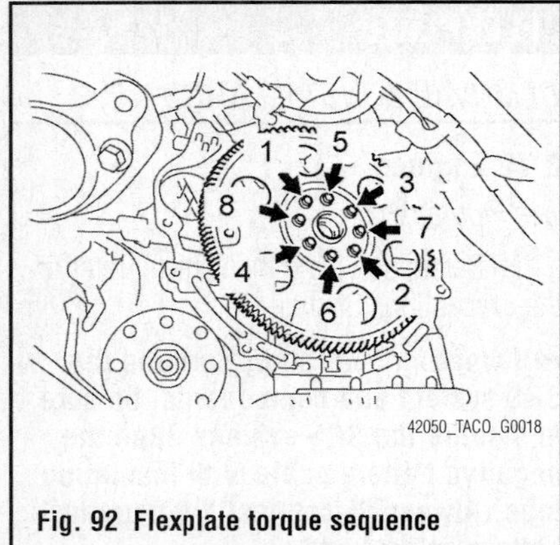

Fig. 92 Flexplate torque sequence

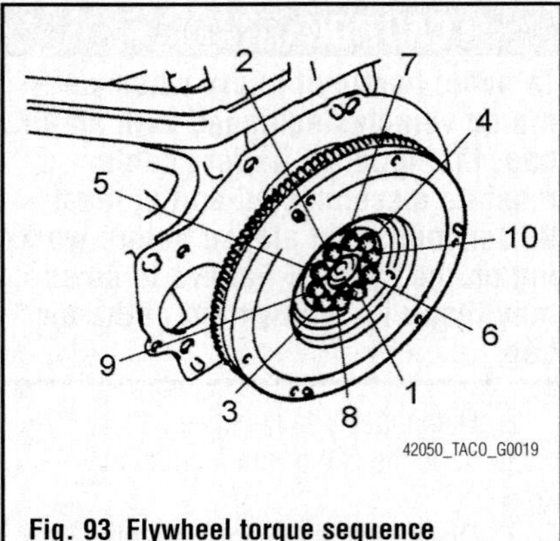

Fig. 93 Flywheel torque sequence

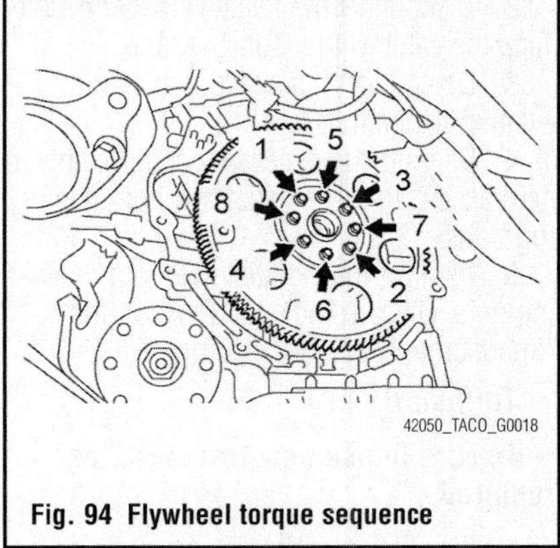

Fig. 94 Flywheel torque sequence

➡If working near and/or around the SRS system and components, be sure to disable the SRS system. Tape the negative battery cable with insulating tape. Always disconnect the negative battery cable first.

⁂ CAUTION

To avoid personal injury when working on vehicles equipped with an air bag, the negative battery cable must be disconnected and at least 90 seconds must elapse before working on the system. Failure to do so may result in deployment of the air bag.

2. Disconnect the negative battery cable. Tape the cable with insulating tape.
3. Remove the transmission.
4. Remove the flexplate.

To install:

➡Be sure to use new fasteners, as required.

5. Apply THREE BOND 1324 or equivalent adhesive to the threads of the mounting bolts.
6. Install the flexplate. Tighten the bolts in several passes in sequence to 83 Nm (61 ft. lbs.).
7. Install the clutch assembly.
8. Install the transmission.

FLYWHEEL

REMOVAL & INSTALLATION

2.7L Engines

See Figure 93.

1. Before servicing the vehicle, refer to the Precautions Section.

➡If working near and/or around the SRS system and components, be sure to disable the SRS system. Tape the negative battery cable with insulating tape. Always disconnect the negative battery cable first.

⁂ CAUTION

To avoid personal injury when working on vehicles equipped with an air bag, the negative battery cable must be disconnected and at least 90 seconds must elapse before working on the system. Failure to do so may result in deployment of the air bag.

2. Disconnect the negative battery cable. Tape the cable with insulating tape.
3. Remove the transmission.
4. Remove the clutch assembly.
5. Remove the flywheel.

To install:

➡Be sure to use new fasteners, as required.

6. Apply THREE BOND 1324 or equivalent adhesive to the threads of the mounting bolts.
7. Install the flywheel. Tighten the bolts in sequence as follows:
   a. Step 1: 27 Nm (20 ft. lbs. (27 Nm).)
   b. Step 2: Plus 90°
8. Install the clutch assembly.
9. Install the transmission.

4.0L Engines

See Figure 94.

1. Before servicing the vehicle, refer to the Precautions Section.

➡If working near and/or around the SRS system and components, be sure to disable the SRS system. Tape the negative battery cable with insulating tape. Always disconnect the negative battery cable first.

⁂ CAUTION

To avoid personal injury when working on vehicles equipped with an air bag, the negative battery cable must be disconnected and at least 90 seconds must elapse before working on the system. Failure to do so may result in deployment of the air bag.

2. Disconnect the negative battery cable. Tape the cable with insulating tape.
3. Remove the transmission.
4. Remove the clutch assembly.
5. Remove the flywheel.

To install:

➡Be sure to use new fasteners, as required.

6. Apply THREE BOND 1324 or equivalent adhesive to the threads of the mounting bolts.
7. Install the flywheel. Tighten the bolts in several passes in sequence to 85 Nm (63 ft. lbs.).
8. Install the clutch assembly.
9. Install the transmission.

INTAKE MANIFOLD

REMOVAL & INSTALLATION

2.7L Engines

See Figure 95.

1. Before servicing the vehicle, refer to the Precautions Section.

➡If working near and/or around the SRS system and components, be sure to disable the SRS system. Tape the negative battery cable with insulating tape. Always disconnect the negative battery cable first.

⁂ CAUTION

To avoid personal injury when working on vehicles equipped with an air bag, the negative battery cable

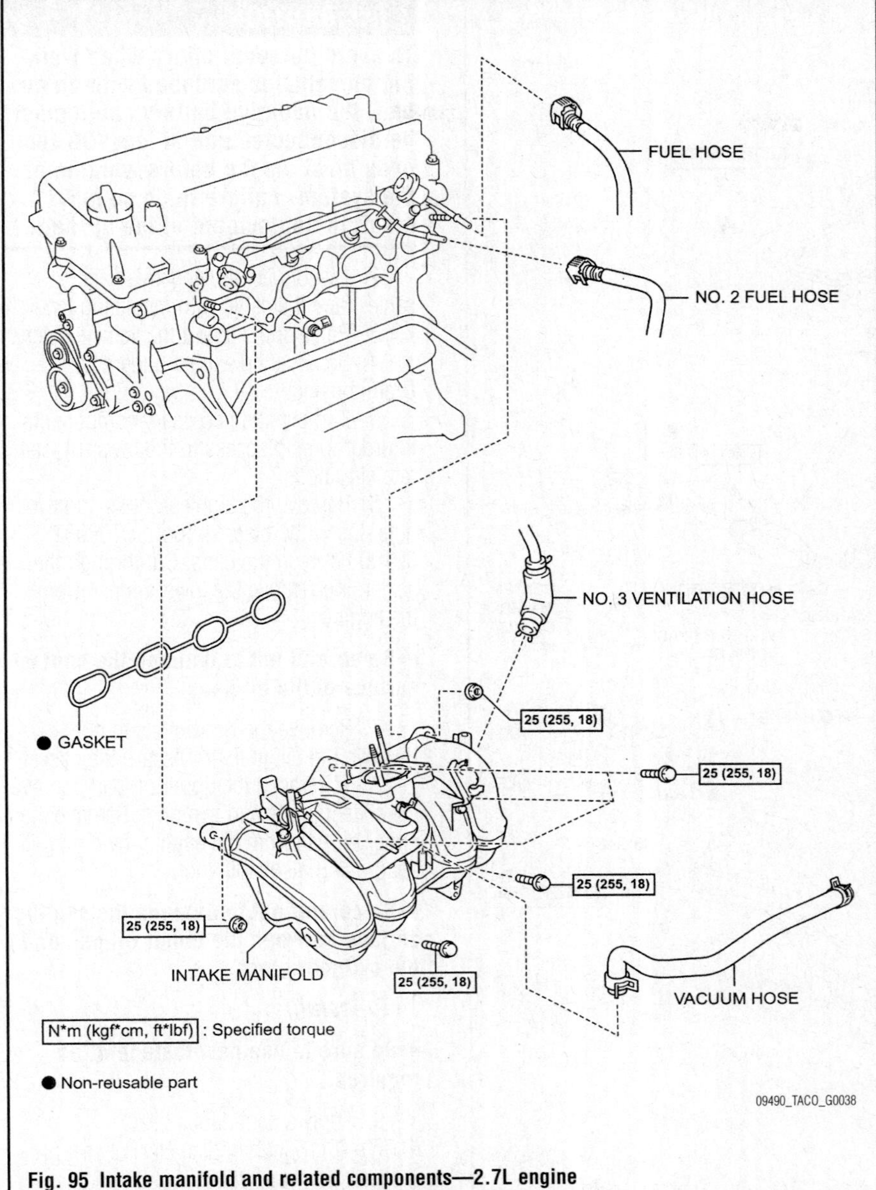

Fig. 95 Intake manifold and related components—2.7L engine

must be disconnected and at least 90 seconds must elapse before working on the system. Failure to do so may result in deployment of the air bag.

2. Disconnect the negative battery cable. Tape the cable with insulating tape.
3. Properly relieve the fuel system pressure.
4. On 4WD and PreRunner, remove the engine undercover subassembly.
5. Drain the engine coolant. Remove the air intake connector.
6. Remove the throttle body and motor assembly. Disconnect the fuel hoses.
7. Disconnect the fuel vapor feed hose from the VSR. Disconnect the vacuum hose. Remove the bolt, and then remove the clamp bracket.
8. Disconnect the number two water bypass hose. Disconnect the number three ventilation hose. Disconnect the VSV connector.
9. Disengage the engine wire harness clamp. Disconnect the air conditioning compressor magnetic clutch connector.
10. Disengage the wire harness clamp. Remove the bolt and harness clamp bracket. Disconnect the three connectors.
11. Remove the retaining nut inside the relay block. Disconnect the engine wire harness from the relay block.
12. Remove the five bolts and two nuts retaining the intake manifold in place. Remove the intake manifold from the engine.

To install:

➡Be sure to use new fasteners, as required.

13. Clean all surfaces.
14. Install a new gasket onto the intake manifold. Position the intake manifold to the engine.
15. Install the retaining bolts and tighten to specification, in an alternating sequence.
16. Continue the installation in the reverse order of the removal procedure.
17. Be sure to fill the cooling system with the proper grade and type engine coolant.
18. Start the engine and check for leaks. Correct as required.

4.0L Engines

See Figure 96.

1. Before servicing the vehicle, refer to the Precautions Section.

➡If working near and/or around the SRS system and components, be sure to disable the SRS system. Tape the negative battery cable with insulating tape. Always disconnect the negative battery cable first.

CAUTION

To avoid personal injury when working on vehicles equipped with an air bag, the negative battery cable must be disconnected and at least 90 seconds must elapse before working on the system. Failure to do so may result in deployment of the air bag.

2. Disconnect the negative battery cable. Tape the cable with insulating tape.
3. Properly relieve the fuel system pressure.
4. Drain the engine coolant. Remove the air cleaner assembly.
5. Disconnect the fuel injector wiring connectors.
6. Remove the necessary components in order to gain access to the intake manifold retaining bolts.
7. Remove the intake manifold retaining bolts. Remove the intake manifold from the engine.

To install:

➡Be sure to use new fasteners, as required.

8. Clean all surfaces.
9. Install a new gasket on each cylinder head.

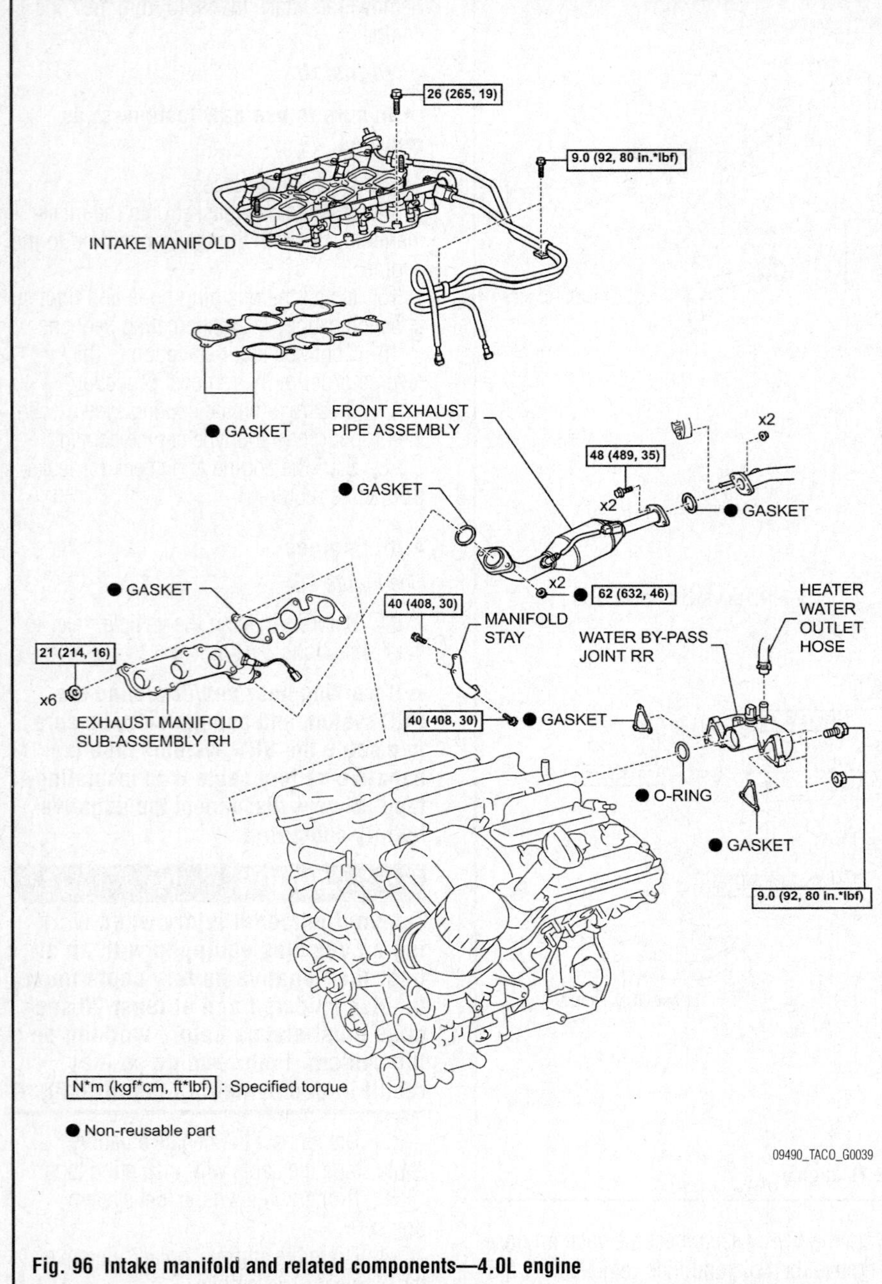

Fig. 96 Intake manifold and related components—4.0L engine

➡Align the ports of the gasket and the cylinder head. Be careful of the installation direction. Position the intake manifold to the engine.

10. Install the retaining bolts and tighten to specification, in an alternating sequence.
11. Continue the installation in the reverse order of the removal procedure.
12. Be sure to fill the cooling system with the proper grade and type engine coolant.
13. Start the engine and check for leaks. Correct as required.

OIL PAN

REMOVAL & INSTALLATION

2.7L Engines

See Figures 97 through 100.

1. Before servicing the vehicle, refer to the Precautions Section.

➡If working near and/or around the SRS system and components, be sure to disable the SRS system. Tape the negative battery cable with insulating tape. Always disconnect the negative battery cable first.

✱✱ CAUTION

To avoid personal injury when working on vehicles equipped with an air bag, the negative battery cable must be disconnected and at least 90 seconds must elapse before working on the system. Failure to do so may result in deployment of the air bag.

2. Disconnect the negative battery cable. Tape the cable with insulating tape.
3. Raise and support the vehicle safely.
4. Remove the engine undercover. Drain the engine oil.
5. Remove the necessary components in order to gain access to the lower oil pan retaining bolts.
6. Remove the eighteen bolts and two nuts. Insert the blade of tool SST09032-00100 between the pans. Cut through the sealer and remove the lower oil pan from the engine.

➡Be careful not to damage the contact surface of the oil pans.

7. Remove the two bolts and nuts. Remove the oil strainer. Discard the gasket.
8. To remove the upper oil pan, remove the sixteen bolts and two nuts. Remove the upper oil pan from the engine, by prying it apart using a suitable tool.

➡Be careful not to damage the sealing surface between the upper oil pan and the cylinder block.

To install:

➡Be sure to use new fasteners, as required.

9. Apply a continuous bead (0.079–0.118 inch in diameter) of seal packing, part number 08826-00080 or equivalent, to the sealing surface of the oil pan.

➡Remove any oil from the contact surface. Install the upper oil pan within three minutes of applying the seal packing. Do not start the engine for at least two hours after the installation of the oil pan.

10. Loosely install the upper oil pan bolts and nuts. Bolt "A" is 0.79 inch long and bolt "B" is 1.57 inch long. Uniformly tighten the bolts to 19 ft. lbs, in the proper sequence.
11. Install the oil strainer assembly. Torque the bolts to 19 ft. lbs.
12. Apply a continuous bead (0.118–0.157 inch in diameter) of seal packing, part number 08826-00080 or equivalent, to the sealing surface of the oil pan.

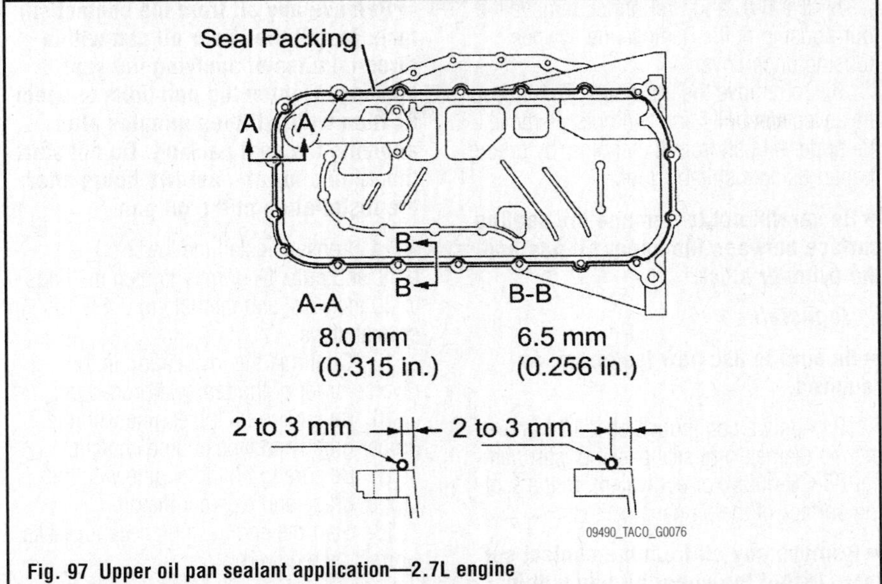

Fig. 97 Upper oil pan sealant application—2.7L engine

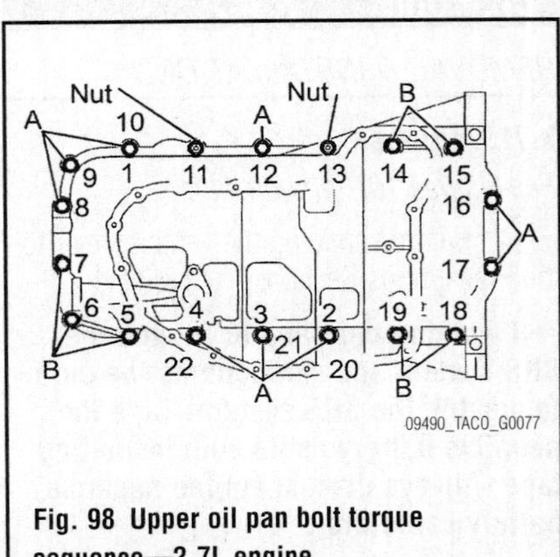

Fig. 98 Upper oil pan bolt torque sequence—2.7L engine

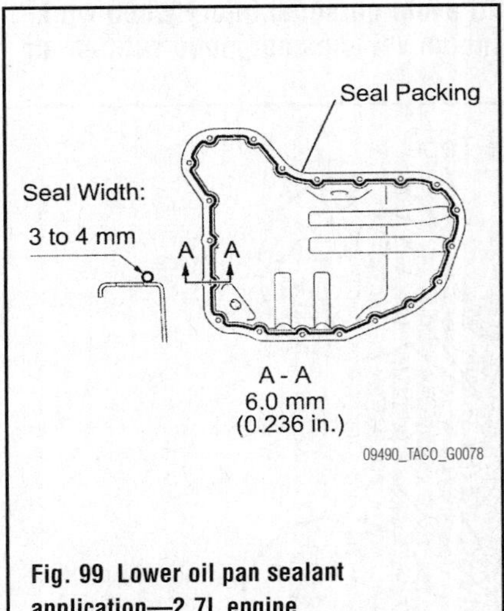

Fig. 99 Lower oil pan sealant application—2.7L engine

➡Remove any oil from the contact surface. Install the lower oil pan within three minutes of applying the seal packing. Do not start the engine for at

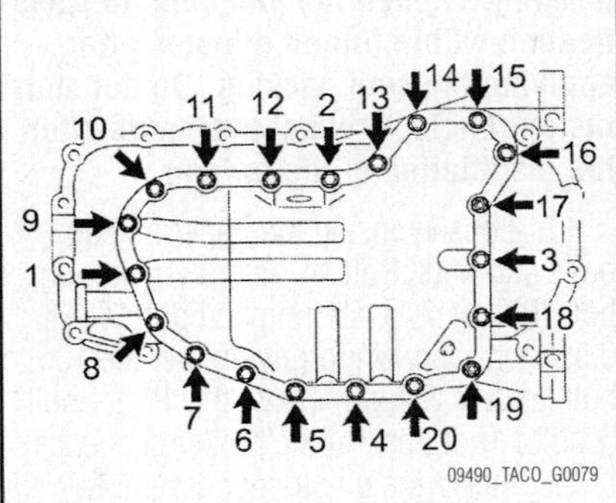

Fig. 100 Lower oil pan bolt torque sequence—2.7L engine

least two hours after the installation of the oil pan.

13. Loosely install the lower oil pan bolts and nuts. Uniformly tighten the bolts to 80 inch lbs, in the proper sequence.
14. Continue the installation in the reverse order of the removal procedure.
15. Be sure to fill the engine with the proper grade and type engine coolant.
16. Be sure to fill the engine with the proper grade and type engine oil.
17. Start the engine and check for leaks. Correct as required.

4.0L Engines

See Figures 101 through 106.

1. Before servicing the vehicle, refer to the Precautions Section.

➡If working near and/or around the SRS system and components, be sure to disable the SRS system. Tape the negative battery cable with insulating tape. Always disconnect the negative battery cable first.

⁂ CAUTION

To avoid personal injury when working on vehicles equipped with an air bag, the negative battery cable must be disconnected and at least 90 seconds must elapse before working on the system. Failure to do so may result in deployment of the air bag.

2. Disconnect the negative battery cable. Tape the cable with insulating tape.
3. Raise and support the vehicle safely.
4. Remove the engine undercover. Drain the engine oil.
5. Remove the necessary components in order to gain access to the lower oil pan retaining bolts.
6. Remove the fifteen bolts and two nuts (2WD vehicles) and ten bolts and two nuts (4WD and PreRunner). Insert the blade of tool SST09032-00100 between the pans. Cut through the sealer and remove the lower oil pan from the engine.

➡Be careful not to damage the contact surface of the oil pans.

7. Remove the two bolts and nuts. Remove the oil strainer. Discard the gasket.

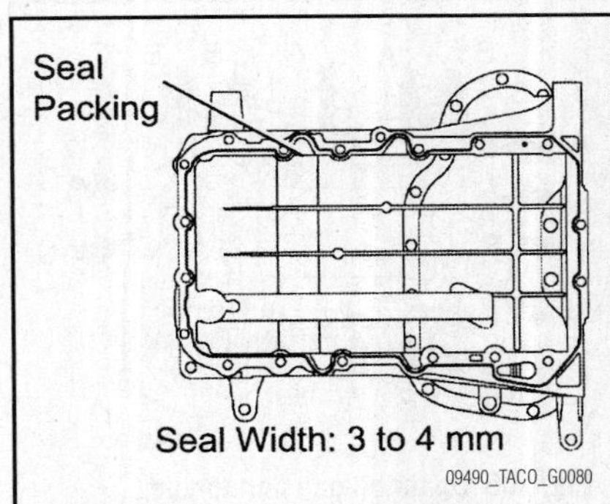

Fig. 101 Upper oil pan sealant application (2WD)—4.0L engine

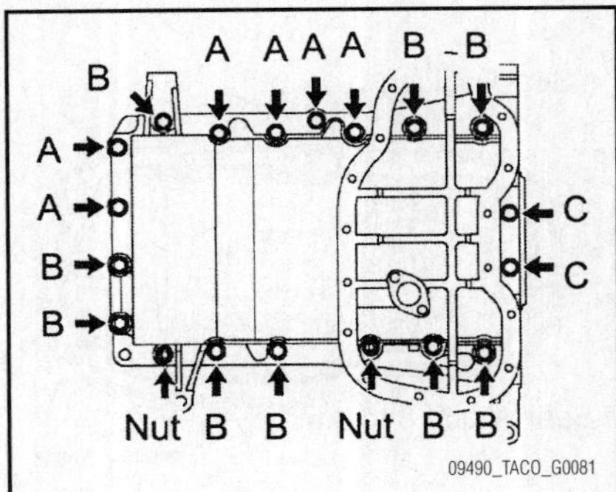

Fig. 102 Upper oil pan bolt torque sequence (2WD)—4.0L engine

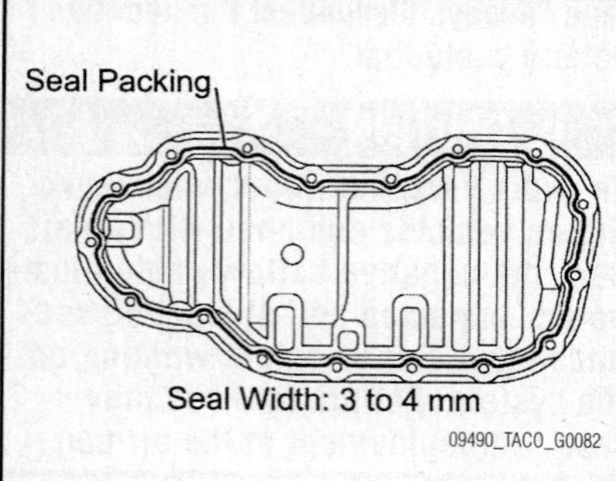

Fig. 103 Lower oil pan sealant application (2WD)—4.0L engine

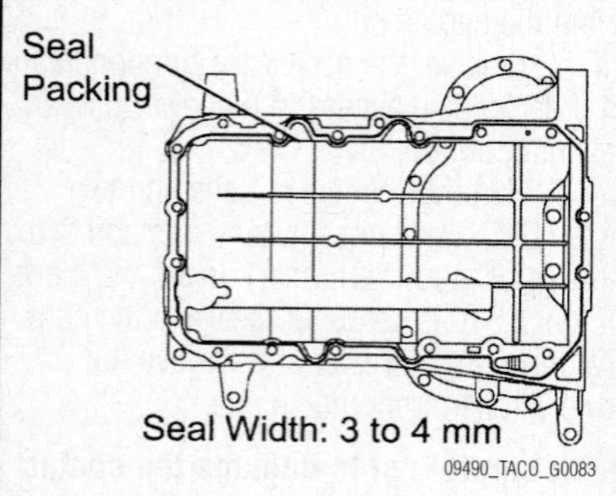

Fig. 104 Upper oil pan sealant application (4WD and PreRunner)—4.0L engine

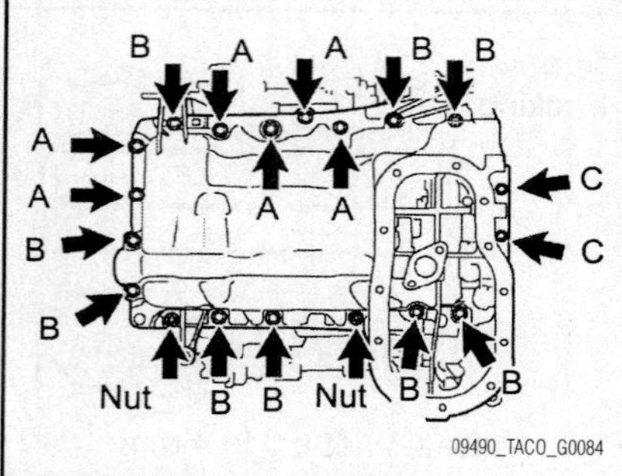

Fig. 105 Upper oil pan bolt torque sequence (4WD and PreRunner)—4.0L engine

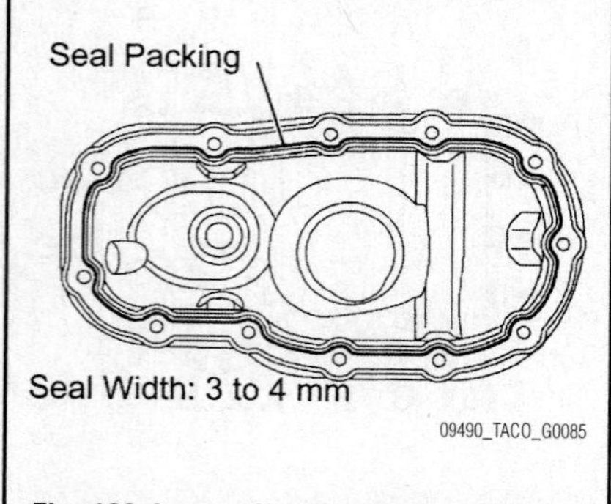

Fig. 106 Lower oil pan sealant application (4WD and PreRunner)—4.0L engine

8. On 4WD and PreRunner, remove the four housing bolts. Remove the flywheel housing undercover.

9. To remove the upper oil pan, remove the seventeen bolts and two nuts. Remove the upper oil pan from the engine, by prying it apart using a suitable tool.

➡**Be careful not to damage the sealing surface between the upper oil pan and the cylinder block.**

To install:

➡**Be sure to use new fasteners, as required.**

10. Apply a continuous bead (0.12–0.16 inch in diameter) of seal packing, part number 08826-00080 or equivalent, to the sealing surface of the oil pan.

➡**Remove any oil from the contact surface. Install the upper oil pan within three minutes of applying the seal packing. Tighten the pan bolts to specification within fifteen minutes after applying the seal packing. Do not start the engine for at least two hours after the installation of the oil pan.**

11. Loosely install the upper oil pan bolts and nuts. Bolt "A" is 0.98 inch long, bolt "B" is 1.77 inch long and bolt "C" is 0.55 inch long. Uniformly tighten the 10mm bolt head to 7.4 ft. lbs, and the 12mm bolt head to 16 ft. lbs., in the proper sequence.

12. Install the oil strainer assembly. Torque the bolts to 80 inch lbs.

13. Apply a continuous bead (0.12–0.16 inch in diameter) of seal packing, part number 08826-00080 or equivalent, to the sealing surface of the oil pan.

➡**Remove any oil from the contact surface. Install the lower oil pan within three minutes of applying the seal packing. Tighten the pan bolts to specification within fifteen minutes after applying the seal packing. Do not start the engine for at least two hours after the installation of the oil pan.**

14. Loosely install the lower oil pan bolts and nuts. Uniformly tighten the bolts to 80 inch lbs, and the nuts to 7.4 ft. lbs., in several steps.

15. Continue the installation in the reverse order of the removal procedure.

16. Be sure to fill the engine with the proper grade and type engine coolant.

17. Be sure to fill the engine with the proper grade and type engine oil.

18. Start the engine and check for leaks. Correct as required.

OIL PUMP

REMOVAL & INSTALLATION

2.7L Engines

See Figures 107 through 111.

1. Before servicing the vehicle, refer to the Precautions Section.

➡**If working near and/or around the SRS system and components, be sure to disable the SRS system. Tape the negative battery cable with insulating tape. Always disconnect the negative battery cable first.**

⁂ CAUTION

To avoid personal injury when working on vehicles equipped with an air

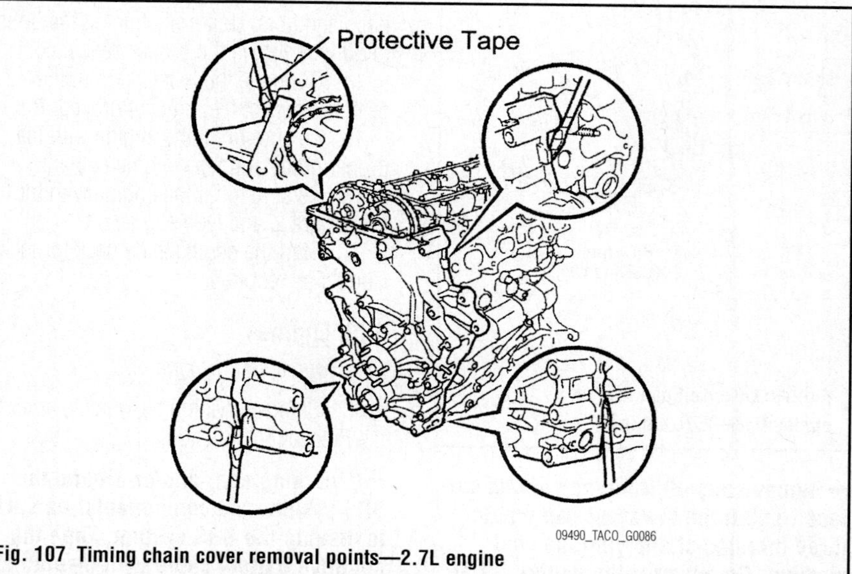

Fig. 107 Timing chain cover removal points—2.7L engine

bag, the negative battery cable must be disconnected and at least 90 seconds must elapse before working on the system. Failure to do so may result in deployment of the air bag.

2. Disconnect the negative battery cable. Tape the cable with insulating tape.
3. Remove the engine and position it in a suitable holding fixture.
4. Remove the air intake connector. Remove the alternator.
5. Remove the belt tensioner assembly. Remove the idler pulley subassembly. Remove the air conditioning idler pulley assembly and bracket.
6. Remove the Crankshaft Position (CKP) sensor. Remove the Camshaft Position (CMP) sensor.
7. Remove the intake manifold. Remove the cylinder head cover.
8. Remove the crankshaft pulley. Remove the oil level gauge subassembly.
9. Remove the lower oil pan. Remove the oil strainer assembly. Remove the upper oil pan.
10. Remove the two nuts and separate the water bypass pipe number one.
11. Remove the nineteen bolts and two nuts retaining the timing chain case cover to its mounting. Remove the timing chain case cover from the engine.

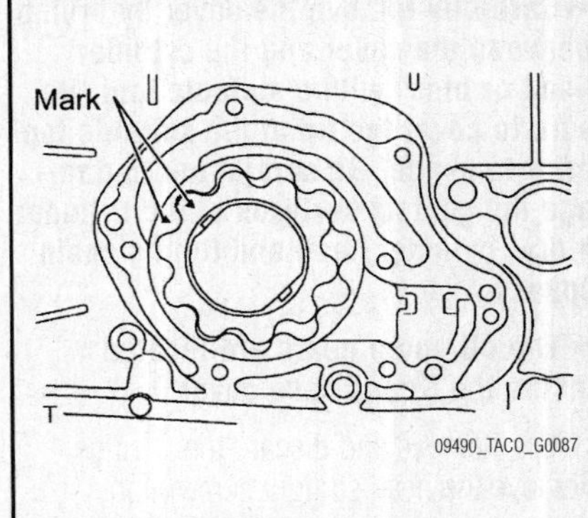

Fig. 108 Oil pump gear alignment marks—2.7L engine

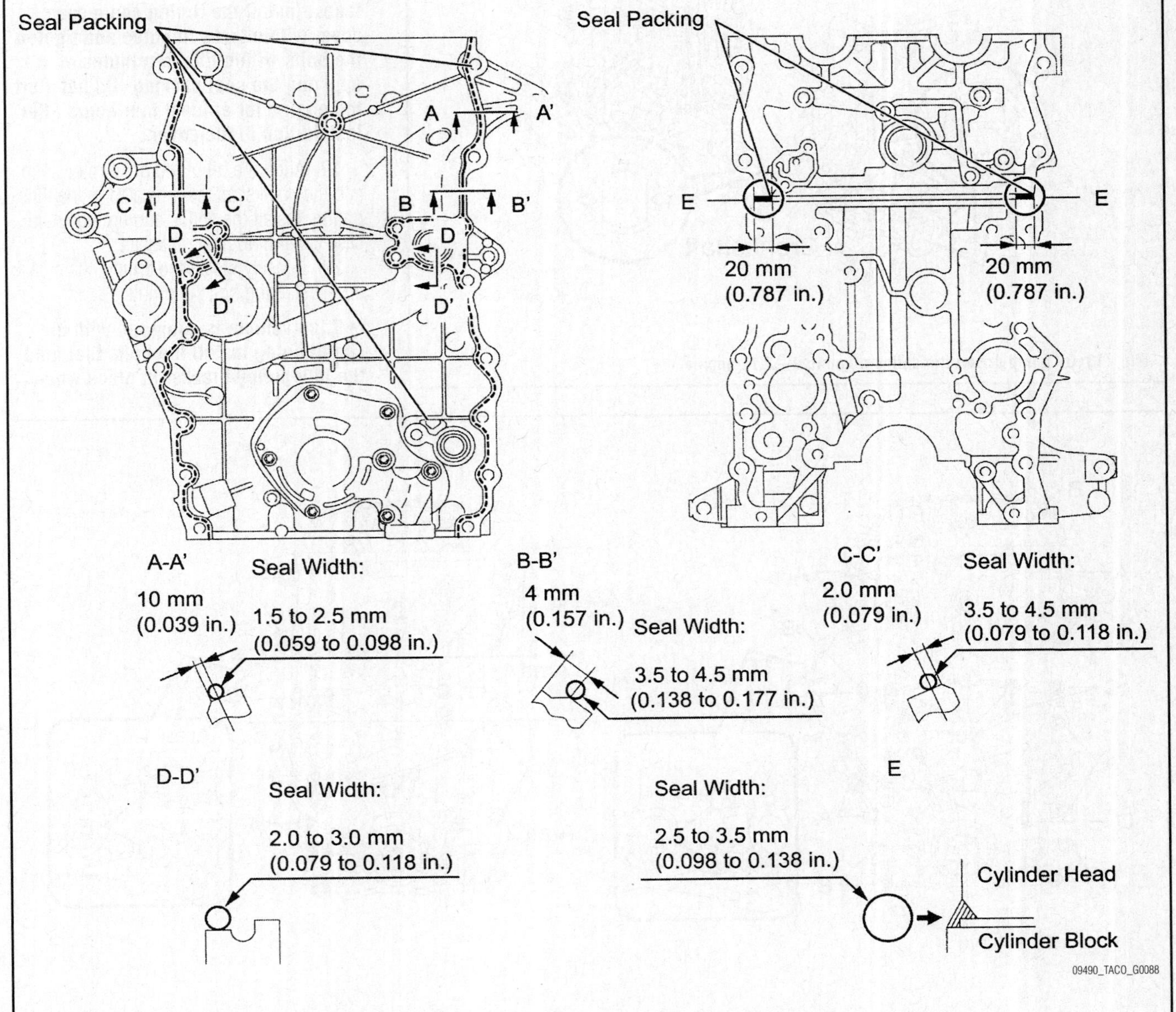

Fig. 109 Timing chain case cover sealant application—2.7L engine

➡**Carefully remove the cover by prying between the cover and the cylinder head or block with a suitable tool. Be sure to cover the tip of the suitable tool prior to usage. Be careful not to damage the contact surfaces of the cylinder block, cylinder head and timing chain cover.**

➡**The oil pump gears are located inside the timing case cover.**

12. Remove and discard the O-rings. Remove the head straight screw plug. Remove the water inlet. Remove the thermostat. Remove the oil seal.

13. Remove the oil pump relief valve. Remove the seven oil pump cover bolts. Remove the gears from the timing chain case cover.

To install:

➡**Be sure to use new fasteners, as required.**

14. Coat the oil pump gears with clean engine oil. Position the gears in the timing chain case cover with the identification marks facing outward. Check that the rotors revolve smoothly. Install the cover. Alternately tighten the bolts to 80 inch lbs.

15. Coat the oil pump relief valve with clean engine oil. Install the plug, using a new gasket and tighten to 36 ft. lbs.

16. Install a new front case oil seal. Install the thermostat. Install the water inlet.

17. Apply adhesive, part number 08833-00070 or equivalent to the head straight screw plug. Install the plug and tighten to 12 ft. lbs. Install four new O-rings onto the timing chain case cover.

18. Apply continuous beads of seal packing, part number 08826-00080 or equivalent as shown in the illustration.

➡**Remove any oil from the contact surfaces. Install the timing chain case cover within three minutes and tighten the bolts within fifteen minutes of applying the seal packing. Do not start the engine for at least four hours after installation of the cover.**

19. Align the oil pump drive rotor spline and the crankshaft, as indicated in the illustration. Install the spline and timing chain case cover onto the crankshaft.

20. Loosely install the timing chain case cover retaining bolts and nuts.

➡**If the vehicle is equipped with air conditioning install the bolts that hold the idle pulley bracket in place when**

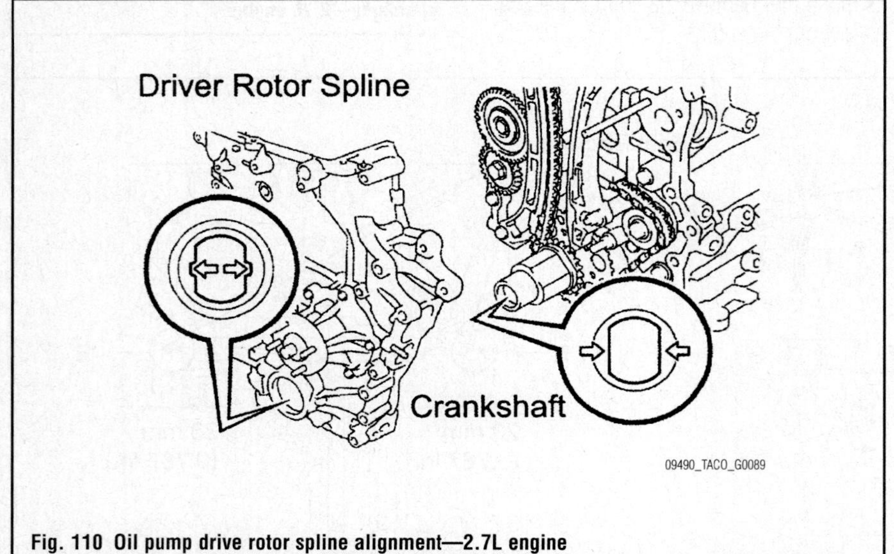

Fig. 110 Oil pump drive rotor spline alignment—2.7L engine

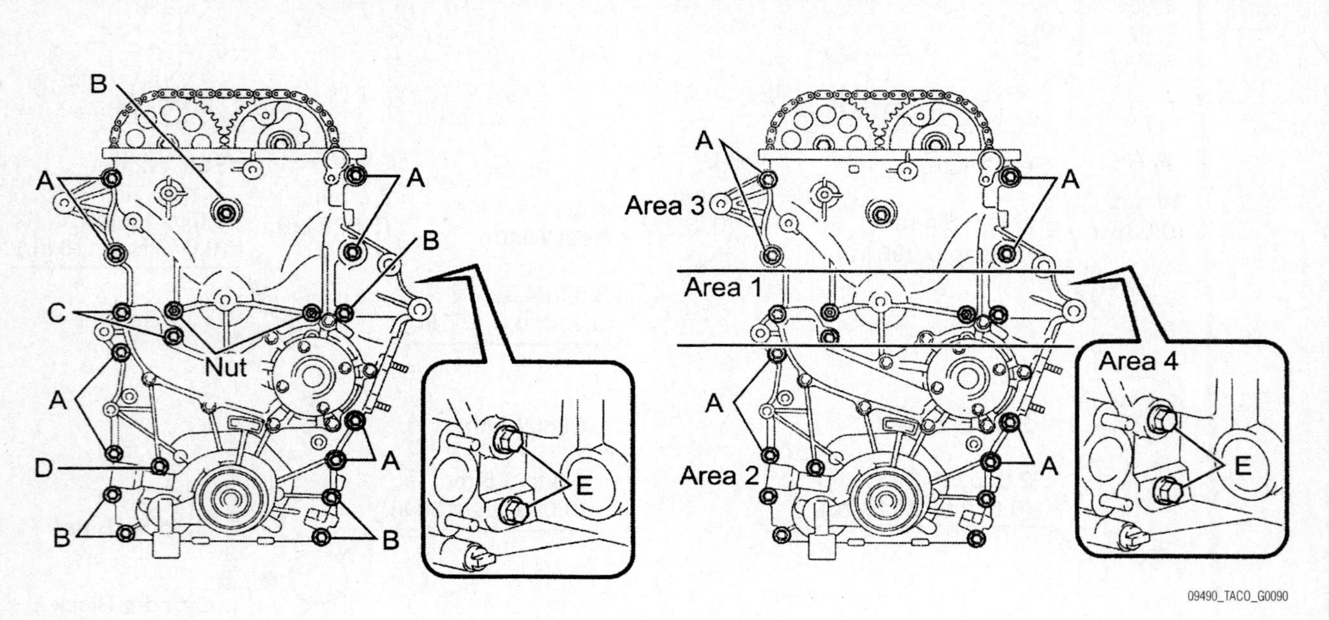

Fig. 111 Timing chain case cover bolt location and torque sequence—2.7L engine

installing the idle pulley, as they are for this purpose.

21. Fully tighten the bolts and nuts, except bolts "A" in the following order: Area 1, Area 3 and then Area 2 to 15 ft. lbs.
22. Fully tighten the bolts "A" in the following order: Area 2 and then Area 3 to 34 ft. lbs.
23. Fully tighten the bolts "E" in Area 4 to 15 ft. lbs.
24. Continue the installation in the reverse order of the removal procedure.
25. When installing the cylinder head cover, apply seal packing, part number 08826-00080 or equivalent, to the cylinder head. Provisionally install the cylinder head cover bolts and nuts. Tighten bolts "A" to 80 inch lbs. Tighten bolts "B" to 80 inch lbs. Retighten bolts "A" to 80 inch lbs.

➡**Be sure to remove any oil from the contact surfaces of the cylinder head cover and the cylinder head. Install the cover within three minutes after applying the seal packing. Do not add engine oil for at least two hours after installing the cover.**

26. Be sure to fill the engine with the proper grade and type engine coolant.
27. Be sure to fill the engine with the proper grade and type engine oil.
28. Start the engine and check for leaks, correct as required.

4.0L Engine

2WD

See Figures 112 through 116.

1. Before servicing the vehicle, refer to the Precautions Section.

➡**If working near and/or around the SRS system and components, be sure to disable the SRS system. Tape the negative battery cable with insulating tape. Always disconnect the negative battery cable first.**

CAUTION

To avoid personal injury when working on vehicles equipped with an air bag, the negative battery cable must be disconnected and at least 90 seconds must elapse before working on the system. Failure to do so may result in deployment of the air bag.

2. Disconnect the negative battery cable. Tape the cable with insulating tape.
3. Properly relieve the fuel system pressure.

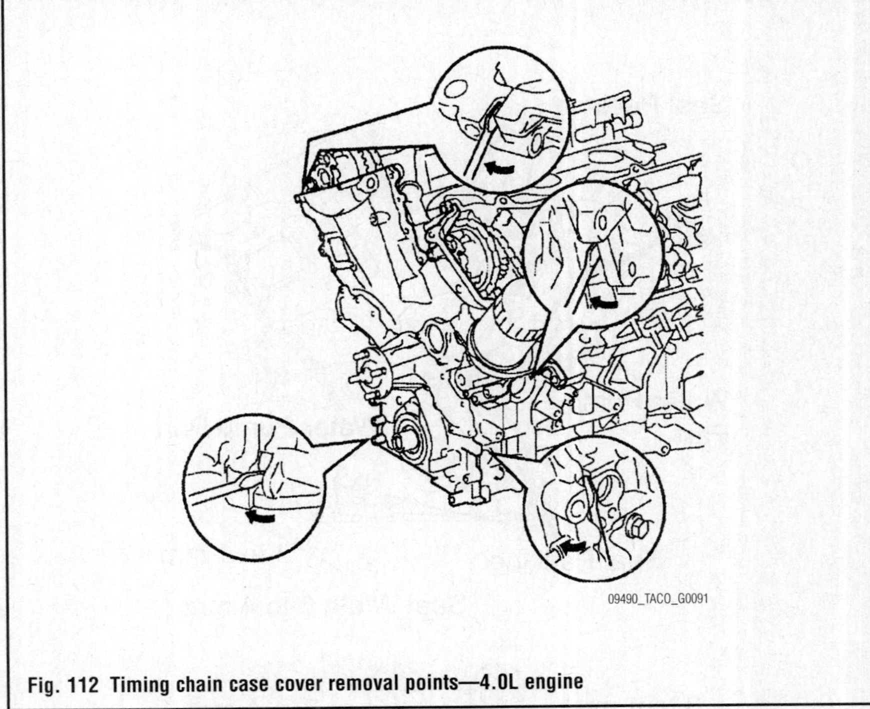

Fig. 112 Timing chain case cover removal points—4.0L engine

4. Disconnect the positive battery cable. Remove the battery.
5. Drain the engine coolant. Drain the engine oil.
6. Remove the engine from the vehicle and position it in a suitable holding fixture.
7. Remove the oil level gauge guide. Remove the water inlet. Remove the belt tensioner.
8. Remove the idler pulley number two subassembly. Remove the idler pulley number one sub assembly. Remove the crankshaft pulley.
9. Remove the lower oil pan. Remove the strainer and pickup tube. Remove the upper oil pan.
10. Remove the ignition coil assembly. Remove the cylinder head cover. Remove the camshaft timing oil control valve assembly.
11. Remove the VVT sensor. Remove the oil filter bracket subassembly.
12. Remove the timing chain case cover retaining bolts. Remove the cover from the engine. Remove the O-ring from the left cylinder head.

➡**Carefully remove the cover by prying between the cover and the cylinder head or block with a suitable tool. Be sure to cover the tip of the suitable tool prior to usage. Be careful not to damage the contact surfaces of the cylinder block, cylinder head and timing chain cover.**

➡**The oil pump gears are located inside the timing case cover.**

13. Remove the three bolts and remove the oil pipe. Remove the two O-rings. Remove the seven oil pump cover bolts. Remove the gears from the timing chain case cover. Remove the oil pump relief valve.

To install:

➡**Be sure to use new fasteners, as required.**

14. Coat the oil pump relief valve with clean engine oil. Install the plug, using a new gasket and tighten to 36 ft. lbs.
15. Coat the oil pump gears with clean engine oil. Position the gears in the timing chain case cover with the identification marks facing oil pump cover side. Install

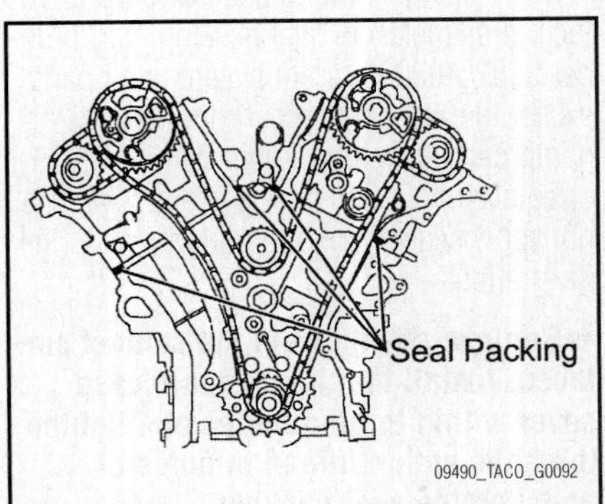

Fig. 113 Timing chain case cover seal packing locating points—4.0L engine

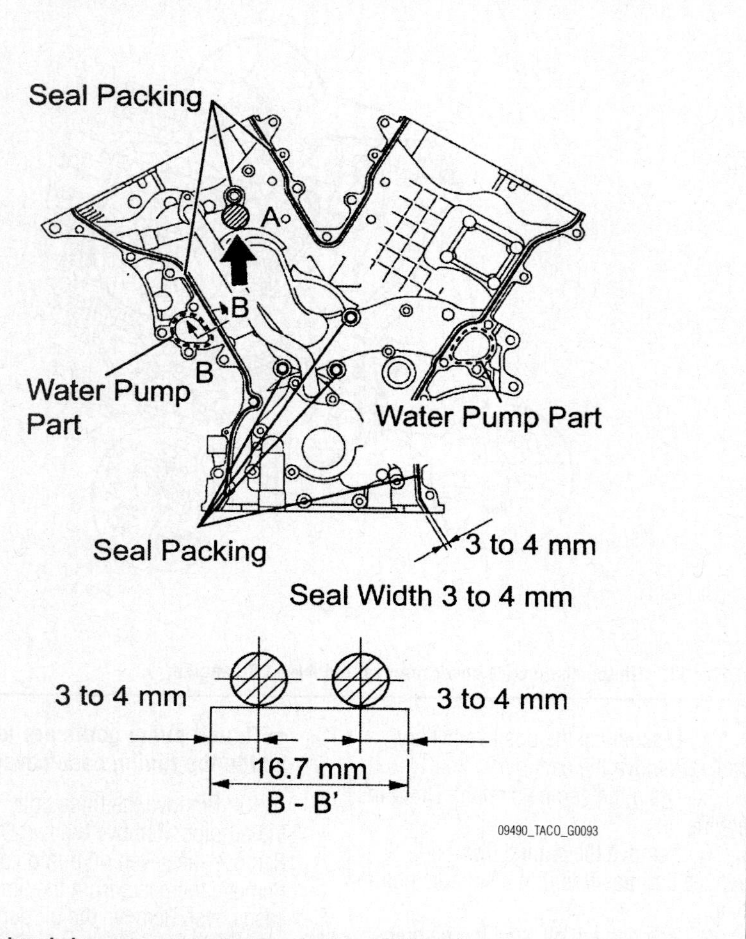

Fig. 114 Timing chain case cover sealant application—4.0L engine

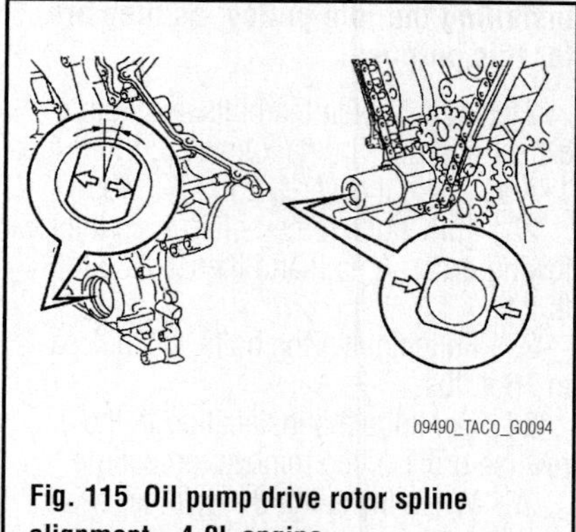

Fig. 115 Oil pump drive rotor spline alignment—4.0L engine

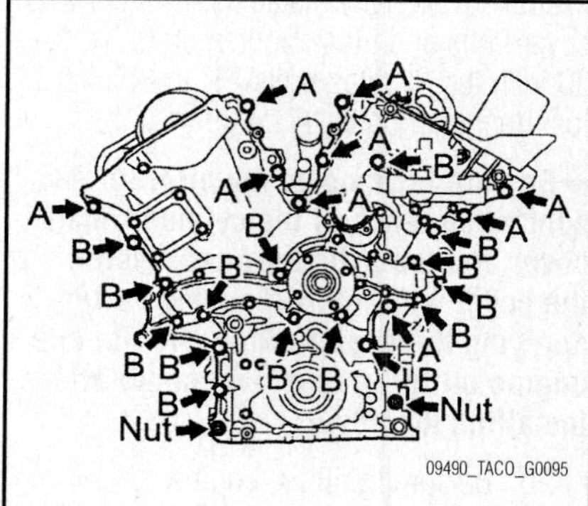

Fig. 116 Timing chain case cover bolt location and torque sequence—4.0L engine

the cover. Alternately tighten the bolts to 80 inch lbs. Install the oil pipe, tighten the bolts to 80 inch lbs.

16. Install a new front case oil seal. Install a new O-ring onto the left cylinder head.

17. Apply continuous beads (0.12–0.16 inch in diameter) of seal packing, part number 08826-00080 or equivalent to the four locations shown in the illustration.

18. Apply continuous beads (0.12–0.16 inch in diameter) of seal packing, part number 08826-00080 or equivalent to all parts except the water pump part: for the water pump part use, part number 08826-00080 or equivalent, to the timing chain cover. Do not apply seal packing to portion "A" in the illustration.

➡Remove any oil from the contact surfaces. Install the timing chain case cover within three minutes and tighten the bolts within fifteen minutes of applying the seal packing.

19. Align the key way of the oil pump drive motor with the rectangular portion of the crankshaft timing gear and slide the timing chain case cover into place.

20. Install the timing chain case cover bolts. Tighten the bolts and nuts uniformly in several steps to 17 ft. lbs.

➡Do not wrap the chain and slipper over the timing chain case cover seal line.

21. Continue the installation in the reverse order of the removal procedure.

22. When installing the cylinder head cover apply a continuous bead (0.08–0.12 inch) of seal packing, part number 08826-00080 or equivalent, to the cylinder head. Install the seal washers onto the bolts. Install the cylinder head cover bolts and nuts. Tighten bolts "A" to 7.4 ft. lbs. Tighten bolts "B" to 80 inch lbs. Tighten nuts to 80 inch lbs.

➡Be sure to remove any oil from the contact surfaces of the cylinder head cover and the cylinder head. Install the cover within three minutes after applying the seal packing. Tighten the bolts to specification within fifteen minutes after installing the cover. Do not add engine oil for at least two hours after installing the cover.

23. Be sure to fill the engine with the proper grade and type engine coolant.

24. Be sure to fill the engine with the proper grade and type engine oil.

25. Start the engine and check for leaks, correct as required.

4WD And PreRunner

1. Before servicing the vehicle, refer to the Precautions Section.

➡If working near and/or around the SRS system and components, be sure to disable the SRS system. Tape the negative battery cable with insulating tape. Always disconnect the negative battery cable first.

CAUTION

To avoid personal injury when working on vehicles equipped with an air bag, the negative battery cable must be disconnected and at least 90 seconds must elapse before work-

ing on the system. Failure to do so may result in deployment of the air bag.

2. Disconnect the negative battery cable. Tape the cable with insulating tape.
3. Properly relieve the fuel system pressure.
4. Disconnect the positive battery cable. Remove the battery.
5. Drain the engine coolant. Drain the engine oil.
6. Remove the power steering gear assembly.
7. If equipped with 4WD, remove the front differential carrier assembly.
8. Remove the V bank cover. Remove the radiator support to frame seal, left side. Remove the fan shroud.
9. Remove the air cleaner assembly. Remove the oil level gauge. Remove the water inlet.
10. Separate the vane pump assembly. Remove the alternator. Remove the air conditioning compressor and position it to the side.
11. Remove the belt tensioner assembly. Remove the idler pulley number two subassembly. Remove the idler pulley number one subassembly. Remove the crankshaft pulley.
12. Remove the lower oil pan. Remove the oil strainer and pickup tube assembly. Remove the upper oil pan.
13. Remove the intake manifold. Remove the cylinder head cover assembly.
14. Remove the camshaft timing oil control valve assembly. Remove the VVT sensor. Remove the oil filter bracket subassembly.

➡Carefully remove the cover by prying between the cover and the cylinder head or block with a suitable tool. Be sure to cover the tip of the suitable tool prior to usage. Be careful not to damage the contact surfaces of the cylinder block, cylinder head and timing chain cover.

➡The oil pump gears are located inside the timing case cover.

15. Remove the three bolts and remove the oil pipe. Remove the two O-rings. Remove the seven oil pump cover bolts. Remove the gears from the timing chain case cover. Remove the oil pump relief valve.

To install:

➡Be sure to use new fasteners, as required.

16. Coat the oil pump relief valve with clean engine oil. Install the plug, using a new gasket and tighten to 36 ft. lbs.
17. Coat the oil pump gears with clean engine oil. Position the gears in the timing chain case cover with the identification marks facing oil pump cover side. Install the cover. Alternately tighten the bolts to 80 inch lbs. Install the oil pipe, tighten the bolts to 80 inch lbs.
18. Install a new front case oil seal. Install a new O-ring onto the left cylinder head.
19. Apply continuous beads (0.12–0.16 inch in diameter) of seal packing, part number 08826-00080 or equivalent to the four locations shown in the illustration.
20. Apply continuous beads (0.12–0.16 inch in diameter) of seal packing, part number 08826-00080 or equivalent to all parts except the water pump part: for the water pump part use, part number 08826-00080 or equivalent, to the timing chain cover. Do not apply seal packing to portion "A" in the illustration.

➡Remove any oil from the contact surfaces. Install the timing chain case cover within three minutes and tighten the bolts within fifteen minutes of applying the seal packing.

21. Align the key way of the oil pump drive motor with the rectangular portion of the crankshaft timing gear and slide the timing chain case cover into place.
22. Install the timing chain case cover bolts. Tighten the bolts and nuts uniformly in several steps to 17 ft. lbs.

➡Do not wrap the chain and slipper over the timing chain case cover seal line.

23. Continue the installation in the reverse order of the removal procedure.
24. When installing the cylinder head cover apply a continuous bead (0.08–0.12 inch) of seal packing, part number 08826-00080 or equivalent, to the cylinder head. Install the seal washers onto the bolts. Install the cylinder head cover bolts and nuts. Tighten bolts "A" to 7.4 ft. lbs. Tighten bolts "B" to 80 inch lbs. Tighten nuts to 80 inch lbs.

➡Be sure to remove any oil from the contact surfaces of the cylinder head cover and the cylinder head. Install the cover within three minutes after applying the seal packing. Tighten the bolts to specification within fifteen minutes after installing the cover. Do not add engine oil for at least two hours after installing the cover.

25. Be sure to fill the engine with the proper grade and type engine coolant.
26. Be sure to fill the engine with the proper grade and type engine oil.
27. Start the engine and check for leaks, correct as required.

PISTON AND RING

POSITIONING

See Figures 117 through 119.

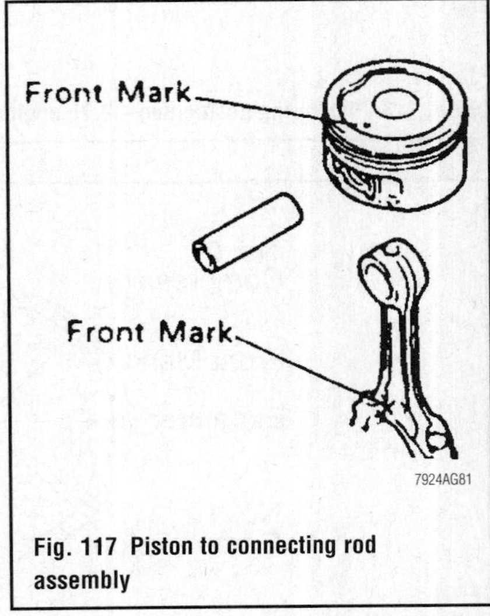

Fig. 117 Piston to connecting rod assembly

REAR MAIN SEAL

REMOVAL & INSTALLATION

2.7L Engines

See Figures 120 through 121.

1. Before servicing the vehicle, refer to the Precautions Section.

➡If working near and/or around the SRS system and components, be sure to disable the SRS system. Tape the negative battery cable with insulating tape. Always disconnect the negative battery cable first.

⁂ CAUTION

To avoid personal injury when working on vehicles equipped with an air bag, the negative battery cable must be disconnected and at least 90 seconds must elapse before working on the system. Failure to do so may result in deployment of the air bag.

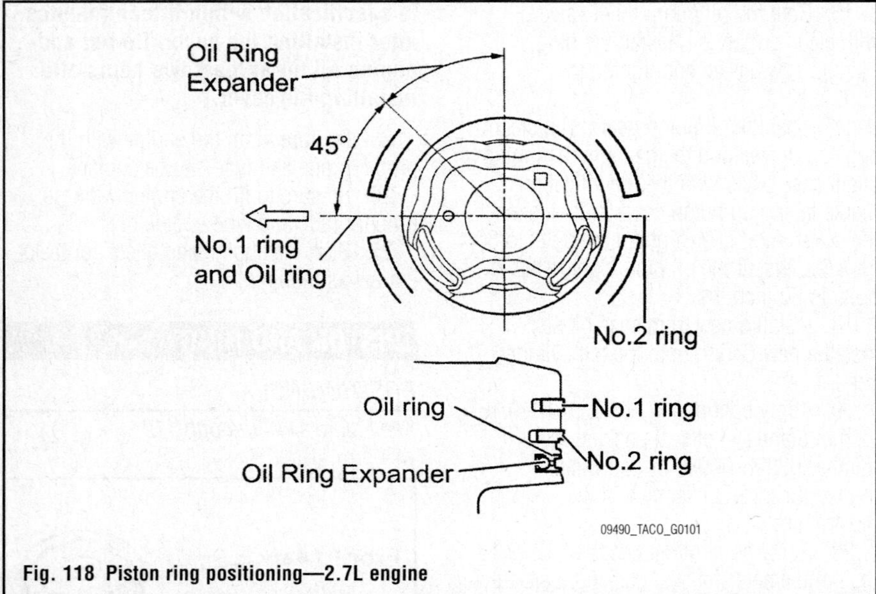

Fig. 118 Piston ring positioning—2.7L engine

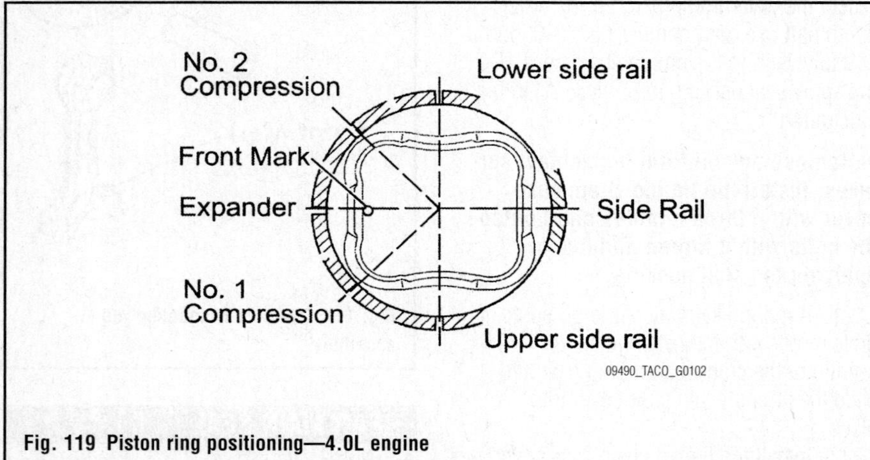

Fig. 119 Piston ring positioning—4.0L engine

2. Disconnect the negative battery cable. Tape the cable with insulating tape.
3. Remove the transmission.
4. Remove the clutch cover assembly and flywheel (manual transmission) or the flexplate and spacers (automatic transmission).
5. Use a small, sharp knife to cut off the lip of the oil seal. Take great care not to score any metal with the knife.
6. Use a small prybar to pry the old seal from the retaining plate. Be careful not to damage the plate. Protect the tip of the tool with tape and pad the fulcrum point with cloth.
7. Inspect the crankshaft and seal lip contact surfaces for any sign of damage.

To install:

➡Be sure to use new fasteners, as required.

8. Apply a light coat of multi-purpose grease to the lip of a new oil seal. Loosely fit the seal into place by hand, making sure it is not crooked.
9. Use a seal driver such as (SST 09223–15030 and 09950–70010) of the correct size to install the seal. Tap it into place until the surface of the seal is flush with the edge of the housing.

➡Use the correct tools. Homemade substitutes may install the seal crooked, resulting in oil leaks and premature seal failure.

10. Installation is the reverse of the removal procedure.
11. Tighten the flywheel/flexplate retaining bolts to specification and in the proper sequence. On 4.0L engine with manual transmission, tighten the flywheel retaining bolts in an alternating sequence.

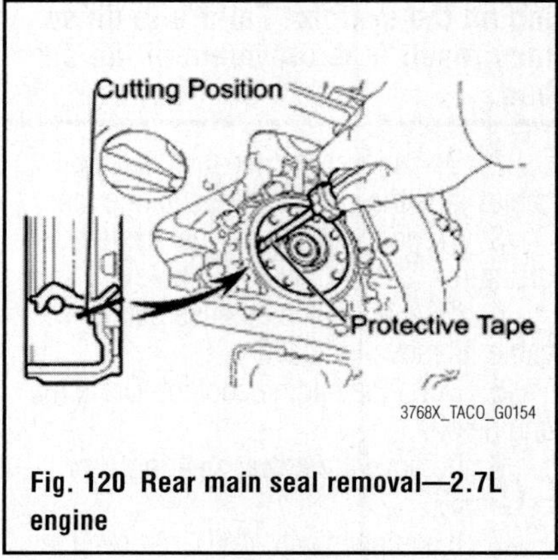

Fig. 120 Rear main seal removal—2.7L engine

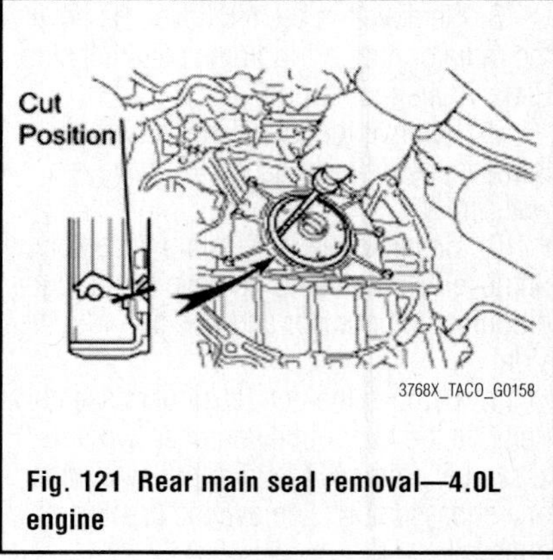

Fig. 121 Rear main seal removal—4.0L engine

TIMING CHAIN FRONT COVER SEAL

REMOVAL & INSTALLATION

Cover Installed

1. Before servicing the vehicle, refer to the Precautions section.
2. Unbolt and remove the oil pump.
3. Using a knife, carefully cut off the oil seal lip. With a flat-bladed tool, (preferably with tape around it) pry the seal from the cover.

To install:

4. Apply multi-purpose grease to the new oil seal lip.
5. Tap the seal into place with SST 09223–50010/60011 or equivalent seal driver, and a hammer. Do this until the seal surface is flush with the cover edge.
6. Install the oil pump with a new O-ring.

Cover Removed

1. Before servicing the vehicle, refer to the Precautions section.
2. Unbolt the timing chain cover assembly. Be careful to loosen only the correct bolts.

3. Pry out the seal from the cover with a flat-bladed tool.

4. It is a good idea to remove the oil pump from the timing cover and replace the O-ring.

To install:

5. Clean and inspect the timing cover area. Install new gaskets around the dowel areas and pump spline.

6. Apply multi-purpose grease to the new oil seal lip.

7. Tap the seal into place with SST 09223–50010/60010 or equivalent, and a hammer. Do this until the seal surface is flush with the cover edge.

8. Install the cover, tighten the bolts as specified for your engine.

9. If the oil pump was removed, install a new O-ring behind the pump prior to installation.

TIMING CHAIN & SPROCKETS

REMOVAL & INSTALLATION

2.7L Engines

See Figures 122 through 125.

1. Before servicing the vehicle, refer to the Precautions Section.

➡If working near and/or around the SRS system and components, be sure to disable the SRS system. Tape the negative battery cable with insulating tape. Always disconnect the negative battery cable first.

⁂ CAUTION

To avoid personal injury when working on vehicles equipped with an air bag, the negative battery cable must be disconnected and at least 90 seconds must elapse before working on the system. Failure to do so may result in deployment of the air bag.

2. Disconnect the negative battery cable. Tape the cable with insulating tape.

3. Remove the engine and position it in a suitable holding fixture.

4. Remove the air intake connector. Remove the alternator.

5. Remove the belt tensioner assembly. Remove the number one idler pulley sub-assembly. Remove the air conditioning idler pulley assembly and bracket.

6. Remove the Crankshaft Position (CKP) sensor. Remove the Camshaft Position (CMP) sensor.

7. Remove the intake manifold. Remove the cylinder head cover.

8. Position the engine at TDC of the compression stroke. Remove the crankshaft pulley.

9. Remove the oil level gauge sub-assembly. Remove the lower oil pan. Remove the oil strainer assembly. Remove the upper oil pan.

10. Remove the two nuts and separate the water bypass pipe number one.

11. Remove the nineteen bolts and two nuts retaining the timing chain case cover to its mounting. Remove the timing chain case cover from the engine.

➡Carefully remove the cover by prying between the cover and the cylinder head or block with a suitable tool. Be sure to cover the tip of the suitable tool prior to usage. Be careful not to damage the contact surfaces of the cylinder block, cylinder head and timing chain cover.

12. Make sure that each matchmark is in the same position as shown in the illustration. Remove the two bolts, timing chain guide and O-ring.

13. Move the stopper plate upward to release the lock, and push the plunger deep into the tensioner.

14. Move the stopper plate downward to set the lock. Insert a 0.118 inch diameter bar into the stopper plate hole. Remove the bolt, nut, number one chain tensioner and gasket.

➡When the number one chain tensioner is removed do not rotate the crankshaft. When the chain is removed and the camshaft needs to be rotated, rotate the crankshaft 90 degrees to the right.

15. Remove the bolt and chain tensioner slipper. Remove the two bolts and remove the number one chain vibration damper. Remove the primary timing chain sub-assembly.

16. Remove the crankshaft timing gear or sprocket. Remove the bolt and remove the number two chain vibration damper.

17. Remove the two bolts and remove the number three chain vibration damper.

18. Remove the nut and the number two chain tensioner assembly. Remove the bolt, balance shaft drive gear shaft and balance shaft drive gear. Remove the crankshaft timing sprocket number two and chain.

To install:

➡Be sure to use new fasteners, as required.

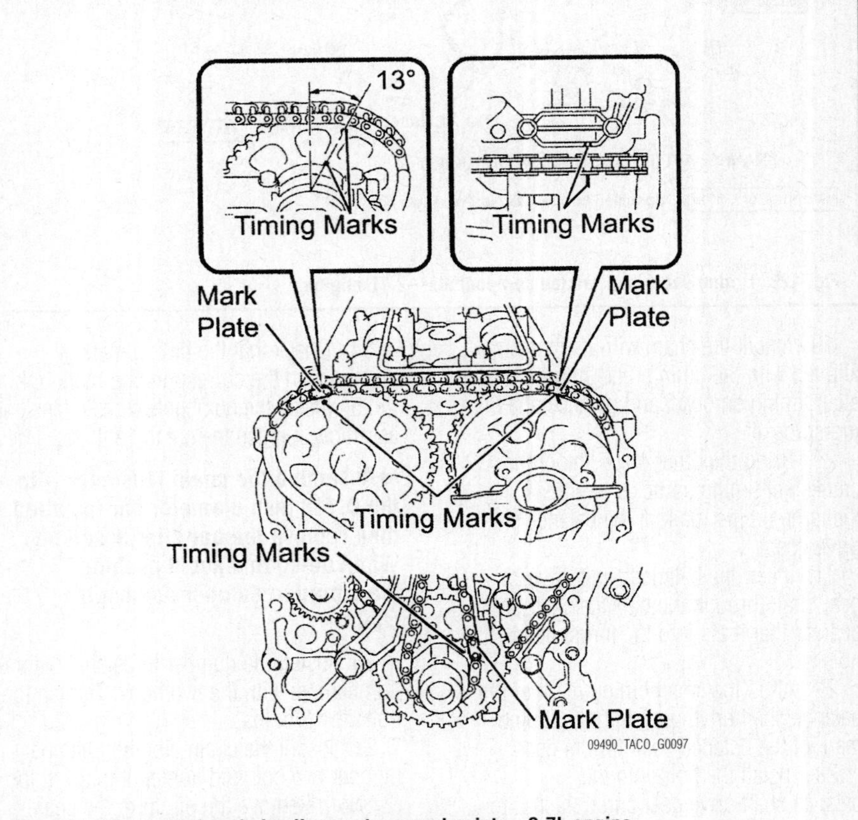

Fig. 122 Primary timing chain alignment removal points—2.7L engine

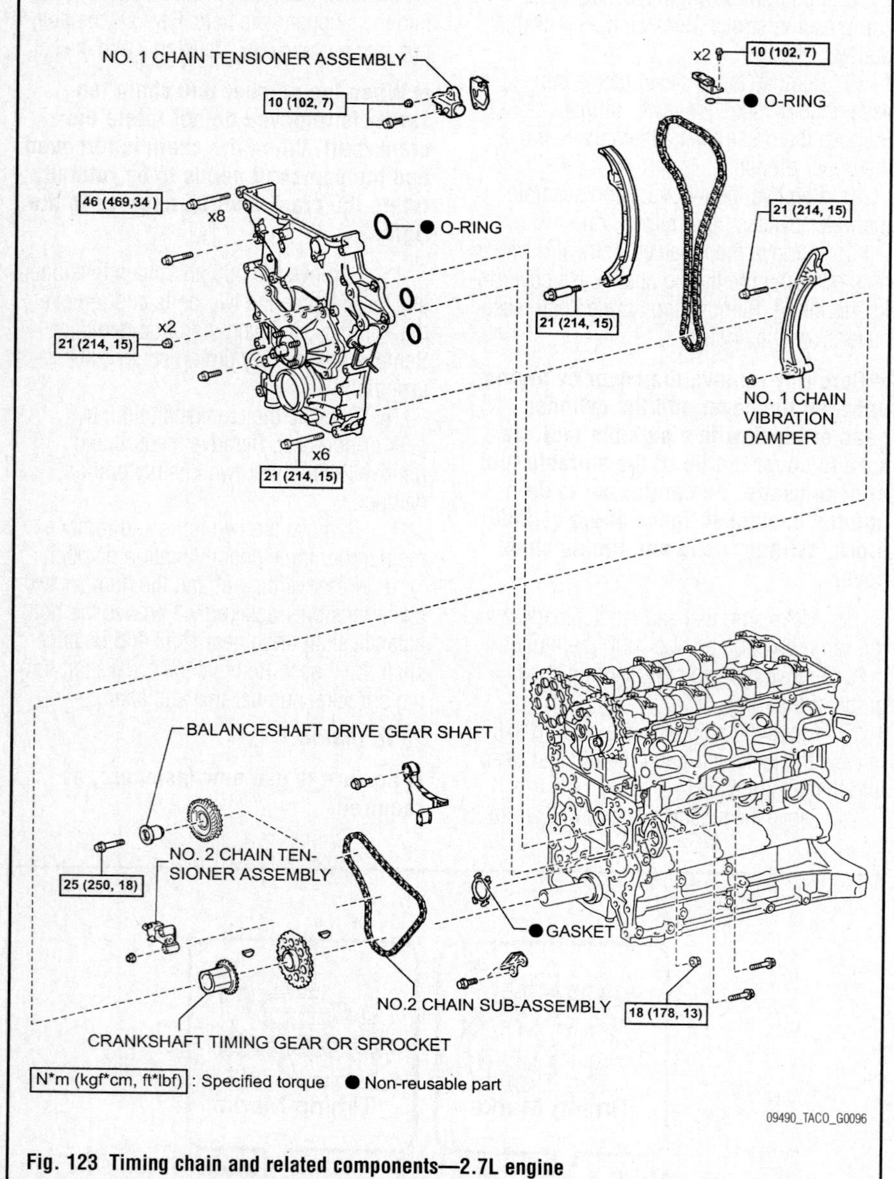

Fig. 123 Timing chain and related components—2.7L engine

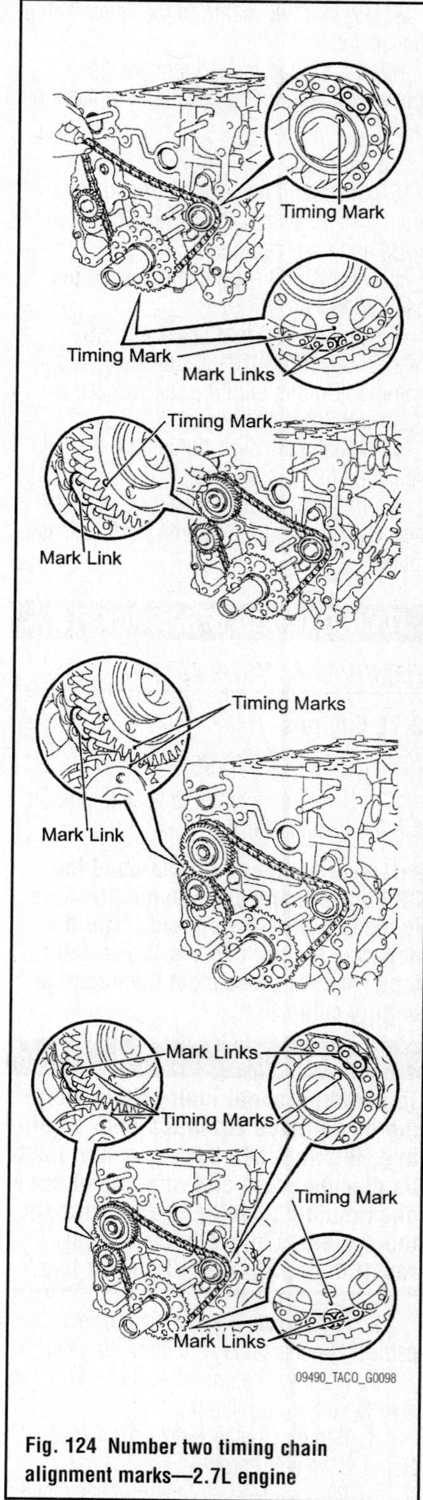

Fig. 124 Number two timing chain alignment marks—2.7L engine

19. Install the chain with its marks aligned with the timing marks on the crankshaft timing sprocket and balance shaft timing sprocket.

20. Bring the other mark link of the crankshaft timing sprocket behind the large timing mark of the balance shaft drive gear.

21. Insert the balance shaft drive gear shaft through the balance shaft drive gear so that it fits into the thrust plate hole.

22. Align the small timing mark of the balance shaft drive gear with the timing mark of the balance shaft timing gear.

23. Install the bolt onto the balance shaft drive gear and tighten it to 18 ft. lbs.

24. Check that the timing mark is aligned with the corresponding mark link.

25. Install the number two chain tensioner assembly. Tighten the nut to 13 ft. lbs.

➡Assemble the chain tensioner with the 0.118 inch diameter bar installed, then remove the bar after assembly. When doing this avoid pushing the vibration damper against the chain.

26. Install the number three chain vibration damper with the two bolts. Tighten the bolts to 13 ft. lbs.

27. Install the chain vibration damper number two bolt and tighten it to 20 ft. lbs. (27 Nm). Remove the pin from the chain tensioner and release the plunger.

28. Install the crankshaft timing gear or sprocket.

29. Install the number one chain vibration damper bolt and nut. Tighten to 15 ft. lbs.

30. Install the primary timing chain onto the sprocket and gear with the painted marks aligned with the timing marks on the sprocket and gear.

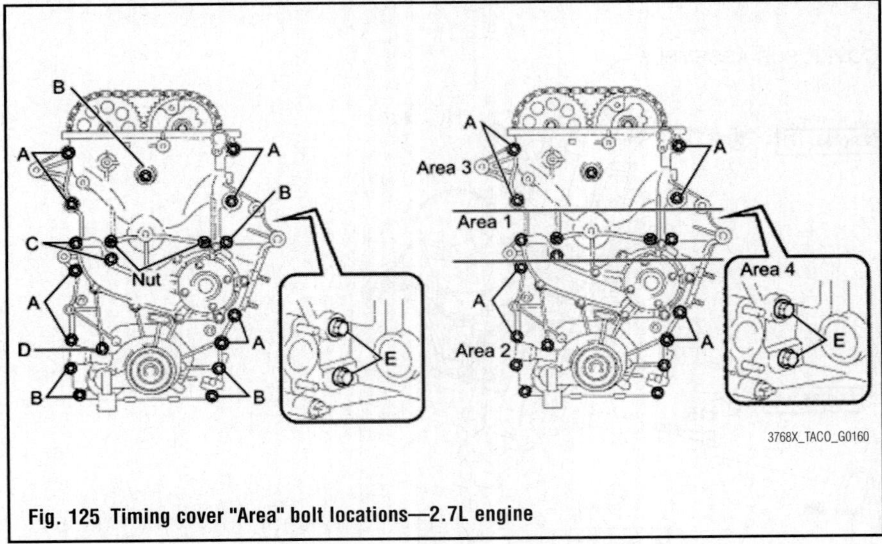

Fig. 125 Timing cover "Area" bolt locations—2.7L engine

➡The camshaft mark plate is orange. The crankshaft mark plate is yellow.

31. Use a rope to tie the chain of the crankshaft timing sprocket. Tie the rope near the sprocket.

➡After the chain tensioner has been installed, remove the rope. The rope is used to prevent gear jumping.

32. Install the tensioner slipper and tighten the bolt to 15 ft. lbs.
33. Install the number one chain tensioner assembly, using a new gasket. Tighten the bolts to 7 ft. lbs.
34. Install a new front case oil seal. Install the thermostat. Install the water inlet.
35. Apply adhesive, part number 08833-00070 or equivalent to the head straight screw plug. Install the plug and tighten to 12 ft. lbs. Install four new O-rings onto the timing chain case cover.
36. Apply continuous beads of seal packing, part number 08826-00080 or equivalent.

➡Remove any oil from the contact surfaces. Install the timing chain case cover within three minutes and tighten the bolts within fifteen minutes of applying the seal packing. Do not start the engine for at least four hours after installation of the cover.

37. Align the oil pump drive rotor spline and the crankshaft. Install the spline and timing chain case cover onto the crankshaft.
38. Loosely install the timing chain case cover retaining bolts and nuts.

➡If the vehicle is equipped with air conditioning install the bolts that hold the idle pulley bracket in place when installing the idle pulley, as they are for this purpose.

39. Fully tighten the bolts and nuts, except bolts "A" in the following order: Area 1, Area 3 and then Area 2 to 15 ft. lbs.
40. Fully tighten the bolts "A" in the following order: Area 2 and then Area 3 to 34 ft. lbs.
41. Fully tighten the bolts "E" in Area 4 to 15 ft. lbs.
42. Continue the installation in the reverse order of the removal procedure.
43. When installing the cylinder head cover, apply seal packing, part number 08826-00080 or equivalent, to the cylinder head. Provisionally install the cylinder head cover bolts and nuts. Tighten bolts "A" to 80 inch lbs. Tighten bolts "B" to 80 inch lbs. Retighten bolts "A" to 80 inch lbs.

➡Be sure to remove any oil from the contact surfaces of the cylinder head cover and the cylinder head. Install the cover within three minutes after applying the seal packing. Do not add engine oil for at least two hours after installing the cover.

44. Be sure to fill the engine with the proper grade and type engine coolant.
45. Be sure to fill the engine with the proper grade and type engine oil.
46. Start the engine and check for leaks, correct as required.

4.0L Engines

2WD

See Figures 126 through 129.

1. Before servicing the vehicle, refer to the Precautions Section.

➡If working near and/or around the SRS system and components, be sure to disable the SRS system. Tape the negative battery cable with insulating tape. Always disconnect the negative battery cable first.

⁂ CAUTION

To avoid personal injury when working on vehicles equipped with an air bag, the negative battery cable must be disconnected and at least 90 seconds must elapse before working on the system. Failure to do so may result in deployment of the air bag.

2. Disconnect the negative battery cable. Tape the cable with insulating tape.
3. Properly relieve the fuel system pressure.
4. Disconnect the positive battery cable. Remove the battery.
5. Drain the engine coolant. Drain the engine oil.
6. Remove the engine from the vehicle and position it in a suitable holding fixture.
7. Remove the oil level gauge guide. Remove the water inlet. Remove the belt tensioner.
8. Remove the idler pulley number two subassembly. Remove the idler pulley number one sub assembly. Remove the crankshaft pulley.
9. Remove the lower oil pan. Remove the strainer and pickup tube. Remove the upper oil pan.
10. Remove the intake manifold. Remove the ignition coil assembly. Remove the cylinder head cover. Remove the camshaft timing oil control valve assembly.
11. Remove the VVT sensor. Remove the oil filter bracket subassembly.
12. Remove the timing chain case cover retaining bolts. Remove the cover from the engine. Remove the O-ring from the left cylinder head.

➡Carefully remove the cover by prying between the cover and the cylinder head or block with a suitable tool. Be sure to cover the tip of the suitable tool prior to usage. Be careful not to damage the contact surfaces of the cylinder block, cylinder head and timing chain cover.

13. Using the crankshaft pulley set bolt, turn the crankshaft to align the crankshaft set key with the timing line of the cylinder block. If not aligned at TDC of the compression stroke, turn the crankshaft one complete revolution, in the direction of rotation
14. While turning the stopper plate of the tensioner upward, push the plunger of the chain tensioner. While turning the stopper

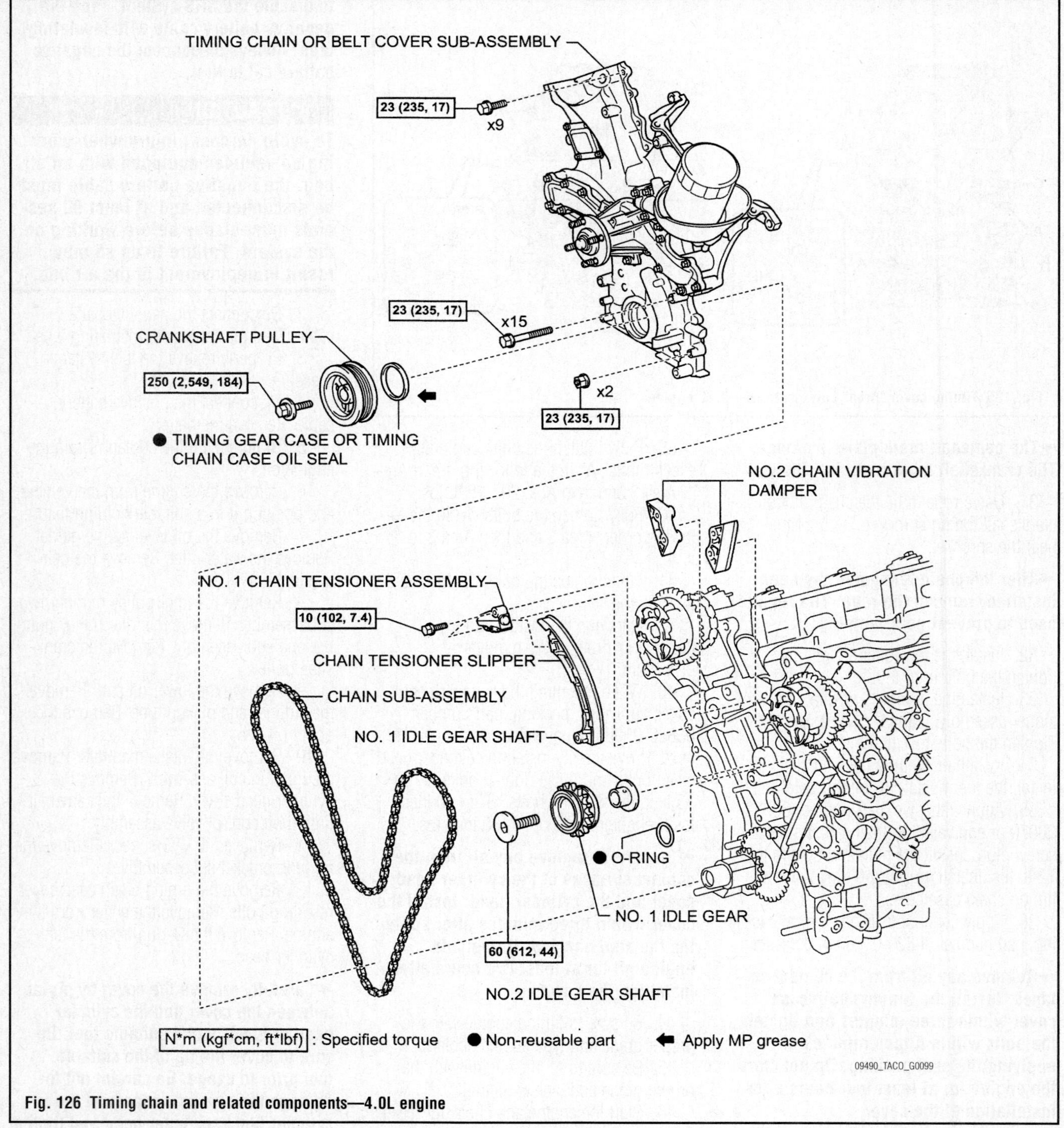

Fig. 126 Timing chain and related components—4.0L engine

plate of the tensioner downward, insert a 0.138 inch diameter bar into the holes in the stopper plate and tensioner to hold the stopper plate.

15. Remove the two bolts, and then remove the chain tensioner.

16. Remove the chain tensioner slipper. Remove the idle gear shaft number two, idle gear number one and idle gear shaft number one.

17. Remove the number two chain vibration damper. Remove the timing chain sub-assembly.

To install:

➡Be sure to use new fasteners, as required.

18. Install the chain tensioner slipper.

19. While turning the stopper plate of the tensioner clockwise, push in the plunger of the chain tensioner. While turning the stopper plate of the tensioner counterclockwise, insert a 0.138 inch diameter bar into the holes in the stopper plate and tensioner to hold the stopper plate.

20. Install the chain tensioner. Tighten the bolts to 7.4 ft. lbs.

21. Position the engine at TDC on the compression stroke. Align the camshaft

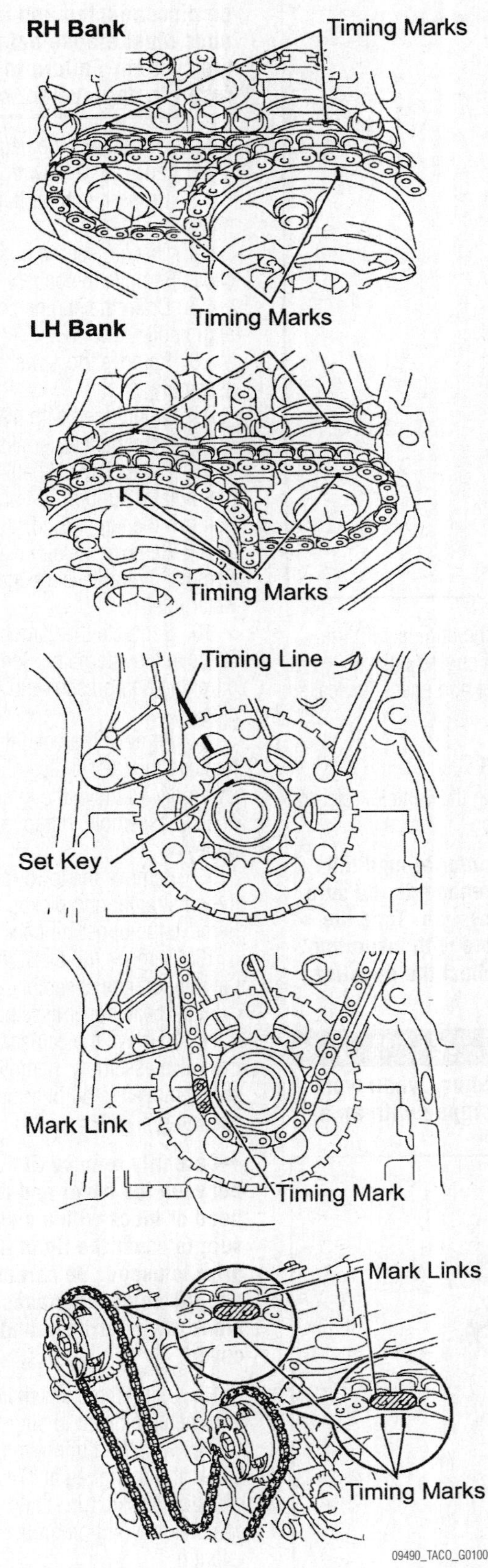

Fig. 127 Timing chain alignment—4.0L engine

timing gears and bearing caps. Using the crankshaft pulley set bolt, align the crankshaft set key with the timing line of the cylinder.

22. Align the yellow mark line with the timing mark of the crankshaft timing link. Align the orange mark links with the timing marks of the camshaft timing gears, and install the chain.

23. Install the number two chain vibration damper.

24. Apply a light coat of clean engine oil to the rotating surface of the idle gear shaft number one.

25. Temporarily install the idle gear shaft number one together with idle gear shaft number two, while aligning the knock pin of idle gear shaft number one with the knock pin groove of the cylinder block.

➡Be care of the idle gear direction.

26. Tighten the idle gear shaft number two to 44 ft. lbs. Remove the bar from the chain tensioner.

27. Install a new front case oil seal. Install a new O-ring onto the left cylinder head.

28. Apply continuous beads (0.12–0.16 inch in diameter) of seal packing, part number 08826-00080 or equivalent to the four locations shown in the illustration.

29. Apply continuous beads (0.12–0.16 inch in diameter) of seal packing, part number 08826-00080 or equivalent to all parts except the water pump part: for the water pump part use, part number 08826-00080 or equivalent, to the timing chain cover. Do not apply seal packing to portion "A".

➡Remove any oil from the contact surfaces. Install the timing chain case cover within three minutes and tighten the bolts within fifteen minutes of applying the seal packing.

30. Align the key way of the oil pump drive motor with the rectangular portion of the crankshaft timing gear and slide the timing chain case cover into place.

31. Install the timing chain case cover bolts. Tighten the bolts and nuts uniformly in several steps to 17 ft. lbs.

➡Do not wrap the chain and slipper over the timing chain case cover seal line.

32. Continue the installation in the reverse order of the removal procedure.

33. When installing the cylinder head cover apply a continuous bead (0.08–0.12 inch) of seal packing, part number 08826-00080 or equivalent, to the cylinder head.

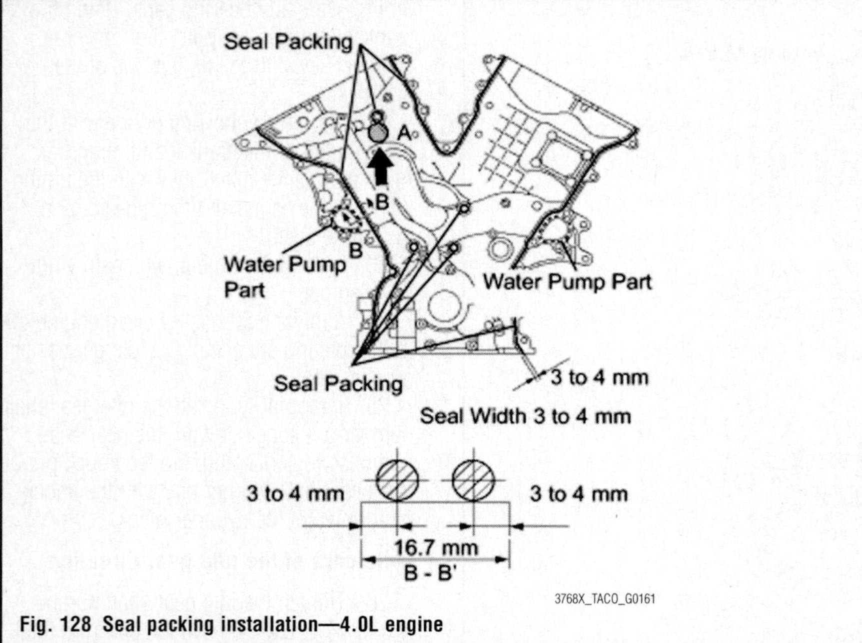

Fig. 128 Seal packing installation—4.0L engine

Install the seal washers onto the bolts. Install the cylinder head cover bolts and nuts. Tighten bolts "A" to 7.4 ft. lbs. Tighten bolts "B" to 80 inch lbs. Tighten nuts to 80 inch lbs.

➡**Be sure to remove any oil from the contact surfaces of the cylinder head cover and the cylinder head. Install the cover within three minutes after applying the seal packing. Tighten the bolts to specification within fifteen minutes after installing the cover. Do not add engine oil for at least two hours after installing the cover.**

34. Be sure to fill the engine with the proper grade and type engine coolant.

35. Be sure to fill the engine with the proper grade and type engine oil.

36. Start the engine and check for leaks, correct as required.

4WD And PreRunner

1. Before servicing the vehicle, refer to the Precautions Section.

➡**If working near and/or around the SRS system and components, be sure to disable the SRS system. Tape the negative battery cable with insulating tape. Always disconnect the negative battery cable first.**

CAUTION

To avoid personal injury when working on vehicles equipped with an air bag, the negative battery cable must be disconnected and at least 90 seconds must elapse before working on the system. Failure to do so may result in deployment of the air bag.

2. Disconnect the negative battery cable. Tape the cable with insulating tape.
3. Properly relieve the fuel system pressure.
4. Disconnect the positive battery cable. Remove the battery.
5. Drain the engine coolant. Drain the engine oil.
6. Remove the power steering gear assembly.
7. If equipped with 4WD, remove the front differential carrier assembly.
8. Remove the V bank cover. Remove the radiator support to frame seal, left side. Remove the fan shroud.
9. Remove the air cleaner assembly. Remove the oil level gauge. Remove the water inlet.
10. Separate the vane pump assembly. Remove the alternator. Remove the air conditioning compressor and position it to the side.
11. Remove the belt tensioner assembly. Remove the idler pulley number two subassembly. Remove the idler pulley number one subassembly. Remove the crankshaft pulley.
12. Remove the lower oil pan. Remove the oil strainer and pickup tube assembly. Remove the upper oil pan.
13. Remove the intake manifold. Remove the ignition coil assembly. Remove the cylinder head cover assembly.
14. Remove the camshaft timing oil control valve assembly. Remove the VVT sensor. Remove the oil filter bracket subassembly.

➡**Carefully remove the cover by prying between the cover and the cylinder head or block with a suitable tool. Be sure to cover the tip of the suitable tool prior to usage. Be careful not to damage the contact surfaces of the cylinder block, cylinder head and timing chain cover.**

15. Using the crankshaft pulley set bolt, turn the crankshaft to align the crankshaft set key with the timing line of the cylinder block. If not aligned at TDC of the compression stroke, turn the crankshaft one complete revolution, in the direction of rotation
16. While turning the stopper plate of the tensioner upward, push the plunger of the chain tensioner. While turning the stopper

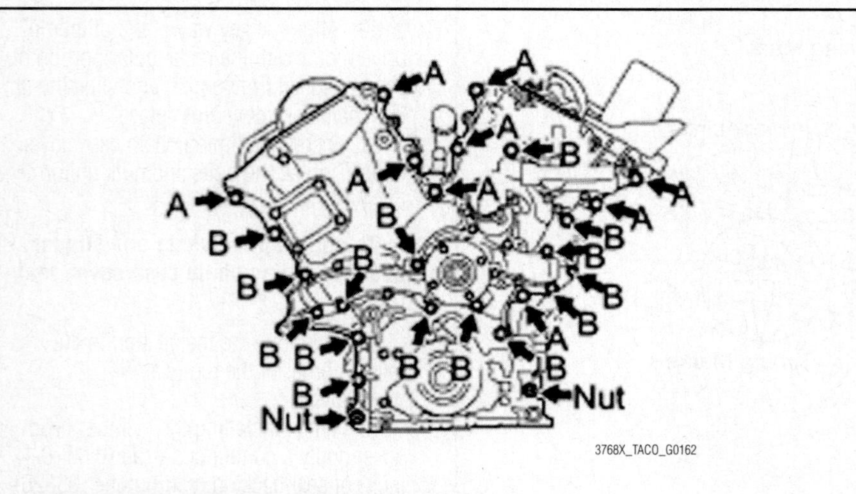

Fig. 129 Timing cover bolt identification—4.0L engine

plate of the tensioner downward, insert a 0.138 inch diameter bar into the holes in the stopper plate and tensioner to hold the stopper plate.

17. Remove the two bolts, and then remove the chain tensioner.

18. Remove the chain tensioner slipper. Remove the idle gear shaft number two, idle gear number one and idle gear shaft number one.

19. Remove the number two chain vibration damper. Remove the timing chain sub-assembly.

To install:

20. Install the chain tensioner slipper.

21. While turning the stopper plate of the tensioner clockwise, push in the plunger of the chain tensioner. While turning the stopper plate of the tensioner counterclockwise, insert a 0.138 inch diameter bar into the holes in the stopper plate and tensioner to hold the stopper plate.

22. Install the chain tensioner. Tighten the bolts to 7.4 ft. lbs.

23. Position the engine at TDC on the compression stroke. Align the camshaft timing gears and bearing caps. Using the crankshaft pulley set bolt, align the crankshaft set key with the timing line of the cylinder.

24. Align the yellow mark line with the timing mark of the crankshaft timing link. Align the orange mark links with the timing marks of the camshaft timing gears, and install the chain.

25. Install the number two chain vibration damper.

26. Apply a light coat of clean engine oil to the rotating surface of the idle gear shaft number one.

27. Temporarily install the idle gear shaft number one together with idle gear shaft number two, while aligning the knock pin of idle gear shaft number one with the knock pin groove of the cylinder block.

➡Be care of the idle gear direction.

28. Tighten the idle gear shaft number two to 44 ft. lbs. Remove the bar from the chain tensioner.

29. Install a new front case oil seal. Install a new O-ring onto the left cylinder head.

30. Apply continuous beads (0.12–0.16 inch in diameter) of seal packing, part number 08826-00080 or equivalent to the four locations shown in the illustration.

31. Apply continuous beads (0.12–0.16 inch in diameter) of seal packing, part number 08826-00080 or equivalent to all parts except the water pump part: for the water pump part use, part number 08826-00080 or equivalent, to the timing chain cover. Do not apply seal packing to portion "A".

➡Remove any oil from the contact surfaces. Install the timing chain case cover within three minutes and tighten the bolts within fifteen minutes of applying the seal packing.

32. Align the key way of the oil pump drive motor with the rectangular portion of the crankshaft timing gear and slide the timing chain case cover into place.

33. Install the timing chain case cover bolts. Tighten the bolts and nuts uniformly in several steps to 17 ft. lbs. (23 Nm).

➡Do not wrap the chain and slipper over the timing chain case cover seal line.

34. Continue the installation in the reverse order of the removal procedure.

35. When installing the cylinder head cover apply a continuous bead (0.08–0.12 inch) of seal packing, part number 08826-00080 or equivalent, to the cylinder head. Install the seal washers onto the bolts. Install the cylinder head cover bolts and nuts. Tighten bolts "A" to 7.4 ft. lbs. (10 Nm). Tighten bolts "B" to 80 inch lbs. 9 Nm). Tighten nuts to 80 inch lbs. (9 Nm).

➡Be sure to remove any oil from the contact surfaces of the cylinder head cover and the cylinder head. Install the cover within three minutes after applying the seal packing. Tighten the bolts to specification within fifteen minutes after installing the cover. Do not add engine oil for at least two hours after installing the cover.

36. Be sure to fill the engine with the proper grade and type engine coolant.

37. Be sure to fill the engine with the proper grade and type engine oil.

38. Start the engine and check for leaks, correct as required.

VALVE LASH

ADJUSTMENT

4.0L Engine

See Figures 130 through 132.

1. Before servicing the vehicle, refer to the Precautions Section.

➡If working near and/or around the SRS system and components, be sure to disable the SRS system. Tape the negative battery cable with insulating tape. Always disconnect the negative battery cable first.

✱✱ CAUTION

To avoid personal injury when working on vehicles equipped with an air bag, the negative battery cable must be disconnected and at least 90 seconds must elapse before working on the system. Failure to do so may result in deployment of the air bag.

2. Disconnect the negative battery cable. Tape the cable with insulating tape.

3. Before servicing the vehicle, refer to the Precautions Section.

4. Disconnect the negative battery cable.

5. Drain the engine coolant. Remove the V bank cover.

6. Remove the air cleaner assembly.

7. Disconnect the two water bypass hoses. Disconnect the fuel vapor feed hose. Disconnect the ventilation hose. Disconnect the VSV connectors.

8. Disconnect the throttle body motor connector. Separate the three wire harness clamps and hose clamp.

9. If equipped with manual transmission, remove the nut, then separate the clutch flexible hose bracket from the surge tank stay.

10. Remove the two bolts and the throttle body bracket. Remove the bolt and the oil baffle plate. Remove the four bolts and the two serge tank stays.

11. Remove the two nuts. Remove the four bolts, intake air surge tank and gasket.

12. Remove the ignition coil assembly.

13. Remove the cylinder head cover retaining bolts. Remove the cylinder head cover.

14. Turn the crankshaft pulley until its groove and the timing mark "0" of the timing chain cover are aligned. If not aligned at TDC of the compression stroke, turn the crankshaft one complete revolution, in the direction of rotation.

15. Using a feeler gauge, check and record the valve clearance on the following valves: right bank, exhaust number three and intake number one, left bank exhaust number two and intake number six.

16. Rotate the crankshaft 240 degrees clockwise and using a feeler gauge, check and record the following valves: right bank, exhaust number five and intake number three, left bank exhaust number four and intake number two.

17. Rotate the crankshaft 240 degrees clockwise and using a feeler gauge, check and record the following valves: right bank,

exhaust number one and intake number five, left bank exhaust number six and intake number four.

18. If adjustment is required, position the engine at TDC on the compression stroke.
19. Place paint marks on the number one chain links corresponding to the timing marks of the camshaft timing gears.
20. Remove the chain tensioner assembly number one. Remove the number two camshaft.
21. Remove the chain tensioner assembly number two. Remove the camshaft.
22. Remove the number four camshaft subassembly. Remove the chain tensioner assembly number three.
23. Remove the number three camshaft subassembly.
24. Remove the valve lifters.
25. Determine the replacement adjusting shim size according to the following formula or use the adjusting shim charts.
26. Using a micrometer, measure the thickness of the removed shim. Calculate the thickness of a new shim so that the valve clearance comes within the specified value.
- T: Thickness of the removed shim
- A: Measured valve clearance
- N: Thickness of the new shim

a. Intake: N = T + A
b. Exhaust: N = T + A

27. Select a new lifter with a thickness as close as possible to the calculated value.
28. Install removed components in the reverse order of the removal procedure.
29. When installing the cylinder head cover, apply a continuous bead (0.08–0.12 inch) of seal packing, part number 08826-00080 or equivalent, to the cylinder head as indicated in the illustration. Install the seal washers onto the bolts. Install the cylinder head cover bolts and nuts. Tighten bolts "A" to 7.4 ft. lbs. Tighten bolts "B" to 80 inch lbs. Tighten nuts to 80 inch lbs.

➡Be sure to remove any oil from the contact surfaces of the cylinder head cover and the cylinder head. Install the cover within three minutes after applying the seal packing. Tighten the bolts to specification within fifteen minutes after installing the cover. Do not add engine oil for at least two hours after installing the cover.

30. Continue the installation in the reverse order of the removal procedure.
31. Be sure to fill the engine with the proper grade and type engine coolant.
32. Be sure to fill the engine with the proper grade and type engine oil.
33. Start the engine and check for leaks. Correct as required.

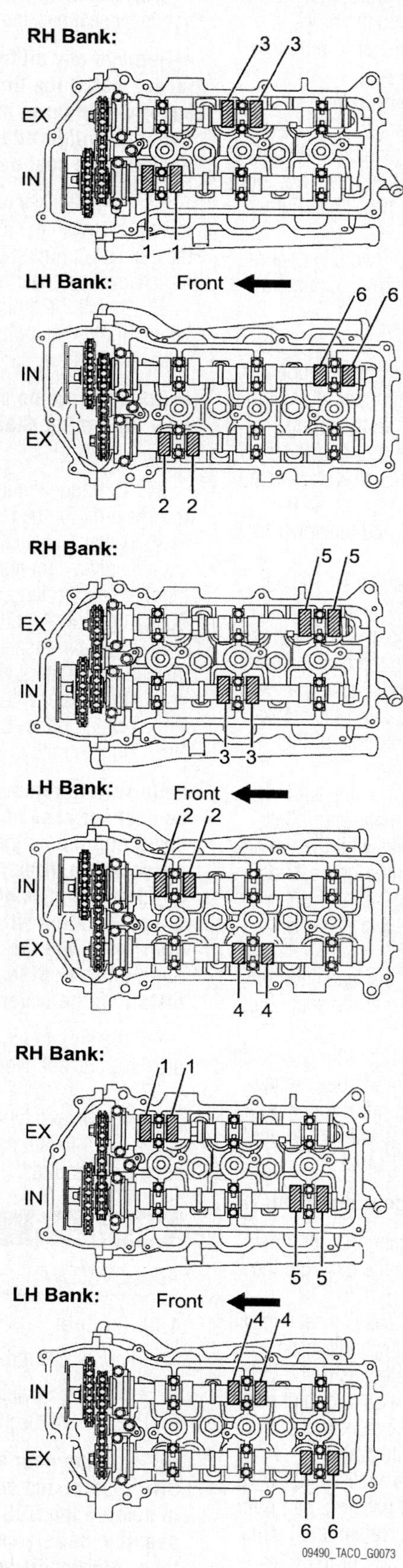

Fig. 130 Valve clearance location—4.0L engine

	5.060 (0.1992)	5.080 (0.2000)	5.100 (0.2008)	5.120 (0.2016)	5.140 (0.2024)	5.160 (0.2031)	5.180 (0.2039)	5.200 (0.2047)	5.210 (0.2051)	5.220 (0.2055)	5.230 (0.2059)	5.240 (0.2063)	5.250 (0.2067)	5.260 (0.2071)	5.270 (0.2075)	5.280 (0.2079)	5.290 (0.2083)	5.300 (0.2087)	5.310 (0.2091)	5.320 (0.2094)	5.330 (0.2098)	5.340 (0.2102)	5.350 (0.2106)	5.360 (0.2110)	5.370 (0.2114)	5.380 (0.2118)	5.390 (0.2122)	5.400 (0.2126)	5.410 (0.2130)	5.420 (0.2134)	5.430 (0.2138)	5.440 (0.2142)	5.450 (0.2146)	5.460 (0.2150)	5.470 (0.2154)	5.480 (0.2157)	5.490 (0.2161)	5.500 (0.2165)	5.510 (0.2169)	5.520 (0.2173)	5.530 (0.2177)	5.540 (0.2181)	5.550 (0.2185)	5.560 (0.2189)	5.570 (0.2193)	5.580 (0.2197)	5.590 (0.2201)	5.600 (0.2205)	5.620 (0.2213)	5.640 (0.2220)	5.660 (0.2228)	5.680 (0.2236)	5.700 (0.2244)	5.720 (0.2252)	5.740 (0.2260)
0.000 - 0.020 (0.0000 - 0.0008)													06	08	08	10	10	12	12	14	14	16	16	18	18	20	20	22	22	24	24	26	26	28	28	30	30	32	32	34	34	36	36	38	38	40	40	42	44	46	48	50	52	54	56
0.021 - 0.040 (0.0008 - 0.0016)											06	08	08	10	10	12	12	14	14	16	16	18	18	20	20	22	22	24	24	26	26	28	28	30	30	32	32	34	34	36	36	38	38	40	40	42	42	44	46	48	50	52	54	56	58
0.041 - 0.060 (0.0016 - 0.0024)									06	08	08	10	10	12	12	14	14	16	16	18	18	20	20	22	22	24	24	26	26	28	28	30	30	32	32	34	34	36	36	38	38	40	40	42	42	44	44	46	48	50	52	54	56	58	60
0.061 - 0.080 (0.0024 - 0.0031)								08	08	10	10	12	12	14	14	16	16	18	18	20	20	22	22	24	24	26	26	28	28	30	30	32	32	34	34	36	36	38	38	40	40	42	42	44	44	46	46	48	50	52	54	56	58	60	62
0.081 - 0.100 (0.0032 - 0.0039)							08	10	10	12	12	14	14	16	16	18	18	20	20	22	22	24	24	26	26	28	28	30	30	32	32	34	34	36	36	38	38	40	40	42	42	44	44	46	46	48	48	50	52	54	56	58	60	62	64
0.101 - 0.120 (0.0040 - 0.0047)						08	10	12	12	14	14	16	16	18	18	20	20	22	22	24	24	26	26	28	28	30	30	32	32	34	34	36	36	38	38	40	40	42	42	44	44	46	46	48	48	50	50	52	54	56	58	60	62	64	66
0.121 - 0.140 (0.0048 - 0.0055)					08	10	12	14	14	16	16	18	18	20	20	22	22	24	24	26	26	28	28	30	30	32	32	34	34	36	36	38	38	40	40	42	42	44	44	46	46	48	48	50	50	52	52	54	56	58	60	62	64	66	68
0.141 - 0.149 (0.0056 - 0.0059)				06	08	10	12	14	16	16	18	18	20	20	22	22	24	24	26	26	28	28	30	30	32	32	34	34	36	36	38	38	40	40	42	42	44	44	46	46	48	48	50	50	52	52	54	54	56	58	60	62	64	66	68
0.251 - 0.270 (0.0099 - 0.0106)	12	14	16	18	20	22	24	26	28	28	30	30	32	32	34	34	36	36	38	38	40	40	42	42	44	44	46	46	48	48	50	50	52	52	54	54	56	56	58	58	60	60	62	62	64	64	66	66	68	70	72				
0.271 - 0.290 (0.0107 - 0.0114)	14	16	18	20	22	24	26	28	30	30	32	32	34	34	36	36	38	38	40	40	42	42	44	44	46	46	48	48	50	50	52	52	54	54	56	56	58	58	60	60	62	62	64	64	66	66	68	68	70	72					
0.291 - 0.310 (0.0115 - 0.0122)	16	18	20	22	24	26	28	30	32	32	34	34	36	36	38	38	40	40	42	42	44	44	46	46	48	48	50	50	52	52	54	54	56	56	58	58	60	60	62	62	64	64	66	66	68	68	70	70	72						
0.311 - 0.330 (0.0122 - 0.0130)	18	20	22	24	26	28	30	32	34	34	36	36	38	38	40	40	42	42	44	44	46	46	48	48	50	50	52	52	54	54	56	56	58	58	60	60	62	62	64	64	66	66	68	68	70	70	72	72							
0.331 - 0.350 (0.0130 - 0.0138)	20	22	24	26	28	30	32	34	36	36	38	38	40	40	42	42	44	44	46	46	48	48	50	50	52	52	54	54	56	56	58	58	60	60	62	62	64	64	66	66	68	68	70	70	72	72	74								
0.351 - 0.370 (0.0138 - 0.0146)	22	24	26	28	30	32	34	36	38	38	40	40	42	42	44	44	46	46	48	48	50	50	52	52	54	54	56	56	58	58	60	60	62	62	64	64	66	66	68	68	70	70	72	72	74										
0.371 - 0.390 (0.0146 - 0.0154)	24	26	28	30	32	34	36	38	40	40	42	42	44	44	46	46	48	48	50	50	52	52	54	54	56	56	58	58	60	60	62	62	64	64	66	66	68	68	70	70	72	72	74												
0.391 - 0.410 (0.0154 - 0.0161)	26	28	30	32	34	36	38	40	42	42	44	44	46	46	48	48	50	50	52	52	54	54	56	56	58	58	60	60	62	62	64	64	66	66	68	68	70	70	72	72	74														
0.411 - 0.430 (0.0162 - 0.0169)	28	30	32	34	36	38	40	42	44	44	46	46	48	48	50	50	52	52	54	54	56	56	58	58	60	60	62	62	64	64	66	66	68	68	70	70	72	72	74																
0.431 - 0.450 (0.0170 - 0.0177)	30	32	34	36	38	40	42	44	46	46	48	48	50	50	52	52	54	54	56	56	58	58	60	60	62	62	64	64	66	66	68	68	70	70	72	72	74																		
0.451 - 0.470 (0.0178 - 0.0185)	32	34	36	38	40	42	44	46	48	48	50	50	52	52	54	54	56	56	58	58	60	60	62	62	64	64	66	66	68	68	70	70	72	72	74																				
0.471 - 0.490 (0.0185 - 0.0193)	34	36	38	40	42	44	46	48	50	50	52	52	54	54	56	56	58	58	60	60	62	62	64	64	66	66	68	68	70	70	72	72	74																						
0.491 - 0.510 (0.0193 - 0.0201)	36	38	40	42	44	46	48	50	52	52	54	54	56	56	58	58	60	60	62	62	64	64	66	66	68	68	70	70	72	72	74																								
0.511 - 0.530 (0.0201 - 0.0209)	38	40	42	44	46	48	50	52	54	54	56	56	58	58	60	60	62	62	64	64	66	66	68	68	70	70	72	72	74																										
0.531 - 0.550 (0.0209 - 0.0217)	40	42	44	46	48	50	52	54	56	56	58	58	60	60	62	62	64	64	66	66	68	68	70	70	72	72	74																												
0.551 - 0.570 (0.0217 - 0.0224)	42	44	46	48	50	52	54	56	58	58	60	60	62	62	64	64	66	66	68	68	70	70	72	72	74																														
0.571 - 0.590 (0.0225 - 0.0232)	44	46	48	50	52	54	56	58	60	60	62	62	64	64	66	66	68	68	70	70	72	72	74																																
0.591 - 0.610 (0.0233 - 0.0240)	46	48	50	52	54	56	58	60	62	62	64	64	66	66	68	68	70	70	72	72	74																																		
0.611 - 0.630 (0.0241 - 0.0248)	48	50	52	54	56	58	60	62	64	64	66	66	68	68	70	70	72	72	74																																				
0.631 - 0.650 (0.0248 - 0.0256)	50	52	54	56	58	60	62	64	66	66	68	68	70	70	72	72	74																																						
0.651 - 0.670 (0.0256 - 0.0264)	52	54	56	58	60	62	64	66	68	68	70	70	72	72	74																																								
0.671 - 0.690 (0.0264 - 0.0272)	54	56	58	60	62	64	66	68	70	70	72	72	74																																										
0.691 - 0.710 (0.0272 - 0.0280)	56	58	60	62	64	66	68	70	72	72	74																																												
0.711 - 0.730 (0.0280 - 0.0287)	58	60	62	64	66	68	70	72	74																																														
0.731 - 0.750 (0.0288 - 0.0295)	60	62	64	66	68	70	72																																																
0.751 - 0.770 (0.0296 - 0.0303)	62	64	66	68	70	72																																																	
0.771 - 0.790 (0.0304 - 0.0311)	64	66	68	70	72																																																		
0.791 - 0.810 (0.0311 - 0.0319)	66	68	70	72																																																			
0.811 - 0.830 (0.0319 - 0.0327)	68	70	72																																																				
0.831 - 0.850 (0.0327 - 0.0335)	70	72																																																					
0.851 - 0.870 (0.0335 - 0.0343)	72																																																						

06	5.060 (0.1992)	30	5.300 (0.2087)	54	5.540 (0.2181)
08	5.080 (0.2000)	32	5.320 (0.2094)	56	5.560 (0.2189)
10	5.100 (0.2008)	34	5.340 (0.2102)	58	5.580 (0.2197)
12	5.120 (0.2016)	36	5.360 (0.2110)	60	5.600 (0.2205)
14	5.140 (0.2024)	38	5.380 (0.2118)	62	5.620 (0.2213)
16	5.160 (0.2031)	40	5.400 (0.2126)	64	5.640 (0.2220)
18	5.180 (0.2039)	42	5.420 (0.2134)	66	5.660 (0.2228)
20	5.200 (0.2047)	44	5.440 (0.2142)	68	5.680 (0.2236)
22	5.220 (0.2055)	46	5.460 (0.2150)	70	5.700 (0.2244)
24	5.240 (0.2063)	48	5.480 (0.2157)	72	5.720 (0.2252)
26	5.260 (0.2071)	50	5.500 (0.2165)	74	5.740 (0.2260)
28	5.280 (0.2079)	52	5.520 (0.2173)		

09490_TACO_G0074

Fig. 131 Shim selection chart, part one—4.0L engine

	5.060 (0.1992)	5.080 (0.2000)	5.100 (0.2008)	5.120 (0.2016)	5.140 (0.2024)	5.160 (0.2031)	5.180 (0.2039)	5.200 (0.2047)	5.210 (0.2051)	5.220 (0.2055)	5.230 (0.2059)	5.240 (0.2063)	5.250 (0.2067)	5.260 (0.2071)	5.270 (0.2075)	5.280 (0.2079)	5.290 (0.2083)	5.300 (0.2087)	5.310 (0.2091)	5.320 (0.2094)	5.330 (0.2098)	5.340 (0.2102)	5.350 (0.2106)	5.360 (0.2110)	5.370 (0.2114)	5.380 (0.2118)	5.390 (0.2122)	5.400 (0.2126)	5.410 (0.2130)	5.420 (0.2134)	5.430 (0.2138)	5.440 (0.2142)	5.450 (0.2146)	5.460 (0.2150)	5.470 (0.2154)	5.480 (0.2157)	5.490 (0.2161)	5.500 (0.2165)	5.510 (0.2169)	5.520 (0.2173)	5.530 (0.2177)	5.540 (0.2181)	5.550 (0.2185)	5.560 (0.2189)	5.570 (0.2193)	5.580 (0.2197)	5.590 (0.2201)	5.600 (0.2205)	5.620 (0.2213)	5.640 (0.2220)	5.660 (0.2228)	5.680 (0.2236)	5.700 (0.2244)	5.720 (0.2252)	5.740 (0.2260)
0.000 - 0.020 (0.0000 - 0.0008)																											06	08	08	10	10	12	12	14	14	16	16	18	18	20	20	22	22	24	24	26	26	28	30	32	34	36	38	40	42
0.021 - 0.040 (0.0008 - 0.0016)																									06	08	08	10	10	12	12	14	14	16	16	18	18	20	20	22	22	24	24	26	26	28	28	30	32	34	36	38	40	42	44
0.041 - 0.060 (0.0016 - 0.0024)																							06	08	08	10	10	12	12	14	14	16	16	18	18	20	20	22	22	24	24	26	26	28	28	30	30	32	34	36	38	40	42	44	46
0.061 - 0.080 (0.0024 - 0.0031)																					06	08	08	10	10	12	12	14	14	16	16	18	18	20	20	22	22	24	24	26	26	28	28	30	30	32	32	34	36	38	40	42	44	46	48
0.081 - 0.100 (0.0032 - 0.0039)																			06	08	08	10	10	12	12	14	14	16	16	18	18	20	20	22	22	24	24	26	26	28	28	30	30	32	32	34	34	36	38	40	42	44	46	48	50
0.101 - 0.120 (0.0040 - 0.0047)																	06	08	08	10	10	12	12	14	14	16	16	18	18	20	20	22	22	24	24	26	26	28	28	30	30	32	32	34	34	36	36	38	40	42	44	46	48	50	52
0.121 - 0.140 (0.0048 - 0.0055)															06	08	08	10	10	12	12	14	14	16	16	18	18	20	20	22	22	24	24	26	26	28	28	30	30	32	32	34	34	36	36	38	38	40	42	44	46	48	50	52	54
0.141 - 0.160 (0.0056 - 0.0063)													06	08	08	10	10	12	12	14	14	16	16	18	18	20	20	22	22	24	24	26	26	28	28	30	30	32	32	34	34	36	36	38	38	40	40	42	44	46	48	50	52	54	56
0.161 - 0.180 (0.0063 - 0.0071)											06	08	08	10	10	12	12	14	14	16	16	18	18	20	20	22	22	24	24	26	26	28	28	30	30	32	32	34	34	36	36	38	38	40	40	42	42	44	46	48	50	52	54	56	58
0.181 - 0.200 (0.0071 - 0.0079)									06	08	08	10	10	12	12	14	14	16	16	18	18	20	20	22	22	24	24	26	26	28	28	30	30	32	32	34	34	36	36	38	38	40	40	42	42	44	44	46	48	50	52	54	56	58	60
0.201 - 0.220 (0.0079 - 0.0087)								08	08	10	10	12	12	14	14	16	16	18	18	20	20	22	22	24	24	26	26	28	28	30	30	32	32	34	34	36	36	38	38	40	40	42	42	44	44	46	46	48	50	52	54	56	58	60	62
0.221 - 0.240 (0.0087 - 0.0094)							08	10	10	12	12	14	14	16	16	18	18	20	20	22	22	24	24	26	26	28	28	30	30	32	32	34	34	36	36	38	38	40	40	42	42	44	44	46	46	48	48	50	52	54	56	58	60	62	64
0.241 - 0.260 (0.0095 - 0.0102)						08	10	12	12	14	14	16	16	18	18	20	20	22	22	24	24	26	26	28	28	30	30	32	32	34	34	36	36	38	38	40	40	42	42	44	44	46	46	48	48	50	50	52	54	56	58	60	62	64	66
0.261 - 0.280 (0.0103 - 0.0110)					08	10	12	14	14	16	16	18	18	20	20	22	22	24	24	26	26	28	28	30	30	32	32	34	34	36	36	38	38	40	40	42	42	44	44	46	46	48	48	50	50	52	52	54	56	58	60	62	64	66	68
0.281 - 0.289 (0.0111 - 0.0114)				06	08	10	12	14	16	16	18	18	20	20	22	22	24	24	26	26	28	28	30	30	32	32	34	34	36	36	38	38	40	40	42	42	44	44	46	46	48	48	50	50	52	52	54	54	56	58	60	62	64	66	68
0.391 - 0.410 (0.0154 - 0.0161)	12	14	16	18	20	22	24	26	28	28	30	30	32	32	34	34	36	36	38	38	40	40	42	42	44	44	46	46	48	48	50	50	52	52	54	54	56	56	58	58	60	60	62	62	64	64	66	66	68	70	72				
0.411 - 0.430 (0.0162 - 0.0169)	14	16	18	20	22	24	26	28	30	30	32	32	34	34	36	36	38	38	40	40	42	42	44	44	46	46	48	48	50	50	52	52	54	54	56	56	58	58	60	60	62	62	64	64	66	66	68	68	70	72					
0.431 - 0.450 (0.0170 - 0.0177)	16	18	20	22	24	26	28	30	32	32	34	34	36	36	38	38	40	40	42	42	44	44	46	46	48	48	50	50	52	52	54	54	56	56	58	58	60	60	62	62	64	64	66	66	68	68	70	70	72						
0.451 - 0.470 (0.0178 - 0.0185)	18	20	22	24	26	28	30	32	34	34	36	36	38	38	40	40	42	42	44	44	46	46	48	48	50	50	52	52	54	54	56	56	58	58	60	60	62	62	64	64	66	66	68	68	70	70	72	72							
0.471 - 0.490 (0.0185 - 0.0193)	20	22	24	26	28	30	32	34	36	36	38	38	40	40	42	42	44	44	46	46	48	48	50	50	52	52	54	54	56	56	58	58	60	60	62	62	64	64	66	66	68	68	70	70	72	72	74								
0.491 - 0.510 (0.0193 - 0.0201)	22	24	26	28	30	32	34	36	38	38	40	40	42	42	44	44	46	46	48	48	50	50	52	52	54	54	56	56	58	58	60	60	62	62	64	64	66	66	68	68	70	70	72	72	74										
0.511 - 0.530 (0.0201 - 0.0209)	24	26	28	30	32	34	36	38	40	40	42	42	44	44	46	46	48	48	50	50	52	52	54	54	56	56	58	58	60	60	62	62	64	64	66	66	68	68	70	70	72	72	74												
0.531 - 0.550 (0.0209 - 0.0217)	26	28	30	32	34	36	38	40	42	42	44	44	46	46	48	48	50	50	52	52	54	54	56	56	58	58	60	60	62	62	64	64	66	66	68	68	70	70	72	72	74														
0.551 - 0.570 (0.0217 - 0.0224)	28	30	32	34	36	38	40	42	44	44	46	46	48	48	50	50	52	52	54	54	56	56	58	58	60	60	62	62	64	64	66	66	68	68	70	70	72	72	74																
0.571 - 0.590 (0.0225 - 0.0232)	30	32	34	36	38	40	42	44	46	46	48	48	50	50	52	52	54	54	56	56	58	58	60	60	62	62	64	64	66	66	68	68	70	70	72	72	74																		
0.591 - 0.610 (0.0233 - 0.0240)	32	34	36	38	40	42	44	46	48	48	50	50	52	52	54	54	56	56	58	58	60	60	62	62	64	64	66	66	68	68	70	70	72	72	74																				
0.611 - 0.630 (0.0241 - 0.0248)	34	36	38	40	42	44	46	48	50	50	52	52	54	54	56	56	58	58	60	60	62	62	64	64	66	66	68	68	70	70	72	72	74																						
0.631 - 0.650 (0.0248 - 0.0256)	36	38	40	42	44	46	48	50	52	52	54	54	56	56	58	58	60	60	62	62	64	64	66	66	68	68	70	70	72	72	74																								
0.651 - 0.670 (0.0256 - 0.0264)	38	40	42	44	46	48	50	52	54	54	56	56	58	58	60	60	62	62	64	64	66	66	68	68	70	70	72	72	74																										
0.671 - 0.690 (0.0264 - 0.0272)	40	42	44	46	48	50	52	54	56	56	58	58	60	60	62	62	64	64	66	66	68	68	70	70	72	72	74																												
0.691 - 0.710 (0.0272 - 0.0280)	42	44	46	48	50	52	54	56	58	58	60	60	62	62	64	64	66	66	68	68	70	70	72	72	74																														
0.711 - 0.730 (0.0280 - 0.0287)	44	46	48	50	52	54	56	58	60	60	62	62	64	64	66	66	68	68	70	70	72	72	74																																
0.731 - 0.750 (0.0288 - 0.0295)	46	48	50	52	54	56	58	60	62	62	64	64	66	66	68	68	70	70	72	72	74																																		
0.751 - 0.770 (0.0296 - 0.0303)	48	50	52	54	56	58	60	62	64	64	66	66	68	68	70	70	72	72	74																																				
0.771 - 0.790 (0.0304 - 0.0311)	50	52	54	56	58	60	62	64	66	66	68	68	70	70	72	72	74																																						
0.791 - 0.810 (0.0311 - 0.0319)	52	54	56	58	60	62	64	66	68	68	70	70	72	72	74																																								
0.811 - 0.830 (0.0319 - 0.0327)	54	56	58	60	62	64	66	68	70	70	72	72	74																																										
0.831 - 0.850 (0.0327 - 0.0335)	56	58	60	62	64	66	68	70	72	72	74																																												
0.851 - 0.870 (0.0335 - 0.0343)	58	60	62	64	66	68	70	72	74																																														
0.871 - 0.890 (0.0343 - 0.0350)	60	62	64	66	68	70	72																																																
0.891 - 0.910 (0.0351 - 0.0358)	62	64	66	68	70	72																																																	
0.911 - 0.930 (0.0359 - 0.0366)	64	66	68	70	72																																																		
0.931 - 0.950 (0.0367 - 0.0374)	66	68	70	72																																																			
0.951 - 0.970 (0.0374 - 0.0382)	68	70	72																																																				
0.971 - 0.990 (0.0382 - 0.0390)	70	72																																																					
0.991 - 1.010 (0.0390 - 0.0398)	72																																																						

06	5.060 (0.1992)	30	5.300 (0.2087)	54	5.540 (0.2181)
08	5.080 (0.2000)	32	5.320 (0.2094)	56	5.560 (0.2189)
10	5.100 (0.2008)	34	5.340 (0.2102)	58	5.580 (0.2197)
12	5.120 (0.2016)	36	5.360 (0.2110)	60	5.600 (0.2205)
14	5.140 (0.2024)	38	5.380 (0.2118)	62	5.620 (0.2213)
16	5.160 (0.2031)	40	5.400 (0.2126)	64	5.640 (0.2220)
18	5.180 (0.2039)	42	5.420 (0.2134)	66	5.660 (0.2228)
20	5.200 (0.2047)	44	5.440 (0.2142)	68	5.680 (0.2236)
22	5.220 (0.2055)	46	5.460 (0.2150)	70	5.700 (0.2244)
24	5.240 (0.2063)	48	5.480 (0.2157)	72	5.720 (0.2252)
26	5.260 (0.2071)	50	5.500 (0.2165)	74	5.740 (0.2260)
28	5.280 (0.2079)	52	5.520 (0.2173)		

09490_TACO_G0075

Fig. 132 Shim selection chart, part two—4.0L engine

ENGINE PERFORMANCE & EMISSION CONTROLS

CAMSHAFT POSITION (CMP) SENSOR

LOCATION

See Figure 133.

At the front of the cylinder head.

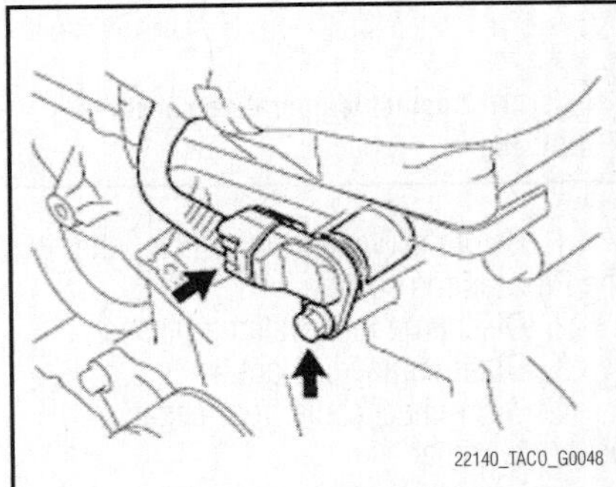

Fig. 133 Camshaft Position (CMP) sensor location

REMOVAL & INSTALLATION

See Figure 133.

1. Before servicing the vehicle, refer to the Precautions Section.

➡If working near and/or around the SRS system and components, be sure to disable the SRS system. Tape the negative battery cable with insulating tape. Always disconnect the negative battery cable first.

⁂ CAUTION

To avoid personal injury when working on vehicles equipped with an air bag, the negative battery cable must be disconnected and at least 90 seconds must elapse before working on the system. Failure to do so may result in deployment of the air bag.

2. Disconnect the negative battery cable. Tape the cable with insulating tape.
3. Disconnect the Camshaft Position (CMP) sensor connector.
4. Remove the bolt, then remove the Camshaft Position (CMP) sensor.
5. Discard the O-ring.

To install:

➡Be sure to use new fasteners, as required.

6. Installation is the reverse of the removal procedure.

CRANKSHAFT POSITION (CKP) SENSOR

LOCATION

2.7L Engines

See Figure 134.

Lower front left of the engine block, under the A/C compressor bracket.

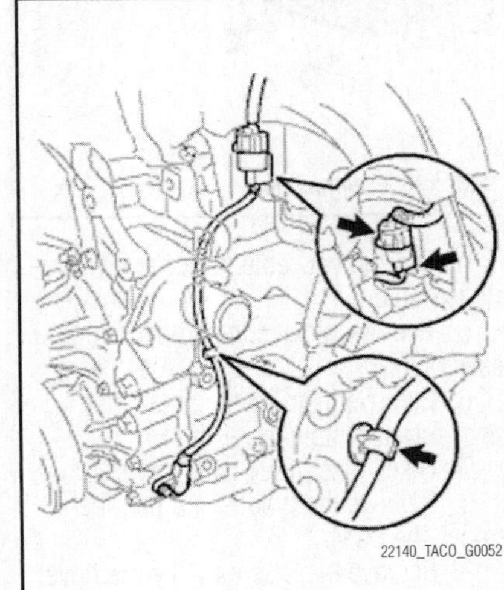

Fig. 134 Crankshaft Position (CKP) sensor location—2.7L engine

4.0L Engines

See Figure 135.

Front left of the engine block, under the A/C compressor.

REMOVAL & INSTALLATION

2.7L Engines

See Figure 134.

1. Before servicing the vehicle, refer to the Precautions Section.

➡If working near and/or around the SRS system and components, be sure to disable the SRS system. Tape the negative battery cable with insulating tape. Always disconnect the negative battery cable first.

⁂ CAUTION

To avoid personal injury when working on vehicles equipped with an air bag, the negative battery cable must be disconnected and at least 90 seconds must elapse before working on the system. Failure to do so may result in deployment of the air bag.

2. Disconnect the negative battery cable. Tape the cable with insulating tape.
3. Remove engine under cover sub-assembly no.1 (for 4wd and pre-runner).
4. Remove fan and generator v belt.
5. Separate compressor and magnetic clutch.
6. Remove idle pulley assembly with bracket.
7. Remove the 5 bolts, then remove the idle pulley with bracket.
8. Disconnect the Crankshaft Position (CKP) sensor connector and the 2 wire harness clamps.
9. Remove the bolt, then remove the Crankshaft Position (CKP) sensor.

To install:

➡Be sure to use new fasteners, as required.

10. Installation is the reverse of the removal procedure.

4.0L Engines

See Figure 135.

1. Before servicing the vehicle, refer to the Precautions Section.

➡If working near and/or around the SRS system and components, be sure to disable the SRS system. Tape the negative battery cable with insulating tape. Always disconnect the negative battery cable first.

⁂ CAUTION

To avoid personal injury when working on vehicles equipped with an air bag, the negative battery cable must be disconnected and at least 90 seconds must elapse before working on the system. Failure to do so may result in deployment of the air bag.

2. Disconnect the negative battery cable. Tape the cable with insulating tape.
3. Remove v-bank cover.
4. Remove fan shroud.
5. Remove generator assembly.
6. Separate cooler compressor assembly.
7. Remove the bolt, then separate the suction hose sub-assembly.
8. Disconnect the cooler compressor assembly connector.
9. Remove the 4 bolts, then separate

Fig. 135 Crankshaft Position (CKP) sensor location—4.0L engine

the cooler compressor assembly from the V-ribbed belt tensioner assembly.

10. Disconnect the Crankshaft Position (CKP) sensor connector.
11. Remove the bolt, then remove the Crankshaft Position (CKP) sensor.

To install:

➡Be sure to use new fasteners, as required.

12. Installation is the reverse of the removal procedure.

ELECTRONIC CONTROL MODULE (ECM)

LOCATION

See Figure 136.

Behind the glove box.

REMOVAL & INSTALLATION

See Figure 136.

1. Before servicing the vehicle, refer to the Precautions Section.

➡If working near and/or around the SRS system and components, be sure to disable the SRS system. Tape the negative battery cable with insulating tape. Always disconnect the negative battery cable first.

CAUTION

To avoid personal injury when working on vehicles equipped with an air bag, the negative battery cable must be disconnected and at least 90 seconds must elapse before working on the system. Failure to do so may result in deployment of the air bag.

2. Disconnect the negative battery cable. Tape the cable with insulating tape.
3. Perform VIN registration when replacing the ECM.

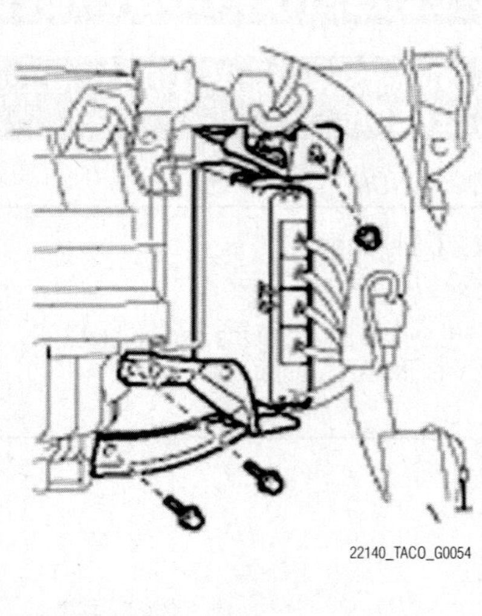

Fig. 136 ECM location

4. Disconnect cable from negative battery terminal
5. Remove glove compartment door assembly
6. Remove instrument panel finish panel sub-assembly lower rh
7. Disconnect the 4 connectors.
8. Remove the 2 bolts and nut, then remove the ECM.
9. Remove the 3 screws, then remove the ECM bracket.
10. Remove the 2 screws, then remove the ECM bracket No. 2.

To install:

➡Be sure to use new fasteners, as required.

11. Installation is the reverse of the removal procedure.
12. If equipped with a 4.0L engine, using the Techstream diagnostic tool, or equivalent perform the RESET MEMORY function.
13. Using the Techstream diagnostic tool, or equivalent perform the REGISTRATION (VIN registration) function.
14. Using the Techstream diagnostic tool, or equivalent perform the throttle body initialization function.

ENGINE COOLANT TEMPERATURE (ECT) SENSOR

LOCATION

Mounted near the thermostat housing.

REMOVAL & INSTALLATION

4.0L Engines

See Figure 137.

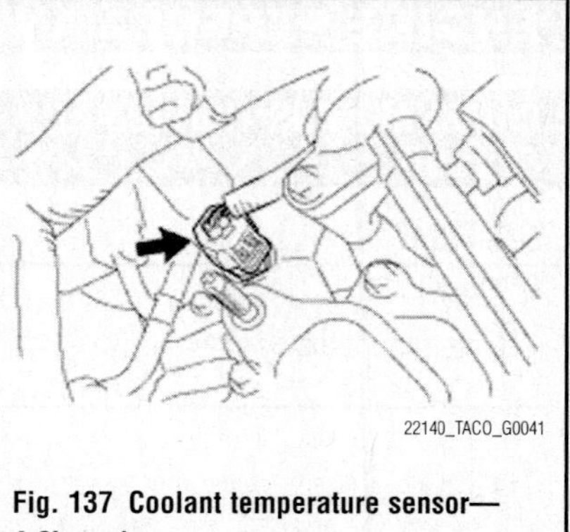

Fig. 137 Coolant temperature sensor—4.0L engine

1. Before servicing the vehicle, refer to the Precautions section.
2. Discharge fuel system pressure.
3. Drain engine coolant.
4. Disconnect cable from negative battery terminal.
5. Remove V-bank cover.
6. Remove air cleaner assembly.
7. Remove intake air surge tank.
8. Disconnect fuel pipe sub-assembly no. 2.
9. Disconnect the connector.
10. Using a 19 mm deep socket wrench, remove the water temperature sensor and gasket.

To install:

11. Using a 19 mm deep socket wrench, install the water temperature sensor with a new gasket and tighten to 15 ft. lbs. (20 Nm).
12. Connect the connector.
13. Connect fuel pipe sub-assembly no. 2.
14. Install intake air surge tank.
15. Install air cleaner assembly.
16. Connect cable to negative battery terminal.
17. Add engine coolant .
18. Check for fuel leakage.
19. Check for engine coolant leakage
20. Install V-bank cover.

HEATED OXYGEN SENSOR (HO2S)

LOCATION

See Figures 138 and 139.

Mounted in the exhaust, between the engine and the catalytic converter(s).

REMOVAL & INSTALLATION

See Figures 154 through 155.

1. Before servicing the vehicle, refer to the Precautions Section.

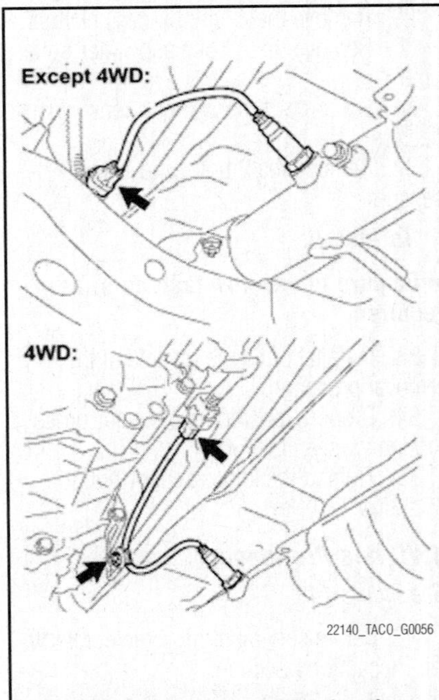

Fig. 138 Heated oxygen sensor location—2.7L engine

➡If working near and/or around the SRS system and components, be sure to disable the SRS system. Tape the negative battery cable with insulating tape. Always disconnect the negative battery cable first.

⁂ CAUTION

To avoid personal injury when working on vehicles equipped with an air bag, the negative battery cable must be disconnected and at least 90 seconds must elapse before working on the system. Failure to do so may result in deployment of the air bag.

2. Disconnect the negative battery cable. Tape the cable with insulating tape.
3. Disconnect the heated oxygen sensor connector.
4. Remove the heated oxygen sensor from the exhaust pipe.

To install:

➡Be sure to use new fasteners, as required.

5. Installation is the reverse of the removal procedure.
6. Tighten the heated oxygen sensor to 33 ft. lbs. (44 Nm).

KNOCK SENSOR (KS)

LOCATION

2.7L Engine

See Figure 140.

On the left side of the engine block, under the intake manifold.

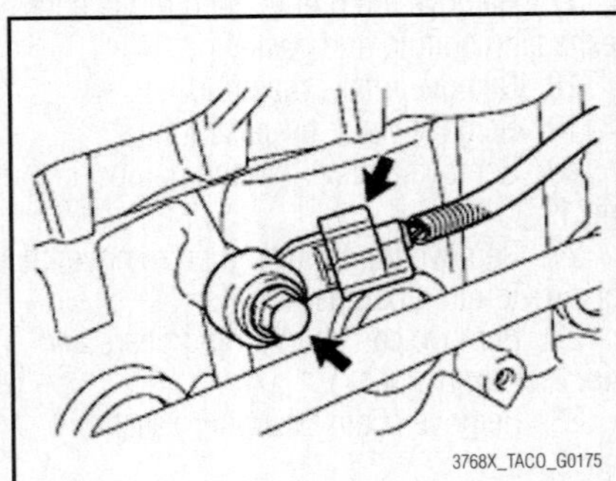

Fig. 140 Knock sensor location—2.7L engine

4.0L Engine

See Figure 141.

Mounted to the top of the engine block under the intake manifold.

REMOVAL & INSTALLATION

2.7L Engine

See Figure 142.

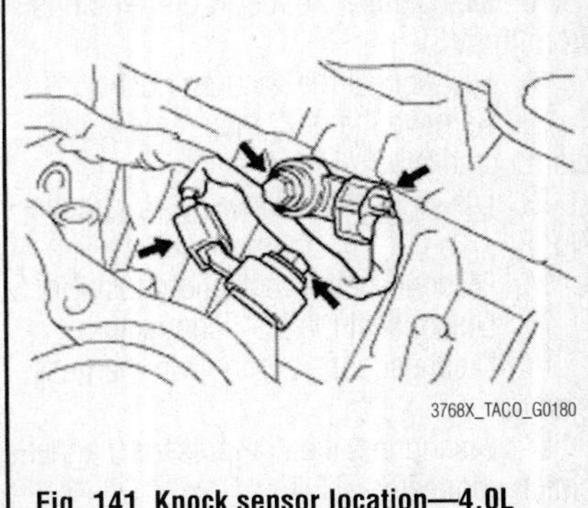

Fig. 141 Knock sensor location—4.0L engine

1. Before servicing the vehicle, refer to the Precautions Section.

➡If working near and/or around the SRS system and components, be sure to disable the SRS system. Tape the negative battery cable with insulating tape. Always disconnect the negative battery cable first.

⁂ CAUTION

To avoid personal injury when working on vehicles equipped with an air bag, the negative battery cable must be disconnected and at least 90 seconds must elapse before working on the system. Failure to do so may result in deployment of the air bag.

2. Disconnect the negative battery cable. Tape the cable with insulating tape.
3. Remove engine under cover sub-assembly.
4. Drain engine coolant.
5. Remove intake air connector.
6. Remove throttle with motor body assembly.
7. Disconnect fuel hose.
8. Disconnect fuel hose no. 2.
9. Remove intake manifold.

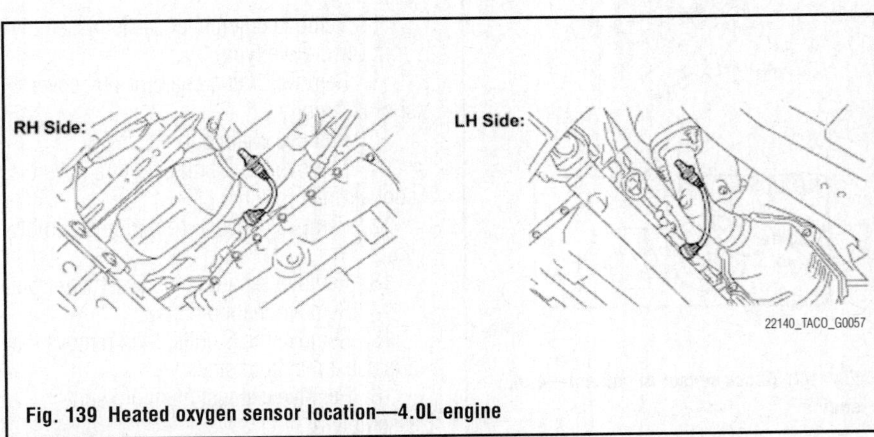

Fig. 139 Heated oxygen sensor location—4.0L engine

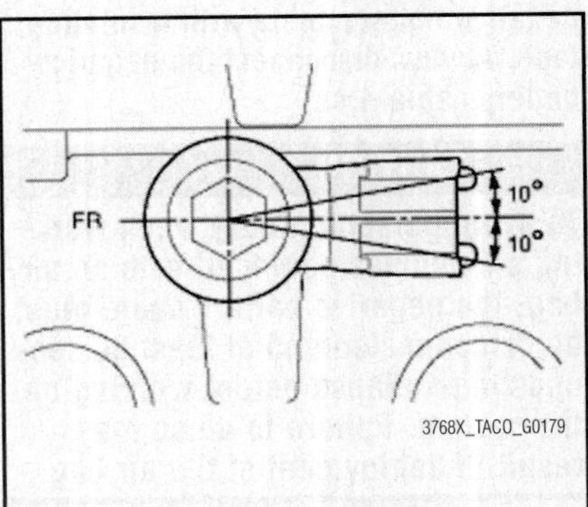

Fig. 142 Knock sensor alignment—2.7L engine

10. Disconnect the fuel vapor feed hose from the VSV.
11. Disconnect the vacuum hose.
12. Remove the bolt, then remove the clamp bracket.
13. Disconnect the water by-pass hose No. 2.
14. Disconnect the ventilation hose No. 3.
15. Disconnect the VSV connector.
16. Disengage the engine wire harness clamp.
17. Disconnect the compressor magnetic clutch connector.
18. Disengage the wire harness clamp.
19. Remove the bolt and harness clamp bracket.
20. Disconnect the 3 connectors.
21. Remove the nut.
22. Disconnect the engine wire harness from the relay block.
23. Remove the 5 bolts and 2 nuts, then remove the intake manifold.
24. Disconnect the knock control sensor connector.
25. Remove the bolt, then remove the knock control sensor.

To install:

➡Be sure to use new fasteners, as required.

26. Installation is the reverse of the removal procedure.
27. Properly position the sensor on its mounting, see illustration.
28. Tighten the knock sensor bolt to 15 ft. lbs. (20 Nm).

4.0L Engine

2WD

See Figure 143.

1. Before servicing the vehicle, refer to the Precautions Section.

➡If working near and/or around the SRS system and components, be sure to disable the SRS system. Tape the negative battery cable with insulating tape. Always disconnect the negative battery cable first.

⁂ CAUTION

To avoid personal injury when working on vehicles equipped with an air bag, the negative battery cable must be disconnected and at least 90 seconds must elapse before working on the system. Failure to do so may result in deployment of the air bag.

2. Disconnect the negative battery cable. Tape the cable with insulating tape.
3. Discharge fuel system pressure.
4. Remove battery.
5. Drain engine coolant.
6. Drain engine oil.
7. Remove engine assembly.
8. Remove timing chain or belt cover sub-assembly.
9. Remove cool air inlet no.1.
10. Remove the 2 bolts, then remove the cool air inlet.
11. Remove exhaust front pipe assembly no. 2.
12. Remove exhaust pipe assembly front.
13. Remove manifold stay.
14. Remove the 3 bolts, then remove the exhaust manifold stay.
15. Remove exhaust manifold sub-assembly rh
16. Disconnect the air fuel ratio sensor connector.
17. Remove the 6 nuts, then remove the exhaust manifold and gasket.
18. Remove intake manifold.
19. Remove water by-pass joint.
20. Remove chain vibration damper no.1.
21. Remove the 2 bolts, then remove the chain vibration damper No. 1.
22. Remove camshaft timing gears and no. 2 chain (rh bank).
23. Remove chain tensioner assembly no. 2.
24. Remove camshafts.
25. Remove camshaft bearing no. 2.
26. Remove cylinder head sub-assembly.
27. Disconnect heater water inlet hose.
28. Remove water outlet pipe no.1.

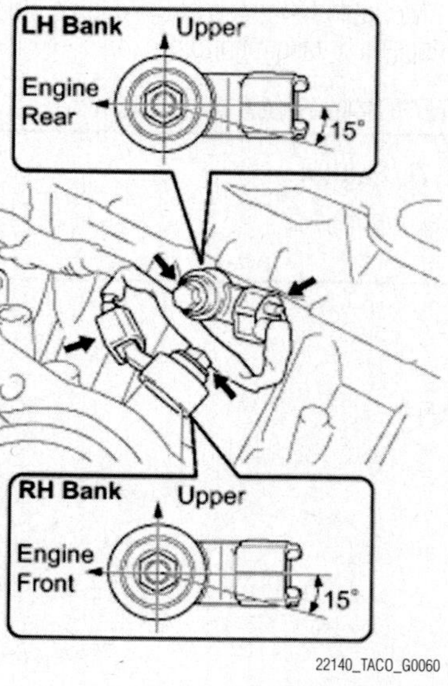

Fig. 143 Knock sensor alignment—4.0L engine

29. Remove the 4 wire harness clamps.
30. Remove the 3 bolts and water outlet pipe.
31. Disconnect the 2 knock sensor connectors.
32. Remove the 2 bolts and 2 knock sensors.

To install:

➡Be sure to use new fasteners, as required.

33. Installation is the reverse of the removal procedure.
34. Properly position the sensor on its mounting, see illustration.
35. Tighten the knock sensor bolt to 15 ft. lbs. (20 Nm).

4WD And PreRunner

See Figure 143.

1. Before servicing the vehicle, refer to the Precautions Section.

➡If working near and/or around the SRS system and components, be sure to disable the SRS system. Tape the negative battery cable with insulating tape. Always disconnect the negative battery cable first.

⁂ CAUTION

To avoid personal injury when working on vehicles equipped with an air bag, the negative battery cable must be disconnected and at least 90 seconds must elapse before working on the system. Failure to do so may result in deployment of the air bag.

2. Disconnect the negative battery cable. Tape the cable with insulating tape.
3. Discharge fuel system pressure.
4. Remove battery.
5. Drain engine coolant.
6. Drain engine oil.
7. Remove power steering link assembly.
8. Remove differential carrier assembly front (4wd drive type).
9. Remove timing chain or belt cover sub-assembly.
10. Remove chain sub-assembly.
11. Remove the 2 bolts, then remove the cool air inlet.
12. Remove exhaust front pipe assembly no. 2.
13. Remove exhaust pipe assembly front.
14. Remove manifold stay.
15. Remove the 6 bolts, then remove the 2 exhaust manifold stays.
16. Remove exhaust manifold sub-assembly rh.

17. Disconnect fuel pipe sub-assembly no.1.
18. Disconnect fuel pipe sub-assembly no. 2.
19. Remove intake manifold.
20. Remove water by-pass joint (right rear).
21. Remove chain vibration damper no.1.
22. Remove camshaft timing gears and no. 2 chain (rh bank).
23. Remove chain tensioner assembly no. 2.
24. Remove camshafts.
25. Remove camshaft bearing no. 2.
26. Remove cylinder head sub-assembly.
27. Disconnect heater water inlet hose.
28. Remove the 4 wire harness clamps.
29. Remove the 3 bolts and water outlet pipe.
30. Disconnect the 2 knock sensor connectors.
31. Remove the 2 bolts and 2 knock sensors.

To install:

➡Be sure to use new fasteners, as required.

32. Installation is the reverse of the removal procedure.
33. Properly position the sensor on its mounting, see illustration.
34. Tighten the knock sensor bolt to 15 ft. lbs. (20 Nm).

MALFUNCTION INDICATOR LIGHT (MIL)

RESET PROCEDURE

Reset the MIL by clearing codes with a scan tool.

MASS AIR FLOW (MAF) SENSOR

LOCATION

See Figure 144.

In the air intake, between the air filter and the throttle body.

REMOVAL & INSTALLATION

See Figure 144.

1. Before servicing the vehicle, refer to the Precautions Section.

➡If working near and/or around the SRS system and components, be sure to disable the SRS system. Tape the negative battery cable with insulating tape. Always disconnect the negative battery cable first.

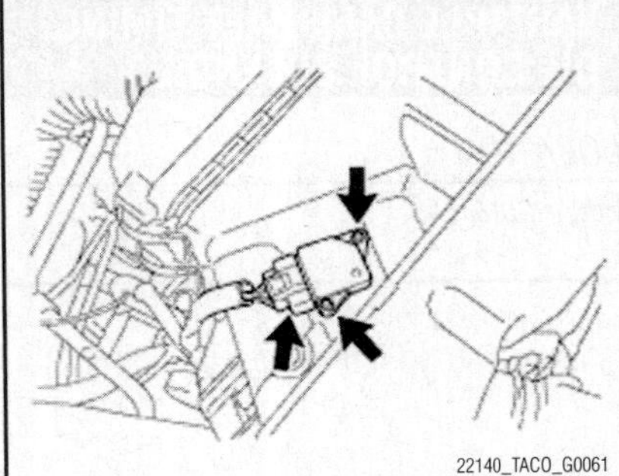

Fig. 144 Mass air flow sensor location—4.0L engine

CAUTION

To avoid personal injury when working on vehicles equipped with an air bag, the negative battery cable must be disconnected and at least 90 seconds must elapse before working on the system. Failure to do so may result in deployment of the air bag.

2. Disconnect the negative battery cable. Tape the cable with insulating tape.
3. Remove v-bank cover.
4. Disconnect the connector.
5. Remove the 2 screws, then remove the mass air flow meter.

To install:

➡Be sure to use new fasteners, as required.

6. Installation is the reverse of the removal procedure.

THROTTLE POSITION SENSOR (TPS)

LOCATION

Part of the throttle body.

REMOVAL & INSTALLATION

2.7L Engines

1. Before servicing the vehicle, refer to the Precautions Section.

➡If working near and/or around the SRS system and components, be sure to disable the SRS system. Tape the negative battery cable with insulating tape. Always disconnect the negative battery cable first.

CAUTION

To avoid personal injury when working on vehicles equipped with an air bag, the negative battery cable must be disconnected and at least 90 seconds must elapse before working on the system. Failure to do so may result in deployment of the air bag.

2. Disconnect the negative battery cable. Tape the cable with insulating tape.
3. Remove no. 1 engine under cover sub-assembly (for 4WD and PreRunner)
4. Drain engine coolant
5. Remove intake air connector
6. Disconnect the pressure sensor connector.
7. Disengage the wire harness clamp.
8. Disconnect the vacuum hose.
9. Disconnect the No. 2 ventilation hose.
10. Disconnect the vacuum hose.
11. Loosen the 2 hose clamp bolts.
12. Remove the 3 bolts, then remove the intake air connector.
13. Disconnect the water by-pass hose.
14. Disconnect the No. 2 water by-pass hose.
15. Disconnect the throttle motor connector.
16. Remove the 2 bolts and 2 nuts, then remove the throttle with motor body.
17. Remove the gasket from the intake manifold.

To install:

➡Be sure to use new fasteners, as required.

18. Install a new gasket onto the intake manifold.
19. Install the throttle with motor body with the 2 bolts and 2 nuts. Tighten to 80 inch lbs. (9 Nm).
20. Connect the water by-pass hose.
21. Connect the water by-pass hose No. 2.
22. Connect the throttle motor connector.
23. Install the intake air connector with the 3 bolts.
24. Tighten the 2 hose clamp bolts.
25. Connect the vacuum hose.
26. Connect the ventilation hose No. 2.
27. Connect the vacuum hose.
28. Engage the wire harness clamp.
29. Connect the pressure sensor connector.
30. Connect cable to negative battery terminal
31. Add engine coolant
32. Check for engine coolant leakage
33. Install engine under cover sub-assembly No.1 (for 4WD and Pre-Runner).

4.0L Engines

1. Before servicing the vehicle, refer to the Precautions Section.

➡If working near and/or around the SRS system and components, be sure to disable the SRS system. Tape the negative battery cable with insulating tape. Always disconnect the negative battery cable first.

⁂ CAUTION

To avoid personal injury when working on vehicles equipped with an air bag, the negative battery cable must be disconnected and at least 90 seconds must elapse before working on the system. Failure to do so may result in deployment of the air bag.

2. Disconnect the negative battery cable. Tape the cable with insulating tape.
3. Drain engine coolant
4. Remove the 2 nuts, then remove the V-bank cover.
5. Disconnect the ventilation hose No. 2.
6. Disconnect the vacuum hose.
7. Disconnect the mass air flow meter connector.
8. Remove the 2 wire harness clamps.
9. Loosen the 2 hose clamps.
10. Remove the 2 bolts, then remove the air cleaner.
11. Disconnect the water by-pass hose No. 5.
12. Disconnect the water by-pass hose No. 4.
13. Disconnect the throttle motor connector.
14. Remove the 4 bolts, then remove the throttle w/ motor body and gasket.

To install:

➡Be sure to use new fasteners, as required.

15. Install a new gasket and the throttle with motor body with the 4 bolts and tighten to 108 inch lbs. (11 Nm).
16. Connect the throttle motor connector.
17. Connect the water by-pass hose No. 4.
18. Connect the water by-pass hose No. 5.
19. Install the air cleaner with the 2 bolts.
20. Connect the ventilation hose No. 2.
21. Connect cable to negative battery terminal
22. Add engine coolant
23. Check for engine coolant leakage
24. Install the V-bank cover with the 2 nuts.

VARIABLE CAMSHAFT TIMING OIL CONTROL SOLENOID

LOCATION

See Figure 145.

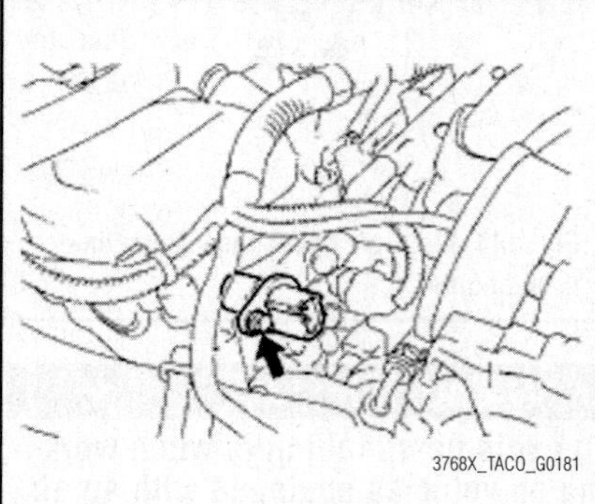

Fig. 145 VVT sensor location—4.0L engine

REMOVAL & INSTALLATION

See Figure 145.

1. Before servicing the vehicle, refer to the Precautions Section.

➡If working near and/or around the SRS system and components, be sure to disable the SRS system. Tape the negative battery cable with insulating tape. Always disconnect the negative battery cable first.

⁂ CAUTION

To avoid personal injury when working on vehicles equipped with an air bag, the negative battery cable must be disconnected and at least 90 seconds must elapse before working on the system. Failure to do so may result in deployment of the air bag.

2. Disconnect the negative battery cable. Tape the cable with insulating tape.
3. Drain engine coolant.
4. Disconnect cable from negative battery terminal.
5. Remove v-bank cover.
6. Remove air cleaner assembly.
7. Remove intake air surge tank.
8. Remove camshaft timing oil control valve assembly.
9. Disconnect the 2 connectors.
10. Remove the 2 bolts, then remove the 2 camshaft timing oil control valves.

To install:

➡Be sure to use new fasteners, as required.

11. Installation is the reverse of the removal procedure.

VEHICLE SPEED SENSOR (VSS)

REMOVAL & INSTALLATION

See Figure 146.

1. Before servicing the vehicle, refer to the Precautions Section.

➡If working near and/or around the SRS system and components, be sure to disable the SRS system. Tape the negative battery cable with insulating tape. Always disconnect the negative battery cable first.

⁂ CAUTION

To avoid personal injury when working on vehicles equipped with an air bag, the negative battery cable must be disconnected and at least 90 seconds must elapse before working on the system. Failure to do so may result in deployment of the air bag.

2. Disconnect the negative battery cable. Tape the cable with insulating tape.
3. Disconnect the speed sensor connector.
4. Remove the bolt and speed sensor.
5. Remove the O-ring from the speed sensor.

To install:

➡Be sure to use new fasteners, as required.

6. Installation is the reverse of the removal procedure.

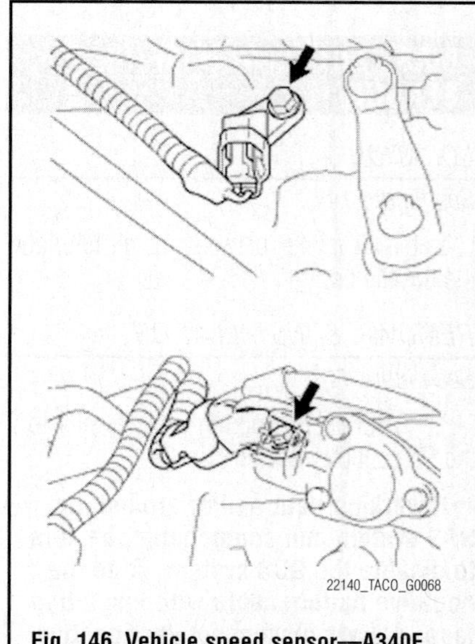

Fig. 146 Vehicle speed sensor—A340E transmission

FUEL SYSTEM SERVICE PRECAUTIONS

Safety is the most important factor when performing not only fuel system maintenance, but any type of maintenance. Failure to conduct maintenance and repairs in a safe manner may result in serious personal injury or death. Work on a vehicle's fuel system components can be accomplished safely and effectively by adhering to the following rules and guidelines.

- To avoid the possibility of fire and personal injury, always disconnect the negative battery cable unless the repair or test procedure requires that battery voltage be applied.
- Always relieve the fuel system pressure prior to disconnecting any fuel system component (injector, fuel rail, pressure regulator, etc.) fitting or fuel line connection. Exercise extreme caution whenever relieving fuel system pressure to avoid exposing skin, face and eyes to fuel spray. Please be advised that fuel under pressure may penetrate the skin or any part of the body that it contacts.
- Always place a shop towel or cloth around the fitting or connection prior to loosening to absorb any excess fuel due to spillage. Ensure that all fuel spillage is quickly removed from engine surfaces. Ensure that all fuel-soaked cloths or towels are deposited into a flame-proof waste container with a lid.
- Always keep a dry chemical (Class B) fire extinguisher near the work area.
- Do not allow fuel spray or fuel vapors to come into contact with a spark or open flame.
- Always use a second wrench when loosening or tightening fuel line connection fittings. This will prevent unnecessary stress and torsion on fuel piping. Always follow the proper torque specifications.
- Always replace worn fuel fitting O-rings with new ones. Do not substitute fuel hose where rigid pipe is installed.

FUEL SYSTEM PRESSURE

RELIEVING

See Figures 147 and 148.

1. Before servicing the vehicle, refer to the Precautions Section.

➡If working near and/or around the SRS system and components, be sure to disable the SRS system. Tape the negative battery cable with insulating tape. Always disconnect the negative battery cable first.

CAUTION

To avoid personal injury when working on vehicles equipped with an air bag, the negative battery cable must be disconnected and at least 90 seconds must elapse before working on the system. Failure to do so may result in deployment of the air bag.

2. Disconnect the negative battery cable. Tape the cable with insulating tape.
3. Remove the engine relay block cover.
4. Remove the circuit opening relay.
5. Connect the negative battery.
6. Start the engine.

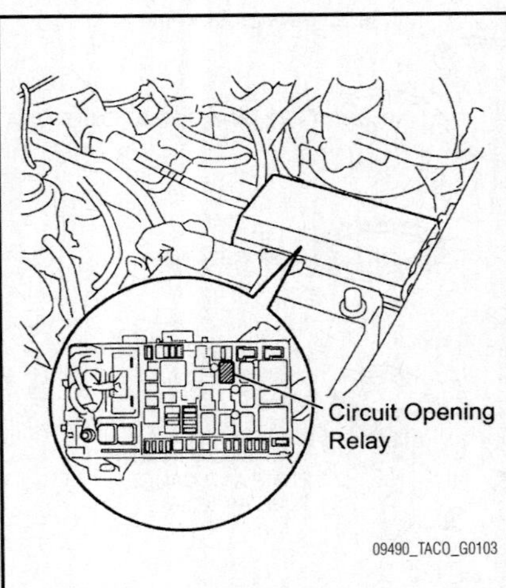

Fig. 147 Circuit opening relay location—2.7L engine

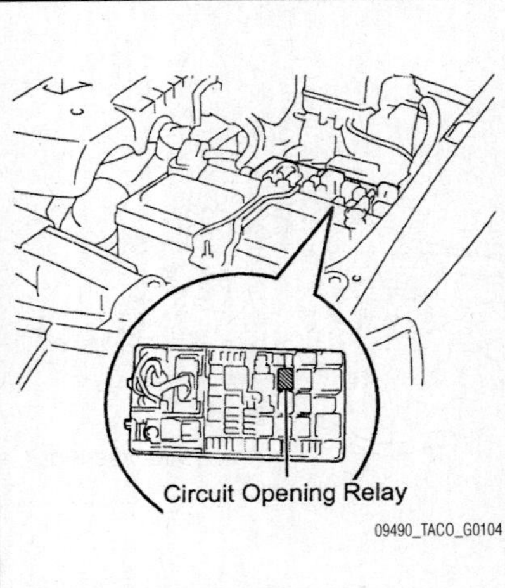

Fig. 148 Circuit opening relay location—4.0L engine

7. Turn the ignition switch "ON" after the engine stops.

➡Code DTC P0171 (system lean) may be present.

8. Crank the engine again. Check that the engine stops.
9. Remove the fuel tank cap and completely discharge the pressure in the fuel tank.

Install the circuit opening relay.

10. Disconnect the negative battery cable.

FUEL FILTER

REMOVAL & INSTALLATION

See Figure 149.

The fuel filter is part of the fuel pump, and is located in the fuel tank.

1. Before servicing the vehicle, refer to the Precautions Section.

➡If working near and/or around the SRS system and components, be sure to disable the SRS system. Tape the negative battery cable with insulating tape. Always disconnect the negative battery cable first.

CAUTION

To avoid personal injury when working on vehicles equipped with an air bag, the negative battery cable must be disconnected and at least 90 seconds must elapse before working on the system. Failure to do so may result in deployment of the air bag.

2. Disconnect the negative battery cable. Tape the cable with insulating tape.
3. Relieve the fuel pressure.
4. Drain the fuel from the fuel tank.
5. On vehicles equipped with off road package and 4.0L engine, remove the number one fuel tank protector subassembly.
6. Remove the fuel tank.
7. Remove the fuel tank main hose and the return hose.
8. Remove the fuel pump and gauge assembly retainer from its mounting, using tool SST09808-14020 or equivalent.
9. Pull the fuel pump and gauge assembly out of the fuel tank. Be careful not to bend the arm of the sender. Remove and discard the gasket.
10. Disconnect the connector from the assembly.
11. Disengage the claw fitting and remove the sender gauge by sliding it forward.

12. Disengage the five claw fittings and remove the fuel pump tank. Separate the connector and disengage the clamp.

13. Disengage the clamp and then disconnect the connector.

14. Disengage the five claw fittings and separate the fuel pump from the fuel pump case. Disconnect the connector from the fuel pump.

15. Remove the fuel filter from the fuel pump.

To install:

➡Be sure to use new fasteners, as required.

16. Installation is the reverse of the removal procedure.

17. Start the engine and check for leaks, correct as required.

FUEL LEVEL SENDING UNIT

REMOVAL & INSTALLATION

See Figure 149.

The fuel filter is part of the fuel pump, and is located in the fuel tank.

1. Before servicing the vehicle, refer to the Precautions Section.

➡If working near and/or around the SRS system and components, be sure to disable the SRS system. Tape the negative battery cable with insulating tape. Always disconnect the negative battery cable first.

CAUTION

To avoid personal injury when working on vehicles equipped with an air bag, the negative battery cable must be disconnected and at least 90 seconds must elapse before working on the system. Failure to do so may result in deployment of the air bag.

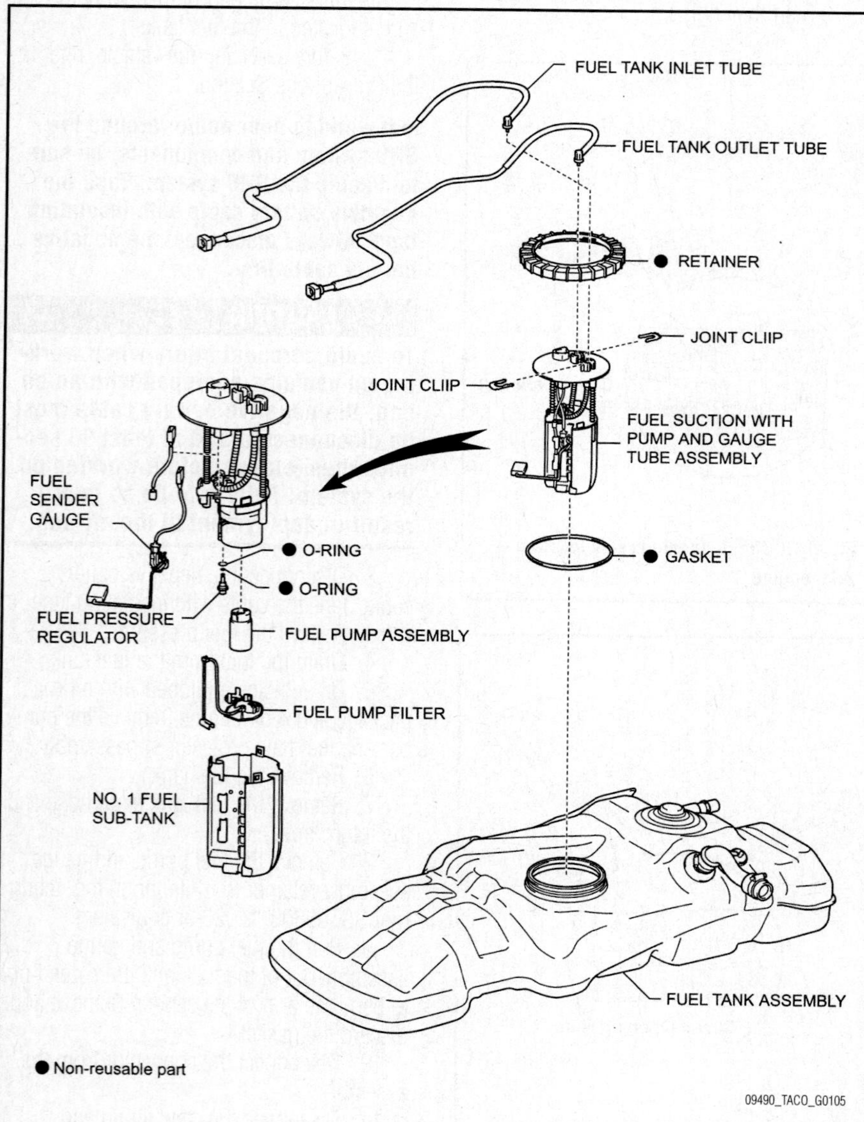

Fig. 149 Fuel pump and related components

2. Disconnect the negative battery cable. Tape the cable with insulating tape.

3. Relieve the fuel pressure.

4. Drain the fuel from the fuel tank.

5. On vehicles equipped with off road package and 4.0L engine, remove the number one fuel tank protector subassembly.

6. Remove the fuel tank.

7. Remove the fuel tank main hose and the return hose.

8. Remove the fuel pump and gauge assembly retainer from its mounting, using tool SST09808-14020 or equivalent.

9. Pull the fuel pump and gauge assembly out of the fuel tank. Be careful not to bend the arm of the sender. Remove and discard the gasket.

10. Disconnect the connector from the assembly.

11. Disengage the claw fitting and remove the sender gauge by sliding it forward.

12. Disengage the five claw fittings and remove the fuel pump tank. Separate the connector and disengage the clamp.

13. Disengage the clamp and then disconnect the connector.

14. Disengage the five claw fittings and separate the fuel pump from the fuel pump case. Disconnect the connector from the fuel pump.

15. Remove the fuel filter from the fuel pump.

To install:

➡Be sure to use new fasteners, as required.

16. Installation is the reverse of the removal procedure.

17. Start the engine and check for leaks, correct as required.

FUEL PUMP

REMOVAL & INSTALLATION

See Figure 149.

1. Before servicing the vehicle, refer to the Precautions Section.

➡If working near and/or around the SRS system and components, be sure to disable the SRS system. Tape the negative battery cable with insulating tape. Always disconnect the negative battery cable first.

⁂ CAUTION

To avoid personal injury when working on vehicles equipped with an air bag, the negative battery cable must be disconnected and at least 90 seconds must elapse before working on the system. Failure to do so may result in deployment of the air bag.

2. Disconnect the negative battery cable. Tape the cable with insulating tape.
3. Relieve the fuel pressure.
4. Drain the fuel from the fuel tank.
5. On vehicles equipped with off road package and 4.0L engine, remove the number one fuel tank protector subassembly.
6. Remove the fuel tank.
7. Remove the fuel tank main hose and the return hose.
8. Remove the fuel pump and gauge assembly retainer from its mounting, using tool SST09808-14020 or equivalent.
9. Pull the fuel pump and gauge assembly out of the fuel tank. Be careful not to bend the arm of the sender. Remove and discard the gasket.
10. Disconnect the connector from the assembly.
11. Disengage the claw fitting and remove the sender gauge by sliding it forward.
12. Disengage the five claw fittings and remove the fuel pump tank. Separate the connector and disengage the clamp.
13. Disengage the clamp and then disconnect the connector.
14. Disengage the five claw fittings and separate the fuel pump from the fuel pump case. Disconnect the connector from the fuel pump.
15. Remove the fuel filter from the fuel pump.

To install:

➡Be sure to use new fasteners, as required.

16. Installation is the reverse of the removal procedure.
17. Start the engine and check for leaks, correct as required.

FUEL RAIL AND INJECTOR

REMOVAL & INSTALLATION

2.7L Engines

See Figure 150.

1. Before servicing the vehicle, refer to the Precautions Section.

➡If working near and/or around the SRS system and components, be sure to disable the SRS system. Tape the negative battery cable with insulating tape. Always disconnect the negative battery cable first.

⁂ CAUTION

To avoid personal injury when working on vehicles equipped with an air bag, the negative battery cable must be disconnected and at least 90 seconds must elapse before working on the system. Failure to do so may result in deployment of the air bag.

2. Disconnect the negative battery cable. Tape the cable with insulating tape.
3. Relieve the fuel system pressure.
4. Remove the engine undercover subassembly.
5. Drain the engine coolant. Remove the intake air connector.
6. Remove the throttle body motor assembly.
7. Disconnect and plug the fuel line hoses.
8. Remove the fuel pressure pulsation damper assembly.
9. Disconnect the fuel injector electrical connectors. Disconnect the VSV connector. Disconnect the engine wiring harness clamp.
10. Disconnect the air conditioning compressor clutch connector. Disconnect the wire harness clamp. Remove the bolt and then remove the harness clamp bracket.
11. Remove the two bolts and remove the fuel rail together with the fuel injectors.
12. Remove the fuel rail number one spacers. Remove the four injector vibration insulators. Remove the four spacers.

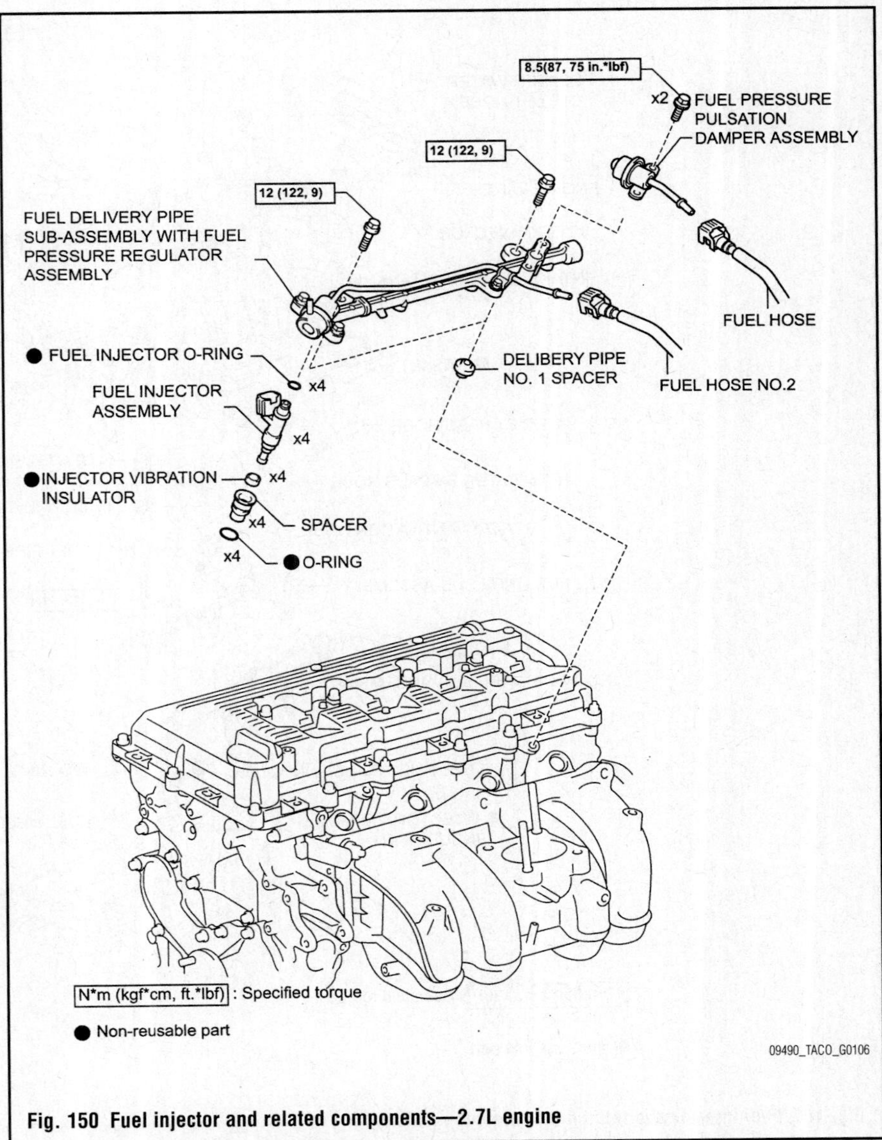

Fig. 150 Fuel injector and related components—2.7L engine

13. Remove the fuel injectors. Discard all gaskets.

To install:

➡Be sure to use new fasteners, as required.

14. Installation is the reverse of the removal procedure.
15. Be sure to use new gaskets and O-rings, as required.
16. Be sure to fill the cooling system with the proper grade and type engine coolant.
17. Start the engine and check for leaks, correct as required.

4.0L Engines

See Figure 151.

1. Before servicing the vehicle, refer to the Precautions Section.

➡If working near and/or around the SRS system and components, be sure to disable the SRS system. Tape the negative battery cable with insulating tape. Always disconnect the negative battery cable first.

CAUTION

To avoid personal injury when working on vehicles equipped with an air bag, the negative battery cable must be disconnected and at least 90 seconds must elapse before working on the system. Failure to do so may result in deployment of the air bag.

2. Disconnect the negative battery cable. Tape the cable with insulating tape.
3. Relieve the fuel system pressure.
4. Drain the engine coolant. Remove the V bank cover.
5. Remove the air cleaner assembly.
6. Remove the intake manifold.
7. Disconnect and plug the fuel line hoses.
8. Disconnect the fuel injector connectors. Remove the six bolts and the fuel rail together with the injectors.
9. Pull the injectors out of the fuel rail.

To install:

➡Be sure to use new fasteners, as required.

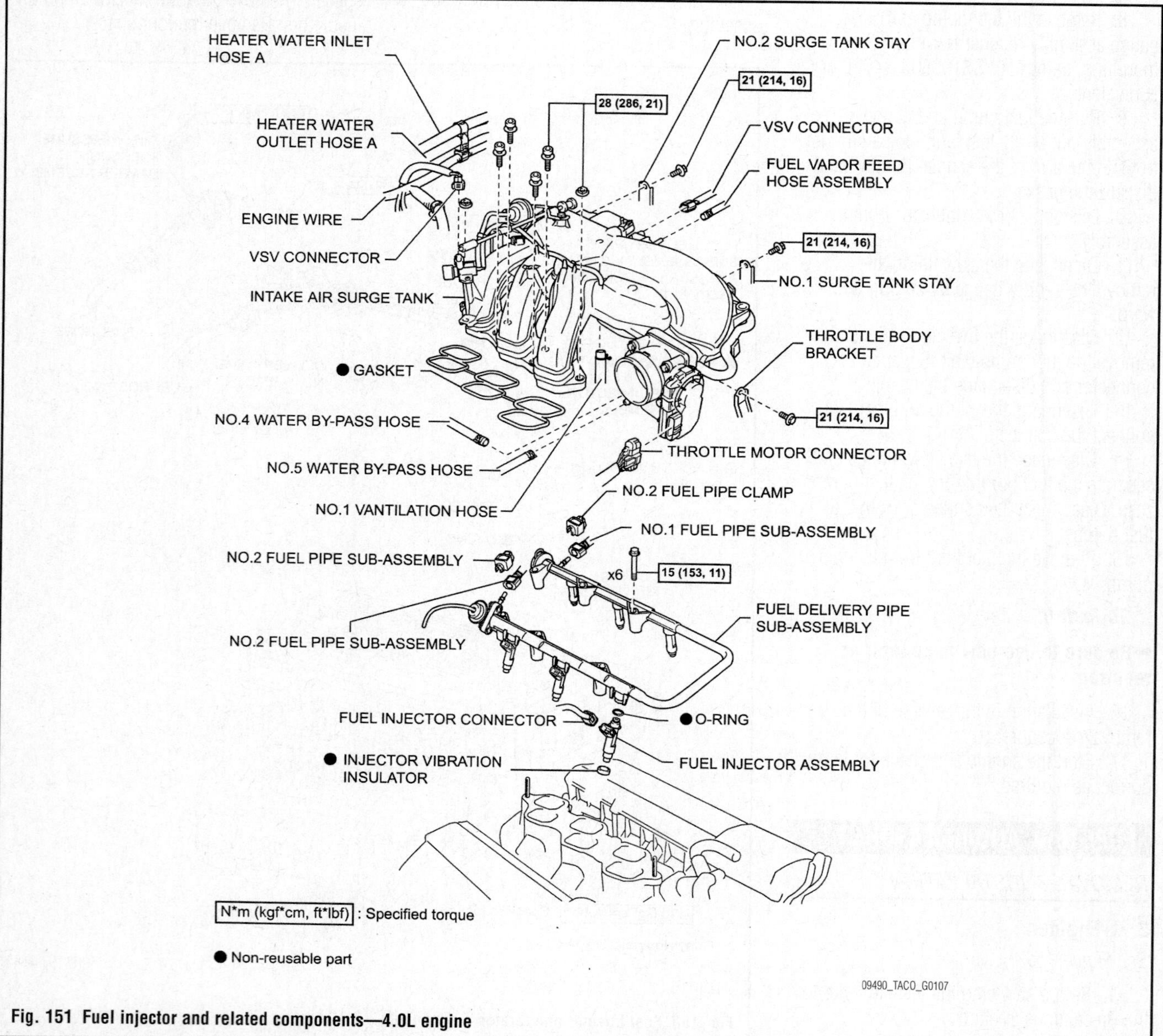

Fig. 151 Fuel injector and related components—4.0L engine

10. Installation is the reverse of the removal procedure.
11. Be sure to use new gaskets and O-rings, as required.
12. Be sure to fill the cooling system with the proper grade and type engine coolant.
13. Start the engine and check for leaks, correct as required.

FUEL TANK

REMOVAL & INSTALLATION

See Figures 152 through 154.

Remove any dirt and foreign objects from the fuel tube connector before performing this work.

Do not allow any scratches or foreign objects on the parts when disconnecting, as the fuel tube connector has the O-ring that seals the pipe.

Perform this work by hand. Do not use any tools.

Do not forcibly bend, twist or turn the nylon tube.

Protect the disconnected part by covering it with a vinyl bag after disconnecting the fuel tube.

If the fuel tube connector and pipe are stuck, push and pull to release them.

1. Before servicing the vehicle, refer to the Precautions Section.

➡If working near and/or around the SRS system and components, be sure to disable the SRS system. Tape the negative battery cable with insulating tape. Always disconnect the negative battery cable first.

✱✱ CAUTION

To avoid personal injury when working on vehicles equipped with an air bag, the negative battery cable must be disconnected and at least 90 seconds must elapse before working on the system. Failure to do so may result in deployment of the air bag.

2. Disconnect the negative battery cable. Tape the cable with insulating tape.
3. Discharge fuel system pressure.
4. Drain the fuel tank to an acceptable level, for tank removal. Be sure to properly dispose of used fuel.
5. Remove no.1 fuel tank protector sub-assembly (with off road package).
6. Disconnect fuel tank main tube and fuel tank return tube.
7. Remove the fuel pipe clamp.

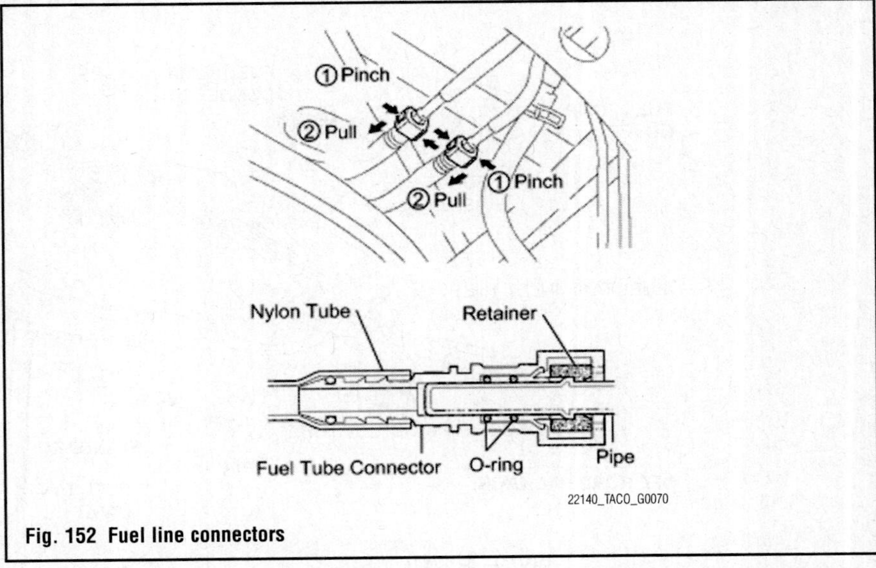

Fig. 152 Fuel line connectors

8. Disconnect the fuel tank main tube and fuel tank return tube.
9. Pinch the retainer as illustrated, then pull the fuel tube connectors out of the pipes.
10. Loosen the clamp bolt.
11. Disconnect the fuel breather tube.
12. Remove the checker of the fuel tube connector from the pipe.
13. Pinch the retainer of the fuel tube connector, then pull the tube connector out of the pipe.
14. Hold the fuel tank using the mission jack.
15. Remove the 2 fuel tank bands.
16. Remove the 2 bolts.
17. Remove the 2 clips and 2 pins, then remove the 2 fuel tank bands.
18. Slightly jack the fuel tank down and separate the fuel tank to filler pipe hose from the inlet pipe.
19. Slowly jack down the fuel tank so as not to tear the wire harness and hose.
20. Remove the fuel pump cover.
21. Disconnect the fuel pump connector.
22. Disconnect the fuel tank vent hose.
23. Pinch the retainer and pull the fuel tank vent hose connector out of the charcoal canister to disconnect the fuel tank vent hose from the charcoal canister.
24. Remove the fuel tank assy.
25. Remove the 2 joint clips, then remove the fuel tank main tube and fuel tank return tube.
26. Remove fuel suction with pump & gauge tube assembly

➡Protect the connector and tube joint with masking tape or equivalent to prevent any foreign objects from sticking to them. Clean any dirt and foreign objects from the fuel suction tube assy before removing.

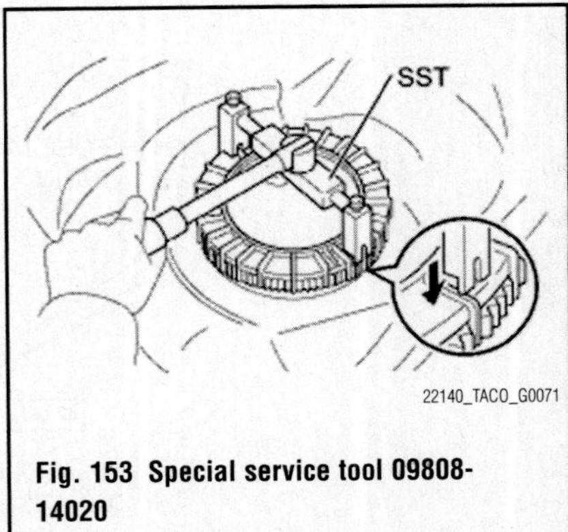

Fig. 153 Special service tool 09808-14020

27. Using SST 09808-14020, loosen the retainer.

➡The ribs on the retainer can be fitted into a tip of the SST.

28. Remove the retainer.
29. Pull the fuel pump assy out of the fuel tank.
30. Be careful not to bend the arm of the sender gauge.
31. Remove the gasket from the fuel tank.
32. Drain fuel.
33. Loosen the clamp bolt, then remove the fuel tank to filler pipe hose.
34. Remove the 4 clips, then remove the fuel tank protector.

To install:

➡Be sure to use new fasteners, as required.

35. Installation is the reverse of the removal procedure.

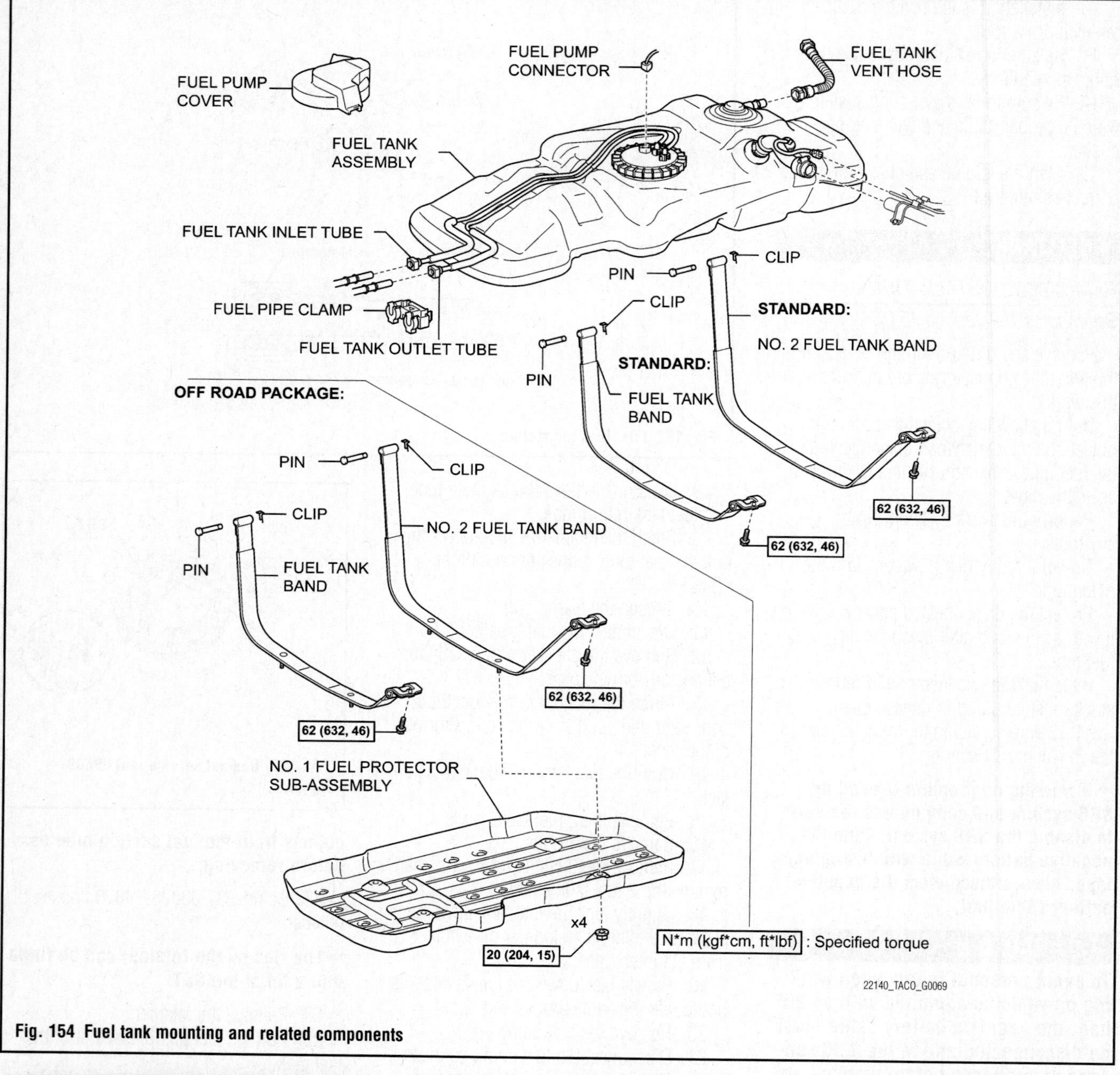

Fig. 154 Fuel tank mounting and related components

IDLE SPEED

ADJUSTMENT

Idle speed is maintained by the Electronic Control Module (ECM). No adjustment is necessary or possible.

THROTTLE BODY

REMOVAL & INSTALLATION

2.7L Engines

See Figure 155.

1. Before servicing the vehicle, refer to the Precautions Section.

➡If working near and/or around the SRS system and components, be sure to disable the SRS system. Tape the negative battery cable with insulating tape. Always disconnect the negative battery cable first.

⁂ CAUTION

To avoid personal injury when working on vehicles equipped with an air bag, the negative battery cable must be disconnected and at least 90 seconds must elapse before working on the system. Failure to do so may result in deployment of the air bag.

2. Disconnect the negative battery cable. Tape the cable with insulating tape.
3. Remove no. 1 engine under cover sub-assembly (for 4wd and pre-runner)
4. Drain engine coolant
5. Remove intake air connector
6. Disconnect the pressure sensor connector.
7. Disengage the wire harness clamp.
8. Disconnect the vacuum hose.
9. Disconnect the No. 2 ventilation hose.
10. Disconnect the vacuum hose.
11. Loosen the 2 hose clamp bolts.
12. Remove the 3 bolts, then remove the intake air connector.

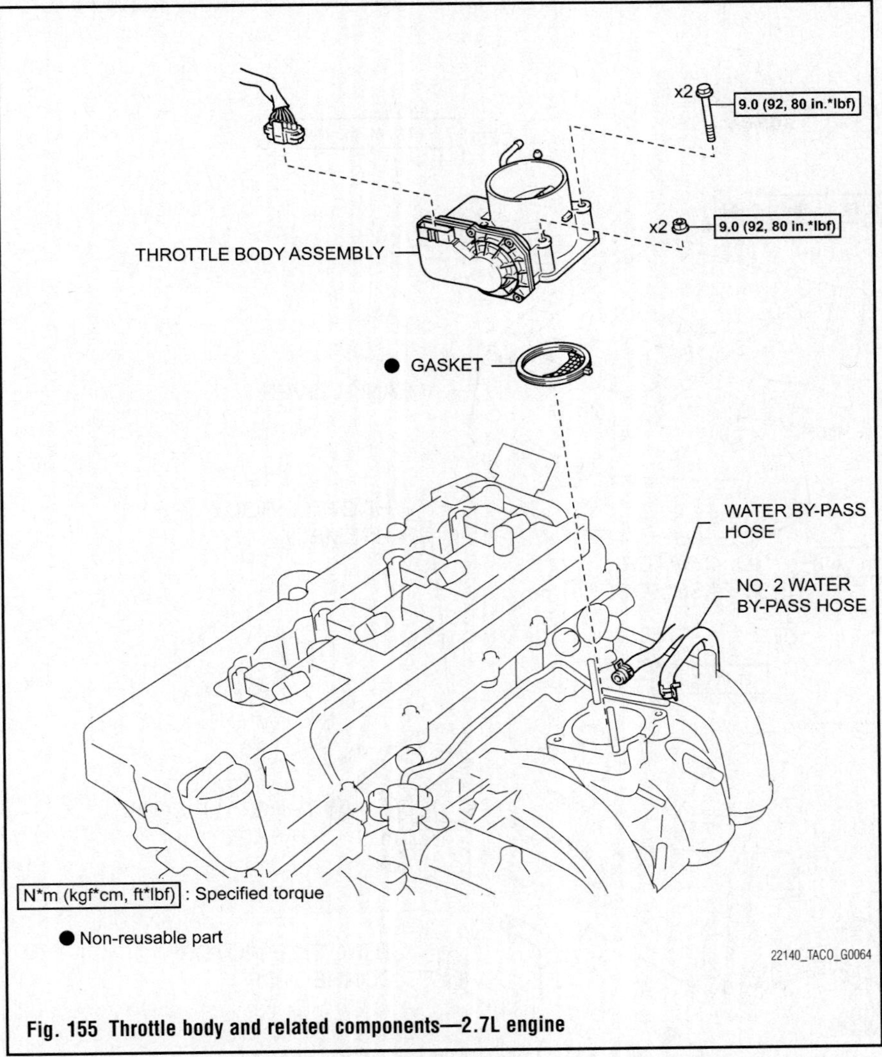

Fig. 155 Throttle body and related components—2.7L engine

13. Disconnect the water by-pass hose.
14. Disconnect the No. 2 water by-pass hose.
15. Disconnect the throttle motor connector.
16. Remove the 2 bolts and 2 nuts, then remove the throttle with motor body.
17. Remove the gasket from the intake manifold.

To install:

➡**Be sure to use new fasteners, as required.**

18. Install a new gasket onto the intake manifold.
19. Install the throttle with motor body with the 2 bolts and 2 nuts. Tighten to 80 inch lbs. (9 Nm).
20. Connect the water by-pass hose.
21. Connect the water by-pass hose No. 2.
22. Connect the throttle motor connector.
23. Install the intake air connector with the 3 bolts.
24. Tighten the 2 hose clamp bolts.
25. Connect the vacuum hose.
26. Connect the ventilation hose No. 2.
27. Connect the vacuum hose.
28. Engage the wire harness clamp.
29. Connect the pressure sensor connector.
30. Connect cable to negative battery terminal
31. Add engine coolant
32. Check for engine coolant leakage
33. Install engine under cover sub-assembly NO.1 (for 4WD and Pre-Runner).
34. Using the Techstream diagnostic tool, or equivalent, perform the initialization procedure.

4.0L Engines

See Figure 156.

1. Before servicing the vehicle, refer to the Precautions Section.

➡**If working near and/or around the SRS system and components, be sure to disable the SRS system. Tape the negative battery cable with insulating tape. Always disconnect the negative battery cable first.**

CAUTION

To avoid personal injury when working on vehicles equipped with an air bag, the negative battery cable must be disconnected and at least 90 seconds must elapse before working on the system. Failure to do so may result in deployment of the air bag.

2. Disconnect the negative battery cable. Tape the cable with insulating tape.
3. Drain engine coolant
4. Remove the 2 nuts, then remove the V-bank cover.
5. Disconnect the ventilation hose No. 2.
6. Disconnect the vacuum hose.
7. Disconnect the mass air flow meter connector.
8. Remove the 2 wire harness clamps.
9. Loosen the 2 hose clamps.
10. Remove the 2 bolts, then remove the air cleaner.
11. Disconnect the water by-pass hose No. 5.
12. Disconnect the water by-pass hose No. 4.
13. Disconnect the throttle motor connector.
14. Remove the 4 bolts, then remove the throttle w/ motor body and gasket.

To install:

➡**Be sure to use new fasteners, as required.**

15. Install a new gasket and the throttle with motor body with the 4 bolts and tighten to 108 inch lbs. (11 Nm).
16. Connect the throttle motor connector.
17. Connect the water by-pass hose No. 4.
18. Connect the water by-pass hose No. 5.
19. Install the air cleaner with the 2 bolts.
20. Connect the ventilation hose No. 2.
21. Connect cable to negative battery terminal
22. Add engine coolant
23. Check for engine coolant leakage
24. Install the V-bank cover with the 2 nuts.
25. Using the Techstream diagnostic tool, or equivalent, perform the initialization procedure.

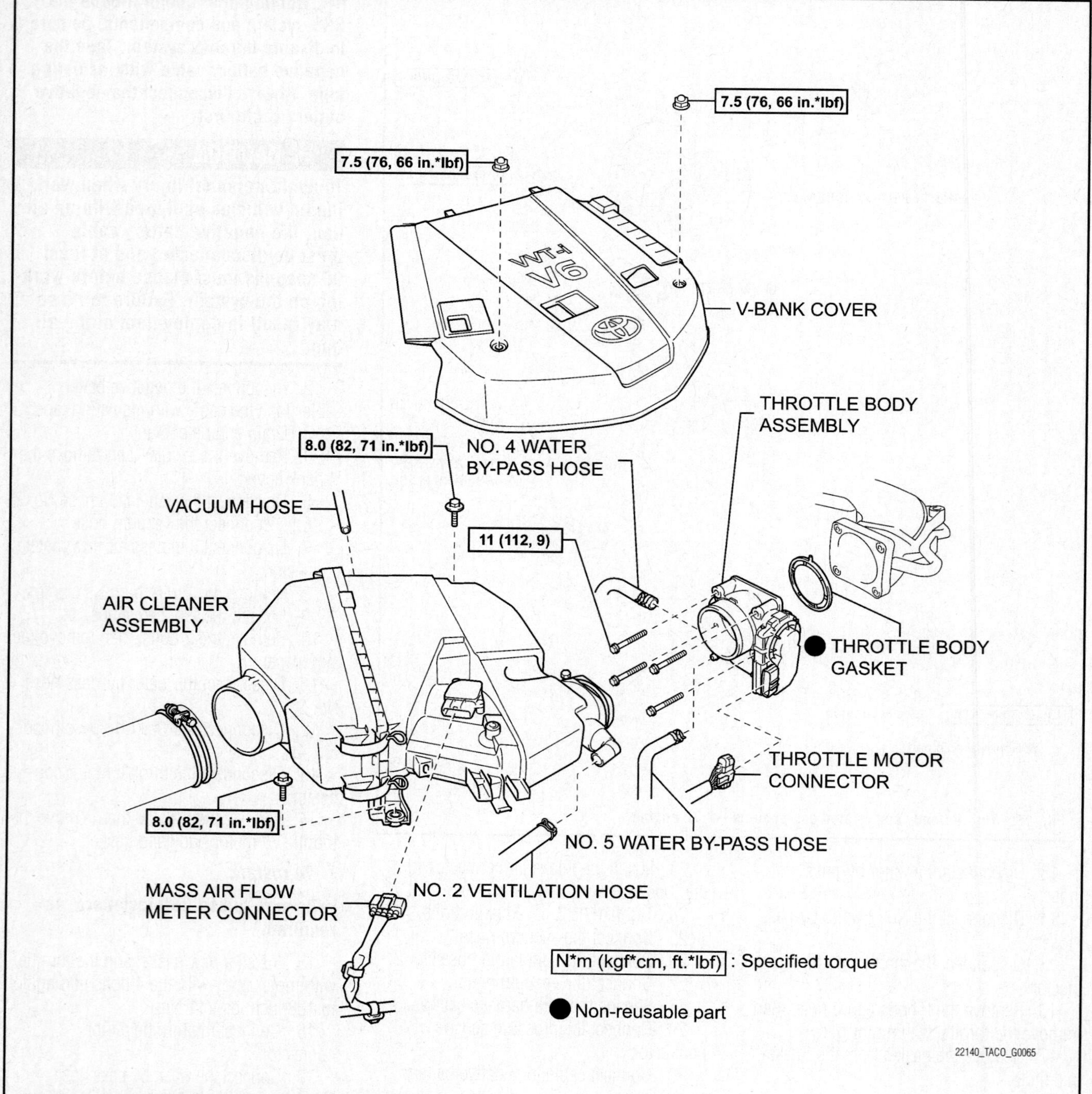

Fig. 156 Throttle body and related components—4.0L engine

HEATING & AIR CONDITIONING SYSTEM

BLOWER MOTOR

REMOVAL & INSTALLATION

See Figure 157.

At this time the manufacturer does not provide removal and installation procedures for this component, refer to the illustration as required.

➡Before servicing the vehicle, refer to the Precautions Section.

➡Disconnect the negative battery cable. Tape the cable with insulating tape.

➡If working near and/or around the SRS system and components, be sure to disable the SRS system. Tape the negative battery cable with insulating tape. Always disconnect the negative battery cable first.

CAUTION

To avoid personal injury when working on vehicles equipped with an air bag, the negative battery cable must be disconnected and at least 90 seconds must elapse before working on the system. Failure to do so may result in deployment of the air bag.

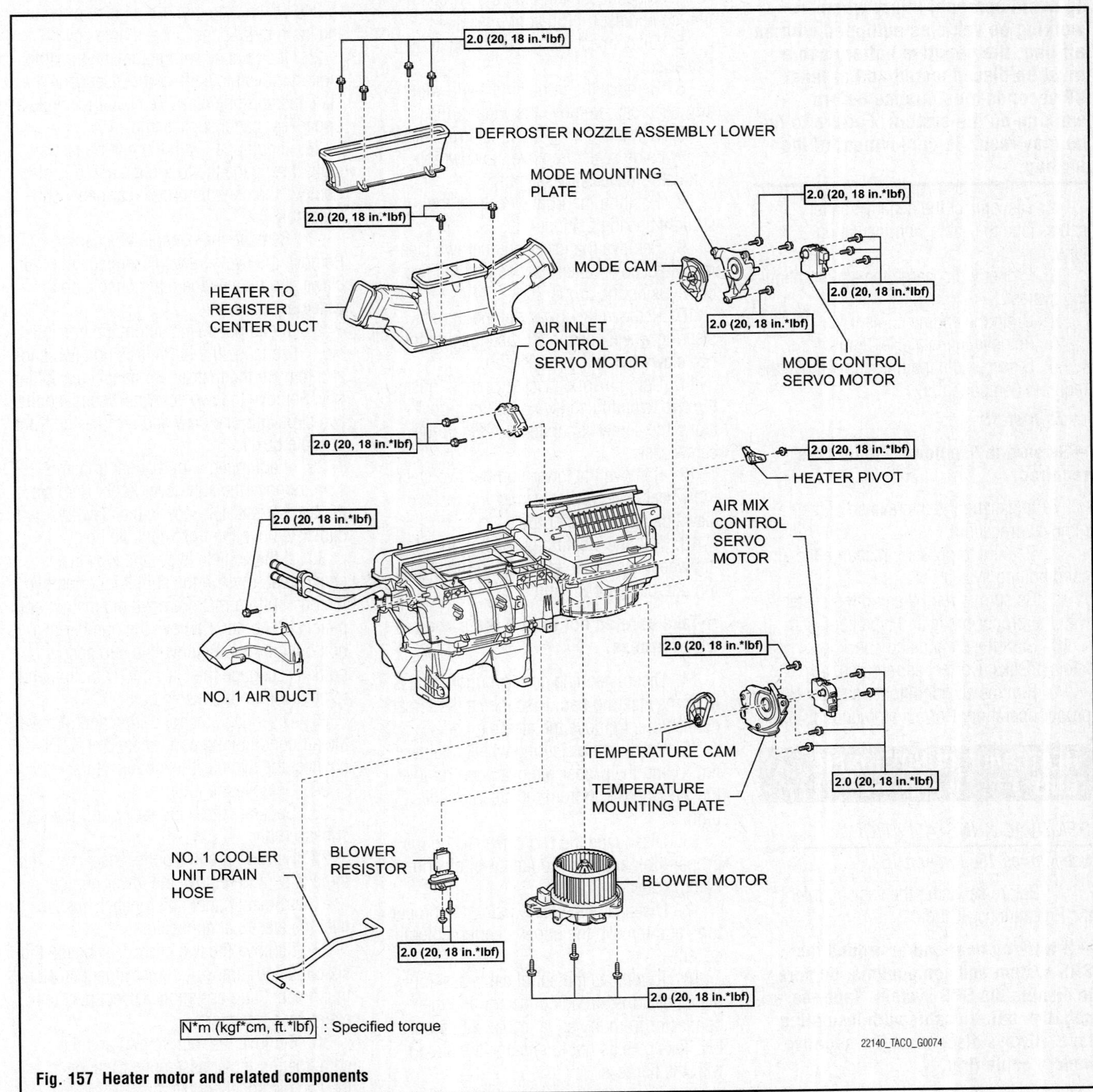

Fig. 157 Heater motor and related components

HEATER CORE

REMOVAL & INSTALLATION

1. Before servicing the vehicle, refer to the Precautions Section.

➡If working near and/or around the SRS system and components, be sure to disable the SRS system. Tape the negative battery cable with insulating tape. Always disconnect the negative battery cable first.

⁂ CAUTION

To avoid personal injury when working on vehicles equipped with an air bag, the negative battery cable must be disconnected and at least 90 seconds must elapse before working on the system. Failure to do so may result in deployment of the air bag.

2. Disconnect the negative battery cable. Tape the cable with insulating tape.
3. Properly discharge the air conditioning system.
4. Drain the engine coolant.
5. Remove the heater/AC unit.
6. Disassemble the unit and remove the required component(s).

To install:

➡Be sure to use new fasteners, as required.

7. Installation is the reverse of the removal procedure.
8. Be sure to properly recharge the air conditioning system.
9. Be sure to refill the cooling system with the proper grade and type coolant.
10. Start the engine and check for coolant leaks. Correct as required.
11. Run the air conditioner and check for proper operation. Correct as required.

HEATER/AIR CONDITIONING UNIT

REMOVAL & INSTALLATION

See Figures 158 through 163.

1. Before servicing the vehicle, refer to the Precautions Section.

➡If working near and/or around the SRS system and components, be sure to disable the SRS system. Tape the negative battery cable with insulating tape. Always disconnect the negative battery cable first.

⁂ CAUTION

To avoid personal injury when working on vehicles equipped with an air bag, the negative battery cable must be disconnected and at least 90 seconds must elapse before working on the system. Failure to do so may result in deployment of the air bag.

2. Disconnect the negative battery cable. Tape the cable with insulating tape.
3. Properly discharge the air conditioning system. Drain the engine coolant.
4. Disconnect and plug the air condition compressor refrigerant lines.
5. Disconnect the heater hoses at the heater core.
6. Remove the front windshield wiper arm head cap. Remove the wiper arms.
7. Remove the right and left front fender to cowl side seals. Remove the cowl top ventilator louver subassembly.
8. Position the front wheels in the straight ahead position.
9. Remove the lower number two steering wheel cover after disengaging the two claws using the proper tool.
10. Remove the lower number three steering wheel cover after disengaging the two claws using the proper tool.
11. Using a Torx® socket wrench loosen the two retaining screws until the groove along the screw circumference fits in the screw case.
12. Remove the lower number two steering wheel cover after disengaging the two claws using the proper tool.
13. Pull the steering pad out of the steering wheel assembly and support the steering pad using your hand.

➡Take care as not to pull the airbag wire harness.

14. Disconnect the horn ground harness from the steering pad. Disconnect the airbag connectors. Remove the steering pad.
15. Remove the steering wheel retainer nut. Using the proper removal tool, remove the steering wheel from the steering column.
16. Disengage the two claws, using the proper tool, and remove the lower steering column cover.
17. Disengage the claw, using the proper tool, and remove the upper steering column cover.
18. To remove the spiral cable assembly, disconnect the airbag connector and the connector from the spiral cable subassembly. Take care as not to damage the airbag with the harness.
19. Disengage the three claws and remove the spiral cable subassembly.
20. Remove the windshield wiper switch assembly.
21. Remove the front floor footrest. Remove the right and left front door scuff plates. Remove the right and left cowl side trim boards.
22. Separate the right and left front door opening trim weatherstrips. Remove the right and left front pillar garnish.
23. Disengage the four clips and two claws and remove the instrument panel under tray.
24. Disengage the four clips and remove the instrument panel access hole cover
25. If equipped with automatic transmission, remove the shift knob. Disengage the six clips and one claw. Remove the console upper rear panel subassembly.
26. If equipped with manual transmission, disengage the four clips and one claw. Remove the console upper rear panel subassembly.
27. Remove the console box carpet. Remove the two screws. Disengage the four claws and remove the rear console box assembly.
28. If equipped with automatic transmission, disengage the four claws and two clips and remove the instrument panel cup holder tray. Remove the two screws. Disengage the two clips and one claw and remove the front console box.
29. If equipped with manual transmission, remove the shift lever knob. Remove the two screws. Disengage the claw and two clips. Remove the front console box.
30. If the vehicle is equipped with a bench seat, remove the shift knob. Remove the box bottom mat. Remove the instrument panel cup holder. Remove the number two box bottom mat. Remove the clip and two bolts. Disengage the claw and two clips and remove the front console box.
31. Disengage the four clips and remove the air conditioning control assembly. Disconnect the two electrical connectors.
32. Remove the radio.
33. Separate the hood lock control lever subassembly.
34. Remove the two bolts. Disengage the three clips and remove the lower number one instrument panel finish panel. Disconnect the electrical connectors.
35. Remove the two clips. Disengage the six clips and remove the instrument cluster finish panel sub assembly. Disconnect the electrical connectors.
36. Remove the four screws and the combination meter assembly. Disconnect the two electrical connectors.

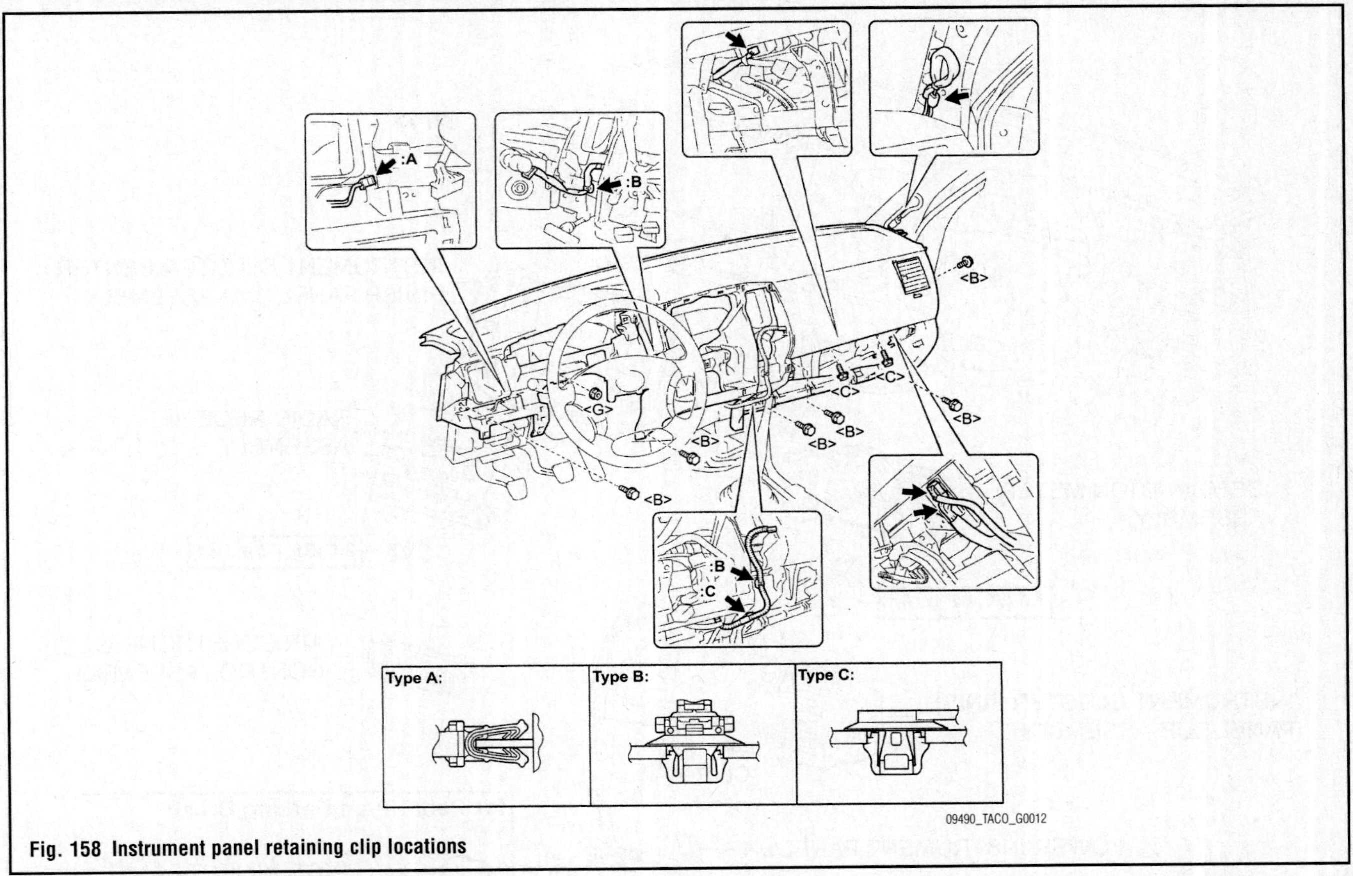

Fig. 158 Instrument panel retaining clip locations

37. Disengage the ten clips and remove the instrument cluster center finish panel subassembly.

38. Disengage the claw and separate the glove box door stopper from the glove box door assembly.

39. Slightly deform the upper part of the glove box door assembly and release the two stoppers and open the door assembly until it becomes horizontal.

40. Pull the glove box door assembly toward the rear of the vehicle and release the two stoppers and open the glove box door until it becomes horizontal.

41. Pull the glove box door assembly toward the rear of the vehicle to release the three hinges and remove the door assembly.

42. Remove the bolt. Disengage the two clips and remove the instrument panel lower finish panel subassembly, right side.

43. Remove the three bolts and the lower instrument panel, left side.

44. Remove the bolt. Disengage the five clips and remove the instrument lower cover subassembly. Disconnect the two connectors.

45. Disengage the three claws and remove the instrument side panel, right side.

46. Disconnect the passenger's side airbag connector. Disengage the ten hooks and remove the front passenger side airbag. Release the front side wall of the airbag door from the other hook and remove the front passenger side airbag assembly.

47. Tape up the steering column cover and upper steering wheel subassembly using protective tape. Disengage the four clamps and disconnect the four connectors. Remove the eight bolts, nut and instrument subassembly.

48. Remove the number one heater to register duct. Remove the number three heater to register duct.

49. On double cab vehicles, remove the number two rear duct and the number one rear duct.

50. Remove the clip. Disengage the four claws and remove the air duct assembly heater, left side.

51. Remove the clip. Disengage the three claws and remove the air duct assembly heater, right side.

52. Remove the bolt, nut and instrument panel brace mounting bracket, left side.

53. Remove the bolt, nut and instrument panel brace mounting bracket, right side.

54. Disengage the four clips and remove the steering column hole cover.

55. Matchmark the steering gear sliding yoke, steering gear intermediate shaft subassembly number two and the steering gear intermediate shaft assembly.

56. Remove bolts "A" and "B" from the steering sliding yoke.

57. Slide the steering yoke up and separate it from the steering intermediate shaft subassembly number two.

58. Pull down the steering sliding yoke from the steering intermediate shaft assembly to remove it.

59. Disconnect the steering column electrical connectors. Remove the steering column retaining bolts. Remove the steering column from the vehicle.

60. Disconnect the connector. Remove the nut and the transponder key ECU assembly.

61. Disconnect the connector. Remove the nut and the air conditioner amplifier assembly.

62. Disengage the four clamps and disconnect the six connectors from the air conditioning unit assembly. Remove the three bolts. Disconnect the connector. Disconnect the two airbag connectors. Disengage the twenty clamps.

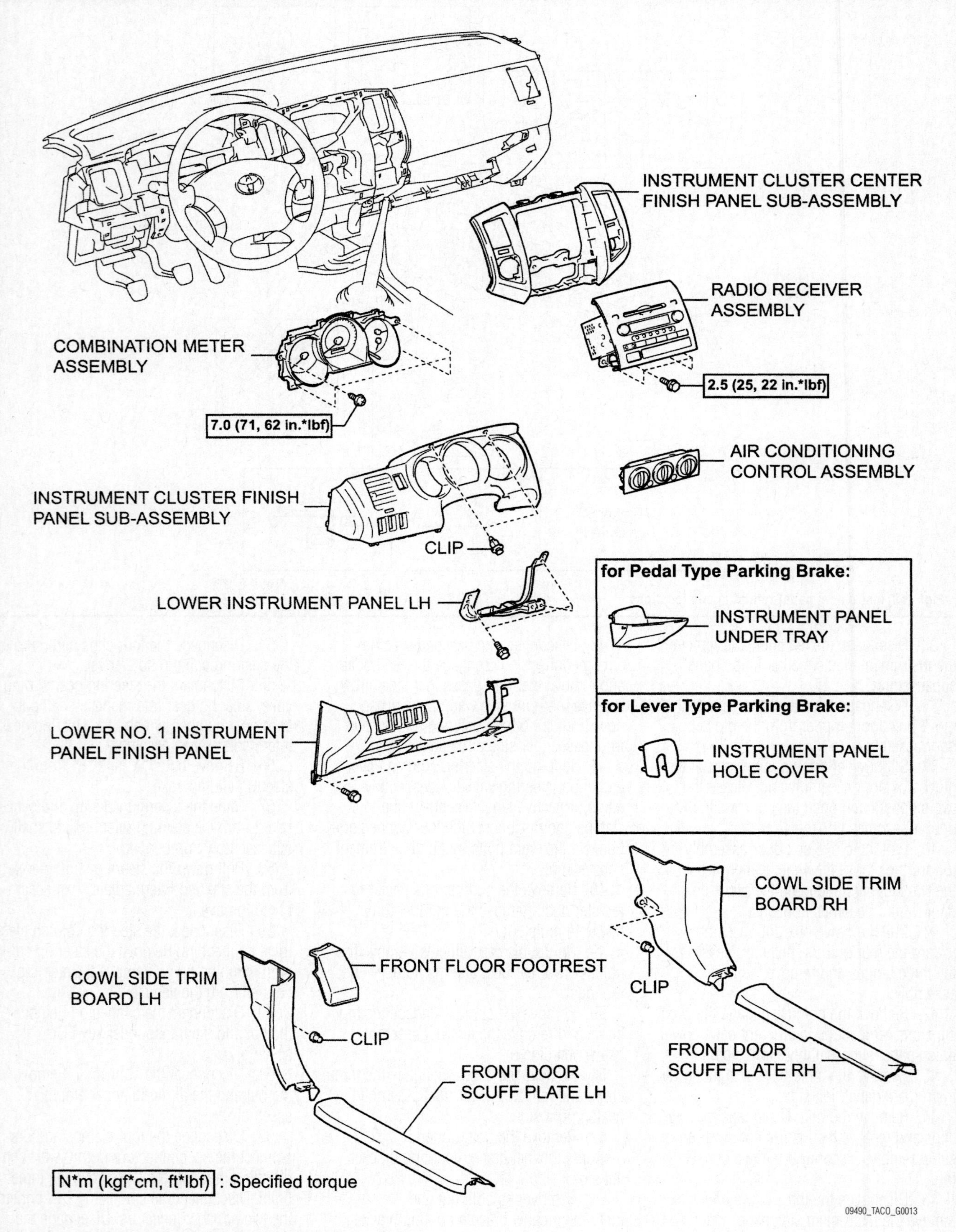

Fig. 159 Instrument panel and related components

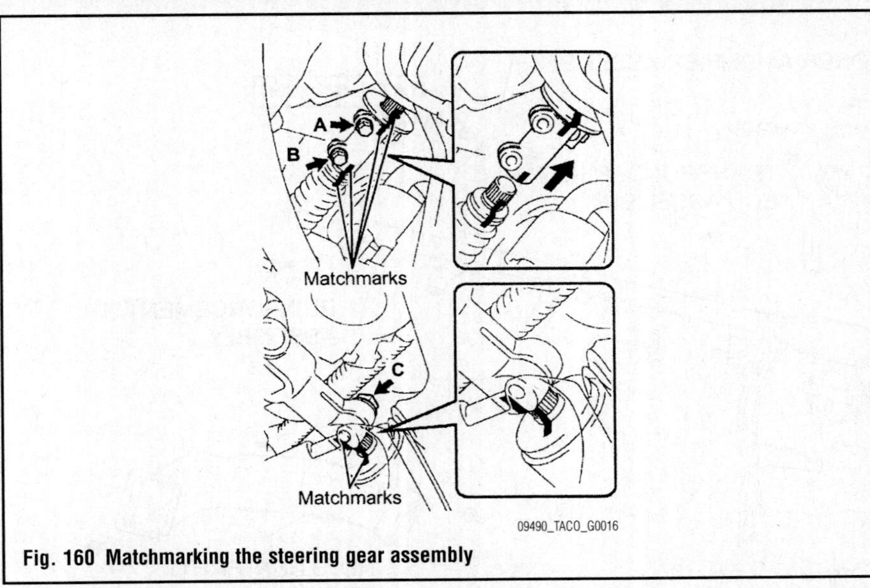

Fig. 160 Matchmarking the steering gear assembly

63. Remove the five bolts and two nuts. Remove the two caps and seven bolts. Disengage the reinforcement hook of the air conditioning unit. Remove the reinforcement.

64. Remove the air conditioning unit assembly from the vehicle.

65. Remove the heater to register center duct. Remove the number one air duct. Remove the air filter. Remove the mode control servo motor.

66. Remove the air mix control servo motor. Remove the air inlet control servo motor. Remove the three heater core assembly retaining bolts and the temperature mounting plate. Remove the temperature cam. Remove the eight bolts and the heater cover case. Remove the bolt and clamp. Remove the heater core from its mounting.

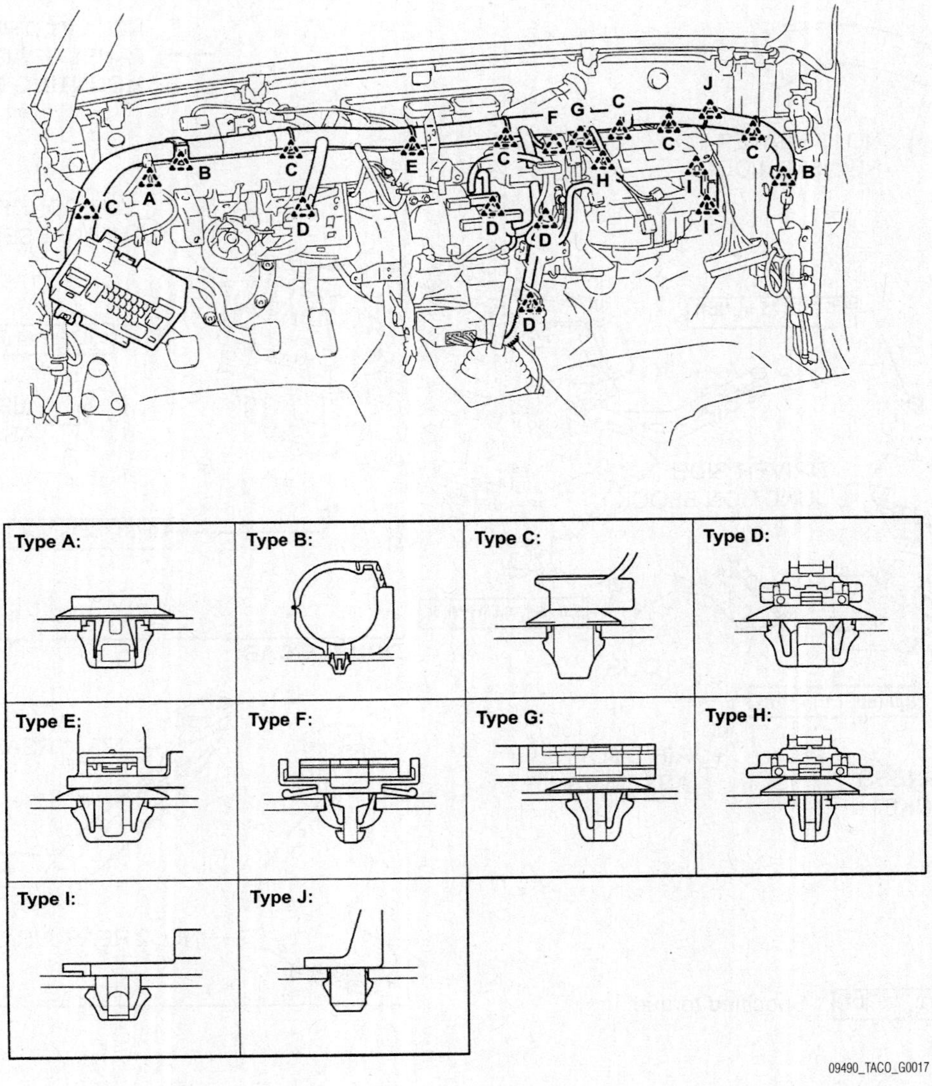

09490_TACO_G0017

Fig. 161 Air conditioning unit clip location

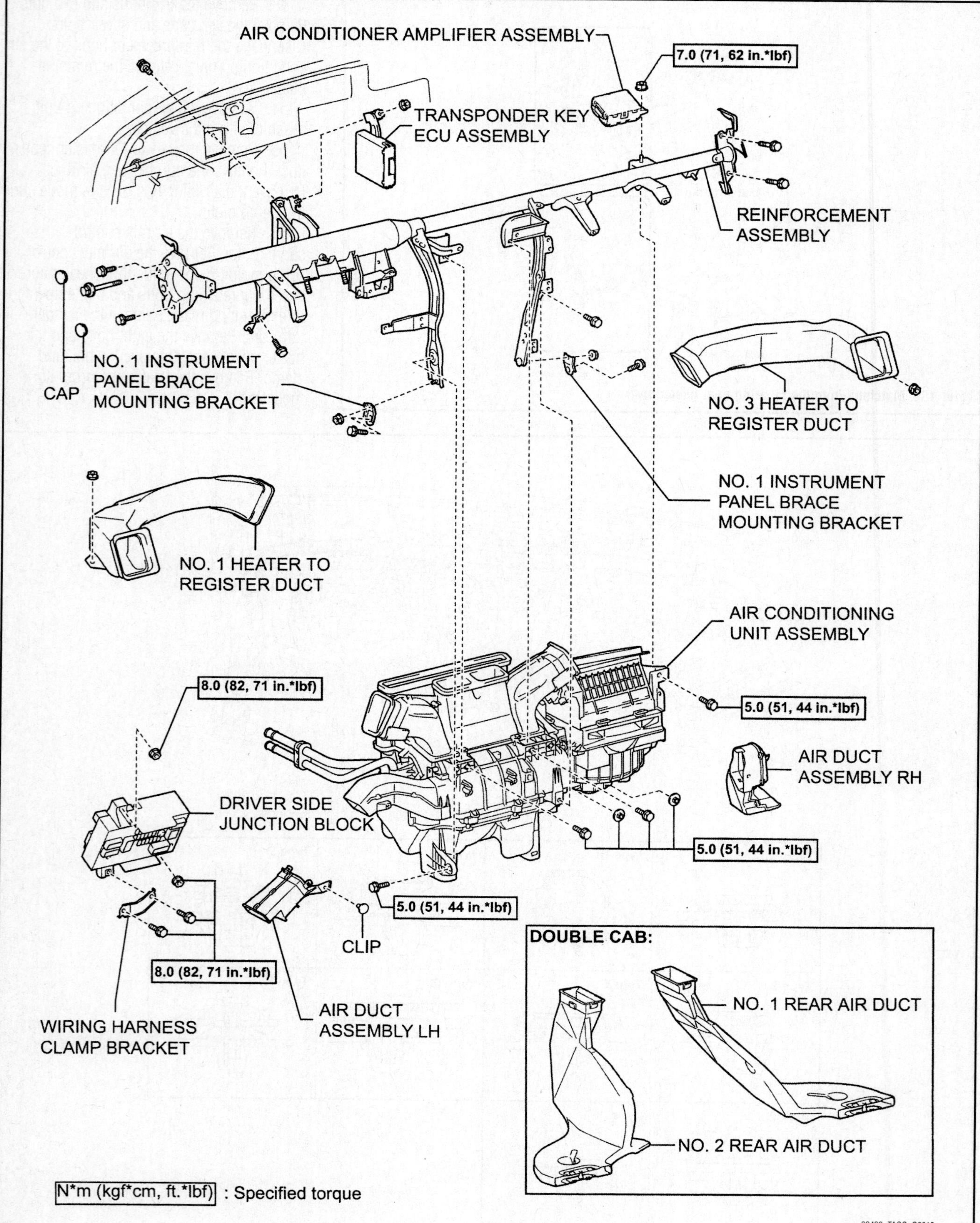

Fig. 162 Reinforcement assembly and related components

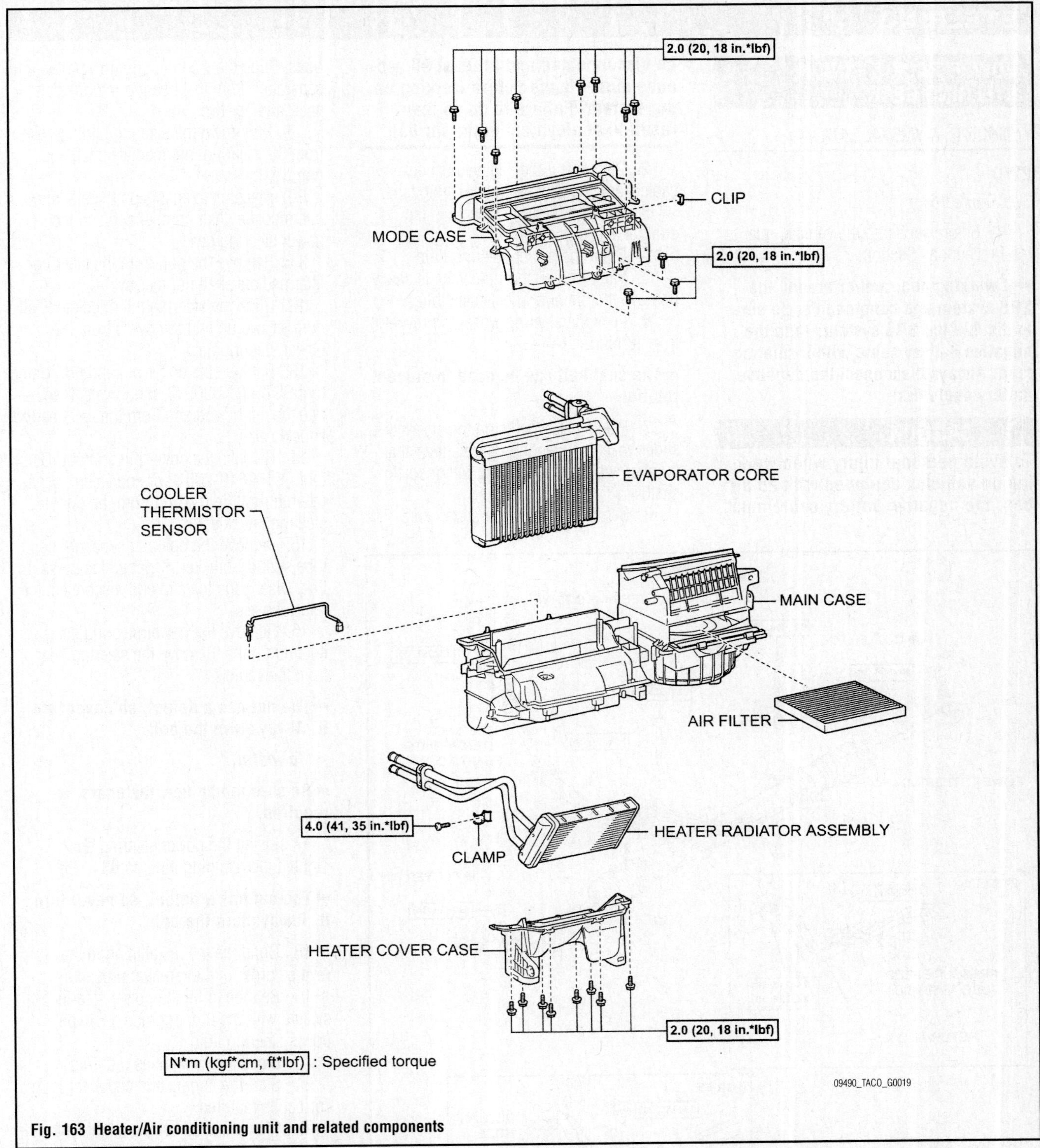

Fig. 163 Heater/Air conditioning unit and related components

To install:

➡Be sure to use new fasteners, as required.

67. Installation is the reverse of the removal procedure.

68. When installing the steering column be sure to align the matchmarks made during the removal procedure. Tighten bolts "A", "B" and "C" to 26 ft. lbs.

69. When installing the spiral cable, check that the ignition switch is in the "OFF" position. Turn the cable counterclockwise by hand until it becomes hard to turn. Turn the cable clockwise about 2 1/2 turns to align the marks.

➡The cable will rotate about 2½ turns both left and right from the center.

70. Refill the cooling system.

71. Evacuate and charge the air conditioning system refrigerant.

72. Run the engine to normal operating temperatures; then, check the climate control operation and check for leaks.

STEERING

POWER RACK & PINION STEERING GEAR

REMOVAL & INSTALLATION

2WD

See Figure 164.

1. Before servicing the vehicle, refer to the Precautions Section.

➡If working near and/or around the SRS system and components, be sure to disable the SRS system. Tape the negative battery cable with insulating tape. Always disconnect the negative battery cable first.

⁂ CAUTION

To avoid personal injury when working on vehicles equipped with an air bag, the negative battery cable must be disconnected and at least 90 seconds must elapse before working on the system. Failure to do so may result in deployment of the air bag.

2. Disconnect the negative battery cable. Tape the cable with insulating tape.
3. Position the front wheels in the straight ahead position.
4. Drain the power steering fluid.
5. Raise and support the vehicle safely. Remove the tire and wheel assemblies.
6. Lock the steering wheel to prevent it from turning.

➡The seat belt can be used to prevent rotation.

7. Place matchmarks on the steering slider yoke, the steering intermediate shaft number two and the steering intermediate shaft.
8. Remove the steering slider yoke bolts. Slide the steering sliding yoke up and separate it from the steering intermediate shaft number two.
9. Pull down the steering sliding yoke from the steering intermediate shaft and remove it.
10. Place matchmarks on the steering intermediate shaft number two and the power steering gear.
11. Remove the bolt from the steering intermediate shaft number two.
12. Slide the steering intermediate shaft number two up and remove it from the power steering gear.
13. Remove the cotter pin and nut. Using tool SST09610-20012 or equivalent, separate the left tie rod end from the left steering knuckle arm.
14. Remove the cotter pin and nut. Using tool SST09610-20012 or equivalent, separate the right tie rod end from the right steering knuckle arm.
15. Remove the bolt and separate the tube support bracket. Separate the pressure line. Disengage the clip and disconnect the return hose.
16. Remove the power steering gear retaining bolts. Remove the steering gear from the vehicle.

➡The nut has a detent, so never turn it. Always turn the bolt.

To install:

➡Be sure to use new fasteners, as required.

17. Install the power steering gear. Tighten the retaining bolts to 68 ft. lbs.

➡The nut has a detent, so never turn it. Always turn the bolt.

18. Continue the installation in the reverse order of the removal procedure.
19. Be sure to fill the power steering system with the proper grade and type power steering fluid.
20. Bleed the system, as required.
21. Start the engine and check for leaks, correct as required.
22. Check and adjust the alignment, as required.

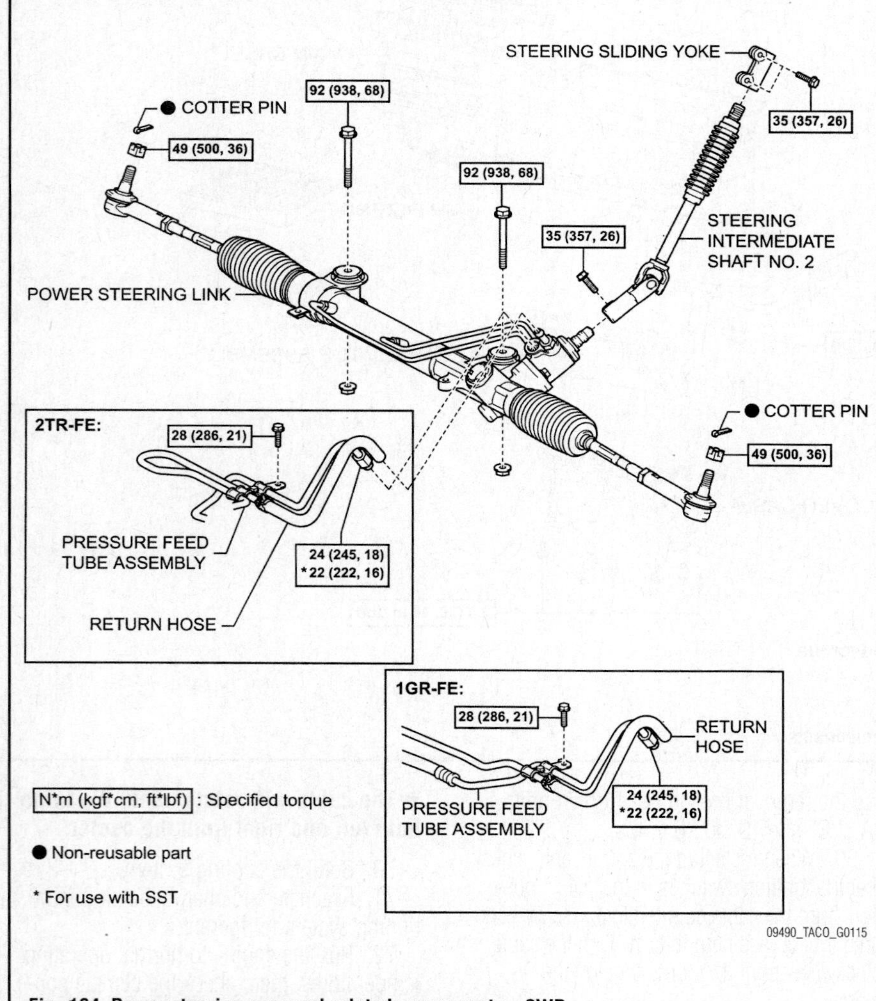

Fig. 164 Power steering gear and related components—2WD

4WD And PreRunner

See Figure 165.

1. Before servicing the vehicle, refer to the Precautions Section.

➡If working near and/or around the SRS system and components, be sure to disable the SRS system. Tape the

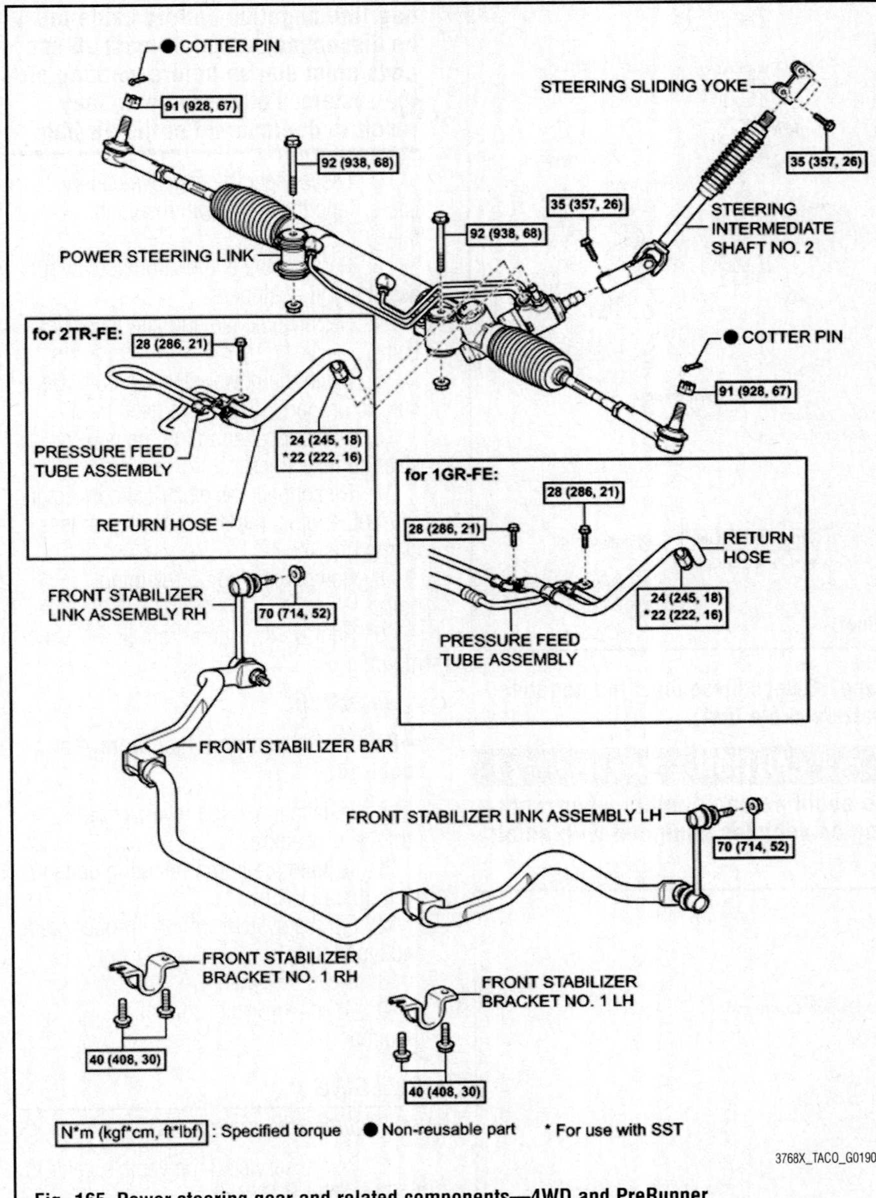

Fig. 165 Power steering gear and related components—4WD and PreRunner

negative battery cable with insulating tape. Always disconnect the negative battery cable first.

CAUTION

To avoid personal injury when working on vehicles equipped with an air bag, the negative battery cable must be disconnected and at least 90 seconds must elapse before working on the system. Failure to do so may result in deployment of the air bag.

2. Disconnect the negative battery cable. Tape the cable with insulating tape.
3. Position the front wheels in the straight ahead position.
4. Drain the power steering fluid.
5. Raise and support the vehicle safely. Remove the tire and wheel assemblies.
6. Remove the number one engine undercover subassembly.
7. Remove the front exhaust pipe assembly. On the 4.0L engine, remove the number two exhaust pipe assembly.
8. Remove the driveshaft. Some vehicles also use a center bearing assembly, remove that too.
9. Remove the frame crossmember subassembly.
10. Remove the stabilizer bar.
11. Lock the steering wheel to prevent it from turning.

➡The seat belt can be used to prevent rotation.

12. Place matchmarks on the steering slider yoke, the steering intermediate shaft number two and the steering intermediate shaft.
13. Remove the steering slider yoke bolts. Slide the steering sliding yoke up and separate it from the steering intermediate shaft number two.
14. Pull down the steering sliding yoke from the steering intermediate shaft and remove it.
15. Place matchmarks on the steering intermediate shaft number two and the power steering gear.
16. Remove the bolt from the steering intermediate shaft number two.
17. Slide the steering intermediate shaft number two up and remove it from the power steering gear.
18. Remove the cotter pin and nut. Using tool SST09610-20011 or equivalent, separate the left tie rod end from the left steering knuckle arm.
19. Remove the cotter pin and nut. Using tool SST09610-20011 or equivalent, separate the right tie rod end from the right steering knuckle arm.
20. Remove the bolt and separate the tube support bracket. Separate the pressure line. Disengage the clip and disconnect the return hose.
21. Remove the power steering gear retaining bolts. Tilt the transmission and remove the steering gear from the vehicle.

➡The nut has a detent, so never turn it. Always turn the bolt.

To install:

➡Be sure to use new fasteners, as required.

22. Install the power steering gear. Tighten the retaining bolts to 68 ft. lbs.

➡The nut has a detent, so never turn it. Always turn the bolt.

23. Continue the installation in the reverse order of the removal procedure.
24. Be sure to fill the power steering system with the proper grade and type power steering fluid.
25. Bleed the system, as required.
26. Start the engine and check for leaks, correct as required.
27. Check and adjust the alignment, as required.

POWER STEERING PUMP

REMOVAL & INSTALLATION

See Figures 166 and 167.

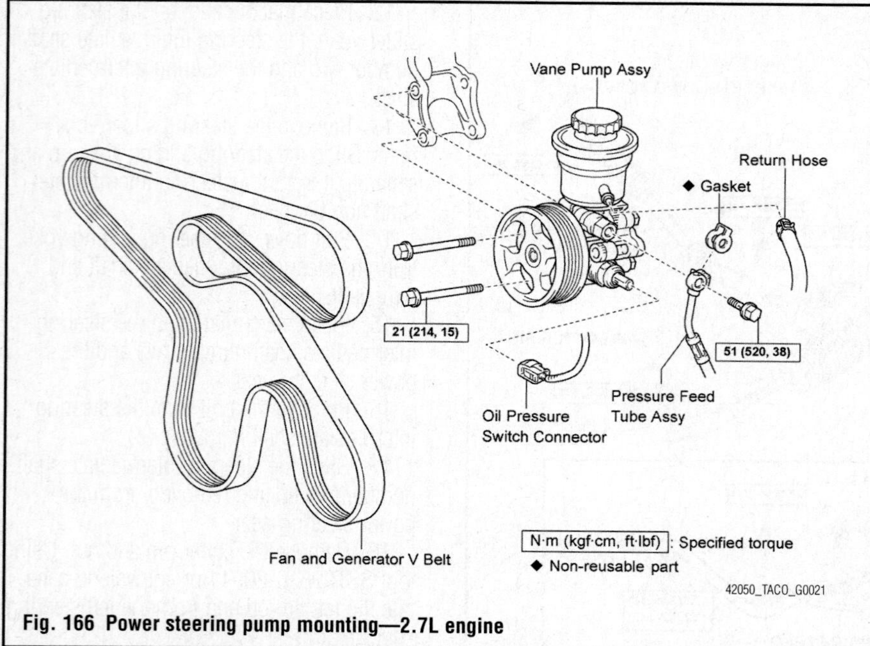

Fig. 166 Power steering pump mounting—2.7L engine

1. Before servicing the vehicle, refer to the Precautions Section.

➡**If working near and/or around the SRS system and components, be sure to disable the SRS system. Tape the negative battery cable with insulating tape. Always disconnect the negative battery cable first.**

⁂ CAUTION

To avoid personal injury when working on vehicles equipped with an air bag, the negative battery cable must be disconnected and at least 90 seconds must elapse before working on the system. Failure to do so may result in deployment of the air bag.

2. Disconnect the negative battery cable. Tape the cable with insulating tape.
3. Remove the engine undercover sub assembly, if equipped.
4. Remove the fan and alternator drive belt.
5. Drain the power steering fluid. Be sure to properly dispose of used fluid.
6. Disconnect and plug the power steering fluid lines.
7. Disconnect the electrical connectors. On 4.0L engine, remove the wire harness clamp bracket.
8. Remove the pump retaining bolts.
9. Remove the pump from its mounting.

To install:

➡**Be sure to use new fasteners, as required.**

10. Installation is the reverse of the removal procedure.
11. Tighten the pump retaining bolts to 15 ft. lbs. (21 Nm).
12. Fill the system with the proper grade and type fluid.
13. Bleed the system.
14. Check for leaks, correct as required.

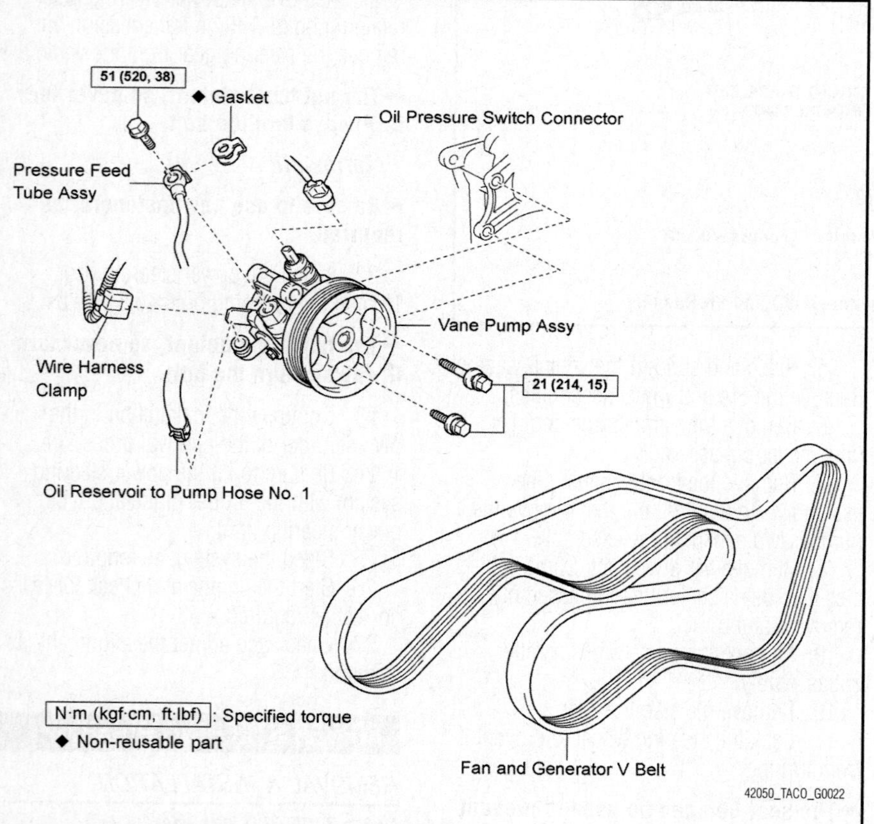

Fig. 167 Power steering pump mounting—4.0L engine

BLEEDING

See Figures 168 through 170.

1. Before servicing the vehicle, refer to the Precautions Section.

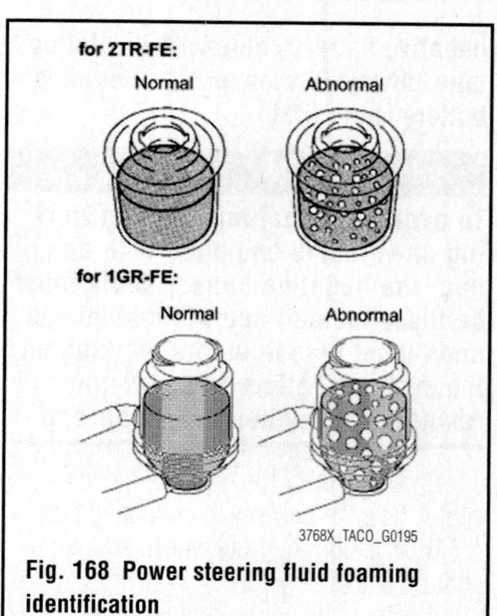

Fig. 168 Power steering fluid foaming identification

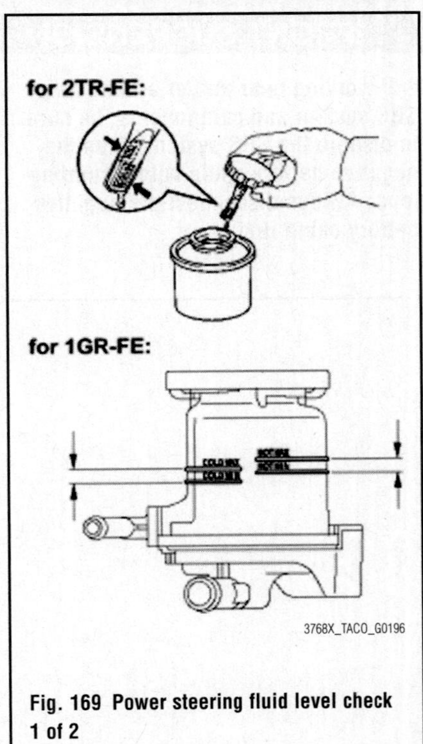

Fig. 169 Power steering fluid level check 1 of 2

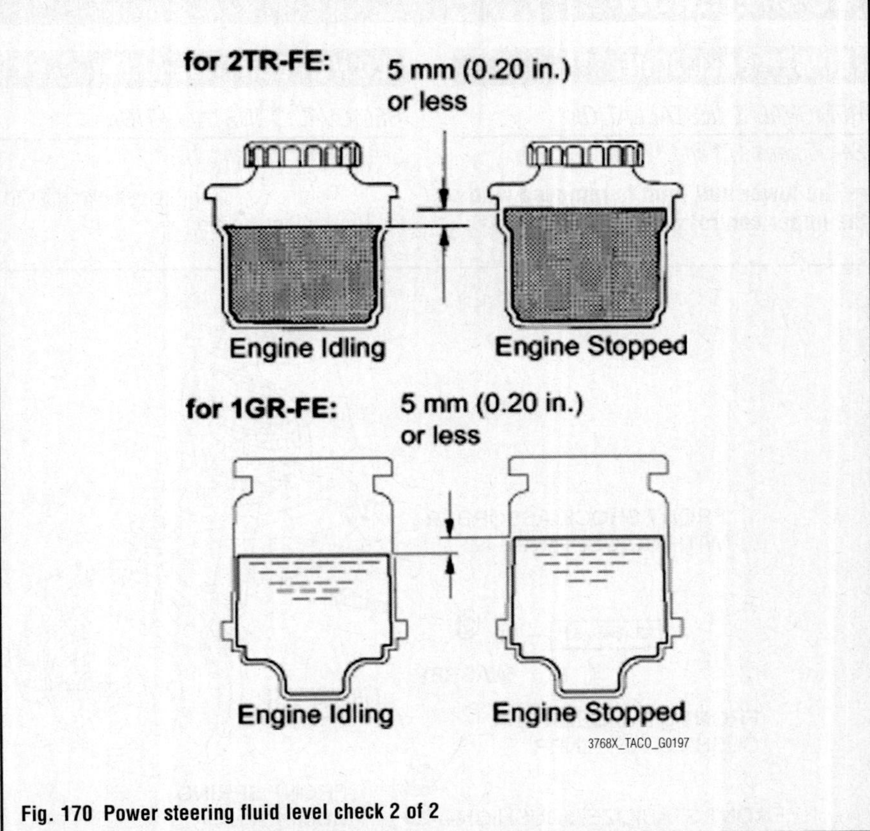

Fig. 170 Power steering fluid level check 2 of 2

➡If working near and/or around the SRS system and components, be sure to disable the SRS system. Tape the negative battery cable with insulating tape. Always disconnect the negative battery cable first.

CAUTION

To avoid personal injury when working on vehicles equipped with an air bag, the negative battery cable must be disconnected and at least 90 seconds must elapse before working on the system. Failure to do so may result in deployment of the air bag.

2. Check the fluid level.
3. Raise and properly support the vehicle.
4. With the engine stopped, turn the wheels slowly from stop to stop several times.
5. Lower the vehicle.
6. Start the engine. Run at idle for a few minutes.
7. With the engine idling turn the steering wheel to the left or right full lock position and keep it there for two or three seconds. Repeat for the other side.
8. Repeat the above step several times.
9. Stop the engine. Check for foaming.
10. If foaming exists, repeat the procedure.
11. With the engine stopped check the fluid level.
12. Adjust as necessary. Be sure to use the proper grade and type fluid.

LOWER BALL JOINT

REMOVAL & INSTALLATION

See Figures 171 and 172.

➡The lower ball joint is removed with the upper control arm.

LOWER CONTROL ARM

REMOVAL & INSTALLATION

See Figures 171 and 172.

1. Before servicing the vehicle, refer to the Precautions Section.

➡If working near and/or around the SRS system and components, be sure to disable the SRS system. Tape the negative battery cable with insulating tape. Always disconnect the negative battery cable first.

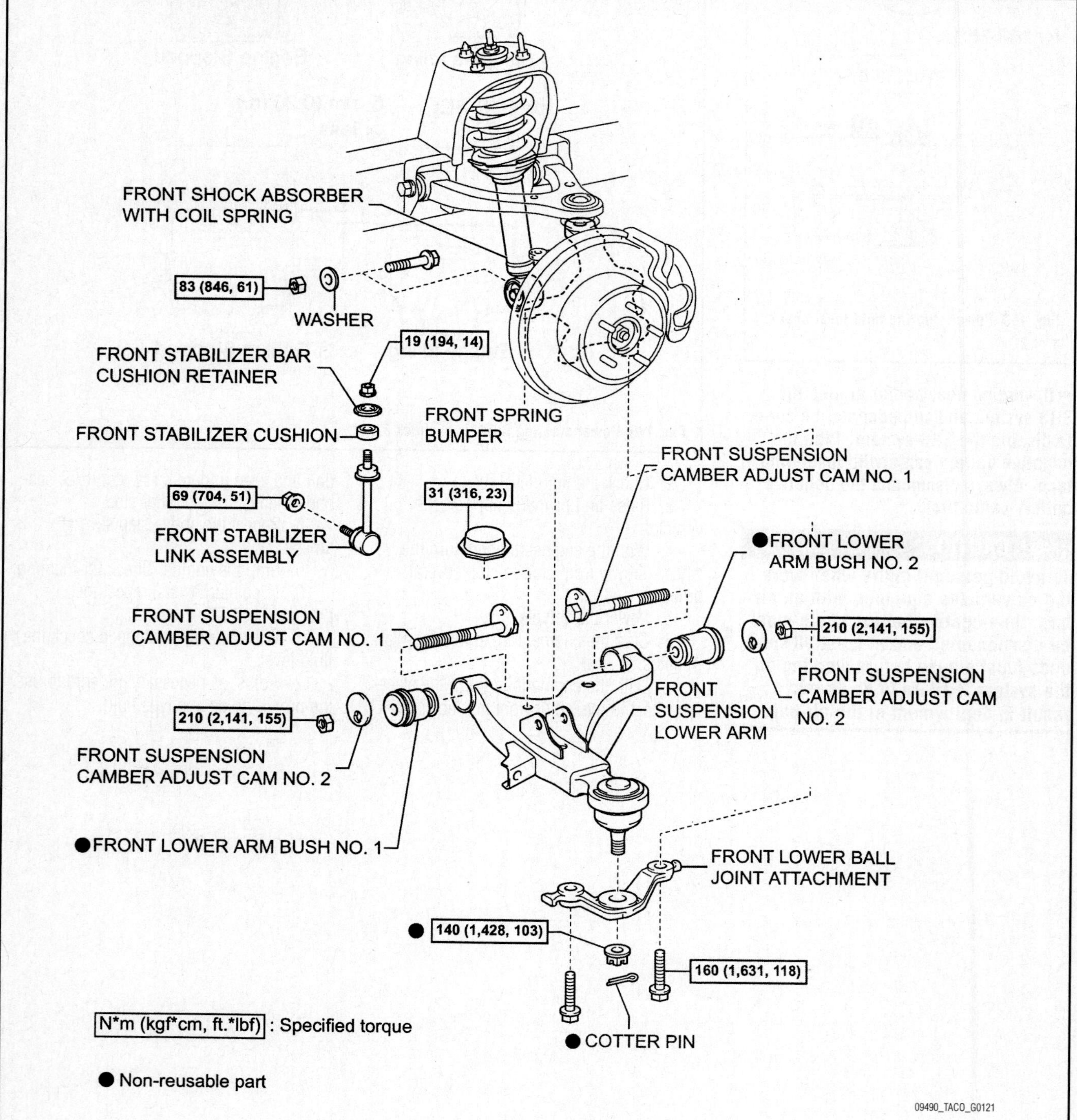

Fig. 171 Lower control arm and related components—2WD

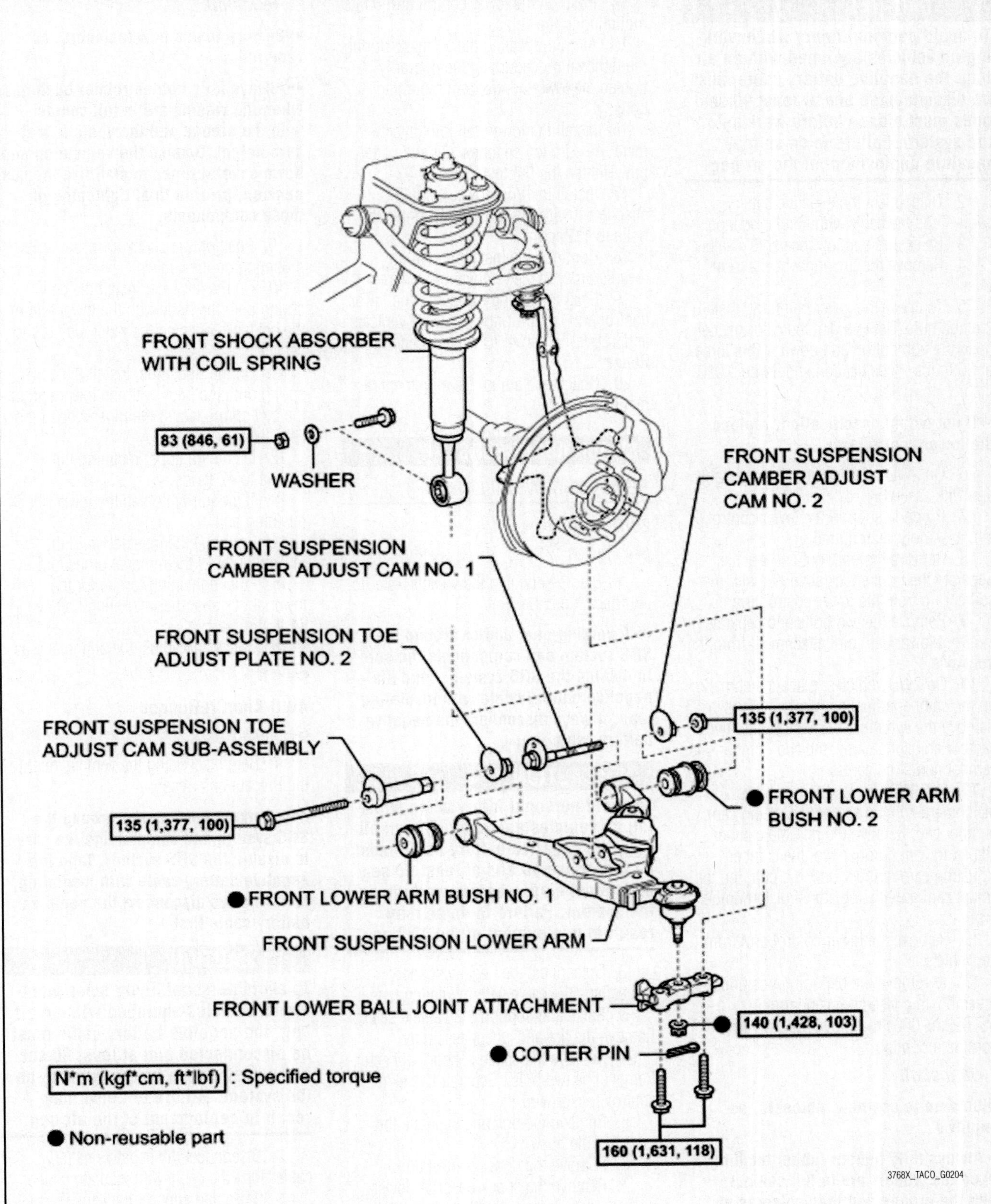

Fig. 172 Lower control arm and related components—4WD

⁂ CAUTION

To avoid personal injury when working on vehicles equipped with an air bag, the negative battery cable must be disconnected and at least 90 seconds must elapse before working on the system. Failure to do so may result in deployment of the air bag.

2. Disconnect the negative battery cable. Tape the cable with insulating tape.
3. Raise and support the vehicle safely.
4. Remove the tire and wheel assemblies.
5. To check the lower ball joint, install the hub nuts. Using a dial indicator gauge push the hub nut up and down with a force of 66 ft. lbs. Specification should be 0.020 inch.

➡If not within specification, replace the lower control arm.

6. On 2WD vehicles, remove the stabilizer link assembly.
7. Properly support the lower control arm assembly, as required.
8. Remove the bolt, nut and washer. Separate the front shock absorber with the coil spring from the lower control arm.
9. Remove the two bolts and separate the front lower ball joint attachment from the front axle.
10. On 2WD vehicles, place matchmarks on the camber adjusting cam number two. Remove the two nuts, the two number two camber adjusting cams, the two number one camber adjusting cams.
11. On 4WD and PreRunner, place matchmarks on the camber adjusting cam number two. Remove the nut, the camber adjusting cam number two, the camber adjusting cam number one, the bolt, the toe adjust cam and the toe adjust plate number two.
12. Remove the lower control arm from the vehicle.
13. To remove the ball joint, position the assembly in a vise and using tool SST09628-00011, remove the ball joint from its mounting.

To install:

➡Be sure to use new fasteners, as required.

➡Always fully tighten rubber bushings when the wheels are in full contact with the ground and the vehicle is at curb height. Bounce the vehicle up and down several times to stabilize the suspension, prior to final tightening of these components.

14. Position the lower control arm on its mounting in the vehicle.
15. Align the mating marks made during the removal procedure and temporarily tighten the lower control arm mounting bolts.
16. Install the lower ball joint attachment. Be sure to use a new nut and cotter pin. Tighten the nut to 103 ft. lbs.
17. Install the front lower ball joint attachment with the two bolts. Tighten the bolts to 118 ft. lbs.
18. Continue the installation in the reverse order of the removal procedure.
19. Final tightening torque for the lower control arm is 155 ft. lbs. on 2WD vehicles and 100 ft. lbs. on 4WD vehicles and PreRunner.
20. Check and adjust the alignment, as required.

SHOCK ABSORBERS

REMOVAL & INSTALLATION

2WD

See Figures 173 and 174.

1. Before servicing the vehicle, refer to the Precautions Section.

➡If working near and/or around the SRS system and components, be sure to disable the SRS system. Tape the negative battery cable with insulating tape. Always disconnect the negative battery cable first.

⁂ CAUTION

To avoid personal injury when working on vehicles equipped with an air bag, the negative battery cable must be disconnected and at least 90 seconds must elapse before working on the system. Failure to do so may result in deployment of the air bag.

2. Disconnect the negative battery cable. Tape the cable with insulating tape.
3. Raise and support the vehicle safely. Remove the tire and wheel assembly.
4. Remove the speed sensor connector. Remove the two bolts. Separate the skid control sensor wire.
5. Remove the bolt and separate the front flexible hose.
6. Remove the upper control arm.
7. Remove the shock absorber lower bolt and nut. Remove the three nuts on the upper side of the shock absorber.
8. Remove the shock absorber and coil spring from the vehicle.

To install:

➡Be sure to use new fasteners, as required.

➡Always fully tighten rubber bushings when the wheels are in full contact with the ground and the vehicle is at curb height. Bounce the vehicle up and down several times to stabilize the suspension, prior to final tightening of these components.

9. Position the shock absorber to its mounting on the vehicle.
10. On the left side, install the coil spring onto the body with the lower end of the coil spring facing the outer side of the vehicle.
11. On the right side, install the coil spring onto the body with the lower end of the coil spring facing the inner side of the vehicle.
12. Install the upper retaining nuts. Torque to 47 ft. lbs.
13. Temporarily tighten the lower shock retaining bolt.
14. Continue the installation in the reverse order of the removal procedure.
15. Final tightening torque for the lower shock absorber mounting bolt is 61 ft. lbs.
16. Check and adjust the alignment, as required.

4WD And PreRunner

See Figure 175.

1. Before servicing the vehicle, refer to the Precautions Section.

➡If working near and/or around the SRS system and components, be sure to disable the SRS system. Tape the negative battery cable with insulating tape. Always disconnect the negative battery cable first.

⁂ CAUTION

To avoid personal injury when working on vehicles equipped with an air bag, the negative battery cable must be disconnected and at least 90 seconds must elapse before working on the system. Failure to do so may result in deployment of the air bag.

2. Disconnect the negative battery cable. Tape the cable with insulating tape.
3. Raise and support the vehicle safely. Remove the tire and wheel assembly.
4. Remove the engine undercover assembly.
5. Remove the stabilizer bar.

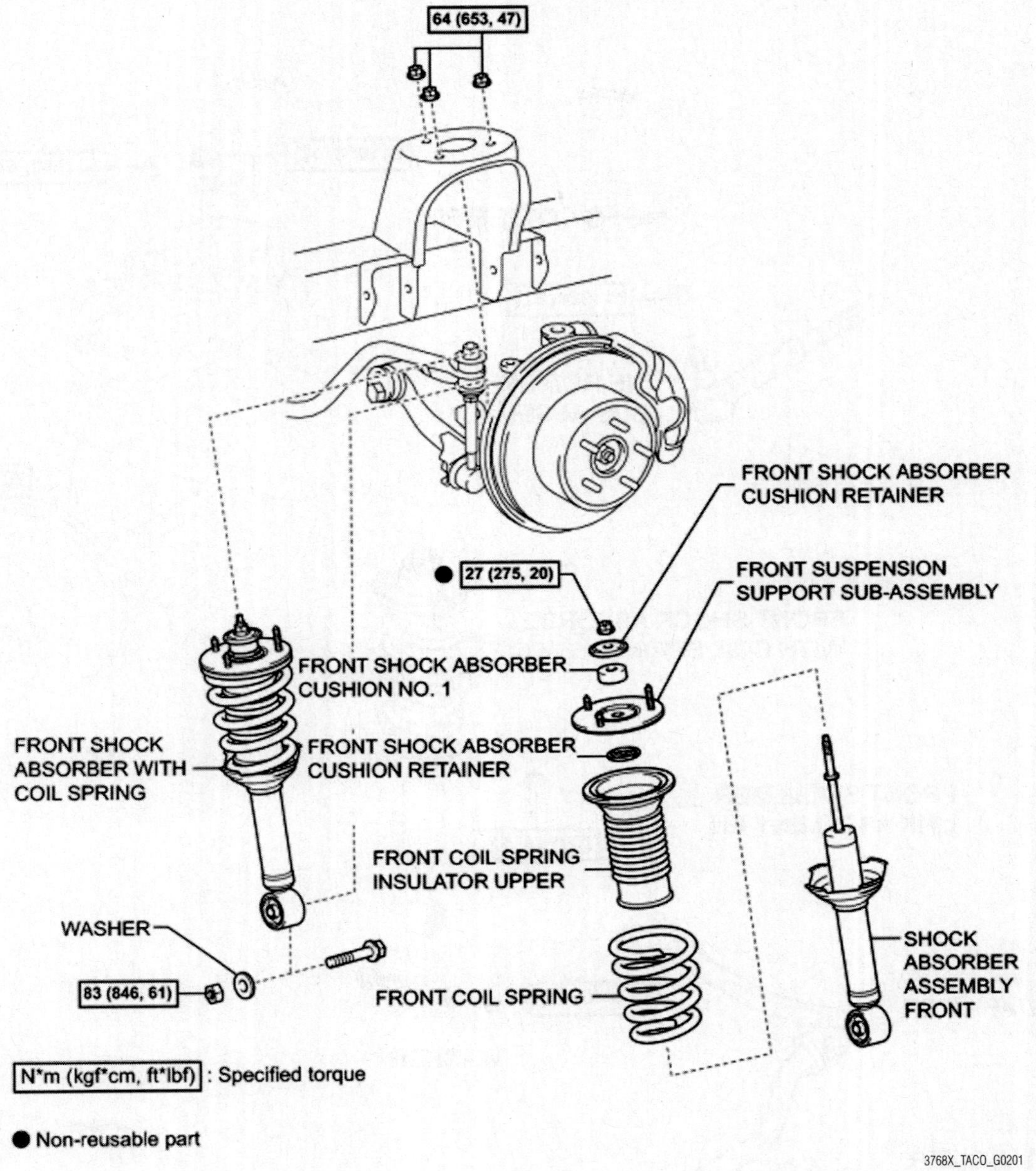

Fig. 173 Front shock absorber and related components—2WD

6. Remove the cotter pin and nut. Using the proper tool, separate the tie rod end from the steering knuckle arm.

7. Remove the shock absorber lower bolt and nut. Remove the three nuts on the upper side of the shock absorber.

8. Remove the shock absorber and coil spring from the vehicle.

To install:

➡**Be sure to use new fasteners, as required.**

➡**Always fully tighten rubber bushings when the wheels are in full contact with the ground and the vehicle is at curb height. Bounce the vehicle up and down several times to stabilize the suspension, prior to final tightening of these components.**

9. Position the shock absorber to its mounting on the vehicle.

10. On the left side, install the coil spring onto the body with the lower end of the coil spring facing the outer side of the vehicle.

11. On the right side, install the coil spring onto the body with the lower

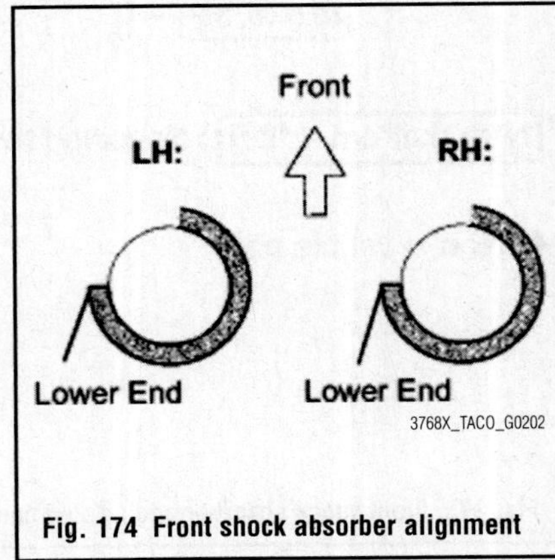

Fig. 174 Front shock absorber alignment

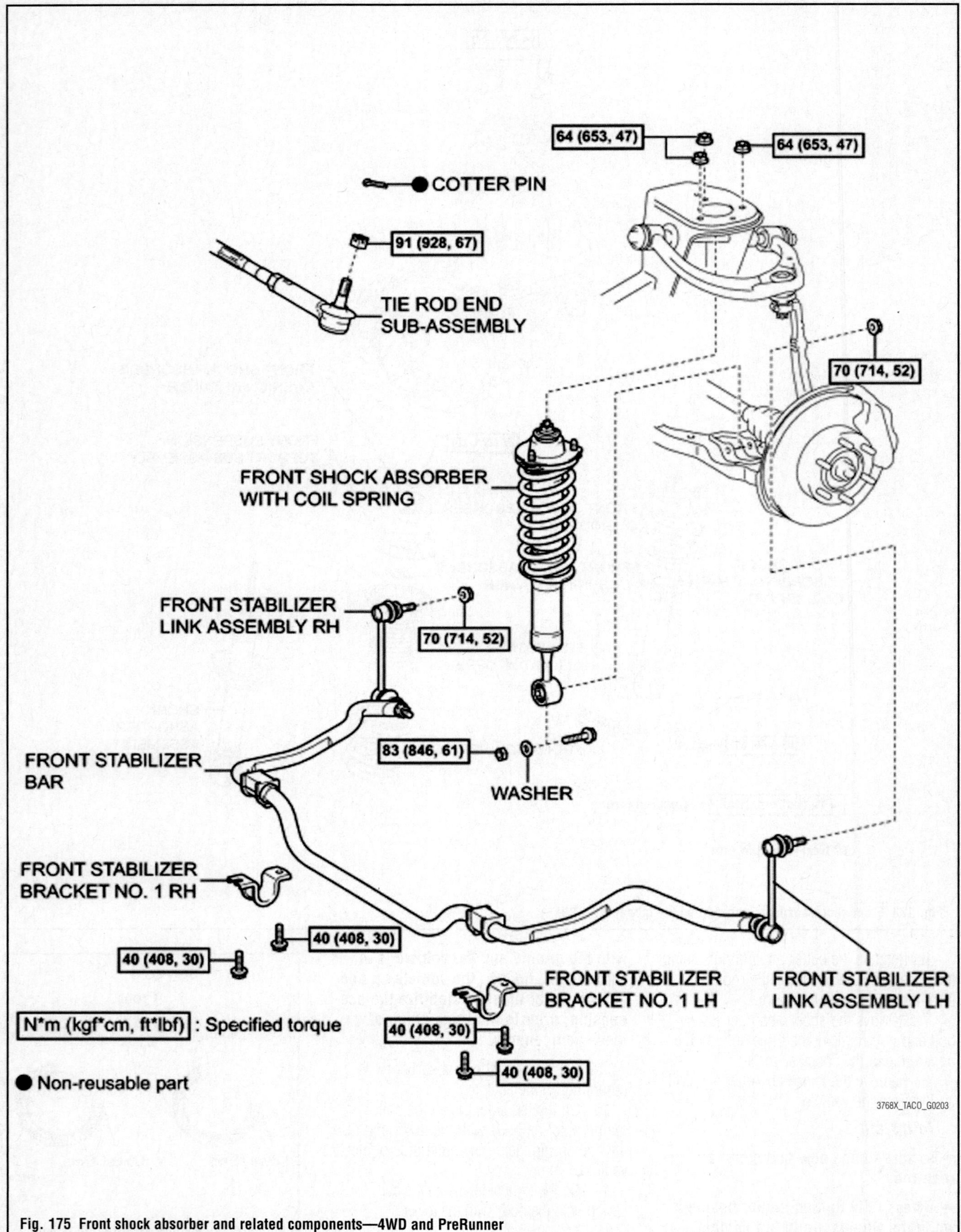

Fig. 175 Front shock absorber and related components—4WD and PreRunner

end of the coil spring facing the inner side of the vehicle.

12. Install the upper retaining nuts. Torque to 47 ft. lbs.

13. Temporarily tighten the lower shock retaining bolt.

14. Continue the installation in the reverse order of the removal procedure.

15. Final tightening torque for the lower shock absorber mounting bolt is 61 ft. lbs.

16. Check and adjust the alignment, as required.

STEERING KNUCKLE

REMOVAL & INSTALLATION

See Figures 176 and 177.

At this time the manufacturer does not provide removal and installation procedures

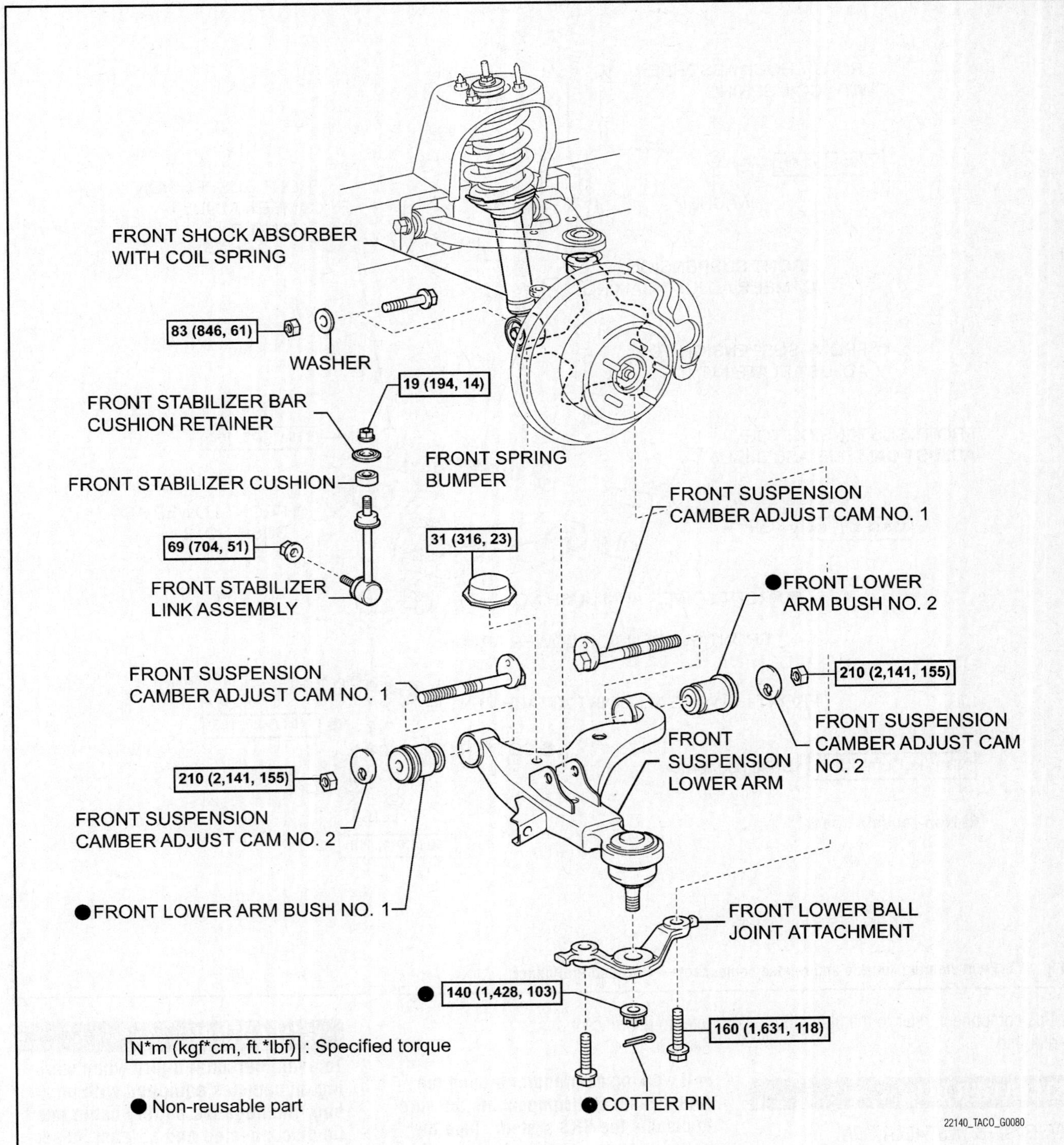

Fig. 176 Front steering knuckle and related components—2WD

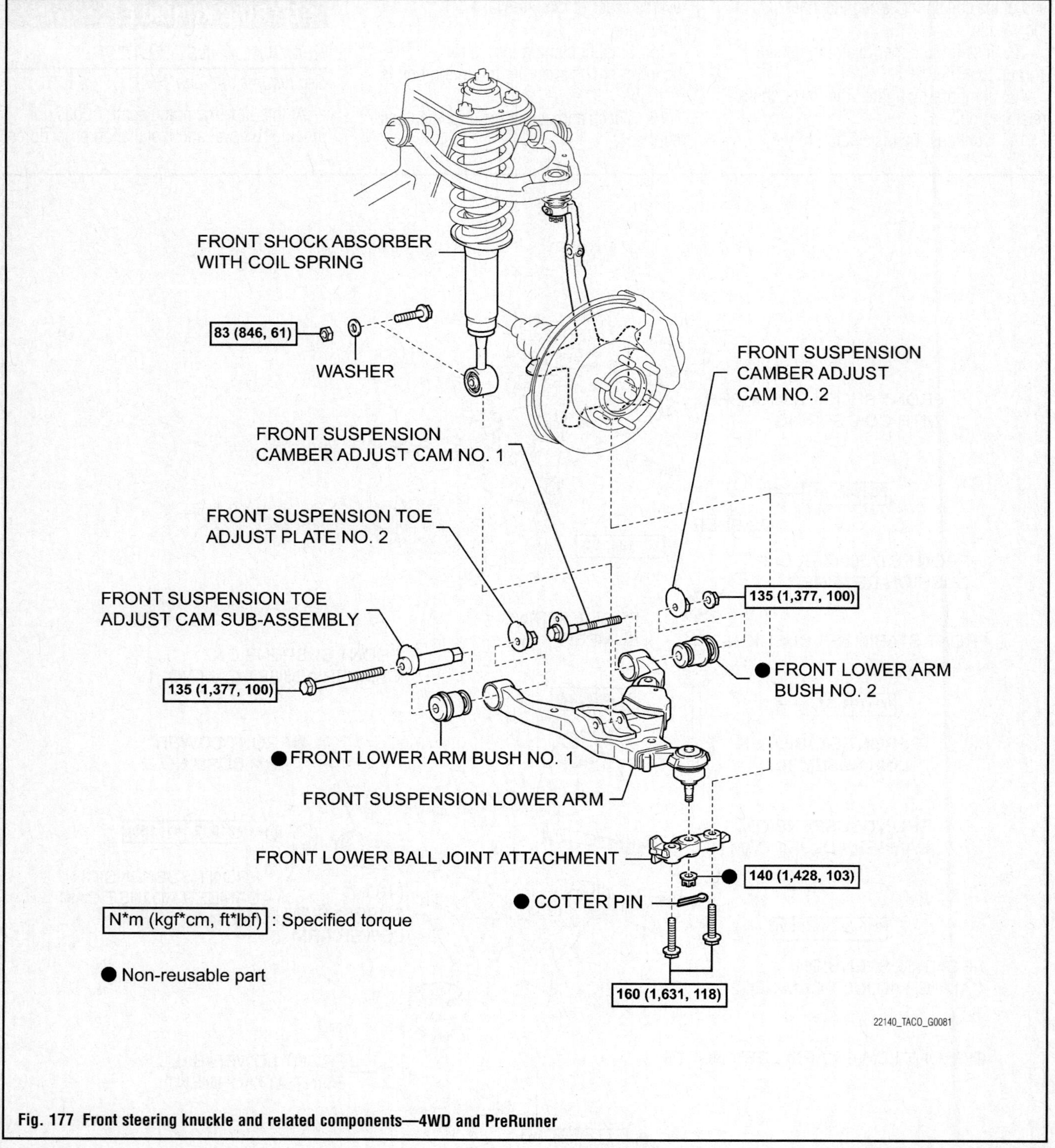

Fig. 177 Front steering knuckle and related components—4WD and PreRunner

for this component, refer to the illustration as required.

STABILIZER BAR

REMOVAL & INSTALLATION

See Figure 178.

1. Before servicing the vehicle, refer to the Precautions Section.

➡If working near and/or around the SRS system and components, be sure to disable the SRS system. Tape the negative battery cable with insulating tape. Always disconnect the negative battery cable first.

⁂ CAUTION

To avoid personal injury when working on vehicles equipped with an air bag, the negative battery cable must be disconnected and at least 90 seconds must elapse before working on the system. Failure to do so may result in deployment of the air bag.

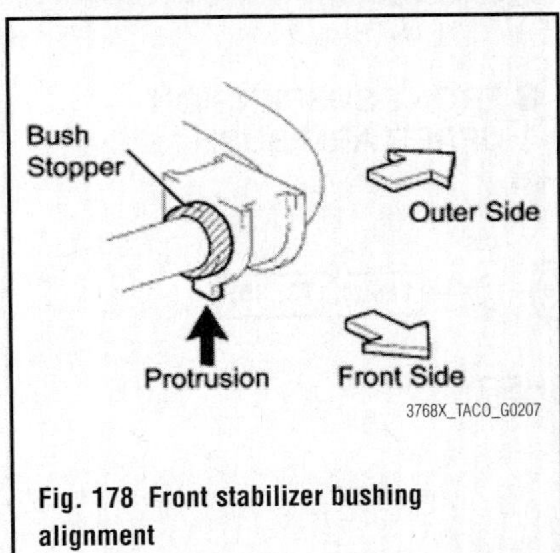

Fig. 178 Front stabilizer bushing alignment

2. Disconnect the negative battery cable. Tape the cable with insulating tape.
3. Raise and support the vehicle safely.
4. Remove the tire and wheel assemblies.
5. On 4WD and PreRunner, remove the engine undercover subassembly.
6. On 2WD (left side), remove the two nuts, stabilizer bar link, two retainers and two bushings.
7. On 2WD (right side), remove the two nuts, stabilizer bar link, two retainers and two bushings.
8. On 4WD and PreRunner (left side), remove the two nuts and stabilizer bar link.
9. On 4WD and PreRunner (right side), remove the two nuts and stabilizer bar link.
10. Remove the four bolts and the stabilizer bar brackets and bushings.
11. Remove the stabilizer bar from the vehicle.

To install:

➡Be sure to use new fasteners, as required.

12. Installation is the reverse of the removal procedure.
13. On 2WD, tighten the stabilizer bar side bushing bolts to 14 ft. lbs. Tighten the lower arm side bushing bolts to 51 ft. lbs. (69 Nm).. Tighten the bushing bracket bolts to 16 ft. lbs.
14. On 4WD and PreRunner, tighten the bar links to 52 ft. lbs. (70 Nm) Tighten the bushing bracket bolts to 30 ft. lbs.
15. Be sure to install the cushion onto the stabilizer bar with its cut line facing the front.

UPPER BALL JOINT

REMOVAL & INSTALLATION

See Figures 179 through 180.

➡The upper ball joint is removed with the upper control arm.

UPPER CONTROL ARM

REMOVAL & INSTALLATION

See Figures 179 and 180.

1. Before servicing the vehicle, refer to the Precautions Section.

➡If working near and/or around the SRS system and components, be sure to disable the SRS system. Tape the negative battery cable with insulating tape. Always disconnect the negative battery cable first.

⁂ CAUTION

To avoid personal injury when working on vehicles equipped with an air bag, the negative battery cable must be disconnected and at least 90 seconds must elapse before working on the system. Failure to do so may result in deployment of the air bag.

2. Disconnect the negative battery cable. Tape the cable with insulating tape.
3. Raise and support the vehicle safely. Remove the tire and wheel assemblies.
4. Remove the speed sensor connector. Remove the two bolts and separate the skid control sensor wire.
5. On 2WD vehicles, remove the bolt and separate the front flexible hose.
6. Support the front suspension lower control arm using a jack.
7. Remove the clip and nut. Using the tool SST09628-62011, or equivalent separate the upper ball joint from the steering knuckle.
8. On 4WD and PreRunner, remove the bolt and bracket.
9. Remove the two upper control arm retaining bolts.
10. Remove the upper control arm from the vehicle.

To install:

➡Be sure to use new fasteners, as required.

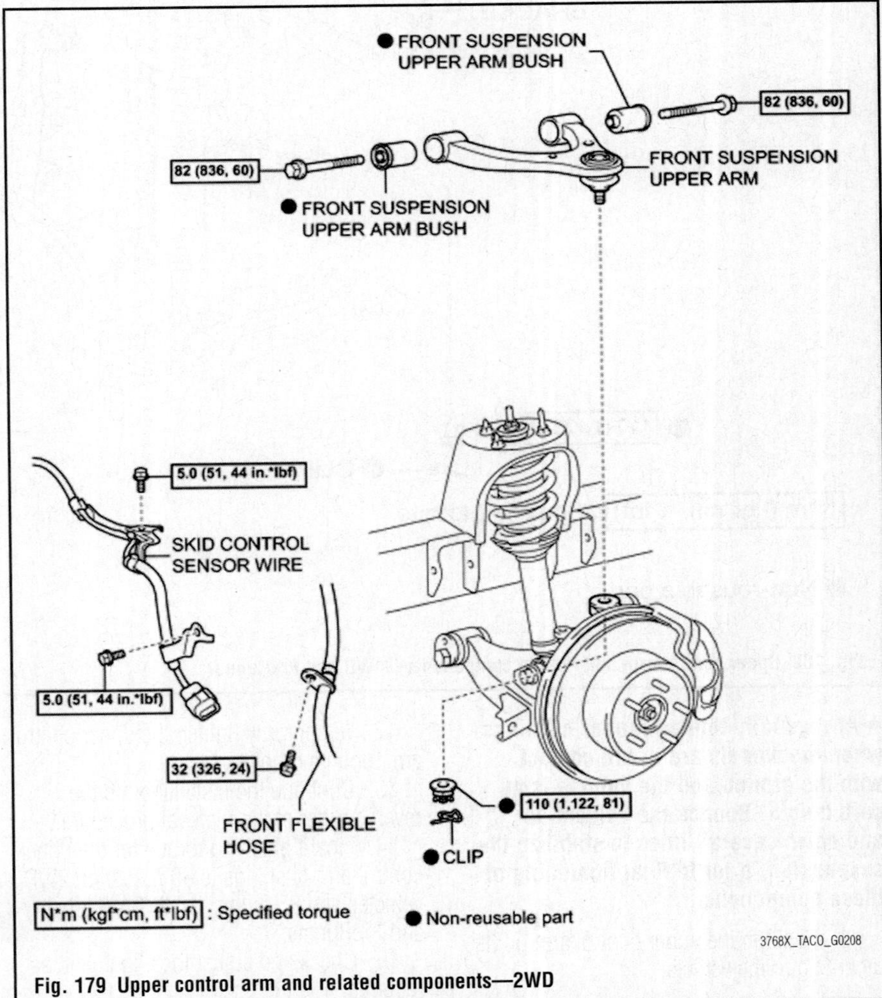

Fig. 179 Upper control arm and related components—2WD

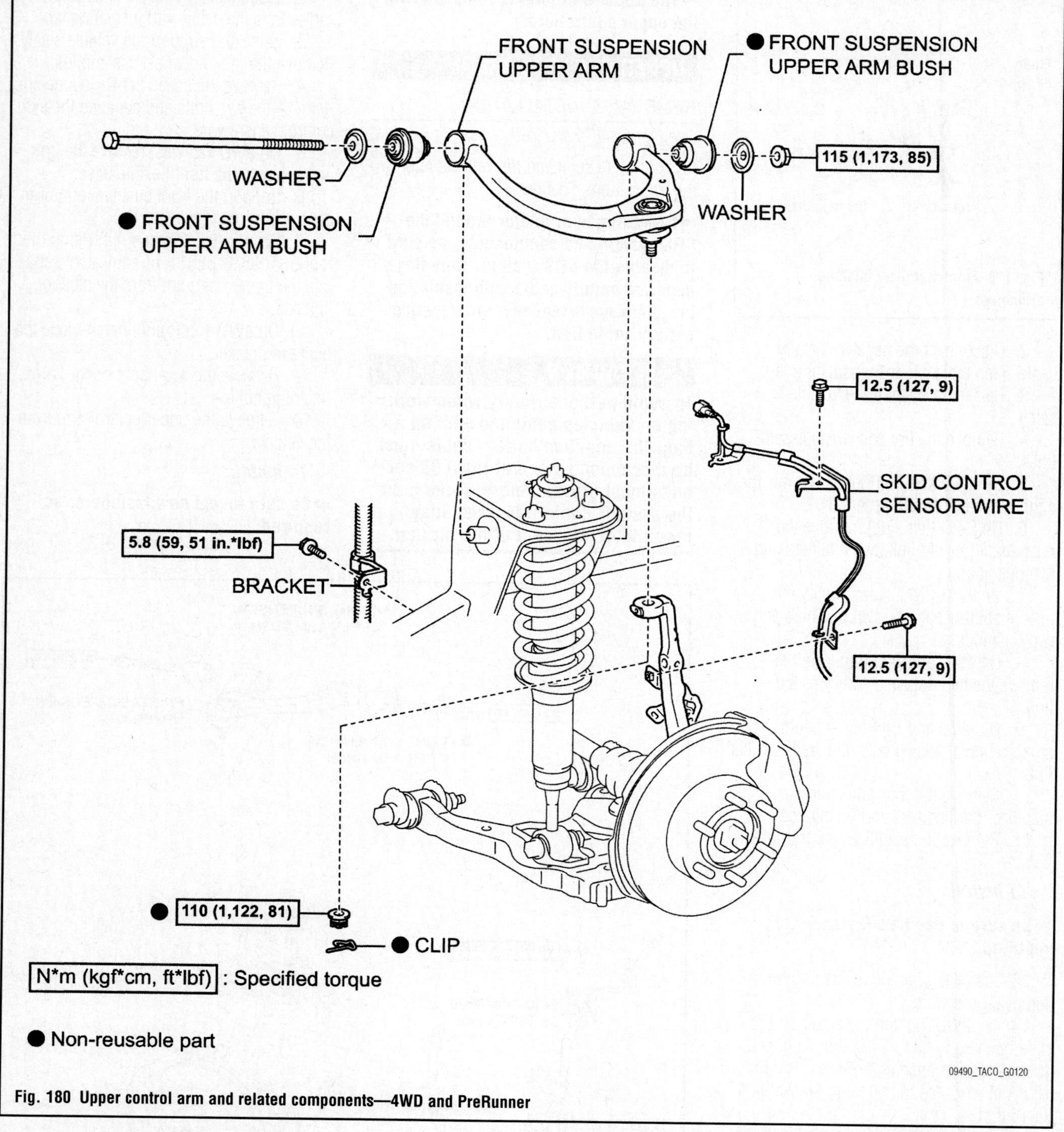

Fig. 180 Upper control arm and related components—4WD and PreRunner

➡Always fully tighten rubber bushings when the wheels are in full contact with the ground and the vehicle is at curb height. Bounce the vehicle up and down several times to stabilize the suspension, prior to final tightening of these components.

11. Position the upper control arm on its mounting in the vehicle.
12. Temporarily tighten the upper control arm mounting bolts.
13. Continue the installation in the reverse order of the removal procedure.
14. Final tightening torque for the upper control arm bushings is 60 ft. lbs. on 2WD vehicles and 85 ft. lbs. on 4WD vehicles and PreRunner.
15. Check and adjust the alignment, as required.

WHEEL HUB & BEARING

REMOVAL & INSTALLATION

2WD

See Figure 181.

1. Before servicing the vehicle, refer to the Precautions Section.

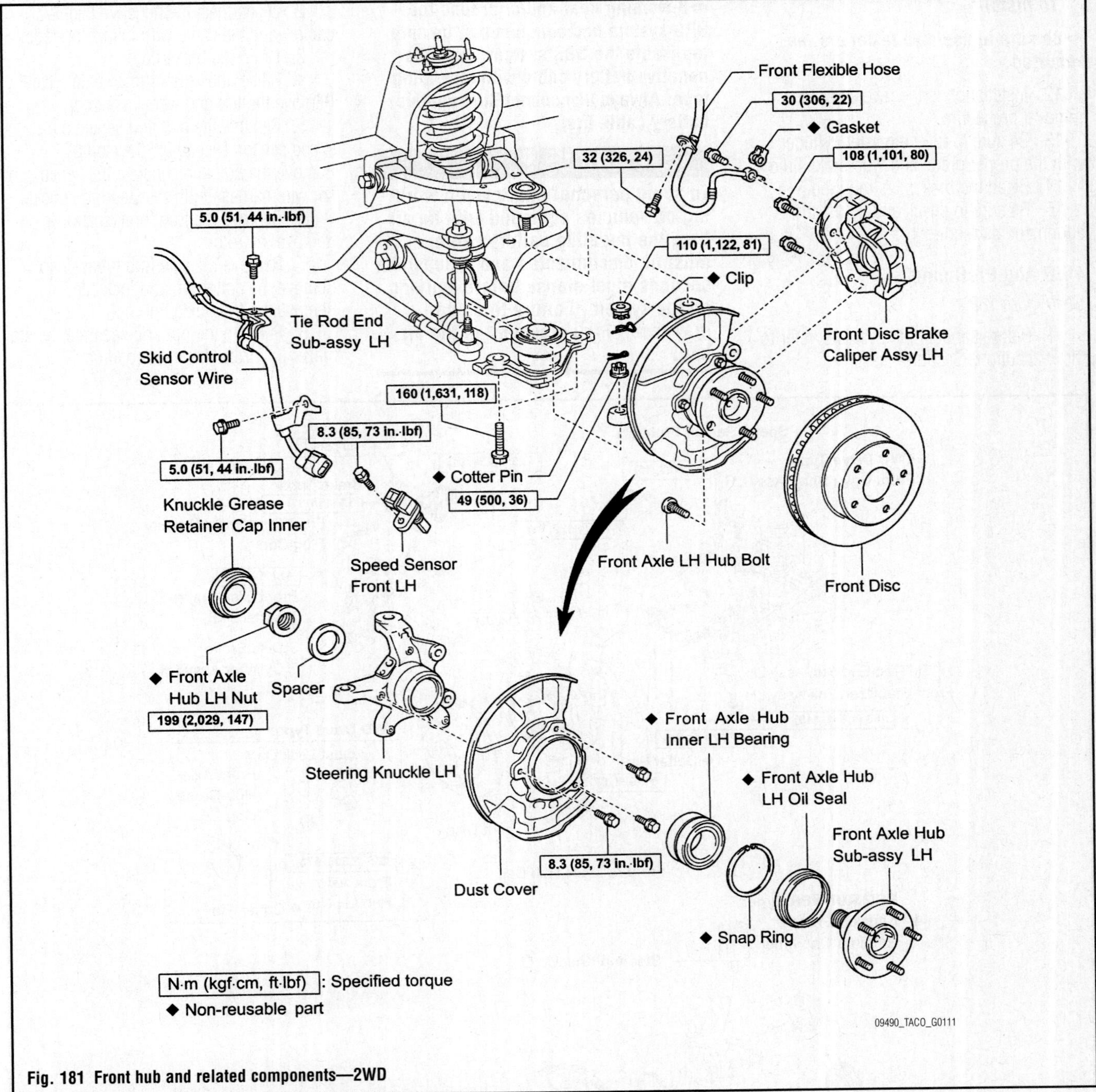

Fig. 181 Front hub and related components—2WD

➡If working near and/or around the SRS system and components, be sure to disable the SRS system. Tape the negative battery cable with insulating tape. Always disconnect the negative battery cable first.

⁂ CAUTION

To avoid personal injury when working on vehicles equipped with an air bag, the negative battery cable must be disconnected and at least 90 seconds must elapse before working on the system. Failure to do so may result in deployment of the air bag.

2. Disconnect the negative battery cable. Tape the cable with insulating tape.
3. Drain the brake fluid.
4. Raise and support the vehicle safely. Remove the tire and wheel assembly.
5. Remove the speed sensor connector. Remove the two bolts and separate the skid control sensor wire.
6. Remove the bolt and separate the front flexible hose.
7. Remove the front brake caliper. Remove the rotor.
8. Remove the cotter pin and nut. Using the proper tool separate the tie rod end from the steering knuckle.
9. Remove the two bolts and separate the front lower ball joint attachment from the front axle.
10. Remove the clip and nut. Using the proper tool, separate the upper ball joint from the steering knuckle.
11. Remove the front axle hub from the vehicle.

To install:

➡**Be sure to use new fasteners, as required.**

12. Installation is the reverse of the removal procedure.
13. Be sure to fill the master cylinder with the proper grade and type brake fluid.
14. Bleed the brakes, as required.
15. Be sure to check and adjust the alignment, as required.

4WD And PreRunner

See Figure 182.

1. Before servicing the vehicle, refer to the Precautions Section.

➡**If working near and/or around the SRS system and components, be sure to disable the SRS system. Tape the negative battery cable with insulating tape. Always disconnect the negative battery cable first.**

CAUTION

To avoid personal injury when working on vehicles equipped with an air bag, the negative battery cable must be disconnected and at least 90 seconds must elapse before working on the system. Failure to do so may result in deployment of the air bag.

2. Disconnect the negative battery cable. Tape the cable with insulating tape.
3. Drain the brake fluid.
4. Raise and support the vehicle safely. Remove the tire and wheel assembly.
5. Remove the bolt and separate the speed sensor. Disengage the clamps. Remove the bolt and separate the speed sensor wire harness from the steering knuckle.
6. Remove the front brake caliper. Remove the rotor.
7. Remove the axle hub grease cap. Remove the cotter pin and lock cap. Remove the front axle hub nut.
8. Remove the nut and separate the stabilizer link from the steering knuckle.

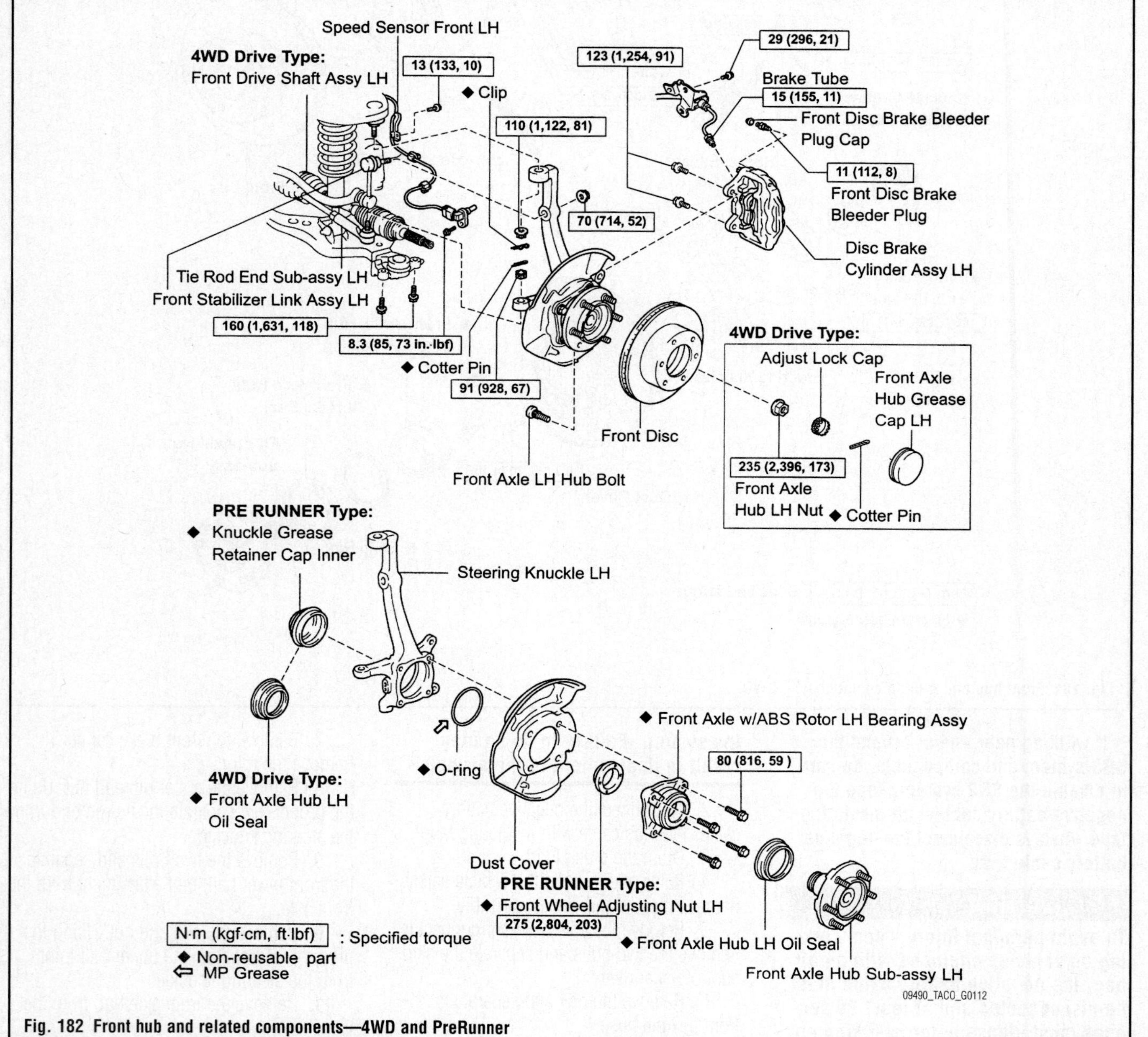

Fig. 182 Front hub and related components—4WD and PreRunner

9. Remove the cotter pin and nut. Using the proper tool separate the tie rod end from the steering knuckle.

10. Remove the two bolts and separate the front suspension lower arm from the front axle.

11. Properly support the front suspension lower arm with a jack. Remove the clip and nut.

12. Using the proper tool separate the steering knuckle from the front suspension upper control arm.

13. On 4WD, use a plastic hammer and separate the front axle hub from the front driveshaft.

14. Remove the front axle hub from the vehicle.

To install:

➡Be sure to use new fasteners, as required.

15. Installation is the reverse of the removal procedure.

16. Be sure to fill the master cylinder with the proper grade and type brake fluid.

17. Bleed the brakes, as required.

18. Be sure to check and adjust the alignment, as required.

ADJUSTMENT

No adjustment necessary.

SUSPENSION

REAR SUSPENSION

LEAF SPRING

REMOVAL & INSTALLATION

See Figures 183 and 184.

1. Before servicing the vehicle, refer to the Precautions Section.

➡If working near and/or around the SRS system and components, be sure to disable the SRS system. Tape the negative battery cable with insulating tape. Always disconnect the negative battery cable first.

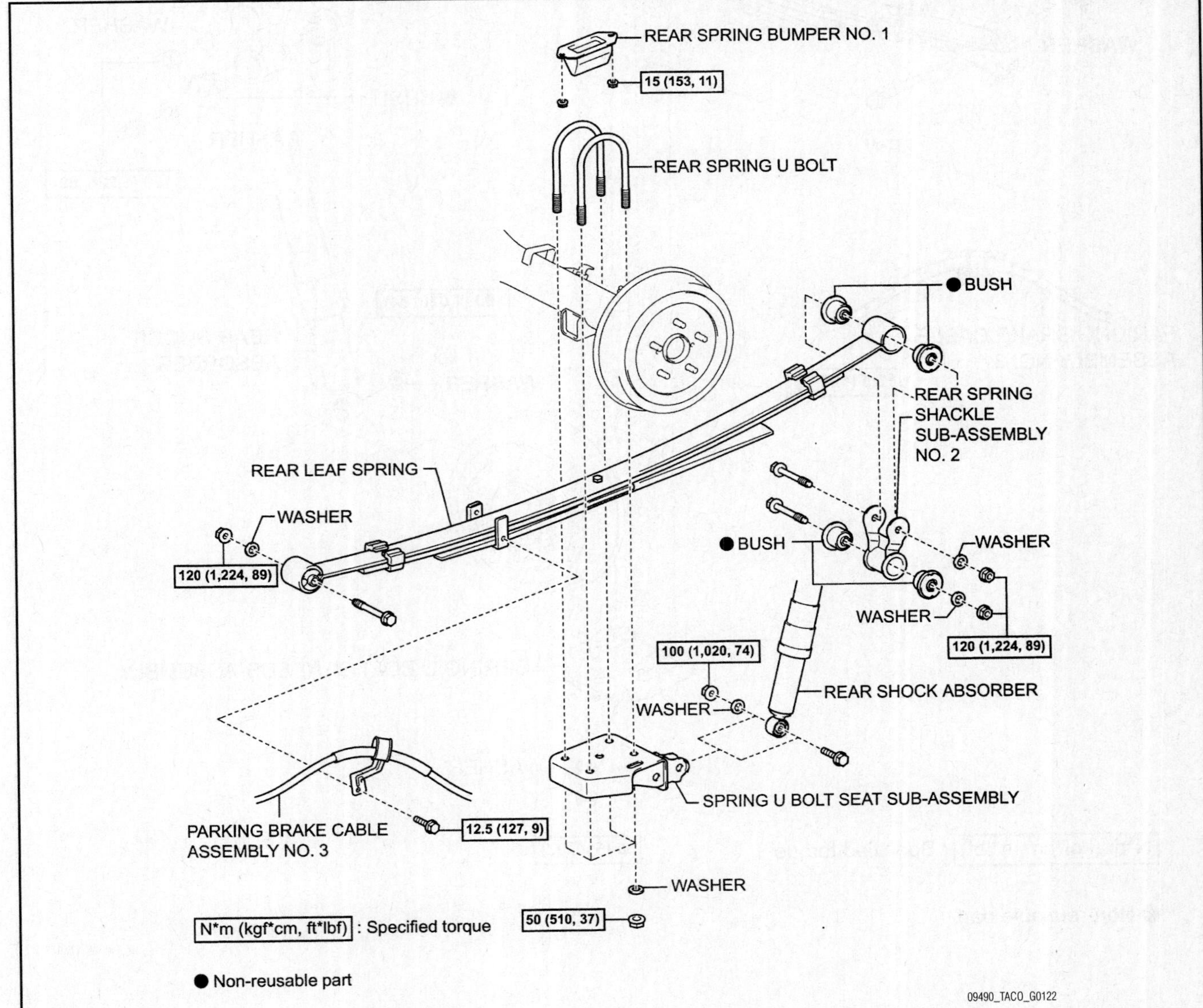

Fig. 183 Rear leaf spring and related components—2WD

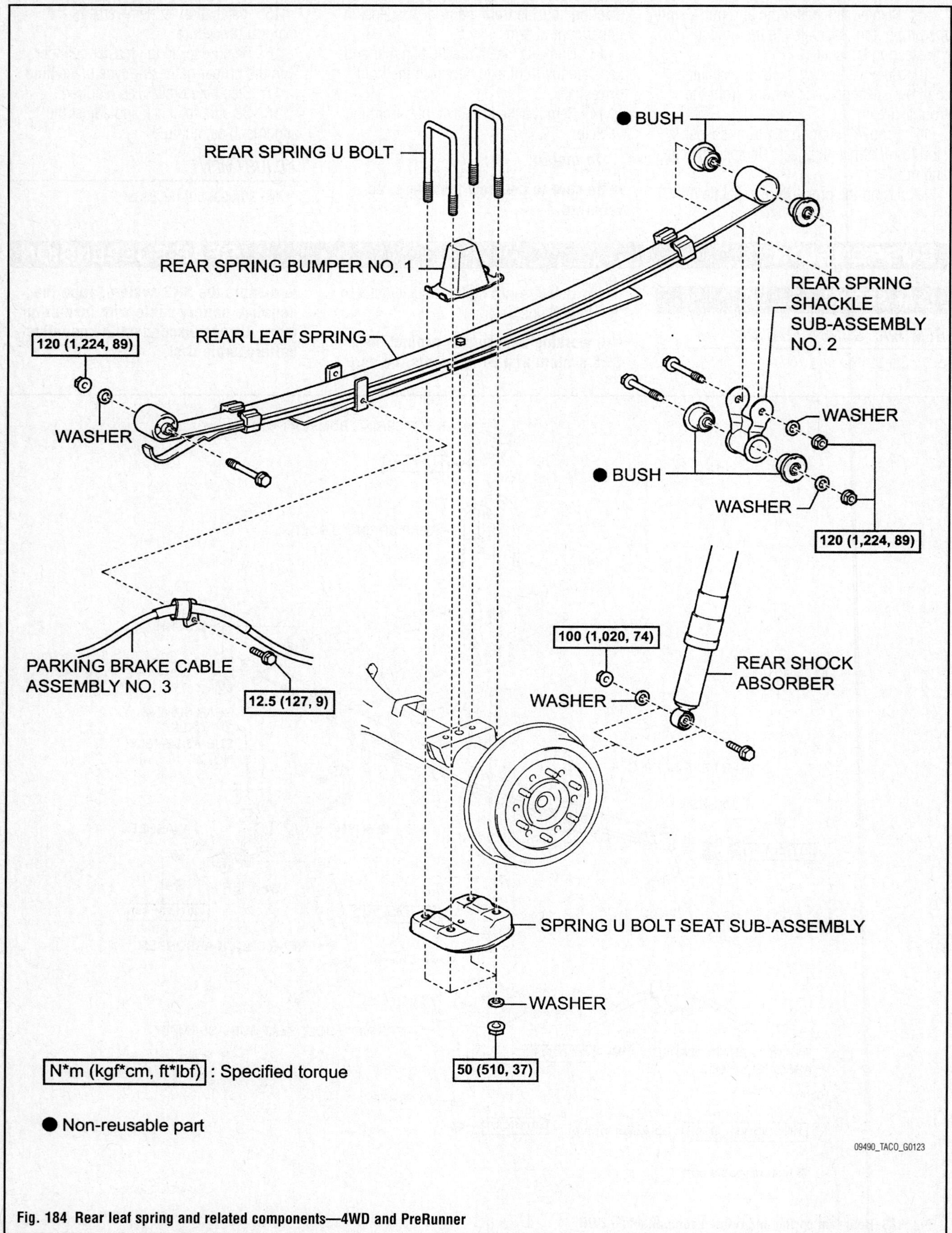

Fig. 184 Rear leaf spring and related components—4WD and PreRunner

⁂ CAUTION

To avoid personal injury when working on vehicles equipped with an air bag, the negative battery cable must be disconnected and at least 90 seconds must elapse before working on the system. Failure to do so may result in deployment of the air bag.

2. Disconnect the negative battery cable. Tape the cable with insulating tape.
3. Raise and support the vehicle safely.
4. Remove the tire and wheel assemblies.
5. On 4WD and PreRunner, remove the spare tire.
6. Support the rear axle housing. Remove the bolt nut and washer.
7. On 2WD separate the shock absorber from the rear spring seat.
8. On 4WD and PreRunner, separate the shock absorber from the rear axle housing.
9. Remove the bolt and then separate the parking brake cable.
10. On 2WD remove the two nuts and spring bumper. Remove the four nuts and four washers. Remove the spring seat and two U-bolts.
11. On 4WD and PreRunner, remove the four nuts and four washers. Remove the spring seat and two U-bolts. Remove the spring bumper.
12. Remove the nut, washer and through bolt. Remove the spring from the vehicle.

➡Be careful not to drop the spring when removing the through bolt.

To install:

➡Be sure to use new fasteners, as required.

➡Always fully tighten rubber bushings when the wheels are in full contact with the ground and the vehicle is at curb height. Bounce the vehicle up and down several times to stabilize the suspension, prior to final tightening of these components.

13. Installation is the reverse of the removal procedure.
14. Tighten the rear spring U-bolts to 37 ft. lbs. (50 Nm).Be sure that the lengths of all of the U-bolts under the spring seat are the same.

SHOCK ABSORBER

REMOVAL & INSTALLATION

See Figure 185.

1. Before servicing the vehicle, refer to the Precautions Section.

➡If working near and/or around the SRS system and components, be sure to disable the SRS system. Tape the negative battery cable with insulating tape. Always disconnect the negative battery cable first.

⁂ CAUTION

To avoid personal injury when working on vehicles equipped with an air bag, the negative battery cable must be disconnected and at least 90 seconds must elapse before working on the system. Failure to do so may result in deployment of the air bag.

2. Disconnect the negative battery cable. Tape the cable with insulating tape.
3. Raise and support the vehicle safely.
4. Remove the tire and wheel assembly.
5. Lower the floor jack to take tension off of the spring.
6. Remove or disconnect the following:
    - Shock absorber from the rear axle housing
    - Nut, retainers and the cushions holding the shock absorber to the frame
    - Shock absorber with the washers and bushings

To install:

➡Be sure to use new fasteners, as required.

➡Always fully tighten rubber bushings when the wheels are in full contact with the ground and the vehicle is at curb height. Bounce the vehicle up and down several times to stabilize the suspension, prior to final tightening of these components.

7. Install the shock absorber to the frame with the washers and bushings.
8. Tighten the shock absorber-to-upper frame nut on to 15 ft. lbs. (20 Nm).
9. Connect the shock absorber to the rear axle housing.
10. Tighten the bolt to 74 ft. lbs. (100 Nm).
11. Install the wheels.

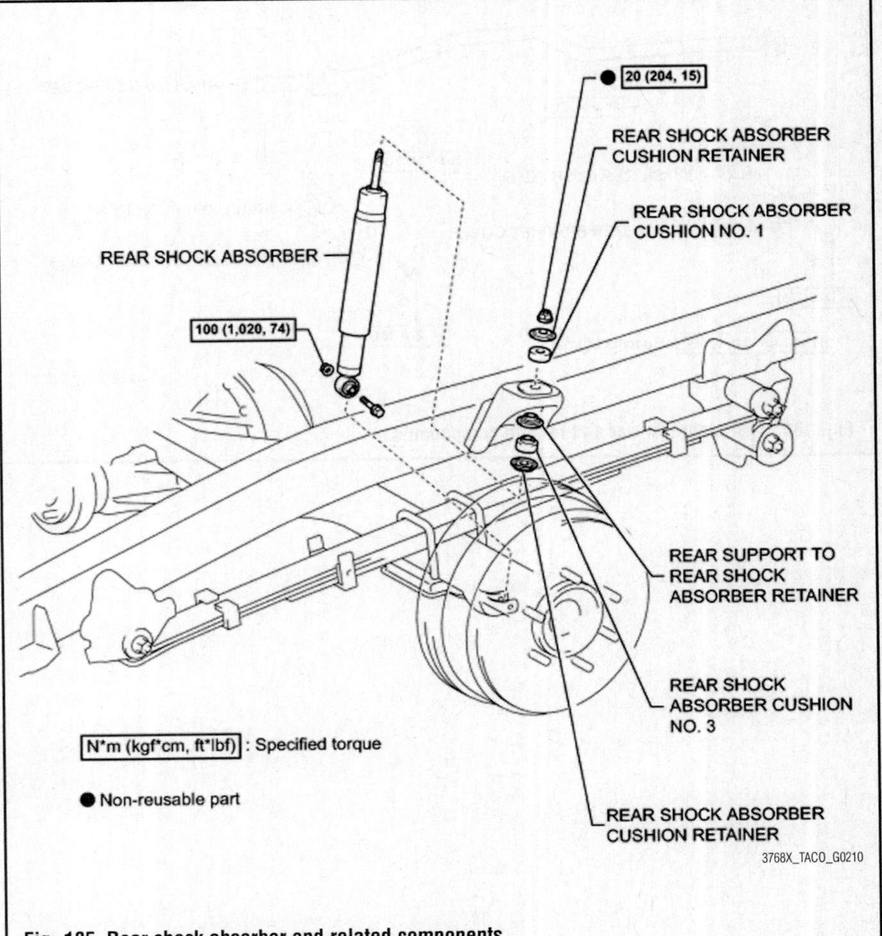

Fig. 185 Rear shock absorber and related components

STABILIZER BAR

REMOVAL & INSTALLATION

See Figures 186 and 187.

1. Before servicing the vehicle, refer to the Precautions Section.

➡If working near and/or around the SRS system and components, be sure to disable the SRS system. Tape the negative battery cable with insulating tape. Always disconnect the negative battery cable first.

CAUTION

To avoid personal injury when working on vehicles equipped with an air bag, the negative battery cable must be disconnected and at least 90 seconds must elapse before working on the system. Failure to do so may result in deployment of the air bag.

2. Disconnect the negative battery cable. Tape the cable with insulating tape.
3. Raise and support the vehicle safely.
4. Remove the tire and wheel assemblies.
5. On the left side, remove the two nuts and stabilizer bar link.
6. On the right side, remove the two nuts and stabilizer bar link.
7. Remove the four bolts and the stabilizer bar brackets and bushings.
8. Remove the stabilizer bar from the vehicle.

To install:

9. Installation is the reverse of the removal procedure.
10. Tighten the stabilizer bar and bushing bracket bolts to 20 ft. lbs. (27 Nm).
11. Tighten the stabilizer link assembly bolts to 51 ft. lbs. (69 Nm).
12. Be sure to install the bushing onto the outer side of the mark on the stabilizer bar.

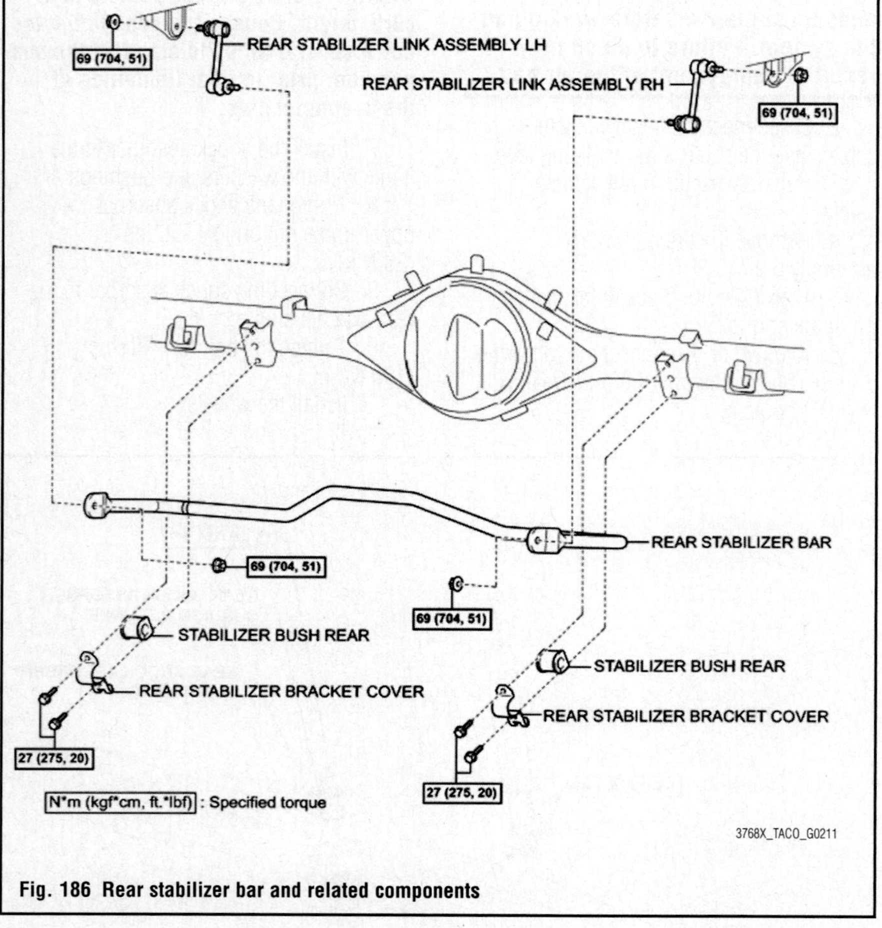

Fig. 186 Rear stabilizer bar and related components

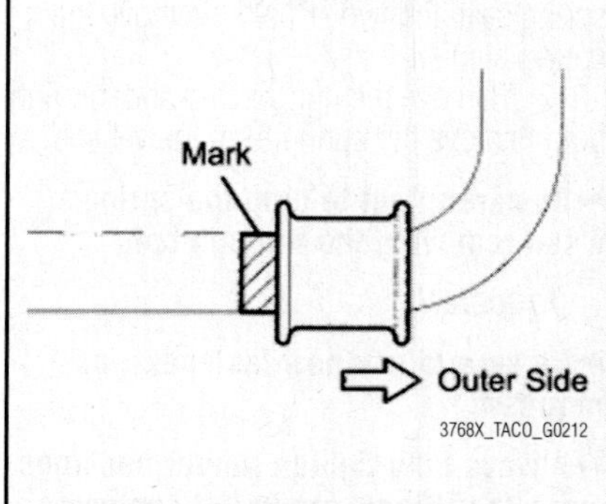

Fig. 187 Rear stabilizer bar bushing alignment

TOYOTA

18

Venza

SPECIFICATIONS AND MAINTENANCE CHARTS

ENGINE AND VEHICLE IDENTIFICATION

Engine							Model Year	
Code/VIN ①	Liters (cc)	Cu. In.	Cyl.	Fuel Sys.	Engine Type	Eng. Mfg.	Code ②	Year
1AR-FE	2.7 (2700)	163	4	SFI	DOHC	Toyota	9	2009
2GR-FE	3.5 (3498)	213	6	SFI	DOHC	Toyota	A	2010

SFI: Sequential Fuel Injection

DOHC: Double Overhead Camshaft

① 5th digit of the vehicle identification number (VIN)

② 10th digit of the Vehicle Identification Number (VIN)

3768X_VENZ_C0001

GENERAL ENGINE SPECIFICATIONS

Year	Model	Engine Displacement Liters	Engine Series Code/VIN	Net Horsepower @ rpm	Net Torque @ rpm (ft. lbs.)	Bore x Stroke (in.)	Com-pression Ratio	Oil Pressure @ rpm
2009	Venza	2.7	1AR-FE	182@5800	182@4200	3.54x4.13	10.0:1	①
		3.5	2GR-FE	269@6200	246@4700	3.70x3.27	10.8:1	②
2010	Venza	2.7	1AR-FE	182@5800	182@4200	3.54x4.13	10.0:1	①
		3.5	2GR-FE	269@6200	246@4700	3.70x3.27	10.8:1	②

NA: Not Available

① 4.3 psi or more at idle

② 11.6 psi or more at idle

3768X_VENZ_C0002

ENGINE TUNE-UP SPECIFICATIONS

Year	Engine Displacement Liters	Engine Code/VIN	Spark Plug Gap (in.)	Ignition Timing (deg.)	Fuel Pump (psi)	Idle Speed (rpm)		Valve Clearance	
						MT	AT	Intake	Exhaust
2009	2.7	1AR-FE	0.043	②	44-50	—	600-700	0.0075-0.0114	0.0150-0.0189
	3.5	2GR-FE	0.043	③	44-50	—	600-700	NA	NA
2010	2.7	1AR-FE	0.043	②	44-50	—	600-700	0.0075-0.0114	0.0150-0.0189
	3.5	2GR-FE	0.043	③	44-50	—	600-700	NA	NA

NOTE: The Vehicle Emission Control Information label often reflects specification changes made during production. The label figures must be used if they differ from those in this chart.

NA: Not Available

① 5-15 degrees BTDC when using intelligent tester tool.

8-12 degrees BTDC at idle. Connect SST when not using intelligent tester tool.

5-15 degrees BTDC at idle. Disconnect SST when not using intelligent tester tool.

② 8-12 degrees BTDC when using intelligent tester tool.

8-12 degrees BTDC at idle. Connect SST when not using intelligent tester tool.

5-15 degrees BTDC at idle. Disconnect SST when not using intelligent tester tool.

3768X_VENZ_C0003

CAPACITIES

Year	Model	Engine Displacement Liters	Engine Code/VIN	Engine Oil with Filter (qts.)	Transmission (pts.) 5-Spd	Transmission (pts.) Auto.*	Transfer Case (pts.)	Drive Axle Front (pts.)	Drive Axle Rear (pts.)	Fuel Tank (gal.)	Cooling System (qts.)
2009	Venza	2.7	1AR-FE	4.5	—	7.4	1.0	—	1.0	15.9	7.2
		3.5	2GR-FE	6.4	—	NA	1.0	—	1.0	15.9	①
2010	Venza	2.7	1AR-FE	4.5	—	7.4	▯	—	1.0	15.9	7.2
		3.5	2GR-FE	6.4	—	NA	1.0	—	1.0	15.9	①

*After draining, add the following amounts, then fill to the cold full line.

NA: Not Available

① STD: 10.4
TWG: 10.6

3768X_VENZ_C0004

FLUID SPECIFICATIONS

Year	Model	Engine Displacement Liters (VIN)	Engine Oil	Auto. Trans.	Drive Axle	Power Steering Fluid	Brake Master Cylinder
2009	Venza	2.7	①	ATF World Standard	②	NA	DOT 3
	Venza	3.5	5W-30	ATF World Standard	③	NA	DOT 3
2010	Venza	2.7	①	ATF World Standard	②	NA	DOT 3
	Venza	3.5	5W-30	ATF World Standard	③	NA	DOT 3

DOT: Department Of Transportation

NA: Not Available

Note: If specification disagrees with specification in owners manual, use specification in owners manaual

① 0W-20 above 40 degrees F. 5W-20 below 40 degrees F

② API GL-5 SAE 90 above 0 degrees F. API GL-5 SAE 80W-90 below 0 degrees

③ API GL-5 SAE 85W-90 above 0 degrees F. API GL-5 SAE 80W-90 below 0 degrees

3768X_VENZ_C0014

VALVE SPECIFICATIONS

Year	Engine Displacement Liters	Engine Code/VIN	Seat Angle (deg.)	Face Angle (deg.)	Spring Test Pressure (lbs. @ in.)	Spring Installed Height (in.)	Stem-to-Guide Clearance (in.) Intake	Stem-to-Guide Clearance (in.) Exhaust	Stem Diameter (in.) Intake	Stem Diameter (in.) Exhaust
2009	2.7	1AR-FE	45	44.5	NA	1.8670	0.0010-0.0024	0.0012-0.0026	0.2154-0.2159	0.2152-0.2158
	3.5	2GR-FE	45	44.5	NA	1.7898	0.0010-0.0024	0.0012-0.0026	0.2154-0.2159	0.2151-0.2157
2010	2.7	1AR-FE	45	44.5	NA	1.8670	0.0010-0.0024	0.0012-0.0026	0.2154-0.2159	0.2152-0.2158
	3.5	2GR-FE	45	44.5	NA	1.7898	0.0010-0.0024	0.0012-0.0026	0.2154-0.2159	0.2151-0.2157

NA: Not Available

3768X_VENZ_C0005

CAMSHAFT SPECIFICATIONS

All measurements in inches unless noted

Year	Engine Displacement Liters	Engine Code/VIN	Journal Dia.	Brg. Oil Clearance	Shaft End-play ①	Circle Runout	Lobe Height Intake	Lobe Height Exhaust
2009	2.7	1AR-FE	②	NA	NA	NA	1.8624-1.8664	1.8104-1.8143
	3.5	2GR-FE	③	④	0.0031-0.0051	0.0016	1.7447-1.7487	1.7426-1.7465
2010	2.7	1AR-FE	②	NA	NA	NA	1.8624-1.8664	1.8104-1.8143
	3.5	2GR-FE	③	④	0.0031-0.0051	0.0016	1.7447-1.7487	1.7426-1.7465

① Thrust clearance

② Mark 1, 2 and 3: 1.4162-1.4167

③ No1: 1.4152-1.4157

All others: 1.0220-1.0226

④ No. 1: 0.0016-0.0031

All others: 0.0010-0.0024

3768X_VENZ_C0006

CRANKSHAFT AND CONNECTING ROD SPECIFICATIONS

All measurements are given in inches.

Year	Engine Displ. Liters	Engine Code/VIN	Crankshaft Main Brg. Journal Dia.	Crankshaft Main Brg. Oil Clearance	Crankshaft Shaft End-play	Crankshaft Thrust on No.	Connecting Rod Journal Diameter	Connecting Rod Oil Clearance	Connecting Rod Side Clearance
2009	2.7	1AR-FE	2.1649-2.1652	0.0006-0.0015	0.0016-0.0095	3	1.8894-1.8898	0.0012-0.0025	0.0063-0.0202
	3.5	2GR-FE	2.4011-2.4016	0.0010-0.0019	0.0016-0.0095	2	2.0863-2.0866	0.0018-0.0026	0.0059-0.0157
2010	2.7	1AR-FE	2.1649-2.1652	0.0006-0.0015	0.0016-0.0095	3	1.8894-1.8898	0.0012-0.0025	0.0063-0.02
	3.5	2GR-FE	2.4011-2.4016	0.0010-0.0019	0.0016-0.0095	2	2.0863-2.0866	0.0018-0.0026	0.0059-0.0157

3768X_VENZ_C0007

PISTON AND RING SPECIFICATIONS

All measurements are given in inches.

Year	Engine Displ. Liters	Engine Code/VIN	Piston Clearance	Ring Gap Top Comp.	Ring Gap Bottom Comp.	Ring Gap Oil Control	Ring Side Clearance Top Comp.	Ring Side Clearance Bottom Comp.	Ring Side Clearance Oil Control
2009	2.7	1AR-FE	NA	0.0086-0.0106	0.0146-0.0165	0.0039-0.0079	0.0008-0.0028	0.0008-0.0024	0.0008-0.0028
	3.5	2GR-FE	NA	0.0098-0.0138	0.0197-0.0236	0.0039-0.0157	0.0008-0.0028	0.0008-0.0024	0.0028-0.0059
2010	2.7	1AR-FE	NA	0.0087-0.0106	0.0146-0.0165	0.0040-0.0079	0.0008-0.0028	0.0008-0.0024	0.0008-0.0028
	3.5	2GR-FE	NA	0.0098-0.0138	0.0197-0.0236	0.0039-0.0157	0.0008-0.0028	0.0008-0.0024	0.0028-0.0059

3768X_VENZ_C0008

TORQUE SPECIFICATIONS

All readings in ft. lbs.

Year	Engine Displacement Liters	Engine Code/VIN	Cylinder Head Bolts	Main Bearing Bolts	Rod Bearing Bolts	Crankshaft Damper Bolts	Flywheel Bolts	Manifold Intake	Manifold Exhaust	Spark Plugs	Oil Pan Drain Plug
2009	2.7	1AR-FE	①	②	③	133	72	22	27	18	30
	3.5	2GR-FE	④	⑤	⑥	184	132	15	15	13	30
2010	2.7	1AR-FE	①	②	③	133	72	22	27	18	30
	3.5	2GR-FE	④	⑤	⑥	184	132	15	15	13	30

① Step 1: 27 ft. lbs.
Step 2: plus 90 degrees

② Step 1: 30 ft. lbs.
Step 2: plus 90 degrees

③ Step 1: 30 ft. lbs.
Step 2: plus 90 degrees

④ Step 1: 27 ft. lbs.
Step 2: plus 90 degrees
Step 3: plus 90 degrees
Bolt should be 22 ft. lbs. on Bank 2

⑤ Step 1: 45 ft. lbs.
Step 2: plus 90 degrees
Main bearing cap bolt: 38 ft. lbs.

⑥ Step 1: 45 ft. lbs.
Step 2: plus 90 degrees
Step 3: 8 side bolts to 38 ft. lbs.

3768X_VENZ_C0009

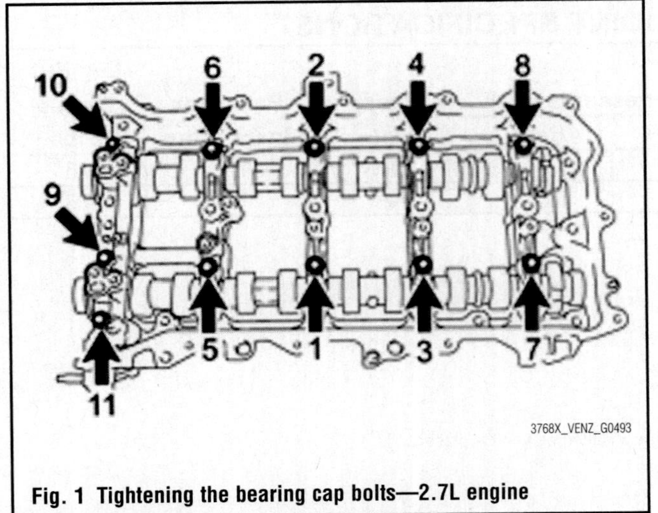

Fig. 1 Tightening the bearing cap bolts—2.7L engine

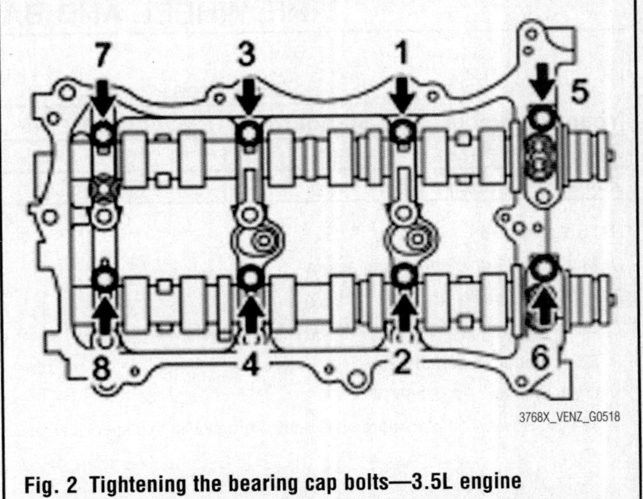

Fig. 2 Tightening the bearing cap bolts—3.5L engine

WHEEL ALIGNMENT

Year	Model	Engine		Caster Range (+/-Deg.)	Caster Preferred Setting (Deg.)	Camber Range (+/-Deg.)	Camber Preferred Setting (Deg.)	Toe-in (in.)	Steering Axis Inclination (Deg.)
2009	Venza 2WD	1AR-FE	F	0.75	+2.65	0.75	-0.58	0.043+/-0.079	10.70+/-0.75
			R	NA	NA	0.75	-1.22	0.150+/-0.079	NA
	Venza AWD	1AR-FE	F	0.75	+2.68	0.75	-0.60	0.043+/-0.079	10.70+/-0.75
			R	NA	NA	0.75	-0.78	0.142+/-0.079	NA
	Venza 2WD	2GR-FE	F	0.75	+2.63	0.75	-0.60	0.043+/-0.079	10.72+/-0.75
			R	NA	NA	0.75	-1.23	0.150+/-0.079	NA
	Venza AWD	2GR-FE	F	0.75	+2.67	0.75	-0.60	0.043+/-0.079	10.72+/-0.75
			R	NA	NA	0.75	-0.80	0.142+/-0.079	NA
2010	Venza 2WD	1AR-FE	F	0.75	+2.65	0.75	-0.58	0.043+/-0.079	10.70+/-0.75
			R	NA	NA	0.75	-1.22	0.150+/-0.079	NA
	Venza AWD	1AR-FE	F	0.75	+2.68	0.75	-0.60	0.043+/-0.079	10.70+/-0.75
			R	NA	NA	0.75	-0.78	0.142+/-0.079	NA
	Venza 2WD	2GR-FE	F	0.75	+2.63	0.75	-0.60	0.043+/-0.079	10.72+/-0.75
			R	NA	NA	0.75	-1.23	0.150+/-0.079	NA
	Venza AWD	2GR-FE	F	0.75	+2.67	0.75	-0.60	0.043+/-0.079	10.72+/-0.75
			R	NA	NA	0.75	-0.80	0.142+/-0.079	NA

3768X_VENZ_C0010

TIRE, WHEEL AND BALL JOINT SPECIFICATIONS

Year	Model	OEM Tires		Tire Pressures (psi)		Wheel	Ball Joint	Lug Nut
		Standard	Optional	Front	Rear	Size	Inspection	Torque (ft. lbs.)
2009	Venza	①	①	②	②	NA	NA	76
2010	Venza	①	①	②	②	NA	NA	76

NA: Not Available

OEM: Original Equipment Manufacturer

PSI: Pounds Per Square Inch

① Base model: P215/70R16, P225/65R17. Optional P235/55R18

Sport model: P235/55R18

Limited model: P225/65R17

② 32 PSI. However if placard on vehicle disagrees with this specification, use the specification on vehicle placard.

3768X_VENZ_C0011

BRAKE SPECIFICATIONS

All measurements in inches unless noted

Year	Model		Brake Disc			Brake Drum Diameter			Minimum	Brake Caliper	
			Original Thickness	Minimum Thickness	Maximum Runout	Original Inside Diameter	Max. Wear Limit	Maximum Machine Diameter	Lining Thickness	Bracket Bolts (ft. lbs.)	Mounting Bolts (ft. lbs.)
2009	Venza	F	1.100	1.020	0.0020	—	—	—	0.039	77	24
		R	0.394	0.354	0.0059	—	—	—	0.039	58	24
2010	Venza	F	1.100	1.020	0.0020	—	—	—	0.039	77	24
		R	0.394	0.354	0.0059	—	—	—	0.039	58	24

F: Front

R: Rear

3768X_VENZ_C0012

SCHEDULED MAINTENANCE INTERVALS

TOYOTA—Venza

TO BE SERVICED	TYPE OF SERVICE	VEHICLE MILEAGE INTERVAL (x1000)																		
		5	10	15	20	25	30	35	40	45	50	55	60	65	70	75	80	85	90	95
Automatic transmission and differential fluid	S/I			✓			✓			✓			✓			✓			✓	
Ball joints and boots	S/I			✓			✓			✓			✓			✓			✓	
Brake system	S/I	✓	✓	✓	✓	✓	✓	✓	✓	✓	✓	✓	✓	✓	✓	✓	✓	✓	✓	✓
Cabin air filter	S/I			✓			✓			✓			✓			✓			✓	
Charcoal canister	S/I												✓							
Drive belts	S/I						✓						✓						✓	
Driveshaft bushing	L						✓						✓						✓	
Engine coolant	R			✓			✓			✓			✓			✓			✓	
Engine oil & filter	R	✓	✓	✓	✓	✓	✓	✓	✓	✓	✓	✓	✓	✓	✓	✓	✓	✓	✓	✓
Exhaust pipes & mounts	S/I			✓			✓			✓			✓			✓			✓	
Fuel tank cap gasket	S/I						✓						✓						✓	
Halfshaft boots & flange bolts	S/I			✓			✓			✓			✓			✓			✓	
Limited slip differential fluid	R						✓						✓						✓	
Manual transmission and differential fluid	S/I						✓						✓						✓	
Platinum spark plugs	R												✓							
Propeller shaft bolts	S/I			✓			✓			✓			✓			✓			✓	
Rack and pinion assy	S/I			✓			✓			✓			✓			✓			✓	
Tires (rotate)	S/I	✓	✓	✓	✓	✓	✓	✓	✓	✓	✓	✓	✓	✓	✓	✓	✓	✓	✓	✓
Transfer case and differential fluid	S/I			✓			✓			✓			✓			✓			✓	
Valves	S/I												✓							

R: Replace S/I: Service or Inspect L: Lubricate

FREQUENT OPERATION MAINTENANCE (SEVERE SERVICE)

If a vehicle is operated under any of the following conditions it is considered severe service:

- Towing a trailer or using a camper or car-top carrier.

- Repeated short trips of less than 5 miles in temperatures below freezing.

- Excessive idling or low-speed driving for long distances as in heavy commercial use, such as delivery, taxi or police cars.

- Operating on rough, muddy or salt-covered roads.

- Operating on unpaved or dusty roads.

Oil filter: service or inspect every 5000 miles or 4 months, whichever occurs first.

Brake linings and discs or drums: service or inspect every 5000 miles or 4 months, whichever occurs first.

Steering linkage: service or inspect every 5000 miles or 4 months, whichever occurs first.

Ball joints and boots: service or inspect every 5000 miles or 4 months, whichever occurs first.

Brake discs & pads (front): service or inspect every 6000 miles.

Halfshaft boots: service or inspect every 5000 miles or 4 months. Retighten the flange bolts, whichever occurs first.

Body chassis bolts and nuts: service or inspect every 5000 miles or 4 months, whichever occurs first.

Transmission and differential fluid: replace every 15,000 miles or 12 months, whichever occurs first.

Transfer case and differential fluid: replace every 15,000 miles or 12 months, whichever occurs first.

Timing belt: replace every 60,000 miles or 48 months, whichever occurs first.

3768X_VENZ_C0013

BRAKES — INFORMATION AND PRECAUTIONS

ANTI-LOCK SYSTEMS

• Certain components within the ABS system are not intended to be serviced or repaired individually.

• Do not use rubber hoses or other parts not specifically specified for and ABS system. When using repair kits, replace all parts included in the kit. Partial or incorrect repair may lead to functional problems and require the replacement of components.

• Lubricate rubber parts with clean, fresh brake fluid to ease assembly. Do not use shop air to clean parts; damage to rubber components may result.

• Use only DOT 3 brake fluid from an unopened container.

• If any hydraulic component or line is removed or replaced, it may be necessary to bleed the entire system.

• A clean repair area is essential. Always clean the reservoir and cap thoroughly before removing the cap. The slightest amount of dirt in the fluid may plug an orifice and impair the system function. Perform repairs after components have been thoroughly cleaned; use only denatured alcohol to clean components. Do not allow ABS components to come into contact with any substance containing mineral oil; this includes used shop rags.

• The Anti-Lock control unit is a microprocessor similar to other computer units in the vehicle. Ensure that the ignition switch is **OFF** before removing or installing controller harnesses. Avoid static electricity discharge at or near the controller.

• If any arc welding is to be done on the vehicle, the control unit should be unplugged before welding operations begin.

DISC AND DRUM SYSTEMS

CAUTION

Dust and dirt accumulating on brake parts during normal use may contain asbestos fibers from production or aftermarket brake linings. Breathing excessive concentrations of asbestos fibers can cause serious bodily harm. Exercise care when servicing brake parts. Do not sand or grind brake lining unless equipment used is designed to contain the dust residue. Do not clean brake parts with compressed air or by dry brushing. Cleaning should be done by dampening the brake components with a fine mist of water, then wiping the brake components clean with a dampened cloth. Dispose of cloth and all residue containing asbestos fibers in an impermeable container with the appropriate label. Follow practices prescribed by the Occupational Safety and Health Administration (OSHA) and the Environmental Protection Agency (EPA) for the handling, processing, and disposing of dust or debris that may contain asbestos fibers.

BRAKES — BLEEDING THE BRAKE SYSTEM

BLEEDING PROCEDURE

See Figure 3.

➡Do not allow brake fluid to adhere to any painted surface such as the vehicle body. If brake fluid leaks onto any painted surface, immediately wash it off.

➡Before bleeding the brake system, confirm that the reservoir located above the master cylinder assembly is filled with brake fluid.

➡If any component of the brake system is removed and reinstalled, or if air in the brake lines is suspected, bleed the brake system.

1. Fill the reservoir with brake fluid.
2. Bleed the brake master cylinder.

➡To prevent brake fluid from damaging painted surfaces cover any surrounding parts with a piece of cloth.

➡If the master cylinder is reinstalled or runs out of brake fluid, bleed the master cylinder.

a. Using a union nut wrench, disconnect the 2 brake tubes from the brake master cylinder assembly.

b. Slowly depress the brake pedal and hold it down.

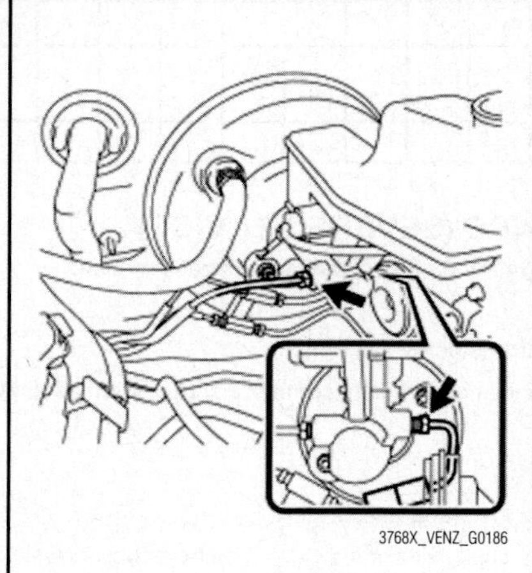

Fig. 3 Disconnecting the 2 brake tubes

c. Cover the 2 tube holes with your fingers and release the brake pedal.

d. Uncover the holes, slowly depress the brake pedal and hold it down. While holding down the brake pedal, cover the tube holes again. Repeat this step 3 or 4 times.

e. Using a union nut wrench, connect the 2 brake tubes to the brake master cylinder assembly.

➡Do not bend or damage the brake lines.

➡Do not allow brake line to twist and interfere with other parts or body during tightening.

➡Do not allow any foreign matter such as dirt or dust to enter the brake line.

➡Use the formula to calculate special torque values for situations where union nut wrench is combined with a torque wrench

3. Bleed the brake line.

➡Bleed the brake line of the wheel farthest from the master cylinder first.

a. Connect a vinyl tube to the bleeder plug.

b. Depress the brake pedal several times, and while holding down the brake pedal, loosen the bleeder plug.

c. When fluid stops coming out, tighten the bleeder plug and release the brake pedal.

d. Repeat previous 2 steps until all the air in the brake fluid is completely bled out and the new brake fluid comes out.

e. Tighten the bleeder plug completely. Tighten to 10 ft. lbs. (13 Nm).

f. Repeat the above steps to replace the brake fluid of the brake lines for each wheel.

4. Inspect for brake fluid leaks.
5. Inspect the fluid level in the reservoir.

WHEEL SPEED SENSORS

REMOVAL & INSTALLATION

Front

See Figures 4 and 5.

➡Use the same procedure for the LH side and RH side.

➡The following procedure is for the LH side.

➡If the sensor rotor needs to be replaced, replace it together with the front driveshaft assembly.

1. Disconnect the cable from the negative battery terminal.

➡When disconnecting the cable, some systems need to be initialized after the cable is reconnected.

2. Remove the front wheel.
3. Remove the front fender outside moulding.
4. Remove the front fender liner.
5. Remove the front speed sensor.
    a. Disconnect the front speed sensor connector and remove the 2 clamps.
    b. Remove the bolt and No. 2 sensor clamp from the body.
    c. Remove the bolt, No. 1 sensor clamp and flexible hose together from the shock absorber assembly.
    d. Remove the bolt, resin clamp and front speed sensor.

➡Prevent foreign matter from attaching to the front speed sensor tip. Clean the speed sensor installation hole and the contact surfaces every time the front speed sensor is removed.

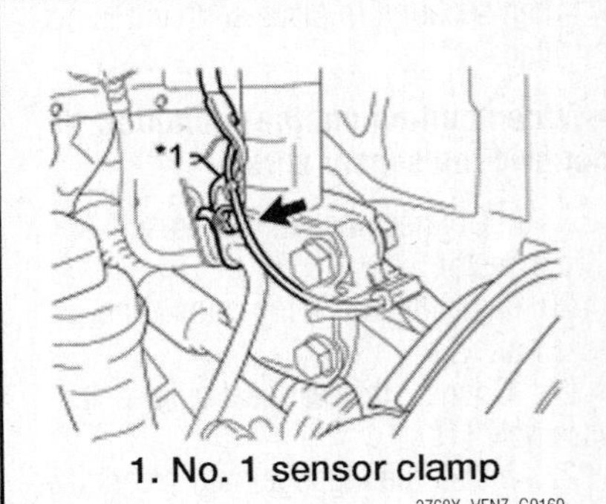

Fig. 4 Removing the No. 1 sensor clamp and hose

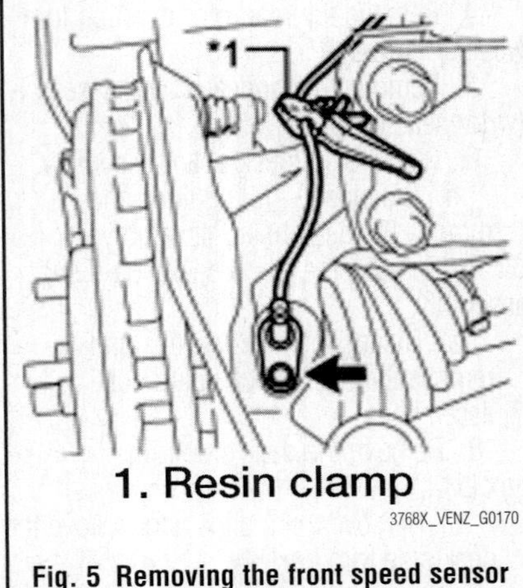

Fig. 5 Removing the front speed sensor

To install:

6. Install the front speed sensor.
    a. Install the resin clamp and front speed senor with the bolt. Tighten to 71 inch lbs. (8 Nm).

➡Prevent foreign matter from attaching to the front speed sensor tip.

➡Firmly insert the front speed sensor body into the knuckle before tightening the bolt.

➡After installing the front speed sensor to the knuckle, make sure that there is no clearance between the front speed sensor stay and knuckle. Also make sure that no foreign matter is stuck between the parts.

➡Before installing the clamp, firmly insert the points of the clamp into the installation holes.

    b. Temporarily install the No. 1 sensor clamp.

➡Be sure to insert the No. 1 sensor clamp claw into the stopper hole while installing the No. 1 sensor clamp.

    c. Install the front brake flexible hose and the No. 1 sensor clamp together to the shock absorber with the bolt. Tighten to 14 ft. lbs. (19 Nm).

➡Do not twist the wire harness for the front speed sensor when installing it.

➡A bolt tightens the brake flexible hose and front speed sensor together. Make sure that the flexible hose is positioned over the front speed sensor.

    d. Install the No. 2 sensor clamp to the body with the bolt. Tighten to 44 inch lbs. (5 Nm).
    e. Install the 2 clamps and connect the front speed sensor connector.
7. Install the front fender liner.
8. Install the front fender outside moulding.
9. Install the front wheel and tighten to 76 ft. lbs. (103 Nm).

➡When disconnecting the cable, some systems need to be initialized after the cable is reconnected.

10. Check for speed sensor signal.

Rear

2WD Vehicles

See Figure 6.

➡Use the same procedure for the RH side and LH side.

➡The following procedure is for the LH side.

➡The rear speed sensor is a component of the rear axle hub and bearing assembly. If the sensor malfunctions, replace the rear axle hub and bearing assembly.

➡If the sensor rotor needs to be replaced, replace it together with the rear axle hub and bearing assembly.

1. Disconnect the cable from the negative battery terminal.

➡When disconnecting the cable, some systems need to be initialized after the cable is reconnected.

2. Remove the rear wheel.
3. Separate the rear speed sensor wire.
    a. Using a screwdriver, disconnect the connector from the rear speed sensor.

➡Be careful not to damage the rear speed sensor.

4. Separate the rear flexible hose.
5. Separate the rear disc brake caliper assembly.
6. Remove the rear disc.
7. Remove the rear axle hub and bearing assembly.

➡The rear speed sensor is a component of the rear axle hub and bearing assembly. If the sensor malfunctions, replace the rear axle hub and bearing assembly.

➡If the sensor rotor needs to be replaced, replace it together with the rear axle hub and bearing assembly.

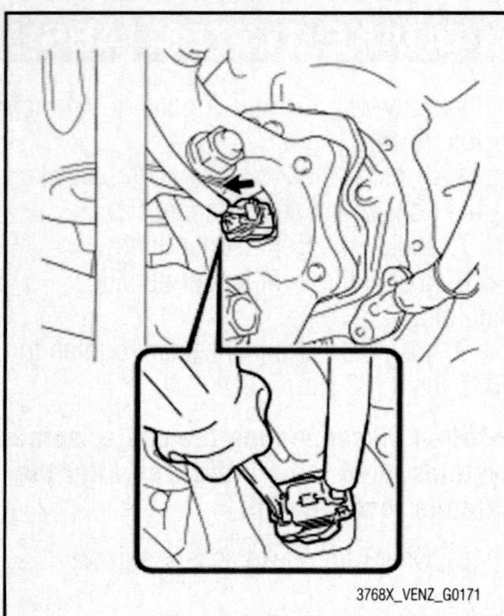
Fig. 6 Separating the rear speed sensor wire (2WD)

To install:

8. Install the rear axle hub and bearing assembly.
9. Inspect the rear axle hub bearing looseness.
10. Inspect the rear axle hub runout.
11. Install the rear disc.
12. Install the rear disc brake caliper assembly.
13. Install the rear flexible hose.
14. Install the rear speed sensor wire.
    a. Connect the connector to the rear speed sensor.
15. Install the rear wheel. Tighten to 76 ft. lbs. (103 Nm).
16. Connect the cable to the negative battery terminal.

➡When disconnecting the cable, some systems need to be initialized after the cable is reconnected.

17. Inspect and adjust the rear wheel alignment.
18. Check for the speed sensor signal.

AWD Vehicles

See Figure 7.

➡Use the same procedure for the LH side and RH side.

➡The following procedure is for the LH side.

➡If the sensor rotor needs to be replaced, replace it together with the rear driveshaft assembly.

1. Disconnect the cable from the negative battery terminal.

➡When disconnecting the cable, some systems need to be initialized after the cable is reconnected.

2. Remove the rear wheel.
3. Remove the rear door scuff plate LH.
4. Remove the rear door opening trim weather-strip LH.
5. Remove the tonneau cover assembly (w/tonneau cover).
6. Remove the deck board assembly.
    a. Disengage the 2 guides and remove the deck board assembly.
7. Remove the No. 3 deck board sub assembly.
    a. Disengage the 2 guides and remove the No. 3 deck board sub-assembly.
8. Remove the deck side trim box LH.
    a. Remove the 3 clips and remove the deck side trim box LH.
9. Remove the No. 2 deck board sub assembly.
    a. Disengage the 2 guides and remove the No. 2 deck board sub-assembly.
10. Remove the deck side trim box RH.
    a. Remove the 4 clips and remove the deck side trim box RH.
11. Remove the No. 1 deck board.
    a. Disengage the 6 clips and remove the No. 1 deck board.
12. Remove the rear seat sub floor panel assembly.
    a. Disengage the 2 claws, 2 guides and 5 clips, and remove the rear seat sub floor panel assembly.
13. Remove the rear floor finish plate.
    a. Disengage the 2 claws and 4 clips, and remove the rear floor finish plate.
14. Remove the reclining remote control bezel LH.
    a. Using a screwdriver, disengage the 3 claws and remove the reclining remote control bezel LH.

➡Tape the screwdriver tip before use.

15. Remove the luggage hold belt striker assembly.
    a. Remove the 2 bolts.
16. Disconnect the rear seat outer belt assembly LH.
17. Remove the deck trim side panel assembly LH.
    a. Remove the bolt and clip.
    b. Disengage the 7 claws and 5 clips.
    c. Disconnect the connector and remove the deck trim side panel assembly LH.
18. Remove the rear speed sensor.
    a. Disconnect the rear speed sensor connector.

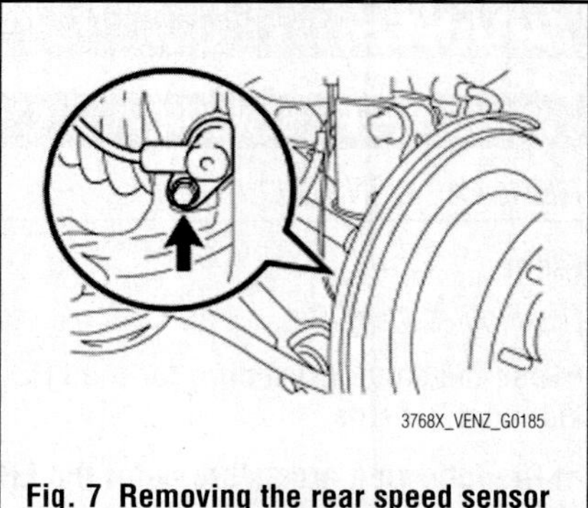
Fig. 7 Removing the rear speed sensor

    b. Disconnect the grommet of the rear speed sensor wire from the hole of the wheel house.
    c. Remove the 2 bolts, No. 1 clamp and No. 2 clamp from the body and absorber.
    d. Remove the bolt and rear speed sensor from the carrier.

➡Keep the sensor tip and rear speed sensor installation hole free from foreign matter.

To install:

19. Install the rear speed sensor with the bolt. Tighten to 71 inch lbs. (8 Nm).

➡Keep the rear speed sensor tip and sensor installation hole free from foreign matter.

    a. Install the No. 1 clamp and No. 2 clamp with the 2 bolts. Tighten to 44 inch lbs. (5 Nm).

➡Do not twist the rear speed sensor wire when installing the clamps.

    b. Insert the connector and grommet to the inside of the vehicle through the passage hole in the wheel well.

➡Make sure that the grommet's band clamp remains on the outside of the vehicle.

    c. Hold the grommet and pull it from the inside to the outside of the vehicle. Then secure it in place so that it is not tilted.

➡When pulling out the grommet, do not grip the sensor wire.

    d. Connect the rear speed sensor connector.
20. Install the deck trim side panel assembly LH.
21. Connect the rear seat outer belt assembly LH.
22. Install the luggage hold belt striker assembly.
23. Install the reclining remote control bezel LH.

24. Install the rear floor finish plate.
25. Install the rear seat sub floor panel assembly.
26. Install the No. 1 deck board.
27. Install the deck side trim box RH.
28. Install the No. 2 deck board sub assembly.
29. Install the deck side trim box LH.
30. Install the No. 3 deck board sub assembly.
31. Install the deck board assembly.
32. Install the tonneau cover assembly (w/tonneau).
33. Install the rear door opening trim weather-strip LH.
34. Install the rear door scuff plate LH.
35. Connect the cable to the negative battery terminal.

➡**When disconnecting the cable, some systems need to be initialized after the cable is reconnected.**

36. Check for speed sensor signal.

BRAKES

FRONT DISC BRAKES

BRAKE CALIPER

REMOVAL & INSTALLATION

See Figures 8 and 9.

➡**Use the same procedure for the LH side and RH side.**

➡**The following procedure listed is for the LH side.**

1. Remove the front wheel.
2. Drain the brake fluid.

➡**If the brake fluid leaks onto any painted surface, immediately wash it off.**

3. Separate the front flexible hose.
   a. Remove the union bolt and gasket, and separate the front flexible hose from the front disc brake caliper assembly.

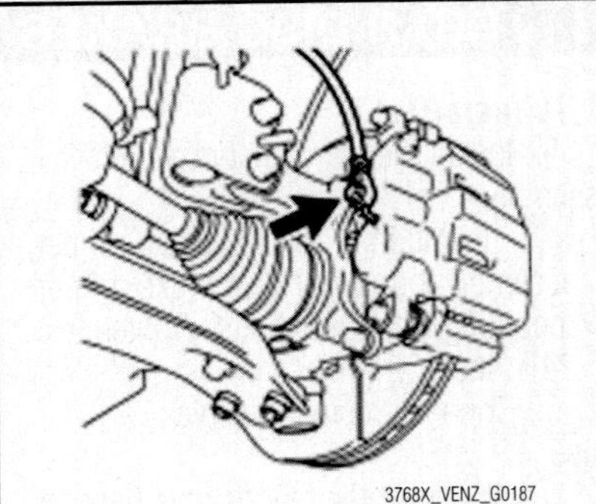

Fig. 8 Separating the front flexible hose

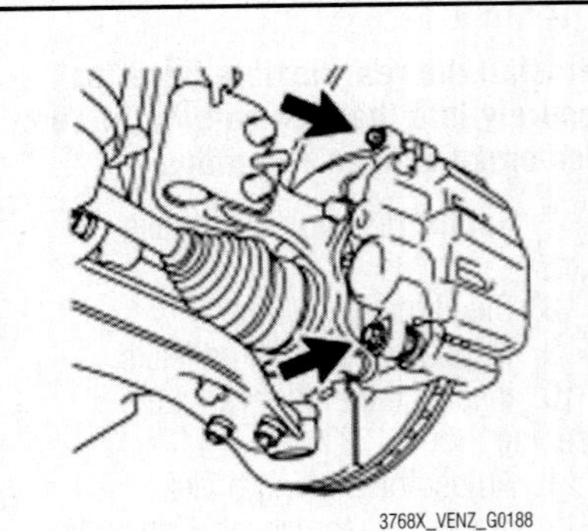

Fig. 9 Removing the brake caliper

4. Remove the front disc brake caliper assembly.
   a. Remove the 2 bolts and the front disc brake caliper assembly from the front disc brake caliper mounting.

To install:

5. Install the front disc brake caliper assembly.
   a. Install the front disc brake caliper assembly to the front disc brake caliper mounting with the 2 bolts. Tighten to 24 ft. lbs. (32 Nm).

➡**Be sure that the anti squeal springs are installed to the front disc brake pads.**

6. Connect the front flexible hose.
   a. Connect the front flexible hose to the front disc brake caliper assembly with a new union bolt and a new gasket. Tighten to 22 ft. lbs. (29 Nm).

➡**Install the front flexible hose lock securely into the lock hole in the front disc brake caliper assembly.**

7. Fill the reservoir with brake fluid.
8. Bleed the brake line.
9. Inspect for brake fluid leaks.
10. Inspect the fluid level in the reservoir.
11. Install the front wheel. Tighten to 76 ft. lbs. (103 Nm).

DISC BRAKE PADS

REMOVAL & INSTALLATION

See Figures 10 through 12.

➡**Use the same procedure for the LH side and RH side.**

➡**The following procedure listed is for the LH side.**

1. Remove the front wheel.
2. Drain the brake fluid.

➡**If the brake fluid leaks onto any painted surface, immediately wash it off.**

3. Remove the brake caliper.
4. Remove the front disc brake pad.

Fig. 10 Removing the anti squeal springs

Fig. 11 Removing the brake pads

   a. Remove the 2 anti-squeal springs from the front disc brake pads.
   b. Using a screwdriver, push the protrusion of the front disc brake pad support plate and remove the 2 front disc brake pads from the front disc brake caliper mounting.

➡**When removing the front disc brake pad, replace the front disc brake pad support plates with new ones.**

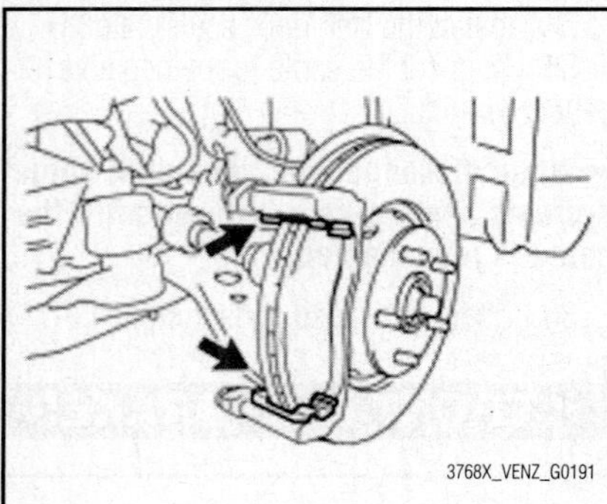

Fig. 12 Removing the front disc brake pad support plates

5. Remove the 2 front anti squeal shims from the front disc brake pads.
6. Remove the 2 front disc brake pad support plates from the front disc brake caliper mounting.

To install:

7. Install the front disc brake pad support plate.
   a. Install 2 new front disc brake pad support plates to the front disc brake caliper mounting.

➡Be sure to install the plate in the correct position and direction.

8. Install the front anti squeal shim.
   a. Apply disc brake grease between the front anti squeal shim and front disc brake pad.

➡Apply 0.03 oz. of disc brake grease to each pad.

   b. Install the 2 front anti squeal shims to the front disc brake pads.

➡When replacing worn pads, the front anti-squeal shims must be replaced together with the pads.

➡Install the front anti-shims in the correct position and direction.

➡Apply disc brake grease to the area that contacts the front anti-squeal shims.

➡Disc brake grease can seep out slightly from the area where the front anti-squeal shim is installed.

➡Make sure that disc brake grease is not applied onto the lining surface.

9. Install the front disc brake pad.
   a. Install the 2 front disc brake pads to the front disc brake caliper mounting.

➡There should be no oil or grease on the friction surfaces of the front disc brake pads or the front disc.

➡Make sure to install the front disc brake pad with pad wear indicator to the inner side of the vehicle.

➡Be sure to engage the protrusion of the pad support plate to the front disc brake pad.

   b. Install the 2 anti-squeal springs to the front disc brake pads.

➡When replacing the front brake pads with new ones, make sure to replace the anti-squeal springs at the same time.

➡Be sure to install the anti-squeal springs into the front disc brake pad installation holes as far as they will go.

10. Install the front disc brake caliper.
11. Fill the reservoir with brake fluid.
12. Bleed the brake line.
13. Inspect for brake fluid leaks.
14. Inspect the fluid level in the reservoir.
15. Install the front wheel. Tighten to 76 ft. lbs. (103 Nm).

BRAKES — REAR DISC BRAKES

BRAKE CALIPER

REMOVAL & INSTALLATION

See Figures 13 and 14.

➡Use the same procedure for the LH side and RH side.

➡The following procedure listed is for the LH side.

1. Remove the rear wheel.
2. Drain the brake fluid.

➡If the brake fluid leaks onto any painted surface, immediately wash it off.

3. Separate the rear flexible hose.
   a. Remove the union bolt and gasket, and disconnect the rear flexible hose from the rear disc brake caliper assembly.
4. Remove the rear disc brake caliper assembly.
   a. Remove the 2 bolts and the rear disc brake caliper assembly from the rear disc brake caliper mounting.

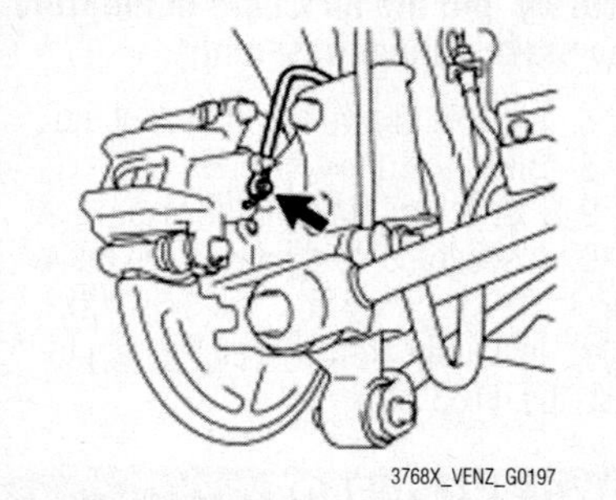

Fig. 13 Separating the rear flexible hose

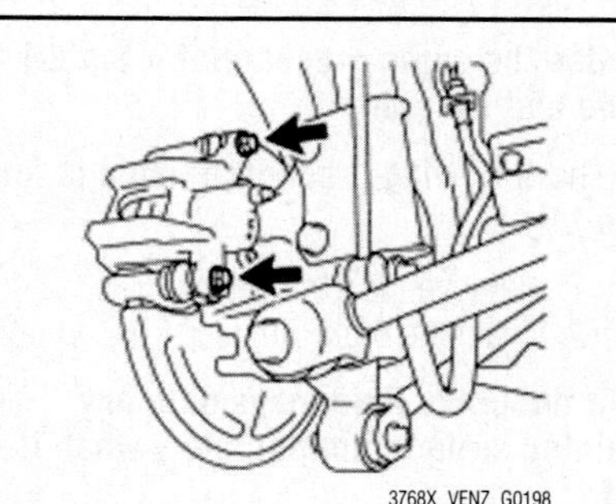

Fig. 14 Removing the rear disc brake caliper

To install:

5. Install the rear disc brake caliper assembly.
   a. Install the rear disc brake caliper assembly to the rear disc brake caliper mounting with the 2 bolts. Tighten to 24 ft. lbs. (32 Nm).
6. Connect the rear flexible hose.
   a. Connect the rear flexible hose to the rear disc brake caliper assembly with a new union bolt and a new gasket. Tighten to 22 ft. lbs. (29 Nm).

➡Install the rear flexible hose lock securely into the lock hole in the rear disc brake caliper assembly.

7. Fill the reservoir with brake fluid.
8. Bleed the brake line.
9. Inspect for brake fluid leaks.
10. Inspect the fluid level in the reservoir.
11. Adjust the parking brake.
12. Install the rear wheel. Tighten to 76 ft. lbs. (103 Nm).

DISC BRAKE PADS

REMOVAL & INSTALLATION

See Figure 15.

➡Use the same procedure for the LH side and RH side.

➡The following procedure listed is for the LH side.

1. Remove the rear wheel.
2. Drain the brake fluid.

➡If the brake fluid leaks onto any painted surface, immediately wash it off.

3. Remove the rear disc brake caliper.
4. Remove the rear disc brake pad.
   a. Remove the 2 rear disc brake pads from the rear disc brake caliper mounting.
5. Remove the rear anti squeal shim.
   a. Remove the 2 rear anti squeal shim and the pad wear indicator from the inner pad.

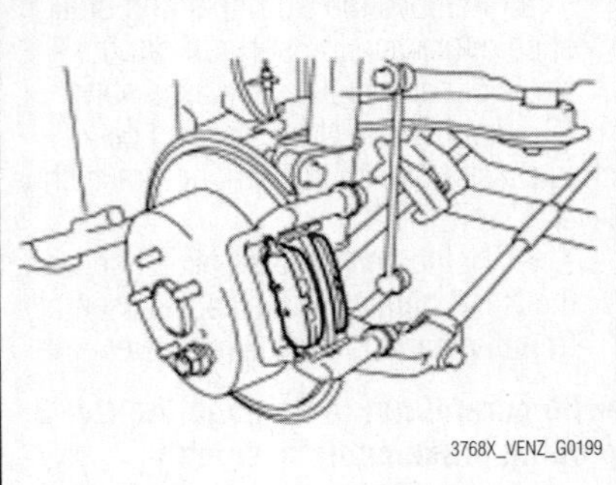

Fig. 15 Removing the rear disc brake pad

To install:

6. Install the rear anti squeal shim.
   a. Install the rear anti squeal shims to the 2 rear brake pads.

➡When replacing worn pads, the rear anti-squeal shims must be replaced together with the pads.

➡Install the shims in the correct positions and directions.

7. Install the rear disc brake pad.
   a. Install the 2 rear disc brake pads to the rear disc brake caliper mounting.

➡Make sure to install the rear disc brake pads with the pad wear indicator to the inner side of the vehicle.

➡There should be no oil or grease on the friction surfaces of the rear disc brake pads or the rear disc.

8. Install the rear disc brake caliper assembly.
9. Fill the reservoir with brake fluid.
10. Bleed the brake line.
11. Inspect for brake fluid leaks.
12. Inspect the fluid level in the reservoir.
13. Adjust the parking brake.
14. Install the rear wheel. Tighten to 76 ft. lbs. (103 Nm).

PARKING BRAKE SHOES

REMOVAL & INSTALLATION

See Figures 16 through 31.

➡Use the same procedure for the RH side and the LH side.

➡The procedure listed below is for the LH side.

1. Remove the rear wheel.
2. Remove the rear axle shaft nut (AWD).

➡Perform this procedure only when the No. 1 parking brake shoe hold down spring pin is replaced.

3. Separated the rear disc brake caliper assembly.
   a. Remove the 2 bolts and rear disc brake caliper assembly.

➡Do not disconnect the rear brake flexible hose from the rear disc brake caliper assembly.

➡Use wire or an equivalent tool to keep the rear disc brake caliper assembly from hanging down by the rear brake flexible hose.

4. Remove the rear disc.
5. Remove the No. 1 parking brake shoe return tension spring.
   a. Remove the 2 No. 1 parking brake shoe return tension springs.

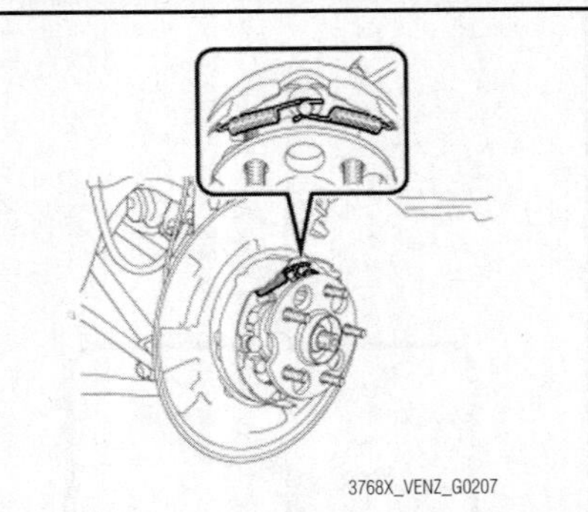

Fig. 16 Removing the No. 1 parking brake shoe return tension springs

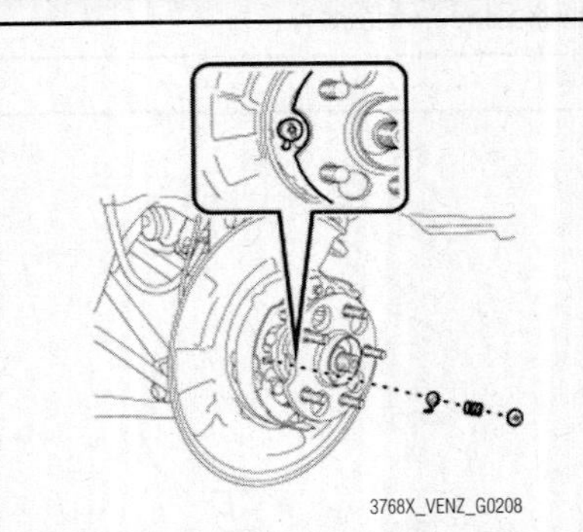

Fig. 17 Separating the No. 1 parking brake shoe assembly

6. Separate the No. 1 parking brake shoe assembly.
   a. Remove the No. 1 parking brake shoe hold down spring cup, parking brake shoe hold down spring and No. 2 parking brake shoe hold down spring cup, and separate the No. 1 parking brake shoe assembly from the backing plate.
7. Remove the parking brake shoe strut.
   a. Remove the parking brake shoe strut and the parking brake shoe strut compression spring.
8. Separate the No. 2 parking brake shoe assembly.
   a. Remove the No. 1 parking brake shoe hold down spring cup, parking brake shoe hold down spring and No. 2

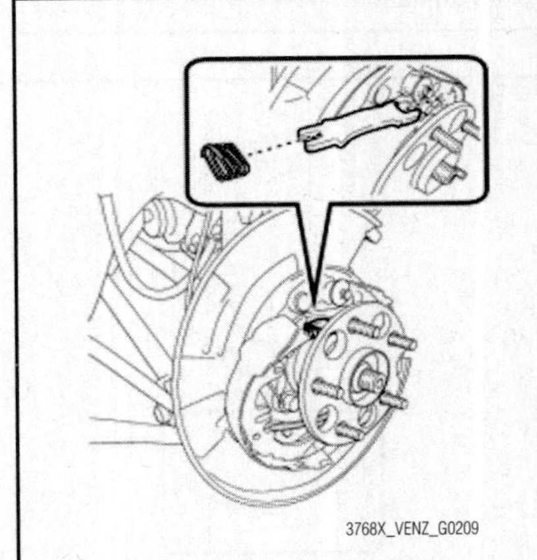

Fig. 18 Removing the parking brake shoe strut

parking brake shoe hold down spring cup, and separate the No. 2 parking brake shoe assembly from the backing plate.

9. Remove the parking brake shoe adjusting screw set.

10. Remove the No. 1 parking brake shoe assembly.

a. Separate the No. 2 parking brake shoe return tension spring and remove the No. 1 parking brake shoe assembly.

11. Remove the No. 2 parking brake shoe return tension spring.

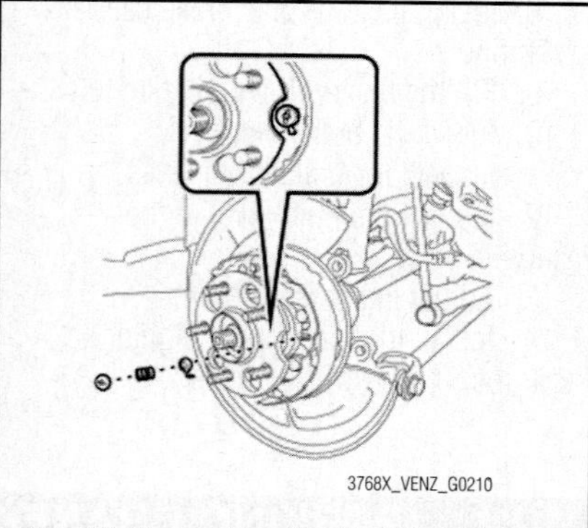

Fig. 19 Separating the No. 2 parking brake shoe assembly

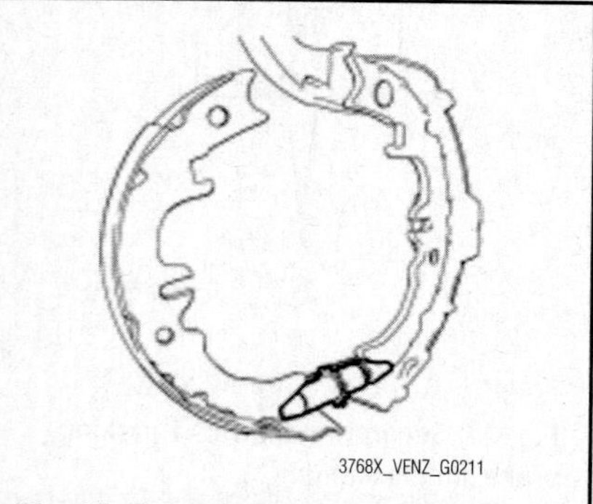

Fig. 20 Removing the parking brake shoe adjusting screw set

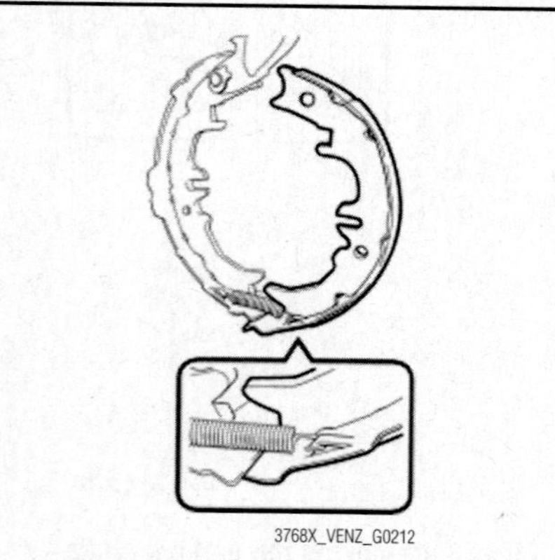

Fig. 21 Removing the No. 1 parking brake shoe assembly

a. Remove the No. 2 parking brake shoe return tension spring from the No. 2 parking brake shoe assembly.

12. Remove the No. 2 parking brake shoe assembly with the parking brake shoe lever.

a. Using needle nose pliers, separate the No. 3 parking brake cable assembly from the parking brake shoe lever.

➡Be careful not to damage the No. 3 parking brake cable assembly.

13. Remove the parking brake shoe lever.

a. Remove the C-washer, shim and the parking brake shoe lever from the No. 2 parking brake shoe assembly.

➡The shim is installed only when a clearance adjustment between the parking brake lever and parking brake shoe C-washer is necessary. Therefore, some models may have no shim.

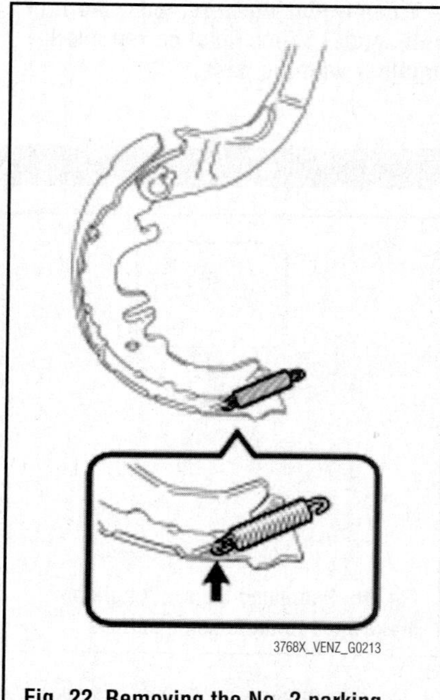

Fig. 22 Removing the No. 2 parking brake shoe return tension spring

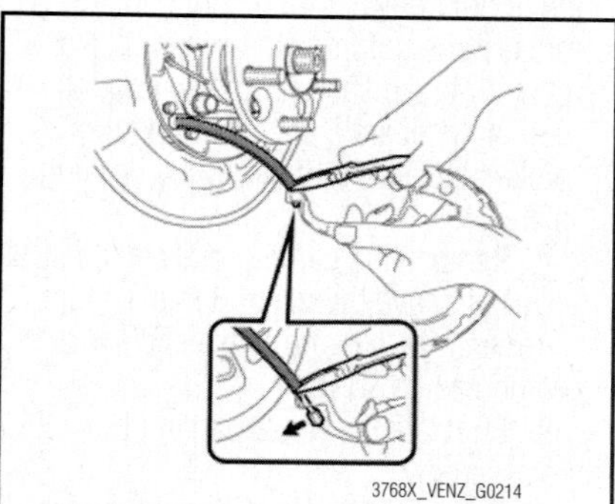

Fig. 23 Removing the No. 2 parking brake shoe assembly with the parking brake shoe lever

14. Remove the parking brake shoe guide plate set bolt and the parking brake shoe guide plate.

15. Remove the No. 1 parking brake shoe hold down spring pin (2WD).

a. Remove the No. 1 parking brake shoe hold down spring pin.

16. Remove the No. 2 parking brake shoe hold down spring pin (AWD).

17. Remove the rear axle hub and bearing assembly (2WD).

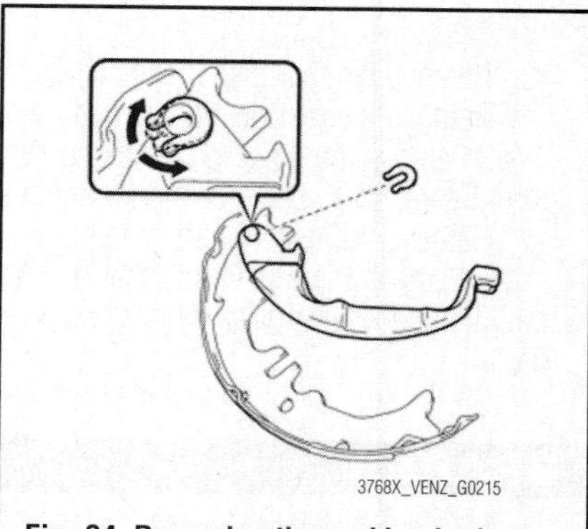

Fig. 24 Removing the parking brake shoe lever

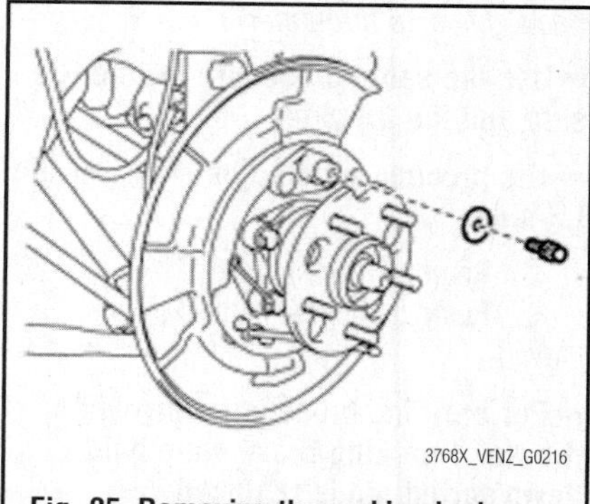

Fig. 25 Removing the parking brake shoe guide plate set bolt

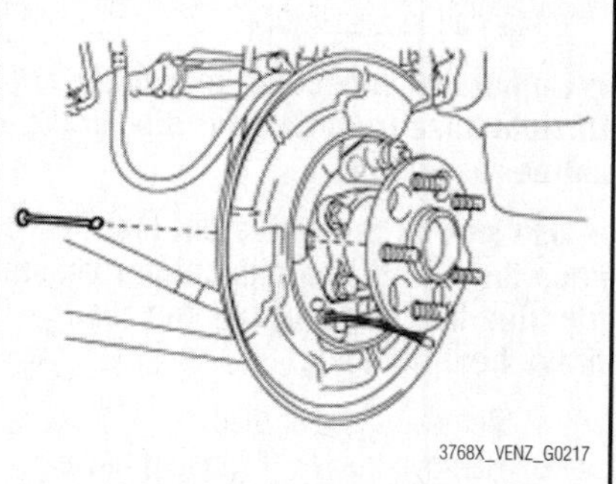

Fig. 26 Removing the No. 1 parking brake shoe hold down spring pin (2WD)

➡**Perform this procedure only when the No. 2 parking brake shoe hold down spring pin is replaced.**

18. Remove the rear axle hub and bearing assembly (AWD).

➡**Perform this procedure only when the No. 1 parking brake shoe hold down spring pin is replaced.**

19. Remove the No. 2 parking brake shoe hold down spring pin (2WD).
20. Remove the No. 1 parking brake shoe hold down spring pin (AWD).

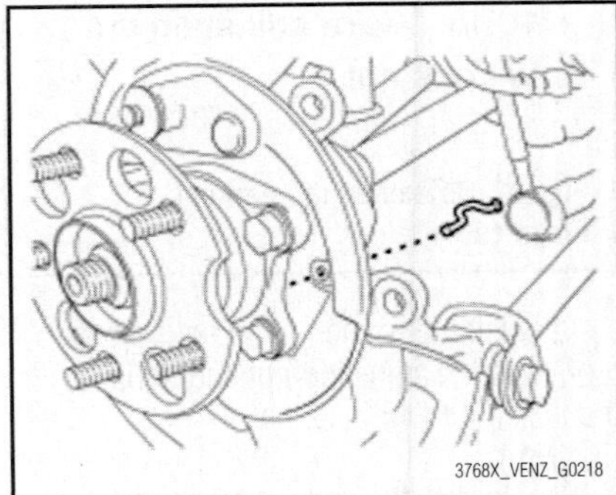

Fig. 27 Removing the No. 2 parking brake shoe hold down spring pin (AWD)

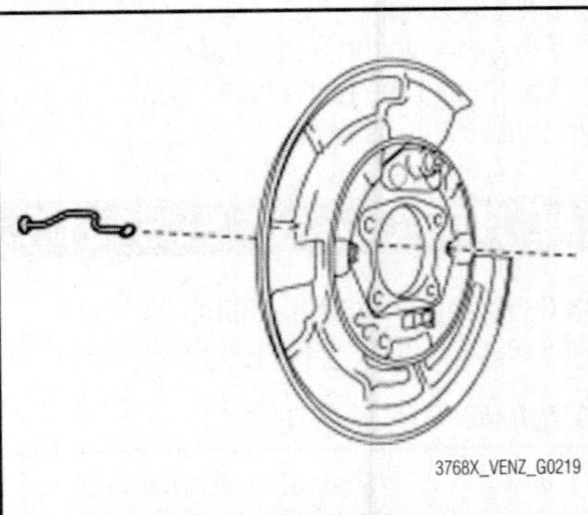

Fig. 28 Removing the No. 2 parking brake shoe hold down spring pin (2WD)

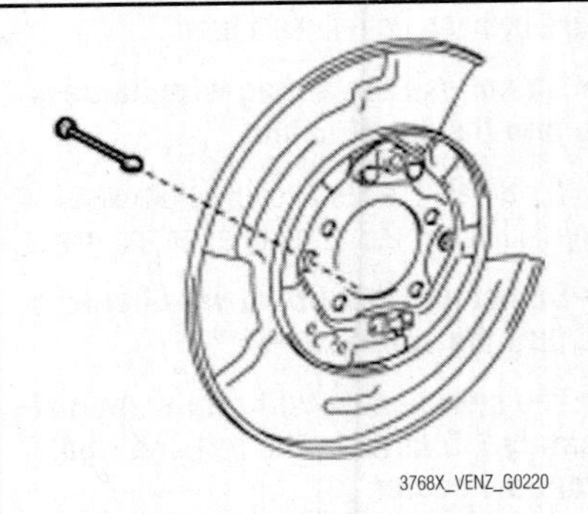

Fig. 29 Removing the No. 1 parking brake show hold down spring pin (AWD)

To install:

21. Install the No. 2 parking brake shoe hold down spring pin (2WD).
22. Install the No. 1 parking brake shoe hold down spring pin (AWD).
23. Install the rear axle hub and bearing assembly (2WD).

➡**Perform this procedure only when the No. 2 parking brake shoe hold down spring pin is replaced.**

24. Install the rear axle hub and bearing assembly (AWD).

➡**Perform this procedure only when the No. 1 parking brake shoe hold down spring pin is replaced.**

25. Install the No. 1 parking brake shoe hold down spring pin (2WD).
26. Install the No. 2 parking brake shoe hold down spring pin (AWD).
27. Apply high temperature grease to the backing plate which makes contact with the shoe.
28. Install the parking brake shoe guide plate set bolt.
    a. Apply adhesive to the threads of the parking brake shoe guide plate set bolt.
    b. Install the parking brake shoe guide plate and the parking brake shoe guide plate set bolt. Tighten to 13 ft. lbs. (18 Nm).
29. Install the parking brake shoe lever.
    a. Apply high temperature grease to the parking brake shoe lever which makes contact with the No. 2 parking brake shoe assembly.
    b. Install the parking brake shoe lever and shim to the No. 2 parking brake shoe assembly with a new C-washer.
    c. Using a feeler gauge, measure the clearance between the No. 2 parking brake shoe assembly and parking brake shoe lever. Standard clearance is less than 0.0138 inch.

➡**If the clearance is not as specified, replace the shim with one of the appropriate size.**

30. Install the No. 2 parking brake shoe assembly with the parking brake shoe lever.
    a. Using needle-nose pliers, connect the No. 3 parking brake cable assembly to the parking brake shoe lever.
31. Install the No. 2 parking brake shoe return tension spring to the No. 2 parking brake shoe assembly.
32. Install the No. 1 parking brake shoe assembly.
    a. Connect the No. 2 parking brake shoe return tension spring to install the No. 1 parking brake shoe assembly.
33. Install the parking brake shoe adjusting screw set.
    a. Apply high temperature grease to the parking brake shoe adjusting screw set.
    b. Install the parking brake shoe adjusting screw set.
34. Install the No. 2 parking brake shoe assembly.
    a. Install the No. 2 parking brake shoe assembly to the backing plate with the No. 1 parking brake shoe hold down spring cup, parking brake shoe hold down spring and No. 2 parking brake shoe hold down spring cup.
35. Install the parking brake shoe strut.
    a. Install the parking brake shoe strut and the parking brake shoe strut compression spring.
36. Install the No. 1 parking brake shoe assembly.
    a. Install the No. 1 parking brake shoe assembly to the backing plate with the No. 1 parking brake shoe hold down spring cup, parking brake shoe hold down spring and No. 2 parking brake shoe hold down spring cup.
37. Install the 2 No. 1 parking brake shoe return tension springs.

➡**First install the front side spring and then the rear side spring.**

38. Check the parking brake installation.
    a. Check that each part is installed properly.

➡**There should be no oil or grease on the friction surfaces of the shoe lining and disc.**

39. Install the rear disc.
40. Install the rear disc brake caliper assembly.
41. Install the rear axle shaft nut (AWD).

➡**Perform this procedure only when the No. 1 parking brake shoe hold down spring pin is replaced.**

42. Adjust the parking brake shoe clearance.
43. Install the rear wheel. Tighten to 76 ft. lbs. (103 Nm).
44. Bed in the parking brake shoes to the discs.
    a. Drive the vehicle at about 31 mph on a safe, level and dry road.
    b. Depress the parking brake pedal with 33.7 lbs. (150 N) of force.
    c. Drive the vehicle for about 0.25 mile (400 m) in this condition.
    d. Repeat this procedure 3 times.

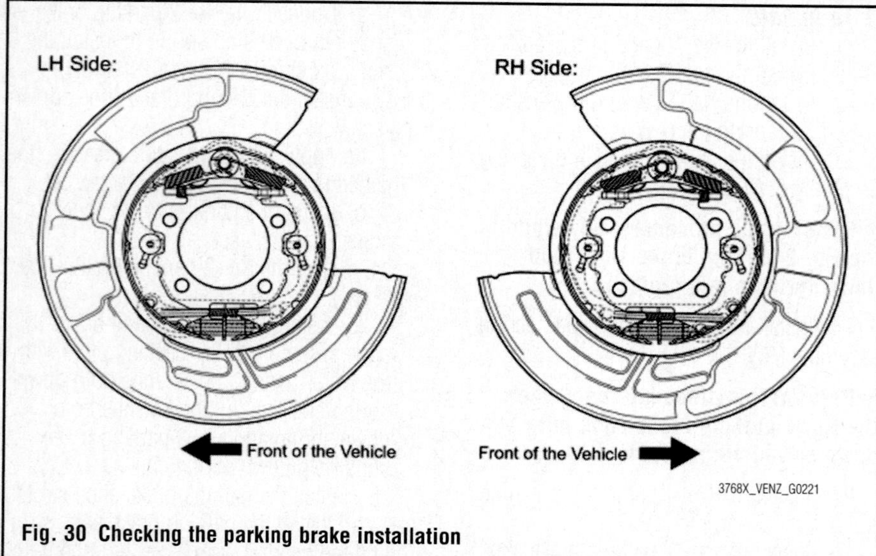

Fig. 30 Checking the parking brake installation

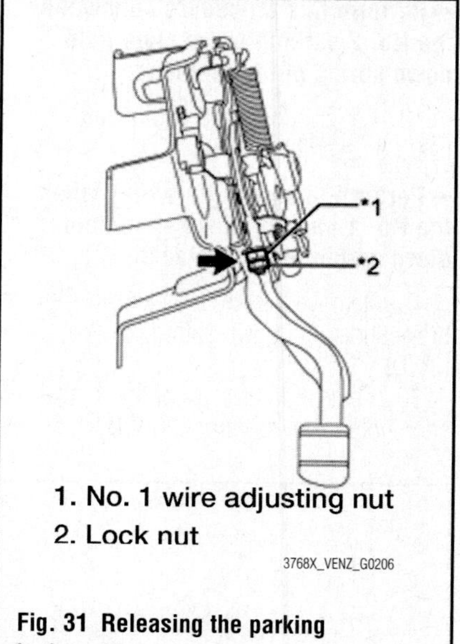

Fig. 31 Releasing the parking brake cable

➡Set a 5 minute interval between each procedure to prevent the brake assembly from overheating.

45. Adjust the parking brake shoe clearance and parking brake pedal travel.

ADJUSTMENT

1. Completely releaser the parking brake pedal.
2. Loosen the lock nut and No. 1 wire adjusting nut to completely release the parking brake cable.
3. Remove the rear wheel.
4. Temporarily install the 5 hub nuts.
5. Remove the shoe adjusting hole plug.
6. Turn the shoe adjuster and expand the shoe until the disc locks.
7. Turn and contract the shoe adjuster until the disc can rotate smoothly. Standard is 8 notches.
8. Check that there is no brake drag against the shoe.
9. Install the shoe adjusting hole plug.
10. Turn the adjusting nut until the parking brake pedal travel is corrected to be within the specified range of 4-6 notches at 67.5 ft. lbs.
11. Using a wrench or an equivalent tool, hold the adjusting nut and tighten the lock nut to 48 inch lbs. (5.4 Nm).
12. Operate the parking brake pedal 3 to 4 times, and check the parking brake pedal travel.
13. Check that there is not brake drag against the shoe.
14. Remove the 5 hub nuts.
15. Install the rear wheel. Tighten to 76 ft. lbs. (103 Nm).

CHASSIS ELECTRICAL

AIR BAG (SUPPLEMENTAL RESTRAINT SYSTEM)

GENERAL INFORMATION

⁂ CAUTION

These vehicles are equipped with an air bag system. The system must be disarmed before performing service on, or around, system components, the steering column, instrument panel components, wiring and sensors. Failure to follow the safety precautions and the disarming procedure could result in accidental air bag deployment, possible injury and unnecessary system repairs.

SERVICE PRECAUTIONS

⁂ CAUTION

Disconnect and isolate the battery negative cable before beginning any airbag system component diagnosis, testing, removal, or installation procedures. Wait at least 90 seconds after the ignition switch is turned off and the negative (-) terminal cable is disconnected from the battery before starting the operation. The SRS is equipped with a backup power source, so if work is started within 90 seconds after disconnecting the negative (-) terminal cable from the battery, the SRS may be deployed. Failure to disable the airbag system may result in accidental airbag deployment, personal injury, or death.

DISARMING THE SYSTEM

To avoid personal injury when working on vehicles equipped with an air bag, the negative battery cable must be disconnected and at least 90 seconds must elapse before working on the system. Failure to do so may result in deployment of the air bag.

ARMING THE SYSTEM

To arm the system after service is finished, connect the negative battery cable.

CLOCKSPRING CENTERING

1. Rotate the spiral cable counterclockwise by hand until it feels firm

➡Do not use the airbag wire harness to turn the spiral cable.

2. Rotate the spiral cable clockwise approximately 2.5 turns to align the marks.

➡Do not use the airbag wire harness to turn the spiral cable.

➡The spiral cable will rotate approximately 2.5 turns to the left and right from the center.

DRIVE TRAIN

FRONT DRIVESHAFT

REMOVAL & INSTALLATION

See Figures 32 through 38.

1. Remove the tail exhaust pipe assembly.
2. Remove the center exhaust pipe assembly.
3. Remove the propeller with the center bearing shaft assembly.
   a. Depress the brake pedal and hold it.
   b. Using a hexagon wrench (6 mm), loosen the cross groove joint set bolts½ turn.

➡Put a piece of cloth or equivalent into the inside of the universal joint cover so that the boot does not touch the inside of the universal joint cover.

➡Do not remove the bolts.

c. Place matchmarks on the rear propeller shaft and electromagnetic control coupling assembly.

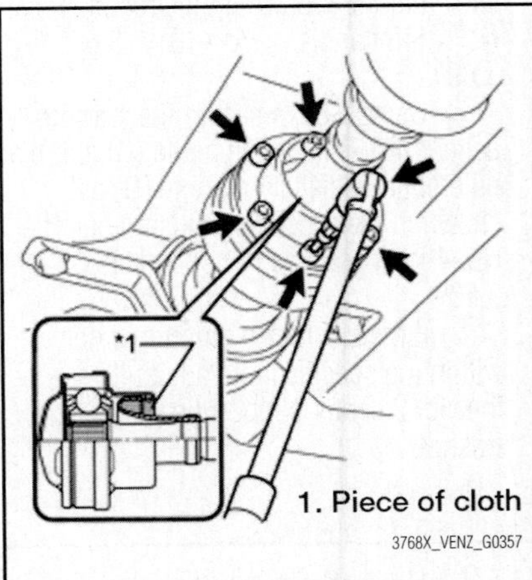

Fig. 32 Loosening the cross groove joint set bolts

d. Remove the 4 nuts and 4 washers.
e. Using a brass bar and a hammer, separate the propeller with center bearing shaft assembly.
f. Remove the 4 bolts, 2 No. 1 center support bearing washers and 2 No. 2 center support bearing washers.

➡When removing the bolts and washers, do not apply excessive force to the universal joint.

g. Pull out the propeller with center bearing shaft assembly from the transfer.

➡When removing the propeller shaft, do not apply excessive force to the universal joint.

➡During and after the removal of the propeller shaft, keep the universal joint angle straight (within 15 degrees).

➡Be careful not to damage the oil seal.

h. Insert the special tool (09325-20010) into the transfer to prevent oil leaks.

➡Be careful not to damage the oil seal.

To install:

4. Temporarily tighten the propeller with the center bearing shaft assembly.
   a. Remove the special tool from the transfer.
   b. Install the propeller with the center bearing shaft assembly.

➡Be careful not to damage the oil seal.

➡Be careful not to damage the universal joint boot when installing the propeller shaft.

c. Align the matchmarks on the rear propeller shaft and electromagnetic control coupling assembly and install the 4 nuts and 4 washers temporarily.

➡Do not allow grease to adhere to the bolts or washers.

d. Temporarily install the propeller with center bearing shaft assembly with the 4 bolts, 2 No. 1 center support bearing washers and 2 No. 2 center support bearing washers.

➡Reuse the washers.

➡Do not allow grease to adhere to be bolts or washers.

e. Fully tighten the 4 nuts. Tighten to 27 ft. lbs. (37 Nm).

5. Fully tighten the propeller with the center bearing shaft assembly.
   a. Remove the piece of cloth or equivalent from the universal joint.
   b. Depress the brake pedal and hold it.
   c. Using a hexagon wrench (6 mm), tighten the 6 bolts. Tighten to 19 ft. lbs. (26 Nm).
   d. With the vehicle unloaded, adjust the dimension between the rear side of the cover and shaft as shown in the illustration. Length A should be 2.579-2.776 inches (65.5-70.5 mm).
   e. With the vehicle unloaded, adjust the front and rear dimensions between the edge surface of the center support bearing and the edge surface of the cushion respectively as shown in the illustration, and then tighten the bolts. Length A should be 0.453-0.532 inch (11.5-13.5 mm).
   f. Check that the center line of the bracket is at a right angle to the shaft axial direction.
   g. Fully tighten the 4 bolts. Tighten to 27 ft. lbs. (37 Nm).
6. Install the center exhaust pipe assembly.

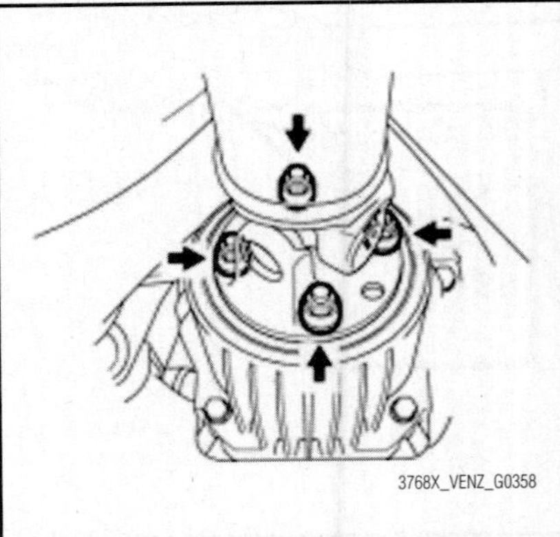

Fig. 33 Removing the nuts and washers

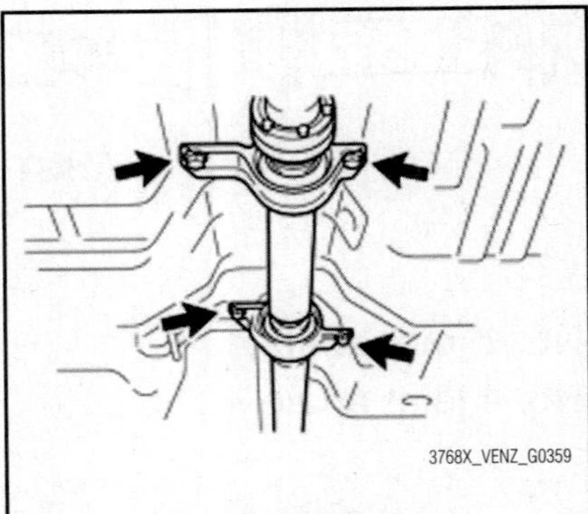

Fig. 34 Removing the bolts and washers

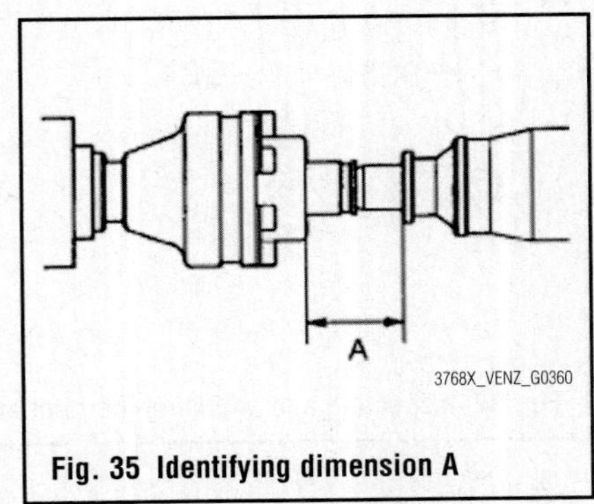

Fig. 35 Identifying dimension A

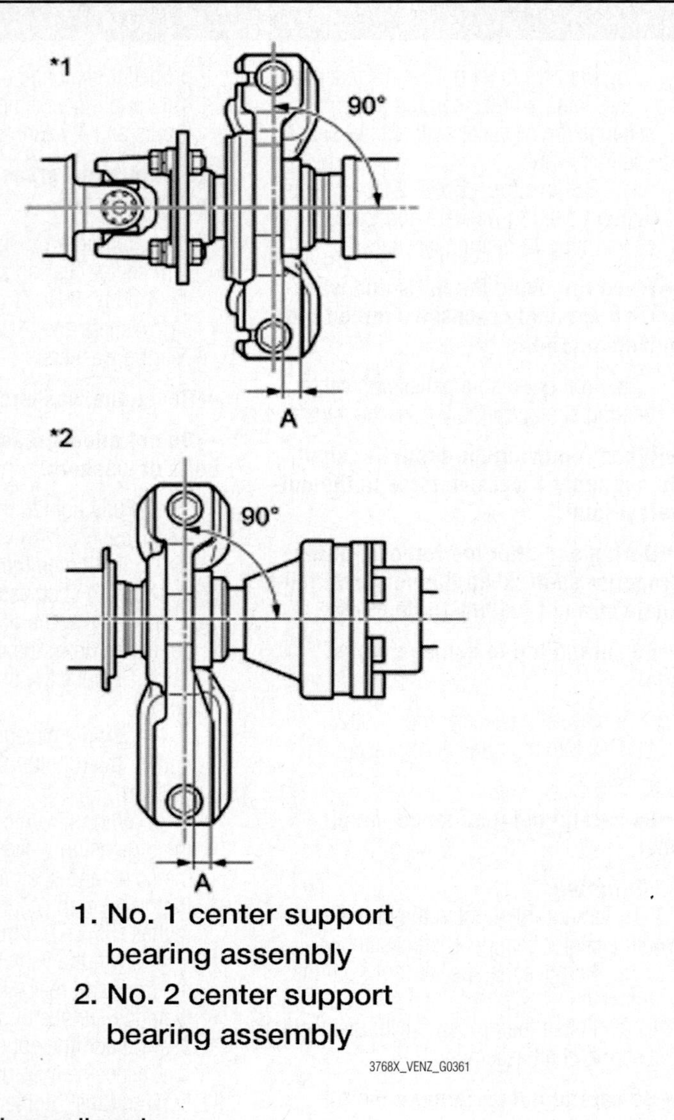

Fig. 36 Adjusting the rear dimensions

7. Install the tail exhaust pipe assembly.
8. Inspect and adjust the transfer oil.
9. Inspect for exhaust gas leaks.
10. Inspect and adjust the joint angle.

If any vibration or noise occurs, perform joint angle check as follows and replace the No. 2 center support bearing washer with a proper one:

a. Turn the propeller shaft several times by hand to stabilize the center support bearings.

b. Using a jack, raise and lower the differential to stabilize the differential mounting cushion.

➡Measure the joint angle while the vehicle is lifted using a 4 pillar lift or while working in a pit.

c. Remove the transfer dynamic damper.

d. Using SST, measure the propeller shaft installation angle (A) and intermediate shaft installation angle (B) as shown in the following illustration. The No. 1 joint angle is (A) - (B)= -3.69° to -1.69°

e. Using SST, measure the rear propeller shaft installation angle (C) and rear differential installation angle (D) as shown in the preceding illustration. The No. 4 joint angle is (C) - (D) = 1.63° to 3.63°

f. If the calculated amount is not within the specification, adjust it with the No. 2 center support bearing washer.

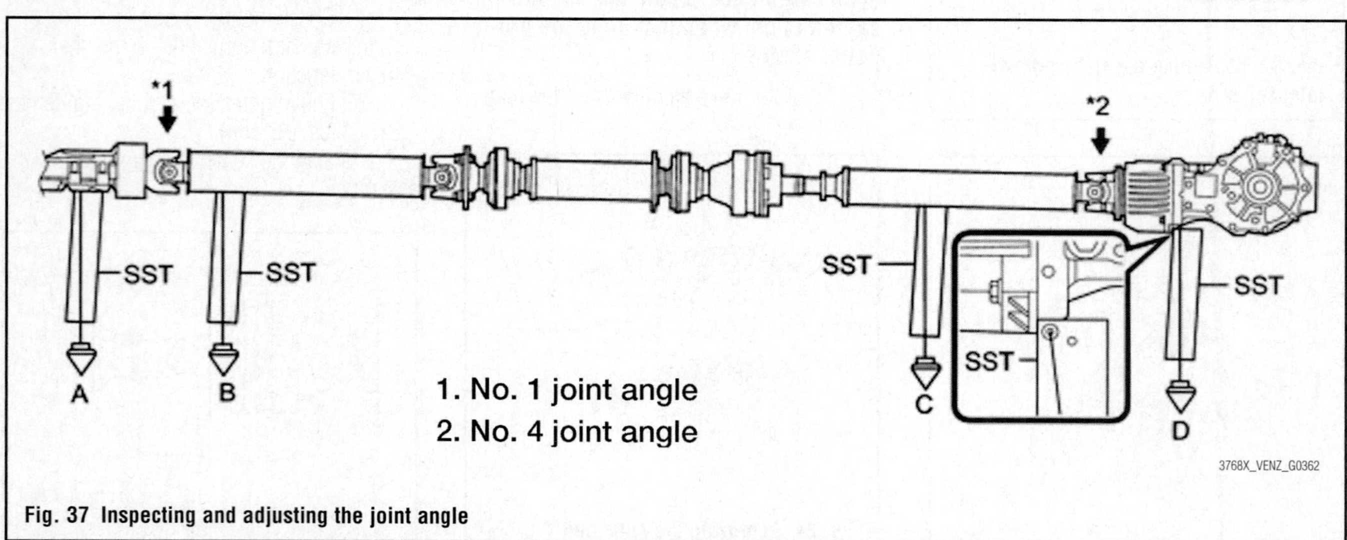

Fig. 37 Inspecting and adjusting the joint angle

Thickness Inch (MM)
0.126 (3.2)
0.177 (4.5)
0.256 (6.5)
0.354 (9.0)

3768X_VENZ_G0363

Fig. 38 Washer thickness

➡Make sure to use a washer of the same thickness on both right and left sides.

➡Do not use 2 or more washers on a bolt.

g. Install the transfer dynamic damper. Tighten to 19 ft. lbs. (26 Nm).

FRONT HALFSHAFT

REMOVAL & INSTALLATION

See Figures 39 through 41.

➡Use the same procedure for the RH side and the LH side.

➡The procedure listed below is for the LH side.

⁂ CAUTION

Refer to Precautions before performing the procedure listed below.

1. Drain the automatic transaxle fluid.
2. Drain the transfer oil.
3. Remove the front wheels.
4. Remove the front axle shaft nut.
   a. Using the special tool (09930-00010) and a hammer, replace the staked part of the front axle shaft nut.

➡Loosen the staked part of the nut completely, otherwise the threads of the driveshaft may be damaged.

   b. While applying the brakes, remove the front axle shaft nut.
5. Separate the front stabilizer link assembly.
6. Separate the front speed sensor.
   a. Remove the bolt and separate the front speed sensor from the steering knuckle.
   b. Remove the bolt and clamp, and separate the front speed sensor and front flexible hose.
7. Separate the tie rod assembly.
8. Separate the front lower suspension arm.
9. Separate the front axle assembly.
10. Remove the front driveshaft assembly LH.
    a. Using the special tool (09520-01010, 09520-24010 or 09520-32040), remove the front driveshaft assembly LH.
11. Remove the front driveshaft assembly RH (2WD).
    a. Remove the bearing bracket hole snap ring from the driveshaft bearing bracket.
    b. Remove the bolt and front driveshaft assembly RH from the driveshaft bearing bracket.

➡Do not damage the boot or oil seal.

12. Remove the front driveshaft assembly RH (AWD).
    a. Remove the bearing bracket hole snap ring from the driveshaft bearing bracket.
    b. Remove the bolt and front driveshaft assembly RH from the driveshaft bearing bracket.

➡Do not damage the boot or oil seal.

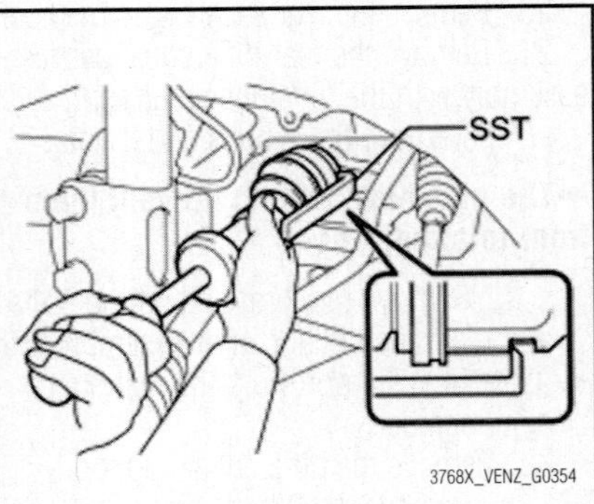

Fig. 39 Removing the front driveshaft assembly LH

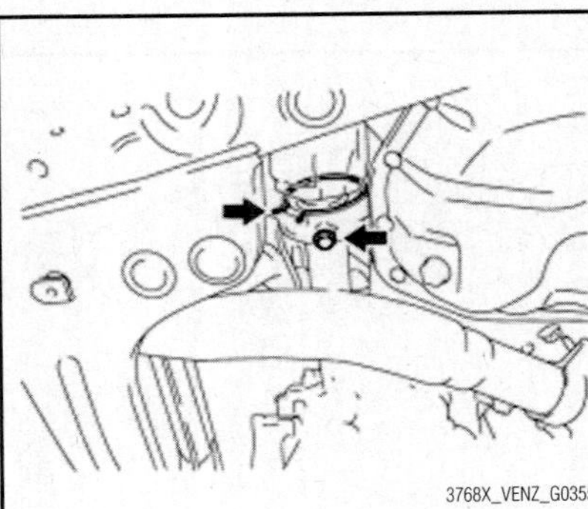

Fig. 40 Removing the front driveshaft assembly RH

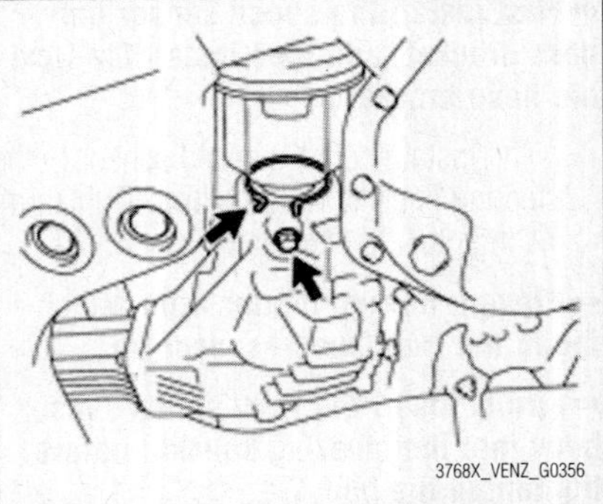

Fig. 41 Removing the front driveshaft assembly RH (AWD)

To install:

13. Install the front driveshaft assembly LH.
    a. Align the splines of the shaft and install the driveshaft assembly LH using a brass bar and a hammer.

➡Set the shaft snap ring with the opening facing down.

➡Be careful not to damage the driveshaft dust cover, boot or oil seal.

➡Move the driveshaft assembly while keeping it level.

14. Install the front driveshaft assembly RH (2WD).
    a. Install the front driveshaft assembly RH.
    b. Install the bearing bracket hole snap ring and a new bolt. Tighten to 24 ft. lbs. (32 Nm).

➡Do not damage the boot or oil seal.

➡Move the driveshaft assembly while keeping it level.

15. Install the front driveshaft assembly RH (AWD).
    a. Install the front driveshaft assembly RH.
    b. Install the bearing bracket hole snap ring and a new bolt. Tighten to 24 ft. lbs. (32 Nm).

➡Do not damage the boot or oil seal.

➡Move the driveshaft assembly while keeping it level.

16. Install the front axle assembly.
17. Install the front lower suspension arm.
18. Connect the tie rod assembly.
19. Install the front speed sensor.
    a. Install the front speed sensor and front flexible hose with the bolt. Tighten to 14 ft. lbs. (19 Nm).

➡Do not twist the front speed sensor when installing it.

➡First install the speed sensor harness bracket, and then install the flexible hose bracket.

b. Install the front speed sensor to the steering knuckle with the bolt. Tighten to 71 inch lbs. (8 Nm).

➡Prevent foreign matter from attaching to the front speed sensor tip.

➡Firmly insert the front speed sensor body into the steering knuckle before tightening the bolt.

➡After installing the front speed sensor to the steering knuckle, make sure that there is no clearance between the front speed sensor stay and steering knuckle. Also make sure that no foreign matter is stuck between the parts.

➡To prevent interference between the front speed sensor and magnetic rotor, do not rotate the front speed sensor body during or after the insertion of the front speed sensor body to the steering knuckle.

20. Install front stabilizer link assembly.
21. Clean the threaded parts on the front driveshaft and a new front axle shaft nut using a non-residue solvent.

➡Be sure to perform this work for a new driveshaft.

➡Keep the threaded parts free of oil and foreign matter.

22. Install the new front axle shaft nut and tighten to 217 ft. lbs. (294 Nm).
    a. Using a chisel and hammer, stake the front axle shaft nut.
23. Install the front wheels. Tighten to 76 ft. lbs. (103 Nm).
24. Add transfer oil.
25. Adjust the transfer oil.
26. Add automatic transaxle fluid.
27. Adjust the front wheel alignment.
28. Check the ABS speed sensor signal.

REAR AXLE HOUSING

REMOVAL & INSTALLATION

See Figures 42 through 48.

1. Drain the differential oil.
   a. Using a hexagon wrench (10 mm), remove the rear differential carrier cover plug and gasket.
   b. Using a hexagon wrench (10 mm), remove the rear differential drain plug and gasket, then drain the differential oil.
2. Remove the rear wheels.
3. Remove the center exhaust pipe assembly.
4. Remove the propeller with the center bearing shaft assembly.
5. Separate the rear speed sensor LH and RH.
6. Remove the rear axle shaft nuts LH and RH.
7. Separate the No. 3 parking brake cable assembly.
8. Separate the No. 2 parking brake cable assembly.
9. Remove the No. 1 floor under cover.
10. Remove the rear strut rod assemblies LH and RH.
11. Remove the rear height control sensor sub assembly (w/HID headlight system).
12. Remove the rear No. 2 suspension arm assemblies LH and RH.
13. Separate the rear No. 1 suspension arm assemblies LH and RH.
14. Remove the rear driveshaft assembly LH.
15. Remove the rear driveshaft snap ring LH.
16. Remove the rear driveshaft assembly RH.
17. Remove the rear driveshaft snap ring RH.
18. Separate the No. 3 floor wire (w/HID headlight system).
19. Separate the frame wire.
20. Remove the rear suspension member.
21. Remove the rear differential carrier assembly with the differential support.
    a. Remove the 2 bolts and 2 nuts.

➡The nuts have tabs to prevent them from rotating.

b. Remove the 3 rear mounting bolts and rear differential carrier assembly with differential support from the rear suspension member.
22. Remove the differential support.
    a. Remove the 3 bolts and differential support.
23. Remove the rear No. 1 differential support.

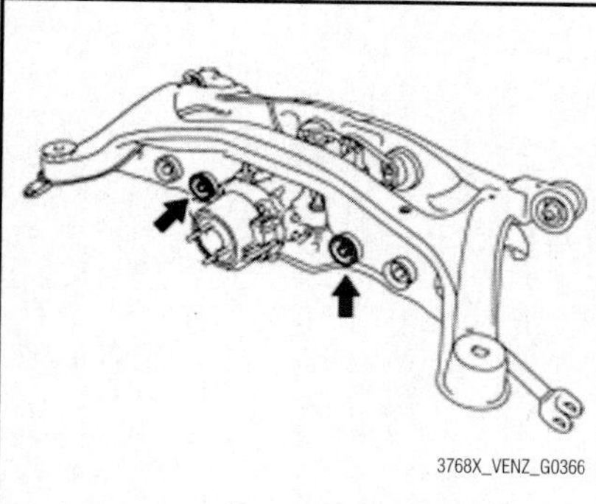

Fig. 42 Removing the differential carrier nuts and bolts

Fig. 43 Removing the rear mounting bolts and carrier assembly

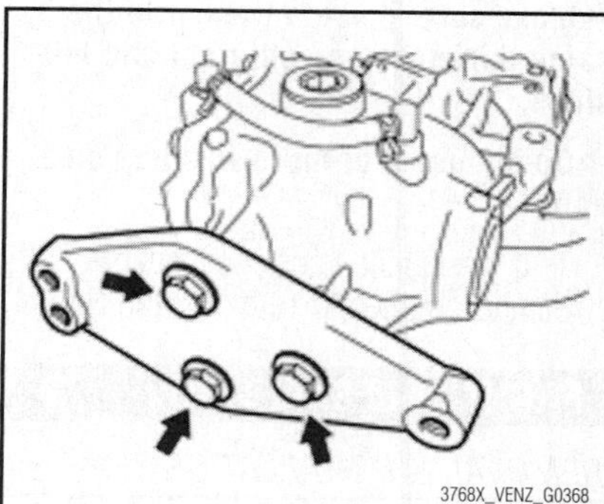

Fig. 44 Removing the differential support

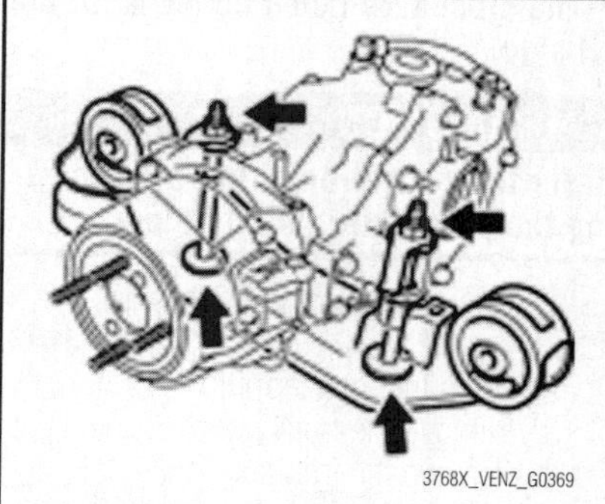

Fig. 45 Removing the rear No. 1 differential support

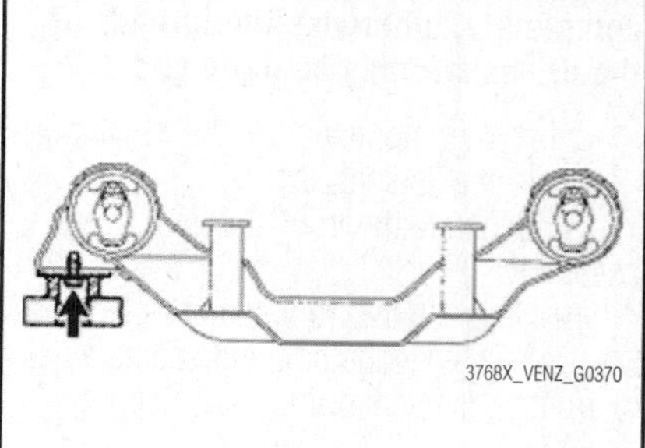

Fig. 46 Removing the rear differential dynamic damper

a. Remove the 2 nuts, 2 bolts and rear No. 1 differential support.

➡**The nuts have tabs to prevent them from rotating.**

24. Remove the rear differential dynamic damper.

➡**This step should be performed only when the rear differential dynamic damper is being replaced.**

a. Remove the bolt and rear differential dynamic damper.

To install:

➡**If installing a new rear differential carrier assembly, remove the 2 differential side seal caps before installing the rear driveshaft assembly.**

25. Install the rear differential dynamic damper.

➡**This step should be performed only when the rear differential dynamic damper is replaced.**

a. Install the rear differential dynamic damper with the bolt. Tighten to 20 ft. lbs. (27 Nm).

26. Temporarily tighten the rear No. 1 differential support.

a. Temporarily install the rear No. 1 differential support to the rear differential carrier assembly with the 2 new bolts and 2 new nuts.

➡**Be sure to install the rear No. 1 differential support facing in the direction shown in the illustration.**

➡**Be sure to install each nut to position (A) shown in the illustration.**

➡**The nuts have tabs to prevent them from rotating.**

27. Install the differential support to the rear differential carrier assembly with the 3 new bolts. Tighten to 123 ft. lbs. (167 Nm).

28. Temporarily tighten the rear differential carrier assembly with the differential support.

a. Temporarily install the rear differential carrier assembly with differential support to the rear side of the rear suspension member assembly with the 3 rear mounting bolts.

b. Temporarily install the rear differential carrier assembly with differential support to the front side of the rear suspension member assembly with the 2 bolts and 2 nuts.

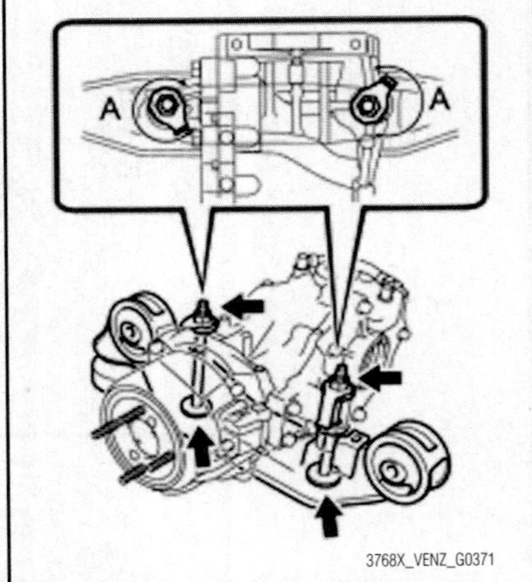

Fig. 47 Temporarily installing the rear No. 1 differential support

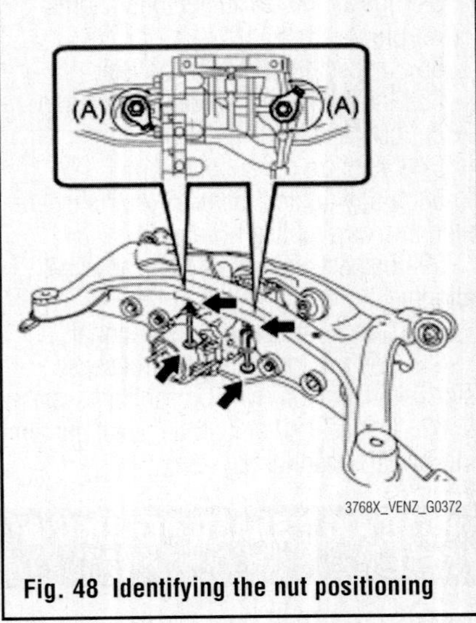

Fig. 48 Identifying the nut positioning

➡**The nuts have tabs to prevent them from rotating.**

29. Fully tighten the rear differential carrier assembly with the differential support.

➡**Do not tighten the bolts with the inner cylinder or rear differential mount cushion tilted.**

a. Install the rear differential carrier assembly with differential support to the rear side of the rear suspension member assembly with the 3 rear mounting bolts. Tighten to 70 ft. lbs. (95 Nm).

b. Install the rear differential carrier assembly with differential support to the front side of the rear suspension member assembly with the 2 bolts and 2 nuts. Tighten to 76 ft. lbs. (103 Nm).

➡**The nuts have tabs to prevent them from rotating.**

30. Fully tighten the rear No. 1 differential support.

a. Install the rear No. 1 differential support to the rear differential carrier assembly with the 2 new bolts and 2 new nuts. Tighten to 63 ft. lbs. (86 Nm).

➡**Make sure that each nut is installed to a position (A) shown in the illustration.**

➡**The nuts have tabs to prevent them from rotating.**

31. Install the rear suspension member.
32. Install the frame wire.
33. Install the No. 3 floor wire (w/HID headlight system).
34. Install the rear driveshaft snap ring LH.
35. Install the rear driveshaft assembly LH.
36. Install the rear driveshaft snap ring RH.
37. Install the rear driveshaft assembly RH.
38. Connect the rear No. 1 suspension arm assembly LH.
39. Connect the rear No. 1 suspension arm assembly RH.
40. Temporarily tighten the rear No. 2 suspension arm assemblies LH and RH.
41. Install the rear strut rod assemblies LH and RH.
42. Install the rear height control sensor sub assembly (w/HID headlight system).
43. Install the No. 3 parking brake cable assembly.
44. Install the No. 2 parking brake cable assembly.
45. Install the rear axle shaft nuts LH and RH.
46. Install the rear speed sensors LH and RH.
47. Temporarily tighten the propeller with the center bearing shaft assembly.
48. Fully tighten the propeller with the center bearing shaft assembly.
49. Install the No. 1 floor under cover.
50. Inspect and adjust the transfer oil.
51. Install the center exhaust pipe assembly.
52. Add differential oil.

a. For a reused rear differential carrier assembly, use a hexagon wrench (10 mm) to install the rear differential drain plug with a new gasket. Tighten to 29 ft. lbs. (39 Nm). Add the differential oil.

b. For a new rear differential carrier assembly, use a hexagon wrench (10 mm) and remove the rear differential carrier cover plug and gasket. Add differential oil.

53. Inspect and adjust the differential oil.

54. Install the rear differential carrier cover plug.
55. Inspect for exhaust gas leak.
56. Install the rear wheels. Tighten to 76 ft. lbs. (105 Nm).
57. Stabilize the suspension.
58. Fully tighten the rear No. 2 suspension arm assemblies LH and RH.
59. Inspect and adjust the rear wheel alignment.
60. Check for speed sensor signal.
61. Perform the height control sensor signal initialization (w/HID headlight system).
62. Inspect and adjust the headlight aiming (w/HID headlight system).

REAR AXLE SHAFT, BEARING & SEAL

REMOVAL & INSTALLATION

2WD Vehicles

See Figure 49.

➡Use the same procedure for the RH side and LH side.

➡The procedure listed below is for the LH side.

1. Remove the rear wheel.
2. Separate the rear flexible hose.
3. Separate the rear disc brake caliper assembly.
4. Remove the rear disc.
5. Separate the rear speed sensor wire.
6. Remove the rear axle hub and bearing assembly.
   a. Remove the 4 bolts and the rear axle hub and bearing assembly.

➡Use wire or an equivalent tool to keep the parking brake assembly from hanging down by the parking brake cable assembly.

To install:

7. Install the rear axle hub and bearing assembly.
   a. Install the parking brake assembly and the rear axle hub and bearing assembly with the 4 bolts. Tighten to 59 ft. lbs. (80 Nm).

➡Do not twist the No. 3 parking brake cable assembly when installing it.

8. Inspect the rear axle hub bearing looseness.
9. Inspect the rear axle hub runout.
10. Install the rear speed sensor wire.
11. Install the rear disc.
12. Install the rear disc brake caliper assembly.
13. Install the rear flexible hose.

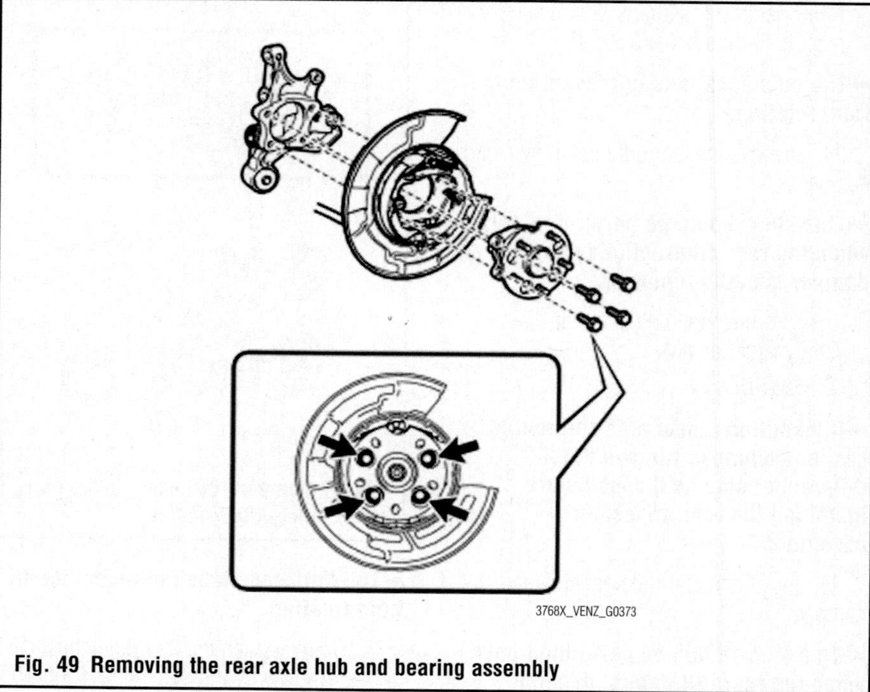

Fig. 49 Removing the rear axle hub and bearing assembly

14. Install the rear wheel. Tighten to 76 ft. lbs. (103 Nm).
15. Check for speed sensor signal.

AWD Vehicles

See Figure 50.

➡Use the same procedure for the RH side and LH side.

➡The procedure listed below is for the LH side.

1. Remove the rear wheel.
2. Separate the rear speed sensor.
   a. Remove the bolt and separate the rear seed sensor from the rear axle carrier sub assembly.

➡Keep the sensor tip and rear speed sensor installation hole free of foreign matter.

3. Remove the rear axle shaft nut.

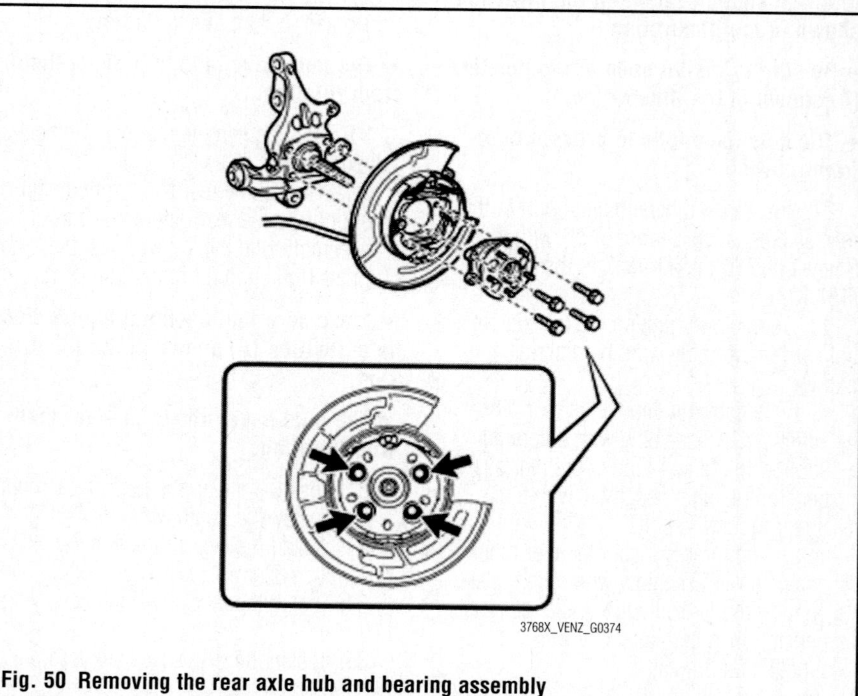

Fig. 50 Removing the rear axle hub and bearing assembly

4. Separate the rear disc brake caliper assembly.
5. Remove the rear disc.
6. Remove the rear axle hub and bearing assembly.
a. Put matchmarks on the rear driveshaft assembly and rear axle hub and bearing assembly.

➡Do not punch the matchmarks.

b. Using a plastic hammer, separate the rear driveshaft assembly from the axle hub and bearing assembly. If it is difficult to separate, tap the end of the rear driveshaft assembly using a brass bar and a hammer.
c. Remove the 4 bolts and the rear axle hub and bearing assembly.

➡Do not rotate the rear driveshaft with the rear axle hub and bearing assembly removed.

➡Use wire or an equivalent tool to keep the parking brake assembly from hanging down by the parking brake cable assembly.

To install:

7. Install the rear axle hub and bearing assembly.
a. Align the matchmarks on the rear driveshaft assembly and the rear axle hub and bearing assembly.

➡Do not rotate the rear driveshaft.

b. Install the parking brake assembly and the rear axle hub and bearing assembly with the 4 bolts. Tighten to 59 ft. lbs. (80 Nm).

➡Do not twist the parking brake cable assembly when installing it.

8. Install the rear axle shaft nut.
a. Clean the threaded parts on the rear driveshaft assembly and a new rear axle shaft nut using a non-residue solvent.

➡Be sure to perform this work for a new rear driveshaft assembly.

➡Keep the threaded parts free of oil and foreign matter.

b. Install the rear disc with the 5 hub nuts.
c. While applying the parking brakes, temporarily install the new rear axle shaft nut. Tighten to 217 ft. lbs. (294 Nm).

➡Stake the nut after inspection for looseness and runout in the following steps.

d. Remove the 5 hub nuts and the rear disc.

9. Install the rear speed sensor.
a. Install the rear speed sensor to the rear axle carrier sub assembly with the bolt. Tighten to 71 inch lbs. (8 Nm).

➡Keep the rear speed sensor tip and sensor installation hole free of foreign matter.

➡Do not twist the rear speed sensor wire when installing it.

10. Inspect the rear axle hub bearing for looseness.
11. Inspect the rear axle hub runout.
12. Install the rear disc.
13. Install the rear disc brake caliper assembly.
14. Stake the rear axle shaft nut.
a. Using a chisel and a hammer, stake the rear axle shaft nut.
15. Install the rear wheel and tighten to 76 ft. lbs. (103 Nm).
16. Check for speed sensor signal.

REAR DRIVESHAFT

REMOVAL & INSTALLATION

See Figure 51.

➡Use the same procedure for the LH side and RH side.

➡The following procedure listed below is for the LH side.

1. Remove the rear wheel.
2. Separate the rear speed sensor.
3. Remove the rear axle shaft nut.
a. Using the special tool (09930-00010) and a hammer, release the staked part of the rear axle shaft nut.

➡Loosen the staked part of the nut completely, otherwise the threads of the driveshaft may be damaged.

b. While applying the brakes, remove the rear axle shaft nut.
4. Separate the rear disc brake caliper assembly.
5. Remove the rear disc.
6. Remove the rear axle hub and bearing assembly.
7. Separate the No. 3 parking brake cable assembly.
8. Remove the rear strut rod assembly.
9. Remove the rear axle carrier sub assembly.
10. Remove the rear driveshaft assembly.
a. Using the special tool (09520-01010, 09520-24010 or 09520-32040), remove the rear driveshaft assembly as shown in the illustration.

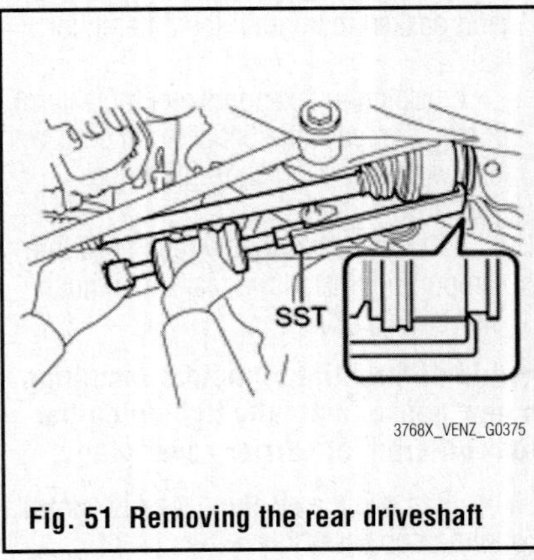

Fig. 51 Removing the rear driveshaft

➡Remove the rear driveshaft assembly while keeping it level.

To install:

11. Install the rear driveshaft assembly.
a. Align the shaft splines and install the rear driveshaft assembly using a screwdriver and hammer.

➡Set the snap ring with the opening facing downward.

➡Be careful not to damage the oil seal, boot or dust cover.

➡Install the driveshaft assembly while keeping it level.

12. Install the rear axle carrier sub assembly.
13. Inspect the rear strut rod assembly.
14. Install the No. 3 parking brake cable assembly.
15. Install the rear axle hub and bearing assembly.
16. Install the rear speed sensor.
17. Install the rear disc.
18. Install the rear disc brake caliper assembly.
19. Install the rear axle shaft nut.
a. Clean the threaded parts on the driveshaft and axle shaft nut using a non residue solvent.

REAR PINION SEAL

REMOVAL & INSTALLATION

See Figures 52 through 57.

1. Remove the center exhaust pipe assembly.
2. Remove the propeller with the center bearing shaft assembly.
3. Drain the differential oil.
a. Using a hexagon wrench (10 mm), remove the rear differential carrier cover plug and gasket.
b. Using a hexagon wrench (10 mm), remove the rear differential drain plug

and gasket, then drain the differential oil.

c. Using a hexagon wrench (10 mm), install the rear differential drain plug with a new gasket. Tighten to 29 ft. lbs. (39 Nm).

d. Using a hexagon wrench (10 mm), temporarily install the rear differential carrier cover plug.

➡Add differential oil before installing a new gasket and fully tightening the rear differential carrier cover plug.

4. Remove the electromagnetic control coupling sub assembly.

a. Disconnect the electromagnetic control coupling sub assembly connector and vacuum hose.

➡Do not damage the electromagnetic control coupling wire harness.

b. Remove the 4 bolts.

c. Using a brass bar and a hammer, tap the electromagnetic control coupling sub assembly to remove the electromagnetic control coupling sub assembly from the rear differential carrier assembly.

➡Do not drop the electromagnetic control coupling sub assembly.

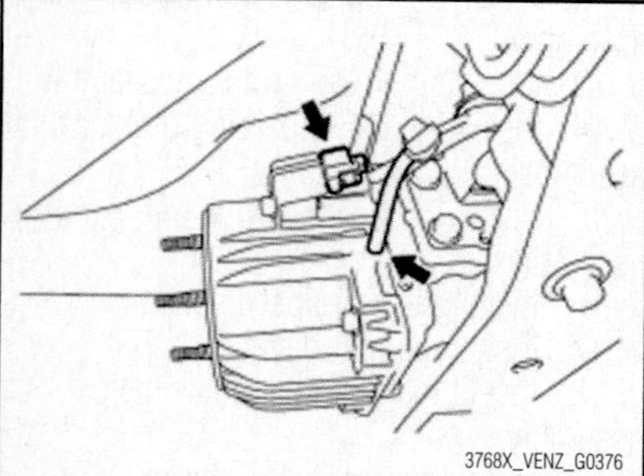

Fig. 52 Disconnecting the electromagnetic control coupling sub assembly connector

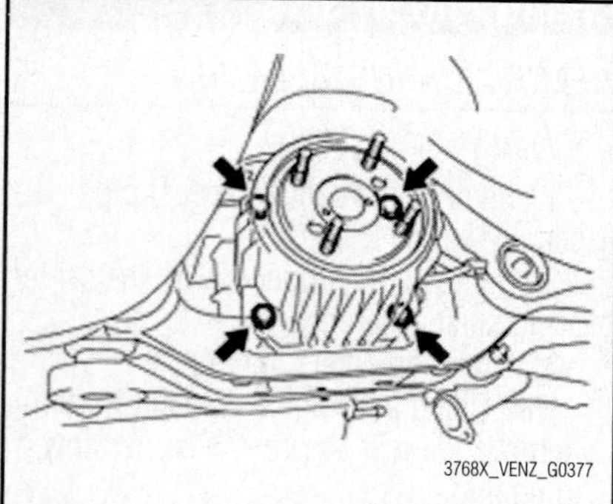

Fig. 53 Removing the electromagnetic control coupling sub assembly

5. Remove the transmission coupling conical spring washer.

a. Remove the transmission coupling conical spring washer from the rear differential carrier assembly.

6. Remove the transmission coupling spacer.

a. Remove the transmission coupling spacer from the rear differential carrier assembly.

7. Remove the diaphragm oil seal.

a. Using the special tool (09308-10010), remove the diaphragm oil seal.

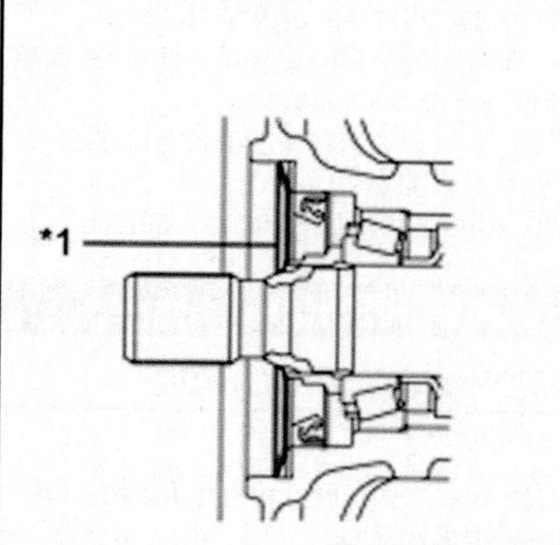

Fig. 54 Removing the transmission coupling conical spring washer

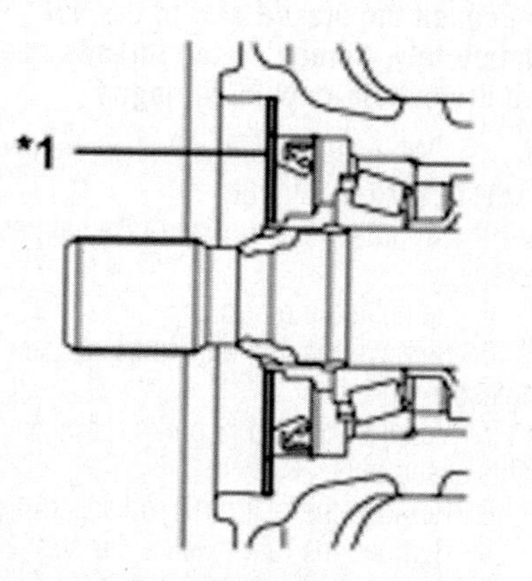

Fig. 55 Removing the transmission coupling spacer

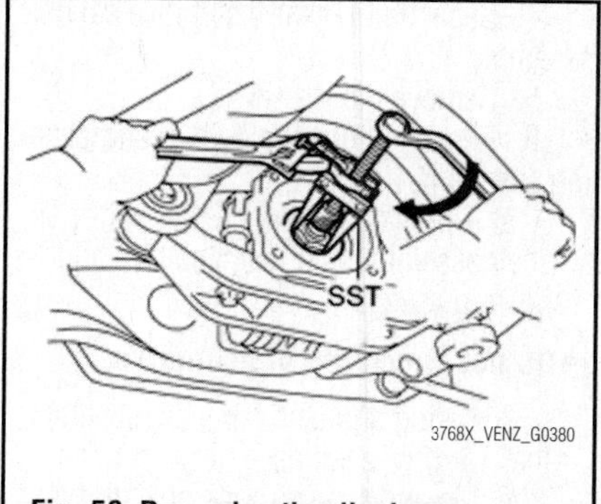

Fig. 56 Removing the diaphragm oil seal

To install:

8. Install the diaphragm oil seal.

a. Using the special tool, install a new diaphragm oil seal. The oil seal driven in depth is 0.0276-0.0512 inch (0.7-1.3 mm).

b. Apply MP grease to the lip of the new diaphragm oil seal.

9. Install the transmission coupling spacer.

a. Install the transmission coupling spacer to the rear differential carrier assembly.

➡Keep the transmission coupling spacer free of oil and foreign matter.

b. Install the transmission coupling conical spring washer to the rear differential carrier assembly.

➡Install the transmission coupling conical spring washer so that the marked surface faces the front of the vehicle (coupling side).

➡Keep the transmission coupling conical spring washer free of oil and foreign matter.

10. Install the electromagnetic control coupling sub assembly.

a. Using a scraper and wire brush, remove the seal packing from the rear differential carrier assembly and electromagnetic control coupling sub-assembly.

➡Do not scratch the installation area.

b. Using a non-residue solvent, remove grease and oil from the contact surfaces of the rear differential carrier assembly and the electromagnetic control coupling sub-assembly.

c. Apply seal packing to the areas indicated in the illustration of the electromagnetic control coupling sub-assembly.

➡Stop applying seal packing after allowing it to overlap with the

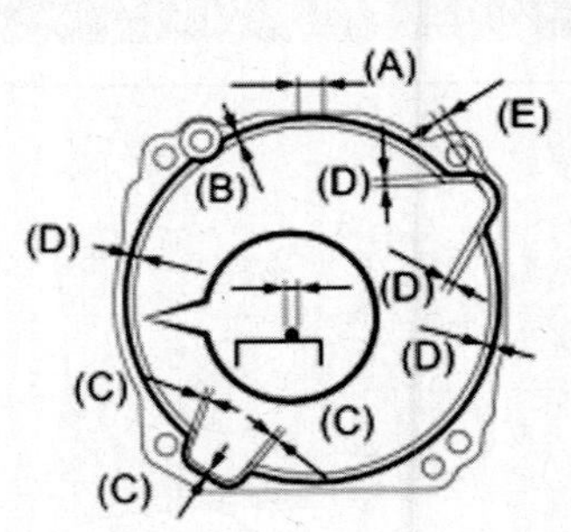

A. 0.394 inch (10 mm)
B. 0.0394 inch (1 mm)
C. 0.0787 inch (2 mm)
D. 0.118 inch (3 mm)
E. 0.197 inch (5 mm)

3768X_VENZ_G0381

Fig. 57 Applying seal packing

beginning of the bead by at least 10 mm (0.394 in.) within range (A) in the illustration.

➡Make sure that the clearance from the center of the bead is within 0.0394 inch (1 mm).

➡Install the electromagnetic control coupling sub-assembly within 10 minutes after applying seal packing.

➡Apply seal packing in a continuous bead 0.0787 to 0.118 inch (2 to 3 mm) in diameter.

d. Apply hypoid gear oil to the splines of the rear drive pinion.

e. Install the electromagnetic control coupling sub-assembly with the 4 bolts. Tighten to 14 ft. lbs. (20 Nm).

➡Do not damage the contact surfaces of the diaphragm oil seal, stud bolt or electromagnetic control coupling sub-assembly.

f. Connect the electromagnetic control coupling sub-assembly connector and vacuum hose.

➡Do not damage the electromagnetic control coupling wire harness.

11. Add differential oil.
    a. Remove the rear differential carrier cover plug.
    b. Add differential oil.
12. Inspect and adjust the differential oil.
13. Install the rear differential carrier cover plug.
14. Inspect for differential oil leak.
15. Temporarily tighten the propeller with the center bearing shaft assembly.
16. Fully tighten the propeller with the center bearing shaft assembly.
17. Install the center exhaust pipe assembly.
18. Inspect and adjust the transfer oil.
19. Inspect for transfer oil leaks.

TRANSFER CASE ASSEMBLY

REMOVAL & INSTALLATION

See Figure 58.

1. Remove the engine.
2. Remove the front No. 1 stabilizer bracket LH and RH.
3. Remove the front stabilizer bar with the front stabilizer link assembly.
4. Remove the steering link assembly.
5. Remove the front frame assembly (3.5L engine).
6. Separate the rear engine mounting insulator assembly (2.7L engine).
7. Remove the front frame assembly (2.7L engine).
8. Remove the starter assembly.
9. Remove the manifold stay (3.7L engine).
10. Remove the transfer stiffener plate RH.
11. Separate the wire harness.
12. Remove the flexible hose bracket sub assembly (3.5L engine).
13. Remove the automatic transaxle.
14. Remove the transfer assembly.
    a. Remove the 2 bolts and 6 nuts.
    b. Using a plastic hammer, remove the transfer assembly from the transaxle assembly.

➡Remove the transfer assembly from the transaxle assembly without tilting it.

➡During removal, do not hold the transfer assembly by the oil seals on either side of the assembly.

To install:

15. Install the transfer assembly.
    a. Install the transfer assembly to the transaxle assembly with 2 new bolts and the 6 nuts. Tighten to 51 ft. lbs. (69 Nm).

➡Install the transfer assembly to the transaxle assembly horizontally.

➡Do not touch the transfer assembly oil seals during installation.

16. Install the automatic transaxle assembly.
17. Install the flexible hose bracket sub assembly (3.5L engine).
18. Install the wire harness.
19. Install the transfer stiffener plate RH.
20. Install the manifold stay (3.5L engine).
21. Install the starter assembly.
22. Install the front frame assembly.
23. Install the rear engine mounting insulator assembly (2.7L engine).
24. Install the steering link assembly.
25. Install the front stabilizer bar with the front stabilizer link assembly.
26. Install the front No. 1 stabilizer bracket LH.
27. Install the front No. 1 stabilizer bracket RH.
28. Install the engine assembly with the transaxle.

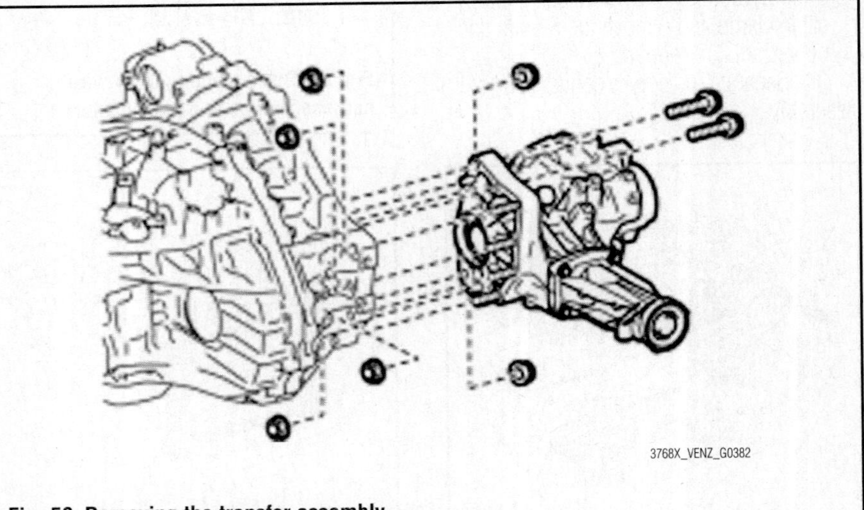

3768X_VENZ_G0382

Fig. 58 Removing the transfer assembly

ENGINE COOLING

ENGINE FAN

REMOVAL & INSTALLATION

1. Remove the radiator removal and installation procedures.

RADIATOR

REMOVAL & INSTALLATION

2.7L Engine

See Figures 59 through 63.

1. Remove the No. 1 engine under cover.
2. Remove the No. 2 engine under cover.
3. Drain the engine coolant.
4. Remove the cool air intake duct seal.
5. Remove the inlet air cleaner assembly.
6. Remove the radiator grille protector.
7. Remove the radiator grille.
8. Remove the low pitched horn assembly.
9. Remove the high pitched horn assembly.
10. Remove the hood lock assembly.
11. Disconnect the outlet reserve tank hose.
    a. Disconnect the outlet reserve tank hose from the radiator.
12. Disconnect the No. 1 radiator hose from the radiator.
13. Remove the No. 2 radiator hose.
14. Disconnect the outlet No. 1 oil cooler hose from the radiator.
15. Remove the upper radiator support.
    a. Disconnect the hood lock control cable clamp and remove the 5 bolts and upper radiator support.
16. Separate the cooler condenser assembly.

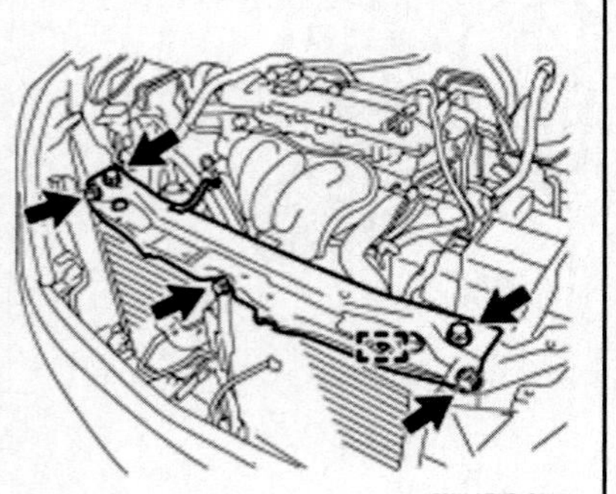

Fig. 59 Removing the upper radiator support

    a. Remove the 4 bolts and separate the cooler condenser assembly.
17. Remove the radiator assembly.
    a. Disconnect the 4 wire harness clamps and 2 connectors.
    b. Remove the radiator assembly and fan assembly with motor.

➡Do not apply any excessive force to the cooler condenser assembly or pipe when removing the radiator assembly.

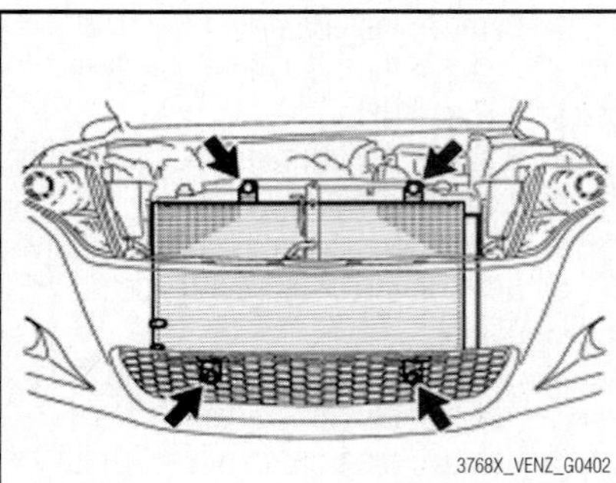

Fig. 60 Separating the cooler condenser assembly

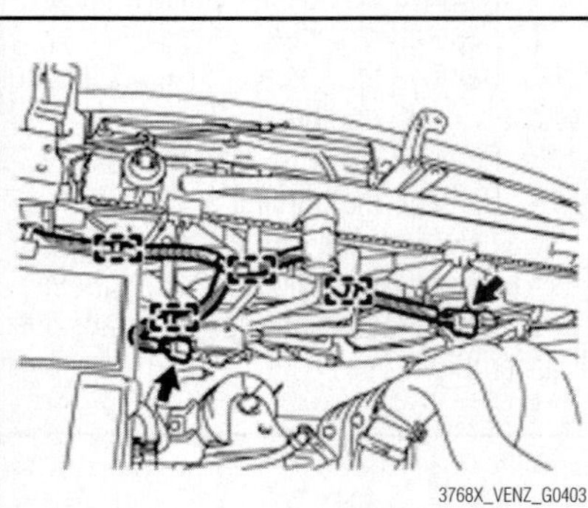

Fig. 61 Disconnecting the 4 wire harness clamps and 2 connectors

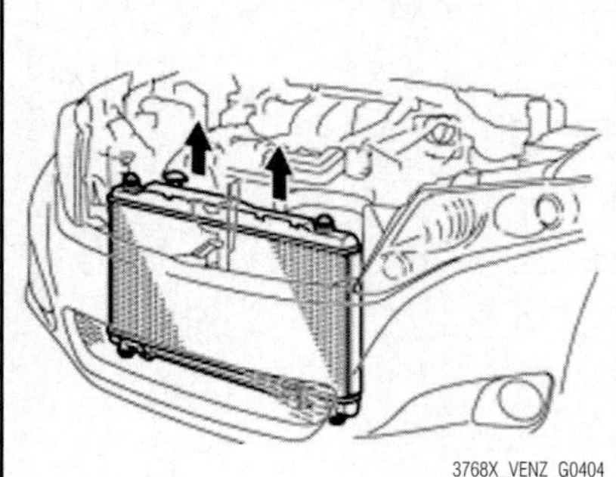

Fig. 62 Removing the radiator assembly and fan assembly with motor

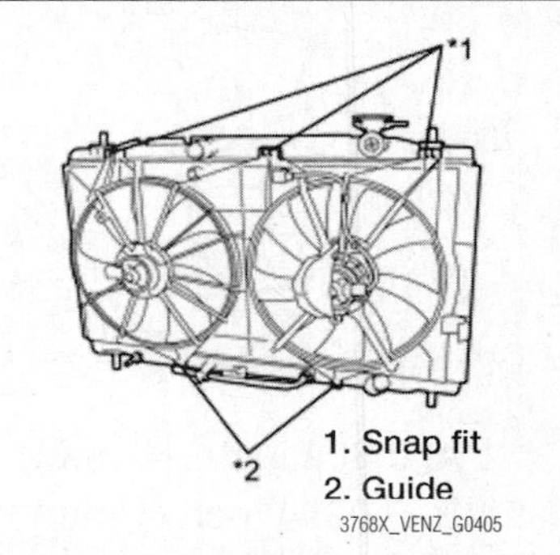

Fig. 63 Removing the fan assembly with the motor

    c. Remove the 2 radiator support cushions and 2 lower radiator supports.
    d. Release the 3 snap fits and pull up the fan assembly with motor from the radiator assembly with the 2 guides to remove the fan assembly with motor.

To install:

18. Install the radiator assembly.
    a. Install the fan assembly with the motor to the radiator with the 2 guides at the bottom and 3 snap fits on the top.
    b. Install the 2 lower radiator supports and 2 radiator support cushions.
    c. Install the radiator assembly and fan assembly with the motor.

➡Do not apply any excessive force to the cooler condenser assembly or pipe when installing the radiator assembly.

    d. Connect the 4 wire harness clamps and 2 connectors.
19. Install the cooler condenser assembly with the 4 bolts. Tighten to 44 inch lbs. (5 Nm).
20. Install the upper radiator support with the 5 bolts and connect the hood lock control cable clamp to the upper radiator support. Tighten to 9 ft. lbs. (12 Nm).
21. Connect the outlet No. 1 oil cooler hose to the radiator.
22. Connect the inlet No. 1 oil cooler hose to the radiator.
23. Install the No. 2 radiator hose to the engine and radiator.
24. Connect the No. 1 radiator hose to the radiator.
25. Connect the outlet reserve tank hose to the radiator.
26. Install the hood lock assembly.
27. Install the low pitched horn assembly.
28. Install the high pitched horn assembly.

29. Install the radiator grille.
30. Install the radiator grille protector.
31. Install the inlet air cleaner assembly.
32. Add engine coolant.
33. Inspect for coolant leaks.
34. Inspect the automatic transaxle fluid.
35. Install the No. 1 engine under cover.
36. Install the No. 2 engine under cover.
37. Install the cool air intake duct seal.

3.5L Engine

See Figures 64 through 66.

1. Remove the No. 1 engine under cover.
2. Remove the No. 2 engine under cover.
3. Drain the engine coolant.
4. Remove the v-bank cover sub assembly.
5. Remove the cool air intake duct seal.
6. Remove the inlet No. 2 air cleaner.
7. Remove the air cleaner cap with the hose.
8. Remove the air cleaner case.
9. Remove the battery.
10. Remove the inlet No. 1 air cleaner.
11. Remove the radiator grille protector.
12. Remove the radiator grille.
13. Remove the low pitched horn assembly.
14. Remove the high pitched horn assembly.
15. Remove the hood lock assembly.
16. Disconnect the outlet reserve tank hose from the radiator.
17. Disconnect the No. 1 radiator hose from the radiator.
18. Remove the No. 2 radiator hose.
19. Disconnect the inlet oil cooler hose from the radiator.
20. Disconnect the outlet oil cooler hose from the radiator.
21. Remove the upper radiator support.

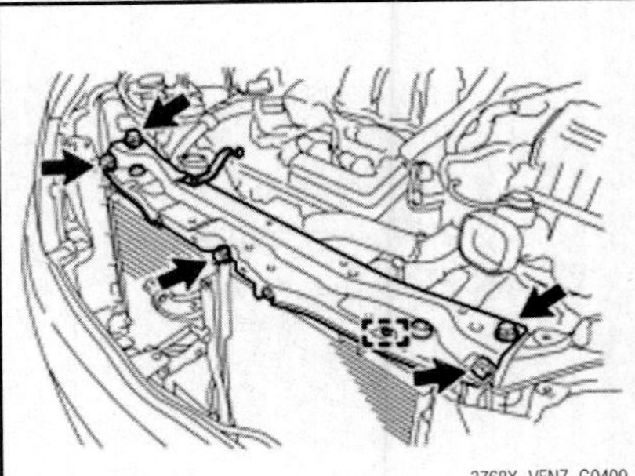

Fig. 64 Removing the upper radiator support

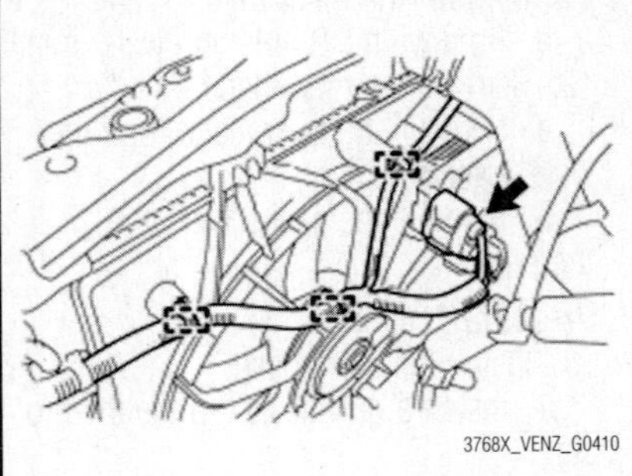

Fig. 65 Disconnecting the clamps and connector

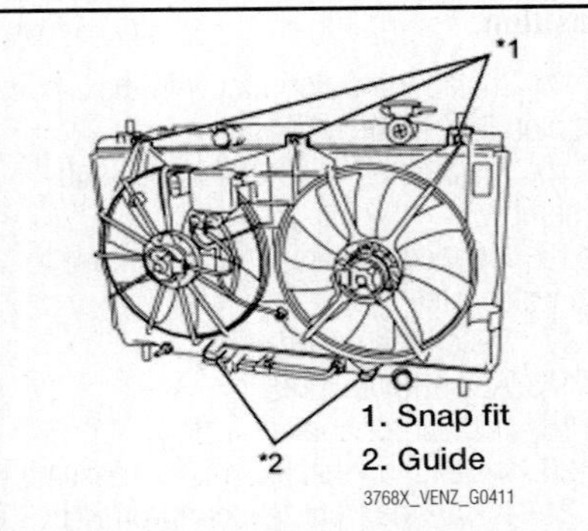

Fig. 66 Removing the fan assembly with motor

a. Disconnect the hood lock control cable clamp and remove the 5 bolts and upper radiator support.

22. Separate the cooler condenser assembly.
    a. Remove the 4 bolts and separate the cooler condenser assembly.
23. Remove the radiator assembly.
    a. Disconnect the 3 wire harness clamps and connector.
    b. Remove the radiator assembly and fan assembly with motor.

➡Do not apply any excessive force to the cooler condenser assembly or pipe when removing the radiator assembly.

c. Remove the 2 radiator support cushions and 2 lower radiator supports.

d. Release the 3 snap fits and pull up the fan assembly with motor from the radiator assembly with the 2 guides to remove the fan assembly with motor.

To install:

24. To install, reverse the removal procedure. Take note to tighten the cooler condenser assembly bolts to 44 inch lbs. (5 Nm). Tighten the radiator support bolts to 9 ft. lbs. (12 Nm).

THERMOSTAT

REMOVAL & INSTALLATION

2.7L Engine

See Figures 67 and 68.

1. Disconnect the cable from the negative battery terminal.

➡When disconnecting the cable, some systems need to be initialized after the cable is reconnected.

2. Remove the No. 1 engine cover sub assembly.
3. Remove the cool air intake duct seal.
4. Remove the No. 1 engine under cover.
5. Remove the No. 2 engine under cover.
6. Drain the engine coolant.
7. Remove the v-ribbed belt.
8. Remove the wire harness clamp bracket.
9. Remove the alternator assembly.
10. Disconnect the No. 2 radiator hose.
11. Remove the water inlet.
    a. Remove the 2 nuts and water inlet.
12. Remove the thermostat. Remove the gasket from the thermostat.

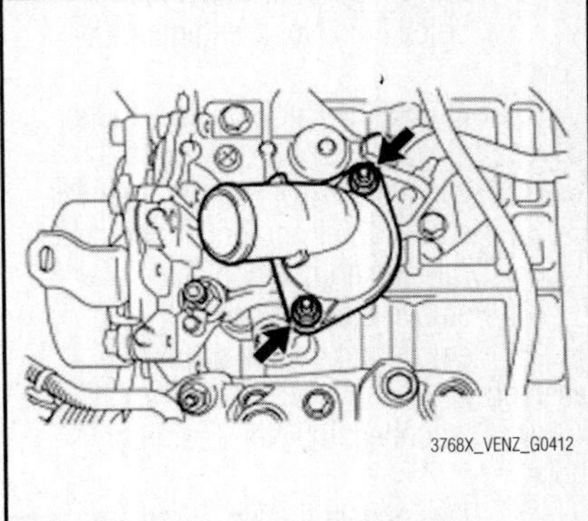

Fig. 67 Removing the water inlet

To install:

13. Install the thermostat.
    a. Install a new gasket to the thermostat.
    b. Install the thermostat with the jiggle valve facing upward.

➡The jiggle valve may be set to within 10°on either side of the prescribed position.

14. Install the water inlet with the 2 nuts. Tighten to 7 ft. lbs. (10 Nm).
15. connect the No. 2 radiator hose.
16. Install the alternator assembly.
17. Install the wire harness clamp bracket.

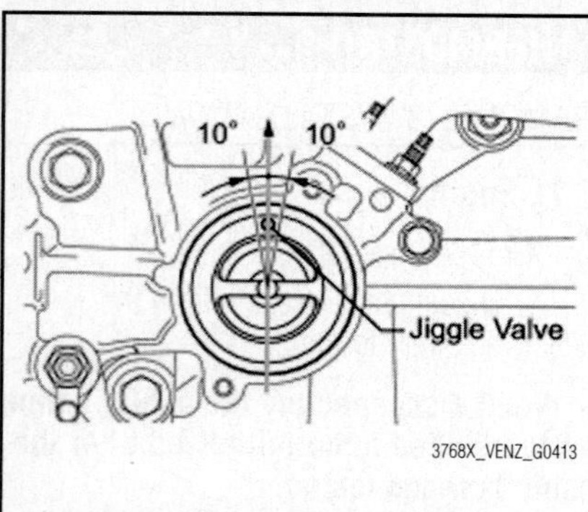

Fig. 68 Installing the thermostat with the jiggle valve facing upward

18. Install the v-ribbed belt.
19. Connect the cable to the negative battery terminal.
20. Add engine coolant.
21. Inspect for coolant leaks.
22. Install the cool air intake duct seal.
23. Install the No. 1 engine under cover.
24. Install the No. 2 engine under cover.
25. Install the No. 1 engine cover sub assembly.

3.5L Engine

See Figures 69 and 70.

1. Remove the front wheel RH.
2. Remove the No. 1 engine under cover.
3. Remove the No. 2 engine under cover.
4. Separate the front fender liner RH.
5. Remove the front fender apron RH.
6. Drain the engine coolant.
7. Remove the v-ribbed belt.
8. Remove the v-bank cover sub assembly.
9. Disconnect the No. 2 radiator hose.
   a. Disconnect the No. 2 radiator hose from the water inlet.
10. Remove the No. 2 idler pulley sub assembly.

Fig. 69 Removing the water inlet

11. Remove the water inlet.
    a. Remove the 2 nuts and water inlet.
12. Remove the thermostat.
    a. Remove the thermostat from the water inlet housing.
    b. Remove the gasket from the thermostat.

To install:

13. Install the thermostat.
    a. Install a new gasket to the thermostat.
    b. Install the thermostat with the jiggle valve facing up.

➡The jiggle valve may be set within 10°on either side of the prescribed position.

14. Install the water inlet with the 2 nuts. Tighten to 7 ft. lbs. (10 Nm).
15. Install the No. 2 idler pulley sub assembly.
16. Connect the No. 2 radiator hose to the water inlet.
17. Install the v-ribbed belt.
18. Add engine coolant.
19. Inspect for coolant leaks.
20. Install the v-bank cover sub assembly.
21. Install the front fender apron RH.
22. Install the front fender liner RH.
23. Install the No. 2 engine under cover.
24. Install the No. 1 engine under cover.
25. Install the front wheel RH.

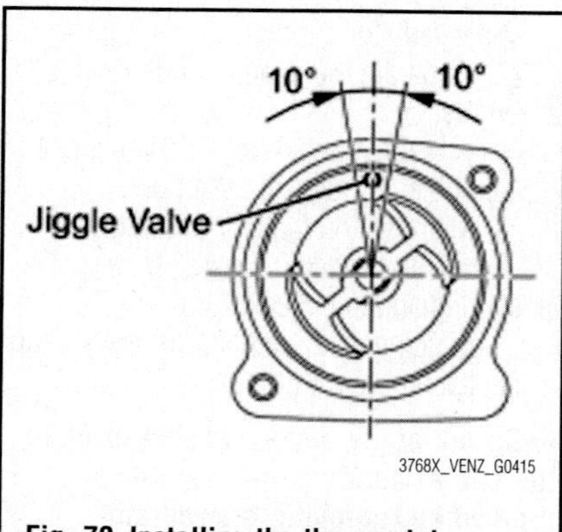

Fig. 70 Installing the thermostat

WATER PUMP

REMOVAL & INSTALLATION

2.7L Engine

See Figure 71.

1. Disconnect the cable from the negative battery terminal.

➡When disconnecting the cable, some systems need to be initialized after the cable is reconnected.

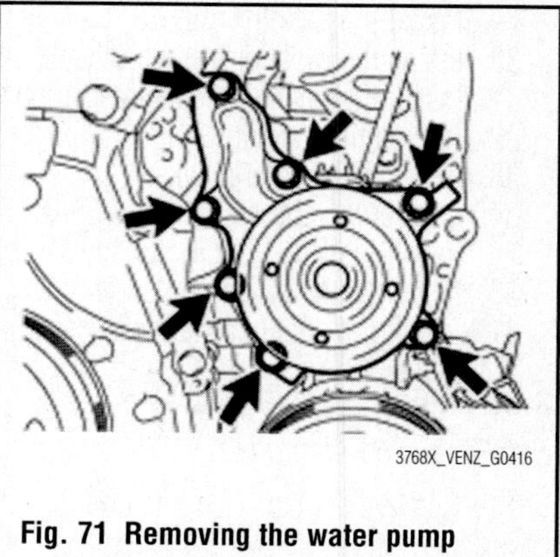

Fig. 71 Removing the water pump

2. Remove the No. 1 engine under cover sub assembly.
3. Remove the cool air intake duct seal.
4. Remove the No. 1 engine under cover.
5. Remove the No. 2 engine under cover.
6. Remove the v-ribbed belt.
7. Remove the wire harness clamp bracket.
8. Remove the alternator assembly.
9. Remove the v-ribbed belt tensioner assembly.
10. Remove the water pump assembly.
    a. Remove the 7 bolts, water pump and water pump gasket.

To install:

11. To install, reverse the removal procedure. Tighten the bolts to 15 ft. lbs. (21 Nm).

3.5L Engine

See Figures 72 through 74.

1. Remove the automatic transaxle assembly.
2. Install the engine on an engine stand.

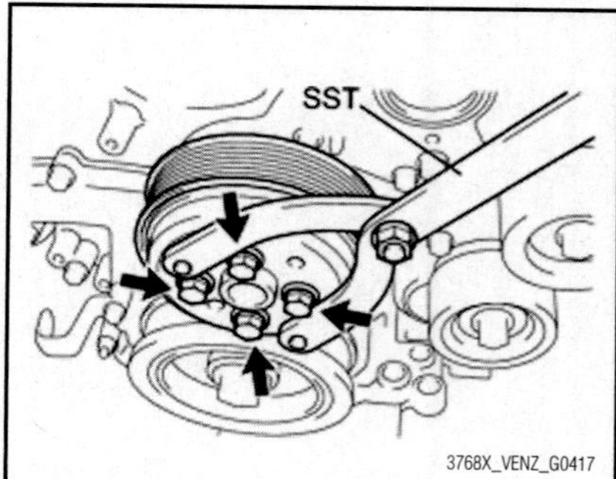

Fig. 72 Removing the water pump pulley

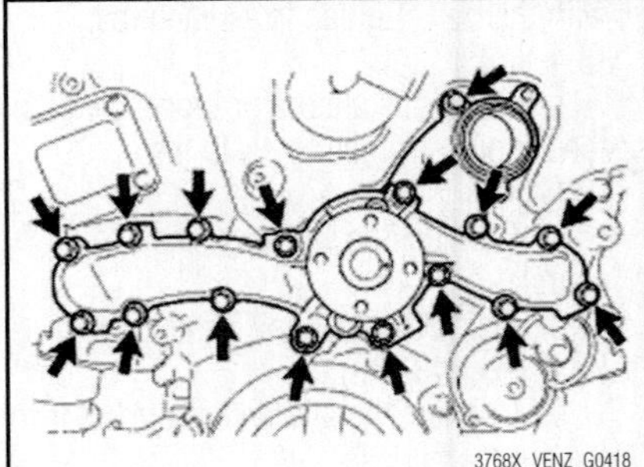

Fig. 73 Removing the water pump assembly

3. Remove the engine hanger.
4. Remove the front No. 1 engine mounting bracket LH.
5. Remove the No. 2 idler pulley sub assembly.
6. Remove the v-ribbed belt tensioner assembly.
7. Remove the water inlet housing.
8. Remove the water pump pulley.
 a. Using the special tool (09960-10010), hold the water pump pulley.
 b. Remove the 4 bolts and water pump pulley.
9. Remove the water pump assembly.
 a. Remove the 16 bolts, water pump assembly and water pump gasket.

To install:

10. Install the water pump assembly.
 a. Install a new water pump gasket and the water pump assembly with the 16 bolts. Tighten bolt A to 15 ft. lbs. (21 Nm). Tighten bolts B and C to 8 ft. lbs. (11 Nm).
11. Install the water pump pulley.
 a. Temporarily install the water pump pulley with the 4 bolts.
 b. Using the special tool, hold the water pump pulley.
 c. Tighten the 4 bolts to 15 ft. lbs. (21 Nm).
12. Install the water inlet housing.
13. Install the v-ribbed belt tensioner assembly.

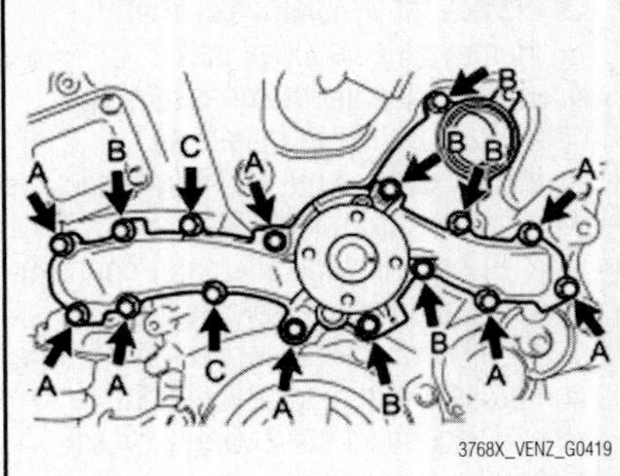

Fig. 74 Installing the water pump assembly

14. Install the No. 2 idler pulley sub assembly.
15. Install the front No. 1 engine mounting bracket LH.
16. Install the engine hanger.
17. Remove the engine from the engine stand.
18. Install the automatic transaxle assembly.

ENGINE ELECTRICAL

CHARGING SYSTEM

ALTERNATOR

REMOVAL & INSTALLATION

2.7L Engine

See Figures 75 and 76.

1. Disconnect the cable from the negative battery terminal.

➡When disconnecting the cable, some systems need to be initialized after the cable is reconnected.

2. Remove the cool air intake duct seal.
3. Remove the No. 1 engine cover sub assembly.
4. Remove the v-ribbed belt.
5. Remove the wire harness clamp bracket.
 a. Detach the wire harness clamp from the clamp bracket.
 b. Remove the bolt and clamp bracket.
6. Remove the alternator assembly.
 a. Disconnect the alternator connector.
 b. Remove the terminal cap.
 c. Remove the nut and disconnect the alternator wire.
 d. Remove the bolt and wire harness clamp bracket.
 e. Remove the 2 bolts and the alternator.

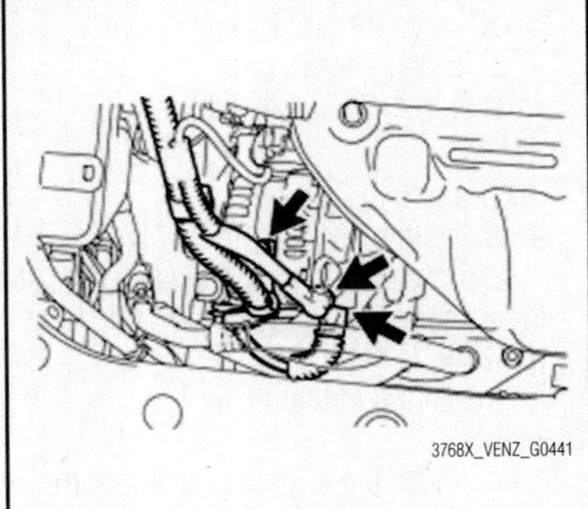

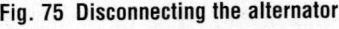

Fig. 75 Disconnecting the alternator

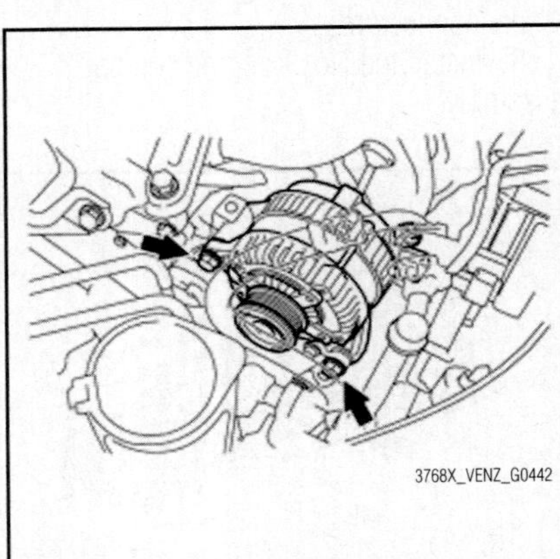

Fig. 76 Removing the alternator

To install:

7. Install the alternator assembly with the 2 bolts. Tighten to 38 ft. lbs. (52 Nm).
 a. Install the wire harness clamp bracket with the bolt. Tighten to 7 ft. lbs. (10 Nm).
 b. Connect the alternator wire with the nut. Tighten to 87 inch lbs. (9.8 Nm).
 c. Install the terminal cap.
 d. Connect the alternator connector.
8. Install the wire harness clamp bracket with the bolt. Tighten to 7 ft. lbs. (10 Nm).
 a. Attach the wire harness clamp to the clamp bracket.
9. Install the v-ribbed belt.
10. Install the No. 1 engine cover sub assembly.
11. Install the cool air intake duct seal.
12. Connect the cable to the negative battery terminal.

➡When disconnecting the cable, some systems need to be initialized after the cable is reconnected.

3.5L Engine

See Figures 77 and 78.

1. Disconnect the cable from the negative battery terminal.

➡When disconnecting the cable, some systems need to be initialized after the cable is reconnected.

2. Remove the radiator assembly.
3. Remove the v-ribbed belt.
4. Remove the alternator assembly.
   a. Remove the terminal cap.
   b. Remove the nut and disconnect the wire harness from terminal B.
   c. Disconnect the alternator connector from the alternator assembly.
   d. Disconnect the connector from the compressor and magnetic clutch.
   e. Disconnect the 2 wire harness clamps.
   f. Remove the 2 bolts.
   g. Remove the bolt and alternator assembly.
   h. Disconnect the wire harness clamp and remove the bolt and alternator bracket.
   i. Remove the bolt and wire harness clamp.

To install:

5. Install the alternator assembly.
   a. Install the wire harness clamp with the bolt. Tighten to 74 inch lbs. (8.4 Nm).
   b. Install the alternator bracket with the 2 bolts. Tighten to 15 ft. lbs. (20 Nm).
   c. Connect the wire harness clamp.
   d. Install the 2 bolts. Tighten to 32 ft. lbs. (43 Nm).
   e. Connect the alternator connector to the alternator assembly.
   f. Install the alternator wire with the nut. Tighten to 87 inch lbs. (9.8 Nm).
   g. Install the terminal cap.
   h. Connect the 2 wire harness clamps.
   i. Connect the magnetic clutch connector to the compressor and magnetic clutch.
6. Install the v-ribbed belt.
7. Install the radiator assembly.
8. Connect the cable to the negative battery terminal.

➡When disconnecting the cable, some systems need to be initialized after the cable is reconnected.

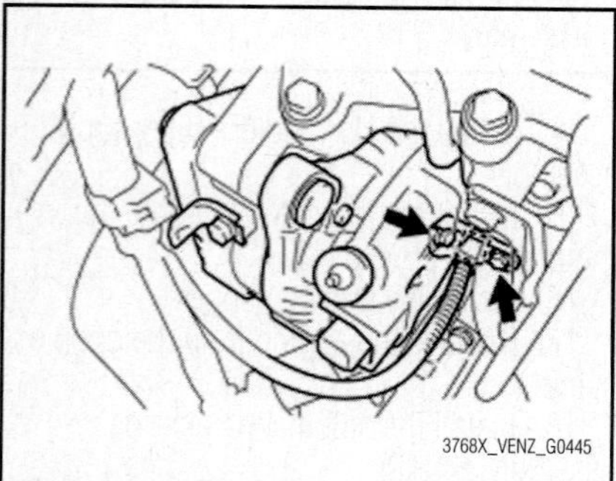

Fig. 77 Removing the bolt and alternator assembly

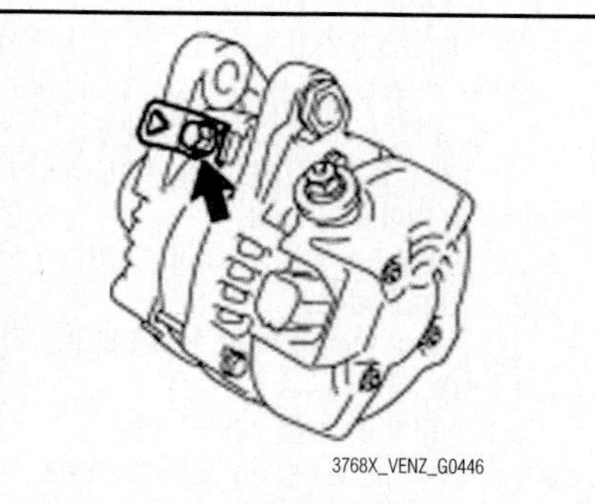

Fig. 78 Removing the bolt and wire harness clamp

ENGINE ELECTRICAL

IGNITION SYSTEM

FIRING ORDER

Firing order for the 2.7L engine: 1–3–4–2.
Firing order for 3.5L engine: 1–2–3–4–5–6

IGNITION COIL

REMOVAL & INSTALLATION

2.7L Engine

See Figures 79 and 80.

1. Remove the No. 1 engine cover sub assembly.
2. Remove the ignition coil assembly.
   a. Disconnect the 4 ignition coil assembly connectors.
   b. Remove the 4 bolts and 4 ignition coil assemblies.
3. Remove the spark plug.
   a. Using a spark plug wrench, remove the 4 spark plugs.

To install:

4. Install the spark plug.
   a. Using a spark plug wrench, install the 4 spark plugs. Tighten to 18 ft. lbs. (25 Nm).
5. Install the ignition coil assembly.
   a. Install the 4 ignition coil assemblies with the 4 bolts. Tighten to 7 ft. lbs. (10 Nm).
   b. Connect the 4 ignition coil assembly connectors.
6. Install the No. 1 engine coil sub assembly.

3.5L Engine

See Figures 81 and 82.

1. Remove the No. 1 engine under cover.
2. Remove the No. 2 engine under cover.
3. Drain the engine coolant.
4. Remove the windshield wiper motor and link assembly.
5. Remove the cowl top panel outer sub assembly.
6. Remove the cool air intake duct seal.
7. Remove the v-bank cover sub assembly.
8. Remove the air cleaner cap with hose.

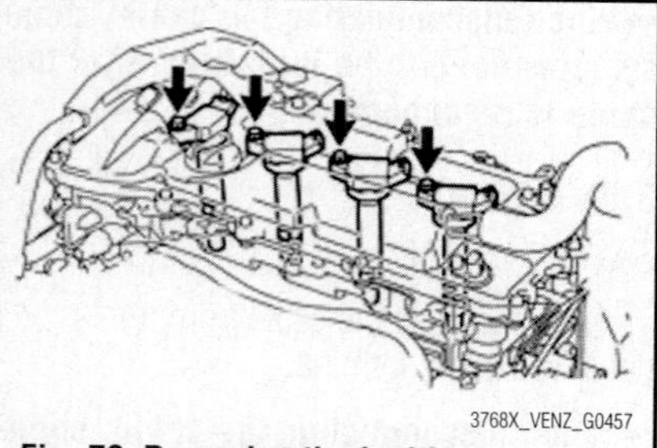

Fig. 79 Removing the ignition coil assembly

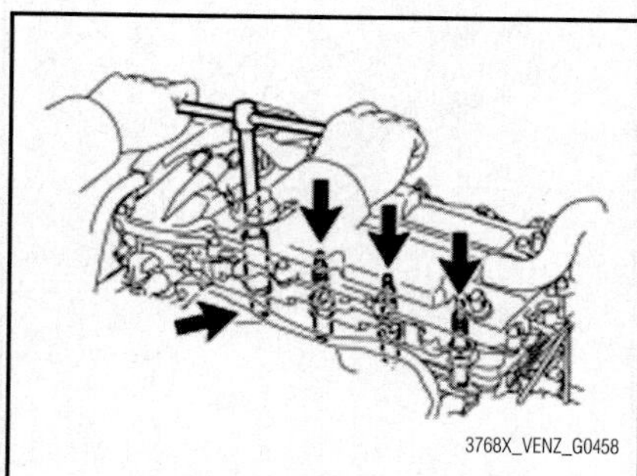

Fig. 80 Removing the spark plugs

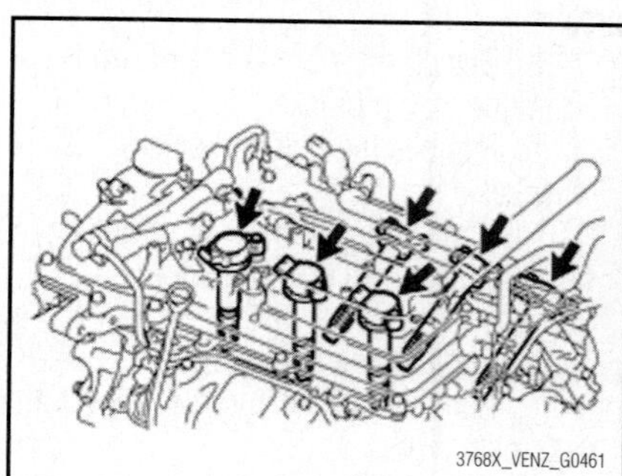

Fig. 81 Removing the 6 bolts and 6 ignition coils

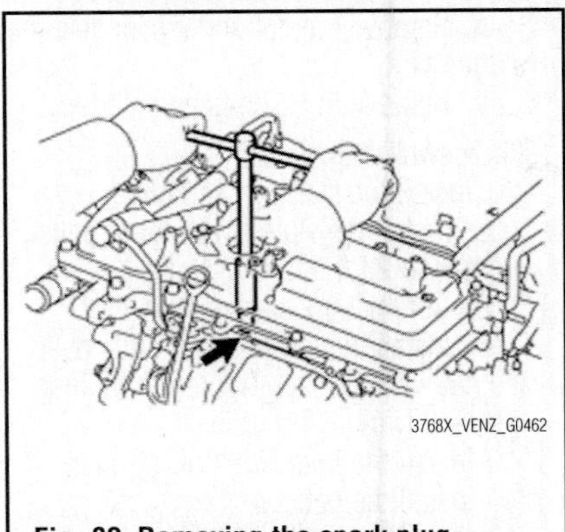

Fig. 82 Removing the spark plug

9. Remove the intake air surge tank assembly.
10. Remove the No. 1 surge tank stay.
11. Remove the ignition coil assembly.
    a. Disconnect the 3 wire harness clamps.
    b. Disconnect the 6 ignition coil connectors.
    c. Remove the 6 bolts and 6 ignition coils.
12. Remove the spark plug.
    a. Remove the 6 spark plugs.

To install:

13. Install the spark plug.
    a. Install the 6 spark plugs and tighten to 13 ft. lbs. (18 Nm).
14. Install the ignition coil assembly.
    a. Install the 6 ignition coils with the 6 bolts. Tighten to 7 ft. lbs. (10 Nm).
    b. Connect the 6 ignition coil connectors.
    c. Connect the 3 wire harness clamps.
15. Install the No. 1 surge tank stay.
16. Install the intake air surge tank assembly.
17. Install the air cleaner cap with the hose.
18. Add engine coolant.
19. Inspect for engine coolant leaks.
20. Install the v-bank cover sub assembly.
21. Install the cool air intake duct seal.
22. Install the cowl top panel outer sub assembly.
23. Install the windshield wiper motor and link assembly.
24. Install the No. 2 engine under cover.
25. Install the No. 1 engine under cover.

IGNITION TIMING

INSPECTION

3.5L Engine

1. Warm up the engine.
2. Using SST: 09843-18040, connect terminals 13 (TC) and 4 (CG) of the DLC3.

WARNING

Confirm the terminal numbers before connecting them. Connecting the wrong terminals can damage the engine.

➡Turn off all electrical systems before connecting the terminals.

➡Perform this inspection after the cooling fan motor is turned off.

3. Remove the v-bank cover.
4. Pull out the red lead wire harness.
5. Connect the tester terminal of the timing light to the red lead wire.

➡Use a timing light which can detect the first signal.

6. Check the ignition timing at idle. Standard ignition timing: 8 to 12°BTDC at idle.

➡When checking the ignition timing, the transmission should be in the neutral position.

➡Run the engine at 1000 to 1300 rpm for 5 seconds, and then check that the engine rpm returns to idle speed.

7. Disconnect terminals 13 (TC) and 4 (CG) of the DLC3.
8. Check the ignition timing at idle. Standard ignition timing: 7 to 24°BTDC at idle.
9. Confirm that the ignition timing advances immediately when the engine rpm is increased.
10. Remove the timing light from the engine.

ADJUSTMENT

All engines are equipped with a Distributorless Ignition System (DIS). No timing adjustment is possible.

SPARK PLUGS

REMOVAL & INSTALLATION

Refer to Ignition Coil.

ENGINE ELECTRICAL

STARTING SYSTEM

STARTER

REMOVAL & INSTALLATION

2.7L Engine

See Figure 83.

1. Disconnect the cable from the negative battery terminal.

➡When disconnecting the cable, some systems need to be initialized after the cable is reconnected.

2. Remove the cool air intake duct seal.
3. Remove the No. 1 engine cover sub assembly.
4. Remove the battery.
5. Remove the inlet air cleaner assembly.
6. Remove the starter assembly.
    a. Disconnect the starter connector.
    b. Open the terminal cap, remove the nut and disconnect the starter wire.
    c. Remove the 2 bolts and starter.

To install:

7. Install the starter.
    a. Install the starter with the bolts. Tighten to 27 ft. lbs. (37 Nm).

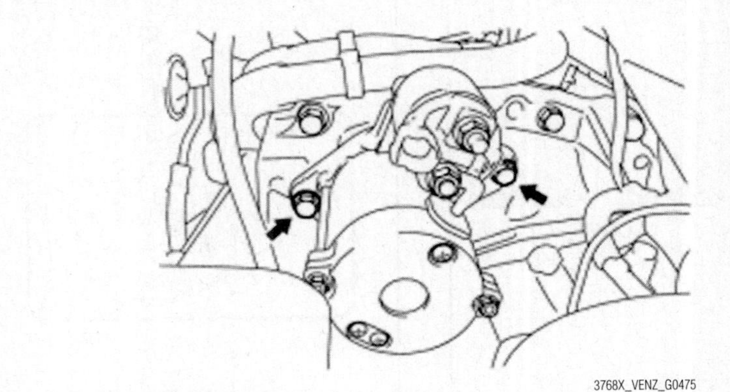

Fig. 83 Removing the starter

b. Connect the starter connector.

c. Install the terminal nut and cover the nut with the cap. Tighten to 87 inch lbs. (9.8 Nm).

8. Install the battery.

9. Install the No. 1 engine cover sub assembly.

10. Install the cool air intake duct seal.

11. Connect the cable to the negative battery terminal.

➡When disconnecting the cable, some systems need to be initialized after the cable is reconnected.

3.5L Engine

See Figure 83.

1. Remove the cool air intake duct seal.

2. Disconnect the cable from the negative battery terminal.

➡When disconnecting the cable, some systems need to be initialized after the cable is reconnected.

3. Remove the v-bank cover sub assembly.

4. Remove the inlet No. 2 air cleaner.

5. Remove the battery.

6. Remove the inlet No. 1 air cleaner.

7. Remove the starter assembly.

a. Disconnect the starter connector.

b. Remove the terminal cap.

c. Remove the nut and disconnect the starter wire.

d. Remove the 2 bolts and starter.

To install:

8. Install the starter assembly.

a. Install the starter with the 2 bolts. Tighten to 27 ft. lbs. (37 Nm).

b. Connect the starter connector.

c. Install the terminal nut and cover the nut with the cap. Tighten to 87 inch lbs. (9.8 Nm).

9. Install the inlet No. 1 air cleaner.

10. Install the battery.

11. Install the inlet No. 2 air cleaner.

12. Install the v-bank cover sub assembly.

13. Connect the cable to the negative battery terminal.

14. Install the cool air intake duct seal.

ENGINE MECHANICAL

ACCESSORY DRIVE BELTS

INSPECTION

See Figure 84.

1. Check the belt for wear, cracks or other signs of damage.

2. If any of the following defects is found; replace the V-ribbed belt:

a. The belt is cracked.

b. The belt is worn out to the extent that the cords are exposed.

c. The belt has chunks missing from the ribs.

3. Check that the belt fits properly in the ribbed grooves.

➡Check with your hand to confirm that the belt has not slipped out of the groove on the bottom of the pulley. If it has slipped out, replace the V-ribbed belt. Install a new V-ribbed belt correctly.

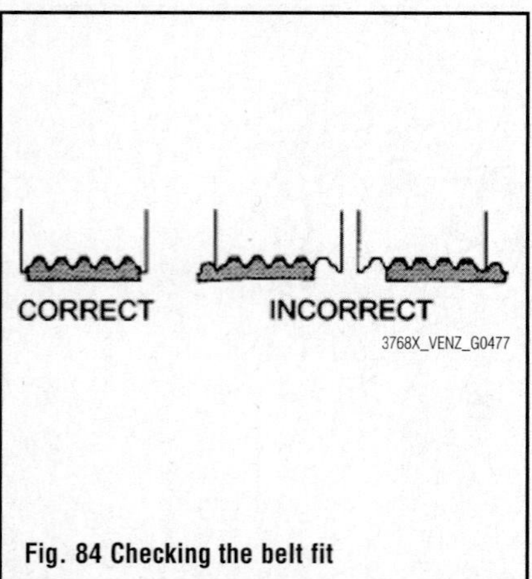

Fig. 84 Checking the belt fit

REMOVAL & INSTALLATION

2.7L Engine

See Figures 85 and 86.

1. Remove the front wheel.

2. Remove the No. 1 engine under cover.

3. Separate the front fender liner RH.

4. Remove the front fender apron seal RH.

5. Remove the v-ribbed belt.

a. Attach a wrench to the hexagonal portion of the belt tensioner as shown in the illustration, rotate the belt tensioner clockwise, and remove the V-ribbed belt.

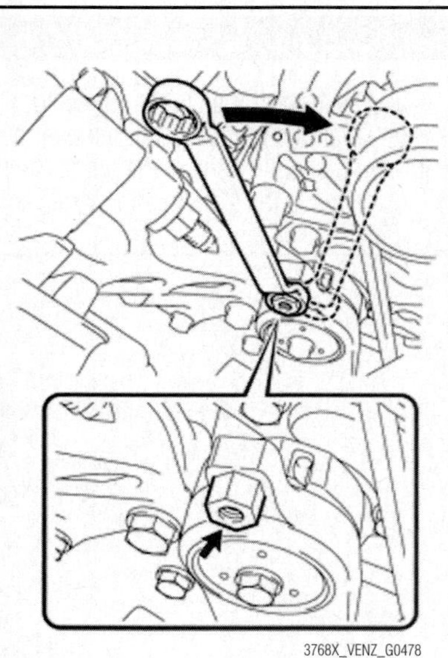

Fig. 85 Removing the drive belt—2.7L engine

To install:

6. Install the v-ribbed belt.

a. Set the V-ribbed belt onto each part as shown in the illustration except the water pump pulley.

b. Loosen the V-ribbed belt by turning the belt tensioner clockwise.

c. Set the V-ribbed belt onto the water pump pulley.

➡Make sure that the belt is attached to each pulley. In particular, make sure that the belt is securely fitted into the grooves of the crankshaft pulley.

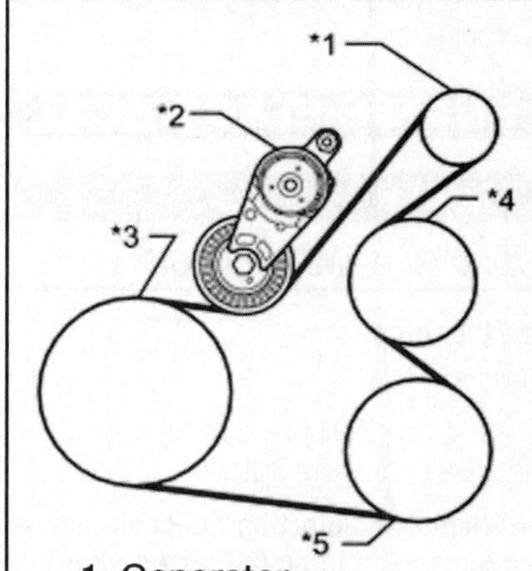

Fig. 86 Installing the drive belt—2.7L engine

7. Install the front fender apron seal RH.
8. Install the front fender liner RH.
9. Install the No. 1 engine under cover.
10. Install the front wheel RH. Tighten to 76 ft. lbs. (103 Nm).

3.5L Engine

See Figures 87 and 88.

1. Remove the front wheel RH.
2. Remove the No. 1 engine under cover.
3. Separate the front fender liner RH.
4. Remove the front fender apron RH.
5. Remove the v-ribbed belt.
   a. Release the V-ribbed belt tension by turning the V-ribbed belt tensioner counterclockwise, and remove the V-ribbed belt from the V-ribbed belt tensioner.
   b. While turning the V-ribbed belt tensioner counterclockwise, aligns with its holes, and then insert a 5 mm hexagon wrench into the holes to fix the V-ribbed belt tensioner.

To install:

6. Install the v-ribbed belt.
   a. Turn the V-ribbed belt tensioner counterclockwise and remove the 5 mm hexagon wrench.
   b. After installing the V-ribbed belt, check that it fits properly in the ribbed grooves. Confirm that the belt has not slipped out of the grooves on the bottom of the crank pulley by hand.
7. Install the front fender apron RH.
8. Install the front fender liner RH.
9. Install the No. 1 engine under cover.
10. Install the front wheel RH. Tighten to 76 ft. lbs. (103 Nm).

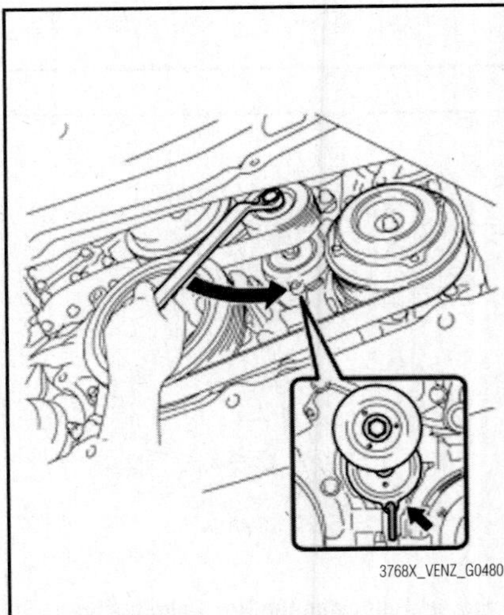

Fig. 87 Releasing the v-ribbed belt tension

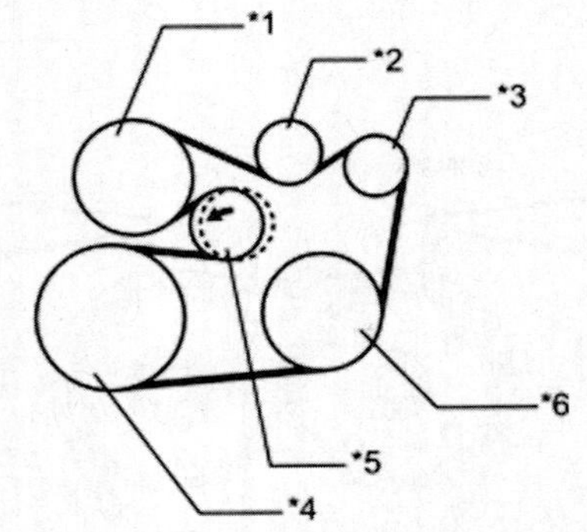

Fig. 88 Belt routing—3.5L engine

BALANCE SHAFT

REMOVAL & INSTALLATION

2.7L Engine

See Figures 89 through 108.

1. Remove the timing chain cover sub assembly.
2. Set the No. 1 cylinder to TDC/compression.
   a. Temporarily install the crankshaft pulley bolt.

➡"A" is not a timing mark.

   b. Rotate the crankshaft clockwise so that the timing marks on the crankshaft timing gear and camshaft timing gears are as shown in the illustration.

➡If the timing marks do not align, rotate the crankshaft clockwise again and align the timing marks.

   c. Remove the crankshaft pulley bolt.
3. Remove the timing chain guide.
4. Remove the No. 1 chain tensioner assembly.
   a. Allow the plunger to extend slightly, and then rotate the stopper plate counterclockwise to release the lock. Once the lock is released, push the plunger into the tensioner.
   b. Move the stopper plate clockwise to set the lock, and insert a pin (*1) into the stopper plate hole.
   c. Remove the 2 bolts, chain tensioner and gasket.
5. Remove the chain tensioner slipper.
   a. Remove the bolt and chain tensioner slipper.
6. Remove the chain sub assembly.
7. Remove the No. 1 chain vibration damper.
   a. Remove the 2 bolts and chain vibration damper.
8. Remove the camshaft timing gear assembly.
   a. Hold the hexagonal portion of the camshaft with a wrench and remove the bolt and camshaft timing gear.

➡Be careful not to damage the cylinder head or spark plug tube with the wrench.

➡Do not disassemble the camshaft timing gear.

9. Remove the camshaft timing exhaust gear assembly.
   a. Hold the hexagonal portion of the camshaft with a wrench and remove the bolt and camshaft timing exhaust gear.

➡Be careful not to damage the cylinder head or spark plug tube with the wrench.

➡Do not disassemble the camshaft timing exhaust gear.

10. Remove the camshaft housing sub assembly.
   a. Uniformly loosen and remove the 20 bearing cap bolts in the sequence shown in the illustration.
   b. Remove the camshaft housing by prying between the cylinder head and camshaft housing with a screwdriver.

➡Tape the screwdriver tip before use.

➡Be careful not to damage the contact surfaces of the cylinder housing.

11. Remove the camshaft bearing cap.
   a. Remove the 11 bearing cap bolts in the sequence shown in the illustration.
   b. Remove the 5 bearing caps.

➡Arrange the removed parts in the correct order.

12. Remove the oil control valve filter.
13. Remove the No. 1 camshaft bearing.
14. Remove the camshaft.
   a. Remove the No. 1 and No. 2 camshafts.
15. Remove the No. 2 camshaft bearing.
16. Remove the No. 1 valve rocker arm sub assembly.
   a. Remove the 16 valve rocker arms from the cylinder head.

➡Arrange the removed parts in the correct order.

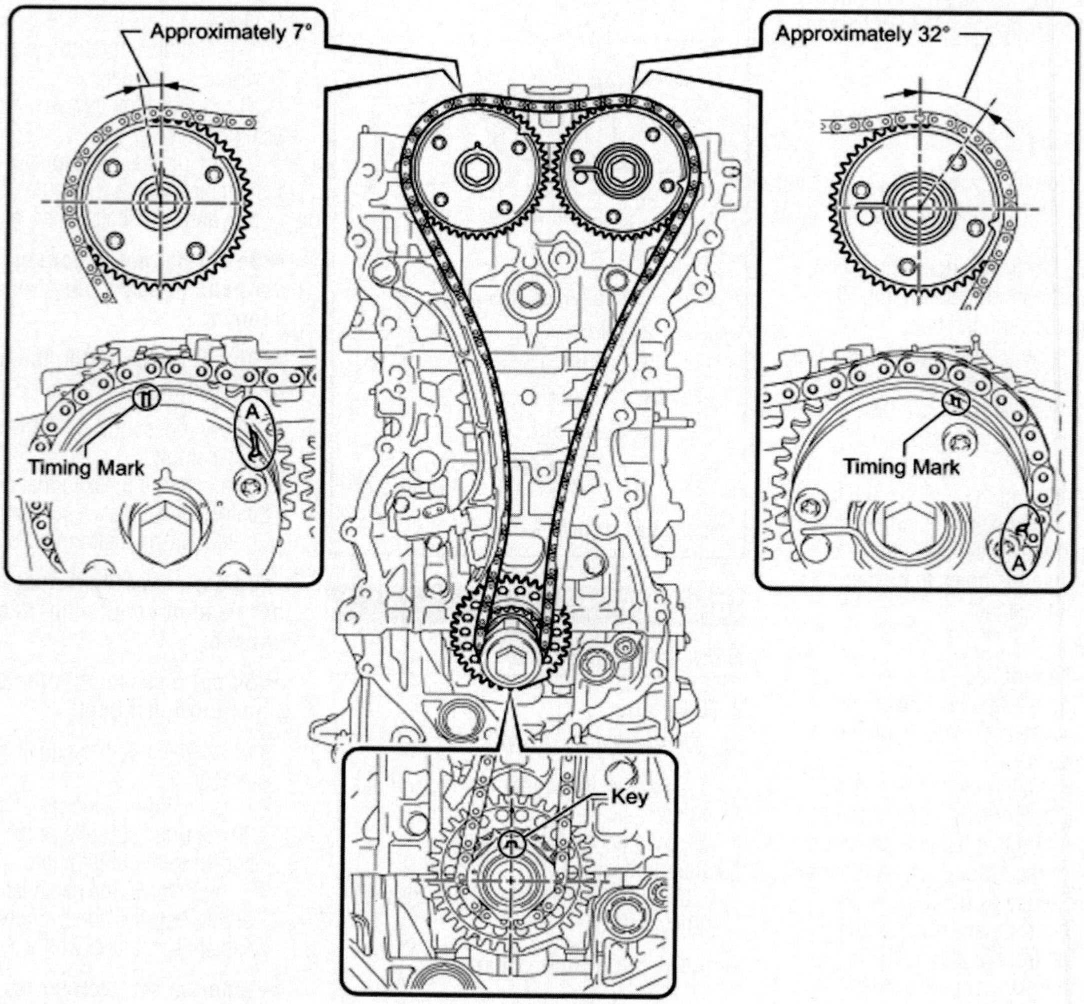

Approximately 7°
Approximately 32°
Timing Mark
A
Timing Mark
A
Key
3768X_VENZ_G0482

Fig. 89 Setting the No. 1 cylinder to TDC/compression

*1
3768X_VENZ_G0484

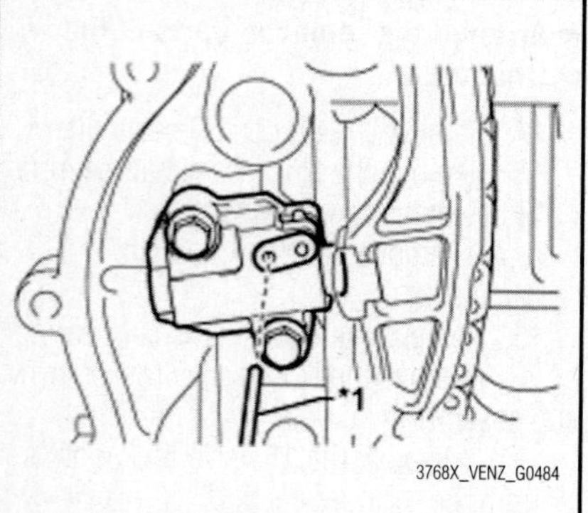

Fig. 90 Inserting the pin into the stopper plate hole

17. Remove the valve lash adjusters from the cylinder head.

➡**Arrange the removed parts in the correct order.**

To install:

18. Set the camshaft gear assembly.

➡**When installing the camshaft timing gear, release the lock pin and set the camshaft timing gear to the advanced position before installation.**

a. Check the camshaft timing gear position.

➡**If the camshaft timing gear is set to the advanced position, do not let the**

3768X_VENZ_G0485

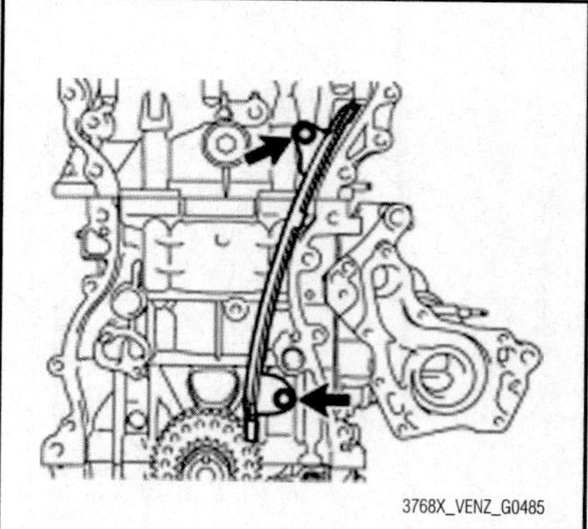

Fig. 91 Removing the No. 1 chain vibration damper

camshaft timing gear rotate clockwise during installation.

➡If the camshaft timing gear rotates to the retarded position, release the lock pin and set the camshaft timing gear to the advanced position.

b. Align and attach the knock pin of the No. 1 camshaft with the pin hole of the camshaft timing gear.

c. Check that there is no clearance between the camshaft timing gear and camshaft flange.

d. Secure the camshaft in place by hand, and then install the installation bolt of the camshaft timing gear by hand.

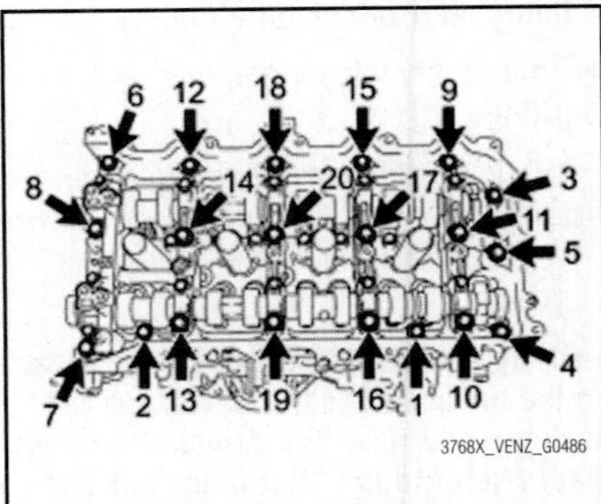

Fig. 92 Loosening the bearing cap bolts

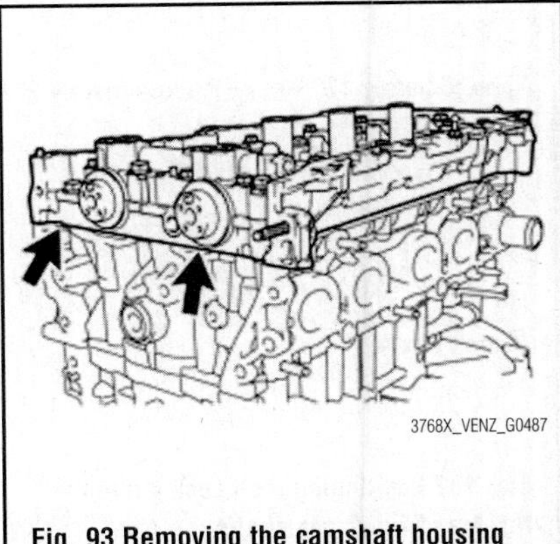

Fig. 93 Removing the camshaft housing

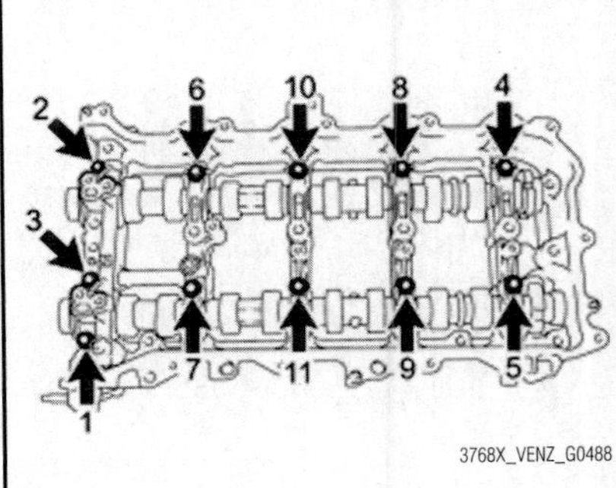

Fig. 94 Removing the camshaft bearing caps

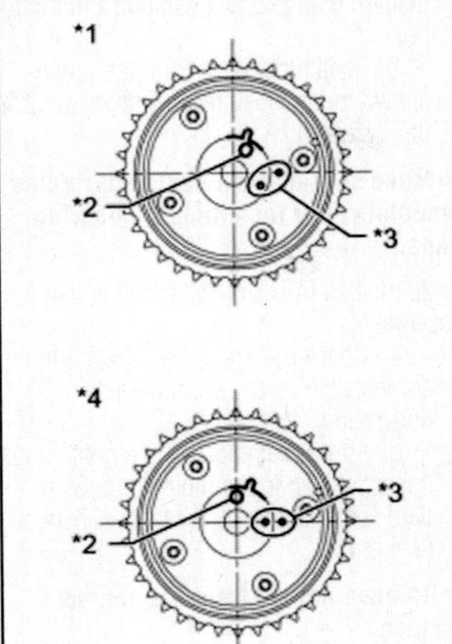

Fig. 95 Checking the camshaft timing gear position

➡Do not use any tools to install the bolt. If the bolt is installed using a tool, the lock pin will be damaged.

e. Release the lock pin.

f. Clean the camshaft journal with non-residue solvent.

g. Cover the 4 oil paths of the cam journal with vinyl tape as shown in the illustration.

➡There are 4 oil paths in the grooves of the camshaft. Plug three of the paths with pieces of rubber.

h. Open a hole at port A shown in the illustration.

i. While applying approximately 29 psi (200 kPa) of air pressure to the oil path, forcibly turn the camshaft timing gear assembly in the advance direction (counterclockwise).

CAUTION

Cover the paths with a piece of cloth when applying pressure to keep oil from splashing.

➡Do not allow the camshaft timing gear assembly to lock. If it locks, release the lock pin again.

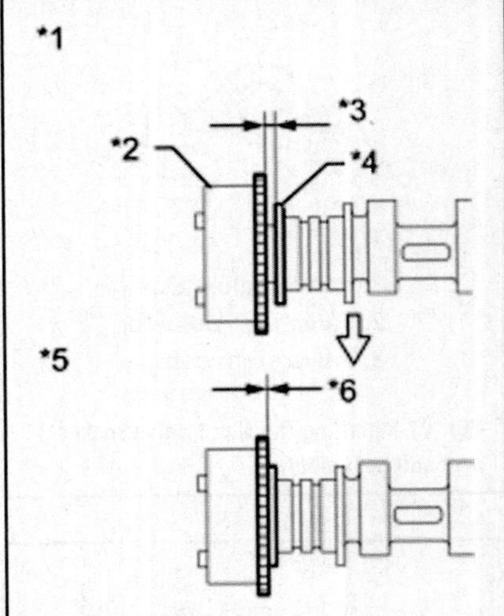

Fig. 96 Checking clearance between the camshaft timing gear and camshaft flange

➡The camshaft timing gear assembly may be turned in the advance direction without applying any force.

➡If enough air pressure cannot be applied because of air leakage from the port, releasing the lock pin may be difficult.

j. Remove the vinyl tape and rubber pieces from the camshaft.

k. Remove the bolt and camshaft timing gear.

➡Do not allow the camshaft timing gear assembly to lock. If it locks, release the lock pin again.

19. Install the valve lash adjuster assembly.

a. Inspect the valve lash adjuster before installing it.

b. Install the 16 lash adjusters to the cylinder head.

➡Install the lash adjuster to the same place it was removed from.

20. Install the No. 1 valve rocker arm sub assembly.

a. Apply engine oil to the lash adjuster tips and valve stem caps.

b. Install the 16 valve rocker arms as shown in the illustration.

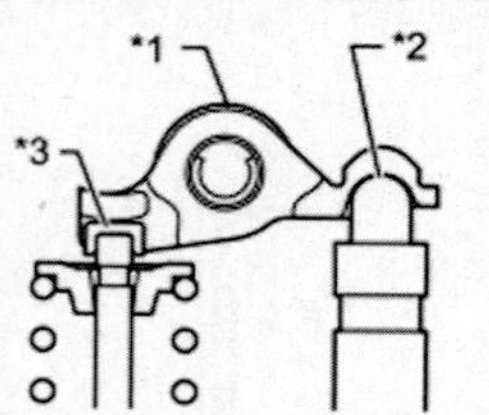

1. Valve rocker arm
2. Valve lash adjuster
3. Valve stem cap

3768X_VENZ_G0491

Fig. 97 Installing the No. 1 valve rocker arm sub assembly

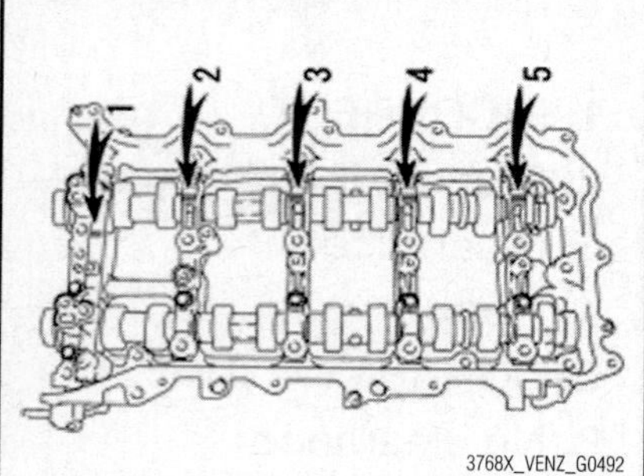

Fig. 98 Confirming marks and numbers on the camshaft bearing caps

21. Install the No. 2 camshaft bearing.
22. Install the No. 1 camshaft bearing.
23. Install the oil control valve filter.
24. Install the camshaft.
    a. Clean the camshaft journals, camshaft housing and bearing caps.
    b. Apply a light coat of engine oil to the camshaft journal, camshaft housing and bearing caps.
    c. Install the No. 1 and No. 2 camshafts to the camshaft housing.
25. Install the camshaft bearing cap.
    a. Confirm the marks and numbers on the camshaft bearing caps and place

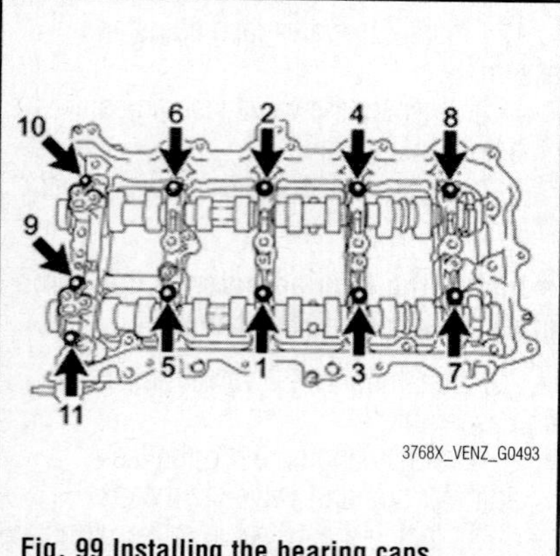

Fig. 99 Installing the bearing caps

them in their proper positions and directions.

b. Install the 11 bolts in the order shown in the illustration. Tighten to 12 ft. lbs. (16 Nm).

➡Make sure that the camshaft rotates smoothly after installing the bearing caps.

26. Install the camshaft housing sub assembly.

a. Check that the valve rocker arms are installed as shown in the illustration.

b. Apply seal packing in a continuous line as shown in the illustration. Standard seal diameter is 0.118-0.157 inch (3-4 mm).

➡Remove any oil from the contact surface.

➡Install the camshaft housing within 3 minutes and tighten the bolts within 10 minutes after applying seal packing.

c. Position the knock pin of the No. 1 and No. 2 camshafts as shown in the illustration.

d. Install the camshaft housing, and then install the 20 bolts in the order shown in the illustration. Tighten to 20 ft. lbs. (27 Nm).

➡Do not apply oil for at least 4 hours after the installation.

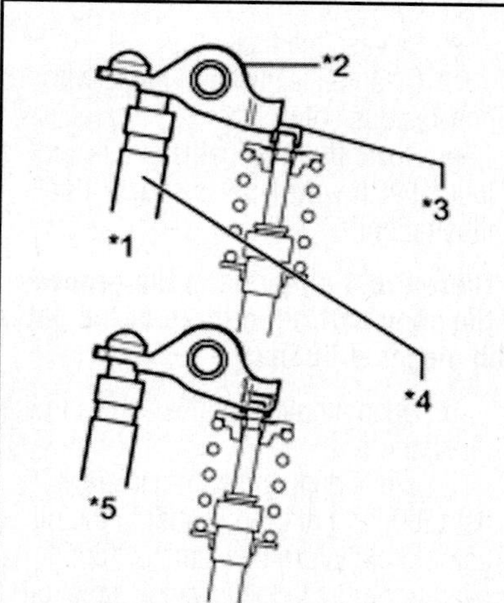

1. INCORRECT
2. Valve rocker arm
3. Valve stem cap
4. Valve lash adjuster
5. CORRECT

3768X_VENZ_G0494

Fig. 100 Checking the valve rocker arm installation

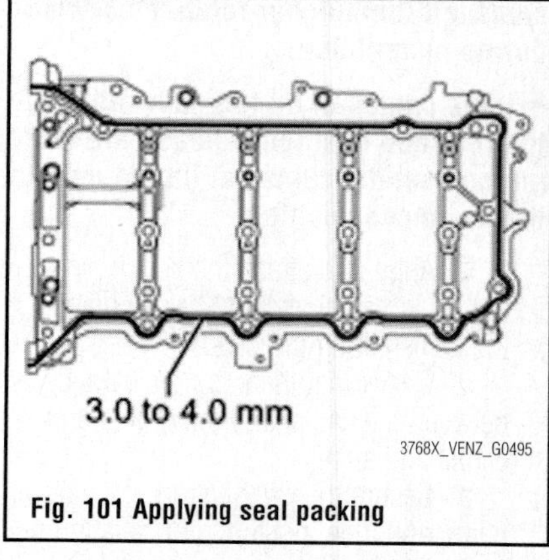

Fig. 101 Applying seal packing

➡Do not start the engine for at least 4 hours after the installation.

➡Thoroughly wipe clean any seal packing.

27. Install the camshaft timing gear assembly.

a. Check the camshaft timing gear position.

➡If the camshaft timing gear is not set to the advanced position, release the lock pin and reset the camshaft timing gear (Refer to the "Set Camshaft Timing Gear Assembly" procedure).

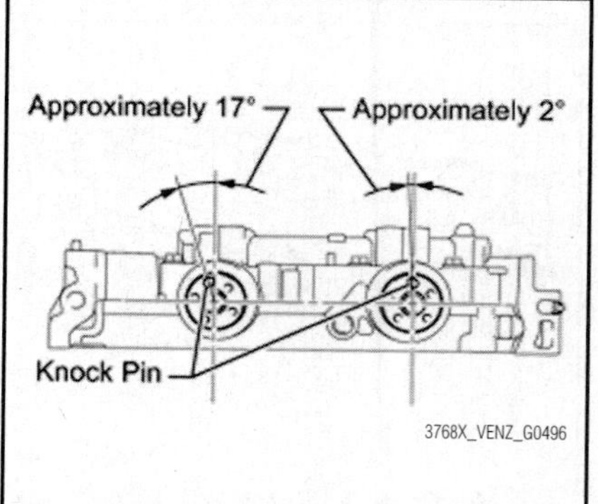

Fig. 102 Positioning the knock pin of the No. 1 and No. 2 camshafts

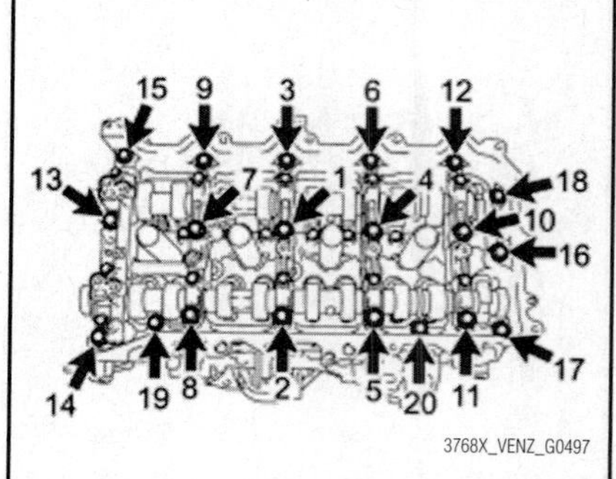

Fig. 103 Installing the camshaft housing

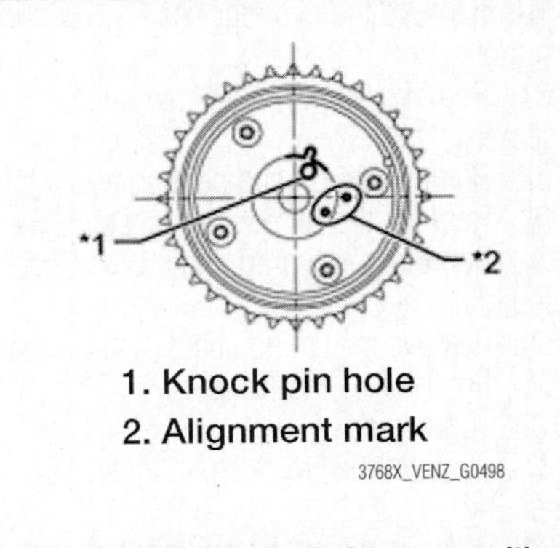

Fig. 104 Checking the timing gear position

b. Align and attach the knock pin of the No. 1 camshaft with the pin hole of the camshaft timing gear.

c. Check that there is no clearance between the camshaft timing gear and camshaft flange.

d. Using a wrench to hold the hexagonal portion of the No. 1 camshaft, install the bolt. Tighten to 63 ft. lbs. (85 Nm).

➡Be careful not to damage the cylinder head or spark plug tube with the wrench.

➡Do not disassemble the camshaft timing gear.

28. Install the camshaft timing exhaust gear assembly.

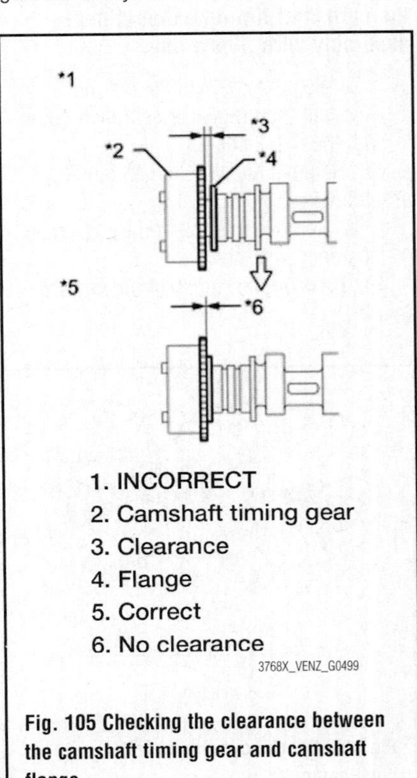

Fig. 105 Checking the clearance between the camshaft timing gear and camshaft flange

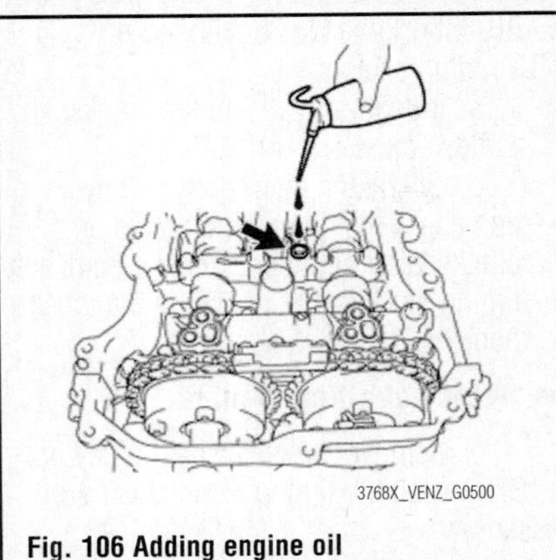

Fig. 106 Adding engine oil

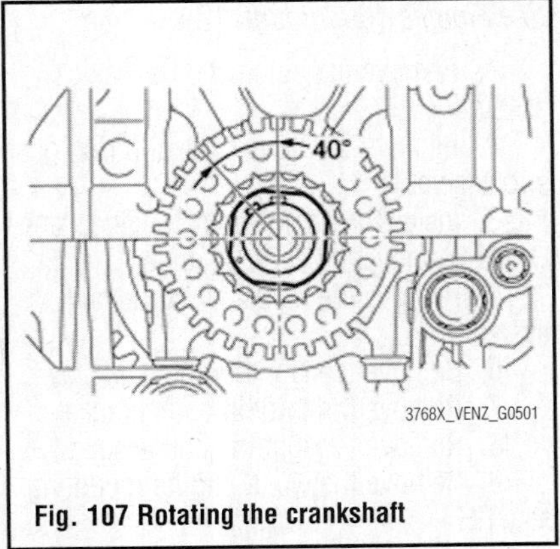

Fig. 107 Rotating the crankshaft

a. Align and attach the knock pin of the No. 2 camshaft with the pin hole of the camshaft timing exhaust gear.

b. Check that there is no clearance between the camshaft timing exhaust gear and camshaft flange.

c. Using a wrench to hold the hexagonal portion of the No. 2 camshaft, install the bolt. Tighten to 63 ft. lbs. (85 Nm).

➡Be careful not to damage the cylinder head or spark plug tube with the wrench.

➡Do not disassemble the camshaft timing exhaust gear.

29. Add engine oil.

a. Add 50 cc (3.1 cu. in) of engine oil into the oil hole.

➡Oil must be added if the lash adjusters were removed.

➡Make sure that the low pressure chamber and oil paths of the lash adjusters are full of engine oil.

30. Set the No. 1 cylinder to TDC/compression.

a. Temporarily install the crankshaft pulley bolt.

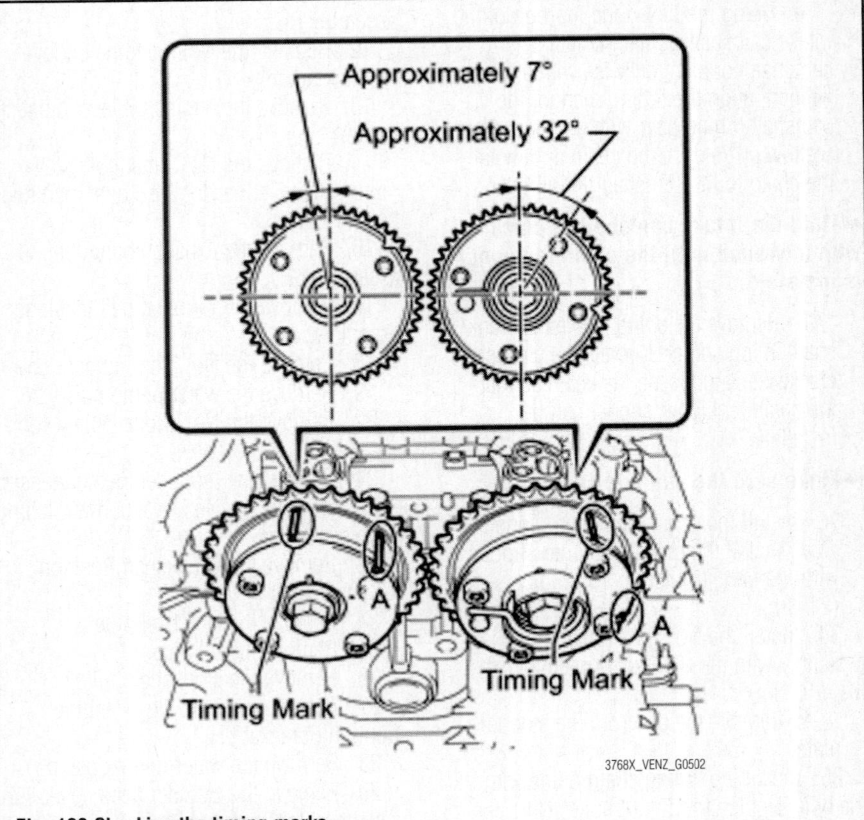

Fig. 108 Checking the timing marks

b. Rotate the crankshaft 40°counter-clockwise to position the crankshaft pulley key as shown in the illustration.

c. Check that the timing marks of the camshaft timing gears are as shown in the illustration.

➡"A" is not a timing mark.

31. Install the No. 1 chain vibration damper with the 2 bolts. Tighten to 15 ft. lbs. (21 Nm).

32. Install the chain sub assembly.

a. Place the chain onto the camshaft timing gears and crankshaft timing sprocket.

➡Make sure the mark plate of the chain faces away from the engine.

➡It is not necessary to install the chain to the teeth of the gears and sprocket.

b. Align the mark plate (yellow or gold) of the chain with the timing mark of the camshaft timing exhaust gear and install the chain to the camshaft timing exhaust gear.

c. Align the mark plate (pink or gold) of the chain with the timing mark of the crankshaft timing sprocket and install the chain to the crankshaft timing sprocket.

d. Tie a string above the crankshaft timing sprocket so that the chain is secure.

e. Using the hexagonal portion of the intake camshaft, rotate the intake camshaft counterclockwise with a wrench, align the timing mark of the camshaft timing gear with the mark plate (yellow or gold) of the chain and install the chain to the camshaft timing gear.

➡Hold the intake camshaft in place with a wrench until the chain tensioner is installed.

f. Remove the string above the crankshaft timing sprocket, rotate the crankshaft clockwise, and loosen the chain so that the chain tensioner slipper can be installed.

➡Make sure the chain is secure.

33. Install the chain tensioner slipper.

a. Install the chain tensioner slipper with the bolt. Tighten to 15 ft. lbs. (21 Nm).

34. Install the No. 1chain tensioner assembly with the 2 bolts. Tighten to 7 ft. lbs. (10 Nm).

a. Remove the pin from the stopper plate.

35. Install the timing chain guide with the bolt. Tighten to 15 ft. lbs. (21 Nm).

36. Check the No. 1 cylinder to TDC/compression.

a. Temporarily install the crankshaft pulley bolt.

b. Rotate the crankshaft clockwise, and check that the timing marks on the crankshaft timing sprocket and camshaft timing gears are as shown in the illustration.

➡"A" is not a timing mark.

c. Remove the crankshaft pulley bolt.

37. Install the timing chain cover sub assembly.

3.5L Engine

See Figures 109 through 127.

1. Remove the automatic transaxle assembly.
2. Remove the drive plate and ring gear sub assembly.
3. Install the engine on an engine stand.
4. Remove the engine hanger.
5. Remove the intake air surge tank assembly.
6. Remove the No. 1 surge tank stay.
7. Remove the throttle body bracket.
8. Remove the ignition coil assembly.
9. Remove the No. 2 engine mounting stay RH.
10. Remove the intake manifold.
11. Remove the exhaust manifold sub assembly RH.
12. Remove the No. 2 engine oil level dipstick guide.
13. Remove the engine oil level dipstick guide.
14. Remove the No. 2 manifold stay.
15. Remove the No. 2 exhaust manifold heat insulator.
16. Remove the exhaust manifold sub assembly LH.
17. Remove the v-ribbed belt tensioner assembly.
18. Remove the No. 2 timing gear cover.
19. Remove the water pump pulley.
20. Remove the No. 2 idler pulley sub assembly.
21. Remove the radio setting condenser.
22. Remove the No. 1 vacuum switching valve assembly.
23. Remove the Crankshaft Position (CKP) sensor.
24. Remove the No. 1 oil pipe.
25. Remove the oil pipe.
26. Remove the crankshaft pulley.
27. Remove the front No. 1 engine mounting bracket LH.
28. Remove the water inlet housing.
29. Remove the cylinder head cover sub assembly (for Bank 1).
30. Remove the cylinder head cover sub assembly LH (for Bank 2).
31. Remove the No. 2 oil pan sub assembly.
32. Remove the oil strainer sub assembly.
33. Remove the oil pan sub assembly.
34. Remove the timing chain cover sub assembly.
35. Remove the timing chain case oil seal.
36. Set the No. 1 cylinder to TDC/compression.
37. Remove the No. 1 chain tensioner assembly.
38. Remove the chain tensioner slipper.
39. Remove the chain sub assembly.
40. Remove the idle sprocket assembly.
41. Remove the camshaft timing gears and No. 2 chain (for Bank 1).

a. While raising the No. 2 chain tensioner assembly, insert a pin of 1.0 mm (0.0394 in.) diameter into the hole to hold the No. 2 chain tensioner assembly.

b. Using SST (09922-10010) to hold the hexagonal portion of each camshaft, loosen the flange bolts of the camshaft timing gear assembly and the camshaft timing exhaust gear assembly RH.

➡Be careful not to damage the cylinder head with SST.

➡Do not loosen the other bolts. If any of the bolts is loosened, replace the camshaft timing gear assembly and/or the camshaft timing exhaust gear assembly with a new one.

c. Remove the 2 bolts and the camshaft timing gear assembly together with the No. 2 chain.

42. Remove the No. 2 chain tensioner assembly.

a. Remove the bolt and No. 2 chain tensioner assembly.

43. Remove the camshaft bearing cap (bank 1).

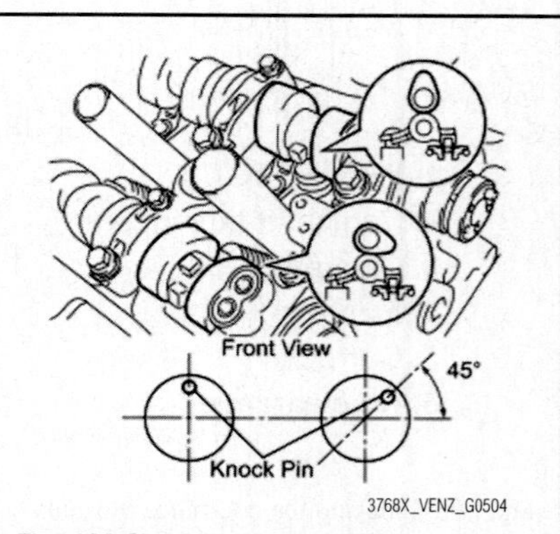

Fig. 109 Checking the camshaft positioning

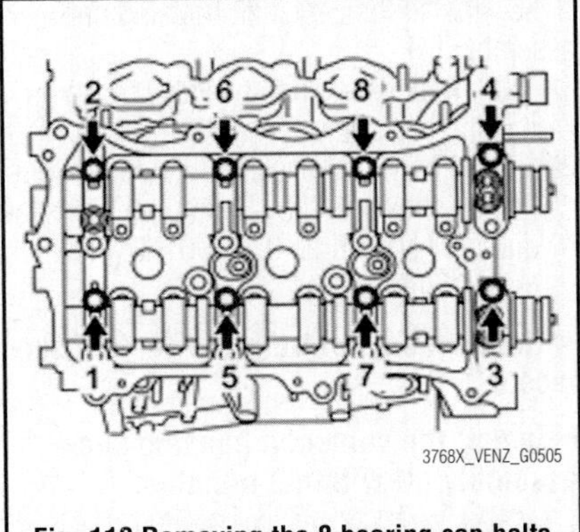

Fig. 110 Removing the 8 bearing cap bolts

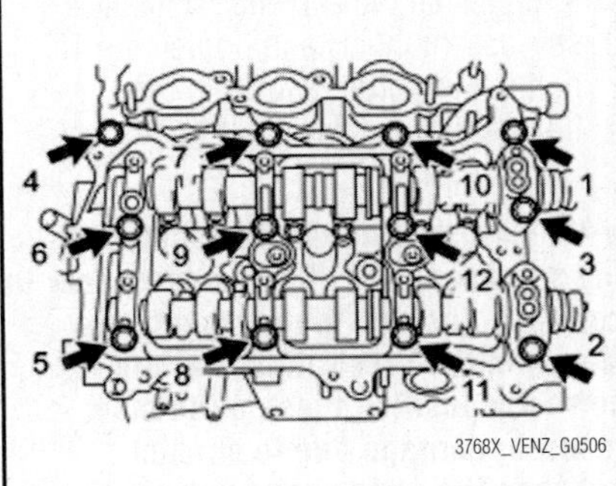

Fig. 111 Removing the 12 bearing cap bolts

a. Check that the camshafts are positioned as shown in the illustration.

b. Uniformly loosen and remove the 8 bearing cap bolts in several steps and in the sequence shown in the illustration.

c. Uniformly loosen and remove the 12 bearing cap bolts in several steps and in the sequence shown in the illustration.

➡Uniformly loosen the bolts while keeping the camshaft level.

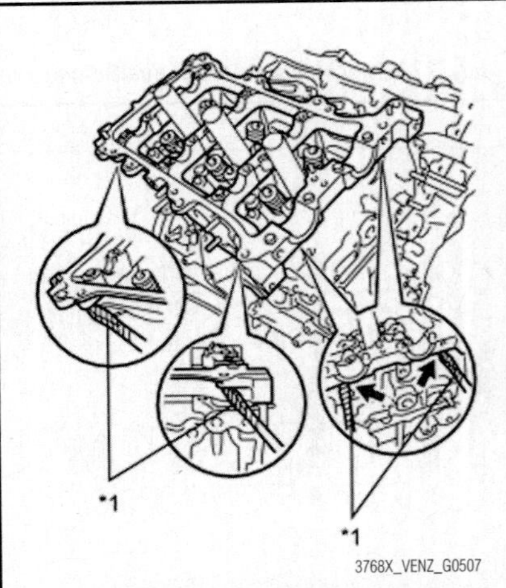

Fig. 112 Removing the camshaft housing sub assembly RH

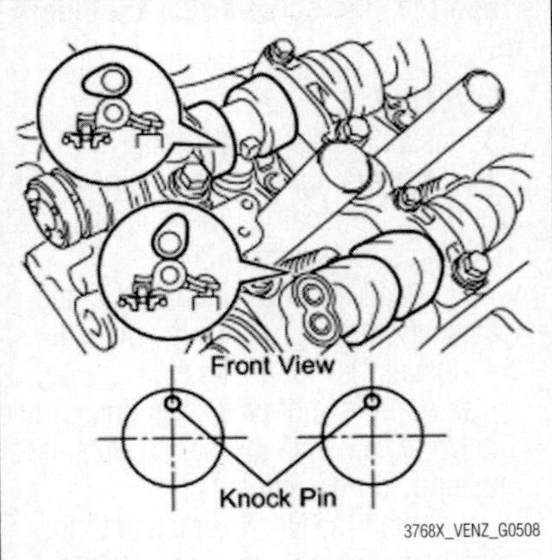

Fig. 113 Checking the camshaft positioning

d. Remove the 5 camshaft bearing caps.

44. Remove the camshaft.

45. Remove the No. 2 camshaft.

46. Remove the camshaft housing sub assembly RH.

a. Remove the camshaft housing sub-assembly RH by prying between the cylinder head and camshaft housing sub-assembly RH with a screwdriver.

➡Be careful not to damage the contact surfaces of the cylinder head and camshaft housing sub-assembly RH.

➡Tape the screwdriver tip before use.

47. Remove the camshaft timing gears and No. 2 chain (bank 2).

a. While pushing down the No. 3 chain tensioner assembly, insert a pin of 1.0 mm (0.0394 in.) diameter into the hole to hold the No. 3 chain tensioner assembly.

b. Using SST (09922-10010) to hold the hexagonal portion of each camshaft, loosen the flange bolts of the camshaft timing gear assembly and the camshaft timing exhaust gear assembly LH.

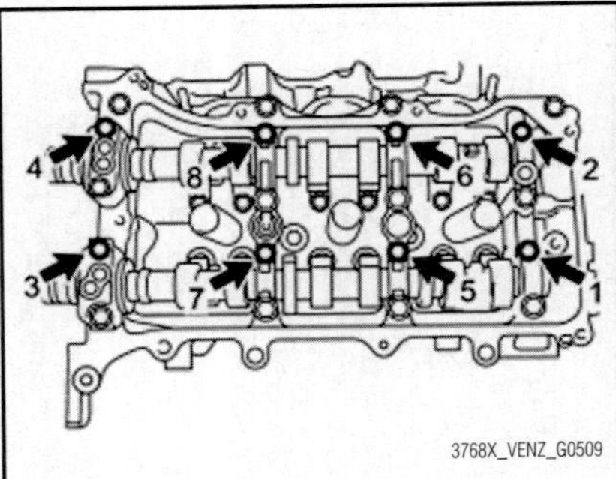

Fig. 114 Removing the 8 bearing cap bolts in sequence

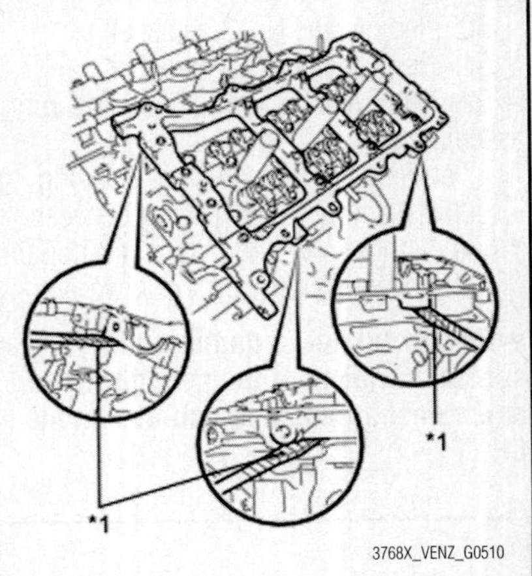

Fig. 115 Removing the camshaft housing sub assembly LH

➡Be careful not to damage the cylinder head with SST.

➡Do not loosen the other bolts. If any of the bolts is loosened, replace the camshaft timing gear assembly and/or the camshaft timing exhaust gear assembly with a new one.

c. Remove the 2 bolts and the camshaft timing gear together with the No. 2 chain.

48. Remove the No. 3 chain tensioner assembly.

a. Remove the bolt and No. 3 chain tensioner assembly.

49. Remove the camshaft bearing cap (bank 2).

a. Check the camshaft are positioned as shown in the illustration.

b. Uniformly loosen and remove the 8 bearing cap bolts in several steps and in the sequence shown in the illustration.

➡Uniformly loosen the bolts while keeping the camshaft level.

c. Remove the 5 camshaft bearing caps.

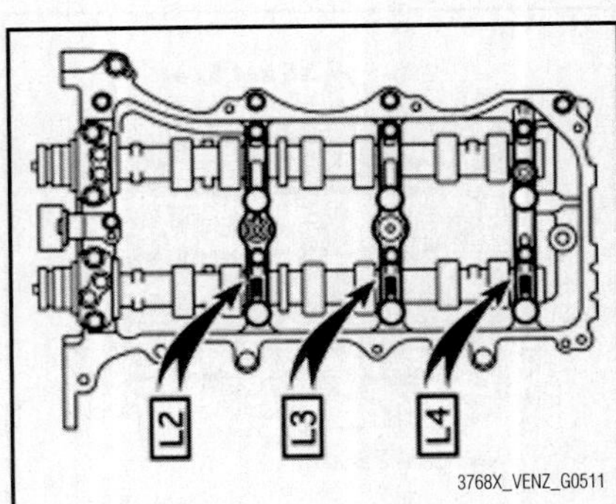

Fig. 116 Checking the camshaft bearing cap positioning

50. Remove the No. 3 camshaft.
51. Remove the No. 4 camshaft.
52. Remove the camshaft housing sub assembly LH.
a. Remove the camshaft housing sub-assembly LH by prying between the cylinder head and camshaft housing sub-assembly LH with a screwdriver.

➡Be careful not to damage the contact surfaces of the cylinder head and camshaft housing sub-assembly LH.

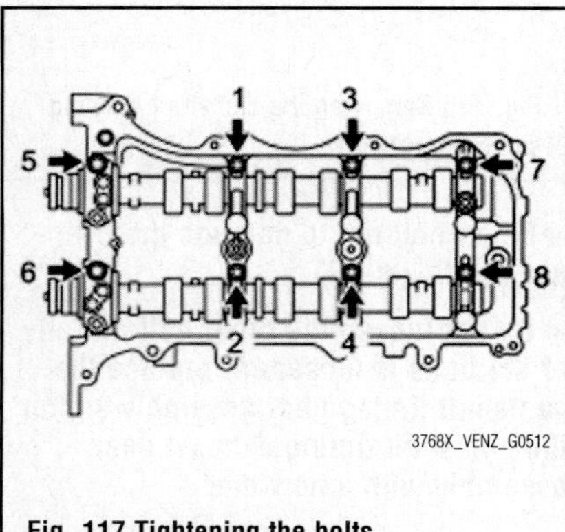

Fig. 117 Tightening the bolts

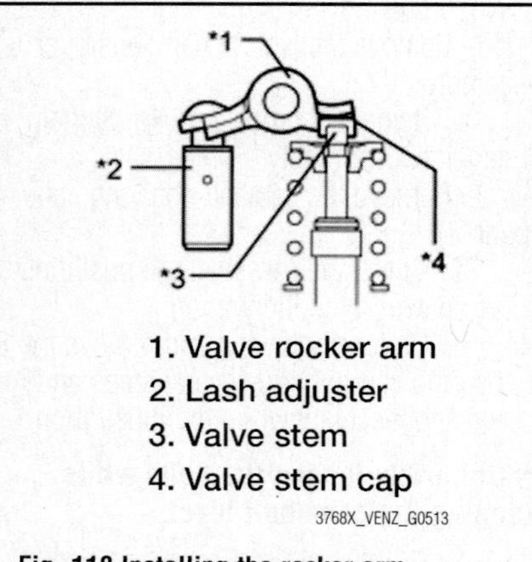

Fig. 118 Installing the rocker arm

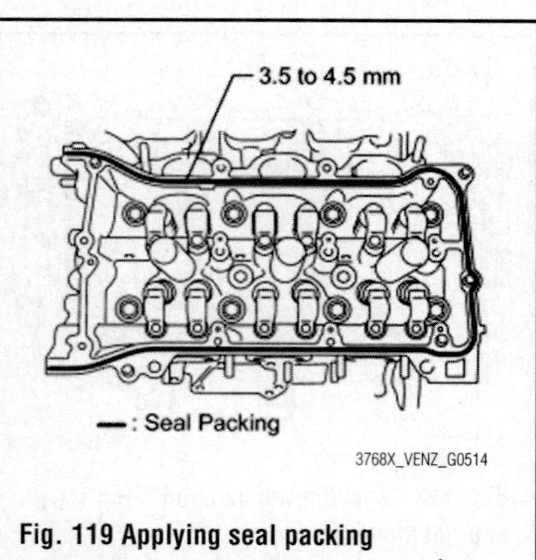

Fig. 119 Applying seal packing

➡Tape (*1) the screwdriver tip before use.

To install:

53. Install the No. 3 camshaft.
a. Apply a light coat of engine oil to the No. 3 camshaft journals and camshaft housing sub-assembly LH.
b. Install the No. 3 camshaft to the camshaft housing sub-assembly LH.
54. Install the No. 4 camshaft.
a. Apply a light coat of engine oil to the No. 4 camshaft journals and camshaft housing sub-assembly LH.
b. Install the No. 4 camshaft to the camshaft housing sub-assembly LH.
55. Install the camshaft bearing cap (bank 2).
a. Apply engine oil to the camshaft bearing caps.
b. Make sure of the marks and numbers on the camshaft bearing caps and place them in each proper position and direction.
c. Temporarily tighten the 8 bolts in the order shown in the illustration. Tighten to 7 ft. lbs. (10 Nm).

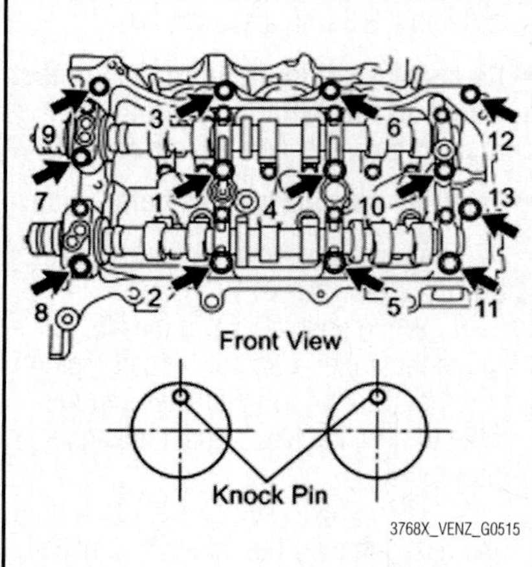

Fig. 120 Installing the camshaft housing sub assembly LH

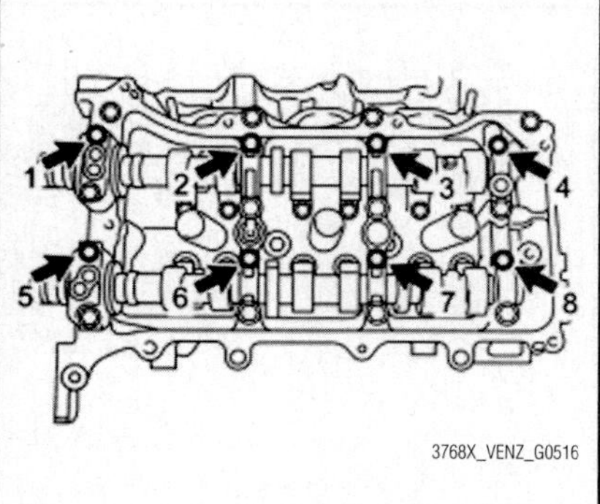

Fig. 121 Tightening the 8 bolts

56. Install the camshaft housing sub assembly LH.
a. Make sure that the valve rocker is installed as shown.
b. Apply seal packing in a continuous line as shown in the illustration. The seal diameter should be 0.138-0.177 inch (3.5-4.5 mm).

➡Remove any oil from the contact surface.

➡Install the camshaft housing sub-assembly LH within 3 minutes.

➡Do not start the engine for at least 2 hours after installing.

c. Install the camshaft housing sub-assembly LH and tighten the 13 bolts in the order shown in the illustration. Tighten to 21 ft. lbs. (28 Nm).

➡When installing the camshaft housing sub-assembly LH, it is necessary to correctly position the camshafts as shown in the illustration. Failure to correctly position these parts may result in damage due to contact between the pistons and valves. If a camshaft is rotated with a piston at TDC, valve contact will occur.

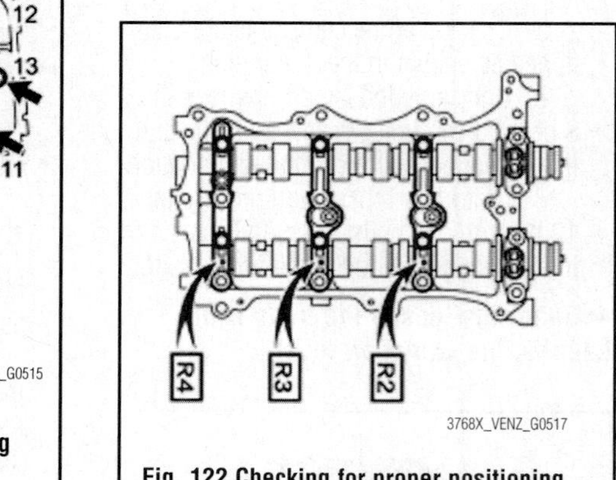

Fig. 122 Checking for proper positioning

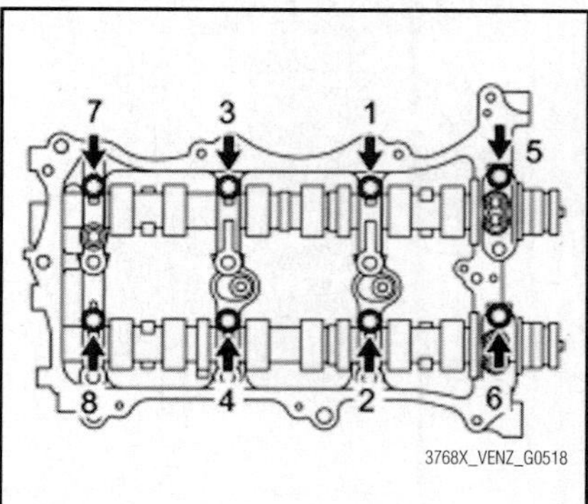

Fig. 123 Tightening the bearing cap bolts

➡If any of the bolts are loosened during installation, remove the camshaft housing sub-assembly LH, clean the installation surfaces, and reapply seal packing.

➡If the camshaft housing sub-assembly LH is removed because any of the bolts are loosened during installation, make sure that the previously applied seal packing does not enter any oil passages.

d. Tighten the 8 bolts in the order shown in the illustration. Tighten to 12 ft. lbs. (16 Nm).

57. Install the camshaft.

a. Apply a light coat of engine oil to the camshaft journals and camshaft housing sub-assembly RH.

b. Install the camshaft to the camshaft housing sub-assembly RH.

58. Install the No. 2 camshaft.

a. Apply a light coat of engine oil to the No. 2 camshaft journals and camshaft housing sub-assembly RH.

b. Install the No. 2 camshaft to the camshaft housing sub-assembly RH.

59. Install the camshaft bearing cap (bank 1).

a. Apply engine oil to the camshaft bearing caps.

b. Make sure of the marks and numbers on the camshaft bearing caps and place them in each proper position and direction.

c. Temporarily tighten the 8 bearing cap bolts in the order shown in the illustration. Tighten to 7 ft. lbs. (10 Nm).

60. Install the camshaft housing sub assembly RH.

a. Make sure that the No. 1 valve rocker arm sub-assembly is installed as shown in the illustration.

b. Apply seal packing in a continuous line as shown in the illustration. The sealant diameter should be 0.138-0.177 inch (3.5-4.5 mm).

➡Remove any oil from the contact surface.

➡Install the camshaft housing sub-assembly RH within 3 minutes.

➡Do not start the engine for at least 2 hours after installing.

c. Install the camshaft housing sub-assembly RH and tighten the 12 bolts in the order shown in the illustration. Tighten to 21 ft. lbs. (28 Nm).

➡When installing the camshaft housing RH, it is necessary to correctly position the camshafts as shown in the illustration. Failure to correctly position these parts may result in damage due to contact between the pistons and valves. If a camshaft is rotated with a piston at TDC, valve contact will occur.

➡If any of the bolts are loosened during installation, remove the camshaft housing sub-assembly RH, clean the installation surfaces, and reapply seal packing.

➡If the camshaft housing sub-assembly RH is removed because any of the bolts are loosened during installation, make sure that the previously applied seal packing does not enter any oil passages.

d. Tighten the 8 bolts in the order shown in the illustration. Tighten to 12 ft. lbs. (16 Nm).

61. Install the No. 3 chain tensioner assembly with the bolt. Tighten to 15 ft. lbs. (21 Nm).

a. While pushing in the tensioner, insert a pin of 0.0394 inch (1 mm) diameter into the hole to hold it.

62. Install the camshaft timing gears and the No. 2 chain (bank 2).

a. Align the mark plates (yellow) with the timing marks of the camshaft timing gear assemblies.

b. Apply a light coat of engine oil to the bolt threads and bolt-seating surface.

c. Align the knock pin of the camshaft with the pin hole of the camshaft timing gear assembly. Install the camshaft timing gear assembly and camshaft timing exhaust gear LH with the No. 2 chain sub-assembly installed.

d. Using SST (09922-10010) to hold the hexagonal portion of each camshaft, tighten the flange bolts of the camshaft timing gear assembly and the camshaft timing exhaust gear assembly LH. Tighten to 74 ft. lbs. (100 Nm).

e. Remove the pin from the No. 3 chain tensioner assembly.

63. Install the No. 2 chain tensioner assembly with the bolt. Tighten to 15 ft. lbs. (21 Nm).

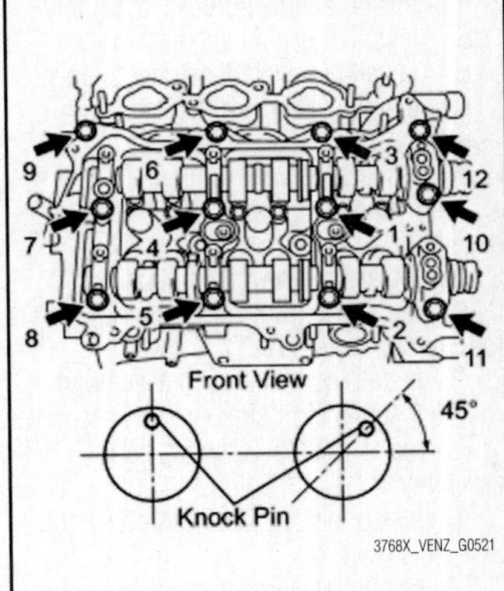

Fig. 126 Installing the camshaft housing sub assembly RH

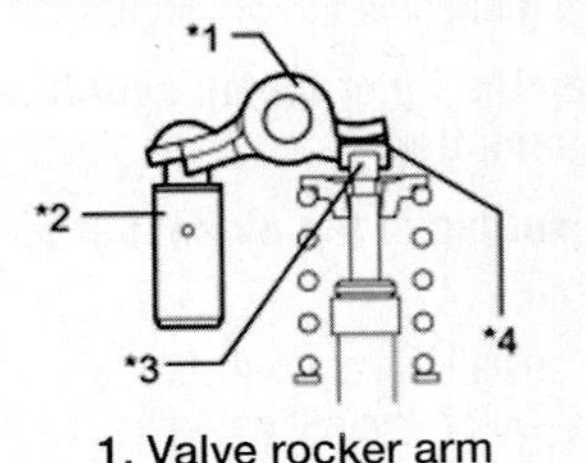

Fig. 124 Checking the No. 1 valve rocker arm sub assembly installation

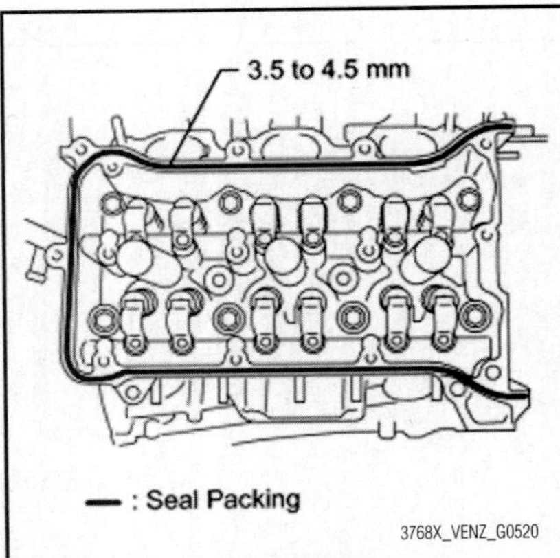

Fig. 125 Applying seal packing

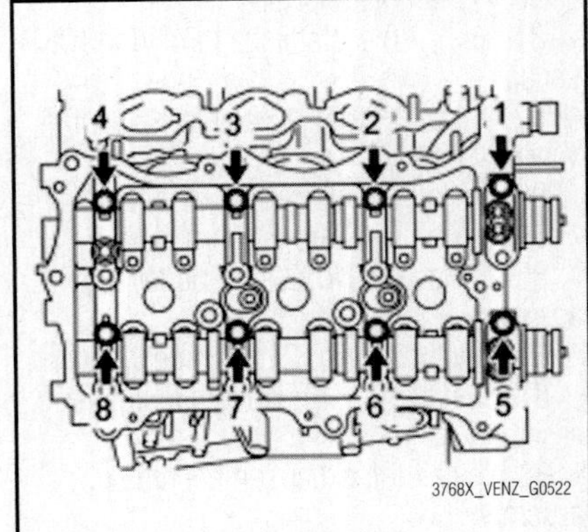

Fig. 127 Tightening the bolts

a. While pushing in the No. 2 chain tensioner assembly, insert a pin of 1.0 mm (0.0394 in.) diameter into the hole to hold it.

64. Install the camshaft timing gears and No. 2 chain (bank 1).

a. Align the mark plates (yellow) with the timing marks of the camshaft timing gear assemblies.

b. Apply a light coat of engine oil to the bolt threads and bolt-seating surface.

c. Align the knock pin of the camshaft with the pin hole of the camshaft timing gear assembly. Install the camshaft timing gear assembly and camshaft timing exhaust gear assembly with the No. 2 chain sub-assembly installed.

d. Using SST (09922-10010) to hold the hexagonal portion of each camshaft, tighten the flange bolts of the camshaft timing gear assembly and the camshaft timing exhaust gear assembly RH. Tighten to 74 ft. lbs. (100 Nm).

e. Remove the pin from the No. 2 chain tensioner assembly.

65. Install the idle sprocket assembly.
66. Install the chain sub assembly.
67. Install the chain tensioner slipper.
68. Install the No. 1 chain tensioner assembly.
69. Inspect the valve timing.
70. Install the timing chain cover sub assembly.
71. Install the water inlet housing.
72. Install the front No. 1 engine mounting bracket LH.
73. Install the oil pan sub assembly.
74. Install the oil strainer sub assembly.
75. Install the No. 2 oil pan sub assembly.
76. Install the cylinder head cover sub assembly (bank 1).
77. Install the cylinder head cover sub assembly (bank 2).
78. Install the crankshaft pulley.
79. Install the No. 1 oil pipe.
80. Install the oil pipe.
81. Install the Crankshaft Position (CKP) sensor.
82. Install the No. 1 vacuum switching valve assembly.
83. Install the radio setting condenser.
84. Install the ware pump pulley.
85. Install the No. 2 idler pulley sub assembly.
86. Install the No. 2 timing gear cover.
87. Install the v-ribbed belt tensioner assembly.
88. Install the exhaust manifold sub assembly LH.
89. Install the No. 2 exhaust manifold sub assembly LH.
90. Install the No. 2 manifold stay.
91. Install the engine oil level dipstick guide.
92. Install the No. 2 engine oil level dipstick guide.
93. Install the exhaust manifold sub assembly RH.
94. Install the intake manifold.
95. Install the No. 2 engine mounting stay RH.
96. Install the ignition coil assembly.
97. Install the throttle body bracket.
98. Install the No. 1 surge tank stay.
99. Install the intake air surge tank assembly.
100. Install the engine hanger.
101. Remove the engine from the engine stand.
102. Install the drive plate and ring gear sub assembly.
103. install the automatic transaxle assembly.

CRANKSHAFT DAMPER

REMOVAL & INSTALLATION

Refer to Crankshaft for the removal and installation procedure.

CRANKSHAFT FRONT SEAL

REMOVAL & INSTALLATION

See Figures 128 and 129.

1. Remove the front wheel RH.
2. Remove the No. 1 engine under cover.
3. Separate the front fender liner RH.
4. Remove the front fender apron seal RH.
5. Remove the v-ribbed belt.
6. Remove the crankshaft pulley.

a. Using the special tool (09213-54015, 09330-00021 or 91551-80650), hold the crankshaft pulley and loosen the pulley bolt. Further loosen the bolt until 2 or 3 threads are screwed into the crankshaft.

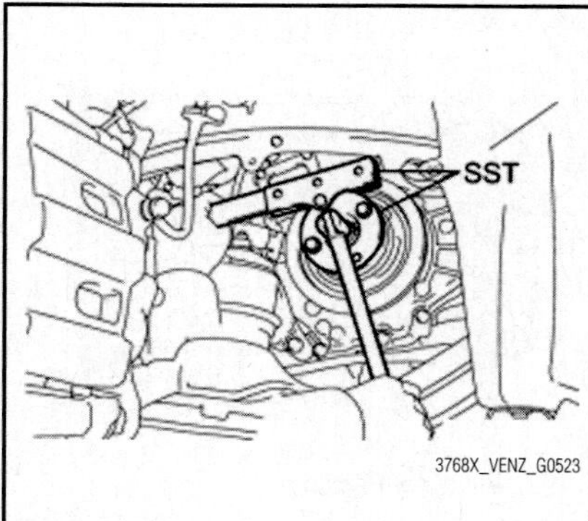

Fig. 128 Removing the crankshaft pulley

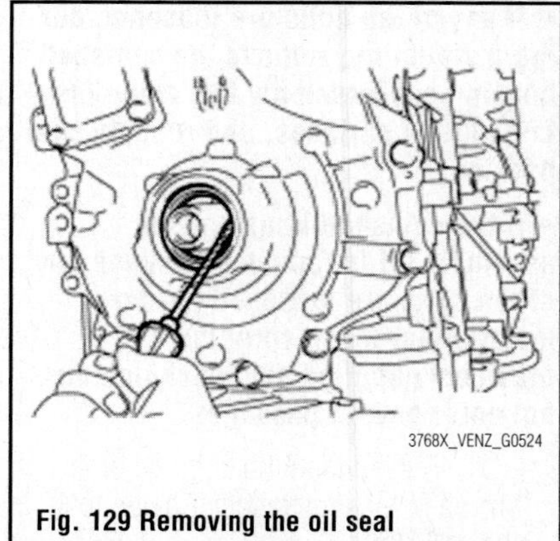

Fig. 129 Removing the oil seal

b. Using SST (09950-50010, 09951-05010, 09952-05010, 09953-05020 or 09954-05011).

➡Apply a lubricant to the threads and end of the SST.

7. Remove the timing chain cover oil seal.

a. For 3.5L engine, using a knife, cut off the timing chain case oil seal lip.

b. Using a screwdriver, pry out the oil seal.

➡Tape the screwdriver tip before use.

➡Do not damage the surface of the seal press fit hole or the crankshaft.

To install:

8. Install the timing chain cover oil seal.

a. Apply MP grease to the lip of a new oil seal.

➡Do not allow foreign matter to contact the lip of the oil seal.

➡Do not allow MP grease to contact the dust seal.

b. Using SST and a hammer, tap in the oil seal until its surface is flush with the timing chain cover edge.

➡Keep the lip of the oil seal free from foreign matter.

➡Do not tap in the oil seal at an angle.

9. Install the crankshaft pulley.

a. Align the pulley set key with the key groove of the crankshaft pulley.

b. Using SST, hold the crankshaft pulley and install the pulley bolt. Tighten to 192 ft. lbs. (260 Nm).

10. Install the v-ribbed belt.
11. Install the front fender apron seal RH.
12. Install the front fender liner RH.

13. Install the No. 1 engine under cover.
14. Install the front wheel RH. Tighten to 76 ft. lbs. (103 Nm).

CYLINDER HEAD

REMOVAL & INSTALLATION

2.7L Engine

See Figures 130 through 132.

1. Remove the engine assembly with the transaxle.
2. Remove the exhaust manifold converter sub assembly.
3. Remove the throttle body assembly.
4. Remove the fuel delivery pipe sub assembly.
5. Disconnect the No. 2 ventilation hose.
6. Remove the intake manifold.
7. Remove the camshaft.
8. Remove the valve stem cap.
    a. Remove the 16 valve stem caps from the cylinder head.

➡Arrange the removed parts in the correct order.

9. Remove the cylinder head sub assembly.
    a. Using a 10 mm bi-hexagon wrench, uniformly loosen the 10 bolts in the sequence shown in the illustration. Remove the 10 cylinder head bolts and plate washers.

➡Be sure to keep the removed parts separate for each installation position.

➡Be careful not to drop washers into the cylinder head.

➡Head warpage or cracking could result from removing bolts in an incorrect order.

    b. Remove the cylinder head.

To install:

10. Inspect the cylinder head.
11. Install the cylinder head sub assembly.
    a. The cylinder head bolts are tightened in 4 progressive steps.
    b. Place the cylinder head on the cylinder block.

➡Ensure that no oil is on the mounting surface of the cylinder head.

➡Place the cylinder head on the cylinder block gently in order not to damage the gasket with the bottom part of the head.

    c. Install the plate washers to the cylinder head bolts.
    d. Apply a light coat of engine oil to the threads and under the heads of the cylinder head bolts.

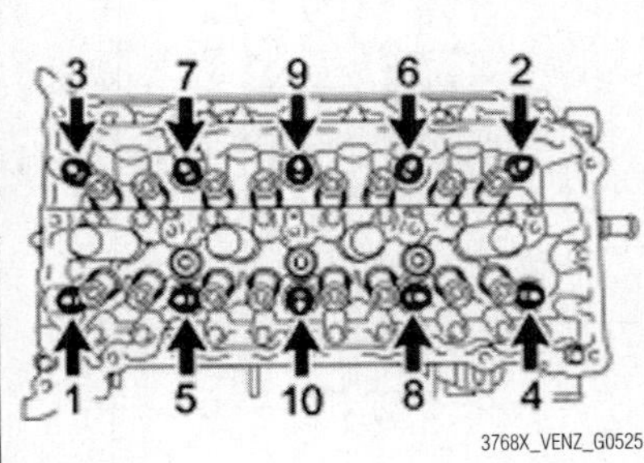

Fig. 130 Removing the cylinder head—2.7L engine

    e. Step 1: Using a 10 mm bi-hexagon wrench, install and uniformly tighten the 10 cylinder head bolts in several steps, in the sequence shown in the illustration. Tighten to 27 ft. lbs. (36 Nm).

➡Do not drop the plate washer for the cylinder head bolt into the cylinder head.

    f. Step 2: Tighten the cylinder head bolts again in the sequence shown in the illustration to make sure that they are tightened to the specified torque. Tighten to 27 ft. lbs. (36 Nm).
    g. Step 3: Mark each cylinder head bolt head with paint as shown in the illustration. Tighten the cylinder head bolts 90°in the sequence shown in Step 1.
    h. Step 4: Tighten the cylinder head bolts another 90°in the sequence shown in Step 1. Check that the painted marks are not facing rearward.

➡Do not apply oil for at least 4 hours after the installation.

➡Do not start the engine for at least 4 hours after the installation.

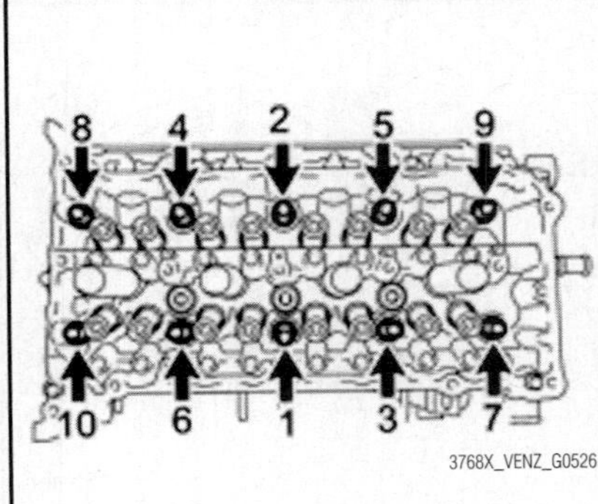

Fig. 131 Tightening the cylinder head bolts

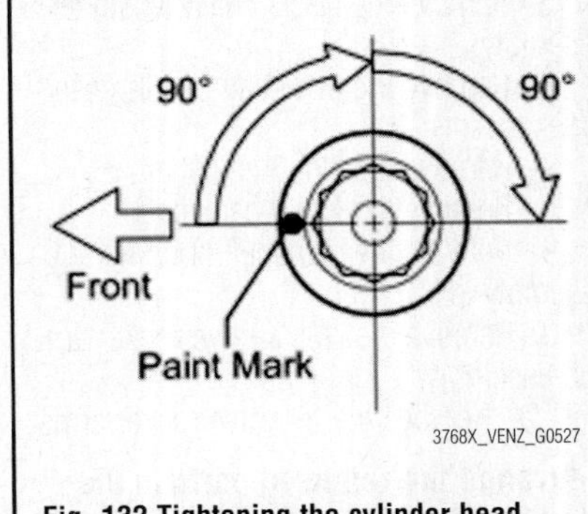

Fig. 132 Tightening the cylinder head bolts 90°

➡After the installation, if the seal packing has seeped out, wipe it off.

12. Install the valve step cap.
    a. Apply a light coat of engine oil to the valve stem ends.
    b. Install the 16 valve stem caps to the cylinder head.

➡Do not drop the valve step caps into the cylinder head.

13. Install the camshaft.
14. Install the intake manifold.
15. Connect the No. 2 ventilation hose.
16. Install the fuel delivery pipe sub assembly.
17. Install the throttle body assembly.
18. Install the exhaust manifold converter sub assembly.
19. Install the engine assembly with the transaxle.

3.5L Engine

See Figures 133 through 143.

1. Remove the camshaft.
2. Remove the No. 2 camshaft.
3. Remove the camshaft housing sub assembly RH.
4. Remove the camshaft timing gears and the No. 2 chain (bank 2).

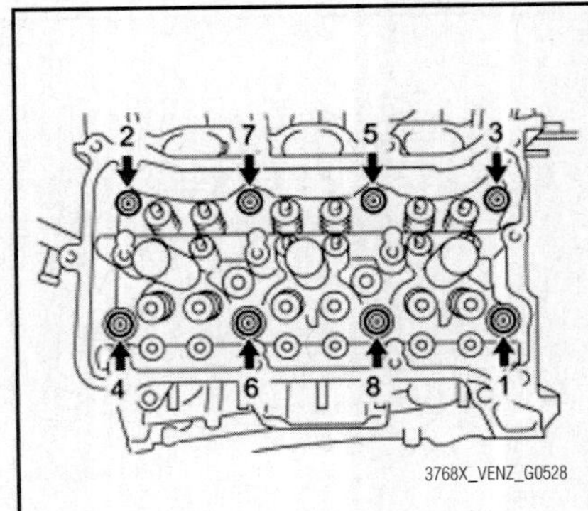

Fig. 133 Removing the cylinder head sub assembly RH

5. Remove the No. 3 chain tensioner assembly.
6. Remove the camshaft bearing cap (bank 2).
7. Remove the No. 3 camshaft.
8. Remove the No. 4 camshaft.
9. Remove the camshaft housing sub assembly LH.
10. Remove the No. 1 valve rocker arm sub assembly.
   a. Remove the 24 valve rocker arms.

➡Arrange the removed parts in the correct order.

11. Remove the valve lash adjuster assembly.
   a. Remove the 24 valve lash adjuster assemblies from the cylinder head.

➡Arrange the removed parts in the correct order.

12. Remove the 24 valve stem caps.
13. Remove the water outlet.
14. Remove the cylinder head sub assembly RH.
   a. Using a 10 mm bi-hexagon wrench, uniformly loosen the 8 cylinder head bolts in the sequence shown in the illustration. Remove the 8 cylinder head bolts and plate washers.

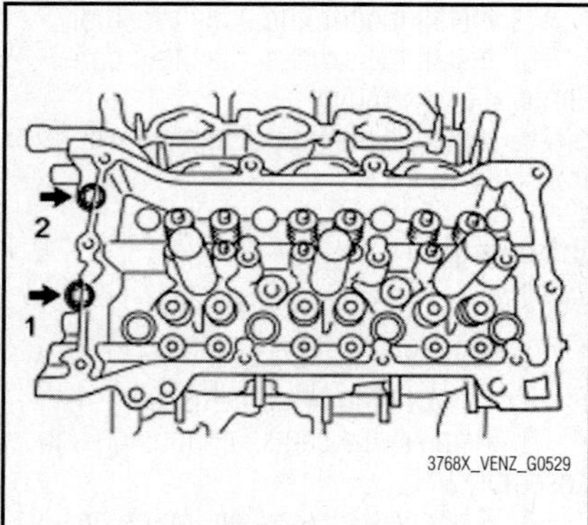

Fig. 134 Removing the set bolts

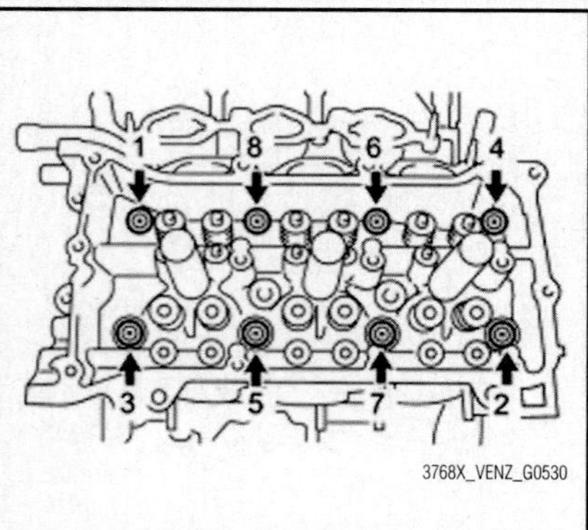

Fig. 135 Removing the cylinder head sub assembly LH

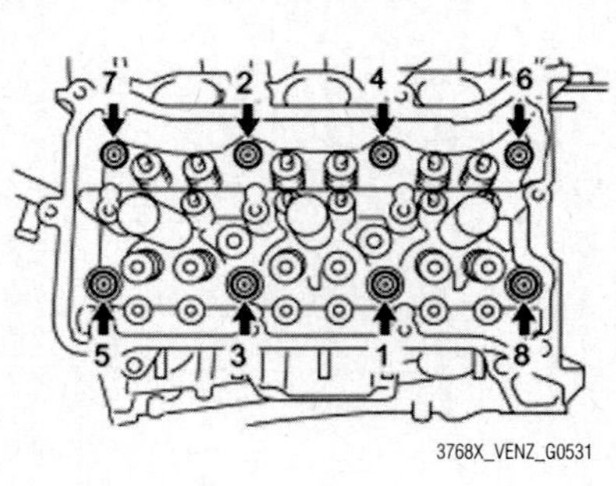

Fig. 136 Tightening the cylinder head bolts

➡Be careful not to drop washers into the cylinder head sub-assembly.

➡Cylinder head warpage or cracking could result from removing bolts in the incorrect order.

➡Arrange the removed parts in the correct order.

   b. Remove the cylinder head sub-assembly RH.

15. Remove the cylinder head gasket RH.
16. Remove the cylinder head sub assembly LH.
   a. Uniformly loosen and remove the 2 cylinder head set bolts in several steps and in the sequence shown in the illustration.
   b. Using a 10 mm bi-hexagon wrench, uniformly loosen the 8 bolts in the sequence shown in the illustration. Remove the 8 cylinder head bolts and plate washers.

➡Be careful not to drop washers into the cylinder head sub-assembly.

➡Cylinder head warpage or cracking could result from removing bolts in the incorrect order.

➡Be sure to keep separate the removed parts for each installation position.

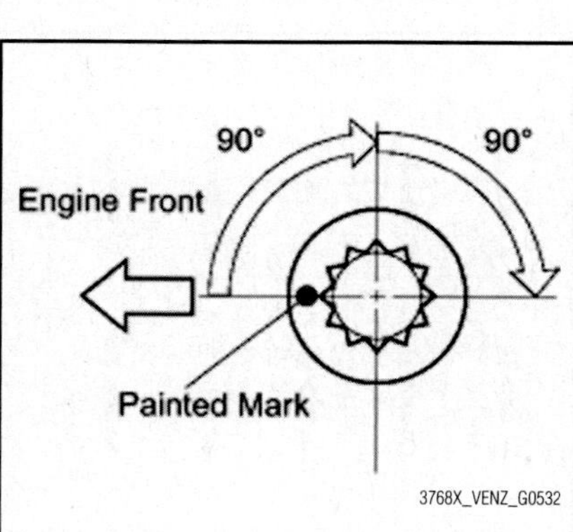

Fig. 137 Marking and tightening the cylinder head bolt

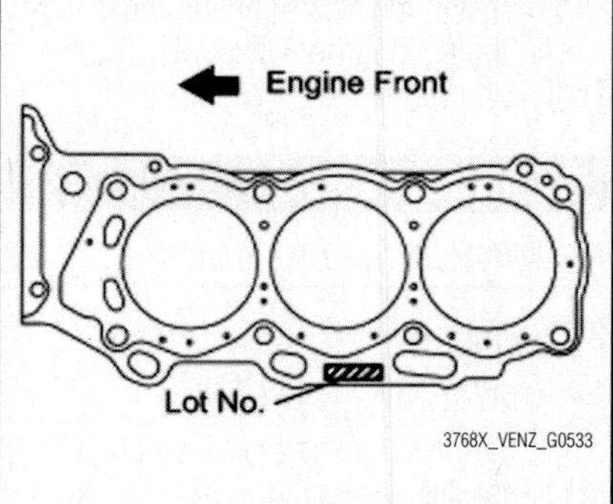

Fig. 138 Installing the cylinder head gasket LH

   c. Remove the cylinder head sub assembly LH.

To install:

17. Install the cylinder head sub assembly RH.
   a. Place the cylinder head on the cylinder block.

➡Be careful not to allow oil to adhere to the bottom part of the cylinder head.

➡The cylinder head bolts are tightened in 3 progressive steps.

   b. Apply a light coat of engine oil to the threads under the heads of the cylinder head bolts.
   c. Step 1: Using a 10 mm bi-hexagon wrench, install and uniformly tighten the 8 cylinder head bolts with the plate washers in several steps and in the sequence shown in the illustration. Tighten to 27 ft. lbs. (36 Nm).
   d. Step 2: Mark the cylinder head bolt head with paint as shown in the illustration. Tighten the cylinder head bolts another 90°.

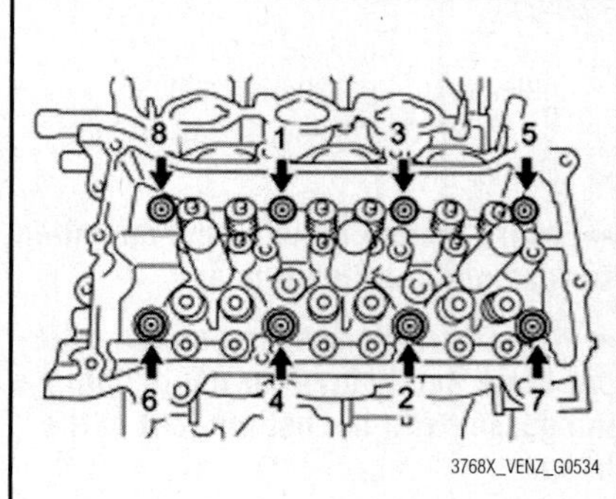

Fig. 139 Tightening the cylinder head bolts

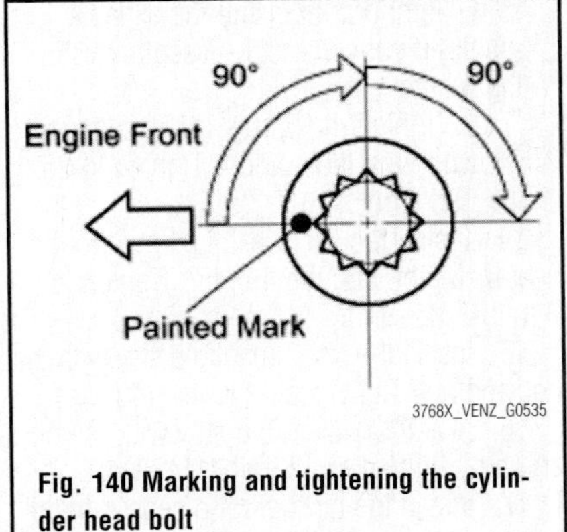

Fig. 140 Marking and tightening the cylinder head bolt

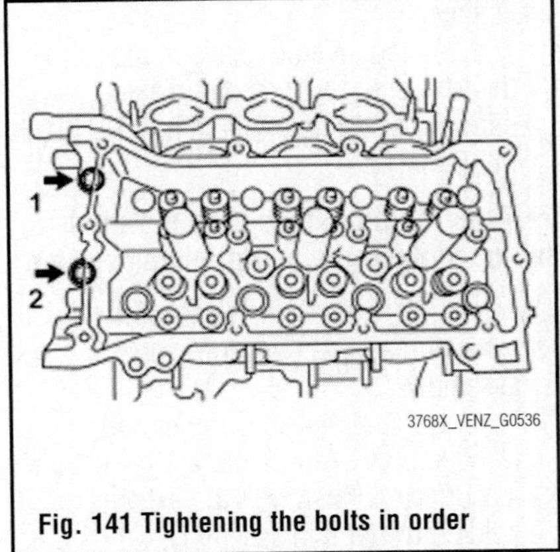

Fig. 141 Tightening the bolts in order

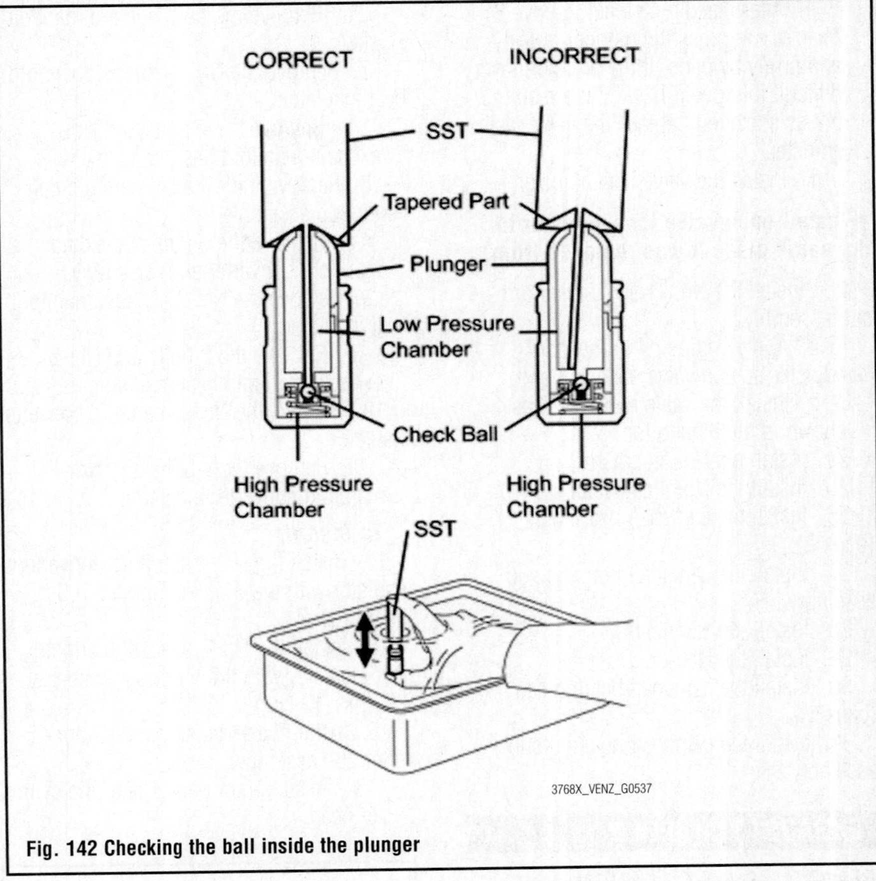

Fig. 142 Checking the ball inside the plunger

e. Step 3: Tighten the cylinder head bolts and additional 90°. Check that the painted mark is now facing rearward.

18. Install the cylinder head gasket LH.

a. Place a new cylinder head gasket LH on the cylinder block surface with the Lot No. stamped upward.

➡**Be careful of the installation direction.**

➡**Gently lower the cylinder head in order not to damage the gasket with the bottom part of the head.**

19. Install the cylinder head sub assembly LH.

a. Place the cylinder head on the cylinder block.

➡**Be careful not to allow oil to adhere to the bottom part of the cylinder head.**

➡**The cylinder head bolts are tightened in 3 progressive steps.**

b. Apply a light coat of engine oil to the threads and under the heads of the cylinder head bolts.

c. Step 1: Using a 10 mm bi-hexagon wrench, install and uniformly tighten the 8 cylinder head bolts with the plate washers in several steps in the sequence shown in the illustration. Tighten to 27 ft. lbs. (36 Nm).

d. Step 2: Mark the cylinder head bolt head with paint as shown in the illustration. Tighten the cylinder head bolts another 90°.

e. Step 3: Tighten the cylinder head bolts an additional 90°. Check that the painted mark is now facing rearward.

f. Tighten the 2 bolts in the order shown in the illustration. Tighten to 22 ft. lbs. (30 Nm).

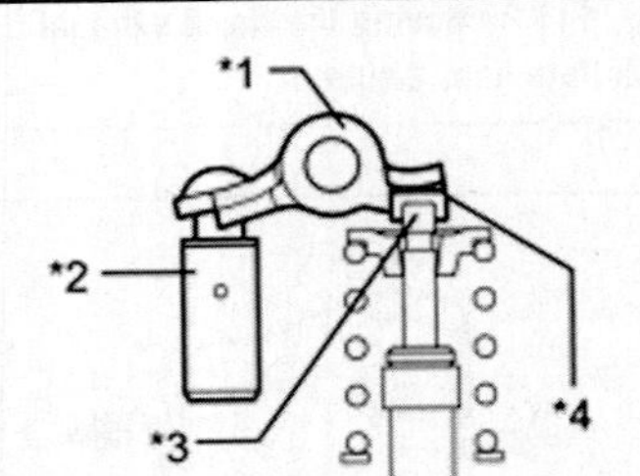

Fig. 143 Installing the No. 1 valve rocker arm sub assembly

20. Install the water outlet.

21. Install the valve stem cap.

a. Install the 24 valve stem caps.

22. Install the valve lash adjuster assembly.

➡**Keep the lash adjuster free of dirt and foreign objects.**

➡**Only use clean engine oil.**

a. Place the lash adjuster into a container filled with engine oil.

b. Insert the tip of SST (09276-75010) into the lash adjuster plunger and use the tip to press down on the check ball inside the plunger.

c. Squeeze SST and lash adjuster together to move the plunger up and down 5 to 6 times.

d. Check the movement of the plunger and bleed the air. If the plunger moves up and down it is OK.

➡**When bleeding air from the high-pressure chamber, make sure that the tip of SST is actually pressing the check ball as shown in the illustration. If the check ball is not pressed, air will not bleed.**

e. After bleeding the air, remove SST. Then, try to press the plunger quickly and firmly by hand. If the plunger is very difficult to move it is ok. If the result is not as specified, replace the valve lash adjuster.

f. Install the valve lash adjuster.

➡Install each valve lash adjuster to the same place it was removed from.

23. Install the No. 1 valve rocker arm sub assembly.

a. Apply engine oil to the lash adjuster tip and valve stem cap end.

b. Install the valve rocker arm as shown in the illustration.

24. Install the No. 3 camshaft.
25. Install the No. 4 camshaft.
26. Install the camshaft bearing cap (bank 2).
27. Install the camshaft housing sub assembly LH.
28. Install the camshaft.
29. Install the No. 2 camshaft.
30. Install the camshaft bearing cap (bank 1).
31. Install the camshaft housing sub assembly RH.

EXHAUST MANIFOLD

REMOVAL & INSTALLATION

2.7L Engine

See Figures 144 through 146.

➡Wear protective gloves when removing the exhaust pipe.

➡The exhaust pipe is extremely hot immediately after the engine has stopped.

➡Confirm that the exhaust pipe has cooled down before removing it.

1. Remove the No. 1 engine under cover.
2. Remove the No. 2 engine under cover.
3. Remove the front exhaust pipe assembly.

a. Disconnect the clamp and connector.

b. Remove the 4 bolts, 2 compression springs and front exhaust pipe assembly.

c. Remove the 2 gaskets from the exhaust manifold converter sub-assembly and center exhaust pipe assembly.

4. Remove the air fuel ration sensor (bank 1 sensor 1).
5. Remove the manifold stay.

a. Remove the bolt, nut and manifold stay.

6. Remove the No. 2 manifold stay.

a. Remove the bolt, nut and manifold stay.

7. Remove the No. 1 exhaust manifold heat insulator.

a. Remove the 4 bolts and No. 1 exhaust manifold heat insulator.

8. Remove the exhaust manifold converter sub assembly.

a. Remove the 5 nuts and exhaust manifold converter sub-assembly.

9. Remove the No. 2 exhaust manifold heat insulator.

a. Remove the 2 bolts and No. 2 exhaust manifold heat insulator.

10. Remove the No. 1 manifold converter insulator.

a. Remove the 4 bolts and No. 1 manifold converter insulator.

To install:

11. Install the No. 1 manifold converter insulator with the 4 bolts. Tighten to 9 ft. lbs. (12 Nm).
12. Install the No. 2 exhaust manifold heat insulator with the 2 bolts. Tighten to 9 ft. lbs. (12 Nm).
13. Install the exhaust manifold converter sub assembly.

a. Install a new gasket onto the cylinder head.

b. Temporarily install the exhaust manifold converter sub-assembly with the 5 nuts.

c. Tighten the 5 nuts in the sequence shown in the illustration. Tighten to 26 ft. lbs. (35 Nm).

14. Install the No. 1 exhaust manifold heat insulator with the 4 bolts. Tighten to 9 ft. lbs. (12 Nm).
15. Install the No. 2 manifold stay with the bolt and nut. Tighten to 32 ft. lbs. (43 Nm).
16. Install the manifold stay with the bolt and nut. Tighten to 32 ft. lbs. (43 Nm).
17. Install the air fuel ratio sensor (bank 1 sensor 1).
18. Install the front exhaust pipe assembly.

a. Using a vernier caliper, measure the free length of the compression spring. The minimum length is 1.64 inch (41.5 mm).

➡If the length is less than the minimum, replace the compression spring.

b. Fully insert a new gasket to the exhaust manifold converter sub assembly.

c. Using a plastic hammer and wooden block, tap in the new gasket until its surface is flush with the exhaust manifold converter sub assembly.

➡Be sure to install the gasket in the correct direction.

➡Do not reuse the gasket.

➡Do not damage the gasket.

➡Do not push in the gasket by using the exhaust pipe when connecting it.

d. Install a new gasket to the center exhaust pipe assembly.

e. Install the front exhaust pipe with the 2 compression springs and 4 bolts. Tighten to 32 ft. lbs. (43 Nm).

f. Connect the clamp and connector.

19. Inspect for exhaust gas leaks.

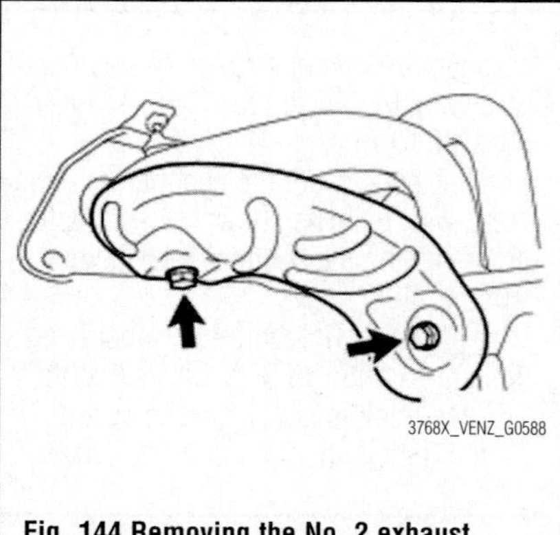

Fig. 144 Removing the No. 2 exhaust manifold heat insulator

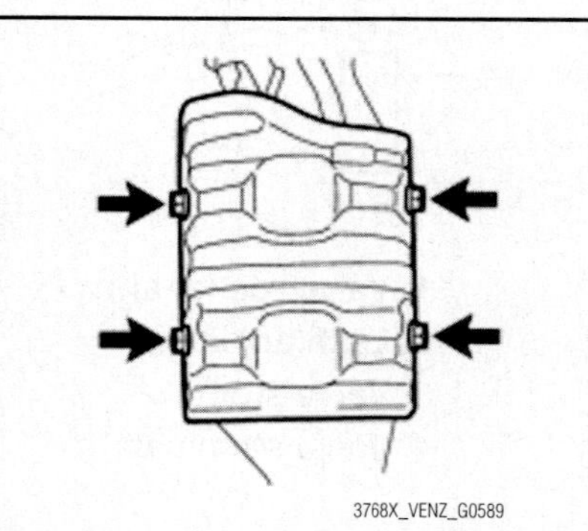

Fig. 145 Removing the No. 1 manifold converter insulator

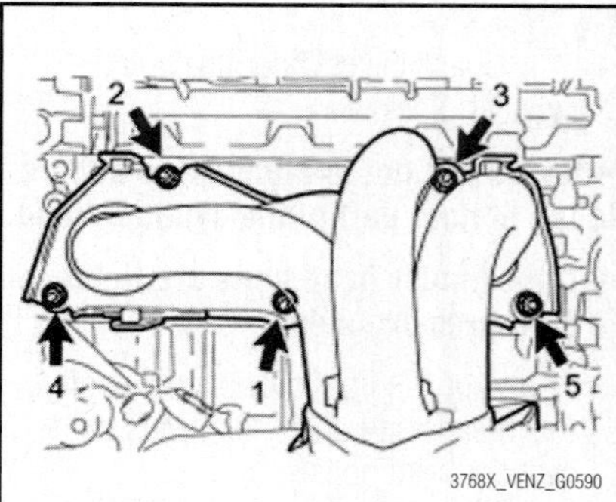

Fig. 146 Installing the exhaust manifold converter sub assembly

20. Install the No. 2 engine under cover.
21. Install the No. 1 engine under cover.

3.5L Engine

See Figures 147 through 153.

1. Remove the front door scuff plate LH.
2. Remove the cowl side trim sub assembly LH.
3. Remove the No. 1 engine under cover.
4. Remove the No. 2 engine under cover.
5. Remove the propeller with the center bearing shaft assembly (AWD).
6. Remove the cool air intake duct seal.
7. Remove the inlet No. 2 air cleaner.
8. Remove the front No. 3 exhaust pipe sub assembly.
    a. Disconnect the 2 clamps and oxygen sensor connector (bank 1 sensor 2).
    b. Remove the grommet.
    c. Remove the 4 bolts, 2 nuts, 2 compression springs and front No. 3 exhaust pipe sub assembly.
    d. Remove the 3 gaskets from the front No. 3 exhaust pipe sub assembly.
9. Remove the manifold stay.
    a. Remove the bolt, nut and manifold stay.
10. Remove the exhaust manifold sub assembly RH.
    a. Disconnect the air fuel ration sensor connector (bank 1 sensor 1).
    b. Using a 12 mm deep socket wrench, remove the 6 nuts and exhaust manifold sub assembly RH.
11. Remove the exhaust manifold to head gasket from the cylinder head sub assembly.
12. Remove the air fuel ratio sensor (bank 1 sensor 1).
13. Remove the front exhaust pipe assembly.
14. Remove the No. 2 manifold stay.
    a. Remove the bolt, nut and No. 2 manifold stay.
15. Remove the No. 2 oil level dipstick guide.
    a. Remove the oil level dipstick.
    b. Remove the bolt and No. 2 engine oil level dipstick guide.
    c. Remove the o-ring from the No. 2 engine oil level dipstick guide.
16. Remove the air fuel ratio sensor (bank 2 sensor 1).
17. Remove the No. 2 exhaust manifold heat insulator.
18. Remove the exhaust manifold sub assembly LH.
    a. Using a 12 mm deep socket wrench, remove the 6 nuts and exhaust manifold sub-assembly LH.

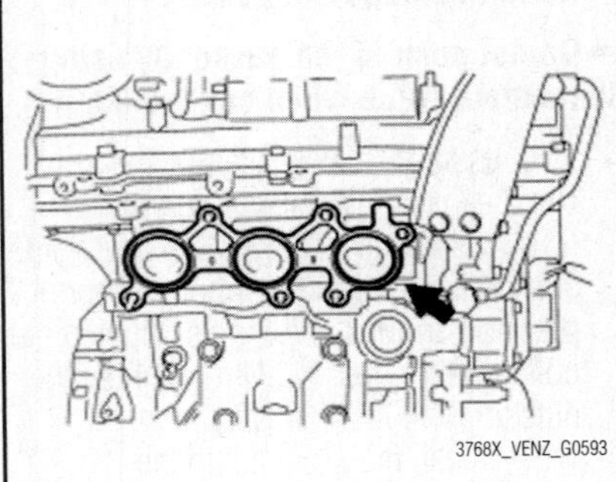

Fig. 147 Removing the exhaust manifold to head gasket

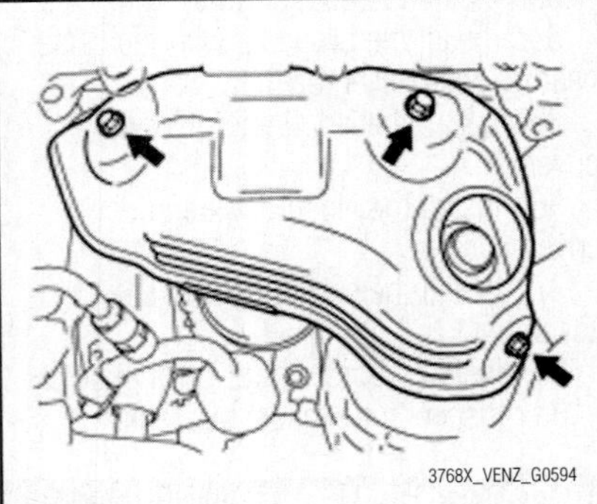

Fig. 148 Removing the No. 2 exhaust manifold heat insulator

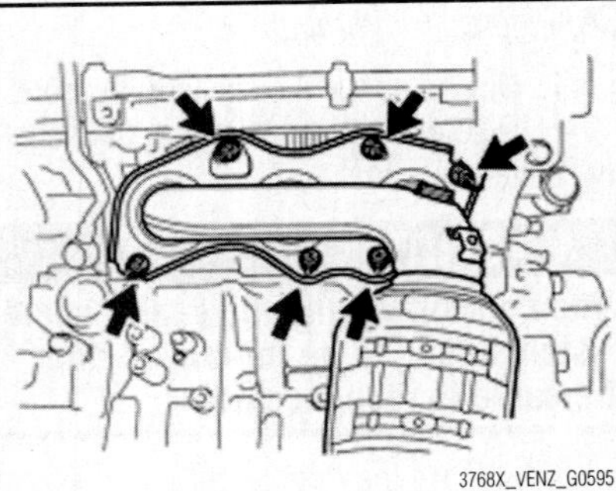

Fig. 149 Removing the exhaust manifold sub assembly LH

Fig. 150 Removing the exhaust manifold to head gasket LH

19. Remove the exhaust manifold to head gasket LH.
    a. Remove the exhaust manifold to head gasket LH from the cylinder head sub assembly.

To install:

20. Install the exhaust manifold to heat gasket LH.
21. Install the exhaust manifold sub assembly LH.
    a. Using a 12 mm deep socket wrench, install the exhaust manifold sub-assembly LH and 6 nuts in the order shown in the illustration. Tighten to 15 ft. lbs. (21 Nm).
22. Install the No. 2 exhaust manifold heat insulator with the 3 bolts. Tighten to 75 inch lbs. (8.5 Nm).
23. Install the No. 2 oil level dipstick guide.
    a. Apply a light coat of engine oil to the o-ring.
    b. Push in the No. 2 oil level dipstick guide end into the oil level dipstick guide.
    c. Install the No. 2 oil level dipstick guide with the bolt. Tighten to 15 ft. lbs. (21 Nm).
    d. Install the oil level dipstick.
24. Install the air fuel ratio sensor (bank 2 sensor 1).
25. Install the No. 2 manifold stay by tightening the bolt and nut in the order shown in the illustration. Tighten to 25 ft. lbs. (34 Nm).
26. Install the front exhaust pipe assembly.
27. Install the air fuel ratio sensor (bank 1 sensor 1).
28. Install the exhaust manifold to head gasket.
29. Install the exhaust manifold sub assembly RH.

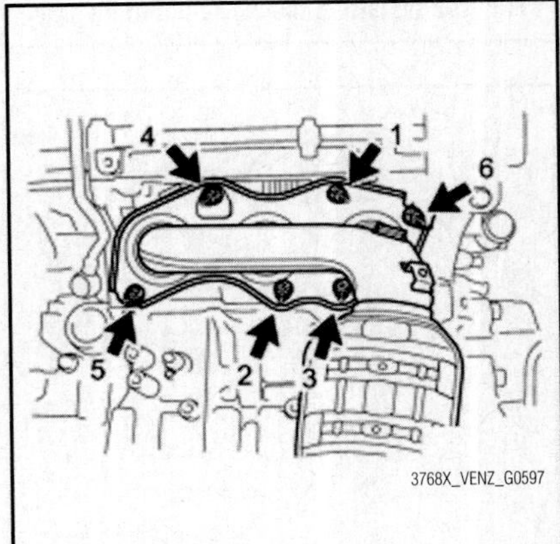

Fig. 151 Installing the exhaust manifold sub assembly LH

a. Using a 12 mm deep socket wrench, install the exhaust manifold sub-assembly RH by tightening the 6 nuts in the order shown in the illustration. Tighten to 15 ft. lbs. (21 Nm).

b. Connect the air fuel ratio sensor connector (bank 1 sensor 1).

30. Install the manifold stay with the bolt and nut. Tighten to 25 ft. lbs. (34 Nm).

31. Install the front No. 3 exhaust pipe sub assembly.

a. Using a vernier caliper, measure the free length of the compression springs. The minimum length is 1.64 inch (41.5 mm).

➡If the free length is less than the minimum, replace the compression spring.

b. Fully insert a new gasket to the front No. 3 exhaust pipe sub-assembly.

c. Using a plastic hammer and wooden block, tap in the new gasket until its surface is flush with the front No. 3 exhaust pipe sub-assembly.

➡Be sure to install the gasket in the correct direction.

➡Do not reuse the gasket.

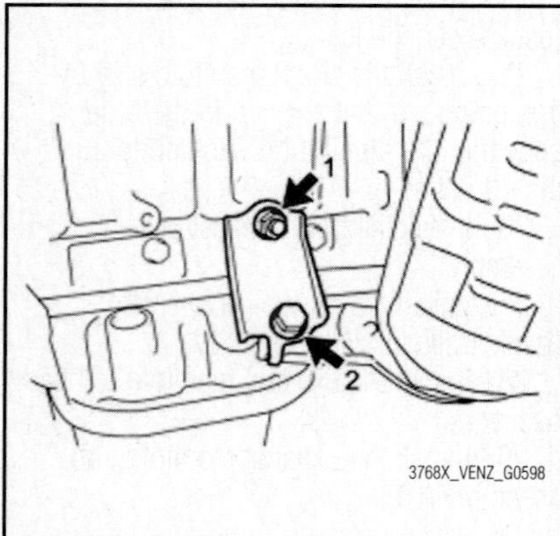

Fig. 152 Installing the No. 2 manifold stay

Fig. 153 Installing the exhaust manifold sub assembly RH

➡Do not damage the gasket.

➡Do not push in the gasket by using the exhaust pipe when connecting it.

d. Install 2 new gaskets to the front No. 3 exhaust pipe sub-assembly.

e. Install the front No. 3 exhaust pipe sub-assembly with the 4 bolts, 2 compression springs and 2 nuts. Tighten the bolts to 32 ft. lbs. (43 Nm). Tighten the nuts to 40 ft. lbs. (55 Nm).

f. Install the grommet to the vehicle.

g. Connect the 2 clamps and oxygen sensor connector (for Bank 1 Sensor 2).

32. Install the inlet No. 2 air cleaner.

33. Install the cool air intake duct seal.

34. Install the propeller with the center bearing shaft assembly (AWD).

35. Install the No. 2 engine under cover.

36. Install the No. 1 engine under cover.

37. Install the cowl side trim sub assembly LH.

38. Install the front door scuff plate LH.

39. Inspect for exhaust gas leaks.

INTAKE MANIFOLD

REMOVAL & INSTALLATION

2.7L Engine

See Figures 154 through 160.

1. Discharge the fuel system pressure.

2. Disconnect the cable from the negative battery terminal.

⁂ CAUTION

When disconnecting the cable, some systems need to be initialized after the cable is reconnected.

3. Remove the throttle body assembly.

4. Remove the vacuum switching valve assembly (ACIS).

a. Disconnect the 2 vacuum hoses, 2 union to connector tube hoses, clamp and connector.

b. Remove the bolt and vacuum switching valve assembly (for ACIS).

5. Disconnect the No. 2 ventilation hose from the intake manifold.

6. Remove the union to connector tube hose form the intake manifold.

7. Remove the fuel delivery pipe sub assembly.

8. Remove the intake manifold.

a. Remove the 2 bolts and 2 wire harness brackets.

b. Disconnect the fuel vapor feed hose, clamp and connector.

c. Remove the bolt and wire harness bracket.

d. Apply battery voltage to the terminals of the connector to close the tumble control valves.

➡If this procedure is not performed, the tumble control valves may be damaged when the intake manifold is removed.

➡Apply battery voltage for 1 to 3 seconds.

➡If battery voltage is applied for more than 3 seconds, the actuator may be damaged.

➡Do not allow the lead wires to contact the other terminals.

e. Remove the bolt.

f. Detach the 2 clamps from the intake manifold and bracket.

g. Disconnect the intake air control valve actuator connector.

h. Remove the 6 bolts and intake manifold.

➡The tumble control valves may be damaged if they are not closed before installing the intake manifold.

➡Connect the battery to the terminals of the actuator to operate the motor and close the valves

i. Remove the intake manifold gasket from the intake manifold.

9. Remove the check valve.

a. Disconnect the 2 vacuum hoses from the intake manifold and remove the check valve.

10. Remove the wiring harness clamp bracket.

11. Remove the engine mounting damper.

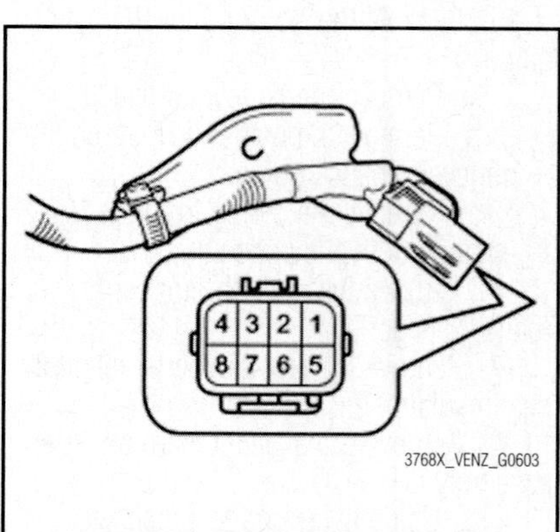

Fig. 154 Identifying the connector terminals

Tester Connection	Specified Condition
Positive (+) battery voltage applied to terminal 8 (M-), and negative (-) battery voltage applied to terminal 4 (M+)	

3768X_VENZ_G0604

Fig. 155 Closing the tumble control valves

a. Remove the 3 bolts and engine mounting damper.

To install:

12. Install the engine mounting damper with the 3 bolts. Tighten to 80 inch lbs. (9 Nm).

13. Install the wiring harness clamp bracket with the bolt. Tighten to 74 inch lbs. (8.4 Nm).

14. Install the check valve.

a. Connect the 2 vacuum hoses to the intake manifold to install the check valve.

b. Check that the check valve is installed as shown in the illustration.

15. Install the intake manifold.

a. Close the tumble control valves.

➡The tumble control valves may be damaged if they are not closed before installing the intake manifold.

➡Connect the battery to the terminals of the actuator to operate the motor and close the valves

b. Install a new gasket to the intake manifold.

c. Install the intake manifold by tightening the 6 bolts in the sequence shown in the illustration. Tighten to 15 ft. lbs. (21 Nm).

d. Connect the intake air control actuator connector.

e. Attach the 2 clamps to the intake manifold and bracket.

f. Install the wire harness with the bolt. Tighten to 74 inch lbs. (8.4 Nm).

g. Install the wire harness bracket with the bolt. Tighten to 74 inch lbs. (8.4 Nm).

h. Connect the fuel vapor feed hose, clamp and connector.

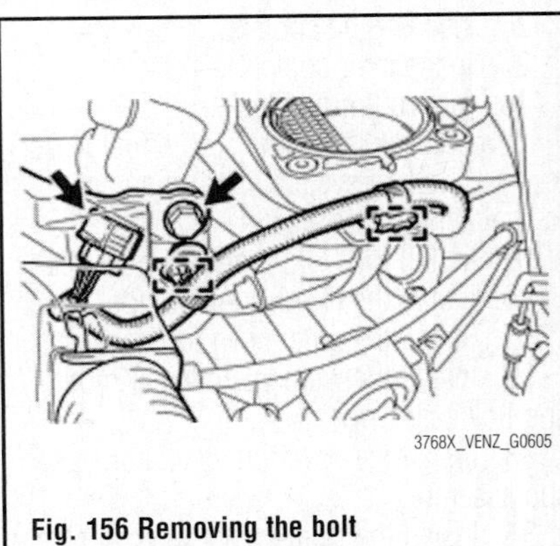

Fig. 156 Removing the bolt

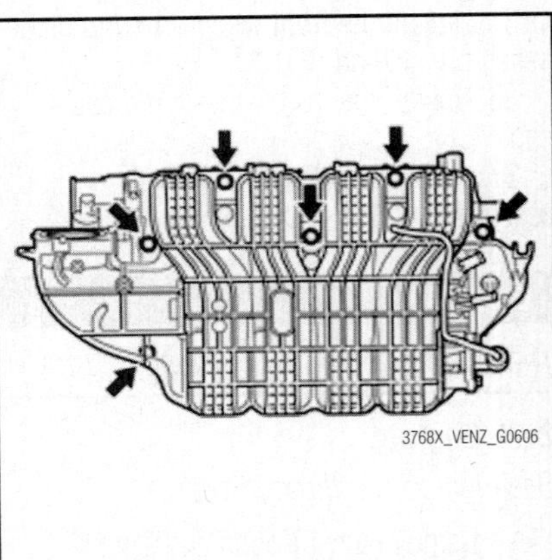

Fig. 157 Removing the intake manifold

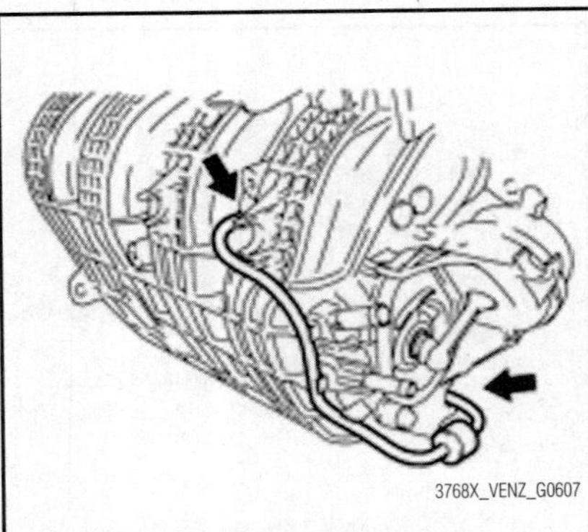

Fig. 158 Removing the check valve

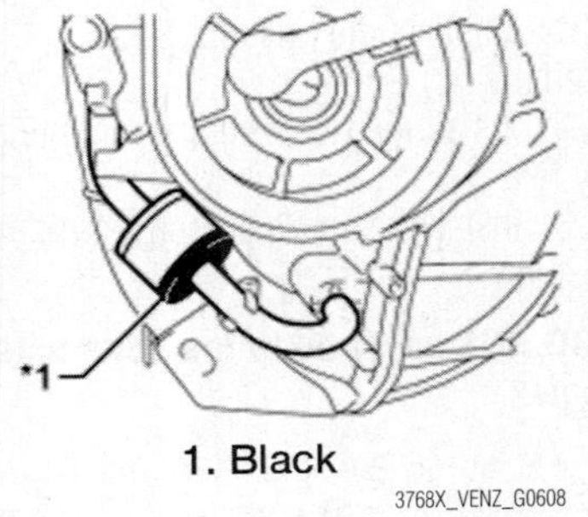

Fig. 159 Checking the check valve installation

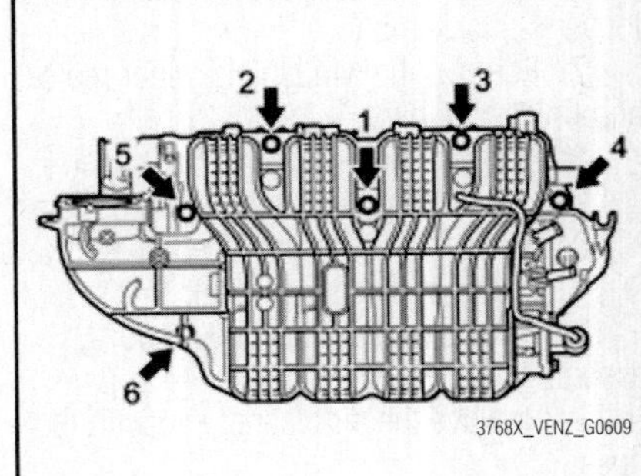

Fig. 160 Installing the intake manifold

i. Install the 2 wire harness brackets with the 2 bolts. Tighten to 74 inch lbs. (8.4 Nm).

16. Install the fuel deliver pipe sub assembly.

17. Install the union to connector tube hose.

a. Install the union to connector tube hose to the intake manifold.

18. Connect the No. 2 ventilation hose to the intake manifold.

19. Install the vacuum switching valve assembly (ACIS).

a. Install the vacuum switching valve (for ACIS) with the bolt. Tighten to 80 inch lbs. (9 Nm).

b. Connect the 2 vacuum hoses, 2 union to connector tube hoses, clamp and connector.

20. Install the throttle body assembly.

21. Connect the cable to the negative battery terminal.

✻✻ CAUTION

When disconnecting the cable, some systems need to be initialized after the cable is reconnected.

22. Inspect for fuel leaks.

3.5L Engine

See Figures 161 through 163.

1. Discharge the fuel system pressure.

2. Disconnect the cable from the negative battery terminal.

➡When disconnecting the cable, some systems need to be initialized after the cable is reconnected.

3. Remove the front wiper arm head cap.

4. Remove the front wiper arm and blade assemblies LH and RH.

5. Remove the front fender to cowl side seals LH and RH.

6. Remove the cowl top ventilator louver sub assembly.
7. Remove the windshield wiper motor and link assembly.
8. Remove the outer cowl top panel.
9. Remove the No. 1 engine under cover.
10. Remove the No. 2 engine under cover.
11. Drain the engine coolant.
12. Remove the v-bank cover sub assembly.
13. Remove the air cleaner cap with the hose.
14. Remove the intake air surge tank assembly.
    a. Disconnect the No. 2 ventilation hose.
    b. Disconnect the union to check vale hose.
    c. Disconnect the throttle body assembly connector and wire harness clamp.
    d. Disconnect the fuel vapor feed hose.
    e. Disconnect the 2 water by pass hoses.
    f. Disconnect the connector from the intake air control valve assembly.
    g. Remove the bolt and separate the No. 1 surge tank stay from the intake air surge tank assembly.
    h. Remove the bolt and separate the throttle body bracket from the intake air surge tank assembly.
    i. Remove the 2 nuts from the intake air surge tank assembly.
    j. Using a 5 mm socket hexagon wrench, remove the 4 bolts.
    k. Remove the intake air surge tank assembly and 3 air surge tank to intake manifold gaskets.
15. Remove the No. 2 engine mounting stay RH.
16. Disconnect the fuel tube sub assembly.
17. Remove the fuel injector assembly.
18. Remove the intake manifold.
    a. Remove the 6 bolts, 4 nuts, intake manifold and 2 intake manifold to head gaskets.

To install:

19. Install the intake manifold.
    a. Set 2 new intake manifold to head gaskets on each cylinder head.

➡Align the port holes of the gaskets and cylinder head.

➡Make sure that the gaskets are installed in the correct direction.

    b. Set the intake manifold on the cylinder head.
    c. Install and tighten the 6 bolts and 4 nuts uniformly in several steps. Tighten to 15 ft. lbs. (21 Nm).

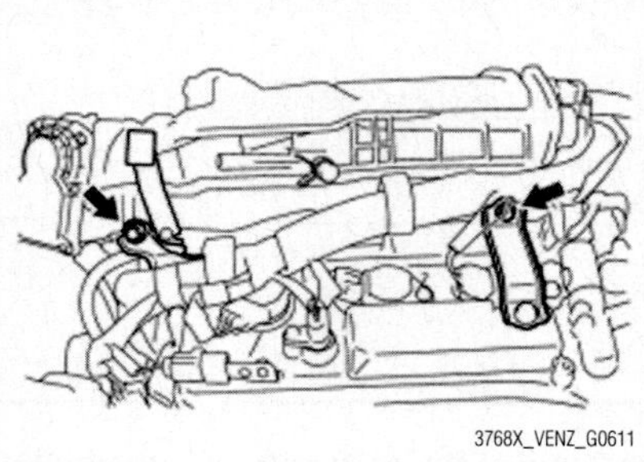

Fig. 161 Removing the bolt and separate the No. 1 surge tank stay from the intake air surge tank assembly

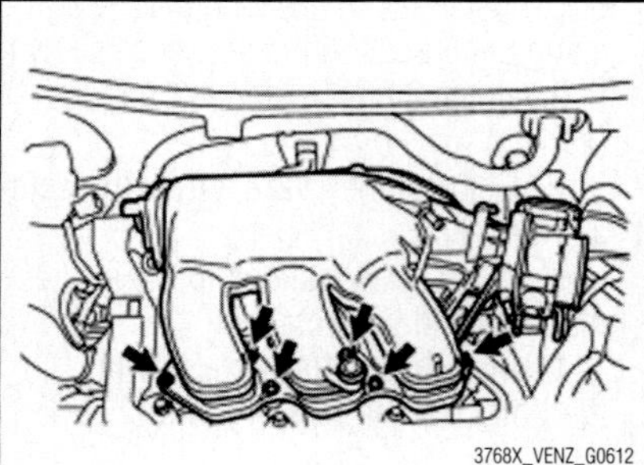

Fig. 162 Removing the intake air surge tank assembly

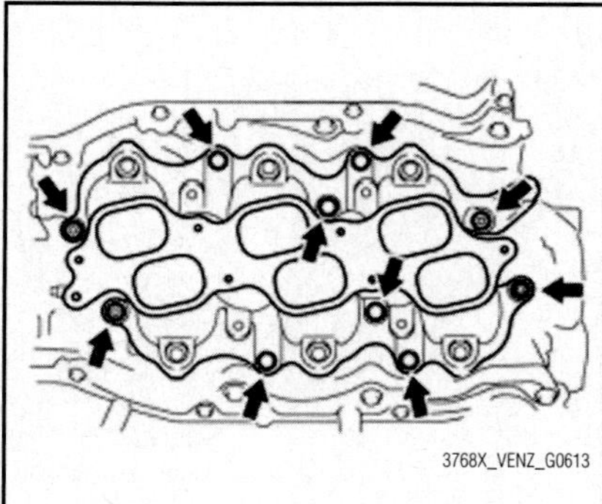

Fig. 163 Removing the intake manifold

20. Install the fuel injector assembly.
21. Connect the fuel tube sub assembly.
22. Install the No. 2 engine mounting stay RH.
23. Install the intake air surge tank assembly.

➡DO NOT apply oil to the bolts listed below:

- Surge tank and intake manifold
- No. 1 surge tank stay and surge tank
- Throttle body bracket and surge tank

    a. Install 3 new air surge tank to intake manifold gaskets to the intake air surge tank.
    b. Using a 5 mm hexagon socket wrench, install the intake air surge tank assembly with the 4 bolts and 2 nuts. Tighten the bolts to 13 ft. lbs. (18 Nm). Tighten the nuts to 12 ft. lbs. (16Nm).
    c. Install the throttle body bracket and No. 1 surge tank stay with the 2 bolts. Tighten to 15 ft. lbs. (21 Nm).
    d. Connect the connector to the intake air control valve assembly.
    e. Connect the 2 water by-pass hoses to the throttle body assembly.
    f. Connect the fuel vapor feed hose to the intake air surge tank assembly.
    g. Connect the throttle body assembly connector and wire harness clamp to the throttle body assembly.
    h. Connect the ventilation hose.
    i. Connect the union to connector tube hose.
24. Install the air cleaner cap with the hose.
25. Connect the cable to the negative battery terminal.

➡When disconnecting the cable, some systems need to be initialized after the cable is reconnected.

26. Add engine coolant.
27. Inspect for coolant leaks.
28. Inspect for fuel leaks.
    a. Check that there are no fuel leaks from the fuel system after doing any maintenance or repairs.
29. Install the No. 2 engine under cover.
30. Install the No. 1 engine under cover.
31. Install the outer cowl top panel.
32. Install the windshield wiper motor and link assembly.
33. Install the cowl top ventilator louver sub assembly.
34. Install the front fender to cowl side seal RH and LH.
35. Install the front wiper arm and blade assembly LH and RH.
36. Install the front wiper arm head cap.
37. Install the v-bank cover sub assembly.

OIL PAN

REMOVAL & INSTALLATION

2.7L Engine

See Figures 164 through 167.

1. Remove the 11 bolts and 2 nuts.
2. Insert the blade of an oil pan seal

cutter between the oil pan and stiffening crankcase, cut off the applied sealer and remove the oil pan.

➡Be careful not to damage the stiffening crankcase contact surface of the oil pan.

➡Be careful not to damage the stiffening crankcase flange.

To install:

3. Apply seal packing in a continuous line as shown in the illustration. The standard seal diameter is 0.0984 to 0.138 inch (2.5-3.5 mm).

➡Remove any oil from the contact surface.

➡Install the oil pan within 3 minutes and tighten the bolts and nuts within 10 minutes after applying seal packing.

➡Do not apply oil for at least 4 hours after the installation.

4. Install the oil pan with the 11 bolts and 2 nuts in several steps, in the sequence shown in the illustration. Tighten to 7 ft. lbs. (10 Nm).

➡Bolt A and nut A are tightened twice.

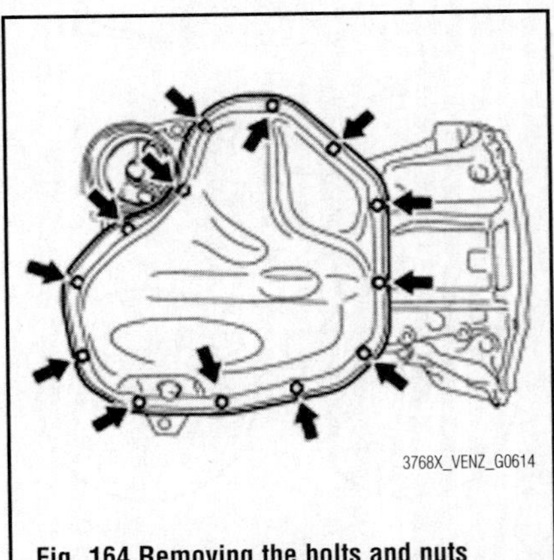

Fig. 164 Removing the bolts and nuts

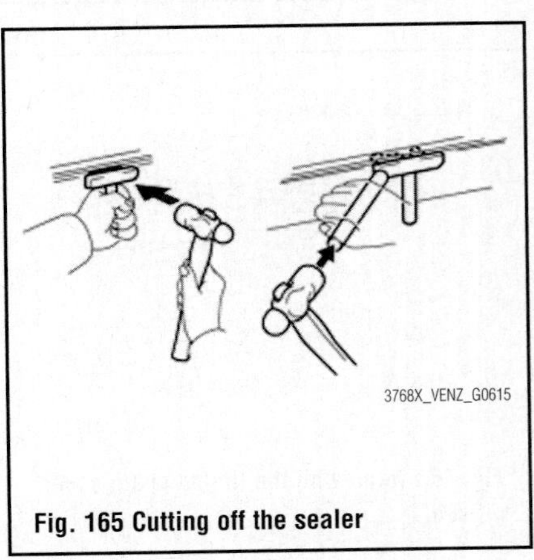

Fig. 165 Cutting off the sealer

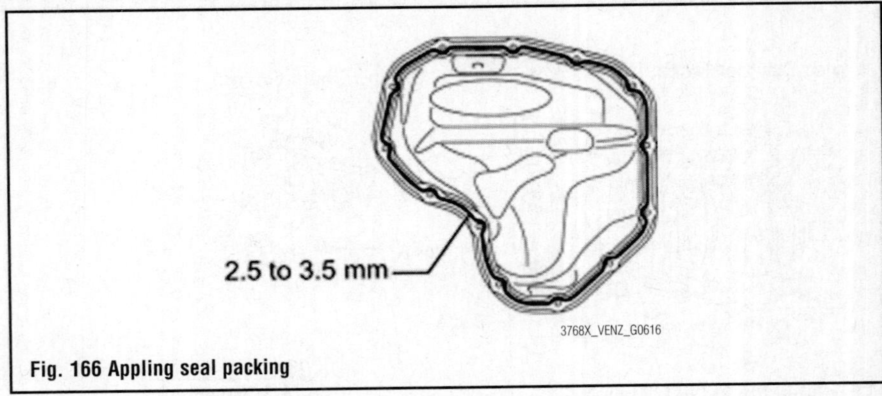

Fig. 166 Appling seal packing

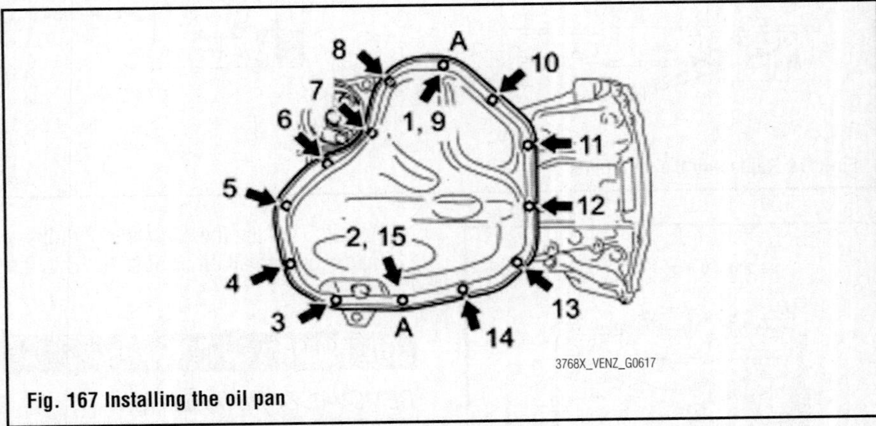

Fig. 167 Installing the oil pan

3.5L Engine

See Figures 168 through 172.

1. Remove the 16 bolts and 2 nuts (*1).
2. Insert the blade of the oil pan seal cutter between the oil pans. Cut through the applied sealer and remove the No. 2 oil pan sub-assembly.

➡Be careful not to damage the contact surfaces of the oil pans.

3. Using an E6 "TORX" socket wrench, remove the 2 stud bolts.

To install:

4. When replacing a stud bolt, install it by using an E8 "TORX" socket wrench. Tighten bolt A to 7 ft. lbs. (10 Nm). Tighten bolt B to 80 inch lbs. (9 Nm).
5. Install 2 new O-rings.
6. Apply seal packing in a continuous line as shown in the illustration. The seal diameter needs to be 0.118-0.157 inch (3-4 mm).

➡Remove any oil from the contact surface.

➡Install the oil pan within 3 minutes after applying seal packing.

➡Do not start the engine for at least 2 hours after installing.

7. Install the oil pan with the 16 bolts

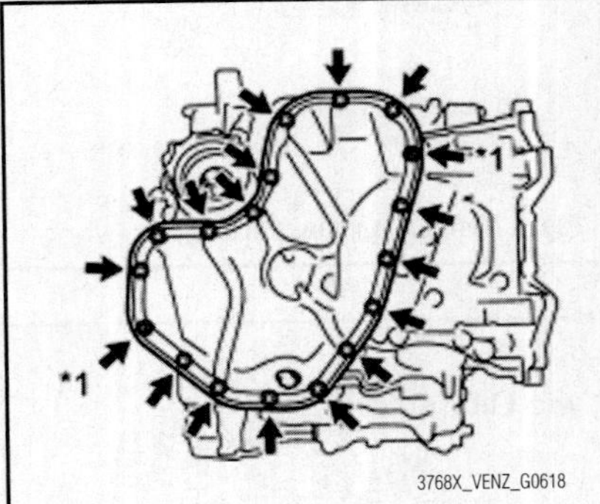

Fig. 168 Removing the oil pan bolts and nuts

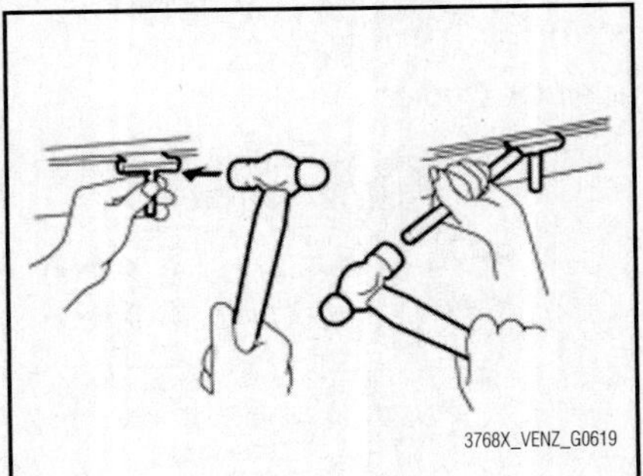

Fig. 169 Removing the oil pan—3.5L engine

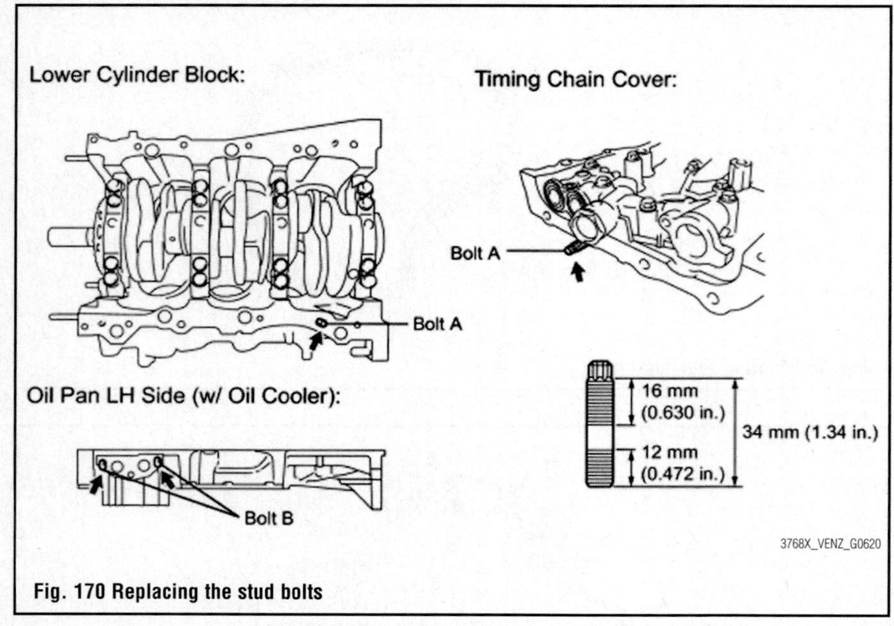

Fig. 170 Replacing the stud bolts

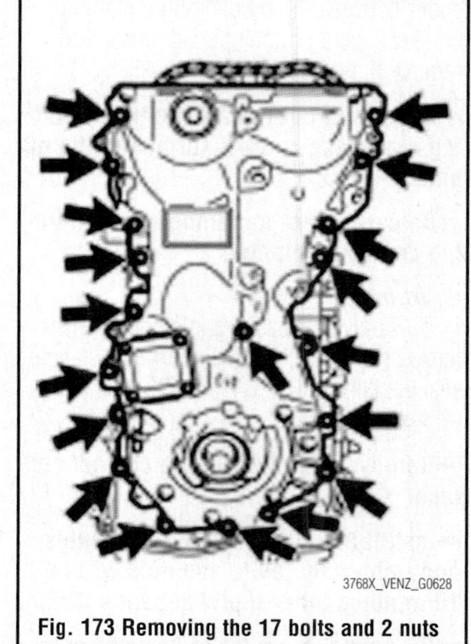

Fig. 173 Removing the 17 bolts and 2 nuts

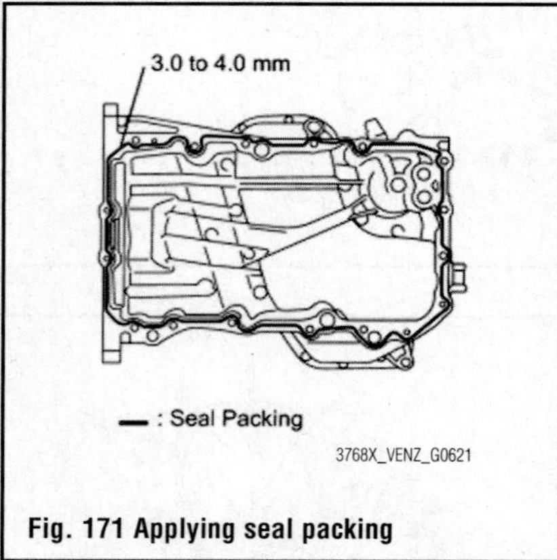

Fig. 171 Applying seal packing

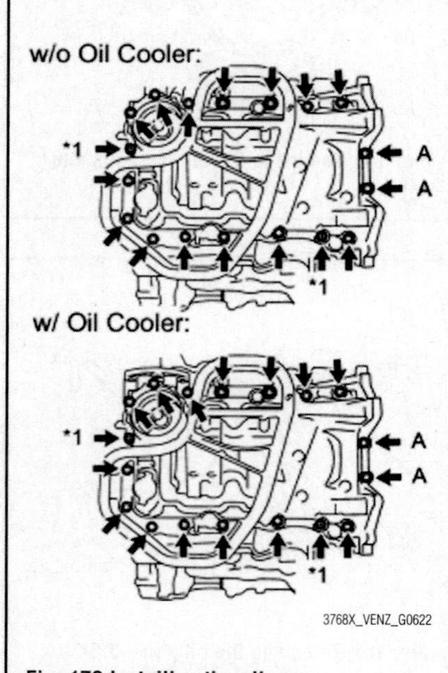

Fig. 172 Installing the oil pan

and 2 nuts. Tighten the A bolts to 7 ft. lbs. (10 Nm). Tighten all other bolts to 15 ft. lbs. (21 Nm).

OIL PUMP

REMOVAL & INSTALLATION

2.7L Engine

See Figures 173 through 181.

1. Remove the engine and transaxle.
2. Remove the engine wire.
3. Remove the ignition coil assembly.
4. Remove the cylinder head cover sub assembly.
5. Remove the Crankshaft Position (CKP) sensor.
6. Remove the crankshaft pulley.
7. Remove the engine mounting bracket RH.
   a. Remove the 5 bolts and engine mounting bracket RH.
8. Remove the v-ribbed belt tensioner assembly.
9. Remove the timing chain cover sub assembly.
   a. Remove the 17 bolts and 2 nuts.
   b. Remove the timing chain cover by prying between the timing chain cover and cylinder head, camshaft housing, cylinder block and stiffening crankcase with a screwdriver as shown in the illustration.

➡**Be careful not to damage the contact surfaces of the cylinder head, camshaft housing, cylinder block, stiffening crankcase and chain cover.**

➡**Tape the screwdriver tip before use.**

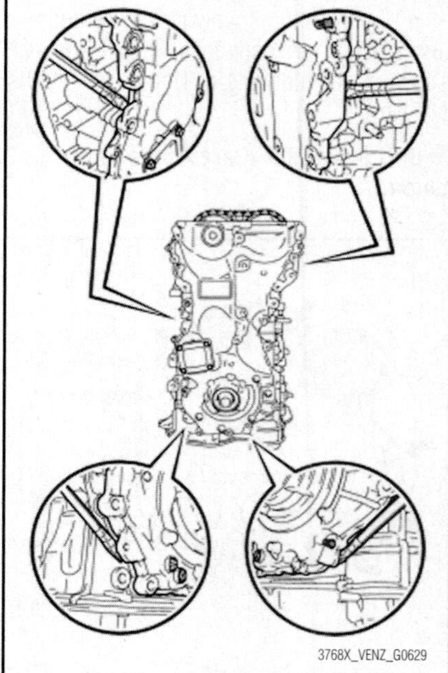

Fig. 174 Removing the timing chain cover

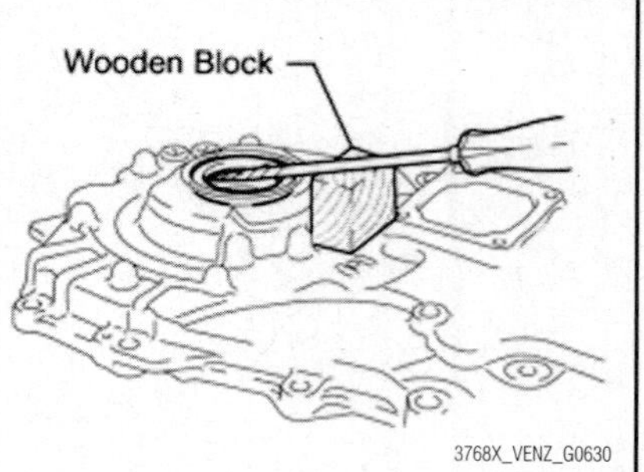

Fig. 175 Removing the timing chain cover oil seal

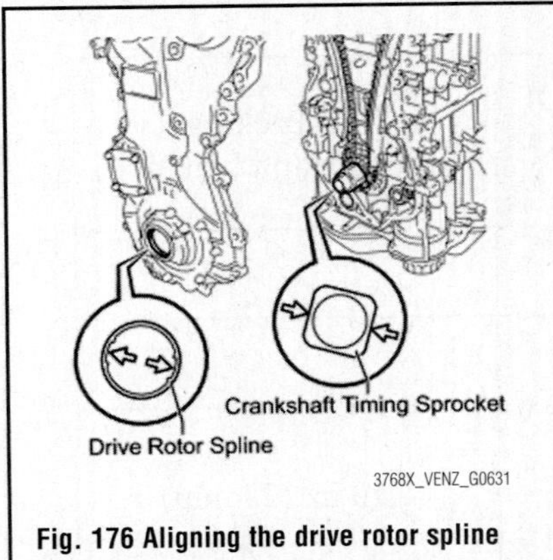

Fig. 176 Aligning the drive rotor spline

c. Remove the 3 gaskets from the stiffening crankcase.

10. Remove the timing chain cover oil seal.

a. Using a screwdriver and wooden block, pry out the oil seal.

➡Do not damage the surface of the oil seal press fit hole.

➡Tape the screwdriver tip before use.

To install:

11. Install the timing chain cover sub assembly.

a. Apply a light coat of engine oil to 3 new gaskets.

b. Install the 3 gaskets to the stiffening crankcase.

c. Align the drive rotor spline and the crankshaft timing sprocket as shown in the illustration.

d. Apply seal packing in a line to the timing chain cover as shown in the following illustration.

➡When the contact surfaces are wet, clean the surfaces with non-residue solvent before applying seal packing.

➡Install the timing chain cover within 3 minutes and tighten the bolts within 10 minutes after applying seal packing.

➡After applying seal packing to the timing chain cover, install the engine mounting bracket within 10 minutes.

➡Do not apply oil for at least 4 hours after the installation.

➡Do not start the engine for at least 4 hours after the installation.

e. Temporarily install the timing chain cover with the 17 bolts and 2 nuts.

➡Make sure there is no oil on the bolts. If there is oil on the bolts, clean them before installing them.

f. Tighten the 17 bolts and 2 nuts in several steps, in the sequence shown in

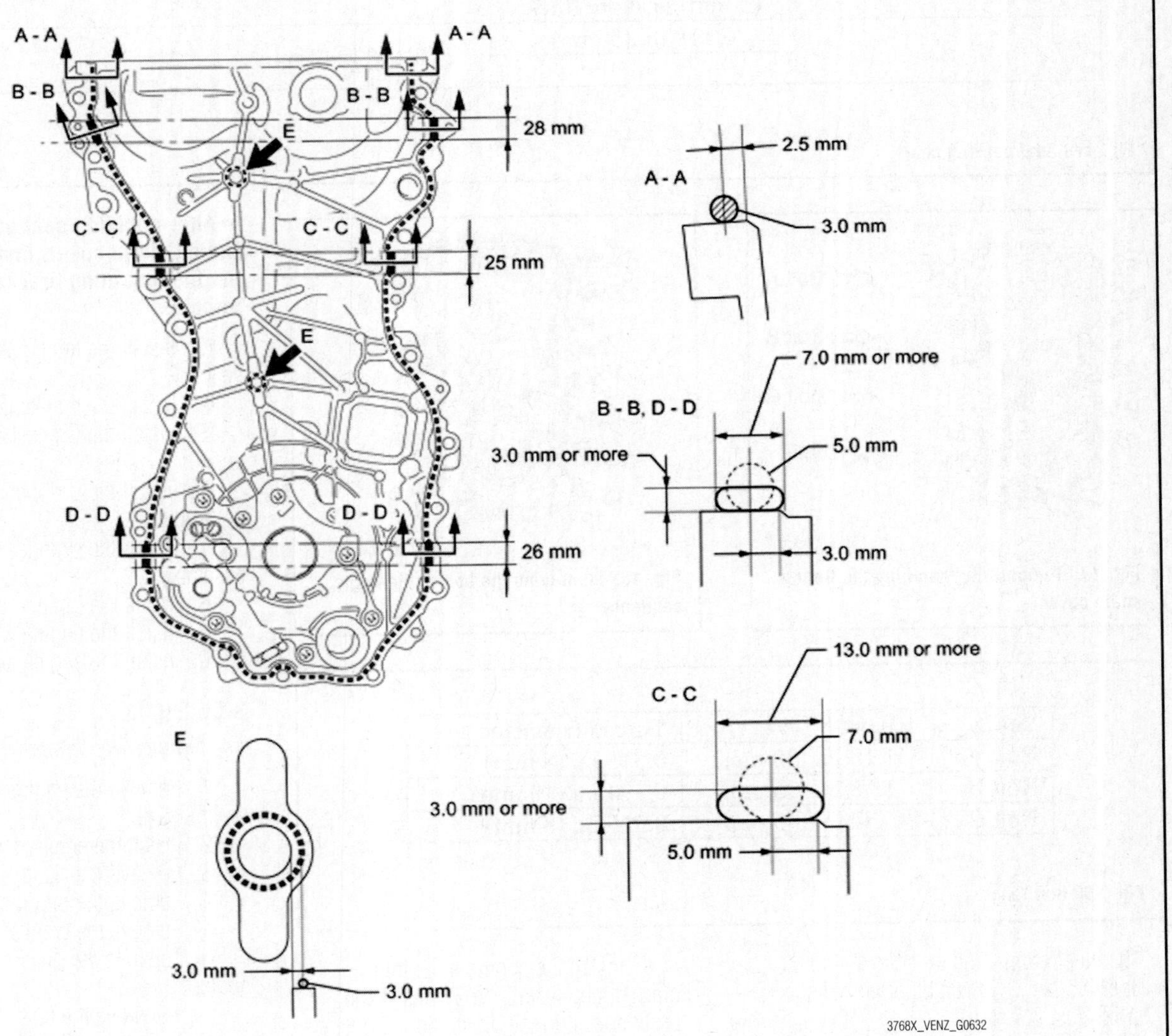

Fig. 177 Applying seal packing

Area	Seal Packing Diameter (Round) / Seal Packing Dimension (Flat)	Distance Front Edge Of Cover To Center Of Seal Packing	Seal Packing Application Length
Dashed Line	0.118 in. (3 mm)	0.0984 in. (2.5 mm)	-
	-		
A - A	0.118 in. (3 mm)	0.0984 in. (2.5 mm)	-
	-		
B - B	0.197 in. (5 mm)	0.118 in (3 mm)	1.10 in. (28 mm)
	0.276 in. (7 mm) or more wide and 0.118 in. (3 mm) or more thick		
C - C	0.276 in. (7 mm)	0.197 in. (5 mm)	0.984 in. (25 mm)
	0.512 in. (13 mm) or more wide and 0.118 in. (3 mm) or more thick		
D - D	0.197 in. (5 mm)	0.118 in (3 mm)	1.02 in. (26 mm)
	0.276 in. (7 mm) or more wide and 0.118 in. (3 mm) or more thick		
E	0.118 in. (3 mm)	0.118 in (3 mm)	-
	-		

3768X_VENZ_G0633

Fig. 178 Seal packing chart

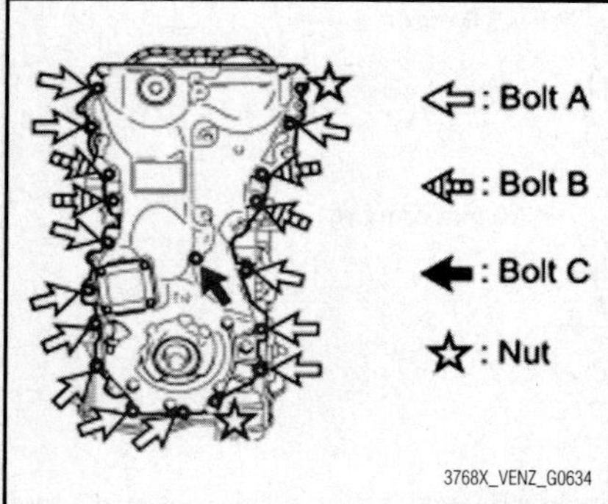

Fig. 179 Temporarily installing the timing chain cover

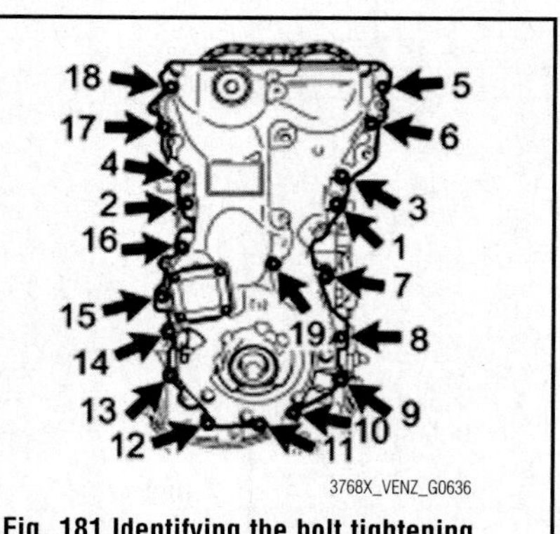

Fig. 181 Identifying the bolt tightening sequence

Item	Length	Thread Diameter
Bolt A	1.18 in. (30 mm)	0.315 in. (8 mm)
Bolt B	1.38 in. (35 mm)	0.394 in. (10 mm)
Bolt C	1.77 in. (45 mm)	0.315 in. (8 mm)

3768X_VENZ_G0635

Fig. 180 Bolt lengths

the illustration. Tighten bolts 1-4 to 41 ft. lbs. (55 Nm). Tighten all other bolts to 15 ft. lbs. (21 Nm).

12. Install the engine mounting bracket RH.

a. Install the engine mounting bracket, and install the 5 bolts in the order shown in the illustration. Tighten bolts 1-3 to 41 ft. lbs. (55 Nm). Tighten bolts 4 and 5 to 15 ft. lbs. (21 Nm).

➡After applying seal packing to the timing chain cover, install the engine mounting bracket within 10 minutes.

13. Install the timing chain cover oil seal.
14. Install the crankshaft pulley.
15. Install the Crankshaft Position (CKP) sensor.
16. Install the v-ribbed belt tensioner assembly.
17. Install the cylinder head cover sub assembly.
18. Install the ignition coil assembly.
19. Install the engine wire.
20. Install the engine and transaxle.

3.5L Engine

See Figures 182 through 190.

1. Remove the engine assembly with the transaxle.
2. Install the engine on an engine stand.
3. Remove the No. 1 oil pipe.
4. Remove the oil pipe.
5. Remove the crankshaft pulley.
6. Separate the oil cooler pipe (w/oil cooler).

a. Remove the bolt and 2 nuts, and disconnect the oil cooler pipe from the oil pan sub-assembly.

b. Remove the gasket from the oil pan sub assembly.

7. Remove the water inlet housing.

8. Remove the cylinder head cover sub assembly (bank 1).

9. Remove the cylinder head cover sub assembly (bank 2).

10. Remove the No. 2 oil pan sub assembly.

11. Remove the oil strainer sub assembly.

12. Remove the oil pan sub assembly.

13. Remove the timing chain cover sub assembly.

a. Remove the 23 bolts and 2 nuts shown in the illustration.

b. Remove the timing chain cover by prying between the timing chain cover and cylinder head or cylinder block with a screwdriver.

➡**Be careful not to damage the contact surfaces of the cylinder head, cylinder block and chain cover.**

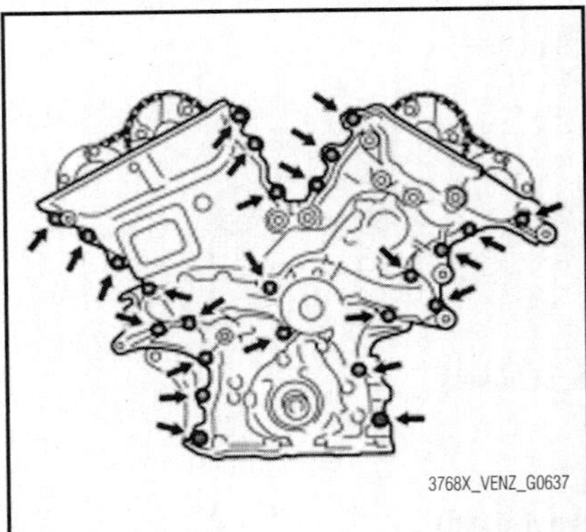

Fig. 182 Removing the 23 bolts and 2 nuts

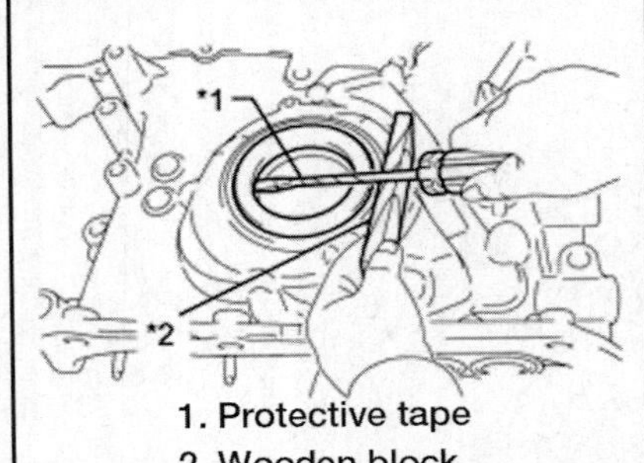

Fig. 184 Removing the timing chain case oil seal

➡**Tape (*1) the screwdriver tip before use.**

c. Remove the gasket.

14. Remove the timing chain case oil seal.

a. Using a screwdriver, pry out the oil seal.

➡**Tape the screwdriver tip before use.**

To install:

15. Install the timing chain case oil seal.

a. Using the special too, tap in the new oil seal until its surface is flush with the timing chain case edge.

➡**Keep the lip free from foreign matter.**

➡**Do not tap on the oil seal at an angle.**

➡**Make sure that the oil seal edge does not stick out of the timing chain case.**

16. Install the timing chain cover sub assembly.

a. Apply seal packing in a continuous line to the engine unit. The seal diameter should be 0.118 inch (3 mm) or more.

➡**Be sure to clean and degrease the contact surfaces, especially the surfaces indicated by C in the illustration.**

➡**If the contact surfaces are wet, wipe them with an oil-free cloth before applying seal packing.**

➡**Install the chain cover sub-assembly within 3 minutes after applying seal packing.**

➡**Do not start the engine for at least 2 hours after installing the chain cover sub-assembly.**

b. Apply seal packing in a line to the timing chain cover as shown in the following illustration.

➡**If the contact surfaces are wet, wipe them with an oil-free cloth before applying seal packing.**

➡**Install the chain cover sub-assembly within 3 minutes and tighten the bolts within 15 minutes after applying seal packing.**

➡**Do not start the engine for at least 2 hours after installing the chain cover sub-assembly.**

c. Install a new gasket.

d. Align the oil pump's drive rotor spline and the crankshaft as shown in the illustration. Install the oil pump and chain cover to the crankshaft.

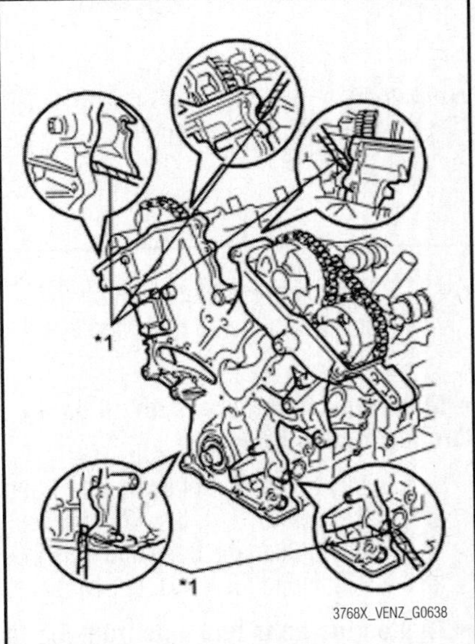

Fig. 183 Removing the timing chain cover

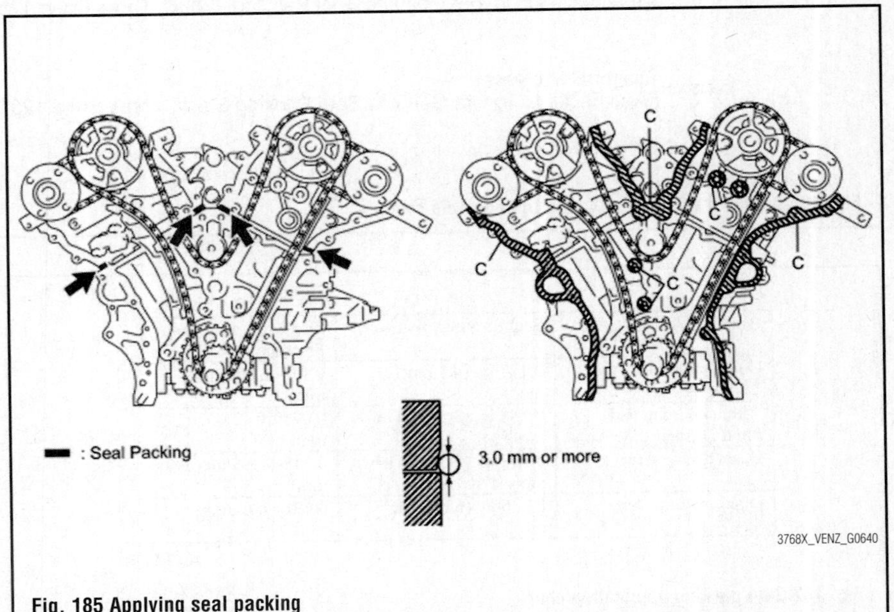

Fig. 185 Applying seal packing

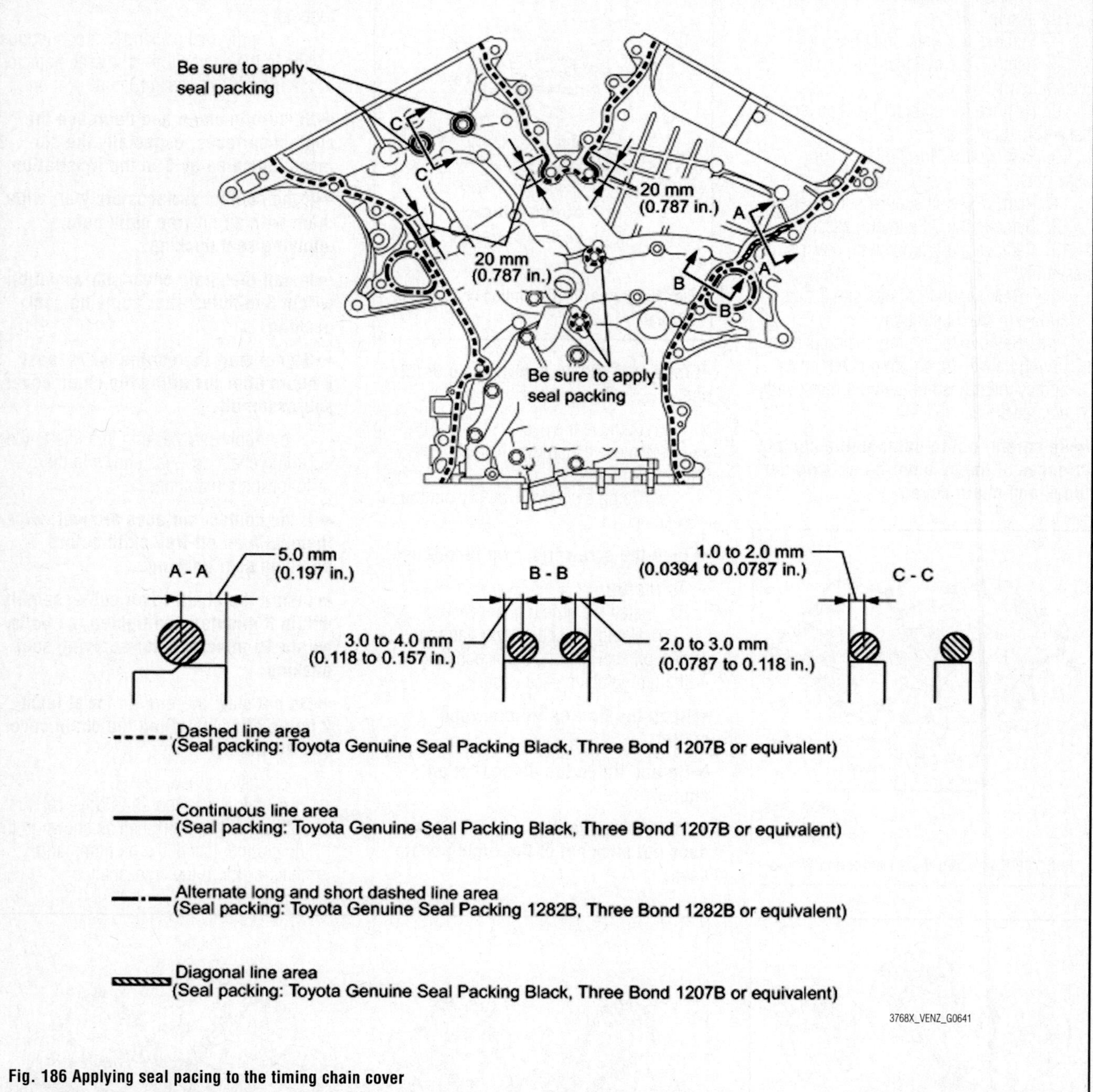

Fig. 186 Applying seal pacing to the timing chain cover

Area	Seal Packing Diameter	Application Position From Inside Seal Line
Continuous Line Area	0.177 in. (4.5 mm) or more	0.118-0.158 in. (3-4 mm)
Alternate Long and Short Dashed Line Area	0.138 in. (3.5 mm) or more	0.0787-0.118 in. (2-3 mm)
Dashed Line Area	0.138 in. (3.5 mm) or more	0.118-0.158 in. (3-4 mm)
Diagonal Line Area	0.236 in. (6 mm) or more	0.197 in. (5 mm)

3768X_VENZ_G0642

Fig. 187 Seal packing application chart

e. Temporarily tighten the timing chain cover with the 23 bolts and 2 nuts.

➡Make sure that there is no oil on the threads of bolt B and C.

f. Fully tighten the bolts in area 1 and area 2. Tighten to 15 ft. lbs. (21 Nm).

g. Fully tighten the bolts and nuts in area 3. Tighten to 15 ft. lbs. (21 Nm).

➡Tighten the bolts and nuts from top to bottom as shown in the illustration.

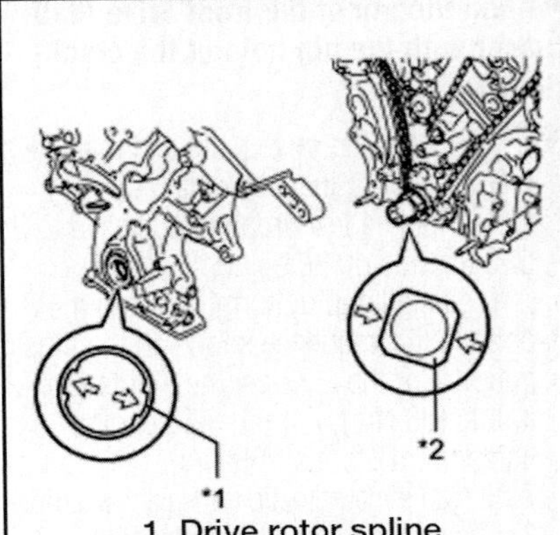

Fig. 188 Aligning the oil pump drive rotor spline

h. Fully tighten the bolts in area 4. Tighten bolt A to 32 ft. lbs. (43 Nm). Tighten all except bolt A to 15 ft. lbs. (21 Nm).

➡Tighten the bolts from the bottom to top as shown in the illustration.

17. Install the oil pan sub assembly.
18. Install the oil strainer sub assembly.
19. Install the No. 2 oil pan sub assembly.
20. Install the cylinder head cover sub assembly (bank 2 and bank 1).
21. Install the water inlet housing.
22. Install the oil cooler pipe (w/oil cooler).
    a. Install a new gasket to the oil pan sub-assembly.
    b. Install the oil cooler pipe with the bolt and 2 nuts. Tighten to 15 ft. lbs. (21 Nm).
23. Install the crankshaft pulley.

Item	Length
Bolt A	1.57 in. (40 mm)
Bolt B	2.17 in. (55 mm)
Bolt C	0.984 in. (25 mm)

3768X_VENZ_G0645

Fig. 190 Bolt length

24. Install the oil pipe.
25. Install the No. 1 oil pipe.
26. Install the engine hangers.
27. Remove the engine stand.
28. Install the engine assembly with the transaxle.

PISTON AND RING

POSITIONING

See Figures 191 through 195.

1. Align the front marks (*1) of the piston and connecting rod, insert the connecting rod into the piston, and then push in the piston pin with your thumb until the pin comes into contact with the snap ring.

➡The piston and pin are a matched set.

2. Using a small screwdriver, install a new snap ring on the other side of the piston pin hole (*1).

➡Be sure that the end gap of the snap ring is not aligned with the service hole cutout portion of the piston.

3. Check the fitting condition between the piston and piston pin.

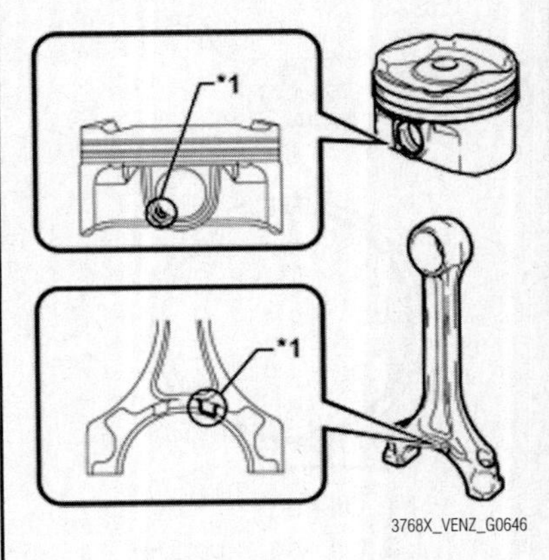

Fig. 191 Aligning the front marks of the piston and connecting rod

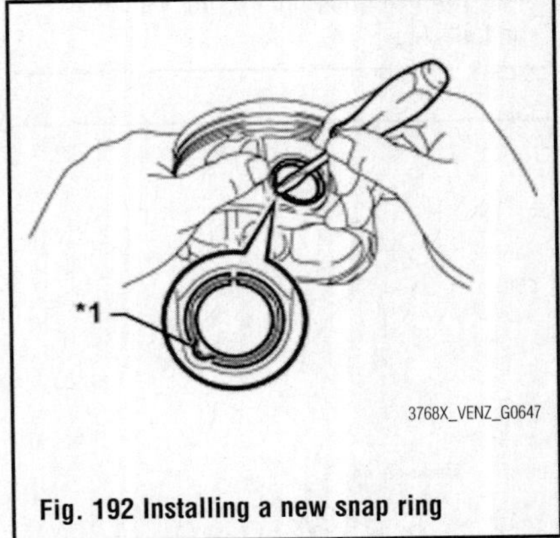

Fig. 192 Installing a new snap ring

a. Move the connecting rod back and forth on the piston pin. Check the fitting condition. If abnormal movement is felt, replace the piston and pin as a set.

b. Rotate the piston back and forth on the piston pin. Check the fitting condition. If abnormal movement is felt, replace the piston and pin as a set.

4. Install the piston ring set.

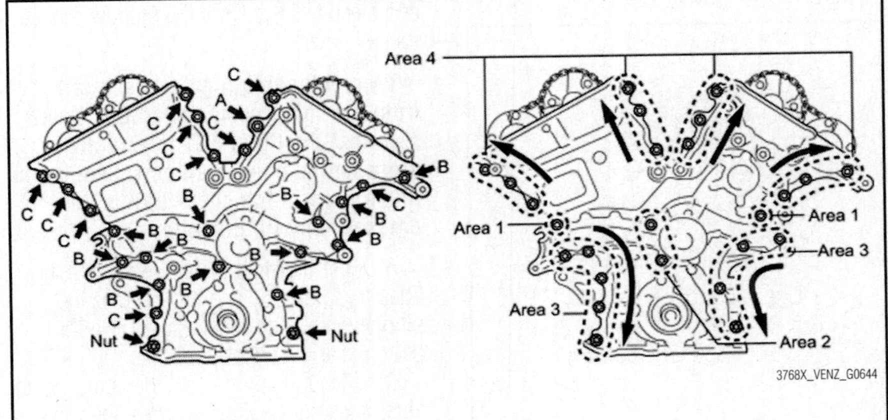

Fig. 189 Temporarily installing the timing chain cover

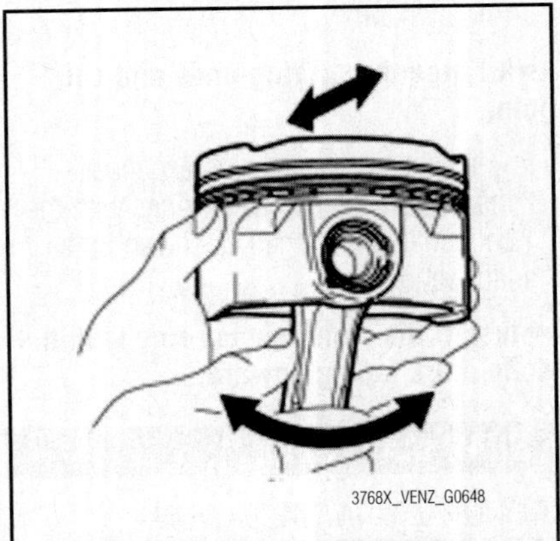

Fig. 193 Checking the fitting condition

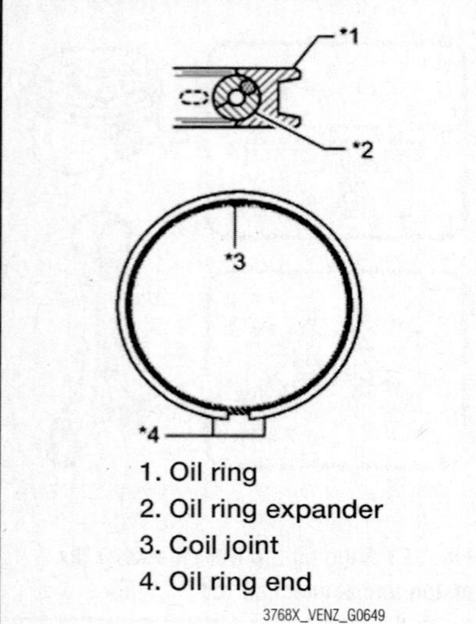

Fig. 194 Installing the oil ring expander and oil ring

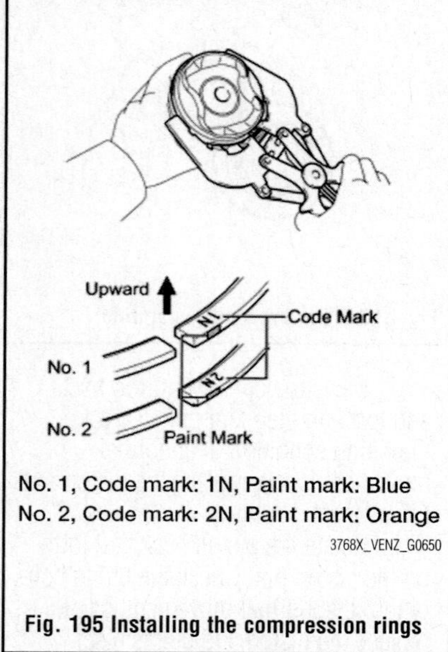

Fig. 195 Installing the compression rings

a. Install the oil ring expander and oil ring by hand.

➡**Arrange the oil ring ends and coil joint.**

b. Using a piston ring expander, install the 2 compression rings with the code mark positioned as shown in the illustration.

➡**Install the compression ring with the code mark facing upward.**

REAR MAIN SEAL

REMOVAL & INSTALLATION

See Figure 196.

1. Remove the engine assembly with the transaxle.
2. Remove the automatic transaxle assembly.
3. Remove the drive plate and ring gear sub assembly.
   a. Using the special tool, hold the crankshaft pulley.
   b. Remove the 8 bolts, front drive plate spacer, drive plate and ring gear sub-assembly, and rear drive plate spacer.
4. Remove the rear engine oil seal.
   a. Using a knife, cut off the lip of the oil seal.
   b. Using a screwdriver, pry out the oil seal.

➡**Tape the screwdriver tip before use.**

➡**Do not damage the surface of the oil seal press fit hole or the crankshaft.**

To install:

5. Install the rear engine oil seal.
   a. Apply MP grease to the lip of a new oil seal.

➡**Do not allow foreign matter to contact the lip of the oil seal.**

➡**Do not allow MP grease to contact the dust seal.**

b. Using the special tool and a hammer, tap in the oil seal until its surface is flush with the edges of the cylinder block and crankcase.

➡**Keep the lip of the oil seal free from foreign matter.**

➡**Do not tap in the oil seal at an angle.**

6. Install the drive plate and ring gear sub assembly.
   a. Using the special too, hold the crankshaft.
   b. Clean the bolts and their installation holes.
   c. Install the front drive plate spacer.

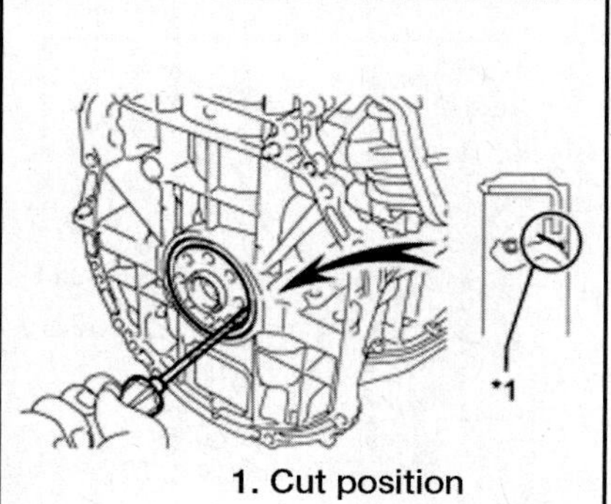

Fig. 196 Removing eh rear engine oil seal

➡**Align the pin of the front drive plate spacer with the pin hole of the crankshaft.**

d. Install the drive plate and rear drive plate spacer to the crankshaft.
e. Apply a few drops of adhesive to 2 or 3 threads of the bolt end.
f. Install and uniformly tighten the 8 bolts in the sequence shown in the illustration. For the 2.7L engine, tighten to 72 ft. lbs. (98 Nm). For the 3.5L engine, tighten to 61 ft. lbs. (83 Nm).

7. Install the automatic transaxle assembly.
8. Install the engine assembly with the transaxle.

TIMING CHAIN FRONT COVER

REMOVAL & INSTALLATION

Refer to Oil Pump for removal and installation procedure.

TIMING CHAIN & SPROCKETS

REMOVAL & INSTALLATION

Refer to Oil Pump for removal and installation procedure.

VALVE LASH

INSPECTION

➡**Keep the adjuster free from dirt and foreign matter.**

➡**Use only clean engine oil.**

1. Place the lash adjuster into a container full of new engine oil.
2. Insert the tip of SST into the lash adjuster plunger and use the tip to press down on the check ball inside the plunger.
3. Squeeze SST and the lash adjuster together to move the plunger up and down 5 to 6 times.
4. Check the movement of the plunger and bleed air. It is OK if the plunger moves up and down.

➡**When bleeding high-pressure air from the compression chamber, make sure that the tip of SST is actually pressing the check ball as shown in the illustration. If the check ball is not pressed, air will not bleed.**

5. After bleeding the air, remove SST. Then try to quickly and firmly press the plunger with your fingers. It is OK if the plunger can be pressed 3 times.
6. If the plunger can still be compressed after pressing it 3 times, replace the valve lash adjuster with a new one.

ENGINE PERFORMANCE & EMISSION CONTROLS

CAMSHAFT POSITION (CMP) SENSOR

LOCATION

See Figures 197 and 198.

Refer to the accompanying illustrations.

REMOVAL & INSTALLATION

2.7L Engine

1. Remove the No. 1 engine cover sub assembly.
2. Remove the Camshaft Position (CMP) sensor (exhaust side).
   a. Disconnect the connector.
   b. Remove the bolt and sensor.
3. Remove the Camshaft Position (CMP) sensor (intake side).
   a. Disconnect the sensor connector.
   b. Remove the bolt and sensor.

To install:

4. Install the Camshaft Position (CMP) sensor (exhaust side).
   a. Apply a light coat of engine oil to the O-ring of the sensor.

➡Make sure that the O-ring is not cracked or does not jump out of position during installation.

   b. Apply adhesive to 2 or 3 threads of the bolt.
   c. Install the sensor with the bolt. Tighten to 7 ft. lbs. (10 Nm).
   d. Connect the sensor connector.
5. Install the Camshaft Position (CMP) sensor (intake side).
   a. Apply a light coat of engine oil to the o-ring of the sensor.

➡Make sure that the o-ring is not cracked or does not jump out of position during installation.

   b. Apply adhesive to 2 or 3 threads of the bolt.
   c. Install the sensor with the bolt. Tighten to 7 ft. lbs. (10 Nm).
   d. Connect the sensor connector.
6. Inspect for oil leaks.

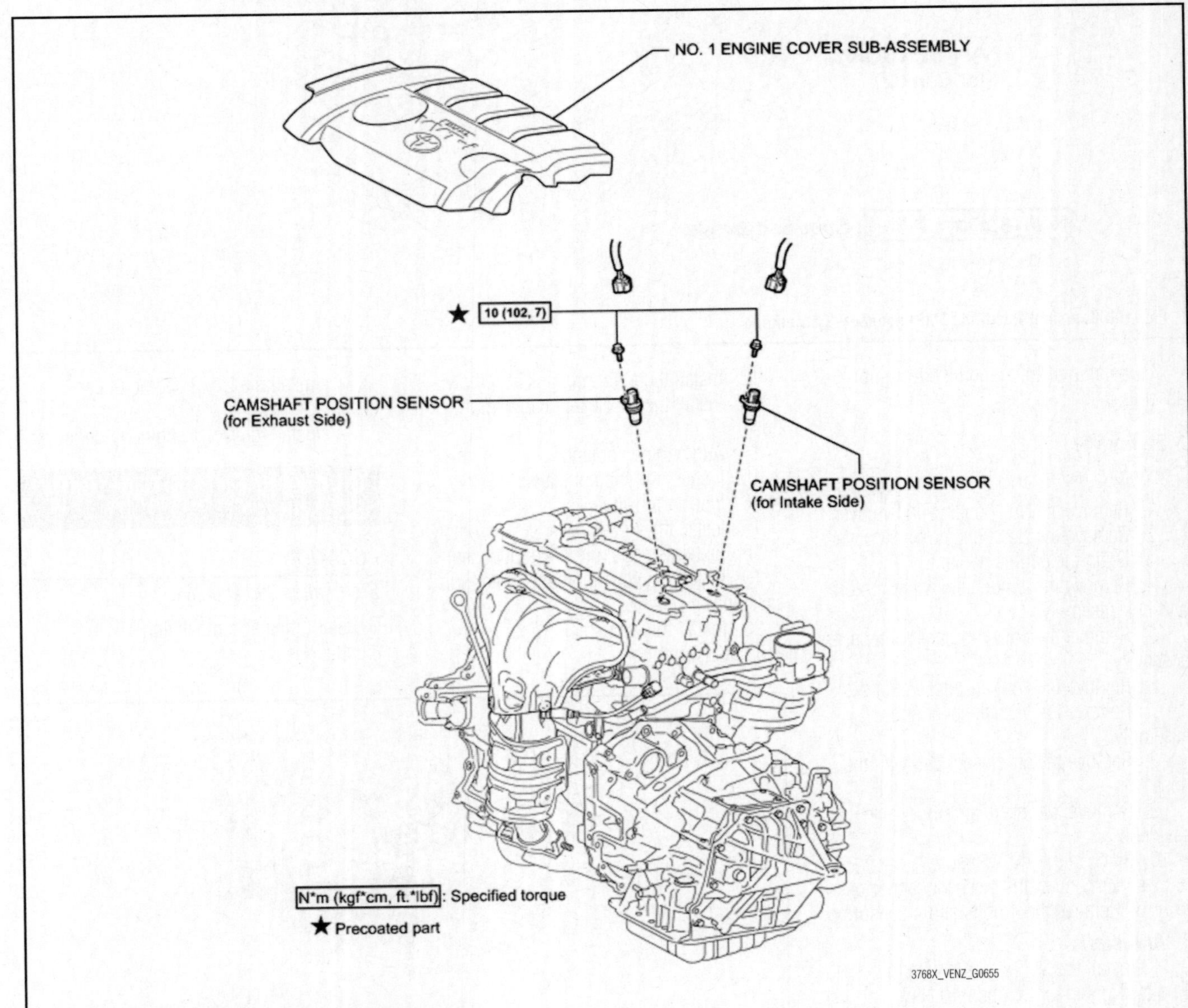

Fig. 197 Camshaft Position (CMP) sensor—2.7L engine

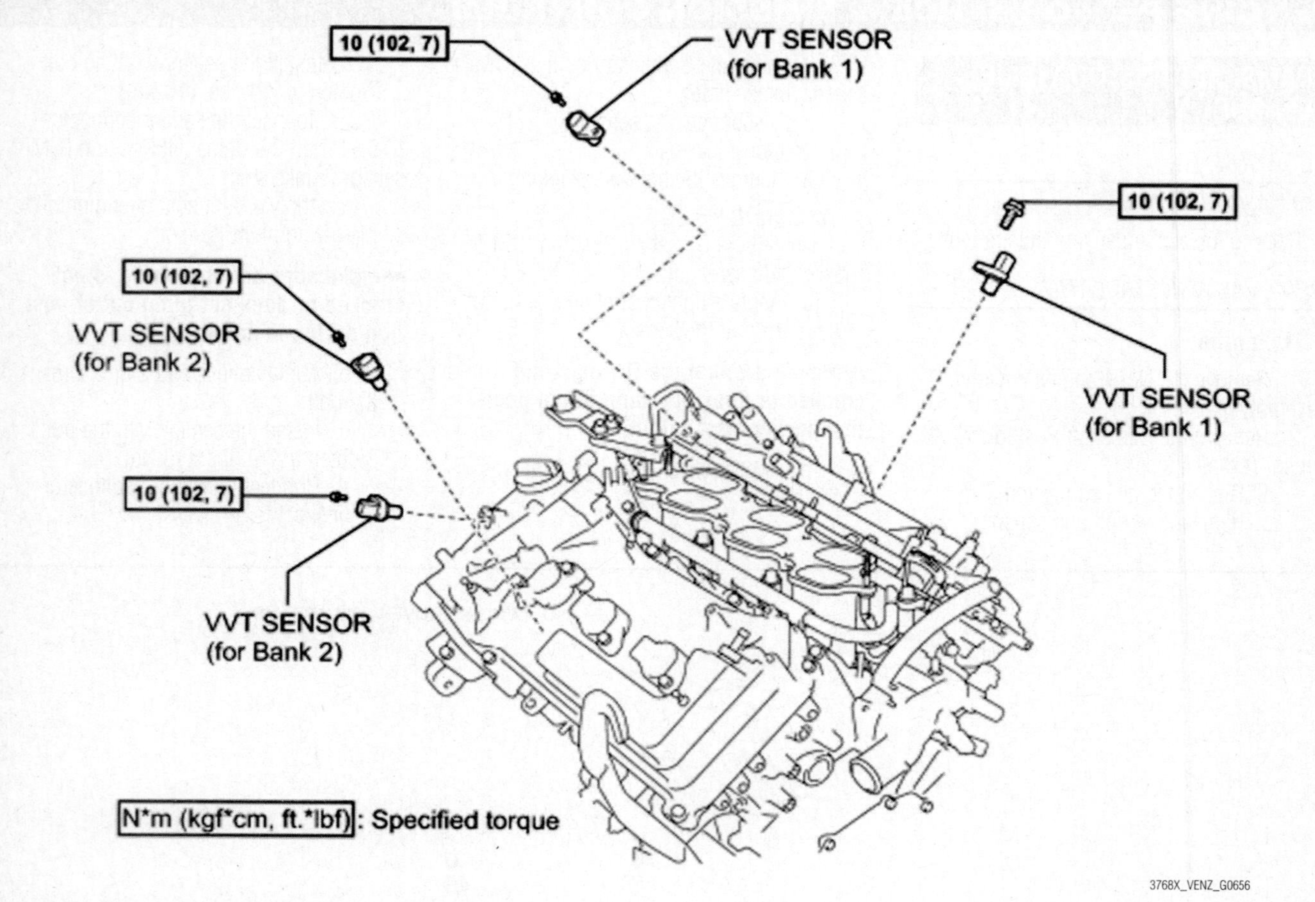

Fig. 198 Camshaft Position (CMP) sensor—3.5L engine

7. Install the No. 1 engine cover sub assembly.

3.5L Engine

See Figure 199.

1. Remove the No. 1 engine under cover.
2. Remove the No. 2 engine under cover.
3. Drain the engine coolant.
4. Remove the windshield wiper motor and link assembly.
5. Remove the cowl top panel outer sub assembly.
6. Remove the cool air intake duct seal.
7. Remove the v-bank cover sub assembly.
8. Remove the air cleaner cap with the hose.
9. Remove the intake air surge tank assembly.
10. Remove the VVT sensor.
    a. Disconnect the 4 sensor connectors.
    b. Remove the 4 bolts and 4 sensors.

To install:

11. Install the VVT sensors with the 4 bolts. Tighten to 7 ft. lbs. (10 Nm).
    a. Connect the 4 sensor connectors.
12. Install the air surge tank assembly.
13. Install the air cleaner cap with the hose.
14. Add engine coolant.
15. Inspect for coolant leaks.
16. Inspect for oil leaks.
17. Install the v-bank cover sub assembly.
18. Install the cool air intake duct seal.
19. Install the cowl top panel outer sub assembly.
20. Install the windshield wiper motor and link assembly.
21. Install the No. 2 engine under cover.
22. Install the No. 1 engine under cover.

CRANKSHAFT POSITION (CKP) SENSOR

LOCATION

See Figures 200 and 201.

Refer to the accompanying illustrations.

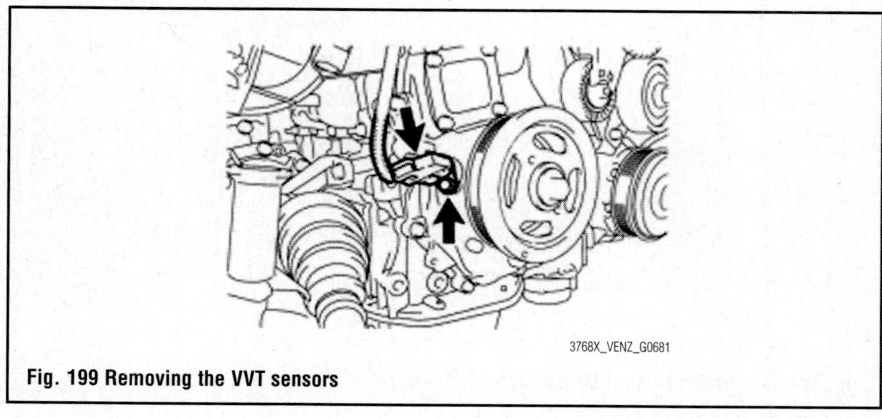

Fig. 199 Removing the VVT sensors

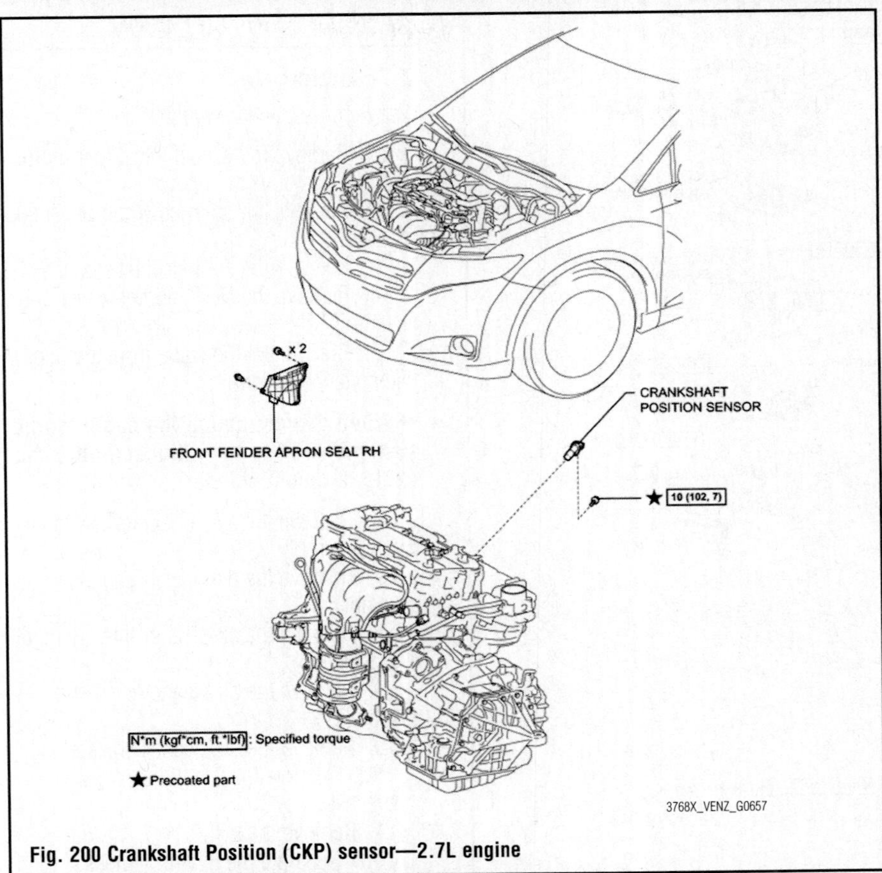

Fig. 200 Crankshaft Position (CKP) sensor—2.7L engine

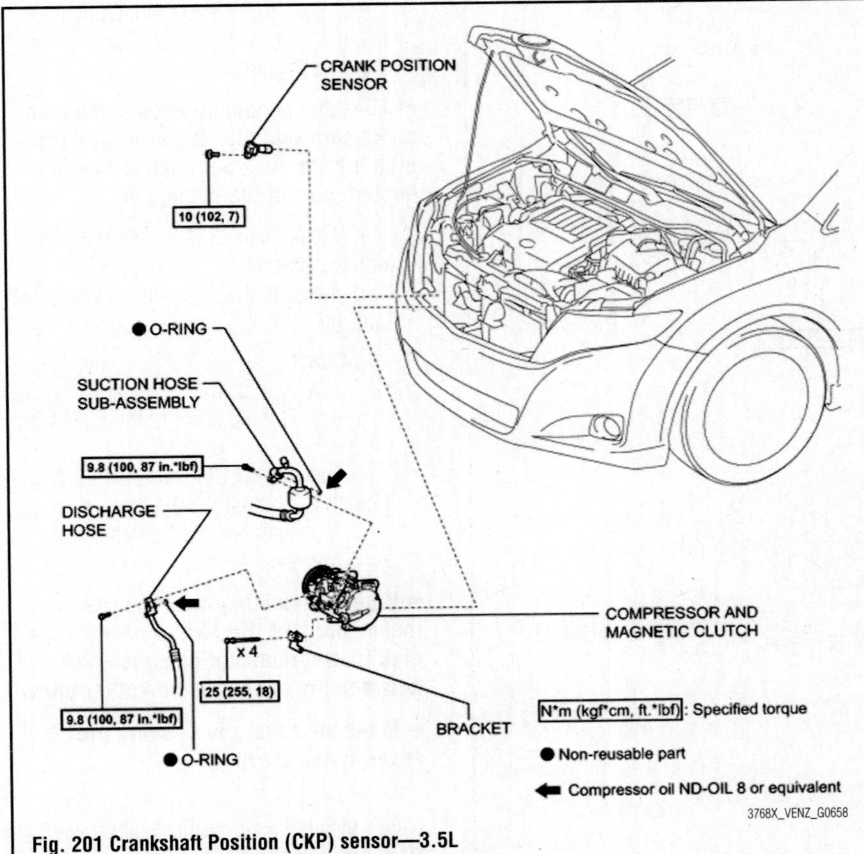

Fig. 201 Crankshaft Position (CKP) sensor—3.5L

REMOVAL & INSTALLATION

2.7L Engine

1. Remove the front fender apron seal RH.
2. Remove the Crankshaft Position (CKP) sensor.
   a. Disconnect the sensor connector.
   b. Remove the bolt and sensor.

To install:

3. Install the Crankshaft Position (CKP) sensor.
   a. Apply a light coat of engine oil to the o-ring of the sensor.

➡Make sure that the o-ring is not cracked or does not jump out of position during installation.

   b. Apply adhesive to 2 or 3 threads of the bolt.
   c. Install the sensor with the bolt. Tighten to 7 ft. lbs. (10 Nm).
   d. Connect the sensor connector.
4. Inspect for oil leaks.
5. Install the front fender apron seal RH.

3.5L Engine

1. Remove the compressor and magnetic clutch.
2. Remove the crank position sensor.
   a. Disconnect the crank position sensor connector.
   b. Remove the bolt and crank position sensor.

To install:

3. Install the crank position sensor.
   a. Apply a light coat of engine oil to the o-ring on the crank position sensor.
   b. Install the cranks position sensor with the bolt. Tighten to 7 ft. lbs. (10 Nm).
   c. Connect the crank position sensor connector.
4. Install the compressor and magnetic clutch.

ELECTRONIC CONTROL MODULE (ECM)

LOCATION

See Figures 202 and 203.

Refer to the accompanying illustrations.

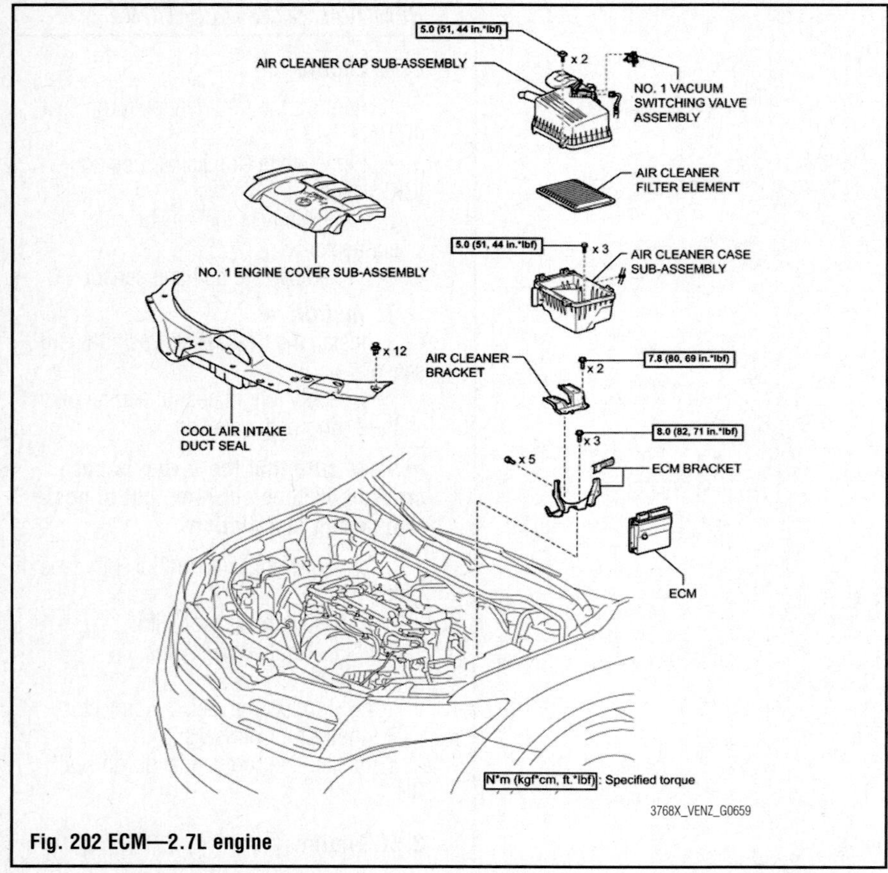

Fig. 202 ECM—2.7L engine

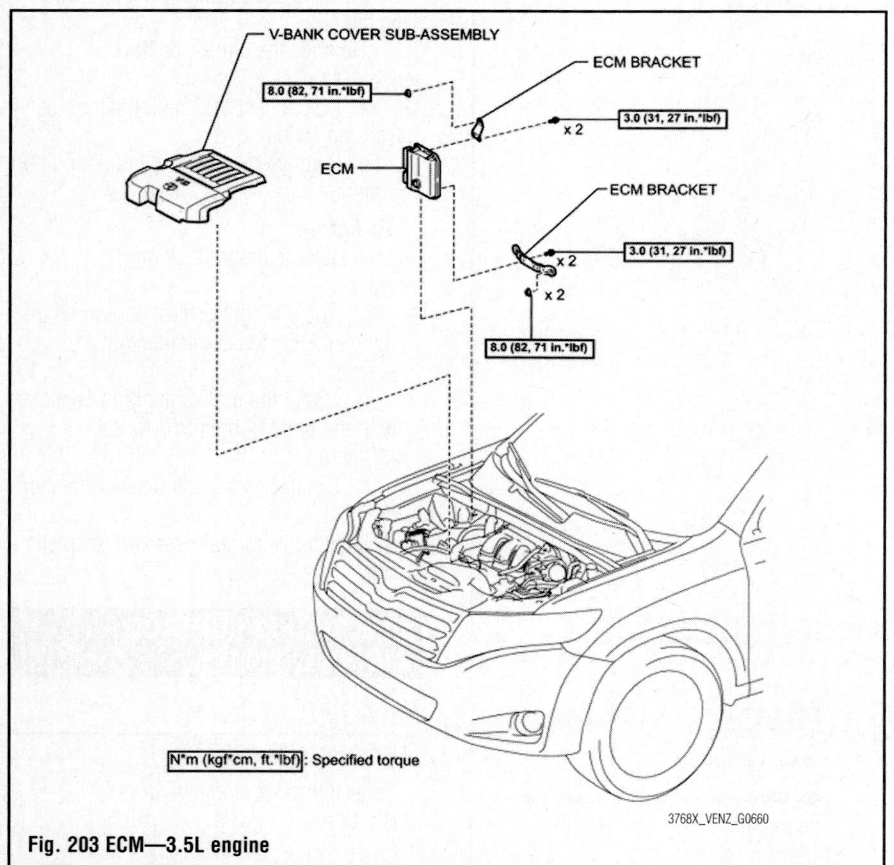

Fig. 203 ECM—3.5L engine

REMOVAL & INSTALLATION

2.7L Engine

See Figures 204 through 208.

1. Remove the windshield wiper motor and link.
2. Remove the outer cowl top panel sub assembly.
3. Remove the cool air intake duct seal.
4. Remove the No. 1 engine cover sub assembly.
5. Disconnect the cable from the negative battery terminal.

➡When disconnecting the cable, some systems need to be initialized after the cable is reconnected.

6. Remove the No. 1 vacuum switching valve assembly.
7. Remove the air cleaner cap sub assembly.
8. Remove the air cleaner filter element sub assembly.
9. Remove the air cleaner case sub assembly.
10. Remove the air cleaner bracket.
    a. Remove the 2 bolts and cleaner bracket.
11. Remove the ECM.
    a. Separate the 3 wire harness clamps.
    b. Raise the 2 levers while pushing the locks on the levers, and disconnect the 2 ECM connectors.

➡After disconnecting each connector, make sure that dirt, water or other foreign matter does not contact the connecting part of the connector.

    c. Remove the 3 bolts and the ECM with the bracket.
    d. Remove the 5 screws and the ECM bracket.

To install:

12. Install the ECM.
    a. Install the bracket to the ECM with the 5 screws.
    b. Install the ECM with the 3 bolts. Tighten to 71 inch lbs. (8 Nm).
    c. Connect the 2 ECM connectors and lower the 2 levers.

➡When connecting the connector, make sure that dirt, water or other foreign matter does not become stuck between the connector and other part.

➡Make sure that the 2 levers are securely lowered.

    d. Install the 3 wire harness clamps.
13. Install the air cleaner bracket with the 2 bolts. Tighten to 69 inch lbs. (7.8 Nm).

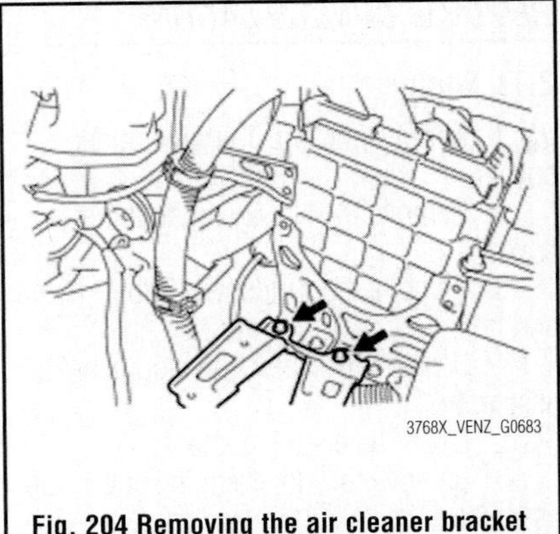

Fig. 204 Removing the air cleaner bracket

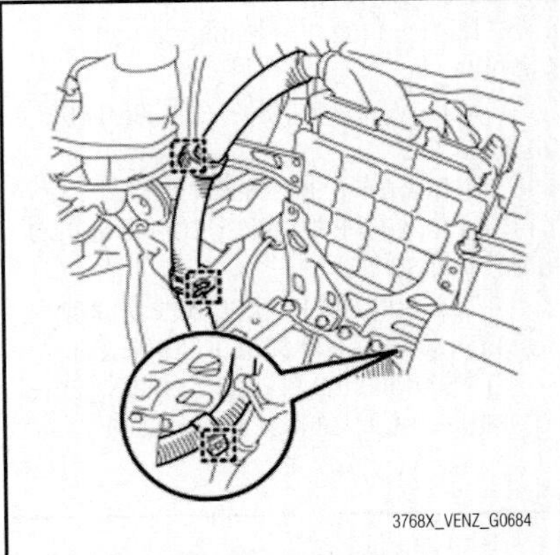

Fig. 205 Separating the wire harness clamps

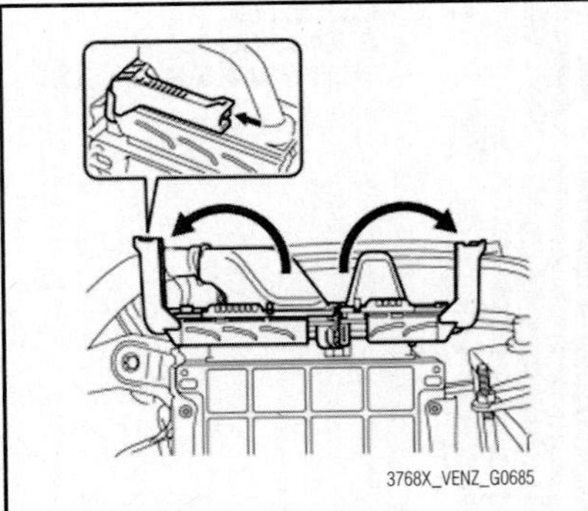

Fig. 206 Disconnecting the ECM connectors

14. Install the air cleaner case sub assembly.
15. Install the air cleaner filter element sub assembly.
16. Install the air cleaner cap sub assembly.
17. Remove the No. 1 vacuum switching valve assembly.
18. Install the No. 1 engine cover sub assembly.
19. Install the outer cowl top panel sub assembly.

Fig. 207 Removing the ECM with the bracket

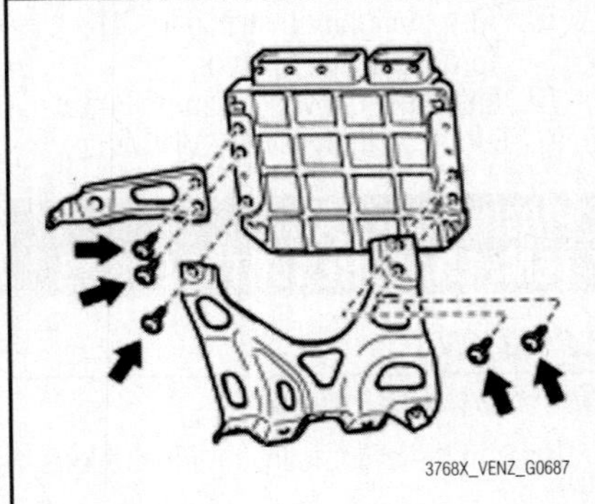

Fig. 208 Removing the ECM bracket

20. Install the windshield wiper motor and link.
21. Install the cool air intake duct seal.
22. Connect the cable to the negative battery terminal.

⁂ CAUTION

When disconnecting the cable, some systems need to be initialized after the cable is reconnected.

23. Perform the registration.
 a. The VIN must be input into a replacement ECM.

3.5L Engine

See Figures 209 through 211.

➡Perform the Vehicle Identification Number (VIN) registration when replace the ECM

1. Disconnect the cable from the negative battery terminal.

➡When disconnecting the cable, some systems need to be initialized after the cable is reconnected.

2. Remove the v-bank cover sub assembly.

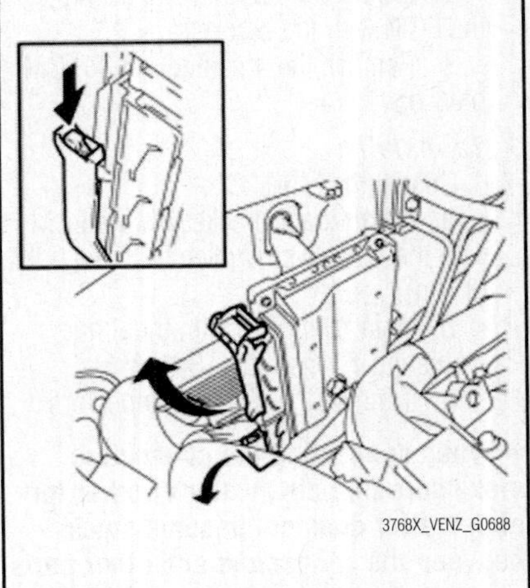

Fig. 209 Disconnecting the ECM connectors

3. Remove the ECM.
 a. Disconnect the 2 ECM connectors and remove the ECM with the brackets.

➡After disconnecting the connectors, make sure that dirt, water or other foreign matter does not contact the connecting part of the connectors.

 b. Push in the locks on the 2 levers, raise the levers, and disconnect the 2 ECM connectors.

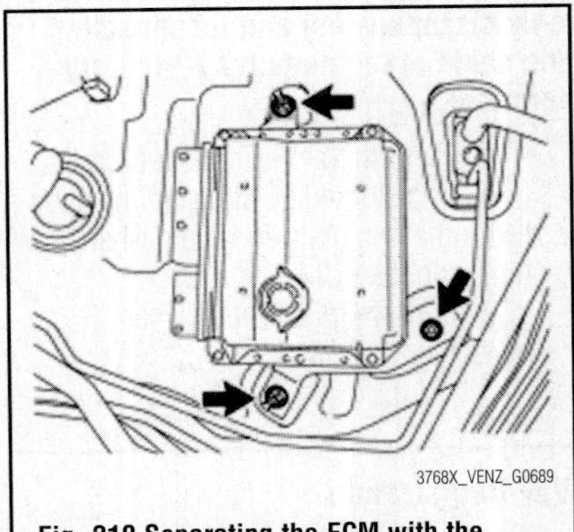

Fig. 210 Separating the ECM with the brackets

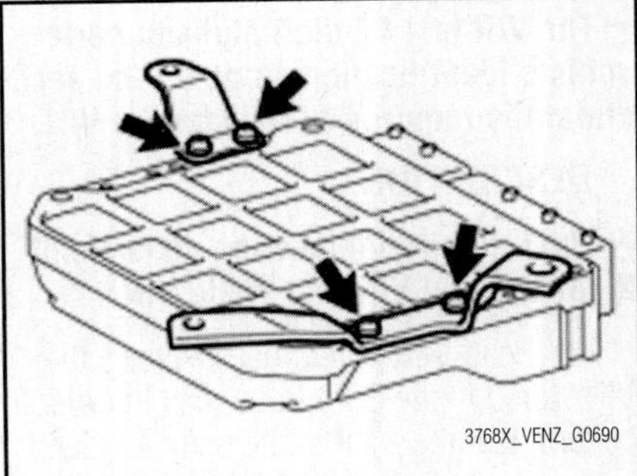

Fig. 211 Removing the ECM brackets

c. Remove the 3 nuts and separate the ECM with the brackets.

d. Remove the 4 screws and 2 ECM brackets.

To install:

4. Install the ECM.

a. Install the 2 brackets to the ECM with the 4 screws. Tighten to 27 inch lbs. (3 Nm).

b. Install the ECM with the 3 nuts. Tighten to 71 inch lbs. (8 Nm).

c. Connect the 2 ECM connectors.

➡When connecting the connectors, make sure that dirt, water or other foreign matter does not become stuck between the connectors and other parts.

d. Connect the 2 ECM connectors and lower the 2 levers.

➡Make sure that the 2 levers are securely locked.

5. Install the v-bank cover sub assembly.

6. Connect the cable to the negative battery terminal.

➡When disconnecting the cable, some systems need to be initialized after the cable is reconnected.

7. Perform the initialization.

➡Initialization cannot be completed by only disconnecting and reconnecting the cable of the negative (-) battery terminal.

a. Perform the VIN registration.

b. Register the ECM communication ID for the immobilizer system when replacing the ECM.

c. Perform initialization when replacing the ECM.

RESET

VIN Registration

➡The Vehicle Identification Number (VIN) must be input into the replacement ECM.

➡The VIN is a 17-digit alphanumeric vehicle identification number. The Techstream is required to register the VIN.

DESCRIPTION

➡This registration section consists of 2 parts: Read VIN and Write VIN.

1. Read VIN: This process allows the VIN stored in the ECM to be read in order to confirm that the 2 VINs, the one provided with the vehicle and stored in the vehicle ECM, are the same.

2. Write VIN: This process allows the VIN to be input into the ECM. If the ECM is changed, or the ECM VIN and vehicle VIN do not match, the VIN can be registered, or overwritten in the ECM by following this procedure.

READ VIN

3. Confirm the vehicle VIN.
4. Connect the Techstream to the DLC3.
5. Turn the ignition switch to ON.
6. Turn the Techstream on.
7. Enter the following menus: Powertrain / Engine / Utility / VIN / VIN Read.

WRITE VIN

8. Confirm the vehicle VIN.
9. Connect the Techstream to the DLC3.
10. Turn the ignition switch to ON.
11. Turn the Techstream on.
12. Enter the following menus: Powertrain / Engine / Utility / VIN / VIN Write.

ENGINE COOLANT TEMPERATURE (ECT) SENSOR

LOCATION

See Figures 212 and 213.

Refer to the accompanying illustrations.

REMOVAL & INSTALLATION

2.7L Engine

1. Remove the No. 1 engine under cover.
2. Remove the No. 2 engine under cover.
3. Remove the windshield wiper motor and link.
4. Remove the outer cowl top panel sub assembly.
5. Drain the engine coolant.
6. Remove the No. 1 engine cover sub assembly.
7. Remove the No. 1 vacuum switching valve assembly.
8. Remove the air cleaner cap sub assembly.
9. Remove the air cleaner filter element sub assembly.
10. Remove the air cleaner case.
11. Remove the Engine Coolant Temperature (ECT) sensor.

a. Disconnect the Engine Coolant Temperature (ECT) sensor connector.

b. Remove the Engine Coolant Temperature (ECT) sensor and gasket.

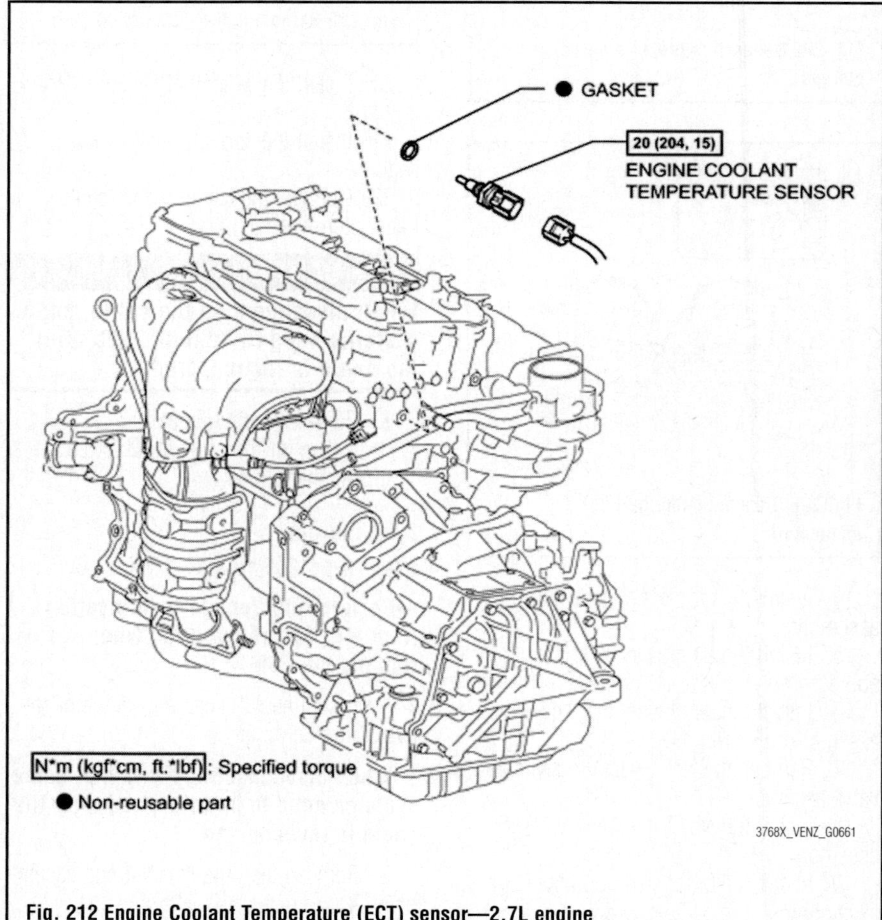

Fig. 212 Engine Coolant Temperature (ECT) sensor—2.7L engine

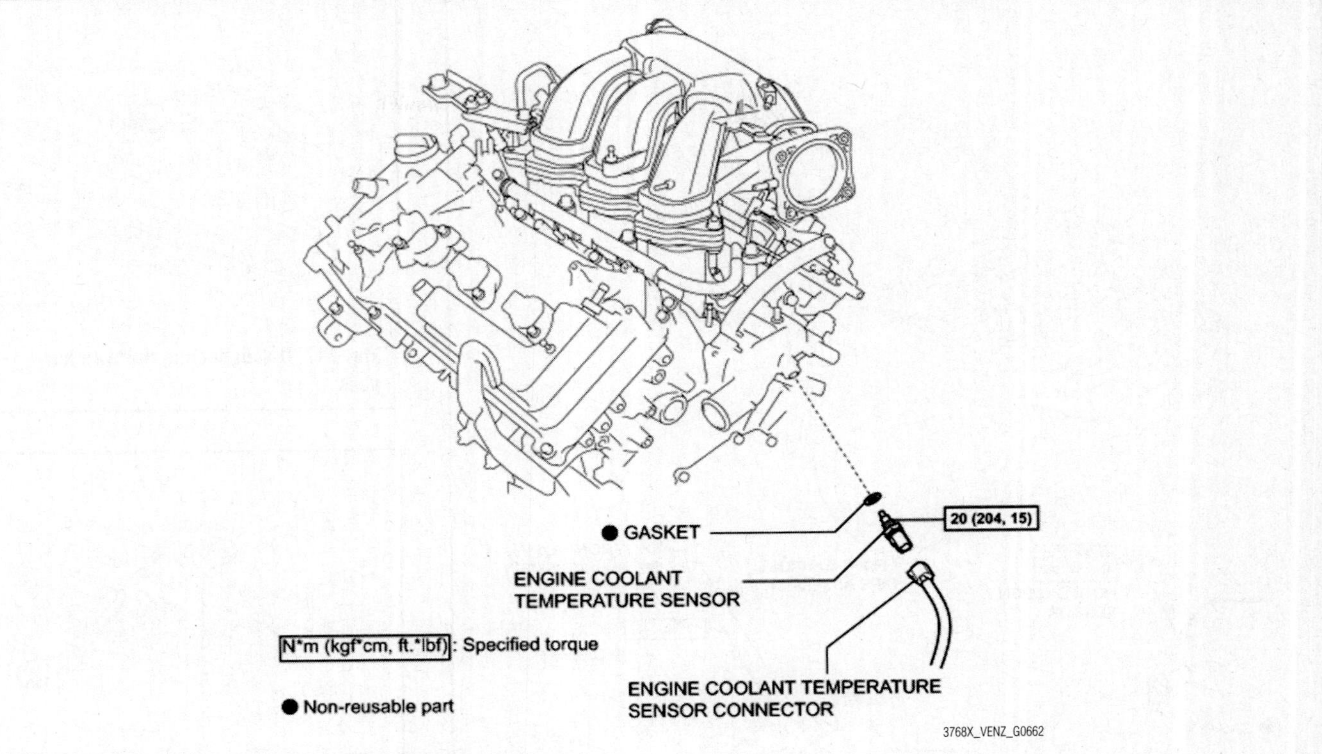

Fig. 213 Engine Coolant Temperature (ECT) sensor—3.5L engine

To install:

12. To install, reverse the removal procedure. Tighten the Engine Coolant Temperature (ECT) sensor to 15 ft. (20 Nm).

3.5L Engine

1. Remove the No. 1 engine under cover.
2. Remove the No. 2 engine under cover.
3. Drain the engine coolant.
4. Remove the v-bank cover sub assembly.
5. Remove the air cleaner cap with the hose.
6. Remove the air cleaner case.
7. Remove the Engine Coolant Temperature (ECT) sensor.
   a. Disconnect the Engine Coolant Temperature (ECT) sensor connector.
   b. Remove the Engine Coolant Temperature (ECT) sensor and gasket.

To install:

8. To install, reverse the removal procedure. Tighten the Engine Coolant Temperature (ECT) sensor to 15 ft. lbs. (20 Nm).

HEATED OXYGEN SENSOR (HO2S)

LOCATION

See Figures 214 and 215.

Refer to the accompanying illustrations.

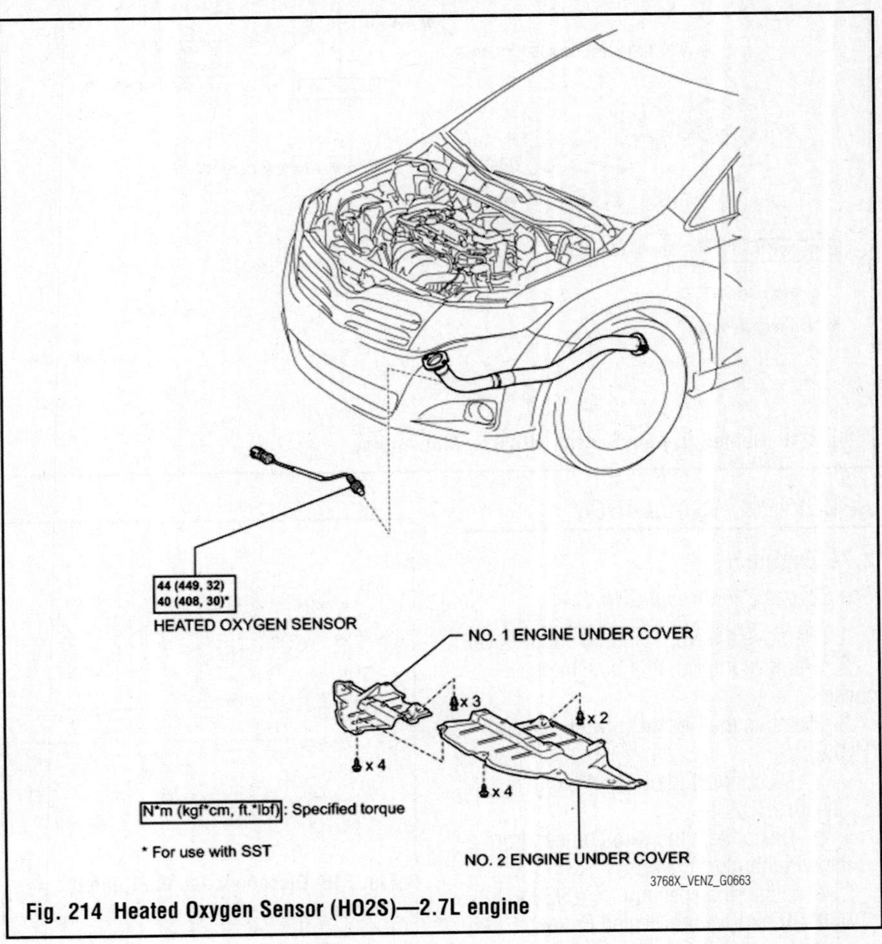

Fig. 214 Heated Oxygen Sensor (HO2S)—2.7L engine

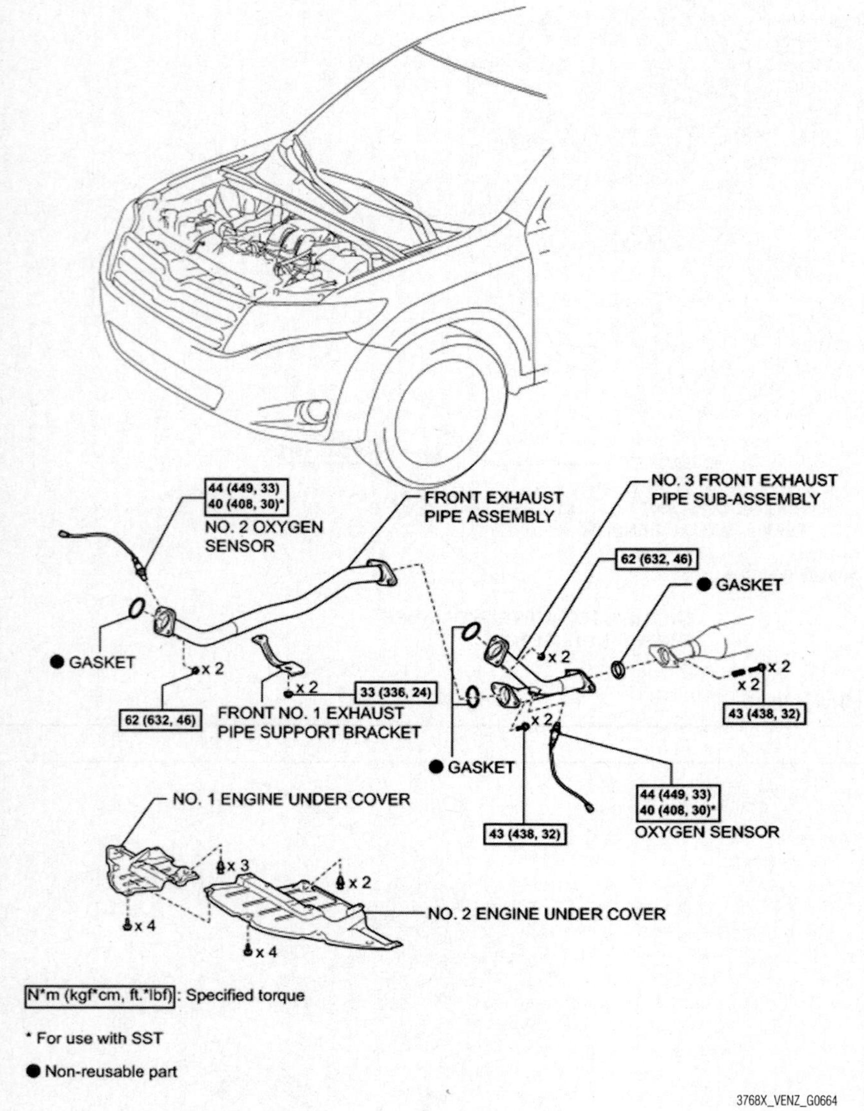

Fig. 215 Heated Oxygen Sensor (HO2S)—3.5L engine

REMOVAL & INSTALLATION

2.7L Engine

See Figures 216 through 218.

1. Remove the No. 1 engine under cover.
2. Remove the No. 2 engine under cover.
3. Remove the Heated Oxygen Sensor (HO2S).
    a. Disconnect the Heated Oxygen Sensor (HO2S).
    b. Disconnect the wire harness from the wire harness clamp.
    c. Using the special tool (09224-00010), remove the Heated Oxygen Sensor (HO2S) from the front exhaust pipe.

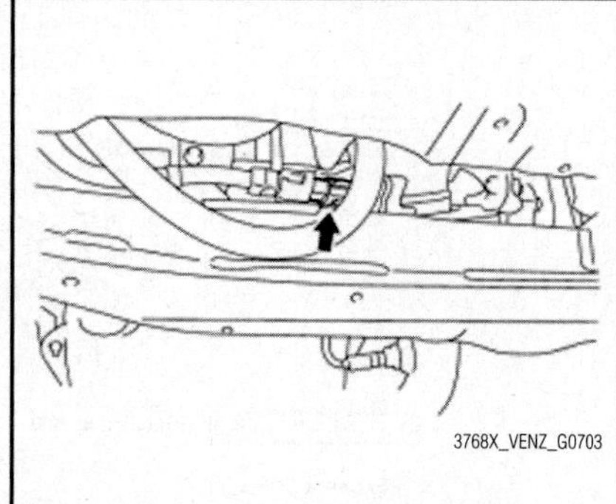

Fig. 216 Disconnecting the Heated Oxygen Sensor (HO2S)

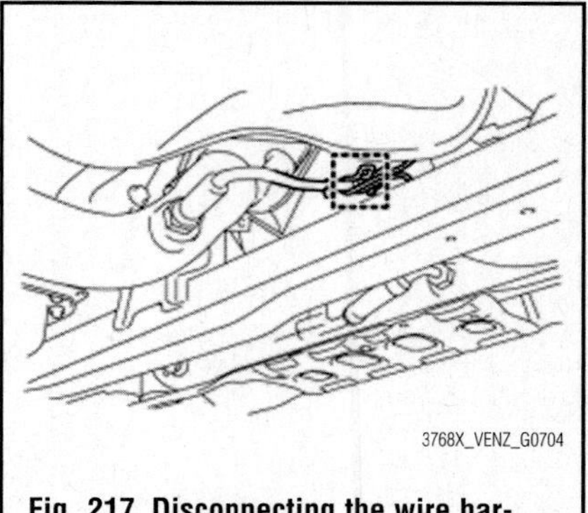

Fig. 217 Disconnecting the wire harness

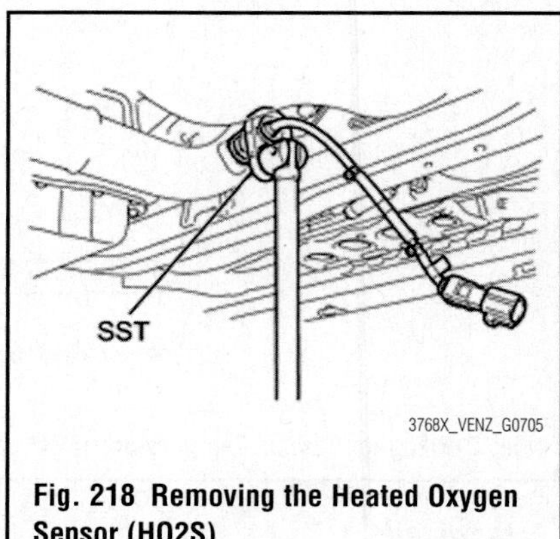

Fig. 218 Removing the Heated Oxygen Sensor (HO2S)

To install:

4. To install, reverse the removal procedure. Tighten the Heated Oxygen Sensor (HO2S) to 32 ft. lbs. (44 Nm). If using the special tool, tighten to 30 ft. lbs. (40 Nm).

3.5L Engine

See Figures 219 and 220.

1. Remove the No. 1 engine under cover.
2. Remove the No. 2 engine under cover.

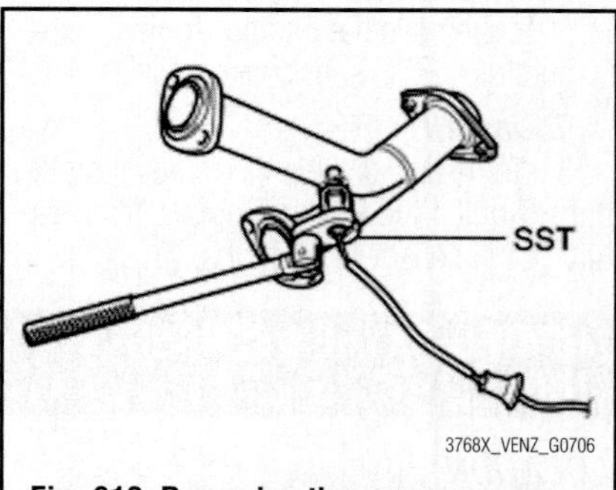

Fig. 219 Removing the oxygen sensor from the No. 3 front exhaust pipe sub assembly

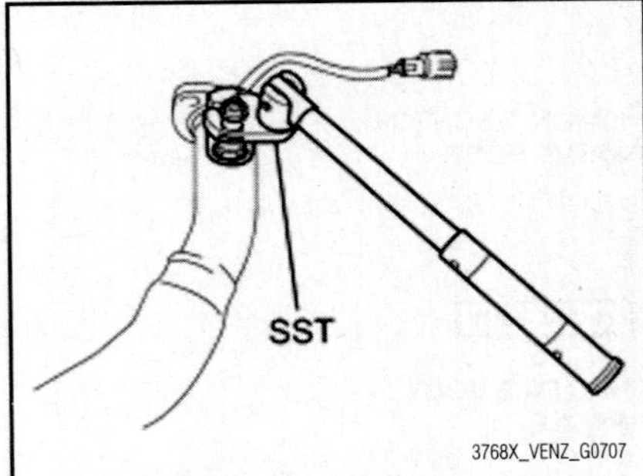

Fig. 220 Removing the No. 2 oxygen sensor from the front exhaust pipe assembly

3. Remove the front door scuff plate LH.
4. Remove the cowl side trim sub assembly LH.
5. Remove the No. 3 front exhaust pipe sub assembly.
6. Remove the oxygen sensor (bank 1 sensor 2).
   a. Using the special tool (09224-00010), remove the oxygen sensor from the No. 3 front exhaust pipe sub assembly.
7. Remove the front exhaust pipe assembly.
8. Remove the No. 2 oxygen sensor (bank 2 sensor 2).
   a. Using the special tool (09224-00010), remove the No. 2 oxygen sensor from the front exhaust pipe assembly.

To install:

9. To install, reverse the removal procedure. Tighten the oxygen sensors to 33 ft. lbs. (44 Nm). If using the special tool, tighten to 30 ft. lbs. (40 Nm).

KNOCK SENSOR (KS)

LOCATION

See Figures 221 and 222.

Refer to the accompanying illustrations.

REMOVAL & INSTALLATION

2.7L Engine

See Figure 223.

1. Remove the intake manifold.
2. Remove the knock sensor.
   a. Disconnect the sensor connector.
   b. Remove the bolt and sensor.

To install:

3. Install the knock sensor.
   a. Install the sensor with the bolt so that the sensor is angled as shown. Tighten to 15 ft. lbs. (20 Nm).

➡The acceptable installation angle of the sensor is between 7°upward and 10°downward from the horizontal position.

   b. Connect the sensor connector.
4. Install the intake manifold.

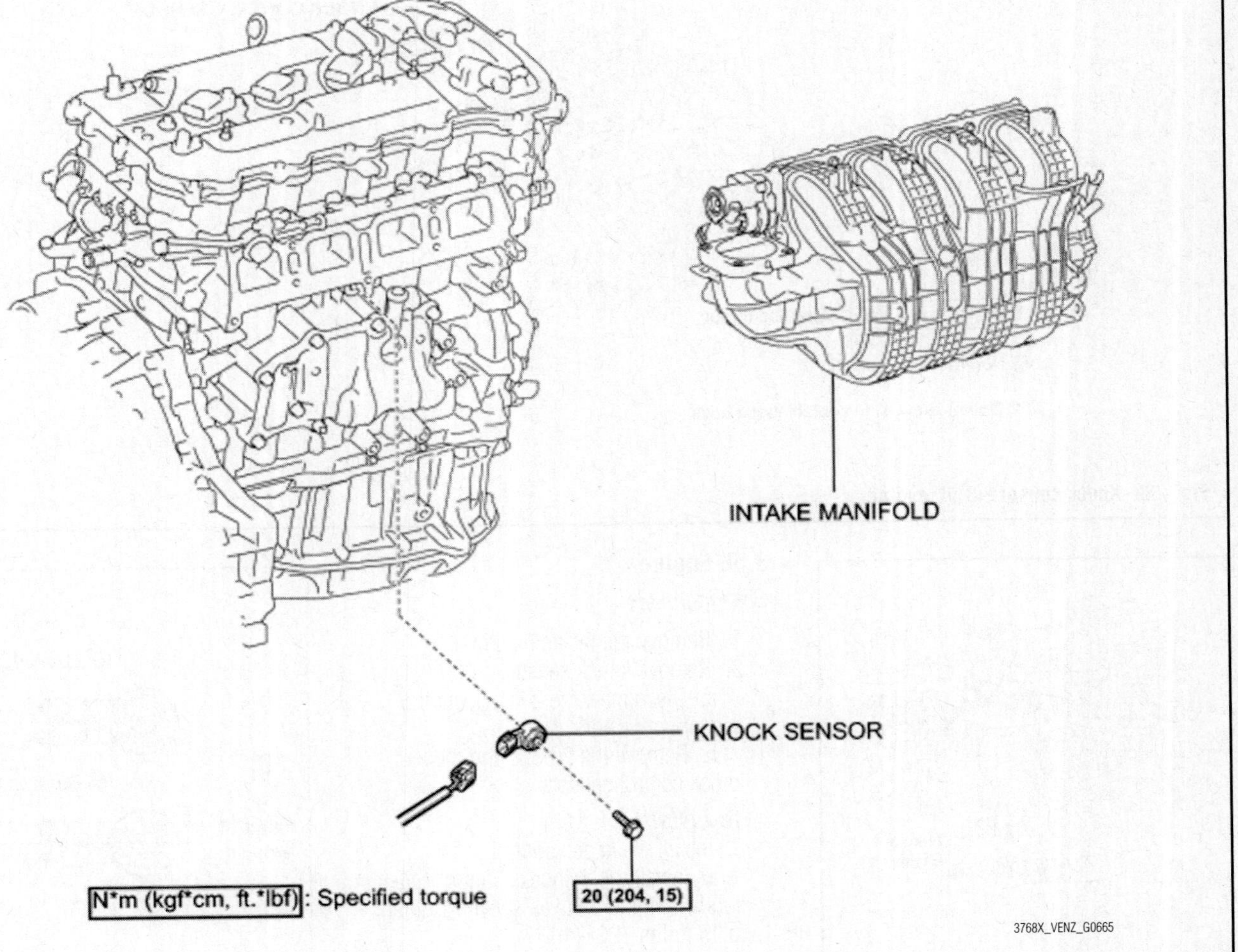

INTAKE MANIFOLD
KNOCK SENSOR
N*m (kgf*cm, ft.*lbf): Specified torque
20 (204, 15)
3768X_VENZ_G0665

Fig. 221 Knock sensor—2.7L engine

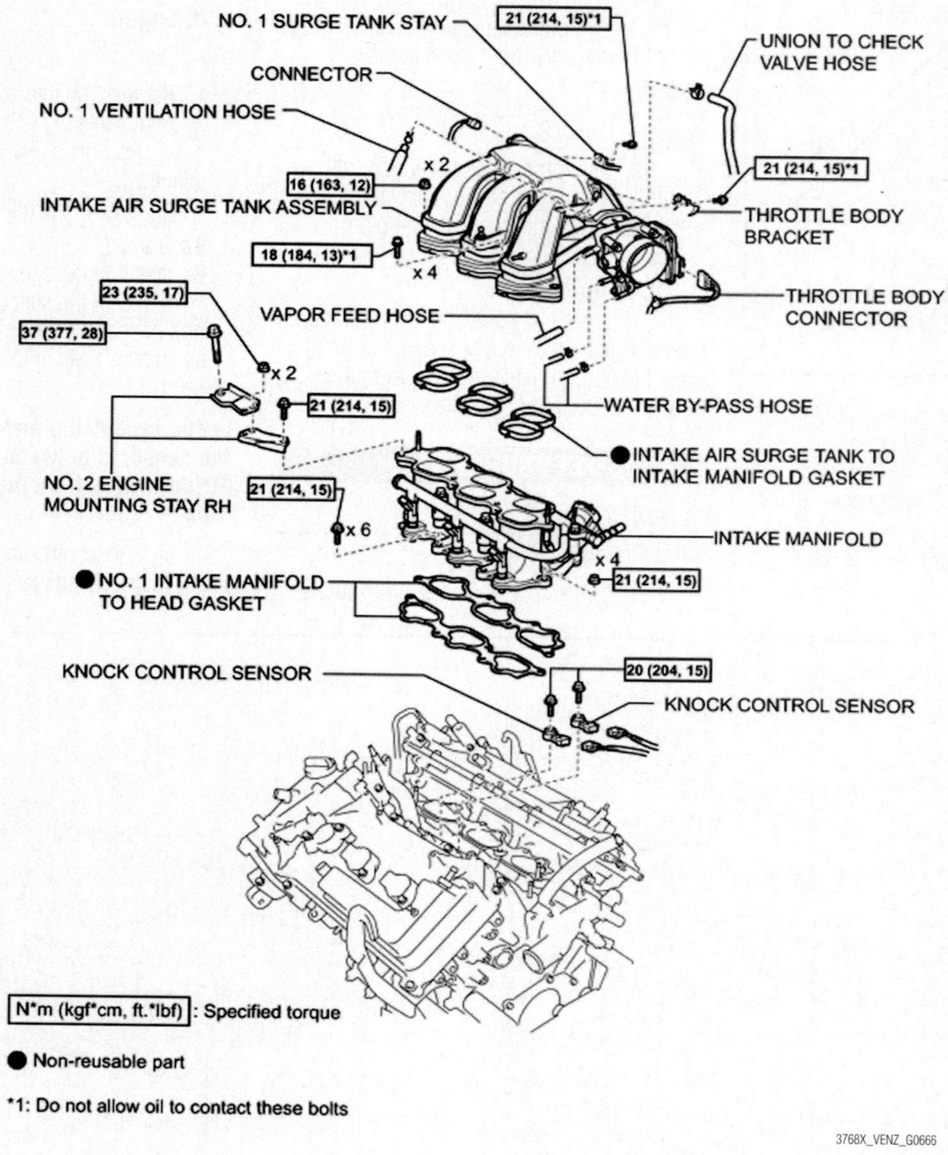

NO. 1 SURGE TANK STAY
21 (214, 15)*1
UNION TO CHECK VALVE HOSE
CONNECTOR
NO. 1 VENTILATION HOSE
x 2
16 (163, 12)
21 (214, 15)*1
INTAKE AIR SURGE TANK ASSEMBLY
THROTTLE BODY BRACKET
18 (184, 13)*1
x 4
23 (235, 17)
THROTTLE BODY CONNECTOR
VAPOR FEED HOSE
37 (377, 28)
x 2
21 (214, 15)
WATER BY-PASS HOSE
● INTAKE AIR SURGE TANK TO INTAKE MANIFOLD GASKET
NO. 2 ENGINE MOUNTING STAY RH
21 (214, 15)
x 6
INTAKE MANIFOLD
x 4
21 (214, 15)
● NO. 1 INTAKE MANIFOLD TO HEAD GASKET
KNOCK CONTROL SENSOR
20 (204, 15)
KNOCK CONTROL SENSOR
N*m (kgf*cm, ft.*lbf) : Specified torque
● Non-reusable part
*1: Do not allow oil to contact these bolts
3768X_VENZ_G0666

Fig. 222 Knock sensor—3.5L engine

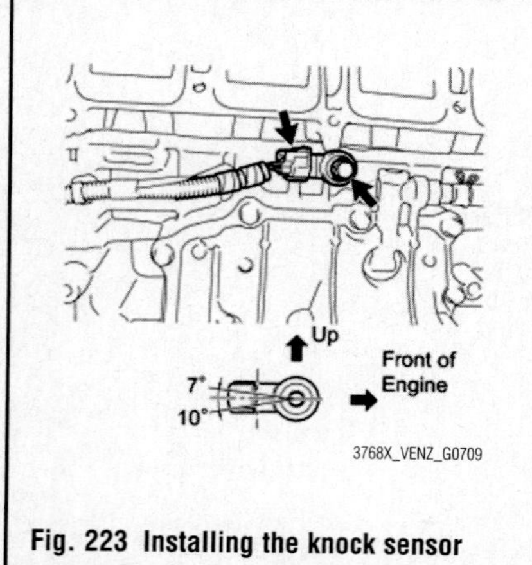

Fig. 223 Installing the knock sensor

3.5L Engine

See Figure 224.

1. Remove the intake manifold.
2. Remove knock sensor.
   a. Disconnect the 2 knock control sensor connectors.
   b. Remove the 2 bolts and the 2 knock control sensors.

To install:

3. Install the knock control sensor.
   a. Install the 2 knocks control sensors with the 2 bolts as shown. Tighten to 15 ft. lbs. (20 Nm).

➡Make sure that each knock control sensor is in the correct position.

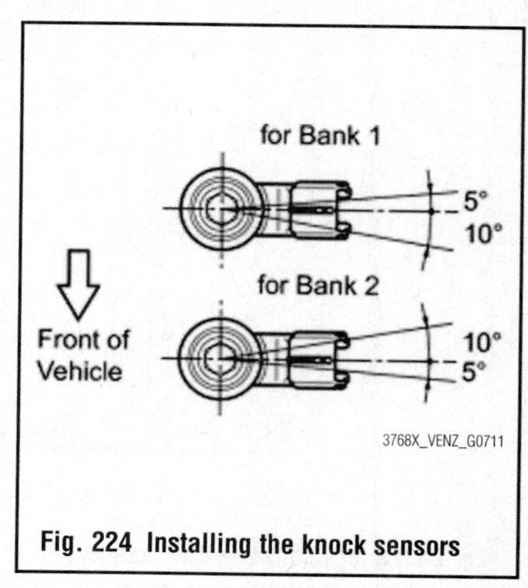

Fig. 224 Installing the knock sensors

b. Connect the 2 knock control sensor connectors.

4. Install the intake manifold.

MASS AIR FLOW (MAF) METER

LOCATION

See Figures 225 and 226.

Refer to the accompanying illustrations.

REMOVAL & INSTALLATION

2.7L Engine

See Figure 227.

1. Remove the Mass Air Flow (MAF) meter sub assembly.
   a. Disconnect the Mass Air Flow (MAF) meter connector.
   b. Remove the 2 screws and Mass Air Flow (MAF) meter.

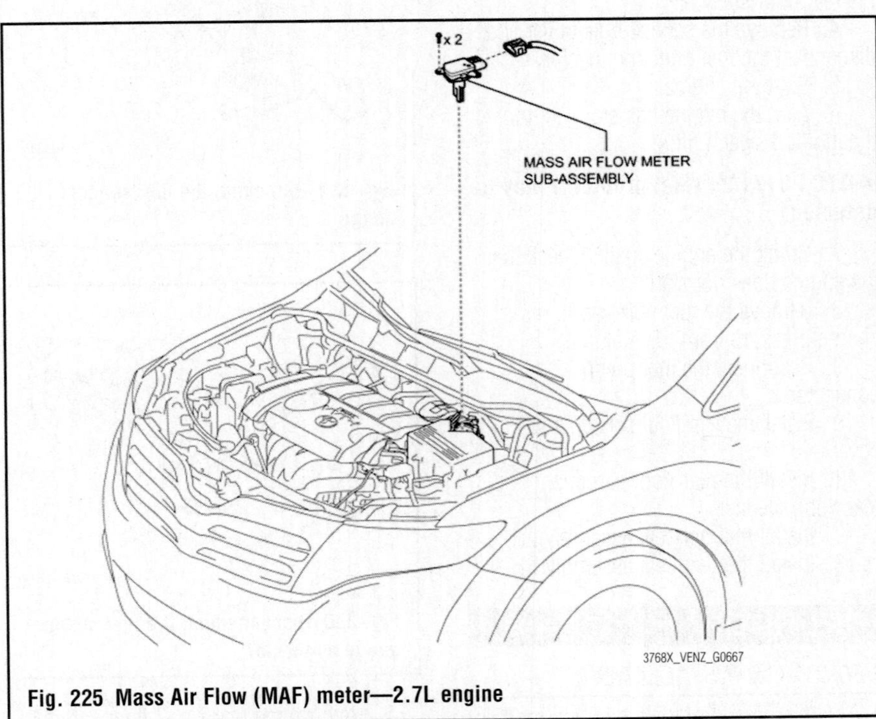

Fig. 225 Mass Air Flow (MAF) meter—2.7L engine

x 2
MASS AIR FLOW METER
3768X_VENZ_G0668

Fig. 226 Mass Air Flow (MAF) meter—3.5L engine

To install:

2. Install the Mass Air Flow (MAF) meter sub assembly.
   a. Install the Mass Air Flow (MAF) meter with the 2 screws.

➡Make sure that the o-ring is not cracked or jammed when installing the Mass Air Flow (MAF) meter.

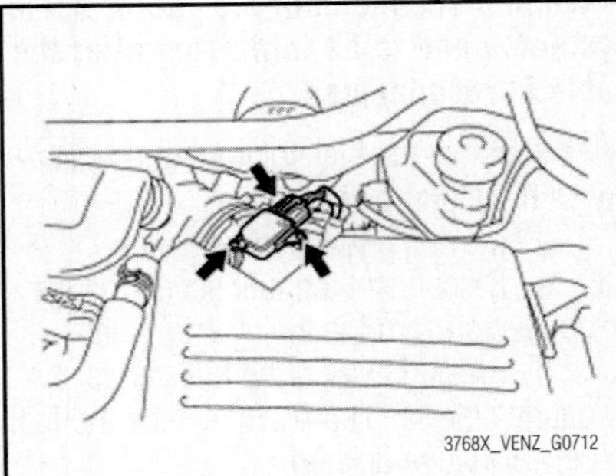

Fig. 227 Removing the Mass Air Flow (MAF) meter sub assembly

b. Connect the Mass Air Flow (MAF) meter connector.

3.5L Engine

1. Remove the Mass Air Flow (MAF) meter.
   a. Separate the Mass Air Flow (MAF) meter connector.
   b. Remove the 2 screws and Mass Air Flow (MAF) meter.

To install:

2. Install the Mass Air Flow (MAF) meter.
   a. Install the Mass Air Flow (MAF) meter with the 2 screws.

➡Make sure that the o-ring is not cracked or does not jump out of position during installation.

b. Connect the Mass Air Flow (MAF) meter connector.

VEHICLE SPEED SENSOR (VSS)

REMOVAL & INSTALLATION

See Figure 228.

1. Remove the automatic transaxle assembly.
2. Remove the automatic transaxle oil pan sub assembly.
3. Remove the valve body oil strainer assembly.
4. Remove the transmission valve body assembly.
5. Remove the speed sensor.
   a. Disconnect the connector.
   b. Remove the 2 bolts and speed sensor from the valve body.

To install:

6. To install, reverse the removal procedure. Tighten the speed sensor to 8 ft. lbs. (11 Nm).

➡Coat the 2 bolts with ATF.

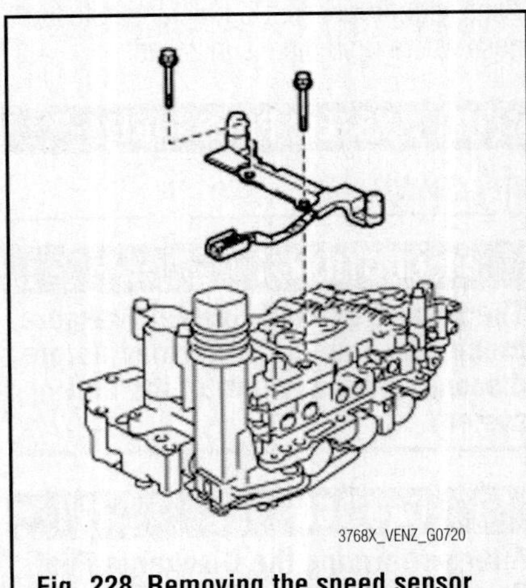

Fig. 228 Removing the speed sensor

FUEL SYSTEM SERVICE PRECAUTIONS

Safety is the most important factor when performing not only fuel system maintenance, but any type of maintenance. Failure to conduct maintenance and repairs in a safe manner may result in serious personal injury or death. Work on a vehicle's fuel system components can be accomplished safely and effectively by adhering to the following rules and guidelines.

- To avoid the possibility of fire and personal injury, always disconnect the negative battery cable unless the repair or test procedure requires that battery voltage be applied.
- Always relieve the fuel system pressure prior to disconnecting any fuel system component (injector, fuel rail, pressure regulator, etc.) fitting or fuel line connection. Exercise extreme caution whenever relieving fuel system pressure to avoid exposing skin, face and eyes to fuel spray. Please be advised that fuel under pressure may penetrate the skin or any part of the body that it contacts.
- Always place a shop towel or cloth around the fitting or connection prior to loosening to absorb any excess fuel due to spillage. Ensure that all fuel spillage is quickly removed from engine surfaces. Ensure that all fuel-soaked cloths or towels are deposited into a flame-proof waste container with a lid.
- Always keep a dry chemical (Class B) fire extinguisher near the work area.
- Do not allow fuel spray or fuel vapors to come into contact with a spark or open flame.
- Always use a second wrench when loosening or tightening fuel line connection fittings. This will prevent unnecessary stress and torsion on fuel piping. Always follow the proper torque specifications.
- Always replace worn fuel fitting O-rings with new ones. Do not substitute fuel hose where rigid pipe is installed.

FUEL SYSTEM PRESSURE

RELIEVING

**** CAUTION**

The Discharge Fuel System Pressure procedure must be performed before disconnecting any part of the fuel system.

**** CAUTION**

After performing the Discharge Fuel System Pressure procedure, pressure will remain in the fuel lines. When disconnecting a fuel line, cover it with a shop rag or a piece of cloth to prevent fuel from spraying or coming out.

1. Remove the rear seat assembly LH.
2. Remove the rear seat assembly RH.
3. Turn over the rear floor carpet and the rear floor silencer.
4. Remove the service hole cover and disconnect the fuel pump connector.
5. Start the engine.
6. After the engine has stopped, turn the ignition switch off.

➡DTC P0171/25 (fuel problem) may be detected.

7. Crank the engine again. Check that the engine does not start.
8. Remove the fuel tank cap to discharge pressure from the fuel tank.
9. Reconnect the fuel pump connector.
10. Install the rear floor service hole cover.
11. Install the rear floor carpet and the rear floor silencer.
12. Install the rear seat assembly LH.
13. Install the rear seat assembly RH.

FUEL FILTER

REMOVAL & INSTALLATION

The filter is part of the fuel pump module and is not normally serviced.

FUEL LEVEL SENDING UNIT

REMOVAL & INSTALLATION

See Figures 229 and 230.

1. Discharge the fuel system pressure.
2. Disconnect the cable from the negative battery terminal.

➡When disconnecting the cable, some systems need to be initialized after the cable is reconnected.

3. Remove the fuel suction tube assembly with the pump and gauge.
4. Remove the fuel sender gauge.
   a. Disconnect the fuel sender gauge connector from the fuel suction plate.
   b. Press down on the fuel sender gauge claw labeled A. Then slide the fuel sender gauge upward.

➡Do not touch the sender resistance plate or contact area.

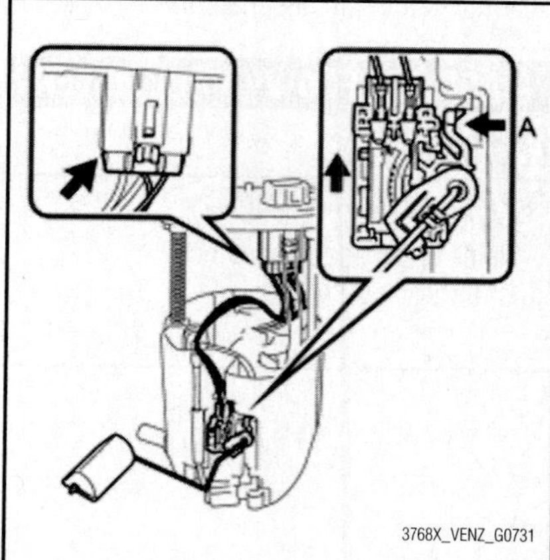

Fig. 229 Removing the fuel sender gauge

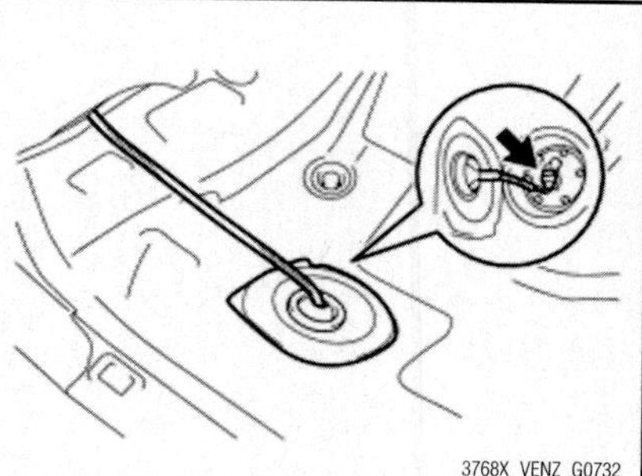

Fig. 230 Disconnecting the fuel sender gauge connector

5. Remove the rear No. 2 floor service hole cover and disconnect the fuel sender gauge connector.
6. Remove the fuel sender gauge assembly.
   a. Remove the 6 bolts and fuel sender gauge assembly from the fuel tank.

➡Be careful not to bend the arm of the fuel sender gauge.

To install:

7. To install, reverse the removal procedure. Tighten the fuel sender gauge assembly bolts to 31 inch lbs. (3.5 Nm).

➡Be careful not to bend the arm of the fuel sender gauge assembly.

FUEL PUMP

REMOVAL & INSTALLATION

See Figures 231 through 235.

1. Discharge the fuel system.
2. Disconnect the cable from the negative battery terminal.

➡When disconnecting the cable, some systems need to be initialized after the cable is reconnected.

3. Remove the rear seats LH and RH.
4. Remove the rear floor service hole cover.
   a. Turn over the rear floor carpet and the rear floor silencer.
   b. Remove the service hole and disconnect the fuel pump connector.
5. Remove the fuel suction tube assembly with the pump and gauge.
   a. Remove the tube joint clip, and pull out the fuel pump tube.

➡Before removing the tube joint clip, check for foreign matter around the clip. Clean it if necessary.

➡Keep the O-rings free of foreign matter, as they become contaminated easily.

➡Do not use any tools in this procedure.

➡Do not forcefully bend or twist the tube.

➡Put the tube in a plastic bag to prevent damage and contamination.

➡If the fuel suction plate and tube are stuck together, pinch the tube and turn it carefully to disconnect it.

➡Be careful not to damage the clip. If the clip is damaged, replace it.

   b. Remove the 8 bolts and the fuel tank vent tube set plate.

➡While holding the fuel suction tube by hand, remove the fuel tank vent tube set plate.

   c. Disconnect the clip and fuel tube.
   d. Remove the fuel suction tube assembly with pump and gauge from the fuel tank.

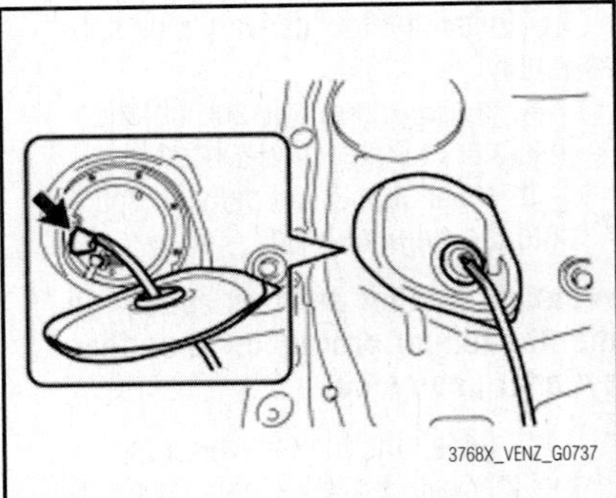

Fig. 231 Disconnecting the fuel pump connector

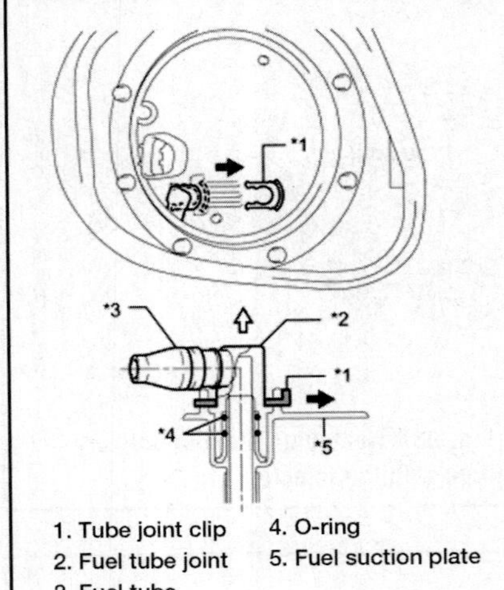

Fig. 232 Removing the fuel suction tube assembly with the pump and gauge

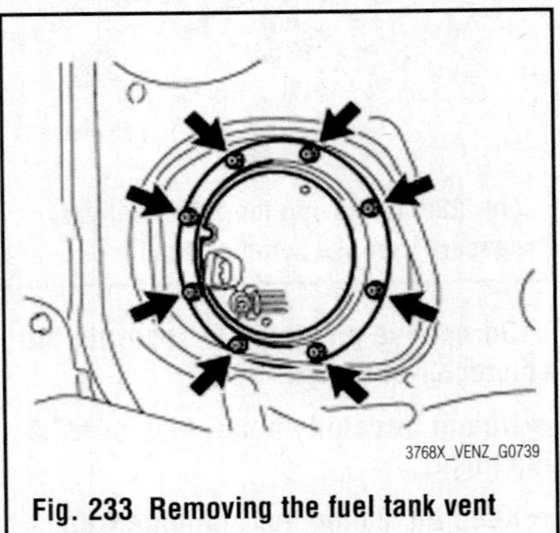

Fig. 233 Removing the fuel tank vent tube set plate

➡Make sure that the sender gauge arm does not bend.

➡Do not damage the fuel suction tube.

   e. Remove the fuel suction tube set gasket from the fuel tank.

To install:

6. Install the fuel suction tube assembly with the pump and gauge.
   a. Install a new fuel suction tube set gasket onto the fuel tank.
   b. Connect the fuel tube with the clip.
   c. Set the fuel suction tube assembly to the fuel tank.

➡Be careful not to bend the arm of the fuel sender gauge.

➡Do not damage the fuel tube.

   d. Align the protrusion of the fuel suction tube assembly with pump and gauge and the cutout of the fuel tank vent tube set plate.

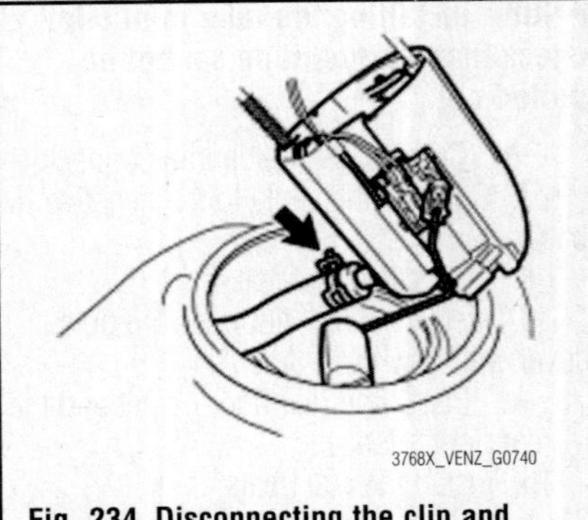

Fig. 234 Disconnecting the clip and fuel tube

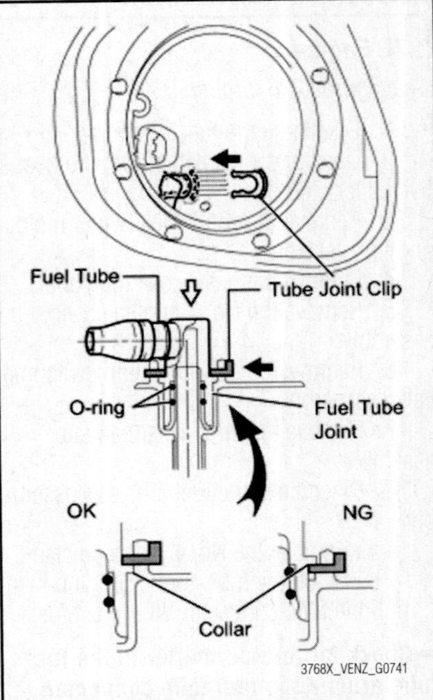

Fig. 235 Installing the fuel tank main tube sub assembly and the tube joint clip

   e. While holding the fuel suction with pump and gauge tube assembly by hand, install the fuel tank vent tube to the fuel tank with the 8 bolts. Tighten to 53 inch lbs. (6 Nm).
   f. Install the fuel tank main tube sub-assembly and the tube joint clip.

➡Check that there are no scratches or foreign matter around the connected part of the fuel tube joint and plug before performing this work.

➡Check that the fuel tube joint is securely inserted.

➡Check that the tube joint clip is on the collar of the fuel tube joint.

➡After installing the tube joint clip, check that the fuel tube cannot be pulled out.

g. Connect the fuel pump connector.

7. Connect the cable to the negative battery terminal.

8. Inspect for fuel leaks.

9. Install the rear floor service hole cover with new butyl tape.

a. Install the rear floor carpet and the rear floor silencer.

10. Install the rear seats.

FUEL RAIL AND INJECTOR

REMOVAL & INSTALLATION

2.7L Engine

See Figures 236 through 239.

1. Discharge the fuel system.
2. Disconnect the cable from the negative battery terminal.
3. Remove the windshield wiper motor and link assembly.
4. Remove the outer cowl top panel.
5. Remove the No. 1 engine cover sub assembly.
6. Remove the No. 1 vacuum switching valve assembly.
7. Remove the air cleaner cap sub assembly.
8. Disconnect the fuel tube sub assembly.

a. Remove the No. 1 fuel pipe clamp.

b. Pinch the tube connector, and then pull the tube connector off of the pipe.

➡Check for foreign matter in the fuel tube around the fuel tube connector. Clean it if necessary. Foreign matter can affect the ability of the O-rings to seal the connector and fuel pipe.

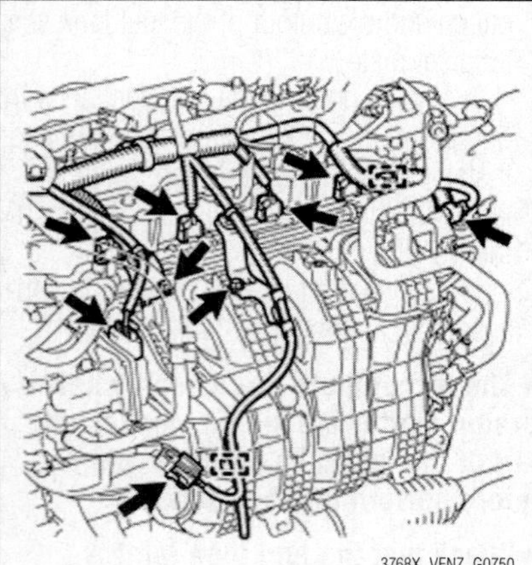

Fig. 236 Disconnecting the wire harness

Fig. 237 Removing the fuel delivery pipe with the injectors

Fig. 238 Removing the 2 fuel delivery spacers from the cylinder head

➡Do not use any tools to separate the connector and pipe.

➡Do not forcefully bend, kink or twist the hose.

➡Keep the connector and pipe free from foreign matter.

➡If the connector and pipe are stuck together, pinch the connector and turn it carefully to disconnect it.

➡Put the connector in a plastic bag to prevent damage and contamination.

c. Remove the fuel tube sub-assembly from the fuel hose clamp.

9. Disconnect the wire harness.

a. Disconnect the 4 fuel injector connectors.

b. Disconnect the 3 connectors.

c. Remove the 2 bolts and 2 wire harness brackets.

d. Detach the 2 clamps to disconnect the wire harness.

10. Remove the vacuum switching valve.

11. Remove the fuel delivery pipe sub assembly.

a. Remove the 2 bolts, and then remove the fuel delivery pipe together with the 4 fuel injectors.

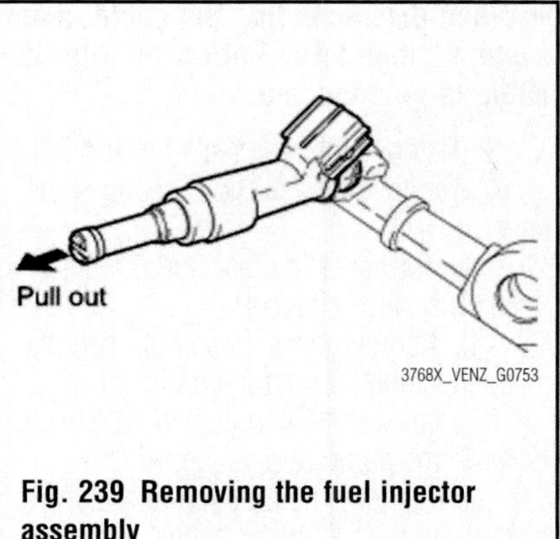

Fig. 239 Removing the fuel injector assembly

➡Be careful not to drop the fuel injectors when removing the fuel delivery pipe.

b. Remove the 2 fuel delivery spacers from the cylinder head.

c. Remove the 4 injector vibration insulators from the cylinder head.

12. Remove the fuel injector assembly.

a. Pull the 4 fuel injectors out of the fuel delivery pipe.

To install:

13. Install the fuel injector assembly.

a. Apply a light coat of gasoline or spindle oil to new O-ring, and then install one onto each fuel injector.

b. Apply a light coat of gasoline or spindle oil to the part of the fuel delivery pipe which comes into contact with the O-ring of the fuel injector.

c. Apply a light coat of gasoline or spindle oil to the O-ring again, and then install the fuel injectors onto the fuel delivery pipe.

➡Make sure that the O-ring is not cracked or jammed when installing the injector.

d. Check that the fuel injector rotates smoothly. If the fuel injector does not rotate, replace the O-ring.

14. Install the fuel delivery pipe sub assembly.

a. Install 4 new injector vibration insulators to the cylinder head.

b. Install the 2 fuel delivery spacers onto the cylinder head.

➡Install the fuel delivery spacer so that the longer protrusion is on the cylinder head side.

c. Install the fuel delivery pipe together with the 4 fuel injectors to the cylinder head, and then temporarily install the 2 bolts.

➡**Be careful not to drop the fuel injectors when installing the fuel delivery pipe.**

d. Check that the fuel injector rotates smoothly. If the fuel injector does not rotate, replace the O-ring.

e. Tighten the 2 bolts to 15 ft. lbs. (21 Nm).

15. Connect the wire harness.

a. Install the 2 wire harness brackets with the 2 bolts. Tighten to 7 ft. lbs. (10 Nm).

b. Connect the 3 connectors.

c. Connect the 4 fuel injector connectors.

d. Attach the 2 clamps to connect the wire harness.

16. Install the vacuum switching valve assembly.

17. Connect the fuel tube sub assembly.

a. Push the tube connector to the pipe until the tube connector makes a "click" sound.

➡**Before connecting the connector and fuel pipe, check that there is no damage or foreign matter on the connecting part of the fuel pipe.**

➡**After connecting the fuel tube connector and pipe, check that they are securely connected by trying to pull them apart.**

b. Install the No. 1 fuel pipe clamp.

c. Install the fuel tube sub-assembly to the fuel hose clamp.

18. Install the air cleaner cap sub assembly.

19. Install the No. 1 vacuum switching valve assembly.

20. Install the No. 1 engine cover sub assembly.

21. Install the outer cowl top panel.

22. Install the windshield wiper motor and link assembly.

23. Connect the negative battery terminal.

24. Inspect for fuel leaks.

3.5L Engine

See Figures 240 through 242.

1. Discharge the fuel system pressure.

2. Disconnect the cable from the negative battery terminal.

➡**When disconnecting the cable, some systems need to be initialized after the cable is reconnected.**

3. Remove the intake air surge tank assembly.

4. Disconnect the fuel tube sub assembly.

a. Remove the No. 2 fuel pipe clamp.

b. Pinch the tube connector and pull out the fuel pipe.

➡**Check that there is no dirt or other foreign objects around the connector when disconnecting it. Clean the connector as necessary.**

➡**It is necessary to prevent dirt or foreign objects from entering the quick connector. If dirt or foreign objects get in the connector, the O-rings may not seal properly.**

➡**Only disconnect the quick connector by hand.**

➡**Do not bend, kink or twist the nylon tubes.**

➡**Protect the connector by covering it with a plastic bag.**

➡**If the pipe and the connector are stuck, carefully try wiggling or pushing and pulling on the connector to release it. Pull the connector off the pipe carefully.**

5. Remove the fuel injector assembly.

a. Disconnect the 6 fuel injector connectors.

b. Remove the 5 bolts and fuel delivery pipe sub-assembly together with the 6 fuel injectors.

➡**Be careful not to drop the fuel injectors when removing the fuel delivery pipe.**

6. Remove the 6 injector vibration insulators from the intake manifold.

7. Pull out the fuel injectors from the fuel delivery pipe.

➡**If the injectors are to be reused, reinstall them to the same cylinder they came from.**

a. Remove the 6 O-rings from the injectors.

To install:

8. Install the fuel injector assembly.

a. Apply a light coat of spindle oil or gasoline to new O-rings, and install them to each injector.

b. Apply a light coat of spindle oil or gasoline where the fuel delivery pipe contacts each O-ring.

c. Push and twist each fuel injector to install them into the fuel delivery pipe.

d. Position each fuel injector connector outward.

➡**Be careful not to twist the O-rings.**

➡**After installing a fuel injector, check that it turns smoothly. If not, reinstall it with a new O-ring.**

e. Install 6 new injector vibration insulators to the intake manifold.

f. Place the fuel delivery pipe with the 6 fuel injectors in position on the intake manifold.

➡**Be careful not to drop the fuel injectors when installing the fuel delivery pipe.**

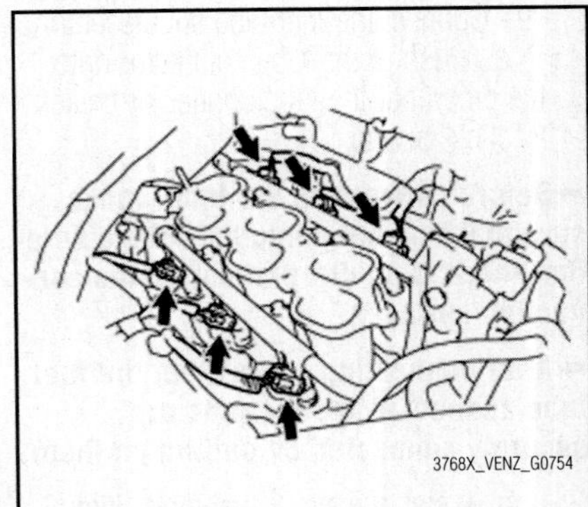

Fig. 240 Disconnecting the 6 fuel injector connectors

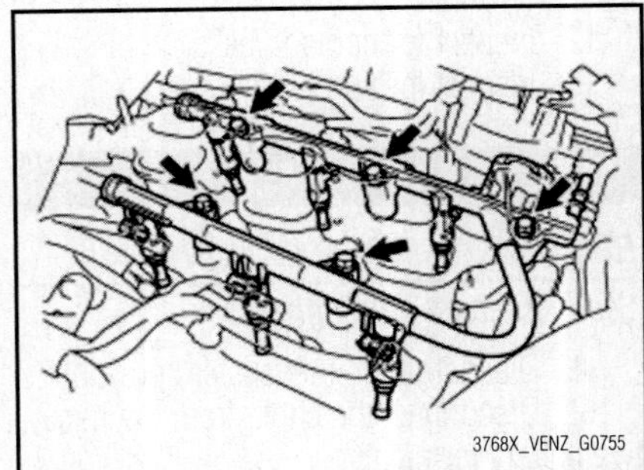

Fig. 241 Removing the 5 bolts and fuel delivery pipe sub-assembly together with the 6 fuel injectors

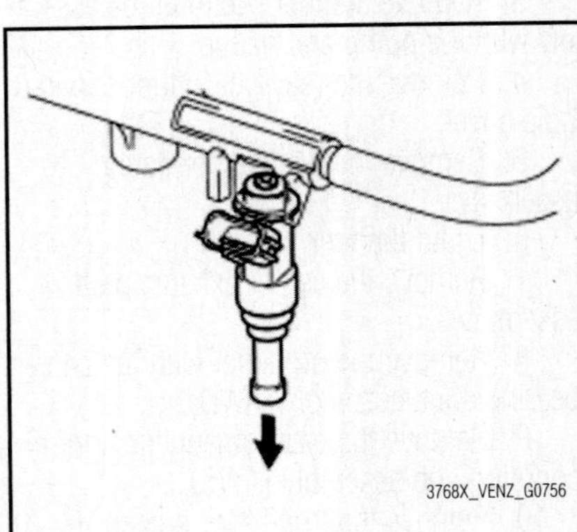

Fig. 242 Pulling out the fuel injectors from the fuel delivery pipe

g. Temporarily install the 5 bolts which are used to hold the fuel delivery pipe to the intake manifold.

➡After installing the fuel injectors, check that they turn smoothly. If not, reinstall the injectors with new O-rings.

h. Tighten the 5 bolts which are used to hold the fuel delivery pipe to the intake manifold. Tighten to 15 ft. lbs. (21 Nm).

i. Connect the 6 fuel injector connectors.

9. Connect the fuel tube sub assembly.

a. Push in the tube connector onto the pipe until the tube connector makes a "click" sound.

➡Before connecting the tube, make sure that it is not damaged. Make sure that there is no dirt present on the connecting surfaces.

➡After connecting, check that the fuel tube connector and the pipe are securely connected by pulling on them.

b. Install the No. 2 fuel pipe clamp.

10. Install the intake air surge tank assembly.

11. Connect the cable to the negative battery terminal.

12. Inspect for coolant leaks.

13. Inspect for fuel leaks.

FUEL TANK

REMOVAL & INSTALLATION

See Figures 243 through 255.

1. Discharge the fuel system pressure.

2. Disconnect the cable from the negative battery terminal.

➡When disconnecting the cable, some systems need to be initialized after the cable is reconnected

3. Remove the fuel suction tube assembly with the pump and gauge.

4. Remove the rear No. 2 floor service hole cover.

5. Remove the fuel sender gauge assembly.

6. Drain the fuel.

7. Remove the center exhaust pipe assembly.

8. Remove the propeller with the center bearing shaft assembly (AWD).

9. Remove the electromagnetic control coupling sub assembly (AWS).

10. Remove the front floor brace LH.

a. Remove the bolt, 2 nuts and front floor brace LH.

11. Remove the front floor brace RH.

12. Remove the bolt, 2 nuts and front floor brace RH.

13. Remove the No. 1 fuel tank protector sub assembly.

14. Separate the No. 2 parking brake cable assembly.

15. Separate the No. 3 parking brake cable assembly.

16. Remove the fuel tank assembly.

a. Disconnect the fuel pump tube. Pinch the tabs of the retainer to release the lock claws and pull it down as shown in the illustration. Pull out the fuel tank main tube.

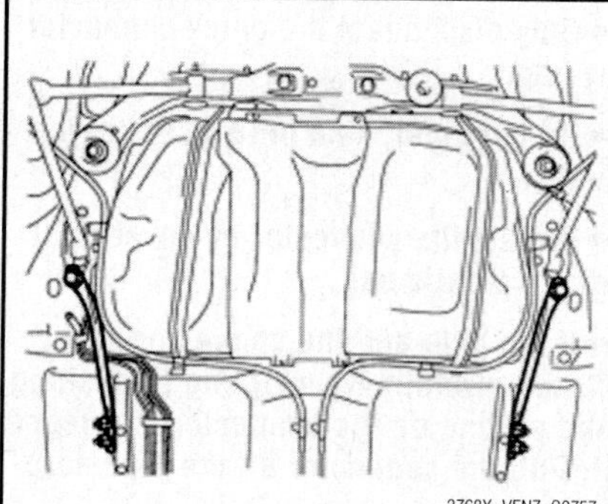

Fig. 243 Removing the front floor brace LH

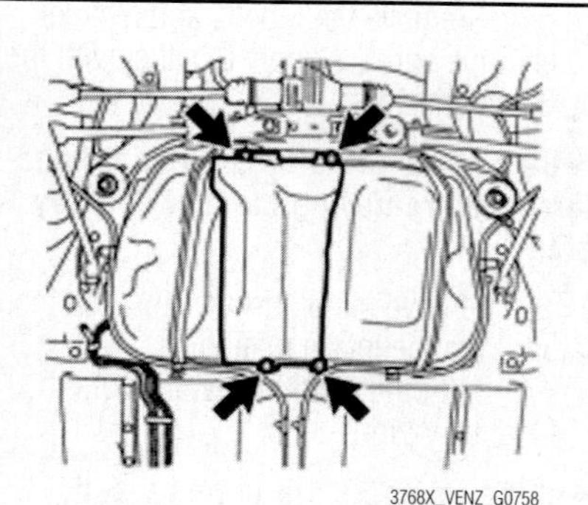

Fig. 244 Removing the No. 1 fuel tank protector sub assembly

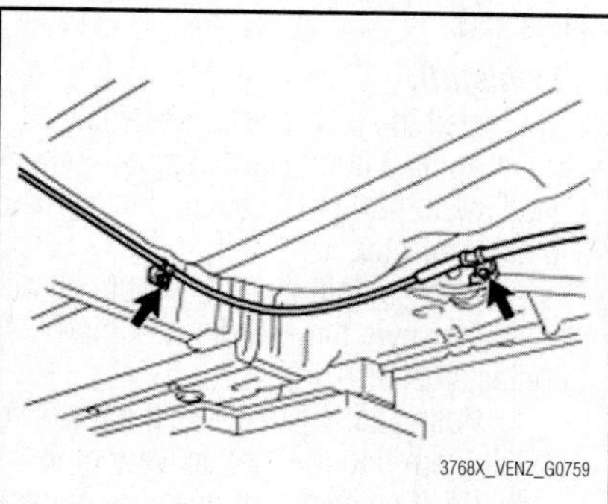

Fig. 245 Separating the No. 2 parking brake cable assembly

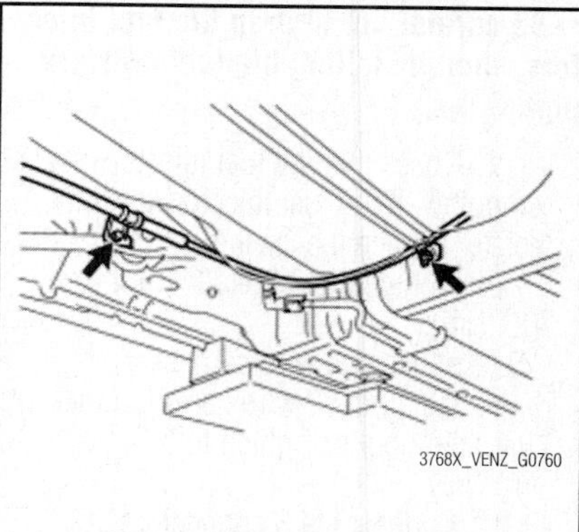

Fig. 246 Separating the No. 3 parking brake cable assembly

➡Check that there is no dirt or other foreign objects around the connector before this operation and clean the connector as necessary.

➡It is necessary to prevent mud or dirt from entering the connector. If mud or dirt gets in the connector, the O-rings may not seal properly.

➡Do not use any tools in this operation.

➡Do not bend, kink or twist the nylon tube. Protect the connector by covering it with a plastic bag.

➡When the pipe and connector are stuck, push and pull on them to release them.

b. Disconnect the fuel tank vent hose sub assembly from the charcoal canister assembly.

c. Set up an engine lifter underneath the fuel tank.

d. Remove the 2 set bolts of the fuel tank bands.

e. Remove the hose clamp and disconnect the fuel tank to filter pipe hose.

f. Remove the fuel tank.

g. Remove the 2 pins and 2 fuel tank bands as shown in the illustration.

h. Remove the 4 clip nuts.

17. remove the fuel tank vent hose sub assembly.

a. Remove the fuel tank vent hose sub-assembly from the fuel tank.

18. Remove the fuel tank to filler pipe hose.

a. Loosen the hose clamp bolt and remove the fuel tank to filler pipe hose from the fuel tank assembly.

19. remove the fuel tank main tube sub assembly.

a. Remove the fuel tank main tube from the fuel main tube support.

20. Remove the fuel main tube support.

21. Remove the No. 1 fuel tank cushion.
   a. Remove the 8 No. 1 fuel tank cushions.

To install:

22. Install the No. 1 fuel tank cushions.
23. Install the fuel main tube support with the bolt. Tighten to 48 inch lbs. (5.4 Nm).

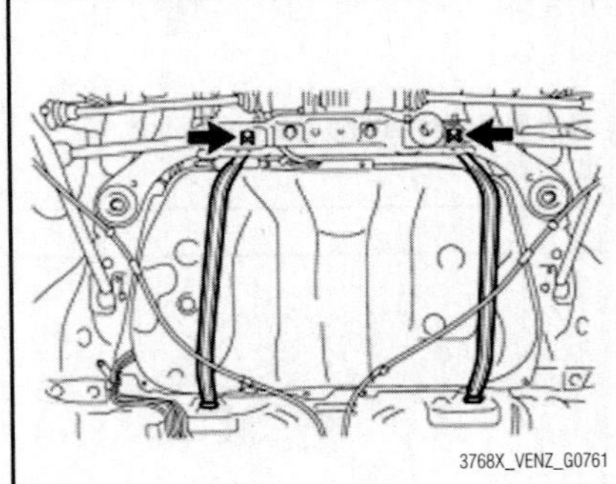

Fig. 247 Removing the 2 set bolts of the fuel tank bands

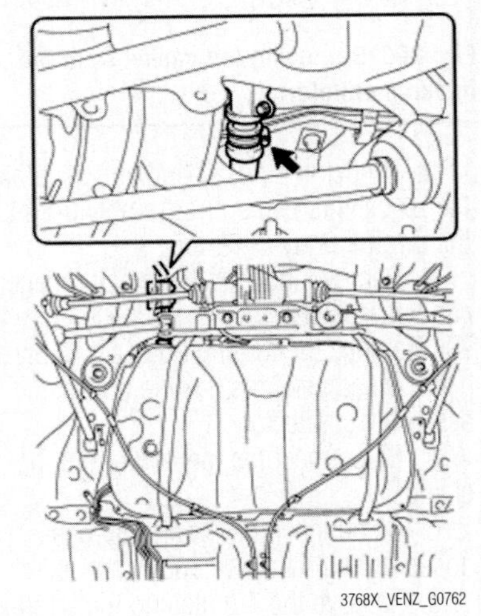

Fig. 248 Removing the hose clamp and disconnecting the fuel tank to filter pipe hose

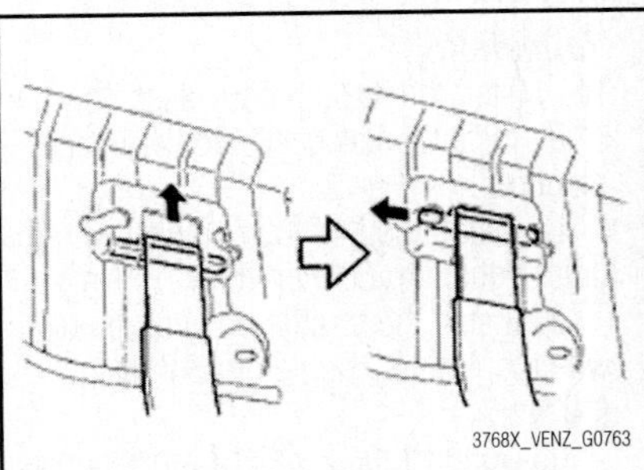

Fig. 249 Removing the 2 pins and 2 fuel tank bands

24. Install the fuel tank main tube sub assembly.
   a. Install the fuel tank main tube to the fuel main tube support.
25. Install the fuel tank to filler pipe hose to the fuel tank assembly with the clamp.
26. Install the fuel tank vent hose sub assembly.
27. Install the fuel sender gauge assembly.
28. Install the fuel tank assembly.
   a. Install the 4 clip nuts.
   b. Install the 2 fuel tank bands with the 2 pins.

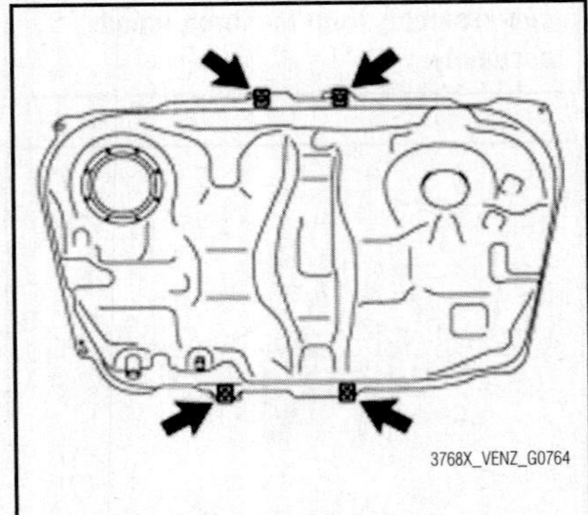

Fig. 250 Removing the 4 clip nuts

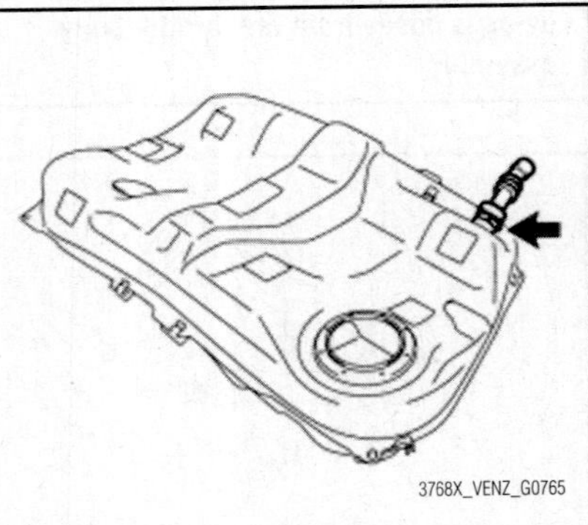

Fig. 251 Loosening the hose clamp bolt and removing the fuel tank to filler pipe hose from the fuel tank assembly

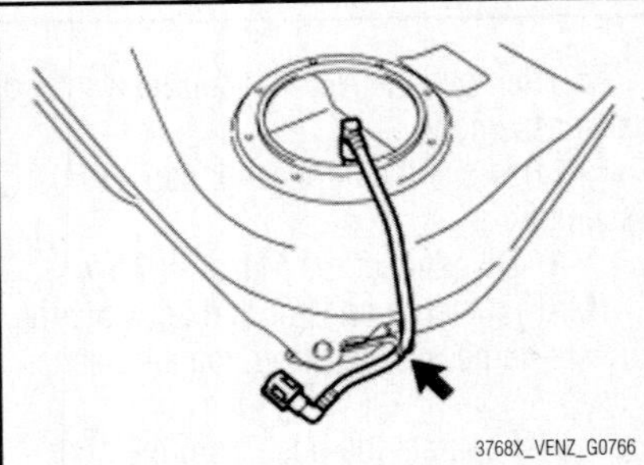

Fig. 252 Removing the fuel tank main tube from the fuel main tube support

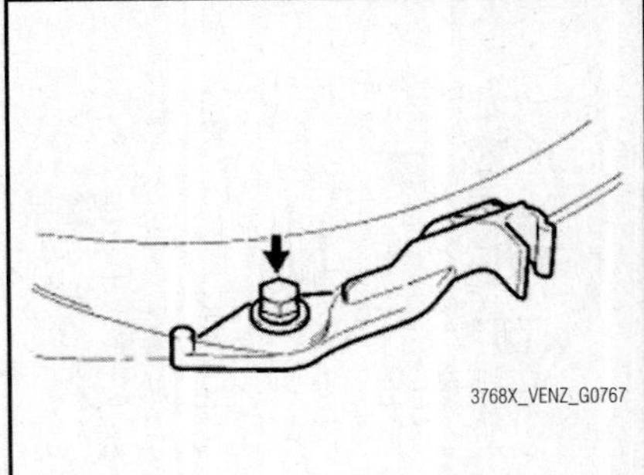

Fig. 253 Removing the fuel main tube support

   c. Set the fuel tank assembly onto the engine lifter.
   d. Lift up the engine lifter.

➡Slowly raise the lifter so as to not drop the fuel tank assembly.

   e. Install the fuel tank assembly with the fuel tank bands and 2 bolts. Tighten to 29 ft. lbs. (39 Nm).
   f. Connect the fuel tank to filler pipe hose with the clamp.
   g. Connect the fuel tank vent hose sub-assembly.
   h. Connect the fuel tank main tube sub-assembly. Push in the fuel pump tube connector to the pipe and push up the retainer so that the claws engage.

➡Check that there is no damage or foreign objects on the connected part.

➡After connecting, check if the fuel tube connector and the pipe are securely connected by trying to pull them apart.

29. Install the No. 3 parking brake cable assembly with the bolt and nut. Tighten to 53 inch lbs. (6 Nm).
30. Install the No. 2 parking brake cable assembly with the bolt and nut. Tighten to 53 inch lbs. (6 Nm).
31. Install the No. 1 fuel tank protector sub assembly with the 4 bolts. Tighten to 48 inch lbs. (5.4 Nm).
32. Install the front floor brace LH.
   a. Install the bolt, 2 nuts and the front floor brace LH. Tighten to 41 ft. lbs. (56 Nm).
33. Install the front floor brace RH.
   a. Install the bolt, 2 nuts and the front floor brace RH. Tighten to 41 ft. lbs. (56 Nm).
34. Install the electromagnetic control coupling sub assembly (AWD).
35. Install the propeller with the center bearing shaft assembly (AWD).
36. Install the center exhaust pipe assembly.

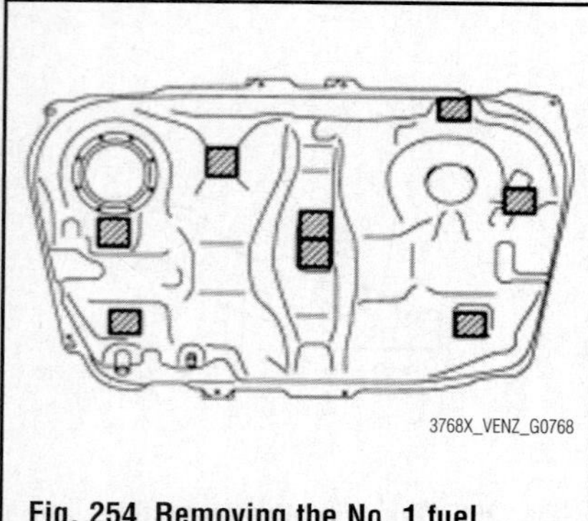

Fig. 254 Removing the No. 1 fuel tank cushions

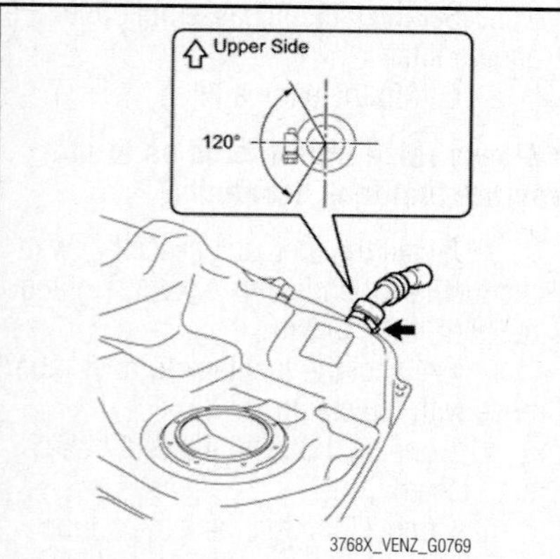

Fig. 255 Installing the fuel tank to filler pipe hose

37. Add fuel.
38. Install the rear No. 2 floor service hole cover.
39. Install the fuel suction tube assembly with the pump and gauge.
40. Connect the cable to the negative battery terminal.
41. Inspect for exhaust gas leak.

THROTTLE BODY

REMOVAL & INSTALLATION

2.7L Engine

See Figures 256 through 260.

1. Remove the windshield wiper motor and link.
2. Remove the outer cowl top panel sub assembly.
3. Remove the No. 1 engine cover sub assembly.
4. Remove the cool air intake duct seal.
5. Remove the No. 1 engine under cover.
6. Remove the No. 2 engine under cover.
7. Drain the engine coolant.

Fig. 256 Unlocking the hose band and separating the air cleaner cap sub-assembly from the throttle body assembly

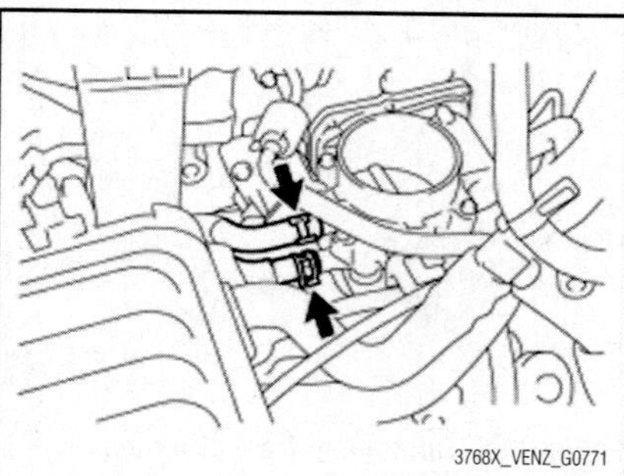

Fig. 257 Disconnecting the 2 water by-pass hoses from the throttle body assembly

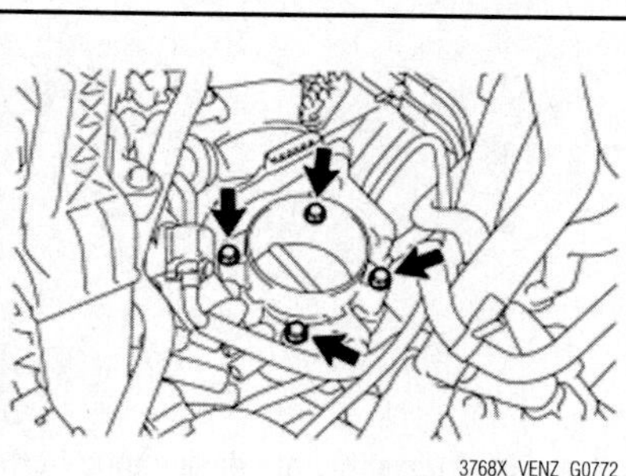

Fig. 258 Removing the 4 bolts and the throttle body assembly with the fuel tube bracket

8. Remove the No. 1 vacuum switching valve assembly.
9. Remove the air cleaner cap sub assembly.
   a. Disconnect the Mass Air Flow (MAF) meter connector and separate the wire harness clamp from the air cleaner cap.
   b. Separate the hose from the hose clamp.
   c. Disconnect the ventilation hose from the cylinder head cover.

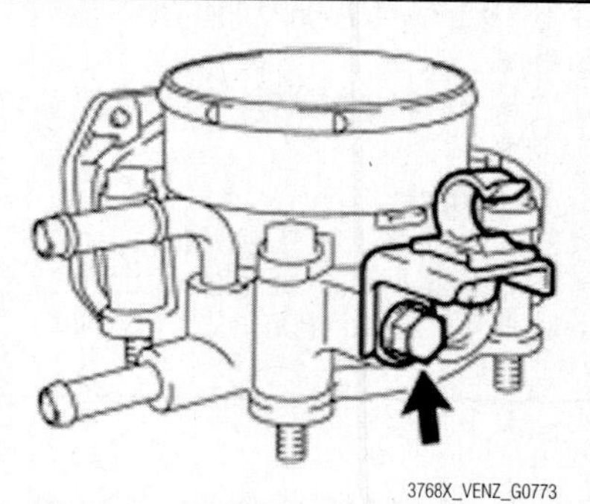

Fig. 259 Removing the fuel tube bracket

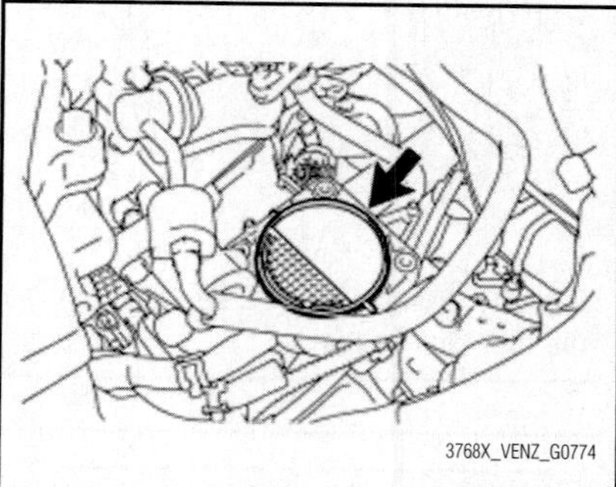

Fig. 260 Removing the gasket from the intake manifold

   d. Unlock the hose band and separate the air cleaner cap sub-assembly from the throttle body assembly.
   e. Remove the 2 bolts and air cleaner cap sub-assembly.
10. Remove the throttle body assembly.
   a. Disconnect the throttle body assembly connector.
   b. Disconnect the fuel tube from the clamp.
   c. Disconnect the 2 water by-pass hoses from the throttle body assembly.
   d. Remove the 4 bolts and the throttle body assembly with the fuel tube bracket.
   e. Remove the bolt and fuel tube bracket.
   f. Remove the gasket from the intake manifold.

To install:

11. Install the throttle body assembly.
   a. Install a new gasket to the intake manifold.
   b. Install the fuel tube bracket with the bolt. Tighten to 66 inch lbs. (7.5 Nm).
   c. Install the throttle body assembly with the 4 bolts. Tighten to 7 ft. lbs. (10 Nm).
   d. Connect the 2 water by-pass hoses to the throttle body.
   e. Connect the throttle body assembly connector.

f. Connect the fuel tube to the clamp.

12. Install the air cleaner cap sub assembly.

a. Connect the air-cleaner cap sub-assembly to the throttle body assembly and lock the hose band.

b. Install the air cleaner cap sub-assembly with the 2 bolts. Tighten to 44 inch lbs. (5 Nm).

c. Connect the ventilation hose to the cylinder head cover.

d. Connect the Mass Air Flow (MAF) meter connector and install the wire harness clamp to the air cleaner cap.

e. Install the hose to the hose clamp.

13. Install the No. 1 vacuum switching valve assembly.

14. Add engine coolant.

15. Inspect for coolant leaks.

16. Install the No. 1 engine cover sub assembly.

17. Install the cool air intake duct seal.

18. Install the outer cowl top panel sub assembly.

19. Install the windshield wiper motor and link.

20. Install the No. 2 engine under cover.

21. Install the No. 1 engine under cover.

22. Perform the initialization.

➡Perform the following procedure after replacing the ECM, throttle body assembly or any throttle body components. The following procedure should also be performed if the throttle body is cleaned.

a. Disconnect the cable from the negative (-) battery terminal. Wait at least 60 seconds and reconnect the cable.

➡When disconnecting the cable, some systems need to be initialized after the cable is reconnected.

b. Turn the ignition switch to ON without operating the accelerator pedal.

➡If the accelerator pedal is operated, perform the above steps again.

c. Connect the Techstream to the DLC3 and clear the DTCs.

d. Start the engine and check that the MIL is not illuminated. After the engine is warmed up, check that the idle speed is within the specified range when the A/C is switched off. The standard idle speed when the A/C is switched off is 600-700 rpm.

➡Be sure to perform this step with all accessories off.

➡Make sure that the shift lever is in N.

e. Enter the following menus: Powertrain / Engine and ECT / Data List / Throttle Sensor Position. Sensor Output. Fully depress the accelerator pedal and check that the value is 60% or more.

f. Perform a road test and confirm that there are no abnormalities.

3.5L Engine

See Figures 261 and 262.

1. Remove the engine No. 1 under cover.

2. Remove the engine No. 2 under cover.

3. Drain the engine coolant.

4. Remove the v-bank cover sub assembly.

5. Remove the air cleaner cap with the hose.

6. Remove the throttle body.

a. Disconnect the 2 water by pass hoses from the throttle body.

b. Disconnect the throttle body connector and wire harness clamp.

c. Remove the 4 bolts, throttle body and wire harness clamp stay.

d. Remove the gasket from the intake air surge tank assembly.

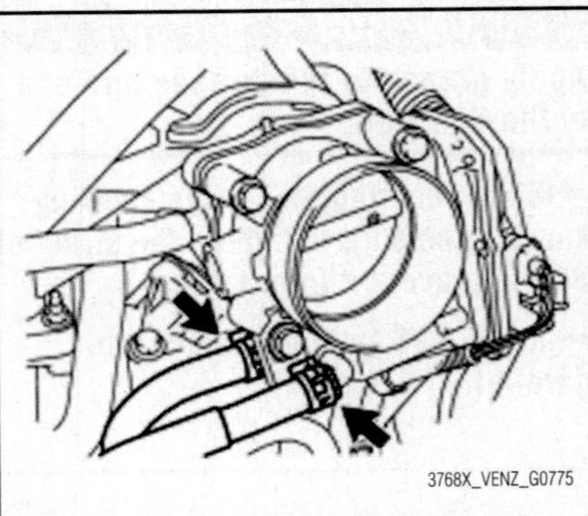

Fig. 261 Disconnecting the 2 water by pass hoses

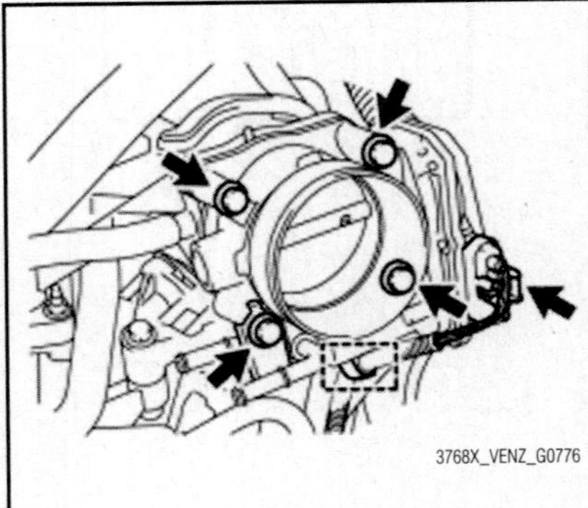

Fig. 262 Removing the throttle body

To install:

7. Install the throttle body.

a. Install a new gasket to the intake air surge tank assembly.

b. Install the throttle body and wire harness clamp stay to the intake air surge tank assembly with the 4 bolts. Tighten to 7 ft. lbs. (10 Nm).

c. Connect the throttle body connector and wire harness clamp.

d. Connect the 2 water by-pass hoses to the throttle body.

8. Install the air cleaner cap with the hose.

9. Add engine coolant.

10. Inspect for engine coolant leaks.

11. Install the v-bank cover sub assembly.

12. Install the No. 2 engine under cover.

13. Install the No. 1 engine under cover.

14. Perform the initialization.

➡Perform the following procedure after replacing the ECM, throttle body assembly or any throttle body components. The following procedure should also be performed if the throttle body is cleaned.

a. Disconnect the cable from the negative (-) battery terminal. Wait at least 60 seconds and reconnect the cable.

➡When disconnecting the cable, some systems need to be initialized after the cable is reconnected.

b. Turn the ignition switch to ON without operating the accelerator pedal.

➡If the accelerator pedal is operated, perform the above steps again.

c. Connect the Techstream to the DLC3 and clear the DTCs.

d. Start the engine and check that the MIL is not illuminated. After the engine is warmed up, check that the idle speed is within the specified range when the A/C is switched off. The standard idle speed when the A/C is switched off is 600-700 rpm.

➡Be sure to perform this step with all accessories off.

➡Make sure that the shift lever is in N.

e. Enter the following menus: Powertrain / Engine and ECT / Data List / Throttle Sensor Position. Sensor Output. Fully depress the accelerator pedal and check that the value is 60% or more.

f. Perform a road test and confirm that there are no abnormalities.

HEATING & AIR CONDITIONING SYSTEM

BLOWER MOTOR

REMOVAL & INSTALLATION

See Figure 263.

1. Disconnect the cable from the negative battery terminal.
2. Remove the No. 2 instrument panel under cover sub assembly.
3. Remove the front blower motor sub assembly.
   a. Disconnect the connector.
   b. Remove the 3 screws and front blower motor sub assembly.

To install:

4. To install, reverse the removal procedure.

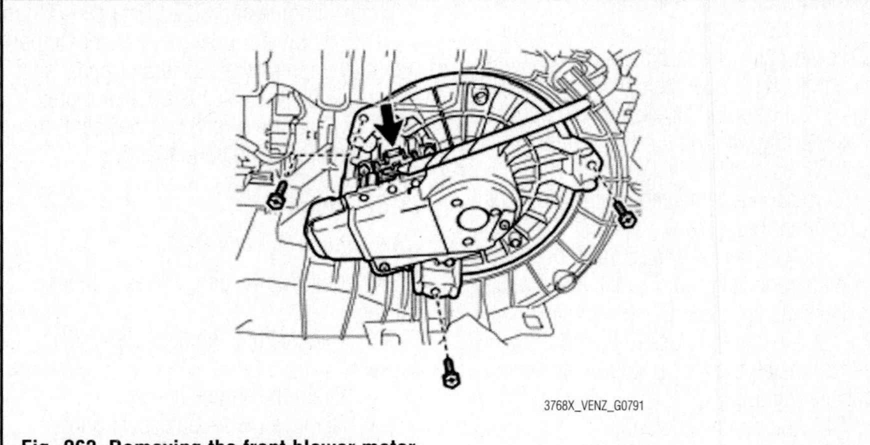

Fig. 263 Removing the front blower motor

STEERING

POWER RACK & PINION STEERING GEAR

REMOVAL & INSTALLATION

2.7L Engine

See Figures 264 through 272.

➡When disconnecting the steering intermediate shaft assembly and pinion shaft of the steering gear assembly, be sure to place matchmarks before servicing.

1. Place the front wheels facing straight ahead (2WD).
2. Secure the steering wheel in order to prevent it from rotating.

➡This operation is necessary to prevent damage to the spiral cable.

3. Remove the front wheels (2WD).
4. Remove the No. 1 engine under cover (2WD).
5. Remove the No. 2 engine under cover (2WD).
6. Remove the No. 3 engine under cover (2WD).
7. Separate the steering intermediate shaft assembly.
   a. Put matchmarks (*1) on the steering intermediate shaft assembly and steering link assembly.
   b. Remove the bolt and slide the steering intermediate shaft assembly.
   c. Separate the steering intermediate shaft assembly from the steering link assembly.
8. Separate the tie rod assembly LH.
   a. Remove the cotter pin and nut.
   b. Install the special tool (09960-20010) to the tie rod end.

➡Make sure that the upper ends of the tie rod ends and the special tool are aligned.

   c. Using the special tool, separate the tie rod end from the steering knuckle.

⁂ CAUTION

Apply grease to the threads and end of the SST bolt.

➡When securing SST to the steering knuckle, be sure to tighten the string of SST to prevent it from falling.

➡Install SST so that A and B are parallel.

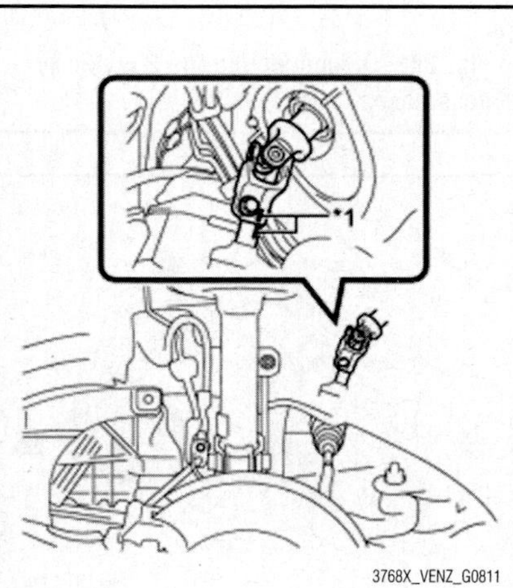

Fig. 264 Putting matchmarks on the steering intermediate shaft assembly

➡Be sure to place a wrench on the part indicated in the illustration.

➡Do not damage the front disc brake dust cover.

➡Do not damage the ball joint dust cover.

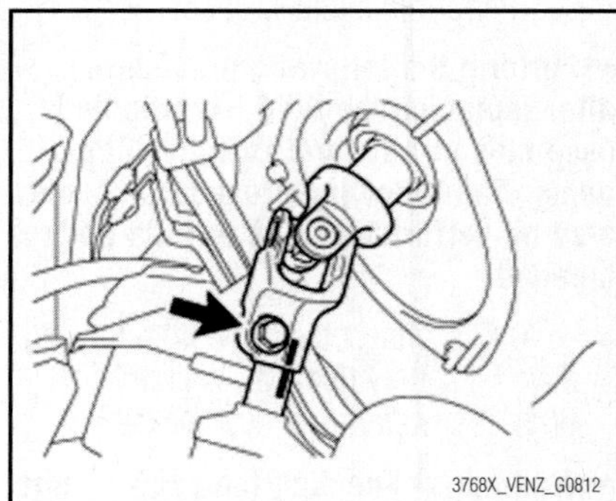

Fig. 265 Removing the bolt and sliding the steering intermediate shaft assembly

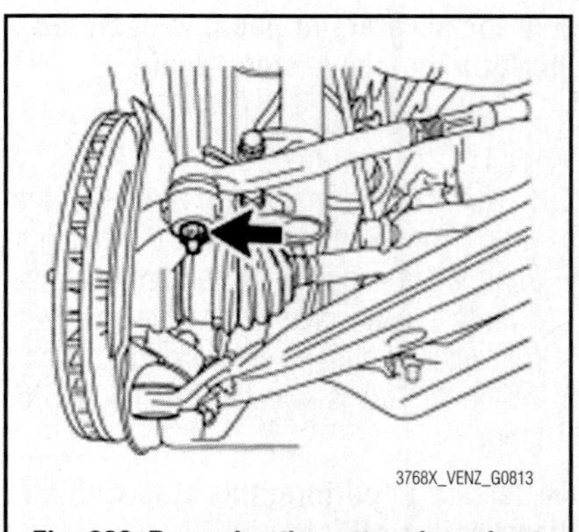

Fig. 266 Removing the cotter pin and nut

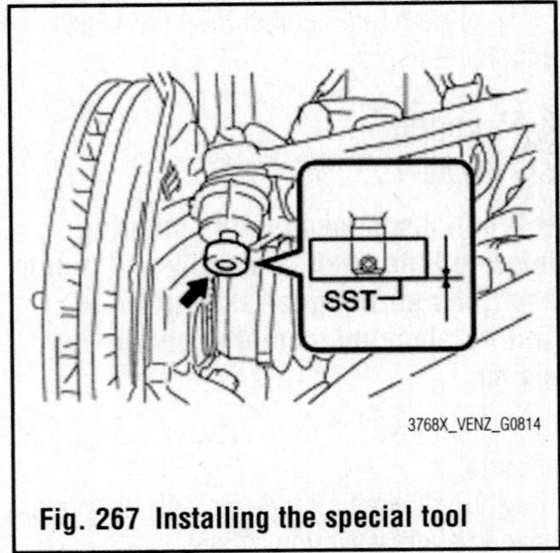

Fig. 267 Installing the special tool

➡Do not damage the steering knuckle.

9. Separate the tie rod assembly RH.

➡Perform the same procedure as for the LH side.

10. Remove the front floor brace (2WD).

11. Remove the front No. 1 stabilizer bracket LH (2WD).

a. Remove the 2 bolts and front No. 1 stabilizer bracket LH.

12. Remove the front No. 1 stabilizer bracket RH (2WD).

➡Perform the same procedure as for the LH side.

13. Separate the front stabilizer bar with the bracket (2WD).

a. Separate the front stabilizer bar with bracket from the front frame assembly.

14. Remove the steering link assembly (2WD).

a. Remove the 2 bolts, 2 nuts and steering link assembly.

➡Because the nut has its own stopper, do not turn the nut. Loosen the bolt with the nut secured.

b. Pull out the steering link assembly towards the left side of the vehicle while lifting the front stabilizer bar with bracket.

15. Remove the engine assembly with the transaxle (AWD).

16. Remove the front No. 1 stabilizer bracket LH (AWD).

17. Remove the front No. 1 stabilizer bracket RH (AWD).

18. Remove the front stabilizer bar with the bracket (AWD).

19. Remove the steering link assembly (AWD).

a. Remove the 2 bolts, 2 nuts and steering link assembly.

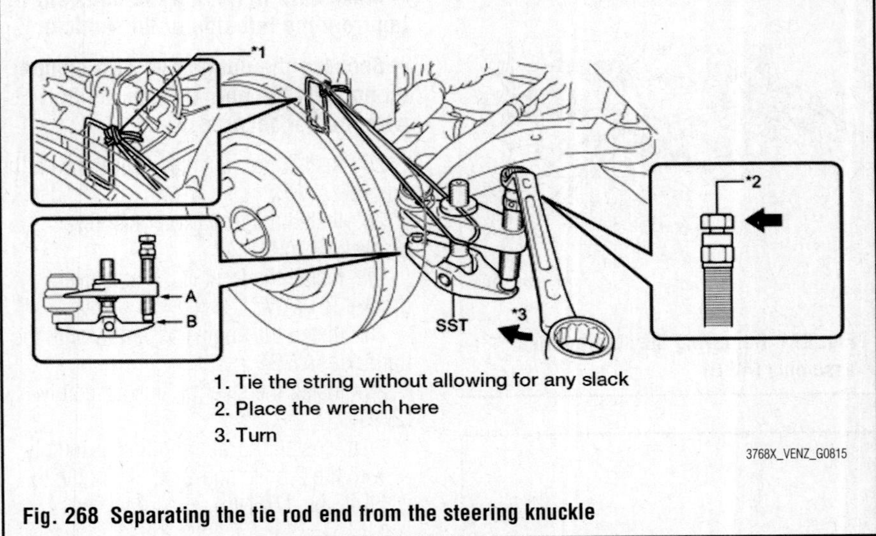

Fig. 268 Separating the tie rod end from the steering knuckle

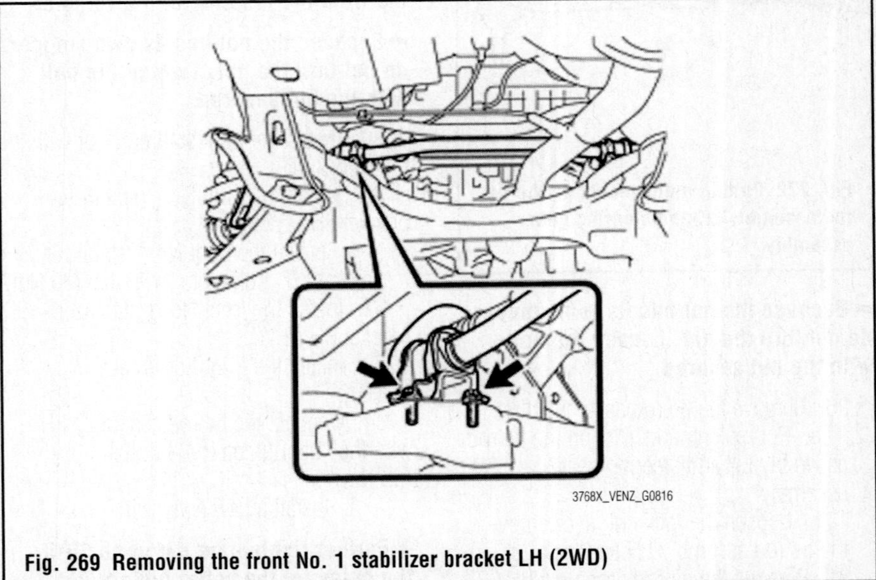

Fig. 269 Removing the front No. 1 stabilizer bracket LH (2WD)

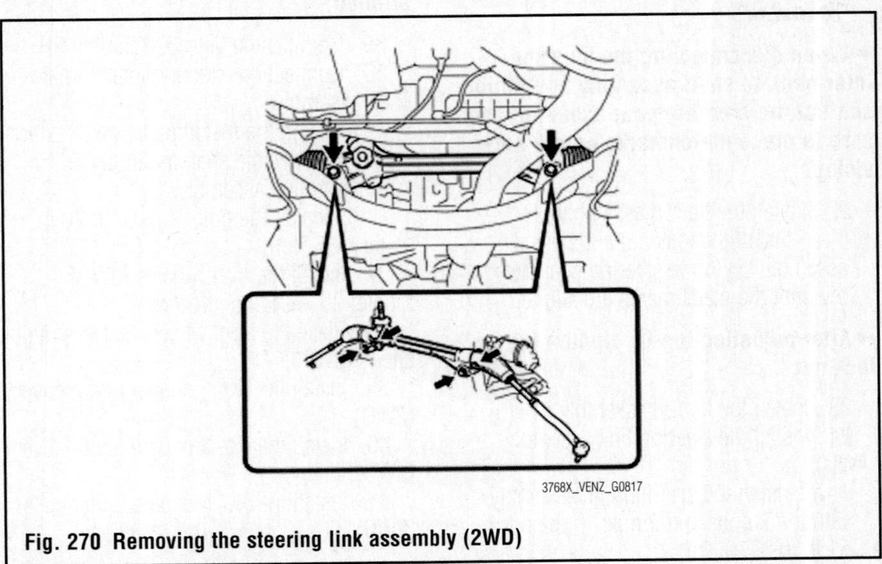

Fig. 270 Removing the steering link assembly (2WD)

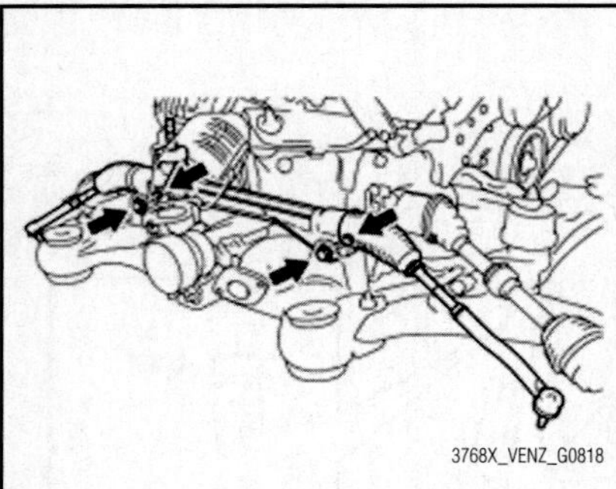

Fig. 271 Removing the steering link assembly (AWD)

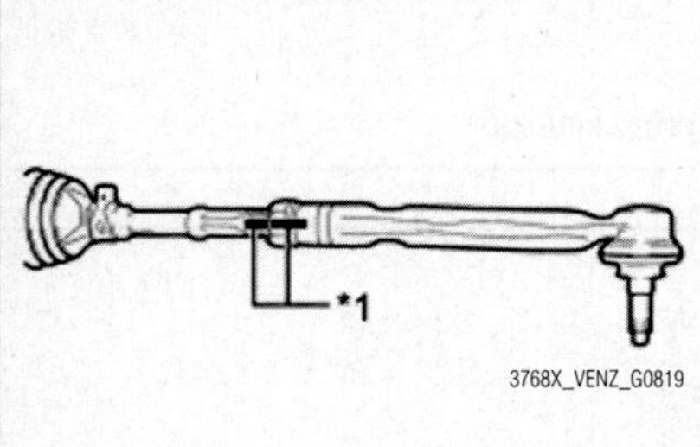

Fig. 272 Putting matchmarks on the tie rod assembly LH and steering gear assembly

➡**Because the nut has its own stopper, do not turn the nut. Loosen the bolt with the nut secured.**

20. Remove the tie rod assembly LH.
    a. Put matchmarks (*1) on the tie rod assembly LH and steering gear assembly.
    b. Loosen the lock nut, and remove the tie rod assembly LH and lock nut.
21. Remove the tie rod assembly RH.

To install:

➡**When disconnecting the steering intermediate shaft assembly and pinion shaft of the steering gear assembly, be sure to place matchmarks before servicing.**

22. Install the tie rod assembly LH.
    a. Install the lock nut and tie rod assembly LH to the steering gear assembly until the matchmarks are aligned.

➡**After adjusting toe-in, tighten the lock nut.**

23. Install the tie rod assembly RH.
24. Install the steering link assembly (AWD).
    a. Install the steering link assembly with the 2 bolts and 2 nuts. Tighten to 51 ft. lbs. (70 Nm).

➡**Make sure to tighten the bolts starting from the left side of the vehicle.**

➡**Because the nut has its own stopper, do not turn the nut. Tighten the bolt with the nut secured.**

25. Install the front stabilizer bar with the bracket (AWD).
26. Install the front No. 1 stabilizer bracket LH (AWD).
27. Install the front No. 1 stabilizer bracket RH (AWD).
28. Install the engine assembly with the transaxle (AWD).
29. Install the steering link assembly (2WD).
    a. Install the steering link assembly with the 2 bolts and 2 nuts. Tighten to 51 ft. lbs. (70 Nm).

➡**Make sure to tighten the bolts starting from the left side of the vehicle.**

➡**Because the nut has its own stopper, do not turn the nut. Tighten the bolt with the nut secured.**

30. Install the front stabilizer bar with the bracket (2WD).
31. Install the front No. 1 stabilizer bracket LH (2WD).
    a. Install the front No. 1 stabilizer bracket LH. Tighten to 21 ft. lbs. (29 Nm).
32. Install the front No. 1 stabilizer bracket RH (2WD).
33. Install the front floor brace (2WD).
34. Connect the tie rod assembly LH with the nut. Tighten to 36 ft. lbs. (49 Nm).
    a. Install a new cotter pin.

➡**Further tighten the nut up to 60°if the holes for the cotter pin are not aligned.**

35. Connect the tie rod assembly RH.
36. Connect the steering intermediate shaft assembly.
    a. Align the matchmarks on the steering intermediate shaft assembly and steering link assembly.
    b. Install the bolt. Tighten to 26 ft. lbs. (35 Nm).
37. Install the front wheels (2WD). Tighten to 76 ft. lbs. (103 Nm).
38. Place the front wheels facing straight ahead (2WD).
39. Install the No. 1 engine under cover (2WD).
40. Install the No. 2 engine under cover (2WD).
41. Install the No. 3 engine under cover (2WD).
42. Inspect and adjust the front wheel alignment (2WD).

3.5L Engine

See Figures 273 through 278.

➡**When disconnecting the steering intermediate shaft assembly and pinion shaft of steering gear assembly, be sure to place matchmarks before servicing.**

1. Remove the front exhaust pipe assembly.
2. Remove the propeller with the center bearing shaft assembly (AWD).
3. Secure the steering wheel in order to prevent it from rotating.

➡**This operation is necessary to prevent damage to the spiral cable.**

4. Separate the steering intermediate shaft assembly.

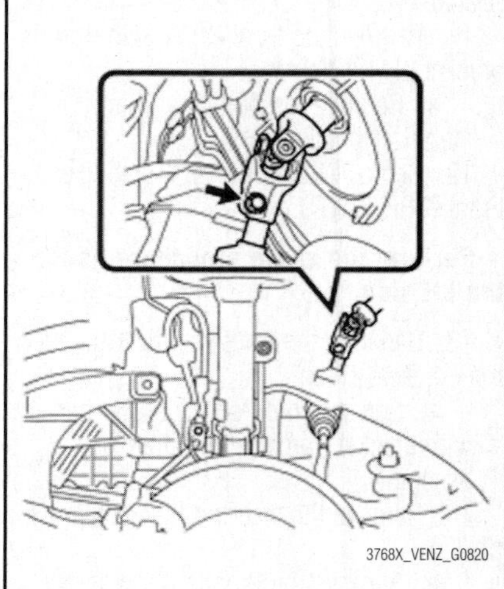

Fig. 273 Removing the bolt and sliding the steering intermediate shaft assembly

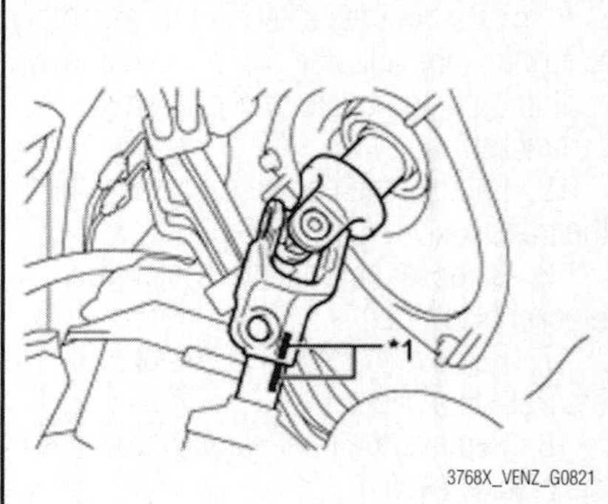

Fig. 274 Putting matchmarks on the steering intermediate shaft assembly and steering link assembly

Fig. 275 Removing the cotter pin and nut

➡When securing SST to the steering knuckle, be sure to tighten SST using a string to prevent it from falling.

➡Install SST so that A and B are parallel.

➡Be sure to place a wrench on the part indicated in the illustration.

➡Do not damage the front disc brake dust cover.

➡Do not damage the ball joint dust cover.

➡Do not damage the steering knuckle.

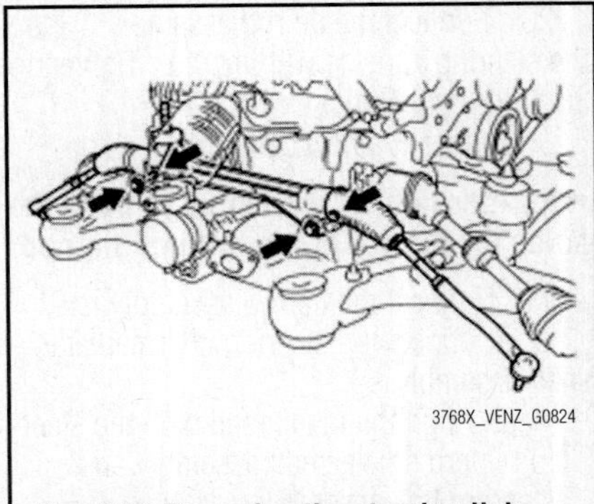

Fig. 277 Removing the steering link assembly

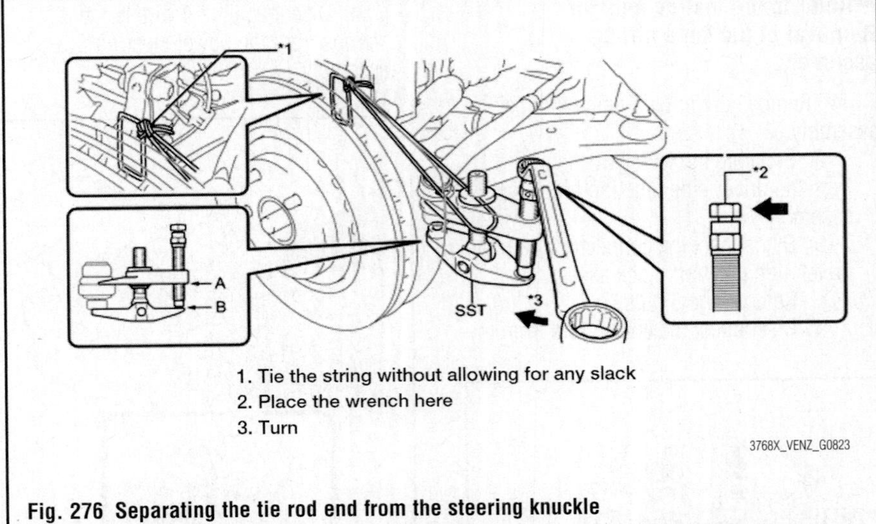
*1
*2
A
B
SST
*3
1. Tie the string without allowing for any slack
2. Place the wrench here
3. Turn

Fig. 276 Separating the tie rod end from the steering knuckle

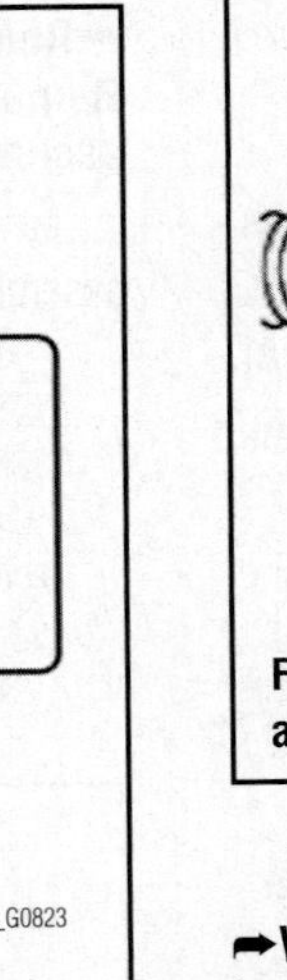
*1

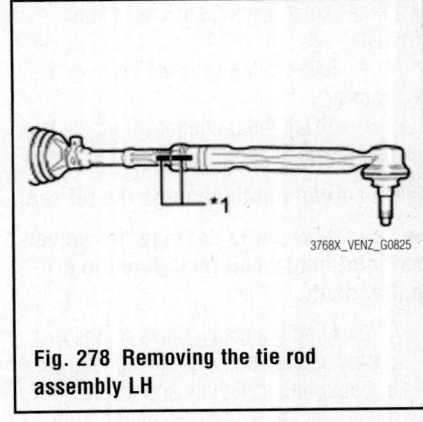

Fig. 278 Removing the tie rod assembly LH

a. Remove the bolt and slide the steering intermediate shaft assembly.

➡**Do not separate the steering intermediate shaft assembly from the steering link assembly.**

b. Put matchmarks on the steering intermediate shaft assembly and steering link assembly.

c. Separate the steering intermediate shaft assembly from the steering link assembly.

5. Separate the tie rod assembly LH.

a. Remove the cotter pin and nut.

b. Install the special tool (09960-20010) to the tie rod end.

➡**Make sure that the upper ends of the tie rod end and special tool are aligned.**

c. Using the special tool, separate the tie rod end from the steering knuckle.

⁂ CAUTION

Apply grease to the threads and end of the SST bolt.

6. Separate the tie rod assembly RH.

7. Remove the engine assembly with the transaxle.

8. Remove the front No. 1 stabilizer bracket LH.

9. Remove the front No. 1 stabilizer bracket RH.

10. Remove the front stabilizer bar.

11. Remove the steering link assembly.

a. Remove the 2 bolts, 2 nuts and steering link assembly.

➡**Because the nut has its own stopper, do not turn the nut. Loosen the bolt with the nut secured.**

12. Remove the tie rod assembly LH.

a. Put matchmarks (*1) on the tie rod assembly LH and steering rack end sub-assembly.

b. Loosen the lock nut, and remove the tie rod assembly LH and lock nut.

13. Remove the tie rod assembly RH.

To install:

➡**When disconnecting the steering intermediate shaft assembly and pinion shaft of steering gear assembly, be sure to place matchmarks before servicing.**

14. Install the tie rod assembly LH.

a. Install the lock nut and tie rod assembly LH to the steering rack end sub-assembly until the matchmarks are aligned.

➡**After adjusting toe-in, tighten the lock nut.**

15. Install the tie rod assembly RH.

16. Install the steering link assembly with the 2 bolts and 2 nuts. Tighten to 51 ft. lbs. (70 Nm).

➡**Make sure to tighten the bolts starting from the left side of the vehicle.**

➡**Because the nut has its own stopper, do not turn the nut. Tighten the bolt with the nut secured.**

17. Install the front stabilizer bar.

18. Install the front No. 1 stabilizer bracket LH and RH.

19. Install the engine assembly with the transaxle.

20. Connect the tie rod assembly LH to the steering knuckle with the nut. Tighten to 36 ft. lbs. (49 Nm).
    a. Install a new cotter pin.

➡Further tighten the nut up to 60°if the holes for the cotter pin are not aligned.

21. Connect the tie rod assembly RH.
22. Connect the steering intermediate shaft assembly.
    a. Align the matchmarks on the steering intermediate shaft assembly and steering link assembly.
    b. Install the bolt. Tighten to 26 ft. lbs. (35 Nm).
23. Temporarily tighten the propeller with the center bearing shaft assembly (AWD).
    a. Remove the special tool from the transfer.
    b. Install the propeller with the center bearing shaft assembly.

➡Be careful not to damage the oil seal.

➡Be careful not to damage the universal joint boot when installing the propeller shaft.

c. Align the matchmarks on the rear propeller shaft and electromagnetic control coupling assembly and install the 4 nuts and 4 washers temporarily.

➡Do not allow grease to adhere to be bolts or washers.

d. Temporarily install the propeller with center bearing shaft assembly with the 4 bolts, 2 No. 1 center support bearing washers and 2 No. 2 center support bearing washers.

➡Reuse the washers.

➡Do not allow grease to adhere to be bolts or washers.

e. Fully tighten the 4 nuts to 27 ft. lbs. (37 Nm).

24. Install the front exhaust pipe assembly.

POWER STEERING ECU

REMOVAL & INSTALLATION

See Figures 279 through 285.

⁂ CAUTION

Some of these service operations affect the SRS airbag system. Read the precautionary notices concerning the SRS airbag system before servicing.

➡Be sure to read "Precaution" thoroughly before servicing.

1. Place front wheels facing straight ahead.
2. Disconnect cable from negative battery terminal.

⁂ CAUTION

Wait at least 90 seconds after disconnecting the cable from the negative (-) battery terminal to disable the SRS system.

➡When disconnecting the cable, some systems need to be initialized after the cable is reconnected.

3. Remove driver side knee airbag assembly.

➡Refer to the instructions for Removal of the knee airbag assembly.

4. Remove driver side junction block assembly
    a. Separate the wire harness clamp from the driver side junction block assembly.
    b. Disconnect the connectors from the driver side junction block assembly.
    c. Remove the 3 nuts.
    d. Disconnect the connectors from the back of the driver side junction block assembly.
    e. Disengage the 2 claws to remove the driver side junction block assembly.
5. Remove the power steering ECU.
    a. Disconnect the 3 connectors from the power steering ECU assembly.
    b. Disconnect the connector from the power steering ECU assembly.

➡Pull out the lock of the lock lever, disengages the claw, and raise the lock lever to disconnect the connector as shown in the illustration.

c. Remove the 3 nuts and power steering ECU assembly.
d. Disengage the 4 wire harness clamps from the power steering ECU assembly

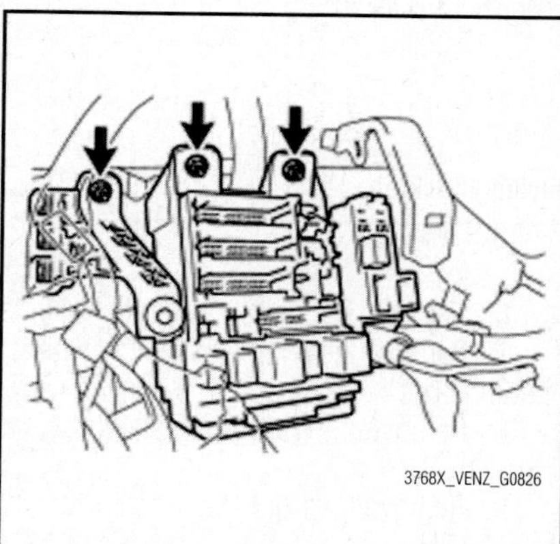

Fig. 279 Removing the 3 nuts

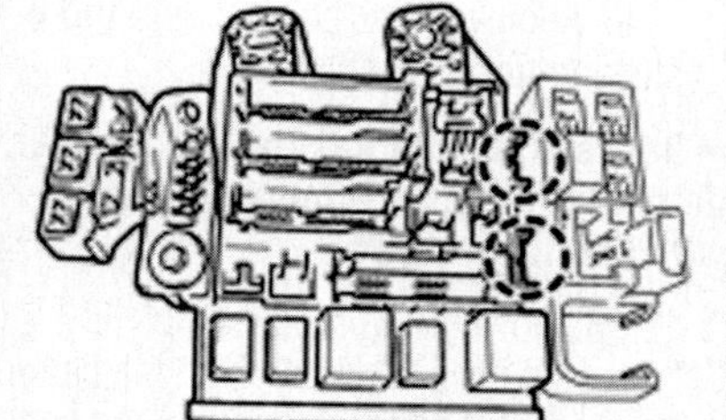

Fig. 280 Disengaging the 2 claws to remove the driver side junction block assembly

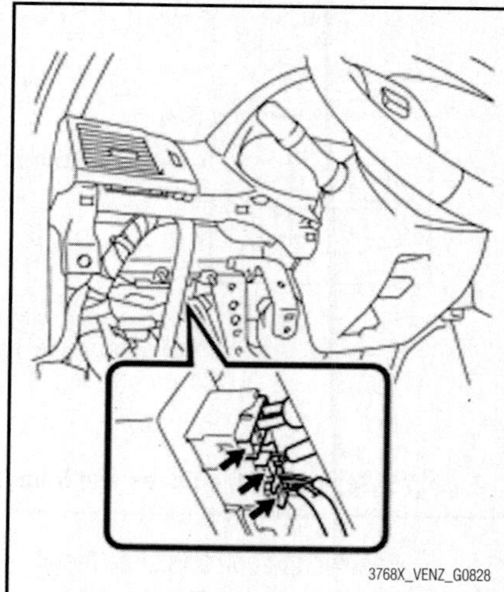

Fig. 281 Disconnecting the 3 connectors from the power steering ECU assembly

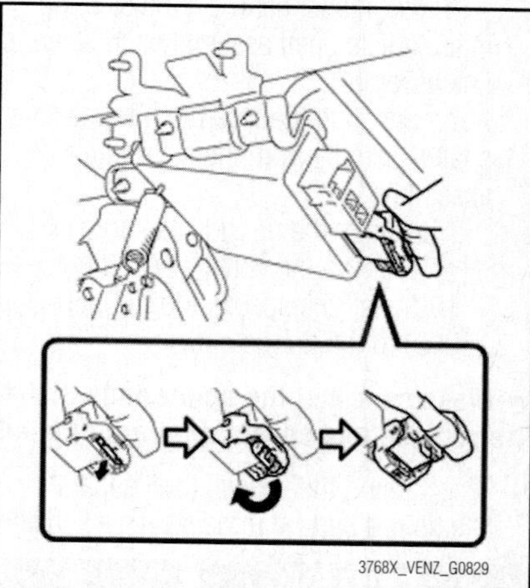

Fig. 282 Disconnecting the connector from the power steering ECU assembly

To install:

6. Install the power steering ECU assembly.

a. Engage the 4 wire harness clamps to the power steering ECU assembly.

b. Install the power steering ECU assembly with the 3 nuts. Tighten to 10 ft. lbs. (14 Nm).

c. Securely connect the connector to the power steering ECU assembly.

➡Return the lock lever to its original position to connect the connector, and then securely push in the lock of the lock lever as shown in the illustration.

d. Connect the 3 connectors to the power steering ECU assembly.

7. Install the driver side junction block assembly.

a. Engage the 2 claws to install the driver side junction block assembly.

b. Connect the connectors to the back of the driver side junction block assembly.

c. Install the driver side junction block assembly with the 3 nuts. Tighten to 71 inch lbs. (8 Nm).

d. Connect the connectors to the driver side junction block assembly.

8. Install the driver side knee airbag assembly.

9. Connect the cable to the negative battery terminal.

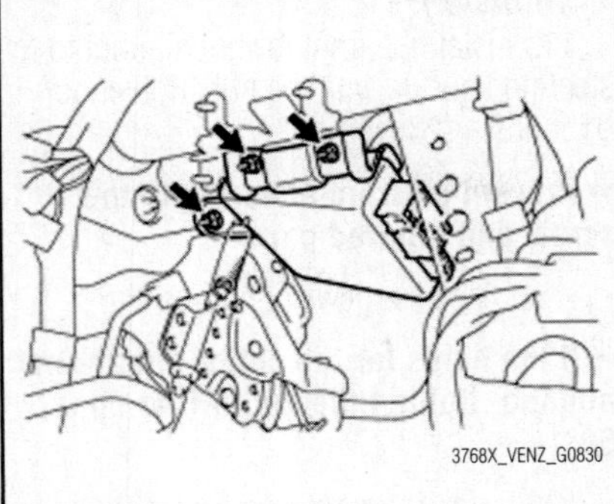

Fig. 283 Removing the 3 nuts and power steering ECU assembly

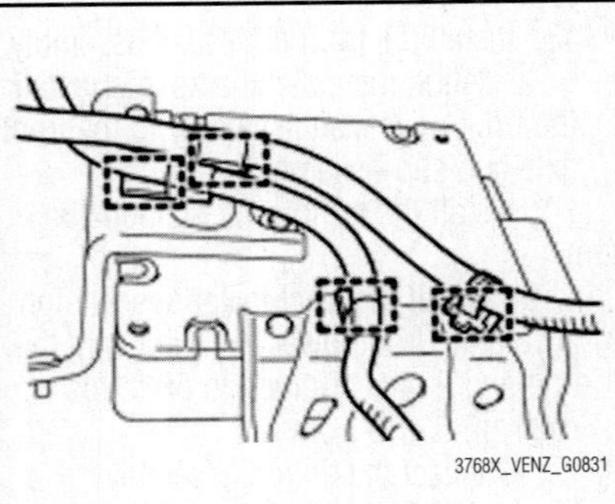

Fig. 284 Disengaging the 4 wire harness clamps from the power steering ECU assembly

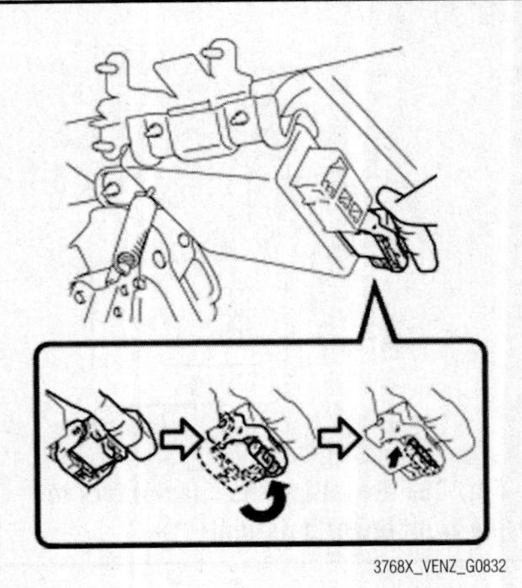

Fig. 285 Connecting the connector to the power steering ECU assembly

➡When disconnecting the cable, some systems need to be initialized after the cable is reconnected.

10. Initialize the rotation angle sensor and calibrate the torque sensor zero point.

a. If replacing the power steering ECU assembly, clear the rotation angle sensor calibration value, initialize the rotation angle sensor, and calibrate the torque sensor zero point.

SUSPENSION

FRONT SUSPENSION

LOWER BALL JOINT

REMOVAL & INSTALLATION

See Figures 286 through 289.

➡Use the same procedure for the LH side and RH side.

➡The following procedure listed is for the LH side.

1. Remove the front wheel.
2. Remove the front axle shaft nut.
3. Separate the front speed sensor.

➡Be sure to completely separate the front speed sensor from the front shock absorber with coil spring.

➡Be careful not to damage the front speed sensor.

➡Clean the speed sensor installation hole and the surfaces every time the speed sensor are removed.

4. Separate the front driveshaft assembly.
5. Separate the front disc brake caliper assembly.
6. Remove the front disc.
7. Separate the tie rod assembly.
8. Separate the front lower suspension arm.

a. Remove the bolt, 2 nuts, and separate the front lower suspension arm from the lower ball joint.

9. Remove the front axle assembly.

a. Remove the 2 bolts, 2 nuts and front axle assembly.

➡When removing the nuts, keep the bolts from rotating.

10. Remove the front lower ball joint.

a. Secure the front axle assembly in a vise using aluminum plates.

➡When using a vise, do not over-tighten it.

b. Remove the cotter pin and nut.

Fig. 286 Separating the front lower suspension arm

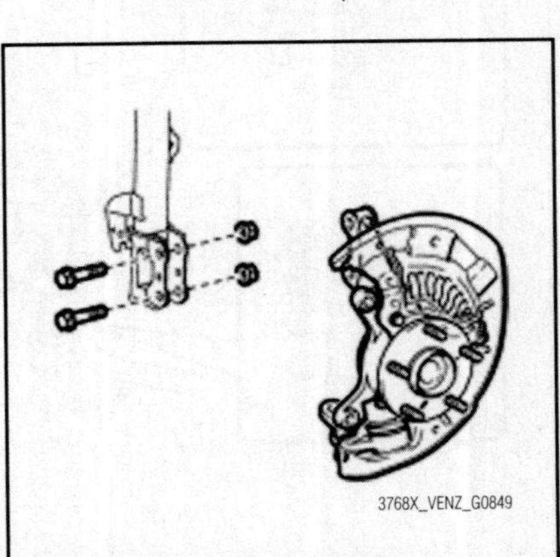

Fig. 287 Removing the front axle assembly

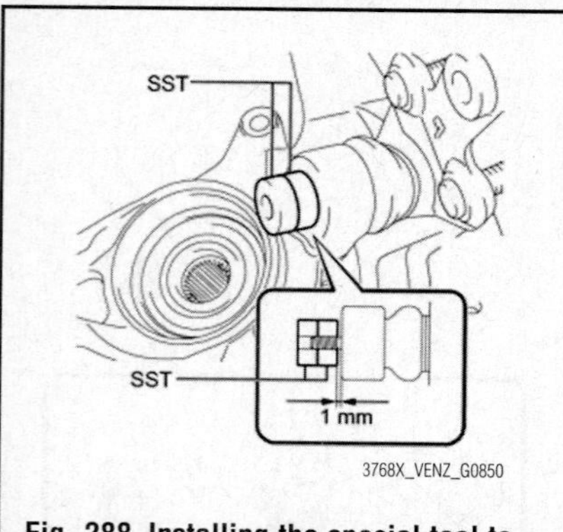

Fig. 288 Installing the special tool to the front lower ball joint

c. Install the special tool (09960-20010) to the front lower ball joint.

➡Check that the clearance measurement between SST and the front axle assembly is 1 mm (0.0394 in.).

d. Using SST, remove the front lower ball joint from the front axle assembly as shown in the illustration.

⁂ CAUTION

Apply grease to the threads and end of the SST bolt.

➡Install SST so that A and B are parallel.

➡Be sure to place a wrench on the part indicated in the illustration.

➡Do not damage the front lower ball joint dust cover, or steering knuckle.

To install:

11. Install the front lower ball joint to the steering knuckle with the nut. Tighten to 91 ft. lbs. (123 Nm).

➡Prevent oil from adhering to the screw and tapered parts.

a. Install a new cotter pin.

➡If the holes for the cotter pin are not aligned, tighten the nut further up to 60°.

12. Install the front axle assembly.
 a. Install the front axle assembly to the front shock absorber with the 2 bolts and 2 nuts. Tighten to 214 ft. lbs. (290 Nm).
13. Install the front driveshaft assembly.
 a. Align the matchmarks and install the front driveshaft assembly to the front axle hub sub-assembly.
14. Install the front lower suspension arm.
 a. Install the front lower suspension arm to the front lower ball joint with the bolt and 2 nuts. Tighten to 68 ft. lbs. (92 Nm).
15. Connect the tie rod assembly.
16. Install the front disc.
17. Install the front disc brake caliper assembly to the steering knuckle with the 2 bolts. Tighten to 77 ft. lbs. (104 Nm).
18. Install the front speed sensor. Tighten the bolt to 71 inch lbs. (8 Nm).

➡Prevent foreign matter from attaching to the sensor tip.

➡Firmly insert the sensor body into the knuckle before tightening the bolt.

➡After installing the sensor to the knuckle, make sure that there is no clearance between the sensor stay and knuckle. Also make sure that no foreign matter is attached between the parts.

➡To prevent interference between the sensor and magnetic rotor, do not rotate the sensor body during or after the insertion of the sensor body to the knuckle.

19. Install the front axle shaft nut.
20. Check the ABS speed sensor signal.
21. Install the front wheel. Tighten to 76 ft. lbs. (103 Nm).
22. Inspect and adjust the front wheel alignment.

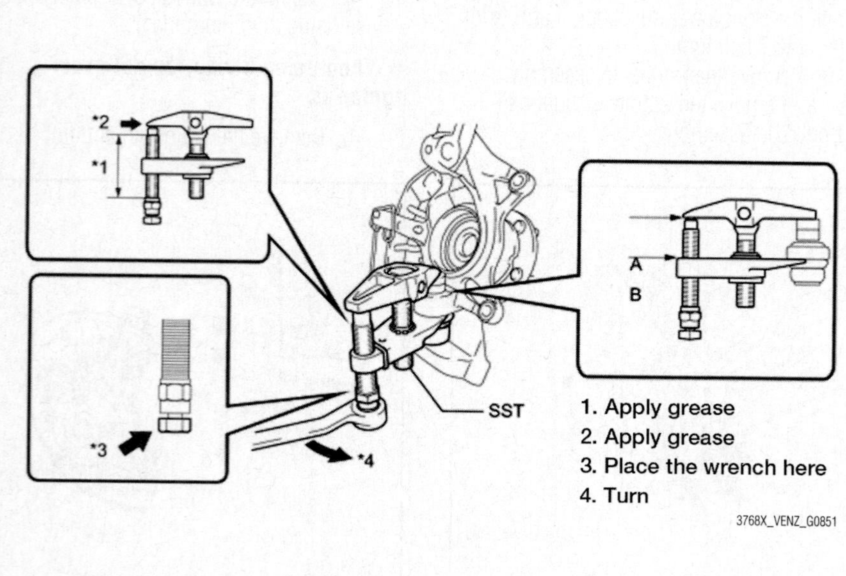

Fig. 289 Removing the front lower ball joint

LOWER CONTROL ARM

REMOVAL & INSTALLATION

See Figures 290 and 291.

1. Remove the engine assembly with the transaxle.
2. Remove the front No. 1 stabilizer bracket LH and RH (AWD).
3. Remove the front stabilizer bar with the front stabilizer link assembly (AWD).
4. Remove the steering link assembly (AWD).
5. Remove the front frame assembly (3.5L engine).
6. Separate the rear engine mounting insulator (2.7L engine).
7. Remove the front frame assembly (2.7L engine).
8. Remove the engine mounting insulator LH.
9. Remove the front lower suspension arm.
 a. Remove the 3 bolts, nut and front lower suspension arm from the front frame assembly.

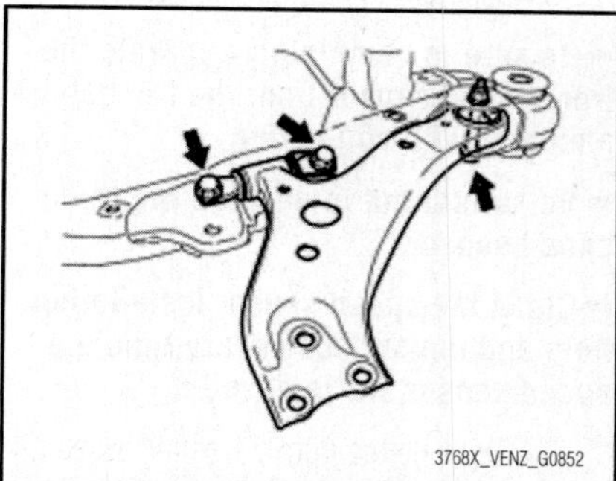

Fig. 290 Removing the front lower suspension arm

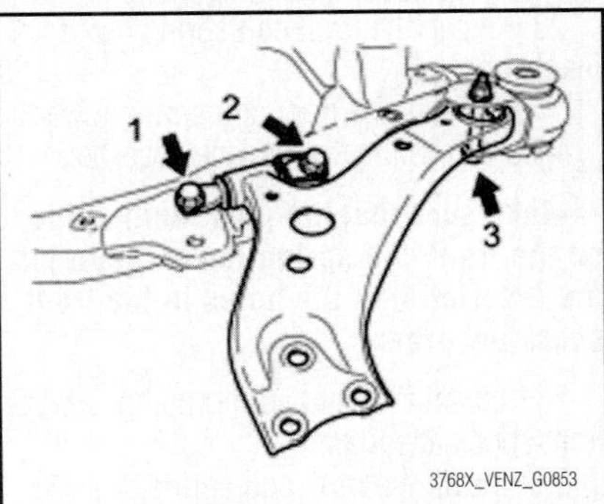

Fig. 291 Installing the front lower suspension arm

b. Remove the front lower arm bushing stopper from the front lower suspension arm.

To install:

10. To install, reverse the removal procedure. Tighten the bolts in the order shown. Tighten bolts 1 and 2 to 147 ft. lbs. (200 Nm). Tighten bolt 3 to 152 ft. lbs. (206 Nm).

➡Start installing the bolts from the front of the vehicle.

SHOCK ABSORBERS

REMOVAL & INSTALLATION

See Figures 292 through 303.

➡Use the same procedure for the LH side and RH side.

➡The following procedure listed below is for the LH side.

1. Remove the front wheel.
2. Remove the front wiper arm head cap.
3. Remove the front wiper arm and blade assemblies LH and RH.
4. Loosen the front suspension support nut.
   a. Loosen the front suspension nut of the front shock absorber.

➡Do not remove the front suspension support nut.

➡Loosen the nut only when the front shock absorber with coil spring needs to be disassembled.

5. Remove front fender to cowl side seal LH
6. Remove front fender to cowl side seal RH.
7. Remove cowl top ventilator louver sub-assembly.
8. Remove windshield wiper motor and link assembly.
9. Remove outer cowl top panel.

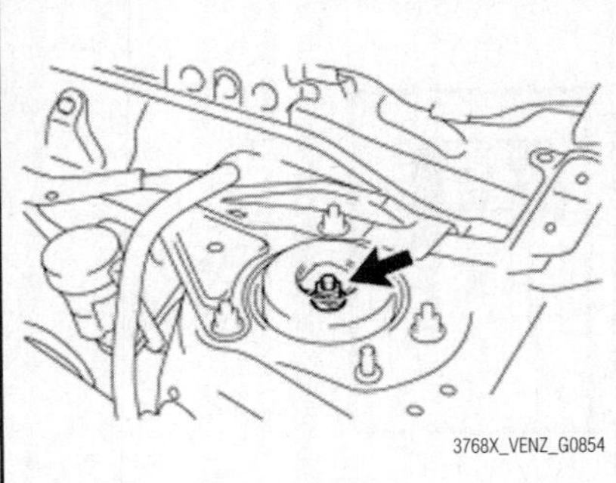

Fig. 292 Locating the front suspension support nut

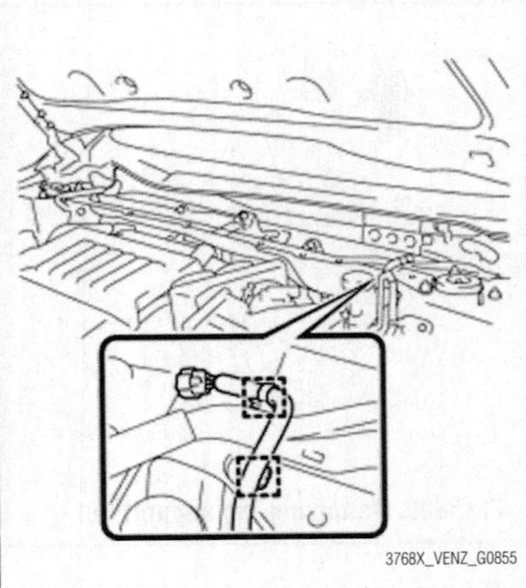

Fig. 293 Disengaging the 2 clamps and separating the wiper wire harness from the outer cowl top panel sub-assembly

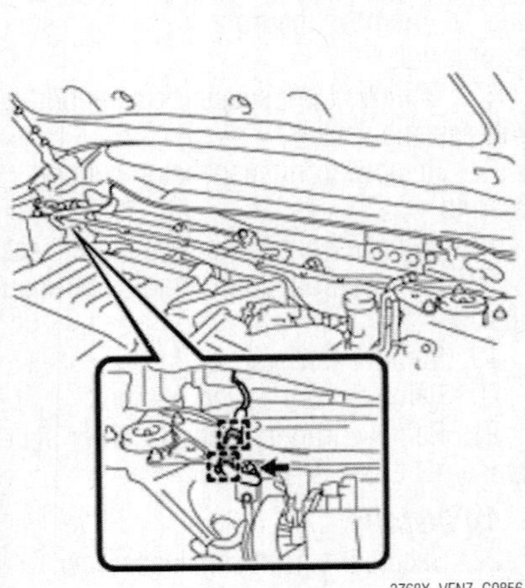

Fig. 294 Disengaging the 2 clamps and connector, and separating the wire harness from the outer cowl top panel sub-assembly (w/ Windshield Deicer)

   a. Disengage the 2 clamps and separate the wiper wire harness from the outer cowl top panel sub-assembly.
   b. Disengage the 2 clamps and connector, and separate the wire harness from the outer cowl top panel sub-assembly (w/ Windshield Deicer).
   c. Remove the 4 bolts, 4 nuts and outer cowl top panel sub-assembly.

10. Separate the front speed sensor.
   a. Remove the bolt and clamp, and separate the front speed sensor and front flexible hose.
11. Separate the front stabilizer link assembly.
   a. Remove the nut and separate the front stabilizer link assembly from the front shock absorber.

➡If the ball joint turns together with the nut, use a hexagon wrench (6 mm) to hold the stud bolt.

12. Remove the front shock absorber with the coil spring.
   a. Support the front axle using a jack and wooden block.
   b. Remove the 2 bolts and 2 nuts, and separate the front shock absorber with coil spring (lower side) from the steering knuckle.

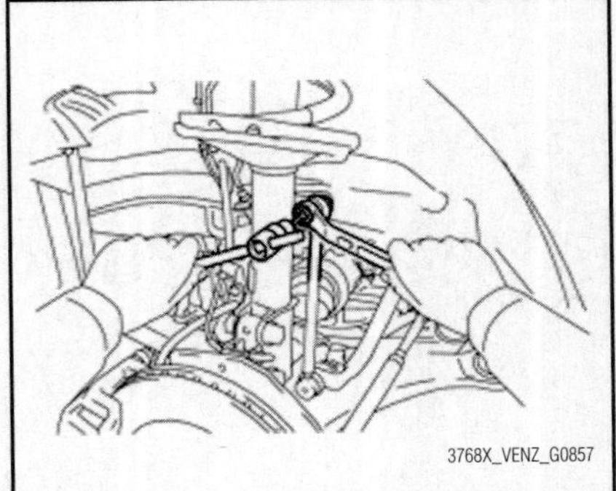

Fig. 295 Separating the front stabilizer link

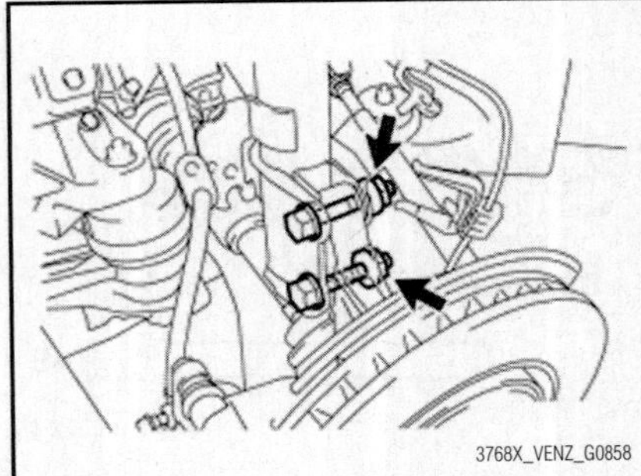

Fig. 296 Separating the front shock absorber with the coil spring from the steering knuckle

➡When removing the nuts, keep the bolts from rotating.

c. Remove the nut (*1) and 2 spacers (*2) on the front shock absorber with coil spring (upper side).

➡Make sure that the front speed sensor is completely separated from the front shock absorber with coil spring.

13. Secure the front shock absorber with the coil spring.

a. Install the special tool (09727-00050) to the front coil spring with the hooks spread as far apart as possible from each other.

➡Make sure that the claws on the hooks are securely engaged to the spring.

b. Install the bolt and nut to the front shock absorber as shown in the illustration and secure the front shock absorber in a vise using aluminum plates.

A: 1.10 inch (28 mm).

14. Remove the front suspension support nut.

a. Using the special tool (09727-00050), compress the front coil spring.

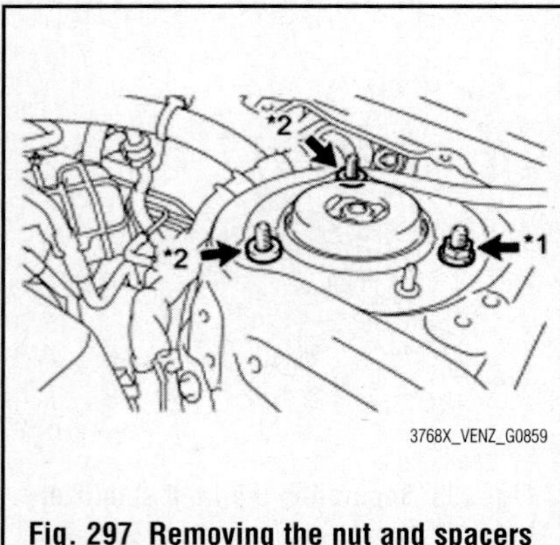

Fig. 297 Removing the nut and spacers

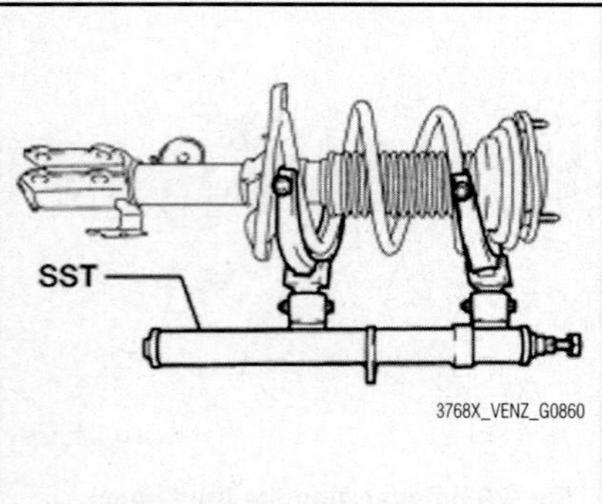

Fig. 298 Securing the front shock absorber

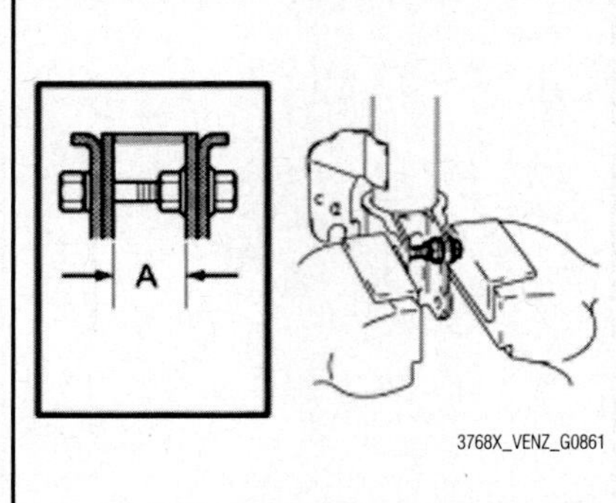

Fig. 299 Installing the bolt and nut to the front shock absorber

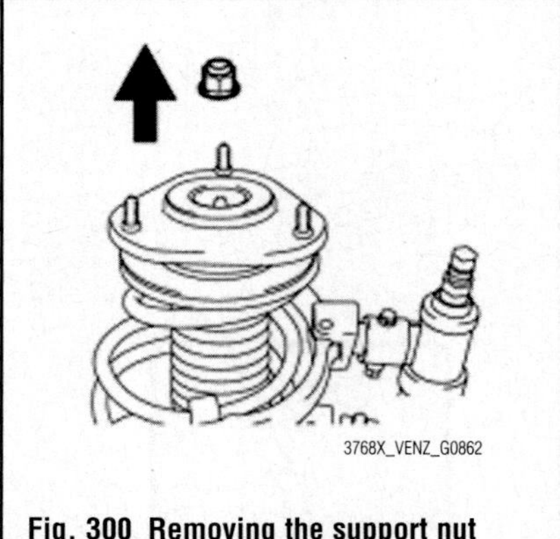

Fig. 300 Removing the support nut

➡Do not use an impact wrench. It will damage the special tool.

➡If the front coil spring is compressed at an angle, using 2 special tools will make the work easier.

b. Check that the front coil spring is fully compressed.

c. Remove the front suspension support nut.

15. Remove front suspension support sub-assembly.

16. Remove front suspension support bearing.

17. Remove front coil spring upper seat.

18. Remove front coil spring upper insulator.

19. Remove front coil spring.

20. Remove front spring bumper.

21. Remove front coil spring lower insulator.

To install:

22. Secure the front shock absorber assembly.

Install the bolt and nut to the front shock absorber assembly as shown in the illustration and secure the front shock absorber assembly in a vise using aluminum plates.

A: 1.10 inch (28 mm).

23. Install the front coil spring lower insulator.

a. Install the front coil spring lower insulator to the front shock absorber.

➡Make sure that the positioning pins on the front coil spring lower insulator are inserted into the holes in the front shock absorber.

24. Install the front spring bumper to the front shock absorber.

25. Install the front coil spring.

a. Install the special tool (09727-00050) to the front coil spring with the hooks spread as far apart as possible from each other.

➡Make sure that the claws on the hooks are securely engaged to the spring.

b. Using the special tool, compress the front coil spring.

➡Do not use an impact wrench. It will damage the special tool.

c. Install the front coil spring to the front shock absorber.

➡Make sure that the end of the fort coil spring is positioned in the depression of the lower spring seat.

26. Install the front coil spring upper insulator.

a. Install the front coil spring upper insulator as shown in the illustration.

➡Any misalignment between the front shock absorber lower bracket and the alignment mark must be +/- 5°.

27. Install the front coil spring upper seat with the mark facing outside of the vehicle.

➡Any misalignment between the front shock absorber lower bracket and the alignment mark must be +/- 5°.

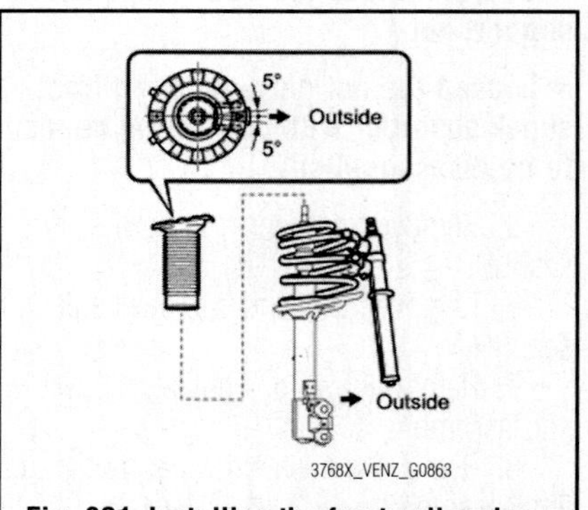

Fig. 301 Installing the front coil spring upper insulator

28. Install the front suspension support bearing.

➡**If there is foreign matter inside the front suspension support bearing, replace it with a new one.**

29. Install the front suspension support sub assembly.
   a. Install the front suspension support sub-assembly as shown in the illustration.

➡**Check that the slots on the piston rod and front suspension support sub-assembly are aligned.**

   b. Temporarily tighten a new front suspension support nut.
30. Install the front shock absorber with the coil spring.
   a. Install the front shock absorber with coil spring (upper side) with the nut and 2 spacers. Tighten to 63 ft. lbs. (85 Nm).
   b. Install the front shock absorber with coil spring (lower side) to the steering knuckle and insert the 2 bolts and 2 nuts. Tighten to 214 ft. lbs. (290 Nm).

➡**When installing the nuts, keep the bolts from rotating.**

31. Install the front stabilizer link assembly to the front shock absorber with the nut. Tighten to 55 ft. lbs. (74 Nm).

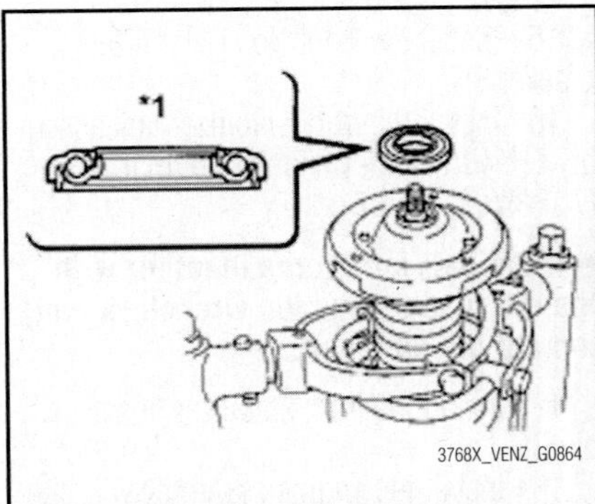

Fig. 302 Installing the front suspension support bearing

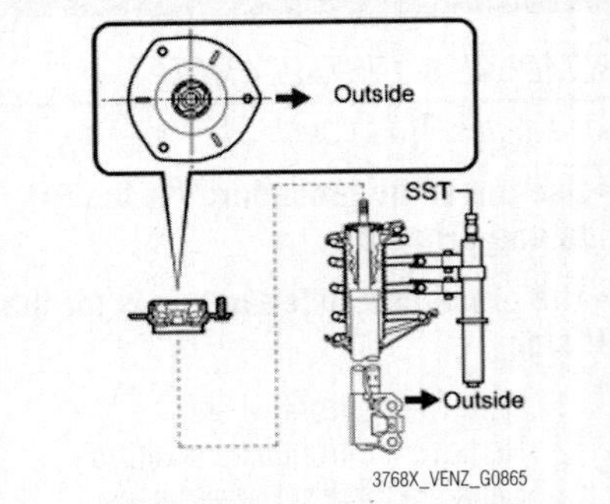

Fig. 303 Installing the front suspension support sub assembly

➡**If the ball joint turns together with the nut, use a hexagon wrench (6 mm) to hold the stud bolt.**

32. Install the front speed sensors. Tighten to 14 ft. lbs. (19 Nm).

➡**Do not twist the front speed sensor when installing it.**

   a. Install the clamp.
33. Install the outer cowl top panel with the 4 bolts and 4 nuts. Tighten the nut to 63 ft. lbs. (85 Nm). Tighten the bolt to 78 inch lbs. (8.8 Nm).
   a. Engage the 2 wire harness clamps to the outer cowl top panel.
   b. Engage the 2 wire harness clamps to the outer cowl top panel and connect the connector (w/ Windshield Deicer).
34. Fully tighten the front suspension support nut. Tighten to 52 ft. lbs. (70 Nm).
35. Install the windshield wiper motor and link assembly.
36. Install the cowl top ventilator louver sub assembly.
37. Install the front fender to cowl side seal LH and RH.
38. Install the front wiper arm and blade assemblies LH and RH.
39. Install the front wiper arm head cap.
40. Install the front wheel. Tighten to 76 ft. lbs. (103 Nm).
41. Inspect and adjust the front wheel alignment.

STABILIZER BAR

REMOVAL & INSTALLATION

2.7L Engine

See Figures 304 through 306.

1. Place the front wheels facing straight ahead.
2. Remove the front wheels.
3. Remove the front stabilizer link assembly LH.
   a. Remove the 2 nuts and separate the front stabilizer link assembly LH.

➡**If the ball joint turns together with the nut, use a hexagon wrench (6 mm) to hold the stud bolt.**

4. Remove the front stabilizer link assembly RH.

➡**Perform the same procedure as the LH side.**

5. Remove the steering link assembly (2WD).

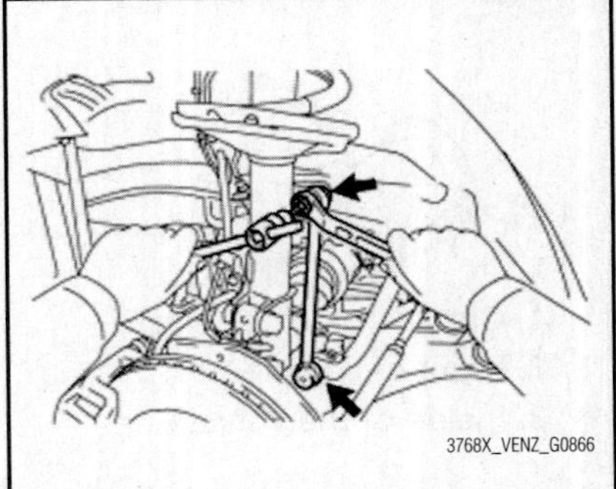

Fig. 304 Removing the front stabilizer link assembly LH

6. Remove the front stabilizer bar (2WD).
   a. Remove the front stabilizer bar from the left side of the vehicle.
7. Remove the engine assembly with the transaxle (AWD).
8. Remove the front No. 1 stabilizer bracket LH (AWD).
   a. Remove the 2 bolts and front No. 1 stabilizer bracket LH from the front frame assembly.
9. Remove the front No. 1 stabilizer bracket RH (AWD).
10. Remove the front stabilizer bar (AWD).
11. Remove the front No. 2 stabilizer bracket LH.
   a. Remove the front No. 2 stabilizer bracket LH from the No. 1 front stabilizer bar bushing.
12. Remove the front No. 2 stabilizer bracket RH.
   a. Remove the front No. 2 stabilizer bracket RH from the No. 1 front stabilizer bar bushing.
13. Remove the front No. 1 stabilizer bar bushings from the front stabilizer bar.

To install:

14. Install the front No. 1 stabilizer bar bushing.

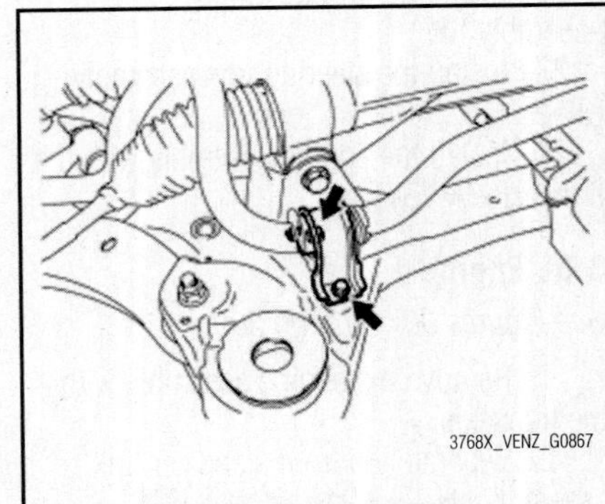

Fig. 305 Removing the front No. 1 stabilizer bracket LH (AWD)

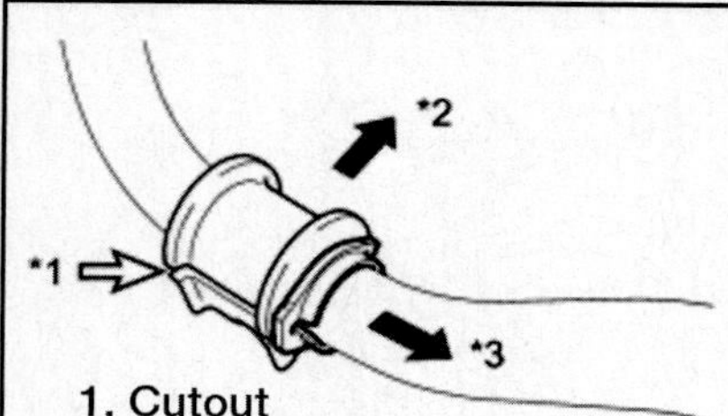

Fig. 306 Installing the front No. 1 stabilizer bar bushing

a. Install the 2 front No. 1 stabilizer bar bushings to the front stabilizer bar as shown in the illustration.

➡When installing the front No. 1 stabilizer bar bushings, make sure that the cutout faces the rear of the vehicle.

15. Install the front No. 2 stabilizer bracket LH to the front No. 1 stabilizer bar bushing.
16. Install the front No. 2 stabilizer bracket RH.
17. Install the front stabilizer bar (AWD).
18. Install the front No. 1 stabilizer bracket LH (AWD).
    a. Install the front No. 1 stabilizer bracket LH to the front frame assembly with the 2 bolts. Tighten to 21 ft. lbs. (29 Nm).
19. Install the front No. 1 stabilizer bracket RH (AWD).
20. Install the front stabilizer bar (2WD).
    a. Install the front stabilizer bar from the left side of the vehicle.
21. Install the front stabilizer link assembly LH with the 2 nuts. Tighten to 55 ft. lbs. (74 Nm).

➡If the ball joint turns together with the nut, use a hexagon wrench (6 mm) to hold the stud bolt.

22. Install the front stabilizer link assembly RH.
23. Install the steering link assembly (2WD).
24. Install the engine assembly with the transaxle (AWD).

3.5L Engine

See Figures 307 through 309.

1. Remove the engine assembly with the transaxle.
2. Separate the front stabilizer link assembly LH.
    a. Remove the nut and separate the front stabilizer link assembly LH.

➡If the ball joint turns together with the nut, use a hexagon wrench (6 mm) to hold the stud bolt.

3. Separate the front stabilizer link assembly RH.
4. Remove the front No. 1 stabilizer bracket LH.
    a. Remove the 2 bolts and No. 1 front stabilizer bracket LH from the front frame assembly.
5. Remove the No. 1 stabilizer bracket RH.
6. Remove the front stabilizer bar.
7. Remove the No. 2 stabilizer bracket LH.
    a. Remove the No. 2 front stabilizer bracket LH from the front stabilizer bar bushing.
8. Remove the front No. 2 stabilizer bracket RH.
9. Remove the front No. 1 stabilizer bar bushings from the front stabilizer bar.

To install:

10. Install the front No. 1 stabilizer bar bushing.
    a. Install the 2 front No. 1 stabilizer bar bushings to the front stabilizer bar as shown in the illustration.

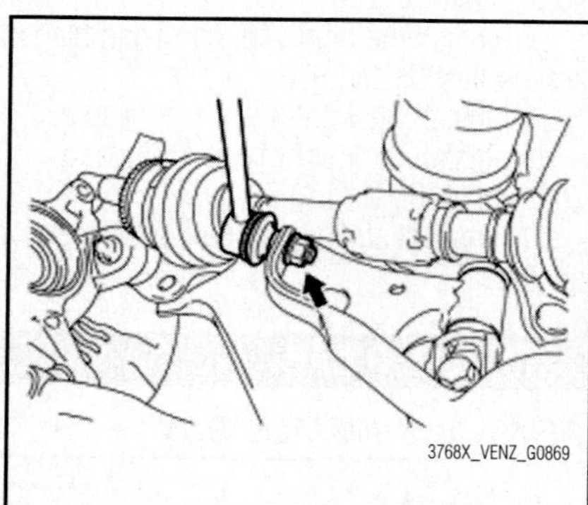

Fig. 307 Separating the front stabilizer link

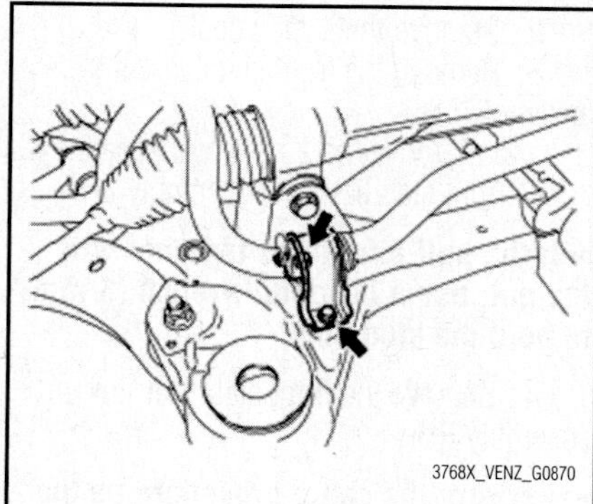

Fig. 308 Removing the front No. 1 stabilizer bracket LH

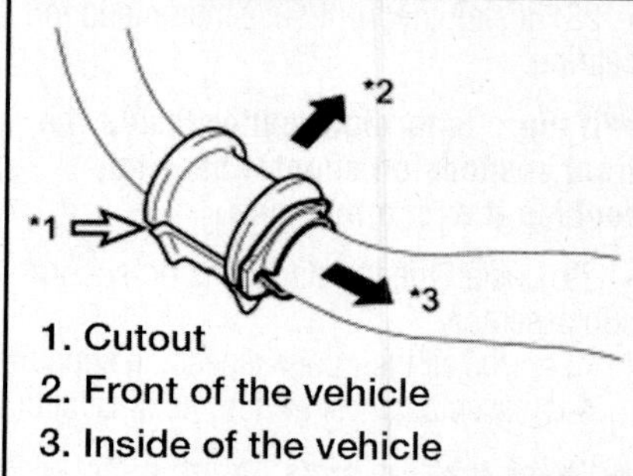

Fig. 309 Installing the front No. 1 stabilizer bar bushing

➡When installing the front No. 1 stabilizer bar bushings, make sure that the cutout faces the rear of the vehicle.

11. Install the front No. 2 stabilizer bracket LH to the front No. 1 stabilizer bar bushing.
12. Install the front No. 2 stabilizer bracket RH.
13. Install the front stabilizer bar to the front frame assembly.
14. Install the front No. 1 stabilizer bracket LH.
    a. Install the front No. 1 stabilizer bracket LH to the front frame assembly with the 2 bolts. Tighten to 21 ft. lbs. (29 Nm).
15. Install the front No. 1 stabilizer bracket RH.
16. Install the front stabilizer link assembly LH with the nut. Tighten to 55 ft. lbs. (74 Nm).

➡If the ball joint turns together with the nut, use a hexagon wrench (6 mm) to hold the stud bolt.

17. Install the front stabilizer link assembly RH.
18. Install the engine assembly with the transaxle.

WHEEL BEARING & HUB ASSEMBLY

REMOVAL & INSTALLATION

See Figures 310 through 317.

➡Use the same procedure for the RH side and LH side.

➡The procedure listed below is for the LH side.

1. Remove the front wheel.
2. Remove the front axle shaft nut.
3. Separate the front speed sensor.
    a. Remove the bolt and resin clamp, and separate the front speed sensor.

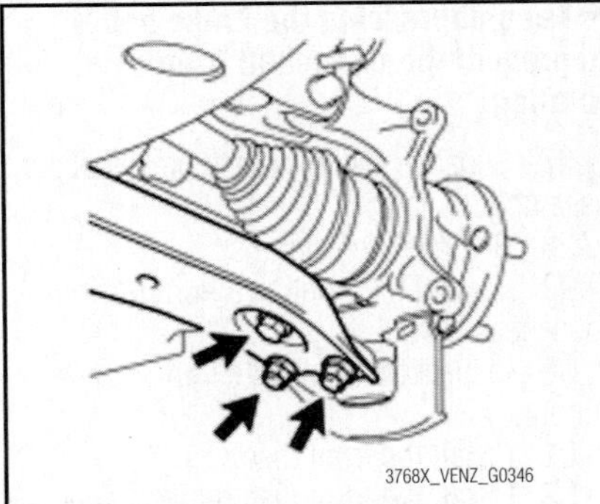

Fig. 310 Separating the front lower suspension arm

➡**Prevent foreign matter from attaching to the front speed sensor tip.**

➡**Clean the speed sensor installation hole and the contact surfaces every time the front speed sensor is removed.**

4. Separate the front disc brake caliper assembly.
5. Remove the front disc.
6. Separate the tie rod assembly.
7. Separate the front lower suspension arm.
 a. Remove the bolt and 2 nuts, and separate the front lower suspension arm from the front lower ball joint.
8. Separating the front driveshaft assembly.
 a. Put matchmarks on the front driveshaft assembly and front axle hub assembly sub assembly.
 b. Using a plastic hammer, separate the front driveshaft assembly from the front axle assembly. If it is difficult to separate, tap the end of the front driveshaft assembly using a brass bar and a hammer.

➡**Be careful not to damage the driveshaft boot and speed sensor rotor.**

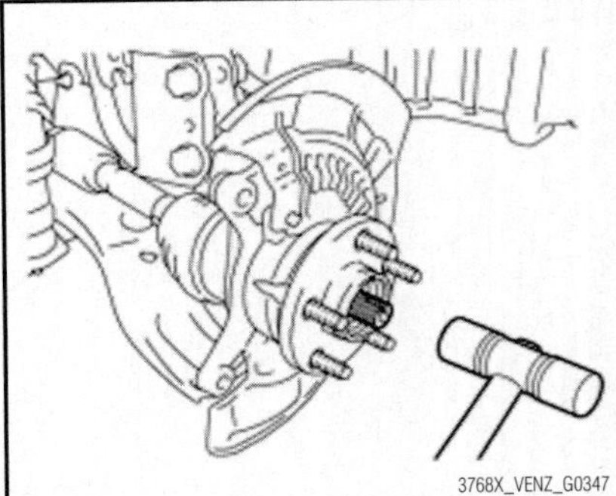

Fig. 311 Separating the front driveshaft assembly from the front axle assembly

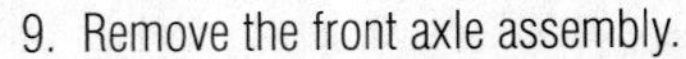

9. Remove the front axle assembly.
 a. Remove the 2 bolts, 2 nuts and the front axle assembly.

➡**When removing the nuts, keep the bolts from rotating.**

➡**Be careful not to damage the driveshaft boot and speed sensor rotor.**

10. Remove the front No. 1 wheel bearing dust deflector.
 a. Using a screwdriver with its tip wrapped with vinyl tape, remove the front No. 1 wheel bearing dust deflector.

➡**Be careful not to damage the steering knuckle.**

11. Remove the front axle hub hole snap ring.
 a. Using snap ring pliers, remove the front axle hub hole snap ring.
12. Remove the front axle hub sub assembly.
 a. Hold the front axle assembly between aluminum plates in a vise.

➡**Do not overtighten the vise.**

 b. Using the special tool (09520-00031).
 c. Using the special tool (09555-55010, 09550-60010, 09951-00430,

Fig. 312 Removing the front axle assembly

Fig. 313 Removing the front No. 1 wheel bearing dust deflector

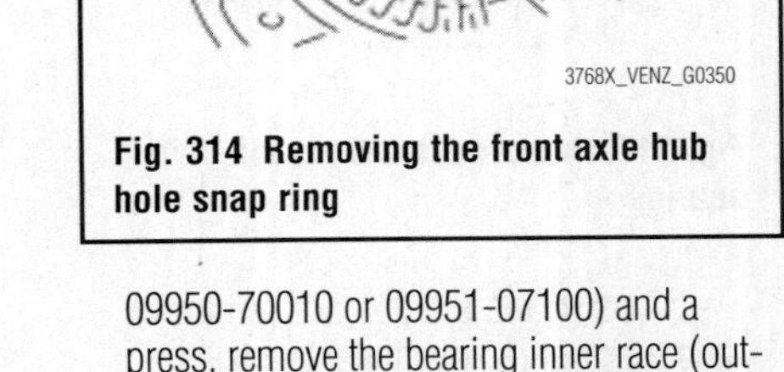

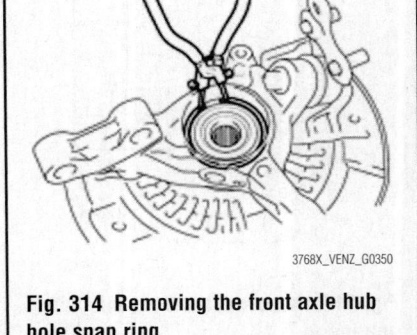

Fig. 314 Removing the front axle hub hole snap ring

09950-70010 or 09951-07100) and a press, remove the bearing inner race (outside) from the front axle hub sub assembly.

➡**Be careful not to drop the front axle hub sub assembly.**

13. Remove the front disc brake dust cover.
 a. Remove the 4 bolts and the front disc brake dust cover from the steering knuckle.
14. Remove the front axle hub bearing.
 a. Place the bearing inner race (outside) on the front axle hub bearing.

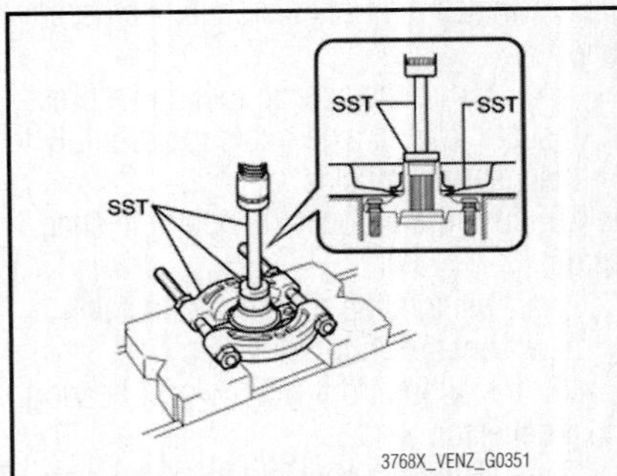

Fig. 315 Removing the bearing inner race (outside) from the front axle hub sub assembly

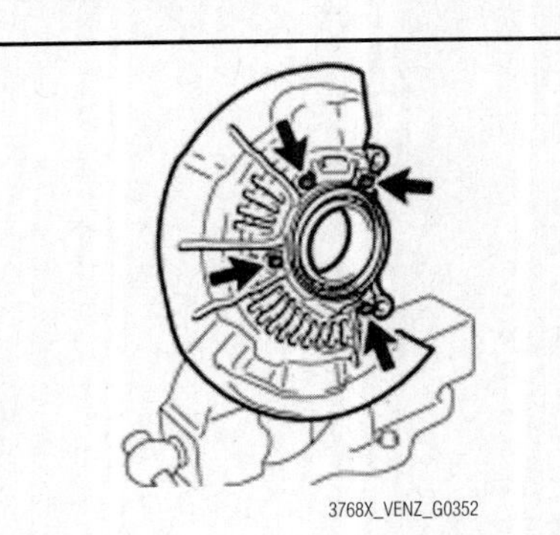

Fig. 316 Removing the front disc brake dust cover

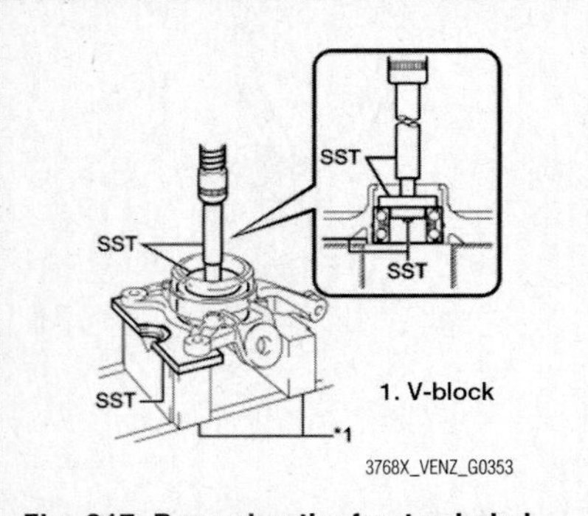

Fig. 317 Removing the front axle hub bearing

b. Using the special tool, V-blocks and a press, remove the front axle hub bearing from the steering knuckle. If the steering knuckle cannot be kept level using the special tool, stabilize the steering knuckle using a washer or an equivalent tool.

To install:

15. Install the front axle hub bearing.
  a. Using the special tool and a press, install a new front axle hub bearing to the steering knuckle.
16. Install the front disc brake dust cover to the steering knuckle with the 4 bolts. Tighten to 73 inch lbs. (8.3 Nm).
17. Install the front axle hub sub assembly.
  a. Using the special tool and a press, install the front axle hub sub assembly to the steering knuckle.
18. Install the front axle hub hole snap ring.
  a. Using snap ring pliers, install a new front axle hub hole snap ring.
19. Install the front No. 1 wheel bearing dust deflector.
  a. Using the special tool and a hammer, install a new front No. 1wheel bearing dust deflector.

➡Align the cutout for the speed sensor in the No. 1 front wheel bearing dust deflector with the hole of the steering knuckle.

20. Install the front axle assembly to the front shock absorber with the 2 bolts and 2 nuts. Tighten to 214 ft. lbs. (290 Nm).

➡When installing the nuts, keep the bolts from rotating.

21. Install the front driveshaft assembly.
  a. Align the matchmarks and install the front driveshaft assembly to the front axle hub sub assembly.
22. Install the front lower suspension arm to the front lower ball joint with the bolt and 2 nuts. Tighten to 68 ft. lbs. (92 Nm).
23. Connect the tie rod assembly.
24. Install the front disc.
25. Install the front disc brake caliper assembly.
26. Install the front axle shaft nut.
  a. Clean the threaded parts on the front driveshaft assembly and a new front axle shaft nut using a non residue solvent.

➡Be sure to perform this work for a new front driveshaft assembly.

➡Keep the threaded parts free of oil and foreign matter.

  b. Using a socket wrench (30 mm), install the new front axle shaft nut. Tighten to 217 ft. lbs. (294 Nm).

➡Stake the front axle shaft nut after inspecting for looseness and runout in the following steps.

➡Keep depressing the brake pedal to prevent the driveshaft from rotating.

27. Separate the front disc brake caliper assembly.
28. Remove the front disc.
29. Inspect the front axle hub bearing looseness.
30. Inspect the front axle hub runout.
31. Install the front disc.
32. Install the front disc brake caliper assembly.
33. Install the front speed sensor with the bolt. Tighten to 71 inch lbs. (8 Nm).

➡Prevent foreign matter from attaching to the front speed sensor tip.

➡Firmly insert the front speed sensor body into the steering knuckle before tightening the bolt.

➡After installing the front speed sensor to the steering knuckle, make sure that there is no clearance between the front speed sensor stay and steering knuckle. Also make sure that no foreign matter is stuck between the parts.

➡Before installing the clamp, firmly insert the points of the clamp into the installation holes.

34. Stake the front axle shaft nut.
  a. Using a chisel and hammer, stake the front axle shaft nut.
35. Install the front wheel. Tighten to 76 ft. lbs. (103 Nm).
36. Inspect and adjust the front wheel alignment.
37. Check for the speed signal.

CONTROL ARMS/LINKS

REMOVAL & INSTALLATION

2WD Vehicles

See Figures 318 through 326.

1. Remove the rear wheels.
2. Separate the rear stabilizer link assembly LH.
    a. Remove the nut and separate the rear stabilizer link assembly LH from the rear stabilizer bar.

➡If the ball joint turns together with the nut, use a hexagon wrench (5 mm) to hold the stud bolt.

3. Separate the rear stabilizer link assembly RH.

➡Perform the same procedure as the LH side.

4. Remove the rear stabilizer bar.

Fig. 318 Separating the rear stabilizer link assembly LH

5. Separate the rear height control sensor sub assembly (w/HID headlight system).
    a. Remove the nut and separate the rear height control sensor sub-assembly from the rear No. 2 suspension arm assembly RH.

➡Use wire or an equivalent tool to keep the rear height control sensor link from hanging down.

6. Remove the rear No. 2 suspension arm assembly LH.
    a. Remove the bolt (A) and the nut, and separate the rear No. 2 suspension arm assembly LH from the rear axle carrier sub-assembly LH.

➡Since a stopper nut is used, loosen the bolt.

    b. Remove the bolt (B) and the rear No. 2 suspension arm assembly LH.

7. Remove the rear No. 1 suspension arm assembly LH.
    a. Remove the bolt and the nut, and separate the rear No. 1 suspension arm assembly LH from the rear axle carrier sub-assembly.

➡Since a stopper nut is used, loosen the bolt.

    b. Remove the bolt and the rear No. 1 suspension arm assembly LH.

8. Remove the rear No. 1 suspension arm assembly RH.

➡Perform the same procedure as the LH side.

To install:

9. Temporarily install the rear No. 1 suspension arm assembly LH to the rear suspension member with the bolt (B).

➡Ensure that the identification mark (*1) faces the rear side of the vehicle.

    a. Temporarily install the rear No. 1 suspension arm assembly LH to the rear axle carrier sub-assembly with the bolt (A) and the nut.

➡Since a stopper nut is used, temporarily tighten the bolt.

10. Temporarily install the rear No. 1 suspension arm assembly RH.

➡Perform the same procedure as the LH side.

11. Temporarily install the rear No. 2 suspension arm assembly LH.
    a. Temporarily install the rear No. 2 suspension arm assembly LH to the rear suspension member with the bolt (B).

➡Ensure that the identification mark (*1) faces the rear side of the vehicle.

    b. Temporarily install the rear No. 2 suspension arm assembly LH to the rear axle carrier sub-assembly LH with the bolt (A) and the nut.

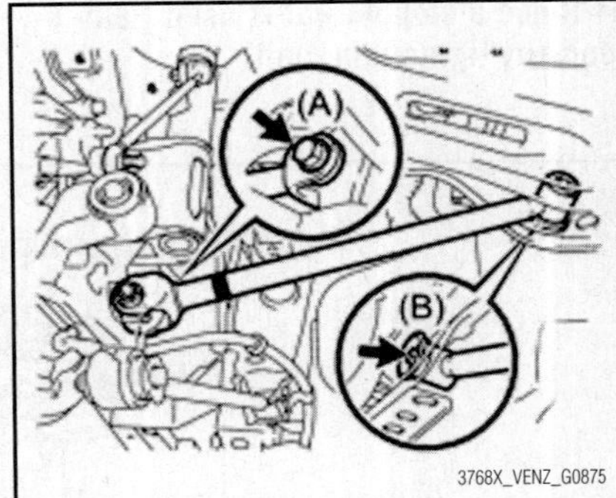

Fig. 321 Removing the rear No. 1 suspension arm assembly LH

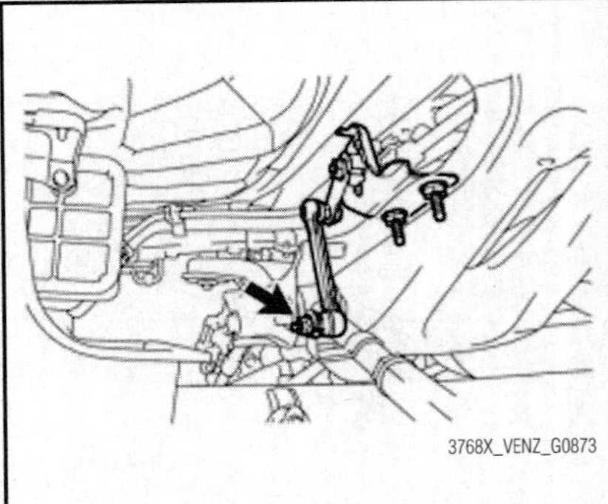

Fig. 319 Separating the rear height control sensor sub assembly

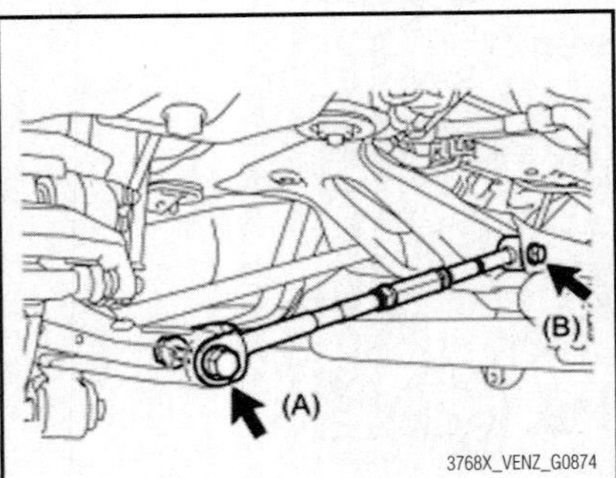

Fig. 320 Removing the rear No. 2 suspension arm assembly LH

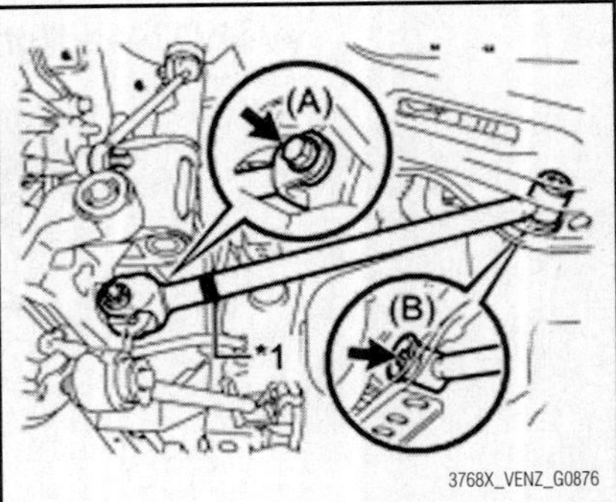

Fig. 322 Temporarily install the rear No. 1 suspension arm assembly LH

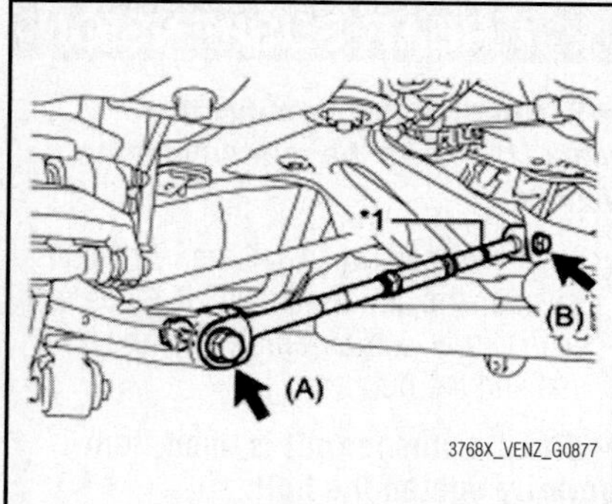

Fig. 323 Temporarily installing the rear No. 2 suspension arm assembly LH

➡Since a stopper nut is used, temporarily tighten the bolt.

12. Temporarily install the rear No. 2 suspension arm assembly RH.

a. Temporarily install the rear No. 2 suspension arm assembly RH to the rear suspension member with the bolt (B).

➡Ensure that the identification mark (*1) faces the rear side of the vehicle.

b. Temporarily install the rear No. 2 suspension arm assembly RH to the rear axle carrier sub-assembly RH with the bolt (B) and the nut.

➡Since a stopper nut is used, temporarily tighten the bolt.

13. Connect the rear height control sensor sub-assembly to the rear No. 2 suspension arm assembly RH with the nut. Tighten to 48 inch lbs. (5.4 Nm).

14. Stabilize the suspension.

a. Install the rear wheels. Tighten to 76 ft. lbs. (103 Nm).

b. Lower the vehicle to the ground.

c. Bounce the vehicle up and down at the corners to stabilize the rear suspension.

d. Remove the rear wheels.

15. Fully tighten the rear No. 1 suspension arm assembly LH.

a. Support the rear axle carrier sub-assembly using a jack and wooden block.

➡Do not bend the brake dust cover.

b. Jack up the rear axle carrier sub-assembly LH to set the rear No. 1 suspension arm assembly LH in the tightening position. Standard angle (A) is 4.6°. Standard length (B) is 1.52 inch (38.6 mm).

CAUTION

Do not jack up the rear axle carrier sub-assembly LH too high as the vehicle may fall.

➡If the rear No. 1 suspension arm assembly LH cannot be positioned as shown in the illustration even when the rear axle carrier sub-assembly LH is jacked up, apply additional load to the vehicle such as by having a person sit in the rear seat.

c. Using the special tool (09961-00950) and a socket wrench (19 mm), fully tighten the bolt in the tightening position. Tighten to 89 ft. lbs. (120 Nm). If using the special tools tighten to 65 ft. lbs. (89 Nm).

➡Use a torque wrench with a fulcrum length of 425 mm (1.39 ft.).

➡This torque value is effective when SST is parallel to the torque wrench.

➡Since a stopper nut is used, fully tighten the bolt.

d. Fully tighten the bolt in the tightening position. Tighten to 83 ft. lbs. (112 Nm).

➡Since a stopper nut is used, fully tighten the bolt.

16. Fully tighten the rear No. 2 suspension arm assembly LH.

a. Support the rear axle carrier sub-

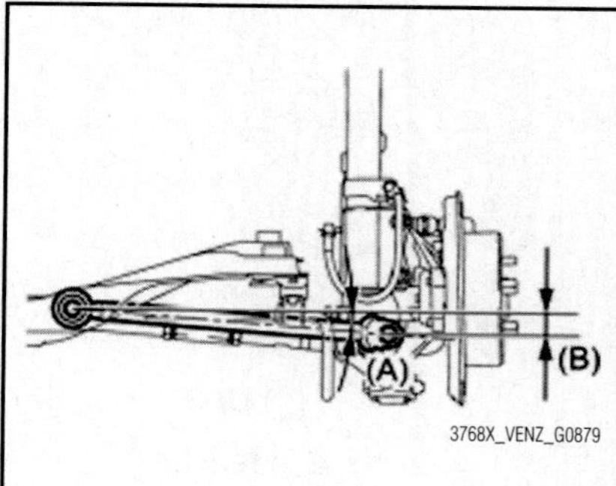

Fig. 325 Identifying the tightening position

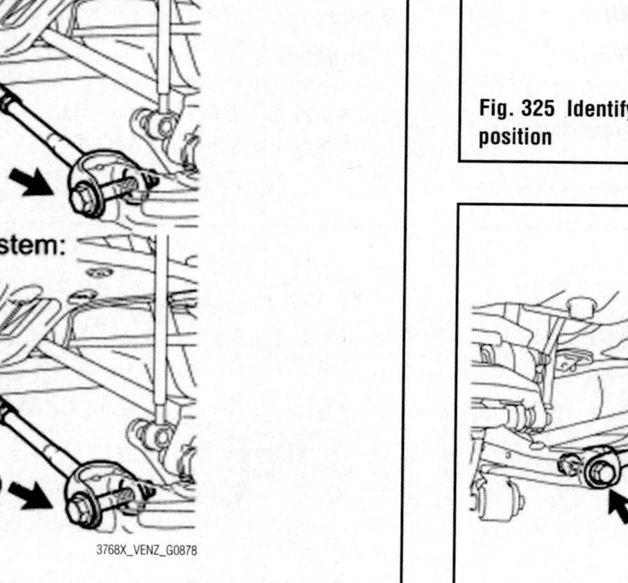

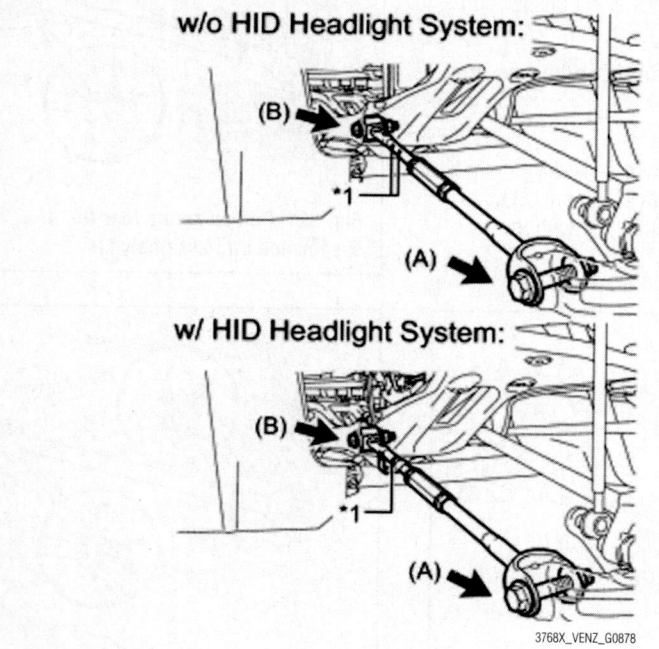

Fig. 324 Temporarily installing the rear No. 2 suspension arm assembly RH

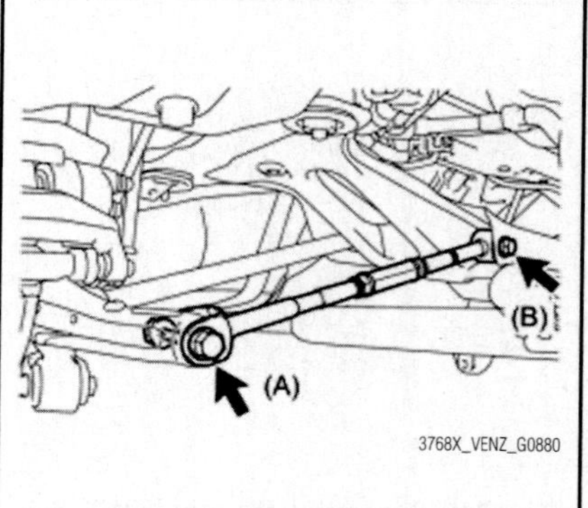

Fig. 326 Tightening the bolts

assembly using a jack and wooden block.

➡Do not bend the brake dust cover.

b. Support the rear axle carrier sub-assembly using a jack and wooden block.

➡Do not bend the brake dust cover.

c. Jack up the rear axle carrier sub-assembly LH to set the rear No. 1 suspension arm assembly LH in the tightening position. Standard angle (A) is 4.6°. Standard length (B) is 1.52 inch (38.6 mm).

⁂ CAUTION

Do not jack up the rear axle carrier sub-assembly LH too high as the vehicle may fall.

➡**If the rear No. 1 suspension arm assembly LH cannot be positioned as shown in the illustration even when the rear axle carrier sub-assembly LH is jacked up, apply additional load to the vehicle such as by having a person sit in the rear seat.**

d. Fully tighten the bolts in the tightening position. Tighten bolt A to 83 ft. lbs. (112 Nm). Tighten bolt B to 89 ft. lbs. (120 Nm).

➡**Since a stopper nut is used, fully tighten the bolt (A) and (B).**

17. Fully tighten the rear No. 2 suspension arm assembly RH.

➡**Perform the same procedure as the LH side.**

18. Install the rear stabilizer bar.
19. Install the rear stabilizer link assembly LH to the rear stabilizer bar with the nut. Tighten to 29 ft. lbs. (39 Nm).

➡**If the ball joint turns together with the nut, use a hexagon wrench (5 mm) to hold the stud bolt.**

20. Install the rear stabilizer link assembly RH.

➡**Perform the same procedure as the LH side.**

21. Install the rear wheel. Tighten to 76 ft. lbs. (103 Nm).
22. Inspect and adjust the rear wheel alignment.
23. Perform the height control sensor signal initialization (w/HID headlight system).
24. Inspect and adjust the headlight aiming (w/HID headlight system).

AWD Vehicles

See Figures 327 through 335.

1. Remove the rear wheels.
2. Remove the center exhaust pipe assembly.
3. Remove the propeller with the center bearing shaft assembly.
4. Separate the rear height control sensor sub assembly (w/HID headlight system).
   a. Remove the nut and separate the rear height control sensor sub-assembly from the rear No. 2 suspension arm assembly RH.

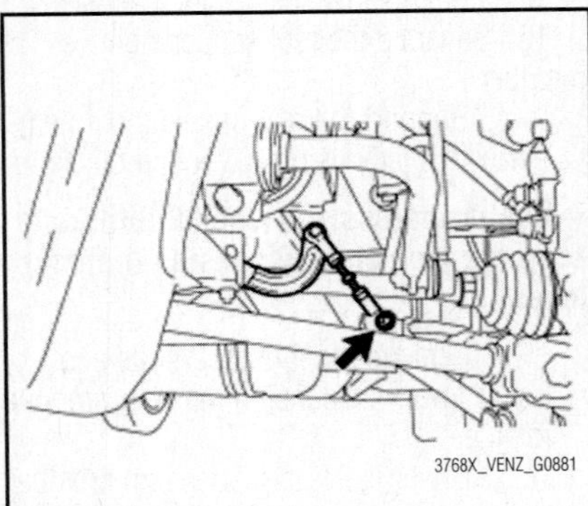

Fig. 327 Separating the rear height control sensor sub assembly

➡**Use wire or an equivalent tool to keep the rear height control sensor link from hanging down.**

5. Separate no. 3 parking brake cable assembly.
6. Separate no. 2 parking brake cable assembly.

➡**Perform the same procedure as the LH side.**

7. Remove rear strut rod assembly LH.
8. Remove rear strut rod assembly RH.

➡**Perform the same procedure as the LH side.**

9. Remove rear No. 2 suspension arm assembly LH.
   a. Put matchmarks (*1) on the adjust cams and the rear suspension member sub-assembly.
   b. Remove the bolt (A) and the nut, and separate the rear No. 2 suspension arm assembly LH from the rear axle carrier sub-assembly LH.

➡**Since a stopper nut is used, loosen the bolt.**

c. Remove the nut (B), the No. 2 camber adjust cam, the rear suspension toe adjust cam sub-assembly, and the rear No. 2 suspension arm assembly LH.

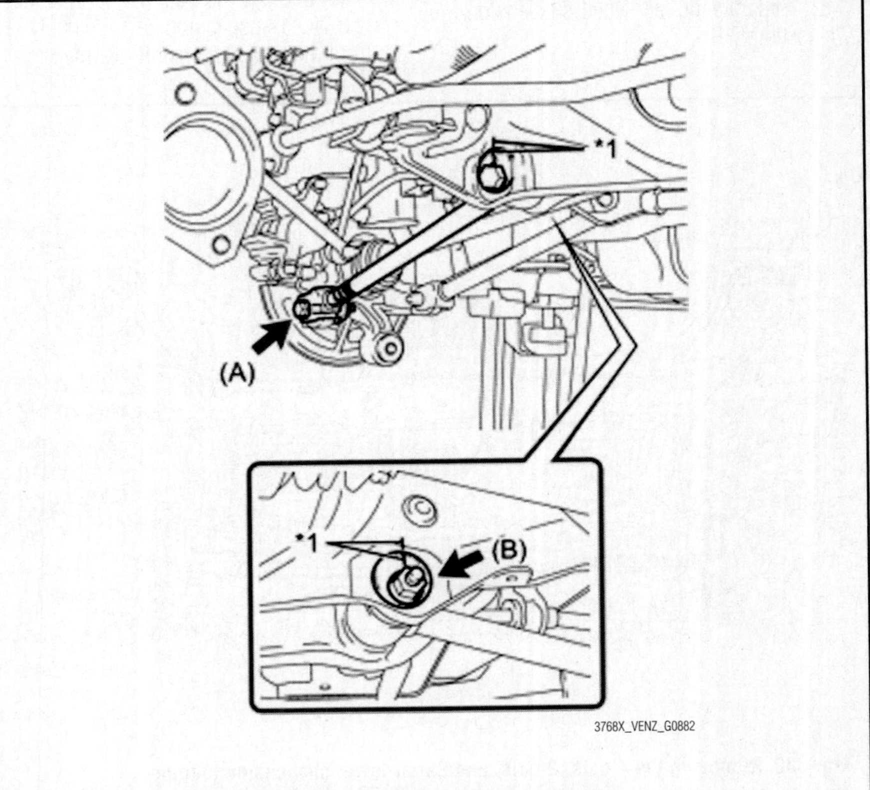

Fig. 328 Removing the rear No. 2 suspension arm assembly LH

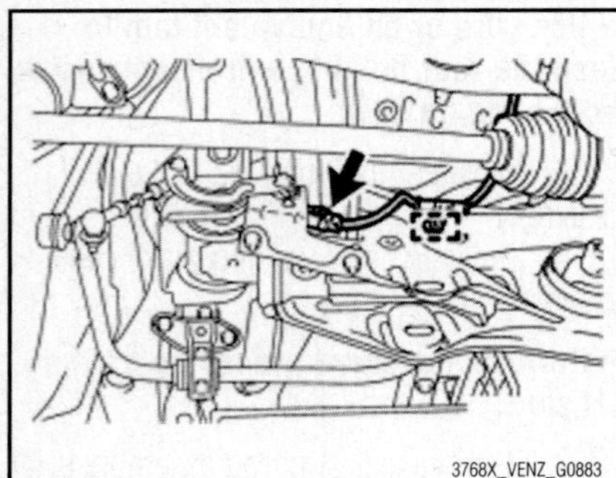

Fig. 329 Disconnecting the No. 3 floor wire

➡**When removing the nut, keep the rear suspension toe adjust cam sub-assembly from rotating.**

10. Remove the rear No. 2 suspension arm assembly RH.

➡**Perform the same procedure as the LH side.**

11. Separate the rear No. 1 suspension arm assembly LH.
    a. Remove the bolt and the nut, and separate the rear No. 1 suspension arm assembly LH from the rear axle carrier sub-assembly LH.

➡**Since a stopper nut is used, loosen the bolt.**

12. Separate the rear No. 1 suspension arm assembly RH.

➡**Perform the same procedure as the LH side.**

13. Remove the No. 1 floor under cover.
    a. Separate the 5 clips to remove the No. 1 floor under cover.

➡**Remove the No. 1 floor under cover with the 5 clips.**

14. Separate the frame wire.
    a. Disengage the 2 clamps to separate the frame wire from the body.
15. Separate the No. 3 floor wire (w/HID headlight system).
    a. Disconnect the connector and disengage the clamp to separate the No. 3 floor wire from the rear suspension member.
16. Separate the rear suspension member.
    a. Support the rear suspension member with a jack using 3 wooden blocks.

➡**Use properly sized wooden blocks to keep the jack and suspension member lever.**

    b. Remove the 4 nuts, 2 bolts and 2 rear lower suspension member stopper retainers.
    c. Lower the rear suspension member to the point shown in the illustration. Length A: 3.94 inch (100 mm).
17. Remove the rear No. 1 suspension arm assembly LH.
    a. Remove the bolt, the nut and the rear No. 1 suspension arm assembly LH from the rear suspension member.

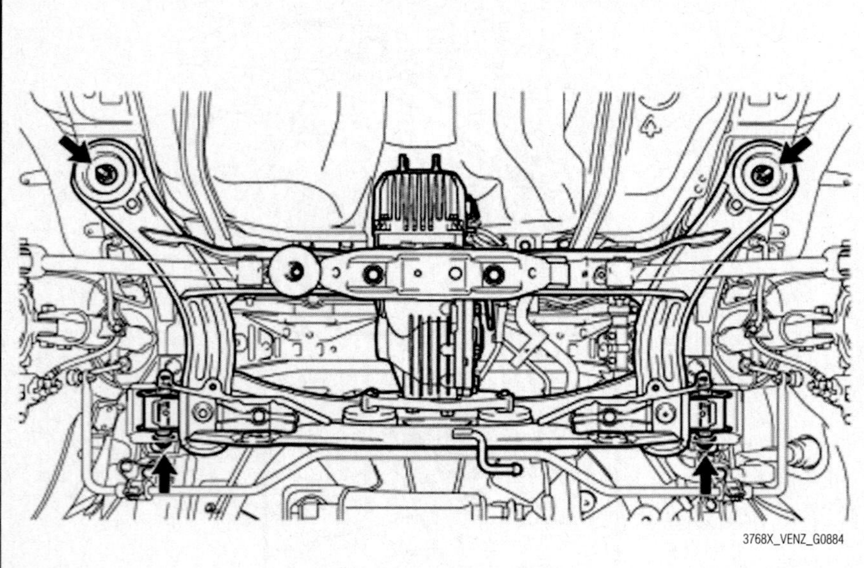

Fig. 330 Removing the 4 nuts, 2 bolts and 2 rear lower suspension member stopper retainers

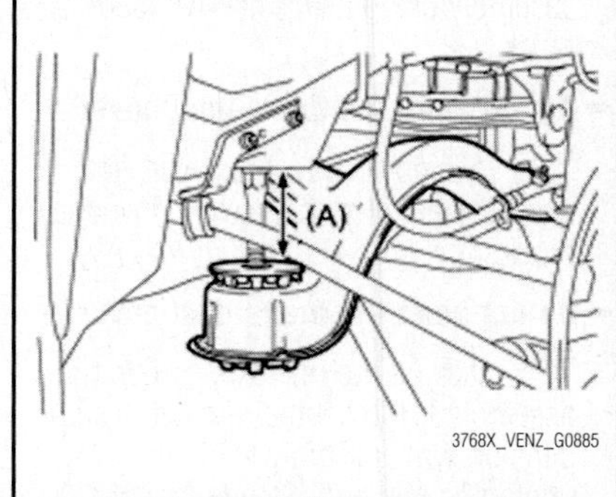

Fig. 331 Lowering the rear suspension member

➡**Since a stopper nut is used, loosen the bolt.**

18. Remove the rear No. 1 suspension arm assembly RH.

➡**Perform the same procedure as the LH.**

To install:

19. Install the rear No. 1 suspension arm assembly LH.
    a. Temporarily install the rear No. 1 suspension arm assembly LH to the rear suspension member with the bolt and the nut.

➡**Ensure that the identification mark faces the rear side of the vehicle.**

➡**Since a stopper nut is used, temporarily tighten the bolt.**

    b. Set the rear No. 1 suspension arm assembly LH in the tightening position shown in the illustration. Standard angle (A): 7.6°. Standard length (B): 1.96 inch (49.7 mm).
    c. Fully tighten the bolt in the tightening position. Tighten to 59 ft. lbs. (80 Nm).

➡**Since a stopper nut is used, fully tighten the bolt.**

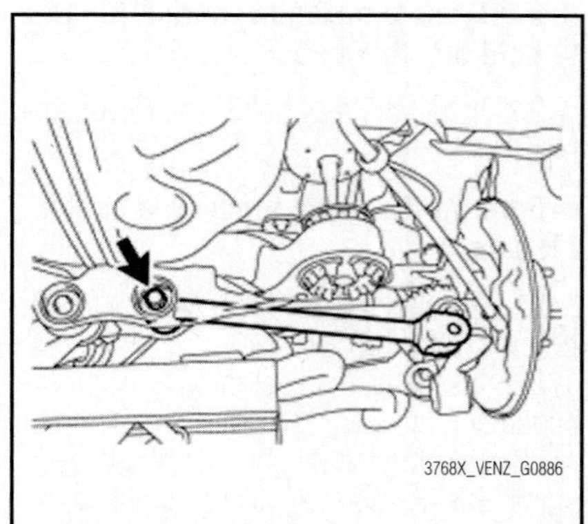

Fig. 332 Removing the rear No. 1 suspension arm assembly LH

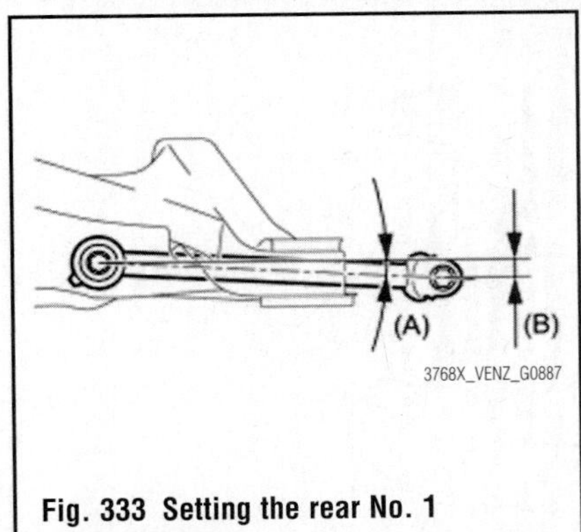

Fig. 333 Setting the rear No. 1 suspension arm

20. Install the rear No. 1 suspension arm assembly RH.

➡**Perform the same procedure as the LH side.**

21. Install the rear suspension member.

a. Raise the rear suspension member with a jack.

b. Temporarily install the rear suspension member, 2 rear upper suspension member stoppers and 2 rear lower suspension member stopper retainers with the 4 nuts and 2 bolts.

➡**Be sure to install the rear suspension member with the rear upper suspension member stopper and the rear lower suspension member stopper retainer in the correct direction as shown in the illustration.**

c. Fully tighten the 2 nuts (A) to 85 ft. lbs. (115 Nm).

d. Using SST and a socket wrench (19 mm), fully tighten the 2 nuts (B). Tighten to 71 ft. lbs. (96 Nm). If using the special tool tighten to 52 ft. lbs. (71 Nm).

➡**Use a torque wrench with a fulcrum length of 425 mm (1.39 ft.).**

➡**This torque value is effective when SST is parallel to the torque wrench.**

22. Connect the rear No. 1 suspension arm assembly LH.

a. Connect the rear No. 1 suspension arm assembly LH to the rear axle carrier sub-assembly LH with the bolt and the nut. Tighten to 83 ft. lbs. (112 Nm).

➡**Since a stopper nut is used, temporarily tighten the bolt.**

23. Connect the rear No. 1 suspension arm assembly RH.

➡**Perform the same procedure as the LH side.**

24. Install the frame wire.

a. Engage the 2 clamps to install the frame wire.

➡**Do not twist the frame wire when installing it.**

25. Install the No. 3 floor wire (w/HID headlight system).

a. Engage the clamp and connect the connector to install the No. 3 floor wire.

➡**Do not twist the No. 3 floor wire when installing it.**

26. Install the No. 1 floor under cover with the 5 clips.

27. Temporarily tighten the rear No. 2 suspension arm assembly LH.

a. Temporarily tighten the rear No. 2 suspension arm assembly LH to the rear suspension member with the rear suspension toe adjust cam sub-assembly, the No. 2 camber adjust cam and the nut (B).

➡**Ensure that the identification mark faces the rear side of the vehicle.**

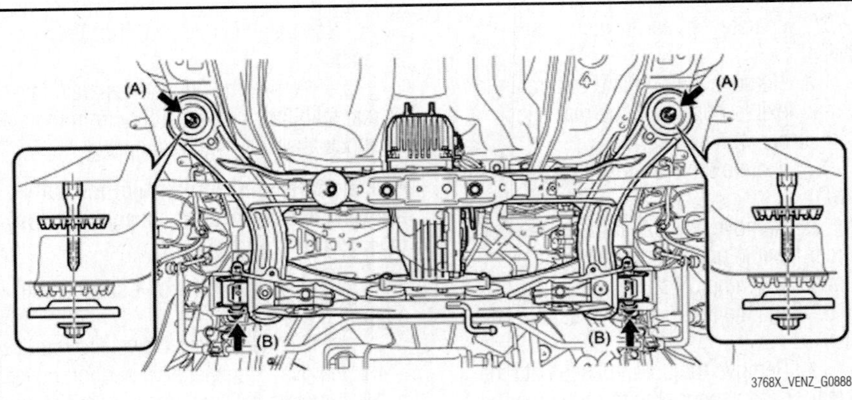

Fig. 334 Installing the rear suspension member with a jack

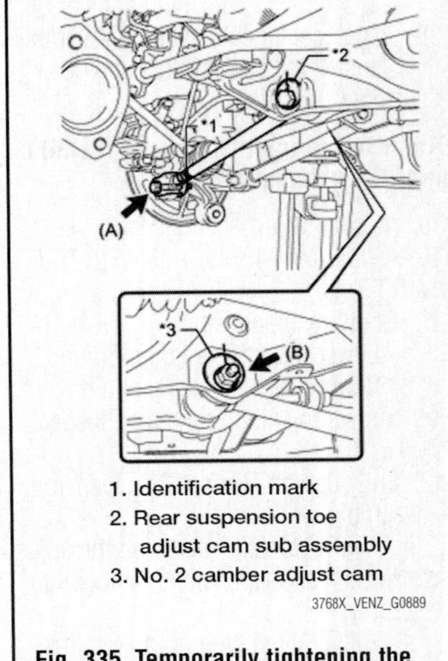

Fig. 335 Temporarily tightening the rear No. 2 suspension arm assembly LH

➡**When temporarily tightening the nut, keep the rear suspension toe adjust cam sub-assembly from rotating.**

b. Fully tighten the rear No. 2 suspension arm assembly LH to the rear axle carrier sub-assembly LH with the bolt (A) and the nut. Tighten to 83 ft. lbs. (112 Nm).

➡**Since a stopper nut is used, fully tighten the bolt.**

28. Temporarily tighten the rear No. 2 suspension arm assembly RH.

➡**Perform the same procedure as the LH side.**

29. Connect the rear height control sensor sub assembly (w/HID headlight system).

a.Connect the rear height control sensor sub-assembly to the rear No. 2 suspension arm assembly

30. RH with the nut. Tighten to 48 inch lbs. (5.4 Nm).

31. Install rear strut rod assembly LH.

32. Install rear strut rod assembly RH.

➡**Perform the same procedure as the LH side.**

33. Install no. 3 parking brake cable assembly.

34. Install no. 2 parking brake cable assembly.

35. Temporarily tighten propeller with center bearing shaft assembly.

36. Fully tighten propeller with center bearing shaft assembly.

37. Inspect and adjust transfer oil.
    a. Inspect and adjust the transfer oil.
38. Install center exhaust pipe assembly.
    a. Install the center exhaust pipe assembly.

➡Refer to the instructions for installation of the exhaust pipe.

39. Inspect for exhaust gas leak.
40. Install rear wheels. Tighten to 76 ft. lbs. (103 Nm).
41. Stabilize suspension.
    a. Lower the vehicle to the ground.
    b. Bounce the vehicle up and down at the corners to stabilize the rear suspension.
42. Fully tighten rear no. 2 suspension arm assembly LH.
    a. Align the matchmarks on the adjust cams and rear suspension member sub-assembly.
    b. Fully tighten the nut to 74 ft. lbs. (100 Nm).

➡The final torque must be applied under standard vehicle height conditions.

➡When fully tightening the nut, keep the rear suspension toe adjust cam sub-assembly from rotating.

43. Fully tighten rear no. 2 suspension arm assembly RH

➡Perform the same procedure as the LH side.

44. Inspect and adjust rear wheel alignment
    a. Inspect and adjust the rear wheel alignment.
45. Height control sensor signal initialization (w/ hid headlight system)
    a. Initialize the height control sensor signal.
46. Inspect and adjust headlight aiming (w/ hid headlight system)
    a. Inspect and adjust the headlight aiming.

SHOCK ABSORBER

REMOVAL & INSTALLATION

See Figures 336 through 341.

➡Use the same procedure for the RH side and LH side.

➡The procedure listed below is for the LH side.

1. Remove the rear wheel.
2. Remove the deck side trim.
    a. Disengage the 5 claws, and then remove the deck side trim.

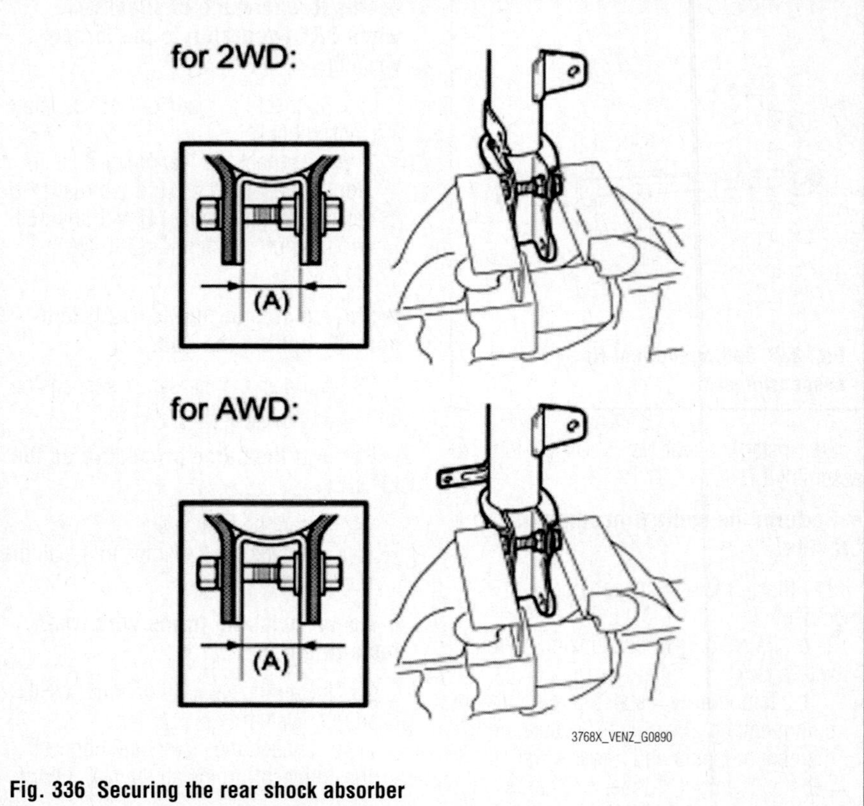

Fig. 336 Securing the rear shock absorber

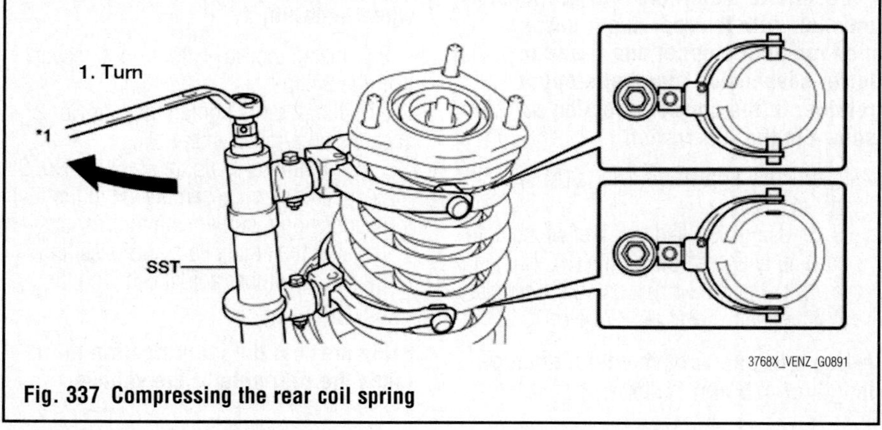

Fig. 337 Compressing the rear coil spring

3. Separate the rear flexible hose (2WD).
    a. Remove the bolt and separate the rear flexible hose from the rear shock absorber with coil spring.
4. Separate the rear flexible hose (AWD).
    a. Remove the bolt and separate the rear flexible hose from the rear shock absorber with coil spring.
5. Separate the rear speed sensor wire.
    a. Remove the bolt and separate the rear speed sensor wire from the rear shock absorber with coil spring.
6. Separate the rear stabilizer link assembly.
    a. Remove the nut and separate the rear stabilizer link assembly from the rear shock absorber with coil spring.

➡If the ball joint turns together with the nut, use a hexagon wrench (5 mm) to hold the stud bolt.

7. Remove the rear No. 1 suspension support cover.
    a. Disengage the 4 claws and remove the rear No. 1 suspension support cover.
8. Remove the rear shock absorber with the coil spring.

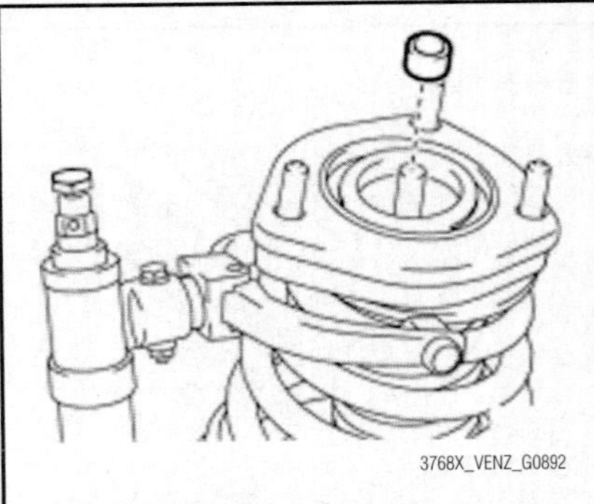

Fig. 338 Removing the rear support to rear shock absorber collar

a. Support the rear axle carrier sub-assembly using a jack and wooden block.

➡**Do not deform the brake dust cover.**

➡**Support the rear axle carrier sub-assembly until reinstallation of the rear shock absorber with coil spring is complete.**

b. Loosen the rear support to rear shock absorber nut.

⁂ CAUTION

Do not remove the rear support to rear shock absorber nut.

➡**Loosen the nut only when the rear shock absorber with coil spring needs to be disassembled.**

c. Remove the 2 bolts and 2 nuts, and separate the rear shock absorber with coil spring from the rear axle carrier sub-assembly.

➡**When removing the nuts, keep the bolts from rotating.**

d. Remove the 3 nuts and rear shock absorber with coil spring.

➡**Make sure that the rear speed sensor and rear flexible hose are disconnected from the rear shock absorber with coil spring.**

9. Secure the rear shock absorber with the coil spring.
   a. Install the bolt and nut to the rear shock absorber with coil spring as shown in the illustration and secure the rear shock absorber with coil spring in a vise. Length (A) is 1.10 inch (28 mm).
10. Remove the rear support to the rear shock absorber nut.
   a. Using the special tool, compress the rear coil spring.

➡**Do not use an impact wrench. It will damage SST.**

   b. Check that the rear coil spring is fully compressed.
   c. Using a screwdriver or an equivalent tool to hold the rear suspension support assembly, remove the rear support to rear shock absorber nut from the rear shock absorber assembly.

➡**Tape the screwdriver tip or the equivalent tool before use.**

11. Remove the rear support to rear shock absorber collar.
   a. Remove the rear support to rear shock absorber collar from the rear shock absorber assembly.
12. Remove the rear suspension support assembly from the rear shock absorber assembly.
13. Remove the rear coil spring together with the special tool from the rear shock absorber assembly.
14. Remove the rear No. 1 spring bumper from the rear shock absorber assembly.
15. Remove the rear lower coil spring insulator from the rear shock absorber assembly.

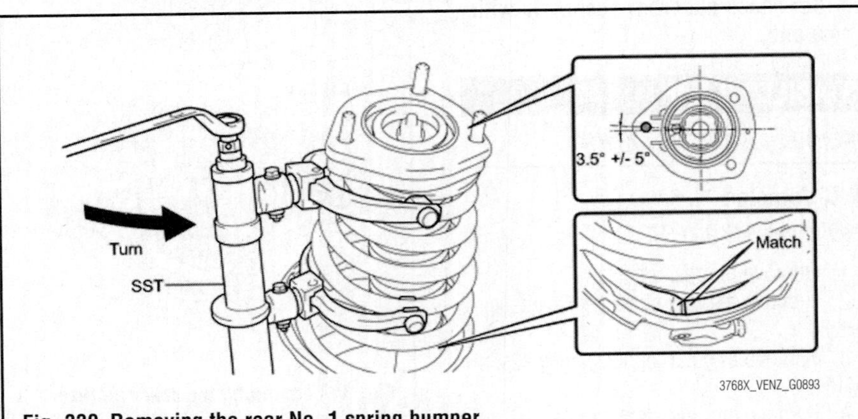

Fig. 339 Removing the rear No. 1 spring bumper

To install:

➡**Use the same procedure for the RH side and LH side.**

➡**The procedure listed below is for the LH side.**

16. Secure the rear shock absorber with the coil spring.
17. Install the rear lower coil spring insulator onto the rear shock absorber assembly.

➡**Fit the recessed part of the rear lower coil spring insulator into the recession on the shock absorber assembly.**

18. Install the rear No. 1 spring bumper to the rear shock absorber assembly.
19. Temporarily install the rear coil spring.
   a. Using the special tool, compress the rear coil spring.

➡**Do not use an impact wrench. It will damage SST.**

   b. Temporarily install the rear coil spring together with SST to the rear shock absorber assembly.
20. Install the rear suspension support assembly to the rear shock absorber assembly.

➡**Align the cutout on the rear shock absorber assembly with the protrusion on the rear suspension support assembly.**

21. Install the rear support to rear shock absorber collar to the rear shock absorber assembly.
22. Temporarily install the rear support to the rear shock absorber nut.
   a. Using a screwdriver or an equivalent tool to hold the rear suspension support assembly, temporarily install a new rear support to rear shock absorber nut to the rear shock absorber assembly.

➡**Do not reuse the old rear support to rear shock absorber nut,**

➡**Use new rear support to rear shock absorber nut.**

➡**Tape the screwdriver or the equivalent tool before use.**

23. Install the rear coil spring.

➡**Do not use an impact wrench. It will damage SST.**

➡**Make sure that the end of the rear coil spring is positioned in the depression of the rear lower coil spring insulator.**

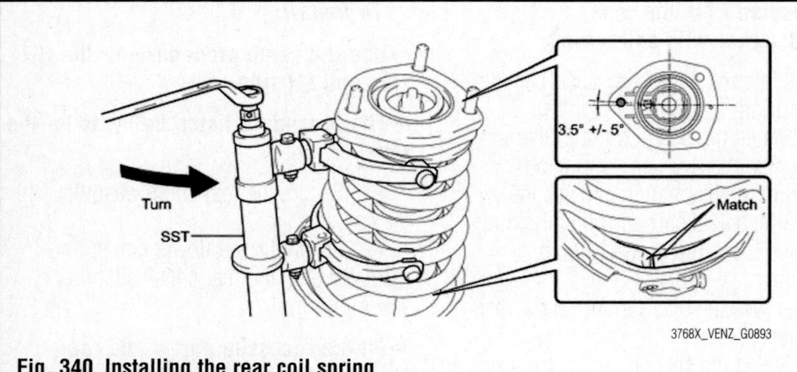

Fig. 340 Installing the rear coil spring

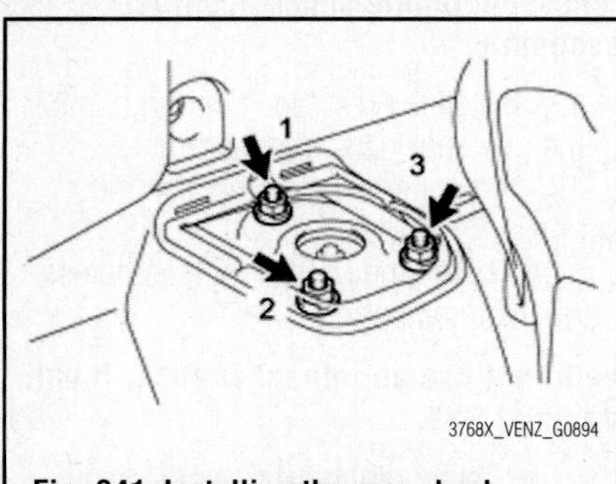

Fig. 341 Installing the rear shock absorber with the coil spring

➡**Ensure the rear lower coil spring insulator is not pinched or folded over and caught by the rear coil spring.**

➡**Ensure that the stud bolt is positioned 3.5°to the outside of the vehicle as shown in the illustration. The deviation should be within +/- 5°.**

24. Install the rear shock absorber with the coil spring
    a. Install the rear shock absorber with coil spring with the 3 nuts in order 1 to 3. Tighten to 43 ft. lbs. (58 Nm).
    b. Install the rear shock absorber with coil spring with the 2 bolts and 2 nuts. Tighten to 214 ft. lbs. (290 Nm).

➡**When installing the nuts, keep the bolts from rotating.**

    c. Fully tighten the rear support to rear shock absorber nut. Tighten to 41 ft. lbs. (55 Nm).
25. Install the rear No. 1 suspension support cover.
26. Install the rear stabilizer link assembly to the rear shock absorber with coil spring with the nut. Tighten to 29 ft. lbs. (39 Nm).

➡**If the ball joint turns together with the nut, use a hexagon wrench (5 mm) to hold the stud bolt.**

27. Install the rear speed sensor wire to the rear shock absorber with coil spring with the bolt. Tighten to 44 inch lbs. (5 Nm).

➡**Do not twist the rear speed sensor wire when installing it.**

28. Install the rear flexible hose to the rear shock absorber with coil spring with the bolt. Tighten to 14 ft. lbs. (19 Nm).

➡**Do not twist the rear flexible hose when installing it.**

29. Install the deck side trim.
    a. Engage the 5 claws to install the deck side trim.
30. Install the rear wheel. Tighten to 76 ft. lbs. (103 Nm).
31. Inspect and adjust the rear wheel alignment.

TESTING

1. Compress and extend the shock absorber rod 4 or more times.

➡**There is no abnormal resistance or sound and operation resistance is normal.**

2. Check for any oil leaks.
3. Check for any damage on the absorber rod.

➡**If there is any abnormality, replace the rear shock absorber assembly with a new one.**

STABILIZER BAR

REMOVAL & INSTALLATION

2WD Vehicles

See Figures 342 and 343.

1. Remove the rear wheels.
2. Remove the rear stabilizer link assembly LH.
3. Remove the rear stabilizer link assembly RH.
4. Remove the rear stabilizer bar.

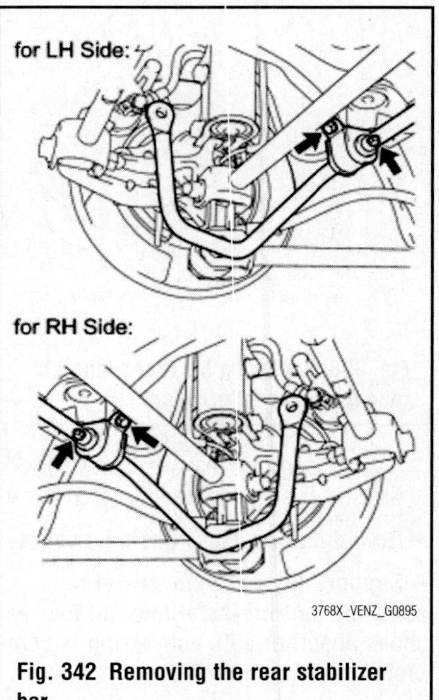

Fig. 342 Removing the rear stabilizer bar

    a. Remove the 4 bolts and the rear stabilizer bar.

To install:

5. Install the rear stabilizer bar with the 4 bolts in order 1 to 2. Tighten to 14 ft. lbs. (19 Nm).
6. Install the rear stabilizer link assembly LH.
7. Install the rear stabilizer link assembly RH.
8. Install the rear wheels. Tighten to 76 ft. lbs. (103 Nm).

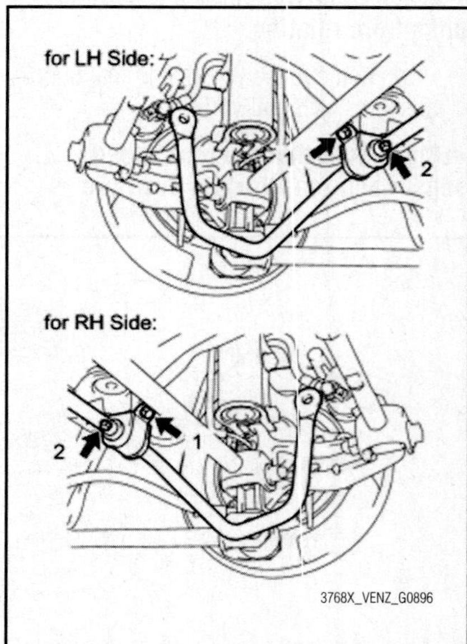

Fig. 343 Installing the rear stabilizer bar

AWD Vehicles

See Figures 344 through 352.

1. Remove the rear wheels.
2. Remove the rear stabilizer link assembly LH.
   a. Remove the nut and separate the rear stabilizer link assembly LH from the rear stabilizer bar.
   b. Remove the nut and the rear stabilizer link assembly LH from the rear shock absorber with coil spring LH.
3. Remove the rear stabilizer link assembly RH.

➡Perform the same procedure as the LH side.

4. Remove the No. 1 floor under cover.
5. Remove the rear lower suspension brace (LH).
   a. Remove the bolt, the nut and the rear lower suspension brace (LH side).
6. Remove the rear lower suspension brace RH.

➡Perform the same procedure as the LH side.

7. Remove the rear stabilizer bar bracket LH (rear side).
   a. Remove the 2 bolts and the rear stabilizer bar bracket LH (rear side).
8. Remove the rear stabilizer bar bracket RH (rear side).
   a. Remove the 2 bolts and the rear stabilizer bar bracket RH (rear side).
9. Remove the rear stabilizer bushing.
10. Remove the rear stabilizer bar bracket (front side).

To install:

11. To install, reverse the removal procedure. Tighten the rear stabilizer bar brackets (rear side) to 14 ft. lbs.

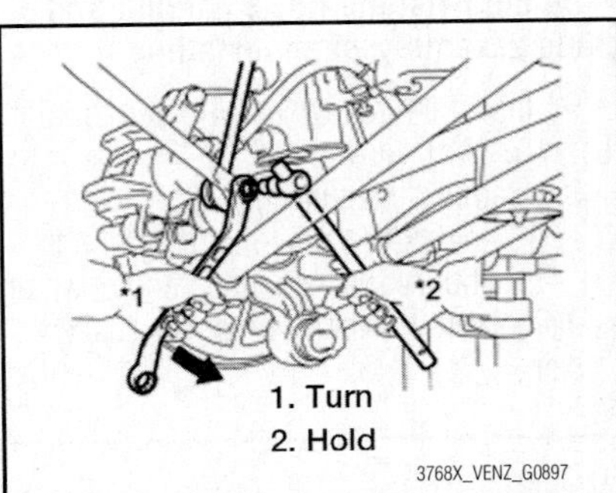

Fig. 344 Separating the rear stabilizer link assembly LH from the rear stabilizer bar

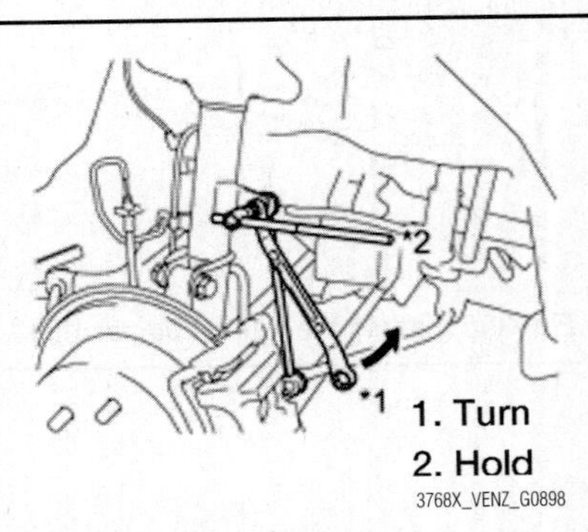

Fig. 345 Removing the stabilizer link assembly

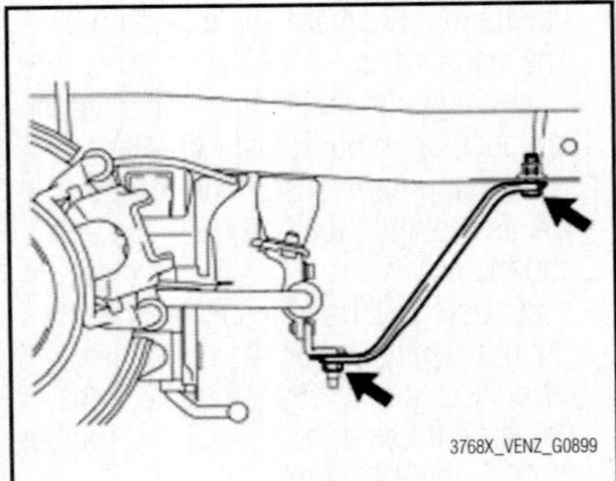

Fig. 346 Removing the rear lower suspension brace LH

Fig. 347 Removing the rear stabilizer bar bracket LH

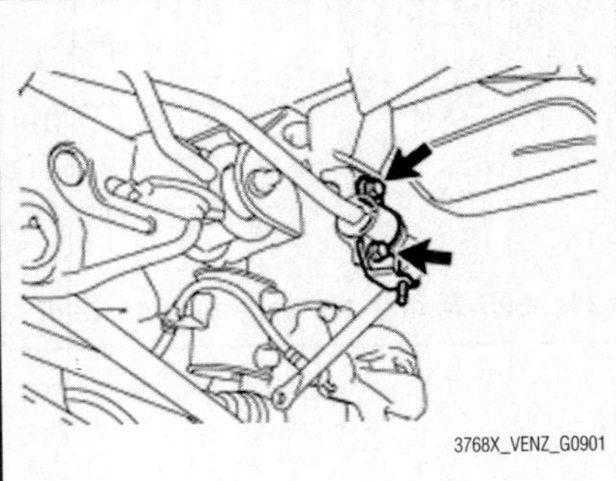

Fig. 348 Removing the rear stabilizer bar bracket RH (rear side)

Fig. 349 Removing the rear stabilizer bushing LH

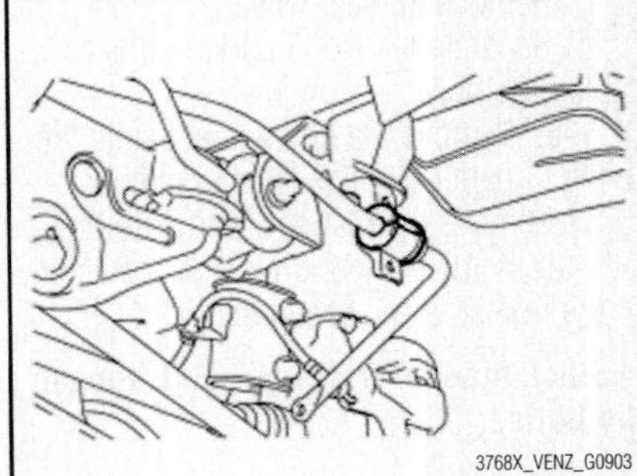

Fig. 350 Removing the rear stabilizer bushing RH

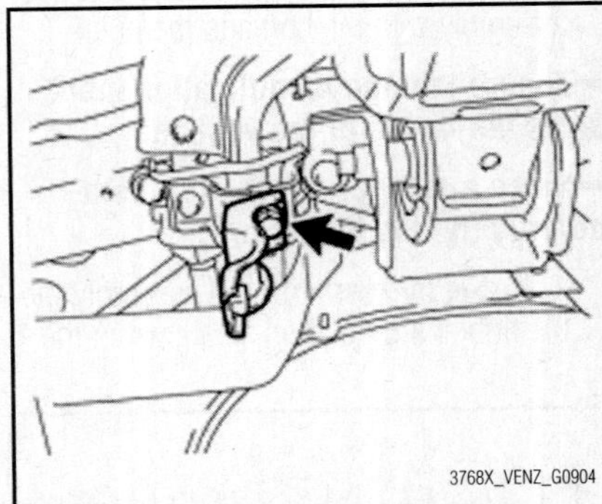

Fig. 351 Removing the rear stabilizer bar bracket (front side)–LH

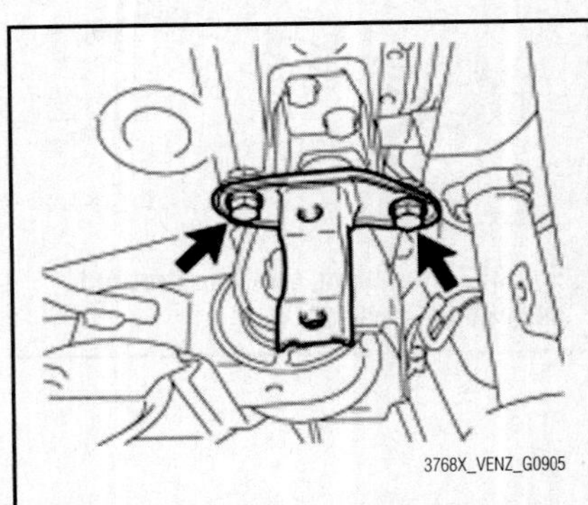

Fig. 352 Removing the rear stabilizer bar bracket (front side)–RH

(19 Nm). Tighten the rear stabilizer bar brackets (front side) to 40 ft. lbs. (54 Nm). Tighten the rear lower suspension brace to 26 ft. lbs. (35 Nm). Install the rear stabilizer link assembly to 29 ft. lbs. (39 Nm).

REAR STRUT ROD

REMOVAL & INSTALLATION

See Figures 353 through 356.

➡Use the same procedure for the RH side and LH side.

➡The procedure listed below is for the LH side.

1. Remove the rear wheel.
2. Separate the No. 3 parking brake cable assembly.
   a. Remove the bolt and separate the No. 3 parking brake cable assembly.
3. Remove the rear strut rod assembly.
   a. Remove the 2 bolts, the 2 nuts and the rear strut rod assembly.

➡Since stopper nuts are used, loosen the bolts.

To install:

4. Install the rear strut rod assembly (2WD).
   a. Temporarily install the rear strut rod assembly to the rear axle carrier sub-assembly with the bolt and the nut.

➡Ensure that the identification mark faces the inside of the vehicle.

➡Since a stopper nut is used, temporarily tighten the bolt.

   b. Set the rear strut rod assembly in the tightening position as shown in the illustration. Standard angle (A) is 9.9°. The standard length is 3.30 inch.
   c. Fully tighten the bolt in the tightening position to 59 ft. lbs. (80 Nm).
   d. Temporarily install the rear strut rod assembly to the body with the bolt and the nut.
   e. Using SST and a socket wrench (17 mm), fully tighten the bolt in the rebound position. Tighten to 59 ft. lbs. (80 Nm). If using the special tool tighten to 44 ft. lbs. (59 Nm).

➡Since the stopper nut is used, fully tighten the bolt.

➡Use a torque wrench with a fulcrum length of 425 mm (1.39 ft.).

➡This torque value is effective when SST is parallel to the torque wrench.

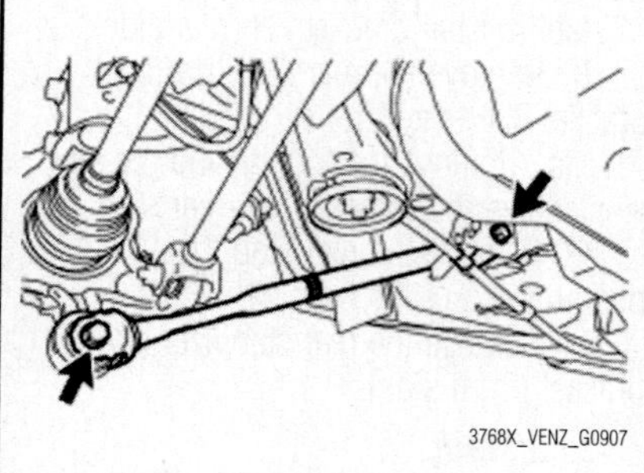

3768X_VENZ_G0907

Fig. 354 Removing the rear strut rod assembly—AWD

5. Install the rear strut rod assembly (AWD).
   a. Temporarily install the rear strut rod assembly to the rear axle carrier sub-assembly with the bolt and the nut.

➡Ensure that the identification mark faces the inside of the vehicle.

➡Since a stopper nut is used, temporarily tighten the bolt.

   b. Set the rear strut rod assembly in the tightening position shown in the illustration. Standard angle (A) is 9.04°. The standard length is 3.08 inch.
   c. Temporarily install the rear strut rod assembly to the body with the bolt and the nut.
   d. Using SST and a socket wrench (17 mm), fully tighten the bolt in the rebound position. Tighten to 59 ft. lbs. (80 Nm). If using the special tool tighten to 44 ft. lbs. (59 Nm).

➡Since a stopper nut is used, fully tighten the bolt.

➡Use a torque wrench with a fulcrum length of 425 mm (1.39 ft.).

➡This torque value is effective when SST is parallel to the torque wrench.

6. Install the No. 3 parking brake cable assembly with the bolt. Tighten to 53 inch lbs. (6 Nm).

➡Do not twist the No. 3 parking brake cable assembly when installing it.

7. Install the rear wheel. Tighten to 76 ft. lbs. (103 Nm).
8. Stabilize the suspension.
   a. Lower the vehicle to the ground.
   b. Bounce the vehicle up and down at the corners to stabilize the rear suspension.

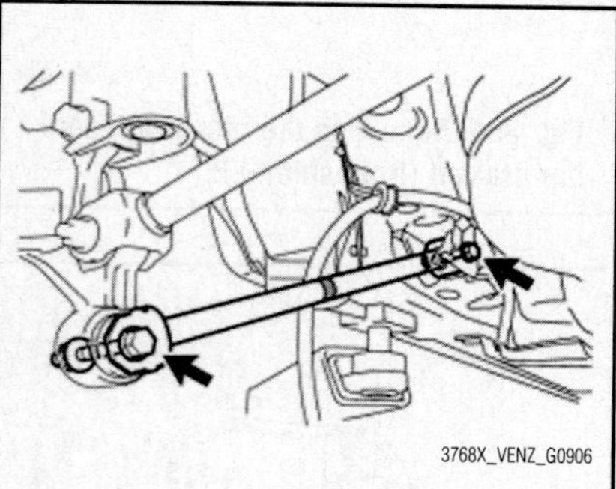

3768X_VENZ_G0906

Fig. 353 Removing the rear strut rod assembly—2WD

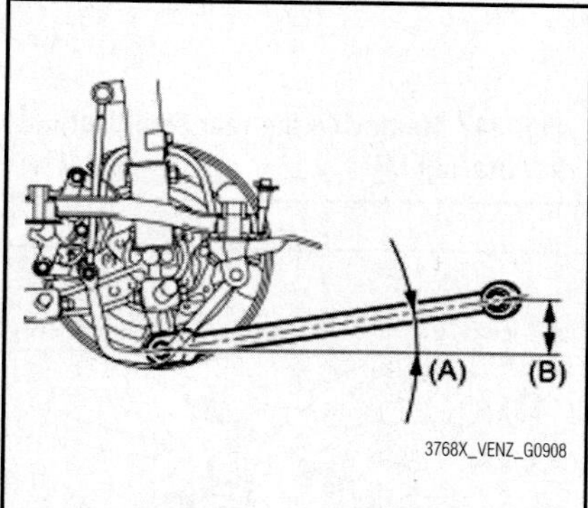

3768X_VENZ_G0908

Fig. 355 Setting the tightening position

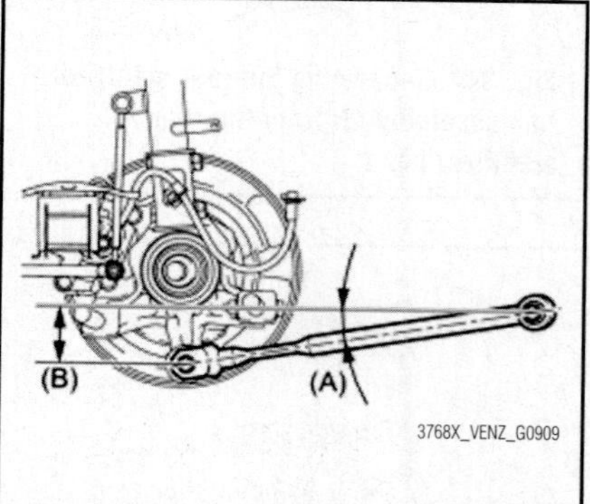

3768X_VENZ_G0909

Fig. 356 Setting the tightening position

TOYOTA

Yaris

19

SPECIFICATIONS AND MAINTENANCE CHARTS

ENGINE AND VEHICLE IDENTIFICATION

Engine						
Code ①	Liters (cc)	Cu. In.	Cyl.	Fuel Sys.	Engine Type	Eng. Mfg.
1NZ-FE	1.5 (1496)	91	4	EFI	DOHC	Toyota

Model Year	
Code ②	Year
8	2008
9	2009
A	2010

EFI: Electronic Fuel Injection

DOHC: Double Overhead Camshaft

① 8th digit of VIN

② 10th digit of VIN

3768X_YARI_C0001

GENERAL ENGINE SPECIFICATIONS

Year	Model	Engine Displacement Liters (cc)	Engine Series (ID/VIN)	Fuel System	Net Horsepower @ rpm	Net Torque @ rpm (ft. lbs.)	Bore x Stroke (in.)	Com-pression Ratio	Oil Pressure @ rpm
2008	Yaris	1.5 (1496)	1NZ-FE	EFI	108@6000	105@4000	2.95x3.33	10.5:1	①
2009	Yaris	1.5 (1496)	1NZ-FE	EFI	108@6000	105@4000	2.95x3.33	10.5:1	①
2010	Yaris	1.5 (1496)	1NZ-FE	EFI	108@6000	105@4000	2.95x3.33	10.5:1	①

EFI: Electronic Fuel Injection

① 4.3 psi @ idle, 22-80 psi @ 3000 rpm

3768X_YARI_C0002

ENGINE TUNE-UP SPECIFICATIONS

Year	Engine Displacement Liters	Engine ID/VIN	Spark Plug Gap (in.)	Ignition Timing (deg.)	Fuel Pump (psi)	Idle Speed (rpm)		Valve Clearance	
						MT	AT	Intake	Exhaust
2008	1.5	1NZ-FE	0.043	8-12 BTDC	44-50	700-800	700-800	0.006-0.010	0.011-0.014
2009	1.5	1NZ-FE	0.043	8-12 BTDC	44-50	700-800	700-800	0.006-0.010	0.011-0.014
2010	1.5	1NZ-FE	0.043	8-12 BTDC	44-50	700-800	700-800	0.006-0.010	0.011-0.014

Note: The Vehicle Emission Control Information label often reflects specification changes made during production. The label figures must be used if they differ from those in this chart.

① With terminal TE1 and E1 connected of DLC1

3768X_YARI_C0003

CAPACITIES

Year	Model	Engine Displacement Liters	Engine ID/VIN	Engine Oil with Filter	Transmission (pts.) 5-Spd	Transmission (pts.) Auto.	Drive Axle Front (pts.)	Fuel Tank (gal.)	Cooling System (qts.)
2008	Yaris	1.5	1NZ-FE	4.0	4.0	6.2	①	11.9	②
2009	Yaris	1.5	1NZ-FE	4.0	4.0	6.2	①	11.9	②
2010	Yaris	1.5	1NZ-FE	4.0	4.0	6.2	①	11.9	②

Note: All capacities are approximate. Add fluid gradually and check to be sure a proper fluid level is obtained.

① Included in transaxle capacity

② w/MT: 4.7
w/AT: 4.5

3768X_YARI_C0005

FLUID SPECIFICATIONS

Year	Model	Engine Displacement Liters (VIN)	Engine Oil	Auto. Trans.	Brake Master Cylinder
2008	Yaris	1.5 (1NZ-FE)	①	Mercon® ATF Fluid	DOT 3
2009	Yaris	1.5 (1NZ-FE)	①	Mercon® ATF Fluid	DOT 3
2010	Yaris	1.5 (1NZ-FE)	①	Mercon® ATF Fluid	DOT 3

DOT: Department Of Transportation

① 5W-20 Premium Synthetic Blend Motor Oil (US) or 5W-20 Super Premium Motor Oil (Canada)

3768X_YARI_C0004

VALVE SPECIFICATIONS

Year	Engine Displacement Liters	Engine ID/VIN	Seat Angle (deg.)	Face Angle (deg.)	Spring Test Pressure (lbs. @ in.)	Spring Installed Height (in.)	Stem-to-Guide Clearance (in.) Intake	Stem-to-Guide Clearance (in.) Exhaust	Stem Diameter (in.) Intake	Stem Diameter (in.) Exhaust
2008	1.5	1NZ-FE	45	44.5	33.5-37@ 1.280	1.280	0.0010-0.0024	0.0012-0.0026	0.1957-0.1963	0.1955-0.1961
2009	1.5	1NZ-FE	45	44.5	33.5-37@ 1.280	1.280	0.0010-0.0024	0.0012-0.0026	0.1957-0.1963	0.1955-0.1961
2010	1.5	1NZ-FE	45	44.5	33.5-37@ 1.280	1.280	0.0010-0.0024	0.0012-0.0026	0.1957-0.1963	0.1955-0.1961

3768X_YARI_C0006

CAMSHAFT AND BEARING SPECIFICATIONS

All measurements are given in inches.

Year	Engine Displacement Liters	Engine VIN	Journal Diameter	Brg. Oil Clearance	Shaft End-play	Runout	Journal Bore	Lobe Lift	
								Intake	Exhaust
2008	1.5	1NZ-FE	2.0505-2.0515	0.0014-0.0028	0.0010-0.0060	0.0012	NA	NA	NA
2009	1.5	1NZ-FE	2.0505-2.0515	0.0014-0.0028	0.0010-0.0060	0.0012	NA	NA	NA
2010	1.5	1NZ-FE	2.0505-2.0515	0.0014-0.0028	0.0010-0.0060	0.0012	NA	NA	NA

NA: Information not available

3768X_YARI_C0009

CRANKSHAFT AND CONNECTING ROD SPECIFICATIONS

All measurements are given in inches.

Year	Engine Displacement Liters	Engine ID/VIN	Crankshaft				Connecting Rod		
			Main Brg. Journal Dia.	Main Brg. Oil Clearance	Shaft End-play	Thrust on No.	Journal Diameter	Oil Clearance	Side Clearance
2008	1.5	1NZ-FE	①	0.0004-0.0009	0.0035-0.0075	3	1.5745-1.5748	0.0006-0.0016	0.0063-0.0142
2009	1.5	1NZ-FE	①	0.0004-0.0009	0.0035-0.0075	3	1.5745-1.5748	0.0006-0.0016	0.0063-0.0142
2010	1.5	1NZ-FE	①	0.0004-0.0009	0.0035-0.0075	3	1.5745-1.5748	0.0006-0.0016	0.0063-0.0142

① Reference mark:
0: 1.81102-1.81110
1: 1.81110-1.81118
2: 1.81118-1.81126
3: 1.81126-1.81133
4: 1.81133-1.81141
5: 1.81141-1.81149

3768X_YARI_C0010

PISTON AND RING SPECIFICATIONS

All measurements are given in inches.

Year	Engine Displacement Liters	Engine ID/VIN	Piston Clearance	Ring Gap			Ring Side Clearance		
				Top Compression	Bottom Compression	Oil Control	Top Compression	Bottom Compression	Oil Control
2008	1.5	1NZ-FE	0.0022-0.0023	0.0098-0.0138	0.0138-0.0197	0.0039-0.0138	0.0012-0.0028	0.0012-0.0028	SNUG
2009	1.5	1NZ-FE	0.0022-0.0023	0.0098-0.0138	0.0138-0.0197	0.0039-0.0138	0.0012-0.0028	0.0012-0.0028	SNUG
2010	1.5	1NZ-FE	0.0022-0.0023	0.0098-0.0138	0.0138-0.0197	0.0039-0.0138	0.0012-0.0028	0.0012-0.0028	SNUG

3768X_YARI_C0008

TORQUE SPECIFICATIONS

All readings in ft. lbs.

Year	Engine Displacement Liters	Engine ID/VIN	Cylinder Head Bolts	Main Bearing Bolts	Rod Bearing Bolts	Crankshaft Damper Bolts	Flywheel Bolts	Manifold Intake	Manifold Exhaust	Spark Plugs	Oil Pan Drain Plug
2008	1.5	1NZ-FE	①	②	③	94	④	22	20	13	28
2009	1.5	1NZ-FE	①	②	③	94	④	22	20	13	28
2010	1.5	1NZ-FE	①	②	③	94	④	22	20	13	28

① Step 1: 22 ft. lbs.
Step 2: 90 degree turn
Step 3: 90 degree turn

② Step 1: 16 ft. lbs.
Step 2: 90 degree turn

③ Step 1: 11 ft. lbs.
Step 2: 90 degree turn

④ M/T: 38 ft. lbs.
A/T: 65 ft. lbs.

3768X_YARI_C0007

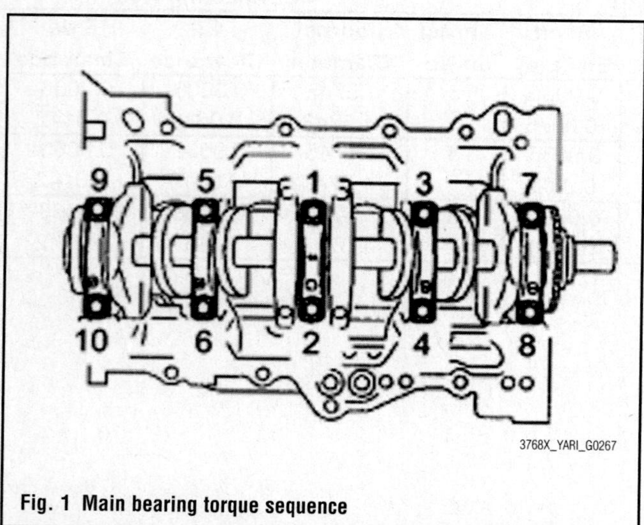

Fig. 1 Main bearing torque sequence

WHEEL ALIGNMENT

Year	Model		Caster Range (+/-Deg.)	Caster Preferred Setting (Deg.)	Camber Range (+/-Deg.)	Camber Preferred Setting (Deg.)	Toe-in (in.)	Steering Axis Inclination (Deg.)
2008	Yaris Sedan	F	0.75	+2.12	0.75	-0.47	0+/-0.08	13.15+/-0.75
		R	—	—	0.75	-1.18	0.12+/-0.08	—
	Yaris Hatchback	F	0.75	+2.02	0.75	-0.42	0+/-0.08	13.07+/-0.75
		R	—	—	0.75	-1.18	0.12+/-0.08	—
2009	Yaris Sedan	F	0.75	+2.12	0.75	-0.47	0+/-0.08	13.15+/-0.75
		R	—	—	0.75	-1.18	0.12+/-0.08	—
	Yaris Hatchback	F	0.75	+2.02	0.75	-0.42	0+/-0.08	13.07+/-0.75
		R	—	—	0.75	-1.18	0.12+/-0.08	—
2010	Yaris Sedan	F	0.75	+2.12	0.75	-0.47	0+/-0.08	13.15+/-0.75
		R	—	—	0.75	-1.18	0.12+/-0.08	—
	Yaris Hatchback	F	0.75	+2.02	0.75	-0.42	0+/-0.08	13.07+/-0.75
		R	—	—	0.75	-1.18	0.12+/-0.08	—

3768X_YARI_C0011

TIRE, WHEEL AND BALL JOINT SPECIFICATIONS

Year	Model	OEM Tires Standard	OEM Tires Optional	Tire Pressures (psi) Front	Tire Pressures (psi) Rear	Wheel Size	Ball Joint Inspection	Lug Nut (ft. lbs.)
2008	Yaris	P175/65R14	P185/60R15	32	32	5.5-JJ	5.2-30 in. ①	76
2009	Yaris	P175/65R14	P185/60R15	32	32	5.5-JJ	5.2-30 in. ①	76
2010	Yaris	P175/65R14	P185/60R15	32	32	5.5-JJ	5.2-30 in. ①	76

OEM: Original Equipment Manufacturer

PSI: Pounds Per Square Inch

① Torque required in inch lbs. to rotate ball joint when removed from the knuckle

3768X_YARI_C0012

BRAKE SPECIFICATIONS

All measurements in inches unless noted

Year	Model	Brake Disc Original Thickness	Brake Disc Minimum Thickness	Brake Disc Maximum Runout	Brake Drum Diameter Original Inside Diameter	Brake Drum Diameter Max. Wear Limit	Brake Drum Diameter Maximum Machine Diameter	Minimum Lining Thickness	Brake Caliper Bracket Bolts (ft. lbs.)	Brake Caliper Mounting Bolts (ft. lbs.)
2008	Yaris	0.709	0.630	0.0020	7.09	NA	7.13	0.039	79	25
2009	Yaris	0.709	0.630	0.0020	7.09	NA	7.13	0.039	79	25
2010	Yaris	0.709	0.630	0.0020	7.09	NA	7.13	0.039	79	25

NA: Not Available

F: Front

R: Rear

3768X_YARI_C0013

SCHEDULED MAINTENANCE INTERVALS
TOYOTA—Yaris

TO BE SERVICED	TYPE OF SERVICE	VEHICLE MILEAGE INTERVAL (x1000)													
		5	10	15	20	25	30	35	40	45	50	55	60	90	120
Engine oil & filter	R	✓	✓	✓	✓	✓	✓	✓	✓	✓	✓	✓	✓	✓	✓
Drive belts	S/I						✓						✓	✓	✓
Automatic transaxle fluid & filter	S/I						✓						✓	✓	✓
Brake line pipes & hoses	S/I	✓	✓	✓	✓	✓	✓	✓	✓	✓	✓	✓	✓	✓	✓
Brake linings & drums	S/I	✓	✓	✓	✓	✓	✓	✓	✓	✓	✓	✓	✓	✓	✓
Brake pads & discs (front & rear if equipped)	S/I	✓	✓	✓	✓	✓	✓	✓	✓	✓	✓	✓	✓	✓	✓
Cabin air filter	R				✓				✓				✓		✓
Differential oil	S/I						✓						✓	✓	✓
Drive shaft boots	S/I	✓	✓	✓	✓	✓	✓	✓	✓	✓	✓	✓	✓	✓	✓
Drive shaft bolt (tighten)	S/I	✓	✓	✓	✓	✓	✓	✓	✓	✓	✓	✓	✓	✓	✓
Engine coolant	S/I			✓			✓			✓			✓	✓	✓
Manual transaxle oil	S/I						✓						✓	✓	✓
Steering gear housing oil	S/I	✓	✓	✓	✓	✓	✓	✓	✓	✓	✓	✓	✓	✓	✓
Steering linkage	S/I	✓	✓	✓	✓	✓	✓	✓	✓	✓	✓	✓	✓	✓	✓
Air filter	R						✓						✓	✓	✓
Rotate tires	S/I	✓	✓	✓	✓	✓	✓	✓	✓	✓	✓	✓	✓	✓	✓
Spark plugs	R									✓				✓	
Fuel lines & connections	S/I						✓			✓			✓	✓	✓
Fuel tank cap gasket	R									✓				✓	
Charcoal canister	S/I									✓				✓	

R: Replace S/I: Service or Inspect

FREQUENT OPERATION MAINTENANCE (SEVERE SERVICE)

If a vehicle is operated under any of the following conditions it is considered severe service:

- Extremely dusty areas.

- 50% or more of the vehicle operation is in 32°C (90°F) or higher temperatures, or constant operation in temperatures below 0°C (32°F).

- Prolonged idling (vehicle operation in stop and go traffic).

- Frequent short running periods (engine does not warm to normal operating temperatures).

- Police, taxi, delivery usage or trailer towing usage.

Oil & oil filter: change every 5000 miles.

Bolts & nuts on chassis & body: tighten every 5000 miles.

Ball joints & dust covers: service or inspect every 5,000 miles.

Drive shaft boots & except Supra): service or inspect every 12,000 miles.

Steering linkage: service or inspect every 12,000 miles.

Air filter: service or inspect every 5,000 miles.

Exhaust system: service or inspect every 15,000 miles.

Timing belt: replace every 60,000 miles.

3768X_YARI_C0014

BRAKES — INFORMATION AND PRECAUTIONS

ANTI-LOCK SYSTEMS

- Certain components within the ABS system are not intended to be serviced or repaired individually.
- Do not use rubber hoses or other parts not specifically specified for and ABS system. When using repair kits, replace all parts included in the kit. Partial or incorrect repair may lead to functional problems and require the replacement of components.
- Lubricate rubber parts with clean, fresh brake fluid to ease assembly. Do not use shop air to clean parts; damage to rubber components may result.
- Use only DOT 3 brake fluid from an unopened container.
- If any hydraulic component or line is removed or replaced, it may be necessary to bleed the entire system.
- A clean repair area is essential. Always clean the reservoir and cap thoroughly before removing the cap. The slightest amount of dirt in the fluid may plug an orifice and impair the system function. Perform repairs after components have been thoroughly cleaned; use only denatured alcohol to clean components. Do not allow ABS components to come into contact with any substance containing mineral oil; this includes used shop rags.
- The Anti-Lock control unit is a microprocessor similar to other computer units in the vehicle. Ensure that the ignition switch is **OFF** before removing or installing controller harnesses. Avoid static electricity discharge at or near the controller.
- If any arc welding is to be done on the vehicle, the control unit should be unplugged before welding operations begin.

DISC AND DRUM SYSTEMS

CAUTION

Dust and dirt accumulating on brake parts during normal use may contain asbestos fibers from production or aftermarket brake linings. Breathing excessive concentrations of asbestos fibers can cause serious bodily harm. Exercise care when servicing brake parts. Do not sand or grind brake lining unless equipment used is designed to contain the dust residue. Do not clean brake parts with compressed air or by dry brushing. Cleaning should be done by dampening the brake components with a fine mist of water, then wiping the brake components clean with a dampened cloth. Dispose of cloth and all residue containing asbestos fibers in an impermeable container with the appropriate label. Follow practices prescribed by the Occupational Safety and Health Administration (OSHA) and the Environmental Protection Agency (EPA) for the handling, processing, and disposing of dust or debris that may contain asbestos fibers.

BRAKES — BLEEDING THE BRAKE SYSTEM

BLEEDING PROCEDURE

➡Immediately wash off any brake fluid that comes into contact with any painted surfaces.

➡If any work is done on the brake system or if air in the brake lines is suspected, bleed the air from the system.

1. Fill reservoir with brake fluid:
   a. Disengage the 3 clips and separate the hood to cowl top seal.
   b. Remove the cowl top ventilator louver.
   c. Set the brake fluid can upside down on the reservoir. Fluid: SAE J1703 or FMVSS No. 116 DOT3.
2. Bleed master cylinder:
   a. Using a union nut wrench, disconnect the brake tubes from the master cylinder.
   b. Slowly depress the brake pedal and hold it there.
   c. Block the outer holes with your fingers, and release the brake pedal.
   d. Repeat 3 or 4 times.
   e. Using a union nut wrench, connect the brake tubes to the master cylinder. Tighten to 10-11 ft. lbs. (144-155 Nm).
3. Bleed brake line:
   a. Connect the vinyl tube to the bleeder plug.
   b. Depress the brake pedal several times, then loosen the bleeder plug with the pedal depressed.
   c. At the point where the fluid stops coming out, tighten the bleeder plug, then release the brake pedal.
   d. Repeat until all the air in the fluid is out.
   e. Tighten the bleeder plug to 73 inch lbs. (8.3 Nm).
   f. Repeat the above procedure to bleed the air out of the brake line for each wheel.

➡After bleeding the air from the brake system, if the height or feel of the brake pedal cannot be obtained, perform air bleeding of the brake actuator with the Techstream by following the procedure below.

4. Bleed brake actuator (with VSC):
   a. Depress the brake pedal more than 20 times with the engine off.
   b. Connect the Techstream to the DLC3, and turn the ignition switch ON.

➡Do not start the engine.

   c. Select "Air Bleeding" on the Techstream.
5. Bleed the air out of the brake line when Step 1: Increase appears on the Techstream display.

➡Bleed the air by following the steps displayed on the Techstream. Make sure that the master cylinder reservoir tank does not become empty.

   a. Connect the vinyl tube to either one of the bleeder plugs.
   b. Depress the brake pedal several times, then loosen the bleeder plug connected to the vinyl tube with the pedal depressed (Step E).
   c. When fluid stops coming out, tighten the bleeder plug and release the brake pedal (Step F).
   - Repeat Steps E and F until all the air in the fluid is completely bled out.
   - Tighten the bleeder plug completely. Tighten front brake to 73 inch lbs. (8.3 Nm). Tighten rear brake to 75 inch lbs. (8.5 Nm).
   - Repeat the above procedure for each wheel to bleed the air out of the brake line.

d. Bleed the air out of the suction line when Step 2: Inhalation appears on the Techstream display.

➡**Bleed the air by following the steps displayed on the Techstream. Make sure that the master cylinder reservoir tank does not become empty.**

- Connect the vinyl tube to the bleeder plug at the right front wheel or the right rear wheel and loosen the bleeder plug.
- Operate the brake actuator to bleed the air using the Techstream (Step G).

➡**Release the brake pedal at this time. This operation stops automatically after 4 seconds.**

- Check if the operation has stopped by referring to the Techstream display and tighten the bleeder plug (Step H).
- Repeat Steps G and H until all the air in the fluid is completely bled out.
- Tighten the bleeder plug. Tighten front brake to 73 inch lbs. (8.3 Nm). Tighten rear brake to 75 inch lbs. (8.5 Nm).
- Repeat the above procedure for the other wheels to bleed the air out of the brake line.

e. Bleed the air out of the pressure reduction line when Step 3: Decrease appears on the Techstream display.

➡**Bleed the air by following the steps displayed on the Techstream. Make sure that the master cylinder reservoir tank does not become empty.**

- Connect a vinyl tube to either one of the bleeder plugs.
- Loosen the bleeder plug (Step I).
- Using the Techstream, operate the brake actuator assembly, completely depress the brake pedal and hold it there (Step J).

➡**During this procedure, the pedal will feel heavy, but completely depress it so that the brake fluid comes out of the bleeder plug. Hold the brake pedal depressed. Do not depress and release the pedal repeatedly.**

➡**The operation stops automatically after 4 seconds. When performing this procedure continuously, set an interval of at least 20 seconds. When the operation is complete, the brake pedal goes down slightly. This is a normal phenomenon caused when the solenoid opens.**

- Tighten the bleeder plug, then release the brake pedal (Step K).
- Repeat Steps I to K until all the air in the fluid is completely bled out.
- Tighten the bleeder plug. Tighten front brake to 73 inch lbs. (8.3 Nm). Tighten rear brake to 75 inch lbs. (8.5 Nm).
- Repeat the above procedure for the other wheels to bleed the air out of the brake line.

f. Bleed the air out of the brake line again when Step 4: Increase appears on the Techstream display.

➡**Bleed the air by following the steps displayed on the Techstream. Make sure that the master cylinder reservoir tank does not become empty.**

- Connect the vinyl tube to either one of the bleeder plugs.
- Depress the brake pedal several times, then loosen the bleeder plug connected to the vinyl tube with the pedal depressed (Step L).
- When fluid stops coming out, tighten the bleeder plug, then release the brake pedal (Step M).
- Repeat Steps L and M until all the air in the fluid is completely bled out.
- Tighten the bleeder plug. Tighten front brake to 73 inch lbs. (8.3 Nm). Tighten rear brake to 75 inch lbs. (8.5 Nm).
- Repeat the above procedure for the other wheels to bleed the air out of the brake line.
- Make sure that the air bleeding is complete by referring to the Techstream display and turn off the Techstream.
- Disconnect the Techstream from the DLC3.
- Turn the ignition switch OFF.

6. Check the fluid level and add fluid if necessary. Fluid: SAE J1703 or FMVSS No. 116 DOT3.

BRAKES — ANTI-LOCK BRAKE SYSTEM (ABS)

WHEEL SPEED SENSORS

REMOVAL & INSTALLATION

Front

See Figure 2.

1. Disconnect the negative battery cable.
2. Remove front wheel.
3. Remove the 3 screws, 6 clips and 4 grommets and remove the front fender liner.
4. Remove the speed sensor clip from the body.
5. Disconnect the speed sensor connector.
6. Remove the 3 clips.
7. Remove the bolt and separate the clamp from the body.
8. Remove the bolt and separate the clamp from the shock absorber.
9. Remove the bolt and remove the speed sensor from the steering knuckle.

➡**Keep the speed sensor tip and installation portion free of foreign matter. Remove the speed sensor without turning it from its original installation angle.**

10. To install, reverse the removal procedure.
11. Observe and use the torque values in the illustration.

Rear

See Figure 3.

➡**The rear wheel speed sensor is integrated with the wheel bearing and hub assembly and must be replaced as a unit.**

1. Disconnect the negative battery cable.
2. Remove front wheel.
3. Remove the rear drum assembly.
4. Using a screwdriver, remove the claw of the connector lock portion and disconnect the skid control sensor wire connector.
5. Remove the rear wheel bearing and hub assembly. Refer to WHEEL BEARING AND HUB ASSEMBLY under REAR SUSPENSION.
6. To install, reverse the removal procedure.

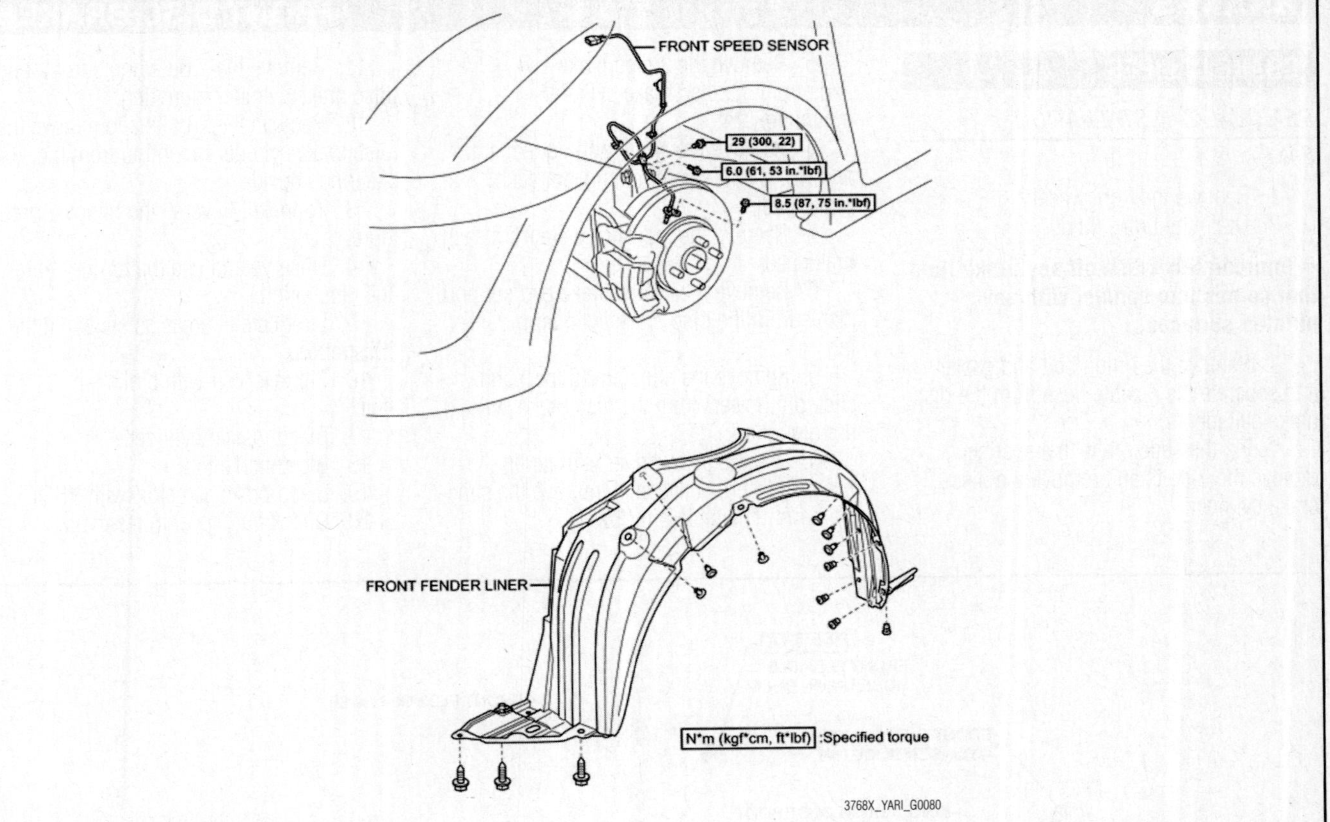

Fig. 2 View of the front speed sensor

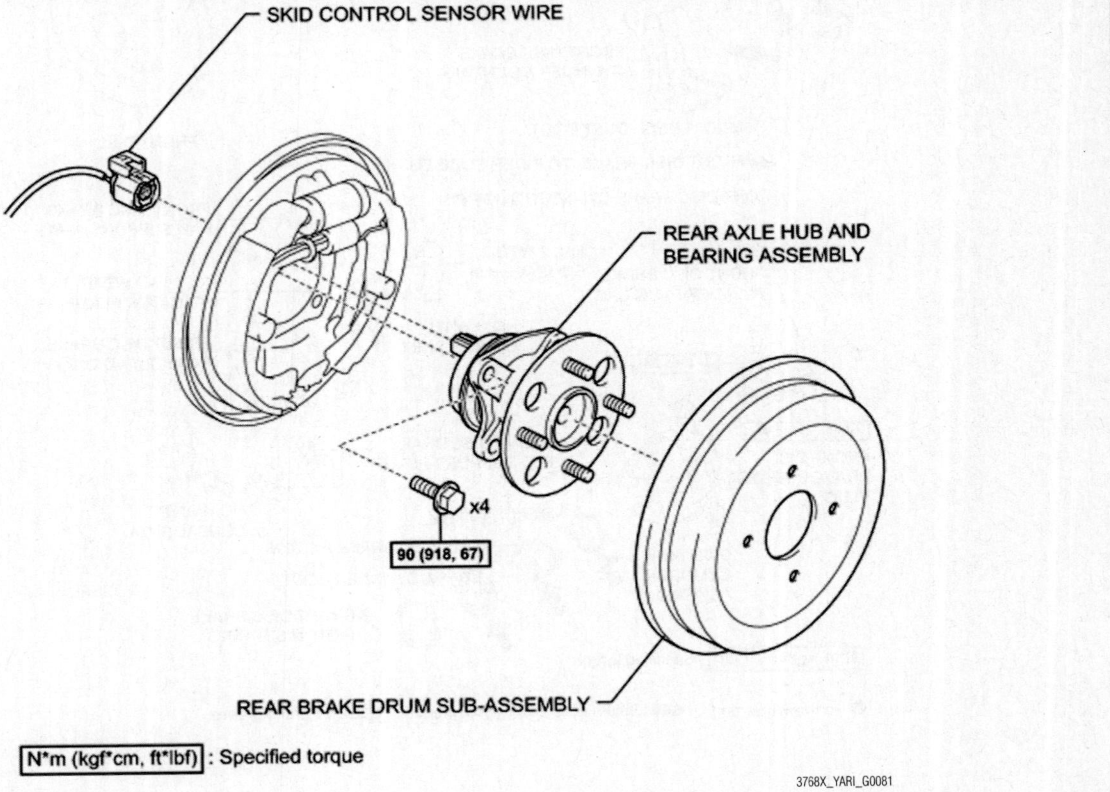

Fig. 3 View of the front speed sensor

BRAKE CALIPER

REMOVAL & INSTALLATION

See Figure 4.

1. Remove the front wheel.
2. Drain the brake fluid.

➡Immediately wash off any brake fluid that comes into contact with any painted surfaces.

3. Remove the union bolt and gasket and separate the flexible hose from the disc brake cylinder.
4. Fix the slide pin with a spanner, remove the 2 bolts and remove the disc brake cylinder.
5. Remove the 2 disc brake pads from the disc brake cylinder mounting.
6. Remove the No. 1 anti squeal shim and No. 2 anti squeal shim from each brake pad.
7. Remove the indicator plate from each brake pad.
8. Remove the 4 disc brake pad support plates from the disc brake cylinder mounting.
9. Remove the slide pin (upper) and slide pin (lower) from the disc brake cylinder mounting.
10. Using a screwdriver with its tip wrapped in protective tape, remove the slide bush from the slide pin (lower).
11. Remove the 2 dust boots from the disc brake cylinder mounting.
12. Remove the 2 bolts and remove the disc brake cylinder mounting from the steering knuckle.
13. To install, reverse the removal procedure.
14. Observe and use the torque values in the illustration.
15. Lubricate in areas as shown in the illustration.
16. Fill reservoir with brake fluid.
17. Bleed master cylinder.
18. Bleed brake line.
19. Bleed brake actuator (with VSC).
20. Check fluid level in reservoir.

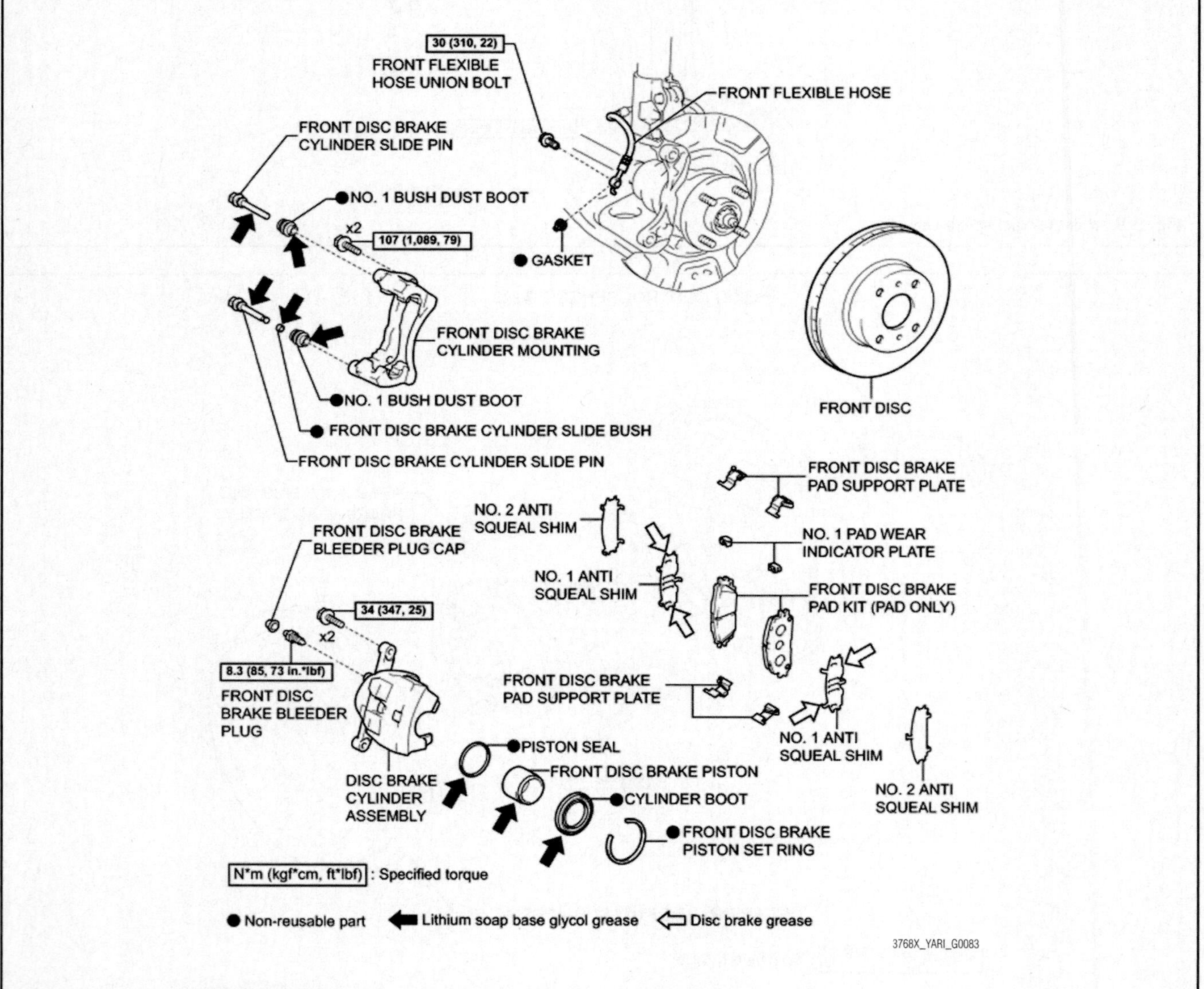

Fig. 4 Exploded view of the front brake assembly

21. Check for brake fluid leakage.
22. Install front wheel.

DISC BRAKE PADS

REMOVAL & INSTALLATION

See Figure 4.

1. Remove the front wheel.
2. Loosen and remove the caliper mounting bolts, then remove the caliper assembly, without disconnecting the brake line. Position it aside.
3. Slide out the old brake pads along with any anti-squeal shims, springs, pad wear indicators and pad support plates.

To install:

4. Install the pad support plates into the torque plate.
5. Install the pad wear indicators onto the pads. Be sure the arrow on the indicator plate is pointing in the direction of rotation.
6. Install the anti-squeal shims on the outside of each pad and then install the pad assemblies into the torque plate.
7. Compress the caliper piston into the bore.
8. Position the caliper back down over the pads.
9. Install and tighten the caliper mounting bolts.
10. Install the wheels. Check the brake fluid level.

BRAKES

REAR DRUM BRAKES

BRAKE DRUM

REMOVAL & INSTALLATION

See Figure 5.

1. Remove the rear wheel.
2. Drain the brake fluid.
3. Release the parking brake and remove the rear brake drum.
4. If the rear brake drum cannot be removed easily, perform the following procedure.:
   a. Remove the hole plug and insert a screwdriver through the hole into the backing plate, and hold the automatic adjust lever away from the adjuster.
   b. Using another screwdriver, contract the brake shoe by turning the adjusting bolt.

To install:

5. Provisionally install the 2 hub nuts.
6. Remove the hole plug, and turn the adjuster to expand the shoe until the drum locks.
7. Using a screwdriver, release the adjuster 12 notches.
8. Install the hole plug.
9. Fill reservoir with brake fluid.
10. Bleed master cylinder.
11. Bleed brake line.
12. Bleed brake actuator (with VSC).
13. Check fluid level in reservoir.
14. Check for brake fluid leakage.
15. Install the rear wheel.
16. Inspect parking brake lever travel.
17. Adjust parking brake lever travel.

BRAKE SHOES

REMOVAL & INSTALLATION

See Figures 5 through 7.

1. Using SST 09703-30010, separate the shoe return spring from the front brake shoe.
2. Using SST 09718-00010, remove the shoe hold down spring cup, shoe hold down spring, pin and front brake shoe.
3. Remove the tension spring.
4. Remove the shoe return spring from the rear brake shoe and remove the parking brake shoe strut set.
5. Using SST 09718-00010, remove the shoe hold down spring cup, shoe hold down spring, pin and rear brake shoe.
6. Using needle-nose pliers, separate the parking brake cable.
7. Remove the automatic adjust lever tension spring and remove the automatic adjust lever.
8. Using a screwdriver, remove the C-washer and remove the parking brake shoe lever.

To install:

➡Refer to the exploded illustration for torque values and grease/lubrications points

9. Using needle-nose pliers, install the parking brake shoe lever with a new C-washer.
10. Install the automatic adjust lever and automatic adjust lever tension spring onto the front brake shoe.
11. Install rear brake shoe kit:
    a. Apply high temperature grease to the surface of the backing plate which is in contact with the shoe.
    b. Using needle-nose pliers, install the parking brake cable onto the parking brake shoe lever.
    c. Using SST 09718-00010, install the rear brake shoe, pin, shoe hold down spring and shoe hold down spring cup.
    d. Apply high temperature grease to the adjusting bolt.
    e. Install the parking brake shoe strut set as shown in the illustration.
    f. Using SST 09718-00010, install the front brake shoe, pin, shoe hold down spring and shoe hold down spring cup.
    g. Using needle-nose pliers, install the tension spring onto the front brake shoe and rear brake shoe.
    h. Using SST 09703-30010, install the shoe return spring onto the front brake shoe.
12. Check that each part is installed properly.
13. Measure the brake drum inner diameter and the diameter of the brake shoes. Check that the difference between the diameters is equal to the specified shoe clearance. Shoe clearance: 0.024 inches. (0.6 mm).

➡There should be no oil or grease adhering to the friction surfaces of the shoe lining or the drum.

ADJUSTMENT

1. Provisionally install the 2 hub nuts.
2. Remove the hole plug, and turn the adjuster to expand the shoe until the drum locks.
3. Using a screwdriver, release the adjuster 12 notches.
4. Install the hole plug.

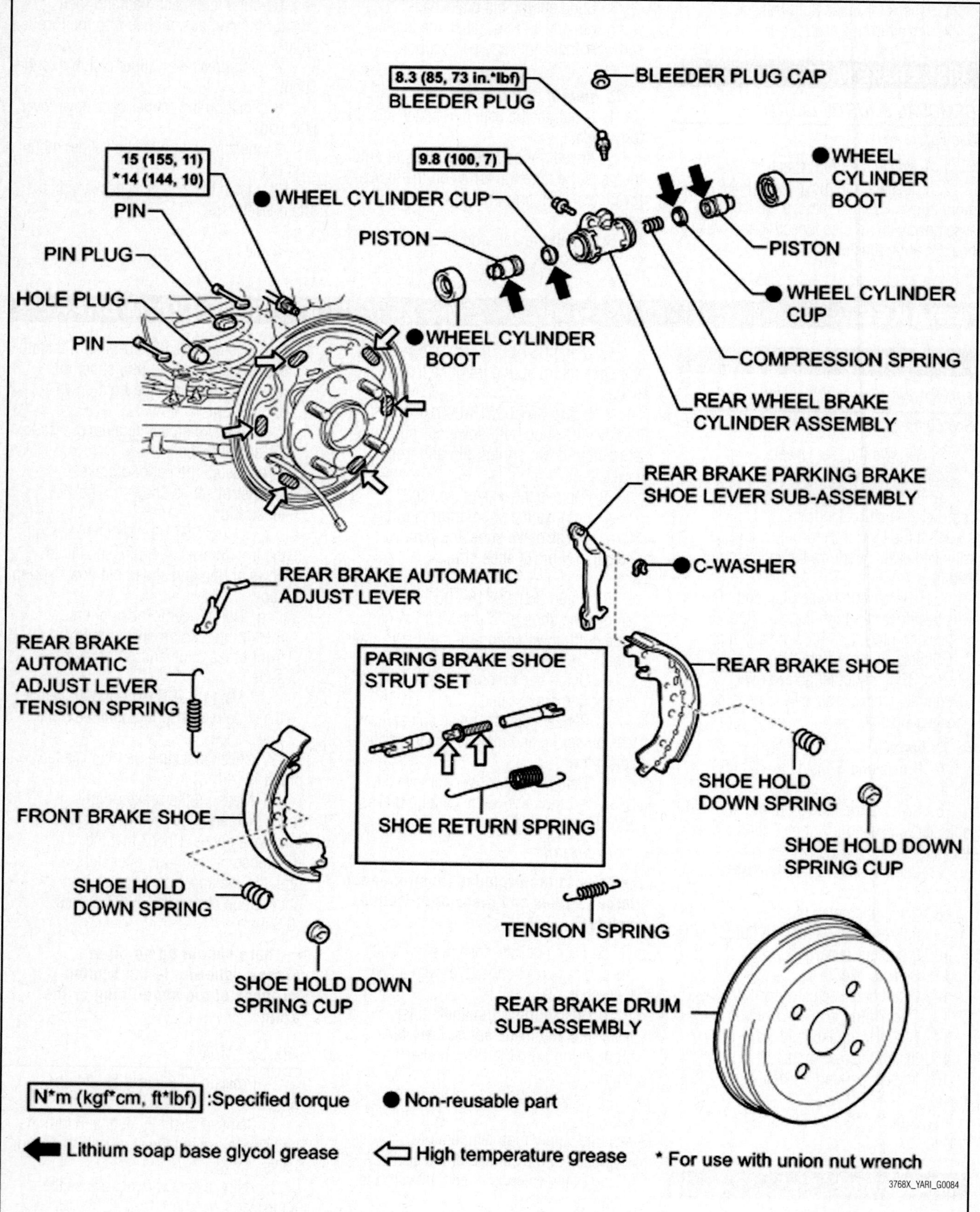

Fig. 5 Exploded view of the rear drum brake and assembly

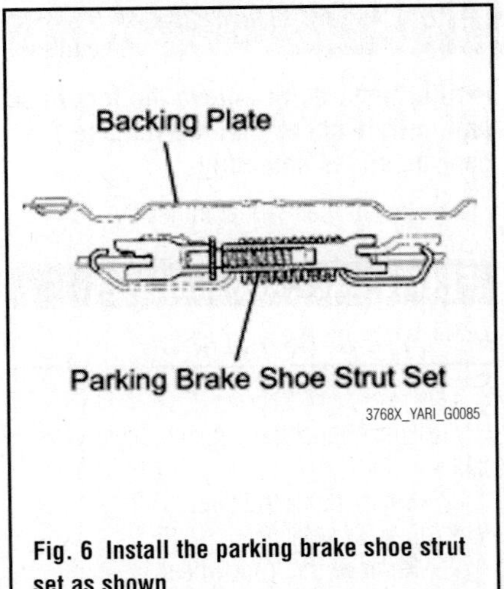

Fig. 6 Install the parking brake shoe strut set as shown

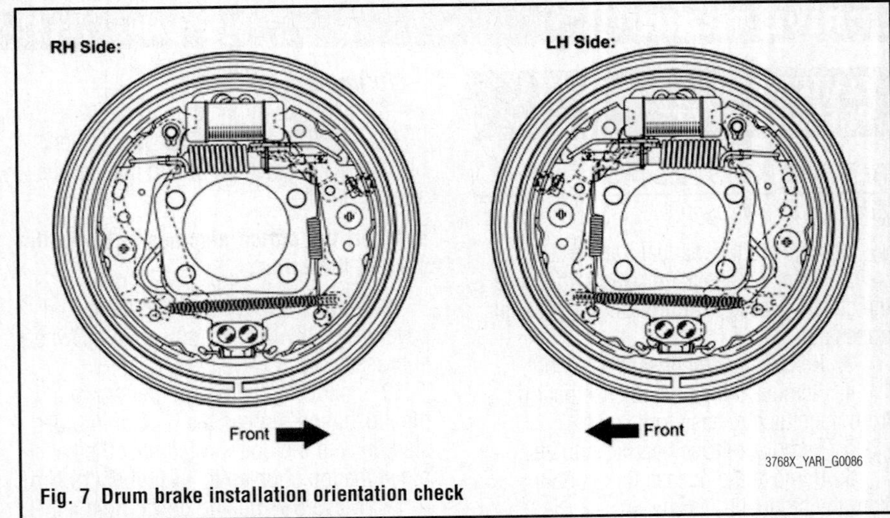

Fig. 7 Drum brake installation orientation check

BRAKES

PARKING BRAKE

PARKING BRAKE CABLES

ADJUSTMENT

See Figure 8.

1. Remove rear console box assembly.
2. Loosen the lock nut and turn the adjusting nut until the parking brake lever travel is corrected to within the specified range. Parking brake lever travel: 6 to 9 clicks (45 lbs.).
3. Tighten the lock nut to 48 inch lbs.
4. Operate the parking brake lever 3 to 4 times, and check the parking brake lever travel.
5. Check whether the parking brake drags or not.
6. When operating the parking brake lever, check that the brake warning light illuminates. Standard: Brake warning light always illuminates at the first click.
7. Reinstall the center console.

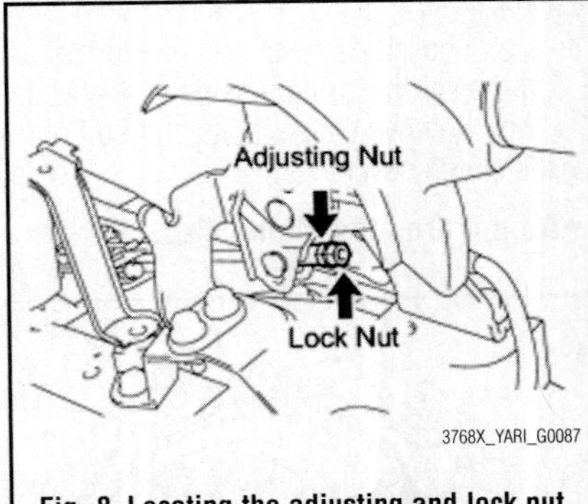

Fig. 8 Locating the adjusting and lock nut

CHASSIS ELECTRICAL

AIR BAG (SUPPLEMENTAL RESTRAINT SYSTEM)

GENERAL INFORMATION

CAUTION

These vehicles are equipped with an air bag system. The system must be disarmed before performing service on, or around, system components, the steering column, instrument panel components, wiring and sensors. Failure to follow the safety precautions and the disarming procedure could result in accidental air bag deployment, possible injury and unnecessary system repairs.

SERVICE PRECAUTIONS

CAUTION

Disconnect and isolate the battery negative cable before beginning any airbag system component diagnosis, testing, removal, or installation procedures. Wait at least 90 seconds after the ignition switch is turned off and the negative (-) terminal cable is disconnected from the battery before starting the operation. The SRS is equipped with a backup power source, so if work is started within 90 seconds after disconnecting the negative (-) terminal cable from the battery, the SRS may be deployed. Failure to disable the airbag system may result in accidental airbag deployment, personal injury, or death.

DISARMING THE SYSTEM

To avoid personal injury when working on vehicles equipped with an air bag, the negative battery cable must be disconnected and at least 90 seconds must elapse before working on the system. Failure to do so may result in deployment of the air bag.

ARMING THE SYSTEM

After vehicle service is completed, reattach the battery cables (positive cable first!) to rearm the air bag system.

CLOCKSPRING CENTERING

1. Check that the ignition switch is off.
2. Check that the battery negative (-) terminal is disconnected.
3. Rotate the spiral cable counterclockwise slowly by hand until it feels firm.

➡Do not turn the spiral cable by the airbag wire harness.

4. Rotate the spiral cable clockwise approximately 2.5 turns to align the marks.

➡Do not turn the spiral cable by the airbag wire harness. The spiral cable will rotate approximately 2.5 turns to both the left and right from the center.

DRIVE TRAIN

CLUTCH DRIVEN DISC & PRESSURE PLATE

REMOVAL & INSTALLATION

See Figures 9 through 11.

1. Remove the manual transaxle.
2. Remove the clutch release fork with the clutch release bearing from the manual transaxle.
3. Remove clutch release fork boot.
4. Remove the clutch release bearing from the clutch release fork.
5. Remove release bearing hub clip.
6. Remove the release fork support from the manual transaxle.
7. Remove clutch cover assembly:
   a. Align the matchmark on the clutch cover assembly with the one on the flywheel.
   b. Loosen each set bolt one turn at a time until the spring tension is released.
   c. Remove the set bolts and pull off the clutch cover.

➡Do not drop the clutch disc.

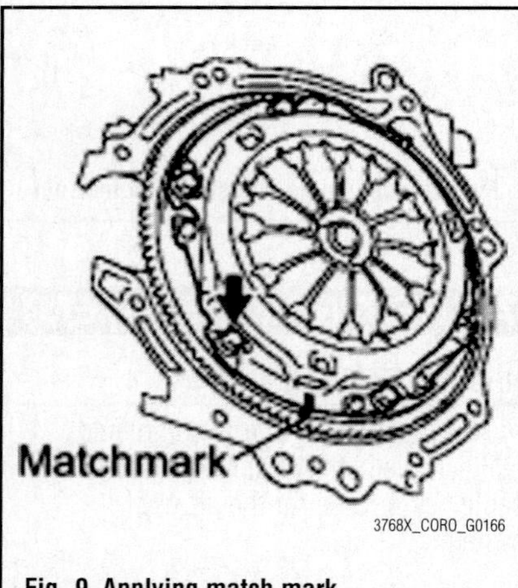

Fig. 9 Applying match mark

To install:

8. Insert the alignment tool (SST: 09301-00110) into the clutch disc assembly, then insert them both into the flywheel sub-assembly.

➡Insert the clutch disc assembly in the correct direction.

9. Align the matchmark on the clutch cover assembly with the one on the flywheel sub-assembly.
10. Following the procedure shown in the illustration, tighten the 6 bolts in order, starting with the bolt located near the knock pin at the top. Tighten to 14 ft. lbs. (19 Nm).
11. Move SST up and down, right and left lightly after checking that the disc is in the center, and tighten the bolts.
12. Using a dial indicator with a roller instrument, check the diaphragm spring tip alignment. Maximum non-alignment: 0.020 inch (0.5 mm)
13. If the alignment is not as specified, adjust the diaphragm spring tip alignment using SST: 09333-00013.
14. Install the release fork support onto the transaxle assembly. Tighten to 27 ft. lbs. (37 Nm).
15. Install the clutch release fork boot to the manual transaxle.
16. Install release bearing hub clip.
17. Apply release hub grease to the contact surfaces between the release fork and release bearing assembly, release fork and push rod, and release fork and fork support. Sealant: Toyota Genuine Release Hub Grease or Equivalent.
18. Install the release fork onto the release bearing assembly with the clip.
19. Apply clutch spline grease to the input shaft spline. DO NOT grease the smooth collar area of the spline.
20. Install the clutch release bearing with the release fork onto the transaxle assembly.

➡After installation, move the fork back and forth to check that the release bearing slides smoothly.

21. Install the manual transaxle.

FRONT AXLE ASSEMBLY

REMOVAL & INSTALLATION

See Figures 12 and 13.

1. Disconnect the negative battery cable.
2. Safely raise and support the vehicle.
3. Remove the front wheel.
4. Separate the front caliper.
5. Remove the front disc.
6. Using SST 09628-10011 and a screwdriver or the equivalent to hold the axle hub, remove the hub bolt.
7. Separate front speed sensor (w/ abs).
8. Separate front stabilizer link assembly.
9. Separate tie rod end sub-assembly.
10. Separate front lower suspension arm.
11. Remove front axle assembly:
    a. Using a plastic hammer, tap the end of the drive shaft and disengage the fitting between the drive shaft and front axle.

➡If it is difficult to disengage the fitting, tap the end of the drive shaft with a brass bar and hammer.

    b. Push the front axle out of the vehicle to remove the drive shaft from the front axle.
    - Do not push the front axle further out of the vehicle than is necessary.
    - Do not damage the speed sensor rotor.
    - Suspend the drive shaft with a piece of string or the equivalent.

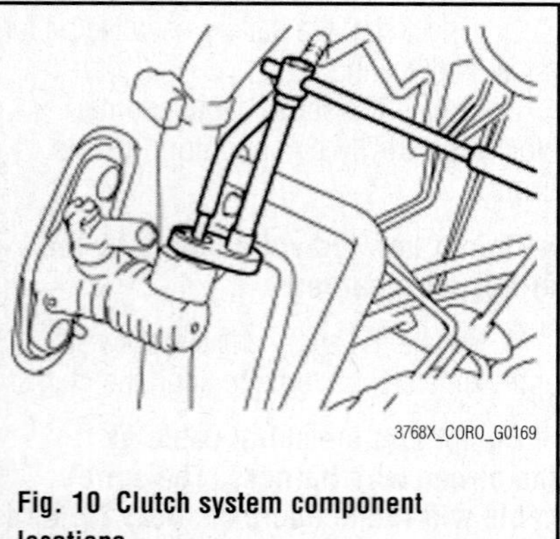

Fig. 10 Clutch system component locations

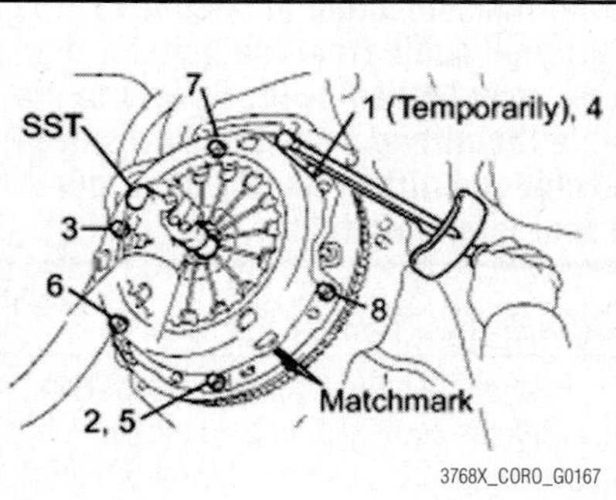

Fig. 11 Clutch assembly tightening sequence

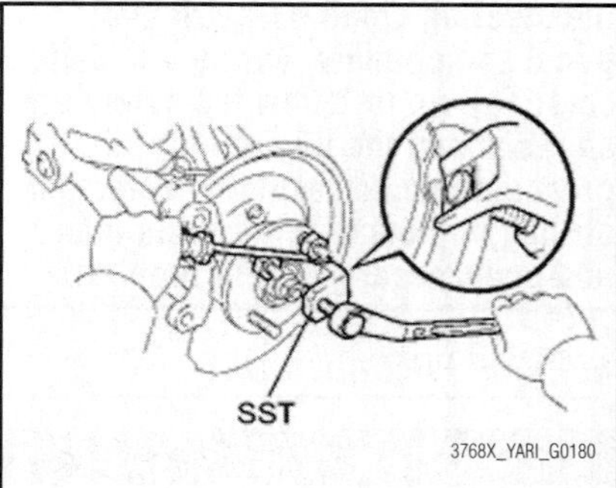

Fig. 12 Using SST 09628-10011 remove the hub bolt

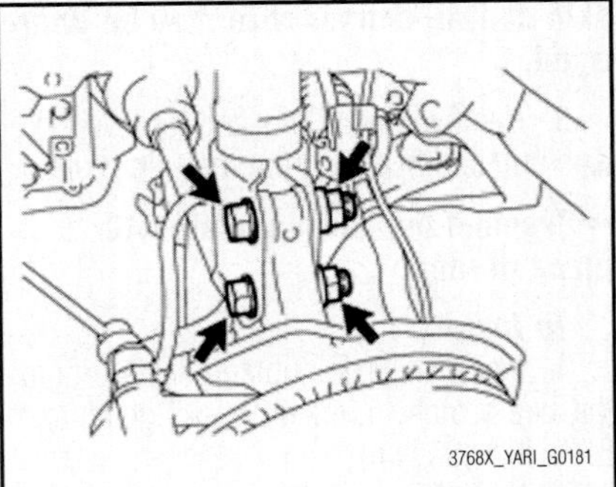

Fig. 13 Remove the 2 nuts and 2 bolts and remove the front axle assembly

c. Remove the 2 nuts and 2 bolts and remove the front axle assembly.

➡Keep the nut from rotating while turning the bolt.

To install:

12. To install, reverse the removal procedure.
13. Tighten the front axle assembly onto the shock absorber to 121 ft. lbs. (164 Nm).
14. Inspect and adjust front wheel alignment.

FRONT HALFSHAFT

REMOVAL & INSTALLATION

See Figures 14 through 17.

1. Disconnect the negative battery cable.
2. Drain automatic transaxle fluid.
3. Drain manual transaxle oil.
4. Remove front wheel.
5. Remove front axle hub nut.
6. Separate front speed sensor (w/ abs).
7. Separate front stabilizer link assembly.

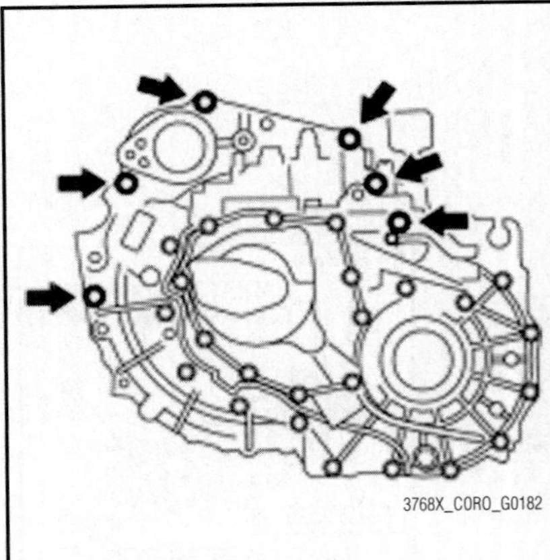

Fig. 14 Using SST: 09520-01010 and 09520-24010 remove the drive shaft

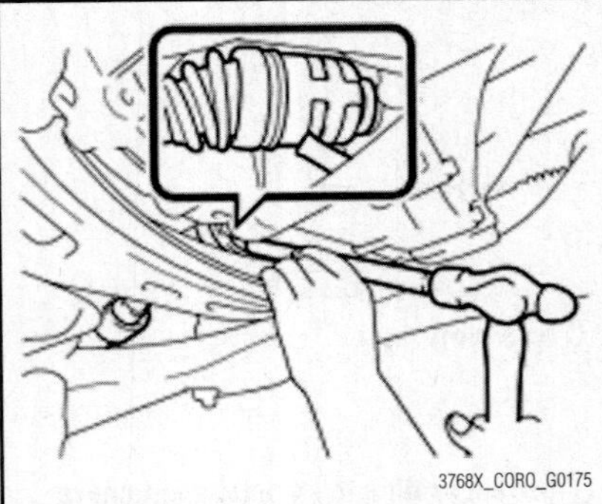

Fig. 15 Removing the RH halfshaft assembly

8. Separate tie rod end sub-assembly.
9. Separate front lower suspension arm.
10. Separate front axle assembly
11. Remove the 2 bolts and transmission case protector.

➡Do not damage the oil seal. Do not damage the inboard joint boot. Do not drop the drive shaft.

12. Remove front halfshaft assembly LH:
    a. Using SST: 09520-01010 and 09520-24010 remove the drive shaft.
13. Using a screwdriver and hammer, remove front drive shaft assembly RH

➡The hub bearing could be damaged if it is subjected to the vehicle's full weight, such as when moving the vehicle with the drive shaft removed. If it is absolutely necessary to place the vehicle's full weight on the hub bearing, first support it with SST 09608-16042.

To install:

14. Install front drive shaft assembly LH and RH:

Fig. 16 Using SST: 09608-16042 to support the hub bearing

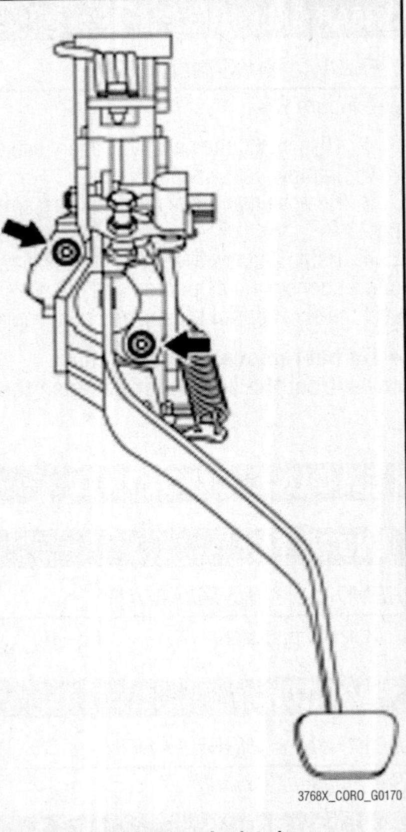

Fig. 17 Halfshafts and related components

a. For Automatic Transaxle: Coat the spline of the inboard joint with ATF.
b. For Manual Transaxle: Coat the spline of the inboard joint with gear oil.
c. Align the inboard joint splines and install the drive shaft with a screwdriver and hammer.

➡

- Face the cut area of the front drive inboard joint hole snap ring downward.
- Do not damage the oil seal.
- Do not damage the inboard joint boot.

➡Confirm whether the drive shaft is securely driven in by checking the reaction force and sound.

15. Install the transmission case protector with the 2 bolts.
16. To complete the installation, reverse the remaining removal procedure.
17. Add automatic transaxle fluid or manual transaxle oil.
18. Inspect the fluid.
19. Check for transmission leakage.
20. Inspect and adjust front wheel alignment.

REAR AXLE HUB

REMOVAL & INSTALLATION

See Figure 18.

1. Disconnect the negative battery cable.
2. Remove rear wheel.
3. Remove rear brake drum sub-assembly.
4. Using a screwdriver, remove the claw of the connector lock portion and disconnect the skid control sensor wire connector.

➡Do not remove the connector cover from the connector because the skid control sensor wire may be damaged.

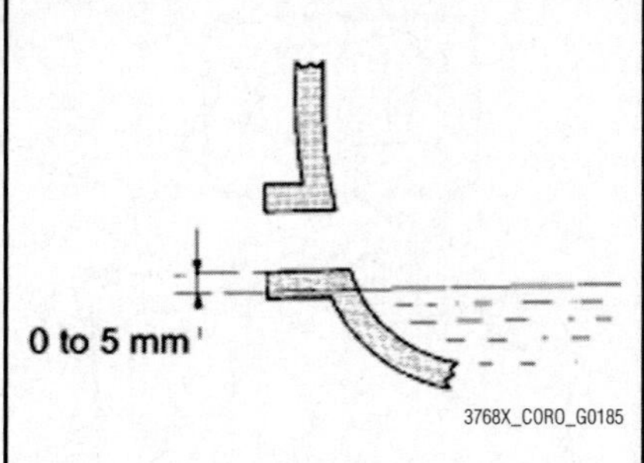

Fig. 18 Locating the 4 bolts and remove the axle hub and bearing from the axle beam

5. Remove the 4 bolts and remove the axle hub and bearing from the axle beam.

➡Suspend the backing plate with a piece of rope.

To install:

6. Install the axle hub and bearing onto the axle beam with the 4 bolts. Tighten to 67 ft. lbs. (90 Nm).
7. Inspect rear axle hub bearing.
8. To complete the installation, reverse the remaining removal procedure.

ENGINE COOLING

ENGINE FAN

REMOVAL & INSTALLATION

Refer to Radiator.

RADIATOR

REMOVAL & INSTALLATION

See Figures 19 through 20.

⁂ CAUTION

Be sure that the ignition is off if you work near the electric cooling fans or radiator grille. With the ignition on the electric cooling fans may automatically start to run if the engine coolant temperature is high and/or the air conditioning is on.

1. Disconnect the negative battery cable.
2. Drain and recycle the engine coolant.
3. Remove the front bumper.
4. Remove air cleaner assembly.
5. For the hatchback, disengage the 6 claws and remove the radiator support absorber upper.
6. If equipped, remove the 2 clips and No. 1 cooler cover.
7. Remove hood lock assembly (w/ theft deterrent system):
   a. Separate the hood lock control cable assembly from the 2 clamps.
   b. Separate the engine hood courtesy switch connector.
   c. Remove the 3 bolts and remove the hood lock assembly.
8. Remove hood lock assembly (w/o theft deterrent system):
   a. Separate the hood lock control cable assembly from the 2 clamps.
   b. Remove the 3 bolts and remove the hood lock assembly.

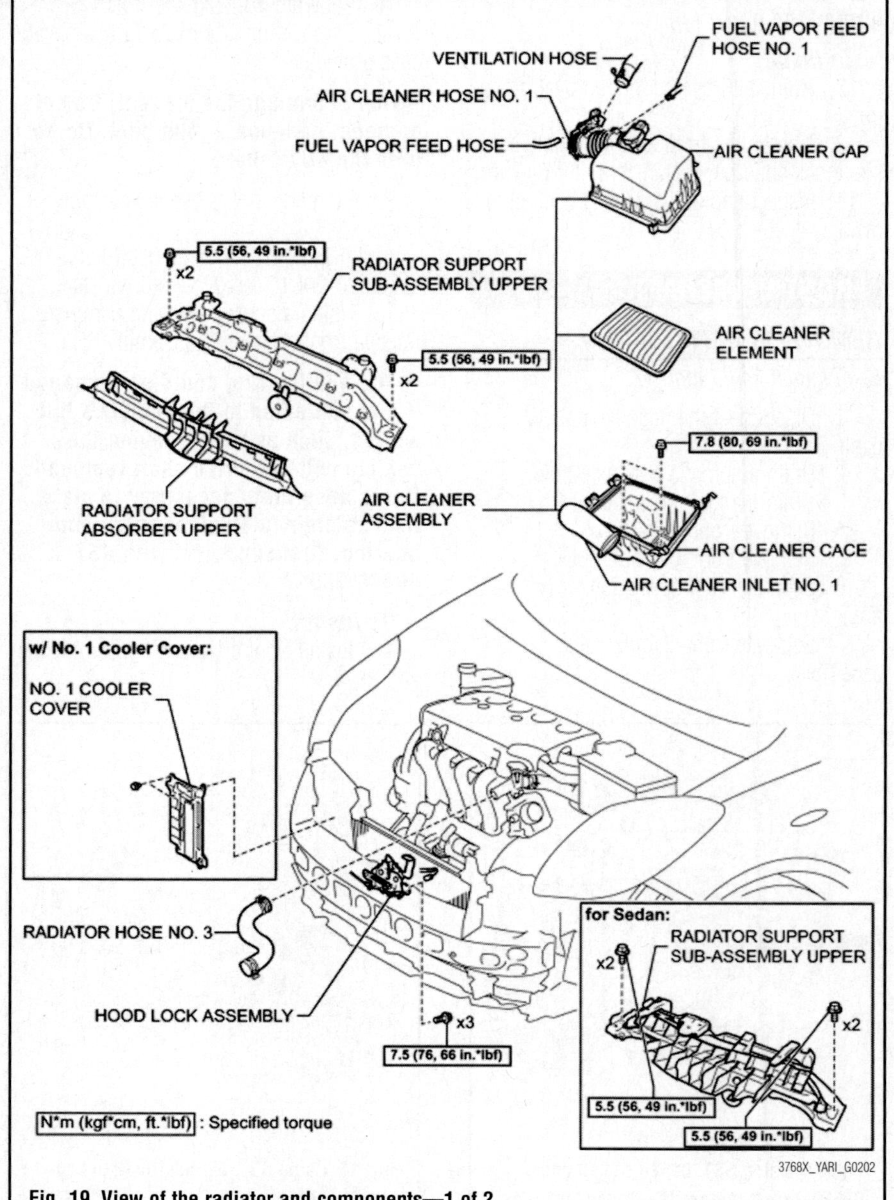

Fig. 19 View of the radiator and components—1 of 2

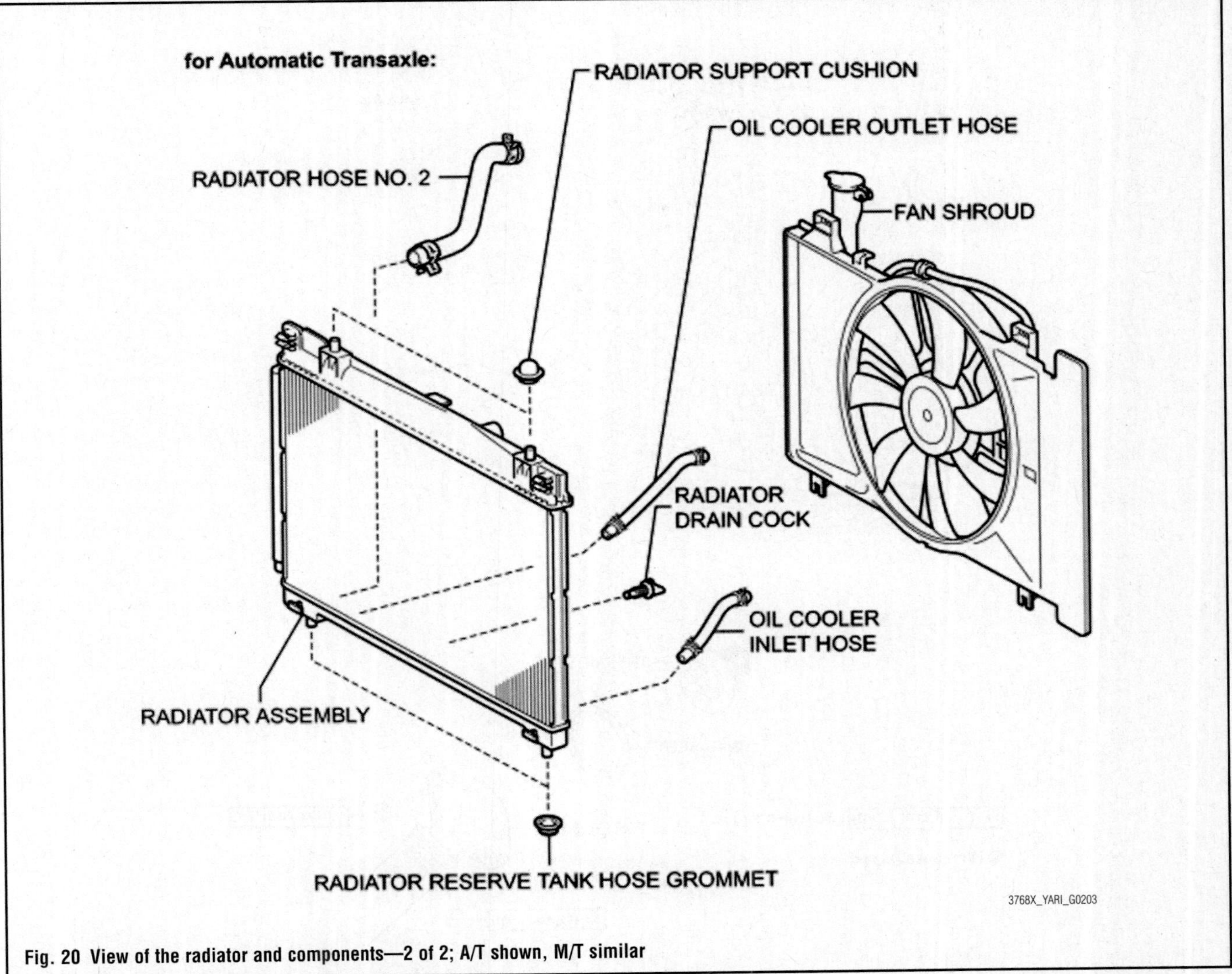

Fig. 20 View of the radiator and components—2 of 2; A/T shown, M/T similar

9. Remove radiator support sub-assembly upper:
   a. Separate the horn assembly connector.
   b. Remove the 4 bolts and remove the radiator support sub-assembly upper.
10. Disconnect the radiator reservoir tank hose from the water filler.
11. Loosen the 2 clips and remove radiator hose No. 3.
12. Loosen the clip and disconnect radiator hose No. 2.
13. Loosen the clip and disconnect the oil cooler outlet hose.
14. Loosen the clip and disconnect the oil cooler inlet hose.
15. Remove radiator assembly (w/ air conditioning system):
    a. Separate the cooling fan motor connector and wire harness clamps.
    b. Disengage the 2 claws and remove the radiator assembly from the vehicle.

➡Do not apply excessive force to the cooler condenser assembly or piping when removing the radiator assembly.

16. Remove radiator assembly (w/o air conditioning system):
    a. Separate the cooling fan motor connector and wire harness clamps.
    b. Remove the radiator assembly from the vehicle.
17. Loosen the clip and remove radiator hose No. 2.
18. Loosen the clip and remove the cooler outlet hose (for A/T).
19. Loosen the clip and remove the cooler inlet hose (for A/T).
20. Disengage the 2 claws and remove the fan shroud.
21. To install, reverse the removal procedure.
22. Refer to the illustrations for torque values.
23. Add coolant and check for leaks.

THERMOSTAT

REMOVAL & INSTALLATION

See Figures 21 and 22.

1. Drain and recycle the engine coolant.
2. Remove the 2 nuts and separate the water inlet with radiator hose from the cylinder block.
3. Remove the thermostat from the cylinder block.
4. Remove the gasket from the thermostat.

To install:

5. Install a new gasket onto the thermostat.
6. Install the thermostat with the jiggle valve facing upward.

➡The jiggle valve may be set within 10° on either side, as shown in the illustration.

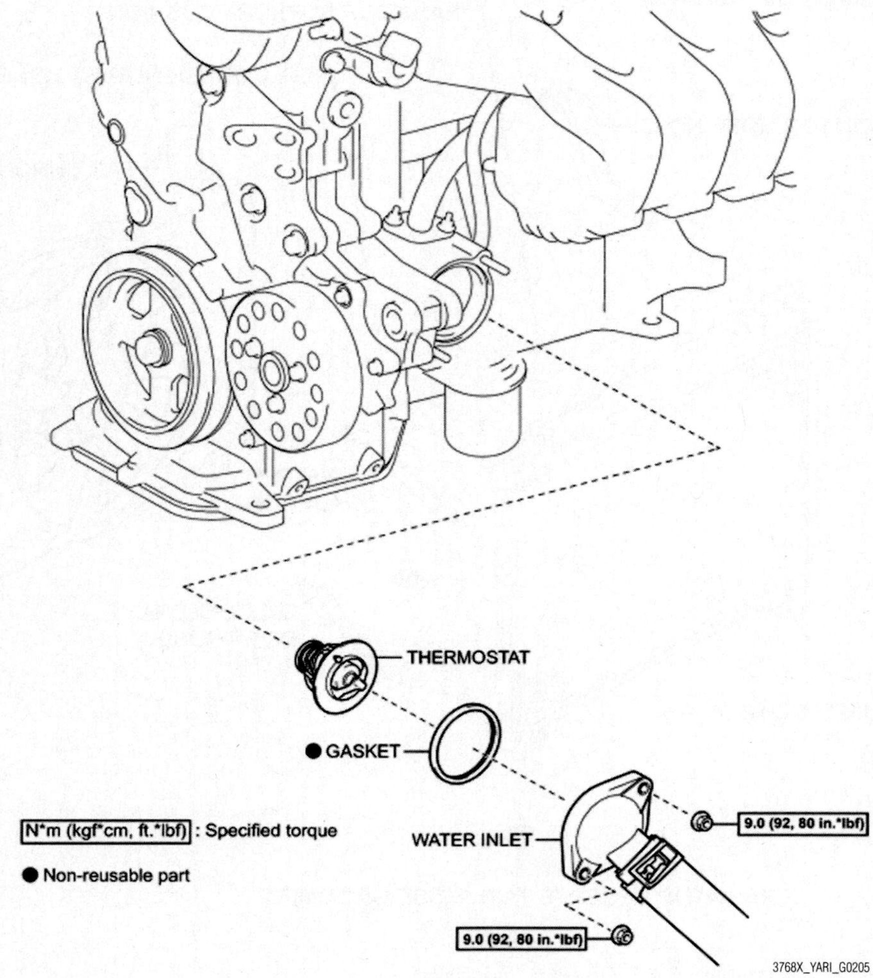

Fig. 21 Removing the thermostat

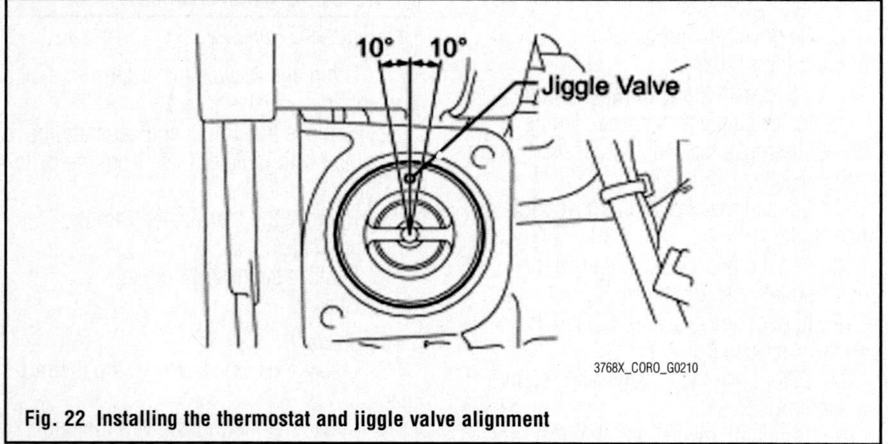

Fig. 22 Installing the thermostat and jiggle valve alignment

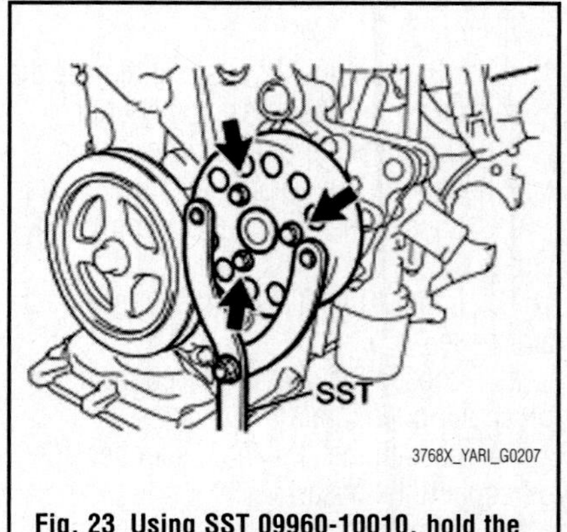

Fig. 23 Using SST 09960-10010, hold the water pump pulley

7. Install the water inlet with radiator hose with the 2 nuts. Tighten to 80 inch lbs. (9 Nm).

8. Add engine coolant and check for leaks.

WATER PUMP

REMOVAL & INSTALLATION

See Figures 23 and 24.

1. Disconnect the negative battery cable.

2. Drain and recycle the engine coolant.

3. Remove accessory drive belt.
4. Remove the alternator.
5. Remove engine mounting insulator sub-assembly RH.
6. Remove water pump pulley:
   a. Using SST 09960-10010, hold the water pump pulley.
   b. Remove the 3 bolts and remove the water pump pulley.
7. Remove the 3 bolts and 2 nuts and remove the water pump assembly and gasket.

To install:

8. To install, reverse the removal procedure.
9. Use a new gasket when installing the water pump. Refer to exploded illustration for torque values.
10. Add coolant and check for leaks.

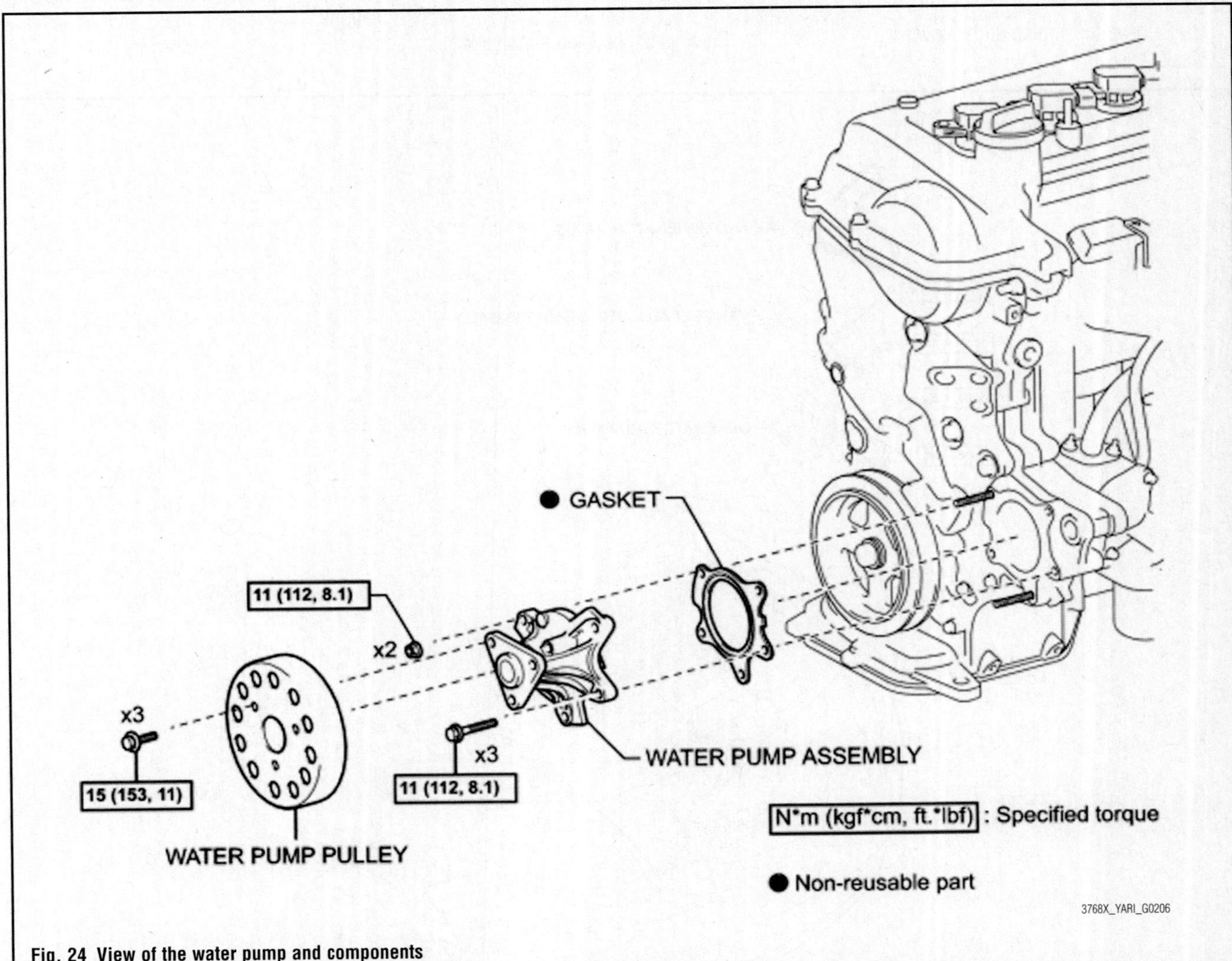

Fig. 24 View of the water pump and components

ALTERNATOR

REMOVAL & INSTALLATION

See Figure 25.

1. Disconnect the negative battery cable.
2. Remove the engine under cover.
3. Remove the accessory drive belt.
4. Remove generator assembly:
    a. Remove the terminal cap.
    b. Separate the connector and the harness clamp.
    c. Remove the nut and remove terminal B.
    d. Remove fan belt adjusting slider fixing bolts and remove the fan belt adjusting slider.
    e. Remove fixing bolt B and remove the generator.
5. To install, reverse the removal procedure. Refer to illustration for torque specifications.

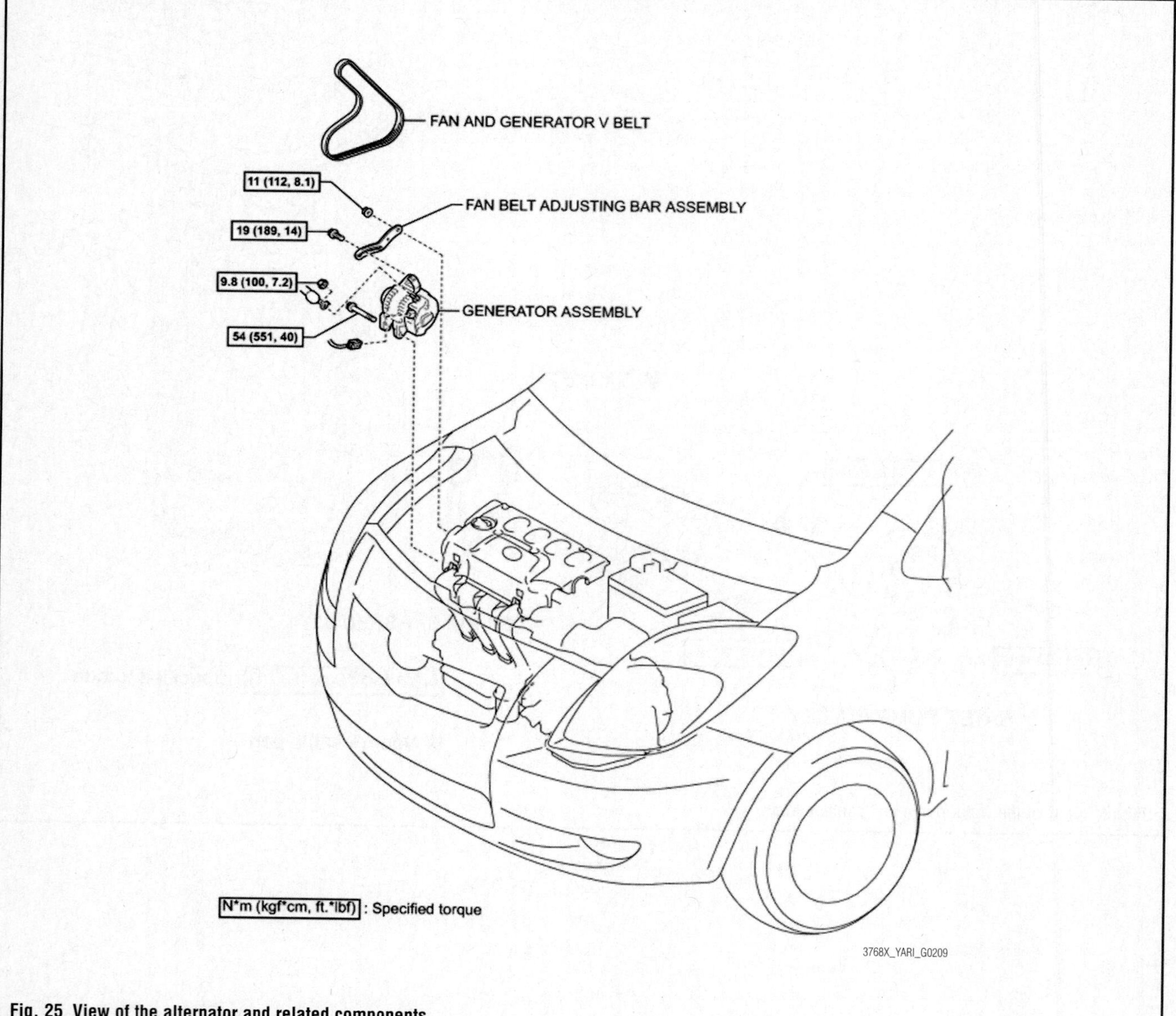

Fig. 25 View of the alternator and related components

FIRING ORDER

See Figure 26.

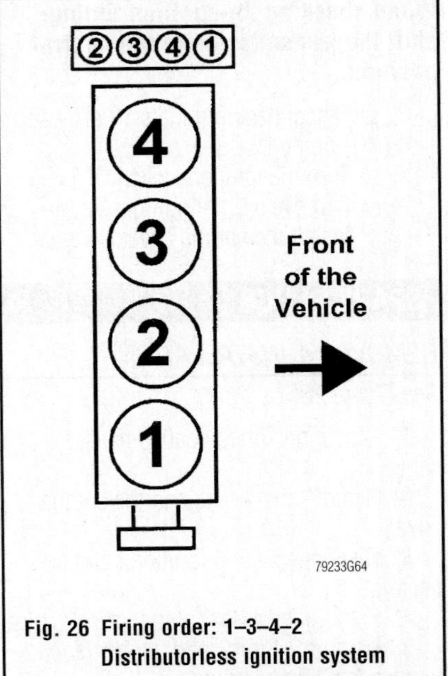

Fig. 26 Firing order: 1–3–4–2 Distributorless ignition system

IGNITION COIL

REMOVAL & INSTALLATION

See Figure 27.

1. Disconnect the negative battery cable.
2. Remove the 4 nuts and the engine cover.
3. Disconnect the 4 ignition coil connectors.
4. Remove the 4 bolts and 4 ignition coils.
5. To install, reverse the removal procedure.

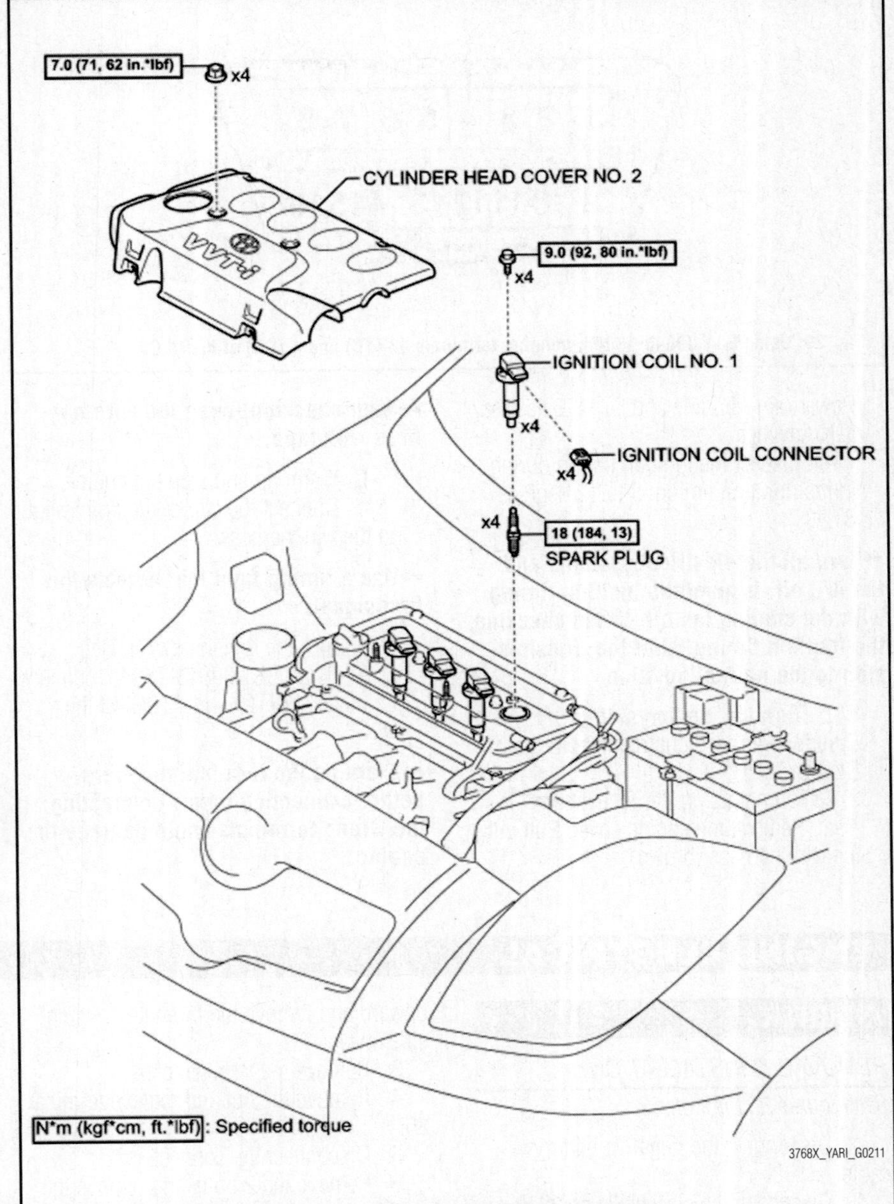

Fig. 27 Removing the ignition coil, spark plugs and related components

IGNITION TIMING

ADJUSTMENT

The ignition timing is controlled by the Powertrain Control Module (PCM). No adjustment is necessary or possible.

INSPECTION

See Figures 28 and 29.

1. Using the Techstream or ODB II scan tool:
   a. Warm up and stop the engine.
   b. Connect the Techstream to the DLC3.
   c. Turn the ignition switch ON.
   d. Select the following menu items:

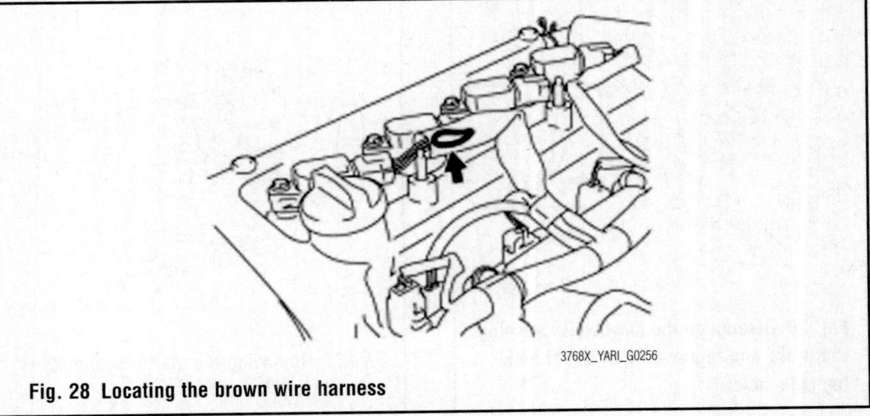

Fig. 28 Locating the brown wire harness

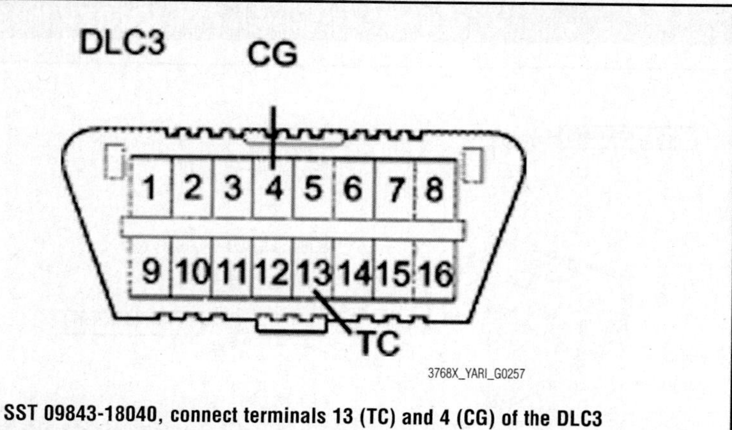

Fig. 29 Using SST 09843-18040, connect terminals 13 (TC) and 4 (CG) of the DLC3

Powertrain / Engine and ECT / Data List / IGN Advance.

e. Inspect the ignition timing during idling. Ignition timing: 8 to 12 degrees BTDC.

➡Turn all the electrical systems and the A/C off. Inspect the ignition timing with the cooling fan off. When checking the ignition timing, shift the transmission to the neutral position.

f. Turn the ignition switch OFF. Disconnect the Techstream from the DLC3.

2. When not using the Techstream:

a. Remove the engine cover. Pull out the wire harness (brown).

➡After checking, wrap the wire harness with tape.

b. Warm up and stop the engine.

c. Connect the clip of the timing light to the wire harness.

➡Use a timing light that detects the first signal.

d. Turn the ignition switch ON.

e. Using SST 09843-18040, connect terminals 13 (TC) and 4 (CG) of the DLC3.

➡Examine the terminal numbers before connecting them. Connecting the wrong terminals could damage the engine.

f. Inspect the ignition timing during idling. Ignition timing: 8 to 12 degrees BTDC.

➡Turn all the electrical systems and the A/C off. Inspect the ignition timing with the cooling fan off. When checking the ignition timing, shift the transmission to the neutral position.

g. Disconnect terminals 13 (TC) and 4 (CG) of the DLC3.

h. Turn the ignition switch OFF.

i. Remove the timing light.

j. Install the engine cover.

SPARK PLUGS

REMOVAL & INSTALLATION

See Figure 28.

1. Disconnect the negative battery cable.
2. Remove the 4 nuts and the engine cover.
3. Disconnect the 4 ignition coil connectors.
4. Remove the 4 bolts and 4 ignition coils.
5. Using a 14 mm spark plug wrench, remove the 4 spark plugs.
6. To install, reverse the removal procedure.
7. Tighten the spark plugs to 13 ft. lbs. (18 Nm).

ENGINE ELECTRICAL

STARTING SYSTEM

STARTER

REMOVAL & INSTALLATION

See Figures 30 through 33.

1. Disconnect the negative battery cable.
2. Disengage the claw while pushing it upward and remove the flywheel housing side cover.
3. Remove the terminal cap.
4. Remove the nut and remove terminal 30.
5. Disconnect the connector.
6. Remove the 2 bolts and remove the starter assembly.
7. To install, reverse the removal procedure.

➡Make sure that the claw makes a click sound, indicating that it fits tightly. Replace the claw with a new one if it does not fit tightly or is deformed.

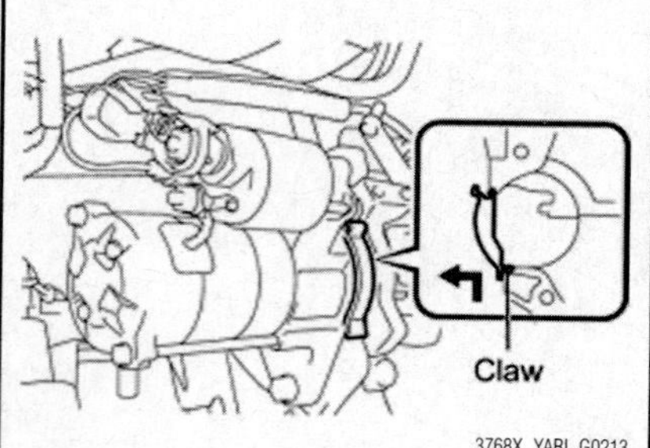

Fig. 30 Disengage the claw while pushing it upward and remove the flywheel housing side cover

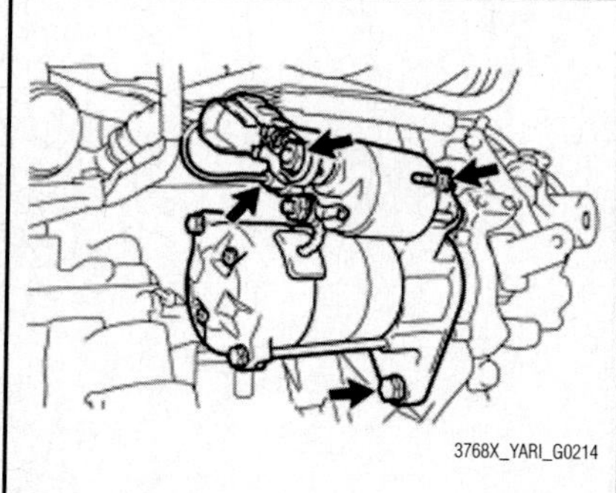

Fig. 31 Removing the starter assembly—0.8 kW type

Fig. 32 Removing the starter assembly—1.6 kW type

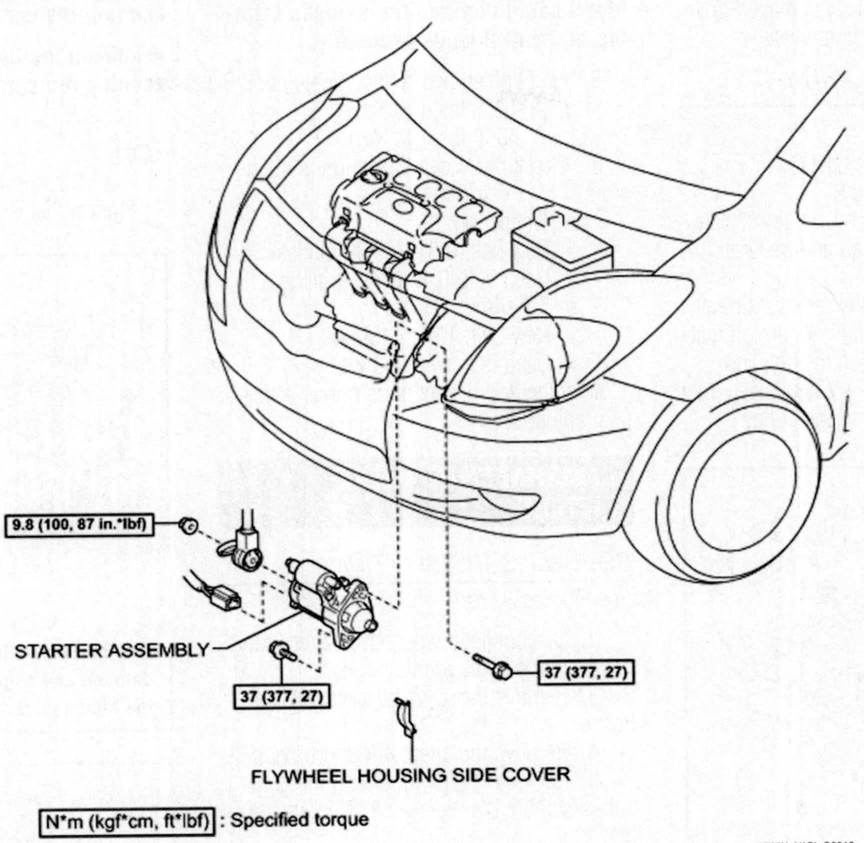

Fig. 33 View of the starter and components

ENGINE MECHANICAL

ACCESSORY DRIVE BELTS

ACCESSORY BELT ROUTING

See Figure 34.

Refer to the accompanying illustration.

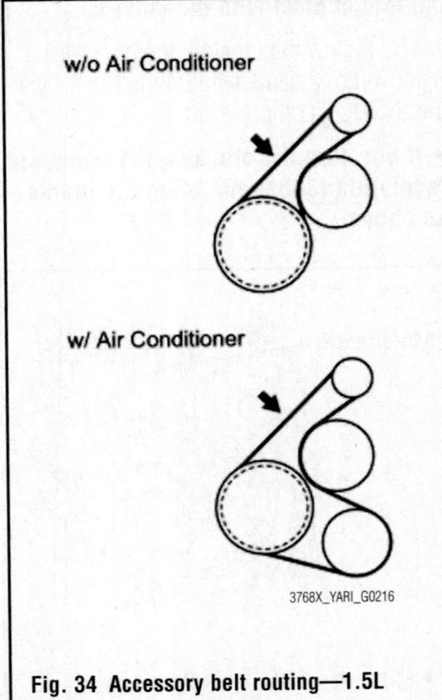

Fig. 34 Accessory belt routing—1.5L Engine

INSPECTION

➡Please note the following:

- Perform the V belt inspection and adjustment while the engine is cold.
- V-ribbed belt tension and deflection should be checked immediately after installation of a new belt, and after cranking the engine when inspecting a used belt.
- Check the V belt deflection at the point between the specified pulleys where the deflection is greatest.
- When installing a new belt, set its tension to the intermediate value of the specification.
- When inspecting a belt which has been used for over 5 minutes, apply the used belt specifications.
- When reinstalling a belt which has been used for over 5 minutes, adjust its deflection and tension to the intermediate values of each used belt specification.
- V-ribbed belt tension and deflection should be checked after 2 revolutions of engine cranking.
- When using a belt tension gauge, confirm its accuracy by using a master gauge first.

Inspect the drive belt for signs of glazing or cracking. A glazed belt will be perfectly smooth from slippage, while a good belt will have a slight texture of fabric visible. Cracks will usually start at the inner edge of the belt and run outward. All worn or damaged drive belts should be replaced immediately.

REMOVAL & INSTALLATION

See Figure 35.

1. Remove the engine cover.
2. Loosen bolts A and B.
3. Release the fan and generator V belt tension and remove the fan and generator V belt.
4. To install, reverse the removal procedure. Make sure that there is no foreign matter or liquid, such as oil, on the belt and pulleys. Make sure that the V belt is securely fitted into the rib grooves of the pulley.

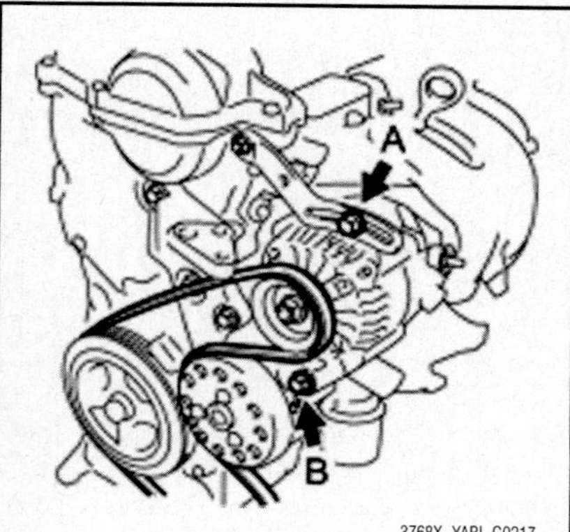

Fig. 35 Removing the accessory drive belt—1.5L Engine

ADJUSTMENT

See Figure 36.

1. Insert an adjusting bar between the engine mounting bracket and generator assembly. Pull the adjusting bar toward the vehicle front to adjust the generator V belt tension.

➡Do not insert the adjusting bar between the camshaft timing oil control valve assembly and generator assembly. It could damage the camshaft timing oil control valve assembly.

2. First tighten bolt A, then tighten bolt B.
   a. A: 14 ft. lbs. (19 Nm).
   b. B: 40 ft. lbs. (54 Nm).
3. Check the V belt deflection and tension:
   a. Deflection:
      - New belt: 0.31 to 0.35 inches
      - Used belt: 0.49 to 0.53 inches
   b. Tension:
      - New belt: 157 to 180 lbs.
      - Used belt: 67 to 90 lbs.
4. If the belt deflection is not as specified, adjust it.

CAMSHAFT AND VALVE LIFTERS

REMOVAL & INSTALLATION

See Figures 37 through 57.

1. Disconnect the negative battery cable.
2. Remove the ignition coils.
3. Remove the valve cover ventilation hoses.
4. Remove the alternator and drive belt.
5. Remove engine mounting insulator sub-assembly RH. Refer to Engine Assembly.

➡When rotating the camshaft with the timing chain removed, rotate the crankshaft damper counterclockwise 40°from the TDC and align its timing notch with the matchmark of the timing chain cover to prevent the pistons from coming into contact with the valves.

6. Follow the illustrations in order. Remove no. 2 camshaft. Set the No. 1 cylinder to TDC / compression:

➡If not, turn the crankshaft 1 complete revolution (360°) and align the marks as above.

➡Loosen the bolts uniformly while keeping the camshaft level.

➡Loosen the bolts uniformly while keeping the camshaft level.

7. Remove the camshaft timing gear assembly:
   a. Clamp the camshaft in a vise and confirm that it is locked.

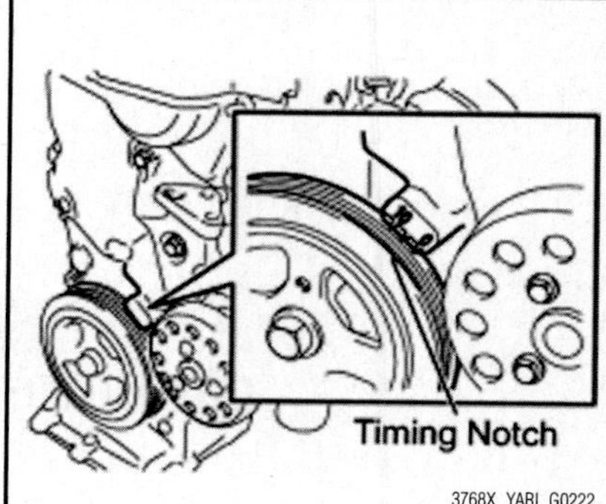

Fig. 38 Step 1: Turn the crankshaft damper, and align its timing notch with the timing mark "0" of the oil pump

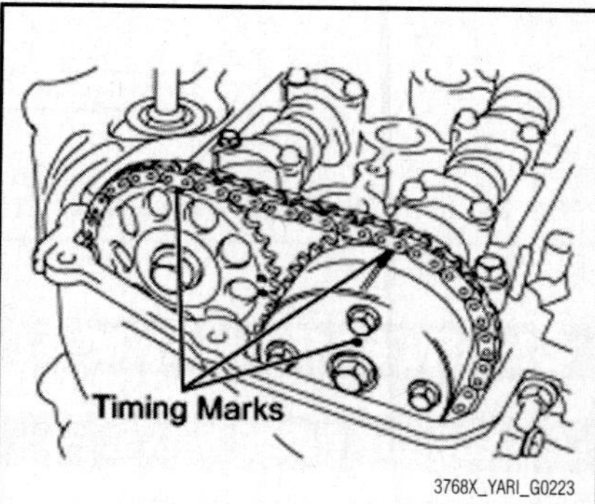

Fig. 39 Step2: Check that the timing marks on both the camshaft timing sprocket and the camshaft timing gear are facing upward

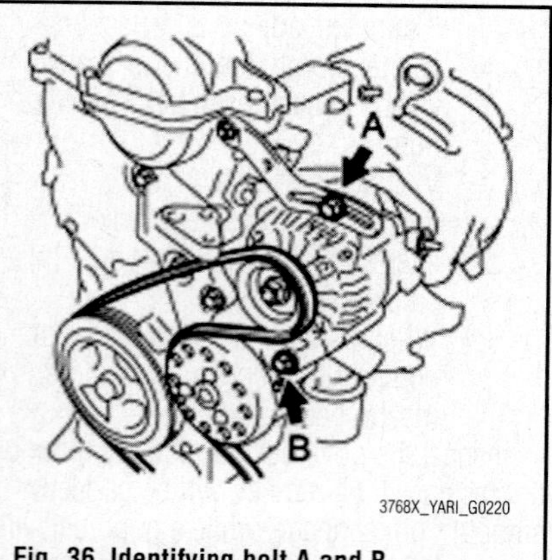

Fig. 36 Identifying bolt A and B

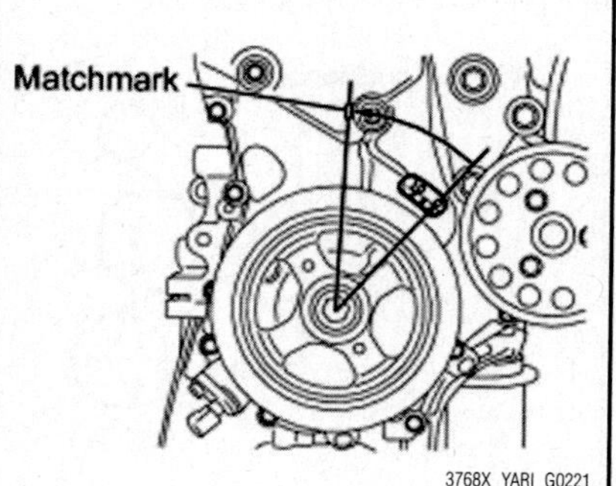

Fig. 37 Rotate the crankshaft damper counterclockwise 40°from the TDC

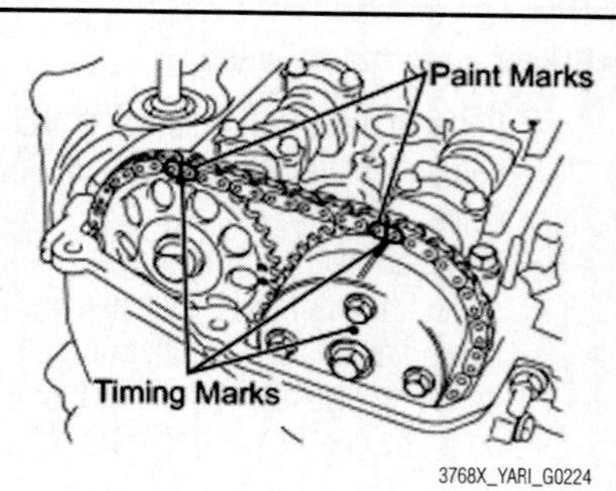

Fig. 40 Step3: Place paint marks on the chain in the places where the timing marks of the camshaft timing sprocket and the camshaft timing gear are located.

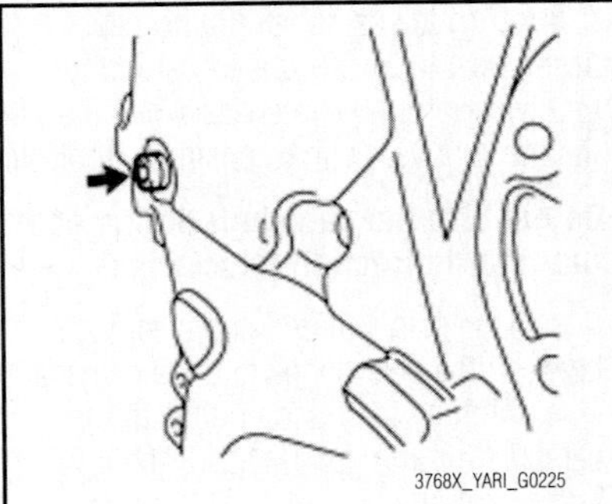

Fig. 41 Step 4: Using an 8 mm hexagon wrench, remove the screw plug

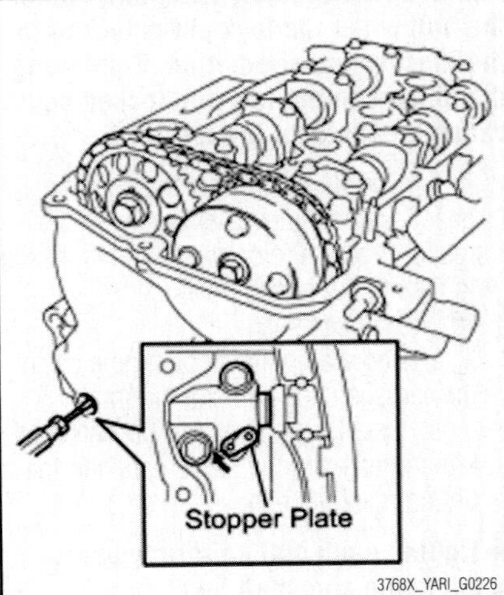

Fig. 42 Step 5: Insert a screwdriver into the service hole in the chain tensioner to pull the stopper plate of the chain tensioner upward

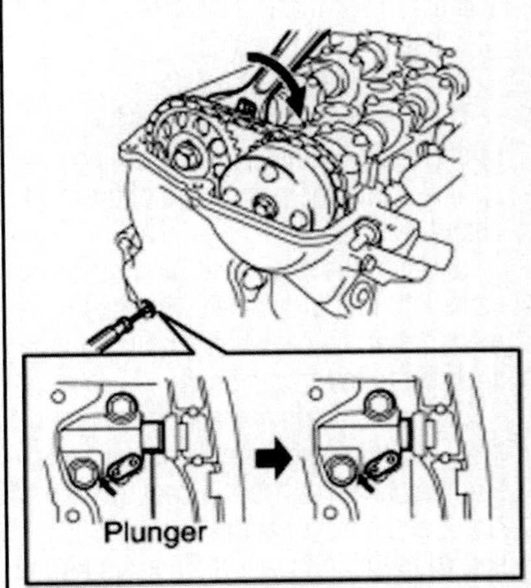

Fig. 43 Step 6: Using a wrench, rotate camshaft No. 2 clockwise to push in the plunger of the chain tensioner

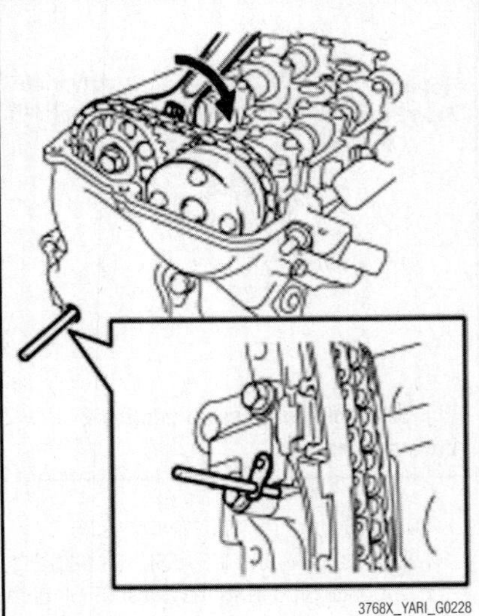

Fig. 44 Step 7: Remove the screwdriver from the service hole, then align the hole in the stopper plate with the service hole and insert a 0.12 inch diameter bar into the holes to hold the stopper plate

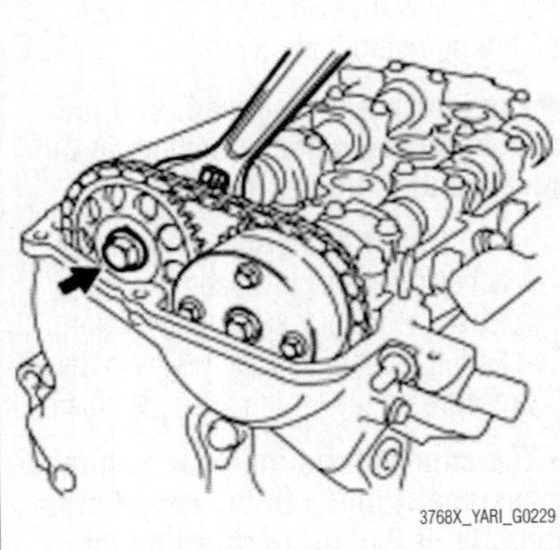

Fig. 45 Step 8: Using a wrench, hold the hexagonal lobe of camshaft No. 2 and remove the flange bolt.

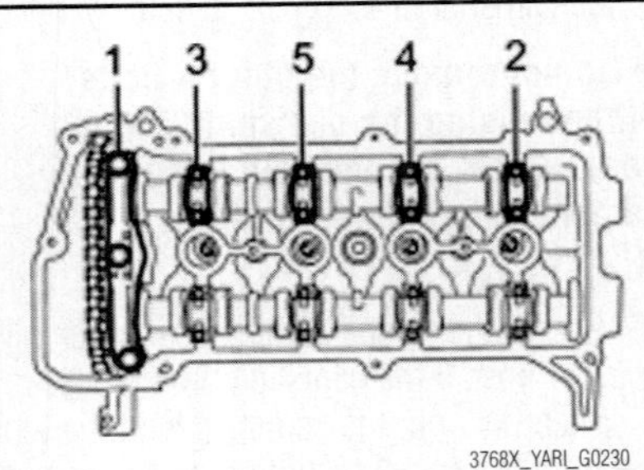

Fig. 46 Step 9: Using several steps, loosen and remove the 11 bearing cap bolts uniformly in the sequence then remove camshaft bearing caps No. 1 and No. 2

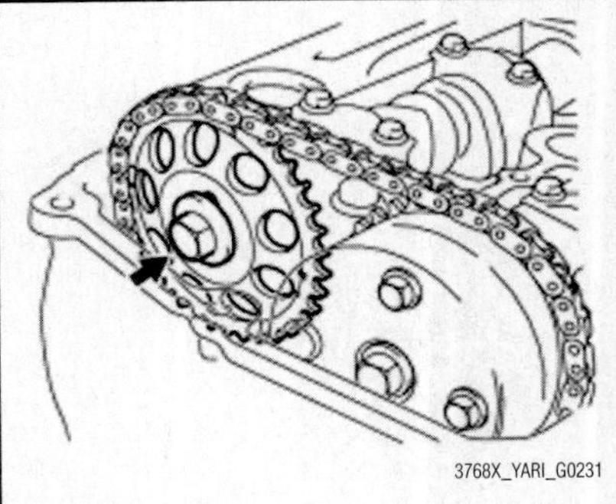

Fig. 47 Step 10: Remove the flange bolt and remove the camshaft timing sprocket

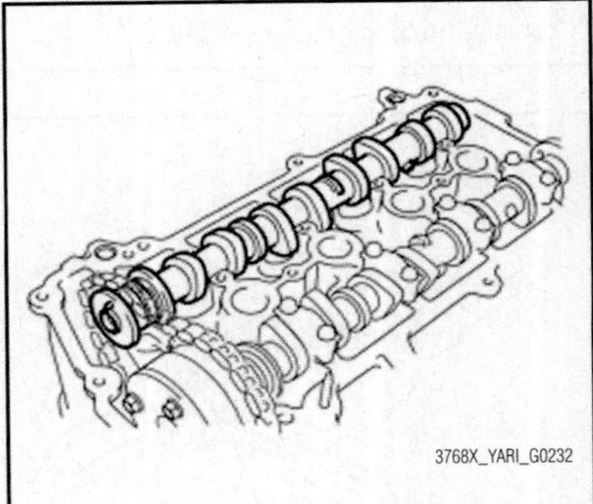

Fig. 48 Step11: Remove camshaft No. 2

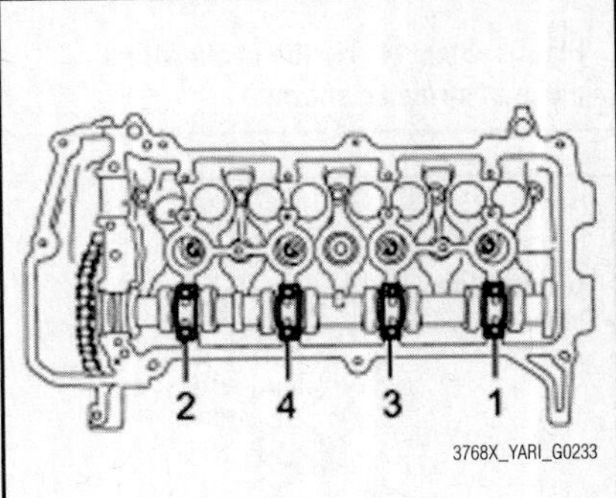

Fig. 49 Step 12: Using several steps, loosen and remove the 8 bearing cap bolts uniformly remove camshaft bearing cap No. 2

➡One of the 2 grooves located on the cam journal is for retarding cam timing (upper) and the other is for advancing cam timing (lower). Each groove has 2 oil paths. Plug one of the oil paths for each groove with a piece of rubber before wrapping the cam journal with the tape.

b. Cover the 4 oil paths of the cam journal with tape as shown in the illustration.

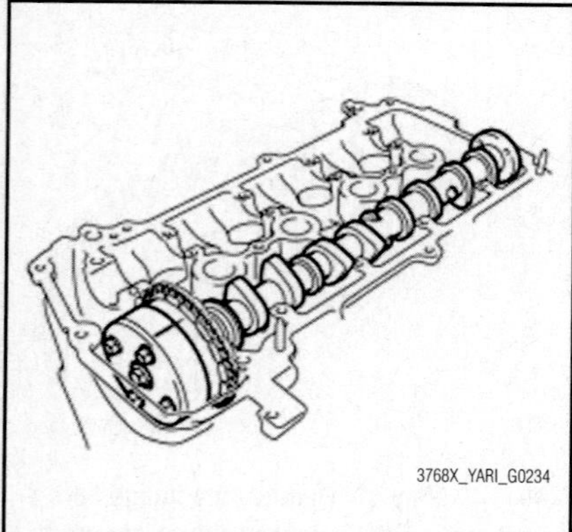

Fig. 50 Step 13: Hold the chain by hand, and remove the camshaft and the camshaft timing gear assembly

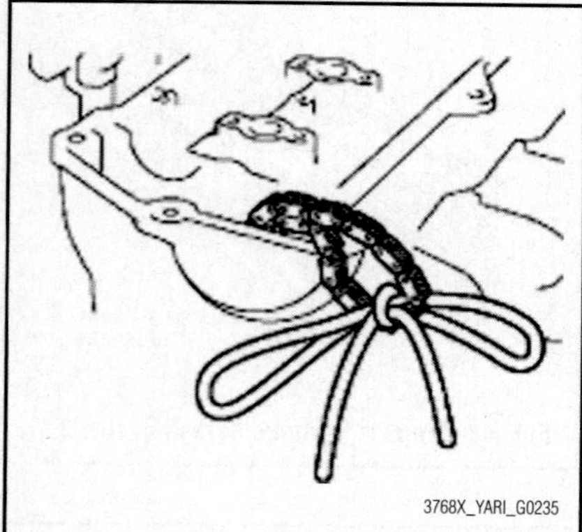

Fig. 51 Step 14: Tie the chain with a piece of string as shown

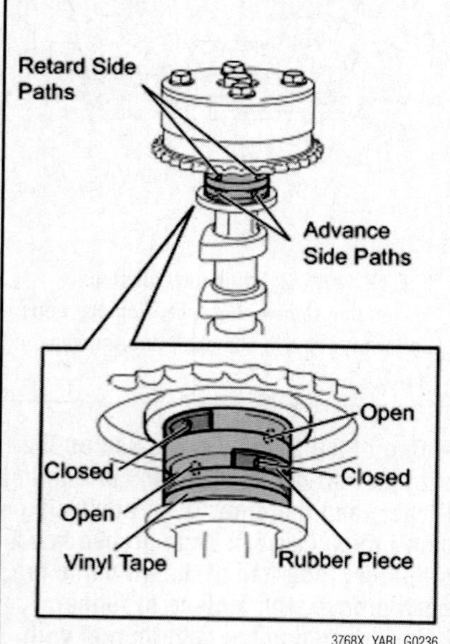

Fig. 52 Cover the 4 oil paths of the cam journal with tape

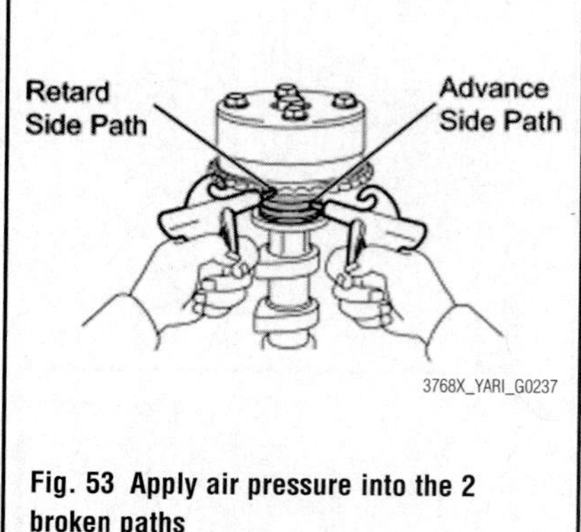

Fig. 53 Apply air pressure into the 2 broken paths

c. Puncture the tape covering the advance oil path and the retard oil path on the opposite side from the advance oil path.

d. Apply air at about 150 kPa (1.5 kgf*cm2) pressure into the 2 broken paths (the advance side path and the retard side path)

➡Cover the paths with a shop rag or piece of cloth to prevent oil splashes.

8. Confirm that the camshaft timing gear assembly revolves in the timing advance direction when the air pressure on the timing retard path is reduced.

➡The lock pin is released, and the camshaft timing gear revolves in the advance direction.

a. When the camshaft timing gear reaches the most advanced position, release the air pressure on the timing retard side path, and then release the air pressure on the timing advance side path.

➡The camshaft timing gear assembly occasionally shifts to the retard side abruptly, if the air pressure on the advance side path is released first. This often results in breakage of the lock pin.

b. Remove the flange bolt and remove the camshaft timing gear assembly.

➡Do not remove the other 4 bolts. When reusing the camshaft timing gear, unlock the lock pin inside the camshaft timing gear first.

To install:

9. Install camshaft timing gear assembly:

a. Install the camshaft timing gear assembly onto the camshaft with the lock pin of the camshaft timing gear assembly released.

b. Put the camshaft timing gear assembly and camshaft together with the straight pin of the groove.

c. Turn the camshaft timing gear assembly clockwise while pushing it gently toward the camshaft. When the pin fits the groove, push to ensure a good fit.

➡Do not turn the camshaft timing gear in the retard direction (clockwise).

d. Check that there is no clearance between the gear flange and the camshaft.

e. Tighten the flange bolt with the camshaft timing gear fixed. Tighten to 47 ft. lbs. (64 Nm).

➡Do not lock the camshaft timing gear assembly when tightening the bolt. Release the lock pin of the camshaft timing gear assembly first, and tighten the bolt when the lock pin is locked in the most retarded position. Tightening the bolts with the lock pin locked could cause breakage of the lock pin.

f. Check that the camshaft timing gear assembly moves smoothly in the retard direction (clockwise) and is locked in the most retarded position.

10. Install camshaft:

a. Apply a light coat of engine oil to the camshaft and camshaft journals.

b. Install the chain onto the camshaft timing gear with the paint mark and the timing mark aligned.

➡Tighten each bolt uniformly while keeping the camshaft level.

c. Examine the front marks and numbers on camshaft bearing cap No. 2 and check that the sequence is as shown in the illustration. Then uniformly tighten the bolts, in several steps, in the sequence shown in the illustration. Tighten to 9.4 ft. lbs. (13 Nm).

11. Install no. 2 camshaft:

a. Install camshaft No. 2.

b. Hold the chain and align the timing mark on the camshaft timing sprocket with the paint mark of the chain.

c. Align the alignment pin hole in the camshaft timing sprocket with the alignment pin of the camshaft, and install the sprocket onto the camshaft.

d. Provisionally install the flange bolt.

e. Examine the front marks and numbers on camshaft bearing caps No. 1 and No. 2 and check that the sequence is as shown in the illustration. Then uniformly tighten the bolts, in several steps, in the sequence shown in the illustration.

- Bearing cap no.2: 9.4 ft. lbs. (13 Nm).
- Bearing cap no. 1: 17 ft. lbs. (23 Nm).

➡**Tighten each bolt uniformly while keeping the camshaft level.**

f. Using a wrench, hold the hexagonal lobe of camshaft No. 2 and install the flange bolt. Tighten to 47 ft. lbs. (64 Nm).

g. Remove the bar from the timing chain tensioner.

h. Turn the crankshaft damper and align its timing notch with the timing mark "0" of the oil pump.

i. Check that all the pairs of timing marks are aligned.

j. Apply adhesive to the end 2 or 3 threads of the screw plug. Adhesive: Toyota Genuine Adhesive 1324, Three Bond 1324 or Equivalent

k. Using an 8 mm hexagon wrench, install the screw plug. Tighten to 11 ft. lbs. (15 Nm).

12. Inspect the valve clearance:

a. Turn the crankshaft damper and align its timing notch with the timing mark "0" of the oil pump.

b. Check that both timing marks on the camshaft timing sprocket and camshaft timing gear are facing upward.

➡**If not, turn the crankshaft 1 complete revolution (360°) and align the marks as above.**

c. Check the valves indicated in the illustration.

- Using a feeler gauge, measure the clearance between the valve lifter and camshaft.
- Valve clearance (cold): Intake: 0.006 to 0.010 inches. Exhaust: 0.010 to 0.014 inches

d. Record any out-of-specification valve clearance measurements. They will be used later to determine the required replacement adjusting shim.

e. Turn the crankshaft 1 complete revolution (360°) and align its timing notch with the timing mark "0" of the oil pump.

f. Check the valves indicated in the illustration.

- Using a feeler gauge, measure the clearance between the valve lifter and camshaft.
- Valve clearance (cold): Intake: 0.006 to 0.010 inches. Exhaust: 0.010 to 0.014 inches

g. Record any out-of-specification valve clearance measurements. They will be used later to determine the required replacement adjusting shim.

13. To complete installation, reverse the remaining removal procedure.

14. Check for engine oil leaks.

CATALYTIC CONVERTER

REMOVAL & INSTALLATION

Refer to Exhaust Manifold in this section.

CRANKSHAFT DAMPER

REMOVAL & INSTALLATION

See Figures 58 through 61.

1. Remove the valve cover.
2. Set cylinder No. 1 to TDC/compression. Turn the crankshaft damper sub-assembly, and align its timing notch with timing mark "0" of the oil pump.
3. Check that the timing marks on the camshaft timing sprocket and the camshaft timing gear are all facing upward. If not, turn the crankshaft 1 complete revolution (360°) and align the marks.
4. Using 2 SSTs, 09213-14010, 09330-00021, loosen the bolt while holding the crankshaft damper sub-assembly.

➡**Check the SST installation positions when installing them, to avoid the SST fixing bolts from coming into contact with the oil pump assembly.**

5. Remove the SSTs and the bolt.

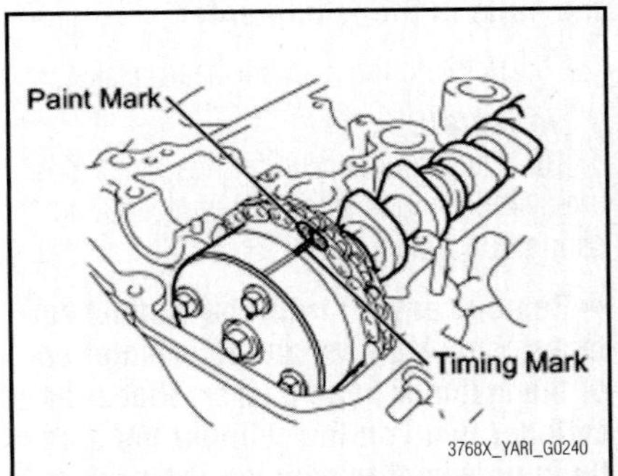

Fig. 54 Install the chain onto the camshaft timing gear with the paint mark and the timing mark aligned

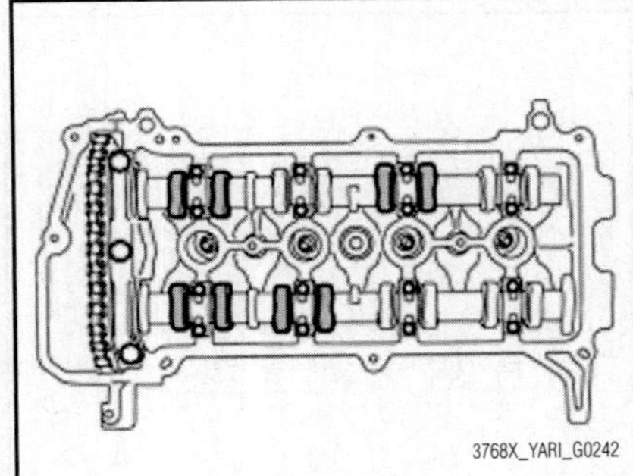

Fig. 56 Check the valves indicated—First set

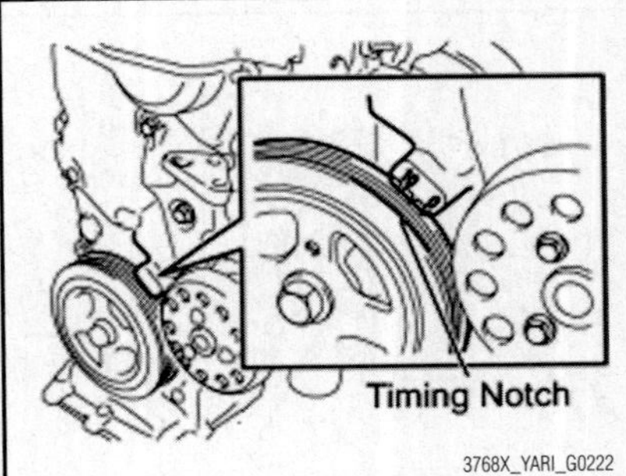

Fig. 58 Turn the crankshaft damper, and align its timing notch with the timing mark "0" of the oil pump

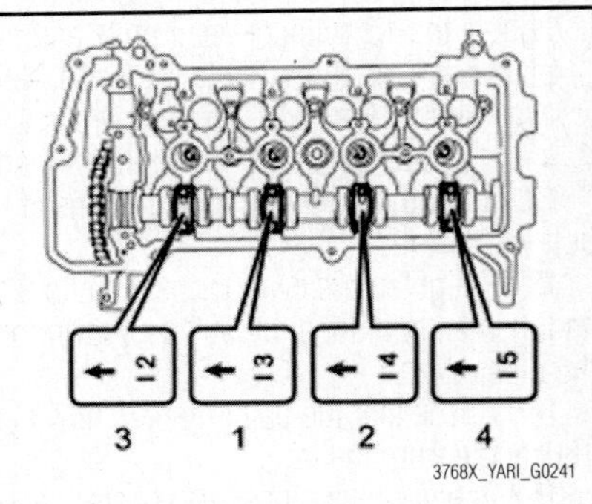

Fig. 55 Camshaft bearing cap No. 2 installation sequence

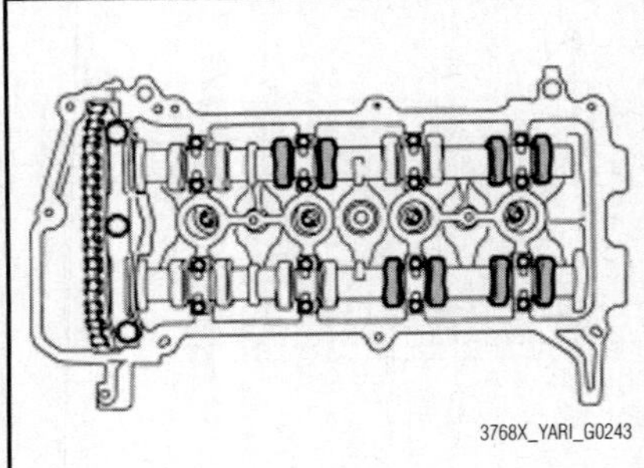

Fig. 57 Check the valves indicated—Second set

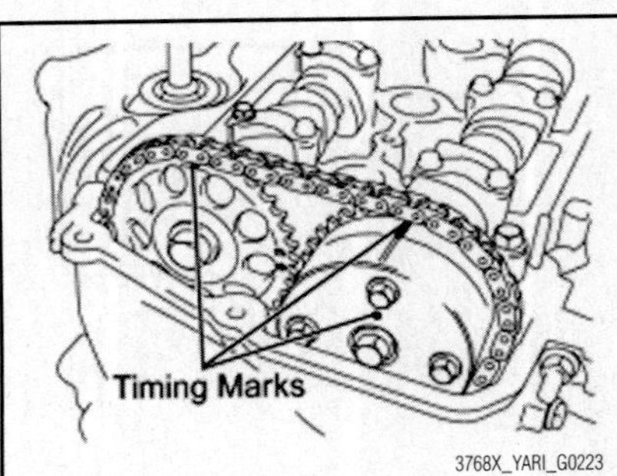

Fig. 59 Check that the timing marks on both the camshaft timing sprocket and the camshaft timing gear are facing upward

6. Remove the crankshaft damper sub-assembly.

7. If necessary, remove timing chain cover front oil seal:

a. Using a knife, cut off the lip of the oil seal.

b. Using a screwdriver with its tip wrapped with tape, pry out the oil seal.

➡After removing, check the crankshaft for damage. If damaged, smooth the surface with 400-grit sandpaper.

To install:

➡Keep the lip free of foreign matter.

8. Apply MP grease to the lip of a new oil seal.

9. Using a seal installer (SST 09223-22010) and a hammer, tap in the oil seal until its surface is flush with the rear oil seal retainer edge.

10. Align the pin hole in the crankshaft damper with the pin position and install the crankshaft damper sub-assembly.

11. Provisionally install the bolt.

12. Using 2 SSTs, tighten the bolt while holding the crankshaft damper sub-assembly. Tighten to 95 ft. lbs. (128 Nm).

➡Check the SST installation positions when installing them, to avoid the SST fixing bolts from coming into contact with the oil pump assembly.

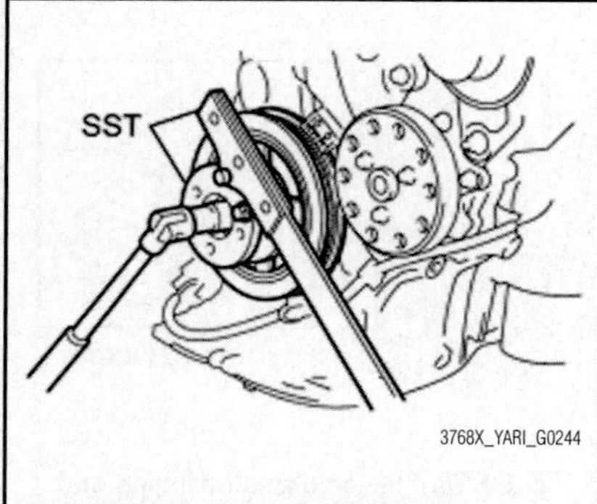

Fig. 60 Using SST's to remove the crankshaft damper

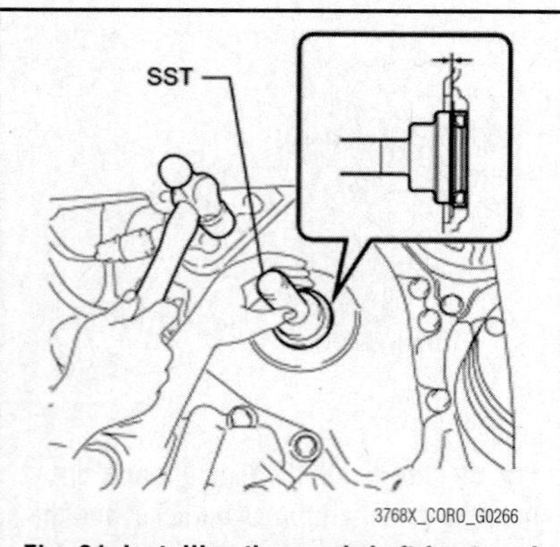

Fig. 61 Installing the crankshaft front seal

CRANKSHAFT FRONT SEAL

REMOVAL & INSTALLATION

Refer to Crankshaft Damper to remove the crankshaft front seal.

CYLINDER HEAD

REMOVAL & INSTALLATION

See Figures 62 through 64.

1. Drain the engine oil.

2. Drain and recycle the engine coolant.

3. Properly relieve the fuel system pressure.

4. Remove the camshaft assembly. Refer to Camshaft in this section.

5. Remove the timing chain sprockets and cover. Refer to Timing Chain And Sprockets in this section.

6. Remove the intake manifold. Refer to Intake Manifold in this section.

7. Remove the exhaust manifold.

8. Remove the cylinder head. Using several steps, uniformly loosen and remove the 10 cylinder head bolts with an 8 mm bi-hexagon wrench in the sequence shown in the illustration. Remove the 10 plate washers.

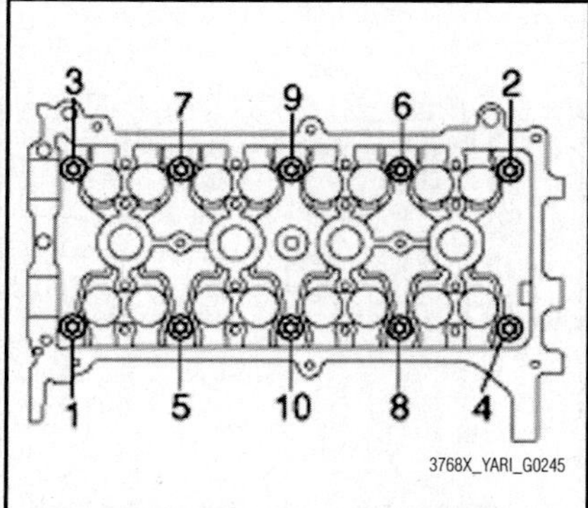

Fig. 62 Cylinder head removal sequence

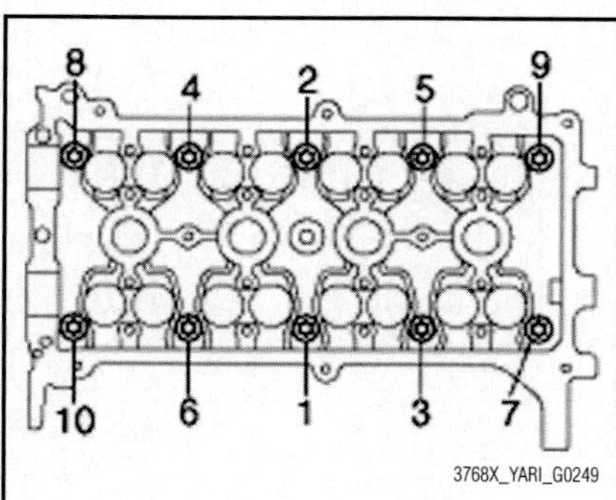

Fig. 63 Cylinder head tightening sequence—First pass

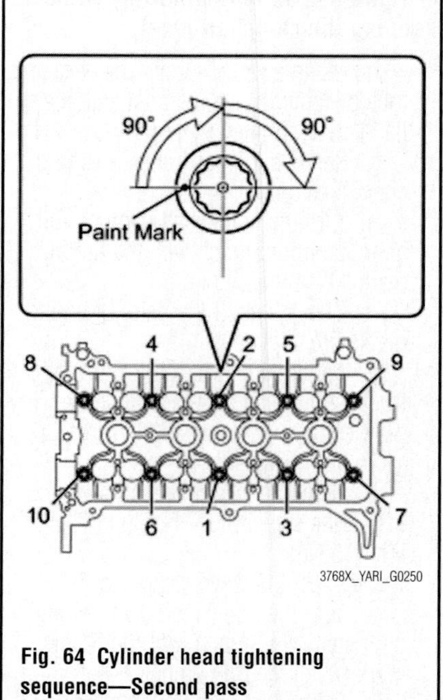

Fig. 64 Cylinder head tightening sequence—Second pass

➡Do not drop the washers into the cylinder head. Head warpage or cracking could result from removing the bolts in the wrong order.

9. Remove the cylinder head gasket.

To install:

10. Place a new cylinder head gasket on the cylinder block with the Lot No. stamp facing upward.

➡Remove any oil from the contact surfaces. Check the mounting orientation of the cylinder head gasket. Place the cylinder head on the cylinder head gently in order not to damage the gasket.

11. Install cylinder head sub-assembly:

a. Apply a light coat of engine oil to the threads of the cylinder head bolts.

b. Using several steps, install and tighten the 10 cylinder head bolts and plate washers uniformly to 22 ft. lbs. (29 Nm) with an 8 mm bi-hexagon wrench, in the sequence shown in the illustration.

12. Mark the front of the cylinder head bolt with paint.

13. Retighten the cylinder head bolts 90° and then an additional by 90° as shown in the illustration.

14. Check that the paint mark is now at a 180°angle from the front.

15. Apply a continuous bead of seal packing (Diameter 4.5 to 5.5 mm (0.177 to 0.217 in.)) as shown in the illustration. Seal Pack-

ing: Toyota Genuine Seal Packing Black, Three Bond 1207B or Equivalent.

➡Remove any oil from the contact surfaces. Install the oil pump assembly within 3 minutes and tighten the bolts within 15 minutes of applying the seal packing.

16. To complete installation, reverse the remaining removal procedure.

EXHAUST MANIFOLD

REMOVAL & INSTALLATION

See Figures 65 and 66.

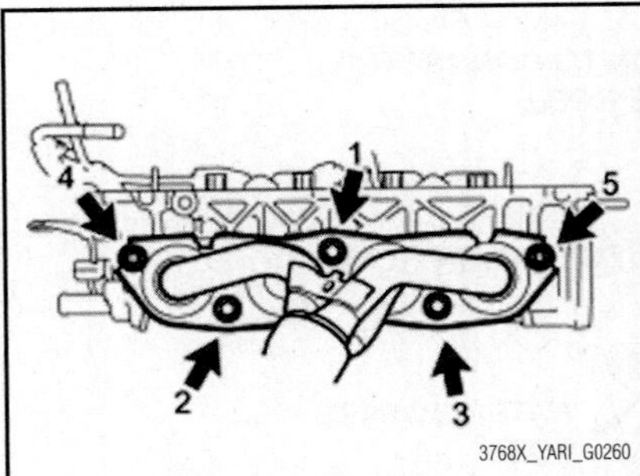

Fig. 65 Exhaust manifold tightening sequence

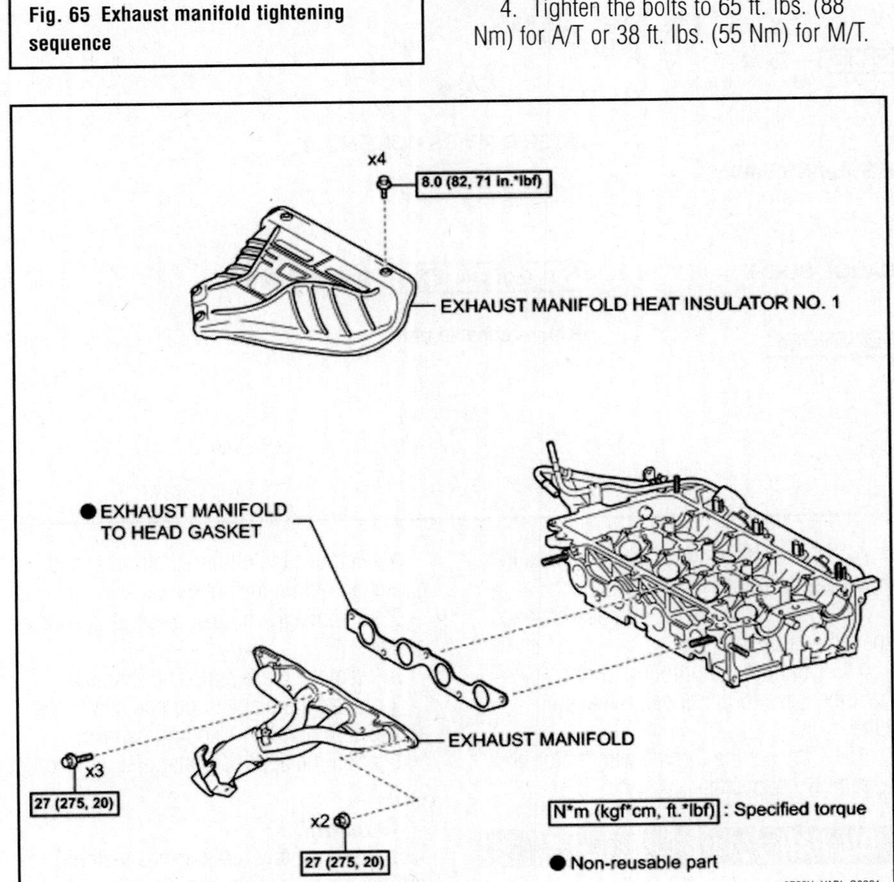

Fig. 66 View of the exhaust manifold and components

1. Remove the 2 bolts and 2 compression springs and separate the exhaust pipe assembly front.\
2. Remove the 3 bolts and remove the manifold support bracket.
3. Remove the exhaust manifold insulator with the 4 bolts.
4. Remove the 2 bolts and 2 compression springs and separate the exhaust pipe assembly front.
5. Remove the 2 nuts and 3 bolts and the exhaust manifold and gasket.
6. To install, tighten the exhaust manifold nuts and bolts, in the order shown in the illustration, through a new gasket.
7. To complete installation, reverse remaining removal procedure.

FLYWHEEL/FLEXPLATE

REMOVAL & INSTALLATION

See Figure 67.

1. Remove the engine from the transaxle.
2. Uniformly remove the flywheel or in the order shown.
3. To install, reverse the removal procedure.
4. Tighten the bolts to 65 ft. lbs. (88 Nm) for A/T or 38 ft. lbs. (55 Nm) for M/T.

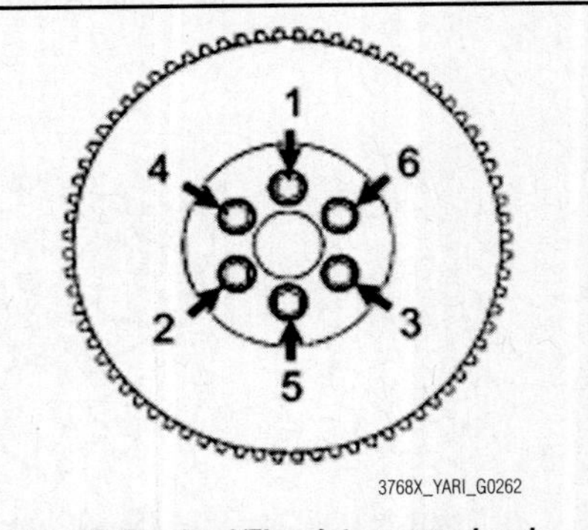

Fig. 67 Flywheel/Flexplate removal and tightening sequence

INTAKE MANIFOLD

REMOVAL & INSTALLATION

See Figures 68 through 70.

1. Disconnect the negative battery cable.
2. Disconnect the union to connector tube hose from the booster vacuum tube.
3. Disconnect the engine wire from the intake manifold.
4. Disconnect the water by-pass hose from the cylinder head.
5. Disconnect the water by-pass hose from water bypass pipe No. 1.
6. Disconnect the throttle with motor body assembly connector.
7. Remove the 3 bolts and 2 nuts in the order shown in the illustration and remove the intake manifold.
8. Remove the gasket from the intake manifold.

To install:

9. Install a new gasket onto the intake manifold.
10. Provisionally tighten the intake manifold nuts and bolts in the order shown in the illustration, and then tighten them to the specified torque. Tighten to 22 ft. lbs. (30 Nm).
11. Connect the engine wire to the intake manifold.

Fig. 68 Remove the 3 bolts and 2 nuts in the order shown

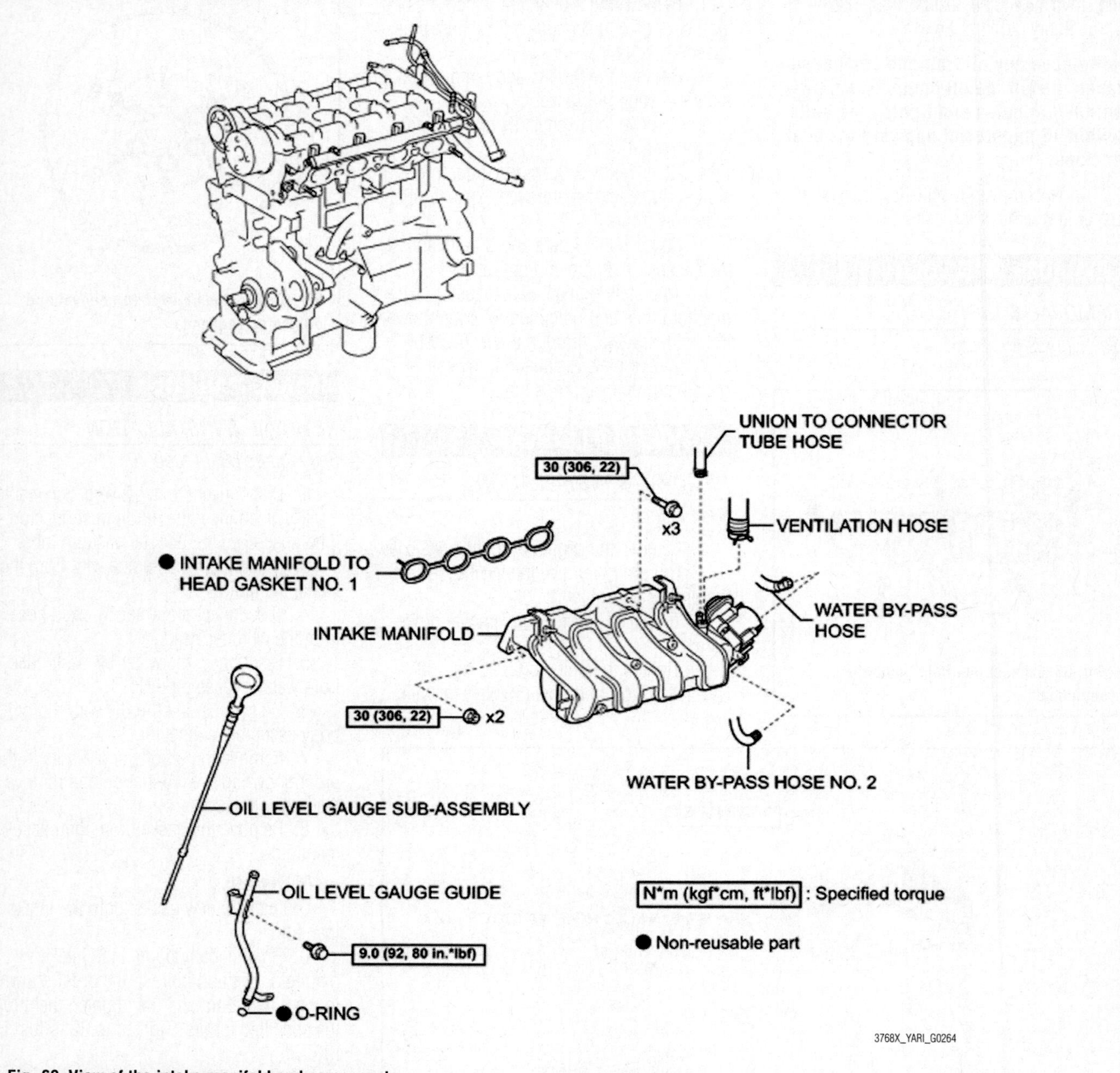

Fig. 69 View of the intake manifold and components

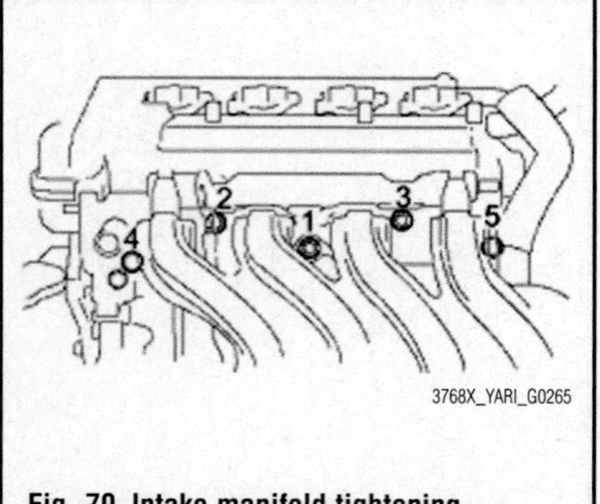

Fig. 70 Intake manifold tightening sequence

12. Connect the water by-pass hose to water bypass pipe No. 1.
13. Connect the water by-pass hose to the cylinder head.
14. Connect the union to connector tube hose to the booster vacuum tube.
15. Connect the throttle with motor body assembly connector.

OIL PAN

REMOVAL & INSTALLATION

See Figures 70 and 71.

1. Remove the oil pan drain plug and gasket and drain the engine oil.
2. Remove the oil pan drain plug and gasket.
3. Remove the 9 bolts and 2 nuts.
4. Insert the blade of oil pan seal cutter between oil pan No. 1 and oil pan No. 2, and cut off the applied sealer and remove oil pan.

To install:

5. Remove any old packing material and be careful not to drop any oil on the contact surfaces of the cylinder block and oil pan.

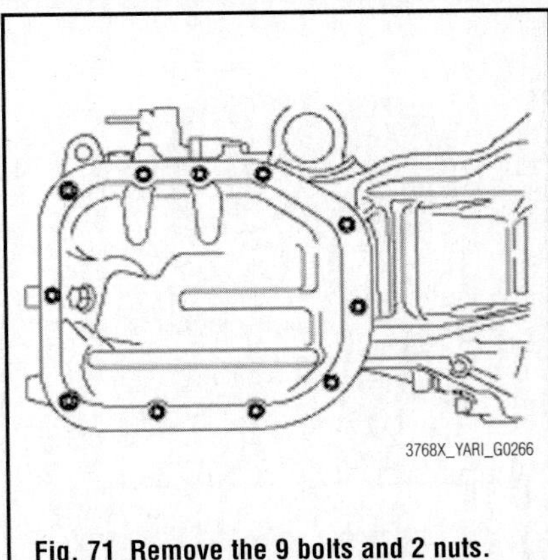

Fig. 71 Remove the 9 bolts and 2 nuts.

6. Apply a continuous bead of seal packing (Diameter 4.0 mm (0.157 in.)). Seal packing: Toyota Genuine Seal Packing Black, Three Bond 1207B or equivalent.

➡Remove any oil from the contact surfaces.

- Install the oil pan within 3 minutes after applying seal packing.
- Do not start the engine for at least 2 hours after installing the oil pan.

7. Install the oil pan with the 9 bolts and 2 nuts. Tighten to 7 ft. lbs. (10 Nm).

OIL PUMP

REMOVAL & INSTALLATION

See Figures 72 through 74.

➡The oil pump is integrated with the timing chain cover.

1. Remove the timing chain cover.
2. Remove the 2 bolts and 3 screws and remove the oil pump cover.
3. Remove the oil pump rotor set.
4. Remove the oil pump relief valve plug, oil pump relief valve spring and oil pump relief valve.

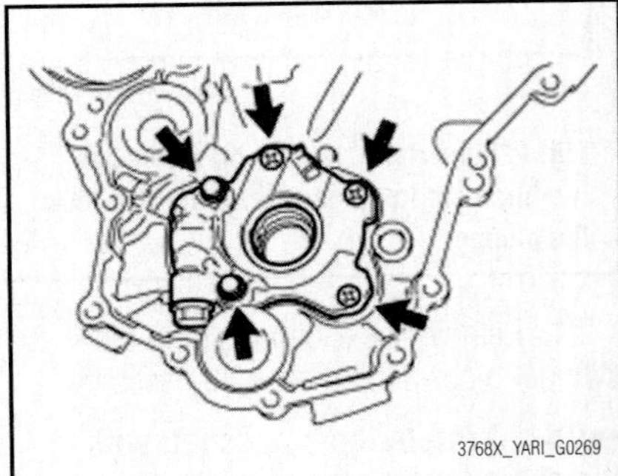

Fig. 72 Remove the 2 bolts and 3 screws and remove the oil pump cover

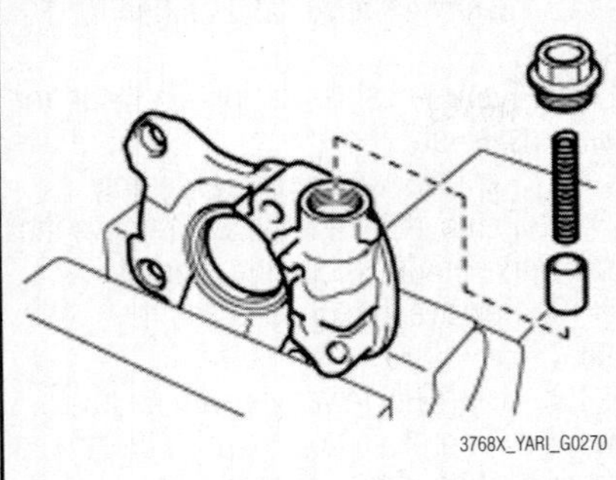

Fig. 73 Remove the oil pump relief valve plug, oil pump relief valve spring and oil pump relief valve

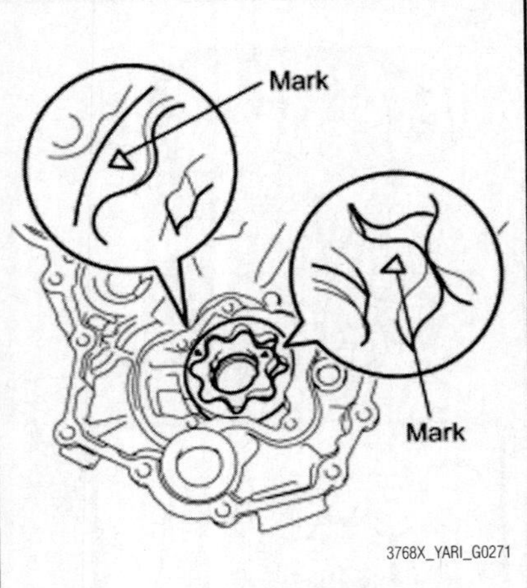

Fig. 74 Oil pump rotor set installation orientation

To install:

5. Coat the oil pump rel ief valve with engine oil and insert the oil pump relief valve and oil pump relief valve spring into the oil pump cover.
6. Install the oil pump relief valve plug. Tighten to 18 ft. lbs. (25 Nm).
7. Coat the oil pump rotor set with engine oil and place it into the oil pump body with the marks facing the oil pump cover side.
8. Install the oil pump cover with the 2 bolts and 3 screws.
   a. Bolt: 78 inch lbs. (8.8 Nm).
   b. Screw: 7.6 ft. lbs. (10 Nm).

PISTON AND RING

POSITIONING

See Figure 75.

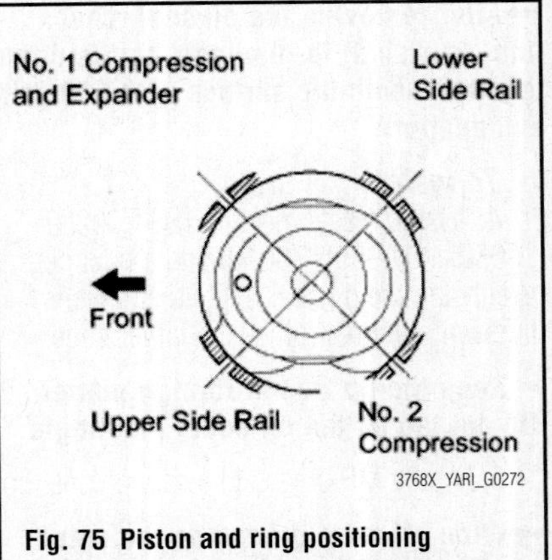

Fig. 75 Piston and ring positioning

REAR MAIN SEAL

REMOVAL & INSTALLATION

See Figures 76 and 77.

1. Remove the appropriate transaxle. Refer to Drivetrain.
2. Remove the flywheel or flexplate. Refer to Drivetrain.
3. Remove rear engine oil seal:
   a. Using a knife, cut off the oil seal lip.
   b. Using a screwdriver with its tip taped, pry out the oil seal.

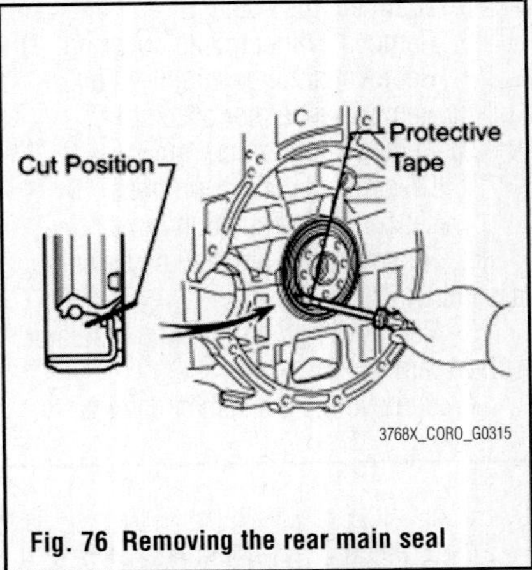

Fig. 76 Removing the rear main seal

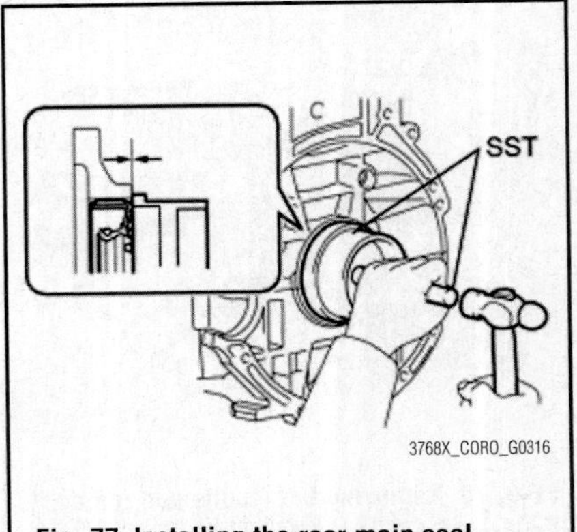

Fig. 77 Installing the rear main seal

➡After removing the oil seal, check the crankshaft for damage. If it is damaged, smooth the surface with 400-grit sandpaper.

To install:

4. Using a seal installer (SST 09223-15030 and 09950-70010) and a hammer, evenly tap the oil seal until its surface is flush with the rear oil seal retainer edge.

➡Keep the lip free of foreign matter. Do not tap on the oil seal at an angle.

5. Apply MP grease to a new oil seal lip.

➡Wipe off extra grease on the crankshaft.

6. To complete installation, reverse remaining removal procedure.

TIMING CHAIN AND FRONT COVER

REMOVAL & INSTALLATION

See Figures 78 through 85.

1. Disconnect cable from negative battery terminal.
2. Remove front wheel RH.
3. Remove engine under cover RH.
4. Drain engine oil.
5. Drain engine coolant.
6. Remove cylinder head cover no. 2.
7. Remove fan & generator v belt.
8. Remove generator assembly.
9. Remove ignition coil no. 1.
10. Disconnect ventilation hose.
11. Disconnect ventilation hose no. 2.
12. Remove cylinder head cover sub-assembly.
13. Remove engine mounting insulator sub-assembly RH.
14. Remove the crankshaft damper.

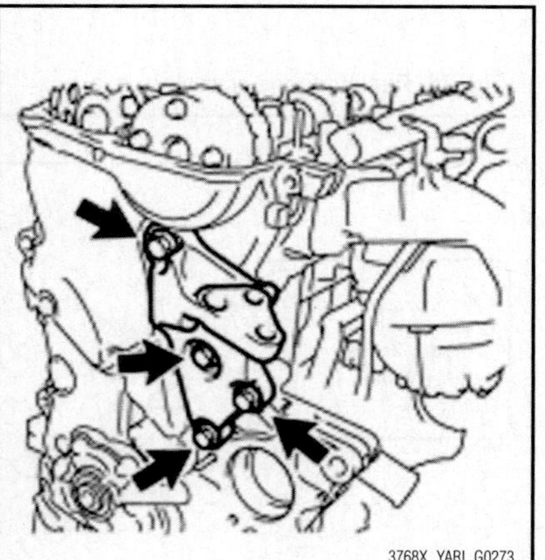

Fig. 78 Removing the 4 bolts and remove the transverse engine mounting bracket

15. Remove Crankshaft Position (CKP) sensor.
16. Remove camshaft timing oil control valve assembly.
17. Remove water pump assembly.
18. Remove the 4 bolts and remove the transverse engine mounting bracket.
19. Remove the 15 bolts and the nut.
20. Using a screwdriver with its tip wrapped in protective tape, prize the timing cover assembly to remove it.

➡Do not damage the contact surfaces of the oil pump assembly or oil pan sub-assembly.

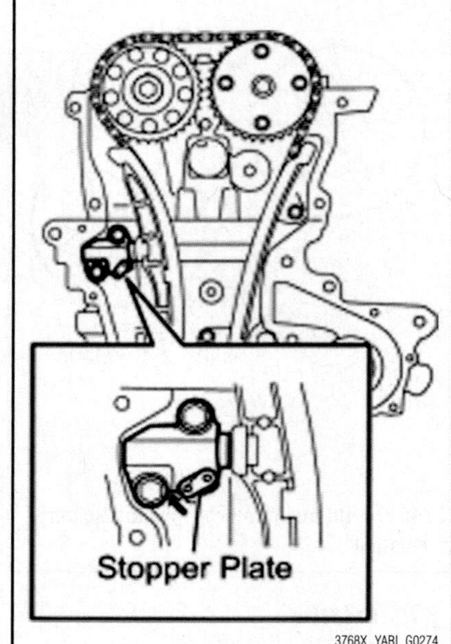

Fig. 79 Pull up the stopper plate and hold it with its lock released

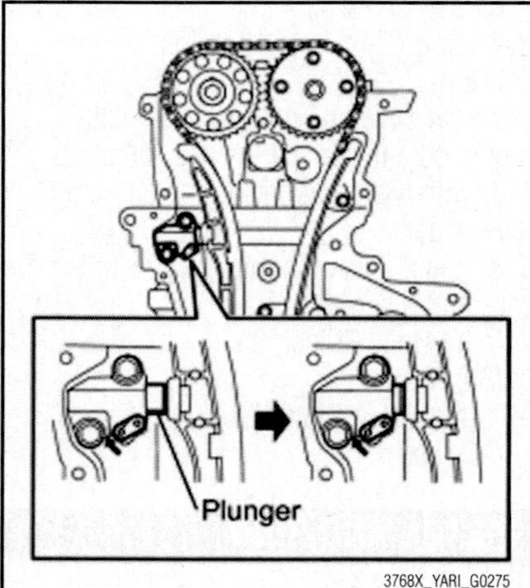

Fig. 80 Unlock the plunger of the tensioner and push it in to the end

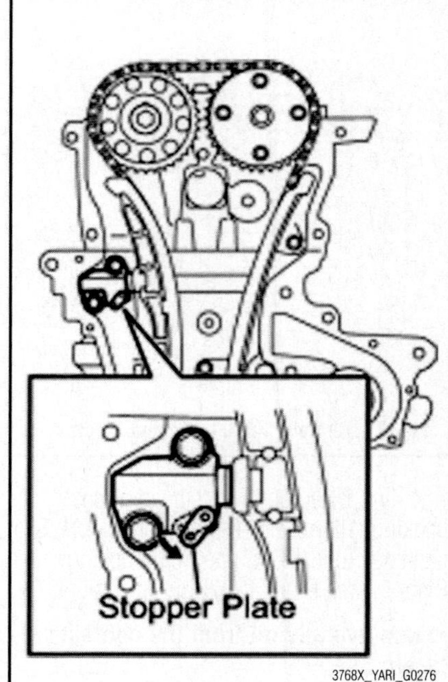

Fig. 81 Pull down the stopper plate with the plunger pushed to the end and lock the plunger

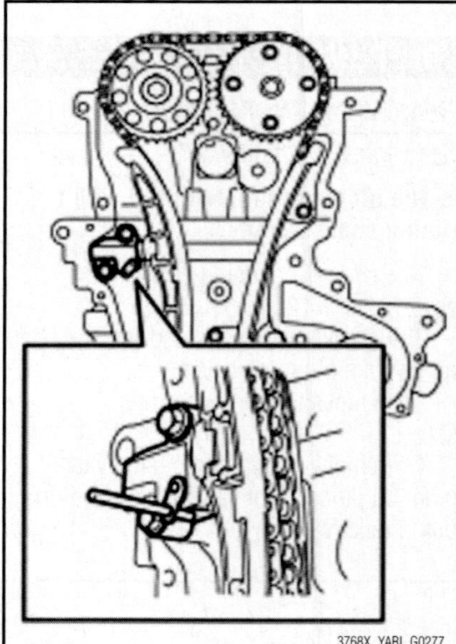

Fig. 82 Insert a 0.12 inch diameter bar into the hole in the stopper plate and lock the plunger

21. Remove the 2 O-rings from the cylinder block and oil pan sub-assembly.

➡Do not rotate the crankshaft with the chain tensioner removed. When rotating the camshaft with the timing chain removed, rotate the crankshaft counterclockwise 40°from the TDC first.

22. Remove chain tensioner assembly no. 1:

a. Pull up the stopper plate and hold it with its lock released.

b. Unlock the plunger of the tensioner and push it in to the end.

c. Pull down the stopper plate with the plunger pushed to the end and lock the plunger.

Fig. 83 Remove the 2 bolts and remove chain tensioner assembly No. 1

d. Insert a 3 mm (0.12 in.) diameter bar into the hole in the stopper plate and lock the plunger.

e. Remove the 2 bolts and remove chain tensioner assembly No. 1.

23. Remove the chain tensioner slipper.

24. Remove the 2 bolts and remove chain vibration damper No. 1.

25. Remove chain sub-assembly.

To install:

26. Make sure that all the timing marks are in the positions (TDC) shown in the illustration.

➡The positions of the timing marks may differ from the predetermined positions due to the force of the valve spring.

27. Set the timing mark of the crankshaft in a position between 40 and 140°ATDC as illustrated.

28. Set the camshaft timing gear and the camshaft timing sprocket in the positions (20°ATDC) shown in the illustration.

29. Set the crankshaft in the position (20°ATDC) shown in the illustration.

30. Install chain vibration damper No. 1 with the 2 bolts.

31. Align the timing marks of the camshaft with the mark plates of the timing chain and install the timing chain.

32. Install the chain tensioner slipper.

33. Install chain tensioner assembly No. 1 with the 2 bolts.

34. Remove the bar from chain tensioner assembly No. 1.

35. Install 2 new O-rings in the 2 locations shown in the illustration.

36. Apply seal packing to the oil pump assembly, cylinder head and cylinder block as shown in the illustration. Seal packing: Water pump part Toyota Genuine Seal Packing 1282B, Three Bond 1282B or Equivalent. Other part Toyota Genuine Seal Packing Black, Three Bond 1207B or Equivalent

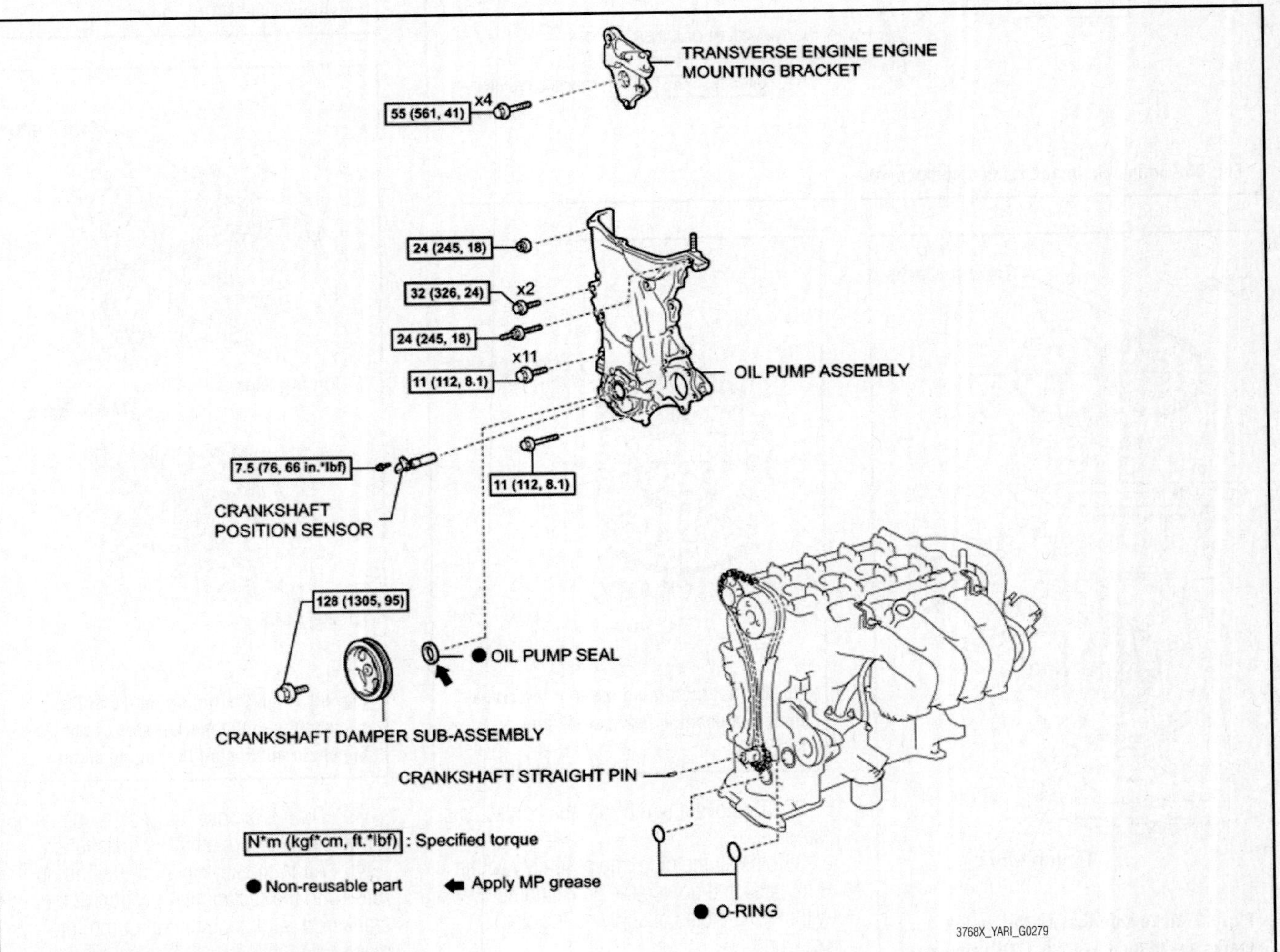

Fig. 84 Timing chain cover and components

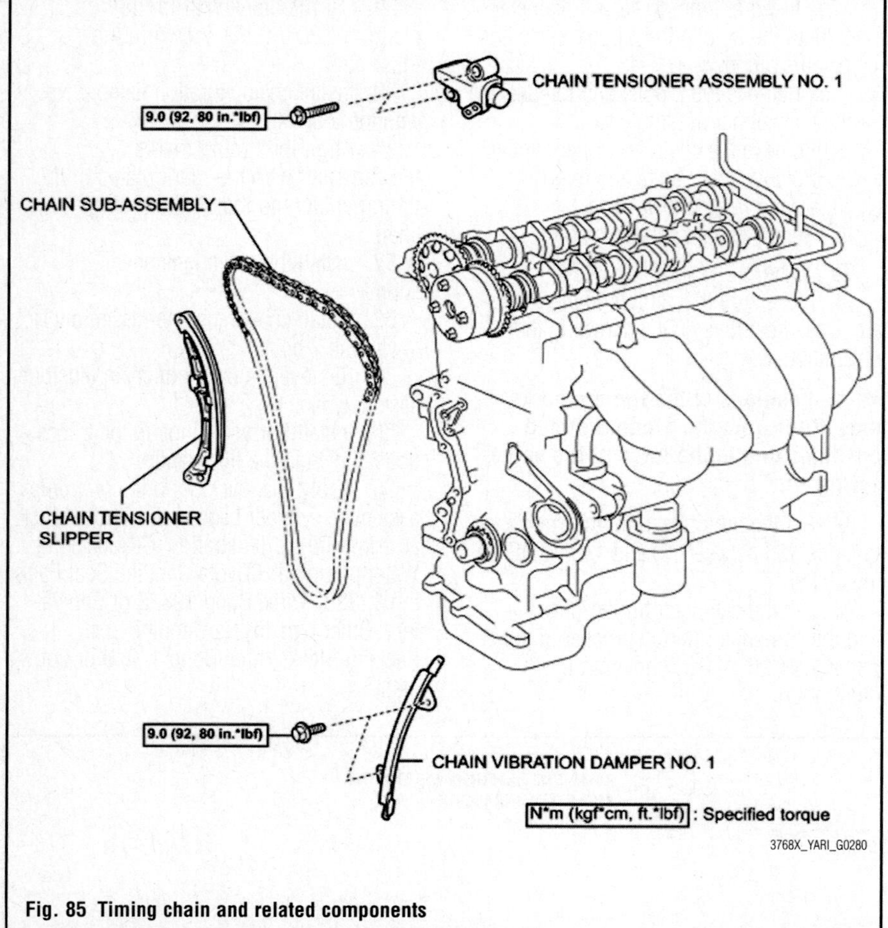

Fig. 85 Timing chain and related components

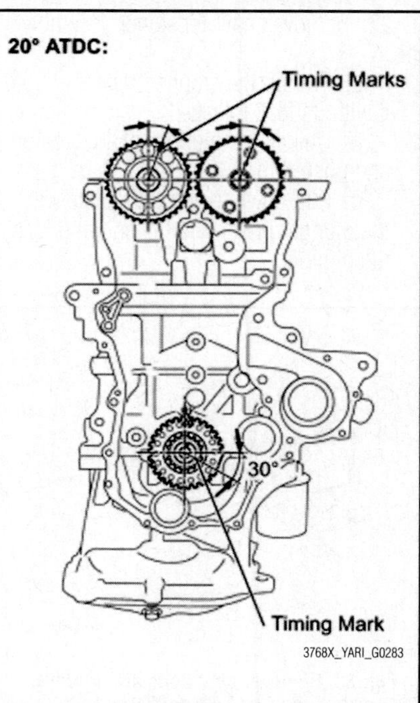

Fig. 88 Set the camshaft timing gear and the camshaft timing sprocket in the positions (20°ATDC) shown

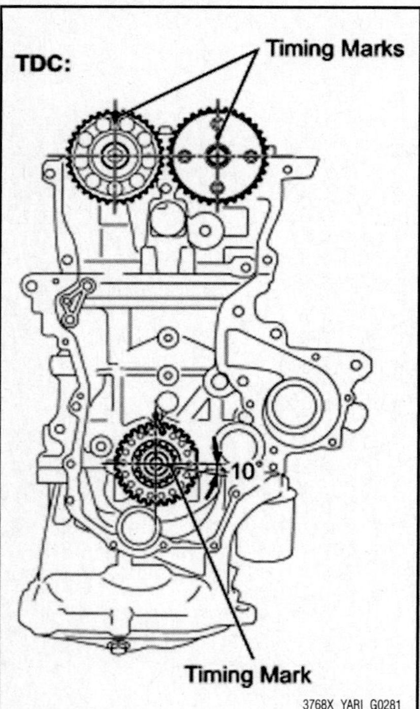

Fig. 86 Make sure that all the timing marks are in the positions (TDC) shown

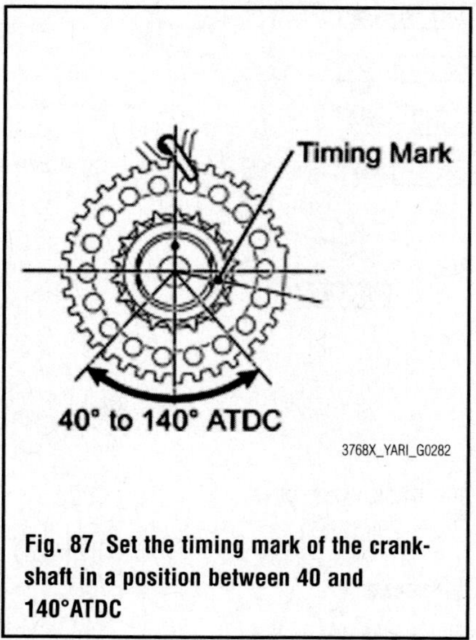

Fig. 87 Set the timing mark of the crankshaft in a position between 40 and 140°ATDC

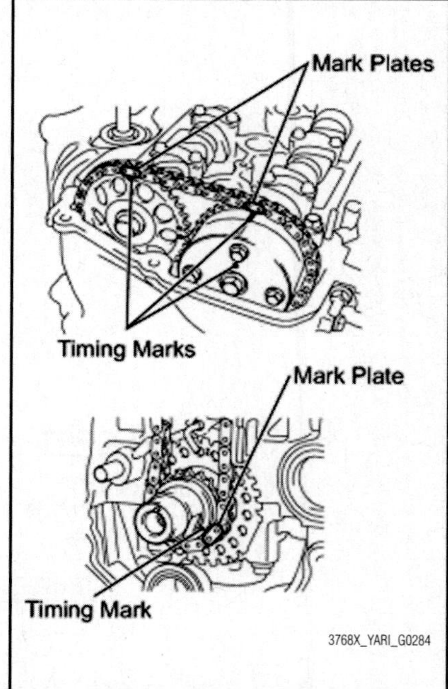

Fig. 89 Align the timing marks of the camshaft with the mark plates of the timing chain and install the timing chain

37. Remove any oil from the contact surfaces.

38. Install the oil pump assembly within 3 minutes and tighten the bolts and nut within 15 minutes of applying the seal packing.

39. Do not expose the seal to engine oil for at least 2 hours after the installation.

40. Align the keyway of the oil pump rotor with the rectangular portion of the crankshaft, and slide the oil pump into place.

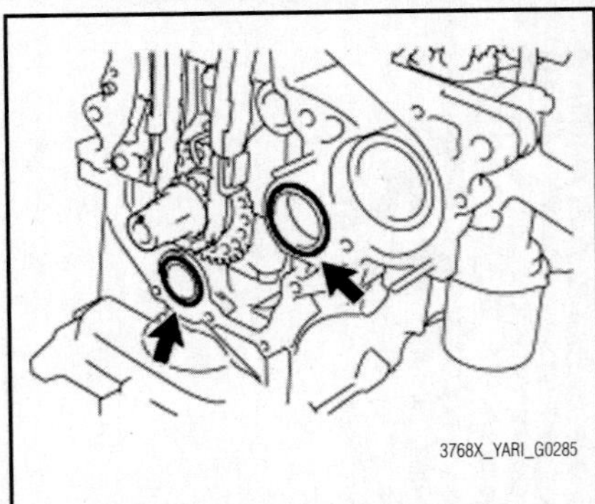

Fig. 90 Install 2 new O-rings in the 2 locations shown

➡After installing the oil pump assembly, install the mounting bracket and water pump within 15 minutes.

41. Install the oil pump assembly with the 15 bolts and the nut. Tighten the bolts and nut uniformly in several steps:
    a. Bolt A: 24 ft. lbs. (32 Nm).
    b. Bolt B: 8 ft. lbs. (11 Nm).
    c. Bolt C: 8 ft. lbs. (11 Nm).
    d. Bolt D: 18 ft. lbs. (24 Nm).
    e. Bolt E: 18 ft. lbs. (24 Nm).

➡After installing the oil pump assembly, install the mounting bracket and water pump within 15 minutes.

42. Install the transverse engine mounting bracket with the 4 bolts. Tighten to 41 ft. lbs. (55 Nm).
43. To complete installation, reverse the remaining removal procedure.
44. Add engine oil.
45. Add coolant.
46. Check for leaks.

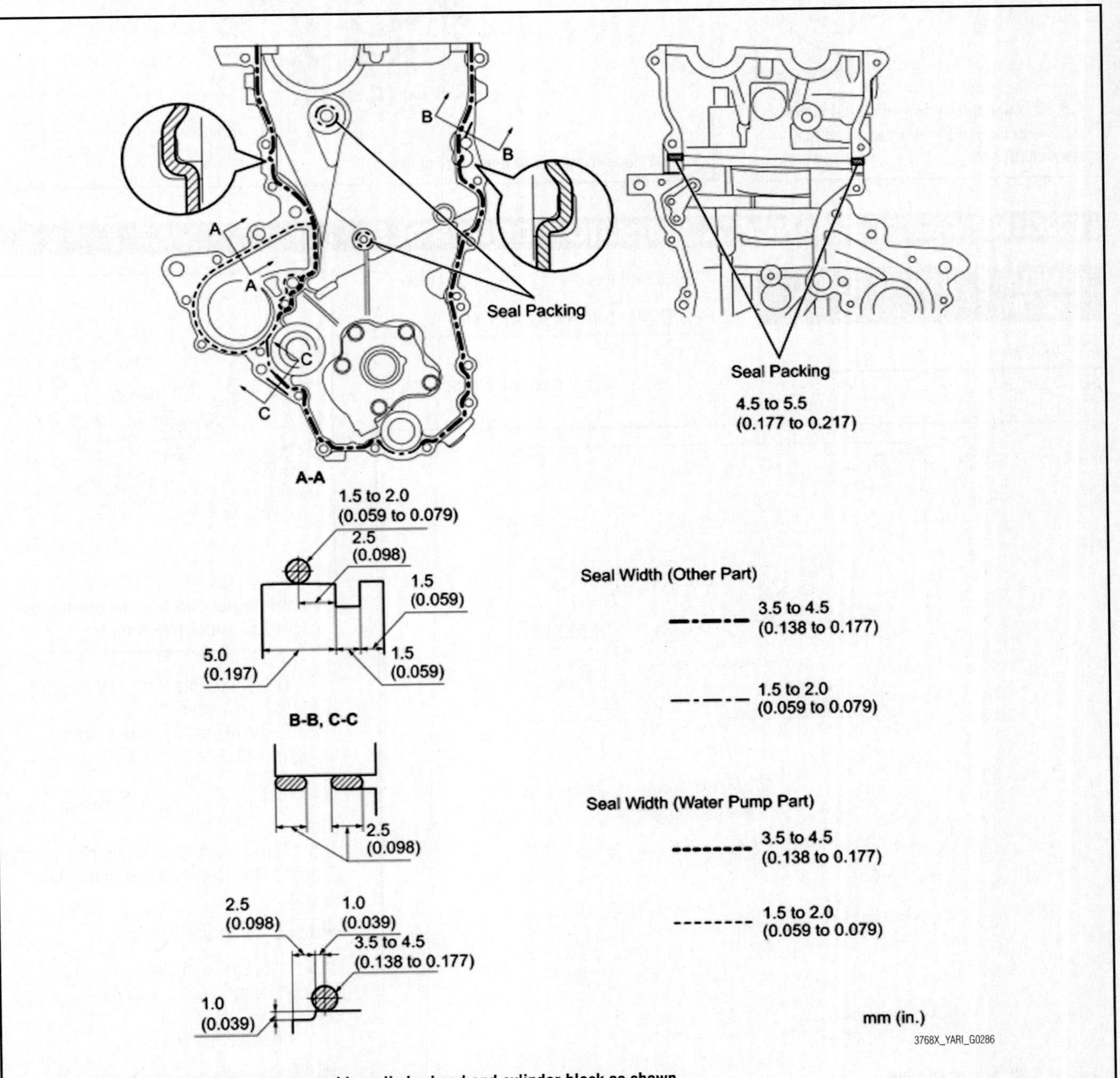

Fig. 91 seal packing to the oil pump assembly, cylinder head and cylinder block as shown

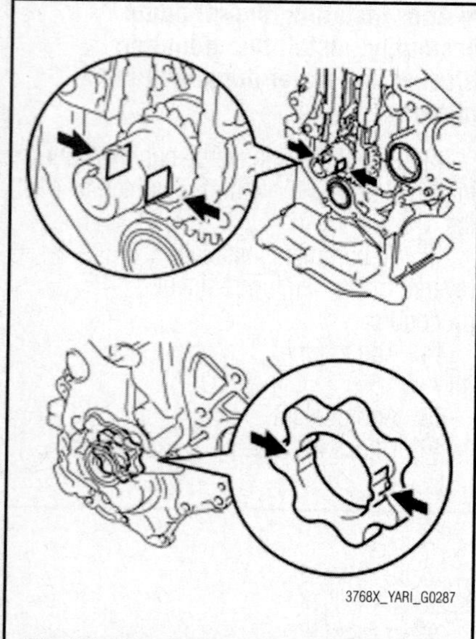

Fig. 92 Align the keyway of the oil pump rotor with the rectangular portion of the crankshaft

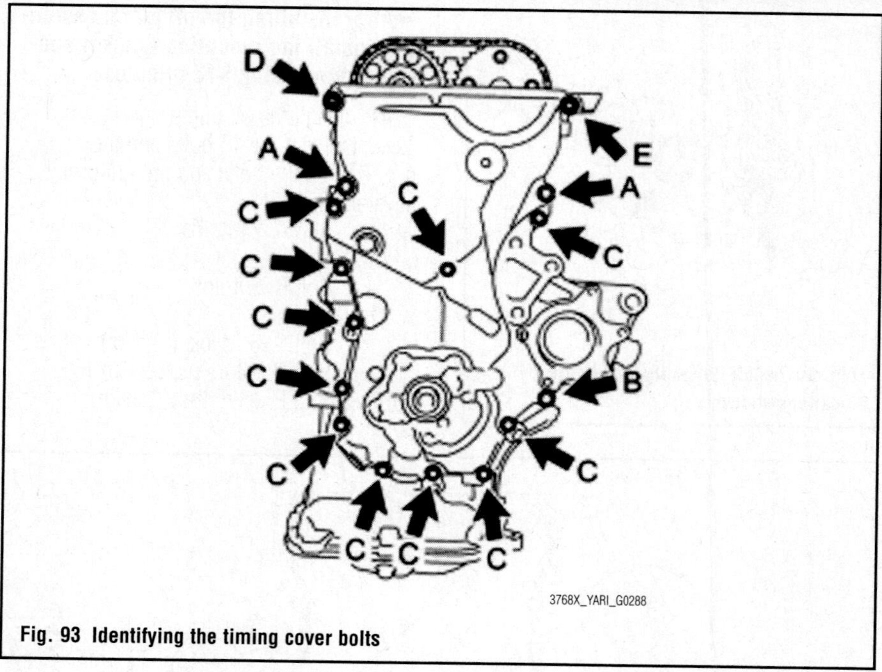

Fig. 93 Identifying the timing cover bolts

ENGINE PERFORMANCE & EMISSION CONTROLS

CAMSHAFT POSITION (CMP) SENSOR

LOCATION

See Figure 94.

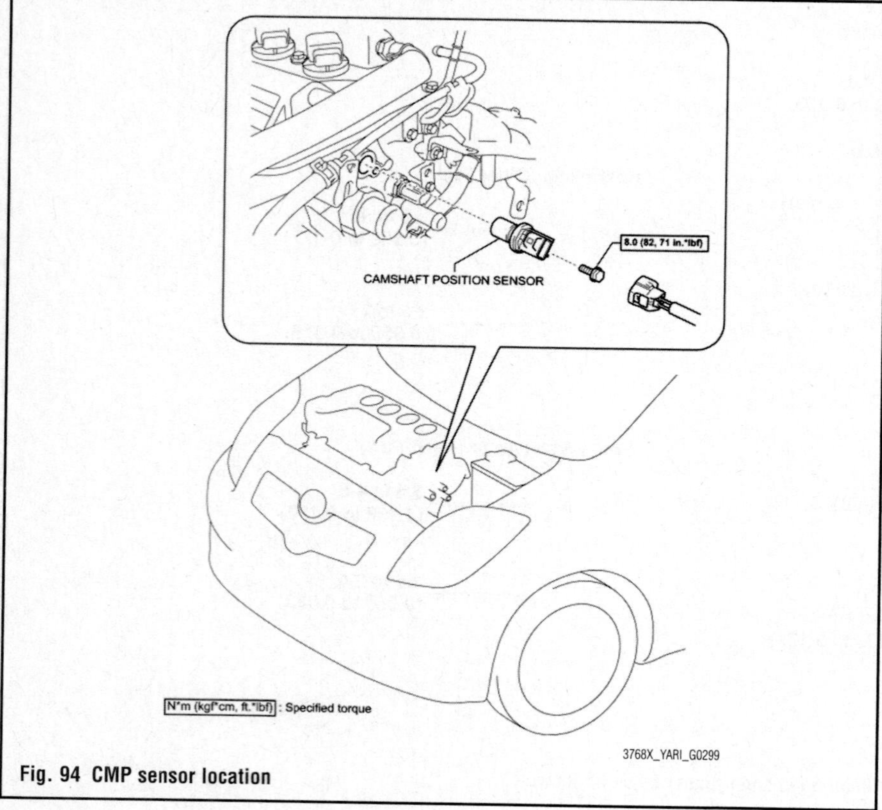

Fig. 94 CMP sensor location

Refer to the accompanying illustration.

REMOVAL & INSTALLATION

See Figure 95.

1. Disconnect the negative battery cable.

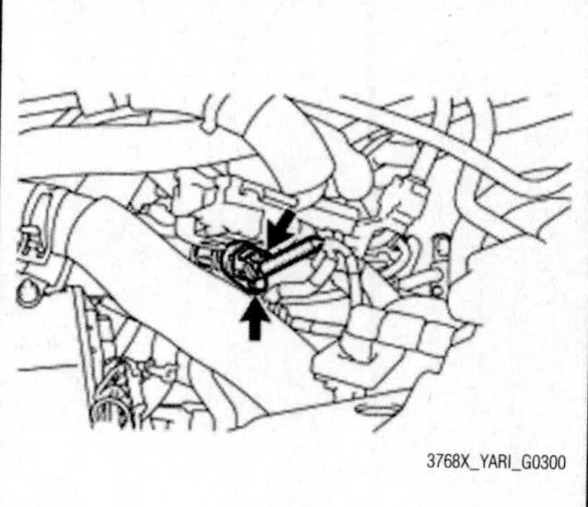

Fig. 95 Remove the bolt and remove the Camshaft Position (CMP) sensor

2. Disconnect the Camshaft Position (CMP) sensor connector.
3. Remove the bolt and remove the Camshaft Position (CMP) sensor.
4. To install, reverse the removal procedure.
5. Apply a light coat of engine oil to the O-ring on the Camshaft Position (CMP) sensor.

➡Do not twist the O-ring.

6. Install the Camshaft Position (CMP) sensor with the bolt.
7. Check for engine oil leakage.

CRANKSHAFT POSITION (CKP) SENSOR

LOCATION

See Figure 96.

Refer to the accompanying illustration.

REMOVAL & INSTALLATION

See Figure 97.

1. Disconnect the negative battery cable.
2. Remove the RH engine under cover.
3. Disconnect the Crankshaft Position (CKP) sensor connector.
4. Remove the bolt and remove the Crankshaft Position (CKP) sensor.

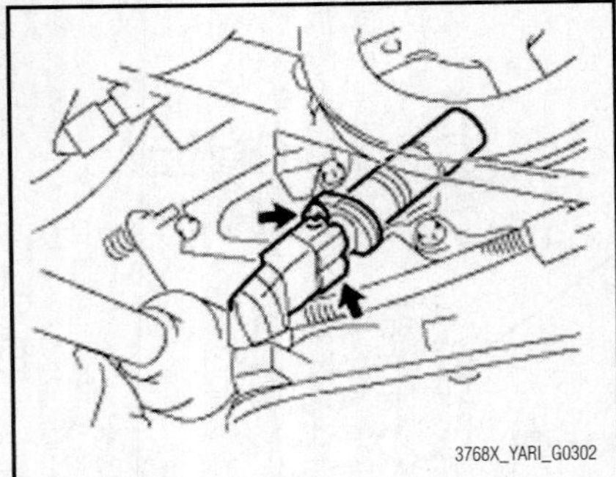

Fig. 97 Remove the bolt and remove the Crankshaft Position (CKP) sensor

ELECTRONIC CONTROL MODULE (ECM)

LOCATION

See Figure 98.

Refer to the accompanying illustration.

REMOVAL & INSTALLATION

See Figures 99 through 101.

1. Remove the wiper motor and link assembly
2. Remove front air shutter seal (for Sedan). Disengage the 3 claws and remove the front air shutter seal.
3. Remove cowl top panel outer (for hatchback):
   a. Disengage the wire harness clamp.
   b. Remove the 9 bolts and remove the cowl top panel outer.
4. Remove cowl top panel outer (for sedan):
   a. Disengage the claw and disconnect the wire harness.

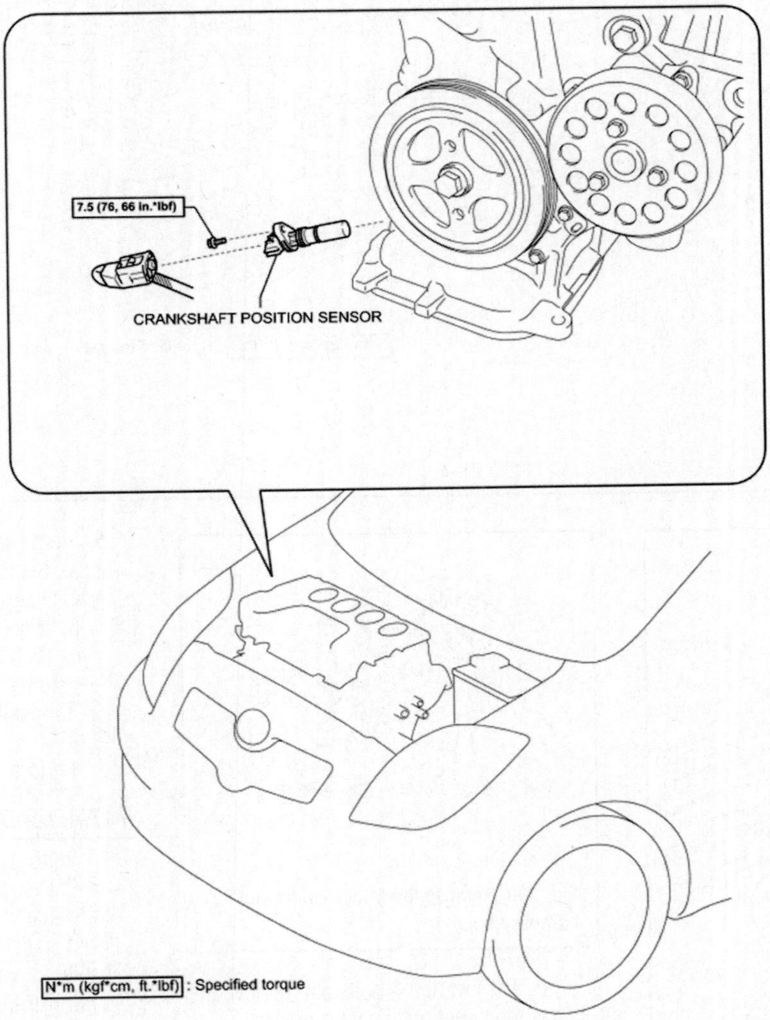

7.5 (76, 66 in.*lbf)

CRANKSHAFT POSITION SENSOR

N*m (kgf*cm, ft.*lbf) : Specified torque

3768X_YARI_G0301

Fig. 96 CKP sensor location

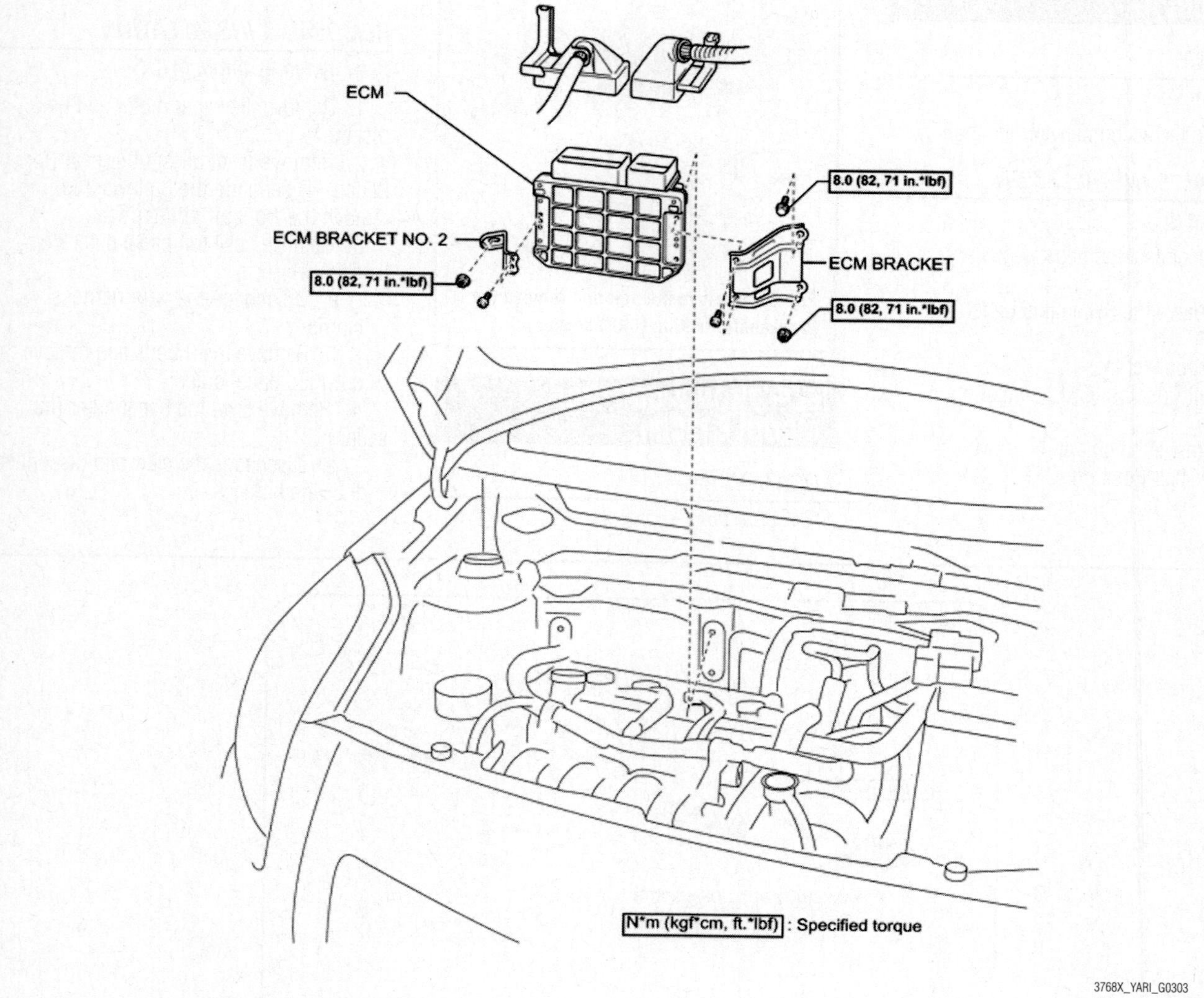

ECM
ECM BRACKET NO. 2
8.0 (82, 71 in.*lbf)
8.0 (82, 71 in.*lbf)
ECM BRACKET
8.0 (82, 71 in.*lbf)
N*m (kgf*cm, ft.*lbf) : Specified torque
3768X_YARI_G0303

Fig. 98 Locating the ECM

3768X_YARI_G0304

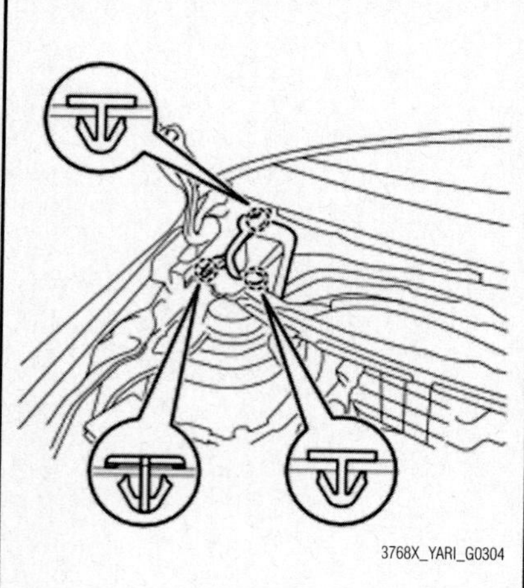

Fig. 99 Disengage the 3 claws and remove the front air shutter seal

3768X_YARI_G0305

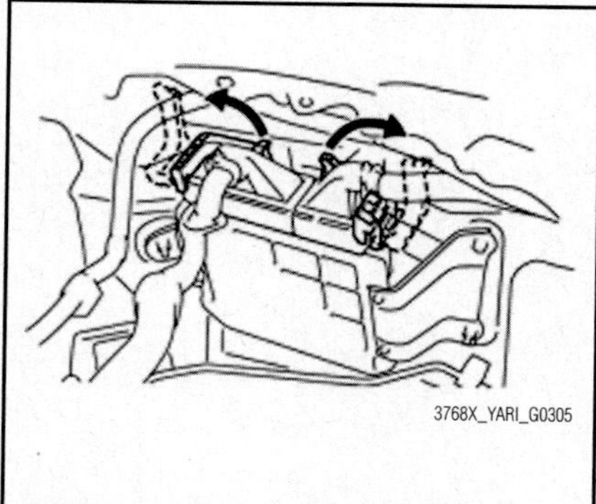

Fig. 100 Remove the 2 lock knobs and harness clamp

b. Remove the 2 bolts and remove the cowl top panel outer center bracket.

c. Remove the 8 bolts and remove the cowl top panel outer.

5. Remove ECM:
   a. Remove the 2 lock knobs and harness clamp.
   b. Disconnect the 2 ECM connectors.
   c. Remove the bolt and 2 nuts and remove the ECM.
6. To install, reverse the removal procedure.
7. Perform initialization.

INITIALIZATION

1. Perform the initialization:
   a. Disconnect the cable from the negative (-) battery terminal. Wait at least 60 seconds and reconnect the cable.
   b. Turn the ignition switch to ON without operating the accelerator pedal.

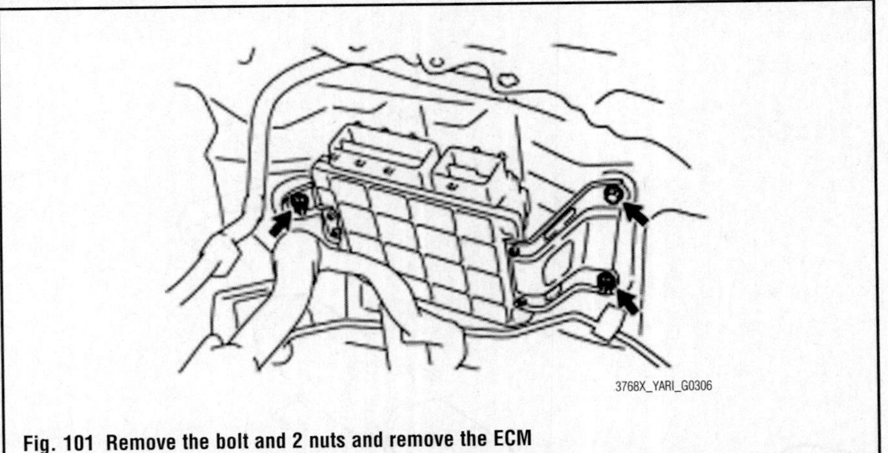

Fig. 101 Remove the bolt and 2 nuts and remove the ECM

➡If the accelerator pedal is operated, perform the above steps again.

c. Connect the Techstream to the DLC3 and clear the DTC's.

d. Start the engine and check that the MIL is not illuminated and that the idle speed is within the specified range when the A/C is switched off after the engine is warmed up.

- Standard (for Automatic Transaxle): With A/C switched off, engine idle speed should be at 650 to 750 RPM.
- Standard (for Manual Transaxle): With A/C switched off, engine idle speed should be at 550 to 650 RPM.

➡Be sure to perform this step with all accessories off. Make sure that the shift lever is in neutral.

e. Enter the following menus: Powertrain / Engine and ECT / Data List / Throttle Sensor Position. Fully depress the accelerator pedal and check that the value is 60% or more.

f. Perform a road test and confirm that there are no abnormalities.

ENGINE COOLANT TEMPERATURE (ECT) SENSOR

LOCATION

See Figure 102.

Refer to the accompanying illustration.

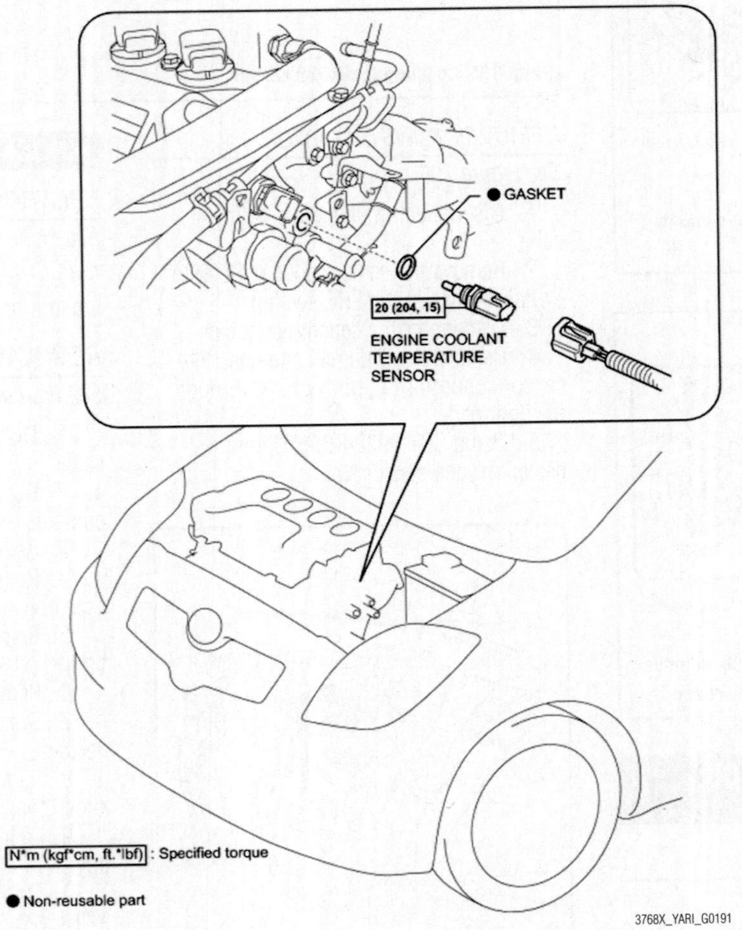

GASKET
20 (204, 15)
ENGINE COOLANT TEMPERATURE SENSOR
N*m (kgf*cm, ft.*lbf) : Specified torque
Non-reusable part
3768X_YARI_G0191

Fig. 102 Locating the Engine Coolant Temperature Sensor

REMOVAL & INSTALLATION

See Figures 103 and 104.

1. Disconnect the negative battery cable.
2. Drain and recycle the engine coolant.
3. Disconnect the engine coolant temperature sensor connector.
4. Using SST 09817-33190, remove the engine coolant temperature sensor.

To install:

5. Provisionally install the engine coolant temperature sensor through a new gasket.
6. Using SST, tighten the engine coolant temperature sensor. Tighten to 15 ft. lbs. (20 Nm).
7. Connect the engine coolant temperature sensor connector.
8. Connect the negative battery cable.
9. Add engine coolant.
10. Check for engine coolant leakage.

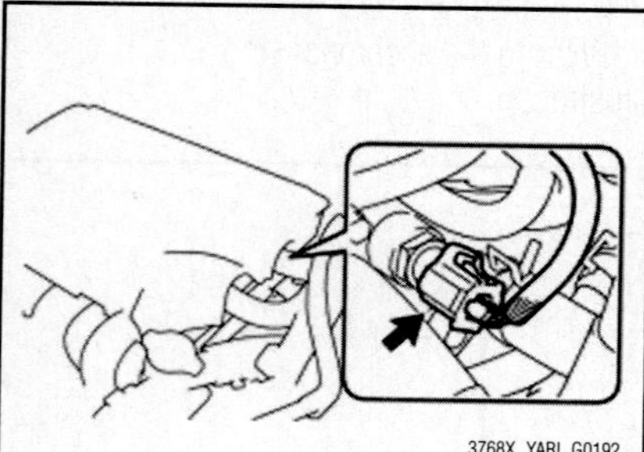

Fig. 103 Disconnect the engine coolant temperature sensor connector

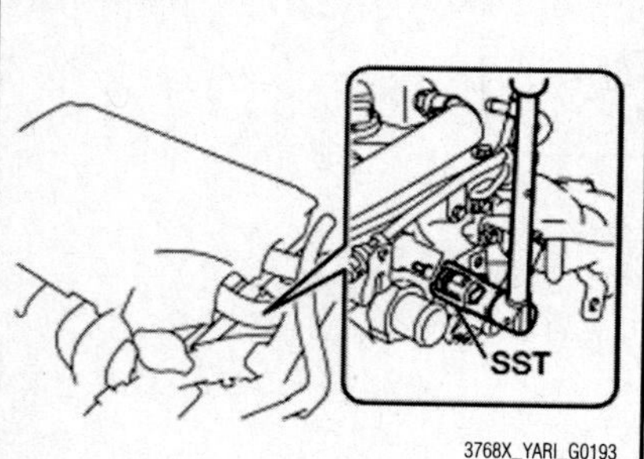

Fig. 104 Using SST 09817-33190, remove the engine coolant temperature sensor

HEATED OXYGEN SENSOR (HO2S)

LOCATION

See Figure 105.

Refer to the accompanying illustration.

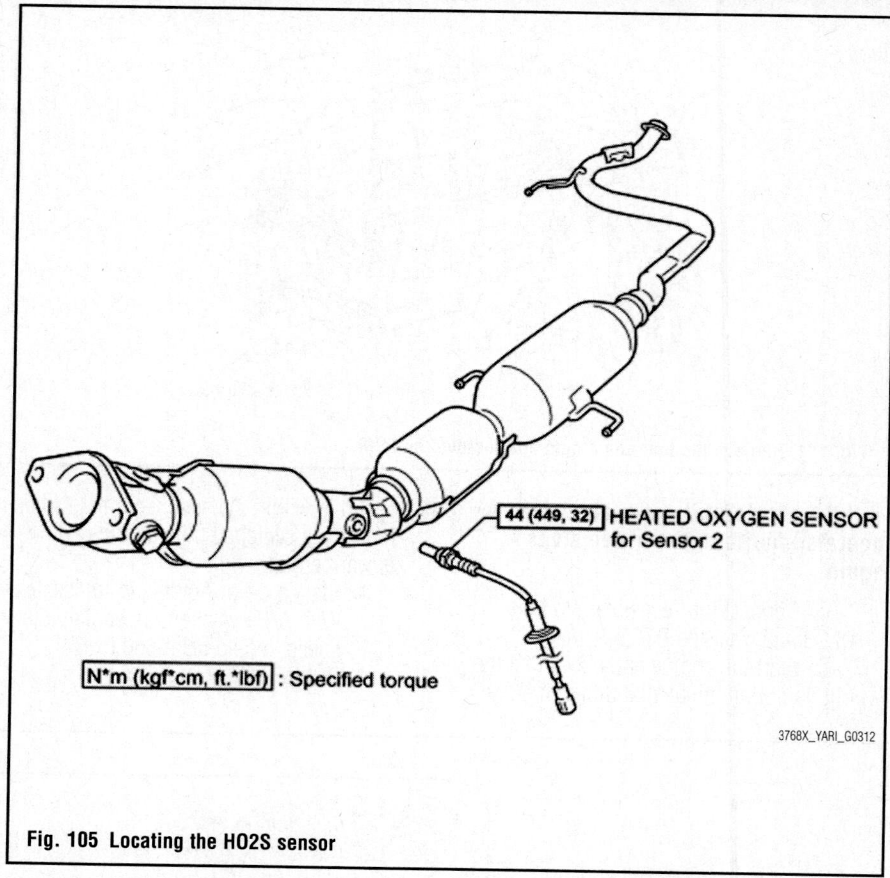

Fig. 105 Locating the HO2S sensor

REMOVAL & INSTALLATION

See Figure 106.

1. Disconnect the negative battery cable.
2. Remove the center console. Refer to Curtain Airbag in the SRS section.
3. Disconnect the sensor connector.
4. Remove the grommet and pass the sensor connector out of the cabin through the floor panel.
5. Using SST 09224-00010, remove heated oxygen sensor No. 2.

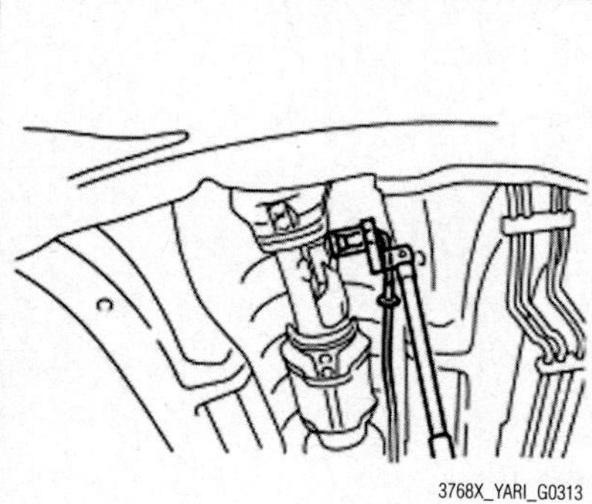

Fig. 106 Using SST 09224-00010, remove heated oxygen sensor No. 2

KNOCK SENSOR (KS)

LOCATION

See Figure 107.

Refer to the accompanying illustration.

REMOVAL & INSTALLATION

See Figure 108.

1. Disconnect the negative battery cable.
2. Drain and recycle the engine coolant.
3. Remove the engine cover.
4. Remove the air cleaner assembly.
5. Remove the throttle body.
6. Remove the intake manifold.
7. Disconnect the knock sensor connector.
8. Remove the nut and remove the knock sensor.
9. To install, reverse the removal procedure.
10. Tighten the knock sensor bolt to 15 ft. lbs. (20 Nm).
11. Make sure the connector is pointing downwards.

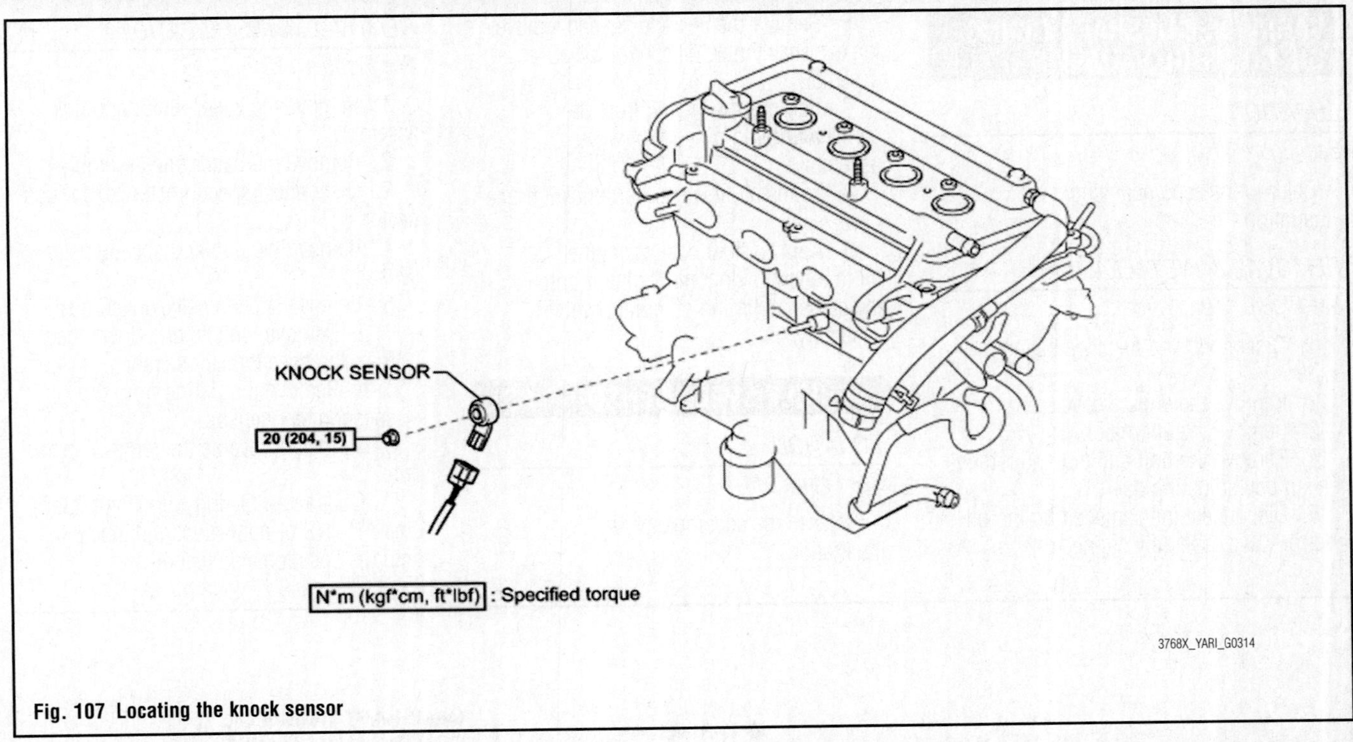

Fig. 107 Locating the knock sensor

Fig. 108 Remove the nut and remove the knock sensor

MALFUNCTION INDICATOR LIGHT (MIL)

RESET PROCEDURE

Use an ODB II scan tool to reset the MIL light.

MASS AIR FLOW (MAF) METER

LOCATION

See Figure 109.

Refer to the accompanying illustration.

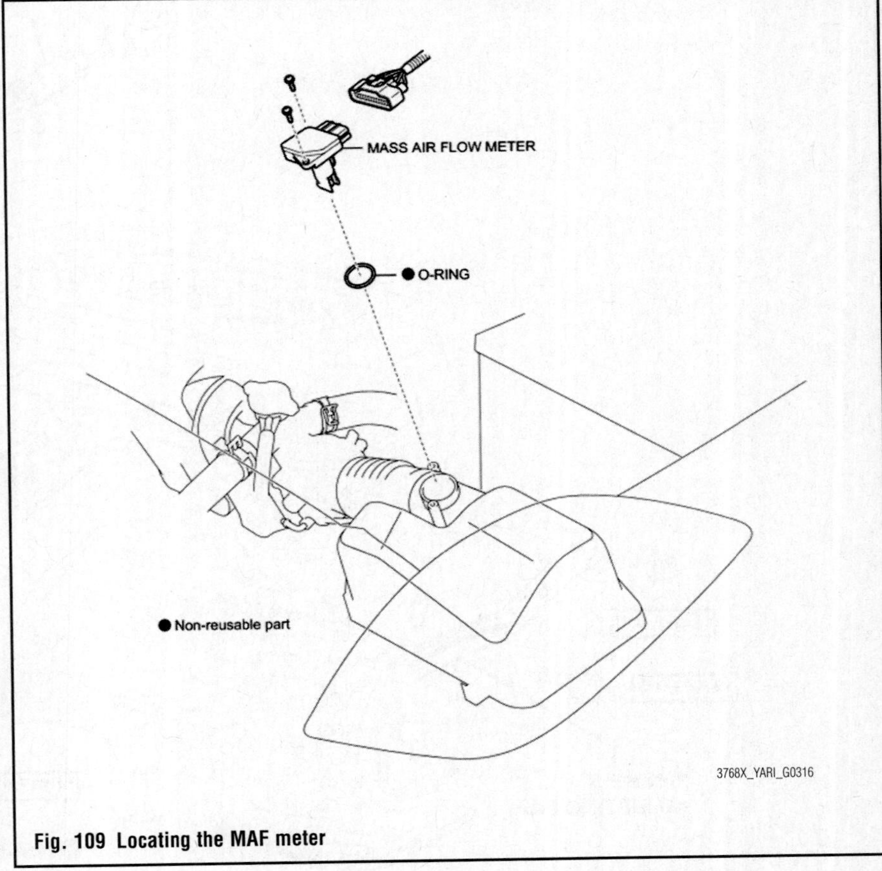

Fig. 109 Locating the MAF meter

REMOVAL & INSTALLATION

See Figure 109.

1. Disconnect the negative battery cable.
2. Disconnect the wire harness clamp and Mass Air Flow (MAF) meter connector.
3. Remove the 2 screws and the Mass Air Flow (MAF) meter.
4. Remove the O-ring from the Mass Air Flow (MAF) meter.
5. To install, reverse the removal procedure.
6. Install a new O-ring on to the MAF sensor.

VARIABLE CAMSHAFT TIMING OIL CONTROL VALVE

LOCATION

See Figure 110.

Refer to the accompanying illustration.

REMOVAL & INSTALLATION

See Figure 110.

1. Disconnect the negative battery cable.
2. Remove the engine cover.
3. Remove the alternator.
4. Remove the bolt and nut and remove the fan belt adjusting bar.
5. Disconnect the camshaft timing oil control valve assembly connector.
6. Remove the bolt and nut and remove the camshaft timing oil control valve assembly.
7. Remove the O-ring from the camshaft timing oil control valve assembly.
8. To install, reverse the removal procedure.
9. Apply a light coat of engine oil to a new O-ring and install it onto the camshaft timing oil control valve assembly.

VEHICLE SPEED SENSOR (VSS)

LOCATION

See Figure 111.

Refer to the accompanying illustration.

REMOVAL & INSTALLATION

See Figures 112 and 113.

1. Remove the wiper motor and link assembly.
2. Remove the batter tray assembly.
3. Disconnect speedometer sensor connector.
4. Remove the bolt and speedometer sensor.
5. Remove speedometer driven gear:
   a. Remove the clip and driven gear from the speedometer sensor.
   b. Remove the O-ring from the speedometer sensor.
6. To install, reverse the removal procedure.
7. Coat a new O-ring with Toyota Genuine ATF WS or equivalent and install it onto the speedometer sensor.

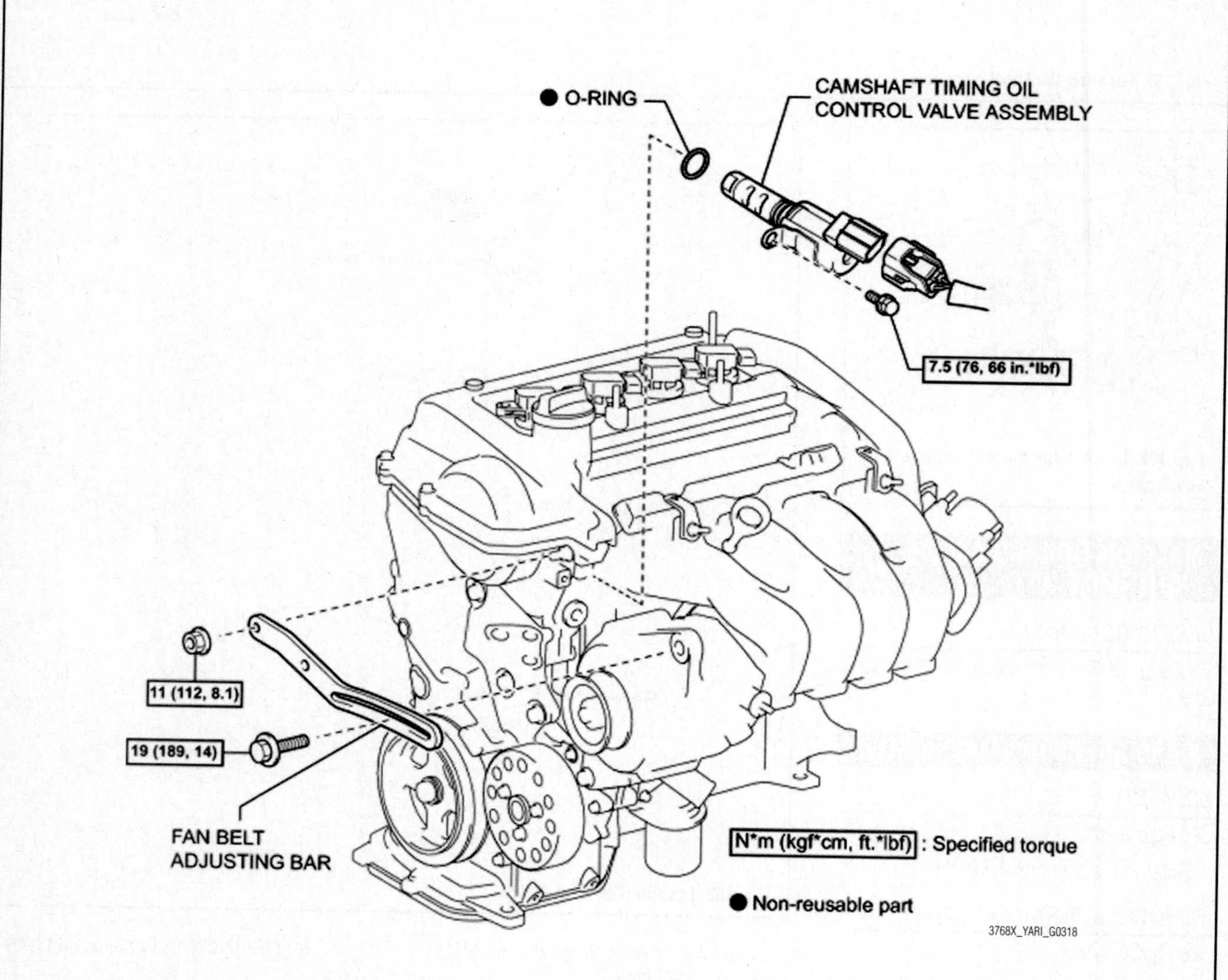

Fig. 110 Locating the Camshaft Timing Oil Control valve

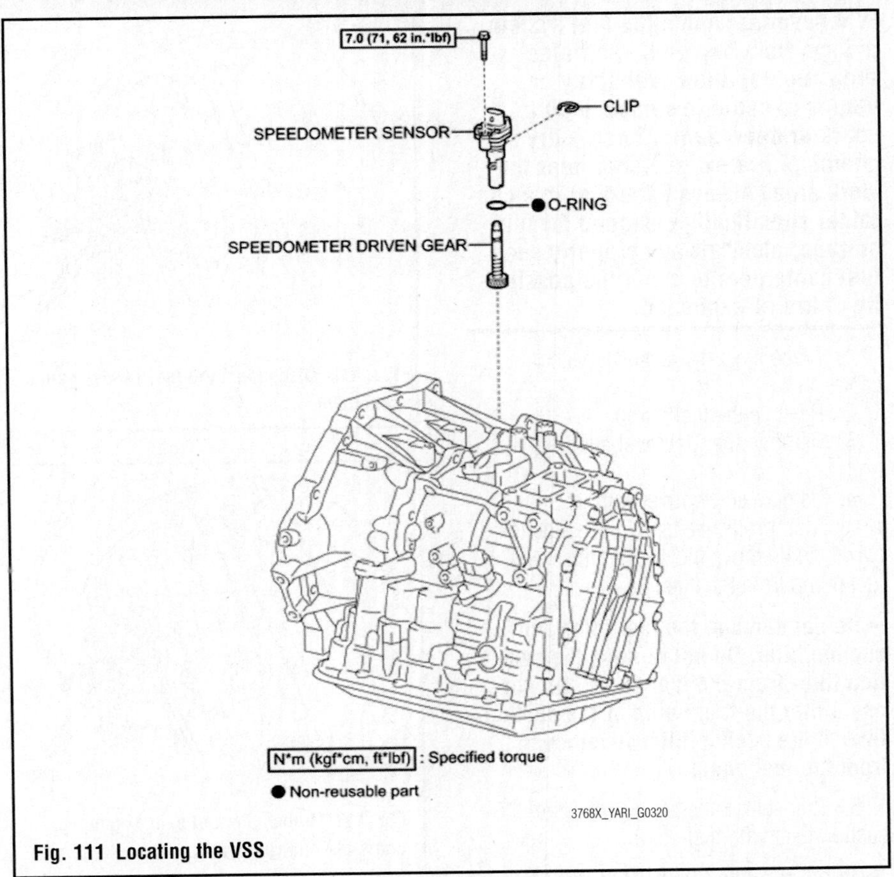

Fig. 111 Locating the VSS

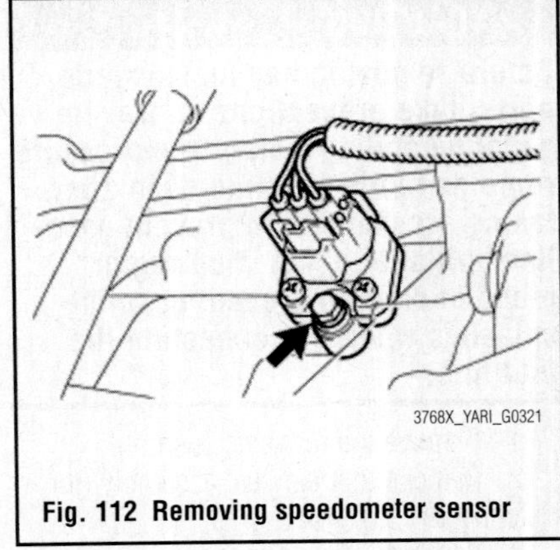

Fig. 112 Removing speedometer sensor

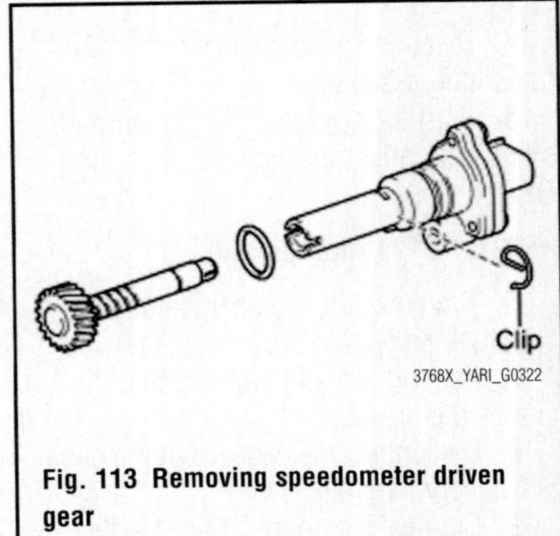

Fig. 113 Removing speedometer driven gear

FUEL

GASOLINE FUEL INJECTION SYSTEM

FUEL SYSTEM SERVICE PRECAUTIONS

Safety is the most important factor when performing not only fuel system maintenance, but any type of maintenance. Failure to conduct maintenance and repairs in a safe manner may result in serious personal injury or death. Work on a vehicle's fuel system components can be accomplished safely and effectively by adhering to the following rules and guidelines.

- To avoid the possibility of fire and personal injury, always disconnect the negative battery cable unless the repair or test procedure requires that battery voltage be applied.
- Always relieve the fuel system pressure prior to disconnecting any fuel system component (injector, fuel rail, pressure regulator, etc.) fitting or fuel line connection. Exercise extreme caution whenever relieving fuel system pressure to avoid exposing skin, face and eyes to fuel spray. Please be advised that fuel under pressure may penetrate the skin or any part of the body that it contacts.
- Always place a shop towel or cloth around the fitting or connection prior to loosening to absorb any excess fuel due to spillage. Ensure that all fuel spillage is quickly removed from engine surfaces. Ensure that all fuel-soaked cloths or towels are deposited into a flame-proof waste container with a lid.
- Always keep a dry chemical (Class B) fire extinguisher near the work area.
- Do not allow fuel spray or fuel vapors to come into contact with a spark or open flame.
- Always use a second wrench when loosening or tightening fuel line connection fittings. This will prevent unnecessary stress and torsion on fuel piping. Always follow the proper torque specifications.
- Always replace worn fuel fitting O-rings with new ones. Do not substitute fuel hose where rigid pipe is installed.

FUEL SYSTEM PRESSURE

RELIEVING

See Figure 114.

⁂ CAUTION

Observe all applicable safety precautions when working around fuel.

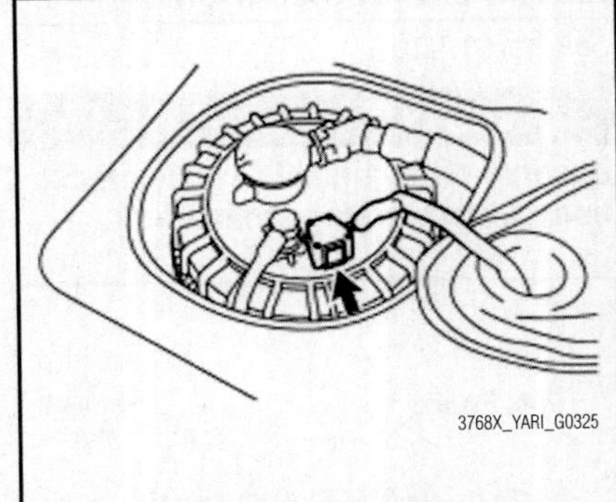

Fig. 114 Disconnect the connector from the fuel pump assembly

Whenever servicing the fuel system, always work in a well ventilated area. Do not allow fuel spray or vapors to come in contact with a spark or open flame. Keep a dry chemical fire extinguisher near the work area. Always keep fuel in a container specifically designed for fuel storage; also, always properly seal fuel containers to avoid the possibility of fire or explosion.

⁂ CAUTION

Before removing any fuel system parts, take precautions to prevent gasoline spillage. As some pressure remains in the fuel line even after taking precautions to prevent gasoline spillage, use a shop rag or piece of cloth to prevent gasoline splashes when disconnecting the fuel line.

1. Remove the rear seat cushion.
2. Remove the rear seat assembly (for Hatchback)
3. Remove the rear floor service hole cover.
4. Disconnect the connector from the fuel pump assembly.
5. Start the engine. After the engine stops naturally, turn the ignition switch OFF.

➡DTC P0171 may be set.

6. Crank the engine again and make sure that the engine does not start.
7. Remove the fuel tank cap and discharge the pressure.
8. Disconnect the cable from the negative battery terminal.
9. Connect the connector of the fuel pump assembly.

FUEL FILTER

REMOVAL & INSTALLATION

See Figure 115.

⁂ CAUTION

Observe all applicable safety precautions when working around fuel.

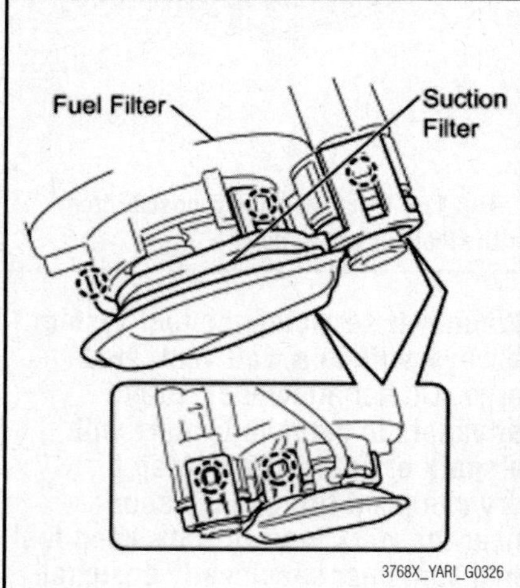

Fig. 115 Identifying the fuel filer and related components

Whenever servicing the fuel system, always work in a well ventilated area. Do not allow fuel spray or vapors to come in contact with a spark or open flame. Keep a dry chemical fire extinguisher near the work area. Always keep fuel in a container specifically designed for fuel storage; also, always properly seal fuel containers to avoid the possibility of fire or explosion.

1. Disconnect the negative battery cable.
2. Remove the fuel pump.
3. Remove the fuel level sending unit.
4. Using a screwdriver with its tip wrapped in protective tape, disengage the 5 claws, and remove the suction filter with fuel pump from the fuel filter.

➡Do not damage the fuel filter and suction filter. Do not remove the suction filter from the fuel pump. Do not use either the fuel pump or the suction filter if the suction filter is removed from the fuel pump.

5. To install, Engage the 5 claws of the suction filter with fuel pump.

➡Do not remove the suction filter from the fuel pump. Do not use either the fuel pump or the suction filter if the suction filter is removed from the fuel pump.

6. Check for fuel leakage:
   a. Connect the Techstream to the DLC3.
   b. Turn the ignition switch to ON and turn the tester ON.

➡Do not start the engine.

   c. Select the following menu items: Powertrain / Engine and ECT / Active Test / Control the Fuel Pump / Speed.
   d. Check that there is no fuel leakage anywhere on the fuel system after doing maintenance.

FUEL LEVEL SENDING UNIT

REMOVAL & INSTALLATION

See Figures 116 and 117.

⁂ CAUTION

Observe all applicable safety precautions when working around fuel. Whenever servicing the fuel system, always work in a well ventilated area. Do not allow fuel spray or vapors to come in contact with a spark or open flame.

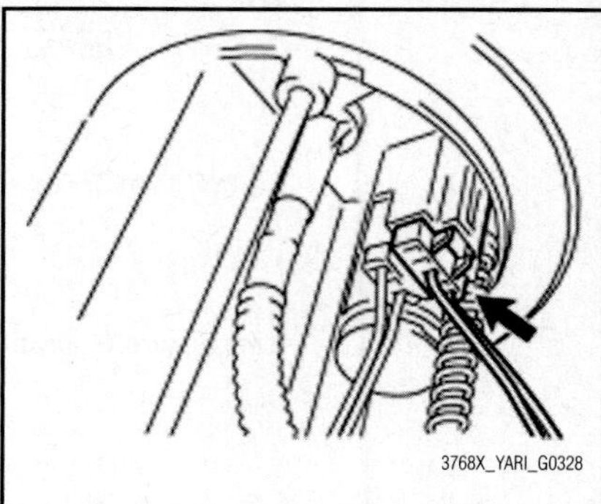

Fig. 116 Disconnect the fuel sender gauge connector

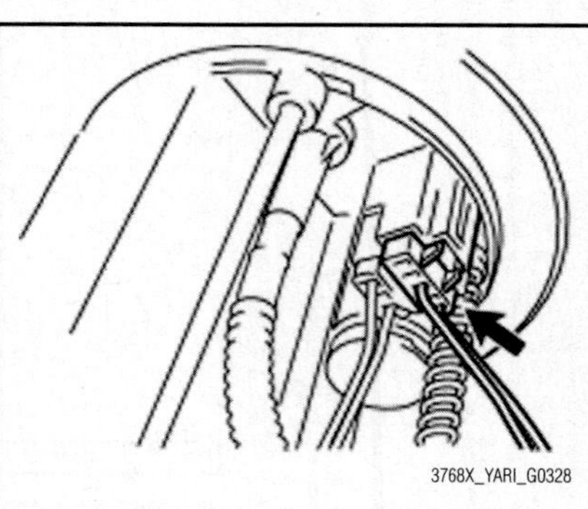

Fig. 117 Unlock the fuel sender gauge, and slide and remove it

Keep a dry chemical fire extinguisher near the work area. Always keep fuel in a container specifically designed for fuel storage; also, always properly seal fuel containers to avoid the possibility of fire or explosion.

1. Disconnect the negative battery cable.
2. Remove the fuel pump.
3. Disconnect the fuel sender gauge connector.
4. Unlock the fuel sender gauge, and slide and remove it.
5. To install, reverse the removal procedure.
6. Check for fuel leakage:
   a. Connect the Techstream to the DLC3.
   b. Turn the ignition switch to ON and turn the tester ON.

➡Do not start the engine.

   c. Select the following menu items: Powertrain / Engine and ECT / Active Test / Control the Fuel Pump / Speed.
   d. Check that there is no fuel leakage anywhere on the fuel system after doing maintenance.

FUEL PUMP & MODULE

REMOVAL & INSTALLATION

See Figures 118 through 123.

CAUTION

Observe all applicable safety precautions when working around fuel. Whenever servicing the fuel system, always work in a well ventilated area. Do not allow fuel spray or vapors to come in contact with a spark or open flame. Keep a dry chemical fire extinguisher near the work area. Always keep fuel in a container specifically designed for fuel storage; also, always properly seal fuel containers to avoid the possibility of fire or explosion.

1. Remove the rear seats and deck board.
2. Remove rear floor service hole cover:
   a. Remove the rear floor service hole cover.

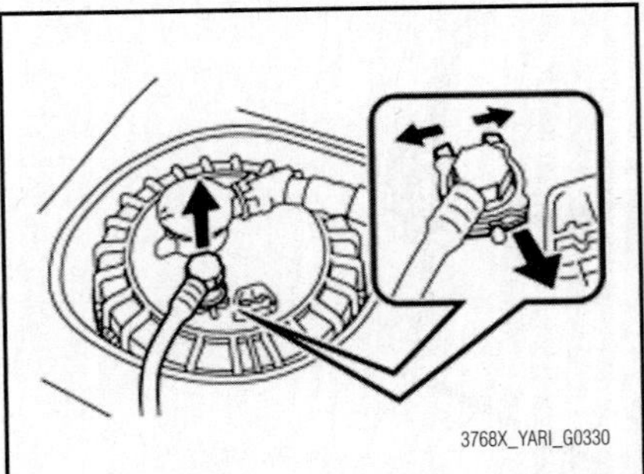

Fig. 118 Disconnect fuel tank main tube sub-assembly

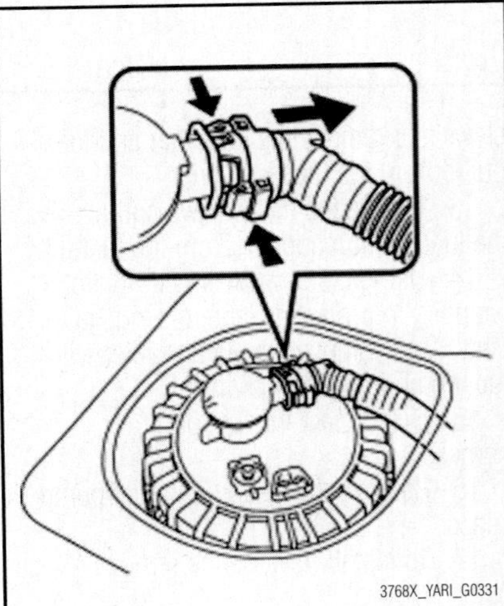

Fig. 119 Disconnect the fuel tank vent hose

   b. Disconnect the fuel pump connector.
3. Properly relieve the fuel system pressure.
4. Disconnect the negative battery cable.
5. Disconnect fuel tank main tube sub-assembly:
   a. Widen the tip of the tube joint clip and pull out the clip in the direction indicated by the arrow.

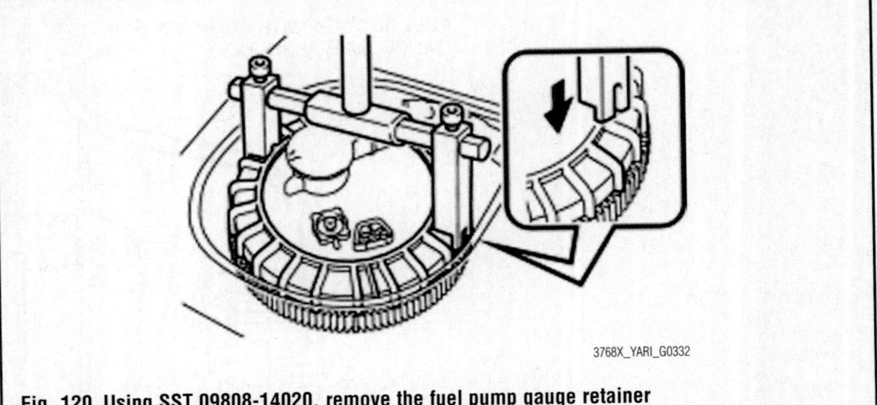

Fig. 120 Using SST 09808-14020, remove the fuel pump gauge retainer

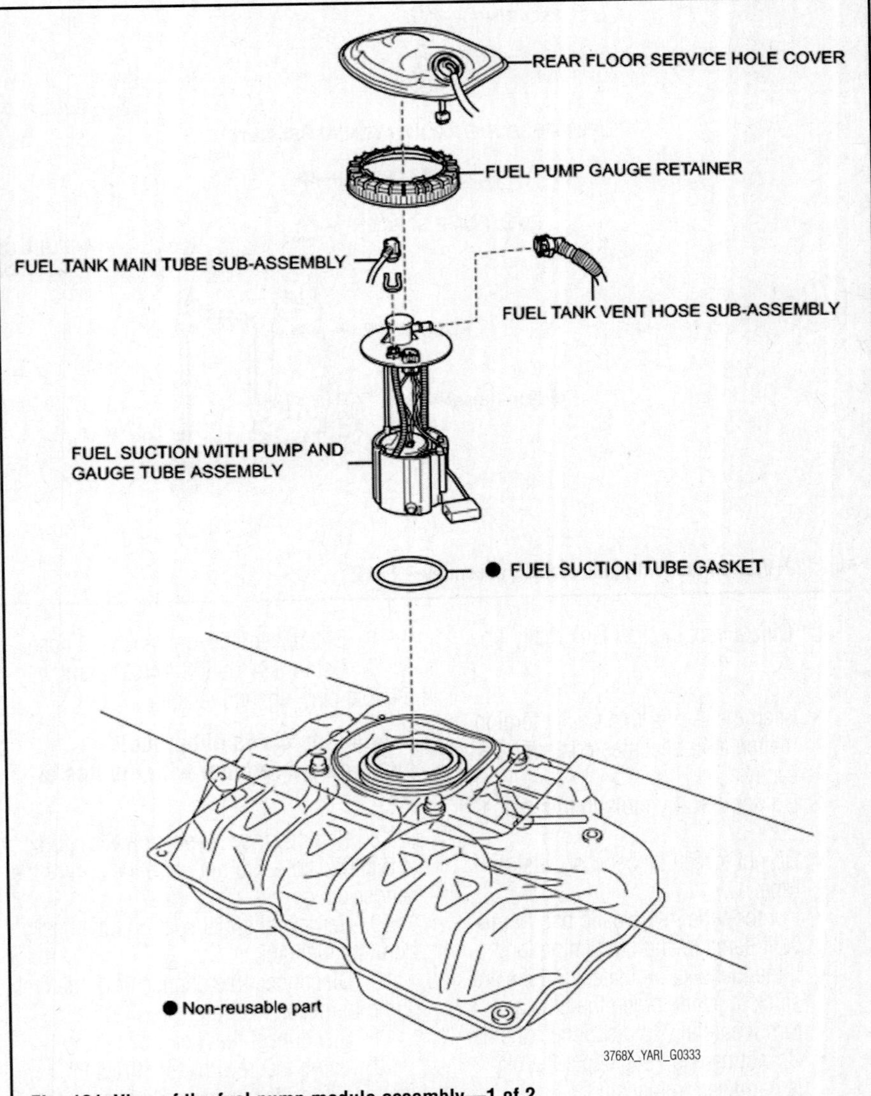

Fig. 121 View of the fuel pump module assembly—1 of 2

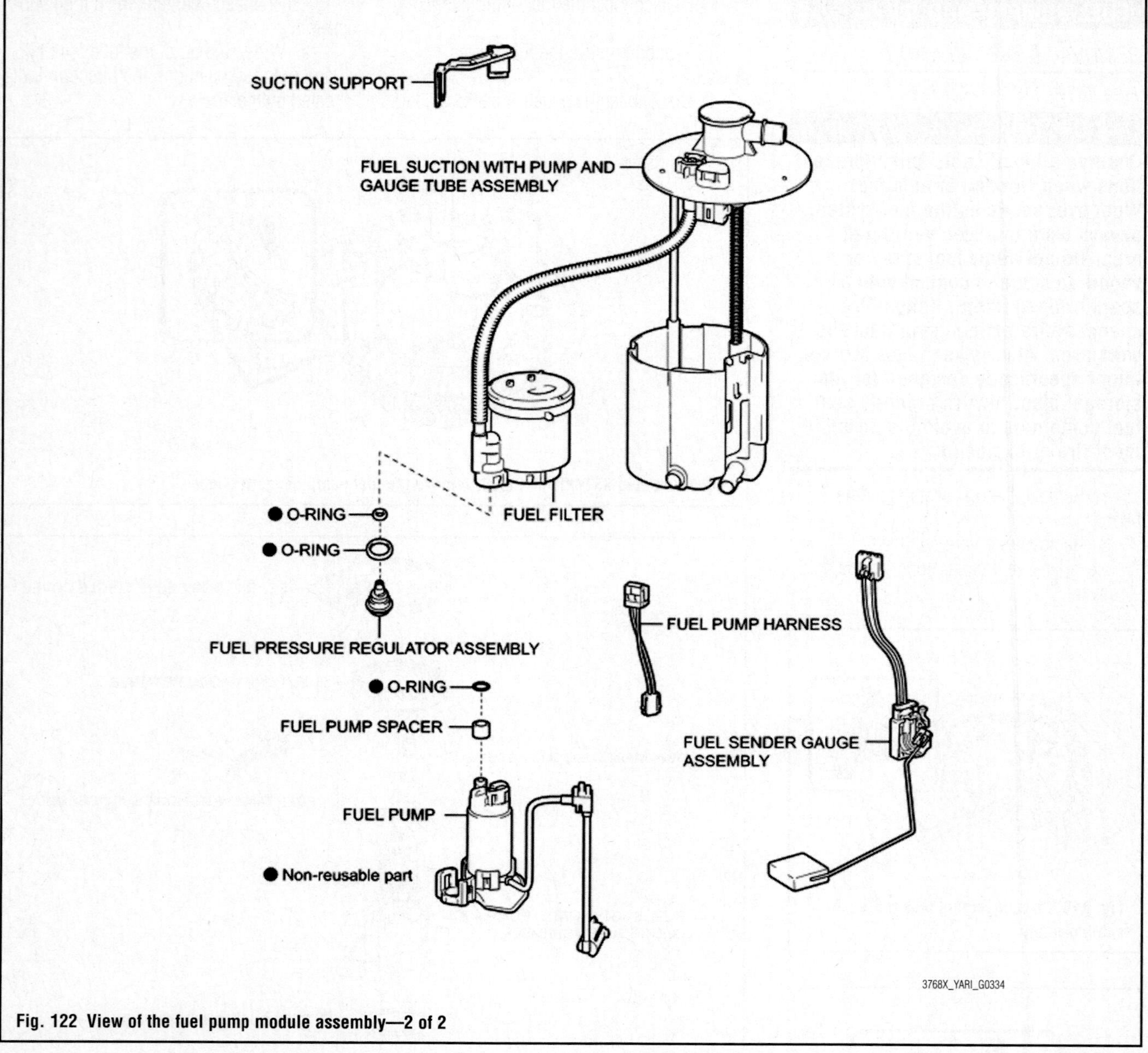

Fig. 122 View of the fuel pump module assembly—2 of 2

b. Disconnect the fuel tank main tube.

➡

- Keep the O-ring free of any foreign matter, as it becomes contaminated easily.
- Do not use any tools in this procedure.
- Do not forcefully bend or twist the tube.
- Put the tube in a plastic bag to prevent damage and contamination.
- If the fuel suction plate and tube are stuck together, pinch the tube and turn it carefully to disconnect them.
- Do not damage any clips. If a clip is damaged, replace it.

6. Disconnect the fuel tank vent hose.
7. Using SST 09808-14020, remove the fuel pump gauge retainer.

➡Align the claws of the fuel pump gauge retainer with the tips of SST.

8. Remove the fuel suction with pump and gauge tube. Do not bend the arm of the sender gauge.
9. Remove the fuel suction tube gasket from the fuel tank.
10. Disconnect the connector of the fuel pump harness.
11. Disconnect the fuel tube.
12. Using a screwdriver with its tip wrapped in protective tape, disengage the 2 claws and remove the fuel filter and fuel pump from the fuel sub-tank.
13. Disengage the 2 claws and remove the fuel suction support from the fuel filter.
14. Using a screwdriver with its tip wrapped in protective tape, disengage the 5 claws, and remove the suction filter with fuel pump from the fuel filter.
15. Disconnect the fuel pump harness.
16. Remove the O-ring and fuel pump spacer.
17. To install, reverse the removal procedure.
18. Apply gasoline to a new O-ring, and then install the fuel pump spacer and new O-ring onto the fuel pump.

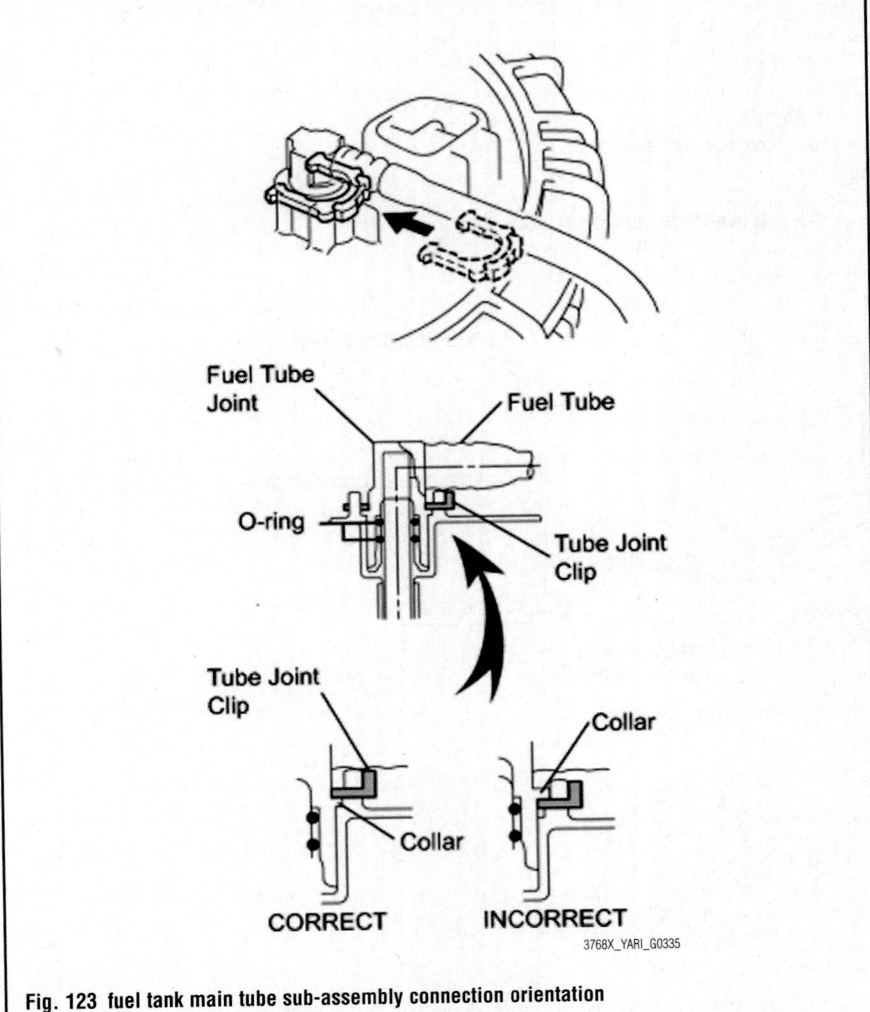

Fig. 123 fuel tank main tube sub-assembly connection orientation

➡Do not disassemble the fuel pump and the suction filter because they are non-reusable parts.

19. Check for fuel leakage:
   a. Connect the Techstream to the DLC3.
   b. Turn the ignition switch to ON and turn the tester ON.

➡Do not start the engine.

   c. Select the following menu items: Powertrain / Engine and ECT / Active Test / Control the Fuel Pump / Speed.
   d. Check that there is no fuel leakage anywhere on the fuel system after doing maintenance.

FUEL RAIL AND INJECTOR

REMOVAL & INSTALLATION

See Figures 124 and 125.

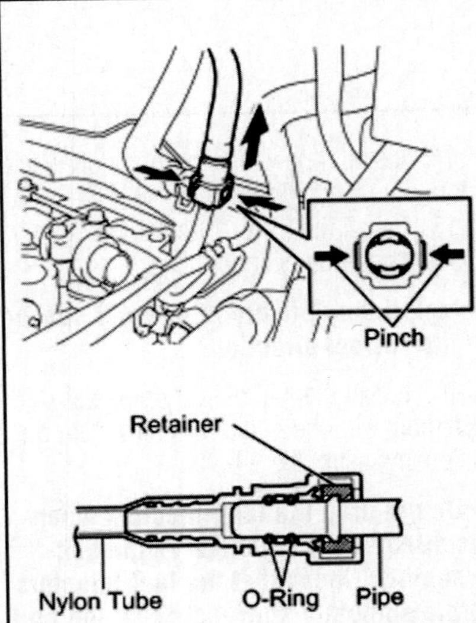

Fig. 124 Pinch the retainer of the fuel tube connector, then pull the fuel tube connector to disconnect the fuel tube out of the fuel delivery pipe sub-assembly

⁂ CAUTION

Observe all applicable safety precautions when working around fuel. Whenever servicing the fuel system, always work in a well ventilated area. Do not allow fuel spray or vapors to come in contact with a spark or open flame. Keep a dry chemical fire extinguisher near the work area. Always keep fuel in a container specifically designed for fuel storage; also, always properly seal fuel containers to avoid the possibility of fire or explosion.

1. Properly relieve the fuel system pressure.
2. Disconnect the negative battery cable.
3. Remove the valve cover.
4. Remove the ignition coils.
5. Remove the engine cover.
6. Disconnect fuel tube sub-assembly:
   a. Remove the fuel pipe clamp.
   b. Pinch the retainer of the fuel tube connector, then pull the fuel tube connector to disconnect the fuel tube out of the fuel delivery pipe sub-assembly.

➡

- Remove any dirt and foreign matter from the fuel tube connector before performing this work.
- Do not allow any scratches or foreign matter on the parts when disconnecting, as the fuel tube connector has the O-ring that seals the pipe.
- Perform this work by hand. Do not use any tools.
- Do not forcibly bend, twist or turn the nylon tube.
- Protect the disconnected parts by covering them with a vinyl bag after disconnecting the fuel tube.
- If the fuel tube connector and pipe are stuck, push and pull to release them.

7. Remove the 3 bolts and remove the fuel delivery pipe sub-assembly with 4 fuel injectors.

➡Do not drop the fuel injectors when removing the fuel delivery pipe sub-assembly.

8. Remove the 2 delivery pipe No. 1 spacers.
9. Remove the 4 injector vibration insulators.

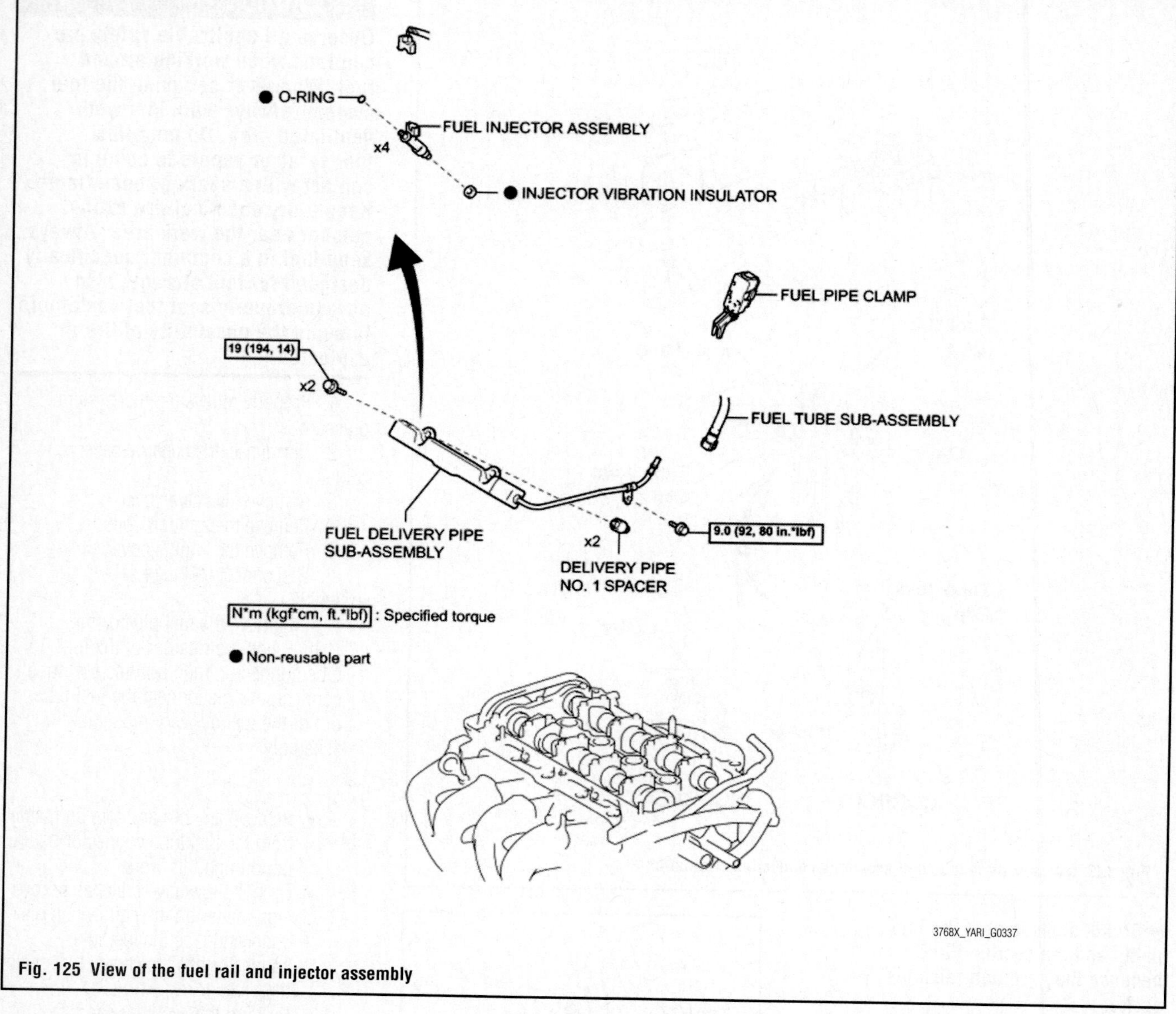

Fig. 125 View of the fuel rail and injector assembly

10. Pull the 4 fuel injector assemblies out of the fuel delivery pipe sub-assembly.

To install:

11. Apply a light coat of gasoline or spindle oil to new O-rings, then install one onto each fuel injector.
12. Apply a light coat of gasoline or spindle oil to the contact surfaces of the fuel delivery pipe and the O-ring of the fuel injector.
13. While turning the fuel injector left and right, install it onto the fuel delivery pipe.

➡Do not twist the O-ring. After installing the fuel injectors, check that they turn smoothly. If not, replace the O-ring with a new one.

14. Install 4 new injector vibration insulators onto the cylinder head.
15. Install the 2 No. 1 delivery pipe spacers onto the cylinder head.

➡Install the delivery pipe No. 1 spacer in the correct direction.

16. Install the fuel delivery pipe sub-assembly with the 4 fuel injectors, then provisionally install the 3 bolts.

➡Do not drop the fuel injectors when installing the fuel delivery pipe sub-assembly. Check that the fuel injectors rotate smoothly after installing the fuel delivery pipe sub-assembly.

17. Connect fuel tube sub-assembly:
    a. Insert the fuel tube connector into the fuel delivery pipe until a click sound can be heard.

➡Check that there are no scratches or foreign matter around the disconnected parts of the fuel tube connector and pipe before performing this work. After connecting the fuel tube, check that the fuel tube connector and pipe are securely connected by pulling them.

18. Install the fuel pipe clamp.
19. To complete installation, reverse the remaining removal procedure.
20. Check for fuel leakage:
    a. Connect the Techstream to the DLC3.
    b. Turn the ignition switch to ON and turn the tester ON.

➡Do not start the engine.

    c. Select the following menu items: Powertrain / Engine and ECT /

Active Test / Control the Fuel Pump / Speed.

d. Check that there is no fuel leakage anywhere on the fuel system after doing maintenance.

21. Check for engine oil leakage.

FUEL TANK

REMOVAL & INSTALLATION

See Figures 126 through 128.

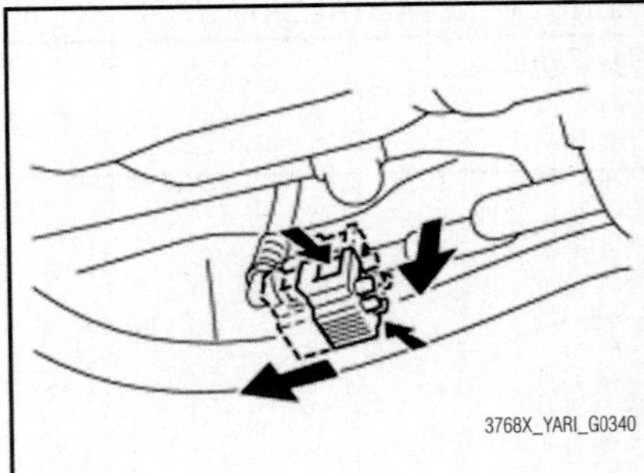

Fig. 126 Release the lock as shown, then pull and remove the fuel tank main tube

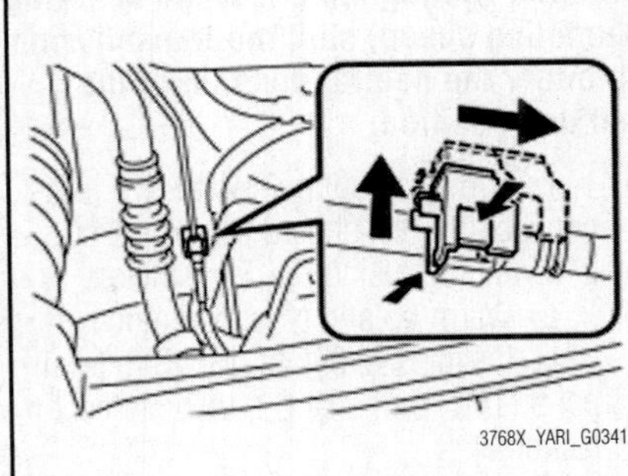

Fig. 127 Disconnect fuel tank breather hose. Release the lock as shown

1. Remove the fuel pump module assembly.
2. Drain the fuel.
3. Disconnect the negative battery cable.
4. Remove front floor heat insulator no. 4. Remove the 2 nuts together with the insulator.
5. Remove fuel tank protector sub-assembly. Remove the bolt together with the clip.
6. Disconnect fuel tank main tube sub-assembly. Release the lock as shown in the illustration, then pull and remove the fuel tank main tube.

➡

- Remove any dirt and foreign matter from the clip before performing this work.
- Avoid any scratches or foreign matter on the parts when disconnecting them, as the quick connector has the O-ring that seals the plug.
- Perform this work by hand. Do not use any tools.
- Do not forcibly bend, twist or turn the nylon tube.
- Protect the disconnected parts by covering them with a plastic bag.
- If the connector and pipe are stuck, disconnect the nylon tube by turning it by hand to release them.

7. Disconnect fuel tank breather hose. Release the lock as shown in the illustration, then pull and remove the fuel tank breather hose.

FUEL TANK FILLER PIPE SUB-ASSEMBLY LOWER
FUEL TANK ASSEMBLY
14 (146, 10)
X4
FUEL TANK MAIN TUBE SUB-ASSEMBLY
5.4 (55, 48 in.*lbf)
FUEL TANK PROTECTOR SUB-ASSEMBLY
5.4 (55, 48 in.*lbf)
5.4 (55, 48 in.*lbf)
FRONT FLOOR HEAT INSULATOR NO. 4
N*m (kgf*cm, ft.*lbf) : Specified torque ● Non-reusable part
3768X_YARI_G0339

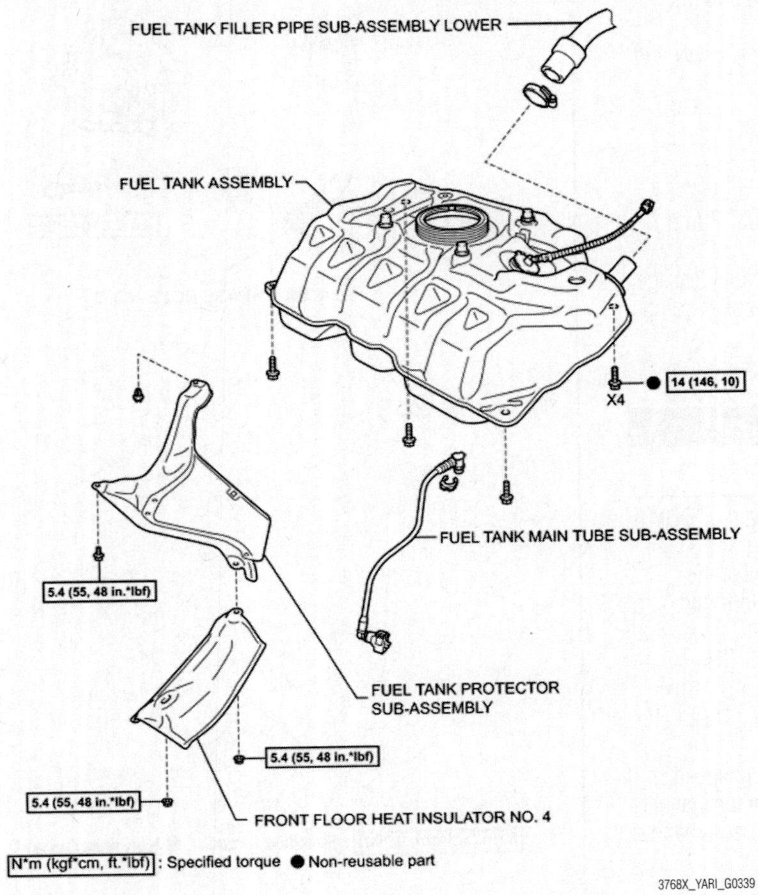

Fig. 126 View of the fuel tank and components

➡

- Remove any dirt and foreign matter from the clip before performing this work.
- Avoid any scratches or foreign matter on the parts when disconnecting them, as the quick connector has the O-ring that seals the plug.
- Perform this work by hand. Do not use any tools.
- Do not forcibly bend, twist or turn the nylon tube.
- Protect the disconnected parts by covering them with a plastic bag.
- If the connector and pipe are stuck, disconnect the nylon tube by turning it by hand to release them.

8. Loosen the clamp and disconnect the fuel tank filler pipe.
9. Remove fuel tank assembly:
   a. Remove the 4 bolts and remove the fuel tank.
   b. Remove the fuel tank main tube from the fuel tank.
10. To install, reverse the removal procedure. Refer to the exploded illustration for torque values.
11. Check for fuel leakage:
   a. Connect the Techstream to the DLC3.
   b. Turn the ignition switch to ON and turn the tester ON.

➡Do not start the engine.

   c. Select the following menu items: Powertrain / Engine and ECT / Active Test / Control the Fuel Pump / Speed.
   d. Check that there is no fuel leakage anywhere on the fuel system after doing maintenance.
12. Check for engine oil leakage.

IDLE SPEED

INSPECTION

1. When using the Techstream or ODB II scan tool:
   a. Warm up and stop the engine.
   b. Connect the Techstream to the DLC3.
   c. Turn the ignition switch ON.
   d. Select the following menu items: Powertrain / Engine and ECT / Data List / Engine Speed.
   e. Inspect the engine idling speed. Idling speed: 550 to 650 rpm for manual transaxle. 650 to 750 rpm for automatic transaxle.

➡Turn all the electrical systems and the A/C off. Inspect the idling speed with the cooling fan off. When checking the idling speed, shift the transmission to either the neutral position or the parking position.

   f. Turn the ignition switch OFF. Disconnect the Techstream from the DLC3.
2. When not using the Techstream:
   a. Warm up and stop the engine.
   b. Install SST 09843-18040 to terminal 9 (TAC) of the DLC3, then connect a tachometer.

➡Examine the terminal numbers before connecting them. Connecting the wrong terminals could damage the engine.

   c. Turn the ignition switch ON.
   d. Inspect the engine idling speed. Idling speed: 550 to 650 rpm for manual transaxle. 650 to 750 rpm for automatic transaxle.
   e. Turn the ignition switch OFF.
   f. Disconnect the tachometer.
   g. Remove SST from terminal 9 (TAC).

THROTTLE BODY

REMOVAL & INSTALLATION

See Figure 129.

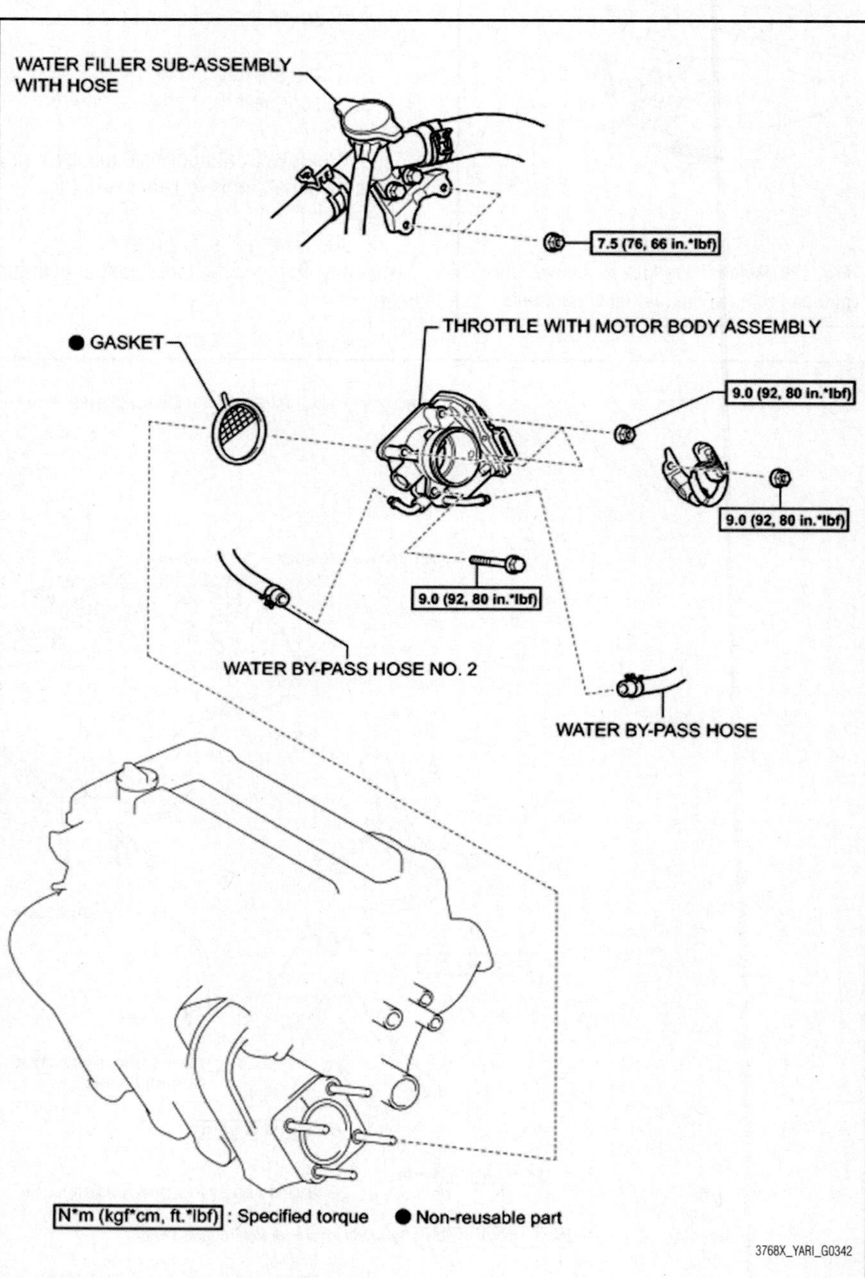

Fig. 129 View of the throttle body and components

1. Disconnect the negative battery cable.
2. Drain and recycle the engine coolant.
3. Remove the engine cover.
4. Remove the air cleaner assembly.
5. Disconnect the water by-pass hose.
6. Disconnect water by-pass hose No. 2.
7. Remove throttle with motor body assembly:
   a. Disconnect the throttle with motor body assembly connector.
   b. Remove the nut and remove the wire harness with bracket.
   c. Remove the 2 nuts and remove the water filler sub-assembly with hose.
   d. Remove the bolt and 3 nuts and remove the throttle with motor body assembly.
   e. Remove the gasket from the intake manifold.
8. To install, reverse the removal procedure.
9. Refer to the illustration for torque values.
10. Perform the initialization:
   a. Disconnect the cable from the negative (-) battery terminal. Wait at least 60 seconds and reconnect the cable.
   b. Turn the ignition switch to ON without operating the accelerator pedal.

➡If the accelerator pedal is operated, perform the above steps again.

   c. Connect the Techstream to the DLC3 and clear the DTC's.
   d. Start the engine and check that the MIL is not illuminated and that the idle speed is within the specified range when the A/C is switched off after the engine is warmed up.
   - Standard (for Automatic Transaxle): With A/C switched off, engine idle speed should be at 650 to 750 RPM.
   - Standard (for Manual Transaxle): With A/C switched off, engine idle speed should be at 550 to 650 RPM.

➡Be sure to perform this step with all accessories off. Make sure that the shift lever is in neutral.

   e. Enter the following menus: Powertrain / Engine and ECT / Data List / Throttle Sensor Position. Fully depress the accelerator pedal and check that the value is 60% or more.
   f. Perform a road test and confirm that there are no abnormalities.

HEATING & AIR CONDITIONING SYSTEM

A/C UNIT

REMOVAL & INSTALLATION

See Figures 130 through 138.

⁂ CAUTION

Some models covered by this manual may be equipped with a Supplemental Restraint System (SRS), which uses an air bag. Whenever working near any of the SRS components, such as the impact sensors, the air bag module, steering column and instrument panel, disable the SRS, as described in Section 6.

1. Disconnect the negative battery cable.

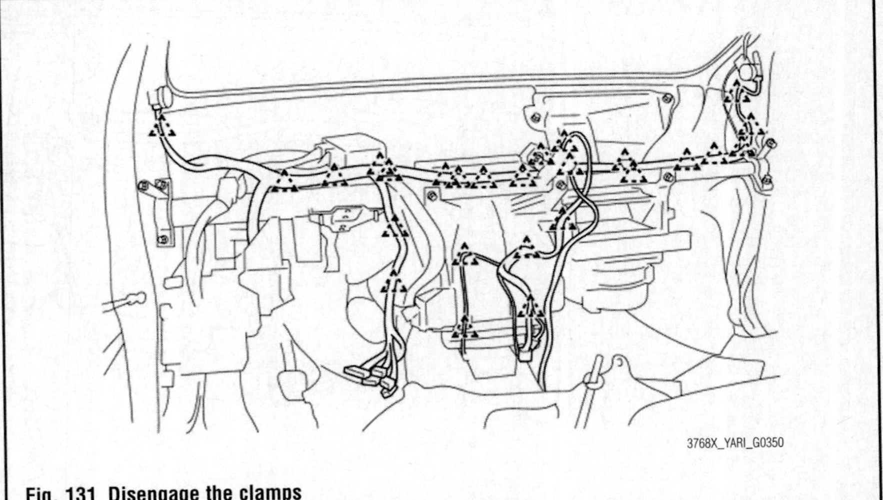

Fig. 131 Disengage the clamps

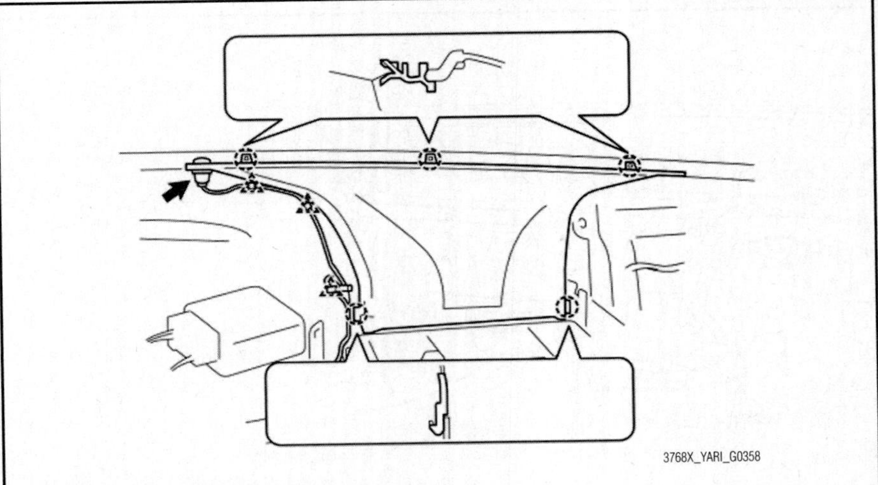

Fig. 130 Remove defroster nozzle assembly

⁂ CAUTION

Wait for at least 90 seconds after disconnecting the cable to prevent the airbag from working.

2. Discharge refrigerant from refrigeration system.
3. Drain and recycle the engine coolant.
4. Remove the wiper arm and link assembly.
5. Remove all cowling.
6. Disconnect suction tube sub-assembly:
   a. Remove the bolt.
   b. Turn the hook type connector clockwise and disconnect the suction tube.
   c. Remove the O-ring from the suction tube.

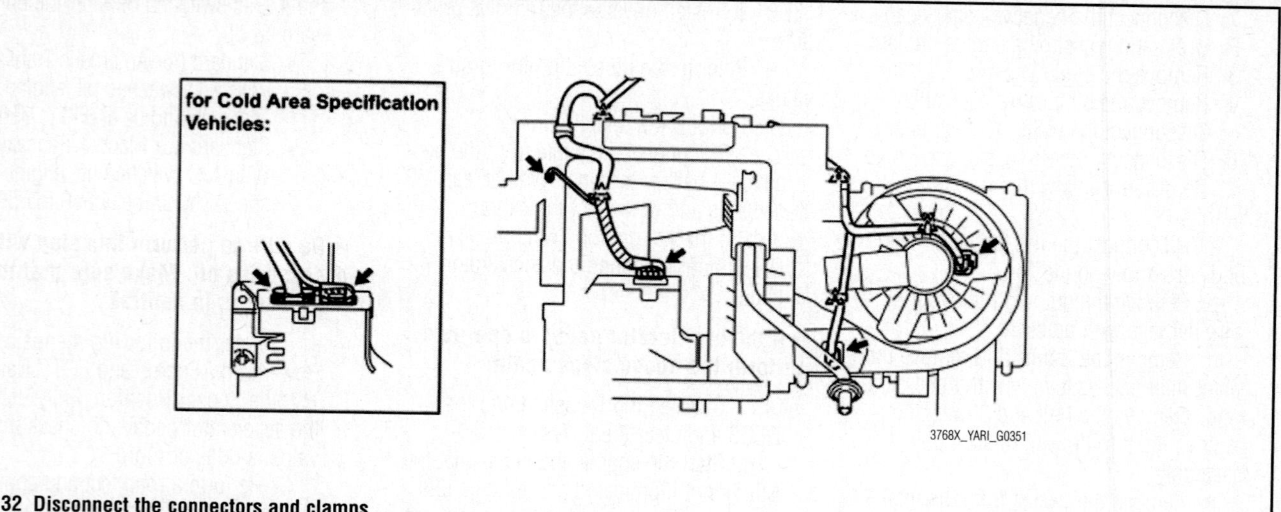

Fig. 132 Disconnect the connectors and clamps

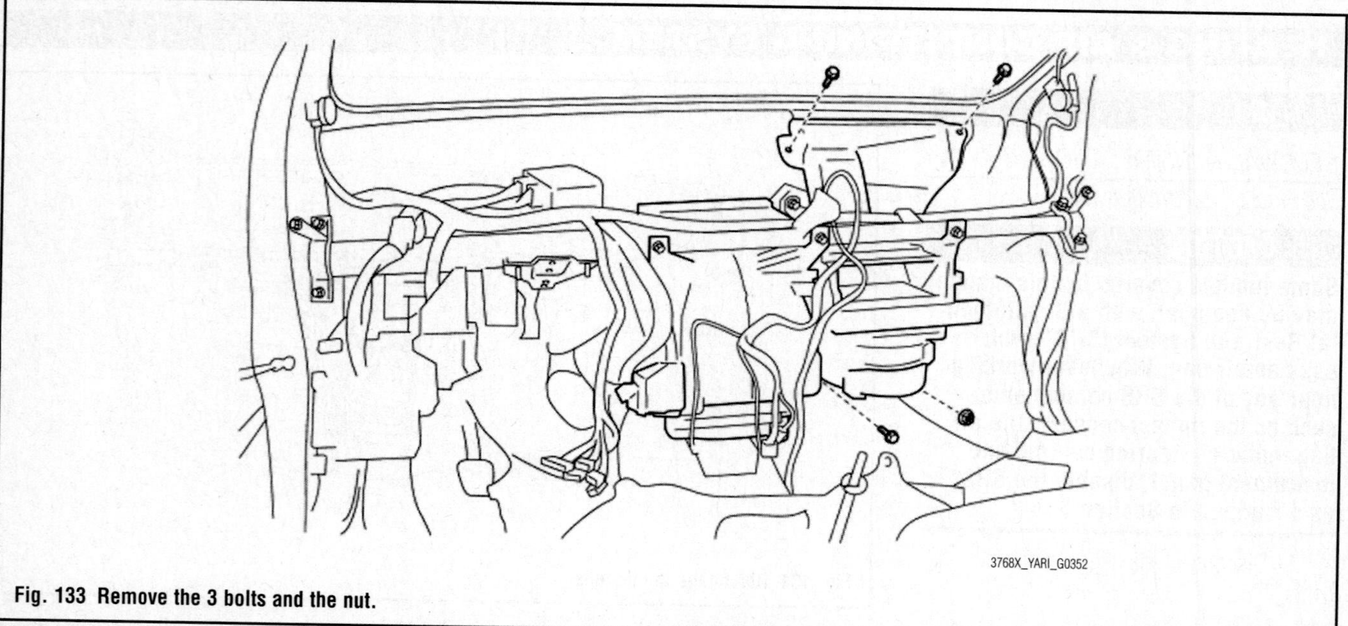

Fig. 133 Remove the 3 bolts and the nut.

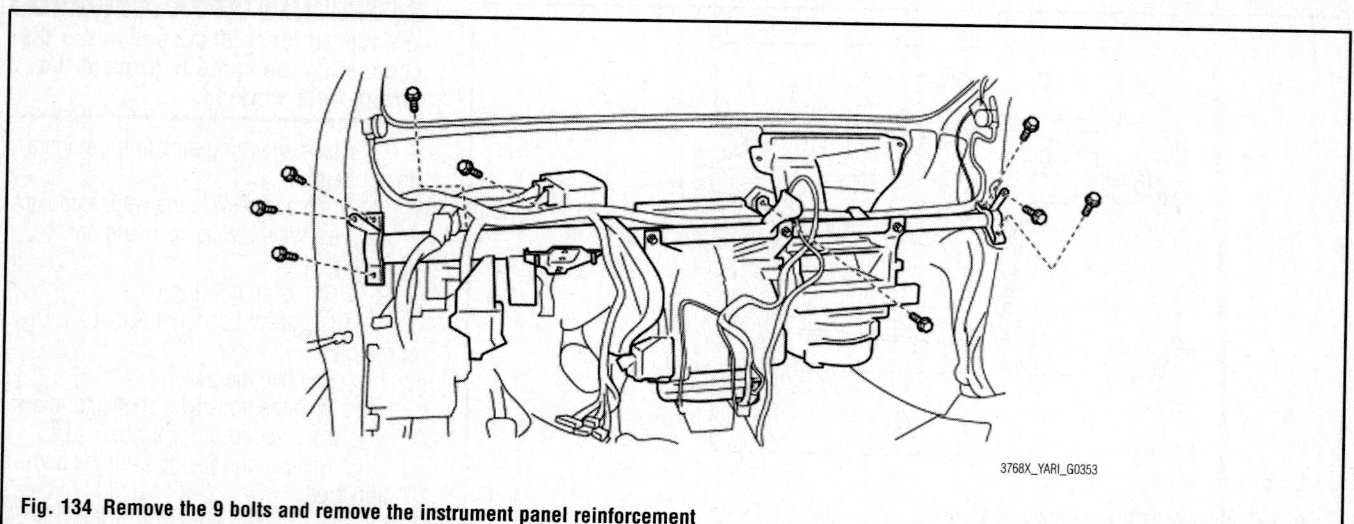

Fig. 134 Remove the 9 bolts and remove the instrument panel reinforcement

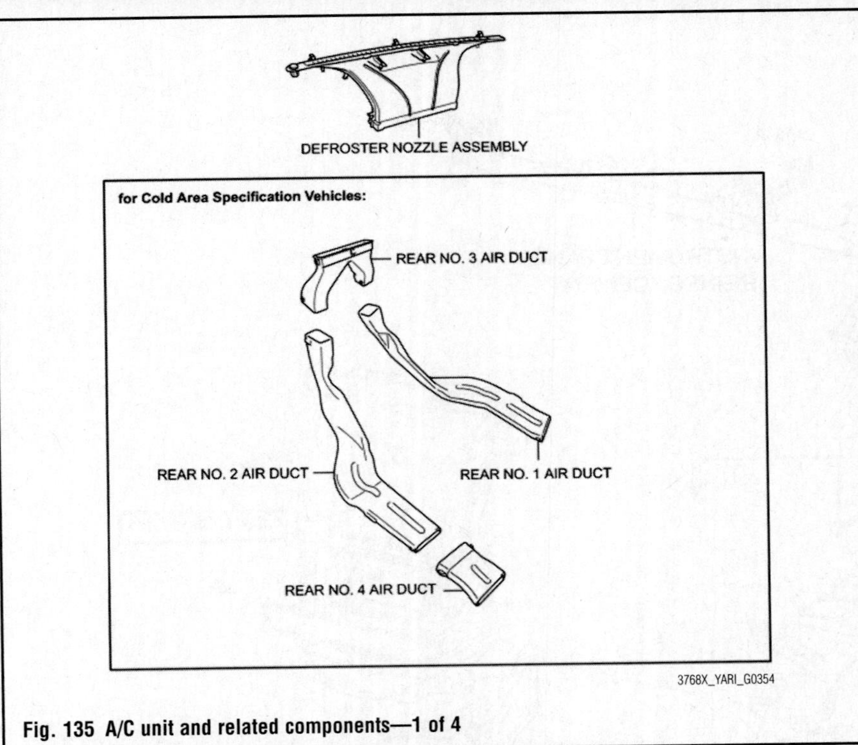

Fig. 135 A/C unit and related components—1 of 4

➡Seal the openings of the disconnected parts using vinyl tape to prevent moisture and foreign matter from entering.

7. Disconnect liquid tube sub-assembly:
   a. Disconnect the liquid tube.
   b. Remove the O-ring from liquid tube.

➡Seal the opening of the disconnected parts using vinyl tape to prevent moisture and foreign matter from entering.

8. Using pliers, grip the claws of the clip, slide the clip and disconnect the heater water outlet hose from the heater unit.
9. Using pliers, grip the claws of the clip, slide the clip and disconnect the heater water inlet hose from the heater unit.
10. Remove the instrument panel.
11. Remove the center console.
12. Position front wheels facing straight ahead.
13. Remove the driver's steering wheel airbag.

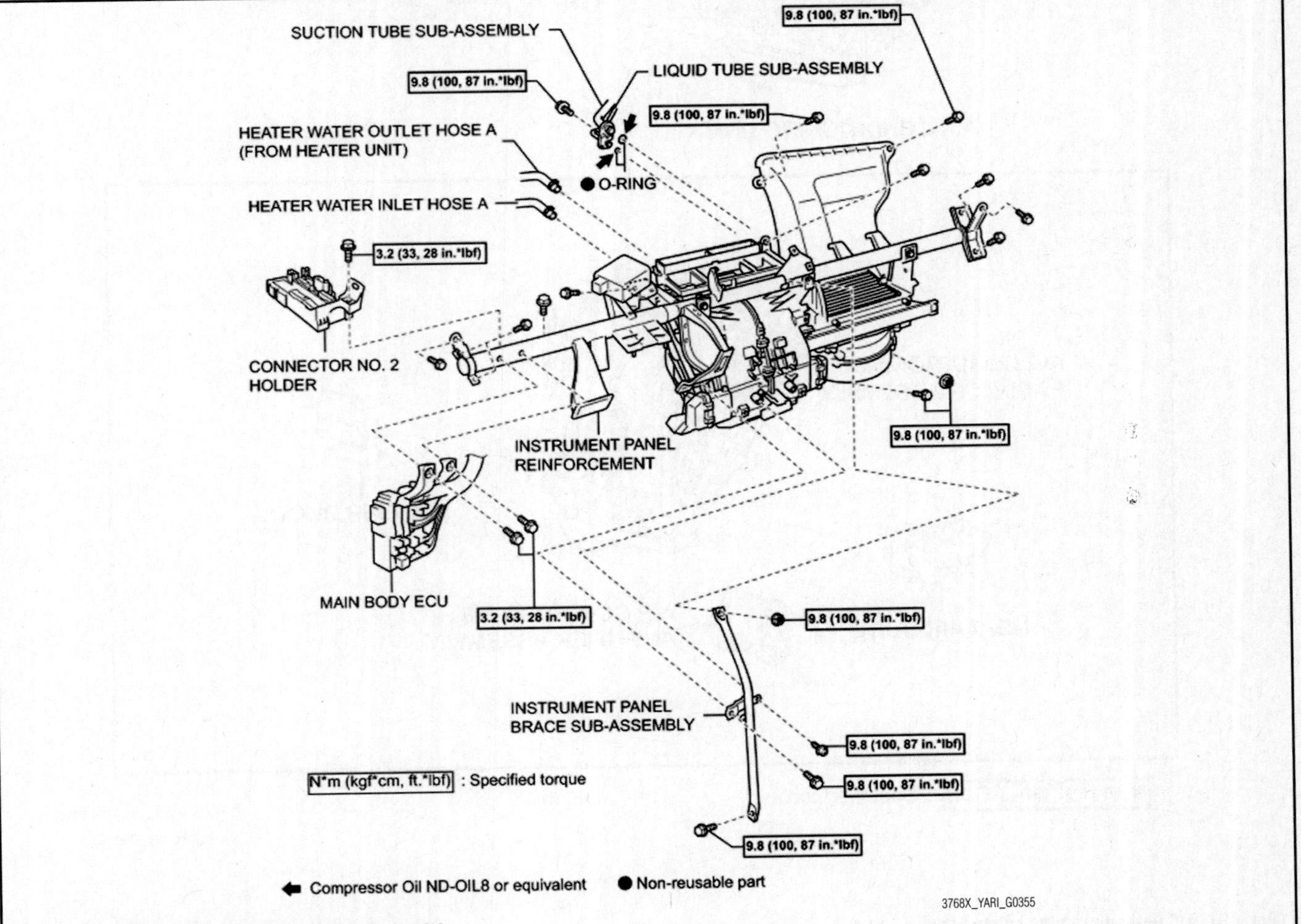

Fig. 136 A/C unit and related components—2 of 4

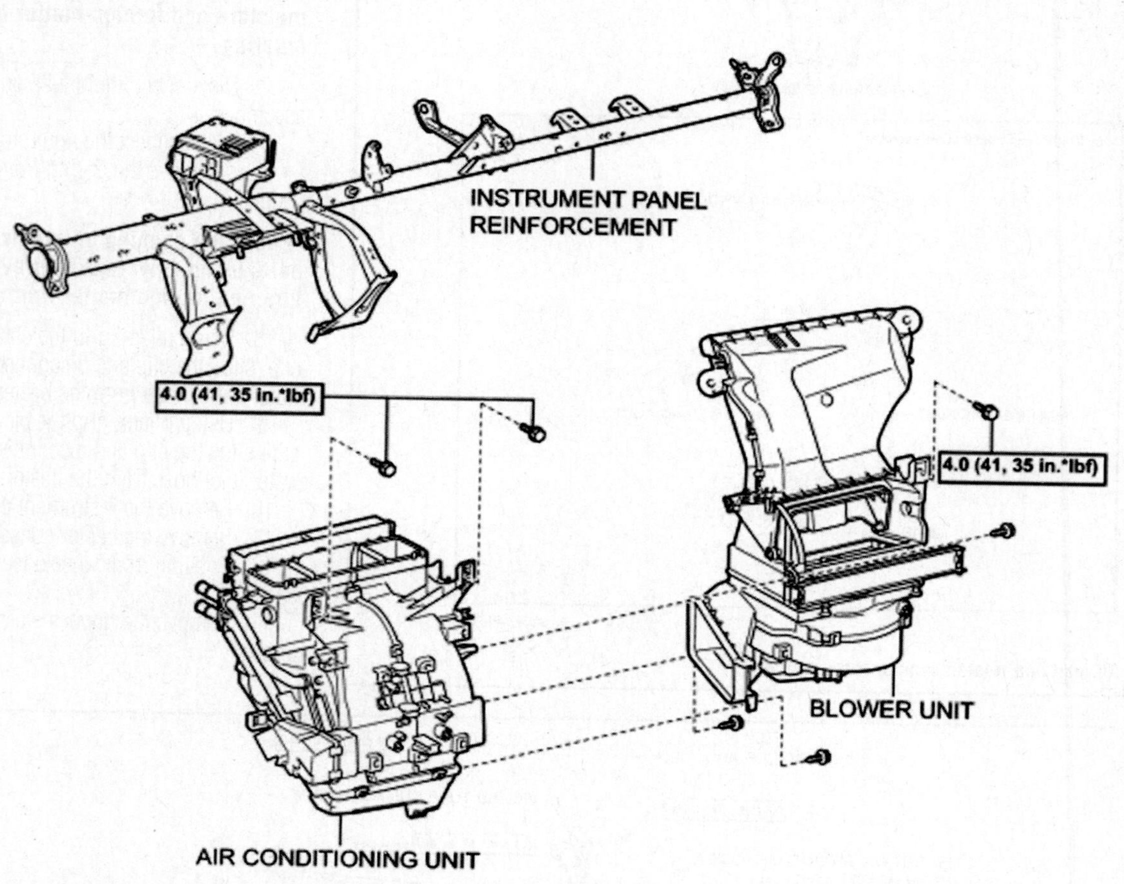

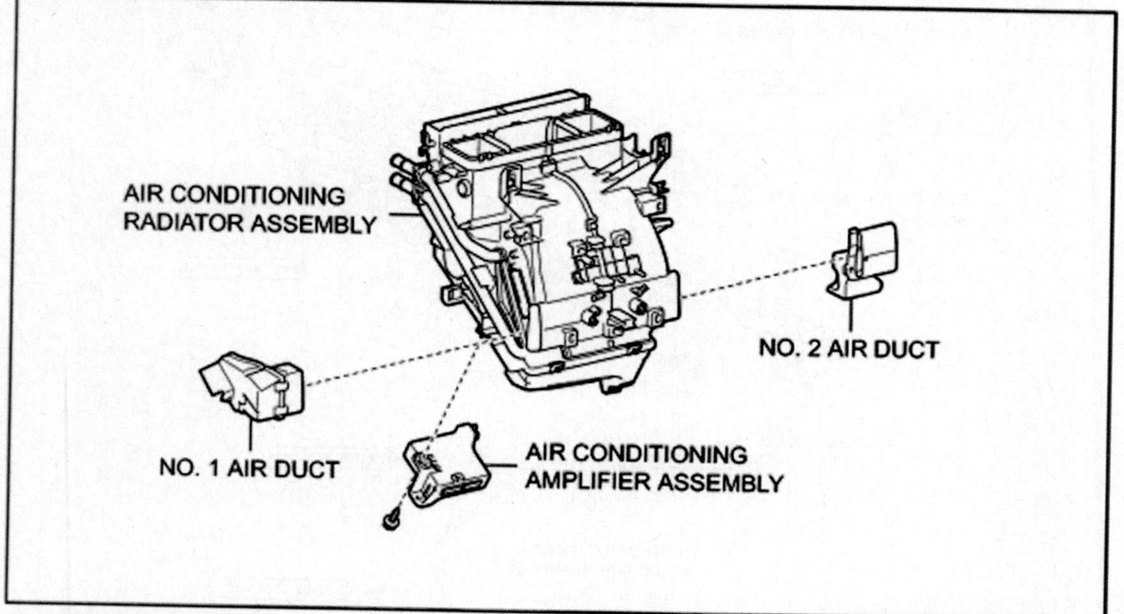

N*m (kgf*cm, ft.*lbf) : Specified torque

3768X_YARI_G0356

Fig. 137 A/C unit and related components—3 of 4

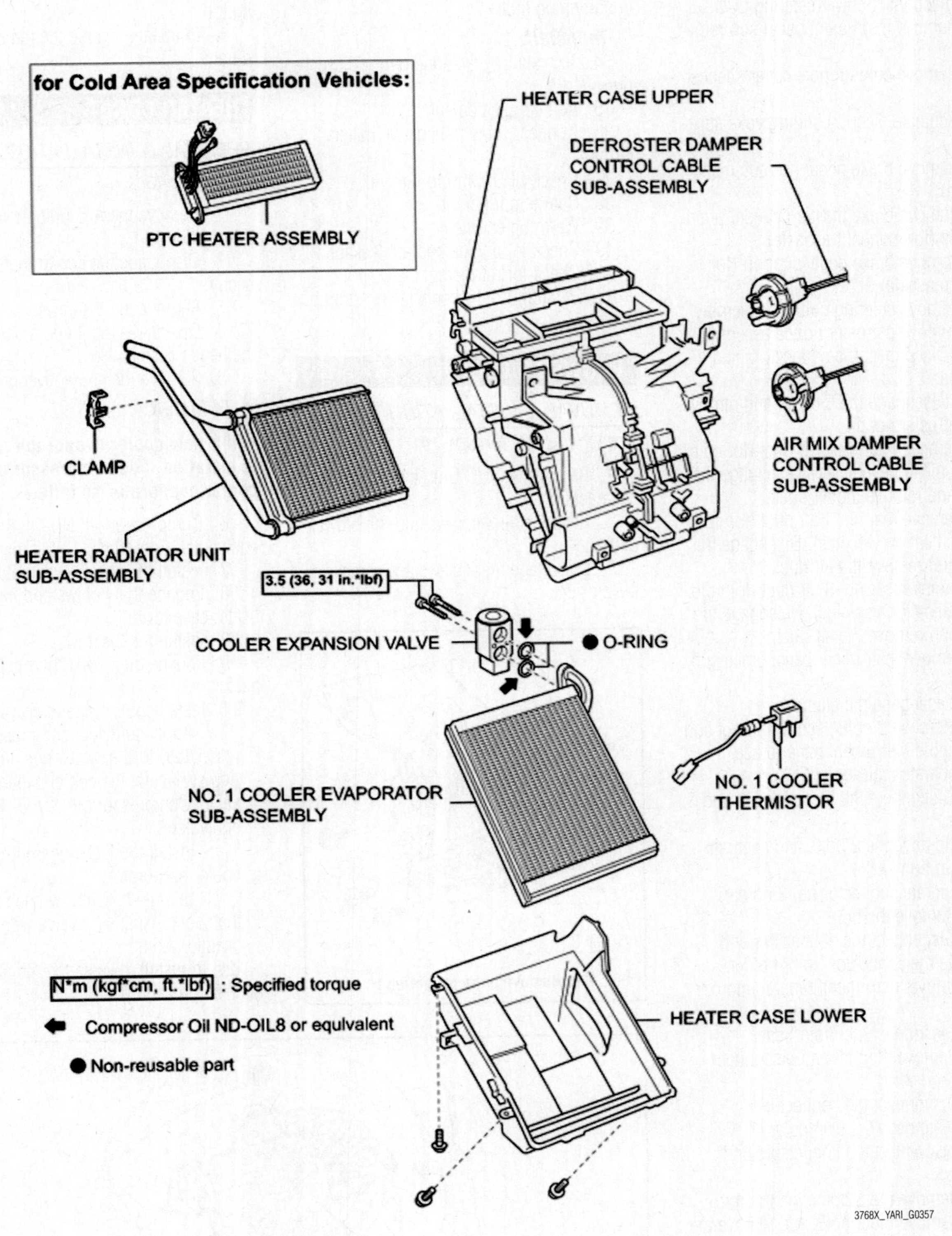

Fig. 138 A/C unit and related components—4 of 4

14. Remove the steering wheel.
15. Remove the combination switch assembly.
16. Disconnect power steering ECU.
17. Remove instrument panel sub reinforcement.
18. Remove column hole cover silencer sheet.
19. Separate steering sliding yoke sub-assembly.
20. Remove brake pedal (for automatic transaxle).
21. Remove brake master cylinder push rod clevis (for manual transaxle).
22. Remove brake pedal support (for manual transaxle) .
23. Remove steering column assembly.
24. Remove defroster nozzle assembly:
    a. Disconnect the connectors and clamps.
    b. Disengage the 5 claws and remove the defroster nozzle.
25. Remove rear no. 2 air duct (for cold area specification vehicles). Disengage the 4 claws and remove the air duct.
26. Remove rear no. 1 air duct (for cold area specification vehicles). Disengage the 2 claws and remove the air duct.
27. Remove rear no. 3 air duct (for cold area specification vehicles). Disengage the 4 claws and remove the air duct.
28. Remove instrument panel brace sub-assembly:
    a. Disengage the clamp.
    b. Remove 2 bolts, screw and nut and remove the instrument panel brace.
29. Separate main body ECU:
    a. Disconnect the 5 connectors and 3 clamps.
    b. Remove the 2 bolts and separate the main body ECU.
30. Separate connector no. 2 holder:
    a. Remove the bolt.
    b. Disconnect the connectors and separate the connector No. 2 holder.
31. Remove instrument panel reinforcement:
    a. Disconnect the drain hose.
    b. Remove the bolt and disconnect the ground wire.
    c. Disconnect the connector.
    d. Disengage the clamps.
    e. Disconnect the connectors and clamps.
    f. Remove the 3 bolts and the nut.
    g. Remove the 9 bolts and remove the instrument panel reinforcement together with the air conditioning unit.
    h. Disconnect the connector and the 3 clamps.
    i. Remove the 3 screws and the air conditioning unit.
32. Remove the screw and the air conditioning amplifier.
33. Remove the 3 screws and the air conditioning unit.

To install:

34. To install, reverse the removal procedure.
35. Add engine coolant.
36. Connect cable to negative battery terminal.
37. Check SRS warning light.
38. Charge refrigerant.
39. Warm up engine.
40. Check for engine coolant leakage.
41. Check for refrigerant leak.
42. Position front wheels facing straight ahead.

BLOWER MOTOR

REMOVAL & INSTALLATION

See Figures 139 through 141.

1. Remove the A/C Unit, as outlined in this section.
2. Remove the 3 screws and the blower unit.
3. Disengage the 2 claws and remove the air duct.

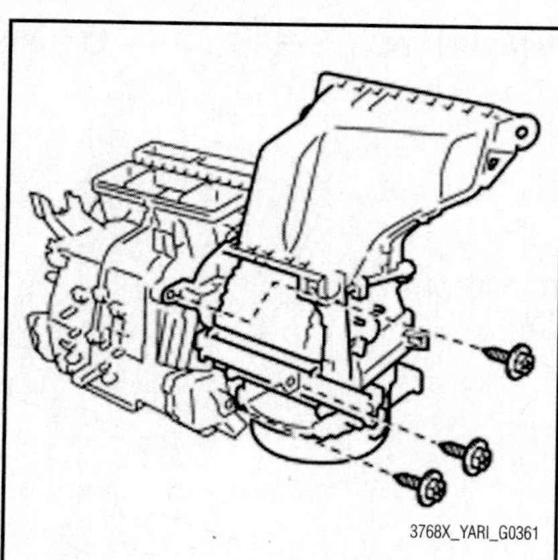

Fig. 139 Removing the blower unit

4. Disengage the 3 claws and remove the air inlet damper control cable.
5. Remove the 3 screws and the blower motor.
6. To install, reverse the removal procedure.

HEATER CORE

REMOVAL & INSTALLATION

See Figure 138.

1. Remove the A/C unit, as outlined in this section.
2. Disengage the cooler thermistor connector.
3. Remove the 3 screws.
4. Disengage the 4 claws and remove the heater case lower.
5. Remove the cooler evaporator.

To install:

➡If a new cooler evaporator is installed, add compressor oil to the cooler evaporator as follows.

6. Compressor oil: ND-OIL8 or the equivalent. Add 40 cc (1.35 fl. oz.)
7. Install the cooler evaporator.
8. Engage the 4 claws and install the heater case lower.
9. Install the 3 screws.
10. Engage the cooler thermistor connector.
11. Install cooler expansion valve:
    a. Apply sufficient compressor oil (ND-OIL8) to 2 new O-rings and the fitting surface of the cooler expansion valve. Compressor oil: ND-OIL8 or the equivalent.
    b. Install the 2 O-rings onto the cooler evaporator.
    c. Using a hexagon wrench 4, install the cooler expansion valve with the 2 hexagon bolts.
12. To install, reverse the remaining removal procedure.

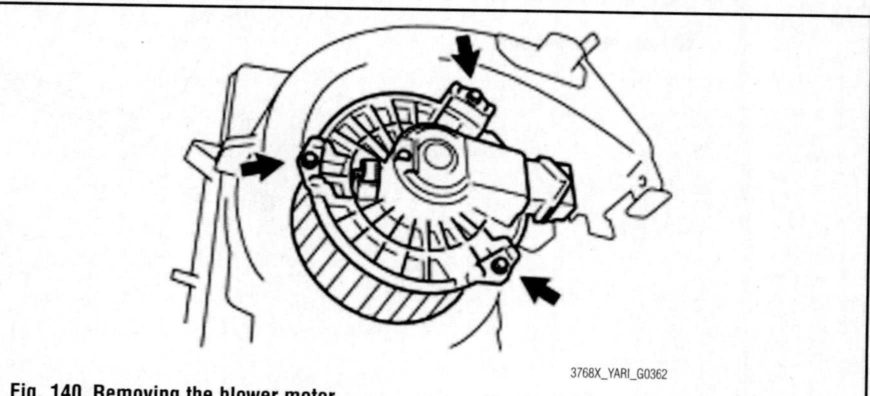

Fig. 140 Removing the blower motor

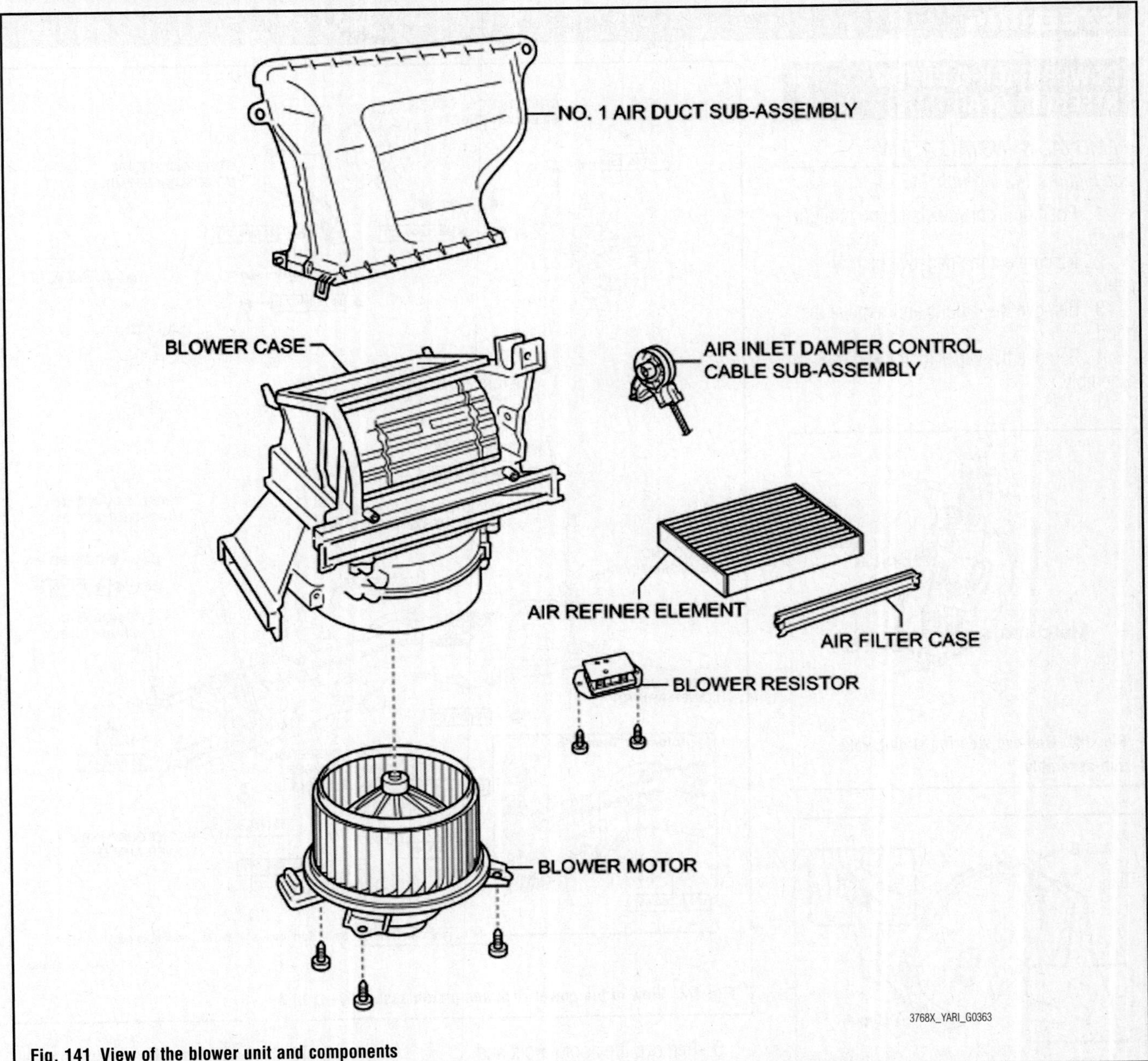

Fig. 141 View of the blower unit and components

STEERING

POWER RACK & PINION STEERING GEAR

REMOVAL & INSTALLATION

See Figures 142 through 146.

1. Position front wheels facing straight ahead.
2. Disconnect the negative battery cable.
3. Remove the 4 bolts and remove the hood.
4. Remove the wiper motor and link assembly.

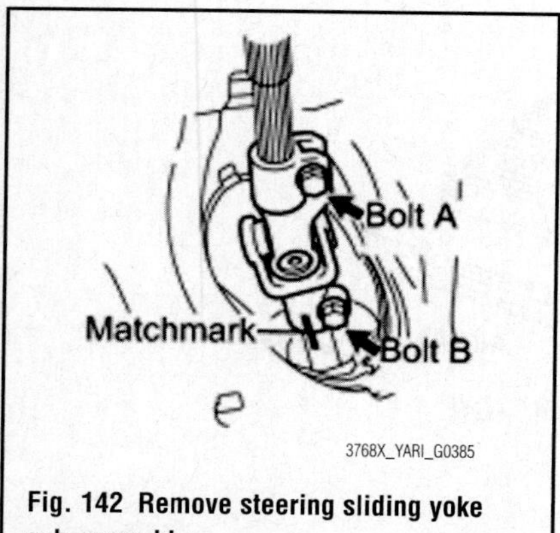

Fig. 142 Remove steering sliding yoke sub-assembly

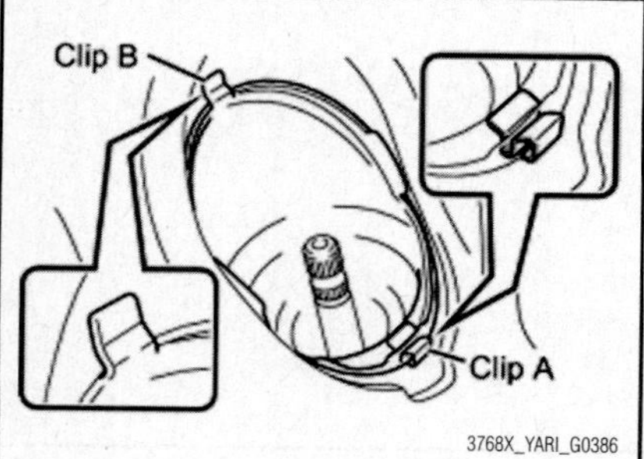

Fig. 143 Remove clip A, separate clip B from the body and separate No. 1 steering column hole cover sub-assembly

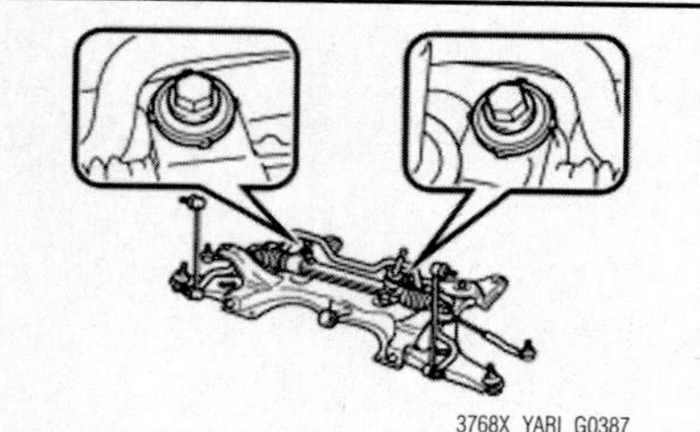

Fig. 144 Removing the and remove the power steering gear from the suspension crossmember

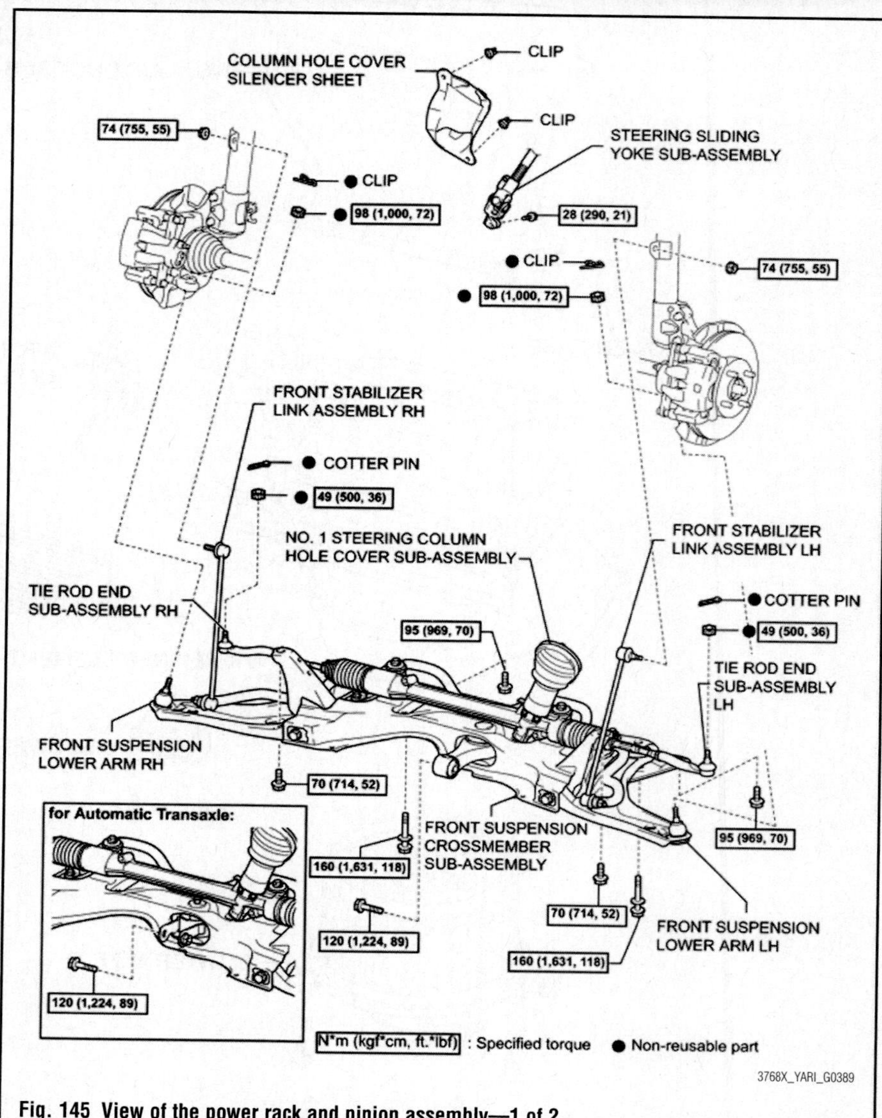

Fig. 145 View of the power rack and pinion assembly—1 of 2

5. Remove the floor carpet and 2 clips and remove the column hole cover silencer.
6. Remove steering sliding yoke sub-assembly:
   a. Use a seat belt to fix the steering wheel assembly, in order to avoid breakage of the spiral cable.
   b. Place the matchmarks on the sliding yoke of the steering intermediate shaft assembly and the power steering.
   c. Loosen bolt A, remove bolt B and separate the steering intermediate shaft assembly.
7. Remove clip A, separate clip B from the body and separate No. 1 steering column hole cover sub-assembly.
8. Remove front wheel.
9. Remove front stabilizer link assembly LH and RH.
10. Remove tie rod end sub-assembly.
11. Remove front suspension lower arm RH and LH.
12. Suspend engine assembly.
13. Remove front suspension crossmember sub-assembly. Refer to Engine Assembly in Engine Mechanical.
14. Remove the 2 bolts and 2 nuts and remove the power steering gear from the suspension crossmember.

➡Keep the nut from rotating while turning the bolt.

15. To install, reverse the removal procedure.

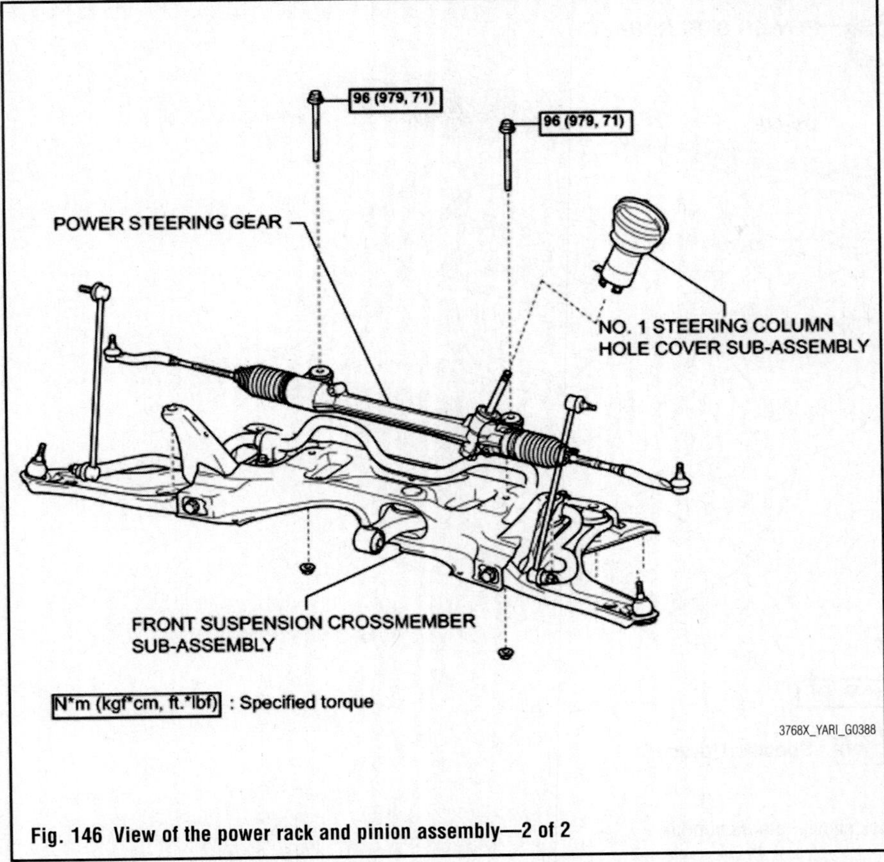

Fig. 146 View of the power rack and pinion assembly—2 of 2

POWER STEERING ECU

REMOVAL & INSTALLATION

See Figures 147 through 149.

1. Place front wheels facing straight ahead.
2. Disconnect the negative battery cable.
3. Remove the upper instrument panel. Refer to Instrument Panel in Body Interior.
4. Separate the wire harness clamp from the power steering ECU assembly.

Fig. 147 Disconnect the 4 connectors from the power steering ECU assembly

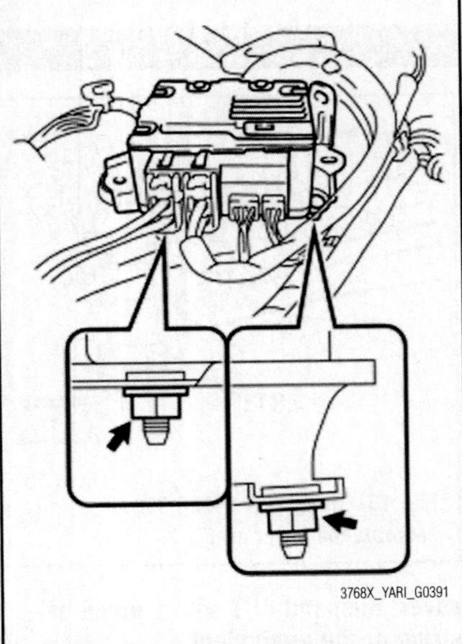

Fig. 148 Remove the bolt, 2 nuts, and the power steering ECU assembly

5. Disconnect the 4 connectors from the power steering ECU assembly.
6. Remove the bolt, 2 nuts, and the power steering ECU assembly.

To install:

7. Install the power steering ECU assembly with the bolt and 2 nuts. Tighten to 71 inch lbs. (8 Nm).
8. Connect the 4 connectors to the power steering ECU assembly.
9. Install the wire harness clamp to the power steering ECU assembly.
10. To complete the installation, reverse remaining removal procedure.
11. Calibrate torque sensor zero point.

TORQUE SENSOR ZERO POINT CALIBRATION AND INITIALIZATION

Torque Sensor Zero Point Calibration (Using Techstream)

➡Perform the torque sensor zero point calibration if any of the following conditions occur:

a. The steering column assembly (containing the torque sensor) has been replaced.
b. The power steering ECU has been replaced.
c. There is a difference in steering effort between turning right and left.

➡When torque sensor zero point calibration is performed, the assist map is written automatically at the same time.

➡If DTC C1516 (Torque Sensor Zero Point Adjustment Incomplete) is stored, the torque sensor zero point cannot be calibrated. Clear the DTC before starting calibration.

➡Do not touch the steering wheel during the torque sensor zero point calibration. Perform the calibration only when the vehicle is stopped.

1. Perform the torque sensor zero point calibration:
   a. Set the steering wheel to the center point and align the front wheels straight ahead.
   b. Turn the ignition switch off.
   c. Connect the Techstream to the DLC3.
   d. Start the engine.
   e. Turn the Techstream on.
   f. Calibrate the power steering ECU. Enter the following menus: Chassis / EMPS / Utility / Torque Sensor Adjustment.
2. Check for DTCs

➡After zero point calibration is completed normally, confirm that a DTC is not output. If DTC C1515, C1516, C1534 or C1581 is output, perform troubleshooting for the corresponding DTC.

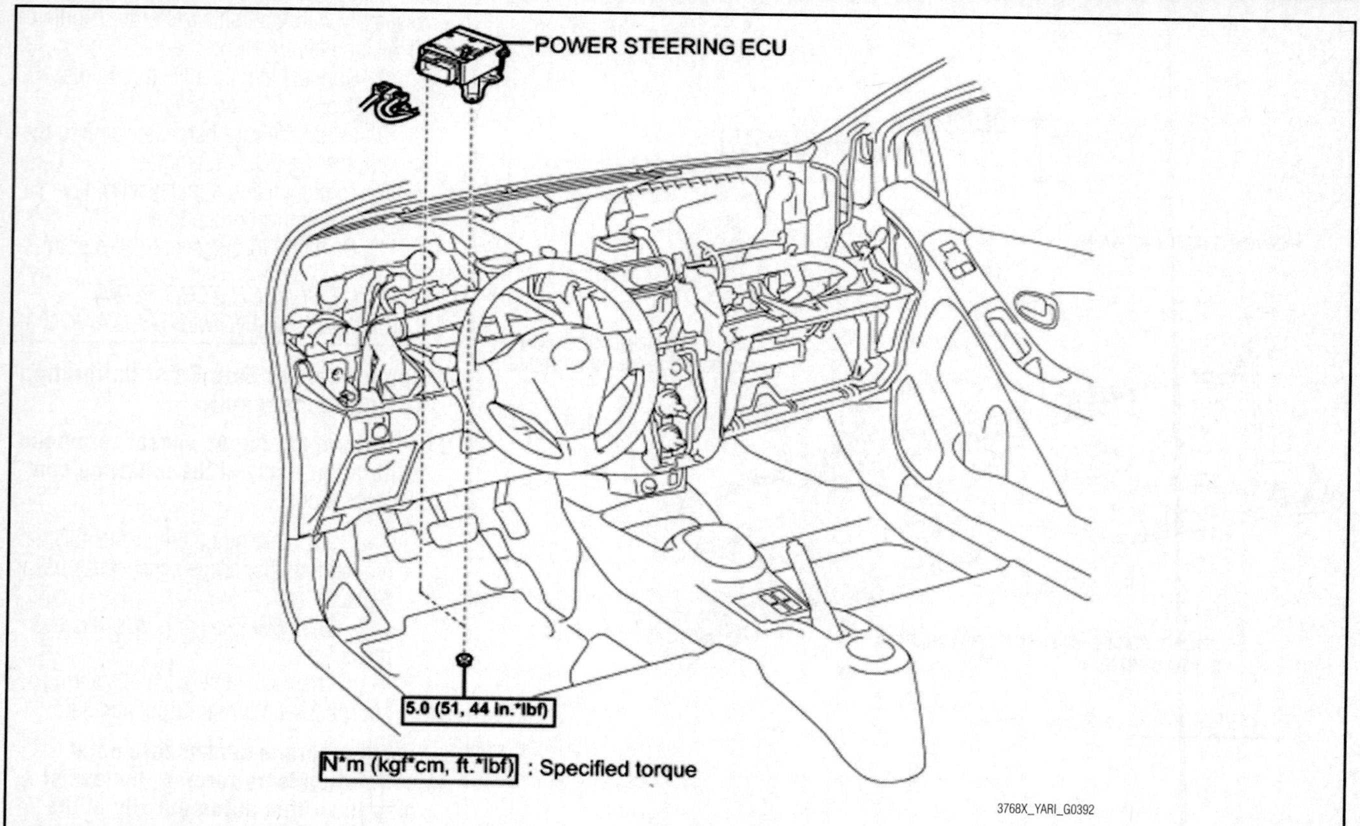

POWER STEERING ECU
5.0 (51, 44 in.*lbf)
N*m (kgf*cm, ft.*lbf) : Specified torque
3768X_YARI_G0392

Fig. 149 View of the power steering ECU—Hatchback shown, others similar

SUSPENSION

FRONT SUSPENSION

LOWER CONTROL ARM

REMOVAL & INSTALLATION

Left Side

See Figures 150 through 153.

1. Remove the front wheel.
2. Remove the clip and castle nut.
3. Using SST 09628-00011, separate the lower arm.

➡**Do not damage the lower arm dust cover. Suspend SST with a piece of string or the equivalent.**

3768X_YARI_G0398

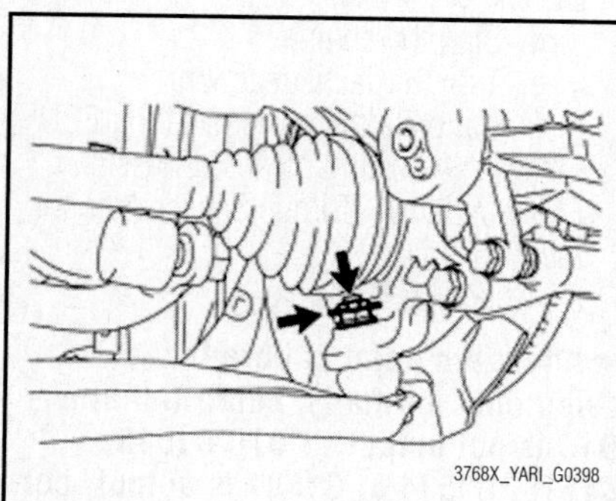

Fig. 150 Remove the clip and castle nut

SST
3768X_YARI_G0399

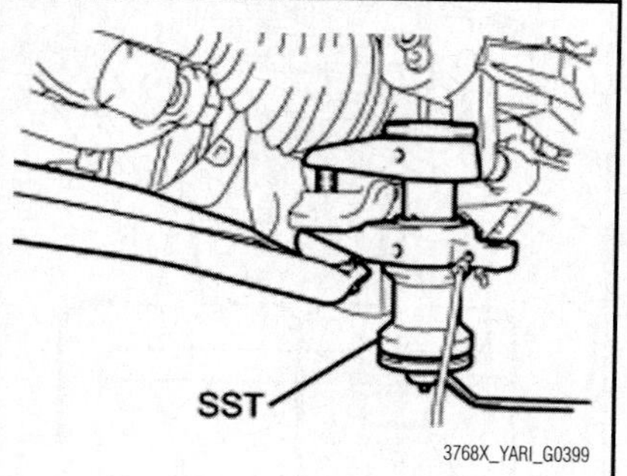

Fig. 151 Using SST 09628-00011, separate the lower arm

4. Remove the 2 bolts and lower arm.

To install:

5. Provisionally tighten the lower arm with the 2 bolts.
6. Install the lower arm onto the steering knuckle with a new castle nut. Tighten to 72 ft. lbs. (98 Nm).

➡**If the holes for the clip are not aligned, tighten the nut by a further turn of up to 60°.**

3768X_YARI_G0400

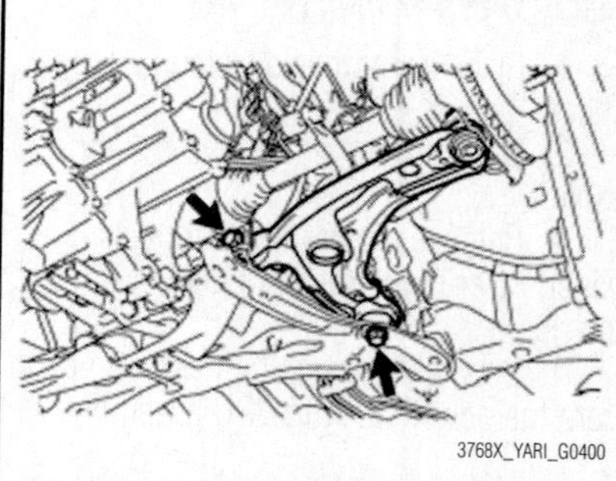

Fig. 152 Remove the 2 bolts and lower arm

7. Install a new clip.
8. Stabilize suspension:
   a. Lower the vehicle from the jack.
   b. Bounce the vehicle up and down several times to stabilize the suspension.
9. Fully tighten the 2 front lower suspension arm bolts:
   a. Bolt A: 101 ft. lbs. (137 Nm).
   b. Bolt B: 118 ft. lbs. (160 Nm).

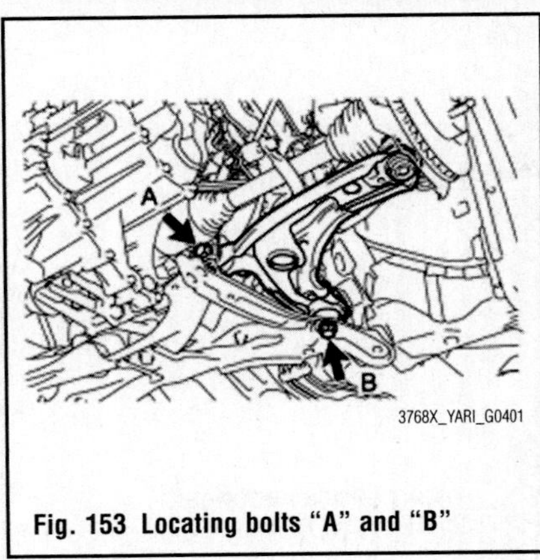

Fig. 153 Locating bolts "A" and "B"

10. Inspect and adjust front wheel alignment.

Right Side

See Figures 154 through 156.

1. Disconnect cable from negative battery terminal.
2. Remove hood sub-assembly.
3. Remove front wiper motor and link.
4. Remove cowl to register duct sub-assembly (for hatchback).
5. Remove front air shutter seal (for sedan) .
6. Remove outer cowl top panel.
7. Position wheels facing straight ahead
8. Remove front wheel.
9. Remove steering column hole cover sub-assembly.
10. Separate front stabilizer link assembly LH and RH.
11. Separate tie rod end sub-assembly LH and RH.
12. Separate front lower suspension arm LH and RH.
13. Suspend engine assembly.
14. Remove the floor carpet and 2 clips and remove the column hole cover silencer.
15. Remove steering sliding yoke sub-assembly:
    a. Use a seat belt to fix the steering wheel assembly, in order to avoid breakage of the spiral cable.
    b. Place the matchmarks on the sliding yoke of the steering intermediate shaft assembly and the power steering.
    c. Loosen bolt A, remove bolt B and separate the steering intermediate shaft assembly.
16. Remove clip A, separate clip B from the body and separate No. 1 steering column hole cover sub-assembly.
17. Remove front wheel.
18. Remove front stabilizer link assembly LH and RH.
19. Remove tie rod end sub-assembly.

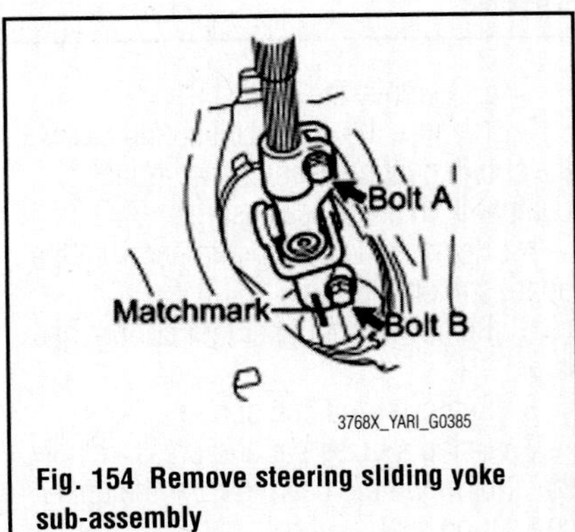

Fig. 154 Remove steering sliding yoke sub-assembly

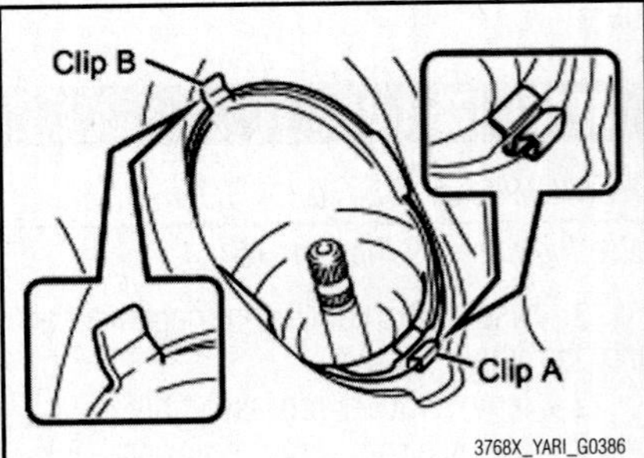

Fig. 155 Remove clip A, separate clip B from the body and separate No. 1 steering column hole cover sub-assembly

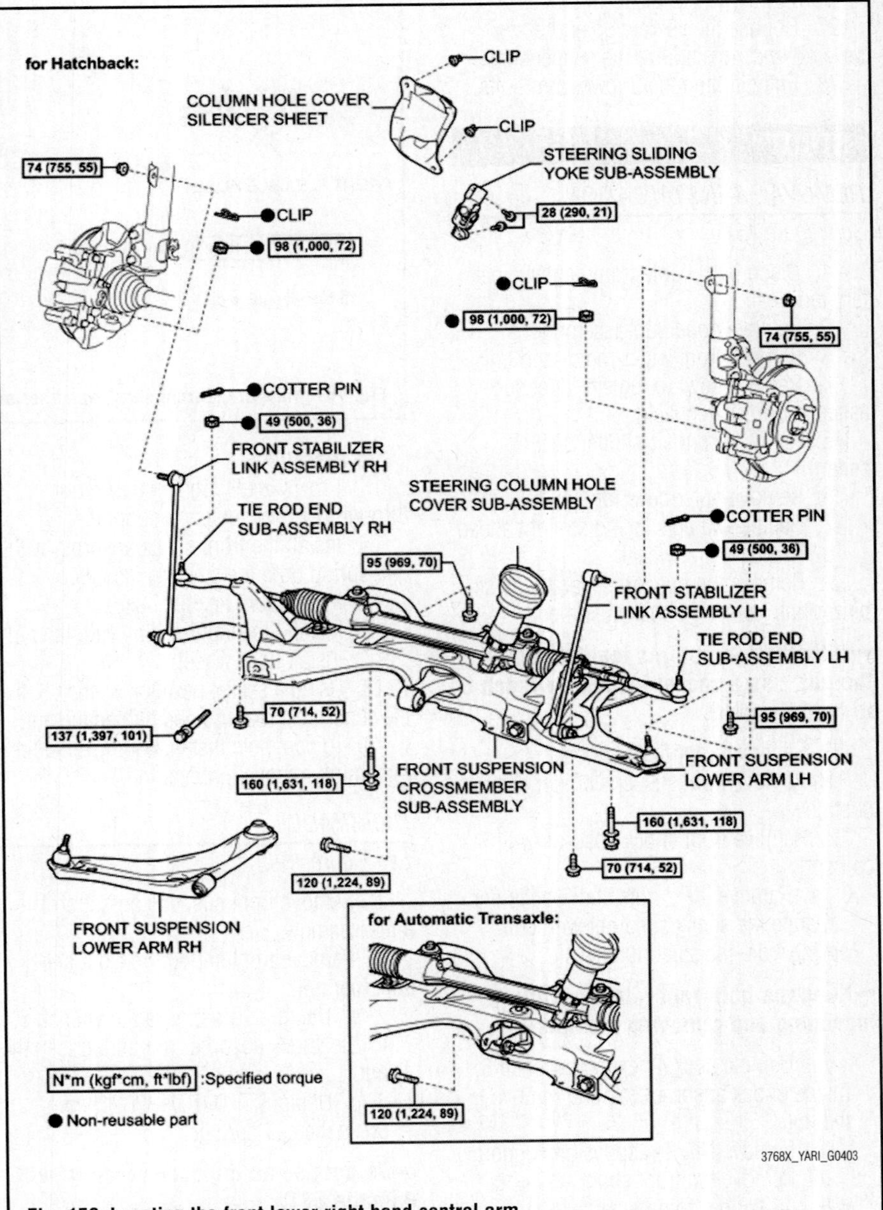

Fig. 156 Locating the front lower right hand control arm

20. Remove front suspension lower arm RH and LH.
21. Suspend engine assembly.
22. Remove front suspension cross-member sub-assembly:
   a. Remove the bolt and separate the engine moving control rod.
   b. Support the front suspension crossmember with a transmission jack.
   c. Remove the 6 bolts and remove the suspension crossmember.
23. Remove front lower suspension arm RH. Remove the bolt and lower arm.
24. To install, install the lower arm onto the crossmember and provisionally tighten the bolt.
25. To complete installation, reverse the remaining removal procedure.
26. Lower the vehicle from the jack.
27. Bounce the vehicle up and down several times to stabilize the suspension.
28. Fully tighten the 2 lower arm bolts.

STRUT

REMOVAL & INSTALLATION

See Figure 157.

1. Disconnect cable from negative battery terminal.
2. Remove hood sub-assembly.
3. Remove front wiper motor and link.
4. Remove cowl to register duct sub-assembly (for hatchback).
5. Remove front air shutter seal (for sedan) .
6. Remove outer cowl top panel.
7. Position wheels facing straight ahead
8. Remove front wheel.
9. Remove the nut and separate the stabilizer link from the shock absorber.

➡If the ball joint turns together with the nut, use a socket hexagon wrench 6 to hold the stud.

10. Separate front flexible hose.
11. Remove front suspension support dust cover.
12. Remove front shock absorber with coil spring:
   a. Remove the 2 nuts and 2 bolts and separate the shock absorber with coil spring from the steering knuckle.

➡Keep the bolt from rotating while loosening and removing the nuts.

   b. Using a socket hexagon wrench 6, fix the shock absorber rod and remove the nut.
   c. Remove No. 2 suspension support.
   d. Remove the front shock absorber with coil spring from the vehicle.

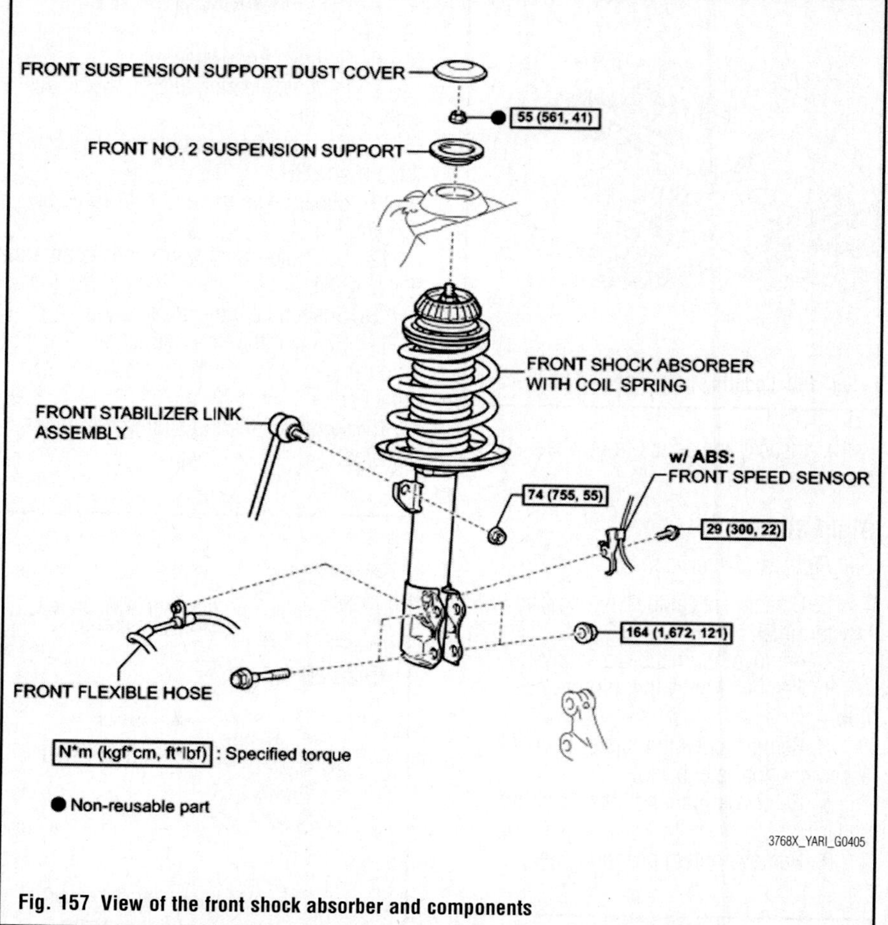

Fig. 157 View of the front shock absorber and components

To install:

13. Provisionally tighten a new nut through No. 2 suspension support.
14. Install the front shock absorber with coil spring onto the steering knuckle.
15. Install front flexible hose.
16. Install the stabilizer link with the nut.
17. Install front wheel.
18. Using a socket hexagon wrench 6, fix the shock absorber rod and tighten the nut.
19. To complete installation, reverse the remaining removal procedure.

OVERHAUL

See Figure 158.

1. Remove front suspension support sub-assembly.
2. Remove front support to front shock absorber nut:
   a. Using a socket hexagon wrench 6, fix the shock absorber rod and loosen the nut.
   b. Using SST 09727-30021, compress the coil spring.

➡Do not use an impact wrench. It will damage SST.

   c. Remove the nut.
3. Remove the upper coil spring seat with the strut mounting bearing and spring bumper from the shock absorber.
4. Remove the spring bumper from the upper coil spring seat.
5. Remove front upper coil spring insulator.
6. Remove front coil spring.
7. Using a brass bar and press, remove the strut mounting bearing from the upper coil spring seat.
8. To install, reverse the removal procedure.

STABILIZER BAR

REMOVAL & INSTALLATION

See Figures 159 through 161.

1. Disconnect cable from negative battery terminal.
2. Remove hood sub-assembly.
3. Remove front wiper motor and link.
4. Remove cowl to register duct sub-assembly (for hatchback).
5. Remove front air shutter seal (for sedan) .

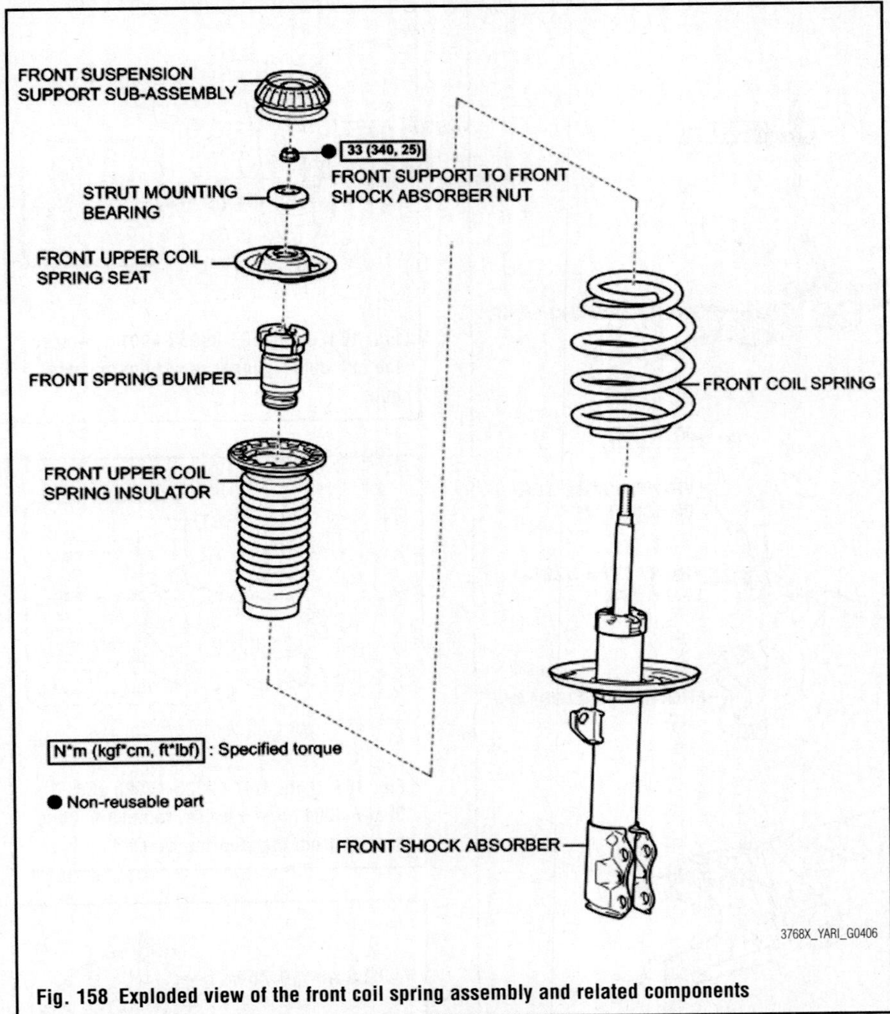

Fig. 158 Exploded view of the front coil spring assembly and related components

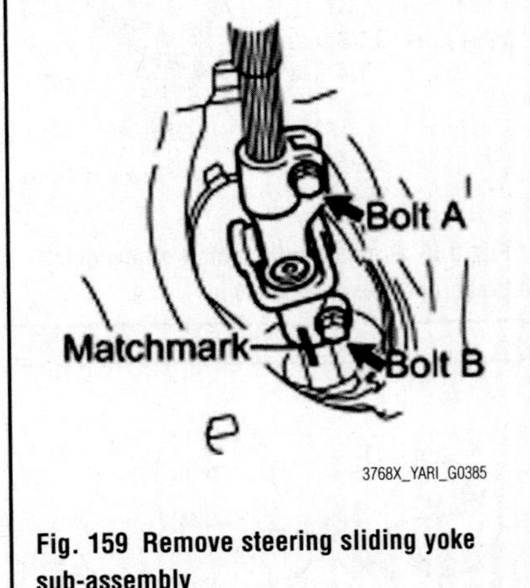

Fig. 159 Remove steering sliding yoke sub-assembly

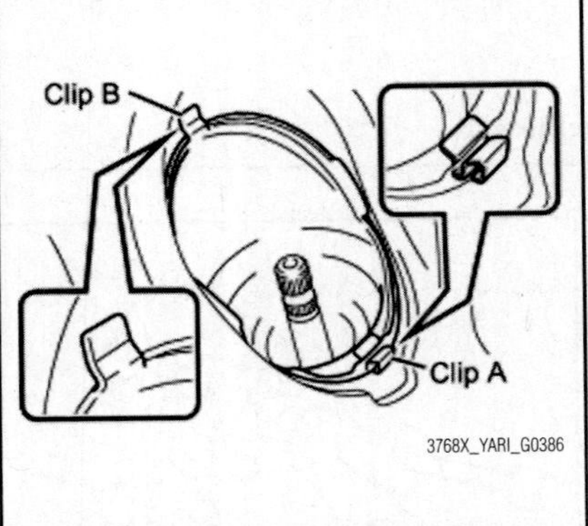

Fig. 160 Remove clip A, separate clip B from the body and separate No. 1 steering column hole cover sub-assembly

6. Remove outer cowl top panel.
7. Position wheels facing straight ahead
8. Remove front wheel.
9. Remove steering column hole cover sub-assembly.
10. Separate front stabilizer link assembly LH and RH.
11. Separate tie rod end sub-assembly LH and RH.
12. Separate front lower suspension arm LH and RH.
13. Suspend engine assembly.
14. Remove the floor carpet and 2 clips and remove the column hole cover silencer.
15. Remove steering sliding yoke sub-assembly:
    a. Use a seat belt to fix the steering wheel assembly, in order to avoid breakage of the spiral cable.
    b. Place the matchmarks on the sliding yoke of the steering intermediate shaft assembly and the power steering.
    c. Loosen bolt A, remove bolt B and separate the steering intermediate shaft assembly.
16. Remove clip A, separate clip B from the body and separate No. 1 steering column hole cover sub-assembly.
17. Remove front wheel.
18. Remove front stabilizer link assembly LH and RH.
19. Remove tie rod end sub-assembly.
20. Remove front suspension lower arm RH and LH.
21. Suspend engine assembly.
22. Remove front suspension crossmember sub-assembly:
    a. Remove the bolt and separate the engine moving control rod.
    b. Support the front suspension crossmember with a transmission jack.
    c. Remove the 6 bolts and remove the suspension crossmember.
23. Remove power steering gear.
24. Remove the 2 bolts and the stabilizer bracket.
25. Remove front stabilizer bar and bushing.
26. To install, reverse the removal procedure.
27. Install the stabilizer onto the crossmember with the paint mark on the left side of the vehicle.
28. Inspect and adjust front wheel alignment.

WHEEL HUB & BEARING

REMOVAL & INSTALLATION

See Figures 162 through 168.

1. Remove the front axle assembly. Refer Axle Assembly under Front Axle.
2. Using snap ring pliers, remove the hole snap ring.

➡When removing the hole snap ring, do not damage the magnetic rotor surface.

3. Remove front axle hub sub-assembly:
    a. Fix the steering knuckle in a vise between aluminum plates.

➡Do not over tighten the vise.

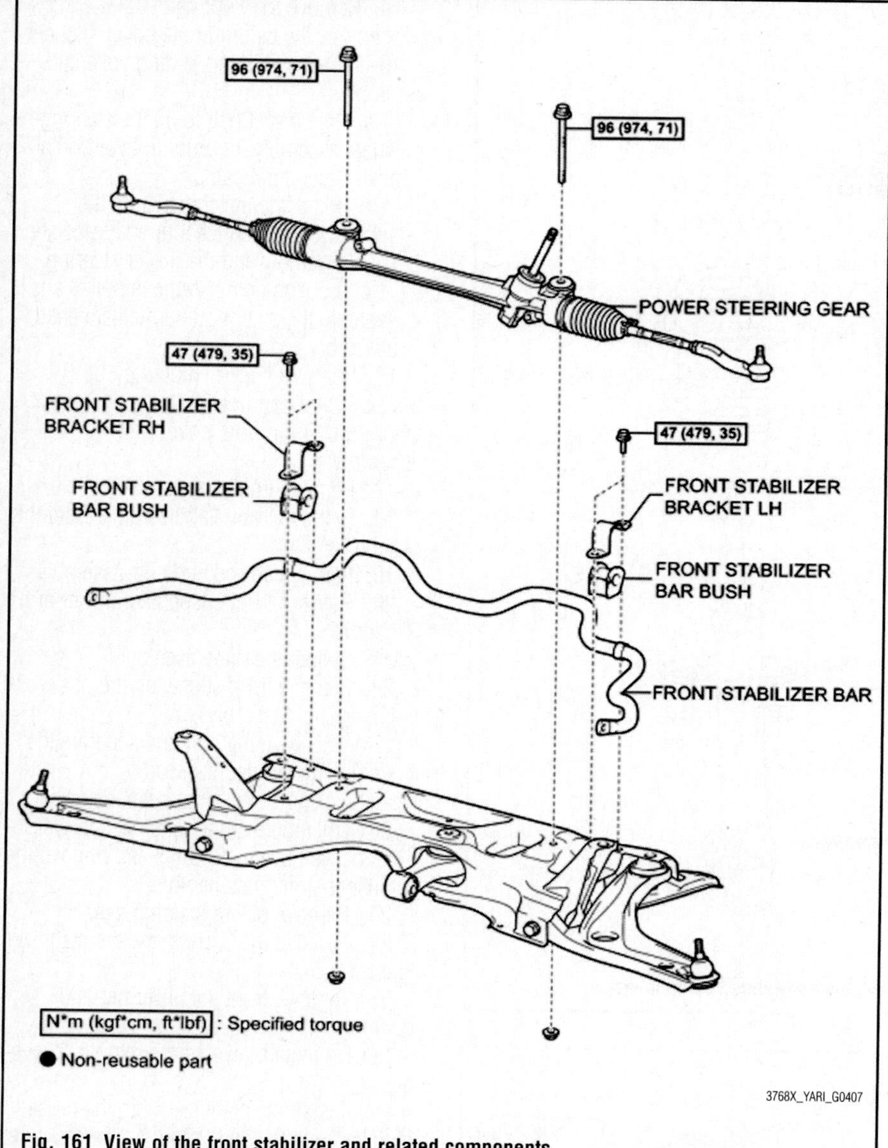

Fig. 161 View of the front stabilizer and related components

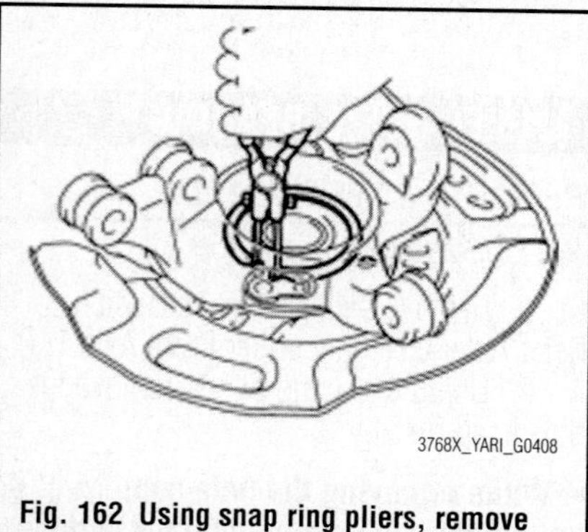

Fig. 162 Using snap ring pliers, remove the hole snap ring

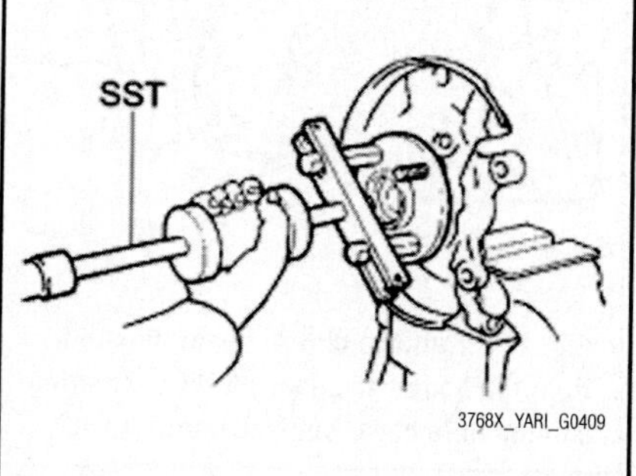

Fig. 163 Using SST 09520-00031, remove the axle hub

b. Using SST 09520-00031, remove the axle hub.

c. Using SST 09950-40011, remove the hub bearing inner race from the axle hub.

4. Remove the dust cover by tapping the back side with a brass bar and hammer.

5. Remove front axle hub bearing:

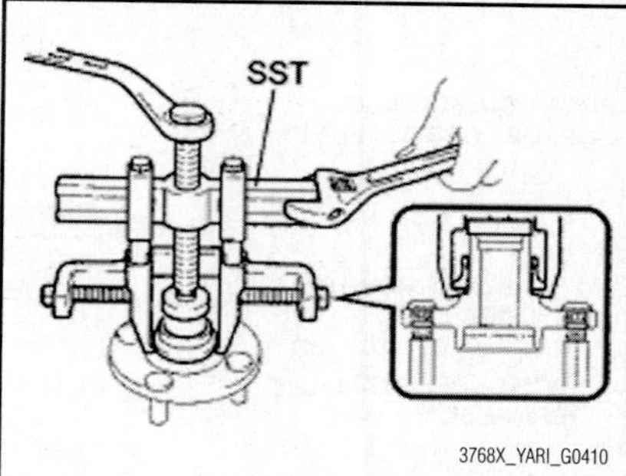

Fig. 164 Using SST 09950-40011, remove the hub bearing inner race from the axle hub

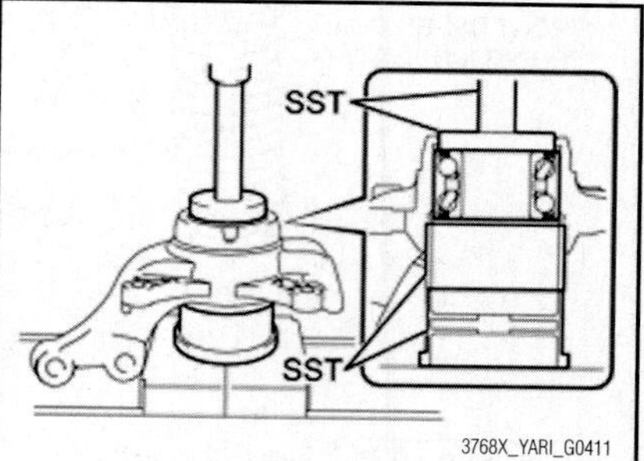

Fig. 165 Using SST 09223-15020 and 09387-00041 and a press, remove the hub bearing from the steering knuckle

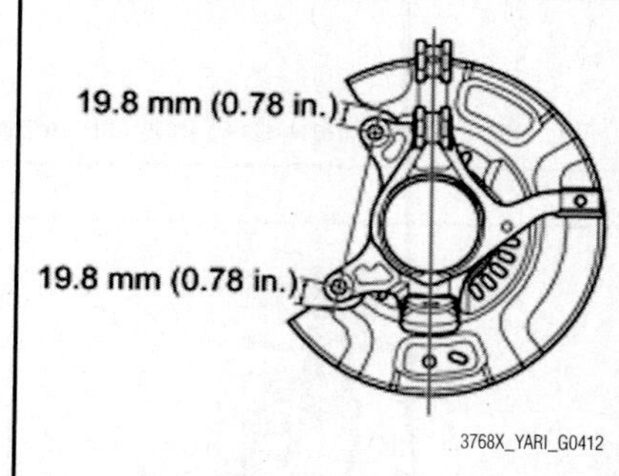

Fig. 166 Provisionally install a new disc brake dust cover as shown

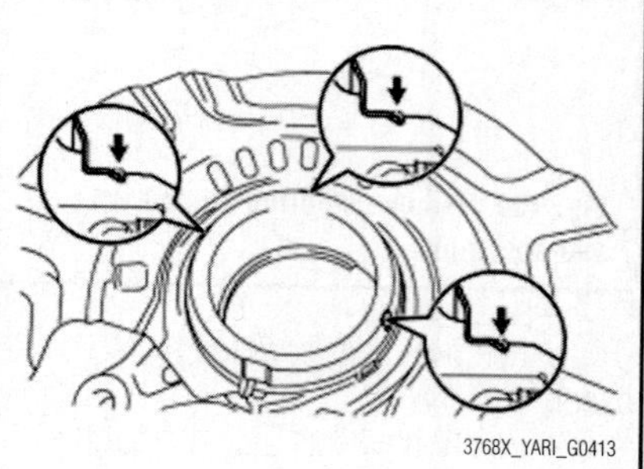

Fig. 167 Using a chisel, fix the 3 points on the circumference

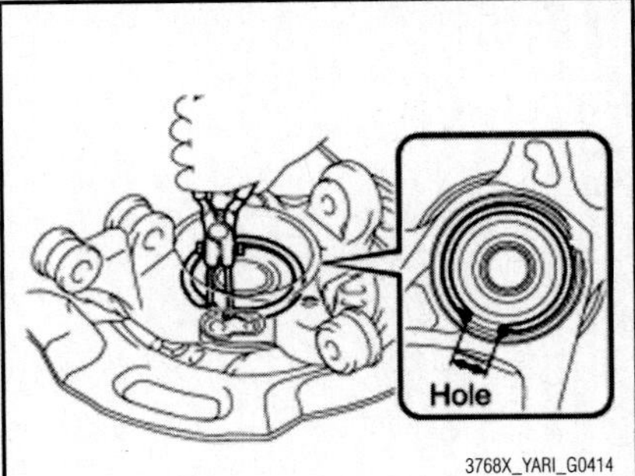

Fig. 168 Snap ring installation orientation

a. Install the removed hub bearing inner race onto the outer side of the hub bearing.

b. Using SST 09223-15020 and 09387-00041 and a press, remove the hub bearing from the steering knuckle.

To install:

6. Using SST 09950-60020 and 09387-00041 and a press, insert a new hub bearing, with its magnetic rotor side facing the inside of the vehicle, until it reaches the end of the steering knuckle.

➡

- Do not remove the inner race because the hub bearing is built into the oil seal.
- Do not use bearings that have been removed.
- Do not wipe off any grease that has been applied to new bearings.
- Do not bring magnets close to the magnetic rotor surface of the bearing.
- Keep the magnetic rotor surface of the bearing free of foreign matter.

7. Provisionally install a new disc brake dust cover.

8. Using SST 09223-56010 and a hammer, install the disc brake dust cover.

➡Uniformly press in the disc brake dust cover while sliding SST slightly. Surely press in the disc brake dust cover until the pressing-in base.

9. Using a chisel, fix the 3 points on the circumference.

➡Securely fold each end into the engagement grooves.

10. Using SST: 09608-32010, 09950-60010 and a press, press the axle hub into the steering knuckle.

11. Using snap ring pliers, install a new hole snap ring, as shown in the illustration.

➡Do not overlap the end of the snap ring and the installation hole in the speed sensor on the knuckle side. Do not damage the magnetic rotor surface of the bearing when installing the snap ring.

SUSPENSION

REAR SUSPENSION

COIL SPRING

REMOVAL & INSTALLATION

See Figure 169.

1. Remove the rear shock absorber.
2. Lower the jacks slowly.
3. Remove the coil spring, coil spring insulator upper and coil spring insulator lower.
4. To install:

a. Install the coil spring insulator lower onto the axle beam.

b. Install the coil spring insulator upper so that its gap fits onto the end of coil spring.

c. Install the coil spring onto the axle beam.

➡The paint mark of the coil spring should be towards the underside and rear side of the vehicle.

5. Install the shock absorber.

SHOCK ABSORBER

REMOVAL & INSTALLATION

See Figure 170.

1. Remove the rear wheel.
2. Remove the rear seat.
3. Remove rear absorber cap (for hatchback).
4. Remove rear shock absorber cap.
5. Remove rear shock absorber:

a. Support the axle beam with a jack. Insert a wooden block between the jack and the rear axle spring seat to prevent damage.

b. Remove the 2 nuts while keeping the piston rod from rotating.

c. Remove the cushion retainer and suspension support.

d. Remove the bolt while keeping the nut from rotating and remove the shock absorber.

➡Remove the nut on the bolt side because the one on the lower side is a jam nut.

6. Remove rear suspension support stopper.
7. Remove rear suspension support assembly.

To install:

8. Install rear suspension support assembly.
9. Install rear suspension support stopper.
10. Temporarily tighten rear shock absorber:

a. Support the axle beam with a jack. Insert a wooden block between the jack and the rear axle spring seat to prevent damage.

b. Jack up the axle beam slowly, and provisionally tighten the shock absorber (lower side) with the bolt and nut to the axle beam.

c. Install the suspension support and cushion retainer.

d. While holding the piston rod, install a new nut (lower nut) to the specified standard. Standard: 0.591 to 0.709 inches.

e. While holding the piston rod, tighten a new nut (upper nut).

11. To complete the installation, reverse the remaining removal procedure.

WHEEL HUB & BEARING

REMOVAL & INSTALLATION

See Figure 171.

1. Disconnect the negative battery cable.
2. Remove rear wheel.
3. Remove rear brake drum sub-assembly.
4. Using a screwdriver, remove the claw of the connector lock portion and disconnect the skid control sensor wire connector.

➡Do not remove the connector cover from the connector because the skid control sensor wire may be damaged.

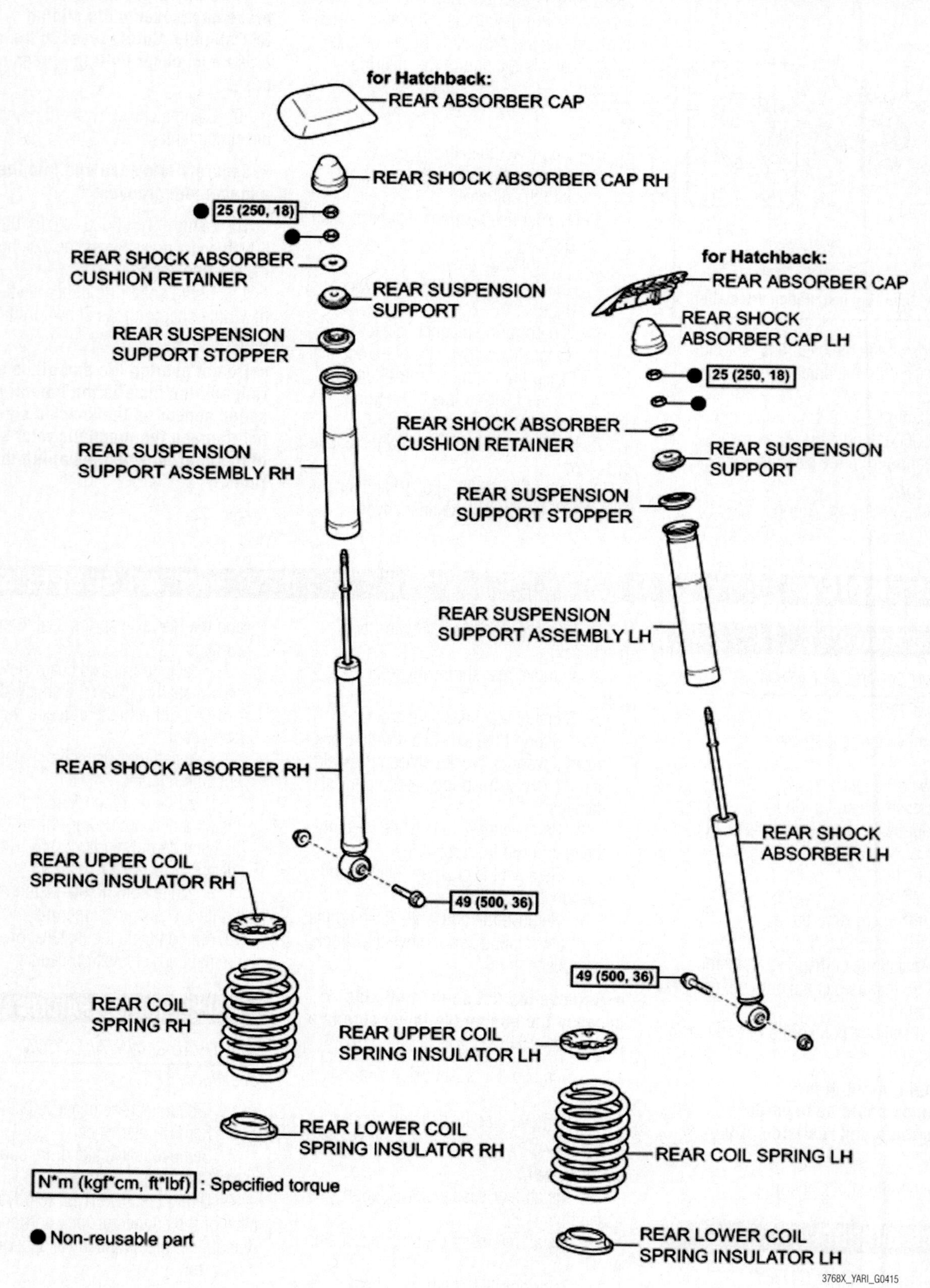

Fig. 169 View of the rear coil spring and components

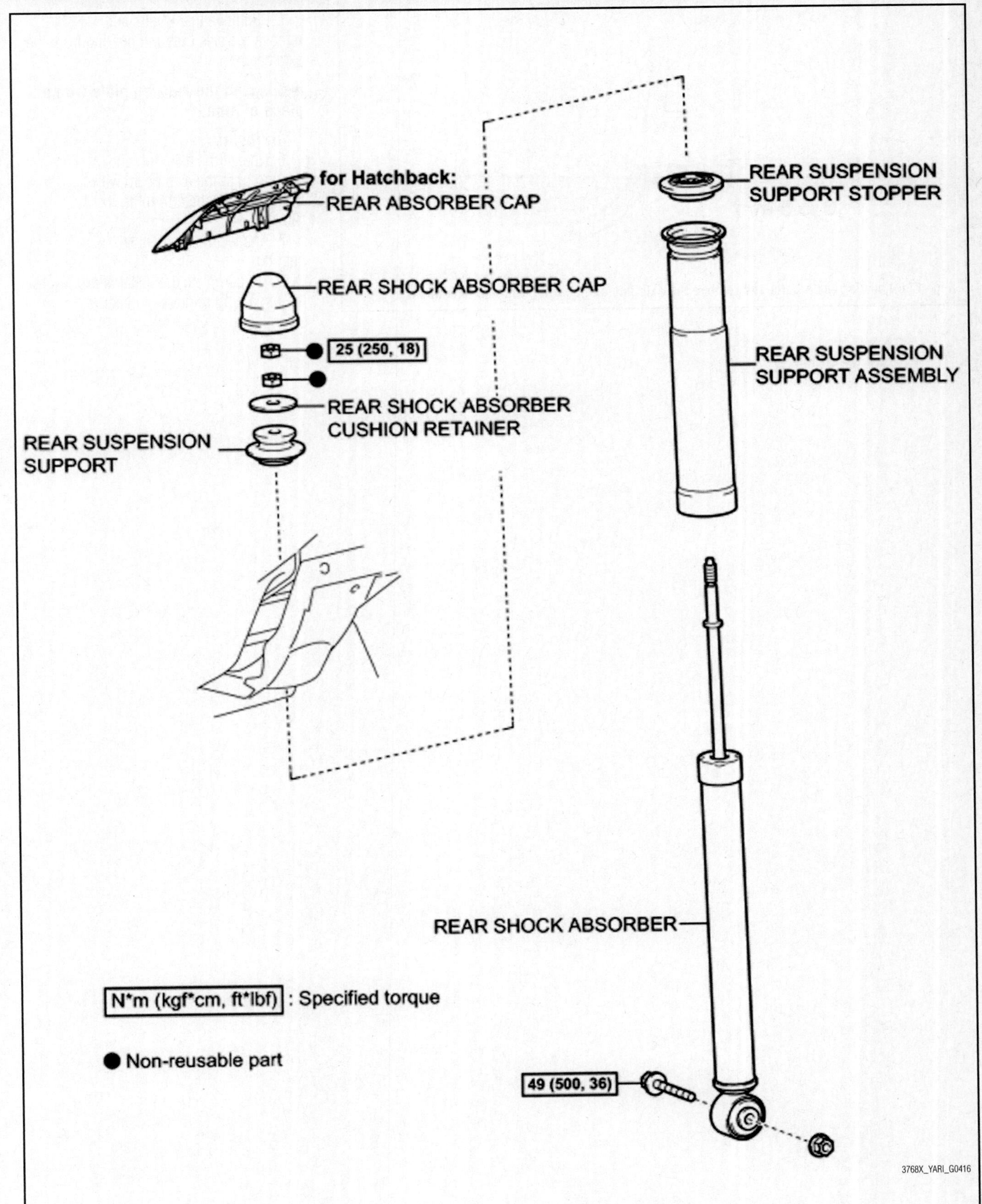

Fig. 170 View of the rear shock absorber components

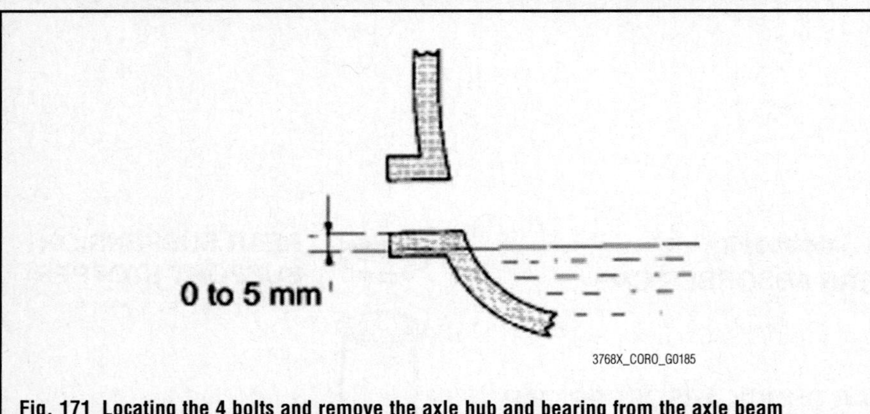

Fig. 171 Locating the 4 bolts and remove the axle hub and bearing from the axle beam

5. Remove the 4 bolts and remove the axle hub and bearing from the axle beam.

➡Suspend the backing plate with a piece of rope.

To install:

6. Install the axle hub and bearing onto the axle beam with the 4 bolts. Tighten to 67 ft. lbs. (90 Nm).
7. Inspect rear axle hub bearing.
8. To complete the installation, reverse the remaining removal procedure.

TOYOTA

Diagnostic Trouble Codes

20

DIAGNOSTIC TROUBLE CODES

OBD II VEHICLE APPLICATIONS

TOYOTA

4Runner
2009–2010
- 4.0L V6 ..... 1GR-FE
- 4.7L V8 ..... 2UZ-FE

Avalon
2009–2010
- 3.5L V6 ..... 2GR-FE

Camry
2009–2010
- 2.4L I4 ..... 2AZ-FE
- 3.5L V6 ..... 2GR-FE

Camry Hybrid
2009–2010
- 2.4L I4 ..... 2AZ-FXE

Corolla
2009–2010
- 1.8L I4 ..... 2ZR-FE
- 2.4L I4 ..... 2AZ-FE

FJ Cruiser
2009–2010
- 4.0L V6 ..... 1GR-FE

Highlander
2009–2010
- 2.7L I4 ..... 1AR-FE
- 3.5L V6 ..... 2GR-FE

Land Cruiser
2009–2010
- 5.7L V8 ..... 3UR-FE

Matrix
2009–2010
- 1.8L I4 ..... 2ZR-FE
- 2.4L I4 ..... 2AZ-FE

RAV 4
2009–2010
- 2.4L I4 ..... 2AZ-FE
- 3.5L V6 ..... 2GR-FE

Sequoia
2009–2010
- 4.6L V8 ..... 1UR-FE
- 4.7L V8 ..... 2UZ-FE
- 5.7L V8 ..... 3UR-FE

Sienna
2009–2010
- 3.5L V6 ..... 2GR-FE

Tacoma
2009–2010
- 2.7L I4 ..... 2TR-FE
- 4.0L V6 ..... 1GR-FE

Tundra
2009–2010
- 4.0L V6 ..... 1GR-FE
- 4.6L V8 ..... 1UR-FE
- 4.7L V8 ..... 2UZ-FE
- 5.7L V8 ..... 3UR-FE

Yaris
2009–2010
- 1.5L I4 ..... 1NZ-FE

OBD II Trouble Code List (P0XXX Codes)

DTC	Trouble Code Title, Conditions & Possible Causes
DTC: P0010 **1T ECM, MIL: Yes** **Year:** 2009, 2010 **Model:** Highlander, Land Cruiser **Engine:** All V6 & V8 **Transmission:** All	**Camshaft Position "A" Actuator Circuit (Bank 1):** Monitor runs whenever following DTCs are not present: None All of the following conditions are met: - Starter: OFF Power switch: ON Time after power switch off to on: 0.5 seconds or more Either of the following conditions is met: - A. All of the following conditions are met: - Battery voltage: 11 V or more and less than 13 V Target duty ratio: Less than 70% Current cut status: Not cut B. All of the following conditions are met: - Battery voltage: 13 V or more Target duty ratio: Less than 80% Current cut status: Not cut **Possible Causes:** • Open or short in oil control valve (bank 1) circuit • Oil control valve (bank 1) • Oil control valve filter • Hybrid vehicle control ECU • ECM
DTC: P0011 **1T ECM, MIL: Yes** **Year:** 2009, 2010 **Model:** Highlander and Land Cruiser **Engine:** All V6 & V8 **Transmission:** All	**Camshaft Position "A" - Timing Over-Advanced or System Performance (Bank 1):** Monitor runs whenever following DTCs are not present: P0010, P0020 (VVT oil control valve bank 1, 2) P0016, P0018 (VVT system bank 1, 2 - misalignment) P0102, P0103 (Mass air flow meter) P0115, O0117, P0118 (Engine coolant temperature sensor) P0125 (insufficient engine coolant temperature for closed loop) P0335 (Crankshaft position sensor) Battery voltage: 11 V or more Engine RPM: 500 to 4000 rpm Engine coolant temperature: 75°C (167°F) to 100°C (212°F) **Possible Causes:** • Valve timing • Oil control valve (bank 1, 2) • Oil control valve filter (bank 1, 2) • Intake camshaft (bank 1, 2) timing gear assembly • Hybrid Vehicle Control ECU • ECM
DTC: P0012 **2T ECT, MIL: Yes** **Year:** 2009, 2010 **Model:** Highlander, Land Cruiser **Engine:** All V6 & V8 **Transmission:** All	**Camshaft Position "A" - Timing Over-Retarded (Bank 1):** Monitor runs whenever following DTCs are not present: P0010, P0020 (VVT oil control valve bank 1, 2) P0016, P0018 (VVT system bank 1, 2 - misalignment) P0102, P0103 (Mass air flow meter) P0115, O0117, P0118 (Engine coolant temperature sensor) P0125 (insufficient engine coolant temperature for closed loop) P0335 (Crankshaft position sensor) Battery voltage: 11 V or more Engine RPM: 500 to 4000 rpm Engine coolant temperature: 75°C (167°F) to 100°C (212°F) **Possible Causes:** • Valve timing • Oil control valve (bank 1, 2) • Oil control valve filter (bank 1, 2) • Intake camshaft (bank 1, 2) timing gear assembly • Hybrid Vehicle Control ECU

<table>
<tr><th>DTC</th><th>Trouble Code Title, Conditions & Possible Causes</th></tr>
<tr><td>DTC: P0013
1T ECM, MIL: Yes
Year: 2010
Model: Camry
Engine: All V6
Transmission: All</td><td>Camshaft Position “B” Actuator Circuit / Open (Bank 1):
Starter: OFF
Engine Switch: On (IG)
Time after turning engine switch off to on (IG): 0.5 seconds or more
Possible Causes:
• Open or short in OCV for exhaust camshaft (bank 1) circuit
• OCV for exhaust camshaft (bank 1)
• ECM</td></tr>
<tr><td>DTC: P0013
1T ECM, MIL: Yes
Year: 2009, 2010
Model: Land Cruiser
Engine: All V8
Transmission: All</td><td>Camshaft Position “B” Actuator Circuit / Open (Bank 1):
With the ignition switch on for at least 5 seconds. An open or short in the Camshaft Oil Control Valve (OCV) for exhaust side (for Bank 1) circuit.
Possible Causes:
• Open or short in Camshaft Oil Control Valve (OCV) (for exhaust side of Bank 1) circuit
• OCV (for exhaust side of Bank 1)
• ECM</td></tr>
<tr><td>DTC: P0014
2T ECM, MIL: Yes
Year: 2009, 2010
Model: Land Cruiser
Engine: All V8
Transmission: All</td><td>Camshaft Position “B” - Timing Over-Advanced or System Performance (Bank 1):
With the engine running at a speed of between 500-4000 RPM, ECT 75 to 100°C (167 to 212°F) and battery voltage 11 V or more.
The exhaust valve timing is not adjusted in the valve timing advance range.
Possible Causes:
• Valve timing
• Oil control valve (OCV) for exhaust camshaft
• OCV filter
• Camshaft timing exhaust gear
• ECM</td></tr>
<tr><td>DTC: P0014
2T ECM, MIL: Yes
Year: 2009, 2010
Model: 4Runner, FJ Cruiser
Engine: 4.0L V6
Transmission: All</td><td>Camshaft Position “B” - Timing Over-Advanced or System Performance (Bank 1):
Monitor runs whenever following DTCs are not present:
P0013, P0023 (Exhaust VVT Oil Control Valve)
P0017, P0019 (Exhaust VVT System - Misalignment)
P0102, P0103 (Mass Air Flow Sensor)
P0115, P0117, P0118 (Engine Coolant Temperature Sensor)
P0125 (Insufficient Coolant Temperature for Closed Loop Fuel Control)
P0335 (Crankshaft Position Sensor)
P0340 (Camshaft Position Sensor)
Battery voltage: 11 V or higher
Engine speed: 500 to 4000 rpm
Engine coolant temperature: 75 to 100°C (167 to 212°F)
Possible Causes:
• Valve timing
• Camshaft timing oil control valve for exhaust camshaft
• Oil control valve filter
• Camshaft timing exhaust gear assembly
• ECM</td></tr>
<tr><td>DTC: P0015
1T ECM, MIL: Yes
Year: 2009, 2010
Model: Land Cruiser
Engine: All V8
Transmission: All</td><td>Camshaft Position “B” - Timing Over-Retarded (Bank 1):
With the engine running at a speed of between 500-4000 RPM, ECT 75 to 100°C (167 to 212°F) and battery voltage 11 V or more.
The exhaust valve timing is not adjusted in the valve timing retard range.
Possible Causes:
• Valve timing
• OCV for exhaust camshaft
• OCV filter
• Camshaft timing exhaust gear
• ECM</td></tr>
</table>

DTC	Trouble Code Title, Conditions & Possible Causes
DTC: P0016 **2T ECM, MIL: Yes** **Year:** 2009, 2010 **Model:** Land Cruiser **Engine:** All V8 **Transmission:** All	**Crankshaft Position - Camshaft Position Correlation (Bank 1 Sensor A):** With the engine running at a speed of between 400-1400 RPM. A deviation in the crankshaft position sensor signal and VVT sensor (for Intake side of Bank 1) signal. **Possible Causes:** • Mechanical system (Timing chain has jumped tooth or chain stretched) • OCV (for intake side of bank 1) • Camshaft timing gear (bank 1) • ECM
DTC: P0016 **2T ECM, MIL: Yes** **Year:** 2009, 2010 **Model:** Highlander **Engine:** All V6 **Transmission:** All	**Crankshaft Position - Camshaft Position Correlation (Bank 1 Sensor A):** Monitor runs whenever following DTCs are not present: P0010, P0020 (VVT oil control valve bank 1, 2) P0102, P0103 (Mass air flow meter) P0115, P0117, P0118 (Engine coolant temperature sensor) P0125 (insufficient Engine coolant temperature for closed loop) P0335 (Crankshaft position sensor) Engine rpm: 900 to 1000 rpm **Possible Causes:** • Valve timing • Oil control valve (Bank 1) • Oil control valve filter (Bank 1) • Intake camshaft (Bank 1) timing gear assembly • Hybrid vehicle control ECU • ECM
DTC: P0017 **2T ECM, MIL: Yes** **Year:** 2009, 2010 **Model:** 4Runner, FJ Cruiser **Engine:** All **Transmission:** All	**Crankshaft Position - Camshaft Position Correlation (Bank 1 Sensor B):** The monitor will run whenever these DTCs are not present: P0013, P0023 (Exhaust VVT Oil Control Valve) P0017, P0019 (Exhaust VVT System - Misalignment) P0102, P0103 (Mass Air Flow Sensor) P0115, P0117, P0118 (Engine Coolant Temperature Sensor) P0125 (Insufficient Coolant Temperature for Closed Loop Fuel Control) P0335 (Crankshaft Position Sensor) P0340 (Camshaft Position Sensor) Engine speed: 500 to 1000 rpm **Possible Causes:** • Valve timing • Camshaft timing oil control valve for exhaust camshaft • Oil control valve filter • Camshaft timing exhaust gear assembly • ECM
DTC: P0018 **2T ECM, MIL: Yes** **Year:** 2009, 2010 **Model:** Land Cruiser **Engine:** All V8 **Transmission:** All	**Crankshaft Position - Camshaft Position Correlation (Bank 2 Sensor A):** With the engine running at a speed of between 400-1400 RPM, a deviation in the crankshaft position sensor signal and VVT sensor (for Intake side of Bank 2) signal. **Possible Causes:** • Mechanical system (Timing chain has jumped tooth or chain stretched) • OCV (for intake side of bank 2) • Camshaft timing gear (bank 2) • ECM

DTC	Trouble Code Title, Conditions & Possible Causes
DTC: P0018 **2T ECM, MIL: Yes** **Year:** 2009, 2010 **Model:** Highlander **Engine:** All V6 **Transmission:** All	**Crankshaft Position - Camshaft Position Correlation (Bank 2 Sensor A):** Monitor runs whenever following DTCs are not present: P0010, P0020 (VVT oil control valve bank 1, 2) P0102, P0103 (Mass air flow meter) P0115, P0117, P0118 (Engine coolant temperature sensor) P0125 (insufficient Engine coolant temperature for closed loop) P0335 (Crankshaft position sensor) Engine rpm: 900 to 1000 rpm **Possible Causes:** • Valve timing • Oil control valve (Bank 2) • Oil control valve filter (Bank 2) • Intake camshaft (Bank 2) timing gear assembly • Hybrid vehicle control ECU • ECM
DTC: P0019 **2T , MIL: Yes** **Year:** 2009, 2010 **Model:** Avalon, Camry **Engine:** All L 4, V6 **Transmission:** All	**Crankshaft Position - Camshaft Position Correlation (Bank 2 Sensor B):** Engine RPM: 500 to 1000 rpm **Possible Causes:** • Valve timing • Camshaft timing oil control valve for exhaust camshaft (bank 1, 2) • Camshaft timing oil control valve filter (bank 1, 2) • Exhaust camshaft timing gear assembly (bank 1, 2) • ECM
DTC: P0019 **2T ECM, MIL: Yes** **Year:** 2009, 2010 **Model:** Land Cruiser **Engine:** All V8 **Transmission:** All	**Crankshaft Position - Camshaft Position Correlation (Bank 2 Sensor B):** With the engine running at a speed of between 400-1400 RPM, a deviation in the crankshaft position sensor signal and VVT sensor (for Exhaust side of Bank 2) signal. **Possible Causes:** • Mechanical system (Timing chain has jumped tooth or chain stretched) • OCV (for exhaust side of bank 2) • Camshaft timing exhaust gear (bank 2) • ECM
DTC: P0020 **1T ECM, MIL: Yes** **Year:** 2009, 2010 **Model:** Land Cruiser **Engine:** All V8 **Transmission:** All	**Camshaft Position "A" Actuator Circuit (Bank 2):** With the ignition switch ON for 5 seconds or more and starter OFF, an open or short is detected in the OCV for intake side (for Bank 2) circuit. **Possible Causes:** • Open or short in OCV (for intake side of Bank 2) circuit • OCV (for intake side of Bank 2) • ECM
DTC: P0020 **1T ECM, MIL: Yes** **Year:** 2009, 2010 **Model:** 4Runner **Engine:** All L4, V6 **Transmission:** All	**Camshaft Position "A" Actuator Circuit (Bank 2):** Monitor runs whenever following DTCs not present: None All of following conditions met: Starter: OFF Ignition switch: ON Time after ignition switch off to ON: 0.5 seconds or more **Possible Causes:** • Open or short in camshaft timing oil control valve for intake camshaft (bank 2) circuit • Camshaft timing oil control valve for intake camshaft (bank 2) • ECM
DTC: P0021 **1T ECM, MIL: Yes** **Year:** 2009, 2010 **Model:** Land Cruiser **Engine:** All V8 **Transmission:** All	**Camshaft Position "A" - Timing Over-Advanced or System Performance (Bank 2):** With the engine running at a speed of between 400-1400 RPM, ECT 75 to 100°C (167 to 212°F) and battery voltage 11 V or more. The intake valve timing is not adjusted in the valve timing advance range. **Possible Causes:** • Valve timing • OCV for intake camshaft • OCV filter • Intake camshaft timing gear • ECM

DTC	Trouble Code Title, Conditions & Possible Causes
DTC: P0021 **1T ECM, MIL: Yes** **Year:** 2009, 2010 **Model:** 4Runner, FJ Cruiser **Engine:** All L4 **Transmission:** All	**Camshaft Position "A" - Timing Over-Advanced or System Performance (Bank 2):** Monitor runs whenever following DTCs not present: P0010, P0020 (VVT Oil Control Valve) P0016, P0018 (VVT System - Misalignment) P0102, P0103 (Mass Air Flow Sensor) P0115, P0117, P0118 (Engine Coolant Temperature Sensor) P0125 (Insufficient Coolant Temperature for Closed Loop Fuel Control) P0335 (Crankshaft Position Sensor) P0340 (Camshaft Position Sensor) Battery voltage: 11 V or higher Engine speed: 400 to 4000 rpm Engine coolant temperature sensor: 75°C (167°F) to 100°C (212°F) **Possible Causes:** • Valve timing • Camshaft timing oil control valve for intake camshaft • Oil control valve filter • Camshaft timing gear assembly • ECM
DTC: P0021 **1T ECM, MIL: Yes** **Year:** 2009, 2010 **Model:** Highlander **Engine:** All V6 **Transmission:** All	**Camshaft Position "A" - Timing Over-Advanced or System Performance (Bank 2):** Monitor runs whenever following DTCs are not present: P0010, P0020 (VVT oil control valve bank 1, 2) P0016, P0018 (VVT system bank 1, 2 - misalignment) P0102, P0103 (Mass air flow meter) P0115, O0117, P0118 (Engine coolant temperature sensor) P0125 (insufficient engine coolant temperature for closed loop) P0335 (Crankshaft position sensor) Battery voltage: 11 V or more Engine RPM: 500 to 4000 rpm Engine coolant temperature: 75°C (167°F) to 100°C (212°F) **Possible Causes:** • Valve timing • Oil control valve (bank 1, 2) • Oil control valve filter (bank 1, 2) • Intake camshaft (bank 1, 2) timing gear assembly • Hybrid Vehicle Control ECU • ECM
DTC: P0022 **2T ECM** **Year:** 2009, 2010 **Model:** Land Cruiser **Engine:** All V8 **Transmission:** All	**Camshaft Position "A" - Timing Over-Retarded (Bank 2):** With the engine running at a speed of between 400-1400 RPM, ECT 75 to 100°C (167 to 212°F) and battery voltage 11 V or more. The intake valve timing is not adjusted in the valve timing retard range. **Possible Causes:** • Valve timing • OCV for intake camshaft • OCV filter • Intake camshaft timing gear • ECM

DTC	Trouble Code Title, Conditions & Possible Causes
DTC: P0022 **2T ECM, MIL: Yes** **Year:** 2009, 2010 **Model:** Highlander **Engine:** All V6 **Transmission:** All	**Camshaft Position "A" - Timing Over-Retarded (Bank 2):** Monitor runs whenever following DTCs are not present: P0010, P0020 (VVT oil control valve bank 1, 2) P0016, P0018 (VVT system bank 1, 2 - misalignment) P0102, P0103 (Mass air flow meter) P0115, O0117, P0118 (Engine coolant temperature sensor) P0125 (insufficient engine coolant temperature for closed loop) P0335 (Crankshaft position sensor) Battery voltage: 11 V or more Engine RPM: 500 to 4000 rpm Engine coolant temperature: 75°C (167°F) to 100°C (212°F) **Possible Causes:** • Valve timing • Oil control valve (bank 1, 2) • Oil control valve filter (bank 1, 2) • Intake camshaft (bank 1, 2) timing gear assembly • Hybrid Vehicle Control ECU • ECM
DTC: P0023 **1T ECM, MIL: Yes** **Year:** 2009, 2010 **Model:** Land Cruiser **Engine:** All V8 **Transmission:** All	**Camshaft Position "B" Actuator Circuit / Open (Bank 2):** With the engine running at a speed of between 400-1400 RPM, ECT 75 to 100°C (167 to 212°F) and battery voltage 11 V or more. An open or short in the OCV for exhaust side (for Bank 2) circuit. **Possible Causes:** • Open or short in OCV for exhaust camshaft (for Bank 2) circuit • OCV for exhaust camshaft (for Bank 2) • ECM
DTC: P0024 **2T ECM, MIL: Yes** **Year:** 2009, 2010 **Model:** Land Cruiser **Engine:** All V8 **Transmission:** All	**Camshaft Position "B" - Timing Over-Advanced or System Performance (Bank 2):** With the engine running at a speed of between 500-4000 RPM, ECT 75 to 100°C (167 to 212°F) and battery voltage 11 V or more. The exhaust valve timing is not adjusted in the valve timing advance range. **Possible Causes:** • Valve timing • Oil control valve (OCV) for exhaust camshaft • OCV filter • Camshaft timing exhaust gear • ECM
DTC: P0025 **1T ECM, MIL: Yes** **Year:** 2009, 2010 **Model:** Land Cruiser **Engine:** All V8 **Transmission:** All	**Camshaft Position "B" - Timing Over-Retarded (Bank 2):** With the engine running at a speed of between 500-4000 RPM, ECT 75 to 100°C (167 to 212°F) and battery voltage 11 V or more. The exhaust valve timing is not adjusted in the valve timing retard range. **Possible Causes:** • Valve timing • OCV for exhaust camshaft • OCV filter • Camshaft timing exhaust gear • ECM
DTC: P0031 **1T ECM, MIL: Yes** **Year:** 2009, 2010 **Model:** Highlander **Engine:** All V6 **Transmission:** All	**Oxygen (A/F) Sensor Heater Control Circuit Low (Bank 1 Sensor 1):** Monitor runs whenever following DTCs are not present: None Battery voltage: 10.5 V or more Air fuel ratio sensor heater duty-cycle ratio: 50% or more Time after engine start: 10 seconds or more **Possible Causes:** • Open in air fuel ratio sensor heater (bank 1, sensor 1) circuit • Air fuel ratio sensor heater (bank 1, sensor 1) • A/F relay • Hybrid vehicle control ECU • ECM

DTC	Trouble Code Title, Conditions & Possible Causes
DTC: P0031 **1T ECM, MIL: Yes** **Year:** 2009, 2010 **Model:** Land Cruiser **Engine:** AllL V8 **Transmission:** All	**Oxygen (A/F) Sensor Heater Control Circuit Low (Bank 1 Sensor 1):** Battery voltage 10.5 V or more, time after engine start 10 seconds or more and heater output duty 50% or more. The Air-Fuel Ratio (A/F) sensor heater current is below 0.8 A. **Possible Causes:** • Open in A/F sensor heater circuit • A/F sensor heater (for Sensor 1) • Integration relay • ECM
DTC: P0032 **1T ECM, MIL: Yes** **Year:** 2009, 2010 **Model:** Land Cruiser **Engine:** All V8 **Transmission:** All	**Oxygen (A/F) Sensor Heater Control Circuit High (Bank 1 Sensor 1):** With the engine running and battery voltage less than 20 V, an Air-Fuel Ratio (A/F) sensor heater current failure was detected. **Possible Causes:** • Short in A/F sensor heater circuit • A/F sensor heater (for Sensor 1) • Integration relay • ECM
DTC: P0032 **1T ECM, MIL: Yes** **Year:** 2009, 2010 **Model:** Matrix **Engine:** All L4 **Transmission:** All	**Oxygen (A/F) Sensor Heater Control Circuit High (Bank 1 Sensor 1):** With battery voltage less than 20 V, 10 seconds after the engine was started. The air fuel ratio sensor heater current failed. **Possible Causes:** • Short in air fuel ratio sensor heater circuit • Air fuel ratio sensor heater (sensor 1) • ECM
DTC: P0032 **1T ECM, MIL: Yes** **Year:** 2009, 2010 **Model:** Highlander **Engine:** All V6 **Transmission:** All	**Oxygen (A/F) Sensor Heater Control Circuit High (Bank 1 Sensor 1):** Monitor runs whenever following DTCs are not present: None Battery voltage: 10.5 V or more Air fuel ratio sensor heater duty-cycle ratio: 50% or more Time after engine start: 10 seconds or more **Possible Causes:** • Short in air fuel ratio sensor heater (bank 1, sensor 1) circuit • Air fuel ratio sensor heater (bank 1, sensor 1) • A/F relay • Hybrid vehicle control ECU • ECM
DTC: P0037 **1T ECM, MIL: Yes** **Year:** 2009, 2010 **Model:** Land Cruiser **Engine:** All V8 **Transmission:** All	**Oxygen Sensor Heater Control Circuit Low (Bank 1 Sensor 2):** With battery voltage 10.5 to 20 V. The Heated Oxygen (HO2) sensor heater current is below 0.3 A. **Possible Causes:** • Open in HO2 sensor heater circuit • HO2 sensor heater (for Sensor 2) • Integration relay • ECM
DTC: P0038 **1T ECM, MIL: Yes** **Year:** 2009, 2010 **Model:** Land Cruiser **Engine:** All V8 **Transmission:** All	**Oxygen Sensor Heater Control Circuit High (Bank 1 Sensor 2):** With the engine running and battery voltage 10.5 V or more. The Heated Oxygen (HO2) sensor heater current is higher than 2 A. **Possible Causes:** • Short in HO2 sensor heater circuit • HO2 sensor heater (for Sensor 2) • Integration relay • ECM

DTC	Trouble Code Title, Conditions & Possible Causes
DTC: P0038 **1T ECM, MIL: Yes** **Year:** 2009, 2010 **Model:** Highlander **Engine:** ALL V6 **Transmission:** All	**Oxygen Sensor Heater Control Circuit High (Bank 1 Sensor 2):** Monitor runs whenever following DTCs are not present: None Case 1: Battery voltage: 10.5 V or more Engine: Running Starter: OFF Case 2: Battery voltage: 10.5 to 20 V **Possible Causes:** • Short in heated oxygen sensor heater circuit • Heated oxygen sensor heater • IGCT relay • Hybrid vehicle control ECU • ECM
DTC: P0051 **T ECM, MIL: Yes** **Year:** 2009, 2010 **Model:** 4Runner, FJ Cruiser **Engine:** All V8 **Transmission:** All	**Oxygen (A/F) Sensor Heater Control Circuit Low (Bank 2 Sensor 1) :** Monitor runs whenever following DTCs not stored: P101D, P103D (Air fuel ratio sensor heater) Battery voltage: 10.5 V or higher Time after engine start: 10 seconds or more Active heater OFF control: Not operating Active heater ON control: Not operating Air fuel ratio sensor heater duty-cycle: 50% or more **Possible Causes:** • Open in air fuel ratio sensor heater circuit • Air fuel ratio sensor heater (sensor 1) • No. 1 integration relay • ECM
DTC: P0051 **1T ECM, MIL: Yes** **Year:** 2009, 2010 **Model:** Highlander **Engine:** All V6 **Transmission:** All	**Oxygen (A/F) Sensor Heater Control Circuit Low (Bank 2 Sensor 1):** Monitor runs whenever following DTCs are not present: None Battery voltage: 10.5 V or more Air fuel ratio sensor heater duty-cycle ratio: 50% or more Time after engine start: 10 seconds or more **Possible Causes:** • Open in air fuel ratio sensor heater (bank 2, sensor 1) circuit • Air fuel ratio sensor heater (bank 2, sensor 1) • A/F relay • Hybrid vehicle control ECU • ECM
DTC: P0052 **1T ECM, MIL: Yes** **Year:** 2009, 2010 **Model:** Highlander **Engine:** All V6 **Transmission:** All	**Oxygen (A/F) Sensor Heater Control Circuit High (Bank 2 Sensor 1):** Monitor runs whenever following DTCs are not present: None Battery voltage: 10.5 V or more Air fuel ratio sensor heater duty-cycle ratio: 50% or more Time after engine start: 10 seconds or more **Possible Causes:** • Short in air fuel ratio sensor heater (bank 2, sensor 1) circuit • Air fuel ratio sensor heater (bank 2, sensor 1) • A/F relay • Hybrid vehicle control ECU • ECM
DTC: P0052 **1T ECM, MIL: Yes** **Year:** 2009, 2010 **Model:** Land Cruiser **Engine:** All V8 **Transmission:** All	**Oxygen (A/F) Sensor Heater Control Circuit High (Bank 2 Sensor 1):** With the engine running and battery voltage less than 20 V, an Air-Fuel Ratio (A/F) sensor heater current failure was detected. **Possible Causes:** • Short in A/F sensor heater circuit • A/F sensor heater (for Sensor 1) • Integration relay • ECM

DTC	Trouble Code Title, Conditions & Possible Causes
DTC: P0057 **1T ECM, MIL: Yes** **Year:** 2009, 2010 **Model:** Highlander **Engine:** All V6 **Transmission:** All	**Oxygen Sensor Heater Control Circuit Low (Bank 2 Sensor 2):** Monitor runs whenever following DTCs are not present: None Battery voltage: 10.5 to 20 V **Possible Causes:** • Open in heated oxygen sensor heater circuit • Heated oxygen sensor heater • IGCT relay • Hybrid vehicle control ECU • ECM
DTC: P0057 **1T ECM, MIL: Yes** **Year:** 2009, 2010 **Model:** Land Cruiser **Engine:** All V8 **Transmission:** All	**Oxygen Sensor Heater Control Circuit Low (Bank 2 Sensor 2):** With battery voltage between 10.5 to 20 V. The Heated Oxygen (HO2) sensor heater current is below 0.3 A. **Possible Causes:** • Open in HO2 sensor heater circuit • HO2 sensor heater (for Sensor 2) • Integration relay • ECM
DTC: P0057 **1T ECM, MIL: Yes** **Year:** 2009, 2010 **Model:** 4Runner, FJ Cruiser **Engine:** All L4 **Transmission:** All	**Oxygen Sensor Heater Control Circuit Low (Bank 2 Sensor 2):** Case 1: Monitor runs whenever following DTCs not stored: P0032, P0052 (Air fuel ratio sensor heater) P0038, P0058 (Rear oxygen sensor heater) Battery voltage: 10.5 V or higher Engine: Running Starter: OFF Catalyst active A/F control: Not operating Time after heater ON: 10 seconds or more Learned heater OFF current operation: Complete Heater OFF current learned value: Acquired Case 2: Monitor runs whenever following DTCs not stored: P0031, P0032, P0051, P0052 (Air fuel ratio sensor heater) P0038, P0058 (Rear oxygen sensor heater) Battery voltage: 10.5 V or higher Engine: Running Starter: OFF Catalyst active A/F control: Not operating Time after heater ON: 10 seconds or more Learned heater OFF current operation: Complete Heated oxygen sensor heater OFF current: Higher than 3.5 A Hybrid IC high current limiter port: Fail **Possible Causes:** • Open in heated oxygen sensor heater circuit • Heated oxygen sensor heater (sensor 2) • No. 1 integration relay (EFI MAIN) • ECM
DTC: P0058 **1T ECM, MIL: Yes** **Year:** 2009, 2010 **Model:** 4Runner, FJ Cruiser **Engine:** All V8 **Transmission:** All	**Oxygen Sensor Heater Control Circuit High (Bank 2 Sensor 2):** Monitor runs whenever following DTCs not stored: None Battery voltage: 10.5 V or higher Engine: Running Starter: OFF Catalyst active A/F control: Not operating Time after heater ON: 10 seconds or more Learned heater OFF current operation: Complete **Possible Causes:** • Short in heated oxygen sensor heater circuit • Heated oxygen sensor heater (sensor 2) • No. 1 integration relay (EFI MAIN) • ECM

DTC	Trouble Code Title, Conditions & Possible Causes
DTC: P0058 **1T ECM, MIL: Yes** **Year:** 2009, 2010 **Model:** Highlander **Engine:** All V6 **Transmission:** All	**Oxygen Sensor Heater Control Circuit High (Bank 2 Sensor 2):** Monitor runs whenever following DTCs are not present: None Case 1: Battery voltage: 10.5 V or more Engine: Running Starter: OFF Case 2: Battery voltage: 10.5 to 20 V **Possible Causes:** • Short in heated oxygen sensor heater circuit • Heated oxygen sensor heater • IGCT relay • Hybrid vehicle control ECU
DTC: P0100 **1T ECM, MIL: Yes** **Year:** 2009 **Model:** Camry **Engine:** All L4 **Transmission:** All	**Mass or Volume Air Flow Circuit:** Monitor runs whenever following DTCs are not present: None MAF meter voltage: Less than 0.2 V, or more than 4.9 V **Possible Causes:** • Open or short in MAF meter circuit • MAF meter • ECM
DTC: P0100 **T ECM, MIL: Yes** **Year:** 2009 **Model:** 4Runner **Engine:** All V8 **Transmission:** All	**Mass Air Flow Circuit Malfunction:** Monitor runs whenever following DTCs not present: None The typical enabling condition is not available: - **Possible Causes:** • Open or short in mass air flow meter circuit • Mass air flow meter • ECM
DTC: P0101 **2T ECM, MIL: Yes** **Year:** 2009, 2010 **Model:** Land Cruiser **Engine:** All V8 **Transmission:** All	**Mass Air Flow Circuit Range / Performance Problem:** Conditions (a), (b), (c), (d) and (e) continue for more than 10 seconds: (a) The engine is running. (b) The engine coolant temperature is 70°C (158°F) or higher. (c) The Throttle Position (TP) sensor voltage is 0.2 V or higher and 3.6 V or less. (d) The average engine load value ratio is less than 0.85, or more than 1.15 (varies with estimated engine load). Average engine load value ratio = Average engine load based on MAF meter output / Average engine load estimated from driving conditions (e) The average air-fuel ratio is less than -15%, or more than 15%. **Possible Causes:** • Mass Air Flow (MAF) meter • Air induction system • PCV hose connections
DTC: P0101 **2T ECM, MIL: Yes** **Year:** 2009, 2010 **Model:** Highlander **Engine:** All V6 **Transmission:** All	**Mass Air Flow Circuit Range / Performance Problem:** Monitor runs whenever following DTCs are not present: None Throttle position (TP sensor voltage): 0.2 to 2 V Time after engine start: 5 seconds or more Battery voltage: 10.5 V or more Engine coolant temperature: 70°C (158°F) or more Intake air temperature sensor circuit: OK Engine coolant temperature sensor circuit: OK Crankshaft position sensor circuit: OK Throttle position sensor circuit: OK Canister pressure sensor circuit: OK EVAP leak detection pump: OK EVAP vent valve: OK **Possible Causes:** • Mass air flow meter • Intake system • PCV hose connections

DTC	Trouble Code Title, Conditions & Possible Causes
DTC: P0102 **1T ECM, MIL: Yes** **Year:** 2009, 2010 **Model:** Land Cruiser **Engine:** All V8 **Transmission:** All	**Mass or Volume Air Flow Circuit Low Input:** The Mass Air Flow (MAF) meter voltage is below 0.2 V for 3 seconds. **Possible Causes:** • Open or short in Mass Air Flow (MAF) meter circuit • MAF meter • ECM
DTC: P0102 **1T ECM, MIL: Yes** **Year:** 2009, 2010 **Model:** Highlander **Engine:** All V6 **Transmission:** All	**Mass or Volume Air Flow Circuit Low Input:** Monitor runs whenever following DTCs are not present: None **Possible Causes:** • Open in mass air flow meter circuit • Short in mass air flow meter circuit • Mass air flow meter • Hybrid vehicle control ECU • ECM
DTC: P0103 **1T ECM, MIL: Yes** **Year:** 2009, 2010 **Model:** Corolla **Engine:** All L4 **Transmission:** All	**Mass or Volume Air Flow Circuit High Input:** Monitor runs whenever following DTCs are not present: None Mass air flow meter voltage more than 4.9 V for 3 seconds. **Possible Causes:** • Short in MAF meter circuit (+B circuit) • MAF meter • ECM
DTC: P0103 **1T ECM, MIL: Yes** **Year:** 2009, 2010 **Model:** Highlander **Engine:** All V6 **Transmission:** All	**Mass or Volume Air Flow Circuit High Input:** Monitor runs whenever following DTCs are not present: None **Possible Causes:** • Short in mass air flow meter circuit (+B circuit) • Mass air flow meter • Hybrid vehicle control ECU • ECM
DTC: P0103 **1T ECM, MIL: Yes** **Year:** 2009, 2010 **Model:** Land Cruiser **Engine:** All V8 **Transmission:** All	**Mass or Volume Air Flow Circuit High Input:** The MAF meter voltage is higher than 4.9 V for 3 seconds. **Possible Causes:** • Open or short in MAF meter circuit • MAF meter • ECM
DTC: P0110 **1T ECM, MIL: Yes** **Year:** 2009, 2010 **Model:** Land Cruiser **Engine:** All V8 **Transmission:** All	**Intake Air Temperature Circuit Malfunction:** An open or short in the Intake Air Temperature (IAT) sensor circuit for 0.5 seconds. **Possible Causes:** • Open or short in Intake Air Temperature (IAT) sensor circuit • IAT sensor (built into Mass Air Flow [MAF] meter) • ECM
DTC: P0110 **1T ECM, MIL: Yes** **Year:** 2009, 2010 **Model:** Avalon, Camry **Engine:** All L4, V6 **Transmission:** All	**Intake Air Temperature Circuit Malfunction:** Open or short in IAT sensor circuit for 0.5 seconds. **Possible Causes:** • Open or short in IAT sensor circuit • IAT sensor (built into MAF meter) • ECM
DTC: P0111 **2T ECM, MIL: Yes** **Year:** 2009, 2010 **Model:** Matrix **Engine:** 1.8L L4 VIN U, 2.4L L4 **Transmission:** All	**Intake Air Temperature Sensor Gradient Too High:** When either of following conditions is met: In duration between engine warmed up and next engine starts, change in intake air temperature sensor output bellow threshold. During engine warming up after cold engine starts, change in intake air temperature sensor output bellow threshold. **Possible Causes:** • Mass air flow meter

DTC	Trouble Code Title, Conditions & Possible Causes
DTC: P0111 **2T ECM, MIL: Yes** **Year:** 2009, 2010 **Model:** Land Cruiser **Engine:** All V8 **Transmission:** All	**Intake Air Temperature Sensor 1 Circuit Range / Performance:** When either of following conditions are met: 1). In duration between engine warmed up and next engine starts, change in Intake Air Temperature (IAT) sensor output below threshold. 2). During engine warming up after cold engine starts, change in IAT sensor output below threshold. **Possible Causes:** • Mass air flow meter
DTC: P0111 **2T ECM, MIL: Yes** **Year:** 2009, 2010 **Model:** Highlander **Engine:** All V6 **Transmission:** All	**Intake Air Temperature Sensor Gradient Too High:** Monitor runs whenever following DTCs are not present: None After engine stop: Time after engine start: 10 seconds or more Battery voltage: 10.5 V or more Engine coolant temperature sensor circuit: OK Accumulated mass air flow amount before engine stop: 3000 g or more Key-off duration: 30 minutes After cold engine start: Key-off duration: 5 hours Time after engine start: 10 seconds or more Engine coolant temperature sensor circuit: OK Engine coolant temperature: 70°C (158°F) or more Accumulated mass air flow amount: 3000 g or more Either of the following conditions 1 or 2 is met: 1. Duration while engine load is low: 120 seconds or more 2. Duration while engine load is high: 10 seconds or more **Possible Causes:** • Mass air flow meter
DTC: P0112 **1T ECM, MIL: Yes** **Year:** 2009, 2010 **Model:** Land Cruiser **Engine:** All V8 **Transmission:** All	**Intake Air Temperature Circuit Low Input:** A short in the IAT sensor circuit for 0.5 seconds. **Possible Causes:** • Short in IAT sensor circuit • IAT sensor (built into MAF meter) • ECM
DTC: P0112 **1T ECM, MIL: Yes** **Year:** 2009, 2010 **Model:** Highlander **Engine:** All V6 **Transmission:** All	**Intake Air Temperature Circuit Low Input:** Monitor runs whenever following DTCs are not present: None Battery voltage: 8 V or more Power switch: On (IG) Starter: OFF **Possible Causes:** • Short in intake air temperature sensor circuit • Intake air temperature sensor (built into mass air flow meter) • Hybrid vehicle control ECU • ECM
DTC: P0113 **1T ECM, MIL: Yes** **Year:** 2009, 2010 **Model:** Land Cruiser **Engine:** All V8 **Transmission:** All	**Intake Air Temperature Circuit High Input:** An open in the IAT sensor circuit for 0.5 seconds. **Possible Causes:** • Open in IAT sensor circuit • IAT sensor (built into MAF meter) • ECM
DTC: P0113 **1T ECM, MIL: Yes** **Year:** 2009, 2010 **Model:** Highlander **Engine:** All V6 **Transmission:** All	**Intake Air Temperature Circuit High Input:** Monitor runs whenever following DTCs are not present: None Battery voltage: 8 V or more Power switch: On (IG) Starter: OFF **Possible Causes:** • Open in intake air temperature sensor circuit • Intake air temperature sensor (built into mass air flow meter) • Hybrid vehicle control ECU • ECM

DTC	Trouble Code Title, Conditions & Possible Causes
DTC: P0115 **1T ECM, MIL: Yes** **Year:** 2009, 2010 **Model:** Land Cruiser **Engine:** All V8 **Transmission:** All	**Engine Coolant Temperature Circuit Malfunction:** An open or short in the Engine Coolant Temperature (ECT) sensor circuit for 0.5 seconds. **Possible Causes:** • Open or short in Engine Coolant Temperature (ECT) sensor circuit • ECT sensor
DTC: P0115 **1T ECM, MIL: Yes** **Year:** 2009, 2010 **Model:** Highlander **Engine:** All V6 **Transmission:** All	**Engine Coolant Temperature Circuit Malfunction:** Monitor runs whenever following DTCs are not present: None **Possible Causes:** • Open or short in engine coolant temperature sensor circuit • Engine coolant temperature sensor • Hybrid vehicle control ECU • ECM
DTC: P0116 **2T ECM, MIL: Yes** **Year:** 2009, 2010 **Model:** Matrix **Engine:** All L4 **Transmission:** All	**Engine Coolant Temperature Circuit Range / Performance Problem:** Case 1: Engine Coolant Temperature (ECT) is between 35°C and 60°C (95°F and 140°F) when engine started, and conditions (a) and (b) are met: (a) Vehicle driven at varying speeds (accelerated and decelerated) (b) ECT change is within 3°C (5.4°F) of initial ECT Case 2: ECT is more than 60°C (140°F) when engine started, and conditions (a) and (b) are met (6 trip detection logic): (a) Vehicle driven at varying speeds (accelerated and decelerated (b) ECT measurements change is within 1°C (1.8°F) of initial ECT on 6 successive occasions **Possible Causes:** • Thermostat • Engine coolant temperature sensor
DTC: P0116 **2T ECM, MIL: Yes** **Year:** 2009, 2010 **Model:** Highlander **Engine:** All V6 **Transmission:** All	**Engine Coolant Temperature Circuit Range / Performance Problem:** Engine coolant temperature sensor cold start monitor: Monitor runs whenever following DTCs are not present: None Battery voltage: 10.5 V or more Time after engine start: 1 second or more Engine coolant temperature at engine start: Less than 60°C (140°F) Intake air temperature sensor circuit: OK Soak time: 0 seconds or more Accumulated mass air flow: 1397 g or more Engine: Running Fuel cut: OFF Difference between engine coolant temperature at engine start and intake air temperature: Less than 40°C (72°F) Engine coolant temperature sensor soak monitor: Monitor runs whenever following DTCs are not present: P0100 to P0103: MAF meter P0110 to P0113: Intake air temperature sensor Battery voltage: 10.5 V or more Engine: Running Soak time: 5 hours or more Either of the following conditions (a) or (b) is met: (a) Engine coolant temperature: 60°C (140°F) or more (b) Accumulated MAF: 2921 g or more **Possible Causes:** • Thermostat • Engine coolant temperature sensor

DTC	Trouble Code Title, Conditions & Possible Causes
DTC: P0116 **2T ECM, MIL: Yes** **Year:** 2009, 2010 **Model:** Land Cruiser **Engine:** All V8 **Transmission:** All	**Engine Coolant Temperature Circuit Range / Performance Problem:** When either of the following conditions is met: 1). When a cold engine is started and the engine is warmed up, the Engine Coolant Temperature (ECT) sensor value does not change. 2). After the warmed up engine is stopped and then the next cold engine start is performed, the ECT sensor value does not change. **Possible Causes:** • Water inlet sub-assembly with thermostat • ECT sensor
DTC: P0117 **1T ECM, MIL: Yes** **Year:** 2009, 2010 **Model:** Corolla, Matrix **Engine:** All L4 **Transmission:** All	**Engine Coolant Temperature Circuit Low Input:** Short in ECT sensor circuit for 0.5 seconds. The engine coolant temperature sensor voltage is less than 0.14 V [More than 140°C (284°F)]. **Possible Causes:** • Short in engine coolant temperature sensor circuit • Engine coolant temperature sensor • ECM
DTC: P0117 **1T ECM, MIL: Yes** **Year:** 2009, 2010 **Model:** Land Cruiser **Engine:** All V8 **Transmission:** All	**Engine Coolant Temperature Circuit Low Input:** A short in the ECT sensor circuit for 0.5 seconds. **Possible Causes:** • Short in ECT sensor • ECT sensor • ECM
DTC: P0117 **1T ECM, MIL: Yes** **Year:** 2009, 2010 **Model:** Highlander **Engine:** All V6 **Transmission:** All	**Engine Coolant Temperature Circuit Low Input:** Monitor runs whenever following DTCs are not present: None **Possible Causes:** • Short in engine coolant temperature sensor • Engine coolant temperature sensor • Hybrid vehicle control ECU • ECM
DTC: P0118 **1T ECM, MIL: Yes** **Year:** 2009, 2010 **Model:** Venza **Engine:** All L4 **Transmission:** All	**Engine Coolant Temperature Circuit High Input:** Monitor runs whenever following DTCs are not present: None **Possible Causes:** • Open in engine coolant temperature sensor circuit • Engine coolant temperature sensor • ECM
DTC: P0118 **1T ECM, MIL: Yes** **Year:** 2009, 2010 **Model:** Land Cruiser **Engine:** All V8 **Transmission:** All	**Engine Coolant Temperature Circuit High Input:** An open in the ECT sensor circuit for 0.5 seconds. **Possible Causes:** • Open in ECT sensor circuit • ECT sensor • ECM
DTC: P0118 **1T ECM** **Year:** 2009, 2010 **Model:** Highlander **Engine:** All V6 **Transmission:** All	**Engine Coolant Temperature Circuit High Input:** Monitor runs whenever following DTCs are not present: None **Possible Causes:** • Open in engine coolant temperature sensor circuit • Engine coolant temperature sensor • Hybrid vehicle control ECU • ECM

DTC	Trouble Code Title, Conditions & Possible Causes
DTC: P011B **2T ECM, MIL: Yes** **Year:** 2009, 2010 **Model:** Land Cruiser **Engine:** All V8 **Transmission:** All	**Engine Coolant Temperature / Intake Air Temperature Correlation:** All of the following conditions are met: a. The battery voltage is 10.5 V or higher. b. 7 hours or more have elapsed from engine stop on the previous trip. c. 15 seconds after a cold engine start. d. The minimum Intake Air Temperature (IAT) after the engine starts is higher than -10°C (14°F). e. The average Engine Coolant Temperature (ECT) before the engine starts is higher than -10°C (14°F). f. The difference between the readings of ECT and IAT is higher than 20°C (36°F). **Possible Causes:** • IAT sensor • ECT sensor • ECM
DTC: P011B **2T ECM, MIL: Yes** **Year:** 2009, 2010 **Model:** Highlander **Engine:** All V6 **Transmission:** All	**Engine Coolant Temperature / Intake Air Temperature Correlation:** The monitor will run whenever these DTCs are not present: None All of the following conditions are met: Conditions 1 and 2 1. All of the following conditions are met: Conditions (a), (b), (c) and (d) (a) After power switch on (IG) and engine not running time: Less than 20 seconds (b) Soak time: 7 hours or more (c) Battery voltage: 10.5 V or more (d) Time after engine start: 76 seconds or more 2. Either of the following conditions is met: Conditions (a) and (b) (a) Minimum intake air temperature after engine start: -10°C (14°F) or more (b) Engine coolant temperature before engine start: -10°C (14°F) or more **Possible Causes:** • Intake air temperature sensor • Engine coolant temperature sensor • Hybrid vehicle control ECU • ECM
DTC: P0120 **1T ECM, MIL: Yes** **Year:** 2009, 2010 **Model:** Corolla **Engine:** All L4 **Transmission:** All	**Throttle / Pedal Position Sensor / Switch "A" Circuit:** Ignition switch ON for 0.012 seconds or more. Electronic throttle actuator power ON. The output voltage of VTA1 quickly fluctuates beyond lower and upper malfunction thresholds for 2 seconds or more when accelerator pedal depressed. **Possible Causes:** • Throttle position sensor (built into throttle body) • ECM
DTC: P0120 **1T ECM, MIL: Yes** **Year:** 2009, 2010 **Model:** Highlander **Engine:** All V6 **Transmission:** All	**Throttle / Pedal Position Sensor / Switch "A" Circuit Malfunction:** Monitor runs whenever following DTCs are not present: None Power switch: ON Electronic throttle actuator power: ON **Possible Causes:** • Throttle position sensor (built into throttle body) • Hybrid vehicle control ECU • ECM
DTC: P0120 **1T ECM, MIL: Yes** **Year:** 2009, 2010 **Model:** Land Cruiser **Engine:** All V8 **Transmission:** All	**Throttle / Pedal Position Sensor / Switch "A" Circuit Malfunction:** With either of following conditions A or B met: A. Engine switch ON. B. Electronic throttle actuator power ON. The output voltage of VTA1 quickly fluctuates beyond lower and upper malfunction thresholds for 2 seconds or more. **Possible Causes:** • Throttle Position (TP) sensor (built into throttle body) • ECM

DTC	Trouble Code Title, Conditions & Possible Causes
DTC: P0121 **1T ECM, MIL: Yes** **Year:** 2009, 2010 **Model:** Corolla **Engine:** All L4 **Transmission:** All	**Throttle / Pedal Position Sensor / Switch "A" Circuit Range / Performance Problem:** With the ignition switch ON, electronic throttle motor power ON. Throttle position sensor malfunction (P0120, P0122, P0123, P0220, P0222, P0223, P2135) is not detected. The difference between VTA1 and VTA2 voltage is less than 0.8 V, or more than 1.6 V for 2 seconds. **Possible Causes:** • Throttle position sensor (built into throttle body) • Throttle position sensor circuit • ECM
DTC: P0121 **1T ECM, MIL: Yes** **Year:** 2009, 2010 **Model:** Highlander **Engine:** All V6 **Transmission:** All	**Throttle / Pedal Position Sensor / Switch "A" Circuit Range / Performance Problem:** Monitor runs whenever following DTCs are not present: None Either of following conditions A or B is met: A. Power switch: ON B. Electric throttle motor power: ON Throttle position sensor malfunction (P0120, P0122, P0123, P0220, P0222, P0223, P2135): Not detected **Possible Causes:** • Throttle position sensor (built into throttle body)
DTC: P0121 **1T ECM, MIL: Yes** **Year:** 2009, 2010 **Model:** Land Cruiser **Engine:** All V8 **Transmission:** All	**Throttle / Pedal Position Sensor / Switch "A" Circuit Range / Performance Problem:** Either of following conditions A or B met: A. Engine switch On (IG). B. Electronic throttle motor power ON, TP sensor malfunction (P0120, P0122, P0123, P0220, P0222, P0223, P2135) Not detected. The difference between the VTA1 and VTA2 voltages is below 0.8 V, or higher than 1.6 V for 2 seconds. **Possible Causes:** • TP sensor (built into throttle body) • TP sensor circuit • ECM
DTC: P0122 **1T ECM, MIL: Yes** **Year:** 2009, 2010 **Model:** Avalon, Camry **Engine:** All L4 **Transmission:** All	**Throttle / Pedal Position Sensor / Switch "A" Circuit Low Input:** Monitor runs whenever following DTCs are not present: None Either of the following conditions A or B is met: A. Engine switch on (IG): 0.012 seconds or more B. Electronic throttle actuator power: ON **Possible Causes:** • Throttle position sensor (built into throttle body) • Short in VTA1 circuit • Open in VC circuit • ECM
DTC: P0122 **1T ECM, MIL: Yes** **Year:** 2009, 2010 **Model:** Highlander **Engine:** All V6 **Transmission:** All	**Throttle / Pedal Position Sensor / Switch "A" Circuit Low Input:** Monitor runs whenever following DTCs are not present: None Power switch: ON Electronic throttle actuator power: ON **Possible Causes:** • Throttle position sensor (built into throttle body) • Short in VTA1 circuit • Open in VC circuit • Hybrid vehicle control ECU • ECM
DTC: P0122 **1T ECM, MIL: Yes** **Year:** 2009, 2010 **Model:** Land Cruiser **Engine:** All V8 **Transmission:** All	**Throttle / Pedal Position Sensor / Switch "A" Circuit Low Input:** Either of following conditions A or B met A. Engine switch ON (IG). B. Electronic throttle actuator power ON. The output voltage of VTA1 is 0.2 V or less for 2 seconds or more. **Possible Causes:** • TP sensor (built into throttle body) • Short in VTA1 circuit • Open in VC circuit • ECM

DTC	Trouble Code Title, Conditions & Possible Causes
DTC: P0123 **1T ECM, MIL: Yes** **Year:** 2009, 2010 **Model:** Highlander **Engine:** All V6 **Transmission:** All	**Throttle / Pedal Position Sensor / Switch "A" Circuit High Input:** Monitor runs whenever following DTCs are not present: None Power switch: ON Electronic throttle actuator power: ON **Possible Causes:** • Throttle position sensor (built into throttle body) • Open in VTA1 circuit • Open in E2 circuit • Short between VC and VTA1 circuits • Hybrid vehicle control ECU • ECM
DTC: P0123 **1T ECM, MIL: Yes** **Year:** 2009, 2010 **Model:** Land Cruiser **Engine:** All V8 **Transmission:** All	**Throttle / Pedal Position Sensor / Switch "A" Circuit High Input:** Either of following conditions A or B met: A. Engine switch ON (IG). B. Electronic throttle actuator power ON. The output voltage of VTA1 is 4.535 V or more for 2 seconds or more. **Possible Causes:** • TP sensor (built into throttle body) • Open in VTA1 circuit • Open in E2 circuit • Short between VC and VTA1 circuits • ECM
DTC: P0125 **2T ECM, MIL: Yes** **Year:** 2009, 2010 **Model:** Highlander **Engine:** All V6 **Transmission:** All	**Insufficient Coolant Temperature for Closed Loop Fuel Control:** Monitor runs whenever following DTCs are not present: None Thermostat: OK Mass air flow meter circuit: OK Intake air temperature sensor circuit: OK Engine coolant temperature sensor circuit: OK **Possible Causes:** • Engine coolant temperature sensor • Cooling system • Thermostat
DTC: P0125 **2T ECM, MIL: Yes** **Year:** 2009, 2010 **Model:** Land Cruiser **Engine:** All V8 **Transmission:** All	**Insufficient Coolant Temperature for Closed Loop Fuel Control:** The Engine Coolant Temperature (ECT) does not reach the closed loop enabling temperature for 20 minutes (this period varies with the engine start ECT). **Possible Causes:** • Engine Coolant Temperature (ECT) sensor • Cooling system • Water inlet sub-assembly with thermostat
DTC: P0128 **2T ECM, MIL: Yes** **Year:** 2009, 2010 **Model:** Avalon, Camry, Matrix **Engine:** All V6 **Transmission:** All	**Coolant Thermostat (Coolant Temperature Below Thermostat Regulating Temperature):** Battery voltage: 11V or more Either following condition 1 or 2 is met: 1. All of the following conditions are met: *ECT at engine start - IAT at engine start: 5-44.6°F (-15-7°C) *ECT at engine start: 14-132.8°F (-10-56°C) *IAT at engine start: 14-132.8°F (-10-56°C) 2. All of following conditions are met: *ECT at engine start - IAT at engine start: More than 44.6°F (7°C) *ECT at engine start: 132.8°F (56°C) or less *IAT at engine start: 14°F (-10°C) or more Accumulated time that vehicle speed is 80 mph (128 km/h) or more speed: Less than 20 seconds **Possible Causes:** • Thermostat • Cooling system • ECT sensor • ECM

DTC	Trouble Code Title, Conditions & Possible Causes
DTC: P0128 **2T ECM, MIL: Yes** **Year:** 2009, 2010 **Model:** Corolla **Engine:** All L4 **Transmission:** All	**Coolant Thermostat (Coolant Temperature Below Thermostat Regulating Temperature):** The following conditions are met for 5 seconds. Cold start Engine is warmed up ECT is less than 75°C (167°F) The engine coolant temperature does not reach 75°C (167°F) despite sufficient engine warm-up time having elapsed. **Possible Causes:** • Thermostat • Cooling system • ECT sensor • ECM
DTC: P0128 **2T ECM, MIL: Yes** **Year:** 2009, 2010 **Model:** Land Cruiser **Engine:** All V8 **Transmission:** All	**Coolant Thermostat (Coolant Temperature Below Thermostat Regulating Temperature):** Conditions (a), (b) and (c) are met for 5 seconds: (a) Cold start. (b) The engine is warmed up. (c) The ECT is below 75°C (167°F). **Possible Causes:** • Water inlet sub-assembly with thermostat • Cooling system • ECT sensor • ECM
DTC: P0136 **2T ECM, MIL: Yes** **Year:** 2009, 2010 **Model:** Matrix **Engine:** All L4 **Transmission:** All	**Oxygen Sensor Circuit Malfunction (Bank 1 Sensor 2):** Abnormal voltage output: During active air fuel ratio control, following conditions (a) and (b) met for certain period of time: (a) Heated oxygen sensor voltage does not decrease to less than 0.21 V (b) Heated oxygen sensor voltage does not increase to more than 0.59 V Low impedance: Sensor impedance less than 5 W for more than 30 seconds when ECM presumes sensor to being warmed up and operating normally. **Possible Causes:** • Open or short in heated oxygen sensor (bank 1 sensor 2) circuit • Heated oxygen sensor (bank 1 sensor 2) • Heated oxygen sensor heater (bank 1 sensor 2) • Air fuel ratio sensor (bank 1 sensor 1) • Gas leaks from exhaust system

DTC	Trouble Code Title, Conditions & Possible Causes
DTC: P0136 **2T ECM, MIL: Yes** **Year:** 2009, 2010 **Model:** Highlander **Engine:** All V6 **Transmission:** All	**Oxygen Sensor Circuit Malfunction (Bank 1 Sensor 2):** Heated oxygen sensor output voltage (Output voltage, High voltage, Low voltage): Active air fuel ratio control: Performing Active air fuel ratio control begins when all of the following conditions are met: Battery voltage: 11 V or more Engine coolant temperature: 75°C (167°F) or more Idle: OFF Engine RPM: Less than 3200 rpm Air fuel ratio sensor status: Activated Fuel system status: Close-loop Fuel-cut: OFF Engine load: 5 to 80% Heated oxygen sensor impedance (Low): Battery voltage: 11 V or more Estimated heated oxygen sensor temperature: Less than 700°C (1292°F) ECM monitor: Completed P0606: Not set Heated oxygen sensor impedance (High): Battery voltage: 11 V or more Estimated heated oxygen sensor temperature: 450°C (842°F) or more P0606: Not set Heated oxygen sensor output voltage (Extremely high): Battery voltage: 11 V or more Time after engine start: 2 seconds or more Heated oxygen sensor voltage during fuel cut: Engine coolant temperature: 70°C (158°F) or more Catalyst temperature: 530°C (986°F) or more Fuel cut: ON **Possible Causes:** • Heated oxygen sensor (bank 1, 2 sensor 2) • Air fuel ratio sensor (bank 1, 2 sensor 2) • Gas leak from exhaust system • Fuel pressure • Fuel injector • PCV valve and hose • Intake system
DTC: P0136 **2T ECM, MIL: Yes** **Year:** 2009, 2010 **Model:** Land Cruiser **Engine:** All V8 **Transmission:** All	**Oxygen Sensor Circuit Malfunction (Bank 1 Sensor 2):** During active air-fuel ratio control, the following conditions (a) and (b) are met for a certain period of time: (a) The Heated Oxygen (HO2) sensor voltage does not decrease to below 0.59 V. (b) The sensor impedance is below 5 W for more than 30 seconds when the ECM presumes the sensor to be warmed up and operating normally. **Possible Causes:** • Open or short in HO2 sensor (for Bank 1) circuit • HO2 sensor (for Bank 1) • HO2 sensor heater (for Bank 1) • Air-fuel Ratio (A/F) sensor (for Bank 1) • Integration relay • Gas leak from exhaust syste

DTC	Trouble Code Title, Conditions & Possible Causes
DTC: P0137 **2T ECM, MIL: Yes** **Year:** 2009, 2010 **Model:** Matrix **Engine:** All L4 **Transmission:** All	**Oxygen Sensor Circuit Low Voltage (Bank 1 Sensor 2):** Low voltage (open): During active air fuel ratio control, following conditions (a) and (b) met for certain period of time: (a) Heated oxygen sensor voltage output less than 0.21 V (b) Target air fuel ratio rich High impedance: Sensor impedance 15 kW or more for more than 90 seconds when ECM presumes sensor to be warmed up and operating normally. **Possible Causes:** • Open in heated oxygen sensor (bank 1 sensor 2) circuit • Heated oxygen sensor (bank 1 sensor 2) • Heated oxygen sensor heater (bank 1 sensor 2) • Gas leaks from exhaust system
DTC: P0137 **2T ECM, MIL: Yes** **Year:** 2009, 2010 **Model:** Highlander **Engine:** All V6 **Transmission:** All	**Oxygen Sensor Circuit Low Voltage (Bank 1 Sensor 2):** Heated oxygen sensor output voltage (Output voltage, High voltage, Low voltage): Active air fuel ratio control: Performing Active air fuel ratio control begins when all of the following conditions are met: Battery voltage: 11 V or more Engine coolant temperature: 75°C (167°F) or more Idle: OFF Engine RPM: Less than 3200 rpm Air fuel ratio sensor status: Activated Fuel system status: Close-loop Fuel-cut: OFF Engine load: 5 to 80% Heated oxygen sensor impedance (Low): Battery voltage: 11 V or more Estimated heated oxygen sensor temperature: Less than 700°C (1292°F) ECM monitor: Completed P0606: Not set Heated oxygen sensor impedance (High): Battery voltage: 11 V or more Estimated heated oxygen sensor temperature: 450°C (842°F) or more P0606: Not set Heated oxygen sensor output voltage (Extremely high): Battery voltage: 11 V or more Time after engine start: 2 seconds or more Heated oxygen sensor voltage during fuel cut: Engine coolant temperature: 70°C (158°F) or more Catalyst temperature: 530°C (986°F) or more Fuel cut: ON **Possible Causes:** • Open or short in heated oxygen sensor (bank1, sensor 2) circuit • Heated oxygen sensor (bank1, sensor 2) • Heated oxygen sensor heater (bank1, sensor 2) • Air fuel ratio sensor (bank1, sensor 1) • Gas leak from exhaust system
DTC: P0137 **2T ECM, MIL: Yes** **Year:** 2009, 2010 **Model:** Land Cruiser **Engine:** All V8 **Transmission:** All	**Oxygen Sensor Circuit Low Voltage (Bank 1 Sensor 2):** During active air-fuel ratio control, the following conditions (a) and (b) are met for a certain period of time: (a) The HO2 sensor voltage output is below 0.21 V. (b) The sensor impedance is 15 kW or higher for more than 90 seconds when the ECM presumes the sensor to be warmed up and operating normally. **Possible Causes:** • Open in HO2 sensor (for Bank 1) circuit • HO2 sensor (for Bank 1) • HO2 sensor heater (for Bank 1) • Integration relay • Gas leak from exhaust system

DTC	Trouble Code Title, Conditions & Possible Causes
DTC: P0138 **2T ECM, MIL: Yes** **Year:** 2009, 2010 **Model:** Avalon, Camry **Engine:** All L4 **Transmission:** All	**Oxygen Sensor Circuit High Voltage (Bank 1 Sensor 2):** Active A/F control: Performing Battery voltage: 11 V or more ECT: 167°F (75°C) or more Idle: OFF Engine rpm: Less than 3200 rpm A/F sensor status: Activatted Fuel system status: Closed loop Fuel cut: OFF Engineload: 10 to 70% Shift position: 4rd or more Battery voltage: 11 V or more Time after engine start: 2 seconds or more **Possible Causes:** • Short in HO2 sensor (sensor 2) circuit • HO2 sensor (sensor 2) • ECM internal circuit malfunction • A/F sensor (sensor 1)
DTC: P0138 **2T ECM, MIL: Yes** **Year:** 2009, 2010 **Model:** Highlander **Engine:** All V6 **Transmission:** All	**Oxygen Sensor Circuit High Voltage (Bank 1 Sensor 2):** Heated oxygen sensor output voltage (Output voltage, High voltage, Low voltage): Active air fuel ratio control: Performing Active air fuel ratio control begins when all of the following conditions are met: Battery voltage: 11 V or more Engine coolant temperature: 75°C (167°F) or more Idle: OFF Engine RPM: Less than 3200 rpm Air fuel ratio sensor status: Activated Fuel system status: Close-loop Fuel-cut: OFF Engine load: 5 to 80% Heated oxygen sensor impedance (Low): Battery voltage: 11 V or more Estimated heated oxygen sensor temperature: Less than 700°C (1292°F) ECM monitor: Completed P0606: Not set Heated oxygen sensor impedance (High): Battery voltage: 11 V or more Estimated heated oxygen sensor temperature: 450°C (842°F) or more P0606: Not set Heated oxygen sensor output voltage (Extremely high): Battery voltage: 11 V or more Time after engine start: 2 seconds or more Heated oxygen sensor voltage during fuel cut: Engine coolant temperature: 70°C (158°F) or more Catalyst temperature: 530°C (986°F) or more Fuel cut: ON **Possible Causes:** • Short in heated oxygen sensor (bank1, 2 sensor 2) circuit • Heated oxygen sensor (bank1, 2 sensor 2) • Hybrid vehicle control ECU internal circuit malfunction • Air fuel ratio sensor (bank1, 2 sensor 1) • ECM

DTC	Trouble Code Title, Conditions & Possible Causes
DTC: P0138 **2T ECM, MIL: Yes** **Year:** 2009, 2010 **Model:** Land Cruiser **Engine:** All V8 **Transmission:** All	**Oxygen Sensor Circuit High Voltage (Bank 1 Sensor 2):** During active air-fuel ratio control, the following conditions (a) and (b) are met for a certain period of time: (a) The HO2 sensor voltage output is higher than 0.59 V. (b) The HO2 sensor voltage output exceeds 1.2 V for more than 10 seconds. **Possible Causes:** • Short in HO2 sensor (for Bank 1) circuit • HO2 sensor (for Bank 1) • Air-Fuel ratio (A/F) sensor (for Bank 1) • ECM
DTC: P0139 **2T ECM, MIL: Yes** **Year:** 2009, 2010 **Model:** Corolla, Matrix **Engine:** 1.8L L4 VIN L, 1.8L L4 VIN U, 2.4L L4 **Transmission:** All	**Oxygen Sensor Circuit Slow Response (Bank 1 Sensor 2):** The Heated oxygen sensor (sensor 2) voltage does not drop to below 0.2 V immediately after fuel cut starts. The heated oxygen sensor (sensor 2) voltage does not drop from 0.35 V to 0.2 V immediately after fuel cut starts. **Possible Causes:** • Short in heated oxygen sensor (bank 1 sensor 2) circuit • Heated oxygen sensor (bank 1 sensor 2) • ECM
DTC: P0139 **2T ECM, MIL: Yes** **Year:** 2009, 2010 **Model:** Land Cruiser **Engine:** All V8 **Transmission:** All	**Oxygen Sensor Circuit Slow Response (Bank 1 Sensor 2):** The HO2 sensor voltage does not drop below 0.2 V immediately after fuel cut starts. The HO2 sensor voltage does not drop from 0.35 V to 0.2 V immediately after fuel cut starts. **Possible Causes:** • Short in HO2 sensor (for Bank 1 Sensor 2) circuit • HO2 sensor (for Bank 1 Sensor 2) • ECM
DTC: P0139 **2T ECM, MIL: Yes** **Year:** 2009, 2010 **Model:** Highlander **Engine:** All V6 **Transmission:** All	**Oxygen Sensor Circuit Slow Response (Bank 1 Sensor 2):** Heated oxygen sensor output voltage (Output voltage, High voltage, Low voltage): Active air fuel ratio control: Performing Active air fuel ratio control begins when all of the following conditions are met: Battery voltage: 11 V or more Engine coolant temperature: 75°C (167°F) or more Idle: OFF Engine RPM: Less than 3200 rpm Air fuel ratio sensor status: Activated Fuel system status: Close-loop Fuel-cut: OFF Engine load: 5 to 80% Heated oxygen sensor impedance (Low): Battery voltage: 11 V or more Estimated heated oxygen sensor temperature: Less than 700°C (1292°F) ECM monitor: Completed P0606: Not set Heated oxygen sensor impedance (High): Battery voltage: 11 V or more Estimated heated oxygen sensor temperature: 450°C (842°F) or more P0606: Not set Heated oxygen sensor output voltage (Extremely high): Battery voltage: 11 V or more Time after engine start: 2 seconds or more Heated oxygen sensor voltage during fuel cut: Engine coolant temperature: 70°C (158°F) or more Catalyst temperature: 530°C (986°F) or more Fuel cut: ON **Possible Causes:** • Short in heated oxygen sensor (bank1, sensor 2) circuit • Heated oxygen sensor (bank1, sensor 2) • Hybrid vehicle control ECU internal circuit malfunction • ECM

DTC	Trouble Code Title, Conditions & Possible Causes
DTC: P0141 **2T ECM, MIL: Yes** **Year:** 2009, 2010 **Model:** Avalon, Camry, Corolla, Matrix **Engine:** All L4 **Transmission:** All	**Oxygen Sensor Heater Circuit Malfunction (Bank 1 Sensor 2):** One of the following conditions is met: Condition A or B A. All of the following conditions are met: Conditions 1, 2, 3, 4 and 5 1. Battery voltage: 10.5 V or more 2. Fuel cut: OFF 3. Time after fuel cut ON to OFF: 30 seconds or more 4. Accumulated heater ON time: 100 seconds or more 5. Learned heater OFF current operation: Completed B. Duration that rear heated oxygen sensor impedance is less than 15 kW: 2 seconds or more **Possible Causes:** • Open or short in HO2 sensor heater circuit • HO2 sensor heater (sensor 2) • EFI relay • ECM
DTC: P0141 **2T ECM, MIL: Yes** **Year:** 2009, 2010 **Model:** Highlander **Engine:** All V6 **Transmission:** All	**Oxygen Sensor Heater Circuit Malfunction (Bank 1 Sensor 2):** Monitor runs whenever following DTCs are not present: None All of the following conditions are met: - Battery voltage: 10.5 V or more Fuel cut: OFF Time after fuel cut ON to OFF: 30 seconds or more Accumulated heater ON time: 100 seconds or more **Possible Causes:** • Open or short in heated oxygen sensor heater circuit • Heated oxygen sensor heater • IGCT relay • Hybrid vehicle control ECU • ECM
DTC: P0141 **2T ECM, MIL: Yes** **Year:** 2009, 2010 **Model:** Land Cruiser **Engine:** All V8 **Transmission:** All	**Oxygen Sensor Heater Circuit Malfunction (Bank 1 Sensor 2):** The battery voltage is 10.5 V or more, fuel cut OFF, time after fuel cut ON to OFF 30 seconds or more. Accumulated heater ON time 100 seconds or more. The Cumulative heater resistance correction value exceeds the threshold, or the duration of the rear heated oxygen sensor impedance is less than 15 kW for 2 seconds or more. **Possible Causes:** • Open or short in HO2 sensor heater circuit • HO2 sensor heater (for Sensor2) • Integration relay • ECM
DTC: P014C **2T ECM, MIL: Yes** **Year:** 2009, 2010 **Model:** 4Runner, FJ Cruiser **Engine:** All L4,V6 **Transmission:** All	**A/F Sensor Slow Response - Rich to Lean Bank 1 Sensor 1:** Monitor runs whenever following DTCs not stored: None Active air fuel ratio control: Performing Active air fuel ratio control is performed when the following conditions met: Battery voltage: 11 V or higher Engine coolant temperature: 75°C (167°F) or higher Idle: OFF Engine speed: 1000 to 4000 rpm Air fuel ratio sensor status: Activated Fuel-cut: OFF Engine load: 10 to 70% Shift position: 2nd or higher Catalyst monitor: Not executed Mass air flow: 6 to 15 g/sec. **Possible Causes:** • Air fuel ratio sensor (bank 1 sensor 1) • Air fuel ratio sensor (bank 1 sensor 1) heater • ECM

DTC	Trouble Code Title, Conditions & Possible Causes
DTC: P014D **2T , MIL: Yes** **Year:** 2010 **Model:** Camry **Engine:** All L4, V6 **Transmission:** All	**A/F Sensor Slow Response - Lean to Rich Bank 1 Sensor 1:** Monitor runs whenever following DTCs are not stored: None Active air fuel ratio control: Performing Active air fuel ratio control is performed when the following conditions are met: Battery voltage: 11 V or higher Engine coolant temperature: 167°F (75°C) or more Idle: OFF Engine speed: 1000-4000 rpm Air fuel ratio sensor status: Activated Fuel cut: OFF Engine load: 10-70% Shift position: 2nd or higher Catalyst monitor: Not yet Mass air flow: 5-15 g/sec. Lean to Rich Response rate deterioration level: -0.038 V or higher **Possible Causes:** • Air fuel ratio sensor • Air fuel ratio sensor heater • ECM
DTC: P014F **2T ECM, MIL: Yes** **Year:** 2009, 2010 **Model:** 4Runner, FJ Cruiser **Engine:** All V6 **Transmission:** All	**A/F Sensor Slow Response - Lean to Rich Bank 2 Sensor 1:** Monitor runs whenever following DTCs not stored: None Active air fuel ratio control: Performing Active air fuel ratio control is performed when the following conditions met: Battery voltage: 11 V or higher Engine coolant temperature: 75°C (167°F) or higher Idle: OFF Engine speed: 1000 to 4000 rpm Air fuel ratio sensor status: Activated Fuel-cut: OFF Engine load: 10 to 70% Shift position: 2nd or higher Catalyst monitor: Not executed Mass air flow: 6 to 15 g/sec. **Possible Causes:** • Air fuel ratio sensor (bank 2 sensor 1) • Air fuel ratio sensor (bank 2 sensor 1) heater • ECM
DTC: P0156 **2T ECM, MIL: Yes** **Year:** 2009, 2010 **Model:** Land Cruiser **Engine:** All **Transmission:** All	**Oxygen Sensor Circuit Malfunction (Bank 2 Sensor 2):** During active air-fuel ratio control, the following conditions (a) and (b) are met for a certain period of time: (a) The Heated Oxygen (HO2) sensor voltage does not decrease to below 0.59 V. (b) The sensor impedance is below 5 W for more than 30 seconds when the ECM presumes the sensor to be warmed up and operating normally.. **Possible Causes:** • Open in HO2 sensor (for Bank 2) circuit • HO2 sensor (for Bank 2) • HO2 sensor heater (for Bank 2) • Integration relay • Gas leak from exhaust system

DTC	Trouble Code Title, Conditions & Possible Causes
DTC: P0157 **2T ECM, MIL: Yes** **Year:** 2009, 2010 **Model:** Land Cruiser **Engine:** All **Transmission:** All	**Oxygen Sensor Circuit Low Voltage (Bank 2 Sensor 2):** During active air-fuel ratio control, the following conditions (a) and (b) are met for a certain period of time: (a) The HO2 sensor voltage output is below 0.21 V. (b) The sensor impedance is 15 kW or higher for more than 90 seconds when the ECM presumes the sensor to be warmed up and operating normally. **Possible Causes:** • Open in HO2 sensor (for Bank 2) circuit • HO2 sensor (for Bank 2) • HO2 sensor heater (for Bank 2) • Integration relay • Gas leak from exhaust system
DTC: P0158 **2T ECM, MIL: Yes** **Year:** 2009, 2010 **Model:** Land Cruiser **Engine:** All **Transmission:** All	**Oxygen Sensor Circuit High Voltage (Bank 2 Sensor 2):** During active air-fuel ratio control, the following conditions (a) and (b) are met for a certain period of time: (a) The HO2 sensor voltage output is higher than 0.59 V. (b) The HO2 sensor voltage output exceeds 1.2 V for more than 10 seconds. **Possible Causes:** • Short in HO2 sensor (for Bank 2) circuit • HO2 sensor (for Bank 2) • Air-Fuel ratio (A/F) sensor (for Bank 2) • ECM
DTC: P0159 **2T ECM, MIL: Yes** **Year:** 2009, 2010 **Model:** Avalon, Camry **Engine:** All **Transmission:** All	**Oxygen Sensor Circuit Slow Response (Bank 2 Sensor 2):** ECT: 167°F (75°C) or more Estimated catalyst temperature: 752°F (400°C) or more Fuel cut: ON **Possible Causes:** • Short in HO2 sensor (bank 1, 2 sensor 2) • HO2 sensor (bank 1, 2 sensor 2) • ECM internal circuit malfunction
DTC: P015A **2T ECM, MIL: Yes** **Year:** 2009, 2010 **Model:** 4Runner, FJ Cruiser **Engine:** All **Transmission:** All	**A/F Sensor Delayed Response - Rich to Lean Bank 1 Sensor 1:** Monitor runs whenever following DTCs not stored: None Active air fuel ratio control: Performing Active air fuel ratio control is performed when the following conditions met: Battery voltage: 11 V or higher Engine coolant temperature: 75°C (167°F) or higher Idle: OFF Engine speed: 1000 to 4000 rpm Air fuel ratio sensor status: Activated Fuel-cut: OFF Engine load: 10 to 70% Shift position: 2nd or higher Catalyst monitor: Not executed Mass air flow: 6 to 15 g/sec. **Possible Causes:** • Air fuel ratio sensor (bank 1 sensor 1) • Air fuel ratio sensor (bank 1 sensor 1) heater • ECM

DTC	Trouble Code Title, Conditions & Possible Causes
DTC: P015B **2T ECM, MIL: Yes** **Year:** 2009, 2010 **Model:** 4Runner, FJ Cruiser **Engine:** All **Transmission:** All	**A/F Sensor Delayed Response - Lean to Rich Bank 1 Sensor 1:** Monitor runs whenever following DTCs not stored: None Active air fuel ratio control: Performing Active air fuel ratio control is performed when the following conditions met: Battery voltage: 11 V or higher Engine coolant temperature: 75°C (167°F) or higher Idle: OFF Engine speed: 1000 to 4000 rpm Air fuel ratio sensor status: Activated Fuel-cut: OFF Engine load: 10 to 70% Shift position: 2nd or higher Catalyst monitor: Not executed Mass air flow: 6 to 15 g/sec. **Possible Causes:** • Air fuel ratio sensor (bank 1 sensor 1) • Air fuel ratio sensor (bank 1 sensor 1) heater • ECM
DTC: P015C **2T ECM, MIL: Yes** **Year:** 2009, 2010 **Model:** 4Runner, FJ Cruiser **Engine:** All **Transmission:** All	**A/F Sensor Delayed Response - Rich to Lean Bank 2 Sensor 1:** Monitor runs whenever following DTCs not stored: None Active air fuel ratio control: Performing Active air fuel ratio control is performed when the following conditions met: Battery voltage: 11 V or higher Engine coolant temperature: 75°C (167°F) or higher Idle: OFF Engine speed: 1000 to 4000 rpm Air fuel ratio sensor status: Activated Fuel-cut: OFF Engine load: 10 to 70% Shift position: 2nd or higher Catalyst monitor: Not executed Mass air flow: 6 to 15 g/sec. **Possible Causes:** • Air fuel ratio sensor (bank 2 sensor 1) • Air fuel ratio sensor (bank 2 sensor 1) heater • ECM
DTC: P015D **2T ECM, MIL: Yes** **Year:** 2009, 2010 **Model:** 4Runner, FJ Cruiser **Engine:** All **Transmission:** All	**A/F Sensor Delayed Response - Lean to Rich Bank 2 Sensor 1:** Monitor runs whenever following DTCs not stored: None Active air fuel ratio control: Performing Active air fuel ratio control is performed when the following conditions met: Battery voltage: 11 V or higher Engine coolant temperature: 75°C (167°F) or higher Idle: OFF Engine speed: 1000 to 4000 rpm Air fuel ratio sensor status: Activated Fuel-cut: OFF Engine load: 10 to 70% Shift position: 2nd or higher Catalyst monitor: Not executed Mass air flow: 6 to 15 g/sec. **Possible Causes:** • Air fuel ratio sensor (bank 2 sensor 1) • Air fuel ratio sensor (bank 2 sensor 1) heater • ECM

DTC	Trouble Code Title, Conditions & Possible Causes
DTC: P0161 **2T ECM, MIL: Yes** **Year:** 2009, 2010 **Model:** Land Cruiser **Engine:** All V8 **Transmission:** All	**Oxygen Sensor Heater Circuit Malfunction (Bank 2 Sensor 2):** The Cumulative heater resistance correction value exceeds the threshold, or the duration of the rear heated oxygen sensor impedance is less than 15 kW for 2 seconds or more. **Possible Causes:** • Open or short in HO2 sensor heater circuit • HO2 sensor heater (for Sensor 2) • Integration relay • ECM
DTC: P0161 **2T ECM, MIL: Yes** **Year:** 2009, 2010 **Model:** Highlander **Engine:** All V6 **Transmission:** All	**Oxygen Sensor Heater Circuit Malfunction (Bank 2 Sensor 2):** Monitor runs whenever following DTCs are not present: None All of the following conditions are met: - Battery voltage: 10.5 V or more Fuel cut: OFF Time after fuel cut ON to OFF: 30 seconds or more Accumulated heater ON time: 100 seconds or more **Possible Causes:** • Open or short in heated oxygen sensor heater circuit • Heated oxygen sensor heater • IGCT relay • Hybrid vehicle control ECU • ECM
DTC: P0171 **2T ECM, MIL: Yes** **Year:** 2009, 2010 **Model:** Highlander **Engine:** All V6 **Transmission:** All	**System Too Lean (Bank 1):** Fuel system status: Closed loop Battery voltage: 11 V or more Either of following conditions 1 or 2 met: 1. Engine RPM: Less than 1100 rpm 2. Intake air amount per revolution: 0.22 g/rev or more Catalyst monitor: Not executed **Possible Causes:** • Intake system • Injector blockage • Mass air flow meter • Engine coolant temperature sensor • Fuel pressure • Gas leakage from exhaust system • Open or short in air fuel ratio sensor (bank 1, 2 sensor 1) circuit • Air fuel ratio sensor (bank 1, 2 sensor 1) • Air fuel ratio sensor heater (bank 1, 2 sensor 1) • A/F relay • Air fuel ratio sensor heater and A/F relay circuits • PCV valve and hose • PCV hose connections • Hybrid vehicle control ECU • ECM

DTC	Trouble Code Title, Conditions & Possible Causes
DTC: P0171 **2T ECM, MIL: Yes** **Year:** 2009, 2010 **Model:** Land Cruiser **Engine:** All V8 **Transmission:** All	**System Too Lean (Bank 1):** Fuel system status is in Closed loop. battery voltage 11 V or higher.Either of following conditions 1 or 2 set 1. Engine speed is less than 1100 rpm. 2. Intake air amount per revolution is 0.36 g/rev or more Catalyst monitor not executed. With a warm engine and stable air-fuel ratio feedback, the fuel trim is considerably in error to the lean side. **Possible Causes:** • Air induction system • Injector blockage • Mass Air Flow (MAF) meter • Engine Coolant Temperature (ECT) sensor • Fuel pressure • Gas leak from exhaust system • Open or short in A/F sensor (for Sensor 1) circuit • A/F sensor (for Sensor 1) • A/F sensor heater (for Sensor 1) • Integration relay • A/F sensor heater and integration relay circuits • PCV valve and hose • PCV hose connections • ECM
DTC: P0172 **2T ECM, MIL: Yes** **Year:** 2010 **Model:** Camry **Engine:** All L4 **Transmission:** All	**System Too Rich (Bank 1):** Fuel system status: Closed loop Battery voltage: 11 V or higher Either of the following conditions is met: 1. Engine speed: Less than 800 rpm 2. Engine load: 11.52% or more Catalyst monitor: Not executed **Possible Causes:** • Injector leak or blockage • Mass air flow meter • Engine coolant temperature sensor • Ignition system • Fuel pressure • Gas leak from exhaust system • Open or short in air fuel ratio sensor (bank 1 sensor 1) circuit • Air fuel ratio sensor (bank 1 sensor 1) • EFI NO. 2 fuse • ECM

DTC	Trouble Code Title, Conditions & Possible Causes
DTC: P0172 **2T ECM, MIL: Yes** **Year:** 2009, 2010 **Model:** Highlander **Engine:** All V6 **Transmission:** All	**System Too Rich (Bank 1):** Monitor runs whenever following DTCs not present: P0010, P0020 (VVT oil control valve Bank 1, 2) P0011, P0021 (VVT System bank 1, 2 - Advance) P0012, P0022 (VVT System bank 1, 2 - Retard) P0016, P0018 (VVT system bank 1, 2 - misalignment) P0031, P0032, P0051, P0052 (Air fuel ratio sensor heater - Sensor 1) P0102, P0103 (Mass air flow meter) P0115, P0117, P0118 (Engine coolant temperature sensor) P0120, P0121, P0122, P0123, P0220, P0222, P0223, P2135 (Throttle position sensor) P0125 (Insufficient engine coolant temperature for closed loop) P0335 (Crankshaft position sensor) P0340, P0342, P0343, P0345, P0347, P0348 (VVT sensor) P0351, P0352, P0353, P0354, P0355, P0356 (Igniter) Fuel system status: Closed loop Battery voltage: 11 V or more Either of following conditions 1 or 2 met: 1. Engine RPM: Less than 1100 rpm 2. Intake air amount per revolution: 0.22 g/rev or more Catalyst monitor: Not executed **Possible Causes:** • Injector leakage or blockage • Mass air flow meter • Engine coolant temperature sensor • Ignition system • Fuel pressure • Gas leakage from exhaust system • Open or short in air fuel ratio sensor (bank 1, 2 sensor 1) circuit • Air fuel ratio sensor (bank 1, 2 sensor 1) • Air fuel ratio sensor heater (bank 1, 2 sensor 1) • A/F relay • Air fuel ratio sensor heater and A/F relay circuits • Hybrid vehicle control ECU • ECM
DTC: P0172 **2T ECM, MIL: Yes** **Year:** 2009, 2010 **Model:** Land Cruiser **Engine:** All V8 **Transmission:** All	**System Too Rich (Bank 1):** With the engine in Closed loop, battery voltage 11 V or higher either of following conditions 1 or 2 set: 1. Engine speed is less than 1100 rpm 2. Intake air amount per revolution 0.36 g/rev or more. Catalyst monitor not executed. With a warm engine and stable air-fuel ratio feedback, the fuel trim is considerably in error to the rich side. **Possible Causes:** • Injector leakage or blockage • MAF meter • ECT sensor • Ignition system • Fuel pressure • Gas leak from exhaust system • Open or short in A/F sensor (for Sensor 1) circuit • A/F sensor (for Sensor 1) • A/F sensor heater (for Sensor 1) • Integration relay • A/F sensor heater and integration relay circuits • ECM

DTC	Trouble Code Title, Conditions & Possible Causes
DTC: P0174 **2T ECM, MIL: Yes** **Year:** 2009, 2010 **Model:** Land Cruiser **Engine:** All V8 **Transmission:** All	**System Too Lean (Bank 2):** Fuel system status is in Closed loop. battery voltage 11 V or higher.Either of following conditions 1 or 2 set 1. Engine speed is less than 1100 rpm. 2. Intake air amount per revolution is 0.36 g/rev or more Catalyst monitor not executed. With a warm engine and stable air-fuel ratio feedback, the fuel trim is considerably in error to the lean side. **Possible Causes:** • Air induction system • Injector blockage • Mass Air Flow (MAF) meter • Engine Coolant Temperature (ECT) sensor • Fuel pressure • Gas leak from exhaust system • Open or short in A/F sensor (for Sensor 1) circuit • A/F sensor (for Sensor 1) • A/F sensor heater (for Sensor 1) • Integration relay • A/F sensor heater and integration relay circuits • PCV valve and hose • PCV hose connections • ECM
DTC: P0174 **2T ECM, MIL: Yes** **Year:** 2009, 2010 **Model:** Highlander **Engine:** All V6 **Transmission:** All	**System Too Lean (Bank 2):** Fuel system status: Closed loop Battery voltage: 11 V or more Either of following conditions 1 or 2 met: 1. Engine RPM: Less than 1100 rpm 2. Intake air amount per revolution: 0.22 g/rev or more Catalyst monitor: Not executed **Possible Causes:** • Intake system • Injector blockage • Mass air flow meter • Engine coolant temperature sensor • Fuel pressure • Gas leakage from exhaust system • Open or short in air fuel ratio sensor (bank 1, 2 sensor 1) circuit • Air fuel ratio sensor (bank 1, 2 sensor 1) • Air fuel ratio sensor heater (bank 1, 2 sensor 1) • A/F relay • Air fuel ratio sensor heater and A/F relay circuits • PCV valve and hose • PCV hose connections • Hybrid vehicle control ECU • ECM

DTC	Trouble Code Title, Conditions & Possible Causes
DTC: P0175 **2T ECM, MIL: Yes** **Year:** 2009, 2010 **Model:** Highlander **Engine:** All V6 **Transmission:** All	**System Too Rich (Bank 2):** Fuel system status: Closed loop Battery voltage: 11 V or more Either of following conditions 1 or 2 met: 1. Engine RPM: Less than 1100 rpm 2. Intake air amount per revolution: 0.22 g/rev or more Catalyst monitor: Not executed **Possible Causes:** • Injector leakage or blockage • Mass air flow meter • Engine coolant temperature sensor • Ignition system • Fuel pressure • Gas leakage from exhaust system • Open or short in air fuel ratio sensor (bank 1, 2 sensor 1) circuit • Air fuel ratio sensor (bank 1, 2 sensor 1) • Air fuel ratio sensor heater (bank 1, 2 sensor 1) • A/F relay • Air fuel ratio sensor heater and A/F relay circuits • Hybrid vehicle control ECU • ECM
DTC: P0175 **2T ECM, MIL: Yes** **Year:** 2009, 2010 **Model:** Land Cruiser **Engine:** All V8 **Transmission:** All	**System Too Rich (Bank 2):** With the engine in Closed loop, battery voltage 11 V or higher either of following conditions 1 or 2 set: 1. Engine speed is less than 1100 rpm 2. Intake air amount per revolution 0.36 g/rev or more. Catalyst monitor not executed. With a warm engine and stable air-fuel ratio feedback, the fuel trim is considerably in error to the rich side. **Possible Causes:** • Injector leakage or blockage • MAF meter • ECT sensor • Ignition system • Fuel pressure • Gas leak from exhaust system • Open or short in A/F sensor (for Sensor 1) circuit • A/F sensor (for Sensor 1) • A/F sensor heater (for Sensor 1) • Integration relay • A/F sensor heater and integration relay circuits • ECM
DTC: P0222 **1T ECM, MIL: Yes** **Year:** 2009, 2010 **Model:** Land Cruiser, Matrix **Engine:** All **Transmission:** All	**Throttle / Pedal Position Sensor / Switch "B" Circuit Low Input:** With the ignition switch ON or electronic throttle actuator power ON, the output voltage of VTA2 is 1.75 V or less for 2 seconds or more. **Possible Causes:** • TP sensor (built into throttle body) • Short in VTA2 circuit • Open in VC circuit • ECM
DTC: P0223 **1T ECM, MIL: Yes** **Year:** 2009, 2010 **Model:** Land Cruiser, Matrix **Engine:** All **Transmission:** All	**Throttle / Pedal Position Sensor / Switch "B" Circuit High Input:** With the ignition switch ON or electronic throttle actuator power ON, the output voltage of VTA2 is 4.8 V or more, and VTA1 is between 0.2 V and 2.02 V for 2 seconds or more. **Possible Causes:** • TP sensor (built into throttle body) • Open in VTA2 circuit • Open in E2 circuit • Short between VC and VTA2 circuits • ECM

DTC	Trouble Code Title, Conditions & Possible Causes
DTC: P0230 **1T ECM, MIL: Yes** **Year:** 2009, 2010 **Model:** Land Cruiser **Engine:** All **Transmission:** All	**Fuel Pump Primary Circuit:** When either condition below is met: (a) When the fuel pump is operating, the remaining fuel is 17 L or more. The DI terminal output is low. (b) When the fuel pump is not operating,the DI terminal output is high. **Possible Causes:** • Open or short in fuel pump circuit • Fuel pump • Fuel pump ECU • ECM
DTC: P0301 **2T ECM, MIL: Yes** **Year:** 2009, 2010 **Model:** Highlander **Engine:** All **Transmission:** All	**Cylinder 1 Misfire Detected:** Battery voltage: 8 V or more VVT system: No operated by scan tool Engine RPM: 750 to 5800 rpm Both of following conditions 1 and 2 met: 1. Engine coolant temperature: -10°C (14°F) or more 2. Either of following conditions (a) or (b) met: - (a) Engine coolant temperature at engine start: More than -7°C (19°F) (b) Engine coolant temperature: More than 20°C (68°F) Fuel cut: OFF Monitor period of emission-related misfire: First 1000 revolutions after engine start, or Check Mode: Crankshaft 1000 revolutions Except above: Crankshaft 1000 revolutions x 4 Monitor period of catalyst-damaging (MIL blinks): All of following conditions 1, 2 and 3 met: Crankshaft 200 revolutions x 3 1. Driving cycles: 1st 2. Check mode: OFF 3. Engine RPM: Less than 2600 rpm Except above: Crankshaft 200 revolutions **Possible Causes:** • Open or short in engine wire harness • Connector connection • Vacuum hose connections • Ignition system • Injector • Fuel pressure • Mass air flow meter • Engine coolant temperature sensor • Compression pressure • Valve clearance • Valve timing • PCV valve and hose • PCV hose connections • EGR valve system • Intake system • Hybrid vehicle control ECU • Tumble Control Valve (TCV) • ECM

DTC	Trouble Code Title, Conditions & Possible Causes
DTC: P0302 **2T ECM, MIL: Yes** **Year:** 2009, 2010 **Model:** Highlander **Engine:** All **Transmission:** All	**Cylinder 2 Misfire Detected:** Battery voltage: 8 V or more VVT system: No operated by scan tool Engine RPM: 750 to 5800 rpm Both of following conditions 1 and 2 met: 1. Engine coolant temperature: -10°C (14°F) or more 2. Either of following conditions (a) or (b) met: - (a) Engine coolant temperature at engine start: More than -7°C (19°F) (b) Engine coolant temperature: More than 20°C (68°F) Fuel cut: OFF Monitor period of emission-related misfire: First 1000 revolutions after engine start, or Check Mode: Crankshaft 1000 revolutions Except above: Crankshaft 1000 revolutions x 4 Monitor period of catalyst-damaging (MIL blinks): All of following conditions 1, 2 and 3 met: Crankshaft 200 revolutions x 3 1. Driving cycles: 1st 2. Check mode: OFF 3. Engine RPM: Less than 2600 rpm Except above: Crankshaft 200 revolutions **Possible Causes:** • Open or short in engine wire harness • Connector connection • Vacuum hose connections • Ignition system • Injector • Fuel pressure • Mass air flow meter • Engine coolant temperature sensor • Compression pressure • Valve clearance • Valve timing • PCV valve and hose • PCV hose connections • EGR valve system • Intake system • Hybrid vehicle control ECU • Tumble Control Valve (TCV) • ECM

DTC	Trouble Code Title, Conditions & Possible Causes
DTC: P0303 **2T ECM, MIL: Yes** **Year:** 2009, 2010 **Model:** Highlander **Engine:** All **Transmission:** All	**Cylinder 3 Misfire Detected:** Battery voltage: 8 V or more VVT system: No operated by scan tool Engine RPM: 750 to 5800 rpm Both of following conditions 1 and 2 met: 1. Engine coolant temperature: -10°C (14°F) or more 2. Either of following conditions (a) or (b) met: - (a) Engine coolant temperature at engine start: More than -7°C (19°F) (b) Engine coolant temperature: More than 20°C (68°F) Fuel cut: OFF Monitor period of emission-related misfire: First 1000 revolutions after engine start, or Check Mode: Crankshaft 1000 revolutions Except above: Crankshaft 1000 revolutions x 4 Monitor period of catalyst-damaging (MIL blinks): All of following conditions 1, 2 and 3 met: Crankshaft 200 revolutions x 3 1. Driving cycles: 1st 2. Check mode: OFF 3. Engine RPM: Less than 2600 rpm Except above: Crankshaft 200 revolutions **Possible Causes:** • Open or short in engine wire harness • Connector connection • Vacuum hose connections • Ignition system • Injector • Fuel pressure • Mass air flow meter • Engine coolant temperature sensor • Compression pressure • Valve clearance • Valve timing • PCV valve and hose • PCV hose connections • EGR valve system • Tumble Control Valve (TCV) • Intake system • Hybrid vehicle control ECU • ECM

DTC	Trouble Code Title, Conditions & Possible Causes
DTC: P0304 **2T ECM, MIL: Yes** **Year:** 2009, 2010 **Model:** Highlander **Engine:** All **Transmission:** All	**Cylinder 4 Misfire Detected:** Battery voltage: 8 V or more VVT system: No operated by scan tool Engine RPM: 750 to 5800 rpm Both of following conditions 1 and 2 met: 1. Engine coolant temperature: -10°C (14°F) or more 2. Either of following conditions (a) or (b) met: - (a) Engine coolant temperature at engine start: More than -7°C (19°F) (b) Engine coolant temperature: More than 20°C (68°F) Fuel cut: OFF Monitor period of emission-related misfire: First 1000 revolutions after engine start, or Check Mode: Crankshaft 1000 revolutions Except above: Crankshaft 1000 revolutions x 4 Monitor period of catalyst-damaging (MIL blinks): All of following conditions 1, 2 and 3 met: Crankshaft 200 revolutions x 3 1. Driving cycles: 1st 2. Check mode: OFF 3. Engine RPM: Less than 2600 rpm Except above: Crankshaft 200 revolutions **Possible Causes:** • Open or short in engine wire harness • Connector connection • Vacuum hose connections • Ignition system • Injector • Fuel pressure • Mass air flow meter • Engine coolant temperature sensor • Compression pressure • Valve clearance • Valve timing • Tumble Control Valve (TCV) • PCV valve and hose • PCV hose connections • EGR valve system • Intake system • Hybrid vehicle control ECU • ECM

DTC	Trouble Code Title, Conditions & Possible Causes
DTC: P0305 **2T ECM, MIL: Yes** **Year:** 2009, 2010 **Model:** Highlander **Engine:** All **Transmission:** All	**Cylinder 5 Misfire Detected:** Battery voltage: 8 V or more VVT system: No operated by scan tool Engine RPM: 750 to 5800 rpm Both of following conditions 1 and 2 met: 1. Engine coolant temperature: -10°C (14°F) or more 2. Either of following conditions (a) or (b) met: - (a) Engine coolant temperature at engine start: More than -7°C (19°F) (b) Engine coolant temperature: More than 20°C (68°F) Fuel cut: OFF Monitor period of emission-related misfire: First 1000 revolutions after engine start, or Check Mode: Crankshaft 1000 revolutions Except above: Crankshaft 1000 revolutions x 4 Monitor period of catalyst-damaging (MIL blinks): All of following conditions 1, 2 and 3 met: Crankshaft 200 revolutions x 3 1. Driving cycles: 1st 2. Check mode: OFF 3. Engine RPM: Less than 2600 rpm Except above: Crankshaft 200 revolutions **Possible Causes:** • Open or short in engine wire harness • Connector connection • Vacuum hose connections • Ignition system • Injector • Fuel pressure • Mass air flow meter • Engine coolant temperature sensor • Compression pressure • Valve clearance • Valve timing • PCV valve and hose • PCV hose connections • EGR valve system • Tumble Control Valve (TCV) • Intake system • Hybrid vehicle control ECU • ECM

DTC	Trouble Code Title, Conditions & Possible Causes
DTC: P0306 **2T ECM, MIL: Yes** **Year:** 2009, 2010 **Model:** Highlander **Engine:** All **Transmission:** All	**Cylinder 6 Misfire Detected:** Misfire: Monitor runs whenever following DTCs not present: P0016, P0018 (VVT system bank 1, 2 - misalignment) P0102, P0103 (Mass air flow meter) P0112, P0113 (Intake air temperature sensor) P0115, P0117, P0118 (Engine coolant temperature sensor) P0120, P0121, P0122, P0123, P0220, P0222, P0223, P2135 (Throttle position sensor) P0125 (Insufficient engine coolant temperature for closed loop) P0327, P0328, P0332, P0333 (Knock sensor) P0335 (Crankshaft position sensor) P0351, P0352, P0353, P0354, P0355, P0356 (Igniter) Battery voltage: 8 V or more VVT system: No operated by scan tool Engine RPM: 750 to 5800 rpm Both of following conditions 1 and 2 met: 1. Engine coolant temperature: -10°C (14°F) or more 2. Either of following conditions (a) or (b) met: - (a) Engine coolant temperature at engine start: More than -7°C (19°F) (b) Engine coolant temperature: More than 20°C (68°F) Fuel cut: OFF Monitor period of emission-related misfire: First 1000 revolutions after engine start, or Check Mode: Crankshaft 1000 revolutions Except above: Crankshaft 1000 revolutions x 4 Monitor period of catalyst-damaging (MIL blinks): All of following conditions 1, 2 and 3 met: Crankshaft 200 revolutions x 3 1. Driving cycles: 1st 2. Check mode: OFF 3. Engine RPM: Less than 2600 rpm Except above: Crankshaft 200 revolutions **Possible Causes:** • Open or short in engine wire harness • Connector connection • Vacuum hose connections • Ignition system • Injector • Fuel pressure • Mass air flow meter • Engine coolant temperature sensor • Compression pressure • Valve clearance • Valve timing • PCV valve and hose • PCV hose connections • EGR valve system • Tumble Control Valve (TCV) • Intake system • Hybrid vehicle control ECU • ECM

DTC	Trouble Code Title, Conditions & Possible Causes
DTC: P0307 **2T ECM, MIL: Yes** **Year:** 2009 **Model:** 4Runner **Engine:** All **Transmission:** All	**Cylinder 7 Misfire Detected:** Battery voltage: 8 V or more Throttle position learning: Completed VVT system: Not operate by scan tool Engine RPM: 400 to 5700 rpm Fuel cut: OFF Either of following conditions is met: Conditions (a) or (b) (a) Engine start ECT: Higher than -7°C (19°F) (b) ECT: Higher than 20°C (68°F) Monitor period of emission-related misfire: First 1,000 revolutions after engine start, or Check Mode: Crankshaft 1000 revolutions Except above: Crankshaft 1000 revolutions x 4 Monitor period of catalyst-damage misfire: All of following conditions 1, 2 and 3 are met: Crankshaft 200 revolutions x 3 1. Driving cycles: 1st 2. Check Mode: OFF 3. RPM: Less than 2800 revolutions Except above: Crankshaft 200 revolutions **Possible Causes:** • Open or short in engine wire harness • Connector connection • Vacuum hose connections • Ignition system • Fuel injector • Fuel pressure • Mass air flow meter • Engine coolant temperature sensor • Compression pressure • Valve clearance • Valve timing • PCV valve and hose • PCV hose connections • EGR valve system • Tumble Control Valve (TCV) • Air induction system • ECM

DTC	Trouble Code Title, Conditions & Possible Causes
DTC: P0308 **2T ECM, MIL: Yes** **Year:** 2009 **Model:** 4Runner **Engine:** All **Transmission:** All	**Cylinder 8 Misfire Detected :** Battery voltage: 8 V or more Throttle position learning: Completed VVT system: Not operate by scan tool Engine RPM: 400 to 5700 rpm Fuel cut: OFF Either of following conditions is met: Conditions (a) or (b) (a) Engine start ECT: Higher than -7°C (19°F) (b) ECT: Higher than 20°C (68°F) Monitor period of emission-related misfire: First 1,000 revolutions after engine start, or Check Mode: Crankshaft 1000 revolutions Except above: Crankshaft 1000 revolutions x 4 Monitor period of catalyst-damage misfire: All of following conditions 1, 2 and 3 are met: Crankshaft 200 revolutions x 3 1. Driving cycles: 1st 2. Check Mode: OFF 3. RPM: Less than 2800 revolutions Except above: Crankshaft 200 revolutions **Possible Causes:** • Open or short in engine wire harness • Connector connection • Vacuum hose connections • Ignition system • Fuel injector • Fuel pressure • Mass air flow meter • Engine coolant temperature sensor • Compression pressure • Valve clearance • Valve timing • PCV valve and hose • PCV hose connections • EGR valve system • Tumble Control Valve (TCV) • Air induction system • ECM
DTC: P0327 **1T ECM, MIL: Yes** **Year:** 2009, 2010 **Model:** 4Runner, FJ Cruiser **Engine:** All **Transmission:** All	**Knock Sensor 1 Circuit Low Input (Bank 1 or Single Sensor):** Monitor runs whenever following DTCs not present: None Battery voltage: 10.5 V or higher Time after engine start: 5 seconds or more **Possible Causes:** • Short in knock sensor circuit • Knock sensor • ECM
DTC: P0328 **1T ECM, MIL: Yes** **Year:** 2009, 2010 **Model:** Corolla **Engine:** All **Transmission:** All	**Knock Sensor 1 Circuit High Input (Bank 1 or Single Sensor):** Battery voltage 10.5 V or more and the vehicle running for at least 5 seconds. The output voltage of knock sensor more than 4.5 V for 1 second or more. **Possible Causes:** • Open in knock sensor circuit • Knock sensor • ECM
DTC: P032D **1T ECM, MIL: Yes** **Year:** 2009, 2010 **Model:** Land Cruiser **Engine:** All V8 **Transmission:** All	**Knock Sensor 3 Circuit High:** With the engine running for at least 5 seconds and battery voltage 10.5 V or more, the output voltage of the knock sensor (for Bank 1 Sensor 2) is 4.5 V or higher. **Possible Causes:** • Open in knock sensor (for Bank 1 Sensor 2) circuit • Knock sensor (for Bank 1 Sensor 2) • ECM

DTC	Trouble Code Title, Conditions & Possible Causes
DTC: P0332 **1T ECM, MIL: Yes** **Year:** 2009, 2010 **Model:** All **Transmission:** All	**Knock Sensor 2 Circuit Low Input (Bank 2):** Monitor runs whenever following DTCs not present: None Battery voltage: 10.5 V or higher Time after engine start: 5 seconds or more **Possible Causes:** • Short in knock sensor circuit • Knock sensor • ECM
DTC: P0333 **1T ECM, MIL: Yes** **Year:** 2009, 2010 **Model:** Avalon, Camry **Engine:** All **Transmission:** All	**Knock Sensor 2 Circuit High Input (Bank 2):** Monitor runs whenever following DTCs are not present: None Battery voltage: 10.5 V or more Time after engine start: 5 seconds or more Engine switch: On (IG) Starter: OFF **Possible Causes:** • Open in knock sensor circuit (bank1, 2) • Knock sensor (bank1, 2) • ECM
DTC: P0335 **2T ECM, MIL: Yes** **Year:** 2009, 2010 **Model:** Corolla **Engine:** All L4 **Transmission:** All	**Crankshaft Position Sensor "A" Circuit:** Case 1: Time after starter OFF to ON 3 seconds or more, The number of camshaft position sensor signal pulse 6 times. Battery voltage 7 V or more, ignition switch ON. Case 2: Starter OFF, engine speed 600 rpm or more. Time after starter from ON to OFF 3 seconds or more. Missing crankshaft position sensor signal despite camshaft position sensor signal inputs normal after engine cranked. No crankshaft position sensor signal to ECM at engine speed of 600 rpm or more. **Possible Causes:** • Open or short in CKP sensor circuit • CKP sensor • Sensor plate (CKP sensor plate) • ECM
DTC: P0335 **1T ECM, MIL: Yes** **Year:** 2009, 2010 **Model:** Land Cruiser **Engine:** All V8 **Transmission:** All	**Crankshaft Position Sensor "A" Circuit:** No Crankshaft Position (CKP) sensor signal is sent to the ECM while cranking. No CKP sensor signal is sent to the ECM at an engine speed of 450 rpm or more. **Possible Causes:** • Open or short in Crankshaft Position (CKP) sensor circuit • CKP sensor • CKP sensor plate • ECM
DTC: P0335 **2T ECM, MIL: Yes** **Year:** 2009, 2010 **Model:** Highlander **Engine:** All V6 **Transmission:** All	**Crankshaft Position Sensor "A" Circuit:** Monitor runs whenever following DTCs are not present: None Power switch: ON Engine rotating signal from hybrid vehicle control ECU: Engine running **Possible Causes:** • Open or short in crankshaft position sensor circuit • Crankshaft position sensor • Crankshaft position sensor plate • Hybrid vehicle control ECU • ECM
DTC: P0337 **1T ECM, MIL: Yes** **Year:** 2009, 2010 **Model:** Land Cruiser **Engine:** 5.7L V8 **Transmission:** All	**Crankshaft Position Sensor "A" Circuit Low Input:** With the ignition switch ON and battery voltage 8.0 V or more, the output voltage of the CKP sensor is 0.3 V or less for 4 seconds. **Possible Causes:** • Open or short in CKP sensor circuit • CKP sensor • CKP sensor plate • ECM

DTC	Trouble Code Title, Conditions & Possible Causes
DTC: P0338 **1T ECM, MIL: Yes** **Year:** 2009, 2010 **Model:** Land Cruiser **Engine:** 5.7L V8 **Transmission:** All	**Crankshaft Position Sensor "A" Circuit High Input:** With the ignition switch ON and battery voltage a minimum of 8 V, the output voltage of the CKP sensor is 4.7 V or higher for 4 seconds. **Possible Causes:** • Open or short in CKP sensor circuit • CKP sensor • CKP sensor plate • ECM
DTC: P0339 **1T ECM, MIL: Yes** **Year:** 2009, 2010 **Model:** Land Cruiser **Engine:** All **Transmission:** All	**Crankshaft Position Sensor "A" Circuit Intermittent:** Under conditions (a), (b) and (c), no CKP sensor signal is sent to the ECM for 0.05 seconds or more: (a) The engine speed is 1000 rpm or more. (b) The starter signal is OFF. (c) 3 seconds or more have elapsed since the starter signal was switched from ON to OFF. **Possible Causes:** • Open or short in CKP sensor circuit • CKP sensor • CKP sensor plate • ECM
DTC: P033C **1T ECM, MIL: Yes** **Year:** 2009, 2010 **Model:** Land Cruiser **Engine:** 5.7L V8 **Transmission:** All	**Knock Sensor 4 Circuit Low Input:** With the engine running for at least 5 seconds and battery voltage 10.5 V or more, the output voltage of the knock sensor (for Bank 2 Sensor 2) is 0.5 V or less. **Possible Causes:** • Short in knock sensor (for Bank 2 Sensor 2) circuit • Knock sensor (for Bank 2 Sensor 2) • ECM
DTC: P033D **1T ECM, MIL: Yes** **Year:** 2009, 2010 **Model:** Land Cruiser **Engine:** 5.7L V8 **Transmission:** All	**Knock Sensor 4 Circuit High Input:** With the engine running for at least 5 seconds and battery voltage 10.5 V or more, the output voltage of the knock sensor (for Bank 2 Sensor 2) is 4.5 V or higher. **Possible Causes:** • Open in knock sensor (for Bank 2 Sensor 2) circuit • Knock sensor (for Bank 2 Sensor 2) • ECM
DTC: P0340 **2T ECM, MIL: Yes** **Year:** 2009, 2010 **Model:** Camry, Matrix **Engine:** All L4 **Transmission:** All	**Camshaft Position Sensor "A" Circuit (Bank 1 or Single Sensor):** Monitor runs whenever following DTCs are not present: None Camshaft Position Sensor Range Check: Stater: ON Minimal battery voltage while starter ON: Less than 11 V Camshaft Position/Crankshaft Position Misalignment: Engine speed: 600 rpm or more Starter: OFF **Possible Causes:** • Open or short in CMP sensor circuit • CMP sensor • Camshaft • Jumped tooth of timing chain • ECM
DTC: P0340 **1T ECM, MIL: Yes** **Year:** 2009, 2010 **Model:** Land Cruiser **Engine:** All V8 **Transmission:** All	**Camshaft Position Sensor Circuit Malfunction:** When either condition below is met: (a) Input voltage to the ECM remains at less than 0.3 V, or more than 4.7 V for 4 seconds when 2 or more seconds have elapsed after turning the engine switch on (IG). (b) No VVT sensor signal to ECM at engine speed of 600 rpm or more. **Possible Causes:** • Open or short in VVT sensor (for intake side) circuit • VVT sensor (for intake side) • Intake camshaft (for Bank 1, 2) • Jumped tooth of timing chain • ECM

DTC	Trouble Code Title, Conditions & Possible Causes
DTC: P0340 **1T ECM, MIL: Yes** **Year:** 2009, 2010 **Model:** Highlander **Engine:** All V6 **Transmission:** All	**Camshaft Position Sensor Circuit Malfunction:** Monitor runs whenever following DTCs are not present: None Engine speed: 600 rpm or more Starter: OFF **Possible Causes:** • Open or short in VVT sensor • VVT sensor • Camshaft timing gear • Hybrid vehicle control ECU • ECM
DTC: P0341 **1T ECM, MIL: Yes** **Year:** 2009, 2010 **Model:** Highlander **Engine:** All V6 **Transmission:** All	**Camshaft Position Sensor "A" Circuit Range / Performance (Bank 1 or Single Sensor):** Monitor runs whenever following DTCs are not present: None Engine revolution angle: 720 °CA **Possible Causes:** • Open or short in VVT sensor • VVT sensor • Camshaft timing gear • Hybrid vehicle control ECU • ECM
DTC: P0342 **1T ECM, MIL: Yes** **Year:** 2009, 2010 **Model:** Avalon, Camry **Engine:** 2.5L L4, 3.5L V6 **Transmission:** All	**Camshaft Position Sensor "A" Circuit Low Input (Bank 1 or Single Sensor):** Starter: OFF Engine switch: On (IG) Time after engine switch off to on (IG): 2 seconds or more VVT sensor verify pulse input fail (P0340): Not detected Battery voltage: 8 V or more **Possible Causes:** • Open or short in VVT sensor for intake side circuit • VVT sensor for intake side • Camshaft timing gear assembly for intake camshaft • ECM
DTC: P0342 **1T ECM, MIL: Yes** **Year:** 2009, 2010 **Model:** Land Cruiser **Engine:** All V8 **Transmission:** All	**Camshaft Position Sensor "A" Circuit Low Input (Bank 1 or Single Sensor):** With the ignition switch ON for at least 2 seconds, the output voltage of the VVT sensor (for intake side of Bank 1, 2) is below 0.3 V for 4 seconds. **Possible Causes:** • Open or short in VVT sensor circuit for intake camshaft • Intake camshaft • Camshaft timing gear for intake camshaft • Jumped tooth of timing chain for intake camshaft • ECM
DTC: P0343 **1T ECM, MIL: Yes** **Year:** 2009, 2010 **Model:** 4Runner, FJ Cruiser **Engine:** 4.0L V6 **Transmission:** All	**Camshaft Position Sensor "A" Circuit High Input (Bank 1 or Single Sensor):** Low Voltage, High Voltage: Monitor runs whenever following DTCs are not present: P0340 (VVT Sensor) Starter: OFF Ignition switch: ON Time after ignition switch off to ON: 2 seconds or more Battery voltage: 8 V or higher **Possible Causes:** • Open or short in VVT sensor circuit for intake camshaft • VVT sensor for intake camshaft • Camshaft timing gear for intake camshaft • Timing chain for intake camshaft has jumped tooth • ECM

DTC	Trouble Code Title, Conditions & Possible Causes
DTC: P0343 **1T ECM, MIL: Yes** **Year:** 2009, 2010 **Model:** Land Cruiser **Engine:** 5.7L V8 **Transmission:** All	**Camshaft Position Sensor "A" Circuit High Input (Bank 1 or Single Sensor):** With the ignition switch ON for at least 2 seconds, the output voltage of the VVT sensor (for intake side of Bank 1, 2) is higher than 4.7 V for 4 seconds. **Possible Causes:** • Open or short in VVT sensor for intake camshaft circuit • VVT sensor for intake camshaft • Intake camshaft • Jumped tooth of timing chain for intake camshaft • ECM
DTC: P0343 **1T ECM, MIL: Yes** **Year:** 2009, 2010 **Model:** Avalon, Camry **Engine:** 2.5L L4, 3.5L V6 **Transmission:** All	**Camshaft Position Sensor "A" Circuit High Input (Bank 1 or Single Sensor):** Starter: OFF Engine switch: On (IG) Time after engine switch off to on (IG): 2 seconds or more VVT sensor verify pulse input fail (P0340): Not detected Battery voltage: 8 V or more **Possible Causes:** • Open or short in VVT sensor for intake side circuit • VVT sensor for intake side • Camshaft timing gear assembly for intake camshaft • ECM
DTC: P0345 **1T ECM, MIL: Yes** **Year:** 2009, 2010 **Model:** 4Runner, FJ Cruiser **Engine:** 4.0L V6, 4.7L V8 **Transmission:** All	**Camshaft Position Sensor "A" Circuit (Bank 2):** Engine Running: Monitor runs whenever following DTCs are not present: P0342, P0343, P0347, P0348 (VVT Sensor) Engine speed: 600 rpm or more Battery voltage: 8 V or higher Starter: OFF Ignition switch: ON VVT sensor voltage: 0.3 to 4.7 V **Possible Causes:** • Open or short in VVT sensor circuit for intake camshaft • VVT sensor for intake camshaft • Camshaft timing gear for intake camshaft • Timing chain for intake camshaft has jumped tooth • ECM
DTC: P0345 **1T ECM, MIL: Yes** **Year:** 2009, 2010 **Model:** Avalon, Camry **Engine:** 3.5L V6 **Transmission:** All	**Camshaft Position Sensor "A" Circuit (Bank 2):** Engine speed: 600 rpm or more Stater: OFF VVT sensor range check fail (P0342, P0343, P0374, P0348): Not detected VVT sensor voltage: 0.3 V or more, and 4.7 V or less Battery voltage: 8 V or more Engine switch: On (IG) Time after engine switch off to on (IG): 0.5 seconds or more **Possible Causes:** • Open or short in VVT sensor for intake side circuit • VVT sensor for intake side • Camshaft timing gear assembly for intake camshaft • ECM
DTC: P0345 **1T ECM, MIL: Yes** **Year:** 2009, 2010 **Model:** Land Cruiser **Engine:** 5.7L V8 **Transmission:** All	**Camshaft Position Sensor "A" Circuit (Bank 2):** When either condition below is met: (a) The input voltage to the ECM remains at below 0.3 V or higher than 4.7 V for 4 seconds when 2 or more seconds have elapsed after turning the engine switch on (IG). (b) No VVT sensor (for intake side of Bank 1, 2) signal is sent to the ECM during cranking. **Possible Causes:** • Open or short in VVT sensor (for intake side) circuit • VVT sensor (for intake side) • Intake camshaft (for Bank 1, 2) • Jumped tooth of timing chain • ECM

DTC	Trouble Code Title, Conditions & Possible Causes
DTC: P0347 **1T ECM, MIL: Yes** **Year:** 2009, 2010 **Model:** Land Cruiser **Engine:** 5.7L V8 **Transmission:** All	**Camshaft Position Sensor "A" Circuit Low Input (Bank 2):** With the ignition switch ON for at least 2 seconds, the output voltage of the VVT sensor (for intake side of Bank 1, 2) is below 0.3 V for 4 seconds. **Possible Causes:** • Open or short in VVT sensor (for intake side) circuit • VVT sensor (for intake side) • Intake camshaft (for Bank 1, 2) • Jumped tooth of timing chain • ECM
DTC: P0347 **1T ECM, MIL: Yes** **Year:** 2009, 2010 **Model:** 4Runner, FJ Cruiser **Engine:** All V6 **Transmission:** All	**Camshaft Position Sensor "A" Circuit Low Input (Bank 2):** Low Voltage, High Voltage: Monitor runs whenever following DTCs are not present: P0340, P0345 (VVT Sensor) Starter: OFF Ignition switch: ON Time after ignition switch off to ON: 2 seconds or more Battery voltage: 8 V or higher **Possible Causes:** • Open or short in VVT sensor circuit for intake camshaft • VVT sensor for intake camshaft • Camshaft timing gear for intake camshaft • Timing chain for intake camshaft has jumped tooth • ECM
DTC: P0348 **1T ECM, MIL: Yes** **Year:** 2009, 2010 **Model:** 4Runner, FJ Cruiser **Engine:** 4.0L V6 **Transmission:** All	**Camshaft Position Sensor "A" Circuit High Input (Bank 2):** Low Voltage, High Voltage: Monitor runs whenever following DTCs are not present: P0340, P0345 (VVT Sensor) Starter: OFF Ignition switch: ON Time after ignition switch off to ON: 2 seconds or more Battery voltage: 8 V or higher **Possible Causes:** • Open or short in VVT sensor circuit for intake camshaft • VVT sensor for intake camshaft • Camshaft timing gear for intake camshaft • Timing chain for intake camshaft has jumped tooth • ECM
DTC: P0348 **1T ECM, MIL: Yes** **Year:** 2009, 2010 **Model:** Land Cruiser **Engine:** 5.7L V8 **Transmission:** All	**Camshaft Position Sensor "A" Circuit High Input (Bank 2):** With the ignition switch ON for at least 2 seconds, the output voltage of the VVT sensor (for intake side of Bank 1, 2) is higher than 4.7 V for 4 seconds. **Possible Causes:** • Open or short in VVT sensor (for intake side) circuit • VVT sensor (for intake side) • Intake camshaft (for Bank 1, 2) • Jumped tooth of timing chain • ECM
DTC: P0348 **1T ECM, MIL: Yes** **Year:** 2009, 2010 **Model:** Avalon, Camry **Engine:** 3.5L V6 **Transmission:** All	**Camshaft Position Sensor "A" Circuit High Input (Bank 2):** Starter: OFF Engine switch: On (IG) Time after engine switch off to on (IG): 2 seconds or more VVT sensor verify pulse input fail (P030, P0345): Not detected Battery voltage: 8 V or more **Possible Causes:** • Open or short in VVT sensor for intake side circuit • VVT sensor for intake side • Camshaft timing gear assembly for intake camshaft • ECM

DTC	Trouble Code Title, Conditions & Possible Causes
DTC: P0351 **1T ECM, MIL: Yes** **Year:** 2009, 2010 **Model:** Land Cruiser **Engine:** All **Transmission:** All	**Ignition Coil "A" Primary / Secondary Circuit:** No IGF signal is sent to the ECM while the engine is running. **Possible Causes:** • Ignition system • Open or short in IGF1 or IGT1 circuit between ignition coil with igniter and ECM • No. 1 ignition coil with igniter • ECM
DTC: P0352 **1T ECM, MIL: Yes** **Year:** 2009, 2010 **Model:** Land Cruiser **Engine:** All **Transmission:** All	**Ignition Coil "B" Primary / Secondary Circuit:** No IGF signal is sent to the ECM while the engine is running. **Possible Causes:** • Ignition system • Open or short in IGF2 or IGT2 circuit between ignition coil with igniter and ECM • No. 2 ignition coil with igniter • ECM
DTC: P0353 **1T ECM, MIL: Yes** **Year:** 2009, 2010 **Model:** Land Cruiser **Engine:** All **Transmission:** All	**Ignition Coil "C" Primary / Secondary Circuit:** No IGF signal is sent to the ECM while the engine is running. **Possible Causes:** • Ignition system • Open or short in IGF2 or IGT3 circuit between ignition coil with igniter and ECM • No. 3 ignition coil with igniter • ECM
DTC: P0354 **1T ECM, MIL: Yes** **Year:** 2009, 2010 **Model:** Land Cruiser **Engine:** All **Transmission:** All	**Ignition Coil "D" Primary / Secondary Circuit:** No IGF signal is sent to the ECM while the engine is running. **Possible Causes:** • Ignition system • Open or short in IGF1 or IGT4 circuit between ignition coil with igniter and ECM • No. 4 ignition coil with igniter • ECM
DTC: P0355 **1T ECM, MIL: Yes** **Year:** 2009, 2010 **Model:** Land Cruiser **Engine:** All **Transmission:** All	**Ignition Coil "E" Primary / Secondary Circuit:** No IGF signal is sent to the ECM while the engine is running. **Possible Causes:** • Ignition system • Open or short in IGF2 or IGT5 circuit between ignition coil with igniter and ECM • No. 5 ignition coil with igniter • ECM
DTC: P0356 **1T ECM, MIL: Yes** **Year:** 2009, 2010 **Model:** Land Cruiser **Engine:** All **Transmission:** All	**Ignition Coil "F" Primary / Secondary Circuit:** No IGF signal is sent to the ECM while the engine is running. **Possible Causes:** • Ignition system • Open or short in IGF1 or IGT6 circuit between ignition coil with igniter and ECM • No. 6 ignition coil with igniter • ECM
DTC: P0357 **1T ECM, MIL: Yes** **Year:** 2009, 2010 **Model:** Land Cruiser **Engine:** All **Transmission:** All	**Ignition Coil "G" Primary / Secondary Circuit:** No IGF signal is sent to the ECM while the engine is running. **Possible Causes:** • Ignition system • Open or short in IGF1 or IGT7 circuit between ignition coil with igniter and ECM • No. 7 ignition coil with igniter • ECM
DTC: P0358 **1T ECM, MIL: Yes** **Year:** 2009, 2010 **Model:** Land Cruiser **Engine:** All **Transmission:** All	**Ignition Coil "H" Primary / Secondary Circuit:** No IGF signal is sent to the ECM while the engine is running. **Possible Causes:** • Ignition system • Open or short in IGF2 or IGT8 circuit between ignition coil with igniter and ECM • No. 8 ignition coil with igniter • ECM

DTC	Trouble Code Title, Conditions & Possible Causes
DTC: P0365 **1T ECM, MIL: Yes** **Year:** 2009, 2010 **Model:** All **Engine:** 2.5L L4, 3.5L V6, 5.7L V8 **Transmission:** All	**Camshaft Position Sensor "B" Circuit (Bank 1):** Monitor runs whenever following DTCs are not present: None Engine speed: 600 rpm or more Starter: OFF Exhaust VVT sensor range check fail (P0367, P0368, P0392, P0393): Not detected Exhaust VVT sensor voltage: 0.3 V or more, and 4.7 V or less Battery voltage: 8 V or more Engine switch: On (IG) Time after engine switch off to on (IG): 0.5 seconds or more **Possible Causes:** • Open or short in VVT sensor for exhaust camshaft circuit • VVT sensor for exhaust camshaft • Exhaust camshaft • ECM
DTC: P0367 **1T ECM, MIL: Yes** **Year:** 2009, 2010 **Model:** Avalon, Camry **Engine:** 2.5L L4 **Transmission:** All	**Camshaft Position Sensor "B" Circuit Low Input (Bank 1):** Monitor runs whenever following DTCs are not present: None Starter: OFF Engine switch: On (IG) Time after engine switch off to on (IG): 2 seconds or more Exhaust VVT sensor verify pulse input fail (P0365, P0390): Not detected Battery voltage: 8 V or more **Possible Causes:** • Open or short in VVT sensor for exhaust camshaft circuit • VVT sensor for exhaust camshaft • Exhaust camshaft • ECM
DTC: P0367 **1T ECM, MIL: Yes** **Year:** 2009, 2010 **Model:** All **Engine:** All V6, V8 **Transmission:** All	**Camshaft Position Sensor "B" Circuit Low Input (Bank 1):** The output voltage of the VVT sensor is below 0.3 V for 4 seconds. **Possible Causes:** • Open or short in VVT sensor (for exhaust side) circuit • VVT sensor (for exhaust side) • Exhaust camshaft • Jumped tooth of timing chain • ECM
DTC: P0368 **1T ECM, MIL: Yes** **Year:** 2009, 2010 **Model:** All **Engine:** V6, V8 **Transmission:** All	**Camshaft Position Sensor "B" Circuit High Input (Bank 1):** The output voltage of the VVT sensor is higher than 4.7 V for 4 seconds. **Possible Causes:** • Open or short in VVT sensor (for exhaust side) circuit • VVT sensor (for exhaust side) • Exhaust camshaft • Jumped tooth of timing chain • ECM
DTC: P0368 **1T ECM, MIL: Yes** **Year:** 2009, 2010 **Model:** Avalon, Camry **Engine:** 2.5L L4 **Transmission:** All	**Camshaft Position Sensor "B" Circuit High Input (Bank 1):** Monitor runs whenever following DTCs are not present: None Starter: OFF Engine switch: On (IG) Time after engine switch off to on (IG): 2 seconds or more Exhaust VVT sensor verify pulse input fail (P0365, P0390): Not detected Battery voltage: 8 V or more **Possible Causes:** • Open or short in VVT sensor for exhaust camshaft circuit • VVT sensor for exhaust camshaft • Exhaust camshaft • ECM

DTC	Trouble Code Title, Conditions & Possible Causes
DTC: P0390 **1T ECM, MIL: Yes** **Year:** 2009, 2010 **Model:** All **Transmission:** All	**Camshaft Position Sensor "B" Circuit (Bank 2):** When either condition below is met: (a) No VVT sensor signal to ECM at engine speed of 600 rpm or more. (b) Input voltage to ECM remains at less than 0.3 V, or more than 4.7 V for more than 4 seconds when 2 or more seconds have elapsed after turning engine switch to on (IG). **Possible Causes:** • Open or short in VVT sensor for exhaust camshaft circuit • VVT sensor for exhaust camshaft • Exhaust camshaft • Jumped tooth of timing chain • ECM
DTC: P0392 **1T ECM, MIL: Yes** **Year:** 2009, 2010 **Model:** All **Engine:** All V6, V8 **Transmission:** All	**Camshaft Position Sensor "B" Circuit Low Input (Bank 2):** Monitor runs whenever following DTCs are not present: None Starter: OFF Ignition switch: ON Time after ignition switch off to on: 2 seconds or more Battery voltage: 8 V or more **Possible Causes:** • Open or short in VVT sensor for exhaust camshaft circuit • VVT sensor for exhaust camshaft • Exhaust camshaft • Jumped tooth of timing chain • ECM
DTC: P0393 **1T ECM, MIL: Yes** **Year:** 2009, 2010 **Model:** All **Engine:** All V6. V8 **Transmission:** All	**Camshaft Position Sensor "B" Circuit High Input (Bank 2):** The output voltage of the VVT sensor is higher than 4.7 V for 4 seconds. **Possible Causes:** • Open or short in VVT sensor for exhaust camshaft circuit • VVT sensor for exhaust camshaft • Exhaust camshaft • Jumped tooth of timing chain • ECM
DTC: P0412 **1T ECM, MIL: Yes** **Year:** 2009, 2010 **Model:** All **Engine:** All **Transmission:** All	**Secondary Air Injection System Switching Valve "A" Circuit:** The monitor will run whenever this DTC is not present: None Case 1: Air pump: Operating Air switching valve: Operating Battery voltage: 8 V or more Ignition switch: ON Starter: OFF Case 2: Air pump: Not operating Air switching valve: Not operating Battery voltage: 8 V or more Ignition switch: ON Starter: OFF **Possible Causes:** • Open in air switching valve drive circuit • Short between air switching valve circuit and +B circuit • Air injection control driver (AID) • Air switching valve (ASV) • ECM

DTC	Trouble Code Title, Conditions & Possible Causes
DTC: P0415 **1T ECM, MIL: Yes** **Year:** 2009, 2010 **Model:** Land Cruiser **Engine:** 5.7L V8 **Transmission:** All	**Secondary Air Injection System Switching Valve "B" Circuit:** After a cold engine start, all of the following conditions are met: 1. The Secondary Air Injection (AIR) system is not operating (air pump OFF/ON, Air Switching Valve [ASV] OFF/ON). 2. The diagnostic signal from the Air Injection Control Driver (AID) is 40%. 3. The battery voltage is 8 V or higher. **Possible Causes:** • Open in Air Switching Valve (ASV) drive circuit • Short between ASV drive circuit and body ground • Air Injection Control Driver (AID) • Air Switching Valve (ASV) • ECM
DTC: P0418 **1T ECM, MIL: Yes** **Year:** 2009, 2010 **Model:** 4Runner **Engine:** 2.7L L4, 4.7L V8 **Transmission:** All	**Secondary Air Injection System Control "A" Circuit:** The monitor will run whenever this DTC is not present: None Case 1: Air pump: Operating Air switching valve: Operating Battery voltage: 8 V or more Ignition switch: ON Starter: OFF Case 2: Air pump: Not operating Air switching valve: Not operating Battery voltage: 8 V or more Ignition switch: ON Starter: OFF **Possible Causes:** • Open in air pump drive circuit • Short between air pump circuit and +B circuit • Air injection control driver (AID) • ECM
DTC: P0418 **1T ECM, MIL: Yes** **Year:** 2009, 2010 **Model:** Land Cruiser **Engine:** 5.7L V8 **Transmission:** All	**Secondary Air Injection System Control "A" Circuit:** After a cold engine start, all of the following conditions are met: 1. Secondary Air Injection (AIR) system not operating (air pump OFF/ON, Air Switching Valve [ASV] OFF/ON) 2. The diagnostic signal from the Air Injection Control Driver (AID) is 20%. 3. The battery voltage is 8 V or higher. **Possible Causes:** • Open in air pump drive circuit • Short between air pump drive circuit and body ground • Air pump • Air injection control driver (AID) • ECM
DTC: P0419 **1T ECM, MIL: Yes** **Year:** 2009, 2010 **Model:** Land Cruiser **Engine:** 5.7L V8 **Transmission:** All	**Secondary Air Injection System Control "B" Circuit:** After a cold engine start, all of the following conditions are met: 1. Secondary Air Injection (AIR) system not operating (air pump OFF/ON, Air Switching Valve [ASV] OFF/ON) 2. The diagnostic signal from the Air Injection Control Driver (AID) is 20%. 3. The battery voltage is 8 V or higher. **Possible Causes:** • Open in air pump drive circuit • Short between air pump drive circuit and body ground • Air pump • Air injection control driver (AID) • ECM

DTC	Trouble Code Title, Conditions & Possible Causes
DTC: P0420 **2T ECM, MIL: Yes** **Year:** 2009, 2010 **Model:** All **Engine:** All V6 **Transmission:** All	**Catalyst System Efficiency Below Threshold (Bank 1):** Battery voltage: 11 V or more Intake air temperature: -10°C (14°F) or more Engine coolant temperature: 75°C (167°F) or more Atmospheric pressure coefficient: 0.75 or more Idling: OFF Engine RPM: Less than 3200 rpm A/F sensor status: Activated Fuel system status: Closed loop Engine load: 5% or more and less than 80% All of the following conditions (a), (b) and (c) are met: (a) Mass air flow rate: 10 to 64 g/sec. (b) Estimated front catalyst temperature: 530 to 660°C (986 to 1220°F) (c) Estimated rear catalyst temperature: 430 to 675°C (806 to 1247°F) **Possible Causes:** • Gas leakage from exhaust system • Exhaust pipe • Air fuel ratio sensor (bank 1 sensor 1) • Heated oxygen sensor (bank 1 sensor 2) • Exhaust manifold sub-assembly RH (TWC: Front catalyst) • Center exhaust pipe assembly (TWC: Rear catalyst)
DTC: P0420 **2T ECM, MIL: Yes** **Year:** 2009, 2010 **Model:** All **Engine:** All L4 **Transmission:** All	**Catalyst System Efficiency Below Threshold (Bank 1):** The oxygen storage capacity value is smaller than the standard value under active air fuel ratio control. **Possible Causes:** • Gas leak from exhaust system • Air fuel ratio sensor (sensor 1) • Heated oxygen sensor (sensor 2) • Exhaust manifold converter sub-assembly (TWC: Front catalyst) • Center manifold converter sub-assembly (TWC: Rear catalyst
DTC: P0420 **2T ECM, MIL: Yes** **Year:** 2009, 2010 **Model:** All **Engine:** All V8 **Transmission:** All	**Catalyst System Efficiency Below Threshold (Bank 1):** Enabling conditions are as follows: Battery voltage 11 V or higher, Intake air temperature -10°C (14°F) or higher, Engine coolant temperature 75°C (167°F) or higher, Atmospheric pressure 76 kPa (570 mmHg) or higher, Idling OFF, Engine speed less than 3200 rpm, A/F sensor status activated, Fuel system status Closed loop, Engine load 10 to 75%. All of following conditions (a), (b) and (c) met (a) Mass air flow rate 7.5 to 75 g/sec. (b) Estimated front catalyst temperature 575 to 820°C (1067 to 1508°F) (c) Estimated rear catalyst temperature 400 to 700°C (752 to 1292°F), Rear HO2 sensor heater monitor Completed, Shift position 4th or higher. The OSC value is less than the standard value under active air-fuel ratio control. **Possible Causes:** • Gas leak from exhaust system • Air-Fuel Ratio (A/F) sensor (for Bank 1 Sensor 1) • Heated Oxygen (HO2) sensor (for Bank 1 Sensor 2) • Front No. 2 exhaust pipe assembly (TWC: Front Catalyst and Rear Catalyst)

DTC	Trouble Code Title, Conditions & Possible Causes
DTC: P0430 **2T ECM, MIL: Yes** **Year:** 2009, 2010 **Model:** All **Engine:** All V8 **Transmission:** All	**Catalyst System Efficiency Below Threshold (Bank 2):** Enabling conditions are as follows: Battery voltage 11 V or higher, Intake air temperature -10°C (14°F) or higher, Engine coolant temperature 75°C (167°F) or higher, Atmospheric pressure 76 kPa (570 mmHg) or higher, Idling OFF, Engine speed less than 3200 rpm, A/F sensor status activated, Fuel system status Closed loop, Engine load 10 to 75%. All of following conditions (a), (b) and (c) met (a) Mass air flow rate 7.5 to 75 g/sec. (b) Estimated front catalyst temperature 575 to 820°C (1067 to 1508°F) (c) Estimated rear catalyst temperature 400 to 700°C (752 to 1292°F), Rear HO2 sensor heater monitor Completed, Shift position 4th or higher. The OSC value is less than the standard value under active air-fuel ratio control. **Possible Causes:** • Gas leak from exhaust system • Air-Fuel Ratio (A/F) sensor (for Bank 2 Sensor 1) • Heated Oxygen (HO2) sensor (for Bank 2 Sensor 2) • Front exhaust pipe assembly (TWC: Front Catalyst and Rear Catalyst)
DTC: P0430 **2T ECM, MIL: Yes** **Year:** 2009, 2010 **Model:** All **Engine:** All V6 **Transmission:** All	**Catalyst System Efficiency Below Threshold (Bank 2):** Battery voltage: 11 V or more IAT: 14°F (-10°C) or more ECT: 167°F (75°C) or more Atmospheric pressure: 570 mmHg (76 kPa) or more Idle: OFF Enigne RPM: Less than 3200 rpm A/F Sensor: Activated Fuel system status: Closed loop Engine load: 10-70% All of the following conditions 1, 2 and 3 are met: 1. MAF: 5-60 g/sec. 2. Front catalyst tempperature (estimated): 1112-1472°F (600-800°C) 3. Rear catalyst temperature (estimated): 212-1652°F (100-900°C) Shift position: 4rd or higher **Possible Causes:** • Gas leakage from exhaust system • A/F sensor (bank 2 sensor 1) • HO2 sensor (bank 2 sensor 2) • Exhaust manifold sub-assembly LH (TWC: Front catalyst) • Front exhaust pipe assembly (TWC: Rear catalyst)
DTC: P043E **2T ECM, MIL: Yes** **Year:** 2009, 2010 **Model:** All **Engine:** All **Transmission:** All	**Evaporative Emission System Reference Orifice Clog Up:** The reference orifice is clogged. Reference orifice high-flow. A leak detection pump OFF malfunction. A leak detection pump ON malfunction. A vent valve ON (close) malfunction. **Possible Causes:** • Canister pump module (Reference orifice, leak detection pump, vent valve) • Connector/wire harness (Canister pump module - ECM) • EVAP system hose (pipe from air inlet port to canister pump module, canister filter, fuel tank vent hose) • ECM

DTC	Trouble Code Title, Conditions & Possible Causes
DTC: P043F **2T ECM, MIL: Yes** **Year:** 2009, 2010 **Model:** All **Engine:** All **Transmission:** All	**Evaporative Emission System Reference Orifice High Flow:** The reference orifice is clogged. Reference orifice high-flow. A leak detection pump OFF malfunction. A leak detection pump ON malfunction. A vent valve ON (close) malfunction. **Possible Causes:** • Canister pump module (Reference orifice, leak detection pump, vent valve) • Connector/wire harness (Canister pump module - ECM) • EVAP system hose (pipe from air inlet port to canister pump module, canister filter, fuel tank vent hose) • ECM
DTC: P0441 **2T ECM, MIL: Yes** **Year:** 2009, 2010 **Model:** All **Engine:** All **Transmission:** All	**Evaporative Emission Control System Incorrect Purge Flow:** The leak detection pump creates negative pressure (Vacuum) in the EVAP system and the EVAP system pressure is measured. The reference pressure is measured at the start and at the end of the leak check. If the stabilized pressure is higher than [second reference pressure x 0.2], the ECM determines that the purge VSV is stuck open. After the EVAP leak check is performed, the purge VSV is turned on (open), and atmospheric air is introduced into the EVAP system. The reference pressure is measured at the start and at the end of the check. If the pressure does not return to near atmospheric pressure, the ECM determines that the purge VSV is stuck closed. While the engine is running, the following conditions are successively met: Negative pressure is not created in the EVAP system when the purge VSV is turned on (open). The EVAP system pressure change is below 0.5 kPa-g (3.75 mmHg-g) when the vent valve is turned on (closed). The atmospheric pressure change before and after the purge flow monitor is below 0.1 kPa-g (0.75 mmHg-g). **Possible Causes:** • Purge VSV • Connector/wire harness (Purge VSV - ECM) • ECM • Canister pump module • Leakage from EVAP system • Leakage from EVAP line (Purge VSV - Intake manifold)
DTC: P0443 **1T ECM, MIL: Yes** **Year:** 2009, 2010 **Model:** All **Engine:** All **Transmission:** All	**Evaporative Emission Control System Purge Control Valve Circuit:** Terminal voltage of ECM output circuit does not correspond with drive signals from ECM to purge VSV. **Possible Causes:** • Open or short in purge VSV circuit • Purge VSV • ECM
DTC: P0450 **1T , MIL: Yes** **Year:** 2009 **Model:**All **Engine:** All **Transmission:** All	**Evaporative Emission Control System Pressure Sensor / Switch:** Monitor runs whenever following DTCs are not present: None Either of following conditions is met: (a) or (b) (a) Ignition switch: ON (b) Soak timer: ON Battery voltage: 8 V or more Starter: OFF **Possible Causes:** • Canister pump module • EVAP system hose (pipe from air inlet port to canister pump module, canister filter, fuel tank vent hose) • ECM

DTC	Trouble Code Title, Conditions & Possible Causes
DTC: P0451 **2T ECM, MIL: Yes** **Year:** 2009, 2010 **Model:** All **Engine:** All **Transmission:** All	**Evaporative Emission Control System Pressure Sensor Range / Performance:** Sensor output voltage fluctuates frequently in certain time period. Sensor output voltage does not vary in certain time period. **Possible Causes:** • Canister pump module • Connector/wire harness (Canister pump module - ECM) • EVAP system hose (pipe from air inlet port to canister pump module, canister filter, fuel tank vent hose) • ECM
DTC: P0452 **2T ECM, MIL: Yes** **Year:** 2009, 2010 **Model:** All **Engine:** All **Transmission:** All	**Evaporative Emission Control System Pressure Sensor / Switch Low Input:** The EVAP pressure is below 42.1 kPa for 0.5 seconds. Enabling conditions are as follows: Battery voltage 8 V or more, Starter OFF, Either of following conditions met (a) Engine switch On (IG) (b) Soak timer ON **Possible Causes:** • Canister pump module • Connector/wire harness (Canister pump module - ECM) • EVAP system hose (pipe from air inlet port to canister pump module, canister filter, fuel tank vent hose) • ECM
DTC: P0453 **1T ECM, MIL: Yes** **Year:** 2009, 2010 **Model:** All **Engine:** All **Transmission:** All	**Evaporative Emission Control System Pressure Sensor / Switch High Input:** The EVAP pressure is higher than 123.8 kPa for 0.5 seconds. Enabling conditions are as follows: Battery voltage 8 V or more, Starter OFF, Either of following conditions met (a) Engine switch On (IG) (b) Soak timer ON **Possible Causes:** • Canister pump module • Connector/wire harness (Canister pump module - ECM) • EVAP system hose (pipe from air inlet port to canister pump module, canister filter, fuel tank vent hose) • ECM
DTC: P0455 **2T ECM, MIL: Yes** **Year:** 2009, 2010 **Model:** All **Engine:** All **Transmission:** All	**Evaporative Emission Control System Leak Detected (Gross Leak):** The leak detection pump creates negative pressure (vacuum) in the EVAP system and the EVAP system pressure is measured. The reference pressure is measured at the start and at the end of the leak check. If the stabilized pressure is higher than [second reference pressure x 0.2], the ECM determines that the EVAP system has a large leak. **Possible Causes:** • Fuel cap (loose) • Leakage from EVAP line (Canister - Fuel tank) • Leakage from EVAP line (Purge VSV - Canister) • Canister pump module • Leakage from fuel tank • Leakage from canister
DTC: P0456 **2T ECM, MIL: Yes** **Year:** 2009, 2010 **Model:** Land Cruiser **Engine:** ALL **Transmission:** All	**Evaporative Emission Control System Leak Detected (Very Small Leak):** The leak detection pump creates negative pressure (vacuum) in the EVAP system and the EVAP system pressure is measured. The reference pressure is measured at the start and at the end of the leak check. If the stabilized pressure is higher than the second reference pressure, the ECM determines that the EVAP system has a small leak. **Possible Causes:** • Fuel cap (loose) • Leakage from EVAP line (Canister - Fuel tank) • Leakage from EVAP line (Purge VSV - Canister) • Canister pump module • Leakage from fuel tank • Leakage from canister

DTC	Trouble Code Title, Conditions & Possible Causes
DTC: P0500 **2T ECM, MIL: Yes** **Year:** 2009, 2010 **Model:** All **Engine:** All **Transmission:** All	**Vehicle Speed Sensor "A":** While the vehicle is being driven, no vehicle speed sensor signal is transmitted to the ECM . **Possible Causes:** • Open or short in speed signal circuit • Wheel speed sensor • Combination meter • ECM • Skid control ECU
DTC: P0503 **T ECM, MIL: Yes** **Year:** 2009, 2010 **Model:** All **Engine:** All L4, V6 **Transmission:** All	**Vehicle Speed Sensor Circuit Malfunction:** Momentary interruption and noise malfunction codes are detected when a rapid change of vehicle speed occurs while the cruise control is in operation. **Possible Causes:** • Combination meter assembly • Vehicle speed sensor • Vehicle speed sensor circuit • ECM
DTC: P0503 **1T ECM, MIL: Yes** **Year:** 2009 **Model:** All **Engine:** All V8 **Transmission:** All	**Vehicle Speed Sensor "A" Intermittent / Erratic / High:** The monitor will run whenever this DTC is not present: P0120 - P0223, P2135 (TP sensor) Time after ignition switch off to ON: 3 seconds or more Vehicle speed: 9 km/h (5.6 mph) or more Battery voltage: 8 V or higher Starter: OFF **Possible Causes:** • Open or short in vehicle speed sensor circuit • Vehicle speed sensor • Combination meter • ECM • Skid control ECU
DTC: P0504 **1T ECM, MIL: Yes** **Year:** 2009, 2010 **Model:** All **Engine:** All **Transmission:** All	**Brake Switch "A" / "B" Correlation:** Conditions (a), (b) and (c) continue for 0.5 seconds or more: (a) The engine switch is on (IG). (b) The brake pedal is released. (c) The STP signal is OFF when the ST1- signal is OFF. **Possible Causes:** • Short in stop light switch signal circuit • STOP fuse • IGN fuse • Stop light switch • ECM
DTC: P0505 **2T ECM, MIL: Yes** **Year:** 2009, 2010 **Model:** All **Engine:** All **Transmission:** All	**Idle Control System Malfunction:** With the engine running the idle speed continues to vary greatly from the target idling speed. **Possible Causes:** • Electronic throttle control system • Intake system • PCV hose connections • ECM
DTC: P050A **2T ECM, MIL: Yes** **Year:** 2009, 2010 **Model:** All **Engine:** All **Transmission:** All	**Cold Start Idle Air Control System Performance:** Battery voltage 8 V or more, time after engine start 3 seconds or more, starter OFF. ECT at engine start -10°C (14°F) or more, ECT -10 to 50°C (14 to 122°F). Engine idling 3 seconds or more, Fuel cut OFF, Vehicle speed less than 3 km/h (1.875 mph). Time after shift position changed 2 seconds or more, Atmospheric pressure 76 kPa (570 mmHg) or higher. Mass air flow is insufficient at cold start. **Possible Causes:** • Throttle body • Mass air flow meter • PCV hose • Air cleaner filter • Air induction system • VVT system • ECM

DTC	Trouble Code Title, Conditions & Possible Causes
DTC: P050B **2T ECM, MIL: Yes** **Year:** 2009, 2010 **Model:** All **Engine:** All **Transmission:** All	**Cold Start Ignition Timing Performance:** Battery voltage 8 V or more, time after engine start 3 seconds or more, starter OFF. ECT at engine start -10°C (14°F) or more, ECT -10 to 50°C (14 to 122°F). Engine idling 3 seconds or more, Fuel cut OFF, Vehicle speed less than 3 km/h (1.875 mph). Time after shift position changed 2 seconds or more, Atmospheric pressure 76 kPa (570 mmHg) or higher. The ignition timing retard is insufficient at cold start. **Possible Causes:** • Throttle body assembly • Mass air flow meter • PCV system • Air cleaner filter element • Air induction system • VVT system • ECM
DTC: P0550 **1T ECM, MIL: Yes** **Year:** 2009, 2010 **Model:** All **Engine:** All **Transmission:** All	**Power Steering Pressure Sensor Circuit Malfunction:** An open or short in the power steering oil pressure sensor circuit is detected for more than 3 seconds after the engine switch is turned on (IG). **Possible Causes:** • Open or short in power steering oil pressure sensor circuit • Power steering oil pressure sensor • ECM
DTC: P0553 **1T ECM, MIL: Yes** **Year:** 2009, 2010 **Model:** All **Engine:** All **Transmission:** All	**Power Steering Pressure Sensor Circuit High Input:** Power steering oil pressure sensor voltage higher than 4.9 V for 0.5 seconds while engine running. **Possible Causes:** • Open or short in power steering oil pressure sensor circuit • Power steering oil pressure sensor • ECM
DTC: P0560 **1T ECM, MIL: Yes** **Year:** 2009, 2010 **Model:** Land Cruiser **Engine:** All **Transmission:** All	**System Voltage:** An open in the ECM back up power source circuit. **Possible Causes:** • Open in back up power source circuit • Battery • Battery terminals • EFI MAIN fuse • ECM
DTC: P0571 **T ECM, MIL: Yes** **Year:** 2009, 2010 **Model:** All **Engine:** All **Transmission:** All	**Brake Switch "A" Circuit:** Stop light switch assembly does not turn off even once the vehicle is driven. **Possible Causes:** • Stop light switch • Stop light switch circuit • ECM
DTC: P0575 **T ECM, MIL: Yes** **Year:** 2009, 2010 **Model:** All **Engine:** All **Transmission:** All	**Cruise Control Input Circuit:** When either of the following conditions is met: * Cruise control input signal abnormal * Stop light switch input signal abnormal **Possible Causes:** • ECM
DTC: P0604 **1T ECM, MIL: Yes** **Year:** 2009, 2010 **Model:** All **Engine:** All **Transmission:** All	**Internal Control Module Random Access Memory (RAM) Error:** The ECM continuously monitors its internal memory status. This self-check ensures that the ECM is functioning properly. It is diagnosed by internal "mirroring" of the main CPU and sub CPU to detect the Random Access Memory (RAM) errors. If outputs from these CPUs are different and deviate from the standards, the ECM will illuminate the MIL and set a DTC immediately. Monitor will run whenever this DTC is not present: None **Possible Causes:** • ECM

DTC	Trouble Code Title, Conditions & Possible Causes
DTC: P0606 **T ECM, MIL: Yes** **Year:** 2009, 2010 **Model:** All **Engine:** All **Transmission:** All	**ECM / PCM Processor:** When either condition below is met: An ECM main CPU error. An ECM sub CPU error. The ECM will illuminate the MIL and store DTC(s) immediately. **Possible Causes:** • ECM
DTC: P0607 **T ECM, MIL: Yes** **Year:** 2009, 2010 **Model:** All **Engine:** All **Transmission:** All	**Input Signal Circuit Malfunction:** The ECM has a supervisory CPU and control ECU inside. When each input STP signal is different for 0.15 sec. or more, this trouble code is output. This trouble code is output after 0.4 sec. has passed from the time the cruise cancel input signal (STP input) is input into the ECM. Hint: When a trouble code is detected, the fail safe must be kept on until the ignition switch is turned off. **Possible Causes:** • ECM
DTC: P0607 **T ECM, MIL: Yes** **Year:** 2009, 2010 **Model:** All **Engine:** All **Transmission:** All	**Control Module Performance:** When one of following conditions is met: The ECM CPUs malfunction. The A/F sensor transistors malfunction. The HO2S transistors malfunction. **Possible Causes:** • Exhaust gas leak • Heated oxygen sensor • ECM
DTC: P060A **T ECM** **Year:** 2009, 2010 **Model:** All **Engine:** All **Transmission:** All	**Internal Control Module Monitoring Processor Performance:** When either condition below is met: An ECM main CPU error. An ECM sub CPU error. An electronic throttle monitoring CPU error. **Possible Causes:** • ECM
DTC: P060B **T ECM** **Year:** 2009, 2010 **Model:** All **Engine:** All **Transmission:** All	**Internal Control Module A/D Processing Performance:** An ECM main CPU communication error. **Possible Causes:** • ECM
DTC: P060D **T ECM** **Year:** 2009, 2010 **Model:** All **Engine:** All **Transmission:** All	**Internal Control Module Accelerator Pedal Position Performance:** When either condition below is met: An ECM main CPU error. An ECM sub CPU error. **Possible Causes:** • ECM
DTC: P060E **T ECM** **Year:** 2009, 2010 **Model:** All **Engine:** All **Transmission:** All	**Internal Control Module Throttle Position Performance:** When either condition below is met: An ECM main CPU error. An ECM sub CPU error. **Possible Causes:** • ECM

DTC	Trouble Code Title, Conditions & Possible Causes
DTC: P0617 **1T ECM, MIL: Yes** **Year:** 2009, 2010 **Model:** All **Engine:** All **Transmission:** All	**Starter Relay Circuit High:** Monitor runs whenever this DTC is not present: None Battery voltage: 10.5 V or more Vehicle speed: 12.43 mph (20 km/h) or more Engine speed: 1000 rpm or more **Possible Causes:** • Park/Neutral Position (PNP) switch • Starter relay circuit • Cranking holding function circuit • ECM
DTC: P062F **T ECM, MIL: Yes** **Year:** 2009, 2010 **Model:** All **Engine:** All **Transmission:** All	**Internal Control Module EEPROM Error:** Time after engine start 10 seconds or more, battery voltage 8 V or more. Engine switch On (IG), starter OFF. ECM internal error (EEPROM). The ECM monitors its internal operation and it will set this DTC when it detects an internal malfunction. **Possible Causes:** • ECM
DTC: P0630 **1T ECM, MIL: Yes** **Year:** 2009, 2010 **Model:** All **Engine:** All **Transmission:** All	**VIN not Programmed or Mismatch - ECM / PCM:** When either condition below is met: The VIN is not stored in the ECM. The input VIN is incorrect. **Possible Causes:** • ECM
DTC: P0657 **T ECM, MIL: Yes** **Year:** 2009, 2010 **Model:** All **Engine:** All **Transmission:** All	**Actuator Supply Voltage Circuit / Open:** With the engine switch ON (IG) to OFF and throttle actuator power supply voltage 7 V or more, a throttle actuator power supply error is detected. The ECM will illuminate the MIL and store DTC when the engine switch is turned on (IG **Possible Causes:** • ECM
DTC: P0705 **2T ECM, MIL: Yes** **Year:** 2009, 2010 **Model:** All **Engine:** All **Transmission:** All	**Transmission Range Sensor Circuit Malfunction (PRNDL Input):** Monitor runs whenever following DTCs are not present: None Ignition switch: ON Battery voltage: 10.5 V or more Starter: OFF Condition B: One of the following conditions is met: Condition (a), (b), (c) or (d) (a) Park/neutral position switch: ON (b) Park position switch: ON (c) Neutral position switch: ON (d) R range position switch: ON **Possible Causes:** • Open or short in park/neutral position switch circuit • Short in park/neutral position switch assembly • Open or short in transmission control switch circuit • Shift lock control unit assembly • ECM
DTC: P0710 **1T TCM, MIL: Yes** **Year:** 2009, 2010 **Model:** Avalon, Camry **Engine:** 2.4L L4, 3.5L V6 **Transmission:** All	**Transmission Fluid Temperature Sensor "A" Circuit:** The monitor will run whenever this DTC is not present: None The typical enablling condition is not available. **Possible Causes:** • Open or short in ATF temperature sensor circuit • Transmission wire • ATF temperature sensor • TCM

DTC	Trouble Code Title, Conditions & Possible Causes
DTC: P0710 **1T TCM, MIL: Yes** **Year:** 2009, 2010 **Model:** Highlander **Engine:** 2.7L L4, 3.5L V6 **Transmission:** All	**Transmission Fluid Temperature Sensor "A" Circuit:** (a) and (b) are detected momentarily within 0.5 sec. when neither P0712 nor P0713 is detected (1-trip detection logic) (a) ATF temperature sensor resistance is less than 79 W. (b) ATF temperature sensor resistance is more than 156 kW. HINT: Within 0.5 sec., the malfunction switches from (a) to (b) or from (b) to (a) **Possible Causes:** • Open or short in ATF temperature sensor circuit • Transmission wire (ATF temperature sensor) • ECM
DTC: P0711 **2T ECM, MIL: Yes** **Year:** 2009, 2010 **Model:** All **Engine:** All **Transmission:** All	**Transmission Fluid Temperature Sensor "A" Performance:** Both (a) and (b) are detected (2-trip detection logic): (a) The intake air and engine coolant temperatures are higher than -10°C (14°F) at engine start. (b) After normal driving for more than 22 min. for 9 km (6 miles) or more, the No. 1 ATF temperature is below 20°C (68°F). **Possible Causes:** • Transmission wire • ATF temperature sensor • TCM/ECM
DTC: P0712 **1T ECM, MIL: Yes** **Year:** 2009, 2010 **Model:** All **Engine:** All **Transmission:** All	**Transmission Fluid Temperature Sensor "A" Circuit Low Input:** The monitor will run whenever this DTC is not present: None The typical enabling condition is not available. **Possible Causes:** • Short in ATF temperature sensor circuit • Transmission wire • ATF temperature sensor • TCM/ECM
DTC: P0713 **1T ECM, MIL: Yes** **Year:** 2009, 2010 **Model:** All **Engine:** All **Transmission:** All	**Transmission Fluid Temperature Sensor "A" Circuit High Input:** The monitor will run whenever this DTC is not present: None The typical enabling condition is not available. **Possible Causes:** • Open in ATF temperature sensor circuit • Transmission wire • ATF temperature sensor • TCM/ECM
DTC: P0715 **T TCM, MIL: Yes** **Year:** 2009, 2010 **Model:** All **Engine:** All **Transmission:** All	**Input / Turbine Speed Sensor Circuit Malfunction:** Battery voltage: 8 V or more Engine switch: ON Starter: OFF **Possible Causes:** • Transmission revolution sensor (speed sensor NT) • TCM
DTC: P0717 **1T TCM, MIL: Yes** **Year:** 2009, 2010 **Model:** 4Runner, FJ Cruiser **Engine:** 4.0L V6, 4.7L V8 **Transmission:** All	**Turbine Speed Sensor Circuit No Signal:** Shift change: Shift change is completed and before starting next shift change operation ECM selected gear: 4th or 5th Output shaft rpm: 1,000 rpm or more R switch: OFF L switch: OFF Engine: Running Ignition switch: ON Starter: OFF Battery voltage: 8 V or more **Possible Causes:** • Open or short in speed sensor (NT) circuit • Speed sensor (NT) • ECM • Automatic transmission (clutch, brake or gear, etc.)

DTC	Trouble Code Title, Conditions & Possible Causes
DTC: P0717 **1T TCM, MIL: Yes** **Year:** 2010 **Model:** All **Engine:** All **Transmission:** All	**Turbine Speed Sensor Circuit No Signal:** Output shaft rpm: 500 rpm or more NSW switch: OFF Battery voltage: 8 V or higher Ignition switch: ON Starter: OFF R switch: OFF Engine: Running **Possible Causes:** • Open or short in speed sensor NCO circuit • Speed sensor NCO • ECM • Automatic transmission (clutch, brake, gear, etc.)
DTC: P0717 **1T TCM, MIL: Yes** **Year:** 2009, 2010 **Model:** All **Engine:** All **Transmission:** All	**Input Speed Sensor Circuit No Signal:** The monitor will run whenever this DTC is not present:P0500 (VSS), P0748, P0778, P0798 (Shift solenoid valve (range)) Shift change: Shift change is completed before starting next shift change operation TCM selected gear: 2nd, 3rd, 4th, 5th or 6th Output shaft rpm: 1,000 rpm or more STAR switch: OFF R switch: OFF Engine: Running Battery voltage: 8 V or more Engine switch: ON Starter: OFF **Possible Causes:** • Transmission revolution sensor (speed sensor NT) • TCM
DTC: P0722 **2T TCM, MIL: Yes** **Year:** 2010 **Model:** All **Engine:** All **Transmission:** All	**Output Speed Sensor Circuit No Signal:** The monitor will run whenever the following DTCs are not stored: None Battery voltage: 8 V or higher Ignition switch: ON Starter: OFF Vehicle speed from vehicle speed sensor: 9 km/h (5.59 mph) or more **Possible Causes:** • Open or short in speed sensor SP2 circuit • Speed sensor SP2 • ECM • Automatic transmission (clutch, brake, gear, etc.)
DTC: P0724 **2T ECM, MIL: Yes** **Year:** 2009, 2010 **Model:** All **Engine:** All **Transmission:** All	**Brake Switch "B" Circuit High:** The stop light switch remains on even when the vehicle is driven in a STOP (less than 3 km/h (2 mph)) and GO (30 km/h (19 mph) or more) pattern 5 times. **Possible Causes:** • Short in stop light switch signal circuit • Stop light switch • ECM
DTC: P0729 **2T ECM, MIL: Yes** **Year:** 2009, 2010 **Model:** Land Cruiser **Engine:** 5.7L V8 **Transmission:** All	**Gear 6 Incorrect Ratio:** A 6th gear shift malfunction: The ECM determines there is a malfunction when both of the following conditions are met: (a) When the ECM directs the transmission to switch to 5th gear, the actual gear is also shifted to 5th. (b) When the ECM directs the transmission to switch to 6th gear, the actual gear is shifted to 4th. **Possible Causes:** • Valve body is blocked up or stuck (sequence valve) • Shift solenoid valve SLT remains open or closed • Automatic transmission (clutch, brake or gear, etc.)

DTC	Trouble Code Title, Conditions & Possible Causes
DTC: P0741 **2T ECM, MIL: Yes** **Year:** 2009, 2010 **Model:** All **Engine:** All **Transmission:** All	**Torque Converter Clutch Solenoid Performance (Shift Solenoid Valve DSL):** Lock-up does not occur when driving in lock-up range. Lock-up remains ON in lock-up OFF range. **Possible Causes:** • Shift solenoid valve DSL remains open or closed • Valve body is blocked • Torque converter clutch • Automatic transaxle (clutch, brake or gear etc.) • Line pressure is too low
DTC: P0746 **2T ECM, MIL: Yes** **Year:** 2009, 2010 **Model:** All **Engine:** All **Transmission:** All	**Pressure Control Solenoid "A" Performance (Shift Solenoid Valve SL1):** Engine coolant temperature 10°C (50°F) or more and transmission range "D". TFT (Transmission fluid temperature) -20°C (-4°F) or more. The gear required by the ECM does not match the actual gear when driving . **Possible Causes:** • Shift solenoid valve SL1 remains open or closed • Valve body is blocked • Automatic transaxle (clutch, brake or gear etc.)
DTC: P0748 **1T ECM, MIL: Yes** **Year:** 2009, 2010 **Model:** All **Engine:** All **Transmission:** All	**Pressure Control Solenoid "A" Electrical (Shift Solenoid Valve SL1):** The ECM checks for an open or short in the shift solenoid valve SL1 circuit while driving and shifting between 4th and 5th gear. The output signal duty equals 100%. The SL1 output signal duty is less than 100% under normal conditions. **Possible Causes:** • Open or short in shift solenoid valve SL1 circuit • Shift solenoid valve SL1 • ECM
DTC: P0751 **2T TCM, MIL: Yes** **Year:** 2009, 2010 **Model:** All **Engine:** ALL **Transmission:** All	**Shift Solenoid "A" Performance (Shift Solenoid Valve S1):** Transmission position: "D" Duration time from shifting "N" to "D": 4 sec. or more ECT: 40°C (104°F) or higher Spark advance from max. retard timing by KCS control: 0° crankshaft angle or more Engine: Starting **Possible Causes:** • Shift solenoid valve remains (open or closed) • Valve body is blocked • Automatic transmission (clutch, brake, gear, etc.)
DTC: P0761 **2T ECM, MIL: Yes** **Year:** 2009, 2010 **Model:** All **Engine:** All **Transmission:** All	**Shift Solenoid "C" Performance (Shift Solenoid Valve S3):** A shift solenoid valve S3 stuck ON malfunction*1: When the ECM directs the transmission to switch to 5th or 6th gear, the engine overruns (clutch slips). The ECM determines there is a malfunction when either of the following conditions is met: (a) When the ECM directs the transmission to switch to 4th gear, the actual gear is shifted to 3rd. (b) When the ECM directs the transmission to switch to 5th gear, the engine overruns (clutch slips). A shift solenoid valve S3 stuck OFF malfunction*2: Shifting to 1st, 2nd, and 3rd gears is impossible. The ECM determines there is a malfunction when the following conditions are both met: (a) When the ECM directs the transmission to switch to 2nd gear, the actual gear is shifted to 4th. (b) When the ECM directs the transmission to switch to 6th gear, the actual gear is shifted to 6th. **Possible Causes:** • Shift solenoid valve S3 remains open • Shift solenoid valve SLT remains open or closed • Valve body is blocked • Automatic transmission (clutch, brake or gear, etc.) • Shift solenoid valve S3 remains closed • Valve body is blocked • Automatic transmission (clutch, brake or gear, etc.)
DTC: P0766 **2T ECM, MIL: Yes** **Year:** 2009, 2010 **Model:** All **Engine:** All **Transmission:** All	**Shift Solenoid "D" Performance (Shift Solenoid Valve S4):** The gear required by the ECM does not match the actual gear when driving. **Possible Causes:** • Shift solenoid valve S4 remains open or closed • Valve body is blocked • Automatic transaxle (clutch, brake or gear etc.)

DTC	Trouble Code Title, Conditions & Possible Causes
DTC: P0771 **2T TCM, MIL: Yes** **Year:** 2009, 2010 **Model:** 4Runner, FJ Cruiser **Engine:** 4.0L V6, 4.7L V8 **Transmission:** All	**Shift Solenoid "E" Performance (Shift Solenoid Valve SR):** ETCS (Electric throttle control system): Not system down Transmission range: "D" Duration time from shifting "N" to "D": 4 sec. or more ECT: 40°C (104°F) or more Spark advance from Max. retard timing by KCS control: 0° CA or more Engine: Starting Malfunction A: ECM selected gear: 5th Vehicle speed: 2 km/h (1.2 mph) or more Throttle valve opening angle: 2.0% or more at 1,000 rpm (Varies with engine speed) Malfunction B: Current ECM selected gear: 5th Last ECM selected gear: 4th Continuous time of ECM selecting 4th gear: 2 sec. or more Malfunction C: Current ECM selected gear: 5th Last ECM selected gear: 4th ON Malfunction: Current ECM selected gear: 2nd Last ECM selected gear: 1st Throttle valve opening angle: (During transition from 1st to 2nd gear) 6.5% or more at 3,000 rpm (Varies with turbine speed) **Possible Causes:** • Shift solenoid valve SR remains open or closed • Shift solenoid valve SL1 remains closed • Valve body is blocked • Automatic transmission (clutch, brake or gear, etc.)

DTC	Trouble Code Title, Conditions & Possible Causes
DTC: P0771 **2T ECM, MIL: Yes** **Year:** 2009, 2010 **Model:** Camry, Corolla **Engine:** 1.8L L4 VIN L, 1.8L L4 VIN U, 2.4L L4 **Transmission:** All	**Shift Solenoid "E" Performance (Shift Solenoid Valve SR):** The monitor will run whenever this DTC is not present: P0115 - P0118 (ECT sensor) P0125 (Insufficient ECT for closed loop) P0500 (VSS) P0748, P0778, P0798 (Shift solenoid valve (range)) ECT (Engine coolant temperature): 10°C (50°F) or more Transmission range: "D" TFT (Transmission fluid temperature): -20°C (-4°F) or more TFT sensor circuit: Not circuit malfunction ECT sensor circuit: Not circuit malfunction Turbine speed sensor circuit: Not circuit malfunction Intermediate shaft speed sensor circuit: Not circuit malfunction Output speed sensor circuit: Not circuit malfunction Shift solenoid valve SL1 circuit: Not circuit malfunction Shift solenoid valve SL2 circuit: Not circuit malfunction Shift solenoid valve SL3 circuit: Not circuit malfunction Shift solenoid valve S4 circuit: Not circuit malfunction Shift solenoid valve SR circuit: Not circuit malfunction Shift solenoid valve DSL circuit: Not circuit malfunction Electronic throttle system: Not circuit malfunction OFF Malfunction (A): ECM selected gear: 5th Throttle valve opening angle: 5% or more Vehicle speed: 10 km/h (6.2 mph) or more OFF Malfunction (B): ECM lock-up command: ON ECM selected gear: 3rd, 4th or 5th Vehicle speed: 25 km/h (15.5 mph) or more ON Malfunction (A): ECM lock-up command: OFF ON Malfunction (B): ECM selected gear: 1st Vehicle speed: Less than 40 km/h (24.9 mph) Throttle valve opening angle: 4.5% or more at engine speed 1,900 rpm (Varies with engine speed) ON Malfunction (C): ECM selected gear: 3rd Throttle valve opening angle: 4.5% or more at engine speed 1,900 rpm (Varies with engine speed) ON Malfunction (D): Duration time from shift command of ECM: 15 sec. or more ECM selected gear: 4th or 5th **Possible Causes:** • Shift solenoid valve SR remains open or closed • Valve body is blocked • Automatic transaxle (clutch, brake or gear etc.)

DTC	Trouble Code Title, Conditions & Possible Causes
DTC: P0771 **2T TCM, MIL: Yes** **Year:** 2009, 2010 **Model:** Highlander **Engine:** 2.7L L4, 3.5L V6 **Transmission:** All	**Shift Solenoid "E" Performance (Shift Solenoid Valve SR):** ECT (Engine coolant temperature): 10°C (50°F) or more Transmission range: "D" TFT (Transmission fluid temperature): -20°C (-4°F) or more OFF Malfunction (A): ECM selected gear: 5th Throttle valve opening angle: 5% or more Vehicle speed: 10 km/h (6 mph) or more OFF Malfunction (B): ECM lock-up command: ON ECM selected gear: 3rd, 4th or 5th Vehicle speed: 25 km/h (16 mph) or more ON Malfunction (A): ECM lock-up command: OFF ON Malfunction (B): ECM selected gear: 1st Vehicle speed: Less than 40 km/h (25 mph) Throttle valve opening angle: 4.5% or more at engine speed 1900 rpm (Varies with engine speed) ON Malfunction (C): ECM selected gear: 3rd Throttle valve opening angle: 4.5% or more at engine speed 1900 rpm (Varies with engine speed) ON Malfunction (D): Duration time from shift command of ECM: 15 sec. or more ECM selected gear: 4th or 5th **Possible Causes:** • Shift solenoid valve SR remains open or closed • Valve body is blocked • Automatic transaxle (clutch, brake or gear, etc.)
DTC: P0776 **2T ECM, MIL: Yes** **Year:** 2009, 2010 **Model:** All **Engine:** All **Transmission:** All	**Pressure Control Solenoid "B" Performance (Shift Solenoid Valve SL2):** The gear required by the ECM does not match the actual gear when driving. **Possible Causes:** • Shift solenoid valve SL2 remains open or closed • Valve body is blocked • Shift solenoid valve SL2 • Automatic transaxle (clutch, brake or gear etc.) • ECM/TCM
DTC: P0778 **1T TCM, MIL: Yes** **Year:** 2009, 2010 **Model:** All **Engine:** All **Transmission:** All	**Pressure Control Solenoid "B" Electrical (Shift Solenoid Valve SL2):** The monitor will run whenever this DTC is not present: None Solenoid current cut status: Not cut Battery voltage: 11 V or more Ignition switch: ON Starter: OFF CPU commanded duty ratio to SL2: 19% or more **Possible Causes:** • Open or short in shift solenoid valve SL2 circuit • Shift solenoid valve SL2 • ECM/TCM
DTC: P0781 **2T ECM, MIL: Yes** **Year:** 2009, 2010 **Model:** All **Engine:** All **Transmission:** All	**1-2 Shift (1-2 Shift Valve):** When conditions (a) and (b), or (a) and (c) are met: (a) When the ECM directs the transmission to switch to 2nd gear, the actual gear is shifted to 1st. (b) When the ECM directs the transmission to switch to 4th gear, the actual gear is shifted to 3rd. (c) When the ECM directs the transmission to switch to 5th gear, the engine overruns (clutch slips). **Possible Causes:** • Valve body is blocked up or stuck (1-2 shift valve) • Shift solenoid valve SLT remains open or closed • Automatic transmission (clutch, brake or gear, etc.) • ECM/TCM

DTC	Trouble Code Title, Conditions & Possible Causes
DTC: P0791 **T TCM, MIL: Yes** **Year:** 2009, 2010 **Model:** All **Engine:** All **Transmission:** All	**Intermediate Shaft Speed Sensor "A" Circuit:** The monitor will run whenever this DTC is not present: P0500 (VSS), P0748, P0778, P0798 (Shift solenoid valve (range)) Vehicle speed: 15.5 mph (25 km/h) or more Battery voltage: 8 V or more Engine switch: ON Starter: OFF **Possible Causes:** • Transmission revolution sensor (speed sensor NC) • ECM/TCM
DTC: P0793 **1T ECM, MIL: Yes** **Year:** 2009, 2010 **Model:** All **Engine:** All **Transmission:** All	**Intermediate Shaft Speed Sensor "A":** ECM detects conditions (a), (b) and (c) continuously for 5 sec. or more: (a) Vehicle speed: 50 km/h (31 mph) or more. (b) Park/neutral position switch (NSW) is OFF. (c) Speed sensor (NC): less than 300 rpm. **Possible Causes:** • Open or short in transmission revolution sensor NC (speed sensor NC) circuit • Transmission revolution sensor NC (speed sensor NC) • ECM/TCM
DTC: P0794 **1T TCM, MIL: Yes** **Year:** 2009, 2010 **Model:** All **Engine:** All **Transmission:** All	**Shift Solenoid "A" Control Circuit High (Shift Solenoid Valve S1):** The monitor will run whenever this DTC is not present: None Shift solenoid valve S1: OFF **Possible Causes:** • Open in shift solenoid valve S1 circuit • Shift solenoid valve S1 • ECM
DTC: P0796 **2T ECM, MIL: Yes** **Year:** 2009, 2010 **Model:** All **Engine:** All **Transmission:** All	**Pressure Control Solenoid "C" Performance (Shift Solenoid Valve SL3):** With the engine coolant temperature at 10°C (50°F) or more and the transmission range in drive, TFT (Transmission fluid temperature) -20°C (-4°F) or more. The gear required by the ECM does not match the actual gear when driving. **Possible Causes:** • Shift solenoid valve SL3 remains open or closed • Valve body is blocked • Automatic transaxle (clutch, brake or gear etc.)
DTC: P0798 **1T ECM, MIL: Yes** **Year:** 2009, 2010 **Model:** All **Engine:** All **Transmission:** All	**Pressure Control Solenoid "C" Electrical (Shift Solenoid Valve SL3) :** The ECM checks for an open or short in the shift solenoid valve SL3 circuit while driving and shifting gears. The output signal duty equals to 100%. **NOTE: SL3 output signal duty is less than 100% under normal condition.** **Possible Causes:** • Open or short in shift solenoid valve SL3 circuit • Shift solenoid valve SL3 • ECM/TCM
DTC: P0818 **2T ECM, MIL: Yes** **Year:** 2009, 2010 **Model:** All **Engine:** All **Transmission:** All	**Driveline Disconnect Switch Input Circuit:** The signal indicating that the transfer is in neutral remains on while the vehicle is running under the following conditions for 30 sec: The vehicle speed is 25 km/h (16 mph) or more. The transfer position switch is in H4. **Possible Causes:** • Short in transfer neutral signal • Transfer shift actuator assembly (transfer system) • ECM
DTC: P0872 **T TCM, MIL: Yes** **Year:** 2009, 2010 **Model:** Avalon, Camry **Engine:** 2.7L L4, 3.5L V6 **Transmission:** All	**Transmission Fluid Pressure Sensor / Switch "C" Circuit Low:** None available. **Possible Causes:** • ATF temperature sensor assembly (ATF pressure switch No. 1) • Transmission wire • TCM

DTC	Trouble Code Title, Conditions & Possible Causes
DTC: P0873 **T TCM, MIL: Yes** **Year:** 2009, 2010 **Model:** Avalon, Camry **Engine:** 2.7L L4, 3.5L V6 **Transmission:** All	**Transmission Fluid Pressure Sensor / Switch "C" Circuit High:** None provided. **Possible Causes:** • ATF temperature sensor assembly (ATF pressure switch No. 1) • Transmission wire • TCM
DTC: P0877 **T TCM, MIL: Yes** **Year:** 2009, 2010 **Model:** Avalon, Camry **Engine:** 2.7L L4, 3.5L V6 **Transmission:** All	**Transmission Fluid Pressure Sensor / Switch "D" Circuit Low:** None provided. **Possible Causes:** • ATF temperature sensor assembly (ATF pressure switch No. 2) • Transmission wire • TCM
DTC: P0878 **T TCM, MIL: Yes** **Year:** 2009, 2010 **Model:** Avalon, Camry **Engine:** 2.7L L4, 3.5L V6 **Transmission:** All	**Transmission Fluid Pressure Sensor / Switch "D" Circuit High:** None provided. **Possible Causes:** • ATF temperature sensor assembly (ATF pressure switch No. 2) • Transmission wire • TCM
DTC: P0894 **2T ECM, MIL: Yes** **Year:** 2009, 2010 **Model:** Land Cruiser **Engine:** 5.7L V8 **Transmission:** All	**Transmission Component Slipping:** The ECM detects a malfunction of the shift solenoid valve SLT, S1, S2, S3, S4, SL2, gear 6 incorrect ratio (sequence valve) or 1-2 shift valves according to the revolution difference of the turbine and output shaft, and also by the oil pressure. **Possible Causes:** • Shift solenoid valve SLT remains open or closed • Shift solenoid valve S1, S2, S3, S4 or SL2 remains open or closed • Gear 6 incorrect ratio (sequence valve) or 1-2 shift valve is stuck • Valve body is blocked • Automatic transmission (clutch, brake, gear, etc.)
DTC: P0973 **1T TCM, MIL: Yes** **Year:** 2010 **Model:** All **Engine:** All **Transmission:** All	**Shift Solenoid "A" Control Circuit Low (Shift Solenoid Valve S1):** The monitor will run whenever the following DTCs are not stored: None Battery voltage: 8 V or higher Ignition switch: ON Starter: OFF Shift solenoid valve S1: ON **Possible Causes:** • Short in shift solenoid valve S1 circuit • Shift solenoid valve S1 • ECM
DTC: P0974 **1T TCM, MIL: Yes** **Year:** 2010 **Model:** All **Engine:** All **Transmission:** All	**Shift Solenoid "A" Control Circuit High (Shift Solenoid Valve S1):** The monitor will run whenever the following DTCs are not stored: None Battery voltage: 8 V or higher Ignition switch: ON Starter: OFF Shift solenoid valve S1: OFF **Possible Causes:** • Open in shift solenoid valve S1 circuit • Shift solenoid valve S1 • ECM
DTC: P0976 **1T ECM, MIL: Yes** **Year:** 2009, 2010 **Model:** Land Cruiser **Engine:** 5.7L V8 **Transmission:** All	**Shift Solenoid "B" Control Circuit Low (Shift Solenoid Valve S2):** The ECM detects a short in the shift solenoid valve S2 circuit 2 times when the shift solenoid valve S2 is operated. **Possible Causes:** • Short in shift solenoid valve S2 circuit • Shift solenoid valve S2 • ECM

DTC	Trouble Code Title, Conditions & Possible Causes
DTC: P0976 **1T TCM, MIL: Yes** **Year:** 2009, 2010 **Model:** All **Engine:** All **Transmission:** All	**Shift Solenoid "B" Control Circuit Low (Shift Solenoid Valve S2):** The monitor will run whenever this DTC is not present: None Shift solenoid valve S2: ON **Possible Causes:** • Short in shift solenoid valve S2 circuit • Shift solenoid valve S2 • ECM
DTC: P0977 **1T TCM, MIL: Yes** **Year:** 2010 **Model:** All **Engine:** All **Transmission:** All	**Shift Solenoid "B" Control Circuit High (Shift Solenoid Valve S2):** The monitor will run whenever the following DTCs are not stored: None Battery voltage: 8 V or higher Ignition switch: ON Starter: OFF Shift solenoid valve S2: OFF **Possible Causes:** • Open in shift solenoid valve S2 circuit • Shift solenoid valve S2 • ECM
DTC: P0977 **1T TCM, MIL: Yes** **Year:** 2009, 2010 **Model:** 4Runner, FJ Cruiser **Engine:** 2.7L L4, 4.0L V6, 4.7L V8 **Transmission:** All	**Shift Solenoid "B" Control Circuit High (Shift Solenoid Valve S2):** The monitor will run whenever this DTC is not present: None Shift solenoid valve S2: OFF **Possible Causes:** • Open in shift solenoid valve S2 circuit • Shift solenoid valve S2 • ECM
DTC: P0979 **1T ECM, MIL: Yes** **Year:** 2009, 2010 **Model:** All **Engine:** All **Transmission:** All	**Shift Solenoid "C" Control Circuit Low (Shift Solenoid Valve S3):** The ECM detects a short in the shift solenoid valve S3 circuit 2 times when the shift solenoid valve S3 is operated. **Possible Causes:** • Short in shift solenoid valve S3 circuit • Shift solenoid valve S3 • ECM
DTC: P0980 **1T ECM, MIL: Yes** **Year:** 2009, 2010 **Model:** All **Engine:** All **Transmission:** All	**Shift Solenoid "C" Control Circuit High (Shift Solenoid Valve S3):** The ECM detects an open in the shift solenoid valve S3 circuit 2 times when the shift solenoid valve S3 is not operated. **Possible Causes:** • Open in shift solenoid valve S3 circuit • Shift solenoid valve S3 • ECM
DTC: P0982 **1T ECM, MIL: Yes** **Year:** 2009, 2010 **Model:** Land Cruiser **Engine:** 5.7L V8 **Transmission:** All	**Shift Solenoid "D" Control Circuit Low (Shift Solenoid Valve S4):** The ECM detects a short in the shift solenoid valve S4 circuit 2 times when the shift solenoid valve S4 is operated. **Possible Causes:** • Short in shift solenoid valve S4 circuit • Shift solenoid valve S4 • ECM
DTC: P0983 **1T TCM, MIL: Yes** **Year:** 2009, 2010 **Model:** Highlander **Engine:** 2.7L L4, 3.5L V6 **Transmission:** All	**Shift Solenoid "D" Control Circuit High (Shift Solenoid Valve S4):** The monitor will run whenever this DTC is not present: None Shift solenoid valve S4: OFF Battery voltage: 8 V or more Ignition switch: ON Starter: OFF **Possible Causes:** • Open in shift solenoid valve S4 circuit • Shift solenoid valve S4 • ECM

DTC	Trouble Code Title, Conditions & Possible Causes
DTC: P0983 **1T ECM, MIL: Yes** **Year:** 2009, 2010 **Model:** All **Engine:** All **Transmission:** All	**Shift Solenoid "D" Control Circuit High (Shift Solenoid Valve S4):** The ECM detects an open in the shift solenoid valve S4 circuit 2 times when the shift solenoid valve S4 is not operated. **Possible Causes:** • Open in shift solenoid valve S4 circuit • Shift solenoid valve S4 • ECM
DTC: P0985 **1T ECM, MIL: Yes** **Year:** 2009, 2010 **Model:** All **Engine:** All **Transmission:** All	**Shift Solenoid "E" Control Circuit Low (Shift Solenoid Valve SR):** The ECM detects a short in the shift solenoid valve SR circuit 2 times when the shift solenoid valve SR is operated. **Possible Causes:** • Short in shift solenoid valve SR circuit • Shift solenoid valve SR • ECM
DTC: P0986 **1T ECM, MIL: Yes** **Year:** 2009, 2010 **Model:** All **Engine:** All **Transmission:** All	**Shift Solenoid "E" Control Circuit High (Shift Solenoid Valve SR):** The ECM detects an open in the shift solenoid valve SR circuit 2 times when the shift solenoid valve SR is not operated. **Possible Causes:** • Open in shift solenoid valve SR circuit • Shift solenoid valve SR • ECM
DTC: P0989 **2T TCM, MIL: Yes** **Year:** 2009, 2010 **Model:** All **Engine:** All **Transmission:** All	**Transmission Fluid Pressure Sensor / Switch "E" Circuit Low:** ECT (Engine coolant temperature): 40°C (104°F) or more Spark advance from Max. retard timing by KCS control: 0°CA or more Transmission range: "D" TFT (Transmission fluid temperature): -10°C (14°F) or more TFT (Transmission fluid temperature) sensor circuit: No circuit malfunction ECT (Engine coolant temperature) sensor circuit: No circuit malfunction Turbine speed sensor circuit: No circuit malfunction Intermediate shaft speed sensor circuit: No circuit malfunction Shift solenoid valve SL1 circuit: No circuit malfunction Shift solenoid valve SL2 circuit: No circuit malfunction Shift solenoid valve SL3 circuit: No circuit malfunction Shift solenoid valve SL4 circuit: No circuit malfunction Shift solenoid valve SLU circuit: No circuit malfunction Shift solenoid valve SL circuit: No circuit malfunction (KCS) Knock control sensor circuit: No circuit malfunction (ETCS) Electronic throttle control system: Not system down CAN communication system: Not system down TCM selected gear: Not 1st Vehicle speed: 15.5 mph (25 km/h) or more Turbine speed/Output speed (NT/NO) with 1st: 3.304 to 7.724 Turbine speed/Output speed (NT/NO) with 2nd: 1.901 to 2.340 Turbine speed/Output speed (NT/NO) with 3rd: 1.399 to 1.649 Turbine speed/Output speed (NT/NO) with 4th: 0.998 to 1.138 Turbine speed/Output speed (NT/NO) with 5th: 0.705 to 0.836 Turbine speed/Output speed (NT/NO) with 6th: 0.568 to 0.695 TCM lock-up command: ON Engine speed - Turbine speed: Less than 35 rpm Throttle valve opening angle: 7% or more Vehicle speed: Less than 62 mph (100 km/h) Shift solenoid valve SLU: Not ON malfunction **Possible Causes:** • ATF temperature sensor assembly (ATF pressure switch No. 3) • Transmission wire • TCM

DTC	Trouble Code Title, Conditions & Possible Causes
DTC: P0990 **2T TCM, MIL: Yes** **Year:** 2009, 2010 **Model:** All **Engine:** All **Transmission:** All	**Transmission Fluid Pressure Sensor / Switch "E" Circuit High:** ECT (Engine coolant temperature): 40°C (104°F) or more Spark advance from Max. retard timing by KCS control: 0°CA or more Transmission range: "D" TFT (Transmission fluid temperature): -10°C (14°F) or more TFT (Transmission fluid temperature) sensor circuit: No circuit malfunction ECT (Engine coolant temperature) sensor circuit: No circuit malfunction Turbine speed sensor circuit: No circuit malfunction Intermediate shaft speed sensor circuit: No circuit malfunction Shift solenoid valve SL1 circuit: No circuit malfunction Shift solenoid valve SL2 circuit: No circuit malfunction Shift solenoid valve SL3 circuit: No circuit malfunction Shift solenoid valve SL4 circuit: No circuit malfunction Shift solenoid valve SLU circuit: No circuit malfunction Shift solenoid valve SL circuit: No circuit malfunction (KCS) Knock control sensor circuit: No circuit malfunction (ETCS) Electronic throttle control system: Not system down CAN communication system: Not system down TCM selected gear: Not 1st Vehicle speed: 15.5 mph (25 km/h) or more Turbine speed/Output speed (NT/NO) with 1st: 3.304 to 7.724 Turbine speed/Output speed (NT/NO) with 2nd: 1.901 to 2.340 Turbine speed/Output speed (NT/NO) with 3rd: 1.399 to 1.649 Turbine speed/Output speed (NT/NO) with 4th: 0.998 to 1.138 Turbine speed/Output speed (NT/NO) with 5th: 0.705 to 0.836 Turbine speed/Output speed (NT/NO) with 6th: 0.568 to 0.695 TCM indicated pressure valve of SLU: Less than 4 kPa TCM lock-up command: OFF Shift solenoid valve SLU: Not malfunction Shift solenoid valve SL: Not OFF malfunction **Possible Causes:** • ATF temperature sensor assembly (ATF pressure switch No. 3) • Transmission wire • TCM

OBD II Trouble Code List (P2XXX Codes)

DTC	Trouble Code Title, Conditions & Possible Causes
DTC: P2004 **2T ECM, MIL: Yes** **Year:** 2009, 2010 **Model:** Camry **Engine:** 2.4L L4, 2.5L L4 **Transmission:** All	**Intake Manifold Runner Control Stuck Open (Bank 1):** Monitor runs whenever following DTCs are not present: P0110 - P0113 (IAT sensor) P0115 - P0118 (ECT sensor) Battery voltage: 8 V or more ECT: -10°C (14°F) or more IAT: -10°C (14°F) or more Ignition switch: ON Command IMRC valve: Closed **Possible Causes:** • IMRV motor circuit • IMRV motor • IMRV position sensor • IMRV • ECM

DTC	Trouble Code Title, Conditions & Possible Causes
DTC: P2006 **2T ECM, MIL: Yes** **Year:** 2009, 2010 **Model:** Camry **Engine:** 2.4L L4, 2.5L L4 **Transmission:** All	**Intake Manifold Runner Control Stuck Closed (Bank 1):** Monitor runs whenever following DTCs are not present: P0110 - P0113 (IAT sensor) P0115 - P0118 (ECT sensor) Battery voltage: 8 V or more Command IMRC valve: Closed **Possible Causes:** • IMRV motor circuit • IMRV motor • IMRV position sensor • IMRV • ECM
DTC: P2014 **1T ECM, MIL: Yes** **Year:** 2009, 2010 **Model:** Camry **Engine:** 2.4L L4, 2.5L L4 **Transmission:** All	**Intake Manifold Runner Position Sensor / Switch Circuit (Bank 1):** Monitor runs whenever following DTCs are not present: None IMRC valve position sensor voltage: Less than 0.2 V, or more than 4.8 V **Possible Causes:** • Open or short in IMRV position sensor circuit • IMRV position sensor • ECM
DTC: P2016 **1T ECM, MIL: Yes** **Year:** 2009, 2010 **Model:** Camry **Engine:** 2.4L L4, 2.5L L4 **Transmission:** All	**Intake Manifold Runner Position Sensor / Switch Circuit Low (Bank 1):** Monitor runs whenever following DTCs are not present: None IMRC valve position sensor voltage: Less than 0.2 V **Possible Causes:** • Open or short in IMRV position sensor circuit • IMRV position sensor • ECM
DTC: P2017 **1T ECM, MIL: Yes** **Year:** 2009, 2010 **Model:** Camry **Engine:** 2.4L L4, 2.5L L4 **Transmission:** All	**Intake Manifold Runner Position Sensor / Switch Circuit High (Bank 1):** Monitor runs whenever following DTCs are not present: None IMRC valve position sensor voltage: More than 4.8 V **Possible Causes:** • Open or short in IMRV position sensor circuit • IMRV position sensor • ECM
DTC: P2102 **1T ECM, MIL: Yes** **Year:** 2009, 2010 **Model:** All **Engine:** All **Transmission:** All	**Throttle Actuator Control Motor Circuit Low:** Conditions (a) and (b) continue for 2.0 seconds: (a) The throttle actuator duty ratio is 80% or more. (b) The throttle actuator current is below 0.5 A. **Possible Causes:** • Open in throttle actuator circuit • Throttle actuator • ECM
DTC: P2103 **1T ECM, MIL: Yes** **Year:** 2009, 2010 **Model:** All **Engine:** All **Transmission:** All	**Throttle Actuator Control Motor Circuit High:** Throttle actuator ON, Duty-cycle ratio to open throttle actuator 80% or more. Throttle actuator power supply 8 V or more, Motor current change during latest 0.016 seconds less than 0.2 A. **Possible Causes:** • Short in throttle actuator circuit • Throttle actuator • Throttle valve • Throttle body • ECM

DTC	Trouble Code Title, Conditions & Possible Causes
DTC: P2109 **5T ECM, MIL: Yes** **Year:** 2009, 2010 **Model:** 4Runner, FJ Cruiser **Engine:** 2.7L L4, 4.0L V6 **Transmission:** All	**Throttle / Pedal Position Sensor "A" Minimum Stop Performance:** Either condition is met: * With atmospheric pressure 85 kPa (638 mmHg) or higher (elevation 1400 m (4592 ft.) or less), when the engine coolant temperature is 45°C (113°F) or less at engine start, the engine is warmed up and conditions for ISC learned are met, the ISC learning value is approximately 3 times larger than normal even though the intake manifold pressure when idling is normal. * With atmospheric pressure 85 kPa (638 mmHg) or higher (elevation 1400 m (4592 ft.) or less), when the ignition switch has been turned to ON for 1 hour or more, the engine is warmed up, conditions for ISC learned are met and the vehicle has been driven at a speed of 30 km/h (19 mph) or more at least once, the ISC learning value is approximately 3 times larger than normal even though the intake manifold pressure when idling is normal. **Possible Causes:** • Throttle body with motor assembly
DTC: P2111 **1T ECM, MIL: Yes** **Year:** 2009, 2010 **Model:** All **Engine:** All **Transmission:** All	**Throttle Actuator Control System - Stuck Open:** System guard is ON when the following conditions set: Throttle actuator, ON Throttle actuator duty calculation, Executing Throttle position sensor fai,l Not detected Throttle actuator current-cut operation, Not executing Throttle actuator power supply, 4 V or higher Throttle actuator fail, Not detected The throttle actuator does not close when signaled by the ECM. (Stuck Open) **Possible Causes:** • Throttle actuator • Throttle body assembly • Throttle valve • ECM
DTC: P2112 **1T ECM, MIL: Yes** **Year:** 2009, 2010 **Model:** All **Engine:** All **Transmission:** All	**Throttle Actuator Control System - Stuck Closed:** Monitor runs whenever following DTCs are not present: None All of the following conditions are met: System guard*:ON Throttle actuator current: 2 A or more Duty cycle to open throttle: 80% or more System guard is ON when the following conditions are met Throttle actuator: ON Throttle actuator duty calculation: Executing Throttle position sensor: Fail determined Throttle actuator current-cut operation: Not executing Throttle actuator power supply: 4 V or more Throttle actuator: Fail determined **Possible Causes:** • Throttle actuator • Throttle body assembly • Throttle valve
DTC: P2118 **1T ECM, MIL: Yes** **Year:** 2009, 2010 **Model:** All **Engine:** All **Transmission:** All	**Throttle Actuator Control Motor Current Range / Performance:** With the electronic throttle actuator power ON and battery voltage 8 V or more. An open in ETCS power source (+BM) circuit was detected. **Possible Causes:** • Open in ETCS power source circuit • Battery • Battery terminals • ETCS fuse • ECM

DTC	Trouble Code Title, Conditions & Possible Causes
DTC: P2119 **1T ECM, MIL: Yes** **Year:** 2009, 2010 **Model:** All **Engine:** All **Transmission:** All	**Throttle Actuator Control Throttle Body Range / Performance:** System guard is ON when the following conditions set: Throttle actuator, ON Throttle actuator duty calculation, Executing Throttle position sensor fail, Not detected Throttle actuator current-cut operation, Not executing Throttle actuator power supply, 4 V or higher Throttle actuator fail, Not detected The throttle valve opening angle continues to vary greatly from the target opening angle. **Possible Causes:** • ETCS • ECM
DTC: P2120 **1T ECM, MIL: Yes** **Year:** 2009, 2010 **Model:** All **Engine:** All **Transmission:** All	**Throttle / Pedal Position Sensor / Switch "D" Circuit:** With the engine switch On (IG) and the throttle actuator power ON, the VPA fluctuates rapidly beyond the upper and lower malfunction thresholds for 0.5 seconds or more. **Possible Causes:** • APP sensor • ECM
DTC: P2121 **1T ECM, MIL: Yes** **Year:** 2009, 2010 **Model:** All **Engine:** All **Transmission:** All	**Throttle / Pedal Position Sensor / Switch "D" Circuit Range / Performance:** The difference between VPA and VPA2 is below 0.4 V, or higher than 1.2 V for 0.5 seconds. **Possible Causes:** • Accelerator Pedal Position (APP) sensor • ECM
DTC: P2122 **1T ECM, MIL: Yes** **Year:** 2009, 2010 **Model:** All **Engine:** All **Transmission:** All	**Throttle / Pedal Position Sensor / Switch "D" Circuit Low Input:** With the engine switch On (IG) and the throttle actuator power ON, the VPA is 0.4 V or less for 0.5 seconds or more when the accelerator pedal is fully released. **Possible Causes:** • APP sensor • Open in VCP1 circuit • Open or ground short in VPA circuit • ECM
DTC: P2123 **1T ECM, MIL: Yes** **Year:** 2009, 2010 **Model:** All **Engine:** All **Transmission:** All	**Throttle / Pedal Position Sensor / Switch "D" Circuit High Input:** With the engine switch On (IG) and the throttle actuator power ON, the VPA is 4.8 V or higher for 2.0 seconds or more. **Possible Causes:** • APP sensor • Open in EPA circuit • ECM
DTC: P2125 **1T ECM, MIL: Yes** **Year:** 2009, 2010 **Model:** All **Engine:** All **Transmission:** All	**Throttle / Pedal Position Sensor / Switch "E" Circuit:** With the engine switch On (IG) and the throttle actuator power ON, the VPA2 fluctuates rapidly beyond the upper and lower malfunction thresholds for 0.5 seconds or more. **Possible Causes:** • APP sensor • ECM
DTC: P2127 **1T ECM, MIL: Yes** **Year:** 2009, 2010 **Model:** All **Engine:** All **Transmission:** All	**Throttle / Pedal Position Sensor / Switch "E" Circuit Low Input:** With the engine switch On (IG) and the throttle actuator power ON, the VPA2 is 1.2 V or less for 0.5 seconds or more when the accelerator pedal is fully released. **Possible Causes:** • APP sensor • Open in VCP2 circuit • Open or ground short in VPA2 circuit • ECM

DTC	Trouble Code Title, Conditions & Possible Causes
DTC: P2128 **1T ECM, MIL: Yes** **Year:** 2009, 2010 **Model:** All **Engine:** All **Transmission:** All	**Throttle / Pedal Position Sensor / Switch "E" Circuit High Input:** With the engine switch On (IG) and the throttle actuator power ON, the Conditions (a) and (b) continue for 2.0 seconds or more: (a) VPA2 is 4.8 V or higher. (b) VPA is between 0.4 V and 3.45 V. **Possible Causes:** • APP sensor • Open in EPA2 circuit • ECM
DTC: P2135 **1T ECM, MIL: Yes** **Year:** 2009, 2010 **Model:** All **Engine:** All **Transmission:** All	**Throttle / Pedal Position Sensor / Switch "A" / "B" Voltage Correlation:** Either condition (a) or (b) is met: (a) The difference between output voltages of VTA1 and VTA2 is 0.02 V or less for 0.5 seconds or more. (b) The output voltage of VTA1 is 0.2 V or less and VTA2 is 1.75 V or less for 0.4 seconds or more. **Possible Causes:** • Short between VTA1 and VTA2 circuits • TP sensor (built into throttle body) • ECM
DTC: P2138 **1T ECM, MIL: Yes** **Year:** 2009, 2010 **Model:** All **Engine:** All **Transmission:** All	**Throttle / Pedal Position Sensor / Switch "D" / "E" Voltage Correlation:** With the engine switch On (IG) and the throttle actuator power ON, the condition (a) or (b) continues for 2.0 seconds or more: (a) The difference between VPA and VPA2 is 0.02 V or less. (b) VPA is 0.4 V or less and VPA2 is 1.2 V or less. **Possible Causes:** • Short between VPA and VPA2 circuits • APP sensor • ECM
DTC: P2195 **2T ECM, MIL: Yes** **Year:** 2009, 2010 **Model:** All **Engine:** All **Transmission:** All	**Oxygen (A/F) Sensor Signal Stuck Lean (Bank 1 Sensor 1):** Conditions (a) and (b) continue for 5 seconds or more: (a) The Air-Fuel Ratio (A/F) sensor voltage is more than 3.8 V for 5 seconds. (b) The Heated Oxygen (HO2) sensor voltage rises from less than 0.21 V to 0.59 V or more. **Possible Causes:** • Open or short in Air-Fuel Ratio (A/F) sensor (for Sensor 1) circuit • A/F sensor (for Sensor 1) • A/F sensor heater (for Sensor 1) • Integration relay • A/F sensor heater and integration relay circuits • Air induction system • Fuel pressure • Injector • ECM
DTC: P2196 **2T ECM, MIL: Yes** **Year:** 2009, 2010 **Model:** All **Engine:** All **Transmission:** All	**Oxygen (A/F) Sensor Signal Stuck Rich (Bank 1 Sensor 1):** Conditions (a) and (b) continue for 5 seconds or more: (a) The A/F sensor voltage is less than 2.8 V for 5 seconds. (b) The HO2 sensor voltage falls from 0.59 V or more to less than 0.21 V. **Possible Causes:** • Open or short in A/F sensor (for Sensor 1) circuit • A/F sensor (for Sensor 1) • A/F sensor heater (for Sensor 1) • Integration relay • A/F sensor heater and integration relay circuits • Air induction system • Fuel pressure • Injector • ECM

DTC	Trouble Code Title, Conditions & Possible Causes
DTC: P2197 **2T ECM, MIL: Yes** **Year:** 2009, 2010 **Model:** Avalon, Camry **Engine:** 3.5L V6 **Transmission:** All	**Oxygen (A/F) Sensor Signal Stuck Lean (Bank 2 Sensor 1):** Sensor voltage detection monitor (Lean side malfunction): Time after engine start: 30 seconds or more Fuel system status: Closed loop Sensor current detection monitor (High and low side malfunction): Battery voltage: 11 V or more Atmospheric pressure: 76 kPa (570 mmHg) or more A/F sensor status: Activated **Possible Causes:** • Open or short in A/F sensor (bank 1, 2 sensor 1) circuit • A/F sensor (bank 1, 2 sensor 1) • A/F sensor (bank 1, 2 sensor 1) heater • A/F sensor heater relay • A/F sensor heater and relay circuits • Air induction system • Injector • ECM
DTC: P2197 **2T ECM, MIL: Yes** **Year:** 2009, 2010 **Model:** All **Engine:** All **Transmission:** All	**Oxygen (A/F) Sensor Signal Stuck Lean (Bank 2 Sensor 1):** Conditions (a) and (b) continue for 5 seconds or more (2 trip detection logic): (a) The Air-Fuel Ratio (A/F) sensor voltage is more than 3.8 V for 5 seconds. (b) The Heated Oxygen (HO2) sensor voltage rises from less than 0.21 V to 0.59 V or more. **Possible Causes:** • Open or short in Air-Fuel Ratio (A/F) sensor (for Sensor 1) circuit • A/F sensor (for Sensor 1) • A/F sensor heater (for Sensor 1) • Integration relay • A/F sensor heater and integration relay circuits • Air induction system • Fuel pressure • Injector • ECM
DTC: P2198 **2T ECM, MIL: Yes** **Year:** 2009, 2010 **Model:** All **Engine:** All **Transmission:** All	**Oxygen (A/F) Sensor Signal Stuck Rich (Bank 2 Sensor 1):** Conditions (a) and (b) continue for 5 seconds or more (2 trip detection logic): (a) The A/F sensor voltage is less than 2.8 V for 5 seconds. (b) The HO2 sensor voltage falls from 0.59 V or more to less than 0.21 V **Possible Causes:** • Open or short in A/F sensor (for Sensor 1) circuit • A/F sensor (for Sensor 1) • A/F sensor heater (for Sensor 1) • Integration relay • A/F sensor heater and integration relay circuits • Air induction system • Fuel pressure • Injector • ECM
DTC: P2237 **2T ECM, MIL: Yes** **Year:** 2009, 2010 **Model:** All **Engine:** All **Transmission:** All	**Oxygen (A/F) Sensor Pumping Current Circuit / Open (Bank 1 Sensor 1):** An open in the circuit between terminals A1A+/A2A+ and A1A-/A2A- of the A/F sensor while the engine is running. **Possible Causes:** • Open in Air Fuel Ratio (A/F) sensor (for Bank 1, 2 Sensor 1) circuit • Air Fuel Ratio (A/F) sensor (for Bank 1, 2 Sensor 1) • ECM
DTC: P2238 **2T ECM, MIL: Yes** **Year:** 2009, 2010 **Model:** All **Engine:** All **Transmission:** All	**Oxygen (A/F) Sensor Pumping Current Circuit Low (Bank 1 Sensor 1):** Case 1: Condition (a) or (b) continues for 5.0 seconds or more: (a) The A1A+/A2A+ voltage is 0.5 V or less. (b) (A1A+/A2A+) - (A1A-/A2A-) = 0.1 V or less. Case 2: The A/F sensor admittance is below 0.022 1/W. **Possible Causes:** • Open or short in A/F sensor (for Bank 1, 2 Sensor 1) circuit • A/F sensor (for Bank 1, 2 Sensor 1) • ECM

DTC	Trouble Code Title, Conditions & Possible Causes
DTC: P2239 **2T ECM, MIL: Yes** **Year:** 2009, 2010 **Model:** All **Engine:** All **Transmission:** All	**Oxygen (A/F) Sensor Pumping Current Circuit High (Bank 1 Sensor 1):** The A1A+/A2A+ voltage is higher than 4.5 V for 5.0 seconds or more. **Possible Causes:** • Open or short in A/F sensor (for Bank 1, 2 Sensor 1) circuit • A/F sensor (for Bank 1, 2 Sensor 1) • ECM
DTC: P2240 **2T ECM, MIL: Yes** **Year:** 2009, 2010 **Model:** All **Engine:** All **Transmission:** All	**Oxygen (A/F) Sensor Pumping Current Circuit / Open (Bank 2 Sensor 1):** An open in the circuit between terminals A1A+/A2A+ and A1A-/A2A- of the A/F sensor while the engine is running. **Possible Causes:** • Open in Air Fuel Ratio (A/F) sensor (for Bank 1, 2 Sensor 1) circuit • Air Fuel Ratio (A/F) sensor (for Bank 1, 2 Sensor 1)* ECM
DTC: P2241 **2T ECM, MIL: Yes** **Year:** 2009, 2010 **Model:** All **Engine:** All **Transmission:** All	**Oxygen (A/F) Sensor Pumping Current Circuit Low (Bank 2 Sensor 1):** Case 1: Condition (a) or (b) continues for 5.0 seconds or more: (a) The A1A+/A2A+ voltage is 0.5 V or less. (b) (A1A+/A2A+) - (A1A-/A2A-) = 0.1 V or less. Case 2: The A/F sensor admittance is below 0.022 1/W. **Possible Causes:** • Open or short in A/F sensor (for Bank 1, 2 Sensor 1) circuit • A/F sensor (for Bank 1, 2 Sensor 1) • ECM
DTC: P2242 **2T ECM, MIL: Yes** **Year:** 2009, 2010 **Model:** All **Engine:** All **Transmission:** All	**Oxygen (A/F) Sensor Pumping Current Circuit High (Bank 2 Sensor 1):** The A1A+/A2A+ voltage is higher than 4.5 V for 5.0 seconds or more. **Possible Causes:** • Open or short in A/F sensor (for Bank 1, 2 Sensor 1) circuit • A/F sensor (for Bank 1, 2 Sensor 1) • ECM
DTC: P2252 **2T ECM, MIL: Yes** **Year:** 2009, 2010 **Model:** All **Engine:** All **Transmission:** All	**Oxygen (A/F) Sensor Reference Ground Circuit Low (Bank 1 Sensor 1):** The A1A-/A2A- voltage is 0.5 V or less for 5.0 seconds or more. **Possible Causes:** • Open or short in A/F sensor (for Bank 1, 2 Sensor 1) circuit • A/F sensor (for Bank 1, 2 Sensor 1) • ECM
DTC: P2253 **2T ECM, MIL: Yes** **Year:** 2009, 2010 **Model:** All **Engine:** All **Transmission:** All	**Oxygen (A/F) Sensor Reference Ground Circuit High (Bank 1 Sensor 1):** The A1A-/A2A- voltage is higher than 4.5 V for 5.0 seconds or more. **Possible Causes:** • Open or short in A/F sensor (for Bank 1, 2 Sensor 1) circuit • A/F sensor (for Bank 1, 2 Sensor 1) • ECM
DTC: P2255 **2T ECM, MIL: Yes** **Year:** 2009, 2010 **Model:** All **Engine:** All **Transmission:** All	**Oxygen (A/F) Sensor Reference Ground Circuit Low (Bank 2 Sensor 1):** The A1A-/A2A- voltage is 0.5 V or less for 5.0 seconds or more. **Possible Causes:** • Open or short in A/F sensor (for Bank 1, 2 Sensor 1) circuit • A/F sensor (for Bank 1, 2 Sensor 1) • ECM
DTC: P2256 **2T ECM, MIL: Yes** **Year:** 2009, 2010 **Model:** All **Engine:** All **Transmission:** All	**Oxygen (A/F) Sensor Reference Ground Circuit High (Bank 2 Sensor 1):** The A1A-/A2A- voltage is higher than 4.5 V for 5.0 seconds or more. **Possible Causes:** • Open or short in A/F sensor (for Bank 1, 2 Sensor 1) circuit • A/F sensor (for Bank 1, 2 Sensor 1) • ECM

DTC	Trouble Code Title, Conditions & Possible Causes
DTC: P2401 **2T ECM, MIL: Yes** **Year:** 2009, 2010 **Model:** All **Engine:** All **Transmission:** All	**Evaporative Emission System Leak Detection Pump Control Circuit Low:** P043E, P043F, P2401, P2402 and P2419 are stored when one of the following conditions is met during the key-off EVAP monitor: The reference orifice is clogged. Reference orifice high-flow. A leak detection pump OFF malfunction. A leak detection pump ON malfunction. A vent valve ON (close) malfunction. **Possible Causes:** • Canister pump module (Reference orifice, leak detection pump, vent valve) • Connector/wire harness (Canister pump module - ECM) • EVAP system hose (pipe from air inlet port to canister pump module, canister filter, fuel tank vent hose) • ECM
DTC: P2402 **2T ECM, MIL: Yes** **Year:** 2009, 2010 **Model:** All **Engine:** All **Transmission:** All	**Evaporative Emission System Leak Detection Pump Control Circuit High:** P043E, P043F, P2401, P2402 and P2419 are stored when one of the following conditions is met during the key-off EVAP monitor: The reference orifice is clogged. Reference orifice high-flow. A leak detection pump OFF malfunction. A leak detection pump ON malfunction. A vent valve ON (close) malfunction. **Possible Causes:** • Canister pump module (Reference orifice, leak detection pump, vent valve) • Connector/wire harness (Canister pump module - ECM) • EVAP system hose (pipe from air inlet port to canister pump module, canister filter, fuel tank vent hose) • ECM
DTC: P2419 **2T ECM, MIL: Yes** **Year:** 2009, 2010 **Model:** All **Engine:** All **Transmission:** All	**Evaporative Emission System Switching Valve Control Circuit Low:** P043E, P043F, P2401, P2402 and P2419 are stored when one of the following conditions is met during the key-off EVAP monitor: The reference orifice is clogged. Reference orifice high-flow. A leak detection pump OFF malfunction. A leak detection pump ON malfunction. A vent valve ON (close) malfunction. **Possible Causes:** • Canister pump module (Reference orifice, leak detection pump, vent valve) • Connector/wire harness (Canister pump module - ECM) • EVAP system hose (pipe from air inlet port to canister pump module, canister filter, fuel tank vent hose) • ECM
DTC: P2420 **2T ECM, MIL: Yes** **Year:** 2009, 2010 **Model:** All **Engine:** All **Transmission:** All	**Evaporative Emission System Switching Valve Control Circuit High:** The following condition is met during the key-off EVAP monitor: The EVAP pressure change when the vent valve is closed (ON) is below 0.3 kPa-g (2.25 mmHg-g). **Possible Causes:** • Canister pump module (Reference orifice, leak detection pump, vent valve) • Connector/wire harness (Canister pump module - ECM) • ECM
DTC: P2430 **1T ECM, MIL: Yes** **Year:** 2009 **Model:** All **Engine:** All **Transmission:** All	**Secondary Air Injection System Air Flow / Pressure Sensor Circuit Bank1:** The monitor will run whenever this DTC is not present: None Starter: OFF Time after starter turned from ON to OFF: 2 seconds or more Battery voltage: 8 V or more Ignition switch: ON **Possible Causes:** • Pressure sensor • Open or short in pressure sensor circuit • ECM

DTC	Trouble Code Title, Conditions & Possible Causes
DTC: P2431 **2T ECM, MIL: Yes** **Year:** 2009, 2010 **Model:** All **Engine:** All **Transmission:** All	**Secondary Air Injection System Air Flow / Pressure Sensor Circuit Range / Performance Bank1:** The pressure sensor indicates below 45.6 kPa (342 mmHg), or higher than 135 kPa (1013 mmHg). **Possible Causes:** • Pressure sensor • Open or short in pressure sensor circuit • ECM
DTC: P2432 **1T ECM, MIL: Yes** **Year:** 2009, 2010 **Model:** All **Engine:** All **Transmission:** All	**Secondary Air Injection System Air Flow / Pressure Sensor Circuit Low Bank1:** While the engine is running, the voltage output of the pressure sensor remains below 0.5 V. **Possible Causes:** • Pressure sensor • Open or short in pressure sensor circuit • ECM
DTC: P2433 **1T ECM, MIL: Yes** **Year:** 2009, 2010 **Model:** All **Engine:** All **Transmission:** All	**Secondary Air Injection System Air Flow / Pressure Sensor Circuit High Bank1:** While the engine is running, the voltage output of the pressure sensor remains higher than 4.5 V. **Possible Causes:** • Pressure sensor • Open or short in pressure sensor circuit • ECM
DTC: P2436 **2T ECM, MIL: Yes** **Year:** 2009, 2010 **Model:** Land Cruiser **Engine:** 5.7L V8 **Transmission:** All	**Secondary Air Injection System Air Flow / Pressure Sensor Circuit Range / Performance Bank 2:** The pressure sensor indicates below 45.6 kPa (342 mmHg), or higher than 135 kPa (1013 mmHg). **Possible Causes:** • Pressure sensor • Open or short in pressure sensor circuit • ECM
DTC: P2437 **1T ECM, MIL: Yes** **Year:** 2009, 2010 **Model:** All **Engine:** All **Transmission:** All	**Secondary Air Injection System Air Flow / Pressure Sensor Circuit Low Bank 2:** While the engine is running, the voltage output of the pressure sensor remains below 0.5 V. **Possible Causes:** • Pressure sensor • Open or short in pressure sensor circuit • ECM
DTC: P2438 **1T ECM, MIL: Yes** **Year:** 2009, 2010 **Model:** All **Engine:** All **Transmission:** All	**Secondary Air Injection System Air Flow / Pressure Sensor Circuit High Bank 2:** While the engine is running, the voltage output of the pressure sensor remains higher than 4.5 V. **Possible Causes:** • Pressure sensor • Open or short in pressure sensor circuit • ECM

DTC	Trouble Code Title, Conditions & Possible Causes
DTC: P2440 **1T ECM, MIL: Yes** **Year:** 2009 **Model:** 4Runner **Engine:** 2.7L L4, 4.7L V8 **Transmission:** All	**Secondary Air Injection System Switching Valve Stuck Open Bank1:** Case 1: Atmospheric pressure: 45 kPa (420 mmHg) or higher Battery voltage: 11.5 V or higher Case 2: Atmospheric pressure: 45 kPa (420 mmHg) or higher Battery voltage: 11.5 V or higher AIR pump: OFF Time after engine start: 10 seconds or more AIR valve bank 1: OFF AIR valve bank 2: OFF AIR status: OFF Engine load: 0% or more Intake air amount: 40 g/sec. or more IAT at engine start: -15°C (5°F) or higher ECT at engine start: Lower than 5°C (41°F) AIR valve: ON Engine RPM: Lower than 3750 rpm Case 3: Cumulative intake air amount: 109 g/sec. or more AIR pump: OFF AIR valve: OFF AIR valve bank 1: OFF AIR valve bank 2: OFF Engine RPM: Lower than 3750 rpm AIR status: OFF **Possible Causes:** • Air switching valve (ASV) • No. 2 air switching valve (Bank 1 and/or 2) • VSV for air injection system (bank 1 and/or 2) • Air injection control driver (AID) • Air injection control driver circuit • ECM
DTC: P2440 **1T ECM, MIL: Yes** **Year:** 2009, 2010 **Model:** Land Cruiser **Engine:** 5.7L V8 **Transmission:** All	**Secondary Air Injection System Switching Valve Stuck Open (Bank 1):** The pressure sensor detects pulsation of the exhaust gas despite the ECM commanding the Air Switching Valve (ASV) to close, while the engine is running. **Possible Causes:** • ASV • Open or short in ASV circuit • Pressure sensor • Pressure sensor circuit • Air Injection Control Driver (AID) • ECM
DTC: P2441 **2T ECM, MIL: Yes** **Year:** 2009, 2010 **Model:** 4Runner **Engine:** 2.7L L4, 4.7L V8 **Transmission:** All	**Secondary Air Injection System Switching Valve Stuck Close Bank1:** Atmospheric pressure: 45 kPa (420 mmHg) or more Battery voltage: 11.5 V or higher **Possible Causes:** • Vacuum hoses (Throttle body - VSVs for air injection system) • Air switching valve • Air injector pipe (No. 2 air switching valve - exhaust manifold) • Air injection hose • No. 2 air switching valve (bank 1 and/or 2) • VSV for air injection system (bank 1 and/or 2) • Air injection control driver • Air injection control driver circuit • ECM

DTC	Trouble Code Title, Conditions & Possible Causes
DTC: P2441 **2T ECM, MIL: Yes** **Year:** 2009, 2010 **Model:** Land Cruiser **Engine:** 5.7L V8 **Transmission:** All	**Secondary Air Injection System Switching Valve Stuck Close (Bank 1):** The pressure sensor detects no pulsation of the exhaust gas despite the ECM commanding the Air Switching Valve (ASV) to open, while the engine is running. **Possible Causes:** • ASV • Open or short in ASV circuit • Vacuum hose (ASV - pressure sensor) • Air injection hose • Pressure sensor • Pressure sensor circuit • Air Injection Control Driver (AID) • ECM
DTC: P2442 **1T ECM, MIL: Yes** **Year:** 2009, 2010 **Model:** Land Cruiser **Engine:** 5.7L V8 **Transmission:** All	**Secondary Air Injection System Switching Valve Stuck Open (Bank 2):** The pressure sensor detects pulsation of the exhaust gas despite the ECM commanding the Air Switching Valve (ASV) to close, while the engine is running. **Possible Causes:** • ASV • Open or short in ASV circuit • Pressure sensor • Pressure sensor circuit • Air Injection Control Driver (AID) • ECM
DTC: P2443 **2T ECM, MIL: Yes** **Year:** 2009, 2010 **Model:** Land Cruiser **Engine:** 5.7L V8 **Transmission:** All	**Secondary Air Injection System Switching Valve Stuck Closed (Bank 2):** The pressure sensor detects no pulsation of the exhaust gas despite the ECM commanding the Air Switching Valve (ASV) to open, while the engine is running. **Possible Causes:** • ASV • Open or short in ASV circuit • Vacuum hose (ASV - pressure sensor) • Air injection hose • Pressure sensor • Pressure sensor circuit • Air Injection Control Driver (AID) • ECM
DTC: P2444 **2T ECM, MIL: Yes** **Year:** 2009, 2010 **Model:** All **Engine:** All **Transmission:** All	**Secondary Air Injection System Pump Stuck On (Bank 1):** The secondary air pressure is higher than 2.4 kPa (18 mmHg) despite the ECM commanding the air pump to turn off. **Possible Causes:** • Short in air pump circuit • Open or short in pressure sensor circuit • Pressure sensor • Air Injection Control Driver (AID) • ECM
DTC: P2445 **2T ECM, MIL: Yes** **Year:** 2009, 2010 **Model:** 4Runner **Engine:** 2.7L L4, 4.7L V8 **Transmission:** All	**Secondary Air Injection System Pump Stuck Off Bank1:** The secondary air pressure is higher than 2.4 kPa (18 mmHg) despite the ECM commanding the air pump to turn off. **Possible Causes:** • Air pump • Open in air pump circuit • Air injection system piping • Vacuum hose (pressure sensor - air switching valve) • Pressure sensor • Open or short in pressure sensor circuit • Air injection control driver • ECM

DTC	Trouble Code Title, Conditions & Possible Causes
DTC: P2445 **2T ECM, MIL: Yes** **Year:** 2009, 2010 **Model:** Land Cruiser **Engine:** 5.7L V8 **Transmission:** All	**Secondary Air Injection System Pump Stuck Off (Bank 1):** The secondary air pressure is below 2.4 kPa (18 mmHg) despite the ECM commanding the air pump to turn on. **Possible Causes:** • Air pump • Open in air pump circuit • Air injection system piping • Vacuum hose (pressure sensor - air switching valve) • Pressure sensor • Open or short in pressure sensor circuit • Air Injection Control Driver (AID) • ECM
DTC: P2446 **2T ECM, MIL: Yes** **Year:** 2009, 2010 **Model:** Land Cruiser **Engine:** 5.7L V8 **Transmission:** All	**Secondary Air Injection System Pump Stuck On (Bank 2):** The secondary air pressure is higher than 2.4 kPa (18 mmHg) despite the ECM commanding the air pump to turn off. **Possible Causes:** • Short in air pump circuit • Open or short in pressure sensor circuit • Pressure sensor • Air Injection Control Driver (AID) • ECM
DTC: P2447 **2T ECM, MIL: Yes** **Year:** 2009, 2010 **Model:** Land Cruiser **Engine:** 5.7L V8 **Transmission:** All	**Secondary Air Injection System Pump Stuck Off (Bank 2):** The secondary air pressure is below 2.4 kPa (18 mmHg) despite the ECM commanding the air pump to turn on. **Possible Causes:** • Air pump • Open in air pump circuit • Air injection system piping • Vacuum hose (pressure sensor - air switching valve) • Pressure sensor • Open or short in pressure sensor circuit • Air Injection Control Driver (AID) • ECM
DTC: P2471 **2T TCM, MIL: Yes** **Year:** 2010 **Model:** 4Runner **Engine:** 2.7L L4 **Transmission:** All	**Transmission Fluid Temperature Sensor "B" Performance:** Condition A: Duration time from engine start: 12 sec. or more Time after engine start: 10 min. Driving distance after engine start: 5 km (3.1 miles) or more ECT (12 sec. after engine start): 15°C (59°F) or higher IAT (12 sec. after engine start): 15°C (59°F) or higher Soak time: 5 hours or more Condition B: Duration time for ECT to reach 60°C (140°F): 10 sec. or more ATF temperature (12 sec. after engine start): 110°C (230°F) or higher ECT (12 sec. after engine start): Below 35°C (95°F) **Possible Causes:** • No. 1 ATF temperature sensor (Transmission wire)
DTC: P2610 **2T ECM, MIL: Yes** **Year:** 2009, 2010 **Model:** All **Engine:** All **Transmission:** All	**ECM / PCM Internal Engine Off Timer Performance:** With the engine switch On (IG) and running, Battery voltage 8 V or more. An ECM internal malfunction has been detected. **Possible Causes:** • ECM
DTC: P2714 **2T ECM, MIL: Yes** **Year:** 2009, 2010 **Model:** All **Engine:** All **Transmission:** All	**Pressure Control Solenoid "D" Performance (Shift Solenoid Valve SLT):** The ECM detects a malfunction of the shift solenoid valve SLT (ON side) according to the revolution difference of the turbine and output shaft, and also by the oil pressure. **Possible Causes:** • Shift solenoid valve SLT remains open or closed • Shift solenoid valve S1, S2, S3, S4 or SL2 remains open or closed • Gear 6 incorrect ratio (sequence valve) or 1-2 shift valve is stuck • Valve body is blocked • Automatic transmission (clutch, brake, gear, etc.)

DTC	Trouble Code Title, Conditions & Possible Causes
DTC: P2716 **1T ECM, MIL: Yes** **Year:** 2009, 2010 **Model:** All **Engine:** All **Transmission:** All	**Pressure Control Solenoid "D" Electrical (Shift Solenoid Valve SLT):** Conditions (a) or (b) below are detected for 1 sec. or more: (a) Shift solenoid valve SLT - terminal: 0 V (b) Shift solenoid valve SLT - terminal: 12 V **Possible Causes:** • Open or short in shift solenoid valve SLT circuit • Shift solenoid valve SLT • ECM
DTC: P2741 **T , MIL: Yes** **Year:** 2009, 2010 **Model:** 4Runner **Engine:** 2.7L L4, 4.7L V8 **Transmission:** All	**Transmission Fluid Temperature Sensor "B" Performance:** Condition A: Duration time from engine start: 12 sec. or more Time after engine start: 10 min. Driving distance after engine start: 5 km (3.1 miles) or more ECT (12 sec. after engine start): 15°C (59°F) or higher IAT (12 sec. after engine start): 15°C (59°F) or higher Soak time: 5 hours or more Condition B: Duration time for ECT to reach 60°C (140°F): 10 sec. or more ATF temperature (12 sec. after engine start): 110°C (230°F) or higher ECT (12 sec. after engine start): Below 35°C (95°F) **Possible Causes:** • ATF level • Transmission wire (No. 1 ATF temperature sensor)
DTC: P2741 00 **1T TCM, MIL: Yes** **Year:** 2009, 2010 **Model:** 4Runner, FJ Cruiser **Engine:** 4.0L V6, 4.7L V8 **Transmission:** All	**Transmission Fluid Temperature Sensor "B" Circuit:** The monitor will run whenever the following DTCs are not present: None The typical enabling condition is not available: - **Possible Causes:** • Open or short in No. 2 ATF temperature sensor circuit • No. 2 ATF temperature sensor • ECM
DTC: P2742 **1T TCM, MIL: Yes** **Year:** 2009, 2010 **Model:** 4Runner, FJ Cruiser **Engine:** 4.0L V6, 4.7L V8 **Transmission:** All	**Transmission Fluid Temperature Sensor "B" Circuit Low Input:** The monitor will run whenever the following DTCs are not present: None The typical enabling condition is not available: - **Possible Causes:** • Short in No. 2 ATF temperature sensor circuit • No. 2 ATF temperature sensor • ECM
DTC: P2742 **1T TCM, MIL: Yes** **Year:** 2010 **Model:** 4Runner **Engine:** 2.7L L4 **Transmission:** All	**Transmission Fluid Temperature Sensor "B" Circuit Low Input:** Battery voltage: 8 V or higher Ignition switch: ON Starter: OFF The monitor will run whenever this DTC is not stored: None **Possible Causes:** • Short in No. 1 ATF temperature sensor circuit • No. 1 ATF temperature sensor (transmission wire) • ECM
DTC: P2742 **1T ECM, MIL: Yes** **Year:** 2009, 2010 **Model:** Land Cruiser **Engine:** 5.7L V8 **Transmission:** All	**Transmission Fluid Temperature Sensor "B" Circuit Low Input:** The No. 2 ATF temperature sensor resistance is below 25 Ohm (0.046 V) for 0.5 sec. or more. **Possible Causes:** • Short in No. 2 ATF temperature sensor circuit • No. 1 transmission wire (No. 2 ATF temperature sensor) • ECM

DTC	Trouble Code Title, Conditions & Possible Causes
DTC: P2743 **1T TCM, MIL: Yes** **Year:** 2010 **Model:** 4Runner **Engine:** 2.7L L4 **Transmission:** All	**Transmission Fluid Temperature Sensor "B" Circuit High Input:** Battery voltage: 8 V or higher Ignition switch: ON Starter: OFF The monitor will run whenever this DTC is not stored: P0115, P0117, P0118 (ECT sensor circuit) P0097, P0098, P0110, P0112, P0113 (IAT sensor circuit) Engine coolant temperature or intake air temperature at engine start: -29.375°C (-20.875°F) or less Time after engine start: 523 sec. or more **Possible Causes:** • Open in No. 1 ATF temperature sensor circuit • No. 1 ATF temperature sensor (transmission wire) • ECM
DTC: P2743 **1T TCM, MIL: Yes** **Year:** 2009, 2010 **Model:** 4Runner, FJ Cruiser **Engine:** 4.0L V6, 4.7L V8 **Transmission:** All	**Transmission Fluid Temperature Sensor "B" Circuit High Input:** The monitor will run whenever the following DTCs are not present: None Time after engine start: 15 minutes or more **Possible Causes:** • Open in No. 2 ATF temperature sensor circuit • No. 2 ATF temperature sensor • ECM
DTC: P2743 **1T ECM, MIL: Yes** **Year:** 2009, 2010 **Model:** Land Cruiser **Engine:** 5.7L V8 **Transmission:** All	**Transmission Fluid Temperature Sensor "B" Circuit High Input:** When 15 min. or more have elapsed after the engine is started, the No. 2 ATF temperature sensor resistance is higher than 156 kOhm (4.915 V) for 0.5 sec. or more. **Possible Causes:** • Open in No. 2 ATF temperature sensor circuit • No. 1 transmission wire (No. 2 ATF temperature sensor) • ECM
DTC: P2757 **2T TCM, MIL: Yes** **Year:** 2009, 2010 **Model:** 4Runner, FJ Cruiser **Engine:** 4.0L V6, 4.7L V8 **Transmission:** All	**Torque Converter Clutch Pressure Control Solenoid Performance (Shift Solenoid Valve SLU):** Transmission range: "D" Duration time from shifting "N" to "D": 4 sec. or more ECT: 40°C (104°F) or more Spark advance from Max. retard timing by KCS control: 0° CA or more Engine: Starting ECM selected gear: 2nd, 3rd, 4th or 5th Vehicle speed: 25 km/h (15.5 mph) or more OFF malfunction (A): ECM lock-up command: ON (SLU pressure: 513 kPa (5.2 kgf/cm(2), 74 psi) or more): Vehicle speed ON malfunction: ECM lock-up command: OFF (SLU pressure: Less than 4 kPa (0.04 kgf/cm2, 0.6 psi)) Throttle valve opening angle: 9% or more Vehicle speed: Less than 120 km/h (74.6 mph) at 2nd gear (varies with ECU selected gear) **Possible Causes:** • Shift solenoid valve SLU remains open or closed • Valve body is blocked • Torque converter clutch • Automatic transmission (clutch, brake or gear, etc.) • Line pressure is too low

DTC	Trouble Code Title, Conditions & Possible Causes
DTC: P2757 **T , MIL: Yes** **Year:** 2010 **Model:** 4Runner **Engine:** 2.7L L4 **Transmission:** All	**Torque Converter Clutch Pressure Control Solenoid Performance (Shift Solenoid Valve SLU):** Transmission position: "D" Duration time from shifting "N" to "D": 4 sec. or more ECT: 40°C (104°F) or higher Spark advance from max. retard timing by KCS control: 0° crankshaft angle or more Engine: Starting ECM selected gear: 3rd or 4th Vehicle speed: 25 km/h (15.5 mph) or more OFF Malfunction (A): ECM lock-up command: ON (SLU pressure: 90% or more) Vehicle speed: Less than 100 km/h (62.1 mph) OFF Malfunction (B): M selected gear: 2nd Vehicle speed: 10 km/h (6.2 mph) or more Output speed: 2nd → 1st down shift point or more Throttle valve opening angle: 6.5% or more at 2000 rpm (Conditions varies with engine speed) ON Malfunction (A) and (B): ECM lock-up command: OFF (SLU pressure: 10% or less) Throttle valve opening angle: 10% or more Vehicle speed: Less than 60 km/h (37.2 mph) **Possible Causes:** • Shift solenoid valve SLU remains open or closed • Valve body is blocked • Shift solenoid valve SLU • Torque converter clutch • Automatic transmission (clutch, brake, gear, etc.) • Line pressure is too low
DTC: P2757 **2T ECM, MIL: Yes** **Year:** 2009, 2010 **Model:** Land Cruiser **Engine:** 5.7L V8 **Transmission:** All	**Torque Converter Clutch Pressure Control Solenoid Performance (Shift Solenoid Valve SLU):** Lock-up does not occur when driving in the lock-up range (normal driving at 80 km/h (50 mph)), or lock-up remains ON in the lock-up OFF range. **Possible Causes:** • Shift solenoid valve SLU remains open or closed • Shift solenoid valve SLU • Valve body is blocked • Torque converter clutch • Automatic transmission (clutch, brake, gear, etc.) • Line pressure is too low

DTC	Trouble Code Title, Conditions & Possible Causes
DTC: P2757 **2T TCM, MIL: Yes** **Year:** 2009, 2010 **Model:** Avalon, Camry **Engine:** 2.5L L4, 2.7L L4, 3.5L V6 **Transmission:** All	**Torque Converter Clutch Pressure Control Solenoid Performance (Shift Solenoid Valve SLU):** ECT (Engine coolant temperature): 40°C (104°F) or more Spark advance from Max. retard timing by KCS control: 0°CA or more Transmission range: "D" TFT (Transmission fluid temperature): -10°C (14°F) or more TFT (Transmission fluid temperature) sensor circuitL: No circuit malfunction ECT (Engine coolant temperature) sensor circuit: No circuit malfunction Turbine speed sensor circuit: No circuit malfunction Intermediate shaft speed sensor circuit: No circuit malfunction Shift solenoid valve SL1 circuit: No circuit malfunction Shift solenoid valve SL2 circuit: No circuit malfunction Shift solenoid valve SL3 circuit:n: No circuit malfunction Shift solenoid valve SL4 circuit: No circuit malfunction Shift solenoid valve SLU circuit: No circuit malfunction Shift solenoid valve SL circuit: No circuit malfunction (KCS) Knock control sensor circuit: No circuit malfunction (ETCS) Electronic throttle control system: Not system down CAN communication system: Not system down TCM selected gear: Not 1st Vehicle speed: 5.5 mph (25 km/h) or more Turbine speed/Output speed (NT/NO) with 1st: 3.304 to 7.724 Turbine speed/Output speed (NT/NO) with 2nd: 1.901 to 2.340 Turbine speed/Output speed (NT/NO) with 3rd: 1.399 to 1.649 Turbine speed/Output speed (NT/NO) with 4th: 0.998 to 1.138 Turbine speed/Output speed (NT/NO) with 5th: 0.705 to 0.836 Turbine speed/Output speed (NT/NO) with 6th: 0.568 to 0.695 **Possible Causes:** • Shift solenoid valve SLU (open or closed) • Valve body (blocked) • Torque converter clutch • Automatic transaxle (clutch, brake or gear, etc.) • Line pressure is too low
DTC: P2759 **1T TCM, MIL: Yes** **Year:** 2009, 2010 **Model:** Avalon, Camry **Engine:** 2.5L L4, 3.5L V6 **Transmission:** All	**Torque Converter Clutch Pressure Control Solenoid Control Circuit Electrical (Shift Solenoid Valve SLU):** The monitor will run whenever this DTC is not present: None Engine switch: ON Starter: OFF Condition (A): Battery voltage: 12 V or more Condition (B): Battery voltage: 10 V or more and less than 12 V Target current: Less than 0.75 A Condition (C): Battery voltage: 8 V or more Target current: 0.25 A or more **Possible Causes:** • Open or short in shift solenoid valve SLU circuit • Shift solenoid valve SLU • TCM

DTC	Trouble Code Title, Conditions & Possible Causes
DTC: P2759 **1T TCM, MIL: Yes** **Year:** 2009, 2010 **Model:** 4Runner, FJ Cruiser **Engine:** 4.0L V6, 4.7L V8 **Transmission:** All	**Torque Converter Clutch Pressure Control Solenoid Control Circuit Electrical (Shift Solenoid Valve SLU):** The monitor will run whenever this DTC is not present: None Ignition switch: ON Starter: OFF Case 1: Solenoid current cut status: Not cut Battery voltage: 11 V or more Case 2: Battery voltage: 8 V or more **Possible Causes:** • Open or short in shift solenoid valve SLU circuit • Shift solenoid valve SLU • ECM
DTC: P2759 **1T TCM, MIL: Yes** **Year:** 2010 **Model:** 4Runner **Engine:** 2.7L L4 **Transmission:** All	**Torque Converter Clutch Pressure Control Solenoid Control Circuit Electrical (Shift Solenoid Valve SLU):** The monitor will run whenever the following DTCs are not stored: None Condition (A): Solenoid current cut status: Not cut Battery voltage: 11 V or higher Ignition switch: ON Starter: OFF Condition (B): Battery voltage: 8 V or higher Ignition switch: ON Starter: OFF **Possible Causes:** • Open or short in shift solenoid valve SLU circuit • Shift solenoid valve SLU • ECM
DTC: P2759 **1T ECM, MIL: Yes** **Year:** 2009, 2010 **Model:** Land Cruiser **Engine:** 5.7L V8 **Transmission:** All	**Torque Converter Clutch Pressure Control Solenoid Control Circuit Electrical (Shift Solenoid Valve SLU):** An open or short is detected in the shift solenoid valve SLU circuit for 1 sec. or more while driving. **Possible Causes:** • Open or short in shift solenoid valve SLU circuit • Shift solenoid valve SLU • ECM
DTC: P2769 **2T TCM, MIL: Yes** **Year:** 2009, 2010 **Model:** Highlander **Engine:** 2.7L L4, 3.5L V6 **Transmission:** All	**Torque Converter Clutch Solenoid Circuit Low (Shift Solenoid Valve DSL):** The monitor will run whenever this DTC is not present: None Shift solenoid valve DSL: ON Solenoid current cut status: Not cut Battery voltage: 8 V or more Ignition switch: ON Starter: OFF **Possible Causes:** • Short in shift solenoid valve DSL circuit • Shift solenoid valve DSL • ECM
DTC: P2769 **2T TCM, MIL: Yes** **Year:** 2009, 2010 **Model:** Avalon, Camry **Engine:** 2.4L L4, 2.5L L4, 2.7L L4, 3.5L V6 **Transmission:** All	**Short in Torque Converter Clutch Solenoid Circuit (Shift Solenoid Valve SL):** The monitor will run whenever the following DTCs are not present: None Solenoid: ON Time after solenoid OFF to ON: More than 0.008 sec. **Possible Causes:** • Short in shift solenoid valve SL circuit • Shift solenoid valve SL • TCM

DTC	Trouble Code Title, Conditions & Possible Causes
DTC: P2769 **2T ECM, MIL: Yes** **Year:** 2009, 2010 **Model:** Corolla, Matrix **Engine:** 1.8L L4 VIN L, 1.8L L4 VIN U, 2.4L L4 **Transmission:** All	**Torque Converter Clutch Solenoid Circuit Low (Shift Solenoid Valve DSL):** Shift solenoid valve DSL is ON, the solenoid current cut status Not cut, battery voltage 8 V or more, ignition switch ON, starter OFF. The ECM detects short in solenoid valve DSL circuit (0.1 sec.) when solenoid valve DSL is operated. **Possible Causes:** • Short in shift solenoid valve DSL circuit • Shift solenoid valve DSL • ECM
DTC: P2770 **2T ECM, MIL: Yes** **Year:** 2009, 2010 **Model:** Corolla, Matrix **Engine:** 1.8L L4 VIN L, 1.8L L4 VIN U, 2.4L L4 **Transmission:** All	**Torque Converter Clutch Solenoid Circuit High (Shift Solenoid Valve DSL) :** The ECM detects open in solenoid valve DSL circuit (0.1 sec.) when solenoid valve DSL is not operated. **Possible Causes:** • Open in shift solenoid valve DSL circuit • Shift solenoid valve DSL • ECM
DTC: P2770 **2T TCM, MIL: Yes** **Year:** 2009, 2010 **Model:** Highlander **Engine:** 2.7L L4, 3.5L V6 **Transmission:** All	**Torque Converter Clutch Solenoid Circuit High (Shift Solenoid Valve DSL):** The monitor will run whenever this DTC is not present: None Shift solenoid valve DSL: ON Battery voltage: 8 V or more Ignition switch: ON Starter: OFF **Possible Causes:** • Open in shift solenoid valve DSL circuit • Shift solenoid valve DSL • ECM
DTC: P2770 **2T TCM, MIL: Yes** **Year:** 2009, 2010 **Model:** Avalon, Camry **Engine:** 2.4L L4, 2.5L L4, 2.7L L4, 3.5L V6 **Transmission:** All	**Open in Torque Converter Clutch Solenoid Circuit (Shift Solenoid Valve SL):** The monitor will run whenever the following DTCs are not present: None Solenoid: ON Time after solenoid ON to OFF: More than 0.008 sec. **Possible Causes:** • Open in shift solenoid valve SL circuit • Shift solenoid valve SL • TCM
DTC: P2772 **1T TCM, MIL: Yes** **Year:** 2009 **Model:** 4Runner **Engine:** 4.7L V8 **Transmission:** All	**Four Wheel Drive (4WD) Low Switch Circuit Range / Performance:** Output speed sensor circuit: Not circuit malfunction Vehicle speed sensor circuit: Not circuit malfunction Transfer neutral position switch: OFF ON malfunction (A): Output speed (Transfer output speed): 1,000 to 3,000 rpm ON malfunction (B): Output speed (Transfer output speed): 143 rpm or more **Possible Causes:** • Short in transfer L4 position switch circuit • Transfer L4 position switch • ECM
DTC: P2772 **1T ECM, MIL: Yes** **Year:** 2009, 2010 **Model:** Land Cruiser **Engine:** 5.7L V8 **Transmission:** All	**Four Wheel Drive (4WD) Low Switch Circuit Range / Performance:** The transfer L4 position switch remains on while the vehicle is running under the following conditions for 1.8 sec. or more: (a) The output shaft speed is between 1000 and 3000 rpm. (b) The transfer position switch is in H4. **Possible Causes:** • Short in transfer L4 position switch circuit • Four wheel drive control ECU • ECM

DTC	Trouble Code Title, Conditions & Possible Causes
DTC: P2808 **2T TCM, MIL: Yes** **Year:** 2009, 2010 **Model:** Avalon, Camry **Engine:** 2.5L L4, 2.7L L4, 3.5L V6 **Transmission:** All	**Pressure Control Solenoid "G" Performance (Shift Solenoid Valve SL4):** Transmission range: "D" TFT (Transmission fluid temperature): -10°C (14°F) or more TFT (Transmission fluid temperature) sensor circuit: No circuit malfunction ECT (Engine coolant temperature) sensor circuit: No circuit malfunction Turbine speed sensor circuit: No circuit malfunction Intermediate shaft speed sensor circuit: No circuit malfunction Shift solenoid valve SL1 circuit: No circuit malfunction Shift solenoid valve SL2 circuit: No circuit malfunction Shift solenoid valve SL3 circuit: No circuit malfunction Shift solenoid valve SL4 circuit: No circuit malfunction (KCS) Knock control sensor circuit: No circuit malfunction (ETCS) Electronic throttle control system: Not system down CAN communication system: Not system down **Possible Causes:** • Shift solenoid valve SL4 (open or closed) • Valve body (blocked) • Automatic transaxle (clutch, brake or gear etc.)
DTC: P2810 **1T TCM, MIL: Yes** **Year:** 2009, 2010 **Model:** Avalon, Camry **Engine:** 2.5L L4, 3.5L V6 **Transmission:** All	**Pressure Control Solenoid "G" Electrical (Shift Solenoid Valve SL4):** The monitor will run whenever this DTC is not present: None Engine switch: ON Starter: OFF Condition (A): Battery voltage: 12 V or more Condition (B): Battery voltage: 10 V or more and less than 12 V Target condition: Less than 0.75 A Condition (C): Battery voltage: 8 V or more Target condition: 0.25 A or more **Possible Causes:** • Open or short in shift solenoid valve SL4 circuit • Shift solenoid valve SL4 • TCM
DTC: P2A00 **T ECM, MIL: Yes** **Year:** 2009 **Model:** 4Runner **Engine:** 4.7L V8 **Transmission:** All	**A/F Sensor Circuit Slow Response (Bank 1 Sensor 1):** P0500 (Vehicle Speed Sensor) P0505 (IAC Valve) Active A/F control: Performing Active A/F control is performed when the following conditions set: Battery Voltage: 11 V or more ECT: 75°C (167°F) or more Idle: OFF Engine RPM: Less than 4000 rpm A/F sensor status: Activated Fuel cut: OFF Engine load: 10 to 70% Shift position: 2nd or more Catalyst monitor: Not executing Intake air amount: 2.5 to 12 g/sec. **Possible Causes:** • Open or short in Air Fuel Ratio (A/F) sensor (sensor 1) circuit • Air Fuel Ratio (A/F) sensor (sensor 1) • ECM

DTC	Trouble Code Title, Conditions & Possible Causes
DTC: P2A00 **2T ECM, MIL: Yes** **Year:** 2009, 2010 **Model:** Avalon, Camry, Corolla **Engine:** 1.8L L4 VIN L, 1.8L L4 VIN U, 2.4L L4, 2.5L L4, 3.5L V6 **Transmission:** All	**A/F Sensor Circuit Slow Response (Bank 1 Sensor 1):** Active A/F control: Performing Battery voltage: 11 V or more ECT: 167°F (75°C) or more Idle: OFF Engine rpm: Less than 4000 rpm A/F sensor status: Activated Fuel cut: OFF Engine load: 10-70% Shift position: 2 or more Catalyst monitor: Not yet MAF: 2.5-15 g/sec **Possible Causes:** • Open or short in A/F sensor (sensor 1) circuit • A/F sensor (sensor 1) • A/F sensor (sensor 1) heater • EFI relay • A/F sensor heater and EFI relay circuits • Air induction system • Fuel pressure • Injector • PCV valve and hose • PCV hose connection • ECM
DTC: P2A00 **2T ECM, MIL: Yes** **Year:** 2009, 2010 **Model:** Matrix **Engine:** 1.8L L4 VIN U, 2.4L L4 **Transmission:** All	**A/F Sensor Circuit Slow Response (Bank 1 Sensor 1):** Calculated value for Air Fuel Ratio (A/F) sensor response rate deterioration level is less than 0.12 V. **Possible Causes:** • Air fuel ratio sensor (sensor 1) • Air fuel ratio sensor (sensor 1) heater • ECM
DTC: P2A00 **2T ECM, MIL: Yes** **Year:** 2009, 2010 **Model:** Land Cruiser **Engine:** 5.7L V8 **Transmission:** All	**A/F Sensor Circuit Slow Response (Bank 1 Sensor 1):** The calculated value of the air-fuel ratio (A/F) sensor response rate deterioration level is less than the threshold. **Possible Causes:** • A/F sensor • A/F sensor heater • ECM
DTC: P2A03 **2T ECM, MIL: Yes** **Year:** 2009, 2010 **Model:** Avalon, Camry **Engine:** 3.5L V6 **Transmission:** All	**A/F Sensor Circuit Slow Response (Bank 2 Sensor 1):** Active A/F control: Performing Battery voltage: 11 V or more ECT: 167°F (75°C) or more Idle: OFF Engine rpm: Less than 4000 rpm A/F sensor status: Activated Fuel cut: OFF Engine load: 10 to 70 % Shift position: 2 or more Catalyst monitor: Not yet MAF: 2.5-15 g/sec **Possible Causes:** • Open or short in A/F sensor (bank 1, 2 sensor 1) circuit • A/F sensor (bank 1, 2 sensor 1) • ECM

DTC	Trouble Code Title, Conditions & Possible Causes
DTC: P2A03 **T ECM, MIL: Yes** **Year:** 2009 **Model:** 4Runner **Engine:** 4.7L V8 **Transmission:** All	**A/F Sensor Circuit Slow Response (Bank 2 Sensor 1):** P0500 (Vehicle Speed Sensor) P0505 (IAC Valve) Active A/F control: Performing Active A/F control is performed when the following conditions set: Battery Voltage: 11 V or more ECT: 75°C (167°F) or more Idle: OFF Engine RPM: Less than 4000 rpm A/F sensor status: Activated Fuel cut: OFF Engine load: 10 to 70% Shift position: 2nd or more Catalyst monitor: Not executing Intake air amount: 2.5 to 12 g/sec. **Possible Causes:** • Open or short in Air Fuel Ratio (A/F) sensor (sensor 1) circuit • Air Fuel Ratio (A/F) sensor (sensor 1) • ECM
DTC: P2A03 **2T ECM, MIL: Yes** **Year:** 2009, 2010 **Model:** Land Cruiser **Engine:** 5.7L V8 **Transmission:** All	**A/F Sensor Circuit Slow Response (Bank 2 Sensor 1):** The calculated value of the air-fuel ratio (A/F) sensor response rate deterioration level is less than the threshold **Possible Causes:** • A/F sensor • A/F sensor heater • ECM

GLOSSARY

ABS: Anti-lock braking system. An electro-mechanical braking system which is designed to minimize or prevent wheel lock-up during braking.

ABSOLUTE PRESSURE: Atmospheric (barometric) pressure plus the pressure gauge reading.

ACCELERATOR PUMP: A small pump located in the carburetor that feeds fuel into the air/fuel mixture during acceleration.

ACCUMULATOR: A device that controls shift quality by cushioning the shock of hydraulic oil pressure being applied to a clutch or band.

ACTUATING MECHANISM: The mechanical output devices of a hydraulic system, for example, clutch pistons and band servos.

ACTUATOR: The output component of a hydraulic or electronic system.

ADVANCE: Setting the ignition timing so that spark occurs earlier before the piston reaches top dead center (TDC).

ADAPTIVE MEMORY (ADAPTIVE STRATEGY): The learning ability of the TCM or PCM to redefine its decision-making process to provide optimum shift quality.

AFTER TOP DEAD CENTER (ATDC): The point after the piston reaches the top of its travel on the compression stroke.

AIR BAG: Device on the inside of the car designed to inflate on impact of crash, protecting the occupants of the car.

AIR CHARGE TEMPERATURE (ACT) SENSOR: The temperature of the airflow into the engine is measured by an ACT sensor, usually located in the lower intake manifold or air cleaner.

AIR CLEANER: An assembly consisting of a housing, filter and any connecting ductwork. The filter element is made up of a porous paper, sometimes with a wire mesh screening, and is designed to prevent airborne particles from entering the engine through the carburetor or throttle body.

AIR INJECTION: One method of reducing harmful exhaust emissions by injecting air into each of the exhaust ports of an engine. The fresh air entering the hot exhaust manifold causes any remaining fuel to be burned before it can exit the tailpipe.

AIR PUMP: An emission control device that supplies fresh air to the exhaust manifold to aid in more completely burning exhaust gases.

AIR/FUEL RATIO: The ratio of air-to-gasoline by weight in the fuel mixture drawn into the engine.

ALDL (assembly line diagnostic link): Electrical connector for scanning ECM/PCM/TCM input and output devices.

ALIGNMENT RACK: A special drive-on vehicle lift apparatus/measuring device used to adjust a vehicle's toe, caster and camber angles.

ALL WHEEL DRIVE: Term used to describe a full time four wheel drive system or any other vehicle drive system that continuously delivers power to all four wheels. This system is found primarily on station wagon vehicles and SUVs not utilized for significant off road use.

ALTERNATING CURRENT (AC): Electric current that flows first in one direction, then in the opposite direction, continually reversing flow.

ALTERNATOR: A device which produces AC (alternating current) which is converted to DC (direct current) to charge the car battery.

AMMETER: An instrument, calibrated in amperes, used to measure the flow of an electrical current in a circuit. Ammeters are always connected in series with the circuit being tested.

AMPERAGE: The total amount of current (amperes) flowing in a circuit.

AMPLIFIER: A device used in an electrical circuit to increase the voltage of an output signal.

AMP/HR. RATING (BATTERY): Measurement of the ability of a battery to deliver a stated amount of current for a stated period of time. The higher the amp/hr. rating, the better the battery.

AMPERE: The rate of flow of electrical current present when one volt of electrical pressure is applied against one ohm of electrical resistance.

ANALOG COMPUTER: Any microprocessor that uses similar (analogous) electrical signals to make its calculations.

ANODIZED: A special coating applied to the surface of aluminum valves for extended service life.

ANTIFREEZE: A substance (ethylene or propylene glycol) added to the coolant to prevent freezing in cold weather.

ANTI-FOAM AGENTS: Minimize fluid foaming from the whipping action encountered in the converter and planetary action.

ANTI-WEAR AGENTS: Zinc agents that control wear on the gears, bushings, and thrust washers.

ANTI-LOCK BRAKING SYSTEM: A supplementary system to the base hydraulic system that prevents sustained lock-up of the wheels during braking as well as automatically controlling wheel slip.

ANTI-ROLL BAR: See stabilizer bar.

ARC: A flow of electricity through the air between two electrodes or contact points that produces a spark.

ARMATURE: A laminated, soft iron core wrapped by a wire that converts electrical energy to mechanical energy as in a motor or relay. When rotated in a magnetic field, it changes mechanical energy into electrical energy as in a generator.

ATDC: After Top Dead Center.

ATF: Automatic transmission fluid.

ATMOSPHERIC PRESSURE: The pressure on the Earth's surface caused by the weight of the air in the atmosphere. At sea level, this pressure is 14.7 psi at 32°F (101 kPa at 0°C).

ATOMIZATION: The breaking down of a liquid into a fine mist that can be suspended in air.

AUXILIARY ADD-ON COOLER: A supplemental transmission fluid cooling device that is installed in series with the heat exchanger (cooler), located inside the radiator, to provide additional support to cool the hot fluid leaving the torque converter.

AUXILIARY PRESSURE: An added fluid pressure that is introduced into a regulator or balanced valve system to control valve movement. The auxiliary pressure itself can be either a fixed or a variable value. (See balanced valve; regulator valve.)

AWD: All wheel drive.

AXIAL FORCE: A side or end thrust force acting in or along the same plane as the power flow.

AXIAL PLAY: Movement parallel to a shaft or bearing bore.

AXLE CAPACITY: The maximum load-carrying capacity of the axle itself, as specified by the manufacturer. This is usually a higher number than the GAWR.

AXLE RATIO: This is a number (3.07:1, 4.56:1, for example) expressing the ratio between driveshaft revolutions and wheel revolutions. A low numerical ratio allows the engine to work easier because it doesn't have to turn as fast. A high numerical ratio means that the engine has to turn more rpm's to move the wheels through the same number of turns.

BACKFIRE: The sudden combustion of gases in the intake or exhaust system that results in a loud explosion.

BACKLASH: The clearance or play between two parts, such as meshed gears.

BACKPRESSURE: Restrictions in the exhaust system that slow the exit of exhaust gases from the combustion chamber.

BAKELITE®: A heat resistant, plastic insulator material commonly used in printed circuit boards and transistorized components.

BALANCED VALVE: A valve that is positioned by opposing auxiliary hydraulic pressures and/or spring force. Examples include mainline regulator, throttle, and governor valves. (See regulator valve.)

BAND: A flexible ring of steel with an inner lining of friction material. When tightened around the outside of a drum, a planetary member is held stationary to the transmission/transaxle case.

BALL BEARING: A bearing made up of hardened inner and outer races between which hardened steel balls roll.

BALL JOINT: A ball and matching socket connecting suspension components (steering knuckle to lower control arms). It permits rotating movement in any direction between the components that are joined.

BARO (BAROMETRIC PRESSURE SENSOR): Measures the change in the intake manifold pressure caused by changes in altitude.

BAROMETRIC MANIFOLD ABSOLUTE PRESSURE (BMAP) SENSOR: Operates similarly to a conventional MAP sensor; reads intake manifold pressure and is also responsible for determining altitude and barometric pressure prior to engine operation.

BAROMETRIC PRESSURE: (See atmospheric pressure.)

BALLAST RESISTOR: A resistor in the primary ignition circuit that lowers voltage after the engine is started to reduce wear on ignition components.

BATTERY: A direct current electrical storage unit, consisting of the basic active materials of lead and sulfuric acid, which converts chemical energy into electrical energy. Used to provide current for the operation of the starter as well as other equipment, such as the radio, lighting, etc.

BEAD: The portion of a tire that holds it on the rim.

BEARING: A friction reducing, supportive device usually located between a stationary part and a moving part.

BEFORE TOP DEAD CENTER (BTDC): The point just before the piston reaches the top of its travel on the compression stroke.

BELTED TIRE: Tire construction similar to bias-ply tires, but using two or more layers of reinforced belts between body plies and the tread.

BEZEL: Piece of metal surrounding radio, headlights, gauges or similar components; sometimes used to hold the glass face of a gauge in the dash.

BIAS-PLY TIRE: Tire construction, using body ply reinforcing cords which run at alternating angles to the center line of the tread.

BI-METAL TEMPERATURE SENSOR: Any sensor or switch made of two dissimilar types of metal that bend when heated or cooled due to the different expansion rates of the alloys. These types of sensors usually function as an on/off switch.

BLOCK: See Engine Block.

BLOW-BY: Combustion gases, composed of water vapor and unburned fuel, that leak past the piston rings into the crankcase during normal engine operation. These gases are removed by the PCV system to prevent the buildup of harmful acids in the crankcase.

BOOK TIME: See Labor Time.

BOOK VALUE: The average value of a car, widely used to determine trade-in and resale value.

BOOST VALVE: Used at the base of the regulator valve to increase mainline pressure.

BORE: Diameter of a cylinder.

BRAKE CALIPER: The housing that fits over the brake disc. The caliper holds the brake pads, which are pressed against the discs by the caliper pistons when the brake pedal is depressed.

BRAKE HORSEPOWER (BHP): The actual horsepower available at the engine flywheel as measured by a dynamometer.

BRAKE FADE: Loss of braking power, usually caused by excessive heat after repeated brake applications.

BRAKE HORSEPOWER: Usable horsepower of an engine measured at the crankshaft.

BRAKE PAD: A brake shoe and lining assembly used with disc brakes.

BRAKE PROPORTIONING VALVE: A valve on the master cylinder which restricts hydraulic brake pressure to the wheels to a specified amount, preventing wheel lock-up.

BREAKAWAY: Often used by Chrysler to identify first-gear operation in D and 2 ranges. In these ranges, first-gear operation depends on a one-way roller clutch that holds on acceleration and releases (breaks away) on deceleration, resulting in a freewheeling coast-down condition.

BRAKE SHOE: The backing for the brake lining. The term is, however, usually applied to the assembly of the brake backing and lining.

BREAKER POINTS: A set of points inside the distributor, operated by a cam, which make and break the ignition circuit.

BRINNELLING: A wear pattern identified by a series of indentations at regular intervals. This condition is caused by a lack of lube, overload situations, and/or vibrations.

BTDC: Before Top Dead Center.

BUMP: Sudden and forceful apply of a clutch or band.

BUSHING: A liner, usually removable, for a bearing; an anti-friction liner used in place of a bearing.

CALIFORNIA ENGINE: An engine certified by the EPA for use in California only; conforms to more stringent emission regulations than Federal engine.

CALIPER: A hydraulically activated device in a disc brake system, which is mounted straddling the brake rotor (disc). The caliper contains at least one piston and two brake pads. Hydraulic pressure on the piston(s) forces the pads against the rotor.

CAPACITY: The quantity of electricity that can be delivered from a unit, as from a battery in ampere-hours, or output, as from a generator.

CAMBER: One of the factors of wheel alignment. Viewed from the front of the car, it is the inward or outward tilt of the wheel. The top of the tire will lean outward (positive camber) or inward (negative camber).

CAMSHAFT: A shaft in the engine on which are the lobes (cams) which operate the valves. The camshaft is driven by the crankshaft, via a belt, chain or gears, at one half the crankshaft speed.

CAPACITOR: A device which stores an electrical charge.

CARBON MONOXIDE (CO): A colorless, odorless gas given off as a normal byproduct of combustion. It is poisonous and extremely dangerous in confined areas, building up slowly to toxic levels without warning if adequate ventilation is not available.

CARBURETOR: A device, usually mounted on the intake manifold of an engine, which mixes the air and fuel in the proper proportion to allow even combustion.

CASTER: The forward or rearward tilt of an imaginary line drawn through the upper ball joint and the center of the wheel. Viewed from the sides, positive caster (forward tilt) lends directional stability, while negative caster (rearward tilt) produces instability.

CATALYTIC CONVERTER: A device installed in the exhaust system, like a muffler, that converts harmful byproducts of combustion into carbon dioxide and water vapor by means of a heat-producing chemical reaction.

CENTRIFUGAL ADVANCE: A mechanical method of advancing the spark timing by using flyweights in the distributor that react to centrifugal force generated by the distributor shaft rotation.

CENTRIFUGAL FORCE: The outward pull of a revolving object, away from the center of revolution. Centrifugal force increases with the speed of rotation.

CETANE RATING: A measure of the ignition value of diesel fuel. The higher the cetane rating, the better the fuel. Diesel fuel cetane rating is roughly comparable to gasoline octane rating.

CHECK VALVE: Any one-way valve installed to permit the flow of air, fuel or vacuum in one direction only.

CHOKE: The valve/plate that restricts the amount of air entering an engine on the induction stroke, thereby enriching the air/fuel ratio.

CHUGGLE: Bucking or jerking condition that may be engine related and may be most noticeable when converter clutch is engaged; similar to the feel of towing a trailer.

CIRCLIP: A split steel snapring that fits into a groove to hold various parts in place.

CIRCUIT BREAKER: A switch which protects an electrical circuit from overload by opening the circuit when the current flow exceeds a pre-determined level. Some circuit breakers must be reset manually, while most reset automatically.

CIRCUIT: Any unbroken path through which an electrical current can flow. Also used to describe fuel flow in some instances.

CIRCUIT, BYPASS: Another circuit in parallel with the major circuit through which power is diverted.

CIRCUIT, CLOSED: An electrical circuit in which there is no interruption of current flow.

CIRCUIT, GROUND: The non-insulated portion of a complete circuit used as a common potential point. In automotive circuits, the ground is composed of metal parts, such as the engine, body sheet metal, and frame and is usually a negative potential.

CIRCUIT, HOT: That portion of a circuit not at ground potential. The hot circuit is usually insulated and is connected to the positive side of the battery.

CIRCUIT, OPEN: A break or lack of contact in an electrical circuit, either intentional (switch) or unintentional (bad connection or broken wire).

CIRCUIT, PARALLEL: A circuit having two or more paths for current flow with common positive and negative tie points. The same voltage is applied to each load device or parallel branch.

CIRCUIT, SERIES: An electrical system in which separate parts are connected end to end, using one wire, to form a single path for current to flow.

CIRCUIT, SHORT: A circuit that is accidentally completed in an electrical path for which it was not intended.

CLAMPING (ISOLATION) DIODES: Diodes positioned in a circuit to prevent self-induction from damaging electronic components.

CLEARCOAT: A transparent layer which, when sprayed over a vehicle's paint job, adds gloss and depth as well as an additional protective coating to the finish.

CLUTCH: Part of the power train used to connect/disconnect power to the rear wheels.

CLUTCH, FLUID: The same as a fluid coupling. A fluid clutch or coupling performs the same function as a friction clutch by utilizing fluid friction and inertia as opposed to solid friction used by a friction clutch. (See fluid coupling.)

CLUTCH, FRICTION: A coupling device that provides a means of smooth and positive engagement and disengagement of engine torque to the vehicle powertrain. Transmission of power through the clutch is accomplished by bringing one or more rotating drive members into contact with complementing driven members.

COAST: Vehicle deceleration caused by engine braking conditions.

COEFFICIENT OF FRICTION: The amount of surface tension between two contacting surfaces; identified by a scientifically calculated number.

COIL: Part of the ignition system that boosts the relatively low voltage supplied by the car's electrical system to the high voltage required to fire the spark plugs.

COMBINATION MANIFOLD: An assembly which includes both the intake and exhaust manifolds in one casting.

COMBINATION VALVE: A device used in some fuel systems that routes fuel vapors to a charcoal storage canister instead of venting them into the atmosphere. The valve relieves fuel tank pressure and allows fresh air into the tank as the fuel level drops to prevent a vapor lock situation.

COMBUSTION CHAMBER: The part of the engine in the cylinder head where combustion takes place.

COMPOUND GEAR: A gear consisting of two or more simple gears with a common shaft.

COMPOUND PLANETARY: A gearset that has more than the three elements found in a simple gearset and is constructed by combining members of two planetary gearsets to create additional gear ratio possibilities.

COMPRESSION CHECK: A test involving removing each spark plug and inserting a gauge. When the engine is cranked, the gauge will record a pressure reading in the individual cylinder. General operating condition can be determined from a compression check.

COMPRESSION RATIO: The ratio of the volume between the piston and cylinder head when the piston is at the bottom of its stroke (bottom dead center) and when the piston is at the top of its stroke (top dead center).

COMPUTER: An electronic control module that correlates input data according to prearranged engineered instructions; used for the management of an actuator system or systems.

CONDENSER: An electrical device which acts to store an electrical charge, preventing voltage surges.

2. A radiator-like device in the air conditioning system in which refrigerant gas condenses into a liquid, giving off heat.

CONDUCTOR: Any material through which an electrical current can be transmitted easily.

CONNECTING ROD: The connecting link between the crankshaft and piston.

CONSTANT VELOCITY JOINT: Type of universal joint in a halfshaft assembly in which the output shaft turns at a constant angular velocity without variation, provided that the speed of the input shaft is constant.

CONTINUITY: Continuous or complete circuit. Can be checked with an ohmmeter.

CONTROL ARM: The upper or lower suspension components which are mounted on the frame and support the ball joints and steering knuckles.

CONVENTIONAL IGNITION: Ignition system which uses breaker points.

CONVERTER: (See torque converter.)

CONVERTER LOCKUP: The switching from hydrodynamic to direct mechanical drive, usually through the application of a friction element called the converter clutch.

COOLANT: Mixture of water and anti-freeze circulated through the engine to carry off heat produced by the engine.

CORROSION INHIBITOR: An inhibitor in ATF that prevents corrosion of bushings, thrust washers, and oil cooler brazed joints.

COUNTERSHAFT: An intermediate shaft which is rotated by a mainshaft and transmits, in turn, that rotation to a working part.

COUPLING PHASE: Occurs when the torque converter is operating at its greatest hydraulic efficiency. The speed differential between the impeller and the turbine is at its minimum. At this point, the stator freewheels, and there is no torque multiplication.

CRANKCASE: The lower part of an engine in which the crankshaft and related parts operate.

CRANKSHAFT: Engine component (connected to pistons by connecting rods) which converts the reciprocating (up and down) motion of pistons to rotary motion used to turn the driveshaft.

CURB WEIGHT: The weight of a vehicle without passengers or payload, but including all fluids (oil, gas, coolant, etc.) and other equipment specified as standard.

CURRENT: The flow (or rate) of electrons moving through a circuit. Current is measured in amperes (amp).

CURRENT FLOW CONVENTIONAL: Current flows through a circuit from the positive terminal of the source to the negative terminal (plus to minus).

CURRENT FLOW, ELECTRON: Current or electrons flow from the negative terminal of the source, through the circuit, to the positive terminal (minus to plus).

CV-JOINT: Constant velocity joint.

CYCLIC VIBRATIONS: The off-center movement of a rotating object that is affected by its initial balance, speed of rotation, and working angles.

CYLINDER BLOCK: See engine block.

CYLINDER HEAD: The detachable portion of the engine, usually fastened to the top of the cylinder block and containing all or most of the combustion chambers. On overhead valve engines, it contains the valves and their operating parts. On overhead cam engines, it contains the camshaft as well.

CYLINDER: In an engine, the round hole in the engine block in which the piston(s) ride.

DATA LINK CONNECTOR (DLC): Current acronym/term applied to the federally mandated, diagnostic junction connector that is used to monitor ECM/PC/TCM inputs, processing strategies, and outputs including diagnostic trouble codes (DTCs).

DEAD CENTER: The extreme top or bottom of the piston stroke.

DECELERATION BUMP: When referring to a torque converter clutch in the applied position, a sudden release of the accelerator pedal causes a forceful reversal of power through the drivetrain (engine braking), just prior to the apply plate actually being released.

DELAYED (LATE OR EXTENDED): Condition where shift is expected but does not occur for a period of time, for example, where clutch or band engagement does not occur as quickly as expected during part throttle or wide open throttle apply of accelerator or when manually downshifting to a lower range.

DETENT: A spring-loaded plunger, pin, ball, or pawl used as a holding device on a ratchet wheel or shaft. In automatic transmissions, a detent mechanism is used for locking the manual valve in place.

DETENT DOWNSHIFT: (See kickdown.)

DETERGENT: An additive in engine oil to improve its operating characteristics.

DETONATION: An unwanted explosion of the air/fuel mixture in the combustion chamber caused by excess heat and compression, advanced timing, or an overly lean mixture. Also referred to as "ping".

DEXRON®: A brand of automatic transmission fluid.

DIAGNOSTIC TROUBLE CODES (DTCs): A digital display from the control module memory that identifies the input, processor, or output device circuit that is related to the powertrain emission/driveability malfunction detected. Diagnostic trouble codes can be read by the MIL to flash any codes or by using a handheld scanner.

DIAPHRAGM: A thin, flexible wall separating two cavities, such as in a vacuum advance unit.

DIESELING: The engine continues to run after the car is shut off; caused by fuel continuing to be burned in the combustion chamber.

DIFFERENTIAL: A geared assembly which allows the transmission of motion between drive axles, giving one axle the ability to rotate faster than the other, as in cornering.

DIFFERENTIAL AREAS: When opposing faces of a spool valve are acted upon by the same pressure but their areas differ in size, the face with the larger area produces the differential force and valve movement. (See spool valve.)

DIFFERENTIAL FORCE: (See differential areas)

DIGITAL READOUT: A display of numbers or a combination of numbers and letters.

DIGITAL VOLT OHMMETER: An electronic diagnostic tool used to measure voltage, ohms and amps as well as several other functions, with the readings displayed on a digital screen in tenths, hundredths and thousandths.

DIODE: An electrical device that will allow current to flow in one direction only.

DIRECT CURRENT (DC): Electrical current that flows in one direction only.

DIRECT DRIVE: The gear ratio is 1:1, with no change occurring in the torque and speed input/output relationship.

DISC BRAKE: A hydraulic braking assembly consisting of a brake disc, or rotor, mounted on an axle shaft, and a caliper assembly containing, usually two brake pads which are activated by hydraulic pressure. The pads are forced against the sides of the disc, creating friction which slows the vehicle.

DISPERSANTS: Suspend dirt and prevent sludge buildup in a liquid, such as engine oil.

DOUBLE BUMP (DOUBLE FEEL): Two sudden and forceful applies of a clutch or band.

DISPLACEMENT: The total volume of air that is displaced by all pistons as the engine turns through one complete revolution.

DISTRIBUTOR: A mechanically driven device on an engine which is responsible for electrically firing the spark plug at a pre-determined point of the piston stroke.

DOHC: Double overhead camshaft.

DOUBLE OVERHEAD CAMSHAFT: The engine utilizes two camshafts mounted in one cylinder head. One camshaft operates the exhaust valves, while the other operates the intake valves.

DOWEL PIN: A pin, inserted in mating holes in two different parts allowing those parts to maintain a fixed relationship.

DRIVELINE: The drive connection between the transmission and the drive wheels.

DRIVE TRAIN: The components that transmit the flow of power from the engine to the wheels. The components include the clutch, transmission, driveshafts (or axle shafts in front wheel drive), U-joints and differential.

DRUM BRAKE: A braking system which consists of two brake shoes and one or two wheel cylinders, mounted on a fixed backing plate, and a brake drum, mounted on an axle, which revolves around the assembly.

DRY CHARGED BATTERY: Battery to which electrolyte is added when the battery is placed in service.

DVOM: Digital volt ohmmeter

DWELL: The rate, measured in degrees of shaft rotation, at which an electrical circuit cycles on and off.

DYNAMIC: An application in which there is rotating or reciprocating motion between the parts.

EARLY: Condition where shift occurs before vehicle has reached proper speed, which tends to labor engine after upshift.

EBCM: See Electronic Control Unit (ECU).

ECM: See Electronic Control Unit (ECU).

ECU: Electronic control unit.

ELECTRODE: Conductor (positive or negative) of electric current.

ELECTROLYSIS: A surface etching or bonding of current conducting transmission/transaxle components that may occur when grounding straps are missing or in poor condition.

ELECTROLYTE: A solution of water and sulfuric acid used to activate the battery. Electrolyte is extremely corrosive.

ELECTROMAGNET: A coil that produces a magnetic field when current flows through its windings.

ELECTROMAGNETIC INDUCTION: A method to create (generate) current flow through the use of magnetism.

ELECTROMAGNETISM: The effects surrounding the relationship between electricity and magnetism.

ELECTROMOTIVE FORCE (EMF): The force or pressure (voltage) that causes current movement in an electrical circuit.

ELECTRONIC CONTROL UNIT: A digital computer that controls engine (and sometimes transmission, brake or other vehicle system) functions based on data received from various sensors. Examples used by some manufacturers include Electronic Brake Control Module (EBCM), Engine Control Module (ECM), Powertrain Control Module (PCM) or Vehicle Control Module (VCM).

ELECTRONIC IGNITION: A system in which the timing and firing of the spark plugs is controlled by an electronic control unit, usually called a module. These systems have no points or condenser.

ELECTRONIC PRESSURE CONTROL (EPC) SOLENOID: A specially designed solenoid containing a spool valve and spring assembly to control fluid mainline pressure. A variable current flow, controlled by the ECM/PCM, varies the internal force of the solenoid on the spool valve and resulting mainline pressure. (See variable force solenoid.)

ELECTRONICS: Miniaturized electrical circuits utilizing semiconductors, solid-state devices, and printed circuits. Electronic circuits utilize small amounts of power.

ELECTRONIFICATION: The application of electronic circuitry to a mechanical device. Regarding automatic transmissions, electrification is incorporated into converter clutch lockup, shift scheduling, and line pressure control systems.

ELECTROSTATIC DISCHARGE (ESD): An unwanted, high-voltage electrical current released by an individual who has taken on a static charge of electricity. Electronic components can be easily damaged by ESD.

ELEMENT: A device within a hydrodynamic drive unit designed with a set of blades to direct fluid flow.

ENAMEL: Type of paint that dries to a smooth, glossy finish.

END BUMP (END FEEL OR SLIP BUMP): Firmer feel at end of shift when compared with feel at start of shift.

END-PLAY: The clearance/gap between two components that allows for expansion of the parts as they warm up, to prevent binding and to allow space for lubrication.

ENERGY: The ability or capacity to do work.

ENGINE: The primary motor or power apparatus of a vehicle, which converts liquid or gas fuel into mechanical energy.

ENGINE BLOCK: The basic engine casting containing the cylinders, the crankshaft main bearings, as well as machined surfaces for the mounting of other components such as the cylinder head, oil pan, transmission, etc.

ENGINE BRAKING: Use of engine to slow vehicle by manually downshifting during zero-throttle coast down.

ENGINE CONTROL MODULE (ECM): Manages the engine and incorporates output control over the torque converter clutch solenoid. (Note: Current designation for the ECM in late model vehicles is PCM.)

ENGINE COOLANT TEMPERATURE (ECT) SENSOR: Prevents converter clutch engagement with a cold engine; also used for shift timing and shift quality.

EP LUBRICANT: EP (extreme pressure) lubricants are specially formulated for use with gears involving heavy loads (transmissions, differentials, etc.).

ETHYL: A substance added to gasoline to improve its resistance to knock, by slowing down the rate of combustion.

ETHYLENE GLYCOL: The base substance of antifreeze.

EXHAUST MANIFOLD: A set of cast passages or pipes which conduct exhaust gases from the engine.

FAIL-SAFE (BACKUP) CONTROL: A substitute value used by the PCM/TCM to replace a faulty signal from an input sensor. The temporary value allows the vehicle to continue to be operated.

FAST IDLE: The speed of the engine when the choke is on. Fast idle speeds engine warm-up.

FEDERAL ENGINE: An engine certified by the EPA for use in any of the 49 states (except California).

FEEDBACK: A circuit malfunction whereby current can find another path to feed load devices.

FEELER GAUGE: A blade, usually metal, of precisely predetermined thickness, used to measure the clearance between two parts.

FILAMENT: The part of a bulb that glows; the filament creates high resistance to current flow and actually glows from the resulting heat.

FINAL DRIVE: An essential part of the axle drive assembly where final gear reduction takes place in the powertrain. In RWD applications and north-south FWD applications, it must also change the power flow direction to the axle shaft by ninety degrees. (Also see axle ratio).

FIRING ORDER: The order in which combustion occurs in the cylinders of an engine. Also the order in which spark is distributed to the plugs by the distributor.

FIRM: A noticeable quick apply of a clutch or band that is considered normal with medium to heavy throttle shift; should not be confused with harsh or rough.

FLAME FRONT: The term used to describe certain aspects of the fuel explosion in the cylinders. The flame front should move in a controlled pattern across the cylinder, rather than simply exploding immediately.

FLARE (SLIPPING): A quick increase in engine rpm accompanied by momentary loss of torque; generally occurs during shift.

FLAT ENGINE: Engine design in which the pistons are horizontally opposed. Porsche, Subaru and some old VW are common examples of flat engines.

FLAT RATE: A dealership term referring to the amount of money paid to a technician for a repair or diagnostic service based on that particular service versus dealership's labor time (NOT based on the actual time the technician spent on the job).

FLAT SPOT: A point during acceleration when the engine seems to lose power for an instant.

FLOODING: The presence of too much fuel in the intake manifold and combustion chamber which prevents the air/fuel mixture from firing, thereby causing a no-start situation.

FLUID: A fluid can be either liquid or gas. In hydraulics, a liquid is used for transmitting force or motion.

FLUID COUPLING: The simplest form of hydrodynamic drive, the fluid coupling consists of two look-alike members with straight radial varies referred to as the impeller (pump) and the turbine. Input torque is always equal to the output torque.

FLUID DRIVE: Either a fluid coupling or a fluid torque converter. (See hydrodynamic drive units.)

FLUID TORQUE CONVERTER: A hydrodynamic drive that has the ability to act both as a torque multiplier and fluid coupling. (See hydrodynamic drive units; torque converter.)

FLUID VISCOSITY: The resistance of a liquid to flow. A cold fluid (oil) has greater viscosity and flows more slowly than a hot fluid (oil).

FLYWHEEL: A heavy disc of metal attached to the rear of the crankshaft. It smoothes the firing impulses of the engine and keeps the crankshaft turning during periods when no firing takes place. The starter also engages the flywheel to start the engine.

FOOT POUND (ft. lbs., lbs. ft. or sometimes, ft. lb.): The amount of energy or work needed to raise an item weighing one pound, a distance of one foot.

FREEZE PLUG: A plug in the engine block which will be pushed out if the coolant freezes. Sometimes called expansion plugs, they protect the block from cracking should the coolant freeze.

FRICTION: The resistance that occurs between contacting surfaces. This relationship is expressed by a ratio called the coefficient of friction (CL).

FRICTION, COEFFICIENT OF: The amount of surface tension between two contacting surfaces; expressed by a scientifically calculated number.

FRONT END ALIGNMENT: A service to set caster, camber and toe-in to the correct specifications. This will ensure that the car steers and handles properly and that the tires wear properly.

FRICTION MODIFIER: Changes the coefficient of friction of the fluid between the mating steel and composition clutch/band surfaces during the engagement process and allows for a certain amount of intentional slipping for a good "shift-feel".

FRONTAL AREA: The total frontal area of a vehicle exposed to air flow.

FUEL FILTER: A component of the fuel system containing a porous paper element used to prevent any impurities from entering the engine through the fuel system. It usually takes the form of a canister-like housing, mounted in-line with the fuel hose, located anywhere on a vehicle between the fuel tank and engine.

FUEL INJECTION: A system replacing the carburetor that sprays fuel into the cylinder through nozzles. The amount of fuel can be more precisely controlled with fuel injection.

FULL FLOATING AXLE: An axle in which the axle housing extends through the wheel giving bearing support on the outside of the housing. The front axle of a four-wheel drive vehicle is usually a full floating axle, as are the rear axles of many larger (1 ton and over) pick-ups and vans.

FULL-TIME FOUR-WHEEL DRIVE: A four-wheel drive system that continuously delivers power to all four wheels. A differential between the front and rear driveshafts permits variations in axle speeds to control gear wind-up without damage.

FULL THROTTLE DETENT DOWNSHIFT: A quick apply of accelerator pedal to its full travel, forcing a downshift.

FUSE: A protective device in a circuit which prevents circuit overload by breaking the circuit when a specific amperage is present. The device is constructed around a strip or wire of a lower amperage rating than the circuit it is designed to protect. When an amperage higher than that stamped on the fuse is present in the circuit, the strip or wire melts, opening the circuit.

FUSIBLE LINK: A piece of wire in a wiring harness that performs the same job as a fuse. If overloaded, the fusible link will melt and interrupt the circuit.

FWD: Front wheel drive.

GAWR: (Gross axle weight rating) the total maximum weight an axle is designed to carry.

GCW: (Gross combined weight) total combined weight of a tow vehicle and trailer.

GARAGE SHIFT: initial engagement feel of transmission, neutral to reverse or neutral to a forward drive.

GARAGE SHIFT FEEL: A quick check of the engagement quality and responsiveness of reverse and forward gears. This test is done with the vehicle stationary.

GEAR: A toothed mechanical device that acts as a rotating lever to transmit power or turning effort from one shaft to another. (See gear ratio.)

GEAR RATIO: A ratio expressing the number of turns a smaller gear will make to turn a larger gear through one revolution. The ratio is found by dividing the number of teeth on the smaller gear into the number of teeth on the larger gear.

GEARBOX: Transmission

GEAR REDUCTION: Torque is multiplied and speed decreased by the factor of the gear ratio. For example, a 3:1 gear ratio changes an input torque of 180 ft. lbs. and an input speed of 2700 rpm to 540 Ft. lbs. and 900 rpm, respectively. (No account is taken of frictional losses, which are always present.)

GEARTRAIN: A succession of intermeshing gears that form an assembly and provide for one or more torque changes as the power input is transmitted to the power output.

GEL COAT: A thin coat of plastic resin covering fiberglass body panels.

GENERATOR: A device which produces direct current (DC) necessary to charge the battery.

GOVERNOR: A device that senses vehicle speed and generates a hydraulic oil pressure. As vehicle speed increases, governor oil pressure rises.

GROUND CIRCUIT: (See circuit, ground.)

GROUND SIDE SWITCHING: The electrical/electronic circuit control switch is located after the circuit load.

GVWR: (Gross vehicle weight rating) total maximum weight a vehicle is designed to carry including the weight of the vehicle, passengers, equipment, gas, oil, etc.

HALOGEN: A special type of lamp known for its quality of brilliant white light. Originally used for fog lights and driving lights.

HARD CODES: DTCs that are present at the time of testing; also called continuous or current codes.

HARSH(ROUGH): An apply of a clutch or band that is more noticeable than a firm one; considered undesirable at any throttle position.

HEADER TANK: An expansion tank for the radiator coolant. It can be located remotely or built into the radiator.

HEAT RANGE: A term used to describe the ability of a spark plug to carry away heat. Plugs with longer nosed insulators take longer to carry heat off effectively.

HEAT RISER: A flapper in the exhaust manifold that is closed when the engine is cold, causing hot exhaust gases to heat the intake manifold providing better cold engine operation. A thermostatic spring opens the flapper when the engine warms up.

HEAVY THROTTLE: Approximately three-fourths of accelerator pedal travel.

HEMI: A name given an engine using hemispherical combustion chambers.

HERTZ (HZ): The international unit of frequency equal to one cycle per second (10,000 Hertz equals 10,000 cycles per second).

HIGH-IMPEDANCE DVOM (DIGITAL VOLT-OHMMETER): This styled device provides a built-in resistance value and is capable of limiting circuit current flow to safe milliamp levels.

HIGH RESISTANCE: Often refers to a circuit where there is an excessive amount of opposition to normal current flow.

HORSEPOWER: A measurement of the amount of work; one horsepower is the amount of work necessary to lift 33,000 lbs. one foot in one minute. Brake horsepower (bhp) is the horsepower delivered by an engine on a dynamometer. Net horsepower is the power remaining (measured at the flywheel of the engine) that can be used to turn the wheels after power is consumed through friction and running the engine accessories (water pump, alternator, air pump, fan etc.)

HOT CIRCUIT: (See circuit, hot; hot lead.)

HOT LEAD: A wire or conductor in the power side of the circuit. (See circuit, hot.)

HOT SIDE SWITCHING: The electrical/electronic circuit control switch is located before the circuit load.

HUB: The center part of a wheel or gear.

HUNTING (BUSYNESS): Repeating quick series of up-shifts and downshifts that causes noticeable change in engine rpm, for example, as in a 4-3-4 shift pattern.

HYDRAULICS: The use of liquid under pressure to transfer force of motion.

HYDROCARBON (HC): Any chemical compound made up of hydrogen and carbon. A major pollutant formed by the engine as a by-product of combustion.

HYDRODYNAMIC DRIVE UNITS: Devices that transmit power solely by the action of a kinetic fluid flow in a closed recirculating path. An impeller energizes the fluid and discharges the high-speed jet stream into the turbine for power output.

HYDROMETER: An instrument used to measure the specific gravity of a solution.

HYDROPLANING: A phenomenon of driving when water builds up under the tire tread, causing it to lose contact with the road. Slowing down will usually restore normal tire contact with the road.

HYPOID GEARSET: The drive pinion gear may be placed below or above the centerline of the driven gear; often used as a final drive gearset.

IDLE MIXTURE: The mixture of air and fuel (usually about 14:1) being fed to the cylinders. The idle mixture screw(s) are sometimes adjusted as part of a tune-up.

IDLER ARM: Component of the steering linkage which is a geometric duplicate of the steering gear arm. It supports the right side of the center steering link.

IMPELLER: Often called a pump, the impeller is the power input (drive) member of a hydrodynamic drive. As part of the torque converter cover, it acts as a centrifugal pump and puts the fluid in motion.

INCH POUND (inch lbs.; sometimes in. lb. or in. lbs.): One twelfth of a foot pound.

INDUCTANCE: The force that produces voltage when a conductor is passed through a magnetic field.

INDUCTION: A means of transferring electrical energy in the form of a magnetic field. Principle used in the ignition coil to increase voltage.

INITIAL FEEL: A distinct firmer feel at start of shift when compared with feel at finish of shift.

INJECTOR: A device which receives metered fuel under relatively low pressure and is activated to inject the fuel into the engine under relatively high pressure at a predetermined time.

INPUT: In an automatic transmission, the source of power from the engine is absorbed by the torque converter, which provides the power input into the transmission. The turbine drives the input(turbine)shaft.

INPUT SHAFT: The shaft to which torque is applied, usually carrying the driving gear or gears.

INTAKE MANIFOLD: A casting of passages or pipes used to conduct air or a fuel/air mixture to the cylinders.

INTERNAL GEAR: The ring-like outer gear of a planetary gearset with the gear teeth cut on the inside of the ring to provide a mesh with the planet pinions.

ISOLATION (CLAMPING) DIODES: Diodes positioned in a circuit to prevent self-induction from damaging electronic components.

IX ROTARY GEAR PUMP: Contains two rotating members, one shaped with internal gear teeth and the other with external gear teeth. As the gears separate, the fluid fills the gaps between gear teeth, is pulled across a crescent-shaped divider, and then is forced to flow through the outlet as the gears mesh.

IX ROTARY LOBE PUMP: Sometimes referred to as a gerotor type pump. Two rotating members, one shaped with internal lobes and the other with external lobes, separate and then mesh to cause fluid to flow.

JOURNAL: The bearing surface within which a shaft operates.

JUMPER CABLES: Two heavy duty wires with large alligator clips used to provide power from a charged battery to a discharged battery mounted in a vehicle.

JUMPSTART: Utilizing the sufficiently charged battery of one vehicle to start the engine of another vehicle with a discharged battery by the use of jumper cables.

KEY: A small block usually fitted in a notch between a shaft and a hub to prevent slippage of the two parts.

KICKDOWN: Detent downshift system; either linkage, cable, or electrically controlled.

KILO: A prefix used in the metric system to indicate one thousand.

KNOCK: Noise which results from the spontaneous ignition of a portion of the air-fuel mixture in the engine cylinder caused by overly advanced ignition timing or use of incorrectly low octane fuel for that engine.

KNOCK SENSOR: An input device that responds to spark knock, caused by over advanced ignition timing.

LABOR TIME: A specific amount of time required to perform a certain repair or diagnostic service as defined by a vehicle or after-market manufacturer .

LACQUER: A quick-drying automotive paint.

LATE: Shift that occurs when engine is at higher than normal rpm for given amount of throttle.

LIGHT-EMITTING DIODE (LED): A semiconductor diode that emits light as electrical current flows through it; used in some electronic display devices to emit a red or other color light.

LIGHT THROTTLE: Approximately one-fourth of accelerator pedal travel.

LIMITED SLIP: A type of differential which transfers driving force to the wheel with the best traction.

LIMP-IN MODE: Electrical shutdown of the transmission/ transaxle output solenoids, allowing only forward and reverse gears that are hydraulically energized by the manual valve. This permits the vehicle to be driven to a service facility for repair.

LIP SEAL: Molded synthetic rubber seal designed with an outer sealing edge (lip) that points into the fluid containing area to be sealed. This type of seal is used where rotational and axial forces are present.

LITHIUM-BASE GREASE: Chassis and wheel bearing grease using lithium as a base. Not compatible with sodium-base grease.

LOAD DEVICE: A circuit's resistance that converts the electrical energy into light, sound, heat, or mechanical movement.

LOAD RANGE: Indicates the number of plies at which a tire is rated. Load range B equals four-ply rating; C equals six-ply rating; and, D equals an eight-ply rating.

LOAD TORQUE: The amount of output torque needed from the transmission/transaxle to overcome the vehicle load.

LOCKING HUBS: Accessories used on part-time four-wheel drive systems that allow the front wheels to be disengaged from the drive train when four-wheel drive is not being used. When four-wheel drive is desired, the hubs are engaged, locking the wheels to the drive train.

LOCKUP CONVERTER: A torque converter that operates hydraulically and mechanically. When an internal apply plate (lockup plate) clamps to the torque converter cover, hydraulic slippage is eliminated.

LOCK RING: See Circlip or Snapring

MAGNET: Any body with the property of attracting iron or steel.

MAGNETIC FIELD: The area surrounding the poles of a magnet that is affected by its attraction or repulsion forces.

MAIN LINE PRESSURE: Often called control pressure or line pressure, it refers to the pressure of the oil leaving the pump and is controlled by the pressure regulator valve.

MALFUNCTION INDICATOR LAMP (MIL): Previously known as a check engine light, the dash-mounted MIL illuminates and signals the driver that an emission or driveability problem with the powertrain has been detected by the ECM/PCM. When this occurs, at least one diagnostic trouble code (DTC) has been stored into the control module memory.

MANIFOLD ABSOLUTE PRESSURE (MAP) SENSOR: Reads the amount of air pressure (vacuum) in the engine's intake manifold system; its signal is used to analyze engine load conditions.

MANIFOLD VACUUM: Low pressure in an engine intake manifold formed just below the throttle plates. Manifold vacuum is highest at idle and drops under acceleration.

MANIFOLD: A casting of passages or set of pipes which connect the cylinders to an inlet or outlet source.

MANUAL LEVER POSITION SWITCH (MLPS): A mechanical switching unit that is typically mounted externally to the transmission/transaxle to inform the PCM/ECM which gear range the driver has selected.

MANUAL VALVE: Located inside the transmission/transaxle, it is directly connected to the driver's shift lever. The position of the manual valve determines which hydraulic circuits will be charged with oil pressure and the operating mode of the transmission.

MANUAL VALVE LEVER POSITION SENSOR (MVLPS): The input from this device tells the TCM what gear range was selected.

MASS AIR FLOW (MAF) SENSOR: Measures the airflow into the engine.

MASTER CYLINDER: The primary fluid pressurizing device in a hydraulic system. In automotive use, it is found in brake and hydraulic clutch systems and is pedal activated, either directly or, in a power brake system, through the power booster.

MacPherson STRUT: A suspension component combining a shock absorber and spring in one unit.

MEDIUM THROTTLE: Approximately one-half of accelerator pedal travel.

MEGA: A metric prefix indicating one million.

MEMBER: An independent component of a hydrodynamic unit such as an impeller, a stator, or a turbine. It may have one or more elements.

MERCON: A fluid developed by Ford Motor Company in 1988. It contains a friction modifier and closely resembles operating characteristics of Dexron.

METAL SEALING RINGS: Made from cast iron or aluminum, their primary application is with dynamic components involving pressure sealing circuits of rotating members. These rings are designed with either butt or hook lock end joints.

METER (ANALOG): A linear-style meter representing data as lengths; a needle-style instrument interfacing with logical numerical increments. This style of electrical meter uses relatively low impedance internal resistance and cannot be used for testing electronic circuitry.

METER (DIGITAL): Uses numbers as a direct readout to show values. Most meters of this style use high impedance internal resistance and must be used for testing low current electronic circuitry.

MICRO: A metric prefix indicating one-millionth (0.000001).

MILLI: A metric prefix indicating one-thousandth (0.001).

MINIMUM THROTTLE: The least amount of throttle opening required for upshift; normally close to zero throttle.

MISFIRE: Condition occurring when the fuel mixture in a cylinder fails to ignite, causing the engine to run roughly.

MODULE: Electronic control unit, amplifier or igniter of solid state or integrated design which controls the current flow in the ignition primary circuit based on input from the pick-up coil. When the module opens the primary circuit, high secondary voltage is induced in the coil.

MODULATED: In an electronic-hydraulic converter clutch system (or shift valve system), the term modulated refers to the pulsing of a solenoid, at a variable rate. This action controls the buildup of oil pressure in the hydraulic circuit to allow a controlled amount of clutch slippage.

MODULATED CONVERTER CLUTCH CONTROL (MCCC): A pulse width duty cycle valve that controls the converter lockup apply pressure and maximizes smoother transitions between lock and unlock conditions.

MODULATOR PRESSURE (THROTTLE PRESSURE): A hydraulic signal oil pressure relating to the amount of engine load, based on either the amount of throttle plate opening or engine vacuum.

MODULATOR VALVE: A regulator valve that is controlled by engine vacuum, providing a hydraulic pressure that varies in relation to engine torque. The hydraulic torque signal functions to delay the shift pattern and provide a line pressure boost. (See throttle valve.)

MOTOR: An electromagnetic device used to convert electrical energy into mechanical energy.

MULTIPLE-DISC CLUTCH: A grouping of steel and friction lined plates that, when compressed together by hydraulic pressure acting upon a piston, lock or unlock a planetary member.

MULTI-WEIGHT: Type of oil that provides adequate lubrication at both high and low temperatures.

needed to move one amp through a resistance of one ohm.

MUSHY: Same as soft; slow and drawn out clutch apply with very little shift feel.

MUTUAL INDUCTION: The generation of current from one wire circuit to another by movement of the magnetic field surrounding a current-carrying circuit as its ampere flow increases or decreases.

NEEDLE BEARING: A bearing which consists of a number (usually a large number) of long, thin rollers.

NITROGEN OXIDE (NOx): One of the three basic pollutants found in the exhaust emission of an internal combustion engine. The amount of NOx usually varies in an inverse proportion to the amount of HC and CO.

NONPOSITIVE SEALING: A sealing method that allows some minor leakage, which normally assists in lubrication.

02 SENSOR: Located in the engine's exhaust system, it is an input device to the ECM/PCM for managing the fuel delivery and ignition system. A scanner can be used to observe the fluctuating voltage readings produced by an 02 sensor as the oxygen content of the exhaust is analyzed.

O-RING SEAL: Molded synthetic rubber seal designed with a circular cross-section. This type of seal is used primarily in static applications.

OBD II (ON-BOARD DIAGNOSTICS, SECOND GENERATION): Refers to the federal law mandating tighter control of 1996 and newer vehicle emissions, active monitoring of related devices, and standardization of terminology, data link connectors, and other technician concerns.

OCTANE RATING: A number, indicating the quality of gasoline based on its ability to resist knock. The higher the number, the better the quality. Higher compression engines require higher octane gas.

OEM: Original Equipment Manufactured. OEM equipment is that furnished standard by the manufacturer.

OFFSET: The distance between the vertical center of the wheel and the mounting surface at the lugs. Offset is positive if the center is outside the lug circle; negative offset puts the center line inside the lug circle.

OHM'S LAW: A law of electricity that states the relationship between voltage, current, and resistance. Volts = amperes x ohms

OHM: The unit used to measure the resistance of conductor-to-electrical flow. One ohm is the amount of resistance that limits current flow to one ampere in a circuit with one volt of pressure.

OHMMETER: An instrument used for measuring the resistance, in ohms, in an electrical circuit.

ONE-WAY CLUTCH: A mechanical clutch of roller or sprag design that resists torque or transmits power in one direction only. It is used to either hold or drive a planetary member.

ONE-WAY ROLLER CLUTCH: A mechanical device that transmits or holds torque in one direction only.

OPEN CIRCUIT: A break or lack of contact in an electrical circuit, either intentional (switch) or unintentional (bad connection or broken wire).

ORIFICE: Located in hydraulic oil circuits, it acts as a restriction. It slows down fluid flow to either create back pressure or delay pressure buildup downstream.

OSCILLOSCOPE: A piece of test equipment that shows electric impulses as a pattern on a screen. Engine performance can be analyzed by interpreting these patterns.

OUTPUT SHAFT: The shaft which transmits torque from a device, such as a transmission.

OUTPUT SPEED SENSOR (OSS): Identifies transmission/transaxle output shaft speed for shift timing and may be used to calculate TCC slip; often functions as the VSS (vehicle speed sensor).

OVERDRIVE: (1.) A device attached to or incorporated in a transmission/transaxle that allows the engine to turn less than one full revolution for every complete revolution of the wheels. The net effect is to reduce engine rpm, thereby using less fuel. A typical overdrive gear ratio would be .87:1, instead of the normal 1:1 in high gear. (2.) A gear assembly which produces more shaft revolutions than that transmitted to it.

OVERDRIVE PLANETARY GEARSET: A single planetary gearset designed to provide a direct drive and overdrive ratio. When coupled to a three-speed transmission/transaxle configuration, a four-speed/overdrive unit is present.

OVERHEAD CAMSHAFT (OHC): An engine configuration in which the camshaft is mounted on top of the cylinder head and operates the valve either directly or by means of rocker arms.

OVERHEAD VALVE (OHV): An engine configuration in which all of the valves are located in the cylinder head and the camshaft is located in the cylinder block. The camshaft operates the valves via lifters and pushrods.

OVERRUNCLUTCH: Another name for a one-way mechanical clutch. Applies to both roller and sprag designs.

OVERSTEER: The tendency of some vehicles, when steering into a turn, to over-respond or steer more than required, which could result in excessive slip of the rear wheels. Opposite of under-steer.

OXIDATION STABILIZERS: Absorb and dissipate heat. Automatic transmission fluid has high resistance to varnish and sludge buildup that occurs from excessive heat that is generated primarily in the torque converter. Local temperatures as high as 6000F (3150C) can occur at the clutch plates during engagement, and this heat must be absorbed and dissipated. If the fluid cannot withstand the heat, it burns or oxidizes, resulting in an almost immediate destruction of friction materials, clogged filter screen and hydraulic passages, and sticky valves.

OXIDES OF NITROGEN: See nitrogen oxide (NOx).

OXYGEN SENSOR: Used with a feedback system to sense the presence of oxygen in the exhaust gas and signal the computer which can use the voltage signal to determine engine operating efficiency and adjust the air/fuel ratio.

PARALLEL CIRCUIT: (See circuit, parallel.)

PARTS WASHER: A basin or tub, usually with a built-in pump mechanism and hose used for circulating chemical solvent for the purpose of cleaning greasy, oily and dirty components.

PART-TIME FOUR WHEEL DRIVE: A system that is normally in the two wheel drive mode and only runs in four-wheel drive when the system is manually engaged because more traction is desired. Two or four wheel drive is normally selected by a lever to engage the front axle, but if locking hubs are used, these must also be manually engaged in the Lock position. Otherwise, the front axle will not drive the front wheels.

PASSIVE RESTRAINT: Safety systems such as air bags or automatic seat belts which operate with no action required on the part of the driver or passenger. Mandated by Federal regulations on all vehicles sold in the U.S. after 1990.

PAYLOAD: The weight the vehicle is capable of carrying in addition to its own weight. Payload includes weight of the driver, passengers and cargo, but not coolant, fuel, lubricant, spare tire, etc.

PCM: Powertrain control module.

PCV VALVE: A valve usually located in the rocker cover that vents crankcase vapors back into the engine to be reburned.

PERCOLATION: A condition in which the fuel actually "boils," due to excessive heat. Percolation prevents proper atomization of the fuel causing rough running.

PICK-UP COIL: The coil in which voltage is induced in an electronic ignition.

PING: A metallic rattling sound produced by the engine during acceleration. It is usually due to incorrect ignition timing or a poor grade of gasoline.

PINION: The smaller of two gears. The rear axle pinion drives the ring gear which transmits motion to the axle shafts.

PINION GEAR: The smallest gear in a drive gear assembly.

PISTON: A disc or cup that fits in a cylinder bore and is free to move. In hydraulics, it provides the means of converting hydraulic pressure into a usable force. Examples of piston applications are found in servo, clutch, and accumulator units.

PISTON RING: An open-ended ring which fits into a groove on the outer diameter of the piston. Its chief function is to form a seal between the piston and cylinder wall. Most automotive pistons have three rings: two for compression sealing; one for oil sealing.

PITMAN ARM: A lever which transmits steering force from the steering gear to the steering linkage.

PLANET CARRIER: A basic member of a planetary gear assembly that carries the pinion gears.

PLANET PINIONS: Gears housed in a planet carrier that are in constant mesh with the sun gear and internal gear. Because they have their own independent rotating centers, the pinions are capable of rotating around the sun gear or the inside of the internal gear.

PLANETARY GEAR RATIO: The reduction or overdrive ratio developed by a planetary gearset.

PLANETARY GEARSET: In its simplest form, it is made up of a basic assembly group containing a sun gear, internal gear, and planet carrier. The gears are always in constant mesh and offer a wide range of gear ratio possibilities.

PLANETARY GEARSET (COMPOUND): Two planetary gearsets combined together.

PLANETARY GEARSET (SIMPLE): An assembly of gears in constant mesh consisting of a sun gear, several pinion gears mounted in a carrier, and a ring gear. It provides gear ratio and direction changes, in addition to a direct drive and a neutral.

PLY RATING: A. rating given a tire which indicates strength (but not necessarily actual plies). A two-ply/four-ply rating has only two plies, but the strength of a four-ply tire.

POLARITY: Indication (positive or negative) of the two poles of a battery.

PORT: An opening for fluid intake or exhaust.

POSITIVE SEALING: A sealing method that completely prevents leakage.

POTENTIAL: Electrical force measured in volts; sometimes used interchangeably with voltage.

POWER: The ability to do work per unit of time, as expressed in horsepower; one horsepower equals 33,000 ft. lbs. of work per minute, or 550 ft. lbs. of work per second.

POWER FLOW: The systematic flow or transmission of power through the gears, from the input shaft to the output shaft.

POWER-TO-WEIGHT RATIO: Ratio of horsepower to weight of car.

POWERTRAIN: See Drivetrain.

POWERTRAIN CONTROL MODULE (PCM): Current designation for the engine control module (ECM). In many cases, late model vehicle control units manage the engine as well as the transmission. In other settings, the PCM controls the engine and is interfaced with a TCM to control transmission functions.

Ppm: Parts per million; unit used to measure exhaust emissions.

PREIGNITION: Early ignition of fuel in the cylinder, sometimes due to glowing carbon deposits in the combustion chamber. Preignition can be damaging since combustion takes place prematurely.

PRELOAD: A predetermined load placed on a bearing during assembly or by adjustment.

PRESS FIT: The mating of two parts under pressure, due to the inner diameter of one being smaller than the outer diameter of the other, or vice versa; an interference fit.

PRESSURE: The amount of force exerted upon a surface area.

PRESSURE CONTROL SOLENOID (PCS): An output device that provides a boost oil pressure to the mainline regulator valve to control line pressure. Its operation is determined by the amount of current sent from the PCM.

PRESSURE GAUGE: An instrument used for measuring the fluid pressure in a hydraulic circuit.

PRESSURE REGULATOR VALVE: In automatic transmissions, its purpose is to regulate the pressure of the pump output and supply the basic fluid pressure necessary to operate the transmission. The regulated fluid pressure may be referred to as mainline pressure, line pressure, or control pressure.

PRESSURE SWITCH ASSEMBLY (PSA): Mounted inside the transmission, it is a grouping of oil pressure switches that inputs to the PCM when certain hydraulic passages are charged with oil pressure.

PRESSURE PLATE: A spring-loaded plate (part of the clutch) that transmits power to the driven (friction) plate when the clutch is engaged.

PRIMARY CIRCUIT: The low voltage side of the ignition system which consists of the ignition switch, ballast resistor or resistance wire, bypass, coil, electronic control unit and pick-up coil as well as the connecting wires and harnesses.

PROFILE: Term used for tire measurement (tire series), which is the ratio of tire height to tread width.

PROM (PROGRAMMABLE READ-ONLY MEMORY): The heart of the computer that compares input data and makes the engineered program or strategy decisions about when to trigger the appropriate output based on stored computer instructions.

PULSE GENERATOR: A two-wire pickup sensor used to produce a fluctuating electrical signal. This changing signal is read by the controller to determine the speed of the object and can be used to measure transmission/transaxle input speed, output speed, and vehicle speed.

PSI: Pounds per square inch; a measurement of pressure.

PULSE WIDTH DUTY CYCLE SOLENOID (PULSE WIDTH MODULATED SOLENOID): A computer-controlled solenoid that turns on and off at a variable rate producing a modulated oil pressure; often referred to as a pulse width modulated (PWM) solenoid. Employed in many electronic automatic transmissions and transaxles, these solenoids are used to manage shift control and converter clutch hydraulic circuits.

PUSHROD: A steel rod between the hydraulic valve lifter and the valve rocker arm in overhead valve (OHV) engines.

PUMP: A mechanical device designed to create fluid flow and pressure buildup in a hydraulic system.

QUARTER PANEL: General term used to refer to a rear fender. Quarter panel is the area from the rear door opening to the tail light area and from rear wheel well to the base of the trunk and roof-line.

RACE: The surface on the inner or outer ring of a bearing on which the balls, needles or rollers move.

RACK AND PINION: A type of automotive steering system using a pinion gear attached to the end of the steering shaft. The pinion meshes with a long rack attached to the steering linkage.

RADIAL TIRE: Tire design which uses body cords running at right angles to the center line of the tire. Two or more belts are used to give tread strength. Radials can be identified by their characteristic sidewall bulge.

RADIATOR: Part of the cooling system for a water-cooled engine, mounted in the front of the vehicle and connected to the engine with rubber hoses. Through the radiator, excess combustion heat is dissipated into the atmosphere through forced convection using a water and glycol based mixture that circulates through, and cools, the engine.

RANGE REFERENCE AND CLUTCH/BAND APPLY CHART: A guide that shows the application of clutches and bands for each gear, within the selector range positions. These charts are extremely useful for understanding how the unit operates and for diagnosing malfunctions.

RAVIGNEAUX GEARSET: A compound planetary gearset that features matched dual planetary pinions (sets of two) mounted in a single planet carrier. Two sun gears and one ring mesh with the carrier pinions.

REACTION MEMBER: The stationary planetary member, in a planetary gearset, that is grounded to the transmission/transaxle case through the use of friction and wedging devices known as bands, disc clutches, and one-way clutches.

REACTION PRESSURE: The fluid pressure that moves a spool valve against an opposing force or forces; the area on which the opposing force acts. The opposing force can be a spring or a combination of spring force and auxiliary hydraulic force.

REACTOR, TORQUE CONVERTER: The reaction member of a fluid torque converter, more commonly called a stator. (See stator.)

REAR MAIN OIL SEAL: A synthetic or rope-type seal that prevents oil from leaking out of the engine past the rear main crankshaft bearing.

RECIRCULATING BALL: Type of steering system in which recirculating steel balls occupy the area between the nut and worm wheel, causing a reduction in friction.

RECTIFIER: A device (used primarily in alternators) that permits electrical current to flow in one direction only.

REDUCTION: (See gear reduction.)

REGULATOR VALVE: A valve that changes the pressure of the oil in a hydraulic circuit as the oil passes through the valve by bleeding off (or exhausting) some of the volume of oil supplied to the valve.

REFRIGERANT 12 (R-12) or 134 (R-134): The generic name of the refrigerant used in automotive air conditioning systems.

REGULATOR: A device which maintains the amperage and/or voltage levels of a circuit at predetermined values.

RELAY: A switch which automatically opens and/or closes a circuit.

RELAY VALVE: A valve that directs flow and pressure. Relay valves simply connect or disconnect interrelated passages without restricting the fluid flow or changing the pressure.

RELIEF VALVE: A spring-loaded, pressure-operated valve that limits oil pressure buildup in a hydraulic circuit to a predetermined maximum value.

RELUCTOR: A wheel that rotates inside the distributor and triggers the release of voltage in an electronic ignition.

RESERVOIR: The storage area for fluid in a hydraulic system; often called a sump.

RESIN: A liquid plastic used in body work.

RESIDUAL MAGNETISM: The magnetic strength stored in a material after a magnetizing field has been removed.

RESISTANCE: The opposition to the flow of current through a circuit or electrical device, and is measured in ohms. Resistance is equal to the voltage divided by the amperage.

RESISTOR SPARK PLUG: A spark plug using a resistor to shorten the spark duration. This suppresses radio interference and lengthens plug life.

RESISTOR: A device, usually made of wire, which offers a preset amount of resistance in an electrical circuit.

RESULTANT FORCE: The single effective directional thrust of the fluid force on the turbine produced by the vortex and rotary forces acting in different planes.

RETARD: Set the ignition timing so that spark occurs later (fewer degrees before TDC).

RHEOSTAT: A device for regulating a current by means of a variable resistance.

RING GEAR: The name given to a ring-shaped gear attached to a differential case, or affixed to a flywheel or as part of a planetary gear set.

ROADLOAD: grade.

ROCKER ARM: A lever which rotates around a shaft pushing down (opening) the valve with an end when the other end is pushed up by the pushrod. Spring pressure will later close the valve.

ROCKER PANEL: The body panel below the doors between the wheel opening.

ROLLER BEARING: A bearing made up of hardened inner and outer races between which hardened steel rollers move.

ROLLER CLUTCH: A type of one-way clutch design using rollers and springs mounted within an inner and outer cam race assembly.

ROTARY FLOW: The path of the fluid trapped between the blades of the members as they revolve with the rotation of the torque converter cover (rotational inertia).

ROTOR: (1.) The disc-shaped part of a disc brake assembly, upon which the brake pads bear; also called, brake disc. (2.) The device mounted atop the distributor shaft, which passes current to the distributor cap tower contacts.

ROTARY ENGINE: See Wankel engine.

RPM: Revolutions per minute (usually indicates engine speed).

RTV: A gasket making compound that cures as it is exposed to the atmosphere. It is used between surfaces that are not perfectly machined to one another, leaving a slight gap that the RTV fills and in which it hardens. The letters RTV represent room temperature vulcanizing.

RUN-ON: Condition when the engine continues to run, even when the key is turned off. See dieseling.

SEALED BEAM: A automotive headlight. The lens, reflector and filament from a single unit.

SEATBELT INTERLOCK: A system whereby the car cannot be started unless the seatbelt is buckled.

SECONDARY CIRCUIT: The high voltage side of the ignition system, usually above 20,000 volts. The secondary includes the ignition coil, coil wire, distributor cap and rotor, spark plug wires and spark plugs.

SELF-INDUCTION: The generation of voltage in a current-carrying wire by changing the amount of current flowing within that wire.

SEMI-CONDUCTOR: A material (silicon or germanium) that is neither a good conductor nor an insulator; used in diodes and transistors.

SEMI-FLOATING AXLE: In this design, a wheel is attached to the axle shaft, which takes both drive and cornering loads. Almost all solid axle passenger cars and light trucks use this design.

SENDING UNIT: A mechanical, electrical, hydraulic or electromagnetic device which transmits information to a gauge.

SENSOR: Any device designed to measure engine operating conditions or ambient pressures and temperatures. Usually electronic in nature and designed to send a voltage signal to an on-board computer, some sensors may operate as a simple on/off switch or they may provide a variable voltage signal (like a potentiometer) as conditions or measured parameters change.

SERIES CIRCUIT: (See circuit, series.)

SERPENTINE BELT: An accessory drive belt, with small multiple v-ribs, routed around most or all of the engine-powered accessories such as the alternator and power steering pump. Usually both the front and the back side of the belt comes into contact with various pulleys.

SERVO: In an automatic transmission, it is a piston in a cylinder assembly that converts hydraulic pressure into mechanical force and movement; used for the application of the bands and clutches.

SHIFT BUSYNESS: When referring to a torque converter clutch, it is the frequent apply and release of the clutch plate due to uncommon driving conditions.

SHIFT VALVE: Classified as a relay valve, it triggers the automatic shift in response to a governor and a throttle signal by directing fluid to the appropriate band and clutch apply combination to cause the shift to occur.

SHIM: Spacers of precise, predetermined thickness used between parts to establish a proper working relationship.

SHIMMY: Vibration (sometimes violent) in the front end caused by misaligned front end, out of balance tires or worn suspension components.

SHORT CIRCUIT: An electrical malfunction where current takes the path of least resistance to ground (usually through damaged insulation). Current flow is excessive from low resistance resulting in a blown fuse.

SHUDDER: Repeated jerking or stick-slip sensation, similar to chuggle but more severe and rapid in nature, that may be most noticeable during certain ranges of vehicle speed; also used to define condition after converter clutch engagement.

SIMPSON GEARSET: A compound planetary gear train that integrates two simple planetary gearsets referred to as the front planetary and the rear planetary.

SINGLE OVERHEAD CAMSHAFT: See overhead camshaft.

SKIDPLATE: A metal plate attached to the underside of the body to protect the fuel tank, transfer case or other vulnerable parts from damage.

SLAVE CYLINDER: In automotive use, a device in the hydraulic clutch system which is activated by hydraulic force, disengaging the clutch.

SLIPPING: Noticeable increase in engine rpm without vehicle speed increase; usually occurs during or after initial clutch or band engagement.

SLUDGE: Thick, black deposits in engine formed from dirt, oil, water, etc. It is usually formed in engines when oil changes are neglected.

SNAP RING: A circular retaining clip used inside or outside a shaft or part to secure a shaft, such as a floating wrist pin.

SOFT: Slow, almost unnoticeable clutch apply with very little shift feel.

SOFTCODES: DTCs that have been set into the PCM memory but are not present at the time of testing; often referred to as history or intermittent codes.

SOHC: Single overhead camshaft.

SOLENOID: An electrically operated, magnetic switching device.

SPALLING: A wear pattern identified by metal chips flaking off the hardened surface. This condition is caused by foreign particles, overloading situations, and/or normal wear.

SPARK PLUG: A device screwed into the combustion chamber of a spark ignition engine. The basic construction is a conductive core inside of a ceramic insulator, mounted in an outer conductive base. An electrical charge from the spark plug wire travels along the conductive core and jumps a preset air gap to a grounding point or points at the end of the conductive base. The resultant spark ignites the fuel/air mixture in the combustion chamber.

SPECIFIC GRAVITY (BATTERY): The relative weight of liquid (battery electrolyte) as compared to the weight of an equal volume of water.

SPLINES: Ridges machined or cast onto the outer diameter of a shaft or inner diameter of a bore to enable parts to mate without rotation.

SPLIT TORQUE DRIVE: In a torque converter, it refers to parallel paths of torque transmission, one of which is mechanical and the other hydraulic.

SPONGY PEDAL: A soft or spongy feeling when the brake pedal is depressed. It is usually due to air in the brake lines.

SPOOLVALVE: A precision-machined, cylindrically shaped valve made up of lands and grooves. Depending on its position in the valve bore, various interconnecting hydraulic circuit passages are either opened or closed.

SPRAG CLUTCH: A type of one-way clutch design using cams or contoured-shaped sprags between inner and outer races. (See one-way clutch.)

SPRUNG WEIGHT: The weight of a car supported by the springs.

SQUARE-CUT SEAL: Molded synthetic rubber seal designed with a square- or rectangular-shaped cross-section. This type of seal is used for both dynamic and static applications.

SRS: Supplemental restraint system

STABILIZER (SWAY) BAR: A bar linking both sides of the suspension. It resists sway on turns by taking some of added load from one wheel and putting it on the other.

STAGE: The number of turbine sets separated by a stator. A turbine set may be made up of one or more turbine members. A three-element converter is classified as a single stage.

STALL: In fluid drive transmission/transaxle applications, stall refers to engine rpm with the transmission/transaxle engaged and the vehicle stationary; throttle valve can be in any position between closed and wide open.

STALL SPEED: In fluid drive transmission/transaxle applications, stall speed refers to the maximum engine rpm with the transmission/transaxle engaged and vehicle stationary, when the throttle valve is wide open. (See stall; stall test.)

STALL TEST: A procedure recommended by many manufacturers to help determine the integrity of an engine, the torque converter stator, and certain clutch and band combinations. With the shift lever in each of the forward and reverse positions and with the brakes firmly applied, the accelerator pedal is momentarily pressed to the wide open throttle (WOT) position. The engine rpm reading at full throttle can provide clues for diagnosing the condition of the items listed above.

STALL TORQUE: The maximum design or engineered torque ratio of a fluid torque converter, produced under stall speed conditions. (See stall speed.)

STARTER: A high-torque electric motor used for the purpose of starting the engine, typically through a high ratio geared drive connected to the flywheel ring gear.

STATIC: A sealing application in which the parts being sealed do not move in relation to each other.

STATOR (REACTOR): The reaction member of a fluid torque converter that changes the direction of the fluid as it leaves the turbine to enter the impeller vanes. During the torque multiplication phase, this action assists the impeller's rotary force and results in an increase in torque.

STEERING GEOMETRY: Combination of various angles of suspension components (caster, camber, toe-in); roughly equivalent to front end alignment.

STRAIGHT WEIGHT: Term designating motor oil as suitable for use within a narrow range of temperatures. Outside the narrow temperature range its flow characteristics will not adequately lubricate.

STROKE: The distance the piston travels from bottom dead center to top dead center.

SUBSTITUTION: Replacing one part suspected of a defect with a like part of known quality.

SUMP: The storage vessel or reservoir that provides a ready source of fluid to the pump. In an automatic transmission, the sump is the oil pan. All fluid eventually returns to the sump for recycling into the hydraulic system.

SUN GEAR: In a planetary gearset, it is the center gear that meshes with a cluster of planet pinions.

SUPERCHARGER: An air pump driven mechanically by the engine through belts, chains, shafts or gears from the crankshaft. Two general types of supercharger are the positive displacement and centrifugal type, which pump air in direct relationship to the speed of the engine.

SUPPLEMENTAL RESTRAINT SYSTEM: See air bag.

SURGE: Repeating engine-related feeling of acceleration and deceleration that is less intense than chuggle.

SWITCH: A device used to open, close, or redirect the current in an electrical circuit.

SYNCHROMESH: A manual transmission/transaxle that is equipped with devices (synchronizers) that match the gear speeds so that the transmission/transaxle can be downshifted without clashing gears.

SYNTHETIC OIL: Non-petroleum based oil.

TACHOMETER: A device used to measure the rotary speed of an engine, shaft, gear, etc., usually in rotations per minute.

TDC: Top dead center. The exact top of the piston's stroke.

TEFLON SEALING RINGS: Teflon is a soft, durable, plastic-like material that is resistant to heat and provides excellent sealing. These rings are designed with either scarf-cut joints or as one-piece rings. Teflon sealing rings have replaced many metal ring applications.

TERMINAL: A device attached to the end of a wire or cable to make an electrical connection.

TEST LIGHT, CIRCUIT-POWERED: Uses available circuit voltage to test circuit continuity.

TEST LIGHT, SELF-POWERED: Uses its own battery source to test circuit continuity.

THERMISTOR: A special resistor used to measure fluid temperature; it decreases its resistance with increases in temperature.

THERMOSTAT: A valve, located in the cooling system of an engine, which is closed when cold and opens gradually in response to engine heating, controlling the temperature of the coolant and rate of coolant flow.

THERMOSTATIC ELEMENT: A heat-sensitive, spring-type device that controls a drain port from the upper sump area to the lower sump. When the transaxle fluid reaches operating temperature, the port is closed and the upper sump fills, thus reducing the fluid level in the lower sump.

THROTTLE POSITION (TP) SENSOR: Reads the degree of throttle opening; its signal is used to analyze engine load conditions. The ECM/PCM decides to apply the TCC, or to disengage it for coast or load conditions that need a converter torque boost.

THROTTLE PRESSURE/MODULATOR PRESSURE: A hydraulic signal oil pressure relating to the amount of engine load, based on either the amount of throttle plate opening or engine vacuum.

THROTTLE VALVE: A regulating or balanced valve that is controlled mechanically by throttle linkage or engine vacuum. It sends a hydraulic signal to the shift valve body to control shift timing and shift quality. (See balanced valve; modulator valve.)

THROW-OUT BEARING: As the clutch pedal is depressed, the throwout bearing moves against the spring fingers of the pressure plate, forcing the pressure plate to disengage from the driven disc.

TIE ROD: A rod connecting the steering arms. Tie rods have threaded ends that are used to adjust toe-in.

TIE-UP: Condition where two opposing clutches are attempting to apply at same time, causing engine to labor with noticeable loss of engine rpm.

TIMING BELT: A square-toothed, reinforced rubber belt that is driven by the crankshaft and operates the camshaft.

TIMING CHAIN: A roller chain that is driven by the crankshaft and operates the camshaft.

TIRE ROTATION: Moving the tires from one position to another to make the tires wear evenly.

TOE-IN (OUT): A term comparing the extreme front and rear of the front tires. Closer together at the front is toe-in; farther apart at the front is toe-out.

TOP DEAD CENTER (TDC): The point at which the piston reaches the top of its travel on the compression stroke.

TORQUE: Measurement of turning or twisting force, expressed as foot-pounds or inch-pounds.

TORQUE CONVERTER: A turbine used to transmit power from a driving member to a driven member via hydraulic action, providing changes in drive ratio and torque. In automotive use, it links the driveplate at the rear of the engine to the automatic transmission.

TORQUE CONVERTER CLUTCH: The apply plate (lockup plate) assembly used for mechanical power flow through the converter.

TORQUE PHASE: Sometimes referred to as slip phase or stall phase, torque multiplication occurs when the turbine is turning at a slower speed than the impeller, and the stator is reactionary (stationary). This sequence generates a boost in output torque.

TORQUE RATING (STALL TORQUE): The maximum torque multiplication that occurs during stall conditions, with the engine at wide open throttle (WOT) and zero turbine speed.

TORQUE RATIO: An expression of the gear ratio factor on torque effect. A 3:1 gear ratio or 3:1 torque ratio increases the torque input by the ratio factor of 3. Input torque (100 ft. lbs.) x 3 = output torque (300 ft. lbs.)

TRACTION: The amount of usable tractive effort before the drive wheels slip on the road contact surface.

TORSION BAR SUSPENSION: Long rods of spring steel which take the place of springs. One end of the bar is anchored and the other arm (attached to the suspension) is free to twist. The bars' resistance to twisting causes springing action.

TRACK: Distance between the centers of the tires where they contact the ground.

TRACTION CONTROL: A control system that prevents the spinning of a vehicle's drive wheels when excess power is applied.

TRACTIVE EFFORT: The amount of force available to the drive wheels, to move the vehicle.

TRANSAXLE: A single housing containing the transmission and differential. Transaxles are usually found on front engine/front wheel drive or rear engine/rear wheel drive cars.

TRANSDUCER: A device that changes energy from one form to another. For example, a transducer in a microphone changes sound energy to electrical energy. In automotive air-conditioning controls used in automatic temperature systems, a transducer changes an electrical signal to a vacuum signal, which operates mechanical doors.

TRANSMISSION: A powertrain component designed to modify torque and speed developed by the engine; also provides direct drive, reverse, and neutral.

TRANSMISSION CONTROL MODULE (TCM): Manages transmission functions. These vary according to the manufacturer's product design but may include converter clutch operation, electronic shift scheduling, and mainline pressure.

TRANSMISSION FLUID TEMPERATURE (TFT) SENSOR: Originally called a transmission oil temperature (TOT) sensor, this input device to the ECM/PCM senses the fluid temperature and provides a resistance value. It operates on the thermistor principle.

TRANSMISSION INPUT SPEED (TIS) SENSOR: Measures turbine shaft (input shaft) rpm's and compares to engine rpm's to determine torque converter slip. When compared to the transmission output speed sensor or VSS, gear ratio and clutch engagement timing can be determined.

TRANSMISSION OIL TEMPERATURE (TOT) SENSOR: (See transmission fluid temperature (TFT) sensor.)

TRANSMISSION RANGE SELECTOR (TRS) SWITCH: Tells the module which gear shift position the driver has chosen.

TRANSFER CASE: A gearbox driven from the transmission that delivers power to both front and rear driveshafts in a four-wheel drive system. Transfer cases usually have a high and low range set of gears, used depending on how much pulling power is needed.

TRANSISTOR: A semi-conductor component which can be actuated by a small voltage to perform an electrical switching function.

TREAD WEAR INDICATOR: Bars molded into the tire at right angles to the tread that appear as horizontal bars when 1⁄16 in. of tread remains.

TREAD WEAR PATTERN: The pattern of wear on tires which can be "read" to diagnose problems in the front suspension.

TUNE-UP: A regular maintenance function, usually associated with the replacement and adjustment of parts and components in the electrical and fuel systems of a vehicle for the purpose of attaining optimum performance.

TURBINE: The output (driven) member of a fluid coupling or fluid torque converter. It is splined to the input (turbine) shaft of the transmission.

TURBOCHARGER: An exhaust driven pump which compresses intake air and forces it into the combustion chambers at higher than atmospheric pressures. The increased air pressure allows more fuel to be burned and results in increased horsepower being produced.

TURBULENCE: The interference of molecules of a fluid (or vapor) with each other in a fluid flow.

TYPE F: Transmission fluid developed and used by Ford Motor Company up to 1982. This fluid type provides a high coefficient of friction.

TYPE 7176: The preferred choice of transmission fluid for Chrysler automatic transmissions and transaxles. Developed in 1986, it closely resembles Dexron and Mercon. Type 7176 is the recommended service fill fluid for all Chrysler products utilizing a lockup torque converter dating back to 1978.

U-JOINT (UNIVERSAL JOINT): A flexible coupling in the drive train that allows the driveshafts or axle shafts to operate at different angles and still transmit rotary power.

UNDERSTEER: The tendency of a car to continue straight ahead while negotiating a turn.

UNIT BODY: Design in which the car body acts as the frame.

UNLEADED FUEL: Fuel which contains no lead (a common gasoline additive). The presence of lead in fuel will destroy the functioning elements of a catalytic converter, making it useless.

UNSPRUNG WEIGHT: The weight of car components not supported by the springs (wheels, tires, brakes, rear axle, control arms, etc.).

UPSHIFT: A shift that results in a decrease in torque ratio and an increase in speed.

VACUUM: A negative pressure; any pressure less than atmospheric pressure.

VACUUM ADVANCE: A device which advances the ignition timing in response to increased engine vacuum.

VACUUM GAUGE: An instrument used for measuring the existing vacuum in a vacuum circuit or chamber. The unit of measure is inches (of mercury in a barometer).

VACUUM MODULATOR: Generates a hydraulic oil pressure in response to the amount of engine vacuum.

VALVES: Devices that can open or close fluid passages in a hydraulic system and are used for directing fluid flow and controlling pressure.

VALVE BODY ASSEMBLY: The main hydraulic control assembly of the transmission/transaxle that contains numerous valves, check balls, and other components to control the distribution of pressurized oil throughout the transmission.

VALVE CLEARANCE: The measured gap between the end of the valve stem and the rocker arm, cam lobe or follower that activates the valve.

VALVE GUIDES: The guide through which the stem of the valve passes. The guide is designed to keep the valve in proper alignment.

VALVE LASH (clearance): The operating clearance in the valve train.

VALVE TRAIN: The system that operates intake and exhaust valves, consisting of camshaft, valves and springs, lifters, pushrods and rocker arms.

VAPOR LOCK: Boiling of the fuel in the fuel lines due to excess heat. This will interfere with the flow of fuel in the lines and can completely stop the flow. Vapor lock normally only occurs in hot weather.

VARIABLE DISPLACEMENT (VARIABLE CAPACITY) VANE PUMP: Slipper-type vanes, mounted in a revolving rotor and contained within the bore of a movable slide, capture and then force fluid to flow. Movement of the slide to various positions changes the size of the vane chambers and the amount of fluid flow. **Note:** GM refers to this pump design as variable displacement, and Ford terms it variable capacity.

VARIABLE FORCE SOLENOID (VFS): Commonly referred to as the electronic pressure control (EPC) solenoid, it replaces the cable/linkage style of TV system control and is integrated with a spool valve and spring assembly to control pressure. A variable computer-controlled current flow varies the internal force of the solenoid on the spool valve and resulting control pressure.

VARIABLE ORIFICE THERMAL VALVE: Temperature-sensitive hydraulic oil control device that adjusts the size of a circuit path opening. By altering the size of the opening, the oil flow rate is adapted for cold to hot oil viscosity changes.

VARNISH: Term applied to the residue formed when gasoline gets old and stale.

VCM: See Electronic Control Unit (ECU).

VEHICLE SPEED SENSOR (VSS): Provides an electrical signal to the computer module, measuring vehicle speed, and affects the torque converter clutch engagement and release.

VESPEL SEALING RINGS: Hard plastic material that produces excellent sealing in dynamic settings. These rings are found in late versions of the 4T60 and in all 4T60-E and 4T80-E transaxles.

VISCOSITY: The ability of a fluid to flow. The lower the viscosity rating, the easier the fluid will flow. 10 weight motor oil will flow much easier than 40 weight motor oil.

VISCOSITY INDEX IMPROVERS: Keeps the viscosity nearly constant with changes in temperature. This is especially important at low temperatures, when the oil needs to be thin to aid in shifting and for cold-weather starting. Yet it must not be so thin that at high temperatures it will cause excessive hydraulic leakage so that pumps are unable to maintain the proper pressures.

VISCOUS CLUTCH: A specially designed torque converter clutch apply plate that, through the use of a silicon fluid, clamps smoothly and absorbs torsional vibrations.

VOLT: Unit used to measure the force or pressure of electricity. It is defined as the pressure needed to move one amp through the resistance of one ohm.

VOLTAGE: The electrical pressure that causes current to flow. Voltage is measured in volts (V).

VOLTAGE, APPLIED: The actual voltage read at a given point in a circuit. It equals the available voltage of the power supply minus the losses in the circuit up to that point.

VOLTAGE DROP: The voltage lost or used in a circuit by normal loads such as a motor or lamp or by abnormal loads such as a poor (high-resistance) lead or terminal connection.

VOLTAGE REGULATOR: A device that controls the current output of the alternator or generator.

VOLTMETER: An instrument used for measuring electrical force in units called volts. Voltmeters are always connected parallel with the circuit being tested.

VORTEX FLOW: The crosswise or circulatory flow of oil between the blades of the members caused by the centrifugal pumping action of the impeller.

WANKEL ENGINE: An engine which uses no pistons. In place of pistons, triangular-shaped rotors revolve in specially shaped housings.

WATER PUMP: A belt driven component of the cooling system that mounts on the engine, circulating the coolant under pressure.

WATT: The unit for measuring electrical power. One watt is the product of one ampere and one volt (watts equals amps times volts). Wattage is the horsepower of electricity (746 watts equal one horsepower).

WHEEL ALIGNMENT: Inclusive term to describe the front end geometry (caster, camber, toe-in/out).

WHEEL CYLINDER: Found in the automotive drum brake assembly, it is a device, actuated by hydraulic pressure, which, through internal pistons, pushes the brake shoes outward against the drums.

WHEEL WEIGHT: Small weights attached to the wheel to balance the wheel and tire assembly. Out-of-balance tires quickly wear out and also give erratic handling when installed on the front.

WHEELBASE: Distance between the center of front wheels and the center of rear wheels.

WIDE OPEN THROTTLE (WOT): Full travel of accelerator pedal.

WORK: The force exerted to move a mass or object. Work involves motion; if a force is exerted and no motion takes place, no work is done. Work per unit of time is called power. Work = force x distance = ft. lbs. 33,000 ft. lbs. in one minute = 1 horsepower

ZERO-THROTTLE COAST DOWN: A full release of accelerator pedal while vehicle is in motion and in drive range.

ENGLISH TO METRIC CONVERSION: TORQUE

To convert foot-pounds (ft. lbs.) to Newton-meters (Nm), multiply the number of ft. lbs. by 1.36
To convert Newton-meters (Nm) to foot-pounds (ft. lbs.), multiply the number of Nm by 0.7376

ft. lbs.	Nm	ft. lbs.	Nm	ft. lbs.	Nm	ft. lbs.	Nm
0.1	0.1	34	46.2	76	103.4	118	160.5
0.2	0.3	35	47.6	77	104.7	119	161.8
0.3	0.4	36	49.0	78	106.1	120	163.2
0.4	0.5	37	50.3	79	107.4	121	164.6
0.5	0.7	38	51.7	80	108.8	122	165.9
0.6	0.8	39	53.0	81	110.2	123	167.3
0.7	1.0	40	54.4	82	111.5	124	168.6
0.8	1.1	41	55.8	83	112.9	125	170.0
0.9	1.2	42	57.1	84	114.2	126	171.4
1	1.4	43	58.5	85	115.6	127	172.7
2	2.7	44	59.8	86	117.0	128	174.1
3	4.1	45	61.2	87	118.3	129	175.4
4	5.4	46	62.6	88	119.7	130	176.8
5	6.8	47	63.9	89	121.0	131	178.2
6	8.2	48	65.3	90	122.4	132	179.5
7	9.5	49	66.6	91	123.8	133	180.9
8	10.9	50	68.0	92	125.1	134	182.2
9	12.2	51	69.4	93	126.5	135	183.6
10	13.6	52	70.7	94	127.8	136	185.0
11	15.0	53	72.1	95	129.2	137	186.3
12	16.3	54	73.4	96	130.6	138	187.7
13	17.7	55	74.8	97	131.9	139	189.0
14	19.0	56	76.2	98	133.3	140	190.4
15	20.4	57	77.5	99	134.6	141	191.8
16	21.8	58	78.9	100	136.0	142	193.1
17	23.1	59	80.2	101	137.4	143	194.5
18	24.5	60	81.6	102	138.7	144	195.8
19	25.8	61	83.0	103	140.1	145	197.2
20	27.2	62	84.3	104	141.4	146	198.6
21	28.6	63	85.7	105	142.8	147	199.9
22	29.9	64	87.0	106	144.2	148	201.3
23	31.3	65	88.4	107	145.5	149	202.6
24	32.6	66	89.8	108	146.9	150	204.0
25	34.0	67	91.1	109	148.2	151	205.4
26	35.4	68	92.5	110	149.6	152	206.7
27	36.7	69	93.8	111	151.0	153	208.1
28	38.1	70	95.2	112	152.3	154	209.4
29	39.4	71	96.6	113	153.7	155	210.8
30	40.8	72	97.9	114	155.0	156	212.2
31	42.2	73	99.3	115	156.4	157	213.5
32	43.5	74	100.6	116	157.8	158	214.9
33	44.9	75	102.0	117	159.1	159	216.2

METRIC TO ENGLISH CONVERSION: TORQUE

To convert foot-pounds (ft. lbs.) to Newton-meters (Nm), multiply the number of ft. lbs. by 1.36
To convert Newton-meters (Nm) to foot-pounds (ft. lbs.), multiply the number of Nm by 0.7376

Nm	ft. lbs.	Nm	ft. lbs.	Nm	ft. lbs.	Nm	ft. lbs.	Nm	ft. lbs.
0.1	0.1	34	25.0	76	55.9	118	86.8	160	117.6
0.2	0.1	35	25.7	77	56.6	119	87.5	161	118.4
0.3	0.2	36	26.5	78	57.4	120	88.2	162	119.1
0.4	0.3	37	27.2	79	58.1	121	89.0	163	119.9
0.5	0.4	38	27.9	80	58.8	122	89.7	164	120.6
0.6	0.4	39	28.7	81	59.6	123	90.4	165	121.3
0.7	0.5	40	29.4	82	60.3	124	91.2	166	122.1
0.8	0.6	41	30.1	83	61.0	125	91.9	167	122.8
0.9	0.7	42	30.9	84	61.8	126	92.6	168	123.5
1	0.7	43	31.6	85	62.5	127	93.4	169	124.3
2	1.5	44	32.4	86	63.2	128	94.1	170	125.0
3	2.2	45	33.1	87	64.0	129	94.9	171	125.7
4	2.9	46	33.8	88	64.7	130	95.6	172	126.5
5	3.7	47	34.6	89	65.4	131	96.3	173	127.2
6	4.4	48	35.3	90	66.2	132	97.1	174	127.9
7	5.1	49	36.0	91	66.9	133	97.8	175	128.7
8	5.9	50	36.8	92	67.6	134	98.5	176	129.4
9	6.6	51	37.5	93	68.4	135	99.3	177	130.1
10	7.4	52	38.2	94	69.1	136	100.0	178	130.9
11	8.1	53	39.0	95	69.9	137	100.7	179	131.6
12	8.8	54	39.7	96	70.6	138	101.5	180	132.4
13	9.6	55	40.4	97	71.3	139	102.2	181	133.1
14	10.3	56	41.2	98	72.1	140	102.9	182	133.8
15	11.0	57	41.9	99	72.8	141	103.7	183	134.6
16	11.8	58	42.6	100	73.5	142	104.4	184	135.3
17	12.5	59	43.4	101	74.3	143	105.1	185	136.0
18	13.2	60	44.1	102	75.0	144	105.9	186	136.8
19	14.0	61	44.9	103	75.7	145	106.6	187	137.5
20	14.7	62	45.6	104	76.5	146	107.4	188	138.2
21	15.4	63	46.3	105	77.2	147	108.1	189	139.0
22	16.2	64	47.1	106	77.9	148	108.8	190	139.7
23	16.9	65	47.8	107	78.7	149	109.6	191	140.4
24	17.6	66	48.5	108	79.4	150	110.3	192	141.2
25	18.4	67	49.3	109	80.1	151	111.0	193	141.9
26	19.1	68	50.0	110	80.9	152	111.8	194	142.6
27	19.9	69	50.7	111	81.6	153	112.5	195	143.4
28	20.6	70	51.5	112	82.4	154	113.2	196	144.1
29	21.3	71	52.2	113	83.1	155	114.0	197	144.9
30	22.1	72	52.9	114	83.8	156	114.7	198	145.6
31	22.8	73	53.7	115	84.6	157	115.4	199	146.3
32	23.5	74	54.4	116	85.3	158	116.2	200	147.1
33	24.3	75	55.1	117	86.0	159	116.9	201	147.8

ENGLISH/METRIC CONVERSION: TEMPERATURE

To convert Fahrenheit (F°) to Celsius (C°), take F° temperature and subtract 32, multiply the result by 5 and divide the result by 9
To convert Celsius (C°) to Fahrenheit (F°), take C° temperature and multiply it by 9, divide the result by 5 and add 32

F°	C°	F°	C°	C°	F°	C°	F°
-40	-40.0	150	65.6	-38	-36.4	46	114.8
-35	-37.2	155	68.3	-36	-32.8	48	118.4
-30	-34.4	160	71.1	-34	-29.2	50	122
-25	-31.7	165	73.9	-32	-25.6	52	125.6
-20	-28.9	170	76.7	-30	-22	54	129.2
-15	-26.1	175	79.4	-28	-18.4	56	132.8
-10	-23.3	180	82.2	-26	-14.8	58	136.4
-5	-20.6	185	85.0	-24	-11.2	60	140
0	-17.8	190	87.8	-22	-7.6	62	143.6
1	-17.2	195	90.6	-20	-4	64	147.2
2	-16.7	200	93.3	-18	-0.4	66	150.8
3	-16.1	205	96.1	-16	3.2	68	154.4
4	-15.6	210	98.9	-14	6.8	70	158
5	-15.0	212	100.0	-12	10.4	72	161.6
10	-12.2	215	101.7	-10	14	74	165.2
15	-9.4	220	104.4	-8	17.6	76	168.8
20	-6.7	225	107.2	-6	21.2	78	172.4
25	-3.9	230	110.0	-4	24.8	80	176
30	-1.1	235	112.8	-2	28.4	82	179.6
35	1.7	240	115.6	0	32	84	183.2
40	4.4	245	118.3	2	35.6	86	186.8
45	7.2	250	121.1	4	39.2	88	190.4
50	10.0	255	123.9	6	42.8	90	194
55	12.8	260	126.7	8	46.4	92	197.6
60	15.6	265	129.4	10	50	94	201.2
65	18.3	270	132.2	12	53.6	96	204.8
70	21.1	275	135.0	14	57.2	98	208.4
75	23.9	280	137.8	16	60.8	100	212
80	26.7	285	140.6	18	64.4	102	215.6
85	29.4	290	143.3	20	68	104	219.2
90	32.2	295	146.1	22	71.6	106	222.8
95	35.0	300	148.9	24	75.2	108	226.4
100	37.8	305	151.7	26	78.8	110	230
105	40.6	310	154.4	28	82.4	112	233.6
110	43.3	315	157.2	30	86	114	237.2
115	46.1	320	160.0	32	89.6	116	240.8
120	48.9	325	162.8	34	93.2	118	244.4
125	51.7	330	165.6	36	96.8	120	248
130	54.4	335	168.3	38	100.4	122	251.6
135	57.2	340	171.1	40	104	124	255.2
140	60.0	345	173.9	42	107.6	126	258.8
145	62.8	350	176.7	44	111.2	128	262.4

LENGTH CONVERSION

To convert inches (in.) to millimeters (mm), multiply the number of inches by 25.4

To convert millimeters (mm) to inches (in.), multiply the number of millimeters by 0.04

Inches	Millimeters	Inches	Millimeters	Inches	Millimeters	Inches	Millimeters
0.0001	0.00254	0.005	0.1270	0.09	2.286	4	101.6
0.0002	0.00508	0.006	0.1524	0.1	2.54	5	127.0
0.0003	0.00762	0.007	0.1778	0.2	5.08	6	152.4
0.0004	0.01016	0.008	0.2032	0.3	7.62	7	177.8
0.0005	0.01270	0.009	0.2286	0.4	10.16	8	203.2
0.0006	0.01524	0.01	0.254	0.5	12.70	9	228.6
0.0007	0.01778	0.02	0.508	0.6	15.24	10	254.0
0.0008	0.02032	0.03	0.762	0.7	17.78	11	279.4
0.0009	0.02286	0.04	1.016	0.8	20.32	12	304.8
0.001	0.0254	0.05	1.270	0.9	22.86	13	330.2
0.002	0.0508	0.06	1.524	1	25.4	14	355.6
0.003	0.0762	0.07	1.778	2	50.8	15	381.0
0.004	0.1016	0.08	2.032	3	76.2	16	406.4

ENGLISH/METRIC CONVERSION: LENGTH

To convert inches (in.) to millimeters (mm), multiply the number of inches by 25.4
To convert millimeters (mm) to inches (in.), multiply the number of millimeters by 0.04

Inches		Millimeters	Inches		Millimeters	Inches		Millimeters
Fraction	Decimal	Decimal	Fraction	Decimal	Decimal	Fraction	Decimal	Decimal
1/64	0.016	0.397	11/32	0.344	8.731	11/16	0.688	17.463
1/32	0.031	0.794	23/64	0.359	9.128	45/64	0.703	17.859
3/64	0.047	1.191	3/8	0.375	9.525	23/32	0.719	18.256
1/16	0.063	1.588	25/64	0.391	9.922	47/64	0.734	18.653
5/64	0.078	1.984	13/32	0.406	10.319	3/4	0.750	19.050
3/32	0.094	2.381	27/64	0.422	10.716	49/64	0.766	19.447
7/64	0.109	2.778	7/16	0.438	11.113	25/32	0.781	19.844
1/8	0.125	3.175	29/64	0.453	11.509	51/64	0.797	20.241
9/64	0.141	3.572	15/32	0.469	11.906	13/16	0.813	20.638
5/32	0.156	3.969	31/64	0.484	12.303	53/64	0.828	21.034
11/64	0.172	4.366	1/2	0.500	12.700	27/32	0.844	21.431
3/16	0.188	4.763	33/64	0.516	13.097	55/64	0.859	21.828
13/64	0.203	5.159	17/32	0.531	13.494	7/8	0.875	22.225
7/32	0.219	5.556	35/64	0.547	13.891	57/64	0.891	22.622
15/64	0.234	5.953	9/16	0.563	14.288	29/32	0.906	23.019
1/4	0.250	6.350	37/64	0.578	14.684	59/64	0.922	23.416
17/64	0.266	6.747	19/32	0.594	15.081	15/16	0.938	23.813
9/32	0.281	7.144	39/64	0.609	15.478	61/64	0.953	24.209
19/64	0.297	7.541	5/8	0.625	15.875	31/32	0.969	24.606
5/16	0.313	7.938	41/64	0.641	16.272	63/64	0.984	25.003
21/64	0.328	8.334	21/32	0.656	16.669	1/1	1.000	25.400
			43/64	0.672	17.066